Gas Engineers Handbook

American Gas Association
Gas Engineers Handbook
Advisory Committee

NAME	TITLE	COMPANY
F. E. Vandaveer (1951–1965)* Chairman	Vice President Research (retired)	Con-Gas Service Corp. Cleveland, Ohio
E. F. Schuldt (1952–1965) Vice Chairman	Assistant Vice President	The Peoples Gas Light & Coke Co. Chicago, Illinois
Guy Corfield (1963–1965)	Utilization Engineer	Southern California Gas Co. Los Angeles, California
Anthony H. Cramer (1959–1965)	Executive Assistant	Michigan Consolidated Gas Co. Detroit, Michigan
Lester J. Eck (1951–1965)	Vice President (retired)	Minneapolis Gas Co. Minneapolis, Minnesota
W. R. Fraser (1951–1958)	Assistant to Manager of Operations (deceased)	Michigan Consolidated Gas Co. Detroit, Michigan
C. S. Goldsmith (1951–1952)	Engineer of Distribution (retired)	Brooklyn Union Gas Co. Brooklyn, New York
Clifford Johnstone (1951–1956)	Managing Director (deceased)	Pacific Coast Gas Association San Francisco, California
Kenneth R. Knapp (1955–1965)	Assistant Director (retired)	A. G. A. Laboratories Cleveland, Ohio
L. E. Knowlton (1951–1958)	Executive Vice President (retired)	Providence Gas Co. Providence, Rhode Island
Harry L. Masser (1952–1965)	Executive Vice President (retired)	Southern California Gas Co. Los Angeles, California
J. C. Stopford (1957–1963)	President (retired)	Honolulu Gas Co. Honolulu, Hawaii
J. Stanford Setchell (1952–1965) Secretary	Secretary, Operating Section	American Gas Association New York, New York

* Years in parentheses indicate term of service on Committee.

Gas Engineers Handbook

FIRST EDITION

FUEL GAS ENGINEERING PRACTICES

INDUSTRIAL PRESS INC.

200 Madison Avenue, New York, N. Y. 10016

Library of Congress Catalog Card Number 65-17328
A.G.A. Catalog Number H20000

Printed in the United States of America.
Manufactured by Quinn Woodbine, Woodbine New Jersey.

15 14 13 12 11

PREFACE

The existence of this edition of the *Gas Engineers Handbook*—not so much a revision of the 1934 edition as a wholly new book—points to the progress of an important, distinctive branch of engineering endeavor. The industry's growth has been impressive, accelerated by the widespread distribution of natural gas, not only in the United States but, in recent years, throughout the world. It is therefore fitting that this book recognize the needs of this profession and present, in convenient and useful form, the basic information required for its practice.

The Table of Contents in the front of the volume lists the 15 Sections of the Handbook. At the head of each Section, a detailed Table of Contents lists the authors, the chapters with their main subject headings, the tables, and the figures in that Section.

To provide maximum usefulness, extra efforts were directed toward preparing a complete subject index. With nearly 8000 entries, the index will provide the best means of locating desired information since related subjects are often discussed in separate locations in the Handbook.

Numbering of pages, tables, and figures throughout the Handbook is by Section. For example, 9/14 indicates page 14 in Section 9 and Table 9-3 is the third table in Section 9.

References in each chapter are indicated by superscript numbers in the text and are listed in numerical order at the end of the chapter. In a few chapters the references have been amplified by a bibliography on the subject in question.

The book is the work of many minds. Authors and contributors to each Section, distinguished authorities in their fields, are listed as a group at the beginning of the Section. Authors are identified individually at the head of each Chapter. The Gas Engineers Handbook Advisory Committee critically reviewed the original manuscripts. When the entire work was completed, it was again reviewed by the able Chairman, Dr. F. E. Vandaveer. Final editing was carried out by C. George Segeler and his associates on the A.G.A. Staff. The gas industry owes thanks to all those who voluntarily gave so much time and effort to bring this work to a fulfillment.

Thanks are also due to the many authors and publishers who have so generously given permission for the reprinting of tables, graphs, and other data and to the Institute of Gas Technology for its invaluable assistance. Underlying all this is the debt of appreciation owed to the research scientists whose painstaking efforts over the years to determine the physical constants and to formulate the generalized principles have made available a system of knowledge that the gas engineer may use with confidence.

It is with appreciation that the editors acknowledge the assistance of numerous correspondents who helped to clarify, verify, and review portions of the manuscript akin to their specialities, especially the following: S. Newton Croll, Philadelphia Gas Works Division, The United Gas Improvement Co.; Robert J. Gustafson, *Assistant Superintendent, Gas Distribution*, Rochester Gas & Electric Corp.; Lawrence T. Johnson, Northwest Natural Gas Co.; G. Russell King, *Staff Engineer*, Philadelphia Electric Co.; Robert Kyle, *Executive Engineer*, Brockton Taunton Gas Co.; J. Henry Long, *Vice President, Gas Operations*, Philadelphia Electric Co.; Patrick B. O'Rourke, *Assistant Vice President*, Northwest Natural Gas Co.; Harry J. Schneider, *Gas Products Manager, Regulators*, Rockwell Manufacturing Co.; Robert D. Sickafoose, *Supervisory Engineer*, The Peoples Gas Light & Coke Co.; Donald A. Vorum, *Consulting Chemical Engineer;* Henry N. Wade, Stearns-Roger Manufacturing Co.

The cooperation of the A.G.A. Headquarters and Laboratories Staffs is also acknowledged, especially the help of Edith Finch and Joann Cataldo, of our Library; E. A. Jahn, *Assistant Director, Utilization Bureau;* Earl J. Weber, *Assistant Chief Research Engineer;* Tom Walsh, *Research Engineer;* and Lester B. Inglis, Jr., *Assistant Secretary, Operating Section.*

Special thanks are due to Kenneth R. Knapp (retired), A.G.A. Laboratories, who performed the preliminary manuscript editing; Robert V. Warrick, former *Assistant Editor;* Roger Sadler, *Staff Artist;* and Don Phelps, who typed the manuscript.

In spite of all care to avoid error, a work of this magnitude cannot be expected to be without flaw. A final debt will be owed to those who discover mistakes and take the trouble to let us know, so that they may be corrected at the earliest opportunity.

C. George Segeler, *Editor-in-Chief*
Marvin D. Ringler, *Assistant Editor*
Evelyn M. Kafka, *Technical Assistant*
Utilization Bureau
American Gas Association, Inc.
New York, N. Y.

ABBREVIATIONS AND SYMBOLS

See Table 1-1 for chemical symbols.

See the end of this listing for abbreviations of organizations, committees, and agencies.

This list contains the abbreviations and symbols frequently used in this handbook and in the gas industry.

A . . . Angstrom unit
abs . . . absolute
a-c . . . alternating-current
amp . . . ampere
amp-hr . . . ampere-hour
atm . . . atmosphere (pressure)
at. wt . . . atomic weight
avg . . . average
bbl . . . barrel
Bé . . . Baumé
bhp . . . brake horsepower
bhp-hr . . . brake horsepower-hour
bmep . . . brake mean effective pressure
bp . . . boiling point
Btu . . . British thermal unit
bu . . . bushel
C . . . degree centigrade
C . . . hundred
c to c . . . center to center
C_p, c_p . . . specific heat at constant pressure
C_v, c_v . . . specific heat at constant volume
cal . . . calorie
CCF . . . hundred cubic feet
CCR . . . Conradson carbon residue
CCR . . . cyclic catalytic reforming
cfd . . . cubic feet per day
cfm . . . cubic feet per minute
cg . . . centigram
cgs . . . centimeter-gram-second (system)
C/H . . . carbon/hydrogen ratio by weight
Chap . . . Chapter
C.I . . . cast iron
cir . . . circular
cir mils . . . circular mils
cl . . . centiliter
Cl . . . percentage clearance
cm . . . centimeter
coef . . . coefficient
const . . . constant
COP . . . coefficient of performance
cos . . . cosine
cp . . . candlepower
cp . . . chemically pure
cps . . . cycles per second
CS . . . Commercial Standard
CTS . . . copper tube size, standard
cu . . . cubic
cu cm . . . cubic centimeter
cu ft . . . cubic foot
cu in . . . cubic inch
cu m . . . cubic meter
cu mm . . . cubic millimeter
cu yd . . . cubic yard
c.w.g . . . carbureted water gas
cyl . . . cylinder
D.A . . . double acting
db . . . decibel
d-c . . . direct-current
DEA . . . diethanolamine
deg, ° . . . degree
°D . . . degree-day
DEG . . . diethylene glycol
diam . . . diameter
D.M.S. . . . Drill Manufacturers Size, equivalent to standard twist drill or steel wire gages
doz . . . dozen
ΔH . . . heat of reaction or of combustion
$+\Delta H$ . . . endothermic reaction
$-\Delta H$ . . . exothermic reaction
eff . . . efficiency
emf . . . electromotive force
emu . . . electromagnetic units
Eq . . . Equation
F . . . degree Fahrenheit
f . . . farad
Fig . . . Figure
fp . . . freezing point
fpm . . . feet per minute
fps . . . feet per second
fps . . . foot-pound-second (system)
ft . . . foot
ft-c . . . foot-candle
ft-L . . . foot-Lambert
ft-lb . . . foot-pound
ft/sec-sec . . . feet per second per second
g . . . gram
gal . . . Galileo (see Table 1-42)
gal . . . gallon
g-cal . . . gram-calorie
gph . . . gallons per hour
gpm . . . gallons per minute
gps . . . gallons per second

h .... henry
ha .... hectare
HC .... hydrocarbons
HHV .... higher heating value
h-p .... high-pressure
hp .... horsepower
hp-hr .... horsepower-hour
hr .... hour
HV .... heating value (gross)
ID .... inside diameter
ihp .... indicated horsepower
ihp-hr .... indicated horsepower-hour
in. .... inch
in.-lb .... inch-pound
in. w.c. .... inches of water column (pressure)
IPS .... iron pipe size
IR .... potential drop across a resistance, where I is the current and R is the resistance
j .... joule
K .... degree Kelvin
kc .... kilocycles per second
kcal .... kilocalorie
kip .... thousand pounds
kg .... kilogram
kg per cu m .... kilograms per cubic meter
kg-cal .... kilogram-calorie
kg-m .... kilogram-meter
km .... kilometer
kv .... kilovolt
kva .... kilovolt-ampere
kvar .... reactive kilovolt-ampere
kw .... kilowatt
kwh .... kilowatt hour
l .... liter
lat .... latitude
lb .... pound
lb-in. .... pound-inch
lb-ft .... pound-foot
lb per bhp-hr .... pounds per brake horsepower-hour
lb per cu ft .... pounds per cubic foot
L.E.L. .... lower explosive limit
LFT .... longitudinal finned tubes
LHV .... lower heating value
LMTD .... logarithmic mean temperature difference
ln .... $\log_e$ or natural logarithm
LP-gas .... liquefied petroleum gas
m .... meter
M .... thousand
ma .... milliampere
max .... maximum
MCF, Mcf .... thousand cubic feet
MEA .... monoethanolamine
mep .... mean effective pressure
mg .... milligram
mh .... millihenry
min .... minimum
min .... minute
m-kg .... meter-kilogram
ml .... milliliter
mm .... millimeter
MM .... million
MMCF .... million cubic feet
MMCFD .... million cubic feet per day
MMCFH .... million cubic feet per hour
mp .... melting point
mph .... miles per hour
MSCFH .... thousand standard cubic feet per hour
mv .... millivolt
mμ .... millimicron
μ .... micron
μa .... microampere
μf .... microfarad
μv .... microvolt
μw .... microwatt
$\mu\mu$ .... micromicron
$\mu\mu$f .... micromicrofarad
N .... normal
nom .... nominal
NS .... nominal size, steel standard
o.c. .... on center
OD .... outside diameter
ohm-cm .... ohm-centimeter
oz .... ounce
oz-in .... ounce-inch
oz-ft .... ounce-foot
p_c .... critical pressure
pf .... power factor
ppm .... parts per million
psf .... pounds per square foot
psi .... pounds per square inch
psia .... pounds per square inch absolute
psig .... pounds per square inch gage
qt .... quart
R .... degree Rankine
rms .... root mean square
rph .... revolutions per hour
rpm .... revolutions per minute
S.A. .... single acting
SCF, scf .... standard cubic foot (dry gas volume at 60 F and 14.73 psia)
SCFD .... standard cubic feet per day
SCFH, scfh .... standard cubic feet per hour
sec .... second
Sec. .... Section
shp .... shaft horsepower
sin .... sine
sp gr .... specific gravity
sp ht .... specific heat
sq .... square
sq cm .... square centimeter
sq ft .... square foot
sq in. .... square inch
sq km .... square kilometer
sq m .... square meter
sq mm .... square millimeter
SSF .... seconds Saybolt Furol
SSU .... seconds Saybolt Universal
std .... standard
STP .... standard temperature and pressure, 60 F and 14.73 psia
t_c .... critical temperature
tan .... tangent
TEA .... triethanolamine
temp .... temperature
v .... volt
va .... volt-ampere
var .... reactive volt-ampere
vol. .... volume
w .... watt
w.c. .... water column
w.g. .... water gas
wt .... weight
yd .... yard
yr .... year

The following organizations, committees, and agencies are frequently referred to in this Handbook and in the literature of the gas industry.

Abbreviation	Organization
AAPG	American Association of Petroleum Geologists
AAR	Association of American Railroads
ACS	American Chemical Society
A.G.A.	American Gas Association
AIA	American Insurance Association (formerly NBFU)
AISI	American Iron and Steel Institute
APGA	American Public Gas Association
API	American Petroleum Institute
ARC	*Approval Requirements Committee
AREA	American Railway Engineering Association
ASA	American Standards Association
ASCE	American Society of Civil Engineers
ASHRAE	American Society of Heating, Refrigerating and Air-Conditioning Engineers
ASME	American Society of Mechanical Engineers
ASTM	American Society for Testing and Materials
AWWA	American Water Works Association
CAA	Civil Aeronautics Administration
C.G.A.	Canadian Gas Association
CGA	Compressed Gas Association
CNGA	California Natural Gasoline Association
EEI	Edison Electric Institute
FIA	Factory Insurance Association
FM	Associated Factory Mutual Fire Insurance Companies
FPC	Federal Power Commission
GAMA	Gas Appliance Manufacturers Association
GID	*Gas Industry Development Committee
IBR	Institute of Boiler and Radiator Manufacturers
ICC	Interstate Commerce Commission
I.G.T.	Institute of Gas Technology
IGU	International Gas Union
I.I.A.	Incinerator Institute of America
INGAA	Independent Natural Gas Association of America
IPAA	Independent Petroleum Association of America
ITAC	*Industry Technical Advisory Committee
I-W-H	Indirect Water Heater Testing and Rating Code
LPGA	LP-Gas Association (changed to NLPGA)
MCOGA	Mid-Continent Oil & Gas Association
NAFOP	*National Appliance Field Observation Program
NARUC	National Association of Railroad and Utilities Commissioners
NBFU	National Board of Fire Underwriters (changed to AIA)
NBS	National Bureau of Standards
NCA	National Coal Association
NEC	National Electric Code
NEGA	New England Gas Association
NEMA	National Electrical Manufacturers Association
NFPA	National Fire Protection Association
NGAA	Natural Gasoline Association of America (changed to NGPA)
NGPA	Natural Gas Processors Association (formerly NGAA)
NLPGA	National LP-Gas Association (formerly LPGA)
NOFI	National Oil Fuel Institute
NSF	National Sanitation Foundation
PAR	*Promotion, Advertising, and Research
PCGA	Pacific Coast Gas Association
PGA	Pennsylvania Gas Association
PUC	Public Utilities Commission
REA	Rural Electrification Administration
REDEX	*Research and Development Executive Committee
SAE	Society of Automotive Engineers
SAMA	Scientific Apparatus Makers Association
SBI	Steel Boiler Institute
SCERA	*Special Committee of Executives on Regulatory Affairs
SGA	Southern Gas Association
SPE	Society of Petroleum Engineers
TVA	Tennessee Valley Authority
UL	Underwriters' Laboratories
UPGRADE	*Utilities for Progress in Gas Residential Air Conditioning Development

* Activity of A.G.A.

SECTION 1

GENERAL TABLES AND CHARTS

Louis Shnidman, *Section Chairman,* Rochester Gas and Electric Corp., Rochester, N. Y.
C. E. Farmer (deceased), United Gas Pipe Line Co., Shreveport, La.
Robert J. Gustafson, Rochester Gas and Electric Corp., Rochester, N. Y.
F. E. Vandaveer (retired), Con-Gas Service Corp., Cleveland, Ohio

CONTENTS

Figures

Tables

Figures

Chapter 1

Properties of Elements, Common Substances, and Materials

by Louis Shnidman, C. E. Farmer, and Robert J. Gustafson

Table 1-1 Physical Properties of Elements[1,2]

Element*	Symbol	Atomic weight	Melting point, °F	Boiling point, °F	Specific gravity‡	Expansion coef§ per °F multiply by 10^{-6}	Specific heat[2] at 68 F, Btu/lb·°F	Heat of fusion,[2] Btu/lb
Aluminum	Al	26.97	1220	3,733	2.70	13.3	0.215	170
Antimony	Sb	121.76	1167	2,516	6.691	4.7–6.0	.049	68.9
Argon	A	39.94	−308.6	−302.3	1.379	§§2500 bp	.125	12
Arsenic, black cryst.	As_4	299.64†	1497 at 36 atm	1,139 (subl.)	5.73 at 57 F	2.6	.082	...
black amor.		...	...	...	4.7	...	...	...
yellow		...	...	...	2.0	...	...	...
Barium	Ba	137.36	1562	2,084	3.5	...	.068	...
Beryllium	Be	9.02	2343	5,013	1.816	6.9 (68–332 F)	.52	470
Bismuth	Bi	209.00	520	2,642	9.80	7.4	.34	22.5
Boron, cryst.	B	10.82	4172	4,622 (subl.)	3.33	4.6 (68–1382 F)	.309	...
amor.		...	...	...	2.32	...	...	...
Bromine	Br_2	159.83†	19	138	3.2	§§600 (32–86 F)	.070	29.2
Cadmium	Cd	112.41	609.6	1,413	8.65	16.6	.055	23.8
Calcium	Ca	40.08	1490	2,138	1.55	12 (32–70 F)	.149	...
Carbon, graphite	C	12.010	6332	7,592 (subl.)	2.255	0.3–2.4 (68–212 F)	.165	...
diamond		...	...	...	3.517	0.49	...	...
Cerium	Ce	140.13	1200	2,552	6.9	...	.042	...
Cesium	Cs	132.91	83.4	1,238	1.873	54 (32 F to mp)	.052	6.8
Chlorine	Cl_2	70.91	−150.9	−30.28	2.49 at 32 F	§§850 bp	.116	38.9
Chromium	Cr	52.01	2939	3,992	7.1	3.4	.11	136
Cobalt	Co	58.94	2696	5,252	8.9	6.8	.099	105
Columbium	Cb (Nb)	92.91	4380	...	8.4	4.0	.065	...
Copper	Cu	63.57	1981	4,172	8.92	9.2	.092	91.1
Fluorine	F_2	38.00†	−369.4	−304.6	1.31	§§1500 (−346 F to bp)	.18	18.2
Gallium	Ga	69.72	85.6	3,760	5.91	10.2 (32–86 F)	.079	34.6
Germanium	Ge	72.60	1757	4,892	5.36	...	.073	...
Gold	Au	197.20	1945	4,712	19.3	7.9	0.031	29.0
Hafnium	Hf	178.6	>3092	>5,792	12.1	...	...	...
Helium	He	4.00	<−458 at 26 atm	−452.0	0.1368	...	1.25	...
Hydrogen	H_2	2.016†	−434.38	−422.9	0.06948	...	3.45	27.0
Indium	In	114.76	311	2,642	7.3	18	0.057	...
Iodine	I_2	253.84†	236.3	363.75	4.93	52 (68–212 F)	.052	25.6
Iridium	Ir	193.10	4262	8,672	22.4	3.8	.031	...
Iron	Fe	55.85	2795	5,432	7.86	6.50	.11	117
Krypton	Kr	83.70	−272.2	−241.2	2.818	...	...	...
Lanthanum	La	138.92	1529	3,272	6.15	...	.045	...
Lead	Pb	207.21	621.5	2,948	11.337	16.3 (68–212 F)	.031	11.3
Lithium	Li	6.94	367	2,437	0.53	31	.79	286
Magnesium	Mg	24.32	1204	2,030	1.74	14 at 104 F	0.25	160

* The following elements are not listed: Actinium, Alabamine, Americium, Astatine, Berkelium, Californium, Curium, Dysprosium, Erbium, Europium, Francium, Gadolinium, Holmium, Lutecium, Neptunium, Plutonium, Polonium, Praseodymium, Prometheum, Protactinium, Samarium, Technitium, Terbium, Thulium, and Ytterbium.

† Molecular weight and formula are given for those elements that do not occur in the atomic form.

‡ Specific gravity at 60 to 68 F, except as noted. Gases with reference to air as 1. Liquids and solids with reference to water as 1.

§ Length per unit length for solids. Cubical coefficient is given for liquids and gases, marked §§. Coefficients given are at 68 F, unless otherwise noted. Values for metallic elements extracted from "Physical Constants of the Principal Alloy-Forming Elements," Mechanical Engineers' Handbook, 6th ed., New York, McGraw-Hill p. 6–68, 1958.

Table 1-1 Physical Properties of Elements (Continued)

Element*	Symbol	Atomic weight	Melting point, °F	Boiling point, °F	Specific gravity‡	Expansion coef§ per °F multiply by 10^{-6}	Specific heat[2] at 68 F, Btu/lb-°F	Heat of fusion,[2] Btu/lb
Manganese	Mn	54.93	2300	3,452	7.2	12	0.115	115
Mercury	Hg	200.61	−38.0	674.4	13.546	§§101	.033	4.9
Molybdenum	Mo	95.95	4748	6,692	10.2	2.7 (68–212 F)	.061	126
Neodymium	Nd	144.27	1544	...	6.9	...	.045	...
Neon	Ne	20.18	−415.61	−410.6	0.674	...	...	...
Nickel	Ni	58.69	2646	5,252	8.90	7.4 (32–212 F)	.105	133
Nitrogen, gas	N_2	28.016†	...	...	0.9674	...	.247	11.2
solid		...	−345.75	...	1.026 at −422.5 F	...	...	...
liquid		...	...	−320.44	0.808 at bp	§§3300 (mp to bp)	...	...
Osmium	Os	190.2	4892	>9,572	22.48	2.6	.031	...
Oxygen, gas	O_2	32.00†	...	...	1.1053	...	.218	5.9
solid		...	−361.1	...	1.426 at −422.5 F	...	...	...
liquid		...	...	−297.4	1.14 at bp	§§2200 (−346 to bp)	...	...
Ozone	O_3	48.00†	−419.8	−169.6	1.658	§§1400 bp	...	...
Palladium	Pd	106.70	2831	3,992	12.0	6.6	.058	61.6
Phosphorus	P_4	123.92†	...	752	2.69	70	.177	9.0
red		...	1094 (43 atm)	1,337	2.20	§§300 (mp to 140 F)	...	...
yellow		...	111.4	536	1.82	...	...	...
Platinum	Pt	195.23	3191	7,772	21.45	4.9	.032	49
Potassium	K	39.10	144.1	1,400	0.86	46	.177	26.1
Radium	Ra	226.05	1760	2,084	5 (?)	...	...	...
Radon	Rn	222.0	−95.8	−79.6	9.73	...	...	...
Rhenium	Re	186.31	6222	...	20.53	...	.033	...
Rhodium	Rh	102.91	3551	>4,532	12.5	4.6 (68–212 F)	.059	...
Rubidium	Rb	85.48	101.3	1,292	1.53	50	.080	11.0
Ruthenium	Ru	101.70	4442	>4,892	12.2	5.1	.057	...
Samarium	Sm	150.43	>2372	...	7.7	...	...	...
Scandium	Sc	45.10	2192	4,352	2.5 (?)	...	...	...
Selenium	Se_8	631.68†	122	1,270	4.26 (77 F)	21	.084	11.9
gray; red		...	428	1,270	4.8; 4.5	...	...	...
steel gray		...	423	1,270	4.8 (77 F)	...	...	...
Silicon	Si	28.06	2588	4,712	2.4	1.6–4.1	.162	607
Silver	Ag	107.88	1761	3,542	10.50	10.9 (32–212 F)	.056	45
Sodium	Na	22.997	207.5	1,616	0.97	39	.295	49.5
Strontium	Sr	87.63	1472	2,102	2.6	...	.176	45
Sulfur, amor.	S	32.06	248	832	2.046	...	.175	16.7
rhombic	S_8	256.48†	235	832	2.07	37.49 (32–212 F)	...	...
monoclinic	S_8	256.48†	246	832	1.96	...	...	...
Tantalum	Ta	180.88	5162	>7,412	16.6	3.6	.036	...
Tellurium, cryst.	Te	127.61	846	2,534	6.24	9.3	.047	13.1
amor.		...	...	...	6.028–6.156	...	...	...
Thallium	Tl	204.39	578.3	3,002	11.85	16	.031	13.0
Thorium	Th	232.12	3353	>5,432	11.2	...	.034	...
Tin, white	Sn	118.70	449.3	4,100	7.31	13	.054	26.1
gray		...	(stable −261 to +64 F)	...	5.750	3 (−261 to 0 F)	...	...
Titanium	Ti	47.90	3272	>5,432	4.50 @ 64 F	4.7	.126	...
Tungsten	W	183.92	6098	10,652	19.3	2.4	.032	79
Uranium	U	238.07	2071	6,332	18.7	...	.028	...
Vanadium	V	50.95	3110	5,432	5.96	4.3	.120	...
Xenon	Xe	131.30	−220	−164.4	4.53	...	...	...
Yttrium	Y	88.92	2714	4,532	5.51	...	...	...
Zinc	Zn	65.38	786.9	1,665	7.140	9.4–22	.0915	43.36
Zirconium	Zr	91.22	3092	>5,252	6.4	3	0.066	...

* The following elements are not listed: Actinium, Alabamine, Americium, Astatine, Berkelium, Californium, Curium, Dysprosium, Erbium, Europium, Francium, Gadolinium, Holmium, Lutecium, Neptunium, Plutonium, Polonium, Praseodymium, Prometheum, Protactinium, Samarium, Technitium, Terbium, Thulium, and Ytterbium.

† Molecular weight and formula are given for those elements that do not occur in the atomic form.

‡ Specific gravity at 60 to 68 F, except as noted. Gases with reference to air as 1. Liquids and solids with reference to water as 1.

§ Length per unit length for solids. Cubical coefficient is given for liquids and gases, marked §§. Coefficients given are at 68 F, unless otherwise noted. Values for metallic elements extracted from "Physical Constants of the Principal Alloy-Forming Elements," Mechanical Engineers' Handbook, 6th ed., New York, McGraw-Hill p. 6–68, 1958.

Table 1-2 Specific Gravity and Specific Heat of Common Solid Substances

Solid	Specific gravity	Specific heat, Btu/lb-°F
Asbestos	2.0 -2.8	0.195 at 68–212 F
Ashes, cinder	.64– .72	.2 at 32–212
Asphalt	.99–1.43	.55
Asphaltum	.87–1.51	.22
Beeswax	.96	.82 at 60–144
Borax	1.70	.238 at 51–208
Brick, soft	1.70–1.89	
common	1.79–2.0	
fire	1.70–2.10	.2–.25 at 64–212
hard	1.89–2.10	
vitrified	2.0 -2.2	
Calcium carbonate	2.71–2.97	.21 at 32–212
Camphor	.99	.44 at 68–353
Caoutchouc	.91– .99	...
Celluloid	1.35–1.39	...
Cellulose	1.51	.32 at 32–212
Celotex	.21– .26	.4
Cement, loose	1.15–1.68 (bulk)	.20 at 68–212
set	2.69–3.0	.20 at 68–212
Cereals	.42– .77	...
Charcoal, pine	.37	.17
oak	.53	...
Clay, dry	1.01	.22 at 68–212
damp	1.76	...
Coal, anthracite	.83– .93 (bulk)	.26–.37
bituminous	.71– .87 (bulk)	...
lignite	1.25	...
peat, dry	.75	...
Coke	.37– .51 (bulk)	.24 at 68–500
Concrete	2.22–2.50	.18–.19 at 72–372
Cork board	.16	about .49
Cotton	1.49	...
Dolomite	2.76–2.95	.22
Dry ice	1.51–1.67	.204 at 75–135
Earth, dry	1.04–1.41	.0014 at 68
moist	1.30–1.60	.0052 at 68
mud	1.28–2.08	.0067 at 68
Emery	3.75–4.34	...
Fat, beef	.91– .98	.52 at 43–79
		.79 at 79–108
		.54 at 151–216
Flour	.45– .75	...
Glass, crown	2.29–2.90	.16 at 50–122
flint	2.90–5.90	.12 at 50–122
jena	2.37–2.58	.20 at 66–212
pyrex	2.24	.196 at 68–212
Granite	2.60–2.72	0.19 at 54–212 F
Graphite	...	0.3–.38 at 70–2200 F
Gypsum	2.31	.259 at 50–212
Hay & straw	.32	...
Ice	.92 at 32 F	.492 at 32
Kaolin	2.40–2.60 (bulk)	.22 at 68–212
Leather, dry	.87	.36
greased	1.03	...
Lime	.85–1.20 (bulk)	...
Limestone	2.28–2.74	.217
Linoleum	1.20	...
Magnesite, refractory	2.90–3.09	.27 at 60–1200
Marble	2.69–2.90	.21 at 32–212
Naphthalene	1.49	.325 at 68–140
Paper	.71–1.15	.349
Paraffin	.87– .95	.7 at 32–68
Pitch, coal tar	1.07–1.15	.45 at 60–212
Plaster of Paris, set	2.31	1.14
Plastics		
acrylonitrile	1.06	.35
butyrate	1.2	.35
polyethylene	.92–1.0	.55
polyvinyl chloride	1.38	.2
Porcelain	2.29–2.50	.26 at 60–1750
Potassium nitrate	2.08	.19 at 59–212
Potatoes	.71	.84
Resin, phenol	1.25	.33–.37 at 167–212
Riprap, limestone	1.28–1.36	...
sandstone	1.44	...
shale	1.68	...
Rubber, India	.91– .93	.48 at 32–212
hard	1.20	.33–.40 at 32–212
Salt, rock	2.15	.219 at 55–113
Sand, dry	1.44–1.92 (bulk)	.195 at 32–212
Sandstone	2.16–2.64	.22
Shale	1.44–2.72	...
Shellac	1.20–1.22	.40 at 60–212
Silicone carbide	...	.23 at 60–950
Slag	2.76	...
Sodium carbonate	2.44	.306
Sodium nitrate	2.24	.231
Starch	1.54	...
Stone	...	about .2
Sugar, cane	1.63	.30 at 68
Tar, coal	.95–1.33	.35–.45 at 68–392 F
Tile	1.79–1.89	.150
Wool	1.30	0.325

Table 1-3 Wood Drying Data[3]

Species	Average moisture* (green wood), per cent	Average specific gravity†	Drying time,‡ days	Species	Average moisture* (green wood), per cent	Average specific gravity†	Drying time,‡ days
Ash, black	95	0.45	10–14	Locust, black	40	0.66	12–16
white	44–46	.55	11–15	Maple, soft	58–97	.44	7–13
Beech	55–72	.56	12–15	hard	65–72	.56	11–15
Birch, yellow	72–74	.55	11–15	Oak	64–83	.51– .81	16–40
Cedar	32–249	.29– .44	3–15	Pine	31–219	.34– .54	3–10
Cherry, black	58	.47	10–14	Poplar, yellow	83–106	.40	6–10
Chestnut	120	.40	8–12	Redwood	86–210	.30– .38	10–24
Douglas-fir	30–154	.40– .45	2–7	Spruce	34–173	.32– .38	3–7
Elm	44–95	.46– .57	10–17	Sweetgum, heartwood	79	.46	15–25
Fir	34–164	.31– .37	3–5	sapwood	137	.46	10–15
Hemlock	85–170	.38	3–5	Walnut, black	73–90	0.51	10–16
Hickory	50–97	0.56– .61	7–15				

Note: Specific heat of wood may be taken as 0.327 (0 to 215 F). Additional heat is required to heat and evaporate water. Because of the many variables that affect kiln-drying, a value of 1½ to 2 boiler horsepower per M board feet capacity is a good rule.

* Average values based on weight of oven-dry wood.

† Based on weight when oven-dry and volume when green.

‡ Time to kiln-dry one-inch stock from the green condition to 6 per cent moisture content. The time required to dry from 20 to 6 per cent is about half the values tabulated.

Table 1-4 Specific Gravity and Specific Heat of Selected Liquids

Liquid	Specific gravity at 60 F/60 F	Specific heat, Btu/lb-°F	Liquid	Specific gravity at 60 F/60 F	Specific heat, Btu/lb.-°F
Acetic acid, glacial	1.055	0.48 at 60 F	Naphtha	0.68–.88	0.493
Acetone	0.800	.52 at 60	Nitric acid, 91%	1.50	
Alcohol, amyl	.8192	.555 at 60	Nonane	.718	.503 at 32–122
ethyl	.7939	.561 at 60	Octane	.706	.505 at 32–122
methyl	.7963	.593 at 60	Oil, castor	.960–.968	
Benzene	.8843	.400 at 60	cottonseed	.922–.925	
Bromine	3.122 at 68 F	.107 at 32–115	creosote	1.04–1.10	
Butane, iso-	.5630	.549	mineral, lubricants	.90–.93	
Calcium chloride, 5.8% solution	1.0469 at 68		linseed	.930–.938	.441–.447
40.9% solution	1.4074 at 68		olive	.915–.920	.471 at 44
Carbon disulfide	1.271	.24 at 60	turpentine	.860–.866	.42 at 68
Carbon tetrabromide	3.42	.268	vegetable	.91–.94	
Carbon tetrachloride	1.602	.20 at 60	Pentane	.634	
Chloroform	1.4998	.23 at 60	Perchlorethylene	1.619	.21
Decane	.730	.588 at 70–309	Petroleum	.88	.511 at 70–135
Ether, diethyl	1.7150	.555 at 60	Phenol	1.072	.561 at 57–79
petroleum	.665		Potassium chloride, 24.3% solution	1.1655	.727 at 68
Ethyl acetate	.899	.457 at 68	Sodium chloride, 20.6% solution	1.1537	.82
Gas oil	.86–.87		Sodium sulfate, 24.0% solution	1.2347	
Gasoline	.66–.69		Sulfuric acid, 87%	1.80	
Glycerine	1.2651	.559 at 60	Toluene	.866	.440 at 54–210
Glycol, mono-	1.116	.55–.75 from 50–356	Trichloroethylene	1.4569 at 77 F	.223 at 68
di-	1.118	.51–.61 from —4 to 212	Turpentine	.864	
tri-	1.126	.48–.7 from 90 to 356	Water, fresh	1.00	1.0 at 60
Heptane	.683	.507 at 32–122	sea	1.02–1.03	
Hexane	.669	.527 at 32–122	Xylene	0.88	0.4 at 32–104 F
Hydrochloric acid, 33.2% solution	1.1699	.591 at 68			
Kerosene	.80–.82				
Lye, soda—66%	1.70	0.78 at 68			

Table 1-5 Strength of Common Building Materials
(in pounds per sq in.)

Building materials	Ultimate average stresses: Com-pression	Tension	Bending	Modulus of elasticity	Safe working stresses: Com-pression	Bearing	Shearing
Steel*							
Shapes, plates, bars	60,000	60,000	60,000	30,000,000			
Stone							
Granite, gneiss, bluestone	12,000	1,200	1,600	7,000,000	1200	1200	200
Limestone, marble	8,000	800	1,500	7,000,000	800	800	150
Sandstone	5,000	150	1,200	3,000,000	500	500	150
Slate	10,000	3,000	5,000	14,000,000	1000	1000	175
Masonry							
Granite	...	...	...	...	420	600	
Limestone, bluestone	...	...	...	...	350	500	
Sandstone	...	...	...	...	280	400	
Rubble	...	...	...	...	140	250	
coursed	...	...	...	...	170	250	
Brick, medium burned	10,000	...	...	...	170	300	
hard burned	15,000	...	...	...	210	300	
pressed, paving brick	6,000						
Terra cotta	5,000						
Cement, Portland							
Neat, 28 days	7,040	740					
90 days	7,350	740					
1:3 sand, 28 days	1,290	320					
90 days	1,490	340					
Concrete, P.C.							
1:1:2							
granite, trap rock	3,300						
furnace slag	3,000						
lime and sandstone, hard	3,000						
lime and sandstone, soft	2,200						
cinders	800						
1:1½:3							
granite, trap rock	2,800						
furnace slag	2,500						
lime and sandstone, hard	2,500						
lime and sandstone, soft	1,800						
cinders	700						
1:2:4							
granite, trap rock	2,200						
furnace slag	2,000						
lime and sandstone, hard	2,000						
lime and sandstone, soft	1,500						
cinders	600						
1:2½:5							
granite, trap rock	1,800						
furnace slag	1,600						
lime and sandstone, hard	1,600						
lime and sandstone, soft	1,200						
cinders	500						
1:3:6							
granite, trap rock	1,400						
furnace slag	1,300						
lime and sandstone, hard	1,300						
lime and sandstone, soft	1,000						
cinders	400						
Miscellaneous							
Glass, common	30,000	3,000					
Plaster	700	70	3,000	8,000,000			

Reinforced Concrete

Modulus of elasticity	
3,000,000	for ultimate compression over 2900
2,500,000	for ultimate compression up to 2900
2,000,000	for ultimate compression up to 2200
750,000	for ultimate compression under 800

Safe working stresses in percentage of ultimate compression

Compression	Plain concrete piers, length 4 diam.	22.5%
	Reinforced columns, length 12 diam.	22.5
	Reinforced beams	32.5
Bearing	Surface twice the loaded area	35.0
Shear and diagonal tension	Horizontal bars, no web reinforcement	2.0
	Horizontal bars, vertical stirrups	4.5
	Bent bars and vertical stirrups	5.0
	Same, securely attached	6.0
Bond stress	Drawn wire	2.0
	Plain reinforcing bars	4.0
	Deformed bars, best type	5.0%

* Elastic limit 30,000 psi, ultimate shear 45,000 psi, elongation 30%.

Table 1-6 Composition of Selected Aluminum Alloys[6]

AA No.	Typical composition, per cent (remainder is aluminum)	Form produced
1100	max: Mn 0.10, Cu 0.20, Zn 0.10, Fe + Si 1.0	Wrought alloy
3003	max: Si 0.60, Mn 1.5, Cu 0.20, Zn 0.10, Fe 0.70 min: Mn 1.0	Wrought alloy
2014	max: Si 1.2, Mn 1.2, Cu 5.0, Mg 0.80, Zn 0.25, Cr 0.10, Fe 1.0 min: Si 0.50, Mn 0.40, Cu 3.9, Mg 0.20	Wrought alloy
2017	max: Si 0.80, Mn 1.0, Cu 4.5, Mg 0.80, Zn 0.25, Cr 0.25, Fe 1.0 min: Mn 0.40, Cu 3.5, Mg 0.20	Wrought alloy
2024	max: Si 0.50, Mn 0.90, Cu 4.9, Mg 1.8, Zn 0.10, Cr 0.25, Fe 0.50 min: Mn 0.30, Cu 3.8, Mg 1.2	Wrought alloy
6061	max: Si 0.80, Mn 0.15, Cu 0.40, Mg 1.2, Ti 0.15, Zn 0.20, Cr 0.35, Fe 0.70 min: Si 0.40, Cu 0.15, Mg 0.80, Cr 0.15	Wrought alloy
6063	max: Si 0.60, Mn 0.10, Cu 0.10, Mg 0.90, Ti 0.10, Zn 0.10, Cr 0.10, Fe 0.35 min: Si 0.20, Mg 0.45	Wrought alloy
7075	max: Si 0.50, Mn 0.30, Cu 2.0, Mg 2.9, Ti 0.20, Zn 6.1, Cr 0.40, Fe 0.70 min: Cu 1.2, Mg 2.1, Zn 5.1, Cr 0.18	Wrought alloy
195	max: Si 1.2, Mn 0.30, Cu 5.0, Mg 0.03, Ti 0.20, Zn 0.30, Fe 1.0 min: Cu 4.0	Sand casting alloy
B195	max: Si 3.0, Mn 0.30, Cu 5.0, Mg 0.03, Zn 0.30, Ti 0.20, Fe 1.0 min: Si 2.0, Cu 4.0	Permanent mold casting alloy
220	max: Si 0.20, Mn 0.10, Cu 0.20, Mg 10.6, Zn 0.10, Fe 0.30 min: Mg 9.5	Sand casting alloy
355	max: Si 5.5, Mn 0.30, Cu 1.5, Mg 0.60, Ti 0.20, Zn 0.20, Fe 0.60* min: Si 4.5, Cu 1.0, Mg 0.40	Sand and permanent mold casting alloy
356	max: Si 7.5, Mn 0.10, Cu 0.20, Mg 0.40, Ti 0.20, Zn 0.20, Fe 0.50 min: Si 6.5, Mg 0.20	Sand and permanent mold casting alloy
A612	max: Si 0.15, Mn 0.05, Cu 0.65, Mg 0.80, Ti 0.20, Zn 7.0, Fe 0.50 min: Cu 0.35, Mg 0.60, Zn 6.0	Sand casting alloy
C612	max: Si 0.30, Mn 0.05, Cu 0.65, Mg 0.45, Ti 0.20, Zn 7.0, Fe 1.4 min: Cu 0.35, Mg 0.25, Zn 6.0	Permanent mold casting alloy
750	max: Si 0.70, Mn 0.10, Cu 1.3, Ni 1.3, Ti 0.20, Sn 7.0, Fe 0.70 min: Cu 0.70, Ni 0.70, Sn 5.5	Permanent mold casting alloy

* When Fe exceeds 0.40 per cent, Mn shall not be less than half of Fe content.

Table 1-7 Composition of Selected Magnesium Alloys[6]

ASTM Designation	Typical composition, per cent (remainder is magnesium)	Form produced
AZ92A	Al 9.0, Mn 0.15, Zn 2.0	Sand and permanent mold castings
AZ63A	Al 6.0, Mn 0.25, Zn 3.0	Sand and permanent mold castings
EK30A	Rare Earths 3.0, Zr 0.55	Sand and permanent mold castings
EZ33A	Zn 3.0, Rare Earths 3.0, Zr 0.7	Sand and permanent mold castings
HZ32A	Zn 2.0, Zr 0.7, Th 3.0	Sand and permanent mold castings
AZ91C	Al 9.0, Mn 0.15, Zn 1.0	Sand and permanent mold castings
AZ31A	Al 3.0, Mn 0.45, Zn 1.0	Extrusions & sheet
ZK60A	Zn 5.7, Zr 0.55	Extrusions

Table 1-8 Composition of Selected Copper Alloys[6]

Name	Nominal composition, per cent (remainder is copper)
Brasses—Nonleaded	
Gilding, 95%	Zn 5
Commercial bronze, 90%	Zn 10
Jewelry bronze, 87.5%	Zn 12.5
Red brass, 85%	Zn 15
Low brass, 80%	Zn 20
Cartridge brass, 70%	Zn 30
Yellow brass	Zn 35
Muntz metal	Zn 40
Brasses—Leaded	
Leaded commercial bronze	Pb 1.75, Zn 9.25
Low-leaded brass	Pb 0.5, Zn 34.5
High-leaded brass	Pb 2.0, Zn 33.0
Leaded Muntz metal	Pb 0.6, Zn 39.4
Forging brass	Pb 2, Zn 38
Architectural bronze	Pb 3, Zn 40
Tin and Aluminum Brasses	
Inhibited Admiralty	Zn 28, Sn 1
Naval brass	Zn 39.25, Sn 0.75
Leaded Naval brass	Pb 1.75, Zn 37.5, Sn 0.75
Manganese bronze	Fe 1, Zn 39.2, Sn 1, Mn 0.3
Aluminum brass	Zn 22, Al 2
Aluminum bronze, 5%	Al 5
Phosphor Bronzes (tin bronzes)	
Phosphor bronze, 10%	Sn 10
Phosphor bronze, 1.25%	Sn 1.25, P Trace
Free-cutting phosphor bronze	Pb 4, Zn 4, Sn 4
Cupro-Nickel and Nickel Silvers	
Cupro-nickel, 30%	Ni 30
Cupro-nickel, 10%	Fe 1.3, Ni 10
Silicon Bronzes (copper-silicon alloys)	
High-silicon bronze	Si 3
Low-silicon bronze	Si 1.5

Table 1-9 Dimensions and Weights* of Steel Pipe[2]

Nom size, in.	OD, in.	Sch No.	Wall thick-ness, in.	ID, in.	Inside sectional area, sq ft	Cir ft or surface, sq ft/ft of length: Out-side	In-side	Wt per ft, lb
⅛	0.405	40†	0.068	0.269	0.00040	0.106	0.0705	0.25
		80‡	.095	.215	.00025	.106	.0563	.32
¼	.540	40†	.088	.364	.00072	.141	.0954	.43
		80‡	.119	.302	.00050	.141	.0792	.54
⅜	.675	40†	.091	.493	.00133	.177	.1293	.57
		80‡	.126	.423	.00098	.177	.1110	.74
½	0.840	40†	.109	.622	.00211	.220	.1630	0.85
		80‡	.147	.546	.00163	.220	.1430	1.09
		160	.187	.466	.00118	.220	.1220	1.31
¾	1.050	40†	.113	.824	.00371	.275	.2158	1.13
		80‡	.154	.742	.00300	.275	.1942	1.48
		160	.218	0.614	.00206	.275	.1610	1.94
1	1.315	40†	.133	1.049	.00600	.344	.2745	1.68
		80‡	.179	0.957	.00499	.344	.2505	2.17
		160	.250	0.815	.00362	.344	.2135	2.85
1¼	1.660	40†	.140	1.380	.01040	.435	.362	2.28
		80‡	.191	1.278	.00891	.435	.335	3.00
		160	.250	1.160	.00734	.435	.304	3.77
1½	1.900	40†	.145	1.610	.01414	.498	.422	2.72
		80‡	.200	1.500	.01225	.498	.393	3.64
		160	.281	1.338	.00976	.498	.350	4.86
2	2.375	40†	.154	2.067	.02330	.622	.542	3.66
		80‡	.218	1.939	.02050	.622	.508	5.03
		160	.343	1.689	.01556	.622	.442	7.45
2½	2.875	40†	.203	2.469	.03322	.753	.647	5.80
		80‡	.276	2.323	.02942	.753	.609	7.67
		160	.375	2.125	.02463	.753	.557	10.0
3	3.500	40†	.216	3.068	.05130	.917	.804	7.58
		80‡	.300	2.900	.04587	.917	.760	10.3
		160	.437	2.626	.03761	0.917	.688	14.3
3½	4.000	40†	.226	3.548	.06870	1.047	.930	9.11
		80‡	.318	3.364	.06170	1.047	0.882	12.5
4	4.500	40†	.237	4.026	.08840	1.178	1.055	10.8
		80‡	.337	3.826	.07986	1.178	1.002	15.0
		120	.437	3.626	.07170	1.178	0.950	19.0
		160	.531	3.438	.06447	1.178	0.901	22.6
5	5.563	40†	.258	5.047	.1390	1.456	1.322	14.7
		80‡	.375	4.813	.1263	1.456	1.263	20.8
		120	.500	4.563	.1136	1.456	1.197	27.1
		160	.625	4.313	.1015	1.456	1.132	33.0
6	6.625	40†	.280	6.065	0.2006	1.734	1.590	19.0
		80‡	.432	5.761	1.1810	1.734	1.510	28.6
		120	.562	5.501	0.1650	1.734	1.445	36.4
		160	.718	5.189	.1469	1.734	1.360	45.3
8	8.625	20	.250	8.125	.3601	2.258	2.130	22.4
		30†	.277	8.071	.3553	2.258	2.115	24.7
		40†	.322	7.981	.3474	2.258	2.090	28.6
		60	.406	7.813	.3329	2.258	2.050	35.7
		80‡	.500	7.625	.3171	2.258	2.000	43.4
		100	.593	7.439	.3018	2.258	1.947	50.9
		120	.718	7.189	.2819	2.258	1.883	60.7
		140	.812	7.001	.2673	2.258	1.835	67.8
		160	.906	6.813	.2532	2.258	1.787	74.7
10	10.75	20	.250	10.250	.5731	2.814	2.685	28.1
		30†	.307	10.136	.5603	2.814	2.655	34.3
		40†	.365	10.020	.5475	2.814	2.620	40.5
		60‡	.500	9.750	.5185	2.814	2.550	54.8
		80	.593	9.564	.4989	2.814	2.503	64.4
		100	.718	9.314	.4732	2.814	2.440	77.0
		120	0.843	9.064	0.4481	2.814	2.373	89.2
10	10.75	140	1.000	8.750	0.4176	2.814	2.290	105.0
		160	1.125	8.500	.3941	2.814	2.230	116.0
12	12.75	20	0.250	12.250	.8185	3.338	3.21	33.4
		30†	.330	12.090	.7972	3.338	3.17	43.8
		40	.406	11.938	.7773	3.338	3.13	53.6
		60	.562	11.626	.7372	3.338	3.05	73.2
		80	.687	11.376	.7058	3.338	2.98	88.6
		100	0.843	11.064	.6677	3.338	2.90	108.0
		120	1.000	10.750	.6303	3.338	2.82	126.0
		140	1.125	10.500	.6013	3.338	2.75	140.0
		160	1.312	10.126	.5592	3.338	2.66	161.0
14	14.0	10	0.250	13.500	.9940	3.665	3.54	36.8
		20	.312	13.376	.9750	3.665	3.51	45.7
		30	.375	13.250	.9575	3.665	3.47	54.6
		40	.437	13.126	.9397	3.665	3.44	63.3
		60	.593	12.814	.8956	3.665	3.36	85.0
		80	.750	12.500	.8522	3.665	3.28	107.0
		100	0.937	12.126	.8020	3.665	3.18	131.0
		120	1.062	11.876	.7693	3.665	3.11	147.0
		140	1.250	11.500	.7213	3.665	3.01	171.0
		160	1.406	11.188	0.6827	3.665	2.93	190.0
16	16.0	10	0.250	15.500	1.3104	4.189	4.06	42.1
		20	.312	15.376	1.2895	4.189	4.03	52.3
		30	.375	15.250	1.2680	4.189	4.00	62.6
		40	.500	15.000	1.2272	4.189	3.93	82.8
		60	.656	14.688	1.1766	4.189	3.85	108.0
		80	0.843	14.314	1.1175	4.189	3.76	137.0
		100	1.031	13.938	1.0596	4.189	3.65	165.0
		120	1.218	13.564	1.0035	4.189	3.56	193.0
		140	1.437	13.126	0.9397	4.189	3.44	224.0
		160	1.562	12.876	0.9043	4.189	3.37	241.0
18	18.0	10	0.250	17.500	1.6703	4.712	4.59	47.4
		20	.312	17.376	1.6468	4.712	4.55	59.0
		30	.437	17.126	1.5993	4.712	4.49	82.0
		40	.562	16.876	1.5533	4.712	4.42	105.0
		60	.718	16.564	1.4964	4.712	4.34	133.0
		80	0.937	16.126	1.4183	4.712	4.23	171.0
		100	1.156	15.688	1.3423	4.712	4.11	208.0
		120	1.343	15.314	1.2791	4.712	4.02	239.0
		140	1.562	14.876	1.2070	4.712	3.90	275.0
		160	1.750	14.500	1.1467	4.712	3.80	304.0
20	20.0	10	0.250	19.500	2.0740	5.236	5.11	52.8
		20	.375	19.250	2.0211	5.236	5.05	78.6
		30	.500	19.000	1.9689	5.236	4.98	105.0
		40	.593	18.814	1.9305	5.236	4.94	123.0
		60	0.812	18.376	1.8417	5.236	4.81	167.0
		80	1.031	17.938	1.7550	5.236	4.70	209.0
		100	1.250	17.500	1.6703	5.236	4.59	251.0
		120	1.500	17.000	1.5762	5.236	4.46	297.0
		140	1.750	16.500	1.4849	5.236	4.32	342.0
		160	1.937	16.126	1.4183	5.236	4.22	374.0
24	24.0	10	0.250	23.500	3.012	6.283	6.16	63.5
		20	.375	23.250	2.948	6.283	6.09	94.7
		30	.562	22.876	2.854	6.283	6.00	141.0
		40	.687	22.626	2.792	6.283	5.94	171.0
		60	0.937	22.126	2.670	6.283	5.80	231.0
		80	1.218	21.564	2.536	6.283	5.65	297.0
		100	1.500	21.000	2.405	6.283	5.50	361.0
		120	1.750	20.500	2.292	6.283	5.37	416.0
		140	2.062	19.876	2.155	6.283	5.21	484.0
		160	2.312	19.376	2.048	6.283	5.08	536.0
30	30.0	10	0.312	29.376	4.707	0.7854	7.69	99.0
		20	.500	29.000	4.587	.7854	7.60	158.0
		30	0.625	28.750	4.508	0.7854	7.53	197.0

* Based on ASA Standard B36.10.
† Designates former "standard" sizes.
‡ Former "extra strong."

Note: The old designation of standard weight corresponds exactly to the Schedule 40 wall thickness in pipe sizes up to 10 in. The extra-strong wall designation corresponds to Schedule 80 in all sizes to 8 in. and to Schedule 60 in the 10-in. size. The old double-extra-strong wall designation has no correspondence with the schedule numbering system. However, pipe manufacturers still use the old designations as well as the new schedule system. So, it is still possible to obtain pipe with the double-extra-strong wall (XXS) in most sizes up to 8 in.

Not all pipe sizes are made regularly in all schedule numbers. The smaller sizes of steel and iron pipe are made only in Schedules 40, 80, and 160, plus the old XXS type. The schedule numbers above 80 are less common on sizes above 12 in.

A further exception should be noted regarding wrought-iron pipe. Outside diameters are the same as steel pipe, and weights per foot are the same as the corresponding sizes of standard, extra-strong, and double-extra-strong steel pipe. However, inside diameters of wrought-iron pipe are slightly smaller to compensate for the difference in density.

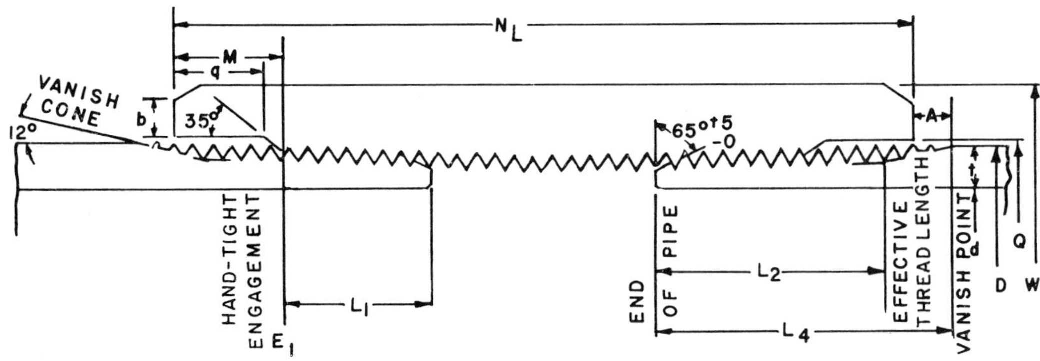

Fig. 1-1 Line pipe hand-tight make-up. (See Tables 1-10 and 1-11.)

Table 1-10 Coupling Dimensions and Tolerances

(All dimensions in inches)
(See Fig. 1-1 and Table 1-11)

(Reproduced by permission from API Std. 5L: Specification for Line Pipe, 20th ed., March 1963)

Size: nominal	Outside diameter of coupling* W	Length N_L	Diameter of recess Q	Width of bearing face b
¾	1.313	2⅛	1.113	1/16
1	1.576	2⅝	1.378	3/32
1¼	2.054	2¾	1.723	3/32
1½	2.200	2¾	1.963	3/32
2	2.875	2⅞	2.469	⅛
2½	3.375	4⅛	2.969	3/16
3	4.000	4¼	3.594	3/16
3½	4.625	4⅜	4.094	3/16
4	5.200	4½	4.594	¼

* Tolerance ± 1 per cent.

Table 1-12 Line Pipe Thread Height Dimensions

(All dimensions in inches)
(See Fig. 1-2)

(Reproduced by permission from API Std. 5B: Specification for Threading, Gaging, and Thread Inspection of Casing, Tubing, and Line Pipe Threads, 5th ed., March 1963)

	Thread element	14 Threads per inch p=0.0714	11½ Threads per inch p=0.0870	8 Threads per inch p=0.1250
$H=$	0.866p	0.0619	0.0753	0.1082
$h_s=h_n=$	.760p	.0543	.0661	.0950
$f_{rs}=f_{rn}=$	.033p	.0024	.0029	.0041
$f_{cs}=f_{cn}=$	0.073p	0.0052	0.0063	0.0091

Table 1-11 Line Pipe Thread Dimensions

(All dimensions in inches)
(See Fig. 1-1 and Table 1-10)
(Included taper on diameter, all sizes, 0.0625 in. per in.)

(Reproduced by permission from API Std. 5B: Specification for Threading, Gaging, and Thread Inspection of Casing, Tubing, and Line Pipe Threads, 5th ed., March 1963)

Size: nominal	No. of threads per inch	Length: end of pipe to hand-tight plane L_1	Length: effective threads L_2	Total length: end of pipe to vanish point L_4	Pitch diameter at hand-tight plane E_1	End of pipe to center of coupling, power-tight make-up	Length: face of coupling to hand-tight plane M	Hand-tight standoff, thread turns A	Depth of recess q
¾	14	0.339	0.5457	0.7935	0.98887	0.2690	0.2403	3	0.1516
1	11½	.400	.6828	0.9845	1.23863	.3280	.3235	3	.2241
1¼	11½	.420	.7068	1.0085	1.58338	.3665	.3275	3	.2279
1½	11½	.420	.7235	1.0252	1.82234	.3498	.3442	3	.2439
2	11½	.436	0.7565	1.0582	2.29627	.3793	.3611	3	.2379
2½	8	.682	1.1375	1.5712	2.76216	.4913	.6392	2	.4915
3	8	.766	1.2000	1.6337	3.38850	.4913	.6177	2	.4710
3½	8	.821	1.2500	1.6837	3.88881	.5038	.6127	2	.4662
4	8	0.844	1.3000	1.7337	4.38713	0.5163	0.6397	2	0.4920

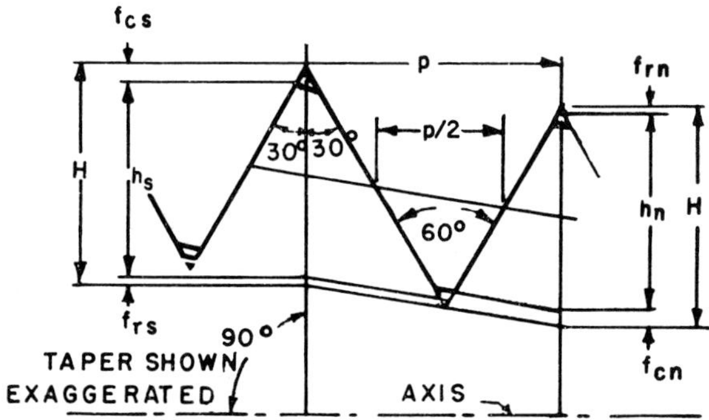

Fig. 1-2 Line pipe thread form. (See Table 1-12.)

Table 1-13 150 Lb Malleable-Iron Screwed Fittings

(all dimensions in inches)

Extracted from American Standard Malleable-Iron Screwed Fittings, 150 Lb (ASA B16.3-1951), with the permission of the publisher, ASME, 345 E. 47th St., New York 17, N. Y.

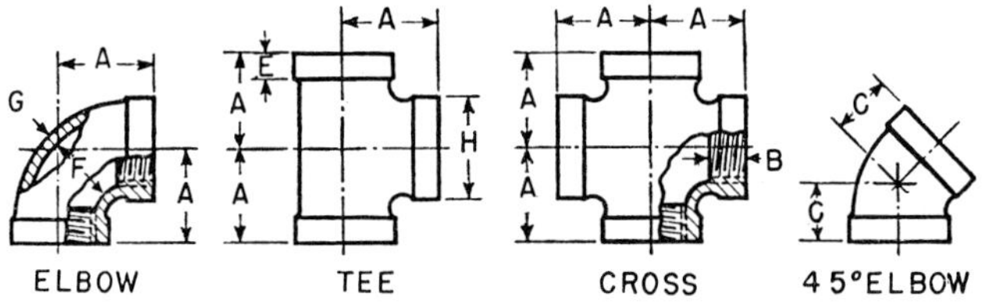

Nominal pipe size	Center-to-end, elbow, tee, and cross A	Center-to-end, 45-deg elbow C	Length of thread, min B	Width of band, min E	Inside diameter of fitting F		Metal thickness* G	Outside diameter of band, min H
					Min	Max		
¾	1.31	0.98	0.50	0.273	1.050	1.107	0.120	1.458
1	1.50	1.12	.58	.302	1.315	1.385	.134	1.771
1¼	1.75	1.29	.67	.341	1.660	1.730	.145	2.153
1½	1.94	1.43	.70	.368	1.900	1.970	.155	2.427
2	2.25	1.68	.75	.422	2.375	2.445	.173	2.963
2½	2.70	1.95	.92	.478	2.875	2.975	.210	3.589
3	3.08	2.17	0.98	.548	3.500	3.600	.231	4.285
3½	3.42	2.39	1.03	.604	4.000	4.100	.248	4.843
4	3.79	2.61	1.08	0.661	4.500	4.600	0.265	5.401

* Patterns shall be designed to produce castings of metal thicknesses given in the table. Metal thickness at no point shall be less than 90 per cent of the thickness given in the table.

Table 1-14 Butt-Welding Steel Fittings

(all dimensions in inches)

Extracted from American Standard Steel Butt-Welding Fittings (ASA B16.9-1958), with the permission of the publisher, ASME, 345 E. 47th St., New York 17, N. Y.

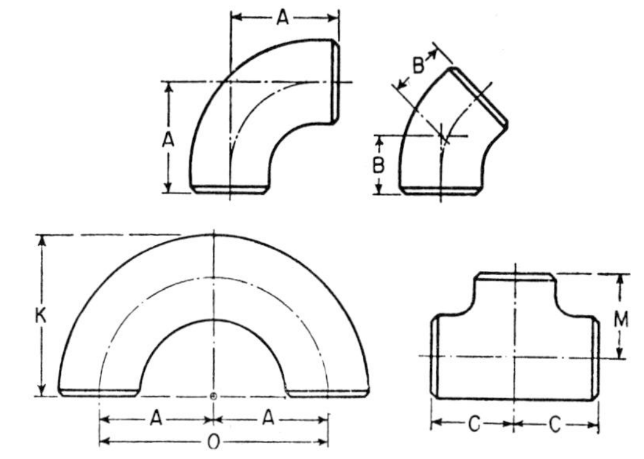

Nominal pipe size	Outside diameter at bevel	Center-to-end				Center-to-center O	Back to face K
		90-deg elbows A	45-deg elbows B	Run C	Outlet M		
1	1.315	1½	⅞	1½	1½	3	2³⁄₁₆
1¼	1.660	1⅞	1	1⅞	1⅞	3¾	2¾
1½	1.900	2¼	1⅛	2¼	2¼	4½	3¼
2	2.375	3	1⅜	2½	2½	6	4³⁄₁₆
2½	2.875	3¾	1¾	3	3	7½	5³⁄₁₆
3	3.500	4½	2	3⅜	3⅜	9	6¼
3½	4.000	5¼	2¼	3¾	3¾	10½	7¼
4	4.500	6	2½	4⅛	4⅛	12	8¼
5	5.563	7½	3⅛	4⅞	4⅞	15	10⁵⁄₁₆
6	6.625	9	3¾	5⅝	5⅝	18	12⁵⁄₁₆
8	8.625	12	5	7	7	24	16⁵⁄₁₆
10	10.750	15	6¼	8½	8½	30	20⅜
12	12.750	18	7½	10	10	36	24⅜
14	14.000	21	8¾	11	11	42	28
16	16.000	24	10	12	12	48	32
18	18.000	27	11¼	13½	13½	54	36
20	20.000	30	12½	15	15	60	40
22	22.000	33	13½	16½	16½	66	44
24	24.000	36	15	17	17	72	48

Table 1-15 Butt-Welding Steel Reducing Tees

(all dimensions in inches)

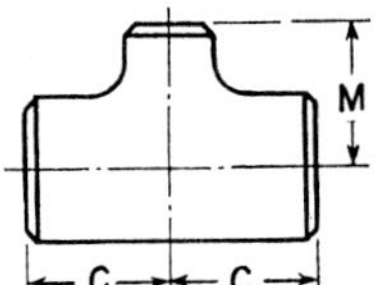

Extracted from American Standard Steel Butt-Welding Fittings (ASA B16.9-1958), with the permission of the publisher, ASME, 345 E. 47th St., New York 17, N. Y.

Nominal pipe size	Outside diameter at bevel		Center-to-end	
	Run	Outlet	Run C	Outlet M
1 × 1 × ¾	1.315	1.050	1½	1½
1 × 1 × ½	1.315	0.840	1½	1½
1¼ × 1¼ × 1	1.660	1.315	1⅞	1⅞
1¼ × 1¼ × ¾	1.660	1.050	1⅞	1⅞
1¼ × 1¼ × ½	1.660	0.840	1⅞	1⅞
1½ × 1½ × 1¼	1.900	1.660	2¼	2¼
1½ × 1½ × 1	1.900	1.315	2¼	2¼
1½ × 1½ × ¾	1.900	1.050	2¼	2¼
1½ × 1½ × ½	1.900	0.840	2¼	2¼
2 × 2 × 1½	2.375	1.900	2½	2⅜
2 × 2 × 1¼	2.375	1.660	2½	2¼
2 × 2 × 1	2.375	1.315	2½	2
2 × 2 × ¾	2.375	1.050	2½	1¾
2½ × 2½ × 2	2.875	2.375	3	2¾
2½ × 2½ × 1½	2.875	1.900	3	2⅝
2½ × 2½ × 1¼	2.875	1.660	3	2½
2½ × 2½ × 1	2.875	1.315	3	2¼
3 × 3 × 2½	3.500	2.875	3⅜	3¼
3 × 3 × 2	3.500	2.375	3⅜	3
3 × 3 × 1½	3.500	1.900	3⅜	2⅞
3 × 3 × 1¼	3.500	1.660	3⅜	2¾
3½ × 3½ × 3	4.000	3.500	3¾	3⅝
3½ × 3½ × 2½	4.000	2.875	3¾	3½
3½ × 3½ × 2	4.000	2.375	3¾	3¼
3½ × 3½ × 1½	4.000	1.900	3¾	3⅛
4 × 4 × 3½	4.500	4.000	4⅛	4
4 × 4 × 3	4.500	3.500	4⅛	3⅞
4 × 4 × 2½	4.500	2.875	4⅛	3¾
4 × 4 × 2	4.500	2.375	4⅛	3½
4 × 4 × 1½	4.500	1.900	4⅛	3⅜
5 × 5 × 4	5.563	4.500	4⅞	4⅝
5 × 5 × 3½	5.563	4.000	4⅞	4½
5 × 5 × 3	5.563	3.500	4⅞	4⅜
5 × 5 × 2½	5.563	2.875	4⅞	4¼
5 × 5 × 2	5.563	2.375	4⅞	4⅛
6 × 6 × 5	6.625	5.563	5⅝	5⅜
6 × 6 × 4	6.625	4.500	5⅝	5⅛
6 × 6 × 3½	6.625	4.000	5⅝	5
6 × 6 × 3	6.625	3.500	5⅝	4⅞
6 × 6 × 2½	6.625	2.875	5⅝	4¾

Nominal pipe size	Outside diameter at bevel		Center-to-end	
	Run	Outlet	Run C	Outlet* M
8 × 8 × 6	8.625	6.625	7	7⅝
8 × 8 × 5	8.625	5.563	7	6⅜
8 × 8 × 4	8.625	4.500	7	6⅛
8 × 8 × 3½	8.625	4.000	7	6
10 × 10 × 8	10.750	8.625	8½	8
10 × 10 × 6	10.750	6.625	8½	7⅝
10 × 10 × 5	10.750	5.563	8½	7½
10 × 10 × 4	10.750	4.500	8½	7¼
12 × 12 × 10	12.750	10.750	10	9½
12 × 12 × 8	12.750	8.625	10	9
12 × 12 × 6	12.750	6.625	10	8⅝
12 × 12 × 5	12.750	5.563	10	8½
14 × 14 × 12	14.000	12.750	11	10⅝
14 × 14 × 10	14.000	10.750	11	10⅛
14 × 14 × 8	14.000	8.625	11	9¾
14 × 14 × 6	14.000	6.625	11	9⅜
16 × 16 × 14	16.000	14.000	12	12
16 × 16 × 12	16.000	12.750	12	11⅝
16 × 16 × 10	16.000	10.750	12	11⅛
16 × 16 × 8	16.000	8.625	12	10¾
16 × 16 × 6	16.000	6.625	12	10⅜
18 × 18 × 16	18.000	16.000	13½	13
18 × 18 × 14	18.000	14.000	13½	13
18 × 18 × 12	18.000	12.750	13½	12⅝
18 × 18 × 10	18.000	10.750	13½	12⅛
18 × 18 × 8	18.000	8.625	13½	11¾
20 × 20 × 18	20.000	18.000	15	14½
20 × 20 × 16	20.000	16.000	15	14
20 × 20 × 14	20.000	14.000	15	14
20 × 20 × 12	20.000	12.750	15	13⅝
20 × 20 × 10	20.000	10.750	15	13⅛
20 × 20 × 8	20.000	8.625	15	12¾
22 × 22 × 20	22.000	20.000	16½	16
22 × 22 × 18	22.000	18.000	16½	15½
22 × 22 × 16	22.000	16.000	16½	15
22 × 22 × 14	22.000	14.000	16½	15
22 × 22 × 12	22.000	12.750	16½	14⅝
22 × 22 × 10	22.000	10.750	16½	14⅛
24 × 24 × 22	24.000	22.000	17	17
24 × 24 × 20	24.000	20.000	17	17
24 × 24 × 18	24.000	18.000	17	16½
24 × 24 × 16	24.000	16.000	17	16
24 × 24 × 14	24.000	14.000	17	16
24 × 24 × 12	24.000	12.750	17	15⅝
24 × 24 × 10	24.000	10.750	17	15⅛

* Outlet dimension "M" for run sizes 14 in. and larger is recommended but not mandatory.

Table 1-16 Butt-Welding Steel Reducers

(All dimensions in inches)

Extracted from American Standard Steel Butt-Welding Fittings (ASA B16.9-1958), with the permission of the publisher, ASME, 345 E. 47th St., New York 17, N. Y.

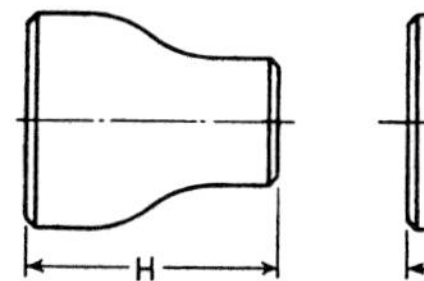

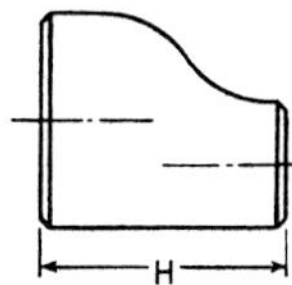

Nominal pipe size	Outside diameter at bevel Large end	Outside diameter at bevel Small end	End-to-end H	Nominal pipe size	Outside diameter at bevel Large end	Outside diameter at bevel Small end	End-to-end H
1 × ¾	1.315	1.050	2	6 × 5	6.625	5.536	5½
1 × ½	1.315	0.840	2	6 × 4	6.625	4.500	5½
1¼ × 1	1.660	1.315	2	6 × 3½	6.625	4.000	5½
1¼ × ¾	1.660	1.050	2	6 × 3	6.625	3.500	5½
1¼ × ½	1.660	0.840	2	6 × 2½	6.625	2.875	5½
1½ × 1¼	1.900	1.660	2½	8 × 6	8.625	6.625	6
1½ × 1	1.900	1.315	2½	8 × 5	8.625	5.563	6
1½ × ¾	1.900	1.050	2½	8 × 4	8.625	4.500	6
1½ × ½	1.900	0.840	2½	8 × 3½	8.625	4.000	6
2 × 1½	2.375	1.900	3	10 × 8	10.750	8.625	7
2 × 1¼	2.375	1.660	3	10 × 6	10.750	6.625	7
2 × 1	2.375	1.315	3	10 × 5	10.750	5.563	7
2 × ¾	2.375	1.050	3	10 × 4	10.750	4.500	7
2½ × 2	2.875	2.375	3½	12 × 10	12.750	10.750	8
2½ × 1½	2.875	1.900	3½	12 × 8	12.750	8.625	8
2½ × 1¼	2.875	1.660	3½	12 × 6	12.750	6.625	8
2½ × 1	2.875	1.315	3½	12 × 5	12.750	5.563	8
3 × 2½	3.500	2.875	3½	14 × 12	14.000	12.750	13
3 × 2	3.500	2.375	3½	14 × 10	14.000	10.750	13
3 × 1½	3.500	1.900	3½	14 × 8	14.000	8.625	13
3 × 1¼	3.500	1.660	3½	14 × 6	14.000	6.625	13

Nominal pipe size	Outside diameter at bevel Large end	Outside diameter at bevel Small end	End-to-end H	Nominal pipe size	Outside diameter at bevel Large end	Outside diameter at bevel Small end	End-to-end H
3½ × 3	4.000	3.500	4	16 × 14	16.000	14.000	14
3½ × 2½	4.000	2.875	4	16 × 12	16.000	12.750	14
3½ × 2	4.000	2.375	4	16 × 10	16.000	10.750	14
3½ × 1½	4.000	1.900	4	16 × 8	16.000	8.625	14
3½ × 1¼	4.000	1.660	4	18 × 16	18.000	16.000	15
4 × 3½	4.500	4.000	4	18 × 14	18.000	14.000	15
4 × 3	4.500	3.500	4	18 × 12	18.000	12.750	15
4 × 2½	4.500	2.875	4	18 × 10	18.000	10.750	15
4 × 2	4.500	2.375	4	20 × 18	20.000	18.000	20
4 × 1½	4.500	1.900	4	20 × 16	20.000	16.000	20
5 × 4	5.563	4.500	5	20 × 14	20.000	14.000	20
5 × 3½	5.563	4.000	5	20 × 12	20.000	12.750	20
5 × 3	5.563	3.500	5	22 × 20	22.000	20.000	20
5 × 2½	5.563	2.875	5	22 × 18	22.000	18.000	20
5 × 2	5.563	2.375	5	22 × 16	22.000	16.000	20
				22 × 14	22.000	14.000	20
				24 × 22	24.000	22.000	20
				24 × 20	24.000	20.000	20
				24 × 18	24.000	18.000	20
				24 × 16	24.000	16.000	20

Table 1-17 Standard Twist Drill Sizes

(Designations are in fractions of an inch, in Standard Twist Drill letters, or in Standard Twist Drill numbers, the latter being approximately the same as Stubs Steel Wire Gage numbers.)

Designation	Diam, in.	Area, sq in.	Designation	Diam, in.	Area, sq in.	Designation	Diam, in.	Area, sq in.
½	0.5000	0.1963	19/64	0.2969	0.06922	8	0.199	0.03110
31/64	.4844	.1843	M	.295	.06835	9	.196	.03017
15/32	.4688	.1726	L	.29	.06605	10	.1935	.02940
29/64	.4531	.1613	9/32	.2813	.06213	11	.191	.02865
7/16	.4375	.1503	K	.281	.06202	12	.189	.02806
27/64	.4219	.1398	J	.277	.06026	3/16	.1875	.02761
Z	.413	.1340	I	.272	.05811	13	.185	.02688
13/32	.4063	.1296	H	.266	.05557	14	.182	.02602
Y	.404	.1282	17/64	.2656	.05542	15	.1800	.02545
X	.397	.1238	G	.261	.05350	16	.1770	.02461
25/64	.3906	.1198	F	.257	.05187	17	.1730	.02351
W	.386	.1170	E-¼	.2500	.04909	11/64	.1719	.02320
V	.377	.1116	D	.246	.04753	18	.1695	.02256
⅜	.375	.1104	C	.242	.04600	19	.1660	.02164
U	.368	.1064	B	.238	.04449	20	.1610	.02036
23/64	.3594	.1014	15/64	.2344	.04314	21	.1590	.01986
T	.358	.1006	A	.234	.04301	22	.1570	.01936
S	.348	.09511	1	.228	.04083	5/32	.1563	.01917
11/32	.3438	.09281	2	.221	.03836	23	.1540	.01863
R	.339	.09026	7/32	.2188	.03758	24	.1520	.01815
Q	.332	.08657	3	.213	.03563	25	.1495	.01755
21/64	.3281	.08456	4	.209	.03431	26	.1470	.01679
P	.323	.08194	5	.2055	.03317	27	.1440	.01629
O	.316	.07843	6	.204	.03269	9/64	.1406	.01553
5/16	.3125	.07670	13/64	.2031	.03241	28	.1405	.01549
N	.302	.07163	7	.201	.03173	29	.1360	.01453

Table 1-17 Standard Twist Drill Sizes (Continued)

Designation	Diam, in.	Area, sq in.	Designation	Diam, in.	Area, sq in.	Designation	Diam, in.	Area, sq in.
30	.1285	.01296	47	.0785	.00484	64	.036	.001018
⅛	.1250	.01227	5/64	.0781	.00479	65	.035	.000962
31	.1200	.01131	48	.0760	.00454	66	.033	.000855
32	.1160	.01057	49	.0730	.00419	67	.032	.000804
33	.1130	.01003	50	.0700	.00385	1/32	.0313	.000765
34	.1110	.00968	51	.0670	.00353	68	.031	.000755
35	.1100	.00950	52	.0635	.00317	69	.0292	.000670
7/64	.1094	.00940	1/16	.0625	.00307	70	.028	.000616
36	.1065	.00891	53	.0595	.00278	71	.026	.000531
37	.1040	.00849	54	.0550	.00238	72	.025	.000491
38	.1015	.00809	55	.0520	.00212	73	.024	.000452
39	.0995	.00778	3/64	.0473	.00173	74	.0225	.000398
40	.0980	.00754	56	.0465	.001698	75	.021	.000346
41	.0960	.00724	57	.0430	.001452	76	.020	.000314
3/32	.0938	.00690	58	.0420	.001385	77	.018	.000254
42	.0935	.00687	59	.0410	.001320	78	.016	.000201
43	.0890	.00622	60	.0400	.001257	1/64	.0156	.000191
44	.0860	.00581	61	.039	.001195	79	.0145	.000165
45	.0820	.00528	62	.038	.001134	80	0.0135	0.000143
46	0.0810	0.00515	63	0.037	0.001075			

Table 1-18 Wire and Sheet Metal Gages[4]

(Diameters and thicknesses in decimal parts of an inch)

Gage no.	American (A.W.G.) wire gage, or Brown & Sharpe (for non-ferrous sheet and wire)	Steel wire gage or Washburn & Moen or Roebling (for steel wire)	Birmingham wire gage (B.W.G.) or Stubs iron wire (for steel rods or sheets)	Stubs steel wire gage	British Imperial standard wire gage (S.W.G.)
0000000	...	0.4900	...	...	0.500
000000	...	.4615	...	...	.464
00000	...	.4305	...	...	.432
0000	0.460	.3938	0.454	...	.400
000	.410	.3625	.425	...	.372
00	.365	.3310	.380	...	.348
0	.325	.3065	.340	...	.324
1	.289	.2830	.300	0.227	.300
2	.258	.2625	.284	.219	.276
3	.229	.2437	.259	.212	.252
4	.204	.2253	.238	.207	.232
5	.182	.2070	.220	.204	.212
6	.162	.1920	.203	.201	.192
7	.144	.1770	.180	.199	.176
8	.128	.1620	.165	.197	.160
9	.114	.1483	.148	.194	.144
10	.102	.1350	.134	.191	.128
11	.091	.1205	.120	.188	.116
12	.081	.1055	.109	.185	.104
13	.072	.0915	.095	.182	.092
14	.064	.0800	.083	.180	.080
15	.057	.0720	.072	.178	.072
16	.051	.0625	.065	.175	.064
17	.045	.0540	.058	.172	.056
18	.040	.0475	.049	.168	.048
19	.036	.0410	.042	.164	.040
20	.032	.0348	.035	.161	.036
21	0.0285	0.0317	0.032	0.157	0.032
22	0.0253	0.0286	0.028	0.155	0.028
23	.0226	.0258	.025	.153	.024
24	.0201	.0230	.022	.151	.022
25	.0179	.0204	.020	.148	.020
26	.0159	.0181	.018	.146	.018
27	.0142	.0173	.016	.143	.0164
28	.0126	.0162	.014	.139	.0148
29	.0113	.0150	.013	.134	.0136
30	.0100	.0140	.012	.127	.0124
31	.0089	.0132	.010	.120	.0116
32	.0080	.0128	.009	.115	.0108
33	.0071	.0118	.008	.112	.0100
34	.0063	.0104	.007	.110	.0092
35	.0056	.0095	.005	.108	.0084
36	.0050	.0090	0.004	.106	.0076
37	.0045	.0085	...	.103	.0068
38	.0040	.0080	...	.101	.0060
39	.0035	.0075	...	.099	.0052
40	0.0031	.0070	...	.097	.0048
41	...	.0066	...	.095	.0044
42	...	.0062	...	.092	.0040
43	...	.0060	...	.088	.0036
44	...	.0058	...	.085	.0032
45	...	.0055	...	.081	.0028
46	...	.0052	...	.079	.0024
47	...	.0050	...	.077	.0020
48	...	.0048	...	.075	.0016
49	...	.0046	...	.072	.0012
50	...	0.0044	...	0.069	0.0010

Table 1-19 Manufacturers' Standard Gage for Uncoated Carbon Steel Sheets[5]

Gage No.	Thickness,* in.	Weight, psf	Thickness† range, in.	Thickness‡ tolerance, ± in.
7	0.1793	7.5000	0.1868–0.1719	0.012
8	.1644	6.8750	.1718– .1570	.012
9	.1495	6.2500	.1569– .1420	.012
10	.1345	5.6250	.1419– .1271	.012
11	.1196	5.0000	.1270– .1121	.012
12	.1046	4.3750	.1120– .0972	.012
13	.0897	3.7500	.0971– .0822	.009
14	.0747	3.1250	.0821– .0710	.007
15	.0673	2.8125	.0709– .0636	.006
16	.0598	2.5000	.0635– .0568	.006
17	.0538	2.2500	.0567– .0509	.006
18	.0478	2.0000	.0508– .0449	.005
19	.0418	1.7500	.0448– .0389	.005
20	.0359	1.5000	.0388– .0344	.004
21	.0329	1.3750	.0343– .0314	.004
22	.0299	1.2500	.0313– .0284	.004
23	.0269	1.1250	.0283– .0255	.004
24	.0239	1.0000	.0254– .0225	.003
25	.0209	0.87500	.0224– .0195	.003
26	.0179	.75000	.0194– .0172	.002
27	.0164	.68750	.0171– .0157	.002
28	.0149	.62500	.0156– .0142	.002
29	.0135	.56250	.0141– .0128	.002
30	0.0120	0.50000	0.0127–0.0113	0.002

* Gage thickness equivalents are based on 41.820 psf per inch thickness.

† Upper and lower limits of thicknesses that may be ordered under the corresponding gage numbers. **These ranges are not tolerances.**

‡ Cold rolled sheets in coils and cut lengths over 12 in. wide. Values given are according to the following schedule for widths:

Width, in.	Thickness, in.
over 90	over 0.0971
to 80	0.0255– .0971
to 60	.0142– .0254
to 48	0.0113–0.0141

Note: Since thickness tolerances increase with width, sheets of progressively smaller widths are held to progressively smaller tolerances.

REFERENCES

1. Perry, John H., ed. *Chemical Engineers' Handbook*, 3d ed. New York, McGraw-Hill, 1950.
2. Mantell, Charles L., ed. *Engineering Materials Handbook.* New York, McGraw-Hill, 1958.
3. U. S. Dept. of Ag. Forest Products Lab. *Dry Kiln Operator's Manual.* Madison, Wisc., 1959.
4. Marks, Lionel S., ed. *Mechanical Engineers' Handbook,* 6th ed. New York, McGraw-Hill, 1958.
5. Am. Iron and Steel Inst. *Steel Products Manual—Carbon Steel Sheets*, New York, n.d.
6. *Steel.* 134: 120e, 120g, 125+, May 10, 1954.

Chapter 2

Properties of Gases, Air, and Water

by Louis Shnidman, C. E. Farmer, Robert J. Gustafson, and F. E. Vandaveer

Table 1-20 Thermodynamic Properties of Saturated Methane[1]

Temp, °F t	Pressure, lb/sq in. abs P	Fugacity / Pressure f/P	Volume: Liquid cu ft/lb v	Volume: Vapor cu ft/lb V	Enthalpy: Liquid Btu/lb h	Enthalpy: Latent Btu/lb L	Enthalpy: Vapor Btu/lb H	Entropy: Liquid Btu/lb-°R s	Entropy: Vapor Btu/lb-°R S
−280	4.90	0.991	0.03635	24.04	0	228.2	228.2	0	1.2699
−275	6.47	.989	.03666	18.64	4.0	226.3	230.3	0.0209	1.2461
−270	8.44	.986	.03698	14.61	8.2	224.1	232.3	.0423	1.2236
−265	10.82	.983	.03732	11.63	12.3	222.1	234.4	.0623	1.2030
−260	13.80	.979	.03766	9.31	16.6	219.8	236.4	.0823	1.1830
−255	17.40	.976	.03801	7.52	20.8	217.6	238.4	.1013	1.1643
−250	21.71	.972	.03839	6.13	25.0	215.3	240.3	.1201	1.1468
−245	26.60	.967	.03876	5.09	29.1	213.1	242.2	.1388	1.1313
−240	32.4	.960	.03915	4.24	33.3	210.6	243.9	.1578	1.1164
−235	39.0	.954	.03956	3.57	37.5	208.1	245.6	.1766	1.1027
−230	46.4	.948	.03999	3.04	42.0	205.3	247.3	.1962	1.0900
−225	54.8	.941	.04045	2.60	46.2	202.6	248.8	.2147	1.0779
−220	64.5	.933	.04092	2.23	50.6	199.6	250.2	.2333	1.0660
−215	75.2	.925	.04141	1.94	55.1	196.4	251.5	.2522	1.0548
−210	87.6	.916	.04193	1.67	59.5	193.3	252.8	.2693	1.0434
−205	101.0	.907	.04248	1.46	64.1	189.7	253.8	.2879	1.0327
−200	115.7	.896	.04306	1.281	68.8	186.0	254.8	.3062	1.0224
−195	132.0	.884	.04366	1.125	73.5	182.1	255.6	.3242	1.0121
−190	150.0	.873	.04431	0.990	78.2	178.0	256.2	.3419	1.0019
−185	170.0	.860	.04501	.874	82.9	173.8	256.7	.3588	0.9915
−180	191.5	.848	.04575	.773	87.8	169.2	257.0	.3767	.9816
−175	214.9	.835	.04656	.687	92.9	164.3	257.2	.3947	.9718
−170	240.0	.822	.04745	.610	98.0	159.2	257.2	.4127	.9622
−165	267.0	.810	.04839	.543	103.3	153.7	257.0	.4308	.9523
−160	297.0	.798	.04944	.483	108.7	147.9	256.5	.4476	.9411
−155	329	.785	.05062	.430	114.4	141.2	255.6	.4655	.9289
−150	364	.772	.05197	.381	120.3	134.1	254.5	.4839	.9169
−145	401	.759	.05347	.339	126.6	126.6	253.0	.5017	.9040
−140	440	.746	.05524	.3008	133.2	118.0	251.2	.5214	.8905
−135	482	.733	.05735	.2654	140.4	108.3	248.7	.5430	.8765
−130	527	.719	.05999	.2318	148.1	97.8	245.9	.5656	.8622
−125	575	.705	.06362	.1986	157.7	83.4	241.1	.5929	.8421
−120	627	.690	.06961	.1613	171.8	59.6	231.4	.6329	.8083
−115.8	673	0.677	0.0983	0.0983	203.4	0	203.4	0.7232	0.7232

Table 1-21 Thermodynamic Properties of Saturated Ethane[2]

			Volume		Heat content above 0°R			Entropy above 0°R	
Temp, °F t	Pressure, lb/sq in. abs P	Fugacity Pressure f/P	Liquid cu ft/lb v	Vapor cu ft/lb V	Liquid Btu/lb h	Latent Btu/lb L	Vapor Btu/lb H	Liquid Btu/lb-°F s	Vapor Btu/lb-°F S
−220	0.27	0.999	0.02669	310.5	117.6	236.3	353.9	0.8249	1.8107
−210	.50	.998	.02694	179.2	123.1	233.7	356.8	.8474	1.7833
−200	0.85	.996	.02720	107.8	128.7	231.0	359.7	.8691	1.7587
−190	1.40	.994	.02746	68.43	134.3	228.3	362.6	.8901	1.7366
−180	2.20	.992	.02774	44.90	139.9	225.6	365.5	.9133	1.7201
−170	3.36	.989	.02802	30.32	145.5	222.8	368.3	.9332	1.7025
−160	4.97	.985	.02831	21.11	151.1	220.1	371.2	.9523	1.6865
−150	7.14	.980	.02861	15.10	156.8	217.1	373.9	.9710	1.6720
−140	9.97	.973	.02893	11.05	162.3	214.3	376.6	.9891	1.6593
−135	11.66	.969	.02908	9.602	165.3	212.7	378.0	0.9981	1.6524
−130	13.68	.965	.02923	8.282	168.3	211.0	379.3	1.0065	1.6464
−127.55	14.696	.963	.02931	7.741	169.9	210.0	379.9	1.0111	1.6435
−125	15.86	.961	.02940	7.302	171.3	209.2	380.5	1.0155	1.6403
−120	18.33	.956	.02957	6.316	174.3	207.5	381.8	1.0240	1.6346
−115	21.08	.952	.02974	5.543	177.3	205.7	383.0	1.0324	1.6293
−110	24.17	.947	.02991	4.876	180.3	204.0	384.3	1.0407	1.6241
−105	27.57	.943	.03010	4.311	183.3	202.2	385.5	1.0489	1.6191
−100	31.32	.938	.03029	3.830	186.2	200.6	386.8	1.0570	1.6143
−95	35.48	.933	.03047	3.404	189.1	198.7	387.8	1.0651	1.6098
−90	39.98	.928	.03067	3.043	192.9	196.9	389.0	1.0731	1.6055
−85	44.96	.923	.03086	2.726	195.1	194.9	390.1	1.0811	1.6014
−80	50.34	.917	.03108	2.451	198.2	193.0	391.2	1.0890	1.5974
−75	56.26	.912	.03129	2.207	201.3	190.9	392.2	1.0969	1.5934
−70	62.63	.906	.03152	1.994	204.4	188.8	393.2	1.1047	1.5896
−65	69.55	.900	.03175	1.805	207.4	186.8	394.2	1.1126	1.5860
−60	77.02	.894	.03199	1.638	210.5	184.7	395.2	1.1205	1.5825
−55	85.06	.887	.03224	1.488	213.6	182.5	396.1	1.1284	1.5790
−50	93.76	.881	.03249	1.355	216.8	180.2	397.0	1.1362	1.5758
−45	103.0	.874	.03276	1.263	219.9	177.9	397.8	1.1441	1.5727
−40	113.1	.867	.03303	1.127	223.2	175.4	398.6	1.1519	1.5697
−35	123.6	.860	.03330	1.032	226.4	173.0	399.4	1.1596	1.5668
−30	135.0	.853	.03359	0.9452	229.7	170.5	400.2	1.1672	1.5639
−25	147.1	.846	.03390	.8675	233.0	168.0	401.0	1.1748	1.5611
−20	159.9	.838	.03422	.7983	236.3	165.4	401.7	1.1824	1.5583
−15	173.7	.830	.03457	.7347	239.7	162.6	402.3	1.1900	1.5556
−10	188.1	.823	.03494	.6775	243.2	159.7	402.9	1.1977	1.5530
−5	203.6	.815	.03530	.6242	246.8	156.6	403.4	1.2054	1.5504
0	219.7	.808	.03570	.5754	250.3	153.6	403.9	1.2132	1.5476
+5	237.0	.800	.03612	.5318	253.9	150.3	404.2	1.2210	1.5449
+10	254.9	.793	.03655	.4909	257.5	147.0	404.5	1.2289	1.5420
15	274.0	.785	.03702	.4541	261.3	143.4	404.7	1.2367	1.5392
20	294.0	.777	.03754	.4198	265.1	139.8	404.9	1.2445	1.5361
25	314.9	.769	.03809	.3886	268.9	136.1	405.0	1.2525	1.5332
30	337.1	.762	.03866	.3595	272.8	132.1	404.9	1.2604	1.5301
35	360.5	.754	.03928	.3320	276.9	127.9	404.8	1.2683	1.5267
40	385.0	.746	.03990	.3062	281.0	123.5	404.5	1.2762	1.5234
45	410.5	.738	.04065	.2822	285.3	118.8	404.1	1.2843	1.5198
50	437.5	.730	.04144	.2596	289.7	113.7	403.4	1.2926	1.5158
55	465.1	.723	.04243	.2375	294.4	108.1	402.5	1.3011	1.5113
60	494.2	.715	.04358	.2164	299.3	102.0	401.3	1.3100	1.5064
65	525.3	.707	.04485	.1968	304.3	95.3	399.6	1.3190	1.5005
70	558.3	.698	.04625	.1795	309.8	87.8	397.6	1.3284	1.4940
75	593.3	.690	.04789	.1622	315.9	78.8	394.7	1.3383	1.4852
80	630.7	.682	.05063	.1411	323.7	67.7	391.4	1.3505	1.4751
85	670.2	.674	.05488	.1184	334.5	49.1	383.6	1.3688	1.4588
90.1	716.0	0.665	0.0755	0.0755	362.0		362.0	1.4234	1.4234

Table 1-22 Vapor Pressure* and Critical Constants of Hydrocarbons[3]

(units: temp, °F; corresponding vapor pressure in atm)

Vapor pressure, atm	Methane CH_4	Ethane C_2H_6	Propylene C_3H_6	Propane C_3H_8	*iso*-Butane C_4H_{10}	*n*-Butane C_4H_{10}	*iso*-Pentane C_5H_{12}	*n*-Pentane C_5H_{12}	Benzene† C_6H_6	Hexane C_6H_{14}	Toluene C_7H_8	Heptane C_7H_{16}	*o*-Xylene C_8H_{10}	Octane C_8H_{18}
0.001	...	...	...	...	...	...	−120	−110	...	−70	−20	−35	19	−2
.003	−332.3‡	...	−198	−192	...	...	−100	−89	...	−48	5	−10	46	25
.01	−322.3‡	−225	−178	−171	−130	−114	−76	−64	...	−21	36	20	76	57
.03	−311.8‡	−212	−156	−148	−106	−89	−49	−36	...	8	69	52	119	91
.1	−298.2‡	−189	−128	−120	−73	−56	−14	0	69	49	113	95	167	137
0.3	−281.7‡	−163	−96	−87	−38	−19	26	42	114	94	163	143	219	189
1	−258.9	−128.0	−53.9	−43.8	10.9	31.1	82.2	96.9	176.2	155.7	231.1	209.2	292.0	258.2
3	−230	−85	−2	8	68	90	148	164	250	230	310	288	379	341
10	−190	−24	70	82	153	176	242	260	357	330	422	396	499	453
30	−140	50	158	173	256	280	361	378	482	459	560	529	...	...
100	−55	175	300	320	412	443	542	560	...	652	...	...	...	...
Critical point														
deg F	−116.3	90.1	196.5	206.3	275	306	369.5	386.5	551.3	455.0	609.1	512	675	565
atm	45.8	48.2	45.4	42.0	36	37.4	32.4	32.6	47.9	29.4	41.6	26.8	37	24.6
cu ft/lb	0.099	0.079	...	0.071	...	...	...	...	0.0526	0.0685	0.055	0.0685	...	0.0685

(For interpolation, use semilog paper and plot pressure on log scale.)

* Although values above critical point are meaningless, they could be used to make rough approximations of fugacity, density, enthalpy above critical temperature.

† Melting point 41.9 F at 0.046 atm.

‡ Vapor pressure above solid except at 0.3 atm. Data interpolated from Ziegler, W. T., Mullins, J. C., and Kirk, B. S. "Calculation of the Vapor Pressure and Heats of Vaporization and Sublimation of Liquids and Solids, Especially below One Atmosphere Pressure. III. Methane," Table XVIII. *Tech. Rept. 3 Eng. Expt. Sta., Georgia Inst. Tech.*, Aug. 31, 1962.

Table 1-23 Bubble and Dew Points of Eight Methane–Ethane Mixtures, °F[4]

Pressure, psia	Mixture (mole %): 97.50% CH_4 2.50% C_2H_6		92.50% CH_4 7.50% C_2H_6		85.16% CH_4 14.84% C_2H_6		70.00% CH_4 30.00% C_2H_6		50.02% CH_4 49.98% C_2H_6		30.02% CH_4 69.98% C_2H_6		14.98% CH_4 85.02% C_2H_6		5.00% CH_4 95.00% C_2H_6	
	B.P.	D.P.	B.P.	D.P.	B.P.	D.P.	B.P.	D.P.	B.P.	D.P.	B.P.	D.P.	B.P.	D.P.	B.P.	D.P.
100	−204.4	−174.2	−202.4	−148.2	−199.5	−128.0	−192.8	−102.2	−181.2	−80.9	−156.7	−65.0	−120.8	−55.2	−75.9	−49.3
150	−189.1	−162.7	−186.3	−136.6	−183.0	−113.9	−175.3	−87.0	−160.8	−62.2	−131.2	−44.9	−92.7	−33.9	−49.0	−26.6
200	−176.9	−153.6	−174.0	−127.2	−170.0	−103.6	−161.2	−75.0	−143.9	−48.4	−112.0	−29.0	−70.2	−17.2	−29.0	−9.3
250	−166.7	−146.3	−163.6	−119.7	−158.9	−95.3	−149.1	−65.1	−130.1	−37.1	−95.4	−16.2	−52.2	−3.1	−12.7	+5.1
300	−157.8	−140.0	−154.7	−113.4	−149.4	−88.7	−138.6	−56.9	−118.0	−27.5	−80.3	−5.2	−36.4	+8.8	+1.2	17.4
350	−149.9	−134.3	−146.4	−108.1	−140.8	−82.8	−129.0	−49.9	−107.0	−19.2	−67.2	+4.2	−22.3	18.8	13.7	28.3
400	−142.9	−129.0	−139.1	−103.7	−132.8	−77.9	−120.4	−44.1	−96.8	−11.8	−55.0	12.6	−9.6	27.8	24.8	38.1
450	−136.4	−124.1	−132.3	−99.8	−125.4	−73.4	−112.5	−39.0	−87.0	−5.7	−43.2	20.1	+1.9	36.1	35.0	46.9
500	−130.5	−119.6	−126.0	−96.2	−118.7	−69.7	−104.7	−34.3	−77.3	0.0	−32.1	26.9	12.1	43.7	44.2	55.0
550	−125.0	−115.4	−120.0	−93.0	−112.1	−66.4	−97.1	−30.2	−68.3	+5.2	−22.0	32.9	21.9	50.4	53.1	62.3
600	−119.6	−111.8	−114.2	−90.2	−105.7	−63.4	−90.0	−26.7	−59.8	9.9	−12.3	38.2	31.2	56.6	61.5	69.3
650	−114.7	−108.8	−108.5	−88.2	−99.5	−61.0	−83.1	−23.6	−51.4	13.8	−3.0	42.9	39.9	62.0	69.3	75.5
700	−109.5	−106.7	−103.0	−86.7	−93.7	−59.1	−76.5	−20.9	−43.2	17.4	+6.0	47.2	48.4	67.2	76.2	81.2
750	...	...	−97.2	−86.0	−87.9	−57.8	−70.0	−18.7	−35.3	20.0	15.0	50.7	56.9	71.2	...	...
800	...	...	...	...	−81.7	−57.0	−63.7	−17.0	−27.3	22.3	23.7	53.3	66.0	72.5	...	...
850	...	...	...	...	−75.0	−57.7	−56.9	−16.1	−19.2	24.1	32.7	54.3	...	...	...	...
900	...	...	...	...	...	...	−50.0	−16.3	−10.7	24.4	44.8	47.0	...	...	...	...
950	...	...	...	...	...	...	−41.6	−18.7	−1.5	22.3	...	...	...	...	...	...

D.P. = Dew Point. B.P. = Bubble Point: Temperature of either completion of vapor condensation or start of liquid boiling.

Table 1-24 Vapor Pressure and Critical Constants of Selected Gases[6]

(units: temp, °F; corresponding vapor pressure in mm Hg below 1 atm and in atm at 1 atm and above)

Gases	Pressure in mm Hg (760 mm Hg = 1 atm): 1 mm	5 mm	10 mm	20 mm	100 mm	200 mm	Pressure in atm: 1 atm	10 atm	30 atm	Critical point: t_c, °F	p_c, atm
Acetylene	−225.2	−207.4	−198.8	−189	−162.2	−148.5	−119.2	−26.9	40.6	96.3	62
Alcohol, ethyl	−24.3	10.4	27.9	46.4	94.8	119.1	173.1	305.2	397.4	469.6	63.1
Alcohol, methyl	−47.2	−13.5	2.8	21.2	70.2	94.6	148.5	280.4	367.7	464.0	78.7
Ammonia	−164.4	−143.5	−133.4	−122.4	−91.1	−70.6	−28.5	78.3	151	270.3	111.5
Carbon dioxide	−209.7	−191.9	−183.1	−173.9	−148.4	−135.4	−108.8	−39.1	22.5	88	73.0
Carbon disulfide	−100.8	−65.7	−48.5	−29.7	22.8	50.7	115.7	277.3	394.7	523.4	72.9
Carbon monoxide	−367.6	−359	−355	−350	−338.3	−330.3	−312.3	−257.8	−223.4	−217.7	34.6
Chlorine	−180.4	−160.1	−150.9	−135.9	−97.1	−76.4	−28.8	96.1	184.6	291.2	76.1
Ethylene	−270.9	−252.9	−243.8	−233.7	−205.2	−190.1	−154.7	−63	6.4	49.3	50.9
Helium	−457.3	−456.7	−456.3	−456	−454.5	−453.6	−451.5	...	...	−450.2	2.26
Hydrogen	−441.9	−439.4	−438.3	−436.7	−432.2	−429.3	−422.5	−403.2	...	−400	12.8
Hydrogen chloride	−239.4	−221.3	−212.1	−202	−173.2	−157.4	−120.6	−25.1	42.6	124.5	81.6
Hydrogen cyanide	−95.8	−67.5	−53.9	−39.5	0	22.5	78.6	216.9	308.8	362.3	50.0
Hydrogen sulfide	−209.7	−188.3	−177.3	−165.5	−132.9	−116.1	−76.7	31.3	107.4	212.5	88.9
Nitric oxide	−300.1	−293.1	−288.8	−283.5	−266.8	−260.1	−241.1	−197.1	−164.2	−135.2	64.6
Nitrogen	−375	−366.3	−362.4	−358.2	−345.5	−338.1	−320.4	−273.6	−234.9	−233	33.5
Nitrous oxide	−226.1	−208.1	−199.7	−191.2	−166.5	−154.5	−127.3	−41.3	24.3	97.7	71.7
Oxygen	−362.4	−352.1	−347.1	−341.5	−325.8	−317.2	−297.6	−243.8	−203.3	−182	49.7
Sulfur dioxide	−139.9	−117.4	−106.2	−93.5	−52.4	−31.7	14	131.9	216.7	315	77.7
Water	0.9	34.2	52	72	124.9	151.7	212	356.9	454.3	705.6	218

(For interpolation, use semilog paper and plot pressure on log scale.)

Table 1-25 Solubility of Selected Gases in Water

α is in lbs of gas soluble in 100 lbs of water when the total gas pressure is 30 in. Hg inclusive of the partial pressure of H_2O vapor.

β is in cu ft of gas soluble in one cu ft of water at 60 F when the total gas pressure is 30 in. Hg inclusive of the partial pressure of H_2O vapor.

Temperature		Air		Ammonia NH_3		Hydrogen sulfide H_2S		Sulfur dioxide SO_2		Carbon dioxide CO_2		Carbon monoxide CO		Methane CH_4		Acetylene C_2H_2		Ethylene C_2H_4		Ethane C_2H_6		Hydrogen H_2		Nitrogen N_2, atmospheric		Oxygen O_2, from air	
°F	°C	$10^3\alpha$	$10^2\beta$	α	β	10α	β	α	β	10α	β	$10^3\alpha$	$10^2\beta$	$10^3\alpha$	10β	10α	β	$10^2\alpha$	10β	$10^2\alpha$	$10^2\beta$	$10^4\alpha$	$10^2\beta$	$10^3\alpha$	$10^2\beta$	$10^3\alpha$	$10^2\beta$
32	0	3.716	3.027	88.93	1215	7.079	4.844	...	...	3.352	1.786	4.404	3.709	3.966	5.827	2.00	1.80	2.81	2.35	1.323	10.27	1.925	2.255	2.948	2.470	6.958	5.128
33.8	1	3.622	2.950	86.31	1179	6.851	4.688	...	...	3.219	1.715	4.301	3.622	3.849	5.655	1.94	1.74	2.72	2.27	1.266	9.830	1.905	2.232	2.875	2.409	6.768	4.988
35.6	2	3.527	2.873	83.81	1145	6.631	4.537	...	...	3.097	1.650	4.199	3.526	3.735	5.488	1.88	1.69	2.63	2.20	1.214	9.421	1.885	2.208	2.803	2.349	6.587	4.855
37.4	3	3.439	2.802	81.43	1112	6.406	4.384	...	...	2.983	1.589	4.099	3.452	3.625	5.326	1.82	1.63	2.54	2.12	1.164	9.035	1.865	2.185	2.734	2.292	6.411	4.725
39.2	4	3.354	2.732	79.16	1081	6.200	4.243	...	...	2.876	1.532	4.004	3.372	3.519	5.170	1.77	1.59	2.45	2.05	1.116	8.666	1.846	2.163	2.668	2.236	6.245	4.603
41.0	5	3.272	2.666	76.99	1052	6.000	4.106	...	...	2.772	1.477	3.911	3.294	3.414	5.016	1.72	1.54	2.37	1.98	1.070	8.310	1.828	2.142	2.604	2.182	6.085	4.485
42.8	6	3.192	2.600	74.92	1023	5.814	3.979	...	...	2.678	1.427	3.820	3.217	3.318	4.875	1.67	1.50	2.29	1.91	1.027	7.971	1.809	2.119	2.543	2.131	5.929	4.370
44.6	7	3.115	2.538	72.93	996	5.634	3.856	...	...	2.586	1.377	3.732	3.143	3.224	4.736	1.63	1.46	2.21	1.85	0.985	7.647	1.792	2.099	2.481	2.079	5.784	4.263
46.4	8	3.042	2.478	71.03	970	5.459	3.738	...	...	2.497	1.330	3.647	3.072	3.132	4.601	1.58	1.42	2.14	1.79	.945	7.338	1.775	2.080	2.424	2.032	5.643	4.159
48.2	9	2.972	2.421	69.20	945	5.289	3.619	...	...	2.410	1.284	3.566	3.003	3.044	4.472	1.54	1.38	2.07	1.73	.907	7.044	1.757	2.058	2.369	1.986	5.509	4.060
50.0	10	2.904	2.366	67.45	921	5.132	3.519	15.24	54.83	2.325	1.239	3.485	2.935	2.967	4.360	1.50	1.35	2.00	1.67	.873	6.776	1.742	2.041	2.317	1.941	5.379	3.964
51.8	11	2.841	2.314	...	...	4.968	3.400	14.65	52.72	2.243	1.195	3.410	2.872	2.885	4.238	1.46	1.31	1.94	1.62	.840	6.520	1.727	2.023	2.268	1.900	5.256	3.874
53.6	12	2.780	2.264	63.91	873	4.823	3.301	14.09	50.71	2.173	1.158	3.338	2.811	2.808	4.126	1.42	1.27	1.88	1.57	.810	6.287	1.713	2.007	2.220	1.860	5.137	3.787
55.4	13	2.720	2.216	...	...	4.679	3.202	13.56	48.78	2.102	1.120	3.268	2.752	2.738	4.022	1.38	1.24	1.82	1.52	.781	6.063	1.698	1.989	2.173	1.821	5.023	3.702
57.2	14	2.665	2.171	60.65	828	4.547	3.112	13.04	46.94	2.033	1.083	3.200	2.695	2.670	3.922	1.35	1.21	1.77	1.48	.754	5.853	1.684	1.973	2.133	1.787	4.915	3.623
59.0	15	2.612	2.128	59.02	806	4.420	3.025	12.56	45.18	1.970	1.049	3.134	2.640	2.604	3.826	1.32	1.18	1.71	1.43	.728	5.652	1.671	1.958	2.089	1.751	4.811	3.546
60.8	16	2.560	2.085	57.46	785	4.299	2.942	12.09	43.50	1.909	1.017	3.071	2.586	2.543	3.736	1.29	1.16	1.67	1.39	.704	5.464	1.657	1.941	2.048	1.717	4.712	3.473
62.6	17	2.510	2.044	55.95	764	4.175	2.857	11.64	41.89	1.850	0.986	3.011	2.536	2.485	3.651	1.26	1.13	1.62	1.35	.681	5.287	1.644	1.926	2.010	1.685	4.616	3.402
64.4	18	2.463	2.006	54.50	744	4.062	2.780	11.21	40.35	1.792	.955	2.952	2.486	2.426	3.564	1.23	1.10	1.58	1.32	.660	5.122	1.631	1.911	1.974	1.654	4.522	3.333
66.2	19	2.415	1.967	53.11	725	3.954	2.706	10.80	38.87	1.737	.925	2.899	2.442	2.373	3.486	1.20	1.08	1.54	1.29	.640	4.968	1.618	1.896	1.938	1.624	4.434	3.268
68.0	20	2.373	1.933	51.76	707	3.849	2.634	10.41	37.46	1.686	.898	2.843	2.395	2.322	3.411	1.17	1.05	1.50	1.25	.622	4.825	1.604	1.879	1.905	1.596	4.348	3.205
69.8	21	2.330	1.898	50.46	689	3.751	2.567	10.03	36.10	1.635	.871	2.793	2.352	2.276	3.343	1.15	1.03	1.46	1.22	.603	4.681	1.591	1.864	1.873	1.570	4.266	3.144
71.6	22	2.288	1.864	49.21	672	3.651	2.498	9.67	34.81	1.585	.844	2.756	2.321	2.226	3.271	1.12	1.00	1.42	1.19	.585	4.542	1.577	1.848	1.842	1.544	4.185	3.085
73.4	23	2.249	1.832	48.00	656	3.555	2.433	9.32	33.55	1.538	.819	2.696	2.271	2.180	3.203	1.10	0.99	1.39	1.16	.568	4.409	1.563	1.831	1.811	1.518	4.106	3.027
75.2	24	2.211	1.801	46.83	640	3.464	2.370	8.99	32.36	1.495	.796	2.651	2.233	2.137	3.139	1.08	0.97	1.36	1.14	.552	4.283	1.549	1.815	1.786	1.497	4.031	2.971
77.0	25	2.171	1.768	45.69	624	3.381	2.314	8.67	31.21	1.451	.773	2.607	2.196	2.095	3.078	1.06	0.95	1.32	1.10	.536	4.161	1.537	1.801	1.755	1.471	3.957	2.916
78.8	26	2.137	1.741	44.59	609	3.296	2.256	8.36	30.10	1.409	.751	2.564	2.159	2.053	3.017	1.03	.93	1.29	1.08	.521	4.046	1.524	1.785	1.726	1.447	3.886	2.864
80.6	27	2.102	1.712	43.53	595	3.214	2.199	8.07	29.04	1.373	.731	2.523	2.125	2.014	2.960	1.01	.91	1.27	1.06	.508	3.940	1.511	1.770	1.700	1.425	3.813	2.811
82.4	28	2.069	1.685	42.50	581	3.138	2.147	7.79	28.02	1.336	.712	2.483	2.091	1.976	2.904	0.99	.89	1.24	1.04	.494	3.836	1.498	1.755	1.675	1.404	3.744	2.759
84.2	29	2.036	1.658	...	...	3.060	2.094	7.51	27.04	1.297	.691	2.445	2.059	1.941	2.851	.97	.87	1.21	1.01	.481	3.733	1.486	1.741	1.650	1.383	3.673	2.707
86.0	30	2.004	1.633	...	...	2.989	2.045	7.28	26.20	1.262	.672	2.409	2.029	1.907	2.801	0.95	0.85	1.18	0.99	.469	3.640	1.474	1.727	1.625	1.362	3.604	2.656
95.0	35	1.852	1.509	...	...	2.653	1.815	6.11	22.00	1.107	.590	2.234	1.882	1.734	2.547	...	...	...	...	.413	3.203	1.428	1.673	1.503	1.260	3.319	2.446
104.0	40	1.719	1.400	...	...	2.365	1.618	5.15	18.52	0.975	.519	2.078	1.750	1.594	2.341	...	...	...	...	.366	2.843	1.385	1.623	1.394	1.168	3.086	2.274
113.0	45	1.604	1.307	...	...	2.114	1.447	4.35	15.64	.863	.460	1.936	1.631	1.468	2.157	...	...	...	...	.327	2.542	1.339	1.569	1.302	1.091	2.863	2.110
122.0	50	1.498	1.220	...	...	1.887	1.291	3.68	12.82	.760	.405	1.799	1.515	1.356	1.996	...	...	...	...	.294	2.285	1.289	1.510	1.219	1.021	2.660	1.961
140.0	60	1.288	1.050	...	...	1.483	1.015	2.62	9.43	0.578	0.308	1.524	1.284	1.145	1.683	...	...	...	...	.238	1.849	1.180	1.382	1.054	0.883	2.278	1.679
158.0	70	1.064	0.867	...	...	1.105	0.756	1.81	6.51	...	...	1.279	1.077	0.929	1.364	...	...	...	...	.186	1.444	1.033	1.210	0.873	.731	1.861	1.371
176.0	80	0.801	.653	...	...	0.751	.514	1.34	4.10	...	...	0.984	0.828	.698	1.026	...	...	...	...	.135	1.049	0.793	0.929	.663	.556	1.384	1.020
194.0	90	.462	.376	...	...	.410	.281	0.56	2.01	...	...	.571	.481	.400	0.588	...	...	...	...	.075	0.586	.463	.542	.384	.322	0.790	0.582
212.0	100	.004	.003	...	...	.003	.021	...	...	...	...	.005	.004	.003	.003	...	...	...	...	0.006	.004	.004	.005	.003	.003	.007	.005
	100.07	0.000	0.000	...	...	0.000	0.000	...	...	...	...	0.000	0.000	0.000	0.000	...	...	...	...	0.000	0.000	0.000	0.000	0.000	0.000	0.000	0.000

The values in this table are computed values based on Henry's law. Basic data for these computations were obtained from the International Critical Tables.

Caution: Since practical problems generally involve mixtures of gases, solubility of individual constituents must be corrected in proportion to partial pressure of the particular constituent under consideration.

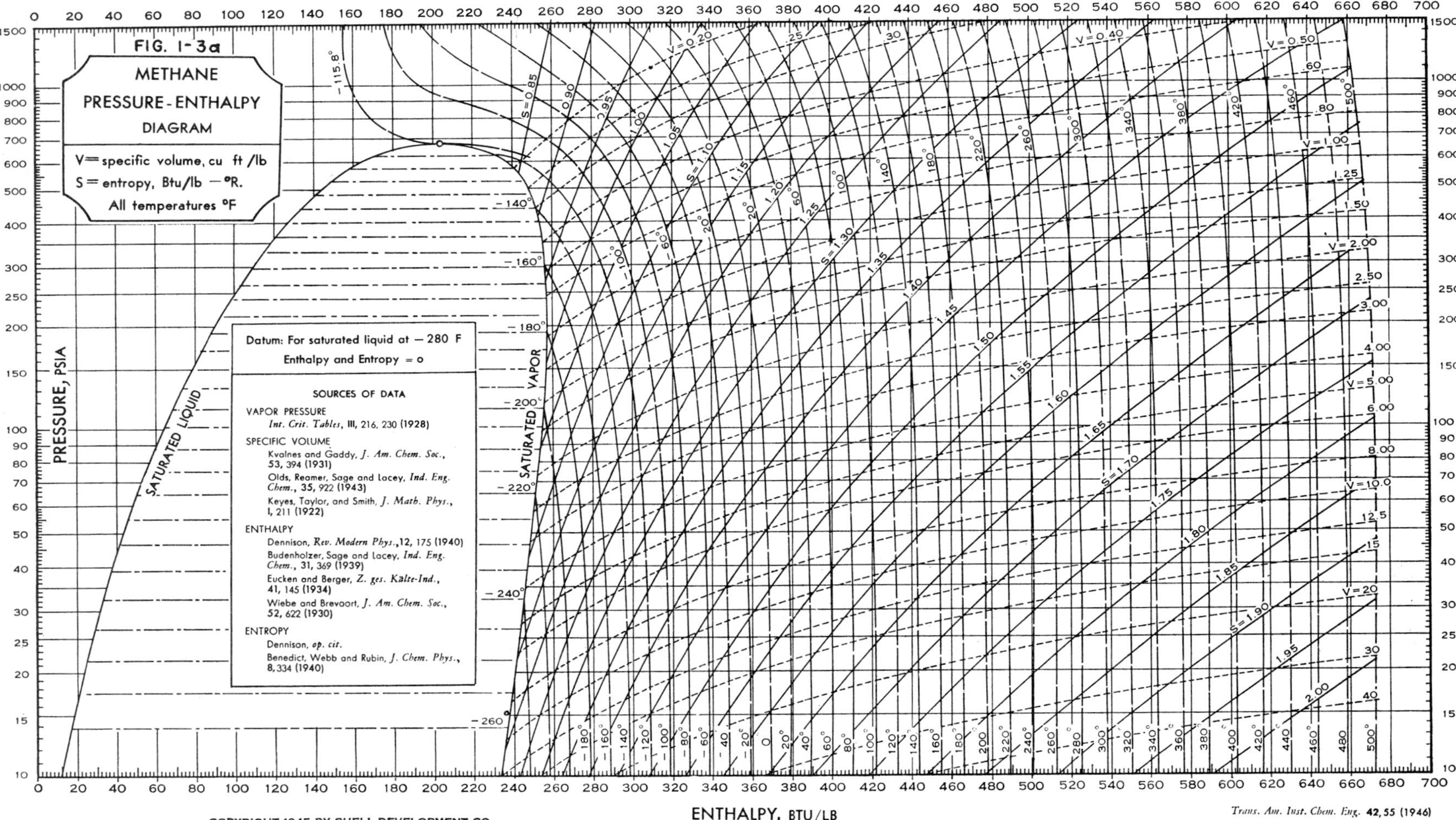

Fig. 1-3a Methane pressure-enthalpy diagram. (Shell Development Co.) (See Fig. 8-43 for Mollier chart.)

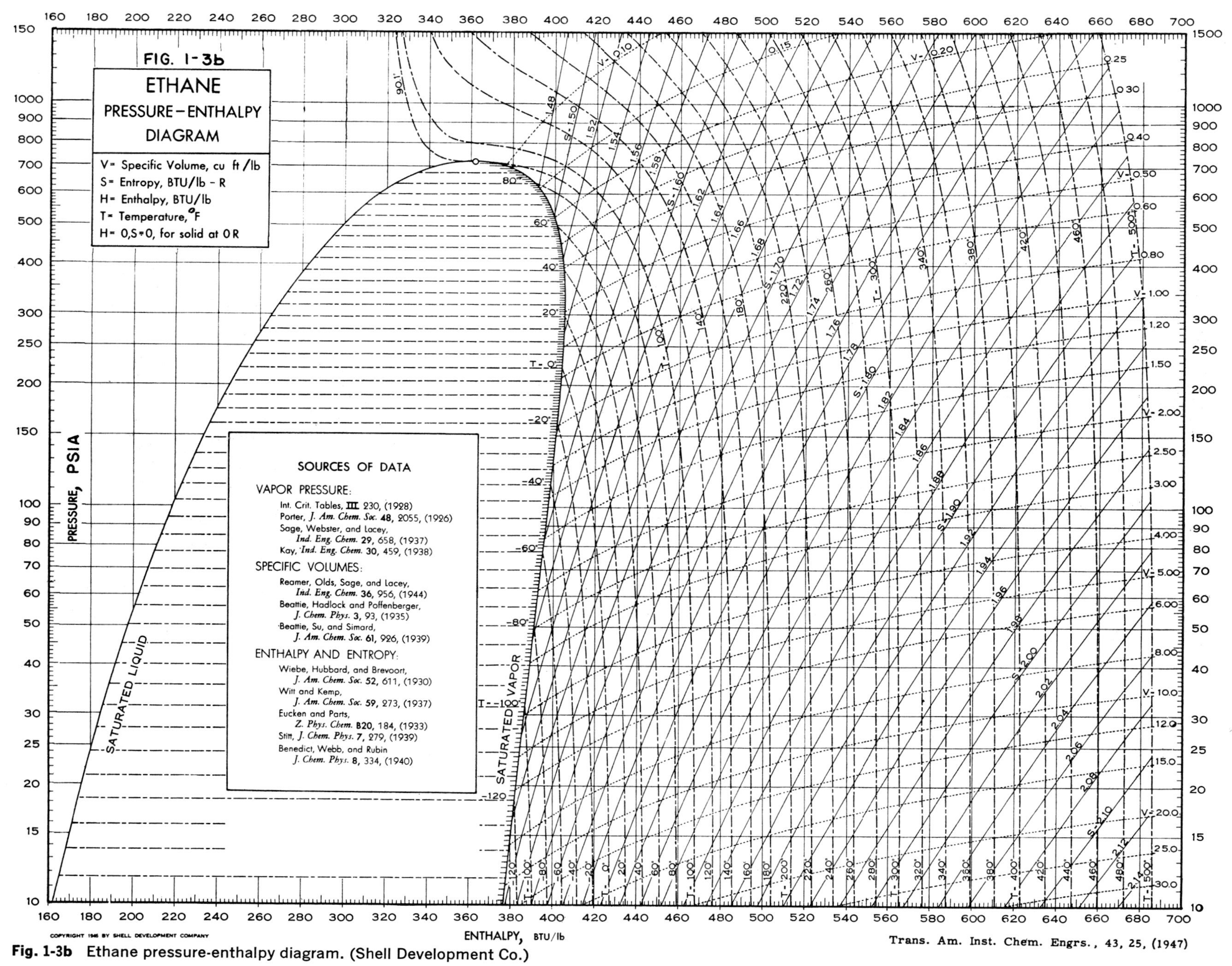

FIG. 1-3b
ETHANE
PRESSURE-ENTHALPY
DIAGRAM
V= Specific Volume, cu ft/lb
S= Entropy, BTU/lb - R
H= Enthalpy, BTU/lb
T= Temperature, °F
H= 0,S=0, for solid at 0 R
SOURCES OF DATA
VAPOR PRESSURE:
Int. Crit. Tables, III 230, (1928)
Porter, J. Am. Chem. Soc. 48, 2055, (1926)
Sage, Webster, and Lacey, Ind. Eng. Chem. 29, 658, (1937)
Kay, Ind. Eng. Chem. 30, 459, (1938)
SPECIFIC VOLUMES:
Reamer, Olds, Sage, and Lacey, Ind. Eng. Chem. 36, 956, (1944)
Beattie, Hadlock and Poffenberger, J. Chem. Phys. 3, 93, (1935)
Beattie, Su, and Simard, J. Am. Chem. Soc. 61, 926, (1939)
ENTHALPY AND ENTROPY:
Wiebe, Hubbard, and Brevoort, J. Am. Chem. Soc. 52, 611, (1930)
Witt and Kemp, J. Am. Chem. Soc. 59, 273, (1937)
Eucken and Parts, Z. Phys. Chem. B20, 184, (1933)
Stitt, J. Chem. Phys. 7, 279, (1939)
Benedict, Webb, and Rubin, J. Chem. Phys. 8, 334, (1940)
PRESSURE, PSIA
ENTHALPY, BTU/lb
SATURATED LIQUID
SATURATED VAPOR
Trans. Am. Inst. Chem. Engrs., 43, 25, (1947)
COPYRIGHT 1946 BY SHELL DEVELOPMENT COMPANY

Fig. 1-3b Ethane pressure-enthalpy diagram. (Shell Development Co.)

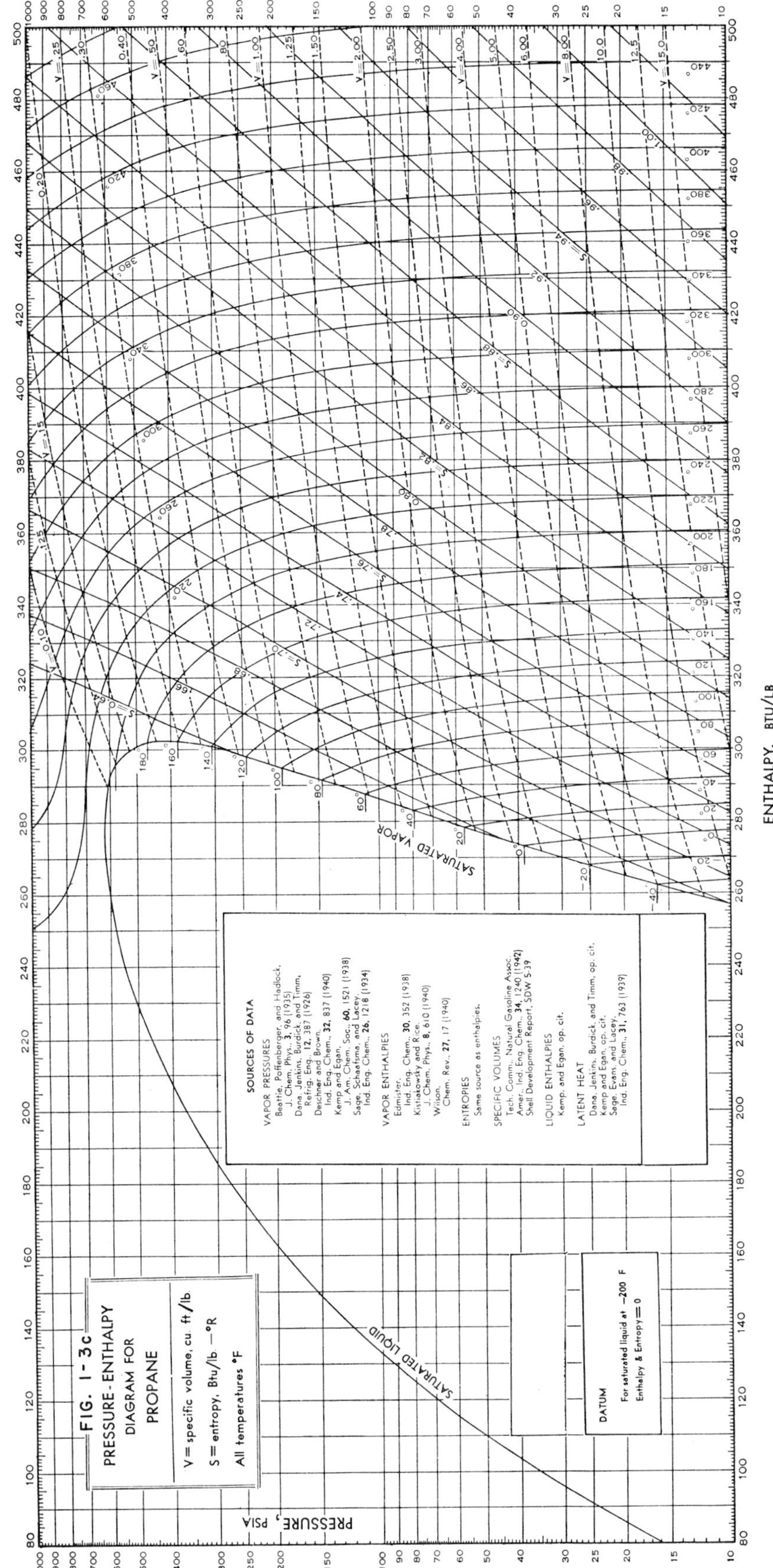

Fig. 1-3c Propane pressure-enthalpy diagram. (Shell Development Co.)

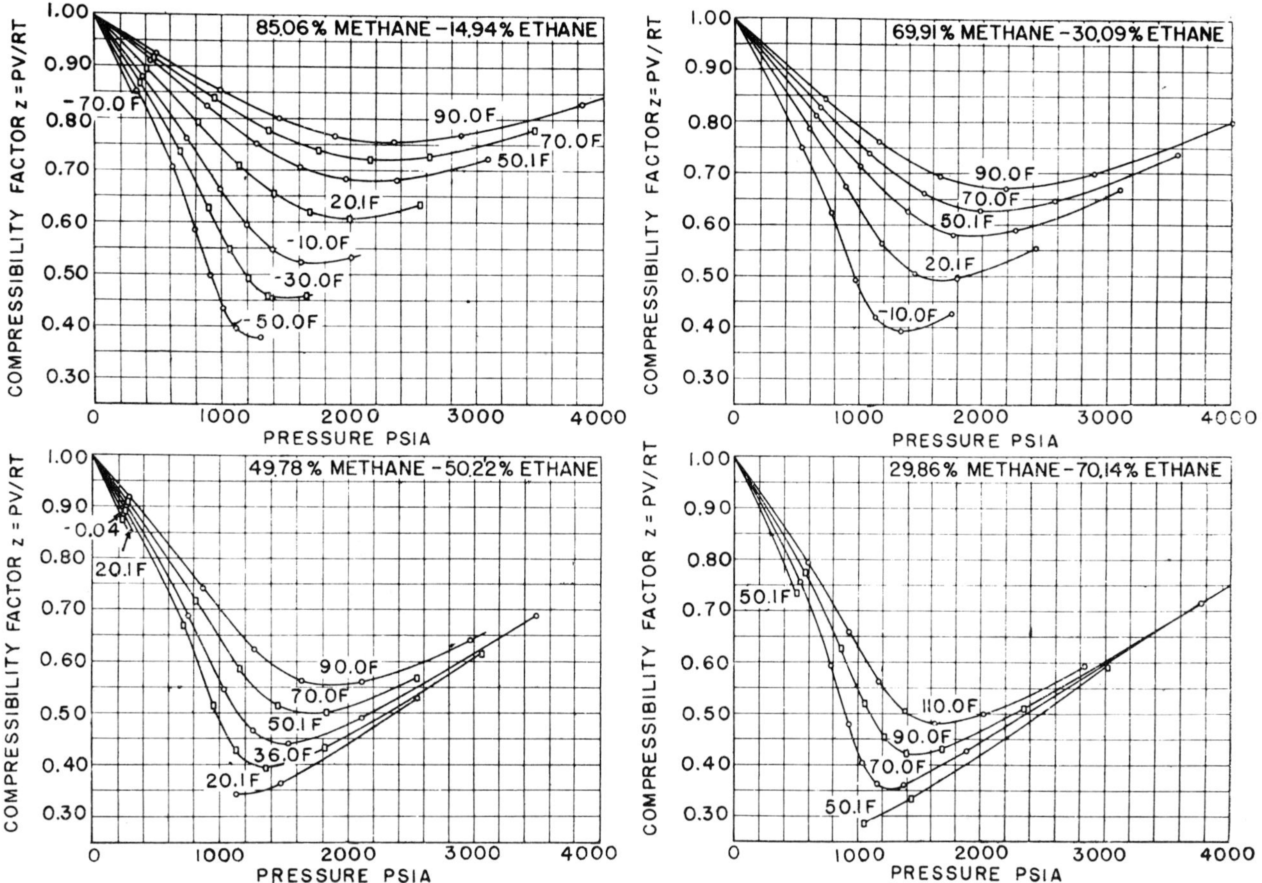

Fig. 1-3d Compressibility charts of methane-ethane mixtures.[4]

where:

z = compressibility factor
P = absolute pressure
R = universal gas law constant, the value of which depends on only the units used to express the pressure, temperature and density, and not on the nature of the gas
T = absolute temperature
ρ = density of the gas in moles per unit volume
V = molal volume = $1/\rho$

The compressibility factor is a measure of the deviation of a gas from the ideal gas law, and is 1.000 for an ideal gas. Its value is readily calculated by applying the above equation to the experimentally measured pressure, temperature and density.

Table 1-26 Weight of Dry Air at Various Pressures and Temperatures*
(Based on 30 in. Hg and 60 F)

Temp, °F	Gage pressures, lb per sq in.												
	0	1	2	3	4	5	10	15	20	30	40	50	60
	Weight, lb per cu ft												
−20	0.09050	0.09664	0.10278	0.10892	0.11506	0.12120	0.15191	0.18262	0.21333	0.27474	0.33616	0.39758	0.45899
−10	.08848	.09449	.10049	.10650	.11250	.11851	.14853	.17856	.20858	.26863	.32868	.38873	.44878
0	.08656	.09243	.09831	.10418	.11006	.11593	.14530	.17467	.20404	.26279	.32153	.38027	.43902
10	.08472	.09046	.09621	.10196	.10771	.11346	.14221	.17095	.19970	.25719	.31468	.37218	.42967
20	.08299	.08858	.09421	.09984	.10547	.11110	.13924	.16739	.19554	.25183	.30812	.36442	.42071
30	.08125	.08677	.09228	.09780	.10331	.10883	.13640	.16397	.19154	.24669	.30183	.35697	.41212
32	.08092	.08642	.09191	.09740	.10289	.10838	.13584	.16330	.19076	.24568	.30060	.35552	.41044
40	.07963	.08503	.09044	.09584	.10124	.10665	.13367	.16069	.18771	.24175	.29579	.34983	.40387
50	.07807	.08336	.08866	.09396	.09926	.10456	.13104	.15753	.18402	.23600	.28998	.34296	.39594
60	.07656	.08176	.08695	.09215	.09735	.10254	.12852	.15450	.18048	.23244	.28440	.33636	.38832
70	.07512	.08022	.08531	.09041	.09551	.10061	.12610	.15159	.17707	.22805	.27903	.33001	.38099
80	.07373	.07873	.08373	.08874	.09374	.09874	.12376	.14878	.17379	.22383	.27386	.32389	.37393
90	0.07238	0.07730	0.08221	0.08712	0.09203	0.09695	0.12151	0.14607	0.17063	0.21975	0.26888	0.31800	0.36712

Table 1-26 Weight of Dry Air at Various Pressures and Temperatures (Continued)

Temp, °F	Gage pressures, lb. per sq in												
	0	1	2	3	4	5	10	15	20	30	40	50	60
	Weight, lb per cu ft												
100	0.07109	0.07591	0.08074	0.08556	0.09039	0.09521	0.11934	0.14346	0.16758	0.21583	0.26407	0.31232	0.36056
110	.06984	.07458	.07932	.08406	.08880	.09354	.11724	.14094	.16464	.21204	.25944	.30683	.35423
120	.06864	.07330	.07795	.08261	.08727	.09193	.11522	.13851	.16180	.20838	.25496	.30154	.34812
130	.06747	.07205	.07663	.08121	.08579	.09037	.11326	.13616	.15905	.20484	.25063	.29633	.34222
140	.06635	.07085	.07535	.07986	.08436	.08886	.11137	.13389	.15640	.20143	.24645	.29148	.33651
150	.06526	.06969	.07412	.07854	.08297	.08740	.10955	.13169	.15384	.19812	.24241	.28670	.33099
175	.06269	.06694	.07120	.07545	.07971	.08396	.10523	.12650	.14777	.19032	.23286	.27540	.31795
200	.06031	.06441	.06850	.07259	.07668	.08078	.10124	.12171	.14217	.18310	.22404	.26497	.30590
225	.05811	.06205	.06600	.06994	.07388	.07783	.09755	.11726	.13698	.17642	.21585	.25529	.29473
250	.05606	.05987	.06367	.06748	.07128	.07509	.09411	.11313	.13216	.17020	.20825	.24630	.28434
275	.05415	.05783	.06150	.06518	.06885	.07253	.09091	.10928	.12766	.16431	.20116	.23791	.27467
300	.05237	.05593	.05948	.06303	.06659	.07014	.08791	.10569	.12346	.15900	.19454	.23008	.26563
350	.04914	.05247	.05581	.05914	.06247	.06581	.08248	.09916	.11583	.14918	.18253	.21587	.24922
400	.04628	.04942	.05256	.05570	.05884	.06198	.07769	.09339	.10909	.14040	.17191	.20332	.23472
450	.04374	.04670	.04967	.05264	.05561	.05858	.07342	.08826	.10310	.13278	.16246	.19214	.22182
500	.04146	.04427	.04708	.04990	.05271	.05552	.06959	.08366	.09772	.12586	.15399	.18213	.21026
550	.03940	.04208	.04475	.04743	.05010	.05277	.06614	.07951	.09289	.11963	.14637	.17311	.19985
600	.03754	.04010	.04264	.04519	.04774	.05028	.06302	.07576	.08850	.11398	.13946	.16494	.19042
650	.03585	.03829	.04071	.04315	.04558	.04802	.06018	.07235	.08451	.10884	.13318	.15751	.18184
700	.03431	.03663	.03896	.04129	.04362	.04595	.05759	.06923	.08087	.10415	.12743	.15072	.17400
800	.03158	.03373	.03587	.03801	.04016	.04230	.05302	.06373	.07445	.09588	.11732	.13875	.16018
900	.02926	.03125	.03323	.03522	.03720	.03919	.04912	.05904	.06897	.08883	.10869	.12854	.14840
1000	.02725	.02910	.03095	.03280	.03465	.03650	.04575	.05500	.06425	.08274	.10124	.11974	.13823
1500	.02030	.02168	.02306	.02443	.02581	.02719	.03408	.04097	.04785	.06163	.07541	.08919	.10296
2000	0.01617	0.01728	0.01837	0.01947	0.02056	0.02166	0.02715	0.03264	0.03813	0.04103	0.06008	0.07106	0.08203

Temp, °F	Gage pressures, lb per sq in.												
	70	80	90	100	120	140	160	175	200	225	250	275	300
	Weight, lb per cu ft												
−20	0.52041	0.58182	0.64324	0.70465	0.82749	0.95032	1.07315	1.16527	1.31881	1.47235	1.62589	1.77932	1.93297
−10	.50883	.56888	.62893	.68898	.80908	.92918	1.04928	1.13935	1.28948	1.43960	1.58973	1.73974	1.88998
0	.49776	.55650	.61525	.67399	.79148	.90896	1.02645	1.11456	1.26142	1.40828	1.55514	1.70189	1.84885
10	.48716	.54465	.60214	.65964	.77462	.88961	1.00459	1.09083	1.23456	1.37829	1.52202	1.66564	1.80948
20	.47700	.53330	.58959	.64588	.75847	.87106	0.98364	1.06808	1.20882	1.34955	1.49028	1.63091	1.77175
30	.46726	.52240	.57755	.63269	.74298	.85326	.96355	1.04627	1.18412	1.32198	1.45984	1.59760	1.73556
32	.46536	.52028	.57520	.63012	.73996	.84979	.95963	1.04201	1.17931	1.31660	1.45390	1.59110	1.72850
40	.45791	.51195	.56599	.62003	.72810	.83618	.94426	1.02532	1.16042	1.29552	1.43062	1.56562	1.70082
50	.44892	.50190	.55488	.60786	.71382	.81977	.92573	1.00520	1.13765	1.27010	1.40255	1.53490	1.66744
60	.44028	.49224	.54420	.59616	.70008	.80400	.90792	0.98586	1.11575	1.24565	1.37555	1.50536	1.63535
70	.43197	.48294	.53392	.58490	.68686	.78882	.89077	.96724	1.09469	1.22213	1.34958	1.47693	1.60447
80	.42396	.47399	.52403	.57406	.67413	.77420	.87426	.94931	1.07440	1.19948	1.32457	1.44956	1.57473
90	.41625	.46537	.51449	.56362	.66186	.76011	.85836	.93204	1.05485	1.17766	1.29046	1.42318	1.54608
100	.40881	.45705	.50530	.55354	.64003	.74653	.84302	.91538	1.03600	1.15661	1.27722	1.39775	1.51845
110	.40163	.44903	.49643	.54383	.63862	.73342	.82822	.89931	1.01781	1.13631	1.25470	1.37321	1.49179
120	.39470	.44128	.48786	.53444	.62760	.72077	.81393	.88380	1.00025	1.11670	1.23315	1.34952	1.46605
130	.38801	.43380	.47959	.52538	.61696	.70854	.80012	.86881	0.98328	1.09776	1.21223	1.32663	1.44119
140	.38153	.42656	.47159	.51662	.60667	.69672	.78678	.85432	.96688	1.07945	1.19202	1.30450	1.41715
150	.37528	.41956	.46385	.50814	.59672	.68529	.77387	.84030	.95102	1.06174	1.17246	1.28310	1.39390
175	.36049	.40303	.44558	.48812	.57321	.65829	.74338	.80720	.91356	1.01991	1.12627	1.23255	1.33899
200	.34683	.38776	.42869	.46962	.55148	.63334	.71521	.77660	.87893	0.98126	1.08358	1.18584	1.28824
225	.33416	.37360	.41303	.45247	.53134	.61021	.68909	.74824	.84683	.94542	1.04401	1.14253	1.24119
250	.32239	.36044	.39848	.43653	.51262	.58872	.66481	.72188	.81700	.91211	1.00723	1.10228	1.19746
275	.31142	.34817	.38481	.42167	.49518	.56868	.64218	.69731	.78919	.88107	0.97295	1.06476	1.15671
300	.30117	.33671	.37225	.40779	.47888	.54996	.62105	.67436	.76322	.85207	.94093	1.02977	1.11864
350	.28257	.31591	.34926	.38261	.44930	.51600	.58269	.63271	.71608	.79945	.88282	0.96612	1.04955
400	.26613	.29754	.32895	.36035	.42317	.48598	.54880	.59591	.67443	.75295	.83146	.90992	0.98850
450	0.25150	0.28118	0.31086	0.34054	0.39991	0.45927	0.51863	0.56315	0.63735	0.71156	0.78576	0.85991	0.93416

Temp, °F	Gage pressures, lb per sq in.												
	70	80	90	100	120	140	160	175	200	225	250	275	300
	Weight, lb per cu ft												
500	0.23840	0.26653	0.29467	0.32280	0.37907	0.43534	0.49161	0.53381	0.60414	0.67448	0.74481	0.81510	0.88549
550	.22659	.25333	.28007	.30681	.36195	.41568	.46726	.50737	.57422	.64108	.70793	.77473	.84163
600	.21590	.24138	.26686	.29233	.34329	.39425	.44521	.48343	.54713	.60082	.67452	.73817	.80192
650	.20617	.23050	.25482	.27916	.32782	.37649	.42515	.46164	.52247	.58330	.64413	.70491	.76578
700	.19728	.22056	.24384	.26712	.31369	.36025	.40682	.44174	.49994	.55815	.61635	.67451	.73276
800	.18162	.20305	.22448	.24592	.28878	.33164	.37452	.40667	.46025	.51384	.56742	.62096	.67459
900	.16826	.18812	.20797	.22783	.26754	.30726	.34697	.37676	.42640	.47604	.52568	.57529	.62497
1000	.15673	.17523	.19362	.21222	.24921	.28621	.32320	.34094	.39719	.44343	.48967	.53588	.58215
1500	.11674	.13052	.14429	.15807	.18562	.21318	.24073	.26140	.29584	.33028	.36473	.39914	.43361
2000	0.09301	0.10398	0.11495	0.12594	0.14789	0.16984	0.19180	0.20826	0.23570	0.26314	0.29058	0.31800	0.34546

* Revised from "Compressed Air," Ingersoll-Rand Company.

Formula: $w = (0.08092396)(P/14.735108)\,[491.58/(459.58 + t)]$
where: t = temperature, °F
P = absolute pressure, lb per sq in.

Table 1-27 Properties of Air*

(Barometer 30 in. Hg)

Temperature		Weight of Mcf dry air, lb	Volume of 1 lb of dry air, cu ft	Mixture of air saturated with water vapor					Weight of water necessary to saturate 100 lb of dry air, lb	Volume of 1 lb dry air and vapor to saturate, cu ft
°F	°F abs, R			Vapor pressure of water vapor, in Hg	Partial pressure of the dry air, in Hg	Weight of dry air, in Mcf of mixture, lb	Weight of water vapor in Mcf of mixture, lb	Weight of Mcf of mixture, lb		
(1)	(2)	(3)	(4)	(5)	(6)	(7)	(8)	(9)	(10)	(11)
0	459.58	86.540	11.555	0.0377	29.962	86.435	0.068	86.503	0.0787	11.569
5	464.58	85.613	11.680	.0488	29.951	85.473	.088	85.561	.1030	11.700
10	469.58	84.702	11.806	.0630	29.937	84.524	.122	84.646	.1443	11.831
15	474.58	83.807	11.932	.0807	29.919	83.582	.141	83.723	.1687	11.964
20	479.58	82.934	12.058	.1028	29.897	82.650	.178	82.828	.2154	12.099
25	484.58	82.077	12.184	.1304	29.870	81.723	.223	81.946	.2729	12.236
30	489.58	81.242	12.309	.1645	29.834	80.791	.278	81.069	.3441	12.378
32	491.58	80.906	12.360	.1803	29.820	80.425	.303	80.728	.3767	12.434
34	493.58	80.584	12.409	.1955	29.804	80.056	.327	80.383	.4085	12.491
36	495.58	80.255	12.460	.2117	29.788	79.690	.353	80.043	.4430	12.549
38	497.58	79.933	12.510	.2292	29.771	79.322	.380	79.702	.4791	12.607
40	499.58	79.612	12.561	.2478	29.752	78.956	.407	79.363	.5155	12.665
42	501.58	79.298	12.611	.2678	29.732	78.588	.441	79.029	.5611	12.724
44	503.58	78.984	12.661	.2891	29.711	78.221	.475	78.696	.6072	12.784
46	505.58	78.670	12.711	0.3120	29.688	77.851	0.510	78.362	0.6551	12.845

Table continues on page 1/26.

Column 3: 1000 × 0.07655 × 519.58/(459.58 + t) = weight of Mcf dry air at t°F and 30 in. Hg, where 0.07655 = weight of 1 cu ft air at 60 F and 30 in. Hg, and t = temperature, °F.

Column 4: Volume of 1 lb dry air at t°F and 30 in. Hg = 1000/wt of Mcf at t°F and 30 in. Hg = 1000/column 3.

Column 5: Authority from 0 to 212 F: I.C.T. *Column 6:* Partial pressure dry air = 30 − vapor pressure water vapor = 30 − column 5.

Column 7: Weight of dry air in Mcf of mixture = 1000/column 11.

Column 8: Weight of water vapor in Mcf of mixture = density of saturated vapor, lb per cu ft, × 1000.

Column 9: Total weight of mixture = sum of columns 7 and 8.

Column 10: Weight of water necessary to saturate 100 lb of dry air = (volume of 1 lb dry air + vapor to saturate) × weight of vapor in 1 cu ft of mixture = column 11 × column 8 ÷ 10.

Column 11: Volume of 1 lb dry air + vapor to saturate. Assume air to follow Boyle's law (correcting wt. of 1 cu ft of air at 60 F and 30 in. Hg for any deviations from Boyle's law)

$$\frac{P_a V_a}{T_a} = K = \frac{30}{(519.58 \times 0.07655)} = 0.7542644. \text{ Volume of 1 lb dry air + vapor to saturate} = \frac{KT_a}{P - P_v},$$

where K = constant; T_a = absolute temperature; P = 30; and P_v = partial pressure of water vapor.

* Revised from Gebhardt, "Steam Power Plant Engineering."

Table 1-27 Properties of Air (Continued)

Temperature		Weight of Mcf dry air, lb	Volume of 1 lb of dry air, cu ft	Mixture of air saturated with water vapor					Weight of water necessary to saturate 100 lb of dry air, lb	Volume of 1 lb dry air and vapor cu ft
°F	°F abs, °R			Vapor pressure of water vapor, in. Hg	Partial pressure of the dry air, in. Hg	Weight of dry air, in Mcf of mixture, lb	Weight of water vapor in Mcf of mixture, lb	Weight of Mcf of mixture, lb		
(1)	(2)	(3)	(4)	(5)	(6)	(7)	(8)	(9)	(10)	(11)
48	507.58	78.357	12.762	0.3364	29.664	77.482	0.547	78.029	0.7060	12.906
50	509.58	78.050	12.812	.3626	29.637	77.108	.587	77.695	.7613	12.969
52	511.58	77.744	12.863	.3904	29.610	76.736	.630	77.366	.8210	13.032
54	513.58	77.446	12.912	.4202	29.580	76.360	.675	77.035	.8840	13.096
56	515.58	77.139	12.964	.4519	29.548	75.981	.724	76.705	0.9529	13.161
58	517.58	76.848	13.013	.4857	29.514	75.601	.775	76.376	1.0251	13.227
60	519.58	76.550	13.063	.5218	29.477	75.215	.829	76.044	1.1022	13.295
62	521.58	76.259	13.113	.5601	29.440	74.833	.886	75.719	1.1840	13.363
64	523.58	75.968	13.163	.6009	29.399	74.443	0.947	75.390	1.2721	13.433
66	525.58	75.677	13.214	.6442	29.356	74.051	1.012	75.063	1.3666	13.504
68	527.58	75.386	13.265	.6904	29.310	73.656	1.080	74.736	1.4663	13.577
70	529.58	75.103	13.315	.7393	29.261	73.255	1.152	74.407	1.5726	13.651
72	531.58	74.820	13.367	.7912	29.209	72.849	1.229	74.078	1.6870	13.727
74	533.58	74.544	13.415	.8463	29.154	72.440	1.309	73.749	1.8208	13.805
76	535.58	74.261	13.466	.9048	29.095	72.023	1.395	73.418	1.9369	13.884
78	537.58	73.986	13.516	0.9666	29.033	71.602	1.484	73.086	2.0726	13.966
80	539.58	73.710	13.567	1.032	28.968	71.177	1.579	72.756	2.2184	14.049
82	541.58	73.442	13.616	1.102	28.898	70.743	1.680	72.423	2.3748	14.136
84	543.58	73.166	13.667	1.175	28.825	70.304	1.785	72.089	2.5390	14.224
86	545.58	72.899	13.718	1.253	28.747	69.857	1.895	71.752	2.7127	14.315
88	547.58	72.638	13.767	1.335	28.665	69.403	2.014	71.417	2.9019	14.408
90	549.58	72.370	13.818	1.422	28.578	69.010	2.137	71.147	3.0998	14.505
92	551.58	72.110	13.868	1.513	28.487	68.472	2.266	70.738	3.3094	14.604
94	553.58	71.850	13.918	1.610	28.390	67.992	2.403	70.395	3.5342	14.707
96	555.58	71.590	13.969	1.712	28.288	67.505	2.545	70.040	3.7701	14.814
98	557.58	71.329	14.019	1.820	28.180	67.005	2.696	69.701	4.0236	14.924
100	559.58	71.077	14.069	1.933	28.067	66.498	2.854	69.352	4.2979	15.038
105	564.58	70.449	14.195	2.243	27.757	65.181	3.283	68.464	5.0367	15.342
110	569.58	69.829	14.321	2.596	27.431	63.850	3.768	67.618	5.9013	15.662
115	574.58	69.224	14.446	2.995	27.005	62.312	4.311	66.623	6.9184	16.048
120	579.58	68.627	14.571	3.446	26.554	60.743	4.919	65.662	8.0981	16.463
125	584.58	68.038	14.698	3.956	26.044	59.066	5.599	64.665	9.4792	16.930
130	589.58	67.463	14.823	4.525	25.475	57.286	6.355	63.641	11.0935	17.456
135	594.58	66.897	14.948	5.165	24.835	55.377	7.19	62.567	12.9837	18.058
140	599.58	66.338	15.074	5.881	24.119	53.332	8.13	61.462	15.2441	18.750
145	604.58	65.787	15.200	6.682	23.318	51.134	9.16	60.294	17.9136	19.556
150	609.58	65.244	15.327	7.572	22.428	48.779	10.30	59.079	21.1155	20.500
155	614.58	64.715	15.452	8.556	21.444	46.260	11.57	57.830	25.0109	21.617
160	619.58	64.195	15.511	9.649	20.351	43.548	12.96	56.508	29.7604	22.963
165	624.58	63.682	15.703	10.86	19.14	40.628	14.48	55.108	35.6400	24.613
170	629.58	63.177	15.828	12.20	17.80	37.628	16.14	53.624	43.0584	26.678
175	634.58	62.679	15.954	13.67	16.33	34.117	17.96	52.077	52.6416	29.310
180	639.58	62.189	16.080	15.29	14.71	30.492	19.94	50.432	65.393	32.795
185	644.58	61.707	16.205	17.07	12.93	26.595	22.10	48.695	83.098	37.601
190	649.58	61.232	16.331	19.02	10.98	22.410	24.44	46.850	109.0574	44.622
195	654.58	60.765	16.457	21.15	8.85	17.925	27.00	44.925	150.0284	55.788
200	659.58	60.298	16.584	23.46	6.54	13.146	29.76	42.906	226.384	76.070
205	664.58	59.847	16.709	25.99	4.01	7.800	32.76	40.760	409.5156	
210	669.58	59.403	16.834	28.75	1.25	2.475	35.96	38.435	633.1861	
212	671.58	59.227	16.884	29.92	0.08	0.158	37.32	37.478		

Footnotes appear on page 1/25.

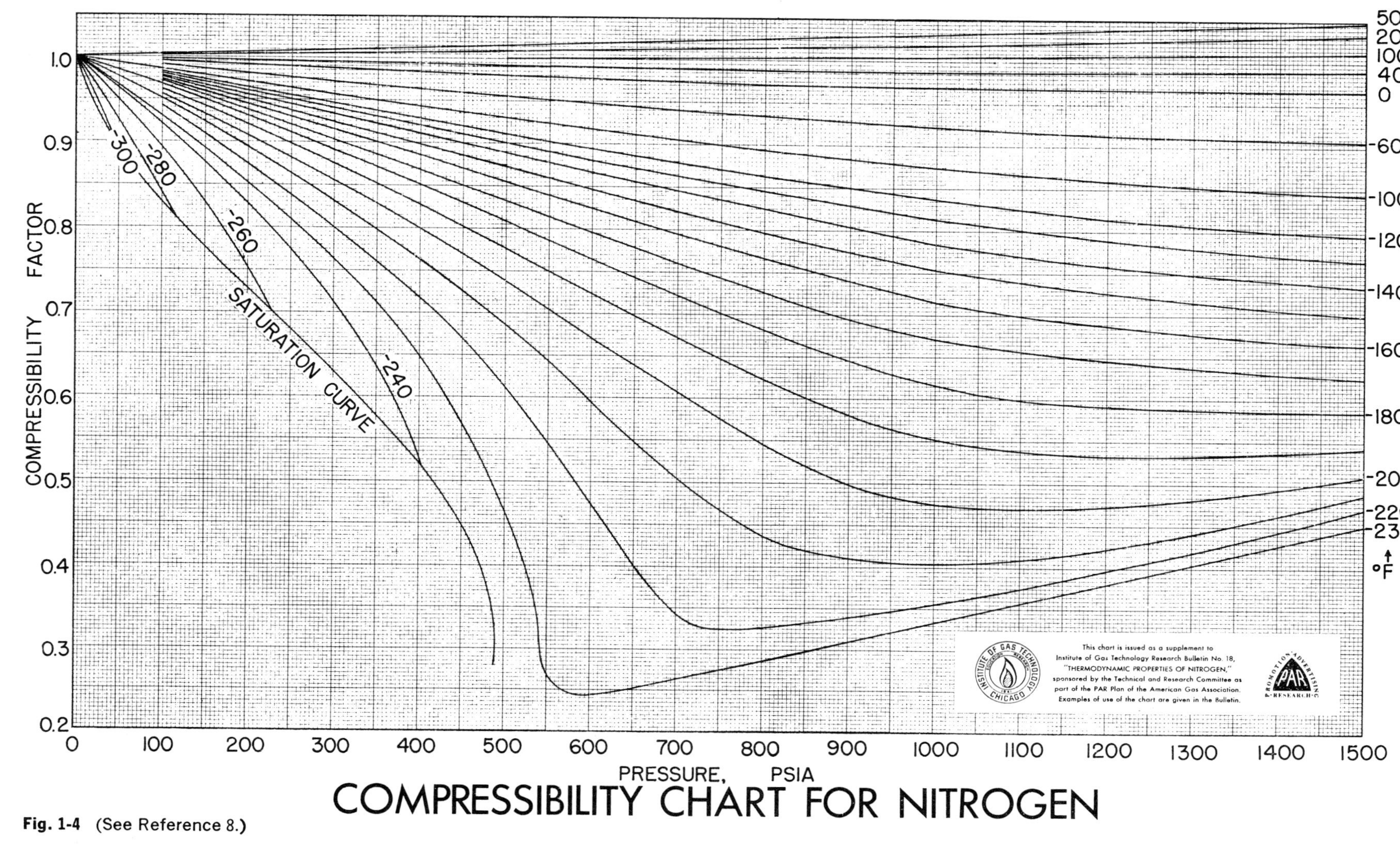

Fig. 1-4 (See Reference 8.)

Table 1-28 Temperature of Dew Point for Air*

Air temp, °F	Depression of wet-bulb thermometer, °F																																		
	1	2	3	4	5	6	7	8	9	10	11	12	13	14	15	16	17	18	19	20	21	22	23	24	25	26	27	28	29	30	31	32	33	34	35
0	−7	−20																																	
5	−1	−9	−24																																
10	5	−2	−10	−27																															
15	11	6	0	−9	−26																														
20	16	12	8	2	−7	−21																													
25	22	19	15	10	5	−3	−15	−51																											
30	27	25	21	18	14	8	2	−7	−25																										
35	33	30	28	25	21	17	13	7	0	−11	−41																								
40	38	35	33	30	28	25	21	18	13	7	−1	−14																							
45	43	41	38	36	34	31	28	25	22	18	13	7	−1	−14																					
50	48	46	44	42	40	37	34	32	29	26	22	18	13	8	0	−13																			
55	53	51	50	48	45	43	41	38	36	33	30	27	24	20	15	9	1	−12	−59																
60	58	57	55	53	51	49	47	45	43	40	38	35	32	29	25	21	17	11	4	−8	−36														
65	63	62	60	59	57	55	53	51	49	47	45	42	40	37	34	31	27	24	19	14	7	−3	−22												
70	69	67	65	64	62	61	59	57	55	53	51	49	47	44	42	39	36	33	30	26	22	17	11	2	−11										
75	74	72	71	69	68	66	64	63	61	59	57	55	54	51	49	47	44	42	39	36	32	29	25	21	15	8	−2	−23							
80	79	77	76	74	73	72	70	68	67	65	63	62	60	58	56	54	52	50	47	44	42	39	36	32	28	24	20	13	6	−7	−53				
85	84	82	81	80	78	77	75	74	72	71	69	68	66	64	62	61	59	57	54	52	50	48	45	42	39	36	32	28	24	19	12	3	−12		
90	89	87	86	85	83	82	81	79	78	76	75	73	72	70	69	67	65	63	61	59	57	55	53	51	48	45	43	39	36	32	28	24	19	11	1
95	94	93	91	90	89	87	86	85	83	82	80	79	78	76	74	73	71	70	68	66	64	62	60	58	56	54	52	49	46	43	40	37	33	29	24
100	99	98	96	95	94	93	91	90	89	87	86	85	83	82	80	79	77	76	74	72	71	69	67	65	63	61	59	57	55	52	50	47	44	41	37
105	104	103	101	100	99	98	96	95	94	93	91	90	89	87	86	84	83	82	80	78	77	75	74	72	70	68	67	65	63	61	58	56	54	51	48
110	109	108	106	105	104	103	102	100	99	98	97	95	94	93	91	90	89	87	86	84	83	81	80	78	77	75	73	72	70	68	66	64	62	60	57
115	114	113	112	110	109	108	107	106	104	103	102	101	99	98	97	96	94	93	92	90	89	87	86	84	83	81	80	78	76	75	73	71	69	67	65
120	119	118	117	115	114	113	112	111	110	108	107	106	105	104	102	101	100	98	97	96	94	93	92	90	89	87	86	84	83	81	80	78	76	75	73
125	124	123	122	121	119	118	117	116	115	114	112	111	110	109	108	106	105	104	103	101	100	99	97	96	95	93	92	90	89	88	86	84	83	81	80
130	129	128	127	126	124	123	122	121	120	119	118	116	115	114	113	112	110	109	108	107	106	104	103	102	100	99	98	96	95	94	92	91	89	88	86

* Southern California Meter Association.

Table 1-29 Per Cent Relative Humidity of Air*

Air temp, °F	Depression of wet-bulb thermometer, °F																																		
	1	2	3	4	5	6	7	8	9	10	11	12	13	14	15	16	17	18	19	20	21	22	23	24	25	26	27	28	29	30	31	32	33	34	35
0	67	33	1																																
5	73	46	20																																
10	78	56	34	13																															
15	82	64	46	29	11																														
20	85	70	55	40	26	12																													
25	87	74	62	49	37	25	13	1																											
30	89	78	67	56	46	36	26	16	6																										
35	91	81	72	63	54	45	36	27	19	10	2																								
40	92	83	75	68	60	52	45	37	29	22	15	7																							
45	93	86	78	71	64	57	51	44	38	31	25	18	12	6																					
50	93	87	80	74	67	61	55	49	43	38	32	27	21	16	10	5																			
55	94	88	82	76	70	65	59	54	49	43	38	33	28	23	19	14	9	5																	
60	94	89	83	78	73	68	63	58	53	48	43	39	34	30	26	21	17	13	9	5	1														
65	95	90	85	80	75	70	66	61	56	52	48	44	39	35	31	27	24	20	16	12	9	5	2												
70	95	90	86	81	77	72	68	64	59	55	51	48	44	40	36	33	29	25	22	19	15	12	9	6	3										
75	96	91	86	82	78	74	70	66	62	58	54	51	47	44	40	37	34	30	27	24	21	18	15	12	9	7	4	1							
80	96	91	87	83	79	75	72	68	64	61	57	54	50	47	44	41	38	35	32	29	26	23	20	18	15	12	10	7	5	3					
85	96	92	88	84	81	77	73	70	66	63	59	57	53	50	47	44	41	38	36	33	30	27	25	22	20	17	15	13	10	8	6	4	2		
90	96	92	89	85	81	78	74	71	68	65	61	58	55	52	49	47	44	41	39	36	34	31	29	26	24	22	19	17	15	13	11	9	7	5	3
95	96	93	89	86	82	79	76	73	69	66	63	61	58	55	52	50	47	44	42	39	37	34	32	30	28	25	23	21	19	17	15	13	11	10	8
100	96	93	89	86	83	80	77	73	70	68	65	62	59	56	54	51	49	46	44	41	39	37	35	33	30	28	26	24	22	21	19	17	15	13	12
105	97	93	90	87	84	81	78	75	72	69	66	64	61	58	56	53	51	49	46	44	42	40	38	36	34	32	30	28	26	24	22	21	19	17	15
110	97	93	90	87	84	81	78	75	73	70	67	65	62	60	57	55	52	50	48	46	44	42	40	38	36	34	32	30	28	26	25	23	21	20	18
115	97	94	91	88	85	82	79	76	74	71	69	66	64	61	59	57	54	52	50	48	46	44	42	40	38	36	34	33	31	29	28	26	25	23	21
120	97	94	91	88	85	82	80	77	74	72	69	67	65	62	60	58	55	53	51	49	47	45	43	41	40	38	36	34	33	31	29	28	26	25	23
125	97	94	91	88	86	83	80	78	75	73	70	68	66	64	61	59	57	55	53	51	49	47	45	44	42	40	38	37	35	33	32	30	29	27	26
130	97	94	91	89	86	83	81	78	76	73	71	69	67	64	62	60	58	56	54	52	50	48	47	45	43	41	40	38	37	35	33	32	30	29	28

* Southern California Meter Association.

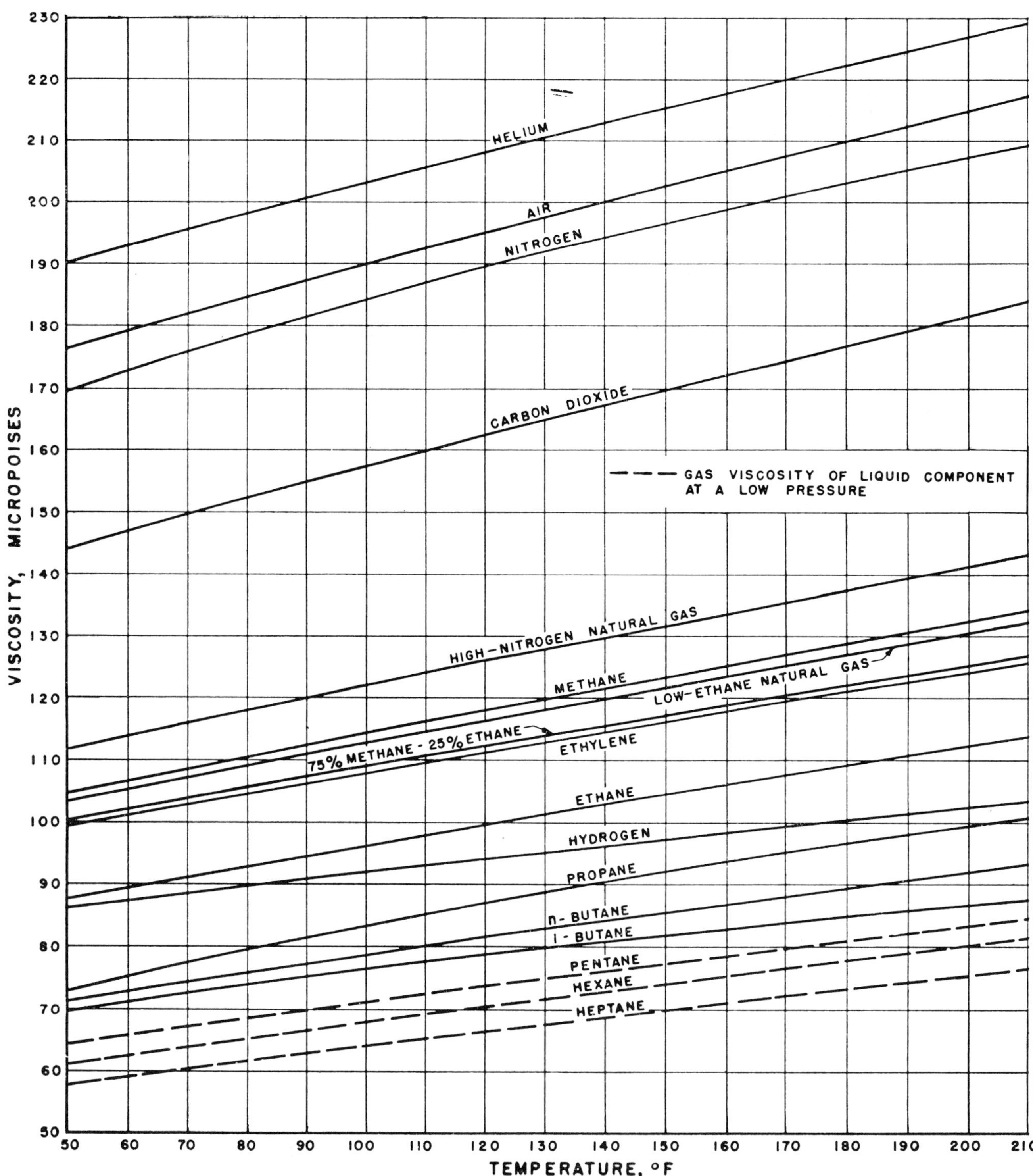

Fig. 1-5 Viscosity vs. temperature at atmospheric pressure for natural gas components and some natural gas mixtures.[5] (The absolute or dynamic viscosities shown here are generally given the symbol μ; see footnotes to Table 1-48.)

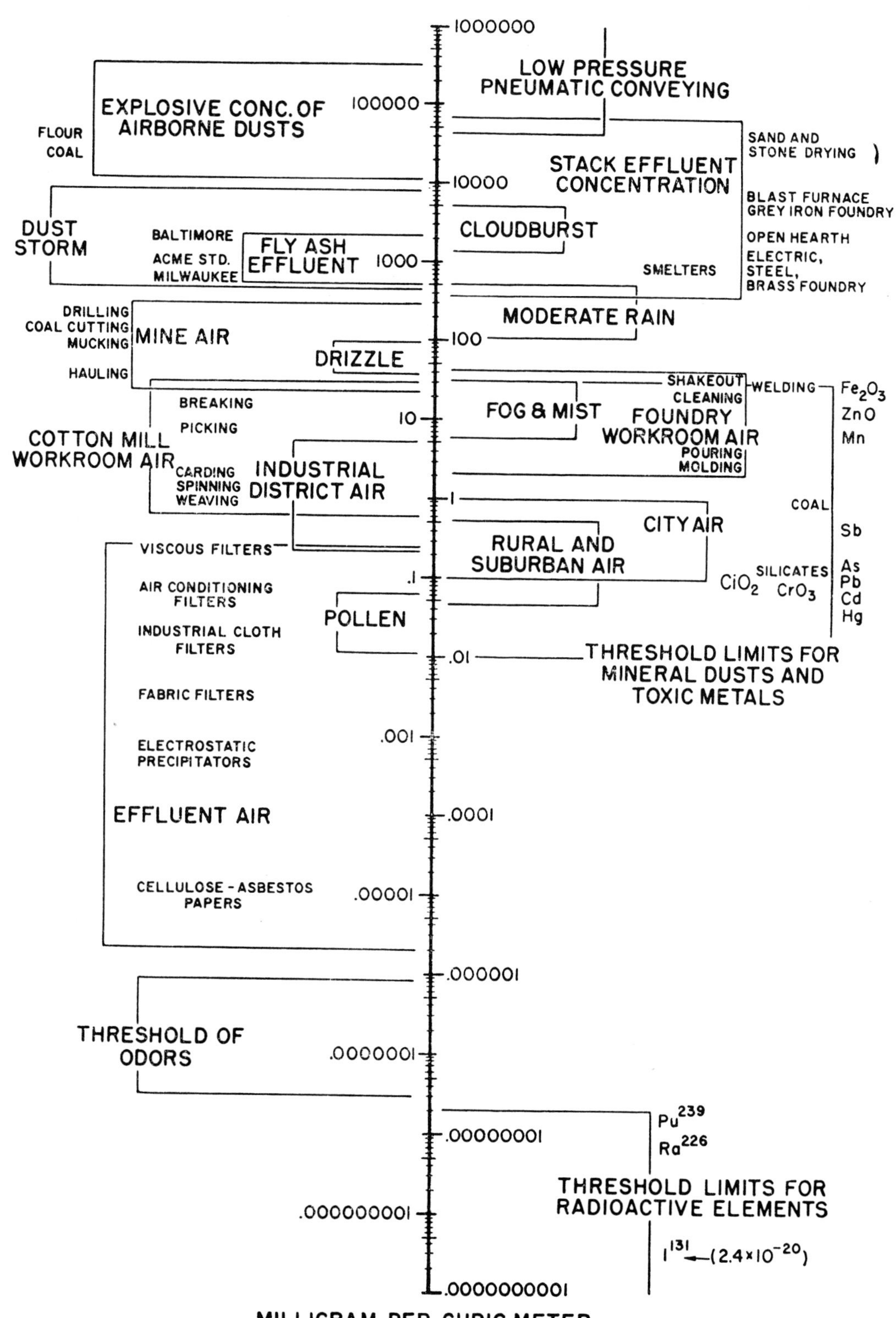

Fig. 1-6a Average concentrations of air-borne particulate matter. (Mine Safety Appliances Co.) (See page 1/35 for footnotes.)

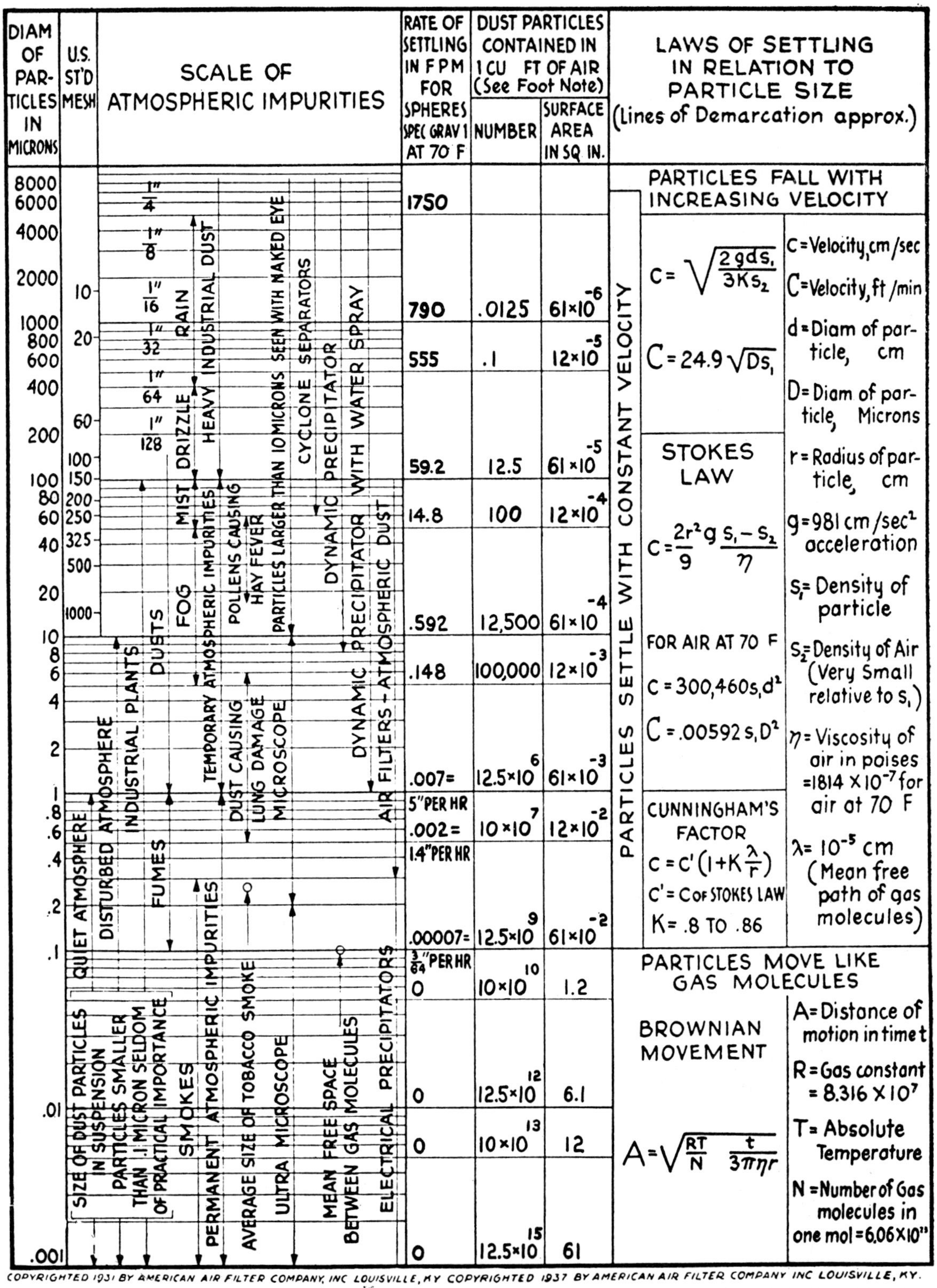

DIAM OF PARTICLES IN MICRONS	U.S. ST'D MESH	RATE OF SETTLING IN FPM FOR SPHERES SPEC GRAV 1 AT 70 F	DUST PARTICLES CONTAINED IN 1 CU FT OF AIR (See Foot Note): NUMBER	SURFACE AREA IN SQ IN.
8000, 6000		1750		
4000, 2000	10	790	.0125	61×10^{-6}
1000, 800, 600	20	555	.1	12×10^{-5}
400, 200	60, 100	59.2	12.5	61×10^{-5}
100, 80, 60	150, 200, 250	14.8	100	12×10^{-4}
40, 20	325, 500, 1000	.592	12,500	61×10^{-4}
10, 8, 6		.148	100,000	12×10^{-3}
4, 2		.007=	12.5×10^{6}	61×10^{-3}
1, .8, .6		5"PER HR .002=	10×10^{7}	12×10^{-2}
.4, .2		1.4"PER HR .00007=	12.5×10^{9}	61×10^{-2}
.1		$\frac{3}{64}$"PER HR 0	10×10^{10}	1.2
		0	12.5×10^{12}	6.1
.01		0	10×10^{13}	12
.001		0	12.5×10^{15}	61

SCALE OF ATMOSPHERIC IMPURITIES

- 1"/4, 1"/8, 1"/16, 1"/32, 1"/64, 1"/128
- RAIN
- MIST, DRIZZLE
- HEAVY INDUSTRIAL DUST
- FOG
- DUSTS
- FUMES
- SMOKES
- TEMPORARY ATMOSPHERIC IMPURITIES
- PERMANENT ATMOSPHERIC IMPURITIES
- POLLENS CAUSING HAY FEVER
- DUST CAUSING LUNG DAMAGE
- AVERAGE SIZE OF TOBACCO SMOKE
- PARTICLES LARGER THAN 10 MICRONS SEEN WITH NAKED EYE
- MICROSCOPE
- ULTRA MICROSCOPE
- CYCLONE SEPARATORS
- DYNAMIC PRECIPITATOR
- DYNAMIC PRECIPITATOR WITH WATER SPRAY
- DYNAMIC PRECIPITATOR
- MEAN FREE SPACE BETWEEN GAS MOLECULES
- AIR FILTERS - ATMOSPHERIC DUST
- ELECTRICAL PRECIPITATORS
- QUIET ATMOSPHERE
- DISTURBED ATMOSPHERE
- INDUSTRIAL PLANTS
- SIZE OF DUST PARTICLES IN SUSPENSION
- PARTICLES SMALLER THAN .1 MICRON SELDOM OF PRACTICAL IMPORTANCE

LAWS OF SETTLING IN RELATION TO PARTICLE SIZE (Lines of Demarcation approx.)

PARTICLES FALL WITH INCREASING VELOCITY

$$C = \sqrt{\frac{2gds_1}{3Ks_2}}$$

$$C = 24.9\sqrt{Ds_1}$$

PARTICLES SETTLE WITH CONSTANT VELOCITY

STOKES LAW

$$C = \frac{2r^2g}{9}\,\frac{s_1 - s_2}{\eta}$$

FOR AIR AT 70 F

$$C = 300{,}460 s_1 d^2$$

$$C = .00592 s_1 D^2$$

CUNNINGHAM'S FACTOR

$$C = C'\left(1 + K\frac{\lambda}{r}\right)$$

C' = C OF STOKES LAW

K = .8 TO .86

c = Velocity, cm/sec
C = Velocity, ft/min
d = Diam of particle, cm
D = Diam of particle, Microns
r = Radius of particle, cm
g = 981 cm/sec² acceleration
s_1 = Density of particle
s_2 = Density of Air (Very Small relative to s_1)
η = Viscosity of air in poises = 1814×10^{-7} for air at 70 F
λ = 10^{-5} cm (Mean free path of gas molecules)

PARTICLES MOVE LIKE GAS MOLECULES

BROWNIAN MOVEMENT

$$A = \sqrt{\frac{RT}{N}\,\frac{t}{3\pi\eta r}}$$

A = Distance of motion in time t
R = Gas constant = 8.316×10^{7}
T = Absolute Temperature
N = Number of Gas molecules in one mol = 6.06×10^{23}

IT IS ASSUMED THAT THE PARTICLES ARE OF UNIFORM SPHERICAL SHAPE HAVING SPECIFIC GRAVITY ONE AND THAT THE DUST CONCENTRATION IS 0.1 GRAINS PER 1000 CU FT OF AIR, THE AVERAGE OF METROPOLITAN DISTRICTS.

Fig. 1-6b Sizes and characteristics of air-borne particulate matter.

Table 1-30 Maximum Allowable Concentration (Threshold Limit*) of Various Substances in Air[7]

Gases and Vapors

Substance	ppm	mg/cu m
Acetaldehyde	200	360
Acetic acid	10	25
Acetic anhydride	5	20
Acetone	1000	2400
Acrolein	0.5	1.2
Acrylonitrile	20	45
Allyl alcohol	5	12
Allyl chloride	5	15
Allyl propyl disulfide	2	12
Ammonia	100	70
Amyl acetate	200	1050
Amyl alcohol (isoamyl alcohol)	100	360
Aniline	5	19
Arsine	0.05	0.2
Benzene (benzol)	25	80
Benzyl chloride	1	5
Bromine	1	7
Butadiene (1,3-butadiene)	1000	2200
Butanone (methyl ethyl ketone)	250	740
Butyl acetate (*n*-butyl acetate)	200	950
Butyl alcohol (*n*-butanol)	100	300
Butylamine	5	15
Butyl cellosolve (2-butoxyethanol)	50	240
Carbon dioxide	5000	9000
Carbon disulfide	20	60
Carbon monoxide	100	110
Carbon tetrachloride	25	160
Cellosolve (2-ethoxyethanol)	200	740
Cellosolve acetate (2-ethoxyethyl acetate)	100	540
Chlorine	1	3
Chlorine trifluoride	0.1	0.4
Chlorobenzene (monochlorobenzene)	75	350
Chloroform (trichloromethane)	100	490
1-Chloro-1-nitropropane	20	100
Chloropicrin	1	7
Chloroprene (2-chloro-1,3-butadiene)	25	90
Cresol (all isomers)	5	22
Cyclohexane	400	1400
Cyclohexanol	100	410
Cyclohexanone	100	400
Cyclohexene	400	1350
Cyclopropane	400	690
Decaborane	0.05	0.3
Diacetone alcohol (4-hydroxy-4-methyl-2-pentanone)	50	240
Diborane	0.1	0.1
o-Dichlorobenzene	50	300
Dichlorodifluoromethane	1000	4950
1,1-Dichloroethane	100	400
1,2-Dichloroethane (ethylene dichloride)	100	400
1,2-Dichloroethylene	200	790
Dichloroethyl ether	15	90
Dichloromonofluoromethane	1000	4200
1,1-Dichloro-1-nitroethane	10	60
Dichlorotetrafluoroethane	1000	7000
Diethylamine	25	75
Difluorodibromomethane	100	860
Diisobutyl ketone	50	290
Dimethylaniline (N-dimethylaniline)	5	25
Dimethylsulfate	1	5
Dioxane (diethylene dioxide)	100	360
Ethyl acetate	400	1400
Ethyl acrylate	25	100
Ethyl alcohol (ethanol)	1000	1900
Ethylamine	25	45
Ethylbenzene	200	870
Ethyl bromide	200	890
Ethyl chloride	1000	2600
Ethyl ether	400	1200
Ethyl formate	100	300
Ethyl silicate	100	850
Ethylene chlorohydrin	5	16
Ethylenediamine	10	30
Ethylene dibromide (1,2-dibromoethane)	25	190
Ethylene imine	5	9
Ethylene oxide	50	90
Fluorine	0.1	0.2
Fluorotrichloromethane	1000	5600
Formaldehyde	5	6
Furfural	5	20
Gasoline	500	2000
Heptane (*n*-heptane)	500	2000
Hexane (*n*-hexane)	500	1800
Hexanone (methyl butyl ketone)	100	410
Hexone (methyl isobutyl ketone)	100	410
Hydrazine	1	1.3
Hydrogen bromide	5	17
Hydrogen chloride	5	7
Hydrogen cyanide	10	11
Hydrogen fluoride	3	2
Hydrogen peroxide, 90%	1	1.4
Hydrogen selenide	0.05	0.2
Hydrogen sulfide	20	30
Iodine	0.1	1
Isophorone	25	140
Isopropylamine	5	12
Mesityl oxide	25	100
Methyl acetate	200	610
Methyl acetylene	1000	1650
Methyl acrylate	10	35
Methyl alcohol (methanol)	200	260
Methyl bromide	20	80
Methyl cellosolve (2-methoxyethanol)	25	80
Methyl cellosolve acetate (ethylene glycol monomethyl ether acetate)	25	120
Methyl chloride	100	210
Methylal (dimethoxymethane)	1000	3100
Methyl chloroform (1,1,1-trichloroethane)	500	2700
Methylcyclohexane	500	2000
Methylcyclohexanol	100	470
Methylcyclohexanone	100	460
Methyl formate	100	250
Methyl isobutyl carbinol (methyl amyl alcohol)	25	100

Continued on page 1/34

* Threshold limits should be used as guides in the control of health hazards and should not be regarded as fine lines between safe and dangerous concentrations. They represent conditions under which it is believed that nearly all workers may be repeatedly exposed, day after day, without adverse effect. The values listed refer to time-weighted average concentrations for a normal work day. The amount by which these figures may be exceeded for short periods without injury to health depends upon a number of factors, such as the nature of the contaminant, whether very high concentrations even for short periods produce acute poisoning, whether the effects are cumulative, the frequency with which high concentrations occur, and the duration of such periods. All must be taken into consideration in arriving at a decision as to whether a hazardous situation exists. Special consideration should be given to the application of these values in the evaluation of the health hazards which may be associated with exposure to combinations of two or more substances.

Table 1–30 Maximum Allowable Concentration (Threshold Limit*) of Various Substances in Air[7] (Continued)

Substance	ppm	mg/cu m
Methylene chloride (dichloromethane)	500	1750
Naphtha (coal tar)	200	800
Naphtha (petroleum)	500	2000
Nickel carbonyl	0.001	0.007
Nitric acid	10	25
p-Nitroaniline	1	6
Nitrobenzene	1	5
Nitroethane	100	310
Nitrogen dioxide	5	9
Nitroglycerin	0.5	5
Nitromethane	100	250
2-Nitropropane	50	180
Nitrotoluene	5	30
Octane	500	2350
Ozone	0.1	0.2
Pentane	1000	2950
Pentanone (methyl propyl ketone)	200	700
Perchlorethylene (tetrachloroethylene)	200	1350
Phenol	5	19
Phenylhydrazine	5	22
Phosgene (carbonyl chloride)	1	4
Phosphine	0.05	0.07
Phosphorus trichloride	0.5	3
Propyl acetate	200	840
Propyl alcohol (isopropyl alcohol)	400	980
Propyl ether (isopropyl ether)	500	2100
Propylene dichloride (1,2-dichloropropane)	75	350
Propylene imine	25	60
Pyridine	10	30
Quinone	0.1	0.4
Stibine	0.1	0.5
Stoddard solvent	500	2900
Styrene monomer (phenylethylene)	100	420
Sulfur dioxide	5	13
Sulfur hexafluoride	1000	6000
Sulfur monochloride	1	6
Sulfur pentafluoride	0.025	0.25
p-Tertiarybutyltoluene	10	60
1,1,2,2-Tetrachloroethane	5	35
Tetrahydrofuran	200	590
Tetranitromethane	1	8
Toluene (toluol)	200	750
o-Toluidine	5	22
Trichloroethylene	200	1050
Trifluoromonobromomethane	1000	6100
Turpentine	100	560
Vinyl chloride (chloroethylene)	500	1300
Xylene (xylol)	200	870

Toxic Dusts, Fumes, and Mists

Substance	mg/cu m
Aldrin (1,2,3,4,10,10-hexachloro-1,4,4a,5,8,8a-hexahydro-1,4,5,8-dimethanonaphthalene)	0.25
Ammate (ammonium sulfamate)	15
Antimony	0.5
ANTU (alpha-naphthyl-thiourea)	0.3
Arsenic	0.5
Barium (soluble compounds)	0.5
Cadmium oxide fume	0.1
Calcium arsenate	0.1
Chlordane (1,2,4,5,6,7,8,8-octachloro-3a,4,7,7a-tetrahydro-4,7-methanoindane)	2
Chlorinated camphene, 60%	0.5
Chlorinated diphenyl oxide	0.5
Chlorodiphenyl (42% chlorine)	1
Chlorodiphenyl (54% chlorine)	0.5
Chromic acid and chromates (as CrO_3)	0.1
Crag herbicide (sodium 2-[2,4-dichlorophenoxy] ethanol hydrogen sulfate)	15
Cyanide (as CN)	5
2,4-D (2,4-dichlorophenoxyacetic acid)	10
DDT (2,2-bis[*p*-chlorophenyl]-1,1,1-trichloroethane)	1
Dieldrin (1,2,3,4,10,10-hexachloro-6,7-epoxy-1,4,4a,5,6,7,8,8a-octahydro-1,4,5,8-dimethanonaphthalene)	0.25
Dinitrobenzene	1
Dinitrotoluene	1.5
Dinitro-*o*-cresol	0.2
EPN (*o*-ethyl *o*-*p*-nitrophenyl thionobenzenephosphonate)	0.5
Ferbam (ferric dimethyl dithiocarbamate)	15
Ferrovanadium dust	1
Fluoride	2.5
Hydroquinone	2
Iron oxide fume	15
Lead	0.2
Lead arsenate	0.15
Lindane (hexachlorocyclohexane, gamma isomer)	0.5
Magnesium oxide fume	15
Malathion (*o,o*-dimethyl dithiophosphate of diethyl mercaptosuccinate)	15
Manganese	6
Mercury	0.1
Mercury (organic compounds)	0.01
Methoxychlor (2,2-di-*p*-methoxyphenyl-1,1,1-trichloroethane)	15
Molybdenum	
(soluble compounds)	5
(insoluble compounds)	15
Nicotine	0.5
Parathion (*o,o*-diethyl *o*-*p*-nitrophenyl thiophosphate)	0.1
Pentachloronaphthalene	0.5
Pentachlorophenol	0.5
Phosphorus (yellow)	0.1
Phosphorus pentachloride	1
Phosphorus pentasulfide	1
Picric acid	0.1
Pyrethrum	2
Rotenone	5
Selenium compounds (as Se)	0.1
Sodium fluoroacetate (1080)	0.1
Sodium hydroxide	2
Strychnine	0.15
Sulfuric acid	1
TEDP (tetraethyl dithionopyrophosphate)	0.2
TEPP (tetraethyl pyrophosphate)	0.05
Tellurium	0.1
Tetryl (2,4,6-trinitrophenylmethylnitramine)	1.5
Thiram (tetramethyl thiuram disulfide)	5
Thallium (soluble compounds)	0.1
Titanium dioxide	15
Trichloronaphthalene	5
Trinitrotoluene	1.5
Uranium	
(soluble compounds)	0.5
(insoluble compounds)	0.25
Vanadium	
(V_2O_5 dust)	0.5
(V_2O_5 fume)	0.1
Warfarin (3-[α acetonylbenzyl]-4-hydroxycoumarin)	0.5
Zinc oxide fumes	15
Zirconium compounds (as Zr)	5

Radioactivity: For permissible concentrations of radioisotopes in air, see "Maximum Permissible Amounts of Radioisotopes in the Human Body and Maximum Permissible Concentration in Air and Water," Handbook 52, U. S. Department of Commerce, National Bureau of Standards, March 1953. In addition, see "Permissible Dose from External Sources of Ionizing Radiation," Handbook 59, U. S. Department of Commerce, National Bureau of Standards, Sept. 24, 1954.

Mineral Dusts

Substance	MPPCF†	Substance	MPPCF†
Aluminum oxide	50	Silica	
Asbestos	5	high (above 50% free SiO_2)	5
Dust (nuisance, no free silica)	50	medium (5 to 50% free SiO_2)	20
Mica (below 5% free silica)	20	low (below 5% free SiO_2)	50
Portland cement	50	Silicon carbide	50
Talc	20	Soapstone (below 5% free SiO_2)	20

Tentative Values‡

Substance	ppm	mg/cu m	Substance	ppm	mg/cu m
Acetylene tetrabromide	1	14	Methyl mercaptan	50	100
Allyl glycidyl ether (AGE)	10	45	α Methyl styrene	100	480
Beryllium		2γ/m³	Monomethyl aniline	2	9
Boron trifluoride	1	3	Paradichlorobenzene	75	450
n-Butyl glycidyl ether (BGE)	50	270	Perchloromethyl mercaptan	0.1	0.8
Butyl mercaptan	10	35	Phenyl glycidyl ether (PGE)	50	310
Chlorine dioxide	0.1	0.3	Phosphoric acid		1
Chloroacetaldehyde	1	3	*n*-Propyl nitrate	25	110
Chlorobromomethane ($ClBrCH_2$)	400	2100	Propylene oxide	100	240
Diglycidyl ether (DGE)	10	55	Tertiary butyl alcohol	100	300
Dimethyl formamide	20	60	Tolylene-2,4-diisocyanate	0.1	0.7
1,1-Dimethyl hydrazine	0.5	1	1,2,3-Trichloropropane	50	300
Dipropyleneglycolmethylether	100	600	Triethyl amine	25	100
Ethyl mercaptan	250	640	Triorthocresyl phosphate		0.1
Furfuryl alcohol	50	200	Vinyl toluene	100	480
Glycidol	50	150	Yttrium and inorganic compounds		5
sec-Hexyl acetate	100	590	Xylidine	5	25
Isopropyl glycidyl ether (IGE)	50	240	Teflon decomposition products	‡	‡
Lithium hydride		25γ/m³	Pentaborane (B_5H_9)	‡	‡

† Millions of particles per cubic foot of air. ‡ Until more data are forthcoming, it is important that atmospheric concentrations of these materials to which workers are exposed must be kept as near 0 as possible.

Footnotes to Fig. 1-6a:

Particulate load is one of the most important factors influencing the selection of air cleaners; shows the average concentrations of air-borne particulate matter under a wide range of industrial conditions.

Particulate loads are commonly classified as:

1. Light Load—0.1 to 1 grain per 1000 cu ft
2. Medium Load—1 to 1000 grains per 1000 cu ft
3. Heavy Load—1000 grains and up per 1000 cu ft; 1 grain per 1000 cu ft = 2.3 mg per cubic meter = 2.3 micrograms per liter. One milligram is equivalent to 0.015432 grains. As a comparison, the concentration of the U. S. Bureau of Mines silica dust test is approximately 22 grains per 1000 cu ft of air.

Normal load ranges may be listed as follows:

1. Country Air—0.2 grains per 1000 cu ft
2. Small Town—0.2 to 0.4 grains per 1000 cu ft
3. City Res.—0.4 to 0.8 grains per 1000 cu ft
4. City Ind.—1.0 to 2.0 grains per 1000 cu ft
5. Ind. Operations—4.0 to 8.0 grains per 1000 cu ft (7000 grains = 1 lb)

In general, the particulate cleaner types can be grouped as follows with respect to load condition:

1. Heavy Loads—gravity separators, cyclone separators, sonic collectors.
2. Medium Loads—all of the above plus bag filters, impingement type box filters, and wet collectors.
3. Light Loads—box filters of both types and electrostatic precipitators.

When MSA **Ultra-Aire** filters are used in conditions of heavy or medium loads, one or more of the other types must be used ahead of it to reduce the load to a practical level.

Table 1-31 Density of Water vs. Temperature*

Temp, °F	Lb/cu ft	Temp, °F	Lb/cu ft
10	62.252	110	61.860
20	62.352	120	61.713
30	62.414	130	61.552
32	62.4184	140	61.380
40	62.4265	150	61.196
50	62.4096	160	61.001
60	62.3668	170	60.796
70	62.3016	180	60.581
80	62.2165	190	60.356
90	62.1134	200	60.121
100	61.9943	212	59.828

* Interpolated from Centigrade values in I.C.T.

References

1. Matthews, C. S. and Hurd, C. O. "Thermodynamic Properties of Methane." *Trans. AIChE*, 42: 55–78, 1009, 1946.
2. Barkelew, C. H. and others. "Thermodynamic Properties of Ethane." *Trans. AIChE*, 43: 25–38, 1947.
3. Maxwell, J. B. *Data Book on Hydrocarbons*. New York, Van Nostrand, 1950.
4. Bloomer, O. T. and others. *Physical-Chemical Properties of Methane-Ethane Mixtures*. (I.G.T. Research Bul. 22) Chicago, I.G.T., 1953.
5. Carr, N. L. *Viscosities of Natural Gas Components and Mixtures*. (I.G.T. Research Bul. 23) Chicago, I.G.T., 1953.
6. Stull, Daniel R. "Vapor Pressure of Pure Substances (Organic and Inorganic Compounds)." *Ind. Eng. Chem.*, 39: 517–550, Ap. 1947; 1684, Dec. 1947.
7. Coleman, Allan L., chm. "Threshhold Limit Values for 1958." A.M.A., *Archives of Ind. Health*, 18: 178–182, Aug. 1958.
8. Bloomer, O. T. and Rao, K. N. *Thermodynamic Properties of Nitrogen*. (I.G.T. Research Bul. 18) Chicago, I.G.T., 1952.

Chapter 3

Mathematical and Conversion Tables

by Louis Shnidman, C. E. Farmer, and Robert J. Gustafson

Table 1-32 Properties of Circles

Diameter	Diameter, decimal equivalent	Circumference	Area	Diameter	Diameter, decimal equivalent	Circumference	Area
1/32	0.031250	0.098175	0.00077	2 5/8	2.62500	8.24668	5.4119
1/16	.062500	.196350	.00307	2 3/4	2.75000	8.63938	5.9396
3/32	.09375	.294524	.00690	2 7/8	2.87500	9.03208	6.4918
1/8	.12500	.392699	.01227	3	3.00000	9.42478	7.0686
3/16	.18750	.589049	.02761	3 1/4	3.25000	10.2102	8.2958
1/4	.25000	.785398	.04909	3 1/2	3.50000	10.9956	9.6211
5/16	.31250	0.981748	.07670	3 3/4	3.75000	11.7810	11.045
3/8	.37500	1.17810	.11045	4	4.00000	12.5664	12.566
7/16	.43750	1.37445	.15033	4 1/4	4.25000	13.3518	14.186
1/2	.50000	1.57080	.19635	4 1/2	4.50000	14.1372	15.904
5/8	.62500	1.96350	.30680	4 3/4	4.75000	14.9226	17.721
3/4	.75000	2.35619	.44179	5	5.00000	15.7080	19.635
7/8	0.87500	2.74889	.60132	5 1/2	5.50000	17.2788	23.758
1	1.00000	3.14159	.78540	6	6.00000	18.8496	28.274
1 1/8	1.12500	3.53429	0.99402	6 1/2	6.50000	20.4204	33.183
1 1/4	1.25000	3.92699	1.2272	7	7.00000	21.9911	38.485
1 3/8	1.37500	4.31969	1.4849	7 1/2	7.50000	23.5619	44.179
1 1/2	1.50000	4.71239	1.7671	8	8.00000	25.1327	50.265
1 5/8	1.62500	5.10509	2.0739	8 1/2	8.50000	26.7035	56.745
1 3/4	1.75000	5.49779	2.4053	9	9.00000	28.2743	63.617
1 7/8	1.87500	5.89049	2.7612	9 1/2	9.50000	29.8451	70.882
2	2.00000	6.28319	3.1416	10	10.00000	31.4159	78.540
2 1/8	2.12500	6.67588	3.5466	10 1/2	10.50000	32.9867	86.590
2 1/4	2.25000	7.06858	3.9761	11	11.00000	34.5575	95.033
2 3/8	2.37500	7.46128	4.4301	11 1/2	11.50000	36.1283	103.87
2 1/2	2.50000	7.85398	4.9087	12	12.00000	37.6991	113.10

Table 1-33 Numerical Constants[1]

Constant	Value	Constant	Value	Constant	Value
π	3.141593	$1/\pi^3$	0.032252	$1/e^2$	0.135335
$1/\pi$	0.318310	$\sqrt[3]{\pi}$	1.464592	$\sqrt{e}$	1.648721
π^2	9.869604	$1/\sqrt[3]{\pi}$	0.682784	1 radian	57.295780 deg
$1/\pi^2$	0.101321	$\log_e \pi$	1.144730	1 radian	3,437.7468 min
$\sqrt{\pi}$	1.772454	$\log_{10} \pi$	0.497150	1 radian	206,264.81 sec
$1/\sqrt{\pi}$	0.564190	e	2.718282	1 deg	0.017453 radian
π^3	31.00628	$1/e$	0.367879	1 min	0.0002909 radian
		e^2	7.389056	1 sec	0.00000485 radian

Table 1-34 Linear Conversion Table

(expressions like 0.0_4, 0.0_5, etc., stand for 0.0000, 0.00000, etc.)

Units	Inches	Feet	Yards	Rods	Chains	Miles	Nautical miles
Millimeter	0.03937	0.003280833	0.001093611	0.0001988383	$0.0_{4}4970957$	$0.0_{6}62137$	$0.0_{6}53959$
Centimeter	0.3937	.03280833	.01093611	.001988383	.0004970957	$.0_{5}62137$	$.0_{5}53959$
Decimeter	3.937	0.3280833	0.1093611	.01988383	.004970957	$.0_{4}62137$	$.0_{4}53959$
Meter	39.37	3.280833	1.093611	0.1988383	.04970957	.00062137	.00053959
Decameter	393.7	32.80833	10.93611	1.988383	0.4970957	.0062137	.0053959
Hectometer	3937.	328.0833	109.3611	19.88383	4.970957	.062137	.053959
Kilometer	39370.	3280.833	1093.611	198.8383	49.70957	.62137	.53959
Inch	1.00	0.083333	0.0277778	0.0050505	0.001262625	$.0_{4}157828$	$.0_{4}13706$
Foot	12.0	1.00	0.333333	.0606061	.0151515	.000189394	.00016447
Yard	36.0	3.0	1.00	0.181818	.0454545	.000568181	.00049339
Rod	198.0	16.5	5.5	1.00	0.25	.003125	.00271365
Chain	792.0	66.0	22.0	4.0	1.00	0.0125	.0108546
Mile	63360.0	5280.0	1760.0	320.0	80.0	1.00	0.868365
Nautical mile	72963.12	6080.26	2026.7533	368.5006	92.12515	1.15156431	1.00

1 Angstrom = 0.0000001 millimeters
1 Micron = 0.001 millimeters = 10,000 Angstroms (Ä or Än)
1 Meter = 1,000 millimeters
1 Kilometer = 1,000 meters
1 Mil = 0.001 inch

Table 1-35 Square-Measure Conversion Table

(expressions like 0.0_4, 0.0_5, etc., stand for 0.0000, 0.00000, etc.)

Units	Square centimeters	Square meters	Square inches	Square feet	Square yards	Square rods	Square chains	Acres	Square miles
Square centimeter	1.00	0.0001	0.15499969	0.001076387	0.0001193985	$0.0_{5}3953670$	$0.0_{6}24710439$	$0.0_{7}24710439$	$0.0_{10}38610061$
Square meter	10,000.00	1.00	1549.9969	10.7638674	1.19398526	0.039536703	0.0024710439	0.00024710439	$.0_{6}38610061$
Hectare	100,000,000.0	10,000.00	15,499,969.00	107,638.674	11,959.8526	395.36703	24.71043930	2.471043930	.0038610061
Square inch	6.45162581	0.0006451628	1.00	0.00694445	0.000771605	$0.0_{4}2550760$	$0.0_{5}15942253$	$0.0_{5}15942253$	$.0_{9}24909767$
Square link	404.687261	.04046873	62.7264	0.4356	.0484	.0016	.0001	.00001	$.0_{7}15625$
Square foot	929.034110	.0929034110	144.00	1.00	0.11111111	.0036730946	.00022956841	$.0_{4}22956841$	$.0_{7}358701$
Square yard	8,361.307	0.8361307	1,296.00	9.00	1.00	0.0330578511	.002066116	.0002066116	$.0_{6}3228306$
Square rod	252,929.537	25.2929537	39,204.00	272.25	30.25	1.00	0.0625	.00625	$.0_{5}9765625$
Square chain	4,046,872.61	404.687261	627,264.00	4,356.00	484.00	16.00	1.00	0.10	.00015625
Acre	40,468,726.10	4,046.872610	6,272,640.00	43,560.00	4,840.00	160.00	10.00	1.00	0.0015625
Square mile	25,899,984,703.2	2,589,998.47	4,014,489,600.00	27,878,400.00	3,097,600.00	102,400.00	6,400.00	640.00	1.00

Table 1-36 Weight Conversion Table

(expressions like 0.0_4, 0.0_5, etc., stand for 0.0000, 0.00000, etc.)

Units	Milligrams	Grams	Metric tons	Grains	Ounces, avoirdupois	Pounds, avoirdupois	Short tons	Long tons
Milligram	1.00	0.001	0.000000001	0.01543236	$0.0_{4}3527396$	$0.0_{5}220462$	$0.0_{8}1102311$	$0.0_{9}984206$
Gram	1,000	1.00	0.000001	15.43236	0.03527396	0.002204622	$0.0_{5}1102311$	$.0_{6}984206$
Metric ton	1,000,000,000	1,000,000	1.00	15,432,360	35,273.96	2,204.622	1.102311	.9842064
Grain	64.79892	0.06479892	$0.0_{7}647989$	1.00	0.002285714	0.0001428571	$0.0_{7}714286$	$.0_{7}637755$
Ounce, avoirdupois	28,349.53	28.34953	$.0_{4}2834953$	437.5	1.00	0.0625	$.0_{4}3125$	$.0_{4}27902$
Pound, avoirdupois	453,592.4	453.5924	.0004535924	7,000	16.00	1.00	0.00050	.00044643
Short ton	907,184,900	907,184.9	0.9071849	14,000,000	32,000	2,000	1.00	0.892857
Long ton	1,016,047,000	1,016,047	1.016047	15,680,000	35,840.00	2,240	1.120	1.00

Dram = 1772 grams

Table 1-37 Liquid and Cubic Measure Conversion Table

(expressions like 0.0_4, 0.0_5, etc., stand for 0.0000, 0.00000, etc.)

Units	Milliliters	Liters	Kiloliters	Fluid ounces	Liquid quarts	Liquid gallons	Cubic inches	Cubic feet	Cubic yards
Milliliter	1.00	0.001	0.000001	0.03381473	0.00105671	0.0002641776	0.0610250	0.0000353154	$0.0_{5}1307978$
Liter	1,000	1.00	0.001	33.81473	1.056710	0.2641776	61.02503	0.03531541	0.001307978
Kiloliter	1,000,000	1000	1.00	33,814.73	1,056.710	264.1776	61,025.03	35.31541	1.307978
Fluid ounce	29.57291	0.02957291	$0.0_{4}2957291$	1.00	0.03125	0.00078125	1.804688	0.00104438	$0.0_{4}386807$
Liquid quart	946.3331	0.9463331	.0009463331	32.00	1.00	0.25	57.75	.03342014	.00123778
Liquid gallon	3,785.332	3.785332	.003785332	128.00	4.00	1.00	231.00	.1336806	.004951132
Cubic inch	16.3872	0.0163872	.0000163872	0.554112	0.017316	0.00432900	1.00	0.000578704	$.0_{4}2143347$
Cubic foot	28,316.25	28.31625	.02831625	957.5055	29.92205	7.480512	1,728	1.00	0.0370370
Cubic yard	764,538.8	764.5388	0.7645388	25,852.65	807.8953	201.9738	46,656	27.00	1.00

Table 1-38 Correction Factors* for Gas Volume[2]

(base: 60 F and 30 in. Hg, saturated)

Temp, °F	Total gas pressure, in. of mercury									
	24.0	24.2	24.4	24.6	24.8	25.0	25.2	25.4	25.6	25.8
46	0.8258	0.8328	0.8398	0.8468	0.8537	0.8607	0.8677	0.8746	0.8816	0.8886
48	.8217	.8287	.8356	.8426	.8495	.8564	.8634	.8703	.8773	.8842
50	.8176	.8245	.8314	.8383	.8453	.8522	.8591	.8660	.8729	.8799
52	.8134	.8203	.8272	.8341	.8410	.8479	.8548	.8617	.8686	.8755
54	.8092	.8161	.8230	.8298	.8367	.8436	.8504	.8573	.8641	.8710
56	.8050	.8119	.8187	.8255	.8324	.8392	.8460	.8529	.8597	.8666
58	.8008	.8076	.8144	.8212	.8280	.8348	.8416	.8484	.8552	.8621
60	.7965	.8032	.8100	.8168	.8236	.8304	.8372	.8439	.8507	.8575
62	.7921	.7989	.8056	.8124	.8191	.8259	.8327	.8394	.8462	.8529
64	.7877	.7944	.8012	.8079	.8146	.8214	.8281	.8348	.8416	.8483
66	.7833	.7900	.7967	.8034	.8101	.8168	.8235	.8302	.8369	.8436
68	.7787	.7854	.7921	.7988	.8055	.8121	.8188	.8255	.8322	.8389
70	.7742	.7808	.7875	.7941	.8008	.8074	.8141	.8208	.8274	.8341
72	.7696	.7762	.7828	.7894	.7961	.8027	.8093	.8160	.8226	.8292
74	.7648	.7715	.7781	.7847	.7913	.7979	.8045	.8111	.8177	.8243
76	.7600	.7666	.7732	.7798	.7864	.7929	.7995	.8061	.8127	.8193
78	.7552	.7618	.7684	.7749	.7815	.7880	.7946	.8011	.8077	.8143
80	.7503	.7568	.7634	.7699	.7764	.7830	.7895	.7960	.8026	.8091
82	.7452	.7518	.7583	.7648	.7713	.7778	.7843	.7908	.7973	.8038
84	.7401	.7466	.7531	.7596	.7661	.7726	.7791	.7855	.7920	.7985
86	.7349	.7414	.7478	.7543	.7608	.7672	.7737	.7801	.7866	.7931
88	.7296	.7360	.7424	.7489	.7553	.7617	.7682	.7746	.7811	.7875
90	.7241	.7305	.7370	.7434	.7498	.7562	.7626	.7690	.7754	.7819
92	.7186	.7250	.7314	.7378	.7441	.7505	.7569	.7633	.7697	.7761
94	.7129	.7193	.7256	.7320	.7384	.7447	.7511	.7575	.7638	.7702
96	.7071	.7135	.7198	.7261	.7325	.7388	.7452	.7515	.7579	.7642
98	.7012	.7075	.7138	.7201	.7265	.7328	.7391	.7454	.7517	.7581
100	0.6951	0.7014	0.7077	0.7140	0.7203	0.7266	0.7329	0.7392	0.7455	0.7518

Temp, °F	Total gas pressure, in. of mercury									
	26.0	26.2	26.4	26.6	26.8	27.0	27.2	27.4	27.6	27.8
46	0.8956	0.9025	0.9095	0.9165	0.9235	0.9304	0.9374	0.9444	0.9513	0.9583
48	.8912	.8981	.9051	.9120	.9189	.9259	.9328	.9398	.9467	.9537
50	.8868	.8937	.9006	.9075	.9144	.9214	.9283	.9352	.9421	.9490
52	.8824	.8892	.8961	.9030	.9099	.9168	.9237	.9306	.9375	.9444
54	.8779	.8847	.8916	.8985	.9053	.9122	.9191	.9259	.9328	.9396
56	.8734	.8802	.8871	.8939	.9007	.9076	.9144	.9212	.9281	.9349
58	.8689	.8757	.8825	.8893	.8961	.9029	.9097	.9165	.9234	.9302
60	.8643	.8711	.8779	.8847	.8914	.8982	.9050	.9118	.9186	.9254
62	.8597	.8664	.8732	.8800	.8867	.8935	.9002	.9070	.9138	.9205
64	.8550	.8618	.8685	.8752	.8820	.8887	.8954	.9022	.9089	.9156
66	.8503	.8570	.8637	.8705	.8772	.8839	.8906	.8973	.9040	.9107
68	.8455	.8522	.8589	.8656	.8723	.8790	.8856	.8923	.8990	.9057
70	.8407	.8474	.8540	.8607	.8674	.8740	.8807	.8873	.8940	.9006
72	.8359	.8425	.8491	.8558	.8624	.8690	.8757	.8823	.8889	.8956
74	.8309	.8375	.8441	.8507	.8573	.8639	.8706	.8772	.8838	.8904
76	.8259	.8324	.8390	.8456	.8522	.8588	.8653	.8719	.8785	.8851
78	.8208	.8274	.8339	.8405	.8471	.8536	.8602	.8667	.8733	.8798
80	.8156	.8222	.8287	.8352	.8418	.8483	.8548	.8614	.8679	.8744
82	.8103	.8168	.8234	.8299	.8364	.8429	.8494	.8559	.8624	.8689
84	.8050	.8115	.8180	.8244	.8309	.8374	.8439	.8504	.8569	.8634
86	.7995	.8060	.8125	.8189	.8254	.8318	.8383	.8448	.8512	.8577
88	.7939	.8004	.8068	.8132	.8197	.8261	.8326	.8390	.8454	.8519
90	.7883	.7947	.8011	.8075	.8139	.8203	.8268	.8332	.8396	.8460
92	.7825	.7889	.7953	.8017	.8081	.8144	.8208	.8272	.8336	.8400
94	.7766	.7829	.7893	.7957	.8020	.8084	.8148	.8212	.8275	.8339
96	.7706	.7769	.7832	.7896	.7959	.8023	.8086	.8150	.8213	.8277
98	.7644	.7707	.7770	.7834	.7897	.7960	.8023	.8086	.8150	.8213
100	0.7581	0.7644	0.7707	0.7770	0.7833	0.7896	0.7959	0.8022	0.8085	0.8148

Temp, °F	Total gas pressure, in. of mercury									
	28.0	28.2	28.4	28.6	28.8	29.0	29.2	29.4	29.6	29.8
46	0.9653	0.9723	0.9792	0.9862	0.9932	1.0001	1.0071	1.0141	1.0211	1.0280
48	.9606	.9676	.9745	.9815	.9884	0.9953	1.0023	1.0092	1.0162	1.0231
50	.9560	.9629	.9698	.9767	.9836	.9905	0.9975	1.0044	1.0113	1.0182
52	.9513	.9582	.9650	.9719	.9788	.9857	.9926	0.9995	1.0064	1.0133
54	.9465	.9534	.9602	.9671	.9740	.9808	.9877	.9946	1.0014	1.0083
56	.9418	.9486	.9554	.9623	.9691	.9759	.9828	.9896	0.9965	1.0033
58	.9370	.9438	.9506	.9574	.9642	.9710	.9778	.9846	.9915	0.9983
60	.9321	.9389	.9457	.9525	.9593	.9661	.9729	.9796	.9864	.9932
62	.9273	.9340	.9408	.9476	.9543	.9611	.9678	.9746	.9813	.9881
64	.9224	.9291	.9358	.9426	.9493	.9560	.9628	.9695	.9762	.9830
66	.9174	.9241	.9308	.9375	.9442	.9509	.9576	.9644	.9711	.9778
68	.9124	.9190	.9257	.9324	.9391	.9458	.9524	.9591	.9658	.9725
70	.9073	.9139	.9206	.9273	.9339	.9406	.9472	.9539	.9605	.9672
72	.9022	.9088	.9154	.9221	.9287	.9353	.9420	.9486	.9552	.9619
74	.8970	.9036	.9102	.9168	.9234	.9300	.9366	.9432	.9498	.9564
76	.8917	.8983	.9048	.9114	.9180	.9246	.9312	.9377	.9443	.9509
78	.8864	.8930	.8995	.9061	.9126	.9192	.9257	.9323	.9389	.9454
80	.8810	.8875	.8940	.9006	.9071	.9136	.9202	.9267	.9332	.9398
82	.8754	.8819	.8884	.8950	.9015	.9080	.9145	.9210	.9275	.9340
84	.8698	.8763	.8828	.8893	.8958	.9023	.9088	.9152	.9217	.9282
86	.8641	.8706	.8771	.8835	.8900	.8965	.9029	.9094	.9158	.9223
88	.8583	.8648	.8712	.8776	.8841	.8905	.8969	.9034	.9098	.9163
90	.8524	.8588	.8652	.8717	.8781	.8845	.8909	.8973	.9037	.9101
92	.8464	.8528	.8592	.8656	.8720	.8784	.8848	.8911	.8975	.9039
94	.8403	.8466	.8530	.8594	.8657	.8721	.8785	.8848	.8912	.8976
96	.8340	.8404	.8467	.8530	.8594	.8657	.8721	.8784	.8848	.8911
98	.8276	.8339	.8403	.8466	.8529	.8592	.8655	.8719	.8782	.8845
100	0.8211	0.8274	0.8337	0.8400	0.8463	0.8526	0.8589	0.8652	0.8715	0.8778

Temp, °F	Total gas pressure, in. of mercury				
	30.0	30.2	30.4	30.6	30.8
46	1.0350	1.0420	1.0490	1.0559	1.0629
48	1.0301	1.0370	1.0440	1.0509	1.0578
50	1.0251	1.0320	1.0390	1.0459	1.0528
52	1.0202	1.0271	1.0340	1.0408	1.0477
54	1.0151	1.0220	1.0289	1.0357	1.0426
56	1.0101	1.0170	1.0238	1.0306	1.0375
58	1.0051	1.0119	1.0187	1.0255	1.0323
60	1.0000	1.0068	1.0136	1.0203	1.0271
62	0.9949	1.0016	1.0084	1.0151	1.0219
64	.9897	0.9964	1.0032	1.0099	1.0166
66	.9845	.9912	0.9979	1.0046	1.0113
68	.9792	.9859	.9925	0.9992	1.0059
70	.9739	.9805	.9872	.9938	1.0005
72	.9685	.9751	.9818	.9884	0.9950
74	.9630	.9697	.9763	.9829	.9895
76	.9575	.9641	.9707	.9772	.9838
78	.9520	.9585	.9651	.9716	.9782
80	.9463	.9528	.9594	.9659	.9724
82	.9405	.9470	.9535	.9600	.9666
84	.9347	.9412	.9477	.9542	.9606
86	.9288	.9352	.9417	.9481	.9546
88	.9227	.9291	.9356	.9420	.9484
90	.9166	.9230	.9294	.9358	.9422
92	.9103	.9167	.9231	.9295	.9359
94	.9039	.9103	.9167	.9230	.9294
96	.8975	.9038	.9102	.9165	.9228
98	.8908	.8972	.9035	.9098	.9161
100	0.8841	0.8904	0.8967	0.9030	0.9093

* Correction factor, $F = \frac{519.7}{459.7 + t} \times \frac{P - w_t}{29.4782}$

where:

P = total gas pressure in inches of mercury at 32F,
w_t = saturated vapor pressure of water, at temperature t, in inches of mercury, and
t = gas temperature in degrees Fahrenheit.

Parallelepiped

S = perimeter, P, perpendicular to sides × lateral length, l: Pl
V = area of base, B × perpendicular height, h: Bh
V = area of section, A, perpendicular to sides × lateral length, l: Al

Prism, Right or Oblique, Regular or Irregular

S = perimeter, P, perpendicular to sides × lateral length, l: Pl
V = area of base, B × perpendicular height, h: Bh
V = area of section, A, perpendicular to sides × lateral length, l: Al

Cylinder, Right or Oblique, Circular or Elliptic, etc.

S = perimeter of base, P × perpendicular height, h: Ph
S = perimeter, P_1, perpendicular to sides × lateral length, l: P_1l
V = area of base, B × perpendicular height, h: Bh
V = area of section, A, perpendicular to sides × lateral length, l: Al

Frustum of any Prism or Cylinder

V = area of base, B × perpendicular distance, h from base to center of gravity of opposite face: Bh
For cylinder: $^1/_2A(l_1 + l_2)$

Pyramid or Cone, Right and Regular

S = perimeter of base, P × $^1/_2$ slant height, l: $^1/_2Pl$
V = area of base, B × $^1/_3$ perpendicular height, h: $^1/_3Bh$

Pyramid or Cone, Right or Oblique, Regular or Irregular

V = area of base, B × $^1/_3$ perpendicular height, h: $^1/_3Bh$
V = $^1/_3$ volume of prism or cylinder of same base and perpendicular height
V = $^1/_2$ volume of hemisphere of same base and perpendicular height

Frustum of Pyramid or Cone, Right and Regular, Parallel Ends

S = (sum of perimeter of base, P, and top, p) × $^1/_2$ slant height, l: $^1/_2l(P + p)$
V = (sum of areas of base, B, and top, b + square root of their products) × $^1/_3$ perpendicular height, h: $^1/_3h(B + b + \sqrt{Bb})$

Frustum of any Pyramid or Cone, Parallel Ends

V = (sum of areas of base, B, and top, b + square root of their products) × $^1/_3$ perpendicular height, h: $^1/_3h(B + b + \sqrt{Bb})$

Wedge, Parallelogram Face

V = $^1/_6$ (sum of three edges, aba × perpendicular height, h × perpendicular width, d): $^1/_6dh(2a + b)$

Prismatoid

V = $^1/_6$ perpendicular height ,h (sum of areas of base, B, and top, b + 4 × area of section, M, parallel to bases and midway between them): $^1/_6h(B + b + 4M)$

The prismatoid formula applies also to any of the foregoing solids with parallel bases, to pyramids, cones, spherical sections, and to many solids with irregular surfaces.

Sphere

$S = 4\pi r^2 = \pi d^2 = 3.14159265d^2$
$V = {}^4/_3\pi r^3 = {}^1/_6\pi d^3 = 0.52359878d^3$

Spherical Sector

$S = {}^1/_2\pi r(4b + c)$
$V = {}^2/_3\pi r^2 b$

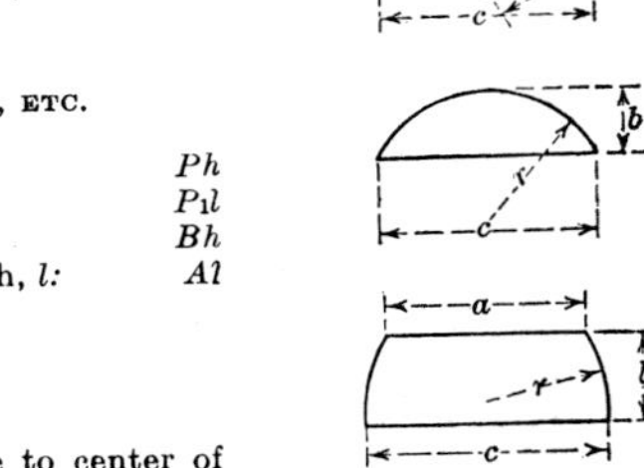

Spherical Segment

$S = 2\pi rb = {}^1/_4\pi(4b^2 + c^2)$
$V = {}^1/_3\pi b^2(3r - b) = {}^1/_{24}\pi b(3c^2 + 4b^2)$

Spherical Zone

$S = 2\pi rb$
$V = {}^1/_{24}\pi b(3a^2 + 3c^2 + 4b^2)$

Circular Ring

$S = 4\pi^2 Rr$
$V = 2\pi^2 Rr^2$

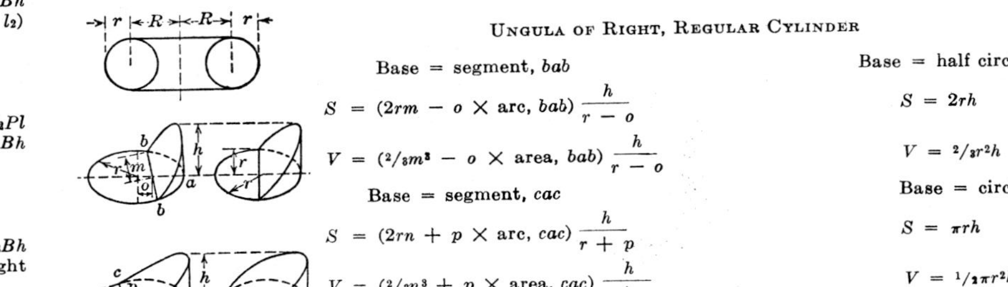

Ungula of Right, Regular Cylinder

Base = segment, bab

$S = (2rm - o \times \text{arc}, bab)\dfrac{h}{r - o}$

$V = ({}^2/_3m^3 - o \times \text{area}, bab)\dfrac{h}{r - o}$

Base = segment, cac

$S = (2rn + p \times \text{arc}, cac)\dfrac{h}{r + p}$

$V = ({}^2/_3n^3 + p \times \text{area}, cac)\dfrac{h}{r + p}$

Base = half circle

$S = 2rh$

$V = {}^2/_3r^2h$

Base = circle

$S = \pi rh$

$V = {}^1/_2\pi r^2h$

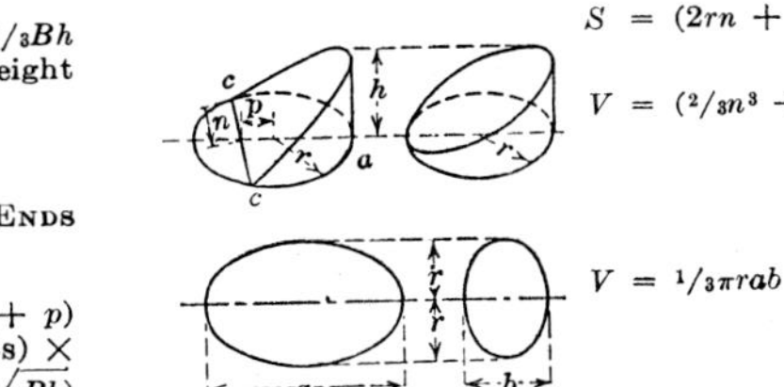

Ellipsoid

$V = {}^1/_3\pi rab$

Paraboloid

$V = {}^1/_2\pi r^2h$

Ratio of corresponding volumes of a cone, paraboloid, sphere, and cylinder of equal height: $^1/_3 : {}^1/_2 : {}^2/_3 : 1$

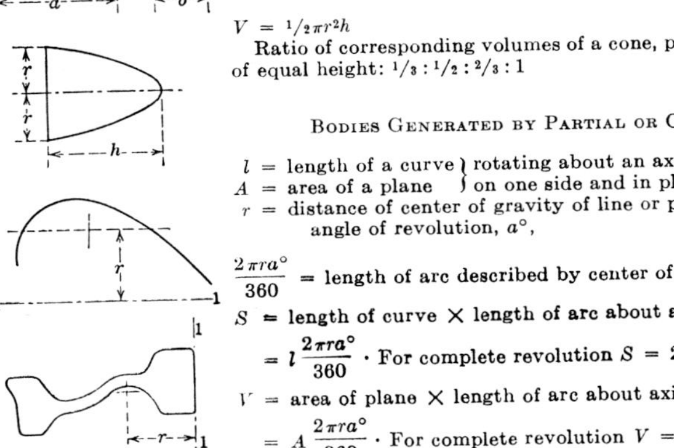

Bodies Generated by Partial or Complete Revolution

l = length of a curve } rotating about an axis 1-1.
A = area of a plane } on one side and in plane of axis
r = distance of center of gravity of line or plane from axis 1-1 and for any angle of revolution, $a°$,

$\dfrac{2\pi ra°}{360}$ = length of arc described by center of gravity

S = length of curve × length of arc about axis

$= l\dfrac{2\pi ra°}{360}$ · For complete revolution $S = 2\pi rl$

V = area of plane × length of arc about axis

$= A\dfrac{2\pi ra°}{360}$ · For complete revolution $V = 2\pi rA$

S = lateral or convex surface. V = volume

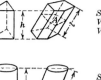

Fig. 1-7 Surface areas and volumes of shapes.

Table 1-39 Pressure Conversion Table

(expressions like 0.0_3 and 0.0_6 stand for 0.000 and 0.000000)

	Centi-meters of water at at 39.2 F	Inches of water at 39.2 F	Feet of water at 39.2 F	Centi-meters of mercury	Inches of mercury at 32 F	Grams per square centimeter	Ounces per square inch	Ounces per square foot	Pounds per square inch	Pounds per square foot	Normal atmos-phere	Dynes per square centimeter
Centimeters of water at 39.2 F	1.000000	0.3937000	0.03280833	0.07355393	0.02895818	0.9999730	0.2275682	32.76983	0.01422301	2.048114	$0.0_3 9678148$	980.6385
Inches of water at 39.2 F	2.540005	1.000000	0.08333333	0.1868273	.07355393	2.539936	0.5780245	83.23552	.03612653	5.202220	.002458255	2490.827
Feet of water at 39.2 F	30.48006	12.00000	1.000000	2.241928	.8826471	30.47924	6.936294	998.8263	.4335183	62.42664	.02949906	29889.92
Centimeters of mercury	13.59547	5.352535	0.4460446	1.000000	0.3937000	13.59510	3.093896	445.5211	.1933685	27.84507	.01315789	13332.24
Inches of mercury at 32 F	34.53255	13.59547	1.132956	2.540005	1.000000	34.53162	7.858513	1131.626	.4911570	70.72661	.03342112	33863.95
Grams per square centimeter	1.000027	0.3937106	0.03280922	0.07355591	0.02895896	1.000000	0.2275744	32.77071	.01422340	2.048169	$.0_3 9678410$	980.6650
Ounces per square inch	4.394286	1.730031	.1441692	.3232170	.1272505	4.394168	1.000000	144.0000	.06250000	9.000000	.004252856	4309.207
Ounces per square foot	0.03051588	0.01201410	0.001001175	0.002244563	$0.0_3 8836843$	0.03051505	0.006944444	1.000000	$0.0_3 4340278$	0.06250000	$.0_4 2953372$	29.92505
Pounds per square inch	70.30858	27.68049	2.306707	5.171473	2.036009	70.30669	16.00000	2304.000	1.000000	144.0000	.06804569	68947.31
Pounds per square foot	0.4882541	0.1922256	0.01601880	0.03591300	0.01413895	0.4882409	0.1111111	16.00000	0.006944444	1.000000	$0.0_3 4725395$	478.8007
Normal atmosphere	1033.255	406.7927	33.89939	76.00000	29.92120	1033.228	235.1361	33859.60	14.69601	2116.225	1.000000	1013250.
Dynes per square centimeter	0.001019744	$0.0_3 4014731$	$0.0_4 3345609$	$0.0_4 7500616$	$0.0_4 2952992$	0.001019716	$0.0_3 2320613$	0.03341682	$0.0_4 1450383$	0.002088552	$0.0_6 9869231$	1.000000

This table is based on the following values and definitions from I.C.T.:

Standard acceleration of gravity, 980.665 cm per second per second; 1 l = 1000.027 cu cm; 1 cm = 0.3937 in.; 1 lb = 453.5924277 g; 1 normal atmosphere = pressure exerted on 1 sq cm by a vertical column of liquid 76 cm in height, of density 13.5951 g per cu cm, at a place where the acceleration of gravity is standard.

1 Torr. = 0.0394 in. Hg = 0.0193 psi.

Table 1-40 Velocity Conversion Table[1]

Cm per sec	Meter per sec	Meter per min	Km per hr	Ft per sec	Ft per min	Miles per hr	Knots (nautical miles per hr)
1	0.01	0.6	0.036	0.03281	1.9685	0.02237	0.01943
100	1	60	3.6	3.281	196.85	2.237	1.943
1.667	0.01667	1	0.06	0.0547	3.281	0.03728	0.03238
27.78	.2778	16.67	1	0.9113	54.68	.6214	.53960
30.48	.3048	18.29	1.097	1	60	.6818	.59209
0.5080	.005080	0.3048	0.01829	0.01667	1	0.01136	.00987
44.70	.4470	26.82	1.609	1.467	88	1	0.86839
51.48	0.5148	30.887	1.8532	1.6889	101.337	1.15155	1

Table 1-41 Power Conversion Table[1]

Hp (550 standard ft-lb per sec)	Metric hp	Kw (1000 joules per sec)	Meter-kg per sec	Ft-lb per sec	K Cal per sec	Btu per sec	Cal per sec
1	1.0138	0.7457	76.04	550	0.1782	0.7068	178.23
0.9863	1	0.7355	75	542.5	.1758	.6971	175.79
1.341	1.3596	1	101.97	737.56	.2390	.9478	239.01
0.01315	0.01333	0.009807	1	7.233	.00234	.009295	2.3438
0.001818	0.001843	0.001356	0.1383	1	0.000324	0.001285	0.3240
5.611	5.689	4.1840	426.7	3086	1	3.966	1000
1.415	1.434	1.055	107.58	778.16	0.2522	1	252.16
0.005611	0.005689	0.004184	0.4267	3.086	0.001	0.00397	1

1 boiler hp = 33,475 Btu per hr.
1 ton refrigeration = 200 Btu per min = 12,000 Btu per hr = 288,000 Btu per day.
1 engine horsepower = 2545 Btu per hr = 42.42 Btu per min.
1 kilowatt = 3413 Btu per hr.

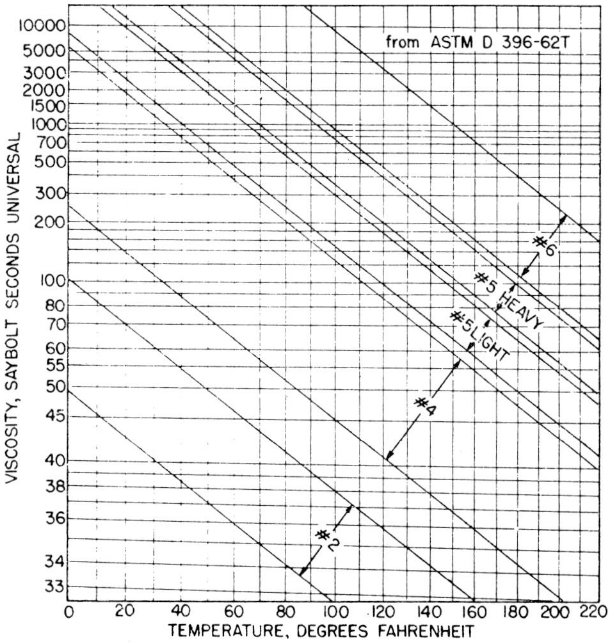

Fig. 1-8 Viscosity-temperature relations for typical fuel oils.

Table 1-42 Theoretical Acceleration of Gravity*

Latitude, $\varphi°$	At 0° long. gal.† = cm/sec-sec	At 0° long. ft/sec-sec	Ratio $\frac{(\varphi, 0)}{(45°, 0)}$
0	978.0	32.088	0.9974
10	978.2	32.093	.9975
20	978.7	32.108	.9980
30	979.3	32.130	.9987
40	980.2	32.158	0.9995
45	980.6	32.173	1.0000
50	981.1	32.188	1.0005
60	981.9	32.215	1.0013
70	982.6	32.238	1.0020
80	983.1	32.253	1.0025
90	983.2	32.258	1.0026

(Courtesy of L. B. Tuckerman, Assistant Chief, Mechanics Division, National Bureau of Standards.)

Actual values for stations within the United States may differ from the values calculated from this table with free air reduction by ±0.3 gal. and for some foreign stations by as much as ±0.6 gal. If more accurate values are needed, inquire of the U. S. Coast and Geodetic Survey, Washington 25, D. C.

Normal or Standard gravity is 980.665 cm/sec-sec or 32.1740 ft/sec-sec according to the National Bureau of Standards.

* Adopted 1930 by the International Geodetic Association. Rounded to 4 and 5 significant figures. Free air reduction for altitude above sea level −0.309 gal. for each 1000 m; and −0.00309 ft/sec-sec for each 1000 ft.

† gal. = contraction of Galileo.

Table 1-43 Energy Conversion Table*

(acceleration of gravity = 980.665 cm/sec-sec)

Units	Btu	P.c.u.†	Calories (large)	Foot-pounds	Foot-tons	Kilogram-meters	Horse-power-hours	Kilowatt-hours	Joules, inter-national	Pounds carbon†	Pounds water†
Btu	1.00000	0.555556	0.251983	777.979	0.388990	107.560	$0.0_3392922$	$0.0_3293000$	1,054.46	$0.0_4687569$	0.0010305
P.c.u.†	1.80000	1.00000	0.453569	1.400.36	0.700181	193.607	$.0_3707260$	$.0_3527400$	1,898.03	$.0_3123762$	.00185490
Calories (large)	3.96852	2.20474	1.00000	3,087.43	1.54372	426.853	.00155932	.00116278	4,184.66	$.0_3272863$	.00408958
Foot-pounds	0.00128538	$0.0_3714101$	$0.0_3323894$	1.00000	$0.0_3500000$	0.138255	$.0_6505051$	$.0_6376617$	1.35539	$.0_7883788$	$.0_5132459$
Foot-tons	2.57076	1.42820	.647788	2,000.00	1.00000	276.510	.00101010	$.0_3753233$	2,710.77	$.0_3176758$	.00264918
Kilogram-meters	0.00929716	0.00516509	0.00234272	7.23300	0.00361650	1.00000	$0.0_5365306$	$.0_5272407$	9.80351	$.0_6639244$	$0.0_5958076$
Horsepower-hours	2,545.03	0.141391	641.304	1,979,980	989.999	273,743	1.00000	0.745694	2,683,640	.174988	2.62266
Kilowatt-hours	3,412.97	$1,896.0_9$	860.009	2,655,220	1,327.61	367,098	1.34103	1.00000	3,598,850	.234665	3.51707
Joules (international)	$0.0_3948350$	$0.0_3526861$	$0.0_3238968$	0.737797	$0.0_3368898$	0.102004	$0.0_6372628$	$0.0_6277867$	1.00000	$0.0_7652056$	$0.0_6977278$
Pounds carbon†	14,544	808.000	3,664.84	11,314,900	5,657.47	1,564,350	5.71466	4.26139	15,336,100	1.00000	14.9876
Pounds water†	970.40	539.111	244.524	754,951	377.476	104,376	0.381292	0.284327	1,023,250	0.0667217	1.00000

This table is based upon the following I.C.T. values:

1 Btu (mean)	= 1054.8	absolute joules
1 g-cal (mean)	= 4.186	absolute joules
1 ft-lb	= 1.35582	absolute joules
1 kg-m	= 9.80665	absolute joules
1 hp-hr	$= 2.6845 \times 10^6$	absolute joules
1 kwh	$= 3.6000 \times 10^6$	absolute joules
1 international joule	= 1.00032	absolute joules

* Adapted from Van Nostrand's "Chemical Annual" and I.C.T.

† P.c.u. refers to pound-Centigrade unit. The ton used is 2000 lb. "Pounds carbon" refers to pounds of carbon oxidized to CO_2 of 100 per cent efficiency equivalent to the corresponding number of heat units. "Pounds water" refers to pounds of water evaporated at 212 F at 100 per cent efficiency.

0.0_3 signifies 0.000.

1 cal (small) per gram = 1.8 Btu per pound.

Table 1-44 Temperature Conversion Formulas

Conversion of deg F to deg C or deg C to deg F requires the following sequence of three steps:

1. Add 40 deg to the value.
2. Multiply result by either 9/5 to convert to deg F or 5/9 to convert to deg C.
3. Subtract 40 from this result.

°R = °F abs = deg Rankine = °F + 459.6
°K = °C abs = deg Kelvin = °C + 273.1
°Celsius* = 0.01 deg + (.9999 × deg C)
°Reaumur = 0.8° Centigrade

* The base of the Celsius scale is the triple point (solid, liquid, and gas in equilibrium) of water which is at 0.01 Centigrade.

Table 1-45 Conversions for Standard Gas Conditions

To convert from	To	Multiply by
Btu per cubic foot, 60 F and 14.73 psia* dry	Calories per cubic meter, 0 C and 760 mm Hg dry	9.377
Btu per cubic foot, 60 F and 30 in. Hg dry	Calories per cubic meter, 0 C and 760 mm Hg dry	9.377
Btu per cubic foot, 60 F and 30 in. Hg saturated	Calories per cubic meter, 0 C and 760 mm Hg dry	9.549
Btu per cubic foot, 60 F and 30 in. Hg saturated	Calories per cubic meter, 0 C and 760 mm Hg saturated	9.490
Calories per cubic meter, 0 C and 760 mm Hg saturated	Btu per cubic foot, 60 F and 30 in. Hg saturated	0.1054
Calories per cubic meter, 0 C and 760 mm Hg dry	Btu per cubic foot, 60 F and 30 in. Hg saturated	.1047
Cubic feet, 60 F and 14.73 psia dry	Cubic meters, 0 C and 760 mm Hg dry	.0283
Cubic feet, 60 F and 30 in. Hg saturated	Cubic meters, 0 C and 760 mm Hg dry	.0264
Cubic feet, 60 F and 30 in. Hg saturated	Cubic meters, 0 C and 760 mm Hg saturated	0.0266
Cubic meters, 0 C and 760 mm Hg dry	Cubic feet, 60 F and 30 in. Hg saturated	37.887
Cubic meters, 0 C and 760 mm Hg saturated	Cubic feet, 60 F and 30 in. Hg saturated	37.656
Btu per pound	Calories per kilogram	0.5556
Calories per kilogram	Btu per pound	1.8000

* Base conditions of temperature and pressure endorsed by A.G.A. in 1963.

Table 1-46 Specific Gravity, Density, and Volume Conversions for Liquids Lighter than Water

(equivalents at 60 F)

°API	Specific gravity*	°Bé	Lb per cu ft	Lb per gal	Lb per bbl (42 gal)
10	1.0000	10.000	62.367	8.337	350.2
11	0.9930	10.989	61.929	8.279	347.7
12	.9861	11.979	61.498	8.221	345.3
13	.9792	12.968	61.072	8.164	342.9
14	.9725	13.958	60.652	8.108	340.5
15	.9659	14.947	60.238	8.053	338.2
16	.9593	15.936	59.830	7.998	335.9
17	.9529	16.926	59.427	7.944	333.7
18	.9465	17.915	59.029	7.891	331.4
19	.9402	18.905	58.637	7.839	329.2
20	.9340	19.894	58.250	7.787	327.0
21	.9279	20.883	57.868	7.736	324.9
22	.9218	21.873	57.491	7.685	322.8
23	.9159	22.862	57.119	7.636	320.7
24	.9100	23.852	56.752	7.587	318.6
25	.9042	24.841	56.389	7.538	316.6
26	.8984	25.830	56.031	7.490	314.6
27	.8927	26.820	55.678	7.443	312.6
28	.8871	27.809	55.329	7.396	310.6
29	.8816	28.799	54.984	7.350	308.7
30	.8762	29.788	54.643	7.305	306.8
31	.8708	30.777	54.307	7.260	304.9
32	.8654	31.767	53.975	7.215	303.0
33	.8602	32.756	53.647	7.172	302.1
34	.8550	33.746	53.323	7.128	299.4
35	.8498	34.735	53.002	7.085	297.6
36	.8448	35.724	52.686	7.043	295.8
37	.8398	36.714	52.373	7.001	294.1
38	.8348	37.703	52.064	6.960	292.3
39	.8299	38.693	51.759	6.919	290.6
40	.8251	39.682	51.457	6.879	288.9
41	.8203	40.671	51.159	6.839	287.2
42	.8156	41.661	50.864	6.800	285.6
43	.8109	42.650	50.572	6.761	283.9
44	.8063	43.640	50.284	6.722	282.3
45	.8017	44.629	49.999	6.684	280.7
46	.7972	45.618	49.718	6.646	279.1
47	.7927	46.608	49.439	6.609	277.6
48	.7883	47.597	49.164	6.572	276.0
49	.7839	48.587	48.891	6.536	274.5
50	.7796	49.576	48.622	6.500	273.0
51	.7753	50.565	48.356	6.464	271.5
52	.7711	51.555	48.092	6.429	270.0
53	.7669	52.544	47.831	6.394	268.6
54	0.7628	53.534	47.574	6.360	267.1
55	0.7587	54.523	47.319	6.326	265.7
56	.7547	55.512	47.066	6.292	264.3
57	.7507	56.502	46.816	8.258	262.9
58	.7467	57.491	46.569	6.225	261.5
59	.7428	58.481	46.325	6.193	260.1
60	.7389	59.470	46.083	6.160	258.7
61	.7351	60.459	45.844	6.128	257.4
62	.7313	61.449	45.607	6.097	256.1
63	.7275	62.438	45.372	6.065	254.7
64	.7238	63.428	45.140	6.034	253.4
65	.7201	64.417	44.910	6.004	252.2
66	.7165	65.406	44.683	5.973	250.9
67	.7128	66.396	44.458	5.943	249.6
68	.7093	67.385	44.235	5.913	248.4
69	.7057	68.375	44.008	5.884	247.1
70	.7022	69.364	43.796	5.855	245.9
71	.6988	70.353	43.580	5.826	244.7
72	.6953	71.343	43.366	5.797	243.5
73	.6919	72.332	43.154	5.769	242.3
74	.6886	73.322	42.944	5.741	241.1
75	.6852	74.311	42.736	5.713	239.9
76	.6819	75.300	42.530	5.685	238.8
77	.6787	76.290	42.326	5.658	237.6
78	.6754	77.279	42.124	5.631	236.5
79	.6722	78.269	41.924	5.604	235.4
80	.6690	79.258	41.725	5.578	234.3
81	.6659	80.247	41.529	5.552	233.2
82	.6628	81.237	41.334	5.526	232.1
83	.6597	82.226	41.142	5.500	231.0
84	.6566	83.216	40.951	5.474	229.9
85	.6536	84.205	40.762	5.449	228.9
86	.6506	85.194	40.574	5.424	227.8
87	.6476	86.184	40.389	5.399	226.8
88	.6446	87.173	40.205	5.375	225.7
89	.6417	88.163	40.022	5.350	224.7
90	.6388	89.152	39.842	5.326	223.7
91	.6360	90.141	39.662	5.302	222.7
92	.6331	91.131	39.485	5.278	221.7
93	.6303	92.120	39.309	5.255	220.7
94	.6275	93.110	39.135	5.232	219.7
95	.6247	94.099	38.962	5.208	218.8
96	.6220	95.088	38.791	5.186	217.8
97	.6193	96.078	38.621	5.163	216.8
98	.6166	97.067	38.453	5.140	215.9
99	.6139	98.057	38.286	5.118	215.0
100	0.6112	99.046	38.121	5.096	214.0

* At 60 F compared with water at 60 F.

$$°\text{Bé} = \frac{140}{\text{sp gr } 60\text{ F}/60\text{ F}} - 130$$

$$°\text{API} = \frac{141.5}{\text{sp gr } 60\text{ F}/60\text{ F}} - 131.5$$

Table 1-47 Hydrometer Conversions for Liquids Heavier than Water[3]

Specific gravity	Deg Bé	Specific gravity	Deg Bé	Specific gravity	Deg Bé	Specific gravity	Deg Bé
1.00	0.00	1.46	45.68	1.23	27.11	1.69	59.20
1.01	1.44	1.47	46.36	1.24	28.06	1.70	59.71
1.02	2.84	1.48	47.03	1.25	29.00	1.71	60.20
1.03	4.22	1.49	47.68	1.26	29.92	1.72	60.70
1.04	5.58	1.50	48.33	1.27	30.83	1.73	61.18
1.05	6.91	1.51	48.97	1.28	31.72	1.74	61.67
1.06	8.21	1.52	49.60	1.29	32.60	1.75	62.14
1.07	9.49	1.53	50.23	1.30	33.46	1.76	62.61
1.08	10.78	1.54	50.84	1.31	34.31	1.77	63.08
1.09	11.97	1.55	51.45	1.32	35.15	1.78	63.54
1.10	13.18	1.56	52.05	1.33	35.98	1.79	63.99
1.11	14.37	1.57	52.64	1.34	36.79	1.80	64.44
1.12	15.54	1.58	53.23	1.35	37.59	1.81	64.89
1.13	16.68	1.59	53.80	1.36	38.38	1.82	65.31
1.14	17.81	1.60	54.38	1.37	39.16	1.83	65.77
1.15	18.91	1.61	54.94	1.38	39.93	1.84	66.20
1.16	20.00	1.62	55.49	1.39	40.68	1.85	66.62
1.17	21.07	1.63	56.04	1.40	41.43	1.86	67.04
1.18	22.12	1.64	56.58	1.41	42.16	1.87	67.46
1.19	23.15	1.65	57.12	1.42	42.89	1.88	67.87
1.20	24.17	1.66	57.65	1.43	43.60	1.89	68.28
1.21	25.16	1.67	58.17	1.44	44.31	1.90	68.68
1.22	26.15	1.68	58.69	1.45	45.00		

$$°\text{Bé} = 145 - \frac{145}{\text{sp gr } 60\ \text{F}/60\ \text{F}}$$

Table 1-48 Approximate Viscosity Conversion Factors* at 68 F

Saybolt Universal time, sec	Engler degrees	Viscosity, centipoises,† also viscosity relative to water at 68 F	Kinematic viscosity, poise-cu cm/g	Redwood No. 1 time, sec	Saybolt Furol time, sec	Redwood Admiralty time, sec	Viscosity, lb/sec-ft, also viscosity relative to water at 68 F
(1)	(2)	(3)	(4)	(5)	(6)	(7)	(8)
32	1.08	1.41	0.0141	...	...	30	0.000948
34	1.14	2.19	.0219	...	...	31	.001472
36	1.19	2.92	.0292	...	...	33	.001962
38	1.25	3.63	.0363	...	...	34	.002439
40	1.31	4.30	.0430	...	...	36	.002890
42	1.37	4.95	.0495	...	...	38	.003326
44	1.43	5.59	.0559	...	...	39	.003756
46	1.48	6.21	.0621	...	...	41	.004173
48	1.53	6.81	.0681	...	...	43	.004576
50	1.58	7.40	.0740	...	...	44	.004973
55	1.73	8.83	.0883	...	...	48	.005934
60	1.88	10.20	.1020	...	...	53	.006854
65	2.03	11.53	.1153	...	...	57	.007748
70	2.17	12.83	.1283	...	...	61	.008622
75	2.31	14.10	.1410	...	...	65	.009475

Table continues on page 1/46.

This table does not take into consideration the usual temperatures at which instruments are used.

Kinematic viscosity ν = absolute viscosity μ divided by mass density ρ; i.e., dimensionally,

$$\frac{\text{lb mass}}{\text{sec-ft}} \bigg/ \frac{\text{lb mass}}{\text{cu ft}} = \frac{\text{sq ft}}{\text{sec}}$$

Kinematic viscosity $= 0.0022t - 1.80/t$, where t = time in Saybolt universal seconds.

Centipoise or viscosity relative to water at 68 F $= 100 \times (0.0022t - 180/t) \times$ (specific gravity)

Viscosity Conversion Factors:

$$1\ \frac{\text{lb mass}}{\text{sec-ft}} = 14.882 \text{ poises} = 14{,}882{,}000 \text{ micropoises} = 0.031081\ \frac{\text{(lb force)(sec)}}{\text{sq ft}}$$

* Sources: Columns 1, 2, 5, 6, and 7, Union Oil Co.
Columns 3, 4, and 8, computed from Formulas in Bureau of Standards, Tech. Papers 100, 112, 210.

† The values of this column must be multiplied by the specific gravity of the fluid at the temperature of the instrument before the conversion is complete.

Table 1-48 Approximate Viscosity Conversion Factors at 68 F (Continued)

Saybolt Universal time, sec	Engler degrees	Viscosity, centipoises,† also viscosity relative to water at 68 F	Kinematic viscosity, poise-cu cm/g	Redwood No. 1 time, sec	Saybolt Furol time, sec	Redwood Admiralty time, sec	Viscosity, lb/sec-ft, also viscosity relative to water at 68 F
(1)	(2)	(3)	(4)	(5)	(6)	(7)	(8)
80	2.46	15.35	.1535	...	...	69	.01032
85	2.59	16.58	.1658	...	...	73	.01114
90	2.74	17.80	.1780	...	...	77	.01196
95	2.88	19.00	.1900	...	...	81	.01277
100	3.02	20.20	.2020	10	15	86	.01357
110	3.31	22.56	.2256	11	16	94	.01516
120	3.60	24.90	.2490	12	17	102	.01673
130	3.69	27.21	.2721	13	18	111	.01829
140	4.19	29.51	.2951	13	18	119	.01983
160	4.77	34.07	.3407	15	20	136	.02290
180	5.35	38.60	.3860	17	22	153	.02594
200	5.92	43.10	.4310	19	23	170	.02896
225	6.64	48.70	.4870	21	26	191	.03273
250	7.35	54.28	.5428	23	28	212	.03648
300	8.79	65.40	.6540	28	32	254	.04395
350	10.25	76.49	.7649	32	37	296	.05140
400	11.68	87.55	.8755	37	42	338	.05883
450	13.00	98.60	0.9860	42	47	380	.06626
500	14.00	109.6	1.096	46	52	422	.07365
550	16.00	120.7	1.207	50	56	465	.08111
600	17.00	131.7	1.317	55	61	507	.08850
650	19.00	142.7	1.427	60	66	549	.09589
700	20.00	153.7	1.537	64	71	591	.1033
800	23.00	175.8	1.758	74	81	676	.1181
900	26.00	197.8	1.978	83	91	760	.1329
1000	29.00	219.8	2.198	92	101	845	.1477
1500	43.00	329.9	3.299	138	150	1267	.2217
2000	58.00	439.9	4.399	184	200	1690	.2956
2500	72.00	549.9	5.499	230	250	2112	.3695
3000	87.00	659.9	6.599	276	300	2535	.4435
3500	101.00	769.9	7.699	322	350	2957	0.5174

Footnotes appear on page 1/45.

Table 1-49 Heat Flow Conversion Table[1]

Cal/sec-sq cm	Cal/hr-sq cm	Btu/hr-sq ft	Btu/day-sq ft	Watts/sq cm
1	3600	13,263	318,322	4.183
0.0002778	1	3.684	88.42	0.001162
.0000754	0.2714	1	24	.0003154
.00000314	0.01131	0.04167	1	0.00001314
0.2390	860.6	3,171	76,094	1

Table 1-50 Thermal Conductivity Conversion Table[1]

Cal/sec-sq cm-°C/cm	International watts/sq cm-°C/cm	Cal/hr-sq cm-°C/cm	Btu/hr-sq ft-°F/in.	Btu/day-sq ft-°F/in.
1	4.1833	3600.	2901.0	69,624
0.2390	1	860.6	693.5	16,643
.000278	0.001162	1	0.8058	19.34
.0003447	.001441	1.241	1	24.0
0.0000144	0.0000601	0.05171	0.04167	1

Table 1-51 Compound Interest Values s^n of 1

	For n years when interest is compounded annually at							
n	2%	3%	4%	4½%	5%	6%	7%	8%
1	1.020	1.030	1.040	1.045	1.050	1.060	1.070	1.080
2	1.040	1.061	1.082	1.092	1.103	1.124	1.145	1.166
3	1.061	1.093	1.125	1.141	1.158	1.191	1.225	1.260
4	1.082	1.126	1.170	1.193	1.216	1.262	1.311	1.360
5	1.104	1.159	1.217	1.246	1.276	1.338	1.403	1.469
6	1.126	1.194	1.265	1.302	1.340	1.419	1.501	1.587
7	1.149	1.230	1.316	1.361	1.407	1.504	1.606	1.714
8	1.172	1.267	1.369	1.422	1.477	1.594	1.718	1.851
9	1.195	1.305	1.423	1.486	1.551	1.689	1.838	1.999
10	**1.219**	**1.344**	**1.480**	**1.553**	**1.629**	**1.791**	**1.967**	**2.159**
11	1.243	1.384	1.539	1.623	1.710	1.898	2.105	2.332
12	1.268	1.426	1.601	1.696	1.796	2.012	2.252	2.518
13	1.294	1.469	1.665	1.772	1.886	2.133	2.410	2.720
14	1.319	1.513	1.732	1.852	1.980	2.261	2.579	2.937
15	1.346	1.558	1.801	1.935	2.079	2.397	2.759	3.172
16	1.373	1.605	1.873	2.022	2.183	2.540	2.952	3.426
17	1.400	1.653	1.948	2.113	2.292	2.693	3.159	3.700
18	1.428	1.702	2.026	2.208	2.407	2.854	3.380	3.996
19	1.457	1.754	2.107	2.308	2.527	3.026	3.617	4.316
20	**1.486**	**1.806**	**2.191**	**2.412**	**2.653**	**3.207**	**3.870**	**4.661**
21	1.516	1.860	2.279	2.520	2.786	3.400	4.141	5.034
22	1.546	1.916	2.370	2.634	2.925	3.604	4.430	5.437
23	1.577	1.974	2.465	2.752	3.072	3.820	4.741	5.871
24	1.608	2.033	2.563	2.876	3.225	4.049	5.072	6.341
25	1.641	2.094	2.666	3.005	3.386	4.292	5.427	6.848
26	1.673	2.157	2.772	3.141	3.556	4.549	5.807	7.396
27	1.707	2.221	2.883	3.282	3.733	4.822	6.214	7.988
28	1.741	2.288	2.999	3.430	3.920	5.112	6.649	8.627
29	1.776	2.357	3.119	3.584	4.116	5.418	7.114	9.317
30	**1.811**	**2.427**	**3.243**	**3.745**	**4.322**	**5.743**	**7.612**	**10.06**
31	1.848	2.500	3.373	3.914	4.538	6.088	8.145	10.87
32	1.885	2.575	3.508	4.090	4.765	6.453	8.715	11.74
33	1.922	2.652	3.648	4.274	5.003	6.841	9.325	12.68
34	1.961	2.732	3.794	4.466	5.253	7.251	9.978	13.69
35	2.000	2.814	3.946	4.667	5.516	7.686	10.68	14.79
36	2.040	2.898	4.104	4.877	5.792	8.147	11.42	15.97
37	2.081	2.985	4.268	5.097	6.081	8.636	12.22	17.25
38	2.122	3.075	4.439	5.326	6.385	9.154	13.08	18.63
39	2.165	3.167	4.616	5.566	6.705	9.704	13.99	20.12
40	**2.208**	**3.262**	**4.801**	**5.816**	**7.040**	**10.29**	**14.97**	**21.72**

Table 1-51 gives the value of s^n (compound amount) for given values of i (interest rate), h (interest periods per year), and n (the number of years).

When interest is compounded annually.

Example 1.—What is the amount of \$1 for 15 years at 7 per cent compounded annually? Under "7 per cent" and opposite "15" of Table 1-51, we find \$2.759, which is the required amount.

Example 2.—What is the amount of \$147.60 for 15 years at 7 per cent compounded annually? As in the example above, we find the amount of \$1 for the given conditions to be \$2.759; and multiplying this by the principal, we obtain $147.60 \times 2.759 = \$407.22$.

When interest is compounded semiannually, or quarterly, etc.

Example 3.—What is the amount of \$1 for 13 years at 4 per cent compounded semiannually? Here, there are 2 interest periods in the year; therefore $h = 2$, $i/h = 0.04/2 = 0.020$, and $nh = 13 \times 2 = 26$. Opposite "26," under "2 per cent," find, in Table 1-51, \$1.673 which is the amount required.

When the number of interest periods is not given in the table.

Example 4.—What is the amount of \$1 for 80 years at 6 per cent compounded annually? In Table 1-51, opposite "40" (which is ½ of 80) and under "6 per cent," we find 10.29. The amount for 80 years is $10.29^2 = \$105.90$. *Ans.* [Explanation: $s^n = (s^{n/2})^2 = (s^{n/3})^3 = (s^{n/4})^4$, etc.]

Example 5.—What is the amount of \$1 for 40 years at 8 per cent compounded quarterly? The number of interest periods is $hn = 4 \times 40 = 160$, which is beyond the reach of the table. $i/h = 0.08/4 = 0.02$. In Table 1-51, opposite "40" and under "2 per cent," find 2.208. The required amount is $2.208^4 = \$22.69$.

Comparisons of alternative investments, e.g., a total energy system vs. a purchased power system, often are made by means of a discounted cash flow analysis. Table 1-52 gives a condensed tabulation of discount factors. In

Table 1-52 Discount Factors for Use in Calculating Discounted Cash Flow

	Rate of interest								
Period	8%	10%	12%	14%	16%	18%	20%	22%	24%
0	1.0000	1.0000	1.0000	1.0000	1.0000	1.0000	1.0000	1.0000	1.0000
1	0.9259	0.9091	0.8928	0.8772	0.8621	0.8475	0.8333	0.8197	0.8065
2	.8573	.8264	.7972	.7695	.7431	.7182	.6944	.6719	.6504
3	.7938	.7513	.7118	.6750	.6407	.6086	.5787	.5507	.5245
4	.7350	.6830	.6355	.5921	.5523	.5158	.4823	.4514	.4230
5	.6806	.6209	.5674	.5194	.4761	.4371	.4019	.3700	.3411
6	.6302	.5645	.5066	.4556	.4104	.3704	.3349	.3033	.2751
7	.5835	.5132	.4523	.3996	.3538	.3139	.2791	.2486	.2218
8	.5403	.4665	.4039	.3506	.3050	.2660	.2326	.2038	.1789
9	.5002	.4241	.3606	.3075	.2630	.2255	.1938	.1670	.1443
10	.4632	.3855	.3220	.2697	.2267	.1911	.1615	.1369	.1164
11	.4289	.3505	.2875	.2366	.1954	.1619	.1346	.1122	.0938
12	.3971	.3186	.2567	.2076	.1685	.1372	.1122	.0920	.757
13	.3677	.2897	.2292	.1821	.1452	.1163	.0935	.0754	.0610
14	.3405	.2633	.2046	.1597	.1252	.0985	.0779	.0618	.0492
15	.3152	.2394	.1827	.1401	.1079	.0835	.0649	.0507	.0397
16	.2919	.2176	.1631	.1229	.0930	.0708	.0541	.0415	.0320
17	.2703	.1978	.1456	.1078	.0802	.0600	.0451	.0340	.0258
18	.2502	.1799	.1300	.0946	.0691	.0508	.0376	.0279	.0208
19	.2317	.1635	.1161	.0829	.0596	.0431	.0313	.0229	.0168
20	.2145	.1486	.1037	.0728	.0514	.0365	.0261	.0187	.0135
21	.1987	.1351	.0926	.0638	.0443	.0309	.0217	.0154	.0109
22	.1839	.1228	.0826	.0560	.0382	.0262	.0181	.0126	.0088
23	.1703	.1117	.0738	.0491	.0329	.0222	.0151	.0103	.0071
24	.1577	.1015	.0659	.0431	.0284	.0188	.0126	.0085	.0057
25	0.1460	0.0923	0.0588	0.0378	0.0245	0.0160	0.0105	0.0069	0.0046

the accompanying example it is assumed that a total energy system costs $254,700 more than its purchased power equivalent.

Example: A discounted cash flow statement[4] using factors from Table 1-52.

		12% Interest rate†		14% Interest rate†	
Year	Cash flow*	Discount factor	Discounted cash flow	Discount factor	Discounted cash flow
0	−$254,700‡	1.0000	−$254,700	1.0000	−$254,700
1	56,299	0.8928	50,263	0.8772	49,385
2	53,849	.7972	42,928	.7695	41,437
3	51,398	.7118	36,585	.6750	34,694
4	48,497	.6355	31,106	.5921	28,982
5	46,497	.5674	26,382	.5194	24,151
6	44,046	.5066	22,314	.4556	20,067
7	41,596	.4523	18,814	.3996	16,622
8	39,145	.4039	15,810	.3506	13,724
9	36,694	.3606	13,232	.3075	11,283
10	34,244	0.3220	11,027	0.2697	9,236
	$198,015		$13,761		−$5,119

Interpolation:

$$\text{Profitability rate} = \left(\frac{13761}{13761 + 5119}\right)(14 - 12) + 12$$

$$= 1.458 + 12$$

$$\text{Profitability} = 13.46\%$$

* Anticipated annual owning and operating cost reduction effected by the total energy system, i.e., income from the incremental initial investment minus income taxes.

† Trial discount rates for trial-and-error solution.

‡ Difference between alternative initial investments (total energy minus purchased power).

Table 1-53 Prefixes for Multiples and Submultiples of Numbers*

Multiples and Submultiples	Prefixes
$1,000,000,000,000 = 10^{12}$	tera
$1,000,000,000 = 10^{9}$	giga
$1,000,000 = 10^{6}$	mega
$1,000 = 10^{3}$	kilo
$100 = 10^{2}$	hecto
$10 = 10$	deka
$0.1 = 10^{-1}$	deci
$0.01 = 10^{-2}$	centi
$0.001 = 10^{-3}$ or 0.0_2	milli
$0.000,001 = 10^{-6}$ or 0.0_5	micro
$0.000,000,001 = 10^{-9}$ or 0.0_8	nano
$0.000,000,000,001 = 10^{-12}$ or 0.0_{11}	pico

* As recommended by the International Committee on Weights and Measures and adopted by the National Bureau of Standards.

References

1. Perry, John H., ed. *Chemical Engineers' Handbook*, 3d ed. New York, McGraw-Hill, 1950.
2. Jessup, Ralph S. and Werner, Elmer R. *Gas Calorimeter Tables.* (Nat. Bur. of Standards, C464) Washington, G.P.O., 1948.
3. Shnidman, Louis, ed. *Gaseous Fuels*, 2nd ed. New York, A.G.A., 1954.

SECTION 2

FUELS, COMBUSTION, AND HEAT TRANSFER

F. E. Vandaveer (retired), *Section Chairman,* and Chapters 1–3, 5, Con-Gas Service Corp., Cleveland, Ohio
Alfred R. Powell (retired), Chapter 4, Koppers Co., Inc. Pittsburgh, Pa.
C. George Segeler, Chapter 5, American Gas Association, New York, N. Y.
Richard L. Stone, Chapter 6, A.G.A. Laboratories, Cleveland, Ohio. Present address: Metalbestos Division, William-Wallace Co., Belmont, Calif.

CONTENTS

Chapter 6—Heat Transfer

Tables

Figures

Chapter 1

Fuel Reserves, Production, Gases Marketed, and Fuels Used by Utilities

by F. E. Vandaveer

Many of the data in this chapter were obtained from "Gas Facts," a statistical record of the gas industry in the United States, published annually since 1946 by A.G.A. This publication and others[1–3] are invaluable for statistics on many phases of the gas industry.

The **natural gas** data in Fig. 2-1, beginning with the year 1946 are prepared annually by the Committee on Natural Gas Reserves of A.G.A. Prior to 1946 the most reliable sources were used.

The favorable ratio of incremental new natural gas reserves to annual production, shown in Table 2-1, greatly minimizes the significance of the "number of years of gas supply available," frequently computed by dividing total proved recoverable reserves by annual withdrawals. As long as new additions exceed production there need be little cause for concern about such a hypothetical ratio. Many geologists and research personnel have made estimates of total ultimate natural gas reserves in the United States. Several such recent analyses have provided estimates of 1400 trillion cu ft, while one specified the possibility of 1700 trillion cu ft.

The ultimate U. S. petroleum reserve including the proved recoverable reserve (Table 2-1) was postulated in 1958 as 200 billion barrels by the A.G.A. Bureau of Statistics. In 1956 the Department of the Interior had estimated the ultimate petroleum reserve at 246 billion barrels.

Dr. A. C. Fieldner, U. S. Bureau of Mines, has estimated an oil reserve from shale of 92 billion barrels, about 2.5 times as much as the 1958 proved liquid hydrocarbon reserves.

An analysis of the consumption of fuel oil and natural gas for the years 1954 and 1955 is given in Table 2-2.

In 1960, estimates of the ultimate reserves of natural gas, petroleum, and coal were made by Lewis G. Weeks, president

Table 2-1 Reserves and Production of Natural Gas and Crude Petroleum and Consumption of Coal in the United States, 1925–1962

	Natural gas*		Crude petroleum*‡		Coal§ (consumption)	
Year	Proved reserves,† billions of cu ft	Net production, billions of cu ft	Proved reserves, millions of 42 gal bbl	Production, millions of 42 gal bbl	Bituminous & lignite, thousands of net tons	Anthracite, thousands of net tons
1925	23,000	1,188	8,500	764		
1930	46,000	1,941	13,600	898		
1934	62,000	1,765	12,177	908	343,814	55,500
1938	70,000	2,960	17,348	1213	336,281	45,200
1940	85,000	2,655	19,025	1352	430,910	49,000
1944	140,900	3,696	20,453	1678	589,599	59,400
1946	160,576	4,943	20,874	1726	500,386	53,900
1948	173,869	6,008	23,280	2002	519,909	50,200
1950	185,593	6,893	25,268	1944	454,202	39,900
1952	199,716	8,640	27,961	2257	418,757	35,300
1954	211,711	9,427	29,561	2257	363,060	26,900
1956	237,775	10,908	30,435	2552	432,858	24,000
1958	254,142	11,485	30,536	2373	366,703	19,000
1960	263,759	13,090	31,613	2471	380,429	17,600
1962	273,766	13,712	31,389	2550	387,774	15,000

* A.G.A.-A.P.I.-C.G.A. *Reports on Proved Reserves of Crude Oil, Natural Gas Liquids, and Natural Gas.* New York, annual. (Excludes allowance for natural gas liquids.)

† A.G.A. Bur. of Statistics. *Historical Statistics of the Gas Industry,* table 7. New York, 1961. (Estimated reserves for 1925–45.)

‡ Data for period 1925–45 includes some natural gas liquids.

§ U. S. Bur. of Mines. *Minerals Yearbook,* vol. II. Washington, G.P.O., annual.

Table 2-2 Consumption of Fuel Oil and Natural Gas in the United States by Major Consumer Groups,[1] 1954–55

(Fuel oils—thousand barrels; natural gas—million cubic feet)

	Railroads	Vessels	Gas and electric power plants	Smelters, mines, and manufactures	Space heating & cooking	Military	Oil company fuel	Miscellaneous	Total
Distillate fuel oil:									
1954	77,389	15,563	6,070	41,589	320,117	8,752	7,699	49,066	526,245
1955	84,668	16,675	5,884	43,606	357,088	10,914	8,597	54,163	581,595
Residual fuel oil:									
1954	16,122	108,790	70,749	160,121	78,845	26,887	52,165	7,035	520,714
1955	15,018	115,128	75,966	173,030	86,282	27,900	53,387	9,804	556,515
Natural gas:									
1954	...	...	1,165,498*	3,903,449	2,479,205	...	2,020,198	...	8,402,852
1955	...	...	1,153,280*	4,184,258	2,753,171	...	2,132,914	...	9,070,343

* Excluded from total, as data includes gas other than natural. Natural gas component included under "Smelters, mines, and manufactures."

Table 2-3 Sales of Liquefied Petroleum Gases and Natural Gasoline and Reserves of Natural Gas Liquids, 1935–1962

Year	Sales		Reserves
	Liquefied petroleum gases,* thousands of gallons	Natural gasoline,*† thousands of gallons	Natural gas liquids reserves,‡ thousands of 42 gallon barrels
1935	76,855	1,651,986	
1938	165,201	2,156,574	
1940	313,456	2,339,400	
1944	1,060,156	2,188,284	
1946	1,704,262	2,691,001	3,163,219
1948	2,736,801	2,979,412	3,540,783
1950	3,482,567	3,228,666	4,267,663
1952	4,477,379	3,665,760	4,996,651
1954	5,125,533	4,104,828	5,244,457
1956	6,635,763	4,438,890	5,902,332
1958	7,462,089	4,355,025	6,204,018
1960	9,544,649	4,479,454	6,816,059
1962	10,729,394	4,772,260	7,311,517

* U. S. Bur. of Mines. *Minerals Yearbook* vol. II. Washington, G.P.O., annual.

† Recovery from natural gas and cycling plant.

‡ A.G.A.-A.P.I.-C.G.A. *Proved Reserves of Crude Oil, Natural Gas Liquids and Natural Gas*. New York, annual. (Data not available prior to 1946.)

Table 2-4 Distribution of LP-Gas Sales, 1956[1]

	% of Total
Gas manufacture	3.2
Domestic & commercial	45.3
Industrial & engine fuel	18.7
Synthetic rubber	6.3
Chemical manufacturing	23.7
Refinery fuel	2.2
Miscellaneous	0.6

Table 2-5 Natural Gas Marketed,* Manufactured Gases Produced and Purchased by Utilities[4]

(Millions of therms)

Year	Natural gas†	Manufactured gas—Total	Carbureted water gas	Coke oven gas	Retort coal gas	Oil & oil refinery gas
1935	21,170	1853	879	774	143	57
1938	25,350	1773	831	727	110	105
1940	29,390	1941	963	795	83	101
1944	41,010	2304	1231	907	82	84
1946	44,640	2690	1543	925	72	150
1948	53,340	2848	1638	979	52	180
1950	67,530	2659	1611	861	21	166
1952	86,140	2009	1160	689	6	155
1954	93,982	1504	969	438	‡	98
1956	108,382	1434				
1958	118,575	1086				
1960	137,288	740	No longer reported			
1962	145,953	731				

* **Marketed production** as tabulated is defined by the Bureau of Mines as useful consumption. It includes: sales by all non-utilities to industrial consumers, field use, the natural gas portion of mixed gases, net change in underground storage, and transmission losses; therefore, the figures given exceed utility sales.

† Note that the Bureau of Mines converts natural gas volume to therms via the factor 1075 Btu per cu ft, which would indicate the presence of natural gas liquids.

‡ Included in coke oven gas.

Table 2-6 U. S. Distribution of Bituminous Coal (and Lignite) to Certain Consumer Groups[1]

(Thousand net tons)

Year	Electric power utilities	Coke	Steel and rolling mills	Cement mills	Other industrials	Retail deliveries	Total (inc. misc.)
1954	115,235	85,391	6983	7924	77,115	51,798	363,060
1955	140,550	107,377	7353	8529	89,611	53,020	423,412
1956	154,983	105,913	7189	9026	93,302	48,667	432,858
1957	157,398	108,020	6938	8633	87,202	35,712	413,668
1958	152,928	76,580	7268	8256	81,372	35,619	366,703
1959	165,788	79,181	6674	8510	73,396	29,138	366,256
1960	173,882	81,015	7378	8216	76,487	30,405	380,429
1961	179,629	73,881	7495	7615	77,280	27,735	374,405

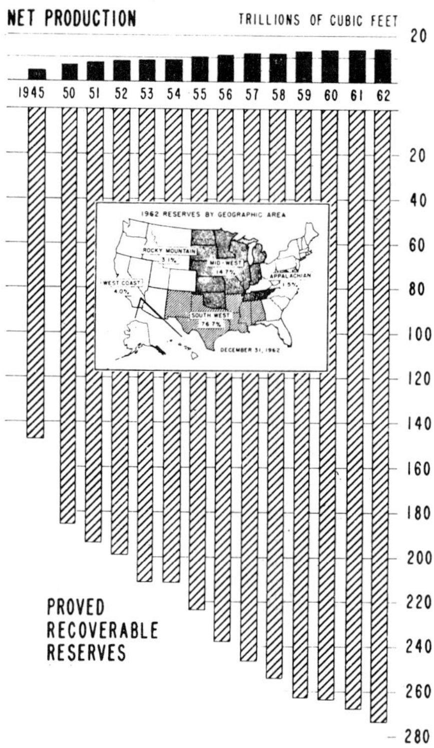

Fig. 2-1 Proved recoverable reserves and net production of natural gas in the United States.

of the American Association of Petroleum Geologists, as follows:

Reserves	United States	Total World
Natural gas,* trillions of cubic feet	1,000	6,000
Oil, billions of barrels		
Conventional recovery	270	2,000
Secondary recovery	190	1,500
Shale oil	2,000	12,000
Coal, billions of tons	2,000	7,000

* Not to be confused with the proved recoverable reserves.

The **liquefied petroleum** and **natural gasoline** data are shown in Table 2-3. Proved recoverable reserves of natural gas liquids are those estimated to be contained in recoverable gas reserves and to be extractable by methods in use. The importance and size of this industry can be shown by converting the LP-gas to equivalent cubic feet of natural gas. For the year 1958 the LP-gases produced were equivalent to 895 billion cu ft of 1000 Btu natural gas or 7.8 per cent as much as the total natural gas produced in 1958 (97,700 Btu/gal was used as an average heating value for LP-gases). In 1958 the natural gasoline produced was equivalent to 432 billion cu ft of 1000 Btu natural gas or equivalent to 3.8 per cent of that year's production of natural gas (107,700 Btu/gal was used in this calculation for natural gasoline). Distribution of LP-gas sales in 1956 is shown in Table 2-4.

Table 2-5 is a summary of natural gas marketed, as defined in the table footnote, and manufactured gases produced and purchased by utilities. If a detailed breakdown

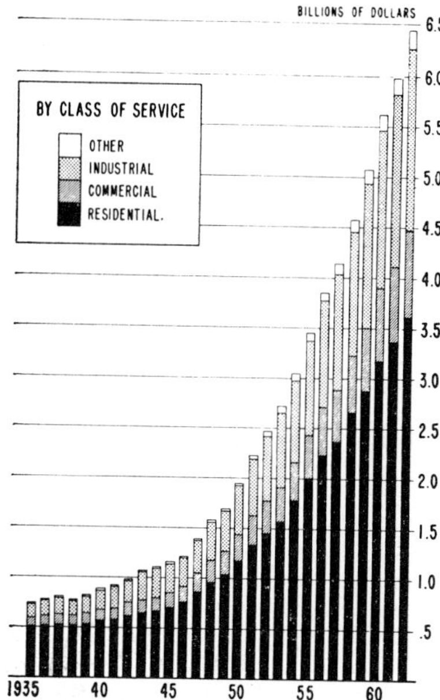

Fig. 2-2 Gas utility revenues by class of service.

is required, e.g., gases marketed by utilities only, the references cited will provide such information. **Marketed production** figures are given for natural gas since they are an accepted standard. The volumes for natural gas may be compared to Table 2-1 which shows net production. All other values have been given in therms, since **heating value** is a more meaningful standard of comparison than volume.

The **coal reserves** of the United States, in the forms of lignite thru bituminous, were estimated at 3.1 trillion tons. A. B. Crichton, in 1948, stated that this figure should be reduced by 92 per cent to an economically recoverable reserve of 233 billion tons with an overall heat content of $6Q$, where $Q = 10^{18}$ Btu.

Estimates of economically recoverable world reserves of coal by Dr. Fieldner were reported by P. C. Putnam, con-

Table 2-7 Solid Fuels and Oil Used by Utilities in the Production of Manufactured Gas, Including Boiler Fuel and Other Uses,[4] 1935–1962

(Includes fuel for peak loads in natural gas systems)

	Thousands of tons			Millions of gallons
Year	Anthracite	Bituminous coal	Coke	Oil
1935	126	7849	2589	510
1938	181	6715	2352	586
1940	214	6882	2599	677
1944	291	7860	3129	837
1946	347	7660	3782	1069
1948	590*	7336	3733	1308*
1950	329	6207	3213	1185
1952	155	4130	1104	467
1954	155	1765	491	301
1956	130	1558	214	246
1958	78	929	210	143
1960	71	575	76	111
1962	78	465	49	117

* Maximums.

Table 2-8 Annual Consumption of Energy Fuels and Energy from Water Power in the United States,[4] 1942–1962
(Trillions of Btu)

Year	Natural gas dry	Natural gas liquids	Crude petroleum and products	Bituminous coal and lignite	Anthracite coal	Water power	Total energy
1942	3,102	367	7,667	14,149	1435	1177	27,897
1944	3,775	442	9,261	15,447	1509	1387	31,821
1946	4,088	493	9,987	13,110	1369	1446	30,493
1948	5,032	619	11,938	13,622	1275	1507	33,993
1950	6,151	783	12,706	12,900	1013	1601	35,154
1952	7,761	954	14,380	10,971	897	1614	36,577
1954	8,553	1042	15,090	9,512	683	1479	36,359
1956	9,834	1209	17,418	11,338	610	1598	42,007
1958	10,995	1240	17,428	9,607	483	1740	41,493
1960	12,736	1427	18,608	9,967	447	1775	44,960
1962	14,036	1598	19,560	10,160	378	1965	47,787*

* Equal to 0.05Q, where $Q = 10^{18}$ Btu.
Note: The heat values employed in this table are: anthracite, 12,700 Btu/lb; bituminous coal and lignite, 13,100 Btu/lb; crude oil, 5,800,000 Btu/barrel; weighted average Btu/barrel on petroleum products by using 5,248,000 for gasoline, 5,670,000 for kerosene, 5,825,000 for distillate, 6,287,000 for residual, 6,064,800 for lubricants, 5,537,280 for wax; 6,636,000 for asphalt, and 5,796,000 for miscellaneous; natural gas dry, 1035 Btu/cu ft; natural gas liquids weighted average Btu based on production; natural gasoline, 110,000 Btu/gal; and LP-gas, 95,500 Btu/gal. Water power is converted to coal equivalent at the prevailing rate of pounds of coal per kwh each year at central electric stations.

sultant to the U. S. Atomic Energy Commission, in *Energy in the Future*, published by Van Nostrand in 1953, as follows:

	Heat content, (Q)
United States	6
Canada	2
United Kingdom	1
China	6
U.S.S.R.	10
Other countries	7
Total ($Q = 10^{18}$ Btu)	32

Note: Underground gasification and/or fuel cells might yield additions.

Data for annual coal consumption is given in Table 2-1. Bituminous coal and lignite coal were distributed to major consumer groups as shown in Table 2-6.

A further indication of the trend in gas production is shown in Table 2-7 which lists the amount of coal and oil used in the production of manufactured gas. Anthracite usage reached a maximum in 1948 and has declined since then; bituminous coal usage was greatest in 1945 and has decreased since then; coke usage reached a peak in 1947 and has decreased to practically nil. Oil for gas manufacture was above a billion gallons per year in 1946 and for five years afterwards. Usage subsequently dropped sharply.

Table 2-8 is included to give the reader an overall grasp of the annual consumption of energy in the United States. Each of the major sources is credited in terms of Btu so that comparisons are facilitated. The heating values employed are given in the footnote.

Figure 2-2 illustrates the steady growth of the gas industry from a revenue standpoint. In 1958, residential revenues represented 58.1 per cent of the total, while industrial and commercial revenues accounted for 26.9 per cent and 12.3 per cent, respectively.

REFERENCES

1. U. S. Bur. of Mines. *Minerals Yearbook.* Washington, G.P.O., annual.
2. Foster, J. F. and Lund, R. J. *Economics of Fuel Gas from Coal.* New York, McGraw-Hill, 1950.
3. U. S. Fed. Power Comm. *Natural Gas Investigation (Docket No. G-580).* Washington, G.P.O., 1948. 2 v.
4. A.G.A. Bur. of Statistics. *Gas Facts.* New York, annual.

Chapter 2

Gaseous Fuels

by F. E. Vandaveer

Fuel gases are listed here in four major groups, namely, natural gases, liquefied petroleum gases, manufactured gases, and mixed gases. In a fifth division a listing of some components of manufactured gases before purification and some trace constituents in city gases are given.

NATURAL GASES

Natural gas, as defined herein,* is a naturally occurring mixture of hydrocarbon and nonhydrocarbon gases found in porous formations beneath the earth's surface, often in association with crude petroleum. The principal constituent of most natural gases is methane. Minor constituents are heavier hydrocarbons and certain nonhydrocarbon gases such as N_2, CO_2, H_2S, and He. Oxygen and argon may be present but the amounts are usually very small. Some natural gases contain major proportions of saturated hydrocarbons heavier than methane. A few are devoid of heavier hydrocarbons, and many contain significant quantities of one or more of the nonhydrocarbons mentioned above. The natural gas of commerce, supplied to fuel gas markets, usually contains from 80 to 95 per cent methane and lesser amounts of ethane and propane. Most of the remainder is nitrogen. The heating value ranges from 900 to 1200 Btu per cu ft and the specific gravity varies from 0.58 to 0.79 (air = 1.00).

Analyses of natural gases distributed in a number of cities in the United States are given in Table 2-9. These gases may be classified in three groups, as shown in Table 2-10.

In Table 2-11 average analyses of natural gas delivered by a number of major transmission lines are given.

As an indication of the general composition of natural gas from some of the major gas fields and a few of the smaller fields in various parts of the country, Table 2-12 is presented. Most of these analyses were made on average samples taken after the processing plant removal of gasoline, liquefied petroleum gas, water vapor, and sometimes sulfur. It was noted that the gasoline content varied from 148 to 1200 gal per MMCF, while some fields had no gasoline at all. Water vapor content varied from a minimum of three to a maximum of 1400 pounds per million cubic feet. In samples where sulfur was found, it varied from 0.02 to 312 grains per CCF. Other fields not listed here have been reported as having as much as 4000 grains of sulfur per CCF. The heating value of the gas is decreased when gasoline and condensable hydrocarbons are removed in the processing plants. These reductions varied in reported cases from 29 to 208 Btu per cu ft. No decrease occurred in the case of a dry gas.

Individual wells in the field may vary appreciably from the average results reported in Table 2-12. A good illustration of this statement may be found in Bulletin 486, Bureau of Mines,[1] where analyses of 2100 natural gas samples from as many wells showed that no two analyses were exactly the same and that wells in the same field may vary appreciably. Natural gas composition for a number of foreign gas fields is shown in Table 2-13.

Natural gas in the United States is the major source of **helium,** although most natural gas contains from zero to less than 0.3 per cent helium. Cady and McFarland[2] discovered helium in a sample of natural gas with a high nitrogen content from a well in Dexter, Kan., in 1905. Over 1.7 billion cu ft of helium were extracted by government plants during the period 1921 thru 1955. In Table 2-14, representative analyses are given of gas from the major United States helium bearing natural gas fields yielding more than one per cent helium. Helium is usually associated with high nitrogen gases. The four government helium plants as of 1955 were at Amarillo and Exel, Tex.; Otis, Kan.; and Shiprock, N. M.

Table 2-15 gives examples of commercial pipeline standards for natural gas in reference to impurities. These data were obtained in a 1957 A.G.A. survey by the Pipeline Research Committee. The information is presented only as an illustration of varying specifications in actual contracts in force in 1957. The actual quality of the gases supplied is likely to be superior to the contractual limits; e.g., actual H_2S and sulfur seldom exceed 0.1 and 1.0 grains per CCF, respectively.

Table 2-16 gives the sulfur content of natural gas in the form of H_2S and mercaptans from various fields, before purification.

Physiologic Effects

Humans and Animals.[18] Following are the results of a study to determine what alterations, however slight, a natural gas might cause in the physiologic processes, even after prolonged applications:

> "Natural gas,† as supplied to domestic consumers in Los Angeles, is without demonstrable effect on human beings in concentrations of 25 per cent.

* Suggested by Petroleum and Natural Gas Branch, U. S. Bureau of Mines, Washington, D. C.

† Composition (per cent by volume): CH_4 (90.00), C_2H_6 (4.63), C_3H_8 (3.91), *iso*-C_4H_{10} (0.47), *n*-C_4H_{10} (0.51), CO_2 (0.4), $C_5H_{12}+$ (0.08), air (trace).

Table 2-9 Natural Gas Distributed in Various Cities in the United States*

(Surveyed by A.G.A. in the fall of 1962)

No.	City	Components of gas, per cent by volume: Methane	Ethane	Propane	Butanes	Pentanes	Hexanes plus	CO_2	N_2	Miscel.	Heat. value,† Btu/cu ft	Sp gr
1	Abilene, Tex.	73.52	13.23	4.35	0.56	0.06	0.11	0.16	8.01	...	1121	0.710
2	Akron, Ohio	93.30	3.49	0.69	.18	.04	.00	0.50	1.80	...	1037	.600
3	Albuquerque, N. M.	86.10	9.49	2.34	.44	.08	.03	1.02	0.50	...	1120	.646
4	Atlanta, Ga.	93.42	2.80	0.65	.33	.12	.10	1.38	1.20	...	1031	.604
5	Baltimore, Md.	94.40	3.40	.60	.50	.00	.00	0.60	0.50	...	1051	.590
6	Birmingham, Ala.	93.14	2.50	.67	.32	.12	.05	1.06	2.14	...	1024	.599
7	Boston, Mass.	93.51	3.82	.93	.28	.07	.06	0.94	0.39	...	1057	.604
8	Brooklyn, N. Y.	94.52	3.29	0.73	.26	.10	.09	0.70	0.31	...	1049	.595
19	Butte, Mont.	87.38	3.02	1.09	.11	.06	.00	1.98	6.36	...	1000	.610
10	Canton, Ohio	93.30	3.49	0.69	.18	.04	.00	0.50	1.80	...	1037	.600
11	Cheyenne, Wyo.	91.00	4.73	1.20	.30	.06	.04	1.86	0.81	...	1060	.610
12	Cincinnati, Ohio	94.25	3.98	0.57	.16	.03	.03	0.68	0.30	...	1031	.591
13	Cleveland, Ohio	93.30	3.49	.69	.18	.04	.00	.50	1.80	...	1037	.600
14	Columbus, Ohio	93.54	3.58	0.66	.22	.06	.03	.85	1.11	...	1028	.597
15	Dallas, Tex.	86.30	7.25	2.78	.48	.07	.02	.63	2.47	...	1093	.641
16	Denver, Colo.	81.11	6.01	2.10	.57	.17	.03	.42	9.19	...	1011	.659
17	Des Moines, Iowa	80.38	6.39	2.46	.61	.08	.03	.20	9.53	0.32 He	1012	.669
18	Detroit, Mich.	89.92	4.21	1.34	.34	.09	.01	.59	3.30	.20 He	1016	.616
19	El Paso, Tex.	86.92	7.95	2.16	.16	.00	.00	.04	2.72	.05 He	1082	.630
20	Ft. Worth, Tex.	85.27	8.43	2.98	.62	.09	.04	.27	2.30	...	1115	.649
21	Houston, Tex.‡	92.50	4.80	2.00	.30	...	...	.27	0.13	...	1031	.623
22	Kansas City, Mo.	72.79	6.42	2.91	.50	.06	Trace	0.22	17.10	...	945	.695
23	Little Rock, Ark.	94.00	3.00	0.50	.20	.20	...	1.00	1.10	...	1035	.590
24	Los Angeles, Calif.	86.50	8.00	1.90	.30	.10	.10	0.50	2.60	...	1084	.638
25	Louisville, Ky.	94.05	3.41	0.40	.13	.05	.09	1.20	0.67	...	1034	.596
26	Memphis, Tenn.	92.50	4.37	0.62	.18	.07	.10	1.60	0.56	...	1044	.608
27	Milwaukee, Wis.	89.01	5.19	1.89	.66	.44	.02	0.00	2.73	.06 He	1051	.627
28	New Orleans, La.	93.75	3.16	1.36	.65	.66	.00	.42	0.00	...	1072	.612
29	New York City	94.52	3.29	0.73	.26	.10	.09	.70	0.31	...	1049	.595
30	Oklahoma City, Okla.	89.57	6.31	1.36	.36	.00	.00	.13	2.06	.21 O_2	1080	.615
31	Omaha, Neb.	80.46	6.30	2.59	.68	.09	.05	.17	9.32	.34 He	1020	.669
32	Parkersburg, W. Va.	94.50	3.39	0.68	.12	.07	.03	.67	0.41	.01 O_2	1049	.592
33	Phoenix, Ariz.	87.37	8.11	2.26	.13	.00	.00	.61	1.37	...	1071	.633
34	Pittsburgh, Pa.	94.03	3.58	0.79	.28	.07	.04	0.80	0.40	.01 O_2	1051	.595
35	Providence, R. I.	93.05	4.01	1.02	.34	.08	.08	1.00	0.42	...	1057	.601
36	Provo, Utah	91.40	3.95	0.84	0.39	.03	.01	0.52	2.86	...	1032	.605
37	Pueblo, Colo.	73.86	5.71	3.20	1.34	.14	.06	.13	15.26	...	980	.706
38	Rapid City, S. D.	90.60	7.20	0.82	0.19	.03	.03	.18	0.93	.02 He	1077	.607
39	St. Louis, Mo.	93.32	4.17	0.69	.19	.05	...	.98	0.61	...	...	...
40	Salt Lake City, Utah	91.17	5.29	1.69	.55	.16	.03	.29	0.82	...	1082	.614
41	San Diego, Calif.	86.85	8.37	1.86	.15	.00	.00	.41	2.32	.04 He	1079	.643
42	San Francisco, Calif.	88.69	7.01	1.93	.28	.03	.00	.62	1.43	.01 He	1086	.624
43	Toledo, Ohio	93.54	3.58	0.66	.22	.06	.03	.85	1.11	...	1028	.597
44	Tulsa, Okla.	86.29	8.36	1.45	.18	.14	.01	0.23	2.95	.39 O_2	1086	.630
45	Waco, Tex.	93.48	2.57	0.89	.43	.17	.11	1.69	0.66	...	1042	.607
46	Washington, D. C.	95.15	2.84	0.63	0.24	.05	.05	0.62	0.42	...	1042	.586
47	Wichita, Kan.	79.62	6.40	1.42	1.12	.48	.14	.10	10.62	0.10 O_2	1051	.660
48	Youngstown, Ohio	93.30	3.49	0.69	0.18	0.04	0.00	0.50	1.80	...	1037	0.600

* Average analyses obtained from the operating utility company(s) supplying the city; the gas supply may vary considerably from these data—especially where more than one pipeline supplies the city. Also, as new supplies may be received from other sources, the analyses may change. Peak shaving (if used) is not accounted for in these data.

† Gross or higher heating value at 30 in. Hg, 60 F, dry. To convert to a saturated basis deduct 1.73 per cent; i.e., 17.3 from 1000, 19 from 1100.

‡ 1954 data.

Table 2-10 Group Classifications of Natural Gases

	Group	Nitrogen, %	Specific gravity	Methane, %	Btu/cu ft, dry
I	High inert type	6.3–16.20	0.660–.708	71.9–83.2	958–1051
II	High methane type	0.1– 2.39	0.59 –.614	87.6–95.7	1008–1071
III	High Btu type	1.2– 7.5	0.62 –.719	85 –90.1	1071–1124

Table 2-11 Natural Gas Delivered by Major Transmission Lines*

No.	Transmission line and major source of supply	Components of gas—per cent by volume									Gross heat value, Btu per cu ft, dry†	Sp gr	Water vapor, lb per MMCF
		Methane	Etnane	Propane	Butane	Pentane	Hexane plus	CO_2	O_2	N_2			
1	Cities Service Gas Co. from Texas Panhandle	73.48	6.86	4.26	2.13	0.63	0.40	0.10	0.32	11.90	1077	0.694	9.80
2	Cities Service Gas Co. from Oklahoma Hugoton	75.28	6.39	3.76	1.45	.29	.29	...	...	12.54	1043	.706	3.50
3	Cities Service Gas Co. from Kansas Hugoton	77.02	3.89	2.58	2.04	.49	0.13	.10	.10	13.65	1005	.698	3.00
4	Colorado Interstate Gas Co. from Kansas Hugoton	72.40	6.12	3.21	1.20	.15	...	...	...	16.92	983	.703	5.80
5	Colorado Interstate Gas Co. from Texas Panhandle	78.76	5.67	2.88	1.06	.10	...	...	...	11.53	1007	.683	8.40
6	El Paso Natural Gas Co. from Permian Basin	81.21	9.42	3.45	0.63	.04	...	...	...	5.25	1097	.665	10.00
7	Kansas Nebraska Natural Gas Co. from Hugoton	71.25	5.69	3.37	.98	.14	.20	.10	...	18.27	956	.7161	...
8	Lone Star Gas Co. from Texas and Oklahoma	85.00	7.10	2.40	.50	.40	...	.60	...	4.00	1069	.650	15.00
9	Michigan-Wisconsin Pipeline Co. from Panhandle	73.10	6.20	4.00	.90	...	...	...	...	15.80	973	.704	3.50
10	Mississippi River Fuel Corp. from Monroe	93.17	4.13	0.75	0.22	.07	.11	1.22	.01	0.32	1049	...	3.0
11	Montana-Dakota Utilities Co. from Montana and the Dakotas	93.00	6.00	0.20	...	...	...	.30	...	0.50	1010	.600	...
12	Natural Gas Pipeline Co. of America from Panhandle	79.00	6.00	3.70	1.00	.10	...	...	...	10.20	1039	.685	8.00
13	Northern Natural Gas Co. from Panhandle	75.78	4.97	3.24	2.12	...	...	.20	.14	13.55	1011	.685	11.08
14	Pacific Gas & Electric Co. 34" from El Paso Nat. Gas Co. System	81.90	9.30	3.30	0.50	...	...	...	...	5.00	1100	.660	Win. 11 Sum. 15
15	Panhandle Eastern Trans. Co. from Panhandle	72.40	15.70	0.10	...	...	...	.20	0.10	11.30	1020	.680	...
16	Southern California Gas Co. from El Paso Nat. Gas Co. System	81.40	8.70	3.60	.60	.10	...	...	...	5.60	1092	.670	10.00
17	Southern Natural Gas Co. from Monroe and others	94.95	1.30	0.33	.11	.09	.14	.70	...	2.39	1008	.590	5.00
18	Tennessee Gas Transmission Co. from Louisiana and Texas	94.61	3.30	.99	.33	.03	.18	.35	...	0.16	1065	.5967	7.28
19	Texas Eastern Transmission Corp. from Louisiana and Texas	92.58	4.27	.97	.21	.04	.06	0.90	...	.95	1051	.606	3.80
20	Texas Gas Transmission Corp. from Louisiana and Texas	92.80	4.20	0.90	.20	...	.10	1.00	...	0.80	1049	.600	...
21	Transcontinental Gas Pipeline Corp. from Louisiana and Texas	93.45	3.59	1.27	.61	.26	.22	0.60	...	...	1085	.6102	3.50
22	United Gas Pipeline Co. from Refugio to Houston	92.61	3.87	1.15	.39	.12	.07	.66	...	1.13	1056	.6049	...
23	United Gas Pipeline Co. from Carthage Longview	91.46	4.18	0.94	.24	.02	.09	.89	...	2.18	1037	.6074	...
24	United Gas Pipeline Co. from Waskom to Dallas	89.55	4.97	.93	.38	.07	.24	0.91	...	2.95	1046	.6219	...
25	United Gas Pipeline Co. from Slico to Shreveport	92.77	3.35	0.83	.45	.18	.06	1.03	...	1.33	1044	.6054	...
26	United Gas Pipeline Co. from Lirette to Mobile	93.32	3.00	1.06	.57	.20	.20	0.49	...	1.16	1062	.6063	...
27	United Gas Pipeline Co. from Carthage to Sterlington	92.82	3.55	0.69	.09	.03	.10	1.01	...	1.71	1029	.5995	...
28	United Gas Pipeline Co. from Agua Dulce to Austin	84.73	7.90	2.49	0.76	0.32	0.25	2.54	...	1.01	1112	0.6695	...

* These are average analyses obtained from the operating companies. As gas from new fields may be acquired or changes made in the processing plants, the analyses may change from time to time.

† 30 in. Hg, 60 F. To convert to a saturated basis deduct 1.73 per cent; i.e., 17.3 from 1000, or 19 from 1100.

Table 2-12 Natural Gas From Various Gas Fields*

(As of November 1951)

No.	Location and field	Components of gas—per cent by volume									Gross heat value, Btu per cu ft, dry†	Sp gr
		Methane	Ethane	Propane	Butane	Pentane	Hexane plus	CO_2	O_2	N_2		
1	California Kettleman North Dome(b)	87.20	5.20	3.50	2.00	1.70	...	0.40	...	...	1212	0.690
2	California, Kettleman North Dome(a)	93.00	4.60	1.80	0.20	...	...	.40	...	...	1080	.602
3	California, Rio Vista	94.20	2.95	0.80	0.15	0.10	...	.30	...	1.50	1038	.590
4	California, Ventura(b)	83.60	5.40	6.10	3.20	1.40	...	.30	...	...	1260	.706
5	California, Ventura(a)	92.70	4.70	2.20	0.10	...	...	0.30	...	...	1083	.604
6	Illinois—Indiana[1]	97.00	...	...	...	...	...	1.00	0.40	1.60	983	...
7	Kansas, Cunningham	62.30	21.20	...	...	...	...	0.20	.30	16.00	1011	...
8	Kansas, Hugoton	77.00	3.90	2.60	2.00	0.50	0.10	.10	.10	13.60	1005	.698
9	Kansas, Otis	71.80	13.90	...	...	...	...	.30	.20	13.80	976	...
10	Kentucky[1]	83.40	13.30	...	...	...	...	.80	.40	2.10	1083	...
11	Louisiana, Monroe	91.28	1.52	0.70	0.41	.19	.15	.30	...	5.45	997	.6075

For footnotes, see end of the table.

Table 2-12 Natural Gas From Various Gas Fields* (Continued)

No.	Location and field	Components of gas—per cent by volume									Gross heat value, Btu per cu ft, dry†	Sp gr
		Methane	Ethane	Propane	Butane	Pentane	Hexane plus	CO_2	O_2	N_2		
12	Louisiana, Paradis	92.18	3.33	1.48	.79	.25	.05	.90	...	1.02	1067	0.6153
13	Michigan[1]	75.80	17.50	...	...	...	...	.10	.30	6.30	1081	...
14	Mississippi, Gwinville	95.74	0.27	0.07	0.10	.03	.18	1.28	...	2.33	989	.5855
15	Ohio, Canton	86.10	10.80	.60	...	...	...	0.10	.40	2.00	1082	.620
16	Ohio, Hinckley-Medina	80.20	16.80	0.20	...	...	...	.20	.10	2.50	1118	.650
17	Oklahoma, Hugoton	75.30	6.40	3.70	1.40	.30	.30	...	...	12.50	1043	.710
18	Oklahoma, Hughes County	79.00	17.50	...	...	...	...	0.20	.30	3.00	1114	...
19	Oklahoma, Keyes	51.50	17.50	...	...	...	...	1.40	.20	29.40	835	...
20	Pennsylvania, Greene County	85.41	8.10	3.34	1.53	.62	...	0.10	...	0.90	1171	.667
21	Pennsylvania, Leidy	96.80	2.50	0.20	...	...	...	.20	...	.30	1031	.5751
22	Pennsylvania, Western Penn.	85.60	13.30	0.30	...	...	...	.10	0.10	.60	1112	.6271
23	Texas, Agua Dulce(b)	90.98	4.56	1.88	1.04	.41	.34	.12	...	0.67	1109	.6307
24	Texas, Agua Dulce(a)	93.00	4.14	1.03	0.19	.04	.07	.49	...	1.04	1051	.5986
25	Texas, Carthage(b)	90.29	4.47	1.45	.87	.38	.39	.88	...	1.27	1093	.634
26	Texas, Carthage(a)	91.73	4.01	0.94	.24	.02	.14	.87	...	2.05	1038	.6073
27	Texas, East Texas	63.28	22.49	11.15	0.09	.08	.19	.67	...	2.05	1336	.7941
28	Texas, Hugoton(b)	76.90	6.07	3.69	1.34	.46	.13	...	...	11.41	1046	.698
29	Texas, Hugoton(a)	79.00	6.31	2.20	0.07	.07	...	...	...	12.35	967	.659
30	Texas, Keystone(a)	86.20	11.90	1.66	0.24	...	...	...	...	...	1133	.630
31	Texas, Keystone(b) (6 formations)	78.77	10.92	6.07	2.38	.70	.43	.40	[H_2S 0.31]	...	1279	.730
32	Texas, Panhandle(b)	81.76	5.63	3.44	1.51	.26	.43	.10	...	6.87	1090	.681
33	Texas, Panhandle(a)	81.50	5.96	3.66	1.23	.05	...	.10	...	7.50	1061	.669
34	Texas, Pledger	94.24	2.85	0.98	0.60	.30	.48	.10	...	0.45	1086	.6132
35	Texas, Tom O'Connor	90.77	4.48	1.88	1.16	0.43	.32	0.11	...	0.85	1117	.6329
36	Texas, Wasson(b)	69.52	9.71	6.48	3.20	1.40	.66	5.60	...	3.14	1238	.805
37	Texas, Wasson(a)	76.87	10.14	2.25	0.12	0.04	0.04	6.02	...	3.54	1030	.684
38	West Virginia, Northern	85.86	8.51	3.15	1.14	0.44	...	0.10	...	0.80	1158	.6475
39	West Virginia, Terra Alta	98.75	1.07	0.04	0.14	...	...	...	...	...	1027	.561
40	West Virginia, Wyoming County	97.75	1.80	0.35	0.10	...	...	...	...	...	1036	0.5683

* Analyses obtained from major companies operating in the field or taking gas from it.
† 30 in. Hg, 60 F. To convert to a saturated basis deduct 1.73 per cent; i.e., 17.3 from 1000, 19 from 1100.
(a) After processing plant. (b) Before processing plant.

Table 2-13 Natural Gas in Foreign Fields

No.	Location	Field	Components of gas—per cent by volume									Gross heat value. Btu per cu ft, dry*	Sp gr
			Methane	Ethane	Propane	Butane	Pentane	Hexane plus	CO_2	O_2	N_2		
1	Argentine	Commodora Rivadavia	95.00	4.00	..	..	..	..	..	..	1.00	1035	0.576
2	Argentina (both to Buenos Aires)	Plaza Huincul	98.00	1.00	..	..	..	..	..	..	1.00	1011	0.561
3	Belgium	21 coal mines	70–100	..	..	..	..	..	..	[30 to 0 air]		680–980	0.570–0.620
4	Canada (Alberta)	Turner Valley Oilfield residue	87.02	9.15	2.78	0.35	..	..	0.24	..	0.46	1126	0.635
5	Canada (Alberta)	Jumping Pound condensate field residue	93.98	3.36	0.79	0.26	..	..	..	..	1.61	1041	0.588
6	Canada (Alberta)	Viking-Kinsella Dry	91.90	2.00	0.90	0.30	..	..	..	..	4.90	1000	0.597
7	Canada (Alberta)	Leduc Oilfield residue	73.59	19.21	3.87	0.70	..	..	..	..	2.63	1208	0.708
8	Canada (Br. Col.)	Fort St. John	88.50	4.31	1.82	0.95	0.42	0.44	0.60	..	2.55	1109	0.646
9	Canada (Alberta & British Columbia)	Pouce Coupe	95.25	0.27	0.10	..	..	..	..	..	4.38	987	0.574
10	Chile (Terra del Fuego)	Chanarcillo	90.03	6.00	2.13	1.05	0.48	0.31	..	..	..	1211	0.647
11	France	St. Marcet											
		Wet gas	89.00	4.50	1.60	0.90	1.10	..	..	..	2.90	1101	0.637
		Dry gas—outlet gasoline plant	91.50	4.60	0.80	0.10	..	..	..	..	3.00	1030	0.598
12	Italy	Cortemaggiore Po Basin	89.87	4.40	1.57	0.77	1.63	..	..	..	1.76	1122	0.684
13	Japan	Kurokawa	96.39	..	..	..	..	..	2.04†	..	1.34	976	0.579
14	Japan	Yabase	50.70	4.50	1.30	..	..	..	29.00	..	..	627	0.818
15	Japan	Nishiyama	22.80	71.30	1.00	..	..	..	0.50	0.50	..	1537	0.908
16	Japan	Ohmo	74.01	16.15	7.31	..	..	..	1.91	0.19	..	1231	0.758
17	Japan	Shikina	94.39	3.28	..	..	..	..	2.16	0.17	..	1012	0.594
18	Netherlands		90.00	2.00	..	..	..	..	[8.00 & N_2]	..	..	950	0.630
19	Poland	Kosow	95.80	1.90	..	..	..	..	0.20	0.30	1.80	1004	0.573
20	Saudi Arabia	Safaniya	56.00	19.00	14.30	5.90	1.80	1.10	1.90	..	..	1502	0.653
21	Saudi Arabia	Abqaiq	55.50	20.40	8.10	3.70	1.40	0.80	7.70	..	[H_2S 2.40]	1373	0.670
22	Saudi Arabia	Ain Dar	27.80	26.10	18.40	8.20	2.60	1.70	9.80	..	[H_2S 5.20]	1717	0.755
23	Saudi Arabia	Uthmaniyah	59.30	17.00	9.40	4.00	1.10	0.40	7.50	..	[H_2S 1.30]	1357	0.679

For footnotes, see end of table.

Table 2-13 Natural Gas in Foreign Fields (Continued)

No.	Location	Field	Components of gas—per cent by volume: Methane	Ethane	Propane	Butane	Pentane	Hexane plus	CO_2	O_2	N_2	Gross heat value, Btu per cu ft, dry*	Sp gr
24	Soviet Russia	Baku Surakhan	88.80	2.20	1.10	0.50	0.50	..	6.90	..	..	1009	0.658
25	Soviet Russia	Kuban—Black Sea Shorakaya Balka	93.30	1.40	0.70	2.10	2.50	..	..	..	..	1158	0.670
26	Soviet Russia	Second Baku Ishimbai	42.40	12.00	20.50	7.10	3.20	..	[1.0 CO_2; 2.8 H_2S]	..	11.00	1543	1.065
27	Soviet Russia	Suretov (Moscow supply) Yelshansk	93.20	0.70	0.60	0.60	0.50	..	..	..	4.40	1006	0.605
28	Soviet Russia	West Ikraine Dashava	97.80	0.50	0.20	0.10	..	..	..	..	1.30	1008	0.568
29	Venezuela	San Joaquin & Santa Rosa—Eastern Venezuela Gas Trans. Co.	76.70	9.79	6.69	3.26	0.94	0.72	1.90	..	..	1307	0.768
30	Venezuela	West Guara	70.62	14.06	7.96	3.54	1.00	0.55	0.21	..	2.06	1346	0.788
31	Venezuela	La Pica	90.20	6.40	..	..	..	..	0.10	0.30	3.00	1065	0.601
32	Venezuela	Quiriquire	63.50	..	..	..	..	..	23.00	1.40	12.10	643	0.834
33	Venezuela	Venezuela Atlantic Gas Trans Co. Guerico & Los Mercedes Field	85.47	3.24	1.52	0.99	0.55	0.38	7.85	..	..	1038	0.695

* 30 in. Hg, 60 F.
† Includes H_2S.

Table 2-14 Major Helium-Bearing Natural Gas Fields in the United States[1]

(as of June 1950)

No.	State	Field	Components of gas—per cent by volume: Methane	Ethane	CO_2	O_2	N_2	He	Gross heat value. Btu per cu ft, dry*
1	Kansas	Otis	71.8	13.9	0.3	0.2	12.18	1.62	976
2	Kansas	Cunningham	62.3	21.2	0.2	0.3	14.77	1.23	1011
3	New Mexico	Rattlesnake	11.5	3.1	2.3	0.1	75.18	7.82	171
4	Oklahoma	Keyes	51.5	17.5	1.4	0.2	27.14	2.26	835
5	Texas	Panhandle	66.3	12.6	0.7	0.2	19.52	1.28	887
6	Utah	Harley Dome	5.1	2.3	1.1	...	84.48	7.02	93

* 30 in. Hg, 60 F.

Table 2-15 Examples of Pipeline Company Purchase Contract Specifications (1957)

Company	Btu per cu ft	H_2O lb per MMCF	H_2S, grain per CCF	S, grain per CCF	CO_2, %	Temp, °F max	O_2, %	Natural gasoline, gal/MCF	Dust, gums, liquids, solids
A	1000	7	1.0	20	3	...	1	0.2	Free
B	1000	6	0.3	9	3	100	1	...	Free
C	1000	None in excess of saturation	Trace to 1	20	..	...	...	...	Free
D	...	7	1	30	3	...	1	...	Free
E	...	..	1	20	..	...	1	...	...
F	1000	7	1	20	2	...	1	0.2	Free
G	950	7	1	20	1	110	...	0.3	Free
H	...	..	Trace	20	..	...	...	...	Free
I	...	7	0.25	5	1	110	0.2	...	Free
J* (Gas wells)	1000	..	0.25	20	..	120	2	...	Free
K (Nat. gas plants)	1000	..	0.25	20	..	120	2	...	Free

* Gas delivered shall be at a pressure which is sufficient to enter buyer's pipeline at the point of delivery against a varying working pressure maintained therein by buyer up to a maximum of 800 psig.

Table 2-16 Sulfur Compounds in Natural Gas at Wells, before Purification[3]

Field	State	Sulfur content, grains S/CCF H_2S	RSH
Casey	Illinois	544–1518	1.7–4.5
Monument	New Mexico	757	1.17
Slaughter	Texas	747	12.6
Monahans	Texas	588	0.70
Panhandle*	Texas	380–505	0.7–1.8
Chatham	Ontario, Canada	400	1.6–2.0
TXL	Texas	396	3.36
Eunice	New Mexico	384	1.09
Fullerton	Texas	199	2.11
Keystone	Texas	195	1.49
Levelland	Texas	182	5.46
Andrews	Texas	171	1.62
Santa Rosa	Texas	165	1.61
Wasson	Texas	158	7.06
Langlie-Mattix*	New Mexico	120	0.78
.....	Alberta, Canada	92	7.8
Panhandle*	Texas	73	0.53
Cooper-Jal	New Mexico	59.5	0.69
Langlie-Mattix*	New Mexico	53	2.99
Benedum	Texas	10.1	0.18

* The two sets of data are from different sources.

"Animals kept in 25 per cent gas for thirty days remain normal in every respect.

"Animals exposed to 80 per cent gas for 8 hr were unaffected."

Plant Life.[19] It has been shown that certain components of manufactured gases (e.g., ethylene and possibly other unsaturated hydrocarbons, as well as carbon monoxide) are harmful to plant life; natural gas, with few qualifications, is not. For example, the roots of maple, oak, and London plane trees, in various states of health, were exposed to natural gas for extended periods without harmful effects. Likewise, concentrations of a mercaptan, at 40 *times* the normal odorant rate, produced no ill effects. It was also shown that there was neither a change in pH nor in the chemical constituents of soil after six weeks' exposure to natural gas. Five hundred dead and dying streetside trees, which for the most part were alleged to have been damaged by gas (since gas leaks were repaired in the vicinities), were later attributed to other factors.[20]

Tomato, sunflower, and geranium plants were not injured by up to 50 per cent natural gas in air for 72 hr.[21] "Concentrations of natural gas in air as high as one per cent were noninjurious to all plants tried [12 species.—Editor] and higher concentrations were injurious only to Bougainvillea and Swainsonia."[22] Experiments with the American elm showed that 2.5 to 4.0 per cent gas in air caused some injury to roots and decreased development of roots; however, the growth of the stem was not influenced.[23]

Analyses of gases sampled from the soil surrounding the roots of dead trees,[24] "which had died for no apparent reason," showed CH_4 concentrations ranging from 13 to 45 per cent, no CO or unsaturated hydrocarbons, and 5 to 14 per cent O_2. Beneath a live tree, 20 per cent O_2 and no CH_4 were found.

Injury to vegetation due to excessively dry soils may also be claimed when large amounts of dry gas (under considerable pressure so as to force the gas thru a large enough soil volume) escape. Gas companies make use of such occurrences to assist in leakage surveys. The exact conditions of pressure and gas flow required to cause such soil dehydration have not been established.

LIQUEFIED PETROLEUM GASES

The term liquefied petroleum gas (LP-gas) is applied to those hydrocarbons the chief components of which consist of propane, propylene, butane, butylene, and *iso*-butane, or mixtures thereof in any ratio or with air. These hydrocarbons can be liquefied under moderate pressure at normal temperatures, but they are gaseous under normal atmospheric conditions. Liquefied petroleum gases derived from natural gas are **saturated*** hydrocarbons, whereas those derived from oil refinery gases may contain varying low amounts of **unsaturated** (olefin) hydrocarbons. Modern refinery operations, such as catalytic polymerization and alkylation of the cracked gases, tend to reduce the per cent of olefins in refinery source LP-gas, since their value as polymer or alkylate base material is considerably higher than their present value in LP-gas. Average content of **propylene** in propane is between five and ten per cent. **Butylene** is used for various chemical purposes and is usually kept under five per cent in commercial butane gas.

Mass spectrometer analysis[4] of propane and butane shows that research and pure grades have less than 0.3 per cent impurities. Commercial butanes, treated by a deisobutanizer, contain little propane and *iso*-butane. In normal samples the ratio of normal butane to *iso*-butane in natural gas is about 2 to 1. Since LP-gas production practices vary widely, there are many combinations of constituents which would be acceptable and yet would vary considerably. Mixtures of propane and butane in various proportions are sold under various trade names for specific conditions.

Industry practice is to use **NGPA** specifications in the east and **CNGA** specifications in the west as shown in Tables 2-17 and 2-18. These specifications do not include a chemical analysis of the gas, but rather the composition is determined indirectly by vapor pressure and weathering limitations. A typical method of reporting LP-gas composition is shown in Table 2-19; any variation in given data of interdependent properties is because the values are extracted from more than one source.

Propane-air and butane-air gases are distributed to communities and small towns and augment natural gas distribution on peak days. They are also used by industries as a standby fuel in case of shortage of natural gas. The composition of mixtures of propane-air and butane-air with Btu values from 500 to 1800 is shown in Table 2-20.

MANUFACTURED GASES

Manufactured gas is defined herein as a combustible gas produced from coal, coke, or oil, or by reforming of natural or liquefied petroleum gases or any mixtures thereof, and in-

* Natural gas from some fields contains *unsaturated* hydrocarbons.

Table 2-17 NGPA* Liquefied Petroleum Gas Definitions and Specifications

(Reproduced by permission from NGAA Publication 2140-62; Liquefied Petroleum Gas Specifications and Test Methods, January 1963)

COMMERCIAL PROPANE

Commercial Propane shall be a hydrocarbon product composed predominantly of propane and/or propylene and shall conform to the following specifications:

Vapor Pressure

The vapor pressure at 100 F as determined by NGPA LPG Vapor Pressure Test shall not be more than 200 pounds per square inch gage pressure.

95 Per Cent Boiling Point

The temperature at which 95 per cent of volume of the product has evaporated shall be —37 F or lower when corrected to a barometric pressure of 760 mm Hg., as determined by the NGPA Weathering Test for Liquefied Petroleum Gases.

Residue

The product shall pass the nonvolatile residue test and shall pass the oil ring test—each as determined by the NGPA Method for Determining Residues in Liquefied Petroleum Gases.

Volatile Sulfur

The unstenched product shall not contain volatile sulfur in excess of fifteen grains per hundred cubic feet as determined by NGPA Volatile Sulfur Test.

Corrosive Compounds

The product shall cause no more discoloration to a polished copper test strip when such product is subjected to the NGPA LPG Corrosion Test than the discoloration of Standard copper strip Classification 1, as described in ASTM Method D 130–56, Table I, Copper Strip Corrosion by Petroleum Products.

Dryness

The product shall be dry as determined by the NGPA Propane Dryness Test (Cobalt Bromide Test).

COMMERCIAL BUTANE

Commercial Butane shall be a hydrocarbon product composed predominantly of butanes and/or butylenes and shall conform to the following specifications:

Vapor Pressure

The vapor pressure at 100 F as determined by NGPA LPG Vapor Pressure Test shall not be more than 70 pounds per square inch gage pressure.

95 Per Cent Boiling Point

The temperature at which 95 per cent of volume of the product has evaporated shall be 36 F or lower when corrected to a barometric pressure of 760 mm Hg., as determined by the NGPA Weathering Test for Liquefied Petroleum Gases.

Volatile Sulfur

The unstenched product shall not contain volatile sulfur in excess of fifteen grains per hundred cubic feet as determined by NGPA volatile Sulfur Test.

Corrosive Compounds

The product shall cause no more discoloration to a polished copper test strip when such product is subjected to the NGPA LPG Corrosion Test than the discoloration of Standard copper strip Classification 1, as described in ASTM Method D 130–56, Table I, Copper Strip Corrosion by Petroleum Products.

Dryness

The product shall not contain free, entrained water.

BUTANE-PROPANE MIXTURES

Butane-Propane mixtures shall be hydrocarbon products composed predominantly of mixtures of butanes and/or butylenes with propane and/or propylene and shall conform to the following specifications:

Vapor Pressure

The vapor pressure at 100 F as determined by NGPA LPG Vapor Pressure Test shall not be more than 200 pounds per square inch gage pressure.

95 Per Cent Boiling Point

The temperature at which 95 per cent of volume of the product has evaporated shall be 36 F or lower when corrected to a barometric pressure of 760 mm Hg., as determined by the NGPA Weathering Test for Liquefied Petroleum Gases.

Volatile Sulfur

The unstenched product shall not contpin volatile sulfur in excess of fifteen grains per hundred cubic feet as determined by NGPA Volatile Sulfur Test.

Corrosive Compounds

The product shall cause no more discoloration to a polished copper test strip when such product is subjected to the NGPA LPG Corrosion Test than the discoloration of Standard copper strip Classification 1, as described in ASTM Method D 130-56, Table I, Copper Strip Corrosion by Petroleum Products.

Dryness

The product shall not contain free, entrained water.

Product Designation

Butane-Propane mixtures shall be designated by the vapor pressure at 100 F in pounds per square inch gage. To comply with the designation the vapor pressure of mixtures shall be within + 0 lb —5 lb of the vapor pressure specified. For example: A product specified as 95 pound LPG shall have a vapor pressure of at least 90 lb but not more than 95 lb, at 100 F.

PROPANE HD 5

Propane HD 5 shall be a special grade of propane for motor fuel and other uses requiring more restrictive specifications than Commercial Propane and shall conform to the following specifications:

Vapor Pressure

The vapor pressure at 100 F as determined by NGPA LPG Vapor Pressure Test shall not be more than 200 pounds per square inch gage pressure.

95 Per Cent Boiling Point

The temperature at which 95 per cent of volume of the product has evaporated shall be —37 F or lower when corrected to a barometric pressure of 760 mm Hg., as determined by NGPA Weathering Test for Liquefied Petroleum Gases.

Residue

The product shall pass the nonvolatile residue test and shall pass the oil ring test—each as determined by the NGPA Method for Determining Residues in Liquefied Petroleum Gases.

Volatile Sulfur

The unstenched product shall not contain volatile sulfur in excess of ten grains per hundred cubic feet as determined by the NGPA Volatile Sulfur Test.

Corrosive Compounds

The product shall cause no more discoloration to a polished copper test strip when such product is subjected to the NGPA LPG Corrosion Test than the discoloration of Standard copper strip Classification 1, as described in ASTM Method D-130-56, Table I, Copper Strip Corrosion by Petroleum Products.

Dryness

The product shall be dry as determined by the NGPA Propane Dryness Test (Cobalt Bromide Test).

Composition

The propylene content of the product shall not exceed five liquid volume per cent and the product shall contain a minimum of ninety liquid volume per cent of propane.

* Formerly named NGAA

Table 2-18 CNGA Liquefied Petroleum Gas Specifications*

	CNGA standard grade					
	A	B	C	D	E	F
Max range of vapor pressures, psi at 100 F†	80	100	125	150	175	200
Range of allowable specific gravities, 60°/60 F	0.585–0.555	0.560–0.545	0.550–0.535	0.540–0.525	0.530–0.510	0.520–0.504
Composition‡	Predominantly butanes	Butane-propane mixtures, largely butane	Butane-propane mixtures, proportions approx. equal	Butane-propane mixture, propane exceeds butane	Propane-butane mixture, largely propane	Predominantly propane
Residue and 90 per cent evaporated temperature	The residue of any liquefied petroleum gas shall not be greater than three per cent. The 90 per cent evaporated temperature shall be stated with an accuracy of 2.5°F.					
Sulfur content	The liquefied petroleum gases shall contain no corrosive form of sulfur such as hydrogen sulfide or highly reactive mercaptans.					
Water and other contaminants	Liquefied petroleum gases shall contain no mechanically entrained water or other contaminants.					
Odorization	Every liquefied petroleum gas sold shall be so odorized by an agent of such offensiveness that the presence of such odorized gas in air shall be positively indicated when its concentration therein exceeds one-fifth of the concentration at the lower limit of flammability of such gas, except that such odorant shall not be required where it will interfere seriously with specific industrial use.					

* Tentative standard test methods covering tests mentioned in the above specifications are available in pamphlet form from the California Natural Gasoline Association, Los Angeles, Calif.

† The actual vapor pressure at 100 F of the product may, in order to provide a manufacturing tolerance, be 5 per cent under the stated pressure, but in no event shall the actual vapor pressure exceed the stated vapor pressure.

‡ When olefin constituents are components of the liquefied petroleum gas mixture, "butanes and butylenes" and "propane and propylene" should be understood, where applicable, instead of "butanes" and "propanes," respectively.

Table 2-19 Properties of Typical Commercial Liquefied Petroleum Gas Products*

	Commercial propane	Commercial butane
Vapor pressure, psig		
at 70 F	124	31
at 100 F	192	59
at 105 F	206	65
at 130 F	286	97
Specific gravity of liquid (60°/60 F)	0.509	0.582
Initial boiling point at 14.7 psia, °F	−51	15
Weight per gallon of liquid at 60 F, lb	4.24	4.84
Dew point at 14.7 psia, °F	−46	24
Specific heat of liquid, Btu/lb-°F at 60 F	0.588	0.549
Cu ft of gas at 60 F, 30 in. Hg per gal of liquid at 60 F	36.28	31.46
Specific volume of gas, cu ft/lb at 60 F, 30 in. Hg	8.55	6.50
Specific heat of gas, c_p, Btu/lb-°F at 60 F	0.404	0.382
Specific gravity of gas (air = 1) at 60 F, 30 in. Hg	1.52	2.01
Ignition temperature in air, °F	920–1020	900–1000
Maximum flame temperature in air, °F	3595	3615
Per cent gas in air for maximum flame temp	4.2–4.5	3.3–3.4
Max rate of flame propagation in 25 mm tube:		
cm per second	84.9	87.1
inches per second	33.4	34.3
Limits of flammability in air, per cent gas in gas-air mixture:		
at lower limit	2.4	1.9
at maximum rate of flame propagation	4.7–5.0	3.7–3.9
at upper limit	9.6	8.6
Latent heat of vaporization at boiling point:		
Btu per pound	185	167
Btu per gallon	785	808
Total heating values (after vaporization):		
Btu per cubic foot	2,522	3,261
Btu per pound (21,370 in 50-50 mixture)	21,560	21,180
Btu per gallon (97,050 in 50-50 mixture)	91,500	102,600

NOTE: The above are typical rather than absolute properties of average products under the corresponding specifications. Blends of one or more of the above products or additional liquefied petroleum gas products having any desired properties can be furnished to customer's specifications.

*Based on Phillips Petroleum Co. Bulletin No. 69, April 1948.

Table 2-20 Propane-Air and Butane-Air Gases

Btu per cu ft	Propane-air*			Butane-air†		
	% gas	% air	Sp gr	% gas	% air	Sp gr
500	19.8	80.2	1.103	15.3	84.7	1.155
600	23.8	76.2	1.124	18.4	81.6	1.186
700	27.8	72.2	1.144	21.5	78.5	1.216
800	31.7	68.3	1.165	24.5	75.5	1.248
900	35.7	64.3	1.185	27.6	72.4	1.278
1000	39.7	60.3	1.206	30.7	69.3	1.310
1100	43.6	56.4	1.227	33.7	66.3	1.341
1200	47.5	52.5	1.248	36.8	63.2	1.372
1300	51.5	48.5	1.268	39.8	60.2	1.402
1400	55.5	44.5	1.288	42.9	57.1	1.433
1500	59.4	40.6	1.309	46.0	54.0	1.464
1600	63.4	36.6	1.330	49.0	51.0	1.495
1700	67.4	32.6	1.350	52.1	47.9	1.526
1800	71.3	28.7	1.371	55.2	44.8	1.557

* Values used for propane-air calculation 2522 Btu per cu ft; 1.52 specific gravity.

† Values used for butane-air calculation 3261 Btu per cu ft; 2.01 specific gravity.

cluding any natural or liquefied petroleum gas if used for enriching.

Table 2-21 includes typical analyses of manufactured fuel gases, a selection of the manufactured gases representative of the process, and in some instances typical variations. Changes in process temperature, pressure, fuel, rate of gas making, introduction of other gases, inclusion of the blowrun gases, and type of generating equipment used, have a bearing on the resultant analysis of the gas made. Some of the gases listed are not made commercially, some were formerly manufactured, and others are not widely used at present but may become a major supply in the future. The 106 gases listed may be divided into about 18 groups or classifications of manufactured fuel gases as follows:

1. **Acetylene**[5] is primarily used for cutting and welding operations requiring high flame temperature. It has been used as an illuminant for lighting. It is made from calcium carbide and water. The crude gas contains the impurities ammonia, hydrogen sulfide, and phosphine, which are removed in the purification process. For transportation, acetylene is dissolved in acetone under pressure and drawn into small containers which are filled with porous material.

2. **Blast furnace gas** is a by-product in the manufacture of pig iron in blast furnaces. It is usually used for heating purposes within the plant, one-third for the blast stove and two-thirds for other purposes. The heating value derived mainly from carbon monoxide is too low for sale by public utilities. The gas is carefully cleaned and washed before use. The low Btu value requires regenerative preheating.

3. **Blue gas,** water gas, or blue water gas,[6] is made by passing steam over hot coke, coal, or other carbonaceous material, and consists essentially of carbon monoxide and hydrogen with varying amounts of carbon dioxide and nitrogen. It burns with a blue flame. Although it is not usually sold as such by public utilities, it serves as a base for carbureted water gas. It is similar to catalytic cracked gas and synthesis gas.

4. **Butane-air** and **propane-air gases** consist of mixtures of butane or propane and air to produce fuel gases with any desired heating values from 450 to 2000 Btu. They are used as the gas supply for small communities and as peak load gas by many natural gas companies.

5. **Carbureted water**[7] **gas** consists of water gas as a base which has been carbureted or enriched with thermally cracked oil, natural gas, or liquefied petroleum gas. In addition to considerable percentages of CO and H_2, there are varying amounts of unsaturated hydrocarbons (illuminants) and saturated hydrocarbons. Nearly any desired heating value from 300 to 1200 Btu may be attained, depending upon the process employed.

6. **Catalytically cracked gas** is made by passing the gas or light hydrocarbon liquid to be cracked over a nickel oxide catalyst maintained at a selected temperature by external heat. Regulated amounts of steam may be introduced. The gas is composed of CO and H_2 with appreciable amounts of N_2 and CO_2. Some of the uncracked gas or liquid vapor may be mixed with the cracked gas to increase the heating value.

7. **Coal gas**[8] is made by the distillation of the volatile matter from coal with some steaming of the coke to produce water gas. This type of gas is generated in so-called retorts. It is high in hydrogen and methane, with lesser amounts of carbon monoxide and illuminants.

8. **Coke oven gas** is made in by-product coke ovens by the distillation of the volatile matter from the coal. Coke is the primary product; gas, tar, and various chemicals are recovered in the process. The gas produced is usually around 500 Btu per cu ft with the combustible constituents consisting of hydrogen, methane, ethane, carbon monoxide, and illuminants. At the end of a coking period the gas is primarily hydrogen; at the beginning it is high in methane. A low temperature coking process will produce a high-Btu gas, as there is little breakdown of the saturated hydrocarbons. Typical analyses are given in Table 2-21. A more complete analysis of unwashed and debenzolized coke oven gas is given in Table 2-22.

9. **Hydrogen** as a fuel is limited to special industrial purposes, such as certain cutting and welding operations. Its primary use is in hydrogenation of fats and oils, in making ammonia, and in production of methanol and other chemicals, and as the reducing agent in the manufacture of certain metals, such as tungsten. It is made by electrolysis of water, by thermal cracking of natural gas and other hydrocarbons, and by the water gas reaction.

10. **Oil gases** are made by thermal decomposition of oils which may vary from naphtha to heavy residuum high carbon oils. Gases with heating values that vary from 300 to 1100 Btu per cu ft may be obtained. Usually gases of about 550 to 1000 Btu per cu ft are made. The lower Btu gases are high in hydrogen and methane and are distributed as manufactured gas by certain utilities, whereas the high-Btu gases are high in illuminants and saturated hydrocarbons and their primary use is for a peak load supplement by natural gas companies.

11. **Pipeline high-Btu gas** from coal, lignite, or oil shale has not as yet been made on a commercial scale, but it may have future application. It would be interchangeable with natural gas. Several processes are being studied, such as gasification to synthesis gas, followed by methanation over catalysts, or hydrogenation, or a combination of these two processes.

Table 2-21 Manufactured Fuel Gases*—Typical Analyses

No.	Type of gas	Source	Components of gas—per cent by volume: CO_2	O_2	Illuminants	CO	H_2	CH_4	C_2H_6	N_2	Other	Sp gr	Btu/cu ft gross	Btu/cu ft net
1	Acetylene	Calcium carbide & water	...	...	[C_2H_2 100]	...	...	...	...	...	...	0.91	1499	1448
2	Blast furnace	By-product in manufacture of pig iron	15.6	...	...	23.4	1.6	0.1	...	59.3	...	1.04	81	80
3	Blast furnace	By-product in manufacture of pig iron	8.7	...	...	32.8	1.8	0.2	...	56.5	...	1.00	111	110
4	Blue	Coke	5.1	...	...	40.2	50.0	0.7	...	4.0	...	0.54	300	273
5	Blue	Bituminous coal	7.0	...	1.0	34.4	48.8	4.8	...	4.0	...	0.55	335	288
6	Blue	Coke, steam, oxygen	16.9	0.1	...	46.6	34.0	0.2	...	2.2	...	0.75	262	244
7	Butane	Natural	...	...	...	...	...	[C_4H_{10} 99.4]	...	...	[C_5H_{12} 0.6]	2.07	3371	3102
8	Butane	Refinery gas	...	...	[C_4H_8 5.0]	...	...	[C_4H_{10} 90.0]	...	...	[C_3H_8 5.0]	2.03	3310	3068
9	Butane-air	Butane, air						See Table 2-20						
10	Carbureted water	Low gravity back run	3.6	0.4	6.1	21.9	49.6	10.9	2.5	5.0	...	0.54	536	461
11	Carbureted water	Normal operation	3.4	1.2	8.4	30.0	31.7	12.2	...	13.1	...	0.64	540	462‡
12	Carbureted water	High-Btu	1.6	0.2	18.9	21.3	28.0	20.7	4.3	5.0	...	0.69	850	791
13	Carbureted water	High-Btu—heavy oil	4.4	1.1	27.4	9.1	19.9	21.8	5.6	10.7	...	0.85	1010	920
14	Catalytic cracked†	Natural gas	3.7	...	...	15.4	43.1	3.8	...	34.0	...	0.59	222	203
15	Catalytic cracked†	Natural gas	4.5	...	...	18.0	49.0	8.5	...	20.0	...	0.50	300	270
16	Catalytic cracked†	Propane	4.6	...	...	21.1	46.3	2.3	...	25.7	...	0.55	230	216
17	Catalytic cracked†	Butane	5.0	...	...	20.5	52.5	6.0	...	16.0	...	0.50	300	265
18	Catalytic cracked†	Natural gasoline	4.4	1.7	2.6	23.8	58.1	3.4	...	6.1	...	0.47	358	319
19	Coal	Contin. vert. retort	3.0	0.2	2.8	10.9	54.5	24.2	...	4.4	...	0.42	532	477
20	Coal	Horizontal retort	2.4	0.8	3.0	7.4	48.0	27.1	...	11.3	...	0.47	542	486
21	Coal	Inclined retort	1.7	0.8	3.4	7.3	49.5	29.2	...	8.1	...	0.47	599	540
22	Coal	Intermittent vert. chamber	2.1	0.4	3.4	13.5	51.9	24.3	...	4.4	...	0.42	520	466
23	Coke oven	By-product	1.8	0.2	3.3	4.5	57.9	30.3	...	2.0	...	0.36	567	505
24	Coke oven	By-product	2.0	0.9	4.0	6.2	53.2	26.7	...	7.0	...	0.40	580	523
25	Coke oven	Sole flue, Ill. coal	4.2	1.0	2.6	14.1	43.5	19.0	...	15.6	...	0.54	435	392
26	Coke oven	Low temp Pitt. Coal	5.1	0.4	2.2	2.9	16.4	70.9	...	2.1	...	0.63	914	838
27	Coke oven	1 hr after charging (11,900 cfh)	1.6	...	4.6	5.3	52.0	34.8	...	2.1	...	0.38	640	570
28	Coke oven	4 hr after charging (13,000 cfh)	1.9	...	3.6	5.3	56.0	31.0	...	2.6	...	0.36	592	536
29	Coke oven	9 hr after charging (13,200 cfh)	1.9	...	1.4	5.3	61.8	25.0	...	3.2	...	0.31	502	443
30	Coke oven	13 hr after charging (below 9000 cfh)	0.3	...	...	5.3	84.0	5.6	...	5.5	...	0.20	337	299
31	Helium-natura	Natural gas						See Table 2-14						
32	Hydrogen	Decom. of hydrocarbons to electrolysis of water	0 to 20	...	...	0 to 2	80 to 99.8	0 to 2	...	0 to 1	...	0.069	325	275
33	Liquefied petroleum	Nat. gas; refinery gas						See Butane, Propane						
34	Marsh	Decayed vegetation						See sewage gas						
35	Mine vent	Ohio coal mine	3.0	0.5	...	...	...	71.0	14.0	11.5	...	0.70	970	...
36	Mine vent	Ohio coal mine	...	...	...	...	...	91.5	2.3	6.2	...	0.60	974	...
37	Mixed							See Table 2-23						
38	Natural							See Tables 2-9 to 2-13, inclusive						
39	Oil	Portland-diesel—1926	0.8	0.3	5.5	7.5	60.1	24.7	...	1.1	...	0.33	559	523
40	Oil	Portland, 1950, straight run residuum	4.1	0.3	4.8	7.3	54.6	28.4	...	3.2	...	0.37	571	528
41	Oil	Portland, 1950, heavy residuum	3.0	0.4	5.3	9.5	46.1	29.9	0.6	5.2	...	0.45	586	544
42	Oil—high-Btu	Blain down blast	1.6	1.2	22.4	0.9	22.9	42.7	...	8.3	...	0.66	970	901
43	Oil—high-Btu	Coke fire—Detroit	1.2	0.8	23.0	8.0	30.0	26.0	4.0	7.0	...	0.68	940	869
44	Oil—high-Btu	Twin gen.—Buffalo	2.7	3.2	29.0	1.8	11.5	31.4	2.2	18.2	...	0.80	1000	939
45	Oil—high-Btu	4 shell—Cambridge New England gas oil	6.0	0.9	24.2	1.5	14.6	33.4	...	19.4	...	0.78	963	831
46	Oil—high-Btu	Hall-Baltimore—0.2 Con. carbon§	4.1	0.2	28.4	2.1	16.0	29.3	4.4	15.5	...	0.89	1096	956
47	Oil—high-Btu	Hall-Balt.—6.02 Con. carbon§	4.9	0.7	33.2	2.2	15.8	25.4	3.8	14.0	...	0.80	1028	1004
48	Oil—high-Btu	Hall-Balt.—13.03 Con. carbon§	5.6	0.1	24.4	3.8	20.1	29.0	3.6	13.4	...	0.82	999	863
49	Oil—high-Btu	Portland test run high carbon heavy oil	4.7	0.3	16.5	8.2	24.6	39.2	4.0	2.5	...	0.64	938	846
50	Pintsch oil		0.5	1.0	31.0	0.2	8.9	55.1	...	3.3	...	0.70	1300	1143.1
51	Pipeline high-Btu	Naphtha	1.5	...	...	...	16.3	82.2	...	...	...	0.59	887	...
52	Pipeline high-Btu	Coal	...	...	...	...	11.1	80.8	...	8.1	...	0.60	855	...
53	Pipeline high-Btu	Coke	2.6	...	...	...	0.4	92.6	...	4.4	...	0.59	941	...
54	Pipeline high-Btu	Coal	0.3	...	7.3	...	1.1	77.9	3.6	5.8	[C_3H_8+ 4.0]	0.70	1124	...
55	Pipeline high-Btu	Low temp carb.						See number 26						
56	Pipeline high-Btu	Coal + hydrogenation	...	...	...	2.0	8.0	57.0	28.0	2.0	[C_3H_8 3.0]	0.70	1190	...
57	Pipeline high-Btu	Coal + Fischer-Tropsch	...	...	...	23.0	23.0	28.0	15.0	9.0	2.0	0.67	650	...
58	Pipeline high-Btu	Coal + Fischer-Tropsch	...	...	...	...	10.9	80.9	3.5	4.7	...	0.60	910	...

59	Pipeline high-Btu	Coal + Fischer-Tropsch from synthesis gas No. 92	...	0.8	0.6	0.8	1.9	90.9	...	5.2	...	0.58	9[illegible]0	...
60	Producer low	Rice anthracite	1.0 to 5.5	0 to 0.3	...	22.0 to 30.0	5.0 to 8.0	1.0 to 3.0	...	45.0 to 60.0	...	0.85	135	125
61	Producer	Bit. coal lumps + 0.32 lb steam	4.3	0.1	...	26.7	13.9	3.1	...	51.5	...	...	169.5	...
62	Producer	Coke 97% thru 1 in., 14% thru ¼ in. + 0.53 lb steam	5.8	0.3	...	26.0	12.1	0.5	...	55.3	...	...	129.4	...
63	Producer	Rice anthracite O_2 + 0.73 lb steam	16.5	0.15	...	40.0	41.0	0.85	...	1.5	...	...	271.0	...
64	Producer	Pea coke O_2 + 1.4 lb steam	15.3	...	...	45.3	38.0	...	...	0.9	...	...	253.0	...
65	Propane	Natural	...	...	...	...	...	[C_3H_8 97.0]	1.1	...	[C_4H_{10} 1.9]	1.56	2593	2397.3
66	Propane	Refinery gas	...	...	[C_3H_6 7.5]	...	...	[C_3H_8 89.7]	2.0	...	[C_4H_{10} 0.8]	1.55	2557	2362
67	Propane-air	Propane, air	See Table 2-20											
68	Refinery oil	Oil refinery low-Btu	0.7	0.1	38.1	1.5	4.2	34.9	17.6	2.9		0.89	1407	1407
69	Refinery oil	Oil refinery high-Btu	0.5	0.2	5.8	0.9	1.0	26.9	21.0	3.4	[C_3H_8+ 40.3]	1.25	2010	...
70	Refinery oil	Average refinery	0.7	0.7	21.4	1.7	5.3	26.0	19.4	2.0	[C_3H_8, 17.8; C_4+, 5.0]	1.02	1734	...
71	Refinery oil	Hammond, Ind.—low	0.6	2.0	9.6	1.2	6.9	42.5	20.9	0.2	[C_3, 11.2; C_4, 4.3; C_5+, 0.6]	...	1424	...
72	Refinery oil	Hammond, Ind.—high	...	...	38.99	0.27	4.37	28.97	23.82	...	[C_3, 1.06; C_4, 0.33; C_5+, 2.19]	0.99	1625	...
73	Refinery oil	Philadelphia—typical refinery dry gas	1.1	...	18.8	5.4	12.7	28.1	17.1	...	[C_3, 14.1; C_4, 0.1; H_2S, 2.6]	0.886	1388	...
74	Reformed natural	Water gas gen. Chicago	4.2	0.6	4.4	17.7	52.3	...	...	20.8	...	0.53	194	...
75	Reformed natural	Water gas gen. Chicago plus natural gas	2.7	0.4	2.7	11.4	33.7	35.5	0.2	13.4	...	0.55	554	...
76	Reformed natural	Water gas gen.	1.2	0.1	0.8	22.3	49.9	21.9	...	3.8	...	0.44	464	424
77	Reformed natural	Natural gas—oxygen	4.1	0.2	1.0	22.2	36.3	5.4	...	30.8	...	0.64	266	...
78	Reformed gas oil	Gas oil on top of fire—back run	4.4	0.3	8.2	16.9	56.6	12.2	...	1.4	...	0.44	537	...
79	Reformed refinery oil	With blow-run gas from sets	6.6	0.3	2.5	16.8	34.8	13.8	...	25.2	...	0.64	361	...
80	Reformed refinery oil	Without blow-run gas from sets	2.4	0.6	2.2	14.6	55.6	16.9	...	7.7	...	0.42	449	...
81	Reformed butane	Butane-water gas gen.	Similar to numbers 76 and 77, also see #16											
82	Reformed propane	Propane-water gas gen.	Similar to numbers 76 and 77, also see #15											
83	Sewage	Chicago, Ill.	14.7	0.5	...	...	...	76.6	...	8.2	...	0.73	775	...
84	Sewage	Decatur, Ill.	22.0	...	...	...	2.0	68.0	...	6.0	2.0	0.79	690	621
85	Sewage	Grand Rapids, Mich.	33.5	...	...	...	...	65.4	...	1.1	...	0.88	660	...
86	Sewage	Rochester, N. Y.	30.0	...	...	...	...	66.0	...	4.0	...	0.86	670	...
87	Sewage	Toronto, Ontario	28.0	1.8	...	...	3.7	58.5	...	8.0	...	0.85	600	...
88	Soil	Decom. of organic matter	9.6	5.9	0.0	0.0	0.0	0.2	...	84.3	...	1.03	2	...
89	Soil	Decom. of organic matter	16.6	1.0	0.3	0.0	0.2	3.2	...	78.7	...	1.05	39.1	...
90	Synthesis	Ruhrchemie commercial scale	14.0	0.2	...	27.5	55.0	0.6	...	2.7	...	0.546	274	...
91	Synthesis	Pulverized coal + O_2 + superheated	18.6	...	...	27.1	48.2	2.3	...	2.8	...	0.62	266	...
92	Synthesis	Naphtha	5.7	...	...	25.8	64.0	4.7	...	...	...	0.41	329	...
93	Synthesis	Coke—Lurgi process crude gas	29.3	...	...	21.9	44.0	3.3	...	1.5	...	0.72	247	
94	Synthesis	Coke—Lurgi process after water wash	1.0	...	...	30.7	61.6	4.6	...	2.1	...	0.40	436	...
95	Synthesis	Pitts. seam West Virginia coal, Lurgi pilot plant, Bureau of Mines	28.6	0.3	0.1	20.1	40.2	7.1	1.0	2.4	[H_2S .2]	0.74	[289 (raw) 393 (washed to 2.5% CO_2)]	...
96	Tail	Coke oven gas after passing ammonia plant	8.0	0.6	6.0	20.0	4.4	50.0	...	11.0	...	0.75	693	...
97	Thermally cracked	Nat. gas, LP-gas, oils	See numbers 39–49, inclusive and 74–82											
98	Thermally cracked	Nat. gas, LP-gas, oils	See numbers 39–49, inclusive and 74–82											
99	Thermally cracked	Natural gasoline 1300 F	...	...	[Unsat. 45.3]	...	10.9	[Sat. 43.3]	...	0.5	...		1608	...
100	Thermally cracked	Natural gasoline 1400 F	...	...	38.0	...	14.6	47.4	...	...	...	0.75	1280	...
101	Thermally cracked	Natural gasoline 1500 F	...	...	25.9	...	22.5	51.5	...	0.1	...	0.59	1051	...
102	Underground gasified coal	Gorgas, Ala., low air input—best	28.3	0.1	0.3	6.3	21.6	4.9	...	38.5	...	0.91	147	...
103	Underground gasified coal	Gorgas, Ala., low air input—average	19.1	...	0.1	1.0	9.5	1.8	...	68.5	...	0.98	54	...
104	Underground gasified coal	Russia—27% O_2	11.0	...	...	25.0	13.5	2.5	...	45.0	...	0.87	150	...
105	Underground gasified coal	Russia—35% O_2	18.0	...	...	15.0	49.0	4.0	...	14.0	...	0.61	247	...
106	Water		See blue gas numbers 4 to 6, inclusive											

These analyses taken from various sources available in the American Gas Association technical library. * Other gases are also included to make this a reasonably complete reference table on gas composition. † Catalytic cracked gas is usually enriched with natural gas or propane to the desired heating value. ‡ 1990 Btu/cu ft used as net heating value of illuminants. § Conradson carbon residue test (ASTM).

Table 2-22 Complete Analyses of Coke Oven Gas from Koppers Ovens[9]

(Both unwashed and debenzolized)

Component	Per cent by volume	
	Unwashed	Debenzolized
Hydrogen sulfide	0.7	0.7
Carbon dioxide	1.7	1.5
Nitrogen	0.9	1.0
Oxygen	0.0	0.0
Hydrogen	56.7	57.2
Carbon monoxide	5.7	5.8
Methane	29.6	29.2
Ethane	1.28	1.35
Ethylene	2.45	2.50
Propylene	0.34	0.29
Propane	.08	.11
Butylene	.16	.18
Butane	.02	.04
Acetylene	.05	.05
Light oil	0.65	0.15

12. **Producer gas**[10,11] is generated when air or oxygen is passed thru a thick bed of hot coal or coke. The products of this process are CO, N_2 (from the use of air), and some CO_2. In actual practice, steam is added to the air to reduce clinker formation and the steam decomposition forms hydrogen in varying quantities. Note items 60 to 64 in Table 2-21.

13. **Refinery oil gas** is produced in oil refineries from two chief sources, namely, evolution of absorbed gases from crudes during the distillation processes and as a by-product of refinery cracking operations. Gas streams thus produced contain considerable quantities of propylenes and butylenes which may be polymerized or alkylated into high and antiknock gasoline components, butanes which are used in vapor pressure gasoline production, and pentanes and heavier components which are economically desirable to include in the gasoline production. Therefore, the refinery usually installs equipment to recover some 75 per cent of the total C_3 fractions, some 95 per cent of the C_4 fractions, and all of the C_5 and heavier fractions normally included in the refinery wet gas production. The usual method of recovery is an absorption system with subsequent stripping of the mentioned fractions from the absorption oils by distillation processes. Thus the wet gas production is dried and the dry gas product is normally used as a fuel. In a light oil refinery equipped with the usual distillation, catalytic and thermal cracking, and catalytic polymerization facilities, a dry gas production of some 16 million SCF per day is obtained when processing some 100,000 barrels of crude per day with crudes from both domestic and foreign sources. The properties of a typical gas are shown in Table 2-21, number 73. Depending upon the amount of hydrocarbons removed, the heating value may vary from 1400 to 2000 Btu per cu ft.

14. **Reformed gases** are usually made by thermally cracking natural gas, propane, butane, or refinery oil gas in water gas generators or similar special equipment. The resultant gas varies appreciably in composition depending upon the equipment used and the percentage of the gas being cracked. Reformed gas contains hydrogen, carbon monoxide, and saturated and unsaturated hydrocarbons.

15. **Sewage gas** is produced from sewage sludge in digesting equipment. It averages between 600 and 700 Btu per cu ft and consists of about two-thirds methane and one-third carbon dioxide.

16. **Synthesis gas**[11,12] is made by various processes from coal, coke, naphtha, or other hydrocarbons, and consists essentially of CO and H_2 in the ratio of 1 to 2. The crude gas also contains high percentages of CO_2.

17. **Thermally cracked gas** is made by decomposition of natural gas, liquefied petroleum gases, or gasoline. It is high in saturated and unsaturated hydrocarbons with some hydrogen.

18. **Underground producer and water gas**[13,14] made in underground coal seams by passing air, with or without supplemental oxygen or steam, through an ignited mass of coal, have been produced experimentally in several countries. The most optimistic claims with regard to the quality of the gases and the economics of the process have come from Russia. English and American experiments have yielded gases which would consistently average around 50 Btu per cu ft, but occasional runs at 100 and even as high as 250 Btu per cu ft have been reported.

The objective of these experiments has been the utilization of inferior or unworkable coal seams with the possible adaptation of these processes to better quality seams as an alternative to coal mining. However, as a result of the low-Btu gases produced and the high costs indicated, experimental work in the United States and England was discontinued in 1958.

19. **Underground oil gases** have been studied in the United States from the viewpoint of generating them via underground atomic bomb explosions in oil shale formations to break up the shale rock and release the kerogen present. Then by retorting in place, petroleum fractions such as a fuel gas could be produced and brought to the surface.

Some Components of Manufactured Gases before Purification and Some Trace Constituents in City Gases

During the manufacture of coal gas or coke oven gas, a large number of chemical constituents are liberated. A list of some 48 constituents, other than the common constituents in

Table 2-23 Constituents of Coal Gas Present in Relatively Small Amounts before Purification

No.	Compound	Relative amounts
Acetylene hydrocarbons (Included as illuminants; not recovered as such.)		
1	Acetylene	Approx. 0.05% by vol.
2	Allylene	Traces only
3	Crotonylene	Traces only
Aromatic hydrocarbons (Constitute about 85% of light oil recovered from coal gas.)		
4	Benzene	Approx. 0.66% by vol.
5	Toluene	Approx. 0.13% by vol.
6	Xylene	Approx. 0.05% by vol. (items 6–14)
7	Ethylbenzene	
8	Propylbenzene	
9	Ethyltoluene	
10	Pseudocumene	
11	Mesitylene	
12	Hemimellitene	
13	Cymene	
14	Durene	
15	Naphthalene	Approx. 0.02% by vol. (items 15–16)
16	Methylnaphthalene	
	Total	approx. 0.91% by vol.

Table 2-23 Constituents of Coal Gas Present in Relatively Small Amounts before Purification (Continued)

No.	Compound	Relative amounts
Miscellaneous unsaturated hydrocarbons (Included as illuminants; partly recovered in light oil forerunnings.)		
17	Butadiene	Up to 0.02% by vol.
18	Cyclopentadiene	Up to 0.006% by vol.
19	Styrene	Up to 0.006% by vol.
20	Indene	Up to 0.013% by vol.
	Total	up to 0.045% by vol.
Nitrogen compounds (Ammonia usually recovered; soluble in water.)		
21	Ammonia	Avg. 1.100% by vol.
22	Hydrocyanic acid	Up to 0.250% by vol.
23	Acetonitrile	Traces
24	Pyridine	Up to 0.004% by vol.
25	Picolines	Traces
26	Lutidines	Traces
	Total	1.354% by vol.
Olefin hydrocarbons (illuminants) (Usually not removed.)		
27	Ethylene	Avg. 2.45% by vol.
28	Propylene	Avg. 0.30% by vol.
29	Butylene	Avg. 0.18% by vol.
30	*iso*-Butene	
	Total	avg. 2.93% by vol.
Oxides (CO and CO_2 are omitted here. Part of water vapor is removed. Nitric oxide can be removed if it causes trouble from gum formation.)		
31	Water vapor	Variable but usually saturated 1.74% at 60 F
32	Nitric oxide	Usual range 0 to 0.0001% by vol.
Oxygen compounds (Tar acids soluble in water and tar oils. Removed by condensation and cooling. Coumarone removed in light oil.)		
33	Phenol	0.004 to 0.006% by vol. tar acids
34	Cresol	
35	Xylenol	
36	Coumarone	Up to 0.013% by vol.
Paraffin hydrocarbons (Methane and ethane omitted here. Usually removed in light oil.)		
37	Propane	Up to 0.10% by vol.
38	*n*-Butane	Up to 0.03% by vol.
39	*iso*-Butane	
40	Higher paraffins	Approx. 0.01% by vol.
Sulfur compounds (95% of sulfur compounds are H_2S which is removed by iron oxide or other processes.)		
41	Hydrogen sulfide	Variable 0.3 to 3% by vol.; 0.6% is typical
42	Carbon disulfide	Variable 0.007 to 0.07% by vol.
43	Carbon oxysulfide	Approx. 0.009% by vol.
44	Methyl mercaptan	Approx. 0.003% by vol.
45	Ethyl mercaptan	
46	Methyl sulfide	Traces
47	Ethyl sulfide	
48	Thiophene	Approx. 0.010% by vol.

the send-out gas, is presented in Table 2-23. Most of these constituents are removed during the purification processes in the plant and either do not exist in the finished gas or might be detected in trace amounts. Some of the constituents reremoved are recovered as by-products.

During the manufacture of oil gas a number of chemical constituents are formed in addition to those gases normally determined in a gas analysis. Most of these constituents are removed during the purification process and either do not exist in the finished gas or are present in minute traces. Some of these constituents were made into commercial products by the Portland Gas & Coke Co. See Table 2-24.

To illustrate the number of gases present in a high-Btu gas made from naphtha, data in Table 2-25 are presented. It is interesting to note the relatively large percentages of the higher unsaturated hydrocarbons, particuarly butylene and butadiene. Other conditions of manufacture and other oils will change the composition of the gas.

Table 2-24 Oil Gas Recovery Products[15]

Using a make oil (API 15 to 17°), Conradson carbon 8 to 10, and 7.7 C/H ratio, the following product recovery was reported:

Lampblack recovery, lb/MCF	22.0
Tar recovery, gal/MCF	0.50
Light oil recovery, gal/MCF	0.53

Products manufactured from straight run oil and heavy fuel oil using gas generators and oil coke ovens:

Oil gas	Solvent naphtha	Soft pitch
Motor fuel	Naphthalene	Hard pitch
Benzol	Tar	Lampblack
Toluol	Creosotes	Briquets
Xylol	Road binder	Coke

Table 2-25 Gas Composition of Thermally Cracked Naphtha[16]
(At about 1350 F)

Component	% by volume	Component	% by volume
Nitrogen	1.6	Propylene	15.8
Oxygen	0.2	Butylene	4.8
Hydrogen	9.6	Pentalene	1.6
Methane	27.6	Hexalene	0.5
Ethane	7.0	Cyclopentene	0.8
Propane	0.8	Butadiene	3.5
Butane	1.0	Cyclopentadiene	0.6
Pentane	1.9	Acetylene	0.6
Hexane	0.8	Benzene	0.6
Ethylene	20.7		

Specific gravity 1.09; heating value 1702 Btu/SCF.

Gum. Gum-forming constituents may be present in manufactured gases. The presence of excess air in the high temperature zones of the manufacturing process tends to produce greater than normal quantities of nitric oxide. The interaction of nitric oxide with unsaturated hydrocarbons and oxygen, when all three are present, may form a vapor phase gum in trace quantities which in turn may polymerize to gummy, semisolid, or solid materials which tend to collect in small gas passages or orifices, as commonly found in gas pilots and parts of gas appliance controls.

Liquid phase gum may be formed by a chemical change of unsaturated hydrocarbons such as indene, styrene, and coumarone. Liquid phase gum is so named because it condenses as a liquid but later may undergo further change to typical gummy materials.

MIXED GASES

Mixed gas as defined for statistical purposes of the American Gas Association[17] is, "a gas in which manufactured gas is *co-mingled* with natural or liquefied petroleum gas (except where the natural or liquefied petroleum gas is used only for

Table 2-26 Typical Mixtures of Manufactured Gas as of 1950

No.*	City	Company	Heat. value, Btu/cu ft	Avg sp gr	Gas mixtures
1	Boston, Mass.	Boston Gas Co.	532	0.56	Water, coke oven, oil refinery, propane-air
2	Brooklyn,† N. Y.	The Brooklyn Union Gas Co.	537	.67	Natural, carb. water, coke oven, refinery oil, producer, oil, LP air
3	Montreal, Canada	Quebec Hydro-Electric Commission	465	.52	...
4	Newark, N. J.	Public Service Electric & Gas Co.	525	.59	Natural, carb. water, coke oven, LP-gas–air, reformed natural
5	New York City†	Consolidated Edison Co.	541	.60	Natural, carb. water, coke oven, natural air jet, reformed natural
6	Philadelphia, Pa.	Philadelphia Gas Works	558	.626	Natural, carb. water, cyclic catalytic reformed, air modified, coke oven
7	Portland,† Ore.	Portland Gas & Coke Co.	572	.45	Oil, oil-coke oven, and auxiliary butane
8	Rochester,† N. Y.	Rochester Gas and Electric Corp.	538	.60	Natural, coke oven, producer, blue, carb. water, oil
9	Seattle,† Wash.	Seattle Gas Co.	503	.55	Water, oil, auxiliary propane-air.
10	Toronto,† Canada	Consumers Gas Co. of Toronto	475	0.475	Coal, carb. water, LP-gas–air auxiliary

* 1, 3, 6–10 *Brown's Directory of American Gas Companies,* 1951–52 ed. New York, Moore Pub. Co., 1951.
2 Hall, A. G. "Utilization of Natural Gas at the Brooklyn Union Gas Co." *A.G.A. Proc.* 1951:690–2.
4 Emanuel, R. S. "Utilization of Natural Gas at Public Service Electric & Gas Co." *A.G.A. Proc.* 1951:693–5.
5 Salzone, V. "Utilization of Natural Gas at Consolidated Edison Co. of New York, Inc." *A.G.A. Proc.* 1951:686–90.
† Converted to natural gas as of 1959 with the exception of some sections of Rochester, New York.

'enriching' or 'reforming') in such a manner that the resulting product has a Btu value higher than that previously produced by the utility prior to the time of the introduction of natural or liquefied petroleum gas." During 1958 total mixed gas utility sales amounted to 2219.7 million therms.

Actually, all fuel gases distributed by utility companies are mixtures of from three to eight or more different chemical constituents. Natural gas is not only a mixture of different chemical constituents, but it is a composite of natural gases of different analyses from many gas wells and possibly from several fields. Manufactured gas always contains from three to eight or more different chemical constituents, and usually a company will mix gases made by two or more processes.

Typical examples of mixtures of manufactured gases as used by certain cities in 1950 are shown in Table 2-26. As many as seven different types of gas are mixed together to form the **send-out** gas in one city, and when it is considered that there are many variations within each process under the control of the operator, the great number of possibilities for gas mixtures may be appreciated. Limitations on resultant mixtures are the heating value and continuity of good gas appliance performance. Mixtures produced must be **interchangeable** so that they do not affect appliance operation.

REFERENCES

1. Anderson, C. C. and Hinson, H. H., *Helium-Bearing Natural Gases of the United States.* (Bur. of Mines, Bul. 486) Washington, G.P.O. 1951.
2. Cady, H. P. and McFarland, D. F. "On the Occurrences of Helium in Natural Gas." *Proc., Am. Chem. Soc.,* 1906.
3. Mason, D. McA. and others. *Identification and Determination of Organic Sulfur in Utility Gases.* (I.G.T. Research Bul. 5) Chicago, 1959.
4. Evans, E. W. (Unpublished communication to Phillips Petroleum Co., Bartlesville, Okla.)
5. Fulweiler, W. H. "Industrial Gases." (In: Rogers, Allen. *Manual of Industrial Chemistry,* 6th ed., New York, Van Nostrand, 1942. Vol. I:760).
6. Morgan, J. J. "Water Gas." (In: Lowry, H. H., ed. *Chemistry of Coal Utilization.* New York, Wiley, 1945. Vol. II, chap. 37).
7. See p. 1721 of reference 6 above.
8. Fulweiler, W. H. "Manufactured Gas" (See p. 578 of reference 5 above).
9. Powell, A. R. "Gas From Coal Carbonization." (See Chap. 25 of reference 6 above) Lowry, H. H., ed.
10. Van der Hoenen, B. J. C. "Producers and Producer Gas." (See Chap. 36 of reference 6 above).
11. Cooperman, J. and others. *Lurgi Process.* (Bur. of Mines, Bul. 498) Washington, G.P.O., 1951.
12. Newman, L. L. "Oxygen in the Production of Hydrogen or Synthesis Gas." *Ind. Eng. Chem.* 40:559–582, April 1948.
13. Powell, A. R. "Future Possibilities in Methods of Gas Manufacture." *A.G.A. Proc.* 1947:631–658.
14. Elder, J. L. and others. *The Second Underground Gasification Experiment at Gorgas, Ala.* (Bur. of Mines, R.I. 4808) Washington, 1951.
15. Bell, J. F. "Portland's Experiments with High-Btu Oil Gas." *Gas* 27: 52–56, Nov. 1951. Also: "The Future of the Gas Industry in the Pacific Northwest." *PCGA Proc.* 42:18–24; 1951.
16. I.G.T. (Unpublished report to the East Ohio Gas Co., January, 1952) Project S-82.
17. A.G.A. *Gas Facts.* New York, annual.
18. Tyler, D. B. and Drury, D. "Natural Gas—Its Physiologic Action." *California & Western Medicine* 47:1, July 1937.
19. Pirone, P. P. "Response of Shade Trees to Natural Gas." *Garden Jnl.* (N. Y. Bot. Garden) 30:25–29, Jan.-Feb. 1960.
20. ——. *Tree Maintenance,* 3rd ed. New York, Oxford Univ. Pr., 1959.
21. Solheim, W. G. and Ames, R. U. "Effects of Some Natural Gases Upon Plants." *Phytopath.* 32:829–30, 1942.
22. Gustafson, F. G. "Is Natural Gas Injurious to Flowering Plants?" *Plant Physiol.* 19:551-8, 1944.
23. ——. "Is the American Elm (Ulmus Americana) Injured by Natural Gas?" *Plant Physiol.* 25:433–40, 1950.
24. Braverman, M. M. and others. "Determining the Cause of Death of Vegetation by Analysis of Soil Gases." *Gas Age* 129: 23–6, April 26, 1962.

Chapter 3

Liquid Fuels

by F. E. Vandaveer

SOURCES AND CHARACTERISTICS OF LIQUID FUELS

The major source for liquid fuels in the U. S. is crude petroleum. Basically, petroleum is a mixture of many hydrocarbons usually containing impurities such as sulfur and nitrogen compounds; vanadium compounds are found in Venezuela oils. Crude oils vary in composition with respect to the **paraffin, naphthene,** and **aromatic groups.** All crude oils are a mixture of these three groups; none has been found that consists purely of one series. Paraffin-base crudes are composed principally of paraffins and some alkyl naphthenes in the heavier fractions; they contain paraffin wax and traces to appreciable amounts of asphalt. Pennsylvania and West Virginia crudes are of this type. Mixed base crudes are lower in paraffins and higher in naphthenes than the paraffin base crudes. Mid-continent crudes are of the mixed base type. Naphthene base crudes contain a high percentage of naphthenes, have a relatively high specific gravity, and very little wax. They occur in the Central, South Central, and Southwestern sections of the United States and around the Caspian Sea. Aromatic crudes contain a relatively high percentage of the lower aromatic hydrocarbons. This type occurs chiefly in California.

Another important source of liquid fuels is from natural gas **condensates.** In 1958 a total of 341,548,000 bbl of these liquids was produced. They are extracted from natural gas by absorption in oil and subsequently fractionated into propane, butane, and natural gasoline of other desired paraffin hydrocarbons.

Coal tar and light oils from coke ovens and gas-making processes are minor liquid fuels. In 1948 about 971 million gallons of coal tar and 256 million gallons of light oil were produced, but these had dropped to a combined total of less than 198 million gallons by 1958.

It is possible to analyze crude oils in terms of the amounts of natural gasoline, naphtha, kerosene, gas oil, residuals, etc., which they contain. However, fuel and oil specifications are such that these natural fractions do not serve particularly useful purposes. This type of analysis, therefore, is of limited use in spite of the probability that fractionation of crudes by distillation continues to be an initial step in refinery operations. It is the subsequent processing which utilizes the potentials of the crude oil and modifies even the *fractions* into marketable products.

The wide range in composition of crude oil is indicated in Table 2-27. Table 2-28 illustrates this wide range in terms of the octane number of some **straight-run** gasolines made from crude oil.

Analyses of petroleum in increasing detail have made possible the identification of many individual constituents and classification of the rest in terms of structure; for example, API Projects 6, 42, 45, 48, and 52. Table 2-29 shows a hydrocarbon type analysis from API Project 6.

Shale oil is different in chemical nature from petroleum, and results from thermal treatment of the shale thru cracking of its organic matter (kerogen). Shale oil has a very high content of sulfur, oxygen, and nitrogen compounds. Table 2-30 gives an analysis of samples of oil recovered by heating. Table 2-31 shows the types of hydrocarbons in the neutral oils from these two shales.

The usefulness of solid fuels for oil making seems much more remote than that of either tar sands or oil shales. Not

Table 2-27 Composition of Typical Crude Oils[1]
(in weight per cent)

	Eastern U. S. paraffinic	California naphthenic	Gulf Coast and California naphthenic-aromatic
Type composition			
Paraffinic hydrocarbons	35–45	10–15	15–25
Naphthenes	40–50	70–80	40–50
Aromatics (sulfonatable)	7–12	8–12	20–25
Resins and asphaltenes	1–3	2–5	10–15
Ring analysis			
Paraffinic side chains	75–80	60–65	45–50
Naphthenic rings	15–20	25–35	25–35
Aromatic rings	4–6	5–10	20–25

Table 2-28 Octane Numbers of Straight-Run Gasolines from Various Crudes[1]

Origin	Type	Octane number of 104–356 F fraction
Michigan	High in n-paraffins	23.1
East Texas	Intermediate	49.1
Winkler, Tex.	High in isoparaffin	60.4
Midway, Calif.	High in cycloparaffin	73.0

Table 2-29 Types of Hydrocarbons in a Crude Oil[1]

	Gasoline	Kerosene	Light gas oil	Heavy gas oil	Lubrication distillate	Residue
Volume, per cent	43.1	12.9	16.8	9	10	8
Boiling range, °F (1 atm)	104–356	356–446	446–572	572–752	752–932	...
Type analysis (vol %)						
n-Paraffins	28	24	22	17	15	...
Isoparaffins	20	16	8	6	...	...
Monocycloparaffins	...	32	29	21	10	...
Alkycyclopentanes	22	...	...	...	...	...
Alkycyclohexanes	20	...	...	...	...	...
Dicycloparaffins	...	11	18	22	26	...
Tricycloparaffins	...	...	3	10	16	...
Mononuclear aromatics	10	15	12	10	9	...
Dinuclear aromatics	...	2	8	10	9	...
Trinuclear aromatics	...	...	...	4	8	...
Residue	...	...	...	...	7	...
Total	100	100	100	100	100	...
Compounds isolated (June 30, 1952)	94	24	7	0	0	...

Note: Analysis done by API Project 6 on Ponca, Okla., petroleum (Continental Oil Co.)

only are mining problems more difficult and costly, but the low H_2 to C ratio will also necessitate the introduction of large quantities of H_2.

Table 2-30 Shale Oils from Fischer Assay Report[1]

	Colorado	Brazil
Specific gravity	0.918	0.888
Sulfur (wt %)	0.64	0.41
Nitrogen (wt %)	1.95	0.98
Distillation yield (vol %)		
Naphtha	13.7	16.4
Light distillate	16.8	19.5
Heavy distillate	35.0	39.1
Residue	34.6	24.6

Table 2-31 Types of Hydrocarbons in Shale Oils[1]
(in volume per cent)

Hydrocarbon type	Naphtha	Light distillate
Brazilian shale		
Paraffins and cycloparaffins	19	18
Olefins and cycloolefins	60	48
Aromatics	21	34
Colorado shale		
Paraffins and cycloparaffins	33	23
Olefins and cycloolefins	56	47
Aromatics	11	30

TRENDS IN PETROLEUM REFINING

The type of crude has little effect on the resultant product. The production of high octane gasoline is expected to be favored into the foreseeable future (to 1975). A 50 per cent yield of 100 research octane gasoline can be made.

Both increased octane and additives are required since engine manufacturers seek to approach 17-to-1 compression ratios in order to maximize thermal efficiency. The increased cost of high octane fuels promises to slow down the "octane race."

"In the U. S., even the 'average' refinery uses catalytic cracking capacity approaching 46 per cent of crude capacity. But in the most modern plants, this level reaches 80 per cent or higher. Such cat crackers handle a high ratio of recycle feed of heavy catalytic gas oil which is processed to extinction."[2]

Catalytic Cracking

The application of heat and pressure to crude oil in the presence of suitable catalysts makes it possible to *steer* chemical reactions so as to obtain a maximum of low-boiling hydrocarbons with high octane rating. The catalyst may be a fixed bed of clay particles as shown in Table 2-32. These clay particles have from 12 to 15 times as much surface area as the original clay.

Oil vapor is forced thru the bed in the **Houdry** process or the catalyst, in the form of small pellets, is dropped thru a tower in the **Thermofor** process. In the fluidized catalyst concept, the moving oil vapor picks up a finely powdered catalyst and whirls it thru the reaction chamber. In all catalytic processes the catalyst is regenerated at short intervals by burning off the carbon that deposits on the catalyst during cracking.

Table 2-33 compares catalytic cracking to thermal cracking. Note the increase in both gasoline yield and octane number.

Table 2-32 Composition of Original Clay and Finished Clay Catalyst

	Per cent by weight			Surface area, sq ft/lb (9000 pellets/lb)
	SiO_2	Al_2O_3	Mg, Ca, Fe (as oxides)	
Clay	65.1	21.7	13.2	7.5–10.0
Catalyst	89.5	10.0	0.5	100–150

Table 2-33 Catalytic vs. Thermal Cracking Yields from Crude Oil

	Catalytic	Thermal*
	(Per cent by volume)	
Gasoline	63	55 (45)
Domestic heating oil	20	0 (20)
Heavy residual oil	7	37 (29)
Octane No. (3 cu cm tetraethyl lead)	**98**	**88**

*By adjusting process to produce 20 per cent as heating oil, other product yields are as shown in parentheses.

Catalytic Reforming

This process is essentially a cracking process except that straight-run gasolines of low octane value are converted into high-octane-number components. Naphthene dehydrogenation to aromatics is considered the backbone of the system. Processes are classified according to catalyst composition; namely, those using platinum-containing catalysts and those employing base metals. Catalytic reforming capacity equaled approximately 16 per cent of the U. S. 1958 total refinery capacity. Three major catalysts are molybdenum oxide on alumina, platinum, and chromium-aluminum oxide. A comparison of gasoline yield and corresponding octane numbers resulting from catalytic reforming versus the earlier thermal reforming method is shown in Table 2-34.

Table 2-34 Catalytic vs. Thermal Reforming Yields from Crude Naphtha

	Catalytic	Thermal
Gasoline, per cent by volume	90	80
Octane No. (3 cu cm of tetraethyl lead)	98	90

Thermal Cracking

This process relies upon heat and pressure to break down the hydrocarbon molecules and reorganize them. Gas oil is generally the raw material; the resultant mixture is fractionated and the heavier fractions are returned to the "cracker." Thermally cracked gasoline came into its own in the late twenties as a product superior in octane rating to straight-run (fractionated from crude oil) gasoline. The advent of **catalytic cracking** during World War II resulted in a decline of **thermal cracking** except for the reforming of straight-run naphthas and other products lighter than gas oil. However, catalytic cracking lost impetus when **catalytic reforming** became a major factor in refining in the late 1940's.

Hydrogen Treating

Hydrogen treatment of finished products and intermediate plant streams removes sulfur, nitrogen, and metal contaminants. Processing capacity exceeded that of catalytic reforming for the first time in 1960. Hydrogen is made in the plant by cracking natural gas or gasoline.

Hydrogenation of heavy residual stocks may find future use. Processes are already developed to convert heavy fractions of oil to light distillates at better than 100 per cent yields. However, the relatively high cost of H_2 may defer wide use of this process.

Alkylation and Isomerization

Alkylation *builds* hydrocarbon molecules in the gasoline range that possess high anti-knock qualities. Low-molecular-weight paraffins (e.g. *iso*-butane) and olefins may be united thermally without a catalyst, or catalytically using sulfuric acid as alkylating agent. Recent developments include alkylating ethylene and propylene with *iso*-butane, using an aluminum chloride catalyst (Phillips Petroleum Co.). Universal Oil Products has a process for alkylating ethylene with aromatics to make petrochemicals.

Isomerization converts *normal* compounds to *iso* compounds. In particular, *iso*-butane is made to meet alkylation requirements. In this case normal butane is brought into contact with aluminum chloride (sometimes antimony trichloride is also added) in the presence of hydrogen chloride. Processes are also available to isomerize pentane, hexane and some heptanes.

Fractionation

The first step in refining is fractional distillation which involves heating the crude and passing it into a fractionating tower. Specific fractions, boiling over particular temperature ranges, collect at different levels of this tower. Naphtha, kerosene, gas, oil, and heavy oil are continuously withdrawn. Gasoline obtained by condensing the vapors leaving the tower is known as **straight-run** gasoline.

FUEL OIL UTILIZATION DATA

Fuel oil for heating follows gasoline in importance as a product of the petroleum industry. Fuel oils are usually produced in compliance with ASTM specifications. Table 2-35 gives the physical properties (adopted from ASTM specifications) for fuel oil numbers 1, 2, 4, 5, and 6. Table 2-36

Table 2-35 Physical Properties of Fuel Oil at 60 F[4]

Fuel oil (CS-12-48) Grade No.	Gravity, °API	Sp gr	Lb per gal	Btu per lb	Net Btu per gal
6	3	1.0520	8.76	18,190	152,100
6	4	1.0443	8.69	18,240	151,300
6	5	1.0366	8.63	18,290	149,400
6	6	1.0291	8.57	18,340	148,800
6	7	1.0217	8.50	18,390	148,100
6	8	1.0143	8.44	18,440	147,500
6	9	1.0071	8.39	18,490	146,900
6	10	1.0000	8.33	18,540	146,200
6	11	0.9930	8.27	18,590	145,600
6	12	.9861	8.22	18,640	144,900
6, 5	14	.9725	8.10	18,740	143,600
6, 5	16	.9593	7.99	18,840	142,300
5	18	.9465	7.89	18,930	140,900
4, 5	20	.9340	7.78	19,020	139,600
4, 5	22	.9218	7.68	19,110	138,300
4, 5	24	.9100	7.58	19,190	137,100
4, 2	26	.8984	7.49	19,270	135,800
4, 2	28	.8871	7.39	19,350	134,600
2	30	.8762	7.30	19,420	133,300
2	32	.8654	7.21	19,490	132,100
2	34	.8550	7.12	19,560	130,900
1, 2	36	.8448	7.04	19,620	129,700
1, 2	38	.8348	6.96	19,680	128,500
1	40	.8251	6.87	19,750	127,300
1	42	0.8156	6.79	19,810	126,200

The relation between specific gravity and degrees API is expressed by the formula:

$$\frac{141.5}{131.5 + °API} = \text{sp gr at 60 F.}$$

For each 10° F above 60 F add 0.7° API.
For each 10° F below 60 F subtract 9.7° AP

Table 2-36 Typical Pacific Coast Specifications for Fuel Oils

	PS-100* Stove distillate or stove oil, (volatile)	PS-200*† Diesel fuel oil or burner oil, (moderately volatile)	PS-300*† Light fuel oil, domestic fuel oil or low viscosity oil residual oil	Portland*† Gas & Coke Co. special straight-run	PS-400*† Industrial fuel oil, heavy fuel oil, high viscosity oil, requires preheating
Equivalent to Eastern	...	No. 3	Between No. 5 & No. 6	...	No. 6 heavy end
Gravity, °API—max	39	31	15	...	17
min	35	28	10	...	7
Characterization "K" factor—min	...	11.3	...	11.3	...
Flash point, °F min	110	150	150	150	150
Water sediment, % max	0.25	0.50	1.00	1.50	2.00
Sulfur, wt % max	...	1.0	1.0	1.2	1.75
Viscosity, sec at 100 F min	...	35	...	...	...
Viscosity, sec at 122 F max	...	...	40	200	200
Distillation, °F—10% min	350	425	...	...	...
Distillation, °F—90% min	450	600	...	...	...
Conradson carbon, % by wt max	...	0.50	...	10.5	19.5
Avg carbon-hydrogen ratio	6.3	7.0	8.0	7.5	8.5
Avg MBtu recovered/gal					
Carb. wate gas	102	90	77	86	70
High-Btu oil gas	90	79	68	75	61
Avg tar & carbon, wt % of oil					
Carb. water gas	21	33	50	43	54
High-Btu oil gas	33	44	58	52	62
Avg oil gas, SCF/gal					
Carb. water gas	73	65	58	62	53
High-Btu oil gas	71	66	62	65	57

*All data *below* Conradson carbon calculated by Henry Linden, Institute of Gas Technology, Chicago.
† Bell, J. F. *Oil Gas Production.* (Report of Portland Gas & Coke Co. to A.G.A. Gas Production Research Com.) 1951.

contains the specifications for Pacific Fuel Oils. Table 2-37 shows the variations in the properties of fuel oils.

Ash and Sulfur. "The ash content of #6 fuel oil, ranging from about 0.01 per cent to 0.50 per cent, is exceedingly low compared to coal. The ash, however, despite this low content, is sometimes responsible for a number of serious operating problems. The sulfur contributes to these problems and is itself the cause of others. Among the operating difficulties due to ash and sulfur are:

1. Fluxing of refractory furnace walls.
2. Slagging of high temperature superheater surfaces with deposits not successfully removed by air or steam soot blowers.
3. Attack on steels including high chromium content alloys when the metal temperatures are something above 1200 F.
4. Coating of boiler, economizer and air heater surfaces or other heat exchangers.
5. Corrosion and plugging of the cold sections of air heaters.

"Experience indicates that the constituents in the ash responsible for the difficulties listed above in items 1 thru 4 are sodium or vanadium or both, particularly when the sulfur content is high. Some of the sulfur in the oil combines with the sodium and usually appears in deposits on heating surfaces as sodium sulfate as analyzed. The vanadium in the oil usually appears in the deposits as V_2O_5 as analyzed. Since both compounds have low melting points, 1625 F for Na_2SO_4 and 1275 F for V_2O_5, their presence tends to lower the fusing temperatures of the deposits, the amount depending on the relative quantities of the more refractory materials such as alumina and silica present."[3]

The API gravity and the Saybolt universal viscosity are two valuable characteristics for identifying a fuel oil. The gravity can be used to convert oil volume to weight, and vice versa. It can also be used to estimate the fuel's heating value by use of the formula:

$$\text{Btu/lb} = 18{,}250 + 40 \times \text{API gravity}.$$

Table 2-35 relates *fuel oil number* to degrees API gravity, Btu/lb, and net Btu/gal. Table 1-46 relates degrees API to specific gravity and various units of density.

The **viscosity** is expressed in seconds of time for a given oil to pass thru a specific opening at a specified temperature. A knowledge of how viscosity varies with temperature is essential for the easy handling and efficient utilization of fuel oil. Figure 1-8 plots viscosity vs. temperature for a number of fuel oils. Note that you may approximate the complete viscosity-temperature relationship for any fuel oil by drawing a line parallel to the lines shown thru the coordinates of any known viscosity-temperature condition.

The heavier oils (Nos. 5 & 6) must be heated to reduce their viscosity before pumping. It is important to hold the upper limit of the oil temperature below the flash point to prevent carbonization of the oil, and to avoid possible loss of burner ignition caused by interruption of the flow of oil by vapor binding of pumps.

Flash Point. Determinations for diesel fuels, distillate and residential fuel oil, kerosene, and illuminating oils may be made by means of the Pensky-Martens closed tester.

Table 2-37 Typical Properties of Commercial Petroleum Products Sold in the East and Midwest[5]

	Gasolines		Premium Fuels			Fuel Oils				
	12 psia Reid vapor pressure natural gasoline	Straight run gasoline	No. 1 fuel, light diesel, or stove oil	Diesel or "gas house" gas oil	Reduced crude, heavy gas oil or premium residuum*	No. 2	No. 3	Cold No. 5	No. 5	Bunker C or No. 6
Gravity, °API	79	63	38–42	35–38	19–22	32–35	28–32	20–25	16–24	6–14
Viscosity	...	...	34–36†	34–36†	15–50‡	34–36†	35–45†	<20‡	20–40‡	50–300‡
Conradson carbon, wt %	None	None	Trace	Trace	3–7	Trace	<0.15	1–2	2–4	6–15
Pour point, °F	<0	<0	<0	<10	...	<10	<20	<15	15–60	>50
Sulfur, wt %	<0.1	<0.2	<0.1	0.1–0.6	<2.0	0.1–0.6	0.2–1.0	0.5–2.0	0.5–2.0	1–4
Water and sed., wt %	None	None	Trace	Trace	<1.0	<0.05	<0.10	0.1–0.5	0.5–1.0	0.5–2.0
Distillation, °F										
10%	110	150	390–410	400–440	550–650	420–440	450–500	<600	<600	600–700
50%	150	230	440–460	490–510	800–900	490–530	540–560	...	...	850–950
90%	235	335	490–520	580–600	...	580–620	600–670	<700	>700	...
End point	315	370	510–560	630–660	...	630–660	650–700	...	...	...
Flash point, °F	<100	<100	100–140	110–170	>150	130–170	>130	>130	>130	>150
Aniline point, °F	145	125	140–160	150–170	...	120–140	120–140	...	...	...
C to H ratio	5.2	5.6	6.1–6.4	6.2–6.5	6.9–7.3	6.6–7.1	7.0–7.3	7.3–7.7	7.3–7.7	7.7–9.0
Average MBtu recovered/gal of oil:										
Carbureted water gas	101	101	102	101	93	91	86	82	83	74
High-Btu oil gas	90	89	90	90	82	80	76	72	73	64
Tar + carbon, wt % of oil										
Carbureted water gas	...	...	20	22	35	31	36	42	42	52
High-Btu oil gas	...	...	32	34	45	42	46	52	52	60

(Data on gasolines and data below C/H ratio not given in reports.)

* The 6 wt % Conradson carbon oil used in the Hall High Btu Oil Gas Tests (A.G.A. Gas Production Research Committee. Hall High Btu Oil Gas Process, New York, 1949.) and "New England Gas Enriching Oil" are typical examples of this group.

† Saybolt Universal seconds at 100 F.

‡ Saybolt Furol seconds at 122 F.

Remarks:

No. 1 Fuel Oil—A distillate oil intended for vaporizing pot-type burners. A volatile fuel.
No. 2 Fuel Oil—For general purpose domestic heating; for use in burners not requiring No. 1 oil. Moderately volatile.
No. 3 Fuel Oil—Formerly a distillate oil for use in burners requiring low viscosity oil. Now incorporated as a part of No. 2.
No. 4 Fuel Oil—For burner installations not equipped with preheating facilities.
No. 5 Fuel Oil—A residual type oil. Requires preheating to 170–220 F.
No. 6 Fuel Oil—Preheating to 220–260 F suggested. A high viscosity oil.

"The sample is heated at a slow constant rate with continual stirring. A small flame is directed into the cup at regular intervals with simultaneous interruption of stirring. The flash point is the lowest temperature at which application of the test flame causes the vapor above the sample to ignite."[10]

The flash points of motor fuels, naphtha, and other hydrocarbon products that flash below 175 F may be determined by means of the Tag closed cup tester[11] or the Tag open cup tester.[12] The apparatus and procedures used are somewhat similar to the Pensky-Martens method.

LUBRICANTS

Motor oil has become a complex fluid, containing as many as five additives, each contributing a special property to the oil. One additive, made from phosphorous pentasulfide, checks an oxidative deterioration; another additive, of a heavy metal sulfonate soap, acts as a detergent to keep the motor clean. High polymer molecules akin to butyl rubber are used to prevent the tendency of natural oil to thicken. An aromatic-wax condensation product lowers the temperature at which the oil will pour. Finally, the whole oil is made alkaline to neutralize the acids produced in the combustion zone. Thus, three to eight per cent concentration of additives in the quality motor oils should be common. Many advantages are claimed for such oils, among them being a five per cent increase in gasoline mileage.

TYPES OF HYDROCARBON LIQUIDS AVAILABLE FOR GAS-MAKING PURPOSES

In reading this material on oils for gas making, the decline in manufactured gas usage should be kept in mind. The average annual use of oils for the purpose has declined from 1182 million (1946–1950) to 153 million gallons in 1957.

Typical Oils Used

For many years the oil used for making carbureted water gas and for other enriching purposes was known as **gas oil.** It has the same characteristics as diesel oil. Usability of this fraction for the manufacture of gasoline and for diesel fuel absorbed its supply.

Many types of hydrocarbon liquids may be used for gas-making purposes. Some of these oils are either unavailable in sufficient quantity or priced too high. Typical oils which have been used are shown in Tables 2-36 and 2-37. They may vary from gaseous propane to very heavy residuum oils or Bunker C oils. On the Pacific Coast a somewhat different designation is used for distillate and residual fuel oils than in the remainder of the United States (Table 2-36). Typical properties of two gasolines and the different grades of fuel oils and premium fuels are listed in Table 2-37. Note that the high-Btu oil gas yield per gallon of oil decreases from about 90 MBtu for No. 1 fuel oil to about 64 MBtu for No. 6 oil. The tar and carbon produced in the high-Btu oil gas process increases as the oil becomes heavier, from 42 per cent of the original weight of oil for No. 2 fuel oil to 60 per cent for No. 6 oil.

Blending and Compatibility of Oils

When changing from one oil or from one source of supply to another, and when storing oils in the same tanks for use in gas making, information on blending and compatibility of the oils should be obtained. Heavy residual oils have a tendency to deposit sludge in storage; this may be aggravated by mixing oils of different character, or when deliveries from two sources go into the same tank. Compatibility of oils may be tested by simply mixing the oils and centrifuging to determine whether the bottom sediment is increased appreciably. To avoid trouble, the following general rules will be helpful:

(a) Straight-run residuals can be mixed with any straight-run product.

(b) Thermal cracked distillates are unstable when blended with straight run and tend to form gum.

(c) The catalytic cracked distillates are stable fuel distillates, either alone or blended with straight-run distillates that do not contain appreciable quantities of active sulfur compounds or organic acids.

(d) Straight-run residuals can be mixed with cracked distillates.

(e) Cracked residuals should be blended with cracked distillates rather than straight-run distillates.

(f) Cracked distillate can be added as a third ingredient to mixtures of all straight-run products.

Factors Affecting Value of an Oil for Gas Making[5,9]

The value of an oil for gas making is normally dependent on: (a) the Btu recovery in the finished make gas; (b) tar and light oil yields and quality; (c) ease of control of gas and tar quality; (d) coke or carbon production; (e) storage, pumping, and general handling costs; (f) the effects of items (a), (c), and (d) on gas-making equipment capacity.

The Btu recovery has the most direct and important effect on the value of a gas-making oil. The other factors listed above being equal or comparable, the oil which provides the higher Btu recovery under the conditions required for its proper gasification is the oil that is worth more. Within the limitations stated, the difference in value would be directly proportional to the difference in Btu recovery in the make gas.

Under specified cracking conditions (as defined by temperature, residence time, and partial pressures) the Btu recoveries *decrease* with an increase in the weight ratio of carbon to hydrogen in the oil. The C/H ratios of oils may be estimated from their mean average boiling points and API gravities by use of the charts shown as Figs. 2-3 and 2-4. In the case of light or distillate oils, the average boiling point is obtained from an ASTM distillation; for residual oils the average boiling point may be estimated from the viscosity and gravity by use of Fig. 2-4.

Since oils are purchased on a *volume* basis, it is usually desired to compare Btu recoveries per gallon rather than per pound. Of two oils having the same C/H ratio, the one having the lower API gravity (the higher specific gravity) will have the greater Btu recovery per gallon. Also, while oils having the same C/H ratio will produce approximately the same percentage by weight of *tar-plus-carbon* under comparable cracking conditions, the ratio of tar to carbon, and the tar properties can vary widely. *Conradson carbon residues* will provide good indications of the ratio of tar to carbon that can be anticipated. The minimum percentage by weight of the oil that will be deposited as carbon or coke in an oil gas generator is roughly proportional to the Conradson carbon residue.

Gas-making oils may be divided into various grades which can be differentiated by simple physical tests. The properties of the various grades of oil commonly used for gasification are summarized in Tables 2-36 and 2-37.

Tar and Coke Credits in Making Oil Gas[5]

Combined tar and coke credits can be estimated by assuming such nongaseous products of cracking to have a value corresponding to their fuel value, or approximately 6.6 cents per gallon in 1950. To estimate tar-plus-carbon yields, the following equation can be used:

$$\text{Lb (T + C) per gal} = 0.07\ (Q - E) - 1.1$$

where Q is the gross heat of combustion of the oil in MBtu per gal, and E is the Btu recovery in the make gas in MBtu per gal. Values of Q, or gross heating value, can be estimated from the API gravities:

°API	Gross heating value, (Q) MBtu gal
0	165
10	155
20	148
30	142
40	136
50	130

Values of E can be obtained in two ways. The value of E is given in MBtu per *lb* in Fig. 2-5. The conversion from a pound to a gallon basis can be obtained in this figure. The

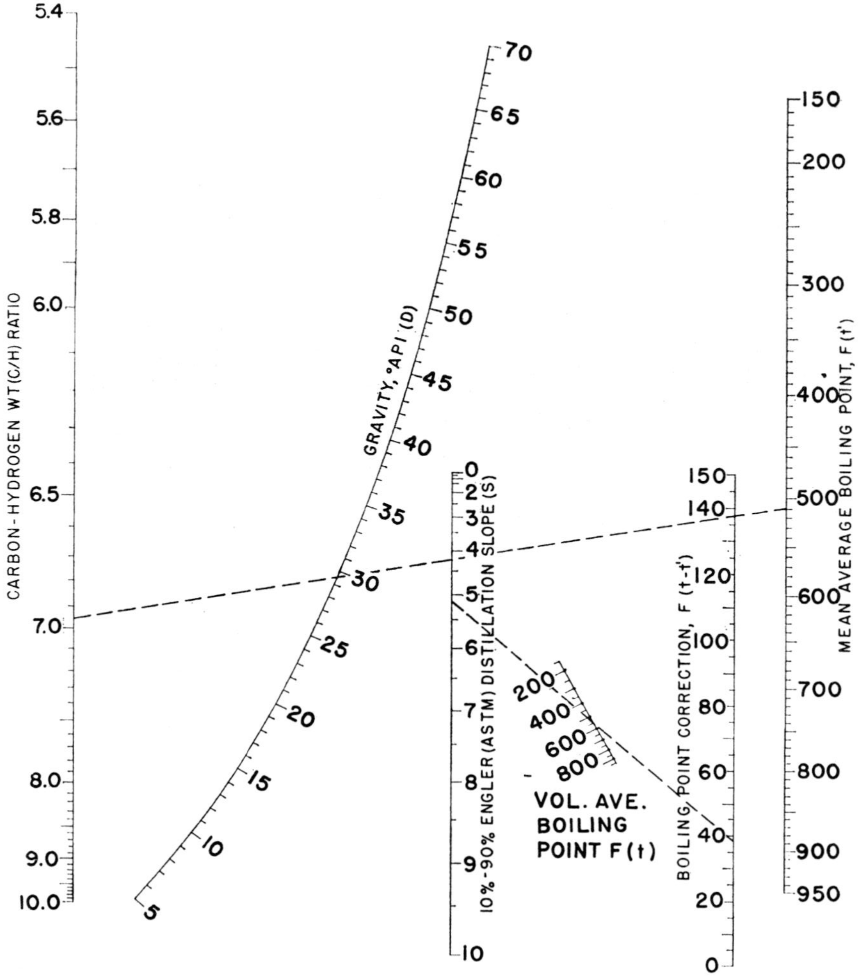

Fig. 2-3 Nomograph for determination of C/H ratios of petroleum fractions.[5] **Example:** Given, an oil with an API gravity of 29.6, a volumetric average boiling point (50 per cent ASTM distillation point) of 550 F, and a 10 to 90 per cent distillation slope of 5.2. Correct boiling point (at lower right) by drawing a line thru the slope on the (S) scale and the volumetric average boiling point on the (t) scale to intersection with the ($t - t'$) scale, which shows a correction of 39 F. The mean average boiling point = 550 — 39 = 511 F. Then locate 511 on the (t') scale and extend a line thru 29.6 on the gravity scale (D) to the C/H scale; read C/H = 6.96. **Note:** In case the necessary distillation data are not available, use Fig. 2-4 to estimate the mean average boiling point from the API gravity and kinematic viscosity.

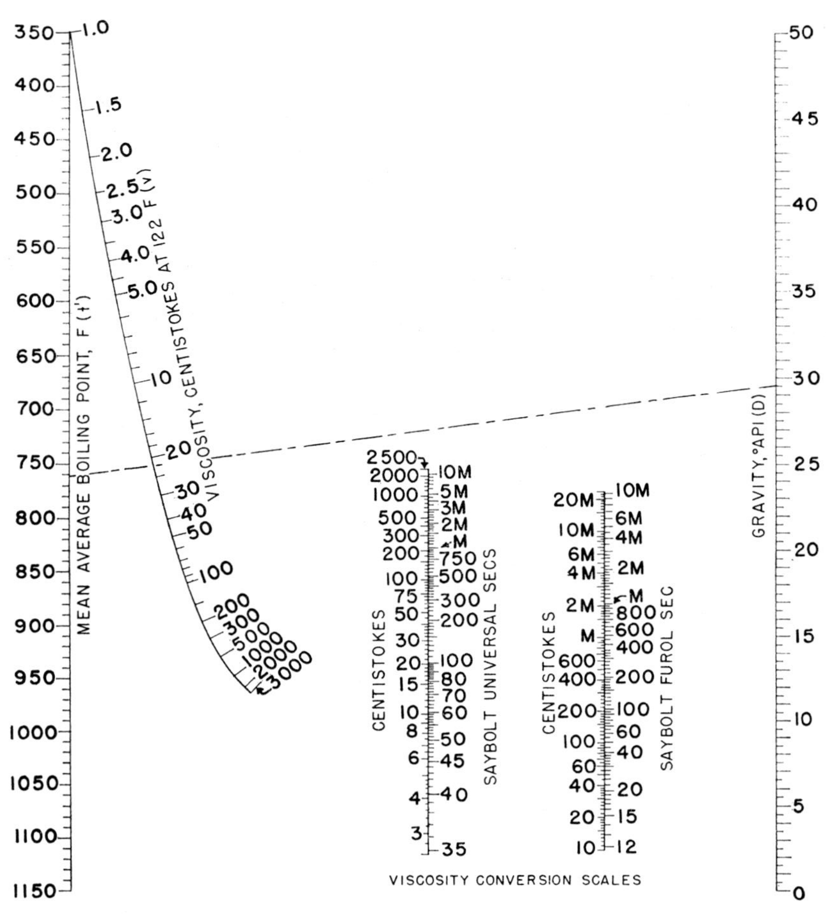

Fig. 2-4 Nomograph for estimating the mean average boiling point of liquid petroleum fractions.[5] **Example:** Given, an oil with a viscosity of 106.2 Saybolt universal seconds at 122 F and an API gravity of 29.5. Convert viscosity to centistoke units by using the proper conversion scale, which in this case gives 22 centistokes. Locate 29.5 on the (D) scale and extend a line from that point thru 22 on the (v) scale to the (t') scale, which shows the mean average boiling point to be 762 F.

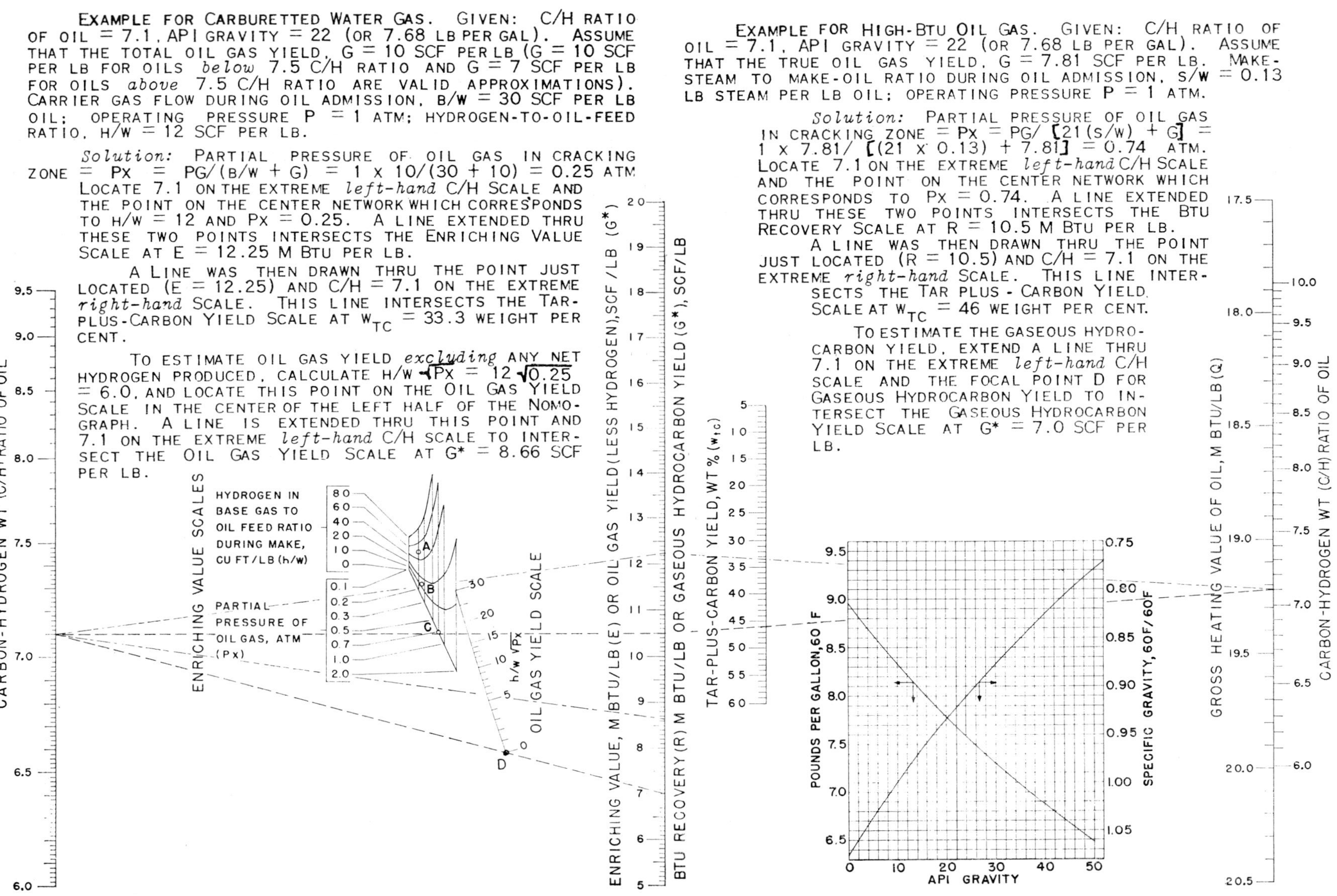

EXAMPLE FOR CARBURETTED WATER GAS. GIVEN: C/H RATIO OF OIL = 7.1, API GRAVITY = 22 (OR 7.68 LB PER GAL). ASSUME THAT THE TOTAL OIL GAS YIELD, G = 10 SCF PER LB (G = 10 SCF PER LB FOR OILS *below* 7.5 C/H RATIO AND G = 7 SCF PER LB FOR OILS *above* 7.5 C/H RATIO ARE VALID APPROXIMATIONS). CARRIER GAS FLOW DURING OIL ADMISSION, B/W = 30 SCF PER LB OIL; OPERATING PRESSURE P = 1 ATM; HYDROGEN-TO-OIL-FEED RATIO, H/W = 12 SCF PER LB.
Solution: PARTIAL PRESSURE OF OIL GAS IN CRACKING ZONE = Px = PG/(B/W + G) = 1 x 10/(30 + 10) = 0.25 ATM
LOCATE 7.1 ON THE EXTREME *left-hand* C/H SCALE AND THE POINT ON THE CENTER NETWORK WHICH CORRESPONDS TO H/W = 12 AND Px = 0.25. A LINE EXTENDED THRU THESE TWO POINTS INTERSECTS THE ENRICHING VALUE SCALE AT E = 12.25 M BTU PER LB.
A LINE WAS THEN DRAWN THRU THE POINT JUST LOCATED (E = 12.25) AND C/H = 7.1 ON THE EXTREME *right-hand* SCALE. THIS LINE INTERSECTS THE TAR-PLUS-CARBON YIELD SCALE AT W_{TC} = 33.3 WEIGHT PER CENT.
TO ESTIMATE OIL GAS YIELD *excluding* ANY NET HYDROGEN PRODUCED, CALCULATE $H/W\sqrt{Px} = 12\sqrt{0.25} = 6.0$, AND LOCATE THIS POINT ON THE OIL GAS YIELD SCALE IN THE CENTER OF THE LEFT HALF OF THE NOMOGRAPH. A LINE IS EXTENDED THRU THIS POINT AND 7.1 ON THE EXTREME *left-hand* C/H SCALE TO INTERSECT THE OIL GAS YIELD SCALE AT G* = 8.66 SCF PER LB.
EXAMPLE FOR HIGH-BTU OIL GAS. GIVEN: C/H RATIO OF OIL = 7.1, API GRAVITY = 22 (OR 7.68 LB PER GAL). ASSUME THAT THE TRUE OIL GAS YIELD, G = 7.81 SCF PER LB. MAKE-STEAM TO MAKE-OIL RATIO DURING OIL ADMISSION, S/W = 0.13 LB STEAM PER LB OIL; OPERATING PRESSURE P = 1 ATM.
Solution: PARTIAL PRESSURE OF OIL GAS IN CRACKING ZONE = Px = PG/[21(S/W) + G] = 1 x 7.81/[(21 x 0.13) + 7.81] = 0.74 ATM.
LOCATE 7.1 ON THE EXTREME *left-hand* C/H SCALE AND THE POINT ON THE CENTER NETWORK WHICH CORRESPONDS TO Px = 0.74. A LINE EXTENDED THRU THESE TWO POINTS INTERSECTS THE BTU RECOVERY SCALE AT R = 10.5 M BTU PER LB.
A LINE WAS THEN DRAWN THRU THE POINT JUST LOCATED (R = 10.5) AND C/H = 7.1 ON THE EXTREME *right-hand* SCALE. THIS LINE INTERSECTS THE TAR PLUS - CARBON YIELD SCALE AT W_{TC} = 46 WEIGHT PER CENT.
TO ESTIMATE THE GASEOUS HYDROCARBON YIELD, EXTEND A LINE THRU 7.1 ON THE EXTREME *left-hand* C/H SCALE AND THE FOCAL POINT D FOR GASEOUS HYDROCARBON YIELD TO INTERSECT THE GASEOUS HYDROCARBON YIELD SCALE AT G* = 7.0 SCF PER LB.
CARBON-HYDROGEN WT (C/H) RATIO OF OIL
9.5
9.0
8.5
8.0
7.5
7.0
6.5
6.0
ENRICHING VALUE SCALES
HYDROGEN IN BASE GAS TO OIL FEED RATIO DURING MAKE, CU FT/LB (h/w)
80
60
40
20
10
0
PARTIAL PRESSURE OF OIL GAS, ATM (Px)
0.1
0.2
0.3
0.5
0.7
1.0
2.0
A
B
C
D
OIL GAS YIELD SCALE
h/w √Px
30
20
15
10
5
0
ENRICHING VALUE, M BTU/LB (E) OR OIL GAS YIELD (LESS HYDROGEN), SCF/LB (G*)
BTU RECOVERY (R) M BTU/LB OR GASEOUS HYDROCARBON YIELD (G*), SCF/LB
20
19
18
17
16
15
14
13
12
11
10
9
8
7
6
5
TAR-PLUS-CARBON YIELD, WT % (w_{tc})
5
10
15
20
25
30
35
40
45
50
55
60
POUNDS PER GALLON, 60 F
9.5
9.0
8.5
8.0
7.5
7.0
6.5
SPECIFIC GRAVITY, 60F/60F
0.75
0.80
0.85
0.90
0.95
1.00
1.05
API GRAVITY
0
10
20
30
40
50
GROSS HEATING VALUE OF OIL, M BTU/LB (Q)
17.5
18.0
18.5
19.0
19.5
20.0
20.5
CARBON-HYDROGEN WT (C/H) RATIO OF OIL
10.0
9.5
9.0
8.5
8.0
7.5
7.0
6.5
6.0

Fig. 2-5 Nomograph for estimating oil gasification yields for high-Btu oil gas and carbureted water gas production.[5] Pivot points: A, laboratory cracking test in hydrogen; B, typical carbureted water gas operation; C, typical high-Btu oil gas operation; D, focal point for gaseous hydrocarbon yield scale.

Table 2-38 Oil Rating Index*

(MBtu per gal)

C/H Ratio	°API 10	15	20	25	30	35	40	45	50	55	60	65	70	75
5.4	...	...	...	...	...	...	...	...	...	109	106	103	101	99[a]
5.6	...	...	...	...	...	...	...	...	108	105	102[b]	99[b]	97[b]	95
5.8	...	...	...	...	...	...	...	107	104	101	98[b]	96[b]	94[b]	91
6.0	...	...	...	...	...	...	105[d]	102[d]	100	97[c]	94[c]	92[c]	90	88
6.2	...	...	...	...	...	104[e]	101[d,e]	98[d]	96	93[c]	91[c]	88[c]	86	84
6.4	...	...	...	...	103	100[e]	97[d,e]	94[d]	91	89	86	84	82	80
6.6	...	...	...	101	98	95[e]	92[e]	90	87	85	83	80	79	77
6.8	...	...	99[g]	96[g]	93[f]	90[f]	88	85	83	81	79	77	75	...
7.0	...	97	94[g]	91[g]	89[f]	86[f]	83	81	79	77	75	73	...	...
7.2	96	92	89[g]	87[g]	84[f]	81[f]	79	77	75	73	71	...	...	...
7.4	91	87	84[h]	82[h]	79	77	75	73	70	69	...	...	...	...
7.6	86	83	80[h]	78[h]	75	73	71	69	67	...	...	...	...	...
7.8	82[j]	79[j]	76[h]	74[h]	71	69	67	64	...	...	...	...	...	...
8.0	78[j]	75[j]	73	70	68	66	64	...	...	...	...	...	...	...
8.2	74[j]	72[j]	70	67	65	63	...	...	...	...	...	...	...	...
8.4	71[j]	69[j]	67	65	63	...	...	...	...	...	...	...	...	...
8.6	69[j]	66[j]	64	62	...	...	...	...	...	...	...	...	...	...
8.8	65[j,k]	63[j,k]	61	...	...	...	...	...	...	...	...	...	...	...
9.0	63[k]	61[k]	...	...	...	...	...	...	...	...	...	...	...	...

* Index = 100 for "gas house" gas oil (C/H = 6.3, gravity = 38° API).

[a] Natural gasoline.

[b] Straight-run gasoline.

[c] Cracked gasoline.

[d] Kerosene, No. 1 fuel oil, stove oil.

[e] "Gas house" gas oil, light gas oil, light diesel oil.

[f] No 2 fuel oil, furnace oil, light cat. cracked distillates.

[g] Premium residuum, heavy gas oil.

[h] Heavy cat. cracked distillates, cycle oils, light cat. cracker residues.

[j] Bunker C fuel oil, No. 6 fuel oil.

[k] Heavy cat. cracker residues and cycle fuel oils.

Btu recovery in MBtu per gal can also be obtained by use of *Index Numbers* (Table 2-38) whose numerical values fall approximately between the enriching values or Btu recoveries obtained in carbureted water gas and in high-Btu oil gas production. On this basis, carbureted water gas operating results will be 5 to 10 MBtu per gal *higher* due to the presence of hydrogen in the carrier gas (**blue gas**) and lower partial pressures of oil gas while high-Btu oil gas results will be 5 to 10 MBtu per gal *lower* due to the absence of hydrogen in the carrier gas (steam in this case) and higher partial pressures of oil gas. A gas plant which has sufficient operating experience for any of the typical oils listed at the bottom of Table 2-38, may wish to modify the Index Number values shown, to portray their own operating conditions more accurately. However, for the purpose of selecting oils, the Index Numbers may be used instead of actual Btu recoveries in terms of MBtu per gal of make oil.

A comparison of the net costs *per therm* of oil gas from typical oils is shown in Table 2-39. This comparison is based only on make oil costs, actual Btu recoveries for 1950 high-Btu oil gas operation, and credits due to the make oil for tar and carbon. It should be noted that to realize "coke credits" from residual oils, that is, the reduction in fuel or heating oil requirements due to deposited carbon, a regenerative process must be used. A value of 0.4 cents per pound of tar plus coke yield was assumed.

Table 2-39 Net Oil Value Per Therm of Oil Gas

	Oil cost, cents/gal	Re-covery, MBtu/gal	Tar plus coke credit, cents/gal	Net value per therm, cents
"Gas house" gas oil	8.0	90	1.0	7.8
No. 2 fuel oil	7.7	80	1.2	8.1
Heavy gas oil (reduced crude)	6.0	82	1.4	5.6
Cold No. 5 fuel oil	5.9	72	1.6	6.0
No. 5 fuel oil	5.2	73	1.6	4.9
No. 6 or Bunker C fuel oil	4.5	64	2.0	3.9

Note: Based on New York posted prices, April 1950.

The values shown in Table 2-39 can only be approached under conditions of continuous operation. Such values cannot be attained when the gas is produced as a *by-product* during emergency or high **sendout** periods.

Procedure for Oil Selection for Gas Production[5]

On the basis of the above ideas and data, it is possible to systematize the selection of oils for carbureted water gas and high-Btu oil gas production, using the following steps:

1. Determine the API gravity, the viscosity, and the *Conradson carbon residue* for residual oils; in the case of distillate or light blended oils, determine the API gravity and mean average boiling point from an ASTM distillation.
2. By use of Figs. 2-3 and 2-4, determine the C/H ratio of the oil from either: (a) the API gravity and the determined mean average boiling point, or (b) the API gravity and the mean average boiling point as estimated from the viscosity and gravity.
3. Determine the grade and Index Number of the oil from Tables 2-37 and 2-38.
4. Estimate tar and coke credits from gross heating value of oil (based on API gravity) and the Index Number. The enriching value (Fig. 2-4*) may be used as an alternative.

* Convert from pound to gallon basis.

5. Calculate the net cost of make oil per therm in the make gas by the formula:

$$J = \frac{100}{I}(O - C)$$

where: J = net make oil cost, cents per therm
I = Index Number of oil from Table 2-38
O = price of oil, cents per gallon
C = tar plus coke credit from step 4

If a plant correlation between the Index Number and actual Btu recoveries is available, the anticipated recovery in MBtu per gal may be used instead of the Index Number. With due regard to expected set capacities and such handling costs as are direct functions of the oil properties, selection of oils can be based on the oil cost per therm of oil gas as determined in step 5.

In this method of selection, observance of the factors listed below is also desirable:

(a) If there is a choice of equally or nearly equally priced oils of a given grade which have the same Index Number, select the one with the *lowest* Conradson carbon residue.

(b) If the oil appears satisfactory but has a 10 per cent distillation point that is considerably below the normal, a poorly blended mixture is indicated. For example, a Bunker C fuel oil with a 10 per cent distillation point below 400 F most probably has been blended with a light distillate to reduce viscosity. A "bumpy" distillation curve will result, and control of operating conditions may be difficult due to nonuniform vaporization rates.

(c) Normal petroleum fractions, or satisfactory blends of fractions, have smooth, slightly *S-shaped,* distillation curves, with nearly uniform slopes between the 10 per cent and 90 per cent points. The method of estimating C/H ratios is most accurate when the distillation slope is normal or uniform.

(d) There is some evidence that oils having very steep distillation slopes (nearly 10°F for each one per cent by volume distilled) may cause trouble by the formation of tars that are difficult to handle (high sulfonation index and high free carbon content). To obtain a distillation curve for the heavier blended oils or residual oils, a rather complicated vacuum distillation would be required.

(e) After a preliminary selection of the preferred oils, the selected oils should be subjected to plant tests on reduced quantities (tank car, or minimum bulk shipments) before a final selection is made to determine whether the tar or tar emulsion produced can be handled in the plant. Variation between plant facilities can lead to trouble if contracts for oil are made before proper plant trials.

Gas-Making Oil Specifications

Approximate specifications for commercial grades of gas-making oils—"gas house" gas oil, No. 2 fuel oil, premium residuum, No. 6 fuel oil or Bunker C fuel oil—are given in Table 2-37.

Generally speaking, the characteristics to avoid in residual oils are a combination of low API gravity with low viscosity, and in distillate oils, the combination of low API gravity with low 50 per cent distillation point. Although a high Conradson carbon residue is usually an objectionable feature, selection of residual oils on this basis alone provides no assurance of satisfactory Btu recovery, and will eventually result in receipts of oils requiring 14–15 gal of make oil per MCF of make gas. If the "gas house" gas oil is assigned an Index Number of 100, a good nonpremium distillate meeting the No. 2 fuel oil specification would have a minimum rating of 90, a premium residue would rate above 85, and a properly selected Bunker C fuel oil above 73.

LIQUID FUELS FOR MOTORS

The gas industry operates a large number of vehicles requiring primarily gasoline, motor benzol, butane, and diesel oil.

Gasoline

The average U. S. octane ratings at the end of 1962 were 99.7 research (90.7 motor) for premium gasolines and 93.0 research (84.9 motor) for regular grades.[9]

Some characteristics of gasolines follow:

The **cold starting** characteristics are measured approximately by the temperature at which 10 per cent is evaporated in the ASTM distillation.

The starting ability is also influenced by the slope of the distillation curve at this 10 per cent point. The *flatter* the curve the lower the starting temperature with a one to one air-fuel ratio carburetor. In general, the following relations hold, although it must be remembered that viscosity of the engine oil and cranking speed greatly influence starting ability:

Temp 10% distilled, °F	Min engine temp to start, °F
110	−20
125	−10
140	0
155	+10

Behavior during *warmup* is thought to be related to the "front-end" volatility, as represented by 15 to 70 per cent on the distillation curve. A relatively flat curve thru this range improves warmup characteristics of a gasoline.

The behavior during *normal operation* is related to overall volatility. This is rather difficult to interpret because complete vaporization is not generally obtained in manifolds. The 90 per cent point on the distillation curve, corrected for loss, is usually relied upon for a general indication of the behavior in normal operation.

The 90 per cent point on the distillation curve is usually taken as a measure of crankcase dilution tendency. With modern thermostats and ventilated crankcases the dilution problem has decreased in importance.

Accelerating ability is generally judged by the 50 per cent point on the distillation curve. The lower the 50 per cent temperature the better the accelerating ability.

The **mileage qualities** of a gasoline are indicated by the "back-ends," or the 70–90 per cent range on the distillation curve. A higher temperature in this region means that a gasoline will provide better mileage per gallon.

The **volatility** of gasolines is varied to suit the weather conditions encountered in each geographical section of the country during the four seasons of the year. It is wise to permit gasoline stocks to decrease to a minimum on the dates when the refiners make these seasonal changes. For example, if the storage tanks were filled with fall-spring blend just before the summer changeover date, this gasoline might cause vapor lock during the summer. Conversely, if the summer blend were carried over into cold weather, starting difficulties might be encountered.

Fresh gasoline does not have an appreciable gum content. A more significant property is the tendency to form gum (**gum stability**). This rate of gum formation will vary with different gasolines. Gum formation also depends upon conditions of storage, such as temperature, access of air, and presence of catalytic materials.

Since most engines have copper fuel lines and copper is a catalyst for gum formation, this gum formation will be greatly accelerated if gasoline remains stagnant in a fuel line. Fuel should be drained from gasoline engines if they are to be idle more than three months. The only alternative for stand-by gasoline-driven equipment is to use straight-run gasoline which, being uncracked, will not form gums.

The **anti-knock rating** (octane number) of gasoline is determined in standard continuously variable compression engines with mixtures of *iso*-octane (100) and normal heptane (0). The ASTM octane number of a fuel is the percentage of *iso*-octane in that mixture of *iso*-octane and heptane which the motor fuel on test matches in knock characteristics when compared by the procedure specified. A common anti-knock test is made on a laboratory test engine under closely controlled conditions and obtains an octane number usually known as the **research octane number.** Another common test is run on a test car engine under road conditions and the result obtained is called the **road** or **motor octane number.** For any given gasoline the research octane number is usually *higher* than the motor octane number.

Although the octane number of a gasoline may be obtained with good precision, this does not necessarily mean that the use of fuel with a high octane number will insure that a particular engine will not knock. **Engine knock** is also influenced by speed, spark timing, mixture ratio, mixture temperature, jacket temperature, and degree of throttle opening.

The corrosive substance which may be present in gasoline is **sulfur.** The best test for corrosion is to determine the discoloration produced when a strip of sheet copper is immersed in gasoline for three hours at 122 F. This test is widely used and is sufficiently severe to afford the degree of protection required by the user. There is no quantitative standard of interpretation.

The **Doctor test** reveals the presence of hydrogen sulfide and mercaptans. No indication of either the total or the free sulfur is given by this test. Since the free sulfur present in a gasoline is usually the major corrosive agent, the Doctor test is not highly regarded. A test for total sulfur combined with the copper strip test is generally believed to yield a better indication of quality, with respect to sulfur content, than does the Doctor test.

Regular gasoline should be used for all motor vehicles except those requiring a premium fuel per manufacturer's instructions. Most lower priced automobiles will perform satisfactorily on regular gasoline. If a knock is found with regular fuel, the cause is usually mechanical maladjustment, such as spark plugs, spark setting, or distributor points.

Motor Benzol

Motor benzol is used to some extent to increase the anti-knock rating of natural and cracked gasolines, and is admixed in quantities up to 50 per cent. In 1940 a total of 32 million gallons of crude and refined benzol was produced, 47 per cent of which went into the making of 101 million gallons of motor benzol. This would indicate about 15 per cent benzol mixed with the gasoline. By 1948 the volume of motor benzol produced had dropped to nine million gallons. Of the total benzol produced in 1948 only 5.7 per cent went into making motor benzol and 94.3 per cent was refined to pure benzol for making phenol, aniline, styrene, and secondary products such as synthetic rubber, nylon, plastics, and some explosives. Inhibited motor benzol, rather than the acid washed benzol, is now being used. Its preparation is very simple, consisting usually of the addition of 0.0165 gram of potassium dichromate, an anti-oxidant, per liter of crude benzol. As shown in Table 2-40, benzol has a blending octane number of 88 which is only slightly above that of some regular gasolines (86) on the market in 1951.

LP-Gases

Propane and butane are proved fuels for internal combustion engines. As shown in Table 2-40 they have high volatility and the highest energy per pound of the liquids listed. They also have fairly good octane ratings. The principal point in their favor is their cleanliness which minimizes engine repair.

Diesel Fuel Oil

Diesel fuel is obtained from crude oil and the cycle stocks from catalytic cracking. Virgin stocks produce oil with higher **cetane** numbers, but this is in competition with domestic heating oil and with cracking units for production of motor gasoline. Use of **cycle stocks** has the advantage of economic supply. Research is directed toward development of **additives for diesel oil** (which may be similar to tetraethyl lead for gasoline) in order to improve ignition characteristics without impairing other properties. This, in turn, might permit decreasing the weight of diesel engines.

Properties of various types of diesel oil are summarized in Table 2-41 for the five U. S. regions: Eastern, Southern, Central, Rocky Mountains, and Western. Data supplied for Type S-M in the Rocky Mountain region reflect one sample only; thus they are not reflected in this general table.

The **cetane** number referred to in Table 2-41 measures ignition quality and is of importance in **cold starting** of diesel engines. It has no quantitative significance; it indicates that one fuel is better than another by marking the ease of self-ignition, whereas the octane number is an indication of resistance to self-ignition. The cetane number can be increased by increasing the average boiling point of the hydrocarbon mixture.

Table 2-40 Comparison of Properties of Various Liquid Fuels for Use in Engines*

No.	Material	Freezing point, °F	Boiling point, °F	Heat of combustion, Btu/lb	Heat of combustion, Btu/gal	Flash point, °F	Octane number
1.	Alcohol, ethyl	−174	173	12,816	82,800	65	90†
2.	Benzol	42	176	17,986	131,140	12.8	88‡
3.	Toluol	−140	231	18,245	131,290	45	102‡
4.	Gasoline C-6§	−60	122–316	20,394	...	0	76†
5.	Propane	−306	−44	21,560	91,500	Gas	(97.1)†† 112**
6.	Butane	−211	31	21,180	102,600	Gas	(92)†† 71‡
7.	Pentane	−202	97	20,808	108,450	0	(58)†† 57.5‡
8.	*iso*-Pentane	−257	82	20,808	107,080	0	(90)†† ...
9.	Hexane	−140	156	20,700	113,510	0	(34)†† 23.5‡
10.	Neohexane (dimethylbutane)	−147	121.5	20,698	111,320	...	95
11.	Heptane	−131	205	20,664	117,190	25	0‡
12.	Octane	−70	258	20,608	120,040	57	−13‡
13.	*iso*-Octane	−161	210–254	20,556	117,990	...	95
14.	Aviation gasoline (70 API)	...	...	21,400	125,000	...	...
15.	U. S. motor gasoline (60 API)	...	...	21,050	129,000	...	...
16.	Kerosene (42 API)	...	...	20,000	135,000	...	...
17.	Automotive diesel oil (35 API)	...	...	19,550	138,500	...	...

* Adopted from: Jacobs, P. B. and Newton, H. P. *Motor Fuels from Farm Products.* (Miscell. Pub 327) Washington, U. S. Dept. of Agriculture, 1938; and Egloff, G. "Petroleum Industry" (In: Rogers, A. *Manual of Industrial Chemistry*, chap. 14. New York, Van Nostrand, 1942).
† CFR motor method.
‡ Blending octane number at 300 F.
§ Standard Oil Development Co.
** Popovich, M. and Hering, C. *Fuels and Lubricants*, p. 108. New York, Wiley, 1959.
†† Indicates motor method octane number.

Table 2-41 Properties of Diesel Fuel Oils[6]

	ASTM test	Types of diesel oils: C-B	T-T	R-R	S-M
Flash point, °F min	D93	125	125	140	162
Viscosity at 100 F max					
Kinematic, centistokes	D445	2.96	4.3	3.92	5.70
Saybolt Universal, sec	D88	35.9	40.1	38.9	44.6
Pour point, °F max	D97	15	20	20	20
Sulfur content, max wt %	D129	0.475	0.78	1.26	0.86
Ramsbottom carbon residue on 10% residuum, max wt %	D524	0.22	0.28	0.44	0.30
Cetane number, min	D613	41	38.0	34.2	34
Initial boiling point, °F max	D86	470	478	478	456

Note: C-B is used for city-bus and similar operations; T-T for trucks, tractors, and similar service; R-R for railroad engines; S-M for stationary and marine engines.

NATURAL GASOLINE

Natural gasoline is defined herein as the gasoline produced from natural gas, as distinguished from gasoline produced from crude oil by distillation and cracking. It is composed chiefly of hydrocarbons heavier than propane, although to meet certain specifications a great part of the butanes may be excluded. It has a high volatility and is used chiefly as a blending agent to raise the volatility of refinery gasoline. Its octane rating is relatively low compared to some blended gasolines (86 octane), varying from about 41 octane rating (**motor method**) for 10 lb gasoline to 72 for 34 lb gasoline. Natural gasoline was formerly called casinghead gas, as the early plants processed the gas flowing between the tubing and the casing in a natural gas well. Natural gasoline is extracted by two systems, compression and absorption, combined with good fractionation.

In 1961 there were 473 million gallons of finished natural gasoline[7] produced in the United States. Reserves[7] as of December 31, 1962 of condensate, natural gasoline, and liquefied petroleum gases were reported to be 292 billion gallons.

Table 2-42 General Specifications for Natural Gasoline

Reid vapor press. at 100 F	Percentage evaporated at 140 F	Percentage evaporated at 275 F	End point, °F	Corrosion	Doctor test	Color (Saybolt)
10–34 psia	25–85	90 min	375 max	Noncorrosive	Negative, sweet	25 min

The **Natural Gasoline Association** has established specifications for natural gasoline based upon the volatility and the **Reid vapor pressure** (at 100 F) of the product. The general specifications are given in Table 2-42.

Natural gasoline is further divided into 24 possible grades on the basis of vapor pressure and percentage evaporated at 140 F. The maximum vapor pressures of the various grades are 14, 18, 22, 26, 30, and 34 psia (**Reid**). Each grade is designated by its *maximum* vapor pressure and its *minimum* percentage evaporated at 140 F as shown in Table 2-43.

Grade 26–70, which is usually taken as a basis for sale purposes, designates a natural gasoline having a maximum vapor pressure (Reid) of 26 psia and a minimum of 70 per cent evaporated at 140 F.

Natural gasoline is stabilized in a distillation tower under pressure to remove the highly volatile constituents such as methane, ethane, propane, and sometimes butane. An illustration of the stabilization process, showing composition of raw gasoline, distillate, and stable gasoline, is given in Table 2-44.

For another example, if a pressure distillate of the composition given in Table 2-45 is to be stabilized so that it has a vapor pressure of 10 psia (Reid), about 20 per cent of the butane and all of the methane, ethane, and propane must be removed.

Table 2-43 Grades of Natural Gasoline

Reid vapor pressure, psia	Percentage evaporated at 140 F			
	From 25% to 40%	From 40% to 55%	From 55% to 70%	From 70% to 85%
34 / 30	Grade 34-25	Grade 34-40	Grade 34-55	Grade 34-70
30 / 26	Grade 30-25	Grade 30-40	Grade 30-55	Grade 30-70
26 / 22	Grade 26-25	Grade 26-40	Grade 26-55	Grade 26-70
22 / 18	Grade 22-25	Grade 22-40	Grade 22-55	Grade 22-70
18 / 14	Grade 18-25	Grade 18-40	Grade 18-55	Grade 18-70
14 / 10	Grade 14-25	Grade 14-40	Grade 14-55	Grade 14-70

Table 2-44 Natural Gasoline Before and After Stabilization[8]

Constituent	Raw gasoline, mole %	Composition of distillate, mole %	Composition of stable gasoline, mole %
Methane	26	42.9	0
Ethane	9	14.9	0
Propane	25	41.2	0.1
Butane	17	1.0	41.6
Pentane	11	...	27.9
Hexane and heavier	12	...	30.4
Octane number (motor calc.)	80	...	65.1

Table 2-45 Example of Stabilizing a Pressure Distillate to 10 psia[8]

Constituent	Pressure distillate, %	Stabilized 10 psia gasoline, %	Composition of vapor removed, %
Methane	0.20	0	2.9
Ethane	0.80	0	11.8
Propane	4.00	0	58.8
Butane	9.00	7.7	26.5
Pentane	10.25	11.0	Trace
Hexane & heavier	75.75	81.3	0

REFERENCES

1. Kobe, K. A. and McKetta, J. J. *Advances in Petroleum Chemistry and Refining*, vol. I. New York, Interscience, 1958.
2. "Refineries Are Geared for Tomorrow's Needs." *Petroleum Panorama*, Oil and Gas J., 57: F-28, Jan. 28, 1959.
3. Babcock and Wilcox Co. *Steam—Its Generation and Use*, 37th ed., p. 3–9. New York, 1955.
4. Hauck Manufacturing Co. *Hauck Industrial Combustion Data*, 3rd ed. Brooklyn, New York, 1953.
5. Pettyjohn, E. S. and Linden, H. R. *Selection of Oils for Carbureted Water Gas.* (I.G.T. Research Bul. 9) Chicago, 1952.
6. "Diesel Fuel Oils, 1962." *Mineral Industry Surveys.* Petroleum Products Survey No. 28, March 1963.
7. A.G.A. *Gas Facts.* New York, annual.
8. Stephens, M. M. and Spencer, O. F. *Petroleum Refining*, vol. III, State College, Pa., Penn State College, 1949.
9. "Gasoline Sales by Grade." *Nat. Pet. News* (*Factbook Issue*) 55: 201, Mid-May 1963.
10. "Standard Method of Test for Flash Point by Pensky-Martens Closed Tester." (ASTM D93–62) *1964 Book of ASTM Standards*, part 17, p. 36–8.
11. "Standard Method of Test for Flash Point by Tag Closed Tester." (ASTM D56–61) *1964 Book of ASTM Standards*, part 17, p. 1–7.
12. "Standard Method of Test for Flash Point of Volatile Flammable Materials by Tag Open-Cup Apparatus." (ASTM D1310–63) *1964 Book of ASTM Standards*, part 17, p. 463–72.

Chapter 4

Solid Fuels

by A. R. Powell

ANALYSIS AND TESTING OF COAL

Authoritative methods for sampling, analyzing, and testing coal and coke are found in **ASTM Standards.**[1]

Proximate analysis (ASTM D271-58) is essential in evaluating coal. It consists of the determination of the percentages of the following four components of the coal sample, the sum of which is 100 per cent. **Moisture** is the loss in weight on heating the coal sample for one hour at a temperature slightly above the boiling point of water. It is an inert constituent of coal since it has no heating value. **Volatile matter** is determined by heating the coal sample out of contact with air for seven minutes at 1742 F. The loss in weight, *minus* the moisture, represents the volatile matter. The percentage of volatile matter is an important, although not wholly determinative, indicator of several useful characteristics of a coal, such as its combustion characteristics and its value for carbonization purposes to make coal gas and coke. **Fixed carbon** is the residue, *minus* the ash content, remaining after the determination of volatile matter. As pointed out later, it is one of the principal factors used in the classification of coals. The **ash** content of a coal is the residue remaining after carefully burning off all combustible matter in the coal sample.

It is sometimes advantageous to express proximate analysis on (1) a *moisture-free* basis, (2) an *ash-free* basis, or (3) a *moisture- and ash-free* basis. Any of these may be easily calculated from the normal four-component analysis.

Besides proximate analysis, the only other tests commonly made on coal are for the heating value, the sulfur content, and the ash-softening temperature.

The **heating value** (Btu per lb), determined in a *bomb-type calorimeter* (ASTM D271-58), is useful in the evaluation of coal for combustion purposes and is also a major factor in the classification of coals. The heating value of most high-rank coals may be calculated within an accuracy of about two or three per cent (if the **ultimate analysis** is available) by the **Dulong formula:**

$$\text{Btu per lb} = 14{,}544\text{C} + 62{,}028\left(\text{H} - \frac{\text{O}}{8}\right) + 4{,}050\text{S}$$

where C, H, O, and S are the fractions by weight of these elements in the coal.

The **sulfur** content of the coal (ASTM D271-58) is an indicator of the sulfur content to be expected in the coke and gas on carbonization of the coal, in the gas from complet gasification of the coal, and in the flue gas product.

The **ash-softening temperature,** determined by heating a small sample of the ash in a laboratory furnace (ASTM D271-58), is an indicator of the tendency of the coal or coke to form **clinker** on combustion of the fuel.

Ultimate analysis is the determination of the percentage of carbon, hydrogen, oxygen, nitrogen, sulfur, and ash in the coal (ASTM D271-58). It is seldom made in connection with industrial uses of coal.

The most common physical test of coal is determination of the size, known as **screen analysis** (ASTM D410-38). A standard procedure for designating the size of coal from its screen analysis has been formulated (ASTM D431-44). A size or **fineness test** for powdered coal is also available (ASTM D197-30). Another size method, known as sieve analysis, has been designed specifically for coarsely crushed bituminous coal, such as is charged into coke ovens (ASTM D311-30). Still another standard size method has been developed for anthracite (ASTM D310-34).

The **bulk density** is the weight per cubic foot of broken coal, including the void spaces between pieces of coal. Since this is of special importance in coal carbonization, a method to determine the bulk density of crushed bituminous coal as charged into coke ovens has been developed (ASTM D291-58T). Bulk density is also of interest in connection with storage and shipping of coal.

The **size stability,** or the ability of coal to withstand breakage during handling or in transit, is measured by the drop-shatter test for coal (ASTM D440-49). The **friability,** or tendency of coal to break into small pieces on repeated handling, is measured by the tumbler test for coal (ASTM D441-45).

The **grindability,** or relative ease of pulverizing coals, is measured by the Hardgrove-machine method (ASTM D409-51). It is useful in determining the capacity of a pulverizer for any given coal. The **dustiness index** (ASTM D547-41) is useful for determining the relative amounts of dust given off in the air when handling coal or coke.

When coal is evaluated for carbonization purposes, not only are some of the general analyses and tests outlined above applied to the coal, but certain special tests may be made. The **free-swelling index** of coal (ASTM D720-57) is sometimes used as an indicator of the coking characteristics when the coal is burned as a fuel, and it is also used in some countries to give an approximate indication of the coking property of the coal related to carbonization. It is determined by heating a small sample in the absence of air to produce a coke *button,* the profile of which is then compared to standard profiles.

The **agglutinating value** of coal is determined by a proposed ASTM method. It gives information regarding coking properties and is an approximate measure of the substance in coal that fuses and becomes plastic on heating. The test is made by carbonizing 20 grams of a mixture of 15 parts of specially prepared silicon carbide to one part of powdered coal. The silicon carbide serves simply as an inert to dilute the coal. The *button* is then crushed in a compression machine and the indicated weight necessary to secure this crushing is called the agglutinating value.

The **plastic properties** of coal may be determined by either one of two proposed ASTM methods.[1] The plastic behavior, as indicated by heating the coal slowly in a specially designed laboratory furnace thru the temperature range where the coal fuses and becomes plastic, is useful in predicting the coking behavior of the coal and in determining the extent of weathering of the coal when results for the fresh coal are known.

The **carbonization pressure** of coal is determined by either one of two proposed **ASTM** methods.[1] This test is intended to indicate the pressure developed against coke-oven walls by the coal during carbonization, so as to evaluate the safety of the coal or blend of coals for this use. The test method most commonly used is that of Russell,[4] which carbonizes about 400 lb of coal in an oven provided with a movable wall and apparatus to measure the pressure exerted against this wall.

Various small-scale **coal-carbonization tests** have been used for evaluating the suitability of coals for making coke of acceptable quality or for indicating the yields of coke, gas, and coal chemicals. In these tests attempts have been made to imitate on a small scale the conditions, especially the rate of heating, that prevail in commercial coke ovens. Probably the most widely known test of this character is that used by the U. S. Bureau of Mines.[2] This test indicates both the yields of products and also the approximate relative coke quality. Another type of carbonization test apparatus that also indicates yields of products and coke quality is that of the Illinois Geological Survey.[3] Probably the most reliable small-scale test for determination of the quality of coke that will be obtained from coal or blend of coals is the **box coking test.**[4] The coal under test, usually about a 60 lb sample, is placed in a steel box and buried in the regular coal charge in the coke oven, so that the test coal is carbonized under approximately the same conditions as it would be if it comprised a full oven charge. All of these small-scale tests for indicating the quality of coke obtainable from any given coal are useful for preliminary survey purposes, but it is always necessary to carbonize full oven charges to determine definitely the actual quality of coke to be expected in plant operation.

In addition to the tests just mentioned, there is a laboratory scale method for determining the approximate yields of coke, gas, tar, ammonia, light oil, etc., to be expected from the usual high-temperature carbonization of a coal. This is the so-called **tube test** of the Chemists Committee of U. S. Steel Corp.[5] This test, which is very useful for yield determination, gives no indication of coke quality to be expected in commercial operation.

CLASSIFICATION OF COAL

Coal ordinarily is classified according to **rank** or according to **grade. Rank** signifies the extent to which the coal has changed since its origin, and is designated by such terms as anthracite, various groups of bituminous coal, lignite, etc. **Grade** refers to the **quality** of the coal as determined by ash and sulfur content, ash-softening temperature, heating value, and size. Details of American coal classification methods will be found in ASTM Standards on Coal and Coke (ASTM D388-38 and D389-37).

Classification by Rank

Classification according to rank is the most commonly used and the most important basis for classifying coals. Ordinarily, the only data required to determine the rank of a coal are the proximate analysis and the heating value. The specific analytical figures used in making this classification are (1) **fixed carbon,** on a moisture-free and mineral-matter-free basis and (2) **heating value,** on a moist, mineral-matter-free basis.

Mineral-matter-free basis is not quite the same as ash-free basis, due to certain changes in weight of the mineral matter of the original coal that take place when the coal sample is burned to ash. For this reason certain special correction formulas must be applied to the figures for fixed carbon and heating value, normally expressed on a moisture- and ash-containing basis, in order to convert them to the proper basis

Table 2-46 Classification of Coals by Rank

Rank	Fixed carbon, per cent (dry, mineral-matter-free)	Heating value, Btu/lb (moist, mineral-matter-free)
Meta-anthracite	98 or more	*
Anthracite	92 to 97.9	*
Semianthracite†	86 to 91.9	*
Low-volatile bituminous coal	78 to 85.9	*
Medium-volatile bituminous coal	69 to 77.9	*
High-volatile A bituminous coal	Less than 69	14,000 or more
High-volatile B bituminous coal	"	13,000 to 13,999
High-volatile C bituminous coal‡	"	11,000 to 12,999
Subbituminous A coal‡	"	11,000 to 12,999
Subbituminous B coal	"	9,500 to 10,999
Subbituminous C coal	"	8,300 to 9,499
Lignite	"	Less than 8,300

* No ASTM limits.

† Must be nonagglomerating. If agglomerating, classify as low-volatile bituminous coal. See note ‡ for meaning of "agglomerating."

‡ High-volatile C bituminous coal must be either agglomerating or nonweathering, whereas subbituminous A coal is both weathering and nonagglomerating. **"Agglomerating"** is the tendency of a coal in the volatile matter determination to produce a button that is either definitely coherent or that shows swelling or cell structure (ASTM D388-38). **"Weathering"** is the tendency of a coal to disintegrate or slack when alternately dried and wetted in the presence of air (ASTM D388-38).

for use in the coal classification table. The following two empirical formulas are used for this purpose:[6]

Dry, mineral-matter-free fixed carbon, per cent =

$$\frac{FC - 0.15S}{100 - (M + 1.08A + 0.55S)} \times 100$$

Moist, mineral-matter-free heating value, Btu per lb =

$$\frac{h - 50S}{100 - (1.08A + 0.55S)} \times 100$$

where: FC = per cent fixed carbon
M = per cent moisture
A = per cent ash
S = per cent sulfur
h = heating value (moist, ash-containing), Btu per lb

The terms "moist" and "per cent moisture" refer to the natural bed moisture, but do not include any visible water on the surface of the coal.

Table 2-46 summarizes the classification according to rank of the coal.

The great majority of commercial coals in the United States fit into this scheme of classification. There are a few exceptional coal types that must be classified outside this scheme and that should be referred to by their type names rather than rank. In general, these less common types of coal lack the banded structure of ordinary coal. Thus, *cannel coal*[7] has a compact, uniform structure, greasy luster, dark gray or black color, and a shell-like fracture. *Boghead coal*[7] is similar to cannel coal in appearance.

Table 2-47 Symbols for Grading Coal According to Ash, Softening Temperature of Ash, and Sulfur

(Based on coal as sampled)

Ash		Softening temp of ash		Sulfur	
Symbol	Per cent	Symbol	Degrees F	Symbol	Per cent
A4	0.0 to 4.0	F28	2800 and higher	S0.7	0.0 to 0.7
A6	4.1 to 6.0	F26	2600 to 2790	S1.0	0.8 to 1.0
A8	6.1 to 8.0	F24	2400 to 2590	S1.3	1.1 to 1.3
A10	8.1 to 10.0	F22	2200 to 2390	S1.6	1.4 to 1.6
A12	10.1 to 12.0	F20	2000 to 2190	S2.0	1.7 to 2.0
A14	12.1 to 14.0	F20—	Less than 2000	S3.0	2.1 to 3.0
A16	14.1 to 16.0			S5.0	3.1 to 5.0
A18	16.1 to 18.0			S5.0+	5.1 and higher
A20	18.1 to 20.0				
A20+	20.1 and higher				

Classification by Grade

ASTM has set up standard specifications for classification of coal according to grade by the use of certain symbols (ASTM D389-37). The symbols for grading according to ash, softening temperature of ash, and sulfur are shown in Table 2-47.

Table 2-48 Typical Analyses of Coals of Various Ranks*

("as received" basis)

Rank	State and county	Proximate analysis, per cent: Moisture	Volatile matter	Fixed carbon	Ash	Ultimate analysis, per cent: S	H_2	C	N_2	O_2	Heating value, Btu/lb	Ash-softening temp, °F
Anthracite	Pa., Schuylkill	4.4	3.4	83.1	9.1	0.7	2.4	81.7	0.9	5.2	12,810	2790
Semianthracite	Va., Montgomery	2.2	12.4	67.4	18.0	0.5	3.6	72.4	0.8	4.7	12,270	2800
Low-volatile bituminous coal	Ark., Sebastian	2.5	16.8	72.8	7.9	0.9	4.4	80.2	1.6	5.0	13,820	2070
	Pa., Somerset	2.6	17.0	73.7	6.7	1.2	4.5	81.7	1.3	4.6	14,240	2740
	W. Va., Raleigh	3.1	17.4	74.8	4.7	0.7	4.7	83.5	1.5	4.9	14,480	2830
Medium-volatile bituminous coal	Ala., Jefferson	2.4	25.9	66.8	4.9	1.5	5.2	81.7	1.5	5.2	14,490	2770
	Pa., Clearfield	2.4	24.3	65.6	7.7	2.1	4.9	79.1	1.2	5.0	14,020	2520
	W. Va., McDowell	1.5	23.6	69.5	5.4	0.5	4.8	83.1	1.2	5.0	14,600	2360
High-volatile A bituminous coal	Ala., Walker	1.4	35.2	56.3	7.1	1.3	5.1	76.3	1.8	8.4	13,800	2440
	Ky., Pike	2.8	33.8	58.8	4.6	0.5	5.3	78.9	1.4	9.3	14,050	2830
	Pa., Fayette	1.8	33.0	58.2	7.0	1.2	5.1	77.7	1.5	7.5	13,920	2660
	Pa., Washington	3.9	37.7	52.4	6.0	1.6	5.9	75.0	1.4	10.1	13,460	2380
	W. Va., Logan	2.8	35.8	56.3	5.1	0.7	5.5	78.8	1.6	8.3	13,980	2880
	W. Va., Marion	1.9	39.2	52.0	6.9	2.4	5.4	76.9	1.5	6.9	13,890	2130
High-volatile B bituminous coal	Ky., Hopkins	7.6	37.8	45.6	9.0	3.5	5.7	67.4	1.3	13.1	12,140	2160
	Utah, Carbon	3.7	43.1	47.8	5.4	0.4	5.7	72.0	1.3	15.2	12,700	2180
High-volatile C bituminous coal	Colo., Routt	9.7	36.4	48.5	5.4	0.5	5.7	66.9	1.3	20.2	11,790	2350
	Ill., Franklin	10.2	33.8	48.2	7.8	1.3	5.4	66.9	1.5	17.1	11,880	2080
Subbituminous coal	Colo., Weld	23.7	28.9	43.1	4.3	0.4	6.2	55.0	1.3	32.8	9,580	2060
	Wyo., Sheridan	23.8	32.4	40.5	3.3	0.4	6.5	55.3	0.9	33.6	9,380	2280
Lignite	N. D., Mercer	36.3	26.2	31.7	5.8	0.7	6.8	42.2	0.6	43.9	7,140	2450
Cannel coal	Ky., Floyd	1.8	44.5	46.8	6.9	0.6	...	...	...	...	14,150	2760

* These analyses are based on data in thousands of published coal analyses in U. S. Bur. of Mines, *Analyses of Coals* (Tech. papers 345, Utah; 347, Ala.; 416, Ark.; 484, Wyo.; 491, Wash.; 574, Colo.; 590, Penna. bituminous; 626, W. Va.; 641, Ill.; 652, Ky.; 656, Va.; 659, Penna. anthracite; 671, Tenn.; 700, Mich., No. Dak., So. Dak., Tex.; and others) Washington, 1925–48. Reference should also be made to Fieldner, A. C. and others, *Typical Analyses of Coals of the United States* (Bur. of Mines Bull. 446) Washington, 1942.

The size designation of coal has been defined in detail (ASTM D431-44) and essentially is the upper and lower limits of screen size between which more than 80 per cent of the coal occurs, in conjunction with other defined percentages over and below these limits. As an example, the size designation or symbol "4 × 2 in." means that more than 80 per cent of the coal passes thru the 4-in. screen and is retained on the 2-in. screen, and that somewhat less than five per cent is retained on the 4-in. screen and somewhat less than 15 per cent passes thru the 2-in. screen.

The heating value symbol used in grade classification is the heating value of the coal in hundreds of Btu to the nearest hundred. For example, heating values of 13,150 to 13,249 Btu per lb would be expressed by the symbol "132."

The classification of coal by grade may be expressed by a combination of the five types of symbols indicated above. As an example,

4 × 2 in., 132—A8—F24—S1.6

indicates a coal of 4 × 2-in. size designation, having a heating value of approximately 13,200 Btu per lb ash content of 6.1 to 8.0 per cent, ash-softening temperature of 2400 to 2590 F, and sulfur content of 1.4 to 1.6 per cent.

If desired, the above use of combinations of symbols may be extended to include the rank of a coal, as well as its grade. For example,

(62–146), 4 × 2 in., 132—A8—F24—S1.6

indicates a high-volatile A bituminous coal analyzing approximately 62 per cent dry, mineral-matter-free fixed carbon, having a moist, mineral-matter-free heating value of approximately 14,600 Btu per lb, and having the same grade classification as illustrated above.

TYPICAL ANALYSES AND MISCELLANEOUS DATA ON COAL

Some typical analyses of coals of different ranks and from various parts of the United States are shown in Table 2-48. Standard specifications for *Pennsylvania* anthracite are summarized in Table 2-49.

Some data on **bulk density** of various coals are shown in Table 2-50. The specific gravity of the coal affects the bulk density to some extent but a major factor is the range in size of the pieces, that is, coal that is not quite uniform in size has a higher bulk density than the same coal in a uniform size. Other factors, such as moisture and degree of settling, also affect this figure. Bulk density as related to coal carbonization is discussed later.

The analyses of some typical samples of coal ash representing ash-softening temperatures of various magnitudes are given in Table 2-51. Barrett[7] has discussed in considerable detail the relationship between ash-softening temperatures, chemical composition of the ash, viscosity of melted ash, clinker formation, and other behavior of coal ash in combustion or in other high-temperature utilization.

Table 2-49 Standard Anthracite Specifications[8]

Size of anthracite	Test mesh round, in. Thru	Test mesh round, in. Retained on	Undersize, %	Ash content, max %	Slate, max %	Bone, max %
Broken	4⅜	3¼–3	15	11	or 1½	2
Egg	3¼–3	2 7/16	15	11	or 1½	2
Stove	2 7/16	1⅝	15	11	or 2	3
Nut	1⅝	13/16	15	11	or 3	4
Pea	13/16	9/16	15	12	or 4	5
Buckwheat	9/16	5/16	15	13		
Rice	5/16	3/16	17	13		

Table 2-50 Weights of Various Coals[9]

Rank	State	Size*	Weight, lb/cu ft
Anthracite	Pa.	Egg	56.0
"	"	Chestnut	53.0
"	"	No. 1 Buckwheat	50.5
L.V. Bit.†	W. Va.	5-10-85	55.5
H.V.A Bit.†	Pa.	90-5-5	49.5
"	"	70-20-10	51.0
"	"	15-15-70	52.0
"	"	0-10-90	51.0
"	"	Lump	46.5
H.V.B Bit.†	Ill.	Lump	49.0
"	"	Run of mine	54.5

* The series numbers indicate percentages of lump, nut, and slack, respectively.
† L.V.—low-volatile; H.V.—high-volatile.

SELECTION OF COALS FOR CARBONIZATION

The selection of coals for carbonization primarily depends on whether gas or coke is the major economic product and, if coke, for what purposes it is to be used.

Coking properties refers to the ability of the coal to produce coke which has satisfactory size, strength, and structure. This ability is the most important factor, even in those plants where the product of primary importance is coal gas.

The rank of a coal usually gives a rough indication of its coking properties; coking coals are found only in the various bituminous ranks, from low-volatile bituminous down to high-volatile C bituminous. All coals that are above or below this middle range of the classification scale are noncoking and therefore not adaptable to carbonization.

Table 2-51 Ash-Softening Temperatures and Ash Composition[10]

Coal	Softening temp, °F	SiO_2	Al_2O_3	Fe_2O_3	TiO_2	CaO	MgO	Na_2O	K_2O	SO_3
		Analyses of ash, per cent								
Subbit., Mont.	2060	30.7	19.6	18.9	1.1	11.3	3.7	1.9	0.5	12.2
H.V. B Bit., Ill.	2320	46.2	22.9	7.7	1.0	10.1	1.6	0.7	0.8	8.9
H.V. A Bit., Pa.	2500	49.7	26.8	11.4	1.2	4.2	0.8	1.6	1.3	2.5
L.V. Bit., W. Va.	2730	51.0	30.9	10.7	1.9	2.1	0.9	1.0	0.4	0.6
H.V. A Bit., Ky.	2900+	58.5	30.6	4.2	1.8	2.0	0.4	0.7	0.9	0.9

L.V.—low-volatile; H.V.—high-volatile.

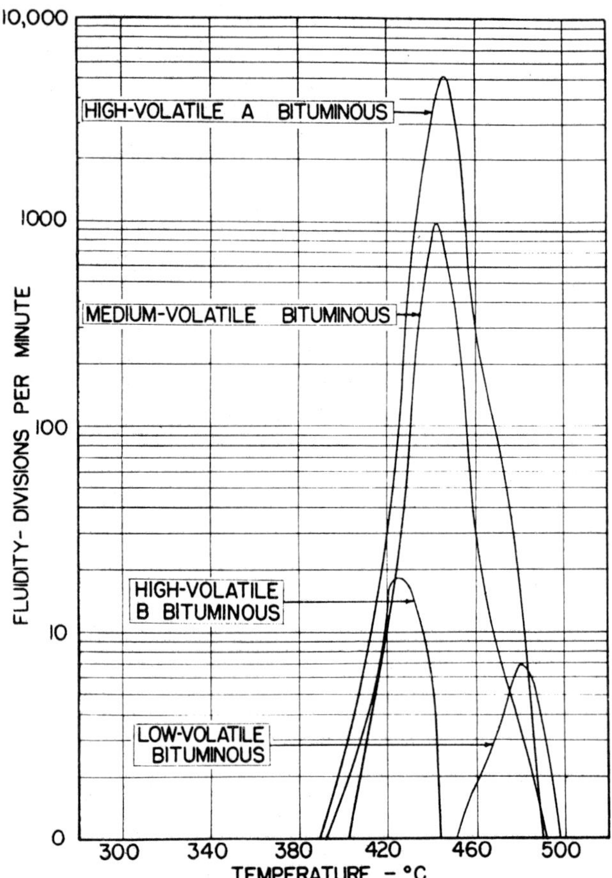

Fig. 2-6 Gieseler fluidity of four typical coals.[13]

Approximately 62 per cent of the coking coal used in the United States is **high-volatile A bituminous.** This rank of coal has good coking properties and gives a good yield of gas and coal chemicals. Coke produced from these coals is inclined to be somewhat small in size when carbonized alone, but despite this fact much high-volatile A coal has been used unmixed with other ranks of coal. It is more common practice to blend these coals with low-volatile bituminous.

Medium-volatile coals account for about 13 per cent of coal carbonized in the U. S., and may be coked straight or mixed with other coals to produce good quality coke. About 22 per cent of coking coal is low-volatile bituminous, which must always be carbonized in mixture with high-volatile coal, due to the dangerously high carbonization pressures encountered when low-volatile bituminous coal is coked alone. A very small amount, less than three per cent of the total, of high-volatile B coal is used in American coke plants, due to its relatively poor coking properties. No high-volatile C bituminous coal is carbonized commercially.

Some laboratory tests are useful in at least preliminary evaluations of coal for coking properties. The appearance of the coke *button* from the volatile-matter determination is useful and in some cases the free-swelling index or the agglutinating value have proved helpful. Published information on the plastic properties of coals is now extensive enough to give an approximate estimation of coking properties.[4] Figure 2-6 shows typical fluidity curves obtained by means of the Gieseler plastometer of the four ranks of coal that are commercially carbonized.

Chemical Composition of Coal

This refers primarily to the ash and sulfur contents and the ash-softening temperature, which are characteristics included in the classification of coal by grade. The American Society for Testing Materials established standard specifications for gas and coking coals in 1924 (ASTM D166-24) but the term "gas coal" referred to therein is based on specifications for coal used in coal-gas retorts rather than coke ovens. Such specifications cannot be regarded as typical practice today.[11]

Since all of the ash and about 60 per cent of the sulfur in the coal remain in the coke and since the softening temperature of the coke ash is the same as that of the ash in the coal from which it is made, the chemical characteristics desired in the coal are largely determined by the use to which the coke is to be put. Therefore, this will be outlined under the types of coke.

Yields of Carbonization Products

Coal may be selected on the basis of coke yield and/or gas yield; the yields of other products do not normally enter the picture. Tables 2-52 and 2-53 show data on some typical yields of coke and gas from a variety of coals.

Experience has shown that it is dangerous to use in coke ovens any coal or coal mixture that produces more than 2.0 psi, and in some cases this should be limited to a maximum of 1.5 psi.[4] The most widely used test apparatus is the **movable-wall** oven. Some typical pressure curves obtained by this apparatus on the four ranks of coking coals are shown in Fig. 2-7. Aside from proper selection of coals, certain operating procedures, modifications of oven design, etc., are available for decreasing carbonization pressures.[12]

Table 2-52 Analyses of Some Coals and Their Estimated Coke and Gas Yields*

Name of coal	Proximate analysis ("as received"), per cent: Moisture	Volatile matter	Fixed carbon	Ash	Btu per lb of dry coal	Estimated coke yield total, % of coal	Estimated gas yield, Btu of gas per lb of coal
W. Va.—Pocahontas†	2.6	16.4	76.4	4.6	15,020	81.0	2650
W. Va.—New River	3.5	24.8	69.0	2.7	15,220	75.0	2750
W. Va.—No. 2 Gas	2.5	34.9	56.4	6.2	14,500	69.0	3300
Pa.—Pittsburgh	3.0	35.2	55.4	6.4	14,230	68.5	3300
Ky.—Elkhorn	2.9	36.3	57.6	3.2	14,650	66.0	3450
Ill.—Franklin Co.	9.9	33.1	48.8	8.2	13,200	61.0	2900

* Prepared by Dr. H. C. Porter, Consulting Engineer, Philadelphia, Pa.
† Calculated yields when blended with other coals; never carbonized by itself.

Following is a brief summary of the qualities desired in various kinds of coke:[4]

Blast-furnace Coke. Uniformity in both physical and chemical characteristics is of prime importance and usually considerable attention is given to size and strength of the coke. Less than one per cent **sulfur** is preferable and ash should not be excessively high. Ash-softening temperature is of no importance in this case, but sometimes phosphorus content is of significance.

Foundry Coke. Strength, large size, and low ash, low sulfur contents are the qualities required. Requirement of large size is regarded so important that slower coking time than normal, employment of special coal mixtures, etc. are usually resorted to.

Domestic Coke. Low ash content, high ash-softening temperature, high density, uniform sizing and minimum of dustiness are important considerations.

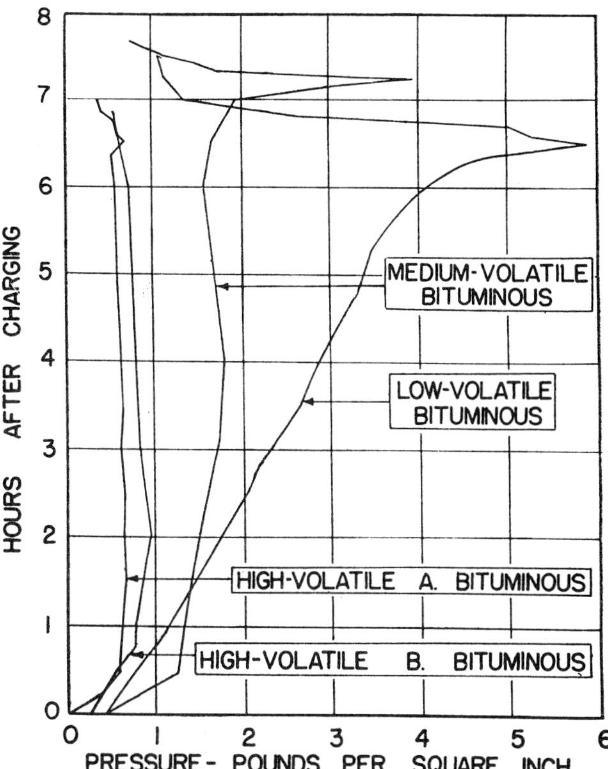

Fig. 2-7 Pressures developed by four typical coals during coking in movable-wall expansion oven.[13]

Table 2-53 Average Yields of Gas from Coals of Various Ranks[11]

Rank of coal	No. of coals tested	Cu ft per ton	Sp gr	Btu per cu ft	Btu per lb of coal
Low-volatile Bituminous	10	10,550	0.281	511	2700
Medium-volatile Bituminous	9	10,400	.308	556	2900
High-volatile A Bituminous	39	10,670	.386	603	3220
High-volatile B Bituminous	2	10,825	.418	580	3145
High-volatile C Bituminous	2	10,500	.421	539	2835
Subbituminous B	2	13,925	.506	421	2905
Lignite	2	17,250	0.561	334	2875

From BM-A.G.A. carbonization tests at 1652 F.

Water-gas Coke. See "Chemical and Physical Properties of Coke."

BLENDING OF COALS FOR CARBONIZATION

The blending of two or more coking coals or the mixing with the coal of other materials is done for the following reasons:[13] (1) improving physical quality and uniformity of coke; (2) controlling carbonization pressures; (3) controlling yields of products; and (4) utilizing inferior coking coals.

The blending of a minor percentage of low-volatile coal with a major percentage of high-volatile coal greatly improves the physical qualities of coke over those secured from high-volatile coal by itself (Fig. 2-8); therefore, this practice is very common in the United States.

Carbonization pressure usually limits the percentage of low-volatile coal that can be blended, and this maximum percentage will vary, depending on the nature of the two coals and other factors entering the carbonization process. Figure 2-9 shows some typical relationships between composition of the blend and pressures as measured by the movable-wall oven.

Addition of small amounts of fine coke breeze[14] or anthracite[15] to coking coals is rather common practice in the U. S., especially in manufacture of foundry coke, where the larger size of the resultant coke and increased resistance to shatter are more important than the resultant decreased resistance to abrasion (tumbler stability factor). Figure 2-10 shows the

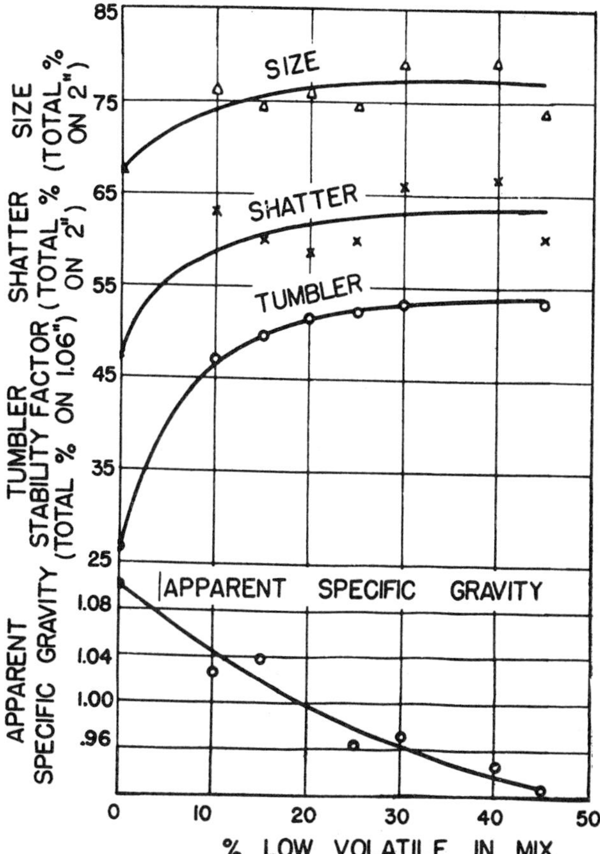

Fig. 2-8 Effect of low-volatile coal on physical properties of cokes.[24] (High-volatile Powellton seam plus low volatile Pocahontas No. 3 seam.)

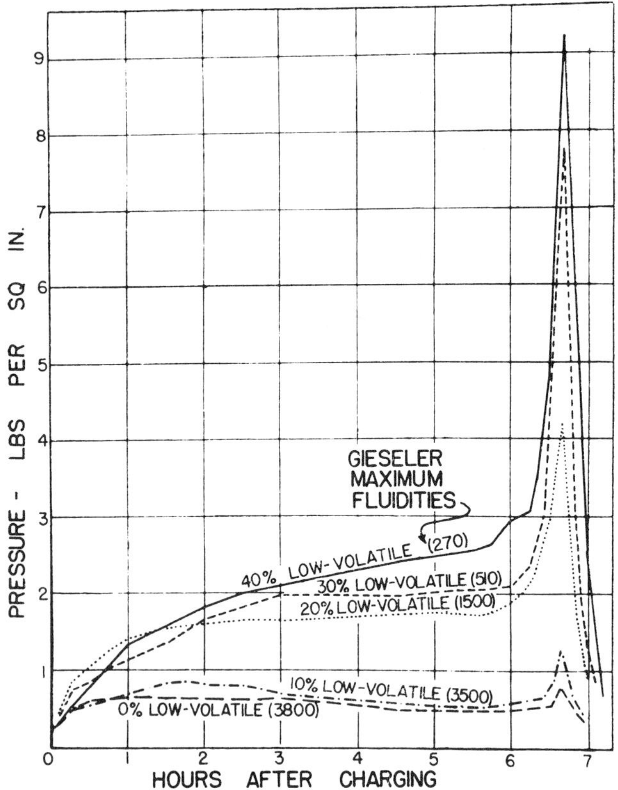

Fig. 2-9 Effect of addition of low-volatile coal to high-volatile coal on pressures developed in movable-wall ovens.[12]

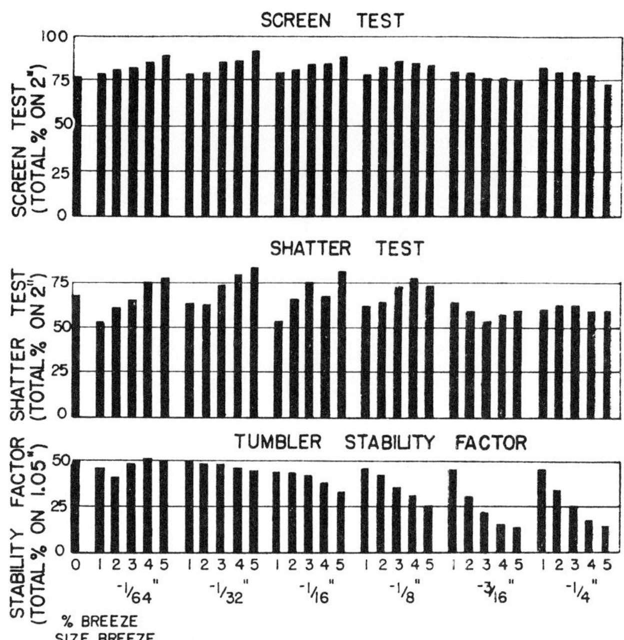

Fig. 2-10 Effect of coke breeze admixture to coal on size, shatter, and tumbler of resultant coke.[14] (Eighty per cent Powellton plus twenty per cent Pocahontas.)

effect on physical properties of the coke of breeze admixture in different sizes of breeze and in different percentages.

A few plants add small amounts of oil to the coal to increase gas yield or to increase bulk density of the coal charge.[16] To increase the gas yield, heavy fuel oil is added in quantities ranging from 2.5 to 8.5 gal per ton of coal (1.0 to 3.5 per cent by weight). The yields of gas and other products per gallon of oil charged in this manner are shown in Table 2-54.

Table 2-54 Distribution of Products from One Gallon of Heavy Oil Carbonized in Coal-Oil Blend

	Btu	Weight, lb
Gas—60 SCF at 1180 Btu per cu ft	70,800	3.2
Tar—0.3 gal	45,000	2.9
Coke—1.7 lb (20% by weight)	24,700	1.7
Unaccounted for	9,500	0.5
TOTAL	150,000	8.3

SELECTION OF COALS FOR DIRECT GASIFICATION

Direct gasification is the process of completely gasifying coal into water gas, synthesis gas, or producer gas. Selection of coals for this purpose is difficult because of the different coal characteristics required by the processes available. Despite these many methods, the background of commercial experience is meager in the U. S. Some of the newer processes for direct gasification of coal that use tonnage oxygen as one of the raw materials have been described.[17]

When **anthracite** is used for water-gas manufacture it is usually broken size or egg size (see Table 2-49). These large lumps should be as free of fines as possible, and the ash content should be low, with a moderately high ash-softening temperature. The analyses of some anthracites that have been used for manufacture of water gas are shown in Table 2-55.

Bituminous coal is used as water-gas fuel when coke is too costly. The chief troubles with bituminous coal in water-gas sets are (1) reduced capacity due to poor fire characteristics, (2) fine fuel carried over into the carburetor and superheater, and (3) smoke during the blow period. Attention to operating procedures[7] will assist in preventing these troubles, but selection of suitable coals is also of major importance. Laboratory testing of bituminous coals as water-gas-generator fuel has shown that chemical analysis is useful only for fixing limits of moisture, sulfur, and ash. A modified shatter test may predict physical degradation on handling; and plasticity and agglutinating tests might be useful in predicting the behavior of coals in the generator[19] but only after further investigations have been made. The behavior of the ash when tested for ash-softening temperature is also important from the standpoint of clinker formation (see "Selection of Coke for Gasification"). In the final analysis, however, actual plant tests have been the only successful means of evaluating bituminous coals for water-gas manufacture.

Table 2-55 shows the chemical and some other characteristics of a variety of solid fuels, including several bituminous coals, that have been successfully used for manufacture of water gas.

In selecting coals for producer-gas manufacture much the same characteristics are sought as for the manufacture of water gas. In one respect, however, there is a difference. In water-gas manufacture, where very high temperatures are obtained on the blow, the ash forms clinkers, and the only control available is to regulate the nature of the clinker so that it can be easily removed. In gas producers (except slagging producers) clinker formation must be prevented, and the

Table 2-55 Analyses of Fuels for Manufacture of Water Gas[18]

Kind of fuel	Per cent moisture as received	Analysis on dry basis, per cent: Volatile matter	Fixed carbon	Ash	Sulfur, per cent	Btu per lb	Ash fusion point, °F	Size of fuel, in.
Anthracite, Pittston broken coal	4.51	6.07	83.94	9.99	0.91	...	2600+	3 7/16 to 4 7/16
Anthracite, broken	3.30	5.23	81.74	13.03	0.91	13,042	...	...
Anthracite	2.77	5.44	84.19	10.37	0.86	12,830	...	...
Water-gas coke	3.56	1.93	89.76	8.31	0.60	...	2300 and above	Over 2 in. grizzly
Horizontal and inclined retort coke	5.88	2.07	90.05	7.88	0.63	13,252	...	...
By-product oven coke	3.13	1.99	89.17	8.84	0.63	13,081	...	...
Water-gas coke	1.67	2.21	87.32	10.47	1.11	13,004	2678	...
Spokane "gas house" coke	1.30	8.3	70.3	21.3	...	11,150	...	...
Denver "gas house" coke	...	2.88	79.58	17.54	0.62	11,899	...	...
Boone-Chilton coal	1.82	36.48	58.73	4.79	0.56	14,380	2825	...
Fairmont gas coal, average	1.06	34.67	58.16	7.17	1.00	...	...	3 to 6
Elkhorn gas coal	1.94	37.77	59.11	3.12	0.54	14,750	...	3 to 6
Perry County, Ill.	7.77	36.72	53.45	9.83	1.40	12,816	...	...
Franklin County, Ill.	7.95	36.08	53.71	10.21	1.31	...	...	6 × 3 lump

characteristics of the ash determine the amount of steam to be added to the blast in order to obtain this result. As with water-gas manufacture, the selection of coals for gas producers must rest on actual plant tests before final decisions can be made.

The various new or proposed processes for the direct gasification of coal can be classified into three general types:

1. **Fixed bed.** The fuel is in the form of large or small lumps and rests in the generator in a relatively static condition while the stream of gasifying agents, such as air, oxygen, and steam, flows thru the bed. The standard water-gas and producer-gas processes, as well as some newer or more uncommon processes are of this type.

2. **Fluidized bed.** The fuel is relatively fine and is lifted bodily by the stream of gasifying agents, causing it to "boil."

3. **Suspension gasification.** Finely pulverized fuel is gasified while being carried in the flowing stream of gasifying agents.

Factors entering into the selection of coal for each of these types of processes are summarized below:

a. **Fixed-bed processes** usually require relatively large lumps of coal in order to permit flow of the large volumes of air, steam, or oxygen required for high capacities. Therefore, size and uniformity of size are of importance. The caking or agglomerating properties of the coal are often of determining influence in selection of coals for most fixed-bed processes, since the lumps of strongly caking coals will fuse together and block the passages necessary for flow of the gasifying agent. The use of noncaking coal has been particularly necessary in the *Lurgi* high-pressure gasification process used commercially in Europe to some extent, as well as in the various European slagging-type gas producers.[17] Highly reactive fuel is desirable for the Lurgi process because of the relatively low temperature at which it operates. Since reactivity usually increases as rank of the coal becomes lower, lignite or other low-rank coal would be preferable for the *Lurgi* process, just as it proved to be in Europe. Since slagging producers operate at very high temperatures, reactivity of the fuel is not important. Ash-softening temperature should be reasonably high for the Lurgi process, although the fuel bed temperature is usually kept below the ash-softening range. For slagging gas generators, ash-softening temperature must be low enough to permit run-off of ash as a liquid; otherwise, a flux must be added.

b. **Fluidized-bed processes** may use a granular fuel (top size ¼ in.) as in the *Winkler* generator,[17] extensively operated in Germany as a commercial process, or finely pulverized fuel as proposed in various methods that have not yet been commercially operated. Fluidization demands **noncaking coal,** since caking causes agglomeration of the particles with consequent loss of fluidity. In order to obtain good gas quality and high capacity, a free-burning, disintegrating, and very reactive fuel, such as lignite or other low-rank coal, or a char made from such coals, is preferable. **Ash-softening temperature** must be higher than the maximum temperature of the fluidized bed to avoid stickiness of the particles which would interfere with good fluidization.

c. **Suspension gasification** may be either slagging or nonslagging. If slagging, ash-softening temperature of the fuel should be low enough to permit liquid run-off from the reacting chamber; if nonslagging, it should be as high as necessary to prevent slagging of the ash in the reactor. Suspension gasification has the advantage over the other two types of processes in that either caking or noncaking coal can be the fuel, since the particles do not come in contact with one another. Coal must *always* be finely pulverized, with a major percentage passing a 200-mesh sieve, and this makes the grindability of the coal a factor in selection.

SELECTION OF COALS FOR COMBUSTION

The selection of acceptable coals for use under boilers or for other heating purposes is usually made primarily on the basis of cost and availability. Most combustion equipment has the flexibility to use various kinds and qualities of coal. However, within reason, certain characteristics of coal are important in considering selection of steam coals. A fairly detailed check list is available to indicate the relative importance of the various factors in coal selection, taking into consideration the type of combustion equipment and the operating conditions.[20] Recently a rather comprehensive survey was made to evaluate the significance of various ASTM laboratory tests of coal and coke for combustion purposes.[21] These evaluations were made for each type of combustion equipment.

Of major importance in selection of **steam** coal is, of course, the heating value. In many types of equipment the rank of the coal influences its behavior on combustion. It is sometimes customary to speak of bituminous coals as **caking** while, in contrast, the ranks above and below the bituminous range, that is, anthracite, subbituminous coals, and lignite, are referred to as **free-burning.** The volatile matter content is also significant in the tendency of coal to produce smoke. Excessive ash is undesirable, but very low ash content (below four per cent) is often undesirable, too, because of increased grate maintenance costs.

Type of combustion equipment has much to do with coal selection. **Spreader stokers** use a very wide range of coals satisfactorily, including all ranks of coal, coals that have relatively high ash and sulfur contents, and coals with ash of low softening temperatures. In pulverized coal installations the grindability index is a major influence on the cost. Also, the volatile matter content influences the **speed of combustion.** As in all other cases of coal selection, the final and deciding criterion for steam coals is the *actual* operating behavior of these coals in the combustion equipment used.

STORAGE OF COAL

The choice of methods used in storage of coal at a plant is subject not only to the ease and economy of placing the coal in storage piles and of later withdrawing it, but also to protection against spontaneous combustion, weathering, and, in some cases, loss of coking properties. To secure this protection for bituminous and subbituminous coals (anthracite is usually not a problem) it is necessary to prevent access of air to the coal pile as completely as possible.[22] Probably the most positive method of doing this is to store under water, but in most cases this is impractical. The practical method of coal storage usually recommended is laying it down in layers two to three feet in thickness, in order to prevent *segregation* of coarse and fine coal, which usually occurs when coal is piled in a high conical-shaped heap. After each layer is completed, it should be *compacted* by means of a road roller, tractor, or even an ordinary truck, before the next layer is started on top of it. Sometimes the pile of coal is further protected from flow of air currents thru it by a covering of fine coal dust, over which a layer of nut-size coal is placed to prevent the coal dust from being blown away. For long-term storage, capping of the pile with heavy tar or oil preparations has been practiced successfully. Local conditions will often indicate other precautions that can be taken to prevent flow of air currents in the coal pile. Incipient spontaneous combustion in the storage pile can be detected by a systematic survey of temperatures at various locations in the interior at regular intervals.

CHEMICAL AND PHYSICAL PROPERTIES OF COKE

The chemical properties of various cokes that have been used for manufacture of water gas are included in Table 2-55. Some of the more important physical properties of various high-temperature cokes are shown in Table 2-56.

A comprehensive study of the physical and chemical properties of cokes from several typical coke plants is the result of the Coke Evaluation Project, sponsored by American Iron and Steel Institute and others.

Table 2-56 Physical Properties of High-Temperature Cokes

	Coal charged			
	100% Pitts-burgh seam high vol. (Pa.)	80% High vol. (W. Va.) 20% low vol. (W. Va.)	75% High vol. (Ky.) 25% low vol. (W. Va.)	90% High vol. (Colo.) 10% non-cok. (Colo.)
Screen test				
On 3-inch	29.9	42.1	...	40.8
On 2-inch	68.4	76.7	64.7	79.6
Tumbler test				
Stability (+1.06 in.)	32.0	50.1	40.9	28.8
Hardness (+0.265 in.)	68.8	67.6	61.0	63.4
Shatter test				
On 2-inch	49.3	66.9	61.2	64.5
On 1½-inch	71.1	84.1	...	83.7
Cell space				
Porosity, %	48.9	51.1	53.1	46.5
App. sp gr	0.96	0.94	...	...
Weight, lb/cu ft	27.4	...	...	...

Table 2-57 Weights and Heating Values of Cordwoods[23]

	Weight per cord,* lb		Available heat per cord,* million Btu		Heat Ratio: cord of wood to ton of coal†	
Wood	Green	Air dried	Green	Air dried	Green	Air dried
Beech	5000	3900	19.7	20.9	0.76	0.80
Birch, yellow	5100	4000	19.4	20.9	.75	.80
Hickory	5700	4600	23.1	24.8	.89	.95
Maple, sugar	5000	3900	20.4	21.8	.78	.84
Oak, red	5800	3900	19.6	21.7	.75	.83
Oak, white	5600	4300	22.4	23.9	.86	.92
Pine, yellow	...	...	21.1	22.0	0.81	0.85

* A cord of wood is a pile 4 × 4 × 8 ft = 128 cu ft.
† Net ton of coal with heating value of 26 million Btu.

High-temperature coke made in chemical-recovery (by-product) coke ovens is the type referred to heretofore. This is the most common kind of coke and the one usually employed for manufacture of water gas, but several other types have been used in the past to a limited extent or could be used under some conditions. **Beehive coke,** made in ovens from which the gas and chemicals are not recovered, is quite similar chemically to the usual by-product coke but differs somewhat in physical properties. **Low-temperature** coke is characterized by a high volatile-matter content (5 to 20 per cent) and is usually used as a **house-heating fuel. Petroleum coke,** the residue from petroleum distillation, has a high-volatile content (5 to 20 per cent), a low ash content (trace to 1.5 per cent), and is largely used for manufacture of **electrode carbon. Pitch coke,** made by the high-temperature carbonization of coal-tar pitch, is also used as a low-ash carbon for manufacture of electrodes.

SELECTION OF COKE FOR GASIFICATION

The extent and nature of **clinker formation** in water-gas generators is the major governing factor in selection of desirable coke. For this reason **low ash** content is of great importance as is also the **ash-softening temperature.** The latter should, in general, be a minimum of 2400 F. Some users

Table 2-58 Proximate Analysis and Heating Value of Miscellaneous Fuels

Kind of fuel	Moisture	Proximate analysis, per cent			Sulfur	Heating value, Btu/lb
		Volatile matter	Fixed carbon	Ash		
Hogged wood fuel: as received	50					
air-dried	7					
moisture-free	...	80	19	1	Trace	9,000
Wood bark: as received	60					
moisture-free	...	73	25	2	Trace	9,000
Peat: as received	90					
air-dried	7					
moisture-free	...	60	33	7	0.5	9,000
Bagasse: as received	50					
moisture-free	...	...	...	2	...	8,500
Charcoal	1	24	72	3	Trace	11,000
Sawdust (pine): air-dried	6					
moisture-free	...	80	19.5	0.5	Trace	9,000

place no maximum on the ash-softening temperature but some other experienced water-gas operators state that very high softening temperatures cause troublesome wall clinker. A short range of temperature between initial softening and fully liquid condition is also desirable, since, if the range is too great, more back steam is required to bring the clinker to the grate, with consequent excessive loss of heat from the carburetor and superheater.

Low sulfur content is desirable, since much of the sulfur in the coke passes into the water gas as hydrogen sulfide.

Preferably, water-gas coke should be uniformly sized, consisting of lumps ranging from two to four inches. Sometimes **run-of-oven** coke, with only the fines (breeze) removed, is used. It is quite essential to remove these fines before the coke is charged into the water-gas generators.[4]

MISCELLANEOUS SOLID FUELS

Wood. Data on the weights and heating values of the more common varieties of **cordwood** are shown in Table 2-57. **Hogged fuel** is a mixture of small pieces of waste wood, bark, and trimmings, with possibly some chips and sawdust. The proximate analysis and heating value of this hogged fuel, as well as the same data for **wood bark** and for **sawdust** are shown in Table 2-58. No wood or wood waste is used directly for gas manufacture in the U. S.

Peat, which is the transition between vegetable matter and lignite, finds little use as fuel in the U. S. As removed from the bog, peat contains about 90 per cent water and must be air-dried before use. See Table 2-58.

Bagasse is the refuse remaining after sugar has been extracted from cane. It contains about 50 per cent moisture and is usually burned as a fuel without air-drying (see Table 2-58).

Charcoal has some limited use as a fuel; it is not used in the gas industry. Table 2-58 includes proximate analysis and heating value of a typical wood charcoal.

Fuel briquettes are manufactured in the U. S. from various solid fuels and binding agents and are of many different shapes and sizes. They are mostly used for house-heating.

REFERENCES

1. A.S.T.M. "Coal and Coke." *Book of ASTM Standards,* 8: 991–111. Philadelphia, 1958.
2. Fieldner, A. C. and Davis, J. D. *Gas-, Coke-, and Byproduct-Making Properties of American Coals and Their Determination.* (Bur. of Mines, Mono. 5) New York, A. G. A., 1934.
3. Reed, F. H. and others. *Ind. Eng. Chem.*, 37: 560–6 1945.
4. Russell, C. C. "Selection of Coals for Manufacture of Coke." *A. G. A. Proc.* 1947: 733–56.
5. U. S. Steel Corp. *Sampling and Analysis of Coal, Coke, and By Products,* 3rd ed. Pittsburgh, 1929.
6. Parr, S. W. (Univ. of Ill. Eng. Exp. Station. Bul. 180) Urbana, Ill., 1928.
7. Lowry, H. H., chm. *Chemistry of Coal Utilization.* New York, Wiley, 1945. 2 vols.
8. Penna. General Assembly. "Anthracite Standards Law, amended Sept. 26, 1951." *Penna. Laws.* Harrisburg, 1951.
9. Flagg, S. B. *Weights of Various Coals.* (Bur. of Mines, Tech. Paper 184). Washington, D. C., 1918.
10. Selvig, W. A. and Fieldner, A. C. *Fusability of Ash from Coals in the United States.* (Bur. of Mines, Bul. 209) Washington, D. C., G. P. O., 1922.
11. Rose, H. J. "Availability of Coals for Gas Manufacture." *A. G. A. Proc.* 1948: 472–86.
12. Russell, C. C. and others. "Blast Furnace, Coke Oven and Raw Materials." *A. I. M. E. Proc.* 8: 32–50, 1949.
13. Powell, A. R. and Russell, C. C. "Coal Blending in the American Coking Industry." *Brit. Coke Res. Assn., Proc. Fourth Conf.*, 1950.
14. Pfluke, F. J. and others. "Coke Strength & Structure As Affected by Coke Breeze Admixture to Coal." *A. G. A. Proc.* 1936: 771–92; 1937: 619–48.
15. Clendenin, J. D. and Kohlberg, J. (Penna. State College. Min. Ind. Expt. Station. Tech. Paper 136) 1948.
16. Ramsburg, C. J. and McGurl, G. V. "Carbonization of Coal-Oil Mixtures." *A. G. A. Proc.* 1940: 666–81.
17. Newman, L. L. *Ind. Eng. Chem.* 40: 568–82, 1948.
18. Morgan, J. J. "Production of Manufactured Gas." *Textbook of American Gas Practice,* 2nd ed., vol. 1. Maplewood, N. J., 1931.
19. Pettyjohn, E. S. "Report of Subcommittee on Evaluation of Bituminous Coals for Water-Gas Use." *A. G. A. Proc.* 1930: 1535–63.
20. A.S.T.M. Tech. Com. on Classification of Coal. *Factors Recommended for Consideration in the Selection of Coal,* no. 21, 2nd ed. New York Natl. Assn. of Purchasing Agents, 1936.
21. Barkley, J. F. *Significance of Laboratory Tests of Coal and Coke for Combustion.* (Bur. of Mines, I.C. 7619) Washington, D. C., 1951.
22. ——. *Storage of Coal.* (Bur. of Mines, I.C. 7235) Washington, D. C., 1943.
23. U. S. Forest Service. *The Use of Wood for Fuel.* (Investigations Bul. 753) Madison, Wis., 1940.
24. Russell, C. C. and Perch, M. "Blast Furnace and Raw Materials." *A. I. M. E. Proc.* 2: 51–69, 1942.

Chapter 5
Combustion

by F. E. Vandaveer and C. George Segeler

CHEMISTRY OF COMBUSTION

Definitions

Combustion as defined herein is the chemical reaction of oxygen with combustible materials, accompanied by evolution of light and rapid production of heat. Gaseous combustible materials with which the gas engineer may be principally concerned consist of hydrocarbons, carbon monoxide, and hydrogen. The combustion of carbon in the form of coal, coke, and charcoal is of interest in manufactured gas and substitute natural gas making processes, steam generation, and general heating. Sulfur in the form of hydrogen sulfide or organic sulfides is a combustible material and, since it may be present in very small amounts in fuel gases, it will be dealt with briefly.

The term **oxidation** is defined herein as the reaction of oxygen and combustible substance which takes place slowly without evolution of light and without rapid generation of heat. Although the total amount of heat produced may be considerable, the slow rate of dissipation precludes the indications of combustion.

Spontaneous combustion occurs when materials which oxidize slowly are so placed that the heat developed is not dissipated; the temperature of the material rises slowly until it reaches the ignition point and the material bursts into flame.

Perfect combustion is the complete burning of a fuel with exactly the theoretical amount of air or oxygen required for the reaction (Table 2-59). The resultant products are essentially carbon dioxide and/or water vapor, and nitrogen (excluding trace compounds).

Complete combustion is the burning of all combustible constituents in a fuel to carbon dioxide or water vapor or both whether or not excess air is present.

Incomplete combustion or **partial combustion** occurs when the burning of the fuel results in such intermediate combustion products as carbon monoxide, hydrogen, and aldehydes in addition to carbon dioxide and water vapor (Tables 2-60 and 2-61). It may be induced by limiting the air or oxygen supply to less than the amount needed for combustion, by chilling the flames as by impingement on a cold surface, or by blowing flames from burner ports.

Normal combustion is the burning of a gas with no sharp or other serious pressure discontinuity at the burning surface.

Detonative combustion[1] is the propagation of a very sharp pressure shock wave thru a gaseous medium; the wave velocity is maintained by utilizing the heat of combustion released in the shock.

Excess air is the quantity of air *above* that required for perfect combustion.

Theories of Combustion

Combustion involves unusually complicated reactions consisting of many steps. Generally speaking, progress of a flame is determined to a greater extent by diffusion, by heat transfer near the flame, and by distortion, disruption, and blending of the flame front by turbulence.

The following theories of combustion are for gaseous hydrocarbons, solid carbon, and carbon monoxide and hydrogen mixtures.

Gaseous Hydrocarbons: Two Accepted Theories. The *free radical reaction mechanism theory*[2,3] for the explanation of the combustion of hydrocarbons is the leading theory. In the slow oxidation of methane, reaction chains are assumed to be initiated at rates proportional to the formaldehyde concentration. The latter undergoes a gradual increase until a steady state value is reached. After that, formaldehyde is both formed and destroyed by the chain reaction at the same rate. The skeletonized reaction mechanisms are:

I $HCHO + O_2 \rightarrow$ free radicals

II $OH + CH_4 \rightarrow H_2O + CH_3$

Hydroxyl radical + methane = water + methyl radical

III $CH_3 + O_2 = HCHO + OH$

IV $OH + HCHO = H_2O + CHO$

V OH diffuses to wall where it may be adsorbed and destroyed

CHO is destroyed

The point of attack, in the oxidation of hydrocarbons, is thru their hydrogen atoms. The initiation is assumed to involve all types of hydrogen: primary, secondary, and tertiary, in the order of their numbers and relative reactivities.

In the *hydroxylation theory* it is assumed that when hydrocarbons burn, there is first an addition or association of oxygen with the hydrocarbon molecule, producing unstable hydroxylated compounds which, in turn, form aldehydes. The aldehydes are oxidized to formaldehyde, to CO and hydrogen or to CO and water, and, in turn, to CO_2 and water. These reactions may be represented as follows:

$$CH_4 + O_2 \rightarrow CH_3OH \rightarrow H_2O + HCHO \rightarrow CO + H_2O \rightarrow CO_2 + H_2O$$

This process is so rapid under favorable conditions that in mixtures of methane and hydrogen or methane and CO, the methane burns faster than either hydrogen or CO.

Solid Carbon. Hot carbon unites with air, forming carbon dioxide, some carbon monoxide, and possibly an intermediate oxide of carbon. The latter oxide breaks down to form CO_2 and CO. At low temperatures considerable CO may be formed, but at high temperatures mostly CO_2 is formed.

Mixtures of CO and Hydrogen. These mixtures burn as if the reactions were:

$$2H_2 + O_2 = 2H_2O$$

$$2CO + O_2 = 2CO_2$$

Heat Sources

Combustion of solid, liquid, or gaseous fuels is the major source of heat in this country. Solid fuels consist primarily of coal, coke, and wood; some liquid fuels are the fuel oils, kerosene, and gasoline; and gaseous fuels may be designated by natural gas, liquefied petroleum gases, and manufactured gases from coal or oil.

Electrical energy can be transformed into heat in a number of ways. Resistance to electrical flow or the so-called **I^2R drop** forms the basis for much of the electrical heating. Alloys of nickel and chromium, or of nickel, chromium, and iron, with *resistivities* as much as *60 times* that of copper are typical resistor materials used for heating. One **kilowatt hour** is equivalent to 3412 Btu. The electric arc produced between carbon electrodes produces heat. Induction heating by setting up induced electric current in metallic bodies that have no direct physical contact with the current carrier is accomplished by making use of the property of metals known as *hysteresis*.

Heat from Exothermic Chemical Reactions. Heat can be generated without combustion. Reactions in which chemical energy is converted into heat are termed **exothermic.** (Those requiring outside heat to keep them going are termed **endothermic.**) A typical exothermic reaction is caused by mixing water with concentrated sulfuric acid. Burning fluorine[4] in hydrogen produces very high temperatures, 7000 F at atmospheric pressure, and 8000 F at five atmospheres. The temperature of the sun is reported to be about 9000 F.

The following two exothermic reactions are potentially useful in the gas industry. To obtain satisfactory reaction rates, proper temperatures, pressures, and catalysts are required. Although heat release is high, the temperatures are relatively low (probably about 800 F), making practical use uncertain.

$$CO + 3H_2 = CH_4 + H_2O^* \qquad \textbf{(a)}$$

$$CO_2 + 4H_2 = CH_4 + 2H_2O \qquad \textbf{(b)}$$

The $\Delta H_{60F,\,1\,atm}$ for Reactions **a** and **b** is $-107{,}600$ and $-108{,}800$ Btu, respectively.

* Liquid.

The Water Gas Reaction. The name is derived from the use of this reaction in the manufacture of gas from coke and steam. However, this reaction is often encountered in the gas industry under circumstances which bear little resemblance to the origin of the term. A few examples are in the decarburization of steel during heat treatment, production and stabilization of special furnace atmospheres, and dissociation of flue gases at flame temperatures.

The reaction in the form

$$CO_2 + H_2 = CO + H_2O$$

is slightly exothermic (-815 Btu per lb-mole) if the water vapor is condensed to water, but endothermic if the water vapor is not condensed ($+17{,}723$ Btu per lb-mole).

Mechanical Sources of Heat. Heat is generated by friction of moving parts, by stressing of metals, and by compression of gases. The mechanical equivalent of one Btu of heat energy is 778 ft-lb.

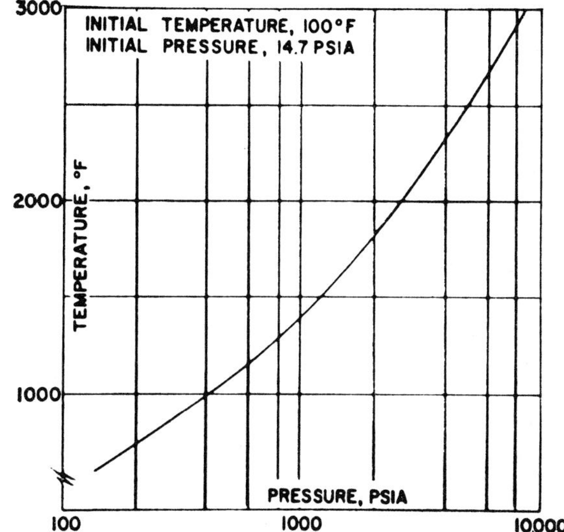

Fig. 2-11a Adiabatic compression temperature vs. air pressure.

Heat of compression of gases[5,6] can be calculated from the equation:

$$T_2/T_1 = (P_2/P_1)^{(n-1)/n} \qquad (1)$$

where:

T_2 and T_1 = absolute outlet and inlet temperatures, respectively, °R

P_2 and P_1 = outlet and inlet pressures, respectively, psia

n = a constant related to the specific heat of the gas

In an **adiabatic compression,** n equals the ratio of specific heat at constant pressure to specific heat at constant volume. However, in practical compressors, some heat is always lost thru the walls of the compressor; this results in a slightly lower value of n. A good *average value* for n for rough determinations is 1.2. The high temperatures resulting from **adiabatic compression** are shown for air in Fig. 2-11a. This curve is a plot of Eq. **1** for $n = k = 1.4$ (see Table 2-74). The various factors affecting the value of n are

Table 2-59 Chemical Reactions and Heats of Combustion of Pure Combustible Materials
[perfect combustion (except No. 1)]

No.	Combustible material	Reaction	Heat of combustion,[a] Btu per lb-mole gross	lb of vapor (except C & S) gross	lb of vapor (except C & S) net[b]	cu ft gross	cu ft net[b]	gal liquid gross	Cu ft gas per gal liquid[c]
1	Carbon (graphite)	$C + 0.5\ O_2 = CO$	47,460	3,950	3,950	...	...		
2	Carbon (coke)	$C + O_2 = CO_2$	174,000	14,500	14,500	...	...		
3	Graphite	$C + O_2 = CO_2$	169,860	14,093	14,093	...	...		
4	Carbon monoxide	$CO + 0.5\ O_2 = CO_2$	122,400	4,347	4,347	321.37[d]	321.37[d]		
5	Hydrogen	$H_2 + 0.5\ O_2 = H_2O$	123,100	61,095	51,623	325.02[d]	274.58[d]		
	Paraffin hydrocarbons								
6	Methane	$CH_4 + 2\ O_2 = CO_2 + 2\ H_2O$	382,980	23,875	21,495	1012.32[d]	911.45[d]	59,755	59.0
7	Ethane	$C_2H_6 + 3.5\ O_2 = 2\ CO_2 + 3\ H_2O$	671,190	22,323	20,418	1773.42[d]	1622.10[d]	74,010	41.3
8	Propane	$C_3H_8 + 5\ O_2 = 3\ CO_2 + 4\ H_2O$	955,430	21,669	19,937	2523.82[d]	2322.01[d]	91,740	35.91
9	*n*-Butane	$C_4H_{10} + 6.5\ O_2 = 4\ CO_2 + 5\ H_2O$	1,239,130	21,321	19,678	3270.69[d]	3018.48[d]	103,787	30.77
10	*iso*-Butane	$C_4H_{10} + 6.5\ O_2 = 4\ CO_2 + 5\ H_2O$	1,236,230	21,271	19,628	3261.17[d]	3008.96[d]	100,176	29.77
11	*n*-Pentane	$C_5H_{12} + 8\ O_2 = 5\ CO_2 + 6\ H_2O$	1,521,880	21,095	19,507	4019.65[d]	3717.15[d]	105,822	26.35
12	*iso*-Pentane	$C_5H_{12} + 8\ O_2 = 5\ CO_2 + 6\ H_2O$	1,518,410	21,047	19,459	4010.71[d]	3708.01[d]	104,863	26.17
13	Neopentane	$C_5H_{12} + 8\ O_2 = 5\ CO_2 + 6\ H_2O$	1,513,440	20,978	19,390	3994	3692	104,603	26.19
14	*n*-Hexane	$C_6H_{14} + 9.5\ O_2 = 6\ CO_2 + 7\ H_2O$	1,806,620	20,966	19,415	4768.27[d]	4415.23[d]	108,806	22.82
15	Neohexane	$C_6H_{14} + 9.5\ O_2 = 6\ CO_2 + 7\ H_2O$	1,803,600	20,931	19,380	4760	4407		
16	*n*-Heptane	$C_7H_{16} + 11\ O_2 = 7\ CO_2 + 8\ H_2O$	2,089,530	20,854	19,329	5459	5056	108,907	19.95
17	Triptane	$C_7H_{16} + 11\ O_2 = 7\ CO_2 + 8\ H_2O$	2,086,520	20,824	19,299	5445	5042		
18	*n*-Octane	$C_8H_{18} + 12.5\ O_2 = 8\ CO_2 + 9\ H_2O$	2,375,400	20,796	19,291	6260	5806	111,240	17.77
19	*iso*-Octane	$C_8H_{18} + 12.5\ O_2 = 8\ CO_2 + 9\ H_2O$	2,372,430	20,770	19,265	6249	5795		
	Olefin series								
20	Ethylene	$C_2H_4 + 3\ O_2 = 2\ CO_2 + 2\ H_2O$	606,910	21,636	20,275	1603.75[d]	1502.87[d]	71,504	44.33
21	Propylene	$C_3H_6 + 4.5\ O_2 = 3\ CO_2 + 3\ H_2O$	885,640	21,048	19,687	2339.70[d]	2188.40[d]	87,390	37.41
22	Butylene (*n*-Butene)	$C_4H_8 + 6\ O_2 = 4\ CO_2 + 4\ H_2O$	1,169,950	20,854	19,493	3084	2885		
23	*iso*-Butene	$C_4H_8 + 6\ O_2 = 4\ CO_2 + 4\ H_2O$	1,163,390	20,737	19,376	3069	2868	102,106	33.27
24	*n*-Pentene	$C_5H_{10} + 7.5\ O_2 = 5\ CO_2 + 5\ H_2O$	1,453,050	20,720	19,359	3837	3585		
	Aromatic series								
25	Benzene	$C_6H_6 + 7.5\ O_2 = 6\ CO_2 + 3\ H_2O$	1,420,300	18,184	17,451	3751.68[d]	3600.52[d]	129,724	34.63
27	Toluene	$C_7H_8 + 9\ O_2 = 7\ CO_2 + 4\ H_2O$	1,704,530	18,501	17,672	4486.44[d]	4284.81[d]	129,003	28.68
27	*p*-Xylene	$C_8H_{10} + 10.5\ O_2 = 8\ CO_2 + 5\ H_2O$	1,978,040	18,633	17,734	5223	4971	127,988	24.50
	Miscellaneous gases								
28	Acetylene	$C_2H_2 + 2.5\ O_2 = 2\ CO_2 + H_2O$	559,830	21,502	20,769	1476.55[d]	1426.17[d]		
29	Naphthalene	$C_{10}H_8 + 12\ O_2 = 10\ CO_2 + 4\ H_2O$	2,217,590	17,303[e]	16,708[e]	5854[e]	5654[e]		
30	Methyl alcohol	$CH_3OH + 1.5\ O_2 = CO + 2\ H_2O$	328,680	10,258	9,066	868	767	66,775	76.93
31	Ethyl alcohol	$C_2H_5OH + 3\ O_2 = 2\ CO_2 + 3\ H_2O$	606,290	13,161	11,917	1600	1449	85,360	53.35
32	Ammonia[f]	$NH_3 + 0.75\ O_2 = 0.5 N_2 + 1.5\ H_2O$	164,640	9,667	7,985	441	364		
33	Sulfur	$S + O_2 = SO_2$	127,800	3,980	3,980	...	...		
34	Hydrogen sulfide	$H_2S + 1.5\ O_2 = SO_2 + H_2O$	241,840	7,097	6,537	646	595	46,822	72.48
35	Formaldehyde[g]	$HCHO + O_2 = CO_2 + H_2O$	209,540	6,980	6,344	643	593		
36	Formic acid[g]	$HCOOH + 0.5\ O_2 = CO_2 + H_2O$	112,700	2,450	2,035	301	251		
37	Acetaldehyde[g]	$CH_3CHO + 2.5\ O_2 = 2\ CO_2 + 2\ H_2O$	501,160	11,390	10,523	1360	1259		
38	Nitric oxide[h]	$NO + 0.5\ O_2 = NO_2$	128,800	4,270	4,270	339[i]	339		
39	Nitrogen tetroxide[h]	$N_2O_4 + 0.5\ O_2 = N_2O_5$	108,000	1,175	1,175	284[i]	284		

[a] All values corrected to 60 F, 30 in. Hg dry. For gases saturated with water vapor at 60 F, deduct 1.74 per cent of the Btu value. Rossini, F. D., and others. *Selected Values of Physical and Thermodynamic Properties of Hydrocarbons and Related Compounds.* (Res. Proj. 44) Pittsburgh, Pa., Am. Petrol. Inst., 1953. Estimated uncertainty of values varies from <0.01 to 0.1 per cent [Rossini, F. D., and others. *Tables of Selected Values of Properties of Hydrocarbons.* (U.S. Nat.Bur. Std. C461), 1947].

[b] Addition from net to gross heating value determined by adding 19,095 Btu per lb-mole of water in the products of combustion. Keenan, J. H., and Keyes, F. G. *Thermodynamic Properties of Steam.* New York, Wiley, 1936.

[c] "Actual" volume of gas at 60 F and 760 mm formed upon vaporization; considered deviation from perfect gas laws. Matteson, R., and Hanna, U. S. "Physical Constants of Low Boiling-Point Hydrocarbons." *Oil & Gas J.* 41: 33–7, 1942.

[d] Mason, D. McA., and Eakin, E. E. "Proposed Standard Method for Calculating Heating Value and Specific Gravity from Gas Composition." *A.G.A. Proc.* 1961: CEP-61-11.

[e] A.G.A. *Combustion,* 3rd ed., rev. New York, A.G.A., 1938.

[f] Does not readily react in air; reaction takes place in pure oxygen. In presence of platinum catalyst: $4\ NH_3 + 5\ O_2 = 4\ NO + 6\ H_2O$.

[g] Calculated from *Handbook of Chemistry and Physics,* 28th ed., p. 1436–44. Cleveland, Ohio, Chemical Rubber Co., 1945.

[h] Shnidman, L., ed. *Gaseous Fuels,* 2nd ed., p. 273. New York, A.G.A., 1954.

[i] Btu per lb-mole divided by 380.

Table 2-60 Chemical Reactions and Heats of Combustion Involved in the Incomplete Combustion of Pure Combustible Materials

Combustible material	Reaction	Heat of combustion,* Btu per lb-mole gross	lb gross	cu ft† gross
Carbon to carbon monoxide	$C + \frac{1}{2}O_2 = CO$	‡	‡	‡
Methane to CO and water	$CH_4 + \frac{3}{2}O_2 = CO + 2H_2O$	260,580	16,245	691.0
Methane to CO and H_2	$CH_4 + \frac{1}{2}O_2 = CO + 2H_2$	14,380	896	41.0
Methane to formaldehyde	$CH_4 + O_2 = HCHO + H_2O$	173,440	10,812	369.8
Methane to formic acid	$CH_4 + \frac{3}{2}O_2 = HCOOH + H_2O$	270,280	16,849	711.8
Methane to methyl alcohol	$CH_4 + \frac{1}{2}O_2 = CH_3OH$	54,300	3,385	144.8
Ethane to CO	$C_2H_6 + \frac{5}{2}O_2 = 2CO + 3H_2O$	426,390	14,181	1148.4
Ethane to acetaldehyde	$C_2H_6 + O_2 = CH_3CHO + H_2O$	170,030	5,655	432.0
Ethane to ethyl alcohol	$C_2H_6 + \frac{1}{2}O_2 = C_2H_5OH$	64,900	2,159	192.0

* Calculated by subtracting the heat of combustion of the unburned product of reaction from the heat of reaction to complete combustion.

† All volumes corrected to 60 F, 30-in. Hg dry.

‡ See Table 2-59.

the temperature and pressure of the gas, composition of the gas, and amount of heat lost thru the compressor walls.

Atomic or Nuclear Energy.[7] The energy released in a nuclear pile appears as the kinetic energy of the fragments of the **fission** of uranium and thorium atoms. The fragments are charged electrically and give off a portion of their energy to the negative particles of each of the atoms thru which they travel. Their kinetic energy is dissipated in speeding up the atoms surrounding them and is converted to heat. It is this heat which is utilized in generating power from nuclear reactions.

Combustion Reactions

Combustion data presented in Tables 2-59, 2-63, and 2-89 are based on perfect combustion. In practical applications, excess air is usually present, and therefore complete combustion is the more appropriate term. Gas appliance performance standards as established in American Standard Approval Requirements, ASA Committee Z21, permit slight but stated departures from "complete combustion." For most appliances, the limit of carbon monoxide which may be present in the dry, air-free flue products is 0.04 per cent. Some types of appliances have limits which vary slightly above or below this figure because of low input rates, intermittent operation, and general operating conditions. These limits for flue gas composition ensure that the maximum amounts of CO which may develop in room air from the use of unvented gas appliances will be far below levels which might be harmful to the occupants. Vented appliances are subject to similar CO limits because of the possibility that the venting system might fail to function properly.

Incomplete or Partial Combustion. In certain industrial operations incomplete or partial combustion of the gas fuel is required; the resultant products of combustion are primarily carbon monoxide and hydrogen.

Table 2-60 presents representative chemical reactions of carbon, methane, and ethane when they are burned *incompletely,* as well as their respective calculated **heats of combustion.** Other hydrocarbons would react in the partial combustion process in a manner similar to methane and ethane. Products of incomplete combustion of carbon monoxide or hydrogen are not found by ordinary methods of analysis. Reactions in the combustion of hydrogen[2,8] involve chain carriers, of which the hydroxyl radical OH is very prominent. How the first OH radical is formed is not known; probably hydrogen peroxide forms on the walls of the containing vessel and then decomposes with subsequent conversion of hydrogen to water as follows:

$$H_2O_2 + M \rightarrow 2\,OH + M$$

where M = any third body, such as a wall or a gas molecule

$$OH + H_2 \rightarrow H_2O + H$$

$$H + O_2 \rightarrow OH + O$$

$$O + H_2 \rightarrow OH + H$$

Table 2-61 Products of Incomplete Combustion[9] of Three Gases With Limited Air Supply, Self-Supporting Flame, and No External Heat

(at one atmosphere pressure)

	Natural	Coke oven	Butane
Heating value of gas, Btu per cu ft	1108	534	3207
Specific gravity of gas	0.63	0.404	2.0
Aeration (100 per cent required for complete combustion)	67.5%	49.5%	49.0%
Products of combustion on dry, air-free basis, per cent			
Orsat analysis:			
Carbon dioxide	3.4	4.6	4.3
Unsaturated hydrocarbons	0.1	0.3	2.0
Carbon monoxide	9.6	11.4	13.5
Hydrogen	9.9	15.4	9.6
Methane	0.3	0.4	3.5
Nitrogen (100 minus sum of above)	76.7	67.9	67.1
Total	100.0	100.0	100.0
Chemical analysis:			
Formaldehyde	0.0133	0.00302	0.0842
Acetaldehyde	.00008	.0002	.0052
Clyoxal	.0005	.00009	.0023
Formic acid	.00282	.00097	.0028
Methyl alcohol	< .0004	< .00025	< .006
Combined nitrogen	0.00106	0.00046	0.00106
Total	0.01816	0.00499	0.11956

The produtcs of **incomplete combustion of hydrogen** (if they could be isolated) would be hydrogen peroxide, hydroxyl radical, atomic hydrogen, and atomic oxygen.

The data in Table 2-61 illustrate what gases may be found in the products of incomplete combustion from a free burning flame supplied with less air than is needed for complete combustion. At least 12 different chemicals are identified. Under the test conditions imposed,[9] the lowest percentage of air which would maintain combustion without external heat or catalysts was found to be 60 per cent for natural gas and butane and 46 per cent for coke oven gas.

The results of partially burning methane with oxygen to form synthesis gas[10] are shown in Table 2-62. Note that both the gas and oxygen were preheated and that the process was carried out at 250 to 300 psig. The reactions are about as follows:

1. $CH_4 + 2O_2 = CO_2 + 2H_2O$
   Part of the methane burns rapidly to completion.
2. $CH_4 + H_2O = CO + 3H_2$
3. $CH_4 + CO_2 = 2CO + 2H_2$
   Part of the methane reacts with the products.
4. $H_2O + CO = H_2 + CO_2$
   Some of the products of combustion react in accordance with the water gas shift reaction. This reaction is used to obtain H_2 for various purposes after the CO_2 is scrubbed out.

The partial combustion process apparently does not take place according to the single step:

5. $CH_4 + 1/2\,O_2 = CO + 2H_2$

Reactions 1 and 5 are *exothermic*, and reactions 2, 3, and 4 are *endothermic*, but the overall reaction is strongly *exothermic*. The partial combustion of methane and oxygen is rate-controlled.

Table 2-62 Partial Combustion of Methane with Oxygen[10]

	Test number			
	23-2A	9-2	6-3	28-4
Gas preheat, °F	1200	963	1200	1200
Oxygen preheat, °F	600	600	600	600
Nitrogen in gas, %	16	16	16	16
Nitrogen in oxygen, %	7.4	7.4	7.4	7.4
Carbon number of feed gas	1.1	1.1	1.07	1.17
Methane (CH_4, commonly called C_1) = 1.0				
Ethane (C_2H_6, commonly called C_2) = 2.0				
90% Methane plus 10% ethane = 1.1				
Psig	267	250	250–300	250–300
Reactor plus burner heat loss, Btu/hr	65,000	65,000	65,000	65,000
Space velocity, cu ft C_1/hr-cu ft of reactor chamber	1900	740–960	740–960	740–960
O_2 to C_1 feed ratio	0.641	0.637	0.601	0.635
Carbon efficiency	2.54	2.34	2.30	2.36
Oxygen efficiency	3.96	3.67	3.92	3.72
Calc. exit temperature, °F	2500	2590	2620	2760
Carbon formation, % of inlet carbon	0.035	...	...	...
Carbon converted to CO or CO_2, % of C_1	97.7	92.9	89.1	94.4

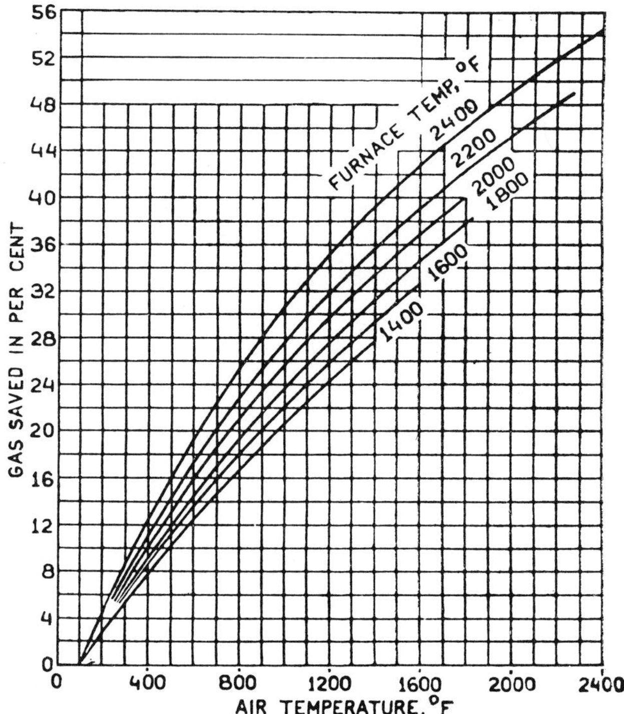

Fig. 2-11b Savings effected by burning natural gas with preheated air (including dissociation effects).[11]

Preheating of Combustion Air. The fuel savings which result from preheated combustion are are shown in Fig. 2-11b; the cost of preheating is assumed to be zero. In many natural gas furnaces, it has been found that premixing the gas and air prior to entry into the furnace permits operation with lower excess air than required in furnace mixing of reactants. Air preheat above 900 F is undesirable for premix, since the temperature of the reactants would favor ignition *within* the burner and the **cracking** of natural gas. Preheating *both* air and fuel is unnecessary because of the high air–fuel ratios.

Combustion Constants and Physical Properties of Pure Combustible Substances, and Supporters of Combustion

Table 2-63, supplemented by preceding Tables 2-59 and 2-60, gives the combustion constants and physical properties of all individual gases and vapors of liquids which may be found in fuel gases or in their combustion reactions.

Composition and Physical Properties of Air

Air takes part in all combustion reactions except those in which oxygen in greater concentration than 21 per cent is used. For normal calculations, air consists of 21 per cent O_2 and 79 per cent N_2. More complete analyses of both *dry* and *moist* air are shown in Tables 2-64 and 2-65. The relative proportions shown remain reasonably constant in all parts of the world, and, except for water vapor, at altitudes up to about 12 miles. Above this point air is rarefied and may show a change in composition. Many gases and dusts may be present in trace amounts in urban areas; in fact, small particles numbering hundreds of thousands per cubic inch can be detected in air. These include carbon particles from smoke and ash, plant spores, microorganisms, and very fine dirt

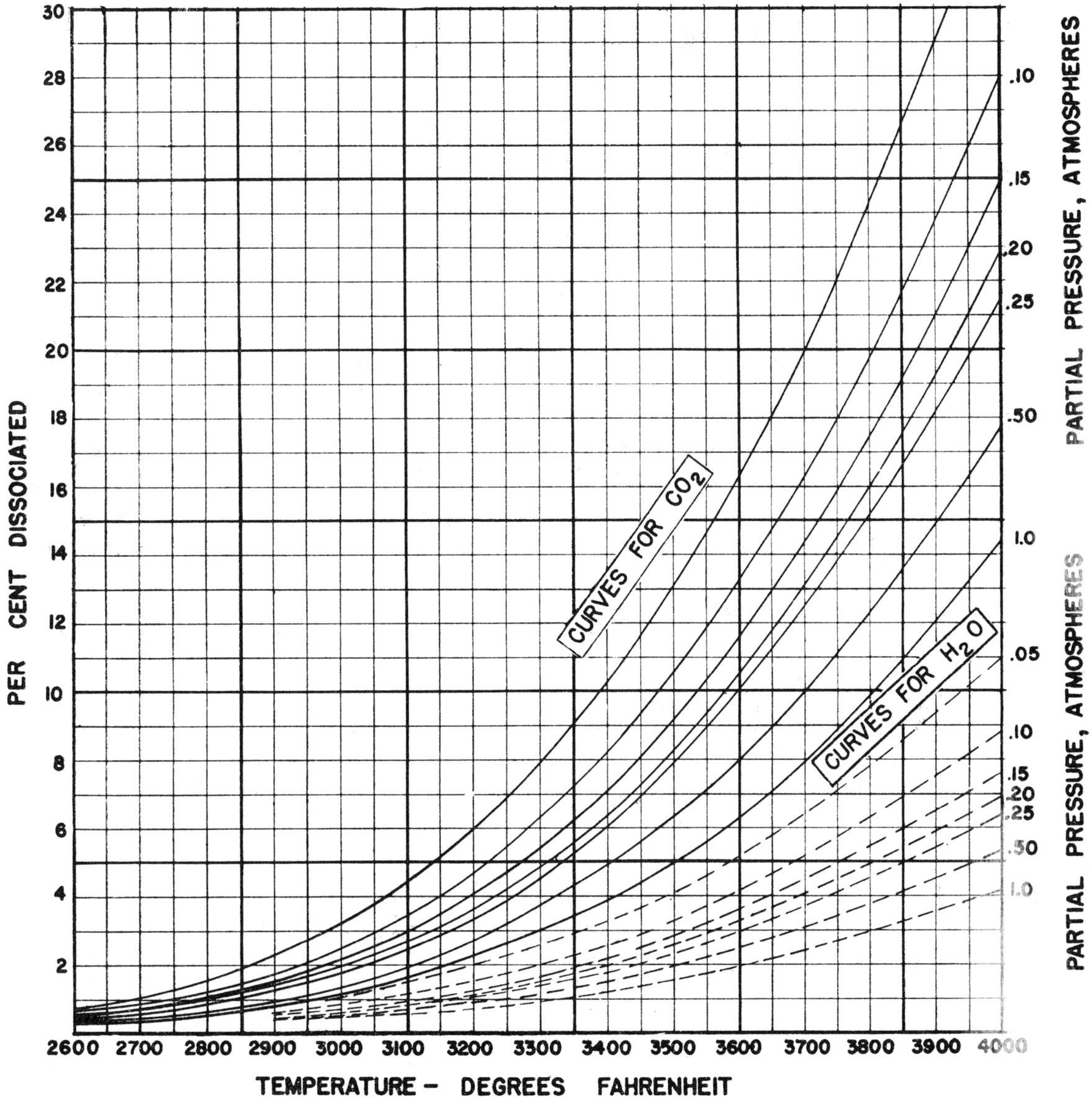

Fig. 2-12 Dissociation of CO_2 and H_2O at various partial pressures in mixtures containing no oxygen.

particles. The volume of these contaminants is so small a portion of the total that they do not affect the combustion reaction.

Table 2-66 presents eight different physical properties of air which may be useful in combustion or flow calculations.

Dissociation, Thermal Cracking, and Pyrolysis of Gases and Vapors

Gases break down or *dissociate* into simpler gases or elements, and molecules of elementary gases break down into atoms by application of heat at elevated temperatures. The reactions are always endothermic and the degree of dissociation is dependent on temperature and contact time. Data on dissociation of flue gases and combustible gases are shown in Tables 2-67 and 2-68a and in Fig. 2-12.

Flue gases CO_2 and H_2O do not dissociate appreciably until a temperature of 3000 F is reached. Dissociation can be neglected in most domestic appliances, but it becomes important in industrial high-temperature operations *above* 3000 F. Carbon monoxide is a very stable gas even at 4500 F, whereas hydrogen dissociates partially at 4000 F. In the presence of copper oxide, both CO and H_2 oxidize readily at the low temperature of 572 F.

Data on the breakdown of hydrocarbons by application of heat are presented in Table 2-67 and in more detail in Table 2-68a. Other information on **thermal cracking** of various grades of oil is included in Chap. 3 of this section. **Methane** is the most resistant to thermal breakdown of any of the hydrocarbons, requiring a temperature of 1445 F to start appreciable decomposition and a temperature of 2120 F to decompose rapidly. Acetylene is also quite resistant to thermal breakdown. From ethane and ethylene on, the higher hydrocarbons in the series break down into numerous gases; the amount of decomposition and the type of gases formed depend upon temperature, time, and pressure, as well as upon the material or catalyst with which they come in contact. Representative products of decomposition are shown in Table 2-68a. Under normal conditions when hydrocarbons are supplied to burners at atmospheric temperature, dissociation is

Table 2-63 Combustion Constants and Physical Properties of Combustible Substances, Supporters of Combustion, and Flue Gases[a]

(gas volumes corrected to 60 F and 30 in. Hg dry, except as indicated)

No.[b]	Substance	Formula	Molecular weight[c]	Density, lb per cu ft[d]	Specific volume, cu ft per lb[d]	Specific gravity[d] (air = 1)	Cu ft per cu ft of combustible required for combustion: O_2	Cu ft per cu ft of combustible required for combustion: Air	Lb per lb of combustible required for combustion: O_2	Lb per lb of combustible required for combustion: Air	Ignition temp, °F	Flash point[e] °F	Limits of flammability,[f] % gas in air: Lower	Limits of flammability,[f] % gas in air: Upper	Flame temp in air calc., °F	Summation factor,[g] $b^{0.5}$
2	Carbon (coke)	C	12.010	0.0668[h]	14.970[h]	0.8200[h]	...	...	2.664[i]	11.482[i]	1220[j] (activated)	...	...	...	...	
4	Carbon monoxide	CO	28.01	.07404	13.506	.9672	0.5	2.382	0.571	2.462	1128[k]	Gas	12.5	74	4475[l]	0.0217
5	Hydrogen	H_2	2.016	.00532	187.970	.06959	0.5	2.382	7.951[i]	34.267[i]	968[m]	Gas	4.0	75.0	4010[l]	
Paraffin hydrocarbons																
6	Methane	CH_4	16.042	.04242	23.574	0.5543	2.0	9.528	4.049[i]	17.195[i]	[n]	Gas	5.0	15.0	3484[i]	0.0436
7	Ethane	C_2H_6	30.068	.08029	12.455	1.0488	3.5	16.675	3.688[i]	15.899[i]	[n]	Gas	3.0	12.5	3540[i]	.0917
8	Propane	C_3H_8	44.094	.1196[o]	8.361[o]	1.5617	5.0	23.821	3.537[i]	15.246[i]	[n]	Gas	2.1	10.1	3573[i]	.1342
9	*n*-Butane	C_4H_{10}	58.12	.1582[o]	6.321[o]	2.0665	6.5	30.967	3.476[i]	14.984[i]	[n]	−76	1.86	8.41	3583[i]	.1841
10	*iso*-Butane	C_4H_{10}	58.12	.1582[o]	6.321[o]	2.0665	6.5	30.967	3.476[i]	14.984[i]	[n]	−117	1.80	8.44	3583	.1723
11	*n*-Pentane	C_5H_{12}	72.146	.1904[o]	5.252[o]	2.4872	8.0	38.114	3.554[i]	15.323[i]	[n]	<−40	1.40	7.80	...	.2377
12	*iso*-Pentane	C_5H_{12}	72.146	.1904[o]	5.252[o]	2.4872	8.0	38.114	3.554[i]	15.323[i]	[n]	<−60	1.32	...	...	.2276
13	Neopentane	C_5H_{12}	72.146	.1904[o]	5.252[o]	2.4872	8.0	38.114	3.554[i]	15.323[i]	842[p]	...	...	...	...	
14	*n*-Hexane	C_6H_{14}	86.172	.2274[o]	4.398[o]	2.9704	9.5	45.260	3.535[i]	15.238[i]	478[p]	−7	1.25	6.90	...	.283
15	Neohexane	C_6H_{14}	86.172	.2274	4.398	2.9704	9.5	45.260	3.535[i]	15.238[i]	797[p]	...	...	...	...	
16	*n*-Heptane	C_7H_{16}	100.198	...	...	3.459	11.0	52.406	...	...	433[p]	25	1.00	6.00	...	
17	Triptane	C_7H_{16}	100.198	...	...	3.459	11.0	52.406	...	...	...	...	...	...	...	
18	*n*-Octane	C_8H_{18}	114.224	...	...	3.943	12.5	59.552	...	...	428[p]	56	0.95	3.20	...	
19	*iso*-Octane	C_8H_{18}	114.224	...	...	3.943	12.5	59.552	...	...	[n]	10	...	...	...	
Olefin series																
20	Ethylene	C_2H_4	28.052	.07456	13.412	0.9740	3.0	14.293	3.422	14.807	914[k]	Gas	2.75	28.6	4250	.0775
21	Propylene	C_3H_6	42.078	.1110[o]	9.009	1.4504	4.5	21.439	3.422	14.807	856[k]	Gas	2.00	11.1	4090	.1269
22	Butylene	C_4H_8	56.104	.1480[o]	6.757[o]	1.9336[o]	6.0	28.585	3.422	14.807	829[k]	Gas	1.98	9.65	4030	
23	*iso*-Butene	C_4H_8	56.104	.1480[o]	6.757[o]	1.9336	6.0	28.585	3.422	14.807	...	...	...	...	...	
24	*n*-Pentene	C_5H_{10}	70.130	.1852[o]	5.400[o]	2.419	7.5	35.732	3.422	14.807	...	...	1.65	7.70	...	
Aromatic series																
25	Benzene	C_6H_6	78.108	.2060[o]	4.854[o]	2.692[o]	7.5	35.732	3.073	13.297	1044[p]	12	1.35	6.75	4110	
26	Toluene	C_7H_8	92.134	.2431[o]	4.114[o]	3.1760[o]	9.0	42.878	3.132[i]	13.503[i]	997[p]	40	1.27	6.75	4050	.311
27	Xylene	C_8H_{10}	106.160	.2803	3.568[o]	3.6618[o]	10.5	50.024	3.170[i]	13.663[i]	867[p]	63	1.00	6.00	4010	
Miscellaneous combustible gases																
28	Acetylene	C_2H_2	26.036	.06971	14.345	0.9107	2.5	11.911	3.073	13.297	[n]	Gas	2.50	81	4770	0.0833
29	Naphthalene	$C_{10}H_8$	128.164	.3384[o]	2.955[o]	4.4208[o]	12.0	57.170	3.000[i]	12.932[i]	959[q]	174	0.90	...	4100	
30	Methyl alcohol	CH_3OH	32.042	.0846[o]	11.820	1.1052	1.5	7.146	1.500[i]	6.466[i]	824[q]	54	6.72	36.50	...	
31	Ethyl alcohol	C_2H_5OH	46.068	.1216[o]	8.224[o]	1.590[o]	3.0	14.293	2.087[i]	8.998[i]	756[q]	55	3.28	18.95	...	
32	Ammonia	NH_3	17.032	.0456[o]	21.930[o]	0.5961[o]	0.75	3.573	1.392[i]	5.998[i]	1204[k]	Gas	15.50	26.60	...	
33	Sulfur	S	32.06	...	...	...	...	...	0.998[i]	4.285[i]	374[j]	...	...	...	...	
34	Hydrogen sulfide	H_2S	34.076	0.09109	10.978[o]	1.1898[o]	1.5	7.146	1.393[i]	6.005[i]	558[k]	Gas	4.3	45.50	...	
35	Formaldehyde	HCHO	30.026	...	...	1.075	1.0	4.764	...	...	...	Gas	7	73	...	
36	Formic acid	HCOOH	46.026	76	0.013	1.59	0.5	2.382	...	...	...	156	...	...	...	
37	Acetaldehyde	CH_3CHO	44.052	48.8	0.0205	1.52	2.5	11.91	...	...	527	−36	4.0	57.0	...	
38	Nitric oxide	NO	30.008	0.0782	12.788	1.018										

Noncombustible gases					
Oxygen	O_2	32.000	.08461	11.819	1.1053
Nitrogen	N_2	28.016	.07439	13.443	0.9718
Carbon dioxide	CO_2	44.01	.1170	8.547	1.5282
Sulfur dioxide	SO_2	64.06	.1733	5.770	2.264
Water vapor	H_2O	18.016	.04758[o]	21.017[o]	0.6215[o]
Air	$O_2 + N_2 +$	28.9	0.07655	13.063	1.00

[a] Heats of combustion are listed in Table 2-59. Except where otherwise indicated, data are from: Shnidman, L., ed. *Gaseous Fuels*, 2nd ed., pp. 34, 118, 427–429. New York, A.G.A., 1954.

[b] Numbers correspond to those in Table 2-59.

[c] Baxter, G. P., and others. "Seventh Report of the Committee on Atomic Weights of the International Union of Chemistry." *J. Am. Chem. Soc.* 59: 225, Feb. 8, 1937.

[d] Densities calculated from values given in grams per liter at 0°C and 760 mm in International Critical Tables, allowing for known deviations from gas laws. Where no densities were available, the volume of the mole was taken as 22.415 liters. The ideal values for specific gravity may be found in CEP 61-11 (see reference g below).

[e] Mine Safety Appliances Co.

[f] Coward, H. F., and Jones, G. W. *Limits of Flammability of Gases and Vapors.* (U. S. Bur. Mines Bull. 503) Washington, D. C., 1952.

[g] At 60 F and 760 mm Hg; used in the calculation of specific gravity and heating value when accurate analyses, for example, by mass spectrometer or gas chromatograph, are available. **Note:** Compressibility factor = $(1 - b)$. Mason, D. McA., and Eakin, B. E. "Proposed Standard Method for Calculating Heating Value and Specific Gravity from Gas Composition." *A.G.A. Proc.* 1961: CEP-61-11.

[h] If carbon could be conceived to exist as a gas under standard conditions, its relative density, weight, and volume would be as shown.

[i] Values have been recalculated.

[j] Hartman, I. "Dust Explosions." *Mechanical Engineers' Handbook*, 6th ed., Sec. 7, p. 41–8. New York, McGraw-Hill, 1958.

[k] Scott, G. S., and others. "Determination of Ignition Temperatures of Combustible Liquids and Gases." *Anal. Chem.* 20: 238–41, Mar. 1948.

[l] Haslam, R. T., and Russell, R. P. *Fuels and Their Combustion*, p. 205. New York, McGraw-Hill, 1926.

[m] Zabetakis, M. G. *Research on the Combustion and Explosion Hazards of Hydrogen–Water Vapor–Air Mixtures.* (U. S. Bur. of Mines, Div. of Explosives Tech., Progr. Rept. No. 1) Washington, D. C., 1956.

[n] See Table 2-77a.

[o] Either the density or the coefficient of expansion has been assumed. Some of the materials cannot exist as gases at 60 F and 30 in. Hg, in which case the values are theoretical ones. These substances exist as gases in a gaseous mixture in quantities proportional to their respective partial pressures.

[p] Zabetakis, M. G., and others. "Minimum Spontaneous Ignition Temperatures of Combustibles in Air." *Ind. Eng. Chem.* 46: 2173–8, Oct. 1954.

[q] Zabetakis, M. G. Unpublished.

Table 2-64 Composition of Dry Air by Volume*

Substance	Per cent	
Oxygen	20.99	
Carbon dioxide	0.03	
Hydrogen	0.01	
Nitrogen	78.03	
Argon	0.94	
Neon	.00123	0.001686† (Neon, Helium, Krypton, Xenon)
Helium	.00040	
Krypton	.00005	
Xenon	0.000006	
Total	100.001686	

* Taken from International Critical Tables.

† Apparently these four items were determined separately and the remaining components were not corrected for their presence.

Table 2-65 Effect of Moisture and Temperature on the Composition of Air*

	Dry bulb temp, °F					
	60			90		
Relative humidity, %	20	80	100	20	80	100
Moisture, lb/lb dry air	0.00218	0.00880	0.01108	0.00597	0.02458	0.03118
Moisture, % by weight	0.217	0.872	1.096	0.593	2.452	3.024
Substance	**Composition by volume, per cent**					
Oxygen	20.917	20.696	20.623	20.790	20.191	19.991
Inert gases	78.733	77.905	77.628	78.258	76.003	75.251
Moisture	0.350	1.399	1.749	0.952	3.806	4.758

* Values calculated from data in ASHRAE Guide, 1960 ed.

Table 2-66 Physical Properties of Air

Temp, °F	Density, lb/cu ft	Specific heat,* Btu/lb-°F const. press.	Specific heat,* Btu/lb-°F const. vol.	Velocity* of sound, ft/sec	Viscosity* × 10^7, lb-mass/sec-ft†	Thermal* conductivity, Btu/hr-ft-°F‡	Heat* content, Btu/lb§
−300	...	0.2392	0.1707	618.7	...	...	38.03
−200	...	.2392	.1707	789.8	...	...	61.96
−100	...	.2393	.1707	929.8	...	...	85.90
0	...	.2394	.1708	1051.2	111	0.0133	109.83
32	0.080710	.2396	.1710	1087.1	117	.0141	117.49
60	.076365	.2397	.1711	1117.4	121	.0148	124.20
90	.072197	.2399	.1713	1149.3	126	.0156	131.39
100	.070907	.2400	.1714	1159.4	128	.0158	133.79
140	.066177	.2403	.1718	1200.0	135	.0168	143.40
200	.060158	.2410	.1724	1257.7	145	.0182	157.84
300	.052238	.2426	.1741	1347.9	160	.0204	182.01
400	.046162	.2448	.1762	1430.9	174	.0227	206.38
500	.041350	.2475	.1789	1508.9	187	.0251	230.98
600	.037448	.2504	.1818	1582.0	200	.027	255.88
700	.034219	.2535	.1849	1651.3	213	.029	281.06
800	0.031502	.2566	.1881	1717.2	225	.031	306.57
900	...	.2598	.1913	1780.1	237	.034	332.39
1000	...	.2630	.1944	1840.6	249	.036	358.55
1200	...	.2687	.2002	1955.2	270	.040	411.74
1400	...	.2740	.2054	2063	289	.044	466.40
1600	...	.2786	.2101	2165	307	.047	521.31
1800	...	.2824	.2139	2263	325	.051	577.43
2000	...	.2857	.2171	2356	343	0.054	634.25
2500	...	.2924	.2238	2576	...	...	778.88
3000	...	.2974	.2288	2778	...	...	926.41
3500	...	.3012	.2326	2966	...	...	1076.10
4000	...	.3044	.2359	3143	...	...	1227.59
5000	...	.3092	.2407	3470	...	...	1534.61
6000	...	0.3130	0.2444	3768	...	...	1845.83

* Keenan, J. H., and Kaye, J. *Gas Tables*, New York, Wiley, 1948. Values interpolated from Tables 1 and 2.
† lb-mass/sec-ft = 14.882 poises (from p. 34 of *Gas Tables*, referenced above*).
‡ Btu/hr-ft-°F = 0.0041339 IT cal/°C-sec-cm (from p. 34 of *Gas Tables*, referenced above*). IT = international temperature.
§ To obtain heat content in Btu per lb-mole, multiply Btu per lb by the molecular weight of air (28.70).

no factor. However, where the gases may become highly heated, as, for example, in a pilot tube or in a manifold to a gas engine, they may break down to carbon and plug an orifice, or to other products and create combustion conditions attributable to the products of decomposition rather than to the original hydrocarbon. In thermal cracking of liquefied petroleum gases and oils it is essential to know the effect of temperature on formation of the products desired.

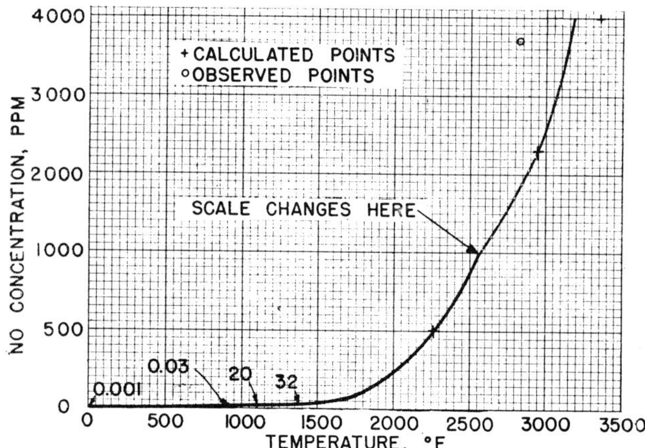

Fig. 2-13a Concentration vs. temperature of $N_2 + O_2 \rightleftharpoons 2NO$ (reference 57 and calculations using data from reference 56).

Nitric Oxide Formation. Figure 2-13a shows the temperature dependence[56] of NO concentration at equilibrium for the reaction $N_2 + O_2 \rightleftharpoons 2NO$. At the lower temperatures (1300–2000 F) the reaction is of zero order and the *decomposition* of NO is catalyzed by the surrounding surfaces. In the 2000–2550 F range, the reaction is both a zero and a second order; above 2500 F, the reaction is solely of second order.

Since the nitric oxide reaction is primarily temperature dependent, it is the rate of quenching[57] which determines the percentage of nitric oxide retained in a gas stream of changing temperature. For example, it has been shown[56] that at temperatures in excess of 2700 F, the gases must be cooled at the rate of 36,000°F per sec to retain their original nitric oxide contents. The phenomenon of an electric spark illustrates the combination of high temperature and subsequent quenching (with good air circulation). Although no determinations have been made for the quenching rates at lower temperatures, it is assumed that the low quench rate would further aid decomposition of NO in conventional domestic appliances.

Table 2-68b shows[58] the equilibrium constants for the formation and decomposition (unquenched) of nitric oxide, including the equilibrium times. The extremely long equilibrium time for the reaction at 1330 F and the very small forward reaction rate constant, K_2, were also noted in Haber's investigation[58] of the effects of activated nitrogen and oxygen atoms upon the nitric oxide reaction. He found that a nitro-

Table 2-67 Dissociation of Gases and Vapor by Heat
(thermal decomposition, pyrolysis, cracking, and reaction with copper oxide)

No.	Gas	Resultant products	Temp, °F	Dissoc., per cent	Energy required for 100% disintegration, Btu/lb-mole	Log_{10} of equilibrium constants*
1	CO_2	$CO + O_2$	2050	0.014	Table 2-59	−5.992
			2800	0.40		−3.632
			3000	1.4		−3.186
			3500	5.0		−2.206
			4000	13.5		−1.478
2	H_2O	$H_2 + O_2$	3190	0.37	Table 2-59	−3.441
			3500	1.5		−2.931
			4000	4.1		−2.259
3	CO	$C + O_2$	4500	Very slightly	Table 2-59	−6.613
CO over copper oxide†		CO_2	293	Begins		...
			572	100		...
4	H_2	2H	2060	0.00029	184,000	−10.627
			4000	6.5		−3.292
			4940	8.31		−1.604
H_2 over copper oxide†		H_2O	338	Begins		...
			572	100		...
5	O_2	2 O	2420	0.0008	211,500	−9.575
			4940	5.95		−1.840
6	CH_4	$C + 2 H_2$	797	$0.0_{12}13$‡	Table 3-36	...
			1067	$0.0_8 33$		...
			1445	8.4		...
			2120	100		...
CH_4 over copper oxide†		$CO_2 + H_2O$	1562	100		...
7	C_2H_6	$C_2 + 3 H_2$	797	$0.0_7 15$	131,600	...
			1067	0.00017		...
			1490	100		...
C_2H_6 over copper oxide†		$CO_2 + H_2O$	1562	Complete		...
8	C_3H_8	$C_3 + 4 H_2$	797	$0.0_6 96$	111,600	...
			1067	.0026		...
9	n-C_4H_{10}	$C_4 + 5 H_2$	797	$0.0_5 15$	105,500	...
			1202	57		...
10	n-C_5H_{12}	$C_5 + 6 H_2$	797	$0.0_5 24$	109,500	...
			1112	30		...
11	*iso*-C_5H_{12}	$C_5 + 6 H_2$	797	$0.0_5 37$	105,400	...
			1067	.0065		...
12	n-C_6H_{14}	$C_6 + 7 H_2$	1148	0.062	...	...
			2192	100		...
13	n-C_7H_{16}	$C_7 + 7 H_2$	1148	0.146	83,500	...
14	n-C_8H_{18}	$C_8 + 9 H_2$	1148	.24	116,700	...
15	*iso*-C_8H_{18}	$C_8 + 9 H_2$	1148	0.24	...	...
16	C_2H_4	$C + H_2$ + other HC	1058	75.6	...	...
			1472	100		...
17	C_3H_6	$C + H_2$ + other HC	1067	75		...
18	C_2H_2	$CH_4 + H_2$	1472	100	...	...
19	C_6H_6	Anthracene and H_2	1697	...	...	...
			1832	100	...	...

Nos. 1, 2: Haslam, R. T., and Russell, R. P. *Fuels and Their Combustion*, p. 160. New York, McGraw-Hill, 1926; and Shnidman, L., ed. *Gaseous Fuels*, 2nd ed., p. 121. New York, A.G.A., 1954.
3: Haslam, R. T., and Russell, R. P. *Fuels and Their Combustion*, p. 168. New York, McGraw-Hill, 1926.
4, 5: Jost, W. *Explosion and Combustion Processes in Gases*, p. 322. New York, McGraw-Hill, 1946.
6–15: *ibid.*, pp. 396–7.
16–19: Hurd, C. D. *Pyrolysis of Carbon Compounds*. (ACS Monograph 50) New York, Chemical Catalog Co., 1929.
* Lewis, B., and von Elbe, G. *Combustion, Flames and Explosions of Gases*, p. 742. New York, Academic Press, 1951.
† Dennis, L. M., and Nichols, M. L. *Gas Analysis*, p. 160. New York, Macmillan, 1929.
‡ $0.0_{12}13 = 1.3 \times 10^{-13}$.

gen gas stream containing an excess of oxygen produced no NO when heated to 1300 F.

Flames

Flame Temperature. Flame temperature is of little significance in domestic applications, since these appliances operate in the relatively low-temperature range of 160 F to 550 F. However, in high-temperature industrial heating, such as melting metal or *burning* lime, the temperature of the flame is of considerable importance. Heat transfer per unit area increases with flame temperature; the rate of heating an object is proportional to the difference in temperature between the flame and the object. Research in high-temperature chemistry was conducted with flame temperatures of 9000 F attained in burning carbon subnitride in oxygen at one at-

Table 2-68a Dissociation (Pyrolysis) of Hydrocarbons by Heat* and by Heat plus Catalytic Cracking

Gas†	Temp decomposition starts, °F	Partial decomposition		Products of decomposition	Complete decomposition temp, °F
		Temp, °F	% Decomposed		
Methane	1202–1292	1445	8.4	C + H_2 + CH_4	2120
Ethane	1067	1067	8.5	C + H_2 + C_2H_4	1490 (1 hr 64% C_2H_4, 36% H_2)
Propane	Below 1067	1067 (6 min)	49.9	CH_4 15.3%	Above 1067
				C_2H_6 3.8	
				C_2H_4 10.3	
				C_3H_6 10.3	
				H_2 8.6	
				C_3H_8 50.1	
Propane‡	About 932	1112	9.9	C_2H_6, H_2, unsat. HC	...
n-Butane	Below 1112	1202	57	C_4H_{10} 43%	Above 1292
				CH_4 24.5	
				Unsat. 26.3	
				H_2 4.5	
n-Butane‡	1022	1112	22.4	H_2 + mixt. of HC	
n-Pentane	Below 1112	1112	30	C_5H_{12} 70%	Above 1112
				C_2H_6, C_3H_6 } 16.5	
				C_3H_8, C_2H_4 } 7.5	
				CH_4, C_4H_8 } 6.0	
n-Pentane‡	About 932	932	5.7		
n-Hexane	1112			CH_4, C_2H_4, C_3H_6	2192 (C + H_2)
Ethylene	Above 651	1058	75.6	C_2H_4 24.4%	1472 (30 min)
				C_2H_2 6.7	
				H_2 27.2	
				CH_4 38.8	
				C_2H_6 2.9	
Propylene	Above 842	1067	75	C_3H_6 25%	Above 1067
				H_2 49.9	
				C_2H_4 6.1	
				CH_4 9.4	
				C_2H_6 1.9	
				C_3H_8 4.3	
				Higher HC 2.4	
Acetylene	About 842	1472	100	CH_4 43%	
				H_2 57	
Benzene	1697	1697	...	$C_{14}H_{10}$ + H_2	1832

* Hurd, C. D. *Pyrolysis of Carbon Compounds.* (ACS Monograph 50) New York, Chemical Catalog Co., 1929.

† Dissociation is by heat except where catalytic cracking is indicated.

‡ By catalytic cracking. Catalyst used = silica–zirconium–alumina. Greensfelder, B. S., and Voge, H. H. "Catalytic Cracking of Pure Hydrocarbons." *Ind. Eng. Chem.* 37: 514–20, June 1945.

Table 2-68b Equilibrium Constants for Formation, K_2, and Decomposition, K_1, of Nitric Oxide, and Time Required for Equilibrium[58]

$$N_2 + O_2 \underset{K_1}{\overset{K_2}{\rightleftharpoons}} 2NO$$

Temp, °F	K_1, 1/(mole-sec)	K_2, 1/(mole-sec)	Time
1330	4.00×10	1.74×10^{-7}	81.6 years
2230	3.78×10^4	2.33×10^{-1}	1.26 days
2950	9.06×10^6	1.20×10^3	2.08 min
3310	1.40×10^8	5.55×10^4	5.06 sec
4030	3.37×10^{10}	6.98×10^7	0.0106 sec

mosphere.[69] Instantaneous high temperatures of 13,000 F were achieved in the detonation of finely sieved cyanogen and liquid oxygen.

Thermodynamic data in graphic form to be used in determining adiabatic flame temperatures at one atmosphere and 2400 to 4000 F have also been presented.[70] Figure 2-13b may be used to obtain a very rough approximation of flame temperature expected with coal, coke, and liquid and gaseous hydrocarbon fuels when burned with 0 to 100 per cent excess air. This figure assumes that a dry fuel at 80 F, containing only carbon and hydrogen, is burned with dry air. The calorific value of the fuel is calculated by the method of W. Boie.[12] The plot shows approximate flame temperature as a function of excess air for fuels of various hydrogen-to-carbon atomic ratios ranging from zero to four. Most coals fall in the range of zero to one ratio, and the ratio of four, represented by methane, is the maximum obtainable with hydrocarbon fuels. Curves for air preheat temperatures of 100 F and 600 F are given.

Note: Flame temperatures under actual (nonadiabatic) conditions are sometimes estimated by assuming that a percentage of the heat energy is lost to surroundings before the heat release process is completed.

Data on flame temperature are given as actual or observed in air, calculated or theoretical in air, and calculated or theo-

Table 2-69 Flame Temperatures in Air and in Oxygen

No.	Gas	Net Btu per cu ft*	Maximum flame temperature in air: Theoretical temp, °F	Observed temp, °F	Gas in air, %	Theoretical flame temp in O_2, °F
1	Carbon monoxide	321.8	4475†	3812	32	...
2	Hydrogen	275	4010†	3713	31.6	5385
3	Methane	913.1	3484	3416	9.4–10.1	...
4	Ethane	1641	3540	3443	5.7–5.9	...
5	Propane	2385	3573	3497	4.0–4.3	...
6	*n*-Butane	3113	3583	3443	3.1–3.4	...
7	*iso*-Butane	3105	3583	3452	3.1–3.2	...
8	Ethylene	1513.2	4250	3587	7.0	...
9	Propylene	2186	4090	3515	4.5	...
10	Butylene	2885	4030	3506	3.4	...
11	Benzene	3601	4110	...	...	...
12	Toluene	4284	4050	...	...	...
13	Acetylene	1448	4770	4207	9.0	5630
14	Natural gas	879	3535	...	...	...
15	Natural gas	904	3565	...	...	5120
16	Natural gas	971	3550	...	...	...
17	Natural gas	1009	3550	...	...	...
18	Natural gas	1021	3562	...	...	5150
19	Carb. water gas	493	3700	...	...	5050
20	Coke oven	514	3610	...	...	...
21	Coal gas	486	3600	...	...	...
22	Oil gas	496	3630	...	...	...
23	Producer gas	153	3175	...	...	...
24	Water gas (coke)	262	3670	...	...	...
25	Gasoline (engine)	...	...	4190	...	...

Nos. 1, 2: Haslam, R. T., and Russell, R. P. *Fuels and Their Combustion,* New York, McGraw-Gill, 1926.
3–24: Shnidman, L., ed. *Gaseous Fuels,* 2nd ed., pp. 34, 128, 129. New York, A.G.A., 1954; and A.G.A. *Combustion,* 3rd ed., rev., p. 95. New York, A.G.A., 1938; and Perry, J. H., ed. *Chemical Engineers Handbook,* 3rd ed., p. 1589. New York, McGraw-Hill, 1950.
25: Jost, W. *Explosion and Combustion Processes in Gases,* p. 223. New York, McGraw-Hill, 1946.
* At 60 F, 30 in. Hg dry.
† Not corrected for dissociation.

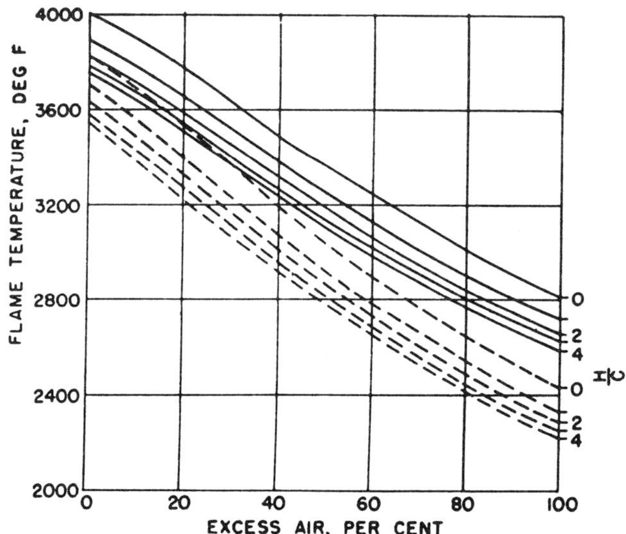

Fig. 2-13b Adiabatic flame temperature as a function of fuel composition and excess air for two different temperatures of combustion air (600 F represented by solid line, 100 F by dotted line). H/C designates hydrogen-to-carbon ratio of fuel.

retical in oxygen (Table 2-69). Differences between calculated and actual flame temperatures are caused by the following factors:

a. Radiation losses; for example, radiation loss from a Bunsen flame may be 12 to 18 per cent of the total heat.

b. All the energy in the fuel is not instantaneously released.

c. Convection losses.

d. Conduction losses; some heat is lost by direct conduction from the burner and furnace.

e. Excess air carries heat away. A slightly gas-rich mixture usually gives the highest temperature in practice; the resultant incomplete reaction results in aldehydes.[13]

f. An object in the flame lowers the flame temperature.

g. Dissociation of diatomic gases at high temperatures.

In practice, flame temperatures may be increased by:

a. Liberating the heat of combustion as fast as possible; for example, by surface catalysis.

b. Using a *minimum* of excess air and intimate mixing of gas and air.

c. Raising the temperature of the gas and air before combustion.

d. Using oxygen or air enriched with oxygen.

e. Selecting combustible gases with high flame temperature or increasing their proportion in the mixture.

f. Superposing electric energy[54] on a natural gas–air flame; this may increase flame temperature by as much as 1000°F.

Maximum flame temperature of any gas may be calculated by a trial-and-error method using the following equation:

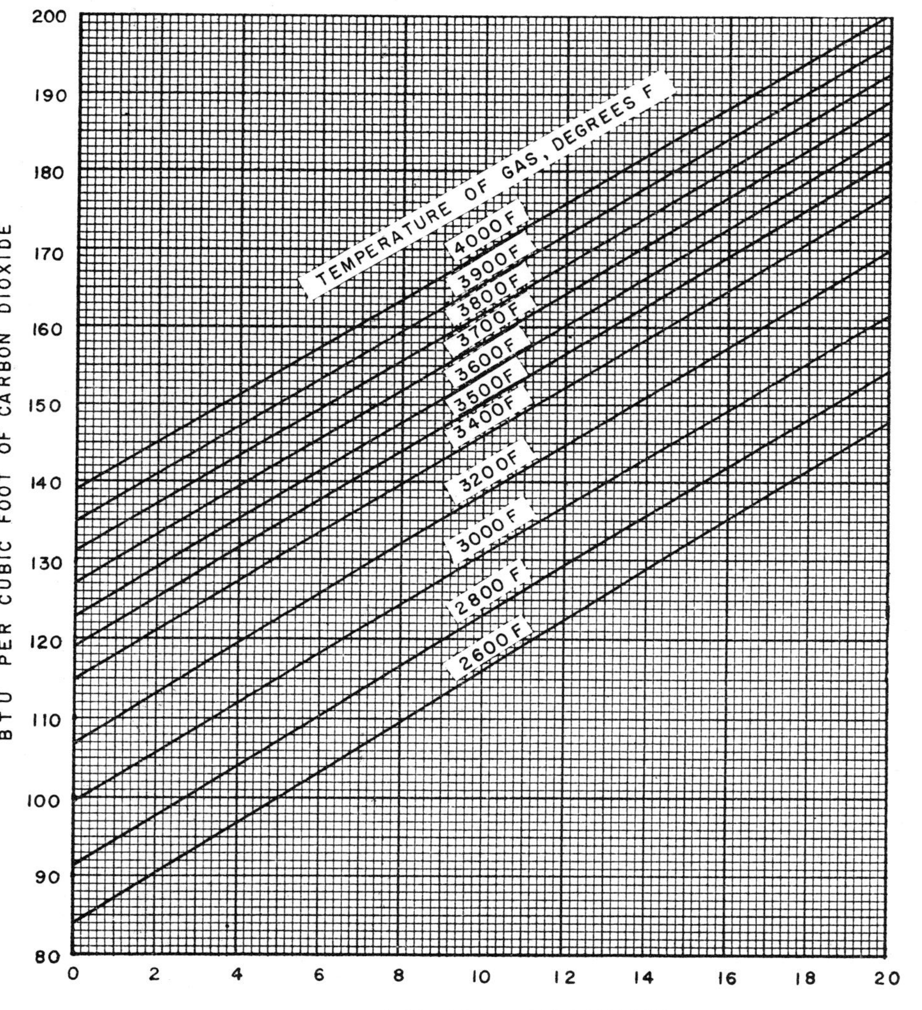

Fig. 2-14 Total heat content of carbon dioxide per cubic foot, including the effects of dissociation.

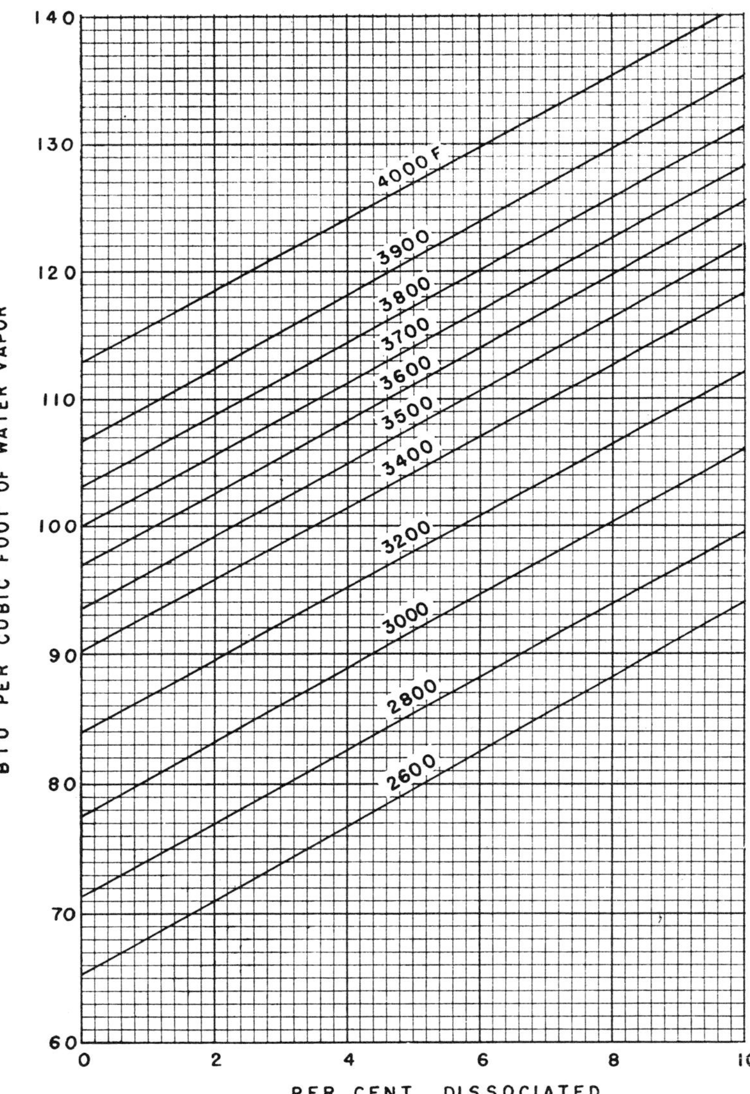

Fig. 2-15 Total heat content of water vapor per cubic foot, including the effects of dissociation.

$$H = a[R_{CO_2}x + 3/2\ R_{O_2}(1-x) + 323.5(1-x)] + b\ [R_{H_2O}y + R_{H_2}(1-y) + 1/2\ R_{O_2}(1-y) + 275.1\ (1-y)] + R_{O_2}(c+d) \quad \text{(2a)}$$

where:

H = net heating value, Btu per cu ft
R = heat content per cu ft of the various gases indicated by subscripts, above 60 F
a = cu ft of CO_2 produced per cu ft of gas burned
b = cu ft of H_2O produced per cu ft of gas burned
c = cu ft of O_2 in the flue gases per cu ft of gas burned
d = cu ft of N_2 in the flue gases per cu ft of gas burned
$(1-x)$ = fraction of CO_2 dissociated, from Fig. 2-12
$(1-y)$ = fraction of H_2O dissociated, from Fig. 2-12

In order to facilitate solution of this equation, the quantities bracketed above have been plotted in Figs. 2-14 and 2-15 for heat contents of CO_2 and H_2O, respectively. As an example of this short solution, methane with a net heating value H of 911.8 Btu per cu ft, burned with the theoretical amount of air, with both air and gas at 60 F, has the following flue gas products per cu ft of gas burned:

		Partial pressures, atm
1 cu ft CO_2	$= a$	1/10.53 = 0.095 for CO_2
2 cu ft H_2O	$= b$	2/10.53 = 0.190 for H_2O
0 cu ft O_2	$= c$	0/10.53 = 0.00 atm for O_2
7.53 cu ft N_2	$= d$	7.53/10.53 = 0.715 for N_2
10.53 cu ft of products		1.0 atm *total* pressure

Then from Fig. 2-12:*		at 3550 F		at 3500 F
$1-x=$		12.1		10.6
$1-y=$		3.0		2.7
Heat content				
CO_2	1 × 158.2†	= 158.2	1 × 152†	= 152
H_2O	2 × 103.7‡	= 207.4	2 × 101.1‡	= 202.2
N_2	7.53 × 73.3§	= 551.9	7.53 × 72.1§	= 542.9
		917.5		897.1

Equation **2a**, therefore, yields a net heating value of 917.5 as against an actual net heating value of 911.8. Hence, 3550 F is too high. Similarly, the heating value of 897.1 at 3500 F is too low. By ***interpolation***, a temperature of 3536 F would be the calculated flame temperature for methane. This is 52° higher than the calculated value for methane in Table 2-69; the negligible difference (1.5 per cent) is mainly due to the inaccuracies in reading the figures.

Flame Length Calculation.[14] The calculation of the length of **free flames** in which the effects of buoyancy are small as a result of high nozzle velocity and small (0.12- to 0.30-in. diam) ports is given by the equation:

$$L/D = \frac{5.3}{C}\left[C\frac{T_f}{AT_n} + (1-C)\frac{M_s}{M_n}\right]^{0.5} \quad \text{(2b)}$$

where:

L = visible flame length, in.
D = nozzle diameter, in.
T_f = adiabatic flame temp, °R
T_n = nozzle fluid temp, °R
M_s, M_n = molecular weight of surrounding and nozzle fluids, respectively; for example, 29 for air, 16 for methane
C = mole fraction of nozzle fluid (which may contain primary air) in the stoichiometric mixture
A = ratio of moles of reactants to moles of products for the stoichiometric mixture; e.g., methane and oxygen: 3/3 = 1

Equation **2b** predicted turbulent flame lengths varying from 40 to 290 nozzle diameters with average and maximum errors of 10 and 20 per cent, respectively. The fuels studied included propane, acetylene, H_2, CO, manufactured gas, CO_2 plus manufactured gas mixtures, and H_2 plus propane mixtures.[14]

Specific Flame Intensity.** The concept of specific flame intensity is useful in characterizing the concentration of heat available from a flame. Flame temperatures are inadequate because they relate only the level of heat and not the rate of its release.

The specific flame intensity, I, is defined as the rate of heat release per unit surface of primary flame (inner cone) in Btu per sec-sq ft. This definition is met by the equation:

$$I = VH/A \quad \text{(3a)}$$

where:

V = volumetric flow of primary gas–air mixture, SCF per sec
H = net heating value of primary *mixture*, Btu per SCF
A = inner cone surface area, sq ft; methods of measurement discussed in the first two references of Table 2-70.

If the flame is *laminar:*

$$A = V/v \quad \text{(3b)}$$

where v = burning velocity of the primary mixture, fps
Thus, for laminar flames:

$$I = vH \quad \text{(3c)}$$

To evaluate H, we must consider four types of flames:

Lean flames.

$$H = H_0X \quad \text{(3d)}$$

where: H_0 = net heating value of *pure fuel*, Btu per SCF
X = fraction of combustible gas in the primary mixture

Stoichiometric flames.

$$H = H_0X_s \quad \text{(3e)}$$

where: X_s = fraction of combustible gas in the primary stoichiometric mixture (see column 4 of Table 2-70 for H value)

* The flame temperature range was narrowed down (3550 F to 3500 F) by trial and error.
† See Fig. 2-14.
‡ See Fig. 2-15.
§ See Fig. 2-18.

** Material submitted by Joseph Grumer, Chief, Flame Research Section, Bureau of Mines, Region V, Pittsburgh, Pa.

Table 2-70 Specific Flame Intensities and Burning Velocities of Various Fuel–Air Mixtures*

Fuel	Stoichiometric mixture				Maximum burning velocity mixture			
	Fuel, per cent	Burning velocity, v, fps	Net heating value, H, Btu/SCF	Specific flame intensity, I, Btu/sec-sq ft	Fuel, per cent	Burning velocity, v, fps	Net heating value, H, Btu/SCF	Specific flame intensity, I, Btu/sec-sq ft
Methane	9.46	1.43	86.5	124	10.14	1.47	85.9	126
Methane†	9.46	1.23	86.5	106	10.10	1.25	86.0	108
Ethane	5.64	1.46	92.4	135	6.38	1.56	91.7	143
Propane	4.02	1.50	95.4	143	4.25	1.52	95.2	145
n-Butane	3.12	1.47	98.0	144	3.21	1.47	98.0	144
n-Pentane	2.54	1.40	99.1	139	2.67	1.41	98.9	140
Acetylene	7.71	4.73	111	525	9.46	5.09	109	554
Ethylene	6.52	2.23	98.7	220	7.31	2.41	97.9	236
Propylene	4.44	1.68	99.2	167	4.44	1.68	99.2	167
Carbon monoxide	29.5	...	94.8	...	46.2	1.71	72.4	124
Hydrogen	29.5	5.58	81.1	452	42.9	10.7	65.7	703
Hydrogen†	29.5	7.36	81.1	597	...	...	...	...
Natural gas‡	8.49	1.05	88.2	93	9.40	1.07	87.3	93
Natural gas	9.0	1.24§	90.0	112	9.5	1.26§	89.6	113
Coke oven gas**	16.9	2.15	86.0	185	17.6	2.20	85.3	188
Blue gas**	32.6	3.55	77.9	277	38.3	3.90	71.4	278
Producer gas**	44.7	0.65	68.4	45	53.6	0.85	57.4	49

* Except where otherwise indicated, data are from: Gibbs, G. J., and Calcote, H. F., "Effect of Molecular Structure on Burning Velocity." *J. Chem. Eng. Data* 4: 226–37, July 1959.
† Ladenburg, R. W., and others, eds. *Physical Measurements in Gas Dynamics and Combustion*, pp. 409–38. Princeton, N. J., Princeton Univ. Press, 1954.
‡ Lewis, B., and von Elbe, G. "Stability and Structure of Burner Flames." *J. Chem. Phys.* 11: 75–97, 1943.
§ Recommended value—average of entries for methane and natural gas.
** Shnidman, L., ed. *Gaseous Fuels*, 2nd ed., p. 32, 118, 185. New York, A.G.A., 1954.

Rich flames.

$$H = H_0X_s(1 - X/1 - X_s) \tag{3f}$$

The above equation is valid only if there is no appreciable diffusion of secondary air into the primary flame and the fuel is completely oxidized. It is reasonable to expect that these conditions will be met by a nearly stoichiometric rich flame, such as one with maximum burning velocity. See column 8 of Table 2-70 for H value.

Very rich flames. Diffusion of secondary air into the primary flame is important and some fuel molecules are partly oxidized and decomposed. The situation is too involved for a simple and general analysis.

Since the most likely use of specific flame intensities will be to compare fuels, the values of v preferably should come from a single source and type of measurement—but such a complete set of measurements is not yet available. Table 2-70 is based on several sources of v values. The minor inconsistencies shown should not be serious when computing flame intensities. Specific flame intensities for both stoichiometric and maximum burning velocity mixtures are given.

Ionization of Flames.[61] Nearly all experimental measurements have involved low-temperature air–gas flames. The limited data on hydrocarbon flames (acetylene, ethylene, and propane) indicate an ion concentration range of about 10^{11} to 10^{12} ions per cu cm. The ionization of CO and H_2 is lower—probably in the range of 10^7 to 2.5×10^9 ions per cu cm. The equilibrium concentration of free electrons in flames has been reported[62] to be of the order of 10^6 electrons per cu cm. While the exact nature of the ions involved has not been ascertained, it is known that the ion concentration is not uniform throughout the flame structure. There is considerable evidence of an abundance of H_3O^+. In any event, few ions survive beyond the reaction zone.

Radiation from Flames. Radiation from gas flames may be of thermal origin or of *chemiluminescent* origin.[15] The *thermal radiation portion* obeys **Kirchhoff's law** and **Planck's law** of radiation. Chemiluminescence is not a temperature function for a given material. Examples of chemiluminescence are the phosphorous vapor flame, the highly attenuated flames of alkali metals and halogens, carbon disulfide with NO or N_2O and carbon monoxide combustion at low pressures. The **inner cone** of the Bunsen flame emits OH, CH, and CC bands;[15] the **aureole** and the **outer cone** emit only OH bands with continuous background. Apparently there is essentially thermal radiation in the outer cone of a Bunsen burner flame.

It has been stated[16] that "in the ordinary Bunsen flame, from 12 to 18 per cent of the total heat of combustion is radiated away. Radiation from **luminous flames** amounts to from 10 to 40 per cent of the potential heat of the gas, depending upon the degree of luminosity and the flame temperature."

Radiation from **luminous flames** is greater than that from the flames of a clear gas which has the same temperature. In tests with natural gas,[17] values as high as 95 per cent black-body radiation were obtained under conditions that were favorable to the *cracking* of methane; but, with ordinary burners, 50 per cent was seldom exceeded at the point of highest luminosity. The value at the end of the combustion varied from about 1.5 *times* the clear gas radiation when 10 per cent excess air was present to about three *times* clear gas radiation with 15 per cent excess gas. The use of only 2 per cent of the black-body radiation is recommended as an average value in spite of the much higher test values.

Luminous flames are produced when gas is burned in a deficiency of air. Luminosity is due to the cracking of hydrocarbons and the resultant liberation of free carbon particles. Nonluminous flames, on the other hand, are produced by burning gases containing no hydrocarbons or by burning gases containing hydrocarbons with an excess of air.

The only clear gases that radiate appreciably are those having three or more atoms per molecule, such as CO_2, H_2O, SO_2, CH_4, C_2H_6, and other hydrocarbons. Carbon monoxide, although diatomic, also gives off some radiation. The diatomic gases O_2, N_2, H_2, and air have negligible radiating power. Gases do not radiate in all wave lengths as do solids; instead each gas radiates only in three or four rather sharply defined bands of wave lengths. Total radiation from clear gases depends not only on their temperature, but also on composition, thickness (up to 40 in.), and shape of the gas layer. Pseudo heat transfer coefficients for radiation from carbon dioxide to a solid and from water vapor to a solid,[17] as well as extensive radiant heat transfer data from both nonluminous gases and luminous flames,[18] are available.

Laws and Equations Employed in Combustion Calculations

The laws and equations most frequently involved follow.

Boyle's Law. The volume occupied by a given mass of gas varies inversely with the absolute pressure if the temperature remains constant.

$$PV = \text{constant} \tag{4a}$$

where: P = absolute pressure

V = volume of gas

Charles' Law. The volume of a given mass of gas is directly proportional to the absolute temperature if the pressure is constant.

$$V/T = \text{constant} \tag{4b}$$

where: V = volume of gas

T = absolute temperature

Avogadro's Law. Under the same condition of temperature and pressure, equal volumes of all gases contain the same number of molecules.

Dalton's Law. Each gas of a mixture exerts a partial pressure which the same mass of the gas would exert if it were present alone in the given space at the same temperature.

$$P_t = P_a + P_b + P_c, \text{ etc.} \tag{4c}$$

where: P_t = total pressure

P_a = partial pressure of gas a

P_b = partial pressure of gas b

P = partial pressure of gas c

$$P_a = P_t X_a \tag{4d}$$

where: P_a = partial pressure of gas a

P_t = total pressure of all gases in the mixture

X_a = volume fraction or mole fraction of gas a

The **pound-mole** is the number of pounds of a given substance equal to the molecular weight of the substance, e.g., 32 lb O_2 = 1 lb-mole O_2; 28 lb N_2 = 1 lb-mole N_2.

$$n = m/M \tag{4e}$$

where: n = number of lb-moles of gas

m = pounds of gas

M = molecular weight of gas in pounds

The **ideal gas law** is the combination of the volume, temperature, and pressure relationships of Charles' and Boyle's laws.

$$PV = nRT \tag{4f}$$

Table 2-71a gives the gas constants, R/M (specific) and R (universal), most useful to the gas engineer. It is worth noting that the *product* R would be a constant for *ideal gases*. The conversions in Table 2-71b facilitate the calculation of R (equal to PV/nT) for different units in the gas equation; e.g., when pressure P is expressed in psia in lieu of in psfa.

Table 2-71a Gas Constants and Volume of the Pound-Mole for Certain Gases

	$R' = R/M$, specific gas constant, ft per °F	M, mole wt, lb per lb-mole	R, universal gas constant, ft-lb per (lb-mole)(°F)	Mv, cu ft per lb-mole*
Hydrogen	767.04	2.016	1546	378.9
Oxygen	48.24	32.000	1544	378.2
Nitrogen	55.13	28.016	1545	378.3
Nitrogen, "atmospheric"†	54.85	28.161	1545	378.6
Air	53.33	28.966	1545	378.5
Water vapor	85.72	18.016	1544	378.6
Carbon dioxide	34.87	44.010	1535	376.2
Carbon monoxide	55.14	28.010	1544	378.3
Hydrogen sulfide	44.79	34.076	1526	374.1
Sulfur dioxide	23.56	64.060	1509	369.6
Ammonia	89.42	17.032	1523	373.5
Methane	96.18	16.042	1543	378.2
Ethane	50.82	30.068	1528	374.5
Propane	34.13	44.094	1505	368.7
n-Butane	25.57	58.120	1486	364.3
iso-Butane	25.79	58.120	1499	367.4
Ethylene	54.70	28.052	1534	376.2
Propylene	36.01	42.078	1515	379.1

* At 60 F, 30 in. Hg, dry.
† Includes other inert gases in trace amounts.

Table 2-71b Multipliers for the Universal Gas Constant* R for Various Gas Equation Units

P	psfa	psia	in. Hg†	in. H_2O†	in. H_2O†	oz/sq in.†	in. Hg†	mm H†
V	cu ft	cu ft	cu ft	cu ft	moles	cu ft	liter	liter
T	°F abs	°F abs	°F abs	°F abs	°F abs	°F abs	°C abs	°C abs
n	lb-mole	lb-mole	lb-mole	lb-mole	lb-mole	lb-mole	g-mole	g-mole
Multiply R by	1	0.00694	0.01415	0.1915	0.000508	0.111	0.00159	0.0403

* Ft-lb per (lb-mole) (°F abs).
† Absolute pressure.

Table 2-72 Coefficients in Equation of State,[73] Eq. 4g

Substance	T_c, °K	P_c, atm	B_0	B_1	B_3	C_0	C_1	C_3'	C_3''	$A_5 \times 10^4$
Nitrogen	126.26	33.54	0.15694	−0.39800	−0.083742	0.02435	−0.018047	0.030135	0.068122	0.93665
Methane	190.7	45.8	.12469	− .34697	− .11609	.028956	− .027045	.038313	.051401	.84333
Ethylene	283.06	50.5	.12107	− .31262	− .15322	.040655	− .052715	.053650	.043633	.96444
Ethane	305.4	48.2	.12073	− .31889	− .14773	.041138	− .050941	.051839	.043646	.88197
Propylene	365.1	45.4	.12891	− .30919	− .16666	.042959	− .059338	.059011	.042004	.94108
Propane	369.97	42.01	.13466	− .31326	− .16926	.043087	− .059781	.059449	.042129	.96188
iso-Butane	408.14	36.0	.14785	− .23844	− .16377	.049035	− .066854	.059241	.039288	.89192
iso-Butylene	419.6	39.7	.13378	− .29984	− .17638	.046289	− .065364	.060297	.039347	.91276
n-Butane	425.17	37.47	.13357	− .31046	− .16908	.046140	− .062238	.057873	.039220	.84921
iso-Pentane	461.0	32.9	.13920	− .29421	− .18893	.050540	− .075114	.065397	.035024	.84012
n-Pentane	469.78	33.31	.13545	− .27302	− .21546	.049889	− .078933	.072340	.035469	.92185
n-Hexane	507.90	29.92	.12766	− .24870	− .22166	.056248	− .088104	.072525	.034370	.91543
n-Heptane	540.16	27.01	0.121272	−0.24089	−0.22363	0.056430	−0.086841	0.070928	0.033422	0.85608

Equation of State.[73] Real gases deviate from the ideal Eq. **4f.** Many equations have been developed which more closely reflect the behavior of real gases. For example, Eq. **4g** follows the *Law of Corresponding States,* which says that all substances behave similarly when at states which bear the same relation to critical temperatures and pressures. Note that critical temperature is an explicit parameter in Eq. **4g.**

Evaluating Z in the manner indicated under this equation results in reasonably accurate determinations for *any* pure gas whose critical temperature is known, as well as for any mixture of gases for which the pseudocritical temperature can be computed. Pseudocritical constants of mixtures may be taken (except near the pseudocritical condition) as the summation of the respective constants for each mixture component in proportion to its mole fraction.

$$Pv = ZRT \tag{4g}$$

where:

P = gas pressure, psfa
v = gas volume, cu ft per lb-mole
R = universal gas constant, ft-lb per (lb-mole) (°F)
T = absolute temperature, °R
Z = compressibility factor = $1 + (A_1/V_R) + (A_2/V_R^2) + (A_4/V_R^4) + (A_5/V_R^5)$
$A_1 = B_0 + (B_1/T_R) + (B_3/T_R^3) + (B_5/T_R^5)$
$A_2 = C_0 + (C_1/T_R) + (C_3/T_R^3) + (C_5/T_R^5)$
$A_4 = (C_3 + C_5)C_3''$
$A_5 = 8.4333 \times 10^{-5}$
V_R = fraction of *ideal* volume calculated at critical pressure and temperature, dimensionless
B_5 = 0.009365, for N_2 only*
$C_3 = C_3' e^{(-C_3''/V_R^2)}$
$C_5 = C_5' e^{(-C_3''/V_R^2)}$
T_R = reduced temperature, dimensionless
e = base of natural logarithms
C_5' = 0.003311, for N_2 only*
Other terms as given in Table 2-72.

Volume and temperature, rather than temperature and pressure, are the independent variables in Eq. **4g.** Equation **4h** may be used in a trial-and-error fashion to approximate volume for substitution into Eq. **4g.**

$$Z = (1/A) + B \tag{4h}$$

where:

$$A = 1.0 + 0.6108A_1 + 1.107A_1^3 + 0.4294A_1^5 - 6.747A_1^7 + 4.211A_1^9 + 43.44A_1^{11} - 52.19A_1^{13} - 40.7A_1^{15} + 57.68A_1^{17}$$

$$A_1 = P_R/T_R^3$$

$$B = (0.2595 - 0.0929B_1 + 0.01437B_1^2)(P_R/T_R)$$

$$B_1 = P_R^{1/3}$$

The foregoing constants may also be used to calculate the entropy, $S - S_0'$,† and enthalpy, $H - H_0'$,† for various gases (Table 2-72) and their mixtures:

$$(S - S_0')/R = [(C_3'/T_R^3) + (2C_5'/T_R^5)] \times (2/C_3'')\,[1 - e^{(-C_3''/V_R^2)}] - [B_0 - (2B_3/T_R^3) - (4B_5/T_R^5)]\,(1/V_R) - \{(C_0/2) + [(C_3'/T_R^3) + (2C_5'/T_R^5)] \times e^{(-C_3''/V_R^2)}\}\,(1/V_R^2) \tag{4i}$$

$$(H - H_0')/RT = [(3C_3'/T_R^3) + (5C_5'/T_R^5)]\,[(1 - e^{(-C_3''/V_R^2)})/C_3''] + [B_0 + (2B_1/T_R) + (4B_3/T_R^3) + (6B_5/T_R^5)]\,(1/V_R) + \{C_0 + (3C_1/2T_R) - [(C_3'/2T_R^3) + (3C_5'/2T_R^5)]e^{(-C_3''/V_R^2)}\}(1/V_R^2) + [(C_3'/T_R^3) + (C_5'/T_R^5)][e^{(-C_3''/V_R^2)}(C_3''/V_R^4)] + (6A_5/5V_R^5T_R) \tag{4j}$$

First Law of Thermodynamics. This law states the principle of conservation of energy, i.e., energy can neither be created nor destroyed. A thermodynamic system has a characteristic property (parameter of state) called its *internal energy.* The energy of a system is increased by the addition of heat and decreased by the work performed by it. *Total amount of energy of an isolated system is constant.*

† S_0' and H_0' are the ideal-gas values of entropy and enthalpy, respectively (at the same temperature and pressure).

$$S_0' = \int_{t_0}^{t} C_p \frac{dT}{T} - R \ln P$$

$$H_0' = \int_{t_0}^{t} C_p\, dT$$

where C_p is the specific heat at zero pressure. To integrate these expressions, define C_p as a function of temperature; for example, a power series in T.

* These values are to be used only in conjunction with Table 2-72 data; zero is to be used for other substances in Table 2-72.

$$dU = dQ - dW \quad (5)$$

where: dU = increase in internal energy of the system
dQ = heat absorbed by the system
dW = work done by the system

Enthalpy. This is a property of the system defined as:

$$H = U + pV \quad (6)$$

where: H = enthalpy of the system
U = internal energy of the system
p = pressure of the system
V = volume of the system
and change in enthalpy,

$$dH = dQ + Vdp \quad (6a)$$

where dp = change in pressure

so that in a constant pressure process the change in enthalpy equals the heat exchanged by the system with its surroundings, i.e., $Vdp = 0$.

Second Law of Thermodynamics. Three statements of the second law of thermodynamics follow:

Kelvin: It is impossible by means of any material agency to derive mechanical effect from any portion of matter by cooling it below the temperature of its surroundings.

Planck: It is impossible to construct an engine which, working in a complete cycle, will produce no effect other than the raising of a weight and the cooling of a heat reservoir.

Kelvin–Planck: It is impossible to construct an engine that, operating in a cycle, will produce no effect other than the extraction of heat from a reservoir and the performance of an equivalent amount of work.

Entropy. There exists a property of state of a thermodynamic system defined as

$$dS = dQ/T \quad (7)$$

where: dS = change in entropy
dQ = heat exchanged by the system
T = absolute temperature

Entropy is calculated by assuming that the state of a system is changed from an arbitrary reference state to the actual state thru a sequence of infinitesimal equilibrium states by summing up the term dQ/T; that is:

$$S_2 - S_1 = \text{entropy change} = \int \frac{dQ}{T} \quad (7a)$$

(for a reversible process)

Carnot cycle. The Carnot cycle consists of two isothermal and two adiabatic processes, and has the maximum efficiency.

$$\text{Efficiency of a Carnot cycle} = \frac{\text{work done}}{\text{heat supplied}} = \frac{T_1 - T_2}{T_1} \quad (8)$$

where: T_2 = final absolute temperature
T_1 = initial absolute temperature

Third Law of Thermodynamics. When the temperature and pressure of the products for a given reaction are the same as those of the reactants, the equation may be written from the definition of free energy (Gibb's function).

$$\Delta G = \Delta H - T\Delta S \quad (9)$$

where: ΔG = change in maximum useful work or free energy
ΔH = change in heat capacity (enthalpy)
ΔS = change in entropy
T = absolute temperature

In order to find the values of ΔH and ΔS of a gas at a temperature, T, and low pressures, the following equations may be used for any substance that normally exists in only one crystalline phase:

$$H = H_0 + \int_0^{T_{mp}} C_p \text{ (crystalline) } dT + \Delta H \text{ (fusion)} + \int_{T_{mp}}^{T_{bp}} C_p \text{ (liquid) } dT + \Delta H \text{ (vaporization)} + \int_{T_{bp}}^{T} C_p \text{ (gas) } dT \quad (10)$$

$$S = S_0 + \int_0^{T_{mp}} C_p \text{ (crystalline) } d(\ln T) + \frac{\Delta H \text{ (fusion)}}{T} + \int_{T_{mp}}^{T_{bp}} C_p \text{ (liquid) } d(\ln T) + \frac{\Delta H \text{ (vaporization)}}{T_{bp}} + \int_{T_{bp}}^{T} C_p \text{ (gas) } d(\ln T) \quad (11)$$

where H_0 and S_0 are the heat content and entropy, respectively, of the substance at absolute zero.

These equations may be altered for substances having *more than one* crystalline phase by adding the terms involving heat or entropy of transition. In similar form, Eqs. **10** and **11** are:

$$H = H_0 + \int_0^T dQ \quad (12)$$

$$S = S_0 + \int_0^T \frac{dQ}{T} \quad (13)$$

where dQ is the heat absorbed by the substance at constant pressure.

Combining Eqs. **9**, **12**, and **13**:

$$\Delta G = \Delta H_0 + \Delta \int_0^T dQ - T\left(\Delta S_0 + \Delta \int_0^T \frac{dQ}{T}\right) \quad (14)$$

or

$$\frac{\Delta G}{T} = \frac{\Delta H_0}{T} + \frac{1}{T} \Delta \int_0^T dQ - \Delta \int_0^T \frac{dQ}{T} - \Delta S_0 \quad (15)$$

Since all the terms on the right side of Eq. **15** may be evaluated except ΔS_0, this can be a powerful equation in evaluating ΔG. As a result of the studies of the problem of evaluating ΔS_0 for chemical reactions, the principle that is today commonly called the *third law of thermodynamics* was evolved. It may be stated as follows:

Planck's statement: As the temperature of a system (in a pure quantum state) approaches absolute zero, its entropy tends to approach a constant value.

Extension of Planck's statement: The entropy of any system vanishes at zero temperature.

Fugacity. Some formulas for real gases can be transcribed to forms having a simplicity resembling that of the perfect gases by introduction of the fictitious pressure called *fugacity*. This term indicates the tendency of a gas to escape from a gas (and/or vapor) mixture, and is defined as:

$$G - G_0 = RT \ln f \qquad \text{(when } T \text{ is constant)} \qquad \textbf{(16)}$$

$$f/p \text{ approaches 1 as } p \text{ approaches 0}$$

where: f = fugacity
p = absolute pressure
G = free energy at absolute temperature, T
G_0 = free energy at absolute zero
R = gas constant
ln = logarithm to the base e

Ultimate CO_2. This is the maximum percentage of CO_2 which can be found in the flue gases under **perfect combustion** conditions. Note illustrative problem at right.

Ultimate % CO_2 =
(no excess air)

$$\frac{\text{cu ft of } CO_2 \text{ per cu ft of gas} \times 100}{\text{total dry cu ft of flue gas}} \qquad \textbf{(17)}$$

$$\underset{\text{(excess air present)}}{\text{Ultimate \% } CO_2} = \frac{\% \, CO_2 \text{ in flue gas sample} \times 100}{100 - (\% \, O_2 \text{ in same sample}/0.21)} \qquad \textbf{(18)}$$

Calculation of Flue Gas Volume.

Volume of flue gases (dry) =

$$\frac{\text{cu ft } CO_2 \text{ produced per cu ft gas burned} \times 100}{\% \, CO_2 \text{ by analysis}} \qquad \textbf{(19)}$$

Excess air = volume flue gases (dry) from Eq. **19** *minus* flue gas volume (dry) for perfect combustion. **(20)**

Calculation of Specific Gravity from the Gas Analysis. Multiply the specific gravity of each component gas from Table 2-63 by the percentage of the component in the mixture and add the results.

Example: Find the specific gravity of the following gas composition—93.09% CH_4, 2.73% C_2H_6, 0.30% C_3H_8, and 3.88% N_2.

$$\begin{aligned} 0.9309 \times 0.5543 &= 0.51600 \\ .0273 \times 1.0488 &= .02863 \\ .0030 \times 1.5617 &= .00047 \\ 0.0388 \times 0.9718 &= 0.03771 \\ \hline \text{Sp gr of dry mixture} &= 0.58281 \end{aligned}$$

Sp gr of saturated mixture =

$$\frac{0.5828 + (0.622 \times 0.0174)}{1.0174} = 0.583$$

Calculation of Gross Heating Value from the Gas Analysis. The gross heating value of a gas is the number of Btu liberated by the **complete** combustion of one cubic foot of gas at a constant pressure of 30 in. Hg at 60 F, with air at the same temperature and pressure and with the water formed by combustion condensed to the liquid state.

Multiply the gross heating value of each component gas from Table 2-59 by the percentage of each component in the mixture and add the results.

Example: Find the gross heating value of the natural gas used in the preceding illustration.

$$\begin{aligned} 0.9309 \times 1012.8 &= 945 \\ .0273 \times 1792 &= 48.4 \\ .0030 \times 2592 &= 7.8 \\ 0.0388 \times 0 &= 0 \\ \hline \text{Gross heating value of dry mixture} &= 1001.2 \end{aligned}$$

On a saturated basis the heating value is 1.74 per cent less than on a dry basis, or 983.8.

Calculation of Net Heating Value from the Gas Analysis. The net heating value of a gas has the same definition as the gross heating value, except that the water vapor formed remains in the *vapor* state.

Multiply the net heating value of each gas from Table 2-59 by the percentage of each gas in the mixture and add the results. The difference between gross and net heating values for most gases is between seven and ten per cent.

The following calculation is offered as a guide for the determination of the *difference* between the *gross* and *net* Btu values of a gas.

By volume: $CH_4 + 2O_2 \rightarrow CO_2 + 2H_2O$
At 60 F (from steam tables):

saturation pressure = 0.2563 psia
latent heat of condensation = 1060 Btu per lb
specific volume of sat. vapor = 1207 cu ft per lb

Therefore, the quantity of heat removed in condensing a cubic foot of water vapor at 60 F and 0.2563 psia = 1060/1207 = 0.88 Btu.

However, since the pressure is usually specified at 30 in. Hg (equal to approximately 14.7 psia): 0.88 × (14.7/0.2563) = 50.4 Btu per cu ft *of water vapor.*

The combustion of 1 cu ft of CH_4 yields 2 cu ft of H_2O. Therefore, 2 × 50.4 = 100.8 Btu "lost"* in the combustion of 1 cu ft of methane. In other words, for methane, the gross Btu value *exceeds* the net Btu value by approximately 100 Btu.

Illustrative Combustion Problem. Calculate the air required for a boiler output of 1000 lb per hr of steam at 30 psig (using water at 60 F) at 4000-ft elevation, 60 F, and 40 per cent relative humidity. Use the natural gas described at left (at standard conditions: 60 F and 30 in. Hg). Assume 25 per cent excess air for boiler and 70 per cent efficiency. Also find the ultimate per cent CO_2.

Barometric pressure at 4000 ft (Fig. 8-40) = 12.65 psi
Output steam pressure = 30 psig + 12.65 psi = 42.65 psia
h_g = enthalpy of saturated steam at 42.65 psia (steam tables) = 1170.5 Btu per lb
h_f = enthalpy of feed water = 28.1 Btu per lb

Net heat transferred to steam = 1170.5 − 28.1 = 1142.4 Btu per lb or a total of 1,142,400 Btu per hr for the 1000-lb requirement.

Input = output/efficiency = 1,142,400/0.7 = 1,632,000 Btu per hr

or 1630 cu ft per hr of 1001 Btu gas (top of column) at 60 F and 30 in. Hg, or X cu ft per hr *actual* gas at 60 F and 4000-ft elevation, where X = 1630 × (30/26) = 1881 cu ft.

* Unavailable for useful purposes unless the water vapor is condensed in the process.

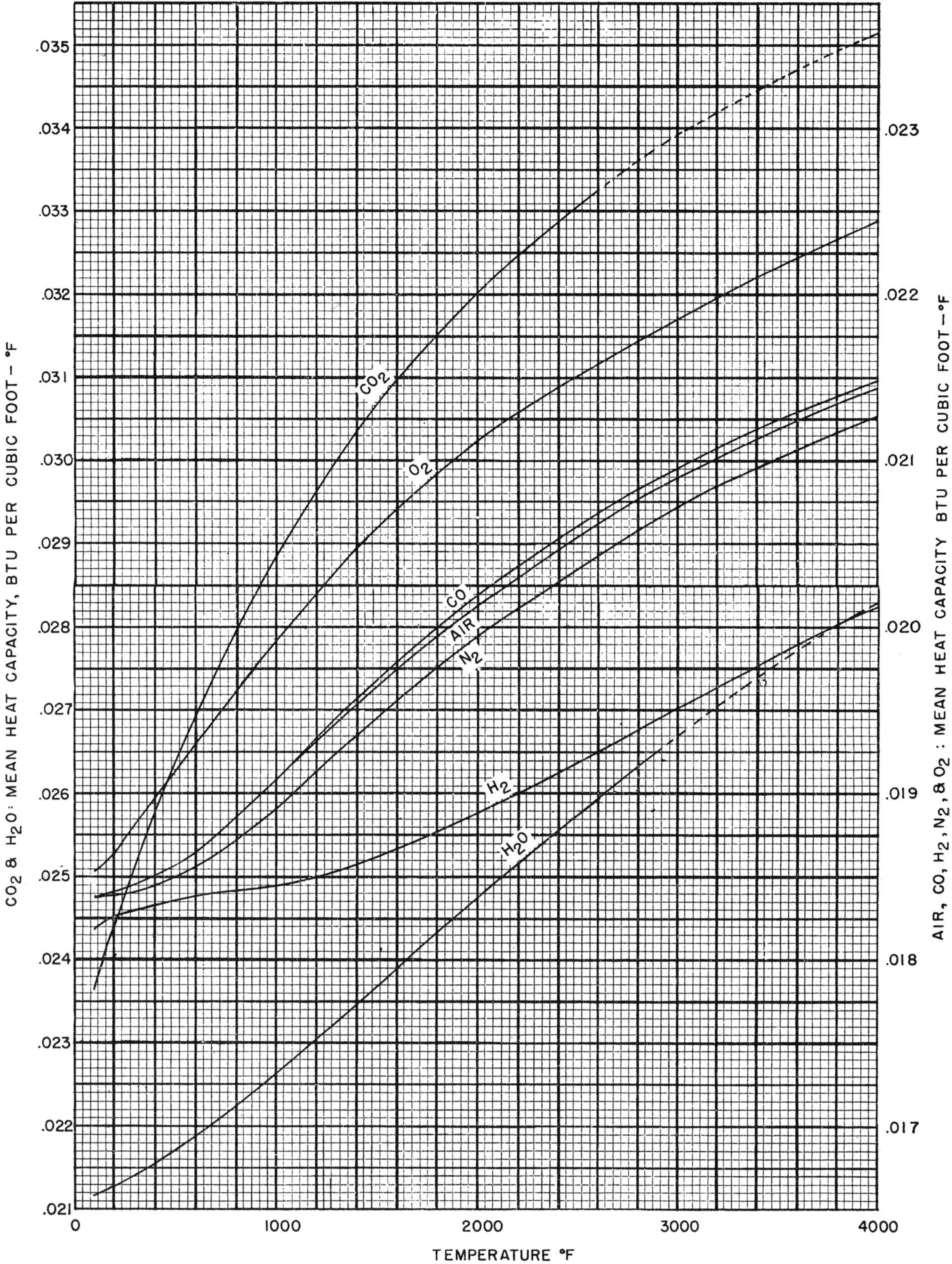

Fig. 2-16 Mean heat capacity of gases found in flue products from 60 to t°F (dashed line indicates dissociation).

Combustion Calculations.

	Mixture before combustion, cu ft			Products of combustion, cu ft			
	Gas	O_2 reqd.	Air reqd.	H_2O	CO_2	N_2	Totals
CH_4	0.931	1.862*	...	1.862	0.931	...	2.793
C_2H_6	.027	0.095*	...	0.081	.054	...	0.135
C_3H_8	.003	0.015*	...	0.012	0.009	...	.021
N_2	0.039	...	...	...	...	0.039	0.039
Dry air	...	...	9.39†	...	...	7.418‡	7.418
H_2O	...	...	0.07§	0.07	...	...	0.07
	1.000	1.972	9.46	2.025	0.994	7.457	10.476

* See Table 2-63.

† Air required = oxygen required/0.21 = 1.972/0.21 = 9.39.

‡ 9.39 − 1.972 = 7.418

Actual air requirement based on 25 per cent excess air = 1881 cu ft gas/hr × 9.46 cu ft air/cu ft gas × 1.25 = 22,260 cu ft of actual air per hr.

Volume of dry products = 10.476 − 2.025 = 8.451 cu ft per cu ft of gas.

Ultimate CO_2 = (0.994/8.451) × 100 = 11.75 per cent.

§ $e = RH \times e_s = 0.40 \times 0.522 = 0.209$ in. Hg

where: e = actual water vapor pressure of air at 4000-ft elevation

RH = relative humidity = 40 per cent

e_s = saturated vapor pressure at 60 F = 0.522 in. Hg

Therefore, the quantity of water vapor per cubic foot of air at 40 per cent RH = (0.209/30) × 9.39 = 0.07 (low-pressure water vapor approximates a perfect gas).

SPECIFIC HEAT

The general definition of the specific heat of a substance is the *amount* of heat required to raise the temperature of a *quantity* of a substance one degree. The units of amount may be in Btu, calories, etc., while quantity may be in terms of pounds, cubic feet, moles, etc. Temperature may be expressed in either °F or °C.

Consideration should be given to the fact that specific heat varies with temperature and pressure. In this regard the two most important specific heats are the specific heat at constant pressure, c_p, and the specific heat at constant volume, c_v.

To avoid confusion between molar specific heats, Btu per (lb-mole)-°F, and the more common specific heats expressed in Btu per lb-°F, this chapter employs the capital letter C for the former and the small c for the latter. Ratios of specific heats may be written either way.

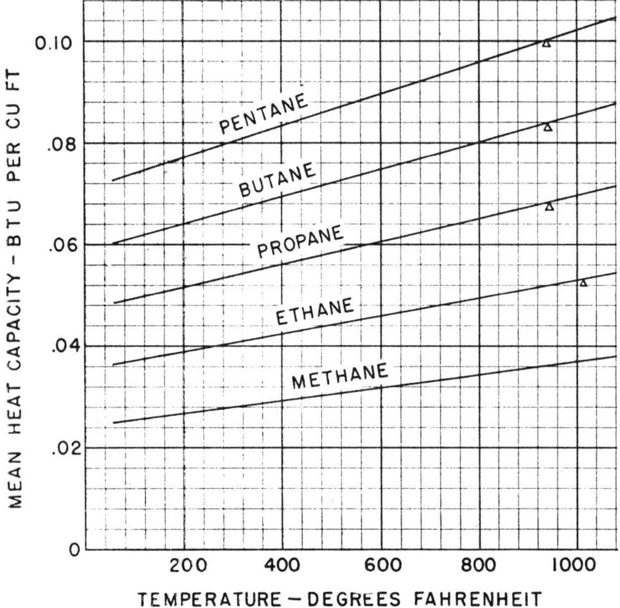

Fig. 2-17 Mean heat capacities of hydrocarbons of the paraffin series, Btu per cu ft-°F. Dissociation not considered; Δ denotes temperature at which dissociation begins.

Mean Heat Capacity

The most subtle pitfall in the subject of specific heat is encountered when the quantity of the gas is expressed in units of volume rather than weight, for while a pound of gas remains a pound at any temperature, a cubic foot of a gas at a low temperature has a larger volume at a higher temperature if the pressure remains constant. This complication is resolved for the common flue gas components and the paraffin hydrocarbons in Figs. 2-16 and 2-17, where the *mean* heat capacity is solved for.

The curves in Fig. 2-16 are based on *Gas Tables* by J. H. Keenan and J. Kaye, New York, Wiley, 1948. The following equations were used for mean heat capacities:

Air: $(28.97h - 3598.074)/378.4(t - 60)$

CO: $(\bar{h} - 3609.1)/(378.3)(t - 60)$

CO_2: $(\bar{h} - 3877.7)/(376.2)(t - 60)$

H_2: $(\bar{h} - 3521.24)/(378.9)(t - 60)$

H_2O: $(\bar{h} - 4119.6)/(378.6)(t - 60)$

N_2: $(\bar{h} - 3609.2)/(378.6)(t - 60)$

O_2: $(\bar{h} - 3604)/(378.2)(t - 60)$

where: h = enthalpy, Btu per lb

$\bar{h}$ = enthalpy, Btu per lb-mole

The mean heat content equals $(t - 60)$ *times* the mean heat capacity.

Mean heat capacity takes account of the variation of specific heat with temperature and is based on a cubic foot of gas at 60 F and one atmosphere. For example, if the temperature of a cubic foot of water vapor at 60 F is raised to 1600 F, the mean heat capacity is 0.02389 Btu per (original cubic foot at 60 F) (Δ°F) from Fig. 2-16, or a heat content of 0.02389 × (1600 − 60) = 36.8 Btu per original cubic foot at 60 F. This

Table 2-73 Specific Heat, C_p, of Various Gases at One Atmosphere in Btu per (Lb-Mole)-°F[20]

°F	Air	CO_2	CO	C_2H_4	CH_4	H_2	N_2	O_2
−103	6.94	7.72	6.96	8.42	8.00	6.53	6.96	6.96
−13	6.94	8.32	6.96	9.35	8.16	6.76	6.96	6.98
32	6.95	8.60	6.96	9.78	8.33	6.84	6.96	7.00
77	6.96	8.89	6.96	10.28	8.52	6.89	6.96	7.02
122	6.96	9.40	6.98	11.43	9.04	6.95	6.97	7.09
212	6.99	9.64	7.00	12.02	9.37	6.97	6.98	7.15
302	7.04	10.09	7.03	13.26	10.07	6.98	7.00	7.25
482	7.17	10.84	7.16	15.34	11.44	7.00	7.10	7.49
932	7.62	12.20	7.59	19.64	14.79	7.07	7.48	8.03
1382	7.92	13.06	7.97	22.60	17.64	7.24	7.85	8.37
1832	8.13	13.64	8.24	24.76	19.54	7.49	8.14	8.59
2282	8.32	14.03	8.44	26.34	20.84	7.74	8.35	8.76
2732	8.47	14.32	8.58	27.52	21.84	7.99	8.50	8.91
3182	8.60	14.52	8.63	28.32	22.64	8.20	8.62	9.05
3632	8.73	14.68	8.75	...	23.24	8.39	8.70	9.19
4082	8.84	14.82	8.82	...	23.64	8.54	8.77	9.33
4532	8.96	14.93	8.86	...	24.00	8.69	8.82	9.45
4982	9.04	15.01	8.91	...	24.26	8.81	8.87	9.53

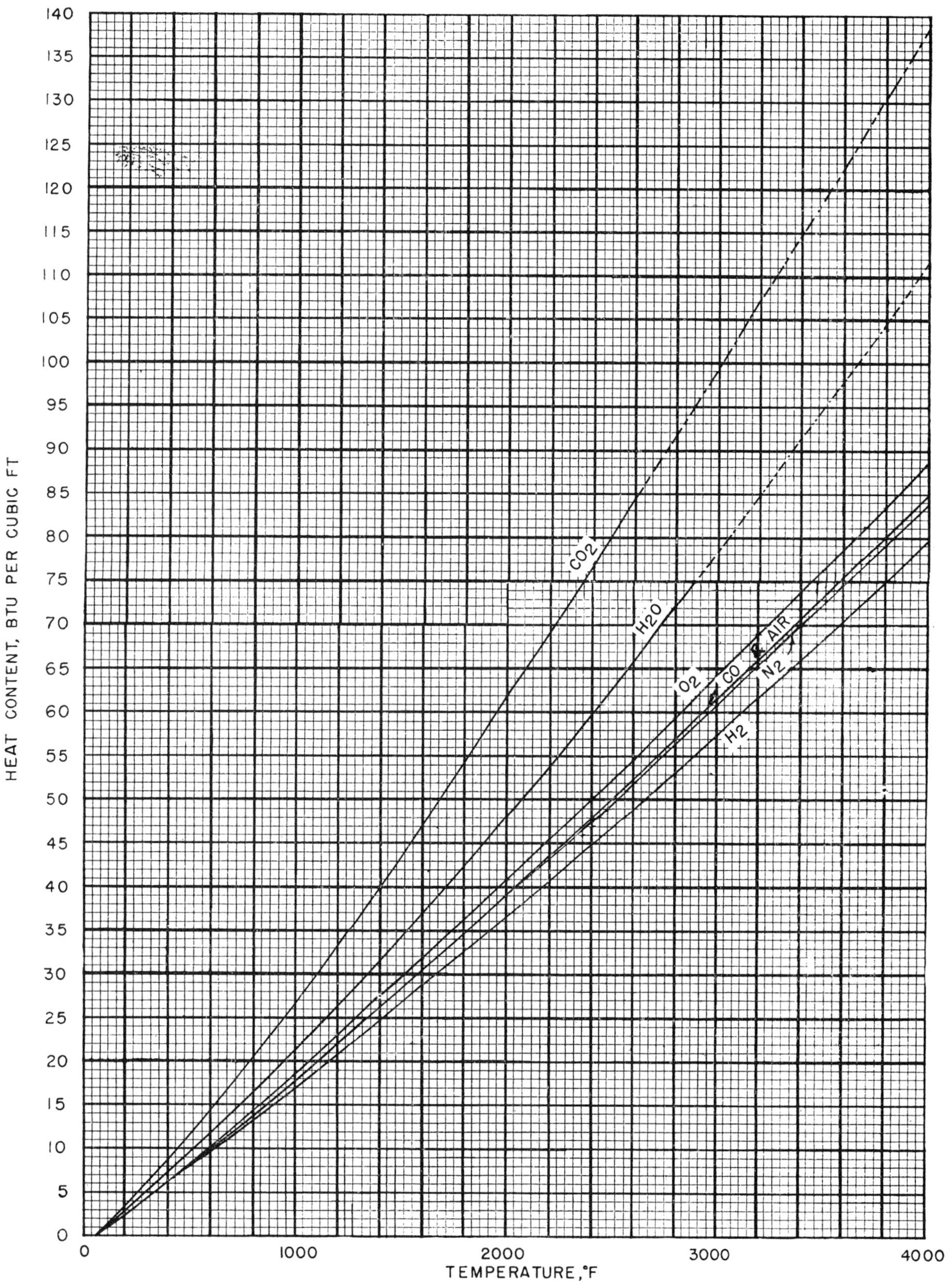

Fig. 2-18 Heat content of gases found in flue products, based on gas volumes at 60 F (dashed line indicates dissociation).

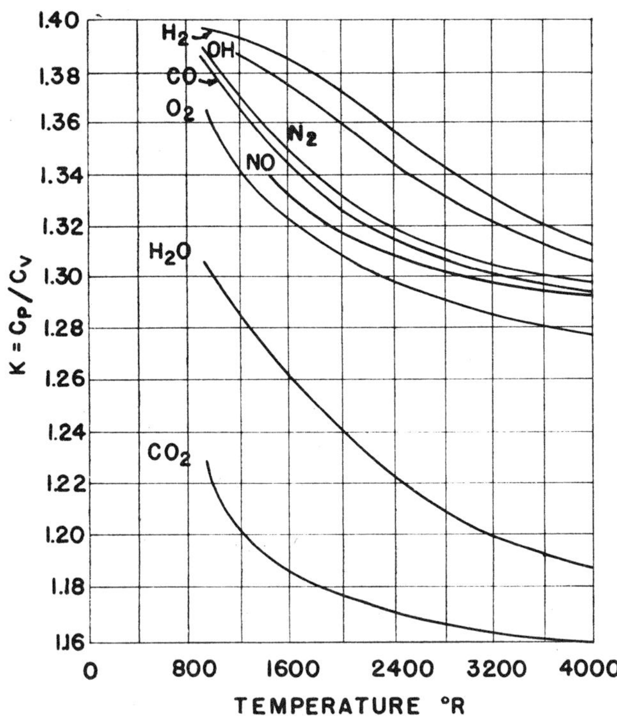

Fig. 2-19 Variation of k with temperature for products of combustion (where $k = c_p/c_v$).

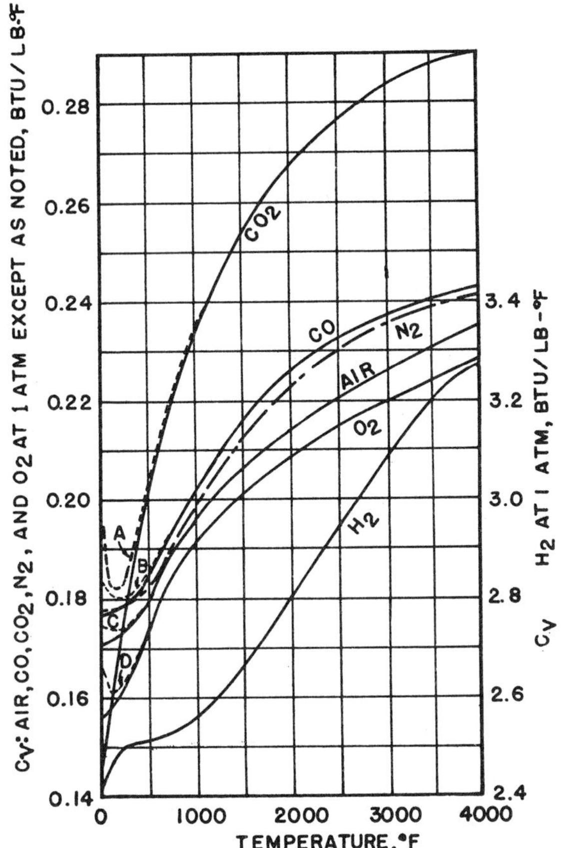

Fig. 2-20 Effect of temperature and pressure on c_v (specific heat at constant volume) of various gases.

answer may be verified by Fig. 2-18, which gives the heat *content* of gases found in flue products based on the mean heat capacity explained above.

Dissociation effects of carbon dioxide and water vapor should be accounted for at temperatures of 2600 F and above by use of Figs. 2-12, 2-14, and 2-15. An example using these figures is under *Flames* in this chapter.

Specific heat data are presented in a number of forms in this handbook in order to minimize the work, and possible errors, experienced in conversion calculations and interpolations.

Molar Specific Heat

Table 2-73, C_p in Btu per (lb-mole)-°F at one atmosphere, is based on spectroscopic data at *zero* pressure which are almost indistinguishable from the values at one atmosphere. Equations **21** thru **29** also give the instantaneous molar specific heats of various gases at one atmosphere in Btu per (lb-mole)-°F.[19] There is a maximum error of 1.8 per cent in the temperature ranges shown.

Temperature range, °R	*Molar specific heat*	
540–5000	C_p for $O_2 = 11.515 - \dfrac{172}{T^{0.5}} + \dfrac{1530}{T}$	**(21)**
540–9000	C_p for $N_2 = 9.47 - \dfrac{3.47 \times 10^3}{T} + \dfrac{1.16 \times 10^6}{T^2}$	**(22)**
540–9000	C_p for $CO = 9.46 - \dfrac{3.29 \times 10^3}{T} + \dfrac{1.07 \times 10^6}{T^2}$	**(23)**
540–4000	C_p for $H_2 = 5.76 + \dfrac{0.578}{1000} T + \dfrac{20}{T^{0.5}}$	**(24)**
540–5400	C_p for $H_2O = 19.86 - \dfrac{597}{T^{0.5}} + \dfrac{7500}{T}$	**(25)**
540–6300	C_p for $CO_2 = 16.2 - \dfrac{6.53 \times 10^3}{T} + \dfrac{1.41 \times 10^6}{T^2}$	**(26)**
540–1500	C_p for $CH_4 = 4.52 + 0.00737\ T$	**(27)**
350–1100	C_p for $C_2H_4 = 4.23 + 0.01177\ T$	**(28)**
400–1100	C_p for $C_2H_6 = 4.01 + 0.01636\ T$	**(29)**

where T = absolute temperature and °R = °F + 460

Ratio of Specific Heats

The specific heat at constant volume equals the specific heat at constant pressure *minus* the *universal gas constant*, 1.9865, or:

$$C_v = C_p - R, \text{ Btu per (lb-mole)-°F} \qquad \textbf{(30)}$$

Table 2-74 Specific Heat, Boiling Point, and Latent Heat of Vaporization of Various Gases and Liquids

No.	Gas or liquid	Boiling point, °F at 1 atm	Specific heat, Btu per lb-°F* — Gas c_p	Gas c_v	Gas k = c_p/c_v	Liquid, c_p	Latent heat of vaporization at bp, Btu/lb	Latent heat of vaporization at 60 F, Btu/lb
2	Carbon (coke)	...	0.2000	...	...	...	...	...
4	Carbon monoxide	−310	0.2484	0.1779	1.395	...	90.6	...
5	Hydrogen	−423	3.4460	2.443	1.412	...	194.5	...
6	Methane	−258.9	0.526	0.400	1.315	...	220.8	...
7	Ethane	−128.2	.409	.347	1.18	0.78	210.5	100†
8	Propane	−43.7	.388	.343	1.13	.58†	183.3	152.6
9	*n*-Butane	31.1	.397	.361	1.10	.55	165.5	162†
10	*iso*-Butane	10.9	.387†	.348	1.11	.56	157.5	143†
11	*n*-Pentane	96.9	.3974	.3699	1.07	.557	153.59	...
12	*iso*-Pentane	82.1	.388	.3605	1.076	.562	145.66	...
13	Neopentane	49.1	.391	.3635	1.076	.543	135.6	...
14	*n*-Hexane	155.7	.3984	.3753	1.062	.536	144.7	...
15	Neohexane	121.5	.3984	.3753	1.062	.511	132.6	...
16	*n*-Heptane	209.17	.3992	.3794	1.052	.525	137.5	...
17	Triptane	177.6	.3992	.3794	1.052	.497	124.6	...
18	*n*-Octane	258.2	.3998	.3824	1.046	.526	131.65	...
19	*iso*-Octane	210.6	.3998	.3824	1.046	.489	116.7	...
20	Ethylene	−154.9	.363	.296	1.22	...	207.6	...
21	Propylene	−53.8	.363	.316	1.15	.57†	188.2	149†
22	Butylene	20.7	.371	.334	1.11	.53	168.0	157.2
23	*iso*-Butene	19.2	.375	.335	1.12	.55†	168.7	156†
24	*n*-Pentene	85.95	.380	.352	1.08	...	...	...
25	Benzene	176.18	.342‡	.317	1.08	.410	169.3	...
26	Toluene	231.1	.347‡	.345	1.06	.404	156.2	...
27	*p*-Xylene	281.03	...	...	...	0.407	146.1	...
28	Acetylene	−128	.689	.547	1.26	...	...	...
29	Naphthalene	420	.325	...	...	64.1 heat of fusion	...	...
30	Methyl alcohol	148	.701	.583	1.203	...	471	...
31	Ethyl alcohol	173	.731	.646	1.13	...	367	...
32	Ammonia	−27	.940	.718	1.310	...	589	...
33	Sulfur	831	.190	...	...	16.9 heat of fusion	...	...
34	Hydrogen sulfide	−81	.456	.345	1.32	...	237	...
36	Formic acid	213.8	...	...	...	...	216	...
37	Acetaldehyde	69.8	...	...	1.14	...	245	...
38	Nitric oxide	−242	.419	.2995	1.40	...	...	...
39	Nitrogen tetroxide	64.4	...	...	...	...	168	...
40	Oxygen	−297	.219	.1565	1.397	...	91.3	...
41	Nitrogen	−319	.2485	.1772	1.400	...	85.7	...
42	Carbon dioxide	−112 subl.	.1989	.1535	1.295	...	157 at −76	...
43	Sulfur dioxide	14	.273	.2115	1.29	...	171	...
44	Water vapor	212	.446	.334	1.335	...	969	...
45	Air	...	0.2397	0.1711	1.400	...	91.6	...

Nos. 4, 5, 40–42, 44, and 45: Keenan, J. H., and Kaye, J., *Gas Tables*, New York, Wiley, 1948.
6–10, 20–23: L. Shnidman, ed. *Gaseous Fuels*, 2nd ed., New York, A.G.A., 1954.
11–19, 24–27: Phillips Petrol. Co. *Hydrocarbons*, 3rd ed. (Bull. 255) Bartlesville, Okla., 1949.
28–39: C. D. Hodgman, ed. *Handbook of Chemistry and Physics*, 41st ed., Cleveland, Ohio, Chemical Rubber Pub. Co., 1959–60.
Numbers correspond to those in Table 2-59.

* At 60 F, 30 in. Hg, dry.
† Estimated value.
‡ At 250 F.

Note that the values for R in Table 2-71 must be divided by 778 ft-lb per Btu if the units of specific heat desired are Btu per (lb-mole)-°F.

The above relationship was used in plotting Fig. 2-19, which shows how the ratio of C_p to C_v is affected by temperature. The ratio of these specific heats is so important in thermodynamic equations that it is given the special symbol k (not to be confused with thermal conductivity). The factor k lies between 1.2 and 1.4 for most gases. Table 2-74 gives this ratio for a number of gases and liquids at 60 F and 30 in. Hg.

Specific Heat at Constant Volume

Figure 2-20 shows the effect of temperature and pressure on c_v for various gases. Equation **30** may be used to derive C_v from C_p.

Mean Specific Heat

Since the specific heat of a substance may vary appreciably with temperature, the *mean specific heat* for the pertinent temperature range should be used. Basically, the mean specific

Table 2-75 Mean Apparent Specific Heats of Steels[21]

(Btu/lb-°F)

Composition	Temperature ranges, °F											
	122 to 212	302 to 392	392 to 482	482 to 572	572 to 662	662 to 752	842 to 932	1022 to 1112	1202 to 1292	1292 to 1382	1382 to 1472	1562 to 1652
Pure iron	0.112	0.122	0.126	0.130	0.135	0.140	0.155	0.175	0.198	0.232	0.218	0.1
Carbon steels												
0.60 C, 0.38 Mn	.115	.124	.128	.132	.137	.142	.158	.180	.207	.264	.209	.202
0.08 C, 0.31 Mn	.115	.125	.130	.133	.136	.142	.158	.177	.205	.272	.229	.195
0.23 C, 0.635 Mn	.116	.124	.127	.133	.137	.143	.158	.179	.202	.342	.227	...
0.415 C, 0.643 Mn	.116	.123	.126	.131	.136	.140	.155	.169	.184	.378	.149	.131
0.80 C, 0.32 Mn	.117	.127	.131	.135	.140	.145	.160	.170	.184	.497	.147	...
1.22 C, 0.35 Mn	.116	.129	.130	.133	.138	.143	.152	.167	.195	.499	.155	...
Alloy steels												
0.23 C, 1.51 Mn, 0.105 Cu	.114	.122	.126	.130	.135	.141	.155	.177	.200	.346	.196	.128
0.325 C, 0.55 Mn, 0.17 Cr, 3.47 Ni*	.115	.125	.128	.131	.136	.141	.158	.179	.391	.228	.144	.153
0.33 C, 0.53 Mn, 0.80 Cr, 3.38 Ni*	.118	.125	.129	.134	.139	.143	.161	.185	.312	.281	.133	.139
0.325 C, 0.55 Mn, 0.71 Cr, 3.41 Ni*	.117	.125	.128	.131	.136	.142	.159	.182	.274	.330	.136	.137
0.34 C, 0.55 Mn, 0.78 Cr, 3.53 Ni, 0.39 Mo*	.116	.125	.129	.133	.139	.145	.160	.184	.251	.397	.152	.152
0.315 C, 0.69 Mn, 1.09 Cr, 0.073 Ni	.118	.125	.128	.132	.137	.142	.157	.177	.200	.358	.223	.137
0.35 C, 0.59 Mn, 0.88 Cr, 0.26 Ni, 0.20 Mo	.114	.123	.126	.130	.136	.142	.157	.176	.197	.386	.211	...
0.485 C, 0.90 Mn, 1.98 Si, 0.637 Cu	.119	.125	.129	.133	.138	.144	.159	.179	.198	.216	.326	...
High-alloy steels												
1.22 C, 13.00 Mn, 0.22 Si*	.124	.133	.136	.139	.141	.143	.138	...	...	...	...	...
0.28 C, 0.89 Mn, 28.37 Ni*	.119	.120	.119	.119	.118	.117	.124	.130	.123	.124	.123	.126
0.08 C, 0.37 Mn, 19.11 Cr, 8.14 Ni*	.122	.127	.129	.131	.134	.136	.142	.155	.149	.150	.153	.153
0.13 C, 0.25 Mn, 12.95 Cr*	.113	.123	.127	.132	.138	.145	.163	.186	.209	.216	.165	.160
0.27 C, 0.28 Mn, 13.69 Cr*	.113	.122	.127	.131	.136	.143	.162	.186	.221	.237	.187	.157
0.715 C, 0.25 Mn, 4.26 Cr, 18.45 W, 1.075 V*	0.098	0.104	0.108	0.111	0.116	0.120	0.132	0.143	0.152	0.171	0.171	0.176

* Steels heated discontinuously; all others heated at steady rate of about 5.4°F per min.

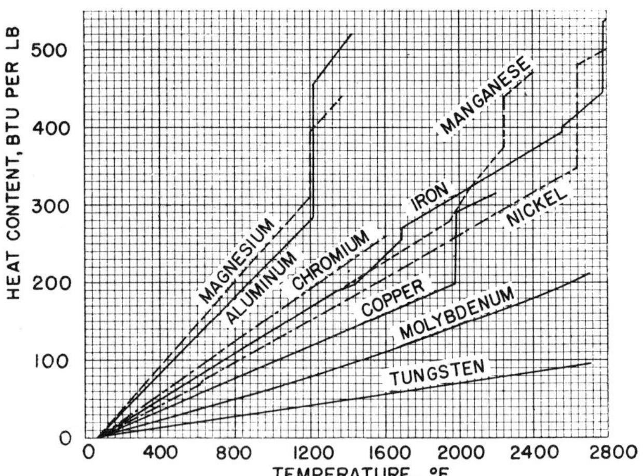

Fig. 2-21 Heat content vs. temperature for pure metals.

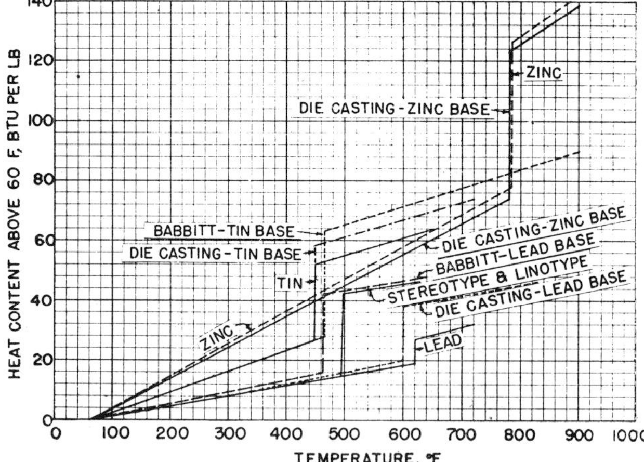

Fig. 2-22 Heat content of zinc, tin, and lead alloys.

heat is calculated by integrating the temperature function for specific heat (instantaneous specific heat) between the temperature limits of the process and then dividing by that temperature differential. Equations **21** thru **29** may be handled in this manner.

Heat Content

The heat content of gases found in flue products at atmospheric pressure may be read directly from Fig. 2-18, except for CO_2 and water vapor above 2600 F, which requires Figs. 2-12, 2-14, and 2-15. Figures 2-36 thru 2-42 give the heat content of a number of commercial fuel gases as a function of temperature and per cent CO_2.

For example, the heat content of 10 cu ft of CO_2 at 1000 F is:

$$10 \times 27.1 = \text{Btu (see Fig. 2-18)}$$

The same problem may be solved using Fig. 2-16. The mean heat capacity of CO_2 for the temperature limits of 60 F to

1000 F is 0.02884 Btu per cu ft. The heat content of 10 cu ft of CO_2 is therefore:

$$10 \times 0.02884 \times (1000 - 60) = 271 \text{ Btu}$$

Specific Heat and Heat Content of Solids

As long as no change in state takes place in solids, such as fusion or internal structural changes, the specific heat will increase slowly with rising temperature. The latent heat of fusion is defined as the amount of heat required to melt one pound of the substance, both the solid and the liquid being at the same temperature. Some solids such as glass and rubber do not have heats of fusion.

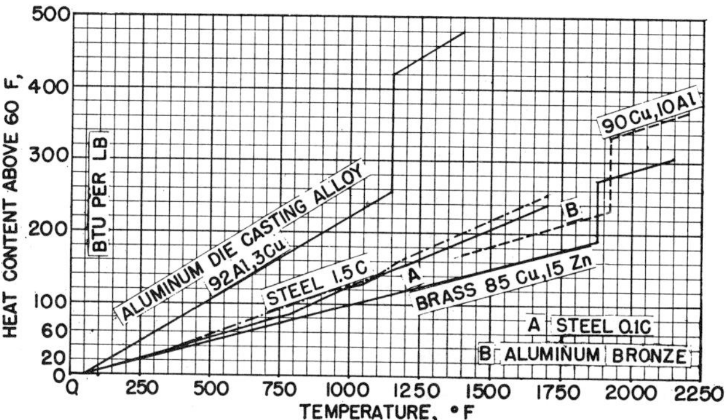

Fig. 2-23 Heat content of steel, brass, and aluminum alloys.

Figures 2-21 thru 2-23, as well as Table 2-75, show the heat content of a number of metals and alloys. The vertical portion of the curves represents the melting of the substance (at constant temperature).

Specific Heat of Liquids

The specific heat of liquids varies widely and irregularly with temperature. In changing from liquid to a gaseous state, liquids require heat. This latent heat of vaporization is the amount of heat required to evaporate a pound of the liquid at its boiling point. Table 2-74 gives the latent heat of vaporization of a number of liquids at their boiling point under 1-atm pressure. Note that water requires 969 Btu per lb, whereas carbon monoxide in the liquid state at −310 F only requires 90.6 Btu per lb. Figure 2-50 gives the specific heat of a variety of liquids as a function of temperature.

COMBUSTION DATA FOR COMMERCIAL FUEL GASES

Combustion Constants and Physical Properties of Typical Fuel Gases

Combustion constants for pure gases are listed in Table 2-63 and may be used for calculating data on mixtures of those gases.

Empirical Combustion Constants for Natural Gas. Natural gas, over a fairly wide range of compositions, requires 9.40 cu ft of *theoretical* air per 1000 gross Btu burned for complete combustion. This number is fairly constant in the range of *heating values* from 950 to 1100 Btu per cu ft.

Occasionally it is convenient to solve gas combustion problems on a mole basis. Since many of the constants in this book are on a cubic foot basis, the following empirical constants on a mole basis may be useful:[22]

To obtain moles of:	*Multiply MBtu burned by factor:*
Theoretical combustion air	0.02485
Wet products formed	.027525
Dry products formed	.0225
H_2O formed	0.005025

Total moles of dry air entering = 0.02485 [1 + (per cent excess air/100)].

Flue products on a *mole basis* are given by:

$$\text{Moles } H_2O = W_{ga} + 0.005025 I_g$$
$$\text{Moles } CO_2 = \%CO_2/100\ (A_d - 0.00235 I_g)$$
$$\text{Moles } (O_2 + N_2) = A_d - 0.00235 I_g - \text{moles } CO_2$$

where: W_{ga} = moles H_2O in entering gas and air
I_g = gross MBtu input
A_d = total moles dry air entering

Effect of Temperature on Heat of Combustion

Table 2-76 may be used to find the variation in the heats of combustion for selected gases when, for example, preheating is effected. Some reactions involving carbon are also included.

Ignition Temperature

Ignition temperature may be defined as the lowest temperature at which heat is generated by combustion faster than heat is lost to the surroundings, and combustion thus becomes *self-propagating*. Below this temperature the gas–air mixture will not burn freely and continuously unless heat is supplied, but chemical reaction between gas and air may take place. Igni-

Table 2-76 Heat of Combustion as a Function of Temperature*,[11]

Reaction	Equation for heat of reaction, Btu per lb-mole; temperature, t, in °F
$CH_4 + 2\,O_2 \rightarrow CO_2 + 2\,H_2O$	$-345{,}220 + 3.432t - (1.348t^2 \times 10^{-3}) - (5.71t^3 \times 10^{-7}) - (5.27t^4 \times 10^{-11})$
$H_2 + 0.5\,O_2 \rightarrow H_2O$	$-103{,}942 - 2.078t - (3.37t^2 \times 10^{-5}) + (2t^3 \times 10^{-7}) - (2.63t^4 \times 10^{-11})$
$CO + 0.5\,O_2 \rightarrow CO_2$	$-122{,}243 - 1.386t + (1.07t^2 \times 10^{-3}) - (1.05t^3 \times 10^{-7})$
$CO_2 + H_2 \rightarrow CO + H_2O$	$+18{,}301 - 0.678t - (1.11t^2 \times 10^{-3}) + (3.06t^3 \times 10^{-7}) - (2.63t^4 \times 10^{-11})$
$C + 0.5\,O_2 \rightarrow CO$	$-47{,}422 + 1.07t - (1.11t^2 \times 10^{-3}) + (1.21t^3 \times 10^{-7})$
$C + O_2 \rightarrow CO_2$	$-169{,}665 - 0.32t - (3.31t^2 \times 10^{-5}) + (1.54t^3 \times 10^{-8})$
$C + CO_2 \rightarrow 2\,CO$	$+74{,}821 + 2.46t - (2.19t^2 \times 10^{-3}) + (2.27t^3 \times 10^{-7})$
$C + H_2O \rightarrow CO + H_2$	$+56{,}520 + 3.148t - (1.076t^2 \times 10^{-3}) - (7.9t^3 \times 10^{-8}) + (2.63t^4 \times 10^{-11})$
$C + 2\,H_2O \rightarrow CO_2 + 2\,H_2$	$+38{,}219 + 3.835t + (3.42t^2 \times 10^{-5}) - (3.85t^3 \times 10^{-7}) + (5.26t^4 \times 10^{-11})$

* The applicable temperature range for the first equation is −189 to 261 F; all other equations are valid from 80 to 4040 F.

Table 2-77a Comparison of Ignition Temperatures of Gases in Air and in Oxygen[16]

Substance	In air, °F	In oxygen, °F
Carbon monoxide	1191–1216	1179–1216
Methane*	1301[55]	*
Ethane	968–1166	968–1166
Propane	871†	914–1058
iso-Butane	864†	...
n-Butane	761†	...
n-Pentane	588†	...
iso-Pentane	788†	...
Ethylene	1008–1018	932–966
Acetylene	763–824	781–824

* See Table 2-77b for effects of excess air and oxygen enrichment.

† Natl. Fire Protection Assn., "Fire-Hazard Properties of Flammable Liquids, Gases and Volatile Solids: Table 6-126." *Fire Protection Handbook*, 12th ed., p. 6–131+, Boston, 1962.

Table 2-77b Ignition Temperatures of Methane for Various Air-Gas Mixtures with and without Oxygen Enrichment[55]

Stoichiometric air used regardless of O_2 content of the air, %*	O_2 in combustion air, %	Cu ft air per cu ft gas	Mixture composition CH_4, %	Mixture composition Air, %	Ignition temp, °F
80	15	10.67	8.5	91.5	1364
	18	8.90	10.0	90.0	1342
(n = 0.8, where n =	21	7.62	11.6	88.4	1323
fraction of stoichi-	25	6.40	13.5	86.5	1306
ometric air)	35	4.57	18.0	82.0	1274
100	15	13.34	7.0	93.0	1328
(n = 1.0)	18	11.12	8.4	91.6	1314
	21	9.52	9.5	90.5	1301
	25	8.00	11.1	88.9	1288
	35	5.71	15.0	85.0	1260
110	21	10.47	9.0	91.0	1297
(n = 1.1)	25	8.00	11.1	88.9	1288
	35	6.28	13.7	86.3	1245
120	21	11.40	8.0	92.0	1288
(n = 1.2)	25	9.60	9.5	90.5	1272
	35	6.85	12.8	87.2	1231

* The oxygen available for combustion under any given percentage category is the same, that is, the product of the second and third columns of the table is constant within any given n category.

tion temperature is affected by a large number of factors in varying degrees. Therefore, it cannot be considered as a fixed property of the gas, and tabulated values are not necessarily valid except for the specific conditions of the test under which they were determined.

However, ignition temperatures are important "constants" of a fuel combustion process along with the limits of flammability. They provide a measure of the tendency of hot objects to ignite a gas, and thus they serve as one basis for technical safety considerations. The data given in Tables 2-77a and 2-77b are valuable also in thermodynamic studies. Oxygen-enriched atmospheres are widely used in industry, particularly in steel plants; thus, data covering such methane–air mixtures are also included in Table 2-77b. Since natural gas contains not only methane, but also hydrocarbons and nitrogen, its ignition temperature will be somewhat lower than that of pure methane. The higher hydrocarbons in natural gas, such as ethane, lower ignition temperature slightly:

C_2H_6 in CH_4–C_2H_6 mixture:	20%	40%	60%	80%
Ignition[66,*] temperature, °F:	2070	1930	1830	1780

Inert gases raise the ignition temperature.

A comparison of the ignition temperatures of methane and a natural gas (CH_4, 88.7%; C_2H_6, 7.4%; C_3H_8, 1.4%; C_4H_{10}, 1.0%; N_2, 1.5%) follows:[67]

Gas in air, %:	4	6	8	10	12	14
For methane, °F:	1179	1175	1188	1198	1218	1238
For natural gas, °F:	1152	1134	1132	1134	1135	1143

The factors which influence the ignition temperature include uniformity of the air–gas mixture, oxygen concentration, velocity (Fig. 2-24a), pressure, materials enclosing the mixture, catalytic materials, ignition induction time, surface conditions, volume of the container, the effects of fluid flow

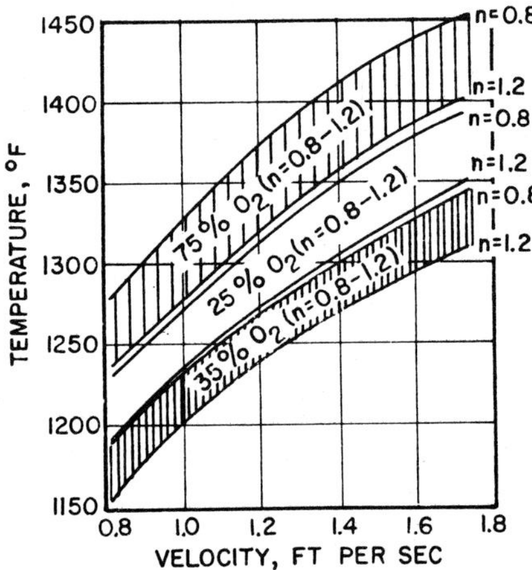

Fig. 2-24a Effect of air-gas velocity on ignition temperatures for various O_2 concentrations and n numbers.[55] See Table 2-77b.

laws, heat capacities, ignition sources, and temperature gradients. Especially important is the method of test used to determine the ignition temperature. There are three major methods:

1. Sudden but uniform heating of the air–gas mixture by adiabatic compression or shock waves.
2. Contact with hot surfaces under various conditions (flow along hot pipe wall, hot bodies introduced into a container, etc.)
3. Preheating of the combustible material under test and introduction of the combustible material into preheated air.

The data in Tables 2-77a and 2-77b were obtained by dynamic methods (item 2 above), furnishing consistent information for the varying parameters shown; other methods would undoubtedly yield somewhat different temperatures. The ignition temperatures as given are the lowest temperatures at which the air–gas mixtures will ignite themselves by contact with a hot surface. A temperature so determined may be initiated neither by spark nor by flame.

The ignition temperature of most substances decreases with rising pressure, which is an important factor in engine and turbine operation. Ignition temperatures are generally

* Ignited by a jet of air.

lower in oxygen than in air, as shown in Table 2-77a, although the values for H_2 are apparently not affected. Ignition temperatures of the paraffin hydrocarbons generally decrease with increasing molecular weight (Table 2-77a). The decrease for paraffin hydrocarbons with increasing pressures is approximately 180°F at pressures up to 150 psi. An increase of as much as 90°F in ignition temperature may be caused by moisture in the air or gas. Catalysts and impurities will lower the temperature of ignition. Addition of antiknock compounds to liquid hydrocarbons tends to raise the ignition temperature appreciably.

Ignition Energy. A match flame will always ignite a flammable gas–air mixture issuing from a burner at atmospheric pressure, but a lighted cigarette or an electric spark may not. A cigarette glowing strongly would not ignite natural gas issuing from a Bunsen burner and small sparks may be passed thru an explosive gas without producing ignition. There is a minimum ignition energy required for given distances between electrodes to prevent quenching of the flame. A mixture of 7.0 per cent methane in oxygen and nitrogen at 1.0 atm, a spark voltage of 10.5 kv, and a capacitance of 10.5 micromicrofarads required a minimum ignition energy of 0.58 millijoules at a quenching distance of 0.109 in. between electrodes.[23] A mixture of 5.6 per cent ethane in oxygen and nitrogen at 1.0 atm, a spark voltage of 8.5 kv, and a capacitance of 8.85 micromicrofarads required a minimum of 0.31 millijoules at a quenching distance of 0.083 in.

A critical review of the literature pertaining to both spontaneous and forced ignition may be found on pages 27 to 37 of Reference 29.

Limits of Flammability

The terms **limits of flammability, explosive limits, limits of inflammability,** and **inflammation limits** mean the same thing. A distinction between explosion and flammability on the basis that one develops a pressure and the other does not is not valid. Violence and pressure developed by a flammable mixture depend upon environment and direction of flame propagation. (**Note:** Avoid using the word "inflammability" since the prefix "in" is somewhat confusing.)

A flammable gas mixture is a mixture of gases thru which flames can propagate. The flame is initiated in the mixture by means of an external source. The limits of flammability may be defined as the limiting composition of a combustible gas and air mixture beyond which the mixture will not ignite and continue to burn. The **lower limit** represents the *smallest* proportion of the gas which, when mixed with air, will burn *without* the continuous application of heat from an external source. Above the **upper limit,** the large amount of gas acts as a diluent and combustion cannot become self-propellant. The limits of flammability of various gases in air are given in Table 2-63.

Qualitative field data are of great value in designing protective systems, for example, to control the programming, for large industrial gas equipment. It was reported[63] that a 55 per cent *air-deficient* gas flame will stay ignited at 1300 F—75 per cent at 1500 F. See Fig. 14-2.

Effect of Initial Temperature. An *increase* in temperature tends to widen the flammability range; i.e., the lower limit is reduced and the upper limit is increased. For the paraffin hydrocarbons,[24] about an 8 per cent decrease in the lower limit for each temperature increase of 212°F is a general rule. Data for a few gases (under stated conditions) are given in Table 2-78.

Table 2-78 Variation of Flammability Limits with Initial Gas Temperature[25]

(conditions: vertical tube, downward flame propagation, atmospheric pressure)

Initial gas temp, °F	Fuel in mixture with air, per cent: Methane	Hydrogen	Carbon monoxide	Ethylene
63	6.3–12.9	9.4–71.5	16.3–70.0	3.45–13.7
212	5.95–13.7	8.8–73.5	14.8–71.5	3.20–14.1
392	5.50–14.6	7.9–76.0	13.5–73.0	2.95–14.9
572	5.10–15.5	7.1–79.0	12.4–75.0	2.75–17.9
752	4.80–16.6	6.3–81.5	11.4–77.5	2.50–...

Effect of Initial Pressure. In general, the lower limit *increases* while the upper limit *decreases* as pressure is reduced below atmospheric. Since this rate of change is small close to atmospheric pressure, the limits of flammability of natural gas–air mixtures and most other fuel–air mixtures are taken at one atmosphere. Below 1.0 psia, carbon monoxide–air mixtures are not flammable.

Most available data above atmospheric pressure were obtained with closed vessels. Since the significance of pressure is strongly affected by the size and shape of the vessel used, there are many differences in these reported data. In general, however, the lower limit of flammability remains relatively constant, while the upper limit rises with increases in the initial pressure. The data in Table 2-79 illustrate how increases in pressure raise the upper flammability limit.

Table 2-79 Effect of Pressure on the Upper Flammability Limit of Gas–Air Mixtures

(per cent gas in gas–air mixture)

Natural gas[59]: Press., psia	Limit, %	Press., psig	Nat. gas, limit,[60] %	Coke oven gas, limit, %	Multipliers*	Natural gas[60]: Press., psig	Limit, %
1.0	†	0	14	30.7	1.0	600	46
1.5	11.5	100	21	38.4	1.3	800	50
2.0	12.5	200	28	54.0	1.9	1000	52
3.0	13.5	300	34	74.9	2.7	1200	54
3.5	13.7	400	40	95.7	3.5‡	1600	57
14.7	13.9	500	44	...	...	2000	59

* For estimating the upper limit of gases other than those given when the upper and lower limits at 0 psig are known, apply the multipliers indicated in the table to the difference between these limits. Add this product to the lower limit to find the new upper limit.

† Curve is tangent to this line.

‡ At 350 psig.

Geometry and Size of the Confining Vessel. Flammability limits are ordinarily determined in vessels of laboratory dimensions. A study[26] in steel tanks of 12½ cu ft capacity with surface discharge spark plugs as the ignition source showed that the upper flammability limit of gasoline–air–nitrogen mixtures was almost 1 per cent greater than the values found by Coward and Jones.[27] The lower flammability limit was not significantly affected by the test in a large volume space. This report does not show whether a similar effect would be obtained with any other fuel than the liquid hydrocarbons.

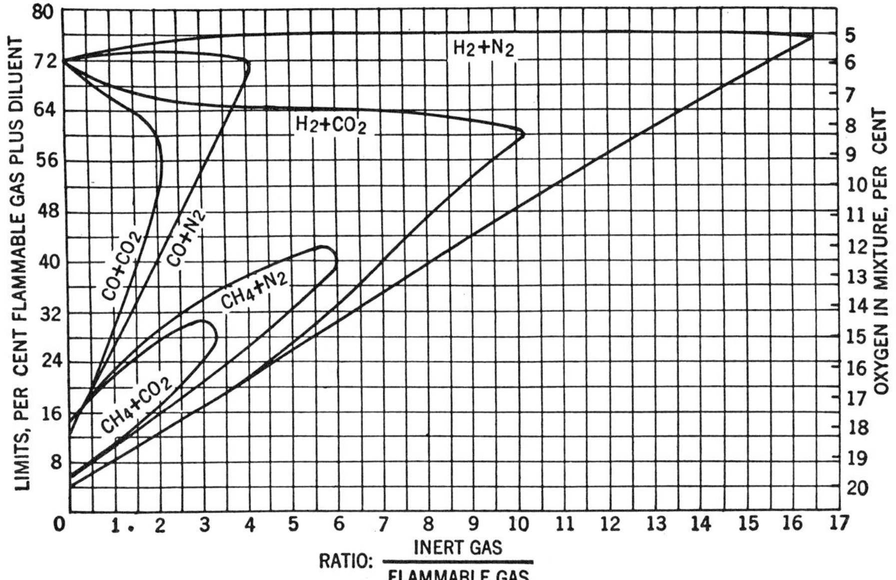

Fig. 2-24b Limits of flammability by volume of hydrogen, carbon monoxide, and methane in air combined with various amounts of carbon dioxide and nitrogen.[27]

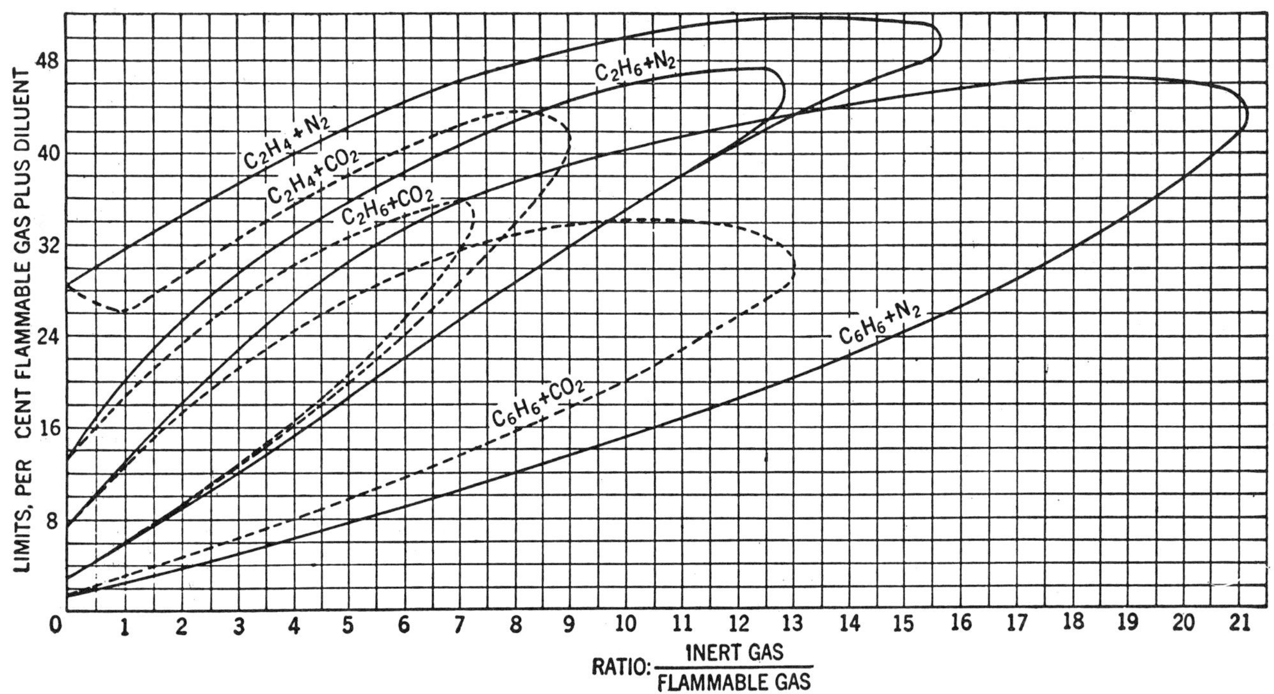

Fig. 2-25 Limits of flammability by volume of ethane, ethylene, and benzene in air combined with various amounts of carbon dioxide and nitrogen.[27]

Even when a gas–air mixture is within the flammable range, its flames will be quenched in tubes of sufficiently small diameter due to the mass and heat-absorbing properties of the tubes. Gas burner port design requires somewhat analogous consideration to prevent flame flashback.[28a–c] Numerical values for the size of the tube would vary, depending upon the gas, the air–gas mixture, and the tube properties. Burning gases containing CO and H_2 have a greater tendency to travel in small-diameter tubes.

Flame arresters are examples of applied flame-quenching technique. Approved flame arresters for flammable mixtures of natural gas, LP-gas, or manufactured gas at pressures not exceeding 5 psig are available for use on gas piping systems. These are intended to prevent flashback thru gas burners from reaching back to equipment in the system which is not strong enough to withstand explosion pressures. They are equipped with special elements for arresting the explosion wave which may already be established in a long pipe. These devices include a thermally actuated, manually resetting shutoff valve. Screens inserted in gas burners for flashback prevention are not recommended because such screens may heat up and thereby ignite the mixture on the far side of the screen.

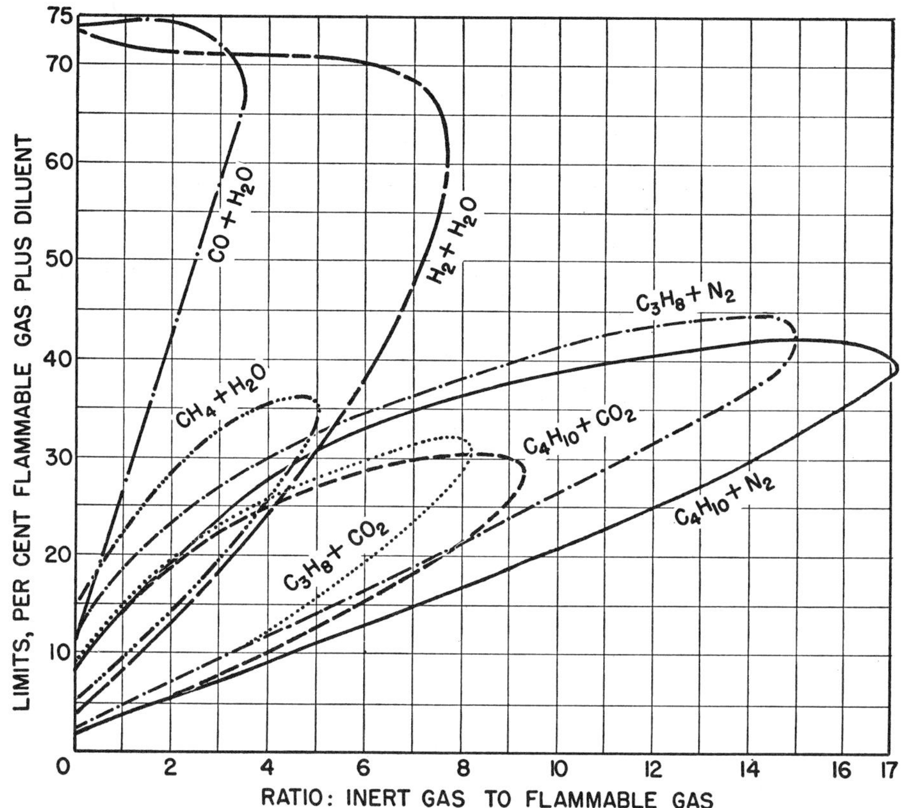

Fig. 2-26 Limits of flammability by volume of methane, carbon monoxide, and hydrogen in air combined with various amounts of water vapor, and propane and butane in air combined with various amounts of carbon dioxide and nitrogen.[27]

Effect of Inerts. The limits of flammability of a gas are changed by the addition of an inert gas. This effect may be seen in Figs. 2-24b, 2-25, and 2-26. For example, the addition of nitrogen to hydrogen results in decreasing the range of flammability limits, for while there is a negligible effect on the upper limit (72 to 76 per cent), the lower limit is raised at a constant rate until an inert gas to flammable gas ratio of approximately 16.5 is reached, at which point the mixture (containing 5.7 per cent H_2 and 94.3 per cent N_2) is nonflammable.

The efficiency of inert gases as flame extinguishers decreases in this order: CO_2, H_2O, N_2, helium, argon.

Effect of Fine Metal Particles.[30] The introduction of very fine metal particles into hydrogen–air mixtures widens the flammability limits and lowers the ignition temperatures. Such fine particles are usually present in industrial flames.

Calculation of Flammability Limits of Mixtures. The limits of flammability for gases of various compositions can be calculated by applying Le Chatelier's equation to data in Figs. 2-24b, 2-25, and 2-26.

$$\text{Limit of flammability} = \frac{100}{\dfrac{a}{A} + \dfrac{b}{B} + \dfrac{c}{C} + \dots}$$

where:

$a, b, c, \dots$ = the proportions of the various constituents of the gas mixture on an air-free basis, expressed as percentages by volume

$A, B, C, \dots$ = the respective flammability limits from Figs. 2-24b, 2-25, and 2-26.

The use of the flammability curves and Le Chatelier's equation will be illustrated in the following two problems:[11]

1. **Determining Whether a Gas is Flammable.** A mine gas has the following composition, percentage by volume: H_2, 4.9; CO, 4.3; CH_4, 3.3; CO_2, 13.8; N_2, 70.9; O_2, 2.8.

Steps in Solution:

A. Recalculate the mixture to an air-free basis.
   a. Percentage of air in mixture = $(2.8 \times 100)/20.9 = 13.4$
   b. Percentage of other gases in mixture = $100 - 13.4 = 86.6$
   c. Percentage of H_2 in air-free mixture = $(4.9/86.6) \times 100 = 5.7$
   d. Repeat step c for CO, CH_4, and CO_2. Air-free nitrogen is by difference from 100.
   e. Therefore, the mixture on an air-free basis, per cent by volume, is: H_2, 5.7; CO, 5.0; CH_4, 3.9; CO_2, 15.9; N_2, 69.6.

B. Determine if there are sufficient percentages of inerts in the air-free mixture to prevent flammability.
   a. Figure 2-24b shows that it requires a N_2 to H_2 ratio of 16.55 for nonflammability.
   b. Repeat step a for combustibles CO and CH_4 in combination with CO_2. The results are given in the third column of Table 2-80.
   c. Multiply the values in the second column by the values in the third column of Table 2-80. Record the results in the appropriate *inerts* column.

d. Total the inerts columns and note that these sums *exceed* the percentage of inerts in the gas; therefore the gas is flammable. Note that a trial-and-error solution would have been necessary if, say, the N_2 had been exceeded (greater than 69.6 per cent) but the CO_2 had not (less than 15.9 per cent).

Table 2-80 Data for Illustrative Problem 1 Relative to Flammability Limits

Combustible gases, per cent		Ratio of inert to flammable gas required for non-flammability	Inerts, per cent	
			N_2	CO_2
H_2	5.7	16.55	94.3	...
CO	5.0	2.20	...	11.0
CH_4	3.8	3.35	...	12.7
Total	14.5	...	94.3	23.7

2. **Estimating the Upper and Lower Limits of Flammability of a Gas.** A natural gas has the following composition, per cent by volume: CH_4, 79; C_2H_6, 17; CO_2, 1; N_2, 3.

Note: If an appreciable percentage of O_2 is present, recalculate to an air-free basis.

Steps in Solution (tabulated in Table 2-81):

A. Dissect the gas mixture into simpler divisions, each of which contains only one combustible gas and *none, part, or all* of one of the inerts. These divisions must be made in flammable ratios as shown in Figs. 2-24b, 2-25, and 2-26.

B. The upper and lower flammability limits for the inert-to-flammable mixtures as well as the limits for the pure gas divisions are read from Figs. 2-24b, 2-25, and 2-26, and tabulated.

C. Substitute the tabulated values in Le Chatelier's equation in order to calculate the lower and upper flammability limits of the mixture:

$$\text{Lower limit of flammability} = \frac{100}{\frac{a}{A}+\frac{b}{B}+\frac{c}{C}+\cdots} = \frac{100}{\frac{6.0}{10.5}+\frac{76.0}{5.0}+\frac{2.0}{6.5}+\frac{16.0}{3.0}} = 4.7$$

$$\text{Upper limit of flammability} = \frac{100}{\frac{a}{A}+\frac{b}{B}+\frac{c}{C}+\cdots} = \frac{100}{\frac{6.0}{23.0}+\frac{76.0}{15.0}+\frac{2.0}{18.5}+\frac{16.0}{12.5}} = 14.9$$

Catalytic Combustion. In order to maintain combustion of a gas by flame, both high-temperature and mixture composition within flammable limits are necessary. Catalytic combustion, on the other hand, takes place without flames, at lower temperatures, and below the lower flammability limit.

Table 2-81 Data for Illustrative Problem 2 Relative to Flammability Limits

Combustible	Ratio of inert to combustible	"Dissection," % CH_4	C_2H_6	N_2	CO_2	Total	Limits of flammability Lower	Upper
CH_4	1.0	3.0	...	3.0	...	6.0	10.5	23.0
CH_4	0.0	76.0	...	...	...	76.0	5.0	15.0
C_2H_6	1.0	...	1.0	...	1.0	2.0	6.5	18.5
C_2H_6	0.0	...	16.0	...	...	16.0	3.0	12.5
	Total	79.0	17.0	3.0	1.0	100.0		

The basic use for this latter combustion phenomenon is the processing of exhaust gases to minimize air pollution. The heat evolved is often a valuable by-product of the process.

The following are the basic features of catalytic combustion:

1. The heat capacity of dry air at 70 F is 0.018 Btu per cu ft-°F. Therefore, one Btu will raise the temperature of a cubic foot of 70 F air about 55°F. The energy in a gas–air mixture at the lower flammability limit is of the order of 52 Btu per cu ft (referred back to 70 F). Let us assume that we wish to oxidize an exhaust gas that contains 13 Btu per cu ft. (Note that the gas assumed is at one-fourth of the lower flammability limit.) We may approximate the temperature rise of the gas as 13 Btu per cu ft *times* 55°F per Btu *equals* 715°F. It is apparent then that process exhaust at 485 F may be recovered at approximately 1200 F via catalytic combustion.

2. The minimum catalytic ignition temperature of virtually all combustible gases falls between the extremes of 70 F (for hydrogen) and 750 F (for methane.) In general, as the minimum catalytic ignition temperature increases, the degree to

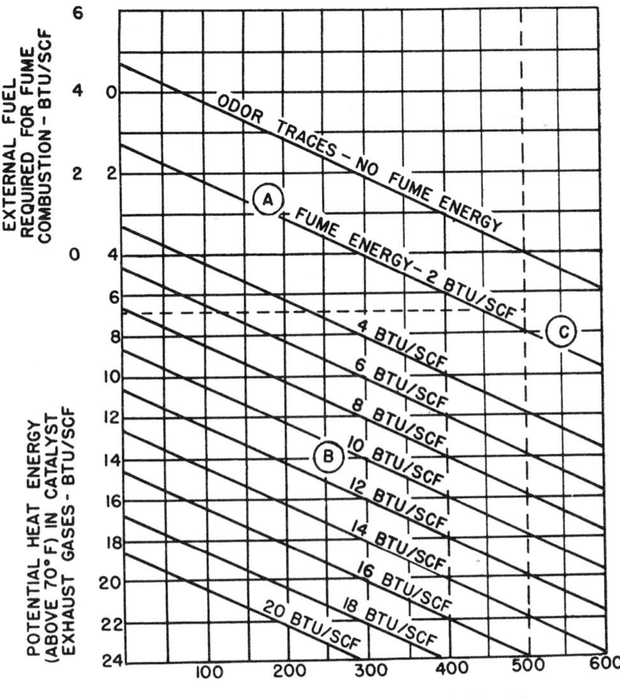

Fig. 2-27 Catalytic combustion of continuous-flow fumes:[32] A, continuous fume preheating required; B, initial preheating only, self-sustaining after start-up; C, no preheating required.

which the catalyst entry temperature can be depressed is reduced.

The ignition (catalyst) temperature for most *organics* ranges from 450 to 650 F.[31] The corresponding discharge temperature ranges from entry temperature for trace concentrations of the combustible to 1200 F or more for large concentrations of combustible gases.

Figure 2-27 gives parameters for fumes in regard to supplementary energy needed either to initiate or to sustain catalytic combustion in low-energy fumes. In area A continuous preheating is required; in area B initial preheating only is required, whereas in area C no preheating is needed. External heat requirements in area A may be as much as four Btu per SCF. In area B the initial addition of one Btu per SCF of fuel may return up to 20 Btu or more to the user. Obviously the energy level of fumes may be increased by decreasing the dilution of the process exhaust.

3. The catalytic *reduction* of nitrogen oxides is accomplished by the addition of a fuel gas. The fuel selected in turn determines the catalyst ignition temperature. The temperature rise thru the catalyst depends on the nitrogen oxides which must be reduced (it may exceed 500°F.) Reductions from 5000 ppm down to 160 ppm residual oxides of nitrogen were reported.

4. The catalysts employed are of two general types:

A. They may be elements similar in appearance to air filter mats, consisting of heat and corrosion resistant metallic strips which are coated with catalytic materials. These materials include nickel, silver, vanadium, and manganese where *partial oxidation* is required. Platinum alloys produce complete oxidation at minimum temperatures. There must be sufficient catalyst surface for the design flow rate and reactant concentration. Normal pressure drops thru these elements are ½-in. w.c.

B. They may be elements consisting of porcelain rods coated with a platinum alloy catalyst. The rods are retained by porcelain end plates. The capacity per cubic foot of these elements ranges from 170 to 670 SCF per minute of exhaust gases.

5. The catalysts must be reactivated periodically, since their activity is decreased by contamination. The minimum catalyst entry temperature of air-borne particles of pitch or tar, for example, must be above the gasification temperature; otherwise, the particles will coat the catalyst.

Rate of Flame Propagation

The terms **rate of flame propagation, ignition velocity,** and **flame speed** are all equivalent to **burning velocity,** which varies for different gases and is also dependent upon the gas-to-air ratio, the size and shape of the container, the test method employed, the temperature of the mixture, and the relation of the flames to energy-absorbing sinks. A discussion of the measurement of burning velocity by **static and dynamic** methods (pages 17–20 of Reference 29) follows.

Table 2-82 Maximum Burning Velocity of Various Gases at Constant Pressure by Different Methods

		Method No. 1, tube-type burner* (0.108 to 0.378 in. diam)		Method No. 2, glass tube† (0.984 in. diam, 22.41 in. long)		Method No. 3, Bunsen-type burner†	
No.	Gas	Gas in air mixture, %	Flame velocity, fps	Gas in air mixture, %	Flame velocity, fps	Primary air portion of theoretical air required for complete comb., %	Flame velocity, fps
1	Hydrogen	42	7.02	38§	16§	55	9.25
2	Carbon monoxide	45	1.405	45§	4.2§	55	1.7
3	Methane	9.8	1.21 1.363**	9.96	1.11	98	0.85
4	Ethane	6.4		6.28	1.316	100	1.06
5	Propane	4.7	1.471	4.54	1.298	90	0.95
6	*n*-Butane	3.65		3.52	1.245	88	1.05
7	*n*-Pentane	2.9	1.143	2.92	1.262		
8	*n*-Hexane			2.51	1.262		
9	*n*-Heptane			2.26	1.265		
10	Ethylene	7.2	2.285	7.40	2.23	90	2.25
11	Propylene			5.04	1.435		
12	Butylene			3.87	1.42		
13	Acetylene	9.5	4.74			85	8.75
14	Benzene			3.34	1.333		
15	Blue water gas					78	4.0
16	Coke oven					90	2.30
17	Carbureted water					90	2.15
18	Natural gas					100	1.0
19	Producer gas			50§	2.2§	90	0.85

* Smith, F. A., and Pickering, S. F. "Measurements of Flame Velocity by a Modified Method." *J. Res. Natl. Bur. Std.* 17, July 1936.

† Gerstein, M., and others. "Fundamental Flame Velocities of Hydrocarbons." *Ind. Eng. Chem.* 43: 2770, Apr. 1951, except where § is indicated.

‡ Corsiglia, J. "New Method for Determining Ignition Velocity of Air and Gas Mixtures." *A.G.A. Monthly* 13: 437–442, Oct. 1931.

§ Haslam, R. T., and Russell, R. P. *Fuels and Their Combustion*, p. 268. New York, McGraw-Hill, 1926.

** Probably the best value to date with controlled moisture content of the gas, controlled gas inlet temperature, and determination of gas velocity thru burner port of a nozzle-type burner. Average of three values. Caldwell, F. R., and others. "Apparatus for Studying Combustion in Bunsen Flames." *Am. Chem. Soc. Div. Petrol. Chem. Joint Symp. Combustion Chem.*, Apr. 9–12, Cleveland, Ohio, 1951.

Static Methods. Here the flame front passes thru an initially nonflowing gas mixture. Tests have been performed on gas–air mixtures in various sizes of tubes and pipes (Table 2-82). Constant volume tests have been run within sealed spheres during which the rate of progress of the flame front was recorded simultaneously with the rate of pressure rise. Thirdly, there are constant pressure methods, wherein a soap or plastic bubble is filled with the gas–air mixture, the mixture is ignited at the center, and the progress of the flame front is recorded by a rotating drum camera. The validity of the results of the latter method is usually questioned, since it is difficult to measure bubble diameters.

Dynamic Methods. In these a constant speed gas mixture flows (laminar) to a fixed flame front. Mixture velocities less than burning velocities cause flashback. Conversely, mixture velocities greater than burning velocities result in flame blowoff and extinction as the mixture becomes diluted in the ambient atmosphere. In other words, flame stabilization requires an equality between the gas–air mixture and the burning velocity that is capable of restoring itself from more or less large imbalances.

Dynamic methods may employ: (1) Bunsen-type burners, (2) nozzle burners, or (3) flat flames produced by screens or other means.

The flames of *Bunsen-type burners* are measured as to size and shape. Burning velocity is then calculated from the formula:

$$V_b = \frac{A_b \times V_a}{A_f}$$

where: V_b = burning velocity in fps
A_b = cross-sectional area of burner, sq ft
V_a = average mixture velocity thru burner, fps
A_f = surface area of flame front, sq ft

The major source of error in this method is the *quenching effect* at the base of the flame which increases with decreasing burner diameter and decreasing flame height.

Lewis and von Elbe used a particle tracking technique. Here the particles of magnesium oxide used may alter the flame front to yield inaccurate results.

Other techniques in the Bunsen method include the *angle method,* which measures the cone angle, and *Dery's method,* which involves the portion of the flame surface resembling the frustum of a cone.

The *nozzle method* improves over the Bunsen method by giving a uniform velocity profile and a more distinguishable flame front at the base of the flame instead of a parabolic profile. This method can produce larger flames without turbulence. The cone angle can thus be measured more accurately;

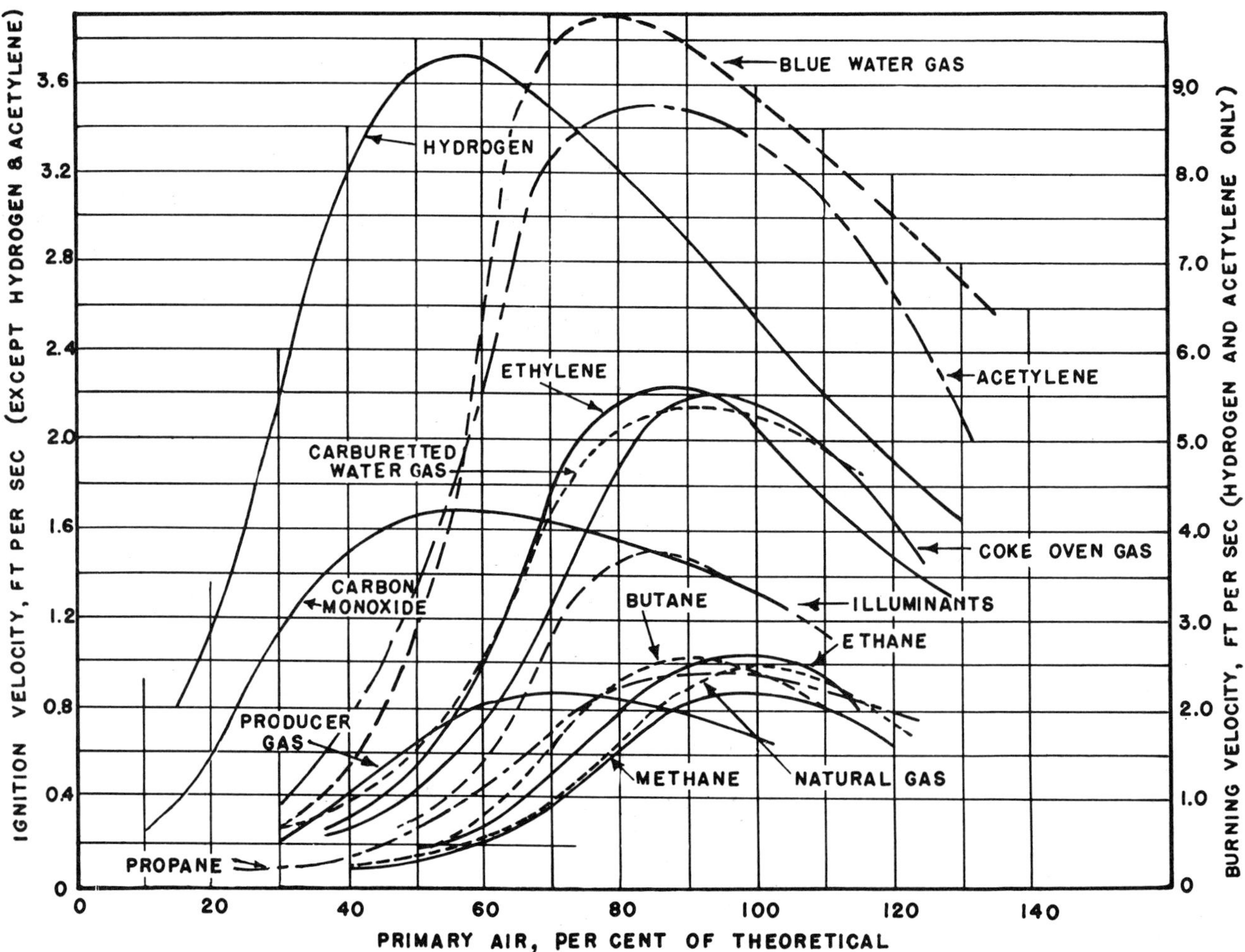

Fig. 2-28 Burning velocity curves for various gases. Ignition velocity and burning velocity have the same meaning.[34]

very tall flames, however, must be avoided in order to prevent the return to a parabolic profile.

In 1958, Carl Halpern[33] published a detailed study of factors to be observed in using the nozzle burner flame method. He obtained the flame speed values shown in Table 2-83.

Table 2-83 Maximum Flame Speeds at Various Methane–Air Ratios

Gas velocity, fps	Fuel–air ratio for max flame speed	Gas temp, °F	Flame speed, fps
6	0.0619	84.4	1.233
6	.0621	100	1.283
6	.0618	120	1.364
4	.0634		2.212
5	.0636	280	2.201
6	.0627		2.164
7	.0628		2.180
5	.0623		2.457
6	.0632		2.400
7	.0627	330	2.426
8	.0630		2.406
9	.0632		2.372
10.2	0.0628		2.360

The *flat flame method* is most useful for determining burning velocities in slow burning mixtures. This method produces a flame in the form of a flat disk slightly larger in diameter than the cross section of a burner. It is extremely difficult to stabilize this sort of flame front with fast flames.

Burning Characteristics of Various Fuel Gases.[28a] Each combustible constituent of a fuel gas has a rather definite burning speed or ignition velocity under given conditions of temperature and aeration. Burning velocity curves for different gases are shown in Fig. 2-28, from which it will be noted that hydrogen is by far the fastest burning gas. Next in order of speed of flame propagation are carbon monoxide and the illuminants (ethylene, benzene, and propylene). Considerable data are available on the ignition velocities of simple gases; however, there is no rational basis by which the ignition speed of commercial fuel gases may be accurately calculated from data on the constituent simple gases. Some data ignition velocities of commercial gases are also presented in Fig. 2-28.

From such data as those shown in Fig. 2-28, *qualitative* estimates of the burning speeds of commercial gases may be made. These burning speeds affect burner design and also affect the heating characteristics of flames. For example, the substitution of a slower burning gas for one having a higher burning speed will change the position of maximum temperature in a gas burning system, moving them further away from the burner.

Flame speed increases with the temperature of the reacting gases. To make use of this effect, it is possible to design burners which incorporate **piloting flames** or hot refractory surfaces that serve to preheat the main flame mixture. Flame speed and flame intensity are thereby increased. Such designs are typically used in glassworking, lamp manufacture, and related high-speed industrial gas applications.

For use in **gas burner design** the values of flame speeds obtained by the burner method are preferable; for **pipeline purging** and where propagation may take place in pipelines or similar vessels, the tube method values are more applicable.

Gas–Air Mixtures: Maximum Burning Velocity vs. Theoretical Combustion. Maximum flame speeds are attained at approximately the percentages of gas in air mixtures listed in Table 2-82. Note that these are slightly larger values for the gas component than would be present in a mixture containing *exact proportions* of gas and air for combustion as illustrated in the following tabulation:

	Gas in maximum speed mixture, %	*Gas for theoretical complete combustion, %*
H_2	42	27.4
CO	45	27.4
CH_4	9.8	9.4
C_3H_8	4.7	4.04
C_2H_4	7.2	6.56
C_2H_2	9.5	7.74

The rate of flame propagation *decreases* as the mixture becomes *leaner* in gas to the lower explosive limit or *richer* in gas to the upper explosive limit. At these limits the flame speed is from $\frac{1}{3}$ to $\frac{1}{2}$ of the value at maximum speed, except in the case of hydrogen, for which it decreases from $\frac{1}{7}$ to $\frac{1}{5}$ of its maximum.

Table 2-84 Effect of Size and Material of Tube or Pipe on Flame Propagation of Methane and Natural Gas

Gas	Material of tube or pipe	Diameter of tube or pipe, in.	Gas in air, %	Maximum flame velocity, fps
Methane*	Glass	1	9.5	2.2
Methane*	Glass	2	9.5	3.1
Methane*	Glass	3.5	9.5	3.5
Methane*		12	9.25	5.5
Methane*		38	10	8.2
Methane†	Glass	1.04	10	2.27
Methane†	Lead	1.04	10	2.12
Methane†	Copper	0.865	10	2.06
Methane†	Iron	1.07	10	2.19
Methane*	Glass	0.142	5–15	Would not propagate
Natural gas‡ (Los Angeles)	Iron	4.03 (327 ft long)	9.2	13.5 maximum pressure developed was 20.5 psi

* Haslam, R. T., and Russell, R. P., *Fuels and Their Combustion*, p. 267, 268. New York, McGraw-Hill, 1926.

† Jost, W., *Explosion and Combustion Processes in Gases*, p. 102. New York, McGraw-Hill, 1946.

‡ Henderson, E., "Combustible Gas Mixtures in Pipe Lines." *P.C.G.A. Proc.* 1941: 98–111.

Effect of Tube Size on Flame Speed. As shown in Table 2-84, the rate of flame propagation increases from 2.2 to 8.2 fps when the tube diameter increases from 1 to 38 in. The material of the tube has little effect on the flame speed. It has been shown that flame will not travel in a glass tube of 0.142-in. diam or less if methane is the combustible gas, but hydrogen will propagate flame thru a tube as small as 0.035-in. diam. Natural gas will have a maximum flame speed of 13.5 fps in a 4-in. pipe, as shown in Table 2-84.

Effect of Preheat on Flame Speed. Data from three investigations are given. It has been stated:[35] "for practical purposes, 'as at present advised,' it may be considered fairly certain that for rich gases the flame speed varies about as the square of the absolute temperature; for a manufactured gas containing 36 per cent methane and 11 per cent H_2, it varies about as $T^{1.8}$; and for dry CO in air, about as $T^{1.6}$."

In investigations[36] of the effect of temperature, t (°F), upon propane–air mixtures, it was found that in the range −117 F to 633 F the burning flame speed, v (ft per sec), increased according to:

$$v = 1.11 + 0.0346(t/100)^2 + 0.329(t/100)$$

Since the experiments were carried to preheat temperatures over 900 F, it was noted[37] that premixed air–gas mixtures had decreased flame velocities. The amount of the decrease varied with the time of contact between hot air and gas because of partial reaction between air and gas. It was also noted[72] that the stability range (defined as the ratio of blowoff to flashback velocity) narrowed with increased initial mixture temperature.

It was found[33] that the flame speed of a 0.062 methane–air mixture from a nozzle burner at 6-fps gas velocity varied approximately as the 1.8 power at the absolute temperature. Experimental results were compared with other predictions;[38,39] the maximum divergence was 5 and 15 per cent, respectively, under the given assumptions.

Catalytic Gas Burners. Some burner materials speed up the combustion process without change to themselves. Available commercial systems may be classified as low temperature—about 800 F or lower (*no* visible light emitted), and high temperature—about 900 F and higher (visible light emitted). The so-called refractory surface catalysis, attributed to many industrial gas burners, is in a separate category (the mechanism involved is probably a preheat effect on flame speeds rather than catalysis).

Low-temperature catalytic combustion results from the appropriate use of platinum, palladium, or other precious metals which are expensive and subject to "poisoning" by sulfur and other undesirable substances. A commercially available burner of this type operates at about 650 F and a loading of 6500 Btu per hr-sq ft.

High-temperature catalytic burners operate with substantially greater loading—12,000 to 15,000 Btu per hr-sq ft—and are intended for various industrial heating applications. Still higher outputs are available from power burners of this class, reaching heat release rates of up to 50 MBtu per hr-sq ft.

Effects of Hot Ceramics on Flame Speed (Catalytic Surface Combustion). Gas burners are available in which turbulent gas–air mixtures burn on ceramic surfaces (cups, tunnels, firebrick, etc.). Another method of utilizing ceramics (as well as stainless steel grids) is to pass the gas thru the material, either thru pores or thru small openings. All of these arrangements produce significant amounts of radiant heat and appear to speed up combustion.

With hot ceramics, for example, the burning rate is apparently increased because of high temperatures. In the case of burners, such as the tunnel type, quenching effects are minimized because of these high temperatures. In another type of burner, the ceramic, in the form of a porous medium, acts as the port surface and provides so-called surface combustion. There is the contention that flame speed is increased with this type of burner by virtue of a catalytic action of the ceramic. This idea was first brought out by Professor W. A. Bone in England in the early 1900's.

Gray,[43] on the other hand, theorized that flameless combustion on incandescent surfaces is due to the high degree of primary aeration. Furthermore, he stated that the flamelessness is also due in part to the detonation type of combustion which results from these conditions. This detonation combustion occurs in the gaseous medium in a thin wave front independent of the nature of the solid surface present. However, the high temperature of the surface sets up and propagates this type of combustion.

There was no agreement as recently as 1959 as to whether or not hot ceramics have any catalytic effects on combustion. Some experiments[44] showed that porous ceramics do *not* have a catalytic effect. Instead, a porous medium splits the flame into a multitude of small ones, increasing the flame cone surface area per unit flame volume. Increase in the apparent burning rate is thus obtained with ceramics in a manner similar to that with turbulent flames. The high temperatures of the ceramics minimize quenching effects.

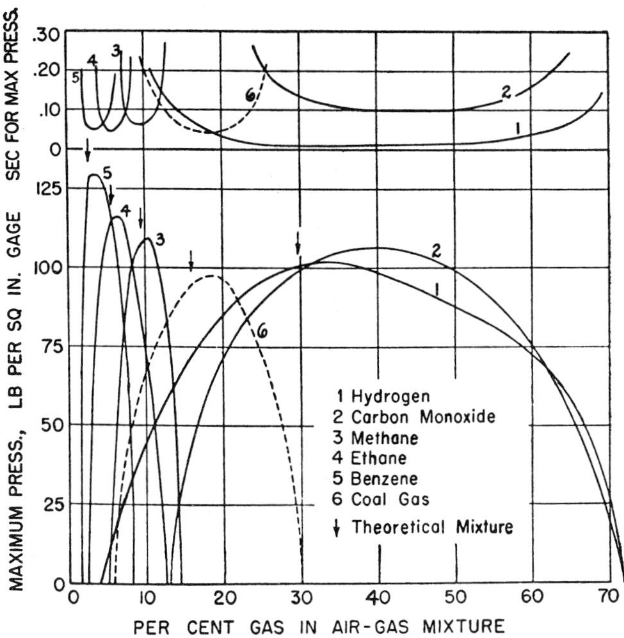

Fig. 2-29 Explosion time and pressure for mixtures of gas and air.[34]

Pressures Developed by Burning Gases. The principal application of gas burning at higher than atmospheric pressure has been in the internal combustion engine, jet engine, and certain boilers (Velox, Lucas-Rotax). In general, increasing the pressure tends to accelerate ignition. Some values for flame speeds of various gases and the pressures developed in pipes, test bombs, and engines are given in Table 2-85. The natural gas values are practical field values. The calculation of maximum combustion pressures and temperatures must take into account heat losses to surroundings during combustion. Furthermore, dissociation and rate of reaction act to reduce practical values below theoretical ones. Figure 2-29 shows the relation of the air–gas mixture to the pressure attained and to the time required to reach the pressure. Note that pressure decreases sharply as you depart from stoichiometric. Data are presented for normal combustion with no detonation. Results shown would in practice be approached only in relatively moderate sized enclosures having comparatively uniform dimensions in all three directions. Therefore, Fig. 2-29 is not applicable to long pipes.

Full-scale tests of the pressures developed in piping have been made by the Factory Mutual Engineering Division, Norwood, Mass., to determine the effects of flashback in piping carrying combustible air–gas mixtures. Results showed flame travel of 10 to 100 fps and pressures under a few hundred pounds per square inch. Others[40,41] also describe tests in which pressures on the order of 300 psig were obtained.

Table 2-85 Some Flame Speeds and Pressures Developed by Various Combustibles in Pipes, Test Bombs, and Engines

Status of gas–air mixture	Initial pressure, psig	Pressure developed, average psig	Pressure developed, maximum psig	Flame speed, fps
Quiescent 9.2% natural gas in air, 4-in. pipe, 337 ft long*	0	10.7	10.7	
	45	45	70	
	75	110	150	
	90	155	190	
Flowing at speed of 259 ft/min, 9.2% natural gas in air, 4-in. pipe*	0	500	1500	5300
	15	1100	3400	
	45	2000	7400	
	90	3100§	8900	
H_2 air 100%, 310-cu in. bomb†	61.6		550	56.1
H_2 air 11.9%, CO air 88.1%, 310-cu in. bomb†	61.6		550	8.65
H_2 air 0.2%, CO air 99.8%, 310-cu in. bomb†	61.6		526	1.895
Gasoline—spark ignition engine‡	103		456	
Diesel oil, Diesel engine†	191		810	

* Henderson, E., "Combustible Gas Mixtures in Pipe Lines." *P.C.G.A. Proc.* 1941: 98–111. Conditions of test: (1) quiescent mixture natural gas (CH_4, 91.0%; C_2H_6, 8.5%; CO_2, 0.5%), pipe ends closed, ignition at one end; (2) turbulent flowing mixture, same natural gas, pipe ends closed, ignition at upstream end. Maximum pressure obtained about 50 ft from ignition end.

† Jost, W., *Explosion and Combustion Processes in Gases*, p. 156. New York, McGraw-Hill, 1946. Initial temperature, 122 F. Time to reach maximum pressure from beginning of perceptible rise in pressure for three tests quoted was 0.0059 sec, 0.0383 sec, and 0.2048 sec, respectively; all tests in cylindrical bomb 7.1 in. diam × 7.87 in. long.

‡ Griswold, J., *Fuels, Combustion, and Furnaces*, p. 211. New York, McGraw-Hill, 1946.

§ Pipe burst open.

A noteworthy observation in the *Fire Protection Handbook*, published by the National Fire Protection Association, is that an 8-in. brick wall will *not* withstand 1 psig. Thus, pressures developed in gas–air explosions in structures are limited.

Pulse Combustion. This method of burning gas is based on the use of essentially self-powered burners capable of producing "resonant" or "pulse" combustion involving relatively small combustion air and flue passages. Extremely high heat releases per cubic foot of combustion space are attainable.

When resonant (pulse) combustion occurs, oscillation takes place in the burner system, analogous to oscillations in an organ pipe. The reported positive pressures in the combustion chambers range as high as 37 psia, with negative pressures of about 7 psia.

This oscillating combustion may impose flue gas velocities of the order of 1000 fps on normal movement. The resultant scrubbing action increases the heat transfer rate thru the flue gas side film.

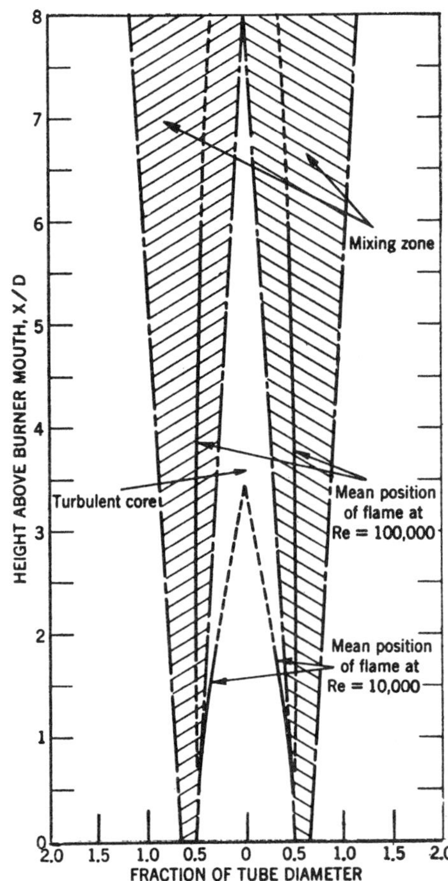

Fig. 2-30 Typical turbulent flames from a 1.25-in. diam Bunsen burner.[65] **Re** = Reynolds number, X = flame height, in., D = 1.25 in.

Turbulent Burning. The higher capacity of industrial burners compared with domestic equipment (close to laminar performance) is well recognized. In certain instances, the turbulent combustion wave is the same phenomenon as in laminar flow except that it is wrinkled.[42] Wrinkling the flame surface increases the flame surface area per unit volume of flame without changing the flame volume. Thus, flame speed on a flame volume basis is higher for turbulent flow compared with laminar flow. The stream velocity fluctuates at random; the flame wrinkles conform to the stream distortions. At some level of turbulence intensity and at some value of the turbulent scale small fragments of flame break away from the main body of turbulent flame and contribute considerably to the overall burning rate.

Studies[64] of turbulent burning (typical of industrial gas burner operation) show that the turbulent flame *brush* is a zone of nonhomogeneous and partly discontinuous burning.

Flamelets within the brush are extinguished continually by flue gases—reignition takes place almost immediately.

There is considerable evidence[65] that premixed open turbulent flames can operate in two distinct modes. The "classical" model of a wrinkled laminar flame appears adequate for the mixing zone (Fig. 2-30); the turbulent core region is far more complex. The flame velocities of turbulent flames are markedly higher than the data given in this chapter, which are for laminar-type flames, indicate.

Another point of interest in connection with flame velocity is the influence of recirculated combustion products. The extent of reduction in velocities and in flame temperatures has been evaluated.[68] The inert dilution undoubtedly requires more combustion space and slows combustion velocities, thereby permitting higher stable firing rates while decreasing flame temperatures.

Detonation

Detonation is the propagation of a sharp pressure shock thru a gaseous medium in which the shock wave velocity is maintained by utilizing the heat of combustion released by detonation. As shown in Table 2-86, some very high flame speeds are attained under these conditions. It has been estimated[1] that at one instant during the thirteenth millisecond of tests the acceleration of the flame on a manufactured gas–air mixture was at the rate of 350,000 fps increase in velocity. The values for flame propagation under detonating conditions (Table 2-86) vary from 1300 to nearly 6000 times as much as those for normal combustion at constant pressure.

The detonation phenomenon in pipes differs from ordinary combustion due to the unexpected effects produced by extreme rate of pressure rise. This causes **impact** rather than expansion effects. J. B. Smith[45] measured axial and radial detonation pressure waves. He found that in detonation explosions in pipes, the pressure exerted along the pipe axis was *greater* than the radial pressure exerted on the wall. In ordinary combustion, the two pressures were about the same. Smith reported that detonation occurred only in pipes exceeding certain length–diameter ratios. Once detonation occurs, the instantaneous or impact pressure is independent of the piping system; that is, of whether the piping is opened or closed at the ends, or of how far beyond the minimum required length it may extend.

Detonation explosions starting with air–gas mixtures at or near atmospheric pressure have occurred only when the shape of the container was long compared with its diameter. Such detonations occur within concentration limits which are much *narrower* than the flammability limits. There are also differences which result from the chemical nature of the gases involved. Thus far not all flammable gases have been found to produce detonation. If initial pressures substantially *above* atmospheric levels are involved, then detonation limits will be *widened* and the length–diameter relationship required to establish detonation will be decreased. It has been shown[46] that natural gas–air mixtures in a 4-in. pipe were incapable of detonation explosions unless the air–gas mixture was flowing at over 250 fpm (i.e., approximately four feet per second) toward the flame. (See Table 2-84 for data on flame speeds in pipes.) It was also found that detonation pressures were unaffected by further increases in the flow rate up to 1100 fpm.

Table 2-86 Flame Velocity of Manufactured Gas under Detonating Conditions[1]

Gas mixture*	Distance from spark ignition, in.	Pressure attained, psia	Flame velocity, fps
Manufactured gas to air ratio = 1 to 4.976	50	36.7	800
	150	117.8	2200
	225	808.0	6800†
	450	235.0	5740†
	850	235.0	5740

* Manufactured gas composition: CH_4, 32.4%; C_2H_4, 3.4%; C_3H_8, 0.8%; CO_2, 1.9%; CO, 7.2%; H_2, 50.0%; O_2, 0.5%; N_2 3.8%. Mixture used was 1 part gas to 4.976 parts air. Total length of 2-in. diam pipe = 74 ft.

† Maximum values attained in 14 milliseconds.

‡ Equilibrium values.

Table 2-87 Minimum Explosive Concentrations of Some Common Dusts

(oz per cubic foot)

Aluminum	0.025
Magnesium	.020
Zirconium	.040
Phenolic resin	.025
Coal	.035
Sulfur	.035
Cornstarch	0.045

Dust Explosions. Whether dusts can explode in air depends on several factors; these include particle size, concentration (see Table 2-87), dispersion uniformity, amounts of impurities, and source of ignition. Generally accepted data for lower explosive limits range from 0.015 oz per cu ft for light, readily dispersed dusts to 0.5 oz per cu ft for heavy powders. Explosion pressures vary widely (approximately 40 to 100 psig, except for the metal dusts, which range from 5 to 100 psig). Generally, it has been noted, dust explosions in industry come in pairs; first, the dust in the air, and second, dust jarred from ledges, cracks, beams, etc., by the first explosion.

Explosion Relief.[47] An ordinary building cannot resist a sustained internal pressure of 1.0 psig. Hence, relief of pressure to atmosphere at a rate which will preclude destructive pressure on the building is necessary. Conservative requirements (F.I.A.) for volumes, in cubic feet, accommodated per square foot of relief area follow:

1. Machines and ovens of light construction and enclosures to 1000 cu ft—10 to 30.
2. Small enclosures, substantially built with reasonably high bursting strength—30.
3. Enclosures of 1000 to 25,000 cu ft—30 to 50.
4. Large rooms and buildings over 25,000 cu ft in which hazard represents small portion of entire volume: heavy reinforced-concrete walls, 80; light reinforced-concrete, brick, or wood, 60 to 80; light construction, e.g., prefabricated, 50 to 60.

Combustion and Physical Properties of Liquid Vapors Which Might Be Found in Sewers

Some of the more combustible liquids with high volatility and low flash point are listed in Table 2-88. Certain properties

Table 2-88 Combustion and Physical Properties of Liquid Vapors Which Might Be Found in Sewers

Vapor	Formula	Vapor in air, %: Explosive limits, Lower	Explosive limits, Upper	For theoretical combustion	Flash point, °F	Liquid density (water = 1.0)	Gaseous density (air = 1.0)
Acetone	C_3H_6O	2.55	12.8	4.97	0	0.792	2.0
Benzene	C_6H_6	1.35	6.75	2.72	12	.879	2.692
Butane	C_4H_{10}	1.86	8.41	3.12	Gas	0.60	2.066
Carbon disulfide	CS_2	1.25	50.00	0.652	22	1.262	2.64
Dichloroethylene	$C_2H_2Cl_2$	7.0	19.5	9.5	57	2.27	...
Divinyl ether	C_4H_6O	1.7	27.0	4.02	—22	...	...
Ethyl acetate	C_4H_8O	2.18	11.4	4.02	28	...	3.04
Ethyl alcohol	C_2H_6O	3.28	18.9	6.52	54	0.789	1.590
Ethyl ether	$C_4H_{10}O$	1.85	36.5	3.37	—49	.774	2.56
Gasoline (regular)	C_8H_{18}	1.4	7.5	...	0	.712	3.943 (*n*-octane)
Methyl alcohol	CH_4O	6.72	36.5	12.24	52	.792	1.1052
Methyl ethyl ketone	C_4H_8O	1.81	9.5	3.67	30	...	...
Naphtha	...	1.1	6.0	...	0–60	.702	3.459 (*n*-heptane)
Pentane	C_5H_{12}	1.4	7.8	2.55	—40	.626	2.487
Petroleum ether	...	1.7	...	...	—69	...	...
Propane	C_3H_8	2.1	10.1	4.02	Gas	.510	1.55
Propyl acetate	$C_5H_{10}O_2$	1.77	8.0	3.12	56	...	...
Propyl alcohol	C_3H_8O	2.15	13.5	4.44	59	.804	...
Stoddard solution	...	...	...	...	117	...	3.9+
Toluene	C_7H_8	1.27	6.75	2.27	40	.866	3.176
Turpentine	$C_{10}H_{16}$	0.80	...	1.47	95	...	...
Vinyl acetate	$C_4H_6O_2$	1.70	...	...	18	...	...
Xylene	C_8H_{10}	1.00	6.0	1.95	63	0.874	3.6618

relating to their combustion and explosion characteristics are included.

FLUE GASES

Composition

Flue gases from the common fuels contain primarily carbon dioxide, water vapor, oxygen, and nitrogen. The term **flue gases** is used interchangeably with **products of combustion.** Flue gases represent the final products of a heating process. As an indication of the quantities of flue gases generated by **perfect combustion** of pure combustibles with *no excess air,* data in Table 2-89 are presented.

When **incomplete combustion** occurs, flue gases may contain many other gases, including carbon monoxide, hydrogen, aldehydes, and unsaturated hydrocarbons, as shown previously under **Incomplete or Partial Combustion,** and in Tables 2-60, 2-61, and 2-62.

Flue gases from either complete or incomplete combustion may also contain traces of oxides of nitrogen, sulfur dioxide, and sulfur trioxide. Nitric oxide is formed in the combustion process in amounts varying up to 0.01 per cent in the flue gases. High temperature and increased pressure cause formation of the larger amounts of nitric oxide. If hydrogen sulfide or organic sulfur is present in the fuel gas, the sulfur will appear in the flue gases as sulfur dioxide and sulfur trioxide. It was found[48] that a gas containing 12 grains of organic sulfur per CCF gave 0.00146 per cent SO_2 and 0.00041 per cent SO_3 in flue gases containing 95 per cent excess air.

In most heating operations there is usually some **excess air** in the flue gases. Domestic heating appliances usually require from 25 to 50 per cent excess air, although some types of equipment may employ as much as 100 per cent excess air. Composition of the flue gases with *known amounts of excess air* can be calculated using Tables 2-89 and 2-63. This calculation is accomplished by dividing the volume of each flue gas constituent by the total volume of flue gases including the excess air. For example, flue gases from methane burned with 50 per cent excess air (4.764 cu ft per cu ft of methane) would contain:

				Per cent in dry flue gas
	CO_2	1.0 ÷ 13.292 × 100	=	7.54
	H_2O	2.0		
	N_2	7.528 ÷ 13.292 × 100	=	84.92
Excess air	N_2	3.764 ÷ 13.292 × 100		
	O_2	1.0 ÷ 13.292 × 100	=	7.54
	Total	15.292		100.00
	Less H_2O	2.0		
		13.292 cu ft dry		

Computations from Flue Gas Analyses

Flue gases are normally analyzed for CO_2, O_2, and CO with an **Orsat** apparatus or thermal conductivity equipment, and for low percentages of CO with an **iodine pentoxide apparatus, colorimetric tubes,** or **heat generation over hopcalite.** The *water vapor content* is not determined by these analyses and it is not normally reported in flue gas analysis. Other methods such as condensation, absorption and weighing of the water vapor, or dew point tests may be used. From these analyses the following calculations can be made:

Ultimate CO_2. This factor may be ascertained under

Table 2-89 Flue Gas Products from Perfect Combustion of Gases
(no excess air)

No.*	Substance	Cu ft per cu ft of combustible: CO_2	H_2O	N_2	Total	Dew point,† °F	Pounds per pound of combustible: CO_2	H_2O	N_2	Total	Ultimate CO_2, %
2	Carbon	...	...	...	...	...	3.664	...	8.863	12.527	29.30
4	Carbon monoxide	1.0	...	1.882	2.882	...	1.571	...	1.900	3.471	34.70
5	Hydrogen	...	1.0	1.882	2.882	162	...	8.937	26.407	35.344	...
Paraffin hydrocarbons											
6	Methane	1.0	2.0	7.528	10.528	139	2.744	2.246	13.275	18.265	11.73
7	Ethane	2.0	3.0	13.175	18.175	134	2.927	1.798	12.394	17.119	13.18
8	Propane	3.0	4.0	18.821	25.821	131	2.994	1.634	12.074	16.702	13.75
9, 10	Butanes‡	4.0	5.0	24.467	33.467	129	3.029	1.550	11.908	16.487	14.05
11–13	Pentanes‡	5.0	6.0	30.114	41.114	128	3.050	1.498	11.805	16.353	14.24
14, 15	Hexanes‡	6.0	7.0	35.760	48.760	128	3.064	1.464	11.738	16.266	14.37
16, 17	Heptanes‡	7.0	8.0	41.404	56.404	128	3.074	1.438	11.688	16.200	14.46
18, 19	Octanes‡	8.0	9.0	47.050	63.050	127	3.082	1.419	11.651	16.152	14.80
Olefin series											
20	Ethylene	2.0	2.0	11.293	15.293	125	3.138	1.285	11.385	15.808	15.05
21	Propylene	3.0	3.0	16.939	22.939	125	3.138	1.285	11.385	15.808	15.05
22, 23	Butylenes‡	4.0	4.0	22.585	30.585	125	3.138	1.285	11.385	15.808	15.05
24	*n*-Pentene	5.0	5.0	28.232	38.232	125	3.138	1.285	11.385	15.808	15.05
Aromatic series											
25	Benzene	6.0	3.0	28.232	37.232	108	3.381	0.692	10.224	14.297	17.53
26	Toluene	7.0	4.0	33.878	44.878	111	3.344	0.782	10.401	14.527	17.12
27	Xylene	8.0	5.0	39.524	52.524	113	3.317	0.849	19.530	14.696	16.83
Miscellaneous gases											
28	Acetylene	2.0	1.0	9.411	12.411	103	3.384	0.692	10.224	14.297	17.53
29	Naphthalene	10.0	4.0	45.170	59.170	101	3.434	0.562	9.968	13.964	18.13
30	Methyl alcohol	1.0	2.0	5.646	8.646	147	1.374	1.125	6.644	9.142	15.05
31	Ethyl alcohol	2.0	3.0	11.293	16.293	138	1.910	1.173	8.088	11.172	15.05
32	Ammonia	...	1.5	3.323	4.823	158	NO = 1.761	1.587	7.815	11.161	...
35	Formaldehyde	1.0	1.0	3.764	5.764	135	1.466	0.600	3.546	5.612	20.99
36	Formic acid	1.0	1.0	1.882	3.882	151	0.956	0.391	1.157	2.504	34.70
37	Acetaldehyde	2.0	2.0	9.410	13.410	130	2.000	0.818	6.049	8.867	17.53

No.	Sulfur-bearing gases	Cu ft per cu ft of combustible: CO_2	H_2O	SO_2	N_2	Total	Dew point, °F	Pounds per pound of combustible: CO_2	H_2O	SO_2	N_2	Total	Ultimate CO_2, %
33	Sulfur	...	...	1.0	3.764	4.764	...	...	...	1.499	1.661	3.160	...
34	Hyd. sulfide	...	1.0	1.0	5.646	7.646	125	...	0.528	2.820	4.686	8.034	...
	Methyl mercaptan	1.0	2.0	1.0	11.292	15.292	125	0.549	0.450	1.199	3.985	6.183	9.55
	Ethyl mercaptan	2.0	3.0	1.0	16.938	22.938	125	0.935	0.574	1.021	5.088	7.618	12.93
	Propyl mercaptan	3.0	4.0	1.0	22.584	30.584	125	1.220	0.666	0.888	5.904	8.678	14.64
	Butyl mercaptan	4.0	5.0	1.0	28.230	38.230	125	1.440	0.737	0.786	6.533	9.46	15.71
	Amyl mercaptan	5.0	6.0	1.0	33.876	45.876	125	1.622	0.794	0.705	7.033	10.254	16.42

* Numbers correspond to those in Table 2-59.
† Calculated and taken from Fig. 2-31.
‡ Flue gas products same for all forms of these gases.

the two possible conditions: no excess air, Eq. **17**; or excess air present, Eq. **18**.

Air-Free Factor. This factor is used to convert actual flue gas analyses to a *standard* undiluted flue analysis. For example, if analysis showed 0.2 per cent CO and 15.7 per cent O_2, and the air-free factor would be 3.96 (by substituting in Eq. **31**) and the *air-free* CO would be 0.79 per cent.

$$\text{Air-free factor} = \frac{21}{21 - \%\ O_2} \tag{31}$$

or

$$\text{Air-free factor} = \frac{\text{Ultimate } \%\ CO_2}{\%\ CO_2 \text{ in flue gas}} \text{ if CO in flue gas is negligible} \tag{32}$$

Flue Loss. (These data are normally taken from alignment charts or flue loss charts based on the percentage of CO_2 in flue gas and flue gas temperature. See Figs. 2-79 and 2-80.)

0.01 (% CO_2) × heat content of CO_2 at flue gas temp from Fig. 2-18 = ______ Btu/cu ft
0.01 (% O_2) × heat content of O_2 at flue gas temp from Fig. 2-18 = ______
0.01 (% N_2) × heat content of N_2 at flue gas temp from Fig. 2-18 = ______
0.01 (% H_2O calc.) × heat content of H_2O at flue gas temp from Fig. 2-18 = ______

Total sensible heat = ______ Btu/cu ft
Latent heat of water vapor = % H_2O vapor/100 × 50.4 Btu/cu ft* = ______ Btu/cu ft

* Heating value at 14.7 psia and 60 F.

$$\frac{\text{Total sensible heat and latent heat in } \textit{flue} \text{ gas} \times 100}{\text{Total gross heat in } \textit{fuel} \text{ gas}} = \% \text{ flue loss} \quad (33)$$

Excess Air. Excess air may be expressed simply as total air in the flue gases *divided* by the air required for combustion, or:

$$\% \text{ Excess air} = \frac{100q(u - C)}{AC} \quad (34)$$

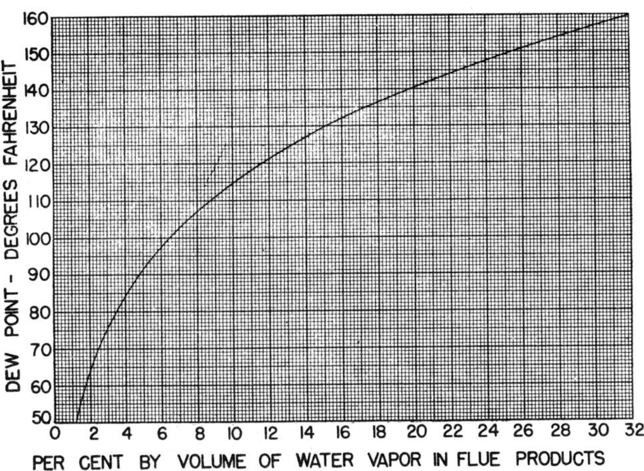

Fig. 2-31 Water vapor and dew point of flue gases.

where:

A = theoretical air required for complete combustion of 1 cu ft of gas, cu ft

q = volume of theoretical flue products formed by combustion of 1 cu ft of gas, cu ft

C = % CO_2 in flue gas sample

u = ultimate % CO_2 in flue gas

$$\% \text{ Excess air} = \frac{100(O_2 - 0.5\ CO)}{0.266\ N_2 - (O_2 - 0.5\ CO)} \quad (35)$$

where the symbols for the gases stand for the percentage of each gas found by Orsat analysis.

Flue Gas Volume.

Total cu ft of flue gas volume =

$$\left[\frac{\text{cu ft } CO_2 \text{ produced per cu ft gas burned} \times 100}{\% \ CO_2 \text{ in flue gas}}\right]^* + \text{total water vapor.}$$

Water Vapor and Dew Point of Flue Gases

Water vapor in flue gases is the total of that contained in the fuel itself, in the air admitted, and resulting from the combustion of hydrogen or hydrogen in hydrocarbons. The amount of water vapor in flue gases may be calculated from the fuel burned using data in Tables 2-63 and 2-89.

* The first term is the volume of dry flue gases.

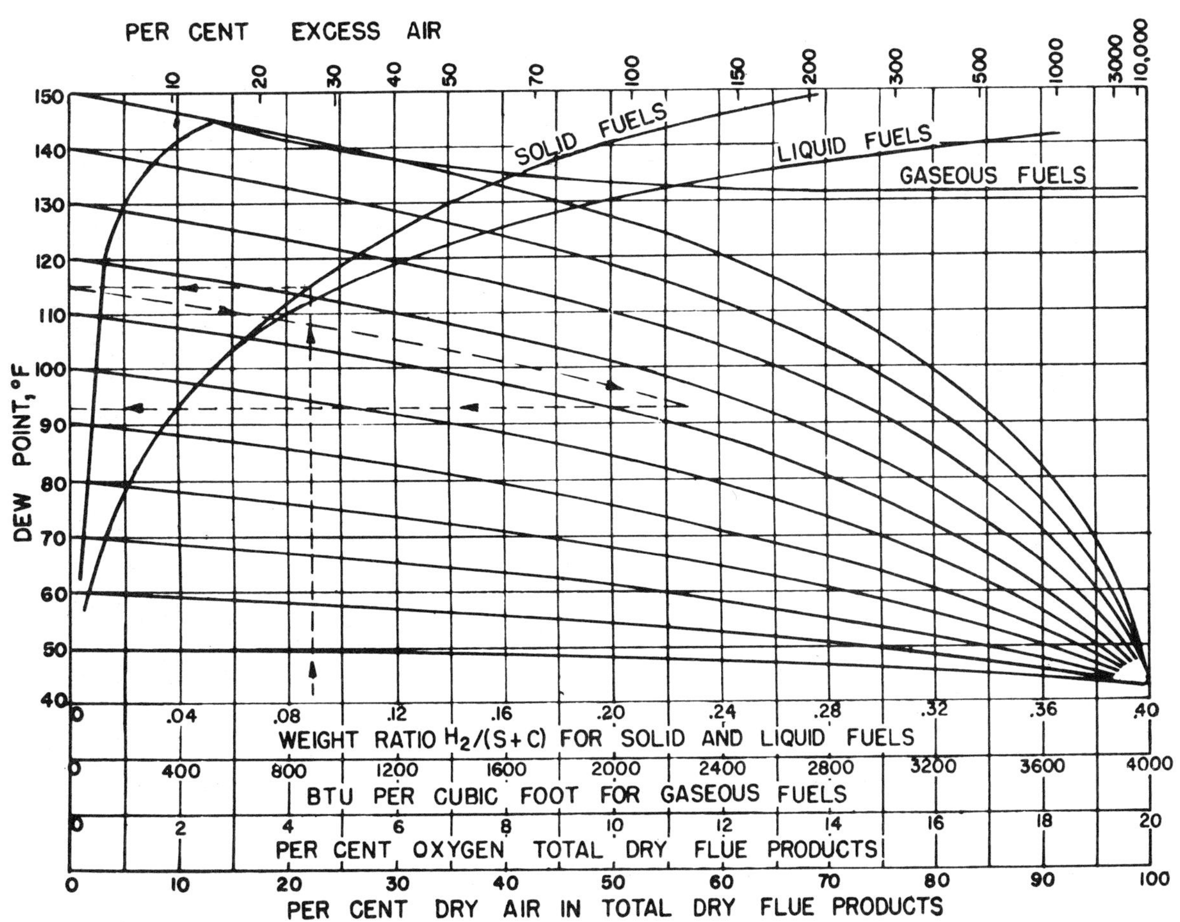

Fig. 2-32 Theoretical dew points of the combustion products of industrial fuels.

Table 2-90 Emissions from Various Industrial and Commercial Gas-Fired Combustion Equipment[71]

		Flue gas data—stack conditions							Air contaminant emissions[a]						
					Orsat analysis (wet basis)				Combustion contaminants[d]			Aldehydes (as formaldehyde)		Oxides of nitrogen (as NO_2)	
Footnote No.	Natural gas burned, SCFH	Vol SCFM[b]	Temp, °F	CO_2, vol %	O_2, vol %	CO, vol %[c]	Excess air, vol %	Moisture, vol %	Grains/ SCF	Grains/ SCF at 12% CO_2	Lb/hr	Ppm	Lb/hr	Ppm	Lb/hr
1	840	400	550	4.8	10.0	0.000	98	9.2	0.0035	0.0088	0.012	3	0.005	33	0.10
2	2,040	720	300	4.5	9.3	.1	93	14.3	.019	.051	.12	4	.02	37	.20
6	7,200	1,800	580	7.4	2.3	.2	13	18.3	.0049	.0079	.08	2	.02	28	.4
7	2,220	2,700	160	1.7	17.2	.00	28	4.3	.0004	.003	.009	3	.05	8	.16
8	4,260	1,700	370	4.8	9.6	.000	94	10.8	.0011	.0028	.016	4	.03	13	.16
9	2,487	1,250	380	3.8	11.3	.000	135	10.8	.0017	.0054	.018	11	.065	16.1	.146
11	4,860	1,200	440	7.8	2.3	.01	13	16.6	.0022	.0034	.02	3	.02	38.2	.334
12	6,120	3,100	330	3.8	12.4	.000	124	8.2	.0079	.0025	.021	4	.06	34.8	0.79
13	10,800	4,100	480	5.0	8.2	.000	72	13.3	.0013	.0031	.046	7	.14	56	1.7
15	15,420	5,400	218	5.2	9.1	.0	85	11.2	.0018	.004	.08	4	.09	39.4	1.55
16	28,800	11,500	600	4.8	9.0	.0	84	10.9	.0033	.0082	.33	4	.2	124	10.4
18	24,000	9,200	480	5.0	8.1	.02	73	14.6	.0022	.0053	.17	10	.5	127	8.5
19	117	81	110	3.4	14.6	.001	203	5.7	.007	.03	.005	2	.002	45.8	0.026
20	130	124	160	2.0	17.7	.0	480	3.8	.001	.01	.001	2	.001	19.0	.016
21	408	155	373	5.0	9.5	.000	99	11.6	.0007	.002	.0009	3	.003	34.1	.038
22	384	230	370	1.7	16.6	.000	437	5.0	.0013	.009	.0026	6	.006	16.0	.026
23	582	310	250	3.5	14.7	.000	38	7.0	.0018	.062	.0048	6	.008	19.5	.045
24	1,146	1,300	330	18.1	1.0	.000	6	20.0	.0069	.075	.073	0	0	102	.16
26	619	240	440	5.6	9.1	.04	67	5.6	.002	.01	.004	2.2	.0025	13.4	.023
27	420	422	360	0.7	19.4	.0	[e]	2.9	.0012	.021	.0043	3.9	.0045	2.8	.0046
28	1,140	450	1040	4.9	10.0	.000	105	11.7	.0021	.0051	.0081	5	.01	66	.22
29	4,679	1,100	590	2.7	15.7	0.000	355	6.1	0.0051	0.023	0.048	6.7	0.035	18.8	0.15

[a] All concentrations expressed at actual stack conditions except where otherwise noted.
[b] Standard conditions are 60°F and 14.7 psia.
[c] All CO measurements below 0.1% were made with Mine Safety Appliances Co. instrument.
[d] "Combustion contaminants" are particulate matter discharged into the atmosphere from the burning of any kind of material containing carbon in a free or combined state. (Rule 2m of the Los Angeles County Air Pollution Control District Rules and Regulations.)
[e] Excess air introduced after combustion zone.

	Equipment	*Type*
No. 1:	Thompson tubeless boiler, 30 hp	Tubeless
2:	Cyclotherm steam generator, 60 hp	Fire-tube
3:	Bryan No. 315, 100 hp	Water-tube
4:	Locomotive-type boiler, 120 hp	Single-pass fire-tube
5:	Pioneer boiler, 125 hp	Scotch marine
6:	Dixon water wall boiler, 150 hp	Scotch marine
7:	Gabriel boiler, 150 hp	Scotch marine
8:	Johnston boiler No. 18, 200 hp	Scotch marine
9:	B & W boiler Model FM-27, 200 hp	Water-tube
10:	Erie City boiler Model 4C-14, 245 hp	Water-tube 3-drum
11:	B & W boiler Type FM-1, 300 hp	Water-tube
12:	Kewanee boiler Model 590, 300 hp	Modified HRT 2 pass-fire-tube
13:	Dixon wet back boiler, 350 hp	Scotch marine
14:	Collins boiler, 425 hp	Water-tube
15:	Springfield boiler, 460 hp	Water-tube
16:	Sterling boiler Model 477-31 (modified), 500 hp	Water-tube 4-drum
No. 17:	Collins boiler, 580 hp	Water-tube
18:	B & W boiler Model FM-9, 870 hp	Water-tube
19:	General Water Heater Corp. Model A-75 H	75 gal water heater
20:	Utility Appliance Corp. Model 125 UF space heater	100,000 Btu/hr, forced air
21:	Vensel bonderizing oven	Indirect fired
21:	Vensel paint bake oven	Indirect fired
23:	Bake oven	Indirect fired
24:	425-lb crucible-type aluminum melting furnace	Indirect fired
25:	Childers oil heater Model D-100	Oil circulating heat exchanger
26:	Prouty custom artware kiln	Indirect fired full muffle
27:	Lehr glassware decorating oven	Indirect fired full muffle
28:	Ceramic tunnel kiln	Indirect fired
29:	Semimuffle dinnerware kiln	Indirect fired

Table 2-91 Emissions from Various Industrial and Commercial Oil-Fired Combustion Equipment[71]

		Flue gas data—stack conditions										Air contaminant emissions[a]										
		Fuel oil analysis					Orsat analysis (wet basis)					Combustion contaminants[c]			Sulfur dioxide		Sulfur trioxide		Aldehydes (as formaldehyde)		Oxides of nitrogen (as NO_2)	
Foot-note No.	Fuel oil burned, gph	API gravity	Sulfur, wt %	Ash, wt %	Vol, SCFM[b]	Temp, °F	CO_2, vol %	O_2, vol %	CO, vol %	Excess air, vol %	Moisture, vol %	Grains/SCF	Grains/SCF at 12% CO_2	Lb/hr	Ppm	Lb/hr	Ppm	Lb/hr	Ppm	Lb/hr	Ppm	Lb/hr
2	9	31.07	1.05	0.02	390	250	7.0	7.9	0.01	65	9.8	0.041	0.069	0.14	355	1.4	1.6	0.0080	9	0.017	47	0.13
3	6.1	28.71	0.71	0	510	290	3.9	15.2	.000	290	4.7	.023	.071	.10	98.2	0.51	1.4	.0092	5	.013	35.8	0.13
4	36.3	16.51	1.0	0	1,800	710	7.0	7.8	.003	68	12.7	.043	.074	.66	414	7.5	4.7	.10	7	.05	368	4.8
5	23.1	11.39	1.78	0.18	1,800	330	5.0	13.3	.000	180	4.8	.0439	.11	.68	264	5.0	3.2	.076	9	.08	128	1.68
7	15.3	40.10	0.09	0	1,700	240	2.7	16.2	.001	150	4.4	.0085	.038	.12	28	0.48	1.7	.036	5	.04	20	0.25
8	21.0	33.82	.97	0	2,000	360	4.3	13.8	.02	210	5.6	.0498	.14	.85	11.2	2.3	5.6	0.00006	52	.50	21	.31
9	9.7	35.09	.55	0	1,290	370	2.8	16.3	.002	370	3.0	.023	.10	0.26	0.2	0.0024	0	0	8	.04	54.9	0.51
10	99.4	11.10	.94	0.13	4,300	540	7.9	6.0	.00	43	10.7	.042	.064	1.55	397	17.3	0.37	0.020	8	.2	387	12.1
11	23.1	33.01	.21	.07	1,300	390	5.5	10.9	.0024	115	6.9	.020	.041	0.22	102	1.35	.5	.008	7	.04	32.8	0.310
12	39.6	34.87	.29	0.01	3,100	320	4.0	13.9	.0	220	6.3	.0472	.142	1.25	7.1	2.24	0.0	0.0	6	.08	14.7	0.33
13	85.5	32.9	0.42	0	4,200	500	6.3	9.8	.000	94	8.2	.0072	.014	0.26	17	7.2	0	0	3	.06	72	2.2
14	160.0	8.0	3.06	0	10,600	630	6.3	10.3	.001	110	10.6	.147	.28	13.4	700	75	6.7	1.2	4	.2	274.9	19.8
15	80.5	12.11	0.78	0.12	4,800	220	5.9	10.7	.0	107	6.6	.0193	.039	0.79	362	17.6	2.2	0.13	7	0.2	199	6.95
16	248.0	15.09	1.39	.04	13,100	560	6.7	9.5	.000	92	9.1	.0249	.0446	2.80	594	79.0	3.6	.6	17	1.0	256	24.5
17	57.5	13.33	1.30	.03	3,200	580	6.4	9.6	.0	95	9.8	.032	.060	0.87	640	21	2.2	0.091	8.5	0.12	205.9	4.3
18	168.0	9.30	1.94	0.03	7,800	530	8.2	8.5	.000	73	7.9	.0653	.096	4.4	344	27.2	1.2	1.2	48	1.8	256	14.6
25	5.2	33.6	0.80	0	290	820	5.4	11.1	.002	120	7.8	.033	.073	0.081	138	0.41	2.8	0.003	11	0.015	33.7	0.065
26	10.0	45.1	Trace	0	230	410	9.7	3.5	.04	21	5.7	.001	.004	.002	0	0	0	0	3.5	.0037	27.1	.045
29	20.0	45.1	Trace	0	1,200	520	3.0	15.9	0.000	373	5.6	0.094	0.038	0.097	0.17	0.0021	0	0	3.4	0.020	19.8	0.17

[a] All concentrations expressed at actual stack conditions except where otherwise noted. [b] Standard conditions are 60°F and 14.7 psia. [c] "Combustion contaminants" are particulate matte discharged into the atmosphere from the burning of any kind of material containing carbon in a free or combined state. (Rule 2m of the Los Angeles County Air Pollution Control District Rules and Regulations.) See numbered footnotes to Table 2-90.

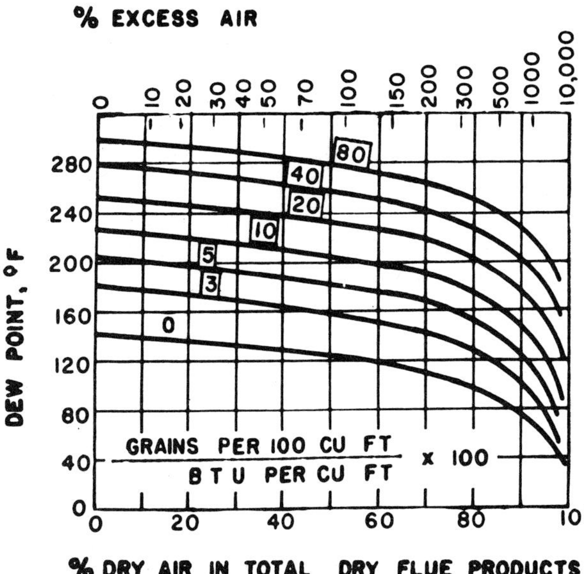

Fig. 2-33 Dew point curves for fuel gases that are sulfur-free (O curve) and for those that contain sulfur (the other curves shown). Sulfur in utility natural gas is so low that no significant dew point increase over theoretical values is involved. As many as 10 grains of sulfur in manufactured gas are needed to cause a 20°F increase in dew point.

The **dew point** is the temperature at which *condensation begins*. With the percentage of water vapor calculated, the dew point of the flue gas can be obtained from Fig. 2-31.

Using Fig. 2-32, the dew points of solid, liquid, or gaseous fuels may be estimated. For example, to find the dew point of flue gases resulting from the combustion of a solid fuel which has a weight ratio (hydrogen to carbon plus sulfur) of 0.088 and excess air in a quantity to give 11.4 per cent oxygen by Orsat analysis in the flue gases, start with the weight ratio 0.088 and proceed vertically to the intersection of the solid fuels curve until the theoretical dew point 115 F is located. (See dotted lines.) The curve fixed by this point is followed to the right to its intersection with the vertical line representing 11.4 per cent oxygen in the flue products. The actual dew point 93 F can then be read on the dew point scale.

The dew point for the flue products of a natural gas of 1020 Btu per cu ft burning to yield flue gases with 6.3 per cent oxygen or 31.4 per cent air may be estimated. The dew point of the flue products of the theoretical air–gas mixture, 139 F, is first located by means of the curve for gaseous fuels. The proper curve sloping to the right is then followed until it intersects the vertical line representing 6.3 per cent oxygen or 31.5 per cent air in the flue gases, and the dew point of this mixture, 127 F, is read on the dew point scale.

The presence of *sulfur dioxide*, and particularly *sulfur trioxide*, influences the *vapor pressure* of condensate in flue gases and the dew point may be *raised* by as much as 25 to 75°F, as shown in Fig. 2-33. To illustrate the use of Fig. 2-33, take a manufactured gas with a heating value of 550 Btu per cu ft containing 15 grains of sulfur per CCF being burned with 40 per cent excess air. The proper curve in Fig. 2-33 is chosen by:

$$\frac{\text{Grains S/100 cu ft}}{\text{Btu/cu ft}} \times 100 \quad \text{or} \quad \frac{15}{550} \times 100 = 2.73$$

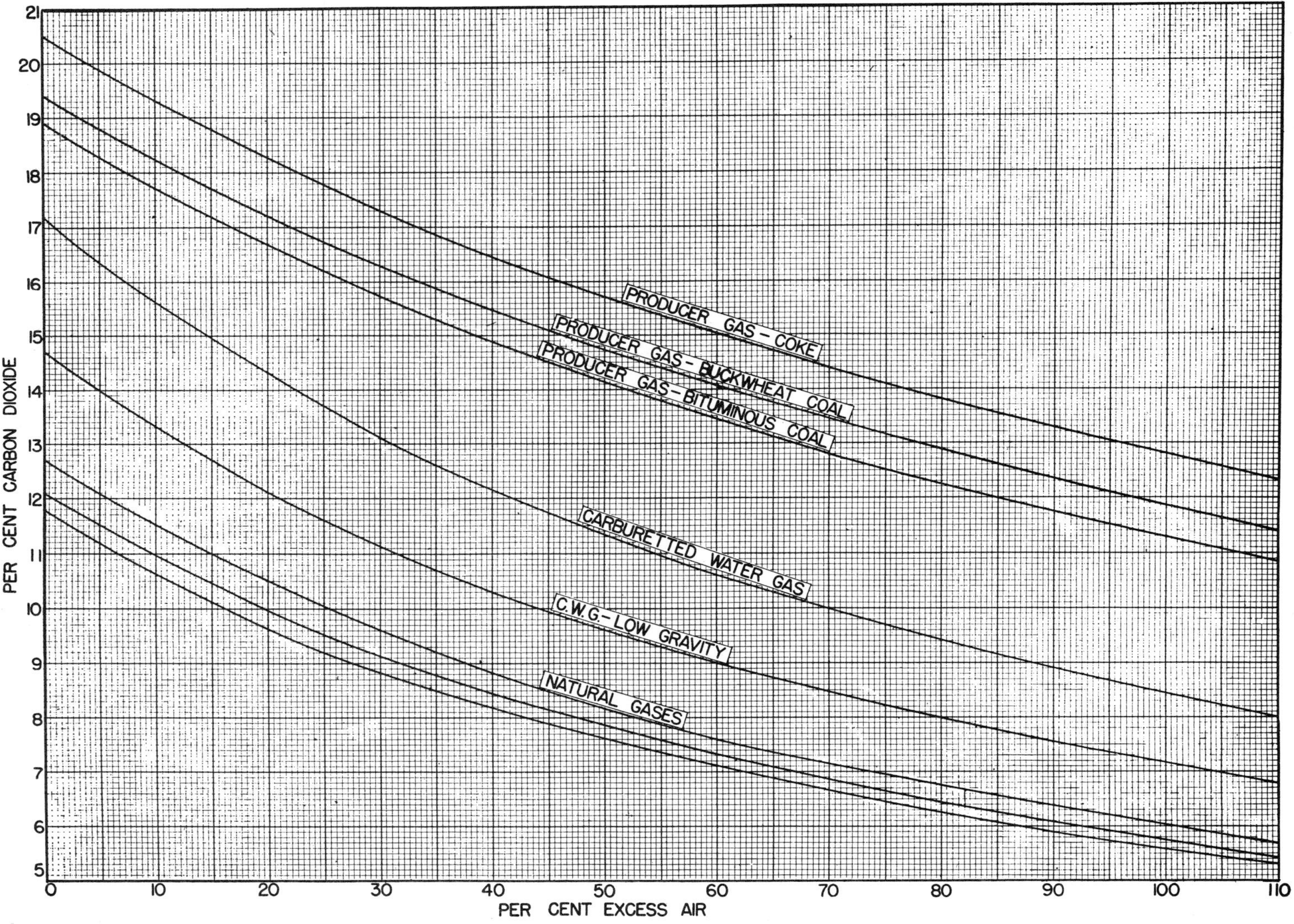

Fig. 2-34 Carbon dioxide vs. excess air in flue products for various typical gases, I.

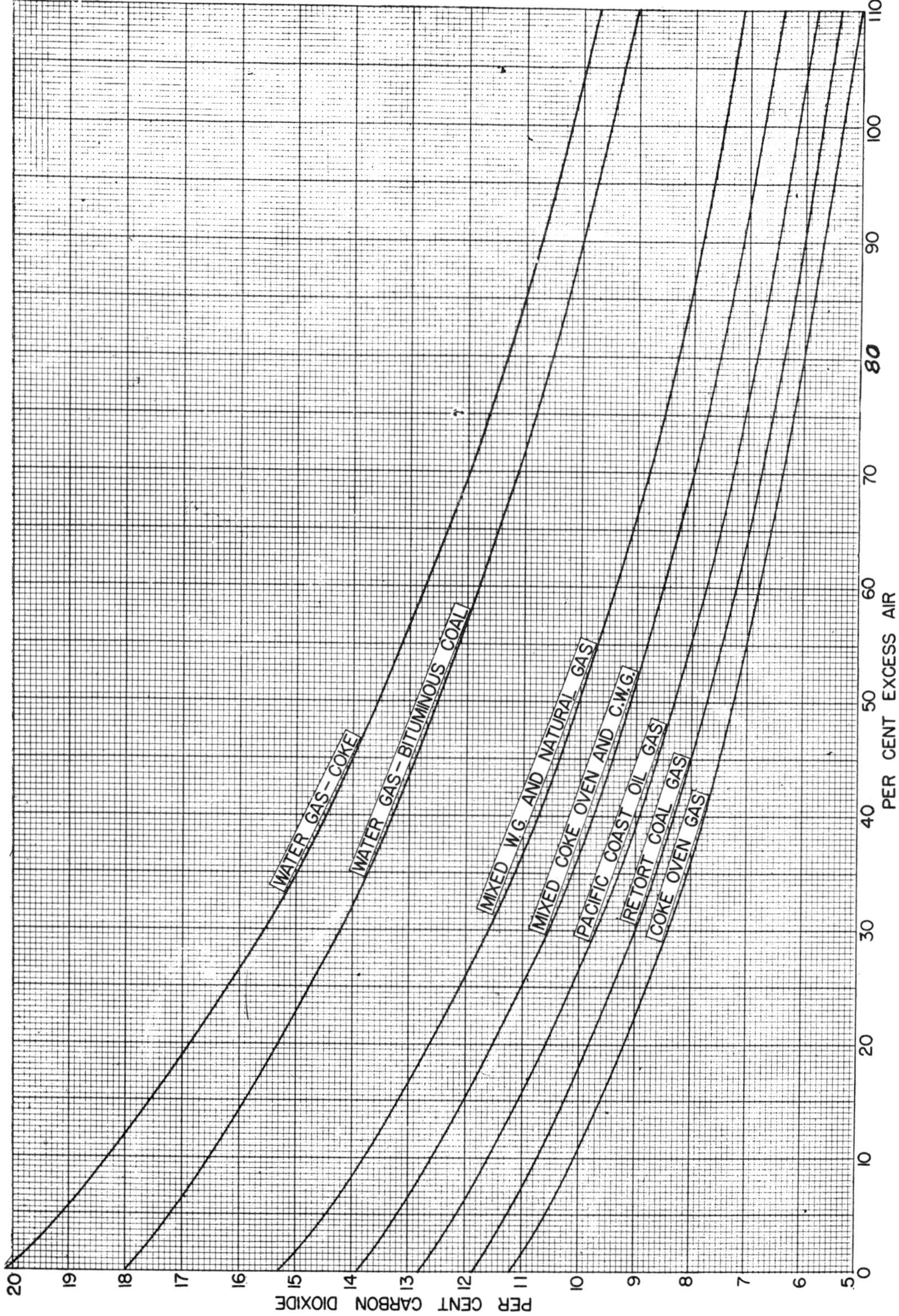

Fig. 2-35 Carbon dioxide vs. excess air in flue products for various typical gases, II.

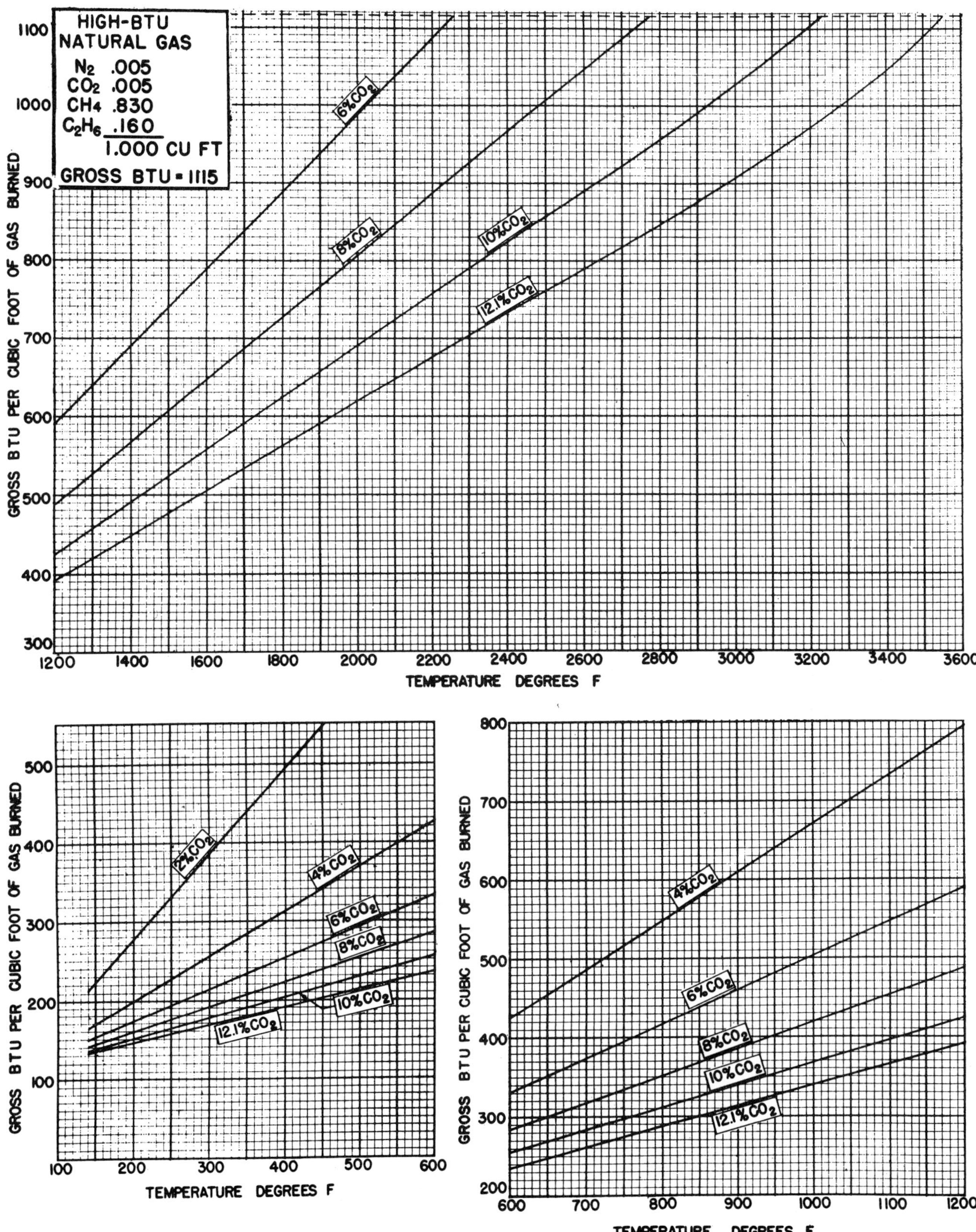

Fig. 2-36 Heat content in combustion products from high-Btu natural gas.

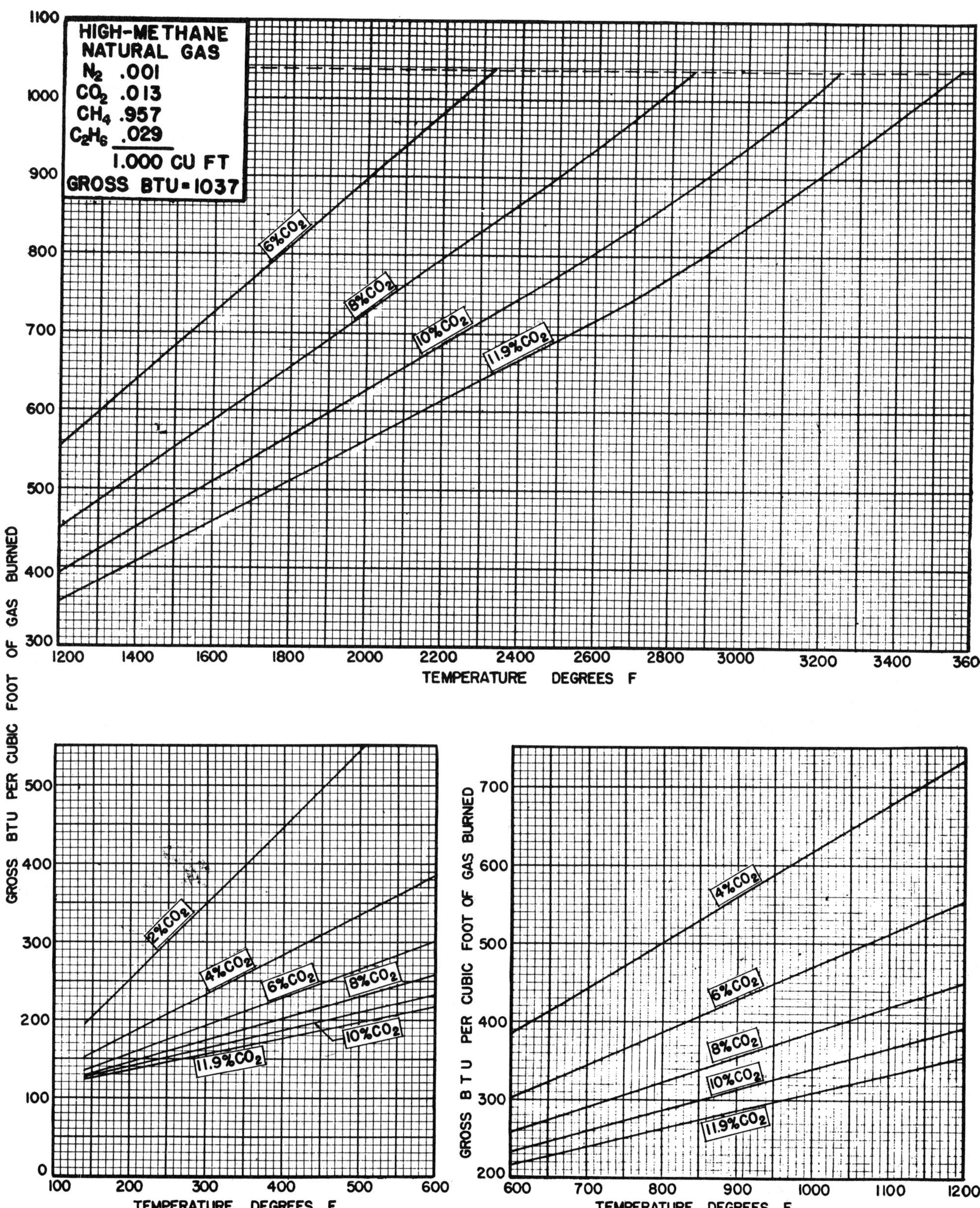

Fig. 2-37 Heat content in combustion products from high-methane natural gas.

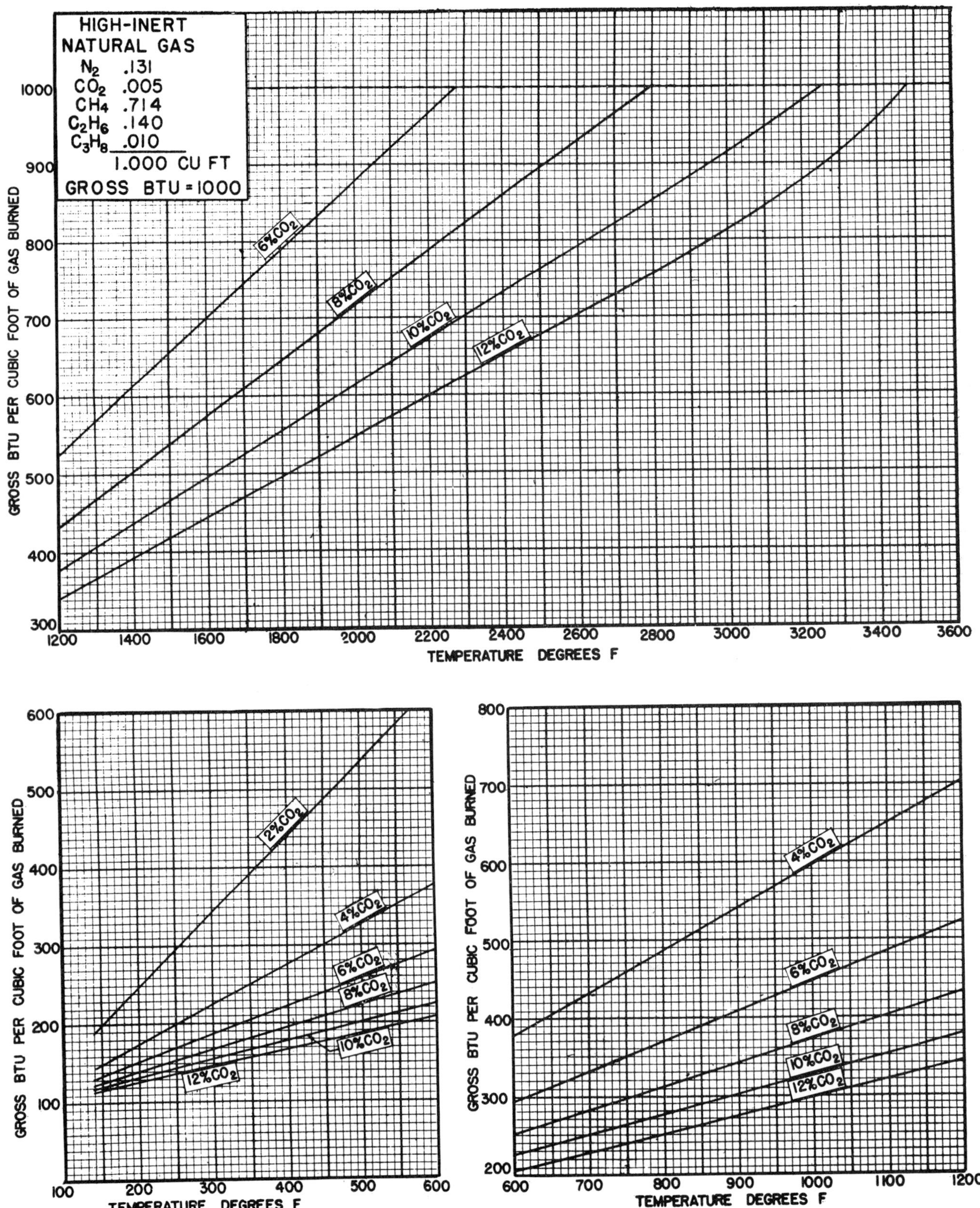

Fig. 2-38 Heat content in combustion products from high-inert natural gas

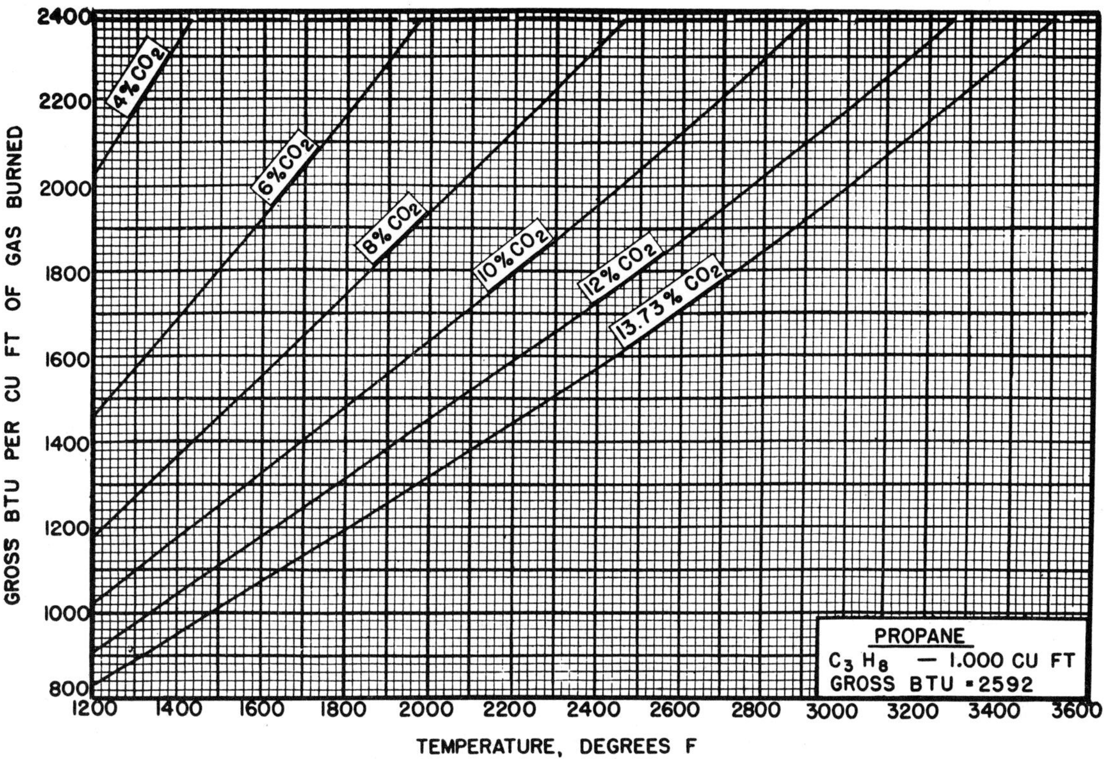

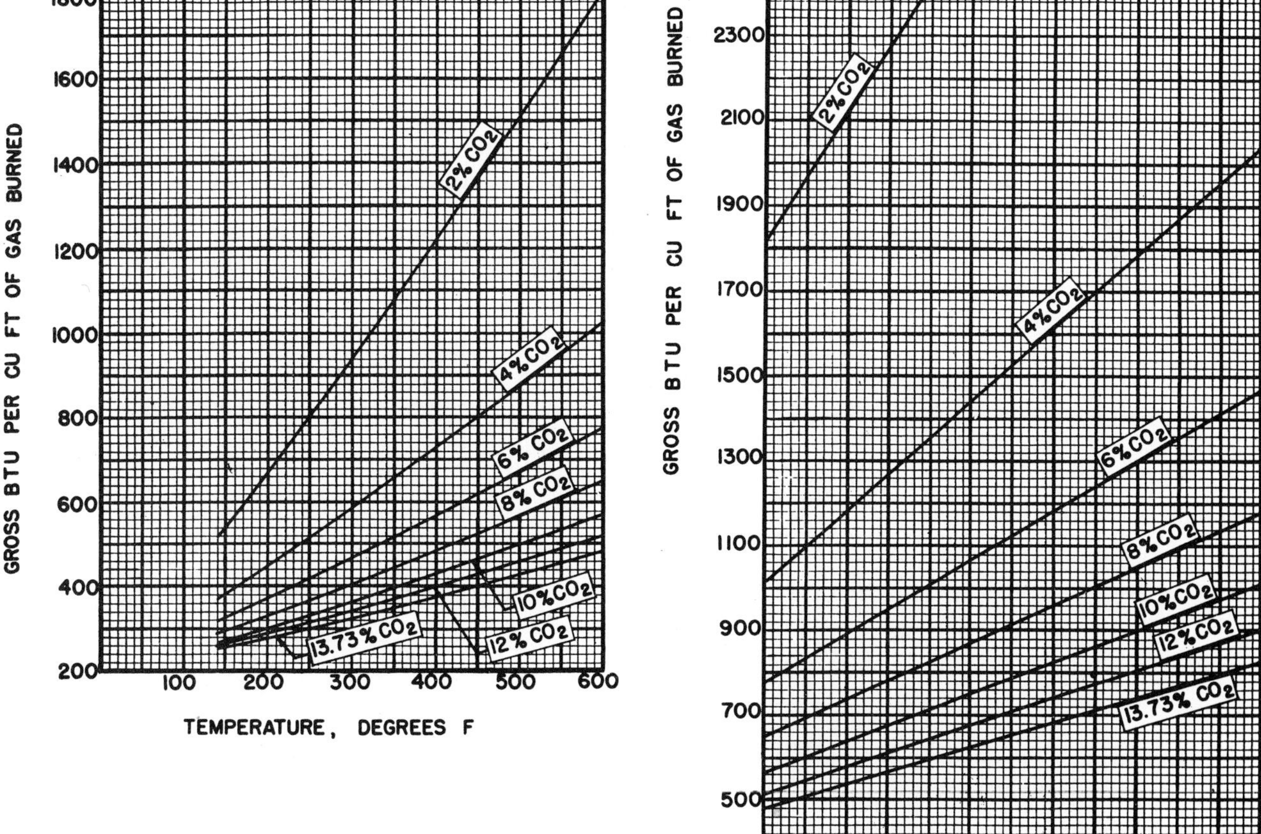

Fig. 2-39 Heat content in combustion products from propane gas.

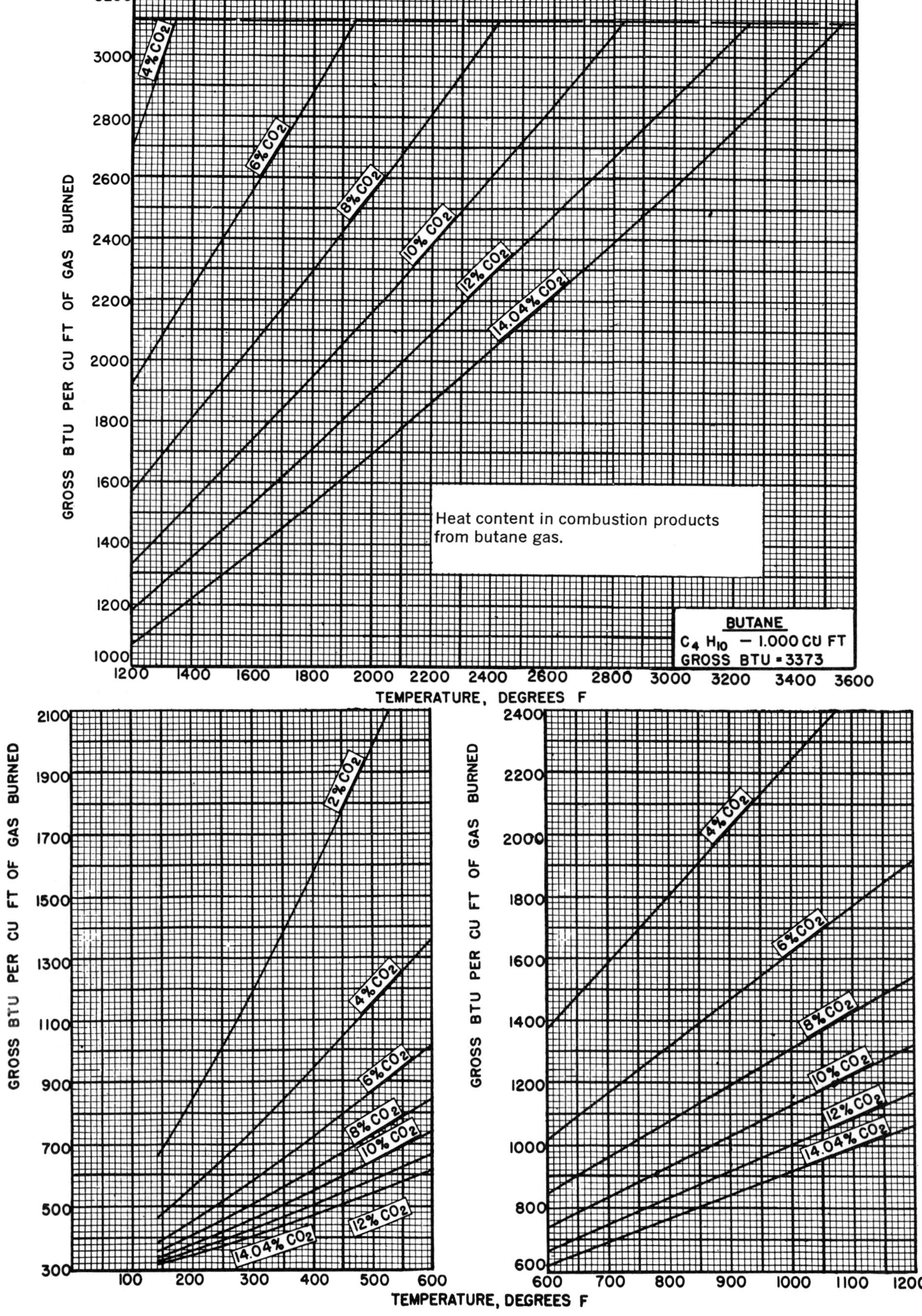

Fig. 2-40 Heat content in combustion products from butane gas.

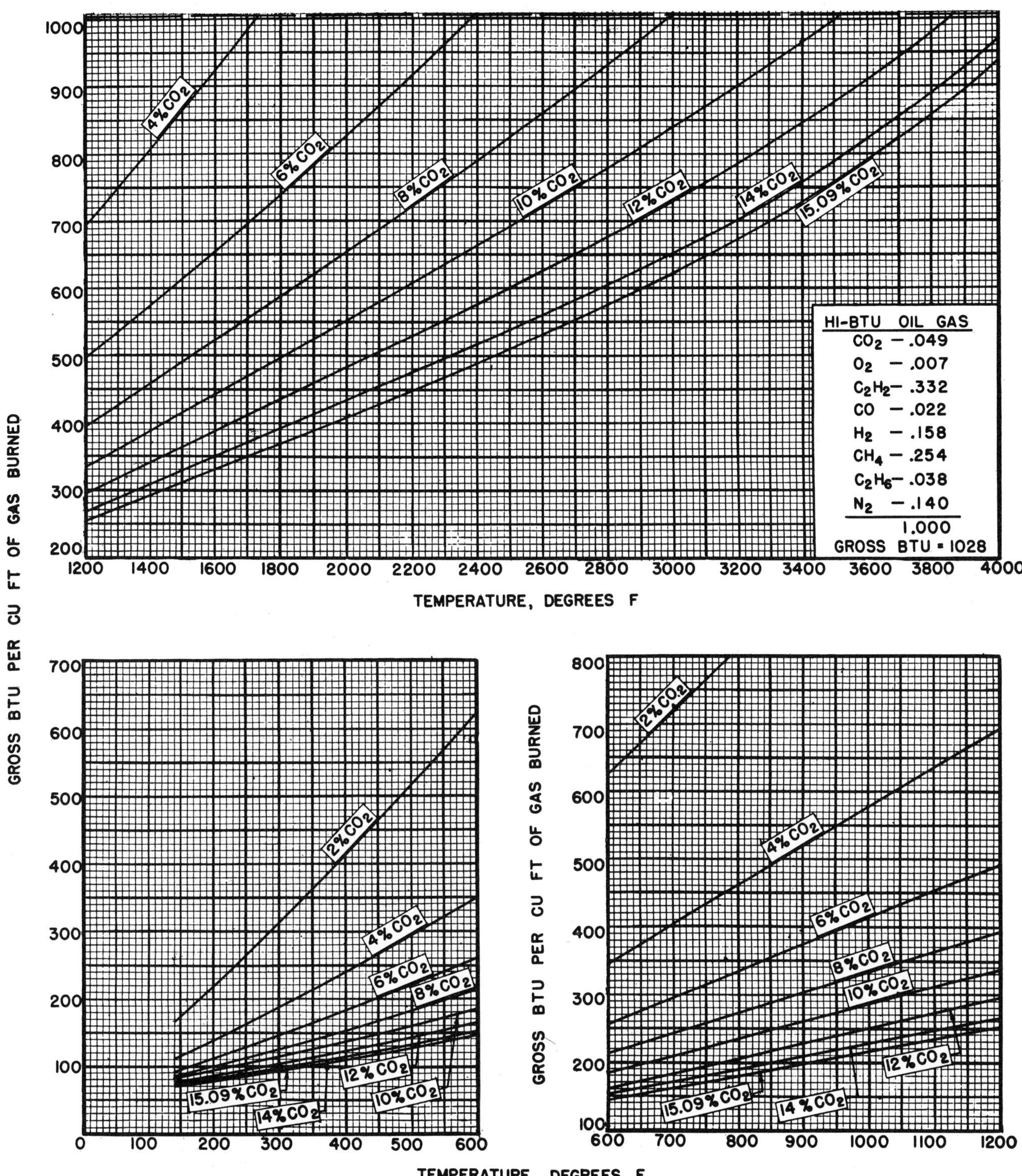

Fig. 2-41 Heat content in combustion products from high-Btu oil gas.

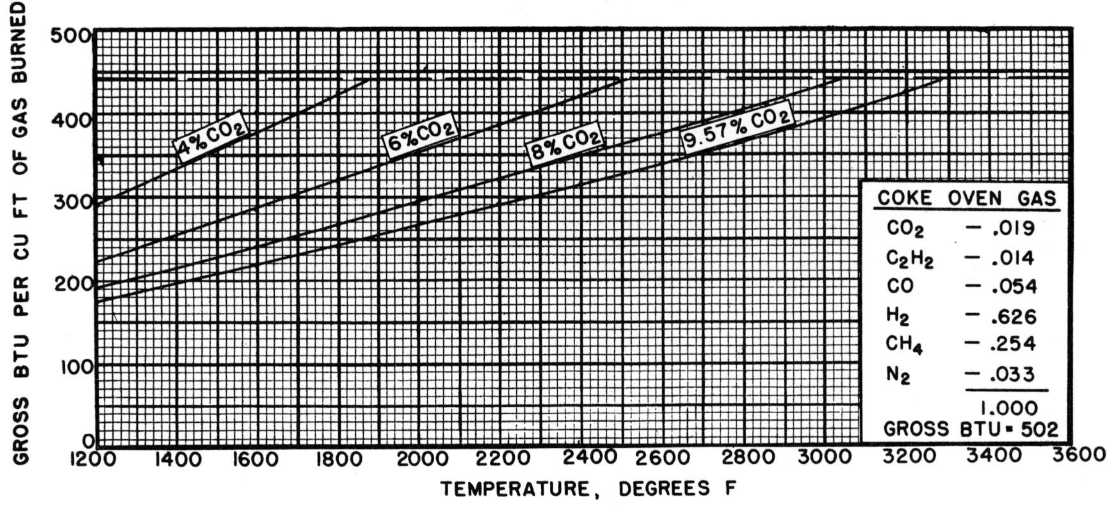

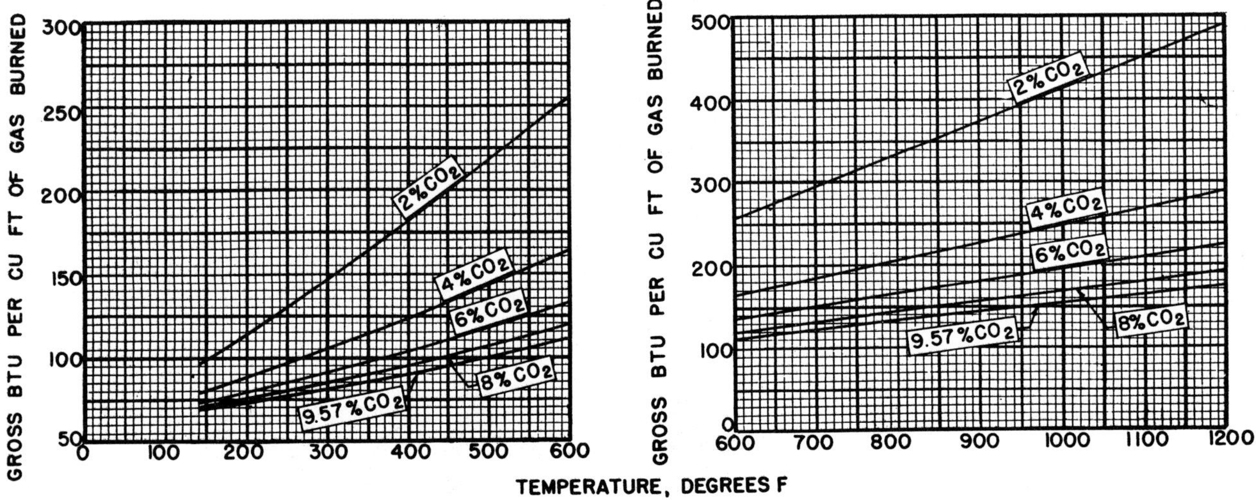

Fig. 2-42 Heat content in combustion products from coke oven gas.

This would be a curve lying between 0 and 3, and is close to the 3 curve. This curve represents the true dew points of all the flue gas mixtures containing any amount of excess air. For a mixture with 40 per cent excess air, the dew point is about 160 F.

Contaminant Emissions

The effluents resulting from combustion were reported (Tables 2-90 and 2-91) for 29 units of combustion equipment, including boilers, heaters, driers, and furnaces. Note that 12 of these were fired with both gas and oil fuels, 10 were tested on natural gas only, and seven were tested on oil only. While these emission data are interesting, they are specifically applicable only to the referenced equipment.

Flue Loss

Loss of heat in flue gases may take the following three forms:

1. **Sensible heat** in flue gases is the sum of the heat in each of the constituents of the flue gases, usually measured above 60 F. When the temperature and flue gas analysis are known, these sensible heat content data may be obtained from Fig. 2-18.

2. **Latent heat** contained in the water vapor in flue gases *cannot* be utilized unless the water vapor is condensed. This loss amounts to 50.4 Btu per cu ft of water vapor on a 60 F, 30-in. Hg basis.

3. **Unburned combustible** loss is due to the presence of such gases as CO, H_2, or hydrocarbons. In domestic gas appliances these gases will not usually be found in measurable quantities, but in industrial furnaces they may be present in appreciable amounts. This loss may be calculated by multiplying the heat value of the unburned gas by the percentage present in the flue gas—on a saturated basis.

Flue Loss Charts and Tables

For most purposes of determining flue losses, flue loss charts or tables are used and any deviation from the conditions listed can, as a rule, be evaluated. For special work a most extensive series of thermodynamic charts for combustion processes was prepared.[49]

To determine the percentage of excess air directly from the percentage of CO_2 in the flue gases when burning a number of typical fuel gases, Figs. 2-34 and 2-35 are presented. **Heat content charts for flue gases** from typical fuel gases are presented in Figs. 2-36 thru 2-42. Flue loss

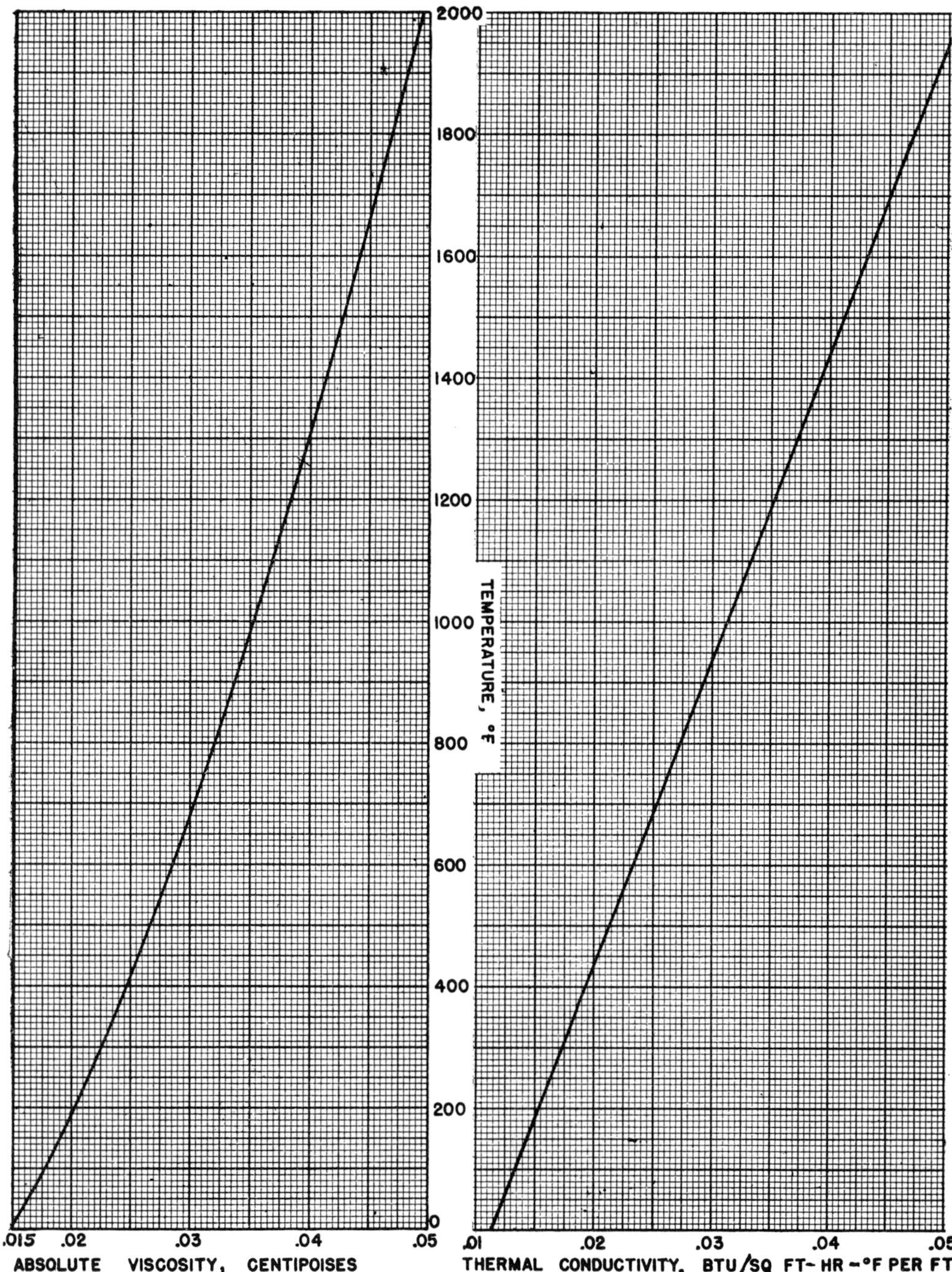

Fig. 2-43 Absolute viscosity of flue gas at one atmosphere.

Fig. 2-44 Thermal conductivity of flue gas.

can also be determined from these charts by *dividing* the heat content of the flue gas by the **gross heating value** of the fuel gas.

In these charts the gross Btu in the flue products is plotted against flue gas temperature, the base temperature being 60 F.

Example: A furnace heated with high-Btu natural gas operates at 1000 F and the flue gas analysis shows 10 per cent CO_2. The flue loss would be 369 Btu per cu ft of gas burned (see Fig. 2-36). Since the gross heating value of the fuel gas is 1115 Btu per cu ft, the loss is $(369/1115) \times 100 = 33.1\%$. The theoretical flame temperature corrected for dissociation is found at the intersection of the CO_2 curves and the broken horizontal line at the top of the charts. The maximum flame temperature for 10 per cent CO_2 on Fig. 2-36 would be 3230 F.

An alignment chart for various gases is presented in Fig. 2-80. The percentage of flue loss can be obtained directly after determining the temperature rise and CO_2 of the flue gases. **Dissociation and flame temperature data** for the gases listed in Figs. 2-36 to 2-42 inclusive are presented in Table 2-92.

Viscosity of Flue Gases from Hydrocarbons

Maxwell[50] has calculated the viscosity of flue gases from the combustion of hydrocarbons in terms of temperature. The

Table 2-92 Dissociation and Flame Temperature Data

High-Btu Natural Gas

	CO_2 by analysis, %, and flame temp, °F			
	12.14% and 3526 F		10% and 2912 F	
	Partial pressures, atm			
	0.098 CO_2	0.189 H_2O	0.084 CO_2	0.162 H_2O
	Dissociation of flue gases, %			
Temp, °F	CO_2	H_2O	CO_2	H_2O
3700	14.8	2.6	...	...
3600	12.0	2.2	...	...
3500	9.5	1.7	4.4	0.8
3400	7.4	1.4	3.0	0.5
3200	4.3	0.8	...	...
3000	2.4	0.5	...	...

High-Methane Natural Gas

	CO_2 by analysis, %, and flame temp, °F			
	11.92% and 3563 F		10% and 2997 F	
	Partial pressures, atm			
	0.096 CO_2	0.194 H_2O	0.083 CO_2	0.168 H_2O
	Dissociation of flue gases, %			
Temp, °F	CO_2	H_2O	CO_2	H_2O
3700	14.9	2.7	...	...
3600	12.0	2.2	...	...
3500	9.6	1.7	4.5	0.7
3400	7.5	1.4	3.2	0.6
3200	4.3	0.8	...	...
3000	2.4	0.5	...	...

High-Inert Natural Gas

	CO_2 by analysis, %, and flame temp, °F			
	12.0% and 3523 F		10% and 2944 F	
	Partial pressures, atm			
	0.098 CO_2	0.186 H_2O	0.084 CO_2	0.160 H_2O
	Dissociation of flue gases, %			
Temp, °F	CO_2	H_2O	CO_2	H_2O
3700	14.9	2.7	...	...
3600	12.0	2.2	...	...
3500	9.5	1.7	4.5	0.8
3400	7.4	1.4	3.1	0.6
3200	4.3	0.8	...	...
3000	2.6	0.5	...	...

Coke Oven Gas

	CO_2 by analysis, %, and flame temp, °F			
	9.57% and 3662 F		8.0% and 3460 F	
	Partial pressures, atm			
	0.073 CO_2	0.241 H_2O	0.063 CO_2	0.210 H_2O
	Dissociation of flue gases, %			
Temp, °F	CO_2	H_2O	CO_2	H_2O
3700	15.3	2.7	...	...
3600	12.4	2.2	6.5	1.1
3500	9.9	1.8	4.7	0.8
3400	7.7	1.4	3.2	0.6
3200	4.5	0.9	...	...
3000	2.5	0.5	...	...

Propane Gas

	CO_2 by analysis, %, and flame temp, °F					
	13.73% and 3670 F		12% and 3485 F		10% and 3127 F	
	Partial pressures, atm					
	0.115 CO_2	0.160 H_2O	0.103 CO_2	0.144 H_2O	0.088 CO_2	0.123 H_2O
	Dissociation of flue gases, %					
Temp, °F	CO_2	H_2O	CO_2	H_2O	CO_2	H_2O
3700	14.4	2.6	...	...	...	...
3600	11.3	2.0	6.8	1.2	...	...
3500	9.2	1.7	4.9	0.7	3.0	0.5
3400	7.1	1.3	3.5	0.6	2.1	0.4
3200	4.2	0.8	...	...	...	...
3000	2.3	0.5	...	...	...	...

Butane Gas

	CO_2 by analysis, %, and flame temp, °F					
	14.04% and 3687 F		12% and 3456 F		10% and 2859 F	
	Partial pressures, atm					
	0.119 CO_2	0.155 H_2O	0.104 CO_2	0.136 H_2O	0.088 CO_2	0.117 H_2O
	Dissociation of flue gases, %					
Temp, °F	CO_2	H_2O	CO_2	H_2O	CO_2	H_2O
3700	14.3	2.5	...	...	...	...
3600	11.5	2.0	6.4	1.1	...	...
3500	9.2	1.7	4.8	0.8	3.5	0.6
3400	7.1	1.3	3.2	0.6	2.4	0.4
3200	4.2	0.8	...	...	...	...
3000	2.3	0.5	...	...	...	...

High-Btu Oil Gas

	CO_2 by analysis, %, and flame temp, °F							
	15.09% and 4350 F*		14% and 4280 F*		12% and 4020 F*		10% and 3661 F	
	Partial pressures, atm							
	0.130 CO_2	0.142 H_2O	0.121 CO_2	0.133 H_2O	0.106 CO_2	0.117 H_2O	0.090 CO_2	0.100 H_2O
	Dissociation of flue gases, %							
Temp, °F	CO_2	H_2O	CO_2	H_2O	CO_2	H_2O	CO_2	H_2O
4200	34.0	6.3	32.0	5.3	...	...	...	...
4000	24.9	4.4	21.5	3.7	16.3	2.8	...	...
3900	20.9	3.6	17.4	3.0	13.0	2.1	...	...
3800	17.2	3.0	13.6	2.3	10.1	1.7	8.4	1.4
3700	14.1	2.5	10.7	1.8	7.6	1.3	6.2	1.0
3600	11.4	2.0	7.9	1.3	5.6	0.9	4.5	0.7
3500	9.0	1.6	5.9	0.9	3.9	0.7	3.2	0.5
3400	7.0	1.3	4.2	0.8	2.8	0.5	2.2	0.4
3200	4.1	0.8	...	...	...	...	...	...
3000	2.3	0.5	...	...	...	...	...	...

* Data estimated.

effect of excess air is almost imperceptible on the viscosity and it has been neglected. These data are reproduced in Fig. 2-43.

Thermal Conductivity of Flue Gases from Hydrocarbons

These data have been calculated by Maxwell[50] and are reproduced in Fig. 2-44. The effect of excess air can be neglected.

Flue Gas Corrosion of Metals

Corrosion of metals by flue gases may be due to one or more of the following conditions:

1. Operation at such low temperature that the moisture in the flue gas *condenses*.
2. Operation at such a high temperature that *oxidation* of the metal occurs.
3. An appreciable content of SO_2 and SO_3 in the flue gases.

The first two conditions may cause rapid corrosion of metals, but it has been demonstrated that **sulfur accelerates corrosion** by flue gases and appears to be the major cause of this condition. Extensive research[51,52] on this subject has been conducted at Battelle Memorial Institute under the sponsorship of the American Gas Association.

Comparative tests[53] for a 15-month period on two external flue-type automatic hot water storage heaters (deoxidized copper tanks and aluminized steel for inner liner of heater jacket) showed that the organic sulfur content of a gas averaging 15.0 grains of sulfur per 100 cu ft produced 40 *times* the weight of corrosion products obtained when the gas burned contained approximately one grain of sulfur per 100 cu ft. The 15-grain sulfur gas caused a pilot stoppage failure in 11 months and severe corrosion was evident on the aluminum pilot tubing and the thermal element The water heater operating on one-grain sulfur gas was free from any corrosive damage at the end of the 15-month test.

REFERENCES

1. Turin, J. J., and Huebler, J. *Advanced Studies in the Combustion of Industrial Gases*, Part II. (Project IGR-59) New York, A.G.A., 1951.
2. Lewis, B., and Von Elbe, G. *Combustion, Flames, and Explosions of Gases*, 2nd ed. New York, Academic Press, 1961.
3. Boord, C. E., and others. "Oxidation Reactions as Related to Hydrocarbon Structure and Engine Knock." *Am. Chem. Soc. Joint Symp. Combustion Chem.*, p. 171, Apr. 9–12, Cleveland, Ohio, 1951.
4. Wilson, R. H., Jr., and others. "Temperature of the Hydrogen-Fluorine Flame." *Energy Transfer in Hot Gases*, p. 111–118. *Natl. Bur. Std. Circ. 523.* 1954.
5. Dunkle, R. V. "Predicting Temperature Distribution along Gas Transmission Pipe Lines." *Gas* 18:31–3, Sept. 1952. (Also: *P.C.G.A. Proc.* 33: 46–8, 1942.)
6. Pfluke, F. J. "Report of Subcommittee on Test Code for Producer and Carbonizing Plants." *A.G.A. Proc.* 1930: 1044–5.
7. Kurz, P. F., and Lund, R. J. "Nuclear Energy as a Source of Electrical Power." In: Foster, J. F., and Lund, R. J. *Economics of Fuel Gas from Coal*, Chap. 14. New York, McGraw-Hill, 1950.
8. Griswold, J. *Fuels, Combustion, and Furnaces*, p. 177. New York, McGraw-Hill, 1946.
9. Batten, J. W., and others. *Combustion of Gas With Limited Air Supply*, p. 97. (A.G.A. Labs. Res. Bull. 15) Cleveland, Ohio, 1942.
10. Mungen, R., and Kratzer, M. B. "Partial Combustion of Methane with Oxygen." *Am. Chem. Soc. Joint Symp. Combustion Chem.*, p. 75–89, Apr. 9–12, Cleveland, Ohio, 1951.
11. A.G.A. *Combustion*, 3rd ed. New York, 1938.
12. Boie, W. "Zur Berechnung von Heizwerten." *Allgem. Waermetech.* 5: 209–13, 1954.
13. Byrne, J. F. "Influence of Atmospheric Oxygen on Bunsen Flames." *Fourth Symposium (International) on Combustion*, p. 345–8. Baltimore, Williams & Wilkins Co., 1953.
14. Hawthorne, W. R., and others. "Mixing and Combustion in Turbulent Gas Jets." *Third Symposium on Combustion, Flame and Explosion Phenomena*, p. 266–88. Baltimore, Williams & Wilkins Co., 1949.
15. Jost, W., and Croft, H. O. *Explosion and Combustion Processes in Gases*, p. 235–9. New York, McGraw-Hill, 1946.
16. Haslam, R. T., and Russell, R. P. *Fuels and Their Combustion.* New York, McGraw-Hill, 1926.
17. Trinks, W. *Industrial Furnaces*, vol. 1, 4th ed. New York, Wiley, 1951.
18. Hottel, H. C. "Radiant Heat Transmission." In: Perry, J. H. *Chemical Engineers' Handbook*, 3rd ed., p. 490–495. New York, McGraw-Hill, 1950.
19. Sweigert, R. L., and Beardsley, M. W. *Empirical Specific Heat Equations Based upon Spectroscopic Data.* (Georgia Inst. Technol. Eng. Exp. Sta. Bull. 2) Atlanta, Ga., 1938.
20. Ellenwood, F. O. *Specific Heats of Certain Gases over Wide Ranges of Pressures and Temperatures.* (Cornell Univ. Eng. Exp. Sta. Bull. 30) Ithaca, N. Y., 1942.
21. Lyman, T., ed. *Metals Handbook*, p. 313. Cleveland, Ohio, Am. Soc. Metals, 1948.
22. A.G.A. Laboratories. *Requirements Committee Investigation of Laboratories Test Gases and Pressures.* (Rept. 847) Cleveland, Ohio, 1938.
23. Lewis, B., and Von Elbe, G. *Combustion, Flames and Explosions of Gases.* New York, Academic Press, 1951.
24. Zabetakis, M. G., and others. "Limits of Flammability of Paraffin Hydrocarbons in Air." *Ind. Eng. Chem.* 43: 2120–24, Sept. 1951.
25. White, A. G. "Limits for the Propagation of Flame in Inflammable Gas–Air Mixtures. Part III: Effects of Temperature on the Limits." *J. Chem. Soc.* 127: 672–84, Jan.-Mar. 1925.
26. Stewart, P. B., and Starkman, E. S. "Flammability Limits for Hydrocarbons at Low Pressures." *Chem. Eng. Progr.* 53: 41–5, Jan. 1957.
27. Coward, H. F., and Jones, G. W. *Limits of Flammability of Gases and Vapors.* (U. S. Bur. Mines Bull. 503) Washington, D. C., 1952.

28a. A.G.A. Laboratories. *Research in Fundamentals of Atmospheric Gas Burner Design.* (Res. Bull. 10) Cleveland, Ohio, 1940.

28b. ——. *Automatic Flash Tube and Pilot Ignition of Oven and Broiler Burners on Manufactured Gas.* (Res. Bull. 17) Cleveland, Ohio, 1943.

28c. Weber, E. J. *Study of Large Single Port Atmospheric Gas Burners—Flashback Characteristics on Ignition.* (A.G.A. Labs. Res. Rept. 1167) Cleveland, Ohio, 1950.

29. Weil, S. A., and others. *Fundamentals of Combustion of Gaseous Fuels.* (IGT Res. Bull. 15) Chicago, Ill., 1957.
30. Burgoyne, J. H., and Thomas, N. "Effects of Very Fine Solid Particles on Flame Propagation." *Nature* (British) 163: 765, May 14, 1949.
31. Ruff, R. J. "Air Pollution Control Economically through Progress in Application of Catalysis." *Ind. Heating* 26: 1187+, June 1959.
32. ——. "Gas Clears the Air." *Gas* 30: 48–51, Aug. 1954. (Modified in accordance with personal communication.)
33. Halpern, C. "Measurement of Flame Speeds by a Nozzle Burner Method." *Nat. Bur. Std. J. Res.* 60: 535–46. June 1958.

34. Shnidman, L., ed. *Gaseous Fuels*, 2nd ed. New York, A.G.A., 1954.
35. Keller, J. D. "Flame Speed as Affected by Preheating of Gaseous Fuels." *Ind. Heating* 17: 780–6, May 1950.
36. Dugger, G. L. "Flame Stability of Preheated Propane–Air Mixtures." *Ind. Eng. Chem.* 47: 109–14, Jan. 1955.
37. ——, and others. "Flame Velocity and Preflame Reaction in Heated Propane–Air Mixtures." *Ind. Eng. Chem.* 47: 114–6, Jan. 1955.
38. Semenov, N. N. *Thermal Theory of Combustion and Explosion.* (Natl. Advisory Comm. Aeron. Tech. Memo. 1026) Washington, D. C., 1942.
39. Tanford, C., and Pease, R. N. "Theory of Burning Velocity." *J. Chem. Phys.* 15: 861–5, Dec. 1947.
40. Grice, C. S. W., and Wheeler, R. V. *Firedamp Explosions within Closed Vessels: "Pressure Piling."* (Min. Fuel Power (Gt. Brit.) Safety Mines Res. Testing Branch Paper 49) Sheffield, England, 1929.
41. Gleim, E. J. *Abnormal Pressures in Explosion-Proof Compartments of Electrical Mining Machines.* (U. S. Bur. Mines Rept. Invest. 2974) Washington, D. C., 1929.
42. Richmond, J. K., and others. "Evidence for the Wrinkled Continuous Laminar Wave Concept of Turbulent Burning." *Jet Propulsion* 28: 393–9, June 1958.
43. Gray, H. H. "Critical Review of Flameless Incandescent Surface Combustion." *J. Gas Lighting* 126: 786–9, June 1914.
44. Korneev, V. L., and Khamalyan, D. M. "Microflame Burning." *Prom. Energ.* 5: 3–7, 1948.
45. Smith, J. B. "Explosion Pressures in Industrial Piping Systems." *Intern. Acetylene Assoc. Official Proc. 1948, 1949, and 1950*, p. 279–93. New York, 1954.
46. Henderson, E. "Combustible Gas Mixtures in Pipe Lines." *P.C.G.A. Proc.* 32: 98–111, 1941.
47. Miller, E. E. "Danger—Explosive Vapors." *Mech. Eng.* 85: 31–3, Dec. 1963.
48. Maconachie, J. E. "Deterioration of Domestic Chimneys." *Can. Chem. Met.* 16: 270+, Nov.-Dec., 1932.
49. Hottel, H. C., and others. *Thermodynamic Charts for Combustion Processes*, 2 parts. New York, Wiley, 1949.
50. Maxwell, J. B. *Data Book on Hydrocarbons*, p. 191–2. Princeton, N. J., Van Nostrand, 1950.
51. Pray, H. A., and others. *Corrosion of Mild Steel by the Products of Combustion of Gaseous Fuels*, p. 46. (Battelle Rept. 2) Columbus, Ohio, Battelle, 1949.
52. ——. *Corrosion of Metals and Materials by the Products of Combustion of Gaseous Fuels.* (Battelle Rept. 3) Columbus, Ohio, Battelle, 1951.
53. Churchill, W. E. *Test on Appliance and Flue Connection.* Unpublished rept. to the A.G.A. Gas Production Research Committee from the Boston Gas Co. New York, 1951.
54. Arthur D. Little, Inc. *Investigation of the Thermodynamic Properties of the Combex-ADL Natural-Gas Burner.* New York, A.G.A., June 15, 1961.
55. Voigt, H. "Ignition Temperatures of Methane," *Wärme* (W. Germany) 68 (2): 45–50, Oct. 1961.
56. Daniels, F. "Nitrogen Oxides and Development of Chemical Kinetics." *Chem. Eng. News* 33: 2370, July 6, 1955.
57. Faith, W. L. "Nitrogen Oxides, A Challenge to Chemical Engineers." *Chem. Eng. Progr.* 52: 342–4, Aug. 1956.
58. Knox, J. *Fixation of Atmospheric Nitrogen*, 2nd ed. New York, Van Nostrand, 1921.
59. Jones, G. W., and Kennedy, R. E. *Inflammability of Natural Gas: Effect of Pressure upon the Limits.* (U. S. Bur. Mines Rept. Invest. 3798), Washington, D. C., 1945.
60. Jones, G. W., and others. *Effect of High Pressures on the Flammability of Natural Gas–Air Mixtures.* (U. S. Bur. Mines Rept. Invest. 4557) Washington, D. C., 1949.
61. Calcote, H. C. "Mechanisms for Formation of Ions in Flames." *Combust. Flame* 1: 385–403, Dec. 1957.
62. Gaydon, A. G., and Wolfhard, H. G. *Flames: Their Structure, Radiation and Temperature*, 2d ed., Chap. 13. New York, Macmillan, 1960.
63. Monroe, E. S. "Needed—Research in Flame Safeguard System Applications." *Combustion* 33: 24–5, Feb. 1962.
64. Grumer, J., and others. "Photographic Studies of Turbulent Flame Structures." *Ind. Eng. Chem.* 49: 305–12, Feb. 1957.
65. Richmond, J. K., and others. "Turbulent Flame Propagation by Large-Scale Wrinkling of a Laminar Flame Front." In: *Seventh Symposium (International) on Combustion*, p. 615–20. New York, Academic, 1959.
66. Wolfhard, H. G., and Vanpée, M. "Ignition of Fuel–Air Mixtures by Hot Gases and Its Relationship to Firedamp Explosions." In: *Seventh Symposium (International) on Combustion*, p. 446–53. New York, Academic, 1959.
67. Coward, H. F., and others. *Explosibility of Methane and Natural Gas.* (Carnegie Inst. Technol. Tech. Bull. 30) Pittsburgh, Pa., Carnegie Inst. Technol., 1926.
68. Karr, C., and Putnam, A. "Influence of Recirculated Combustion Products on Burning Velocity." *ASHRAE J.* 4: 43–7, Sept. 1962.
69. "Putting on the Heat: Temperatures above 5000°K." *Chem. Eng. News* 34: 3442–5, July 16, 1956.
70. Myers, J. W., and others. "Calculation of Theoretical Flame Temperatures in Furnaces." *Trans. ASME* 80: 202–16, 1958.
71. Chass, R. L., and George, R. E. "Contaminant Emissions from the Combustion of Fuels." *J. Air Pollution Control Assoc.* 10: 34–43, Feb. 1960.
72. A.G.A. Laboratories. *Heating Element Temperatures of Warm-Air Furnaces as Related to Corrosion.* (Rept. 914) Cleveland, Ohio, 1940.

Chapter 6
Heat Transfer

by Richard L. Stone

GENERAL

The energy transport resulting from a temperature gradient in a system or from a difference of temperature between two systems is known as *heat transfer*. There are three modes of heat transfer: *conduction*, *convection*, and *radiation*.

Conduction

Conduction is the transfer of heat thru a substance without any appreciable displacement of any of its molecules. Although the thermal conductivity of a fluid is an important property, pure conduction thru fluids is seldom encountered. The law of conduction states that the heat flow per unit time is directly proportional to the area thru which the heat is flowing, to the temperature gradient in the direction of flow, and to the thermal conductivity of the material.

A temperature gradient is equal to the temperature difference across the heat flow path divided by the thickness of the path. For one-dimensional heat flow in a single layer of a homogeneous substance:

$$q = kA(t_1 - t_2)/L \qquad (1)$$

where: q = heat conduction, Btu per hr
k = thermal conductivity, Btu per hr-sq ft-°F per in.
A = area thru which heat is conducted, sq ft
t_1 = higher temperature, °F
t_2 = lower temperature, °F
L = distance separating t_1 and t_2, in.

Note: Values of k are given in Tables 2-93 to 2-96. Conversion factors for k are given following Table 2-93 and in Table 1-50.

Temperature affects the conductivity of substances, so that the mean thermal conductivity should be used where an appreciable temperature difference exists.

For conduction thru multiple parallel layers of different substances for an overall temperature differential of $(t_1 - t_2)$:

$$q = \frac{(t_1 - t_2)}{\dfrac{L_1}{Ak_1} + \dfrac{L_2}{Ak_2} + \dfrac{L_3}{Ak_3} + \ldots \text{ etc.}} \qquad (2)$$

where:
$L_1, L_2, \ldots$ etc. refers to thickness of each layer, in.
$k_1, k_2, \ldots$ etc. refers to thermal conductivity of respective layers
L/Ak is the thermal resistance of a layer

Where thermal contact between adjacent layers is poor, allowance must be made for the intervening substance.

Where heat is being conducted thru a substance of varying area, a mean area must be employed. For circular lagged pipes, the **logarithmic mean** area is:

$$A_m = \frac{A_2 - A_1}{2.3 \log (A_2/A_1)} \qquad (3)$$

where: A_2 = outside area, sq ft
A_1 = inside area, sq ft

When A_2/A_1 does not exceed 2, the **arithmetic mean,** A_a, gives results within 4 per cent of the logarithmic mean:

$$A_a = (A_1 + A_2)/2 \qquad (4)$$

The significance of this error is minimized by inaccuracy in the other terms of Eq. **1**.

Additional thicknesses of insulation always decrease heat losses from flat surfaces, but when a curved surface, such as a pipe, is insulated, the heat loss may *increase* because of greater outside surface area. For pipes, the **critical outside radius** at which maximum heat transfer occurs is equal to the conductivity of the insulation divided by the surface coefficient of heat transfer.

$$r_c = k/h_a \qquad (5)$$

where:
r_c = critical outside radius, in.
h_a = surface coefficient of heat transfer, Btu per hr-sq ft-°F
k = thermal conductivity of insulation (neglect pipe resistance), Btu per hr-sq ft-°F per in.

For cases in which the inside radius of the pipe is larger than r_c, as a result of the lowest possible k, the addition of insulation will always reduce the rate of heat transfer.

The proper thickness of insulation to employ when heat loss *prevention* is the main consideration may be determined by an economic balance in which the decreasing value of heat lost must be added to the increasing fixed charges resulting from

Table 2-93 Thermal Conductivity, *k*, of Miscellaneous Solid Substances*

(Btu/hr-sq ft-°F/in.)

Material	Apparent density, lb/cu ft at room temp	t, °F	*k*
Aerogel, silica, opacified	8.5	248	0.156
		554	0.312
Asbestos-cement boards	120	68	5.16
Asbestos sheets	55.5	124	1.152
Asbestos slate	112	32	1.044
	112	140	1.368
Asbestos	29.3	−328	0.516
	29.3	32	1.08
	36	32	1.044
	36	212	1.332
	36	392	1.440
	36	752	1.548
	43.5	−328	1.080
	43.5	32	1.620
Air spaces (¾ in.) faced with Al foil	...	100	0.300
Ashes, wood	...	32–212	0.492
Asphalt	132	68	5.16
Bricks	115	2012	7.56
Building brick work	...	68	4.8
Carbon	96.7		36.0
Diatomaceous earth, natural, across strata†	27.7	399	0.612
	27.7	1600	.924
Diatomaceous earth, natural, parallel to strata†	27.7	399	0.972
	27.7	1600	1.272
Diatomaceous earth, molded and fired†	38	399	1.68
	38	1600	2.16
Diatomaceous earth and clay, molded and fired†	42.3	399	1.68
	42.3	1600	2.28
Diatomaceous earth, high burn, large pores‡	37	392	1.56
	37	1832	4.08
Calcium carbonate, natural	162	86	15.6
White marble	175	...	20.4
Chalk	96	...	4.8
Calcium sulfate ($4H_2O$), artificial	84.6	104	2.64
Plaster, artificial	132	167	5.16
building	77.9	77	3.00
Cambric, varnished	...	100	1.092
Carbon, gas	...	32–212	24.0
Carbon stock	94	−300	6.60
		32	43.2
Cardboard, corrugated	...	...	0.444
Cattle hair, felted	11–13	70	0.264
Celluloid	87.3	86	1.44
Charcoal flakes	11.9	176	0.516
	15	176	.612
Charcoal, 6 mesh	15.2	70	0.372
Clinker, granular	...	32–1292	3.24
Coke, petroleum	...	212	40.8
		932	34.8
Coke petroleum, 20–100 mesh	62	752	6.60
Coke, powdered	...	32–212	1.32
Concrete, cinder	...	...	2.40
stone	...	...	6.48
1:4 dry	...	...	5.28
Cotton wool	5	86	0.288
Cork board	7.0	70	.270
	10	86	.30
Cork, regranulated	8.1	86	.312
ground	9.4	86	.30
Diatomaceous earth powder coarse†	20.0	100	.432
	20.0	1600	.984
fine†	17.2	399	.480
	17.2	1600	.888
molded pipe covering†	26.0	399	0.612
	26.0	1600	1.56
4 vol. calcined earth and 1 vol. cement, poured and fired†	61.8	399	1.92
	61.8	1600	2.76
Dolomite	167	122	12.0
Ebonite	77	...	1.2
Eel grass between paper	3.4–4.6	70	0.252–0.264
Enamel, silicate	38	...	6.0–9.0
Felt, wool	20.6	86	0.36
Fiber insulating board	14.8	70	0.336
Fiber, red	80.5	68	3.24
with binder, baked	...	68–207	1.164
Flax fibers between paper	4.9	70	0.276
Flax fiber, sheets	13.0	70	0.312
Gas carbon	...	32–212	24.0
Glass	150–175	...	2.4–8.76
Borosilicate type	139	86–167	7.56
Window glass	...	...	3.6–7.32
Soda glass	...	...	3.6–5.28
Glass wool, curled	4–10	70	0.288
Granite	168	...	12.0–27.6
Graphite, longitudinal	...	68	1140.0
powdered, thru 100 mesh	30	104	1.248
Gypsum, molded and dry	78	68	3.0
Gypsum, cellular	8	70	0.348
	12	70	.444
	18	70	.588
	24	70	.768
	30	70	0.996
Gypsum, powdered form	26–34	70	0.516–0.60
Hair felt, perpendicular to fibers	17	86	0.252
Hair and asbestos fibers, felted	7.8	70	0.276
Ice	57.5	32	15.6
Infusorial earth—see diatomaceous earth	...	...	...
Insulating hair and jute	6.1–6.3	70	0.264–0.276
Jute and asbestos fibers, felted	10.0	70	0.372
Kapok	0.88	68	.240
Lampblack	10	104	0.456
Lava	...	...	5.88
Leather, sole	62.4	...	1.104
Limestone (15.3 vol. % H_2O)	103	75	6.48
Linen	...	86	0.60
Magnesia, powdered	49.7	117	4.20
Magnesia, light carbonate	13	70	0.408
Magnesium oxide, compressed	49.9	68	3.84
Marble	...	...	14.4–20.4
Mica, perpendicular to planes	...	122	3.00
Mill shavings	...	...	0.396–0.60
Mineral wool	19.7	86	0.288
Mineral wool, fibrous	6	70	.2604
	10	70	.270
	14	70	.2796
	18	70	.2904
Mineral wool, block, with binder	16.7	70	.372
Paper	...	...	0.900
Paraffin wax	...	32	1.68
Petroleum coke	...	212	40.8
	...	932	34.9

Table 2-93 Thermal Conductivity, k, of Miscellaneous Solid Substances (Continued)

Material	Apparent density, lb cu ft at room temp	t, °F	k
Porcelain	...	392	10.56
Portland cement (see concrete)	...	194	2.04
Pumice stone	...	70–151	1.68
Pyroxylin plastics	...	...	0.90
Rubber, hard	74.8	32	1.044
para	...	70	1.308
soft	...	70	0.90–1.104
Sand, dry	94.6	68	2.28
Sandstone	140	104	12.72
Sawdust	12	70	0.36
Silk	6.3	...	0.312
varnished	...	100	1.152
Slag, blast furnace	...	75–261	0.768
Slag wool	12	86	0.264
Slate	...	201	10.32
Snow	34.7	32	3.24
Sulfur, monoclinic	...	212	1.08–1.164
rhombic	...	70	1.92

Material	Apparent density, lb/cu ft at room temp	t, °F	k
Sugar-cane fiber, stiff sheets	13.2–14.8	70	0.336
Wallboard, insulating type	14.8	70	.336
Wallboard, stiff pasteboard	43	86	.480
Wood fiber, chemically treated	2.2	70	.276
Wood pulp, stiff sheets	16.2–16.9	70	.336
Wood shavings	8.8	86	0.408
Wood, across grain			
Balsa	7–8	86	0.30–0.36
Oak	51.5	59	1.44
Maple	44.7	122	1.32
Pine, white	34.0	59	1.044
Teak	40.0	59	1.2
White fir	28.1	140	0.744
Wood, parallel to grain			
Pine	34.4	70	2.40
Wool, animal	6.9	86	0.252

* Compiled from Marks, L. S., ed. *Mechanical Engineers' Handbook*, 4th ed. New York, McGraw-Hill, 1941; and *International Critical Tables*. New York, McGraw-Hill, 1929.

† Townshend, B. and Williams, E. R. "Heat Insulation Developed for Every Purpose." *Chem. Met. Eng.* 39:219, Apr. 1932.

‡ Norton, F. H. *Refractories*, 2nd ed. New York, McGraw-Hill, 1942.

Note: When k is expressed in units other than Btu per hr-sq ft-°F per in., the units of corresponding terms of Eq. 1 must be changed to agree. (See adjacent column.)

To obtain k in terms of:	*Multiply by:*
Btu/hr-sq ft-°F/ft	1/12
p.c.u./hr-sq ft-°C/in.	1
g-cal/sec-sq cm-°C/cm	0.000344
kilo-ergs/sec-sq cm-°C/cm	14.4
watts/sq cm-°C/cm	0.00144
kg-cal/hr-sq m-°C/m	0.124

p.c.u. = pound-centigrade unit (the sensible heat required to raise the temperature of one pound of water one degree centigrade) = 1.8 Btu = pound calorie or centigrade heat unit

the use of thicker insulation. When it is necessary to maintain the exterior surface of a furnace or oven at some limiting temperature, this temperature, rather than the cost, will determine the thickness.

Convection

Convection is the transfer of heat by moving and mixing masses of fluid. Natural convection occurs when an otherwise stagnant fluid moves and mixes because of density differences caused by temperature differentials. **Forced convection** occurs when a fluid intermixes because of outside forces which are independent of temperature differences in the system itself. Heat transfer from solid surface to fluid in convective processes is handled by assuming the existence of a film, known as a *temperature boundary layer*, between the bulk of the fluid and the solid surface. Nearly all of the temperature drop between surface and fluid is assumed to take place in the film. This permits use of a relationship analogous to the conduction relationship (Eq. **1**) to compute convected heat transfer between surfaces and fluids. The constant, h, in Eq. **6**, which is determined by flow conditions, fluid properties, and the shape of the surface involved, is called the **film coefficient** of heat transfer,

$$q = hA(t_1 - t_2) \tag{6}$$

where: q = heat conduction, Btu per hr
h = film coefficient, Btu per hr-sq ft-°F
A = area of convection, sq ft

Thus, when a quantity of heat flows from surface to fluid, t_1 is the surface and t_2 the bulk temperature of the fluid.

Specific heats, viscosity, conductivity, and mass rate of flow are temperature and pressure dependent. These are calculated either at *film temperature*, which is an arithmetic mean of wall temperature and fluid temperature, or bulk temperature as defined below. If a fluid flowing thru a tube is collected in a container and allowed to mix thoroughly, the temperature of the fluid in the container is known as its *bulk temperature*.

Natural or Free Convection. Heat transfer by natural convection occurs from the jackets of furnaces to still air, from immersed surfaces to nonagitated liquids, and to objects being heated in low temperature ovens and furnaces.

The dimensional relationship of **natural convection** factors is:

$$\left[\frac{h_c D_0}{k}\right] = C\left[\frac{D_0^3 d^2 g V \Delta t}{z^2}\right]^a \left[\frac{c_p z}{k}\right]^b \tag{7}$$

Note: All properties are evaluated at the mean of the surface and bulk temperatures.

where:

h_c = film coefficient for natural convection, Btu per hr-sq ft-°F
D_0 = outside diameter of pipe (or shape factor), ft
k = thermal conductivity of the fluid at film temperature, Btu per hr-sq ft-°F per ft
C = a constant
d = fluid density at film temperature, lb per cu ft
g = gravitational constant, ft per sec-sec
V = coefficient of volumetric expansion; for perfect gases, $V = 1/T$, dimensions = 1/°R

Table 2-94 Thermal Conductivity, *k*, of Liquids[1]

(Btu/hr-sq ft-°F/in.)

Liquid	*t*, °F	*k**
Acetic acid, 100%	68	1.188
50%	68	2.40
Acetone	86	1.224
	167	1.14
Allyl alcohol	77–86†	1.248
Ammonia	5–86†	3.48
Ammonia, aqueous, 26%	68	3.132
	140	3.48
Amyl acetate	50	0.996
alcohol (*n*-)	86	1.128
	212	1.068
(*iso*-)	86	1.056
	167	1.044
Aniline	32–68†	1.20
Benzene	86	1.104
	140	1.044
Bromobenzene	86	0.888
	212	0.840
Butyl acetate (*n*-)	77–86†	1.02
alcohol (*n*-)	86	1.164
	167	1.140
(*iso*-)	50	1.092
Calcium chloride brine, 30%	86	3.84
15%	86	4.08
Carbon disulfide	86	1.116
	167	1.056
tetrachloride	32	1.284
	154	1.128
Chlorobenzene	50	0.996
Chloroform	86	.960
Cymene (*para*-)	86	.936
	140	0.948
Decane (*n*-)	86	1.02
	140	0.996
Dichlorodifluoromethane	20	.684
	60	.636
	100	.576
	140	.516
	180	.456
Dichloroethane	122	0.984
Dichloromethane	5	1.332
	86	1.152
Ethyl acetate	68	1.212
alcohol, 100%	68	1.260
80%	68	1.644
60%	68	2.112
40%	68	2.688
20%	68	3.372
100%	122	1.044
benzene	86	1.032
	140	0.984
bromide	68	.840
ether	86	.960
	167	.936
iodide	104	.768
	167	0.756
Ethylene glycol	32	1.836
Gasoline	86	0.936
Glycerol, 100%	68	1.968
80%	68	2.268
60%	68	2.640
40%	68	3.108
20%	68	3.336
100%	212	1.968
Heptane (*n*-)	86	0.972
	140	0.948
Hexane (*n*-)	86	0.960
	140	0.936
Heptyl alcohol (*n*-)	86	1.128
	167	1.092
Hexyl alcohol (*n*-)	86	1.116
	167	1.080
Kerosene	68	1.032
	167	0.972
Mercury	82	57.96
Methyl alcohol, 100%	68	1.488
80%	68	1.848
60%	68	2.280
40%	68	2.808
20%	68	3.408
100%	122	1.368
chloride	5	1.332
	86	1.068
Nitrobenzene	86	1.140
	212	1.056
Nitromethane	86	1.500
	140	1.440
Nonane (*n*-)	86	1.008
	140	0.984
Octane (*n*-)	86	.996
	140	.972
Oils	86	0.948
Oils, castor	68	1.248
	212	1.200
Oils, olive	68	1.164
	212	1.140
Paraldehyde	86	1.008
	212	0.936
Pentane (*n*-)	86	.936
	167	0.888
Perchlorethylene	122	1.104
Petroleum ether	86	0.900
	167	0.876
Propyl alcohol (*n*-)	86	1.188
	167	1.140
(*iso*-)	86	1.092
	140	1.080
Sodium	212	588
	410	552
chloride brine, 25.0%	86	3.96
12.5%	86	4.08
Sulfuric acid, 90%	86	2.52
60%	86	3.00
30%	86	3.60
Sulfur dioxide	5	1.536
	86	1.332
Toluene	86	1.032
	167	1.008
β-Trichlorethane	122	0.924
Trichlorethylene	122	.960
Turpentine	59	0.888
Vaseline	59	1.272
Water	32	4.116
	100	4.356
	200	4.716
	300	4.740
	420	4.512
	620	3.300
Xylene (*ortho*-)	68	1.080
(*meta*-)	68	1.080

* See Tables 2-93 and 1-50 for conversion factors to change *k* values to other systems of measurement.

† A linear variation with temperature may be assumed. The extreme values given constitute the temperature limits over which the data are recommended.

Table 2-95 Thermal Conductivity, k, of Gases and Vapors[1]

(Btu/hr-sq ft-°F/in.)

For extrapolation to other temperatures, it is suggested that the data given be plotted as log k vs. log $(t + 460)$, or that use be made of the assumption that the ratio $c_p z/k$ (see Eq. **7**) is practically independent of temperature (or of pressure, within moderate limits*).

Substance	t, °F	k
Acetone	32	0.0684
	115	.0888
	212	.1188
	363	.1764
Acetylene	−103	.0816
	32	.1296
	122	.1680
	212	.2064
Air	−148	.1140
	32	.1680
	212	.2196
	392	.2712
	572	.3180
Ammonia	−76	.1140
	32	.1536
	122	.1884
	212	.2220
Benzene	32	.0624
	115	.0876
	212	.1236
	363	.1824
	413	.2112
Butane (n-)	32	.0936
	212	.1620
(iso-)	32	.0960
	212	.1668
Carbon dioxide	−58	.0816
	32	.1020
	212	.1596
	392	.2172
	572	.2736
disulfide	32	.0480
	45	.0504
monoxide	−312	.0492
	−294	.0552
	32	.1620
tetrachloride	115	.0492
	212	.0624
	363	.0780
Chlorine	32	.0516
Chloroform	32	.0456
	115	.0552
	212	.0696
	363	.0924
Cyclohexane	216	.1140
Dichlorodifluoro-methane	32	.0576
	122	.0768
	212	.0960
	302	.1164
Ethane	−94	.0792
	−29	.1032
	32	.1272
	212	.2100
Ethyl acetate	115	.0864
	212	.1152
	363	.1692
alcohol	68	.1068
	212	.1488
chloride	32	.0660
	212	.1140
	363	.1620
	413	.1824
ether	32	.0924
	115	.1188
	212	.1572
	363	.2268
	413	0.2508
Ethylene	−96	.0768
	32	.1212
	122	.1572
	212	.1932
Heptane (n-)	392	.1344
	212	0.1236
Hexane (n-)	32	0.2064
	68	.0960
Hexene	32	.0732
	212	.1308
Hydrogen	−148	.780
	−58	0.996
	32	1.200
	122	1.380
	212	1.548
	572	2.136
Hydrogen and carbon dioxide		
20% H_2	32	0.2180
40%	32	.324
60%	32	.492
80%	32	0.744
100%	32	1.200
Hydrogen and nitrogen		
20% H_2	32	0.2544
40%	32	.3756
60%	32	.5256
80%	32	.7620
Hydrogen and nitrous oxide		
20% H_2	32	.2040
40%	32	.3240
60%	32	.4920
80%	32	.780
Hydrogen sulfide	32	.0912
Mercury	392	.2364
Methane	−148	.120
	−58	.1740
	32	.2100
	122	.2580
Methyl alcohol	32	.0996
	212	.1536
acetate	32	.0708
	68	.0816
Methyl chloride	32	.0636
	115	.0864
	212	.1128
	363	.1560
	413	.1776
Methylene chloride	32	.0468
	115	.0588
	212	.0756
	413	.1140
Nitric oxide	−94	.1236
	32	.1656
Nitrogen	−148	.1140
	32	.1680
	122	.1920
	212	.2160
Nitrous oxide	−98	.0804
	32	.1044
	212	.1536
Oxygen	−148	.1140
	−58	.1428
	32	.1704
	122	.1968
	212	.2220
Pentane (n-)	32	.0888
	68	0.0996

Table 2-95 Thermal Conductivity, *k*, of Gases and Vapors (Continued)

Substance	*t*, °F	*k*	Substance	*t*, °F	*k*
Pentane (*iso-*)	32	0.0864	Water vapor, 1 atm†	32	0.1584
	212	.1524		200	.1908
Propane	32	.1044		400	.2388
	212	.1812		600	.3072
Sulfur dioxide	32	.060		800	.3672
	212	0.0828		1000	0.5940

* Thermal conductivities for methane, ethane, nitrogen, and argon at pressures up to 200 atm are given in: Lenoir, J. M., Junk, W. A. and Comings, E. W. "Measurement and Correlation of Thermal Conductivities of Gases at High Pressure." *Chem. Eng. Progr.*, 49: 539–542, Oct. 1953.

† For saturated water vapor:

psia	250	500	1000	1500	2000
t, °F	401	467	545	596	636
k	0.2976	0.3588	0.4740	0.5832	0.6936

Note: See Tables 2-93 and 1-50 for conversion factors to change *k* values to other systems of measurement. See Fig. 2-44 for thermal conductivity of flue gas.

Table 2-96 Effect of Temperature Upon Thermal Conductivity of Metals and Alloys*

(Btu/hr-sq ft-°F/in.)

		32 F	212 F	392 F	572 F	752 F	932 F	1112 F	1472 F	2192 F
Aluminum		1404	1428	1488	1596	1728	1860	...	...	...
Brass (70-30)		672	720	756	792	804	...	...	...	...
Copper, pure		2688	2592	2580	2544	2520	2484	2448	...	...
Iron, pure		...	439.2	...	...	...	...	...	...	...
Lead		240	228	216	216	...	...	...	...	...
Magnesium		1104	1104	...	...	...	...	...	...	...
Nickel		432	408	396	384	...	...	...	...	...
Cast iron	3.16TC, 1.54Si, 0.57Mn, 0.22P, 0.11S	325.14	319.33	...	...	284.49	...	...	...	...
	3.93TC, 3.34GC, 0.59CC, 1.40Si, 0.63Mn, 0.134P, 0.077S	...	380.30	...	339.66	...	299.01	...	...	...
Carbon steels	0.08C, 0.31Mn, 0.045Cr, 0.07Ni, 0.020Mo, trace Cu	412.23	400.62	368.69	342.56	316.43	284.49	255.47	197.41	206.12
	0.23C, 0.635Mn, trace Cr, 0.074Ni, 0.13Cu	359.98	354.17	339.66	319.33	296.11	272.89	246.76	179.99	206.12
	0.80C, 0.32Mn, 0.11Cr, 0.13Ni, 0.01Mo, 0.07Cu	345.46	333.85	313.53	287.40	264.18	243.85	226.43	168.38	209.02
	1.22C, 0.35Mn, 0.11Cr, 0.13Ni, 0.01Mo, 0.08Cu	313.53	310.63	301.92	284.49	267.08	249.66	232.24	165.47	197.41
Alloy steels	0.34C, 0.55Mn, 0.78Cr, 3.53Ni, 0.39Mo, 0.05Cu	229.34	235.15	243.85	246.76	246.76	232.24	211.92	185.79	209.02
	0.315C, 0.69Mn, 1.09Cr, 0.07Ni, 0.012Mo, 0.07Cu	336.76	322.24	307.72	293.21	267.08	246.76	220.63	179.99	209.02
Stainless and heat resisting steels	5Cr, 0.50Mo,	...	255.47	249.66	243.85	240.95	235.15	229.34	...	...
	410 12Cr	...	171.28	179.99	185.79	191.60	200.31	206.12	...	...
	430 17Cr	...	179.99	179.99	179.99	179.99	179.99	179.99	...	...
	446 27Cr	...	145.15	150.96	156.76	162.57	168.38	174.18	...	...
	304 18Cr, 8Ni	...	113.22	121.93	130.64	139.35	148.06	156.76	...	...
	310 25Cr, 20Ni	...	87.09	95.80	104.51	113.22	121.93	127.73	...	...
High alloy steels	1.22C, 0.22Si, 13.00Mn, 0.03Cr, 0.07Ni, 0.07Cu	89.99	95.80	113.22	124.83	133.54	142.24	150.96	162.57	194.50
	0.08C, 0.68Si, 0.37Mn, 19.11Cr, 8.14Ni, 0.60W, 0.03Cu	110.32	113.22	119.02	124.83	136.44	148.06	159.67	179.99	206.12
	0.13C, 0.17Si, 0.25Mn, 12.95Cr, 0.14Ni, 0.12V, 0.06Cu	185.79	191.60	191.60	194.50	191.60	188.70	182.89	174.18	211.92
	0.27C, 0.28Mn, 13.69Cr, 0.20Ni, 0.25W, 0.022V	174.18	182.89	188.70	191.60	191.60	188.70	185.79	174.18	209.12
	0.715C, 0.25Mn, 4.26Cr, 0.067Ni, 18.45W, 1.075V	168.38	179.99	188.70	194.50	197.41	194.50	188.70	179.99	203.21

* Data for nonferrous metals based on: Perry, J. H., ed. *Chemical Engineers' Handbook*, 3rd ed. New York, McGraw-Hill, 1950; ferrous metals data calculated from: Lyman, T., ed. *Metals Handbook*, p. 314. Cleveland, Ohio, Amer. Soc. for Metals, 1948.

Note: See Tables 2-93 and 1-50 for conversion factors to change *k* values to other systems of measurement.

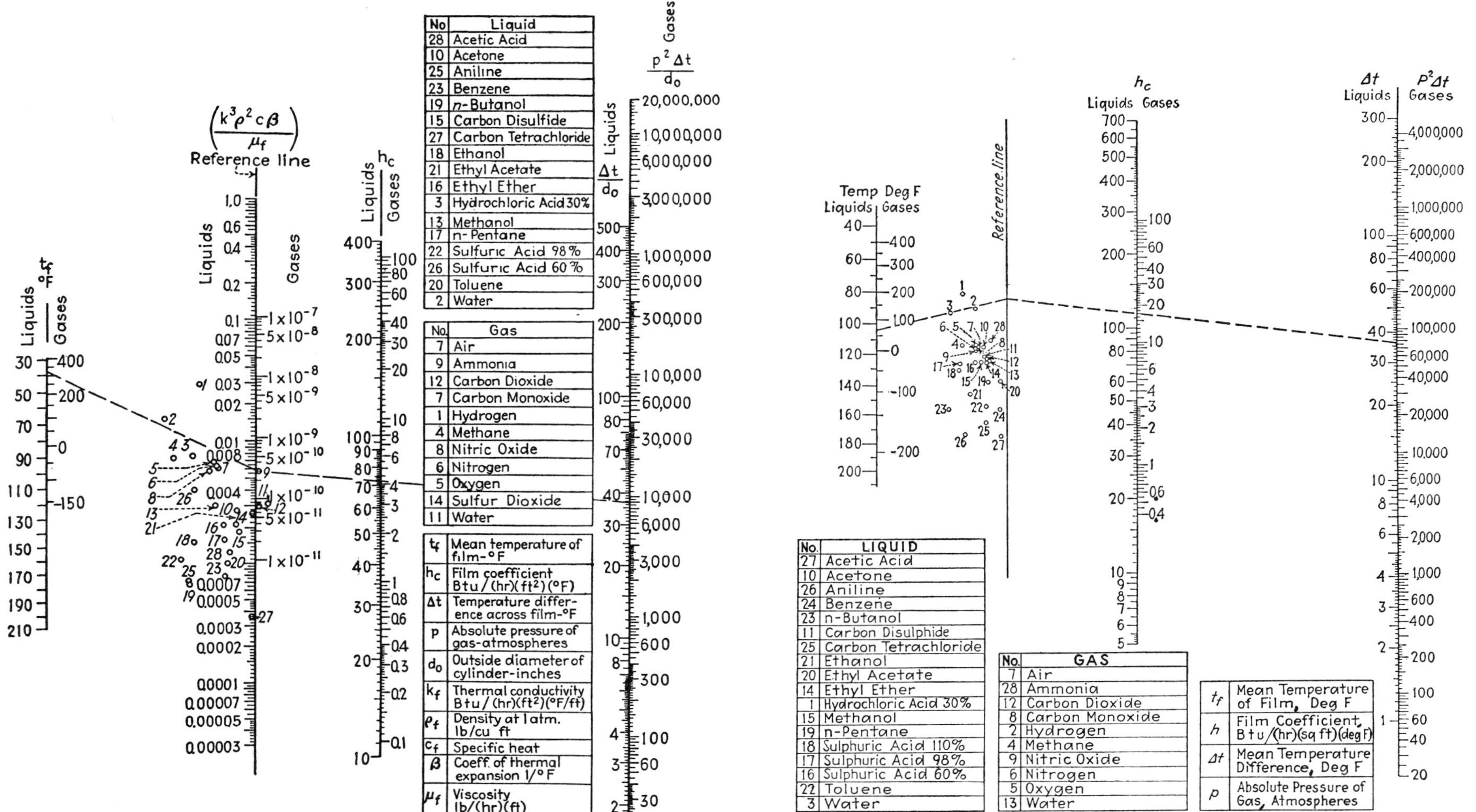

Fig. 2-45 Free convection outside horizontal pipes and tubes.[3]

Fig. 2-46 Coeffient h_c for a turbulent boundary layer of air or liquid on a vertical surface[3] for products of Grashof and Prandtl numbers from 10^8 to 10^{12}.

Δt = temperature difference between surface and ambient fluid, °F
z = dynamic viscosity of fluid at film temperature, lb per ft-hr
c_p = specific heat of fluid at constant pressure, Btu per lb-°F
a,b = exponents

Each of the bracketed dimensionless terms in Eq. **7** has been given a special symbol, that is,

$$(\mathbf{Nu}) = C(\mathbf{Gr})^a(\mathbf{Pr})^b \tag{7a}$$

where: **Nu** is the Nusselt number
Gr is the Grashof number
Pr is the Prandtl number

Compare with Eq. **15** for **forced convection.** The difference is the substitution of the Reynolds number for the Grashof number.

Equations **7** and **7a** may be evaluated for heat transfer from the outside of a **single horizontal pipe** by use of an alignment chart (Fig. 2-45). This chart may be used with negligible error for the calculations of free convection outside banks of tubes[2] when the tubes are located about one or more diameters from each other and several diameters from shell surface so that excessive interference with the convection currents does not occur. Figure 2-45 is based on fluid properties evaluated at film temperature.

The dotted lines on Fig. 2-45 illustrate the solution for h_c between ammonia gas (point number 9) at 10 atm and 392 F, and a 2-in. cylinder at 212 F. Therefore,

$$t_f = \frac{392 + 212}{2} = 302 \text{ F}$$

$$\frac{p^2 \Delta t}{d_0} = 10^2 \frac{(392 - 212)}{2} = 9000$$

Read h_c = 4.3 Btu per hr-sq ft-°F.

Figure 2-46 gives a rapid method of determining the film coefficient for natural convection thru **vertical tubes.**

For heat transfer by **natural convection between solid surfaces and air,** McAdams[4] recommends the following simplified equations for the film coefficient, h_c, at ordinary temperatures and pressures, as a function of the temperature difference between surface and air, Δt, °F:

Heated plates facing up, or cooled plates facing down:

$$\textit{Turbulent range:} \quad h_c = 0.22\ \Delta t^{0.333} \tag{8}$$

for **Gr** from 2×10^7 to 3×10^{10}

$$\textit{Laminar range:} \quad h_c = 0.27(\Delta t/L)^{0.25} \tag{9}$$

for **Gr** from 10^5 to 2×10^7

Cooled plates facing up, or heated plates facing down:

$$\textit{Laminar range:} \quad h_c = 0.12(\Delta t/L)^{0.25} \tag{10}$$

for **Gr** from 3×10^5 to 3×10^{10}

Horizontal pipes (flow outside):

$$X \text{ from } 10^3 \text{ to } 10^9: \quad h_c = 0.27(\Delta t/D_0)^{0.25} \tag{11}$$

$$X \text{ from } 10^9 \text{ to } 10^{12}: \quad h_c = 0.18\ \Delta t^{0.333} \tag{12}$$

where: X = the product of Grashof and Prandtl numbers

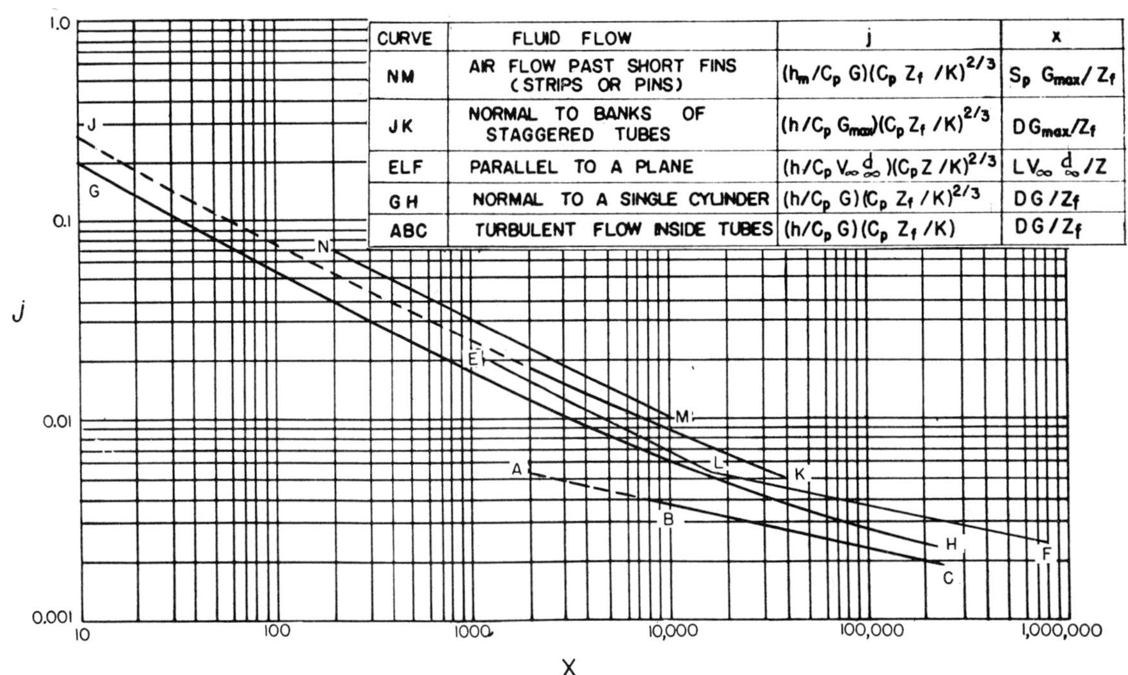

CURVE	FLUID FLOW	j	x
NM	AIR FLOW PAST SHORT FINS (STRIPS OR PINS)	$(h_m/C_p G)(C_p Z_f/K)^{2/3}$	$S_p G_{max}/Z_f$
JK	NORMAL TO BANKS OF STAGGERED TUBES	$(h/C_p G_{max})(C_p Z_f/K)^{2/3}$	$D G_{max}/Z_f$
ELF	PARALLEL TO A PLANE	$(h/C_p V_\infty d_\infty)(C_p Z/K)^{2/3}$	$L V_\infty d_\infty/Z$
GH	NORMAL TO A SINGLE CYLINDER	$(h/C_p G)(C_p Z_f/K)^{2/3}$	DG/Z_f
ABC	TURBULENT FLOW INSIDE TUBES	$(h/C_p G)(C_p Z_f/K)$	DG/Z_f

1.0
0.1
0.01
0.001
j
10
100
1000
10,000
100,000
1,000,000
X
J
G
N
E
M
L
K
A
B
H
C
F

Fig. 2-47 Heating and cooling of fluids by forced convection.[1] $j = (\mathbf{Nu/RePr})\mathbf{Pr}^{1/3} = C_f/2$ where C_f is the average friction or drag coefficient; X is the Reynolds number.

Vertical surfaces for X from 10^4 to 10^9:

$$h_c = 0.29(\Delta t/L)^{0.25} \quad (13)$$

Gr = Grashof number
L = characteristic length, ft
D_0 = outside diameter, ft

Equations **8** thru **13** have been found to have many useful gas applications. For example, heat transfer by convection to foods in a domestic gas range oven follows a law with the constant C and exponent a dependent on the size, nature, location, and number of loads placed in the oven:

$$h_c = C(\Delta t)^a \quad (14)$$

For a single cake in the center of an oven, average convection heat transfer data were found to fit the equation $h_c = 0.0058(\Delta t)^2$, where Δt is the difference between oven and cake surface temperature.[5]

Forced Convection. The motion of a fluid in forced convection is not caused by a density differential as in free convection, but is due to external forces. The mixing of fluid particles is greater (depends on the Reynolds number) and, consequently, the coefficient of heat transfer for forced convection is usually much larger. It can be shown by dimensional analysis that:

$$\mathbf{Nu} = a\mathbf{Re}^b\mathbf{Pr}^c \quad (15)$$

where: **Nu** = Nusselt number
Re = Reynolds number
Pr = Prandtl number

The values of constants a, b, and c follow:

Constant	*Laminar flow*	*Turbulent flow*
a	0.664	0.036
b	.500	.800
c	0.333	0.333

The flow is laminar for Reynolds numbers below 2100 and turbulent for Reynolds numbers above 10,000. To determine the convective heat transfer coefficient:

1. Determine Reynolds number and pick constants a, b, and c, depending on flow.
2. Find Prandtl number from fluid properties.
3. Determine h from the relation $\mathbf{Nu} = hL/k$, where L is the length or diameter involved.

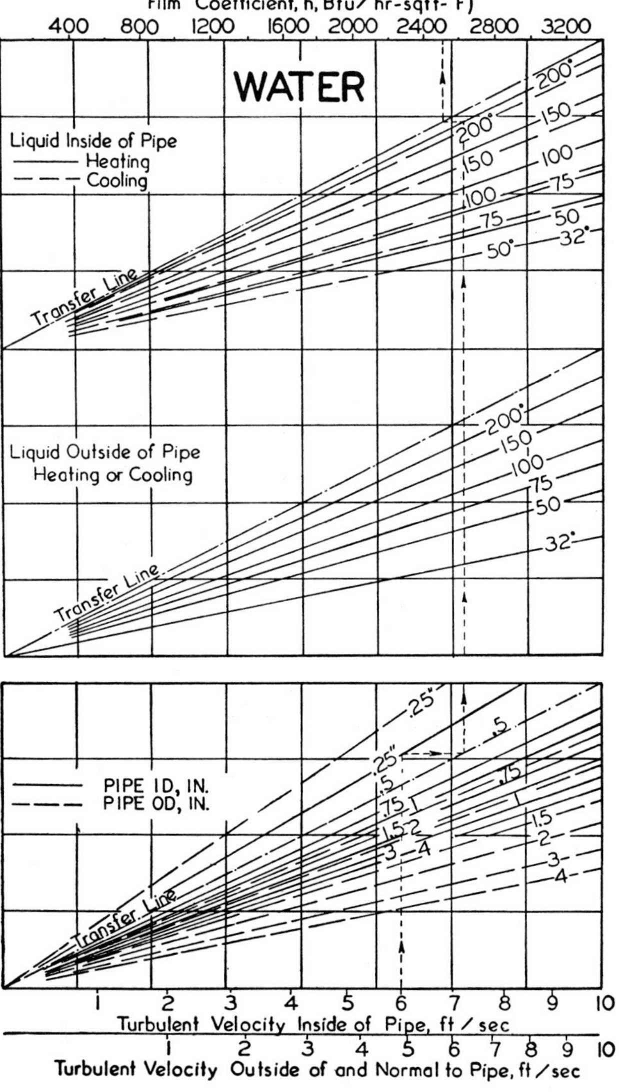

Fig. 2-48 Film coefficient for water heating or cooling in turbulent flow (over 1 fps) within a tube and in turbulent flow (over 1 fps unless Reynolds number exceeds 100) outside and normal to a single tube.[6]

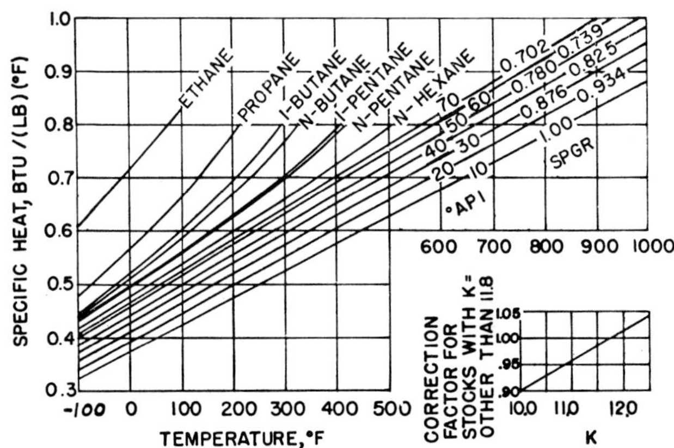

Fig. 2-49 Specific heats of hydrocarbon liquids.[7] K = correction or characterization factor = $(T_B)^{0.333}$/(sp gr), where T_B = cubic average boiling point of stock, °R, and sp gr = specific gravity at 60/60 F.

A general method of obtaining **forced convection film coefficients** applied to various heat transfer situations involves a plot of **dimensionless ratios** such as that given in Fig. 2-47. Definitions of the terms indicated in Fig. 2-47 are:

C_p = specific heat at constant pressure, Btu per lb-°F· see Figs. 2-49, 2-50, 2-51, Table 2-74, etc.
D = diameter, ft
G = mass velocity, lb-mass per hr-sq ft of cross section
G_{max} = mass velocity thru minimum free area in a row of pipes normal to the fluid stream, lb-mass per hr-sq ft
h = local individual coefficient of heat transfer, Btu per hr-sq ft-°F

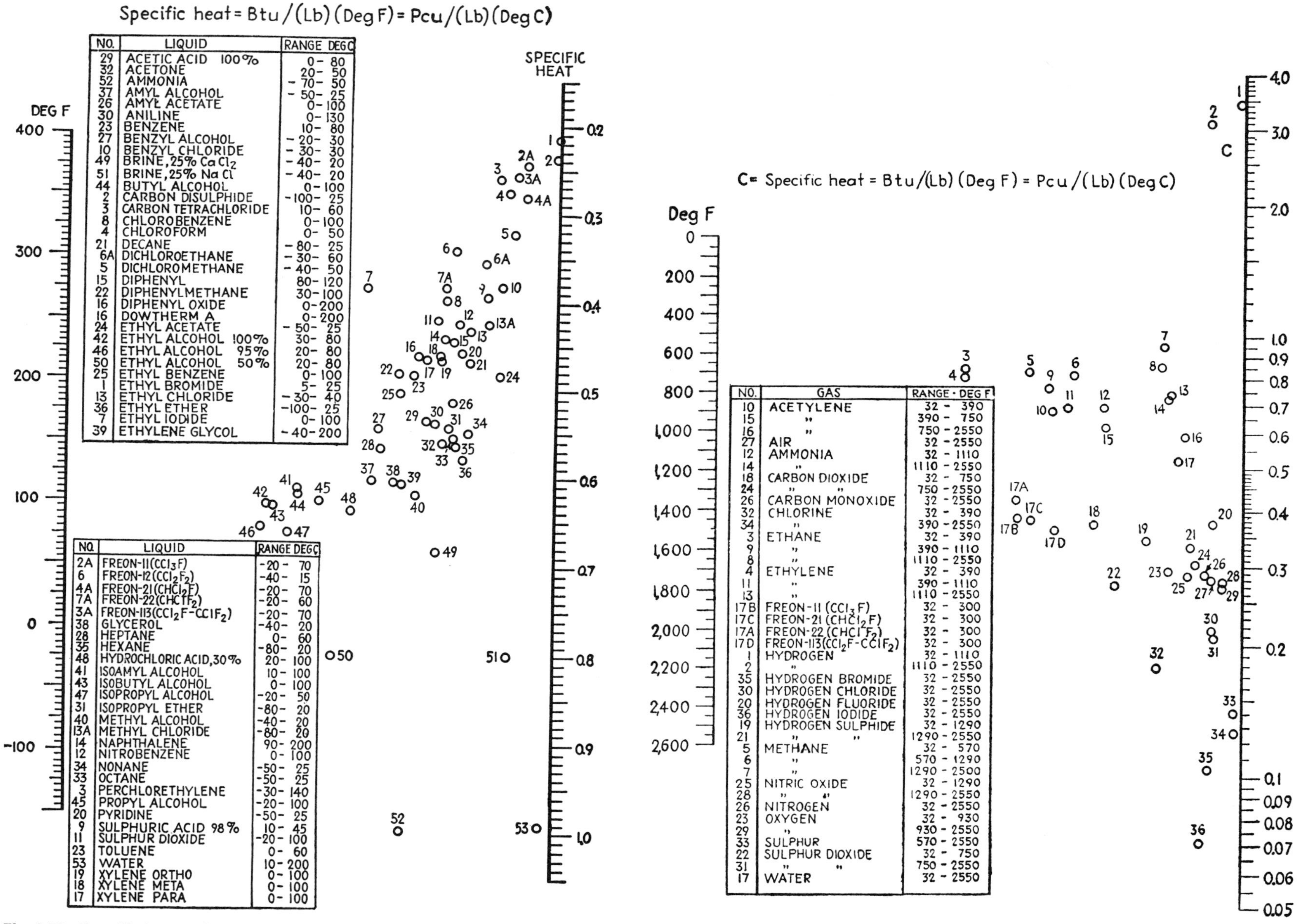

NO.	LIQUID	RANGE DEG C
29	ACETIC ACID 100%	0 - 80
32	ACETONE	20 - 50
52	AMMONIA	- 70 - 50
37	AMYL ALCOHOL	- 50 - 25
26	AMYL ACETATE	0 - 100
30	ANILINE	0 - 130
23	BENZENE	10 - 80
27	BENZYL ALCOHOL	- 20 - 30
10	BENZYL CHLORIDE	- 30 - 30
49	BRINE, 25% $CaCl_2$	- 40 - 20
51	BRINE, 25% NaCl	- 40 - 20
44	BUTYL ALCOHOL	0 - 100
2	CARBON DISULPHIDE	-100 - 25
3	CARBON TETRACHLORIDE	10 - 60
8	CHLOROBENZENE	0 - 100
4	CHLOROFORM	0 - 50
21	DECANE	- 80 - 25
6A	DICHLOROETHANE	- 30 - 60
5	DICHLOROMETHANE	- 40 - 50
15	DIPHENYL	80 - 120
22	DIPHENYLMETHANE	30 - 100
16	DIPHENYL OXIDE	0 - 200
16	DOWTHERM A	0 - 200
24	ETHYL ACETATE	- 50 - 25
42	ETHYL ALCOHOL 100%	30 - 80
46	ETHYL ALCOHOL 95%	20 - 80
50	ETHYL ALCOHOL 50%	20 - 80
25	ETHYL BENZENE	0 - 100
1	ETHYL BROMIDE	5 - 25
13	ETHYL CHLORIDE	- 30 - 40
36	ETHYL ETHER	-100 - 25
7	ETHYL IODIDE	0 - 100
39	ETHYLENE GLYCOL	- 40 - 200

NO.	LIQUID	RANGE DEG C
2A	FREON-11 (CCl_3F)	-20 - 70
6	FREON-12 (CCl_2F_2)	-40 - 15
4A	FREON-21 ($CHCl_2F$)	-20 - 70
7A	FREON-22 ($CHClF_2$)	-20 - 60
3A	FREON-113 (CCl_2F-$CClF_2$)	-20 - 70
38	GLYCEROL	-40 - 20
28	HEPTANE	0 - 60
35	HEXANE	-80 - 20
48	HYDROCHLORIC ACID, 30%	20 - 100
41	ISOAMYL ALCOHOL	10 - 100
43	ISOBUTYL ALCOHOL	0 - 100
47	ISOPROPYL ALCOHOL	-20 - 50
31	ISOPROPYL ETHER	-80 - 20
40	METHYL ALCOHOL	-40 - 20
13A	METHYL CHLORIDE	-80 - 20
14	NAPHTHALENE	90 - 200
12	NITROBENZENE	0 - 100
34	NONANE	-50 - 25
33	OCTANE	-50 - 25
3	PERCHLORETHYLENE	-30 - 140
45	PROPYL ALCOHOL	-20 - 100
20	PYRIDINE	-50 - 25
9	SULPHURIC ACID 98%	10 - 45
11	SULPHUR DIOXIDE	-20 - 100
23	TOLUENE	0 - 60
53	WATER	10 - 200
19	XYLENE ORTHO	0 - 100
18	XYLENE META	0 - 100
17	XYLENE PARA	0 - 100

Fig. 2-50 Specific heats of liquids.[4]

NO.	GAS	RANGE - DEG F
10	ACETYLENE	32 - 390
15	"	390 - 750
16	"	750 - 2550
27	AIR	32 - 2550
12	AMMONIA	32 - 1110
14	"	1110 - 2550
18	CARBON DIOXIDE	32 - 750
24	" "	750 - 2550
26	CARBON MONOXIDE	32 - 2550
32	CHLORINE	32 - 390
34	"	390 - 2550
3	ETHANE	32 - 390
9	"	390 - 1110
8	"	1110 - 2550
4	ETHYLENE	32 - 390
11	"	390 - 1110
13	"	1110 - 2550
17B	FREON-11 (CCl_3F)	32 - 300
17C	FREON-21 ($CHCl_2F$)	32 - 300
17A	FREON-22 ($CHClF_2$)	32 - 300
17D	FREON-113 (CCl_2F-$CClF_2$)	32 - 300
1	HYDROGEN	32 - 1110
2	"	1110 - 2550
35	HYDROGEN BROMIDE	32 - 2550
30	HYDROGEN CHLORIDE	32 - 2550
20	HYDROGEN FLUORIDE	32 - 2550
36	HYDROGEN IODIDE	32 - 2550
19	HYDROGEN SULPHIDE	32 - 1290
21	" "	1290 - 2550
5	METHANE	32 - 570
6	"	570 - 1290
7	"	1290 - 2500
25	NITRIC OXIDE	32 - 1290
28	" "	1290 - 2550
26	NITROGEN	32 - 2550
23	OXYGEN	32 - 930
29	"	930 - 2550
33	SULPHUR	570 - 2550
22	SULPHUR DIOXIDE	32 - 750
31	" "	750 - 2550
17	WATER	32 - 2550

Fig. 2-51 Specific heats of gases and vapors at 1 atm.[4]

Gas	*X*	*Y*
Acetic acid	7.7	14.3
Acetone	8.9	13.0
Acetylene	9.8	14.9
Air	11.0	20.0
Ammonia	8.4	16.0
Argon	10.5	22.4
Benzene	8.5	13.2
Bromine	8.9	19.2
Butene	9.2	13.7
Butylene	8.9	13.0
Carbon dioxide	9.5	18.7
Carbon disulfide	8.0	16.0
Carbon monoxide	11.0	20.0
Chlorine	9.0	18.4
Chloroform	8.9	15.7
Cyanogen	9.2	15.2
Cyclohexane	9.2	12.0
Ethane	9.1	14.5
Ethyl acetate	8.5	13.2
Ethyl alcohol	9.2	14.2
Ethyl chloride	8.5	15.6
Ethyl ether	8.9	13.0
Ethylene	9.5	15.1
Fluorine	7.3	23.8
Freon-11	10.6	15.1
Freon-12	11.1	16.0
Freon-21	10.8	15.3
Freon-22	10.1	17.0
Freon-113	11.3	14.0
Helium	10.9	20.5
Hexane	8.6	11.8
Hydrogen	11.2	12.4
$3H_2 + 1N_2$	11.2	17.2
Hydrogen bromide	8.8	20.9
Hydrogen chloride	8.8	18.7
Hydrogen cyanide	9.8	14.9
Hydrogen iodide	9.0	21.3
Hydrogen sulfide	8.6	18.0
Iodine	9.0	18.4
Mercury	5.3	22.9
Methane	9.9	15.5
Methyl alcohol	8.5	15.6
Nitric oxide	10.9	20.5
Nitrogen	10.6	20.0
Nitrosyl chloride	8.0	17.6
Nitrous oxide	8.8	19.0
Oxygen	11.0	21.3
Pentane	7.0	12.8
Propane	9.7	12.9
Propyl alcohol	8.4	13.4
Propylene	9.0	13.8
Sulfur dioxide	9.6	17.0
Toluene	8.6	12.4
2,3,3-Trimethyl-butane	9.5	10.5
Water	8.0	16.0
Xenon	9.3	23.0

Fig. 2-52 Viscosities of gases at 1 atm.[1] To convert centipoises to lb-mass/hr-ft, multiply centipoises by 2.42.

Liquid	*X*	*Y*
Acetaldehyde	15.2	4.8
Acetic acid, 100%	12.1	14.2
Acetic acid, 70%	9.5	17.0
Acetic anhydride	12.7	12.8
Acetone, 100%	14.5	7.2
Acetone, 35%	7.9	15.0
Allyl alcohol	10.2	14.3
Ammonia, 100%	12.6	2.0
Ammonia, 26%	10.1	13.9
Amyl acetate	11.8	12.5
Amyl alcohol	7.5	18.4
Aniline	8.1	18.7
Anisole	12.3	13.5
Arsenic trichloride	13.9	14.5
Benzene	12.5	10.9
Brine, $CaCl_2$, 25%	6.6	15.9
Brine, NaCl, 25%	10.2	16.6
Bromine	14.2	13.2
Bromotoluene	20.0	15.9
Butyl acetate	12.3	11.0
Butyl alcohol	8.6	17.2
Butyric acid	12.1	15.3
Carbon dioxide	11.6	0.3
Carbon disulfide	16.1	7.5
Carbon tetrachloride	12.7	13.1
Chlorobenzene	12.3	12.4
Chloroform	14.4	10.2
Chlorosulfonic acid	11.2	18.1
Chlorotoluene, ortho	13.0	13.3
Chlorotoluene, meta	13.3	12.5
Chlorotoluene, para	13.3	12.5
Cresol, meta	2.5	20.8
Cyclohexanol	2.9	24.3
Dibromoethane	12.7	15.8
Dichloroethane	13.2	12.2
Dichloromethane	14.6	8.9
Diethyl oxalate	11.0	16.4
Dimethyl oxalate	12.3	15.8
Diphenyl	12.0	18.3
Dipropyl oxalate	10.3	17.7
Ethyl acetate	13.7	9.1
Ethyl alcohol, 100%	10.5	13.8
Ethyl alcohol, 95%	9.8	14.3
Ethyl alcohol, 40%	6.5	16.6
Ethyl benzene	13.2	11.5
Ethyl bromide	14.5	8.1
Ethyl chloride	14.8	6.0
Ethyl ether	14.5	5.3
Ethyl formate	14.2	8.4
Ethyl iodide	14.7	10.3
Ethylene glycol	6.0	23.6
Formic acid	10.7	15.8
Freon-11	14.4	9.0
Freon-12	16.8	5.6
Freon-21	15.7	7.5
Freon-22	17.2	4.7
Freon-113	12.5	11.4
Glycerol, 100%	2.0	30.0
Glycerol, 50%	6.9	19.6
Heptene	14.1	8.4
Hexane	14.7	7.0
Hydrochloric acid, 31.5%	13.0	16.6
Isobutyl alcohol	7.1	18.0
Isobutyric acid	12.2	14.4
Isopropyl alcohol	8.2	16.0
Kerosene	10.2	16.9
Linseed oil, raw	7.5	27.2
Mercury	18.4	16.4
Methanol, 100%	12.4	10.5
Methanol, 90%	12.3	11.8
Methanol, 40%	7.8	15.5
Methyl acetate	14.2	8.2
Methyl chloride	15.0	3.8
Methyl ethyl ketone	13.9	8.6
Naphthalene	7.9	18.1
Nitric acid, 95%	12.8	13.8
Nitric acid, 60%	10.8	17.0
Nitrobenzene	10.6	16.2
Nitrotoluene	11.0	17.0
Octane	13.7	10.0
Octyl alcohol	6.6	21.1
Pentachloroethane	10.9	17.3
Pentane	14.9	5.2
Phenol	6.9	20.8
Phosphorus tri-bromide	13.8	16.7
Phosphorus tri-chloride	16.2	10.9
Propionic acid	12.8	13.8
Propyl alcohol	9.1	16.5
Propyl bromide	14.5	9.6
Propyl chloride	14.4	7.5
Propyl iodide	14.1	11.6
Sodium	16.4	13.9
Sodium hydroxide, 50%	3.2	25.8
Stannic chloride	13.5	12.8
Sulfur dioxide	15.2	7.1
Sulfuric acid, 110%	7.2	27.4
Sulfuric acid, 98%	7.0	24.8
Sulfuric acid, 60%	10.2	21.3
Sulfuryl chloride	15.2	12.4
Tetrachloroethane	11.9	15.7
Tetrachloroethylene	14.2	12.7
Titanium tetrachloride	14.4	12.3
Toluene	13.7	10.4
Trichloroethylene	14.8	10.5
Turpentine	11.5	14.9
Vinyl acetate	14.0	8.8
Water	10.2	13.0
Xylene, ortho	13.5	12.1
Xylene, meta	13.9	10.6
Xylene, para	13.9	10.9

Fig. 2-53 Viscosities of liquids at 1 atm.[1] To convert centipoises to lb-mass/hr-ft, multiply centipoises by 2.42.

h_m = mean value of h for entire apparatus, based on true average (mean length) value of the terminal temperature differences

K = thermal conductivity of fluid, Btu per hr-sq ft-°F per in; note that this K is per *foot* of thickness; see Tables 2-94 or 2-95

K_f = thermal conductivity of fluid at film temperature t_f; $t_f = \frac{1}{2}$ (fluid temperature + wall temperature); note that this K is per *foot* of thickness; see Tables 2-94 and 2-95

L = heated length of heat transfer surface, ft

V_∞ = velocity of stream at great depth, ft per hr

S_p = *twice* the perimeter over which the fluid flows in passing fin, ft

Z = absolute viscosity at bulk fluid temperature, lb-mass per hr-ft = 2.42 × centipoises, = 105,800 × (lb-force)-sec per sq ft; see Figs. 2-52 and 2-53

Z_f = absolute dynamic viscosity at arithmetic mean of wall and fluid temperature, lb-mass per hr-ft; see Figs. 2-52 and 2-53

d_∞ = density of stream at great depth, lb-mass per cu ft

In Fig. 2-47 the Prandtl number, C_pZ/K, which appears in all j groups is nearly independent of temperature; and for air, oxygen, nitrogen, hydrogen, and carbon monoxide has an average value of 0.74. For other gases its average values are: methane, 0.78; carbon dioxide, 0.79; ethylene, 0.81; and low pressure steam, 0.78. Flue products from the combustion of common fuel gases have a Prandtl number value of 0.74. The film coefficients given by Fig. 2-47 *do not include* transfer by gaseous radiation which must be accounted for separately at high gas temperature.

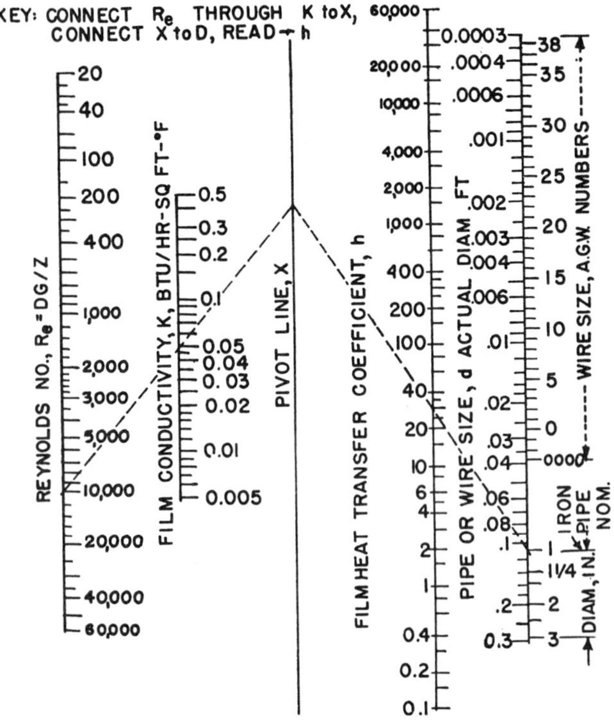

Fig. 2-54 Forced convection heat transfer coefficients for pipe and wire.[9] In example shown, Reynolds number = 10,000; K = 0.05; d = 0.1095 ft (OD of 1 in. IPS pipe). Therefore, h = 29.

A graphic means of evaluating the heat transfer film coefficient for two common cases of flowing water is given in Fig. 2-48. The example shown by dotted lines is for water in turbulent flow at 6 fps thru a tube of 0.25 ID. The fluid is heating at a mean temperature of 200 F. The inside tube coefficient is read as approximately 2500 Btu per hr-sq ft-°F.

The following equation is recommended by McAdams for general use for **turbulent flow** of gases in straight tubes.[1]

$$h = 16.6c_p(G')^{0.8}/(D_i')^{0.2} \qquad (16)$$

where:

h = coefficient of heat transfer, Btu per hr-sq ft-°F

c_p = specific heat, Btu per lb-°F

G' = mass velocity, lb of gas per sec-sq ft cross section

D_i' = actual inside diameter, in.

Equation **16** gives conservative values for film coefficients of flames and gaseous combustion products at temperatures up to 3000 F in medium and high temperature gas glow tubes.[8] For low flow velocities in tubes, it is preferable to use Fig. 2-46 for the film coefficient.

Film coefficients for flow over **small pipes and wires** in forced convection may be found in Fig. 2-54.

The Reynolds number shown in Fig. 2-54 is a dimensionless factor which is defined by the equation:

$$\mathbf{Re} = LVp/Z, \quad \text{or, for circular pipes,} \quad \mathbf{Re} = DG/Z$$

where:

L = characteristic linear dimension of the apparatus, ft

V = linear velocity, ft per hr

p = density, lb-mass per cu ft

Z = absolute viscosity, lb-mass per hr-ft

D = pipe diameter, ft

G = mass velocity, lb-mass per hr-sq ft of cross-sectional area

In computing a Reynolds number, it is imperative that a consistent system of units be used to ensure the **dimensionless** character of the factor.

Radiation

Radiation, in contrast to both conduction and convection, involves no material means of heat transfer. Infra-red electromagnetic waves, similar to light waves, are the carriers of heat. The quantity of heat available from radiation *is proportional to the fourth power of the absolute temperature of the heat source*, whereas conduction and convection are dependent on a temperature difference as the "driving force" for heat transfer.

When radiation falls on any body a fraction, A, is *absorbed;* a fraction, B, is *reflected;* and a fraction, C, is *transmitted* thru the body. Transmission is practically negligible for opaque materials—metals, refractories, etc. For any body,

$$A + B + C = 1 \qquad (17)$$

A *black body* absorbs all incident radiation; i.e., it reflects and transmits none. Actually, there is no material that will absorb all the radiant energy that is imposed on it. Hence, the absorptivity of any real body is *always less than unity.*

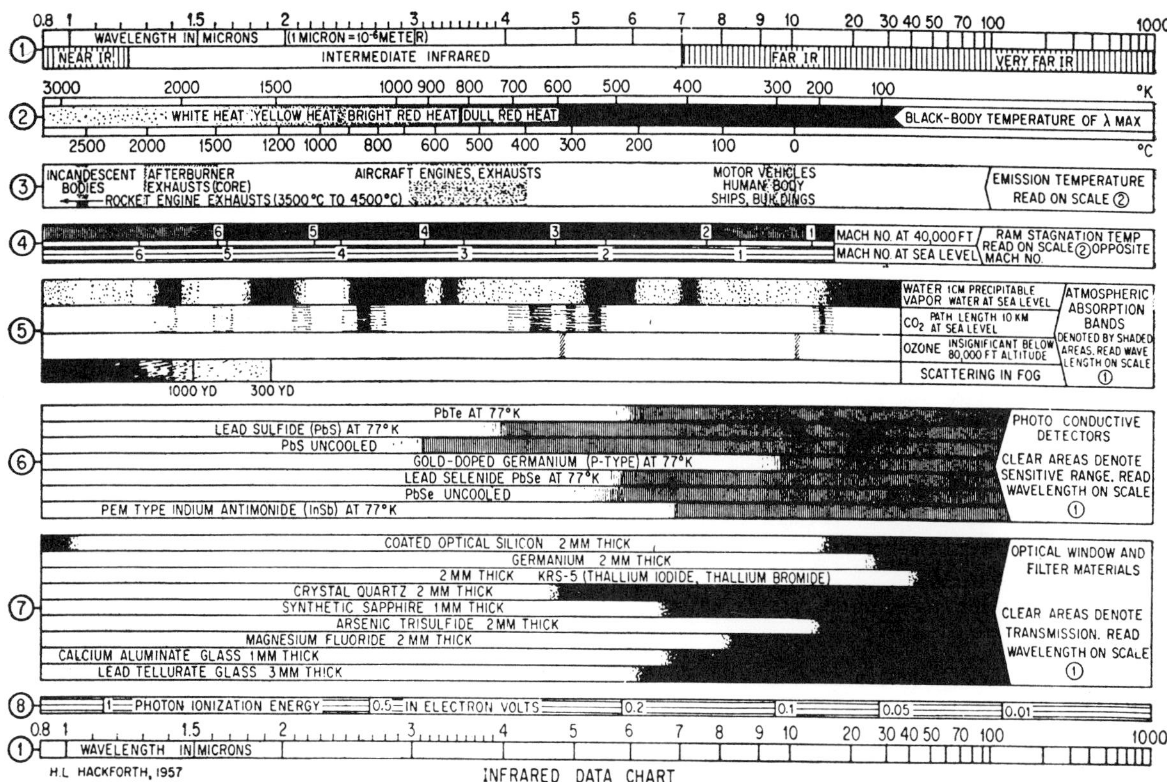

Fig. 2-55a Infra-red data chart.[10]

Figure 2-55a was designed as a ready reference of approximate values and of conversions for common parameters used in infra-red work.

Scale **(2)** is related to scale **(1)** by Wien's law, $\lambda_{max}\,T$ = constant. λ_{max} is the wave length of maximum intensity for a radiating black body at T degrees absolute temperature. Scales **(3)** and **(4)** are referenced to scale **(2)**. Scales **(5)**, **(6)**, **(7)**, and **(8)** are referenced to scale **(1)**.

Example of Use: The IR (infra-red) search system is required for detection of a 600 C source.

1. Scale **(2)** indicates detection is required in the neighborhood of 3.2μ. Lay slider vertically across extreme scales **(1)** at this figure.

2. Scale **(5)** indicates a water vapor absorption band from 2.5 to 3μ and atmosphere "windows" of good transmission from 2 to 2.5 and 3 to 5μ.

3. Scale **(6)** shows that a liquid nitrogen-cooled PbS, PbTe, or PbSe detector is required to take advantage of the atmosphere windows indicated in 2 above. Scale **(5)** shows that radiation cutoff below 2μ is desirable for suppression of scattered sunlight. This can be effected by a germanium filter, scale **(7)**, which also indicates suitable optical material.

4. Scale **(8)** shows that if a semiconductor-type photoconductive detector is used, a photon-ionization energy of about 0.5 ev is required.

5. Scale **(3)** indicates types of sources radiating at this temperature. Scale **(4)** shows that the nose of an aircraft flying at Mach 3.25 at sea level or Mach 3.9 at 40,000 ft altitude would achieve a ram-stagnation temperature of this value.

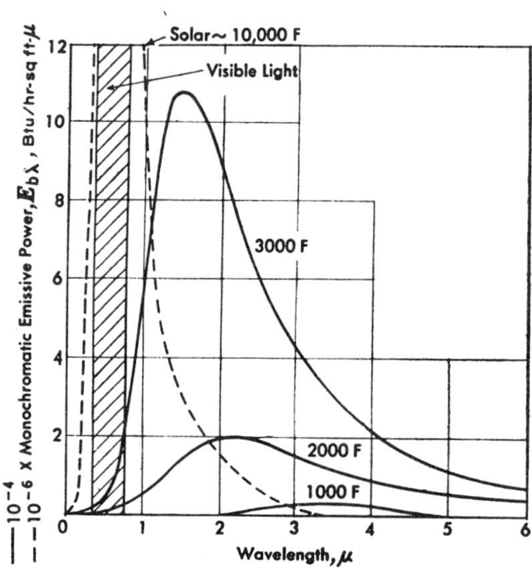

Fig. 2-55b Spectral distribution of monochromatic emissive power for ideal radiator at various temperatures.[11]

Black-Body Radiation Laws. There are three fundamental laws which can be conveniently illustrated by Fig. 2-55b. In all cases, the radiation is measured on the *inner surface of a hemispherical shell* resting on the plane of the heat source.

Stefan-Boltzmann Law. The total area under any of the curves in Fig. 2-55b is equal to the total radiant energy available (emissive power) at the given temperature. This energy is a function of temperature, or:

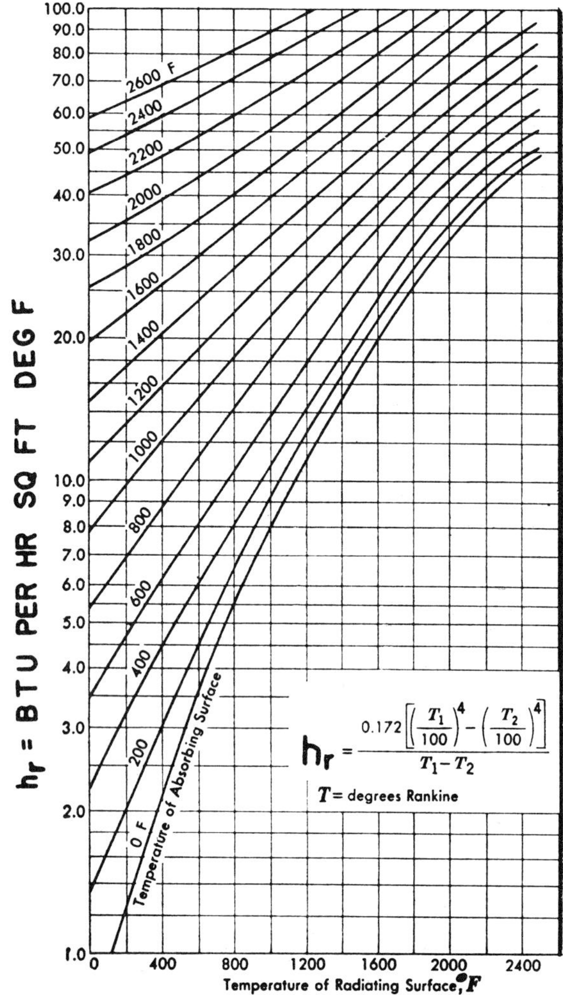

Fig. 2-56 Radiation coefficient of heat transfer, h_r, as a function of temperature, °F. (Ref. 11.)

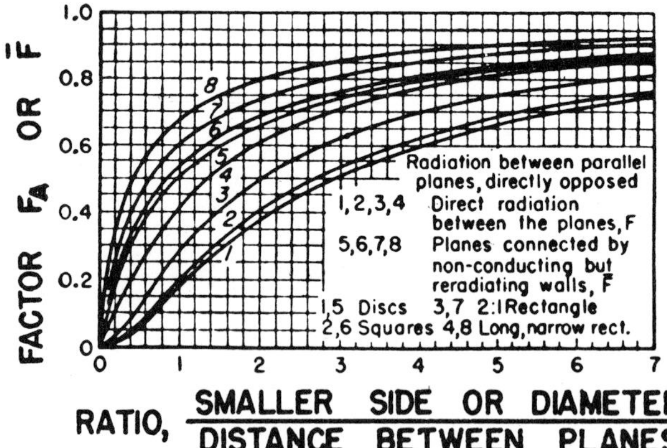

Fig. 2-57 F_A in case 5, Table 2-97, and $\bar{F}$ in Eq. **18b.** (Ref. 12.)

$$E_b = 0.1713(T/100)^4 \tag{18}$$

where:

E_b = total quantity of heat radiated by a black body over the *entire* wave-length spectrum, Btu per hr-sq ft

T = absolute temperature, °R

Example: Find the radiant energy from a black body at 3000 F.

Solution: Integrate the area under the 3000 F curve in Fig. 2-55b, or, more simply:

$$E_b = 0.1713\left(\frac{3000 + 460}{100}\right)^4 = 247{,}000 \text{ Btu per hr-sq ft}$$

Planck's Law. Figure 2-55b is a plot of Planck's law, which may be expressed as:

$$E_{b\lambda} = \frac{A\lambda^{-5}}{e^{B/\lambda T} - 1} \tag{19}$$

or

$$E_{b\lambda} = T^5 f(\lambda T) \tag{19a}$$

where:

$E_{b\lambda}$ = quantity of heat available from a black body, in Btu per hr-sq ft at a specified wave-length band width and temperature *divided* by the width of the wave-length band in microns—1 micron = $1\mu = 10^{-6}$ meters (**Note:** Most texts define $E_{b\lambda}$ as the *monochromatic* emissive power of a black body.)

λ = wave length, μ

T = temperature of body, °F abs

e = Napierian base of logarithms

A = 1.1870×10^8, Btu μ^4 per sq ft-hr

B = 2.5896×10^4, °F abs-μ

Example: Find the radiant energy available from a black body at 3000 F in the wave-length band from 3μ to 3.5μ.

Solution:

$$E_{b\lambda}\,(\text{avg}) = \frac{4.1 + 3.0}{2} \times 10^4 = 3.55 \times 10^4 \text{ Btu/hr-sq ft-}\mu$$

(See Fig. 2-55b.)

Width of wave-length band = $4\mu - 3.5\mu = 0.5\mu$.

Therefore, the energy available = $(3.55 \times 10^4)(0.5)$ = 17,750 Btu per hr-sq ft.

Wien's Displacement Law. The correlation between the wave length λ_{max} in microns (at which $E_{b\lambda}$ is a maximum) and the temperature, °R, is:

$$\lambda_{max}T = 5215.6 \tag{20}$$

Example: Find the *instantaneous* wave length for 3000 F at which the radiant energy from a black body is maximum.

Solution:

$$\lambda_{max} = \frac{5215.6}{3000 + 460} = 1.5$$

(Check this value on Fig. 2-55b.)

Radiation from Real Surfaces. Radiation and absorption of radiant energy by real surfaces depart significantly from black-body values. For heat transfer, surfaces are regarded as **gray bodies,** even though they only approximate this concept. A gray body is one whose characteristic radiation effects are proportional to (but lower than) those of a black body over the entire wave-length spectrum. The ratio of the emissive power of an actual surface to that of a black body is called the emissivity, e, of the surface.

The absorptivity and emissivity of any body can be considered equal at the same temperature and wave length. Nevertheless, in real situations a body emits radiation at wave lengths which are usually different from those at which the body accepts radiation. Therefore, for best results, choose the absorptivity of the receiving substance at the temperature (or wave length) of the body emitting radiation. When dealing with radiation emission, the emissivity value corresponding to the actual temperature of the emitter should be used. It is essential that the *average* emissivity or absorptivity be used for the wave-length band in which the bulk of the radiation is emitted or absorbed.

In many important cases of thermal radiation between surfaces of solids, separated by a nonabsorbing (transparent to radiation) medium, the *net rate* of direct radiant heat transfer may be approximated by the following relationship:

$$q_r = \sigma A_1 F_A F_e \left[\left(\frac{T_1}{100}\right)^4 - \left(\frac{T_2}{100}\right)^4\right] \quad \textbf{(18a)}$$

where:

σ = Stefan-Boltzmann radiation constant = $0.1713 \dfrac{\text{Btu}}{\text{hr}(^\circ\text{F abs}/100)^4 \text{ sq ft}}$

A_1 = area of radiating body, sq ft
F_A = configuration or "shape" factor
F_e = emissivity factor (Table 2-97)
T = absolute temperature, °R
q_r = net rate of direct radiant heat transfer, Btu per hr

In many cases it is convenient to express radiation heat transfer by the equation:

$$q_r = h_r A_1 (T_1 - T_2) \quad \textbf{(21)}$$

where:

$$h_r = \frac{(T_1^2 + T_2^2)(T_1 + T_2) F_A F_e}{5.78 \times 10^8} \quad \textbf{(22)}$$

Values of h_r are given in Fig. 2-56, based on Eq. 22 for $F_A = F_e = 1$.

Values of F_A and F_e, for use in Eq. 22, are given in Table 2-97. In cases 5 and 7 in this table, F_A is determined by following the procedures shown on Figs. 2-57 and 2-58, respectively. Figure 2-59 shows the method for obtaining F_A for a surface element, dA, and a rectangle in a parallel plane.

The following relations indicate that the number of unique F_A required to describe a system is $\frac{1}{2}n(n - 1)$ if all the surfaces are curved so that they "see" one another (n is the number of "isothermal" zones in a system). If all surface elements are flat, they number only $\frac{1}{2}(n - 1)(n - 2)$ unique F_A:

$$A_1 F_{12} = A_2 F_{21} \quad \textbf{(23)}$$

$$F_{11} + F_{12} + F_{13} + \ldots = 1 \quad \textbf{(23a)}$$

$$F_{11} = 0 \text{ (when } A_1 \text{ can "see" no part of itself)}$$

Note: F_{12} is the shape factor to allow for direct interchange from surface 1 to surface 2 based on A_1; F_{21} from surface 2 to surface 1 based on A_2, where A_2 is the area of a second radiating body.

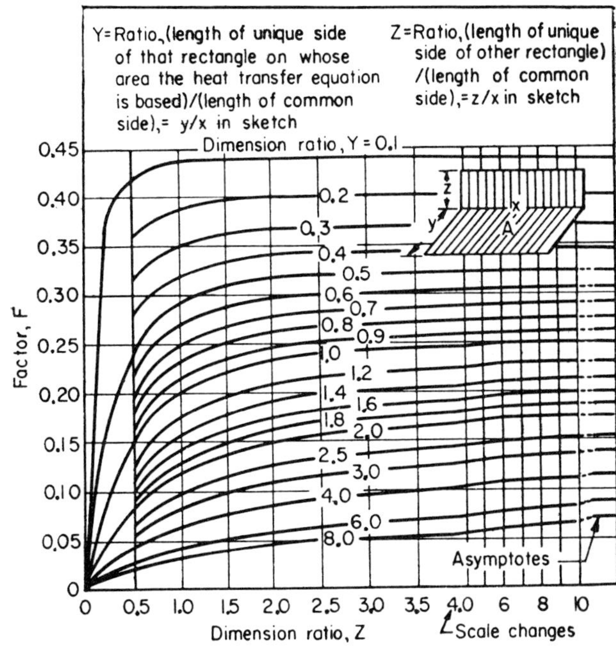

Fig. 2-58 Radiation between adjacent rectangles in perpendicular planes and F_A in case 7, Table 2-97. (Ref. 12.)

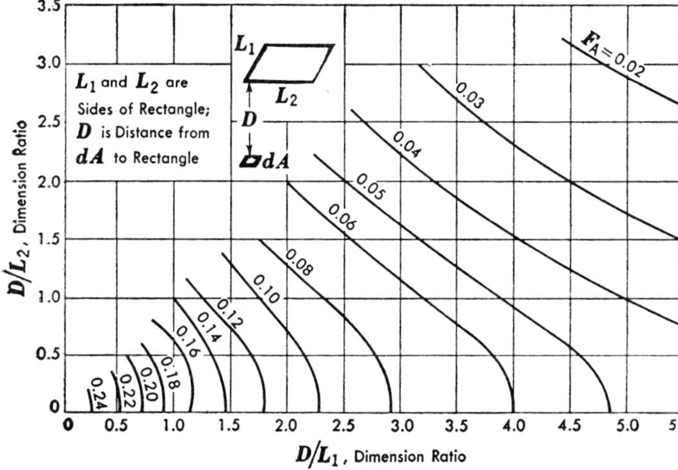

Fig. 2-59 Radiation between a plane element, dA, and a rectangle above and parallel to it[12] (one corner of rectangle contained in normal to dA).

In many cases, the complete enclosure consists in part of black heat sources and sinks and in part of refractory surfaces. The following equation expresses the net radiant heat flux from A_1 to A_2 by the combined mechanisms of direct radiation and reradiation from the refractory surfaces:

$$q_{12} = \sigma A_1 \bar{F}_{12} \left[\left(\frac{T_1}{100}\right)^4 - \left(\frac{T_2}{100}\right)^4\right] \quad \textbf{(18b)}$$

Note: $A_1\bar{F}_{12} = A_2\bar{F}_{21}$ where $\bar{F}$ is this geometrical factor to allow for net radiation between black surfaces, *including* the effect of refractory surfaces. Values of $\bar{F}$ are given in Fig. 2-57.

Emissivities. Emissivities (and absorptivities) of materials vary from almost zero to slightly less than one, depending on the nature of the material, its surface finish, and its

Table 2-97 Configuration and Emissivity Factors[12]
(e = emissivity of material; subscripts 1 and 2 refer to the two bodies involved)

Case	Surface relations	Area used, A	Configuration factor, F_A	Emissivity factor, F_e
1	Infinite parallel planes	Either plane	1	$\dfrac{1}{\dfrac{1}{e_1}+\dfrac{1}{e_2}-1}$
2	Completely enclosed body, 1 (without depressions) small compared with enclosure 2	A_1	1	e_1
3	Completely enclosed body, 1 large compared with enclosure 2	A_1	1	$\dfrac{1}{\dfrac{1}{e_1}+\dfrac{1}{e_2}-1}$
4	Concentric spheres or infinite concentric cylinders	A_1	1	$\dfrac{1}{\dfrac{1}{e_1}+\dfrac{A_1}{A_2}\left(\dfrac{1}{e_2}-1\right)}$
5	Direct radiation between parallel and equal squares or disks of width D, and distance between, L	Either plane	See Fig. 2-57	If A is small compared with L, use e_1e_2; if A is large compared with L, use F_e, case 3
6	Two equal rectangles in parallel planes directly opposite each other	Either plane	$(F_A'F_A'')^{0.5}$ $F_A' = F_A$ for squares equivalent to smaller side of rectangle; $F_A'' = F_A$ for squares equivalent to larger side of rectangle. See Fig. 2-57	
7	Two rectangles with a common side in perpendicular planes	Either	See Fig. 2-58	e_1e_2

temperature. Polished metal surfaces have low emissivities, while oxidized surfaces and nonmetals generally approach a value of one (Table 2-98).

The directional emissivity, which is the ratio of radiating powers in a direction making a given angle with the normal to the radiating surface, is of significance for well-polished surfaces. In these cases increase the pertinent values in Table 2-98 by 15 to 20 per cent.

Gaseous Radiation. Many gases such as O_2, N_2, H_2, and dry air are practically transparent to thermal radiation. Conversely, gases such as CO_2, H_2O, SO_2, CO, NH_3, hydrocarbons, and alcohols absorb and emit radiation. Whereas solids radiate at all wave lengths, gases emit and absorb radiation in specific wave-length bands.

Discussion is restricted to H_2O and CO_2 since they are the most important gases in furnaces. Figure 2-60 shows the emissivity curve produced by burning a natural gas containing about 88 per cent methane in a Bunsen burner as obtained with a NaCl prism in the region from 1 to 5μ. At 4.36μ the maximum emission of CO_2 occurs. The small bands in the region of 1.8 to 2μ are produced by water vapor and CO_2. The band at 2.49μ is probably due to water vapor, and the bands in the region of 2.7μ are attributed to CO_2.

At low inputs, radiation maxima of 59 per cent have been reported.[14] Once the limiting temperature of the burner head is approached, the percentage of radiated heat decreases with further increases in heat input to the burner.

Flames may be either luminous or nonluminous. The maximum amount of thermal radiation is obtained from the luminous type of flame. Under proper burning conditions this flame can be made to fill the entire combustion space. The flame radiates heat directly to the surface of the material at a lower temperature than a nonluminous flame, allowing for the penetration of heat into the interior of the material without high surface temperatures. Such a flame, however, is sometimes undesirable because of its low temperature and reducing character.

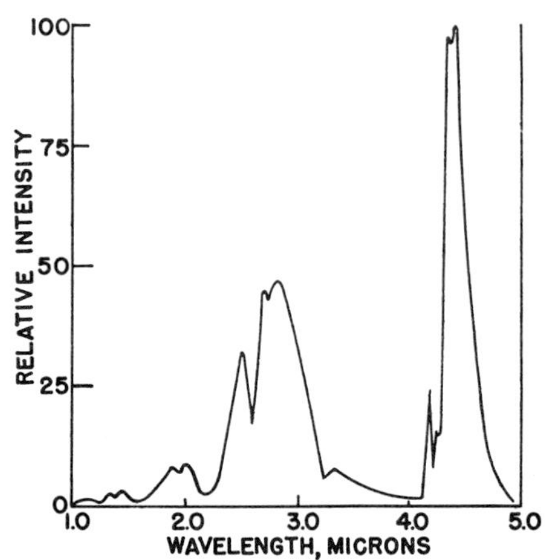

Fig. 2-60 Infra-red emission spectrum of a natural gas flame.[13]

Table 2-98 Emissivities of Various Surfaces

Material	Wave length and average temperature				
	9.3 μ 100 F	5.4 μ 500 F	3.6 μ 1000 F	1.8 μ 2500 F	0.6 μ Solar
Metals					
Aluminum					
Polished	0.04	0.05	0.08	0.19	~0.3
Oxidized	.11	.12	.18	...	...
24-ST weathered	.4	.32	.27	...	...
Surface roofing	.22	...	...	...	...
Anodized (at 1000 F)	.94	.42	.60	.34	...
Brass					
Polished	.10	.10	...	...	...
Oxidized	.61	...	...	...	...
Chromium, polished	.08	.17	.26	.40	.49
Copper					
Polished	.04	.05	.18	.17	...
Oxidized	.87	.83	.77	...	...
Iron					
Polished	.06	.08	.13	.25	.45
Cast, oxidized	.63	.66	.76	...	...
Galvanized, new	.23	...	...	.42	.66
Galvanized, dirty	.28	...	...	.90	.89
Steel plate, rough	.94	.97	.98	...	...
Oxide	.96	...	.85	...	.74
Molten	...	...	...	.3–.4	...
Magnesium	.07	.13	.18	.24	.30
Molybdenum filament	...	...	~ .09	~ .15	~ .2*
Silver, polished	.01	.02	.03	...	.11
Stainless steel					
18-8, polished	.15	.18	.22	...	...
18-8, weathered	.85	.85	.85	...	...
Steel tube, oxidized	...	.80	...	...	...
Tungsten filament	.03	...	...	~ .18	.35†
Zinc					
Polished	.02	.03	.04	.06	.46
Galvanized sheet	~ .25	...	...	...	...
Building and insulating materials					
Asbestos paper	.93	.93	...	...	...
Asphalt	.93	...	.9	...	.93
Brick					
Red	.93	...	...	...	.7
Fire clay	.9	...	~ .7	~ .75	...
Silica	.9	...	~ .75	~ .84	...
Magnesite refractory	.9	...	...	~ .4	...
Enamel, white	.9	...	...	...	...
Marble, white	.95	...	.93	...	.47
Paper, white	.95	...	.82	.25	.28
Plaster	.91	...	...	...	...
Roofing board	.93	...	...	...	...
Enameled steel, white	...	...	...	.65	.47
Asbestos cement, red	...	...	...	.67	.66
Paints					
Aluminized lacquer	.65	.65	...	...	...
Cream paints	.95	.88	.70	.42	.35
Lacquer, black	.96	.98	...	...	...
Lampblack paint	.96	.97	...	.97	.97
Red paint	.96	...	...	...	.74
Yellow paint	.95	...	.5	...	.30
Oil paints (all colors)	~ .94	~ .9	...	...	...
White (ZnO)	.95	...	.91	...	0.18
Miscellaneous					
Ice	~ .97†	...	...	...	...
Water	~ .96	...	...	...	...
Carbon	...	...	...	...	...
T-carbon, 0.9% ash	.82	0.80	0.79	...	...
Filament	~ .72	...	...	0.53	...
Wood	~ .93	...	...	...	...
Glass	0.90	...	...	...	(Low)

* At 5000 F.
† At 6000 F.
‡ At 32 F.

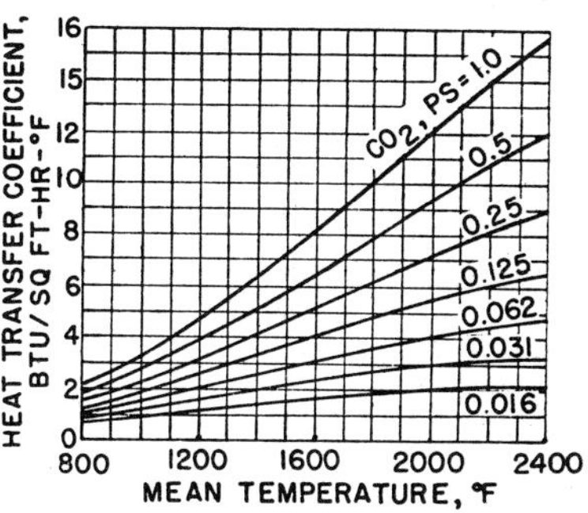

Fig. 2-61 Radiation heat transfer coefficients for gas layers containing CO_2 at partial pressure, P, in atmospheres and thickness, S, in feet at the mean temperature of the gas and the bounding solid.[16]

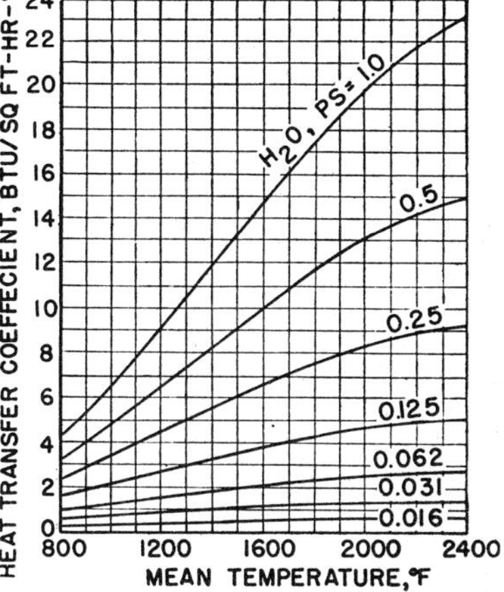

Fig. 2-62 Radiation heat transfer coefficients for gas layers containing H_2O at partial pressure, P, in atmospheres and thickness, S, in feet at the mean temperature of the gas and the bounding solid.[16]

A **nonluminous flame** is practically invisible, or blue in color. The heat in the combustion products is transferred mainly by convection. The combustion chamber walls thus may become incandescent and radiate heat to the product being heated, whereas with a **luminous flame,** which is relatively opaque to radiation, the flame itself tends to block wall radiation from the product. A nonluminous flame transfers about 10 per cent of its heat by radiation; a luminous flame generally produces 20 to 120 per cent more radiation. The radiation from a luminous flame may range from 10 to 40 per cent of its potential heat, depending on the luminosity and the flame temperature.[15]

H. C. Hottel noted that the overall opacity of a flame determines the quantity of radiation. In **small systems,** when the flame does not radiate strongly, the radiation absorbed by a solid body is proportional to the flame volume, which is evaluated in terms of the temperature, concentration of the heat-radiating gases CO_2 and H_2O, and geometric shape of the gas mass.

Radiations from the nonluminous gases CO_2 and H_2O are given in Figs. 2-61 and 2-62, respectively. The mean temperature in these figures is defined as the average of the gas temperature and the temperature of the bounding solid. The parameter curves are for varying values of *PS*, which equals the partial pressure of the gas in atmospheres *times* the thickness of the gas layer in feet. The following example illustrates the use of these curves.

Example: Find the heat transfer due to radiation for the volumetric flue gas analysis: 9.5 per cent CO_2, 19.0 per cent H_2O, and 71.5 per cent N_2. Resultant partial pressures are: 0.095 atm, 0.19 atm, 0.715 atm, respectively. Assuming the

Table 2-99a Radiation Data for Gas-Fired Infra-Red Burners[44]

Type	Emitting face material of construction	Red brightness temp, °F	Total normal radiation, Btu/hr-sq ft	GIR factor	Spectral distribution of energy at maximum temperature in selected wave-length spans (μ), per cent					
					0.75–2	2–3	3–4	4–5	5–6	6–16
Atmospheric	Ceramic tile	1420	18,400	0.85						
		1500	22,420	.87						
		1675	29,920	.84	14.6	25.3	20.0	14.7	6.7	18.7
Atmospheric	Ceramic tile	1400	14,000	.76						
		1490	17,820	.71	8.2	20.8	14.6	23.0	10.4	23.0
Atmospheric	Ceramic tile	1420	17,800	.83						
		1500	21,450	.84						
		1590	27,650	.94	10.6	23.3	19.9	17.1	9.2	19.9
Atmospheric	Inconel	1480	19,000	.77						
		1550	24,000	.85	12.7	26.7	20.7	16.7	7.0	16.1
Atmospheric	Inconel	1400	18,720	0.90						
		1540	28,200	1.02						
		1595	32,600	1.06	14.1	28.2	21.2	14.0	8.5	14.0
Atmospheric	Cercor glass	1550	14,380	0.51						
		1710	23,200	.60	8.2	20.9	15.8	21.9	10.8	22.3
Catalytic	Glass wool	600	800	.37						
		700	1,120	.36						
		850	3,300	.65	1.2	4.3	13.7	17.5	15.3	48.0
Powered	Cordurite	1500	22,250	.87						
		1800	32,700	.73						
		2000	44,400	.69	17.4	19.0	14.3	19.0	8.6	21.7
Powered	Low density refractory	1500	17,700	.79						
		1665	23,500	.66						
		1720	29,100	.74	14.9	23.9	21.1	14.1	8.4	17.6
Powered	Refractory brick	1800	38,800	.86						
		2000	51,500	.81						
		2300	62,800	.63	26.0	24.5	15.3	12.9	7.1	14.1
Powered	Refractory brick	1600	21,650	.69						
		1800	34,600	.77						
		2000	58,000	.91	20.2	26.0	14.2	13.0	7.8	18.2
Powered	Inconel	1660	25,650	.73						
		1790	36,700	.83						
		1900	46,600	0.85	17.8	29.4	20.3	12.8	6.8	12.9
Powered	Silicon carbide	1450	24,550	1.06						
		1525	30,900	1.16						
		1610	37,150	1.17	11.5	25.4	21.8	14.9	9.5	16.7

shape of the gas mass to be a cylinder of 3.17 ft diam, the product PS equals 0.095 × 3.17 = 0.3 atm-ft for CO_2 and 0.19 × 3.17 = 0.61 atm-ft for H_2O. Assuming the products of combustion at 2300 F and the wall or charge at 2000 F, the mean temperature equals (2300 + 2000)/2 = 2150 F. The intersection of mean temperature and parameter PS yields 8.2 for CO_2 and 14.7 for H_2O. At temperatures of 1200 F and above, which are common in furnace practice, it is unnecessary to consider the fact that radiation of water vapor is not a direct function of the product PS.

The overall heat transfer coefficient for CO_2 and H_2O, then, is 8.2 + 14.7 = 22.9 Btu per sq ft-hr-°F. However, an average value of five per cent should be deducted from this figure to account for the fact that the emission bands of CO_2 and water vapor overlap at some wave lengths. If an average value of the emissivity of the furnace walls and charge is taken as 0.9, the actual heat transfer coefficient is 0.9 × 0.95 × 22.9 = 19.6. The resultant $Q/A = h\Delta t$ = 19.6 × 300 = 5880 Btu per sq ft-hr.

If the shape of the gas mass is other than a cylinder, the following multipliers[17] should be used to correct the rate of heat transfer: sphere—⅔ diam; space between infinite parallel planes—1.8 × distance between planes; cube—⅔ × side; space outside infinite bank of tubes with centers on equilateral triangles (tube diameter equals clearance)—2.8 × clearance; same as latter, except tube diameter = ½ clearance—3.8 × clearance.

The method given is accurate to within 10 per cent of the true value. Further refinement in calculations is unnecessary since the data used (namely, the furnace gas temperature, distribution of the temperature in the gas layer, large variations in furnace wall emissivity, etc.) cannot be evaluated accurately.

In many heat applications it is desirable to produce higher percentages of radiated heat from the gas burner flames than are obtainable under normal circumstances. To accomplish this, the following means are available: (1) air–gas mixture is directed along the surface of an incandescent refractory, where it is ignited; (2) air–gas mixture is burned at the surface of a perforated or porous refractory used as the burner head; and (3) burner flame is directed against a ceramic or wire screen surface.

Hot surfaces influence combustion speed, making it possible to produce such short flames that for all practical purposes combustion is considered "flameless." It is probable that this phenomenon results from a preheat effect rather than from catalysis. The relation between percentage radiation from such equipment and the total gas input varies over wide ranges, since many units combine radiant emission with convective heat transfer means. The reasons for selecting gas burning equipment with high radiant effectiveness include the following:

1. Radiant burners can be made for directed heat transmittance.
2. Radiant energy travels with the velocity of light, and it is often possible to increase heating speeds or product output by increasing radiant heat transfer.
3. Heating can be accomplished in the open without heating the intervening air space.
4. Heat may be transferred without the products of combustion coming in contact with the charge.

Yellow flames, in which submicroscopic carbon particles contribute to radiation, have a higher radiating power than blue flames. It is difficult to secure accurate data on the effects of **luminosity.** It has been observed in experimental domestic gas appliances in which the proportion of radiation heat transfer was about 15 per cent of the total that an increase in luminosity was offset by a decrease in flame turbulence and a reduction in the convection heat transfer coefficient.

In **industrial furnaces,**[1,2,18] a luminous gas flame is formed by non-premix types of burners, so that the gas and air layers are distinct and must diffuse into each other to support combustion. In this way, a large area of more highly radiating yellow flame is produced. While the theoretical flame temperature may be the same as with a blue flame, the higher emissivity and greater area of a yellow flame result in improved radiation heat transfer.

Table 2-99b Results of Gas-Fired Burner Radiation Studies[44]

Type of infra-red source*	Operating red brightness temperature, °F	Total normal radiation, Btu/hr-sq ft†	GIR factor	Portion of energy radiated between 2–6 μ, per cent
Gas-fired				
Atmospheric	1400–1650	17,800–32,600	0.51–1.06	69.2–71.5
Powered	1500–2300	18,000–62,800	.63–1.17	49.5–61.2
Catalytic	600–800	800– 3,300	0.36–0.65	36.0–50.8
Electric				
Heat lamp	2950–3150	12,000–32,000	...	36.26–48.10
Quartz lamp	3000–3500	40,560–63,089	...	29.17–39.28
Quartz tube	1400–1600	10,580–23,650	...	64.00–67.00

* The gas-fired radiant tube and the electric resistance heater are not included, since the quality and quantity of the energy radiated by them are functions of emissivity and temperature only.

† The sum of the energy radiated at all wave lengths between 14 and 16μ in a direction normal to the burner surface.

Table 2-100 Absorption Wave Lengths of Some Common Loads[44]

Product	Major absorption wave lengths, μ
Paints	2.8–3.5, 5.7–6.2, 7.5–10.0
Printing inks	3.5, 5.8, 6.8–10.0, 13.75
Oils	3.5, 5.75, 8.0–9.0, 13.80
Meats	2.8–3.4, 5.7–6.8, 7.8–9.0, 13.5
Vegetables	1.9, 2.8–3.4, 5.5–6.5, 8.5–10.0

Gas Infra-Red Radiation (GIR) Factor.[44] Essentially a burner emissivity factor, the GIR factor can exceed unity since it combines (1) the emissivity of the burner material of construction, (2) the emissivity of the layer of hot gases on the surface of the burner (5 to 18 per cent of energy input) and the port area and configuration of the burner, and (3) a factor accounting for some simplifying assumptions made in deriving the GIR factor. Table 2-99a gives radiation data for 13 gas-fired infra-red burners. Table 2-99b summarizes the data of Table 2-99a and compares them to the output of electrical units. Table 2-100 details the major absorption wave lengths of some common loads. The study[44] also showed the effect of different burner-coating materials on energy emission.

Condensation

Pure Vapors. Heat transfer by condensation of vapors at constant temperature occurs by recovery of the latent heat of vaporization of those vapors. Condensation in the absence of radiation can only take place if the wall temperature is less than the saturation temperature of the vapor at the prevailing pressure. Two important types of condensation are *film-type*, in which the liquid wets the entire surface, and *dropwise*, in which the liquid forms into drops leaving a bare surface between drops. For film-type condensation, the rate of heat transfer is given by:

$$q = h_m A_w (t_v - t_w) \qquad (24)$$

where:

q = heat transferred, Btu per hr
h_m = average film coefficient between vapor and wall, Btu per hr-sq ft-°F
A_w = wall area, sq ft
t_v = saturation temperature of vapors, °F
t_w = surface temperature of wall, °F

Film coefficients of 200 to 400 are recommended for the following organic vapors condensing at one-atmosphere pressure with cooling water at ordinary temperatures:[1] benzene, carbon tetrachloride, dichloromethane, dichlorodifluoromethane, diphenyl, ethyl alcohol, heptane, hexane, methyl alcohol, octane, toluene, and xylene.

Ammonia condenses with a mean film coefficient h_m = 1000. Mixtures of organic vapors and steam forming immiscible condensates give h_m of 250 to 750, increasing with an increasing proportion of steam. A linear variation from organic film coefficient to that of pure steam may be assumed. For film-type condensation of steam on clean tubes the value of h_m ranges from 1000 to 3000. The use of traces of organic compounds[1] to promote dropwise condensation of steam gives values of h_m up to, and sometimes greater than, 12,000. Because of the difficulty in maintaining clean surfaces in plant service, the use of these high coefficients is not advisable for general design work. To determine the latent heats for condensation of vapors, Fig. 2-63 may be employed.

It has been suggested that neither true laminar nor fully developed turbulent flow exists in film. See Reference 42 for graphic relationships for the calculation of heat transfer coefficients by this theory.

Vapor in Presence of Noncondensable Gas. The presence of a noncondensable gas in a condensing mixture

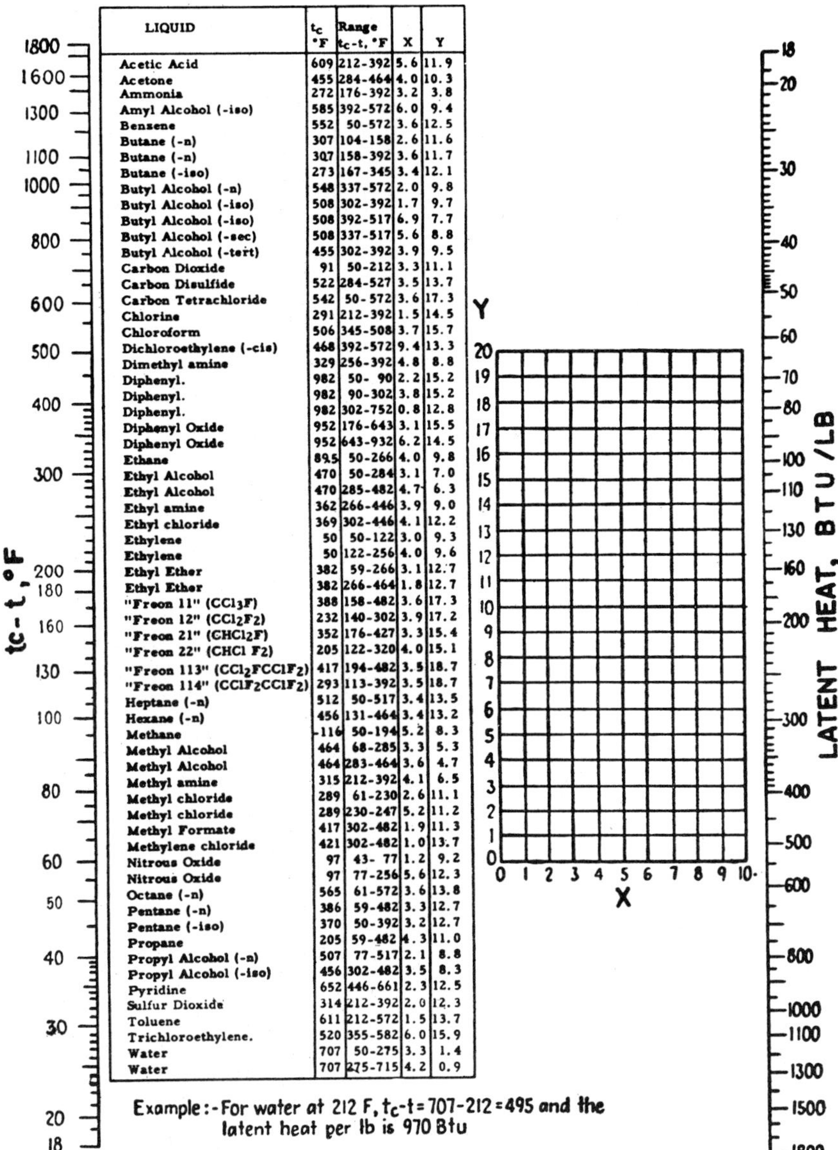

LIQUID	t_c °F	Range $t_c - t$, °F	X	Y
Acetic Acid	609	212-392	5.6	11.9
Acetone	455	284-464	4.0	10.3
Ammonia	272	176-392	3.2	3.8
Amyl Alcohol (-iso)	585	392-572	6.0	9.4
Benzene	552	50-572	3.6	12.5
Butane (-n)	307	104-158	2.6	11.6
Butane (-n)	307	158-392	3.6	11.7
Butane (-iso)	273	167-345	3.4	12.1
Butyl Alcohol (-n)	548	337-572	2.0	9.8
Butyl Alcohol (-iso)	508	302-392	1.7	9.7
Butyl Alcohol (-iso)	508	392-517	6.9	7.7
Butyl Alcohol (-sec)	508	337-517	5.6	8.8
Butyl Alcohol (-tert)	455	302-392	3.9	9.5
Carbon Dioxide	91	50-212	3.3	11.1
Carbon Disulfide	522	284-527	3.5	13.7
Carbon Tetrachloride	542	50-572	3.6	17.3
Chlorine	291	212-392	1.5	14.5
Chloroform	506	345-508	3.7	15.7
Dichloroethylene (-cis)	468	392-572	9.4	13.3
Dimethyl amine	329	256-392	4.8	8.8
Diphenyl.	982	50-90	2.2	15.2
Diphenyl.	982	90-302	3.8	15.2
Diphenyl.	982	302-752	0.8	12.8
Diphenyl Oxide	952	176-643	3.1	15.5
Diphenyl Oxide	952	643-932	6.2	14.5
Ethane	89.5	50-266	4.0	9.8
Ethyl Alcohol	470	50-284	3.1	7.0
Ethyl Alcohol	470	285-482	4.7	6.3
Ethyl amine	362	266-446	3.9	9.0
Ethyl chloride	369	302-446	4.1	12.2
Ethylene	50	50-122	3.0	9.3
Ethylene	50	122-256	4.0	9.6
Ethyl Ether	382	59-266	3.1	12.7
Ethyl Ether	382	266-464	1.8	12.7
"Freon 11" (CCl_3F)	388	158-482	3.6	17.3
"Freon 12" (CCl_2F_2)	232	140-302	3.9	17.2
"Freon 21" ($CHCl_2F$)	352	176-427	3.3	15.4
"Freon 22" ($CHClF_2$)	205	122-320	4.0	15.1
"Freon 113" (CCl_2FCClF_2)	417	194-482	3.5	18.7
"Freon 114" ($CClF_2CClF_2$)	293	113-392	3.5	18.7
Heptane (-n)	512	50-517	3.4	13.5
Hexane (-n)	456	131-464	3.4	13.2
Methane	-116	50-194	5.2	8.3
Methyl Alcohol	464	68-285	3.3	5.3
Methyl Alcohol	464	283-464	3.6	4.7
Methyl amine	315	212-392	4.1	6.5
Methyl chloride	289	61-230	2.6	11.1
Methyl chloride	289	230-247	5.2	11.2
Methyl Formate	417	302-482	1.9	11.3
Methylene chloride	421	302-482	1.0	13.7
Nitrous Oxide	97	43-77	1.2	9.2
Nitrous Oxide	97	77-256	5.6	12.3
Octane (-n)	565	61-572	3.6	13.8
Pentane (-n)	386	59-482	3.3	12.7
Pentane (-iso)	370	50-392	3.2	12.7
Propane	205	59-482	4.3	11.0
Propyl Alcohol (-n)	507	77-517	2.1	8.8
Propyl Alcohol (-iso)	456	302-482	3.5	8.3
Pyridine	652	446-661	2.3	12.5
Sulfur Dioxide	314	212-392	2.0	12.3
Toluene	611	212-572	1.5	13.7
Trichloroethylene.	520	355-582	6.0	15.9
Water	707	50-275	3.3	1.4
Water	707	275-715	4.2	0.9

Fig. 2-63 Latent heats of vaporization.[4]

reduces the film coefficient by reducing the concentration of vapor in the gas film adjacent to the liquid film of condensate. Common examples of this situation are the condensation of air–steam or gas–steam mixtures, and the mixing of hydrocarbons on surfaces at a temperature which will remove only the heavier constituents. Coefficients are shown in Table 2-101 for **condensing hydrocarbon vapors** outside tubes. In general, the condensation to mixed vapors is a complex problem for which reliance must be placed on the organization furnishing the condenser or similar item of equipment.

Direct Contact of Liquid and Gas. The problems of fouling factors and low heat transfer rates encountered in the design of tubular condensers may be eliminated by direct contact condensing or scrubbing of gases. In this method, sprays of cold liquid are introduced into the hot gas stream to cool the gas by direct contact. The condensed components of the gas stream may be soluble or insoluble in the liquid, depending on the nature of the two streams. In this process, very little of the cooling liquid employed will be lost if the dew point of the liquid in the gas at all points in the tower is higher than the liquid temperature. Under these conditions, most of the heat transfer is by convection.

Table 2-101 Film-Transfer Coefficients for Condensing Oil Vapor Outside Tubes

(Btu/hr-sq ft-°F)

Material	No steam	With steam, but steam does not condense with oil
Natural gasoline	210–250	170–210
Gasoline	180–220	130–170
Kerosene	150–190	100–120
Gas oil	130–180	60–70
Wax distillate	110–160	50–60

Boiling Liquids

Heat transfer coefficients in pure liquids can be estimated by the laws of free convection if the temperature of the hot surface is less than the boiling point of the liquid. When bubbles are formed on the solid surface, caused either by dissolved gas or by liquid vapor, the free convection equations do not apply.

An increase in temperature difference between surface and boiling liquid increases the film coefficient, slowly at first and then faster as agitation becomes more violent. It reaches a maximum and then decreases as the heat transfer surface becomes vapor-bound or insulated with a continuous layer of vapor which prevents contact of the liquid. The boiling film coefficient improves if the liquid can *wet* the surface, and may be higher for rough surfaces than for smooth. A small amount of scale can cause an appreciable decrease in the boiling film coefficient.

Flux is defined as the overall heat transfer coefficient *times* the temperature difference, Btu per hr-sq ft on the boiling side. For toluene the flux is 26,500 Btu per hr-sq ft at a Δt of 53°F, and for heptane boiling on a copper surface at atmospheric pressure the maximum flux is 53,000 at a Δt of 55°F.[4]

In the design of vaporizers and reboilers, use a maximum allowable flux of 20 MBtu per hr-sq ft for forced circulation, and 12 MBtu for vaporizing organics by natural circulation.[2] For water or low concentration aqueous solution, the maximum allowable flux for either forced or natural circulation is 30,000. Maximum allowable film coefficient for boiling organics is 300 Btu per hr-sq ft-°F, and 1000 Btu for water or aqueous solutions. Both values apply to either type of circulation and should also apply to LP-gas vaporizers.

GENERAL DESIGN CONSIDERATIONS FOR A HEAT TRANSFER APPARATUS

Suggestions for routing of fluids[20] and a procedure for shell and tube exchanger design[21] were reported. Short-cut methods, which give practical answers and eliminate much tedious calculation, are frequently possible. Some of the general considerations follow.

Steady State Heat Transfer

At any given point in an exchanger, the local temperature gradient controls local heat transfer; but when an entire unit containing flowing fluids must be analyzed, the temperature and temperature difference of both fluids may vary continuously with position. The heat transferred in an entire heat exchanger may be expressed as:

$$Q = UA\Delta t \tag{25}$$

where: Q = heat transferred, Btu per hr
A = heat transfer surface area, sq ft
U = overall heat transfer coefficient
Δt = mean overall temperature difference

Neither the overall coefficient nor the temperature difference is entirely independent, because the values of one may depend on the assumptions made about the other. In a simple form of heat exchanger with two fluids in **pure parallel flow or counterflow,** the **logarithmic mean temperature difference LMTD,** is applicable:

$$\Delta t_0 \text{ or LMTD} = \left(\frac{\Delta t_1 - \Delta t_2}{\ln \dfrac{\Delta t_1}{\Delta t_2}}\right) = \left(\frac{\Delta t_1 - \Delta t_2}{2.3 \log_{10} \dfrac{\Delta t_1}{\Delta t_2}}\right) \tag{26}$$

In the above equation Δt_1 and Δt_2 are the higher and lower terminal temperature differences, respectively. The difference between counterflow and parallel flow may be seen from Fig. 2-64, in which T refers to the hotter fluid, t to the colder fluid, and L to the length of heat transfer surface. For many types of heat transfer, a counterflow exchanger is preferred because it permits the recovery of greater amounts of heat from the hot fluid. When one of the fluids is isothermal—for example, condensing steam—the LMTD is the same for either flow assumption. When there is a large temperature drop in a small amount of gas and the weight of gas is small relative to the weight of other fluid, there is little difference between parallel flow and counterflow.

These principles are applied to equipment heated by gas flames, such as domestic gas heating furnaces, by reducing the theoretical flame temperature to account for radiation and convection losses from the flame. This corrected temperature can then be used in calculating the LMTD.

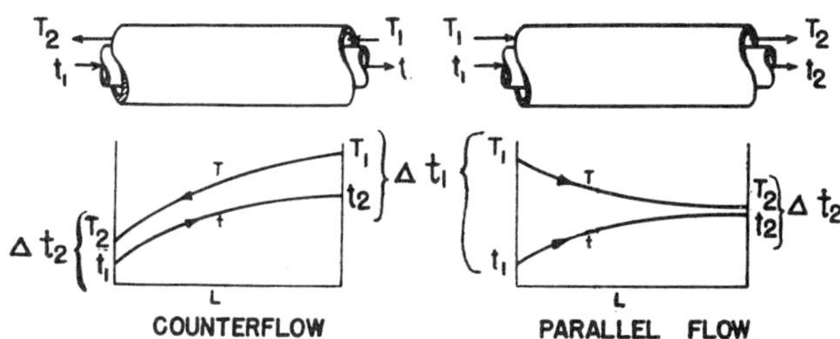

Fig. 2-64 Comparison of counterflow and parallel flow heat transfer.

Example: A warm air heating furnace burns gas at a corrected flame temperature of 2400 F, which cools to 500 F at the flue outlet while heating air from 70 F to 150 F. Compute the LMTD in parallel and counterflow:

(*a*) *Parallel flow*

$$\text{LMTD} = \frac{(2400 - 70) - (500 - 150)}{2.3 \log \dfrac{(2400 - 70)}{(500 - 150)}} = 1043 \text{ F}$$

(*b*) *Counterflow*

$$\text{LMTD} = \frac{(2400 - 150) - (500 - 70)}{2.3 \log \dfrac{(2400 - 150)}{(500 - 70)}} = 1100 \text{ F}$$

Although, as this example shows, the LMTD in a warm air gas furnace is sensitive (five per cent variation) to the type of flow, many designs use parallel flow to avoid continuous cooling of moist combustion gases below their dew point at the flue gas outlet.

LL Slide Rule Computation of LMTD:

Set the difference ($\Delta t_1 - \Delta t_2$) on the C scale opposite ($\Delta t_1/\Delta t_2$) on the proper LL scale and read the LMTD on the C scale over e on the LL scale. Thus, to find the logarithmic mean of 2000 and 50, set 2000 − 50, or 1950, on the C scale opposite 2000/50, or 40, on the LL scale, and over e on the LL scale read the logarithmic mean on the C scale as 529.

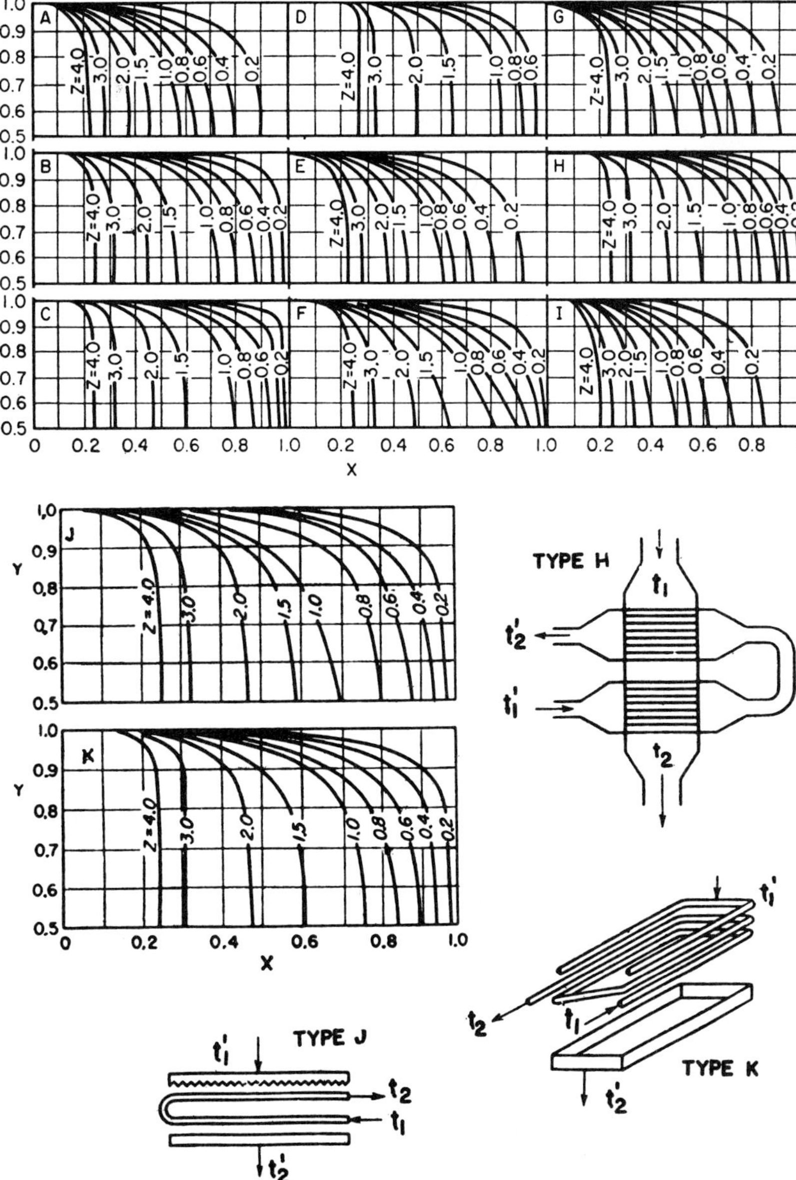

Fig. 2-65 Correction factor Y, applied to counterflow heat exchanger LMTD for the following arrangements:[23] A, one shell pass and two* tube passes; B, two shell passes and four* tube passes; C, three shell passes and six* tube passes; D, four shell passes and eight* tube passes; E, crossflow, one shell pass and one tube pass; F, crossflow, both fluids unmixed, single pass; G, crossflow, shell fluid mixed, tube fluid unmixed, single pass; H, crossflow, shell fluid mixed, tube fluid unmixed, shell fluid flows across second and first passes in series; I, same as H, but shell fluid crosses first and second passes in series; J, crossflow (drip type), two horizontal passes with U-bend connections (trombone type); K, crossflow (drip type), helical coils with two turns.

* Or multiples thereof.

If one terminal temperature difference is not more than twice the other terminal temperature difference, the arithmetic mean $(\Delta t_1 + \Delta t_2)/2$ may be used, because it will be within four per cent of the LMTD.

Heat Transfer with Variable Coefficient

The use of the LMTD as indicated above is predicated on the assumption of a constant value of U, the overall heat transfer coefficient. When U varies appreciably from one end of the exchanger to the other in parallel or counterflow, the heat transferred in Btu per hr is calculated using the following equation:[22]

$$Q = A_m \left(\frac{\Delta t_1 U_2 - \Delta t_2 U_1}{2.3 \log \dfrac{\Delta t_1 U_2}{\Delta t_2 U_1}} \right) \tag{27}$$

where:
Q = heat transferred, Btu per hr
A_m = mean heat transfer surface area, sq ft
Δt_1 = temperature difference at inlet
Δt_2 = temperature difference at outlet
U_1 = overall heat transfer coefficient at inlet
U_2 = overall heat transfer coefficient at outlet

In practice, many shell and tube and other heat exchanger layouts are neither pure parallel nor pure counterflow, but mixtures of the two. Correction factors, Y, to obtain the true temperature difference from the LMTD for counterflow are given in Fig. 2-65. These charts are applied by means of the equation:

$$Q = U_m A_m \Delta t_m = U_m A_m Y \Delta t_0 \tag{28}$$

or

$$\Delta t_m = Y \Delta t_0 \tag{29}$$

where:

Q = heat transferred, Btu per hr
U_m = mean (arithmetic or logarithmic) overall heat transfer coefficient, Btu per hr-sq ft-°F
Δt_m = true mean overall temperature difference, °F
A_m = mean heat transfer surface area, sq ft (see Eq. **3**)
Δt_0 = counterflow LMTD, °F (from Eq. **26**)

For example, assume a type C reversed current exchanger with three shell passes and six tube passes. The hot fluid enters at 300 F and leaves at 150 F, while the cold fluid enters at 100 F and leaves at 200 F. In counterflow, the LMTD is (from Eq. **26**):

$$\Delta t_0 = \frac{(300 - 200) - (150 - 100)}{\ln [(300 - 200)/(150 - 100)]} = 72 \text{ F}$$

The correct temperature difference to employ is found as follows:

$$X = \frac{200 - 100}{300 - 100} = 0.5$$

$$Z = \frac{300 - 150}{200 - 100} = 1.5$$

From type C in Fig. 2-65, $Y = 0.93$. Therefore, the true mean temperature difference, Δt_m, is:

$$Y\Delta t_0 = 0.93 \times 72 = 67 \text{ F}$$

Heat Transfer for Unknown LMTD

When the overall coefficient U is known or can be estimated, the trial-and-error procedure necessary to determine outlet temperatures and heat flow rate can be circumvented by use of charts found in *Compact Heat Exchangers* by W. M. Kays and A. L. London, published by McGraw-Hill, New York, 1958. These charts enable the exploration of a great variety of heat transfer apparatus without resort to their LMTD.

Overall Heat Transfer Coefficients

The overall heat transfer coefficient indicates the net effect of all resistances to heat flow in an exchanger. In general, when two fluids are flowing on either side of a wall, the overall coefficient is determined from the equation:

$$\frac{1}{U} = \frac{1}{h_i} + \frac{x}{k} + \frac{1}{h_o} \tag{30}$$

where:

U = overall heat transfer coefficient, Btu per hr-sq ft-°F
h_i = film coefficient for inside of wall, Btu per hr-sq ft-°F
h_o = film coefficient for outside of wall, Btu per hr-sq ft-°F
x = thickness of wall, ft
k = thermal conductivity coefficient for wall, Btu per hr-sq ft-°F per ft

This equation applies also to clean tubes with thin walls in which the difference in areas is negligible. However, in the case of two fluids flowing on either side of a thick tube wall on which appreciable dirt or scale has accumulated, the overall coefficient and area must be computed together:

$$\frac{1}{UA} = \frac{1}{h_iA_i} + \frac{x}{kA_{avg}} + \frac{1}{h_oA_o}$$

where: A = area of heat transfer surface, sq ft
A_i = inside area of tubes, sq ft
A_{avg} = mean area of tubes, sq ft
A_o = outside area of tubes, sq ft

For a gas-to-gas heat exchanger or for any exchanger in which the conductivity of the heat transfer surface wall is large relative to that of the fluid films:

$$U = \frac{1}{(1/h_o) + (1/h_i)} = \frac{h_oh_i}{h_o + h_i} \tag{32}$$

This relationship applies also to gravity and forced air gas furnaces, and to all similar cases in which **film coefficients** are low relative to the conductivity of the metal. When the resistance, x/k, of the metal wall is low and one film coefficient is large relative to the other, as is true for boiling water with a flame or for a hot gas stream, the overall coefficient approaches the value of the smaller film coefficient. This effect makes it simple to obtain rough estimates of overall coefficients in many types of flame or gas heated exchangers, because the value will be approximately equal to the gas film coefficient.

Fouling Resistances. The overall coefficient of a heat exchanger for fluids which may cause corrosion, scaling, fouling, or dirt deposition on a surface must take account of these effects or the equipment will be undersized for continuous maintenance-free service. Fouling resistances for various fluids are combined with the film coefficients in the expression for the overall coefficient. The deposition of dirt on the in-

Table 2-102 Overall Coefficients for Coils Immersed in Liquids[1]

(Btu/hr-sq ft-°F)

Substance inside coil	Substance outside coil	Coil material	Agitation	U
Steam	Water	Lead	Agitated	70
Steam	Sugar and molasses solutions	Copper	None	50–240
Steam	Boiling aqueous solution	...	...	600
Cold water	Dilute organic dye, intermediate	Lead	Turboagitator at 95 rpm	300
Cold water	Warm water	Wrought iron	Air bubbled into water surrounding coil	150–300
Cold water	Hot water	Lead	0.40 rpm, paddle stirrer	90–360
Brine	Amino acids	...	30 rpm	100
Cold water	25% oleum at 140 F	Wrought iron	Agitated	20
Water	Aqueous solution	Lead	500 rpm, sleeve propeller	250
Water	8% NaOH	...	22 rpm	155
Steam	Fatty acid	Copper (pancake)	None	96–100
Milk	Water	...	Agitation	300
Cold water	Hot water	Copper	None	105–180
60 F water	50% aqueous sugar solution	Lead	Mild	50–60
Steam and hydrogen at 1500 psi	60 F water	Steel	...	100–165

Note: Chilton, T. H., Drew, T. B., and Jebens, R. H. *Ind. Eng. Chem.* 36:510, 1944. Gives film coefficients for heating and cooling agitated fluids using a coil in a jacketed vessel.

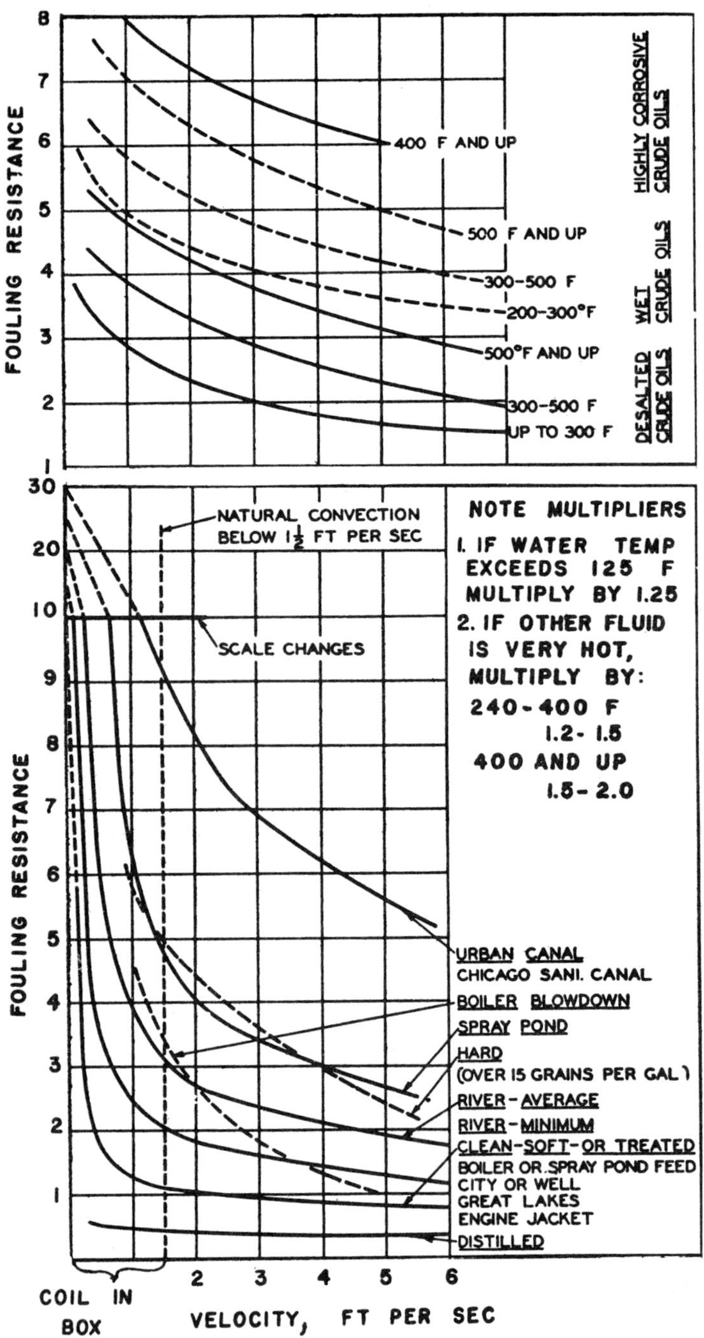

Fig. 2-66 (left) Fouling resistances for crude oils and water[25] in 1000 hr-sq ft-°F per Btu. To use in Eq. 33, divide values shown by 1000.

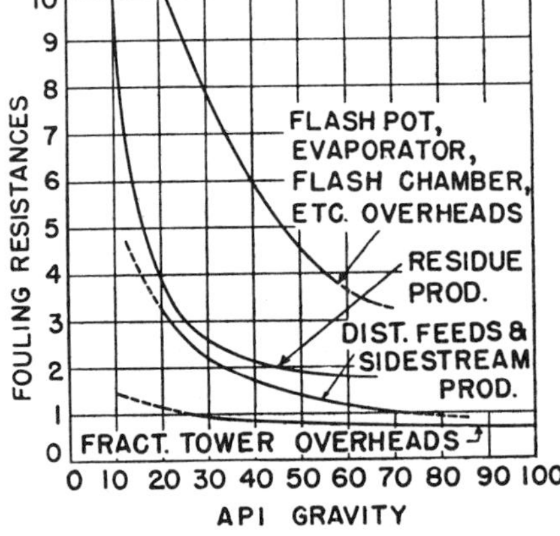

Fig. 2-67 (above) Fouling resistances of petroleum and natural gas products[25] in 1000 hr-sq ft-°F per Btu (topping, vacuum rerun, treating, and cracking units). To use in Eq. 33, divide values shown by 1000.

side or outside of tube surfaces may be treated by use of appropriate data for the fluids involved.

$$\frac{1}{U_d} = \frac{1}{U_c} + R_{di} + R_{do} \tag{33}$$

where U_d is the overall coefficient for dirty tubes, U_c is the coefficient for the clean condition, and R_{di} and R_{do} are fouling resistances for the inside and outside surfaces of the tubes. Dirt resistances chosen to make 12 to 18 months of continuous heat transfer service possible without cleaning are given in Table 2-107. When dissolved solids are present in the fluid streams, tube surface temperatures likely to cause precipitation, decomposition, or scale deposition should be avoided. Severe corrosion, deposits of coke, salt, or soot, and extremely reactive stocks such as severely cracked stocks may cause fouling resistances more than *twice* as large as those shown in Figs. 2-66 and 2-67.

In cooling a wax-bearing stock the temperature at the surface may be lower than the pour point of the stock even when the main body temperature of the waxy stock is quite high. In such a situation, wax is deposited on the surface, and the fouling resistance may be 25 or larger.[25]

A somewhat similar situation may arise with hard water. Calcium sulfate is most soluble at about 100 F, and at about 145 F a rapid decrease in its solubility occurs. Thus, if the film temperature of the hard water exceeds 145 F, scale will be

Table 2-103 Variations* in Overall Heat Transfer Coefficients
(Btu/hr-sq ft-°F)

Type of heat exchanger	State of controlling resistance: Free convection, U	Forced convection, U	Typical fluid	Typical apparatus
Liquid to liquid	25–60	150–300	Water	Liquid-to-liquid heat exchangers
Liquid to liquid	5–10	20–50	Oil	
Liquid to gas (1 atm)	1–3	2–10	...	Hot-water radiators
Liquid to boiling liquid	20–60	50–150	Water	Brine coolers
Liquid to boiling liquid	5–20	25–60	Oil	
Gas (1 atm) to liquid	1–3	2–10	...	Air coolers, economizers
Gas (1 atm) to gas	0.6–2	2–6	...	Steam superheaters
Gas (1 atm) to boiling liquid	1–3	2–10	...	Steam boilers
Condensing vapor to liquid	50–200	150–800	Steam–water	Liquid heaters and condensers
Condensing vapor to liquid	10–30	20–60	Steam–oil	
Condensing vapor to liquid	40–80	60–300	Organic vapor–water	
Condensing vapor to liquid	...	15–300	Steam–gas mixture	
Condensing vapor to gas (1 atm)	1–3	6–16	...	Steam pipes in air, air heaters
Condensing vapor to boiling liquid	40–100	...	...	Scale-forming evaporators
Condensing vapor to boiling liquid	300–800	...	Steam–water	
Condensing vapor to boiling liquid	50–150	...	Steam–oil	
Condensing vapor to boiling liquid	...	50–400	Steam–organic liquid	Steam-jacketed tubes

* Modified from Lucke, C. E. *Engineering Thermodynamics*, p. 550. New York, McGraw-Hill, 1912. Higher or lower values may be realized under some conditions.

Table 2-104 Overall Coefficients for Heat Exchangers in Petroleum Service
(Btu/hr-sq ft-°F)

Service of exchanger	Fluids: Tubes	Shell	Velocity in tubes, fps	Δt_m, °F	U
Stabilizer reflux condensers	Water	Condensing vapors + residual gas	3.0	13.5	94
			5.0	22	145
		108°–118° API	0.3–0.6	...	55–67*
			0.7	...	98–125*
Partial condensers	39° API crude	58° API gasoline	2.4	147	24
	55° API crude	62° API naphtha	4.4	80	37
	55° API crude	62° API naphtha	6.8	90	48
Stabilizer reboiler	Steam	58° API naphtha	...	33.5	42
Absorber reboiler	Steam	37° API oil	...	41.4	45
Stabilizer reboiler	Steam	67–74° API	...	...	43–183*
Oil preheater	42° API oil	Steam	1.4	32	108
Exchangers	60° API	58° API	1.4	65	74
	63° API	57° API	4.6	69	139
	70–82° API	64–74° API	0.3–0.7	...	18–37*
	70–82° API	67–74° API	0.3–0.7	...	35–45*
	43° API	37° API	1.6	59	33
	39° API crude	13° API residue	3.9	262	19
Coolers	Water	57° API	...	40	52
	Water	44° API	...	97	40
	Water	67–74° API	0.2	...	20*
	Water	67–74° API	0.4–0.7	...	51–53*

* McGiffin, J. Q. *Am. Inst. Chem. Engrs. Trans.* 38:761–790, 1942. All other data are from Higgins, M. B. "Transfer Rates on Heat Exchange Services" (In: ASME *Heat Transfer*, pp. 56–60, New York, 1936).

Δt_m = true mean overall temperature differences, °F.

produced which greatly increases the fouling resistance. This accounts for the fact that water is seldom heated to temperatures exceeding about 120 F, particularly if the other fluid is very hot (produces a high temperature in the water film).[25]

Overall Coefficients for Exchangers

The five tables, Tables 2-102 thru 2-106, contain heat transfer coefficients for a number of combination fluids under a variety of situations. The effects of varying numbers of baffles on heat transfer coefficients in turbine-agitated 30-gallon tanks were studied and found to be negligible.[24]

Overall coefficients for ferrous and nonferrous tubes up to ½ in. thick for film coefficients up to 1000 Btu per hr-sq ft-°F are charted by the International Nickel Co. in *Heat Transfer through Metallic Walls.*

Double-Pipe Exchangers.[22] This design (two concentric pipes) is suitable for small installations where large

Table 2-105 Overall Heat Transfer Coefficients for Miscellaneous Equipment and Materials[1]

(Btu/hr-sq ft-°F)

Type of equipment	Hot material	Cold material	U	Remarks
High-pressure boiler	Molten salt	Boiling water	100–150	
Tubular exchanger	Molten salt	Oil	52–80	
Steam superheater	Molten salt	Steam	70	
Air heater	Molten salt	Air	6	
Catalyst case	Gas	Molten salt	6	Fins on outside of tube
Double-pipe Karbate exchanger	Water	Water	300–500	
Karbate trombone cooler	20° Bé HCl	Water	300	Water $W_1 = 1750$
Karbate tube reboiler	Steam	20% HCl	136	Vertical thermosiphon reboiler
	Steam	35% HCl	472–575	
Double-pipe Pyrex glass exchanger	Air–water vapor	Water	25–75	Cooling water in annulus
using heat exchanger tubing	Water	Water	80–110	
	Condensing steam	Water	100–125	
Glass trombone cooler	50% sugar solution	60 F water	50–60	Sugar solution inside pipe
Glass pipe in trough	20° Bé HCl	Water	25	
Votator	Water	Water	520–1120	Rotor velocity = 300–1900 rpm
Pebble heater	Solid pebbles	Air	4	Heating gases to 1900°F using ½-in. pebbles
		Methane	9	
		Hydrogen	22	
Long-tube vertical evaporator	Condensing steam	Water	300–1200	
Falling-film condenser	Condensing steam	Water	574–2300	Water W_2 = 400 to 21,000 inside tubes
Stainless steel conveyor belt	Molten TNT	50 F air	5–7	Air blowing under and over belt
Partial condenser	Hydrocarbons and chlorinated hydrocarbons	Boiling propane	55–76	Refrigerated condenser
Shell and tube reboiler	Hot water	Hydrocarbons	42–88	Hot water in tubes
Reboiler	Steam	Chlorinated hydrocarbons	67	Clean reboiler, $\Delta t = 12$ F
			20	Same reboiler after several months service, $\Delta t = 96$ F

W_1 = lb/hr-ft of pipe length for each side of pipe; W_2 = lb/hr-ft of periphery.

Table 2-106 Overall Heat Transfer Coefficients for Miscellaneous Substances[1]

(Btu/hr-sq ft-°F)

Fluid inside jacket	Fluid in vessel	Wall material	Agitation	U
Steam	Water	Enameled C. I.*	0–400 rpm	96–120
Steam	Milk	Enameled C. I.	None	200
Steam	Milk	Enameled C. I.	Stirring	300
Steam	Milk boiling	Enameled C. I.	None	500
Steam	Milk	Enameled C. I.	200 rpm	86
Steam	Fruit slurry	Enameled C. I.	None	33–90
Steam	Fruit slurry	Enameled C. I.	Stirring	154
Steam	Water	C. I. and loose lead lining	Agitated	4–9
Steam	Water	C. I. and loose lead lining	None	3
Steam	Boiling SO_2	Steel	None	60
Steam	Boiling water	Steel	None	187
Hot water	Warm water	Enameled C. I.	None	70
Cold water	Cold water	Enameled C. I.	None	43
Ice water	Cold water	Stoneware	Agitated	7
Ice water	Cold water	Stoneware	None	5
Brine, low velocity	Nitration slurry	...	35–58 rpm	32–60
Water	Sodium alcoholate solution	"Frederking" (cast-in-coil)	Agitated, baffled	80
Steam	Evaporating water	Copper	...	381
Steam	Evaporating water	Enamelware	...	36.7
Steam	Water	Copper	None	148
Steam	Water	Copper	Simple stirring	244
Steam	Boiling water	Copper	None	250
Steam	Paraffin wax	Copper	None	27.4
Steam	Paraffin wax	Cast iron	Scraper	107
Water	Paraffin wax	Copper	None	24.4
Water	Paraffin wax	Cast iron	Scraper	72.3
Steam	Solution	Cast iron	Double scrapers	175–210
Steam	Slurry	Cast iron	Double scrapers	160–175
Steam	Paste	Cast iron	Double scrapers	125–140
Steam	Lumpy mass	Cast iron	Double scrapers	75–96
Steam	Powder (5% moisture)	Cast iron	Double scrapers	41–51

* C. I. = cast iron.

Table 2-107 Fouling Resistances, R
(hr-°F-sq ft of outside surface/Btu)

Water

	Water velocity, fps*		Water velocity, fps†	
	Up to 3	Over 3	Up to 3	Over 3
Sea water	0.0005	0.0005	0.001	0.001
Brackish water	.002	.001	.003	.002
Cooling tower and artificial spray pond				
Treated make-up	.001	.001	.002	.002
Untreated	.003	.003	.005	.004
City or well water, Great Lakes	.001	.001	.002	.002
River water				
Minimum	.002	.001	.003	.002
Mississippi, Delaware, Schuylkill, East River, and New York Bay	.003	.002	.004	.003
Chicago Sanitary Canal	.008	.006	.010	.008
Muddy or silty	.003	.002	.004	.003
Hard (over 15 grains/gal)	.003	.003	.005	.005
Engine jacket	.001	.001	.001	.001
Distilled	.0005	.0005	.0005	.0005
Treated boiler feedwater	.001	.0005	.001	.001
Boiler blowdown	0.002	0.002	0.002	0.002

For inside surface fouling, multiply these values by the outside to inside surface ratio.

Industrial Fluids

Oils		Gases and vapors	
Fuel oil	0.005	Mfd. gas, engine exhaust	0.01
Engine lube and transformer oil	.001	Steam (nonoil-bearing)	.0005
Quench oil	.004	Exhaust steam (oil-bearing)	.001
Liquids		Refrigerant vapors (oil-bearing)	.002
Refrigerants, hydraulic fluid	.001	Compressed air	.002
Organic heat transfer media	.001	Organic heat transfer media	0.001
Molten heat transfer salts	0.0005		

Chemical Processing Streams

Acid gases, solvent vapors, stable overhead products	0.001	Stable side draw and bottom liquids	0.001
MEA, DEA, DEG, and TEG solutions	0.002	Caustic solutions	.002
		Vegetable oils	0.003

Natural Gas–Gasoline Processing Streams

Natural gas and overhead products	0.001	Rich oil	0.001
Lean oil	0.002	Nat. gasoline and LP-gases	0.001

Oil Refinery Streams

Atmospheric tower overhead vapors, gasoline, naphtha, light distillates, kerosene	0.001	Vacuum overhead vapors, light gas oil	0.002
		Heavy gas oil	.003
		Heavy fuel oils	.005
		Asphalt and residuum	0.010

Cracking and Coking Unit Streams

Overhead vapors, light cycle oil	0.002	Heavy coker gas oil	0.004
Heavy cycle oil, light coker gas oil	0.003	Bottoms slurry oil (4.5 fps min)	.003
		Light liquid products	0.002

Catalytic Reforming and Hydrodesulfurization Streams

Reformer charge	0.002	Overhead vapors	0.001
Reformer effluent	.001	Liquid product over 50°API	.001
Hydrodesulfurization charge and effluent	0.002	Liquid product 30–50°API	0.002

Light Ends Processing Streams

Overhead vapors and gases	0.001	Alkylation trace acid streams	0.002
Liquid products	.001	Reboiler streams	0.003
Absorption oils	0.002		

Lube Oil Processing Streams

Stock and solvent mix feeds	0.002	Extract and wax slurries	0.003
Solvent and raffinate	0.001	Asphalt	.005
		Refined lube oil	0.001

Crude Oil Streams

	0–199 F		200–299 F		300–499F		500 F and over	
Velocity, fps	Under 2	4 and over	Under 2	4 and over	Under 2	4 and over	Under 2	4 and over
Dry	0.003	0.002	0.003	0.002	0.004	0.002	0.005	0.003
Salt	0.003	0.002	0.005	0.004	0.006	0.004	0.007	0.005

* At heating medium temperatures up to 240 F and water temperatures of 125 F or less.
† At heating medium temperatures from 240 F to 400 F and water temperatures over 125 F. If the heating medium temperature is over 400 F and the cooling medium is known to cause scaling, these ratings should be modified accordingly.

quantities of heat are not required. Maximum flow limits are dictated by pressure drop and power consumption, and minimum flow limits by the desirability of having turbulent flow. Linear velocities for gases should be in the range of 20 to 100 fps at pressures close to atmospheric. Liquid linear velocities should be 3 to 6 fps.

For convenience, a 1-in. pipe will be used as a datum. Film coefficients for 1-in. pipe with 3 fps flow in Btu per hr-sq ft-°F are approximately:

Water	600
Saturated brine	500
98% sulfuric acid	110
Light oils	150
Alcohols and light organic liquids	200

Figures 2-68 and 2-69 show the effects of changes in velocity and pipe diameter. The coefficient for the annulus is based on its *equivalent* diameter. The effect of temperature is neglected for the sake of simplicity except at low temperatures, where it may be very important.

Gases close to atmospheric pressure with a velocity of 20 fps in 1-in. pipe have a film coefficient of 5 to 8 for the molecular weight range from 2 to 70. **Hydrogen** is a special case, requiring a velocity of about 100 fps to assure turbulence.

Temperature changes reduce the film coefficient about 10 per cent for every 100 F above a datum of 100 F. The reverse is true for temperatures below 100 F.

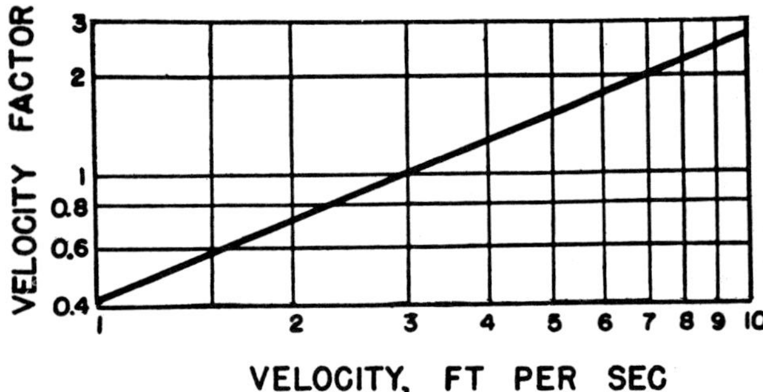

Fig. 2-68 Velocity correction factor for double-pipe exchanger. Apply to estimated film coefficient when velocity differs greatly from 3.0 fps.

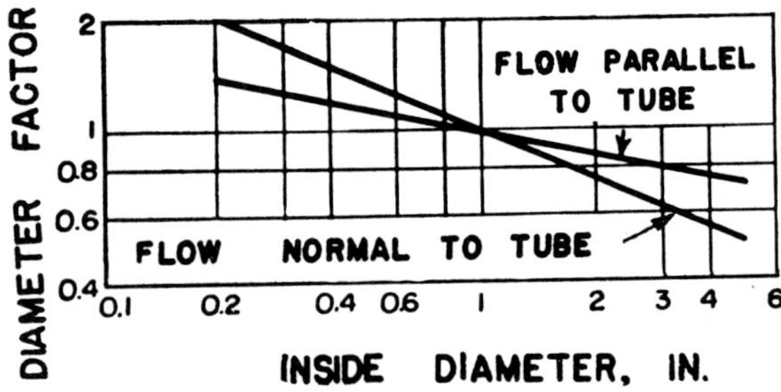

Fig. 2-69 Diameter correction factor for double-pipe exchanger. Apply to estimated film coefficient when diameter differs greatly from 1.0 in.

Shell and Tube Exchangers. Ease of removal and distortion at high temperatures limit tube lengths. The ratio of tube length to shell diameter is usually approximately 6 to 1, although considerable leeway may be given. Baffles give desired velocity distribution. Although shell-side fluid velocities will be reasonable, they will be lower than tube-side velocities, and pressure drop will have to be taken into account.

Allowing for fouling resistances from Fig. 2-66 and Table 2-107, calculate the overall coefficient from Eq. 33. Typical examples of overall coefficients in exchangers of this type are:

Water to water	100 to 150
Gas to gas	2 to 4
Gas to water	20 to 40
Water to organic liquid	59 to 100

Cascade coolers consist of banks of horizontal pipes with water dripping evenly from pipe to pipe. The cooling is accomplished by convection or, in the cases of liquids or gases at high temperatures, by evaporation. Pipe size is usually 2 to 4 in. Rate of water flow to the top pipe should be between 1 and 6 gpm per ft of pipe for convection cooling. For evaporative cooling, 0.2 gpm per ft of pipe is usually adequate.

Heat transfer coefficient inside the pipe is the same as for a double-pipe exchanger. Increases in both evaporative cooling and pipe diameter tend to decrease the coefficient.

Regardless of individual coefficients, an overall coefficient of 30 to 50 should be used for preliminary design of liquid cascade coolers, because experience has shown that most coolers of this type operate in this range of coefficients. For evaporative cooling of gases, an overall coefficient of 4 to 10 may be used.

Jacketed vessels find most use when only a moderate amount of heat exchange is required.

Immersed Coils (except gas fired). Immersed coils are relatively inexpensive and are particularly suited to a batch process or a reaction vessel. A coil of surface equal to the jacketed vessel surface will increase the total heat transfer by about 125 per cent. Tube sizes most frequently used for coils are in the range of ¾ to 2 in.

For a liquid flowing thru a coil the approximate value of the inside film coefficient will be about 20 per cent higher than the value of the coefficient for flow in a straight pipe.

With moderate agitation and a linear velocity of about 2 fps, the film coefficient will be about 600 for water and about 200 for most organic liquids. Pipe coils are often used for steam heating of a liquid. When they are, most of the resistance to heat transfer is in the outside liquid and dirt films.

Heat Transfer to Boiling Liquids. Design of equipment for boiling liquids introduces the factor of the critical temperature across the boiling film. For most liquids, the drop is between 70 and 100 F.

For clean horizontal pipes or plates an overall coefficient of about 250 would be reasonable for liquids such as benzene or alcohol when the overall temperature difference is 50 to 70°F. Allowing for fouling, the coefficient would be 50 to 100. Avoid using temperature differentials of less than 50°F.

Boiling coefficients for cylinders will run about 25 per cent higher than those for flat plates and coils. Coefficients for forced circulation evaporators and boilers parallel those for liquids flowing thru pipes at high velocities and may be estimated on the same basis.

Another factor which should not be overlooked is the effect of temperature on the boiling coefficient. This may be of importance in designing for vacuum operation. Coefficients given in the literature usually refer to liquids boiling at atmospheric pressure. For most liquids, the film coefficient for a fixed temperature difference will double for each 10°F above the normal boiling point. A decrease of 10°F below the normal boiling point halves the coefficient. This relationship gives a straight line on a semilog plot.[26]

Condensers. Some typical film coefficients for vapors condensing on horizontal tubes, measured in Btu per hr-sq ft-°F, are: most organic vapors, 200 to 400; ammonia, 1000; and steam, 1000 to 3000. Condensation inside tubes appears to give coefficients of the same order of magnitude as condensation outside tubes, but the drawbacks of liquid hold-up will be a factor.

In many condensers the **water-side film** will be the controlling factor. Attention should be directed to achieving a good water velocity and keeping the tubes free from dirt. Water-side coefficients may be estimated by the methods previously outlined.

The condensation of mixed vapors of immiscible liquids, as occurs in steam distillation, will show coefficients which vary almost linearly from organic to steam coefficients, depending on the relative proportions of the two phases.

The presence of a noncondensable gas in the vapors passing thru a condenser will have a profound effect on the overall heat transfer coefficients at every point in the condenser. The design of a condenser for this case is quite complicated there seems to be no short method for approximating the surface requirements.

Gas Heaters with Tube Banks. When gas pressures are sharply reduced, it is necessary to raise the temperature of the expanded gas to prevent regulator freeze-ups or hydrate formation. For rough approximations, 100-psi pressure drop will cause 7°F natural gas temperature drop. A common method for gas heating is to pass the gas across a number of rows of heated pipes or tubes. Most of the resistance to heat transfer is in the gas film, and the overall coefficient observed will not differ very much from the gas film coefficient. The number of rows of pipes will influence this coefficient to some extent since it affects the amount of **turbulence.** Any advantage caused by increase in the number of rows is said to level out at four rows.

For air and 1-in. tubes, a velocity of 10 fps past the tube will give a coefficient of about 8; at 60 fps, the coefficient increases to 20. Four rows in the bank will have little effect at low velocity, but at 50 to 100 fps the coefficient may be about 50 per cent greater. Coefficients will be better for gases lighter than air and not as good for gases denser than air. Figure 2-69 may be used to estimate the effect of changing tube diameter.

Since higher gas temperatures mean poorer coefficients, it is suggested that for waste heat boilers the overall coefficient be reduced about 40 per cent. An overall coefficient of 6 to 10 would be reasonable for flue gas velocities of 50 to 100 fps.

Economics of Pressure Drop

High film coefficients are characteristic of turbulent flow; baffling on the shell sides of tubes minimizes channeling and promotes turbulence and mixing. Increased cost of pumping, however, must be balanced against the improvement in the overall heat transfer coefficient and the decrease in the size of the exchanger before determining the optimum velocity and pressure drops.

Use of Longitudinal Finned Tubes (LFT)

When double-pipe exchangers are to be used for two fluids and the ratio of film coefficients is 3 to 1 or greater, the use of tubes with external longitudinal fins will reduce the size or increase the efficiency of the unit. The benefit is most pronounced if both coefficients are low. To obtain the advantages of these tubes, the fluid with the low coefficient must be passed over the fins. Overall coefficients of heat transfer for typical externally finned LFT applications are given in Table 2-108.

The coefficients given in Table 2-108 apply to the external finned surface of the tube. Thus, the entire outside tube area, fins and all, must be used as the area in the relationship $Q = UA\,\Delta t$.

Table 2-108 Typical Coefficients for LFT Sectional Exchangers[27]
(Btu/hr-sq ft external surface-°F)

Residuum to crude oil	10–12
Bunker C fuel oil to steam	12–15
Vegetable oil to water	13
Paraffin wax to water	18
Vegetable oil to steam	17–22
Heavy gas oil to heavy gas oil	20
Crude oil to light gas oil	29
90% Ethyl alcohol to water	29
Gas oil to steam	30–40
50% Caustic solution to water	35
Naphtha or kerosene to water	38
Light hydrocarbons to steam	40–80
Natural gas to kerosene	5
Air to water	10–15
Air to steam	15–20
Hydrogen to water	15–20
Natural gas to water	25–30
Bunker C fuel oil to steam (tank heater)	8–10
Light lube oil to steam (tank heater)	14

Heating of Solids

Heat transfer to solids is not readily calculated by direct methods, particularly for complicated shapes. For heating or cooling various geometrical shapes, Grober[28] offers the following method of determining temperatures at different time intervals. In Parts a and b of Fig. 2-70, the abscissas are $(f/k)L$, or YL, and in Parts c to f inclusive, $(f/k)R$, or YR, where f = overall coefficient of heat transfer between surrounding medium and body, Btu per hr-sq ft-°F; k = thermal conductivity, Btu per hr-sq ft-°F per ft; R = radius, ft; L = thickness, if heat is flowing thru one side only, or ½ thickness, if heat is flowing thru both sides, ft; and $Y = f/k$. The parameter of the curves is hF/R^2 or hF/L^2, where h = diffusivity, sq ft per hr, = k/(specific heat)(density); and F = time, hr.

The ordinates represent the difference in temperature between the body and the medium which it is approaching as a limit, expressed as a fraction of the initial temperature difference.

Example: An 8-in. steel sphere at a temperature, T_0, of 550 F is quenched in an oil bath, which is at T_∞ = 85 F. What is the temperature, T, of the surface of the sphere at the end of three minutes, if f = 100 Btu per hr-sq ft-°F?

Solution:

$$R = 0.333 \text{ ft} \qquad YR = 100 \times 0.333 \div 34 = 0.98$$

$$hF/R^2 = 0.42 \times (3/60) \div (0.333)^2 = 0.189$$

From Fig. 2-70, Part c, the factor is $0.53T_c$. Substituting this factor into the equation:

$$(T_0 - T_\infty)T_c = T - T_\infty$$

results in $0.53(550 - 85) = T - 85$.

Therefore, T = 331 F.

HEAT TRANSFER APPLICATIONS IN MANUFACTURED GAS PRACTICE

Gas Cooling and Scrubbing

After leaving the water seal and wash box of a gas-making set, the gases may contain up to 50 per cent water vapor at a temperature of up to 185 F. Some cooling takes place in the piping and relief holder, but most of the latent heat of the water vapor must be removed in a scrubber type of heat exchanger designed for that purpose. Two main methods of scrubbing the gases are in use. The first method makes use of a water spray in various types of packed or empty columns to

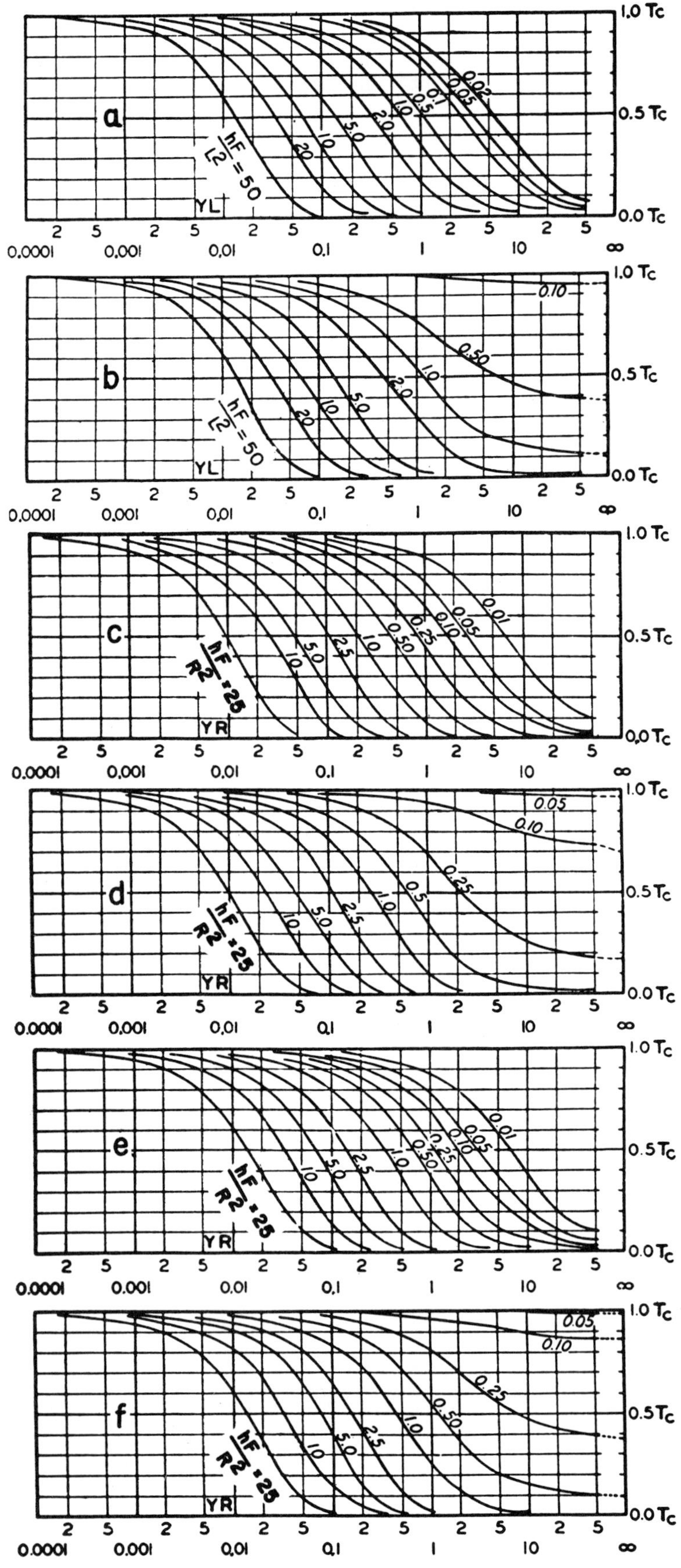

a. Difference between temperature of the exposed surface of a flat plate of thickness L (other surface insulated), or $2L$ (both surfaces exposed), and temperature of the surrounding medium, expressed as a fraction of the initial temperature difference, T_c.

b. Difference between temperature of the insulated surface of a flat plate of thickness L, or at the center of a flat plate (both sides exposed) of thickness $2L$, and temperature of the surrounding medium, expressed as a fraction of the initial temperature difference, T_c.

c. Difference between surface temperature of a sphere of radius R and temperature of the surrounding medium, expressed as a fraction of the initial temperature difference, T_c.

d. Difference between temperature at center of a sphere of radius R and temperature of the surrounding medium, expressed as a fraction of the initial temperature difference, T_c.

e. Difference between surface temperature of a long cylinder of radius R and temperature of the surrounding medium, expressed as a fraction of the initial temperature difference, T_c.

f. Difference between temperature at the axis of a long cylinder of radius R and temperature of the surrounding medium, expressed as a fraction of the initial temperature difference, T_c.

Fig. 2-70 Transient heat flow to solid shapes. $T_c = T - T_\infty/T_0 - T_\infty$; see illustrative example for meaning and use of terms.

cool the gases and remove the latent heat. The second method requires a tubular condenser in which the hot gases impinge on the outer surface of water-cooled tubes, with resulting condensation of the moisture and of some of the tars. A third method which is sometimes used is forced air circulation.

In any scrubbing system, **low pressure drop** keeps gas pumping costs low. Consideration of this factor eliminates the use of bubble cap columns and many types of packings in column types of scrubbers, and also influences the tube spacing in tubular condensers.

Tubular condensers are subject to continuous fouling because of the accumulation of tarry condensate, which greatly increases the resistance to heat flow. The overall coefficient of heat transfer must, therefore, make adequate allowance for the tar accumulation. In a vertical tube condenser, the tar condensing at the top runs down the tube in an accumulating layer, increasing the fouling effects greatly at the lower end of the tubes.

Although higher heat transfer coefficients should be possible by using crossflow in horizontal tubular condensers, the tubes may become fouled if tar accumulations do not drop free. Therefore, horizontal types of condensers are generally not selected.

Spray-Packed Tower Scrubbers. In spray-packed tower scrubbers, gases enter at the bottom and leave at the top. The tower has partial *countercurrent* operation with several water sprays, one above the other. Use of sprays in this manner permits the entire tower to be filled with well-dispersed small droplets of water which ensure good contact with, and cooling of, the gas.

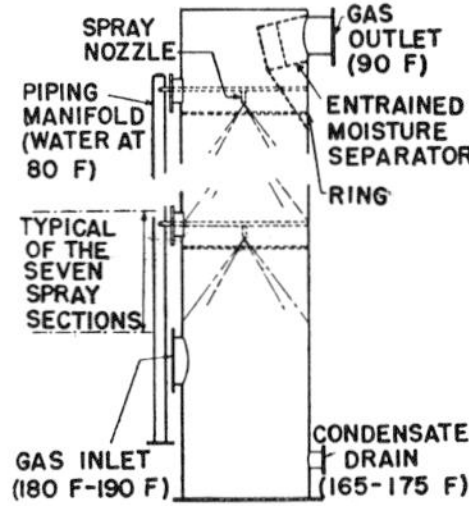

Fig. 2-71 Spray-packed scrubbing cooling column. (Courtesy of Gas Machinery Co.)

The design of such a tower is shown schematically in Fig. 2-71. Design factors furnished by the Gas Machinery Co. are as follows:

1. Gas velocity in empty column, 2.6 to 2.7 fps (corresponds to item 5).
2. Spray height, 30 to 35 ft.
3. About six sprays, each located at the point of impingement on the wall of the spray above.
4. Five feet allowed for top and bottom sections.
5. Ten MCF per hr-sq ft of tower cross section.
6. Gas inlet temperature, 180 to 190 F.
7. Gas to water temperature difference at top of column, 10°F (water enters at 80 F, gas leaves at 90 F).
8. Gas to water temperature differences at bottom of column, 15° F (water leaves bottom of column at 165 to 175 F)
9. Water rates, 16 gal per sq ft of cross section, maximum; 9 gal per sq ft of cross section, minimum.
10. All sprays adjusted at the same pressure, about 25 to 30 psig, using Sprayco or equivalent nozzles.
11. Annular ledges or rings inside column used to break up water film on walls, located just below spray impingement points.

Spray tower scrubbers require the use of an auxiliary heat exchanger to cool the spray water. This cooler may be any suitable type of shell and tube cooler, with overall coefficients chosen to compensate for fouling caused by tar residue and water supply.

When the gases contain water-soluble constituents, suitable provisions must be made to prevent excessive build-up in the spray water. Water spray towers cannot be expected to remove naphthalene and similar constituents from gases. The removal of these components may be accomplished by oil scrubbing in a grid-packed tower.

Grid-Packed Scrubber Designs. Older types of water and oil scrubbers contain wooden grids placed horizontally edgewise, each grid at right angles* to adjacent grids, with grids staggered over the open space of the parallel grid below. For scrubber design, the following methods and data may be used.[29]

Formulas for computing tray areas and other data:

$$E = \frac{24}{x + a} \tag{34}$$

$$S = \frac{24}{b}\left[\frac{b + a}{x + a} - \frac{a^2}{(x + a)^2}\right] \tag{35}$$

$$V = 0.7854\, D^2h \tag{36}$$

$$A = VS \tag{37}$$

where: x = tray spacing, in.
a = thickness of board, in.
b = width of board, in.
h = height of space in scrubber occupied by trays, ft
D = diameter, ft
E = total linear feet of vertical surface edge per square foot of scrubber cross section in top tray
S = net exposed surface per cubic foot of tray volume, sq ft (deducting for intersections and omitting the top edge of boards)
A = total tray surface in scrubber, sq ft
V = volume of trays in scrubber, cu ft

Tray spacing, x, closer than ½-in. tends to cause flooding. The selection of spacing depends on the character of the substances being separated from the gas.

The *critical velocity* for the tray spacing selected may be determined from Fig. 2-72. The maximum gas rate (SCF) should *never* be less than 1.5 *times* the critical velocity. Note that inlet volume is twice outlet volume.

Water or Solvent Rates. The amount of water or solvent used should fall within the range of $0.2E$ to $0.6E$ gpm. Less than $0.2E$ does not wet the tray surface sufficiently, and more than $0.6E$ does not show a corresponding increase in absorption.

The exact amount depends on the gas velocities and the amount of heat or soluble constituent that is to be removed.

* Heavy oils clog this arrangement.

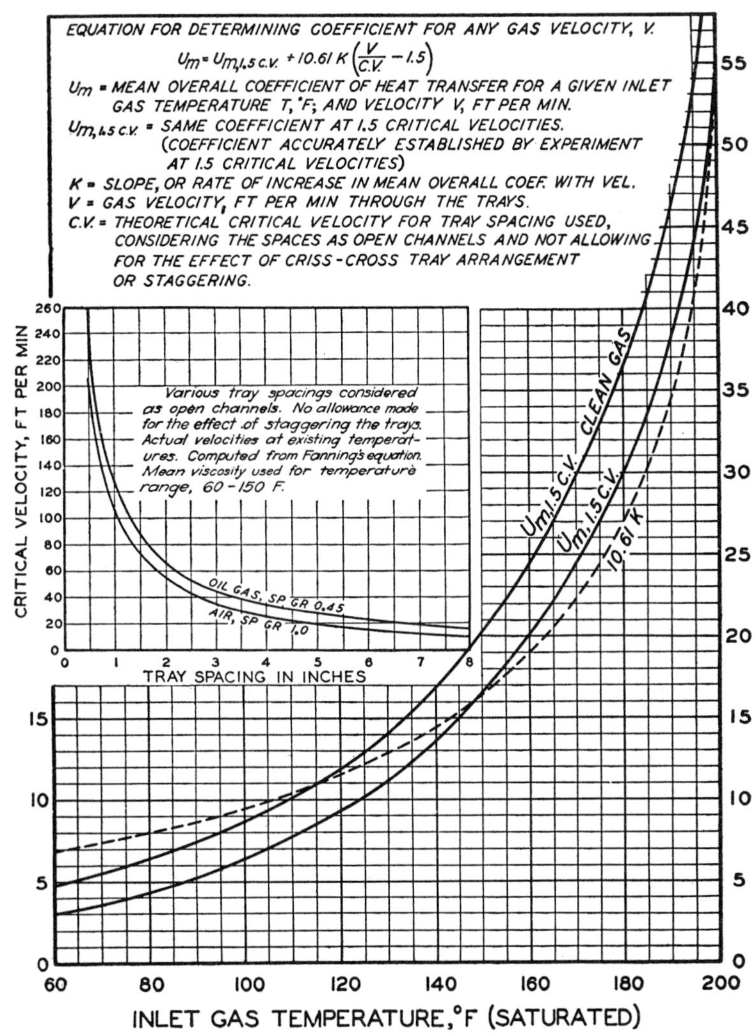

Fig. 2-72 Mean overall coefficient of heat transfer for grid-packed scrubbers (gas cooled to 60 F).

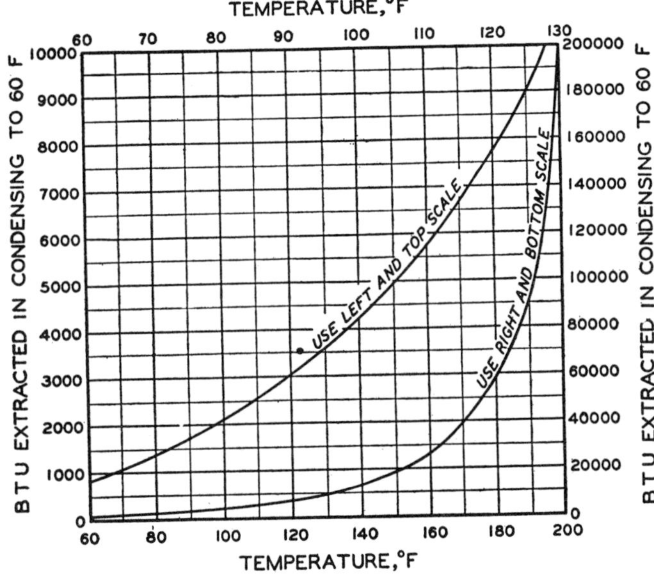

Fig. 2-73 Total heat content of water-saturated oil gas above 60 F, in Btu per MCF (gas corrected to 60 F and 30 in. Hg).

Heat Content. From Fig. 2-73, the heat content above 60 F of MSCF of water-saturated oil gas may be found.

Heat Transfer.

$$Q = U_m A \Delta t$$

where: Q = rate of heat flow from gas to water, Btu per hr; see Fig. 2-73

U_m = mean overall coefficient of heat transfer, Btu per hr-sq ft-°F

A = total tray surface in scrubber, sq ft (includes *only* the sides and bottom-exposed edges of trays, deducting for intersections)

Δt = mean temperature difference

The determination of the mean temperature difference in this equation is a subject of some controversy. Use of the LMTD gives a poor approximation. The Haug and Mason[30] method is sufficiently accurate under some conditions, especially when it may be assumed that the cooling water absorbs heat uniformly. A case in point is when the gas passageways are reduced to compensate for changing volumes resulting from cooling and condensation. An example using this method follows. The Colburn and Hougen[30,31] method, however, represents the best design practice for computing the rate of condensation of a vapor from a noncondensable gas.

Table 2-109 Data on Various Trays for Scrubbers

Conventional tray arrangement, crisscross (90°) and staggered; serrated edges on lowest tray of each section except bottom. Areas include sides and bottom edges only of boards, deducting for intersections. All lumber is dressed.

Tray spacing, x, in.	Percentage of "free" space of cross-sectional area	E*	S*	Board ft of lumber per cu ft of trays	Water rate per cu ft of cross section: Min to wet, $0.2E$ gpm	Water rate per cu ft of cross section: Max permissible, $0.6E$ gpm
		½ × 3¾-in. boards†				
0.5	50.0	24.00	25.60	12.00	4.80	14.40
0.75	60.0	19.20	20.75	9.60	3.84	11.52
1.0	66.6	16.00	17.40	8.00	3.20	9.60
1.5	75.0	12.00	13.23	6.00	2.40	7.20
2.0	80.0	9.60	10.62	4.80	1.92	5.76
		½ × 5¾-in. boards				
0.5	50.0	24.00	25.02	12.00	4.80	14.40
0.75	60.0	19.20	20.20	9.60	3.84	11.52
1.0	66.6	16.00	17.00	8.00	3.20	9.60
1.5	75.0	12.00	12.82	6.00	2.40	7.20
2.0	80.0	9.60	10.62	4.80	1.92	5.76
		13⁄16 × 3¾-in. boards†				
0.5	38.0	18.30	20.00	9.15	3.66	10.98
0.812	50.0	14.80	16.40	7.40	2.96	8.88
1.0	55.1	13.23	14.85	6.61	2.646	7.938
1.5	64.8	10.40	11.82	5.20	2.08	6.24
2.0	71.1	8.54	9.85	4.27	1.708	5.124
		13⁄16 × 5¾-in. boards				
0.5	38.0	18.30	19.30	9.15	3.66	10.98
0.812	50.0	14.80	15.85	7.40	2.96	8.88
1.0	55.1	13.23	14.30	6.61	2.646	7.938
1.5	64.8	10.40	13.80	5.20	2.08	6.24
2.0	71.1	8.54	9.40	4.27	1.708	5.124

* See Eqs. 34 and 35.

† 3¾ in. width gives 3.2 changes in direction of flow per vertical foot compared with 2.8 changes for 5¾ in. boards. The greater number of changes in direction per foot increases turbulence and raises the rate of absorption. The height of trays required must be calculated for each case, as most scrubber problems involve widely different conditions.

Vertical Feet of Trays Required [sample calculations for a scrubber (oil gas)]. Assume saturated inlet gas at 180 F to be cooled to 60 F, trays 13⁄16 by 5¾ in., spaced 13⁄16 in. apart. From Table 2-109, free space = 50 per cent; linear feet of vertical surface edge per square foot of scrubber cross section = 14.80; square feet of surface per square foot of tray volume = 15.85.

For the special case of primary scrubbers, divide the hourly rate of the generator, cubic feet, by the time of make, minutes. This will give a figure for the average minute. Increase this by 50 per cent for the maximum minute. For the trays selected, the maximum minute should correspond to 1.5 *times* the critical velocity. From the insert of Fig. 2-72, the critical velocity for 13⁄16-in. tray spacing is 160 fpm. The actual volume flowing per maximum minute per square foot of scrubber cross section is 0.5 × 160 × 1.5 = 120 cu ft.

Assuming that the generator rate is 100 MCF per hr and the actual time of make is 24 min, the capacity of the scrubber must be 1.5 × 100,000 ÷ 24 = 6250 cfm. The cross-sectional area required for the scrubber is 6250 ÷ 120 = 52.08 sq ft (diameter = 8.15 ft)

The heat content above 60 F for 180 F saturated gas per MCF is 58,500 Btu from Fig. 2-73. Assuming that the temperature rise of the water is 0.96 of the maximum (Δt inlet *minus* Δt outlet for the gas) and the initial water temperature is 60 F, the outlet water temperature during the maximum minute will be 0.95(180 − 60) + 60 = 174 F.

See tabulation of determination of Δt.

Determination of Δt Using the Method of Haug and Mason and the Trapezoidal Rule

Water temp	Btu to water	Btu in gas	Gas temp (Fig. 2-73)	Temp difference	Averages
60	0	880	60	0 × 0.5	0
88.5	14,625	15,505	145	56.5 × 1.0	56.5
117	29,250	30,130	165	48.0 × 1.0	48.0
145.5	43,875	44,755	175	29.5 × 1.0	29.5
174	58,500	59,380	180	6.0 × 0.5	3.0
Total	...	...	...	...	137.0
Δt = total/4	...	...	...	...	34.25

The heat transfer coefficient, U_m, at 1.5 *times* the critical velocity for 180 F gas is 30, from Fig. 2-72. The cooling surface required per square foot of scrubber cross section is:

$$A = \frac{Q}{U_m \Delta t} = \frac{120 \times 58{,}500 \times 60}{1000 \times 30 \times 34.25} = 410 \text{ sq ft}$$

The vertical height of trays required *theoretically* is 410 ÷ 15.85 = 25.9 ft. This assumes ideal water distribution and wetting. Good practice indicates that this should be increased about 50 per cent as a safety factor, which would give a height of about 40 ft.

Since the maximum amount of heat removed per minute per square foot of cross section is 120 × 58,500 ÷ 1000 = 7020 Btu, the water required is 7020 ÷ (174 − 60)(8.33) = 7.4 gpm per sq ft of scrubber cross section.

The water per linear foot of vertical surface edge would be 7.4 ÷ 14.8 = 0.50 gpm (14.8 from Table 2-109). This rate is satisfactory, as it is between the allowable minimum (0.2) and maximum (0.7) rates.

The empirical design of a gas condenser may be determined as follows:

1. Heat load is the difference between heat content at inlet and outlet. The assumption made with regard to gas inlet temperature is most important. The following are suggested guides: 180–185 F for carbureted water gas; 190–195 F for high-Btu oil gas. Assumption of temperatures above 195 F is generally invalid. Note the large variation in heat content that exists between 180 and 190 F (Fig. 2-73).

2. Cooling water and/or liquor maximum exit temperature, 140 F; inlet liquor at minimum of 10° below gas outlet temperature.

3. Assume 30 ft grid height and a rate of 7500 SCF of gas per hr-sq ft of washer-cooler cross section to calculate scrubber diameter.

4. The cooling liquor rate should be within the range of 6 to 13 gpm per sq ft. Consider cooling water temperature in both winter and summer.

5. Additional heat load allowance for tar content ranges between 5 and 10 MBtu per MSCF; i.e., a full cooling job (180 F inlet to 85 F outlet) requires an allowance of 10 MBtu for tar.

Pressure Drop of Grid-Packed Towers. Johnstone pressure drop data on air flow thru wood grid packings as given by Sherwood[32] may be used to estimate the pressure drop thru grid-packed scrubbers. Johnstone's work applies only to ¼-in.-thick boards, and does not apply exactly to the grid arrangements of the previous section. The results of his work may be summarized by the equation:

$$\Delta h = fG^{1.8} \tag{38}$$

where: Δh = pressure drop, in. of water per ft of tower height
f = pressure drop constant depending on grid shape
G = mass velocity of the gas in lb per hr-sq ft of total cross section

The data obtained cover the range of G from 700 to 4000 for air flowing at 55 to 85 F, over grids wet with water circulated at 0.08 to 0.136 gpm per ft of wetted perimeter. Values of f are given in Table 2-110. *Pitch* refers to horizontal clearance between parallel grids. *Height* is the height of each individual grid.

Table 2-110 Johnstone Pressure Drop Data for Air Flow thru Various Packings[32]

Type of packing	Height, in.	Pitch, in.	Arrangement	$f \times 10^8$ (see Eq. 42)
Wood grids	1	0.625	Staggered	21.4
	1	0.625	Nonstaggered	18.9
	2	0.625	Staggered	14.6
	2	0.625	Nonstaggered	13.7
	4	0.625	Staggered	10.2
	4	0.625	Nonstaggered	8.3
	8	0.625	Staggered	6.4
	8	0.625	Nonstaggered	5.7
	4	1.25	Nonstaggered	2.43
	8	1.25	Nonstaggered	1.79
	4	1.75	Staggered	1.75
	4	1.75	Nonstaggered	1.73
	12	1.25	Nonstaggered	1.20
	16	1.25	Nonstaggered	1.17
	8	1.75	Nonstaggered	1.16
	8	1.75	Staggered	1.04
	12	1.75	Staggered	0.99
	12	1.75	Nonstaggered	.95
	16	1.75	Nonstaggered	.87
	16	1.75	Staggered	.82
	4	2.25	Nonstaggered	.87
	8	2.25	Nonstaggered	.68
	12	2.25	Nonstaggered	.63
	16	2.25	Nonstaggered	.61
Smooth sheets	96	1.0	Parallel	.21
	96	2.0	Parallel	0.124
1¼-in. corrugated sheets	96	1.0	Parallel	2.11
2⅝-in. corrugated sheets	96	1.0	Parallel	1.88
	96	2.0	Parallel	1.07
	96	3.0	Parallel	0.56

HEAT TRANSFER APPLICATIONS IN NATURAL GAS PRACTICE

Use of the Gas Stream as a Heat Transfer Fluid

Use of the cold gas stream as a heat-absorbing medium for compressor oil cooling, for lowering gas inlet temperatures to glycol dehydration plants, and for cooling the overhead of a desulfurization plant has been described as a method of eliminating cooling towers at compressor stations and obtaining more automatic operation.[33]

In a gas-to-gas exchanger serving a **glycol dehydration plant** field gas entered and absorbed heat on the shell side of a shell and tube exchanger. Discharge gas passed thru the tubes, was cooled, and entered the contactors at a lower temperature. Young indicated that the expected equilibrium dew point of wet gas received at 85 F would be 37 F with a 95 per cent glycol solution, and that the use of gas-to-gas heat exchangers might readily lower the contact temperature to 70 F, corresponding to a 24 F dew point. These heat exchangers ironed out the diurnal temperature fluctuations commonly experienced with cooling towers, and permitted better control of dehydration. Little fouling was experienced and the performance of all units closely checked with the design figures.

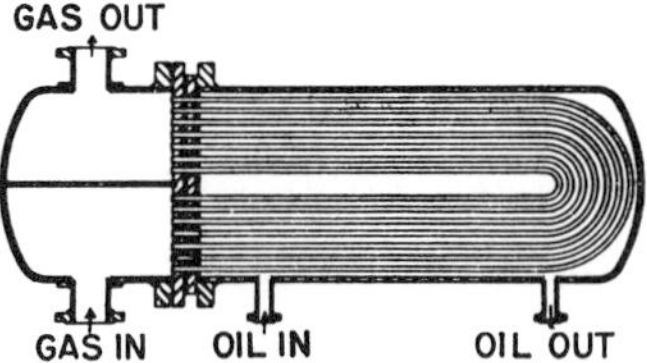

Fig. 2-74 Sectional view of oil-to-gas heat exchanger.

For **engine oil cooling,** an oil-to-gas exchanger, Fig. 2-74, was employed. This exchanger, installed on the suction side of the compressor, dissipated its heat by raising inlet temperature slightly. It was constructed with a double tube sheet to prevent gas leakage into the circulating engine oil.

Using the type of cooler shown in Fig. 2-74 for a main line station, the cooler had 0.7 sq ft of surface area per horsepower served, and the gas-side pressure drop was 1.7 psi at full capacity. Careful analysis of the increase in power due to increased compressor inlet temperature showed an overall loss of 0.6 to 0.7 per cent.

For the desulfurization plant, a small shell and tube heat exchanger was installed on top of the still, and cooled with a side stream of sweet gas, which eventually released its heat to the ground.

Cooling units equipped with roll-type steel curtains so that the cooling can be reduced during cold weather make it unnecessary to cut coolers out of operation and eliminate the necessity of draining them.

Air-Cooled Heat Exchangers[34]

In a *dry-surface cooler* (noncontact convection-type heat exchanger) the fluid flowing thru finned-tube sections is cooled by fan-circulated air. This type of exchanger may be specified where water is scarce or unsuitable for cooling or where small compact units are desired. The air-cooled exchangers are easy to start and easier to operate in cold weather. Their usage is for "high-level" heat removal (e.g., above 130 F referred to 100 F dry-bulb air).

When selecting the **design air temperature** for dry-surface coolers it is generally acceptable to choose a *dry-bulb* temperature which will not be exceeded more than five per cent of the hours between noon and midnight during the months of June to September, inclusive. (Consult local U. S. Weather Bureau Statistics.) Generally used dry-bulb design temperatures are 95 to 110 F.

For the air-cooled exchanger, **cooling range** is the number of degrees that the fluid is cooled in the exchanger. **Approach** is the difference between the temperature of the cold fluid leaving the dry cooler and the dry-bulb temperature of the ambient air.

Water-Cooling Towers[34]

In a mechanical-draft tower, water drops thru upward-moving air, is circulated by fans, and falls into the cold water collecting basin. As the water falls from level to level over the wood filling, it splashes into fine droplets and spreads out over the wood slats in thin films, thus providing additional water surface to the upward-moving air.

The water-cooling tower would generally be used in preference to the dry-cooling unit when cold water temperatures are less than 130 F and approach within a few degrees of the *wet-bulb* temperature of the air (e.g., 85 F cold water to 80 F wet-bulb).

Cooling range is the temperature difference between hot water entering the tower and cooler water leaving. *Approach* is the difference between cooled water leaving the tower and the wet-bulb temperature of the ambient air.

Cooling tower design and performance guarantees are based on: (1) a given approach to a specified wet-bulb temperature, (2) a stated cooling range, and (3) a given heat load. It is common practice to select a wet-bulb temperature which will not be exceeded more than five per cent of the time. (Consult local U. S. Weather Bureau Statistics.) Wet-bulb design temperatures of 70 to 80 F are ordinarily used.

Average **evaporation losses** will run from 0.75 to 1.0 per cent of the water circulated for each 10°F cooling range. Drift loss may run at about 0.1 per cent of the water circulated. Hardness of circulating water, treatment employed, and amount of drift loss determine the amount of **blowdown.** The ratio of make-up to blowdown indicates the number of concentrations in the system, since the amount of solids introduced by make-up water equals the total removed by blowdown.

Heat Exchangers (Immersion Coil Type)

Important design considerations are *low gas pressure drop* thru the heat exchanger, 2 to 5 psi at maximum capacity; adequate heat input; and flexibility to accommodate future loads. One design, shown in Fig. 2-75, uses Griscom-Russell Bentube sections for the gas-heating elements, immersed in a water tank which is heated by four double-pass immersion tubes fired with high-pressure burners. The heater was originally built with only one of the two Bentube units, and

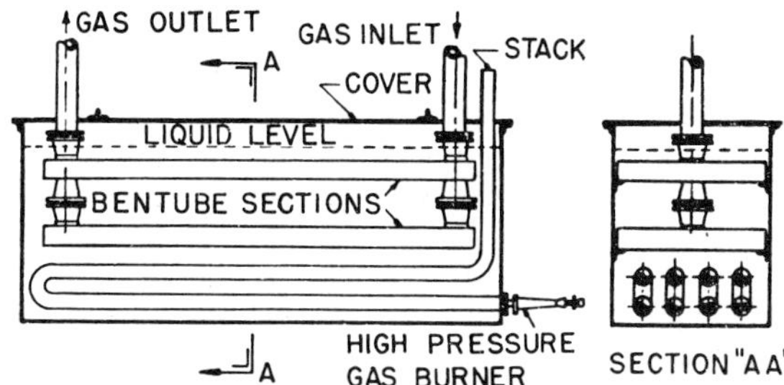

Fig. 2-75 Immersion-fired heat exchanger.[35]

was designed to heat 5,000,000 ft of gas per day at 600 psi thru a total temperature rise of 50°F with a 3-psi drop. A second Bentube unit was to be installed later to increase the capacity to 10,000,000 ft per day. The general design arrangement shown in Fig. 2-75 may be extended to heat exchangers of any desired capacity.

Heaters must be properly sized to meet heat loads based on obtainable heat transfer rates. Heat loads may be calculated from entropy–enthalpy diagrams of the gas. However, as a preliminary approximation, a specific heat of 0.5 Btu per lb-°F may be assumed together with a 7°F cooling effect for every 100-psi drop thru the regulator.

The heater is installed in the line ahead of the regulator or control valve thru which gas is fed from the high-pressure lines, and is regulated so that the heated gas will leave the regulator at a temperature of 32 F. The thermostat which controls the on–off burners is located on the outlet side of the regulator.

HEAT TRANSFER IN GAS UTILIZATION

Use of gas as a fuel requires that it be burned in one or more stages. The amount of air employed in each stage as primary or secondary air influences the shape, color, chemical composition, and temperature of the flame. However, the total amount of air employed is the principal factor in the quantity of heat transferred in most appliances having confined gas flames. Most domestic gas appliances, small boilers, and small commercial cooking appliances fall into this category.

Heating Vessels with Flames

Heating a fluid contained in a light gage metal container thru a small (less than 400°F) temperature rise with a flame usually yields time–temperature heating curves which are almost straight lines. Situations of this kind include heating water to boiling in vessels over open burners, preheating deep fat fryers, and so forth. Actually, the heating-up curve is *exponential*, approaching a limiting temperature which is determined by the point at which heat input balances heat losses from the vessel. When boiling takes place, the temperature limit is the boiling point of that liquid. For fats in fryers, a much higher temperature asymptote exists, which is well above the flash point of the fat; thus, the maximum temperature must be thermostatically limited. Tests of heating speeds of a group of commercial deep fat fryers, in which the heat capacity of the kettle was relatively small compared with that of the fat, yielded the following expression for heating time:

$$S = \frac{5500M(t_2 - t_1)}{IE_c} \tag{39}$$

where:

S = heating time, min
M = weight of heated fat in the fryer, lb
t_1 and
t_2 = initial and final fat temperatures, °F
I = gross heat input rate, Btu per hr
E_c = combustion efficiency, per cent, measured at a fat temperature between 300–350 F, = 100 minus per cent flue loss

A similar relationship applies to heating pans of water on range open-top burners. Thermal efficiency of such heating operations is determined by the formula:

$$E_t = \frac{(W + W_v)(t_2 - t_1)}{(\text{HV})(\text{SCF})} \times 100 \tag{40}$$

where:

E_t = thermal efficiency, per cent
W = weight of water in pan, lb
W_v = water equivalent of pan (weight times specific heat), lb
t_1 and
t_2 = initial and final temperatures (no boiling), °F
HV = heating value of gas, gross Btu per cu ft
SCF = standard cubic feet of gas consumed

For a steady rate of gas consumption of I Btu per hr, the time T in hours to heat the vessel thru a rise of $(t_2 - t_1)$ is:

$$T = \frac{(W + W_v)(t_2 - t_1)}{IE_t} \times 100 \tag{41}$$

In all these relationships, the liquids are assumed to be intimately mixed, and do not undergo any change of state involving an appreciable amount of latent heat.

Excess Aeration Effects on Heat Transfer

All other factors remaining constant, **excess air** supplied to a gas appliance always reduces the rate of heat transfer at a given input. This is because the effects of reduced flame temperature and increased heat content carried off in the flue products are *always greater* than the effect of increasing the heat transfer coefficient. Figure 2-76, based on experimental data taken from actual appliances, illustrates how these points can be evaluated.

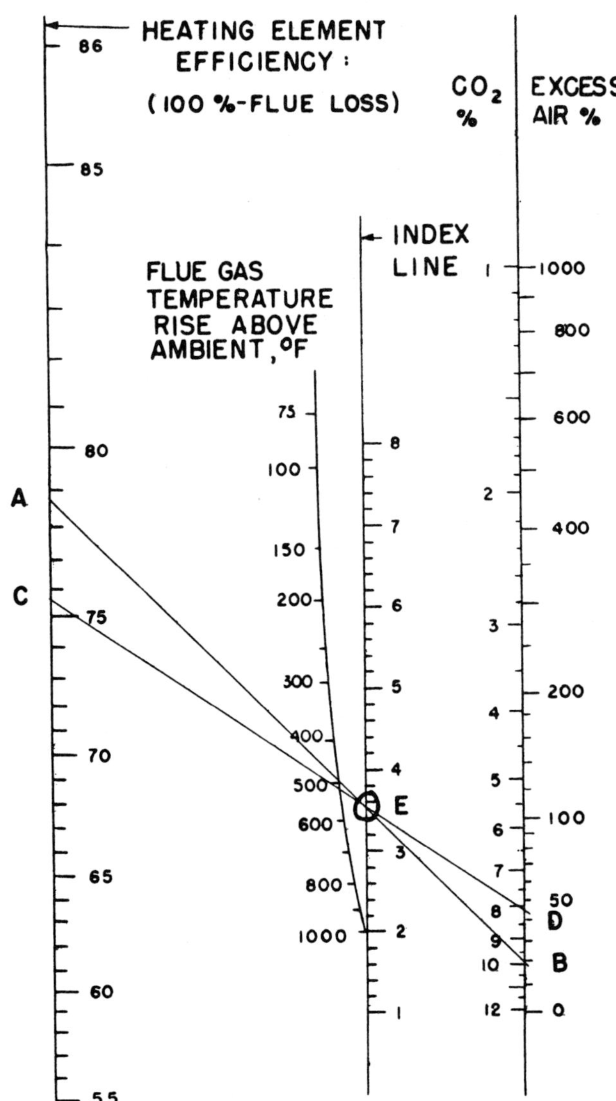

Fig. 2-76 Chart for determination of effect of a change in per cent CO_2 on efficiency at constant heat input.[36]

Example: Suppose that the draft hood of a furnace is raised until the percentage of CO_2 in the flue gases is reduced from 9.9 per cent to 8.1 per cent. What is the effect on element efficiency and flue gas temperature? Flue gas temperature at 9.9 per cent CO_2 is 489°F above room temperature.

Solution:

Construct line AB for the 9.9 per cent CO_2 and 489°F temperature rise condition.

Thru the intersection of AB with the index line at 3.5, construct line CD thru 8.1 per cent CO_2.

Line CD gives these results:

1. Although excess air has been increased from 19 per cent to 45 per cent, the flue gas temperature has *increased*.

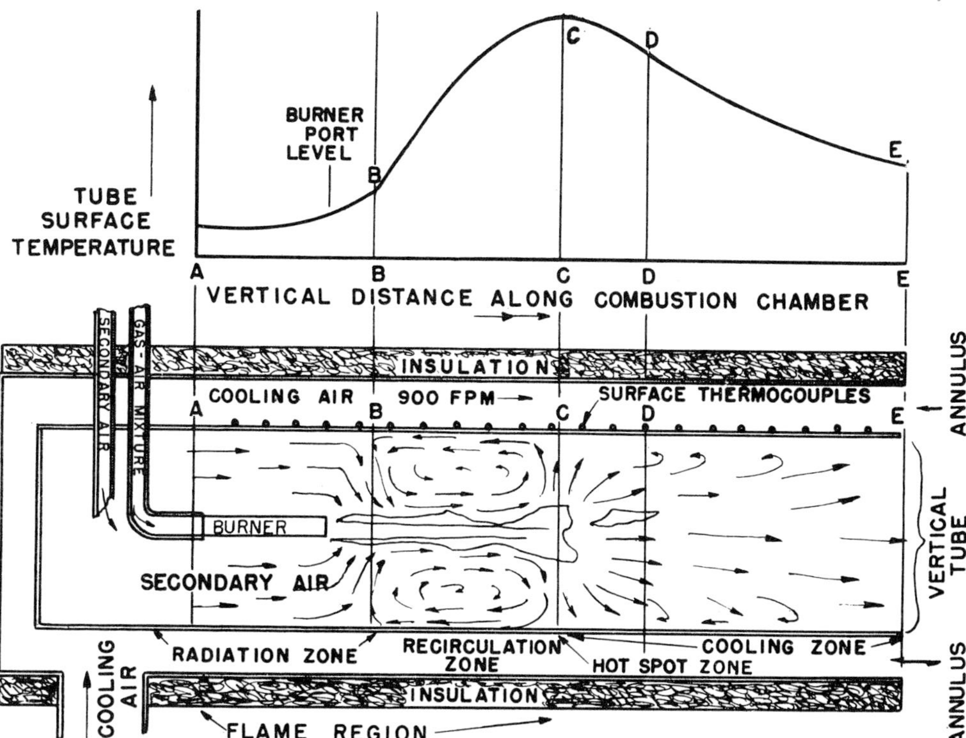

Fig. 2-77 Zones and wall temperatures in a gas flame combustion chamber.[36]

2. The combination of a larger flue gas volume and a higher flue gas temperature, however, caused the efficiency to *drop* from 78.7 per cent to 75.6 per cent.

Zones of the Combustion Chamber

Observations[36] of temperature and flow of burning gases in vertical tubular heat exchangers, air-cooled in parallel flow, showed that this heating process could be divided into distinct zones. The lowest zone, around the burner, termed the **radiation zone,** was caused by a layer of cold secondary air between the flame and exchanger walls. The next highest zone, termed the **recirculation zone,** contained the flame plus a vortex or recirculation eddy consisting of a mixture of air and combustion gases and created by velocity and buoyancy effects of the flame. At the uppermost limit of the recirculation zone, a **hot spot zone** was formed on the wall, due to the turbulence at this point. The position of this hot spot was fixed by the velocity breakdown of the jet of gases forming the flame, and not by the location of the visible flame mantle. Thus, hot spots of this type can be located above or below the end of the visible flame. The zone beyond the end of the flame is termed the **cooling zone.** Here the turbulence introduced by the burner and the combustion process diminishes, and the resemblance to conventional heat exchangers begins.

A diagrammatic representation of these zones is shown in Fig. 2-77, along with the accompanying wall temperatures by means of which the zones are defined.

Flame Temperatures in Heat Transfer Calculations

Calculate the **combustion temperature** from a heat balance equation involving the volumes of the flue gas constituents, their heat contents, and the net available heat. Then determine the maximum temperature or corrected flame temperature by deducting the heat transferred in the zones upstream of the hot spot (Fig. 2-77) from the net heat available in the gas. The corrected flame temperature should then

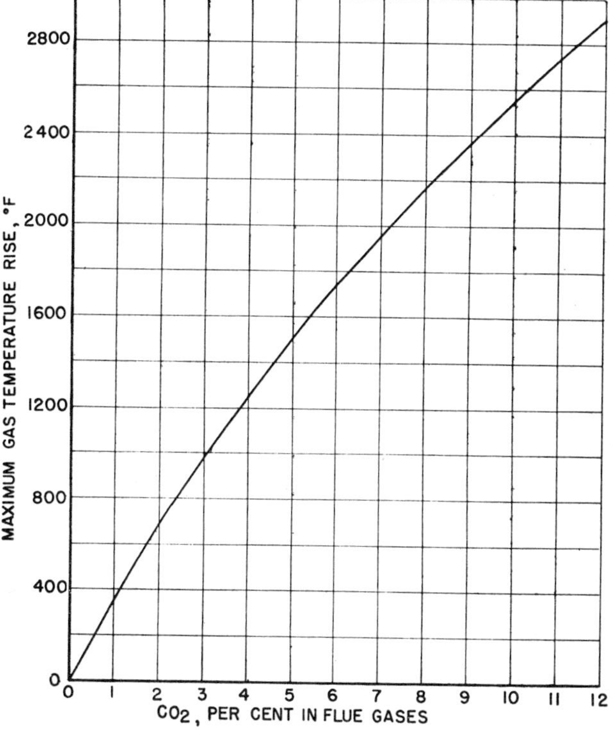

Fig. 2-78 Maximum gas temperature rise (compensated for 20 per cent heat losses in flame region) as a function of CO_2 in flue gas.[36] Note limitations in text.

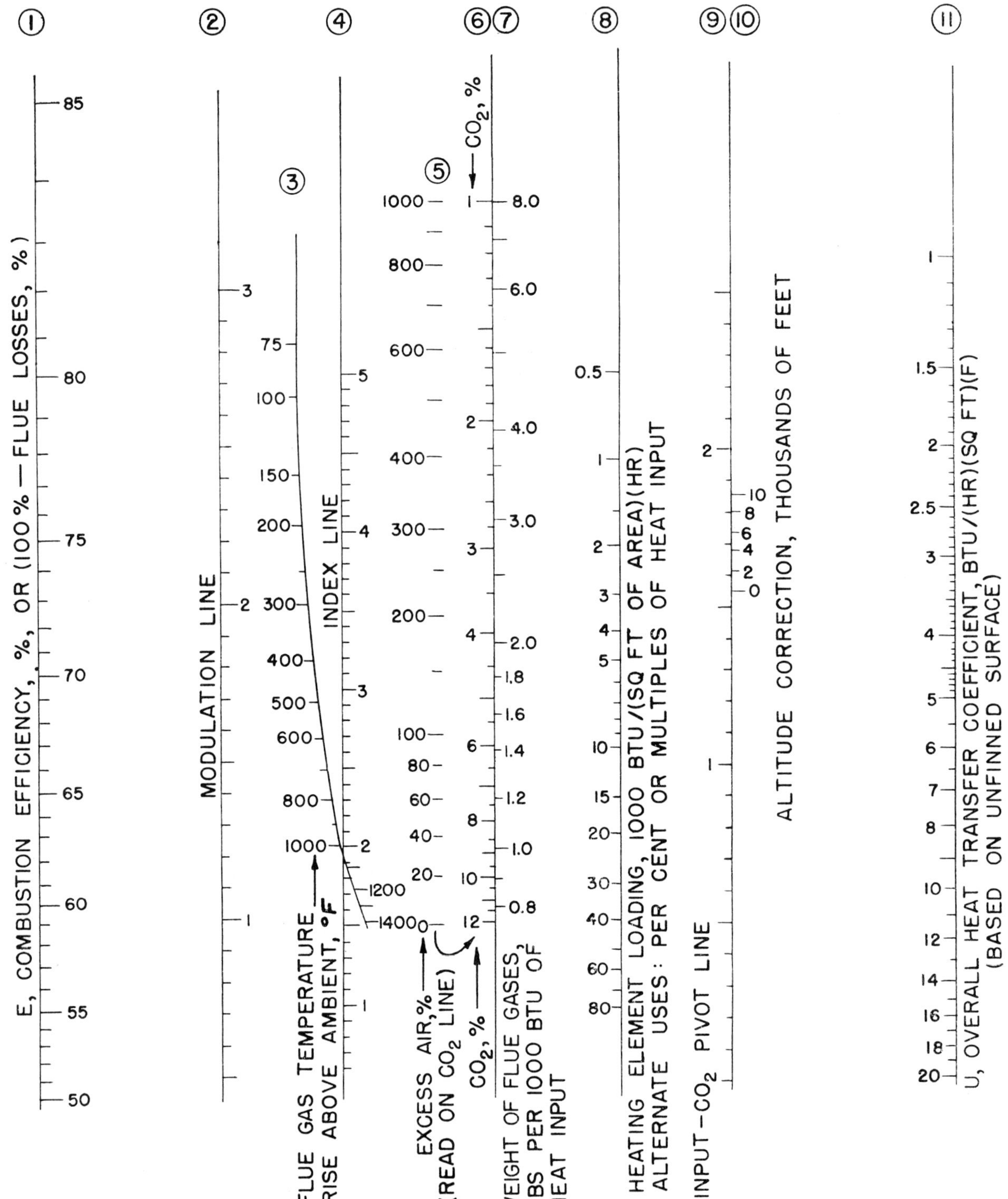

Fig. 2-79 Gas appliance performance chart. (Data obtained in part from A.G.A. Laboratories Res. Bull. 63, 1951.)

be used together with the flue gas outlet temperature and the wall temperatures to establish the LMTD which is needed to calculate heat transfer.

The amount of heat lost by the flame before its maximum temperature is reached ranges from 15 to 25 per cent of the gross heating value of the gas burned. A graph showing the maximum temperature rise of the combustion gases, assuming a 20 per cent heat loss, vs. the percentage of CO_2 in the burned mixture is given in Fig. 2-78. The temperature rise values are entirely dependent on the ratio of the total air supplied to the volume of gas burned. No compensation is needed for the relative amounts of primary air and secondary air. The graph applies only to small domestic and commercial types of appliances using atmospheric burners with natural draft secondary air supply. Power burners with turbulet flames constitute an exception to this rule, for which the true theoretical flame temperature will give more satisfactory results.

The outlet flue gas temperature theoretically requires a correction along with that for the maximum flame temperature, in order to obtain the true LMTD. Because the gas-side and other film coefficients in a gas appliance are approximated by making the assumptions just described, the use of outlet temperature, in general, yields fair results; however, the use of the **gas appliance performance chart** (Fig. 2-79) is generally preferable.

Gas-Side Film Coefficients

In domestic gas appliances, the gases formed by the flame may be initially turbulent, although the computed Reynolds number of the flame gases frequently indicates laminar flow. Thus, considerable difficulty is encountered in developing theoretically valid methods for determining gas-side film coefficients. Inside a 4-in. diam steel tube, and in most domestic gas appliances, combustion flow rates must be limited to laminar flows (low Reynolds numbers) to prevent excessive tube wall temperatures. For these conditions, it was found that the *cooling zone gas film coefficient* in a 4-in. tube, at all heat inputs greater than 10,000 Btu per hr, could be correlated by the following equation:

$$h = (1.2 + 0.25x)e^{0.067w/d} \qquad (42)$$

where:

h = gas film coefficient in cooling zone, Btu per hr-sq ft-°F
x = fraction of excess air in total air–gas mixture
w = weight of unburned gases issuing from burner port, lb per hr
d = equivalent port diameter = $(4A/\pi)^{0.5}$, in. (applies to both single and multiport burners; A is port area in square inches)
e = 2.718, base of natural logarithms

The fraction w/d is approximately proportional to the Reynolds number of the air–gas mixture as it issues from the burner ports.

Since several possible appliance adjustments will change the values of x, w, and d, it follows that the value of the gas-side film coefficient, h, will change accordingly. The weight of gases issuing from the burner can be raised by (a) increasing the gas input, (b) opening the primary air shutter, (c) cooling the burner head. The use of a smaller equivalent port diameter will raise the value of the exponent. The excess air, x, can be increased by increasing the secondary air supply or increasing the appliance draft. All of these tend to raise the value of h.

The maximum observed value of gas film coefficient was 3.85, while the minimum (which was not correlated by Eq. **42**) was about 0.65. It is considered that heat transfer from combustion gases in laminar flow follows the rules for that type of flow, particularly because the equation indicates the low value of 1.2 for the film coefficient after burner turbulence effects have died out in flames of more than 10,000 Btu per hr in the 4-in. tube.

Because of diminishing gas turbulence, the use of Eq. **42** for estimating gas film coefficients is only recommended for tubes between 3 and 4 ft long.

Radiation Effects from Gas Flames in Domestic Gas Appliances

Radiation of heat from typical **domestic appliance flames** is known to contribute a maximum of 12 per cent of the gross heating value of the fuel. Most of this heat is transferred by flame radiation prior to flue gas entry into the cooling zone.

Surveys of the flames of burners tested in both open air and tubes showed that the heat radiation originated mostly in the visible portion of the flame. The amount of radiation was proportional to the flame surface area. At 50 per cent primary air and 120 per cent excess air, heat radiation was found to be 13.4 Btu per sq in. of flame surface for all single port burners. This small amount of radiation means that heat transfer in small gas appliances may be treated primarily as a convective process.

The Gas Appliance Performance Chart

To simplify the process of analyzing heat transfer and other phenomena in domestic gas appliances, the relationships for heat transfer thru the wall were combined with those for heat balances in the two streams of fluids for the case of a *parallel flow* gas-to-warm-air heat exchanger, such as a **forced warm air furnace.** The resulting equations permitted development of the nomogram, Fig. 2-79, which is based on the equation:

$$A = \frac{1}{U}\left[\log_e\left(\frac{0.87}{0.87 - E}\right)\right]\left[\frac{96}{CO_2} + 4.17\right] \qquad (43)$$

where:

A = mean area of heat transfer surface, sq ft
U = overall coefficient of heat transfer, Btu per hr-sq ft-°F
E = combustion efficiency, = 100 per cent *minus* latent and sensible heat losses in flue gases. The value of E must be greater than 50 per cent for the relationship and graph to apply accurately.
CO_2 = CO_2 content in flue gases after complete combustion, per cent for natural gas

A discussion of stoichiometry is available which considers the principles upon which Fig. 2-79 is based from the appliance designer's point of view.[37]

Overall Heat Transfer Coefficients for Gas Appliances. Rather than analyze the heat transfer process on both sides

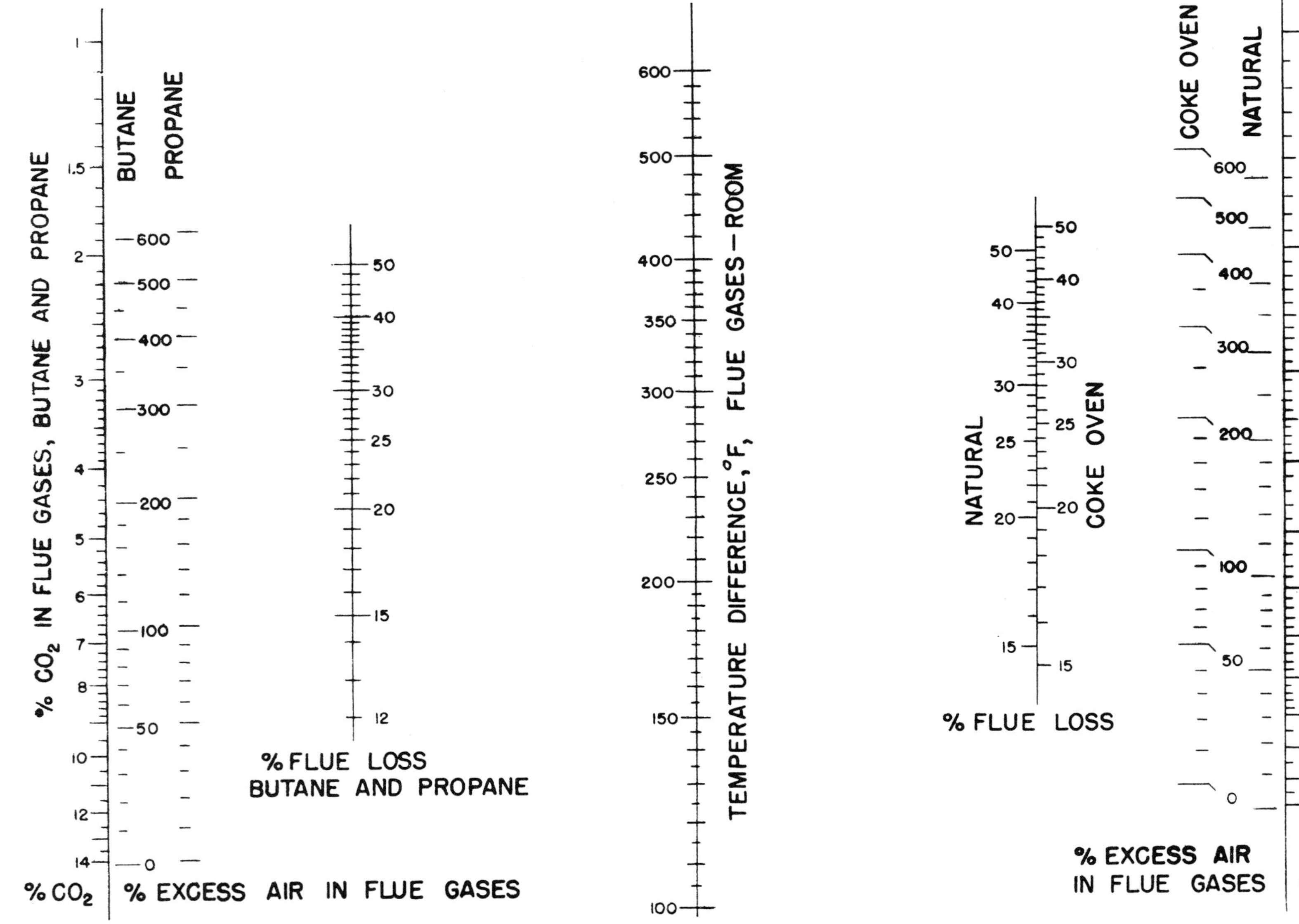

Fig. 2-80 Alignment chart for calculation of flue losses for butane, propane, and coke oven and natural gases. (Adapted from A.G.A. Laboratories' flue loss charts.)

of the surface separating the gas from the other fluids, it is preferable to assign overall coefficients which can be used in conjunction with the chart. For the types of appliances which have been analyzed to date, overall heat transfer coefficients observed are shown in Table 2-111.

Table 2-111 Overall Heat Transfer Coefficients, U, for Gas Appliances

(Btu/hr-sq ft-°F)

Appliance	U
Forced warm air furnaces	2.0–2.7
Gravity furnaces, recessed heaters	2.0–2.2
Underfired tank-type water heaters (30 gal and above)	
Internal flue	5.0–6.4
External flue	3.3–4.0
Deep fat fryers	
Externally heated	6.0–7.0
Immersion tubes (loose internal baffles)	6.4
Finned on gas side (U, based on fat-side area; 1-in. wide 18-gage fins, ½ in. apart)	11–12

Existing ASA requirements governing appliance efficiency, maximum and minimum heating element surface temperatures, and flue gas outlet temperatures often combine to restrict the permissible design factors of the appliance within narrow limits. Thus, most forced warm air furnaces using plain sheet metal (unfinned) construction have hourly heating element loadings of 3500 to 4000 Btu per sq ft of heating surface area. Combustion efficiencies greater than 80 per cent are obtained only with difficulty because of the likelihood of **cold spots** (less than 108 F *above* room temperature) which may cause *moisture condensation* if lower loadings are used.

Applications of the Appliance Performance Chart (Fig. 2-79).

Example 1: Find the combustion efficiency of an appliance in a room whose air temperature is 70 F. The appliance produces gases containing 7 per cent CO_2 at a temperature of 470 F. The combustion efficiency is found by drawing a straight line thru the 7 per cent CO_2 point on scale **(6)**, thru the 400°F rise point on the flue gas line **(3)** (470 − 70), and over to the combustion efficiency line **(1)**—read 77.5 per cent. The code for this operation is **(6) (3)→(1)**.

Example 2: This problem involves the use of *pivot* scale, as when computing the heating element loading at which a predetermined efficiency will be obtained.

			Scale
Known:	CO_2	= 8 per cent	**(6)**
	Flue temperature rise	= 500°F	**(3)**
	U	= 3.0 Btu per hr-sq ft-°F	**(11)**
Find:	The heating element loading to satisfy these conditions.		**(8)**

Solution: Connect **(6)** and **(3)** with a straight line thru the index line, to locate a *pivot point* at 2.8 on scale **(4)**.

Connect the point on scale **(4)** with the value of U on scale **(11)** to find the intersection on scale **(8)** at a loading equal to 4500 Btu per sq ft-hr.

The code for this operation is:

(6) (3)→(4) (2.8)
‖
(4) (11)→(8)

Here the vertical *equal sign* between the two **(4)** symbols denotes that the point on that scale is used for subsequent operations. Its numerical value is indicated in the lightface parentheses.

For problems involving changes in operating altitude, the CO_2 line, scale **(6)** is a pivot line for heat input variations; similarly, the heat input line **(8)** is a pivot line for CO_2 changes. In either case, the numerical value of the point on the pivot scale is for reference only and is not to be considered an answer.

The CO_2 scale can be used to indicate the percentage of **excess air** for natural gas and the weight of flue products per 1000 Btu of natural gas burned. Flue gas temperature rise above ambient conditions has been added by graphic methods to permit rapid estimation of flue losses. The input–CO_2 pivot line **(9)** is used to relate changes in heat input to changes in CO_2. The modulation line is employed to correlate efficiency and flue outlet temperature changes during variations of heat input to the appliance.

A review of Table 2-112, which summarizes the eight types of calculation, will show solutions to nearly all common problems of appliance heat transfer and internal draft.

Figure 2-79 was computed for any natural gas with a heating value between 950 and 1100 gross Btu per standard cubic foot. It will apply within 1 per cent to A.G.A. Laboratories' manufactured test gas or any manufactured, mixed, or LP-enriched gas having a similar carbon-to-hydrogen ratio. The relative accuracy with which the chart may be applied to other gases may be estimated from Fig. 2-80, which compares **flue losses** of butane, propane, and natural and coke oven gases. The results on the CO_2 and flue temperature scales are very comparable. Considerable differences exist, however, in the percentage of CO_2 for the same proportion of excess air.

The classes referred to in Table 2-112 are as follows:

1. Forced air furnaces:
   Parallel flow, Counterflow, Crossflow } with { updraft, downdraft, steel or cast iron
2. Gravity furnaces; floor furnaces; room, space, and recessed heaters
3. Water heaters and boilers
4. Deep fat fryers
5. Gas range ovens (for input–CO_2 relationships and flue losses)

Heat Losses from Closed Tanks and Closed Vats[38]

The following procedure can be used to give close estimates of heat losses from tanks and similar closed vessels:

1. Estimate the surface temperature of the tank or insulation surrounding it. In the case of *metallic, uninsulated* tanks, this temperature will be practically the same as the inside temperature.

2. Obtain the surface emissivity, e, from Table 2-98.

3. Obtain the radiation coefficient, h_r, from Fig. 2-81 and the emissivity.

Table 2-112 Applications of the Gas Appliance Performance Chart

(see Fig. 2-79)

Problems involving	Known factors	Find	Procedure	Additional steps or remarks	Answer	Applicable appliance classes*	Applicable fuel gases†
Efficiency, CO_2, flue gas temperature	Efficiency **(1)**, CO_2 **(6)**	Flue gas temp rise **(3)**	**(1) (6)→(3)**			1, 2, 3, 4, 5	N, T, M
	Flue gas temp rise **(3)**, CO_2 **(6)**	Combustion efficiency **(1)**	**(3) (6)→(1)**				
	Efficiency **(1)**, flue gas temp rise **(3)**	CO_2 **(6)**	**(1) (3)→(6)**				
Efficiency, CO_2, overall heat transfer coefficient, area, heat input	CO_2 = 8% **(6)**, flue gas temp rise = 300 F **(3)**, loading = 3000 Btu/sq ft-hr **(8)**	*U:* overall heat transfer coefficient **(11)**	**(6) (3)→(4)** (3.27) ‖ **(4) (8)→(11)**	Variations of this procedure permit solving for flue temperature, CO_2, etc.	$U = 2.75$ Btu/hr-sq ft-°F	1, 2, 3, 4	N, T, M
	Estimated $U = 6.0$ **(11)**, input = 72,000 Btu/hr, area = 4 sq ft, loading = 72,000/4 = 18,000 **(8)**, operating CO_2 = 7% **(6)**	*E:* predicted combustion efficiency **(1)**	**(11) (8)→(4)** (1.82) ‖ **(4) (6)→(1)**		E = 52.5%		
Input and CO_2 (no changes in appliance geometry)	CO_2 = 8.5% **(6)**, input = 80,000 Btu/hr, i.e., 8.0 on **(8)** scale	CO_2 **(6)** at: (a) 100,000 input, (b) 50,000 input	**(6) (8)→(9)** (1.47) ‖ **(9) (8)$_a$→(6)$_a$** ‖ **(9) (8)$_b$→(6)$_b$**	Applies from 40 to 125% of normal input to all appliances with enclosed flames	CO_2: (a) 11%, (b) 5.4%	1, 2, 3, 4, 5	N, T, M C, B, P
Efficiency variations for excess air changes at constant heat input	Input = 60,000 Btu/hr **(8)**, CO_2 = 9% **(6)**, E = 70% **(1)**	Efficiency at 6% CO_2 and 60,000 Btu/hr input	**(1) (6)→(4)** (2.34) ‖ **(4) (6)′→(1)′**	Total air may be changed by flue outlet or combustion air inlet baffles, or by different draft hood	E = 61%	1, 2, 3, 4	N, T, M
Effect of altitude changes on heat input and CO_2 in flue products	Input = 50,000 Btu/hr, i.e., 5.0 on **(8)** scale, at sea level **(10)**, CO_2 value not reqd.	Input at 6000 ft **(10)′** to obtain same CO_2 as that at sea level (without altering appliance)	**(10) (8)→(6)** (5.8) ‖ **(6) (10)′→(8)′**	Altitude changes follow rule of 4% reduction in input or in weight flow/1000 ft alt. increase‡	Input = 39,000 Btu/hr	1, 2, 3, 4, 5	N, T, M C, B, P
	CO_2 = 7.5% **(6)** at sea level **(10)**	CO_2 at same input and 4000 ft alt. **(10)** (without altering appliance)	**(10) (6)→(8)** (6.5) ‖ **(8) (10)′→(6)′**	Value on **(8)** acts as pivot only	CO_2 = 9.0%		
	CO_2 = 7.5% **(6)**, input = 50,000 Btu/hr, i.e., 5.0 on **(8)′** scale, at sea level **(10)**	CO_2 at 30,000 Btu/hr input, i.e., 3.0 on **(8)″** scale, and 5000 ft alt. **(10)′**	(a) CO_2 **(6)′** at 5000 ft and 50,000 input: **(10) (6)→(8)** (6.5) ‖ **(8) (10)′→(6)′** (9.4)	(b) CO_2 **(6)″** at 30,000 input: **(6)′ (8)′→(9)** (1.9) ‖ **(9) (8)″→(6)″**	CO_2 = 5.5%		
Weight of flue products	Input = 150,000 Btu/hr, CO_2 = 7.9% **(6)**	Flue gas rate	**(6)→(7)** (1.12 lb/1000 Btu)		(1.12 × 115,000)/1000 = 129 lb/hr	1, 2, 3, 4, 5	N
Effect of modulated heat input on appliance efficiency (from 50 to 125% of normal heat input)	Input = 170,000 Btu/hr, i.e., 17 on **(8)** scale, CO_2 = 7% **(6)**, efficiency = 80% **(1)**	Efficiency **(1)′** at 120,000 Btu/hr, i.e., 12 on **(8)′** scale (without altering appliance)	**(6) (8)→(9)** (0.78) ‖ **(9) (8)′→(6)′** CO_2 at 120,000 input = 4.8%	**(1) (6)→(2)** (2.2) ‖ **(2) (6)′→(1)**	E = 78.3%	1, 2, 3, 4	N, T, M

Table 2-112 Applications of the Gas Appliance Performance Chart (Continued)

Problems involving	Known factors	Find	Procedure	Additional steps or remarks	Answer	Appli. cable appli- ance classes*	Appli- cable fuel gases‡
		Efficiency **(1)″** at 220,000 Btu/hr, i.e., 22 on **(8)″** scale	Using points **(9)** and **(2)** as before: CO_2 at 220,000 input = 9.2% and **(2) (6)″→(1)″**	For precise work, the modulation point should be found by operating at two different heat inputs	E = 81.3%		
Temperature of flue gases after dilution at draft hood	Flue gas temp rise = 450°F **(3)** CO_2 = 8% **(6)** (before dilution)	Temperature at 3% CO_2 **(6)′** after dilution with room air	**(6) (3)→(1)** (77.5) ‖ **(1) (6)′→(3)′**	Assume no heat loss at draft hood; thus, E is constant	Temp = 190°F above room	1, 2, 3, 4	N, T, M C, B, P
	Flue gas temp = 800 F at 8.0% CO_2 **(6)** in recessed heater. Dilution air is heated to 300 F, CO_2 after dilution is 2.5% **(6)′**	Gas temperature downstream of draft hood	Temp rise entering hood = 800 − 300 or 500°F **(3)** above dilution air temp **(6) (3)→(1)** (76.0) ‖ **(1) (6)′→(3)′**	Dilution air temperature is treated as ambient air temperature for this calculation	Temp = 175°F above 300, or 475 F		

* See text.

† Code for fuel gases: N = natural; T = manufactured, 525 Btu, 0.42 sp gr; M = mixture, natural and manufactured; C = coke oven; B = butane and butane-air; P = propane and propane-air.

‡ This rule as subsequently revised states that appliances do not have to be derated at elevations up to 2000 ft; for elevations above 2000 tf, ratings should be reduced at the rate of 4 per cent for each 1000 ft above sea level.

4. Obtain the convection coefficient, h_c, from Fig. 2-82.

5. Obtain the thermal conductivity of the insulation (if present) from Tables 2-93 to 2-97 in Btu per hr-sq ft-°F per in.

6. Evaluate the ratio of the thickness of insulation in inches to the thermal conductivity of the insulation, L_1/K_1.

7. Evaluate U, overall heat transfer coefficient, from either of the following equations:

For *insulated* tanks:

$$U = \frac{1}{\frac{L_1}{K_1} + \frac{1}{h_c + h_r}} \tag{44}$$

For *uninsulated* tanks:

$$U = h_c + h_r$$

8. Calculate the surface temperature from the following equation:

$$t_s = t_i - [(t_i - t_a)(UL_1/K_i)] \tag{45}$$

where: t_s = surface temperature of insulation, °F
t_i = temperature of tank contents, °F
t_a = surrounding temperature, °F

If initial estimate of surface temperature is in error, re-evaluate U.

9. Calculate overall heat loss from the equation:

$$q = UA\Delta t$$

where: q = overall heat loss, Btu per hr
A = area exposed to the surrounding air, sq ft
Δt = difference in temperature between tank contents and surrounding air

Heat Losses from Open Vats and Kettles

The following procedure can be used to give estimates of heat losses from open vats or kettles.[38]

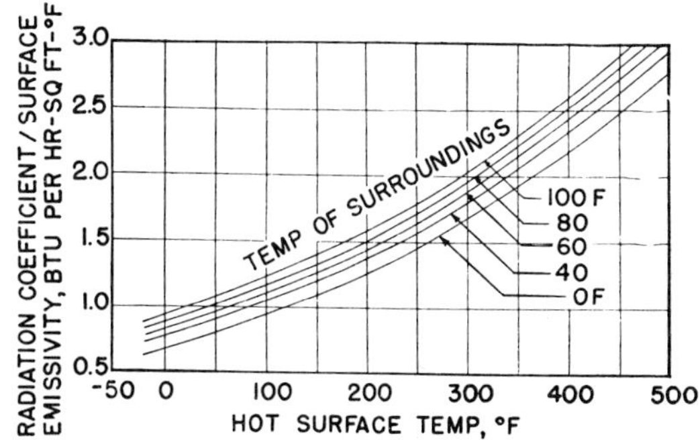

Fig. 2-81 Radiation heat transfer coefficients.[39]

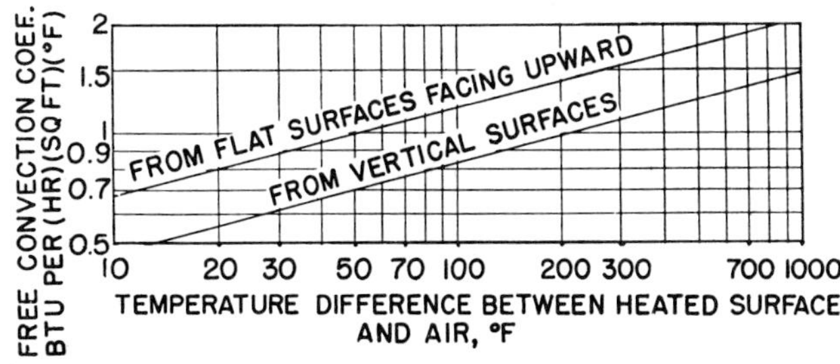

Fig. 2-82 Free convection coefficients from flat surfaces to air.[39]

1. Knowing the surface and air temperatures, obtain the coefficient of convection, h_c, from Fig. 2-82.

2. Obtain the coefficient of radiation, h_r, from Fig. 2-81, assuming e = 0.90 to 0.98 (e = emissivity).

3. (a) For *still air*, calculate the evaporation rate from the equation:

$$E = 0.00084\,(D_v/z)^{0.25}(p_v - p_a)^{1.25}(M_a - M_v)^{0.25}(M_v) \tag{46}$$

where:

E = evaporation rate, lb per hr-sq ft

Table 2-113 Heat Loss from Water Surface at Various Air and Water Temperatures and Air Velocities

(Btu/hr-sq ft; relative humidity 70 per cent)

Water temp, °F	Air velocity, ft/sec: 0	0.5	1	2	5	10	20	50
Air temp, 100 F								
100	35	70	82	105	130	200	330	700
110	105	170	200	250	315	480	800	1,600
120	200	295	350	420	550	810	1,380	2,750
130	325	460	520	650	820	1,250	2,050	4,200
140	500	670	800	950	1230	1,800	3,000	6,200
150	730	930	1100	1300	1700	2,600	4,300	8,700
160	1000	1290	1500	1800	2400	3,600	5,900	12,100
170	1380	1750	2050	2450	3300	4,950	8,200	17,000
180	1830	2450	2800	3300	4600	7,000	11,500	24,000
190	2400	3500	4050	4800	6600	10,000	16,500	34,000
200	3100	5100	6000	7300	9400	14,300	24,000	48,000
Air temp, 80 F								
80	17	34	40	49	66	106	185	330
100	58	96	113	135	185	280	460	920
110	210	285	320	380	540	780	1,250	2,550
120	315	420	470	560	760	1,150	1,850	3,800
130	460	580	660	780	1090	1,620	2,600	5,200
140	640	780	900	1060	1480	2,200	3,600	7,200
150	860	1020	1200	1420	1970	2,950	4,800	9,700
160	1150	1330	1590	1900	2550	3,800	6,300	13,000
170	1500	1730	2050	2500	3400	5,200	8,500	17,300
180	2000	2400	2750	3400	4700	7,100	12,000	24,000
190	2550	3400	3900	4800	6600	10,000	17,000	34,000
200	3300	4900	5500	6900	9500	14,000	24,000	50,000
Air temp, 60 F								
60	7	18	21	26	34	56	95	180
70	33	52	60	78	105	170	270	580
80	78	110	125	150	210	315	510	1,050
90	130	180	210	240	330	490	780	1,600
100	205	270	320	350	480	710	1,150	2,300
110	290	370	420	480	670	1,000	1,600	3,300
120	400	500	570	660	930	1,350	2,200	4,400
130	550	780	750	890	1240	1,800	3,000	6,000
140	710	870	970	1150	1600	2,400	4,000	8,000
150	950	1130	1260	1510	2100	3,200	5,300	10,500
160	1230	1450	1600	2000	2700	4,050	6,900	13,700
170	1600	1900	2100	2600	3700	5,200	9,100	18,500
180	2050	2600	2900	3550	5000	7,200	12,500	25,000
190	2600	3550	4000	4950	6900	10,300	17,500	35,000
200	3300	5200	5700	7000	9700	14,600	25,000	50,000
Air temp, 40 F								
40	3	8	10	13	17	26	46	85
50	25	36	42	47	60	100	160	320
60	58	75	86	94	130	190	310	620
70	90	115	130	150	210	300	490	970
80	135	170	195	225	310	450	720	1,460
90	190	230	265	310	440	640	1,030	2,100
100	260	315	360	420	600	860	1,400	2,800
110	350	420	470	550	790	1,130	1,880	3,700
120	460	550	610	720	1020	1,500	2,500	5,000
130	600	700	770	930	1300	1,950	3,150	6,400
140	800	880	970	1200	1760	2,500	4,200	8,300
150	1020	1130	1430	1530	2150	3,200	5,300	10,600
160	1300	1450	1570	2000	2800	4,200	7,000	14,000
170	1700	1900	2100	2600	3700	5,500	9,400	18,800
180	2200	2550	2750	3500	4900	7,400	12,500	25,000
190	2750	3400	3750	4700	6800	10,300	17,000	34,000
200	3400	4800	5500	6700	9700	15,000	25,000	50,000

D_v = diffusion coefficient of the evaporating vapor in air at the *average* of the air and liquid temperatures, sq ft per hr

z = viscosity of the air–vapor mixture at the average on the air and liquid temperatures, lb per ft-hr

p_v = vapor pressure of the evaporating liquid at the temperature of the liquid, in. Hg

p_a = partial pressure of the evaporating liquid in the air, in. Hg

M_a = molecular weight of the air (28.8)

M_v = molecular weight of the evaporating liquid

Note: If p_a is larger than p_v, condensation occurs.

(b) For *moving air*, calculate E from the equation:

$$E = FM_vD_v^{0.667}\left(\frac{z}{d}\right)^{0.333} \log\left(\frac{30 - p_a}{30 - p_v}\right) \tag{47}$$

where:

$$F \begin{cases} = \left(0.46 + 0.0693\,\dfrac{Vd}{z}\right) & \text{when } \dfrac{Vd}{z} \text{ is less than 30} \\ = 16.2\left(\dfrac{Vd}{z}\right)^{0.78} & \text{when } \dfrac{Vd}{z} \text{ is greater than 30} \end{cases}$$

For the air moving over the surface:

V = velocity, ft per sec

d = density, lb per cu ft

z = viscosity, lb per ft-hr

Other values are defined above.

4. Calculate overall heat loss from the equation:

$$q = (h_c + h_r)A(\Delta t) + EbA \tag{48}$$

where:

q = overall heat loss, Btu per hr

A = area exposed to air, sq ft

Δt = difference in temperature between tank contents and surrounding air, °F

b = latent heat of vaporization at the temperature of the liquid, Btu per lb

The h_c, h_r, and E are as previously defined.

Note: If condensation occurs, E should have a negative value.

5. If the liquid in the open container is *not* water and its temperature is *below* the surrounding air temperature, calculate the condensation rate of water from the applicable equation given in step 3.

Since *water* is probably the fluid most often encountered as the liquid in open vessels, the heat loss may be obtained directly from Table 2-113 if the water temperature and air velocity are known. These data[38] are based on a 70 per cent relative humidity. At water temperatures *above* 150 F, varying humidity does not seriously affect the results. However, at water temperatures *below* 150 F, where humidity varies significantly, serious errors will be avoided by the calculation of heat loss with the foregoing equations.

References 39 and 40 are excellent sources for further study of the evaporation from a free water surface.

Calculated comparisons of heating requirements for **bare open tanks** and **insulated tanks with covers** are given in the

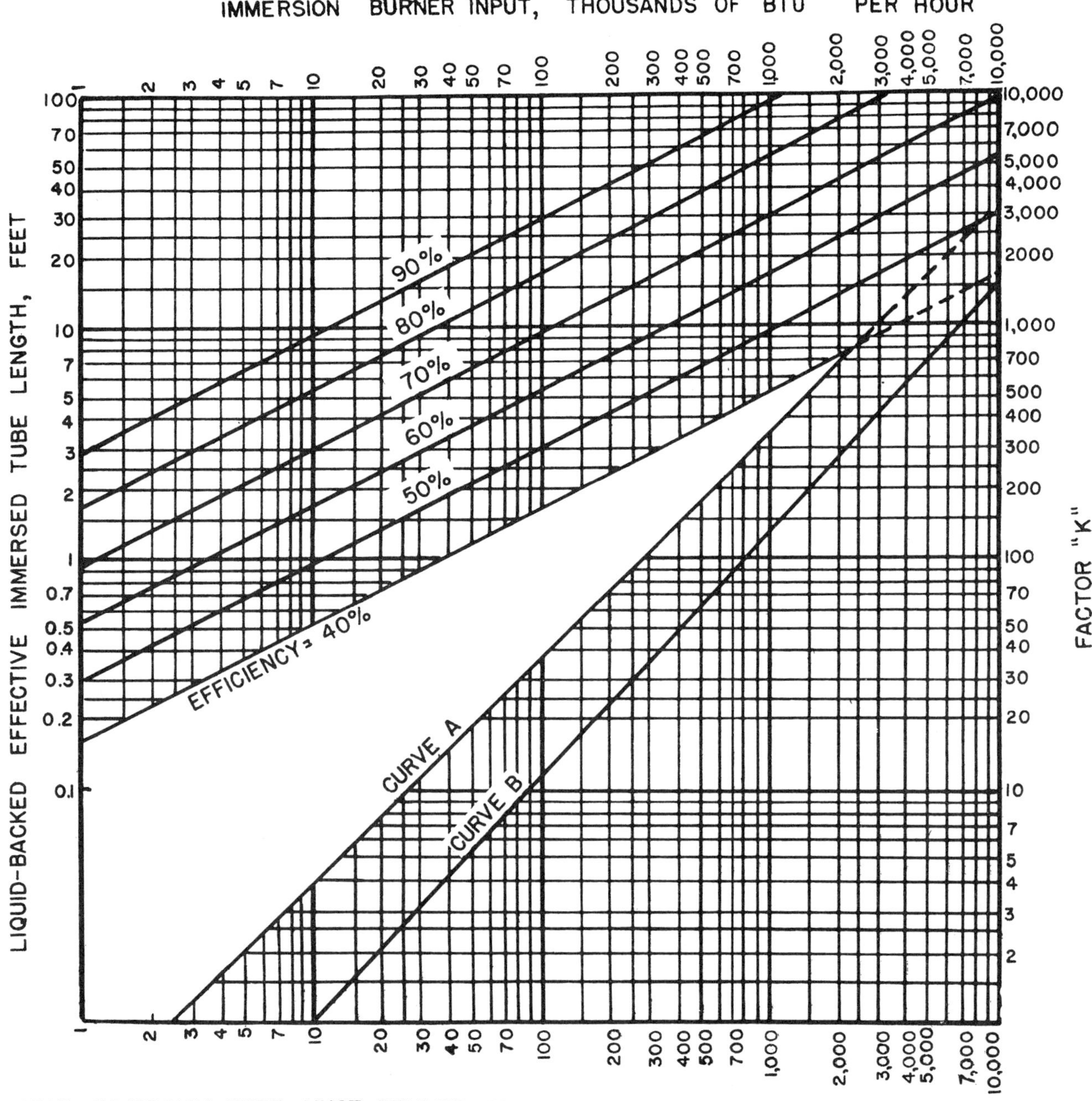

Fig. 2-83 Sizing chart for immersion tube heaters. **Note:** Efficiency = burner input minus flue loss. Curve A is for estimating tank heat loss and curve B is for estimating heat requirements to heat tank and contents in a specific amount of time.[41]

article, "Heating Vats and Tanks by Gas," in the April 1943 issue of *Gas in Industry*.

Immersion Tube Design

An equation for the sizing of gas-fired immersion tube tank heaters was developed in A.G.A. Laboratories Research Bulletin No. 24.

$$E = 20 \log (L^2/R) + 71 \qquad (49)$$

where:

E = efficiency, input minus flue losses, per cent

L = equivalent centerline length of immersed tube including fittings, plus 1.1 ft per 90 deg elbow and return bend, ft

R = burner input rate, MBtu per hr

The tube diameter is not needed in the above equation, since this dimension has a negligible effect on efficiency.

In the absence of manufacturers' recommendations, the following conservative relationships may be used to determine input rates per square inch of cross-sectional area:

With atmospheric burners:

$$5000(10/L)^{0.5}$$

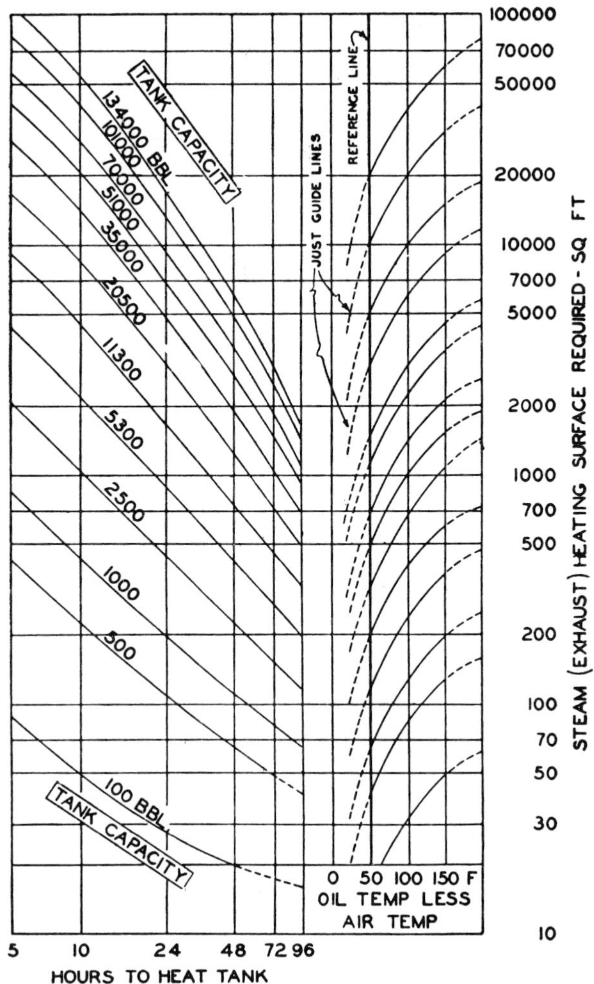

Fig. 2-84 Steam coils for gas oils and crude oils.

With open-type pressure burners:

$$15{,}000(10/L)^{0.5}$$

Equation **49** is charted in the upper section of Fig. 2-83. The lower section of this figure determines the heat that must be delivered to a water solution in a *bare steel tank:*

1. Hourly heat loss with negligible surface agitation.

$$K = A(\Delta t/100)^2 \tag{50}$$

2. Heating-up requirement.

$$K = V\Delta t/100H \tag{51}$$

where: K = factor required in Fig. 2-83
A = surface area, sq ft
Δt = solution temperature − ambient, °F
V = solution volume, cu ft
H = allowable heating time, hr

Oil Tank Heaters

Factors[20] involved in design are: time of heating, pour point of oil or oil temperature required, air temperature, wind velocity, viscosity of oil, dirtying or fouling of the steam coil, steam temperatures, and capacity and shape of tank.

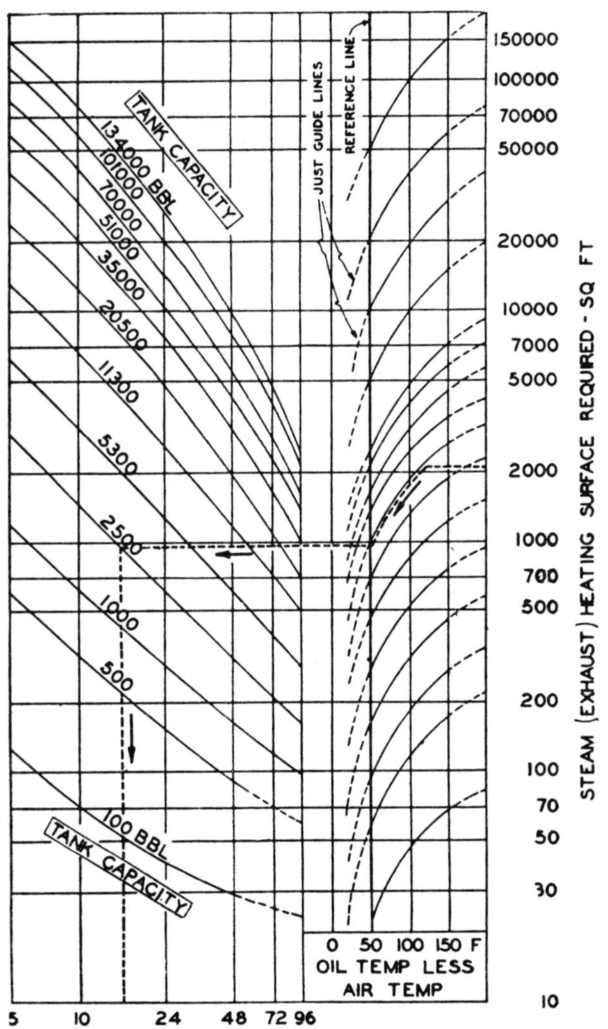

Fig. 2-85 Steam coils for lubricating oils.

Three charts are given, Figs. 2-84, 2-85, and 2-86, for (1) gas oils and crude oils, (2) lubricating oils, and (3) semisolids oils such as Bunker C. These charts assume no wind or rain, and are based on exhaust steam at 212 F. For other pressures the required surface may be *reduced* by the following multipliers:

Pressure, psig	Temp, °F	Factor
Exhaust	212	1.0
21	260	0.67
63	310	0.50
138	360	0.33

Examples of the use of these charts are as follows:

Example 1: A 250-bbl tank of Bunker C fuel oil is to be heated in 7 hr. The air and tank are at 50 F, and it is necessary to hold the oil at 120 F. Thus, the difference in temperature between air and oil will be 70°F. The dotted line in Fig. 2-86 shows the solution. Move upward on the 7-hr line to the 250-bbl capacity and follow horizontally to the right to the 50°F temperature-difference line. Follow a path first between the guidelines at the right to a temperature difference of 70°F and then horizontally to the right, finding a surface of

Table 2-114 Comparison of Heat-Absorption Rates in Fire Tests[43]

No.	Tests	Wetted surface, sq ft	Avg absorp. rate on wetted surface, Btu/hr-sq ft	Total absorp. rate on vessel, Btu/hr
1	Hottel, avg. 36 tests	296	12,700	3,760,000
2	Hottel, avg. 13 tests	296	7,226	2,139,000
3	Standard Oil Co., Calif.	29	16,000	416,000
4	Standard Oil Co., Calif.	105	32,000	3,370,000
5	Underwriters' Labs.	24	32,500	780,000
6	Rubber Reserve Corp. Test #17	400	23,200	9,280,000
7	Rubber Reserve Corp. Test #17	400	21,000	8,400,000
8	Rubber Reserve Corp. Test #17	9.0	30,400	274,000
9	API Project Test #1	6.1	15,700	95,800
10	API Project Test #2	6.1	16,800	102,500
11	Report to API on a 48-ft butane sphere	4,363	5,400	23,560,000
12	Union Carbide Corp.*	1,480	28,800	42,600,000

* U.C.C. Fire Research Lab Report #62.

520 sq ft. If 120-lb steam is used rather than exhaust steam, only about 40 per cent as much surface is required, or 210 sq ft.

Example 2: A 2500-bbl tank of cylinder stock must be heated to 130 F for pumping. The air temperature is 10 F, and the tank has 2100 sq ft of coil in it. Can we begin pumping in one day?

Use Fig. 2-85. Locate 2100 on the sq ft scale on the right. Move horizontally to a temperature difference of 120 (130 less 10) and move downward to the left between the guidelines to the 50°F reference line. Move horizontally to the 2500-bbl line and read 15 hr on the bottom scale.

The steam (exhaust) required for heating will be about 14 lb per 1000 bbl per degree difference in air and oil temperature. The heated oil temperature is 130 F and the air is at 10 F. During heating the steam consumption is about 120 × 14 × 2500/1000, or 4200 lb per hr. Less than half as much steam is required to maintain the tank at the high temperature.

Fire Exposure of Liquid Fuel Storage Tanks

Liquid storage containers exposed to external fire are subject to unusual heat absorption conditions because completely enveloping fires are all but impossible except for the smallest containers. Furthermore, there are winds (natural and induced by the fire itself) to consider. The heat absorption rate is important in determining such features as relief valve capacity, cooling requirements, and insulation for protection. Table 2-114 shows the heat absorption rates obtained in a series of diverse fire tests.

Flames of open fires generally have a core of flammable vapors which are largely unburned. Combustion takes place on the external envelope of this core, accompanied by considerable cracking of the vapors. Thus, there is generally a pall of black smoke, which in turn influences the heat transfer rates. In extensive refinery fires flames are turbulent and, as masses of burning vapor tumble and bellow, the smoky mantle is displaced. The bright flames, which show intermittently, are not white as in furnaces, but red or orange in color, indicating that lower flame temperatures are involved.

Although there is much emphasis on relief valve sizing, the major hazard resulting from fires applied to pressure containers is that of weakening of the container metal strength as a result of the rise in temperature. Figure 2-87 should serve to give an approximate idea of the drop in tensile strength of steel when steel is heated. In this connection it should be remembered that the internal pressure is held constant by the setting of the relief valves—provided they are properly

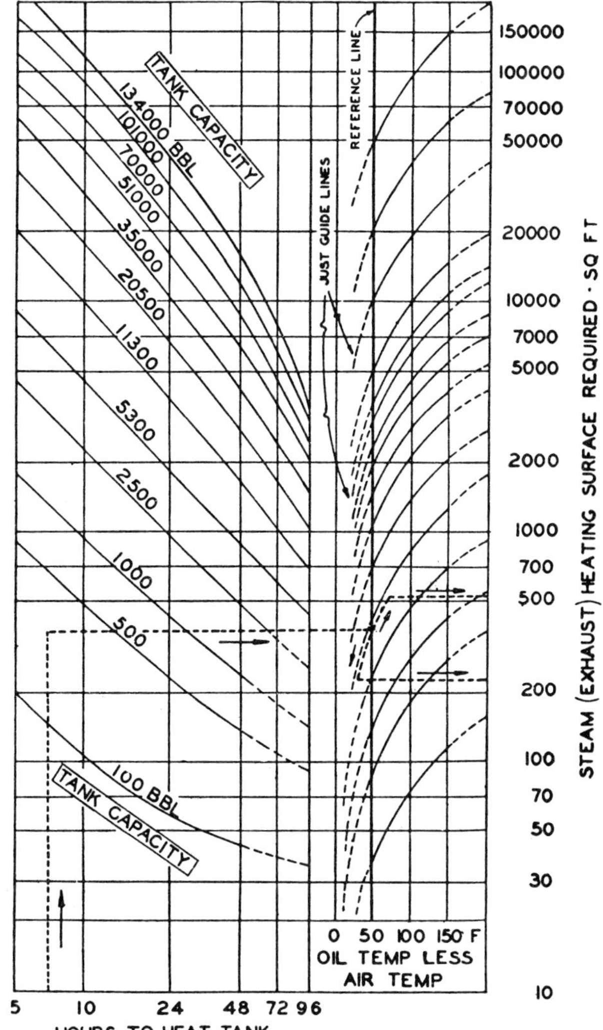

Fig. 2-86 Steam coils for semisolid oils.

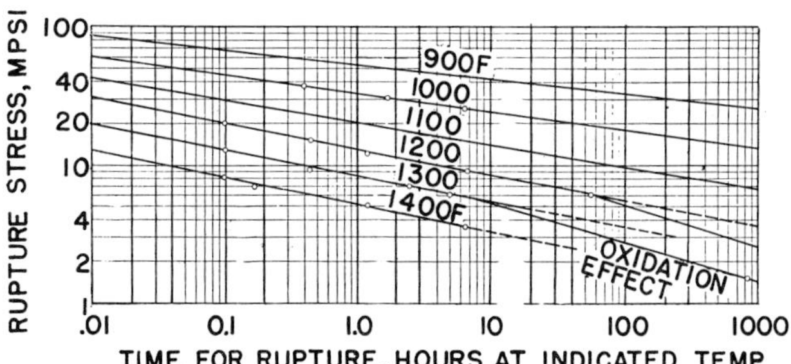

Fig. 2-87 Effect of overheating steel.[43]

sized to discharge all evaporated liquid. Thus, the container metal stress remains substantially unchanged while the metal's strength is reduced as it is heated.

REFERENCES

1. Perry, J. H., ed. *Chemical Engineers' Handbook,* 3d ed. New York, McGraw-Hill, 1950.
2. Tubular Exchanger Manufacturers Assn. *Standards,* 4th ed., p. 57. New York, 1959.
3. Chilton, T. H., and others. "Heat-Transfer Design Data and Alignment Charts." *Trans. ASME.* 1933.
4. McAdams, W. H. *Heat Transmission,* 3rd ed. New York, McGraw-Hill, 1954.
5. A.G.A. Laboratories. *Comparative Study of Various Methods of Cooking; Part II—Heat Application in Gas Oven Cookery.* (Res. Rept. 1182) Cleveland, Ohio, 1952.
6. Hutchinson, F. W. *Industrial Heat Transfer.* New York, Industrial Press, 1952.
7. Holcomb, D. E., and Brown, G. G. "Thermodynamic Properties of Light Hydrocarbons." *Ind. Eng. Chem.* 34: 591, May 1942.
8. Badger, K. L. *A Study of Small Diameter Gas Glow Tubes.* (A.G.A. Laboratories Res. Bull. 56) Cleveland, Ohio, 1950.
9. Hooks, I. J., and Kerze, F., Jr. "Heat Transfer Coefficients for Pipe and Wire." *Chem. & Met. Eng.* 52: 117, Nov. 1945.
10. Hackforth, H. L. *Infrared Radiation,* p. 190. New York, McGraw-Hill, 1960.
11. Kreith, F. *Principles of Heat Transfer.* Scranton, Pa., International Text Book Co., 1958.
12. Hottel, H. C. "Radiant Heat Transmission." *Mech. Eng.* 52: 699–704, July 1930.
13. Plyler, E. K. "Infrared Radiation from a Bunsen Flame." *J. Res. Natl. Bur. Std.* (RP 1860) 40: 113, 1948.
14. Keller, J. D. *Radiant Gas Burners.* (Paper 50-A-59) New York, ASME, 1950.
15. Haslam, R. T., and Russell, R. P. *Fuels and Their Combustion.* New York, McGraw-Hill, 1926.
16. Trinks, W. "Simplified Calculation of Radiation from Non-Luminous Furnace Gases." *Ind. Heating,* 14: 40–6, Jan. 1947.
17. Hottel, H. C., and Egbert, R. B. "Radiant Heat Transmission from Water Vapor." *Am. Inst. Chem. Engrs. Trans.* 38: 531–65, 1942.
18. Trinks, W. *Industrial Furnaces,* vol. 1, 4th ed. New York, Wiley, 1951.
19. Nelson, W. L. "Heat Rates outside Tubes." *Oil Gas J.* 45: 123–4, May 11, 1946.
20. Nelson, W. L. "Heating Coils in Tanks." *Oil Gas J.* 43: 22, Oct. 7, 1944.
21. Davies, G. F. "Quick Estimation Method for Heat Exchanger Dimensions." *Chem. Eng.* 59: 170–1, Mar. 1952.
22. Davies, G. F. "Short Cuts to Heat Exchanger Coefficients." *Chem. Eng.* 58: 122–4, Aug. 1951.
23. Bowman, R. A., and others. "Mean Temperature Difference in Design." *Trans. ASME.* 62: 283–93, 1940.
24. Brooks, G., and Su, G. J. "Heat Transfer in Agitated Kettles." *Chem. Eng. Progr.* 55: 54–7, Oct. 1959.
25. Nelson, W. L. "Fouling Factors." (Refiners' Notebook, Nos. 94, 95) *Oil Gas J.* 45: 141, May 25, 1946; 95, June 1, 1946.
26. Cryder, D. S., and Finalborgo, A. C. "Heat Transmission from Metal Surfaces to Boiling Liquids." *Am. Inst. Chem. Engrs. Trans.* 33: 346–62, 1937.
27. Johnston, L. C. "Finned Tubes Find Process Uses." *Chem. Eng.* 59: 148–9, Jan. 1952.
28. Grober, H. "Heating and Cooling of Bodies of Simple Geometrical Shapes." *Z. Ver. Deut. Ing.* May 23, 1925. Abstracted in *Fuels and Furnaces* 3: 807–12, Aug. 1925.
29. Rosebaugh, T. "Heat Transfer in Grid-Packed Scrubbers Used for Cooling Oil Gas Saturated With Water Vapor." *Pacific Coast Gas Assoc. Trans.* 18: 322–39, 1927.
30. Colburn, A. P., and Hougen, O. A. "Design of Cooler Condensers for Mixtures of Vapors With Noncondensing Gases." *Ind. Eng. Chem.* 26: 117–82, Nov. 1934.
31. Revilock, J. F., and others. "Heat and Mass Transfer Analogy." *Chem. Eng. Progr.* 55: 40–4, Oct. 1959.
32. Sherwood, T. K., and Pigford, R. L. *Absorption and Extraction,* 2nd ed. New York, McGraw-Hill, 1952.
33. Young, F. S. "Aspects of Gas Heat Exchangers as Applied to Transmission System Problems." *Gas Age* 105: 40–4, Feb. 16, 1950.
34. Degler, H. E. "Water Cooling Towers and Air Cooled Heat Exchangers." *Oil Gas J.* 50: 139, Feb. 11; 159, Feb. 18; 181, Feb. 25, 1952.
35. Lauderbaugh, A. B. "Immersion Fired Heat Exchangers at Regulator Stations." *Gas Age* 106: 68–70, Sept. 28, 1950.
36. Stone, R. L. *Fundamentals of Heat Transfer in Domestic Gas Furnaces.* (A.G.A. Laboratories Res. Bull. 63) Cleveland, Ohio, 1951.
37. Mueller, J. C. "Stoichiometry—A Tool in Gas Appliance Development." *Gas* 37: 94–7, Oct. 1961.
38. "Heat Loss from Water Surface in Tanks." *Heating and Ventilating* 46: 88–9, Apr. 1949.
39. Friedman, S. J. "Heat Losses from Tanks, Vats and Kettles." *Heating and Ventilating* 45: 94–108, Apr. 1948.
40. Lurie, M., and Michailoff, N. "Evaporation from a Free Water Surface." *Ind. Eng. Chem.* 28: 345–9, Mar. 1936.
41. Eeles, C. C. "Heating by Immersion and Submerged Combustion." *Am. Gas J.* 165: 14–6, Sept. 1946.
42. Dukler, A. E. "Dynamics of Vertical Falling Film Systems." *Chem. Eng. Progr.* 55: 62–7, Oct. 1959.
43. Am. Petrol. Inst. *Recommended Practice for the Design and Construction of Pressure-Relieving Systems in Refineries.* (RP 520) New York, 1955.
44. DeWerth, D. W. *A Study of Infra-Red Energy Generated by Radiant Gas Burners.* (A.G.A. Laboratories Res. Bull. 92) Cleveland, Ohio, 1962.

SECTION 3

PRODUCTION AND CONDITIONING OF MANUFACTURED GAS

R. E. Kruger (retired), *Section Chairman*, Rochester Gas & Electric Corp., Rochester, N. Y.
E. G. Boyer (retired), *Editorial Reviewer*, Philadelphia Electric Co., Philadelphia, Pa.
James F. Bell, Chapter 7, Portland Gas & Coke Co., Portland, Ore.
Mark G. Eilers, Chapter 4, Rochester Gas & Electric Corp., Rochester, N. Y.
E. D. Freas, Chapter 2, Philadelphia Electric Co., Philadelphia, Pa.
C. B. Glover, Chapter 5, United Engineers & Constructors, Philadelphia, Pa.
H. A. Gollmar, Chapter 8, Koppers Co., Inc., Pittsburgh, Pa.
W. J. Huff, Chapter 8, University of Maryland, College Park, Md.
W. A. Leech, Jr., Chapter 1, Koppers Co., Inc., Pittsburgh, Pa.
H. R. Linden, Chapter 7, Institute of Gas Technology, Chicago, Ill.
Charles R. Locke, Chapter 4, The Peoples Gas Light & Coke Co., Chicago, Ill.
Carl J. Lyons, Chapter 4, Battelle Memorial Institute, Columbus, Ohio
S. A. Petrino, Chapter 2, Kings County Lighting Co., Brooklyn, N. Y.
C. U. Pittman, Chapter 2, Koppers Co., Inc. Verona, Pa.
J. M. Reid, Chapter 7, Institute of Gas Technology, Chicago, Ill.
C. C. Russell, Chapter 1, Koppers Co., Inc., Pittsburgh, Pa.
S. C. Schwarz, Chapters 2 and 7, Consulting Chemical Engineer (formerly with Portland Gas & Coke Co.), Portland, Ore.
E. F. Searight, Chapter 7, Institute of Gas Technology, Chicago, Ill.
J. P. Templin, Chapter 4, Philadelphia Coke Co., Philadelphia, Pa.
N. C. Updegraff, Chapter 8, Girdler Construction Div., Chemetron Corp. Louisville, Ky.
F. E. Vandaveer (retired), Chapters 3, 4, and 9, Con-Gas Service Corp., Cleveland, Ohio
C. G. von Fredersdorff (deceased), Chapter 9, Institute of Gas Technology, Chicago, Ill.
D. A. Vorum, Chapter 6, Consulting Chemical Engineer (formerly with M. W. Kellogg Co.), New York, N. Y.

CONTENTS

Chapter 3—Manufactured Gas Conditioning

Chapter 4—Producer Gas

Chapter 5—Blue Gas and Carbureted Water Gas

Chapter 6—Fuel and Synthesis Gases from Gaseous and Light Liquid Hydrocarbons

Chapter 7—Oil Gas

Chapter 8—Manufactured Gas Purification

Chapter 9—Substitute Natural Gas from Coal

Tables

Figures

Chapter 1
Coal Carbonization

by C. C. Russell and W. A. Leech, Jr.

NATURE OF COAL CARBONIZATION

Commercial coke and coal gas are made by heating bituminous coals in coke ovens or retorts. Only the bituminous class of coals is used since anthracite and subbituminous coals, lignites, and peat do not exhibit fusion with subsequent solidification when heated. Definitions of the various classes of coals by rank will be found in **ASTM Standards.**[1] When bituminous coals are heated in the *absence* of air, softening first appears between 575 and 825 F (depending upon the type). At or near the temperature at which coal begins to fuse, gases of decomposition first appear in appreciable quantities; as the temperature rises, the quantities increase. After the initial fusion stage the degree of **fluidity** increases with increasing temperature to a maximum, after which it decreases. Complete rigidity of the mass is attained between 840 and 930 F, after which its structure continues to change due to further devolatilization and shrinkage.

The degree of fluidity attained by a coal is dependent upon its rank, as indicated by its fixed carbon content[1] or, more commonly, by its volatile matter content. A detailed discussion of the mechanics of coal carbonization was presented in 1947.[2] Brewer[3] described many different methods for measuring plastic properties of coals during heating. The Gieseler method[4] has found wide interest in the United States.

When *coking* coal is charged into coke ovens, softening and gas evolution occur. The oven shape and the physical condition of the coal influence this process. The **chemical recovery** (by-product) **oven** is a narrow, slot-type chamber. The oven wall temperature at the time of charging ranges from 1500 to 2125 F, depending on the operating rate and other conditions. At a *normal* operating rate the oven wall temperature is about 2000 F. When the coal comes in contact with the hot walls it softens immediately. As the heat penetrates into the charge, two layers of softened coal are formed adjacent to the heating walls. As carbonization progresses, these *plastic zones* advance toward the oven center, leaving coke of varying temperatures between them and the oven walls. Coke temperatures[5] will vary from wall temperature down to about 930 F. Temperatures in the plastic zone will vary from 930 to 660 F, while uncarbonized coal at the oven center will be at about 212 F.

Most of the gases and vapors are liberated on the hot side of the plastic zone and in the adjacent coke. They then move out toward the walls thru shrinkage cracks in the coke. Small quantities of gas are present in the uncarbonized coal but are a minor portion of the total. Rich vapors passing thru the hot coke are partially decomposed and deposit a silvery carbon film on the coke surface. The coking process in the chemical recovery oven is not completed until the two plastic zones have met at the center, the final plastic mass has been carbonized, and the coke has shrunk. Any attempt to push out coke before these processes are complete will result in its sticking to the oven walls.

All coals charged into chemical recovery ovens exert some pressure against the oven walls during carbonization. This pressure depends on the kind of coal and conditions of coking. Aside from the coal characteristics, *bulk density* is the most important factor affecting carbonization pressure. The higher this density, the higher the pressure is. Some coal charges may develop pressures high enough to damage the oven walls.

Determination of the pressure developed by coals, sources of this pressure, and methods of control have been extensively investigated. Brewer[3] reviews methods developed up to 1942. In addition, methods for reducing coking pressure have been reported.[6]

Pressures exerted by coals against the coke oven walls prior to the meeting of the two plastic zones at the oven center have been shown to be on the order of 0.2 to 1.0 psi for high-volatile coals or mixtures of high- and low-volatile coals most generally used. This pressure is fairly constant throughout the coking period after the plastic zones are established. Certain coals or coal mixtures will exert a peak pressure at the time the two plastic zones meet. Even though the fairly constant pressure established previous to this occurrence is less than 1.0 psi, the peak pressure may be as high as 3.0 to 5.0 psi. Such a peak is followed by a precipitous drop due to release of gases between the two plastic zones and also to coke shrinkage. Figure 3-1 illustrates this behavior.

It is generally accepted that in movable-wall ovens carbonization pressures must not exceed 2.0 psi, and in ovens over 10 ft high the maximum pressure should not exceed 1.5 psi, in order to prevent distortion or damage to walls. A coal can exert excessive pressure against the oven walls during coking even though the coke shrinks sufficiently at the end of the coking period to facilitate pushing it out easily.

The commercial value of coke depends on its purity and on such factors as size, strength, and structure. Its physical characteristics are determined not only by the kind of coal and/or the proportions of various coals in the mixture but also by the degree of pulverization, moisture content, bulk density in the ovens, presence and size of impurities such as slate, temperature of oven walls, rate of heating, and other variables.[2]

Approximately 94 per cent of the coal carbonized in the United States is processed in *slot-type coke ovens*, and the remainder in *beehive ovens*. Formerly a small amount of coal was carbonized in *gas retorts*, but this process is no longer used in the United States. The fundamentals of coal carbonization in *beehive ovens* are substantially the same as described above, but heat application is different. *No volatile products are recovered* from beehive ovens; coke is the only product.

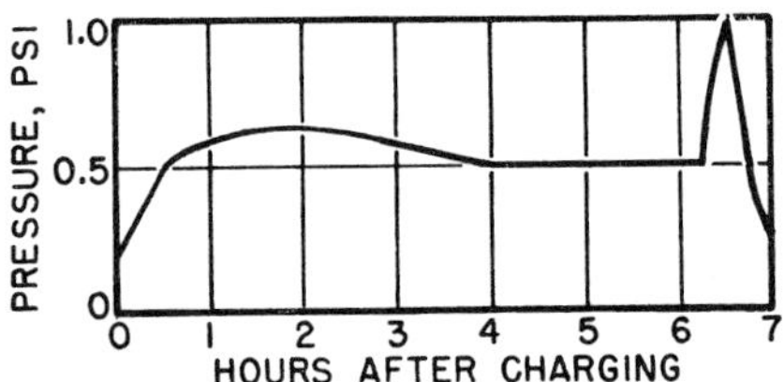

Fig. 3-1 Typical pressures in a movable wall oven with typical coal mix.

80% high volatile coal	Avg flue temp, 2530 F
20% low volatile coal	Moisture, 0.6%
Bulk density, 51.6 lb/cu ft	Coal size (0–⅛ in.), 71.2%

PREPARATION OF COAL FOR CARBONIZATION

Coal used in carbonization passes thru a number of stages of preparation from mine to coke oven.[7] Its selection is most generally based on the quality of coke desired. The yield of coal chemicals is frequently of secondary importance.

Coke oven plant operators customarily keep a 40 to 60 days' supply of coal on hand to avoid shutting down ovens in case of interruption of shipments. Coal, after storage[8] for extended periods, may show a decrease in its coking properties due to oxidation.

Coal is carried on conveyor belts (usually rubber) to the crusher building, where a *breaker* reduces it to pieces less than 1½ in. in size. From the breaker, the coal passes either to a crusher or to a mixer bin. However, before charging, it should be pulverized so that about 80 per cent is less than ⅛ in. in size.

There is a mixer bin for each kind or type of coal. Proper mixing of the various coals for charging is accomplished by regulating rate of flow from the various mixer bins onto the conveyor belts that carry the coals to a mechanical mixer or pulverizer en route to the oven charging bin.

A coal handling system usually includes (1) track hoppers; (2) coal bridge and associated equipment; (3) conveyor belts; (4) a primary crusher, a Bradford breaker; (5) mixing equipment and bins; (6) a hammer mill to pulverize about 80 per cent of the coal to less than ⅛ in. in size; and (7) an oven charging bin.

Coal handling in coke oven plants was described[9] in 1924. Subsequent changes are matters of detail rather than of principle. One important improvement is the use of *weightometers* or similar equipment so that coals are blended by weight rather than by volume.

Accurate control[10] of **bulk density** can be maintained by adding water or as little as 0.1 gallon of oil per ton before the coal reaches the hammer mills. Automatic equipment for the purpose is used in some plants.

HIGH-TEMPERATURE CARBONIZATION

Coal carbonization processes are divided into *three* groups according to carbonization temperature. Low-temperature processes operate in the range of 930 to 1380 F. Medium-temperature carbonization takes place in the range of 1380 to 1650 F. High-temperature carbonization is carried out above 1650 F, but most generally at 1832 F. Temperatures in the oven chamber should not exceed 2150 F because of the tendency of the silica brick to react with the iron of the coal ash above that temperature.

Substantially all carbonization of coal in the United States falls into the high-temperature category. Approximately 85 per cent of the coke produced is used for smelting iron in blast furnaces. No medium-temperature carbonization processes operate in the United States, except for some foundry cokes. Only two low-temperature processes are in operation.

High-temperature carbonization is carried out in **beehive** and **slot-type** ovens. Beehive ovens, because of their low investment cost and ease of shutting down and reactivating, are used when coke demand is high. They are generally inactive in periods of low industrial activity, and are frequently located at or near coal mines.

Many designs of **slot-type coke ovens** have been developed, designated by the name of the designer or that of the company constructing them. At the end of 1959, 15,993 slot-type ovens were in operation. An historical account of their development has been given.[8]

The **beehive oven** is generally a circular chamber with an arched roof and flat floor. After the coal is charged, air input is regulated thru an opening near the top of the door to support combustion of evolved gases which supply the heat needed for the coking operation. Carbonization proceeds downward and requires 48 to 72 hr for its completion.[9] Recent beehive ovens have rectangular rather than circular chambers. The coke is discharged by pushing rather than by pulling.

FUELS FOR HEATING COKE OVENS

The fuel commonly used for heating coke ovens is coke oven gas. From 35 to 40 per cent of the total gas produced is required to fire the coke oven. Other gaseous fuels may be used satisfactorily where conditions are favorable for the sale of coke oven gas. They include low heating value fuels such as producer gas and blast furnace gas or, in certain cases, high heating value gases such as natural gas, refinery oil-still gas, and liquefied petroleum gases.

The time required to carbonize a coal charge when operating at full capacity is a function of the oven width. Coking rates are commonly described in *inches per hour*, but are not a straight line function of oven width. One frequently used formula, assuming wall temperatures are the same, is:

$$\frac{T_1}{T_2} = \left(\frac{W_1}{W_2}\right)^{1.6}$$

where: T_1 and T_2 = coking times, hr

W_1 and W_2 = oven widths, in.

Although it is possible to adjust the combustion in individual flues in each wall and also to adjust the individual walls for proper heating, many other factors affect heat flow into the coal charge. These include: (1) sequence of pushing ovens; (2) procedure used in charging; (3) moisture content of the coal; (4) bulk density of the coal (largely controlled by the moisture, pulverization, and addition of oil); and (5) composition of the coal blend.

The sequence in which the ovens in a battery are charged and discharged profoundly affects the heat distribution of adjacent ovens. It is customary to arrange the pushing sequence so that when an oven charge is completely carbonized, those in adjacent ovens will be approximately 50 per cent carbonized. One plan in common use involves numbering of the ovens in the battery in sequence but omitting the numbers divisible by 10. The ovens are thus numbered 1 to 9, 11 to 19, etc. They are pushed as follows: 1, 11, 21, 31, etc.; 3, 13, 23, 33, etc.; 5, 15, 25, 36, etc.; until all *odd-numbered* ovens have been pushed. Then the *even-numbered* ovens are pushed.

Table 3-1 Heat Balance of Combustion for Oven Heating per Pound of Coal

	W. C. Rueckel*		J. K. Munster	
Sensible heat introduced, Btu				
By coal		1		1.6
By air		...		...
By fuel gas		1		1.4
By combustion of fuel gas		1038		1344.6
Total		1040		1347.6
Sensible heat in products, Btu				
In coke	(1945 F)	484.6	(2000 F)	552
In gas yield		158.6		195
In water vapor	(1220 F)	128.9	(1200 F)	130.5
In tar and light oil		40.7		8.5
In stack gas	(515 F)	226.7	(650 F)	490
In radiation and convection		195.4		178
Total		1234.9		1554.0
Exothermic heat of carbonization		195		206.4

* Tests for Koppers Co., blast furnace underfiring.

Heat requirements are generally expressed in terms of gross Btu required to carbonize one pound. Depending on type of oven, kind of coal being carbonized, physical conditions of the battery, and coking time, heat required varies from 1000 to 1300 or more Btu per pound of coal. The carbonization process is generally considered slightly exothermic. Laboratory studies of heat of carbonization and data obtained from heat balances of coke ovens are available.[8] Table 3-1 shows data as determined by two authorities.

CHEMICAL RECOVERY COKE OVENS

General

A typical modern oven has a chamber about 40 ft long, 13 ft high, and 18 in. (average) wide. Actual width of the oven may taper from 16½ in. on the pusher side to 19½ in. on the coke side to facilitate pushing. Allowing a one-foot gas space at the top, such an oven has a capacity of 720 cu ft of coal. For calculating capacity, assume coal density as 50 lb per cu ft. The typical oven would then hold 18 net tons of coal per charge. Coking rates vary with coals used and with the end use of coke, but for standard purposes a rate of one inch of average width per hour is common. The daily coal carbonizing capacity of the oven described is then 24 net tons, or one ton per oven per hour.

Ovens are built in groups or batteries generally ranging from 25 to 80 ovens. Small batteries of only 5 to 15 ovens have been built in specialized cases and for research purposes. Large batteries of 106 ovens have been built with bulkheads dividing them into two groups of 53 each.

Fuel gas is burned in the walls between adjacent oven chambers as well as in the walls at each end of the battery. The hot gaseous products of combustion heat the coal in the ovens. All chemical recovery coke ovens built in the 1950's had vertical heating flues to the exclusion of horizontal heating flues. In a vertical-flued oven battery, at any given time gas and preheated air burn in the bottoms of one-half of the flues, the hot products of combustion rising thru them and then descending thru the remaining half of the flues. From the base of the downflow flues the combustion gases enter the regenerators located beneath the ovens, giving up much of their remaining heat to the checker-brick in the regenerators before entering the stack. Periodically, usually every 20 to 30 min, the flow of gases is reversed, so that air is preheated in the regenerators that formerly recovered heat from the hot combustion products; the flues which carried gases downward carry gases upward and vice versa; and the regenerators that formerly preheated air recover heat from the hot combustion gases before their discharge.

Instead of using a part of the coke oven gas for underfiring, many batteries use blast furnace or producer gas which, in addition to the air, is preheated in the regenerators.

In one type of battery, three adjacent sets of regenerators heat the blast furnace gas and air for combustion in the flues in two oven walls. One regenerator preheats the gas for the two walls and the regenerators on either side of it, which are only one-half its width, heat the air. The next group of three regenerators are simultaneously being heated by the waste products of combustion. Because the gas and air in the regenerators are flowing in the same direction and at the same pressure, there is almost no tendency toward cross leakage between the air and gas regenerators.

Slot Type of Oven

All modern types of chemical recovery ovens use the general method of heating described. However, the design and operation of individual heating systems differ sharply. Three principal heating flue systems for ovens built in the United States during the years 1950 to 1958 are described here.

Koppers Ovens. These ovens may be about 11 ft high, 40 ft long, and 18 in. (average) wide, with a coal capacity of about 600 cu ft or 15 net tons per charge. Vertical flues and regenerators are divided into two groups at the centerline of the battery. During any one reversing period, fuel gas enters thru the so-called gun flues in one side of the battery. These flues are built into the brickwork of each heating wall at about oven-floor level. There are two for each heating wall, one in each side of the battery. Fuel gas is apportioned among the flues in that half of the battery in which gas is entering by ceramic gas nozzles of selected sizes inserted in the top of the gun flues.

Just above the gas nozzles the gas meets air previously preheated in the group of regenerators lying under that half of the battery, and burns in the bottom of the flues. Hot combustion gases rise in these heating flues and pass thru ports, partially closed by sliding brick dampers, into the horizontal flues thru which they flow to the other side of the battery at about the level of the top of the coal charge. The combustion gases are then redistributed to the vertical flues in the other half of the battery, thru which they descend and enter the opposite regenerators. After half of the 40- or 60-min cycle has elapsed, direction of flow in the horizontal flues is reversed, and gas is burned in the opposite half of each heating wall of the battery.

Koppers-Becker Gun Flue Ovens. These ovens differ fundamentally from Koppers ovens. The flue system and regenerators are not divided along the centerline of the battery, and reversals of combustion gas flow are between one entire heating wall and an entire adjacent heating wall instead of between halves of the same wall.

To accomplish this, single gun flues supply gas to all vertical flues in one of each pair of alternate oven walls at the same time. Combustion gases from every group of, say, four or six vertical flues are collected in short bus flues at about the level of the top of the charge and caused to flow thru crossover flues above the oven to similar bus flues in the companion heating wall. Here the gases are redistributed to an equal number of vertical flues thru which they descend into the off-regenerator and thence thru individual waste gas-reversing valves to the stack. On reversal, the flow of gases in the entire wall changes from downflow to upflow and vice versa. Direction of flow thru the crossover flues above the oven also is reversed.

The principal advantage of this oven is that the cross-sectional area of its horizontal flue can be reduced by one-half or more, permitting construction of higher ovens with more uniform heat distribution and increased wall strength. Additional advantages include ease of inspection of regenerators, since they are undivided, and lower differential pressure across regenerator walls, since no sliding brick and no horizontal flues are required. Ovens of this type have been built up to 14 ft in height, thus holding more coal per charge than the Koppers ovens.

Koppers-Becker Underjet Ovens. These ovens differ from gun flue ovens in the method of introduction of fuel gas. The underjet battery has no gun flues, these being replaced by steel pipe gas manifolds parallel to each wall, located in a basement below the regenerators. Connected to these manifolds are small stainless steel gas nozzles with ceramic tips bored to close tolerances that permit closely controlled feeding of fuel gas to each vertical flue. Gas leaving these nozzles forms a jet which aspirates a volume of combustion products from the bottom of the corresponding heating flue in the companion "off" wall, as shown in Fig. 3-2. The mixture of fuel and combustion gas flows up thru a duct in the regenerator wall to the bottom of the vertical flue, where it meets preheated air from the regenerator and burns. This diluted fuel gas has the advantage of burning with an

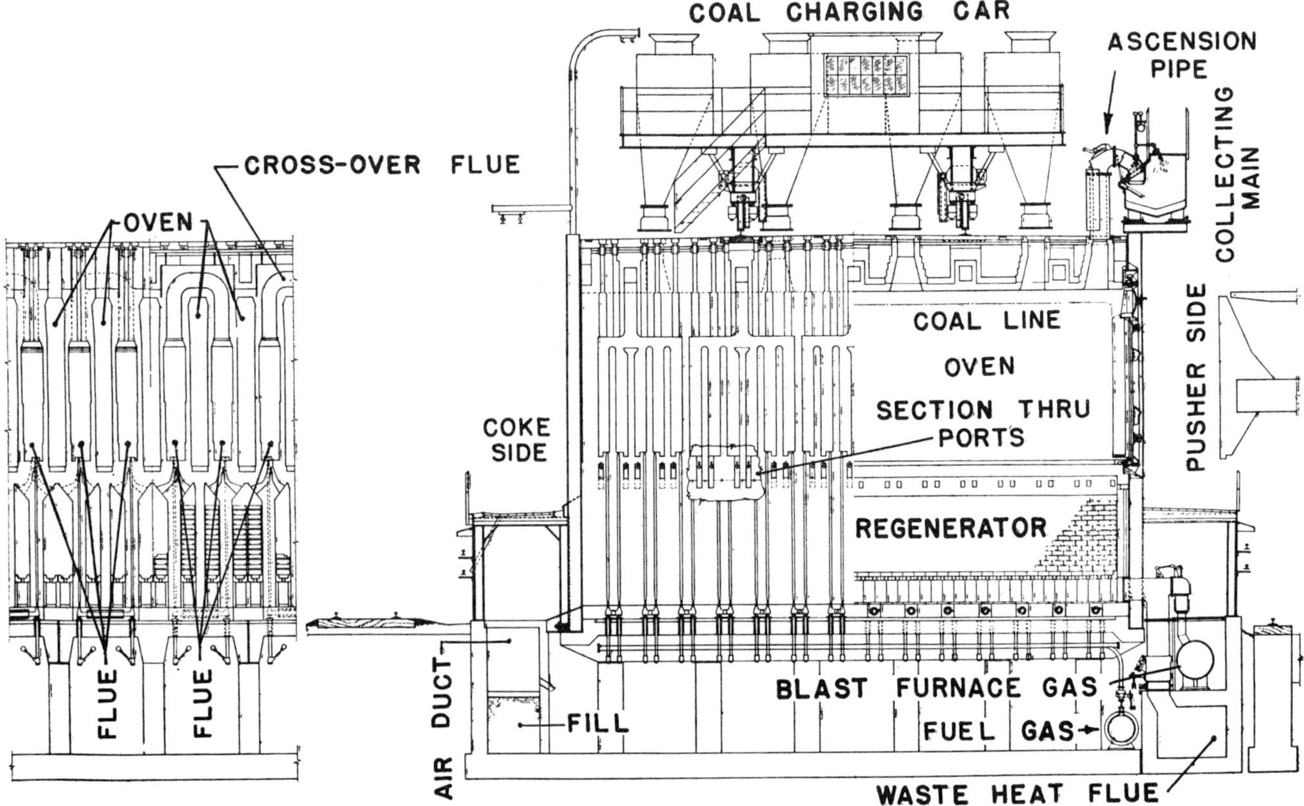

Fig. 3-2 Koppers-Becker underjet low-differential combination coke oven with waste gas recirculation. (Koppers Company, Inc.)

elongated flame of low luminosity, permitting use of ovens 15 ft high without difficulty in heating them uniformly. Admixture of combustion products with the fuel gas also adds sufficient water vapor and carbon dioxide to inhibit formation of carbon in ducts and flues, eliminating the need for conventional decarbonizing by air.

This heating system can be arranged to provide high coking speeds using mostly 85–90 Btu blast furnace gas. About 90 per cent of the underfiring heat is provided by this gas passed thru the regenerator, supplemented by ten per cent of the heat from a 250 Btu mixture of blast furnace and coke oven gases introduced thru the regular underjet system. At the other extreme, the underjet system has used natural gas or refinery gas for underfiring with no carbon trouble in ducts or heating flues.

Wilputte Ovens.* These ovens use a heating system similar in design to the Koppers oven. However, each horizontal flue is divided into two equal parts, so that only half the combustion gases are carried in each, decreasing the cross section required. Under each heating wall there are four regenerators in line, each about one-quarter the length of the heating wall.

Assuming coke oven gas as underfiring fuel, air will be preheated, say, in the two outer regenerators at the ends of the heating wall. Combustion will occur at the bottoms of the vertical heating flues directly above these regenerators. Combustion gases will rise in these flues, flow toward the center of the battery thru the shortened horizontal flues, and then down thru the two inner zones of vertical flues, finally passing thru the two inner zone regenerators. On reversal, air enters the two inner regenerators and combustion gases flow in the reverse direction thru the heating system. Control of vertical heating is secured by a system of high and low burners which are placed close to the oven floor level in alternate flues but some distance above it in the others. Sliding brick dampers are used to control flow of combustion gases from the vertical heating flues into the horizontal heating flues.

For underfiring with blast furnace gas there are two sets of regenerators, one for air, the other for the lean gas. Coke oven gas may be introduced by the underjet method. Controls for feed regulation to each individual vertical flue are located in a basement under the battery foundation pad.

Other Types of Ovens. The ovens which have been described are the most common types. Many other types have been built at various times. These include the Foundation, Simon Carves, and Otto.

COAL CHARGING AND COKE PUSHING

After proper preparation, coal is conveyed to the oven charging bin, usually capable of holding one day's supply. From this bin, located above and beyond the end of the oven battery, sufficient coal for one oven charge is withdrawn into a larry car. After weighing, the car is moved over the battery top to fill the proper oven. Cars may fill an oven either by gravity only or by means of rotating table or screw-type feeders. There are usually two to four oven charging holes per oven.

Upon expiration of the selected coking time, the oven is disconnected from the gas-collecting main, usually by water-sealed cup dampers. The charging hole lids and oven doors are removed, and the coke guide and quenching car are made ready to receive the coke. The pusher ram then shoves the glowing mass of coke out into the quenching car for transport to the quenching station.

* Wilputte Coke Oven Div. of Allied Chemical Corp.

COKE HANDLING

Quenching

At the quenching station, a number of sprays quench the coke with about 10,000 gallons of water for an 18-ton oven in about one to two minutes, reducing its temperature from 1800 F to about 350 F. Most quenching stations recover the unevaporated quenching water for reuse on the succeeding quench. Some plants dispose of contaminated water by adding it to the quenching station hot well.

Wharf and Conveyors

Cooled coke is discharged from the quenching car onto a brick-paved inclined (about 27 degrees) wharf. It is built in multiples of the length of the quenching car, usually 40 ft, and is frequently 160 ft long. Hand-operated gates at the bottom of the slope permit loading of the coke gently on the first belt of the conveyor system. Hand-operated water hoses are used to spot quench any coke hot enough to damage the belt. The wharf also provides storage to allow shutdown of the coke handling and screening system for maintenance.

Conveyors usually consist of rubber belts from 18 to 48 in. wide operating at speeds of from 100 to 300 fpm. Conveyor systems are designed with some excess capacity to allow for repairs and replacements.

Screening, Crushing, and Storing

Rotating multiple-disk "grizzlies" screen the large-sized foundry and furnace coke from the smaller sizes. Shaking or vibrating screens serve to separate the various domestic and producer fuel sizes and to remove the fine coke breeze for use as boiler or sinter fuel.

Crushing is sometimes necessary to reduce the top-sized coke to acceptable maximum size. Special crushers keep the amount of breeze produced to a minimum.

Coke is preferably stored in concrete bins provided with brick-lined spiral lowering chutes. Large stocks are commonly stored on the ground and reclaimed by clamshell buckets operated by locomotive cranes or storage bridges.

Dustproofing is desirable for domestic coke to reduce dispersal of fine dust. It may be accomplished by spraying with oils or emulsions which cause the dust to adhere to the coke surface. Calcium chloride and other moisture gathering chemicals are also used.

COKE OVEN GAS HANDLING

Figure 3-3 is a diagram of a coal chemical recovery plant. Collecting mains operating at a closely controlled pressure receive the coke oven gas produced in the ovens. Liquid-sealed valves are closed to disconnect the ovens during pushing and charging. Double mains may be provided for smoke control while charging and for recirculating raw gas across oven tops to control oven top temperature. Sprays of circulating ammoniacal flushing liquor reduce gas temperature and remove much of the pitch and tar. A so-called "down comer" (vertical runoff pipe) in the gas suction main system provides primary separation by gravity of tar and flushing liquor from the gas.

Indirect tubular or direct contact spray-type primary coolers cool the gas further and condense additional water vapor and tar before they enter the exhausters.

Exhausters are usually steam turbine-driven centrifugal gas pumps. They pull the gas from the ovens thru the primary coolers and then force it on thru the coal chemical recovery apparatus and gas purification plant into a gas holder. Older plants and some smaller new plants use positive displacement-type blowers for this purpose.

Electrical precipitators operating on high-voltage, unidirectional current are commonly used to remove final traces of tar. The corona discharge they provide causes tar particles in the gas passing thru it to collect on electrodes and flow down to a sealed outlet.

Ammonia Absorbers

Spray-type ammonia absorbers are a comparatively recent development in ammonia recovery. The ammonia-bearing coke oven gas passes thru a chamber where it is sprayed with a solution of sulfuric acid and ammonium sulfate to remove ammonia as ammonium sulfate. The supersaturated solution from the spray chamber is circulated thru a bed of seed crystals on which the supersaturation is deposited, forming larger crystals of ammonium sulfate. The solution may also be circulated thru a vacuum evaporator for controlling the degree of supersaturation.

A packed tower may be utilized in place of a spray chamber in installations provided with a vacuum evaporator. In that case the circulating solution must be maintained in an unsaturated condition to prevent plugging of the tower due to crystallization on packing.

Reheater

Water present in the diluted sulfuric solution added to the absorber circulatory liquor must be evaporated. The heat required for evaporation may be supplied by reheating the coke oven gas so that it enters the absorber in unsaturated condition. On many installations, necessary heat is added to the sulfate liquor instead of to the gas.

Final Cooler

Heat is added to the coke oven gas both by the reheater and by exothermic reaction in the ammonia absorber. The gas leaving the absorber at 52 to 58 C is practically saturated with water vapor. Gas cooling to between 20 and 30 C is accomplished in a direct, tower-type, wood grid packed or spray-type final cooler, where the gas contacts a countercurrent flow of water. One type of cooler has a lower section in which a counterflow stream of coal tar is used for the absorption of **naphthalene** produced during the gas cooling.

Benzol Washers

Crude light oil components are removed from the coke oven gas by direct absorption in a stream of petroleum wash oil in

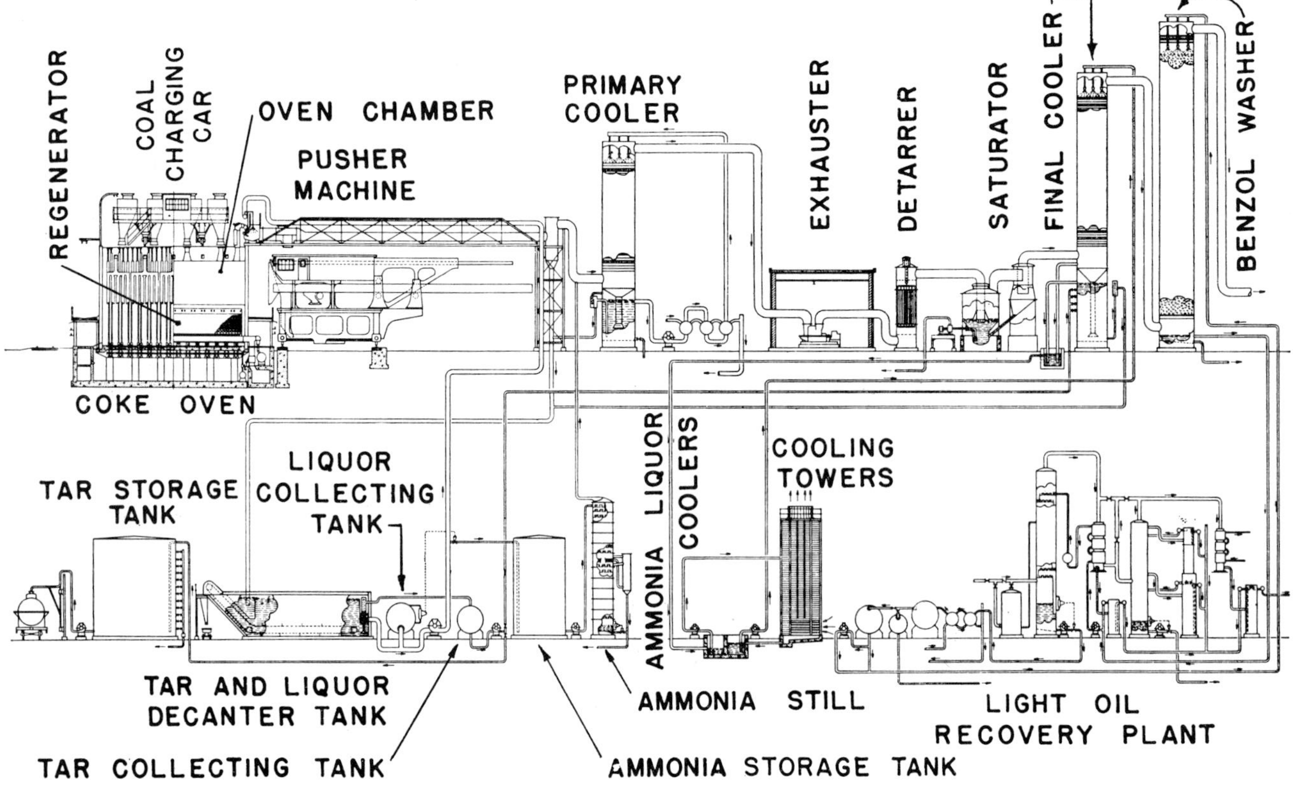

Fig. 3-3 Diagram of coal chemical recovery plant.

a tower packed with spiral steel strip packing or other suitable packing material. The crude light oil contains benzene, toluene, xylene, and solvent naphtha, together with some unsaturated hydrocarbons. Spray-type towers are also used.

Final Treatment

Heating value control of coke oven gas may be obtained by addition of producer gas. The producer gas may be admitted to the collecting main and allowed to pass thru the recovery and purification apparatus with the coke oven gas.

COAL NITROGEN

Determination of nitrogen in coal is time-consuming. It is generally agreed that bituminous coals contain 0.6 to 2.8 per cent nitrogen, while American gas coals have a practically constant 1.4 per cent nitrogen content.

Many authorities believe that the nitrogen in coal is linked as amino or substituted amino groups and is given off on distillation as ammonia, all other reactions being secondary. Evidence indicates that amino groups are not present as such in coal but appear on primary distillation.

Coals of high rank which give high coke yield will give low yields of volatile nitrogen and vice versa.

Nitrogen distribution in products of coal carbonization on the basis of nitrogen in coal follows:[9]

	J. Morgan	*H. C. Porter*
Coke	46.3	50.0
Tar	...	3.3
Cyanides	1.4	1.2
Ammonia	13.1	18.0
Free nitrogen (gas)	39.2	27.5

In America most ammonia is recovered as *sulfate* (used for fertilizer); therefore, few figures are quoted for free ammonia.

The water in coal has a definite protective influence against decomposition of ammonia, since slowly evolved moisture from the coal reacts with glowing coke and converts some coke nitrogen into ammonia. The principal ammonia production occurs only after completion of the coal fusion process (1475* to 1650 F), with 85 per cent of the ammonia evolved after the coke has set. In general, increasing carbonizing temperature from 1475 to 1830 F decreases ammonia yield by 33 per cent.

* Optimum yield.

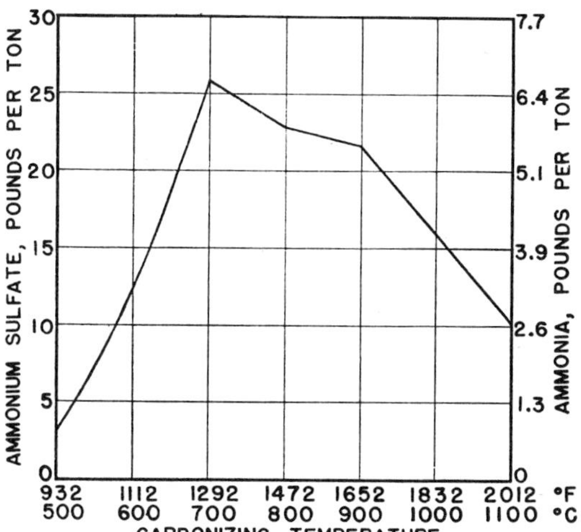

Fig. 3-4 Yield of NH_3 and $(NH_3)_2SO_4$ vs. carbonizing temperature for Pratt coal.[11]

Suggestions[9] for increasing ammonia yields include: (1) Keep oven tops cool. (2) Maintain positive pressure in oven chamber. (3) Avoid too rapid coking rate. (4) Use coals with minimum of iron in ash. (5) Use full charges (little space left above charge). (6) Heat oven uniformly. (7) Steam charge during last ⅕ of the carbonizing period (after raw coal has passed its fusion stage). (8) Use high-oxygen coals.

Ammonia yields in chemical recovery coke plants range from five to six pounds per ton of coal carbonized. Very little ammonia is produced in low-temperature carbonization. Relations between coking temperature and ammonium sulfate and ammonia yields[11] are shown in Fig. 3-4.

Ammonia Liquors

Salable ammonia liquors are divided into three classes.

1. "A" liquor must contain at least 29.4 per cent by weight ammonia, be *water white* in color, be free of H_2S, have a maximum organic titer of 50 by a specified *permanganate test,* and have a maximum of one gram per liter pyridine equivalent content.
2. "B" liquor must contain at least 25 per cent ammonia, have only a slight yellow color, be free of H_2S, have a maximum organic titer of 100 (permanganate test), and have a maximum of two grams per liter pyridine equivalent content.
3. "C" liquor must contain at least 15 per cent ammonia and be as free of tar as possible.

Since synthetic ammonia has become plentiful, "A" and "B" liquors have been priced about the same; therefore, little "A" liquor is produced. Sale of "C" liquor is confined to soda ash plants.

Ammonia Recovery Processes

Three principal processes by which ammonia has been recovered from coke oven gas are **semidirect, direct,** and **indirect.**

In the direct and semidirect processes, ammonia is recovered as ammonium sulfate. Technological developments during 1956 have fostered a changeover to production of diammonium phosphate. The direct process is no longer used in America.

In the indirect process, recovery is as a weak ammonia liquor, from which ammonia is stripped and converted to salable products such as ammonium sulfate or concentrated "A," "B," or "C" liquors.

Semidirect Process. Ammonia is recovered from the gas by direct contact with dilute sulfuric acid according to the reaction:

$$2NH_3 + H_2SO_4 \rightarrow (NH_4)_2SO_4$$

After leaving the primary cooler and exhauster, gas enters the reheater, where its temperature is raised to 120 to 140 F. Reheating is necessary to keep the volume of *mother liquor* in the saturator at a minimum. Gas then enters the saturator thru a distributor pipe, where it is broken up into many small streams which bubble up thru the saturator liquor.

The saturator is partly filled with a saturated aqueous solution of ammonium sulfate termed *mother liquor* which contains four to eight per cent free sulfuric acid. The reaction pre-

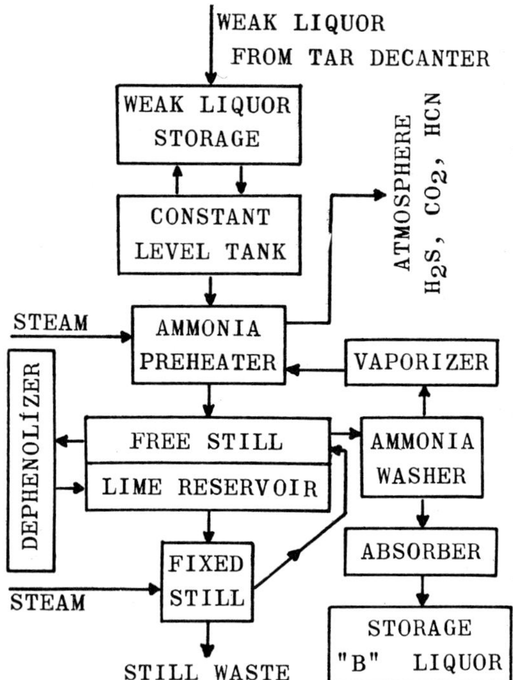

Fig. 3-5 Flow sheet of ammonia "B" liquor plant.

cipitates small ammonium sulfate crystals from the saturated solution. These crystals are pumped from the bottom of the saturator to a drain table, where excess mother liquor is removed, and thence to a centrifugal dryer. Acid spray droplets in the gas stream are removed in the separator by impingement or by centrifugal force. Separated acid returns thru a seal to the saturator.

Approximately one pound of acid (1.706 sp gr) is consumed per pound of sulfate produced. Commercial ammonium sulfate made in a saturator contains 25.0 to 25.7 per cent ammonia, is white or gray in color, is free-flowing, and has a maximum moisture content of about 2 per cent by weight, with 0.0 to 0.15 per cent free sulfuric acid and organic matter (tar).

Indirect Process. In this process about 98 per cent of the ammonia is recovered from the gas as weak liquor by scrubbing with water. The liquor is distilled and either the stripped ammonia gas is passed to the saturator, where it is converted to sulfate, or the ammonia gas, along with H_2S, CO_2, and HCN, is water-washed to produce "C" grade ammonia liquor.

"B" Ammonia Liquor Recovery. Ammonia is recovered from the gas in the form of a weak liquor, which is subsequently distilled, washed, and reabsorbed in fresh water to a resultant strength of 25 to 30 per cent ammonia. The weak liquor distilled is a combination of "flushing" liquor and condensate from the primary coolers and ammonia scrubbers. Ammonia concentration of the resultant weak liquor is maximized by passing effluent from the ammonia scrubber thru the second stage of the primary gas cooler.

Sixty to ninety gallons of weak liquor containing free and fixed ammonium salts are produced per ton of coal carbonized. *Free salts* are those which are decomposed by heat:

$$(NH_4)_2CO_3 + \text{heat} \rightarrow 2NH_3 + CO_2 + H_2O$$

Fixed salts are those which are decomposed by heat in the presence of an alkali, such as lime:

$$2NH_4Cl + Ca(OH)_2 + \text{heat} \rightarrow CaCl_2 + 2NH_3 + 2H_2O$$

The principal free and fixed salts present in the weak liquors are:

Free Salts	*Fixed Salts*
Ammonium carbonate	Ammonium chloride
Ammonium bicarbonate	Ammonium ferrocyanide
Ammonium sulfide	Ammonium thiocyanate
Ammonium bisulfide	Ammonium thiosulfate
Ammonium carbamate	Ammonium sulfate
Ammonium cyanide	
Ammonium hydroxide	

In addition to these salts, weak ammonia liquor contains low concentrations of phenol, pyridine bases, neutral oils, and empyreumatic compound.

In the indirect recovery process the saturators in Fig. 3-3 are replaced by gas washer–coolers, in which incoming gas is cooled from 180 to 80 F by a countercurrent stream of cool, recirculated weak liquor. Large quantities of condensed water, the majority of the tar, and some ammonia are removed by this process. The heat of compression of the gas exhauster is usually removed by a secondary cooler, with additional tar and ammonia removal. The gas is then purified of suspended matter by electrostatic precipitators and finally washed free of ammonia with fresh water in the intensive washer scrubber.

Overflows from the cooling and washing equipment are pumped to a tar decanter where tar is mechanically separated from the weak liquor. Ammonia is then recovered from the liquor by the process shown in Fig. 3-5.

This process treats weak ammonia liquor, containing about ten grams per liter of ammonia, by pumping it to an overhead constant level tank from which it flows at a uniform rate to the preheater. Here the descending liquor is both heated and stripped of acidic gases by ascending steam. Acidic gases and ammonia are then passed thru a cooler which refluxes the ammonia back to the system and allows the impurities to escape to the atmosphere.

The hot liquor is then pumped from the base of the preheater to the *free* still, where ascending steam distills out all free ammonia. Liquor enters the *lime leg,* where the fixed salt is converted to free hydroxide, and goes from there to the fixed still, where the free hydroxide is stripped from it by steam. The alkaline liquor may require further treatment before disposal.

From the top of the free still the mixture of steam and ammonia vapor from the free ammonium salts enters a cooler, where a large part of the steam is refluxed to the *free* still. The proper heat balance must be maintained here so as to reflux water and very little ammonia (200 *down* to 175 F). Ammonia vapor then enters the washer, in which the last traces of acidic gases, pyridine, etc., are removed by water in a short bubble plate column. Ammonia in water causes a strongly exothermic reaction, and sufficient cooling surface must be provided in the washer to remove the heat of solution as well as to condense the water vapor in the incoming vapor mixture. The strong ammonia wash water is removed at the bottom of the washer, vaporized, and returned to the preheater system.

Pure ammonia vapor enters the absorber, where sufficient

water is added to absorb all ammonia and produce the highest possible concentration. Internal cooling is necessary to remove the heat of absorption. The finished "B" liquor is then returned to storage.

In the event that "C" liquor is salable, the preheater and washer are eliminated, thus producing a crude yellow liquor containing 15 to 25 per cent ammonia combined as sulfide, carbonate, carbamate, and cyanide, along with organic matter and other impurities.

Ammonia and the Future

In the mid-1950's a buyers' market for ammonium sulfate and concentrated ammonia liquor developed. This change was due principally to the construction of a large number of new synthetic ammonia plants capable of exceeding the production required by industry and agriculture. The shift affects the economics of the recovery and sale of coke plant ammonia, particularly in the smaller coke plants, making it economically desirable, in certain plants, to remove the ammonia from the coke oven gas produced and destroy it by combustion (in lieu of making "B" liquor). Economics will indicate the most profitable procedure to follow.

LIGHT OIL RECOVERY

The benzolized wash oil from the benzol washers is stripped of its light oil content in a steam-stripping wash oil still. Operating in conjunction with this still is a gravity-fed short column for wash oil purification. Total steam for the *wash oil still* first passes thru the wash oil purifier and then leaves overhead with the wash oil vapors which enter the bottom of the wash oil still. The purifier bottoms are discarded as sludge and muck. The overhead from the wash oil still is crude light oil. It is *rectified* in a continuous column to produce an *overhead product* with a closely controlled dry point and a *bottom product* usually consisting of solvent naphtha, naphthalene, and higher boiling material. The stripped wash oil, after separation from water, is cooled and continuously recirculated thru the benzol washers.

Light oil refining plants may be classified as batch, semicontinous, or continuous. Batch plants represent early practice. They are no longer built except for very small throughputs. A **continuous plant** must have about eight distillation units and can only be justified for large coke plants. A **semicontinuous plant** offers a mid-ground for usual American practice. Such a plant, with a batch still for pure products, is particularly economical because of its adaptability to constantly changing specifications for end products.

In a semicontinuous light oil refining plant, crude light oil is first run in a continuous carbon disulfide column. The overhead from this column contains hydrogen sulfide, carbon disulfide, methyl mercaptan, butadiene, butylenes, and other low boiling constituents. These are bled continuously to the fuel gas system. *Bottoms* from this column, which are free from low boiling impurities, may be acid washed in an agitator, neutralized, and fed to the continuous benzol column. Nitration grade benzol is produced continuously as the *overhead product.* Bottoms from the continuous benzol column, comprising toluene and xylene fractions, are pumped as feed to the *pure still kettle.* This still is run batch-wise to produce pure toluene and xylene. The *overhead* may be neutralized before condensation in a vapor neutralizer containing bubble-cap trays, by aqueous caustic soda solution, with complete removal of all acid constituents. Suitable receivers are provided for the various pure products and for the intermediates, which are run off to storage until enough have accumulated for a rerun batch. Normal end products are nitration grade benzene, toluene, and xylene.

The *bottom fraction* from the light oil rectifier may be sold without further processing. Depending on market conditions, naphtha and naphthalene may be distilled in a separate *batch-type still* for unwashed light oil products. If market conditions do not justify production of nitration grade benzol, toluol, and xylol, the *bottoms* from the carbon disulfide column can be fed directly, without acid washing, to the continuous benzol column for conversion of as much of the light oil as desired to motor fuel. In this case the addition of an antioxidant inhibitor is required to prevent gum formation. Also, the bottoms from the continuous benzol column are acid washed and charged to the *batch pure still* for conversion into pure products.

YIELD OF CARBONIZATION PRODUCTS

The primary products customarily recovered at coke oven plants include coke, tar, ammonia, light oil, and gas. Their yields depend principally on the coal used and the temperature of carbonization. Average yields for the primary products of carbonization in the United States for the year 1959 are tabulated below:[12]

Coke	70.49 per cent of coal
Breeze	4.78 per cent of coal
Tar	8.41 gal per ton coal
Ammonium sulfate equiv.	18.90 lb per ton coal
Light oil	2.81 gal per ton coal
Gas, total	10,350 cu ft per ton coal

Other products recovered from the carbonization process in some plants include H_2S, HCN, naphthalene, and pyridine.

Coke

Coke is the primary product in coal carbonization. Substantially all coke produced in the United States is used as fuel. The iron blast furnace is by far its largest consumer. Of a total of 45.8 million tons of coke produced in both chemical recovery and beehive ovens in 1959, 90 per cent was consumed in iron blast furnaces. Foundries were the next largest consumer, followed by industrial users, water gas manufacture, house heating, and producer gas.

One of the chief attributes of coke is that its properties can be controlled to some extent by coal selection and oven operation. Coke producers select the coals economically available and blend them to approach the coke properties best suited to usage conditions. Although it is common plant practice to measure the size, strength, and density of the coke produced, no precise relation exists between the measured physical properties of coke and its performance. It is generally accepted, however, that uniformity of properties of coke, both physical and chemical, is highly important.

Coke *breeze* (also called braize) is the small (under ½ in.) size portion of the coke. Many coke oven plants burn it on chain grate stokers for steam raising. Some breeze is sold for

Table 3-2 Variation of Gas Analyses during Coking Period*

Hours after charging	CO_2 (+H_2S, etc.)	Illuminants	CO	CH_4	H_2	N_2	Gas produced, cu ft per hr
	Per cent by volume of moisture-free gas						
1	1.6	4.6	5.3	34.8	52.0	2.1	11,900
2	1.7	4.4	5.3	33.6	53.4	2.3	10,600
3	1.8	4.0	5.3	32.4	54.8	2.5	13,800
4	1.9	3.6	5.3	31.0	56.0	2.6	13,000
5	2.0	3.2	5.3	30.0	57.2	2.7	13,000
6	2.1	2.6	5.3	28.8	58.4	2.8	13,600
7	2.2	2.2	5.3	27.6	59.5	3.0	14,000
8	2.1	1.8	5.3	26.4	60.6	3.1	13,600
9	1.9	1.4	5.3	25.0	61.8	3.2	13,200
10	1.4	1.0	5.3	22.0	66.0	3.4	14,800
11	0.8	0.4	5.3	16.2	72.0	3.6	9,100
12	0.4	...	5.3	11.0	78.0	4.1	...
13	0.3	...	5.3	5.6	84.0	5.5	...

* Based on coking 13.4 tons (dry basis), 16 in. average oven width, 14 hr coking time.

various fuel purposes. Sump breeze is the fine coke that is washed out during quenching. It is generally drained and added to the screenings.

Tar and Light Oil

Tar yields depend upon both the coking rate and the coals used. High oven top temperatures cause decreased tar yields. Coal that has deteriorated because of poor storage practice can also account for low tar yields. Tar is generally sold to tar refiners for further processing.

Light oil yields are dependent upon both the coal used and the temperature of carbonization. Light oils scrubbed from coke oven gas contain benzene, toluene, xylenes, and a number of other higher benzene ring compounds. The products obtained from refining crude light oil (2.81 gal per ton of coal coked) in U. S. coke oven plants in 1959 were:[12]

Products	*Volume, per cent*
Benzene, motor	0.3
Benzene, all other grades	60.4
Toluene	13.1
Xylene	3.8
Solvent naphtha	2.0
Other light oil products	1.6

Gas

Coke oven gas yields are principally dependent on the coal used, but the temperature of carbonization is also an important factor. Yield of gas increases while its heating value decreases with increasing temperature. Rate of formation and composition of coke oven gas change continuously throughout the coking period. Coke oven gas during coking is analyzed in Table 3-2.[13]

Many coke oven plants use part of the gas produced to underfire the ovens. Approximately 35 to 40 per cent of the gas is required to do this. The balance of the gas is generally used for heating purposes in the various steel plant operations or sold to public utilities. Some coke oven plants have been operated independently of steel plants.

LOW-TEMPERATURE CARBONIZATION

Low-temperature coal carbonization processes operate in the temperature range of 930 to 1380 F. Interest in them is based on their **high tar yields,** together with their production of a smokeless fuel. Only two commercial low-temperature processes have been successful in the United States. Shortcomings[8] of low-temperature processes are as follows:

1. It has not been proved that low-temperature coke is desirable in metallurgical practice.
2. There is doubt that low-temperature coke has any outstanding advantages either as a stoker or as a hand-fired fuel over high-temperature coke.
3. The economics of the process are questionable due to low gas yield and difficulty in refining the tar.
4. No satisfactory mechanical system has been devised to produce low-temperature coke on a large scale under American conditions. Plant capacities are low, resulting in high investment cost, and great difficulties are encountered in handling a moving mass of coal in the plastic state. The **Wisner process,** developed and modified by the Pittsburgh Consolidation Coal Co., has been the most satisfactory American method.

There is some interest in production of a substitute for low-volatile coal by low-temperature carbonization, especially in the Western United States, where low-volatile coals are not readily accessible to coke oven plants. No commercial scale plant had been constructed as of 1959 for that purpose. There has been some experimental work in low-temperature carbonization of lignite for use as a power plant fuel.

Low-temperature carbonization processes have been reported.[14,15]

Table 3-3 Value of Coal Chemical Materials, Coke, and Breeze per Ton of Coal Carbonized in the U. S.[12]

Products	1947–49 (average)	1959
Ammonia and its compounds	$ 0.356	$ 0.331
Light oil and its derivatives	0.451	0.620
Surplus gas sold or used	1.291	1.584
Tar and its derivatives sold	0.501	0.776
Tar burned by producers	0.228	0.388
Miscellaneous products	0.020	0.008
Total	$ 2.847	$ 3.707
Coke produced	$ 8.488	$12.562
Breeze produced	0.191	0.329
Grand Total	$11.526	$16.598
Value of coal at the ovens	7.79	9.88

ECONOMICS OF COAL CARBONIZATION

The U. S. Department of the Interior, Bureau of Mines, publishes data on the quantities of coke and coal chemicals produced in the United States each year together with the value of the products sold.[16] The various unit values so published are an average for the entire country and do not represent any one plant or location. Some products are consumed within the producing plant so that the values are not necessarily market prices.

Table 3-3 shows the value of various carbonization products per ton of coal carbonized in the United States.

REFERENCES

1. A.S.T.M. "Classification of Coals by Rank." (Spec. D388-38) *A.S.T.M. Standards 1958*, p. 1078–83. Philadelphia, 1959.
2. Russell, C. C. "Selection of Coals for Manufacture of Coke." *A.G.A. Proc.* 1947: 733.
3. Brewer, R. E. *Plastic and Swelling Properties of Bituminous Coking Coals.* (Bur. of Mines Bull. 445) Washington, D. C., 1942.
4. Soth, G. C., and Russell, C. C. "Gieseler Method for Measurement of the Plastic Characteristics of Coal." *A.S.T.M. Proc.* 43: 1176, 1943.
5. Wilson, P. J., Jr., and Wells, J. H. *Coal, Coke, and Coal Chemicals*, p. 194. New York, McGraw-Hill, 1950.
6. Russell, C. C., and others. "Reducing Coal Expansion Pressure." *A.I.M.E. Blast Furnace, Coke Oven and Raw Materials Proc.* 8: 32–50, 1949.
7. Mitchell, D. R. "Coal Preparation Plant Control in United States." Intern. Conf. on Coal Prep., Paris, 1950. *Revue de l'Industrie Minerale*, special issue, Nov. 1950, p. 73–83.
8. Lowry, H. H., ed. *Chemistry of Coal Utilization*, vol. I, New York, Wiley, 1945.
9. Porter, H. C. *Coal Carbonization.* New York, Chemical Catalog Co., 1924.
10. Russell, C. C. *Control of the Bulk Density of Coal at Coke Oven Plants.* Pittsburgh, Pa., Koppers Co., 1951.
11. Fisher, C. H. *Composition of Coal Tar and Light Oil*, p. 12. (Bur. of Mines Bull. 412) Washington, D. C., 1938.
12. U. S. Bur. of Mines. *Minerals Yearbook 1957; Vol. II: Fuels.* Washington, D. C., 1959.
13. Lowry, H. H., ed. *Chemistry of Coal Utilization*, vol. II, p. 921–46. New York, Wiley, 1945.
14. Gentry, F. M. *Technology of Low Temperature Carbonization.* Baltimore, Williams & Wilkins, 1938.
15. Utah. Conservation and Research Foundation. *Low-Temperature Carbonization of Utah Coals.* Report to the Governor and Legislature. Salt Lake City, Utah, 1939.
16. Otero, M. M., and others. *Coke and Coal Chemicals in 1952.* (Minl. Market Rept. MMS2186) Washington, D. C., Bur. of Mines, 1953.

BIBLIOGRAPHY

Gluud, W. *International Handbook of the By-Product Coke Industry.* New York, Chemical Catalog Co., 1932.

Morgan, J. J. *American Gas Practice*, 2nd ed., 2 vols. Maplewood, N. J., J. J. Morgan, 1931.

Chapter 2

Gas Tars and Pitches

by S. A. Petrino, C. U. Pittman, S. C. Schwarz, and E. D. Freas

In the production of gas from oil and of gas and/or coke from coal by destructive distillation or cracking, a black to dark brown bituminous condensate known as tar[1] is obtained. The tar[2] yield from coal ranges up to twelve per cent. On the other hand, oil yields highly variable amounts of tar depending on the cracking temperature level. Tars vary greatly in composition and properties depending on the parent material and manner of production.

Pitches,[1] produced by the partial evaporation or fractional distillation of tars, also vary widely by reason of the composition of the parent tar and the treatment given prior to and during their production.

TAR CHARACTERISTICS

Tars are colloidal systems consisting of disperse particles or micelles in the dispersion medium of tar oils. Composition of these micelles may differ considerably from tar to tar.[3–7] Temperature of carbonization and nature of the coal coked also affect composition.

Table 3-4 Analyses and Properties of Typical Coal Tars[15]

	Types of carbonizing equipment											
					By-product coke ovens (Koppers)							
	Rotary retort, Disco process		Sole-flue oven (Curran–Knowles)		Vertical retort		Average		Heavy		Horizontal retort	
Specific gravity at 15.5/15.5 C	1.141		1.113		1.103		1.180		1.226		1.249	
Absolute viscosity, centistokes												
At 20 C	3545		253		1493		1931		...		...	
At 35 C	530		63		237		316		2850		14090	
At 50 C	...		...		...		...		493		2065	
Viscosity-temp susceptibility, S	1.19		1.28		1.29		1.21		1.16		1.04	
Carbon I (nitrobenzene-insoluble)	12.3		2.1		2.2		2.5		10.8		17.3	
Carbon II (acetone-insoluble, tar-soluble)	0.2		1.1		2.6		8.7		10.8		12.3	
Benzene-insoluble	13.6		3.4		3.5		4.6		12.0		21.0	
Distillation, % by weight												
To 170 C	1.8	...	2.1	...	1.2	...	0.7	...	0.4	...	1.5	...
To 200 C	3.3	...	4.3	...	2.6	...	1.1	...	0.7	...	1.7	...
To 210 C	5.0	...	7.4	...	4.0	...	1.8	...	1.3	...	2.0	...
To 235 C	13.2	...	22.3	...	11.4	...	7.1	...	6.1	...	4.3	...
To 270 C	25.8	...	33.8	...	21.2	...	18.2	...	15.4	...	11.0	...
To 300 C	35.4	33.9	41.8	41.7	28.8	28.2	26.3	24.9	20.9	20.1	17.5	15.2
To 315 C	...	39.2	...	45.6	...	32.7	...	28.3	...	21.8	...	17.6
To 335 C	...	44.2	...	53.8	...	40.6	...	33.4	...	25.8	...	21.0
To 355 C	...	55.8	...	61.6	...	49.6	...	41.9	...	34.0	...	30.3
Residue	62.4	42.4	57.6	37.7	70.8	49.7	73.7	57.6	78.2	65.9	82.3	69.6
Softening point of distillation residue (R & B method)												
residue at 270 C	77.5 C		37.2 C		30.9 C		41.5 C		41.6 C		56 C	
residue at 300 C	93.5 C		45.5 C		37.2 C		48.5 C		53.9 C		59.2 C	
residue at 355 C	209.0 C		94.2 C		80.0 C		89.8 C		80.4 C		108.5 C	
Tar acids in 0–270 fraction, % by volume of tar	17.5		18.5		7.36		2.01		2.2		1.21	
Naphthalene in 170–235 fraction, % by volume of tar	Trace		Trace		0.55		7.45		6.3		5.31	
Sulfonation factors												
0–300 C fraction	1.27		1.79		3.11		...		0.04		0.42	
300–355 C fraction	...		Trace		1.86		...		0.03		0.12	

In American practice, high-temperature tars are the most important. They contain large proportions of aromatic compounds, phenols, cresols, naphthalene, benzene and its homologues, and anthracene. They are produced by destructive distillation of coal in the absence of air and can therefore be defined as pyrogenous distillates.[8]

Low-temperature carbonization yields tars containing paraffins, phenols, cresols, xylenols, and alkylated naphthalene. Variations in composition of such tars are much greater than in those produced at high temperatures, because of the degree of cracking of primary compounds.[8]

In general, maximum tar yield from coal is obtained at 500 to 600 C. Pitch increases and tar oils decrease with increasing temperature. Neutral oil and tar-acid contents decrease with increase in temperature, while free carbon and specific gravity increase.[9] Tars produced at temperatures of up to about 750 C generally are considered primary or low-temperature tars. They are of considerable importance in Europe.

For additional data see References 9 and 10.

Coal tars[11] and their distillates differ in many important respects from oil gas tars; therefore, their analytic procedures are not necessarily interchangeable. Phenolic derivatives, characteristic of coal tar distillates, as well as almost all oxygenated and nitrogenous hydrocarbon derivatives, are wholly lacking in oil gas tars.[12–14] Table 3-4 gives analyses and properties of various coal tars. Table 3-5 shows products from distillation of crude coal tar together with their composition. Properties of various coal tar pitches are given in Table 3-6.

Tars from the cracking of petroleum distillates may be divided into two groups: water gas and oil gas tars.

The so-called water gas tars result from the cracking of petroleum oils used to enrich the heating value of water gas. Since the water gas reaction yields no tar, the water gas tars are basically low-temperature oil gas tars.

The most characteristic difference between water gas and conventional oil gas tars is the higher amount of "tar insolubles" (quinoline insolubles) in the latter. Water gas tars are lower in filterable carbon or carbon disulphide insolubles because of the lower cracking temperatures. Consequently, there is usually a high sulfonation residue content in water gas tars.

The correlation of yield and quality of tars from oil gas with the Btu level of the gas is unsatisfactory. This can be attributed largely to the fact that all commercial oil gases contain more or less air, combustion or oxidation* products, and/or blow-run producer gas. A more realistic index is the heating value of the *true* oil gas consisting only of the hydrocarbon and free hydrogen portion of the mixture, excepting when water gas has been enriched with oil gas. In the latter case, the water gas reaction contributes free hydrogen which can be approximated by the Wills formula.

When petroleum oils are cracked to produce true oil gases having heating values below about 900 Btu, the resulting large proportion of filterable carbon can absorb most or all of the tar hydrocarbons to make a solid or semisolid effluent in the initial quenching apparatus or wash box.[16] This product, commonly referred to as lampblack, contains variable amounts of tar hydrocarbon, depending on the mechanical efficiency of the wash box as well as on the temperature of the water used for quenching the off-take gas. This mixture of filterable carbon and liquid hydrocarbons usually becomes pumpable when made at cracking levels to produce true oil gases of greater than about 900 Btu. The product then is commonly classified as "tar" even though it may contain up to 40 per cent suspended solids or filterable carbon. As the depth or severity of cracking is reduced to produce oil gases of still higher Btu levels, the suspended solids content of the tar phase gradually diminishes practically to the vanishing point, as in tars from water gas carburetors.

The common practice of identifying the nonpumpable effluent from oil gas operations as "lampblack" and the pumpable portion as "tar," even though both are mixtures of filterable carbon and liquid hydrocarbons, has led to poor correlation of their quality and yield from various operations. To be more consistent and realistic, such comparisons may be made on the more fundamental basis of yield of "quinoline" insolubles, or filterable carbon, and "quinoline" solubles, or tar hydrocarbon content of both lampblack and tar. These data may then be compared with the true oil gas Btu level or severity of cracking in the operation.

Another factor influencing the quality and yield of effluents from oil gas operations is the uniformity of cracking.[13,14]

Table 3-5 Crude Coal Tar Distillates Obtained in Practice, and Products Derived Therefrom[9]

(percentages are based on the original tar)

Light oil, up to 200 C (392 F)	5.0
Benzene	0.1
Toluene	0.2
Xylene	1.0
Heavy solvent naphtha	1.5
Middle oil, 200–250 C (392–482 F)	17.0
Tar acids	2.5
Phenol	0.7
Cresols	1.1
Xylenols	0.2
Higher tar acids	0.5
Tar bases	2.0
Pyridine	0.1
Heavy bases	1.9
Naphthalene	10.9
Unidentified	1.7
Heavy oil, 250–300 C (482–572 F)	7.0
Methylnaphthalenes	2.5
Dimethylnaphthalenes	3.4
Acenaphthene	1.4
Unidentified	1.0
Anthracene oil, 300–350 C (572–662 F)	9.0
Fluorene	1.6
Phenanthrene	4.0
Anthracene	1.1
Carbazole	1.1
Unidentified	1.2
Pitch	62.0
Gas	2.0
Heavy oil	21.8
Red wax	7.0
Carbon	32.0

* Commercial oil gas operating temperatures have been proved to be too low to produce significant volumes of water gas. Investigations at Portland (Oregon) Gas & Coke Co. have proved that the excessive CO and CO_2 can be attributed to oxidation of carbon and/or hydrocarbons by the reduction of metallic oxides in the checker slag. These oxides absorb oxygen during the heating cycle and release it during the make cycle.

Table 3-6 Properties of Gas Works Coal Tar Pitches[10]

Test No.		Horizontal retorts	Inclined retorts	Vertical retorts	Coke oven coal tar pitch	Blast furnace coke tar pitch	Gas producer coal tar pitch	Low-temperature coal tar pitch
1	Color in mass	Black	Black	Black	Black	Black	Black	Black
2a	Homogeneity to the eye at 25 C	Variable	Variable	Variable	Variable	Variable	Variable	Variable
2b	Appearance under microscope	Lumpy	Lumpy	Sl. Lumpy	Sl. Lumpy	Lumpy	Lumpy	Sl. Lumpy
3	Appearance surface on aging	Unchanged	Unchanged	Unchanged	Unchanged	Dull	Unchanged	Dull
4	Fracture	Conchoidal	Conchoidal	Conchoidal	Conchoidal	Conchoidal	Conchoidal	Conchoidal
5	Luster	Variable	Variable	Variable	Variable	Variable	Variable	Variable
6	Streak on porcelain	Black	Black	Black	Black	Black	Black	Black
7	Specific gravity at 25 C	1.25–1.40	1.25–1.35	1.15–1.30	1.20–1.35	1.20–1.30	1.20–1.35	1.10–1.26
9b	Hardness at 25 C (needle penetrometer)	0–100	0–100	0–100	0–100	0–100	0–100	0–100
9d	Susceptibility index	100	100	100	100	100	100	100
10	Ductility at 25 C, cm	Variable	Variable	Variable	Variable	Variable	Variable	Variable
15c	Fusing point (cube method), °F	80–212*	80–212*	80–300*	80–300*	80–212*	80–212*	80–200*
16	Volatile at 163 C for 5 hr, %	1–10	1–10	1–10	1–10	1–10	1–10	1–10
	Volatile at 260 C for 5 hr, %	3–20	3–20	3–20	3–20	3–20	3–20	3–20
17b	Flash point (open tester), °C	121–232	121–232	121–232	121–232	121–232	121–232	121–232
19	Fixed carbon, %	35–65	30–45	15–40	17–60	10–30	25–45	8–22
21	Soluble in carbon disulfide, %	45–70	63–78	70–94	50–92	45–75	60–85	85–98
	Nonmineral matter insoluble	30–55	28–37	6–30	8–50	15–35	15–40	2–15
	Mineral matter, %	0–½	0–½	0–½	0–½	10–20	0–2	0–3
22	Carbenes, %	2–10	2–10	2–10	2–10	1–5	2–10	1–5
23	Soluble in 88° petroleum naphtha, %	10–20	15–30	20–40	10–30	5–25	10–30	25–60
24	Free carbon, %	30–55	25–40	5–35	6–40	15–35	15–40	2–25
28	Sulfur, %	½–1	½–1	½–1	½–1	½–1	½–1	½–1
31	Tar acids (distillate to coke), %	5–20	10–15	20–30	1–12	20–30	10–15	25–50
32	Naphthalene, %	Trace–2	Trace–2	0–Trace	1–2½	0–Trace	0–Trace	0–Trace
33	Solid paraffins, %	0	0	0–Trace	0	2–5	0–Trace	2–5
34a	Soluble in conc. H_2SO_4, %	95–100	90–95	90–95	95–100	85–100	90–95	85–90
34b	Sulfonation residue of distillate at 235–315 C, %	0–3	2–6	4–6	0–5	15–25	0–5	5–20
37c	Saponifiable matter, %	Trace–1	Trace–1	Trace–1	Trace–1	Trace–1	Trace–1	Trace–1
39	Diazo reaction	Yes	Yes	Yes	Yes	Yes	Yes	Yes
40	Anthraquinone reaction	Yes	Yes	Yes	Yes	No	Yes	No

* Figures apply to commercial grades of pitch. It should be noted, however, that coal tar pitches can be produced with fusing points as high as 450 F.

The gaseous phase in such operations has been shown to attain its "pseudoequilibrium" composition quite rapidly,[17,18] whereas the slower condensation and dehydrocyclization reactions which govern the quality of the liquid phase products have been shown to require more time of contact at a given temperature level to become thoroughly "aromatized" or free of unsulfonatable products.[14,19–21] Consequently, the quality of the tar, as well as of light oil, may vary widely at a given gas phase Btu level due to lack of uniformity as well as to severity of cracking.

The effect of feed stock quality on the characteristics and yield of high Btu oil gas tars has been studied in connection with the selection of oils, but no generalizations were apparent.[18] Because, as has been shown, the Conradson carbon portion of heavy feed stocks remains largely as coke or checker deposit during the initial vaporization phase of oil gas operations,[22] the true feed stock for the subsequent cracking reactions is not the original feed, but rather the volatile portion therefrom.

Tables 3-7 and 3-8 list properties of various water gas, oil gas, and coke oven tars. Table 3-9 gives properties of pitches from such sources.

Table 3-8 indicates the trend of gas Btu level, feed oil, and uniformity of cracking on tar quality. In the 570 Btu operation most of the filterable carbon formed had been removed from the operation as lampblack. Sample 2 represents an improved feed and cracking condition over Sample 3. Sample 4 resulted from better uniformity of cracking. The above samples were taken with reference to quality for the production of electrode pitch. See Table 3-10.

Table 3-7 Properties of Various Gas Tars*

	Heavy water gas and oil gas tars	Light water gas and oil gas tars	Coke oven tars
Sp gr at 25/25 C	1.16–1.24	1.02–1.15	1.13–1.24
Consistency			
Engler specific viscosity at 40 C	...	1.0–3.0	...
Engler specific viscosity at 50 C	...	...	2.0–120.0
Float test at 32 C	20–300	...	–300 max
CS_2 insoluble	1.0–20	0.5–5.0	1.5–16.0
Quinoline insoluble	0.5–15	0.5–5.0	0.6–11.0
ASTM D-20 distillation, %			
0–170 C	0–4	0–7	0–1.5
200	0–7	5–15	0–4.0
235	2–10	15–30	0.9–21.0
270	5–15	25–40	5.6–32.0
300	15–25	35–50	11.0–37.0
355	25–40	50–75	26.0–54.0
Softening point (R & B method), °C			
Residue to 300 C	50–70	25–60	31–53
Residue to 355 C	100–140	100–140	58–92
Sulfonation index			
Distillate to 300 C	0–2.5	5–10.0	0–0.5
Distillate, 300–355 C	0–1.5	0.5–2.0	Trace

* Source: C. U. Pittman, Tar Products Div., Koppers Co., Inc., Pittsburgh, Pa.

Table 3-8 Properties of Various Oil Gas Tars

	Sample			
	1 Tar from 570 Btu gas	2 Tar from 1000 Btu gas	3 Tar from 1000 Btu gas	4 Tar from high-Btu gas
Sp gr at 60 F	1.164	1.156	1.102	1.162
Carbon disulfide insol., %	7.7		10.9	12.5
CI_{NB},* %	4.6			
CI_Q,† %		8.6	10.9	10.8
Distillation, % by wt				
to 190 C	0.7	0.2	0.7	
to 300 C	25.8	5.8	32.6	26.5
Naphthalene, %	9.2	1.1	2.7	
S.P. c/w °F of dist. residue	133		109	
Unsulfonated residue of distillate to 300 C, %	4.5	6.5	20.0	1.5
Sulfonation index	1.2		6.5	

Data from Northwest Natural Gas Co., Portland, Ore.
* Nitrobenzene insolubles.
† Quinoline insolubles.

TAR DISTILLATION

A large proportion of the coke oven tar produced in this country is at least topped or distilled to remove the oils boiling below 270 C for the recovery of chemicals. This operation is usually conducted continuously, with the tar fed to a column in which it is stripped of the desired distillate in the lower part. The distillate is then rectified in the section above the feed tray and taken overhead to a condenser. The distillation is often performed under vacuum. Batch distillation, usually without a column, is still sometimes used for topping.

Table 3-9 Properties of Water Gas and Oil Gas Tar Pitches[10]

Test No.		Water gas tar pitch	Oil gas tar pitch
1	Color in mass	Black	Black
2a	Homogeneity to the eye	Uniform	Uniform
2b	Homogeneity under the microscope	Small amount of carbon visible	
3	Aged surface	Variable	Variable
4	Fracture	Conchoidal	Conchoidal
5	Luster	Bright	Bright
6	Streak	Black	Black
7	Specific gravity at 25 C	1.10–1.25	1.15–1.35
9b	Penetration at 25 C	0–100	0–100
9d	Susceptibility index	>100	>100
10	Ductility	Variable	Variable
11	Tensile strength at 25 C	Variable	Variable
15a	Fusing point (K & S method), °C	27–135	27–135
15b	Fusing point (R & B method), °C	38–149	38–149
15c	Fusing point (cube method), °C	43–160	43–160
16	Volatile matter at 260 C for 5 hr, %	5–15	5–15
17a	Flash point, °C	149–204	149–204
19	Fixed carbon, %	25–45	20–35
21	Solubility in carbon disulfide, %	75–98	70–98
	Nonmineral matter insoluble, %	2–25	2–30
	Mineral matter, %	0–1/2	0–1/2
22	Carbenes, %	5–10	5–10
23	Solubility in 88° petroleum naphtha, %	50–70	60–80
28	Sulfur, %	<4	<4
30	Oxygen, %	0–2	0–2
33	Solid paraffins, %	0–5	0–5
34b	Sulfonation residue, %	0–15	20–40
37e	Saponifiable matter, %	0–1	0–1
39	Diazo reaction	Yes*	Yes*
40	Anthraquinone reaction	Yes	Yes

* Slight; contains only a trace of phenols.

The distillate collected is further processed to recover chemicals such as tar acids, tar bases, and naphthalene.

Batch stills, except for the production of special products or high softening point pitches, are gradually giving way to continuous stills.

If the tar is to be processed to a pitch residue, a larger proportion of distillate may be taken off as described above and its higher boiling portion removed as a side stream. However, it is more usual to feed the residue from the topping operation continuously to a vacuum flash unit or another still. In the vacuum flash unit the feed temperature and the vacuum used determine the consistency of the residue. The distillate taken off may be separated by fractional condensation into creosote and heavy creosote oil fractions, depending upon the softening point of its residue. When the topped tar is fed to another still, vacuum or steam agitation may be used to reduce the temperature required for distillation. If a column is used, creosote oil may be taken overhead and heavy creosote oil taken as a side stream.

A continuous tar distillation unit was operated by the Portland Gas & Coke Co. for the production of electrode pitch from oil gas tar. By selection of feed stocks and control

Table 3-10 Electrode Pitches from Coal Tar and Oil Gas Tar

	Sample source		
	1 Oil gas tar 570 Btu gas	2 Oil gas tar high-Btu gas	3 Coal tar pitch
Softening point cube in air, °C	112	108	104
CI_Q,* %	11.6	17.5	7.3
Coking value, %	57.7	58.9	55.5
Net coking value,† %	52.1	50.2	52.0
Benzol insoluble, %	26.0	27.9	29.5
Beta resin, % (Benzene insoluble, CI_Q)	14.4	10.4	22.2
Ultimate analysis, % C	92.77		92.57
Ultimate analysis, % H	4.68		4.37
C/H ratio	19.82	18.89	21.18
Ash, %		0.05	
Sulfur, %		0.94	

Data from Northwest Natural Gas Co., Portland, Ore.
* Bitumens insoluble in quinoline.
† Coking value (C.V.) of quinoline insoluble free bitumen; C.V. net = (C.V. gross − CI_Q)/(100 − CI_Q).

of both depth and uniformity of cracking, acceptable soft and hard electrode pitches and binders for other purposes were made. Table 3-10 shows a comparison of the quality of such pitches from coal and oil gas tars. The advent of natural gas in this area ended the operation.

Special materials have almost entirely eliminated corrosion and erosion in stills. Free hydrochloric acid from decomposition of ammonium chloride and tar acids are not particularly troublesome since the advent of stainless steels.

TAR EMULSIONS

As established by common industrial usage, a tar containing more than five per cent water is considered an emulsion. Practically all oil gas operations use water for quenching the off-take gas, which invariably results in the formation of emulsions that are difficult to break. Many methods have been developed for breaking such preformed emulsions.

The Ripley-Schwarz (R-S) process was one of the first to break high-temperature oil gas emulsions by chemical treatment successfully. It consisted of heating the emulsion in a closed vessel to 75 to 100 psi after addition of a small proportion of caustic soda, followed by a cooling and settling period.[23] Some emulsions, especially low-temperature oil gas emulsions, did not react satisfactorily with this method. Schwarz[23] later prevented the formation of stable emulsions from both high- and low-temperature oil gas tars by using surface tension depressants or wetting agents in the quenching liquor. This method has become the most common practice in recent years.

Undesirable stable emulsions, with water in the disperse phase, are not formed in the absence of an emulsifying agent or protective colloid. The latter hinders the desired coalescence of the emulsified water particles.[3]

Filterable carbon plays an important role as an emulsifying agent. It has been proved that in its absence tar does not form a stable emulsion with water.

Plant tests using anionic and cationic surface active agents have shown that the cationic agents (concentration of 0.02%) have a larger degree of effectiveness. It was not known (in 1959) whether these agents prevent formation of emulsions or merely reduce their stability.

Heating, settling, mechanical processes, flash distillation, and the use of chemical additives, either singly or in combination, are the present means of dehydration.

Heating (200 to 250 F) and settling are performed at atmospheric pressure or in pressure vessesls with the disadvantage that polymerization, decomposition, and volatilization result.

Chaney[14] points to the desirability of recovering and processing tars with a minimum time–temperature effect to preserve the high boiling monomer oils. Continuous vacuum flash distillation with simultaneous removal of water and light oil and heavy oil distillates from the residual pitch is the preferred tar processing method when heat sensitive monomers are to be preserved for conversion to salable resins.

Mechanical processes include the **Sharples Nozljector Centrifuge**[5] in which tars of satisfactory specific gravity (normally over 1.10), emulsified with not over about an equal volume of water, can be processed to two per cent or less moisture content, providing the emulsion also possesses satisfactory fluidity. To ensure reasonably continuous operation with such centrifuges, solid particles larger than the size of the orifices must be removed, usually by filtering thru a Purolator continuous filter. Experience indicates that addition of soda ash or caustic soda to maintain a pH of 7.5–8.5 in the tar liquor tends to produce emulsions that break more easily.

It was reported that heating tar emulsions containing as much as 90 per cent water to 170 to 180 F and discharging them at 200 psi into a storage tank reduced moisture to between 11 and 20 per cent.[20]

With use of alkali metal hydroxides or carbonate treatment,[4] deemulsification proceeds at a very low rate. It can be catalyzed and improved by an additive and heat, followed by a cooling and settling period. Additives include Paromate JC, Sterox 510, Amine 220, Dox Antifoam 200, Fatchemco, Dri-Tar, and the sodium sulfonates.

USES AND PRODUCTS

The light water gas tars produced from lower boiling petroleum distillates were used principally as fluxes in making low-viscosity road tars from high-viscosity coal tars. The heavier water gas tars and most of the oil gas tars are too high in viscosity for use as fluxes. Their principal outlet is in road tars, where they are used alone or in combination with coal tars.

In the Pacific Northwest, because of the proximity to aluminum reduction plants, oil gas tar has been distilled to pitch which has many uses, such as:

1. Electrode pitch: a binder for electrode carbon aggregate.
2. Pipe enamel: for corrosion abatement.
3. Road tars: softer grades for this use.
4. Binders: for briquetting operations.
5. Roofing: a binder and waterproofing agent.

Uses of coke oven tar pitches also include the production of pitch coke and binders in foundry molds and cores.

Distillates from oil gas tars are not used for the production of chemicals and solvents because of their nature.

The chemicals recovered from the distillate obtained in topping coke oven tars include crude solvent naphtha, a raw material for coumarone-indene resin production, tar acids, which include phenols, cresols, and xylenols, tar bases such as quinoline, and naphthalene.

An appreciable amount of coke oven tar is used in the production of road tars. Also, a large amount of tar is still used as fuel in steel plants, usually after topping to remove the chemical oil fraction.

REFERENCES

1. Rhodes, E. O. "Tar and Pitch." *Encyclopedia of Chemical Technology,* vol. 13, p. 614–31. New York, Wiley, 1951.
2. Willien, L. J. "Experiments to Determine Steam–Carbon Reaction in an Oil Gas Generator." *P.C.G.A. Proc.* 20: 556, 1929.
3. Porter, S. C. "Water-in-Tar Emulsions." (I.G.E. Communication 408) *Inst. Gas Engrs. Trans.* 101: 798–809, 1951–52.
4. Petrino, S. A. "Chemical Treatment of Water Gas Tar Emulsions." *A.G.A. Proc.* 1947: 608–14.
5. Sweeney, J. C., Chm. "Report of Joint Subcommittee on Tar Dehydration" (PC-50-25) *A.G.A. Proc.* 1950: 448–451.
6. Laudani, H., Chm. "Report of Subcommittee on Tar Dehydration" (PC-52-23) *A.G.A. Proc.* 1952: 324–5, 845–7.
7. Rhodes, E. O. *German High-Temperature Coal-Tar Industry.* (Bur. of Mines I.C. 7409) Washington, D. C., 1947.
8. Wilson, P. J., Jr., and Wells, J. H. *Coal, Coke, and Coal Chemicals,* p. 372–99. New York, McGraw-Hill, 1950.
9. Fisher, C. H. *Composition of Coal Tar and Light Oil.* (Bur. of Mines, Bull. 412) Washington, D. C., 1938.
10. Abraham, H. *Asphalts and Allied Substances,* 5th ed. New York, Van Nostrand, 1945. 2 vols.
11. Volkmann, E. W. "Structure of Coke Oven Tar." *Fuel* (London) 38: 445, 1959.
12. Kinney, C. R. "Hidden Values in Gas Tar." *A.G.A. Proc.* 1949: 737–42.
13. A.G.A. Gas Prod. Res. Comm. *Characterization of Water Gas Tars* (Research Bull. 2, Proj. PSC-1) New York, 1949.
14. Chaney, N. K., and others. "Developments in High-Temperature Pyrolysis." *Chem. Eng. Progr.* 45: 71–80, Jan. 1949.
15. Lowry, H. H., ed. *Chemistry of Coal Utilization,* vol. II, p. 1287–1370. New York, Wiley, 1945.
16. Schwarz, S. C. "Committee Report on Tar and Lampblack Removal." *P.C.G.A. Proc.* 16: 329–44, 1925; and "Condensing and Scrubbing" *Ibid.* 18: 270–322, 1927.
17. Pettyjohn, E. S., and Linden, H. R. *Selection of Oils for Carbureted Water Gas.* (I.G.T. Research Bull. 9) Chicago, 1952.
18. Linden, H. R., and Pettyjohn, E. S. *Selection of Oils for High-Btu Oil Gas.* (I.G.T. Research Bull. 12) Chicago, 1952.
19. Weizmann, C., and others. "Aromatizing Cracking of Hydrocarbon Oils." *Ind. Eng. Chem.* 43: 2312–8, Oct. 1951.
20. Geniesse, J. C., and Reuther, R. "Effect of Time and Temperature on the Cracking of Oils." *Ind. Eng. Chem.* 22: 1274–9, 1930. Also their: "Reaction-Velocity Constants of Oil Cracking." *Ibid.* 24: 219–22, 1932.
21. Groll, H. P. A. "Vapor-Phase Cracking." *Ind. Eng. Chem.* 25: 784–98, 1933.
22. Hull, W. Q., and Kohlhoff, W. A. "Oil Gas Manufacture." *Ind. Eng. Chem.* 44: 948, May 1952.
23. Linden, H. R., and Parker, R. *Prevention and Resolution of Tar Emulsions in High-Btu Oil Gas Production.* (I.G.T. Interim Rept.) Chicago, 1953. Also their: "Control and Resolution of High-Btu Oil Gas Tar Emulsions." *A.G.A. Proc.* 1953: 740–50.

Chapter 3
Manufactured Gas Conditioning

by F. E. Vandaveer

Manufactured gas conditioning may encompass all production plant processes for treating manufactured gas for distribution and sale to customers. Processes included are generally designated as purification, cleaning, scrubbing, precipitation, and condensing. Conditioning may be carried out for one or more of the following reasons:

1. To separate and recover certain chemicals.
2. To avoid trouble with stoppages, corrosion, or "dust storms" in distribution systems.
3. To avoid trouble with pilot outages and burner performance on customer's appliances.
4. To permit distribution of a uniform product.
5. To remove potential sources of trouble.

Extent and thoroughness of gas conditioning employed will vary considerably, depending upon:

1. Kind of manufactured gases involved.
2. Pipe materials and joints in distribution systems.
3. Economics of manufacture and sale.
4. Manufacturing and purification processes employed.

CONSTITUENTS IN MANUFACTURED GASES AS PRODUCED

Table 3-11 lists a number of manufactured gas constituents that may be removed, greatly reduced, or controlled by one or more processes. Reasons for control or removal, and processes generally used to do so are included. For details on removal processes for various constituents listed, see References 2 thru 6.

CONSTITUENTS REMOVED, CONTROLLED, OR ADDED IN CONDITIONING FOR DISTRIBUTION

Major items in various manufactured gases requiring removal to an acceptable minimum, continuous control within acceptable limits, or addition for customer protection and satisfactory appliance performance are listed in Table 3-12. Natural gas is also included, since it is sometimes used for carburetion of manufactured gases and, when mixed with air, for supplementing manufactured gases. Items shown are of more concern to distributing utilities than the list given in Table 3-11.

Water Vapor

Gases manufactured from coal or oil, such as oil gas, carbureted water gas, coke oven gas, or reformed gases in which steam is used, are practically saturated with water vapor during manufacturing or treating processes. Water vapor content is largely determined by temperature, as shown in Table 4-33. For example, a gas saturated with water vapor at 60 F would contain 834 lb per MMCF, or 1.74 per cent by volume, whereas at 132 F it would contain 7580 lb per MMCF, or 16 per cent by volume. In practically all instances manufactured gas is cooled by condensers and scrubbers to water or ground temperature before distribution to avoid excess condensate and excess gas temperature in the system. It will be saturated at ground temperature unless dehydration equipment is used. Certain utilities prefer to distribute a saturated gas and may humidify the natural gas used. Others prefer to distribute a dry gas.

Condensers. Crude coal gas leaving the hydraulic main at 130 to 150 F is partially saturated (60 to 95 per cent). This gas also contains tar, ammonia, hydrogen sulfide, organic sulfur compounds, naphthalene, and cyanogen which must be removed before distribution. Average content of these impurities is given in Table 3-13.

Condensers of various types are used to reduce gas temperature and to remove or reduce tar, naphthalene, water vapor, and certain high boiling hydrocarbons such as styrene and indene. If gas left the hydraulic main at 140 F saturated with water vapor, its total heat above 60 F would be about 15,000 Btu per MCF measured at 60 F, made up as follows: gas, 1600 Btu; tar, 1700 Btu; water, 11,700 Btu—*total,* 15,000 Btu.

This heat must be removed by the condenser, 1.28 gal of water per MCF being condensed and removed. Gas would remain saturated at this temperature (60 F) or at the condenser outlet temperature.

Two types of condensers are used for the aforementioned purposes. One type is tubular, arranged to secure high velocities of both gas and water to drop temperature, after which tar can be completely removed by extractors of the Pelouze and Audouin or static type. The other system lowers temperature and removes tar and some ammonia by intensively scrubbing crude gas with ammonia liquor sprayed in large quantities against a rising gas stream in grid-filled towers. Hot liquor is cooled in cooling coils or in refrigerating coils in a closed system to prevent ammonia loss.

Table 3-11 Manufactured Gas Constituents That May Be Removed or Reduced in Gas Conditioning

Class	Constituent	Formula	Reasons for removal	General process employed
Element	Hydrogen	H_2	Used in hydrogenation processes. To increase hydrocarbon content of remaining gas.	Refrigerate or liquefy other gases present; thermally decompose other gases and remove CO_2 by absorption, leaving H_2.
Element	Oxygen	O_2	Seldom removed as it is present in small amount. Contributes to corrosion of pipe interiors and enters into reaction to form vapor phase gum.	Keep air infiltration into generators at minimum. May be reduced by passing gas over hot copper or hot iron.
Element	Nitrogen	N_2	Seldom removed as such, but is removed in form of ammonia, cyanogen, and tar. Causes increase in gas specific gravity, decrease in heating value, and increase in distribution costs.	Never removed from manufactured gas.
Element	Lampblack	C	Causes deposit trouble in distribution system.	Keep gas-set temperatures low; oil fog at exhauster inlet with low-viscosity oil; Cottrell precipitator; iron oxide purifier boxes.
Oxide	Carbon monoxide	CO	Seldom removed from fuel gases as it contributes to heating value. Can be removed to reduce toxicity of supply or to purify hydrogen for hydrogenation purposes.	Convert to CO_2 by water gas shift reaction with steam and remove CO_2 by water absorption. Pass over calcium oxide at 900–1000 F. Absorb in cuprous chloride at high pressures.
Oxide	Carbon dioxide	CO_2	To reduce gas specific gravity and increase heating value. To reduce corrosivity of gas.	Very soluble in water and alkali. Removed in water scrubbers and in ammonia separators.*
Oxide	Water	H_2O	To reduce internal corrosion of mains; to reduce restrictions and stoppages in mains and house piping and to avoid ice stoppages in exposed piping.	Treat with hygroscopic substances such as calcium chloride or diethylene glycol; cool with water in indirect-type coolers to below dew point of distribution system; absorb in silica gel or activated alumina; compress and cool.
Oxide	Nitric oxide	NO	Reacts with unsaturated hydrocarbons and oxygen to form vapor phase gums. Not over one gram NO per MMCF tolerable under some conditions; 30 grams per MMCF tolerable under others.	Reduce air inleakage on coke ovens; exclude blast purge cycles and as much blow run as possible on water gas sets. Remove by chemical reaction with fouled iron oxide or by an electrical treater.*
Oxide	Iron oxide	Fe_2O_3	To prevent distribution stoppages and improper functioning of controls. Formed by rusting of main interiors and carry-over from purifier boxes.	Reduce constituents in gas causing rusting. These are O_2, H_2O, CO_2, S, and CN. Install oil scrubbers or dry filters to scrub out rust particles. Keep rust wetted with oil fog or moisture.
Oxide	Silicon dioxide, sand	SiO_2 and mixture of many substances.	To prevent main and meter stoppages and improper functioning of controls.	Install oil scrubbers or dry filters to remove dust. Reduce gas velocity, allowing particles to drop out.*
Unsaturated hydrocarbons Olefin and acetylene groups	Ethylene Propylene Butylene Acetylene Allylene	C_2H_4 C_3H_6 C_4H_8 C_2H_2 C_3H_4	Valuable compounds. To reduce yellow tips of gas flames, carbon deposits, and gum formation. To reduce solvent action on meter diaphragm dressing oil and hardening of meter and regulator diaphragms.	Autohydrogenate with H_2 and catalysts to form paraffin hydrocarbons. Direct absorption by sulfuric acid, activated charcoal, or other absorbents. Concentrate by fractional distillation at low temperatures.
Miscellaneous unsaturated	Butadiene Cyclopentadiene Styrene Indene	C_4H_6 C_5H_6 C_8H_8 C_9H_8	To prevent their polymerization and avoid liquid phase gum.	Oil scrub, using a greater volume of fresh feed than for removal of naphthalene.
Aromatic hydrocarbons	Benzene Toluene Xylenes	C_6H_6 C_7H_8 C_8H_{10}	Valuable chemicals. To reduce yellow tips on gas flames and carbon deposits. To reduce solvent action on meter and regulator diaphragms.	Recover in light oil fraction by (a) scrubbing gas with high boiling petroleum or creosote solvents, (b) adsorption on beds of granular active solids, or (c) compression and cooling.
Aromatic hydrocarbons	Naphthalene	$C_{10}H_8$	Solidifies in distribution system. Removal down to one to three grains per CCF may be necessary for such prevention.	Remove with tar and light oil. Further removal is possible by condensation, cooling, or absorption in petroleum oils.

Table 3-11 Manufactured Gas Constituents That May Be Removed or Reduced in Gas Conditioning (Continued)

Class	Constituent	Formula	Reasons for removal	General process employed
Oxygen compounds	Tar acids: Phenol Cresols Xylenols Coumarone	 C_6H_6O C_7H_8O $C_8H_{10}O$ C_8H_6O	To prevent trouble in distribution system and recover these substances for sale. Coumarone forms a gum or resin on solidification.	Condense and cool as a part of the tar. Coumarone is removed with light oil and appears in solvent naphtha fraction on distillation.
Nitrogen compounds	Ammonia Hydrocyanic acid Pyridine Picoline	NH_3 HCN C_5H_5N C_6H_7N	Ammonia may be a valuable by-product. To avoid trouble with brass parts of meters and fittings and leather diaphragms. To comply with state regulations. HCN corrodes mains and appliances.	Ammonia removed by water scrubbing or by combination with SO_2. HCN removed with ammonia, but removal is not normally practiced by any method. Pyridine and picoline are removed along with tar and ammonia.
Sulfur compounds	Hydrogen sulfide Carbon disulfide Carbon oxysulfide Mercaptans Organic sulfides Thiophene	H_2S CS_2 COS C_2H_6S $C_4H_{10}S$ C_4H_4S	H_2S removed to prevent excessive corrosion of iron and copper in system, reduce stench, comply with state regulations. Other organic sulfur compounds removed for same reasons as H_2S, but slightly larger amounts are tolerated.	H_2S removal: iron oxide processes; activated carbon; silica gel; oxidize to SO_2 by activated carbon at 660 F or nickel oxide at 750 F. Ferrox sulfur recovery—iron suspension in alkaline solution. Nickel sulfur recovery—nickel in alkaline solution. Thylox process—neutral thioarsenate salt solution. Seaboard soda liquid purification—soda ash alkaline solution. Vacuum carbonate process. Girbitol process using ethanolamines. Feld process, using ammonium thiosulfate and polythionates. Burkheiser process—H_2S burned to SO_2 and combined with ammonia to form ammonium sulfide. Gluud's process to form ammonium sulfate.* Organic sulfur compounds removal: absorbent oil circulated at higher rates than those used for light oil plus addition of more steam; activated carbon; silica gel; convert to H_2S or SO_2 by passing gas over hot checker-brick at 650 F. Refrigerate and condense.
Paraffin hydrocarbons	Methane Ethane Propane Butane Higher hydrocarbons	CH_4 C_2H_6 C_3H_8 C_4H_{10} C_5^+	Methane seldom removed. Ethane and higher hydrocarbons seldom removed, but may be reduced if recovery for sale or control of heating value desired.	Ethane and higher hydrocarbons reduced by absorption in heavy paraffin oil during light oil recovery, or by methods similar to those for absorption of LP-gases from natural gas.

* In 1963 the following relative absorption rates were reported[1] for foam methods and conventional methods of gas cleaning:

Process	For foam-type equipment	For other types of equipment
Absorption of N_2O_3	10,000–100,000	100–150 (packed tower)
Removal of H_2SO_4 mist by water	3,000–6,000	150 (packed tower)
Removal of dust and volatile salts	30,000–70,000	10,000 (electrostatic precipitator)
Absorption of CO_2 in ammonic brine	12,000	100–115 (bubble-cap tower)

Foam is produced on a perforated plate or in venturi constrictions and maintained by control of aerodynamic and hydrodynamic conditions. Foaming agents are not used, since unstable foam is desired.

In washer–scrubber condenser systems, condensers consist of one or more relatively narrow and high towers packed with wooden grids ½ in. by 6 in. deep, spaced ½ in. apart. Condensation of constituents in crude carbureted water gas follows the same general lines as in coal gas, except that no ammonia is present. Important impurities are tar, hydrogen sulfide, and naphthalene. Gas leaves the wash box outlet of the superheater at 190 F and is nearly saturated with water. At 90 per cent saturation about 86,600 Btu per MCF are removed to bring it to 80 F; about 8.70 gal of water are condensed and removed. Condensers of the direct or tubular type are used. As no ammonia is present in water gas, use of the direct contact type of scrubber condenser is quite general. In this type of condenser, gas is scrubbed with water in a tall grid-filled tower. Water is cooled in coils and recirculated. Condensate mingles with recirculated liquor and reduces tendency to lose gas constituents by absorption. Cooling 1 MCF of saturated gas from 170 to 100 F gives us approximately 3.4 gal of condensate. This condensate is largely water of saturation, together with condensed hydrocarbon vapors.

Table 3-12 Contaminants Occurring in Gases Naturally and from Distribution Operations

Contaminants	Sources*	Contaminants	Sources*
Water vapor	a, b, e, k	Nitric oxide	a, b, c, e
Humidification	f	Vapor phase gum	a, b, c, e, h
Lampblack	a, b, c	Liquid phase gum	a, b, e, g
Naphthalene	a, b, c, d, e	Silica	b, c, f
Light Oil	a, b, e	Iron oxide	f, k
Tar	a, b, e	Dirt	k
Sulfides	a, b, e	Oil	i, j
Ammonia	b, e	Odorization	c, d, f

* The presence of particles or the need to add them for odorization and humidification is shown in the body of the table by the letters a to k, signifying the following: a, oil gas; b, carbureted water gas; c, reformed natural gas fuel bed; d, cyclic catalytic reforming; e, coke oven gas, f, natural gas, g, overflowed drips; h, main and holder deposits; i, oil fogging; j, compressor oil; and k, mains and services.

Table 3-13 Impurities in Crude Coal Gas[2]

Impurity	Percentage by volume	Grains per CCF
Hydrogen sulfide	0.4–1.6	250–1,000, sometimes 3000
Ammonia	.48–1.26	150–400, occasionally 550
Cyanogen	.05–0.135	25–65
Organic sulfur	.0085–0.046	12–65
Naphthalene	0.084–0.210	200–500
Nitric oxide	...	0.015–0.3

The condensing apparatus should be located upstream of the relief holder to avoid considerable loss of relief holder capacity in terms of corrected gas, excessive pressure losses in relief holder inlet connections, and pressure losses from relief holder outlet to exhausters. When the condensing apparatus is located between the relief holder and the exhausters, the loss in pressure may be enough to produce a vacuum at the exhauster inlet. This can be hazardous, especially with electric tar precipitators, which may cause ignition of air–gas mixtures produced when air has been drawn in thru broken seals.

Observation of a great many installations over a considerable period indicates that the saturation temperature of water gas rarely exceeds 185 F.

With ¼ in. grids spaced ½ in. apart, the minimum cross-sectional area of a washer–cooler tower should be 1 sq ft for each 8.3 MCF passed per hr. A vertical grid height of 30 ft is about the minimum for complete heat transfer between cooling liquor and gas. These figures do not apply to flows under 0.5 MMCF per 24 hr or to extreme temperature ranges.

The best spray header design distributes liquor over the entire condenser cross section as nearly as possible in the form of rain, the larger the drops the better. Nozzles delivering liquor should be placed below gas exit level if possible. A heat transfer coefficient of 1 Btu per min–sq ft–degree mean temperature difference is a safe figure for commercial operation.

Countercurrent heat exchange gives greater water economy. It also raises cooling water to a higher temperature and cools liquor to a lower temperature than parallel flow.

The single pipe system with liquor inside and cooling water outside is more accessible for cleaning.

Figure 3-6 shows test results on a combined washer–cooler and cooling coils.

Water-cooled tubular condensers[7] in best practice circulate water thru tubes with gas flowing countercurrent (preferably downward) in a baffled course back and forth across tube exteriors. A high heat transfer requires high water velocity and small diameter tubes which permit maximum cooling surface. Reduction of tube size reaches a limit where cost per square foot of cooling surface increases and probability of stoppage with mud requires too frequent cleaning. Too great a ratio of length to diameter also increases the likelihood of buckling the tubes.

Gas passages should provide for as high a velocity as allowable back pressure will permit. A reasonable resistance should be expected since a low gas velocity will permit only a low heat transfer rate.

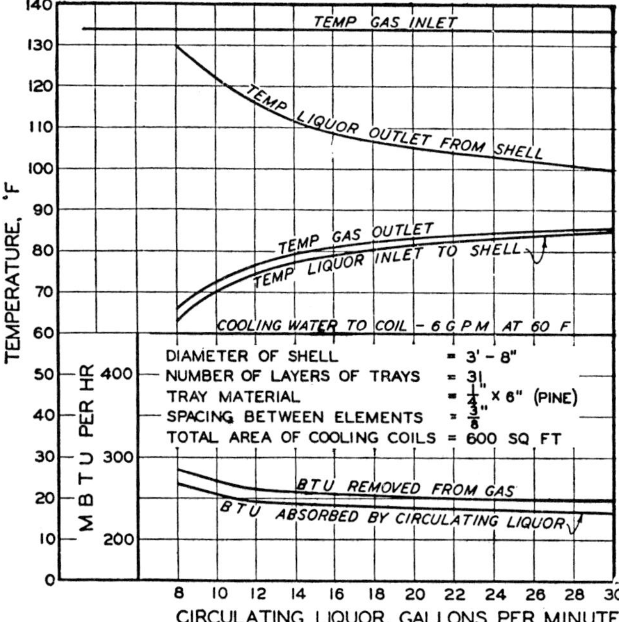

Fig. 3-6 Combined washer-cooler and cooling coils test results.

The cost of water is by far the highest item in the condensing costs. There has been a temptation, in a few cases, to reduce the volume of water by using an excessive amount of cooling surface to raise the exit water temperature. Such attempts have proved ill-fated. The condensing water carries an appreciable amount of dissolved gases, including oxygen and carbon dioxide. At increased water temperatures, more and more of these gases are released. They tend to accumulate in bubbles on the water side of the tubes at the hot end of the condenser. Since they are corrosive, they cause deterioration wherever they accumulate, especially to ferrous tubes. In one case, the tubes were so badly corroded that they required renewal after only one year's service. Waters vary considerably in their corrosive tendencies but, in general, a water outlet temperature of 120 F is a safe maximum limit.

Entraining gas in water leaving the scrubber at its seal pipes can be avoided either by maintaining the liquid level above the seal pipe top by some float control or by making the seal pipe large enough to allow a considerable gas space above the liquid.

Table 3-14 gives calculated diameters of seal pipes for various liquid flow rates and three different stream depths in pipe. If maximum depth is used, and there is little chance of exceeding it, Case I may be assumed. For average conditions it is better to use Case II. With this in mind, choose actual pipe diameter just above or just below calculated diameter, depending upon safety factor desired.

Condensation and purification of oil gas closely follow practices described for carbureted water gas, except that removal of final traces of lampblack and small quantities of tar requires special attention.

Dehydration of Manufactured Gas. From the preceding it is apparent that gas entering the distribution system at 50 to 100 F is essentially saturated with water. Any lowering of gas temperature will cause condensation and leave gas saturated at that lower temperature.

Table 3-14 Calculated Seal Pipe Diameters for Various Flows

Liquid rate, gal per min, where:	Diam *D*, in.		
	Case I	Case II	Case III
Depth of stream in pipe →	1/2 *D*	1/3 *D*	1/4 *D*
Level of liquid in scrubber above bottom of pipe→	0.71 *D*	0.47 *D*	0.35 *D*
25	3.3	4.4	5.5
50	4.3	5.9	7.2
100	5.7	7.7	9.6
150	6.7	9.1	11.2
200	7.5	10.2	12.6
300	8.8	12.0	14.8
400	9.9	13.5	16.6
600	11.6	15.8	19.6
800	13.1	17.8	21.9
1000	14.3	19.4	24.0
1500	16.8	22.8	28.2
2000	18.8	25.6	31.6

Of the four methods available for removal of water vapor from manufactured gas,[8] only two have been used as shown:

1. Compression. Not used; appears too costly.
2. Refrigeration. Partial dehydration used by one company.
3. Hygroscopic substances. Limited use in England.
4. Solid absorbents. Not used to date.

Partial dehydration[9] was used in 1939 at Poughkeepsie, New York. Water cooling by means of vacuum pumps was used in indirect-type gas coolers to reduce gas dew point below lowest distribution system temperature. Shively[9] has also described a calcium chloride solution system of absorbing water vapor. Glycerin has been used in England for dehydration in several plants.[10] Diethylene glycol, silica gel, activated alumina, and fluorite have not been used on manufactured gas, but are used extensively to dry natural gas under pressure to a very dry basis, possibly 7 lb of water vapor per MMCF or less.

Humidification. Addition of water vapor to manufactured gas is unnecessary because the gas is saturated when it enters the distribution system. However, humidification has been used by certain utilities when natural gas, natural gas–air mixtures, LP-air mixtures, or certain reformed natural gases are mixed with manufactured gases and distributed thru cast iron mains with jute-packed bell and spigot joints. One company[6] supplying a mixture of gases, including natural gas, admits steam at the mixing point in excess of that needed for saturation at existing temperature. In addition, it uses a portable steamer at 20 system locations as needed. Steam is injected into mains at rates of 50 to 300 lb per hr.

Tar

Most of the tar in crude manufactured gas has been condensed and removed along with the water in initial condensers; water and tar are separated later by gravity difference. The tar that remains in the form of mist or fog is removed by additional treatment, such as electrical precipitation, impingement, adsorption, or other methods suitable for removal of dispersoids (5 microns diam or less).

Electrical precipitation is widely used in this country[11,12] for removing suspended tar and lampblack. Koppers Co.[13] (Fig. 3-7), Cottrell, and others have developed electrostatic precipitators.

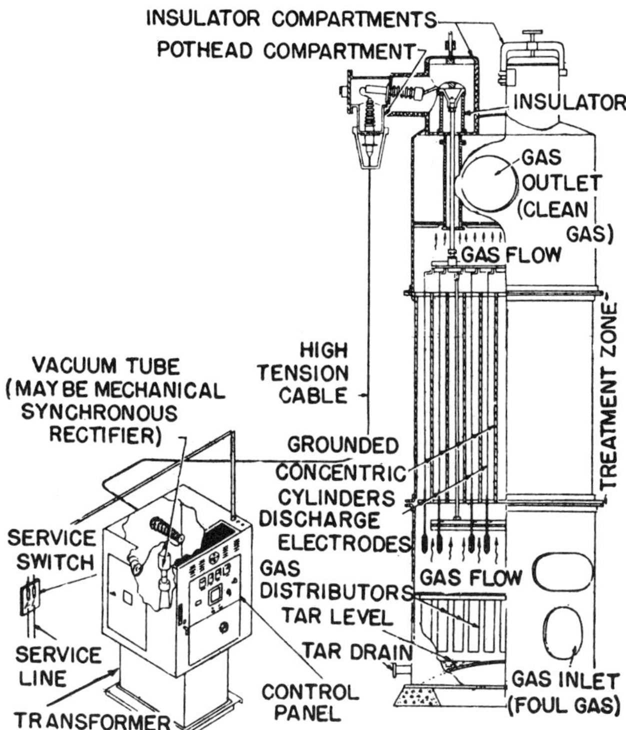

Fig. 3-7 Koppers vertical-flow electrostatic precipitator or detarrer.

Simplicity, inherent low draft loss (0.4 H_2O), and low operating costs (5 kwh per MMCF of gas to be cleaned) have gained acceptance for cleaning large volumes of gas with electrostatic equipment. The vertical flow detarrer pictured in Fig. 3-7 utilizes its entire cross-sectional area for precipitation, effecting lower gas velocities and reducing space requirements.

The Pelouze and Audouin (P & A) tar extractor is an example of the impingement principle of removing suspended matter. Gas passes thru small orifices in a drum and impinges on a surface, causing separation of a large percentage of particles. This extractor is seldom used in large plants and should not be used where appreciable amounts of lampblack may be associated with tar, as in heavy oil operations. A more recent design for a gas cleaner, based on the impingement principle, is the "Tracyfier" made by Blaw-Knox Co., Pittsburgh. It is reported to have high removal efficiency.

These tar extractors do not remove all tar mist. Properly operated electrical precipitators will remove from 95 to 99 per cent of suspended material.[2] Probable concentrations of tar likely to be found in outlet gases of Cottrell precipitators and Pelouze and Audouin extractors are as follows:[14]

	Cottrell	*P & A*
Grams per MCF	1.6–3.0	5–30
Grains per 100 cu ft	2.5–4.6	7.7–46.2
Gallons per MMCF	0.38–0.72	1.2–7.2

If tar extractors are operated at too high a temperature or too low a voltage or are poorly maintained, larger amounts of tar fog will pass thru.

Various washer–scrubbers bringing gas and a scrubbing liquid in intimate contact have been used in this country for tar mist extraction. Blaw-Knox[15] scrubbers are of this type.

An example of equipment for removing tar mist by adsorption is the shavings scrubber.[16] Wood shavings present a very large surface to the gas and a large portion of tar particles adhere to them. Iron oxide purifier boxes are also used to remove tar mist. However, the tar fog must be kept below 0.5 gal per MMCF to prevent trouble in both iron oxide boxes and liquid purification plants.

Although no published data on the composition of suspensoids in manufactured gases as they leave manufacturing plants indicate that tar fog contributes to these suspensoids, it probably does. Precipitation and scrubbing equipment normally used as described above does not remove all tar fog, and during periods of improper operation considerable fog may flow into the distribution system. Because mist particles are extremely small, it is very difficult to trap them in any kind of scrubber. It is believed that dry filters with fibers as small as or smaller than the diameter of the mist particles to be trapped might remove all tar fog particles.

Tar Fog Detection and Measurement. There is no specific test method for tar fog alone. Several methods may be used to determine total suspensoids trapped by a filter, including all solid and liquid particles. If sufficient material is retained on the filtering media, it may be analyzed chemically to determine the tar content.

The filter[6] used by the Philadelphia Gas Works Co. is a glass tube partly filled with glass wool weighed before and after gas is passed thru the filter. In plant operation these filters are placed in the: (1) inlet to condensers, (2) outlet of condensers, (3) outlet of naphthalene scrubbers, (4) outlet of relief holders, (5) outlet of Cottrell precipitators, and (6) outlet of dry box purifiers. These filters are also installed at various selected places throughout the system.

Ammonia

Crude coal gas and crude coke oven gas may contain from 1.0 to 1.5 per cent ammonia by volume. There is essentially no ammonia in crude carbureted water gas, oil gas, reformed natural gas, or LP-gas.

Conditioning of gas to control ammonia content at considerably less than one grain per 100 cu ft appears possible and desirable for several reasons:

1. Ammonia is an irritant and its concentration in air must be kept below 100 ppm for prolonged exposure; 2500 to 6500 ppm is dangerous for even a short exposure (½ hr).[17]
2. Some state regulations limit ammonia content in gas to not more than five grains per 100 cu ft (0.016 per cent by volume). Others limit it to ten grains per 100 cu ft.
3. Ammonia in aqueous solution is corrosive to brass and copper.
4. Ammonia reacts with carbon dioxide[18] in the presence of water vapor to form ammonium carbonate, a solid. When water evaporates, pipe stoppages may occur.
5. Ammonia travels throughout the distribution system.[12] Ammonia and ammonium thiocyanate were found in drip water at points throughout a large system.
6. Ammonia may affect leather meter diaphragms.[19]

From coal gases a small volume of ammonia is absorbed by water condensed in the initial condenser. Ammonia is very soluble in water and in smaller plants is removed by water scrubbing in a tower scrubber. Usually more than one tower is used. In the first tower in a series, a weak ammonia solution is circulated by centrifugal pumps, building up its strength to about 1½ per cent ammonia. In the final tower, fresh water is used so that ammonia is reduced to less than one grain per 100 cu ft. This may require from 10 to 15 gal of fresh water per ton of coal. Some removal of hydrogen sulfide and carbon dioxide occurs during this stage.

Ammonia recovered as condensate and as a weak liquor in ammonia scrubbers will average from one to two per cent by weight. It exists principally as an acid carbonate and chloride, together with small quantities of several other ammonia compounds.

In coke oven plants, ammonia is usually recovered as ammonium sulfate by scrubbing with a weak solution of sulfuric acid. In the Koppers system, gas is separated from ammonia liquors as it enters the collection main from the oven. Crude gas is then cooled in the primary cooler, where tar and most ammonia are condensed and washed out. Gas is exhausted thru tar extractors, where remaining tar is removed. The gas is reheated and forced thru saturators, where it meets a dilute sulfuric acid solution, causing ammonium sulfate to be precipitated as a solid salt. The gas must then be reheated to enable it to carry off the water formed in the reaction between sulfuric acid and ammonia so as to prevent accumulation of excess liquid. Sulfate formed in the saturator is dried in a centrifugal dryer.

Liquor collected at various points, principally at primary coolers, is pumped to a storage tank. The ammonia vapor recovered by having been driven off in the stills is added to the gas just before the gas enters the saturators.

After leaving the saturators, the gas is passed to a final cooler, where it is often washed with water. In some cases it passes to a naphthalene scrubber so that its naphthalene content may be controlled.

Naphthalene

Naphthalene is found in all types of manufactured gases in varying amounts from 5 to 40 grains per CCF at scrubber inlets. Catalytically cracked natural gas generally has about one grain or less, reflecting the lower temperatures at which this process operates. Vertical retort gas may average two pounds of naphthalene per ton of coal gasified; coke ovens will give about six pounds; and horizontal retorts at high temperature may give over eleven pounds per ton of coal. Lower fuel-bed temperatures tend to reduce naphthalene content, but this creates other difficulties as well as inefficient operation.

At ordinary temperatures naphthalene is a white solid with a melting point of 176.4 F and a boiling point of 424.4 F. It is probably formed by pyrolysis of other hydrocarbons, since it is absent in both gas oil and primary distillation products from coal. When gas is subjected to high temperatures, large concentrations of naphthalene are formed; at lower temperatures, lesser amounts are formed. Most of it condenses out together with other hydrocarbons, including tar, in initial condensers; possibly ten per cent of the tar is naphthalene. Upon further cooling, more naphthalene will condense and form solid deposits unless other liquid hydrocarbons condense simultaneously in sufficient volume to keep it in solution. With carbureted water gas, sufficient other hydrocarbons are usually present so that condensed naphthalene is largely dissolved in the condensate, termed **water gas drip.**

Table 3-15 Weight of Naphthalene Needed to Saturate Gas at Various Temperatures*

Gas saturation temp, °F	Vapor pressure of naphthalene in dry coal gas, in. Hg	Grains of naphthalene to saturate 100 cu ft of coal gas†
32	0.000220	1.71
41	.0003976	3.08
50	.0006929	5.38
59	.001185	9.19
68	.001992	15.5
77	.003291	25.5
86	.005354	41.5
95	.008583	66.6
104	.01350	105
113	.02098	163
122	.03213	250
131	.04842	376
140	0.07244	563

* Recalculated from Thomas, J. S. G., "Evaporation of Naphthalene in Dry and in Moist Coal Gas." *J. Soc. Chem. Ind.*, 35: 506–13, May 15, 1916.

† At 60F, 30 in. Hg saturated with water vapor. This column calculated from

$$\text{Grains of } C_{10}H_8 \text{ per CCF} = 232{,}746.1078 \frac{F_d}{30.00 - F_d}$$

where F_d = vapor pressure of naphthalene in dry coal gas, in. Hg.

Gas during its manufacture is ultimately cooled to about 75 to 85 F. During this cooling most of the naphthalene is removed. Table 3-15 shows amounts that can be retained as vapor by gas at different temperatures. To reduce naphthalene content lower than can be accomplished by cooling, a removal process is necessary.

The comparatively high vapor pressure of naphthalene at ordinary temperatures and its ability to crystallize from the gaseous phase on cooling and to resublime on warming are principal causes for naphthalene troubles in distribution systems. If it is allowed to enter mains in any large concentration, stoppage of small mains and customers' services will follow. Accumulated deposits not large enough to affect large mains will be sublimed by warm gas and carried forward to crystallize in cool, contracted sections with resulting loss of service. This ready formation of solid crystals is a most objectionable property, since the crystals serve as nuclei for deposition of more naphthalene if conditions are otherwise favorable. Foreign material such as rust, dust, or gum may also combine with the naphthalene deposition.

Deposition of naphthalene on dry surfaces, when not associated with oily or similar solvent matter, is a simple vapor pressure phenomenon. When the naphthalene content is under 5.38 grains per 100 cu ft (0.0023 per cent by volume at standard conditions), naphthalene should not deposit until the temperature drops below 50 F (see Table 3-15). No simple reasoning can be applied when naphthalene solvents, such

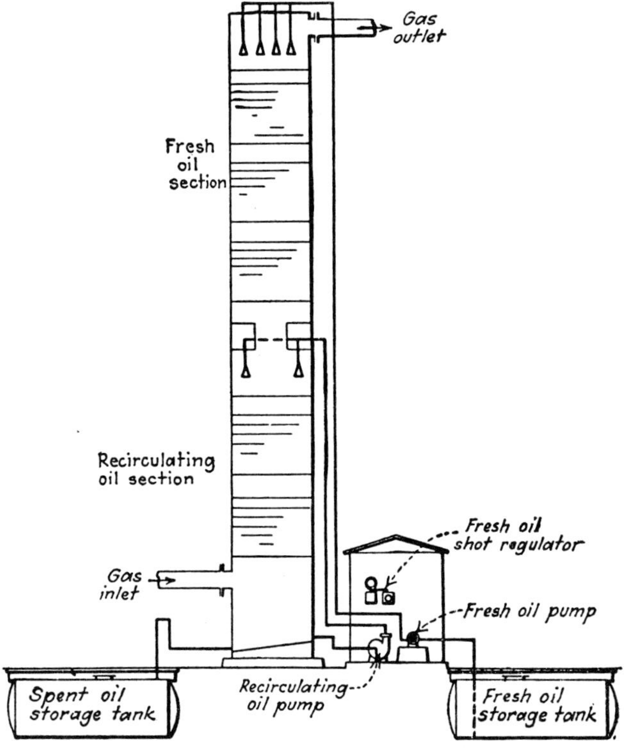

Fig. 3-8 Koppers naphthalene scrubber.

as oil or tarry or gummy material, are present because these act to aid its removal and tend to keep it in liquid solution. Ground temperature and depth of cover must be considered before fixing limits below which naphthalene stoppages may be expected not to occur.

Naphthalene Scrubbers. Special oil scrubbers (see Fig. 3-8) scrub the gas with fresh gas oil to absorb naphthalene.[3,20] Spent gas oil can be used instead of fresh gas oil for water gas manufacture. The spent gas oil can also be sold as fuel oil. The amount of oil required depends on gas temperature and on desired naphthalene content of finished gas. Between 15 and 50 gal of oil per MMCF of gas is adequate in temperature ranges usually encountered to reduce naphthalene content to two grains per CCF. The liquid gum-forming constituents, indene and styrene, can also be removed in this process by means of increased oil circulation rates.

Other Naphthalene Removal Methods. If equipment for light oil recovery is used, it can be operated so that efficient

Table 3-16 Typical Light Oil Yields

Light oil products	Horiz. retorts,[2] gal/ton	Vert. retorts,[2] gal/ton	Coke ovens,[2] gal/ton	Carb. water gas[2] (approx. 550 Btu), % of gas oil used	All oil water gas[2] (approx. 550 Btu), % of gas oil used	Oil gas[22] (970 Btu), gal/MCF of gas produced	Oil gas[23] (1021 Btu; 0.78 sp gr), gal/MCF of gas produced
Benzol*	1.62	1.81	1.782	4.5	3.3	..	..
Toluol*	0.41	0.43	0.402	2.2	0.2	..	..
Refined naphthas	.30	.41	.216	1.0	.03	..	..
Unsaturated hydrocarbons	0.36	0.43	0.297	3.1	0.60	..	..
Total	2.69	3.08	2.697†	10.8	4.13	0.81	0.72

* Chemically pure. † 2.94 in 1957 (ref. 24).

naphthalene removal is obtained. Otherwise, an oil washer designed for naphthalene removal can be used.

Ordinary light oil recovery equipment will remove naphthalene efficiently if the naphthalene is thoroughly stripped from the **absorbent oil.** Otherwise naphthalene in recirculated absorbent oil may accumulate to such a concentration that its removal is unsatisfactory. Use of additional steam in wash oil stripping stills will strip naphthalene from wash oil more thoroughly and consequently result in its more efficient removal by light oil equipment.

In some instances distribution lines have been used as naphthalene scrubbers by the injection of solvents such as tar oils, kerosene, or tetraline.[3,20] This practice has not been widely used in this country. Solid adsorbents[3,20] such as activated charcoal and silica gel are reported to remove naphthalene, as well as other material, completely.

Removal of H_2S by iron oxide obviously requires contact between the oxide and H_2S. Anything that hinders such contact clearly will reduce the efficiency of this process. Gas may become supersaturated with naphthalene due to temperature drop. The resulting contamination has an adverse effect on purification results. Note, however, that naphthalene is *chemically* inert at temperatures encountered in ordinary purification work. Other hydrocarbons present usually condense along with naphthalene, adding to the film deposited on the oxide. Obviously, naphthalene removal before purification results in a very distinct increase in purification efficiency.

Reduction in purification efficiency as described may be *minimized* in several ways:

1. Conditions of gas manufacture may be adjusted to produce a minimum of naphthalene whenever economically practical.
2. Low condenser temperatures will remove a large part of the naphthalene before purification.
3. An oil scrubber may be used to remove naphthalene prior to purification.
4. The temperature of the purification system may be maintained at a high enough level to avoid naphthalene deposition (Table 3-15).

Test Methods. Experience has indicated the desirability of maintaining the naphthalene content at less than one grain (0.00044 per cent by volume) per 100 cu ft.[6] Test for naphthalene and other suspended matter with an impingement device, a glass wool filter, a row of Rutz pilot burners, and Arthur D. Little absolute filters.

The method commonly used to determine the amount of naphthalene in gas is based on the reaction of naphthalene with picric acid solution. Details of various methods are available.[21]

Light Oil

Light oil is produced in manufacturing coal gas, coke oven gas, and oil gas in amounts approximately as given in Table 3-16. Production depends upon coal or oil used, type of manufacturing plant, and recovery or scrubbing system employed.

Light oil contains benzene (C_6H_6; bp, 176 F), toluene (C_7H_8; bp, 231.8 F), and xylene (C_8H_{10}; bp, 280.4 to 291.4 F) and other unsaturated hydrocarbons. The composition of crude light oils is given in Tables 3-17 and 3-18.

Table 3-19 analyzes oil gas,[22] approximately 970 Btu, produced in twin generator back blast machines before and after scrubbing with wash oil.

A recovery method in general use involves scrubbing or washing with a petroleum oil of high boiling point (low vapor tension) until 2.5 to 3.0 per cent of the light hydrocarbons are absorbed. So-called benzolized oil is then heated to 240 to 280 F and distilled with live steam, liberating low-boiling benzol homologues. Debenzolized wash oil is then cooled by heat exchangers and water coolers and returned to the cycle.

It is evident that, given a wash oil of low vapor tension, extraction will depend upon relative amount of low-boiling vapors present and saturation of benzolized wash oil (volume of wash oil per gas unit volume and temperature). As temperature *increases*, the volume of wash oil must be increased to decrease the percentage of its saturation and obtain equal efficiency of extraction. The highest commercial efficiency of extraction is between 92 to 95 per cent, while the average probably does not exceed 90 per cent. Wash oil used is generally known as "absorbent" or "straw" oil.

Table 3-17 Composition of Crude Light Oils from Various Processes

(expressed as percentage of light oil)

Gas-making process	Benzene	Toluene	Xylene	Solvent naphtha, wash oil, and naphthalene	Paraffins
Horizontal retort coal gas	38	16	...	46	<2
Coke oven gas[24]	62.5	13.1	3.7	20.7	...
Continuous vertical retort coal gas	15	15	...	58	12
Carbureted water gas	42	25	...	33	<2
Oil gas (970 Btu)	45.5	22.5	8.0	24	...

Table 3-18 Analysis of a Typical Crude Coke Oven Light Oil

	Percentage by volume
Forerunnings:	
Cyclopentadiene	0.5
Carbon disulfide	0.5
Amylenes and unidentified	1.0
Crude benzol:	
Benzene	57.0
Thiophene	0.2
Saturated nonaromatic hydrocarbons, unidentified	0.2
Unsaturates, unidentified	3.0
Crude toluol:	
Toluene	13.0
Saturated nonaromatic hydrocarbons, unidentified	0.1
Unsaturates, unidentified	1.0
Crude light solvent:	
Xylenes	5.0
Ethyl benzene	0.4
Styrene	0.8
Saturated nonaromatic hydrocarbons	0.3
Unsaturates, unidentified	1.0
Crude heavy solvent:	
Coumarone, indene, dicyclopentadiene	5.0
Polyalkyl benzenes, hydrindene, etc.	4.0
Naphthalene	1.0
Unidentified "heavy oils"	1.0
Wash oil	5.0*
Total	100.0

* The amount of wash oil present depends greatly on the performance and design of the debenzolization apparatus as well as on the nature of the wash oil employed.

Table 3-19 Analysis of High-Btu Oil Gas before and after Light Oil Scrubbing[22]

	Unscrubbed	Scrubbed
Nitrogen	13.5	13.6
Carbon monoxide	0.3	0.0
Carbon dioxide	3.1	3.1
Hydrogen	17.0	16.9
Methane	37.3	37.9
Ethane	3.1	3.3
Propane	0.1	0.4
Ethylene	19.1	18.9
Propylene	3.0	3.2
C_4 compounds	0.4	0.4
1,3-Butadiene	1.1	1.1
Cyclopentadiene	0.3	0.2
Acetylene	0.3	0.4
Benzene	1.3	0.6
Toluene	0.1	Trace
Total	100.0	100.0

Benzene in unscrubbed gas is probably too low.
Xylene, not shown, was possibly 0.05 per cent.

Recovery apparatus is usually installed in coal gas plants alongside ammonia washers in which gas is cooled first. The temperature for highest extraction is rarely allowed to exceed 75 F. In water gas plants and oil gas plants recovery apparatus is usually installed next to purifiers, which are used immediately before the recovery apparatus.

The main features of recovery apparatus are the same in nearly all modern systems. Gas is brought into contact with wash oil in two tall towers filled with layers of boards on edge or ceramic forms to break up the gas stream and expose the maximum surface of oil to the gas. Oil is sprayed down the towers against the rising gas stream. Separators similar to steam separators in gas outlet pipes prevent loss of wash oil in the form of mist or spray while bottoms of towers and external separators remove any water deposited. The wash oil now contains from 2.5 to 3.0 per cent crude benzol and is at a temperature of possibly 75 F. It then passes to an oil heat exchanger, where it is heated to 150 F by latent heat of debenzolizing *still* vapors, then thru an oil-to-oil heat exchanger, where it is heated to 220 F by hot wash oil from the *still* bottom. From there it goes to a steam preheater, where high-pressure steam brings temperatures up to 270 to 290 F.

In coke oven plants in which complete recovery is the goal, temperatures are usually held below 75 F. Sufficient wash oil is used to yield saturation of the benzolized oil of not over 2.5 per cent. Using 2.0 gal of light oil and 10,800 cu ft of gas per ton of coal requires 7.22 gal of wash oil per MCF. With higher temperatures and greater quantities of light oils present, wash oil used may run up to 14 to 16 gal per MCF.

Practically complete elimination of light oil from benzolized oil is important in securing high recovery efficiencies. Elimination of the light oil is largely accomplished by a uniformly high oil temperature entering the debenzolizing column; it is difficult to achieve with a temperature of less than 220 F. Recovery is complete, steam required is minimum, and operation is most nearly uniform at 240 to 280 F. Live steam required varies from 70 to 100 per cent by volume of light oil produced.

An apparatus for removing light oil from oil gas (970 Btu) is shown in Fig. 3-9. It consists essentially of a scrubber with wood grids, condensers, heat exchangers, a still, and storage tanks. Gas at about 90 F enters a two-pass scrubber, where it is scrubbed with benzolized wash oil in both passes and 20 gpm of fresh wash oil which is added at the outlet of the second gas pass. Scrubbing rates on the benzolized oil in the first gas pass are 125 gpm and in the second, 60 gpm. After scrubbing, benzolized wash oil is pumped from the scrubber bottom at 20 gpm to a heat exchanger, where it is heated by light oil vapors from the top of the still at from 90 to 125 F. It then enters the oil-to-oil heat exchanger, where it is again heated by wash oil, this time from the still bottom at from 125 to 170 F. Benzolized oil then enters the third and last preheater and is heated to 270 F. From there it goes to the still for final distillation.

Efficient removal of light oil is desirable in conditioning processes for several reasons:

1. Recover salable by-products: benzene, toluene, and xylene.
2. Decrease oil drippage in distribution system.
3. Decrease oil deposits in services, regulators, meters, and piping.
4. Reduce customer complaints of odor from range burners, pilots, and very small leaks.
5. Prevent yellow-tipped flames and carbon deposits on cooking utensils and heating surfaces.
6. Reduce trouble with meter diaphragms.
7. Reduce pilot outage.

Tests to determine the amount of light oil in gas are usually made to estimate the efficiency of recovery. These tests can show *both* the amount and kinds of oils present. Outlet gases containing predominantly benzene and the lighter hydro-

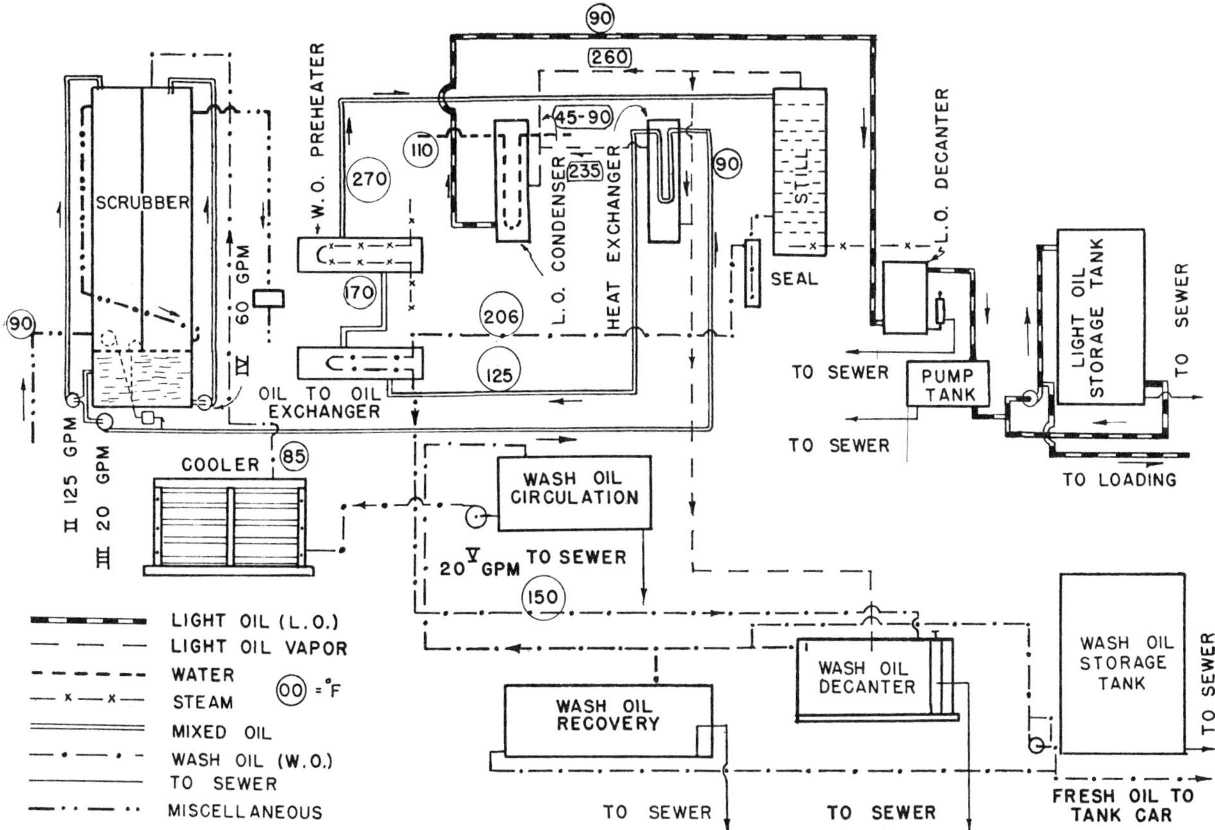

Fig. 3-9 Light oil recovery and scrubbing plant for high-Btu oil gas.[22]

carbons may indicate that insufficient absorbent oil or too high a temperature was used for scrubbing. The presence of heavy constituents in light oil, such as xylol, solvent naphtha, and naphthalene, indicates the incomplete stripping of light oil in wash oil still.

Three test methods[3,25] commonly used for the determination of light oil in gas are:

1. Cool gas to a very low temperature and measure amount of light oil condensed.
2. Pass gas thru wash oil and measure amount of light oil *absorbed.*
3. Pass gas thru activated carbon and measure amount of light oil *adsorbed.*

Light oils resulting from processes other than the coking of coal may present a disposal problem. These oils are difficult to sell because of the presence of large amounts of unsaturates, paraffins, and sulfur.

Styrene, Indene, and Liquid Phase Gum

Styrene (C_8H_8) and indene (C_9H_8) are liquid unsaturated hydrocarbons occurring in the vapor state as minor constituents in coal gas, carbureted water gas, and oil gas. Their boiling points are 293 and 352.4 F, respectively. Coal gases from various sources show styrene contents ranging from 0.002 to 0.106 per cent by volume, and indene contents from 0.004 to 0.013 per cent by volume.[26] Carbureted water gas and oil gas before purification may contain similar percentages.

These unsaturated hydrocarbons are parent substances of liquid phase gum. To form gum, however, they must be condensed to the liquid state or be dissolved in a liquid such as oil in the presence of oxygen. Then they polymerize and combine with oxygen to form liquid phase gum. Analyses[12] of liquid phase gums taken from mains, governors, and meters showed a carbon content of 58.11 to 75.42 per cent, a hydrogen content of 4.73 to 6.68 per cent, an oxygen content of 11.22 to 23.21 per cent, and an ash content of 2.08 to 16.08 per cent *by weight.* These analyses show that nitrogen is not an important constituent, that sulfur is probably of significance, and that gums are oxidized products and not merely polymers of hydrocarbons. Organic sulfur compounds have no effect on oxidation of styrene and indene,[26] but traces of **mercaptans** have a very strong catalytic effect.

Liquid phase gum is a neutral or slightly acidic mixture, ranging from a viscous sticky mass to hard, easily powdered, solid deposits. This type of gum may interfere with the operation of compressors, governors, and meters and may cause stoppages in services, house piping, pilot orifices, and automatic appliance controls.

Liquid phase gum may be removed or controlled, if necessary, by various procedures.[26,27] Powell[27] stated that styrene and indene are not removed from coal gas because they are largely removed in light oil recovery, with the remainder in a vapor state in which it is harmless. If coal gas is to be mixed with carbureted water gas, styrene and indene may be removed by oil scrubbing. Equipment and method of operation are similar to those for naphthalene removal except that a greater volume of fresh scrubbing oil must be fed. Efficient removal of liquid phase gum formers calls for 0.10 to 0.50 gal of oil per MCF of gas, considerably less than used for light oil recovery. Keep oxygen content of gas low to reduce gum formation. Better condensation and removal of liquids will also decrease traces of gum formers and reduce gum formation.

The presence of liquid phase gum in distribution systems is generally indicative of improper condensing facilities.[2] Gas should enter mains with a vapor dew point *below* the lowest temperature it may later encounter. Pumping drips regularly so that they do not overflow will also reduce gum formation. Use of solvents such as benzol and acetone is effective in clearing gum deposits. Stoppages in pilot orifices may be prevented by using gum filters.

To determine the amount of resin-forming hydrocarbons in gas, condense the gas at various temperatures, fractionate the light oil, and determine the styrene and indene by titration with bromine. The unsaturation as determined by bromine titration can be confirmed by polymerization with sulfuric acid. The apparatus and methods have been described in detail.[26]

A complete microscopic technique for separation of gum deposits on small samples (naphthalene was distinguished by a polarizing microscope) has been reported.[28]

Nitric Oxide and Vapor Phase Gum

Nitric oxide may be present in manufactured gases in extremely small amounts, as indicated in Table 3-20. Conversion factors used in computing these values are given in Table 3-21.

Nitric oxide is one of the constituents which reacts to form vapor phase gum*—a most undesirable compound. *Four conditions are necessary to form vapor phase gum.*

1. Nitric oxide must be present in amounts greater than those given in Table 3-20.
2. Unsaturated hydrocarbons must be present.
3. Oxygen must be present.
4. There must be time for the reaction to take place. The amount of time varies with temperature and amounts of above three constituents present. Probably an hour or more is required for reaction in gas holders, street mains, and services.

Vapor phase gum is a strongly *acidic* substance, soluble in aqueous alkalies, and in organic solvents containing oxygen, such as acetone. It decomposes on heating, forming nitrogen oxide, carbon monoxide, carbon dioxide, amines, and small quantities of oil. Some samples of gum, molasses-like in consistency, harden rapidly on exposure to air. This gum is unstable at room temperature, and the rate of decomposition increases rapidly with heat, with the reaction becoming violent at about 350 F.

An average ultimate analysis of vapor phase gum is given in Table 3-22. Vapor phase gum may form at the plant or any place in the distribution system. The actual quantity produced is very small, of the order of 0.0154 to 0.769 grain per CCF or 0.000021 to 0.00091 per cent by volume. In this finely divided state, the gum is like a mist which remains suspended in the gas.

In gas giving objectionable but not excessive gum trouble, the number of *particles* ranges from 20,000 to 100,000 per cu cm, or a maximum of nearly 3 billion per cu ft. The diameter of the smallest particle visible under an ultramicroscope is

* Vapor phase gum should be distinguished from liquid phase gum.

Table 3-20 Nitric Oxide Limitations in Sendout of Various Manufactured Gases

Kind of gas	Grams/ MMCF	Grains, CCF	Per cent by volume
Catalytic cracked[6] natural gas	1	0.00154	0.0000027
Fuel bed[6] reformed natural gas	30	.0462	.000080
Carbureted water gas	...	...	...
Fuel bed reformed[6] natural (90%) and carbureted water (10%) gas	10	.0154	.000027
Fuel bed reformed[6] natural (60%) and carbureted water (40%) gas	1	.00154	.0000027
CCR naphtha[29] reforming	0	0	0
CCR kerosene[29] reforming	1	0.00154	0.0000027
Coke oven gas[3]		0–0.055	0–0.0001
Coal gas[2]	10	0.015	0.000027

Table 3-21 Conversion Factors for Nitric Oxide Limitations

Grains per CCF	1
Per cent by volume	0.00183
Parts per million	18.32
Milligrams per cubic meter	24.56
Grams per MMCF	648.1

Table 3-22 Ultimate Analysis of Vapor Phase Gum as Received

Carbon, per cent	63.36
Hydrogen, per cent	6.71
Nitrogen, per cent	3.81
Oxygen, per cent	23.51
Sulfur, per cent	0.82
Ash, per cent	1.79
Simplest formula	$C_{19}H_{24}NO_5$
Calculated molecular weight	346
Observed molecular weight	381

approximately 0.000008 (eight millionths) in. That of larger particles is between about 0.00004 (40 millionths) in. and 0.00008 in. The average diameter is 0.00001 in. At this size, *without coalescence*, it is estimated that about three months would be required for particles to drop 100 ft by gravity in perfectly still gas. Hence, they could never be expected to drop out of gas before appliances are reached, since gas "life" is only a matter of hours.

The amount of gum for stoppage of a needle pilot adjustment device is about 0.0_63 to 0.0_67 (three to seven ten-millionths) ounce.

Various methods for controlling or preventing vapor phase gum deposits have been used. Removal of oxygen or the more reactive unsaturated hydrocarbons has not been successful to date and it appears simpler to control nitric oxide content.

Removal by Iron Sulfide. Iron oxide boxes which have been heavily sulfided or are used to remove large quantities of H_2S will remove a large portion of the NO. In some instances, however, a special NO box is added ahead of other purifying boxes. In one English installation, a special box *decreased* NO from 0.066 grain per CCF at the inlet to 0.00218 grain at the outlet. The foul box technique has also been used, as described in U. S. Patent 1976704, assigned to the United Gas Improvement Co.

Removal by Treatment with Electric Corona Discharge.[30] In this treatment, electric discharge energizes reactants and accelerates the reaction to such a degree that an exposure of ¾ sec is sufficient to remove 97 per cent or more of the NO present.

The electric treater is *not* a precipitator. Very little of the gum formed deposits in the treater, but the discharge produces an aggregation of fine gum particles that drop out in any downstream apparatus. For example, a naphthalene removal scrubber downstream from one particular installation removes almost all of the gum mist.

Essentially, the electric treater consists of a pair of electrodes, one a flat plate and the other a plate studded with sharp points. Gas passes between these electrodes, which are at a potential difference of about 10,000 v. Energy consumption is approximately 20 kwh per MMCF of gas treated, or about four times the energy consumed by a tar precipitator. It has a *lagging* power factor of about 90 per cent.

The first treater built at the Philadelphia Coke Co. plant in 1935 has removed 96 to 98 per cent of the NO from *sendout* gas since that time. Maintenance consists of cleaning the electrodes once a year and occasionally replacing an insulator. Electrode points have shown no deterioration or dulling either by corrosion or by electrical erosion. Very little attention is required of operators, about the same as that required for an electric precipitator.

Gum Filters. Vapor phase gum continues to form downstream of *any* filter under readily attainable conditions. Hence, it is advisable to filter only the small quantities of gas required for appliance pilots. Effective vapor phase gum filters[31] are available from a number of manufacturers.

Many thousands of these filters have been installed since 1935 on pilots and other small flows. They have given excellent service. A.G.A.-approved gum protective devices effectively protect pilots from outages which are caused by accumulations of vapor phase gum and dust on the adjusting needles. When the nitric oxide content of manufactured gas is kept at low concentration by known operational and purification techniques, pilot filters will give effective protection against outages for many years, usually for the useful life of the appliance.

Analytical Methods for Nitric Oxide and Vapor Phase Gum.[32] For water gas, mixed gases (containing appreciable amounts of water gas or refinery still gas), etc., either the Shaw-Schuftan[33] or modified United Gas Improvement (U.G.I.)[34] apparatus can be employed. The suggested method employs the U.G.I. practice of removing catalytic unsaturates with maleic anhydride, then adding butadiene. Either Griess or metaphenylenediamine reagent can be used with both types of absorbers.

This so-called *universal method* may be modified when analyzing producer gas, blue gas, and combustion gases by omitting the maleic anhydride scrubber and adding butadiene.

For coke oven or retort gas, both the maleic anhydride scrubber and butadiene catalyst are omitted.

The Shaw conversion factor of 2 should always be used rather than the original Schuftan factors, which were based upon NO in nitrogen in the absence of catalysts and are in error for manufactured gas.

A fast but sensitive test for detecting vapor phase gum in a deposit removed from a gas control involves heating a sample of the deposit in a test tube with a piece of filter paper moistened with Griess reagent. If NO is present the characteristic pink color is observed. As an alternative, a needle valve may be heated directly underneath a piece of filter paper moistened with the same reagent. When the deposit contains considerable oil, the test may be inconclusive as the result of oil vapors bleaching the dye. A less sensitive test using starch iodide paper, is carried out in a similar manner.

Bayer[28] has proposed a method for the quantitative analysis of resins from distribution system deposits by the selective solubility of its components, e.g., vapor phase gum, liquid phase gum, naphthalene, oil, and iron rust.

For *quantitative* determination of suspended gum in gases, a modification of the **Owens jet method** as described by the English Air Pollution Committee is one of the simplest available. The gum is collected on a microscope cover glass and weighed, preferably on a microbalance.

The efficiency of this method depends largely upon three important factors:

1. Velocity of jet. High velocity is necessary for efficient deposition.
2. Humidity. The higher the humidity the greater the efficiency.
3. Particle size of suspensoids. Impingement methods become very inefficient with particles smaller than 0.5 micron diameter.

Methods have been described[35,36] which use this principle for the determination of suspended material in gas.

Table 3-23 Grains/CCF of Mercaptans in Crude Manufactured Gases

Oil gas (1032 Btu)		CCR process gas	
Diesel oil	H.E. oil*	Naphtha	Kerosene
70	450	4	1

* "Heavy ends" or Bunker C.

Table 3-24 Organic Sulfur Compounds in Various Gases as Manufactured[3]

Type of retort or gas manufacturing equipment	Total organic sulfur, grains/100 cu ft	Distribution of organic sulfur among different compounds (in per cent of total organic sulfur): Mercaptans	Thiophene	CS_2	Volatile sulfur compounds*
Vertical retorts, continuous	31.7	4.4	12.0	66.3	17.3
	23.6	4.2	15.7	61.5	18.6
	29.4	5.4	17.4	59.0	18.3
Vertical retorts, intermittent	20.1	6.0	16.4	53.6	24.0
Inclined retorts	31.2	3.5	26.6	54.2	15.7
Horizontal retorts	34.5	4.1	18.3	67.2	10.4
Horizontal retorts, steamed	35.6	6.5	22.0	56.3	15.2
Coke ovens	25.2	...	...	...	13.5
Carbureted water gas	11.7	3.4	30.8	17.1	48.7

* "Volatile sulfur compounds" were described by the authors as compounds not removable by oil washing. Since carbon oxysulfide is the only sulfur compound (other than hydrogen sulfide) that has been identified in coal gas and that is not removable by oil washing, it can be assumed that this is actually carbon oxysulfide.

Sulfides and Organic Sulfur

Sulfur compounds appear in crude gases manufactured from coal or oil in varying amounts, depending on those originally present in the coal or oil and the process employed. The per cent by volume of various sulfur compounds in coal gas are as follows: H_2S, 0.3–3.0 (0.6 avg); mercaptans, 0.003; thiophene, 0.010; carbon disulfide, 0.016; and carbon oxysulfide, 0.009. Other typical amounts are given in Tables 3-23 and 3-24.

Regulations in most states limit the maximum amount of hydrogen sulfide in purified gas to 0.0005 per cent by volume (0.32 grain per CCF). State regulations often restrict the content of organic sulfur in purified city gas to 0.025 per cent by volume or to 30 grains per CCF calculated as volume of carbon disulfide. Organic sulfur seldom exceeds 15 to 20 grains per CCF in purified manufactured gases.

Sulfur compounds are reduced in *sendout* gas by purification processes to:

1. Prevent excessive corrosion of gas lines.
2. Reduce corrosion of appliance heating elements, flue pipes, and flue ways by sulfur dioxide in flue gases.
3. Decrease tarnishing of silverware and other metals by unburned gas.
4. Control objectionable odor of gas to within reasonable limits.
5. Comply with state and city regulations.
6. Prevent lowering of dew point with resultant condensation of moisture from flue gases, causing accelerated corrosion.

Organic sulfur can also be removed from gas to very low concentrations, especially by absorbent oil at high rates plus steam, and by activated charcoal or silica gel. Removal of organic sulfur compounds from gas has been considered less important than removal of H_2S. State regulations permit 50 *times* as much organic sulfur as H_2S. However, reducing organic sulfur content to a very low level is desirable for domestic use and for many industrial gas uses. Whenever possible, the total sulfur content in city gas should be kept under 0.5 grain per CCF.

Hydrogen Cyanide

Hydrogen cyanide, HCN, also known as hydrocyanic acid and prussic acid, may be present in unpurified coal gas in amounts varying from 0.12 to 0.24 per cent by volume or 60 to 120 grains per CCF.

Processes that remove H_2S, i.e., iron oxide and liquid purification processes, also remove HCN. Ammonium polysulfide has been used successfully at Rochester for removal of HCN. With the iron oxide system of purification, the decrease in hydrocyanic acid content may only be slight, but with liquid purification processes, removal may be 90 per cent or more.

The HCN in city gas is very corrosive to copper and galvanized iron and forms deposits of Prussian blue in distribution systems. HCN is very toxic. Not over 0.005 per cent in air can be breathed for one hour without serious effects, and 0.3 per cent is rapidly fatal. Altieri gives methods for determining HCN in gas.[21]

Some larger gas plants formerly recovered HCN as ferroferricyanide of ammonia, but this recovery has generally been discontinued. Spent oxide from coal gas plants in which cyanogen is not extracted will contain practically all hydrocyanic acid as Prussian blue together with some ammonia salts. In ordinary oxide, Prussian blue will range from 7 to 13 per cent.

Lampblack

Lampblack, or finely divided carbon, may be produced in any thermal reforming process of oils or gases. It is also produced in carbureted water gas and oil gas processes. In these latter two processes, formation of lampblack presents no particular problem, since it is removed with tar and light oil and seldom gets beyond **Cottrell** precipitators. The last traces would be caught in iron oxide purification boxes.

In 550 Btu oil gas manufacture, about 12 lb of lampblack is produced per MCF of gas in the **Jones** apparatus and about 23 lb per MCF in the straight shot set of a single shell. Water from wash boxes of these sets is run into a lampblack separator, where lampblack is removed, either by settling or by vacuum filters.

Considerable lampblack may be formed in a fuel bed reforming apparatus unless bed temperature is controlled by proper proportioning of primary air, steam, and natural gas. Secondary air is not used; *uprun* steam is used as a final bed temperature control.

An operating technique found desirable in fuel bed reforming operations is a weekly routine of rodding foul gas mains and inspecting scrubber and condenser seal pots, purifier drops, Cottrell runoffs, and exhauster impellers and casings. Such roddings and inspections provide an excellent operating guide and a most important indication of lampblack or emulsion accumulations that might occur. It has been found that entrained lampblack might drop out and accumulate in sufficient quantity to cause *back* pressure problems in foul gas mains or operating difficulties with exhausters.

In spite of process controls, some lampblack will be formed in reforming processes. Because of this, even if the H_2S content is low, the practice of passing manufactured gas thru purifiers, using these as final filters to assure substantially lampblack-free gas, is advisable.

In the **cyclic catalytic reforming** of natural gas, proper control of the heating and reforming part of the cycle is necessary to prevent lampblack formation. When lampblack is produced, its particle size is extremely small—from about 0.5 to 4.0 microns.[37] A water seal and water spray will *not* remove such particles. If the purification system for sulfur, naphthalene, ammonia, and light oil is omitted, lampblack will flow thru the distribution system, causing **pilot outages** and other troubles. A continuous recording instrument was devised[37] to determine the presence of lampblack at the gas plant. It is based on principles of filtering any lampblack in suspended form out of a gas stream continuously and measuring its quantity by photoelectric means. Good normal operation of a gas set should produce no filter paper stain in 30 minutes' time. Optical density readings of 0.02 to 0.04, equivalent to very faint stains, were considered the maximum acceptable for a 30 min test run.

Silica and Fly Ash

Silica and fly ash may originate in a fuel bed reforming of natural gas. To prevent formation of lampblack, bed temperatures are kept at a low of 2500 F and fly ash and silica may be carried over.[6] These dispersoids have an average diameter of about 5 microns. Filtering thru iron oxide beds or dry filters, or oil scrubbing would be necessary for their removal.

Odorization

As a general rule, gas produced from coal or oil has high odor intensity and can be easily detected by smell when less than one per cent gas is present in air. In fact, some odor contained in illuminants, in trace compounds such as sulfides and organic sulfur, and in oil vapors could be removed and still leave an adequate odor intensity.

Reformed natural gases and natural gas may not have adequate odor intensity. Addition of odorants to condition them for ready detection by smell may be necessary.

QUALITY CONTROL PROBLEMS

Preceding portions of this chapter clearly indicate that manufactured gas plants are potential sources of many chemical compounds. These compounds are made in varying amounts, depending upon the process used. From the tar fraction of coal carbonization, at least 348 chemical compounds[3] have been identified. Most of these liquid and solid chemicals and some of the gases are considered undesirable impurities in the gas supply.

Table 3-25 Quantities of Constituents Involved in Gas Conditioning

Constituent	Amount present in crude gas	Maximum permissible in sendout gas leaving the plant for excellent results
Water vapor	1.74% at 60 F	If saturated gas is desired at 40 F = 0.820%
Water vapor	15.1% at 130 F	If dry gas is desired (7 lb/MMCF) = 0.024%
Tar	0 to 8.4% of coal weight	Possibly below 0.5 gal/MMCF or less than 3.5 grain/CCF
Ammonia	0 to 1.5% by volume	Less than 1 grain/CCF
Naphthalene	1 to 500 grains/CCF	Less than 1 grain/CCF (0.0044% by volume)
Styrene, indene	0 to 0.119% by volume	Prevent liquefaction in distribution system. Keep dew point below lowest system temperature.
Light oil	0.7 gal/MCF of gas to 3.0 gal/ton of coal used	90–95% removal.
Nitric oxide	0 to 30+ grains/CCF	Varies with gas. Less than 1 grain is indicated in many instances.
Sulfides and organic sulfur	1 to 3000 grains/CCF	H_2S less than 0.3 grain/CCF (0.0005% by volume). Organic sulfur 30 grains per CCF (0.025%). If possible, 0.3 grain or less total sulfur/CCF.
Hydrogen cyanide	60 to 120 grains/CCF	90% removal or more.
Lampblack	0 to 23 lb/MCF	No stain on filter paper in 30 min.
Silica, fly ash, iron oxide, dust	0 to a dust storm	0 on an absolute filter.

Purification processes normally used may remove major portions of undesirable constituents. Quality control problems then become those of removal of traces of these constituents. Such traces in terms of parts per 100,000, parts per million, or less are involved. The amounts involved, indicated in preceding portions of this chapter, are summarized in Table 3-25.

Failure to remove these constituents to the degree indicated may result in plant and distribution system difficulties. Economics and the standard of quality control desired will then determine how thorough a gas conditioning process is installed and maintained. Certain control tests before and after each conditioning process, at the plant outlet, and possibly at selected places in the distribution system, are necessary to maintain desired gas quality. Such test procedures have been listed earlier in this chapter.

Methods employed for monitoring and pinpointing abnormal system conditions have been described.[6] At selected places, a **Rutz** lighter bank is installed, consisting of six Rutz lighters with a 1¼ in. pipe cap over each lighter and an iron-constantan thermocouple inserted thru a hole in each cap. Flames are set to maintain a temperature of 1500 F. Partial stoppages which reduce thermocouple temperatures below 500 F are investigated. The sources of the troubles are determined and corrected. In addition to the Rutz lighter bank, a Flowrator device, an impingement device, microbalance filters, Arthur D. Little absolute filters, Tyndall beam apparatus, recording dew point apparatus, and a nitric oxide analyzer are used as continuous checks on manufacturing and purification processes.

REFERENCES

1. Sherwood, P. W. *Gas J.* 315: 76, July 17, 1963.
2. Furnas, C. C., ed. *Rogers' Industrial Chemistry,* 6th ed., vol. I, chap. 15, p. 19. New York, Van Nostrand, 1942.
3. Lowry, H. H., ed. *Chemistry of Coal Utilization,* vol. II. New York, Wiley, 1945.
4. McGurl, G. V., and Lusby, D. W. "Bibliography on Gas Conditioning." (CEP-55-1) *A.G.A. Proc.* 1955: 684–751.
5. Van der Pyl, L. M. "Bibliography on Gas Conditioning." (CEP-56-7) *A.G.A. Proc.* 1956: 517–34.
6. Symnoski, S. C., and others. "Undesirable Deposits in a Utility Distribution System." (CEP-56-5) *A.G.A. Proc.* 1956: 486–504.
7. Haug, J. S. "Economic Factors in Gas Condensing and their Influence on Condenser Design." *A.G.A. Proc.* 1925:1165.
8. Sperr, F. W., Jr. "Dehydration of Manufactured Gas." *A.G.A. Proc.* 1926: 1250–73.
9. Shively, W. L. "Ten Years of Gas Dehydration." *A.G.A. Proc.* 1939: 501–15.
10. Tupholme, C. H. S. "Glycerin Process for Dehydration." *Gas Age Record* 63: 311–3, Mar. 9, 1929.
11. Sultzer, N. W. "Progress Made in Detarring of Gas With Cottrell Equipment." *Gas Age Record* 57: 319, Mar. 6, 1926.
12. Shnidman, L., ed. *Gaseous Fuels,* 2nd ed. New York, A.G.A., 1954.
13. Vollmer, R. W., and Pfaff, G. C. "Koppers Company Expansion in Gas Industry." (CEP-55-3, p. 32) *A.G.A. Proc.* 1955: 787.
14. Gluud, W. *International Handbook of the By-Product Coke Industry.* New York, Chemical Catalog Co., 1932.
15. Judd, W. F. "Cleaning Coke Oven and Water Gas." *Am. Gas J.* 146: 51–2, Feb. 1937.

16. Morgan, J. J. *Textbook of American Gas Practice*, vol. I. Maplewood, N. J., 1931.
17. Henderson, Y., and Haggard, H. W. *Noxious Gases*, 2nd ed., p. 126. (A.C.S. Mono. 35) New York, Reinhold, 1943.
18. Hammerschmidt, E. G. "Preventing and Removing Gas Hydrate Formations in Natural Gas Pipe Lines." *A.G.A. Proc. Natl. Gas* 1939: 223–33.
19. Anthes, J. F. "Study of the Premature Deterioration of Gas Meter Diaphragms." *A.G.A. Proc.* 1928: 1584.
20. Powell, A. R. "Selective Removal of Liquid-Phase Gum-Formers and Naphthalene by Oil Scrubbing." *A.G.A. Proc.* 1935: 699.
21. Altieri, V. J. *Gas Analysis & Testing of Gaseous Materials*, p. 284, 151. New York, A.G.A., 1945.
22. Sholar, J. O. "Effects on Scrubbing High Btu Gases" (PC-52-9) *A.G.A. Proc.* 1952: 664.
23. Ceccarelli, F. E. "Oil Gas Manufacture at the Hunts Point Plant." (CEP-56-51) *A.G.A. Proc.* 1956: 640–2.
24. DeCarlo, J. A., and Otero, M. M. *Coal-Chemical Materials Produced at Coke Plants in the United States 1948–57.* (Bur. of Mines I.C. 7925) Washington, D. C., 1959.
25. Altieri, V. J. *Gas Chemists' Book of Standards for Light Oils and Light Oil Products.* New York, A.G.A., 1943.
26. Ward, A. L., and others. "Gum Deposits in Gas-Distribution Systems. I, Liquid-Phase Gum." *Ind. Eng. Chem.* 24: 969–77, 1932; 25: 1224–34, 1933.
27. Powell, A. R. "A Primer on Deposits in Gas Pipes and Appliances." *A.G.A. Proc.* 1938: 577–92.
28. Bayer, A. R. "Quantitative Analysis of Resins from Deposits Found in Manufactured Gas Distribution Systems." *A.G.A. Proc.* 1936: 714. Also his: "Quantitative Analysis of Deposits in Manufactured Gas Distribution Systems." *Am. Gas J.* 144: 22, June, 1936.
29. Smoker, E. H. "Experiences in the Production of High Btu Gas from Naphtha and Kerosene Using the UGI Cyclic Catalytic Reforming Process." *A.G.A. Proc.* 1953:852.
30. Fulweiler, W. H. "Discussion." *A.G.A. Proc.* 1931:770. Also: Shively, W. L., and Harlow, E. V. "Electrical Removal of Gum-Forming Constituents from Manufactured Gas." *Electrochem. Soc. Trans.* 69: 495–517, 1936.
31. Vandaveer, F. E. "Gum Protective Devices for Gas Appliances." *A.G.A. Proc.* 1943: 270.
32. Huff, W. J. "Determination of Nitrogen Oxides in Gas." *A.G.A. Proc.* 1938: 625.
33. Shaw, J. A. "Determination of Nitric Oxide in Coke-Oven Gas." *Ind. Eng. Chem., Anal. Edition* 8: 162, May 15, 1936.
34. Fulweiler, W. H. "Analytical Methods for Determining Nitric Oxide." *Gas Age Record* 75: 586, June 15, 1935.
35. Wilson, C. W. "Method for the Determination of the Quantity of Suspended Material in Gas." *A.G.A. Monthly* 24: 325, Sept. 1942.
36. Shnidman, L. "'Nitric Oxide' Control." *A.G.A. Proc* 1942: 272.
37. Gilkinson, R. W. "Continuous Determination of Suspended Solids in Fuel Gas as an Aid in Control of Production Equipment." *A.G.A. Proc.* 1954: 606.

Chapter 4
Producer Gas

by Mark G. Eilers, Charles R. Locke, Carl J. Lyons, J. P. Templin, and F. E. Vandaveer

USES

Producer gas may be used hot and without purification, as in open hearth furnaces, or after cleaning (tar and dust removed) and cooling, as in heating coke ovens.[1]

Hot uncleaned gas is preferable from a heat utilization standpoint. Its sensible heat is utilized and the tar, partly in vapor form and partly suspended, is burned, with direct recovery of its heating content. Further, cost of cleaning and cooling installations is avoided and preheating in regenerators is partially or completely eliminated. Disadvantages of uncleaned gas include rapid contamination of pipelines (which should be short) and contamination of direct-fired products such as glass and ceramic ware by ash precipitation, sulfur contamination, and deposition of ashes in regenerators. A rather elaborate installation is required to produce gas from fuels of high volatile content such as bituminous coal and wood. Removal of small amounts of tar fog is a difficult cleaning problem. Low-volatile fuels such as anthracite and coke are used in the manufacture of cold, clean producer gas. The status of the gas producer in the iron and steel industry,[2] the use of producer gas in glass-making furnaces,[3,4] and its use for underfiring coke oven batteries and gasworks furnaces[5] have been reported.

With the availability of natural gas, gasoline, and fuel oil, producer gas is rarely used as engine fuel, except in England, France, and Germany. Individual producer engines are used on automotive vehicles such as trucks. Cold, clean gas is essential and a number of cleaning devices for gas have been built. The development of power producers has been reported.[6]

PRODUCER GAS THEORY

A gas producer converts solid fuel into a combustible gas by continuous reaction with an oxygen-carrying gasifying agent. Producer gas may have a wide range of heating values (50 to 500 Btu per cu ft), depending primarily on the fuel used, and also on the oxygen and steam concentrations and ratio in the blast or gasifying agent.

Gas producers using solid fuels are classified according to the type of fuel bed employed:

1. Fixed or relatively static beds with fuel supported on a grate.
2. Suspended or fluidized beds with fuel supported by the gaseous blast.
3. Entrained or pulverized coal systems in which the solid fuel is introduced *concurrently* with the blast and moves with it thru the reactor. Operation may be at atmospheric or elevated pressures.

Oxygen-blown units are sometimes referred to as synthesis gas generators, since the scrubbed gas is primarily a mixture of carbon monoxide and hydrogen. Obviously, N_2 is not present when O_2 is used in lieu of air.

Heats of Reaction

The reactions of interest in gasification processes all involve heat. Heat-liberating reactions are termed ***exothermic;*** conversely, heat-absorbing reactions are termed *endothermic.* In this book, $-\Delta H$ signifies that the reaction is exothermic and $+\Delta H$ that it is endothermic.

An example will illustrate the principles involved.

Assume that the heat of reaction accompanying the reduction of CO_2 to CO, that is, $C + CO_2 = 2CO$, is unknown, and that the heats of reaction of two more basic reactions (lines *a* and *b*) are known, namely:

Line	Reaction	Heat of reaction, $\Delta H_{60F,\,1\,atm}$
a	$C + O_2 = CO_2$	−169,288 Btu/lb-mole C
b	$2CO + O_2 = 2CO_2$	−243,448 Btu/2 lb-mole CO
c	$C + CO_2 = 2CO$	+74,160 Btu/lb-mole C

Line *c* (the reaction desired) is the algebraic difference between line *b* and line *a*. Note that:

1. The heat of reaction is at a specified temperature (60 F) and pressure (1 atm).
2. Both lines *a* and *b* are exothermic, as shown by the *minus* signs.
3. The heat of reaction of line *b* reflects the quantities involved in the reaction by being given for 2 lb-moles (*not* 1 lb-mole) of CO.
4. Line *c* is endothermic, as ΔH is positive; that is, $(-169{,}288) - (-243{,}448) = (+74{,}160)$.
5. Since both these and other reactions of interest are reversible, line *a* may be rewritten as $CO_2 = C + O_2 +$ 169,288 Btu per lb-mole C.

Since these reactions are reversible, some means is required of determining the relative degrees of their forward and reverse directions. For example, in the reversible reaction of

$C + CO_2 = 2CO$, there are 0.44 moles of CO present per mole of CO_2 at 1161 F.

Equilibrium Constants

The **law of mass action,** or, more accurately, the criterion of equilibrium, may be expressed as follows: At equilibrium and at any one constant temperature, for any one reversible chemical reaction, *the product of the pressures (concentrations) of the substances on the right, each raised to a power equal to the number of moles reacting (according to the chemical reaction), divided by the product of the pressures (concentration) of the left, each raised to a corresponding power, is a constant.* Since this derivation assumes the accuracy of the perfect gas laws, it follows that it is only rigidly applicable when the constituents obey these laws.

Table 3-36 shows that the equilibrium constant for the reaction $C + CO_2 = 2CO$ is $K_p = (CO)^2/CO_2 = 0.011$ at 980 F. Note that the solid, carbon, is excluded from the equilibrium constant since its volume is relatively small. Also note that the exponent, 2, is derived from the balanced reaction. In general, for the reaction $aA + bB = cC + dD$, the equilibrium constant is $K_p = (C)^c(D)^d/(A)^a(B)^b$.

Chemical Reactions

The main reactions in a producer are those of carbon with O_2, CO_2, and water vapor to form CO and H_2. These are also intermediate reactions in complete combustion of carbon where sufficient oxygen is used with the carbon and hydrogen to give carbon dioxide and water vapor as end products.

Reaction

Oxidation of carbon by oxygen:

$$C + O_2 \rightleftharpoons CO_2 \qquad (1)$$

The Boudouard reaction or the reduction of carbon dioxide by carbon:

$$C + CO_2 \rightleftharpoons 2CO \qquad (2)$$

The water gas reaction or the reduction of water by carbon:

$$C + H_2O^* \rightleftharpoons H_2 + CO \qquad (3)$$

Reactions 2 and 3 are related by the water gas shift equation:

$$CO + H_2O^* \rightleftharpoons H_2 + CO_2 \qquad (4)$$

Methane formation, represented most simply by:

$$C + 2H_2 \rightleftharpoons CH_4 \qquad (5)$$

All five basic reactions are reversible. Since Reactions **2** and **3** are related by the water gas shift, Reaction **4**, the reactants and products tend to approach an equilibrium composition dictated by the temperatures and pressures within the producer. Values for the heats of reaction[7] and equilibrium constants, K_p,† of the five reactions at various temperatures may be obtained from Table 3-36. The following equations[8] can also be used:

* Steam.

† Expressed as a quotient of partial pressures of the gas constituents.

For Reaction 2:

$$\log K_p = 3.26730 - (8820.69/T) - 0.001208714T + 0.153734 \times 10^{-6}T^2 + 2.295483 \log T$$

For Reaction 3 (see Fig. 3-21 for approximate values):

$$\log K_p = -33.45778 - (4825.986/T) - 0.005671122T + 0.8255484 \times 10^{-6}T^2 + 14.515760 \log T$$

For Reaction 4 (see Fig. 3-23 for approximate values of reciprocals):

$$\log K_p = 36.72508 - (3994.704/T) + 0.004462408T - 0.671814 \times 10^{-6}T^2 - 12.220277 \log T$$

For Reaction 5:

$$\log K_p = -13.06361 + (4662.80/T) - 2.09594 \times 10^{-3}T + 0.38620 \times 10^{-6}T^2 + 3.034338 \log T$$

In the above equations, T is the temperature in degrees Kelvin.

Figure 3-10 is a plot of log K_p for Reactions **2**, **3**, and **5**, from which approximate values may be obtained.

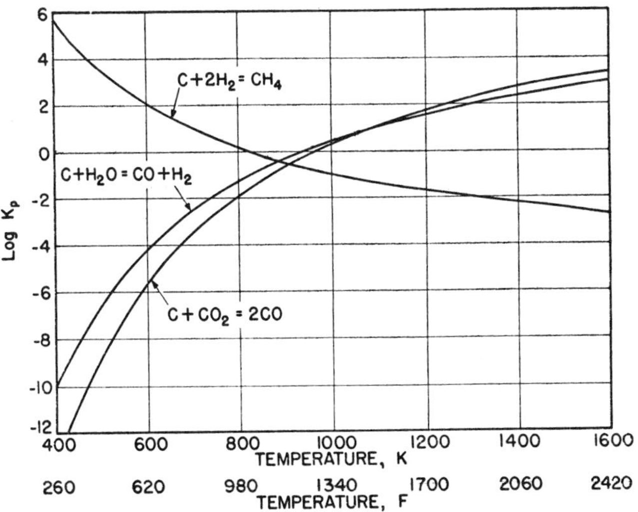

Fig. 3-10 Variation of log K_p with temperature for gasification reactions.

The precise effects of temperature,[9] pressure,[9] and blast composition on the products of the gasification reactions are complicated, but it is generally recognized that elevated pressures favor the formation of CH_4, but not of CO and H_2.

Mechanism of Gasification

The three essential steps in gasification of solid fuels are: (1) oxidation of carbon to produce the heat required to maintain a temperature at which reactions proceed at an adequate rate; (2) transfer of heat by radiation and convection from the point of oxidation to the pile of carbon present; and (3) further reaction of the blast and gaseous products of step 1 with the heated carbon to produce CO, H_2, and CH_4. The sequence of reactions occurring in various types of producers differs widely.

Fixed Fuel Beds. Here the *upward* flowing blast and reactant gases meet the slowly *descending* fuel. A producer with a fixed fuel bed is characterized by a deep fuel bed with a large inventory of carbon and a fairly well-defined sequence of zones. In the direction of gas flow, these are: (1) the ash zone, (2) the oxidation zone, (3) the reduction zone, and (4) the preheating and devolatilization zone.

Within the *oxidation zone,* the oxygen burns a portion of the carbon until all oxygen is exhausted; either CO or CO_2 is formed. Below about 1350 F, the oxidation rate is controlled by the chemical reaction on the carbon surface. At higher temperatures the reaction rate appears to be limited by the rate at which the oxygen can diffuse to the carbon surface. CO is the primary product on the carbon surface and, in the presence of oxygen in the atmosphere surrounding the coal particle, the carbon monoxide is oxidized by the reaction $CO + \frac{1}{2} O_2 = CO_2$ (see Ref. 10). However, for practical purposes the oxygen reacts with carbon to form CO_2.

The heat liberated in the oxidation zone is conveyed by radiation and as sensible heat in the gases to the *reduction zone,* where Reactions **2**, **3**, and **4** occur in the *endothermic* direction. The reactions of carbon with CO_2 and H_2O are slower than that of carbon with O_2. At temperatures above 2100 F, under usual conditions, reaction rates are probably controlled by a mass transfer or diffusion process. Principal products of the gasification reactions, CO and H_2, approach an equilibrium with the reactants, CO_2 and H_2O, thru the water gas shift reaction: $CO + H_2O \rightleftharpoons H_2 + CO_2$. Temperature determines the equilibrium composition.

The consumption of carbon by the gasification reactions absorbs heat, so that a falling temperature gradient exists as the gases leave the oxidation zone. Reaction of steam or carbon dioxide with solid carbon becomes negligible in the range of falling temperature from 1600 to 1300 F. The remaining sensible heat in the gases is transferred to the top layer of fresh fuel, preheating it and driving off any volatile matter present.

The composition of the final gas depends on: (1) blast composition, which fixes the relative amounts of H_2 and combined O_2 in the products and also affects the bed temperature reached; (2) approach toward equilibrium of the reactions; (3) system pressure; and (4) contribution of volatile matter which mixes with the product gas.

Fuel for fixed bed gasification should preferably be uniformly sized and nonagglomerating, and should contain a minimum of *fines* to ensure even flow of the blast thru the bed with little resistance. The gasification rate *increases* with an increase in the flow of reactants thru the bed and with an increase in the gasification temperature. Therefore, the **limits on gasification rate** are fixed by the maximum rate of blast which can be used without lifting the fines from the bed, and, in dry ash removal, by the oxygen-to-steam ratio that will keep the maximum bed temperature below the ash-softening temperature.

Fixed bed producers have also been operated with a *concurrent* downward flow of both blast and coal. Thus, without preheating of fuel by the hot gasification products, the volatile matter and the moisture evolved flow thru the high-temperature zone in the reaction bed.

Another fixed bed configuration is the *crosscurrent* type, in which the blast is directed into the fuel bed at 90° to the direction of fuel flow.

Fluidized Beds. Here the solid particles are in random motion and exhibit the liquid-like characteristics of mobility, hydrostatic pressure, and a surface or boundary zone, in which particle concentration decreases rapidly. The main characteristic of a reactor with a fluidized bed is the virtual elimination of temperature zones corresponding to predominantly exothermic or endothermic reactions. The net effect is that the entire bed approaches a temperature dictated by the relative rates of the combustion and gasification reactions. Because of the relatively low overall reaction temperature, 1800 to 2000 F, the chemical reaction rates control the rate of gasification of carbon.

Because of the low temperature involved, fluidized bed units have been successful only with highly reactive fuels such as lignite and char of relatively small top size. Another limiting factor is the ash-softening temperature. As soon as slag particles start to soften and agglomerate, fluidization is lost.

Entrained or Pulverized Coal Systems. These processes are characterized by a short residence time in the reactor and by conditions not favoring rapid transport of the oxygen to the surface of the "floating" particle of fuel, because there is little *relative* movement of the particle and its surrounding gas. They have the advantage of using almost any fuel, provided it is pulverized.

Almost all entrained or pulverized coal gasification units have been oxygen blown. With air, the high temperatures required for rapid reactions of the blast constituents with the carbon cannot be reached, since the nitrogen present absorbs heat.

A comprehensive study has been made[11] of the mechanism of gasification in entrained systems in terms of: (1) a preheat period; (2) a period of predominantly exothermic oxidation reactions; and (3) a period of endothermic gasification reactions.

In the first stage of pulverized coal gasification, fuel particles are carried into the reactor with the oxygen-containing blast and immediately subjected to radiation from the walls and flame. The radiant heat drives off the volatile matter and elevates the particle temperature to the ignition point. The volatile matter probably burns first and may exhaust all available oxygen before oxidation of the carbon particle can begin.

During the combustion period, oxygen diffuses to the carbon surface, where carbon monoxide is probably formed as the primary product by the reaction $2C + O_2 \leftrightarrows 2CO$. The CO will be further oxidized to CO_2 upon leaving the carbon surface as long as O_2 is present. Steam diffusing to the carbon surface is decomposed by the reaction $C + H_2O \leftrightarrows CO + H_2$. Upon entering the gas phase, these products may be oxidized to CO_2 and H_2O. Concentrations of the gaseous products CO, H_2, H_2O, and CO_2 are probably determined by the gas-phase temperature and the water gas shift equilibrium $CO + H_2O \rightleftharpoons H_2 + CO_2$. Maximum flame temperature occurs when the oxygen is exhausted.

After the period of oxidation, the carbon is gasified by the endothermic versions of Reactions **2**, **3**, and **4**, in which steam and carbon dioxide diffuse to the carbon surface and react. The heat absorbed by these reactions causes a drop in the temperature with a progressive drop in the rate of gasification.

The ratio of oxygen to carbon, residence time in the reactor, oxygen-to-steam ratio of the blast, size of the original coal

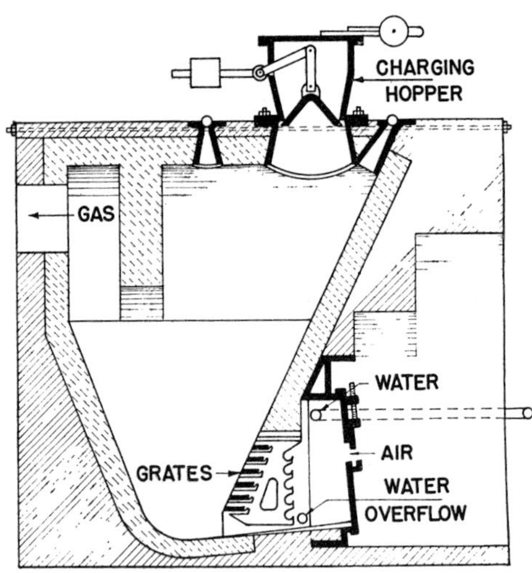

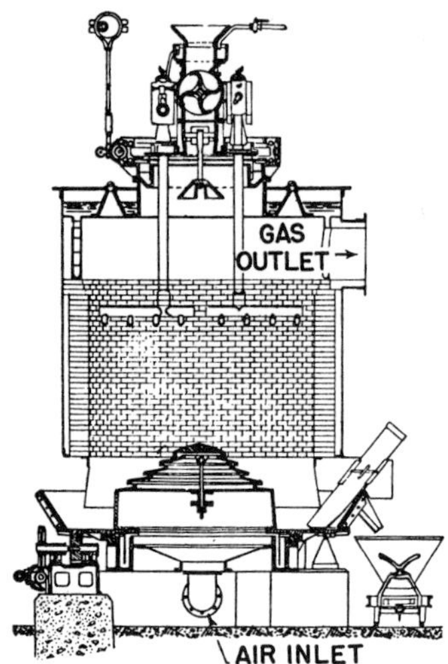

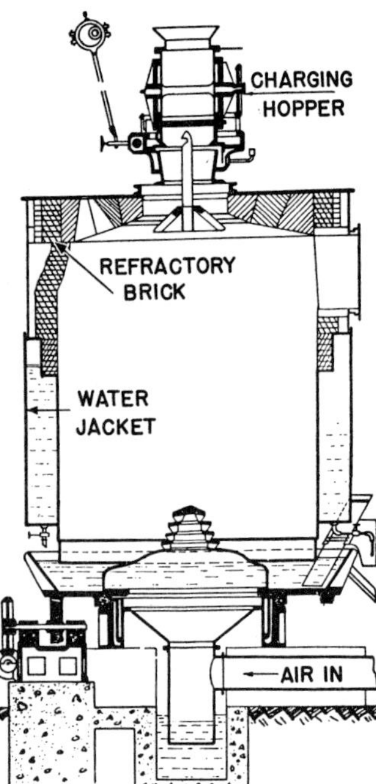

Fig. 3-11 (left) Siemens coal fueled gas producer.
Fig. 3-12 (center) Heurtey coal fueled gas producer.
Fig. 3-13 (right) Heurtey coke fueled gas producer.

particles, and heat supplied by preheated reactants largely control the gas composition and yield.

Methods of Calculating Gasification Results

There are no simple methods for predicting accurately the results which will be obtained with a given set of operating conditions. In general there are two methods of calculation, those based on *thermodynamic* considerations alone, and those based on *kinetic* theories. The latter are by far the more complex because of the necessity for taking into account rates of reaction and rates of heat and material transfer, as well as the usual material and heat balances. With the thermodynamic type of analysis, heat and material balances are used, and an assumption is made as to the approach to equilibrium at the temperature indicated by the heat balances.

For **fixed bed producers,**[8] a material balance is combined with a heat balance, with the assumption that all reactions are at equilibrium except that of methane formation. This assumption is probably valid because of the large excess of carbon present. With some modification[12] based on assumed reaction mechanisms, a similar method has been proposed for entrained and suspended gasification systems.

For **pulverized coal-fired producers** there are two outstanding methods of calculating expected results: thermodynamic and kinetic. Charts (thermodynamic method) have been developed[13] from which results of gasifying a wide range of coals can be predicted, assuming the water gas shift reaction is in equilibrium. With these charts, an assumption of only the blast composition, the temperature of reaction, or the heat loss from the unit need be made.

A kinetic method[11] for calculating results from pulverized coal gasification has been developed. This method, while too involved for ordinary usage, includes graphs from which close approximations of results can be predicted. Essentially the method gives gasification results as a function of time.

The high-pressure gasification of low-grade Scottish coal to produce 30 MMCF gas daily has been reported.[14] The Lurgi gasifiers, 118 in. in diameter and 28 ft high, were designed to operate at 1200 F and 370 psig. Each unit requires 220 tons of steam and 60 tons of O_2 to produce 7.5 MMCF of 410 Btu gas daily. After purification, the gas is enriched with butane to 450 Btu.

GAS PRODUCERS

Early gas producers were generally found in the iron and steel industry; ash was removed as a liquid slag. The first gas producer built in France, about 1840, by Ebelmen, was similar to a blast furnace shaft. Air was introduced at the base and slag was removed from a tapping hole at the bottom of the hearth. Centrally in the producer a retort was suspended which also served as a crude fuel valve. The fuel used was charcoal to which iron furnace slag or lime was added as a slagging medium to obtain an easily fusible slag. The producer was worked with and without steam.

The analysis of the gas obtained with a *dry* blast was: CO_2, 0.5 per cent; CO, 33.3; H_2, 2.8; and N_2, 63.4 per cent.

One early gas producer in which the ash was removed as a solid was developed by the Siemens brothers in connection with their open hearth furnace about 1861. The air for combustion in these first producers was obtained by the natural draft in the chimney of the furnace, coal retort, etc., that they served. These producers were generally 10 to 12 ft on a side, and were 10 ft high or higher (Fig. 3-11). Their brick shell enclosed a refractory brick lining, and the unit was held together with tie rods or steel sections. In the original installations no water saturation of the air was attempted. Later, the air was saturated by passing thru a water screen.

This type of producer gasifies about five tons of coke per day. Its maximum rate is generally reported at 8.2 to 8.5 lb per sq ft of grate area per hr. Figure 3-11 shows a horizontal step grate. A vertical grate bar arrangement is common.

In place of the natural draft method described, an improvement consists of blowing air thru the producers. Another improvement is the use of hollow grate bars with perforations, permitting blowing of the air and its saturation with water vapor. This grate is known as the *Sauvageot grate.*

The Siemens producer has a thermal efficiency of approximately 80 per cent and produces about 65.5 cu ft of gas per lb of coke. Gas heating value is 112 Btu per cu ft. Average gas composition (per cent) is: CO_2, 6.4; CO, 25.6; H_2, 8.2; and N_2, 59.8. Because the ashes are removed by hand, with resulting high labor costs and varying gas composition and heating value, this type of producer possesses distinct disadvantages. Siemens producers are located in the Eastern Hemisphere in gasworks with coke as fuel, and in steel, glass, and ceramic industries with coal as fuel.

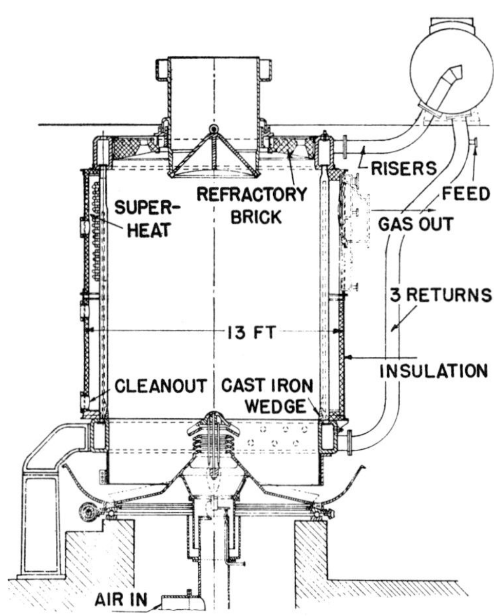

Fig. 3-14 Marischka coke fueled gas producer.

Difficulty with hand cleaning led to development of mechanical rotary grate type producers. These may be separated into:

1. Producers lined with refractories, Heurtey type, coal fuel (Fig. 3-12).
2. Producers with water jackets, Heurtey type, coke fuel (Fig. 3-13).
3. Producers with water tubes, Marischka type, coke fuel (Fig. 3-14).

These three types of producers are *top fed,* either thru the center or around the periphery. Furthermore, they have a conical grate traveling eccentrically. Ashes are removed thru a *water seal* by a plow, star wheel, or paddle, or, in a few cases, in dry form. Air is saturated and blown or blasted into the producer thru the grate. Gas is removed thru a side opening or, when there is a peripheral feed arrangement, sometimes taken off the top.

Refractory-lined producers are available in a wide range, from 3 to 12 ft in diameter, and with from 40 lb to 5 tons per hr gasification capacity. The refractory lining deteriorates rapidly in the heat zone. Ashes become fused to the lining and cause a bridging over of the ash zone with its attendant difficulties.

Mechanical producers are generally eight feet or more in diameter, gasifying 1.5 to 5 tons per hr. Water jackets are usually constructed as low-pressure boilers. The upper portion is lined with refractory brick. In some cases the area in which the fuel first enters is lined with a hard floor of abrasion-resisting brick. Hot water, if available, is fed to the water jacket, to supply the total vapor necessary for air saturation.

To reduce labor costs, the Wellman Co. designed a gas producer (Fig. 3-15) which feeds anthracite, bituminous coal, or coke from a bin holding a day or more's supply. Ash is removed dry from a hopper.

The third or Marischka type of gas producer is found only with diameters of from 8 to 12 ft. Water tubes are connected to top and bottom headers and may have a superheating section, as shown in Fig. 3-14. This type of producer combines two units in one because occasionally the sensible heat of the gas made in the preceding classes of producers is recovered in a fire tube boiler. The capacity is similar to that of the preceding type. The steam pressure can reach as high as 240 psig, and one pound of steam is produced per pound of carbon gasified.

All three classes of producers show comparable results, have thermal efficiencies ranging from 85 to 90 per cent, and produce a gas with a heating value of from 120 to 150 Btu per cu ft and the following analysis range (per cent): CO_2, 7 to 4; CO, 25 to 30; H_2, 12 to 14; CH_4, 0 to 2; and N_2, 56 to 50.

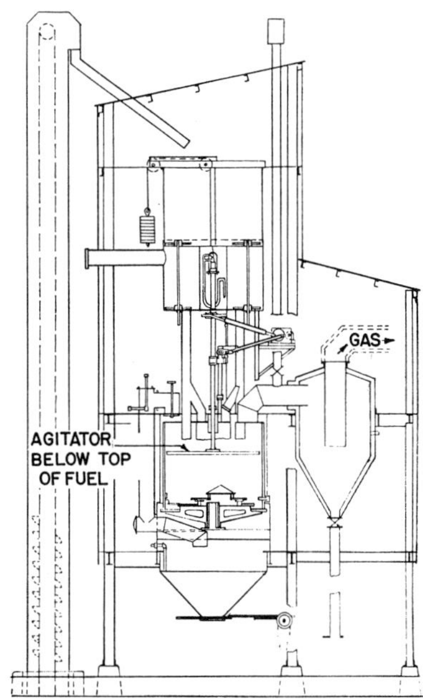

Fig. 3-15 Wellman–Galusha agitator gas producer, bituminous coal, anthracite, or coke fueled.

The mechanical rotary grate type of producer is used in chemical, iron and steel, gas, ceramics, and other industries. Fuels gasified are coke, anthracite, semibituminous coal, bituminous coal, and lignite. Occasionally, attempts are made to gasify such waste products as wood chips, but indications are that it is necessary to blend higher grade fuels with these products.

Newer developments in the gas producer field involve automatic fuel charging and ash removal by means of probing devices.

A new French development is the gasification of pulverized fuel. Figure 3-16 shows the arrangement of equipment in the **Panindco** process. Briefly, a mixture of air and water vapor superheated to a temperature of 1800 to 2200 F meets a stream of coal pneumatically inserted at the top of a tall tower. Gasification takes place immediately in the upper portion, the lower portion serving essentially to cool the ashes. The regenerators or superheaters operate cyclically, being heated by a portion of the gas produced. The gas produced (based on reports of operation with lignite and coal) has the following range of composition (per cent): CO_2, 9 to 14.4; O_2, 0.1 to 0.4; CO, 12.1 to 20.8; H_2 16.5 to 22.8, CH_4, 0.4 to 0.8; and N_2, balance. Heating values range from 114 to 124 Btu per cu ft.

A second new development relates to gasification of oil and tar. The equipment differs from that of the customary producer, but is included here because of the similarity of the gas produced. The operation is basically controlled combustion.

Figure 3-17 shows the equipment at the gasworks in Andresy, France. Fuel oil No. 2, tar, and creosote oil have been used with the following gas composition (per cent): CO_2, 4.8 to 5.9; O_2, 0.2 to 0.3; CO, 12.7 to 14.4; H_2, 13.6 to 17.8; CH_4, 1.3 to 2.4; and N_2, balance. The gas produced has a heating value of 108 to 128 Btu per cu ft.

Saturation temperature control prevents ash fusion with its accompanying erratic operating conditions. Saturation temperature is generally controlled by varying the amount of low-pressure steam added to the air blast. However, the amount of steam added must vary with the set operating conditions.

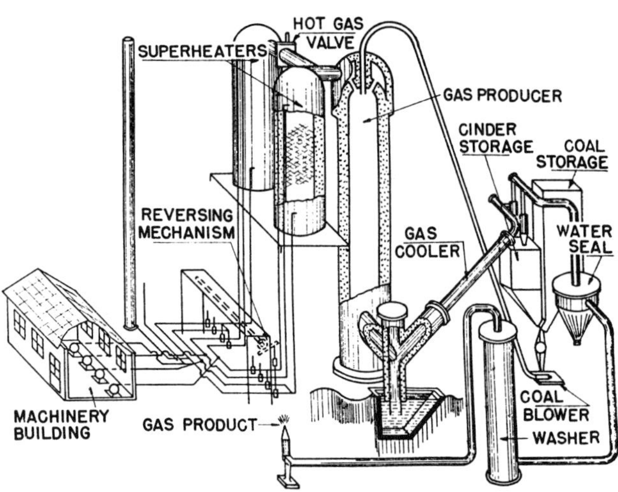

Fig. 3-16 Panindco process, coal fueled.

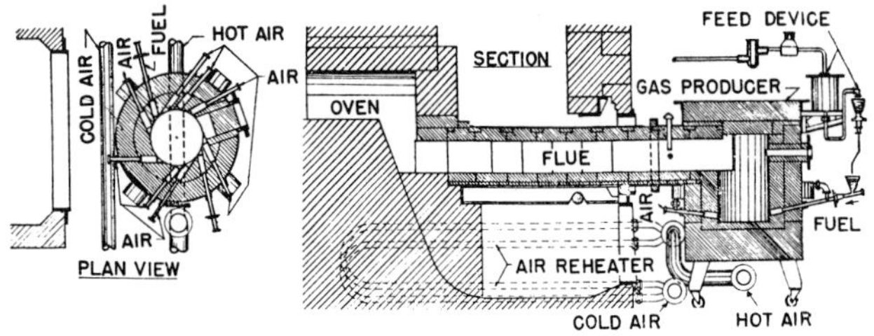

Fig. 3-17 L'office central de chauffe rationnelle, oil and tar fueled.

OPERATING RESULTS

Reviews of producer practices and a general discussion of the processes are available.[15,16] Also, heat balances and test data have been reported.[17,18] Standards for testing gas producers have been set up by the American Gas Association.[19]

Typical operating data on large mechanical *updraft* producers using coal or coke are given in Tables 3-26 and 3-27.

Typical heat balances of a coal producer making a hot uncleaned gas and a cold clean gas are given in Table 3-28.

Operating results using oxygen instead of air and anthracite and coke are given in Table 3-29. These tests were made under regular operating conditions in producers generating

Table 3-26 Typical Operating Figures on Large Mechanical Updraft Producers[1]

	Fuel		
	Coal lumps	Coal slack	Coke
Fuel			
Size	2 to 4 in.	70% thru ½ in.	97% thru 1 in., 14% thru ¼ in.
Moisture, per cent	1.3	9.5	10.9
Volatile matter, per cent	35.0	38.0	1.9
Fixed carbon, per cent	55.5	44.4	88.3
Ash, per cent	8.2	8.2	9.9
Heat value (dry basis), Btu per lb	15,296	13,833	21,605
Ash fusion temperature, °F	2,780	...	2,550
Throughput dry fuel, lb per hr	3,890	4,040	3,200
Blast, air–steam			
Air volume, cu ft per lb dry fuel	49.0	42.8	51.6
Saturation temperature, °F	126	138	137
Steam, lb per lb dry fuel	0.316	0.43	0.53
Gas composition, per cent			
CO_2	4.3	4.8	5.8
Illuminants	0.4	0.5	0.0
O_2	0.1	0.2	0.3
CO	26.7	25.0	26.0
CH_4	3.1	2.8	0.5
H_2	13.9	12.8	12.1
N_2	51.5	53.9	55.3
Heat value (gross), Btu per cu ft	169.5	160.2	129.4
Temperature, outlet producer, °F	1,400	1,146	1,148
Gas volume, cu ft per lb dry fuel	68.6	63.1	74.1

Table 3-27 Typical Results of Operation of American Gas Producers Using Various American Fuels[21]

	Fuel		
	Bitu-minous coal	Anthra-cite	Coke breeze
Fuel gasified, lb per sq ft per hr	30–70	15–30	15–30
Volume of gas at 60 F, cu ft per lb coal	60	65	68
Temp of gas leaving producer, °F	1250–1500	550–650	550–650
Lower or net heating values, Btu			
Coal or coke used per lb	13,500	12,500	12,000
Steam, 0.4 lb per lb fuel (anthracite and coke producers make own steam)	450	(450)	(450)
Heat input to producer per lb fuel	13,950	12,500	12,000
Raw producer gas per cu ft at 60 F	140	130	120
Sensible heat of gas leaving producer per cu ft	23	10	10
Tar in hot gas leaving producer (0.0006 lb tar per cu ft gas at 15,000 Btu)	9	...	...
Heat output from producer			
per cu ft	172	140	130
per lb fuel	10,320	9,100	8,840
Efficiency of producer not including losses in boiler or gas lines, per cent	73.9	72.7	73.6
Typical analysis of gas, per cent			
CO_2	7.4	9.0	7.6
CO	22.5	24.8	26.0
H_2	12.5	15.0	11.0
CH_4 (and other hydrocarbons)	3.1	0.9	0.7
N_2	54.5	50.8	54.7
Air for perfect combustion, cu ft per cu ft gas	1.17	1.04	0.95
Additional air for combustion of 0.0006 lb tar per cu ft hot gas from bituminous coal	0.11	...	...
Total air, cu ft per cu ft gas	1.28	1.04	0.95
Products of perfect combustion, cu ft per cu ft gas	2.14	1.83	1.76
Weight of producer gas, lb per cu ft	0.069	0.066	0.069
Weight of air for perfect combustion, lb per cu ft gas	0.098	0.080	0.073
Weight of products of perfect combustion, lb per cu ft producer gas	0.167	0.146	0.142
Theoretical flame temperature based upon gas at 60 F, °F	3,170	3,150	3,000

synthesis gas for ammonia production. Test results on a Kerpely producer using ¼ to ¾ in. pea coke with various oxygen concentrations are given in Table 3-30.

Operating data on modified producers for making synthesis gas, such as in Winkler and Lurgi generators and Leauna slagging generators, have been reported.[20]

UNDERGROUND GASIFICATION OF COAL

Gasification of coal in place underground by its partial burning with air yields a type of producer gas. Large-scale experiments have been conducted by Russia, Great Britain, and the United States.

Extensive Russian experiments covered five experimental and one industrial plant. One plant[1] operated for 18 months

Table 3-28 Typical Heat Balances of Coal Producers Making Hot Uncleaned Gas and Cold Clean Gas[1]

	Coal, per cent	Coke, per cent
Input		
Heating value in fuel	97.6	96.2
Sensible heat in steam	2.4	3.3
in air	...	0.5
Total	100.0	100.0
	Hot, unclean	Cold, clean
Output		
Heating value in cold gas	74.3	76.0
Sensible heat in gas	12.1	5.7
Heating value in tar	5.1	...
Sensible heat in tar	0.2	...
Heat in steam from waste-heat boiler	...	7.3
from jacket	...	5.1
Heat loss in tar, dust, etc.	2.2	...
in cooling water	0.3	...
in carbon in ash	0.4	1.0
in water vapor	1.9	3.8
in radiation	3.5	1.1
Total	100.0	100.0
Cold efficiency (based on gross heating value of cold gas)	76.1	78.9
Hot efficiency (based on useful heat in gas and tar)	93.9	86.5

(1938), producing seven million cubic meters of 115 Btu per cu ft gas and two million cubic meters of 210 Btu per cu ft gas. Four methods[23] of underground gasification were developed:

1. **Chamber Method.** Chambers were constructed underground and filled with broken coal. The coal was gasified as in a gas producer above ground. This method was abandoned because it required much underground labor and proved costly.

2. **Stream Method.** Devised in 1934 at the Donets Institute of Coal Chemistry, this method has been used at Gorlovka since 1935 on an experimental scale and since 1938 on an industrial scale with considerable success. First, the gasification area is marked on three sides. Two parallel galleries are connected by a third passage, the "fire-face," in which fire is started and the coal panel burnt from the lower end toward the surface. Combustion is maintained by a flow of air or air enriched with oxygen. This method requires very little underground work and no additional drilling after the coal is ignited.

3. **Borehole Method.** Two or more vertical boreholes are drilled into the coal seam and cross-connected. Coal is ignited in one borehole and gasification occurs in the cross-connection. As a result of combustion, the holes gradually increase in diameter, making continuous operation difficult.

4. **Percolation Method.** A number of boreholes are drilled from the surface into a coal bed, usually in a radial pattern. Coal is ignited at the bottom of one borehole and combustion continues until the fire reaches an adjacent borehole, after which another borehole is put into operation. The chief difficulty of this method is in making connections between boreholes at a reasonable distance.

Table 3-29 Oxygen-Blown Gas Producers Using Anthracite or Coke

	Kind of generator				
	Wellman-Galusha, 10 ft ID, at Trail, British Columbia				U.G.I. generator, converted to continuous operation, 9 ft ID, at Belle, West Virginia
	Anthracite rice		Coke		Coke
Nominal size	5/16 in. × 3/16 in. round hole screen		2 in. × 1.8 in. square hole screen; 2 5/8 in. × 5/16 in. round hole screen		Run of oven
Proximate analysis (as fired), per cent					
Moisture	6.4		5.0		
Volatile matter	4.5		2.1		
Fixed carbon	78.9		81.0		
Ash	10.2		11.9		
Estimate analysis (as fired), per cent					
C	78.7		82.3		
H_2	2.4		0.9		
O_2	7.4		4.7		
N_2	0.7		0.5		
S	0.6		0.5		
Ash	10.2		11.1		
Heating value, Btu per lb	12,250		12,000		
Ash fusion temp, °F					
Initial deformation	2,100		...		
Softening temp	2,810		2,200		
Fluid temp	2,940		...		
Duration of test, hr	25		29		2,890
Fuel consumed, lb	69,280		72,100		23,687,000
Pure oxygen flow, cfm	400		378		1,400
Oxygen purity, per cent	96.8		97.5		95.2
Total steam used, lb per hr	3,875		2,520		11,220
Steam generated in producer jacket, lb per hr	615		695		...
Net addition of steam to producer, lb per hr	3,260		1,825		...
Offtake temperature, °F	879		508		1,210
Estimated combustion zone depth, in.	7.10		20		...
Combustible in dry refuse, per cent	2.6		2.0		...
Fuel gasified, lb per hr	2,771		2,490		8,200
Fuel gasified, lb per hr per sq ft	33.0		31.7		128.9
Oxygen (100% pure) used, cu ft per lb fuel	8.67		9.20		10.2
Steam used, lb steam per lb fuel	1.40		1.01		1.4
Steam decomposed, lb steam per lb fuel	0.73		0.62		...
Total fuel, lb per MCF gas	23.6		24.6		23.7
Total steam, lb per MCF gas	33.0		24.8		32.4
Total oxygen (100% pure), cu ft per MCF gas	204.2		227		243
Gas analysis, per cent					
CO_2	17.70		11.3		16.9
O_2	0.15		0.6		0.1
H_2	40.55		31.0		34.0
CO	39.75		54.1		46.6
CH_4	0.85		0.4		0.2
N_2	1.00		2.6		2.2
Heat balance basis: 1 lb of fuel as fired at 60 F					
Input	Btu	Per cent	Btu	Per cent	
Calorific value of fuel	12,250	89.8	12,000	92.9	
Sensible heat of fuel; assumed entering at 60 F	0	0	0	0	
Total heat of water supplied; assumed entering at 60 F	0	0	0	0	
Total heat of steam entering	1379	10.1	898	7.0	
Sensible heat of oxygen entering	4	0	3	0	
Total heat of water in oxygen entering	18	0.1	13	0.1	
Total	13,651	100.0	12,914	100.0	
Output	Btu	Per cent	Btu	Per cent	
Calorific value of gas	11,376	83.4	11,173	86.5	
Sensible heat of dry gas	715	5.2	492	3.8	
Total heat of water vapor in gas	1006	7.4	487	3.8	
Sensible heat of ashes	4	0	0	0.1	
Heat lost in jacket overflow water	458	3.4	752	5.8	
Calorific value of the ashes	40	0.3	39	0.3	
Radiation and convection	31	0.2	21	0.2	
Errors and unaccounted for	11	0.1	58	0.5	
Total	13,651	100.0	12,914	100.0	

Table 3-30 Effect of Various Concentrations of Oxygen in a Kerpely* Producer Using Pea Coke[22]

Coke, lb per hr	1257	1578	1803	1553
Lb steam per lb coke	0.61	1.11	1.30	1.43
Blast rate, MSCFH	61.55	28.94	21.30	13.89
Oxygen, per cent	21	50.7	76.1	98.4
Make gas, MSCFH	95.15	82.60	85.60	69.20
$CO + H_2$ MSCFH	40.81	57.95	67.60	57.60
Make gas analysis, per cent				
CO_2	5.9	12.2	13.8	15.3
CO	28.8	40.3	44.1	45.3
H_2	14.2	29.9	34.8	38.0
N_2	50.3	17.0	6.7	0.9
Heat value	133	214	240	253
Steam decomposition, per cent	77.1	63.6	57.6	53.4

* Similar to the Heurtey producer (Fig. 3-13).

A British experiment was also reported.[23]

American experiments have been conducted by the U. S. Bureau of Mines and the Alabama Power Corp. at Gorgas, Ala.[24,25] in the first experiment, 1946–1947, some 220 tons of coal were gasified, the gas having a heating value of 47 Btu per cu ft with air alone, and 135 Btu per cu ft with oxygen–steam blasts. The roof rock softened, expanded, and filled the space occupied by the consumed coal. The second experiment,[25] 1949–1951, used the stream method. It ran for 22½ months, gasifying 10,846 tons of coal. The gas had a calorific value of from 42 to 147 Btu per cu ft.

The analysis (per cent volume) of dry effluent gas was: 19 to 28 CO_2, 0 to 0.5 O_2, 7 to 21.6 H_2, 0.7 to 6.3 CO, 0.9 to 4.9 CH_4, and 70 to 38.5 N_2. Heat loss to underground strata varied from 20 to 36 per cent. Leakage from the underground system increased to approximately 50 per cent of the total fluid input. The overall thermal efficiency, electric power generated compared with the heating value of the coal consumed, was 15 to 20 per cent. A steam generating power plant using mined coal may have an overall efficiency of approximately 12.5 per cent.

Further laboratory and field-scale experiments were carried out at the Missouri School of Mines at Hume, Missouri, to establish a gas passage thru a coal bed by an electric method. The method developed consists of three phases:

1. Electrolinking. Electrodes were placed in solid coal thru boreholes. Current passing between them carbonized the coal and established good electrical contact thru it.
2. Electrocarbonization. This served to free the volatile constituents of coal, leaving the coke underground.
3. Gasification of the coke and delivery of the energy above ground.

A third experiment at Gorgas, Ala., was effected using the electrocarbonization method. The technique required no horizontal mining or boring, and the porous coke bed provided excellent distribution of the air blast to the "fire-face." Gas produced had a calorific value of about 100 Btu per cu ft.

Field-scale tests at Gorgas conducted by the Bureau of Mines from 1951 to 1954 were reported[26] in 1957. Reaction paths were formed by electrolinking carbonization, using various electrode spacings up to 155 ft. Coal along these paths was gasified with air, oxygen, and steam, separately or in various combinations. Location and approximate area of the reaction zones were later determined by a drilling survey.

During peak operation of one system with air, a gas of the following average analysis was produced continuously for 21 days:

Analysis	Per cent
CO_2	9.7
Illuminants	0.3
O_2	0.5
H_2	9.2
CO	13.1
CH_4	1.5
N_2	65.7
Sp gr	0.94
Btu (at 60 F, 30 in. Hg dry)	93

During this 21-day period a gas production rate of 2160 SCFM was maintained at an average input air pressure of 27 psig. This operation represented the best obtained using air with paths prepared by electrolinking carbonization. Moreover, gas produced from electrolinked systems was much superior to that obtained from mined passages.[26]

With oxygen used as a gas-making fluid, a gas of 195 Btu per cu ft was obtained. Oxygen usage was much higher than oxygen required in gas producers.[26]

Tests also have been undertaken by the Bureau of Mines of hydraulic fracturing of the coal followed by back burning for seam preparation; the preliminary results[27] were encouraging. Over 2000 tons of coal were gasified with an average recovery of about 5500 Btu per lb of coal. The heating value ranged from 80 to 150 Btu per SCF. Several 20-day runs and one 75-day run were made.

During one stage of the work at Gorgas, the energy in the

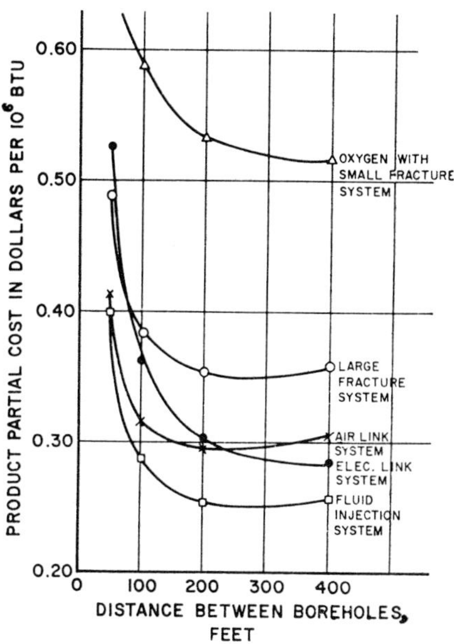

Fig. 3-18 Comparison of process costs for different systems.[28] Included are costs of boreholes, linking, gasification energy, and operation.

product gases was used to run a two-stage gas turbine, which drove an air compressor and supplied air to the gasification galleries.

A 1960 economic analysis, based on experimental Bureau of Mines results, indicated that a gas suitable for use in a power plant might be produced for approximately 62 to 77 cents per million Btu.[24] Figure 3-18, prepared earlier, compares *partial* costs of different systems.[28]

REFERENCES

1. Lowry, H. H., ed. *Chemistry of Coal Utilization,* Vol. II, Chap. 36. New York, Wiley, 1945.
2. Slottman, G. V. "Changing Status of the Gas Producer in the Iron and Steel Industry." *3rd Intern. Conf. on Bituminous Coal Proc.,* 1: 866–73, 1931.
3. Romig, J. W. "Economics of Producer Gas, Fuel Oil and Natural Gas Compared." *Glass Ind.* 10: 283–5, Dec. 1929.
4. Gauger, A. W. "Physical Chemistry of Gas-Producer Reactions in Relation to Ceramic Firing." *Am. Ceramic Soc. Bull.* 19: 365–8, Oct. 1940.
5. Rambush, N. E., and Townend, F. S. "Producer Gas Practice from the Point of View of the Carbonizing Industries with Special Reference to Low Priced Fuels." *Second World Power Conf. Trans.* 2: 165–74, 1930.
6. Goldman, B., and Clarke-Jones, N. "Modern Portable Gas-Producer." *J. Inst. Fuel* 63: 103–40, Feb. 1939.
7. Wagman, D. D., and others. *Heat, Free Energies, and Equilibrium Constants of Some Reactions Involving O_2, H_2, H_2O, C, CO, CO_2, and CH_4.* (Natl. Bur. of Standards R.P. 1634) Washington, D. C., 1945.
8. Gumz, W. *Gas Producers and Blast Furnaces.* New York, Wiley, 1950.
9. Parent, J. D., and Katz, S. *Equilibrium Compositions and Enthalpy Changes for the Reactions of Carbon, Oxygen and Steam.* (I.G.T. Research Bull. 2) Chicago, Ill., 1948.
10. Arthur, J. R., and others. "Combustion in Fuel Beds." *J. Soc. Chem. Ind.* 68: 1–6, Jan. 1949.
11. Batchelder, H. R., and others. "Kinetics of Coal Gasification." *Ind. Eng. Chem.* 45: 1856–78, Sept. 1953.
12. Gumz, W., and Foster, J. F. *Critical Survey of Methods of Making a High Btu Gas from Coal.* (A.G.A. Gas Prod. Res. Comm. Research Bull. 6) New York, 1953.
13. Edmister, W. C., and others. "Thermodynamics of Gasification of Coal with Oxygen and Steam." *ASME Trans.* 74: 621–36, July 1952.
14. "Gas-Making Plant to Use Low-Grade Coal." *Oil & Gas J.* 57: 142 Nov. 16, 1959.
15. Denig, F. "Coke-Fired Gas Producer Operation." *A.G.A. Proc.* 1927: 1210–35.
16. Slottman, G. V. "Factors Influencing Gas Producer Operation." *A.G.A. Proc.* 1930: 978–85.
17. Morris, W. R. "Carbonization of Coal with Blue Gas and Producer Gas." *A.G.A. Proc.* 1922: 39.
18. Romig, J. W. "Gas Producer Heat Balance." *Am. Gas J.* 138: 17–20, May 1933.
19. Pfluke, F. J. "Report of Subcommittee on Test Code for Producer and Carbonizing Plants." *A.G.A. Proc.* 1930: 1042–95.
20. Newman, L. L. "Oxygen in Production of Hydrogen or Synthesis Gas." *Ind. Eng. Chem.* 40: 559–82, April 1948.
21. Hendryx, D. B. "Utilization of Producer Gas in Industrial Furnaces." *ASME Trans.* 68: 877–82, Nov. 1946.
22. Batchelder, H. R., and others. "Operation of Kerpely Producer with Oxygen-Enriched Blast." *A.G.A. Proc.* 1950: 348.
23. Machelson, S. G. "Underground Gasification of Coal." *Gas Age* 115: 31, Feb. 10, 1955.
24. Katell, S., and Faber, J. H. "Estimated Costs of Gasifying Coal in Place." *Gas Age* 128: 40–4, Sept. 14, 1961.
25. Elder, J. L., and others. *Second Underground Gasification Experiment at Gorgas, Ala.* (Bur. of Mines R.I. 4808) Washington, D. C., 1951.
26. Elder, J. L., and others. *Field-Scale Experiments in Underground Gasification of Coal at Gorgas, Ala.* (Bur. of Mines R.I. 5367) Washington, D. C., 1957.
27. Carman, E. P., and others. *Report of Bureau of Mines Research and Technologic Work on Coal and Related Investigations, 1955,* p. 77. (Bur. of Mines I.C. 7794) Washington, D. C., 1957.
28. Pears, C. D., and others. "Energy and Economic Evaluation of the Parameters of Several Underground Gasification Processes." *Am. Power Conf. Proc.* 20: 335–45, 1958.

Chapter 5
Blue Gas and Carbureted Water Gas

by C. B. Glover

BLUE GAS (WATER GAS)

Modern blue gas production apparatus consists of a **generator,** usually equipped with mechanical grates, an **igniter** for preheating the *down-run* steam, and a waste heat boiler thru which both the blast gas and *up-run* blue gas are passed for maximum recovery of sensible heat. This apparatus, operating on a short cycle and employing high rates of air and steam, tends to produce the best overall economy, efficiency, and capacity.

A typical blue gas cycle consists in attaining the vigorous combustion of coke or coal by blowing in air (blast) during the first part of the cycle. The resulting gas, mostly CO_2 and N_2, is vented to the atmosphere. Then steam is substituted for the air, and the resulting blue gas is collected. Since the production of blue gas involves absorption of heat, the coal or coke soon becomes too cool to support the blue gas reactions. It is then necessary to turn the steam off and to turn the air on again.

Steam–Carbon Reactions

Production of blue gas from coke, bituminous coal, and anthracite coal requires the cyclic storage of large quantities of heat in the fuel bed to permit the chemical reaction (Table 3-36):

$$2H_2O^* + C \rightleftharpoons 2H_2 + CO_2 \qquad (6)$$

as well as Reactions **2**, **3**, and **4** of Table 3-36. Commercial blue gas always contains about one per cent methane. Reaction **5**, $C + 2H_2 = CH_4$, at temperatures up to 2700 F is slow. These reactions are of vital importance in the manufacture of producer, blue, carbureted water, and reformed natural gases or LP-gases.

Mechanism of Steam–Carbon Reactions and Reaction Rates

The blue gas reactions have not been completely understood. However, several groups of investigators have agreed that the aforementioned Reactions **2**, **3**, and **6** occur on the surface of the carbon and Reaction **4** (Table 3-36) is a gas phase equilibrium reaction occurring either in the gas film immediately above the carbon surface or in the main gas stream. Reactions **3** and **6** may be regarded as *primary* and **2** as *secondary,* while **4** may not be of importance when time of contact is short.

Early investigators,[1] using a 10-in. bed of 5 to 8 mm ground coke and electric arc light carbon, varied the pressure and used times of contact of both 0.113 and 0.226 sec.

Figure 3-19 shows the variation in wet water gas composition for various percentages of steam decomposition. Figure 3-20 gives the relation between the ratio of CO to CO_2 and the same base.

In one experiment using a 3-in. fuel bed 1 in. in diameter with particle sizes from 0.1 to 0.2 in., steam partial pressure (by dilution with nitrogen) and time of contact were varied over wide ranges.[2] The three fuels, medium-temperature (1652 F) coke, high-temperature (2318 F) coke, and metallurgical (by-product) coke, each required a different time of contact for equilibrium (Reaction **4**) to be complete. General conclusions were that the effectiveness of steam decomposition in a fuel bed of fixed dimensions and given temperature is determined by the steam supply rate *divided* by the volume of free space in the fuel bed, *independently of whether or not* the steam supply is diluted with nitrogen. Thus, the partial pressure of

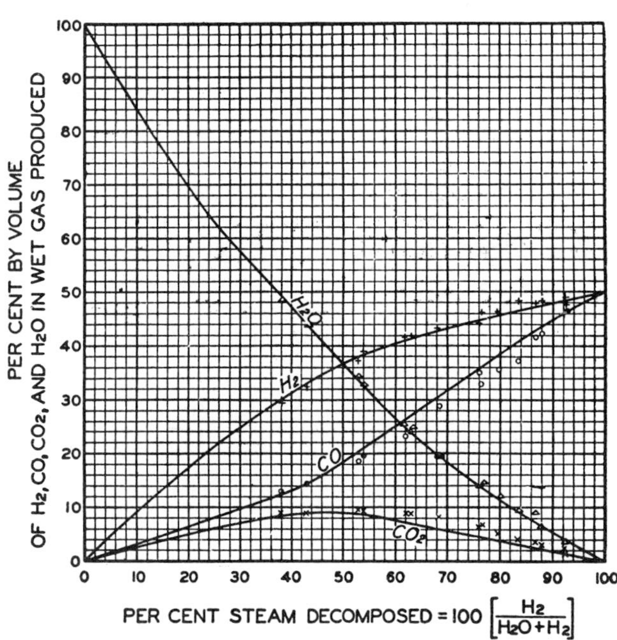

Fig. 3-19 Variation in composition of wet water gas with percentage of steam decomposed.[1]

* Gas.

the steam has no effect upon the steam decomposition. Also, it was evident that differently prepared cokes from the same coal do not react at the same rate with steam.

It was also found[2] that the 2318 F coke reacted with steam at 1832 F two-thirds as fast as the by-product coke. The hourly rate of gasification of the by-product coke at 1832 F was four *times* as fast as at 1652 F.

Failure of large-scale operations to match laboratory tests at a fair time of contact is due to the larger voids letting thru more steam without penetrating the gas film on the carbon surface, where equilibrium is established in a very short time. Making good quality water gas on a large scale depends on equilibrium *not* being attained. The final product is really steam mixed with good quality gas. Obviously gas production, not steam decomposition, is the goal of the operation.[2]

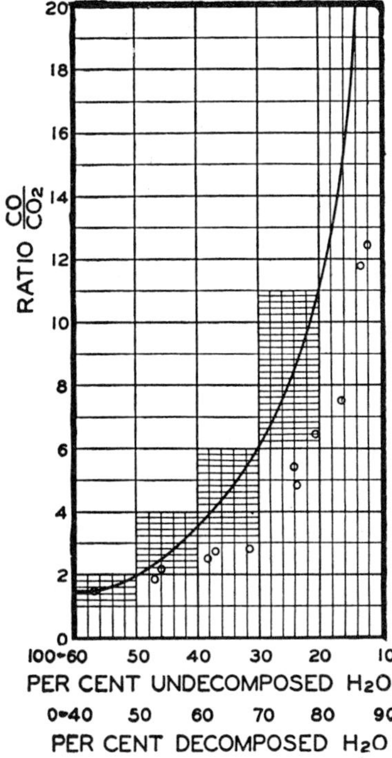

Fig. 3-20 CO/CO_2 ratio vs. decomposed and undecomposed steam.[1]

In an experiment,[3] four different kinds of carbon were used (activated, natural graphite, arc-electrode, and retort) and the pressure was varied. The steam was also diluted with nitrogen. Three of the fuels indicated a zero order of reaction, which checked with earlier work.[2] The retort carbon actually showed a "negative" order of reaction, i.e., the reaction rate increased with a reduction in pressure.

The reactions may be[3]

$$C + 2H_2O = 2H_2 + C_xO_y \quad \textbf{(a)}$$

and

$$C_xO_y = aC + bCO_2 \quad \textbf{(b)}$$

or, alternately,

$$C + H_2O = H_2 + CO \quad (3)$$

Reaction **b** is assumed to be slower than **a** so that **b** controls the decomposition rate of steam by Reactions **a** and **b**, while 3 is supposed in many cases to be relatively less important than **a**, because most of the surface is covered with the complex C_xO_y.

If **3** is entirely negligible, **b** predominates, and the reaction will be of the zero order with respect to steam. This is substantially the case with three of the fuels and with the retort carbon below 1850 F.

During these experiments, it was found that between 1850 and 2057 F, the rate of reaction with retort carbon *doubles* for each 45 to 54°F rise in temperature.

Since the latter two investigations found zero reaction order on all fuels tested save one, and the exception occurred below 1850 F, it seems practical to consider ordinary fuels and operating conditions as falling within this class.

In Figs. 3-19 and 3-20 points have been plotted to show the values obtained by Pexton and Cobb when gasifying by-product coke at 1832 F.

An interesting series of experiments was conducted on charcoal, three British coals, three German coals, and retort carbon using a wide range of temperatures.[4] Steam decomposition for the four kinds of fuel commenced at 1067, 1292, 1517, and 1580 F, respectively. At this temperature each fuel produced a gas of 33.3 per cent CO_2 and 66.6 per cent H_2. This is interpreted to support the view that the water gas reactions are:

$$C + 2H_2O = CO_2 + 2H_2 \quad (6)$$

$$C + CO_2 = 2CO \quad (2)$$

See Table 3-36.

Various equilibrium vs. temperature relationships are shown in Figs. 3-21, 3-22, and 3-23 and in Table 3-36.

Later Studies. Principal stoichiometric relations involved in water gas reactions have been presented above. A study of the mechanism of these reactions and their rates, sponsored by A.G.A., was made by the Institute of Gas Technology and the Battelle Memorial Institute. It consisted of a review of all previous investigations and intensive experimental work. The entire project covered a period of about eight years. Results were presented in I.G.T. Research Bulletin No. 19[6] and a Battelle Memorial Institute report.[7] They are summarized below:

1. A brief summary of I.G.T. Bulletin 19[6] shows that while any two of four reactions can account for all of the composi ion changes that occur in the carbon–steam reactions, the two principal reactions are:

Reaction **3**: $C + H_2O = CO + H_2$ (predominating)
Reaction **4**: $CO + H_2O \rightleftharpoons CO_2 + H_2$ (water gas shift reaction accompany ng Reaction **3**)

Reaction 2, $CO_2 + C = 2CO$, is slow compared with Reaction **3**. Reaction **6**, $C + 2H_2O = CO_2 + 2H_2$, is practically nonexistent.

2. Wet gas compositions above 1900 F have been shown to depend almost entirely upon the amount of steam decomposi ion, and are practically independent of temperature. In the lower operating range of 1700–1900 F, wet gas compositions seem to depend to a very slight extent upon operating temperature.

3. Product gas compositions also appear to be inde endent of operating pressures in the range of 1.0–3.5 atm.

4. Reaction rate equations and related velocity and mass transfer coefficients were developed, *applicable only to the two fuels used.* These were the −10 to +20 U. S. sieve particle size range of Koppers pitch coke, used by I.G.T., and ¾ × 1¼ in. high-temperature oven coke (Philadelphia Coke Co.), used by

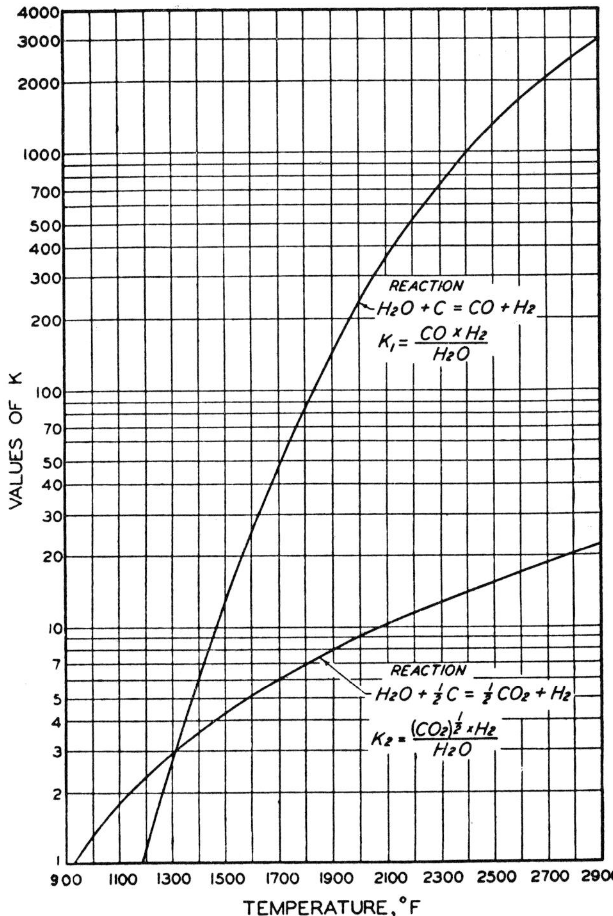

Fig. 3-21 Equilibrium constants (Reactions 3 and 6) vs. temperature.[5]

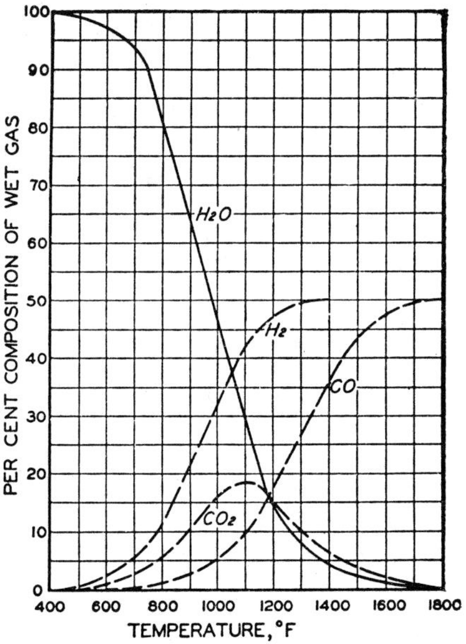

Fig. 3-22 Equilibrium composition of wet water gas at atmospheric pressure and varying temperature.[5]

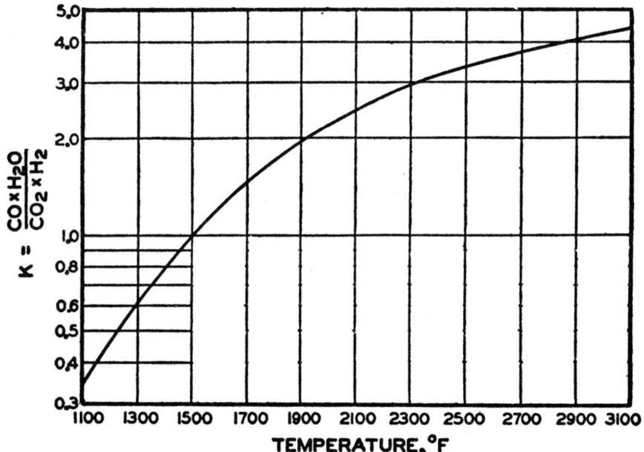

Fig. 3-23 Equilibrium constant for $CO_2 + H_2 = CO + H_2O$ vs. temperature.

Battelle Memorial Institute. In the 1700 to 2300 F temperature range, reaction rates for the cyclic water gas generator with oven coke appeared to be from 5 to 50 *times* greater than I.G.T. values obtained with pitch coke. These differences may be partly due to a more reactive coke, errors in temperature and composition measurement, and the eight- to fortyfold higher steam flow rates employed in the cyclic water gas generator.

Factors Influencing Reactivity of Coke

Many attempts have been made to increase the reactivity of coke used in the production of blue gas. The aim has been to increase both the percentage of steam decomposition and the synthesis of methane.

Special cokes made by adding five per cent of such oxides as silica, alumina, lime, and ferric oxide and equivalent quantities of calcium carbonate and sodium carbonate to the original coal were steamed.[8] The following data concern the decomposition of steam in coke beds thus impregnated and tests on decomposing carbon dioxide made by Branson.

	"Pure" coke	Calcium oxide coke	Iron oxide coke	Sodium carbonate coke
Steam decomposed	61	82	91	98
CO_2 in gas	9.2	5.4	2.6	0.4
CO_2 decomposed	6.6	29.9	45.6	89.0

Effects of the presence of lime in coke on the formation of methane have been studied.[9] Below 1500 F, formation of calcium carbonate is a factor in the synthesis of methane; above 1600 F, calcium is formed and acts catalytically, favoring the following three reactions:

$$4CO + 2H_2O = 3CO_2 + CH_4 \quad \text{(c)}$$

$$2CO + 2H_2 = CO_2 + CH_4 \quad \text{(7)}$$

$$CO_2 + 4H_2 = 2H_2O + CH_4 \quad \text{(d)}$$

In spite of these promising results, no practical application has developed, probably because the minerals having such a marked effect on the reactivity of coke are fluxes to the refractories used in gas-making equipment.

Coke porosity is of far greater influence than the catalytic effect of these minerals. A coke of light structure and high porosity produces a high degree of steam decomposition and high set capacities.

Table 3-31 Operating Results for 14-Day Continuous Run of a Nine Ft ID U.G.I.* Mechanical Generator Blue Gas Set†

Make per set day, MCF							5096
Make per running hour, MCF							216
Make per run, MCF							10.5
Fuel used as charged, lb/MCF							32.61
Moisture in fuel, %							2.24
Combustible in fuel, %							87.94
Dry fuel used, lb/MCF							31.88
Combustible used, lb/MCF							28.64
Btu per cu ft							292
Steam used, lb/MCF							56.5
Generator air used, cu ft/MCF							1700
Analysis	Ill.	CO	H_2	CH_4	CO_2	O_2	N_2
Blue gas	0.02	38.46	52.16	0.38	5.21	0.08	3.69
Generator blast gas	0	5.88	1.18	0	15.84	0.06	77.04
Carbon balance							Lb/MCF
Carbon in blue gas							13.76
Carbon in blast gas							12.46
Carbon in clinker and ash							0.34
Carbon in blown-over dust							0.38
Total carbon accounted for							26.94
Accounted for by carbon balance, %							94

* United Gas Improvement Corp.
† Courtesy of E. I. du Pont de Nemours & Co., March 1930.

Table 3-32 Operating Data for Standard Blue Gas Sets*

Generator size, ft	Generator ID, in.	Grate area, sq ft	Make per day, MCF	Make per cycle, MCF	Fuel per MCF, continuous operation, lb	Air per minute to generator, cu ft
6†	50	13.64	740	1.83	37	2,740
7†	62	20.97	1160	2.90	37	4,360
8†	74	29.87	1690	4.12	36	6,190
9†	86	40.34	2160	5.55	36	8,370
10†	98	52.38	2690	7.23	35	10,900
11†	108	63.62	3460	9.20	35	13,850
12†	120	78.54	4320	11.55	35	17,370
Mechanical generator	108	78.54	6100	13.10	33	20,000

* Courtesy of United Engineers & Constructors, Inc., March 1953.
† Hand-clinkered sets.
Notes: All data are on the basis of back-run operation. Blue gas, 292 Btu; average make per sq ft grate area per day, 54.3 MCF; generator air/MCF = 1800 cu ft; air-to-steam ratio, 30; cycle, 3.00 min; 40% blow, 60% run; operating time per hour, 56 min. For specific conditions, capacity will vary with material flow rates.

Table 3-33 Calculated Heat Balance of a Nine Ft ID Blue Gas Set

(based on production of 1000 cu ft blue gas; for operating results see Table 3-31)

	Heat supplied		Heat accounted for	
	Btu	%	Btu	%
Coke consumed during blow	207,000	44.2		
Coke consumed during run	198,000	42.4		
Exhaust steam	62,700	13.4		
Potential heat of blue gas			292,400	62.5
Jacket steam			27,900	6.0
Sensible heat of blue gas			15,900	3.4
Sensible heat of blast gas			40,200	8.6
Excess steam			50,900	10.8
Unaccounted-for losses			40,400	8.7
Total	467,700	100.0	467,700	100.0
Heat recoverable in waste heat boiler from blast gas, excess steam, and up-run blue gas			30,300	6.4

Table 3-34 Heat Balances of Blue Gas Sets[5]

	Fuel Research Board,[10] %	Inst. of Gas Engrs.,[11] %
Heat input:		
Coke	92.52	93.41
Steam	7.35	6.50
Air	0.13	0.09
Total	100.00	100.00
Heat output:		
Water gas, heat of combustion	57.81	52.19
sensible heat	3.23	2.04
Blow gas, heat of combustion	18.95	22.59
sensible heat	9.33	5.30
Clinker and unburned coke, heat of combustion	4.54	6.94
Dust, heat of combustion	0.33	6.94
Sensible heat of clinker, unburned coke and dust	0.33	0.42
Uncondensed steam, sensible and latent heat	2.20	2.32
Radiation and unaccounted-for losses	3.28	8.20
Total	100.00	100.00
Approximate average temperature of blow gas	1280 F	625 F
Approximate average temperature of water gas	935 F	690 F

Operating Results

Table 3-31 shows the operating results of a blue gas set under specific conditions. Table 3-32 lists operating data for standard blue gas sets.

Heat Balances

Heat balance data for blue gas operations are shown in Tables 3-33 and 3-34.

Use of Oxygen in Blue Gas Manufacture

The low average efficiency of 62 per cent common to cyclic blue gas production is due to alternate blasting of the fuel bed with air to supply the heat storage necessary for the steam–carbon reaction during the *make* portion of the cycle. By using oxygen instead of air, an efficiency of 90 per cent might be realized, since the process would be continuous. Oxygen and steam are introduced into the fuel bed continuously, so that the exothermic reaction between oxygen and carbon is carried on simultaneously with the endothermic reaction between steam and carbon.

The reaction of oxygen and carbon to form carbon monoxide is the *controlling* reaction when carbon, oxygen, and steam are used for making blue gas. This reaction produces heat for decomposition of the steam and yields a combustible gas at the same time.

The reactions between carbon and oxygen are:

$$2C + O_2 = 2CO \quad (e)$$

$$2CO + O_2 = 2CO_2 \quad (f)$$

$$C + O_2 = CO_2 \quad (1)$$

$$CO_2 + C = 2CO \quad (2)$$

The $\Delta H_{60F,\,1\,atm}$ for Reactions **e** and **f** are −95,128 and −243,448 Btu per lb-mole of O_2, respectively.

$$H_2O + C = CO + H_2 \quad (3)$$

$$2H_2O + C = CO_2 + 2H_2 \quad (6)$$

See Table 3-36 for ΔH on Reactions **1** to **3** and **6**.

If Reaction **e** is tripled and added to Reaction **3**, theoretically enough heat will be produced to keep the carbon–oxygen–steam reactions continuous:

$$7C + 3O_2 + H_2O = 7CO + H_2 \quad (g)$$

In Reaction **g**, 228,947 Btu per lb-mole of water is given off.

Since original experiments[12] in 1925 in a model water gas set, industrial applications of the process have been made in the United States, in Canada,[13] and in Europe.[14]

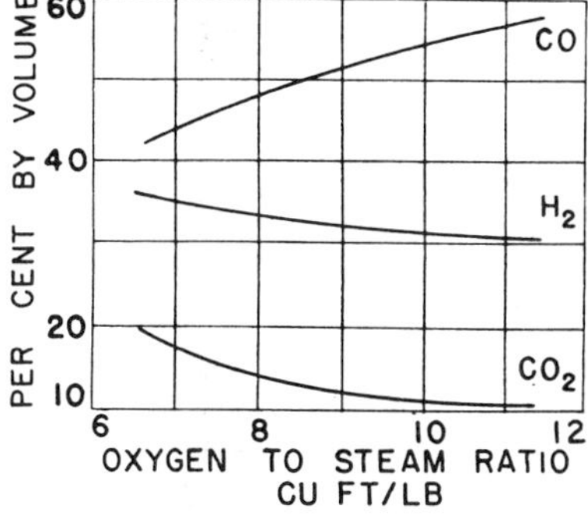

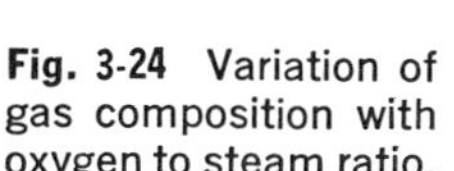
Fig. 3-24 Variation of gas composition with oxygen to steam ratio.

Reports on conversion of several standard U.G.I. mechanical blue gas sets for oxygen service, including erection of an oxygen plant of about 360 tons daily capacity, are available.[15] Conversion of the cyclic generator was actually a simplification of the apparatus, with its final arrangement essentially that of an updraft producer.

Due to new developments for producing low-cost oxygen,[16] its use in blue gas manufacture is becoming more feasible. General data on plant sizes and costs (1956) follow:

Plant size, tons/day	50	100	200	350	500
Plant cost, $1000*	690	1080	1560	1070	2520
O_2 cost, $/ton†	14.40	9.72	7.28	5.92	5.32

* Exclusive of building and site.

† Includes plant cost given above, plus 20 per cent for site preparation amortized over 15 years; two operators per shift at $2.00 per hr; maintenance plus taxes and insurance at four per cent of investment; power at one cent per kwh; 400 hp-hr per ton O_2; and overhead taken equal to operating payroll.

The O_2 requirements for MMCF per day of pipeline gas are 20 and 44 tons per day for the Lurgi process and suspension gasification, respectively.

Cost of O_2, $/ton	O_2 consumed, SCF/MSCF pipeline gas	Cost of O_2 per MSCF pipeline gas, cents
4	300–1300	5.18–22.35*
5	300–1300	6.64–28.22
6	300–1300	7.74–34.64

* Linear interpolation permissible; e.g., 12 cents worth of O_2 per MSCF pipeline gas for a consumption of 700 SCF O_2 per MSCF pipeline gas.

Operating Results. Table 3-35 and Fig. 3-24 give operating data on the use of oxygen to produce blue gas.

Table 3-35 Operating Results Using Oxygen in Blue Gas Production[15] with U.G.I. Generators

	Average operating results using air	High gasification rate using O_2
Set operation, hr	2,890	44.7
Oxygen flow, cfm (32 F, 1 atm)	1,400	1,957
Oxygen temp, °F	210	208
Oxygen purity, %	95.2	95.0
Steam flow, lb/min	187	257
Oxygen/steam ratio, cu ft/lb	7.5	7.6
Grate speed, rph	0.68	0.78
Ashpit pressure, in. H_2O	33	41
Generator differential, in. H_2O	4	9
Offtake gas temp, °F	1,209	1,211
Steam decomposition, %	53.5	51.0
Gas analysis, % CO_2	16.9	16.6
O_2	0.1	0.0
CO	46.6	47.8
H_2	34.0	33.6
CH_4	0.2	0.2
N_2	2.2	1.8
Coke used, lb/MCF (32 F, 1 atm)	23.7	24.1
Oxygen used, cu ft/MCF (32 F, 1 atm)	243	248
Steam used, lb/MCF (1 atm)	32.4	32.6
Gas made/24-hr set day, MCF	8,300	11,400
Gas heating value, Btu (60 F, 30 in. Hg, sat.)	258	261

CARBURETED WATER GAS

General

The basic carbureted water gas* process has not changed much since its development by Prof. T. S. C. Lowe in 1875. Carbureted water gas, long made in *sets* consisting of a generator (for blue gas making), a carburetor (for oil gas making), and a superheater (for final cracking of oil vapors), is currently (1960) made by modification of the process to permit use of oil, natural gas, or LP-gas addition to the generator and other parts, reducing the amount of coke required. Essentially, the process involves intermittent blow-runs to heat the coke (various coals may also be used), followed by make-runs in which gas is made by a combination of the water gas reaction and thermal cracking of hydrocarbons.

Manufacture of carbureted water gas is a compromise between the blue gas and oil gas processes in that generator efficiency is sacrificed so that heat may be stored in the carburetor and superheater for the purpose of gasifying enriching oil. Unlike the oil gas process, oil is then cracked in the presence of hydrogen and superheated steam.

* Also known as carbureted blue gas.

Table 3-36 Quantitative Study of Equilibria in Selected Reactions[17]

(see Table 12-119 for more data on selected reactions; Table 3-42 gives an application of these data)

No.*	Reaction	Heat of reaction ΔH, Btu/lb-mole at 298.16 K	Volume changes	Equilibrium constant† K_p	K_p at 800 K (980 F)	K_p at 1000 K (1340 F)	K_p at 1200 K (1700 F)
1	$C + O_2 \rightarrow CO_2$	−169,288	1 to 1	$\frac{(CO_2)}{(O_2)}$	6.709×10^{25}	4.751×10^{20}	1.738×10^{17}
2	$C + CO_2 \rightarrow 2CO$‡	+74,160	1 to 2	$\frac{(CO)^2}{(CO_2)}$	1.098×10^{-2}	1.900	57.09
3	$C + H_2O \rightarrow H_2 + CO$	+56,437	1 to 2	$\frac{(H_2)(CO)}{(H_2O)}$	4.399×10^{-2}	2.609	39.77
4	$CO + H_2O \rightarrow H_2 + CO_2$‡	−17,723	2 to 2	$\frac{(H_2)(CO_2)}{(CO)(H_2O)}$	4.038	1.374	0.6966
5	$C + 2H_2 \rightarrow CH_4$‡	−32,198	2 to 1	$\frac{(CH_4)}{(H_2)^2}$	1.411	9.829×10^{-2}	1.608×10^{-2}
6	$C + 2H_2O \rightarrow 2H_2 + CO_2$	+38,714	2 to 3	$\frac{(H_2)^2(CO_2)}{(H_2O)^2}$	0.1777	3.608	28.01
7	$CH_4 + CO_2 \rightarrow 2CO + 2H_2$	+106,358	2 to 4	$\frac{(CO)^2(H_2)^2}{(CH_4)(CO_2)}$	7.722×10^{-3}	19.32	3.548×10^{3}
8	$2C + H_2 \rightarrow C_2H_2$	+97,485	1 to 1	$\frac{(C_2H_2)}{(H_2)}$	1.602×10^{-12}	1.337×10^{-9}	1.156×10^{-7}
9	$C_2H_6 \rightarrow C_2H_4 + H_2$	+58,879	1 to 2	$\frac{(C_2H_4)(H_2)}{(C_2H_6)}$	4.557×10^{-3}	0.3443	6.224
10	$2CH_4 \rightarrow C_2H_2 + 3H_2$	+161,840	2 to 4	$\frac{(C_2H_2)(H_2)^3}{(CH_4)^2}$	8.017×10^{-13}	1.384×10^{-7}	4.474×10^{-4}
11	$CH_4 + H_2O \rightarrow CO + 3H_2$	+88,635	2 to 4	$\frac{(CO)(H_2)^3}{(CH_4)(H_2O)}$	3.120×10^{-2}	26.56	2.473×10^{3}
12	$CH_4 + 2H_2O \rightarrow CO_2 + 4H_2$	+70,912	3 to 5	$\frac{(CO_2)(H_2)^4}{(CH_4)(H_2O)^2}$	0.1260	36.49	1.723×10^{3}
13	$2CH_4 \rightarrow C_2H_4 + 2H_2$	+86,836	2 to 3	$\frac{(C_2H_4)(H_2)^2}{(CH_4)^2}$	1.021×10^{-7}	6.939×10^{-5}	5.540×10^{-3}
14	$C_2H_2 + O_2 \rightarrow 2CO + H_2$	−192,520	2 to 3	$\frac{(CO)^2(H_2)}{(C_2H_2)(O_2)}$	4.563×10^{35}	6.747×10^{28}	8.562×10^{25}

* Equation numbers outside of this chapter do not necessarily correspond.
† (CO), (H_2), etc., are the *partial pressures* per atmosphere; see illustration in Table 3-42.
‡ Expanded data given in Table 12-119.

The presence of hydrogen during decomposition of gas oil under conditions prevailing in the carburetor and superheater (1) diminishes partial pressures of all reacting gases; (2) decreases contact time; and (3) causes hydrogenation of hydrocarbons. Volumes of methane and ethane produced per unit quantity of oil are higher when cracked in hydrogen.

This process is basically a thermochemical operation, but equilibrium conditions never prevail. Many variables control the quantity and quality of the product, including the kind and size of fuel, analyses of oil, depth and condition of fire, rate of air admission, distribution and rate of steam admission, rate and design of oil admission, and temperature.

Carbureted water gas reactions and their equilibria are shown in Table 3-36.

Evaluation of Carbureted Water Gas Results

Plant scale tests by H. G. Terzian resulted in development of the "Terzian factor"* and the "Terzian plant constant"* as items in evaluation of carbureted water gas results. Reasons for these tests as set forth by Terzian may be summarized as follows:

* Although these concepts were not technically correct, they constituted an indicator of relative day-to-day efficiency.

> In plants operating with gas oil and without blow-run, the usual procedure is to determine oil efficiency or oil enriching value in Btu per gallon when calculating operating results. For this purpose the following enriching value formula is generally used:
>
> $$E = \frac{1000\,(A - B)}{C} + BD$$
>
> where: E = oil efficiency or enriching value, Btu per gal
> A = Btu per cu ft make gas
> B = Btu per cu ft blue gas
> C = oil used, gal per MCF
> D = cu ft of oil gas per gal
>
> The value of this formula depends primarily on the accuracy of determination of the blue gas calorific value. This is generally assumed to be 300 Btu per cu ft, and the oil gas yield is assumed to be 65 to 70 cu ft per gal. When operating without blow-run, using coke as generator fuel and gas oil, it is possible to obtain a good quality blue gas of about 300 Btu per cu ft. Hence, the

above formula can be used successfully to determine oil enriching value.

Since the advent of blow-run, it has been very difficult to determine the calorific value of the blue gas (including the blow-run gas). Therefore, the enriching value as determined by the above formula may be inaccurate and somewhat misleading when blow-run is used.

Use of heavy oil in the water gas set, not only for enriching the blue gas, but also for reforming the oil in the generator fuel bed, made it just as difficult to determine the blue gas Btu, and also added another variable, i.e., variation in generator fuel due to amount of reforming. Heavy oil operating technique has so changed set operation that a set can produce a finished gas of varying specific gravity and inert content. In producing this gas, it is possible to operate with generator fuel ranging from 10 to 25 lb per MCF, and with oil varying from 3 to 6 gal per MCF. Development of the *heavy oil process* necessitated plant scale tests under varied operating conditions.

Terzian Factor.[18] This term was developed for comparing operating results. It has been successfully used during the last few years.

$$\text{Terzian factor} = \frac{10\,F + 100\,C}{A}$$

where: F = generator fuel used, lb per MCF
C = oil used, gal per MCF
A = Btu per cu ft make gas

When a fairly uniform fuel and oil are used, operating results obtained under extremely varying conditions can be compared successfully by use of this formula; *the smaller its numerical value, the better the gasification efficiencies.*

Tests using various kinds of heavy oils showed that the Terzian factor is practical for comparing plant operating results. Figures 3-25, 3-26, and 3-27 show variations of gas- and tar-making qualities of heavy oils with respect to this factor. The factor gives an approximation of the enriching value and amount of tar produced per MCF when a certain grade of heavy oil is used. For practically the same grade of oil, the Terzian factor apparently stays *constant* whether or not operations are conducted with reforming. Similarly, variation in amount of blow-run does not seem to affect this factor.

Terzian Plant Constant. Plant scale tests showed[18] that energy introduced into a water gas set in the form of coke and oil bears a practically constant ratio to that removed in the gas and tar. This is illustrated by the relative stability of a factor, developed thru tests and termed the *Terzian plant constant*, over a wide range of operating conditions, including reforming, Btu value, and oil quality.

$$\text{Terzian plant constant} = \frac{10\,F + 100\,(C - T)}{A}$$

where: F = generator fuel used, lb per MCF
C = oil used, gal per MCF
T = tar made, gal per MCF
A = Btu per cu ft make gas

This constant varies between 0.88 and 0.92, primarily due to the quality of the heavy oil used. Oils of high enriching value give a low constant. Those of low enriching value give the higher figure. Slight variations also exist between different plants, even with oil of practically the same quality. This variation is due to certain plant characteristics such as quality of fuel used, operating technique, miscellaneous plant equipment, etc.

A study of heavy oil operating results of various plants for recent years indicates that the Terzian plant constant[18] may be used as a means of checking plant operating results, including tar production.

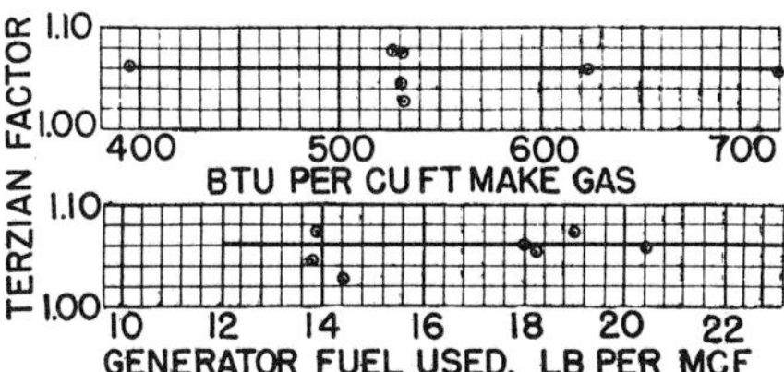

Fig. 3-25 Relations between Terzian factor and make gas Btu and generator fuel, using uniform grade heavy oil.

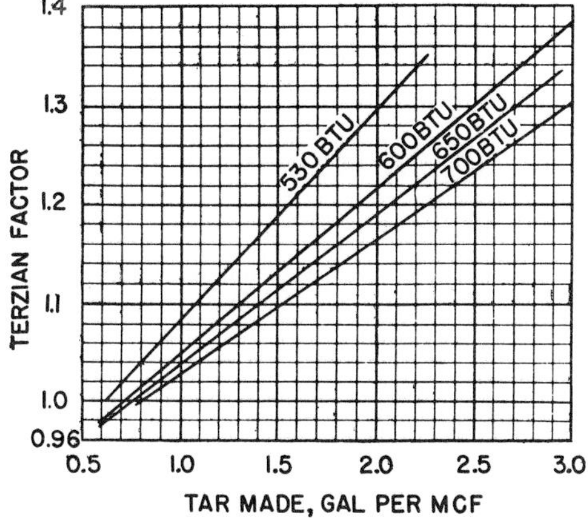

Fig. 3-26 Terzian factor vs. tar production, using heavy oil of various enrichment values.

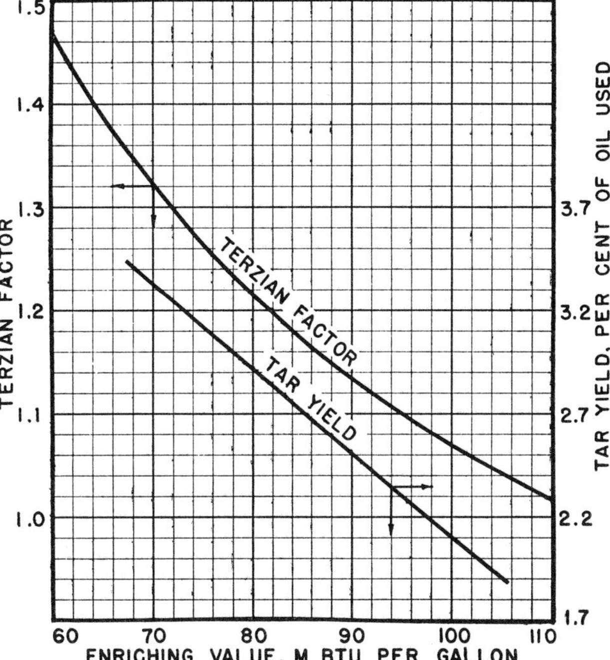

Fig. 3-27 Terzian factor vs. heavy oil enrichment values.

Analysis of Carbureted Water Gas Operation

The following analysis is based on operating data* using a representative gas analysis.

Assumptions.

1. N_2 and O_2 contents of the solid and liquid fuels are negligible.
2. Any N_2 in finished gas comes from air or its products. Any O_2 or oxides in the gas resulting from air is accompanied by 3.8 moles of N_2 per mole of O_2.
3. CO and CO_2 in the gas come from carbon–steam and carbon–air reactions. The CO and CO_2 from carbon–air reactions are accompanied by 3.8 moles of N_2 per mole of O_2, the same as is true for any free O_2 in the gas.
4. From the water gas reactions:

$$C + H_2O = CO + H_2 \quad (3)$$

$$C + 2H_2O = CO_2 + 2H_2 \quad (6)$$

there will be two moles of H_2 per mole of O_2 involved. This ratio, however, does not hold if CH_4 is formed, which does happen to a slight extent in water gas production. Where the ratio of CH_4 to H_2 in the water gas is known, it can be shown that the H_2/O_2 ratio equals $2/(1 + 2R)$, where R = vol. CH_4/vol. H_2. Samples of straight blue gas from a plant for which operations are analyzed indicate that R is about 0.02. Therefore, the multiplying factor for finding the H_2 should be $2/[1 + (2 \times 0.02)] = 1.92$ instead of 2.0. The CH_4 formed would then be 0.02 × per cent H_2. This does not apply to gas from fuels having an appreciable volatile content, as CH_4 from that source would not affect the H_2/O_2 ratio.
5. Analysis of fuel (coke) as fired should be known, as its per cent carbon is an important factor.
6. Ultimate analysis or C/H_2 ratio of the oil and its specific gravity must be known.
7. Tar yield must be determined either from reported results or estimated from the reported Terzian factor.
8. Blue gas is defined as a mixture of blast gas from the air–carbon reactions and water gas from the carbon–steam reactions, possibly including small quantities of air.
9. Reforming is considered in these calculations as complete hydrocarbon decomposition to elemental carbon and hydrogen.

Reported Operating Results.

Representative carbureted water gas analysis, %		Operating results, 24-hr period	
Ill.	9.7	Fuel/MCF, lb	14.76
C_2H_6	3.0	Oil/MCF, gal	4.43
CH_4	7.4	Btu/cu ft	536
H_2	38.7	Sp gr of gas	0.661
CO	23.2	Terzian factor	1.106
CO_2	4.0	Tar/MCF, gal	1.15
O_2	1.1	Steam/MCF, lb	29.3
N_2	12.9	Gen. air/MCF, cu ft	990
Total	100.0	Air/steam ratio, cu ft/lb	33.7
		Oil nominally reformed, %	26.0

* Fuel analysis: 12% ash and moisture, 88% carbon. Oil analysis: 11% H_2, 89% carbon, 15.8° API. Tar analysis: 1.08 sp gr, 12% free carbon.

* Courtesy Philadelphia Gas Works Division, United Gas Improvement Co.

Calculations.

Reforming.

Oxygen balance		Moles O_2
O_2 from CO	$= 23.2 \times ½$	= 11.6
O_2 from CO_2	$= 4.0 \times 1$	= 4.0
Free O_2	$= 1.1 \times 1$	= 1.1
Total O_2		= 16.7 moles O_2/100 moles gas

$$O_2 \text{ accompanying } N_2 = \frac{12.9}{3.8} = \frac{-3.4 \text{ moles } O_2 \text{ from air}}{13.3 \text{ moles } O_2 \text{ from w.g. reaction}}$$

Moles H_2 due to water gas reaction = 13.3 moles O_2 from w.g. × 1.92 Terzian factor = 25.5

Moles H_2 due to oil decomposition = 38.7 H_2 in gas − 25.5 moles H_2 from w.g. reaction = 13.2 moles H_2 from reforming, or 132 cu ft H_2/MCF gas

H_2 available in oil = 8.0 lb/gal (15.8° API) × 11 per cent H_2 = 0.88 lb H_2/gal or 0.88 lb H_2 × 190 cu ft H_2/lb = 167 cu ft H_2/gal

$$\text{Oil reformed, gal/MCF} = \frac{132 \text{ cu ft } H_2 \text{ from oil}}{167 \text{ cu ft } H_2\text{/gal}} = 0.79$$

$$\text{Actual oil reforming, \%} = \frac{0.79 \text{ gal oil reformed}}{4.43 \text{ gal oil/MCF}} = 17.8$$

$$\text{Reforming efficiency} = \frac{17.9\% \text{ actual reforming}}{26.0\% \text{ nominal reforming}} \times 100 = 68.5\%$$

Coke Equivalent Deposited in Fuel Bed Due to Reforming
From oil analysis, one gallon of oil contains:

8.0 lb/gal × 89% carbon = 7.12 lb carbon

$$\frac{7.12 \text{ lb carbon}}{88\% \text{ carbon in coke}} = 8.10 \text{ lb coke equivalent/gal of oil}$$

0.79 gal oil reformed × 8.10 lb coke equivalent/gal = 6.4 lb coke equivalent deposited in fuel bed due to reforming

Blue Gas Composition.

The blue gas contains, per cent:

CO_2	4.0	from original analysis
O_2	1.1	from original analysis
N_2	12.9	from original analysis
CO	23.2	from original analysis
H_2	25.5	from oxygen balance calculation
CH_4	0.5	from assumption 4 = 25.5% × 0.02.
Total	67.2%	blue gas, or 672 cu ft blue gas/MCF.

Btu Contributed by Blue Gas.

CO	232 cu ft/MCF × 317 Btu/cu ft	= 74,000
H_2	255 cu ft/MCF × 321 Btu/cu ft	= 82,000
CH_4	5 cu ft/MCF × 1005 Btu/cu ft	= 5,000
Total		= 161,000 gross Btu/MCF finished gas from the carbon-air-steam reactions

$$\text{Btu/cu ft of blue gas} = \frac{161{,}000 \text{ Btu/MCF}}{672 \text{ cu ft blue gas/MCF}} = 239 \text{ Btu/cu ft blue gas}$$

Solid Fuel Efficiency.

Solid fuel lost in tar due to free carbon = 1.15 gal tar/MCF × 1.08 sp gr × 8.33 lb/gal H_2O × 12% free carbon = 1.24 lb carbon†

† An approximation, since free carbon in tar is not all elemental carbon.

Equivalent coke lost in tar in free carbon =

$$\frac{1.24 \text{ lb carbon in tar}}{88\% \text{ carbon in coke}} = 1.41 \text{ lb}$$

Coke Balance.

6.40 lb coke equivalent from oil reformed
−1.41 lb coke equivalent out in tar

Net 4.99 lb coke equivalent due to reforming
14.76 lb coke/MCF charged thru coaling door

Total 19.75 lb coke/MCF reacted by carbon–steam–air reactions

Btu in final gas due to blue gas = 239 Btu/cu ft × 672 cu ft blue gas/MCF = 161,000

Btu/lb fuel obtained in final gas due to blue gas =

$$\frac{161{,}000 \text{ Btu}}{19.75 \text{ lb coke}} = 8150 \text{ Btu/lb}$$

$$\text{Fuel efficiency, \%} = \frac{8150 \text{ Btu/lb coke}}{14{,}500 \times 88\% \text{ carbon}} = 63.8$$

Btu Due to Oil.

Btu/MCF finished gas = 1000 × 536 Btu/cu ft = 536,000
Btu/MCF finished gas due to coke = 8150 Btu/lb × 14.76 lb/MCF = 120,000

Total Btu/MCF finished gas due to oil = 416,000

Total Btu in finished gas per gallon of oil used =

$$\frac{416{,}000 \text{ Btu}}{4.43 \text{ gal oil/MCF}} = 94{,}000$$

Oil Gas Data.

Total Btu in final gas per gallon of oil is the result of:

1. Oil to oil gas and tar reactions.
2. Oil to carbon and hydrogen and utilization of net carbon deposited in fuel bed by air–carbon and air–steam or water gas reactions.

This total is a proportionate mixture of heat contributed by oil to oil gas, and oil to H_2 and carbon, the carbon being converted to blue gas. Oil to oil gas reactions occur mainly on the up-run portion of the cycle and the remainder on the back-run part.

Estimation of Allocation.

Btu contributed by reforming
(Btu from H_2) 132 cu ft H_2/MCF from oil × 321 Btu/cu ft = 42,300 Btu
8150 Btu/lb coke to blue gas × 4.99 lb coke eq. from oil = 40,700 Btu

Total Btu/MCF contributed by reforming = 83,000 Btu/MCF

Gasification value obtained per gallon oil reformed =

$$\frac{83{,}000 \text{ Btu/MCF}}{0.79 \text{ gal/MCF actually reformed}} = 105{,}000 \text{ Btu/gal}$$

Btu contributed from oil due to oil gas reaction = 416,000 total Btu/MCF due to oil − 83,000 Btu/MCF due to reforming = 333,000 Btu

Gasification value of oil to oil gas =

$$\frac{333{,}000 \text{ Btu from oil gas reactions}}{4.43 \text{ total oil/MCF} - 0.79 \text{ gal/MCF reformed}} = 91{,}500 \text{ Btu/gal}$$

Oil Gas Constituent Values

From gas analysis, illuminants	= 9.7%	=	97 cu ft/MCF	49.5%
From gas analysis, C_2H_6	= 3.0	=	30	15.3
CH_4 = 7.4% − 0.5% from w.g. reactions	= 6.9	=	69	35.2
Oil gas	= 19.6%	=	196 cu ft/MCF	100.0%

$$\text{Btu/cu ft of oil gas} = \frac{333{,}000 \text{ Btu from oil gas reactions}}{196 \text{ cu ft oil gas}} = 1695$$

Cu ft oil gas/gal =

$$\frac{196 \text{ cu ft/MCF}}{4.43 \text{ total oil/MCF} - 0.79 \text{ gal/MCF reformed}} = 53.8$$

Btu/Cu Ft of Illuminants

Btu due to C_2H_6 = 30 cu ft/MCF × 1764 Btu/cu ft = 53,000 Btu
Btu due to net CH_4 = 69 cu ft/MCF × 1005 Btu/cu ft = 69,300 Btu

Total Btu due to saturates/MCF = 122,300 Btu

Total Btu due to hydrocarbon gases = 333,000
Minus Btu due to saturate gases = 122,300

Btu due to illuminants = 210,700

$$\text{Btu/cu ft of illuminants*} = \frac{210{,}700 \text{ Btu}}{97 \text{ cu ft of illuminants}} = 2175$$

Steam Decomposition.

Total H_2 produced from w.g. reactions = 255 cu ft/MCF
5 cu ft CH_4/MCF × 2 = 10 cu ft/MCF

Total H_2 = 265 cu ft/MCF

Pounds of steam decomposed =

$$\frac{265 \text{ cu ft } H_2\text{/MCF}}{21.1 \text{ cu ft } H_2 \text{ per lb steam decomposed}} = 12.55 \text{ lb/MCF}$$

$$\text{Steam decomposition, \%} = \frac{12.55 \text{ lb/MCF decomposed}}{29.3 \text{ lb steam used/MCF}} = 42.9$$

Composition of Water Gas, Blow-Run Gas, and Final Gas.

It can be shown from stoichiometrical relations that the per cent CO and per cent H_2 composition of a theoretical water gas, when no air or blast products are included, is a function of the per cent CO_2 as shown below:

$$\% \text{ CO} = \frac{100 - 3 \times \% CO_2}{2}$$

$$\% \text{ H}_2 = \frac{100 + \% CO_2}{2}$$

Where CH_4 is formed simultaneously with the water gas reactions it can be shown that:

$$\% \text{ CO} = \frac{100\,(1 + 2R) - \%\, CO_2\,(3 + 4R)}{2 + 3R}$$

$$\% \text{ H}_2 = \frac{100 + \%\, CO_2}{2 + 3R}$$

where $R = \%\, CH_4/\%\, H_2$

Assuming 4% CO_2 and $R = 0.02$ in the water gas formed,

$$\% \text{ CO} = \frac{100\,(1 + 2 \times 0.02) - 4\,(3 + 4 \times 0.02)}{2 + 3 \times 0.02} = 44.5$$

$$\% \text{ H}_2 = \frac{100 + 4.0}{2 + 3 \times 0.02} = 50.5$$

$$\% \text{ CH}_4 = 50.5\% \times 0.02 = 1.01$$

* Mainly ethylene.

Table 3-37 Summary of Results Calculated from Operating Data

Net % H_2 in finished gas due to reforming	13.2
Gallons oil/MCF reformed	0.79
Total oil reformed, %	17.9
Reforming efficiency, %	68.8
Overall gasification value obtained, Btu/gal	94,000
Gasification value obtained from reforming, Btu/gal	105,000
Gasification value from oil to oil gas, Btu/gal	91,500
Cu ft oil gas/gal, oil to oil gas	53.8
Btu/cu ft oil gas	1695
Btu/cu ft of illuminants	2175
Steam decomposition, %	42.9
Btu contributed/lb of coke	8150
Fuel efficiency, %	63.8

	Blow-run gas analysis	Water gas analysis	Average blast gas analysis
CO_2	17.5%	4.0%	15.3%
CO	6.1	44.5	9.2
N_2	76.4	50.5	75.5
CH_4	...	1.0	...
Total	100.0%	100.0%	100.0%
Btu/cu ft	19.0	313	29

Oil gas analysis (hydrocarbons)		Composition of final gas	
Ill.	49.4%	Air	5.3%
CH_4	35.3	Blast gas	11.4
C_2H_6	15.3	Water gas	50.5
		Oil gas	19.6
		Excess H_2 (from reforming)	13.2
Total	100.0%	Total	100.0%
Btu/cu ft	1695	Btu/cu ft	536

Additional fuel required for oil gasification: Coke, lb/MCF	1.45
Air/MCF 239-Btu blue gas, cu ft	1470
Equivalent air/MCF 295-Btu blue gas, cu ft	1820

Table 3-38 Operating Data for Standard Hand-Clinkered Water Gas Sets*

Generator OD, ft	Generator ID, in.	Grate area, sq ft	Capacity, MCF/day	Generator air rate, cfm	Fuel, continuous operation, lb/MCF
4	28	4.28	260	735	29.0
5	38	7.88	550	1,575	29,0
6	50	13.64	950	2,740	28.0
7	62	20.97	1500	4,360	28.0
8	74	29.87	2100	6,190	28.0
9	86	40.34	2800	8,370	27.0
10	98	52.38	3600	10,900	27.0
11	108	63.62	4500	13,850	27.0
12	120	78.54	5600	17,370	27.0

* Courtesy United Engineers & Constructors, Inc.

Note: The above is on the basis of a 530 Btu and 0.65 sp gr make gas, using a good gas oil with a gasification value of 105,000 Btu/gal when used in the sets. Fuel used: Coke, containing 90 per cent combustible as charged.

Table 3-39 Heat Balance of 11 Ft Diam Hand-Clinkered Carbureted Water Gas Set with Waste Heat Boiler*

	Heavy oil operation, 1934, Btu/MCF	%	Gas oil operation, 1928, Btu/MCF	%
Heat Input				
Generator fuel				
Heat of combustion	164,000	19.7	323,000	39.4
Sensible heat	50	0.0	100	0.0
Enriching oil				
Heat of combustion	639,000	76.7	435,000	53.1
Sensible heat	2,700	0.3	0	0.0
Steam				
Total heat	22,800	2.7	58,500	7.1
Blast air				
Sensible heat in dry air	1,230	0.2	0	0.0
Sensible heat in moisture	30	0.0	0	0.0
Feed water				
Total heat	3,100	0.4	3,000	0.4
Total Input	832,910	100.0	819,600	100.0
Heat Output				
In gas				
Heat of combustion	528,700	63.5	530,000	64.7
Sensible heat	23,800	2.9	23,000	2.8
In tar				
Heat of combustion	133,000	16.0	100,000	12.3
Latent heat of vaporization	1,000	0.1	900	0.1
Sensible heat	5,600	0.7	4,000	0.5
In drip oil				
Heat of combustion	10,500	1.3	8,700	1.1
Latent heat of vaporization	100	0.0	100	0.0
Sensible heat	600	.0	200	.0
In blast products				
Heat of combustion for unburned stack gases	2,900	0.3	3,000	0.4
Sensible heat	18,100	2.2	14,500	1.8
In waste heat boiler				
Total heat of steam	24,100	2.9	24,100	2.9
In undecomposed steam				
Total heat	13,000	1.6	43,200	5.3
In water vapor				
Moisture in generator fuel	400	0.1	600	0.0
Moisture in blast air	490	.1	400	.0
Combustion of hydrogen from fuel	440	.0	3,900	0.5
In refuse				
Heat of combustion for unburned carbon	2,430	.3	14,600	1.8
Sensible heat of refuse	170	0.0	1,000	0.1
Radiation and unaccounted for	67,580	8.1	46,900	5.7
	832,910	100.0	819,600	100.0
Gas conversion efficiency		63.5		64.7
Total efficiency (with waste heat boiler)†		83.6		80.9

* Courtesy Public Service Electric & Gas Co.

† Total of heats of combustion in gas, tar, and drip oil, and total heat of steam in waste heat boiler.

Water gas composition will then be:

CO_2	4.0%	
CO	44.5 × 317 Btu/cu ft	= 141 Btu
H_2	50.5 × 321 Btu/cu ft	= 162
CH_4	1.0 × 1005 Btu/cu ft	= 10
	100.0%	313 Btu/cu ft

Relation of Theoretical Water Gas Analysis to Actual Blue Gas Analysis.

Actual H_2 in blue gas analysis = 25.5%
H_2 in theoretical w.g. = 50.5%
Relating factor = 25.5%/50.5% = 0.505

Amount from theoretical water gas analysis	Original blue gas analysis	Difference due to air and producer air gas*
CO_2 4.0% × 0.505 = 2.0%	4.0%	CO_2 2.0%
CO 44.5 × .505 = 22.5	23.2	CO 0.7
H_2 50.5 × .505 = 25.5	25.5	H_2 0.0
CH_4 1.0% × 0.505 = 0.5	0.5	CH_4 0.0
Total water gas = 50.5%		
O_2	1.1	O_2 1.1
N_2	12.9	N_2 12.9%
	67.2%	

* Same as blow-run gas.

Air and N_2 in Producer Air Gas* Mixture
1.1% O_2 × 4.8 = 5.28% Air
1.1% O_2 × 3.8 = 4.18% N_2
Producer Air Gas*
Air will be:

			Analysis, %	
CO_2	2.0%	= 2.0%	CO_2	17.5%
O_2	1.1 − 1.1%	= 0.0	O_2	0.0
CO	0.7	= 0.7	CO	6.1
N_2	12.9% − 4.18% N_2	= 8.7	N_2	76.4
		11.4%		100.0%

Blow-Run Gas.

Btu/cu ft = 6.1% CO × 317 Btu/cu ft = 19 Btu/cu ft

Final Gas Composition.

Air	5.3%	
Blow-run gas	11.4 × 19 Btu/cu ft	= 2 Btu
Water gas	50.5 × 313	= 158
Oil gas	19.6 × 1695	= 332
Excess H_2 (from reforming)	13.2 × 321 Btu/cu ft	= 43
	100.0%	535 Btu Total

Carbon Balance from Blue-Gas Composition.

	Moles	
CO_2	4.0 × 12	= 48 lb
CO	23.2 × 12	= 278
H_2	25.5	
CH_4	0.5 × 12	= 6
O_2	1.1	
N_2	12.9	
	67.2 moles	332 lb carbon/67.2 moles of blue gas

* Same as blow-run gas.

67.2 moles × 380 cu ft/mole = 25.6 MCF

Carbon/MCF of blue gas = 332 lb/25.6 MCF = 13.0 lb

Coke equivalent in blue gas $= \frac{13.0 \text{ lb carbon} \times 672 \text{ MCF blue gas}}{88\% \text{ carbon in coke}} = 9.9$ lb

Total fuel = 19.75 lb coke equivalent
−9.90 lb coke equivalent in blue gas

9.85 lb/MCF

Difference of 9.85 lb/MCF must be accounted for in the blast gas, lost in clinker, blown over and not recovered from carburetor and superheater, or blown out of set as fines.

Fuel Required for Oil Gasification.

Good blue water gas production should require about 100 lb of combustible per MMBtu produced. Therefore, a 239-Btu water gas would require 23.9 lb of carbon, or 23.9 lb÷

Table 3-40 Steam Consumption in Water Gas Plant[21] with Accumulator

Unit	Consumption, lb/MCF
High-Pressure Steam Requirements	
Gas compressors, including extraction*	18.25
Exhausters	6.40
Blowers	19.20
Circulating pump	1.87
Hydraulic pump	1.27
Forced-draft fan	2.56
Boiler-feed pump	2.66
Stoker engines	0.393
Soot blowers	0.040
Salt-water condenser pumps	9.75
Jacket water pump	0.788
Gas oil pump	1.18
Tar pump	0.393
Low-pressure fans	3.35
Solution pumps	3.15
Auxiliary fire pump	0.098
Tar still	1.08
Conditioner	0.640
Heating holders	0.049
Live steam to accumulators	1.97
Total accounted for	75.10
Radiation, trap losses, etc.	7.20
Total	82.30
Low-Pressure Steam Requirements	
Basis: 24,400 MCF/day production and sendout.	
Low-Pressure Steam Produced	
From prime movers and water jackets	63.73
Live steam make-up	1.97
Total low-pressure steam	65.70
Low-Pressure Steam Usage	
Steam to sets	46.90
Liquid purification solution heater	2.95
Boiler feed water heater	12.88
Building heating	1.00
Radiation (trap losses) and unaccounted for	1.97
Total low-pressure steam used	65.70

* All gas is compressed to 3 to 15 psig on leaving plant. Low-pressure steam is extracted from low-pressure stages of turbo-compressors to supply make-up steam to accumulators automatically.

88% carbon = 27.1 lb coke equivalent per MCF 239-Btu blue gas. Since blue gas comprised 67.2 per cent of the total gas, fuel for its production would be 67.2% × 27.1 lb = 18.2 lb coke.

Allowing 0.5 lb fuel lost in clinker, blown over material, etc., coke to be accounted for would be 19.75 − 0.5 = 19.25 lb. Since coke required for blue gas production is 18.2 lb, the difference, or 1.05 lb, is required to supply additional heat for oil gasification.

Average Blast Composition.

Stoichiometrically, it can be shown that in the carbon–air reaction the carbon consumed per MCF of reacting air is expressed as a function of the per cent CO_2 in the blast gas as:

$$\text{Lb C/MCF air} = 13.15 - 0.315 \times \%\ CO_2$$

or

$$\%\ CO_2 = \frac{13.15 - \text{lb C/MCF air}}{0.315}$$

Since 990 cu ft air was reported in the results and we have (9.85 − 0.5) lb of coke × 88% carbon in fuel = 8.22 lb carbon to be accounted for in the air–carbon reaction, the CO_2 percentage will be:

$$\frac{13.15 - (8.22/0.990)}{0.315} = 15.3\%\ CO_2$$

The per cent CO is derived from the stoichiometrical relation for the air–carbon reaction as:

$$\%\ CO = 34.45 - 1.65 \times \%\ CO_2$$

$$= 34.45 - 1.65 \times 15.3 = 9.2$$

% N_2 is obtained by difference, or 75.5% N_2

The average blast gas analysis will then be:

CO_2	15.3%
CO	9.2
N_2	75.5
	100.0%

The average blast gas analysis differs from that of the blast gas included in the blue gas, because that blast gas was obtained mostly from the air purge portion of the cycle when fire temperature is lowest.

Air/MCF of 239 Btu blue gas =

$$\frac{\text{990 cu ft/MCF finished gas}}{\text{67.2\% blue gas}} = \text{1470 cu ft}$$

The equivalent air/MCF of 295-Btu blue gas would be:

$$1470 \times (\text{295 Btu/239 Btu}) = \text{1820 cu ft}$$

For a summary of the results obtained from operating data, see Table 3-37.

Thermal Decomposition of Hydrocarbons in Carbureted Water Gas Production

At temperatures prevailing in a water gas set, carbureting oils are split into olefins and paraffins* of lower molecular weight.[19] Chain molecules are cracked at each link, several splits occurring simultaneously with subsequent reactions which depend entirely upon the prevailing atmosphere. A high hydrogen concentration leads to hydrogenation of unsaturated hydrocarbons.

According to I.G.T. Bulletin No. 9,[20] regardless of the character of the oil used, the gas as derived will consist largely of methane, ethane, ethylene, and propylene, with varying quantities of hydrogen, depending on composition of the cracking atmosphere and severity of cracking.

Except for benzene and toluene, which are present in the gas phase in only limited quantities, all these gaseous hydrocarbons have low carbon-to-hydrogen weight ratios; any hydrocarbons having relatively higher C/H ratios appear as condensables and in the tar. This recovery of low molecular weight, low C/H ratio, stable, gaseous hydrocarbons in the finished gas is the enriching value measure of the carbureting oil used. Due to the severity of cracking conditions, distribution of products between gas and tar depends largely on the C/H ratio of the oil and on the composition and pressure of the cracking atmosphere.[5]

Studies of effects of cracking temperature and residence time on oil gasification for five oils (having C/H ratios ranging from 6.39 to 8.47) in a blue gas atmosphere resulted in these conclusions:[5]

1. There is a residence time for optimum enriching value of a given oil at otherwise constant cracking conditions.
2. Optimum enriching values for higher aromaticity oils are obtained at higher temperatures. The C/H ratio alone may not characterize this trend over a wide range of oil types. Other factors, such as boiling range and molecular weight, may need consideration.
3. Oil gas yields follow the enriching values approximately, except in the range of 1-sec residence time. This is due to production in this region of very low paraffin/unsaturates ratio (high heating value) oil gases which give relatively high thermal but low volume yields.
4. Carbon formation in excess of minimum carbon deposition corresponding to the Conradson carbon residue of an oil is related to dehydrogenation reactions. When net hydrogen produced from the oil increases, carbon is formed. For all oils some irreversible dehydrogenation forms carbon under severe cracking conditions, even if sufficiently high partial pressures of hydrogen exist in the cracking atmospheres. Some net hydrogen formation thus occurs.

Selection of Oils for Carburetion

When enriching oil is admitted during the up-run to the top of the generator or is sprayed onto the generator fire, the carbon or coke formation will exceed the *Conradson carbon residue*, the amount of excess depending on fuel bed temperature and residence time. This carbon, cracked out of the oil, replaces an equivalent portion of solid fuel. Thus, when heavy oils are used, generator fuel requirements may be reduced substantially by admitting the oil to the top of the generator. The relationship of enriching value and C/H ratio still holds under these conditions. In reforming, when enriching oil is admitted during the back- or down-run to the top of the generator or is sprayed onto the generator fire, the amounts of carbon deposited approach the total in the oil, this deposited carbon again acting as a substitute for generator fuel. The enriching value of the oil is distorted by the partial recovery of carbon deposited in the generator fire as blue gas. The C/H ratio retains its value as a measure of the product distribu-

* Appreciable amounts of aromatics are also formed, especially when the proportion of olefins is high.

Table 3-41 Characteristics of Typical Carbureting Oils and Some Practical Operating Results for Various Types of Sets

	Hand clinkered							Mechanical grate						Hand clinkered		Mechanical grate	
	11 ft OD					12 ft OD		10 ft ID				9 ft ID					
	Sun Oil Co. Bunker C (1)	Sun Oil Co. Bunker C (2)	Sun Oil Co. Bunker C (3)	Atlantic Ref. Co. (4)	Tide-water Bunker C (5)	Esso Bunker C (6)	Cycle gas oil (7)	(8)	(9)	Caripito Enrich. (10)	(11)	(12)	Atlantic Ref. Co. (13)	7 ft 10 in. ID, (14)*	11 ft OD, Sun Oil Co. Bunker C (15)*	9 ft ID, Atlantic Ref. Co. (16)*	10 ft ID, Residuum (17)*
Sp gr at 60 F	0.968	0.9346	0.9619	0.967	0.9545	0.9600	0.8807	...	0.9642	0.9315	0.845	0.9749	0.846	...	0.969	0.876	0.9196
API gravity at 60 F	14.7	19.9	15.6	14.8	16.75	15.9	29.2	22.3	...	20.4	...	...	35.4	14.6	14.5	30.0	...
Visc. sec. Say. Univ. at 100 F	...	...	...	...	2875	...	40	...	18 at 122 F	...	...	...	42 at 70 F	...	...	70	31 at 122 F
Visc. sec. Say. Fur. at 122 F	291	47	359	40	122	154	...	22.8	...	50	...	118.5	...	...	298	11	...
Carbon residue, wt %	7.0	4.6	7.9	6.7	4.36	8.49	0.33	6	7.16	4.65	...	9.2	...	5.3	8.0	0.06	3.4
Sulfur, % by wt	0.65	0.55	0.81	0.63	0.53	2.19	1.09	...	...	1.63	...	0.62	...	0.67	0.53	0.50	0.88
Ash content, % by wt	...	0.58	0.68	...	...	0.09	...	...	...	0.03	0.001	...	...	0.03	0.69	Trace	...
Pour point, °F	...	35	35	20	50	...	−30	...	...	80	...	...	...	10	35	50	...
Flash point, °F	296	258	202	174	265	200+	180	...	...	200+	...	...	...	...	266	220	...
Distillation, % by vol.																	
Initial boiling point, °F	...	375	450	370	466	...	380	...	...	...	...	...	325	...	475	490	...
300–400 F	...	0.1	Nil	...	Nil	...	0.9	...	...	...	...	...	9.00	...	Nil	Nil	...
400–500 F	...	0.9	1.5	...	Nil	...	36.5	...	...	...	...	...	57.5	...	1.2	0.5	...
500–600 F	...	11.0	42.0	...	0.8	...	43.2	...	...	...	...	...	28.5	...	57.0	33.4	...
600–700 F	...	80.0	42.9	85.5	7.3	...	16.6	...	...	...	...	...	3.5	...	26.3	58.0	...
Residuum	...	7.0	10.0	14.5	8.3	...	2.8	...	...	...	...	...	1.5	...	7.5	0.1	...
Maximum temp, °F	...	720	720	658	943	...	684	...	...	...	...	...	640	...	725	725	...
Cracking test																	
Cracking temp	...	1500	1500	1500	1500	1500	1500	...	...	1500	...	1500	1350	...	1500	1500	1500
MBtu/gal oil	...	100.0	97.0	97.5	101.0	100.2	100.7	98–106	100	108.2	114.5	103.4	105.9	94.4	101.8	105.0	110.2
Btu/cu ft oil gas	...	1630	1650	2060	2263	2412	2222	...	...	2185	...	1484	1960	...	1475	1681	221
Cu ft oil gas/gal oil	...	59	56	47.5	44.64	41.6	45.2	...	...	49.5	...	69.68	60	...	69	65	49.79
Practical results																	
Make/running hr, MCF	108	250	235	242.0	215.0	225.1	151.2	415	320	587.9	278.7	330.9	308.5	210.9	153	386	331.3
Htg. value, Btu/cu ft	535	536	535	532	539	540	545	500	540	539	521	539	528	579	716	680	567.6
Sp gr of gas	0.671	0.66	0.68	0.697	0.718	0.73	0.70	0.7	0.67	0.74	0.612	0.695	0.65	0.71	0.691	0.696	0.643
H_2S, grain/CCF	120	120	150	107	153	228	100	...	...	200	...	...	90	...	120	100	...
Kind of fuel	Coke	Coke	Coke	Coke	Coke	Coke	Coke	Coke	Coke	Coke	Anth.	Coke	Coke	Coke	Coke	Coke	Coke
Fuel, lb/MCF	10.3	14.76	13.7	15.5	20.83	20.9	30.0	15.0	20.7	16.6	31.26	20.0	27.0	15.5	17.3	16.8	25.4
Oil, gal/MCF	4.42	4.43	4.34	4.64	3.80	4.40	3.79	4.5	4.25	4.34	2.50	3.81	2.85	5.27	5.75	5.68	3.19
Tar, gal/MCF	...	1.15	0.91	1.34	0.94	1.25	0.86	1.0	1.05	1.13	0.225	...	0.40	1.28	1.48	1.33	0.96
Light oil, gal/MCF	...	0.10	0.08	...	...	0.22	...	...	...	0.21	...	...	...	...	0.12	0.11	0.25
Oil for reforming, % of total	46	26	25	7.5	6	0	0	10–15	0	0	0	0	0	22.7	5	15	0
Gas analysis																	
CO_2	...	4.0	3.7	4.8	6.1	4.4	4.0	5.8	5.0	4.5	4.4	6.6	4.8	5.3	4.2	2.7	5.6
Ill.	...	9.7	9.3	9.6	9.6	9.5	9.1	9.2	9.1	10.1	8.4	8.6	9.1	9.9	17.8	13.5	9.5
O_2	...	1.1	0.8	0.3	0.6	0.9	0.9	1.1	1.0	1.3	0.5	0.6	0.7	0.7	0.9	0.6	0.1
CO	...	23.2	21.2	25.3	23.2	25.8	28.3	17.0	24.7	22.5	31.9	27.0	31.2	20.4	22.5	19.7	38.6
H_2	...	38.7	38.5	26.4	30.6	27.4	27.1	30.1	31.0	25.1	40.2	32.7	35.1	27.5	33.0	29.8	25.7
CH_4	...	7.4	11.0	16.1	12.9	14.2	14.3	15.3	13.6	13.4	9.7	13.2	10.0	17.9	12.6	20.5	12.8
N_2	...	12.9	13.6	17.5	17.0	17.8	16.3	21.5	15.6	23.1	4.3	11.3	9.1	18.3	5.6	11.2	7.7
C_2H_6	...	3.0	1.9	...	...	...	...	...	...	...	0.6	...	...	...	3.4	2.0	...

* Above normal Btu.

Koppers reverse air blast process: (5), (8), and (9). Semet-Solvay reverse flow process: (11), (12), (14), and (17). U.G.I. heavy oil process: (1), (2), (3), (4), (10), (15), and (16). Operation from up- and down-run sets: (7). Back-run operation: (6) and (13).

tion in terms of solid carbon deposited in the fuel bed and of reformed oil gas addition to the make gas.[20]

Sulfur Tolerance in Gas Oils*

The amount of sulfur that may be tolerated in oils for manufacture of carbureted water gas depends on the amount of hydrogen sulfide absorption by the tar, the sealing and scrubbing water, and the plant equipment design.

A formula used successfully for approximating the H_2S content of crude gas quickly when oil sulfur content by weight is known follows:

$$\text{Grains } H_2S/\text{CCF make gas} = (\%\ S \text{ in oil} \times 100) + 53$$

For most plants, this formula will be accurate within ±15 to 20 grains. For closer results, a constant representing a plant design and operation factor may be used.

The formula assumes that 90 grains of H_2S result from blue gas operation. As blue gas is approximately 59 per cent of the finished carbureted water gas (Pacific Coast method, not including combustion products and air), the final gas contains 53 grains H_2S per CCF.

An approximate equation for each plant may be computed by plotting known results of sulfur against hydrogen sulfide.

X axis = $100 \times \%\ S$
Y axis = grains S/CCF make gas
Using the formula for a straight line,

$$Y = MX + A$$

where: M = slope of line = $(Y_2 - Y_1)/(X_2 - X_1)$

A = Y intercept at $X = 0$

Miscellaneous Operating Data

Tables 3-38 to 3-41 show miscellaneous operating data covering: (a) set capacity of various size sets, (b) heat balance for a water gas set equipped with waste heat boiler, (c) steam consumption in a water gas plant, and (d) characteristics of typical carbureting oils and some practical operating results.

REFERENCES

1. Haslam, R. T., and others. "Water-Gas Reactions." *Ind. Eng. Chem.* 15: 115, Feb. 1923.
2. Pexton, S., and Cobb, J. W. "Gasification of Coal in Steam." *Gas J.* 167: 161, July 2, 1924.
3. Thiele, E. W., and Haslam, R. T. "The Mechanism of the Steam-Carbon Reactions." *Ind. Eng. Chem.* 19: 882–7, Aug. 1927.
4. Brender a Brandis, G. A., and Le Nobel, J. W. "The Reactivity of Coke." *Gas, The Hague* 47: 37–47, 1927.
5. Haslam, R. T., and Russell, R. P. *Fuels and Their Combustion,* p. 162, 611. New York, McGraw-Hill, 1926. Also: Emmett, P. H., and Schultz, J. F. "Equilibrium in the System Co-H_2O-CoO-H_2. Free Energy Changes for the Reaction CoO + H_2 = Co + H_2O and the Reaction Co + $\frac{1}{2}O_2$ = CoO." *J. Am. Chem. Soc.* 51: 3249, Nov. 1929. Also: Emmett, P. H., and Schultz, J. F. "Equilibrium in the System Co-CO_2-CoO-CO. Indirect Calculation of the Water Gas Equilibrium Constant." *J. Am. Chem. Soc.* 52: 1782, May 1930.
6. Von Fredersdorf, C. G. *Reactions of Carbon with Carbon Dioxide and with Steam.* (I.G.T. Research Bull. 19) Chicago, Ill., I.G.T., 1955.
7. Foster, J. F., and Vorum, D. A. *Mechanism of the Water-Gas Reaction.* (Battelle Rept.) New York, A.G.A., 1951.
8. Marson, C. B., and Cobb, J. W. "Influence of the Ash Constituents in the Carbonization and Gasification of Coal, Part II: Gasification of Special Cokes in Steam." *Gas J.* 175: 882, Sept. 29, 1926.
9. Vignon, L. "Water-Gas." *Ann. Chim.* 15: 42–60, 1921.
10. King, J. G., and Shaw, R. N. *Comparisons of Some Methods of Running Water-Gas Plant.* (Fuel Research Board Tech. Paper 6) London, Gt. Brit. Scientific and Ind. Res. Dept., 1923.
11. Inst. of Gas Engrs. "Sixth Report of the Research Sub-Committee of the Gas Investigation Committee." *Gas J.* 154: 619, June 15, 1921.
12. Vandaveer, F. E., and Parr, S. W. "Use of Oxygen in the Manufacture of Water Gas." *Ind. Eng. Chem.* 17: 1123, Nov. 1925.
13. Mitchell, R. F. "Oxygen-Steam Producer Blast." *Can. Chem. & Proc. Ind.* 30: 34–42, Aug. 1946.
14. Newman, L. L. "Recent European Developments in the Use of Oxygen in Gas Manufacture." (PC-51-1) *A.G.A. Proc.* 1951: 585–90.
15. Blatchford, J. W. "Production of Water Gas With Tonnage Oxygen." (PC-50-16) *A.G.A. Proc.* 1950: 652.
16. Lyons, C. J., and Batchelder, H. R. "Tonnage Oxygen Production, 1956." (CEP-56-16) *A.G.A. Proc.* 1956: 606–21.
17. Wagman, D. D., and others. *Heats, Free Energies, and Equilibrium Constants of Some Reactions Involving O_2, H_2, H_2O, C, CO, CO_2, and CH_4.* (Natl. Bur. of Standards R.P. 1634) Washington, D. C., 1945.
18. Terzian, H. G. "Method of Determining the Relation of Generator Fuel, Oil and Tar in the Evaluation of Heavy Oil for Carbureted Water Gas." *A.G.A. Proc.* 1936: 848.
19. Whitaker, M. C., and Rittman, W. F. "Thermal Reactions in Carbureting Water Gas." *Ind. Eng. Chem.* 6: 383, May 1914; 472, June 1914.
20. Pettyjohn, E. S., and Linden, H. R. *Selection of Oils for Carbureted Water Gas.* (I.G.T. Research Bull. 9) Chicago, Ill., 1952.
21. Murdock, W. J. "Water Gas." *A.G.A. Proc.* 1928: 1301.

* Courtesy Public Service Electric & Gas Co., N. J.

Chapter 6

Fuel and Synthesis Gases from Gaseous and Light Liquid Hydrocarbons

by Donald A. Vorum

This chapter is concerned essentially with the use of light hydrocarbons, essentially paraffinic, ranging from methane thru petroleum naphtha, as feed stocks for production of fuel and synthesis gases. There was little application and little immediate future prospect in 1964 for methane decomposition processes for American gas utility companies, and beyond petrochemicals, little immediate future application was anticipated for CH_4 decomposition. However, in the more distant future, there is likely to be increased U. S. interest in processes for converting hydrocarbons heavier than methane into acceptable substitutes for or supplements to natural gas. Abroad, the situation is quite different, in that the increased use of either domestic or imported hydrocarbons will call for more and more conversion systems as long as the standard heating value of the order of 500 Btu per cu ft of product gas is maintained.

PROCESSES FOR CONVERSION OF HYDROCARBONS

Hydrocarbons are converted to fuel and synthesis gases by one of the following three general processes or a combination thereof:

1. Cracking, or thermal decomposition
2. Reforming, or reaction with steam
3. Partial combustion, or reaction with oxygen as such or as contained in air

Cracking. Thermal decomposition changes feed stock hydrocarbon molecules into product molecules of lower average molecular weight, saturated and unsaturated (with H_2), plus usually appreciable amount of hydrogen and carbon. Thermal decomposition is appreciable at temperatures as low as 1000 F, but commercial operation is more likely to be at 1400 F or higher. The extent of cracking and product molecule distribution varies with system temperature; composition of the feed stock, presence of steam, if any, time of exposure, and type of equipment. In simplest form, or when carried to extreme limits, this process can be represented as:

$$C_nH_{2n+2} \rightarrow nC + (n + 1)H_2 \qquad (1)$$

The reaction system is quite complex, passing thru various stages with many intermediate short-lived products, and with paraffinic feed stocks, ultimately resulting in other paraffinic compounds, olefins, diolefins, and other hydrocarbons, together with C and H_2. For details of the possibilities, see Fig. 3-28, which summarizes mechanisms and products of cracking in tubular furnaces.[1]

Much of the cracking of hydrocarbons for commercial-scale gas utility work has been conducted in equipment which operates cyclically, although work has been done in recent years on continuous processes.[1]

Reforming.* In this gas-making process a hydrocarbon reacts with steam, usually in the presence of a catalyst. The reaction is a relatively high-temperature one, usually requiring about 1000 F for attainment of significant rates, and often being finished at temperatures of the order of 1800 F. With the introduction and decomposition of steam, new molecular species and oxides of carbon are formed (rather than free carbon), and the number of chemical reactions which can at least be presumed grows. There has not been unanimity on the nature or extent of reactions, but those commonly considered for reforming of saturated hydrocarbons are:

$$C_nH_{2n+2} + nH_2O \rightleftharpoons nCO + (2n + 1)H_2 \qquad (2)$$

$$C_nH_{2n+2} + 2nH_2O \rightleftharpoons nCO_2 + (3n + 2)H_2 \qquad (3)$$

The presence of steam, oxides of carbon, hydrogen, and frequently solid carbon forces consideration of the following additional reactions:

$$CO + H_2O \rightleftharpoons CO_2 + H_2 \qquad (4)$$

$$CO_2 + CH_4 \rightleftharpoons 2CO + 2H_2 \qquad (5)$$

$$2CO \rightleftharpoons CO_2 + C \qquad (6)$$

$$C + H_2O \rightleftharpoons H_2 + CO \qquad (7)$$

$$C + 2H_2O \rightleftharpoons 2H_2 + CO_2 \qquad (7a)$$

Further, there is the prospect that the CO and H_2 formed in decomposition of heavier hydrocarbons react to form, for example, methane, i.e.,

$$CO + 3H_2 \rightleftharpoons CH_4 + H_2O \qquad (8)$$

Partial Combustion. Here hydrocarbons react with enough oxygen as such, or with that of air, to produce exten-

* The term *reforming* is also used in the petroleum refining field to describe reaction of a hydrocarbon with additional hydrogen, and has been used in the gas industry to refer to thermal cracking.

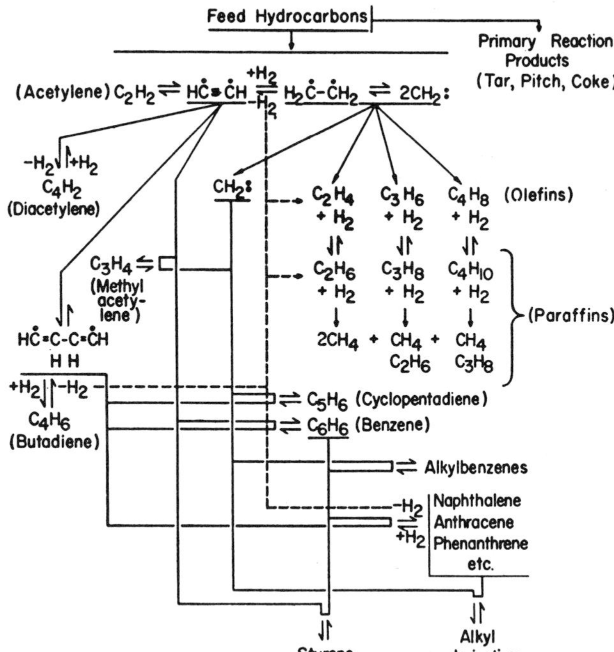

Fig. 3-28 General mechanism of high-temperature vapor phase cracking of hydrocarbons.[1] Dots above C indicate free radicals.

sive composition change, but insufficiently to give complete conversion to the end products CO_2 and H_2O. In such a process, the heat to support the reaction is supplied by combustion of only a part of the feed stock. Steam is commonly present either as a supplementary starting material or as a product. The basic reactions, at least in molecular terms, for saturated hydrocarbons are:

$$C_nH_{2n+2} + \frac{n}{2} O_2 \rightleftharpoons NCO + (n + 1)H_2 \qquad \textbf{(9)}$$

and

$$C_nH_{2n+2} + \left(\frac{3n + 1}{2}\right)O_2 \rightleftharpoons nCO_2 + (n + 1)H_2O \qquad \textbf{(10)}$$

The products are divided among those shown according to circumstances of (1) system temperature, (2) the amount of oxygen, (3) the addition of steam, if any, as well as (4) catalytic surfaces. These again are relatively high temperature reactions, usually the highest of any considered here, and usually resulting in the formation of solid carbon and, at least temporarily, of the other products of cracking of hydrocarbons. In summary, in all but the simplest systems there will usually be a considerable variety of products formed:

1. Carbon, largely as a result of cracking reactions;
2. Hydrocarbons, representing the results of cracking of larger molecules, residual unreacted feed stock, or the results of recombinations;
3. Oxides of carbon, resulting either from steam reforming reactions, or from partial combustion;
4. Hydrogen, resulting from any one or all of the above reactions; and
5. Steam, either as an unreacted feed material or as a product of combustion.

The foregoing reactions refer to molecular species only. There is some cracking also to form numerous short-lived free radicals. Full consideration of these latter products becomes very complicated, and beyond the scope of this handbook.

Discussion of Process Reactions

The preceding paragraphs indicated three major routes to conversion of hydrocarbons, five groups of products, and a number of major chemical reactions or types of reactions by which these products can be obtained.

There is a wide variety of considerations which can be applied to these reactions. For example, the list of reactions is deficient in that only end products are shown and only a moderate indication of equilibrium dissociation is given. Conversely, from some viewpoints, there is duplication, because stoichiometrically Reaction **3**, for example, represents a summation of Reactions **2** and **4**. Kinetically, however, the situation can be quite different. If, for example, Reaction **3** should prove to be very slow, its status in an actual system as summarizing the other reactions may not be important. It is thus necessary to consider to a reasonable degree all reactions which appear likely to represent a process system.

Temperature Considerations. Processes in this field are carried out at relatively high temperatures, although temperature ranges used may vary widely. The cracking of some petroleum fractions (Reactions **1**, **1a**, **1b**) can become appreciable at 680 to 700 F. Time requirements at low temperatures can, however, be prohibitive industrially. As an indication of this, Nelson found that in industrial processing butane decomposition at 1000 F requires 380 sec, but if the temperature is raised to 1100 F, the time requirement falls to 40 seconds.[2] Nelson showed further that the rate of cracking increases sharply with temperature, doubling at intervals of 20 to 60°F, depending on the temperature level.[3]

Reaction **4**, the "water gas shift," has long been used effectively to adjust gas compositions in the petrochemical industries at temperatures as low as 600 to 800 F (as low as 400 F by new developments). The steam hydrocarbon or reforming reactions, Reactions **2** and **3**, become appreciable at 1100 F, and in industrial practice in fired tubes are commonly carried out at temperatures of the order of 1500 to 1800 F.

Partial combustion conditions are favored at the upper end of the temperature range, approaching 3000 F.

Pressure Considerations. Conversion processes in the utility gas industry have traditionally been carried out at approximately atmospheric pressure, in keeping with the conditions of local distribution of the product. With the advent of high-pressure transmission and distribution, however, the interest in higher pressure operation has grown. Petrochemical technology, in which pressure levels have increased steadily into the range of several hundred pounds per square inch, has also influenced the trend to higher pressures. Economic advantages accrue from this development in terms of decreases in equipment size or, conversely, increased rates of throughput, and savings in equipment and energy for compression. Of general interest also is the fact that the effectiveness of a large number of the gas separation and purification processes increases with pressure.

Concerning Reactions **1** through **10**, the effect of pressure, at least on the degree of conversion feasible, depends qualitatively on the relationship between the volumes of feed materials and products. Processes which result in an increase in the total gaseous volume of the system are hindered by an increase in pressure; conversely, processes which result in a decrease in the total gaseous volume are aided by an increase in pressure.

Thermodynamics of Reactions. The science of thermodynamics has many aspects; here it will be applied to consideration of the heat effects of chemical reactions and the chemical composition and stability of systems under varying conditions of temperature and pressure. Chemical reactions result in transformations of matter from one chemical composition to another, and often indirectly from one physical state to another. These transformations often either consume or release large amounts of energy as heat which must either be supplied to (endothermic reaction) or removed from (exothermic reaction) the system, depending on the reaction(s) involved. Among the steps involving the largest heat effects are the forming or decomposing of water and carbon dioxide molecules from and to their respective elements. In each case, formation of the molecule releases heat, and decomposition requires heat. Formation of saturated hydrocarbons from their elements is likewise exothermic, but the formation of unsaturated hydrocarbons is endothermic. A distinction, however, is that the heat effects involved in forming H_2O and CO_2 molecules from the elements are generally far greater than those involved in forming hydrocarbons. For example, the heats of formation of water as steam and carbon dioxide are −104,010* and −169,288 (Table 3-36) Btu per lb-mole, respectively, while the corresponding values for methane, ethane, and propane are −32,198, −36,344, and −44,550 Btu per lb-mole, respectively. Against this background, it is possible to appraise the heat effects of the reactions discussed earlier, and of new reactions with which one may be confronted. Thus, for the following chemical transformations:

1. Process steps based on reactions such as **1** and **1a** require that moderate amounts of heat be supplied; 32,157 Btu per lb-mole in the case of decomposition of methane to carbon and hydrogen.
2. Reaction **2**, breaking both water and methane molecules, requires the supplying of 88,667 Btu per lb-mole.
3. Combustion reactions, notably exothermic, require in the case of Reaction **10** the removal of as many as 34 MBtu per lb-mole of methane.
4. The "water gas shift," Reaction **4,** is very nearly neutral thermally, releasing only 17,800 Btu when moving to the right. The reason is that water and carbon monoxide molecules are decomposed, while carbon dioxide is formed, the opposing effects nearly balancing each other.

Equilibrium Considerations. Equilibrium refers to a stable state reached by a chemical system under given process conditions when allowed unlimited time. In other terms, it is concerned with the ultimate composition of the mixture of reactants and products, again under specified conditions, particularly of pressure and temperature. The composition of the mixture at equilibrium may be described by the ratio of a function of the concentrations of the products by a function of the concentrations of the reactants. These concentrations may be in various terms, but in industrial work they are usually in "partial pressures" of the constituents. The partial pressure of a constituent is its contribution to the total system pressure; if a system at 100 psia contains 50 per cent hydrogen, the hydrogen partial pressure is 50 psia.

The above ratio is usually designated K_p, and is a constant at a given temperature; hence the designation "equilibrium constant." For Reaction 5:

$$CO_2 + CH_4 \rightleftharpoons 2CO + 2H_2$$

and

$$K_p = \frac{P_{CO}^2 \times P_{H_2}^2}{P_{CO_2} \times P_{CH_4}}$$

Kinetics of Reactions. Kinetics refers to the rates at which chemical process steps take place. The term "process steps" is used here as distinguished from "chemical reactions" to emphasize the fact that the rate of the chemical reaction as such is not always the governing aspect in the question of how rapidly a given process can be accomplished. Among the impediments to a process proceeding as rapidly as the strictly chemical aspect might permit are:

1. Rates of heat transfer thru confining surfaces such as furnace tube walls;
2. Rates of gaseous diffusion in space or to and from solid surfaces;
3. Rates of diffusion into and out of the pore structure of catalysts; and
4. Rates of adsorption onto, and desorption from, catalytic surfaces.

Kinetic data are expressed in different forms:

1. Basic data for the Arrhenius equation;
2. In terms of rate equations, such as

$$v = F(dX/dV_R)$$

in which
v = the rate of conversion of material
F = the rate at which feed is supplied to the reactor
dX/dV_R = the differential extent of conversion per unit of reactor

3. The more empirical expressions in terms of "space velocity," i.e., the amount of material which can be processed or produced per hour to obtain a given result under given conditions;
4. Data which give required contact or residence times to obtain similar results, particularly in the case of catalytic systems.

Use of expressions of the rate equation type is beyond the scope of this handbook, but is treated in a number of texts. Expressions and data for the space velocity and contact time types will be applied in subsequent paragraphs.

Those interested in further study of principles and fundamentals of systems are referred to Hougen and Watson,[4] Dodge,[5] Smith,[6] Montgomery,[7] Mayland and Hays,[8] Mungen and Kratzer,[9] and Akers and Camp.[10]

Process Data and Calculations. Most of the materials, reactions, and equilibrium data discussed here are pre-

* Minus sign signifies heat removed (exothermic reaction).

sented either in this handbook or in other engineering handbooks[11] and the current literature.[12] It is expected that the routine thermal and thermodynamic calculations of material and heat requirements and heat effects of process steps and reactions can be made from these sources. The following paragraphs will therefore be confined to the application of equilibrium considerations and such kinetic data as are available.

Equilibrium Considerations. **The law of mass action** is a powerful tool for analyzing the performance of a given system or for predicting the performance of a new system under consideration. This is because at a given temperature, pressure, and specific feed composition, there is a particular mixture composition and maximum conversion which will eventually be attained.

Equilibria and heats of reaction for many reactions of interest are given in Table 3-36. Figure 3-23 plots the equilibrium constant for the water gas shift reaction. Note that the data plotted in Fig. 3-23 are the reciprocals of the data given for the water gas shift (Reaction 4) in Table 3-36.

Example. To illustrate the use of these data, let it be assumed that a mixture of three volumes of steam and one of methane is passed over a catalyst at ten atmospheres absolute pressure (i.e., 132 psig). What degree of conversion can be expected? This problem is approached by assumptions which are reviewed and adjusted in the light of related data. For this problem the following assumptions are made:

1. The water gas shift reaction, Reaction 4, is assumed to be at equilibrium. This may not always be the case, but is a reasonable point of departure.

2. Successive assumptions are made of the degree of methane conversion until one is found which produces a gas composition fitting the equilibrium data of Table 3-36.

Table 3-42 shows the results of successive assumptions in terms of the calculated equilibrium constants compared with equilibrium data in Table 3-36. The input column shows the composition of the incoming mixture, and the succeeding columns show the probable compositions at increasing levels of conversion. For each degree of conversion there is a calculated equilibrium constant, K_p' which increases with the degree of conversion, finally greatly exceeding the actual equilibrium constant, K_p, at 98 per cent conversion. Inspection shows that the two values are very close together at 96 per cent, indicating that this column represents closely the eventual degree of conversion and product composition for the stated feed mixture and process conditions. It should be emphasized that this is the degree of conversion and composition reached eventually. In actual practice methane conversion usually falls somewhat short of this, the product composition representing a temperature somewhat below the defined outlet temperature, in this case 1600 F. The difference between this latter value as a measured outlet temperature and the temperature to which the product mixture corresponds is called the "approach to equilibrium." This approach varies widely with the process, catalyst, and conditions of operation. Evaluation or prediction of this approach to equilibrium is quite an empirical problem and is probably best undertaken in cooperation with an equipment designer or catalyst maker.

Table 3-42 Equilibrium Conversion of Methane by Reforming with Steam at 1600 F and 10 Atmospheres Absolute Pressure for Four Degrees of Conversion

	Mole per cent				
	0% (Input)	80%	90%	96%	98%
Mixture composition					
H_2	0.0	48.0	51.7	53.54	54.24
CO	0.0	9.1	10.3	11.30	11.50
CO_2	0.0	5.2	5.2	4.92	4.90
CH_4	25.0	3.6	1.7	0.67	0.34
H_2O	75.0	34.1	31.1	29.57	29.02
Total	100.0	100.0	100.0	100.00	100.00
Calculated equilibrium constant K_p'	...	82	268	873*	1860
Tabulated equilibrium constant K_p†	850	850	850	850	850

$$* \; K_p' = \frac{[(11.30 \times 10)/100] \times [(53.54 \times 10)/100]^3}{[(0.67 \times 10)/100] \times [(29.57 \times 10)/100]} = 873.$$

† Interpolated from a semilog plot of Reaction 11 in Table 3-36. $K_p = [(CO)(H_2)^3]/[(CH_4)(H_2O)]$, where the terms in parentheses are the respective partial pressures in atmospheres; Reaction 11 of Table 3-36 gives K_p at three temperatures.

These equilibrium calculations can also be used for predicting or avoiding process difficulties, particularly deposition of solid carbon. The subject is treated by Hougen and Watson.[4] It should be noted that a prediction of carbon deposition by these thermodynamic equilibria does not always mean that it will occur. Thermodynamic calculations again represent limit situations, and some aspect of process rate may govern the situation and prevent deposition.

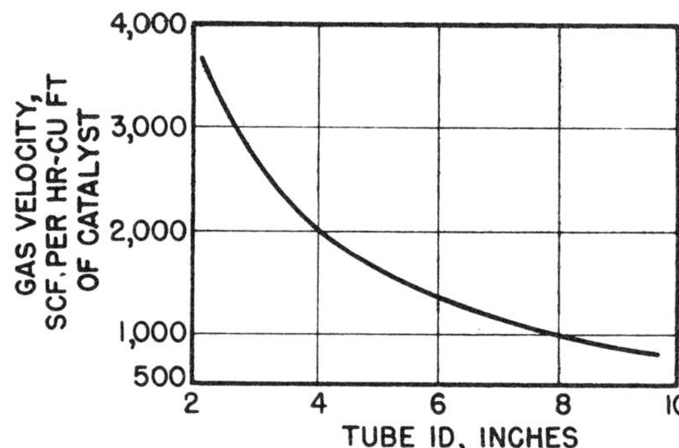

Fig. 3-29 Theoretical hydrogen space velocity vs. reformer tube diameter. (Catalysts and Chemicals, Inc., Fig. 5 of Bull. C11-26160.)

Process Rates. Kinetics, as noted earlier, refers to consideration of rate of reaction or other process steps. This science is relatively new, the data much less plentiful, and the practice less exact than in the case of thermodynamics.

For thermal cracking operations, data already cited in works such as that by Nelson[2] give ranges of time requirement. Data and references on cracking practice over a considerable range of materials are given in Table 3-53 and elsewhere in the following chapter. From photographs of the Hasche process units (discussed later), it appears that residence time is very low, of the order of a few seconds.

Kinetic considerations for fired-tube reforming installations are quite complex,[10] because they involve not only the process mixture and conditions, but also turbulence of flow, catalyst type and particle size, and diameter of the tubes containing the catalyst. Involved in all this is the fact touched on earlier that physical considerations such as the rate of heat transfer thru confining surfaces such as tube walls may actually govern the rate of an overall process, rather than the chemical step. Very helpful sources of information are the Bulletins of the catalyst manufacturers, such as Catalysts and Chemicals, Inc.* Its Bulletin C11-26160 deals with kinetics in terms of the "Theoretical Hydrogen Space Velocity, Volumes Per Volume Per Hour." "Theoretical hydrogen" refers to the maximum amount of hydrogen which can be produced by reaction of a hydrocarbon with steam and subsequent shift of all the CO formed to H_2 as well.

In the case of methane,

$$CH_4 + H_2O \rightarrow CO + 3H_2$$

and

$$CO + H_2O \rightarrow CO_2 + \underline{H_2}$$

$$\text{Total } 4H_2$$

The theoretical hydrogen is thus equal to four. As a general guide, Fig. 3-29 shows a relationship between theoretical hydrogen space velocity and tube diameter, without reference to other variables.

Note the rather marked influence of tube diameter on allowable space velocity (Fig. 3-29). By way of interpretation, a four-inch tube would allow production of 2000 SCF H_2 per hour per cubic foot of catalyst. In the case of methane, this is equivalent to reforming 500 cu ft of feed gas per hour per cubic foot of catalyst. As with other aspects of the rapidly developing field of catalysis, it appears that these values are now quite conservative. It is indicated[13] that:

1. One reforming unit is running at an outlet temperature of 1850 to 1900 F.
2. Theoretical space velocities in practice have gone beyond 5000.
3. Heat transfer rates thru tube walls as high as 26,000 Btu per sq ft-hr have been observed (15,000 to 20,000 is common, depending on the pressure of the system).

Kinetic data on partial combustion in practice are not known, but from the general behavior of oxygen it is expected that rates are very high and residence times low, conceivably fractions of a second.

INDUSTRIAL APPLICATIONS OF PROCESSES

Industrial applications of the various approaches to gas production have been numerous and varied, with two or more of the approaches often combined to the point of near inseparability. The general classification into cracking, reforming, and partial combustion is generally adhered to, however, with exceptions noted as necessary.

Cracking

Cracking or thermal decomposition goes far back into the history of the gas industry. Many of the systems employed are used to make oil gas. Cracking also played a part, probably along with some reforming, in the widespread use of water gas carburetion. As fluid fuels became available, the practice of "fuel-bed reforming" arose, in which natural gas, for example, was reformed and probably cracked by passing with steam thru the hot coke bed of a water gas generator. The gas substantially replaced the coke required; reviews of the practice are available.[14]

* 1230 South 12th Street, Louisville, Ky.

Hasche Process.[15] This cyclic, usually noncatalytic, process is used to crack or thermally decompose hydrocarbons with the heat supplied by combustion within the cracking zone. The process is carried out in an ovenlike structure built of and filled with ceramic shapes, all enclosed in a steel shell. Figure 3-30 shows the system schematically in its basic form; feed materials and products pass alternately in one direction and then the other thru the identical regenerative sections and the center combustion zone. Assuming stable operation, feeds may enter the left side of the oven, be heated in the tile mass, ignite and react in the combustion zone, finish reacting in the right tile mass, and give up heat there en route to storage. When the left tile mass cools and the right reaches an optimum temperature, the flow is reversed.

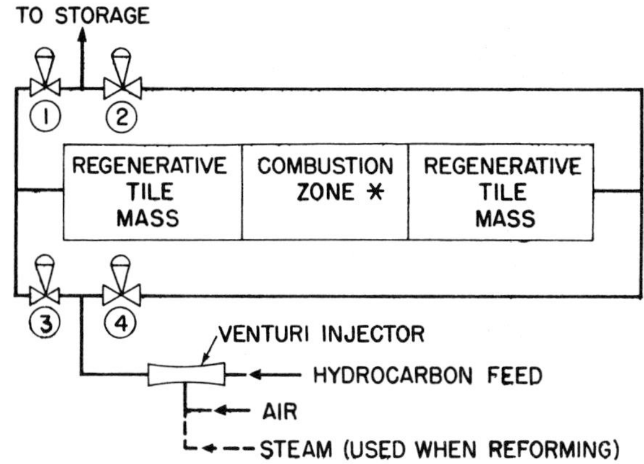

Fig. 3-30 Hasche cyclic cracking unit.

Partial Combustion Cycle	Valves open	Valves closed
Make to right	3.2	4.1
Make to left	4.1	3.2

* May be void or packed with catalyst, depending on type of gas desired.

There are two principal modes of operation in commercial use, "partial combustion" and "cyclic cracking." The latter uses less air; normally neither requires steam. From the product composition it appears that both represent thermal cracking supported by partial combustion. A third mode of operation using steam to produce synthesis gas over a catalyst is also described.[15] There have been some 75 installations in the United States and eight foreign countries.

Table 3-43 shows typical operating data. The results of simply cracking naphtha with internal heating (called reformed naphtha) and the effects of enriching the gas by addition of propane are shown. The first gas is used as a substitute for manufactured gas and the second as a supplement for natural gas.

The last two columns show the results of restricting or eliminating the air during part or all of the gas-making period and of pilot plant work on producing synthesis gas or hydrogen by cyclic reforming over a catalyst. Additional but less complete data have been published for operation on natural gas.[15] The vendor reports a range of investment costs for the

Table 3-43 The Hasche Process Operating Data[15]

	Partial Combustion		Cyclic Cracking	
	Re-formed naphtha	Propane enriched	Re-formed naphtha	Synthesis gas (pilot plant)
Product gas composition, mole %				
Hydrogen	3.6	5.4	22.0	69.5
Methane	13.0	10.3	35.4	8.8
Ethane	0.9	1.0	3.3	0
Other paraffins	0.5	23.9	2.9	0
Ethylene	15.5	14.7	17.8	0.2
Other olefins	1.5	5.9	7.3	0
Butadiene	0	0.9	0.9	0
Acetylene	0	0	0	0.2
Benzene	0	0.2	0.2	0
Nitrogen	52.6	32.9	7.3	1.5
Carbon monoxide	7.8	2.3	2.0	15.0
Carbon dioxide	4.6	2.5	0.9	4.8
Total	100.0	100.0	100.0	100.0
Feed stock				
Type	Pentane–octane	Propane & naphtha	Pentane–octane	Methane & steam
Heat. value MBtu/gal	119	92 & 119	119	...
Consumption, gal/MCF	4.2–6.5	3.7 & 5.6 8.6 & 4.5	13.4	*
Efficiency, Btu out/Btu in, %	96	98	70	83
Pressure, psig	35	35	30	5
Temperature, °F	400	20	450	0 & 300
Product				
Heat. value, Btu/cu ft	480–740	980–1300	1120–1140	370
Specific gravity (air = 1)	0.92–0.99	1.06–1.16	0.72–0.76	0.33
Tar, gal/MCF	Trace	Trace	2	...
Auxiliaries				
Wash water, gal/MCF	40	35	55	...
Electricity, kwh/MCF	0.08	0.07	0.52	0.44
Steam, lb/MCF	None	None	12	22

* 412 cu ft of methane at 1013 Btu per cu ft.

chamber without accessories of $15,000 to $25,000 per MM-SCFD of gas capacity.

Catalytic Reforming

Reforming, or the reaction of hydrocarbons with steam, results essentially in the formation of oxides of carbon and hydrogen. Depending on process conditions, "fragments" of the original hydrocarbon also result from thermal decomposition. The general process is therefore represented principally by Reactions **2** and **3**, incidentally by Eq. **1**, and with the product composition markedly influenced by Reaction **4**.

Reforming is highly endothermic. The method of delivering the requisite heat to the immediate reaction area not only constitutes one of the biggest single problems in conducting the process, but also provides the basis for classifying the various systems which have been developed. The systems can be described as:

1. Externally heated and continuous;
2. Internally heated and either continuous or cyclic.

The appreciable crossing of classification lines and combining of process features in practical application are noted in the following.

General Process Considerations. The reactions of steam and hydrocarbons are by no means as rapid as those of oxygen and hydrocarbons. The difference is so marked, in fact, that the use of catalysts is necessary to attain rates of commercial interest. Any great increase in the rate of reaction consumes heat at a proportionately higher rate. This heat is usually available only by virtue of storage in or transfer thru solid materials, most often in limited quantities. Therefore, the process is inherently self-damping and relatively easy to control. Combustion, on the other hand, usually proceeds readily.

Process Requirements. These include a supply of hydrocarbon and steam and a catalyst. Hydrocarbons discussed in this Chapter range from methane to petroleum naphtha. The problems of operation increase with the molecular weight of the feed stock, particularly in terms of the amount of steam which must be mixed with the hydrocarbon and the likelihood of carbon formation. The hydrocarbon should be virtually free from sulfur, to the order of a very few molar parts per million, in order to assure selectivity and activity of the catalyst and freedom from corrosion of equipment. Gas purification process data are available[16] for both feed stock and raw product gases.

Another desirable feature is a high degree of saturation of the hydrocarbon molecules. Unsaturated feed molecules are relatively prone to cracking and carbon deposition. In petrochemical plants, a feed hydrogenation step has sometimes been introduced to minimize subsequent cracking in the reforming step.

Catalytic Reforming, Externally Heated and Continuous. The furnaces used contain metal tubes filled with catalyst particles. Limited use of silicon carbide tubes has been reported.[17] The extent of the reaction of a mixture of steam and hydrocarbons within the tubes is governed by temperature, pressure, and the activity of the catalyst. Catalytic reforming systems have been used for a number of years to produce both public utility gas and chemical synthesis gas.[18,19] The advent of natural gas has dimmed current interest in conversion processes for U. S. utility gas purposes. Development in the chemical field, however, has been rapid; much of the equipment used is also applicable to utility gas production.

Process Conditions. As noted earlier, temperatures range from 1000 to 1800 F and pressures from atmospheric to 275 psig. High-pressure developments came from the petrochemical industry requirements to reduce equipment costs and compression work before subsequent synthesis steps.

Equipment. The central piece of equipment is a furnace consisting of a refractory-lined shell containing catalyst-filled tubes. Fuel is burned in the shell at slightly over atmospheric pressure, while mixtures of steam and gas and sometimes air pass over the catalyst at pressures appropriate to the application and within the limits of the technology and economics of tube design. Air may be injected with the process gas to burn inside the tubes and help supply the endothermic heat of reaction. Both round and rectangular furnaces have been built. Tubes are suspended around the periphery or in diametrical rows in circular furnaces and in banks across rectangular furnaces. Tubes vary from four to eight inches in diameter, and up to 32 ft in length. The burners may be at the top or bottom of the furnace or in the form of radiant cups in the side walls.[19] Figures 3-31 and 3-32 show approximate extremes in design variations. The latter shows a complete furnace with the various aspects which have been developed to meet recent process requirements.

Catalysts. Catalysis is not only a large and complicated subject, but also one in which specific data become obsolete rather rapidly. For detailed study, reference should be made to the various texts in the field and to the manufacturers' technical literature. Some basic catalyst information follows:

1. The principal catalytic agent in use for reforming is nickel deposited in extended form on various supports. Chromia, a relatively mild catalyst, has also been used.
2. There has been appreciable development of catalyst supports to increase porosity and nickel carrying capacity, and hence capacity for reforming service. Resistance to process conditions, particularly high temperature, has also been improved.
3. The principal enemies of activity and longevity are carbon deposition, excessive temperature, and sulfur. Degree of removal of sulfur is receiving increased attention, with even one-half part per million of sulfur in the feed gas considered significant at times.

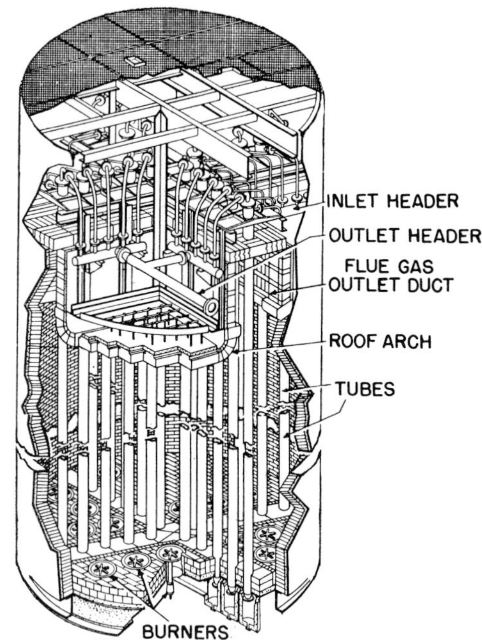

Fig. 3-31 Chemico reforming heater. (Chemical Construction Co.)

Table 3-44 Results of Catalytic Steam–Hydrocarbon Reforming

	Feed gas		
	Natural gas	Natural gas[18]	Butane[20] (contains 18% propane)
Higher heating value, Btu/cu ft	1048	1060	3211
Feed mixture, mole %			
Gas	31.0	22.5	13.4
Steam	69.0	22.0	60.0
Air		55.5	26.6
	100.0	100.0	100.0
Product comp., mole %			
N_2	2.0	22.5	11.7
H_2	70.5	52.3	57.0
CO	16.0	18.0	23.0
CO_2	6.9	3.7	5.0
CH_4	4.6	3.5	3.0
C_2			0.1
O_2			0.2
	100.0	100.0	100.0
Higher heating value, Btu/cu ft	327*	264*	292*
Requirement per MCF of reformed product, cu ft			
Process gas	270	224	81.5
Fuel gas	125	106	31.7
	395	330	113.2
Cold gas efficiency, %	79	76	80

* All subsequently enriched to distribution value of approximately 500 Btu/cu ft.

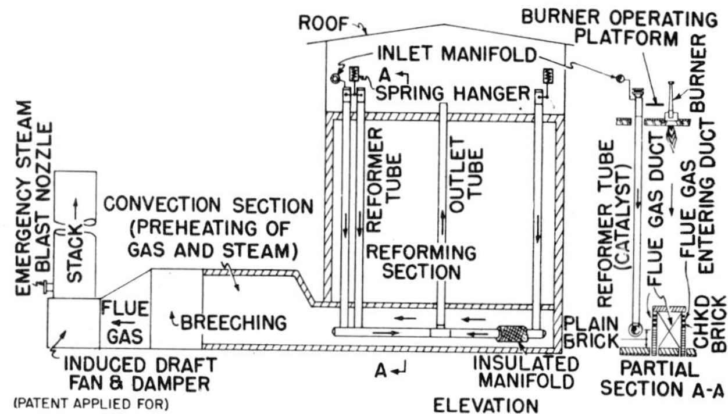

Fig. 3-32 Kellogg rectangular reforming furnace. (M. W. Kellogg Co.)

Operating Results. The gas mixtures resulting from hydrocarbon–steam reforming usually resemble those already presented for varying degrees of conversion in the discussion of thermodynamic equilibrium. Those made in industrial practice usually have a low heating value, requiring enrichment before use in any utility gas system. Of course, this may not be a significant disadvantage if an appropriate enriching gas is available.

The use of published operating data is difficult without a complete description of the plant and every aspect of the operation. This is true because of the wide disparity not only in basic designs but also in the significant effects of auxiliaries such as heaters and waste heat boilers.

The data in Table 3-44 indicate what has been done; however, they are not necessarily typical or limiting, because plant design and the objective of operation can be very influential. The table indicates the amounts of principal process materials and furnace fuel required per unit of product and compares reforming with and without air* in the process mixture. In these two cases, at least, the air resulted in an expected reduction in product gas heating value, in total natural gas

* Air may be admitted into the reforming tubes with the gas and steam or it may be added to a partially reformed gas for reaction in a separate vessel.

consumption, and in cold gas efficiency. If the gas composition is acceptable, however, the large volume produced per volume consumed may outweigh the reduced cold gas efficiency. Note that the data in Table 3-44 refer entirely to reforming essentially gaseous materials in fired tubes; in 1962, there was marked interest in processing liquid feed stocks by this means.[19]

An externally fired generator containing carbide tubes is in operation in an incandescent lamp manufacturing plant.[17] Average life of tubes and catalyst over a 12-year period is six months. A 50 per cent increase in life would be likely if reliability of continuous operation of equipment (except for semiannual shutdowns) were not such a critical factor. Product gas composition is given in Table 3-45a.

Table 3-45a Reformed Gases Made in Silicon Carbide Tubes[17]

Product composition, vol. %	Feed: Propane	Feed: Gasoline*
Carbon dioxide	1.3	2.0
Illuminants	12.7	11.6
Oxygen	0.6	0.4
Hydrogen	39.8	33.7
Carbon monoxide	14.5	11.9
Paraffins†	14.0	16.3
Nitrogen	17.1	24.1
Total	100.0	100.0

* 120 MBtu per gal, 73 API, 0.69 sp gr.
† Mainly methane.

Small-scale reforming equipment for process applications is available.[21] The A.G.A. Laboratories in Cleveland make both manufactured and mixed test gases by natural gas enrichment of the reformed gas made (Table 3-45b). The A.G.A. unit, containing six 35 per cent nickel–15 per cent chromium alloy tubes, is charged with equal volumes of gas and steam. The catalyst in the tubes consists of Grade A pure nickel wire (0.02 to 0.08 in. diam).

Catalytic Reforming, Internally Heated and Continuous. Here the endothermic heat of steam–hydrocarbon reactions is supplied by the controlled admission of air or oxygen with the hydrocarbon. The steam for reforming is usually also admitted with the hydrocarbon, although there are cases of apparent dependence on the steam formed by combustion. Sufficient air or oxygen is admitted to supply the endothermic heat plus the sensible heat consumption and losses, but it is markedly insufficient for complete combustion.

The reactors used are comparatively large, upright, cylindrical vessels, lined with brick or poured refractory material, and filled with one or more beds of catalyst. Reactants are commonly preheated to about 1000 F before they enter the reaction chamber, and products leave at 1500 to 1800 F. Pressures in the earliest reactors were only sufficient for entry into gasholders, but with the advent of gas grids and newer synthesis systems, the pressures have mounted into the hundreds of pounds. Applications include public utility gases, chemical synthetic gases, and metallurgical process gases.

Table 3-45b Composition of A.G.A. Laboratories Reformed and Test Gases

(Cleveland, Ohio; 1962)

	Feedstock (nat gas)	Product (reformed)	Test gas B*	Test gas C†
Composition, vol. %				
Methane	93.2	1.9	27.5	63
Ethane	3.4	...	2.0	3.4
Carbon monoxide	...	23.8	16.6	7.6
Hydrogen	...	72.0	51.5	23.2
Carbon dioxide	0.6	1.1	1.0	0.7
Oxygen	...	...	...	...
Propane	0.83	...	...	0.5
Butane	0.41	...	...	0.2
Nitrogen	1.60	1.2	1.4	1.4
Heating value, Btu/cu ft sat.	1024	330	526	803
Specific gravity	0.593	0.32	0.408	0.51
Flow rate, cfh	250	1000		

* Manufactured gas.
† Mixed gas.

Table 3-45c Surface Combustion Light Hydrocarbon Processes

	Endothermic reactions: With air and steam	Endothermic reactions: With air	Adiabatic with air	Pre-heated hydro-carbon and air
Product composition, mole %				
Hydrogen	69.9	40.7	30.0	37.80
Methane	1.5	0.13	1.8	0.04
Nitrogen	4.5	38.3	49.0	41.63
Carbon monoxide	20.6	20.90	16.6	19.70
Carbon dioxide	3.5	0.00*	2.6	0.83
Total	100.0	100.03	100.0	100.00
Feedstock per MCF of dry product gas				
Natural gas (C_1 thru C_3) for all, 1000 Btu/cu ft				
Consumption, MBtu	250†	232†	200	200
Temperature, °F	1800	1800	1450	1750
Product				
Heat. value Btu/cu ft	310	201	151	187
Auxiliaries per MCF of dry product gas				
Fuel input MBtu	235	206	...	...
Reaction air, SCF	57	525	610	520
Steam, lb	20	None		
Cooling water, gal	127‡	80‡	120	2

* After CO_2 absorption.
† Reaction heating fuel given separately below.
‡ 30 °F temperature rise.

Low-Pressure Operation. An early and very fundamental use was in public utility gas production at Toulouse, France.[22] Here natural gas and air were the only reactants entering the system. The methane was converted to a fairly rich producer gas with a heating value of 200 Btu per cu ft at a thermal efficiency of 90 per cent. The gas was then enriched with natural gas to 500 Btu per cu ft. The extent of reaction with steam is not certain, but the ability to maintain operation without carbon blockage is a tribute to either equipment design or catalyst. The Surface Combustion Co. has described similar equipment for producing reducing gases in the metallurgical field.[23] See Table 3-45c.

The Haldor Topsoe Co. installed a plant in Copenhagen for reforming refinery gases with internal heating by partial com-

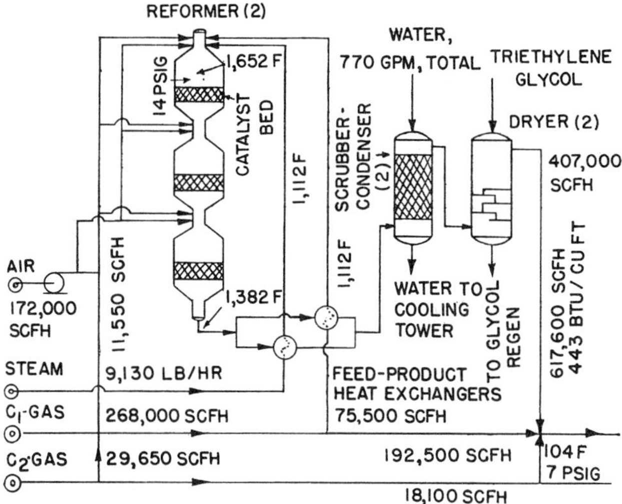

Fig. 3-33 The Topsoe reforming system for public utility gas.[24]

bustion with air (Fig. 3-33). The heart of the system is a pair of upright three-chamber reactors with burners and catalyst beds in series.

Feed material consists of air, steam, and two hydrocarbon streams. Part of each stream is reformed and part is used for enriching the reformed product. Steam and part of the lighter hydrocarbon stream are preheated by recuperative heat exchange with the hot product gases and pass to the tops of the chambers for reaction with air. Portions of the heavier hydrocarbon stream and additional air are added at the intermediate burners. The hot gas–steam mixture leaves the bottom of the reactors and gives up heat to the incoming feed. It is then washed and cooled with water, dried with ethylene glycol, enriched by blending with the mixed gas stream, and passed to the city network.

Compositions of the various gas streams are shown in Table 3-46a; rates of flow, heating values, and efficiencies are shown in Table 3-46b. A favorable aspect of the system is freedom from carbon deposition.

High-Pressure Operation. The foregoing discussed relatively low-pressure systems. Various aspects of demand have called for higher system pressures and the use of steam in the chemical field. The M. W. Kellogg Co. installed the Sasol plant in South Africa in 1955, using oxygen and steam at a

Table 3-46a Topsoe Utility Gas System Stream Compositions[25]

(in mole per cent; see Fig. 3-33)

Component	C_1 gas	C_2 gas	Reformer effluents: Wet	Reformer effluents: Dry	Mixed product gas (reformed plus by-passed)
H_2	38.4	...	34.2	44.8	41.5
CO	6.5	...	8.9	10.9	9.2
CO_2	...	...	7.3	9.6	6.3
CH_4	31.0	4.1	0.7	0.9	10.4
C_2H_4	16.4	64.6	...	...	7.0
C_2H_6	6.6	31.3	...	...	3.0
C_3H_6	1.1	...	...	...	0.3
N_2	...	...	25.3	33.2	21.9
H_2O	...	...	23.6	0.6	0.4
Total	100.0	100.0	100.0	100.0	100.0

Table 3-46b Topsoe Utility Gas System Material Flow Rates and Efficiencies[25]

Material flows, two reactors	Rate	High heating value: Btu/cu ft	High heating value: MBtu/hr	Cold gas efficiency, %
Feed streams				
Air to reformers	172,000 scfh			
Steam to reformers	9,130 lb/hr			
C_1 gas				
To reformers	75,500 scfh	870	65,700	
By-pass to blending	192,500 scfh	870	167,500	
Total	268,000 scfh	870	233,200	
C_2 gas				
To reformers	11,550 scfh	1,646	19,020	
By-pass to blending	18,100 scfh	1,646	29,800	
	29,650 scfh	1,646	48,820	
Product streams				
Reformed product gases (dry)	407,000 scfh	190	77,400	91.3
By-passed gases	210,600 scfh	937	197,800	
	617,600 scfh	443	275,200	97.5
Water flows				
Quench coolers, total	770 gpm			
Other services	95 gpm			
Total cooling tower load (between 122 and 85 F)	865 gpm			
Triethylene glycol dryer data not available				

pressure of 320 psig, and producing 200 MMCFD of synthesis gas.[26] The firms of Haldor Topsoe and Société Belge de l'Azote have announced a similar system, using air, steam, and hydrocarbons at 10 to 14 atm, with applications to the production of H_2 and synthesis gas. Material and cost details for producing hydrogen and ammonia synthesis gas have been reported.[27] The use of air for partial combustion above atmospheric pressure for the final or "secondary" step in producing ammonia synthesis gas is well known, and any of these pressurized systems can be applied to the production of public utility gas.

Catalytic Reforming, Internally Heated, Cyclic. In these systems, the endothermic heat required for the reactions of hydrocarbons and steam is alternately stored in and withdrawn from refractory masses and catalyst beds during a repetitive sequence of phases of operation. The two major phases of the typical cycle are:

1. The heating phase, during which a fuel is burned essentially to completion, with the heat stored in checker-brick and catalyst beds;
2. The gas-making phase, in which hydrocarbons and steam pass over the checker-brick and react over the catalyst bed, absorbing the necessary heat of reaction from both.

In the latter phase, inert nitrogen of the heating phase is excluded from the product gas. Brief phases of purging with

Table 3-47 Operating Results of Cyclic Catalytic Reforming

	Feed stock and unit					
	Natural gas, 1048 Btu/cu ft, in United Engrs. two-shell CCR,[28] 12-ft OD shells	Butane, 3200 Btu/cu ft, in United Engrs. two-shell CCR[28] 12-ft OD shells	Gasoline, in ONIA-GEGI,[29] Cahors, France		Light virgin naphtha (butane–hexane), boiling range 74–161 F, in ONIA-GEGI, Cork, Ireland,[30] 9-ft single shell	Light distillate, boiling range 102–302 F, in South Eastern Gas Board (Segas) three-shell,[31] 11-ft 9-in. ID
Product composition, mole %						
N_2	19.4	12.6	0.9	1.4	4.7	5.4
H_2	53.2	56.7	55.0	24.4	47.0	52.5
CO	17.7	16.8	21.4	7.2	22.2	14.9
CO_2	5.1	7.2	6.1	4.3	6.5	4.1
CH_4	4.5	6.1	0.0	0.0	11.9	17.7
C_nH_{2n}	0.0	0.0	3.7	26.5	0.0	3.0
C_nH_{2n+2}	.0	.0	12.7	35.6	0.0	1.7
Hvy H.C.	.0	.0	0.0	0.0	7.5*	0.0
Ill.	.0	.2	.0	.0	0.0	.0
O_2	0.1	0.4	0.2	0.6	0.2	0.7
Total	100.0	100.0	100.0	100.0	100.0	100.0
Product heating value, Btu/cu ft	279†	298†	456	1065	508	502
	Cu ft/MCF Prod.	Gal/MCF Prod.			Lb/MCF Prod.	
Feed requirement	...	...	3.69	9.88	21.7	33.7
Process heat req.	...	...	1.72	1.73	7.04	2.3
Total	321	3.62	5.41	11.61	28.74 (+0.36 gas oil)	36.0
Cold gas eff. (calc.), %	82.6	82.0	75.5	82.0	83.8	69.5

* C_nH_m.

† Subsequently enriched to 500 Btu.

steam are used between major phases to prevent the formation of explosive mixtures. The phases often overlap appreciably either for control of product composition or continuity of operation. In the case of extreme overlap, the process may become virtually continuous reforming. The overall thermal efficiency may be raised by this means, but the heating value of the product gas falls, frequently below an acceptable minimum.

The system is often used to process liquid fuels in equipment originally designed for solid fuels, e.g., converted water gas sets. Limited application of such systems is made in the United States. With increasing availability of fluid fuels, however, such installations are becoming more and more common abroad. A 500-Btu per cu ft gas is usually made.

Plant designs vary appreciably, but each contains one or

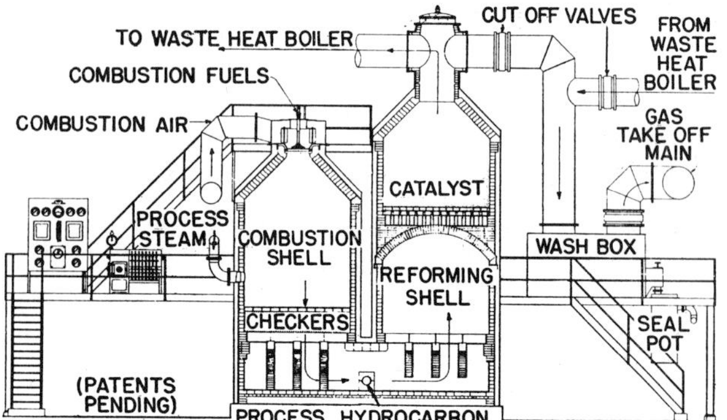

Fig. 3-34 The U.G.I. cyclic catalytic reforming apparatus. (United Engineers & Constructors, Inc.)

more upright refractory-lined vessels called shells. These house a combustion space, some heat storage material, and reforming space filled with catalyst. A two-shell design (Fig. 3-34) is frequently used. In Europe, designers such as the ONIA-GEGI organization of France and the South Eastern Gas Board of England tend to incorporate virtually all equipment features into a single shell. The checker-brick is either eliminated or combined with the catalyst bed, and the space above the catalyst is used for combustion. There is undoubtedly an investment saving, but possibly at the expense of less heat exchange and less mixing of materials.

In two-shell designs (Fig. 3-34), combustion air and fuel enter the top of the left shell during the heating phase, burn, and pass thru the system, storing heat in checker-brick, catalyst, and the considerable volume of refractory. The flue gases go to the waste heat boiler and then either to the stack or, during appropriate portions of the cycle, back to the wash box and to the gas main. During the gas-making phase, process steam enters at mid-level in the left shell, is preheated in the checker-brick, and is joined by the process hydrocarbon in the lower duct, or crossover. The mixture rises into the catalyst bed for reforming to hydrogen and oxides of carbon; the product gases pass to the offtake main and gas holder. A short steam purge clears the system of product gas before the next combustion phase. The cycle lengths are of the order of four minutes, of which approximately half is devoted to the steam–hydrocarbon phase proper. For continuity, the system is controlled by a cycle timer, with particularly detailed interlocks against the formation of explosive mixtures. This type of plant presents problems in high-pressure systems because of

Table 3-48 Performance Data for the Texaco Partial Oxidation System[33]

	Fuel used					
	Natural gas	Propane	64° API naphtha	9.6° API fuel oil	9.7° API fuel oil	−11.4° API coal tar
Fuel composition, wt %						
Carbon	73.40	81.69	8.38	87.2	87.2	88.1
Hydrogen	22.76	18.31	16.2	9.9	9.9	5.7
Oxygen	0.76	...	...	0.8	0.8	4.4
Nitrogen	3.08	...	...	0.7	0.7	0.9
Sulfur	...	...	...	1.4	1.4	0.8
Ash	100.00	100.00	100.00	100.0	100.0	100.0
Gross heating value, Btu/lb	22,630	21,662	20,300	18,200	18,200	15,690
Flow rates, per MMCF of dry product gas						
Fuel, lb	16,354	17,969	18,524	20,139	19,486	22,196
Steam, lb	None	None	4,625	9,261	11,043	11,612
Oxygen, MCF	248	270	239	240	240	243
Net carbon produced, lb	None	123	112	639	None*	379
Composition of product gas, mole %						
Hydrogen	61.1	54.0	51.2	45.9	45.8	38.9
Carbon monoxide	35.0	43.7	45.3	48.5	47.5	54.3
Carbon dioxide	2.6	2.1	2.7	4.6	5.7	5.7
Nitrogen	1.0	0.1	0.1	0.7	0.2	0.8
Methane	0.3	0.1	0.7	.2	.5	.1
Hydrogen sulfide	...	...	...	.1	.3	.2
Carbonyl sulfide	...	...	...	0.0	0.0	0.0
Performance data						
Oxygen consumed, cu ft/MCF of $H_2 + CO$	255	276	248	254	258	259
Cold gas efficiency, gross heating value of $H_2 + CO$, % of fuel used	83.8	80.9	82.7	83.1	84.7	86.4
Operating conditions						
Pressure, psig	340	247	350	350	350	347
Fuel and steam preheat, °F	900	376	665	691	630	750
Oxygen feed temperature, °F	260	44	105	71	72	64

* 1034 lb recycled.

the frequent valve and stack lid movements and the apparent need of a gas holder or multiple units for continuity of output. However, it is noted for flexibility in operation and use of feed stocks, and there have been scores of installations in recent years.

Table 3-47 shows some operating results for the system using various fuels; these data indicate what has been done, and not necessarily the optimum or limit of what can be done. Note that in one of the ONIA-GEGI operations, a gas of 1065 Btu per cu ft was apparently made directly from the unlikely fuel, gasoline. Also note that the direct production of 700–800 Btu gas has been obtained in the United States for relatively brief peak load periods.[32]

Table 3-47 indicates a general cold gas efficiency of about 82 per cent for all but the heaviest fuel but, as indicated earlier, firm conclusions should be reached only after study of the design and complete data of the plants involved.

Partial Oxidation

This incomplete combustion process yields a mixture of CO and H_2 and, usually, residual hydrocarbons, as well as some of the end products of combustion (CO_2 and H_2O). The residual hydrocarbons may consist of a considerable amount of "fragments" from thermal cracking or traces of methane only, depending on the extent of combustion practiced, the temperature attained, and the residence time at high temperature. Therefore, the general process is represented to varying degrees by Eqs. **1**, **9**, and **10**.

Partial combustion or partial oxidation enters to varying degrees into a variety of utility gas-making processes. For example, it is used to supply the heat necessary within the system to carry on the thermal decomposition (endothermic reaction) of the Hasche process as well as the endothermic steam–hydrocarbon reforming processes. However, the term *partial oxidation* as used in the following paragraphs refers essentially to the continuous process wherein raw hydrocarbon feed material is reacted with air, with air enriched with O_2, or with O_2 as such, without the aid of catalysis. Steam may or may not be added, depending on the process feed.

The two major commercial systems in use (1962) are those of the Texaco Development Corp.[33] (see Fig. 3-35) and The Shell Development Co.[34,35] Basically, these systems involve partial combustion and some thermal decomposition in an empty, refractory-lined vessel. Necessary features are supplies of preheated hydrocarbons, air or oxygen, and equipment for cleaning and extracting heat from or quenching the product gas. The oxygen and hydrocarbon are preheated separately, and steam may or may not be added to the latter. Steam is not ordinarily used with natural gas, but it is used in increasing proportions with heavier feed stocks. Mixing and combustion take place promptly in the upper portion of the chamber. The products are quenched, washed, cooled, and filtered.

Table 3-48 presents process data; oxygen was the combus-

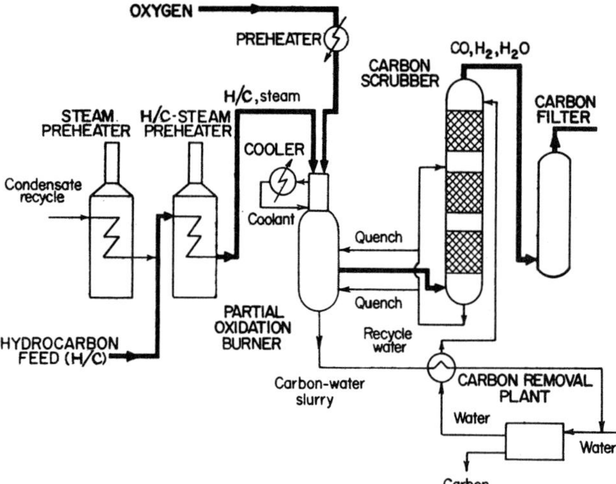

Fig. 3-35 The Texaco partial oxidation system.

tion agent in all cases. Notable features are the wide range in feed stocks and the selective use of steam. The product gas is essentially a blue gas, as the reported operating temperatures of 2000 to 2700 F largely preclude the existence of hydrocarbons in the product. The gas can, of course, be enriched if a suitable material is available. The system pressure, of the order of 350 psig, is compatible with high-pressure pipelines. A Shell system has been installed for the South Eastern Gas Board and one utility gas system of the Texaco type exists in Europe.[33]

NEW PROCESSES UNDER DEVELOPMENT

Developments not yet in commercial practice, but with interesting aspects for the future, include:

1. Catalytic steam reforming of liquid hydrocarbons.[18]
2. Simultaneous or stepwise reforming and hydrogenation, the exothermic heat of the latter helping to supply the endothermic heat of the former.[36]
3. Extensive laboratory and pilot plant work on the high-pressure hydrogenation of oils directly to gas.[37]
4. New oil hydrogenation process available for production of a high heating value gas for enriching other gases.[33]

Synthesis gas has been produced in a single-cylinder high-speed compression ignition engine with power produced simultaneously. Input was a mixture of washed sludge or sewage gas (90.7 CH_4, 6.3 CO_2, 0.8 O_2, and 2.2 N_2) and oxygen-enriched air.[38]

Recent British Developments

The Imperial Chemical Industries Co. of England developed a process for the manufacture of hydrogen for chemical purposes. Raw materials were light hydrocarbon liquids with low unsaturated sulfur contents. The hydrogen so made is enriched with an external source of other hydrocarbon gases, such as natural gas or propane-butane mixtures, to produce a mixed gas of 500 Btu per cu ft free from sulfur and carbon monoxide. This gas is used in England for the replacement of and as an addition to traditional coal gas.

The Midland Research Station of the Gas Council has taken this process a step further. Light hydrocarbon liquids are converted directly to a low sulfur, low CO city gas in one stage. For this purpose two methods are employed:

1. Catalytic hydrogenation (Bromley Works of the North Thames Gas Board), and
2. Recycle hydrogenation which works on slightly different principles (Seabanks Works,[39] Avon Mouth, Bristol of the Southwestern Gas Board).

The Seabanks Works expects to produce a final gas (500 Btu per cu ft, 0.508 sp gr) having the following analysis (per cent): H_2(55.10), CO(4.09), CH_4(15.80), C_2H_6(7.65), C_2H_4 (0.20), C_3H_6(0.72), CO_2(16.44).

It remains to be seen whether this type of gas manufacturing process will have an important economic influence on the manufactured gas industry. In any event, a number of processes of this general type are being tried at laboratory and pilot stages in various parts of the world.

REFERENCES

1. Reid, J. M., and others. "Production of Petrochemicals from Natural Gas and Natural Gas Liquids." (CEP-55-17) *A.G.A. Proc.* 1955: 931–45.
2. Nelson, W. L. *Petroleum Refinery Engineering,* 3rd ed., p. 586. New York, McGraw-Hill, 1940.
3. *Ibid.*, p. 593.
4. Hougen, O. A., and Watson, K. M. *Chemical Process Principles,* 3 vols. New York, Wiley, 1947.
5. Dodge, B. F. *Chemical Engineering Thermodynamics.* New York, McGraw-Hill, 1944.
6. Smith, J. M. *Chemical Engineering Kinetics.* New York, McGraw-Hill, 1956.
7. Montgomery, C. W., and others. "Thermodynamics and Stoichiometry of Synthesis Gas Production." *Ind. Eng. Chem.* 40: 601–7, Apr. 1948.
8. Mayland, R. J., and Hays, G. E. "Thermodynamic Study of Synthesis Gas Production from Methane." *Chem. Eng. Progr.* 45: 452–8, July 1949.
9. Mungen, R., and Kratzer, M. B. "Partial Combustion of Methane with Oxygen." *Ind. Eng. Chem.* 43: 2782–7, Dec. 1951.
10. Akers, W. W., and Camp, D. P. "Kinetics of Methane–Steam Reaction." *Am. Inst. Chem. Engrs. J.* 1: 471–4, Dec. 1955.
11. Shnidman, L. *Gaseous Fuels,* 2nd ed. New York, A.G.A., 1954.
12. Kobe, K. A., and others. "Thermo Data for Petrochemicals." *Petrol. Refiner* 28–36, Jan. 1949–July 1958.
13. Habermehl, R. Private communication. Louisville, Ky., Catalysts & Chemicals, Inc.
14. Carroll, J. W., and Paquette, R. B. "Review of Reforming of Hydrocarbons in Water Gas Sets." *A.G.A. Proc.* 1948:613–22.
15. Clarke, R. P. "Recent Developments in the Hasche Process." *Gas Age* 125: 31–2, June 23, 1960. Also, *ibid.* Private communications, April 3 and July 23, 1962. Johnson City, Tenn., Hasche Process Co.
16. Kohl, A. L., and Riesenfeld, F. C. *Gas Purification,* New York, McGraw-Hill, 1960.
17. Krause, A. W. "Continuous Catalytic Generator Produces Special Purpose Fuel Gas." *Ind. Heating* 18: 620–6, Apr. 1951.
18. Horsfield, S. W. "Reforming of Hydrocarbons as Experienced by the Long Island Lighting Company." (PC-50-38) *A.G.A. Proc.* 1950: 778–80. Also as: "Continuous Catalytic Cracking Process Used to Reform Natural Gas." *Am. Gas J.* 175: 17–20, Sept. 1951.

19. Kenard, R. J., Jr. "Steam Methane Reforming for Hydrogen Production." *World Petrol.* 33: 60+, Mar.; 48–50, Apr. 1962.
20. Vane, S. G., and Morris, J. E. *Reflections on the Continuous Catalytic Reforming of Butane by the Otto Process.* (C.P. 613) London, I.G.E., 1962.
21. Arnold, M. R., and others. "Nickel Catalysts for Hydrocarbon–Steam Reaction." *Ind. Eng. Chem.* 44: 999–1003, May 1952.
22. Brunelli, R. "L'Installation de Cracking du Gaz Naturel par Catalyse à l'Air à l'Usine à Gaz de Toulouse." *Compt. Rend. Congr. Ind. Gaz* 54: 329–42, 1947.
23. Huebler, J. "Gas Reactions Play Key Role in Iron-Pellet Reduction." *Iron Age* 186: 104–6, Sept. 22, 1960.
24. Jensen, H. "Copenhagen's City Reforming Plant in Service." (Transl. of title) *Ingenioeren* 68: 133–8, Feb. 15, 1959.
25. Barry, M. J. Private communications. New York, Haldor Topsoe Inc.
26. Garrett, L. W., Jr. "Gasoline from Coal via the Synthol Process." *Chem. Eng. Progr.* 56: 39–43, Apr. 1960.
27. "Hydrogen Process Broadens Feedstock Range." *Chem. Eng.* 69: 88+, July 9, 1962.
28. United Engineers and Constructors. "Developments in the U.G.I. Cyclic Catalytic Reforming Process," Table II. *Tech. Bull.* Oct. 1952.
29. Calderwood, G. L. "The 'ONIA-GEGI' Process for Cyclic Catalytic Cracking of Liquid Hydrocarbons." (CEP-58-11) *A.G.A. Operating Sec. Proc.* 1958: 89–97.
30. Dunphy, T. E. *O.N.I.A.-G.E.G.I. Plant in Cork.* (C.P. 615) London, I.G.E., 1962.
31. Stott, C. *Gasification at the Isle of Grain.* (C.P. 568) London, I.G.E., 1960.
32. Milburn, C. G. Private communication. Philadelphia, Pa., United Engineers and Constructors.
33. Texaco Development Corp. Private communication, Aug. 2, 1962.
34. Van Amstel, A. P. "New Data on Shell's Synthesis Gas Process." *Petrol. Refiner* 39: 151–2, Mar. 1960. Also in: *Oil Gas J.* 57: 174–5, Dec. 28, 1959.
35. Singer, S. C., Jr., and ter Haar, L. W. "Reducing Gases by Partial Oxidation of Hydrocarbons." *Chem. Eng. Progr.* 57: 68+, July 1961.
36. Cockerham, R. G., and Percival, G. *Experiments on the Production of Peak-Load Gas from Methanol and Petroleum Distillate.* (GC 41) London, Gas Council, 1957.
37. Schultz, E. B., Jr., and Linden, H. R. *High-Pressure Hydrogasification of Petroleum Oils.* (Res. Bull. 29) Chicago, I.G.T., 1960.
38. Karim, G. A., and Moore, N. P. W. "Production of Synthesis Gas and Power in a Compression Ignition Engine." *J. Inst. Fuel* 36: 98–105, Mar. 1963.
39. Olive, G. F. and Olden, J. F. *Development of Production in the Southwest.* (Pub 657) London, Inst. Gas Engrs., 1964.

Chapter 7
Oil Gas

by James F. Bell, Henry R. Linden, J. M. Reid, S. C. Schwarz, and E. F. Searight

Oil gas is manufactured by oil "cracking" or pyrolysis. Cyclic thermal processes in which cracking is done on heated refractory checkerwork used to be the most important commercially.

The first commercial oil gas production in the United States employed the Pintsch process (1873). In 1889 the L. P. Lowe patents[1] were issued for an oil gas set with "chambers lined with refractory material and each having an openwork filling of refractory material." The set somewhat resembled a carbureted water gas generator.

The production of oil in California resulted in a decline in gases made from coal and coke on the West Coast and a corresponding increase in oil gas production. By 1925 most oil gas generators in operation were "straight shot" or Jones generators.

The term "Pacific Coast process" was adopted to cover the general group of processes of that area for production of oil gas of from 500 to 570 Btu per cubic foot with lampblack as a by-product.

In sections of the East and Middle West, 1000 Btu oil gas augmented natural gas supplies during peak periods. High-Btu oil gas is suitable for peak shaving because it provides flexibility of operation and is interchangeable to a large extent with natural gas. It is particularly important where distributing companies lack other suitable peak shaving means, such as underground storage or LP-gas facilities.

In 1932 extensive tests were made[2] in California on Jones-type double-shell and straight shot generators producing 1000 Btu gas. In the same year, use of the "refractory screen process"[3] was reported. The conversions involved moderate capital investment and simplicity of both design and operation. These conversions developed maximum capacity, using up to three per cent Conradson carbon oils, and were nonregenerative, that is, without provision for back blast.

In general, **nonregenerative conversions** are best suited to operation with distillate feed stocks and lighter residuums. Some nonregenerative conversions are: refractory screen processes; single generator oil gas sets; twin generators; solid fuel fired oil gas generators; single burner series generators; and the U.G.I. oil gas process.

In 1948, a regenerative oil gas[4,5] set was tested at Baltimore under plant conditions. This Hall set achieved greater thermal efficiency and higher gas production rates than previous processes. It also made use of more readily available, low cost, and higher (0.2 to 13.0 per cent) Conradson carbon oils. The Hall four-shell unit was able to operate on residuums having up to 13 per cent Conradson carbon without down time for carbon scurfing. Regeneration was inherent in the Hall design because of its four-shell construction and special method of blasting. A similar effect has been obtained in other conversions by including back blast, which transfers some heat from the superheater back to the generators and provides better thermal efficiency and scurfing of deposited carbon.

Regenerative conversions generally operated well on gas oil; they also allowed use of residuums containing up to 6 per cent Conradson carbon with only small loss of capacity. Some regenerative conversions are: the reverse flow high-Btu oil gas process; the Hall four-shell set; the Hall two-shell inverted "U" generator (Henry design); and the twin generator (with back blast).

FUNDAMENTALS OF OIL GASIFICATION

Variables and Tests

In oil gasification, a limited number of variables will determine quantity and composition of the diluent-free oil gas (Fig. 2-5) and quantities of carbon or coke produced. Physical properties of the tar, as made, are largely a function of the following variables:

1. Oil characteristics as determined by (a) carbon to hydrogen weight (C/H) ratio, and (b) Conradson carbon residue;
2. Cracking temperature;
3. Residence time;
4. Partial pressure of oil gas leaving the cracking zone; and
5. Quantities of reactive constituents in carrier gas per unit quantity of make oil.

Oil characteristics can be determined by standard test methods. However, operating variables can only be approximated, particularly in cyclic cracking operations. For example, **average cracking temperatures** are greatly affected by factors other than the true vapor stream temperatures and normally serve only as control measurements for a particular gas-making unit. **Residence times** can be approximated on the basis of total make gas volume, including steam in vapor form, free space within heated zone, and average cracking temperatures and pressures. Volume contributions of tar and light oil vapors can be neglected because of their high relative molecular weights. This repre-

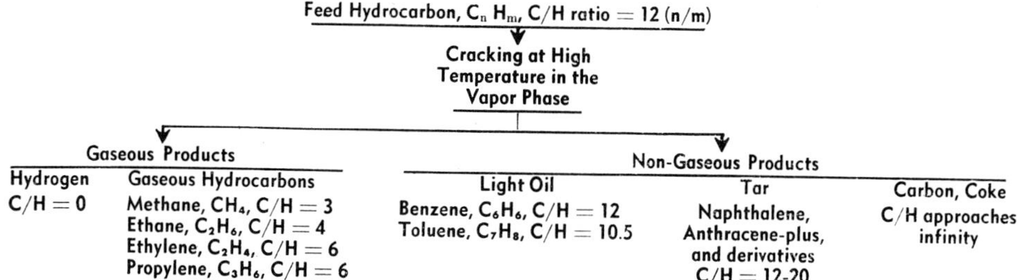

Fig. 3-36 Schematic product distribution in high-temperature vapor phase cracking.[7]

sents a considerable idealization, since it assumes instantaneous cracking of the oil to the final product volume as well as average gas and vapor temperatures corresponding to indicated temperatures. Similar approximations must be made in determining the partial pressure of oil gas.

For simplification, all data reported herein are based on results obtained under controlled conditions at the Institute of Gas Technology[6,7] in an empty, electrically heated, 2½ in. ID cracking tube of 0.115 cu ft free space. The tar and light oil recovery system simulated plant systems. However, since oil introduction was continuous, without intermittent burnoff, carbon or coke formation could be determined by direct measurement. Material balances averaging over 94 per cent were obtained.

Gas analyses and heating values were based on unscrubbed gases stabilized for 16 hr in a water-sealed holder at room temperature. Laboratory results obtained checked with plant data within anticipated limits of experimental errors.[6–8]

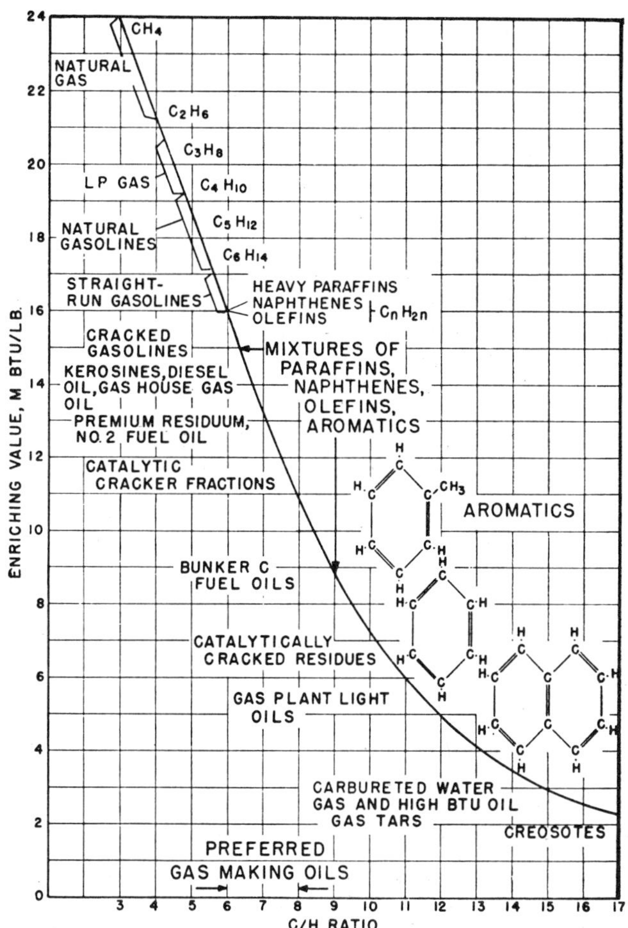

Fig. 3-37 Relationship of C/H ratio and Btu recovery.[7]

Correlation of Oil Properties and Operating Conditions with Gasification Yields

Effects of Carbon–Hydrogen Weight Ratio. The C/H ratio facilitates comparison of gasification data since it largely determines how much oil will be converted to gas. When petroleum oils are cracked in gas-making operations, the gaseous hydrocarbons formed (light paraffins and unsaturates) are normally limited to C/H ratios from 3 to 6, while such ratios of liquid products (aromatics) are normally between 10 and 20. Thus, the *lower* the C/H ratio of the hydrocarbon cracked, the higher its potential gas yield and the lower its potential tar or aromatics yield will be. Formation of free hydrogen and carbon does not alter this relationship. Figure 3-36 shows distribution of the cracking products between gaseous products of low and nongaseous products of high C/H ratios.

Figure 3-37 shows the relationship between Btu recovery per pound of various oils under typical operating conditions and C/H ratio.

Normal gas-making oils are in the 6 to 9 C/H ratio range (Fig. 3-37), corresponding to a variation of Btu recovery of 86 per cent. Since these ratios and, therefore, Btu recoveries

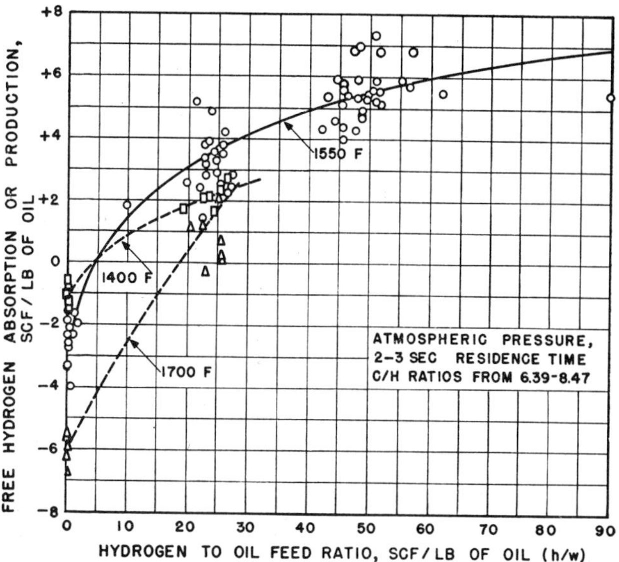

Fig. 3-38 Effects of cracking temperature and hydrogen-to-oil feed ratio on absorption or production of free hydrogen.[7] Positive values—hydrogen absorption; negative values—hydrogen production.

of oils, can vary widely within some commercial grades, buying gas-making oils based on grade designation alone is unwise.

Effects of Partial Pressures of Reactive Constituents. Among the carrier gases, such as blue gas or steam, hydrogen appears to be their only reactive constituent which becomes reactive with steam only above 1700 F. Other components, including steam and unreacted hydrogen, act as diluents with an effect comparable to that of reduced pressure.

To reduce the number of variables, effects of total pressure, carrier gas hydrogen concentration, and oil gas concentration on gasification yields can be combined empirically in the following terms:[7,9]

1. Partial pressure of oil gas, Px, as the product of total tube pressure in atmospheres, P, and the fraction, x, of oil gas in the gas leaving the cracking zone (steam, if any, in gaseous form). Oil gas is considered to be the net production of gaseous paraffins, unsaturates, and hydrogen as calculated from the quantities of carrier and make gas and their analyses.

2. Hydrogen-to-oil feed ratio, h/w, in SCF of carrier gas hydrogen per pound of oil. The partial pressure of carrier gas hydrogen in the cracking zone can be approximated by a function of $Px \times h/w$.

Net hydrogen production from liquid petroleum fractions is normally suppressed if the hydrogen-to-oil feed ratio, h/w, is *above* a critical value characteristic of the cracking temperature and, to a lesser extent, of the residence time and partial pressure of oil gas (Fig. 3-38). These values are approximately 5 SCF per lb at both 1400 and 1550 F, and 19 SCF per lb at 1700 F, at 2 to 3 sec residence time and atmospheric pressure. Therefore, no significant net hydrogen production from the oil occurs in carbureted water gas production except when the incandescent solid fuel bed is used for reforming the normal cracking products. Suppression of free hydrogen formation makes more hydrogen available for formation of low molecular weight gaseous hydrocarbons.

The lower Btu recoveries in high-Btu oil gas production compared with those in carbureted water gas are partially due to the absence of free hydrogen in the carrier gas (steam). Effects of increasing hydrogen-to-oil feed ratios are particularly apparent when Btu recoveries reported for the Dick cracking test[10] ($h/w \approx 50$ SCF per lb, $Px \approx 0.2$ atm) are compared with those for high-Btu oil gas plant results when $h/w = 0$.

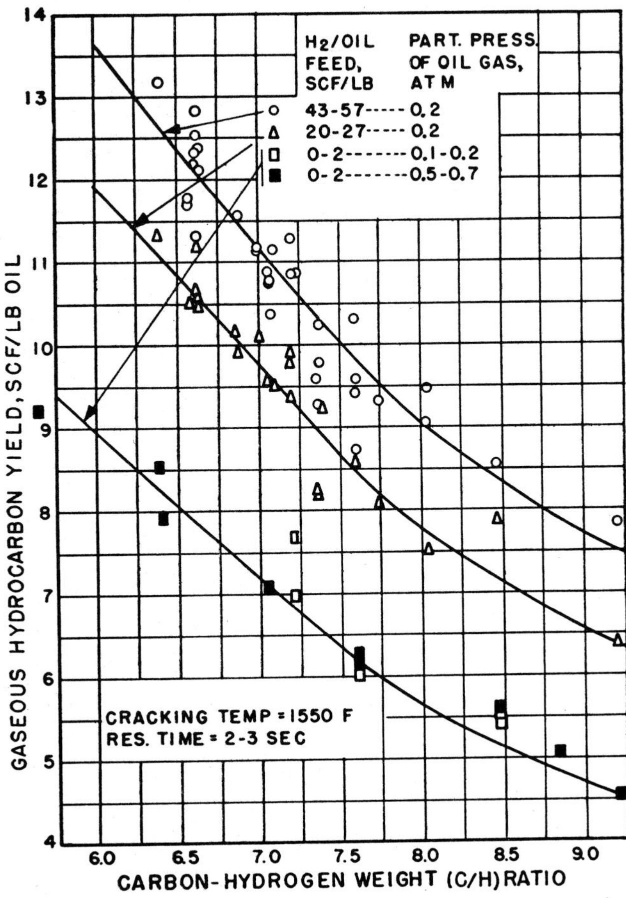

Fig. 3-40 Effects of C/H ratio, hydrogen-to-oil feed ratio, and partial pressure on gaseous hydrocarbon yield.[7]

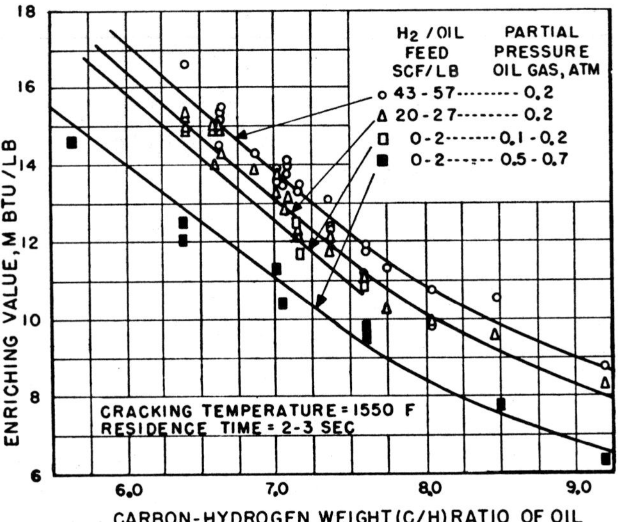

Fig. 3-39 Effects of C/H ratio, hydrogen-to-oil feed ratio, and partial pressure of oil gas on Btu recovery.[7]

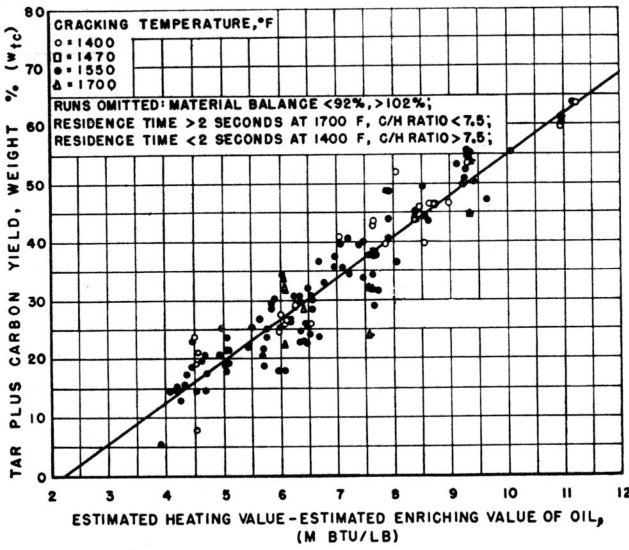

Fig. 3-41 Correlation of tar-plus-carbon yield with estimated gross heating value minus enriching value of oil.

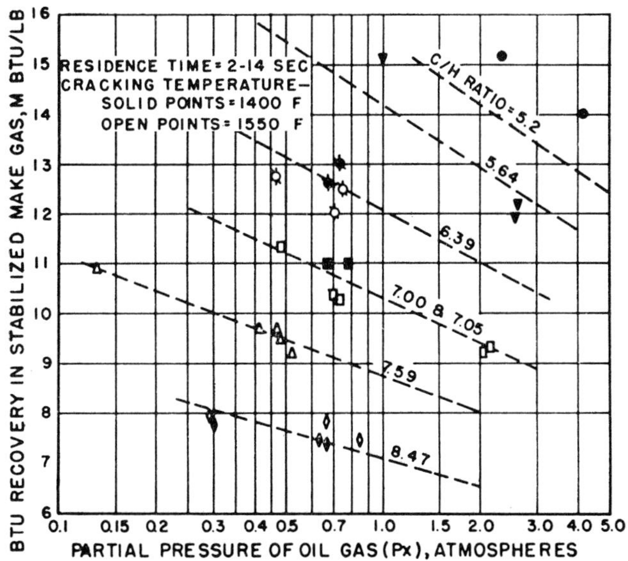

Fig. 3-42 Effects of C/H ratio of oil and partial pressure of oil gas on Btu recovery.[1 3]

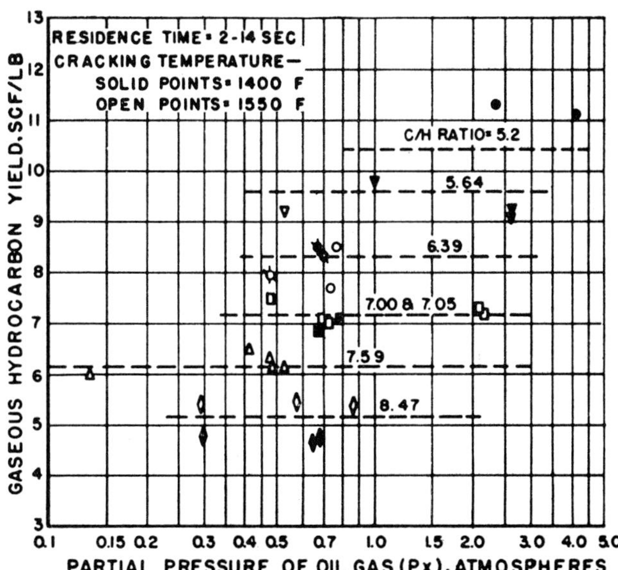

Fig. 3-43 Effects of C/H ratio of oil and partial pressure of oil gas on gaseous hydrocarbon yield.[13]

Increasing the pressure or partial pressure at constant temperature favors a reaction in the direction of volume decrease. However, when the total pressure or partial pressure of oil gas is increased in the presence of feed (carrier gas) hydrogen, increased rates of the hydrogenation reactions may more than compensate for the pressure effects (Figs. 3-39 and 3-40).

The tar-plus-carbon yields, w_{tc}, in weight per cent of feed oil were correlated with the gross heating value *less* estimated recovery in the make gas expressed in Btu per lb (Fig. 3-41). Heats of combustion of the oils were estimated from published correlations.[11]

Equations **a** and **b**, applicable to oil gas production when the carrier gas contains no hydrogen ($h/w = 0$), are given below:[6–9,12]

Btu recovery

$$\text{Btu/lb} = [0.830 - 0.228 \log Px]\,[36{,}400 - 3{,}410\ \text{C/H}] \quad \textbf{(a)}$$

at 5.2 *to* 7.5 C/H *ratios;*

$$\text{Btu/lb} = [0.830 - 0.228 \log Px]\,[39{,}800/(\text{C/H} - 3.82)] \quad \textbf{(b)}$$

at 7.5 *to* 9.2 C/H *ratios.*

Gaseous hydrocarbon yield

$$\text{SCF/lb} = 19.50 - 1.76\ (\text{C/H}) \quad \text{(c)}$$

at 5.2 *to* 7.5 C/H *ratios;*

$$\text{SCF/lb} = [20.4/(\text{C/H} - 3.82)] + 0.76 \quad \text{(d)}$$

at 7.5 *to* 9.2 C/H *ratios.*

In the foregoing, Px is the partial pressure of the oil gas leaving the cracking zone (product of total pressure in atmospheres, P, and the net sum, x, of volume fraction yields of paraffins, unsaturates, and hydrogen). The gas volume upon which these fractions are based must include the carrier steam but no diluents added subsequent to the cracking step. Effects of the tar vapor volume can be neglected as insignificant.

The accuracy of these equations in predicting Btu recoveries and gaseous hydrocarbon yields for these cracking conditions is shown in Figs. 3-42 and 3-43.

Effects of Cracking Temperature and Residence Time. Figure 3-44 shows effects of cracking temperature on Btu recovery for oils ranging from gasolines to Bunker C fuel oils. Btu recoveries are inversely proportional to C/H ratios.

Effects of residence time on Btu recovery are relatively small in most cases. Combined effects of cracking temperature and residence time in a given partial pressure range and in the absence of hydrogen-containing carrier gases on cracking product yields and compositions are best correlated by means of the diluent-free or true heating value of the stabilized prod-

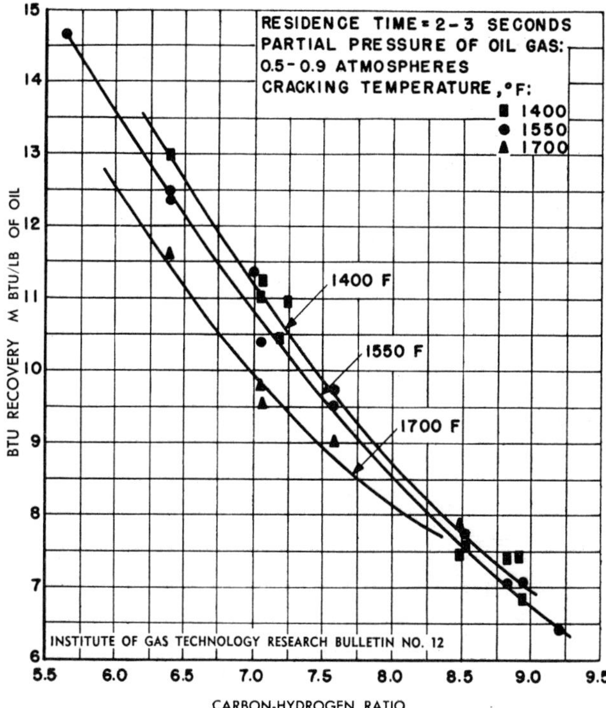

Fig. 3-44 Effects of oil properties and cracking temperature on Btu recovery.[13]

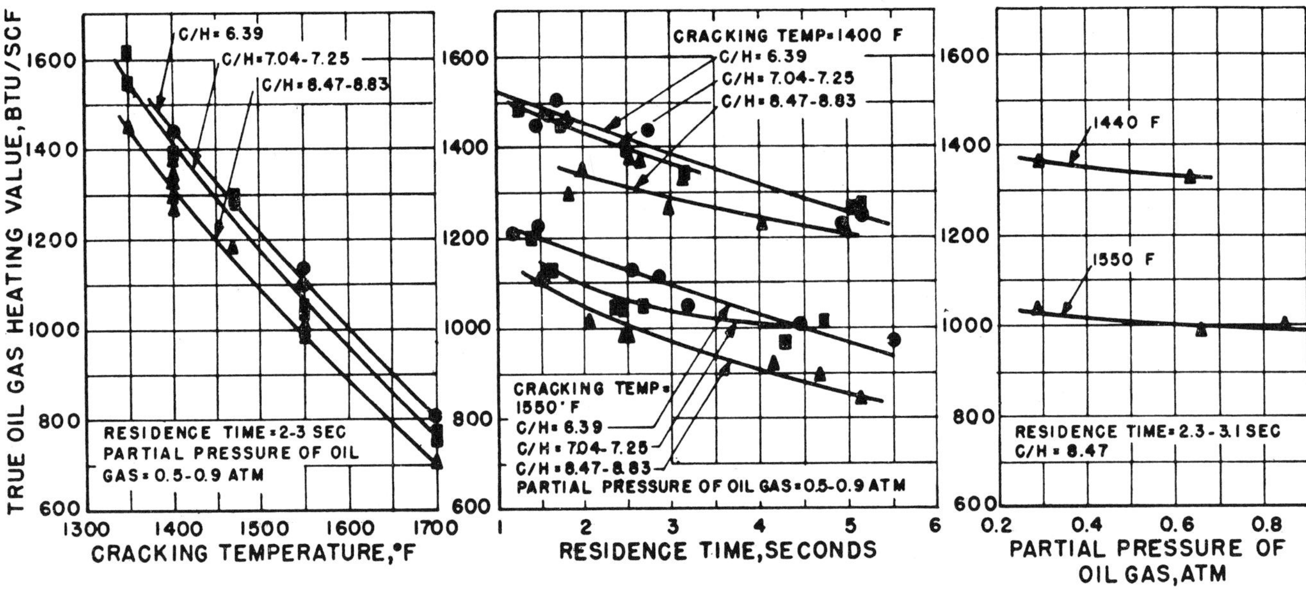

Fig. 3-45 Effects of oil properties and operating variables on true oil gas heating value.[6]

DIESEL OIL, 37.3°API, 6.39 C/H RATIO, 0 WT % CONRADSON CARBON RESIDUE

HEAVY GAS OIL (ARUBA), 22.2°API, 7.05 C/H RATIO, 6 WT % CONRADSON CARBON RESIDUE

BUNKER "C" FUEL OIL (TEXAS CRUDE) 9.9° API, 8.51 C/H RATIO, 11 WT % CONRADSON CARBON RESIDUE

SCF OIL GAS/GALLON OF OIL

GAL OIL/MCF OIL GAS

H_2

PARAFFINS

C_2H_4

C_3H_6+

TRUE OIL GAS HEATING VALUE BTU/SCF

LBS OF LIQUID OR SOLID PRODUCTS/GAL OF OIL

DIST 0-200 C

DIST 200-355 C

PITCH

CARBON

TRUE OIL GAS HEATING VALUE BTU/SCF

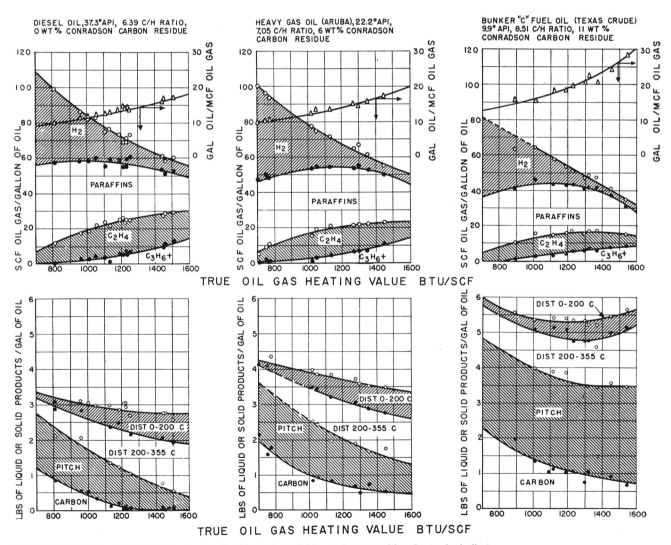

Fig. 3-46 Make oil requirements, gas and liquid product yields, and composition for typical oils.[6]

A. Bunker "C" (West Coast)
C/H = 8.83, CCR = 7 wt %

B. Heavy Gas Oil (Aruba)
C/H = 7.05, CCR = 6 wt %
New Eng. Gas Enrich. Oil
C/H = 7.25, CCR = 6 wt %

C. Bunker "C" (Texas Crude)
C/H = 8.51, CCR = 11 wt %

D. Reduced Crude
C/H = 7.04, CCR = 4 wt %

E. Catalytic Cracker Residue Blend
C/H = 8.47, CCR = 2 wt %

F. Diesel Oil
C/H = 6.39, CCR = 0 wt %

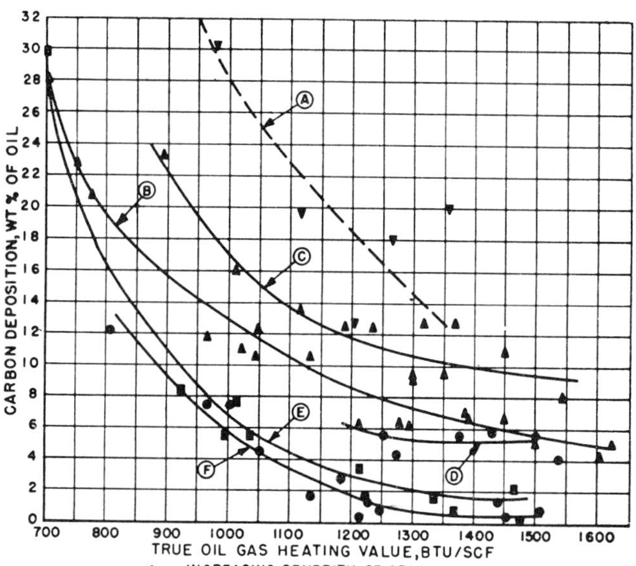

Fig. 3-47 Variation of carbon formation with true oil gas heating value for typical oils.[6]

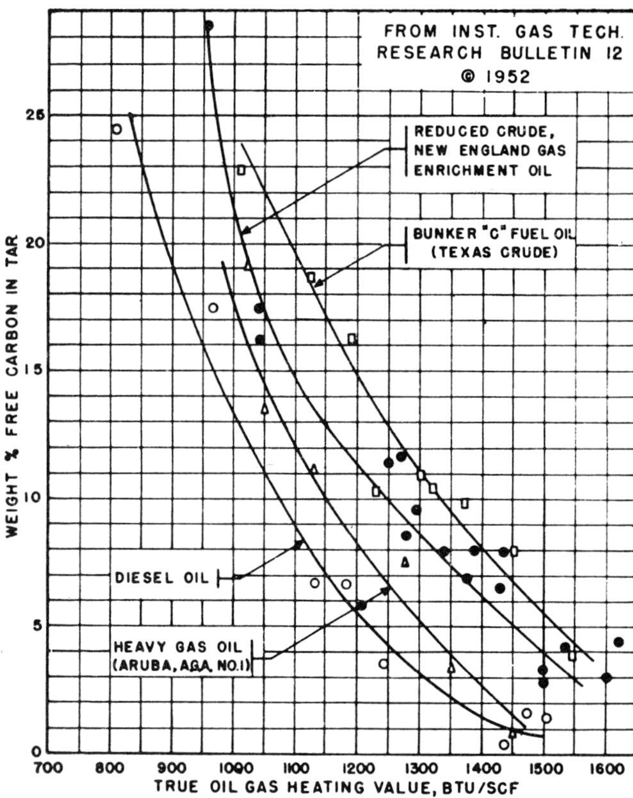

Fig. 3-48 Effects of oil type and severity of cracking on free carbon content of tar.[6]

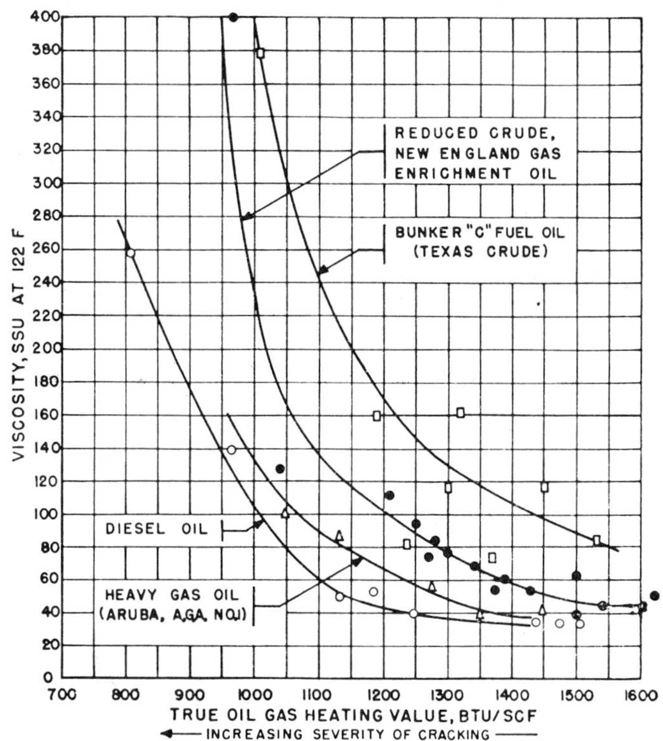

Fig. 3-49 Effects of oil type and severity of cracking on tar viscosity.[6]

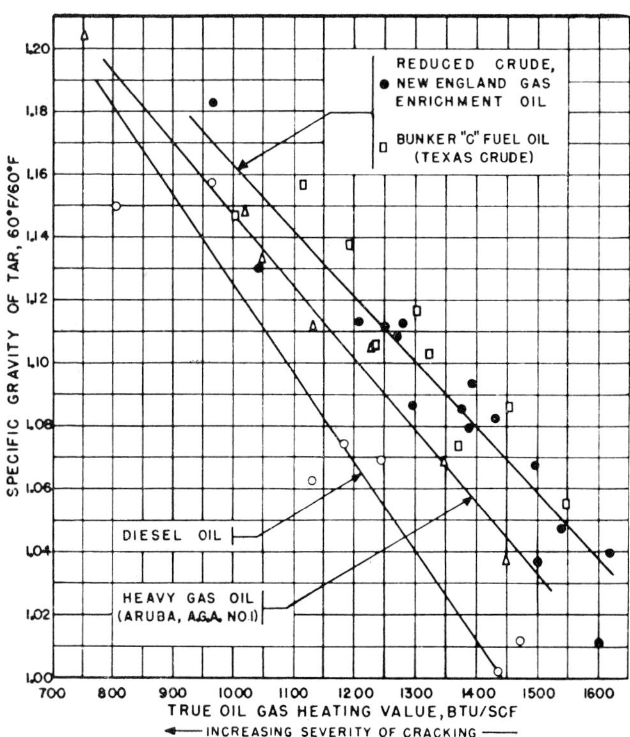

Fig. 3-50 Effects of oil type and severity of cracking on specific gravity of tar.[6]

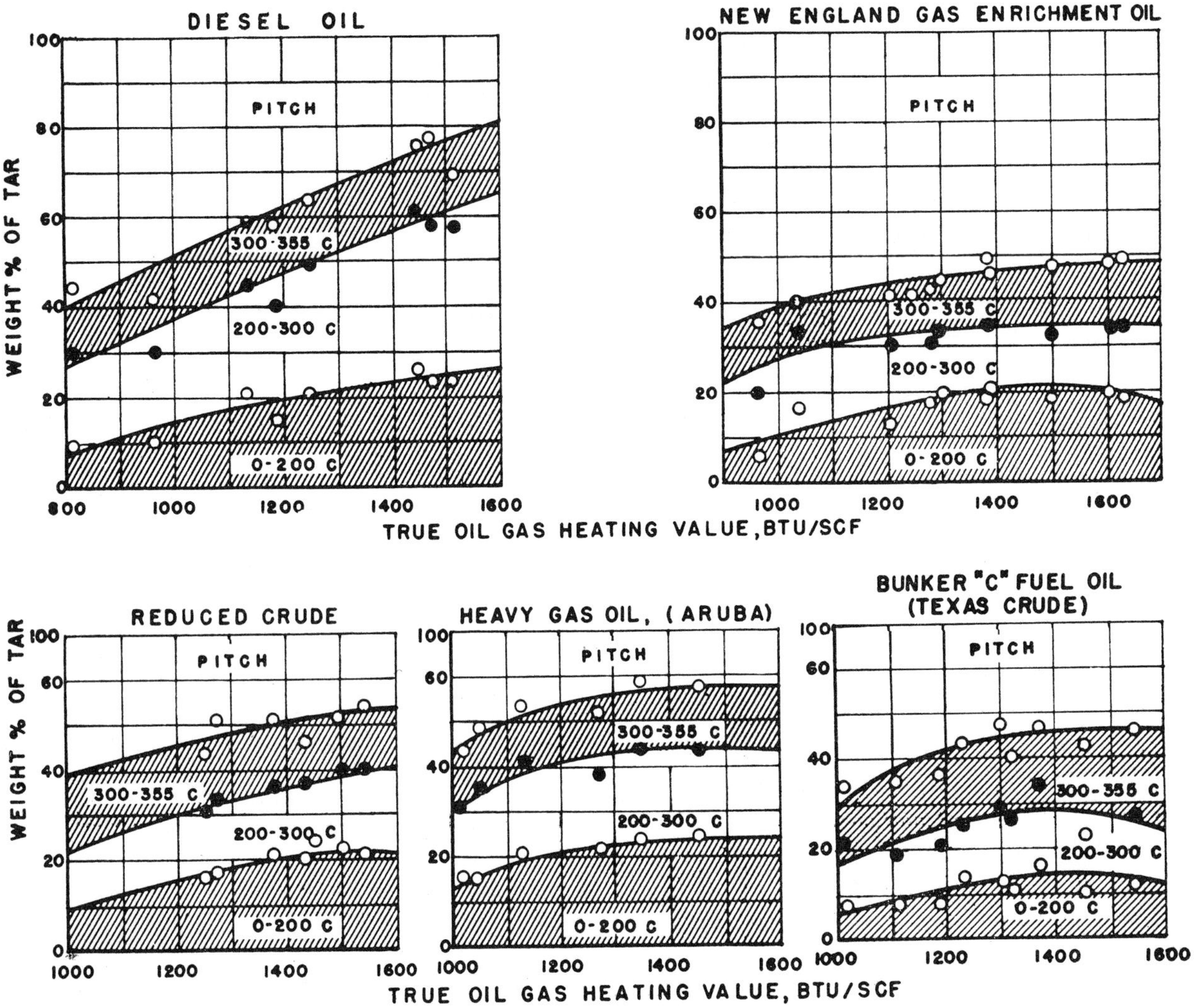

Fig 3-51 Effect of severity of cracking on distribution of tar fractions.[6]

uct gas.[8] Relationship of the oil gas heating value to the C/H ratio, cracking temperature, residence time, and oil gas partial pressure is given in Fig. 3-45.

Figure 3-46 shows variations of oil gas yields and compositions, make oil requirements, tar yields and compositions, and carbon depositions for three typical gas-making oils within the normal range of severities of cracking.

The weight per cent of the oil which forms carbon or coke is related to the heating value of the true oil gas and to the severity of cracking, as shown in Fig. 3-47. Increase of carbon formation, with decrease in oil gas heating value, is a C/H ratio function. High C/H ratio oils show a more rapid increase in carbon deposition.

Effects of Oil Properties and Operating Conditions on Products. *Liquid Products.* Effects of severity of cracking and of type of gas-making oil on tar products are shown in Figs. 3-48, 3-49, and 3-50. Gravities, viscosities, and free carbon contents of the tars as made increase rapidly with increasing severity of cracking and with increasing C/H ratios and Conradson carbon residues.

Viscosities of tars from the laboratory cracking unit presented in Fig. 3-49 are generally lower than those of dehydrated plant tars from cracking similar oils. This is because of prolonged exposure at elevated temperatures in the plant condensing and separating systems and less light oil in the plant tars.

Weight distribution (Fig. 3-51) of tar distillates among the 0 to 200 C, 200 to 300 C, and 300 to 355 C fractions for the five oils reflects the trends of other tar properties at various true oil gas heating values. The distillate to 300 C is consistently higher for the diesel oil and consistently lower for the Bunker C fuel oil (Texas crude) at comparable oil gas heating values than for the three intermediate heavy gas oils. The opposite is true for the residues above 355 C (pitch), which increase with increasing severity of cracking.

Gaseous Products. An approximate relationship exists between percentages of paraffins, unsaturates, and hydrogen and the heating value of a true oil gas cracked within a limited range of partial pressures.[6,8] This has been found to be true for diluent-free oil gases from all types of liquid petroleum fractions (*excluding* natural gas liquids) over a wide range of cracking conditions. Results obtained at one and three atmospheres are shown in Fig. 3-52.

There is a less empirical method for estimating gaseous product distribution based on a partial pressure *product constant* for the major gaseous product constituents.[13] This constant, K_p, *similar to an equilibrium* constant in that it is a function of cracking temperature only, is expressed as:

$$K_p = \frac{(H_2)^2(U)}{100\ (S)^2} \qquad \text{(e)}$$

where U and S are the mole per cents of *unsaturates* (olefins or illuminants) and *saturates* (paraffins), respectively. Ideal gas behavior is assumed.

Experimental partial pressure products of the gaseous paraffins, unsaturates, and hydrogen formed in oil gasification have been published[8,13] for a wide range of cracking conditions. The concept of the partial pressure product constant appears valid for oil gas produced from all types of petroleum oils under sufficiently severe cracking conditions to give essentially complete conversion, as in typical oil gas production.

A comparison of laboratory cracking tube and estimated gaseous product distributions for operating conditions typical for high-Btu oil gas production is shown below for a common gas-making oil (C/H = 7.05).

Average maximum cracking temp, °F	Res. time, sec	Oil gas composition, vol. %					
		Hydrogen		Paraffins		Unsaturates	
		Actual	Est.	Actual	Est.	Actual	Est.
1400	1.70	12.5	14	44.2	46	43.3	40
1400	3.13	15.2	14	45.6	47	39.2	39
1400	5.05	16.0	16	50.6	50	33.4	34
1550	1.62	25.0	26	45.2	47	29.8	27
1550	2.67	27.1	27	46.8	48	26.1	25
1550	4.71	30.6	32	49.0	49	20.4	19

An improvement of this method permitted estimates of the hydrogen, methane, ethane, ethylene, and propylene concentrations in the product gas.[14]

Mechanism of Oil Gasification

If exponents of mole percentages (Eq. e) are chosen to correspond to the partial pressure products characterizing the methane–ethylene–hydrogen equilibrium, the following is obtained:

$$K_p' = P\,\frac{(H_2)^2\,(C_2H_4)}{100\ (CH_4)^2}$$

The assumption that the paraffins are equivalent to methane and the unsaturates to ethylene closely approximates actual results, average carbon numbers generally being below 1.3 for paraffins and 2.6 for unsaturates in oil gases.

Thermal cracking data[15,16] for C_2 to C_4 paraffins and olefins agree moderately well with Eq. e. Partial pressure product constants calculated from other oil cracking data also show good agreement.[17]

Other developments in thermal cracking of petroleum fractions included the identification[18] of acetylene as a basic intermediate in hydrocarbon pyrolysis. The free radical mechanism was originally proposed[19] on the assumption that methyl ($CH_3\cdot$), methylene ($CH_2{:}$), and methyne ($CH{:}$) radicals exist as intermediates in such pyrolysis. Existence of such free radicals at temperatures of 1500 to 1600 F was later reported.[20,21] For a survey of the literature up to 1938, see Ref. 22.

Hydrocarbons of equal carbon–hydrogen ratio produce approximately the same quantities of gaseous products of similar composition when cracked under comparable conditions above 1400 F and above one sec residence time. These products are primarily methane, ethylene, and also hydrogen if insufficient hydrogen is present in the cracking atmosphere to suppress hydrogen liberation from the hydrocarbon feed. Most data available are for normally liquid hydrocarbons, but correlations for estimating thermal gasification yields from liquid petroleum fraction data are also found to be applicable to high-temperature pyrolysis for gaseous hydrocarbons.

During severe cracking, all types of hydrocarbons form similar fragments containing one or two carbon atoms. The formation of these fragments or free radicals—methyl ($CH_3\cdot$),

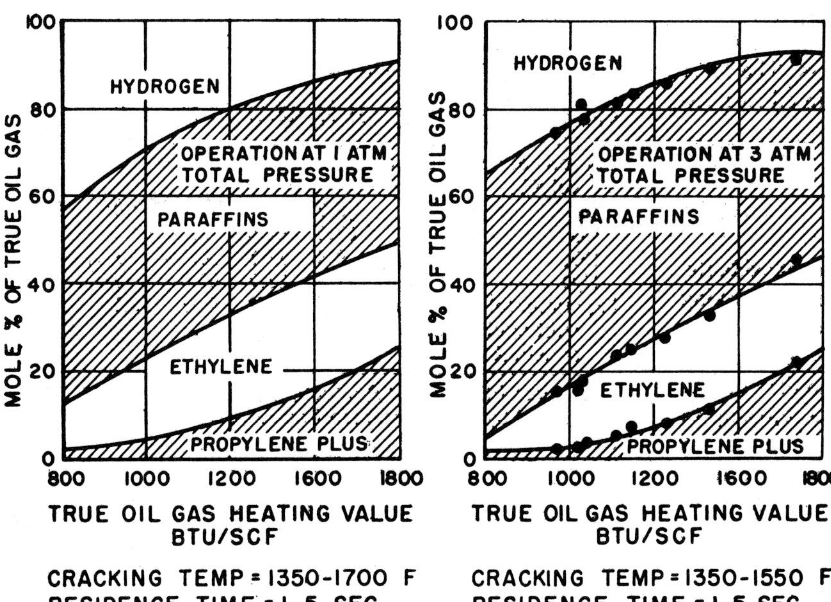

Fig. 3-52 Effects of severity of cracking and pressure on true oil gas composition.[6]

methylene (CH_2:), methyne ($CH\vdots$), and dimethyne ($\cdot HC = CH\cdot$)—is supported by the extremely short residence times (0.001–0.1 sec) and low pressures or low partial pressures. The cracking products under these conditions contain large quantities of ethylene, acetylene,[22,23] and direct dimerization products of methylene and methyne radicals. At higher pressures and longer residence times, increased condensation of these radicals to aromatics apparently takes place. Studies of acetylene production also indicate that for short residence times there is a tendency toward production of ethane rather than methane from the ethylene hydrogenation reaction.[23,24]

Carbon may be formed by secondary dehydrogenation reactions when free hydrogen is required to establish methane–ethylene–hydrogen partial pressure relationships, or whenever cracking conditions are so severe that localized irreversible dehydrogenation to carbon and hydrogen occurs. Increased pressures should favor condensation reactions and therefore give higher aromatic (tar) yields. They also cause higher gaseous paraffin concentrations. Published experimental data[7,25] support these deductions.

Effects of introducing gaseous hydrogen, namely, reduction of carbon and aromatics formation and increase in the paraffin-unsaturates ratio,[7,9,26] also stem from the partial pressure product relationship. At sufficiently high hydrogen concentrations in the cracking atmosphere no dehydrogenation is necessary, since the hydrogen partial pressure requirement is already met. Amounts of hydrogen liberated during cracking with no free hydrogen in the carrier gas are primarily a function of cracking temperature. They vary from 1 to 2 SCF per lb of oil or hydrocarbon feed at 1400 F to 2 to 4 SCF per lb at 1550 F, and 6 to 8 SCF per lb at 1700 F over a wide range of hydrocarbon types, pressures, and residence times.[9,15–17]

The proposed mechanism is in agreement with the experimental observation that the C/H ratio of the hydrocarbon cracked largely determines the relative distribution of gas and tar. Relative quantities of methylene and methyne or dimethyne radicals formed initially should depend upon this ratio. The methyne or dimethyne radicals (C/H = 12) would then be the principal tar formers and the methylene radicals (C/H = 6) would be the principal gas formers. The C/H ratio should therefore be the primary variable in oil gasification yield correlations.

Oxidation Reactions

When hydrocarbons are cracked in the presence of oxidants, CO and CO_2 are formed in addition to the normal cracking products. In the absence of catalysts, steam becomes an effective oxidant only when very severe cracking is involved. For example, when a 37.3° API, 6.37 C/H ratio diesel oil in an empty stainless steel cracking tube at temperatures up to 1770 F, 5 sec residence time, and a steam concentration of 25 mole per cent was cracked, no significant steam decomposition was observed. However, in catalytic reforming of the lower molecular weight hydrocarbons over nickel catalysts, the following equilibria are approached at reaction temperatures of 1200 F and residence times of <1 sec.[27] See Table 3-36.

$$CO + H_2O \rightleftharpoons CO_2 + H_2 \quad (4)$$

$$CH_4 + H_2O \rightleftharpoons CO + 3H_2 \quad (11)$$

Table 3-49 Effects of Feed Air on Oil Gasification

	Oil type				
	Reduced crude, 21.4° API, 7.0 C/H ratio, 4.4 wt % Conradson carbon residue				Diesel oil, 34.6° API, 6.76 C/H ratio, 0.0 wt % CCR
I.G.T. run No.*	400	395	398	399	404
Operating conditions					
Dry air feed, SCF/gal	27.06	31.49	22.94	18.31	22.98†
Steam feed, SCF/gal	0.81	0.91	0.63	0.69	0.51
Total pressure, atm	1.02	1.03	1.04	1.03	1.01
Partial pressure, atm	0.743	0.682	0.732	0.753	0.553
Cracking temp, °F	1710	1560	1470	1400	1410
Residence time, sec	2.45	2.57	2.55	2.43	2.38
Gal oil/cu ft free space-hr	2.98	3.45	4.23	5.21	4.64
Operating results					
Btu recovery, MBtu/gal‡	75.31	77.80	83.10	85.53	74.55
Oil gas yield, SCF/gal	89.50	72.01	66.55	60.82	54.66
Tar yield, wt %§	30.4	40.8	37.5	40.5	50.9
Light oil yield, wt %	...	5.480	5.097	4.578	3.308
Carbon yield, wt %	22.5	8.6	5.2	5.1	0.2
True oil gas composition, vol. %					
H_2	45.9	25.1	16.7	13.0	17.3
CH_4	39.8	43.8	40.7	37.1	37.0
C_2H_6	0.8	2.6	4.9	6.9	4.2
C_3H_8	...	...	0.3	0.5	...
Butanes	...	0.1	...	...	0.1
Pentanes	...	...	...	...	...
C_2H_4	10.6	23.7	26.3	26.2	27.0
C_3H_6	0.1	1.3	5.8	10.8	8.0
Butenes	...	0.1	0.7	1.6	1.0
Pentenes	...	...	...	0.1	0.1
Acetylene	0.5	0.4	0.3	0.1	0.1
Butadienes	0.3	0.7	1.4	2.0	2.0
Pentadienes	...	0.1	0.4	0.5	0.6
Benzene	1.9	2.0	2.2	1.1	2.2
Toluene	0.1	0.1	0.3	0.1	0.4
Total	100.0	100.0	100.0	100.0	100.0
True oil gas heating value, Btu/SCF	815	1053	1228	1382	1275
Oxygen gas yield, SCF/gal**	11.53	12.19	9.76	7.87	31.81
Oxygen gas composition, vol. %					
O_2	5.2	3.4	...	2.0	0.5
CO	63.6	51.8	43.4	50.3	48.5
CO_2	9.4	12.1	4.6	4.1	7.3
H_2O	21.8	32.7	52.0	43.6	43.7
Experimental water gas shift equilibrium constant	0.417	0.952	4.27	5.41	9.80

* Project PB-8, 1952, "Production and Utilization of Oil Gas," sponsored by Gas Production Research Committee of A.G.A. at I.G.T.

† Oxygen-enriched air, 73.5% oxygen.

‡ For stable, unscrubbed gas.

§ Corrected to 100% material balance.

** Includes carrier steam plus H_2O formed as calculated from oxygen balance.

In oil gas production such catalytic activity was unusual. Only small percentages of CO and CO_2 were formed by steam decomposition. This was confirmed by plant scale tests using Pacific Coast (low-Btu) oil gas sets.[28,29] Product gases from these sets contained substantial amounts of CO and CO_2, originally thought to be due primarily to steam decomposition at the high cracking severities employed. However, CO and CO_2 production persisted after conversion to high-Btu oil gas operation. Further tests in the absence of steam indicated no decrease in "steam gas" production. Oxygen appeared to be released from the multivalent metal oxides in the heavy feed oil ash depositing as slag on the refractory surfaces. Their reduction from higher to lower stages of oxidation during oil admission and reoxidation during the blast period appeared feasible. Oxygen deficiencies were observed[29] in the stack gases during the blast period equivalent to approximately twice the oxygen accounted for in the CO and CO_2 content of the make gas. The remainder of the oxygen released during oil admission probably formed water vapor.

Table 3-49 gives laboratory data for gasification of a reduced crude oil and a diesel oil in the presence of air and oxygen-enriched air. These results are similar to the plant data discussed above, although the portion of the feed oxygen forming steam seems considerably less than half the total, particularly at higher cracking temperatures.

A number of processes in development stages employ air, oxygen, or oxygen-enriched air mixtures in hydrocarbon gasification. Since oxidation reactions are highly exothermic, enough heat can be liberated by partial combustion to satisfy the process requirements. An additional objective in most partial combustion processes is the reduction of nongaseous product formation, particularly of carbon lay-down, to permit continuous operation. However, except for the lowest molecular weight, low C/H ratio feed stocks, complete gasification appears attainable only at very high oxygen-to-oil ratios, giving product gases of low heating value and high specific gravity. In the range in which complete gasification is not reached, oxygen does not appear to alter the true oil gas composition (Table 3-49).

Tar composition was similarly determined by oil type and true oil gas heating value. Thus, to gasify a proportionately higher percentage of the feed oil thru carbon oxidation to CO and CO_2, reduction of steam formation from the hydrogen content of the oil is necessary. If no feed oil hydrogen is lost, all CO and CO_2 must originate from hydrocarbon fragments normally forming tar and carbon.

METHODS OF OIL GAS MANUFACTURE

Gas Machinery Oil Gas Set

This equipment consists of two shells (Fig. 3-53), a generator and superheater, similar to the carburetor and superheater of a conventional water gas set. The generator top is equipped with a tee connection to accommodate steam and air blast lines, sight cocks, a lighter opening, and an oil burner and make oil spray either singly or in combination.

Suitable meters, strainers, and pressure regulators are installed in both the burner oil and make oil systems. Steam and air meters are also included. A six-lever automatic control is connected to purge steam supply to make oil spray, make oil, air blast, stack valve, burner oil, and main or sweep steam. A rearrangement of the purge steam control can be made, if necessary, to start and stop an individual air blower. Checkerbrick spacing is usually a $1\frac{1}{4}$ in. staggered flue in the generator and a $2\frac{1}{2}$ in. staggered flue in the superheater.

Converting a standard three-shell *water gas machine* to a single generator oil gas set requires disconnecting the original generator by removing the generator–carburetor connection and the dust trap. The carburetor inlet connection is then bricked up and blanked off. Back-run lines and valves are removed and a tee connection is installed on top of the former carburetor. A converted machine at one installation produces 250 MCFH of 1050 Btu gas (0.74 to 0.76 sp gr), using No. 2 oil as feed stock.[30]

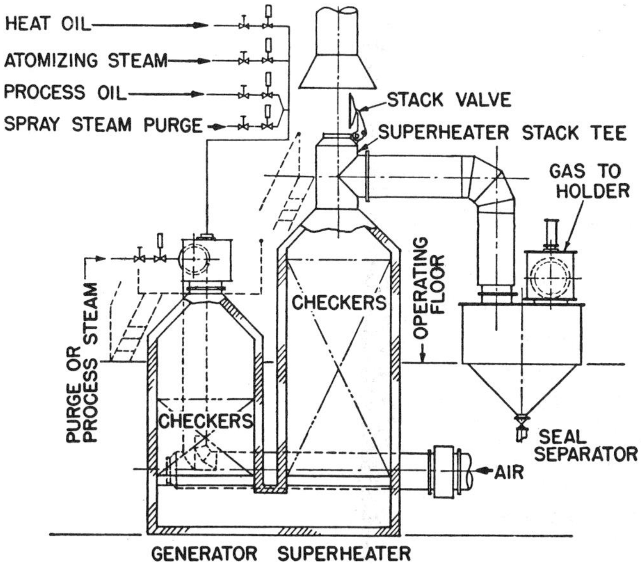

Fig. 3-53 Single generator high-Btu oil gas process for light oil operation. (Patented, Gas Machinery Co.)

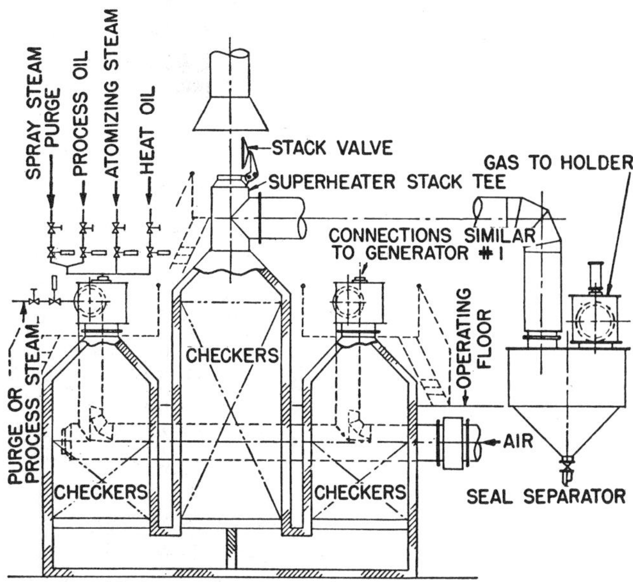

Fig. 3-54 Twin generator high-Btu oil gas set for light oil operation. (Patented, Gas Machinery Co.)

An 11 ft single generator oil gas set will produce between three and four MMCF of gas per day, depending upon the kind of oil used. Make will increase with API gravity of the oil. High production rates can be expected from naphthas and lower rates from 30° API gas oils.

Twin Generator High-Btu Oil Gas Set

The 1960 set shown in Fig. 3-54 differs from the earlier arrangements in which the superheater is located between the two generators. It is claimed that there is better gas flow thru a centrally located superheater. The feed stocks for the machine shown are light distillates up to and including No. 2 oil. A suggested three-minute operating cycle follows: make, 65 sec; steam purges, 13 sec; blow purge, 9 sec.; heat period, 86 sec; and valve changes, 7 sec.

Back Blast. Two general arrangements were used for fitting the twin generator for back blast.[31] In both, back blast air entered at or near the superheater top and passed in a reverse direction downward through it, then divided and passed upward thru the twin generators. Combustion products then united in a common back blast purge pipe and went from there to a stack. The three-way valve shut off the back blast from the wash box and the back-run pipe was connected to the back blast purge pipe near its top, serving both as a vent connection, if necessary, and as a means for a back purge. In another arrangement, an angle valve at the wash box top is used to control the direction of the back blast.

In operation, the back blast serves both to burn carbon deposits from the checkers of the generator and superheater and to carry heat from the superheater back to the generators.

Semet-Solvay Regenerative Reverse Flow High-Btu Oil Gas Generator

The Semet-Solvay high-Btu oil gas processes incorporate the principle of counterflow of atomized oil and carrier gas in a checkerless reactor. This principle previously proved successful in Semet's reverse flow carbureted water gas process. The generator is a two-shell four-chamber unit making gas in alternate or reverse directions during each cycle (Fig. 3-55). Process features follow:

1. Make oil is injected downward in the reactor, countercurrent to an ascending flow of carrier gas which enters the checkerless space thru an aperture. The velocity developed here prevents the passage of oil vapors or carbon into the checkered chamber space below.

2. Vaporized oil and make gas pass to the other reactor and descend thru the adjoining checker chamber. Controlled heating and vaporization of the make oil are thus induced in the reactor chambers and fixing of the oil gas vapors occurs in the outlet checkered chamber.

3. During the alternating make period carbonaceous material is deposited on the walls of both reactors and fixing chamber checkers. During the alternating reverse flow heating cycle both carbon and fuel oil are burned, and heat is absorbed in both reactors and fixing chambers. This heat is subsequently transferred either to up-run steam or carrier gas or to down-run oil vapors and make gas.

4. The products of combustion are forced to *descend* thru the reactors and checkered chambers, while the gases requiring preheat are forced to *ascend*.

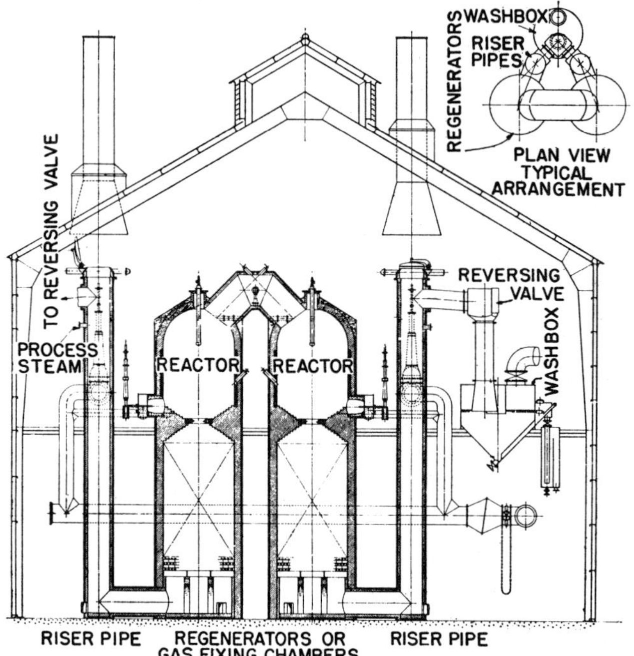

Fig. 3-55 Regenerative reverse flow oil gas machine for all grades of oil. Recycle gas line connects riser pipes to reactor manifold (not shown). (Semet-Solvay)

Operating experience has demonstrated that mixtures containing over 75 per cent oil gas from these generators and natural gas are fully interchangeable with natural gas. Advantages claimed for this process include: easy control of gas composition; complete control of gas gravity; smokeless operation; no objectionable odors; tar produced without stable emulsifications and easily dehydrated; and dependable, uniform day-to-day operation.

Production capacity of this unit is claimed to approximate 3 MSCF of 1000 Btu gas per sq ft of reactor cross-sectional area per hour. It is claimed that total consumption of an oil with a recoverable Btu value for gas enrichment of not less than 100,000 Btu per gal and a Conradson carbon content of not more than five per cent will not exceed 12.5 gal per MCF of 1000 Btu gas produced. Heavier oils with Conradson carbon content of not more than nine per cent can be used with a relative reduction in production capacity and a slight increase in oil consumption.

Poughkeepsie Process—Semet-Solvay Two-Shell Single Run Oil Gas Machine

This unit (Fig. 3-56) uses lighter grades of make oil, such as gas oil, diesel, H-fuel, and gasoline. The design is adaptable to any type of reverse flow or standard water gas machine. A very simple up-run process is utilized.

The burner nozzle inducing the heating medium and burner air is located above a firebrick constructed aperture in the reactor shell. At this point the combustion is completed by the main air, which enters thru the aperture. The combustion products then radiate upward into the fixing chamber to the open stack.

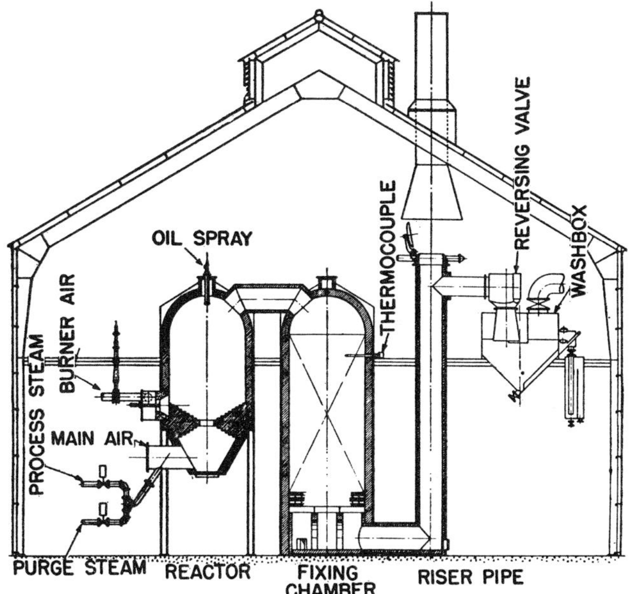

Fig. 3-56 Poughkeepsie process—two-shell oil gas machine for light oil. (Semet-Solvay)

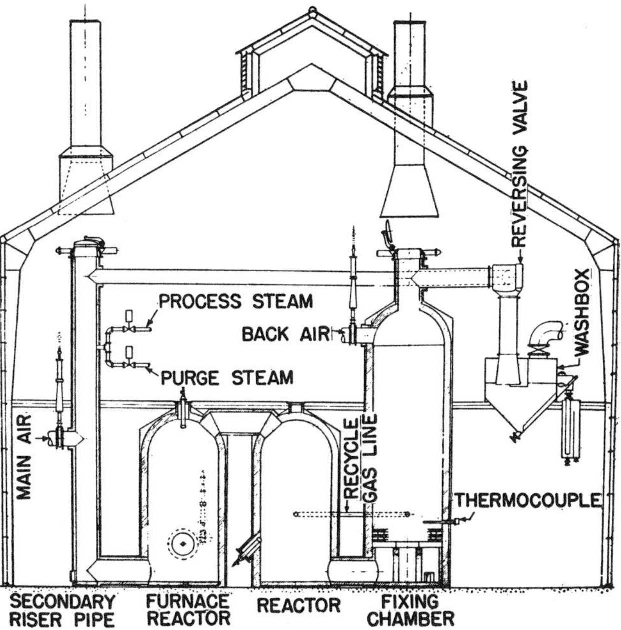

Fig. 3-57 Poughkeepsie process—three-shell regenerating oil gas machine. (Semet-Solvay)

The temperature-controlling thermocouple is located about eight courses down in the checker-brick area of the fixing chamber. When the temperature reaches a set controlling point, make oil under pressure is sprayed into the reactor thru a steam atomizing oil spray, countercurrent to an ascending flow of carrier steam. This steam enters thru the aperture, which develops enough velocity to prevent passage of oil vapors below it.

The make period is followed by an increased amount of steam for purging. Alternatively, this purge period can be followed by an air purge which creates a lean combustion gas. Some of this lean gas can be passed into the gas system, but the amount must be controlled to avoid adverse Btu, specific gravity, or chemical composition changes.

After the purge period a new cycle starts. The cycle is approximately three minutes long, consisting of a 45 per cent heat period and a 55 per cent make and purge period.

Poughkeepsie Process—Semet-Solvay Three-Shell Regenerating Oil Gas Machine

This unit (Fig. 3-57) uses all light or heavy grades of oil, with up to a nine per cent Conradson carbon content and without cycle changes. The design is adaptable to any type of reverse flow or standard water gas machine embodying the principle of a checkerless counterflow gasification chamber.

The burner nozzle inducing the heating medium and burner air is located in the furnace reactor. Additional air for total combustion enters below the burner nozzle. The combustion products then flow upward thru the connecting reactor and fixing chamber to the open stack.

The temperature-controlling thermocouple is located about eight courses up in the checker-brick area of the fixing chamber. When the temperature reaches a set controlling point, about 25 per cent of the make oil is injected downward into the furnace reactor, while the remainder is simultaneously injected upward into the reactor. Carrier steam enters the furnace reactor, counterflows, and motivates the gasified particles thru the machine into the conditioning system.

The make period is followed by an increased amount of steam for purging. This is followed by a back-air purge. In heavier oil operation some carbon is freed and remains in the producing unit. By use of this back-air purge period, deposits may be kept to a minimum. The first part of the air stream reacts with the carbon, creating a lean combustion gas which is purged into the gas system. The secondary stack valve then opens, and carbon is burned to the atmosphere.

After this back-air purge period a new cycle starts. The cycle is approximately three minutes long. It consists of: heating, 72 sec; making, 72 sec; steam purging, 27 sec; and back blow, 9 sec.

Hall High-Btu Oil Gas Process (Four-Shell)

Based on a design by Edwin L. Hall, this process[32] utilizes high-carbon low-cost oils for efficient production of high-Btu gas. A regenerative cycle is used to maintain relatively low temperatures in the superheater tops and relatively high temperatures in the generators, giving high thermal efficiency and providing maximum heat storage in the generators for oil vaporization. These conditions are accomplished by admitting air into the superheater tops to cool their checker-brick and to preheat the air required to burn carbon from the checker-brick in the bases of the superheaters and in the generators, where carbon deposits are heaviest. Introducing steam into the superheaters during the make period results in a superheated steam atmosphere, which aids oil vaporization and cracking.

The cycle used was four minutes for the blow and make periods in one direction, or a total of eight minutes for the complete regenerative cycle. It was later reduced to a total of six minutes, three in each direction, to reduce temperature

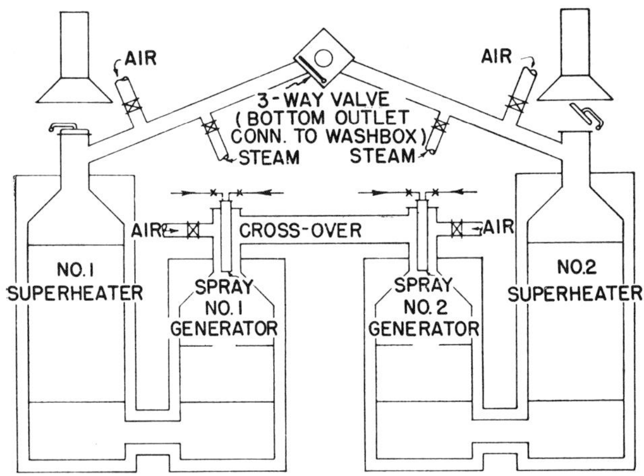

Fig. 3-58 Hall oil gas process.

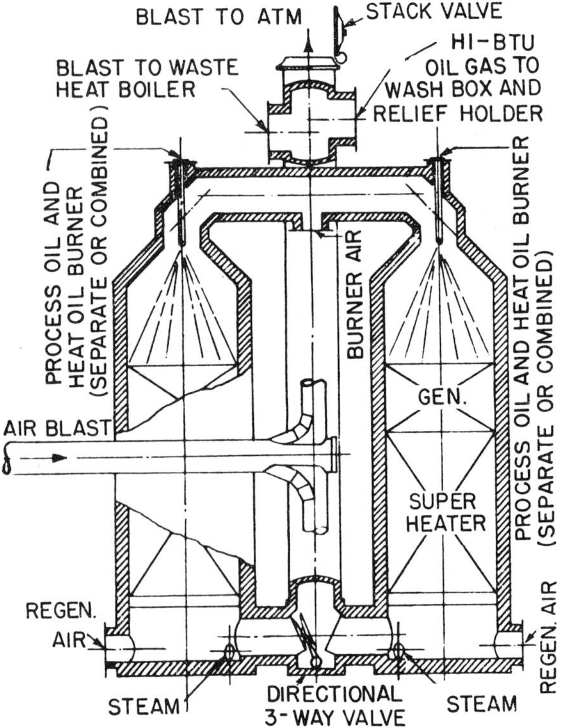

Fig. 3-59 Two-shell regenerative oil gas set; Hall process, Henry design. (Gas Machinery Co.)

variation in the checker-brick during the cycle, thereby obtaining more uniform cracking conditions and reducing deterioration of the generator checker-brick because of spalling (Fig. 3-58).

Silicon carbide checkers are usually used in the generators to provide a more rapid heat release for oil vaporization and thereby increase gas-making capacity. Generator checkers are replaced more frequently than superheater checkers because of spalling. The large amount of heat required from the generator checkers to vaporize the oil causes enormous changes in their temperature, particularly in the top courses, causing the brick to break into small cubes. The rate of spalling depends upon the magnitude of temperature changes which, in turn, is affected by the cycle length and the amount of oil vaporized per cycle. Checker life of from 1000 to 3500 hr has been reported, depending upon operating conditions and type of brick.

Rapid ash accumulation on the checkers occurs from oils of high ash content. Oils of less than 0.05 per cent ash can be used without accumulating serious ash deposits in the checker-brick which would reduce its normal life.

In addition to the original four-shell design, many variations have been constructed. The most familiar is the two-shell inverted "U" type with generators built on top of the superheaters in the same shells.

Two-Shell Oil Gas Set (Hall Process, Henry Design)

This set (Fig. 3-59) is essentially the same as the four-shell except for greater length to accommodate both a generator and a superheater in one shell. It is termed an "inverted U" because the superheaters are on the bottom of the shells and the crossover at their tops.

U.G.I. Oil Gas Process

This *water gas set conversion* requires a minimum of expenditures (Fig. 3-60). The generator becomes the primary vaporizer, with oil burners and wind boxes mounted on its side, a process oil spray at the top to spray downward, and branches for steam admission. The carburetor becomes the secondary vaporizer, with an oil spray at its top and full checkering. The superheater becomes the fixing chamber and is also checkered.

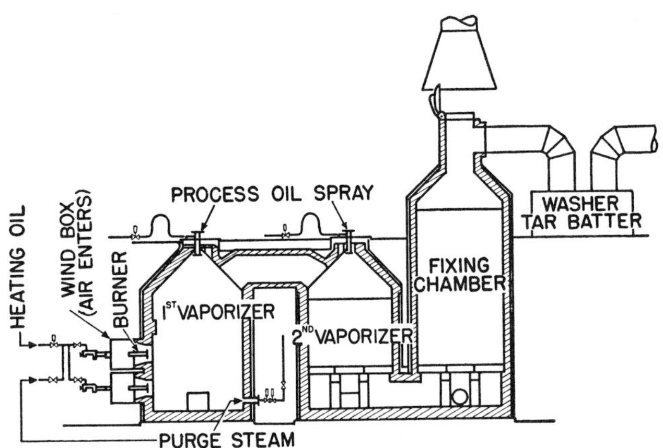

Fig. 3-60 U.G.I. high-Btu oil gas process. (Applied to existing three-shell carbureted water gas sets.)

During the blow period, heating oil and air are introduced into the bottom of the primary vaporizer. During the run period, process oil is sprayed into the top of the primary vaporizer, steam is admitted into its bottom, and additional oil is sprayed into the secondary vaporizer.

Honolulu Oil Gas Process

The Honolulu Gas Co. utilizes two types of generator sets, one of which is a modified two-shell Jones regenerative set (Fig. 3-61), and the other a unique arrangement of a four-shell set. Both types of generators have two sets of generator checkers and two sets of regenerator checkers. Only the generator checkers are contacted by the make oil. The regenerator checkers absorb and store flue gas waste heat and use it to preheat the purge and process steam and heat air. Generator cycles are the same, with modified steam, air, and oil flow rates.

One-half generator cycle follows the make, purge, heat, purge, and reverse pattern. During the make period, 180 gal of 200 penetration grade asphalt at 320 F is sprayed on the upstream make checkers. Process steam, preheated by the regenerator, is introduced during the gas-making period at a *decreasing* rate (230 to 30 lb per min). Vaporizing and thermal cracking of the oil occur as the oil passes downward thru generator "A," across the throat, upward thru generator "B," and off to the wash box thru the side offtake nozzle. The oil vapors completely by-pass the regenerator checkers.

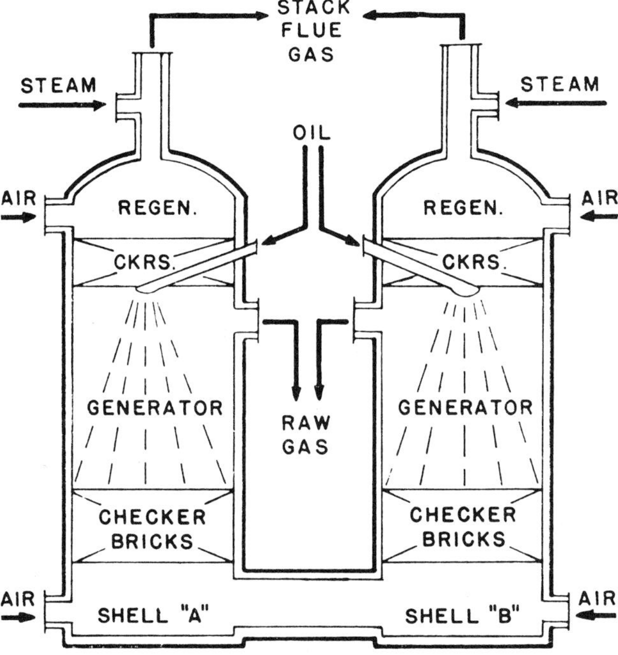

Fig. 3-61 Jones two-shell regenerative oil gas set.

After the make oil period, purge steam and then purge air introduced thru the "A" regenerator checkers rid the generator checkers of oil and gas vapors. As the purge air reaches the offtake to the wash box, the stack valve on side "B" opens, releasing the flue gases to the atmosphere. The heat and blow period continues, with some low-pressure purge steam being introduced into the offtake, regenerator, and throat. During the latter part of the heat period, throat air is admitted into the base of the generator shells as the regenerator air is closed to burn out the carbon deposits at the bottom. Normally, no heat oil is required to maintain proper cracking temperature. Combustion of the residual carbon provides sufficient heat, and the necessary temperature control is provided by varying flows of air and process steam.

The steam purge thru regenerator "A" follows the heat period flow pattern out thru shell "B," completing one-half of the cycle. *The direction of flow is reversed on the subsequent make period.*

The Ranarex gravity of the wash box raw gas and the quality of tar in the wash box are used as the primary guides to cracking severity. The production of a 1260 Btu per cu ft lean true oil gas from asphalt feed stock yields a wash box tar (viscosity: 1500 SSF at 180 F) which can be transferred by pumps and used as boiler fuel. A blend of one part wash box tar and three parts scrubber tar (200 SSF at 122 F) is sold to a manufacturer of electrode pitch.

An alternating water-sealed wash box[33] successfully handles the pastelike tars.* Eighty-five per cent of the heat content of the asphalt cracked is recovered in the gas, light oil, and tar products.

Refinery Oil Gas

Chief sources of gas production in a refinery are dissolved gases from distillation of crudes and by-products of refinery cracking operations. Gases from these sources contain: (1) considerable quantities of propylenes and butylenes which may be polymerized or alkylated into high antiknock gasoline components; (2) butanes which are used to increase the vapor pressure of gasoline; and (3) pentanes and heavier components economically desirable for inclusion in gasoline production. Refineries therefore usually install equipment to recover about 75 per cent of the total C_3 fractions, about 95 per cent of the C_4 fractions, and all C_5 and heavier fractions normally included in the refinery *wet* gas production. Usually, recovery is by means of an absorption system with subsequent stripping of these fractions from the absorption oils by distillation. The resultant *dry* gas product is normally used as a fuel.

In a light oil refinery equipped with usual distillation, catalytic and thermal cracking, and catalytic polymerization facilities, 160 cu ft of dry gas is obtained from processing a barrel of crude oil. Properties of this gas follow:

Typical Refinery Dry Gas

(0.886 sp gr, 1388 Btu per cu ft)

Component	Mole %
H_2	12.7
Methane	28.1
Ethylene	11.6
Ethane	17.1
Propylene	6.6
Propane	11.9
iso-Butane	2.2
Butene	0.6
Butane	0.1
H_2S	2.6
CO	5.4
CO_2	1.1
Total	100.0

The H_2S and CO_2 above can be removed by many gas purification processes. The H_2S is due to breakdown of sulfur-containing hydrocarbons in the distillation and cracking processes. CO and CO_2 originate primarily from adsorption on, and subsequent evolution from, circulating regenerative cracking catalysts.

Analysis of refinery oil gas shows it to be similar to high-Btu oil gas. It can be used as an enricher for low-Btu gases, or as a fuel bed reforming material.

* Viscosity at 210 F, 400 SSF.

Table 3-50 Diesel Oil Gasification by Fluid Cracking

°API	41.4	41.4	41.4
C/H ratio	6.23	6.23	6.23
Cracking temp, °F	1228	1250	1298
Oil gas res. time, sec	11.1	11.1	12.4
Oil gas partial pressure	0.89	0.89	0.84
Solid/oil ratio	5.9	7.5	7.6
Btu recovery/gal	90,450	86,910	84,090
Oil, gal/MMBtu	11.06	11.5	11.9
Gas, SCF/gal	61.6	62.7	66.9
Tar, wt %	46.1	42.1	47.5
Coke, wt %	0.6	1.1	1.0
Heating value, Btu/SCF	1469	1387	1257
Sp gr (air = 1)	0.87	0.81	0.73
Gas composition, vol. %			
H_2	9.9	11.0	13.1
Paraffins	48.8	51.2	53.9
C_2H_4	23.6	23.7	24.4
C_3H_6+	17.7	14.1	8.6

Ugite Process

The Ugite process involves control of cracking conditions to a degree not usually attained in normal oil gas operation.[34]

The objective of this process is to reduce oil gas costs by increasing the by-product credits. Since a gas utility is primarily concerned with base and peak gas loads, the handling of a complex pyrolytic hydrocarbon by-product appeared to be too far afield. A noteworthy solution involves close cooperation between a utility and a chemical manufacturer. Decreased idle time on peak load facilities would be a major advantage to the participating utility.

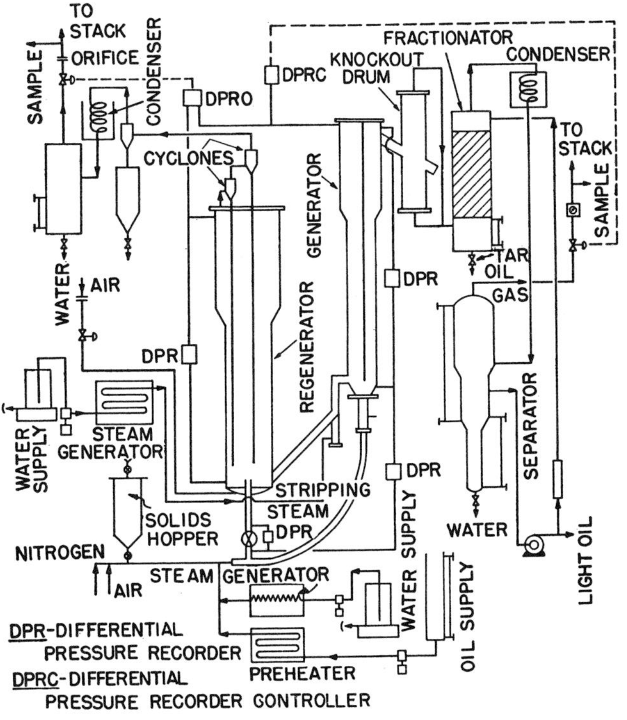

Fig. 3-62 Flow diagram of oil gasification fluid pilot plant.

Table 3-51 Operating Data from Various Oil Gas Plants and Processes

	Process and generator diam							
	4-Shell Hall regenerative (Cambridge)		2-Shell Henry design Hall regenerative (Cambridge)		Semet–Solvay regenerative reverse flow, high-Btu oil gas		Regen. thermal cracking (Los Angeles), 27 ft	UGI oil gas
					11 ft conversion	13 ft × 58 ft		
	1	2	3	4	5	6	7	8
Fuels used: Designation	N.E. No. 6	Aruba	N.E. No. 6	Aruba	Gas oil, naphtha,	All light oils	Naphtha	
Grade	(Gulf)	(Esso)	(Gulf)	(Esso)	gasoline, diesel,	plus Bunker	54.6	
°API	18–20	18–20	18–20	18–20	etc.	C	...	
% Conradson carbon	6.4	6–7	6.4	6–7		Up to 13	...	
Cycle: minutes, total	7		7		3.5	3.5	5 min, 52 sec	
make, %	30		33		40	35	42	
heat, %	50	↑	43	↑	40	35	50	
blow, %	St. purge 8		St. purge 9		5	12	Valving 3	
purge, %	8		9		15	18	5	
Operating variables								
Temp, °F, Shell	Bot. 1,400		Bot. 1,450		1,600	1,580	1,310	1,465
Shell	Top 1,200		Top 1,250		...	...	...	
Shell		Same as Column 1		Same as Column 3	1,460 bot. checkers	1,380 bot. checkers	...	
							...	
Make oil	200		200		60–70	235	572	
Steam	15 psig		15 psig				610	
Rates: Blast air, CFM	9,000–16,000		12,000–16,000		34,000	32,000	13,350	
Purge air, CFM	12,000		6,000–10,000		3,500	12,500	...	
Steam, lb/hr, Make	100/min		125/min		2,600 max	3,100 max	34,315	
Purge	275/min		275/min		2,600 max	3,100 max	34,315	
Make oil, gal/cycle	150		175		110–150	140–180	372	
gpm	150	↓	175	↓	80–110	120–150	151	
Residence time, sec	...		...		1.5	1.5	1.5	...
Total pressure, atm	...		...		1.1	1.2	1.12	37 in.
Capacity: MCF/hr	180	185	220	222	190–260	190–260	265	145
MCF/day	4,000	4,070	5,280	5,328	4,500–6,000	4,500–6,000*	6,354	3,480
Hours operation per day	22	22	24	24	23	23	24	
Thermal yield, MBtu/gal	88	...	93	...	100	83.5	106	
Make oil usage, gal/MCF	11.45	12.08	10.52	12.60	10	11.5	13.5	8.63
Heat oil usage, gal/MCF	0.67	0.78	1.13	0.68	1.5	1.7	Gas 0.244	1.26
Gas quality, Btu/cu ft light oil free						1,270		
Raw gas (generator offtake)	1,010	1,000	1,000	1,010	1,040	1,040	Outlet purif. 1,240	810
True oil gas	1,308	1,385	...	...	1,152	1,152	1,430	
Tar yield, gal/MCF dry	2.86	3.02	2.63	3.15	2.0	3.0	0.80	

* Based on 1000-Btu gas.

Fluid Cracking

Oil gasification in a fluidized bed type of cracking unit similar to cracking equipment used in the oil industry has been carried out in a pilot plant by I.G.T. under Gas Production Research Committee sponsorship. Figure 3-62 indicates basic flow of materials thru component pilot plant equipment used.

Essentially, oil contacts hot, finely divided solid particles in a bed which is maintained in the fluidized state by upward flow of steam and oil vapors. Thermal cracking takes place on the hot solids, producing oil gas and vaporized tars, and coke which is deposited on the solids. Oil gas, tars, and light oil vapors are withdrawn and recovered conventionally. Coke-covered solids are continuously transferred from the reaction zone to a regenerator adjacent to the generator. In it carbon is burned off with air, raising the temperature and heat content of the solids to furnish heat for feed oil vaporization and cracking. Regenerated solids are transferred back to the generator to be contacted with fresh oil feed, forming a continuous gasification process. Solids conveyance is accomplished pneumatically, by gravity, or by a combination of these methods.

These solids include 30 mesh silica gel or other materials which have good resistance to attrition, such as granular refractories, silica–alumina, and high-temperature coke breeze. Unlike the gasoline fluid processes, *no catalytic effect* is produced by the heat transfer solids at the elevated temperatures encountered in oil gasification units.

Premium distillate and residual feed stocks can be utilized with high C/H ratio residual oils after further development of the oil feed system to prevent coagulation of heat transfer solids from excessive pitch and coke formation. Oil gases produced can be varied from 1050 to 1500 Btu per SCF and from 0.6 to 0.9 specific gravity, depending upon operating conditions and make oil properties. Under normal conditions, high-severity cracking can be obtained at low reaction temperatures because of rapid heat transfer and long oil gas residence times.

Fluid gasification has the advantages of continuous coke re-

Table 3-52 Product Analyses from Various Oil Gas Processes

	4-Shell Hall regenerative		2-Shell Henry design Hall regenerative		Semet-Solvay regenerative reverse flow		Regenerative thermal cracking	UGI oil gas
	N.E. Sp No. 6	Aruba 270	N.E. Sp No. 6	Aruba 270	Conv. w.g. set	High-Btu oil gas		
	1	2	3	4	5	6	7	8
Raw gas analysis—light oil free, vol. %								
H_2	15.1	11.6	16.6	10.4	18.3	14.5	15.6	19.2
Methane, CH_4	34.6	33.9	33.1	36.6	36.3	33.4	28.0	34.8
Paraffins, C_2+	...	...	...	...	4.1	6.9	6.0	1.2
Ethene, C_2H_4	...	...	...	...	24.8	24.6	23.8	
Propene, C_3H_6	...	...	...	...	2.2	9.7	9.6	19.4
Unsaturates, C_4+	III. 25.1	26.7	27.1	27.6	1.3	5.5	5.6	
CO	1.6	2.0	1.9	1.8	2.6	1.0	1.0	1.6
CO_2	6.2	6.5	3.8	5.1	1.6	0.4	1.9	2.4
O_2	1.0	0.6	1.6	0.8	0.5	1.0	0.0	2.4
N_2	16.4	18.7	15.9	16.7	8.3	3.0	8.5	19.0
Heating value calc. dry, Btu/cu ft	...	...	...	...	...	1264	1240	875
Heating value sat. at 60°F, Btu/cu ft	1001	1000	1005	1010	1045	1020	1220	810
Sp gr, calc. dry	0.84*	0.86	0.84	0.886	0.725	0.78	0.86	0.69
Raw gas composition, vol. %								
Air					2.5	5.0	0.0	
Combustion products					...		2.0	
Oxidation gas					...		9.4	
True oil gas					87.0	94.6	88.6	
True oil gas analysis, vol. %								
H_2	20.2	16.0	21.6	13.9	21.1	15.3	17.6	
Methane, CH_4	46.2	46.9	43.2	48.9	41.7	35.3	31.6	
Paraffins, C_2+	33.6				4.7	7.3	6.8	
Ethene, C_2H_4	33.6				28.5	26.0	26.9	
Propene, C_3H_6	33.6	37.1	35.2	37.2	2.5	10.3	10.8	
Unsaturates, C_4+	33.6				1.5	5.8	6.3	
Heating value calc. dry, Btu/cu ft	1338	1385	1322	1354	...	1418	1430	
Sp gr, calc. dry	...	...	...	...	...	0.822	0.83	
Tar: sp gr at 77/77 F	...	1.167	...	...	1.197 at 15.5 F	1.227 at 15 F	1.09	
sp viscosity, Engler 50 cc, 40 C	...	...	...	...			203 SUS at 100°F	
Float test at 89.6 F, sec	...	63.8	...	...				
Insolubles, in carbon bisulfide, wt %	...	12.7			20.0	18.8	...	
Tar distillation to 338 F	...	1.4					2	
to 392 F	...	2.3					5	
to 455 F	...	3.5					19	
to 518 F	...	10.9					39	
to 572 F	...	16.7				18.4	64	
Light oil and carbon omitted; data incomplete.								

* By Ranarex.

Table 3-53 Comparison of Plant Results and Operating

Type of plant process	Make oil data									Cracking temp, °F			Operating
			Distillation, °F			Viscosity at							
	Type	°API	10%	50%	90%	100 F	122 F	Aniline point, °F	Est. C/H ratio	Carburetor or oil gas generator	Superheater	Gal make oil/MCF	Steam dur. oil adm., lb/gal make oil
Hall	A.G.A. No. 3†	30.2	480	575	682	5.31 CSKS		146.5	6.86‡	1461–1689	1080–1663	11.50§	0.49
	A.G.A. No. 2†	19.0	552	801	...		157.5 SSU	151.0	7.35‡	1440–1627	1038–1574	11.46§	0.71
	A.G.A. No. 1†	22.2	545	903	...		193.2 SSU	179.7	7.05‡	1433–1569	1043–1602	10.67§	1.04
	A.G.A. No. 4†	14.2	650	985	...		163.8 SSF		7.73‡	1441–1563	1024–1588	11.69§	1.47
	A.G.A. No. 5†	12.0	688	972	...		153.4 SSF		8.03‡	1549–1653	1034–1595	13.31§	1.58
Twin Gen.	A.G.A. No. 3†	30.2	480	575	682	5.31 CSKS		146.5	6.86‡	1715–1766	1526–1570	11.42§	2**
	A.G.A. No. 2†	19.0	552	801			157.5 SSU	151.0	7.35‡	1312–1552	1567–1652	12.70§	2**
	A.G.A. No. 1†	22.2	545	903			193.2 SSU	179.7	7.05‡	1499–1570	1542–1574	12.39§	2**
Single Gen.	Gas Oil	35.2	451	517	612	36 SSU			6.51	1560–1380	1500–1350	12.52	2.03
Single Gen.	Gas Oil†	35.7	299	552	646			144.3	6.53	1320–1700	1530–1455	11.05	1.67
Twin Gen.	Gas Oil†	36.2	503	584	677	42.2 SSU		170	6.30	1630–1650	1595–1545	9.43	1.78
Twin Gen.	Louisiana Crude†	25.4	550	795	...		79.1 SSU		6.86	1510	1470	11.56	0.44
Twin Gen.	No. 2 Fuel Oil	33.3	...	490	603	38 SSU			6.67	...	...	11.93	1.1
Twin Gen.	Special No. 2 Fuel Oil	37.2				34.9 SSU			6.39	...	...	10.26	0.6
	Offshore Crude	25.1				114.8 SSU			6.88	...	...	11.43	1††
Jones Gen.	Diesel Oil	31.2	466	550	675				6.76	...	1400	11.78	2

* Oil gas is considered to consist of the gaseous hydrocarbons plus H_2 produced from the make oil. All H_2 is assumed to be produced from oil, since there is no free H_2 in carrier gas (steam), and since temperatures are too low for significant steam decomposition ($h/w = 0$). Values in parentheses are based only on hydrocarbon constituents of oil gas.

† Inst. of Gas Technology oil analysis.

‡ Actual C/H ratio.

§ 1000 Btu basis.

** Estimated.

†† Assumed.

moval and continuous production of high-Btu gas free from flue gas dilution. More rapid heat transfer to the oil vapor is accomplished with a bed of hot fluidized solids than by any other commercial process. Uniform cracking temperatures thus produced allow excellent control of oil gas composition.

Results from typical gasification runs on diesel oil are presented in Table 3-50.

OPERATING DATA AND PRODUCT ANALYSES

Detailed operating data and analyses of products from various oil gas plants and processes are given in Tables 3-51 and 3-52.

EVALUATION OF OILS FOR GAS MAKING

Most methods were developed as an adjunct to carbureted water gas production; five classifications of methods follow:

1. Methods for determining percentages of the major classes of hydrocarbons and calculating oil enriching values from carbureting efficiency values assigned each class of hydrocarbons—these were developed[35] and subsequently modified.[36,37]
2. Methods based on available hydrogen content, i.e., total hydrogen content *less* one-eighth of oxygen content.[38]
3. Evaluation by gravity determination of a key fraction distilling within a specified temperature range.[39]
4. Methods employing specific oil dispersion* in conjunction with gravity–boiling point and gravity–iodine number relationships.[40–42]
5. Empirical methods using two or more standard physical tests, such as viscosity–gravity or average boiling point–gravity for establishing oil paraffinicity.[43]

Standardized laboratory cracking tests were also developed.[10,41,44,45] By long comparisons of laboratory and plant results, purely empirical relationships were developed between laboratory and plant data. Relationships between C/H ratio and Btu recovery in high-Btu oil gas production are summarized in Fig. 3-63 for normal limits of oil specific gravity and partial oil gas pressure, and for cracking temperature and residence time variations required to produce true heating value oil gases of 900 to 1500 Btu per SCF.

In carbureted water gas production, hydrogen absorption from blue gas is normally greater than hydrogen production from oil cracking. Thus, the (true) oil gas produced consists primarily of paraffinic and unsaturated hydrocarbons.[7]

In oil gas production, oil is cracked into hydrogen as well as into paraffinic and unsaturated hydrocarbons, since the atmosphere contains no hydrogen. Production of hydrogen from the cracking and the relatively small dilution with carrier

* These methods are based on the refractive index, a characteristic property that may be used to identify a material.

Estimated Results for High-Btu Oil Gas Operation[6]

conditions		Make gas data, per cent							Operating results est. from Fig. 2-5		Deviations	
Carrier gas H_2/lb oil, h/w in SCF/lb make oil	Partial pressure oil gas (Px), atm	H. V., Btu/SCF	H_2	Paraff.	Olef.	Oil gas*	Cu ft of oil gas/gal oil*	Btu recov./gal oil	Btu recov./gal oil	Cu ft of oil gas/gal oil*	Btu recov., MBtu/gal	Gaseous hydroc. yield, SCF/gal
1.4	0.87	1046	16.1	35.2	26.2	77.5(61.4)	64.5(51.0)	87,000	79,800	−(54.3)	−7.2	+3.3
1.9	.84	1006	20.0	32.6	22.4	75.0(55.0)	65.1(47.7)	87,300	76,000	−(51.4)	−11.3	+3.7
2.9	.78	1047	19.4	34.7	23.4	77.5(58.1)	69.4(52.1)	93,700	80,900	−(54.5)	−12.8	+2.4
3.8	.72	966	17.1	31.6	23.6	72.3(55.2)	64.0(48.8)	85,500	70,700	−(48.0)	−14.8	−0.8
4.1	.71	974	21.3	28.6	24.3	74.2(52.9)	57.1(40.8)	75,100	66,300	−(45.8)	−8.8	+5.0
6	.6	1088	17.2	39.7	27.1	84.0(66.8)	67.6(53.7)	87,600	83,800	−(54.3)	−3.8	+0.6
6	.6	900	19.0	31.6	21.0	71.6(52.6)	62.6(46.0)	78,700	78,200	−(51.4)	−0.5	+5.4
6	.6	970	13.4	33.3	25.9	72.6(59.2)	60.5(49.2)	80,800	83,200	−(54.5)	+2.4	+5.3
6.1	.62	1046	16.1	42.3	28.9	87.3(71.2)	69.7(56.9)	83,600	87,900	−(56.9)	+4.3	0.0
5.0	.67	911	11.2	32.8	24.5	68.5(57.3)	62.0(51.9)	82,500	86,000	−(56.6)	+3.5	+4.7
5.3	.65	917	20.4	29.5	25.8	75.7(55.3)	80.3(58.7)	97,200	91,400	−(59.3)	−5.8	+0.6
1.3	.89	975	16.9	43.2	22.8	82.9(66.0)	71.7(57.1)	84,300	81,700	−(54.3)	−2.6	−2.8
3.2	.76	972	21	36	26	83(62)	70(52)	81,500	83,200	−(56.6)	+1.7	+4.6
1.7	.85	966	14	35	23	72(58)	70(57)	94,200	86,500	−(57.8)	−7.7	+0.8
2	.8	993	...	...	...	...	...	86,900	82,900	...	−4.0	...
6	0.6	975	21.7	48.0	20.7	90.4(68.7)	76.8(58.3)	82,580	85,000	−(55.2)	+2.4	−3.1

steam (high partial oil gas pressure) result in reduction of Btu recoveries as compared with blue gas enriching values and in formation of proportionately greater amounts of tar and carbon.

Table 3-53 gives 16 sets of operating data and results in several plants. Operating results estimated from Fig. 2-5 are also included. Comparison of actual and estimated Btu recoveries may readily be made. This shows that, although values estimated from the nomographs average 4 MBtu per gal lower than plant figures because of more effective removal of lower aromatics in the laboratory-produced oil gases, a conclusive correlation is established.

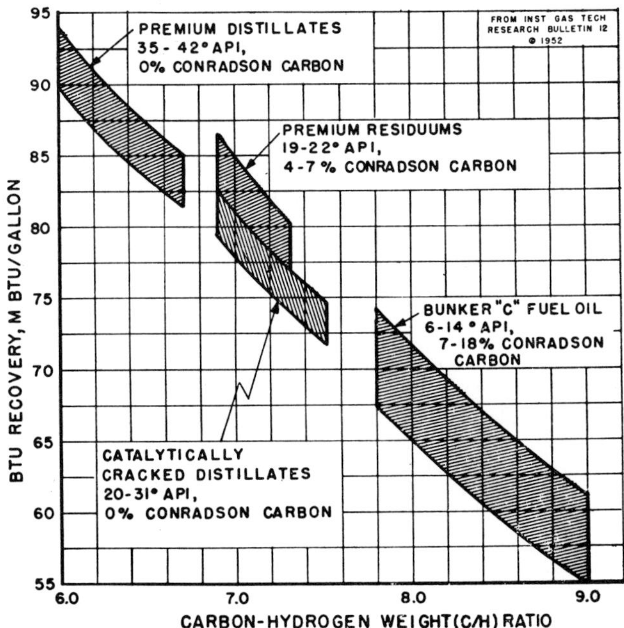

Fig. 3-63 Btu recoveries in unscrubbed oil gas for high-Btu oil gas production.[6]

PEAK SHAVING OIL GAS PRODUCTION SURVEY

Peak shaving oil gas production is often used to maintain an economical load factor on natural gas pipelines. An investigation of equipment and practice in this area was made by means of a detailed questionnaire to about 40 companies (24 replied) making oil gas.

The survey was designed to find:

(1) Operating guides for existing oil gas sets;

(2) Changes in design to increase the peak load capacity of existing sets at minimum capital costs;

(3) Data for the design of new sets with low investment and operating costs; and

(4) Oil gas production operating problems and answers.

Results of Survey

Capacity of Sets. Factors which may be involved include: (1) blower capacity; (2) type and arrangement of checker-brick in the reformer (superheater); (3) combustion system design; (4) use of specific gravity recorder on individual sets as primary operating control; (5) gas handling train design; (6) feed stock type and quality; and (7) type of set.

(1) *Blower Capacity.* Oil gas set capacity using *distillate* feed stocks appears to vary directly with blower capacity. For example, doubling the air rate from, say, 300 to 600 cfm per sq ft of reformer area doubles the set capacity for a kerosene feed stock (from 800 to 1600 therms per day-sq ft of reformer area). The data received on No. 4 and No. 6 fuel oils

were insufficient to develop a firm correlation between combustion air rate and set capacity.

(2) *Reformer Checker-Brick Considerations.* The optimum type of reformer checker-brick arrangement gives maximum surface with minimum pressure drop. Several plants increased set capacities 20 to 50 per cent by using special checker-brick to increase the reformer checker-brick surface about 100 per cent without increasing the pressure drop with increased air rates.

(3) *Combustion System Design.* This factor affects set capacity, efficiency of fuel utilization, and degree of smoke formation. Efficiently designed combustion systems give smoke-free operation with a stack blast gas analysis of one to two per cent oxygen.

(4) *Use of Specific Gravity Recorder as Primary Operating Control.* The use of this instrument assists in obtaining optimum quality oil gas, minimizes carbon deposition in the set, and minimizes tar emulsion troubles in the separator and dehydration systems. Furthermore, it facilitates the production of optimum quality tar with minimum carbon, sulfonation residue, and water content.

(5) *Gas Handling Train Design.* In many cases in which carbureted water gas plants have been converted to oil gas, the most serious bottlenecks to increased production occur from the gas offtake thru the condensers, because of the fact that this part of the gas handling train was designed for carbureted water gas saturated at 180 F leaving the wash box. The temperature of oil gas runs between 195 and 200 F. This 20°F increase in temperature means that the volume of vapor leaving the wash box is more than doubled; thus, the load on the condensing system is increased 200 to 300 per cent. These bottlenecks cause back pressure on the set, limiting the allowable make oil rate to the set. The effect of make oil on set capacity is appreciable. For example, doubling the make oil rate from, say, 2.3 to 4.6 gpm per sq ft of reformer area doubles the set capacity for a kerosene feed stock (from 800 to 1600 therms per day-sq ft of reformer area).

(6) *Feed Stock Type and Quality.*[6] The type and quality of the feed stock have an important effect on the capacity of an oil gas set. Table 3-54 shows the typical variations in the unscrubbed oil efficiencies with various feed stocks.

Table 3-54 Unscrubbed Oil Efficiency and Tar Produced for Various Feed Stocks

Type of oil	API gravity	MBtu/ gal	MBtu to gas/gal	Oil eff., %	Tar, gal/MCF
H-fuel	71.5	119.0	103.5	87	0.6
G-fuel	53.8	130.0	96.5	74	...
Kerosene	42.0	134.0	91.7	68	2.5
Diesel oil	39.8	135.0	90.3	67	3.0
No. 2 fuel oil	36.8	138.0	81.5	59	2.5–3.0
No. 4 fuel oil	20.2	145.0	80.0	55	3.0–4.0
No. 6 fuel oil	11.9	150.0	70.0	47	4.8

The effect of feed stock quality on Btu recovery (as measured by the carbon–hydrogen ratio) was considerably less than that shown in Fig. 3-37. For example:

Feed stock	C/H ratio	Recovery, therms/gal
H-fuel	5.3	0.98
No. 6 oil	8.1	0.70

The type and quality of feed stock also affect the capacities of the tar separator and tar dehydration systems, as shown by the variation in the quantity of tar produced with various feed stocks; see Table 3-54.

The effect of the type and quality of feed stock on the purifier system is approximated by the equation $G = 219S$, where G = the grains of H_2S per 100 cu ft of oil gas and S = the per cent of sulfur in the feed stock.

(7) *Type of Set.* An evaluation of this effect is beyond the scope of this handbook. However, the variation in total feed stock (heat oil plus make oil) required for various sets and fuels is given in Table 3-55.

Table 3-55 Oil Gas Feed Stock Requirements for Various Sets and Fuels

Type of oil	Type of set	Total oil, gal per therm*
H-fuel	Twin generator	1.175–1.295
H-fuel	3-Shell	1.130–1.500
G-fuel	3-Shell	1.185–1.303
Kerosene	Twin generator	1.255–1.385
Diesel	Twin generator	1.233–1.397
Diesel	2-Shell	1.395–1.533
Diesel	3-Shell	1.331–1.358
Diesel	Hall	1.285
No. 2 fuel oil	Twin generator	1.451–1.581
No. 2 fuel oil	3-Shell	1.235–1.450
No. 2 fuel oil	Hall	1.324
No. 4 fuel oil	Twin generator	1.483
No. 4 fuel oil	Hall	1.289–1.370
Aruba oil	Twin back blast	1.36
No. 6 Bunker C	Hall	1.361–1.463

* 100 cu ft of 1000 Btu per cu ft scrubbed oil gas.

Note: Factors other than the type of set, such as quality of the feed stock or burner design, probably account, at least in part, for the differences in total oil required per MCF.

Operating Problems. Major problems reported were: (1) manpower requirements and training; (2) net operating time of oil gas sets not permitting sufficient down time; (3) carbon deposition in set; (4) pitch deposition in the wash box; (5) tar separator operation; (6) circulating water treatment; (7) effluent water treatment; (8) gas scrubbing; (9) safety features; and (10) stand-by procedure. Details are given in Ref. 46.

REFERENCES

1. Lowe, L. P. *Apparatus for the Manufacture of Hydrocarbon Gas.* (U. S. Patent No. 405,426) Washington, D. C., June 18, 1889.
2. Harritt, J. A., and others. "High Heating Value Oil Gas as Standby for Natural Gas." *P.C.G.A. Proc.* 23: 262–88, 1932.
3. Johnson, A. "Refractory Screen Oil Gas Process." *Gas Age Record* 69: 737+, June 18, 1932.
4. Utermohle, C. E. "Hall High B.t.u. Oil Gas Process." *A.G.A. Monthly* 30: 27–8, Nov. 1948.
5. A.G.A. Gas Production Research Comm. *Hall High B.T.U. Oil Gas Process.* (Research Bull. Proj. HB-1) New York, 1948.
6. Linden, H. R., and Pettyjohn, E. S. *Selection of Oils for High-BTU Oil Gas.* (Research Bull. 12) Chicago, I.G.T., 1952.
7. Pettyjohn, E. S., and Linden, H. R. *Selection of Oils for Carbureted Water Gas.* (Research Bull. 9) Chicago, I.G.T., 1952.
8. Linden, H. R., and Pettyjohn, E. S. "Process Variables and Oil Selection for High Btu Oil Gas Production." *A.G.A. Proc.* 1951: 553–75.

9. Linden, H. R. "Methods of Estimating Oil Gasification Yields." *A.G.A. Proc.* 1949: 755–80.
10. Dick, I. B. "Laboratory Cracking Furnace for Heavy Oils and Gas Oils." *Am. Gas J.* 138: 32–4, June 1933. Discussion in *A.G.A. Operating Sec. Joint Prod. and Chem. Conf.* 1933: 121–6.
11. Linden, H. R., and Othmer, D. F. "Combustion Calculation for Hydrocarbon Fuels." *Chem. Eng.* 54: 115–9, April 1947.
12. Pettyjohn, E. S., and Linden, H. R. "Estimating Potential Gasification Value and Yields of Oils for Gas Plants." *Am. Gas J.* 171: 26+, Oct.; 31+, Nov. 1949.
13. Linden, H. R. "Thermal Cracking Effects of Process Variables and Feedstock Properties on the Yields and Compositions of the Products." *Petrol. Process.* 6: 1389–96, Dec. 1951.
14. Linden, H. R., and Peck, R. E. "Gaseous Product Distribution in Hydrocarbon Pyrolysis." *Ind. Eng. Chem.* 47: 2470–4, Dec. 1955.
15. Hague, E. N., and Wheeler, R. V. "Mechanism of the Thermal Decomposition of the Normal Paraffins." *J. Chem. Soc.* 132: 378–93, 1929.
16. Wheeler, R. V., and Wood, W. L. "Mechanism of Thermal Decomposition of the Normal Olefins." *J. Chem. Soc.* 135: 1819–28, 1930.
17. White, C. E. "Yield of Tar in the Manufacture of High B.t.u. Oil Gas." *Gas* 13: 25–6, Sept. 1937.
18. Berthelot, M. "Chimie Organique—Action de la Chaleur sur Quelques Carbures d'Hydrogène." (Part I) *Compt. Rend.* 62: 905–9, Jan.–June 1866.
19. Bone, W. A., and Coward, H. F. "Thermal Decomposition of Hydrocarbons." (Part I) *J. Chem. Soc.* 93: 1197–1225, July 1908.
20. Rice, F. O., and others. "Thermal Decomposition of Organic Compounds from the Standpoint of Free Radicals. II, Experimental Evidence." *J. Am. Chem. Soc.* 54: 3529–43, 1932.
21. Rice, F. O., and Dooley, M. D. "Thermal Decomposition of Organic Compounds from the Standpoint of Free Radicals. IV, The Dehydrogenation of Paraffin Hydrocarbons and the Strength of the C–C Bond." *J. Am. Chem. Soc.* 55: 4245–7, 1933.
22. Dunstan, A. E., Ed. *The Science of Petroleum,* vol. III, p. 1994–2075. London, Oxford Univ. Press, 1938.
23. Hasche, R. L. "Production of Acetylene by Thermal Cracking of Petroleum Hydrocarbons." *Chem. Met. Eng.* 49: 78–83, July 1942.
24. Tropsch, H., and others. "High-Temperature Pyrolysis of Gaseous Olefins." *Ind. Eng. Chem.* 28: 581–6, May 1936.
25. Linden, H. R., and others. "Production of Oil Gases as Substitutes for Natural Gases." (PC-52-13) *A.G.A. Proc.* 1952: 702–15.
26. Dent, F. J. "Third Report on the Use of Creosote in the Manufacture of Carburetted Water Gas: Detailed Description of the Investigation." (Comm. 37) *I.G.E. Trans.* 81: 309–49, 1931.
27. Sebastian, J. J. S., and Riesz, C. H. "Sulfur-Resistant Catalysts for Reforming Propane." *Ind. Eng. Chem.* 43: 860–6, 1951.
28. Hull, W. Q., and Kohlhoff, W. A. "Oil Gas Manufacture." *Ind. Eng. Chem.* 44: 936–48, May 1952.
29. Kohlhoff, W. A. Private communication. Portland Gas and Coke Co., 1952.
30. Middleton, A. J. "Increasing Capacity of Oil–Gas or Converting Water–Gas Machines." *Gas Age* 129: 17–8, Mar. 1, 1962.
31. Morgan, J. J. "Gasification of Hydrocarbons." *Gas Age* 110: 38–44, Oct. 9, 1952.
32. Utermohle, C. E. "Hall High B.t.u. Oil Gas Process." *A.G.A. Monthly* 30: 27+, Nov. 1948.
33. Joy, P. C. "Advances in High Btu Oil–Gas Manufacturing in Hawaii." (CEP-58-12) *A.G.A. Operating Sec. Proc.* 1958: 99–104.
34. Chaney, N. K., and others. "Developments in High-Temperature Pyrolysis: Control Methods and Product Separations." *Chem. Eng. Progr.* 45 (1): 71–80, 1949.
35. Mighill, T. A. "Evaluation of Gas Oils." *A.G.A. Proc.* 1927: 1454–63.
36. Griffith, R. H., and Holliday, G. C. "Determination of Iron Carbonyl." *Chem. & Ind.* (*J. Soc. Chem. Ind.*) 47: 311t–312t, Oct. 19, 1928.
37. Murphy, E. J. "Report of Subcommittee on Evaluation of Gas Oils." *A.G.A. Proc.* 1930: 1189–1200.
38. Merkus, P. J., and White, A. H. "Evaluation of Oils for the Manufacture of Carburetted Water Gas by their Available Hydrogen Content." *A.G.A. Proc.* 1934: 986–91.
39. Ferguson, H. E. "Gas Oil Value and Supply." *Gas Age Record* 64: 235+, Aug. 24, 1929.
40. Holmes, A. "Carbureting Values of Gas Oils and a New Method for their Evaluation." *Ind. Eng. Chem.* 24: 325–8, Mar. 1932. Also in: *Gas J.* 199: 459–60, Aug. 31; 505–6, Sept. 7, 1932.
41. Kugel, W. F., and Bliss, E. M. *Empirical Relation between the Physical Characteristics of an Oil and its Yield as Determined by a Cracking Test.* A.G.A. Prod. and Chem. Comm. Conf. 1940. Abstract in *A.G.A. Proc.* 1940: 735.
42. Kugel, W. F. "Gas-Making Qualities of Oils." *A.G.A. Monthly* 29: 297–303, June 1947.
43. Cauley, S. P., and Delgass, E. B. "Gas Enriching Value of Oils." *A.G.A. Monthly* 27: 494–9, Nov. 1945.
44. Downing, R. C., and Pohlmann, E. F. "A Study of Gas Oils." *Proc. Am. Gas Inst.* 11: 587–636, 1916.
45. Murphy, E. J. "Evaluation of Gas Oil by the Laboratory Cracking Method." *Gas Age Record* 65: 389+, Mar. 22, 1930.
46. Milbourne, C. G. "High Btu Oil Gas—Progress Report, Oil Gas Production Task Group." *A.G.A. Operating Sec. Proc.* 1962: CEP-62-1.

Chapter 8

Manufactured Gas Purification

by H. A. Gollmar, W. J. Huff, and N. C. Updegraff

GENERAL

Manufactured gas purification refers to the removal of sulfur compounds, mainly H_2S. Carbon disulfide, carbon oxysulfide, and other organic sulfur compounds are also present. Impurities that do *not* contain sulfur are largely removed by scrubbing (dry or wet) or condensing.

Hydrogen sulfide is removed by dry box or liquid purification. Dry box purification provides reliable and effective H_2S removal, without continuous operation of moving parts. It is well suited to the needs of small gas manufacturing plants. Its space requirements, however, may be excessive, and the operation produces a considerable quantity of spent oxide (when revivification is not effective) which must be discarded.

Liquid purification[1] is sometimes the only practical means for H_2S removal, because of limited ground space. A system of this type exerts low back pressure on the gas train. It is adapted to high-pressure gas purification. Recovery of sulfur as a salable product also may be accomplished by some liquid purification processes.

DRY BOX PURIFICATION

In dry box purification, hydrogen sulfide is removed by bringing the gas into contact with hydrated colloidal ferric oxide or colloidal ferric hydrate.[2] The oxide may have a variety of compositions and structures. Although the chemical reactions of fouling and revivification are complex, the following equations seem to apply:

Fouling:

$$Fe_2O_3{\cdot}(H_2O)_x + 3H_2S = Fe_2S_3 + 3H_2O + xH_2O + \text{heat} \qquad \textbf{(a)}$$

$$Fe_2O_3{\cdot}(H_2O)_x + 3H_2S = 2FeS + S + 3H_2O + xH_2O + \text{heat} \qquad \textbf{(b)}$$

Revivifying:

$$2Fe_2S_3 + 3O_2 + xH_2O = 2Fe_2O_3{\cdot}(H_2O)_x + 6S + \text{heat} \qquad \textbf{(c)}$$

$$4FeS + 3O_2 + xH_2O = 2Fe_2O_3{\cdot}(H_2O)_x + 4S + \text{heat} \qquad \textbf{(d)}$$

The overall transformation of hydrogen sulfide is then:

$$2H_2S + O_2 = 2H_2O + 2S + \text{heat} \qquad \textbf{(e)}$$

Equation **e** is the sum of Eqs. **a** and **c** or **b** and **d**.

The reactions of sulfiding and revivifying in total give about 3300 Btu per lb of H_2S removed.[3] About one-third of this is given off in the sulfiding stage and about two-thirds in that of revivifying. There are several intermediate reactions[4] under usual box operating conditions. The bulk of the H_2S is absorbed per Eq. **a** and perhaps one-fifth per Eq. **b**. Alkaline[5] conditions favor the formation of *ferric* sulfide; acid conditions favor *ferrous* sulfide and ferrous disulfide, F_2S_2.

Purifying Materials

The colloidal hydrated iron oxides used for purification are prepared from metallic iron, bauxite residues, or high iron bog ores. These prepared oxides have characteristic properties, depending upon their source and method of preparation. Activity and capacity are important. Activity of an oxide is a measure of its ability to remove H_2S rapidly and completely. Capacity of an oxide is a measure of the total weight of H_2S that it is capable of absorbing. Chemically prepared oxides are usually more active, and they are used when trace removal is important. High iron content bog ores have a high capacity and are used when loads are heavy but complete removal is unnecessary.

Granular oxides and bog ores are sometimes used for high-capacity purification. Generally, the finely divided oxide is mixed with a fluffing agent, such as wood shavings (most commonly used), crushed slag, or ground corn cobs. When the oxide is prepared from iron borings, usual practice has been to add the wood shavings during the oxidation period, coating them with oxide as the borings are turned. To mix a prepared oxide with wood shavings, it is preferable to mix the oxide with water to form a slurry and then to add the slurry with constant mechanical stirring to the shavings. The oxide and wood shavings mixture is termed **sponge.** Weight of oxide used per bushel of sponge depends upon the characteristics of the oxide, the H_2S content, and the rate of gas flow. Sponge mixtures containing from 6 to 12 lb of Fe_2O_3 per bushel are satisfactory in most cases. Mixes containing 20 to 25 lb of oxide per bushel have been used efficiently. Too large a quantity of oxide must not be used, because the gas only penetrate a definite thickness, and any excess oxide is inert. "Back pressure" troubles, as well as caking of the sponge after sulfiding, making its removal difficult, may also result from too much oxide.

General Operation of Purifier Boxes

Hydrogen sulfide removal in dry box purification is a surface reaction.[6] Light oils, naphthalene, and tars should be removed before purification, since these materials coat the sponge and render it unreactive. When their removal before purification is not complete, a wood shaving scrubber or filter may be used ahead of the iron oxide boxes. This filter may be a separate box or a layer of coarse wood shavings a foot or more thick at the inlet of the box.

In filling the purifying boxes, it is essential that the sponge be packed as uniformly as possible to prevent channeling. A mechanical distributor or other device should be used to distribute the sponge as the box is filled. Men forking or raking should stand on planks. The sponge should be carefully tamped for a distance of six in. from all metal surfaces.

It is preferable to keep the box temperature[7] between 85 and 100 F. Each box should be equipped with recording thermometers showing inlet and outlet gas temperatures on the same chart. Convergence or divergence of the temperature curves indicates if the reaction is causing the box to heat up or if the gas flow and water evaporation are causing a temperature drop. It is good practice to have steam available at each box inlet so that gas temperature can be regulated.

Measurement of gas humidity as it enters and leaves the purifier is important. The gas should not carry away more water than it introduces. The box should not become dry. Sponge does not function properly outside the limits of 17 to 55 per cent moisture. A dry gas stream will remove moisture, while one supersaturated with moisture favors its presence in the sponge. However, the presence of varying amounts of deliquescent impurities in the sponge and its varying colloidal surface properties make it impossible to derive a simple quantitative correlation between the sponge moisture content and the humidity of the gas stream.[8]

Moisture must be present in the gas stream for colloidal hydrated iron oxide to remove H_2S. Dry oxide will not interact with H_2S in a dry gas stream and dry iron sulfide will not react with the oxygen in dry air. Therefore, humidity of the gas stream is an important factor in the purification process. Sulfiding of the sponge reaches its maximum rate at an intermediate humidity, but revivification of the iron sulfide reaches its maximum at or near saturation. Since revivification is a much slower reaction, both revivification and sulfiding should proceed simultaneously at or near saturated humidities.[8]

Water also furnishes enough drips to dissolve and remove from the box soluble salts formed during purification. If not removed, these salts crystallize on the iron oxide surface and prevent contact with the gas. Water may be added by introducing steam or by a spray system in each box. Absence of drippage from the purifiers indicates insufficient moisture.[8]

Presence of free alkalies or ammonia in the sponge increases H_2S removal from the gas stream. The higher pH due to alkalies also favors the removal process. The pH of the drip water is generally considered an indication of alkalinity or acidity of the box. Presence of an easily available potential alkali in a box containing new oxide is essential to neutralize any acid formed from various acids of sulfur or cyanogen. Acid solutions are poor preferential solvents for H_2S. In a partly fouled box, presence of Fe_2S_3 precludes the possibility of an acid box. Acidity in drips from partly fouled boxes is an after-purification development, usually because of oxidation of some salts in solution. Any acidity formed in a partly fouled box will be neutralized by the ferric sulfide, with evolution of H_2S. Therefore, it is usually advantageous to increase the alkalinity of the moisture in the box by adding soda ash, lime, or ammonia. It is good practice to allow a concentration of one or two grains per CCF of ammonia in the gas entering the boxes. Ammonia in excess of four grains per CCF of gas will tend to cause formation of undesirable sulfites, thiocyanates, and other salts.

Gas to the purifiers may contain cyanogen or hydrogen cyanide. Both have strong affinities for iron and iron salts, forming ferrous thiocyanate, which in solution is highly dissociated and tends to produce an acid condition. When purifying material is required to remove cyanogen as well as hydrogen sulfide, additional ammonia (four to six grains per CCF of gas) may be required to maintain good purification.[5]

Disposing of spent sponge is sometimes a problem. Processes for recovery of elemental sulfur by steaming or solvent extraction have proven uneconomical. In some locations, the used sponge has been burned, recovering the sulfur oxides for sulfuric acid production, with the high iron ash going to blast furnaces. Care must be taken that drainage from dumped or buried sponge does not contaminate water supplies.

When boxes exhibit traces of fungus growth, they may be sprayed before refilling with a dilute solution of an inhibitor such as Santobrite.*[,9,10] Spraying should include all surfaces within the box and trays. A thin layer of this inhibitor may also be spread over the top layer of sponge.

Revivification of Oxide

Revivification is the oxidation of ferric and ferrous sulfides formed during fouling to ferric oxide and sulfur. It is accomplished by three general methods. These involve the mixing of air with gas, an operation which should be attempted only on plant scale by men experienced in methods for avoiding fire or explosion that may result.[11,12]

Passing Air thru Box with Gas. When a regulated amount of air is mixed with the gas, practice has shown that the oxygen content must exceed 0.4 to 1.0 per cent for revivification reactions to occur. The active sponge must also be at least 15 per cent fouled. With this revivification method, fouling and revivification occur simultaneously, with the oxide never becoming completely fouled. The purifying material is not removed from service until it has become so contaminated with sulfur that its active surface and porosity are greatly reduced. Controlled introduction of air provides one-half cubic foot of O_2 per cubic foot of H_2S in the gas, *plus* the air required to provide the minimum concentration at which the reaction will occur. Removal of one pound of sulfur per pound of iron oxide in the sponge is typical of the oxide capacity when used in this manner. Air addition may be split up to each purifying box or the admission point rotated, since a considerable rise in temperature may occur, causing removal of moisture from the zone of reaction. Air must be added cautiously to a heavily fouled box until equilibrium is established, or overheating will result.

Passing Air thru Box with Gas Off. Where blowing of the fouled oxide with a controlled amount of air is practiced, the purifier is by-passed and then air is admitted in such a manner that it prevents overheating, combustion, or

* 78.6 per cent sodium pentachlorophenate and 11.4 per cent sodium tetrachlorophenate.

formation of an explosive mixture in the purifier. Several methods[7] may be used. In one, air under pressure is admitted to the box with the outlet valve throttled to maintain a constant five psig pressure. If this pressure is exceeded, or if the temperature rises above 115 F, the blower is shut off until conditions are corrected. Revivification takes place at low oxygen concentration, and the increased pressure gives good penetration and revivification of any lumps in the box.

A second method is to circulate the gas in the box by means of a blower, with a valve on its suction side to control air admission to the system. This valve is gradually opened until a concentration of five per cent O_2 is obtained. The pressure and temperature limits of the first method apply.

A third method consists of circulating air and gas as described above, with the additional step of allowing gas to escape to the atmosphere with gradually increasing oxygen concentrations and reduced circulation, until air is being blown directly thru the box to the atmosphere. The box should then be purged with inert gas before it is returned to service. In using the second and third methods, it is recommended that the boxes be equipped with water sprays to add moisture to the sponge if necessary. In some plants, the combustible gas is displaced with inert gas before either of these methods is employed.

Removal of Fouled Oxide for Exposure to Air. Where space and labor costs permit, revivification by removing the purifying material from the box is practiced. The fouled material is turned repeatedly, or passed thru a mechanical oxide conditioner which breaks the lumps and discharges the material thru the air. Change of color from the black sulfide to the red or brown oxide indicates progress of revivification. For maximum activity, the material should not be allowed to heat unduly. Water sprays should be available to control temperature or combustion. Excessive moisture will retard revivification. This method requires considerable handling and competent attention, but it has the advantage that all lumps formed during fouling are broken up, eliminating channeling from this source.

The Honolulu Gas Co. mixes soda ash and wood shavings with its fouled oxide and effects such a high degree of revivification that oxide activity and capacity are claimed equal to that of new oxide. There is no danger of overheating or combustion, because the box is removed in its most highly revived condition (the fourth or last position in the four-box purifier system).

Intermittent vs. Continuous Process. Revivification rates increase rapidly with increasing humidity. However, simultaneous revivification and purification are *not* advisable at high humidities if deposition of water is permitted. This condensation of water greatly lowers the purification efficiency.[8]

Purifier Box Rotation

The theory is that gas very low in H_2S and containing O_2 will revivify a badly fouled box. However, there is a chance of poor trace removal. The Honolulu Gas Co., after years of operation involving the placing of the cleanest box in the catch-box, or last-stage, position and moving it to the head of the series as it became more fouled, had tremendous success in 1958 with the opposite process. Here, the cleanest box was placed at the head of the series and moved in steps to the catch-box position. The length of time in each position was 24 hours. Improved scrubbing, which virtually eliminated the carry-over of tars and resins into the lead purifier box, was a prerequisite to the change in operation.

The lead box consistently removed at least 99 per cent of the 400 grains of H_2S per CCF of raw gas throughout its 24-hr tour. The gas leaving the second-stage box is always "white" when given the standard lead-acetate test. Boxes in the second, third, and fourth stages are reviving continuously thru the action of 1.0–1.5 per cent by volume of oxygen added in air ahead of the lead box.

A wood-shaving sponge containing four pounds of iron oxide per cu ft is used in the ratio of 2.5 to 3.0 bu/MCF-day of gas. The oxide sponge was formerly changed after four months of service; it appears that changes will now come at eight-month intervals. When back pressure builds up because of sulfur deposits, the sponge is removed and crushed. New shavings and soda ash are mixed in and the reworked sponge returned to the purifier box. Activity of the reworked sponge is equal to or greater than that of new sponge. The sulfur content of sponge is carried as high as 70 per cent by weight; formerly, 55 per cent was normal (an increase of 91 per cent in the weight of sulfur removed per cubic foot of sponge).

Purifier Box Design[6]

Specific operating characteristics of a dry box purifying system are functions of sponge activity and capacity, rate of flow, H_2S concentration, preconditioning of the gas, and physical properties of the box and oxide bed. One design with three layers of oxide and a dry gas seal is shown in Fig. 3-64.

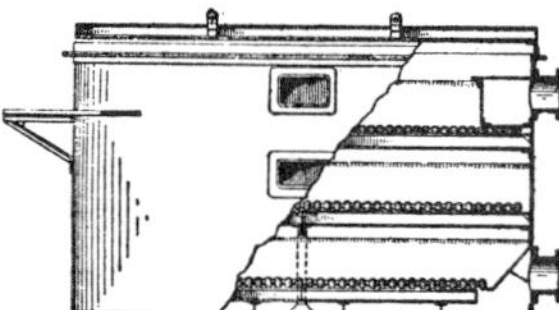

Fig. 3-64 Deep circular box-type iron oxide purifier with dry gas seal.

Five bushels of sponge per MCF of gas passed per day is usual, although efficient operation is possible with as little as one bushel.[4] The latter ratio requires more frequent box changes; oxide cost per MCF of gas is the same. Labor costs decrease as the ratio of bushels of sponge to gas flow rate increases up to ten bushels per MCF per day, above which labor costs remain constant.

Duplex boxes (two layers of sponge) five feet deep on wooden trays are common. Downward flow is preferred by some, who reason that concurrent gas and moisture flow results in less back pressure and better moisture distribution thru the sponge. Others favor upward flow, since gas passing countercurrent to water flow receives better scrubbing with better trace removal. Downflow boxes are recommended if the gas contains naphthalene or suspensoids, since accumulation occurs on the top layer and can be removed without the entire box being dumped.

Round boxes of steel or concrete with a smooth interior are more easily packed than angular boxes. They may have oxide depth of 10 to 15 ft, allowing 300 cu ft of gas per cu ft of oxide per 24 hr.[13] Boxes should be connected in ways that permit rearrangement. Both upward and downward flow should be possible.

Three duplex boxes in series are common for H_2S concentrations up to 200 grains per 100 cu ft. For higher concentrations, four boxes in series are usually used.

Design Formula

American oxide box design is ordinarily based on the Steere formula given below. This formula is based on the *maximum hourly rate* of gas to be purified, not the average daily flow. Steere formula:

$$A = \frac{GS}{3000\,(D + C)}$$

where: A = cross-sectional area of oxide through which gas passes in any one box in a series of a set of boxes (Note 1), sq ft

G = maximum amount of gas to be purified, cu ft per hr at 60 F

S = factor for grains of H_2S per CCF of unpurified gas (Note 2)

D = total depth of oxide thru which gas passes consecutively, ft (equals depth of such oxide per box *times* number of boxes in series). A single "catch box" used for two or more sets may be disregarded (Note 1)

C = 4 for two-box, 8 for three-box, and 10 for four-box series, respectively (Note 3)

Note 1: Duplex boxes, with two layers of oxide each and divided flow, whereby half the gas passes through each layer, present a combined area of both layers but the depth of only one to gas passage. Area A of a duplex box is the sum of the areas of the two layers of oxide, or double the box cross section. Depth D is the depth of one layer of oxide per box multiplied by the number of boxes, in series, in the set.

Note 2:

Grains of H_2S per CCF of unpurified gas	S factor
≥1000	720
800	675
600	600
400	525
≤ 200	480

The direct use of the number of grains of H_2S per CCF of gas for the S factor would result in absurdity, except within very small limits A purifier set with a capacity based upon 600 grains of H_2S would not have three times that capacity for gas containing 200 grains. Values for S have been selected, as were those for C, on the basis of both experience and theory.

Note 3: Factor C is the number of boxes in use, in series, at one time, disregarding any one box of the set out of use at times for purposes of oxide changing or revivifying. Factor C is based on reversing gas flow during purification; otherwise, the number of boxes used for determining the factor C must be reduced by one.

Two-box sets should never be installed without means for reversing the flow within the boxes, unless a catch box is provided for each two sets. Each set can then be regarded as a two-box series with $C = 4$. If such two-box sets with a catch box for each two sets are arranged throughout for reversal of flow, they can be regarded as two-box series with $C = 6$.

Evaluation of Purifying Materials

The following list indicates the determinations usually made on samples of gas oxide purifying materials submitted for laboratory examination.[7,14]

1. Oxides—moisture, ferric oxide, calcium oxide, active alkalinity, total alkalinity, carbon dioxide, total sodium as sodium oxide, total sulfur, activity.
2. Unused sponge—moisture, ferric oxide, percentage of oxide in the sponge.
3. Used sponge—moisture plus light oils, ferric oxide, total sulfur, free sulfur and tar, light oils, naphthalene, thiocyanates, ferrocyanides, organic sulfur, vapor phase gums, pyridine.

LIQUID PURIFICATION

Liquid purification processes employ absorbing solutions as their active medium to remove H_2S from manufactured gas. Solutions of relatively inexpensive alkalies, such as soda ash and ammonia, are commonly used by three general types of such processes.

In the first type, H_2S is removed by transfer to a relatively large volume of air and the mixture is vented to the atmosphere or burned. In the second type, H_2S is converted to precipitated sulfur (Seaboard process, Fig. 4-67). In the third type, H_2S is recovered in concentrated gaseous form, which may be used in chemical manufacturing.

Girbotol Process.[15] Since this aqueous ethanolamine process removes CO_2 as well as H_2S, it may not be economical for manufactured gases containing substantial amounts of CO_2. Pretreatment to remove tars, cyanogen, and oxygen may also be necessary.

Thylox Process

This process (Fig. 3-65) removes H_2S and recovers it as precipitated sulfur. A mildly alkaline solution of arsenic trioxide (As_2O_3) in soda ash is used as the absorbing medium, which is then regenerated by oxidation with air.[16,17] The oxidation reaction causes precipitation of elemental sulfur, which is recovered in a flotation froth.

The Thylox solution originally contains nearly equal parts of soda ash and arsenic trioxide, but normal operation converts it into one of the sulfur-containing sodium arsenates, known as thioarsenates. This solution is almost neutral but must be maintained very slightly alkaline (pH about 7.7 to 8.0). Chemically, there are several sodium thioarsenates that differ in the number of their sulfur atoms. When the solution absorbs H_2S from the gas, the sodium thioarsenate is converted into another sodium thioarsenate that contains more sulfur. Both thioarsenate solutions have almost no H_2S vapor pressure For this reason, almost complete H_2S removal is possible, and the air leaving the aerator has no appreciable H_2S content. The total arsenic content of the Thylox solution is usually maintained at about 0.5 per cent As_2O_3, but lower concentrations are used in plants operating below design capacity.

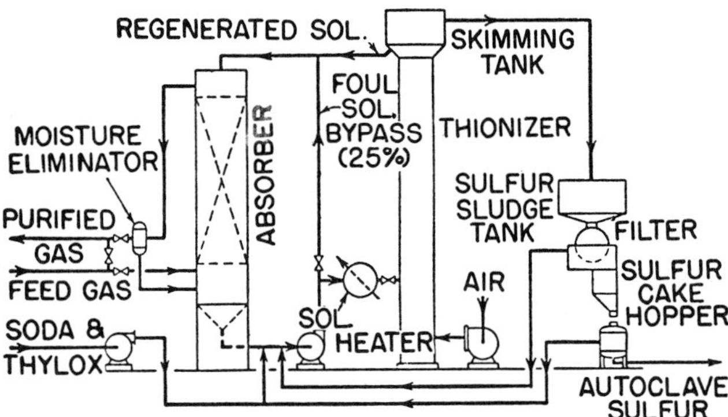

Fig. 3-65 Thylox process for gas purification and recovery of precipitated sulfur.[1]

The aerator has two functions: (1) to regenerate the solution by oxidation, thereby precipitating sulfur; and (2) to separate the precipitated sulfur from the solution by flotation. The aerator must be operated at about 100 F to function efficiently. Sulfur-bearing froth accumulates in the aerator top, overflows into a sulfur slurry tank, and is then filtered and washed. The filter cake is a wet sulfur paste that is sold without further drying to the agricultural insecticide trade as "Thylox flotation sulfur." This paste can also be melted in an autoclave and cast in cakes, but such a sulfur product has less value.

The Thylox process also includes chemical reactions that consume alkali, so that soda ash must be added to maintain the solution in operating condition. The more vigorous aeration required by this process causes oxidation of part of the sulfur to sodium thiosulfate and conversion of all the HCN absorbed from the gas to sodium thiocyanate. These reactions usually consume about 0.4 lb soda ash per pound of H_2S removed. In general, about one-fourth of the sulfur is lost in these reactions and about three-fourths of the H_2S removed from the gas is recovered as salable sulfur. Figures on sulfur recovery and soda ash consumption vary somewhat with operating conditions and the HCN content of the gas.

For efficient operation, the solution temperature must be kept at about 100 F. Higher temperatures unnecessarily increase the reaction rates and waste alkali. The compressed air rate to the thionizer must be high enough to maintain a sulfur-bearing froth in the aerator top for overflow to the sulfur slurry tank. Sometimes, especially when the gas contains larger amounts of HCN or the sulfur load on the plant is heavy, it is advantageous to add a small amount of ferrous sulfate or other iron salt to the solution as an oxidation catalyst. Too high a concentration of salts in the solution can reduce efficiency, and it is considered inadvisable to allow the specific gravity of the solution (a measure of its total salt content) to exceed 1.18. Obviously, the necessary As_2O_3 content of the solution and the proper alkalinity must be maintained. Too high an alkalinity interferes with oxidation and the pH should be kept between about 7.7 and 8.0 throughout.

The higher cost of operation of the Thylox process is offset by the higher value of its precipitated sulfur. This process is especially suitable in areas where the market for agricultural sulfur is assured.

Before development of the Thylox process, other processes had been used to produce precipitate sulfur. For example, the Ferrox and nickel processes used soda ash solutions to which iron oxide or nickel sulfate, respectively, was added as an oxidation catalyst. These processes required practically the same equipment as the Thylox process, but they consumed more soda ash, produced less sulfur, and were more expensive to operate.

Vacuum Actification (Carbonate) Process

In this process, H_2S is recovered in sufficient concentration to be used for production of sulfuric acid or elemental sulfur. The process uses a soda ash solution and is similar in principle to the Seaboard process, except that water vapor under vacuum is passed thru the actifier instead of air. Sufficient heat is added to the actifier bottom to produce the required volume of water vapor, which is condensed from the vapors (containing H_2S) leaving the tower top. A vacuum pump maintains the system vacuum, and pumps the H_2S gases to their utilization point (Fig. 3-66).

The soda ash solution used in this process is somewhat stronger than in the Seaboard process, and its pumping rate is smaller. The actifier operates at the boiling point of the soda ash solution under vacuum, about 140 F or less. Because of this higher temperature, less vapor volume is required than air volume in the Seaboard process.

The recovered H_2S gases will contain some CO_2 and HCN if these are present in the unscrubbed gas. A vacuum artification process designed to recover 90 per cent of the H_2S from coal gas will also remove about 80 per cent of the HCN, but only less than six per cent of the CO_2. Less soda ash is required to maintain alkalinity in the vacuum process, since very little oxidation of the H_2S or HCN occurs. HCN may also be recovered as an additional product.

It is important to supply an adequate amount of heat to the actifier bottom. Where heat is obtained from a heat exchanger or flushing liquor in coke plants, recording temperature and flowmeters are recommended on both solution and flushing liquor flows. It is also important that the unscrubbed gas does not contain excessive amounts of naphthalene and ammonia, because these can cause obstructions to form in equipment handling concentrated H_2S gases.

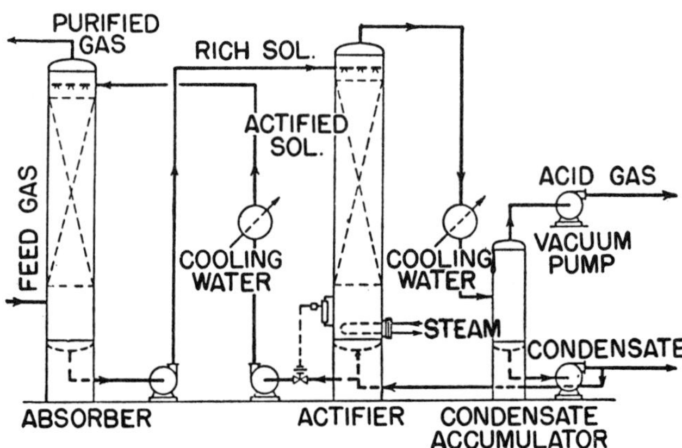

Fig. 3-66 Vacuum actification process for gas purification and recovery of concentrated H_2S.[1]

Where large tonnages of H_2S must be removed, processes that recover it in a form convertible to sulfur or sulfuric acid may be preferable. The vacuum artification process, even when treating gases of high CO_2 content, recovers H_2S of adequate purity to produce these products.

REMOVAL OF SULFUR COMPOUNDS OTHER THAN H_2S

Organic sulfur compounds present in coal distillation gases include carbon disulfide, thiophene, carbon oxysulfide, mercaptans, and trace compounds such as thioethers.[18] These constituents vary in volatility from the fixed gas carbon oxysulfide to thiophene, which boils at 183 F. Carbon disulfide may account for about two-thirds of the total organic sulfur content. Processes for total organic sulfur removal must be effective for carbon disulfide. For very low total sulfur content of the purified gas, they must also be effective in removing thiophene, carbon oxysulfide, and mercaptans.

Suitable choice of low sulfur gas-making materials is largely responsible for low total sulfur in sendout. Certain special gas applications, however, demand practically complete absence of all forms of sulfur in the fuel. Examples are found in optical glass manufacture, catalytic production of synthetic fuels, and some metal treating processes.

The principal types of process that have been used for organic sulfur removal follow:[18]

1. *Absorption by Liquids and Desorption.* A light oil plant as generally used for benzene recovery may be modified for organic sulfur removal as well. Removal of 73 per cent of total organic sulfur by a plant of this type has been reported.

2. *Conversion of Organic Sulfur to H_2S or SO_2 at High Temperature.* Various metallic catalysts were first employed to decompose organic sulfur compounds to H_2S for subsequent removal. Other metallic catalysts have been used more recently for absorbing organic sulfur and H_2S with regeneration of the catalyst by burning off the sulfur with air.

3. *Adsorption by Solids and Desorption.* Activated carbon and silica gel serve as adsorbing mediums. Plants using these materials for light oil recovery also remove most organic sulfur present in the gas. For successful use of activated carbon for light oil removal, gas must be free of H_2S and heavy condensable hydrocarbons. Because of accumulation of tar and resinous compounds from many manufactured gases, the activated carbon process has not generally been used with manufactured gas.

REFERENCES

1. Kohl, A. L., and Riesenfeld, F. C. *Gas Purification.* New York, McGraw-Hill, 1960.
2. Dotterweich, F. H., and Huff, W. J. "Colloidal Properties of Iron Oxide in the Removal of Hydrogen Sulphide from Gas." *A.G.A. Proc.* 1938: 699–717.
3. Evans, O. B. "Revivification in Place." *A.G.A. Tech. Sec. Proc.* 1919: 205–239.
4. Meade, A. *Modern Gasworks Practice,* 2d ed. London, Benn Bros., 1921.
5. Murphy, E. J. "The Control of Alkalinity in Dry Box Purification." *A.G.A. Proc.* 1935: 701–2.
6. "Proteus." "Design of Purifiers for the Removal of Hydrogen Sulphide from Town Gas by Iron Oxide." *Gas J.* 202: 922–4, June 28; 203: 51–2, July 5, 1933.
7. Seil, G. E. *Gas Chemists' Manual of Dry Box Purification of Gas.* New York, A.G.A., 1943.
8. Huff, W. J., and Milbourne, C. G. "Humidity Effects in the Iron Oxide Process for the Removal of Hydrogen Sulphide from Gas." *A.G.A. Proc.* 1930: 856–86. [Note: Figs. 6(a) and 6(b) are for air.] Same in: *Ind. Eng. Chem.* 22: 1213–24, Nov. 1930.
9. Guba, E. F., and Seeler, E. V., Jr. "A Troublesome Mold and Its Control in Gas-Purifying Sponge." *Economic Botany* 2: 170–7, 1948.
10. *Studies on the Identity and Control of Stilbaceous Mold in Gas Purifying Sponge.* A.G.A. Tech. Sec. Joint Prod. & Chem. Comm. Conf., 1944.
11. Huff, W. J. "Report of Subcommittee on Chemical Aspects of Safety and Accident Prevention." *A.G.A. Proc.* 1945: 207–9.
12. *Ibid. A.G.A. Proc.* 1946: 510–26.
13. Fulweiler, W. H. "Manufactured City Gas." (In: Furnas, C. C., Ed. *Rogers' Manual of Industrial Chemistry,* 6th ed., vol. 2, Chap. 15. New York, Van Nostrand, 1942.
14. Altieri, V. J. *Gas Analysis and Testing of Gaseous Materials.* New York, A.G.A., 1945.
15. Bottoms, R. R. "Organic Bases for Gas Purification." *Ind. Eng. Chem.* 23: 501–4, May 1931.
16. Jacobson, D. L. "The Thylox Process of Liquid Purification." *Gas Age-Record* 63: 597–600, May 4, 1929.
17. Gollmar, H. A. "Chemistry of the Thylox Gas-Purification Process." *Ind. Eng. Chem.* 26: 130–2, Feb. 1934.
18. ——. "Removal of Sulfur Compounds from Coal Gas." (In: Lowry, H. H. *Chemistry of Coal Utilization,* vol. II, Chap. 26. New York, Wiley, 1945.

BIBLIOGRAPHY

Sperr, F. W., Jr. "Gas Purification in Relation to Coal Sulphur." *Intern. Conf. on Bituminous Coal* II: 37–64, 1928.

Powell, A. R. "Recovery of Sulfur from Fuel Gases." *Ind. Eng. Chem.* 31: 789–96, July 1939.

Leech, W. A., Jr., and Schreiber, F. D. "Sulphuric Acid from Coke-Oven Gas." *Iron Steel Engr.* 23: 93–101, Dec. 1946.

Kastens, M. L., and Barraclough, R. "Cyanides from the Coke Oven." *Ind. Eng. Chem.* 43: 1882–92, Sept. 1951.

Chapter 9

Substitute Natural Gas from Coal

By C. G. von Fredersdorff and F. E. Vandaveer

REASONS FOR DEVELOPMENT FROM COAL

Sources of substitute natural gas are being considered in long-range planning, even though large natural gas reserves are assured. Any process for large-scale manufacture of substitute natural gas indicates use of coal because of the abundance of this resource. Advantageous location of large coal beds in areas traversed by natural gas transmission lines and increases in production and processing costs of natural gas have been contributing factors to extensive research in making high-Btu gas from coal.

The principal economic factor that would permit the use of a substitute natural gas over other forms of energy is cost at the point of consumption. This cost comprises manufacturing, transmission, and distribution costs, plus the return on plant investment. Manufacturing costs are determined by raw material, labor, and fixed plant charges. These costs depend directly upon the complexity of operations for converting coal into gas. The low hydrogen-to-carbon ratio of coal, as compared with that of natural gas, gives a measure of this complexity. Since large capital investments are required, processes utilizing coal for producing methane are suitable only for high load factor (base load) operations.

Transmission cost per unit of energy for natural gas in pipelines is the lowest for any mode of long distance transmission of any fuel, including electricity. Obviously, if this substitute gas is produced near existing pipeline facilities, close to major market areas, transmission charges will be minimized. Distribution costs of this gas should not differ significantly from corresponding costs for natural gas.

Production costs could not be calculated with accuracy in 1965 because integrated processes are still in developmental stages. However, the technology of certain component steps, such as oxygen and hydrogen production, gas purification, catalytic water gas shift, and carbon dioxide removal, has been developed commercially. Pipeline gas production costs have been estimated as between \$0.60 and \$1.20 per MCF, depending upon assumptions made on capital, operating, and raw material unit costs. On the basis of equivalent energy, if the production cost of pipeline gas is assumed to be \$1.00 per MCF or approximately 11 cents per therm, the corresponding electric generation cost is 3.8 mills per kwh. Electric generating costs in 1959 ranged from 2.5 to 7 mills per kwh, or 7.3 to 20.5 cents per therm for large coal-fired power stations. If a gas-from-coal program should become established as early as 1975, projections of electric statistics indicate that approximately 15 per cent of all new electric power installations would have generating costs competitive on an equivalent therm basis with pipeline gas production costs arbitrarily taken at \$1.00 per MCF. However, electric transmission plus distribution costs range typically from 20 to 30 cents per therm, while gas transmission (over considerably longer distance than normal electric transmission) plus distribution costs were typically six to eight cents per therm in 1959.

Substitute natural gas made from oil was used by some companies to meet peak loads and emergencies, and to avoid demand charges by pipeline companies. How soon substitute natural gas made from coal will be needed is a matter of conjecture. That need may be determined by factors other than availability of natural gas in the field. These factors include cost of natural gas at point of consumption, financing problems, coal costs, and production and transmission costs of the substitute gas. Figures 2-1 and 4-1 give an indication of increasing reserves compared with production.

DESIRABLE SUBSTITUTE NATURAL GAS CHARACTERISTICS

The following characteristics of a substitute natural gas made from coal are desirable, although recognized as difficult to attain:

1. Complete interchangeability with natural gas insofar as burning characteristics in domestic and industrial appliances are concerned.
2. Heating value not less than 950 Btu per cu ft on a dry basis.
3. Chemical composition essentially methane with other saturated hydrocarbons and possibly some inert gases.
4. No carbon monoxide (less than 0.1 per cent) to avoid field asphyxiation troubles.
5. Very little unsaturated hydrocarbon to avoid gum, meter, and regulator diaphragm troubles
6. Cost close to that of natural gas as distributed.

METHODS OF PRODUCING SUBSTITUTE NATURAL GAS FROM COAL

Basic Methods

In principle, three basic methods for producing high methane content gases from coal are: (1) pyrolysis, (2) hydrogenation, and (3) gasification–methanation.

Pyrolysis, or direct application of heat to the coal, drives off the volatile matter, leaving a high carbon content residue.

As exemplified in a coke oven, pyrolysis represents production of a small amount of gas, 15 to 30 per cent of the energy of bituminous coal, at the expense of large rejected amounts of carbon. Gas quantity increases in proportion to the hydrogen content of the coal. Since gases from coal pyrolysis contain a large proportion of hydrogen and some carbon monoxide, they cannot serve as a natural gas substitute without treatment. Because of the large proportion of carbon rejected as coke or breeze, partial gasification of coal by pyrolysis alone is not an economical approach to large-volume synthetic methane production.

In **hydrogenation** or hydrogasification, externally produced hydrogen is added directly to the coal substance under moderate temperatures (1200–1350 F) and elevated pressures (50–200 atm) to form methane and possibly other hydrocarbons.

Hydrogenation is an upgrading process which utilizes both hydrogen addition and carbon rejection for methane production. Only part (about 50–60 per cent by processes known in 1958) of the coal can be hydrogenated within economic limits of time, temperature, and pressures. The coal rejected may be used to produce the required hydrogen. External hydrogen is obtained from decomposition of steam by one of the following methods:

1. Complete gasification with oxygen and steam of either the carbon residue or part of the coal feed, followed by catalytic water gas shift with excess steam, and removal of CO_2.
2. Partial combustion of the carbon residue with air and steam in gas producers to form CO and H_2 and subsequent conversion by water gas shift of CO to H_2.
3. Catalytic reforming in presence of steam of part of the product methane, followed by catalytic water gas shift with excess steam, and removal of CO_2.
4. Passing steam over hot iron[1] to form H_2 and iron oxide.

Gasification–methanation is a dual-step process: making synthesis gas (H_2–CO mixture) and catalytically upgrading it to CH_4.

The ideal high-Btu gas process may be considered as one that yields only methane or methane and ethane. Accordingly, the basic requirements for methane production from coal are either hydrogen addition or excess carbon rejection.

Gasification and catalytic methanation are methods of methane synthesis by indirect addition of hydrogen to carbon. External hydrogen is obtained from steam or hydrocarbon decomposition, and carbon is rejected as carbon dioxide in the accompanying chemical reactions.

Each method of methane synthesis from coal is characterized by heat requirements which vary with coal composition and the process. The manner of supplying heat depends in part upon the amount of heat required, the nature of the chemical reactions, and the cost. In general, coal pyrolysis requires heat, usually by indirect transfer, to initiate the reactions. Coal hydrogenation liberates enough heat to maintain the system in thermal balance. Provision for heat transfer, except to initiate the reactions, is usually unnecessary. Reactions occur only at high pressure. Hydrogenation production through decomposition of steam with carbon, or with the product methane, is highly endothermic. It is usually achieved thru partial combustion of the fuel with oxygen, although in catalytic steam reforming of methane, heat requirements would preferably be supplied by external firing of catalyst tubes. Production of additional hydrogen in the water gas shift reaction is sufficiently exothermic to obviate external heat supply.

In complete gasification of coal for synthesis gas production, large heat requirements for steam decomposition are usually best fulfilled thru partial combustion of part of the coal with oxygen. Alternatives are using tonnage oxygen, thus making the gasification process continuous, or using air in intermittent or cyclic blasting of the fuel bed. The choice between a continuous or an intermittent process depends primarily upon fuel characteristics, desired operating pressures of the system, and cost. In methanation of synthesis gas, large amounts of heat liberated are usually best removed by indirect transfer to a cooling medium surrounding the reactor or by heat exchangers within it.

Approximate Methane Yields and Heat Requirements

The amount of methane-produced coal depends upon coal composition, type of process, and method of heat supply.

Carbon, heat requirements, and thermal efficiencies of methane production by gasification-methanation, by partial hydrogenation–residual carbon gasification, and by complete hydrogenation–catalytic reforming are compared on a theoretical basis for idealized processes[2] in Table 3-56 and Fig. 3-67. There appears to be a 12 to 15 per cent greater thermal efficiency in favor of hydrogenation.

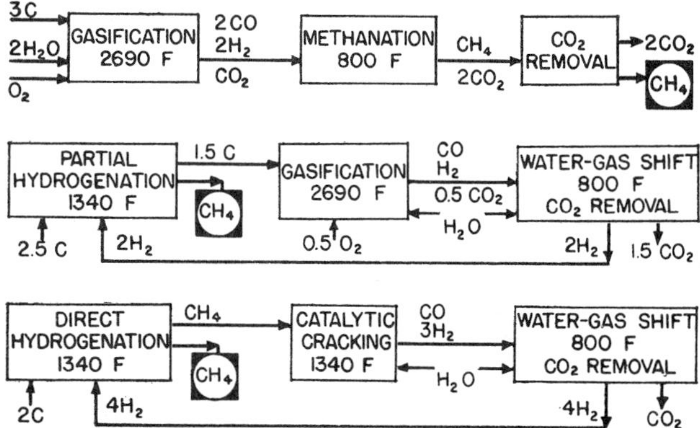

Fig. 3-67 Theoretical process requirements for conversion of carbon to methane.[2]

Process Steps and Equipment

Processes for methane production from coal are in various stages of experimental development. Figure 3-68 shows major steps associated with two approaches (hydrogenation and gasification–catalytic methanation) for making pipeline gas from coal. Comments relative to the state of the art are included. For the hydrogenation process, a coal pretreatment step may be necessary for caking or coking coals to improve the fuel bed performance in a pressure hydrogenation reactor. This step may not be necessary for noncoking coal and lignites. Pretreatment methods by low-temperature partial devolatilization or by treatment at low temperatures with air,

Table 3-56 Heat Balances and Efficiencies of Ideal Processes for Converting Carbon to Methane

(based on enthalpies and standard heats of formation of graphite and ideal gases H_2O, O_2, H_2, CO, CO_2, and CH_4, and reference conditions of 77 F, 1 atm, and liquid water.[3])

Process*	3 moles C per mole CH_4 I	2½ moles C per mole CH_4 II	2 moles C per mole CH_4 III
Btu per lb-mole of product methane			
Heat of combustion of reactant carbon	507,879	423,233	338,587
Heat requirements:			
Sensible heats	131,380	89,635	84,769
Endothermic heats of reaction	...	...	96,977
Latent heats	37,870	37,870	37,870
Total input	677,129	550,738	558,203
Heat of combustion of product methane	383,040	383,040	383,040
Heat available from sensible heat content of final and intermediate products	127,985	85,334	81,697
Exothermic heats of reaction	166,104	82,364	93,466
Total output	677,129	550,738	558,203
Thermal efficiency, per cent			
Without heat recovery†	56.6	69.6	68.6
With recovery of sensible and latent heat requirements‡	75.4	90.5	87.9§

* See Fig. 3-67.

† Heat of combustion of product CH_4/(heat of combustion of reactant carbon + total heat requirements).

‡ Heat of combustion of product CH_4/(heat of combustion of reactant carbon + net heat requirements).

§ Endothermic heat of reaction assumed to be supplied from external source at 1340 F.

steam, nitrogen, or CO_2 to diminish or destroy coking properties were investigated. Pretreatment schemes may also include the low-temperature distillation and carbonization techniques as practiced in the Carbolux,[4] Carbocite,[5] KSG,[6] and McEwen-Runge[7] processes on a plant scale in Europe and the United States. The partial devolatilization method employed in the Disco process[8,9] yields a reactive coal char which appears suitable for direct hydrogenation.[7]

The pressure hydrogenation reactor could employ coarse fuel particles in a fixed bed or finely divided particles in a fluidized bed. Both systems are under investigation to determine their feasibility for intermittent or continuous operation. Although not employed presently as a high-Btu reactor, the Lurgi pressure gasifier[10,11] typifies equipment for pipeline gas production by means of direct coal hydrogenation in a continuous fixed-bed process. The Winkler water gas generator[11] employing oxygen–steam blast is typical of the large-scale equipment required for coal hydrogenation in a fluid-bed operation.

Experimental results indicate that the volatile matter portion of coal is its most readily hydrogenated component.[2] Part of the carbonaceous residue is generally very unreactive, and difficult to hydrogenate. A favored experimental development comprises partial hydrogenation and generation of the required hydrogen from the coal residue. Alternatively, the residue could be employed as boiler fuel, the hydrogen being produced from raw coal or the product methane.

Methane produced from direct coal hydrogenation contains small quantities of higher hydrocarbons, N_2, CO, CO_2, sulfur compounds, and varying amounts of unreacted hydrogen.[2] Ordinarily, it requires purification for removal of sulfur compounds and condensable hydrocarbons. The quantity of unreacted hydrogen (10–30 per cent in raw product gas) depends primarily upon the reactor operating conditions and the reactivity of the fuel to hydrogenation. Although the product gas should have satisfactory combustion characteristics in appliances adjusted to natural gas, the disadvantage of the process is that unreacted hydrogen represents a loss in methane production potential per unit quantity of feed hydrogen to the hydrogenation reactor. Accordingly, a step-like methane purification by liquefaction, and recovery and reuse of unreacted hydrogen, might have merit in ensuring maximum heating value of the product gas and full use of all potential hydrogen available to the process.

Pipeline gas production by the dual process (coal gasification–catalytic methanation) requires generation of a synthesis gas from complete gasification of coal, adjustment of the H_2 to CO ratio of the synthesis gas over a water gas shift catalyst, removal of sulfur compounds and CO_2 from the adjusted synthesis gas, conversion of CO plus H_2 to CH_4 over a suitable methanation catalyst and, finally, removal of CO_2. This results in nearly complete conversion of the synthesis gas to a gas having 900 Btu per SCF heating value.

Purification methods for removal of sulfur-bearing compounds from synthesis gas have been used on a commercial scale. Fixed-bed dry processes include use of iron oxide for H_2S removal and activated carbon for carbon disulfide and carbon oxysulfide removal. Selective H_2S removal is accomplished in the Seaboard and Thylox processes. Other liquid-contacting regenerative processes include: (1) oil scrubbing for carbon disulfide removal, (2) scrubbing with glycol-amine or di- and tri-ethanolamine solutions for H_2S and CO_2 removal, and (3) use of refrigerated methanol (Rectisol process) for H_2S, organic sulfur, and CO_2 removal.

Methanation of synthesis gas with nickel catalysts has been accomplished on a pilot plant scale. Satisfactory catalyst activity was maintained over periods of 1500 hours, for the catalyst lifetime, producing 2500 lb of methane per lb of catalyst. However, further catalyst development to improve chemical composition, physical and chemical stability, and resistance to sulfur poisoning is necessary.

Basic advantages of coal hydrogenation include:

1. No need for sulfur removal except in the product gas.
2. Significantly lower coal and oxygen requirements per unit of product methane.
3. Considerably higher potential thermal efficiency.

Among the disadvantages are:

1. Requirement of hydrogen-resistant high-pressure vessels operating at 50 to 200 atm pressure.
2. Dilution of product gas with some unreacted hydrogen, unless a methane separation or methanation step is provided.

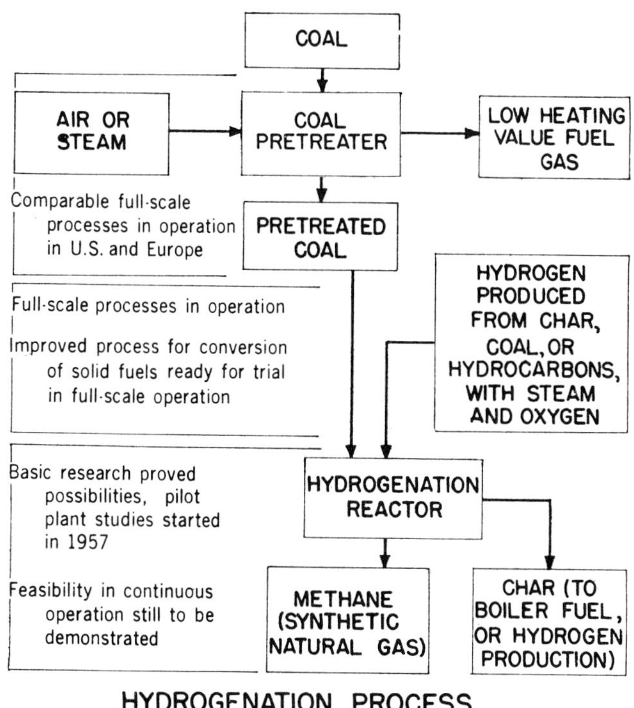

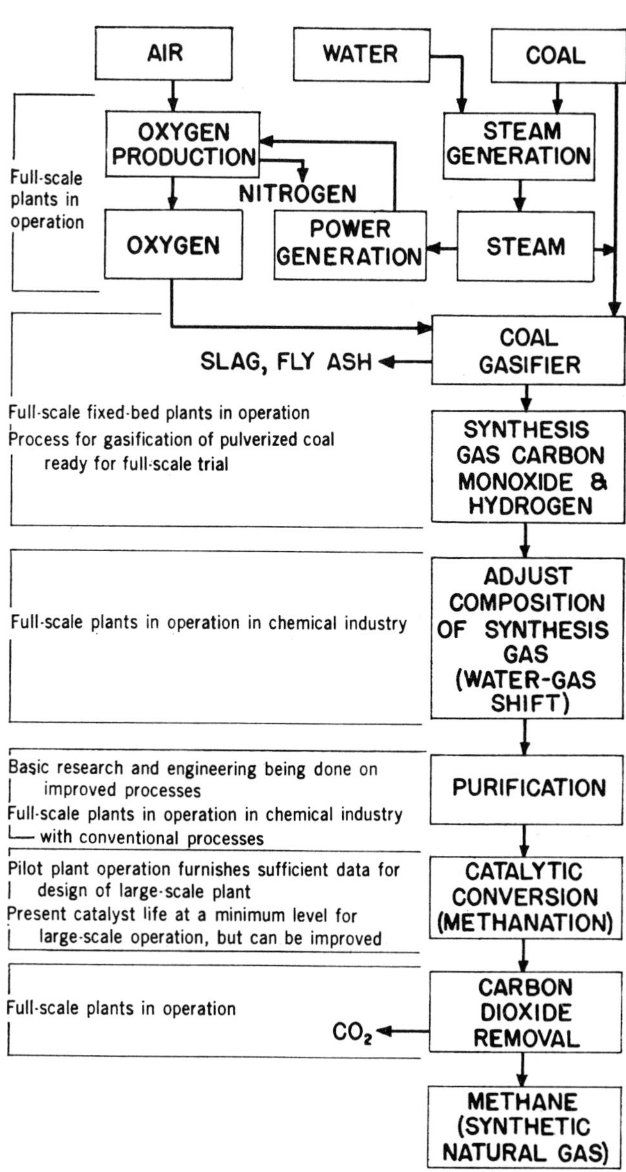

Fig 3-68 (above and right) Schematic diagrams of high-Btu gas production from coal.

3. Low gas production per unit volume of reactor equipment, since coal hydrogenation reaction rates are comparatively slow.

Principal advantages of coal gasification–methanation are:

1. Adaptability to a wide variety of coal.
2. A minimum of coal pretreatment required with respect to agglomerating characteristics.
3. High rates of methane production per unit volume of reactor equipment.
4. Probable operation at any pressure level, from one atm to pipeline pressure.

Disadvantages include:

1. High degree of sulfur removal from synthesis gas required to prevent early loss of activity of methanation (nickel) catalyst.
2. Rapid heat removal required from the highly exothermic methanation reactions near reaction point.
3. Low thermal efficiency of conversion of coal to methane.

PRODUCTION OF SYNTHESIS GAS

General Considerations

The first step in making methane by the process of coal gasification and subsequent methanation is to produce a synthesis gas, consisting of CO and H_2. Coal characteristics greatly influence the type of process and equipment to be used. They include caking or agglomerating characteristics, friability or resistance to degradation ash properties and content (e.g., fusibility), and cost.

Coal gasification processes may be grouped according to:

1. Type of fuel bed: fixed, fluidized, dilute phase, or suspended bed.

2. Gas flow: continuous or intermittent (cyclic).
3. Coal movement: continuous or intermittent feeding.
4. Heat transfer: direct or indirect.
5. Operating pressure: atmospheric or higher.
6. Temperature of ash removal region: slagging or nonslagging.

Fixed-Bed Gasification. Principal conditions for efficient operation of an atmospheric pressure, fixed-bed cyclic process, such as the water gas generator, are: (1) uniformity in coal size, (2) absence of agglomeration, (3) low pressure drop thru fuel bed, and (4) absence of clinker formation and continuity of ash removal.

The most efficient conversion of coal into gas is usually accomplished in continuous processes utilizing reasonably deep fixed beds, as exemplified in the conventional gas producer or the Lurgi generator,[11] which operates with oxygen–steam blast at pressures up to 30 atm. The fuel bed is characterized by zones of different temperature. They may be designated as: (1) ash zone at the generator bottom, (2) oxidation zone or region of heat supply, (3) reduction zone or region of steam decomposition, and (4) preheat zone, where the incoming fuel charge is dried and heated to operating temperature by upflowing hot make gases.

Fuels for fixed-bed gasification may be lump coke, screened anthracite, noncaking or weakly caking bituminous coals, briquetted brown coals, or lignite. These same fuel limitations apply to fixed-bed continuous processes operating under ash slagging conditions. Slagging generators are simpler in construction, requiring no grate to support the fuel bed. They can be of considerably larger diameter. Gasification rates under slagging conditions in excess of 500 lb fuel per hr per sq ft of hearth area have been obtained, compared with 50 to 200 lb fuel per hr per sq ft of grate area for nonslagging processes.

Fluid-Bed Gasification. Gasification of finely divided coal in a fluidized bed requires a noncaking coal for effective operation. Weakly caking coals may be accommodated by preatment or special mixing processes. Fluid-bed gasification is limited in capacity by restrictions on maximum allowable velocity of the fluidizing gases to prevent excessive bed carry-over. Operating temperature is normally below that of ash-softening. Objectionable features are: (1) generally high unreacted carbon losses, necessitating recycling of entrained

Table 3-57 Features of Selected Gasification Processes[62]

Process	State of bed and particles		Gasification pressure, atm	Gasification media	State of ash leaving reaction zone	Approx. fuel throughputs, lb/hr-sq ft
	Bed*	Particles†				
Commercial						
Marischka producer	Fixed	Static	1	Air and steam	Solid	15–100
Power-Gas B.W.T. Integrale, etc.	Fixed	Static	1	Air and steam	Solid	15–100
Kerpely producer	Fixed	Static	1	Air, O_2, and steam	Solid	60–100
Leuna slagging producers	Fixed	Static	1	Air, O_2, and steam or O_2 and steam	Solid	150–300
Lurgi	Fixed	Static	10–30	O_2 and steam	Solid	100–400
Winkler	Fixed	Fluidized (mainly)	1	Air, O_2 and steam or O_2 and steam	Solid	200–450
Koppers–Totzek	No bed	Suspended	1	O_2 and steam	Solid	50‡ 600§
Ruhrgas A.G. cyclonic	No bed	Suspended	1	Air	Slag	250–300
Experimental						
Flesch Winkler B.A.S.F. Flesch-Demag	Fixed	Static, fluidized	1	Air and steam or O_2 and steam	Solid	70–200
Lurgi Ruhrgas	Moving (heat carrier)	Suspended (fuel)	1	Air and steam	Solid	400–450 (20–70% ungasified)
Panindco	No bed	Suspended	1	O_2 and steam	Solid (mainly)	40–50
Texaco	No bed	Suspended	14	O_2 and steam	Slag	900–2500
U.S. Bureau of Mines	No bed	Suspended	7–20	O_2 and steam	Slag	900–2500

* **Gasifying beds,** composed of catalyst, heat storing material, or solid fuel to be gasified, are defined as follows:

1. Fixed beds—the bed as a whole is stationary. It may be of static or of circulating fluidized particles. In the case of the latter, if they are composed of the fuel being gasified, they will pass out in suspension in the gas stream when burned down below a critical size.

2 Moving beds—the whole mass of catalyst, heat carrier, or fuel, is circulated between the gassifer and some other vessel in which it is reheated or recooled or otherwise treated. These beds may also be composed of either static or fluidized particles, which in the latter case would be circulating also within the moving bed.

† **Particle definitions:**

1. Static or fixed—lumps or particles (usually 8–75 mm or more) resting upon each other, i.e., motionless relative to each other.
2. Fluidized—particles (usually 0.1–8 mm) torn apart by gas velocity in turbulent motion relative to each other, but in a clearly defined bed of, for example, roughly two-thirds the bulk density of the material when static.
3. In suspension—particles carried in a gas stream, in which case there is no clearly defined bed and gasification takes place in a chamber or shaft (usually up to 0.1 mm but it is practical to suspend particles up to 2 mm). Density of the material may be from 1–20% of the static bulk density.

‡ Of gasifier vessel cross section.

§ Of gasifying head cross section.

solids back to the fluidized bed, and (2) difficulty in collecting ash at the generator bottom, necessitating its removal mainly as fly ash from the make gases. Fluid-bed gasification is potentially more versatile than fixed-bed gasification in respect to arrangement and design of equipment, but its cost is potentially higher.

Suspension or Dilute Phase Gasification. Gasification of finely divided coal by the suspension or dilute phase technique tends to overcome some of operating difficulties with weakly caking coals in fixed-bed or fluidized-bed processes. Pulverized coal is entrained in a gasifying medium consisting of steam and oxygen-enriched air, or steam and high-purity oxygen. The mixture is blown continuously into a refractory-lined gasifier operating at atmospheric or higher pressures. A sequence of oxidation and steam decomposition reactions occurs. The product gases generated range in quality from producer gas to water or synthesis gas, depending upon oxygen concentration of the gasifying medium. One attractive feature of suspension gasification is the relatively wide spatial separation of the individual coal particles. This permits each particle to swell, soften, decrepitate, and even slag without interfering with others and without altering the gas flow pattern or fuel inventory of the system. Effective operation requires high temperatures, normally above the ash fusion point. Slagging conditions are readily obtained thru use of steam and commercial grade oxygen as the gasifying medium. Thus, by reason of dispersion of fuel to minimize contact between particles, and use of slagging conditions to remove a large portion of the ash in fused form, this process would be applicable to a wide range of solid fuel types, including coking coal. Experimental investigations have verified the operability of this process with coking and noncoking coals.[12,13]

Process Features. Table 3-57 summarizes the features of a number of commercial and experimental processes. The table gives a rough indication of the variations in pressure and ash slagging, and in the use of oxygen.

European Developments

In Europe several gasification processes have been developed for converting coal into low-Btu city gas or synthesis gas for production of ammonia and synthetic liquid fuels by the Fischer-Tropsch process.

Lurgi Process. During the 1930–1940 decade, a continuous pressure gasification process was developed capable of utilizing coal inferior to high-grade bituminous grades. The Lurgi process[11,14] for production of 430–450 Btu methane-rich water gas (after CO_2 removal) has the following characteristics:

1. Fixed-bed complete gasification of noncoking brown coal, lignite, and subbituminous coals in particle sizes of $\frac{1}{4}$ in. or less.
2. Continuous gasification at nonslagging conditions and 10–30 atm operating pressure assured thru use of oxygen for internal heating.
3. Methane formation favored in upper portion of fuel bed.

Table 3-58a Operating Results of Large Lurgi Plants[15]

	Plant									
	Morewell-Australia pipeline gas, 6 generators, 2.6 m diam		Sasol-South Africa Fischer-Tropsch synthesis, 10 generators, 3.7 m diam				Ruhrgas Dorsten pipeline gas, 6 generators, 2.6 m diam			
Fuel										
Type	Brown coal		Bituminous coal				Gas coal			
Size	Briquettes		7–20 mm or 20–60 mm				6–30 mm			
Moisture, wt %	15.6		9.7				5.0			
Ash, wt %	1.6		26.4				22.0			
Fixed carbon, wt %	42.2		42.8				42.0			
Volatile matter, wt %	40.6		21.1				31.0			
Tar content (Fischer–Hempel), wt %	5.6		2.1				10.5			
Gas composition, vol %	Crude	Purified	Crude	Purified	Crude	Purified	Crude	Purified	Crude	Purified
$CO_2 + H_2S$	34.0	4.5	28.0	0.8	32.3	0.9	27.0	2.0	30.5	2.0
O_2	0.1	0.1	0.0	...	0.01	...	0.0	...	0.0	...
C_nH_m	0.4	0.5	0.2	0.2	0.2	0.2	0.4	0.5	0.4	0.5
CO	14.5	20.7	22.4	31.5	16.4	24.8	21.0	28.5	17.0	24.1
H_2	35.8	52.6	38.0	53.3	39.4	59.3	40.5	54.1	40.5	57.3
CH_4	13.5	19.4	10.9	13.4	11.3	14.0	10.1	13.6	10.7	14.9
N_2	1.7	2.2	0.5	0.8	0.4	0.8	1.0	1.3	0.9	1.2
Gross heating value, purified gas, Btu/SCF		457		421		424		415		428
Raw gas per generator, MCF/hr	260–372		930–1190				559–596			
Oxygen used, SCF/SCF	0.121	0.178	0.139	0.213	...	...	0.17	0.23	0.16	...
Steam used for gasification from outside sources, lb/MCF ranges	51	...	35.5	...	...	...	41.5	...	56.3	...
Tar and light oil yield, wt %	3.5		1.6		...	...	7.5			
Refined light oil yield, wt %	0.75		0.4		...	...	1.67			
Ammonium sulfate yield, wt %	...		5.2		...	...	2.98			
Low-pressure steam produced per SCF raw gas, lb/MCF	62.1		33.2		...	...	35.5			
Efficiency referred to raw gas, %	Gross heating value (gas + by-products)/Gross heating value of coal = 88–92%									

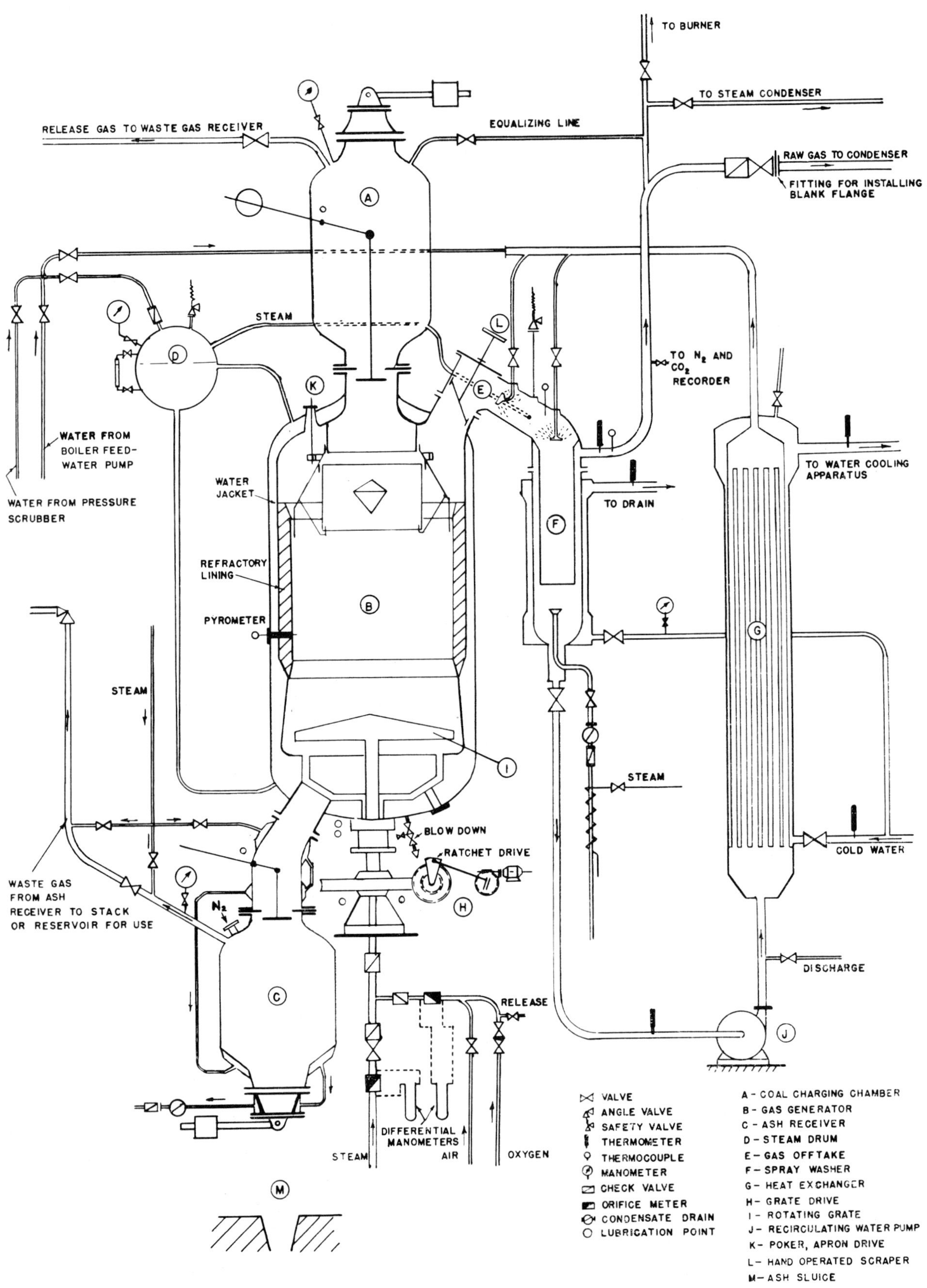

Fig. 3-69 Lurgi fixed-bed generator. (Bureau of Mines)

A typical generator (Fig. 3-69) consists of a water-jacketed vertical pressure shell equipped with a rotating grate, pressure fuel charging bell, and ash discharge hopper.

The Lurgi process is a successful gasification method, as demonstrated by installations for city gas and pipeline gas production in Germany, Czechoslovakia, Australia, and South Africa (Table 3-58a). Use of American coals in Lurgi pressure gasification has been tested experimentally.[10,16] Results from a 4-in. and 6-in. diam generator indicated: (1) caking coals cannot be gasified satisfactorily, (2) weakly caking or noncaking coals may exhibit caking behavior under pressure gasification conditions, thus becoming useless without prior treatment, and (3) noncaking fuels, such as anthracite and low- and high-temperature cokes, can be completely gasified. Reactive coal chars for Lurgi gasification may be prepared in rotary drum low-temperature carbonization equipment. The Lurgi pressure gasifier is of interest because its gasification process is relatively economical in oxygen consumption, and gasification can be conducted at relatively high pressure.

Winkler Process. The desirability of gasifying large quantities of fines from brown coal char produced in Germany resulted in development of the Winkler process[14] for treatment of this material in a "fluidized bed" at essentially atmospheric pressure. Designed originally for air–steam blast and later modified for oxygen–steam operation for hydrogen and synthesis gas generation, this process was adapted primarily to predried brown-coal and pulverized brown-coal coke (3 to 65 mesh) for producing a 200–270 Btu water gas. A typical Winkler generator (Fig. 3-70) consists of a tall insulated vertical shell, equipped with a screw feeder and a fixed grate thru which steam and air or oxygen are blown at high rates to agitate the fuel bed. High capacities (3,500 to 14,000 SCF make gas per hr per sq ft of grate area) were achieved. Because of the high gas velocities employed, 20 to 30 per cent of the fuel bed was eventually blown from the generator thru the gas offtake into the ash collecting system. Carbon gasification efficiency was therefore unsatisfactorily low for a single pass operation. Recent designs return and gasify this material. The entrained unreacted material can also be utilized in external steam generating equipment.

To overcome fuel limitations a new technique embodying a combined fixed-bed and fluid-bed operation has been under development. Originally consisting of a single generator (Flesch–Winkler process)[17] and later modified with two generators in parallel (Flesch–Demag process),[17] it achieves gasification of various grades of finely divided fuel in a cyclic system.

Thyssen–Galocsy Process. Another development was the application of oxygen–steam blast for continuous low-pressure fixed-bed gasification of coke and low-grade, high-ash fuels under slagging conditions in grateless producers. The Thyssen-Galocsy[14] slagging producer (Fig. 3-71) is representative. It is shaped like a blast furnace, its lower conical part tapering down to the oxygen tuyeres. A second set of tuyeres provides steam and additional oxygen admission. The lower part of the generator is cooled by circulating water in cooling boxes, while molten slag is withdrawn thru a tap hole in the earth. A centrally located large-diameter gas outlet tube serves to prevent excessive and uneven gasification at the generator rim. Any noncoking coal or coke of high-ash content may be employed. Weakly caking coals may be accommodated if sufficiently mixed with coarse iron ore or broken slag prior to charging. Due to high fuel-bed temperatures, the synthesis gases contained up to 86 per cent combustibles (320 to 330 Btu per SCF).

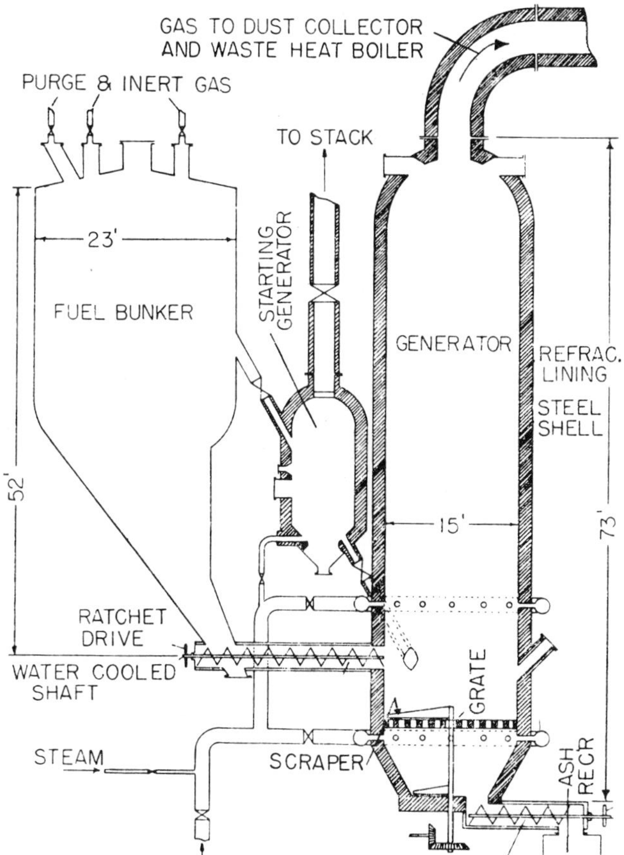

Fig. 3-70 Winkler fluid-bed generator. (Bureau of Mines)

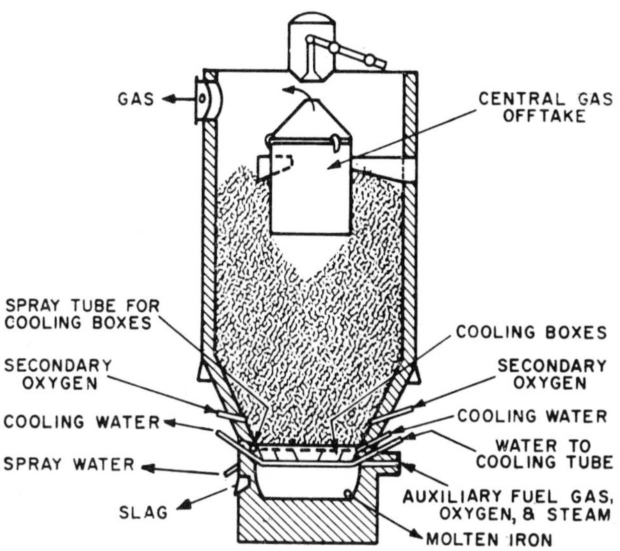

Fig. 3-71 Thyssen–Galocsy fixed-bed slagging generator.

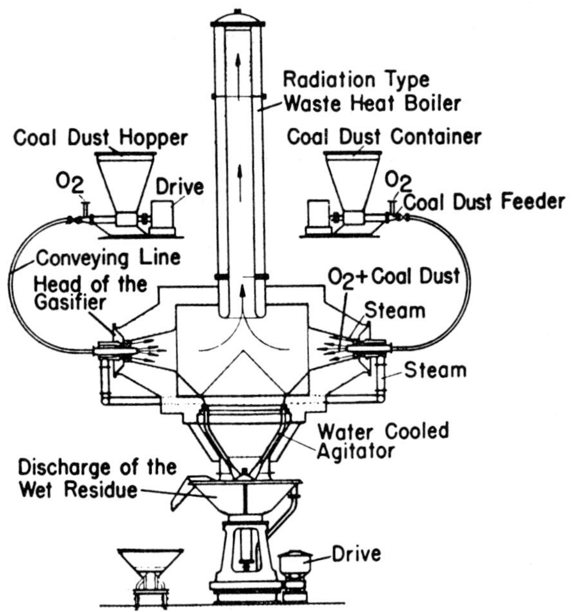

Fig. 3-72 Koppers–Totzek pulverized coal gasifier.

Koppers Dust Gasification Unit. This unit has oxygen–coal suspensions introduced thru two opposing nozzles to form a turbulent reaction mixture. Superheated steam at about 2200 F, introduced with the oxygen–coal suspension, gave a reported 94 to 95 per cent carbon gasification in producing a 250 and 270 Btu water gas from coal and bituminous coal, respectively.[18] Molten ash particles in the combustion zone should cool and solidify before reaching the reactor wall. If not, and appreciable slagging occurs over the ash collecting zone, the unit on prolonged operation would probably become plugged with solid particles imbedded in a viscous slag layer. Three modified Koppers–Totzek generators (Fig. 3-72) installed in Finland have operated continuously since July 1952, gasifying 50 tons of bituminous coal dust per day in production of ammonia synthesis gas.[19,20]

Table 3-58b Performance Data for a Koppers–Totzek Bituminous Coal Gasifier
(Oulu, Finland)[15]

Input		Hydrogen/carbon monoxide	
Fuel, lb/hr	4280	ratio	0.67
therms/hr	517	Gross heating value, Btu/SCF	271
Oxygen, lb/hr	3352	Materials requirements/MCF	
Steam, lb/hr	2315	$CO + H_2$	
Steam temperature, °C	128	Fuel, lb	39.5
		Fuel, therms	4.77
Gas composition, vol %		Oxygen (97%), lb	31.4
CO_2	12.6	Steam, lb	21.3
C_n–H_m	0	Steam generated/MCF $CO + H_2$, lb	42.4
O_2	0.1		
CO	51.1	Performance	
H_2	34.0	Specific gasification rate,	
CH_4	0.1	therms/cu ft-hr*	1.1
N_2	1.9	Steam decomposition, %†	29.7
		Carbon conversion, %‡	88.6

* Height of shaft below heat exchangers only used in calculating the volume.
† Steam decomposition equals

$$\frac{H_2 \text{ in gas} - H_2 \text{ in coal} + (O_2 \text{ in coal}/8)}{H_2 \text{ in (steam blast} + \text{moisture from coal)}}.$$

‡ Carbon in gases/carbon in input fuel.

Performance data for a Koppers-Totzek gasifier operating at 1.0 atm are given in Table 3-58b. The virtual absence of hydrocarbons and inerts in the product gas facilitates chemical synthesis.

American Developments

Continuous Gasification at Atmospheric Pressure with Oxygen and Steam. United States and Canadian developments have been directed primarily toward continuous complete coal gasification with oxygen–steam blast in fixed-bed units at nonslagging conditions, and in suspension gasifiers at ash slagging temperatures.

Wellman–Galusha Producer. Pilot plant scale tests were conducted on continuous gasification of coke, bituminous coal, and peat at atmospheric pressure in a 5-ft diam fixed-grate blue gas generator at the Leaside Station of the Hydro-electric Power Commission of Ontario. Full-scale tests on continuous gasification of rice and barley size anthracite with oxygen and steam have been conducted in 10-ft diam Wellman–Galusha producers[21,22] (Fig. 3-73 and Table 3-29) at Consolidated Mining and Smelting Co., Trail, British Columbia. They are operated normally with 1/8 to 2½ in. coke for producing ammonia synthesis gas. Tests with anthracite indicated: (1) satisfactory gas quality for synthesis purposes, (2) oxygen-steam gasification rates double those with air–steam blast, and (3) thermal efficiencies above 85 per cent based on total heat input.

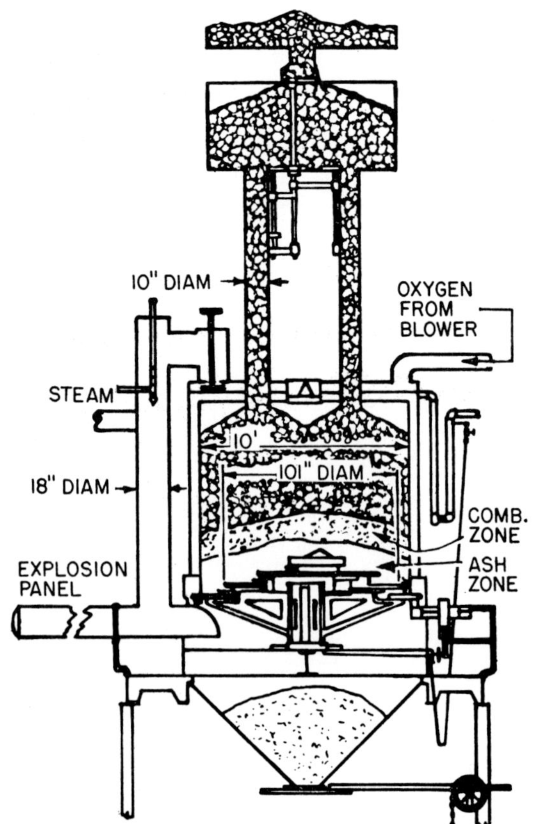

Fig. 3-73 Wellman–Galusha oxygen-steam blown producer.

Kerpely Producer. Full-scale use of oxygen and oxygen-enriched air with steam for continuous gasification of high-temperature coke and low-temperature coal char prepared by the Disco process has been demonstrated in a 7-ft diam Kerpely producer in conjunction with gas synthesis tests at Louisiana, Missouri.[23,24] The producer (Fig. 3-74) was suitable for this operation with substantially no modification of its rotary grate or other appurtenances. Results indicated satisfactory routine performance with any blast oxygen concentration or with any of the fuels investigated (Table 3-30).

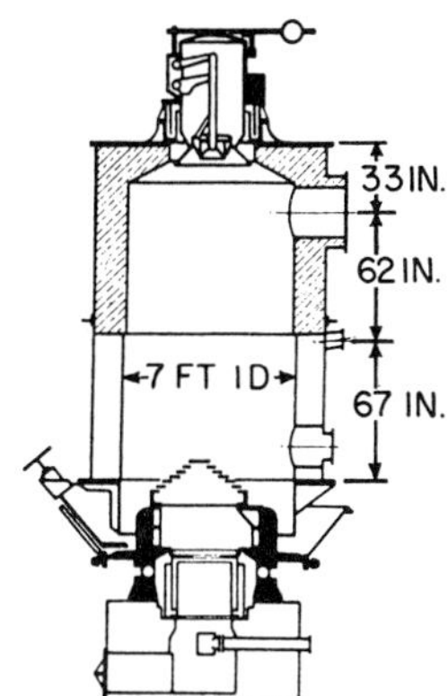

Fig. 3-74 Kerpely producer.

U. S. Bureau of Mines Gasifier. Carbon gasification efficiencies from 65 to nearly 100 per cent were reported[25] for a Bureau of Mines pilot unit upflow gasifier operated at atmospheric pressure. The process yielded gas (230 to 300 Btu per SCF) at thermal efficiencies of 50 to 75 per cent, and thus appeared applicable for synthesis gas production on a large scale. A larger unit (Fig. 3-75) was equipped with a water-filled slag removal catch pot and an outlet gas heat exchanger. Pulverized coal was fed at uniform rates by discharging the coal feed from a fluidized bed under pressure thru small-diameter transfer lines.

Another unit designed for coal feed rates up to 2000 lb per hr introduces coal, steam, and oxygen thru a single combination burner.[24] The unit was originally operated with noncoking coal at carbon conversions of 78 to 95 per cent and, after installation of the combination burner, with Illinois No. 6 low ash fusion coal at carbon conversions of about 80 per cent, with satisfactory slag tapping.

A semiplant scale pulverized coal gasifier (Fig. 3-76) similar in design to the Bureau of Mines unit has been operated since 1951 for synthesis gas production at the E. I. du Pont Co., Belle Works, W. Va.[26] Designed for 3000 lb per hr coal feed rates, it has temperatures in the lower gasifier zone high enough to melt the coal ash. The molten slag is continuously tapped by permitting it to drop into a water tank, where it solidifies and fractures into small particles which are removed by an underwater conveyor. The gasifier provides enough residence time to complete gasification reactions up to about 85 per cent carbon conversion. Anthracite, several bituminous coals, and lignite, representing ash fusion temperatures of 2000 to 3000 F, have been successfully gasified at thermal efficiencies near 70 per cent. The only apparent difficulty has been some erosion of the high-temperature refractory lining of the combustion zone. A full-scale atmospheric pressure gasifier of similar design and of 17 tons of coal per hr rated capacity, equipped with waste heat recovery and continuous slag removal, has been placed in operation at the same plant.

The Babcock and Wilcox System. The Babcock and Wilcox system is based on development work done in cooperation with the U. S. Bureau of Mines and a subsequent large-scale installation at the Belle, W. Va., works of the du Pont Co. This process was also based on the gasification of pulverized, coking coal at atmospheric pressure, in suspension, in a chamber fed by a number of nozzles sloping downward. The suspension was in a state of violent agitation and at such temperature that the ash separated and dropped out as a molten slag, while the product gas rose upward thru a heat recovery boiler. The gas was then cleaned and purified. This gas (Table 3-59) is low in hydrocarbons and inerts, although the carbon dioxide content appears high. It is particularly suited to chemical synthesis or conversion to hydrogen.

Table 3-59 Performance Data for Babcock and Wilcox-duPont Coal Gasifier (1955)*

Ultimate analysis	
C	81.09
H_2	5.53
O_2 (by diff.)	6.33
N_2	1.6
S	0.77
Ash	4.68
Coal rate, lb/hr	31,600
Oxygen rate, cfh	301,000
Steam rate, lb/hr	27,400
Oxygen–coal ratio, cu ft/lb	9.5
Steam–coal ratio, lb/lb	0.86
Oxygen inlet temp, °F	80
Steam inlet temp, °F	...
Coal inlet temp, °F	80
Steam production, lb/hr	64,500
Temperature, °F	805
Pressure, psig	456
Feed water entering economizer, °F	235
Product gas leaving economizer, °F	565
Gas production rate, MSCF/day	26,800
Gas analysis, vol % dry	
CO_2	19
CO	40
H_2	38
CH_4 and N_2	3

* Coal fired, No. 2 gas seam, 3 per cent moisture, 70–200 mesh, Kanawha County, W. Va.

Gasification of Pulverized Coal at High Pressure with Oxygen and Steam. *U. S. Bureau of Mines.* Experimental studies on a pilot scale pulverized coal downflow gasifier designed for 500 to 1000 lb coal per hr have indicated advantages of operating pressures up to 30 atm in increasing gasification capacity per unit of reactor volume.[12,27] At 30 atm, up to 800 lb coal per hr per cu ft was gasified with over 90 per cent carbon conversion efficiency as against the 20 to 30 lb coal per hr per cu ft reactor volume and approximately 85 per cent carbon conversion efficiency achieved at atmospheric pressure.

An early design version is shown in Fig. 3-77. Tangential firing resulted in rapid erosion and fusion of refractories in the high-temperature zone. The down-jet single nozzle firing system lessened but did not eliminate erosion. Stabilization of the refractory wall dimensions required use of studded

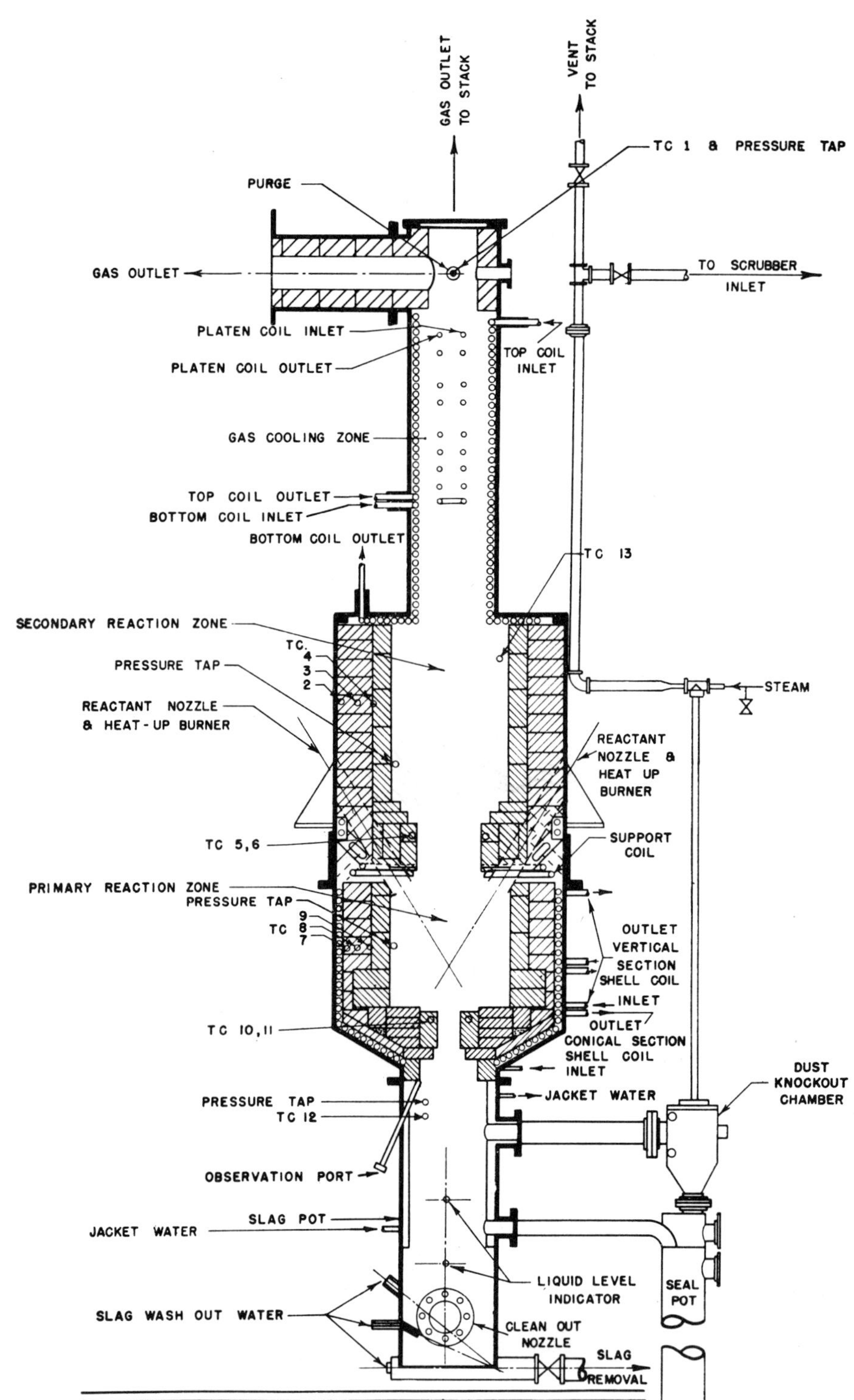

Fig. 3-75 Atmospheric pressure pulverized coal gasifier. (Bureau of Mines)

water wall construction. In latest modifications, a cooling coil was inserted within the original 12 in. ID reaction space and a castable refractory installed over the studded surface.

Operating results showed that a thin "equilibrium layer" of fused ash mixed with unreacted carbon could be maintained over a thin layer of refractory adjacent to the cooling coil. Thus, the solidified slag layer provided some insulation of the cooling coil against ash slagging temperatures. This represented an improvement, since maintaining high-temperature refractories under oxidizing and reducing conditions combined with molten coal ash is difficult at best. Heat losses per unit weight of coal gasified in a water-cooled pressure gasifier are small, particularly in a large unit, since the gasification rate per unit volume of reaction space is high.

Effects of operating variables like pressure, oxygen–steam–coal ratio, and coal feed rate on gas composition and carbon gasification efficiency have been investigated with anthracite, bituminous, and subbituminous coals. The synthesis gas and

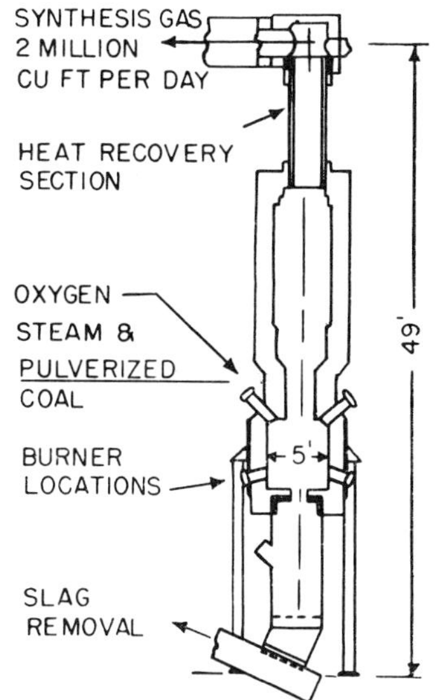

Fig. 3-76 Du Pont atmospheric pressure pulverized coal gasifier.

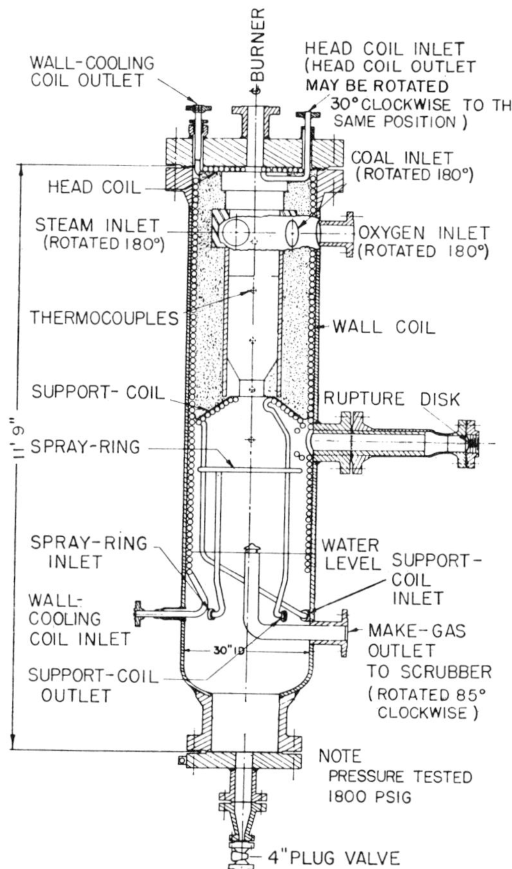

Fig. 3-77 Pressure gasifier for pulverized coal (early design). (Bureau of Mines)

slag may be quenched by water spray in the lower section of the reactor. The slag stream is broken into fragments that fall into a slag chamber filled with water and can then be intermittently discharged by a letdown valve. This system does not permit sensible heat recovery from the hot synthesis gas, since all available heat is removed by large quantities of quench water. Controlled quenching, however, will produce usable steam.

Institute of Gas Technology.—Further pilot plant scale studies of pulverized coal gasification at pressures up to 7 atm in a downflow system designed for 500 lb of coal per hr have been conducted at I.G.T.[13] Gasification equipment (Fig. 3-78) consisted of a refractory-lined, noncooled reactor, insulated slag catch pot, and refractory-lined heat exchanger for sensible heat recovery from the synthesis gas. Coarsely pulverized coal (less than 20 mesh) from a hammer mill, with the additional fineness imparted by flowing pulverization in the nozzle system, was gasified successfully at rates approaching 100 lb of coal per hr per cu ft reactor volume at 5 atm gasification pressure, with carbon conversion efficiencies exceeding 90 per cent. This system offers a potential saving in pulverization costs, since the accepted coal fineness for other suspension gasification units previously discussed has been from 70 to 90 per cent of 200 mesh or finer.

Texaco Pilot Unit. A pressure gasification pilot unit for synthesis gas production from pulverized coals ranging from lignites to anthracite has been developed at Montebello, Calif. A suspension of coal and steam heated to 100 F and introduced thru jets at the reactor top is subsequently mixed with oxygen in the downflow generator operating at ash slagging temperatures. Synthesis gas is withdrawn near the bottom, while molten slag, flowing into a water reservoir, disintegrates into small particles and is removed in water suspensions. Carbon gasification efficiencies as high as 95 per cent are reported.

METHANATION—CONVERSION OF SYNTHESIS GAS TO METHANE

Overall Reactions and Equilibrium Compositions

Production of methane and higher hydrocarbons from carbon monoxide and hydrogen proceeds over catalytic surfaces at temperatures normally not above 1000 F and at atmospheric or higher pressures. Depending upon feed gas composition, type of catalyst, catalyst activity, space velocity, and pressure–temperature conditions, reaction products of varying quantity and type can be obtained. They range from liquid paraffins, alcohols, aldehydes, ketones, and acids to gaseous hydrocarbons such as methane, ethane, propane, and small amounts of light olefins. Formation of liquid products is favored at low temperatures, 400–600 F; of gaseous products, at high temperatures, above 500 F.

For the synthesis primarily of methane, it is necessary to promote the gas-forming reactions. In their simplest form* these may be expressed as:

$$CO + 3H_2 \rightarrow CH_4 + H_2O \qquad (11)$$

$$2CO + 2H_2 \rightarrow CH_4 + CO_2 \qquad (7)$$

$$CO + H_2O \rightarrow H_2 + CO_2 \qquad (4)$$

In the presence of active catalysts and with a sufficiently high synthesis gas H_2-to-CO ratio, three to one and higher, Reaction **11** predominates. At intermediate or low H_2-to-CO ratios, a combination of overall Reactions **11** and **7** occurs.

* As numbered in Table 3-36.

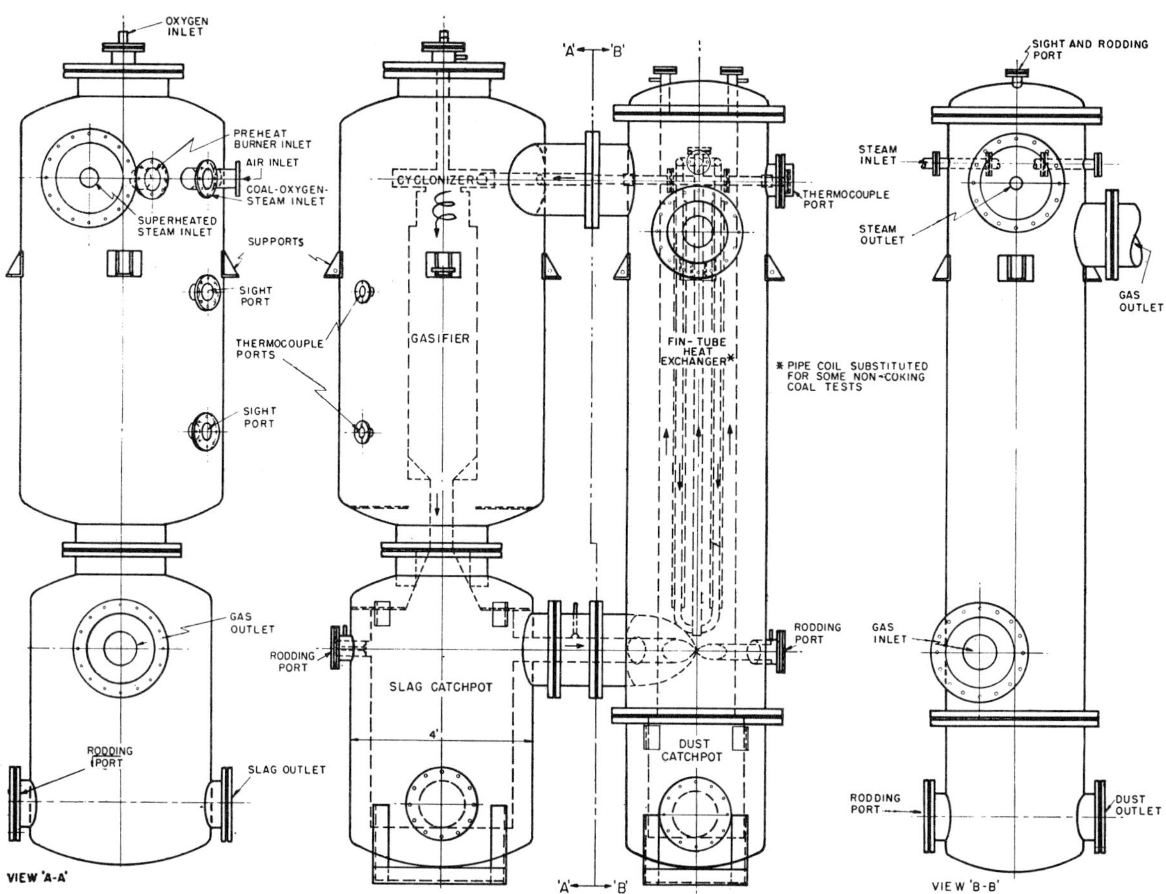

Fig. 3-78 I.G.T. downflow pressure gasifier for pulverized coal.[13]

It has been demonstrated that the water gas shift, Reaction **4**, is approximately in thermodynamic equilibrium at methanation conditions. Thus, a plausible course of the methane formation would be thru Reaction **11**, followed by Reaction **4** in practical equilibrium; the sum of Reactions **11** and **4** being equivalent to Reaction **7**. An overall representation of the reaction for various synthesis gas H_2-to-CO ratios is, therefore:

$$nH_2 + CO \rightarrow \frac{1+n}{4} CH_4 + \frac{n-1}{2} H_2O + \frac{3-n}{4} CO_2 \quad \textbf{(a)}$$

where n is the H_2-to-CO ratio.

Other possible reactions of the methane synthesis include:

1. Hydrogenation of carbon dioxide:*

$$CO_2 + 4H_2 \rightarrow CH_4 + 2H_2O \quad (12)$$

2. Formation of hydrocarbons higher in molecular weight than methane:

$$(2n+1)H_2 + nCO \rightarrow C_nH_{2n+2} + nH_2O \quad \textbf{(b)}$$

$$2nH_2 + nCO \rightarrow C_nH_{2n} + nH_2O \quad \textbf{(c)}$$

3. Decomposition reactions which result in carbon deposition:*

$$2CO \rightarrow CO_2 + C \quad (2)$$

$$CO + H_2 \rightarrow H_2O + C \quad (3)$$

$$CH_4 \rightarrow 2H_2 + C \quad (5)$$

4. Removal of portions of the deposited carbon by the reverse of Reactions **2** to **5**, and

5. Reforming of the product methane with steam and carbon dioxide thru the reverse of Reactions **11** and **7**.

Reaction **12**, which may be considered as a combination of Reactions **11** and **4**, usually occurs to a negligible extent in the presence of CO; however, it can proceed to a small degree if the synthesis gas already has appreciable concentrations, up to 30 per cent, of CO_2.[28] Reactions leading to the formation of ethane, ethylene, and higher saturated and unsaturated hydrocarbons (Reactions **b** and **c**) occur only to a minor extent under the pressure–temperature conditions for methanation with a nickel catalyst. For example, maximum concentrations of ethylene and ethane at equilibrium with synthesis gas of a one-to-one H_2-to-CO ratio are only approximately one per cent of the equilibrium methane concentration at 600 F, provided no carbon deposition or liquid hydrocarbon formation occurs. Some of the factors involved in carbon deposition (Reactions **2**, **3**, and **5**) are discussed later.

Maximum conversion of CO and H_2 to CH_4 would occur at thermodynamic equilibrium of the methane-forming reactions. A number of thermodynamic analyses of this reaction system have been made to define the effects of the major

* As numbered in Table 3-36.

operating variables (pressure, temperature, and H_2-to-CO ratio) on the theoretically attainable conversions of synthesis gas to methane, water vapor, CO_2, and carbon. In equilibrium product distribution calculations, it is only necessary to consider any two of Reactions **11**, **7**, and **4** and any one of Reactions **2**, **3**, and **5**. If carbon formation is not taken into consideration, Reactions **11** and **4** suffice to define the system at equilibrium. Results of equilibrium computations show that in the absence of carbon deposition nearly complete conversion of one-to-one or three-to-one H_2-to-CO ratio synthesis gases can be attained at 600–700 F and 25 atm or at 600 F and 1 atm.

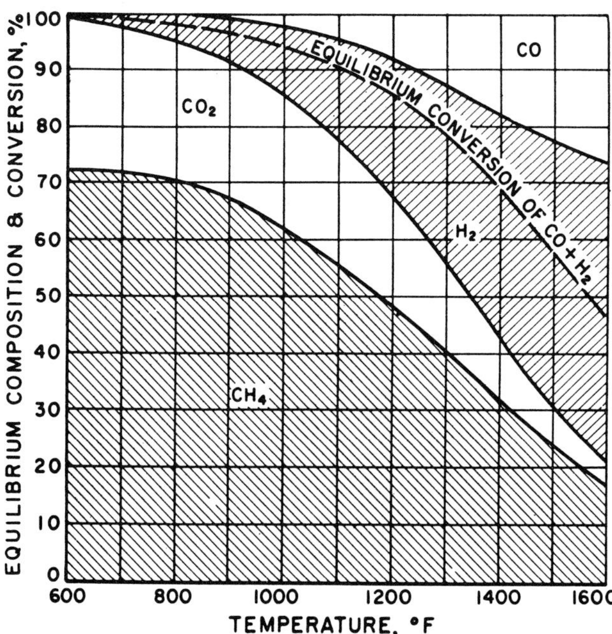

Fig. 3-79 Equilibrium gas compositions and conversions for methanation process 25. Ratio of H_2 to CO is two to one.

Typical equilibrium dry gas composition and equilibrium CO plus H_2 conversions at 25 atm are shown in Fig. 3-79 for two-to-one H_2-to-CO ratio synthesis gas. The theoretical methane yield decreases with increase in temperature, but remains reasonably high below 900 F and 1 to 25 atm. The equilibrium conversion of CO + H_2 in the synthesis gas is in the range of 86 to 96 per cent at these conditions. On a CO_2-free basis, the equilibrium methane concentration aproaches 95 per cent at 800 F and 25 atm. In the absence of undesirable side reactions, such as carbon deposition, increasing pressure permits using a higher operating temperature for a given methane content, with the possible advantage of increased reaction rates.

Process Efficiency

Production of methane from synthesis gas involves considerable heat release with attendant loss in process efficiency, maximum expected thermal efficiencies for complete conversion being about 70 to 75 per cent expressed as Btu in methane to Btu in synthesis gas. Generation of synthesis gas from coal, as by continuous gasification with oxygen and steam, occurs at a maximum efficiency of 75 to 80 per cent. Thus, an overall process efficiency of 52 to 60 per cent for conversion of coal to methane may be anticipated.

Sufficient excess heat at a relatively high temperature (1500 F and above) is normally available in coal gasification to generate the process steam required. Even greater amounts of excess heat are available, but at a lower temperature, in the methanation step. Efficient utilization of waste heat for process purposes requires heat exchange with water or other cooling medium for feed water preheating or for steam generation.

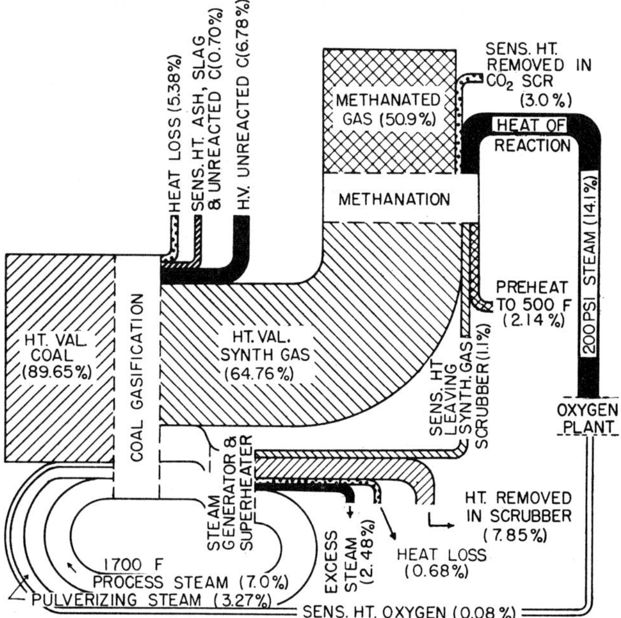

Fig. 3-80 Heat distribution for coal gasification and methanation.

The maximum heat of reaction for methane using synthesis gas from coal is equivalent to approximately 62 lb of 200 psig saturated steam per MCF of CO + H_2 converted. Figure 3-80 shows a typical heat distribution outline for a combined coal gasification and methanation process. It is based on I.G.T. pilot plant data on continuous pressure gasification of coal in suspension with oxygen and steam and on thermal calculations for the methanation reactions.

Experimental Developments

Historical. Discovery of catalytic synthesis of methane is credited to Sabatier and Sendersens[29–33] who, in 1897, obtained practically stoichiometric conversion of a three-to-one H_2-to-CO ratio pure synthesis gas at approximately 550 F in the presence of nickel. These investigators found no appreciable methane formation with other catalysts such as platinum, palladium, iron, and copper. These experimental results were confirmed by Mayer *et al.*[34] in 1910.

Although numerous early laboratory scale studies on methanation were conducted, Sabatier's discovery remained a scientific curiosity until about 1910. A scheme proposed by Cedford Gas Process Co. of Great Britain[35] for the catalytic enrichment of water gas was then put into operation briefly on a small scale.

Since the discovery in 1923 by Fischer and Tropsch that reactions of CO and hydrogen at high pressures could be catalyzed by iron to produce oxygenated liquid products, extensive experimental work in Europe has been directed toward development of improved catalysts and pressure reactors for production of synthetic gasoline and other hydrocarbons. The Fischer–Tropsch and related synthesis methods, such as the Oxo and Synol processes, were widely utilized in Germany during World War II. The Fischer–Tropsch synthesis employed improved cobalt–kieselguhr catalysts promoted with thoria and magnesia, producing small percentages of methane and higher gaseous hydrocarbons with the liquid products. The Synol process, which employed a fused-iron synthetic ammonia type of catalyst, produced higher proportions of alcohols and olefins than the Fischer–Tropsch process. The Oxo process was conducted in two steps: (1) addition of CO and hydrogen to an olefin in the presence of cobalt catalyst to form aldehydes, and (2) hydrogenation of the aldehydes to primary alcohols. Nickel catalysts were not employed in the European commercial scale synthetic liquid fuels program.

Later Investigations. Since 1940, most developments have been made by the British Fuel and Gas Research Boards, the U. S. Bureau of Mines, and thru projects sponsored by A.G.A. at the Institute of Gas Technology. Only laboratory scale and small pilot plant scale reactors have been employed at atmospheric and higher pressures.

British Gas and Fuel Research Boards. In the Gas Research Board[36] work, synthesis of methane from coke-produced, sulfur-purified water gas was conducted in small pilot plant fixed-bed reactors operated at atmospheric pressure and 20 atm; see Table 3-60 for test conditions. Gas recirculation and auxiliary cooling were employed for temperature control of the catalyst bed. Figure 3-81 shows equipment for the 20 atm operation. The equipment for atmospheric operation was similar; its catalyst chamber consisted of a fixed bed 5.75 in. deep and 12.5 in. in diam.

Table 3-60 Gas Research Board Tests on Methanation of Synthesis Gas at Atmospheric Pressure, with Temperature Control by Gas Recirculation

Catalyst bed volume, cu ft	0.41
Synthesis gas feed rate, SCF/hr	820
Synthesis gas H_2/CO ratio	1.07
Space velocity, SCF/hr-cu ft catalyst	2000
Inlet gas temperature, °F	572–626
Outlet gas temperature, avg, °F	932
Gas recirculation ratio, approx.	5
Synthesis gas sulfur content, grains/CCF	0.002
Steam/synthesis gas, volume ratio	0.20
Pressure drop across catalyst, in. H_2O	3–5

Gas compositions, per cent by volume:

Component	Avg. synthesis gas	Product gas, first stage, hours from start: 250	1000	1700	2000
CO_2	0.5	38.5	33.4	35.7	34.8
CO	46.0	5.8	3.5	7.2	7.6
H_2	49.0	24.3	30.2	25.8	25.9
CH_4	0.5	24.1	25.1	22.5	23.4
N_2	4.0	7.3	7.8	8.8	8.3
Product gas composition, first stage, with CO_2 removal to 3 per cent:					
CO_2		3.0	3.0	3.0	3.0
CO		9.1	5.1	10.9	11.3
H_2		38.4	44.0	38.9	38.5
CH_4		38.0	36.5	34.0	34.8
N_2		11.5	11.4	13.2	12.4
Heating value, Btu/SCF		529	520	497	505

Product gas composition, second stage:

	As made	After CO_2 removal
CO_2	44.4	3.0
CO	0.15	0.3
H_2	9.85	17.2
CH_4	37.7	65.7
N_2	7.9	13.8
Heating value, Btu/SCF	407	710

Tests at atmospheric pressure (Table 3-60) produced approximately 500 Btu per SCF methanated gas (after CO_2 removal) at the outlet of the first stage catalyst bed. The heating value was increased to about 700 Btu per SCF (after CO_2 removal) by using a second stage catalyst bed. About 20 per cent steam was added to the synthesis gas feed to avoid carbon deposition. Tests at 20 atm (Table 3-61) were conducted with a single catalyst bed 10 in. deep and 9.5 in. in diam. Heating value of the methanated gas averaged above 700 Btu per SCF when corrected to a two per cent nitrogen content of the synthesis gas.

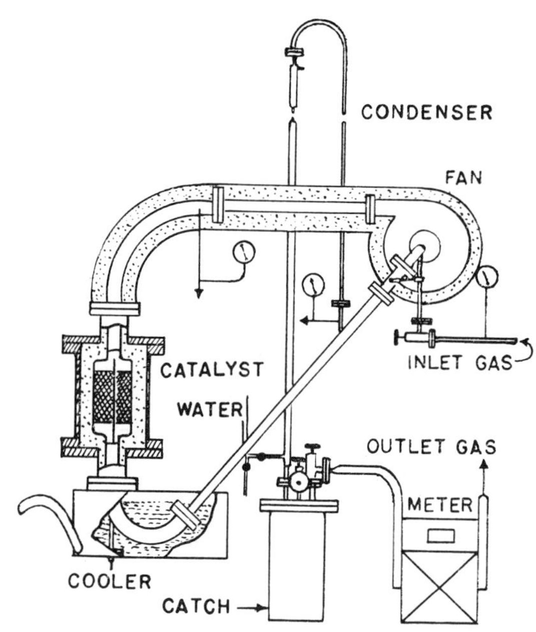

Fig. 3-81 Gas Research Board pilot plant methanation unit employing gas recirculation at 20 atm.[36]

The catalyst for both tests consisted of nickel and aluminum on chain clay prepared by precipitating 100 parts nickel nitrate and 300 parts aluminum nitrate on 25 parts of the support material. Maintenance of catalyst activity required synthesis gas purification to a limiting content of 0.002 grain sulfur per 100 cu ft. This was accomplished by removing H_2S thru use of a copper–chromium catalyst at 520 F. Residual organic sulfur was removed by reaction at 390 F with nickel hydroxide on chain clay. The nickel methanation catalyst probably could have been operated for a longer time than indicated in Tables 3-60 and 3-61, since it still exhibited sufficient activity at the end of the test.

Studies of the British Fuel Research Board[28] with laboratory scale fixed-bed methanation reactors at atmospheric pressure demonstrated that: (1) product gas compositions obtained with catalysts at 662 F exit temperature from the reactor and inlet space velocity of 3700 SCF synthesis gas per hr per cu ft of catalyst agreed generally with the thermodynamic equilibrium compositions over a 1.3 to 3.0 range of H_2-to-CO ratios of the synthesis gas, (2) carbon deposition ranged from six per cent of the carbon content of the inlet gases at a unity H_2-to-CO ratio to a negligible 0.003 per cent at a three-to-one H_2-to-CO, (3) carbon deposition at fixed conditions tended to increase with pressure, and (4) with a three-to-one H_2-to-CO ratio, conversion of CO averaged 99 per cent and that of H_2 averaged 82 per cent, provided the synthesis gas sulfur content was less than 0.004 grain per CCF. Also, by use of 10 to 13 per cent steam addition with a H_2-to-CO ratio of three-to-two synthesis gas at an inlet space velocity of 3700 SCF per hr per cu ft catalyst and 662 F exit temperature, carbon deposition could be decreased to little or nothing. Best results obtained with the laboratory scale fixed beds containing 0.49 cu in. of catalyst showed that catalyst life was about 2500 hr. This corresponded to a production of approximately 2000 lb methane per lb catalyst with an average conversion of CO above 90 per cent. Several small-scale pilot reactors were constructed and operated at atmospheric pressure with approximately 200 SCF per hr of a H_2-to-CO ratio of three-to-two synthesis gas, confirming laboratory results.

Table 3-61 Gas Research Board Tests on Methanation of Synthesis Gas at 20 Atm with Temperature Control by Gas Recirculation

Catalyst bed volume, cu ft	0.43
Synthesis gas feed rate, SCF/hr	830
Synthesis gas H_2/CO ratio	3.6
Space velocity, SCF/hr-cu ft catalyst	2000
Inlet gas temperature, °F	590–626
Outlet gas temperature, avg, °F	932
Gas recirculation ratio, approx.	6
Synthesis gas sulfur content, grains/CCF	0.002

Gas compositions, per cent by volume:

Component	Avg. synthesis gas	Product gas, hours from start: 300	1100	1900	2700	3500
CO_2	2.4	3.1	2.5	3.9	5.6	4.5
CO	19.6	0.3	0.2	0.3	1.2	0.8
H_2	71.4	21.6	24.0	24.3	20.4	22.2
CH_4	0.0	57.7	56.5	61.0	55.0	54.7
N_2	6.6	17.3	16.8	10.5	17.8	17.8
Product gas composition corrected to 2 per cent N_2 content of synthesis gas:						
CO_2		3.5	2.8	4.1	6.5	5.2
CO		0.4	0.3	0.4	1.4	0.9
H_2		24.4	27.1	25.4	23.2	25.3
CH_4		65.3	64.0	63.9	62.7	62.4
N_2		6.4	5.8	6.2	6.2	6.2
Heating value, Btu/SCF		730	724	719	703	705

U. S. Bureau of Mines. A large fixed-bed reactor with 92 tubes of ½ in. ID in a 12-in. diam Dowtherm jacket was constructed for methanating Lurgi gas at 300 psig.[10] This gas, produced by gasification of coal char with oxygen and steam in an experimental pressure generator, contained 25–29 per cent CO_2 and approximately 350 grains sulfur as H_2S plus ten grains sulfur as organic sulfur per 100 cu ft. Before methanation it was: (1) washed with water at 300 psig to reduce its CO_2 content to approximately three per cent and H_2S content to 6–10 grains per CCF, and (2) treated with alkalyzed iron at 350–420 F to reduce its final sulfur content to approximately 0.3 grain per CCF. Although methanation catalyst life tests were not made, this amount of sulfur in the synthesis gas was believed too high to permit an extended life of about 2000 hr for nickel catalysts. Typical gas compositions obtained are summarized in Table 3-62. Methanated gas heating values were above 900 Btu per SCF on a CO_2-free basis.

Sulfur tolerance of the nickel–manganese–alumina–china clay catalyst developed by the British Gas Research Board was determined by the U. S. Bureau of Mines using a laboratory scale fixed-bed reactor with a ¾-in. diam and 6-in. long catalyst chamber.[37,38] Heat of reaction was removed by circulation of 12 volumes of product gas per volume of feed gas. Fresh gas space velocity was 2000 SCF per hr per cu ft catalyst. Acceptable catalyst activity was arbitrarily defined as that giving a minimum of 97 per cent conversion of H_2 plus CO, corresponding to methanated gas heating values of 930 to 940 Btu per SCF on a CO_2-free basis. Tests showed that the time over which the 97 per cent minimum conversion was maintained depended almost inversely on the amount of sulfur as COS added to the synthesis gas (Table 3-63). These data indicate that, under operation at constant space velocity, the product of sulfur concentration, S, in grains per CCF and hours of maintained 97 per cent minimum conversion, H, was practically a constant, as shown by:

$$SH = 12.6 \pm 0.3$$

Initial methanation studies were limited to the use of sulfur-sensitive nickel catalysts in fixed beds. These required:

1. Either highly purified synthesis gas or frequent catalyst regeneration for an extended catalyst life, and

2. Complex heat exchange equipment or intricate catalyst bed design when a liquid coolant is used for temperature control, or

3. High product gas circulation rates for temperature control without a liquid coolant.

Table 3-62 Typical Gas Compositions in Bureau of Mines Methanation of Lurgi Gas at 300 Psig

Component	Gas composition, per cent by volume: Raw Lurgi gas	After water scrubbing	After alkalyzed iron purifier	After methanation, CO_2-free
CO_2	27.8	2.9	12.3	...
Ill.	0.3	0.3	0.4	0.4
O_2	0.2	0.5	0.1	0.2
H_2	41.9	55.8	50.6	0.9
CO	18.3	26.3	15.2	0.4
CH_4	6.3	8.9	17.7	92.8
N_2	5.2	5.3	3.7	5.3

Table 3-63 Sulfur Tolerance of Nickel–Manganese–Alumina–China Clay Catalyst at 2000 Synthesis Gas Space Velocity

Synthesis gas sulfur content, grains per CCF	Life of catalyst at a 97% minimum gas conversion, hr
0.04	320
0.12	104
0.19	68
1.53	8

Table 3-64 Experimental Data on Production of Pipeline Gas from Various Synthesis Gases Using Nickel Catalysts in Bench Scale Fixed- and Fluid-Bed Reactors*

(I. G. T. data)

	Fixed-bed operation						Fluid-bed operation			
Catalyst										
Composition	5% Ni		15% Ni		42% Ni, 58% Al		16.6% Ni		42% Ni, 58% Al	
Support	Kieselguhr		Kieselguhr		...		Kieselguhr		...	
Preparation	Impregnated		Impregnated		Caustic extracted		Precipitated		Caustic extracted	
Size	¼ in.		¼ in.		4–8 mesh		40–140 mesh		—40 mesh	
Bed height, in.	9		2¼		2¼		16		20	
Volume, cc	226		56		56		200		100	
Feed gas H_2/CO ratio	1.26		3.01		2.76		1.28		3.43	
Pressure, psig	300		27		27		Atmos.		150	
Temperature, °F										
Top	735		745		755		710		745	
Middle	740		775		790		700		680	
Bottom	730		775		725		430		670	
Feed gas space velocity, SCF/cu ft cat.-hr	803		465		1503		287		15316	
Feed gas rate										
SCF/hr-sq ft	588		84		275		...		...	
Ft/sec	...		...		...		0.21		1.35	
SCF product gas/SCF feed gas	0.438		0.411		0.323		0.476		0.287	
Gas composition, mole %	In	Out	In	Out	In	Out	In	Out	In	Out
CO_2	0.2	42.4	20.7	49.0	0.9	14.7	0.4	40.8	1.0	4.8
CO	44.2	...	19.8	...	26.1	...	43.7	1.1	21.7	...
H_2	55.6	1.4	59.5	7.2	72.0	17.0	55.9	2.9	74.4	8.0
CH_4	...	55.2	...	41.0	...	65.8	...	52.3	0.9	80.7
C_2H_6	...	...	...	1.4	...	...	...	...	...	0.1
N_2†	...	1.0	...	1.4	1.0	2.5	...	2.9	2.0	6.4
Total	100.0	100.0	100.0	100.0	100.0	100.0	100.0	100.0	100.0	100.0
Inert-free product gas heating value, Btu/SCF	980		920		858		948		937	
Inert-free product gas sp gr (air = 1)	0.542		0.498		0.455		0.537		0.511	
Water formation, mole/mole feed gas										
By hydrogen balance	.0662		.2115		.2406		.0502		.2660	
By oxygen balance	0.0745		0.2094		0.1841		0.0482		0.2094	
H_2–CO conversion, %	97		91		87		100		93	
Net methane-equivalent space-time yield, SCF/cu ft catalyst-hr	194		84		319		71		3420	

* Gas volumes in standard cubic feet (SCF) at 60 F, 30 in. Hg, and saturated with water vapor. H_2–CO conversion is defined as the percentage of complete H_2–CO conversion represented by the actual methane-equivalent $[CH_4 + 2(C_2H_6) + \ldots + n\,(C_n\text{–}H_m)]$ yield:

$$\%\ H_2\text{–CO conversion} = (400)\,\frac{\left(\dfrac{\text{Moles dry product gas}}{\text{Moles dry feed gas}}\right)(n)\left(\begin{array}{c}\text{\% Hydrocarbon in}\\ \text{dry product gas}\end{array}\right) - (n')\left(\begin{array}{c}\text{\% Hydrocarbon}\\ \text{in dry feed gas}\end{array}\right)}{\%\ CO + \%\ H_2\ \text{in dry feed gas}}$$

where n and n' are the average carbon numbers of the hydrocarbons present in the product and feed gases, respectively.

† May include some CO not distinguishable from N_2 by mass spectrometer analysis when present in small concentrations.

Further research at the Bureau of Mines and I.G.T. has been devoted to:

1. Evaluation of nickel catalysts in fixed-bed reactors.
2. Testing and development of fluid-bed reactors capable of conducting methanation at high throughput rates.
3. Development of active nickel catalysts capable of withstanding abrasion in a fluid bed.
4. Testing of sulfur-resistant iron catalysts in fixed and fluid beds.

Since catalyst preparation and evaluation work is largely empirical, any one of several steps involved in catalyst precipitation or impregnation, drying, extraction, and reduction may be critical in making the difference between an acceptable or a poorly performing final product.

Institute of Gas Technology. Other phases of methanation development work in a gas research program at I.G.T.[39–41] showed that 900 Btu per SCF gases on an inert-free basis can be produced without significant carbon deposition and without gas recycle in laboratory scale fixed-bed and fluid-bed reactors operated at reasonably high gas flow rates with several commercial nickel catalysts. Typical results under various operating conditions are summarized in Table 3-64. Under conditions of complete CO conversion, a low-activity 15 per cent nickel-impregnated kieselguhr catalyst showed a tendency toward incomplete conversion of hydrogen at low pressures with three-to-one H_2-to-CO ratio synthesis gas. Since this result could be attributed to the relatively higher rate of the CO_2-forming methanation reaction as compared with the water-forming methanation reaction, use of 20 to 40 per cent CO_2 concentration in the synthesis gas seemed beneficial in suppressing additional CO_2 formation, with resultant improved hydrogen utilization.

Synthesis gas purification to a sulfur content of about 0.01

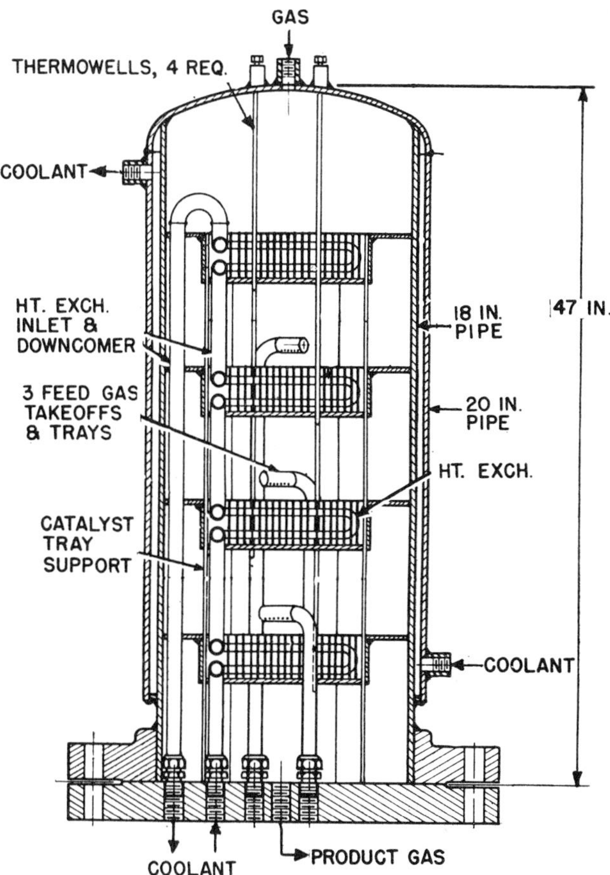

Fig. 3-82 I.G.T. pilot plant fixed-bed methanation reactor.

grain or less per 100 cu ft appeared too rigorous a requirement in the fixed-bed tests with nickel–kieselguhr catalysts. Except when extremely deactivated by sulfur, fixed-bed catalysts could be regenerated to substantially their initial activity by burn-off with air or oxygen followed by reduction with hydrogen.

Raney nickel catalysts prepared from a 42 per cent nickel—58 per cent aluminum alloy, with the aluminum partly extracted by sodium hydroxide, gave the highest methanated gas production capacities in laboratory scale fixed and fluid beds (Table 3-64). Production capacities up to 3500 SCF methane per hr per cu ft catalyst, corresponding to synthesis gas space velocities up to 15,000 SCF per hr per cu ft catalyst, were attained with 40 to 200 mesh Raney nickel at 150 psig, 660 to 745 F, using substantially sulfur-free three-to-one H_2-to-CO ratio synthesis gas. A life test of the catalyst showed a total methane yield of 2800 lb per lb nickel for an operating period of over 1700 hr, during which the methanated gas (CO_2-free) heating value averaged 800 Btu per SCF at an average synthesis gas space velocity of 10,000 SCF per hr per cu ft catalyst. However, experience with Raney nickel catalysts indicates that they may be critically dependent upon alloy composition and method and extent of extraction of the aluminum. Their sulfur tolerance appeared to be equivalent to a total exposure of 0.5 lb sulfur in synthesis gas per lb catalyst nickel.

Pilot plant scale methanation studies are being conducted at I.G.T. with a multiple-tray fixed-bed reactor design for 100 to 300 SCF synthesis per hr per tray with a 6-in. diam fluid-bed reactor equipped with three-point synthesis gas introduction at rates of 500 SCF per hr and above. The fixed-bed reactor consists of an 18-in. diam steel shell (Fig. 3-82) with a cooling jacket and four catalyst trays, each having 7⅜-in. diam hairpin finned tubes containing Dowtherm cooling fluid. The catalyst, in pellet form, is placed to a depth of about 3 in. around each finned tube, each pellet being in contact with a cooling surface.

Tests with active Raney nickel catalyst in the 6-in. diam fluid-bed reactor indicated production of 800 to 900 Btu per SCF methanated gases on an inert-free basis at 100 to 190 psig, 600 to 900 F, and at synthesis gas space velocities up to 6000 SCF per hr per cu ft catalyst, without product gas recycling. The synthesis feed gas consisted of substantially sulfur-free three-to-one H_2-to-CO mixtures obtained from catalytic reforming of natural gas. Although one extended test of approximately 175 hr duration was completed, it has not been possible to duplicate this result of the methane production rates and catalyst life obtained in earlier laboratory scale fluid-bed experiments with Raney nickel satisfactorily.

HYDROGENATION—DIRECT COMBINATION OF COAL AND HYDROGEN TO METHANE

Hydrogenation is the chemical term for directly combining hydrogen with coal to form methane or other gaseous or liquid products. The process may be called "hydrocarbonization," "hydrogenolysis," or "hydrogasification."

Experimental Developments

Historical. Hydrogenation of coal for production of motor fuels and chemicals has been practiced commercially for over 30 years. A vast literature has accumulated on the subject. In Bureau of Mines Bulletin 455, published in three volumes in 1952, Wiley and Anderson cited 2503 articles and 3569 patents before 1950. Many more have appeared since then. Hydrogenation processes, including coal hydrogenation, are reviewed by Groggins.[42]

The literature on coal hydrogenation for producing high-Btu pipeline gas is much smaller than that for producing liquids. In general, gas production requires higher temperatures and lower pressures than the Bergius process for making liquids.

Much information on the production of gas by direct hydrogenation has been published by Dent[43–47] and his associates in England, who first reported in 1936. In this country, Bray[48–50] and his associates have published information on laboratory scale experiments, beginning in 1943. The U. S. Bureau of Mines has also done considerable work, published by Symonds *et al*,[51] Newman and Pipilen,[52] and Hiteshue, Anderson, and Schlesinger.[53] In 1955, the A.G.A. Gas Operations Research Committee initiated a research project. Two papers by Channabasappa and Linden[2,54] have been published.

Later Investigations. *The Gas Council.* During laboratory investigations 8- to 16-mesh coke prepared at 840 to 930 F from a strongly caking Yorkshire coal was heated at atmospheric pressure in nitrogen and hydrogen. The fuel

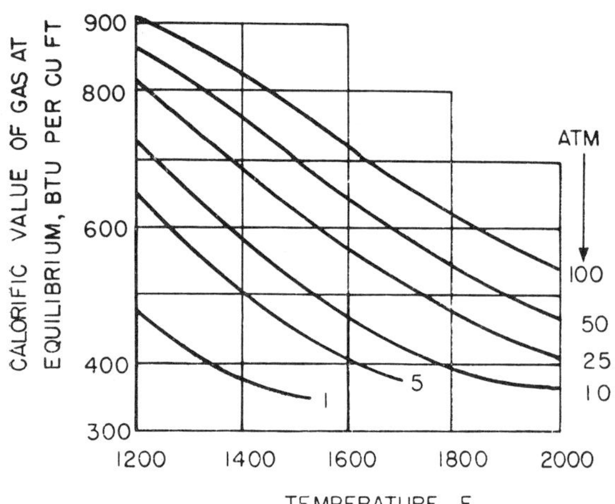

Fig. 3-83 Calorific value of methane and hydrogen mixture in equilibrium with coke at various temperatures and pressures (atm).

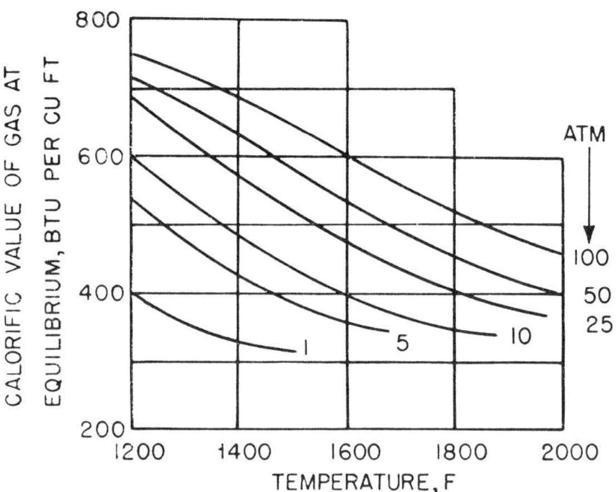

Fig. 3-84 Calorific value of methane and hydrogen mixture in equilibrium with coke at various temperatures and pressures (atm). (Gases at equilibrium diluted with 15 per cent CO, 3 per cent N_2, and 2 per cent CO_2.)

column was heated at the rate of 2.5 degrees per minute to a final temperature between 1470 and 1740 F. Gas mass flow was the same for all pressures. At one atmosphere the contact time was 1.3 seconds, increasing linearly with the pressure.

In nitrogen, the yields per ton of coal treated were: methane, 15.3 therms; ethane, 1.9 therms; and unsaturated hydrocarbons, 0.7 therm, making a hydrocarbon total of 17.9 therms; in hydrogen, the yield was 26.1 therms.

In hydrogen, the coke evolved less hydrogen, and considerably more methane was obtained. Coke yielded only 1.3 therms in hydrogen, but 33.8 therms in gaseous hydrocarbons per ton. The higher hydrocarbon yield was the result of the coke being hydrogenated, even at atmospheric pressure, with absorption of the hydrogen supplied up to 1200 F.

By heating the coke above atmospheric pressure, hydrogenation became much more pronounced, with remarkably high yields of hydrocarbons. Absorption began with the appearance of methane at about 930 F and continued beyond 1470 F.

Alkalies and alkaline earths were found to catalyze the hydrogenation reaction. The gasification rate increased with the amount of sodium carbonate until a ten per cent addition gave 95 per cent gasification before the outlet gas calorific value fell below 500 Btu per cu ft.

In Fig. 3-83 calorific values are shown for mixtures of methane and hydrogen in equilibrium with coke at various temperatures and pressures. In practice hydrogen is probably diluted with some CO_2, CO, and N_2; see Fig. 3-84. Methane formation is greatest at 1110 F rather than at higher temperatures, and increases with higher pressures.

From laboratory operations the scheme shown in Fig. 3-85 was developed for employing hydrogenation in combination with gasification.

Tests were next conducted in successively larger apparatus to determine if the heat of the reaction—some 37,000 Btu per lb-mole of carbon reacted—was enough to maintain it in industrial equipment. The final unit held a static fuel bed 12 in. in diam and 8 ft deep.

The smaller units demonstrated that once reaction had been initiated, no further heat needed to be supplied. In the larger units, however, thermal conditions were found to be markedly different, since the reaction ceased with a considerable proportion of the charge ungasified, and both the yield and the concentration of methane were much lower. These conditions resulted from having too much reaction heat available and from difficulties in controlling fuel-bed temperature. There was also difficulty from agglomeration with some coals.

In a static bed, the outlet gas consists essentially of methane diluted by the volume of unreacted hydrogen necessary to drive the reaction thru the charge. In such a bed it is difficult to dissipate the heat liberated, which limits the methane

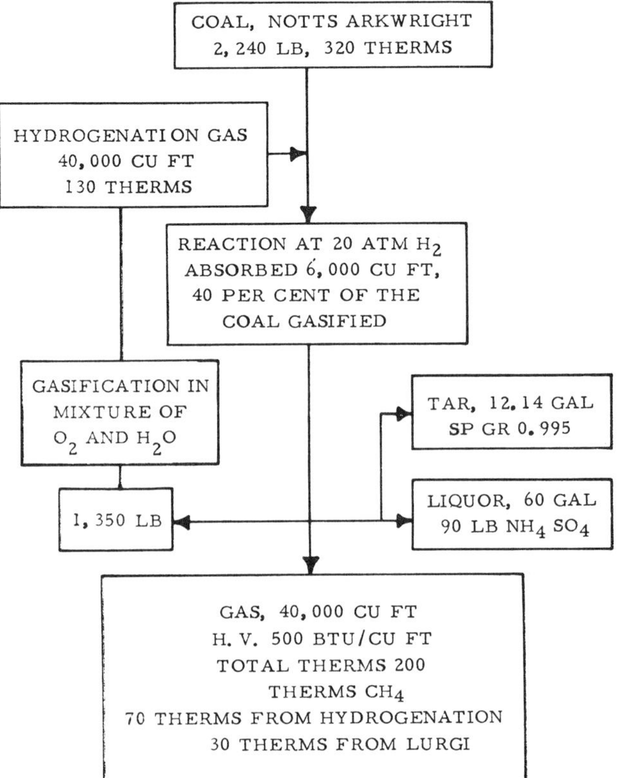

Fig. 3-85 Flow of materials in Gas Research Board hydrogenation and gasification system.

yield. The limitation is not great enough to prevent production of 500 Btu gas at 25 atm, but attaining higher calorific values or using a lower pressure is desirable.

Fluidization was therefore adopted in a laboratory scale unit. There was no difficulty in maintaining temperature at any chosen value within the optimum range of 1470 to 1650 F. Even though there was some coal agglomeration, reaction with hydrogen was rapid when the coal was undergoing thermal decomposition and slower when the hydrogen was attacking the residue. Total methane yields were higher than in the static bed.

These results warranted larger scale pilot plant investigations. Accordingly, the pilot plant used earlier for static-bed experiments was modified for fluidized operation, as shown in Fig. 3-86. Its reaction vessel was 12 in. in diam and 9 ft deep. Disappointing test results were attributed in part to introduction of the fuel with hydrogen at the reactor base. The plant was therefore altered to feed fuel from the top, but no further tests were made due to interruption of the program.

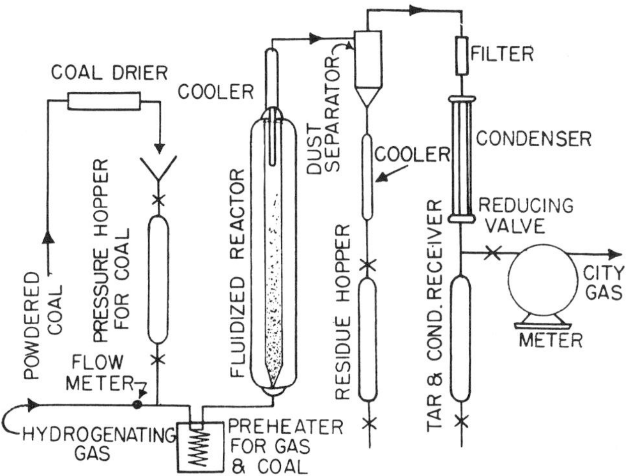

Fig. 3-86 Flow sheet of bottom feed fluidized hydrogenation plant.

Methane is not produced in adequate concentration or yield using a single fluidized reaction vessel at one temperature.

At high temperature, the reaction proceeds at a sufficient rate for a high yield, but equilibrium limits methane concentration unless pressure is also high. While higher concentrations are theoretically possible at lower temperature, both concentration and yield may be difficult to attain because of reduced reaction velocity. This difficulty is more severe because, with fluidization, gas velocities must be within the range for preserving turbulent motion, and prolonging contact time to counteract effects of a slow reaction may require an undesirably deep fuel bed. Furthermore, granular fuel attains the temperature of a fluidized bed almost immediately on entrance, since there is no progressive temperature rise like that during which most methane is obtained in the laboratory.

Possibly fluidization potentialities will be fully realized only if a graduated temperature is maintained. Also, heating in steps results in a free-flowing char. Accordingly, a new reactor of one-half ton coal capacity per hour was designed for installation in the Midlands Research Station at Solihull, England. Its five fluidized beds in series will permit operation with stepwise temperature increases.

U. S. Bureau of Mines. Bench scale tests on hydrocarbonization of coal at 500 psig are in progress at their Bruceton Station. The apparatus is especially designed to simulate rapid heating of coal, presumed to be an essential condition of continuous operation.

Test objectives are to determine the causes of agglomeration and to eliminate their resulting difficulties in earlier pilot plant operations. The reactor tube is of 2-in. pipe containing a porous stainless steel bed support and a stainless steel exit filter 34 inches above it. Gas velocity chosen is such that the fuel bed is expanded with a minimum of entrainment.

It was found that rapid heating (167 °F per minute to temperatures above 930 F up to 1290 F) produced weak to strong agglomeration. Slow heating (46°F per minute) of Rock Springs, Wyo., coal to 1290 F did not produce agglomeration. Heating in steps at 500 psig to 1290 F produced a free-flowing (no agglomeration) char like that with slow heating to the same temperatures. Stepwise heating was obtained by dropping the coal into the reactor at 750 F, holding this temperature for one hour, then heating successively to 930 F, 1110 F, and 1290 F, holding at each temperature for one hour, with a ten minute interval between temperature levels. Hydrogen velocity in the reactor was 0.06 ft per sec.

Use of a catalyst (molybdenum) was found to cause agglomeration at 840 F and above. Large particle size (4 to 6 mesh) coal agglomerated more readily than small particles (100 to 200 mesh). Texas lignite did not agglomerate when rapidly heated to 1110 F at a particle size of either 100 to 200 mesh or 4 to 6 mesh.

In an inert atmosphere (helium), agglomeration of Rock Springs, Wyoming, coal was more severe than in hydrogen. A possible explanation is that the asphaltic oils wet the particles and are coked, thereby bonding them.

In general, conclusions were that it would be difficult to produce high-Btu gas at high reaction rates with simultaneous high carbon conversion. Therefore, to obtain an acceptable production rate, hydrogenation limited to the primary products of distillation and subsequent gasification of the residual char were indicated.

The Bureau of Mines also began a study of hydrogenation of coal reactions at high temperatures to determine if appreciable quantities of gaseous hydrocarbons and low-boiling aromatics could be produced from coal. Its first phase was designed to elucidate effects of short residence time at 1470 F and 6000 psi on hydrogenation of a bituminous *C* coal from Wyoming, using ammonium molybdate plus sulfuric acid as a catalyst. The following results were obtained.

With zero residence time, 6000 psi and 1470 F, 65 to 68 per cent of the coal was converted to liquids and gases. After 15 minutes, conversion was 90 per cent. With zero residence, 38 per cent of the coal was converted to gaseous hydrocarbons. After 15 minutes, yield rose to 69 per cent, with 90 volume per cent methane and about 10 per cent ethane. Yield of oil based on coal rose rapidly from five per cent at zero residence to nine per cent at one minute. At lower temperatures (750 to 896 F), the mechanism of coal hydrogenation reaction differed. An active catalyst can prevent polymerization of primary products to char. Stabilization resulted in high yields (60 per cent) of oil and low yields (20 per cent) of hydrocarbon gases.

Institute of Gas Technology. 1. Pretreatment of bituminous coals and lignite: A nonagglomerating reactive coal, lignite, or char appears necessary for hydrogenation if this process is to be a continuous fluid-bed operation. Effects of temperature, residence time, and atmosphere on quantity and composition of gas evolved in distillation of low-rank coals have been extensively investigated by others. In general, as the temperature is raised, oxygen-containing substances break down to form water, CO_2, and CO. The highest CO_2 evolution occurs between 400 and 600 F. With further temperature increase, bituminous and subbituminous coals evolve mainly hydrocarbons and hydrogen; lignites evolve mainly carbon oxides.

Channabasappa and Linden[54] studied pretreatment of coal and lignite in a laboratory unit consisting of a glass fluid bed retort heated by two electric furnaces. Powdered coal (¾ lb) dried to 230 F was charged to the reactor. Measured volumes of nitrogen, air, CO_2, or steam at rates to keep the bed fluidized were preheated to the desired temperature and introduced at retort bottom.

With increases in temperature, the greatest agglomeration of bituminous coal was found in steam atmospheres; a considerable tendency to agglomerate, in CO_2 atmospheres; and very little agglomeration, in air or nitrogen atmospheres. Subbituminous coal and lignite exhibited little tendency to agglomerate except at higher temperatures and in steam and CO_2 atmospheres. About 2.5 to 9.9 per cent of coal by weight was converted to gas in the pretreatment runs. Steam runs produced gases with heating values of from 200 to 1000 Btu.

2. Batch hydrogenation of pretreated coal: Reactivities of the chars in respect to gaseous hydrocarbon formation were evaluated in batch hydrogasification tests of approximately 60 gram sample charges at 1350 F maximum reactor temperature and a 25 minute run at maximum temperature. These charge quantities corresponded to a hydrogen-to-coal ratio of 17 SCF per lb and gave maximum reactor pressures of 2900 to 3300 psig (Table 3-65.) The char charge to the reactor was obtained by sampling residual chars from pretreatment runs with a small riffle sampler and crushing to less than 60 mesh.

Threshold temperatures in Table 3-65 corresponds to points

Table 3-65 Hydrogasification Characteristics of Pretreated Coals

[pretreated charge: 60 gram (0.132 lb)—maximum reactor temperature: 1348–1350 F]

Coal	High volatile B bituminous				Subbituminous				Lignite			
Run No.	128	95	106	112	108	102	104	113	127	129	121	110
Pretreatment conditions												
Feed gas type	N_2	Air	CO_2	Steam	N_2	Air	CO_2	Steam	N_2	Air	CO_2	Steam
Maximum temp, F	610	610	605	620	620	610	620	620	510	520	510	520
Operating conditions												
H_2/coal ratio, SCF/lb*	17.13	16.94	20.09	16.77	16.88	16.94	16.88	16.85	16.97	17.03	16.93	16.93
Threshold temp, F	996	930	1011	977	1002	965	956	964	980	962	1002	994
Reactor pressure, psig†	2970	2940	3310	2990	3140	3070	3030	3120	3190	3180	3160	3210
Operating results†												
Net Btu recovery, MBtu/lb	6.945	6.991	6.738	7.179	7.576	7.740	7.104	7.514	7.801	7.524	6.872	7.899
Product gas yield, total SCF/lb	14.61	14.46	16.28	14.70	15.45	15.08	14.90	15.33	15.68	15.63	15.53	15.77
Gaseous hydrocarbon yield, SCF/lb	11.4	11.0	11.3	11.4	11.6	11.5	11.3	11.8	12.0	11.9	11.0	11.9
Net moisture-free ash-free coal charge hydrogasified, wt %‡	51.0	50.4	48.9	50.3	53.0	54.6	50.3	49.3	54.8	53.9	54.5	56.5
Material balance, %	100.6	98.4	99.3	99.2	98.9	98.0	96.2	96.1	101.8	100.5	98.5	97.5
Product gas properties												
Composition, mole %												
N_2 + CO	3.3	3.3	2.2	2.2	4.8	3.1	3.9	3.7	3.1	2.4	4.5	4.6
CO_2	1.5	0.7	0.2	0.6	1.0	2.0	1.2	0.3	1.3	1.7	2.5	1.0
H_2	17.1	19.8	28.4	19.6	19.4	18.3	18.9	19.2	18.9	19.9	21.9	18.7
H_2S	...	...	...	...	...	0.2	0.3	0.1	...	...	...	...
CH_4	77.3	75.2	67.6	76.4	73.6	74.6	74.9	76.3	75.6	75.2	70.3	74.8
C_2H_6	0.6	0.2	0.9	0.8	0.7	0.4	0.3	0.2	1.0	0.6	0.4	0.1
C_3H_8	...	...	...	...	...	...	...	...	...	...	...	...
Butanes	...	...	...	...	...	0.2	0.1	...	...	...	...	...
Pentanes +	...	...	...	...	...	0.1	...	...	...	...	...	...
Olefins	...	0.5	0.2	...	...	0.3	0.2	0.1	...	...	...	0.1
Benzene	0.2	0.3	0.5	0.4	0.5	0.6	0.2	0.1	0.1	0.2	0.4	0.5
Toluene +	...	...	...	...	...	0.2	...	...	...	...	...	0.2
Total	100.0	100.0	100.0	100.0	100.0	100.0	100.0	100.0	100.0	100.0	100.0	100.0
Heating value, Btu/SCF	850	858	808	853	840	872	838	841	843	829	791	844
Specific gravity, air = 1	0.5056	0.5008	0.4438	0.4849	0.5028	0.5277	0.4984	0.4818	0.4935	0.4901	0.5004	0.5073
Residue properties												
Screen analysis, wt %												
+ 40 mesh	9.5	11.1	33.1	20.5	12.2	6.7	11.4	9.8	2.6	2.6	6.2	8.8
+ 100 mesh	11.8	8.4	14.3	7.8	13.4	10.4	14.4	15.4	5.7	5.0	7.3	13.0
+ 200 mesh	30.6	15.0	19.7	22.8	16.2	4.9	18.5	10.5	18.9	5.9	24.2	13.2
− 200 mesh	48.1	65.5	32.9	48.9	58.2	80.0	55.7	64.3	72.8	86.5	62.3	65.0

* Initial H_2 pressure, 1000–1010 psig, except for run 106, 1200 psig.

† At the time maximum reactor temperature was reached.

$$‡\ 100\left[\frac{\text{wt of product gas} - \text{wt of hydrogen in it}}{\text{wt of moisture-free, ash-free pretreated coal charge}}\right].$$

Table 3-66 Operating Results for Hydrogasification of Low-Temperature Char in Semicontinuous Pilot Plant Reactor

Coal	Bituminous
Source	Consolidation Coal Co.
Sieve size	140/325
Run No.	7
Duration of test, hr	4
Steady state period, min*	174–243
Operating conditions	
Standpipe height, ft	5
Reactor pressure, psig	1027
Bed temperatures, °F	
Bottom	1290
8 in. height	1300
18 in. height	1405
28 in. height	1430
42 in. height	1420
54 in. height	1130
Average	1330
Coal rate, lb/hr	3.22
Hydrogen rate, SCF/hr	65.19
Hydrogen/coal ratio, SCF/lb	20.22
Coal bed pressure differential, in. H_2O	23.8
Coal space velocity, lb/cu ft-hr	39.4
Coal residence time, min†	...
Hydrogen residence time, min‡	1.55
Superficial hydrogen velocity, ft/sec§	0.0539
Operating results	
Product gas rate, SCF/hr	50.31
Net Btu recovery, MBtu/lb	3.530
Product gas yield, total SCF/lb	15.60
Gaseous hydrocarbon space–time yield, SCF/cu ft-hr	281.1
Residue, lb/lb coal	0.568
Liquid products, lb/lb coal	0.0020
Net MAF coal hydrogasified, wt %**	31.7
Overall material balance, %	94.6
Product gas properties	Composite
Time spot sample taken, min*	174–243
Gas composition, mole %	
N_2	0.9
CO	2.0
CO_2	0.2
H_2	51.2
H_2S	...
CH_4	44.8
C_2H_6	0.5
C_3H_8	...
Butanes	...
Pentanes +	...
Olefins	...
Benzene	0.4
Toluene +	...
Total	100.0
Heating value, Btu/SCF††	640
Specific gravity, air = 1	0.331

* From start of coal feed.

† $\frac{\text{Wt coal recovered from reactor}}{\text{Coal feed rate}}$.

‡ $\frac{\text{Reactor vol. at standpipe height}}{\text{Cu ft/min } H_2 \text{ at reactor press. and temp}}$.

§ $\frac{\text{Cu ft/sec } H_2 \text{ at reactor press. and temp}}{\text{Cross-sectional area of reactor}}$.

** $100\left[\frac{\text{Wt of product gas} - \text{wt hydrogen in it}}{\text{Wt of moisture-free ash-free coal}}\right]$.

†† Gross, gas saturated at 60 F, 30 in. Hg pressure. Reported data based on dry coal and residue. Liquid products and overall material balance include moisture.

at which the number of moles of gas in the reactor passed thru a maximum due to initiation of hydrogenolysis. Reactor pressures increased nearly linearly with temperature increases up to threshold temperatures. As the reactor attained the maximum run temperature, pressures decreased rapidly and continued to decrease at a lower rate during the remainder of the run.

These data indicate that optimum pretreatment conditions for hydrogasification yields result from two factors. (1) increase in reactivity of the char to a maximum as pretreatment temperature is increased, followed by decrease as low-temperature carbonization temperatures are approached, and (2) continuous decrease in materials relatively easily hydrogenated as pretreatment temperature is increased. The increased coal conversion to liquid products with increases in pretreatment temperature indicates loss of the more reactive coal constituents, which eventually offsets any general reactivity increase due to pretreatment.

3. Fluid bed hydrogenation of coal: A small pilot plant for fluid bed hydrogenation of coal has been placed in operation at I.G.T. Its reactor is 2 in. ID, 5 in. OD, and 113 in. long for a working pressure of 3500 psi at 1400 F. The coal feeder is 5 in. ID and 60 in. long for a working pressure of 3500 psi at 600 F.

Table 3-66 shows initial test results at a reactor pressure of 1000 psig and average maximum temperature of 1430 F. Only 31.7 per cent of the coal char was gasified, and 51.2 per cent hydrogen remained in the product gas. If this hydrogen were separated for reuse, the resultant product gas would have an acceptable heating value. The two per cent CO could be methanized or separated.

HYDROGENATION OF OIL SHALE

Conversion of oil shale[55] to high heating value gases by direct hydrogenation was investigated by I.G.T. to determine whether such production of pipeline gas was feasible and whether it offered potential advantages over alternate methods for utilization of large reserves of this fossil fuel. Data on batch hydrogenolysis of a 22.9 gal per ton Colorado oil shale in three particle size ranges were obtained at a maximum reactor temperature of 1300 F, maximum pressures of 1200 to 5700 psig, and hydrogen-to-shale ratios equivalent to 50 to 200 per cent of stoichiometric requirements for complete conversion of organic carbon plus hydrogen content to methane. Nearly complete conversion of organic carbon and hydrogen to a fuel gas with a heating value of over 800 Btu per cu ft was obtained in relatively short residence times at temperatures of 1200 to 1300 F, with only little information of carbon oxides from mineral carbonate decomposition. Since material costs are relatively low, these results indicate that consideration can be given to supplementing future natural gas supplies with synthetic high heating value gas from oil shale, particularly in areas served by long-distance transmission lines passing near Colorado oil shale deposits.

METHANE DRAINAGE FROM COAL MINES

Natural gas is often compressed at pressures of 10 to 20 atm in the porous structure of coal and surrounding rocks. Its drainage thru 2½ in. diam bore holes at selected spacings 20 to 30 yards apart and its collection in a gathering pipeline have been practiced in Europe[56,57] and England[58] for over 15 years. Dangerous gassy seams of coal have been made safe, coal production has been increased by 50 per cent, and utility gas and gas boiler fuel equivalent to 14,000 tons of coal per year have been recovered from the Haig Mine[58] in England. Experiments were conducted in Ohio in 1947 and 1948 on collecting gas from an unused coal mine. Over 500,000 cu ft of gas per day was developed, with an average heating value of 970 Btu per cu ft and 0.705 sp gr. Its analysis in per cent by volume was: CO_2, 3.0; O_2, 0.5; CH_4, 71.0: C_2H_6, 14.0; N_2, 11.5; and S, 0.0. No commercial use of gas from coal or coal mines has been made in the United States, but over 28 MMCF per day is collected in England, Germany, Belgium, the Saar, and France. It is estimated[56] that the potential gas volume from such sources is seven and one-half times the natural gas reserves for the year 1951.

BIOLOGICAL FORMATION OF METHANE

Methane is a common and abundant product of bacterial decomposition of organic materials under anaerobic conditions.[59,60] Where plants die and decompose under water, methane, or swamp gas which is largely methane and CO_2, is formed. Methane is also formed in the digestive tracts of cows in volumes of up to 100 to 500 liters of methane per day per animal; 5 to 10 per cent of the calorific value of the feed may be lost as methane. Sewage plants using the fermentation process produce considerable volumes of methane,[61] which is used in internal combustion engines and for boiler fuel or, after purification, is sold to local gas companies. Considerable information is given in *Industrial Fermentations*, Chap. 14, Chemical Publishing Co., New York, 1954, on this subject. Buswell[61] states that.

1. Methane fermentation may be aided by mixed or enriched cultures.
2. Fermentation can be applied to any type of substrate; lignin and mineral oil are not fermentable.
3. Reaction is quantitative, converting entire substrate to CO_2 and CH_4.
4. No temperature limitation is observed. Reaction proceeds at temperatures of from 32 to 131 F, but a sudden temperature change should be avoided.
5. Addition of inert solid material, such as straw or sawdust, is important.
6. Decomposition of organic compounds passes thru the lower aliphatic acids, characteristically acetic. They are precursors of CO_2 and CH_4.

REFERENCES

1. Fairclough, H. "What Price Hydrogen." *Petrol. Refiner* 35: 333–5, Sept. 1956.
2. Channabasappa, K. C., and Linden, H. R. "Hydrogenolysis of Bituminous Coal." *Ind. Eng. Chem.* 48: 900–5, May 1956.
3. Rossini, F. D., and others. *Selected Values of Physical and Thermodynamic Properties of Hydrocarbons and Related Compounds.* Pittsburgh, Carnegie Press, 1953.
4. Kopper, H. "The Effect of Moisture Content Upon Heat Consumption During Carbonization of Coal." *Fuel* 12: 139–43, 1933.
5. Wisner, C. B. "The Missing Link in Low Temperature Carbonization." *Intern. Conf. Bituminous Coal Proc.* 1926: 800–11.
6. Brownlee, D. "Low Temperature Carbonization of Bituminous Coal." *Engineering* 124: 36–8, July 8, 1927.
7. Zielke, C. W., and Gorin, E. "Kinetics of Carbon Gasification." *Ind. Eng. Chem.* 47: 820–5, Apr. 1955; 49: 396–403, Mar. 1957.
8. Lesher, C. E. "Production of Low Temperature Coke by the Disco Process." *A.I.M.E. Trans.* 139: 328; *Combustion* 11: 39, Apr. 1940.
9. Lesher, C. E., and Zimmerman, R. E. "325 Tons Daily and Treated at Disco Plant of Pittsburgh Coal Carbonization Co." *Coal Age* 44: 45–9, Mar. 1939.
10. Cooperman, J., and others. *Lurgi Process.* (Bur. of Mines Bull. 498) Washington, D. C., 1951.
11. Odell, W. W. *Gasification of Solid Fuels in Germany by the Lurgi, Winkler, and Leuna Slagging-Type Gas-Producer Processes.* (IC-7415) Washington, D. C., Bur. of Mines, 1947.
12. Strimbeck, G. R., and others. "Recent Work by the Bureau of Mines on the Gasification of Pulverized Coal." (CEP-55-27) *A.G.A. Proc.* 1955: 1006–37.
13. von Fredersdorff, C. G., and others. *Gasification of Pulverized Coal in Suspension.* (Research Bull. 7) Chicago, I.G.T., 1957.
14. I.G.T. *Gas Making Processes.* Chicago, 1945.
15. Lowry, H. H. *Chemistry of Coal Utilization,* Suppl. Vol. New York, Wiley, 1963.
16. Hubman, O.: "Advances in the Technique of Gasifying Fine, High-Ash Coal under Pressure." (In: *Intern. Conf. Complete Gasification of Mined Coal,* p. 169–77. Liège, May 1954.)
17. Newman, L. L. "Recent European Developments in the Use of Oxygen in Gas Manufacture." *A.G.A. Proc.* 1951: 585–90.
18. Atwell, H. V. *Koppers Powdered Coal Gasification Process.* (FIAT Final Rept. 1303, PBL 85165) Washington, D. C., Office of Technical Services, 1947.
19. Totzek, F. "Synthesis Gas from the Koppers–Totzek Gasifier." *Chem. Eng. Progr.* 50: 182–7, Apr. 1954.
20. Totzek, F. "Total Gasification of Coal Dust." *Coke Gas* 16: 89–96, Mar. 1954.
21. Wright, C. C., and Newman, L. L. "Oxygen Gasification of Anthracite in the Wellman–Galusha Producer." *A.G.A. Proc.* 1947: 701–12.
22. Wright, C. C., and others. "Production of Hydrogen and Synthesis Gas by the Oxygen Gasification of Solid Fuel." *Ind. Eng. Chem.* 40: 592–600, Apr. 1948.
23. Batchelder, H. R., and others. "Operation of Kerpely Producer with Oxygen-Enriched Blast." *A.G.A. Proc.* 1950: 348–54.
24. Batchelder, H. R., and Hirst, L. L. "Coal Gasification at Louisiana, Missouri." *Ind. Eng. Chem.* 47: 1522–8, Aug. 1953.
25. Strimbeck, G. R., and others. "Pilot-Plant Gasification of Pulverized Coal with Oxygen and Highly Superheated Steam." *A.G.A. Proc.* 1950: 501–63; (R.I. 4733) Washington, D. C., Bur. of Mines, 1950.
26. Grossman, P. R., and Curtis, R. W. "Pulverized-Coal-Fired Gasifier for Production of Carbon Monoxide and Hydrogen." *ASME Trans.* 76: 689–95, May 1954.
27. Strimbeck, G. R., and others. *Progress Report on Operation of Pressure-Gasification Pilot-Plant Utilizing Pulverized Coal and Oxygen.* (R.I. 4971) Washington, D. C., Bur. of Mines, 1953.
28. Great Britain Fuel Research Station, Dept. of Sci. & Ind. Research. *Catalytic Enrichment of Industrial Gases by the Synthesis of Methane.* (Tech. Paper No. 7) London, H.M.S.O., 1953.
29. Sabatier, P. *Catalysis in Organic Chemistry,* New York, Van Nostrand, 1922.
30. Sabatier, P., and Sendersens, J. B. "Action of Nickel on Ethylene; Synthesis of Ethane." *Compt. Rend.* 124: 1358–61, June 14, 1897.
31. ——. "Direct Hydrogenation of Carbon Monoxide in the Presence of Finely-Divided Metals." *Compt. Rend.* 134: 689–91, Mar. 24, 1902.
32. ——. "New General Methods of Hydrogenation and of Molecular Reactions Based on the Use of Finely-Divided Metals." *Ann. Chim. Phys.*, viii, 4: 319–432, 1905; abstracted in *J. Chem. Soc.* 88: 333–4, 1905.
33. ——. "New Synthesis of Methane." *Compt. Rend.* 134: 514–6 Mar. 3, 1902.

34. Mayer, M., and others. "Gas Reactions." *J. Gasbelsuchtung* 52: 166–94, 238–82, 305–26, 1910.
35. Erdmann, E. "Synthetic Production of Methane." *Chem. Trade J.* 49: 175, 1911.
36. Dent, F. J., and others. "An Investigation into the Catalytic Synthesis of Methane for Town Gas Manufacture." (GRB 20) *Inst. Gas Engrs. Trans.* 95: 602–709, 1945.
37. Wainwright, H. W., and others. *Laboratory-Scale Investigation of Catalytic Conversion of Synthesis Gas to Methane* (R.I. 5046) Washington, D. C., Bur. of Mines, 1954.
38. ——. *Removal of Hydrogen Sulfide and Carbon Dioxide from Synthesis Gas Using Di- and Tri-Ethanolamine.* (R.I. 4891) Washington, D. C., Bur. of Mines, 1952.
39. Dirksen, H. A., and others. *Pipeline Gas from Coal: Methanation.* (I.G.T. Rept. No. 2, Project PB-19, to A.G.A.) Chicago, I.G.T., Mar. 1957.
40. ——. *Catalytic Conversion of Synthesis Gas from Coal to High Btu Pipeline Gas.* New York, A.G.A. Chem. Eng. and Manufactured Gas Proc. Conf., May 23–25, 1955.
41. Dirksen, H. A., and Linden, H. R. *Pipeline Gas from Coal: Methanation.* (I.G.T. Rept. No. 1, Project PB-19, to A.G.A.) Chicago, I.G.T., May 1956.
42. Groggins, P. H. *Unit Processes in Organic Synthesis,* 5th ed. New York, McGraw-Hill, 1958.
43. Dent, F. J., and others. *39th Report of the Joint Research Committee of the Institution and Leeds University: The Investigation of the Use of Oxygen and High Pressure in Complete Gasification—Part I, Gasification with Oxygen.* (Communication 141) London, Inst. of Gas Engrs., 1936. Also in *I.G.E. Trans.* 86: 118–201, 1936.
44. *Ibid. 41st Report—Part II, Synthesis of Gaseous Hydrocarbons at High Pressure.* (Communication 167) London, Inst. of Gas Engrs., 1937. Also in *I.G.E. Trans.* 87: 231–87, 1937.
45. *Ibid. 43rd Report—Part III, Synthesis of Gaseous Hydrocarbons at High Pressure.* (Communication 190) London, Inst. of Gas Engrs., 1938. Also in *I.G.E. Trans.* 88: 150–217, 1938.
46. Dent, F. J., and others. Private communication, 1950.
47. Dent, F. J. "The Hydrogenation of Coal to Gaseous Hydrocarbons." (In: *Intern. Conf. Complete Gasification of Mined Coal,* p. 113–24, Liège, May 1954.)
48. Bray, J. L., and Howard, R. E. *The Hydrogenation of Coal at High Temperatures (Rept. No. 1).* (Research Series 90) (Engineering Bull. XXVII, No. 5, Sept. 1943) Lafayette, Ind., Purdue Univ., 1943.
49. Bray, J. L., and Morgal, P. W. *The Hydrogenation of Coal at High Temperatures (Rept. No. 2).* (Research Series 93) (Engineering Bull. XXVIII, No. 4, July 1944) Lafayette, Ind., Purdue Univ., 1944.
50. Stockman, C. H., and Bray, J. L. *The Hydrogenation of Coal at High Temperatures (Rept. No. 3).* (Research Series 111) (Engineering Bull. XXXIV, No. 6, Nov. 1950) Lafayette, Ind., Purdue Univ., 1950.
51. Symonds, F. L., and others. "Public Utility Gas as a By-Product of Synthetic Liquid Fuels Production." *A.G.A. Proc.* 1949: 789–95.
52. Newman, L. L., and Pipilen, A. P. "Hydro-Carbonization of Coal for High Btu Gas." *Gas Age* 119: 16–21, May 16; 18+, May 30, 1957.
53. Hiteshue, R. W., and others. "Hydrogenating Coal at 800°C." *Ind. Eng. Chem.* 49: 2008, Dec. 1957.
54. Channabasappa, K. C., and Linden, H. R. "Fluid-Bed Pretreatment of Bituminous Coals and Lignite." *Ind. Eng. Chem.* 50: 637–44, Apr. 1958.
55. Schultz, E. B., Jr., and Linden, H. R. "Production of Pipeline Gas by Batch Hydrogenolysis of Oil Shale." *Amer. Chem. Soc. Div. of Gas & Fuel Chem. Preprints* 1958: 26–37.
56. Minchin, L. T. "Fuel Hungry Europe Is Recovering Natural Gas From Coal Mines." *Gas Age* 108: 31+, Dec. 6, 1951.
57. "Methane from Coal Mines—The Present Position." *Coke Gas* 16: 357–62, Sept. 1954. "Utilization of Methane from Coal Mines." *Coke Gas* 17: 33–5, Jan. 1955.
58. Cawley, R. L., and Jones, J. H. "Methane Drainage: An Account of the Work Carried Out at Haig Pit, Whitehaven." *J. Inst. Fuels (British)* 28: 366–82, Aug. 1955.
59. Barker, H. A. "Biological Formation of Methane." *Ind. Eng. Chem.* 48: 1438–42, Sept. 1956.
60. Davis, J. B. "Microbial Decomposition of Hydrocarbons." *Ind. Eng. Chem.* 48: 1444–8, Sept. 1956.
61. Buswell, A. M. "Discussion, Biological Formation of Methane." *Ind. Eng. Chem.* 48: 1443, Sept. 1956.
62. British Petroleum Co., Ltd. *Gasmaking.* London, circa 1960.

SECTION 4

PRODUCTION, GATHERING, AND CONDITIONING OF NATURAL GAS

Thomas S. Bacon, *Section Chairman,* Lone Star Gas Co., Dallas, Texas
Warren L. Baker, Chapter 2, American Association of Oilwell Drilling Contractors, Dallas, Texas
J. S. Connors, Chapter 9, Phillips Petroleum Co., Bartlesville, Okla.
J. L. Eakin, Chapter 4, Bartlesville Petroleum Research Center, Bureau of Mines, Bartlesville, Okla.
A. W. Francis, Jr., Chapter 6, National Tank Co., Tulsa, Okla.
E. G. Hammerschmidt, Chapter 8, Natural Gas Pipeline Company of America, Chicago, Ill.
John W. Harrington, Chapter 3, Southern Methodist University, Dallas, Texas
Wm. Hinchliffe, Chapter 5, Halliburton Co., Dallas, Texas
Donald E. Kliewer, reviewed Chapter 2, World Oil, Houston, Texas
K. R. Knapp (retired), Chapter 8, American Gas Association Laboratories, Cleveland, Ohio
George W. McKinley, Chapter 7, Consolidated Natural Gas System, Pittsburgh, Pa.
J. S. Miller, Chapter 4, Bartlesville Petroleum Research Center, Bureau of Mines, Bartlesville, Okla.
Russell L. Morris, Chapter 6, National Tank Co., Tulsa, Okla.
P. V. Mullins, Chapter 10, Bureau of Mines, Amarillo, Texas
Neal Neece, Jr., Chapter 3, Southern Methodist University, Dallas, Texas
C. L. Perkins, Chapters 8 and 9, El Paso Natural Gas Co., El Paso, Texas
J. E. Schaefer, Chapter 3, The East Ohio Gas Co., Cleveland, Ohio
R. V. Smith, Chapter 4, Phillips Petroleum Co., Bartlesville, Okla.
Oscar F. Spencer, Chapter 1, The Pennsylvania State University, State College, Pa.
F. E. Vandaveer (retired), Chapters 3, 7, and 9, Con-Gas Service Corp., Cleveland, Ohio

CONTENTS

Chapter 5—Well Completion Practices

Chapter 6—Gas Separators, Heaters, and Cleaners

Chapter 7—Natural Gas Gathering Systems

Chapter 8—Gas Hydrates and Gas Dehydration

Chapter 9—Removal of Sulfur Compounds and Carbon Dioxide from Sour Natural Gas

Chapter 10—Nitrogen Removal from Natural Gas

Tables

Figures

Chapter 1

History of the Natural Gas Industry in the United States

by Oscar F. Spencer

Natural gas was first discovered in the United States in 1775 in the Kanawha Valley, W. Va. Since 1900 rapid growth has taken place in the natural gas industry, with the principal increase having occurred since 1948 (see Table 4-1). Marketed production of natural gas in 1900 accounted for 3.2 per cent of the total energy produced in the United States; by 1962 this figure had risen to 34.2 per cent, with marketed production 13.9 trillion cu ft (Fig. 4-1).

Marketed production of natural gas in countries outside the Western hemisphere has been relatively small to date, accounting for only 18 per cent of the world marketed production in 1961. In that same year, in the United States alone, nearly 75 per cent of the world production was marketed. United States *net* marketed production was 83 per cent of the total United States marketed production, as shown in Table 4-1.

The natural gas utility and pipeline industry in the United States in 1963 served 33,940,200 customers[2] or about 95 per cent of all U. S. gas users. Over 1,266,600 new natural gas customers were added during 1963. At the end of 1963 natural gas was supplied thru 694,620 miles of main (including field and gathering lines, 199,630 miles of transmission mains, and 434,010 miles of distribution lines, but excluding service pipe). Employees of natural gas utility companies totaled 196,200 or almost 96 per cent of those employed by all gas utilities. Total gas utility and pipeline construction expenditures for 1963 were $1.56 billion.

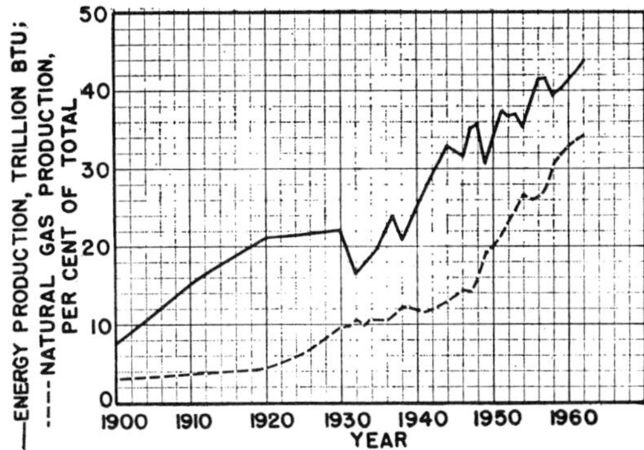

Fig. 4-1 U. S. total production of energy and marketed production of wet natural gas as a per cent of this total.[1,3]

Table 4-1 Production and Disposition of Natural Gas in the United States[1]

Note: Includes allowances for natural gas liquids in the natural gas and therefore differs from data originating with A. G. A.

(Billions of cu ft)

Year	Gross production from gas & oil wells*	Repres-suring	Net production†	Losses & waste*	Marketed production‡	Field use	Net change in under-ground storage	Lost in trans-mission	Net marketed production
1936	2,645	74	2,571	346	2,225	618	11	47	1,549
1940	3,694	363	3,331	597	2,734	712	15	59	1,948
1944	5,614	883	4,731	798	3,815	855	10	94	2,856
1948	7,179	1,221	5,958	810	5,148	1,022	57	127	3,942
1952	10,273	1,411	8,862	849	8,013	1,484	177	203	6,149
1956	12,373	1,427	10,946	864	10,082	1,421	136	213	8,312
1960	15,088	1,754	13,334	563	12,771	1,780	132	274	10,585
1961	15,460	1,682	13,778	524	13,254	1,881	146	235	10,992
1962	16,039	1,736	14,303	426	13,877	1,993	86	286	11,512
1963	16,973	1,843	15,130	383	14,747	2,081	131	365	12,170

* Includes gas blown to air but not direct waste on producing properties except where data are available.

† See Table 2-1 for net production of *dry* gas.

‡ *Marketed production,* as tabulated, is defined by the Bureau of Mines as useful consumption, and includes sales by all non-utilities to industrial consumers, field use, the natural gas portion of mixed gases, net change in underground storage, and transmission losses; therefore, the figures given exceed utility sales.

NATURAL GAS PRODUCTION

Production of natural gas in the United States may be divided into five geographic areas:

1. **Appalachian,** including all Atlantic seaboard states, and Pennsylvania, Vermont, and West Virginia. Producing states are Florida, Maryland, New York, Pennsylvania, Virginia, and West Virginia.
2. **Midwest**, including all states between the Appalachian and Rocky Mountain areas north of the Gulf Coast states. Producing states are Illinois, Indiana, Kansas, Kentucky, Michigan, Missouri, Nebraska, North Dakota, Ohio, Oklahoma, South Dakota, and Tennessee.
3. **Southwest,** including Gulf Coast states (except Florida), and Arkansas and New Mexico. All states of this area, Alabama, Arkansas, Louisiana, Mississippi, New Mexico, and Texas, are natural gas producers.
4. **Rocky Mountain,** including producing states of Colorado, Montana, Utah, and Wyoming.
5. **Pacific,** including states bordering the Pacific, and Arizona and Nevada. California is the only producing state.

Discovery of natural gas and its first production occurred in the Appalachian area. This area held the lead in production until 1926, when it shifted to the Southwest area, where it has remained. The six largest natural gas producing states, accounting for 92 per cent of the total marketed production, are Texas, Louisiana, Oklahoma, California, Kansas, and New Mexico.

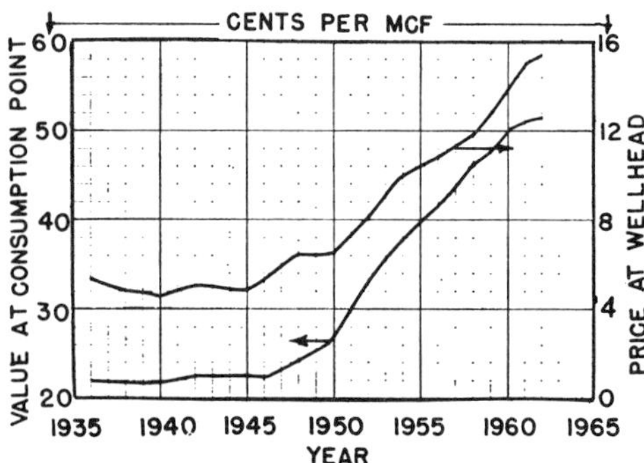

Fig. 4-2 U. S. average wellhead price and value at point of consumption (i.e., revenue from customers/sales volume).[1,3]

The largest United States gas-producing fields are located within the states of Texas, Louisiana, Oklahoma, and Kansas. The *Hugoton field*, covering an estimated 4,000,000 acres, extends 150 miles from southwestern Kansas across the Oklahoma Panhandle, terminating in the Texas Panhandle. The *Panhandle field* in northern Texas contains nearly 2,000,000 acres. Two other large fields, approximately 300,000 acres each, are the *Carthage field* in eastern Texas and the *Monroe field* in northern Louisiana.

Natural gas and crude oil are often found together in the same type of reservoirs. Search for crude oil has led to the discovery of vast proved recoverable gas reserves.

Dry gas from the well presents few processing problems, since it can be admitted to a pipeline with little or no treatment. Much natural gas, however, is produced together with crude oil as a *wet* gas containing heavier, easily liquefiable hydrocarbons. These natural gas liquids (for example, propane and butane) are usually removed in processing plants before delivery of the gas for transmission.

Table 4-1 shows gross, net, and marketed production of natural gas in the United States since 1936. Figure 4-2 presents a graph of average wellhead price and value at consumption point of marketed production for the period. Rise in marketed production has been particularly rapid since 1942, increasing from about 3.1 trillion cu ft that year to 13.9 trillion cu ft in 1963.

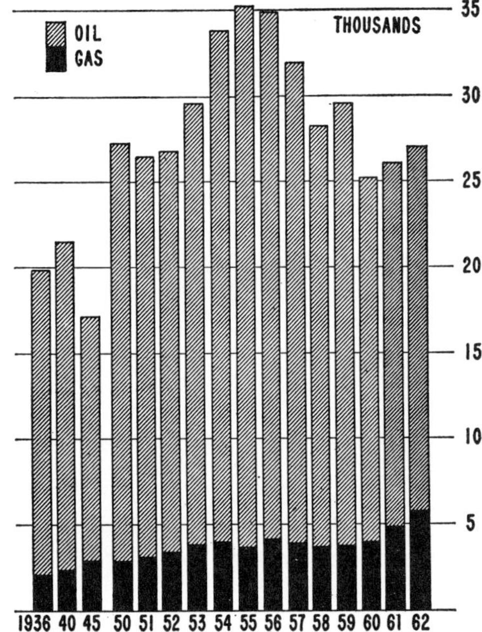

Fig. 4-3 Total gas and oil new well completions.[2]

Figure 4-3 shows total new oil and gas well completions yearly since 1936; many oil wells also produce gas. The number of dry gas and condensate-producing wells at the end of 1962 was 100,267 compared with 60,660 in 1945, an increase of over 65 per cent. In 1963, 4751 new gas wells (including condensate wells) were completed.

NATURAL GAS TRANSMISSION

Early production of natural gas was utilized near the fields owing to a lack of suitable lines for its distant transportation. Pipeline developments during the 1920 decade extended natural gas utilization more than 500 miles from its source of production. Manufacture of large-diameter, thin-walled, welded, seamless steel pipe capable of withstanding high pressures facilitated the construction of long-distance lines that now extend some 2000 miles from the wells. Large producing fields such as the Hugoton, Panhandle, Monroe, and Carthage* are input terminals for many of the larger pipeline systems.† Table 2-9 lists 48 United States cities in which natural gas is distributed.

Figure 4-4 shows total natural gas main mileage, marketed production, and interstate transmission since 1935.

Several developments have helped greatly in increasing efficiency of natural gas transmission. These include gas tur-

* See Table 2-12 for others.

† See Table 2-11 for major transmission lines.

Table 4-2 Industrial Consumption of Natural Gas in the United States by Type of Industry[2]

Note: Industrial consumption as reported by the Bureau of Mines includes sales by nonutility producers and others, and natural gas mixed with manufactured gas.

(Millions of cubic feet)

Year	Total	Field use	Carbon black plants	Petroleum refineries	Portland cement plants	Electric public utility power plants*	Other industrial*
1945	3,062,980	916,952	431,830	338,458	38,349	326,190	1,011,201
1950	4,440,197	1,187,473	410,852	455,096	96,986	628,919	1,660,871
1951	5,163,528	1,441,870	426,423	537,774	102,508	763,898	1,891,055
1952	5,475,843	1,483,754	368,399	536,402	111,479	910,117	2,065,692
1953	5,763,185	1,471,085	300,942	558,695	115,039	1,034,172	2,283,152
1954	5,923,647	1,456,883	251,176	563,315	125,257	1,165,498	2,361,518
1955	6,317,172	1,507,671	244,794	625,243	131,400	1,153,280	2,654,784
1956	6,662,443	1,420,550	242,598	679,343	144,192	1,239,311	2,936,449
1957	7,003,590	1,479,720	233,788	678,810	146,000	1,338,079	3,127,193
1958	7,174,623	1,604,104	211,048	681,912	164,000	1,372,853	3,140,706
1959	7,931,930	1,737,402	214,612	752,239	188,000	1,627,097	3,412,580
1960	8,386,038	1,779,671	197,628	775,154	171,000	1,724,763	3,737,822
1961	8,756,287	1,881,208	161,377	772,028	180,000	1,825,341	3,936,333
1962	9,204,898	1,993,128	133,302	789,877	188,000	1,965,590	4,135,001†

* Consumption by electric public utility power plants includes small quantities of gas other than natural, impossible to segregate. To this extent consumption by other industrials is understated.

† Includes Portland cement plants.

bine-driven compressors, automatic instrument-controlled compressor stations, microwave radio-relay systems, and airplane reconnaissance of pipeline rights of way for detection of breaks. Advances in storage techniques have permitted gas flow thru transmission lines to remain as near as possible to maximum rate throughout the year, in spite of fluctuating consumption.

NATURAL GAS UTILIZATION

Ultimate consumers may be classified into three main groups by type of service: (1) residential or domestic, (2) commercial, and (3) industrial. The first and third groups are self-defining. Commercial consumers include those engaged in selling or distributing a commodity, in some business or profession, or in other forms of economic or social activity, for example, offices, hotels, restaurants, hospitals, and other institutions. Sales of gas to consumers by gas utility companies according to major types of usage are represented in Fig. 4-5; about 98 per cent of these sales is natural gas.

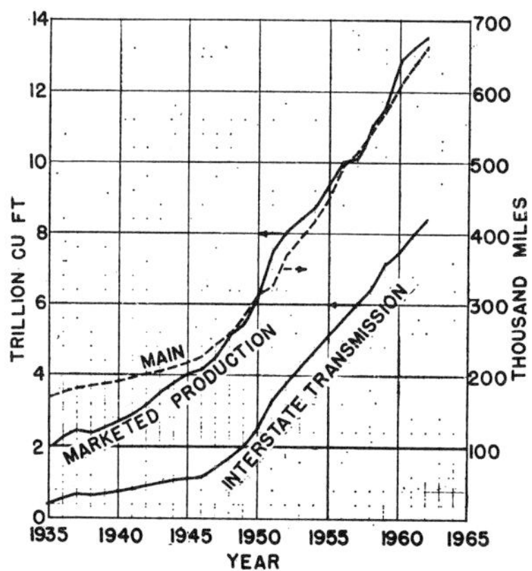

Fig. 4-4 U. S. natural gas main mileage and interstate transmission (including field, gathering, transmission and distribution main, but excluding service pipe).[1,3]

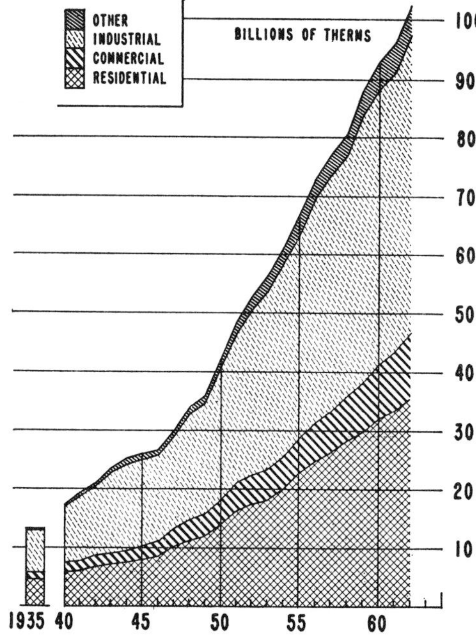

Fig. 4-5 Total gas utility sales by type of service.[2]

Consumption of natural gas since 1945 by various classes of industrial customers is given in Table 4-2. This table represents all natural gas sales to industrial users by distribution companies, pipelines, and producers.

REFERENCES

1. U.S. Bur. of Mines. *Minerals Yearbook.* Washington, D. C., G.P.O., annual.
2. A. G. A. *Gas Facts.* New York, annual.
3. A. G. A. *Historical Statistics of the Gas Industry.* New York.

Chapter 2
World Natural Gas Reserve and Production

*by Warren L. Baker**

The world's proved reserves and production of natural gas, by country, are shown in Table 4-3; the 13 major gas areas are shown in Table 4-4. Reserve data for many countries are approximations. For example, the Middle East reserve of 200,000 billion cu ft is simply an estimate of 1000 cu ft of gas per barrel of crude oil reserve. Estimates for other major reserves include Western Europe, over 21,000 billion cu ft,[1] and North Africa, over 30,000 billion cu ft.

A brief description of the reserve and production status is given for a number of countries.

UNITED STATES

The United States is the world's greatest user of natural gas. During 1961 it produced about 75 per cent of the total world production; the second largest producer, the Soviet Union, produced only 11.6 per cent. Table 4-5 gives the United States gas reserve data by states.[2]

Most oil-producing states also produce natural gas. In 1962, Texas, the largest producer, marketed 6.1 trillion cu ft of natural gas, almost 44 per cent of the country's total. The second largest producing state was Louisiana, with 3.5 trillion cu ft. Texas also has nearly 43 per cent of United States reserves and Louisiana more than 27 per cent.

At the end of 1963, more than 31 million homes in the United States were served with natural gas by utilities and pipeline companies, about four times the 1945 figure. Commercial natural gas customers totaled more than 2.540 million in 1963, and industrial natural gas customers had risen to 154,600, about five times the 1945 figure.

CANADA[6]

Tremendous strides have been made in recent years by Canada's expanding natural gas industry. Gross marketed production of natural gas in 1963 rose 17.9 per cent over the previous year to reach a new production estimate of 1105.7 billion cu ft. Alberta, with a marketed production of 933 billion cu ft, accounted for 84.4 per cent of the total; British Columbia was next with 117.0 billion cu ft, representing 10.6 per cent of the 1963 total estimated gross production.

Natural gas utility sales in 1963, up 9.6 per cent over 1962, also reflect the industry's growth. A new high of 1.4 million customers in Canada purchased 452 billion cu ft of natural gas and increased 1963 revenues 11.6 per cent over the previous year, to $287.6 million.[6]

In 1958, it was estimated that 30 trillion cu ft would be required over the next 30 years.[11] Future discoveries in Alberta to 1987 were estimated to be 51 trillion cu ft. Proved reserves were 37 trillion cu ft in 1963 and ultimate recovery was estimated as high as 300 trillion cu ft.

SOUTH AMERICA AND THE CARIBBEAN

Although the South American and Caribbean area has large reserves (mainly in Mexico, Venezuela, and Argentina), few markets are available to utilize the resources.

Mexico.[12] The main gas producing areas lie along the coastal plain of the Gulf of Mexico, where the principal sedimentary basins occur. In the north, the Reynosa district produced about 300 MMCF per day in 1960 and contained proved reserves of 3.15 trillion cu ft. The Macuspana district at the southern end of the area also produced about 300 MMCF per day and contained reserves of 2.8 trillion cu ft.

Two major projects were completed in 1960: a $40 million, 491-mile long, 500 MMCFD capacity gas line from the Tabasco fields[9] to Mexico City; and a 220-mile long, 40 MMCFD capacity, 16-in. diam gas line running from Monterrey to Torreon. A 1200-mile long, 34-in. diam line was under construction in 1961 from Reynosa to Mexicali (the line is to be extended to Los Angeles).

The major portion of the gas produced in Mexico is used for pressure maintenance and secondary recovery from oil fields.

Venezuela. Venezuela has the largest unexploited South American gas resources. The Venezuelan government prohibits flaring except for test purposes and when it is uneconomical to sell gas from gas condensate wells. Much of the gas is used for repressuring; very little is consumed.

Argentina. The production of gas in 1961 amounted to 83.3 billion cu ft, most of it coming from the Comodoro Rivadavia area which had 500 to 875 billion cu ft reserves at the end of 1952.

Argentina has the longest natural gas transmission line outside North America. The Comodoro Rivadavia fields, along the Atlantic Coast in southern Argentina, are connected with Buenos Aires, La Plata, and intermediate cities by an 1100-mile long line with a capacity of 35 MMCF per day. A 62-mile

* Subsequently reviewed by Donald E. Kliewer, Editorial Director of *World Oil*.

long gas line from Caleta Olivia to Comodoro also supplies gas for the Buenos Aires system. A line from the Plaza Huincul group of fields on the western edge of central Argentina, 310 miles long, with a capacity of 17.5 MMCF per day, connects with the Comodoro Rivadavia-Buenos Aires system.

Other South American Countries. For production and reserves of Brazil, Chile, Colombia, Peru, and Trinidad, see Table 4-3.

WESTERN EUROPE

Before World War II annual natural gas production in Western Europe amounted to only 1.77 billion cu ft. Subsequently, production rose to 14.7, 271.8, and 495 billion cu ft in 1946, 1958, and 1961, respectively. England[13] was expected to import 90 MMCFD of gas in 1964.

Italy. Italy has large reserves (Table 4-3) and the

Table 4-3 World Proved Reserves, Marketed Production, and Cumulative Production of Natural Gas

Note: 1 billion = 10^9

Country	Reserve			Production* (1961), billions of cu ft	Cumulative production†	
	Billions of cu ft	As of end of‡	Reference		Billions of cu ft	Period
North America						
Canada	35,436.9	1962	2	655.7	4,269.0	1889–1957
United States	273,765.6	1962	2	13,254.0[6]	193,000	1858–1960
South America						
Argentina	7,415	1960	7	83.3	...	...
Bolivia	...	...	...	0.35	...	...
Brazil	380	1957	7	18.7	17.66	1942–1957
Chile	2,825	1961	7	37.4	...	...
Colombia	...	...	...	14.8	...	...
Mexico	8,121	1960	7	306.5	...	...
Peru	1000–2000	1962	8	1.8	...	...
Trinidad & Tobago	722	1960	7	29.3	§	1909–1956
Venezuela	33,637	1960	7	170.23	...	1937–1954
Western Europe						
Austria	925	1962	1	55.1	208.3	1939–1957
Belgium, Luxembourg	§	1955	4	2.5	7.252	1951–1954
France (Lacq)	5,190	1962	1	141.6	118.6	1941–1957
Italy	2,800**	1960	9	242.3	...	...
Netherlands	14,050	1962	...	17.3	25.1	1947–1957
United Kingdom	0	1957	4	2.4	...	...
West Germany	1,158	1962	...	32.9	...	...
Far East						
Brunei	...	...	...	2.8	...	...
Burma	...	...	...	0.35	...	...
China (Taiwan)	...	...	...	1.4	...	...
Indonesia	...	...	...	90.4	...	...
Japan	5,031	1957	4	33.5	96.11	1915–1957
Pakistan	20,459	1960–61	7	34.6	27	1955–1958
Africa						
Algeria	26,000	1960	9	8.1	...	...
Nigeria	2,648	1961	7	...	...	...
Morocco	28	1960	9	0.35	...	...
Tunisia	...	....	...	0.35	...	...
French Sahara	42,372	1961	7	...	...	...
European Communist Countries						
Czechoslovakia	...	1957	4	53.0	...	...
East Germany	...	...	...	2.1	...	...
Hungary	318	1960	9	11.3	225.7	1936–1957
Poland	...	1960	...	25.8	...	...
Rumania	...	...	...	341	...	...
U. S. S. R.	78,300	1962	1	2,081	...	...
Yugoslavia	141.24	1957	4	2.4	9.339	1925–1957
Middle East						
Iran	3,107	1956	7	34.6	...	...
Iraq	2,966	1956	7	22.6	...	...
Saudi Arabia	9,181	1956	7	...	...	...
Israel	...	...	...	1.4	...	...
Kuwait	9,887	1956	7	...	...	...

* Data taken from Ref. 3 unless otherwise indicated.
† Data taken from Ref. 4, except for United States.
‡ Ref. 5 may be used to update a number of these.
§ Unestimated.
** Low limit; high limit, 3180.

Table 4-4 Thirteen Major Gas Areas of the World
(based on estimated remaining reserve)[10]

Rank	Field	Location	Year discovered	Original reserve (Trillions of cu ft)	Cumulative prod, 1-1-60 (Trillions of cu ft)	Est remaining reserve (Trillions of cu ft)	Productive gas wells drilled, 1-1-60	Producing formation: Age	Producing formation: Avg depth, ft
1	Panhandle–Hugoton	Kansas–Oklahoma–Texas	1918	70.0	33.0	37.0	7783	Permian	1400–3800
2	Hassi R'Mel	Algeria	1956	35.3	0.0	35.3	9 (shut in)	Triassic	6880
3	Slochteren	Netherlands	1962	39.0	0.0	39.0	...	Permian	9,840–13,120
4	Lacq	France	1951	14.0	0.07	13.93	35	Cretaceous	12,000
5	San Juan (all fields)	New Mexico–Colorado	1927	15.0	1.8	13.2	4260	Cretaceous	5500
6	Puckett	West Texas	1952	6.5	0.3	6.2	40	Permian, Devonian, Ordovician	10,000–15,000
7	Sui	West Pakistan	1952	4.0	0.05	6.0	...	Eocene	4000 (?)
8	Katy	Texas coast	1935	6.0	0.5	5.5	40	Eocene	6250–7450
9	Jalmat-Eumont	New Mexico	1927	8.1	4.1	4.0	791	Permian	3700
10	Old Ocean	Texas coast	1934	5.0	1.0	4.0	35	Oligocene	10,500
11	Big Piney	Wyoming	1922	2.8	0.144	2.656	160	Cretaceous-Tertiary	900–9000
12	Peace River region	British Columbia	1943	2.5	0.141	2.359	193	Triassic, Cretaceous, Permian, Pennsylvanian, Mississippian	2,100–11,000
13	Chocolate Bayou	West Texas	1952	2.5	0.2	2.3	57	Oligocene	9,600–12,800

largest gas pipeline system in Western Europe. The Ente Nazionale Idrocarburi (ENI) network (2700 miles in 1959) accounted for 11 per cent of the total Italian energy consumption in 1956. It was reported in 1964 that natural gas was in danger of running low by 1970. The low reserve level prompted Italian interest in trans-Mediterranean pipeline and liquefaction plans.

France. The gas industry (Régie Autonome des Pétrôles), nationalized in 1946, had as its only domestic source of natural gas in 1954 the St. Marcet field, which produced 9.45 billion cu ft. However, by 1959 the total national output was 92 billion cu ft, of which 81 billion cu ft came from Lacq field. Lacq field has a reserve of about 7 trillion cu ft, with a 15 per cent content of H_2S. The French have also been studying the techniques of handling liquefied natural gas from the Sahara. A pipeline supply has been negotiated with the Dutch. Trans-Mediterranean pipeline schemes are also under consideration.[14]

Netherlands. The Nederlandse Aardolie Maatschappij (NAM), the only producer of natural gas, sells gas to the government, which resells to consumers thru state-owned pipelines. In 1956 natural gas accounted for 7 per cent of the total gas used and 0.8 per cent of the total Netherlands energy requirements. The discovery of the Slochteren field (Table 4-4) promises to change this picture. Holland plans to replace all manufactured gas with natural gas by 1970. Export plans include sending 150 billion cu ft to France per annum.[15]

Germany.[17,18] Natural gas accounted for only 0.7 per cent of the 1956 mixed gas output. Germany, however, is actively exploring for natural gas and doubled its gross 1954 natural gas production in 1958.

Austria.[3] Natural gas production has been rising steadily. The 1957 production was 26.8 billion cu ft (about 1.8 per cent more than 1956); 1958 production, 29.0 billion cu ft; and 1962 production, 62 billion cu ft.

Under the North Sea. The Continental Shelf, which is of the same sedimentary composition[16] as the Dutch Slochteren field, was intensively surveyed by the British and the Dutch in 1963, using seismic techniques. The ratified 1958 Geneva Convention on Offshore Rights established the median line between maritime nations as the division for exploiting natural resources.

FAR EAST

Indonesia, Pakistan, and Japan produce appreciable amounts of natural gas. Other countries have either small reserves or few explorations. Japan has large gas requirements which are met mainly by manufactured gas.[19]

Pakistan.[20] The natural gas industry started in 1955 after the discovery of gas at Sui, in West Pakistan, in 1952. The 1960 reserves at Sui were 6.5 trillion cu ft and at Mari 3.5 trillion cu ft. The gas was used mainly for industrial application of electric power generation, fertilizer, and chemicals. Annual Sui gas consumption is expected to be about 1 trillion cu ft by 1965.

EUROPEAN COMMUNIST COUNTRIES

Yugoslavia. Oil and gas production has increased rapidly,[9] from 1765 MMCF of natural gas in 1959 to 1925 MMCF in 1960. Gas was discovered at Boca, Plantiste, and Begejac.

Table 4-5 Estimated Proved Recoverable Reserves of Natural Gas, by State, 1962–1963

Note: Volumes are calculated at a pressure base of 14.73 psia and at a standard temperature of 60 F. Reserves of dissolved gas were estimated jointly with the API Committee on Petroleum Reserves. Shrinkage caused by natural gas liquids recovery excluded.

(Millions of cubic feet)

State	Reserves* as of December 31, 1962	Reserves* as of December 31, 1963				
		Total	Non-associated†	Associated‡	Dissolved§	Underground storage**
Alaska	1,634,321	1,690,724	1,626,290	0	64,434	0
Arkansas	1,643,669	1,792,644	1,331,319	266,282	178,455	16,588
California††	9,121,385	8,865,726	3,140,246	1,708,976	3,825,882	190,622
Colorado	2,204,777	1,876,057	1,509,657	99,939	262,209	4,252
Illinois	158,213	168,595	80	0	41,434	127,081
Indiana	49,638	60,180	696	686	17,665	41,133
Kansas	18,567,174	17,994,235	17,241,316	494,664	168,013	90,242
Kentucky	1,108,942	1,085,236	979,238	0	70,352	35,646
Louisiana††	71,544,088	75,364,992	61,759,903	8,643,963	4,960,535	591
Michigan	647,039	722,812	112,403	80,472	58,452	471,485
Mississippi	2,735,845	2,481,627	1,988,655	195,205	291,795	5,972
Montana	600,171	598,131	399,266	22,624	80,934	95,307
Nebraska	100,194	100,042	65,074	6,797	17,828	10,343
New Mexico	14,112,734	15,037,822	10,868,323	2,326,626	1,816,952	25,921
New York	131,196	132,285	40,869	0	31	91,385
North Dakota	962,709	1,119,575	6,955	338,531	774,089	0
Ohio	727,912	748,187	273,761	0	98,426	376,000
Oklahoma	18,259,036	19,138,820	14,301,295	2,681,100	2,023,291	133,134
Pennsylvania	1,175,083	1,214,498	704,026	0	19,241	491,231
Texas††	118,854,773	117,809,376	78,449,748	25,771,799	13,515,098	72,731
Utah	1,786,366	1,638,324	978,846	404,312	254,378	788
Virginia	33,045	31,303	31,303	0	0	0
West Virginia	2,025,999	2,311,164	1,930,259	0	60,802	320,103
Wyoming	3,931,224	3,988,546	3,441,002	138,922	387,207	21,415
Miscellaneous‡‡	163,325	180,332	39,119	0	18,530	122,683
Total	272,278,858	276,151,233	201,219,649	43,180,898	29,006,033	2,744,653

* Excludes gas loss due to natural gas liquids recovery.
† Non-associated gas is free gas not in contact with crude oil in the reservoir, and free gas in contact with oil where the production of such gas is not significantly affected by the production of crude oil.
‡ Associated gas is free gas in contact with crude oil in the reservoir where the production of such gas is significantly affected by the production of crude oil.
§ Dissolved gas is gas in solution with crude oil in the reservoirs.
** Gas held in underground reservoirs (including native and net injected gas) for storage purposes.
†† Includes off-shore reserves.
‡‡ Includes Alabama, Arizona, Florida, Iowa, Maryland, Missouri, and Tennessee.

U. S. S. R.[21] The U. S. S. R. is the world's second largest producer of natural gas (Table 4-3). Although the Russian gas industry is expanding, it is expected[22,23] that in 1965 it will reach only the 1944 United States level (5.6 trillion cu ft gross production). Gas accounted for 5.4 per cent (980 billion cu ft) of the total Soviet energy production in 1958, equivalent to one-fifth of the crude oil and one-tenth of the coal consumption for that year. In 1960, 1.8 trillion cu ft at wellhead provided for 10.3 per cent of the total fuel needs. The 1965 production is expected to provide for 17.5 per cent of the total; these increases are expected to come chiefly from the Northern Caucasus and Central Asia.

Transmission lines totaled 7500 miles in 1959; construction of 16,000 additional miles is planned by 1965. The principal gas markets are the industrial areas of Moscow, Leningrad, the Ukraine, and the Ural Mountains. Before 1956 pipeline diameters did not exceed 20 in.; in 1959 40-in. pipe was used for the first time.

A breakdown of Soviet consumption by class of service follows:

	Per cent of total	
	1958	1965 (planned)
Chemicals, carbon black, and natural gasoline	6.3	5.7
Domestic	10.4	8.3*
Steel	5.3	18.3
Cement	7.2	5.4
Power stations	33.5	20.0
Others	34.1	39.8
Pipeline losses	3.2	2.5

* Later reported as 5 per cent.[22]

Rumania. Out of 212 billion cu ft of natural gas produced in 1959,[24] about three-quarters was used by industry and the remainder by domestic consumers. The existing 2190 miles of pipeline had a capacity of about 350 billion cu ft per year. Gas constituted 19 per cent of the overall energy consumption.

Other Countries. During 1959 Rumania supplied 7.4 billion cu ft of natural gas to Hungary by means of a 231-

mile long, 11-in. diam pipeline. Production in Hungary was about 4 billion cu ft. Czechoslovakia and Poland share the Malaky and Cerkytesin fields. Although a common oil pipe network is planned among European Communist countries, no plan for a gas network exists.[25]

REFERENCES

1. U. N. Econ. Commission for Europe. Committee on Gas. *Report of the Meeting of Rapporteurs, May 1962.* Geneva, 1962.
2. A. G. A. *Reports on Proved Reserves of Crude Oil, Natural Gas Liquids, and Natural Gas.* (Vol. 17) New York, 1962.
3. U. N. Statistical Office. *World Energy Supplies, 1958–1961.* (Stat. Papers, J, No. 6) New York, 1963. (Figures for net production)
4. World Power Conf. *Statistical Year-Book, No. 9.* London, Percy Lund, 1960.
5. Weeks, L. G. *World Gas Reserves, Occurrence, Production.* Dallas, Institute on the Economics of the Gas Industry, 1962.
6. A. G. A. *Gas Facts.* New York, annual.
7. World Power Conf. *Survey of Energy Resources, 1962.* London, Percy Lund, 1962.
8. "Large Gas Field Is Found in Peru." *N. Y. Times,* Apr. 15, 1962.
9. "World Petroleum Report." *World Petrol.,* Feb. 15, 1961.
10. "Major Gas Areas of the World." *Oil & Gas J.* 59: 106, Apr 24, 1961.
11. Canada. Royal Commission on Energy. *First Report.* Ottawa, 1958.
12. Rojas, A. G. "Growth of the Mexican Gas Industry." *A. G. A. Monthly* 42: 28–31, Apr. 1960.
13. "Algerian Gas to Enter England by '64." *Oil & Gas J.* 60: 84, Sept. 3, 1962.
14. Rose, Stephane. "Europe's Energy Market Is Target for Both Sahara and Dutch Gas." *World Petrol.* 35: 48+, June 1964.
15. Haffner, A. E. *Gas in Europe.* (Paper given to 57th Annual Meeting, CGA) Don Mills, Ontario, Canadian Gas Assoc., 1964.
16. "Wild Well Spurs North Sea Play." *Oil & Gas J.* 62: 59–61, July 6, 1964.
17. O.E.E.C. *Gas in Europe.* (C(58)78) Paris, 1958.
18. —— (GA(60)3) Paris, 1960.
19. Japan Gas Assn. *Gas Industry in Japan.* Tokyo, 1962.
20. Pakistan. Planning Commission. *Second Five Year Plan (1960–1965),* p. 269–71. Karachi, 1960.
21. Shabad, T. "Russia's Present and Future in Natural Gas." *Gas Age* 124: 49, Dec. 10, 1959.
22. "American Gas Industry Group Reports on Russian Tour." *A. G. A. Monthly* 43: 35, Sept. 1961.
23. A. G. A. *Historical Statistics of the Gas Industry.* New York, 1961.
24. "World Petroleum Report." *World Petrol.,* Feb. 15, 1960.
25. "U.S.S.R.–East Europe Pipeline System." *Pet. Press Service* 28: 203, June 1961.

Chapter 3

Prospecting and Drilling for Natural Gas

by John W. Harrington, Neal Neece, Jr., J. E. Schaefer, and F. E. Vandaveer

PROSPECTING

Common Geological Terms[1]

Rock Types.

Agglomerate. Contemporaneous pyroclastic rock containing a predominance of rounded or subangular fragments larger than 32 mm diam.

Igneous. Formed by solidification of hot mobile material termed magma.

Metamorphic. All rocks formed in the solid state in response to pronounced changes of temperature, pressure, and chemical environment, taking place generally below shells of weathering and cementation.

Sedimentary. Rocks formed by accumulation of sediment in water (aqueous deposits) or from air (eolian deposits). Sediment may consist of rock fragments or particles of various sizes (conglomerate, sandstone, shale); of remains or products of animals or plants (certain limestones and coal); of products of chemical action or of evaporation (salt, gypsum, etc.); or of mixtures of these materials.

a. Limestone. Bedded sedimentary deposit consisting chiefly of calcium carbonate ($CaCO_3$) yielding lime when burned. Limestone is the most important and widely distributed of carbonate rocks. It is the consolidated equivalent of limy mud, calcareous sand, or shell fragments.

b. Sandstone. Cemented or otherwise compacted detrital sediment composed predominantly of quartz grains, their grades being those of sand.

c. Shale. Laminated sediment, in which constituent particles are predominantly of clay grade.

Structural Terms.

Anticline. Beds arched so as to incline away from each other.

Fault. A fracture or fracture zone along which displacement of two sides, relative to one another and parallel to fracture, has occurred. Displacement may be a few inches or many miles.

Isopach.[2] Line drawn thru points of equal rock thickness and entered on a map. Sedimentary basins are identified and studied by means of such maps.

Monocline. Strata that dip for an indefinite or unknown length in one direction, and which apparently do not form sides of ascertained anticlines or synclines.

Nonconformity, unconformity.[2] (1) Discordance in attitude with underlining rocks, due to overlap or deposition lapse, during which rocks beneath were deformed, eroded, or both. (2) Contact surface between unconformable strata and rocks beneath them.

Plunge, pitch, rake.[2] (1) In surveying, to set horizontal cross wire of a theodolite in direction of a grade. (2) To turn transit telescope on its horizontal transverse axis. (3) Vertical angle between a horizontal plane and line of maximum elongation of a body.

Salt dome. A structure resulting from upward movement of a salt mass, with which oil and gas fields are frequently associated. In the Gulf Coast area, salt is in form of a roughly cylindrical plug of relatively narrow diameter, but often several thousand feet in depth.

Sand lens. A body of sand in general form of a lens, thick in central part and thinning toward edges.

Stratigraphic trap. A trap resulting from variation in lithology of reservoir rock, a termination of reservoir (usually on updip extension), or other interruption of continuity.

Structural contour. A contour line drawn thru points of equal elevation on a stratum, key bed, or horizon, for depicting attitude of rocks.

Structural trap. Where entrapment results from folding, faulting, or their combination.

Structure section. Diagrams to show observed geological structure on vertical faces or, more commonly, inferred geological structure as it would appear on sides of a vertical trench.

Syncline. A fold in rocks in which strata dip inward from both sides toward axis. The opposite of anticline.

General Terms.

Casinghead gas. Unprocessed natural gas produced from a reservoir containing oil. Such gas contains gasoline vapors and is so called because it is usually produced under low pressure thru an oil well casing head.

Chert. Cryptocrystalline varieties of silica of any color, composed mainly of petrographically microscopic chalcedony and/or quartz particles with outlines ranging from easily resolvable to nonresolvable under binocular microscope at magnifications ordinarily used. Particles rarely exceed 0.5 mm diam.

Connate water. Water entrapped in interstices of a sedimentary rock at time it was deposited.

Table 4-6 The Geologic Time Scale[4]

Era	North America			Europe			Age, years
	Period	Epoch	Age	Age	Epoch	Period	
	Quaternary	Recent pleistocene			Pleistocene		10,000
		Pliocene			Pliocene		
			Upper			Neogene	
Cenozoic		Miocene	Middle		Miocene		
			Lower				1 million
	Tertiary	Oligocene			Oligocene		
			Jackson				
		Eocene	Claiborne		Eocene	Paleogene	
			Wilcox				
		Paleocene	Midway		Paleocene		70 million
			Montanan	Danian			
		Upper		Senonian	Upper		
			Coloradoan	Turonian			
				Cenomanian			
	Cretaceous					Cretaceous	
			Comanchean	Albian			
		Lower		Aptian	Lower		
			Coahuilian	Barremian			
				Neocomian			
		Upper		Malm	Upper		
Mesozoic	Jurassic	Middle		Dogger	Middle	Jurassic	
		Lower		Lias	Lower		
		Upper		Keuper			
	Triassic	Middle		Muschelkalk		Triassic	
		Lower		Bunter			260 million
		Upper	Ochoa	Thuringian			
	Permian	Middle	Guadalupe	(Artinskian)		Permian	
		Lower	Leonard	Saxonian			
			Wolfcamp	(Sakmarian)			
			Virgil				
		Upper	Missouri	Stephanian			
	Pennsylvanian	Middle	Des Moines	Westphalian			
		Lower	Morrow	Dinanthian			
						Carboniferous	
		Upper	Chester				
				Visean			
	Mississippian	Middle	Osage				
				Tournasian			
		Lower	Kinderhook				
		Upper	Chautauquan	Famennian	Upper		
			Senecan				
	Devonian	Middle	Erian		Middle	Devonian	
			Ulsterian				
		Lower	Oriskanian	Gedinnian	Lower		
			Helderbergian				
Paleozoic							
		Upper	Cayugan	Downtonian			
				Ludlovian			
	Silurian	Middle	Niagaran	Wenlockian	Gotlandian		
		Lower	Alexandrian	Llandoverian			
						Silurian	
		Upper	Cincinnatian				
	Ordovician	Middle	Mohawkian		Ordovician		
			Chazyan				
		Lower	Canadian				
		Upper	St. Croixan		Upper		
	Cambrian	Middle	Albertan		Middle	Cambrian	
		Lower	Waucoban		Lower		500 million
	Proterozoic						
Pre-Cambrian							
	Archeozoic						2–3 billion

Dip. Angle at which a stratum or any planar feature is inclined from horizontal. Dip is at right angle to strike.

Horizon. Surface separating two beds, hence having no thickness.

Lithology.[2] Branch of geology dealing with study of rocks. (Data from microscopic examinations of rock sections are as valid for most purposes as data from chemical analysis.)

Migration. Movement of oil, gas, or water thru porous and permeable rock.

Open flow. Rate of flow of a gas well into atmosphere, unrestricted by any other pressure. It is usually stated in units of cubic feet per 24 hr.

Permeability. Capacity of rock for transmitting a fluid. Degree of permeability depends upon size and shape of pores, and size, shape, and extent of their interconnections. It is measured by rate at which a fluid of standard viscosity can move a given distance in a given interval of time under a given pressure difference. Unit of permeability is the Darcy.

Pool. Figurative term for an underground accumulation of oil or gas in a single and separate natural reservoir (ordinarily of porous sandstone or limestone), characterized by a single natural-pressure system and bounded by geologic barriers, such as structural conditions, impermeable strata, changes in porosity, or water in formations.

Porosity. Ratio of aggregate volume of interstices in a rock or soil to its total volume, usually stated as a percentage.

Reservoir. A natural underground container of fluids, such as oil, water, and gases. In general, such reservoirs were formed by local deformation of strata, by changes of porosity and by intrusions.

Reservoir energy. Energy within an oil or gas reservoir which causes oil, gas, and water to flow into a well.

Reservoir pressure, rock pressure, static-fluid pressure, bottom-hole pressure. Terms applied to pressure of liquids or gases confined in rock, usually referring to original pressure in reservoir. (The most accepted theory holds that this pressure derives from hydrostatic pressure of the water underlying the gas accumulation. If the reservoir is continuously permeable to its outcrop, then the hydrostatic pressure will be roughly equivalent to that of a vertical column of water extending from the oil-water interface to the ground-water table, and reservoir pressure will generally increase with the depth in an almost linear manner. Approximate pressure in psi may be predicted by multiplying the depth in feet from the ground water table to the oil-water interface by the factor 0.435. There are, however, many exceptions to this rule, particularly below 7000 ft depth.)[3]

Strike. Course or bearing of outcrop of an inclined bed or structure on a level surface; direction or bearing of a horizontal line in plane of an inclined stratum, joint, fault, cleavage plane, or other structural plane. Strike is perpendicular to direction of dip.

Geologic Age of Reservoir Rock[4]

This means of classifying fields and provinces, Table 4-6, may also serve to classify the characteristics and productivity of the petroleum contained therein (Table 4-7). Of course, the petroleum may have accumulated in a pool subsequent to the original reservoir rock formation.

Table 4-7 Relations of Geologic Age, Gravity, Number of Pools, and Oil Reserves in the United States*

Age	API gravity	Oil pools: Number	Oil pools: Per cent of total	Past production plus proved reserves: Million barrels	Past production plus proved reserves: Per cent of total
Pliocene	22	70	.9	5,218	8.6
Miocene	26	410	5.4	7,337	12.2
Oligocene	30	410	5.4	6,048	10.0
Eocene	36	420	5.5	2,532	4.2
U. Cretaceous	33	440	5.8	8,518	14.1
L. Cretaceous	34	290	3.8	1,546	2.6
Jurassic-Triassic	38	116	1.5	799	1.3
Permian	33	400	5.3	7,904	13.1
Pennsylvanian	38	1960	25.7	10,097	16.7
Mississippian	38	1610	21.7	3,377	5.6
Devonian and Silurian	39	615	8.1	2,674	4.4
Ordovician	40	825	10.8	4,267	7.1
Cambrian	33	55	.7	59	.1
Pre-Cambrian		3	—	3	—

* Adapted from: Hopkins, G. R. "Projection of Oil Discovery 1949–1965." J. Petrol. Technol., Vol. 2, June 1950, p. 7, Table I.

Surface Geology

Techniques of oil and natural gas discovery in early days consisted chiefly of hunting for oil or gas seepages; for outcrops of bituminous rocks or of fissures filled with congealed hydrocarbons; for natural paraffin, called ozocerite; and for mud volcanoes, dikes, salt, and sulfur water seepages. Wildcat wells were then located as near to such favorable occurrences as possible. The long list of productive fields discovered is eloquent testimony to the success of these methods, but there were many instances of failure also. In recent petroleum geology practice, considerable importance has been attached to surface indications of oil and gas, especially when occurring in regions of relatively young rocks, i.e., Tertiary age.

Surface geology produces the most detailed understanding of exposed parts of structures by the very nature of continuity of observation. A geologic map synthesized from these data is in reality a cross-sectional diagram made possible by truncating effects of erosion. Thicknesses are only apparent since

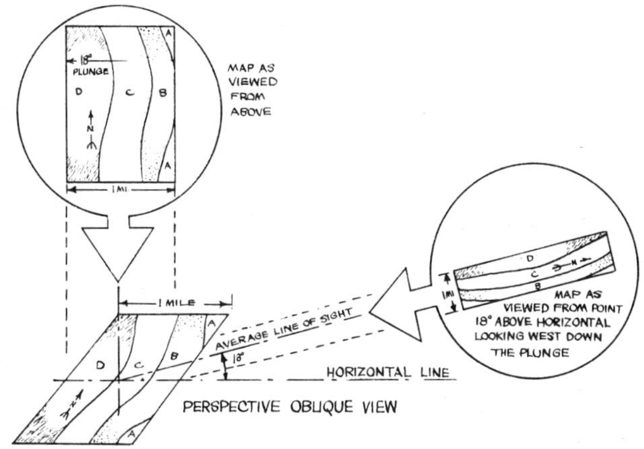

Fig. 4-6 How to read a geologic map as an approximate structure section.

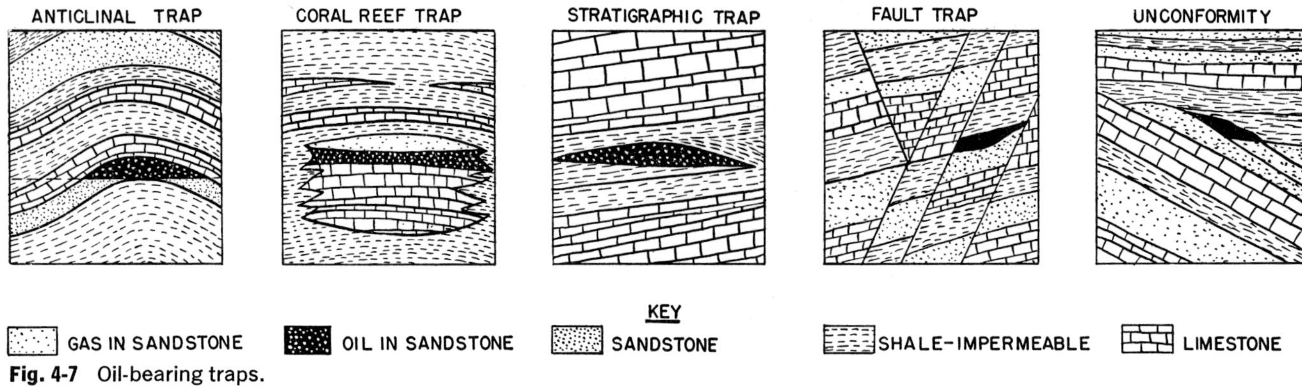

Fig. 4-7 Oil-bearing traps.

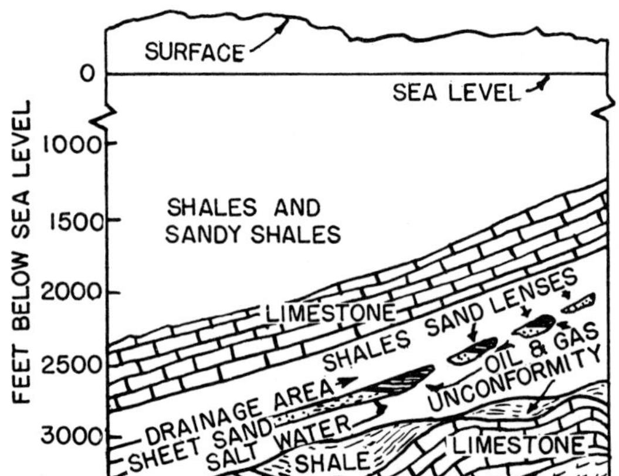

Fig. 4-8 Oil and gas accumulation in a monocline. (The reservoir is a sheet sand pinching updip. Note the sand lenses which represent irregular shore line deposits.)

topographic slope and angle of beveling impose artificial values; i.e., true thicknesses would be evident if the topographic surface were horizontal and the beds dipped vertically. A relatively true scale section may be seen by orienting the map so that the observer looks directly down the dip, or preferably the plunge of the folds. Figure 4-6 illustrates this analytical procedure.

Surface geology may often be extrapolated into the subsurface on the basis of dip projection. The difficulty is that minor variations of true dips from assumed values are sources of considerable error. Furthermore, existence of an unconformity or a fault (as shown in Fig. 4-7) not recognized on the surface may completely change the situation. Oil or gas may be trapped as illustrated in Fig. 4-8.

Aerial photography is another method of recording geologic data. As far as rock exposures themselves are concerned, they must be considered as a base for a surface geologic map and subject to some limitations. Photographs taken from a somewhat removed vantage point are capable of showing details of structure, e.g., significant faults and minor faults not evident to the land-bound observer.

Geochemical Exploration Methods

Geochemical methods of exploring for gas or oil consist of making chemical tests for hydrocarbons in the soil and mapping the results. These methods are based on the theory that gases associated with oil and gas deposits will penetrate and spread upward thru overlying rock formations and remain in soil near the surface. Soil or gas samples are taken from holes in a selected area. Holes are usually spaced about 1000 ft apart, or closer when advisable.

In the gas analysis method, gas samples are collected from holes 10 to 30 ft deep, and analyzed for methane, ethane, and heavier hydrocarbon content. Accurate and sensitive analytical procedures involving infra-red mass spectroscopes, low temperature fractionation, or other micro-analytical devices must be used.

In soil analysis, soil from depths of several feet is analyzed for hydrocarbon content. Since gas may be found in parts per million by weight, great care must be taken in this analytical procedure.

Percentages of hydrocarbons found are plotted on maps, using a single map for each type of hydrocarbon. From these plots the location of commercial gas and oil deposits may often be determined. A *halo* may be observed on the maps, with maximum concentration around the periphery of the deposit. This method, although reported to be widely accepted in Russia, has appeared to yield poor results in the United States. Unreliability seems due to varying soil capacities for gas retention and to formation of hydrocarbons by organic soil material.

Geophysical Prospecting Methods

The following methods, which depend upon variations in physical properties between reservoir rocks and neighboring rocks, have been widely used. They locate rock structure and formations favorable to containment of gas or oil.

Gravitational Method. Slight variation in the force of gravity due to difference in density of matter in porous reservoir rock and dense basement rock can be detected with a gravity meter (Fig. 4-9). Modern gravity meters are sensitive to two or three parts per 100 million in measurement of the force of gravity. The unit commonly employed is the milligal., 0.001 gal. A gal. (named after Galileo) is a unit of acceleration in the cgs system, i.e., 1 cm per sec-sec.

A torsion type gravity meter was originally used, but nearly all now in use are adaptations of the unstable, spring-supported principle (Fig. 4-10). These instruments require less time to set up and the data are easier to interpret than readings obtained with torsion balances.

Magnetic Method. This method depends upon variations in the magnetic susceptibility of different rock types. The magnitude of the magnetic field intensity is directly proportional to the susceptibility contrast between rocks. A salt

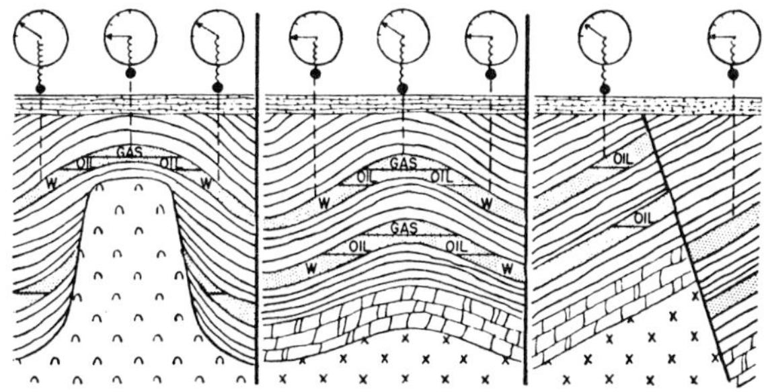

Fig. 4-9 Use of the gravity meter. Geological structures cause slight variations in the force of gravity. The salt dome, shown at left, is a prolific source of oil accumulation along the Gulf Coast of Texas and Louisiana.[7]

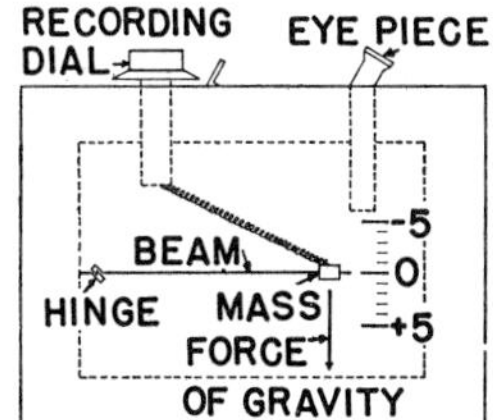

Fig. 4-10 Gravity meter.

dome has a very small negative susceptibility contrast with the surrounding sediments, and therefore provides a poor target for detection by magnetic methods, unless it occurs at shallow depths of the order of 1000 to 2000 ft.

The instrument employed for detection of changes in the earth's magnetic field is called a magnetometer. Instrument sensitivity varies, depending upon the type, from about 0.01 to 10 gammas, and the instrument can be airborne.[5,6]

This method is quite useful for determining such things as the areal extent and configuration of a sedimentary basin, major faults involving the basement, and regional basement uplifts. Although the method does not always provide a positive answer to a given question, it may yield a positive answer when used with other methods, such as gravity.

Seismic Methods. Reflection and refraction seismic methods, although expensive, are the best known as well as most popular and productive of all geophysical techniques They have been instrumental in discovery of five billion barrels of oil.

When a charge is set off, three types of shock waves are set up: longitudinal, **P**, transverse, **S**, and surface, **Rayleigh**, waves. The last-named travel comparatively slowly along the surface and at boundaries between rock strata. The former types are partially transmitted thru strata and partially reflected back toward the surface from strata boundaries, **P** waves being faster than **S** waves. Wave velocity varies considerably with rock type and increases with depth. These principles are the bases of seismic methods. Velocity of transmission of elastic waves is 6000 to 13,000 ft per sec in shales, 15,000 to 21,000 ft per sec in limestones, 8000 to 14,000 ft per sec in sandstones, and 15,000 to 17,000 ft per sec in salt.

In the **refraction technique** only longitudinal waves are used. Stations one-half mile apart are established, the farthest from three to six miles from the explosion of a buried dynamite charge. Station spacing is usually at least four times the depth of the strata to be studied. First longitudinal wave signals come from the denser, deeper formations, where velocity is greater. Wave paths are slightly curved because velocity varies with depth. By making a series of observations, a contour map of the higher-speed strata can be prepared. Depth of penetration of refraction surveys is limited in most areas to a few thousand feet.

Reflection technique has largely superseded that of refraction, because it is superior in depth of penetration, accuracy, and speed. Strata as deep as 15,000 ft have been studied. Location of strata boundaries and contours are determined by the time required for shock waves, set up by explosion of a buried dynamite charge, to travel down to and back from such boundaries. Pickup stations which record the waves are spaced from 100 to 300 ft apart. Several receivers in a line may be used to differentiate between reflected and refracted waves (Fig. 4-11). The optical principle of equality of angles of incidence and reflection is applied in the calculations.

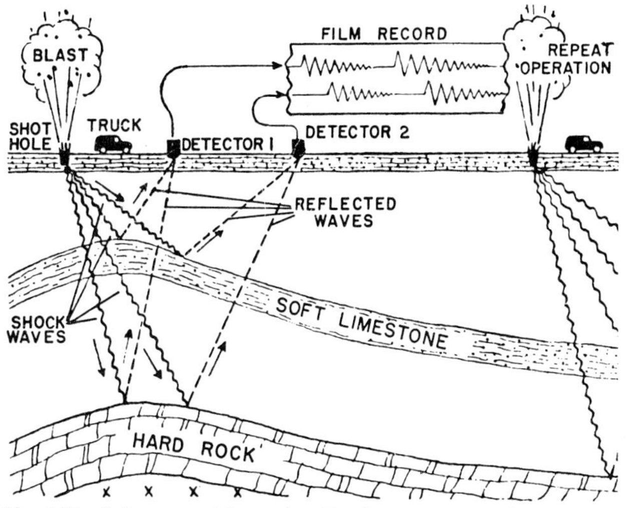

Fig. 4-11 Seismographic exploration.[7]

Time required for waves to reach pickup stations is determined by means of a recording device sensitive to vibrations. Its recording chart is mounted on a moving drum, and 0.001 sec intervals are marked off by means of a tuning fork. Character of disturbances and actual time lapses are evaluated from the chart tracing, called a seismogram.

Vibroseis is a technique involving a surface vibrator system that generates controlled seismic waves for recording.[8]

Electrical Resistivity Method. This was one of the earliest methods of studying formations of subsurface stratum used by the mining industry and to a lesser extent by the oil industry. An electric current is introduced into the earth and the potential difference produced between separated electrodes is measured. A change of resistance caused by the subsurface formation can be calculated using the following three experimental results: (1) area traversed, (2) amount of current injected, and (3) difference in potential experienced.

This method requires from six to twelve men to operate and is fairly expensive. Its use as a geophysical tool for discovery of oil and gas has been small, due to its limited measuring depth.

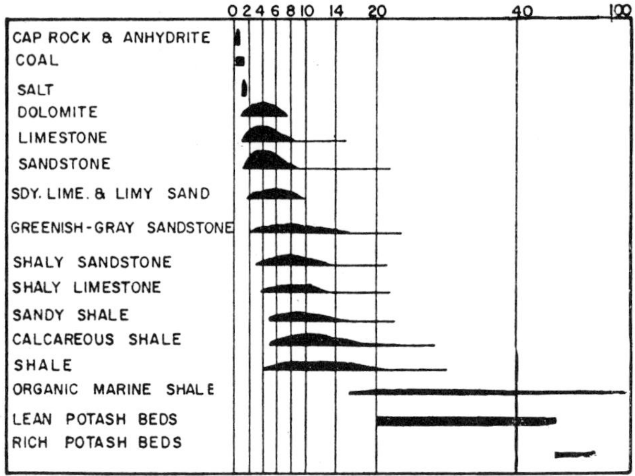

Fig. 4-12 Relative radioactivity of various rocks.[9,12]

Radioactive Method.[9] Percolation of ground waters or gases tends to bring radioactive materials to the surface. By measuring radioactivity of soil gas, it was expected that zones of greater circulation could be detected and that the method would be particularly useful in locating faults. Considerable testing of this idea was carried out by passing gases, drawn from shallow holes, thru an electroscope or a scintillometer, and by testing soil samples in a special electroscope. In both cases, variation in discharge rate of the electroscope indicated variation in amount of radioactive material (i.e., the radioactive gas radon) in air or soil.

Numerous investigators have undertaken the task of measuring the radioactivity of different rocks and formations and classifying them. Such activity has been expressed generally in terms of the radium (or thorium) content of the rocks, per gram of rock (Fig. 4-12). The average radium and thorium content of rocks of the upper crust is possibly equivalent to 2×10^{-12} grams of radium per gram of rock.

Measuring the radioactivity of the rocks at the surface constitutes a method of geophysical exploration. Two types of measurements are used for the purpose. One procedure measures the radiation emitted by the ground; the other, the activity of the air pumped from the soil. An exhaustive discussion of the processes (ionometric method), including a number of references, is available.[10]

In addition to the surface survey made with gamma ray techniques, airplane surveys using scintillation counters are employed. There are thus several procedures which have adapted radioactive measurements for location of oil and gas. Considering that these are also a form of geochemical prospecting, it may be well to point out that all such methods are directly aimed at finding hydrocarbons; all other methods are intended to locate stratigraphic traps. However, the various "soil analysis" methods need considerable refinement to make them more workable. Even then, they would seem to be suitable primarily for rocks that are not too indurated. Considerably greater success with these methods was obtained in their application to the mapping of lithologic contacts.[11] This makes possible the recognition of hidden faults and the preparation of structural and areal geological maps.

Interpretation of Geophysical Measurements. Reliability and precision of geophysical measurements are high

Table 4-8 Relationships Between Geologic and Geophysical Measurements

Method	Units	Dimensions	Means of usual conversion to useful geologic measurements
Seismic	millisecond	$\frac{\text{sec}}{1000}$	Distance = Velocity × Time. In both refraction and reflection methods, time is measured between instant of energy release and first detected arrival. Vertical depths to key horizons are computed from trigonometric versions of the basic equation.
Gravitational	milligal.	$\frac{\text{cm} \times 10^{-3}}{\text{sec}^2}$	Localization by symmetry of anomaly field as shown in map view. Effect of anomalous structure is inversely proportional to square of distance of its center of gravity from measuring device. No unique solution is possible, but with geologically reasonable assumptions some definite structural conclusions may be justified.
Magnetic	gamma*	oersteds $\times 10^{-5}$	Localization is by symmetry of map view of anomaly field. The effect of a buried dipole with a moment given by the product of the pole strength (m) and the distance between poles (l) is inversely proportional to the square of the distance from the measuring device (R), when l is very much smaller than R. No unique solution is possible, but with application of geologically reasonable assumptions, some structural conclusions may be drawn, particularly with reference to basement features.
Resistivity & self potential logging	ohmmeter & millivolt	ohmmeter & volt $\times 10^{-3}$	Comparison is made of relative values for similar and dissimilar lithologies. Depth is established by vertical position of device in drill hole. Instrument is especially sensitive to distinction between impervious shales and other fluid-bearing rocks where higher potentials are generated.
Radioactivity logging	older logs uncalibrated, new logs—microroentgen per hr	$\frac{83.8 \text{ ergs} \times 10^{-8}}{36 \text{ gr (of air) sec}}$	Comparison is made of relative values for similar and dissimilar lithologies. In earth formations, the natural gamma radiating materials—uranium, thorium, and potassium—normally are concentrated more in shales than sands and, as a result, the shale sections are more radioactive. In carbonates the concentration, and therefore radioactivity, is variable.

* Strictly speaking, magnetic field intensity is measured in oersteds, and magnetic flux density in gauss. In free space and in the cgs system of units, the flux density is numerically equal to the field intensity. Among professional geophysicists, the unit of measure of field intensity has been commonly called a gauss.

within limits of proper usage. Yet it is necessary to recognize that geophysical tools do not give direct readings—note units listed in Table 4-8. The basic geologic unit is the formation defined as a mapable entity such as a black shale, a red sandstone, an intricately bedded limestone suite, a metamorphic complex, or a granite mass. Physical measurements are not the same as those considered diagnostic in field mapping. Integration of the two systems will necessarily remain a somewhat qualitative transposition until an equation may be written to convert *black shale* (for example) into ohm-meters or milliseconds.

Applied to a problem of interpretation, proper use of geophysics may be considered to be controlled by the following principal limits:

1. Geologically, it is a last resort employed in those areas where there is insufficient knowledge.
2. Conversion from physical to geologic units requires thorough regional knowledge.
3. Ability to execute surface and subsurface geological mapping.
4. Geophysical systems, the **magnetic method** excepted, are quite responsive to shale thru one or another of its physical properties. Therefore analysis should be restricted solely to spatial dimensions of length, width, and depth defining forms and shapes.

Although geologists classify sediments on completely different criteria (limestones on composition, sandstones on grain size, shales on composition, thin bedding on the ability to be split), origin of most shale is recognizably unique. Sandstones and limestones generally have much larger grain sizes and were deposited in spite of rather high water velocities. Shales, on the other hand, consist almost entirely of grain sizes much nearer colloidal range. They are deposited only in those areas where water velocities are quite low. Thus separate depositional environments can be recognized even with geophysical tools.

5. **Logging methods** produce some of the most reliable interpretations because measuring devices have a limited range and are thus in proximity to materials tested.

Subsurface Methods of Well Logging

Subsurface geology depends upon information obtained from wells drilled for oil, gas, water, or essentially for geologic data. It consists of recognizing and following beds or series of beds from well to well, or from well to outcrop, and of determining and interpreting the instructure. Basic data are obtained as follows:

Drillers' Logs. Drillers keep records of depths, type of rocks, fluids, and other features encountered in drilling. Good drillers recognize and report key beds.

Sample Logs, Lithologic Logs, or Rock Sample Logs. Well cuttings, rock fragments, cores, and other samples from each formation are set aside for identification by geologists.

Time Logs. In rotary drilling, time records may be kept. It is assumed that under similar drilling practices the same beds will be drilled at comparable speeds in adjacent wells.

Micropaleontology Analysis. Study of fossil remains in rocks of different periods helps to identify strata. Foraminifera, algae, diatoms, spores, pollen, fish scales, and other fossil remains are associated with rocks of certain ages.

Electrical Resistivity Logs. This method is based on the fact that rocks vary in electrical properties, particularly if they are porous and contain fluids.[13] Underground water is normally saline and therefore quite conductive. Both compact rock and porous rock containing oil have high resistivities (except where the latter contains some salt water).

Electric logging has as its primary purposes the location of porous formations and indication of the nature of their fluid contents. A secondary, but extremely valuable use is the characterization and correlation of such data from well to well. In contrast to limestone and sandy shales, clays and shales yield fairly uniform records which are quite useful in correlation work.

Usually, three electrodes of known area and fixed separation are attached to insulated electric wire cables and lowered into the hole containing mud. A known and fixed current is applied, and a resistivity log is traced as apparatus is raised thru the hole. Depth is measured as cable is raised, rather than when lowered, to avoid error due to slack. One cable is connected to a source of current, the other pole of which is grounded near the well. The other cables are connected to a potentiometer at the surface. In accordance with Ohm's law, measured voltages for the condition of fixed current reflect apparent resistances. Actual resistances depend upon well diameter, bed thickness, mud resistivity, and electrode area and spacing. High resistance may mean a compact rock, a fresh water sand, or an oil sand. A moderate resistance means a loosely consolidated rock, while a very low resistance means a porous rock or sand containing salt water. The resistivity range encountered is roughly 100 to 10,000,000 ohm-cm.

Very close electrode spacing, as used in the *microlog*, gives more detail, but wider spacing gives truer values of wall rock character, since influence of the mud itself is less. It is common to record two or three resistivity curves corresponding to different spacing.

Gamma Ray Logs. The gamma ray log,[14] neutron log, and temperature survey are usually a part of one survey of a newly drilled well. Generally the bottom 200 to 300 ft of the hole in the area of expected gas sands are surveyed, but the entire well may be tested if desired. Gamma ray log and neutron log are run at the same time, followed by a temperature survey.

The gamma ray log shows natural radioactivity vs. depth. Use is made of the fact that all rocks contain some material with unstable atoms which undergo breakdown into simpler particles while emitting radiation similar to ordinary X-rays. Atomic breakdown is termed radioactivity. The slow natural gamma or X-ray emission of rock substance in different strata is measured by an instrument with surface leads as it is raised in the drilling hole. Success of gamma ray logging depends on difference in concentrations of radioactive material in different rock types. Pure limestones and sandstones are low in radioactivity, while certain dark shales are relatively high. A great advantage is the ability to log thru steel casing, since X-rays can pass thru steel to some extent. A typical gamma ray-neutron-temperature survey of an Ohio well is shown in Fig. 4-13.

Neutron Logs. While radioactivity is inherent in some elementary substances, such as radium and uranium, it can be induced in others by neutron bombardment. A neutron

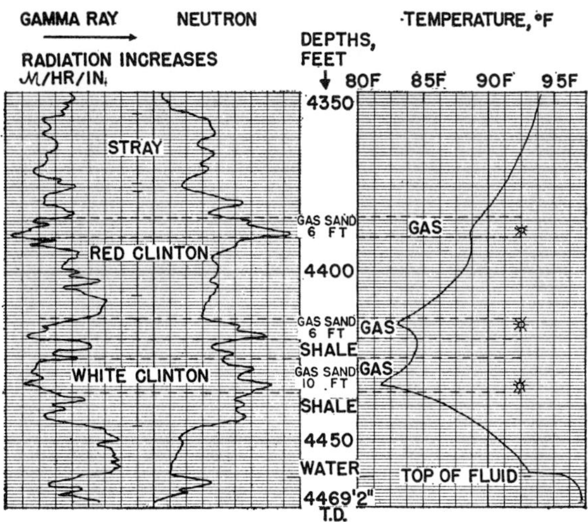

Fig. 4-13 Gamma ray-neutron-temperature survey of a newly drilled gas well showing presence of gas in three places at depths below 4350 ft.

is an electrically neutral atomic fragment, of about unit atomic weight, produced by certain atomic disintegration processes. It possesses considerable penetrating power. In practice, neutrons from a capsule containing a radium-beryllium alloy are allowed to bombard the rock substance, which then emits induced as well as natural radiation. Radiation is picked up by a probing instrument, which is raised in the hole at a rate of about 5 fpm. Well fluids containing hydrogen (water and hydrocarbons) are readily detected, because hydrogen is very effective in slowing down or stopping neutrons, without inducing artificial radioactivity. A neutron log of one Ohio well is shown in the center curve of Fig. 4-13.

Temperature Surveys. These measurements of temperature at various depths are of great value in locating entry of gas or oil into the wellbore, in locating leaks in tubing or casing, and in detecting migration of fluids back of casing. Fluids from different horizons have measurably different temperatures, so circulation effects can be detected. Expanding gas undergoes self-cooling, thus permitting location of escaping gas. Three instances of drop in temperature appear on the right curve of Fig. 4-13, denoting gas flow. In another well, with a very large gas flow from 1400 psi rock pressure, temperature opposite the gas sand dropped from 100 F to 16 F in the drill hole.

The last three methods described have been very valuable in the Appalachian area in finding both the thickness and depth of gas sands, as well as layers of shale and gas sand, difficult if not impossible to evaluate by other methods. A few instances have occurred, however, where such logs have indicated no gas flow, whereas after hydraulic fracturing, entry into the gas area was made and an acceptable flow produced.

Reliability Limits in Subsurface Correlation

Data on subsurface geology are admittedly very meager. Even in well-drilled areas observation points are a quarter mile or more apart. Subsurface mapping from such scattered observations is not fully reliable. Rock records usually consist of small chips secured from an 8-in. hole and floated to the surface. Samples must be taken regularly or the accumulated record may be misleading.[15] Cores are usually quite scarce and rarely oriented. This is unfortunate because cores furnish reliable data about primary structures destroyed in drilling.

In all but two respects well sampling leaves much to be desired and compares unfavorably with surface information. It does, however, provide an insight into distribution of rock units in space and permits direct observation of fluid content of rocks from drill stem tests and core analyses.[16–18]

Subsurface Structural Contouring

Figure 4-14a shows that there should be a crowding of contours at the inflection point and a spreading out toward crest and trough. This means that equispaced contours are not natural and any map using them as a final interpretation will be in error to the extent of variation between actual and assumed curves. The curve of Fig. 4-14a is much steeper than those of any but Appalachian type folds. Nevertheless, the much gentler anticlinical flexures of the Mid-Continent region will still follow the general contour pattern.

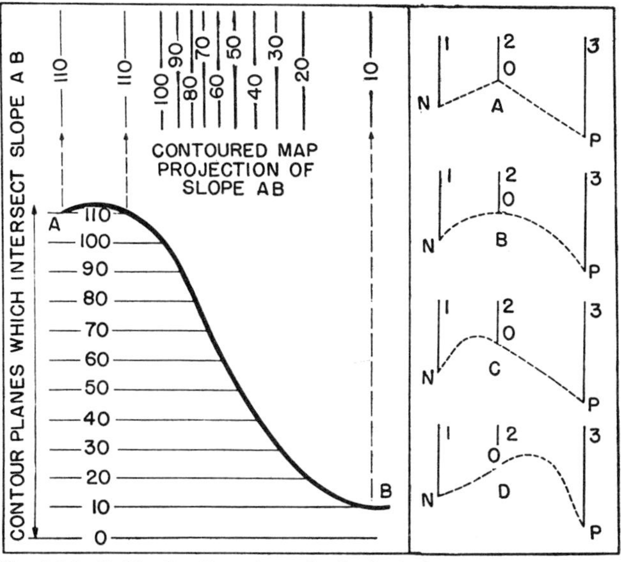

Fig. 4-14a (left) Section view of a typical slope curve.
Fig. 4-14b (right) Interpretations of structure section based on the same three data points.

Figure 4-14b shows three wells, 1, 2, and 3, topping a given horizon at points N, O, and P. Incompleteness of data requires some interpretation of the position of the crest of a structure since the most correct shape cannot be drawn by equal division downward from the highest known point.

Diagram **A** may be ruled out as inaccurate. Actual choice must be made between diagrams **B**, **C**, and **D**. If in reality point O is the crest, diagram **B** is the best. If O is not the crest, choice of either **C** or **D** represents an assignment of extra value to land area on one side and the condemnation of that on the other. It is quite obvious from the Figure that the crest would be expected to lie on the side opposite the flatter slope; i.e., the greatest differential curvature is on the crestal side. There is no truly quantitative way to make this interpretation with physical certainty.

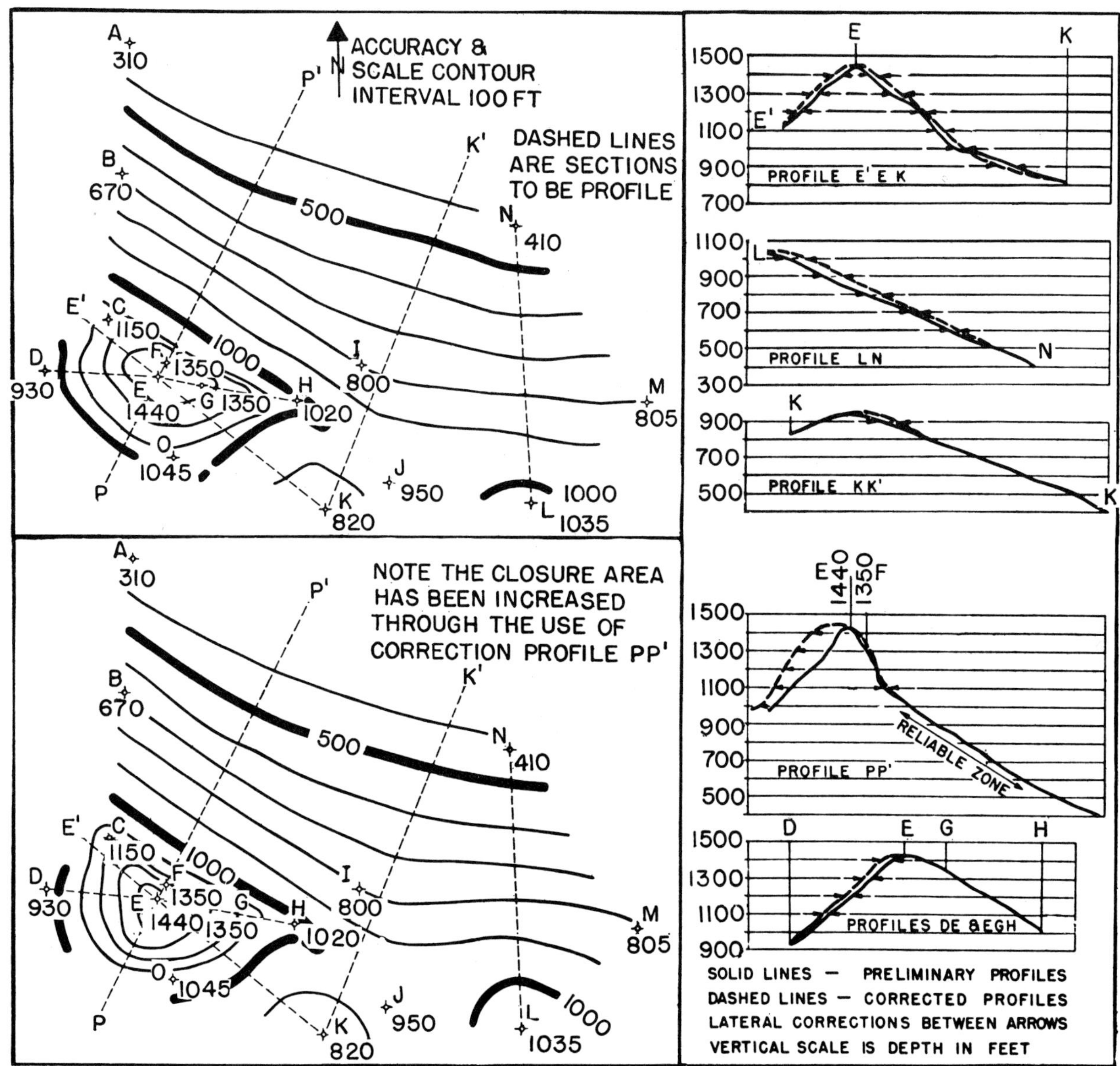

Fig. 4-15 (upper left) Preliminary subsurface structure contour map based on mechanical division.
Fig. 4-16 (right) Correction profiles used to prepare the final subsurface structure contour map.
Fig. 4-17 (lower left) Final subsurface structure contour map.

If a simple map is drawn, using mechanical division and equal spacing between data points, cross section profiles thru the surface will show a composite of straight line segments. It is possible to superimpose on these segments reasonable curved line relationships, keeping the regional character and maximum slopes in agreement. Positions of contour lines may then be shifted from a preliminary determination on the straight line segments to corresponding positions on the curves. Figures 4-15, 4-16, and 4-17 illustrate just such a process of preliminary contouring, profiling, adjustment, and recontouring.[19] Discussions concerning contouring and presentation of all subsurface data are given by Low.[20]

Types of Traps

A system of classifying various types of petroleum traps, useful in estimating size and shape of a partially proved reservoir, was reported.[21] The system lists elements of data necessary for recognition of each type. An outline of this system of classification of reservoirs is given by Table 4-9 and Fig. 4-18.

Table 4-9 Classification of Petroleum Reservoirs[21]

Trap or type of Reservoir*	Lettered diagram, Fig. 4-18	Typical examples
Convex	a and b	Anticlinical structures, convex thickening of beds, reefs.
Permeability	c	Variations of permeability in the same lithologic unit. Facies changes to different lithologies in same stratigraphic unit.
Convergence	d and e	Depositional pinchouts and truncational pinchouts.
Fault	f	Trapping by interruption of lateral permeability thru faulting other than piercement.
Piercement	g	Salt dome and dip piercement.

* Combinations of types shown may be formed by coexistence of two or more conditions.

Analysis of Data

In Fig. 4-18 each solid vertical line under a drilling rig represents a single random discovery thru the particular trap structure shown. Each diagram is distinctive because there are characteristic correlation limits and angular relationships between the beds themselves and between the beds and the drilled hole. If it were possible to read the same angle values at the position of the drill hole, at least part of each structure could be reconstructed. Devices are available which can perform this function. However, the dips measured in small holes may vary considerably from the average for the bed. A small diameter hole cannot show more than a local value unless the local and broader area values are the same.

For the reason given, the fact of discovery alone does not define the shape, size, character, or value of the reservoir. Other data from sources already described must be integrated before these properties can be determined.

Dashed lines of Fig. 4-18 under the dry hole symbol represent old subsurface control points thru areas east of the discoveries. Correlation is the first step toward integrating the old and new data. Beds must be identified. Continuity of the producing bed in both wells will limit trap possibilities to those of diagrams **a**, **b**, and **f**, or a porosity pinchout in the type **c** trap. Here the mechanics of subsurface structural contouring is the key. Accurate contour maps with cross section profiles will show whether curvatures are too extreme or quite expectable. If they are extreme and far beyond the regional range, suspect a fault, or miscorrelation. If they are too flat to fit drill stem test data, which may require two water levels, then greater curvature, separate flexures, and consequent loss of reservoir size are implied.

The ability to draw truly accurate structural maps is a powerful tool if used analytically and critically, but such maps cannot be drawn from the knowledge of only two points. Distant controls are of limited importance, as shown previously under limits of reliability.

Surface mapping may locate the anticlinal crest thru slight changes in the degree of dip, even if not in actual reversal. This latter relationship is quite common in minor flexuring accompanying compaction effects at great depth and somewhat suggestive of the trap in diagram **b**. Surface expression of faulting shown in aerial photographs is a possible key to the trap of diagram **f**. Geophysical tools have usually been applied by the time the discovery is made. Since they enjoy their greatest sensitivity in resolution of variations in vertical position, these data may serve to define the trap. The reflection seismic method is recognized as the most precise in this respect.

Definition of other types of traps is perhaps even more dependent on a knowledge of the regional geology. Regional evidence alone may determine presence of a permeability trap, but other data are needed for selection among those of diagrams **c**, **d**, and **e**. Since their fundamental element is the convergence, regional isopachous maps may indicate existence of these types. More data of a strictly local nature are always helpful. Reworking of the seismic interpretation on the basis of correlations obtained in the discovery well may serve to distinguish between these various trap types.

The real problem is that of estimating the updip and downdip extent beyond the discovery. Truly accurate cross sections drawn on the line of the dip will evaluate traps **d** and **e**. The facies change is beyond analysis in that manner. Until complete basic research involving variations of depositional velocities with grain size and lateral position in a bed have been made, solution will remain a matter of opinion and dependent upon more drilling. There are several partial approaches which should suggest reasonable interpretations.[22–25]

Specific solution for a type **g** trap flanking a salt dome may best be made with the gravity meter.

No panacea to geologic exploration exists and each case requires its individual solution. Basic principles herein described have proved useful and may be expected to develop reasonable approximations. This is possible only if the individual applying these principles is able to recognize and isolate the correct anomaly with incomplete information.

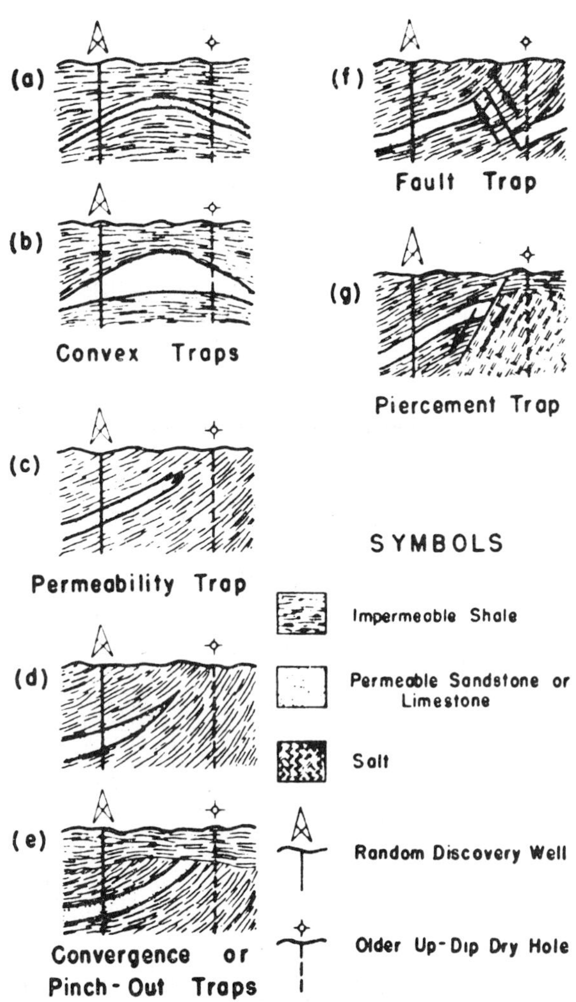

Fig. 4-18 Essential types of traps—after Wilhelm.

DRILLING[17,26]

Only by drilling can it be determined whether a favorable location actually contains oil or gas.

A drilling rig consists of a derrick, machinery, and a housing to protect equipment and workers. Nature of rig and drilling tools depends largely upon character of rock formations involved, and depth of gas- or oil-bearing formation. In the Appalachian district older percussion-type cable tools are normally used for depths down to 5000 ft, either cable or rotary for 5000 to 7000 ft, and rotary for over 7000 ft. In other sections, both types are employed; to keep the sand face clean, a rotary drill is used almost to the gas-bearing formation, and cable tools may be used for "tailing-in" operations. Generally, cable tools are used to make shallow holes while rotary tools are used for intermediate to deep holes.

Land and Leases

Except where otherwise stated by law, ownership of land includes the subsurface minerals. The owner has the right to dispose of these, provided he does not waste natural resources or become a public nuisance. In some cases, the owner has leased coal rights to one company, oil rights to another, and gas rights to a third. Federal and state governments regulate production and handling of oil and gas for conservation purposes and for protection of correlative rights of land owners.

Almost all natural gas comes from lands in which the producing company has no property rights other than to explore for and produce gas and oil. The latter rights include not only removal of oil and gas from the premises, but also placing, constructing, and maintaining machinery, structures, and pipelines necessary for production and operation. After the company geologist has determined what places are likely to be productive, it is the duty of the company's land or lease department to obtain rights to drill and produce, since it is usually more economical to lease exploration and production rights than to buy the land outright.

A lease is a contract between the owner (lessor) and operator (lessee), which confers on the operator rights to drill for and produce oil and gas, in exchange for some form of compensation (fee or royalty). Set forms of leases are employed. Provision is made that royalty begins with production. Although its amount may be fixed on any mutually agreeable basis, it is generally a portion, commonly ⅛ of production, or the equivalent in cash. It has been said that ⅛ of production corresponds roughly to the limit at which the lessee can make a reasonable profit. It is usually agreed that the lease shall remain in operation for a stated period of time, say five to ten years, and as long thereafter as production is maintained. Provision is commonly made for its expiration unless a well is drilled within the stated time, although delay clauses may be included.

Sometimes a fee or bonus is paid the lessor on signing of oil and gas leases. Certain other restrictions or liabilities are often imposed on the lessee, such as not drilling too close to buildings, and granting free use of gas by the lessor, up to a certain amount.

A company will normally have a considerable amount of unoperated land on long-term leases, since acreage beyond that immediately required is needed for future operations. Such acreage is often maintained to prevent others from drilling on the edge of a block (thereby spoiling the spacing pattern) and to have production rights in the event that a field proves larger than originally estimated.

Unitization

When more than one group has rights in a given field, unitized operation is a virtual necessity for fair profits. Under a unitization system, each operator and royalty-owner receives a fixed fraction of the revenue from the entire field. In reaching this formula, an effort is made to consider fairly all significant factors, such as the reserves and deliverability of each section. Commonly, a single operator, in the interest of economy, is designated to process all gas from the field. Unitization can be required in some states under certain conditions. It is almost a necessity when gas injection is practiced, as in storage, pressure maintenance, or repressuring.

Cable Drilling

Tools. *Standard* or churn drilling tools have been in use for several hundred years in drilling for fresh water and brine. Rock is fractured by pounding with a bit, and fragments, suspended in water, are removed by a bailing device. *Percussion-type* tools now used consist of a drilling bit, stem jars, and rope socket. These devices, called the *string*, are joined in the order named. The string is attached to a wire cable, by which it is raised and lowered in the well.

The *bit* is usually 6 to 8 ft long, irregular in cross section, and somewhat wider than thick. It is available in several styles and widths, and must be reshaped as it wears. It must be light for ease of handling, but heavy enough to permit drilling at a reasonable rate. Necessary weight is supplied by a steel bar called the *stem*, which also functions as a guide. The stem varies from 6 to 42 ft in length and from 2½ to 6 in. in diameter. *Jars* are bar links, which are useful for loosening jammed bits. They are normally extended, due to line tension, but may open when the tool sticks. The drilling line is fastened to the rope socket, which permits tools to rotate without twisting the line. Devices are available for removing tools if the cable snaps.

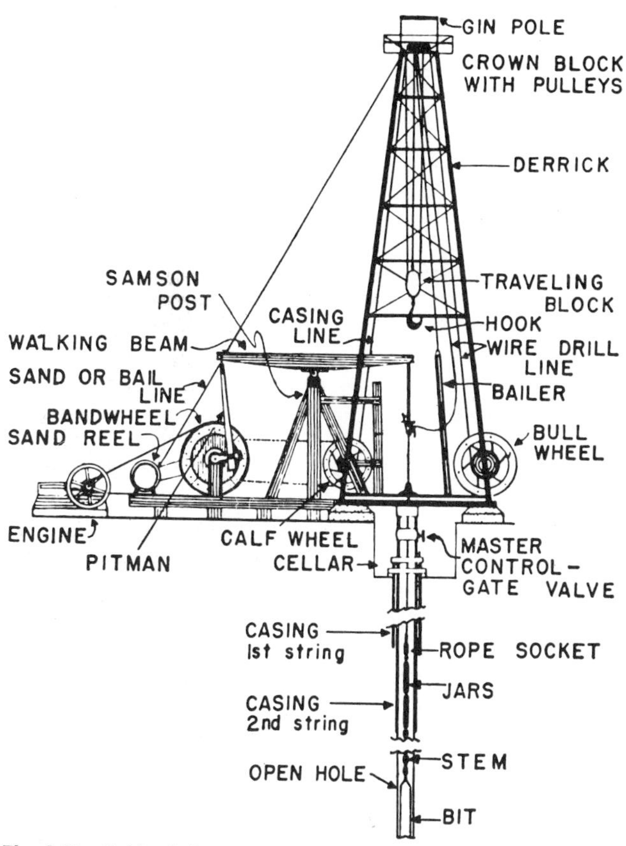

Fig. 4-19 Cable rig.[7]

Rig. The cable rig consists of a prime mover, band wheel, sand reel, calf wheel, bull wheel, walking beam, derrick, and pulleys or sheaves (Fig. 4-19). The prime mover rotates the band wheel. A pitman connects one end of the walking beam, which is simply a long member pivoted in the center, to an eccentric on the band wheel shaft. The up-and-down motion thus supplied is communicated to the drilling tools hung from the temper screw attached to the opposite end of the walking beam. As the bit works its way down into the hole, the drilling line connecting the tools to the temper screw is lengthened by *playing out* the adjustable temper screw. The screw can be reset by raising and reclamping. This operation may be repeated several times before it is necessary to increase the line length.

One end of the drilling line is spooled. The line passes from this spool over a sheave on the crown block at the top of the derrick and down to the temper screw. The bull wheel is driven by lines from the band wheel. The bull wheel is also used for raising the tools, as is necessary, for example, when the bit must be reshaped. The calf wheel, also driven from the band wheel, is used for handling and supporting casing. The casing line is spooled on it and passes thru the crown and traveling blocks. A pipe clamp is provided for the hook on the bottom of the traveling block. The sand reel, upon which the sand line is spooled, is used for lowering and raising the bailer. It is driven from the band wheel. Various modifications of the arrangements described may also be employed.

Procedure. Before the hole is deep enough, say 60 to 100 ft, to permit use of the walking beam and regular tools, it must be *spudded in* by use of a smaller bit and a jerk line with one end attached to a crank on the band wheel, and the other end attached to the cable near the bull wheel. The jerking action causes the raising and lowering of the bit.

One or two barrels of water are introduced into the hole to suspend finely ground rock. After a certain amount of drilling, the tool is removed and a bailer is lowered into the hole. This device admits the suspension thru a bottom check valve, which opens when resting on the bottom of the hole, and closes on raising. The bailer is then withdrawn from the hole, and suspension is removed. From time to time some rock particles are retained as samples for the drilling or lithologic log. The log can then be studied for information on formations encountered. When the drilling tools are not in the hole, a core or solid cylindrical section can be removed by a device, commonly known as a core barrel, which cuts along only the periphery or circumference. The core is then used for study of the rock formations penetrated.

As drilling proceeds, wells are protected from caving and seepage by insertion of steel casing. A conductor or drive pipe is used from the surface to the first solid formation. Normally the surface string is set thru all potable water formations. Some uncased length of hole is necessary, however, to prevent the tools from hitting the casing. A considerable run of unprotected (open) hole is often used in hard formations, but only 20 to 30 ft of unprotected hole may be used where caving is serious. Cases have occurred where sealing off a series of formations with successively smaller casing led to such decrease in ultimate diameter that *running out of hole* was experienced.

Drilling is normally continued until a water-bearing formation is encountered. Casing is then set and cemented into place to seal off water from the hole, and drilling is resumed with a narrower bit until water is again encountered, at which point a narrower string of casing is cemented into place.

Sometimes a gas-bearing formation is sealed off because it is not economically productive. If one is encountered which is productive enough for commercial purposes, but is not the lowest formation producing gas, the well may be dually completed. In that case, gas from the lower formation is brought thru a tubing set on a packer, while gas from the higher formation is conducted thru the annular space between this tubing and casing or next wider tubing. Such separation is normally required in dual completions, because of pressure difference at the two levels. An effort is made to provide a tight seal between casing and hole by cementing this space. This provides added strength, increased corrosion resistance, and protection from blow-outs.

Rotary Drilling

Tools. Rotary drilling was first used in Texas in 1901, where poorly consolidated sands were encountered. These sands caused caving-in of formations when cable tools were used. Rotary drilling proved very effective in formations which were easily cut. It is usable for most types of rock, due largely to development of hard-rock bits.

Selection of bits depends mainly upon type of rock encountered. Fishtail-shaped bits are used in soft formations; drag bits, which depend on dragging an abrasive substance on the bit end over the rock surface, are useful in shale and soft sandstone. In all kinds of rock, various types of *rock bits* are best. They crush and grind their way thru by action of revolving-toothed cones or disks.

The bit is fitted with several lengths of extra-heavy drill pipe, known as collars, and is attached to the regular-weight drill pipe. Collars provide weight, strength, and stiffness, and tend to keep the drill pipe in tension, thereby helping to keep the hole straight. The drill pipe is rotated, and the bit is allowed to press against the formation. The swivel has the dual function of supporting the weight of the entire drill string and providing a pressure-tight seal under rotation for the drilling fluid. It remains stationary to accommodate the mud hose, except for the lower part, to which the drill stem is attached. This section rotates on bearings, and may turn at speeds of 50 to 300 rpm or more.

The bit is wider than the pipe to which it is attached, and is perforated. Water or mud is forced by pumps thru a hose and swivel joint, down thru the rotating drill pipe, and out thru ports in the bit, returning to surface by passing up annular space between drill pipe and formation. The fluid serves to cool and lubricate the bit, cool the formation, wash abrasive cuttings from bit and convey them to surface, and seal the sand face against intrusion or extrusion of water and gas, depending on conditions, until casing is set.

Rig. The rig consists essentially of a derrick, draw works, hoisting system, rotary table, and mud-circulation system (Fig. 4-20). The derrick is usually from 84 to 176 ft high, has a cross-sectional area at crown of $5\frac{1}{2}$ sq ft, and a working floor area at base of 20 to 32 sq ft. Draw works is a collective term for the hoisting drum and other operating machinery. The prime mover can be a steam engine, internal-combustion engine, diesel-electric, or turbine depending on local costs and availability of water and fuels. One hundred horsepower for every 1000 ft of depth has been suggested as an estimate of power required.

Hoisting and lowering are accomplished by means of a cable, spooled on a drum in the draw works. The cable is passed thru sheaves of crown block and traveling block. The rotary table, which receives its power from the draw works, rotates the drill pipe. This table may be driven by the same prime mover that is used for hoisting, but more flexible operation results from a separate source of power. Turning is accomplished by use of a square drive bushing in the table, thru which passes a length of square or hexagonal drill pipe, called the grief stem or kelly joint. Fit is loose enough to permit downward movement of drill stem as hole is made, and to allow an even pressure on bit. The regular drill pipe is joined to the kelly, so rotary motion of the table is communicated to pipe, thereby rotating bit.

Procedure. Successive sections of drill pipe, usually in 30 ft lengths, are coupled by disconnecting the kelly joint. These lengths, usually stored in a *mousehole* in the derrick floor, are joined to the top joint of the drill pipe in the hole. The kelly is then reconnected and drilling resumed. The derrickman on the elevated platform latches and unlatches elevators and guides the upper end of pipe. If a bit is replaced, the drill stem is hoisted in steps and disconnected at every second, third, or even fourth joint, as it is being withdrawn. Pipe sections are stacked, and the kelly is set in the *rathole*. Reassembly is done in reverse manner. Pipe may also be withdrawn for coring or in case of accident.

When making or breaking joints during running or pulling, the *string* of pipe is supported from the surface by means of a spider or forged steel ring with a conical opening and two lugs at opposite sides. Slips with horizontal grooves fit into the opening. When pipe is to be suspended, slips are dropped into place and the weight of pipe causes them to be tightly gripped.

Drilling speed and bit behavior are good rough guides as to type of rock being drilled. The operator also has available a rotary table tachometer or speed-of-rotation indicator, rotary table torque gage, mud-line pressure gage, and either a manual or automatic mud density indicator.

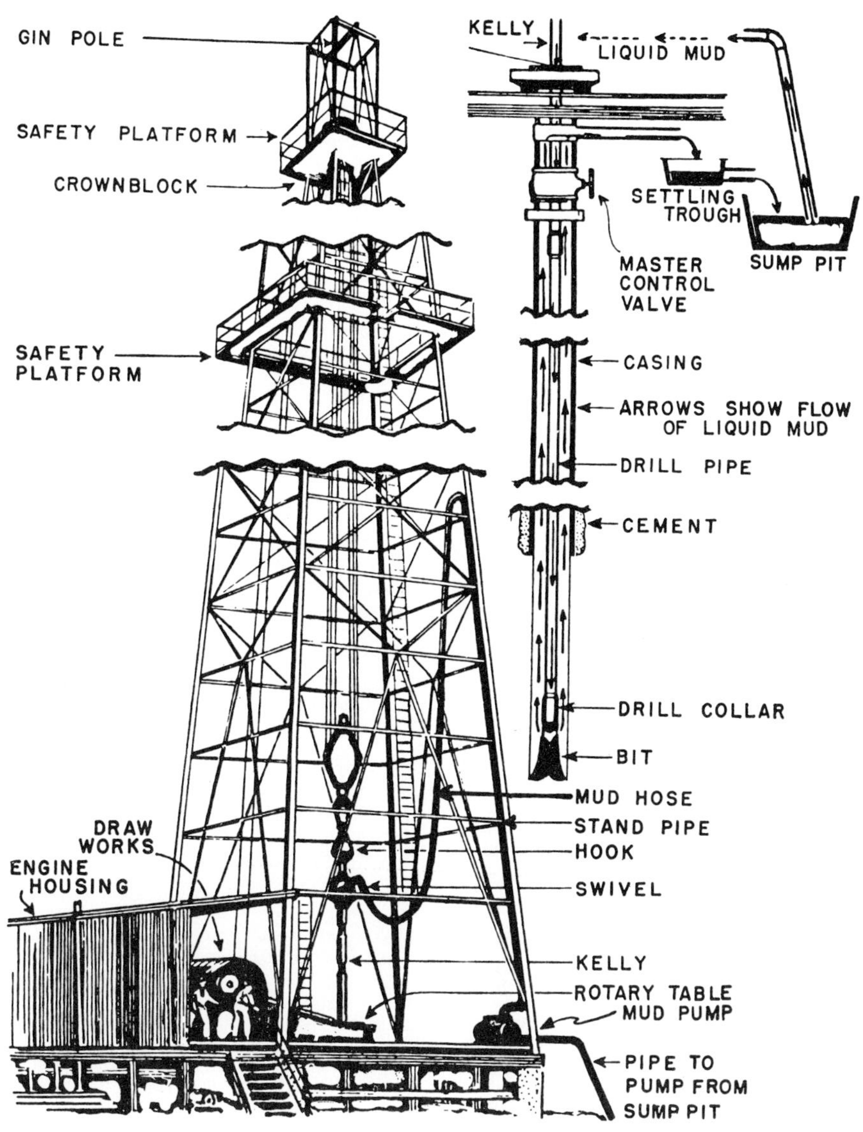

Fig. 4-20 Rotary rig.[7]

Mud and Mud Control. In rotary drilling, the ability of mud to seal water-bearing and porous formations permits the completion of a hole with only a few casing sizes, so that the minimum hole diameter exceeds that possible with cable tools. Mud is usually made from Wyoming-type bentonite, a clay that swells readily in water and has properties easily controlled by carefully selected additives. A typical mud may be 65 to 98 per cent water, 2 to 30 per cent clay, 0 to 35 per cent barytes (a dense mineral used for increasing density of suspension), and 0 to 10 per cent materials retained from the drilling operation. Shale shakers (vibrating screens) are used to remove cuttings from the circulating mud stream. The mud then passes into either steel tanks or earthen pits before it is returned to the hole.

Muds of other compositions are used for different applications. For example, oil base muds are useful when hydratable, swelling clays are encountered; sealing of the formation by infiltrating water is thus prevented. Long cleanup periods may also be avoided, although the fluid is flammable and disagreeable to handle.

Drilling muds are *thixotropic*, i.e., they tend to thin out and flow more freely when agitated, but thicken or even congeal when left standing. Thixotropic behavior is important. Mud should be viscous enough to keep cuttings suspended at a reasonable mud-circulation rate, but it should not be so viscous as to overload the *slush* (mud) pumps. Mud velocity should not exceed 8 to 11 ft per sec in the drill pipe, and 2 to 4 ft per sec in the annular space. Mud-circulation is said to consume 70 to 85 per cent of the energy required for drilling. Various additives, such as sodium tannate, serve to minimize the need for viscous mud. Both mud flow rate and viscosity should be kept as low as possible, to minimize friction losses.

In general, mud should be dense enough to prevent blowouts due to gas pressure. Mud pressure must exceed gas pressure to control it. Mud pressure is the product of depth of the mud column in the well and its density (more properly, specific weight). Mud should form a deposit on formation walls thick enough for sealing and prevention of caving. However, it should not form so thick a filter cake as to interfere with rotation of the drill pipe, because this can result in shearing the stem. In this connection, special tools are available for fishing jobs, but it is often cheaper to drill directionally around the obstruction.

Mud specific gravities of 1.05 to 1.32 are used. This range corresponds to $8\frac{3}{4}$ to 11 lb per gal, or 65.5 to 82.4 lb per cu ft. A density of about $9\frac{1}{2}$ lb per gal is usually satisfactory, but up to 19 lb per gal have been used in special cases. Mud properties checked may include: weight (expressed in lb per gal or lb per cu ft), Marsh funnel viscosity (in sec per qt), water loss (e.g., in cu cm per 30 min), pH, sand content, plastic viscosity, yield point, alkalinity (when drilling with special muds), and oil, salt, and solids content.

Rotary Drilling with Air and Natural Gas.[27,28] Use of air and natural gas as the circulating medium, to cool the bit and remove water and solids during drilling and completion operations, has proved successful in the Appalachian area. For drilling down to within approximately 50 ft of the expected gas sand, compressed air from three 600 cfm, 200 psig air compressors, is used. To avoid explosions, the last 50 ft or so are drilled with natural gas circulation, 400 to 800 psig, and gas flow rates of 2000 to 4000 cfm. Cuttings are discharged thru a 7-in. line, 200 to 300 ft, and 300 to 400 ft from the rig for air and gas, respectively. Where such a process is applicable, claimed advantages from field tests are:[27]

1. No risk of *mudding off* pay zone as when mud is used. Previous practice was to remove rotary rig when pay zone was approached and *drill in* with a cable tool spudder.
2. Total time of drilling a well is reduced.
3. Average footage drilled per bit is reduced.
4. Consequently, drilling cost is greatly reduced.

Rotary drilling with air or gas[29] has been highly successful in four wells in extremely hard-rock area of eastern Oklahoma and northwestern Arkansas. Advantages include: increased penetration rate, compared to mud-drilled holes, increased footage per bit, and low optimum rotary speeds. However, air velocities below 2000 fpm were insufficient to remove cuttings at depths below 4700 ft.

Possible fields for improvement in air-gas drilling include new types of rock bits, reverse circulation, and additives to gas or air such as foaming agents, lubricants, penetration stimulants, and well-sealing compounds.

Directional Drilling.[17] Controlled steering of the wellbore has been made possible thru development of reliable well-surveying instruments for checking correct setting of the deflecting tool and progress of the course of the directed hole to its objective. It has been used extensively on wells drilled with rotary equipment. Objectives of directional drilling include:

1. To straighten crooked holes, i.e., bring them back to vertical from a point of excessive angle of deviation.
2. To sidetrack lost tools and junk and thus avoid interferences to the well.
3. To accomplish a specific purpose; e.g., to maintain course of hole so that it will bottom at a predetermined location, either under derrick or offshore, to an oil sand location beneath water (Fig. 4-21a); to avoid trespass; to deflect a hole to underside of a salt dome or to a point of inaccessible location; to drill several wells from one location; or to drill a relief well to kill cratered or burning wells (Fig. 4-21b).

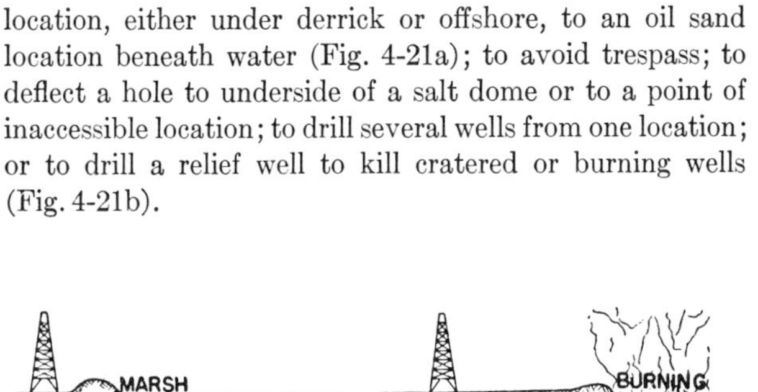

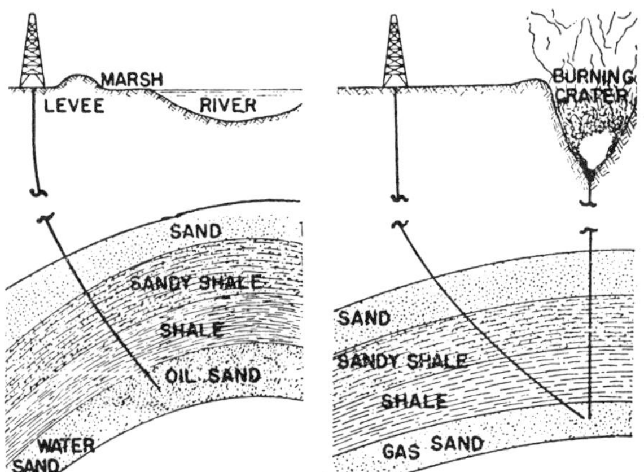

Fig. 4-21a (left) Directional drilling to inaccessible locations.
Fig. 4-21b (right) Controlling a burning well by directional drilling.
(Courtesy of Eastman Well Survey Corp.)

Turbodrilling

This modification of rotary techniques converts hydraulic to mechanical power at the bottom of the hole at high speeds (approaching 1000 rpm), resulting in higher bit horsepowers (say 450 hp) than are obtained with rotary drilling. Bit loading up to 35 tons is possible.

The turbodrill consists of stators mounted in a nonrotating casing containing a rotating shaft with complimenting rotors; up to 100 turbine stages have been used. Downward flowing mud drives the turbine in addition to lubricating and cooling the bearings.

Evaluation.[30] Extensive U. S. field tests conducted during 1958–59 showed that although effective in certain applications, turbodrilling cannot compete economically with good rotary drilling practices. Test results included poor performance with rock bits—penetration rate increase not offset by reduced (30 per cent of rotary) bit footage. However, turbodrilling was advantageous where effective bit weight is limited, e.g., crooked hole, directional drilling, and small hole size. Many performance inconsistencies were attributed to lack of knowledge of design parameters.

The following criteria should offer the best applications for turbodrilling: (1) formations conducive to rotary diamond bits, such as sandstone, limestone, and dolomites (avoid chert, quartzite, and soft shale); (2) mud with minimum amount of lost circulation material, with low abrasive solids (above one per cent sand), low aniline point oil (under 160 F), and temperatures below 180 F; (3) adequate rig hydraulics where the pumps can be operated at rated loads without excessive downtime, (4) high rig operating cost, (5) rotary trip time exceeding 5 hr; (6) rotary rock bits for drilling under 70 ft; (7) rotary penetration rate below 7 ft per hr; (8) rotary rock bit life under 15 hr; and (9) hole size 9 in. or less.

Evaluation of Types of Drilling

Rotary drilling is generally considered faster and more economical than cable tool drilling. It can be used in both hard rock and unconsolidated sands. Gas pressure can be readily controlled, and loss of diameter with depth is less than with cable tool drilling.

Cable tools cannot readily be adapted to drilling at depths much in excess of 10,000 ft, and cannot drill effectively in unconsolidated sands. However, lithologic logs can be readily kept, the sand face is not contaminated by mud used in the drilling process, and gas sands are not easily missed. Necessity of sealing off water and nonproducing porous horizons during drilling is a real drawback, but great difficulty may be encountered in rotary drilling in some instances with mud loss into fissures.

Cable tools are considered preferable in the Appalachian area for depths down to 7000 ft where unconsolidated sands and undue trouble from water or porous formations are not normally encountered. Much of the rock encountered in that area was formerly regarded as too hard for rotary tools. However, experience has proved quite favorable with rotary tools and drilling mud as well as with air and gas as circulating fluids.

REFERENCES

1. Am. Geol. Inst. *Glossary of Geology and Related Sciences.* (Pub. 501) Washington, 1957.
2. Fay, A. H. *Glossary of the Mining and Mineral Industry.* (Bur. of Mines Bul. 95) Washington, G.P.O., 1948.
3. Pirson, S. J. *Oil Reservoir Engineering,* 2d ed., p. 358. New York, McGraw-Hill, 1958.
4. Levorsen, A. I. *Geology of Petroleum.* San Francisco, Freeman, 1954.
5. Hoylman, H. W. "Airborne Magnetometer Profiles." *Oil Gas J.* 48: 55, Dec. 29, 1949.
6. Balsley, J. R. "Aeromagnetics Surveying." *Advan. Geophys.* I: 313. New York, Academic Pr., 1952.
7. Am. Pet. Inst. *Petroleum Discovery and Production.* New York, 1956.
8. "Vibroseis Hailed as Major Tool." *Oil Gas J.* 59: 64, Feb. 27, 1961.
9. Am. Pet. Inst. *History of Petroleum Engineering.* New York, 1961.
10. Rothé, E. and Rothé, J. P. *Prospection Géophysique.* Paris, Gauthier-Villars, 1950–52. 2v.
11. Landes, K. K. *Petroleum Geology,* 2d ed. New York, Wiley, 1959.
12. Russell, W. L. "Well Logging by Radioactivity." *Am. Assoc. Petrol. Geologists Bull.* 25: 1768, Sept. 1941.
13. Doll, H. G. and others. "True Resistivity Determination from the Electric Log." A.P.I. *Drilling and Production Practice* 1947: 215.
14. Bush, R. E. and Mardock, E. S. "Some Preliminary Investigations of Quantitative Interpretations of Radioactive Logs." *J. Petrol. Technol.* 2: 19, Jan. 1950 (Trans.).
15. Hills, J. M. "Sampling and Examination of Well Cuttings." *Am. Assoc. Petrol. Geologists Bull.* 33: 73, 1949.
16. LeRoy, L. W. and Crain, H. M., eds. "Subsurface Geologic Methods: a Symposium." *Quart., Colo. School Mines* 44, July 1949.
17. Stephens, M. M. and Spencer, C. F. *Petroleum and Natural Gas Production.* University Park, Pennsylvania State College, 1946.
18. Nettleton, L. L. *Geophysical Prospecting for Oil.* New York, McGraw-Hill, 1940.
19. Harrington, J. W. "Elementary Theory of Subsurface Structural Contouring." *Am. Geophys. Union Trans.* 32: 77–80, Feb. 1951.
20. Low, J. W. "Subsurface Maps and Illustrations." Chap. VII in reference 16 above.
21. Wilhelm, O. "Classification of Petroleum Reservoirs." *Am. Assoc. Petrol. Geologists Bull.* 29: 1537–80, Nov. 1945.
22. Barrell, J. "Rhythms and the Measurements of Geologic Time." *Geol. Soc. Am. Bull.* 28: 745–904, Dec. 4, 1917.
23. Krumbein, W. C. "Lithofacies Maps and Regional Sedimentary-Stratigraphic Analysis." *Am. Assoc. Petrol. Geologists Bull.* 32: 1909–23, Oct. 1948.
24. Dapples, E. C. and others. "Tectonic Control of Lithologic Associations." *Am. Assoc. Petrol. Geologists Bull.* 32: 1924–47, Oct. 1948.
25. Rich, J. L. "Three Critical Environments of Deposition, and Criteria for Recognition of Rocks Deposited in Each of Them." *Bull. Geol. Soc. Am.* 62: 1–13, Jan. 1951.
26. Condensed from: I.G.T. *Natural Gas Production and Transmission Home Study Course,* chap. V, VI, Chic., 1952, plus additions by authors.
27. Henrickson, J. W. "New York State Natural Gas Corporation's Rotary Air and Natural Gas Drilling in the Benezette Oriskany Gas Field." *Producers Monthly* 19: 38–41, Dec. 1954.
28. Magner, H. J. and Bono, T. J. "Air Drilling in Appalachian Basin." Schramm, Inc. *Applications.* West Chester, Pa., 1954.
29. Adams, J. H. "Air and Gas Drilling Gain Prestige." *Oil Gas J.* 55: 123–6, Aug. 12, 1957.
30. Bingman, W. E. "Turbodrilling: What Is Its Status Today?" *World Oil* 150: 111–5, May 1960.

Chapter 4
Estimation of Well Capacities and Gas Reserves*

by J. L. Eakin, R. V. Smith, and J. S. Miller

BACK-PRESSURE TESTING OF NATURAL GAS WELLS

General

The back-pressure methods for testing gas wells help to predict pipeline delivery rates and to avoid both possible damage to the reservoir and waste of gas in flowing wells wide open during open-flow tests. Relatively little information about the ability of a well to produce gas into a pipeline under pressure is gained from open-flow tests. Formerly the approximate open-flow capacity of a gas well was gaged with a Pitot tube. Later the capacity was gaged more accurately with a critical flow prover.

The volume of condensate and water produced from a gas well is expressed here as barrels per million cubic feet of gas corrected to standard conditions* (14.65 psia and 60 F).

Back-pressure testing of gas wells has been widely used in the natural gas industry since publication of the method in Monograph 7 in 1935.[1] Although it has been adopted by several states for **proration** purposes, the test procedures described here are not necessarily those approved by the various state regulatory bodies. Information regarding gas well testing in a particular state should be obtained from the test manual of that state.[2,3] Usually the official bodies require a multipoint test of short duration. In addition, some states require a single rate of flow for a period of 24 to 72 hr.

Natural gas production problems solved by back-pressure methods include: effects on delivery capacities of liquids, e.g., water, brine, condensate, and crude oil in the wellbore and in the producing formation; effects of the formation of gas hydrates; effects of casing and tubing leaks; effects of treating wells with acid; possible effects of shooting or fracturing; effects of the accumulation of salts, cavings, "mud" and foreign objects in the wellbore; and effects of changes in producing characteristics of a well. It is useful in solving problems of natural gas storage in depleted reservoirs and in formulation of a rate of development for a given gas productive area.

The **multipoint** (flow-after-flow) back-pressure method[1] consists of multiple rates of flow from a well, each successively greater than the preceding one. The first and all succeeding rates should be continued until flowing pressure stabilization is approached. The usual procedure is to measure or calculate the pressure at the sand face after the well has been shut in and allowed to stabilize. Measurements are then recorded for a series of gas producing rates and corresponding pressures, either at predetermined time intervals or at stabilized conditions. Enough points, generally four, should be obtained to permit a proper correlation of pressures and rates.

The **isochronal** back-pressure method[4] is readily adapted to testing wells that stabilize slowly. The prescribed method is to open the well to flow from a shut-in condition at a given rate for a specific period of time. Then the well is shut in, allowing the pressure to attain the original magnitude after which it is opened at a different flow rate. Each flow period starts from a comparable shut-in condition and the reservoir drainage area is maintained at a simple pressure gradient during testing. Accurate estimation of pipeline deliveries for wells producing from reservoirs with low permeability is practically impossible without isochronal test data.

Permeability Considerations. The empirical relationships for porous media have received extensive laboratory study. Work on gas flow thru cylindrical tubes filled with sand tends to verify the back-pressure method[5,6] (provided transient pressure conditions in the reservoir do not interfere). Either the multipoint or isochronal test is satisfactory for wells producing gas from reservoirs with high permeability. However, analyses of isochronal-type tests are necessary for an accurate interpretation of producing problems for wells producing gas from low-permeability reservoirs.

Reservoirs of extremely low permeability require special procedures to establish delivery capacity under stabilized conditions. Procedures for estimating stabilized delivery capacity of gas wells in the Hugoton field from test flow rates of short duration have been published.[7]

Basic Equations. Studies have shown a consistent relationship between rates of gas delivery and corresponding pressures in the productive formation for normal gas wells tested by either multipoint or isochronal back-pressure methods. Results of tests on 582 wells show that when delivery rate, Q, is plotted on logarithmic paper against $(P_f^2 - P_s^2)$,† the respective differences of squares of formation pressure, P_f, and pressure at the sand face, P_s, the relationship is represented empirically by a straight line.[1]

Bottom-hole curve,[8] Eq. **1**, useful in the study of reservoir performance, indicates the capacity of a well to deliver gas into a wellbore.

* This chapter uses 14.65 psia as the pressure base because of the historical use of this level in many reserve and well capacity studies. A.G.A. recommends 14.73 psia, which is used elsewhere in this Handbook.

† Both in absolute units.

$$Q = C(P_f^2 - P_s^2)^n \tag{1}$$

where: C = performance coefficient
n = a constant

Top hole curve (or multipoint curve), Eq. **2,** useful in estimating the capacity of a well to deliver gas into a pipeline under specified conditions, indicates the capacity of the well to produce gas at the wellhead.

$$Q = C(P_c^2 - P_t^2)^n \tag{2}$$

Parameters, C, the performance coefficient of the back-pressure equation, and n, the exponent, differ from those appearing in Eq. **1**. The shut-in and flowing pressure in psia at the wellhead are shown as P_c and P_t, respectively.

Although a determination of the value of the performance coefficient, C, is not always necessary for a practical analysis of a back-pressure test, C values may be determined by substituting in, say, Eq. **1**, predetermined values of n and corresponding sets of values of Q and $P_f^2 - P_s^2$. The value of C may be determined graphically by extending the straight line relationship to $P_f^2 - P_s^2 = 1$ and reading the corresponding Q; i.e., C equals Q when $P_f^2 - P_s^2 = 1$—true also in Eq. **2**.

Effects of liquids in the well on multipoint testing have been described in detail.[1] The value of the exponent, n, for both the multipoint test and the isochronal test ranges from 0.5 to 1.0. Multipoint tests run in decreasing rate sequence may have exponents greater than 1.0 for wells producing gas from reservoirs that stabilize slowly. Either the accumulation of liquids in the wellbore or the slow stabilization characteristics of a reservoir are reflected in multipoint tests by exponents less than 0.5. Conversely, tests resulting in exponents greater than 1.0 may indicate that during the tests liquids or restrictions caused by drilling mud or well stimulation liquids were removed from the formation around the wellbore. An erratic alignment of data points resulting from either method of testing usually indicates changes during the test period in the capacity of the well to produce gas. Usually such changes are effected either by cleaning the sand face or by the accumulation of liquids in the wellbore, resulting in erratic exponents from isochronal testing. Tests resulting in exponents with values either less than 0.5 or more than 1.0 should be rerun.

Points for the back-pressure curve should be plotted on equal-scale log paper. As straight a line as possible should be drawn thru three of the points.

Values of the exponent, n, in the basic back-pressure Eq. **1**, may be determined by substituting values of Q read directly from the straight line relationship, rather than data points, and corresponding values of $P_f^2 - P_s^2$ in:

$$n = \frac{\log Q_2 - \log Q_1}{\log[(P_f)^2 - (P_{s2})^2] - \log[(P_f)^2 - (P_{s1})^2]} \tag{3}$$

where:

Q_1 or Q_2 = low and high rates of flow, respectively, MCFD
P_f = subsurface shut-in pressure in the reservoir at vertical depth, H, psia
P_{s1} or P_{s2} = subsurface flowing pressure, corresponding to Q_1 or Q_2 in the wellbore at vertical depth, H, psia
n = a constant

Subscripts 1 and 2 refer to points on the straight line.

Cleaning Wells. Before the well is shut in for a back-pressure test, it should be cleaned of liquids that may have accumulated in its bore and in the immediate surrounding formation. If the well is equipped with tubing open opposite the producing formation, cleaning may be accomplished with relatively low flow rates thru the tubing. Liquids may be removed from gas wells by a foaming technique.[9] Removal of liquids from wells without tubing or foaming agents requires high producing rates to develop sufficient velocities. Where pressures range from 1200 to 1300 psig, a flow rate as large as 20 MMCF per 24 hr is necessary to clean a well thru 7-in., 24-lb per ft pipe.[10] Linear velocities of 6 to 10 ft per sec in the flow string are sufficient to remove liquids.[11] Equation **4** computes the average linear gas velocity, V, in ft per sec, at any point in the flow string.

$$V = 0.06QZT/Pd^2 \tag{4}$$

where:

Q = flow rate (at 14.65 psia and 60 F), MCFD
Z = compressibility factor, dimensionless
T = temperature, °F abs
P = pressure, psia
d = pipe ID, in.

In wells completed with casing set at the top of the producing formation, cavings may accumulate in the open-hole wellbores, with material decrease in productive capacity. Presence of cavings may be determined by measuring apparent well depth with a weighted wire line and subsequent comparison with the drilled depth. Cavings may be removed in some cases by blowing, but usually a work-over operation is required.

Well Temperatures and Depths. Accurate temperature measurements are essential. Thermometers in liquid-filled wells at the wellhead, and other measurement equipment, are used to take temperature measurements. Usually bottom-hole temperature is taken either from a continuous survey or from a maximum-registering thermometer run in conjunction with a bottom-hole pressure gage. The temperature at a depth midway in a well is estimated as an arithmetic average of the wellhead and bottom-hole temperatures.

Depth at which the subsurface pressure at the sand face is calculated or measured is usually considered as the distance between control wellhead valve and a point midway between top and bottom of the producing formation, for a uniform or relatively thin productive stratum. For wells producing from two or more closely-spaced sands in the same producing horizon, or from a thick stratum not uniformly productive, depth can be approximated by averaging the distance between control valve and points midway between top and bottom of each producing stratum, as indicated by drilling and log records. If a well has tubing set with a packer, the pressures are determined at the entrance to the tubing.

Determination of Specific Gravity of Flowing Fluid. This information for the two-phase flow produced from a gas or gas-condensate well is essential in calculating either the shut-in or flowing pressures. The specific gravity of the fluid flowing in the wellbore may be calculated[12] by Eq. **5**.

$$G_m = \frac{G_g + Bd_L/76{,}130}{1 + BV_L/1{,}000{,}000} \tag{5}$$

where:

G_m = specific gravity of flowing fluid—mixture of gas and gas condensate (air = 1.000)

G_g = specific gravity of separator gas (air = 1.000)

76,130 = weight of dry air (at 60 F and 14.65 psia), lb per MMCF

B = liquid to gas ratio, bbl per MMCF

d_L = liquid density, Table 4-10, lb per bbl

V_L = vapor volume equivalent of d_L, Table 4-10, cu ft per bbl

Equation **5** was used to compute specific gravities of flowing fluids from 48 wells. The average deviation between the computed specific gravity and that measured experimentally was sufficiently small on recombined samples to indicate the adequacy of Eq. **5** and data from Table 4-10 for gas-condensate wells.

Table 4-10 Approximate Relationship Between Gravity of Gas-Condensate Liquids and Equivalent Vapor Volume

Gravity, °API	Density, lb per bbl, d_L	Vapor volume, cu ft per bbl, V_L
40	288.5	635
45	280.4	676
50	272.6	721
55	265.3	765
60	258.3	813
65	251.7	863
70	245.5	916
75	239.6	970

Shut-In Pressure. After the well has been cleaned of liquids, it is shut in at the wellhead and, after pressure in the well and reservoir in its vicinity becomes stabilized, wellhead shut-in pressure is measured with a piston gage (dead-weight tester or gage). Spring gages are not accurate enough for back-pressure tests. It is generally recommended that the well be shut in for at least 24 hr. The degree of pressure stabilization is followed during shut-in time by periodic measurement of wellhead pressure to obtain the rate of pressure buildup. After the pressure reaches satisfactory stabilization, either the isochronal or multipoint test may be started. In general, a greater degree of stabilization is required for shut-in pressures than for flowing wellhead pressures.

Surface Pressure Measurement. All wellhead pressure measurements in back-pressure tests should be taken with a piston gage to a precision of 1 part in 5000 or 0.1 psi in 500 psi. Such precision is easily attained with commercial piston gages.

Open-Flow Potential. The volume of gas that would be produced in 24 hr from a well with only atmospheric pressure against the sand face is commonly termed the *calculated absolute open flow* or the open-flow potential. By plotting different rates of flow, Q, against the corresponding values of $(P_f^2 - P_s^2)$ and extending this straight line relationship to a point where P_s is equivalent to atmospheric pressure, then the value of Q, which is the absolute open-flow potential of a well, can be read directly from the chart in cubic feet per 24 hr. Corrections for atmospheric pressure may be neglected where P_f values are 100 psia or more. The use of plotted relationships for reading open-flow potentials is recommended in preference to the cumbersome computations involved with Eq. **1**.

It is important that the length of time and type of test be carefully recorded when measuring the open-flow potential of wells, particularly those wells producing from reservoirs with low permeability. The open-flow potential values are different for a specific well for either a 2-hr multipoint, a 3-hr multipoint, or a 3-hr isochronal test. Cullender[4] gives examples showing the relationship between open-flow values and type of test.

Gas Measurements with Critical Flow Prover. When gas wells are not equipped with orifice meters, open flow tests may be made with what is essentially a steel tube 12 in. long, with provision for installing an orifice plate at one end. Gas vented to the atmosphere during any type of operation in the vicinity of a well should be blown vertically as a safety precaution to keep gas away from the working area. Consequently, provers should be installed vertically either at the wellhead or downstream from the separator.

Equation **6** gives a modification of the Rawlins and Schellhardt[1] method for the computation of flow rates for critical flow provers.

$$Q = F_p F_{tf} F_g F_{pv} P_m \tag{6}$$

where:

Q = flow rate (at 14.65 psia and 60 F), MCFD

F_p = basic orifice factor for critical-flow prover, Table 4-11 (at 14.65 psia, 60 F, and 1.000 sp gr), MCFD

F_g = specific gravity factor = (1 ÷ flowing gas specific gravity)$^{0.5}$—Table 15 of Sec. 7, Chap. 2

F_{tf} = flowing temperature factor = (520 ÷ flowing gas temp in °R)$^{0.5}$—Table 14 of Sec. 7, Chap. 2, for flowing temperatures from 1 to 150 F, and Table 4-12 for 150 to 249 F

F_{pv} = supercompressibility factor = (1 ÷ compressibility factor)$^{0.5}$—Table 16-A of Sec. 7, Chap. 2

P_m = static pressure on critical-flow prover, psia

Table 4-11 Basic Orifice Factors, F_p, for Critical-Flow Prover[1] (Bureau of Mines plate design)

2-In. Prover		4-In. Prover	
Orifice diam, in.	F_p, MCFD	Orifice diam, in.	F_p, MCFD
1/16	0.06569	1/4	1.074
3/32	0.1446	3/8	2.414
1/8	0.2716	1/2	4.319
3/16	0.6237	5/8	6.729
7/32	0.8608	3/4	9.643
1/4	1.115	7/8	13.11
5/16	1.714	1	17.08
3/8	2.439	1 1/8	21.52
7/16	3.495	1 1/4	26.57
1/2	4.388	1 3/8	31.99
5/8	6.638	1 1/2	38.12
3/4	9.694	1 3/4	52.07
7/8	13.33	2	68.80
1	17.53	2 1/4	88.19
1 1/8	22.45	2 1/2	110.6
1 1/4	28.34	2 3/4	136.9
1 3/8	34.82	3	168.3
1 1/2	43.19		

Base temperature = flowing temperature = 60 F, base pressure = 14.65 psia, and specific gravity = 1.000.

Table 4-12 Flowing Temperature Factor, F_{tf}

Observed temp, °F	0	1	2	3	4	5	6	7	8	9
150	0.9233	0.9225	0.9217	0.9210	0.9202	0.9195	0.9187	0.9180	0.9173	0.9165
160	.9158	.9150	.9143	.9135	.9128	.9121	.9112	.9106	.9099	.9092
170	.9085	.9077	.9069	.9063	.9055	.9048	.9042	.9035	.9028	.9020
180	.9014	.9007	.9000	.8992	.8985	.8979	.8972	.8965	.8958	.8951
190	.8944	.8937	.8931	.8923	.8916	.8910	.8903	.8896	.8889	.8882
200	.8876	.8870	.8863	.8856	.8849	.8843	.8836	.8830	.8823	.8816
210	.8810	.8803	.8797	.8790	.8784	.8777	.8770	.8764	.8758	.8751
220	.8745	.8738	.8732	.8725	.8719	.8713	.8706	.8700	.8694	.8687
230	.8681	.8675	.8668	.8662	.8656	.8650	.8644	.8637	.8631	.8625
240	0.8619	0.8613	0.8606	0.8600	0.8594	0.8588	0.8582	0.8576	0.8570	0.8564

Note: For lower flowing temperature, $T_f = 1$ thru 150 F, see Table 14 of Sec. 7, Chap. 2.
$F_{tf} = (520/T_f)^{0.5}$; T_f in °R.

Table 4-13a Basic Orifice Factors, F_b, for Flange Taps in MCFD[3]

Pipe sizes—nominal and published diameters, in.

Orifice diam, in.	2 std., 2.067	3 std., 3.068	4 std., 4.026	6 std., 6.065	8 std., 8.071
0.250	0.3067	0.3066	0.3061	...	...
.375	.6860	.6848	.6841	...	...
.500	1.219	1.214	1.212	1.211	...
.625	1.914	1.897	1.892	1.889	...
.750	2.779	2.740	2.729	2.720	...
0.875	3.824	3.744	3.723	3.706	3.700
1.000	5.073	4.912	4.874	4.847	4.836
1.125	6.557	6.251	6.186	6.143	6.126
1.250	8.329	7.771	7.661	7.595	7.571
1.375	10.46	9.486	9.303	9.203	9.171
1.500	13.09	11.41	11.12	10.97	10.93
1.625	...	13.58	13.12	12.90	12.84
1.750	...	16.01	15.31	14.98	14.90
1.875	...	18.75	17.71	17.23	17.13
2.000	...	21.86	20.32	19.65	19.51
2.125	...	25.40	23.18	22.24	22.05
2.250	...	29.52	26.30	25.01	24.75
2.375	...	...	29.72	27.95	27.61
2.500	...	...	33.48	31.09	30.64
2.625	...	...	37.60	34.41	33.84
2.750	...	...	42.15	37.94	37.21
2.875	...	...	47.19	41.69	40.74
3.000	...	...	52.97	45.66	44.46
3.125	...	...	...	49.86	48.36
3.250	...	...	...	54.32	52.44
3.375	...	...	...	59.05	56.71
3.500	...	...	...	64.07	61.17
3.625	...	...	...	69.40	65.85
3.750	...	...	...	75.08	70.73
3.875	...	...	...	81.12	75.83
4.000	...	...	...	87.56	81.15
4.250	...	...	...	101.8	92.53
4.500	...	...	...	118.3	105.0
4.750	...	...	...	...	118.5
5.000	...	...	...	...	133.4
5.250	...	...	...	...	149.8
5.500	...	...	...	...	167.8
5.750	...	...	...	...	187.7
6.000	...	...	...	...	210.1

Base temperature = flowing temperature = 60 F, base pressure = 14.65 psia, and specific gravity = 1.000.

Note: When inside diameter of pipe varies from standard by more than tolerances recommended by the A.G.A., coefficient should be calculated as shown in Reference 13.

Table 4-13b Basic Orifice Factors, F_b, for Pipe Taps in MCFD[3]

Pipe sizes—nominal and published diameters, in.

Orifice diam, in.	2 std., 2.067	3 std., 3.068	4 std., 4.026	6 std., 6.065	8 std., 8.071
0.250	0.3089	0.3076	0.3070	...	...
.375	.6999	.6921	.6898	...	...
.500	1.266	1.237	1.228	1.222	...
.625	2.029	1.951	1.927	1.910	...
.750	3.016	2.848	2.793	2.756	...
0.875	4.273	3.941	3.836	3.762	3.739
1.000	5.871	5.249	5.066	4.933	4.893
1.125	7.915	6.797	6.494	6.274	6.207
1.250	10.57	8.618	8.134	7.791	7.685
1.375	14.09	10.76	10.00	9.492	9.330
1.500	...	13.27	12.12	11.38	11.15
1.625	...	16.24	14.52	13.47	13.14
1.750	...	19.77	17.23	15.77	15.32
1.875	...	23.99	20.30	18.28	17.68
2.000	...	29.09	23.77	21.02	20.23
2.125	...	35.36	27.71	24.00	22.98
2.250	...	...	32.20	27.24	25.94
2.375	...	...	37.34	30.76	29.11
2.500	...	...	43.25	34.57	32.49
2.625	...	...	50.10	38.71	36.11
2.750	...	...	58.09	43.20	39.97
2.875	...	...	...	48.08	44.07
3.000	...	...	...	53.37	48.44
3.125	...	...	...	59.13	53.09
3.250	...	...	...	65.40	58.03
3.375	...	...	...	72.23	63.28
3.500	...	...	...	79.71	68.87
3.625	...	...	...	87.90	74.80
3.750	...	...	...	96.89	81.11
3.875	...	...	...	106.8	87.82
4.000	...	...	...	117.7	94.96
4.250	...	...	...	...	110.7
4.500	...	...	...	...	128.6
4.750	...	...	...	...	149.0
5.000	...	...	...	...	172.5
5.250	...	...	...	...	199.7
5.500	...	...	...	...	231.3

Base temperature = flowing temperature = 60 F, base pressure = 14.65 psia, and specific gravity = 1.000.

Note: When inside diameter of pipe varies from standard by more than tolerances recommended by the A.G.A., coefficient should be calculated as shown in Reference 13.

Specific gravity of gas produced from the well may be determined accurately with a gas gravity balance. Usually gas specific gravity for a given well is constant for various flow rates, but for wells producing hydrocarbon liquids it may change with flow rate. Consequently, specific gravities should be determined frequently during testing until behavior of the particular well is known; data reliability beyond two significant figures is questionable.

Gas Measurements with Orifice Meters. Gas wells are usually fitted with such equipment. Specifications for orifice meters and methods for computing flow rates should be in accordance with Gas Measurement Committee Report No. 3.[13] Basic orifice factors in this Report are for a pressure base of 14.73 psia. Some authorities calculate flow using other pressure bases, for example: a pressure base of 14.65 psia[3,14] is used.* Table 12 of Sec. 7, Chap. 2, may be used to change from a pressure base of 14.73 psia to other pressure bases.

Basic Flow Rate Calculations. Recommended equations and factors are given on page 7/21. An example which employs the various factors in these equations is given on page 7/23. Some authorities recommend an abbreviated version of the equation on page 7/21, i.e., they neglect factors F_r, Y, and F_m.[3] Tables 4-13a and 4-13b give basic orifice factors, F_b, for flange taps and pipe taps consistent with the development of this chapter, i.e., flow in MCFD at 14.65 psia.

Multipoint Back-Pressure Testing and Examples

Procedure. After the well has been cleaned, shut in, stabilized, and necessary pressure measurements have been taken, it is allowed to produce gas at a high back pressure—about 95 per cent of shut-in pressure. When flow conditions again become stabilized, observations are made of wellhead pressure and gas delivery rate. Back pressure is then decreased, and another set of observations is taken of wellhead pressure and delivery rate. Except for very high-volume wells, four sets of observations for back pressures ranging from about 95 to 70 or 75 per cent of shut-in pressure usually furnish enough data to establish the relationship between Q and $(P_c^2 - P_t^2)$. Sequence of flow rates should be from lower to higher rates. Maintain each flow rate until stabilization is approached. Some authorities recommend flow rates of at least 2 hr even though "satisfactory" stabilization is obtained in less time.

Examples of Multipoint Tests. Table 4-14a is an actual copy of the field data sheet for a multipoint back-pressure test.[15] The "remarks" column is for recording such items as the results of the specific gravity measurement and the condition of the flow with regard to whether the well is producing water. All observations shown are necessary for accurate analysis of test results.

Computation of the results of a back-pressure test on a gas well involve the following steps:

1. Compute rates of flow and pressures at the face of the producing formation from pressure and volume observations made at the wellhead.
2. Determine values of $(P_c^2 - P_t^2)$ and $(P_f^2 - P_s^2)$ and rates of flow corresponding to these pressure factors. P_f and P_s are calculated at the midpoint of the sand face in wells without tubing. If the well has tubing, these pressures are determined at the entrance to the tubing.
3. Plot values of Q and corresponding values of $(P_f^2 - P_s^2)$ and $(P_c^2 - P_t^2)$ on logarithmic coordinates.
4. Determine values of the exponent n and the performance coefficient C in the flow Eqs. **1** and **2**. For most routine analyses of back-pressure tests, determination of the value of C is not necessary.
5. Determine the calculated absolute open flow.

A convenient form for reporting the results of a multipoint test is shown in Table 4-14b for the test data given for Well A in Table 4-14a. Table 4-14b also shows general well information, a summary of test data, calculation of rates of flow, data for determining compressibility, and the difference of squares of pressures for wellhead and bottom-hole conditions. The calculated open flow potential of 5200 MCFD was determined in Fig. 4-22a. The data points were connected by a straight line and extrapolated to a value of $(P_f^2 - P_s^2)$, where P_s is equivalent to atmospheric pressure. In this case, the value is 1,387,000. The corresponding rate of flow is 5200 MCFD. Actually, atmospheric pressure was neglected. The

* Also used in this Section of the Handbook.

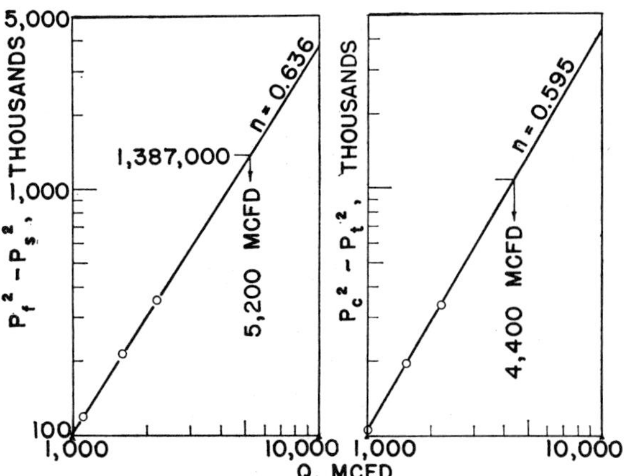

Fig. 4-22a (left) Multipoint test showing bottom-hole performance of Well A;[15] Q (in cubic feet per day) $= 652(P_f^2 - P_s^2)^{0.636}$
Fig. 4-22b (right) Multipoint test showing wellhead performance of Well A;[15] Q (in cubic feet per day) $= 1161(P_c^2 - P_t^2)^{0.595}$

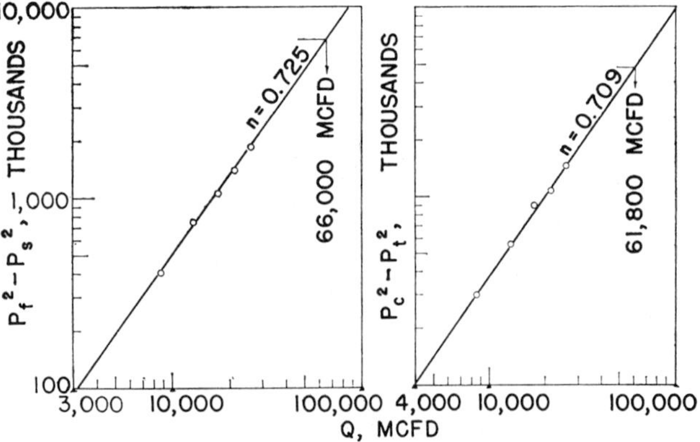

Fig. 4-23a (left) Multipoint test showing bottom-hole performance of Well B.[15]
Fig. 4-23b (right) Multipoint test showing wellhead performance of Well B.[15]

Table 4-14a Field Data Sheet for Multipoint Test of Well A.[15]

Type Test: ☐ Initial ☐ Annual ☒ Special			Test Date	Lease No. or Serial No.
Company		Connection		Allottee
Field	Reservoir: Mayflower		Location: Kay County, Oklahoma	Unit
Completion Date	Total Depth	Plug Back TD	Elevation	Farm or Lease Name
Csg. Size / Wt / d	Set At	Perforations: From	To	Well No.
Tbg. Size: 2" / Wt: 4.70# / d: 1.995	Set At	Perforations: From	To	Sec. - - Twp - Blk Rge - Sur - - -
Type Completion (Describe)			Packer Set At	County or Parish
Producing Thru: Tubing	Reservoir Temp, °F @	Mean Ground Temp, °F: 60	Baro. Press., P_a: 14.2	State
L: 4,530 / H: 4,530	G_g: 0.661	% CO_2 / % N_2	% H_2S	Prover - - Meter Run Taps - - - 4 inch

DATE Time of Reading	ELAP. TIME hr	WELLHEAD WORKING PRESSURE Tbg., psig	Csg., psig	Temp, °F	METER OR PROVER Pressure, psig	Diff.	Temp, °F	Orifice	REMARKS (Include liquid production data: Type - API Gravity - Amount)
		Well shut in for 3 days							
7:55		1,027.0							S.I.P.
8:00	0	1,027.0	- -						S.I.P.
8:15	.25	994.0	- -	53	222	8.0	53	2 in.	
8:30	.50	987.7	- -	54	222	8.0	54		
8:45	.75	984.7	- -	55	220	7.9	55		
9:00	1.00	981.6	- -	56	219	8.0	56		
9:30	1.50	977.5	- -	57	219	8.0	57		
10:00	2.00	974.5	- -	57	217	8.0	57		
		Rate changed on meter							
10:15	.25	941.9	- -	59	222	16.6	59	2 in.	
10:30	.50	938.9	- -	60	223	16.6	60		
10:45	.75	936.7	- -	60	224	16.4	60		
11:00	1.00	934.4	- -	60	225	16.2	60		
11:30	1.50	930.5	- -	60	226	16.2	60		
12:00	2.00	927.6	- -	60	224	16.2	60		
		Rate changed on meter							
12:15	.25	870.5	- -	61	229	31.8	61	2 in.	
12:30	.50	865.5	- -	62	229	31.7	62		
12:45	.75	861.6	- -	62	228	31.5	62		
1:00	1.00	858.9	- -	63	228	31.2	63		
1:30	1.50	853.0	- -	63	227	31.0	63		
2:00	2.00	848.7	- -	63	227	30.9	63		

PAGE 1 OF ____ DATA BY ____________

absolute open flow of 5200 MCFD for Well A is for a 2-hr, 3-point test.

The exponent n and the performance coefficient C on all back-pressure calculations made in this chapter were determined by the method of least squares.[16]

The C value of 652 may be checked by extrapolating the straight line on Fig. 4-22a to $(P_f^2 - P_s^2) = 1$ and reading the corresponding value of Q. It should be noted that the value of $C = 652$ is for Q in units of CFD.

The wellhead performance of Well A, as determined by the test results given in Table 4-14b, is illustrated in Fig. 4-22b where Q and corresponding $(P_c^2 - P_t^2)$ values are plotted. The straight line has been extended to show a wellhead open-flow potential of 4400 MCFD.

The wellhead performance equation given under Fig. 4-22b measures the ability of Well A to deliver gas at the wellhead thru 2-in. tubing as indicated by the multipoint test given in Table 4-14b. The relationship is influenced by the size of the flow string and weight of the gas column, as well as the productive capacity of the well.

Another example of the bottom-hole performance as indicated by a multipoint test is given in Fig. 4-23a for a large capacity well. Well B was dually completed with 6.214-in. ID casing and 2.875-in. OD tubing with a packer. The pro-

Table 4-14b Results of Multipoint Back-Pressure Test of Well A.[15]

Type Test: ☐ Initial ☐ Annual ☒ Special		Test Date	Lease No. or Serial No.
Company	Connection		Allottee
Field	Reservoir Mayflower	Location Kay County, Okla.	Unit
Completion Date / Total Depth	Plug Back TD	Elevation	Farm or Lease Name
Csg. Size / Wt. / d	Set At	Perforations: From / To	Well No.
Tbg. Size 2 in. / Wt. 4.70# / d 1.995	Set At	Perforations: From / To	Sec. --- Twp – Blk Rge – Sur ---
Type Completion (Describe)		Packer Set At	County or Parish
Producing Thru Tubing	Reservoir Temp, °F @ 128 F	Mean Ground Temp, °F 60 F / Baro. Press. – P_a 14.2	State
L 4,530 / H 4,530	G_g 0.661	% CO_2 / % N_2 / % H_2S	Prover -- Meter Run Taps --- 4 in.

	FLOW DATA					TUBING DATA			CASING DATA	
NO.	Prover Line Size	X Choke Orifice Size	Press., psig	Diff., h_w	Temp, °F	Press., psig	Temp, °F	Press., psig	Temp, °F	Duration of Flow, hr
SI	4"	2"	217	8	57	974.5	57	--	--	2
1.		2"	224	16.2	60	927.6	60	--	--	2
2.		2"	227	30.9	63	848.7	63	- -	- -	2

NO.	Coefficient (24–Hour)	$\sqrt{h_w P_m}$	Pressure, P_m	Flow Temp. Factor, F_t	Gravity Factor, F_g	Super Compress. Factor, F_{pv}	Rate of Flow Q, Mcfd
1.	30.32	43.01	231.2	1.003	1.23	1.025	1,105
2.	20.32	62.12	238.2	1.000	1.23	1.023	1,588
3.	30.32	86.33	241.2	.9971	1.23	1.023	2,201

NO.	P_r	Temp, °R	T_r
1.	0.32	517	1.37
2.	.33	520	1.38
3.	.34	523	1.38
4.			

Gas Liquid Hydrocarbon Ratio		Mcf/bbl
API Gravity of Liquid Hydrocarbons		deg
Specific Gravity Separator Gas		x x x x x x x x x x x
Specific Gravity Flowing Fluid	x x	0.661
Critical Pressure	670 psia	psia
Critical Temperature	378 °R	°R

P_c 1,041.2 P_c^2 1,084,097 P_f 1,177.9 P_f^2 1,387,448

NO.	P_t	P_t^2	$P_c^2-P_t^2$	P_w	P_w^2	$P_c^2-P_w^2$	P_s	P_s^2	$P_f^2-P_s^2$
1.	988.7	977,527	106,570				1,126.1	1,268,101	119,347
2.	941.8	886,987	197,110				1,083.8	1,174,622	212,826
3.	862.9	744,596	339,501				1,017.3	1,034,899	352,549

AOF 5,200 Mcfd

n = 0.636

Commission ______

Company ______

Others ______

ductivity tests were based upon flow from the upper formation produced through the annulus. Well B had a shut-in pressure, P_f, of 2610 psia at a depth of 8072 ft and a wellhead pressure of 2191 psia. The calculated open-flow potential at bottom-hole conditions was 66,000 MCFD. The corresponding wellhead performance for Well B, produced thru the annulus, is illustrated in Fig. 4-23b. The wellhead open-flow potential of Well B was 61,800 MCFD.

Isochronal Test and Example

The term isochronal was adopted because only those conditions extending as a result of a single disturbance of constant duration are considered related to each other by Eqs. **1** and **2**. The expression "single disturbance of constant duration" is defined as that condition existing around a well as a result of a constant flow rate for a specific period of time from shut-in conditions. Under actual test conditions this requirement is

Table 4-15a Field Data Sheet for Isochronal Test of Well C.[15]

Type Test: ☐ Initial	☐ Annual	☒ Special	Test Date	Lease No. or Serial No.
Company: Lone Star		Connection		Allottee
Field: San Carlos	Reservoir: Hidalgo County, Texas		Location	Unit
Completion Date	Total Depth: 7,500	Plug Back TD: 7,467	Elevation: Casing	Farm or Lease Name
Csg. Size: 7 in. · Wt · d	Set At: 7,500	Perforations: From	To	Well No.
Tbg. Size: 2-7/8" · Wt: 6.5# · d	Set At: 7,345	Perforations: From	To	Sec. - - Twp – Blk Rge – Sur - - -
Type Completion (Describe)			Packer Set At	County or Parish
Producing Thru: Tubing	Reservoir Temp, °F: @ 240	Mean Ground Temp, °F: 75	Baro. Press, – P_a: 14.66	State
L: 7334	H: 7388	G_g: 0.621	% CO_2 · % N_2: 2 · % H_2S	Prover - - - Meter Run Taps - - - 4.026

DATE Time of Reading	ELAP. TIME hr	WELLHEAD WORKING PRESSURE Tbg. Psig	Csg, Psig	Temp, °F	METER OR PROVER Pressure, Psig	Diff.	Temp, °F	Orifice	REMARKS (Include liquid production data: Type – API Gravity – Amount)
			WELL SHUT-IN FOR 72 HOURS						
4-28-59									
8:00	0	2,325	2,337	72					S.I.P.
8:15	0								Open to Flow
8:30	.25	2,272	--	79.5	7.0	2.0	78	1-5/8	Meter Chart
8:45	.50	2,252	2,265	83.5	7.1	2.1	75.5		Gas Producing Condensate
9:00	.75	2,244	2,259	85.0	7.1	2.2	76.0		
9:13	1.00	2,237	2,252	87.0	7.1	2.2	78.5		
9:15									Shut-in
11:13		2,317	2,330						S.I.P.
11:15	0							1-5/8	Open to Flow
11:30	.25	2,220	2,242	93	6.99	4.55	84		Gas Producing Condensate
11:45	.50	2,170	2,195	94	6.99	4.41	82		
12:00	.75	2,141	2,172	96	6.99	4.39	81		
12:13	1.00	2,123	2,153	97	6.98	4.38	80		
12;15									Shut-in
2;13		2,312	2,327						S.I.P.
2:15	0							1-5/8	Open to Flow
2:30	.25	2,132	2,180	100	6.95	6.71	81.5		Gas Producing Condensate
2:45	.50	2,050	2,112	100	6.95	6.50	79.5		
3:00	.75	2,032	2,075	101	6.95	6.30	79.5		
3:13	1.00	2,012	2,059	102	6.97	6.30	78.5		
3:15									Shut-in
5:13		2,313	2,326						S.I.P.
5:15	0							1-5/8	Open to Flow
5:30	.25	2,050	2,112	95.5	7.00	9.50	78.0		Gas Producing Condensate
5:45	.50	1,943	2,001	96.0	7.00	8.98	81.5		
6:00	.75	1,873	1,937	97.0	7.01	8.75	81.5		
6:13	1.00	1,843	1,908	97.0	7.02	8.61	81.0		
6:15									Shut-in

rarely satisfied. However, the condition may be approximated by starting a well on production and allowing it to produce without further outside or mechanical adjustments in rate of flow.

The performance coefficient, C, in Eqs. **1** and **2** is considered a variable with respect to the passage of time, but a constant with respect to a specific time.[4] Thus, the back-pressure performance of a well producing from a reservoir with low permeability is a series of parallel curves. Each curve represents the performance of the well at the end of a given time interval. Isochronal performance curves are closely spaced for wells producing from reservoirs with relatively higher permeability.

Procedure. The isochronal method of testing permits the determination of the true exponent, n, of the performance curve for a given gas well. This is accomplished by establishing a simple pressure gradient around a producing well during the test period, which prevents the variation of the performance coefficient with time from obscuring the true value of the exponent. The determination of the relationship between performance coefficient and time permits the estimation of the rate of flow of a given well into a pipeline over long periods of time.

The presentation of isochronal test data as a series of parallel curves with a constant exponent, n, and a constant performance coefficient, C, for a specific time interval involves certain assumptions. The exponent of the performance curves for a gas well is assumed to be independent of the drainage area. It is established immediately after the well is opened. The variation of the performance coefficient with respect to time is believed to be independent of the rate of flow and the pressure level under simple gradient conditions.

The procedure is to open a well from shut-in conditions and obtain rate of flow and pressure data at specific time intervals during the flow period without disturbing the rate of flow. After data have been obtained, the well is shut in and allowed to return to a shut-in condition comparable to that existing when the well was first opened. The well is then reopened at a different rate of flow, and data are obtained for the same time intervals as before, to obtain the desired number of data points.

With the exception of starting each rate of flow from shut-in conditions, the procedure for running isochronal tests is the same as that for the multipoint test. The necessity for cleaning the well, calibrating the gas measuring equipment, and accurately measuring pressures and temperatures remains the same. At least four rates of flow should be taken. The lowest rate should reduce the pressure at the wellhead about five per cent, and the pressure reduction for the highest rate of flow should be about 25 per cent.

Examples of Isochronal Tests. Field data and corresponding results for Well C are given in Tables 4-15a and 4-15b. This well produced approximately 22 bbl of condensate per day. Four rates of flow of 1-hr duration were used, with each flow starting from shut-in conditions. Reported shut-in pressures varied from 2325 psig (before the first rate of flow) to 2314 psig (just previous to the fourth rate of flow). The difference in shut-in pressure between the first rate of flow and the second rate of flow was due to the liquid or condensate column in the tubing. Bottom-hole and wellhead performance curves are given in Figs. 4-24a and 4-24b.

Table 4-15b Results of Isochronal Tests of Well C[15]

P_c, psia	Duration of flow, hr	Q, MCFD	$(P_f^2 - P_s^2)$, thousands	$(P_c^2 - P_t^2)$, thousands
2332	0.25	800	313	208
	0.50	855	453	299
	0.75	895	507	335
	1.00	891	566	367
2332	0.25	1792	667	443
	0.50	1742	981	664
	0.75	1737	1161	790
	1.00	1733	1278	867
2332	0.25	2631	1181	829
	0.50	2556	1666	1174
	0.75	2478	1783	1248
	1.00	2490	1899	1329
2332	0.25	3774	1568	1174
	0.50	3550	2193	1604
	0.75	3462	2584	1874
	1.00	3418	2741	1986

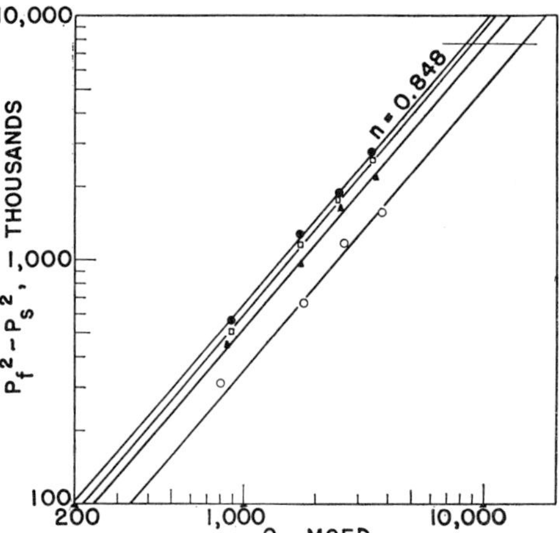

Fig. 4-24a Isochronal test showing bottom-hole performance of Well C.[15]

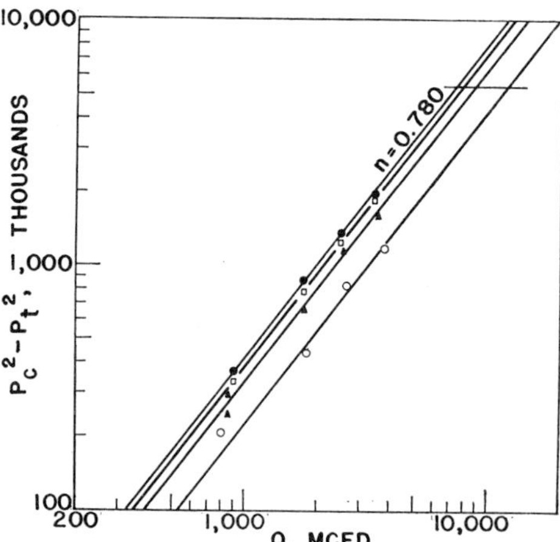

Fig. 4-24b Isochronal test showing wellhead performance of Well C.[15]

Figure 4-24a shows that the calculated open flow potential for a bottom-hole pressure of 2771 psia was 14,300, 10,100, 8900, and 8400 MCFD at the end of 0.25, 0.50, 0.75, and 1.0 hr, respectively. Thus, the calculated potential at the end of 1 hr of flow was only 59 per cent of the potential after 0.25 hr of flow. If Well C were opened into a pipeline with a constant pressure, the rate of flow at the end of 1 hr would be 61 per cent of the rate of flow at 0.25 hr. Data (not given here) showed that the production at the end of 52 hr decreased to about 55 per cent of that at 0.25 hr. The foregoing data showing change of performance characteristics with time illustrate the need for isochronal test data in estimating the delivery from a particular well into a pipeline.

Examination of the field notes under the "remarks" column on Table 4-15a indicates that Well C was producing liquid during the entire flow test. The effect of liquid production on well performance is illustrated by the irregularities in the corresponding data on Figs. 4-24a and 4-24b, especially on the first points on the 0.25 and 0.50-hr curves. Liquid production of any kind or the accumulation of liquids in the wellbore causes the performance characteristics of a well to deteriorate.

Comparison of Multipoint and Isochronal Tests (for Well A)

The results of the 2-hr multipoint test and the 2-hr isochronal test on Well A are shown together in Fig. 4-25 as wellhead performance curves. Note that in this Figure and in general, exponents of multipoint curves are less than those for isochronal curves for the same well. The first data point on the multipoint test (Q = 1105 MCFD) is on the isochronal curve (Fig. 4-25) because the first rate of flow for the multipoint curve was started from shut-in conditions. Thereafter, the position of each succeeding point of the multipoint test is influenced not only by the rate of flow, but also by each preceding point.

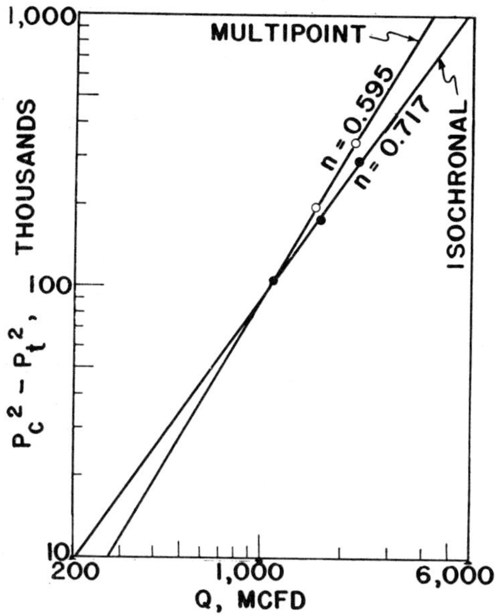

Fig. 4-25 Comparison of a multipoint and an isochronal test of Well A for wellhead performance.[15]

The initial points of each multipoint test on wells producing from reservoirs with low permeability represent the *formation* characteristics, while other points represent complex conditions which are almost impossible to interpret. The characteristic exponent of the isochronal curve still applies to the complex points, with the only difference in performance being in the performance coefficient C. If the exponent 0.717 is applied to the complex points of the multipoint test (Fig. 4-25), it can be seen that the coefficient obtained in each case can be considered as the result of an "effective" time, which has no permanent significance because it is neither equal to the total elapsed time nor the elapsed time since the last change in flow rate.

Equations for Computing Subsurface Pressures

Equations[17] for the practical solution of gas-flow problems have been adopted by the major conservation agencies. Outlined here are the steps best suited for direct calculation of subsurface pressures and equivalent static wellhead pressures. Those methods outlined for **low-pressure wells** are based on the assumption that both temperature and compressibility can be considered constant below 2000 psig. Shut-in pressures—above 2000 psig—use a numerical procedure in steps.

The general flow equation for gas wells is:

$$\frac{1000G_mL}{53.33} = \int_{P_2}^{P_1} \frac{P/TZ}{(F_rQ_m)^2 + (H/L)(\bar{P}/TZ)^2}\,dP \qquad (7)$$

where:

G_m = specific gravity of fluid flowing in well (air = 1.00)
L = length of flow string in well, corresponding to H, ft
53.33 = specific gas constant for air, ft per °F
P_1 = subsurface pressure at $H/2$, psia
P_2 = subsurface pressure at H, psia
T = temperature, °R
Z = compressibility factor
dP = the mathematical term "differential of P"
F_r = friction factor = $(2.6665f/d^5)^{0.5}$
  where: f = coefficient of friction
  d = internal diameter, in.
Q_m = flow rate, MMCFD
H = vertical depth in a well, ft. In *untubed wells*, H is the vertical depth to the midpoint of the productive formation. In *tubed wells*, H is the vertical depth to the entrance to the tubing.
$\bar{P} = P/\sqrt{1000}$

For positive (upward) inclined flow, Eq. **7** becomes:

$$\frac{1000G_mH}{53.33} = \int_{P_t}^{P_s} \frac{P/TZ}{(F_rQ_m)^2(L/H) + (\bar{P}/TZ)^2}\,dP \qquad (8)$$

where:

P_s = subsurface flowing pressure in the wellbore at vertical depth H, psia
P_t = flowing pressure at wellhead measured on a flowing column of gas, psia

Other terms are defined under Eq. **7**.

In numerical calculations, Eq. **8** is reduced to:

$$37.5G_mH = (P_1 - P_t)(I_1 + I_t) + (P_2 - P_1)(I_2 + I_1) \qquad (9)$$

where:

$$I_1 = \frac{P_1/T_1Z_1}{\frac{L}{H}F^2 + (\bar{P}_1/T_1Z_1)^2}$$

$$I_t = \frac{P_t/T_tZ_t}{\frac{L}{H}F^2 + (\bar{P}_t/T_tZ_t)^2}$$

$$I_2 = \frac{P_2/T_2Z_2}{\frac{L}{H}F^2 + (\bar{P}_2/T_2Z_2)^2}$$

$$F^2 = (F_rQ_m)^2$$

F_r = friction characteristic of flow string (Table 4-16) or annuli (Table 4-17); for values not included in these tables see Table 4-18; subscripts t, 1, and 2 indicate conditions at wellhead: when flowing, at $H/2$, and H, respectively.

Other terms are defined under Eqs. **7** and **8**.

Table 4–16 Friction Factors, F_r, for Various Flow Strings[17]

Nominal size, in.	OD, in.	Weight, lb/ft	ID, in.	F_r
1	1.315	1.80	1.049	0.095288
1¼	1.660	2.40	1.380	.046552
1½	1.990	2.75	1.610	.031122
2	2.375	4.70	1.995	.017777
2½	2.875	6.50	2.441	.010495
3	3.500	9.30	2.992	.006167
3½	4.000	11.00	3.476	.004169
4	4.500	12.70	3.958	.002970
4½	4.750	16.25	4.082	.002740
	4.750	18.00	4.000	.002889
4¾	5.000	18.00	4.276	.002427
	5.000	21.00	4.154	.002617
	5.000	13.00	4.494	.0021345
	5.000	15.00	4.408	.0022437
5³⁄₁₆	5.500	14.00	5.012	.0016105
	5.500	15.00	4.976	.0016408
	5.500	15.50	4.950	.0016631
	5.500	17.00	4.892	.0017145
	5.500	20.00	4.778	.0018221
	5.500	23.00	4.670	.0019329
	5.500	25.00	4.580	.0020325
5⅝	6.000	15.00	5.524	.0012528
	6.000	17.00	5.450	.0012972
	6.000	20.00	5.352	.0013595
	6.000	23.00	5.240	.0014358
	6.000	26.00	5.140	.0015090
6¼	6.625	20.00	6.049	.00099103
	6.625	20.00	5.989	.0010169
	6.625	24.00	5.921	.0010473
	6.625	26.00	5.855	.0010781
	6.625	28.00	5.791	.0011091
	6.625	31.80	5.675	.0011686
	6.625	34.00	5.595	.0012122
6⅝	7.000	20.00	6.456	.00083766
	7.000	22.00	6.398	.00085741
	7.000	23.00	6.366	.00086858
	7.000	24.00	6.336	.00087924
	7.000	26.00	6.276	.00090111
	7.000	28.00	6.214	.00092451
	7.000	30.00	6.154	.00094795
	7.000	40.00	5.836	.0010871
7¼	7.625	26.40	6.969	.00068759
	7.625	29.70	6.875	.00071213
	7.625	33.70	6.765	.00074241
	7.625	38.70	6.625	.00078360
	7.625	45.00	6.445	.00084136
	8.000	26.00	7.386	.00059178
7⅝	8.125	28.00	7.485	.00057179
	8.125	32.00	7.385	.00059199
	8.125	35.50	7.285	.00061320
	8.125	39.50	7.185	.00063548
8¼	8.625	17.50	8.249	.00044488
	8.625	20.00	8.191	.00045306
	8.625	24.00	8.097	.00046677
	8.625	28.00	8.003	.00048106
	8.625	32.00	7.907	.00049623
	8.625	36.00	7.825	.00050982
	8.625	38.00	7.775	.00051833
	8.625	43.00	7.651	.00054030
8⅝	9.000	34.00	8.290	.00043922
	9.000	38.00	8.196	.00045235
	9.000	40.00	8.150	.00045897
	9.000	45.00	8.032	.00047658
9	9.625	36.00	8.921	.00036342
	9.625	40.00	8.835	.00037264
	9.625	43.50	8.755	.00038149
	9.625	47.00	8.681	.00038995
	9.625	53.50	8.535	.00040741
	9.625	58.00	8.435	.00042000
9⅝	10.000	33.00	9.384	.00031893
	10.000	55.50	8.908	.00036481
	10.000	61.20	8.790	.00037759
10	10.750	32.75	10.192	.00025767
	10.750	35.75	10.136	.00026136
	10.750	40.00	10.050	.00026718
	10.750	45.50	9.950	.00027417
	10.750	48.00	9.902	.00027761
	10.750	54.00	9.784	0.00028634

The final evaluation of P_s, the flowing pressure at a vertical depth of H in the well, is computed by Eq. **10**.

$$P_s = P_t + \Delta P \qquad (10)$$

where:

$$\Delta P = \frac{3(37.5G_mH)}{I_t + 4I_1 + I_2}$$

Other terms are defined under Eqs. **7**, **8**, and **9**.

The details of computations of a flowing bottom-hole pressure by Eqs. **9** and **10** are illustrated by **Example 4.**

For a shut-in gas well, $(Q_m = 0)$, Eq. **7** becomes:

$$\frac{G_mH}{53.33} = \int_{P_c}^{P_f} \frac{TZ}{P}\,dP \qquad (11)$$

where:

P_f = subsurface shut-in pressure in the reservoir at vertical depth, H, psia

P_c = shut-in pressure at wellhead, psia

Other terms are defined under Eq. **7**.

Or for numerical computations:

$$0.0375G_mH = (P_1 - P_c)(I_1 + I_c) + (P_2 - P_1)(I_2 + I_1) \qquad (12)$$

where: $I_c = \dfrac{T_cZ_c}{P_c}$; etc.

Other terms are defined under Eqs. **7** and **11**.

The final evaluation of P_f, the shut-in pressure at a vertical depth of H in the well, is computed by Eq. **13**.

$$P_f = P_c + \Delta P \qquad (13)$$

where: $\Delta P = \dfrac{3(0.0375G_mH)}{I_c + 4I_1 + I_2}$

Other terms are defined under Eqs. **7**, **9**, **11**, and **12**.

The details of computations of a shut-in bottom-hole pressure by Eqs. **12** and **13** are given by **Example 3.**

In **low-pressure wells** (shut-in wellhead pressures less than 2000 psia) temperature, T, and compressibility, Z,

Table 4-17 Friction Factors, F_r, for Various Annuli[17]

(F_r values within table*)

Casing ID, in.	Tubing OD, in. 1.990	2.375	2.875	3.500	4.000	4.500	4.750	5.000
4.154	0.0050631	0.0065339	0.010331					
4.276	.0045445	.0057636	.0087675					
4.408	.0040656	.0050718	.0074456					
4.494	.0037919	.0046847	.0067389					
4.580	.0035440	.0043394	.0061280	0.011804				
4.670	.0033088	.0040164	.0055730	.010260				
4.778	.0030551	.0036732	.0050005	.0087831				
4.892	.0028166	.0033552	.0044860	.0075490				
4.950	.0027053	.0032087	.0042539	.0070197				
4.976	.0026575	.0031459	.0041555	.0068006				
5.012	.0025933	.0030620	.0040248	.0065138	0.011753			
5.140	.0023604	.0027884	.0036060	.0056294	.0095637			
5.240	.0022154	.0025987	.0033220	.0050586	.0082610			
5.352	.0020678	.0023816	.0030413	.0045167	.0071001			
5.450	.0019499	.0022343	.0028232	.0041110	.0062769			
5.524	.0018671	.0021316	.0026734	.0038397	.0057479	0.010367		
5.595	.0017923	.0020394	.0025404	.0036037	.0053016	.0092398		
5.675	.0017131	.0019423	.0023715	.0033629	.0048589	.0081822		
5.791	.0016068	.0018128	.0021942	.0030541	.0043101	.0069508	0.0096004	
5.836	.0015681	.0017660	.0021306	.0029455	.0041221	.0065486	.0089292	
5.855	.0015521	.0017467	.0021046	.0029014	.0040465	.0063895	.0086675	
5.921	.0014985	.0016821	.0020177	.0027555	.0037993	.0058805	.0078448	
5.989	.0014460	.0016190	.0019335	.0026163	.0035676	.0054183	.0071172	
6.049	.0014018	.0015661	.0018633	.0025019	.0033804	.0050551	.0065582	0.0090467
6.154	.0013289	.0014794	.0017493	.0022860	.0030866	.0045031	.0057301	.0076842
6.214	.0012897	.0014329	.0016886	.0021929	.0029355	.0042279	.0053270	.0070434
6.276	.0012508	.0013870	.0016290	.0021025	.0027907	.0039696	.0049545	.0064642
6.336	.0012148	.0013446	.0015742	.0020202	.0026606	.0037417	.0046306	.0059709
6.366	.0011973	.0013241	.0015478	.0019808	.0025988	.0036352	.0044807	.0057458
6.398	.0011791	.0013027	.0015204	.0019400	.0025353	.0035265	.0043288	.0055197
6.445	.0011530	.0012722	.0014813	.0018823	.0024462	.0033756	.0041196	.0052117
6.456	.0011470	.0012652	.0014724	.0018692	.0023861	.0033417	.0040729	.0051434
6.625	.0010603	.0011642	.0013445	.0016832	.0021135	.0028794	.0034457	.0042463
6.765	.00099541	.0010891	.0012503	.0015491	.0019214	.0025657	.0030307	.0036725
6.875	.00094827	.0010348	.0011829	.0014544	.0017881	.0023128	.0027555	.0033006
6.969	.00091045	.00099146	.0011293	.0013800	.0016848	.0021573	.0025492	.0030264
7.185	.00083125	.00090112	.0010187	.0012289	.0014786	.0018543	.0021190	.0025135
7.285	.00079786	.00086323	.00097275	.0011670	.0013955	.0017352	.0019718	.0022765
7.385	.00076633	.00082756	.00092969	.0011096	.0013192	.0016272	.0018394	.0021101
7.386	0.00076602	0.00082722	0.00092928	0.0011090	0.0013184	0.0016262	0.0018382	0.0021086

* For explanation of broken horizontal line, see footnote, Table 4-18.

may be considered constant. Thus, static column Eq. **13** may be simplified to:

$$P_f^2 = e^s P_c^2 \quad (14)$$

where: $e = 2.7183$

$s^* = 0.0375 G_m H / TZ$

Other terms are defined under Eqs. **7** and **11**.

And flowing column Eq. **10** may be simplified to:

$$P_s^2 = e^s P_t^2 + L/H(F_r Q_m TZ)^2 (e^s - 1) \quad (15)$$

Terms are defined under Eqs. **7, 8, 9, 11**, and **14**.

* When used as an exponent only.

The use of Eqs. **14** and **15** is illustrated in **Examples 1** and **2**, respectively.

Packers. In computing subsurface pressures in wells equipped with tubing set without a packer, the preferred practice is to calculate the flowing subsurface pressure from the wellhead pressure, measured on the static gas column by means of the static column equations (e.g., Eq. **14**). If the well contains a packer, it is necessary to calculate the flowing subsurface pressure by means of the equations for flowing gas columns (e.g., Eq. **15**).

Depths for calculating or measuring subsurface pressures in wells are determined in practice according to the equipment installed in the well. Where a well is equipped without

Table 4-18 Formulas for F_r, Friction Factors[3]

1. For internal diameters less than 4.277 in.:

$$F_r = \frac{0.10797}{d^{2.612}}$$

2. For internal diameters greater than 4.277 in.:

$$F_r = \frac{0.10337}{d^{2.582}}$$

3. For *annulus,* where F_r is greater than 0.0024278:*

$$F_r = \frac{0.10797}{(d_1 + d_2)(d_1 - d_2)^{1.612}}$$

4. For *annulus,* where F_r is less than 0.0024278:*

$$F_r = \frac{0.10337}{(d_1 + d_2)(d_1 - d_2)^{1.582}}$$

where: d_1 = ID of casing, in.
d_2 = OD of tubing, in.

Note: Use these formulas for wells not covered by Tables 4-16 and 4-17.

*The broken horizontal lines in Table 4-17 are the break-even lines for $F_r = 0.0024278$ and may be used to select between Formulas 3 and 4. For example: use Formula 4 to find the F_r between 5½ in. ID casing and 1¼ in. tubing, where the tubing has an OD of 1.660 in. However, the friction factor for flow between 5-in. ID casing and 2-in. tubing with an OD of 2.375 in. would be calculated by Formula 3.

TABLE 4-19 **WORK SHEET FOR CALCULATION OF SUBSURFACE PRESSURES P_f, and P_s BY EQUATIONS 14 and 15**

COMPANY Consolidated Gas Utilities Corporation LEASE N.W. Mayflower WELL NO. A DATE 2-18-56

L 4,530 H 4,530 L/H 1.000 G_m 0.661 % CO_2 0 % N_2 0 % H_2S 0

d 1.995 F_r 0.0178 G_mH 2994 P_{cr} 670 T_{cr} 378

LINE		EXAMPLE 1 S.I.P.		EXAMPLE 2 1st Flow		2nd Flow		3rd Flow	
		1st Trial	2nd Trial	1st Trial	2nd Trial	1st Trial	2nd Trial	1st Trial	2nd Trial
1	Q_m	0	0	1.105	1.105	1.588	1.588	2.201	2.201
2	T_w	512	512	517	517	520	520	523	523
3	T_s	588	588	588	588	588	588	588	588
4	T	550	550	553	553	554	554	556	556
5	Z	0.836	0.827	0.847	0.838	0.855	0.846	0.866	0.855
6	TZ	459.8	454.9	468.4	463.4	473.7	468.7	481.5	475.4
7	G_mH/TZ	6.512	6.582	6.392	6.461	6.320	6.388	6.218	6.298
8	e^s	1.277	1.280	1.271	1.274	1.267	1.271	1.263	1.266
9	P_c or P_t	1,041.2	1,041.2	988.7	988.7	941.8	941.8	862.9	862.9
10	$\bar{P}_c^2$ or $\bar{P}_t^2$	1,084.0	1,084.0	977.5	977.5	886.9	886.9	744.6	744.6
11	$e^s(\bar{P}_c^2$ or $\bar{P}_t^2)$	1,384.3	1,387.5	1,242.4	1,245.3	1,123.7	1,127.2	940.4	942.7
12	F_r	---	---	0.0178	0.0178	0.0178	0.0178	0.0178	0.0178
13	$F_c = F_rTZ$	---	---	8.338	8.249	8.432	8.343	8.571	8.462
14	F_cQ_m	---	---	9.213	9.115	13.390	13.249	18.865	18.625
15	$L/H\ (F_cQ_m)^2$	---	---	84.879	83.083	179.292	175.536	355.888	346.891
16	$F_s = L/H(F_cQ_m)^2(e^s - 1)$	---	---	23.002	22.765	47.871	47.570	93.599	92.273
17	$P_s^2 = e^sP_t^2 + F_s$	---	---	1,265.4	1,268.1	1,171.6	1,174.6	1,034.0	1,035.0
18	P_f or P_s	1,176.1	1,177.9	1,124.8	1,126.1	1,082.4	1,083.8	1,016.9	1,017.3
19	P	1,108.7	1,109.5	1,056.7	1,057.4	1,012.1	1,012.8	939.9	940.1
20	P_r	1.65	1.66	1.58	1.58	1.51	1.51	1.40	1.40
21	T_r	1.45	1.46	1.46	1.46	1.47	1.47	1.47	1.47
22	Z	0.827	0.827	0.838	0.838	0.846	0.846	0.855	0.855

tubing or with tubing set without a packer, the proper depth for pressure determinations is the distance to the midpoint of the productive sand face. If the well has tubing set with a packer, the pressures are determined at the entrance to the tubing.

Example 1—Low-Pressure Well (Static Column). The shut-in subsurface pressure, P_f, is calculated by means of Eq. **14**, as illustrated under "S.I.P." in Table 4-19 for Well A. This well (Tables 4-14a and 4-14b) had a wellhead shut-in pressure, P_c, of 1041.2 psia, $H = L = 4530$ ft, $G_m = 0.661$, $CO_2 = 0$ per cent, $N_2 = 0$ per cent, and wellhead temperature = 52 F. Calculations for the required shut-in subsurface pressure, P_f, follow:

First Trial of S.I.P. (Table 4-19)*

Enter well information as shown at top of Table 4-19.

Determine H, the vertical distance from the midpoint of the flow string to the wellhead, for the well. Calculate L/H, the length of the flow string divided by the vertical distance. In most gas wells L/H is unity (L/H is more than unity for directionally drilled wells).

$G_mH = 0.661 \times 4530 = 2994.$

Line 1. Q_m = rate of flow, MMCFD = 0.

Line 2. T_w = wellhead temperature, °R = 52 + 460 = 512.

Line 3. T_s = bottom-hole temperature, °R = 128 + 460 = 588 (measured or estimated from reliable data on other wells in the area—usually given on electric logs).

Line 4. $T = (T_w + T_s)/2$, °R = (512 + 588)/2 = 550.

Line 5. Z = estimated effective compressibility = 0.836 (at $T_r = T_w/T_{cr}$ and $P_r = P_w/P_{cr}$).

Line 6. $TZ = 550 \times 0.836 = 459.8.$

Line 7. $G_mH/TZ = 2994/459.8 = 6.512$; therefore, $s = 0.0375G_mH/TZ = 0.0375 \times 6.512 = 0.2442.$

Line 8. $e^s = 2.7183^{0.2442} = 1.277.$

Line 9. P_c = wellhead shut-in pressure, psia = 1041.2.

Line 10. $\bar{P}_c^2 = (1041.2)^2/1000 = 1084.0.$

Line 11. $e^s\bar{P}_c^2 = 1.277 \times 1084.0 = 1384.3.$

Line 12 thru 17. Not used in the static column calculation.

Line 18. $P_f = (e^sP_c^2)^{0.5} = (1384.3)^{0.5} = 1176.1.$

Line 19. $P = (P_c + P_f)/2 = (1041.2 + 1176.1)/2 = 1108.7.$

Line 20. $P_r = P/P_{cr} = 1108.7/670 = 1.65.$

Line 21. $T_r = T/T_{cr} = 550/378 = 1.45.$

Line 22. Z = compressibility (at $P_r = 1.65$ and $T_r = 1.45$) = 0.827.

Second Trial of S.I.P. (Table 4-19)

Line 5. Since, in the *First Trial*, Z (*Line 22*) is not equal to Z (*Line 5*), enter $Z = 0.827$.

Line 6 thru 22. Calculate as per *First Trial.*

Since the final value of Z (*Line 22* of *Second Trial*) is equal to the assumed value of Z (*Line 5* of *Second Trial*), the value of $\bar{P}_f^2$ = 1387.5 and P_f = 1177.9 is used in the back-pressure computations of Table 4-14b for Well A.

Example 2—Low-Pressure Well (Flowing Column). The flowing subsurface pressure, P_s, is calculated by means of Eq. **15,** as illustrated under "1st Flow" and "2nd Flow" in Table 4-19 for Well A (Tables 4-14a and 4-14b). Calculations for the required flowing subsurface pressure, P_s, follow:

First Trial of First Flow (Table 4-19)

Line 1. Q_m = rate of flow, MMCFD = 1.105.

Line 2. T_w = wellhead temperature, °R = 57 + 460 = 517 (from Tables 4-14a and 4-14b).

Line 3. T_s = bottom-hole temperature, °R = 128 + 460 = 588.

* Shut-in pressure.

Line 4. $T = (T_w + T_s)/2$, °R = (517 + 588)/2 = 553.

Line 5. Z = estimated effective compressibility = 0.847.

Line 6. $TZ = 553 \times 0.847 = 468.4.$

Line 7. $G_mH/TZ = 2994/468.4 = 6.392$; therefore, $s = 0.0375G_mH/TZ = 0.0375 \times 6.392 = 0.2397.$

Line 8. $e^s = 2.7183^{0.2397} = 1.271.$

Line 9. P_t = wellhead working pressure, psia = 988.7.

Line 10. $\bar{P}_t^2 = (988.7)^2/1000 = 977.5.$

Line 11. $e^s\bar{P}_t^2 = 1.271 \times 977.5 = 1242.4.$

Line 12. F_r = 0.0178 (for flow string with d = 1.995 in. ID from Table 4-16).

Line 13. $F_c = F_rTZ = 0.0178 \times 468.4 = 8.338.$

Line 14. $F_cQ_m = 8.338 \times 1.105 = 9.213.$

Line 15. $(L/H)(F_cQ_m)^2 = 1.000 \times 9.213^2 = 84.879.$

Line 16. $F_s = (L/H)(F_cQ_m)^2(e^s - 1) = 84.879 \times 0.271 = 23.002.$

Line 17. $\bar{P}_s^2 = es\bar{P}_t^2 + F_s = 1242.4 + 23.002 = 1265.4.$

Line 18. $P_s = (P_s^2)^{0.5} = (1265.4)^{0.5} = 1124.8.$

Line 19. $P = (P_t + P_s)/2 = (988.7 + 1124.8)/2 = 1056.7.$

Line 20. $P_r = P/P_{cr} = 1056.7/670 = 1.58.$

Line 21. $T_r = T/T_{cr} = 553/378 = 1.46.$

Line 22. Z = compressibility (at $P_r = 1.58$ and $T_r = 1.46$) = 0.838.

Second Trial of First Flow (Table 4-19)

Line 5. Since, in the *First Trial*, Z (*Line 22*) is not equal to Z (*Line 5*), enter $Z = 0.838$.

Lines 6 thru 22. Calculate as per *First Trial.*

Since the final value of Z (*Line 22* of *Second Trial*) is equal to the assumed value of Z (*Line 5* of *Second Trial*), the value of $\bar{P}_s^2$ = 1268.1, and P_s = 1126.1 is then used in the back-pressure computations of Table 4-14b for Well A.

Example 3—High-Pressure Well (Static Column). The shut-in subsurface pressure, P_f, at H = 8072 ft is calculated by means of Eqs. **12** and **13**, as illustrated in Table 4-20 for Well B. This well had a wellhead shut-in pressure, P_c, of 2191 psia, $G_m = 0.642$ and $CO_2 = 0$ per cent. Temperature at string midpoint of $H/2$ ft may be estimated as the arithmetic average of subsurface or bottom-hole temperature (at H = 8072 ft) and wellhead temperature (just before shutting in the well), or (240 + 130)/2 = 185 F at a depth of 4036 (8072/2) ft. Calculation details follow:

Line 1 (Table 4-20)

Col. 1. H = depth, ft = 0.

Col. 2. P_c = wellhead shut-in pressure, psia = 2191.

Col. 3. P_r = reduced pressure = P_c/P_{cr} = 2191/670 = 3.27.

Col. 4. T = wellhead temperature, °R = 130 + 460 = 590.

Col. 5. T_r = reduced temperature = T/T_r = 590/373 = 1.58.

Col. 6. Z = compressibility (at $P_r = 3.27$ and $T_r = 1.58$) = 0.811.

Col. 7. $TZ = 590 \times 0.811 = 478.490.$

Col. 8. $I_c = TZ/P = 478/2191 = 0.218389.$

Line 2 (Table 4-20)

Col. 1. First trial calculation for the pressure at $H/2$ = 8072/2 = 4036 ft follows:

Col. 13. $0.0375G_mH = 0.0375 \times 0.642 \times 4036 = 97.167.$

Col. 9. $M = 0.0375G_mH/2I_c = 97.17/(2 \times 0.218) = 222.$

Col. 2. Add M to P_c = 2191 + 222 = 2413.

Cols. 3 thru 8. Calculate as per *Line 1.* (Note that I_1 is found in *Col. 8.*)

Col. 10. $N = I_c$ plus trial I_1 = 0.218389 + 0.231751 = 0.450140.

Check Col. 9. $0.0375G_mH/N = 97.167/0.450140 = 216 \neq 222$; (if M has been estimated correctly, the value just determined would equal 222).

Line 3 (Table 4-20)

Col. 9. Enter 216 (just determined).

Cols. 2 thru 10. Calculate as per *Lines 1* and *2* until correct value of M is determined (per the "*Check Col. 9*").

Cols. 11 and 12. Final $MN = 216 \times 0.450718 = 97.355$.

Line 4 (Table 4-20)

Col. 1. First trial calculation for the pressure at $H = 8072$ ft follows:

Col. 13. $0.0375 G_m H = 0.0375 \times 0.642 \times 8072 = 194.333$.

Col. 9. M: Estimate value by dividing N (*Line 3, Col. 10*) into the difference, $0.0375 G_m H$ (*Col. 13*), minus ΣMN (*Line 3, Col. 12*). Thus, $(194.333 - 97.355)/0.450718 = 215$.

Col. 2. Add M to $P_1 = 2407 + 215 = 2622$.

Cols. 3 thru 8. Calculate as per *Lines 1* and *2*. (Note that I_2 is found in *Col. 8*.)

Col. 10. N: Add final I_1 and trial $I_2 = 0.232329 + 0.243478 = 0.475807$.

Check Col. 9. M: Divide N (*Col. 10*) into the difference, $0.0375 G_m H$ (*Col. 13*), minus ΣMN (*Line 3, Col. 12*). Thus, $(194.333 - 97.355)/0.475807 = 204 \neq 215$; (if M had been estimated correctly, the value just determined would equal 215).

Lines 5 and 6 (Table 4-20)

Col. 9. Enter 204 (just determined).

Cols. 2 thru 11. Calculate as per *Lines 1, 2,* and *3* until correct value of M is determined (per the "*Check Col. 9*").

Col. 12. Add MN (*Line 3*) and MN (*Line 6*) $= 97.355 + 96.816 = 194.171$.

Line 7 (Table 4-20)

Using Eq. **13**, first calculate ΔP by substituting $I_c = 0.218389$ (*Line 1, Col. 8*), $I_1 = 0.232329$ (*Line 3, Col. 8*), and $I_2 = 0.244598$ (*Line 6, Col. 8*); then substitute the value of ΔP, 419 psi, into Eq. **13**, thus:

$$P_f = 2191 + 419 = 2610 \text{ psia, and } \bar{P}_f{}^2 = 6812.$$

In this particular case P_f by Eq. **13** is equal to P_2 determined by Eq. **12** (as shown under P_n in *Line 6, Col. 2* of Table 4-20). However, all calculations do not agree in this manner with the result that the final step using Eq. **13** is necessary.

The value of shut-in pressure, P_f, for Well B was used to compute the back-pressure test illustrated in Fig. 4-23a.

Example 4—High-Pressure Well (Flowing Column). The flowing subsurface pressure, P_s, is calculated by means of Eqs. **9** and **10** as illustrated in Table 4-21 for Well B. This well had a wellhead flowing pressure, $P_t = 2122$ psia at a flow rate of 8.737 MMCFD. The testing was done

Table 4-20 Calculation Sheet for Static Column Pressures, P_f, of Well B

Company: Lone Star; Lease: Opelika; Date: 2-6-57; $G_m = 0.642$; $\%CO_2 = 0$; $\%N_2 = 0$; $P_{cr} = 670$; $T_{cr} = 373$.

	(1)	(2)	(3)	(4)	(5)	(6)	(7)	(8)	(9)	(10)	(11)	(12)	(13)
								I	M	N			
Line	H	P_n	P_r	T	T_r	Z	TZ	TZ/P	$P_n - P_{n-1}$	$I_n + I_{n-1}$	MN	ΣMN	$0.0375 G_m H$
1	0	2191	3.27	590	1.58	0.811	478.490	0.218389					0
2	4036	2413	3.60	645	1.73	.867	559.215	.231751	222	0.450140			97.167
3		2407	3.59	645	1.73	.867	559.215	.232329	216	.450718	97.355	97.355	
4	8072	2622	3.91	700	1.88	.912	638.400	.243478	215	.475807			194.333
5		2611	3.90	700	1.88	.912	638.400	.244504	204	.476833			
6		2610	3.90	700	1.88	0.912	638.400	0.244598	203	0.476927	96.816	194.171	
7	Substituting into Eq. **13**:												

$$\Delta P = \frac{3 \times 194.333}{0.218389 + 4(0.232329) + 0.244598} = \frac{582.999}{1.39230} = 419 \text{ psi}$$

$$P_f = 2191 + 419 = 2610 \text{ psia}$$

Table 4-21 Calculation Sheet for Subsurface Flowing Pressures, P_s, of Well B

Company: Lone Star; Lease: Opelika; Data: 2-6-57; Casing ID = 6.214 in.; Tubing OD = 2.875 in.; $F_r = 0.00169$; $Q_m = 8.737$; MMCFD = 0.642; $\%CO_2 = 0$; $\%N_2 = 0$; $P_{cr} = 670$; $T_{cr} = 373$.

	(1)	(2)	(3)	(4)	(5)	(6)	(7)	(8)	(9)	(10)	(11)	(12)	(13)	(14)	(15)	(16)	(17)
								A	B	C		I_n	M	N			
Line	H	P_n	P_r	T	T_r	Z	P/Z	P/TZ	$\frac{(P/TZ)^2}{1000}$	$(L/H)F^2$	$B + C$	$\frac{A}{B + C}$	$P_n - P_{n-1}$	$I_n + I_{n-1}$	MN	ΣMN	$37.5 G_m H$
1	0	2122	3.17	592	1.59	0.817	2597	4.3868	0.019244	0.000218	0.019462	225.403					0
2	4036	2338	3.49	646	1.73	.864	2706	4.1889	.017547	.000218	.017765	235.795	216	461.198			97,167
3		2333	3.48	646	1.73	.865	2697	4.1749	.017430	.000218	.017648	236.565	211	461.968			
4		2332	3.48	646	1.73	.865	2696	4.1734	.017417	.000218	.017635	236.654	210	462.057	97,032	97,032	
5	8072	2543	3.80	700	1.88	.911	2791	3.9871	.015897	.000218	.016115	247.415	211	484.069			194,333
6		2533	3.78	700	1.88	0.911	2780	3.9714	0.015772	0.000218	0.015990	248.368	201	485.022	97,489	194,521	
7	Substituting into Eq. **10**:																

$$\Delta P = \frac{3 \times 194{,}333}{225.403 + 4(236.654) + 248.368} = \frac{582{,}999}{1420.387} = 410 \text{ psi}$$

$$P_s = 2122 + 410 = 2532 \text{ psia}$$

with flow thru the annulus from the upper formation. Gas properties are given in foregoing **Example 3.**

The flow string was made up of 8400 ft of 7.00-in. OD (6.214-in. ID) casing set thru the Rodessa formation and a perforated liner extended thru the Travis Peak formation. Gas was produced from the Rodessa formation thru perforations in the casing from depth H at 8072 ft thru the annulus. Gas from the lower formation was produced thru 2.875-in. OD tubing on a packer set at 8870 ft.

Computations given in Table 4-21 solve Eq. **9** by trial and error. The steps in the computations follow:

F_r = 0.00169 (for annular flow through a 6.214-in. ID casing, 2.875-in. OD tubing combination [from Table 4-17]).

Temperature at string midpoint of $H/2$ ft may be estimated as the arithmetic average of subsurface or bottom-hole temperature (at H = 8072 ft) and wellhead temperature (just before shutting in the well) or (240 + 132)/2 = 186 F at a depth of 4036 (8072/2) ft. Calculation details follow:

Line 1 (Table 4-21)

Col. 1. H = depth, ft = 0.
Col. 2. P_t = wellhead flowing pressure, psia = 2122.
Col. 3. P_r = reduced pressure = P_c/P_{cr} = 2122/670 = 3.17.
Col. 4. T = wellhead temperature, °R = 132 + 460 = 592 R.
Col. 5. T_r = reduced temperature = T/T_{cr} = 592/373 = 1.59.
Col. 6. Z = compressibility (at P_r = 3.17 and T_r = 1.59) = 0.817.
Col. 7. P/Z = 2122/0.817 = 2597.
Col. 8. $A = P/TZ$ = 2597/592 = 4.3868.
Col. 9. $B = (P/TZ)^2/1000$ = $(4.3868)^2/1000$ = 0.019244.
Col. 10. $C = (L/H)F^2 = (L/H)(F_rQ_m)^2$ = $(1.000)(0.00169 \times 8.737)^2$ = 0.000218.
Col. 11. $B + C$ = 0.019244 + 0.000218 = 0.019462.
Col. 12. $I_t = A/(B + C)$ = 4.3868/0.019462 = 225.403.
Col. 17. $37.5G_mH$ = 37.5 × 0.642 × 0 = 0.

Line 2 (Table 4-21)

Col. 1. First trial calculation for the pressure at $H/2$ = 8072/2 = 4036 ft, follows:
Col. 17. $37.5G_mH$ = 37.5 × 0.642 × 4036 = 97,167.
Col. 13. Estimate $M = 37.5G_mH/2\,I_w$ = 97,167/(2 × 225.403) = 216.
Col. 2. Add M to P_w = 2122 + 216 = 2338.
Cols. 3 thru 12. Calculated as per *Line 1.*
Col. 14. $N = I_t$ plus trial I_1 = 225.403 + 235.795 = 461.198.
Check Col. 13. $37.5G_mH/N$ = 97,167/461.198 = 211 ≠ 216; (if M has been estimated correctly, the value just determined would equal 216).

Lines 3 and 4 (Table 4-21)

Col. 13. Enter 211 (just determined).
Cols. 2 thru 14. Calculate as per *Lines 1* and *2* until correct value of M is determined (per the "*Check Col. 13*").
Cols. 15 and 16. MN = 210 × 462.057 = 97,032.

Line 5 (Table 4-21)

Col. 1. First trial calculation for the pressure at H = 8072 ft (bottom of the producing formation) follows:
Col. 17. $37.5G_mH$ = 37.5 × 0.642 × 8072 = 194,333.
Col. 13. M: Estimate value by dividing N (*Line 4, Col. 14*) into the difference, $37.5G_mH$ (*Col. 17*), minus ΣMN (*Line 4, Col. 16*). Thus, (194,333 − 97,032)/462.057 = 211.
Col. 2. Add M to P_1 = 2332 + 211 = 2543.
Cols. 3 thru 12. Calculate as per *Lines 1, 2,* and *3.*
Col. 14. N: Add final I_1 and trial I_2 = 236.654 + 247.415 = 484.069.

Check Col. 13. M: Divide N (*Col. 14*) into the difference 37.5 G_mH (*Col. 17*), minus Σ MN (*Line 4, Col. 16*), thus (194,333 − 97,032)/484.069 = 201 ≠ 211; (if M has been estimated correctly, the value just determined would equal 211).

Line 6 (Table 4-21)

Col. 13. Enter 201 (just determined).
Cols. 2 thru 14. Calculate as per *Lines 1, 2, and 3* until correct value of M is determined.
Col. 16. Add MN (*Line 6*), and ΣMN (*Line 4, Col. 16*) = 97,489 + 97,032 = 194,521.

Line 7 (Table 4-21)

Using Eq. 10, first calculate ΔP by substituting I_t = 225.4 (*Line 1, Col. 12*), I_1 = 236.654 (*Line 4, Col. 12*), and I_2 = 248.368 (*Line 6, Col. 12*). Then, substitute the value of ΔP, 410 psi, into Eq. 10, thus: P_s = 2122 + 410 = 2532 psia at 8072 ft. The foregoing procedure must be carried out for each flow rate. The value of the flowing pressures, P_s, was used with the shut-in pressure of 2610 psia to compute the results of the back-pressure test for Well B, which are illustrated in Fig. 4-23a.

ESTIMATION OF GAS RESERVES

Methods used for estimating quantities of natural gas in underground reservoirs have been discussed thoroughly in technical literature.[18] Although methods have not been changed in recent years, means of obtaining data for reserve estimation have been improved greatly, and new tools for obtaining information have been developed.

Natural gas reserves have been classified by the Committee on Natural Gas Reserves of A.G.A., according to the nature of their occurrence, as follows:

1. **Non-associated gas** is free gas not in contact with crude oil in the reservoir.
2. **Associated gas** is free gas in contact with crude oil in the reservoir.
3. **Dissolved gas** is gas in solution with crude oil in the reservoir.

Methods of Estimating Non-Associated Gas Reserves

Comparison with other fields, volumetric estimation, and decline curve are methods which may be used to estimate gas reserves in place in the reservoir, but in actual practice **recoverable reserves** are of greatest interest. Their estimation requires prediction of an abandonment pressure at which further production from the wells will no longer be profitable. The abandonment pressure is determined principally by economic conditions such as future market value of gas, cost of operating and maintaining wells, and cost of compressing and transporting gas to consumers. Since these factors are quite variable, this discussion of estimation methods *is confined to reserves in place in the reservoir* and their recovery efficiencies.

Comparison with Other Fields. Judging the applicability of similar fields with known producing histories depends upon the experience of the estimator. When reserve estimates are required early in the producing life of a reservoir, for example after completion of one or two wells when practically no gas has been produced, experience in similar fields may be the only guide to reserve estimation.

Volumetric Method. Here pore space volume in the reservoir containing gas is converted to gas volume at standard conditions. Net volume of reservoir rock containing the gas reserve is determined by geological information based on cores, electric or radioactivity logs, drilling records, and drill stem and production tests. Reservoir rock volume is usually obtained by planimetering isopachous maps of productive reservoir rock or by the polygon method[19] for computing volumes.

Gas in place per acre-foot of reservoir volume may be calculated by the following equation:

$$V_{af} = 43{,}560\phi(1 - S_w)\frac{PT_b}{ZP_bT_r} \tag{16}$$

where:

V_{af} = volume of gas in place at P_b and T_b, cu ft per acre-ft
43,560 = cu ft per acre-ft
ϕ = average effective porosity, decimal fraction
S_w = connate water saturation in pore space, decimal fraction
P = reservoir pressure (assuming complete stabilization), psia
T_b = temperature base of gas measurement, °F abs
Z = compressibility factor for gas at P and T_r
P_b = pressure base of gas measurement, psia
T_r = reservoir temperature, °F abs

Although Eq. **16** may be used for any reservoir pressure, the development of natural water drive will reduce both reservoir pressure and efficiency of recovery. Results from studies of the efficiency of gas displacement from porous media by liquid flooding indicate that a residual gas saturation, ranging from 15 to 50 per cent of pore space, remains in the sand after water flooding.[20] This work indicates that recovery factors ranging from 80 to 95 per cent of gas in place, used by earlier estimators, are too high. Experience gained thru studies of similar fields depleted with natural water drives is helpful in determining recovery factors.

Rate of advance of water into a reservoir is related to pressure decrease created by production of gas, permeability of porous media, and area thru which water moves. Water influx rate may be estimated by material balance computations or by methods of Hurst.[21]

Because presence or absence of a water drive is important both in estimation of recoverable gas reserves and in performance of a gas field, pressure measurements with wells shut in and cumulative production data may be used to determine whether the field has an active water drive. Calculations may be made by the following:

$$AF = \frac{Q_1}{V_{af} - V_{af1}} \tag{17}$$

where:

AF = volume of void space in reservoir containing gas, acre-ft
Q_1 = cumulative production to reservoir pressure of P_1, cu ft at standard conditions
V_{af} = volume of gas in place initially, cu ft per acre-ft at standard conditions
V_{af1} = volume of gas in place at reservoir pressure of P_1, cu ft per acre-ft at standard conditions

In absence of a water drive, successive calculations of reservoir void volume by Eqs. **16** and **17** will show a constant value. Under water drive, each successive void volume will be apparently larger than the previous one. However, the conclusion that a particular field is subject to a water drive should be drawn only after study of all factors involved, such as rates of water production in wells. Substantial quantities of gas migrating to wells from an undeveloped part of reservoir, not considered in use of Eqs. **16** and **17**, will cause the same effect as a water drive.

Porosity. Average effective porosity of the reservoir rock may best be determined from study of core analyses. The porosity of clean sandstones may be calculated from electric logs if the resistivity of the sandstone filled with formation water, R_0, and if the resistivity of the formation water, R_w, are known. In limestone reservoirs, core analyses may be supplemented by porosity determinations from electric logging information. Average porosities may be determined with a fair degree of accuracy for homogeneous sandstones, but for limestones and heterogeneous sandstones estimation of an average porosity is at best uncertain. Also, in estimating net effective thickness of reservoir rock from porosity and permeability information, estimators must set arbitrary lower limits below which the rock is considered nonproductive. There is little agreement on those limits, and no standard figures can be set. In general, limits are lower for gas than for oil.

Connate-Water Saturation. Many early estimates of gas reserves were in error because they were made without considering space occupied by connate water. Even though determination of connate-water saturation has received considerable attention recently, such data may be lacking entirely for many older gas reservoirs and must be estimated. Connate-water saturations may be determined from electric logging information or by laboratory determinations run on cores by restored state, evaporation mercury injection, or centrifuge methods. Connate-water saturation may range as high as 50 per cent in some gas reservoirs.[22]

A detailed discussion and comparison of seven methods for determining the water content of reservoir rocks has been published.[23] The results of this study indicated that the best water content values were obtained from cores cut with a nonaqueous drilling fluid in the hole.

Reservoir Pressure and Temperature. Both may be measured directly with subsurface gages or reservoir pressure may be calculated from wellhead pressure by Eqs. **7**, **8**, etc. In using Eq. **16** to estimate gas reserves, it is assumed that the reservoir pressure used is completely stabilized. After a considerable quantity of gas has been produced from reservoirs with relatively low permeability, pressures stabilize very slowly, often requiring several days (or longer). Transient pressure conditions may exist for years in parts of the Monroe gas field.[24] Consequently, estimators should study pressure build-up characteristics of individual wells as well as pressure distribution throughout the reservoir. Where pressure varies over the reservoir, the average should be determined by volumetrically weighting individual pressures.

Example 5.[25] Estimate gas in place in a reservoir with a total volume of 127,000 acre-ft, average porosity of 21 per cent, connate-water saturation of 15 per cent, reservoir temperature of 186 F, initial reservoir pressure of 2651 psia, and reservoir gas compressibility factor of 0.880 at 186 F and 2651 psia. Using Eq. **16**:

$$V_{af} = 43{,}560\phi(1 - S_w)\,\frac{PT_b}{ZP_bT_r}$$

$$V_{af} = 43{,}560 \times 0.21 \times (1 - 0.15) \times \frac{2651 \times (460 + 60)}{0.880 \times 14.65 \times (460 + 186)}$$

$$V_{af} = 1{,}287{,}000 \text{ cu ft per acre-ft}$$

Gas in place = 1.287 MMCF per acre-ft × 127,000 acre-ft = 163,000 MMCF.

Decline Curve Methods. These techniques plot reservoir pressure (may be corrected for compressibility) against cumulative production and extrapolate the curve to an abandonment pressure to arrive at recoverable gas reserves. These methods and their variations are based on material balance concepts and the gas laws.

By means of the following assumptions, symbols, and gas laws, basic equations for decline curve methods may be developed:

1. Volume of pore space containing gas remains constant throughout productive life of reservoir (i.e., no water drive and no oil intrusion),
2. Shut-in reservoir pressures are representative of reservoir pressures,
3. Gas production data include all gas produced from reservoir, and
4. No gas is added to reserve by evolution of solution gas from oil.

V_0 = initial gas volume, SCF
V_1 = volume of gas remaining in reservoir after Q_1 cu ft of gas have been produced, SCF
Q_1 = cumulative volume of gas produced at a given time from reservoir, SCF
V_r = volume of reservoir pore space containing gas, cu ft
P_0 = initial reservoir pressure, psia
P_1 = reservoir pressure after Q_1 cu ft of gas have been produced, psia
P_b = pressure base for gas measurement, psia
t_b = temperature base for gas measurement, °F
T_b = t_b + 460, °F abs
T_f = reservoir temperature, °F abs
Z_0 = compressibility factor for gas at P_0 and T_f
Z_1 = compressibility factor for gas at P_1 and T_f
Z_b = 1.000

Material balance equation is:

$$Q_1 = V_0 - V_1 \tag{18}$$

Gas laws are:

$$\frac{P_0V_r}{Z_0T_f} = \frac{P_bV_0}{T_b};\quad \frac{P_1V_r}{Z_1T_f} = \frac{P_bV_1}{T_b}$$

Then:

$$\frac{P_1}{Z_1} = -\frac{T_fP_b}{V_rT_b}Q_1 + \frac{P_0}{Z_0} \tag{19}$$

$$V_0 = \frac{P_0/Z_0}{P_0/Z_0 - P_1/Z_1}Q_1 \tag{20}$$

$$\log(P_0/Z_0 - P_1/Z_1) = \log Q_1 + \log\frac{(P_0/Z_0)}{V_0} \tag{21}$$

Equation **19**, the basis of "reservoir pressure corrected for compressibility decline curve method," states that P_1/Z_1 is a straight line function of Q_1, cumulative production. When plotted on regular coordinate paper, a straight line results with a slope of $-T_fP_b/V_rT_b$. This straight line may then be extrapolated to an abandonment P/Z in obtaining recoverable reserve. If gas compressibility is appreciable, plots of reservoir pressure against cumulative production will form a curved line.

Equation **20**, derived directly from Eq. **19**, is basis of the widely used "pressure-volume" and "equal pound loss" methods. It merely considers two points on the decline curve to calculate reserve initially in place and, depending on accuracy of individual data points, may lead to conflicting results.

Equation **21**, derived directly from Eq. **20**, is basis of method in which cumulative pressure drop is plotted against cumulative production on logarithmic coordinates. It states that difference in initial P/Z and successive values plotted against corresponding cumulative production on logarithmic coordinates forms a straight line at a 45 degree angle with either coordinate. Possible interpretations of causes of slopes other than 1.0 have been discussed.[26]

Table 4-22 Reservoir Pressures and Cumulative Production from a Gas Reservoir[27]

Reservoir pressure, psia	Compressibility factor, Z, dimensionless	P/Z	Cumulative production at 14.4 psia and 60 F, billion cu ft
2080	0.759	2740	0
1885	.767	2458	6.873
1620	.787	2058	14.002
1205	.828	1455	23.687
888	.866	1025	31.009
645	0.900	717	36.207

Although these three equations apparently present different methods of estimating gas reserves, they are essentially identical. Since each has individual advantages in interpretation of results, these will be discussed in connection with examples.

Example 6. Use pressures and cumulative production data in Table 4-22 to estimate recoverable gas reserves initially in place in gas reservoir.

Data presented in Table 4-22 are plotted in Fig. 4-26, where a straight line (1) has been drawn thru a plot of P/Z against cumulative production and extrapolated to zero pressure, indicating an initial reserve of 48.3 billion cu ft of gas in reservoir. Line 1 also shows that there must have been an error in either initial reservoir pressure or cumulative production to first interval. Actually, cumulative production includes estimated production from a wild well. Study of Fig. 4-26 indicates this estimated production may have been too large. Curve 2, a plot of reservoir pressure against cumulative production, shows curvature resulting from neglect of gas compressibility. Curve 3 illustrates a possible erroneous

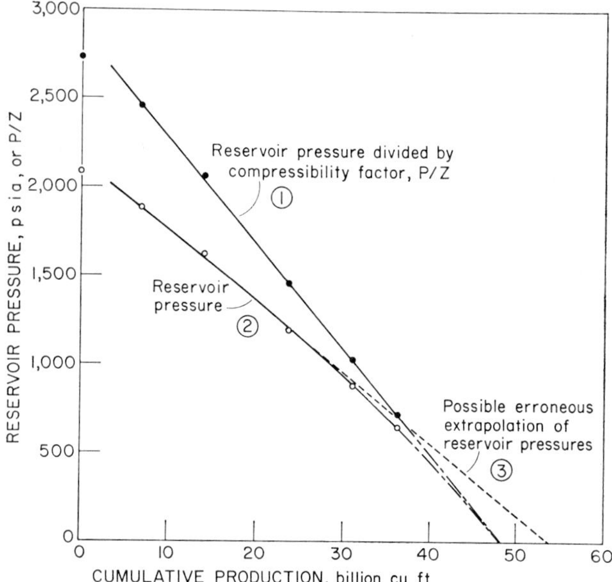

Fig. 4-26 Pressure-decline curves for a natural gas reservoir.[27]

extrapolation of reservoir pressures taken during early field life, neglecting gas compressibility.

Use of Eq. **20** for estimating gas reserves initially in place is illustrated in Table 4-23. Quantities given in Columns 1 and 2 were substituted into Eq. **20**, to compute initial gas reserve in place in reservoir, given in Column 3 for successive values of P/Z. At each successive calculation, apparent initial reserve in place decreased as shown in Column 3. Previously (see Fig. 4-26), a discrepancy was noted in these data, so initial reservoir pressure and cumulative production to the first pressure after initial were disregarded (Column 4), to compute an adjusted reserve shown in Column 5. Initial reserve in place is then computed by adding disregarded cumulative production of 6.9 billion cu ft to each adjusted reserve estimate in Column 5 to obtain Column 6. These reserve values in Column 6 are consistent and identical with value of 48.3 billion cu ft indicated by Fig. 4-26, where P/Z was plotted against cumulative production. If a water drive were present, the initial reserve estimate would apparently increase with each successive computation.

Data necessary for use in Eq. **21** in estimating recoverable reserves for the gas reservoir described in **Example 6** are given in Table 4-24. The actual method is illustrated in

Table 4-23 Estimation of Initial Gas Reserves in Place for Example 6 Using Eq. 20

Volumes in billions of cubic feet

(1)	(2)	(3)	(4)	(5)	(6)
Cumulative production	P/Z	Initial reserve in place	Adjusted cumulative production	Adjusted reserve in place	(5) + 6.9, Initial reserve in place
0	2740	..	...	..	..
6.873	2458	66.8	0	..	..
14.002	2058	56.3	7.129	43.8	50.7
23.687	1455	50.5	16.814	41.2	48.1
31.009	1025	49.5	24.136	41.4	48.3
36.207	717	49.0	29.334	41.4	48.3

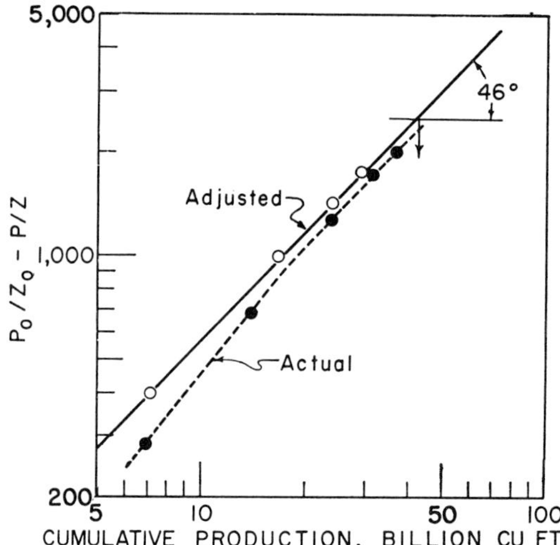

Fig. 4-27 Change in P/Z with cumulative production for natural gas reservoir.

Fig. 4-27. Values in Column 3 of Table 4-24, the change in P/Z, or $(P_0/Z_0 - P/Z)$, are plotted on logarithmic coordinates with cumulative gas production from the reservoir as abscissa to give the dashed curve of Fig. 4-27. The curvature, contrary to Eq. **21**, results from data discrepancy previously explained. By neglecting initial reservoir pressure and cumulative production to the first pressure after initial, this discrepancy may be eliminated. Adjusted values of cumulative production and change in P/Z are given in Columns 4 and 5 of Table 4-24. These data are shown by the solid line of Fig. 4-27 that forms an angle of 46 degrees with the abscissa. The reserve in place when P/Z was 2458 is 41.4 billion cu ft as read from the abscissa where the change in P/Z is 2458. Adding the neglected production of 6.9 billion cu ft, the initial reserve in place is 48.3 billion cu ft. Thus, Eqs. **19**, **20**, and **21** can be used to obtain identical results.

Table 4-24 Data for Estimation of Initial Gas Reserves in Place for Example 6 Using Eq. 21

(1)	(2)	(3)	(4)	(5)
Cumulative production, billion cu ft	P/Z	$\frac{P_0}{Z_0} - \frac{P_1}{Z_1}$	Adjusted cumulative production, billion cu ft	Adjusted $\frac{P_0}{Z_0} - \frac{P_1}{Z_1}$
0	2740	..	...	..
6.873	2458	282	0	..
14.002	2058	682	7.129	400
23.687	1455	1285	16.814	1003
31.009	1025	1715	24.136	1433
36.207	717	2023	29.334	1741

Methods of Estimating Associated Gas Reserves

The initial gas reserve in contact with crude oil in the reservoir or the initial quantity of gas in a gas cap can be readily determined by the volumetric method (Eq. **16**), providing the gas-oil contact position can be fixed. Subsequent estimations of gas reserve in the cap are difficult to make after production of oil and gas from the reservoir. Gas from the gas cap may have migrated into the oil zone, dissolved gas may have migrated into the cap, and some wells may have produced a mixture of associated and dissolved gas with crude oil. For these reasons Gruy and Crichton[18] recommend making no effort to distinguish between associated and dissolved gas. The total initial gas reserve may be computed and cumulative production of associated and dissolved gas subtracted to obtain total remaining at any time. It was also stated that decline methods for estimating gas remaining in a gas cap are inaccurate because volume of reservoir occupied by the cap will remain constant only under exceptional conditions.[18]

Methods of Estimating Dissolved Gas Reserves

The initial dissolved gas reserve in a reservoir is computed by estimating initial reserve of stock-tank oil and by multiplying oil volume by the initial solution gas-oil ratio. Initial oil reserve is estimated by the volumetric method. Recoverable dissolved gas reserve is determined by application of a recovery factor which varies according to reservoir characteristics and mechanism by which the reservoir is produced.

In complete water-drive reservoirs, where pressures remain above saturation pressure or bubble point of the oil throughout producing life of the reservoir, recovery factor for dissolved gas will be the same as that for oil. If reservoir pressure is reduced below saturation pressure of reservoir oil, recovery factor for dissolved gas will be greater than for oil. This results from increasing relative permeability of porous media to gas as gas saturation increases and from decreasing relative permeability to oil.

References

1. Rawlins, E. L. and Schellhardt, M. A. *Back-Pressure Data on Natural Gas Wells and Their Application to Production Practices.* (Bur. of Mines Mono. 7) Baltimore, Lord Baltimore Press, 1935.
2. Texas Railroad Commission. *Back-Pressure Test for Natural Gas Wells.* Austin, 1950.
3. Kansas State Corporation Commission. *Manual of Back Pressure Testing of Gas Wells.* Topeka, 1959.
4. Cullender, M. H. "Isochronal Performance Method of Determining Flow Characteristics of Gas Wells." *J. Petrol. Technol.* 7: 137–42, Sept. 1955 (Trans.)
5. Johnson, T. W. and Taliaferro, D. B. *Flow of Air and Natural Gas through Porous Media.* (Bur. of Mines TP 592) Washington, G.P.O., 1938.
6. Hetherington, C. R. and others. "Unsteady Flow of Gas through Porous Media." *A.I.M.E. Trans. (Pet. Dev. & Tech.)* 146: 166–74, 1942.
7. Haymaker, E. R. and others. "Method of Establishing Stabilized Back-Pressure Curve for Gas Wells Producing from Reservoirs of Extremely Low Permeability." *J. Petrol. Technol.* 1: 71–82, Mar. 1949 (Trans.)
8. Pierce, H. R. and Rawlins, E. L. *Study of a Fundamental Basis for Controlling and Gaging Natural-Gas Wells.* (Bur. of Mines RI 2929 and 2930) Washington, 1929.
9. Dunning, H. N. and others. *Using Foaming Agents to Remove Liquids from Gas Wells.* (Bur. of Mines Mono. 11) New York, A.G.A., 1961.
10. Capshaw, E. "Fundamentals of Gas Well Back-Pressure Testing." *Oil Weekly* 125: 56+, Mar. 10, 1947.
11. Vitter, A. L., Jr., "Back-Pressure Tests on Gas-Condensate Wells." A.P.I. *Drilling and Production Practice* 1942: 79–87.
12. Frick, T. C., ed. *Petroleum Production Handbook,* New York, McGraw-Hill, 1962. 2 v.
13. A.G.A. Gas Measurement Com. *Orifice Metering of Natural Gas.* (Rept. 3) New York, 1955.
14. Interstate Oil Compact Commission. *Manual of Back-Pressure Testing of Gas Wells.* Oklahoma City, 1961.
15. Walker, C. J. and others. *Back-Pressure Tests on Gas-Storage Projects.* (Bur. of Mines RI 5606) Washington, 1960.
16. Dunning, H. N. and others. "Non-Graphical Solution of Back-Pressure Tests on Gas Wells." *Petrol. Engr.* 30: B77+, Jan. 1958.
17. Cullender, M. H. and Smith, R. V. "Practical Solution of Gas-Flow Equations for Wells and Pipelines with Large Temperature Gradients." *J. Petrol. Technol.* 8: 281–7, Dec. 1956 (Trans.)
18. Gruy, H. J. and Crichton, J. A. "Critical Review of Methods Used in Estimation of Natural Gas Reserves." *J. Petrol. Technol.* 11: TP-2402, July 1948. Also in: *Petrol. Engr.* 20: 208+, Oct. 1948.
19. Anderson, K. F. "Volumetrically Weighting Reservoir Data." *Oil Gas J.* 51: 202–3, May 26, 1952.
20. Geffen, T. M. and others. "Efficiency of Gas Displacement from Porous Media by Liquid Flooding." *J. Petrol. Technol.* 4: 29–38, Feb. 1952 (Trans.)
21. Hurst, W. *Water Influx Into Reservoir and Its Application to Equation of Volumetric Balance.* (T.P. 1473) New York, A.I.M.E., 1942. Also in: *Petrol. Technol.,* May 1942.
22. Hill, H. B. and F. A. Vogel, Jr. *Petroleum-Engineering Study of the Lake Creek Field, Montgomery County, Tex.* (Bur. of Mines RI 4319) Washington, 1948.
23. Caraway, W. H. and Gates, G. L. *Methods for Determining Water Contents of Oil-Bearing Formations.* (Bur. of Mines R.I. 5451) Washington, G.P.O., 1959.
24. MacRoberts, D. T. "Effects of Transient Conditions in Gas Reservoirs." *J. Petrol. Technol.* 1:36–8, Feb. 1949 (Trans.)
25. Wilhelm, C. J. and others. *Petroleum-Engineering Study of the Carthage Gas Field, Panola County, Tex.* (Bur. of Mines RI 4698) Washington, 1950.
26. Miller, H. C. *Oil-Reservoir Behavior Based Upon Pressure-Production Data.* (Bur. of Mines RI 3634) Washington, 1942.
27. Heithecker, R. E. *Estimate of Natural-Gas Reserves from the Layton, Oolitic, and Oswego-Prue Horizons in the Oklahoma City Fields.* (Bur. of Mines RI 3338) Washington, 1937.

Chapter 5
Well Completion Practices

by Wm. Hinchliffe

CASING

After a drill stem test has demonstrated probable presence of oil or gas in commercial quantites in a well, running and cementing of casing at or near the bottom of the well are usually the next steps. In some instances, the casing is swung above the producing formation, and the well is produced from open hole. This casing string is referred to as the *flow string, oil string,* or *long string. Casing string* is defined as the total footage of casing run in the well at one time. The following principal types are used in rotary drilling: (1) surface string, which protects fresh-water sands and prevents cave-ins, (2) intermediate or protection string, which is set inside the surface pipe to protect the hole (one or more such strings may be set), and (3) oil string, which completes well as a producer or for test.

A master gate or its equivalent may be put in place before last lengths of casing are set for control of high gas pressure. Drilling may be completed thru a drilling valve or stuffing box, and the well can be shut in until connected with the transmission system. In rotary drilling, mud of suitable density can control production until the drill pipe has been removed and controls set in place. This mud must then be washed away, to avoid interference with gas flow into the wellbore.

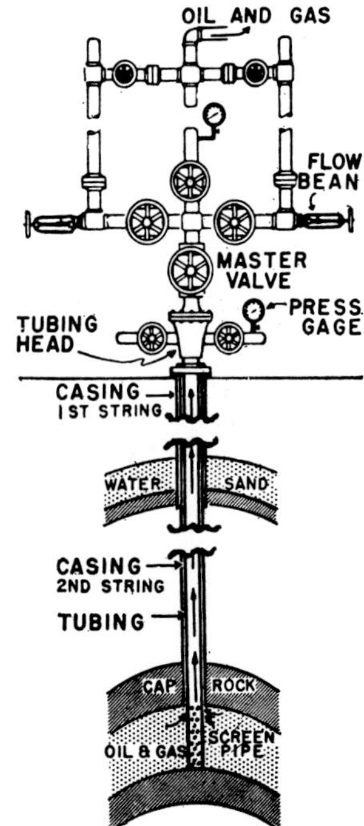

Fig. 4-28 A Christmas tree installation.[1]

A *Christmas tree* (Fig. 4-28) is mounted at the wellhead to control flow from tubing and from the annular space between tubing and casing. Adjustable flow beans (equivalent to needle valves), several fixed-size beans, or their equivalents are included for production control. A Christmas tree usually includes a master gate valve capable of withstanding highest shut-in wellhead pressure, a tee or cross for horizontal connection with valves for flow lines, and a blow-off gate, bull plug, and pressure gage connection at the top (Fig. 4-28).

Chokes, placed in the well bottom for flow control, decrease pressures on tubing, and the gas, having expanded to a greater volume, moves at greater velocity and carries fluids out with less slippage. Bottom hole chokes seem to give good control of production and help prevent *freezing* at wellhead controls.

During well productive life, casing keeps the hole open, prevents entry of water, confines gas and oil (thus preventing them from passing into other sands), and conducts them to the surface. It can perform these functions only when surrounded by cement, with a tight seal formed between casing and hole wall.

Standard Grades

The American Petroleum Institute cooperating with pipe manufacturers has standardized physical properties of casings according to grades. Table 4-25 shows three principal grades and their physical properties.

Yield strength in API specifications is the load in pounds per square inch of cross-sectional area required to stretch the metal 0.5 per cent of its length. Tensile strength, similarly ex-

Table 4-25 Physical Properties of Three Principal Grades of Casing[2]

(API Standard 5A)

Property	Grade		
	H-40	J-55	N-80
Yield strength (min), 1000 psi	40	55	80
Tensile strength (min), 1000 psi	60	75	100
Elongation in 2 in. (min), per cent			
Strip specimens	27	20	16
Full-section specimens	32	25	18

Table 4-26 Dimensions and Strengths of Casing[4]

Size, OD, in.	Nom. wt thread & coupling, lb/ft	Grade per API Std 5A	ID,* in.	Collapse resistance, psi	Internal yield press., psi	Joint strength, 1000 lb†	
						Short threads	Long threads
4½	9.50	H-40	4.090	2550	3,190	96	...
	9.50	J-55	4.090	3320	4,380	128	...
	11.60	J-55	4.000	4540	5,350	159	189
	11.60	N-80	4.000	5930	7,780	...	220
	13.50	N-80	3.920	7350	9,020	...	258
5	11.50	J-55	4.560	3130	4,240	152	...
	13.00	J-55	4.494	3930	4,870	178	210
	15.00	J-55	4.408	4980	5,700	210	247
	15.00	N-80	4.408	6520	8,290	...	288
	18.00	N-80	4.276	8550	10,140	...	354
5½	14.00	H-40	5.012	2440	3,110	139	...
	14.00	J-55	5.012	3170	4,270	186	...
	15.50	J-55	4.950	3860	4,810	211	247
	17.00	J-55	4.892	4500	5,320	234	275
	17.00	N-80	4.892	5890	7,740	...	320
	20.00	N-80	4.778	7580	9,190	...	382
	23.00	N-80	4.670	8900	10,560	...	440
6	18.00	H-40	5.424	2780	3,360	179	...
	18.00	J-55	5.424	3620	4,620	239	278
	18.00	N-80	5.424	4740	6,720	...	323
	20.00	N-80	5.352	5690	7,560	...	366
	23.00	N-80	5.240	7180	8,870	...	432
6⅝	20.00	H-40	6.049	2360	3,040	195	...
	20.00	J-55	6.049	3060	4,180	259	299
	24.00	J-55	5.921	4250	5,110	320	370
	24.00	N-80	5.921	5550	7,440	...	430
	28.00	N-80	5.791	7110	8,810	...	511
	32.00	N-80	5.675	8490	10,040	...	582
7	17.00	H-40	6.538	1370	2,310	160	...
	20.00	H-40	6.456	1920	2,720	191	...
	20.00	J-55	6.456	2500	3,730	254	...
	23.00	J-55	6.366	3290	4,360	300	344
	26.00	J-55	6.276	4060	4,980	345	395
	23.00	N-80	6.366	4300	6,340	...	400
	26.00	N-80	6.276	5320	7,240	...	460
	29.00	N-80	6.184	6370	8,160	...	520
	32.00	N-80	6.094	7400	9,060	...	578
	35.00	N-80	6.004	8420	9,960	...	635
	38.00	N-80	5.920	9080	10,800	...	688
7⅝	24.00	H-40	7.025	1970	2,750	227	...
	26.40	J-55	6.969	3010	4,140	333	378
	26.40	N-80	6.969	3930	6,020	...	439
	29.70	N-80	6.875	4910	6,890	...	505
	33.70	N-80	6.765	6070	7,890	...	581
	39.00	N-80	6.625	7530	9,180	...	676
8⅝	28.00	H-40	8.017	1580	2,470	252	...
	32.00	H-40	7.921	2110	2,860	295	...
	24.00	J-55	8.097	1430‡	2,950	288	...
	32.00	J-55	7.921	2740	3,930	393	437
	36.00	J-55	7.825	3420	4,460	448	499
	36.00	N-80	7.825	4470	6,490	...	581
	40.00	N-80	7.725	5390	7,300	...	655
	44.00	N-80	7.625	6320	8,120	...	729
	49.00	N-80	7.511	7370	9,040	...	812
9⅝	32.30	H-40	9.001	1320	2,270	279	...
	36.00	H-40	8.921	1710	2,560	318	...
	36.00	J-55	8.921	2220	3,520	422	462
	40.00	J-55	8.835	2770	3,950	477	521
	40.00	N-80	8.835	3530‡	5,750	...	606
	43.50	N-80	8.755	4280	6,330	...	670
	47.00	N-80	8.681	4900	6,870	...	727
	53.50	N-80	8.535	6110	7,930	...	841

Table 4-26 Dimensions and Strengths of Casing[4] (Continued)

Size, OD, in.	Nom. wt. thread & coupling, lb/ft	Grade per API Std 5A	ID,* in.	Collapse resistance, psi	Internal yield press., psi	Joint strength, 1000 lb†	
						Short threads	Long threads
10¾	32.75	H-40	10.192	830	1,820	265	...
	40.50	H-40	10.050	1340	2,280	338	...
	40.50	J-55	10.050	1730	3,130	450	...
	45.50	J-55	9.950	2300	3,580	518	...
	51.00	J-55	9.850	2870	4,030	585	...
	51.00	N-80	9.850	3750	5,860	680	...
	55.50	N-80	9.760	4420	6,450	750	...
11¾	42.00	H-40	11.084	940	1,980	336	...
	47.00	J-55	11.000	1630‡	3,070	507	...
	54.00	J-55	10.880	2270	3,560	593	...
	60.00	J-55	10.772	2840	4,010	668	...
	60.00	N-80	10.772	3680‡	5,830	778	...
13⅜	48.00	H-40	12.715	740‡	1,730	352	...
	54.50	J-55	12.615	1140‡	2,730	545	...
	61.00	J-55	12.515	1670‡	3,090	613	...
	68.00	J-55	12.415	2140	3,450	695	...
	72.00	N-80	12.347	2880‡	5,380	868	...
16	65.00	H 40	15.250	640‡	1,640	423	...
	75.00	J-55	15.124	1010‡	2,630	662	...
	84.00	J-55	15.010	1480‡	2,980	753	...
20	94.00	H-40	19.124	520‡	1,530	487	...

(Reproduced by permission).
* Drift diam = ID − 0.125 for OD up to 8⅝ in.
Drift diam = ID − 0.156 for OD from 9⅝ to 13⅜ in.
Drift diam = ID − 0.188 for OD from 16 to 20 in.
† Joint strength in feet = joint strength in lb divided by nominal weight from column 2.
‡ Collapse resistance values calculated by the elastic formula.

pressed, is the load required to part the metal. API casing must conform to tensile properties listed in Table 4-25.

Steel casing pipe may be butt welded, lap welded, electric welded, or seamless. Butt welded pipe is usually made in small sizes only. Lap welded pipe may be employed for shallow wells. Steel of a higher carbon content may be used for electric welded pipe than is used for butt or lap welded. Seamless casing is required for the deepest wells.

Conductor pipe may be used to start the hole vertically and to shut off surface material and water. It is often made by bending sheet steel into a cylindrical form and riveting, this being termed *stove pipe*. These cylindrical forms are usually joined by welding.

The production of very *sour natural gas* requires increased protection against rupture or leaks in production casing and wellhead, and the prevention of the release of H_2S into the air. Tubing materials less subject to hydrogen embrittlement (e.g., API grade J-55) are used. Integral joint tubing appears more dependable than collared tubing for this reason.[3]

Size and Range

Casing sizes up to 14 in. formerly were designated like ordinary pipe by nominal inch diameter, and 14 in. and larger sizes by outside diameter. API has standardized sizes, weights, and couplings, as well as materials. All casings meeting these API standards are interchangeable. They are designated by external diameter and weight per foot.

Table 4-26 shows dimensions and strengths of various standard grades of casing. Corresponding API casing thread specifications call for eight threads per inch tapered ¾ in. per ft on diameter. See Table 1-12 and Fig. 1-2 for thread dimensions.

Casing lengths are joined by external couplings and threaded with standardized tools. Increase of thickness can be obtained only by decrease of internal diameter. A particular size casing may be made in different thicknesses and, therefore, may have different weights (Table 4-26). Thickness is important in determining clearance when one string is run inside another.

Length of sections or joints of casing is known as *range*. API casing is available in the following ranges:

Range I—16 to 25 ft
Range II—25 to 34 ft
Range III—34 ft and longer

Joints

Except in special cases, joints are made with external screw couplings. Threads are so cut that pipe ends do not butt together in collars except for drive pipe. Casing is shipped with a coupling collar and coupling protector on one end and a thread protector on the other end, commonly called the field end. Joints are designed to be as strong as the remainder of the pipe when properly made up.

Figure 4-29 shows eight types of casing joints. Coupled API round-thread casing shown as **A** is at present most common for gas and oil use (sharp V threads are standard in 16 and 20 in. sizes).

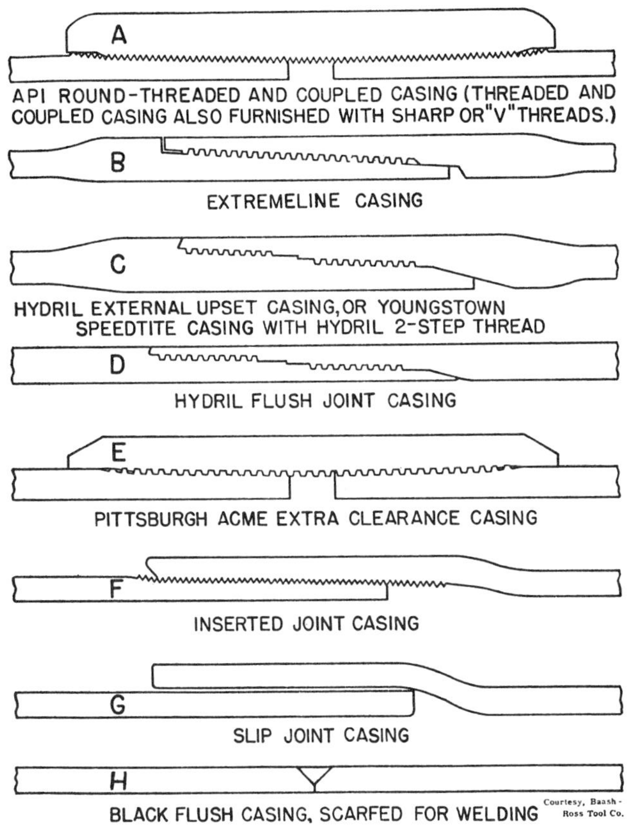

Fig. 4-29 Eight types of casing joints.

Some types have advantage of smaller overall diameters at joints than coupled casing **A**. This is true in casings **B**, **C**, **D**, and **F**. Because of the increased clearance, these types are run inside other strings of casing and are sometimes used in liner jobs. Note that OD of joint shown as **D** is same as casing OD.

To insure quality of material for high pressure wells, casing is often ordered tested at the mill by the field test method. A plug is screwed into the coupling and a cap placed on the threaded field end. The pipe is then tested to pressures considerably beyond the range of ordinary field testing equipment. To insure tight joints, couplings are usually screwed on the casing to a three-turn power make-up for these tests.

Casing Stresses

Casing design for a well is based on certain assumed loads, with provision for additional safety factors. Safety factor may be regarded as the ratio between strength and load. Formulas for joint strength are based upon a large number of joint tests, summarized as a straight line expression for joint efficiency (ratio of load causing failure to load calculated from net area and strength of material). In accordance with values listed in API Bul 5C2 (Ref. 4), strength requirements were obtained as follows:

Minimum collapse resistance = 75 per cent of average collapse resistance values.

Minimum joint strength = 80 per cent of average ultimate joint strength values.

Internal yield pressure strength = 87.5 per cent of average values of internal pressure at minimum yield point.

Safety factors given in tables for minimum properties are, therefore, lower than average values. A safety factor against collapse of 1.5 for average values is equal to $1.5 \times 0.75 = 1.125$ when based on minimum values. A safety factor against joint failure in tension of 2.5 for average values is equal to $2.5 \times 0.80 = 2.0$ when based on minimum values. A safety factor of 1.5 based on average values thus gives the same margin of resistance to failure by collapse as one of 1.125 based on minimum values.

It is not possible to know in advance exact loads which casings must bear. Most important are those due to weight, outside pressure, bending, and to what is called column loading, which occurs when the casing bottom supports part or all of the string weight. It is not intended that the latter condition shall occur, except in short strings or lower ends of long strings.

Shock caused by sudden application of loads may result in rupture of a casing which would be strong enough if the load were applied gradually. Such shocks can occur in many ways. Depth of a well determines casing tension, since each foot of depth adds its proportional load. Collapsing pressure is proportional to depth of submergence, but this depth cannot always be determined exactly. Properly manufactured casing, correctly made up, is more likely to fail from tension or collapsing pressure than from any other cause. Tables to aid proper selection are included in several steel mill handbooks and in API Bul 5C2.[4]

Running Casing

Selection. Relationship of hole sizes to casing is important. Generally, casing diameter should be from 2 to 4 in. less than hole size, thus leaving an annular space of 1 to 2 in. For economic and other practical reasons, a hole should only be enough larger than the casing to be run to permit easy entry of pipe. Excess hole diameters tend to make it more difficult to develop desired flow rates during cementing, to centralize pipe, and to clean filter cake from the hole.

Selection of casing is governed not only by well depth and expected pressure but also by clearance necessary between the inside of one string and the outside of the next smaller one. Clearance depends upon internal and external diameter of the larger pipe, and external diameter of couplings of the smaller pipe. Published tables, e.g., API Bul 5C2, give dimensions and weights for various casing grades,[4] so that casing programs can be selected and adapted to local conditions. Within districts, casing programs become somewhat standardized.

Before installation, engineering design has designated grade, weight, and length of casing for the entire well depth. Figure 4-30 shows a sample casing tally sheet.

Preparation and Inspection.[5] Casing, whether new, used, or reconditioned, should always be handled with protectors in place.

Slip elevators are recommended for long strings. Both spider and elevator slips should fit properly and be clean, sharp, and extra long for heavy strings.

If collar-pull elevators are used, bearing surface should be carefully inspected for uneven wear, which may produce a side lift with danger of jumping off the coupling, and for uniform load distribution over coupling bearing face.

CASING TALLY SHEET PAGE____OF____PAGES

FIELD________PROPERTY________WELL No.____

SIZE____O D WT____ #/FT THREADS______GRADE______RANGE____

MAKE________ TYPE______MIN I D____ INTERNAL UPSET NOT UPSET____

COUPLINGS: TYPE________ MAX O D________

DELIVERED: JTS.______FT OVERALL: USED______JTS.: LEFT ON RACK______JOINTS

TYPE OF SHOE________ TYPE FLOAT COLLAR________

STARTED CASING A.M. P.M.______DATE______CASING ON BOTTOM A.M. P.M.______DATE______

JOINT	MEASUREMENTS				REMARKS	JOINT	MEASUREMENTS				REMARKS
	RACK-LENGTH		MADE UP-LENGTH		CHECK WHERE		RACK-LENGTH		MADE UP-LENGTH		CHECK WHERE
	FEET	TENTHS	FEET	TENTHS	MEAS. DIFFER		FEET	TENTHS	FEET	TENTHS	MEAS. DIFFER
1						1					
0						0					
TOTAL A						TOTAL C					
1						1					
0						0					
TOTAL B						TOTAL D					

TOTALS	TOTAL THIS TYPE CASING		REMARKS	TOTAL	TOTAL ALL CASING		ADDED	CHECKED
	RACK	MADE UP			RACK	MADE UP	BY	BY
TOTAL A				BRT. FWD.				
TOTAL B				PAGE TOTAL				
TOTAL C				GR. TOTAL				
TOTAL D				REMARKS				
PAGE TOTAL								
BRT. FWD.								
TOTAL								

NOTE: START NEW SHEET FOR EACH DIFFERENT GRADE OR WEIGHT OF CASING

Fig. 4-30 Sample casing tally sheet.

Spider and elevator slips should be examined and watched to see that all lower together. If they lower unevenly, there is danger of denting or badly slipcutting the pipe.

Care must be exercised, particularly with long casing strings, to insure that the slip bushing or bowl is in good condition.

The following precautions should be taken in preparation of casing threads:

1. Immediately before running casing, remove protectors from both field and coupling ends of several lengths and clean threads thoroughly.
2. Inspect threads (API Std 5B). Any even slightly damaged should not be used, unless satisfactory means are available for correcting damage.
3. When threads are thoroughly dry, apply thread compound over entire surface of both internal and external threads, and reapply clean protectors on field ends.

Each joint of casing should be lowered or rolled carefully to the walk without dropping, using rope snubber if necessary.

For mixed or unmarked strings, a drift or *jackrabbit* should be run thru each casing length. This avoids running a heavier length or one with a lesser inside diameter than called for in the casing string.

Stabbing, Making Up, and Lowering. Protector should not be removed from field end of casing until ready to stab. If necessary, thread compound should be applied over entire surface of threads just before stabbing, keeping application free of foreign matter.

Stabbing should be done vertically, preferably with assistance of a man on the stabbing board. If the casing stand tilts to one side after stabbing, it should be lifted, any damaged thread cleaned and repaired, any filings carefully removed, and compound reapplied. After stabbing, if spinning line or rotary table make-up post is used, casing should be rotated very slowly at first to insure that threads are engaging properly and are not crossthreading. The spinning line should pull close to the coupling. Tongs should be used for proper tightening. The joint should be made up at least three turns ($3\frac{1}{2}$ turns for sizes $7\frac{5}{8}$ in. and larger) beyond handtight position. Final make-up should be more than three turns if there are sufficient indications that the joint is still not tight.

Joints of questionable tightness should be unscrewed and, if practical, the casing should be laid down for inspection and repair when the mating coupling should be carefully examined for damaged threads. Parted joints should never be reused without shopping or regaging, even though they may show little evidence of damage.

If casing has a tendency to wobble unduly at its upper end when making up, indicating that threads may not be axial, speed of rotation should be decreased to prevent thread galling. If wobbling persists despite reduced rotational speed, the casing should be laid down for inspection. Great care should be exercised in using such casing where a heavy tensile load is imposed.

Casing strings should be picked up and lowered carefully, and care exercised in setting slips, to avoid shock loads. Dropping a string even a short distance may loosen couplings at the bottom of the string. Care should be exercised in setting casing down on bottom, or otherwise placing it under compression, because of danger of buckling, particularly where hole enlargement has occurred.

Definite string design data should be available, including proper location of various grades of steel, weights of casing, and types of joint. Care should be exercised to run the string in exactly the order designed. If any length cannot be clearly identified, it should be laid aside until its grade, weight, or type of joint can be positively identified.

Immediately after each length is made up, the string lowered, and slips set, the length should be filled with mud, using a conveniently located hose of adequate size for filling within time required for preparing next length. A quick opening and closing plug valve on mud hose facilitates operations and prevents overflow. A copper nipple may be used to prevent coupling thread damage during filling operations.

Landing Procedure. It is common practice when casing is set thru the formation to land it on the bottom until the weight indicator shows that it is taking weight. This point should be marked on the casing with relation to the rotary table or some other type surface equipment and the casing raised a minimum of 3 to 4 ft or more, using this marker as the indicator. Common practice is to raise casing until free circulation is established. With casing raised off bottom, its full weight is on the top joint of casing string. Natural stretch in pipe places its lower end closer to bottom of hole than is indicated by the distance that it has been raised above the marked point.

Definite instructions should be available on proper string tension, also on proper landing procedure, in order to apply tension after cement has set. The purpose is to avoid critical stresses or any excessive, unsafe tensile stresses during life of the well. In arriving at proper tension and landing procedures all factors should be considered, such as well temperature and pressure, temperature developed by cement hydration, mud temperature, and temperature changes during producing operations. Adequacy of the original tension safety factor of the string as designed also influences landing procedure. If special landing procedure instructions are believed unnecessary, as is probably true with most wells drilled, casing should be landed in the casing head at exactly the position in which it was hanging when the cement plug reached its lowest point.

Drill pipe should be equipped with suitable protectors when it is run inside casing.

Use of Packers

Proper well completion involves preventing undesirable matter from entering the gas or oil stream. Rubber packers, which are essentially ring-shaped rubber sleeves about 1 ft long, can be used to seal space between tubing and next larger size of casing or formation, to keep water and cavings from entering the production string. In operation, weight of tubing or casing flattens rings sealing the space. Packers can also be used during drilling, to exclude water and to prevent gas from escaping between casing and formation. They may be used in dual completions to direct gas flow into the tubing. Packers are not as lasting as cement, and replacements may be required during life of the well.

WELL CEMENTING

Well cementing in the petroleum and natural gas industries is the equivalent of pressure grouting in mining and construction. Cementation of casing produces a number of important results in drilling and completion of wells. Those most generally recognized are:

1. Prevention of vertical migration of formation fluids from one zone to another, commonly termed segregating the formations.
2. Exclusion of undesirable fluids from the well.
3. Protection of casing.
4. Sealing off of formations with troublesome zones, to allow deeper drilling and help prevent blowouts.
5. Preservation of natural permeabilities and porosities of formations.
6. Avoidance of loss of circulating fluids.

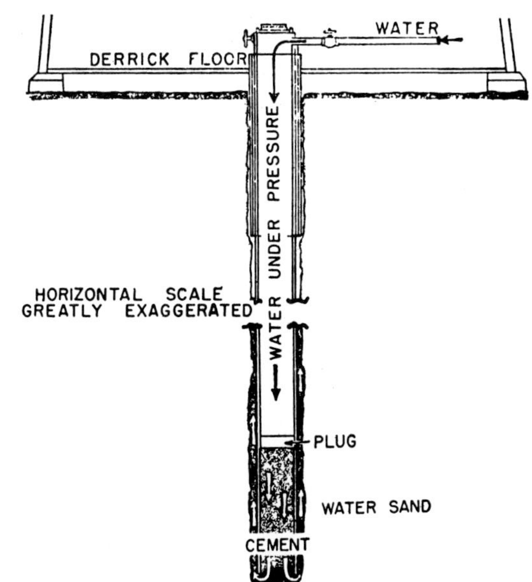

Fig. 4-31 Cementing casing in place.

Before reaching the producing zone most wells are drilled through permeable formations which carry fresh or salt water, or gas. Cement must be placed between casing and wellbore wall to prevent migration of fluid into producing zones and contamination of fresh-water strata as shown in Fig. 4-31. Surface and shallow waters must be protected from contamination if they are used for irrigation or for municipal water supply.

Coal and many other mineral strata must be protected in certain areas. It is also necessary to seal the producing formation so that it will not lose oil or gas into other permeable zones where pressures are less. Formation production tests are impossible in some instances, especially where soft formations prevail, until casing has been cemented into or thru the zone to be tested. Likewise, troublesome formations, which often make deep drilling extremely difficult, must be cased off so that drilling may proceed.

For example, circulation of drilling fluid, mud, or oil, may be *lost*, i.e., fluid fails to return to surface; the formation itself may slough off into the open hole, a condition frequently described as *heaving;* or drilling fluid may be contaminated by salt water, anhydrite, or corrosive waters.

Equipment and Materials

Cementing trucks usually contain one or more high pressure pumps for deep well cementing and possibly one low pressure mixing pump for shallow wells or surface pipe. Some present-day equipment is capable of 12,000 psi pumping pressures and uses or generates as much as 300 hp for pumping. Most cementing slurries can be mixed continuously at rates of 10 to 40 sacks per minute.

Most cement used in wells is shipped from mills in enclosed hopper-bottom railroad cars to bulk cement plants of service companies for blending with required additives. Blended cement is then carried to wells and discharged into jet mixers as cementing proceeds.

Cementing Surface Strings

Usually surface holes are drilled with lightweight slurries of low viscosity. Comparatively little trouble is to be expected from commingling and gelation of mud and cement. If it is suspected that cement-contaminated mud may interfere with placement of cement, 20 or 30 bbl of water can often be used to minimize this condition. Only under extreme conditions should this practice cause any ill effects such as heaving. To prevent lubrication or by-passing of cement into mud below it while cement is still in the pipe, both a bottom and top plug should be run.

During no cementing operation is it more important to consider rate of flow than during cementing of surface strings. Flow properties of a plastic fluid are independent of pipe diameter, and such flow divides itself into three broad flow regions: plug, laminar, and turbulent. Tests indicate that flow rate at which cement slurry displaces circulatable mud in a well greatly affects percentage of mud displaced. In plug-flow regions cement slurry displacement rates seem to reach approximately 60 per cent of circulatable mud, in laminar-flow regions about 90 per cent, and in turbulent-flow regions more than 95 per cent.[6]

Flow rate corresponding to *turbulent flow* for a $10\frac{3}{4}$ in. casing set in a 15 in. hole was found to be approximately 50 bbl per min. For middle of the *laminar-flow region,* 19 bbl per min are required for same pipe and hole sizes. Rates of 19 bbl per min are well within capacity of many rigs and can probably be maintained without ill effect to equipment or hole. Laminar flow is much more desirable than plug flow, which is often in effect during surface pipe cementing. It would be necessary to exercise precaution and reduce pump speed near end of job in order to land the top cement plug safely.

Cementing Protection Strings

Procedures for intermediate or protection strings are quite similar to those for production strings. Protection strings are generally set wholly for expediting deeper drilling, and therefore, actual success of cementing often cannot be known. Protection strings, like surface strings, must withstand stresses resulting from drilling thru and below. Again, it is essential that bottom joints be as near perfectly cemented as possible. Corrosive fluids often occur in quantities that demand extensive coverage of pipe by cement for its protection. In such cases, cement can be brought up the annulus to the desired height by using lightweight slurries or multistage cementing equipment.

It is sometimes desirable to place and utilize a multistage cement tool to supplement a deliberately shortened surface string. Centralizers and wall cleaners should be planned on the same basis as long strings if future production is planned, and similar to surface strings if no production is planned. Pump rates should be selected that will, where possible, create turbulent flow during cement placement.

Cementing Production Strings

Because of increased depths with correspondingly greater formation pressures, it becomes increasingly difficult to isolate penetrated formations from each other. Often, shale beds separating two or more fluid-bearing sand bodies are extremely thin. Occurrence of two or more different fluids within a single sand body becomes more prevalent and creates a more difficult isolation problem. Abnormal pressures in certain formations penetrated call for a drilling fluid with column weight exceeding tensile strengths of other formations exposed. Often a different situation exists in which it is possible to drill with fluids of weights so low that it is impractical to match them with weight of cement slurry to be run, yet greater columnar weight created by cement causes loss of circulation during cementing. Higher differential pressures and higher temperatures tend to develop thicker and tougher filter cakes. These same pressures and temperatures encourage use of drilling fluids extensively treated with chemicals that may interfere with proper setting of cement slurries.

For best results and minimum damage to formations, consideration should be given the following:

1. Electric logs and caliper logs may be consulted where possible, to determine location for centralizers and wall cleaners. If no caliper log is available, assume hole to be near bit size in sand sections and washed out to a greater diameter in shale sections. Centralizers should be positioned in sections expected to be near bit size. Wall cleaners should be placed as recommended by the manufacturer.
2. If multistage equipment is to be run, it should be placed in string according to purpose intended.
3. It has been found beneficial to plan and execute several periods of circulation during running in of casing. Each is continued until fluid has been returned from bottom and/or cuttings from wellbore cease to return. This practice does much to reduce final circulating pressure and hazard of lost circulation.
4. When pipe has been landed and circulation established, the latter should be maintained until filter cake and cuttings displaced by wall cleaners have been returned to surface. If circulation and returns are normal, cementing may be done as soon as assured cuttings have been received from bottom.

Two methods of procedure are most common: (a) if pumping pressure is normal, cementing may be undertaken in usual manner, or (b) content of lost circulation material in drilling mud may be built up and re-establishment of circulation attempted before cementing.

If casing becomes stuck at or near landing point and circulation is lost, it is often necessary to *nipple up* and go into hole with a squeeze tool and squeeze cement thru casing shoe. Before doing so, an attempt should be made to determine depth at which pipe is stuck, as an aid in determining where casing should be perforated for cementing of annulus above section creating restriction.

5. Choice of cement calls for consideration of depth of casing, temperature of bottom of section, length of time needed to mix and land required volume of cement, nature of formations to be covered, and capacity of formations affected by cement column to withstand pressure.

6. Plugs are now available that can safely be run ahead of cement on prime jobs, to prevent its contamination with drilling fluids while they are still in the casing and simultaneously to wipe inside wall of casing clean. Their use has become almost universal. A bottom plug can prevent contamination of cement coming to rest around bottom joint or joints, and in area between guide or float shoe and float collar. This zone of contamination may be of major importance.

7. Surface cementing equipment, such as plug containers, should be chosen with thought of minimizing or eliminating stoppage of fluid travel, once circulation has been established ahead of cement. Caution should be used to insure that top and bottom plugs are not interchanged. Movement of casing is often impossible after circulation is allowed to stop during cementing.

8. Results definitely indicate that better cementing is obtained where casing is moved during cementing. Type of movement, spudding or rotation, depends upon type of wall cleaning equipment or discretion of operator if no such equipment is used.

9. Once cement has started out of bottom of pipe and up annulus, far better results will probably be obtained if circulation rates in the turbulent-flow region are developed. Such flow can be developed with typical slow-set cement as follows:

Hole size, in.	Pipe size, in.	Flow rate, bbl/min
7⅞	5½	8.5
9⅞	7	13.0

10. Practice of leaving any designated amount of pressure on casing after plugs are landed is becoming less common. If floating equipment is holding after plug is down, many operators remove cementing head immediately after completing job.

11. If well being cemented is a wildcat, or if any unusual effects are noticed in connection with returns during cement, valuable information can be obtained by running a temperature survey. Any further cementing should be accomplished as soon as possible before annulus mud becomes gelled. A successful temperature survey can now be run much sooner than previously thought possible. Period at which maximum heat of hydration occurs is influenced by native temperature of well, and possibly by pressure applied to cement and by water-cement mixing ratio.

Table 4-27 shows the minimum waiting time for temperature surveying various brands of cement.

Table 4-27 Earliest Temperature Surveying Times for Various Cements

(Ratios are by volume; data taken at temperatures and pressures shown)

Brand of cement	Time	Temp, °F	Pressure, psi
Lone Star Common	2 hr, 40 min	120	1750
(water-cement ratio: 0.46)	1 hr, 50 min	140	2700
	1 hr, 40 min	170	4000
	1 hr, 30 min	200	5500
Longhorn Slo-Set	6 hr	140	2700
(water-cement ratio: 0.40)	7 hr	170	4000
	7 hr, 30 min	200	5500
Trinity Inferno	6 hr	140	2700
(water-cement ratio: 0.40)	7 hr, 30 min	170	4000
	10 hr	200	5500
Univ. Atlas Unaflo	6 hr	140	2700
(water-cement ratio: 0.40)	6 hr, 30 min	170	4000
	8 hr, 30 min	200	5500
Lone Star Starcor	4 hr	140	2700
	2 hr, 45 min	170	4000
	3 hr	200	5500

Squeeze Cementing

Squeeze cementing is the process of applying hydraulic pressure to force a cement slurry into permeable space of an exposed formation or thru perforation of casing or liner. Many problems in producing wells are solved by means of proper selection of pressures, prepared oil well cements, water-cement ratios, various admixes, and squeeze cement processes.

Objects of squeeze cementing include: (1) correcting well defects, such as high gas–oil ratios and excessive water, (2) repairing casing leaks or perforations improperly placed, (3) isolating producing zone before perforating for production, (4) remedial or secondary cementing to correct a defective condition, (5) sealing off low pressure formation that *engulfs* oil, gas, or drilling fluids, and (6) abandoning depleted producing zones to prevent migration of fluids and to eliminate possibilities of contaminating other zones or wells.

Cementing procedure can not be established until initial pumping of fluid into the formation indicates breakdown pressures and well behavior. Basic items that must be considered even before pumping into the formation to be squeezed include:

1. Dimensions and strengths of casing, tubing, or drill pipe for performing squeeze job. Maximum permissible collapse strength and internal yield pressures should be determined. These pressures will determine annulus pressure required to keep differential within safe limits.

2. Condition of primary cementing as indicated by perforating and testing, temperature survey, or well behavior, as well as probable fill-up behind casing, either calculated or result of temperature survey.

3. Top hole connections: (a) Bradenhead with stripper rubber should be installed to withstand maximum pressures to be allowed on casing; (b) pressure manifold to control annulus pressure and facilitate washing out surplus cement should always be available.

4. Special tools: (a) high pressure pumps; (b) type squeeze retainer or packer, whether drillable or retrievable, as packer setting point and well conditions dictate.

5. Cements: (a) type based on well depth, temperature, breakdown pressure, and other well conditions; (b) amount on hand dictated by type of job. Amount used is determined after initial formation breakdown and well behavior.

6. Admixes: expanded perlite, graded granular materials, cellophane flakes, fibrous materials, gypsum cements, resin cements, Bentonite, calcium chloride, and many others. Need for some type of admix may be indicated. Well condition and breakdown pressures will dictate amount and kind.

7. Breakdown fluid: (a) if drilling mud is to be used, sufficient water should be on hand to mix slurry plus additional water to run ahead and behind it; amounts are dictated by well condition and behavior; (b) breakdown with water is usually more desirable than with mud and gives better results; (c) mud cleanout and acid wash are good breakdown fluids run ahead of cement slurry, following a breakdown with mud or water.

Hazards. These include:

1. *Collapse of Casing*
   (a) Above packer, with no cement, or possible channel in cement, outside of casing opposite packer seat. May be minimized by keeping safe differential pressures across casing by means of pressure on annulus between tubing and casing.
   (b) Below bridge plug, if set below perforations with channel or no cement outside of casing. There are no means of controlling differential pressures; therefore, pressure must be limited to minimum collapse strength of casing below bridge plug until cement is squeezed downward and allowed to set.
2. *Necessity of Setting Packer Between Perforations*
   (a) Annulus pressure is limited to breakdown pressure of formation opposite upper set of perforations.
   (b) Should squeeze be accomplished with tubing or drill pipe full of cement, and should formation opposite perforations above retainer break down below pressure required to reverse surplus cement out of tubing, cement must be washed out "the long way" thru annulus or dumped into casing. Squeeze retainer, drill pipe, or tubing is immediately pulled out of hole.
   (c) Communication between perforations may cause sand and shale to stick to packer or tubing.
3. *Squeezing Opposite Pay Zone*
   (a) Low pressure producing zones should be protected by covering with river sand, cal-seal, or cement if possible.
   (b) Production from fractures may be lost completely, following squeeze cementing.
   (c) Cement should be spotted near packer with circulating valve open, to minimize amount of fluid pumped into producing formation ahead of cement.
4. *Failure of Packer*
   (a) Packer may fail if set in section of casing where casing scraper was not used following drilling out of cement.
   (b) Size and weight of casing at point of packer setting must be known.
5. *Suitability of Cement*
   (a) High temperature and deep wells demand proper type of cement for long pumping period, together with a safety factor if some mechanical failure occurs.
6. *Leakage Under Pressure*
   (a) Tubing, when possible, should be tested for leaks and at maximum pressure required to accomplish squeeze.

Bottom Hole Squeeze Pressure. Overburden pressure gradient of earth formations is roughly 1.0 psi per ft of depth. Squeeze cementing, hydraulic fracturing, and acidizing records on a number of wells show that bottom hole pressure gradients in wells intentionally fractured in treatment varied from 0.50 to 1.13 psi per ft of depth. Wide variations were found in different fields and in different wells in the same field.

For correcting well defects like high gas–oil ratio or excessive water, final bottom hole squeeze pressures on an initial job may be approximated as follows:

Minimum Gas Squeeze Pressure (G.S.P.), psi = well depth in feet + 1000

Minimum Water Squeeze Pressure (W.S.P.), psi = well depth in feet + 2000

Surface Gage Pressure (G.S.P. or W.S.P.), psi = hydrostatic head.

Another formula used is:

Surface Gage Pressure for G.S.P. and W.S.P.,
psi = 0.4 well depth in feet + 1000
with water in tubing, psi = 0.4 well depth in feet − 500

For **shallow wells,** the following formula may be used:

G.S.P., psi = 0.6 well depth in feet + 4000
W.S.P., psi = 0.6 well depth in feet + 5000
Surface Gage Pressure, psi = (G.S.P. or W.S.P.) − hydrostatic head in psi

WELL CEMENTING MATERIALS

Two stages are recognized in setting of cement; (1) initial set, when the paste slurry begins to harden or offers resistance to change of form; and (2) final set, when the mix cannot be appreciably distorted without rupture. Laboratory instruments definitely determine these two stages. Tests of tensile and compression strength are made by rupturing standard size molded samples in laboratory equipment. Thickening time or pumping time is determined by special cement consistometers which test ease of pumping. There are API Standards for testing procedures.[7]

Primary Properties

In addition to several secondary properties, cement must possess the following primary properties:

1. **Thickening Time.** A cement must possess enough thickening time for it to be pumped into position with a reasonable factor of safety. Thickening time requirements will vary greatly, depending upon such factors as well depth, bottom hole temperature, bottom hole pressure, and amount of cement.

2. **Setting Time.** Cement must set within a reasonable time after it is placed in the well.

3. **Strength.** Cement must develop sufficient strength to permit resumption of operations within a relatively short time, usually 24 hr or less.

Cement Types

Types used for cementing wells include:

ASTM Type I—Common Construction: satisfactory for cementing wells to depths of 6000 ft provided there are no temperature anomalies.

ASTM Type III—High Early-Strength: used in some areas for cementing surface pipe or shallow wells having low temperatures.

Special Oil-Well Cements. (a) Slow-setting cements have modified clinker composition. They are usually safe for wells up to approximately 12,000 ft in depth. (b) Retarded oil-well cements have a modified clinker composition and also contain a chemical set retarder. They are usually satisfactory for well depths up to 14,000 to 16,000 ft. (c) Sulfate resisting cements resist disintegrating action of sulfate waters. Some slow-setting and retarded oil-well cements are more sulfate-resistant than ASTM Type I cements.

Expanded Perlite, made by expansion of certain volcanic rock, forms a low density cellular product. When a mixture of Perlite and cement is added to water, the Perlite tends to float. Addition of from two to six per cent bentonite greatly reduces flotation and produces a slurry in which Perlite remains much better dispersed. Such low density slurry also yields more volume per sack of cement. It also sets with lower strength than does neat cement. Improved penetration is thus afforded with less shattering as a result of gun perforation.

Gel Cement—Modified Low Strength. A modified cementing composition may be prepared from normal Portland cement by addition of bentonite clay and a suitable agent for dispersing and controlling the set slurry. Light weight of such slurries allows higher placement behind casing in some areas. Such low strength cements also have better gun perforating qualities.

Modified cement is defined as one containing Portland cement, 0.10 to 1 per cent lignin dispersing agent, and 5 to 35 per cent bentonite or other colloidal clays. Gel cement with a range of four to five per cent bentonite (or other colloidal clays) is generally used with Portland cement or high early-strength cement containing no dispersing agent.

Pozzolan Cement. Pozzolans include any siliceous materials, either natural or artificial, processed or unprocessed, which in presence of lime and water will develop cement-like qualities. Natural pozzolans are mostly of volcanic origin, but also include certain diatomaceous earths. Artificial pozzolans are mainly products of heat treatment of natural materials, such as clays, shales, and siliceous rocks.

Lower strength cements have recently been accepted by the industry. Some engineers believe that approximately 500 psi compressive strength, as tested according to API Code 32, should be considered the minimum strength requirement. Others consider 200 psi compressive strength wholly adequate.

When oil well cementing was first introduced, ten days were usually allowed for setting. Recently, few operators have required more than 48 hr, and petroleum engineers have advocated still shorter times.

They claim that it is necessary to wait only until tensile strength of a set cement is sufficient to withstand weight of pipe plus closed-in wellhead pressure (usually 8 to 12 hr). Considering all these points, the properties that appear ideal for a cementing material include: less time for reaching proper strength, lighter slurry weights, better perforating qualities, more volume per sack of cement, and better control of lost circulation.

A 1:1 ratio of pozzolanic material and Portland cement, commonly spoken of as Pozmix cement, covers these ideal properties of a cement material. While other proportions have been used, 1:1 is now most common.

It has also been observed that pozzolan cement possesses positive expansion under atmospheric conditions. Confirmation has been given by laboratory tests. This characteristic is quite valuable in a well-cementing material.

Diesel Oil Cement. Field use of this type of mixture is comparatively recent. A survey of many jobs indicates that a marked improvement has resulted in quality of remedial work. Diesel oil cement is a mixture of dry cement, kerosene or diesel oil, and a surfactant (surface tension reducer). Diesel oil and kerosene must be water-free, and no water can be mixed with this type of cement slurry before it starts down the hole. The surfactant material is an extremely sensitive ingredient which promotes rapid and uniform dispersion of formation water throughout the slurry as the latter reaches a water-bearing zone. Bentonite should never be added to diesel oil cement, since it tends to absorb the wetting agent present and renders the surfactant ineffective.

Points particularly important in using this material for remedial work are:

1. By using a surfactant, twice as much dry cement can be mixed with same amount of diesel oil or kerosene. A weight of 16.2 lb per gal has been obtained using this mixture. Slurry of this weight cannot be mixed without a surfactant.
2. Mixture is placed in much the same manner as in any other conventional squeeze job using either a drillable or retrievable squeeze tool or one used on various types of Bradenhead squeeze jobs.
3. Because cement is squeezed into producing formation against a pressure equivalent to hydrostatic pressure plus pump pressure, slurry is somewhat compressed before any water enters it.
4. As water from producing formation penetrates slurry, it becomes equally distributed thru it due to surfactant action. This produces a hard, dense slurry and also ultimately results in an exceptionally tight, compact cement seal.

Cement Additives

Thru modern bulk cement blending equipment, facilitating close control of materials used, a cement can be prepared to fit practically any given well condition. Many types of additives are used in preparing blended cements.

Common additives used today, including those for preventing lost circulation, are cellophane flakes, mica flakes, expanded perlite, granulated nut shells, and shredded fibrous materials. All can be blended in any practical proportion with bulk cement and other types and so delivered.

Accelerators: calcium chloride and sodium chloride.
Retarders and Dispersants: *HR-4*, *Kembreak*, and *Lignicite*.
Bulk Additives: bentonite, pozzolan, and expanded perlite.

A retarding agent, e.g., Halliburton Retarder No. 4 (HR-4), can be used with both Portland cement and other types. Added to Portland, it makes a slow setting cement under certain temperatures and pressures. Tests with oil well slow setting cements show that this retardant can be used at 400 F with a 3-hr period in which it can be pumped. It is in powdered form that can be blended with cement, and the mixture can be hauled in bulk. It has been used with Pozmix cement where a retarded Pozmix cement is required.

LANDING EQUIPMENT

Floating equipment generally consists of a guide shoe and a float collar. The guide shoe, a round nose device that screws on the casing bottom, both guides the casing as run and prevents it from hanging on the well side. The float collar is placed in the casing string, usually one to three joints (30 to 90 ft) above its bottom. Various designs are available, all essentially back-pressure valves that allow fluid to be pumped downward but prevent reverse fluid movement. They prevent drilling fluid from entering casing as it descends and produce a floating action which takes considerable load off the derrick on a long string of casing. They also prevent flow-back of mud or cement into casing and act as stops for plugs used to separate cement and other fluids inside casing during cementing.

A modified form of float equipment is used in some instances where the casing is permitted to fill up as it enters the hole, but the back-pressure or float valve can be activated merely by circulating drilling fluid thru the casing. Circulation trips a mechanism at any chosen point, after which the back-pressure or float valve functions as described.

Plug containers are often used on top of casing so that **cementing plugs** may be released at the proper time without opening it. A bottom cementing plug is placed ahead of cement, and a top plug on top of cement column in casing. Drilling mud or other fluids are then pumped behind this plug which, as it moves downward, forces cement slurry out around the shoe and up the annular space, where it is allowed to set between casing and formation. Bottom plug contains a diaphragm that ruptures or shears under pressure when it comes to rest on the float collar. The solid top plug is so constructed that it serves the dual purpose of separating the cement from the displacing fluid and indicating that the cement job is complete. This prevents over-displacement of cement slurry and insures good quality cement around casing, where a good seal is usually desired.

Cementing plugs now most commonly used, both top and bottom, consist of a five wiper design. They are made of rubber molded around a core of hard material.

Casing centralizers located in the hole at strategic points so that cement may fill annular space around it, prevent pipe contact against the side of the hole. Otherwise, cement slurry may channel thru larger spaces but will not fill entire annulus or effectively seal against fluid migration. Centralizers are run with and without wall cleaning devices.

Wall cleaners are of various types, rotational and reciprocating. All are similar in having cleaning spikes of spring steel wire that produce abrasive action, thus removing mud filter cake or other loose materials from wall while cement fills the annular space around the pipe. With rotating type cleaning devices, casing is placed at the proper casing point, and there is no reciprocating action. This keeps bottom of pipe at the casing point at all times. Thus, during cementing, casing is at bottom at correct depth, minimizing danger of landing it above or below desired point.

Multiple stage cementing, using the *two-stage method*, makes it possible to do a cement job at the bottom thru the casing shoe and then thru the special two-stage tool, placed at any predetermined point in the casing string, for completion of second stage. Cementing can be done as one continuous operation, or it is possible to cement bottom portion, then open the tool and circulate, allowing the first cement to set before cementing thru the multiple stage tool.

Advantages of this arrangement are full depth cementing, formation shutoffs at any point, reduced pump pressures, and prevention of cement loss to thieving formations. These tools have been used to overcome special problems in many fields and at all depths. After cementing has been completed with two-stage tool, a special plug closes it by operation of a built-in sleeve. Different type plugs are used to adapt the two-stage tool to various requirements.

Multiple stage cementing with the *three-stage method* makes it possible to cement thru the casing shoe and also to perform two other cementing jobs at predetermined depths. Two multiple stage cementers, a two-stage, and a three-stage tool, are placed at the designated casing point. The first two jobs can run as one continuous operation, after which a bomb type plug is used to open the third-stage tool.

Float shoes equipped with a packer set by pump pressure are now in common use. After the packer is set, circulation takes place thru ports above it, permitting cementation above packer and avoiding excess pressure on formations below. This equipment is also available in collar form to be run in the casing string with a perforated liner below, so that the cement job can be above both liner and packer. The same size liner can be used as the production string. Many applications can be made for this type of tool. Its principal purpose is to facilitate placement of the packer below the opening thru which cementing will be performed, and to prevent cement from entering the producing formation.

WELL STIMULATION METHODS

After waiting required time for setting, cementing plugs and cement are drilled out from inside casing, if casing was set above producing zone and well is to be completed in what is known as an *open hole completion*. Cement and plugs will be drilled out with either cable or rotary tools. Both methods have certain advantages and limitations. Advantages of drilling-in with cable tools include: less harm to mud-sensitive zones, lower cost, ease of deepening and casing swabbing, and effectiveness in lost circulation zones.

Greatest limitation of this type of drilling-in operation is that there is practically no control of the well in high pressure areas.

Advantages of rotary tools in drilling-in operations include: well pressure control, faster completions, and better cores and electric logs. Greatest limitation is damage to permeability caused by drilling fluid.

If casing was set thru the producing formation, it will be necessary only to perforate producing zone, with no drilling-in operation usually required. The open hole method is frequently used in formations where poor permeability of zones renders proper drainage by perforations unlikely. Setting of casing thru producing zone is indicated for caving formations, for thin producing zones where gas cap or water table requires careful segregation, for alternate shale or sand stringers, and for multiple zone work. After casing has been either perforated or drilled out, basic methods for stimulating production are: nitro-shooting, acidizing, hydraulic fracturing, and use of surface active agents.

The greatest advantage of **nitro-shooting** is combined effect of bore hole enlargement with fracturing. Greatest disadvantage is its limitation to open hole completions only, along with a very expensive cleanout program.

Acidizing

Increased production from oil and gas wells was first accomplished by use of acid. Hydrochloric acid is most suitable for limestone and dolomite well treating. These formations are carbonates, and wells producing from them are subject to acid action. Drilling muds also contain carbonate particles which react similarly with acid treating mixtures.

Chemical reactions are as follows:

Limestone: $CaCO_3 + 2HCl = CaCl_2 + H_2O + CO_2$
Dolomite: $CaMg(CO_3)_2 + 4HCl = CaCl_2{\cdot}MgCl_2 + 2H_2O + 2CO_2$

Example. If limestone density is taken as 169 lb per cu ft and dolomite as 184 lb per cu ft, 1000 gal of 15 per cent acid will consume and produce the following products of reaction:

Limestone consumed: 11.0 cu ft
$CaCl_2$ solution: 20 per cent at 9.8 lb per gal
CO_2 (free gas): 7000 cu ft at 68 F
Dolomite consumed: 9.25 cu ft
$CaCl_2{\cdot}MgCl_2$ solution: 19.6 per cent at 9.7 lb per gal
CO_2: 7000 cu ft at 68 F

CO_2 will stay in solution at pressures above 800 psi. Not all formations are completely soluble in acid, because of presence of other materials besides carbonates; however, a proportionate amount of solids in suspension should leave the well with spent acid. Visual observations can indicate whether enough carbonate is present to warrant use of acid to increase production by dissolving action alone. It is believed that at least 25 per cent solubility is necessary, although lower ranges can be treated in other ways.

Other reactions are possible with acid, such as with iron oxide scale and iron sulfide. Both produce iron chloride solutions, but the latter produces hydrogen sulfide gas, which is poisonous in sufficient concentration. Ordinarily this reaction is always at or near well bottom, so that no generation of gas is to be feared during treatment with the well full of fluid. Caution should be observed with H_2S where it is produced naturally from wells.

Permeability is a term used to evaluate flowing capacity from or thru a porous medium. Porosity is the void space in a formation containing fluids. A formation can have a high porosity, but if connecting channels to wellbore and some of their tributaries are plugged or quite small, a drilled well can be a small producer. Thus *effective porosity* may be considered proportional to permeability in referring to pores capable of giving up their contents. Acid, by chemical action, dissolves restrictions and enlarges channel or crevice walls, thus increasing permeability and allowing greater flow.

Limestone and dolomite usually have a variable assortment of fissures or crevices. Soft streaks resembling sand can occur, or the whole body can be quite hard but contain enough capillarity (hairlike cracks) and irregular small holes or vugs to hold large amounts of oil or gas, especially in a thick zone.

Ability for flow or well production depends upon crevice sizes, number of crevices, their interconnection, formation pressure, and the physical characteristics of the formation fluids. In addition to main drainage channels, the formation matrix can have a porosity similar to sand particles with varying degrees of permeability or numerous very fine cracks that act similarly to hold fluids.

Propulsive forces that make producing wells can be internal energy from dissolved gas at reservoir pressure, possibly with a gas cap at top of zone; from underlying or frontal water that has forced the oil or gas into a stratigraphic trap under hydraulic pressure; from a combination of above, or from simple drainage by gravity flow. Surface energy of reservoir fluids as a result of capillary forces is another distinct production process; it is similar to rise of oil thru a lamp wick.

Acidizing methods can use packers and special tools to segregate and direct treatment into proper part of the wellbore, to avoid increasing flow of unwanted fluids as far as possible.

Large crevices are not necessary to deliver large volumes of fluid to a wellbore. A single crevice 0.04 in. wide connected to tributaries can drain all of the oil from a section 40 ft thick.* A condition of such uniform permeability is not common, but it can be seen that a very little widening of numerous leading channels or crevices by chemical action can increase production substantially, if enough oil or gas is present with sufficient formation pressure to move it to the wellbore.

Smaller crevices and flow channels in a formation are more easily plugged by mixtures of fluids and gases, sometimes to an extent that formation pressure cannot clean them out. Lowering the surface tension of an acid treating solution, by addition of a suitable wetting agent, reduces the necessary pressure. This can be important in low pressure formations and will aid in faster cleanup and production from those with higher pressures.

Some dolomitic formations contain gypsum. Gypsum reacts very slowly with HCl to form sulfates. Addition of polyphosphates (retarding agents) further slows this action and helps to keep the sulfates in suspension, thereby permitting them to be ejected with the spent acid and acid products.

Emulsions with acid, spent acid, and oil have caused a decrease in production from an acid treatment because high viscosity emulsions are too thick for removal by existing formation pressure. In other cases, even though emulsions will not affect production, they can cause production of a large amount of contaminated oil which requires treatment.

These emulsifying tendencies can be found and prevented before treatment is made. The selection of a nonemulsifying agent is a trial-and-error process. No single known agent prevents emulsion with acid and all oils.

Special retarding agents are available to thicken or gel the acid solution. The vehicle also provides better penetration of the acid into the formation. Heat breaks down this gel and permits this chemical action to progress. While this material starts to react immediately on contact, it takes place so slowly that after 30 min under pressure, reaction is 50 per cent complete. From this point 3 to 4 hr are required for completion. Thus, a comparative volume of thickened acid by proper manipulation may cover from four to nine times the drainage area of that of ordinary acid. Gelled acid can be made up to contain as high as 20 per cent HCl in order to give more limestone consumption to compensate partially for greater area travel.

---

* According to Morris Muskat, recognized authority on fluid flow thru porous media.

Hydraulic Fracturing

In March, 1949, this process, developed by the Stanolind Oil and Gas Co., was commercially introduced for treating wells to revive old production or to complete them for new production. Fluid product from a formation can be increased by injecting a fluid, in which sand is suspended, into the producing formation at a sufficiently rapid rate, thereby making fractures.

Hydraulic fracturing is believed to create in a formation a crack or fracture which can be used as a drainage channel for oil or gas to enter a wellbore at a more rapid rate. Fractures are created by applying sufficient hydraulic pressure in a well to rupture the formation surrounding its bore. These fractures may then be extended some distance away by injecting fluids at a higher rate than that at which fluid will be lost. The viscous fluid carries suspended quantities of graded sand grains added during injection. When pressure is released after treatment, treating fluids return as production, leaving sand within the created fracture to serve as a propping agent to hold it open.

Within a few years hydraulic fracturing has become an accepted means of production stimulation for both old and new oil wells. Conservative estimates indicate that three out of every four wells treated have resulted in profitable production increases, which in turn have raised potential reserves tremendously. Many fields are in existence today solely because of these techniques.

Fracturing Fluids. Since hydraulic fracturing of hydrocarbon-bearing formations was introduced, many changes have been made in fracturing fluids. All have been steps toward development of an ideal fracturing fluid. The latest version is obtained by adding relatively inexpensive chemicals to native crude oils, kerosene, diesel fuel, or other petroleum products, primarily to give them gel strength and increase their sand-carrying ability, and also to improve certain other properties.

Ideal properties of a fracturing fluid include: No free water, low pressure drop due to fluid friction during pumping in tubing or casing, high gel strength, low fluid loss, good viscosity index, stability to agitation, compatibility with native crude oil, minimum closed-in time after treatment, resistance to emulsification, and economy.

A fluid offering *low frictional resistance* allows more power to be applied to the formation, resulting in higher injection rates or using fewer pumps for same rates.

Gel strength is not essential when a fracturing fluid moves at relatively high velocity, but if movement rate is slowed or completely stopped, sand may drop out rapidly unless fluid possesses gel strength to support or hold it.

A *low fluid loss* fracturing fluid will theoretically produce a greater fracture area per unit volume of fluid than one with a high leak-off rate. The less fluid lost, the higher velocity will be at fracture extremities. These two factors will result in placing more sand in a fracture without screen-out.

A *good viscosity index* means that viscosity changes in relation to temperature are relatively small, especially when compared with many refined or native fracturing oils. It has been difficult to fracture many deep wells with elevated bottom hole temperature, because of high surface viscosities necessary for adequate properties at the formation. A gel having relatively little change in viscosity characteristics with temperature is quite useful for these problem wells.

Stability to agitation is important in any fracturing fluid because increasingly higher injection rates are being demanded. Many *Big-Frac* treatments are conducted thru small perforations, which subject the fluid to high shear and velocity.

A fracturing fluid must be *compatible* with oil produced from an interval to be treated; i.e., no sludge, residue, or viscosity increase must occur with intimate mixing. Otherwise, plugging may occur within the formation matrix.

Since some fracturing fluids require considerable time to revert to a low viscosity before they can be produced from the formation, much costly rig-waiting time is necessary. A *minimum closed-in time* for the formation to reach equilibrium and settle on the propping agent is desirable.

Hydraulic fracturing produces either a horizontal or vertical fracture. Type of fracture created in open holes is usually determined by type of breakdown fluid.

Horizontal Fractures. When breakdown fluids with a high fluid loss, such as kerosene, diesel fuel, water, lease crudes, emulsions, gelled crudes, and conventional fracturing oils, are used with relatively low pumping rates, horizontal fractures usually result. Part of fluid invades formation, entering some zones more rapidly than others because of variations in vertical permeability. Application of breakdown pressure to fluid in the wellbore imposes pressure on fluid in the formation. This pressure is exerted uniformly in all directions, with pressure promoting a horizontal fracture distributed over a much greater area than pressure promoting a vertical fracture. Since fracturing force is *product* of pressure times area on which it acts, force tending to fracture formation horizontally is thus usually far greater than any tending to fracture vertically. A horizontal fracture results in overlying and underlying strata compressed sufficiently to accommodate it.

Vertical Fractures. Formation of vertical fractures depends upon confining the pressure to the wellbore. To meet this requirement, a breakdown fluid of low fluid loss and containing lost circulation material is used. Lost circulation material is used solely to bridge any incipient formation fractures. Application of low pressures to fluid in the wellbore causes part to filter into the formation, depositing a filter cake on the wellbore wall. Continuation of this operation results in formation of a substantially impermeable filter cake. At this point, hydraulic pressure against the formation is abruptly increased. Pressure is confined to the wellbore by the filter cake, and any force acting within the formation to effect a horizontal fracture is eliminated. Formation pressure is thus resolved into a single force which will fracture the formation vertically.

A vertical fracture is desired in sands laminated with shale, sands separated by large shale breaks, and pure sands of widely varying permeability. This procedure has been applied only to open hole completions.

Where active water or gas zones are present, vertical fracturing is not recommended. Possible fracturing into such zones might result in excessive water or gas production. Fractures originally existing in formations adjacent to the wellbore usually will be extended by either of the foregoing techniques.

Sand Use in Hydraulic Fracturing. The sand (propping agent) is an important constituent. Rounded silica sand grains averaging approximately 0.026 in. diam are normally used, although both larger and smaller particles have been employed. Close tolerance in grading is desirable, and cleanliness is essential. Angular and subangular grains have performed satisfactorily in selected formations but have not proved acceptable for a wide range of conditions. Sand is transported either in bags or in bulk conveyed to fields in specially constructed hauling equipment. Special equipment mixes it with fracturing fluids at the well in various proportions, as desired.

Although types of fluid may be varied for hydraulic fracturing, experience has shown that a propping agent must be employed for successful use. Only in rare cases have viscous fluids alone resulted in sustained production increases when injected into formations to create fractures. Some treatments without a propping agent have failed to give lasting results, while applications in same well and formation using same fluid with a propping agent have been highly successful.

Sand Characteristics. First propping agent utilized in fracture treatments was ordinary screened river sand. Difficulty was experienced when this material refused to enter many formations and caused bridges from within the hole or in subsurface tools. Because trouble existed even with improved forms of screening, use of angular type sands was discontinued in favor of clean, carefully graded, rounded grain silica.

Benefits derived from sand used in hydraulic fracturing are believed to be results of propping within a fracture to provide and preserve a more permeable path for fluid flow. No evidence supports the theory that injected sand has a cutting action. Measurement of ability of a fracture to convey flow is termed *fracture flow capacity,* and is expressed as product of fracture permeability *times* width. Type of sand most capable of accomplishing required propping action should be chosen because a high capacity value is always desirable and even necessary in many types of producing formations.

Laboratory studies of flow capacities thru sands under various compressive loads have shown that the larger the grains the greater the flow capacity will be under corresponding loads. Also, flow capacities thru rounded grains of limited particle size range tend to decrease more or less uniformly as compressive load increases, while sands having a wider range exhibit good capacity under low pressure but a sharp decrease as load is raised. Additional tests have shown that angular type sands under compression will permit as much as 20 per cent greater reduction in permeability than round grains.

The most efficient propping agent is one in which the particles are: rounded in shape, of minimum variation in size, as large in diameter as can be placed practically within formation being fractured, of high compressive strength, and free of dust or silt.

Proportioning and Mixing. Methods of adding propping agent (sand) to fluid were rather crude in early fracturing treatments. Lack of control allowed sand to enter in uneven proportions and created slugs of high concentrations which promoted injection refusal. Two general mixing methods are now used. One is a batch type operation in which sand and fluid are blended in relatively large portable tanks by special self-contained mixers. By this process, mixing is thoroughly accomplished, but sand-to-fluid ratios are fixed. Continuous blending allows flexibility in these ratios. Considerable equipment is required where treatment volumes are large.

By the second method, sand and fluid, admitted at controlled rates into a relatively small tank containing a mechanical agitator, are intimately mixed and then pumped to well. This method permits continuous operation, limited only by availability of fracturing ingredients. It maintains any desired constant sand-to-fluid ratio within narrow limits, although mixing rates may vary. Ratios may be changed by simple control manipulation as desired. This last feature is very important, since it is often desirable to begin an injection with a small sand concentration, which is gradually increased as treatment progresses. For example, some formations will plug if fluid containing 1 to 2 lb of sand per gallon is used at first. By starting with 0.5 lb per gal and slowly increasing sand proportion as pumping proceeds, 3 to 4 lb per gal can be injected.

Surface Active Agents

Drilling into an oil-producing formation can result in impairment of normal permeability and create restrictions to flow of fluids to the wellbore. This has been demonstrated many times where formation characteristics, disclosed by cores, drill stem tests, and electric logs, indicate potential production not developed in a completed well. Many wells have been abandoned which, in view of presently known facts, might have been made into satisfactory producers. Older producing wells, once good producers, are frequently damaged during performance of various kinds of remedial work and prove difficult to restore to the previous production rate.

Phenomenal results from relatively small acid treatments in limestone formations or shooting in sandstone formations have demonstrated that both rotary and cable-tool drilled wells have been plugged too close to their bores (*skin effect*). It should not be limited to a small radius around the wellbore since blocking materials can penetrate highly permeable formations, crevices, and fractures to a considerable distance. Causes of blocking action vary from field to field.

Unless a well is drilled-in with oil or an oil-base mud, presence of water is necessary. Unavoidable filtrate from a water-base mud can act in several ways to reduce permeability. Drilling fluids can penetrate appreciable distances into oil-bearing formations. Mud losses barely noticeable while drilling or running casing can be sufficient to plug the well. Excessive pressures created by drilling and running of casing may force large quantities of drilling fluids into the formation. Even minute losses may be sufficient to plug the well within a short radius. Fresh or salt water introduced during remedial operations penetrates adjacent formation and can act to reduce oil flow. In some cases, oil-water emulsions are developed and form efficient flow blocks. End effects are often noticed where formation fines and interstitial water accumulate near the wellbore, reducing permeability of the producing formation. Some formations contain varying quantities of bentonite, which swells by adsorption of water and thus greatly reduces flow channel diameters in the formation.

Water and/or bentonite are chief causes of permeability reduction. Removal of water or bentonitic drilling mud would restore original conditions existing before penetration of the drill and permit effective flow of well fluids to the bore. Several procedures for physical removal of blocking materials were considered, but all were abandoned in favor of a chemical method. Such a method is designed to break water blocks, disperse muds, reduce size of water-swollen bentonite particles, and destroy emulsions near wellbore. Actual dissolving of siliceous and bentonitic materials cannot be achieved in appreciable amounts under actual well conditions. A chemical action must produce a series of physical changes to accomplish desired results.

Laboratory studies show that a mixture of dilute hydrochloric acid and a surfactant of the aromatic sulfonic acid class has a marked effect in increasing effective permeability of bentonite sands. Water adsorption or swelling properties of bentonite depend mainly upon surface characteristics of individual particles. A mixture of dilute hydrochloric acid and aromatic sulfonic acid has a powerful effect on surface characteristics of bentonite. Hydrogen ions from hydrochloric acid will replace predominant sodium ions attached to swollen bentonite, thus causing formation of hydrogen bentonite. This change causes bentonite particles to lose most of their affinity for water. The system, bentonite and water, is reduced in size, and free water is released from each nucleus.

Such a mixture will penetrate mud filter cakes and bentonitic sand formations effectively, thus contacting a large quantity of bentonite. Efficiency of ion exchange is increased by this action, and the released water, having a low surface tension along with the original solution, is easily recovered from the treated horizon. New wells drilled-in with oil or otherwise containing no bentonitic material and old wells of the same nature which are making water are often benefited by a simple treatment with a chemical surfactant.

Interstitial waters are usually present in most formations. This water can cause a partial or complete block of oil flow to the wellbore and reduce effective permeability of sand to oil. Theory holds that continued oil production will cause a gradual movement of connate water, further reducing this effective permeability.

Many crude oils and various waters are capable of forming emulsions. These are almost always more viscous than crude oil and form an even more effective block to oil flow than water alone. Emulsions, in addition to their high viscosity, often exhibit thixotropic properties. Thixotropic emulsions may be so thick that they resemble an elastic solid. When stirred or caused to flow, their thixotropic structure is partially destroyed, but it slowly heals on standing. Both water blocks and water-oil emulsions can be present, often fairly close to the wellbore, causing more resistance to flow than formation pressure can overcome. Emulsions can result in an increased fluid viscosity of from 10 to 2000 times that of the oil in place. Either breaking or preventing them will be of great benefit in increasing productive flow toward the wellbore.

Salt and Water Removal

Salt and water are removed from a well when the preceding stimulation methods are used. In the Appalachian area it is also common practice to water down the well to remove salt, and to *bail out* the water. A siphon is commonly employed for removing water from the well bottom. This is a small pipe, of say ½ to 2 in. diam, running from top to bottom of the well. Bottom end is sealed, but lowest length of tubing is slotted. Water invading the well bottom can be forced thru the slots into and up the pipe, together with some gas, and thus removed by gas pressure. It is not unusual for a well to become so clogged during production that its output is unduly decreased. Well cleaning of this sort may be required periodically. Pumping and bailing are also used. Bailing is quite effective where cavings are to be removed, but well must be open during this operation, with considerable gas lost. Salt may deposit on tubing walls, requiring wash-down and removal. Such salt deposition can result from evaporation of water from bottom-hole brine.

REFERENCES

1. Am. Pet. Inst. *Petroleum Discovery and Production.* New York, 1946. (chart—folder).
2. ——. *Specification for Casing, Tubing and Drill Pipe,* 26th ed. (Std 5A) Dallas, 1963.
3. "Some Natural Gas Problems Are Under Attack." *Petrol. Week* 10: 350, June 24, 1960.
4. Am. Pet. Inst. *Bulletin on Performance Properties of Casing and Tubing,* 7th ed. (Bul. 5C2) Dallas, 1957. Supplement 1961.
5. ——. *Recommended Practice for Care and Use of Casing, Tubing and Drill Pipe,* 8th ed. (RP 5C1), Dallas, 1963.
6. Howard, G. C. and Clark, J. B. "Factors to be Considered in Obtaining Proper Cementing of Casing." *Oil & Gas J.* 47: 243+, Nov. 11, 1948.
7. Am. Pet. Inst. *Specification for Oil-Well Cements and Cement Additives,* 7th ed. (Std 10A) Also: *Recommended Practice for Testing Oil-Well Cements and Cement Additives,* 10th ed. (RP 10B) Dallas, 1960 and 1961.

Chapter 6

Gas Separators, Heaters, and Cleaners

by A. W. Francis, Jr., and Russell L. Morris

GAS SEPARATORS

Gas and liquid separators,* widely used in the production and transmission of natural gas, include gas field production separators, transmission line scrubbers,* and plant separators and scrubbers. By far the largest number of separators is in field production systems. Problems are varied because of complexities of mixtures, foaming conditions, presence of two liquid phases, solid corrosion products in the wellstream, drilling mud, formation solids, extreme paraffin conditions, and other less common factors.

Gas and liquid separators must comply with pressure vessel code requirements. The usual separator has automatic controls for unattended operation.

Removal of liquid can be accomplished because: (1) density of liquid droplets is greater than that of the gas stream, and (2) force of surface tension of the liquid film wetting a properly designed collecting surface is greater than forces tending to disperse liquid back into the gas stream.

Care must also be taken to allow all gas entrained in the oil to escape back into the gas stream. Allowing the liquid to come to true equilibrium at separation temperature and pressure can be the controlling efficiency factor of a separator. Actual time required will vary from a very short interval for light hydrocarbon liquids, like wellstream distillates and plant gasoline fractions, to many minutes for heavy, high viscosity crude oils. Viscosity and surface tension of the liquid are the physical characteristics governing required retention time. The only other factor to establish besides retention time is required gas–liquid interfacial area in the separator. Some oils show degassing rates nearly proportional to interfacial area.

Types

A simple and widely used type is the **labyrinth** unit, which directs the stream thru a maze of either a rigid fabricated pattern or a random packing of a ceramic material used in fractionation towers. Principle of operation is that high turbulence or abrupt directional changes of the gas stream will *throw out* liquid droplets onto obstructing surfaces because of their greater density and inability to change direction. This type operates between both upper and lower velocity limits. Liquid droplets will meander thru passages available when velocity is too low to deposit them on a collection surface. When velocity is excessive, extreme turbulence will re-entrain liquid, causing rapid carry-over, increasing very rapidly to actual flooding of the separator.

This type separator is rather sensitive to overloading with liquid. Amount being separated will have a significant effect on gas capacity. Also, its labyrinth construction or packing makes it susceptible to plugging and collecting of solids encountered in field production. These solids are generally composed of drilling mud, paraffin, or both. Cleaning such a separator is more difficult than cleaning other general types.

Centrifugal separation is usually accomplished in a vertical cylindrical vessel with a side inlet opening. The entering stream is diverted to a tangential flow which thereby spins the gas around the shell interior. Heavier components tend to fall from the inlet level, while lighter components tend to rise toward the gas outlet in the vessel top. Centrifugal action accelerates classification of liquid to the outermost gas film, where it wets the vessel wall and runs downward. Liquid level is maintained in some kind of quieting chamber for degassing. Gas passes thru a final scrubbing element, usually a combination centrifugal and labyrinth unit.

Centrifugal type separators are very difficult to overload with liquid because essentially everything is separated in a relatively large chamber, with the final scrubbing element doing only the last cleaning job. Regardless of the gas–liquid ratio at separator inlet, the final scrubbing element will always have a very small liquid load on it.

The **horizontal type** of separator makes use of the normal trajectory of a liquid particle in a horizontally moving gas stream. This type always embodies preliminary impingement and deflection surfaces to knock down liquid entering as *slugs* or large drops, allowing only droplets small enough to remain entrained in the gas to move past obstructions into the main separation chamber. This chamber is a horizontal cylinder filled with closely spaced parallel plates, which are oriented parallel to its axis and at sufficient angle with the horizontal for rapid runoff of collected liquid. Plates are spaced closely enough that a droplet of predetermined size will have sufficient time during its travel to fall in a normal trajectory and adhere to a plate. Design is based on gas velocity limit for a given shell length. Small plate indentations normal to gas flow provide shelter for liquid drops being swept along in direction of gas flow. Liquid entering them is free to run down the plate to the vessel wall and thence to a quieting chamber or degassing space.

* Distinctions may be made between separators and scrubbers; the former do not use auxiliary media, e.g., scrubbing oil, while the latter do.

The horizontal separator has the advantage of lowest overall height of any type. It fits under low overhead obstructions and is more readily skid mounted, prepiped, and packaged, because of lower center of gravity and overall height. It is manufactured in both one and two-barrel designs. The two-barrel design accomplishes separation and liquid quieting functions in separate cylinders. Thus, smaller diameter cylinders can be used than when separation and liquid quieting are accomplished in one cylinder.

Vertical shell separators as shown in Fig. 4-32 are probably most commonly used. They can conform most easily to space limitations. Two or more separators can be made within one shell if space is at a high premium.

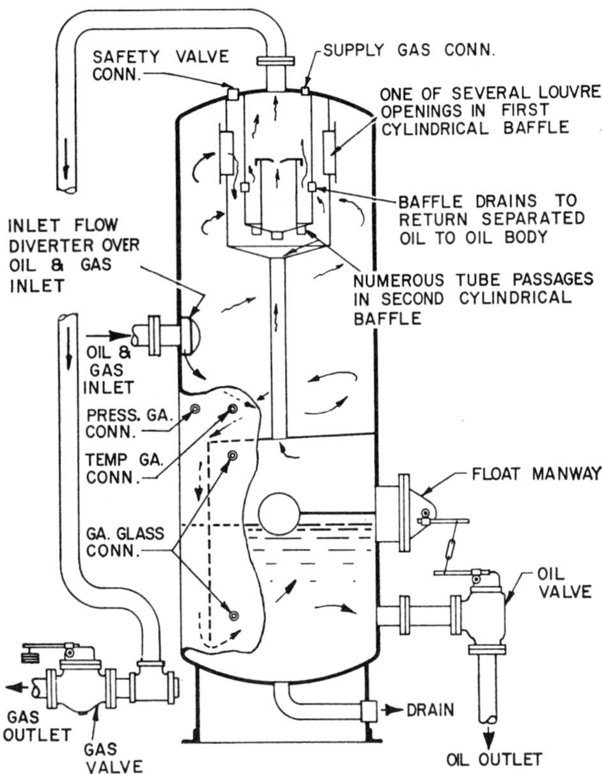

Fig. 4-32 Low pressure vertical separator. (National Tank Co.)

Attention must be given in separator design to the general range of gas-to-liquid ratio. This must be done to insure sufficient cross-sectional area for separation and sufficient liquid quieting volume with required gas–liquid interface for degassing. For small gas and large liquid volumes, liquid space obviously should predominate, while for large gas and small liquid volumes liquid space should be small compared to gas separation space.

Accessories for gas–liquid separators can be significant design features, particularly for liquid level controls, liquid discharge valves, and gas back-pressure valves if required. Factors that can cause trouble include presence of abrasive particles, extreme pressure drops, very low and widely variable flow rates, and other less common considerations. Control manufacturers have developed on-off or intermittent controls to minimize the erosive action of liquids, extreme pressure drops, or abrasive particles across the inner valve in a control valve. Valves are manufactured with special, hard inner valve trim to further minimize erosion for use in conjunction with intermittent controls where conditions merit its use. Special stem packing problems exist in some extreme pressure applications that demand special attention. Full investigation of any separation problem will allow reasonable prediction of conditions which should be met.

Stage Separation

Individual units of separation equipment may be combined to do a given job. The simplest flow pattern is thru a single unit, as usually required for plant inlet and transmission line scrubbing, large enough to separate the entire stream but containing negligible amounts of liquid compared to value of that recovered. Wellstream production, on the other hand, usually contains a sufficient volume of recoverable liquid hydrocarbons to make the use of a more elaborate installation attractive because of additional liquid recovery that can be attained with the added separation equipment. Considerable study is involved in determining the optimum or economic limit of equipment expenditure.

Stage separation, generally referred to as *differential liberation,* has been evolved as an economical approach to the ultimate in separation. Differential liberation may be defined as separation of the wellstream into gas and liquid at optimum separation pressure, followed by an *infinite* number of liquid pressure reductions, each in its turn followed by removal of gas liberated, as illustrated in Fig. 4-33. This process will yield maximum volume of liquid obtainable by separation at storage pressure, usually atmospheric. This ideal separation is a *batch* type process, and the practical approach is to use a *series of incremental pressure reductions.* The problem is to find, by calculation or experiment, the number of increments economically justified by added increase in recovered volume thru use of each successive increment. The usual domestic gas well will justify three stages of separation, including the storage tank as the final one. When large quantities of oil

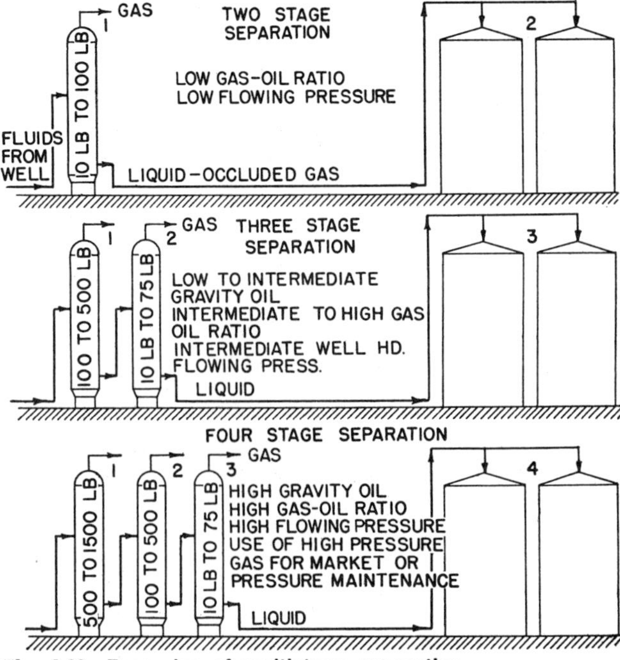

Fig. 4-33 Examples of multistage separation.

production are involved, four or more stages of separation may be justified.

Low Temperature Separation

Recent advances in separation methods take advantage of pressure differential between wellhead flowing pressure and transmission line pressure, to allow a constant enthalpy or Joule–Thomson expansion into the separator to obtain cooling effect. This is coupled with heat exchange precooling to achieve rather low temperatures. Equilibrium constants for the greater molecular weight fractions of the mixture decrease much more rapidly with decreasing temperatures than do those for lighter components which should leave the separator in the gas phase. Depending upon wellstream composition, this effect can yield a substantial increase in recovered liquid in return for a small reduction in volume of gas sold from the high pressure separation vessel. Usually the hydrocarbon material has a much greater value as a liquid than it does as a gas. Therefore this process yields an attractive increase in revenue from a given unit of produced well fluid. The amount of cooling available to this process is sharply defined by available pressure drop.

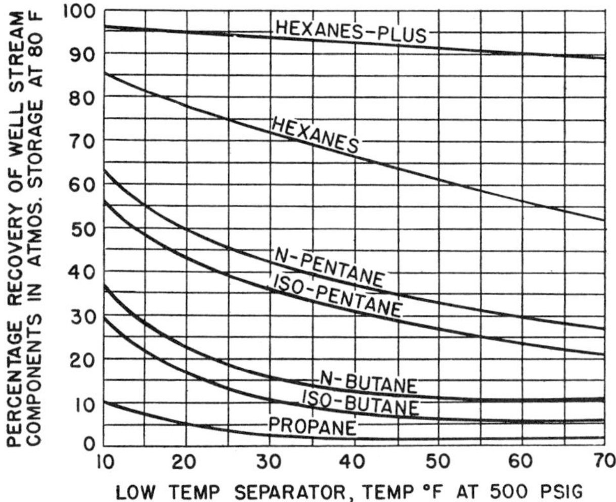

Fig. 4-34 Percentage wellstream components recovery from low temperature separator at constant pressure.

Wellstream Composition

	Mole, %		Mole, %
Methane	97.71	*iso*-Pentane	0.12
Ethane	4.44	*n*-Pentane	.20
Propane	1.50	Hexane	.34
iso-Butane	0.30	Heptanes	0.83
n-Butane	0.56		

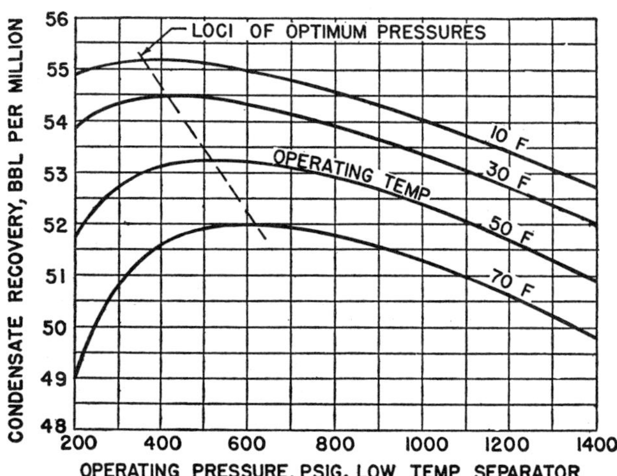

Fig. 4-35 Condensate recovery from low temperature separator at varying pressures.

Figures 4-34 and 4-35 show liquid recovery from a low temperature separator operating under conditions set forth.

Hydrate Handling. Gaseous hydrocarbons form hydrates at elevated pressures and at temperatures considerably above the actual freezing point of liquid water. Since gas production is usually saturated with water at reservoir pressure and temperature, provision must be made to cope with hydrates forming with cooling, or to prevent their formation.

One approach is to allow gas hydrates to form, with provision in the separator for melting them into their hydrocarbon and liquid water components. Some of these hydrocarbons will be gaseous and some liquid at the separation pressure. Therefore, a three-phase separation problem exists after melting the hydrates; that of gas, condensate or distillate, and liquid water. Hydrate melting is accomplished by using a heating coil low enough in the vessel base to be in the liquid water phase. Heat is usually furnished by

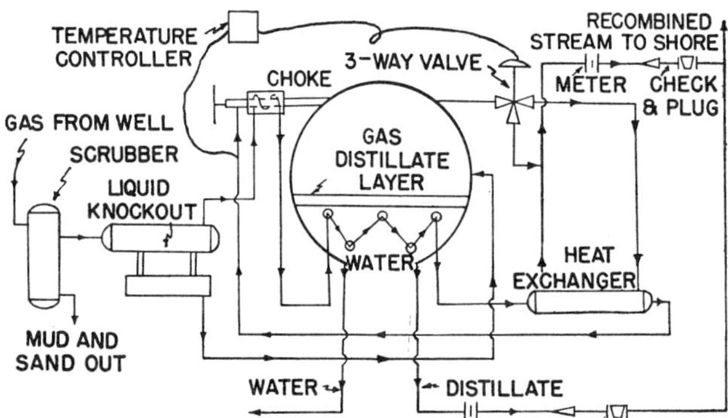

Fig. 4-36 Schematic diagram of low temperature extraction equipment and free water knockout.

the incoming wellstream but may be from an outside source. The slight amount of heat required for this melting operation has a beneficial stabilizing effect on the condensate, thereby reducing solution gas content of the liquid. This increases gas volume slightly and reduces flash losses when stage separating the resultant liquid to the storage tanks. Figure 4-36 shows a typical flow diagram.

Another approach is to inject a hydrate inhibitor into the gas stream to prevent hydrate formation. This requires mechanical pump equipment, with attendant cost and maintenance. This approach can use either an expendable inhibitor or a recoverable type such as mono-, di-, or triethylene glycol, with which collection and reconcentration equipment must be provided. The average case in which liquid water is produced from the well with the gas presents a further problem. Such water must be removed before the inhibitor is injected, to prevent inhibitor dilution to such an extent that it would not be effective. Produced water often contains minerals that may ruin a reconcentrator rapidly if allowed to deposit. Use of a nonrecoverable inhibitor like methanol eliminates recovery and reconcentration problems, but this may become so costly as to be prohibitive.

Some types of hydrate melting units utilize an inhibitor to prevent hydrate formation in the precooling exchanger *before* expansion takes place. In this way utility of the unit is extended when available, and pressure drop declines with declining well pressure to a point where desired temperature is no longer obtainable. When heat is supplied to the liquid storage portion of the recovery vessel, required inhibitor concentration is much less than for a total hydrate inhibition problem. Attendant inhibitor losses are lower at the lower concentrations because inhibition is required only in precooling before expansion, instead of at low temperature after expansion. The inhibitor system need not be installed until actual need arises in the hydrate melting system.

Efficiency

Vertical separator efficiency has been measured in terms of carry-over. This has been correlated to a derivation of *Stokes' law* for spherical particles (see Fig. 1–6b), with empirical constants substituted for theoretical factors. This correlation is based on 10 ft shell height (values for use with a 5 ft shell height are 52 per cent thereof). The basic relation is expressed as a maximum upward velocity as follows:

$$V_{max} = K\sqrt{\frac{P_1 - P_g}{P_g}}$$

where:

V_{max} = velocity, fpm
P_1 = density of liquid, lb per cu ft
P_g = density of gas, lb per cu ft
K = 0.17 for max allowable carry-over of 1.0 gal per MMCF gas; 0.11 for max allowable carry-over of 0.1 gal per MMCF gas

GAS HEATERS

Gas heaters may be required to heat a gas stream and keep its temperature above that of hydrate formation, particularly when pressure reduction occurs. Gathering systems may also require heat to permit transmission to a central separation and dehydration location. Regeneration heat to solid desiccant dehydration plants, special process streams for high temperatures, and other special or local problems throughout the system may require the application of heat. Flame arrestors may be used to preclude the possibility of any flame propagation outside heater fire tubes.

Types

These heaters are manufactured in several easily recognizable types. One of the most popular, an **indirect heater,** allows unattended use even in remote locations (Fig. 4-37). Its heating element, usually a fire tube, and coils containing gas to be heated share the same water bath. Heat transfer is from fire tube to water to coils. Some coils might be exposed with decrease of available heat transfer area, should the water level become too low. Fuel gas flow is thermostatically controlled, thereby maintaining desired water-bath temperature (maximum—the boiling point of water). However, gas temperatures above 100 to 150 F are seldom required.

Internal construction of an indirect heater varies for different manufacturers. Some embody patented features. Varieties include natural convection (limited by fire tube percolation) and specially baffled thermosiphon convection circulation. Outside film heat transfer coefficient will be in the range of 200 to 250 Btu per hr-sq ft-°F for thermosiphon action. Heat losses between fuel gas and heated gas will include those from the fire tube and heater shell. Such loss will be present in all gas-fired heaters. The limiting heat flux from metal to water will be in the range of 10,000 Btu per hr-sq ft.

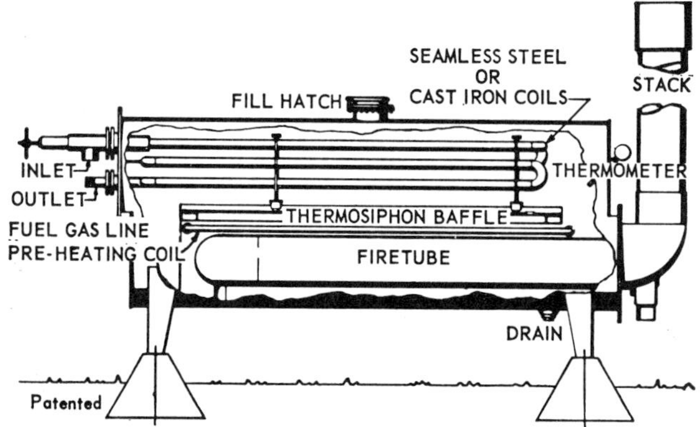

Fig. 4-37 Indirect heater—oil field type. (National Tank Co.)

Another type is the **14.5 psig nonboiler code steam generator** with coils installed in the upper part of the vessel. This heater features ease of operation and temperature levels up to the condensing temperature of 14.5 psig steam. Outside film heat transfer coefficients on the coils range upward to 1500 Btu per hr-sq ft-°F, with the inside film transfer coefficient thus controlling overall heat transfer rate. The unit has thermostatically controlled heat input and should embody a low water shutoff.

This general type may be made for operation at any steam pressure. It should meet existing boiler codes for pressures over 14.5 psig.

When higher temperatures for gas heating are required, a **salt bath heater** may be used. Selection of salt should be made on basis of melting and boiling points, toxicity, stability, corrosive nature when molten, cost, and availability. Salt mixtures adequate for usual applications are available. The salt is melted into a shell containing the fire tube and heating coils. Its specific heat makes intermittent operation somewhat difficult, but correct process design can minimize this effect. Operating temperatures for this type unit range from 350 to 500 F. Above 500 F fire tube duty presents a maintenance problem. Also salt dissociation starts. Salt bath film heat transfer coefficients are roughly two-thirds that of water.

Special plant processes may require direct-fired tubes; each application should be engineered individually.

GAS CLEANERS

Gas cleaners are usually required and are generally located upstream of gas transmission compression equipment, metering equipment, pressure reduction stations, city trunk lines, and certain types of processing plants.

Efficiencies of dust removal equipment are being improved steadily and can be considered satisfactory for many applications. Their success is best measured by the amount of dust and other solids remaining in the system.

Solid materials present in natural gas systems originate from the following principal sources: (1) construction dirt and welding residues, (2) pipe mill scale, (3) pipeline corrosion inside the line, and (4) solids produced from gas wells (latest gas production practices have reduced these to an insignificant proportion).

The amount of dust to be removed from a pipeline generally will be the greatest immediately after it is put into service. At that time the cleaning equipment may prove undersized because dust volume will probably exceed the amount normally expected. Usually new lines are not started up under full capacity, and since dust moving in the line increases with increasing flow, construction dust can be expected to be present for a considerable time. A gradual change in dust composition occurs with a diminishing of construction dust and an increase in erosion products. Amount of dust varies with flow rate. During low demand seasons, the gas may appear almost dust free. As high demand seasons approach, dust which has formed or been generated during low demand periods will be picked up and appear in the cleaning equipment.

Sizing cleaning equipment must generally be done on the basis of gas volume with reference to operating pressure. It is impossible to predict dust loads except from experience, until sufficient time has elapsed to obtain samples and analyze them for particle size, amount, and sometimes composition.

Types

Filters can be designed with very low resistance to flow because they can be constructed with extremely large filter medium area for space occupied. This allows velocities of about 5 to 25 fpm across filter media, with resultant small pressure drops. Filters have been designed to remove dust particles as small as about two microns (can be designed for 0.3 microns), and to minimize frequency of element changing. Since filters afford a means of dry scrubbing, there is no carry-over of scrubbing liquids. Filters are generally offered in working pressures up to 2500 psi. They have the advantage in certain installations of extreme compactness. Their initial and operating costs are relatively low.

In **liquid bath scrubbers** gas passes thru an oil bath and is subjected to sufficient turbulence to insure good mixing, thereby wetting all dust particles. Gas then passes thru a primary separation section, to remove large liquid drops and surges containing dust particles. Next it passes thru a mist extraction stage to remove last traces of droplets of scrubber oil. Oil losses as low as 0.03 gal per MMCF have been claimed for these units. Oil and dust thus separated return to an oil and dust settling reservoir.

The contamination rate of oil in this type of unit is governed by the foreign material in the gas and may vary from a few weeks to a year. When contaminated, oil can be circulated thru a diatomaceous earth filter or withdrawn and allowed to settle long enough for its cleaning. Cleaners of this type are generally considered low in operating cost, simple to operate and maintain, and capable of removing dust without reference to particle size. Design of this type of scrubber should be based on minimum pressure and maximum volume.

Cleaners of this type usually employ a scrubbing oil having the following approximate specifications:

1. 25–35° API gravity at 60 F.
2. Initial boiling point above 500 F.
3. Distillation end point below 850 F.
4. Viscosity about 100 Saybolt seconds at 100 F.
5. Pour point between −30 F and 0 F.

A **rotating vane type scrubber,** usually a horizontal vessel of two compartments, can remove liquids and/or solids without making any changes or adjustments. Its first compartment is a straightforward primary separation chamber, while the second is a revolving labyrinth for final scrubbing. The labyrinth revolves thru an oil bath, which serves to maintain an oil film on the scrubber element vanes at all times and provides a washing action for all material trapped in this oil film from the flowing gas stream. A clean oil film is thus presented at all times. These scrubbers have some features in common with oil bath scrubbers. Gas does not actually pass thru the oil bath itself, thus reducing both oil carry-over tendency and gas pressure drop. Specifications may call for no oil carry-over within rated capacity. If they are used for scrubbing liquids from the gas stream, liquid level controls would be required.

Cyclone or dry type scrubbers have been used successfully in numerous gas installations. Circular motion is imparted to the gas; centrifugal force drives all material of greater density to the enclosing circular wall. The more successful types employ a tube within a tube, with gas traveling downward between them and discharging the dust downward. Cleaned gas then reverses direction and discharges thru the inner tube. Removed dust travels downward to a suitable collector and is withdrawn. Sizes and capacities of these scrubbers may cover a very wide range. Units of 100 MMCF daily capacity are now in use, and can be manifolded in parallel to obtain multiples of this capacity. They can be purchased for design pressures within the approximate range of 250 to 1500 psig. Greatest efficiency is usually obtained where 2 to 5 psi pressure drop is maintained across scrubber tubes.

Since these units are dry scrubbers there are no scrubbing oils to maintain or carry-over with the gas. They are also able to separate both liquids and solids indiscriminately. A disadvantage is that larger particles are separated more easily than smaller ones. Therefore, loss of efficiency due to operating below the design range causes undesirable losses of smaller particles. These scrubbers are generally quite compact except for rather low operating pressures. Theoretically, since the separating force for a given cyclone scrubber is proportional to entering velocity *squared,* operation at $\frac{1}{5}$ of rated capacity would provide only $\frac{1}{25}$ of separating power at rated capacity. This would entail a serious loss of particle cleaning in the very smallest sizes (0–10 microns). However, overall particle cleaning efficiency might not diminish greatly, because the smallest particles generally do not constitute a very large proportion of total dust. Dust cleaning efficiencies have been reported as high as 99.8 per cent under ideal conditions and can generally be expected to be 99 per cent.

Chapter 7
Natural Gas Gathering Systems

by George W. McKinley and F. E. Vandaveer

Systems usually consist of the piping and processing equipment between individual wells in a gas field or fields and the compressor station at the inlet of transmission or distribution lines. System size and complexity may vary both within a gas producing area and from one area to another. The smallest gathering system consists simply of two or more gas wells interconnected by piping and tied directly into a distribution system. Wells of a small gas field may be connected to a common line, connected in turn directly to a transmission line. A *dry* gas field may have no processing equipment except drips at each well and in low places in the line. For a *wet* gas field, however, drips, separators, heaters, and a gasoline plant may be required. For large fields and for several interconnected fields, involving hundreds of miles of piping, gathering systems may include, besides piping and valves, such equipment as drips, separators, meters, heaters, dehydrators, gasoline plant, sulfur plant, cleaners, and compressors.

PIPING SYSTEM

Piping system design for a given field will be influenced by the shape of the field, the anticipated future developments, and the roughness of the terrain.

One or a combination of the following general plans is usually selected:

1. Main gathering lines laid in a *loop* (Fig. 4-38) about the producing field, lines from each well running to these main lines, with the compressor station in the loop. Such a system, while not economical with respect to piping, permits system operation while a part of the loop is being repaired.
2. Main gathering lines laid radiating out from a compressor station near the field center (Fig. 4-39).

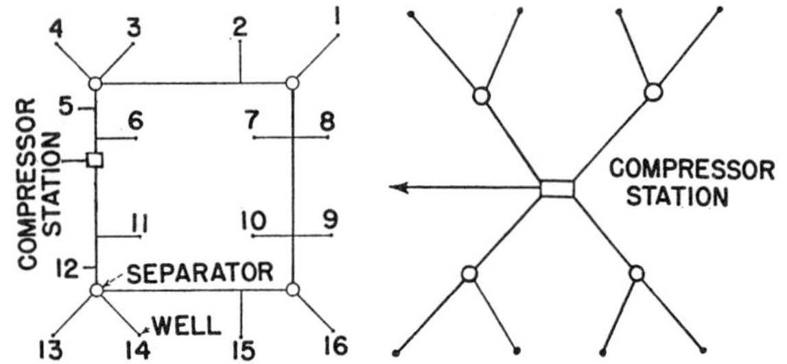

Fig. 4-38 (left) Representative loop gathering system with four oil-gas separators.
Fig. 4-39 (right) Representative radiating line gathering system.

3. The main line run directly thru the field with laterals to wells taken off each side (Fig. 4-40).

Sizing of main gathering lines depends upon the following factors: (1) number and design of compressor stations (i.e., pipeline diameter varies inversely with number of compressor stations required); (2) anticipated peak day demands; (3) safety at maximum pressures; (4) initial reservoir pressure of wells; (5) expected production of wells; (6) whether liquid condensate, water, and hydrate may be present; (7) line pressures to be maintained as well as future lower pressure; (8) allowable pressure drop when producing the field with compressors; and (9) final abandonment pressure.

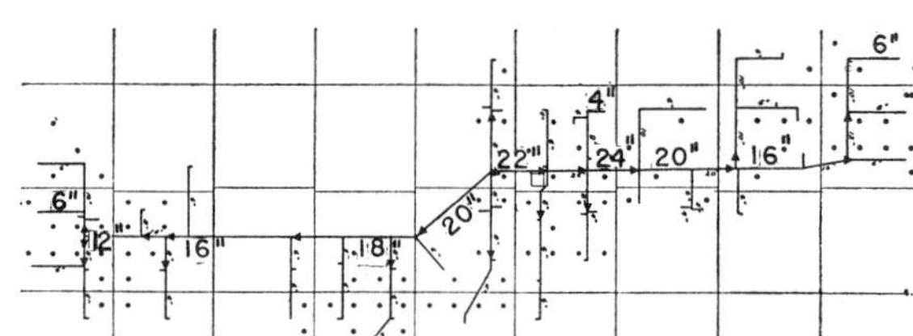

Fig. 4-40 Representative main gathering line directly thru the field with laterals.[1]

Determination of size of main gathering lines and laterals can be made with any of the commonly used pipeline flow formulas. These formulas will not be accurate for short lines, three to five miles long, if large quantities of liquid are present to restrict gas flow. An experience factor must be used to determine pressure drop. This factor may vary from 1.10 to 1.5 or greater, when large quantities of liquid are expected to accumulate in the gathering system.

Table 4-28 gives a comparison of relative carrying capacities of various diameters of pipe when using **Weymouth's** formula. An experience factor should be applied to these values, in line with the general use of this formula.

Electronic computers[1] may be used in calculating gathering system pipe sizes.

DRIPS AND SEPARATORS

Removal of liquid water and liquid hydrocarbons, like condensate, gasoline, and oil, soon after gas leaves the wellhead is essential in reducing restriction to flow thru the gathering system, and in avoiding ice and hydrate formation within it. This is usually effected to an acceptable level by means of a

Table 4-28 Miles of Pipe of Various Diameters Having Delivery Capacity Equivalent to One Mile of a Diameter Given in Extreme Left Column[2]

Internal diam, in.	2.067	3.068	4.026	6.065	8.071	10.025	12.125
2.067	1.000	8.218	35.01	311.40	1429.4	...	...
3.068	0.1217	1.000	4.260	37.89	173.9	622.2	...
4.026	.1286	0.2347	1.000	8.895	40.83	146.1	357.8
6.065	0.0032	0.0264	0.1124	1.000	4.590	16.42	40.23
8.071			0.0245	0.2178	1.000	3.577	8.763
10.250				.0609	0.2795	1.000	2.450
12.125				.0249	.1141	0.4082	1.000
13.375				0.0147	.0676	.2419	0.5925
15.375					0.0322	.1151	.2818
17.375						0.0599	0.1468

See Eq. 6 of Chap. 2 in Sec. 8 for formula used.
Example: Capacity of one mile of 8.071 in. pipe is equivalent to 0.0245 miles of 4.026 in. or 3.577 miles of 10.025 in. pipe for same pressure conditions.

gas drip and/or gas separators at the wellhead and by gas drips at various low places in the line.

Gas drip designs usually include one or more of the following features:

1. Velocity reduction. Gas stream velocity is reduced, usually by means of an enlarged pipe or cylinder, to a point where entrained liquids will be separated by gravity.

2. Impingement and agglomeration. The gas stream strikes or passes over a baffle surface where entrained liquid accumulates into drops which are separated by gravity. Baffle surface may be a plate or plates, a packed section, a porous stone element, or a porous plastic element.

3. "Stepladder" type. As used at various system low points, this type consists of an enlarged section in the line, to which a reservoir is connected by means of two or more pipes. Such a reservoir is placed directly beneath enlarged section so that any liquids may flow into it thru two or more pipes by gravity.

4. Temperature reduction. Enough heat is removed to lower gas temperature to the dew point and cause deposition of droplets of water on walls of drips. Gas temperature at bottom of a well 8000 ft deep will be approximately 150 F if top of well temperature is 45 F. Temperature increases about 1 °F for each 80 ft in depth in Pennsylvania wells, and for each 76 ft in depth in Alberta fields. Gas produced thru tubing is likely to be quite warm and thus carry considerable water, while that produced thru casing or open hole is likely to be cooled somewhat in the well. As much as 20 gal or more of water per MMCF of gas may be brought to the surface under deep well gas production conditions. Provisions for avoidance or removal of hydrate are necessary where hydrate forming temperature may be reached within the well tubing.

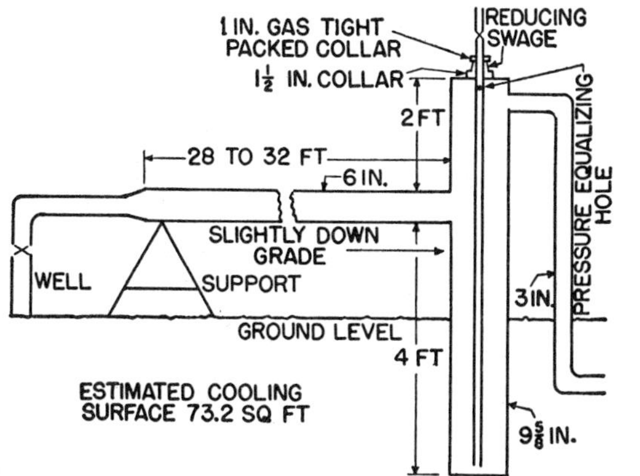

Fig. 4-41 Gas drip in common use.[3] Turbulent flow may be established in the horizontal section by inserting baffles or fins.

Figure 4-41 shows a drip based upon fundamentals of heat exchange. The horizontal section is a 6 or 8 in. casing, its length depending upon amount of cooling surface desired. A well with a 100 psig head pressure, 50 psig line pressure, 25 MCF daily flow, a well temperature of 40 F, and an outside temperature of 20 to 25 F, requires a 6 in. by 30 ft long casing. The horizontal section slopes downward into the sump, which must be large enough to hold water accumulated between *blowings*. Water in the sump must not rise to height of gas inlet. The sump should be buried deep enough so that water accumulation is protected from freezing. A 1 in. OD pipe with valve provides for water removal when drip is blown.

GAS MEASUREMENT

Gas produced by each well, if individually owned, or by a group of wells under same ownership, must be measured for payment of royalties as well as for company record purposes. For large volumes of gas under high pressure, orifice meters are installed. For relatively small volumes at low pressure, large displacement meters may be used. In warm climates these meters are set outside with some protection from sun and rain. In cold climates meters may be heated by a variety of devices. Indirect type gas heaters may be used for maintaining gas temperature above that of hydrate or ice formation.

DEHYDRATORS—HYDRATES

Dehydrators used to decrease moisture content so that hydrates will not form are usually installed at some central point in a gathering system, such as at a gasoline removal station or at a compressor station. Occasionally, they may be installed at a wellhead or for a group of wells. Expense of such equipment and care of operation may preclude their use for individual wells unless they are large producers.[4,5]

GASOLINE—LP-GAS RECOVERY

Wet natural gas containing considerable gasoline (150 gal per MMCF or more), gas from oil wells, gas having a heating value of 1100 Btu per cu ft or more, and gas from condensate fields are all rich enough in gasoline and LP-gases to recover these liquid products for sale. It is also necessary to remove them before the gas is highly compressed for cross-country transmission. Some fields of dry gas of less than 1100 Btu per cu ft contain little or no recoverable liquids. Where recovery is economical, gasoline—LP-gas plants are installed, usually ahead of the compressing station.

SULFUR REMOVAL PLANTS

Where the field gas contains sulfur in form of hydrogen sulfide, mercaptans, organic sulfides, or a combination of these, it is usual practice to install purification plants to remove these gases and recover the sulfur for sale.

COMPRESSORS FOR FIELD PRODUCTION

When field rock pressure decreases to the point where gas will not flow into the piping systems, the field must be abandoned or produced with compressors. Abandonment of wells with reservoir pressures as high as 250 psig has occurred in some areas. Where compressors are used, however, gas can be removed from the field down to pressures of a few pounds. Normally, the first step in using compressors is the installation of gas engine (Fig. 4-42) or motor-driven reciprocating units. For final depletion of the field, the use of **rotary compressors** may be advantageous.

Compressors for field production are manufactured in a wide variety of sizes and types, ranging from 10 to 3000 bhp and operating at from 100 to 600 rpm. Two cycle and four cycle engines cover this range of sizes and speeds between them. Many engines are available in packaged portable units, which have the advantages of flexibility and ease of installation and removal. These units are skid-mounted complete with all necessary auxiliaries, such as starting air compressor and storage tank, jacket-water cooler, over-speed trip, and gages. Other automatic equipment can be added for the temperature control of jacket-water, gas, and lubricating oil by use of by-pass valves or regulation of radiator shutters. Control of compressor capacity can be obtained, within limits, by control of clearance and suction valves.

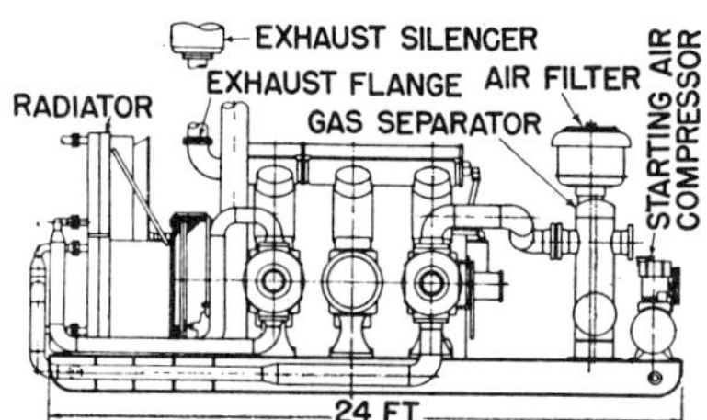

Fig. 4-42 Typical semiportable six cylinder gas engine.

Fans and air cooled jacket-water and gas coolers can be substituted for the more conventional water cooled units. Thus a permanent installation can be made at any site where sufficient water is available for whatever leakage may occur and the small amount evaporated. Electric motor-driven reciprocating compressor units are available in sizes ranging to 3000 bhp, with the small units portable.

Rotary Compressors

In final depletion of a gas field, rotary compressors (see Fig. 9-6 and Table 4-29) have proved quite adaptable. They are available in sizes from 5 to 350 bhp, both electric and gas-driven. Rotary compressors are limited to approximately 50 psi pressure differential across their blades. This limits their use to final clean-up, where gas is pumped into the suction of larger and more permanent field stations. Inlet gas must be

Table 4-29 Capacity and Horsepower Requirements of Rotary Compressors

Speed, rpm	Actual free air delivery, cfm				Rating of nearest commercial size squirrel cage motor, hp				Approx. overall dimensions incl. motor		Ship wt. (minus motor), lb
	Discharge pressure, psig				Discharge pressure, psig						
	20	30	40	50	20	30	40	50	Length	Width	
1160	32	31	...	...	5	5	...	...	3′11″	2′1″	490
1160	44	42	...	...	5	7.5	...	...	4′0″	2′1″	510
1160	52	50	...	...	5	7.5	...	...	4′0″	2′1″	515
1160	76	74	72	70	7.5	10	15	15	4′3″	2′2″	680
1160	83	82	80	78	7.5	10	15	15	4′3″	2′2″	700
1160	112	109	107	105	10	15	20	20	5′5″	2′3″	840
1160	129	127	124	122	15	15	20	25	5′6″	2′3″	870
1160	154	152	149	146	15	20	25	25	6′0″	2′8″	1090
1160	197	194	190	186	20	25	30	30	6′4″	2′8″	1120
1160	232	228	224	220	20	25	30	40	6′6″	2′8″	1150
870	284	280	275	270	25	30	40	50	7′3″	3′2″	2000
870	339	334	329	324	30	40	50	50	7′3″	3′2″	2150
870	377	370	365	360	30	40	50	60	7′11″	3′4″	2350
870	403	396	390	385	40	50	60	60	7′11″	3′4″	2450
870	482	473	467	460	40	50	75	75	8′3″	3′4″	2550
690	534	526	519	512	50	60	75	100	8′6″	3′8″	3100
690	607	598	592	585	50	75	100	100	8′6″	3′8″	3300
690	685	675	665	656	60	75	100	100	9′6″	3′8″	3950
690	773	763	754	745	75	100	100	125	9′7″	3′8″	4200
575	890	878	866	855	75	100	125	150	10′7″	4′4″	5600
575	1050	1037	1023	1010	100	125	150	150	10′7″	4′4″	5800
575	1410	1392	1374	1355	125	150	200	200	12′0″	4′4″	7100
575	1610	1592	1572	1554	125	200	200	250	12′9″	4′4″	7450

cleaned by an efficient scrubber, since pipeline dust and other abrasives will cause excessive blade wear. Advantages of rotary compressors include their mobility and small space and foundation requirements. Some of the smaller sizes can be anchored to heavy timbers buried in the ground. Automatic controls can be supplied for unattended operation.

Table 4-29 gives capacities and horsepower requirements for various rotary compressors.

FIELD AND FARM TAPS

Supplying gas to the owner of the farm where a well is located may be required by contract. Such taps and connections are made either at the well or from gathering lines. Reduction of gas pressure without freeze-up is a major problem in some areas.

ABANDONMENT

No specific minimum production level can be set as an index for abandonment of a gas well. Some wells may flood with water and must be abandoned. In other cases, reservoir pressure drops so low that gas will not flow against the receiving pipeline pressure. For a high pressure transmission line, abandonment may occur at several hundred pounds reservoir pressure; otherwise, compressors must be installed to overcome line pressure. When well reservoir pressure drops to a few pounds, compressor operation may not be profitable. After the field is abandoned for commercial use, a number of wells may still remain to produce gas for the landowner's use for many years.

A well should be plugged after it is no longer usable. Methods for accomplishing this follow:[6]

(a) The bottom of the hole should be filled to the top of each producing formation, or a bridge placed at the top of each producing formation. In either event, a cement plug not less than 15 ft in length should be placed immediately above each producing formation whenever possible.

(b) A cement plug not less than 15 ft long should likewise be placed approximately 50 ft below all fresh water bearing strata.

(c) A plug should be placed at the surface of the ground in each hole plugged so that it does not interfere with soil cultivation.

(d) The interval between plugs should be filled with an approved, heavy mud-laden fluid.

(e) An uncased rotary drilled hole should be plugged with approved heavy mud up to the base of the surface string, where a plug of not less than 15 ft of cement should be placed. The hole should also be capped as are other abandoned holes.

(f) The operator may place cement in the hole by (1) dump bailer, (2) pumping through tubing, (3) pump and plug, or (4) other approved methods.

REFERENCES

1. Hass, A. E. and Benear, J. B. "Gas-Gathering-System Design is 'Natural' for Electronic Computer." *Oil Gas J.* 55: 132–134, May 13, 1957.
2. Johnson, T. W. and Berwald. *Flow of Natural Gas Through High-Pressure Transmission Lines.* (U. S. Bur. of Mines Mono. 6) New York, A.G.A., 1935.
3. Stephans, M. M. and Spencer, O. F. *Natural Gas Engineering,* 3rd ed. University Park, Pennsylvania State University, 1954.
4. Schuster, R. A. "Producing and Gathering High-Pressure Gas." *Oil Gas J.* 50: 186+, May 5, 1952. Also in *Gas* 28: 102+, July 1952.
5. Hetherington, C. R. "5 Trillion Cubic Feet of Gas Ready for Westcoast." *Oil Gas J.* 55, Aug. 1957, "Canada Oil and Gas Unlimited."
6. Interstate Oil Compact Commission. *Suggested Form of General Rules and Regulations for the Conservation of Oil and Gas.* Oklahoma City, 1960.

Chapter 8
Gas Hydrates and Gas Dehydration

by E. G. Hammerschmidt, K. R. Knapp, and C. L. Perkins

GAS HYDRATES

Water vapor is present in practically all natural gas as it comes from the well. In many locations liquid water is also entrained. Amounts of water vapor in gas at equilibrium conditions depend upon pressure and temperature as shown in Table 4-33.

Water in natural gas represents a potential source of stoppage troubles in high pressure pipelines. Solid compounds known as hydrates may be formed by chemical combinations of hydrocarbons and water under pressure, resulting in partial or complete blocking of lines. Application of the phase rule to methane–propane water data has proved that hydrates behave as solid solutions.[1]

Natural gas hydrates are one form of the chemical inclusion compounds, since evidence suggests that molecules of natural gas occupy space in between the water molecules. In these natural gas complexes (clathrates), the H_2O molecules are linked so that they form roughly spherical capacities which trap the natural gas. There is no chemical bond between the H_2O and the natural gas, since the latter occupies available structural voids. The combination is in the ratio of 46 water molecules to six gas molecules.[32]

Many gases form solid hydrates that crystallize in the system, usually at elevated pressures and possibly at temperatures considerably *above the normal freezing point of water* (32 F). The chemical nature of hydrate-forming gases may be very diverse. Such gases include inert gases and both saturated and unsaturated hydrocarbons. Substances easily soluble in water do *not* form hydrates. Otherwise, determining factors are size and shape of molecules. Among the paraffin hydrocarbons, methane, ethane, propane, and *iso*-butane form hydrates. Normal butane may take part in hydrate formation only in a mixture with smaller molecules. The larger molecules usually form more stable hydrates. Thus, those of propane and *iso*-butane are more stable than those of ethylene, ethane, and acetylene, which in turn are more stable than methane hydrate.

Freezing of natural gas pipelines was long assumed to result from ice formation. If ground temperatures were above 32 F, freezing was assumed to be due to reduction of gas temperature to 32 F or lower by gas throttling action thru some restriction. Later, it was generally recognized that such freezing was often caused by formation of gas hydrates.

A clear realization that natural gas pipeline freezing was caused by hydrate formation resulted from the publication of tests[2] conducted at the Texoma Gas Co. gas cooling plant at Fritch, Texas. This plant was built to cool (to 40 F) natural gas for pipeline transmission to Chicago. Cooling to that temperature, which was below any probable line temperature, was expected to reduce the gas *dew point* sufficiently to prevent any condensation in the line. Stoppage difficulties, however, occurred soon after the plant began operations. Gas passages were found coated with ice even though gas temperature had been maintained around 40 F and local cooling was insufficient to produce temperatures as low as 32 F.

Gas hydrate usually obtained from pipelines looks much like packed snow. It is very light and porous. Relatively large volumes of gas are given off from freshly obtained hydrate at atmospheric pressure. On ignition it burns quietly, leaving a small water residue; some water resulting from decomposition probably evaporates during burning.

Behavior of hydrates under conditions similar to those in natural gas pipelines has been reported.[3] The formation of hydrates was observed thru a glass window in a high pressure chamber under favorable pressure and temperature conditions. These hydrates melted under appropriate pressure reduction and/or temperature increases. Hydrate formed without agitation of water present was in the form of transparent crystals. When water was agitated, however, a snowy porous hydrate formed, resembling that usually found in lines.

Solubility of Methane in Water. Table 1-25 covers the range from 32 to 212 F at methane partial pressures of 30 in. Hg. Values for methane partial pressures up to 500 psia are given in Table 4-30, expressed in moles of gas per mole of water. For purposes of conversion, note that there are 18.016

Table 4-30 Solubility of Methane in Water at Various Partial Pressures of Methane Vapor[5] in Moles per Mole × 10[4]*

Pressure, psia	Temperature, °F					
	100	130	160	190	220	250
50	0.70	0.53	0.46	0.44	0.39	0.24
100	1.39	1.08	0.98	0.99	0.98	0.85
150	2.09	1.63	1.49	1.53	1.57	1.46
200	2.79	2.19	2.01	2.09	2.16	2.07
250	3.49	2.74	2.53	2.64	2.75	2.68
300	4.18	3.29	3.04	3.19	3.34	3.29
350	4.88	3.87	3.56	3.74	3.93	3.90
400	5.58	4.40	4.07	4.29	4.52	4.51
450	6.27	4.95	4.49	4.84	5.11	5.12
500	6.97	5.50	5.10	5.39	5.70	5.73

* Values were multiplied by 10,000 for greater readability.

Table 4-31 Values of X in Formulas of Gas Hydrate Composition Determined by Different Investigators[3]

Investigator	Methane $CH_4 \cdot XH_2O$	Ethane $C_2H_6 \cdot XH_2O$	Propane $C_3H_8 \cdot XH_2O$	Carbon dioxide $CO_2 \cdot XH_2O$	Natural gas $\cdot XH_2O$
DeForcrand	6	7		6	
Villard				6	
Hammerschmidt	6	6	7		
Roberts, Brownscombe, and Howe	7	7			
Miller and Strong			6		
Deaton and Frost	7	8	18	7	9

and 16.032 lb per lb-mole for H_2O and CH_4, respectively. Data for pressures up to 10,000 psia are also available.[4]

Chemical Composition

Gas hydrates were discovered[3] in 1810. Little additional information about them was obtained until late in the century, when hydrates of carbon dioxide, methane, ethane, ethylene, acetylene, and propane were discovered. Gas hydrates are chemical compounds of gas and water similar to hydrated salts of inorganic compounds. Copper sulfate ($CuSO_4 \cdot 5H_2O$) and calcium sulfate ($CaSO_4 \cdot 2H_2O$) are examples of such salts.

Table 4-31 shows composition of hydrates of several constituent gases of natural gas. Values of **X**, representing water content as obtained by different investigators, are included.

Formation Characteristics

Hydrate forming characteristics of several natural gases are shown in Fig. 4-43, which gives equilibrium curves for these gases as well as for methane and ethane. These curves fall between those for methane and ethane. The curves also indicate that natural gases with *low* inert content and *high* content of ethane and heavier gases form hydrates at lower pressures. Generally, high-Btu natural gases will form hydrates more readily than those of lower heating value.

There is no certainty that hydrates will form immediately upon reaching equilibrium conditions. It is sometimes necessary to change these conditions in order to start formation, or to inoculate with a small crystal of hydrate.

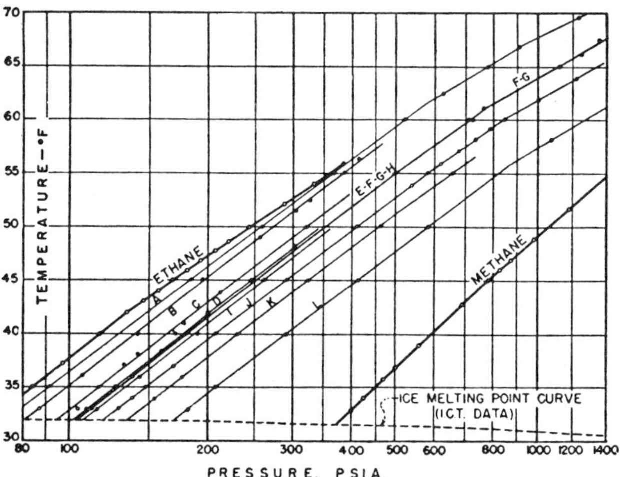

Fig. 4-43 Gas hydrate equilibrium curves for natural gases.[3] (See Table 4-32 for analyses of gases shown by letters.)

Hydrate equilibrium curves for pure paraffin hydrocarbon gases above and below 32 F were reported.[3] Several mixtures of hydrocarbon gases were also investigated. It was found that a small content of the heavier hydrocarbons would materially shift the hydrate curve for the mixture. Addition of *iso*- or *n*-butane to methane changed hydrate formation considerably, although butane used alone did not form hydrates.

Prediction of Hydrate Formation

No simple relationship has been found that can permit accurate calculation of hydrate characteristics of a gas from its analysis.[3] If analysis of a natural gas of unknown hydrate-forming characteristics is known, it may be compared with those natural gases of known analyses listed in Table 4-32, with hydrate curves shown in Fig. 4-43. A fairly dependable prediction of its hydrate characteristics should then be possible. When gas composition is known, a method for calculating a natural gas hydrate line is available[1] using equilibrium constants. Although this has been useful in some cases, it is not known whether the method is applicable to all gas mixtures.

Early experimental work indicated that methane hydrate could *not* be formed at temperatures above 71 F, but apparatus then available used pressures limited to 4410 psi. Later experiments, using a three-phase, methane hydrate water-rich liquid-vapor system showed that the hydrate temperature of decomposition at 11,240 psia was 84 F. It was also shown[6] that data could be extrapolated so that at 40,000 psia methane hydrate would be expected to form at 100 F. Other investi-

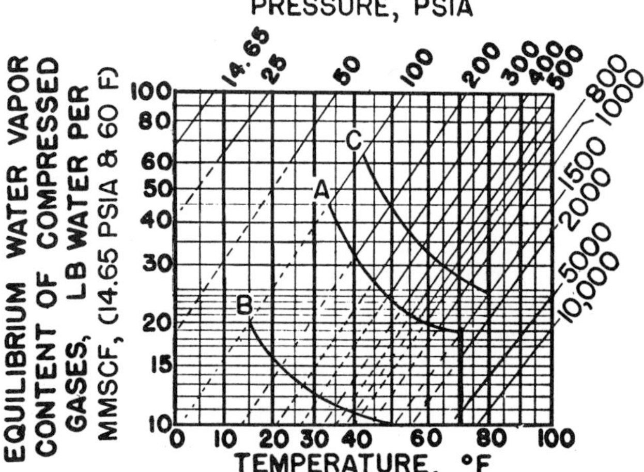

Fig. 4-44 Hydrate lines for compressed natural gases. (Pressure, temperature, and moisture parameters beyond the range of this Figure are given in Table 4-33.) Curves: A, Avg (79% CH_4, 10% N_2, 6% C_2H_6); B, Min (high-CH_4); C, Max (high-C_2H_6 and C_3H_8).

Table 4-32 Analyses of Gases[3] Represented by Curves of Fig. 4-43

Component	Ethane	Methane	Natural gases A	B	C	D	E	F	G	H	I	J	K	L
CO_2	0.0	0.3	0.2	0.2	0.2	0.3	3.25	0.4	0.0	0.2	0.0	0.9	0.8	0.6
H_2S	..	..	..	..	..	..	0.25	..	..	..	..	..	..	..
N_2	..	..	7.7	1.1	9.4	9.5	1.1	0.3	1.0	14.3	3.4	1.2	25.0	0.2
CH_4	0.8	99.7	65.4	87.9	78.4	79.4	87.8	91.0	90.8	75.2	88.5	90.6	67.4	96.5
C_2H_6	97.1	..	12.7	4.4	6.0	5.8	4.0	3.2	3.0	5.9	4.3	3.8	3.7	0.9
C_3H_8	2.1	..	10.3	4.9	3.6	3.6	2.1	2.0	2.1	3.3	2.0	1.5	1.9	1.8
C_4H_{10} & up	..	..	3.7	1.5	2.4	1.4	1.5	3.1	3.2	1.1	1.7	2.0	1.2	..
Heating value, calc., Btu/cu ft			1283	1147	1077	1049	1067	1138	1135	990	1086	1095	839	1056

gators[7] determined that the hydrate decomposition temperature of three different natural gases at pressures up to 4000 psia ranged from 71 to 75 F.

Figure 4-44 shows as Curve **A** the hydrate line of an average natural gas.

Hydrate Stoppage Prevention or Removal

One or a combination of the following procedures may be used:

1. Introduce inhibitors (e.g., methanol) to lower hydrate and ice forming temperature.
2. Reduce line pressure on both sides of hydrate plug, thus upsetting hydrate equilibrium and permitting evaporation.
3. Raise temperature above that of formation of hydrate. Gas in reservoir sands is often at a temperature far above that of formation of any known hydrate. Because of expansion and contact with well piping, the gas may be rapidly cooled below the hydrate forming temperature.
4. Dehydrate to a *dew point* below any temperature the gas may encounter in transmission or distribution. Required amount of dehydration may be found by referring to Table 4-33.

Lowering of Formation Temperature by Inhibitors. Addition of chemicals that lower the freezing point of water exerts a *similar reduction* in temperatures at which hydrates will form at a particular pressure. *Antifreezes* such as methanol, ethanol, propanol, or glycol may thus be used to prevent hydrate stoppage of gas lines where dehumidifying the gas is not practical. Field gathering lines, small distribution systems, and systems where hydrate formation is infrequent or of short duration represent examples where inhibitors may be used. Apparatus for introducing inhibitors is similar to that for injecting odorant into natural gas.

Lowering of hydrate freezing point in a natural gas system by an *antifreeze* compound may be calculated by the following equation:[8]

$$d = \frac{KW}{M(100 - W)} \tag{1}$$

where:

K = 2335 for methanol* (M = 32.04), ethanol (M = 46.07), isopropanol (M = 60.10), and ammonia; 2200 for ethylene glycol (M = 62.07), urea, and sodium chloride; 3590 for propylene glycol (M = 76.10—calculated from data on glycols published by Carbide & Carbon Chemicals Corp.); 4367 for diethylene glycol, (calculated values ranged from 4080 to 4780—M = 106.10)

d = depression of gas hydrate freezing point, °F

M = molecular weight of solute inhibitor

W = solute inhibitor in aqueous solution, weight per cent

* An alternate method for methanol is presented in the next paragraph.

Figure 4-45a shows freezing point lowering of hydrates in **methanol solution** of varying concentration. Since amounts of water and hydrate present are usually unknown, additional information is needed for practical application of Fig. 4-45a.

Table 4-33 Equilibrium Moisture

°F	psia 14.7	100	200	300	400	500	600	700	800	900	1000
−40	9.1	1.5	0.88	0.66	0.55	0.49	0.44	0.41	0.39	0.37	0.36
−36	11.5	1.9	1.1	.80	.68	.59	.54	.50	.47	.45	.43
−32	14.4	2.4	1.3	0.99	0.82	.72	.65	.60	.57	.54	.51
−28	17.8	2.9	1.6	1.2	1.0	0.87	.79	.72	.68	.64	.61
−24	22.0	3.6	2.0	1.5	1.2	1.1	0.95	0.87	.81	.77	.73
−20	27.0	4.4	2.4	1.8	1.5	1.3	1.1	1.0	0.97	0.92	0.87
−16	33.1	5.4	3.0	2.2	1.8	1.5	1.4	1.2	1.2	1.1	1.0
−12	40.5	6.5	3.6	2.6	2.1	1.8	1.6	1.5	1.4	1.3	1.2
− 8	49.3	7.9	4.3	3.1	2.5	2.2	1.9	1.8	1.6	1.5	1.5
− 4	59.8	9.5	5.2	3.7	3.0	2.6	2.3	2.1	1.9	1.8	1.7
0	72.1	11.4	6.2	4.5	3.6	3.1	2.7	2.5	2.3	2.1	2.0
4	86.8	13.7	7.4	5.3	4.3	3.6	3.2	2.9	2.7	2.5	2.4
8	104	16.4	8.8	6.3	5.1	4.3	3.8	3.4	3.2	3.0	2.8
12	124	19.5	10.5	7.5	6.0	5.1	4.5	4.0	3.7	3.5	3.3
16	148	23.2	12.4	8.8	7.0	5.9	5.2	4.7	4.3	4.0	3.8
20	176	27.4	14.6	10.4	8.2	7.0	6.1	5.5	5.1	4.7	4.4
24	208	32.4	17.2	12.2	9.7	8.2	7.2	6.4	5.9	5.5	5.1
28	246	38.1	20.2	14.3	11.3	9.5	8.3	7.5	6.8	6.3	5.9
32	289	44.7	23.7	16.7	13.2	11.1	9.7	8.7	7.9	7.3	6.9
34	313	48.4	25.6	18.0	14.2	11.9	10.4	9.3	8.5	7.9	7.4
36	339	52.4	27.7	19.4	15.3	12.9	11.2	10.0	9.2	8.5	7.9
38	367	56.6	29.9	20.1	16.5	13.9	12.1	10.8	9.8	9.1	8.5
40	396	61.1	32.2	22.6	17.8	14.9	13.0	11.6	10.6	9.8	9.1
42	428	66.0	34.8	24.4	19.2	16.0	13.9	12.5	11.3	10.5	9.8
44	462	71.2	37.5	26.2	20.6	17.2	15.0	13.4	12.2	11.2	10.5
46	499	76.7	40.3	28.2	22.2	18.5	16.1	14.4	13.1	12.0	11.2
48	538	82.6	43.4	30.3	23.8	19.9	17.3	15.4	14.0	12.9	12.0
50	580	89.0	46.7	32.6	25.6	21.3	18.5	16.5	15.0	13.8	12.9
52	624	95.7	50.2	35.0	27.4	22.9	19.8	17.7	16.1	14.8	13.8
54	672	103	54.0	37.6	29.4	24.5	21.3	18.9	17.2	15.8	14.7
56	721	111	57.9	40.3	31.5	26.7	22.8	20.3	18.3	16.9	15.7
58	776	119	62.1	43.2	33.8	28.1	24.4	21.7	19.0	18.0	16.5
60	834	128	66.6	46.3	36.2	30.1	26.1	23.2	21.0	19.3	17.9
62	895	137	71.4	49.6	38.7	32.2	27.9	24.7	22.4	20.6	19.1
64	960	147	76.5	53.1	41.4	34.4	29.8	26.4	23.9	22.0	20.4
66	1030	157	81.8	56.8	44.3	36.8	31.8	28.2	25.5	23.4	21.8
68	1100	168	87.6	60.7	47.3	39.3	33.9	30.1	27.2	25.0	23.2
70	1180	180	93.7	65.0	50.6	42.0	36.2	32.1	29.0	26.6	24.7
72	1260	192	100	69.4	54.0	44.8	38.6	34.2	30.9	28.4	26.3
74	1350	206	107	74.0	57.6	47.7	41.1	36.4	32.9	30.2	28.0
76	1440	220	114	79.0	61.4	50.9	43.8	38.8	35.0	32.1	29.8
78	1540	235	122	84.2	65.5	54.2	46.7	41.3	37.3	34.2	31.7
80	1650	250	130	89.8	69.7	57.7	49.7	44.0	39.7	36.3	33.6
82	1760	267	138	95.6	74.2	61.4	52.8	46.7	42.1	38.6	35.7

* Eight subsequent corrections incorporated.

This is supplied by Fig. 4-45b, which shows experimentally determined equilibrium conditions between *methanol vapor in the gas stream* and that in fluid at various pressures and temperatures.

Example. Determine amount of methanol which must be introduced into gas J (Table 4-32) to prevent formation of hydrate of 400 psia and 40 F.

From Fig. 4-43, gas J at 400 psia has a hydrate temperature of 50 F. Therefore, the required freezing point lowering = 50 − 40 = 10 °F. Enter Fig. 4-45a at 10°F and read 12.2 per cent methanol (by weight). Now enter Fig. 4-45b at 400 psia and 40 F, and read C = 1.7. Thus, if methanol content of gas is adjusted to contain 12.2 × 1.7 = 20.7 lb methanol per MMCF, hydrates and water vapor in line will absorb methanol vapor until resulting fluid contains 12.2 per cent methanol.

Ammonia may also be used as a hydrate inhibitor in gas transmission lines.[3] While more effective than methanol, it has some serious drawbacks. In presence of water ammonia reacts with carbon dioxide (usually present in small quantities in natural gases) to form ammonium bicarbonate, a relatively poor inhibitor. This solid compound has been reported to cause difficulties equal to or greater than the hydrates eliminated by the ammonia. Ammonia also attacks brass.

Line Pressure Reduction. Tests[3] have shown that hydrates begin to break down almost at once when pressure is reduced below that of equilibrium. Where conditions permit, reduction of operating pressure has proved effective in removing hydrate stoppages. For best results, reduction of pressure on both sides of the stoppage is necessary. Reduction of pressure on only one side of an extensive stoppage may result in freezing of water, due to lowering of temperature produced by breaking down of hydrates. It is desirable to maintain gas flow in the line and to remove all water released from hydrates. By maintaining flow of gas, its heat will assist in hydrate decomposition. Removal of released water also prevents the re-formation of hydrates upon resumption of normal pressure and flow.

It has been noted that when line pressure is lowered quickly, temperature drops considerably, due to expansion of the gas. Because of this drop, pressure should be lowered to a point somewhat below what would be necessary if line remained at normal operating temperature, or reduced operating pressure should be maintained until temperature returns to normal.

Heat is absorbed in decomposition of hydrates. Where a considerable mass has accumulated in a line, its breakdown on reduction of pressure may lower its temperature to 32 F. Further breakdown at 32 F is accompanied by conversion of water to ice. This condition accounts for presence of ice that may often be ejected from a pipeline when it is depressured and blown.

Contents of Natural Gases Above the Critical Temperature[17]* (pounds per MMCF where P_b = 14.7 psia, t_b = 60F)

	psia																	
°F	14.7	100	200	300	400	500	600	700	800	900	1000	1500	2000	2500	3000	3500	4000	4500	5000
84	1870	285	148	102	79.0	65.3	56.2	49.7	44.8	41.0	37.9								
86	2000	303	157	108	84.1	69.5	59.7	52.8	47.6	43.5	40.3								
88	2130	323	167	115	89.4	73.8	63.5	56.1	50.5	46.2	42.7								
90	2270	344	178	123	95.0	78.5	67.4	59.5	53.6	49.0	45.3								
92	2410	366	189	130	101	83.3	71.5	63.1	56.8	51.9	48.0								
94	2570	389	201	138	107	88.4	75.9	67.0	60.3	55.0	50.9								
96	2730	413	214	147	114	93.8	80.5	71.0	63.9	58.3	53.9								
98	2900	439	227	156	121	99.5	85.3	75.2	67.6	61.8	57.0								
100	3080	466	241	166	128	105	90.4	79.7	71.6	65.4	60.4	45.4	37.9	33.3	30.3	28.2	26.6	25.3	24.3
102	3270	495	256	176	136	112	95.8	84.4	75.9	69.2	63.9	47.9	40.0	35.5	32.0	29.7	28.0	26.6	25.6
104	3470	525	271	186	144	118	101	89.3	80.2	73.1	67.5	50.6	42.1	37.0	33.6	31.2	29.4	28.0	26.9
106	3680	557	287	197	152	125	107	94.5	84.9	77.4	71.4	53.4	44.5	39.1	35.5	32.9	31.0	29.5	28.3
108	3900	589	304	209	161	133	114	99.9	89.7	81.7	75.4	56.4	46.9	41.1	37.3	34.6	32.6	31.0	29.7
110	4130	624	322	221	170	140	120	106	94.7	86.3	79.6	59.4	49.4	43.3	39.3	36.4	34.2	32.5	31.2
112	4380	661	341	234	180	148	127	112	100	91.2	84.1	62.7	52.1	45.6	41.4	38.3	36.0	34.2	32.8
114	4640	700	360	247	191	157	134	118	106	96.2	88.7	66.1	54.8	48.0	43.4	40.2	37.8	35.9	34.4
116	4910	740	381	261	201	165	142	124	112	102	93.6	69.7	57.7	50.5	45.7	42.3	39.8	37.8	36.2
118	5190	783	403	276	213	175	149	131	118	107	98.7	73.4	60.7	53.1	48.0	44.4	41.7	39.6	37.9
120	5490	828	426	292	225	185	158	139	124	113	104	77.3	63.9	55.9	50.5	46.7	43.8	41.6	39.8
124	6130	923	474	325	250	205	175	154	138	125	116	85.6	70.7	61.7	55.7	51.4	48.2	45.7	43.7
128	6830	1030	528	361	278	228	195	171	153	139	128	94.7	78.0	68.0	61.3	56.6	53.0	50.2	48.0
132	7580	1140	585	400	308	252	215	189	169	154	141	104	85.8	74.7	67.3	62.0	58.1	55.0	52.5
136	8470	1270	653	446	343	281	240	210	188	171	157	116	94.9	82.5	74.2	68.3	63.9	60.3	57.7
140	9360	1410	721	492	378	310	264	231	207	188	173	127	104	90.4	81.3	74.7	69.9	66.0	63.0
144	10,400	1560	799	545	419	343	292	256	229	207	191	140	115	99.3	89.2	81.9	76.5	72.3	68.9
148	11,500	1720	882	602	462	378	322	282	252	229	210	154	126	109	97.6	89.6	83.6	78.9	75.6
152	12,700	1910	975	665	510	417	355	311	277	252	231	169	138	119	107	98.0	91.4	86.2	82.1
156	14,000	2100	1070	732	561	458	390	341	305	276	253	185	151	130	117	107	100	94.0	89.4
160	15,400	2300	1180	802	615	502	427	374	333	302	277	202	165	142	127	116	108	102	97.1
164	...	2540	1300	883	676	552	469	410	366	332	304	221	180	155	139	127	118	111	106
168	...	2780	1420	967	740	604	514	449	400	363	332	242	196	169	151	138	128	121	115
172	...	3040	1550	1060	810	661	562	491	437	396	363	263	214	184	164	150	139	131	124
176	...	3330	1700	1160	885	722	613	535	477	432	396	287	233	200	178	163	151	142	135
180	...	3640	1860	1260	967	789	670	585	521	471	432	313	253	217	194	177	164	154	146
184	...	3980	2030	1380	1060	860	730	637	567	513	470	340	275	236	210	191	177	167	158
188	...	4340	2210	1500	1150	936	794	693	617	558	511	369	298	256	227	207	192	180	171
192	...	4720	2410	1630	1250	1020	863	753	670	606	554	400	323	277	246	224	207	194	184
196	...	5140	2620	1780	1360	1110	938	818	728	658	602	434	350	299	266	242	224	210	199
200	...	5570	2840	1930	1470	1200	1020	885	788	712	651	469	378	323	286	260	241	226	213
212	...	...	3620	2450	1870	1520	1290	1120	999	902	824	591	475	405	359	325	301	281	266
220	...	...	4220	2860	2180	1780	1500	1310	1160	1050	959	687	551	469	415	376	347	324	306
230	...	...	5100	3460	2630	2140	1810	1580	1400	1260	1150	824	660	561	495	448	413	385	363
240	...	...	...	4160	3170	2570	2180	1890	1680	1510	1380	985	787	668	589	532	490	456	430
250	...	...	...	...	3770	3060	2590	2250	2000	1800	1640	1170	932	790	695	628	577	538	506

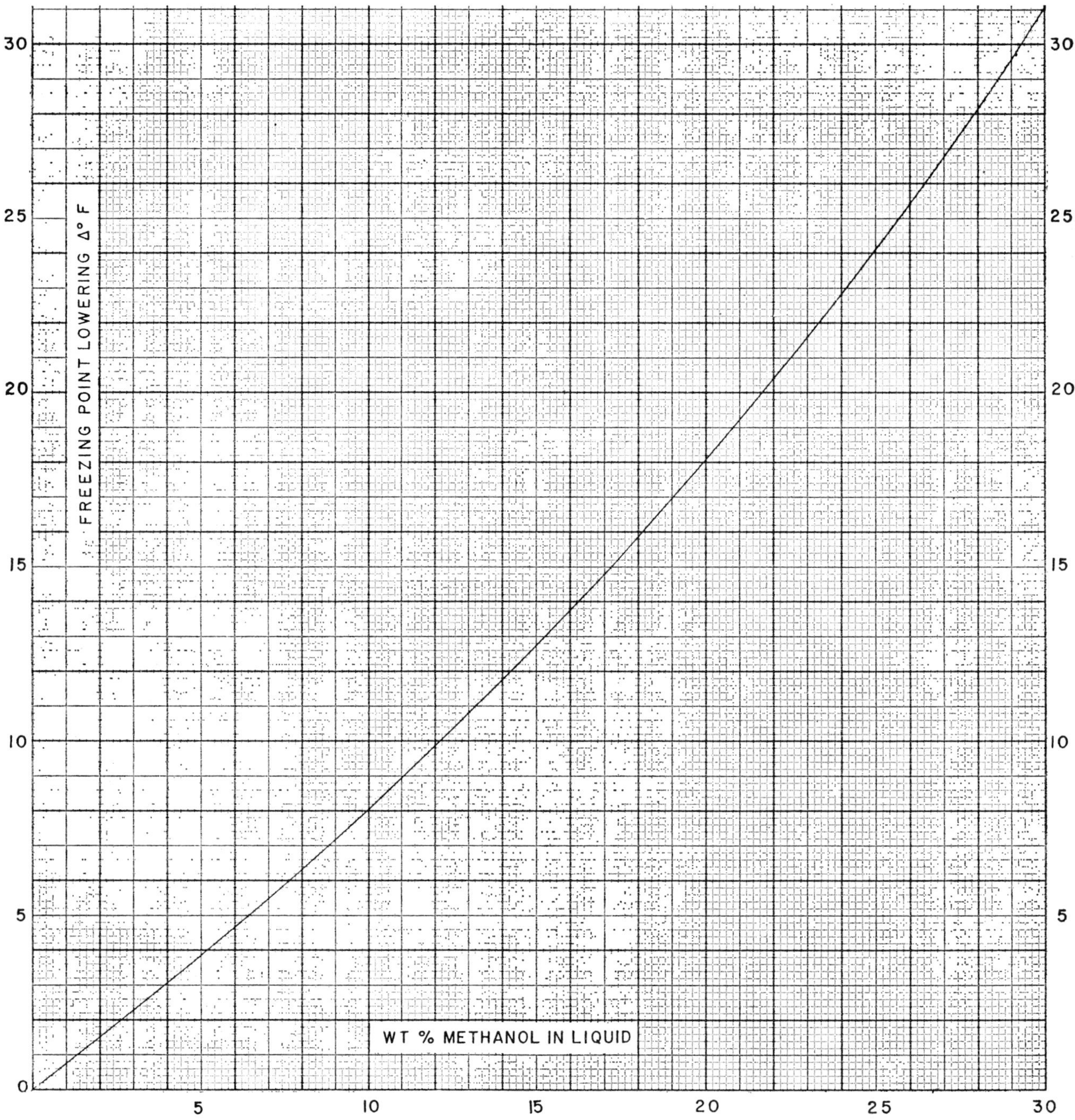

Fig. 4-45a Gas hydrate freezing point depression vs. weight per cent methanol in liquid.

Heat Application. Several factors must be carefully considered before attempting use of heat for hydrate removal. Unless exact stoppage location is known, it is difficult to secure sufficient heat absorption by the gas without resorting to elaborate heating arrangements. If heating is attempted at some distance from the stoppage, much heat is lost before it reaches the stoppage. If stoppage is nearly complete, reduced gas flow increases difficulty of transmitting heat to the point where it is needed. Even if hydrates present are melted, water formed must be quickly removed. Otherwise, hydrates may again be formed when the line returns to its normal temperature, with a recurrence of stoppage.

For safety reasons direct heating, as by open flames, is inadvisable. When heaters are used for prevention of hydrate formation, those designed for that purpose should be employed.

GAS DEHYDRATION

Principal reasons[9] for dehydration of natural gas for long distance transmission include the prevention of: (1) freeze-ups due to ice of hydrate formation, (2) corrosion, with subsequent dirt troubles, (3) condensation—requiring removal, as by blowing of drips, and (4) reduced line capacity due to water or dust accumulations.

Dehydration is the best method of hydrate prevention in a natural gas pipeline. To be effective, water must be removed

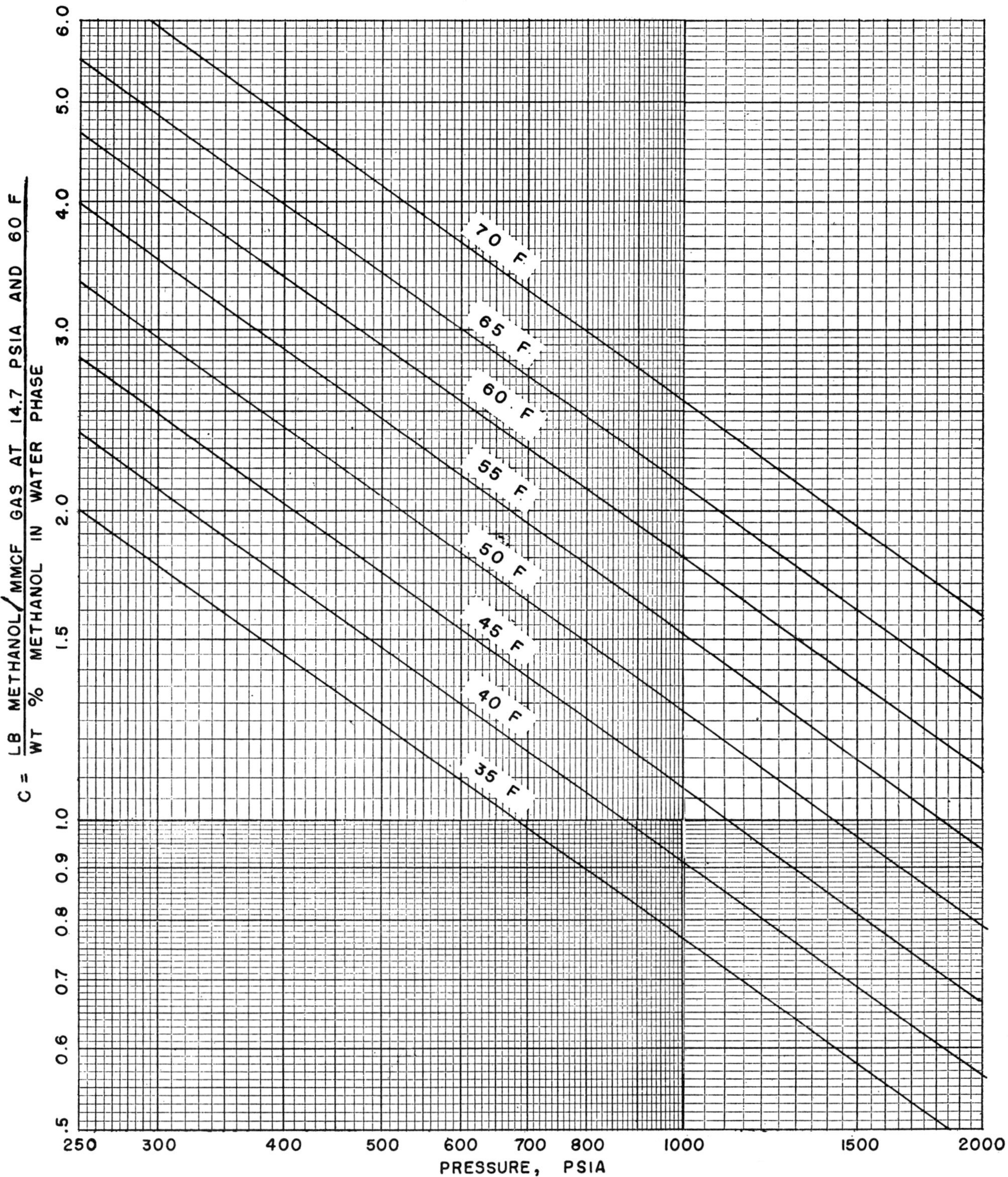

Fig. 4-45b Ratio of methanol vapor to liquid composition vs. pressure at various temperatures.

to the point where the *dew point* will *not* be reached at the *lowest* operating temperature and *maximum* pressure of the system. This condition assumes special significance where gas is transported long distance to a cold climate.

Water vapor may be removed from natural gas by several methods.[10] Simple compression and cooling represent one of the earliest methods. Its effectiveness was limited, because of hydrate formation with resulting flow stoppage. This method has been modified and improved by using it in combination with various hygroscopic solutions. Low temperature extraction units[11,12] are another modification. They are sometimes used where high wellhead pressures are available for expansion cooling. Operation is continuous with utilization of hydrates formed to lower water vapor content of effluent gas.

Various liquid absorption mediums have been used for dehydrating purposes. Calcium chloride brine was an early

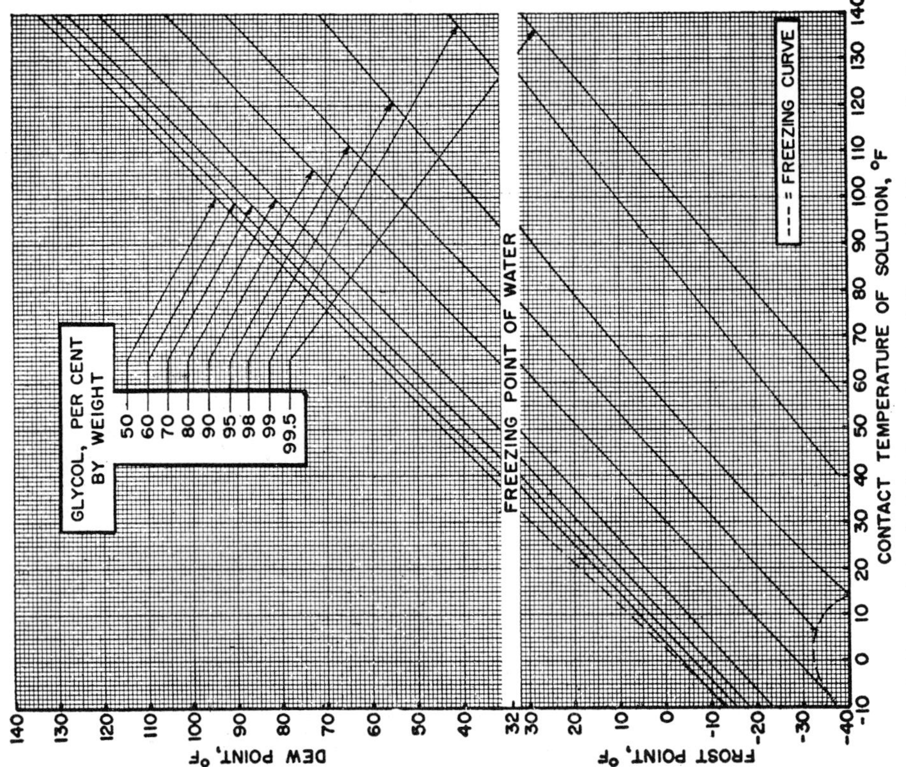

Fig. 4-47 Dew points of triethylene glycol solutions at various contact temperatures.[10]

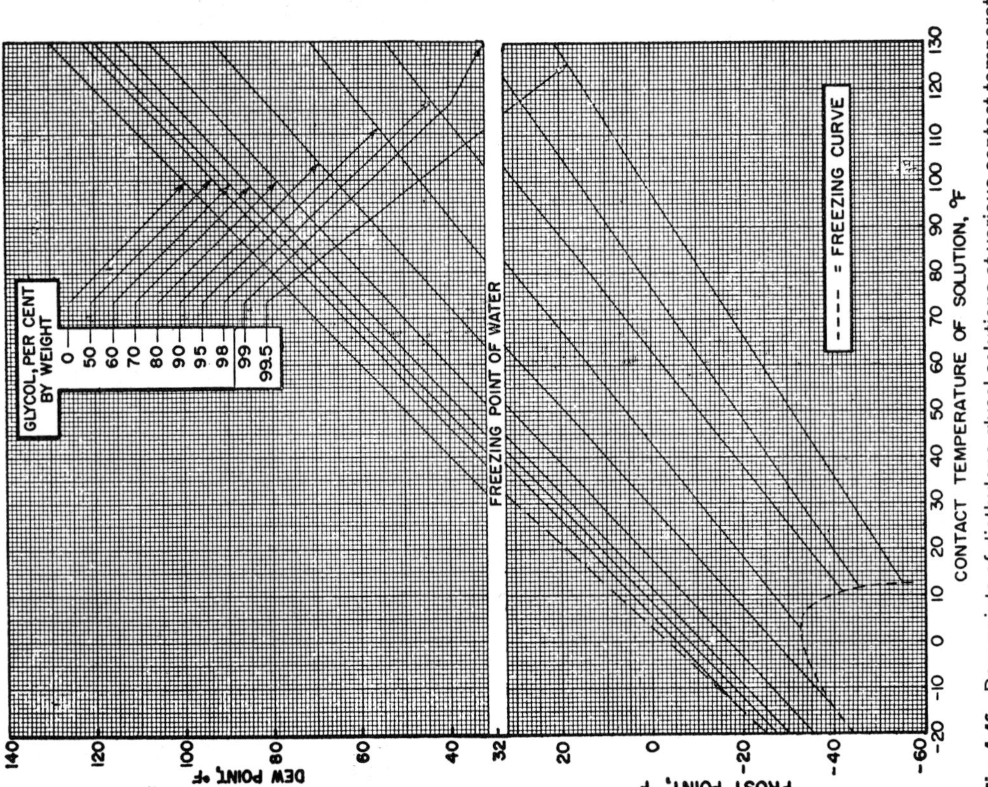

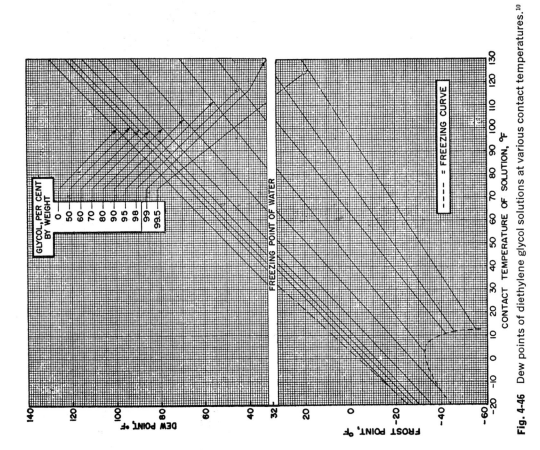

Fig. 4-46 Dew points of diethylene glycol solutions at various contact temperatures.[10]

example.[13] Refrigerated brine was used in towers to chill natural gas and depress its equilibrium *dew point*. Operating difficulties were the limited dew point depression obtainable, need for refrigeration and corrosion from the brine. Development and use of ethylene glycol as a moisture absorbent led to its gradual adoption as the liquid absorption medium for many gas dehydration plants.

Hot potash–amine treatment has been used to remove acid components. The hot potash, which removes most of the acid gases, is followed by an amine treatment for final clean-up. The stream is then passed over a solid adsorbent bed for water vapor removal. This process was utilized to treat a feed-stream with a 6.5 per cent CO_2 and a 2.0 per cent H_2S content (i.e., 1268 grains per CCF). Advantages include low fuel requirements, and low investment and maintenance costs.[14]

Glycol may be added directly to gas within pipelines, as well as in absorption plants. The former method requires approximately 34.5 gal of glycol per inch-mile of pipeline, since this inhibitor characteristically covers piping with a thin coating.[8] In costs, a 16 in. diam pipeline, 10 miles long, would require 5520 gal, which at, say, $1.50 per gal would cost $8280. Recovery, reconcentration, return of the glycol to the point of injection, and making up of glycol losses would be additional considerations. The coating decreases effective pipe diameter and probably decreases transmission efficiency. Partial blockages, if any, are indicated by unexpected pressure drops. A description of glycol use and recovery in a low temperature system is available.[15]

Solid calcium chloride was used in individual wellhead dehydration units at Rocky Mountain locations. Conditions encountered included: shut-in pressures ranging from 200 to 3000 psi, low average producing temperatures, distillate contents ranging from small to 30 bbl per MMCF of gas, and flow rates of 0.1 to 6.0 MMCF per day.[25]

Beds containing solid desiccants like potash or $CaCl_2$, and a series of trays containing brine were used. Gas, partially dehydrated by passage thru them, was further treated by a bed of solid $CaCl_2$ pellets.[16]

Solid absorbents like activated alumina or silica gel have also come into extensive use for natural gas dehydration.

Most high pressure natural gas transmission lines carry gas previously dehydrated to a specified water vapor content, usually in the range of 5 to 7 lb per MMCF.

Water Content of Natural Gas Under Pressure

Equilibrium moisture contents of natural gases are given in Table 4-33. These data which are also part of ASTM D1142-58 (Ref. 18), are considered sufficiently accurate for gas industry needs, except for unusual cases where the *dew point* is measured at conditions close to the critical temperature of the gas.

Glycol Absorption Plants

Plants using diethylene glycol (*DEG*) as the absorbing medium have been in operation for about 20 years. More recently, triethylene glycol (*TEG*) has been employed. Quantity of water vapor removable from treated gas in well designed glycol plants depends upon concentration of the lean glycol solution and absorber contact temperature. Figures 4-46 and 4-47 are equilibrium dew point charts for aqueous glycol solutions. A reasonable approach to equilibrium probably can be obtained with glycol circulation rates of two to three gallons per pound of water removed.

Use of *TEG* has been favored where highly concentrated solutions, with resulting low dew points, were desired. As indicated below, the boiling point of *TEG*, as well as its point of initial thermal decomposition, is considerably higher than for *DEG*. Thus vaporization losses are lower for *TEG*, and it can be heated to a higher reboiler temperature without decomposition.

	Normal boiling point, °F	Initial thermal decomposition temperature, °F
Diethylene glycol	473	328
Triethylene glycol	549	404

Normal glycol losses from a glycol dehydration system are usually about 0.1 gal per MMSCF of gas passing thru the absorber.[19]

In conventional *DEG* plants, dew-point depressions of 40–50°F can be obtained at regenerator atmospheric pressures when reboiler temperatures range between 290 and 320 F. In similar operation *TEG*, at reboiler temperatures of 350 to 375 F, can produce dew-point depressions ranging from 60 to 75°F. High concentration and correspondingly high dew-point depressions of either *DEG* or *TEG* can be obtained without thermal decomposition by distillation under partial vacuum, or by using natural gas as a stripping agent. Such concentrations permit correspondingly high dew-point depressions.

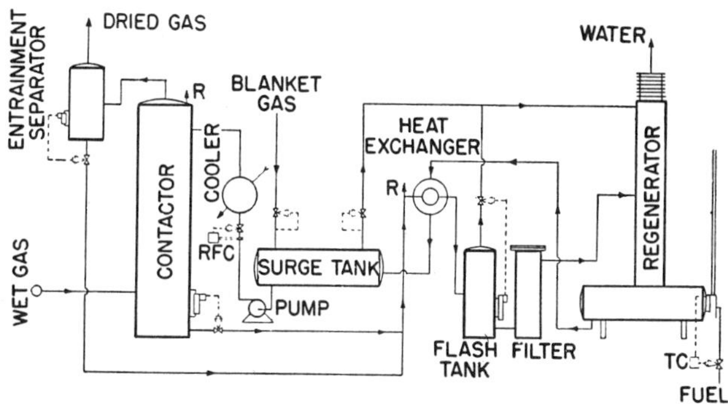

Fig. 4-48 Flow diagram for glycol absorption unit for natural gas dehydration.[20]

R: relief RFC: recording flow control
TC: temperature control

Operation. Figure 4-48 shows a typical flow diagram for a dehydration process using either *DEG* or *TEG*.

Concentrated glycol solution flows down thru the contactor countercurrent to the stream of gas to be dried. It then passes thru the heat exchanger to the regenerator, where water is removed by distillation. Regenerated glycol solution is then returned to the contactor, and the cycle is repeated.

A filter may be used as shown for removal of any solid material accumulated by corrosion or decomposition. The flash tank improves filtering by separating out natural gas which may contain corrosive acid gases.

Calculated dew-point depression of dried gas at various

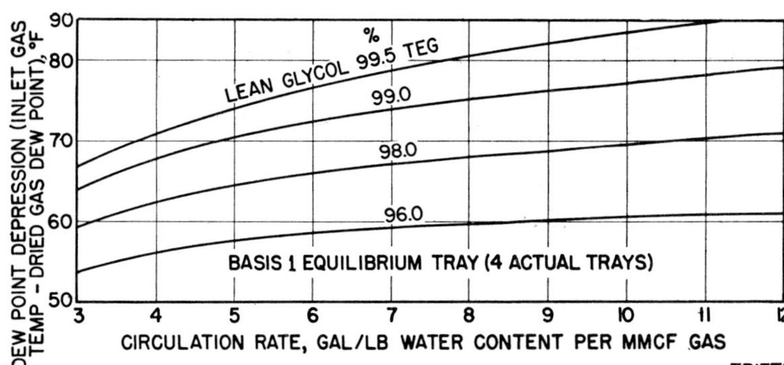

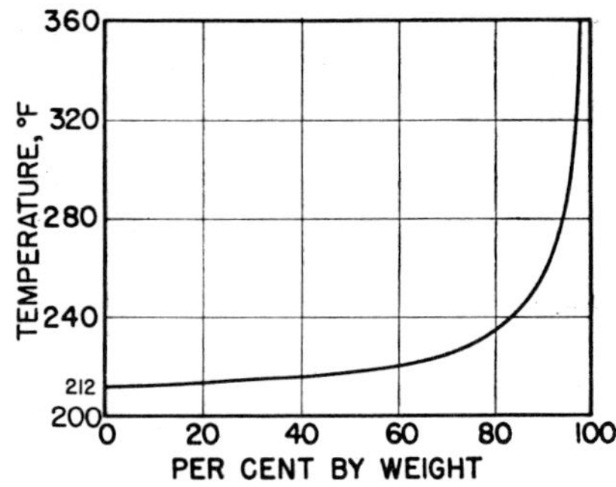

Fig. 4-49a (left) Dew point depression of *TEG* solutions at various circulation rates.[20]
Fig. 4-49b (right) Boiling point of *TEG* solutions at atmospheric pressure.[20]

circulation rates of *TEG* solution is shown by Fig. 4-49a. Dew-point depression is difference between inlet gas temperature and dew point of dried gas, both at same contactor pressure, and assuming incoming gas is saturated. It is apparent that dew-point depression varies materially with small changes in water content of lean glycol solution. If glycol could be regenerated to zero water content, its drying ability would be greatly increased.

Figure 4-49b shows temperature at which a glycol still reboiler must be carried to obtain a lean glycol of a particular water content. Since temperatures above 400 F are not generally used, because of possibility of glycol decomposition, approximately a 99 per cent glycol solution represents about the maximum practical, unless vacuum regeneration permits a higher concentration.

Example. Assume specifications require that gas contain not more than 7.0 lb water vapor per MMCF, with the operating pressure fixed at 500 psia. The maximum permissible contact temperature in the glycol absorber may be determined as follows:

From Table 4-33, the *dew point* of gas at the specified moisture content and pressure is 20 F. Figure 4-49a shows the dew point depression at 98 per cent. *TEG* concentration, at three gallons per pound of water circulation rate, is about 59 F. Hence the contact temperature of the 98 per cent glycol–gas mixture must not be greater than 59 + 20 = 79 F.

If the glycol concentration is increased to 99 per cent at the same circulation rate, a depression of 64 F can be obtained. The allowable contact temperature could then be increased to 64 + 20 = 84 F. The same results could also be obtained by increasing the circulation rate of the 98 per cent glycol from 3.0 to 4.75 gallons per pound of water. Again, this results in a depression of 64 F with a corresponding contact temperature of 84 F.

Advantages Over Solid Adsorbent Plants. These include:[10,21]

1. Initial equipment costs are lower as shown in Fig. 4-50. Data given by Campbell[22] have been increased by 10 per cent to reflect higher construction costs.

2. Low pressure drop across absorption towers saves power.

3. Operation is continuous; this is generally preferred to batch operation.

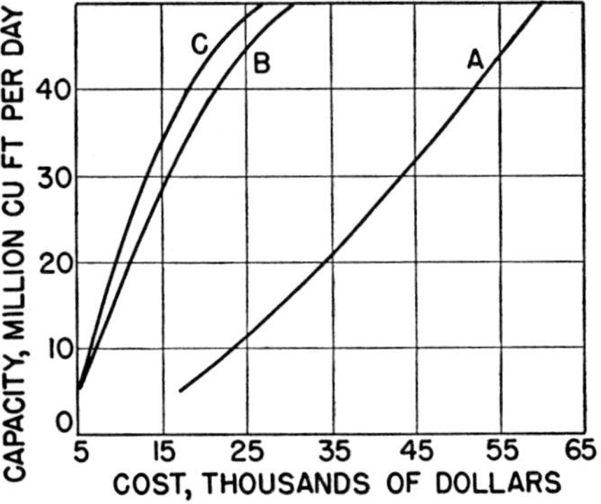

Fig. 4-50 Comparative initial equipment costs of glycol absorption and solid adsorbent dehydration plants.[10]

Curve A — Dry desiccant at 500 and 1000 psig
B — Glycol at 500 psig
C — Glycol at 1000 psig

4. Make-up requirements may be added readily; recharging of towers presents no problems.

5. Plant may be used satisfactorily in presence of materials which would cause fouling of some solid adsorbents.

Operating Problems. These include:[10]

1. Suspended foreign matter, such as dirt, scale, and iron oxide, may contaminate glycol solutions. Also, overheating of solution may produce both low and high boiling decomposition products. Resultant sludge may collect on heating surfaces, causing some loss in efficiency, or, in severe cases, complete flow stoppage. Placing by-pass mechanical filter ahead of solution pump usually prevents such troubles.

When both oxygen and hydrogen sulfide are present, corrosion may become a problem because of formation of acid material in glycol solution.

2. Liquids (e.g., water, light hydrocarbons, or lubricating oils) in inlet gas may require installation of an efficient separator ahead of absorber. Highly mineralized water entering system with inlet gas may, over long periods, crystallize and fill reboiler with solid salts.

3. Foaming of solution may occur with resultant carry-over of liquid. Addition of a small quantity of antifoam compound usually remedies this trouble.

4. Some leakage around packing glands of pumps may be permitted since excessive tightening of packing may result in scoring of rods. This leakage is collected and periodically returned to system.

5. Highly concentrated glycol solutions tend to become viscous at low temperatures and, therefore, are hard to pump. Glycol lines may solidify completely at low temperatures when the plant is not operating. In cold weather continuous circulation of part of solution thru heater may be advisable. This practice can also prevent freezing in water coolers.

6. In starting plant, all absorber trays must be filled with glycol before good contact of gas and liquid can be expected. This may also become a problem at low circulation rates because weep holes on trays may drain solution as rapidly as it is introduced.

7. Sudden surges should be avoided in starting and shutting down plant. Otherwise, large carry-over losses of solution may occur.

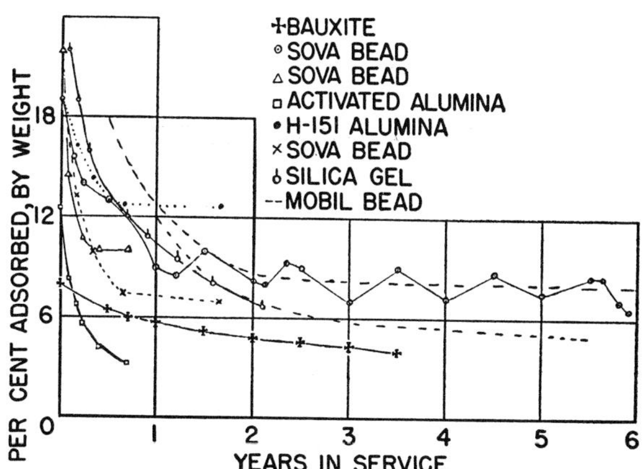

Fig. 4-51 Adsorptive capacity of dry desiccants at several natural gas dehydration plants.[10]

Solid Adsorbent Plants

Adsorption is defined as the ability of a substance to hold gases or liquids on its surface. This property occurs to a greater or lesser extent on all surfaces.[23]

Dehydration plants using solid adsorbents are capable of removing practically all water from natural gas. They can be used with temperatures higher than those for which glycol plants are satisfactory. Because of their great drying ability, solid adsorbents may be employed where higher efficiencies are required.

Various forms of alumina or silica gel represent solid types of commercial adsorbents usually used in natural gas dehydration plants. Bauxite and other native alumina ores carrying trade names of *Florite* and *Driocel*[24,25] have been used. Their adsorptive capacity as shown in Fig. 4-51 is somewhat lower than silica gel compounds and refined activated alumina. *Sovabead,*[26] now known as *Mobilbead,* is a siliceous material in form of small beads. Use of **molecular sieves,** mainly in trimmer beds of conventional adsorption units, was reported.[27] This material was relatively high in cost in 1961.

Capacity of these solid adsorbents decreases with use. Ultimately a point is reached when a tower on dehydrating cycle becomes exhausted before an alternate tower can be regenerated. Recharging with fresh desiccant is then required. Causes for decline in capacity have not yet been satisfactorily explained, but it has been shown that adsorption capacity is diminished by combined effects of adsorbed water and heat during desorption cycle.

Operation. Figure 4-52 is a flow diagram of a typical solid adsorption dehydration unit. It includes complete instrumentation for automatic switching by a time-cycle controller.

Necessity of handling solid adsorbents in batches requires a minimum of two separate charges of desiccant. While one is drying gas, the other is being regenerated for use when the first becomes exhausted.

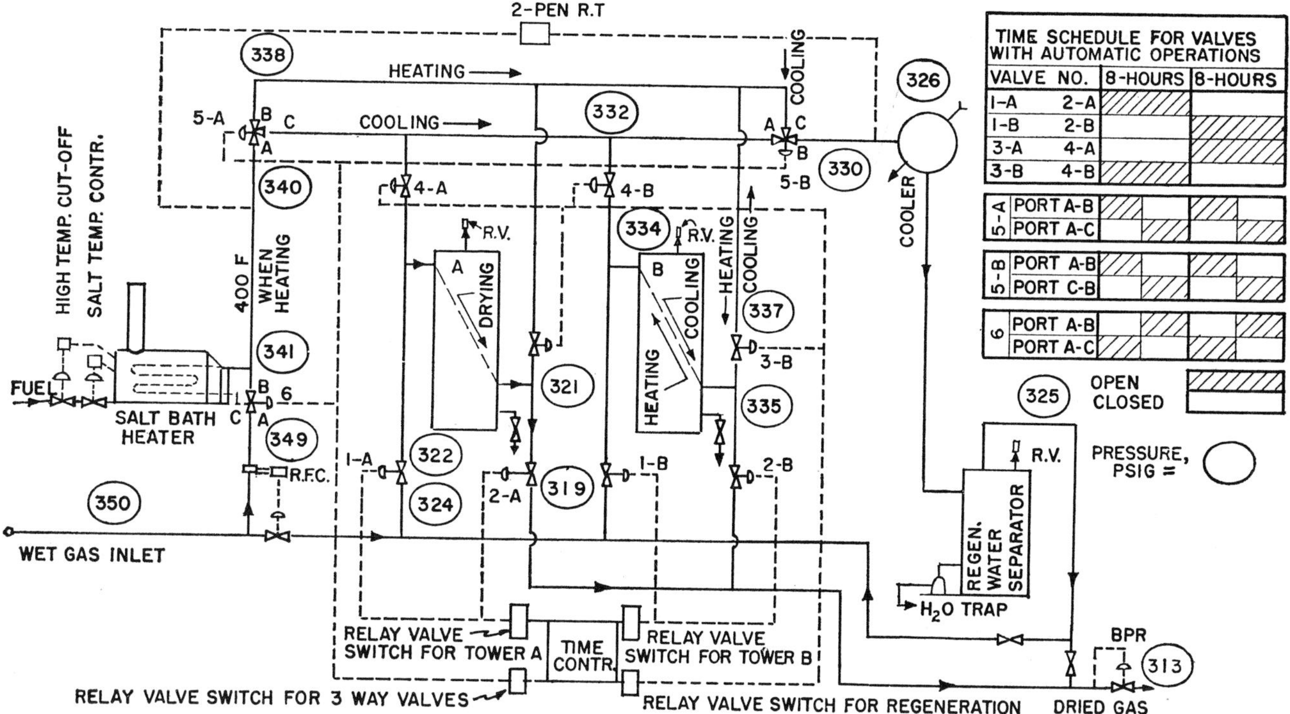

Fig. 4-52 Flow diagram of gas dehydration unit using solid desiccant.[20]

Regeneration is accomplished by by-passing part of wet gas thru a heater and then up thru desiccant tower, in direction opposite to that used for drying. Heated gas, saturated at ordinary temperatures, is capable of regenerating desiccant because its relative humidity has been greatly reduced by heating.

The heater is then turned off, and the desiccant bed is cooled by same gas stream used for heating. To minimize dewpoint effects of this cooling gas, cooling cycle is stopped when temperature of cooling gas from bed drops to about 30°F above the desired operating temperature.

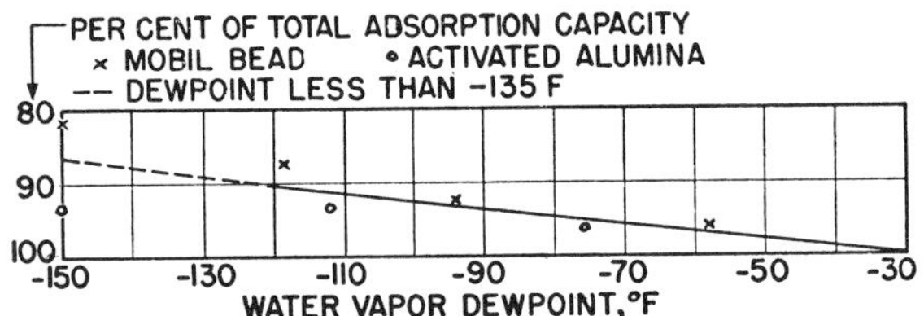

Fig. 4-53 Dew points of gas dried by two solid adsorbents vs. per cent of total adsorber capacity.[20] Process stream being dried on solid desiccant; adsorber pressure, 275 psig; regeneration gas inlet temperature, 350 F; maximum regeneration gas exit temperature, 275 F.

Gas used for regeneration is cooled to condense out the water picked up during regeneration cycle. This cooled gas then has about the same water content as inlet gas and joins with it for dehydration, or, when specifications permit, it may be mixed with the dry gas from the on-stream tower.

Figure 4-53 shows dew point of gas dried by *F-3* activated alumina and *Mobilbead* in a plant recovering natural gasoline by a refrigerating process. Little difference is evident in dew points obtainable.

Advantages over Glycol Absorption Plants. These include:[10,21]

1. Lower dew points are obtainable over a wide range of operating conditions.
2. Essentially dry gas (moisture content less than 1.0 lb per MMCF) can be produced.
3. Higher contact temperatures can be tolerated with some adsorbents.
4. Greater adaptability is offered to sudden load changes, especially on starting up. Plant may be put in operation more quickly after a shutdown. In some cases adsorption temperatures up to 125 F have been reported for silica gel adsorbents.
5. Adaptability for recovery of certain liquid hydrocarbons in addition to dehydration functions.

Operating Problems. These include:[21]

1. Space adsorbents degenerate with use and eventually require replacement.
2. Amount of water vapor adsorbed per regeneration decreases with continued use. Loss in capacity may be accelerated by collection of contaminants like compressor cylinder oil, deposited in desiccant beds.
3. A tower must be regenerated, cooled, and readied for operation as another tower approaches exhaustion. With more than two towers involved, this operation is relatively complicated. If operation is on a time cycle, maximum allowable time on dehydration gradually shortens because desiccant loses capacity with use.
4. Unloading towers and recharging them with new desiccant should be completed well ahead of the operating season. In interest of maintaining continuous operation when most needed, this may require discard of desiccant before end of its normal operating life.

To conserve material, inlet part of tower may be recharged and remainder of desiccant retained since it may still possess some useful life. Additional service life may be obtained if direction of gas flow is reversed at time when tower would normally be recharged.

5. Sudden pressure surges should be avoided. They may result in upsetting the desiccant bed and channeling the gas stream with poor dehydration. If plant is operated above its rated capacity, pressure loss will increase, and some attrition may occur. Attrition causes fines which may in turn cause excessive pressure loss with resulting loss in capacity.
6. If water cooling is used during reactivating cycle, provision must be made against freezing in water coolers. This problem generally occurs only during shutdown of coolers.

Small adsorption units, termed dynamic or short cycle adsorbers,[28] have been developed for dehydration and partial hydrocarbon extraction at wellhead. Pressure vessel size and amount of desiccant (about one-fifth that used by conventional plants) have been greatly reduced. This has been made possible by more frequent reactivation at higher temperatures. Reactivation is performed at 600 F, using only two per cent of maximum throughout for reactivation gas. Units are air-cooled.

These units, according to their manufacturer, can be sized for each well and selected to provide exact dehydration capacity required for a given situation.

COMBINED DEHYDRATION AND DESULFURIZATION

Glycol–Amine Gas Treating Plants

In special cases where both gas desulfurization and dehydration are required, the glycol–amine process is attractive; it may also be used economically for desulfurization alone. The process is described in a U. S. patent[29] issued in 1939.

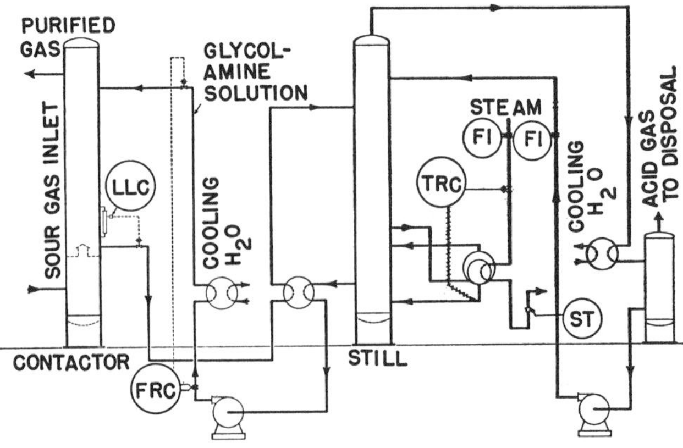

Fig. 4-54 Flow diagram of typical glycol–amine gas treating plant. (Courtesy of the Fluor Corp.) Legend: ST—Steam trap, FI—Flow indicator, LLC—Liquid level controller, FRC—Flow recorder controller, TRC—Temperature recorder controller

Recent improvements in equipment and process design have materially reduced operating costs.

Operation. Figure 4-54 shows a flow diagram of a typical glycol–amine gas treating plant. Natural gas containing H_2S, CO_2, and water enters the bottom of a bubble tray column called the contactor. This sour gas comes in contact with a solution normally containing diethylene glycol, monoethanolamine, and a small proportion of water. The amine, RNH_2, reacts with CO_2 and H_2S according to the following (where $R = HOCH_2CH_2$):

$$2RNH_2 + H_2S = (RNH_3)_2S, \text{ (sulfide)}$$
$$(RNH_3)_2S + H_2S = 2RNH_3HS, \text{ (bisulfide)}$$
$$2RNH_2 + CO_2 = RNHCOONH_3R, \text{ (carbamate)}$$
$$2RNH_2 + CO_2 + H_2O = (RNH_3)_2CO_3, \text{ (carbonate)}$$
$$(RNH_3)_2CO_3 + CO_2 + H_2O = 2RNH_3HCO_3, \text{ (bicarbonate)}$$

"As carbon dioxide is the anhydride of an acid, it cannot react to form the carbonate or bicarbonate unless water is present. In the absence of water, however, it may react with monoethanolamine directly to form a carbamate. As only a small amount of water is present in the glycol–amine solution, most of the carbon dioxide is believed to go to the carbamate . . ."[30]

In typical installations, purified gas leaves the contactor top containing considerably less than 0.25 grain of H_2S per CCF and substantially free of CO_2. Its water vapor content will depend upon solution, composition, and temperature and pressure conditions. For example, the purified gas effluent from a plant operating at 550 psia and 80 F and using a solution containing five per cent water contains about nine pounds of water per MMCF, compared to dehydration possible with straight glycol solution. Reaction of amine with CO_2 and H_2S gives off heat as shown by an increased solution and/or gas temperature. Figure 4-55a shows effects of this heat of reaction on solution temperature in various contactor trays when treating a gas of relatively high CO_2 and H_2S content. As shown, CO_2 is absorbed in the lower trays. Heat of its reaction with amine is sufficient to strip the solution of some H_2S, which is again picked up in upper trays. This H_2S is recycled with the gas to a point higher in the column, where the solution is cool enough to reabsorb it. Figure 4-55b shows that this recycle of H_2S may not be present in plants treating gases containing less CO_2 and H_2S.

The solution leaves the contactor and flows thru a solution heat exchanger, where it is heated to a relatively high temperature, and then flows into the still. The still is usually

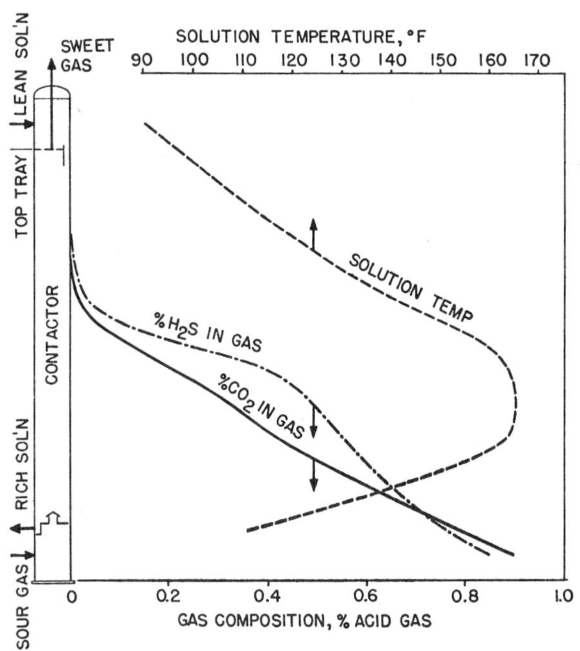

Fig. 4-55a Gas composition and solution temperature in a contactor (relatively high CO_2 and H_2S content).

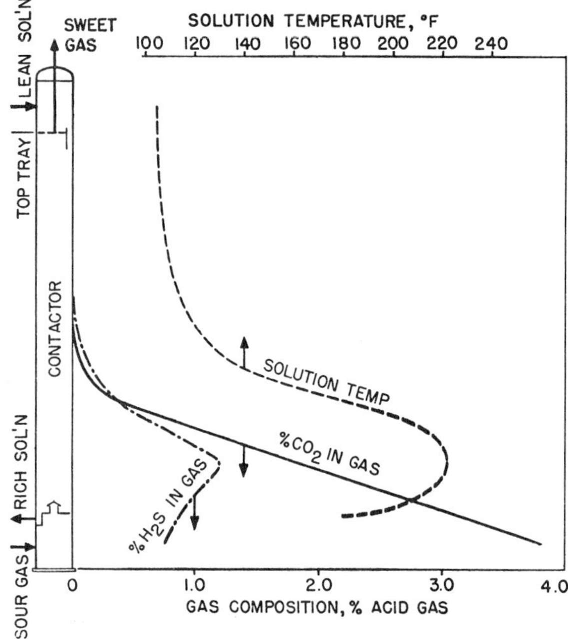

Fig. 4-55b Gas composition and solution temperature in a contactor (relatively low CO_2 and H_2S content).

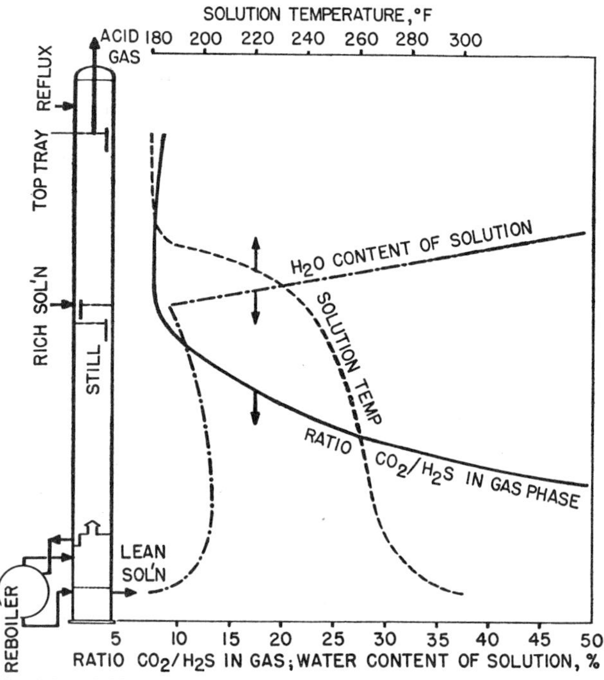

Fig. 4-56 Still and reboiler temperatures and compositions in a plant.

a bubble tray column, equipped with a reboiler and a source of reflux. In the still the solution is subjected to combined effects of relatively high temperatures and stripping action of steam rising from the reboiler and leaving with the acid gas. The reflux, composed mainly of water, enters the top tray and absorbs any glycol or amine vapors in acid gas flowing in the upper column section. It is also a source of the stripping steam.

Figure 4-56 shows graphically operation of the still. The water content curve of the solution in various parts of the still indicates that some steam rising from the reboiler is condensed and recycled. Its heat of condensation is given up, to raise solution temperature and to supply heat of reaction necessary to decompose the amine–acid gas compounds. The curve showing ratio of CO_2 to H_2S in the gas phase indicates relatively little H_2S in the solution entering or leaving the reboiler. Actually, the entering solution contains only a very small amount of either H_2S or CO_2. If temperature of the solution leaving the reboiler is lowered, thereby lowering that in the column stripping section and increasing water content of the solution, the acid gas content of the solution entering the reboiler is increased. As this reboiler temperature is increased, acid gas content in the solution going to the reboiler decreases. This acid gas content can also be reduced by increasing amount of reflux, thus increasing amount of stripping vapor. Conversely, lowering the reflux increases acid gas content.

Figure 4-57 shows effects of reboiler temperature and reflux rate on solution flowing from still to reboiler. Increasing still pressure raises solution boiling point, thereby increasing temperatures throughout the column. The increased temperature overrides effects of pressure on stripping of acid gases from the solution, resulting in a lower acid gas content in solutions to and from the reboiler.

Operating data from 11 plants are presented in Table 4-34.

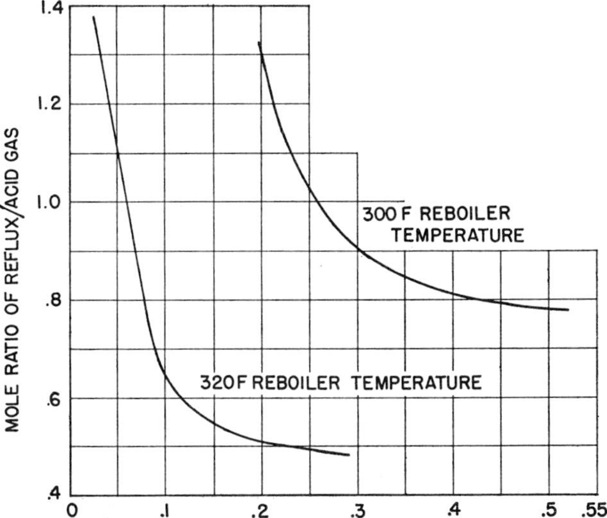

Fig. 4-57 Effects of reboiler temperature and reflux rate on solution entering reboiler from still.

Basic Process Variations. Several variations in this basic process have been made. One is to increase water content. The typical glycol–amine treating plant solution contains five to eight per cent water. Where dehydration requirements are not stringent, solution water content has been increased to about 15 or 20 per cent without changing major characteristics of the process. Some plants are operating with solutions containing 50 to 60 per cent water. These higher water contents make the process more similar to aqueous amine treating plants in chemical properties and operating characteristics. Approximately 20 per cent glycol is used in these solutions, to reduce foaming problems which sometimes exist with aqueous amine solutions. A few plants now operat-

Table 4-34 Average Yearly Operating Conditions at Various Glycol–Amine Plants
(data circa 1949)

Plant	Contactor pressure, psig	Gas analysis		Lean solution analysis†		*MEA*/A.G.* in rich solution, mole ratio	Still feed temp, °F	Reboiler temp, °F	Reflux/ A.G.*, mole ratio	Still pressure, psig	Cu ft of A.G.* per gal of solution to reboiler
		% H_2S	% CO_2	% *MEA*	% *DEG*						
A	520	0.48	1.68	21.0	73.7	3.80	237	300	1.35	3	0.07
	525	0.47	3.43	21.6	70.7	3.10	235	300	1.35	..	.11
B	520	0.75	3.90	23.0	69.0	3.12	245	300	1.31	6	..
C	47	0.66	4.40	18.5	28.5	3.10	212	270	3.30	18	.94
D	583	0.28	0.61	10.5	82.2	6.66	240	300	4.00	1.5	.03
	556	0.15	0.73	10.5	82.2	6.99	210	300	3.39	3.5	.01
F	510	0.12	0.42	17.2	76.3	3.33	190	300	1.22	2.5	..
	510	0.07	0.63	14.6	79.6	3.45	215	300	2.07	..	.02
G	525	0.10	0.18	16.2	72.8	3.00	210	300	..	12	..
	507	0.07	0.40	18.3	73.4	6.19	220	300	4.25	..	.04
H	510	0.22	0.48	14.5	78.5	6.04	190	250	..	3	.08
	..	0.13	0.80	16.1	75.4	6.70	210	300	3.58	3	.03
J	545	0.86	0.90	13.0	81.1	..	232	300	1.50	6.5	.16
K	580	..	..	7.0	87.2	6.25	200	300	..	..	.04
	..	0.03	0.07	10.5	82.7	12.03	190	303	18.6	1.4	.03
L	35	8.80	3.07	27.7	35.5	3.00	200	260	..	..	.73
N	..	0.31	6.03	18.2	69.2	2.41	220	285	1.60	4.0	.15
	165	0.22	5.80	16.0	59.3	2.40	205	244	..	3.5	0.43

* A.G. = acid gas.
† % *MEA* + % *DEG* + % H_2O = 100.

ing remove major H_2S and CO_2 portions at a low pressure with solutions of relatively high water content, the remainder being removed at a high pressure with a solution of low water concentration. Dehydration also occurs at the higher pressure.

Monoethanolamine (*MEA*) is normally used with diethylene glycol (*DEG*) in gas treating plant solutions. Diethanolamine (*DEA*) has also been used, and several others have been proposed. Some available amines[31] will remove substantially all H_2S, but only some CO_2.

REFERENCES

1. Carson, D. B. and Katz, D. L. "Natural Gas Hydrates." (T.P. 1371) *A.I.M.E. Trans.* (*Pet. Dev. & Tech.*) 146: 150–8, 1942. Also: *Gas Age* 89: 15+, Feb. 26, 1942.
2. Hammerschmidt, E. G. "Formation of Gas Hydrates in Natural Gas Transmission Lines." *Ind. Eng. Chem.* 26: 851–5, Aug. 1934.
3. Deaton, W. M. and Frost, E. M., Jr. *Gas Hydrates and Their Relation to the Operation of Natural-Gas Pipe Lines.* (Bur. of Mines Mono. 8) New York, A.G.A., 1948.
4. Culberson, O. L. and McKetta, J. J., Jr. "Solubility of Methane in Water at Pressures to 10,000 Psi." *J. Petrol. Technol.* 3: 223–6, Aug. 1951.
5. Davis, J. E. and McKetta, J. J. "Solubility of Methane in Water." *Petrol. Refiner* 39: 205–6, Mar. 1960.
6. Kobayashi, R. and Katz, D. L. "Methane Hydrate at High Pressure." *J. Petrol. Technol.* 1: 66–70, Mar. 1949 (Trans.)
7. Wilcox, W. I. and others. "Natural Gas Hydrates." *Ind. Eng. Chem.* 33: 662–5, May 1941.
8. Connealy, L. E. "Controlling Gas Hydrates." *Gas* 36: 101–8, Nov. 1960.
9. Shively, W. L. "Ten Years of Gas Dehydration." *A.G.A. Proc.* 1939: 501–15.
10. Hammerschmidt, E. G. "Natural Gas Dehydration Practices." *A.G.A. Proc., Operating Sec.* 1957: CEP-57-7.
11. Boston, F. C. "Production of Marketable Natural Gas with Low Temperature Well-Head Units." *A.G.A. Proc.* 1951: 576–84.
12. Records, L. R. and Seely, D. H., Jr. "Low Temperature Dehydration of Natural Gas." (A.I.M.E. Tech. pub. 3022) *J. Petrol. Technol.* 3: 61–6, Feb. 1951 (Trans.)
13. Spangler, C. V. "Solid Adsorption-Type Natural Gas Dehydration Plants." *A.G.A. Proc. Nat. Gas Dept.* 1946: 56–72.
14. "Gas Treating by New Method." *Petrol. Engr.* 30: D34, Feb. 1958.
15. Sullivan, J. H. "Dehydration of Natural Gas by Glycol Injection." *Oil Gas J.* 50: 70–1, Mar. 3, 1952.
16. Moore, E. R. and Cutler, W. G. "What It Costs Pacific Northwest Pipeline to Dehydrate with Calcium Chloride." *Oil Gas J.* 57: 166–9, July 27, 1959.
17. Bukacek, R. F. *Equilibrium Moisture Content of Natural Gases*, p. 21+ (charts). (Research bul. 8) Chicago, I.G.T., 1955.
18. A.S.T.M. "Standard Method of Test for Water Vapor Content of Gaseous Fuels by Measurement of Dew-Point Temperature." (Standard D-1142–58) *ASTM Standards* 8: 1491–1503. Philadelphia, 1961.
19. Laurence, L. L. and Worley, M. S. "Gas Dehydration and Desulfurization Operating Problems." (GSTS-58-17) *A.G.A. Proc. Operating Sec.* 1958: T95–103.
20. Swerdloff, W. "What We've Learned in 20 Years about Gas Dehydrators." *Oil Gas J.* 55: 122–9, Ap. 29, 1957.
21. Weil, A. H. "Economics of Modern Gas Dehydration." *Gas* 30: 110–3, Nov. 1954.
22. Campbell, J. M. "Design and Choice of Equipment for Gas Dehydration." *Chem. Eng. Progr.* 48: 440–8, Sept. 1952. ——. "Methods for Gas Dehydration for 1954." *Petrol. Engr.* 26: C11+, Sept. 1954.
23. Mantell, C. L. *Adsorption*, 2nd ed. New York, McGraw-Hill, 1951.
24. Amero, R. C. and others. "Design and Use of Adsorptive Drying Units." *Chem. Eng. Progr.* 43: 349–70, July 1947.
25. LaLande, W. A., Jr., and others. "Bauxite as a Drying Adsorbent." *Ind. Eng. Chem.* 36: 99–109, Feb. 1944.
26. Wilkinson, E. P. and Sterk, B. J. "New Data on Use of Solid Adsorbents in Natural Gas Drying." *A.G.A. Proc. Nat. Gas Dept.* 1950: 57–61.
27. Clark, E. L. "How To Use Molecular Sieves for Natural-Gas Treating." *Oil Gas J.* 57: 120–3, Ap. 27, 1959.
28. Dow, W. M. and Parks, A. S. "Development of Wellhead Dynamic Adsorption Equipment." Oklahoma Univ. *Proc. Gas Conditioning Conf.*, 1956: 1–28.
29. Hutchinson, A. J. L. *Process for Treating Gases.* (Patent 2,177,068) Washington, U. S. Patent Off., 1939.
30. Kohl, A. L. and Blohm, C. L. "Technical Aspects of Glycol-Amine Gas Treating." *Petrol. Engr.* 22: C37+, June 1950.
31. Kohl, A. L. "Selective H_2S Absorption." *Petrol Process.* 6: 26–31, Jan. 1951.
32. Brown, J. F., Jr. "Inclusion Compounds." *Sci. Am.* 207: 82–92, July 1962.

Chapter 9

Removal of Sulfur Compounds and Carbon Dioxide from Sour Natural Gas

by J. S. Connors, C. L. Perkins, and F. E. Vandaveer

OCCURRENCE OF SULFUR AND CARBON DIOXIDE IN NATURAL GAS

Gas from many natural gas fields contains no sulfur compounds and very little carbon dioxide. Such gas is designated *sweet* and presents no purification problem. That from other fields, including some very large ones in the U. S. such as the Permian Basin and Panhandle in Texas, and the McCamey Field in Arkansas, and the Southern Alberta[1] in Canada, may contain from a trace up to 46 per cent by volume of sulfur compounds (mainly H_2S plus mercaptans and traces of other organic sulfides). The Lacq field in southern France contains up to 18 per cent. Gases from these fields are designated *sour* and must be purified before distribution to domestic and industrial customers. One definition of a sour gas is one with a *hydrogen sulfide* content above 1.5 grains per CCF, or *total sulfur* content above 30 grains per CCF (7000 grains = 1 lb; 635 grains per CCF = 1 per cent by volume). For good distribution practice total sulfur content below one grain per CCF is desirable. In many systems, particularly when supplying brass manufacturing concerns, total sulfur content below 0.25 grain is desired.

Carbon dioxide is present in most natural gases in amounts varying from 0.1 to 6.0 per cent by volume. A few wells in the Rocky Mountain area produce almost 100 per cent CO_2. In small amounts, CO_2 is not harmful to the distribution system. However, it reduces gas heating value and causes corrosion of tubing near the surface of gas-condensate high pressure wells.

REMOVAL OF HYDROGEN SULFIDE

Aqueous Amine Process

General. This process is the most widely used in the natural gas industry for separation of acid constituents, e.g., H_2S and CO_2, from gaseous mixtures. It is employed for removal of both H_2S and CO_2 from natural gas and also from refinery vapors. The process, discovered by R. R. Bottoms, was granted U. S. Patent No. 1,783,901 in 1930. In early development stages, preferred extractive agents were 50 per cent water solutions of triethanolamine or diethanolamine. At present the preferred solutions are usually 15 to 20 per cent monoethanolamine or 20 to 30 per cent diethanolamine.

Because natural gas very seldom contains carbonyl sulfide (COS) the more highly active monoethanolamine solutions are usually employed, and H_2S content of treated gas will not exceed 0.25 grain per CCF. If desired, it can be reduced to 0.05 grain per CCF. If carbonyl sulfide is present, diethanolamine solutions must be used, since carbonyl sulfide reacts with the primary amine to form a stable compound which cannot be regenerated. Gas treated by diethanolamine solutions will seldom have an H_2S content lower than 0.5 grain per CCF.

The principle of this process rests on reaction of water solutions of aliphatic alkanolamines with acid gases like H_2S and CO_2 around atmospheric temperatures. The equilibrium decreases rapidly at slightly elevated temperatures, with resulting release of such gases. This is actually a reversible chemical reaction in which chemical bonds are so loosely attached that process behavior is similar to that of absorption. This reaction may be expressed by the following equations where RNH_2* represents monoethanolamine (*MEA*):

$$1.\ 2RNH_2 + H_2S \underset{240\text{ F}}{\overset{100\text{ F}}{\rightleftharpoons}} (RNH_2)_2 \cdot H_2S$$

$$2.\ 2RNH_2 + CO_2 + H_2O \underset{300\text{ F}}{\overset{120\text{ F}}{\rightleftharpoons}} (RNH_2)_2 \cdot H_2CO_3$$

or

$$3.\ 2RNH_2 + H_2S \underset{240\text{ F}}{\overset{100\text{ F}}{\rightleftharpoons}} (RNH_3)_2S$$

$$4.\ 2RNH_2 + CO_2 + H_2O \underset{300\text{ F}}{\overset{120\text{ F}}{\rightleftharpoons}} (RNH_3)_2CO_3$$

A similar group of equations can be set up for diethanolamine and triethanolamine. It should be noted that reverse reaction is initiated between H_2S and amine at a lower temperature than for CO_2. Therefore, contact temperatures for H_2S removal should be 100 F or lower. Also, dissociation temperature for CO_2 is considerably higher than for H_2S. Therefore, while virtually complete stripping of H_2S is possible from amine solutions at 240 F, approximately 0.5 to 1.0 cu ft CO_2 per gallon will remain in reactivated solutions, while at temperatures of 300 F nearly all will be stripped out.

Operation. Since reactions are chemical in nature rather than physical, increasing solution strength or rate of flow increases capacity to remove acid gases. However,

* Where R = $HOCH_2CH_2$.

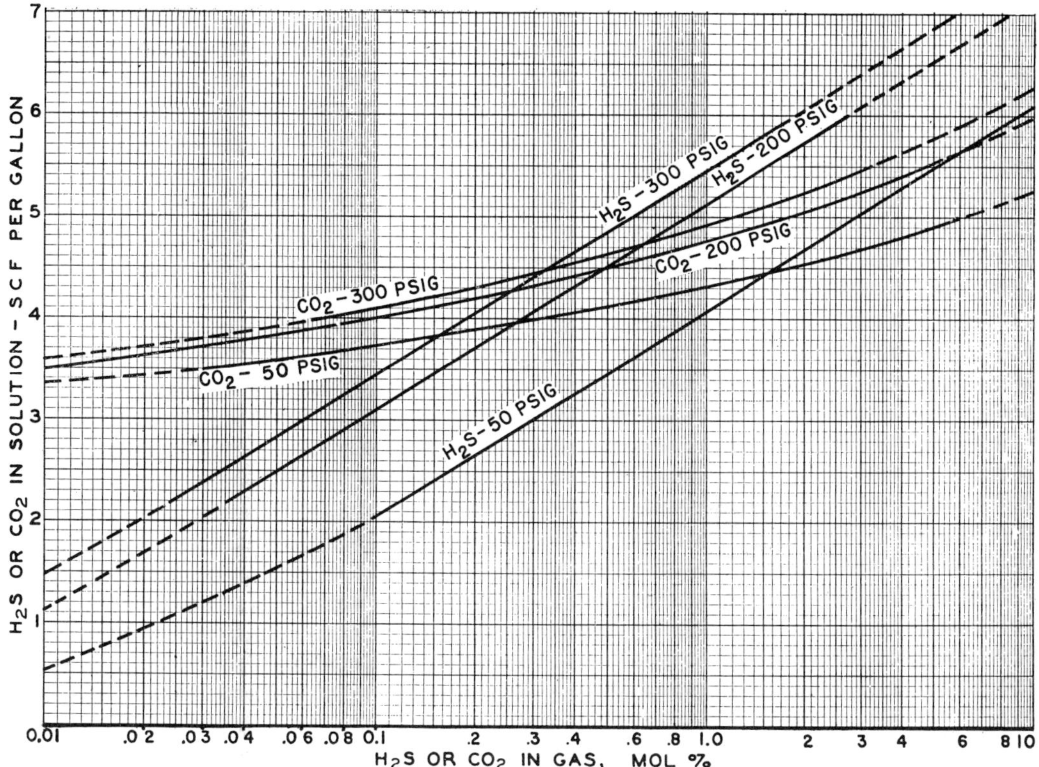

Fig. 4-58 H_2S and CO_2 equilibrium over 15 per cent *MEA* solution at 100 F (recalculated from Girdler data).

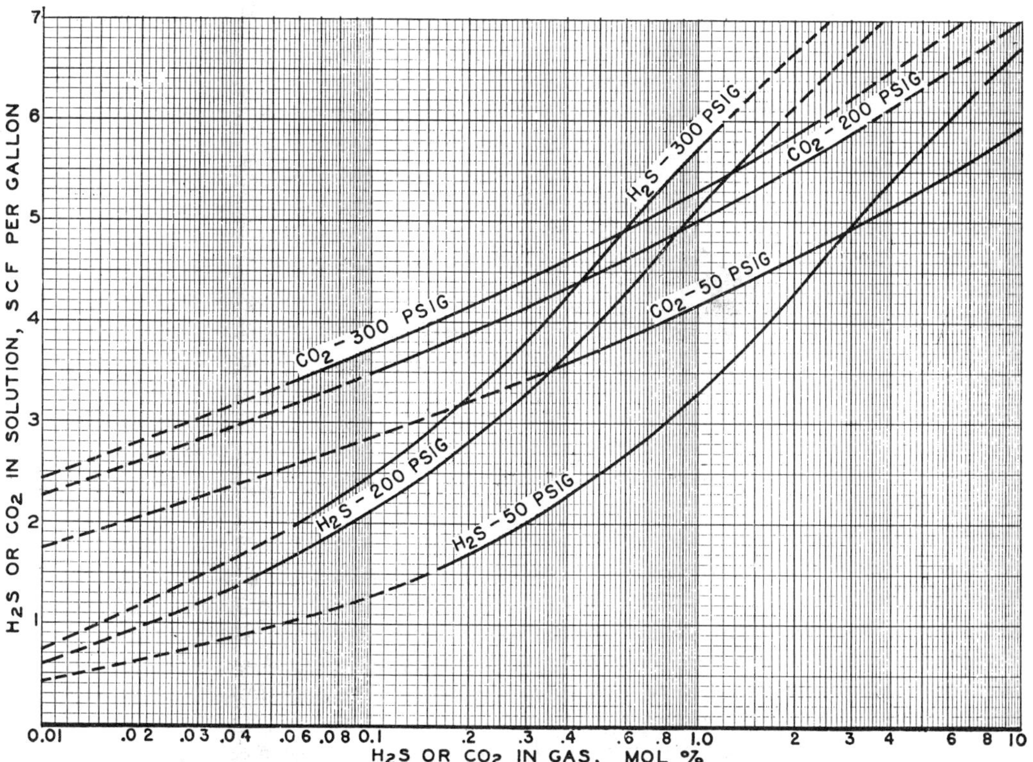

Fig. 4-59 H_2S and CO_2 equilibrium over 30 per cent *DEA* solution at 100 F (recalculated from Girdler data).

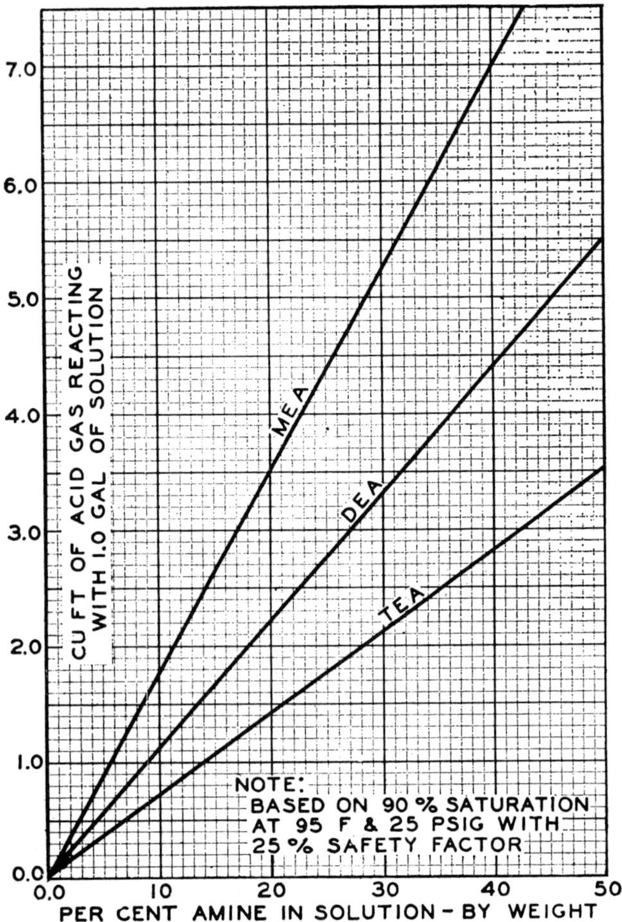

Fig. 4-60 Acid gas pickup with varying strength amine solutions.

such increases must be evaluated with care because stronger solutions or increased flow rates will result in decreased regeneration, and thus amine consumption will be higher. For plants using 15–20 per cent solutions of *MEA*, consumption will vary from approximately 0.6 to 1.5 lb per MMCF of gas processed because contact pressures vary from 200 to 50 psig, where the acid gas content of the gas treated is below three per cent.

Equilibrium of H_2S and CO_2 over 15 per cent *MEA* and 30 per cent diethanolamine (*DEA*) solutions are shown in Figs. 4-58 and 4-59, respectively. Since the process is based on chemical rather than physical equilibrium, total loading of the foul solution should never exceed amine reacting values for H_2S and CO_2 as shown by preceding equations.

Figure 4-60 shows allowable pickup of acid gas based on reacting values with allowance for acid gas left in solutions after regeneration; i.e., a specific job in a specific unit would require about 1.6 *times* as much *DEA* compared to *MEA* circulation. These curves apparently apply for a wide range of pressures and acid gas contents, as shown by data in Table 4-35 obtained from commercial operating units. Solution loading was obtained by reducing solution circulation until the treated effluent gas reached the maximum H_2S content permissible with operating requirements.

The operating cycle and equipment required for this type of amine gas purifier are quite similar to those for the absorption and stripping process of a natural gasoline plant. Figure 4-61 shows a typical setup; input gas flows upward thru a bubble tower contactor countercurrent to an aqueous amine solution at 80 to 100 F, which enters top of tower and flows from tray to tray picking up acid gas. Purified gas leaves tower at top. Foul amine solution saturated with acid gas

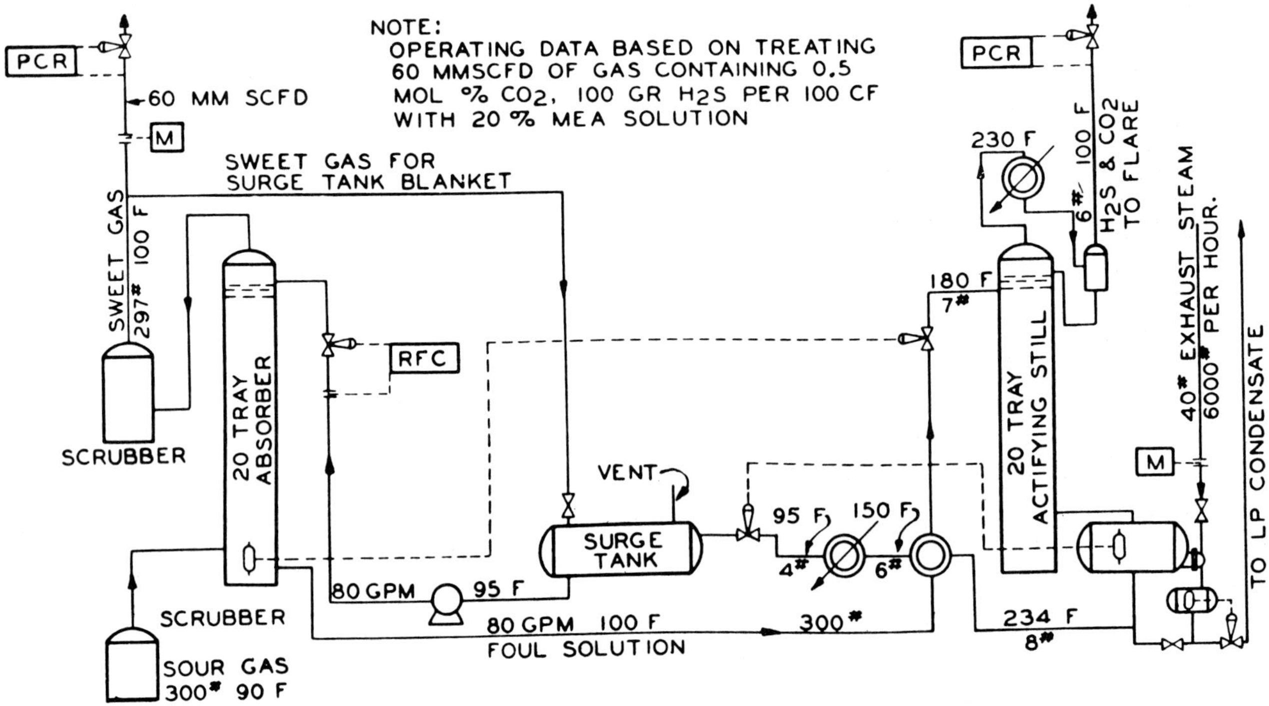

Fig. 4-61 Amine gas treater arrangement. (See also Fig. 5-9.)

Table 4-35 Acid Gas Pickup By Aqueous *MEA* Solutions

	Pressure of absorbers, psig				
	50	50	200	200	200
MEA solution, %	14.97	19.12	20.1	18.2	19.6
H_2S in gas to absorbers, grains per CCF	399	472	8.0	72.5	92.4
CO_2 in gas to absorbers, mole %	4.967	4.451	0.105	0.6	0.63
H_2S in treated gas, grains per CCF	1.415	2.17	0.1	0.1	0.07
CO_2 in treated gas, mole %	0.555	0.61	0.03	0.062	0.03
Total acid gas in foul solution, cu ft per gal	3.954	5.077	4.5	4.07	4.40
Total acid gas in lean solution, cu ft per gal	0.782	1.333	1.2	0.58	1.39
Acid gas pickup by solution balance,* cu ft per gal	3.172	3.744	3.3	3.49	3.01
Acid gas pickup by gas balance,* cu ft per gal	3.59	3.31	3.53	3.35	3.79

* Difference between gas and solution balance is due to analysis and flow measurement differences.

leaves bottom of tower and passes thru heat exchangers, where its temperature is raised to 180–200 F, into top tray of stripper or reactivating still. This stripper, which is also a bubble tower, contains in its base or in an external kettle a tubular heating element or reboiler. At top of stripper are ondensers and a dephlegmator or water separator. Amine solution flowing downward thru stripper to kettle element or reboiler is heated to approximately 230–240 F by vapor rising up from boiling solution in kettle section. Acid gas liberated from amine solution and some internally generated stripping steam are carried overhead from top of the stripper thru condensers where steam is condensed and cooled. Condensed steam and acid gases are separated in dephlegmator. Condensate is returned to top of stripper as cold reflux and acid gases are burned in a flare or sent to a processing unit.

Hot lean solution, stripped of acid gas flows from kettle to heat exchangers where heat is exchanged in heating up cold foul solution, and then thru water coolers where it is cooled to approximately atmospheric temperature. From coolers, cold lean solution goes to a surge tank from which it is pumped to contactor, thereby completing cycle.

Because of fairly rapid rate of reaction in the stripper, virtually complete stripping of H_2S from solution can be accomplished in a single tower using 1.2 pounds of steam (exhaust at 40 psig) per gallon of solution. Since aqueous amine solutions have very little affinity for hydrocarbon constituents of gas, this process is applicable to purification of field gases before gasoline extraction as well as to residue gas.

Guide Features for Design and Operation. While aqueous amine solutions used in H_2S and CO_2 removal are

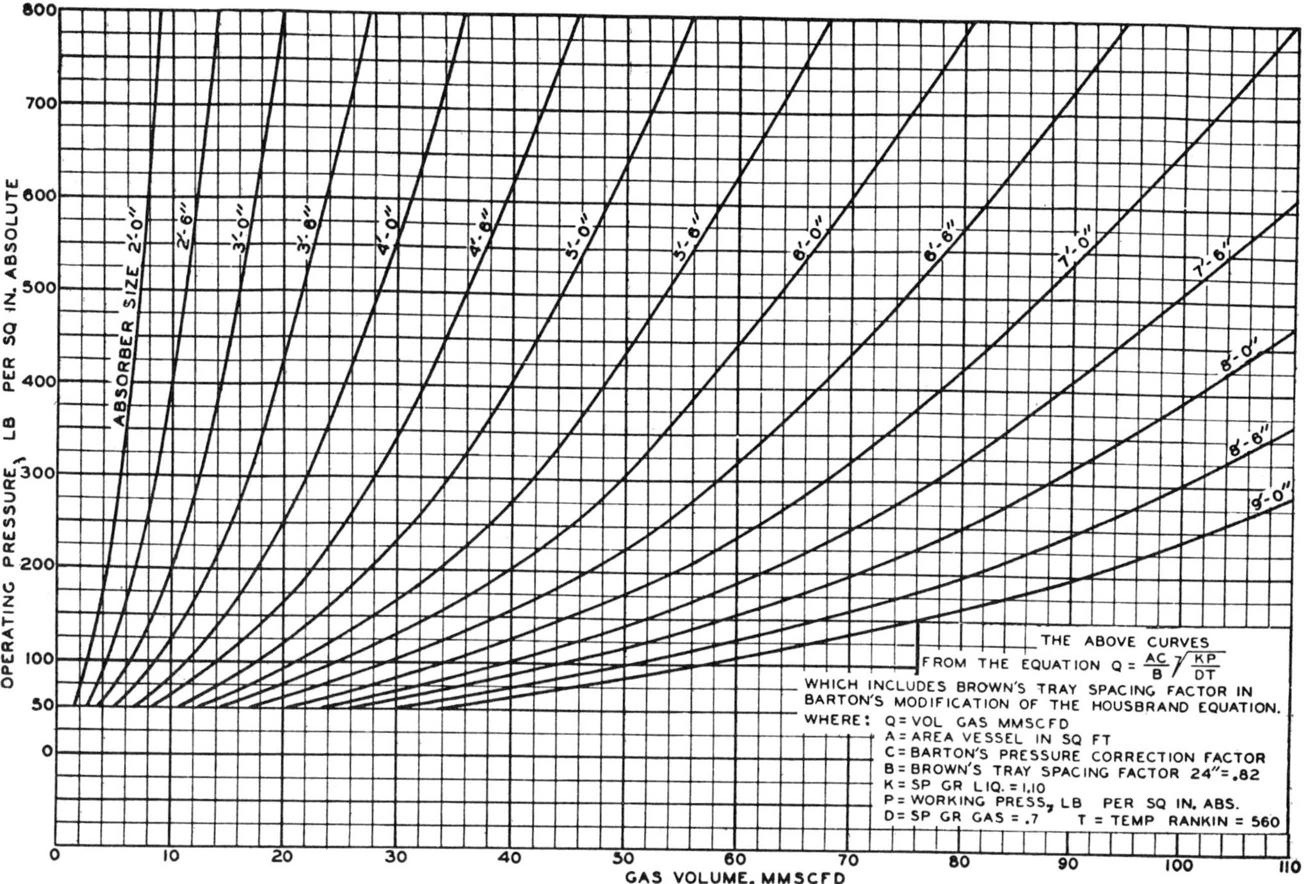

Fig. 4-62 Curves for estimating absorber size in gas treating.

generally considered only mildly corrosive to carbon steel, their susceptibility to decomposition[2,3] at elevated temperatures and their tendency to form glycine in presence of oxygen make extreme care mandatory in plant design and operation. Amine solutions also tend to foam, especially when purifying wet field gas. Considering these properties, the following may serve as a general guide for operators and designers.

In design of contactors, chemical equilibrium curves are not always reliable. Experience has shown that while only three or four theoretical stages of contact are required by calculations, actual number of bubble plates required may be 16 to 20. This is apparently due to difficulty of securing adequate contact when removing last traces of H_2S as well as to the reaction time element. The contactor should always be stress relieved to minimize the ever-present danger of stress corrosion cracking. The following formula[4] has been found to give excellent results for vessel sizing, subject to some consideration of maximum slot mass velocity:

$$A = \frac{BQ}{C}\sqrt{\frac{DT}{KP}}$$

where:

A = tower cross-sectional area (excludes downcomer area), sq ft
B = Brown bubble plate tray spacing factor (see Table 4-36)
C = Barton bubble plate absorber pressure correction factor (Table 4-37)
D = specific gravity of gas, referred to air ($29D$ = mole weight of gas)
K = specific gravity of liquid at T°, referred to water at 60 F
P = gas pressure, psia
Q = gas volume, MMSCF per day
T = temperature, °R or (°F + 460)

Table 4-36 Brown Tray Spacing Factors

Tray spacing, in.	Factor, B
18	1.00
24	0.82
30	0.75

Table 4-37 Barton Pressure Correction Factors

Pressure range, psig	Factor, C
0–25	0.78
25–50	1.12
50–100	1.36
100–400	1.56

Figure 4-62 shows a chart of tower sizing for various conditions as determined on basis of foregoing formula.

Carbon steel heat exchangers are generally satisfactory as long as solutions have no aggravated tendency toward corrosion and if linear velocities do not exceed about 2 fps. Figure 4-63a shows the effect of linear velocity on overall **K*** rate, Btu per hr-sq ft-°F, assuming that exchanger is designed with equal shell and tube velocities.

* $1/U = \Sigma 1/h + 1/\mathbf{K}$, where U = overall heat transfer coefficient.

Figures 4-63b and 4-63c show the variation of overall **K** rates for solution coolers for varying velocities on shell and tube sides, with different fouling factors introduced for water-side fouling.

Since amines decompose at elevated temperatures, 250 F may be regarded as maximum allowable kettle temperature, with steam not exceeding 300 F as sole kettle heating medium. Tube spacing should be arranged to provide adequate vapor relieving capacity to prevent vapor binding and local overheating. For ¾ in. OD tubes on triangular setting, a 1¼ in. center-to-center spacing is recommended. Stripper and external kettle shell (where used) should always be stress-relieved to minimize danger of stress corrosion cracking. It is difficult to calculate theoretical number of trays required and to correlate tray efficiencies, contact and reaction times; most strippers contain approximately 20 bubble trays.

Figure 4-64 shows a chart of stripping still sizes which have been found adequate. Brown-Souders formula[5] was used in computing overall tower cross-sectional area, with modifications for slot velocity.

If *MEA* solutions become corrosive from either oxidation or high temperature breakdown, they can be purified by simple distillation using a semicontinuous process as indicated by the vapor–liquid compositions shown in Fig. 4-65. Figure 4-66 is a sketch of a redistillation process where amine gravity feed is a side stream from the operating still. Tendency of solution to corrode can be approximated from its iron capacity under set conditions of test. Such a test is qualitative. A better test is a measure of build-up rate of nitrogen system compounds other than *MEA*, as determined by Kjeldahl and Van Slyke procedures.[6]

Seaboard Process—Sodium Carbonate Solution

This process (Fig. 4-67), used mainly for manufactured gas purification, was the first to commercially employ a liquid. Normally, the absorber operates at a very low pressure, and approximately 85 to 95 per cent of H_2S is removed there. For several years this process was used to purify natural gas at approximately 500 psig. Gas leaving the absorber in this case is much purer, containing from 1 to 5 grains of H_2S per CCF. Solution composition is normally 3 to 3½ per cent sodium carbonate. Acid gases react with sodium carbonate as follows:

$$H_2S + Na_2CO_3 \rightarrow NaHS + NaHCO_3$$

$$CO_2 + Na_2CO_3 + H_2O \rightarrow 2NaHCO_3$$

Absorption of H_2S causes the solution to develop a definite H_2S vapor pressure. Theoretically, it can continue to absorb H_2S until its H_2S vapor pressure equals the H_2S partial pressure in the inlet gas. In practice, however, it is not possible to saturate the solution to this degree, and it is necessary to circulate somewhat more solution than is theoretically required. The solution flow rate is between 60 and 150 gal per MCF of gas depending upon H_2S and CO_2 concentration.[8]

For a typical coke oven gas containing three to five grains H_2S per SCF, and 1.5 to 2.0 per cent CO_2, the solution flow rate will be 60 to 80 gal per MSCF—a carrying capacity of 50 grains per gal.

Solution leaving the absorber is heated with exhaust steam

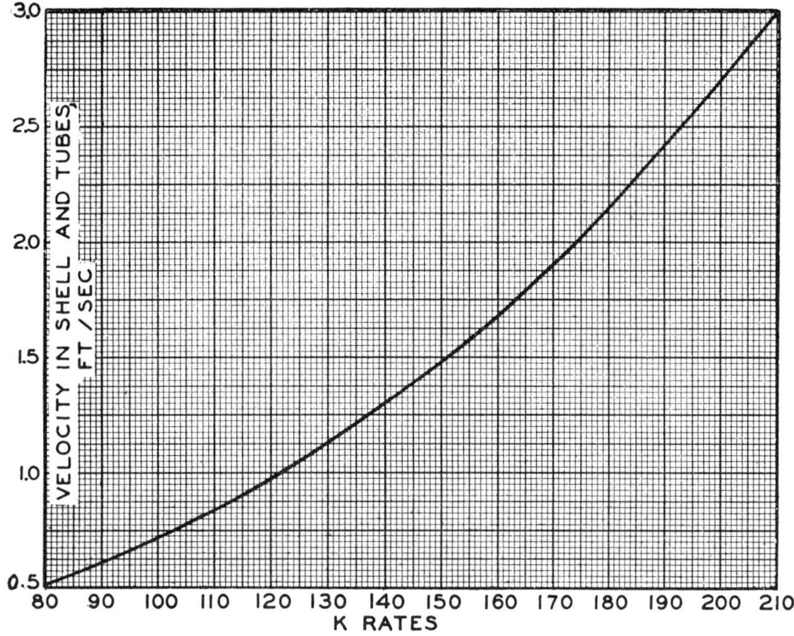

Fig. 4-63a K rates, Btu per hr-sq ft-°F, for amine heat exchangers (fouling factor of 2).

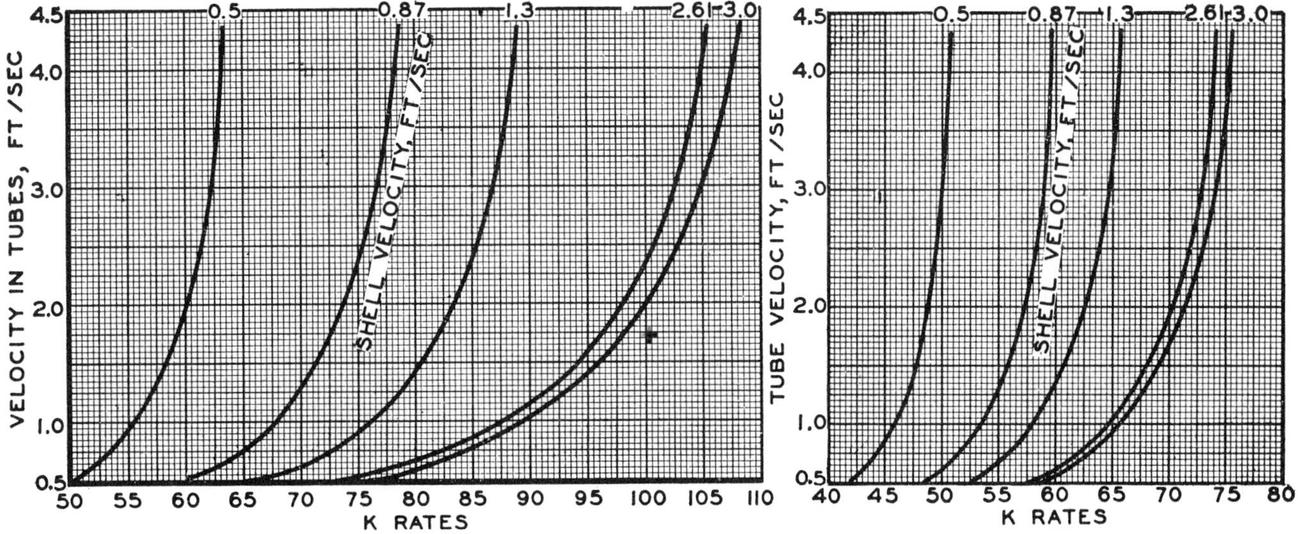

Fig. 4-63b (left) K rates, Btu per hr-sq ft-°F, for amine solution coolers (fouling factor of 6).
Fig. 4-63c (right) K rates, Btu per hr-sq ft-°F, for amine solution coolers (fouling factor of 10).

and flows to the top of an aerating tower or *actifier*, where reactions shown above are reversed. These reactions are due to effects of high temperature and stripping action of air blown upward thru the tower. Acid gases pass to atmosphere along with reactivating air, or in a few cases are piped to a boiler and burned along with additional fuel.

The absorber is normally a bubble tray or packed column with solution entering at top and sour gas at bottom. When used for natural gas purification at 500 psig, absorber design is somewhat different. Both gas and solution enter absorber at the top and flow concurrently thru two groups of short tubes. Solution and gas are separated at absorber bottom, gas passing into the pipeline. Spent solution passes thru heat exchangers to top of a wooden aerating tower. In some plants the absorber is located directly above the actifier in the same vessel, usually when gas is purified at low pressure.

Formation of thiosulfate takes place in some cases because of oxidation of the solution. This thiosulfate cannot be regenerated; therefore, some solution must be discarded and fresh solution added to keep thiosulfate concentration relatively low. If gas being treated contains a relatively high CO_2 concentration, the solution must be heated to a higher temperature to effect reactivation. This high temperature increases water loss and requires addition of relatively large quantities of make-up water. If this water contains any dissolved solids, they will concentrate in the solution necessitating its being discarded. If care is not taken to control concentration of these solids they will precipitate in aerating towers or absorber and heat exchangers, causing plugging or reduced capacity.

A practical test in this regard is to chill a sample of the solution to the minimum plant temperature and then add

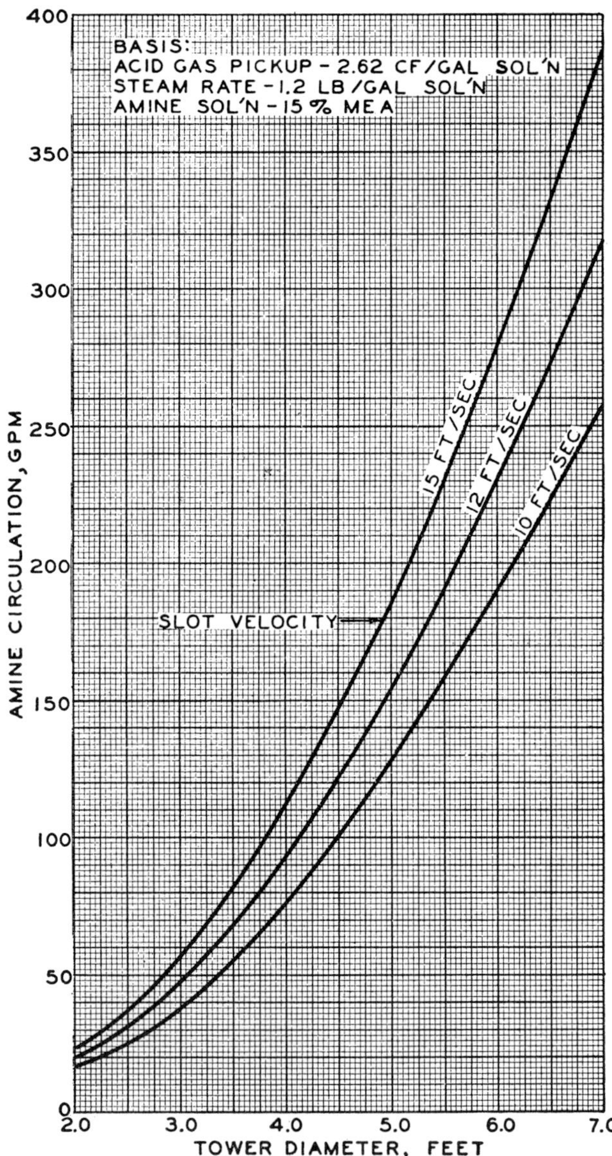

Fig. 4-64 Curves for estimating reactifier size for amine gas treater; pressure 7 psig, bubble cap slot area 12.2 sq in.

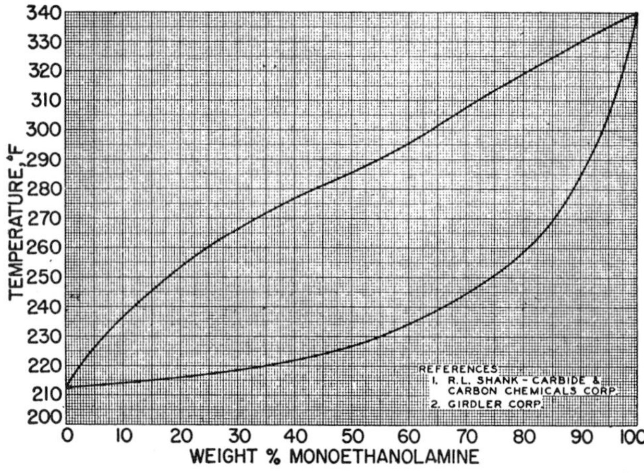

Fig. 4-65 Vapor-liquid composition boiling point curve for aqueous *MEA* solutions.

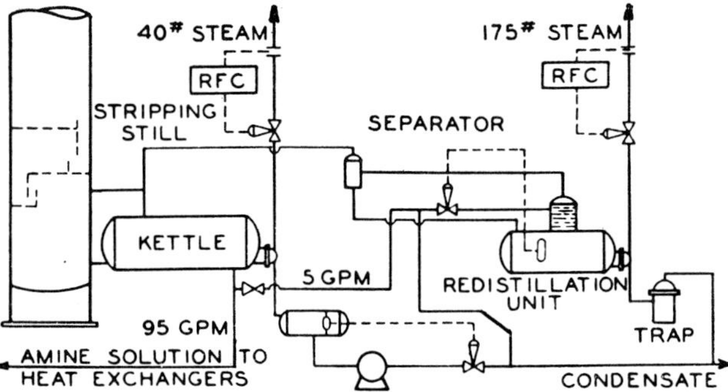

Fig. 4-66 Semicontinuous *MEA* purification system.

some solid sodium bicarbonate. The solution is then tested to see whether the alkalinity was thereby increased. An increase shows that the solution has actually dissolved some sodium bicarbonate and consequently will not deposit solid sodium bicarbonate in the cold locations. If such test shows no increase in alkalinity, the conditions in the plant should be corrected by increasing the temperature or by diluting the solution to reduce the concentration of thiosalts.

Main advantages of this process are small plant space requirements, cheapness of reagent used, and simplicity of operation. Consumption of sodium carbonate in plants treating natural gas at 500 psig was from 0.01 to 0.05 lb per MCF. Operating costs of treating natural gases containing less than one per cent CO_2 are low. Above one per cent, however, costs may be excessive.

To reduce formation of thiosulfates due to oxidation, a modification of the Seaboard Process was developed, called the **Vacuum Carbonate Process.** In its first development steam was used instead of air for stripping carbonate solution of acid gases. Steam requirements, when stripping at atmospheric pressure or above, are relatively high and can be reduced considerably by regeneration under vacuum. A higher degree of regeneration is obtainable with this modification than with the Seaboard Process. This makes possible removal of a larger percentage of acid gases or reduction of solution rates.

Other Sulfur Removal Processes

Other processes for removal of sulfur from manufactured gas may be applicable. Processes that use amine in conjunction with hot potash or glycol may also be used.

Potassium Carbonate Process[9,10]—Bureau of Mines. Hot potassium carbonate absorbs CO_2 and H_2S. These are subsequently released by additional steam heat. It is claimed that less steam is needed for regeneration of the carbonate than for amine solutions.

Iron Oxide Process.[11,12] Iron oxide (5 to 10 lb per cu ft) mixed with wood shavings will remove H_2S from a gas stream as shown in Fig. 4-68. Revivifying is done with air, with oxide in place or taken outside the box and aerated. Removal of H_2S is more efficient at high than at low pressure.[12]

Applicable equations follow:

$$Fe_2O_3 + 3H_2S = Fe_2S_3 + 3H_2O$$

$$2Fe_2S_3 + 3O_2 = 2Fe_2O_3 + 6S$$

$$2H_2S + O_2 = 2H_2O + 2S$$

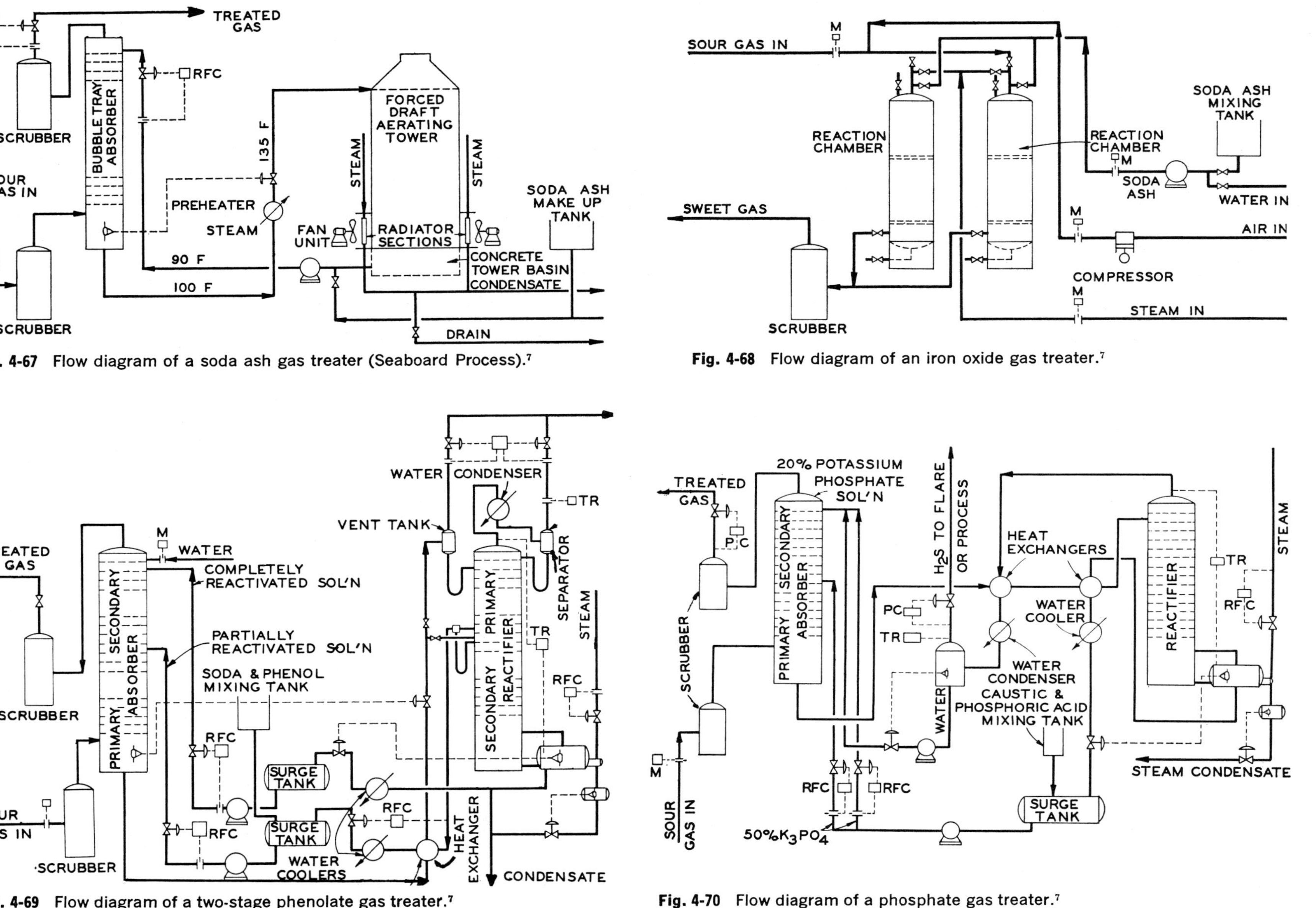

Fig. 4-67 Flow diagram of a soda ash gas treater (Seaboard Process).[7]

Fig. 4-68 Flow diagram of an iron oxide gas treater.[7]

Fig. 4-69 Flow diagram of a two-stage phenolate gas treater.[7]

Fig. 4-70 Flow diagram of a phosphate gas treater.[7]

Thylox Process[11]—Sodium or Ammonium Thioarsenate. It is seldom used for natural gas because operating costs are high, and it is a rather complex process.

Haines Process.[13] Alkali-metal alumin-osilicates (synthetic zeolites) can simultaneously dehydrate and sweeten natural gas, and convert the adsorbed H_2S to elemental sulfur. Regeneration media reported include air containing various proportions of oxygen and sulfur-burner gas with both 3 and 100 per cent excess air.

Sodium Phenolate Process.[11] This process (Fig. 4-69) is seldom used on natural gas because efficiency of H_2S removal is generally low (90 per cent) and steam consumption is very high.

Applicable equations follow:

$$NaOC_6H_5 + H_2S = NaHS + C_6H_5OH$$

$$NaOC_6H_5 + CO_2 + H_2O = NaHCO_3 + C_6H_5OH$$

Tripotassium Phosphate Process.[11] It is used primarily in treating refinery gas (Fig. 4-70). H_2S content may be reduced from 3000 to 15 grains per CCF with a solution containing about 32 per cent K_3PO_4. It is not competitive with amine systems.

Applicable equations follow:

$$K_3PO_4 + H_2S = KHS + K_2HPO_4$$

$$K_3PO_4 + CO_2 + H_2O = KHCO_3 + K_2HPO_4$$

Small Scale Sulfur Removal Processes. When an industrial process requires that the sulfur content of the gas used be below that of the gas as distributed, the following techniques may be considered:

1. Activated charcoal filtering—carbon grain size, bed thickness, and flow rate are important variables. Use of two or more beds permits reactivation of charcoal outside the process setup.
2. Oil scrubbing towers—designed to accommodate a particular application.
3. Catalyst processes—e.g., the United Gas Improvement Co. made a patented catalyst available. This catalyst was successfully used for 17 years to reduce the sulfur content of carbureted blue from 15 to less than 3 grains per MCF at under 10 psig. The catalyst, consisting of magnesium sulfate and zinc oxide, lasts about 6 months at gas stream rates of 2.5 MCFH (max 4 MCFH) with weekly catalyst regenerations.

Stretford Process.[18] It was claimed that this process, which is used in England to remove H_2S and recover pure sulfur from manufactured gases containing up to 700 grains of H_2S per CCF of gas, is capable of treating gases containing as much as 50 per cent CO_2 and H_2S. The process involves washing the gas with a quinone solution which reacts with H_2S to form hydroquinone and sulfur. The sulfur is filtered out and the reagent is reconstituted. Pilot plant operation at 300 psig (on 500 grains per CCF gas) was also reported.[19]

REMOVAL OF CARBON DIOXIDE

Carbon dioxide is mainly removed from natural gas as a result of treatment for hydrogen sulfide. A weak acid gas, like H_2S, it is absorbed in amine or other caustic solutions (notably hot potassium carbonate) employed to remove H_2S. The Giammarco-Vetrocoke[14] and the Fluor Solvent processes are two recent developments. Depending upon the efficiency of the purification process, usually from 0.1 to 1.0 per cent CO_2 remains in the gas. Processes for H_2S removal are applicable to CO_2 also. In only three instances (up to 1961) were natural gases (overall CO_2 range: 5 to 57 per cent) treated for removal of CO_2 only.

Fluor Solvent Process[15,16]

This organic solvent process, applicable to high pressure (e.g., 300 to 1000 psig) lean natural and synthetic gases containing high CO_2 concentrations, removes CO_2 solely by absorption; i.e., there is no chemical reaction between the CO_2 and the absorbents (propylene carbonate, glycerol triacetate, butoxy diethylene glycol acetate, and methoxy triethylene glycol acetate). Some dehydration of the natural gas is also achieved by the solvent; dew point depressions of up to 50°F are sometimes obtained.

A plant designed to reduce the CO_2 concentration of 220 MMSCFD of natural gas from 53 to 2 per cent has been placed in operation using propylene carbonate as the solvent. Solvent is regenerated in successive flash operations. Dissolved hydrocarbons are flashed from the rich solvent in the first stage of pressure reduction. The flash gas is recompressed and combined with the feed. The solvent then undergoes a second pressure reduction in which considerable quantities of waste CO_2 at elevated pressures are evolved. This stream is used in gas expansion turbines. The solvent is then flashed to low pressure. Sufficient CO_2 is desorbed to reduce its equilibrium partial pressure in the solvent, to permit the production of 2 per cent CO_2 residue gas. Hydraulic turbines are used to recover power from the pressure letdown of the rich solvent. Lean solvent is then pumped back to the absorbers by multistage centrifugal pumps.

Virtually all the power needed for recirculating solvent, except during start-up periods, is derived from the expansion of CO_2 in the feed gas from field pressure to atmospheric pressure via gas turbines. A part of the energy required is in turn recovered by hydraulic turbines. In addition, a substantial amount of refrigeration is also obtained from these expansions, thus lowering both the solvent temperature and the solution circulation rate.

REMOVAL OF MERCAPTANS

Alkaline solutions, like monoethanolamine, have little effect on removing mercaptans from sour natural gas because these compounds are not appreciably acidic. Iron oxide does not remove mercaptans satisfactorily. Typical data on a Girbotol amine–iron oxide-activated charcoal plant follow:

	Grains of sulfur/CCF
Inlet of Girbotol amine	30.75
Outlet of Girbotol amine	2.75
Outlet of iron oxide	2.25
Outlet of charcoal tanks	0.16

Practically all the mercaptan content of the gas passed thru the amine and iron oxide, but it was removed by activated charcoal.

Mercaptans[17] are removed from natural gas by the same process used for gasoline and LP-gas removal, namely, contact with absorption oils like kerosene, naphtha, or mineral seal oil. They are absorbed to the same extent as butanes.

Solid adsorbents[17] have the ability to remove mercaptans selectively without reducing gas heating value or specific gravity, as measured by laboratory and pilot plant studies. These adsorbents are: activated bauxite, Fuller's earth, Sovabead (form of silica gel), silica gel, and molecular sieves (synthetic zeolites).

Activated charcoal[12] removes mercaptans very effectively, but it also absorbs gasoline and LP-gases and decreases gas heating value and specific gravity to extent of hydrocarbon removal.

Mercaptans can be converted[17] to less odorous disulfides by oxidation with air and a suitable catalyst, such as active iron oxide.

Gasoline or LP-gases containing mercaptans removed from natural gas can be sweetened by a caustic wash followed by Perco copper sweetening or other similar processes.

PRODUCTION OF SULFUR

Chemicals made from gas include large quantities of sulfur produced each year from hydrogen sulfide mercaptans removed from sour natural gas. In 1957, there were 41 plants at work or under construction to perform this operation.

A Texas Gulf Sulphur Co. plant in Alberta Province was designed to handle 30 MMCF of raw gas per day, yielding 370 long tons of sulfur and 12.5 MMCF of residue gas per day.[1]

H_2S is converted to sulfur in two steps:

$$H_2S + 3/2\ O_2 \rightarrow SO_2 + HO_2$$

$$2H_2S + SO_2 \rightarrow 3S + 2H_2O$$

The basic overall reaction is $2H_2S + O_2 \rightarrow 2S + 2H_2O$.

REFERENCES

1. "Sour Gas Problems Are Under Attack." *Petrol. Week* 10: 35–6, June 24, 1960.
2. Bottoms, R. R. "Organic Bases for Gas Purification." *Ind. Eng. Chem.* 23: 501–4, May 1931.
3. Connors, J. S. and Miller, A. J. "Operating Problems in Gas Treating for Hydrogen Sulfide Removal." *Petrol. Process.* 5: 29–31, 1950.
4. Reed, R. M. and Wood, W. R. "Recent Design Developments in Amine Gas Purification Plants." *A.I.M.E. Trans.* 37: 363–84, 1941.
5. Souders, M., Jr. and Brown, G. G. "Design of Fractionating Columns." *Ind. Eng. Chem.* 26: 98–103, Jan. 1934.
6. Fluor Corp. *Total Nitrogen in Glycol-Amine Solutions.* (Lab. Test Method GA-9-51) and *Primary Amine Nitrogen in Glycol-Amine Solutions.* (Lab. Test Method GA-10-51).
7. Culbertson, L. and Connors, J. S. "Desulfurization and Dehydration of Natural Gas." *Petrol. Engr.* 24: D10+, Sept. 1952. Also in: *Oil Gas J.* 51: 114+, Aug. 11, 1952; 114+, Aug. 18, 1952.
8. Kohl, A. L. and Riesenfeld, F. C. *Gas Purification.* New York, McGraw-Hill, 1960.
9. Benson, H. E. and others. "Improved Process for CO_2 Absorption Uses Hot Carbonate Solutions." *Chem. Eng. Progr.* 52: 433–8, Oct. 1956.
10. Crow, J. H. and Dungan, J. R. "Hot Potash Process." *Oil Gas J.* 56: 97–9, Nov. 24, 1958.
11. Kohl, A. L. "Selective H_2S Absorption." *Petrol. Process.* 6: 26–31, Jan. 1951.
12. Vandaveer, F. E. "A Natural Gas Purification and Gasoline Plant." *A.G.A. Proc.* 1951: 514–26.
13. Haines, H. W., Jr. and others. "Regenerating Zeolites Used for Sulfur Removal." *Oil Gas J.* 59: 78–80, May 22, 1961.
14. Riesenfeld, F. C. and Mullowney, J. F. "Giammarco-Vetrocoke Process for Acid-Gas Removal." *Oil Gas J.* 57: 86–9, May 11, 1959. Also: *Petrol. Refiner* 38: 161–8, May 1959.
15. Kohl, A. L. and Miller, F. E. *Organic Carbonate Process for Carbon Dioxide.* (Patent 2,926,751) Washington, U. S. Patent Off., 1960.
16. Kohl, A. L. and Buckingham, P. A. "Fluor Solvent CO_2-Removal Process." *Oil Gas J.* 58: 146–56, May 9, 1960.
17. Oberseider, J. L. "Removal of Traces of Sulfur from Natural Gas." *A.G.A. Proc. Nat. Gas Dept.* 1950: 62–77.
18. "New Process Developed by Britons Produces Sulphur from Sour Gas for Less Than 1½ cents per M.c.f." *Oil Gas J.* 59: 94, Jan. 23, 1961.
19. Ryder, C. and Smith, A. V. *The Application of the Stretford Process to the Removal of Hydrogen Sulphide at High Pressure.* (Institution of Gas Engineers. Comm. 624) London, 1962.

Chapter 10

Nitrogen Removal from Natural Gas

by P. V. Mullins

NITROGEN CONTENT OF NATURAL GASES

Natural gas from some sources contains appreciable amounts of incombustible constituents—carbon dioxide, nitrogen, and helium. Of these, only nitrogen is present in large amounts of about 10 per cent or more.

Analyses[1] by the Bureau of Mines of natural gases from producing fields in the United States show widespread occurrence of nitrogen. A majority of the proved gas reserves contain less than 5 per cent nitrogen. Such a content is not objectionable in usual gas transmission and utilization. However, extensive gas reserves in the Texas Panhandle and Hugoton fields of Texas, Oklahoma, and Kansas contain large amounts of nitrogen. Another notable example is the Keyes field, Cimarron County, Okla.

According to Bureau of Mines analyses, gas from wells in the Texas Panhandle field contains an average of about 11.3 per cent N_2 (1.5 min and 33.6 max). Gas from wells in the Hugoton field contains an average of about 16.3 per cent N_2 (7.3 min and 33.6 max). These two fields are the sources of gas for several long cross-country pipelines which supply large industrial and domestic markets.

The Keyes field of Oklahoma, with reserves estimated at over half a trillion cubic feet, produces gas containing about 28 per cent N_2. Some small but appreciable gas reserves contain 30 to 80 per cent N_2 and have little or no market value.

Natural gas with high nitrogen content has the following disadvantages compared to low nitrogen gas:

1. Relatively low heating value, limiting utilization or requiring mixing with natural gas of higher heating value to obtain the necessary blended heating value.
2. Reduced thermal deliverability of a transporting pipeline because of appreciable volume of accompanying noncombustibles and less favorable flow characteristics of gas containing excess nitrogen.
3. Limitation on extraction of heavy hydrocarbons (gasoline and LP-gas) for separate sale because such extraction causes further heating value reduction.

BENEFITS FROM NITROGEN REMOVAL

A cooperative project[1] between the American Gas Association and the Bureau of Mines yielded an analysis and discussion of physical benefits from removing nitrogen and from removing both nitrogen and heavy hydrocarbons (propane plus). Table 4-38 shows such an analysis for a natural gas containing 15 per cent nitrogen.

Figure 4-71 (from the same report) shows the relationship between *Btu deliverability* increases of a pipeline, as a result of removing all nitrogen from gas having a composition of **Y** per cent nitrogen, (88 − **Y**) per cent methane, 6 per cent ethane, 4 per cent propane, and 2 per cent butane (plus). When **Y** equals 15 per cent, gas composition is identical to that indicated in Table 4-38. Also shown in Fig. 4-71 is the relationship between Btu deliverability increases as a result of removing propane and heavier hydrocarbons, in addition to nitrogen. Obviously, a wide variety of effects is possible, depending on gas composition and degree of removal of the specified constituents. Economic benefits that might be obtained depend on individual conditions—principally, amount of nitrogen removed, distance gas is transported, and ability to purchase and sell the fuel on a suitable basis.

PROSPECTIVE METHODS FOR NITROGEN REMOVAL

This cooperative work included reviews and reports on prospective gas separation methods for possible application to nitrogen removal.[2] Wade[3] also reviewed and discussed

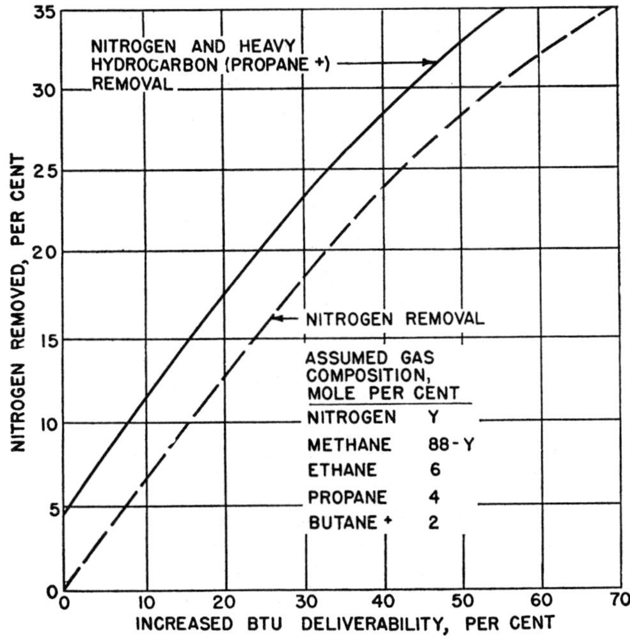

Fig. 4-71 Pipeline deliverability increase by nitrogen and heavy hydrocarbon removal.

Table 4-38 Computed Specific Gravities, Heats of Combustion, and Pipeline Deliverabilities of a Nitrogen-Bearing Natural Gas[1]

(unprocessed and processed)

			Before nitrogen removal			After nitrogen removal			After nitrogen and heavy hydrocarbon removal		
Constituent	Sp gr	Heat of combustion, Btu/cu ft	Composition,* %	G_B, sp gr	Heat of combustion, Btu/cu ft	Composition, %	G_A, sp gr	Heat of combustion, Btu/cu ft	Composition, %	G_F, sp gr	Heat of combustion, Btu/cu ft
Nitrogen	0.967	0	15.0	0.145	0	0	0	0	0	0	0
Methane	0.554	1013	73.0	.404	739.5	85.9	0.476	870.2	92.4	0.512	936.0
Ethane	1.038	1792	6.0	.062	107.5	7.0	.073	125.4	7.6	0.079	136.2
Propane	1.522	2590	4.0	.061	103.6	4.7	.072	121.7	0	0	0
Butane†	2.006	3370	2.0	0.040	67.4	2.4	0.048	80.9	0	0	0
Totals			**100.0**	**0.712**	**1018.0**	**100.0**	**0.669**	**1198.2**	**100.0**	**0.591**	**1072.2**

	After nitrogen removal		After nitrogen and heavy hydrocarbon removal	
(a) Increased heat of combustion, per cent	17.70		5.32	
(b) Decreased specific gravity, per cent	6.04		16.99	
(c) Decreased $(1-\sqrt{G_2/G_1})100$, per cent	3.07	(where $G_2 = G_A$ and $G_1 = G_B$)	8.89	(where $G_2 = G_F$ and $G_1 = G_B$)
Increased pipeline deliverability (volumetric):				
(d) Increase from reduced specific gravity, $(\sqrt{G_1/G_2}-1)100$, per cent	3.16		9.76	
(e) Increase from reduced friction (est.), per cent	0.40		0.70	
(f) Increase from reduced compressor fuel, per cent†	1.10		0.30	
(g) Total increase: $[(100 + \mathbf{d})(100 + \mathbf{e})/100] + \mathbf{f} - 100$, per cent	4.67		10.83	
Increased pipeline deliverability (Btu):				
(h) $[(100 + \mathbf{a})(100 + \mathbf{g})/100] - 100$, per cent	23.19		16.73	
Recovery of hydrocarbons, gal per MCF of gas processed:				
(i) Propane	...		1.16 gal	
(j) Butane†	...		0.65 gal	

* The possible presence of small quantities of carbon dioxide, hydrogen sulfide, and helium is disregarded.
† Overall fuel requirements before nitrogen removal are assumed to be 8 per cent.

prospective methods. To be satisfactory, a method must be practical and economical in processing for transmission (preferably at elevated pressures) large volumes of nitrogen-bearing gas to the extent of several hundred MMCF per day. Further, for efficiency and economy, a highly effective process, capable of removing virtually all nitrogen, is desirable.

A desirable choice, if available, would be use of an absorbent or chemical reactant that would permit selective absorption and removal of nitrogen. Unfortunately, there are very few such possibilities because few substances afford the necessary affinity for nitrogen under practical application conditions. A patent was issued[4] on a method using lithium–amalgam to react selectively and remove nitrogen and then provide recovery of both lithium and nitrogen in a continuous regenerative process. Liquid sulfur dioxide was reported[5] to have a relatively high absorptive capacity for nitrogen, and a patent was issued[6] for removing nitrogen by selective absorption in sulfur dioxide. However, later investigations indicated a very low absorptive capacity of sulfur dioxide for nitrogen.[7] A patent was also issued[8] for a method in which nitrogen-bearing natural gas is scrubbed with ammonia, and the nitrogen selectively absorbed and removed in a continuous regenerative process.

A promising gas-separation method, possibly suitable for nitrogen removal, has been developed using adsorption and fractionation of a gas mixture on a moving bed of activated carbon in a continuous regenerative cycle. Information is available[9–17] on the *Hypersorption* process, a method of this kind. Similar processes using the same general method have been proposed.[18–22]

LOW TEMPERATURE PROCESSING

The low-temperature separation method, involving liquefaction and fractionation, has received great attention and most favorable consideration. Fundamental data and other information on low-temperature processing cycles for separation of gaseous mixtures have been presented.[23,24] The low-temperature processes used for extracting helium from natural gas,[25,26] a natural gas liquefaction plant,[27] an ethane-extraction plant,[28] and tonnage oxygen plants[29–32] illustrate large-scale industrial applications of low-temperature processing. Numerous patents have been issued on liquefaction and fractionation methods, mostly for oxygen production from air. References to many such patents are listed in bibliographies of the articles in References 1 and 2.

A number of low-temperature processing cycles for nitrogen removal from natural gas have been proposed.[33–40] Virtually all are modifications and refinements of a conventional basic cycle (Fig. 4-72) for low temperature gas separation. Compressed nitrogen-bearing natural gas, at about 600 psig, is cooled and liquefied by heat interchange with regasifying natural gas, or components, previously liquefied in the cycle. The feed gas is then fractionated. Most cycles contemplate reduction of pressure of the feed stream to 200–400 psig in order to: (1) accommodate favorable fractionating conditions; (2) secure Joule–Thomson cooling to establish suitable temperature differentials for satisfactory heat exchange; and (3) provide a regenerative cycle which supplies part of refrigeration needed to support it. Additional low-level refrigeration is required for condensation of needed nitrogen reflux for frac-

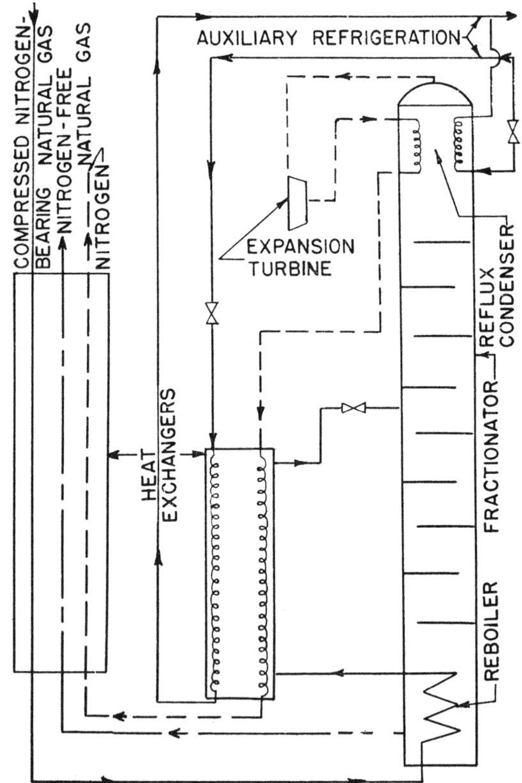

Fig. 4-72 Basic low temperature nitrogen removal cycle. (Note that the two upper horizontal lines are the outlet and inlet, respectively, for auxiliary refrigeration.)

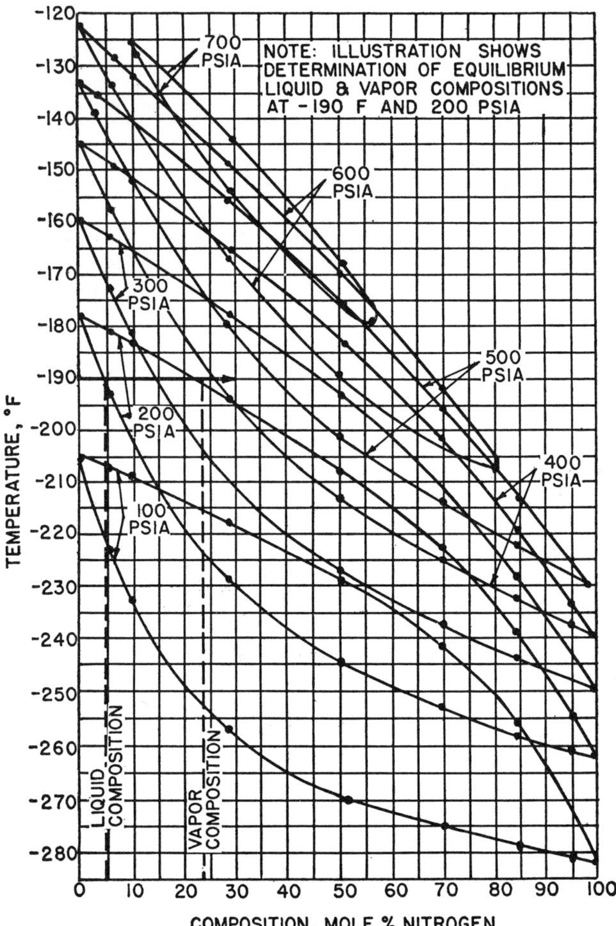

Fig. 4-73 Isobaric temperature-composition diagrams, methane–nitrogen system.

tionation, and for final cooling of the feed gas before it enters the fractionator. This additional or auxiliary refrigeration may be obtained by one of the following means, depending upon conditions (given in order of probable preference and economy):

1. Isentropic expansion of the overhead nitrogen removed from the fractionator, possibly supplemented by an auxiliary cycle also using isentropic expansion of compressed nitrogen.[37,40] (This method is especially favorable when the amount of nitrogen removed is relatively large.)
2. Throttling to low pressure and gasifying part of the liquid bottom product from the fractionator, chiefly methane, to secure cooling and condensation of liquid nitrogen for reflux.[38] (Use of this method depends upon obtaining satisfactory fractionation at 300–400 psig to permit reflux condensation with methane at low pressure.)
3. Direct provision of supplemental refrigeration by a conventional cascade system.[23]
4. Some combination of 1, 2, and 3.

Suitable modifications and additions to the cycle may permit separation and recovery of other selected constituents such as helium, ethane, and LP-gas products. Use of restricted amounts of reflux would tend to convert the fractionator to a simple stripping column, and reduce refrigeration and power requirements—but it would result in a lower purity (methane-bearing) overhead nitrogen.

Fundamental data on physical–chemical properties of methane–nitrogen mixtures at low temperatures[41] are valu-

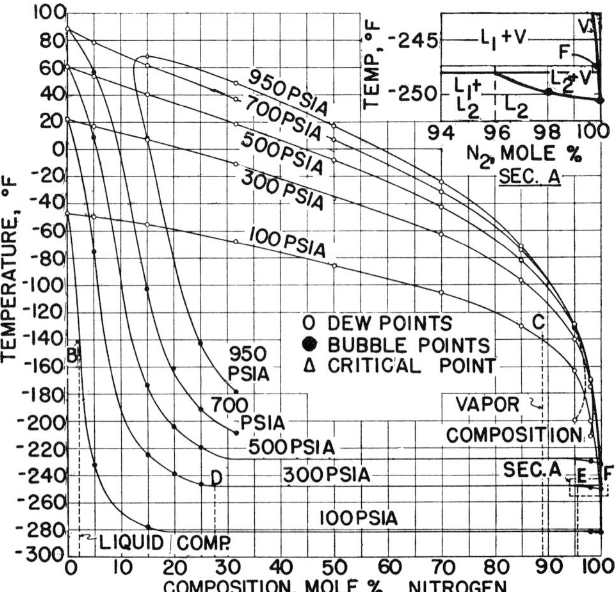

Fig. 4-74 Isobaric temperature-composition diagrams, ethane–nitrogen system.

able in process calculations and equipment design, insofar as the separation required is primarily between methane and nitrogen, chief constituents of typical nitrogen-bearing natural gas. Figure 4-73 shows isobaric temperature-composition diagrams for the methane–nitrogen system as developed from experimental work.

Mollier chart for nitrogen will be found in Fig. 8-44; compressibility chart[42] for nitrogen in Fig. 1-4; and isobaric temperature-composition diagrams[43] for an ethane–nitrogen system in Fig. 4-74.

Cost estimates for N_2 removal plants for gas containing 16 to 27 per cent N_2 by low temperature fractionation are available.[44] For a five-year payout, raw gas price would have to be 7.2 cents per MCF—far below 1961 field prices.

REFERENCES

1. Mullins, P. V. and Wilson, R. W. "Prospective Benefits from Removing Excess Nitrogen from Natural Gas." *A.G.A. Proc.* 1948: 601–12.
2. ——. *Prospective Methods and Estimated Costs for Removing Excess Nitrogen from Natural Gas.* (Project NGD-6) New York, A.G.A., 1952.
3. Wade, H. N. "Removal of Nitrogen from Natural Gas." *Oil Gas J.* 48: 195–200, June 23, 1949. Also: *Mines Mag.* 38: 65+, Dec. 1948.
4. Rohrman, F. A. *Removal of Nitrogen from Mixtures of Combustible Gases.* (Patent 2,660,514) Washington, U. S. Patent Off., 1953.
5. Dornte, R. W. and Ferguson, C. V. "Solubility of Nitrogen and Oxygen in Liquid Sulfur Dioxide." *Ind. Eng. Chem.* 31: 112–3, Jan. 1939.
6. Latchum, J. W., Jr. *Method for Removing Noncombustibles from Fuel Gas.* (Patent 2,448,719) Washington, U. S. Patent Off., 1948.
7. Dean, M. R. and Walls, W. S. "Solubility of Nitrogen and Methane in Sulfur Dioxide." *Ind. Eng. Chem.* 39: 1049–51, Aug. 1947.
8. Latchum, J. W., Jr. *Absorption of Nitrogen by Liquid Ammonia.* (Patent 2,521,233) Washington, U. S. Patent Off., 1950.
9. Berg, C. "Hypersorption: A Process for Separation of Light Gases." *Gas* 23: 32–7, Jan. 1947.
10. Berg, C. and others. "Hypersorption: Process for Separation of Gases and Vapor." *Oil Gas J.* 47: 95+ Ap. 28, 1949.
11. *Ibid. Petrol. Refiner* 28: 113–20, Nov. 1949.
12. Berg, C. H. O. *Adsorption Process.* (Patent 2,519,344) Washington, U. S. Patent Off., 1950.
13. *Ibid.* (Patent 2,519,342) Washington, U. S. Patent Off., 1950.
14. Berg, C. H. O. *Adsorption Process and Apparatus.* (Patent 2,519,343) Washington, U. S. Patent Off., 1950.
15. —— and others. *Adsorption Process.* (Patent 2,545,067) Washington, U. S. Patent Off., 1951.
16. Imhoff, D. H. *Adsorption Process and Apparatus.* (Patent 2,545,850) Washington, U. S. Patent Off., 1951.
17. Berg, C. H. O. *Adsorption Process and Apparatus.* (Patent 2,548,192) Washington, U. S. Patent Off., 1951.
18. Gilliland, R. R. *Fractionation of Gases with Adsorbents.* (Patent 2,495,842) Washington, U. S. Patent Off., 1950.
19. Robinson, S. P. *Separation of Gases.* (Patent 2,527,964) Washington, U. S. Patent Off., 1950.
20. Small, J. K. *Fractionation with Solid Adsorbents in a Single Column.* (Patent 2,548,502) Washington, U. S. Patent Off., 1951.
21. Brandt, P. L. *Separation by Adsorption.* (Patent 2,493,911) Washington, U. S. Patent Off., 1950.
22. Groebe, J. L. and Karbosky, J. T. *Removal of Nitrogen from Hydrocarbon Gases.* (Patent 2,598,785) Washington, U. S. Patent Off., 1952.
23. Ruhemann, M. *Separation of Gases,* 2nd ed. London, Oxford U. Pr., 1949.
24. Davies, M. *Physical Principles of Gas Liquefaction and Low Temperature Rectification,* p. 205. London, Longmans Green, 1949.
25. Mullins, P. V. "Helium Production Process." *Chem. Eng. Progr.* 44: 567–72, July 1948.
26. Cattell, R. A. and Wheeler, H. P., Jr. "Growing Demand Is Seen for Helium." *Petrol. Refiner* 30: 91–4, Mar. 1951.
27. Clark, J. A. and Miller, R. W. "Liquefaction, Storage and Regasification of Natural Gas." *Oil Gas J.* 39: 48+, Oct. 17, 1940. Also: *Am. Gas J.* 153: 52–5, Nov. 1940. Also: *Gas Age* 86: 46–50, Oct. 24, 1940.
28. King, J. J. and Mertz, R. V. "Tennessee Gas Begins Operation of Refrigerated Extraction Plant." *Petrol Refiner* 31: 118–24, Mar. 1952. Also: *Oil Gas J.* 50: 95+, Mar. 10, 1952. Also: *Petrol. Engr.* 24: D3+, Mar. 1952. Also: *Gas* 28: 128+, Ap. 1952.
29. Lobo, W. E. "Low Pressure Oxygen Process." *Petrol. Engr.* 18: 120+, May 1947. Also: *Iron Age* 160: 49–53, July 17, 1947.
30. Sherwood, P. W. "Tonnage Oxygen Today." *Chem. Eng.* 56: 97–100, Dec. 1949.
31. Roberts, I. "Economics of Tonnage Oxygen Production." *Chem. Eng. Progr.* 46: 79–88, Feb. 1950.
32. ——. "Tonnage Oxygen Plants." *Refrig. Eng.* 60: 950+, Sept. 1952.
33. Claude, G. and others. *Treatment of Natural Gases.* (Patent 1,497,546) Washington, U. S. Patent Off., 1924.
34. Deschner, W. W. and Bodle, W. W. "Nitrogen Removal from Natural Gas." *Oil Gas J.* 46: 76+, Ap. 15, 1948; 92+, Ap. 22.
35. Gilmore, F. R. *Treatment of Natural Gas.* (Patent 2,457,957) Washington, U. S. Patent Off., 1949.
36. Bodle, W. W. aud Deschner, W. W. *Method of Treating Natural Gas.* (Patent 2,557,171) U. S. Patent Off., 1951.
37. Cost, J. L. *Separation of Natural Gas Mixtures (Removal of Nitrogen)* (Patent 2,583,090) Washington, U. S. Patent Off., 1952.
38. Mullins, P. V. *Separating Gaseous Mixtures.* (Patent 2,595,-284) Washington, U. S. Patent Off., 1952.
39. Gilmore, F. E. *Separation of Petroleum Well Gases.* (Patent 2,603,310) Washington, U. S. Patent Off., 1952.
40. Miller, B. *Transportation of Natural Gas.* (Patent 2,658,360) Washington, U. S. Patent Off., 1953.
41. Bloomer, O. T. and Parent, J. D. *Physical-Chemical Properties of Methane-Nitrogen Mixtures.* (Research Bul. 17) Chicago, I.G.T., 1952.
42. Bloomer, O. T. and Rao, K. N. *Thermodynamic Properties of Nitrogen.* (Research Bull. 18) Chicago, I.G.T., 1952.
43. Eakin, B. E. and others. *Physical-Chemical Properties of Ethane-Nitrogen Mixtures.* (Research Bull. 26) Chicago, I.G.T., 1955.
44. Burnham, J. G. "Which Nitrogen-Removal Method is Best?" *Oil Gas J.* 55: 143, July 8; 120, July 15; 129, Aug. 12, 1957.

SECTION 5

PRODUCTION AND HANDLING OF LIQUEFIED PETROLEUM GASES

H. Emerson Thomas, *Section Chairman,* Chapter 5, Thomas Associates, Inc., Westfield, N. J.
E. Martin Anderson, Chapter 4, MECAW Industries, Portland, Me.
E. W. Evans, Chapters 2 and 3, Phillips Petroleum Co., Bartlesville, Okla.
M. G. Farrar (retired), Chapter 1, Union Carbide and Carbon Corp., New York, N. Y.
W. R. Fraser (deceased), Chapter 4, Michigan Consolidated Gas Co., Detroit, Mich.
Walter H. Johnson, Chapter 4, National LP-Gas Association, Chicago, Ill.
R. W. Miller (retired), Chapter 3, Consolidated Natural Gas Co., Pittsburgh, Pa.
C. George Segeler, Chapter 4, American Gas Association, New York, N. Y.
John L. Turnan, Chapter 6, Worcester Gas Light Co., Worcester, Mass.
F. E. Vandaveer (retired), Chapters 4 and 6, Con-Gas Service Corp., Cleveland, Ohio
A. E. Wastie, Chapter 5, Drake & Townsend, Inc., New York, N. Y.
A. H. Withrow, Chapter 4, The Verkamp Corp., Cincinnati, Ohio

CONTENTS

Chapter 5—Design of LP-Gas Plants for Utility Companies

Chapter 6—Operation and Safety of Utility LP-Gas Plants

Tables

Figures

Chapter 1
History of the LP-Gas Industry

by M. G. Farrar

Liquefied petroleum gas is essentially a twentieth century development. Long before that time, however, use had been made of gas compressed in containers. Limited amounts of this product were sold in cylinders in England[1] as early as 1810. In 1870 Pintsch gas was developed and later used in railway car lighting. Around 1907, **blaugas**, formed by cracking oil, was liquefied by compressing it to about 1800 psig. Several companies were formed for distributing this gas in containers in the United States, despite attendant difficulties such as high transportation costs and complexity of equipment needed for its utilization.

Experience with these early fuels compressed in containers indicated that full development of such a product depended on its compliance with the following essential requirements: operation at comparatively low pressure, constancy of chemical composition, availability in large quantities, and reasonable price.

Total industry sales as first reported by the Bureau of Mines[2,3] for 1922 were 223,600 gal. Table 5-1 gives distribution of U. S. sales for major uses and Table 5-2 gives distribution of LP-gas sales among propane, butane, and propane–butane mixtures.

DEVELOPMENT PERIOD

Standards for handling LP-gases were first published in 1932. Requirements are available (NFPA No. 58)[5] for all LP-gas distribution, except use by utility companies which is covered in NFPA No. 59.[4] The latter standard was first issued in 1949 as a result of the cooperative efforts of the NFPA Committee on Gases and the A.G.A. LP-Gas Utility Code Committee.

Odorization of LP-gas began in 1934 as the result of industry research conducted with assistance of the U. S. Bureau of Mines. Ethyl mercaptan was used as the odorizing agent. LP-gas ordinarily used for domestic and commercial service is now odorized as required by NFPA Standard 58:[5]

> B.1 Odorizing Gases
>
> (a) All liquefied petroleum gases shall be effectively odorized by an approved agent of such character as to indicate positively, by distinct odor, the presence of gas down to concentration in air of not over one-fifth the lower limit of flammability. Odorization, however, is not required if harmful in the use or further processing of the liquefied petroleum gas, or if odorization will serve no useful purpose as a warning agent in such use or further processing.
>
> (b) The odorization requirement of B.1 (a) shall be considered to be met by the use of 1.0 lb of ethyl mercaptan, 1.0 lb of thiophane, or 1.4 lb of amyl mercaptan per 10,000 gal of LP-gas. However, this listing of odorants and quantities shall not exclude the use of other odorants that meet the odorization requirements of B.1 (a).

Appliances

Domestic appliances for city gas (natural and manufactured) which comply with national safety standards, as evidenced by display of A.G.A. Laboratories Approval Seal, have been available since 1925.

Almost from the start of the Approval Requirements program, the A.G.A. Laboratories have tested and certified equipment for LP-gas in the same categories as for natural gas.

Utility Uses

The first recorded replacement of manufactured gas by LP-gas–air was in Linton, Ind., in 1928. Numerous other systems, replacements and new, diluted LP-gas and undiluted, were installed in the next thirty years (Table 5-3). Only the rapid growth of natural gas reversed this trend. Other utility

Table 5-1 LP-Gas Sales in United States[2]

(millions of gallons)

Year	Total	Domestic and commercial	Gas manufacturing	Industrial	Synthetic rubber	Chemical plants	Internal combustion	All others
1950*	3483	2022	252	217	228	624	130	9
1955	6123	2801	214	423	406	1493	652	32
1960	9545	4225	157	439	539	3019	897	58
1961	9798	4318	169	402	520	3239	880	51
1962	10729	4713	173	425	587	3571	932	55

* See Table 2-3 for data on earlier years.

Table 5-2 Butane, Propane, and Butane–Propane Mixture Sales in United States[2]

(millions of gallons)

Year	Butane	Propane	Butane–propane mixtures
1940	77	109	123
1950	568	1938	976
1955	724	3261	1429
1960	1100	5744	1094
1961	1066	5936	1107
1962	1443	6464	1075

Table 5-3 LP-Gas Customers of Utilities by Class of Service*

(yearly averages, in thousands)

Year	Total	Resi-dential	Com-mercial	Indus-trial	Other
1946–1950 (average)	270	246	23	1	†
1955	241	219	22	†	‡
1960	125	113	12	‡	‡
1961	98	88	10	‡	‡
1962	88	79	9	‡	‡
1963(prelim)	72	65	7	‡	‡

* A. G. A. *Gas Facts.* New York, annual. Excludes data for Alaska; includes data for Hawaii after 1959.

† Less than 500 customers.

‡ Less than 50 customers.

applications included use as a substitute for gas oil, for cold enrichment of water gas, for catalytic cracking or reforming to make manufactured gas, for stand-by use in peak load shaving on natural gas systems, and for augmenting or replacing normal gas supplies by other means.

REFERENCES

1. *Handbook, Butane-Propane Gases,* 3rd ed. Los Angeles, Western Business Papers, 1942.
2. U. S. Bur. of Mines. *Minerals Yearbook,* Washington, G.P.O. (annual).
3. Coumbe, A. T. and Avery, I. F. *A Third of a Century of LP-Gas Sales, 1922–52.* Washington, U. S. Bur. of Mines, 1954. (I.C. 7684).
4. National Fire Protection Assn. *Standard for the Storage and Handling of Liquefied Petroleum Gases at Utility Gas Plants.* (NFPA No. 59) Boston, 1963.
5. ——. *Standard for the Storage and Handling of Liquefied Petroleum Gases.* (NFPA No. 58) Boston, 1963.

Chapter 2
Liquid Hydrocarbon Production

by E. W. Evans

About 75 per cent of the LP-gas marketed is extracted from natural gas, and the remainder comes from refinery gas and liquid streams resulting from processing of crude oil. Liquid hydrocarbons recovered from natural gas consist primarily of liquefied petroleum gases (mainly propane and butanes) and natural gasoline (mainly pentanes thru octanes). *Natural gasoline* is so termed to distinguish it from *gasoline* distilled or cracked from crude oil. If the liquefiable hydrocarbon content of a natural gas exceeds 0.3 gal per MCF, it is both necessary and worthwhile to extract this fraction; extractions of contents under 0.1 gal per MCF are generally not profitable.

Some typical analyses of Appalachian field natural gasolines are:

	Reid vapor pressure, psia			
	26	22	18	14
Gravity at 60 F, °API	90.0	88.0	86.0	83.5
Butanes	36	25	15	4
Pentanes	29	35	40	45
Hexanes	18	21	24	27
Heptanes	12	13	14	16
Octanes	5	6	7	8
Total, %	100.0	100.0	100.0	100.0

The following covers natural gas with respect to its propane, butane, and higher hydrocarbon contents: *dry*—less than one gallon per Mcf; *condensate*—usually contains 0.7 to 1.2 gal per Mcf; and *wet*—may contain in excess of ten gallons per Mcf.

TESTS FOR GAS HYDROCARBON CONTENT

Liquid hydrocarbons in gas are usually reported in **GPM**—the gallons of natural gasoline per thousand standard cubic feet of natural gas.

The **Standard Compression test,**[1] also known as the **test car method,** is the oldest to have any widespread use. Here the gas sample was compressed to around 250 psig and cooled with water or in an ice bath. Any liquid so condensed was measured, tested for specific gravity, and sometimes specially examined in other ways. The results gave a rough indication of the amount and quality of products which might be recovered. This method is now used only where required by old contracts.

The **charcoal adsorption test**[1 2] is commonly used for proration of product among various sources feeding a natural gasoline extraction plant, and for rough preliminary determinations of potential production of liquids from new gas sources. The equipment used is in principle simply a miniature charcoal adsorption plant. A measured volume of gas is passed at a controlled rate thru a standard sized bed of coconut charcoal. The gasoline adsorbed is distilled from the charcoal by heating, and the vapors are largely condensed in an iced condenser. The quantity and quality of the liquid recovered give a fairly good indication of the value of the liquefiable content of the gas. Specifications for testing and reactivating charcoal and determining the water vapor content of the gas are also available.[2]

Low temperature fractional distillation[3] (also known as **fundamental analysis**) is a precision method which cannot compete favorably with the charcoal test because of the expensive apparatus required for the distillation. Gas sample is fed into the bottom of a tiny glass fractionating column, heated at the base by an electric heater, and refluxed from the top by a condenser cooled by liquid air (or more often by liquid oxygen because of its availability). By accurately controlling the reflux condenser temperature, one component after another is drawn out of the column by vacuum and measured by the rise in pressure in the evacuated receivers.

A **mass spectrometer**[4–8] ionizes gas molecules at low pressure and sorts them according to their mass. It is an extremely fast method requiring only a very small gas sample (0.2 cu cm at 1 atm). However, equipment cost (over $30,000 in 1961) precludes its widespread use.

Chromatographic analysis became a commercially accepted means of evaluating gas sources and apportioning royalties around 1960.[9]

LP-GASES FROM REFINERY SOURCES

Oil refinery operations like crude oil distillation, thermal and catalytic cracking, and thermal and catalytic reforming,[10] produce some lighter hydrocarbons, including the saturated group—propane and butanes, and the unsaturates—propylene and butylenes. The presence of unsaturated hydrocarbons in an LP-gas (principally a small amount of propylene) is an indication that the gas originated in a refinery. About 25 per cent of LP-gases sold is obtained from refineries. Although the total LP-gas yield may run 15 per cent of the crude oil run,

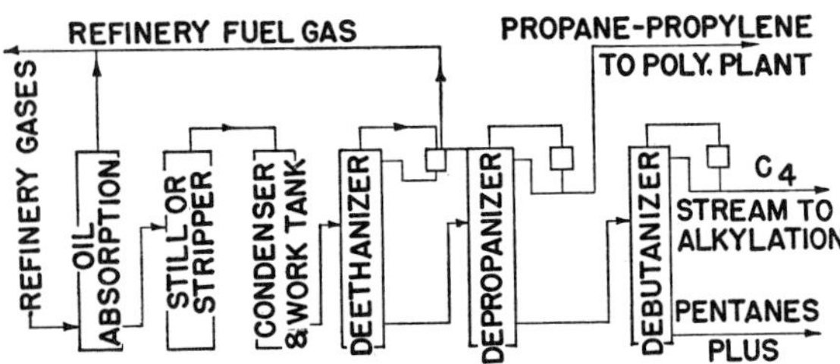

Fig. 5-1 Flow chart for typical refinery gas plant.

less than 4 per cent of the crude becomes available for the LP-gas market. Recovery employs the same processes used in natural gasoline plants, usually separation of components by oil absorption (Fig. 5-1). Since the refinery gases are richer in LP-gas than are natural gases, lower absorber pressures are used.

ECONOMICS OF LP-GAS RECOVERY

The recovery of LP-gas, natural gasoline, gasoline from oil, and other refined oil products, and in many cases the requirements of the natural or refinery gas from which they are extracted, should all be coordinated. It is economically desirable to retain the butanes as much as possible in natural gasoline and refinery gasoline, since their value in motor fuel exceeds that in the LP-gas market. During World War II, large volumes of normal butane were isomerized to *iso*-butane, to be used in making alkylates for aviation gasoline. Butylenes in refinery gases were the other major feed to alkylation processes, although propylene and other unsaturates were also alkylated with *iso*-butane. Catalytic polymerization of butylenes and propylene was also used to form aviation gasoline components. During the same period, a new industry—synthetic rubber—required dehydrogenation of normal butane or butylenes to make butadiene. As a result of higher value uses for propylene, butanes, and butylenes, practically all LP-gas from refineries consisted of propane containing a small percentage of propylene which was not reacted in polymerization.

Most of the recently built natural gasoline plants have provided equipment for deep extraction of propane. Since they deliver residue gas to pipelines at fairly high pressures—500 to 1000 psi—their absorbers can be operated at these pressures without additional compression costs. Partly as a result of tremendous increases in demand for natural gas, LP-gas recovery capacity has increased faster than the rapidly growing LP-gas market. Since additional large volumes could be extracted by existing plants, there appears little chance of a shortage of LP-gas in the foreseeable future.

RECOVERY PROCESSES

By 1957, the absorption process was in almost universal use, although the compression and adsorption processes were also used. The latter processes are not as selective as absorption and cannot be used on sour gases without extensive treatment before processing.

Retrograde condensate, liquid formed by isothermal expansion of a single-phase fluid,[11] occurs at high underground pressures (e.g., 2000 psi). To avoid loss of these hydrocarbon condensates to the formation, residue gas is returned to the field from the cycling plant to drive the retrograde to the producing wells. Figure 5-2 shows the extraction of LP-gas and heavier hydrocarbons from condensate by means of flashing, as well as the preparation of the residue gas for recycling.

The **phase relations of gas condensate fluids** "can be useful in most recovery operations that require mixture of natural gas or processed gas with reservoir oil. These correlations can be used for computing the compressibility of gases, the pressure and temperature of the critical state,

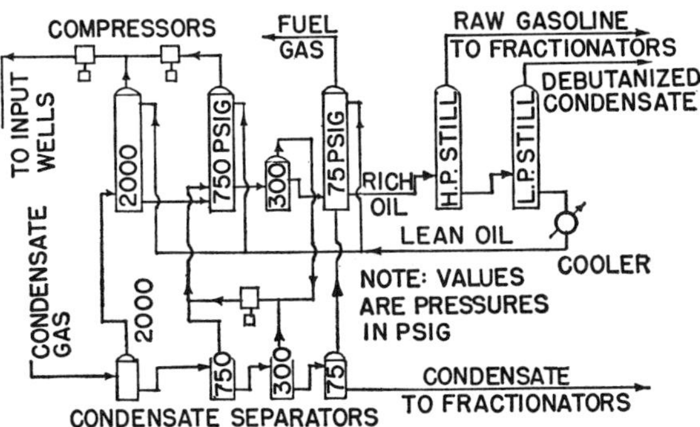

Fig. 5-2 Flow chart for LP-gas recovery section of condensate field cycling plant.

and the dew points and liquid–gas ratios of phase diagrams. These properties are important in an operation that involves the return of gas to a reservoir to obtain gaseous solutions or mobile liquids that will flow thru the formation to producing wells. As gas-condensate fluids represent one-fourth or more of the domestic reserves of gas and as nearly one-half of the oil discovered each year is likely to be abandoned ultimately unless improved means are found to move it from its place in the formation to producing wells, the results of research presented in the monograph (Bureau of Mines Monograph 10) can have many applications."[12]

Compression and Cooling

In 1905 an operator, Mayberg, collected liquefiable hydrocarbons, then called **drip gasoline** or **casinghead gasoline,** by compressing natural gas to 60 psig and cooling it in pipe coils. This process developed rapidly (Fig. 5-3)

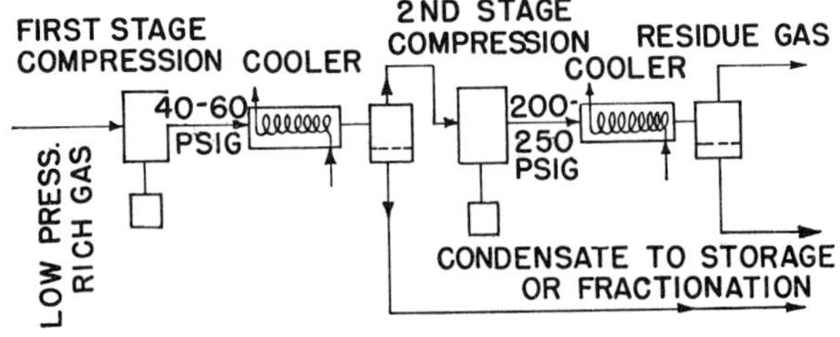

Fig. 5-3 Recovery of natural gas liquids by compression.

and was fairly efficient with very *wet* gases at pressures up to and over 250 psig. In some cases, cooling by means of spray water over coils was augmented by refrigeration. One plant, processing pipeline natural gas to recover ethane and heavier hydrocarbons, employed a pressure of 530 psig and a temperature of −105 F. To avoid ice formation in its heat exchangers, incoming gas to this plant was thoroughly dried with a solid desiccant. Conventional fractionation then separated the recovered liquids into ethane for petrochemical plant feed stock, propane for LP-gas, and butane for LP-gas or motor fuel blending. With this noteworthy exception, compression plants play a very minor part in LP-gas production. The more efficient absorption process displaced recovery by compression and cooling during the 1920's.

Adsorption by Charcoal

In 1918, experimental gasoline extraction plants were built utilizing the ability of activated charcoal and similar materials to selectively adsorb relatively heavy vapors from mixtures of gases and vapors. Figure 5-4 shows recovery of LP-gas in a continuous process. Most earlier plants consisted essentially of three vessels, each containing a bed of granulated coconut charcoal, apparatus to generate and superheat steam for distilling adsorbed material from the charcoal, and condensers to liquefy adsorbed materials. Raw gas was passed thru one cold charcoal bed, while a second bed was being cooled with treated gas and a third was being distilled (reactivated).

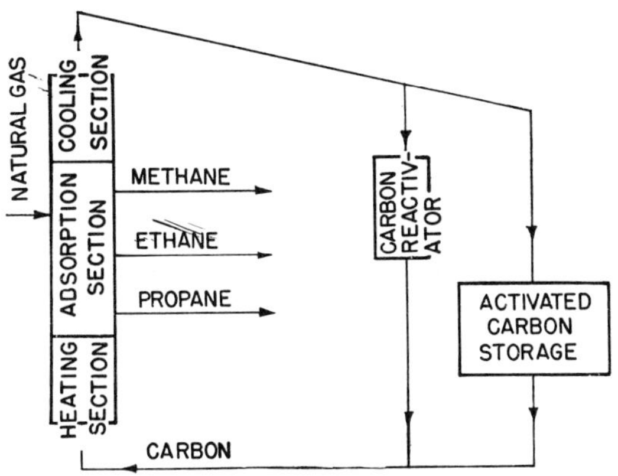

Fig. 5-4 Recovery of LP-gas by adsorption.

This type of plant was quite efficient and mechanically simple (few moving parts). However, charcoal depreciated in service—apparently because of plugging of its pores by heavy hydrocarbons and sulfur compounds. Since charcoal was expensive, overall operating costs of adsorption plants, including periodic charcoal renewal, were generally high. Nevertheless, at least one company[13] has successfully used the same batch of charcoal for 19 years without deterioration. Attempts to substitute silica gel for coconut charcoal were unsuccessful because the gel tended to break down mechanically in service.

In the *Hypersorption* process (Union Oil Co. of California) charcoal is circulated thru a continuous adsorption and distillation process, instead of remaining in stationary beds. Although this process affords extremely high extraction efficiencies at moderate pressures (up to perhaps 100 psig), high fuel and charcoal costs generally have kept this *process noncompetitive in the natural gasoline industry.* Attempts to operate at pressures above about 200 psi make thorough stripping of the charcoal difficult. At usual pipeline pressures of 450 to 800 psi, clean stripping is impossible, and the method cannot be used for efficient recovery of hydrocarbons heavier than butane. However, Hypersorption has been used to a considerable extent in extraction of certain gases, notably ethylene, from off-gas produced by cracking and other types of chemical conversion. More than 40 charcoal plants were in operation in this country in 1957.

Absorption by Oil

This process extracts liquid hydrocarbons from natural gas by absorbing them in an oil stream, then distilling them from the oil by heating and finally condensing them. The fractions having the lowest vapor pressures and the highest molecular weights are the most readily absorbed. Used as early as 1875 to recover benzene and similar materials from manufactured gas made in coke ovens or shale retorts, an absorption process was not applied to natural gas until the Hope Natural Gas Co. built an absorption plant at Hastings, W. Va., in 1913. This recovery process has survived competition.

Oil absorption is subject to many variations. Most of its complications arise in the techniques for retaining desirable components after their extraction in the primary absorbers.

Inlet pressure of field gas to absorber is determined by compression costs, absorber rating, required residue or field maintenance gas pressure, and the desired extent of recovery of liquids. Absorption of LP-gas components is materially increased as pressure is increased, but so is compression cost.

Since compression increases gas temperature, and the absorption efficiency varies inversely with temperature, the compressed gas is cooled either by water or by refrigeration before entering the absorber. Any hydrocarbon liquids condensed by cooling are usually sent directly to the fractionating unit, by-passing the absorber. If the feed gas is cooled by refrigeration below 60 F, it may be necessary to dehydrate it to avoid hydrate or ice formation.

Compressed and cooled field gas enters the low portion of the absorber column and passes upward thru a descending flow of absorber oil. The column contains from 20 to 30 trays or bubble plates (spaced 20 to 24 in. apart) to insure intimate gas-oil contact. It may also contain intercoolers to remove heat of absorption which would otherwise reduce absorption efficiency. Intercoolers are usually placed in the upper portion of the column.

The following factors favor absorption of propane and higher hydrocarbon vapors by oil:

Higher Oil Circulation Rate. This requires extra capacity and fuel for pumping, stripping, and cooling. However, it also results in a greater than proportionate increase in extraction of methane and ethane. These light hydrocarbons, if removed in the fractionation unit, often carry with them part of the increased propane extracted.

Lower Absorption Temperature. This is accomplished by adding more lean oil cooling coils or by supplementing water cooling of the absorption oil with refrigeration, and results in the selective absorption of propane as compared to lighter components. Cooling may be applied directly to the lean oil feed, to the absorber, or thru intercoolers located in its upper portion. Lowering the lean oil temperature results in a lower residue gas temperature. Since a 20°F reduction in residue gas temperature would approximately halve its moisture content, investment, and operating expenses of a gas dehydrator would be minimal if the residue gas meets pipeline moisture limitations. If temperatures are sufficiently low, gas dehydration is necessary to prevent freezing and hydrate formation in absorbers.

If it is not beneficial to lower the residue gas temperature, the intercooler method is more efficient than refrigerating the lean oil because it directs most of the cooling to lowering the effective absorption temperature with less loss of refrigeration to the residue gas. At high absorber pressures, most of the absorption oil temperature rise occurs near the absorber top.

Lower Molecular Weight Absorption Oil. This results in more oil molecules per gallon, and thus proportionately increases the absorption coefficient and conversely lowers the stripping coefficient. Other considerations such as oil entrainment, oil retrograde losses, and costs also influence oil selection. The maintenance of optimum molecular weight requires an efficient oil reclaimer. Low absorber temperatures permit use of oil having molecular weights too volatile for use at ordinary temperatures.

Higher Absorption Pressure. Higher compression is usually not the most economical way to obtain increased

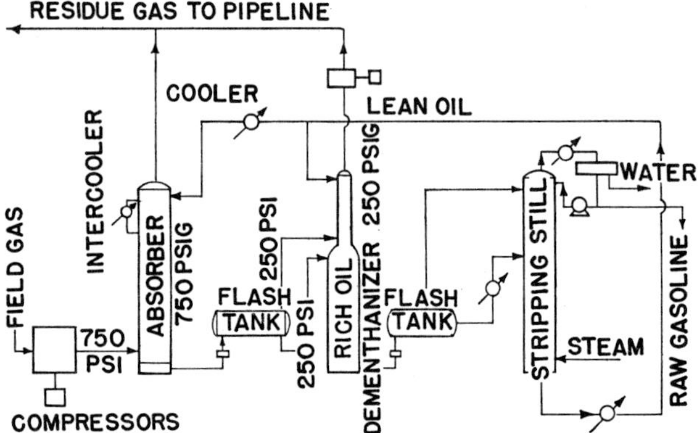

Fig. 5-5 High propane recovery from natural gas by means of absorption and distillation.

propane recovery. However, since treated gas is ordinarily delivered into lines operating at 400 to 1000 psig, economic studies have usually favored compressing *wet* gas to line pressure before absorption. Figure 5-5 illustrates use of high absorption pressure, low absorption temperature, and a fractionating absorber for high propane recovery from natural gas. A low molecular weight oil may also be used in this system.

Some Complete Absorption Systems

Plants must strip and cool the absorber outlet (rich) oil before recycling it to the absorber. Use of a **reabsorber** (Fig. 5-6) is one method of increasing recovery. Rich oil from the main absorber is reduced in pressure at a flash tank. Vapors coming out of solution at the lower pressure are fed to a reabsorber, in which most of the propane and heavier hydrocarbons in the flashed vapors are reabsorbed. Several reabsorbers may be used in series at successively lower operating pressure when the main absorber pressure is quite high.

Figure 5-7 shows a second method for obtaining higher propane recovery without increasing the load on the distillation unit. Rich oil, after being flashed at a reduced pressure and partly denuded of propane and lighter hydrocarbons, is pumped into the midsection of the main absorber where it absorbs additional propane. This type of multiple absorption has been termed *flash flood* because the oil required for high absorption of LP-gas is circulated at a high rate to the absorber midsection and is partially denuded by cold flashing rather than by steam stripping.

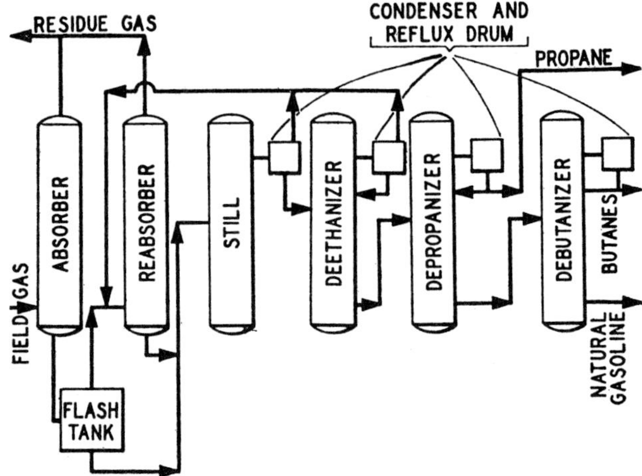

Fig. 5-6 Natural gasoline plant LP-gas recovery system.

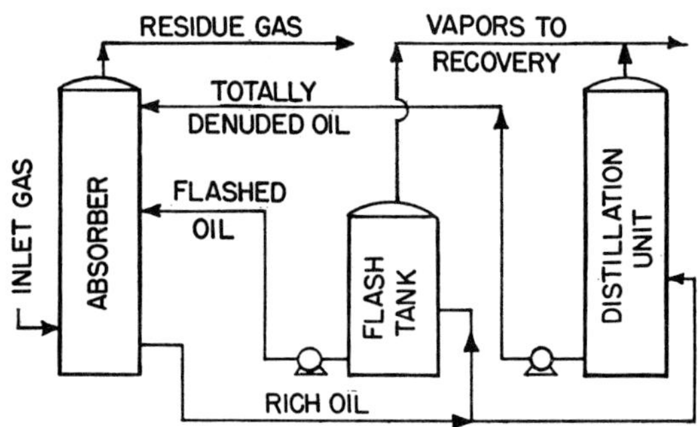

Fig. 5-7 Flash-flood LP-gas recovery system.

Flash tanks like those shown in Figs. 5-6 and 5-7 afford only a single plate or equilibrium-flash separation. More efficient recovery of propane, along with elimination of excess light components, may be obtained by **bubble-plate separation methods:** (1) rich oil de-ethanization, (2) two-stage rich oil stripping at high and low pressure, (3) rich oil desorption, and (4) lean oil presaturation.

Reabsorbers and Fractionating Absorbers. Rich oil from the main absorber and its auxiliary reabsorbers and fractionating absorbers, as shown in Fig. 5-4, are then heated and fed to a still in which absorbed hydrocarbons are steam-stripped from the absorption oil. The still is a bubble-plate column with the reverse function of the absorber. It is also referred to as a *stripper* or *evaporator*. Its bubble plates provide a means of contact between the oil to be stripped and the superheated steam used as the stripping medium.

Superheated steam, injected into the still bottom, flows upward countercurrent to the oil flow, leaving the still top with the stripped gases. This mixture enters a dephlegmator or knock-out tower (if not built into the still), where steam is condensed and withdrawn. Condensed liquid hydrocarbons are then pumped to the fractionation unit.

When stripping an oil containing primarily propane and butane, two stills in series, the first at 100 psig and the second at about atmospheric pressure, were effective.

FRACTIONATION

Raw natural gasoline condensed from the stripping still is separated into commercial grades of LP-gas and natural gasoline by fractionation (Fig. 5-8).

This process consists of a series of distillation and condensation operations conducted at various temperatures within a single apparatus usually containing 30 or more trays or plates.

Heat is applied near the column base and extracted near its top. Extraction is accomplished by reflux of part of the condensed and cooled overhead product to the top column tray. The reflux ratio, or ratio of reflux volume to volume of that part of the overhead removed as a gas or commercial product, varies widely with individual fractionators. It depends upon several factors, such as desired overhead product purity, amount of column cooling required, temperature of available condenser cooling water or refrigerant, feed composition, and column pressure.

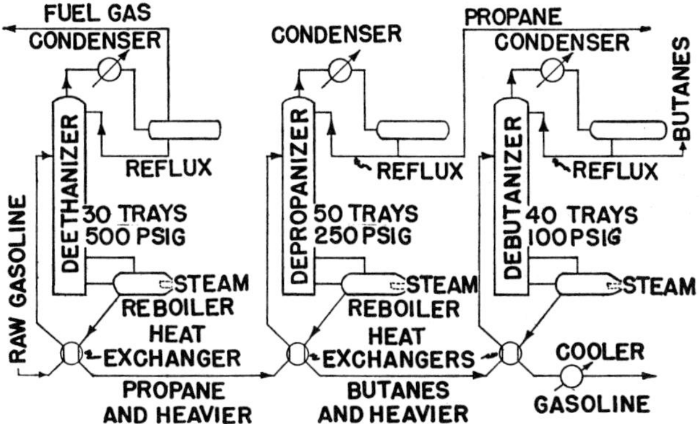

Fig. 5-8 Typical natural gasoline fractionation unit.

Pressure in columns must be high enough to liquefy a major portion of the overhead gases at condenser temperature. Use of a refrigerant in the condenser system permits lower column pressure and usually produces higher separation efficiency. **De-ethanizers,** in particular, benefit from lower condenser temperature because they permit operation at pressures and overhead temperatures substantially below the critical point of ethane (90.1 F at 48.2 atm).

Pressure drop in a fractionator is usually at the rate of not more than 0.1 psi per tray. Practically speaking, a column is at a single pressure at any particular instant. De-ethanizers usually operate between 450 and 550 psig, **de-propanizers** between 225 and 275 psig, and **de-butanizers** between 75 and 125 psig.

Column operation depends upon maintenance of a suitable **temperature gradient.** This gradient varies widely but is usually within a 100 to 200°F range. Composition of the feedstream is the major factor contributing to selection of a suitable temperature gradient.

A **stabilizer** is a type of fractionator employed in a simple absorption plant to remove propane and excess butane from a commercial grade of natural gasoline.

The feed to a fractionating column enters at or near the tray on which the liquid composition approaches that of the feedstream. Some columns have optional feed inlets to permit handling of feed stocks varying slightly in composition.

Number of trays required is a function of difference in boiling points of overhead and bottom products. Separation of two components of relative close boiling points, as normal and *iso*-butane, requires considerably more trays than noted in Fig. 5-8.

That portion of the fractionating column above the feed inlet tray is usually referred to as the *absorbing section* and the lower portion as the *stripping section.* Both absorption and stripping occur in each tray. Absorption predominates at the lower temperatures above the feed inlet, and stripping predominates at the higher temperatures below it.

Low Temperature Fractionation

Complete liquefaction of natural gas from wells is made possible by subjecting the gas to a sufficiently low temperature. Liquid components may then be separated by fractionation. To avoid freezing troubles, such an operation requires complete gas dehydration before processing, as well as removal of carbon dioxide if present in large amounts.

Since natural gas may contain 90 per cent or more methane, this theoretically simple process would require condensation and re-evaporation of a tremendous quantity of material, under very difficult operating conditions, to extract a small percentage of liquid products; i.e., it would be necessary to operate the primary fractionating column at a temperature at least as low as −150 F, and at a pressure below about 500 psig, to produce a reflux of methane and ethane only.

If the untreated gas has a high propane content, and only moderate propane recovery efficiency is all that is desired, primary fractionator conditions may be moderated in some cases to make the process economic. One refrigeration plant designed to extract propane and ethane from 750,000 MCF of feed gas per day has been operating successfully for several years.

TREATMENT OF LP-GAS

Following fractionation, separated LP-gas must be treated to meet corrosion and dryness specifications.

Sulfur and Carbon Dioxide Removal

If H_2S and/or CO_2 are present in the source streams in appreciable quantities, they should be substantially removed by an amine process like that shown in Fig. 5-9.

Where necessary, final removal of H_2S and reduction of mercaptans to acceptable concentrations (usually less than two grains per CCF of flash-vaporized LP-gas) is accomplished by caustic washing. Since by this method H_2S is removed in a nonreversible reaction, the caustic must be periodically replaced. Mercaptans are removed by absorption, and can be steam-stripped from the caustic solution. Figure 5-10 is a refinery schematic.

The strength of the caustic solution depends on relative

quantities of H_2S and mercaptans to be removed. A 20 per cent (about 16° Baumé) solution of sodium hydroxide in water is best suited for removal of mercaptans. A weaker (10 to 15 per cent) solution does a better job of removing H_2S.

LP-gas is usually mixed with the caustic solution before entering the treating or settling tank. Either an open impeller type centrifugal mixing pump or an eductor may be used. The ratio of gas to caustic should be between 1:1 and 3:1.

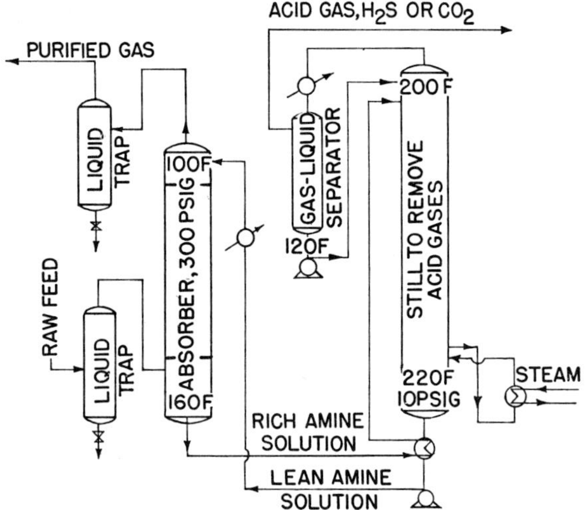

Fig. 5-9 Amine process for removal of acidic gases from feedstreams. (See also Fig. 4-61.)

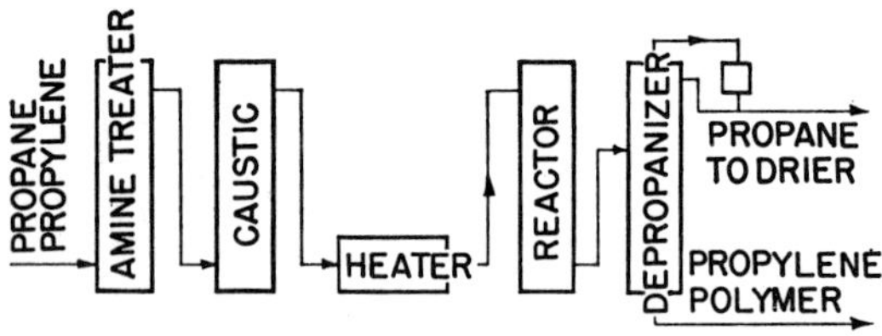

Fig. 5-10 Recovery of propane at refinery polymerization plant.

The caustic settling tank should have a capacity of at least three *times* the hourly charge rate. Since caustic must be replaced when its strength is so reduced that H_2S is no longer satisfactorily removed, two units are usually operated in series. These alternate in position relative to gas flow, so that the downstream tank is always the one last recharged with fresh caustic. The tank being recharged must be removed from service, the gas receiving only one-stage treatment during recharging.

LP-gas from the final settling tank top should go thru a gravel or salt tower to remove entrained caustic. This tower should have a space velocity not exceeding four volumes of charge per hour per volume of gravel or salt. Downward flow is advisable, with LP-gas removed a foot or so above the tank bottom. Caustic is drained from the bottom outlet.

Dehydration

Caustic washing results in LP-gas becoming water saturated. Activated alumina or bauxite is usually employed to dehydrate it following caustic treatment. Commercial propane must be dried to pass the **cobalt bromide test,** or to a *dew point* of −15 F or lower. Two or more drying towers are required, so that at least one unit is operable at all times. Each unit must be regenerated when its effluent shows excessive moisture content. Regeneration is accomplished by passing gas (natural or LP), preheated to 350 to 400 F, thru the dryer bed after the unit has been taken out of service. Regeneration is complete when the outlet gas temperature approaches that of the inlet, indicating that no more water is being driven off. The dryer unit should then be allowed to cool to approximately atmospheric temperature before being returned to service.

Following caustic treatment and dehydration, LP-gas is ready for storage and marketing, providing that it meets all sales specifications.

REFERENCES

1. A.G.A.-N.G.A.A. *Standard Compression and Charcoal Tests for Determining the Natural Gasoline Content of Natural Gas.* (Code 101-43) New York, 1943.
2. Calif. Natural Gasoline Assn. *Procedure for the Charcoal Test for the Determination o^f^ the Gasoline Content of Natural Gas.* (Bul. TS-351) Los Angeles, 1947.
3. ——. *Tentative Standard Method for Analysis of Natural Gas and Gasoline by Fractional Distillation.* (Bul. TS-411) Los Angeles, 1957.
4. Washburn, H. W. and others. "Mass Spectrometry." *Ind. Eng. Chem. Anal. ed.* 17: 74–81, Feb. 1945.
5. Shepherd, M. "Cooperative Analysis of a Standard Sample of Natural Gas with the Mass Spectrometer." *J. Res. Natl. Bur. Stds.* 38: 491-8, May 1947.
6. Barnes, R. B. and others. "Infrared Spectroscopy." *Ind. Eng. Chem. Anal. ed.* 15: 659–709, Nov. 15, 1943.
7. Willard, H. H. and others. *Instrumental Methods of Analysis,* 3rd ed. New York, Van Nostrand, 1958.
8. Barnard, G. P. *Modern Mass Spectrometry.* London, Inst. of Physics, 1953.
9. Calif. Natural Gasoline Assn. *Tentative Standard Chromatographic Test Procedure for the Determination of Propane, Butanes, and Pentane-plus in Natural Gas.* (Bul. TS-611) Los Angles, 1961.
10. *Ibid.*, ref. 1 above.
11. Katz, D. L., and others. *Handbook of Natural Gas Engineering.* New York, McGraw-Hill, 1959.
12. Eilerts, C. K. and others. *Phase Relations of Gas-Condensate Fluids.* (Bur. of Mines Mono. 10), New York, A.G.A., 1957–59. 2v.
13. Vandaveer, F. E. "A Natural Gas Purification and Gasoline Plant." *A.G.A. Proc.* 1951: 514–26.

Chapter 3

Testing and Properties of LP-Gases

by E. W. Evans and R. W. Miller

LP-gases are usually shipped and stored as liquids, but they are utilized in the gaseous state. Test methods covering both physical states are, therefore, necessary. Specifications and test methods for these gases have been developed by NGPA[1] (previously NGAA) and CNGA.[2] ASTM[3–5] has also prepared several test methods. NGPA specifications (Table 2-17) cover four types of LP-gases: commercial propane, commercial butane, butane–propane mixtures, and propane HD-5.

Tests are run to determine such properties as dryness, composition, corrosivity, residue (end point index),[6] sulfur content, specific gravity, and vapor pressure.

Interstate Commerce Commission regulations on movement of LP-gases require determination of vapor pressure at designated temperatures and of the specific gravity of the liquid at 60 F. Most states and many communities also specify both vapor pressure limitations for various types of containers and maximum permissible filling densities, based on specific gravity of the liquid.

Table 5-4 Physical Properties* of LP-Gases[9]

(all values at 60 F and 14.696 psia unless otherwise stated)

	Propane	*iso*-Butane	Butane
Molecular weight	44.09	58.12	58.12
Boiling point, °F	−43.7	+10.9	+31.1
Boiling point, °C	−42.1	−11.7	−0.5
Freezing point, °F	−305.8	−255.0	−216.9
Density of liquid			
Specific gravity, 60 F/60 F	0.508	0.563	0.584
Degrees, API	147.2	119.8	110.6
Lb per gallon	4.23	4.69	4.87
Density of vapor (ideal gas)			
Specific gravity, (air = 1)	1.522	2.006	2.006
Cu ft gas per lb	8.607	6.53	6.53
Cu ft gas per gal of liquid	36.45	30.65	31.8
Lb gas per 1000 cu ft	116.2	153.1	153.1
Total heating value (after vaporization)			
Btu per cu ft	2,563	3,369	3,390
Btu per lb	21,663	21,258	21,308
Btu per gal of liquid	91,740	99,790	103,830
Critical constants			
Pressure, psia	617.4	537.0	550.1
Temperature, °F	206.2	272.7	306.0
Specific heat, Btu/lb-°F			
c_p, vapor	0.388	0.387	0.397
c_v, vapor	0.343	0.348	0.361
c_p/c_v	1.13	1.11	1.10
c_p, liquid 60 F	0.58	0.56	0.55
Latent heat of vaporization at boiling point, Btu per lb	183.3	157.5	165.6
Vapor pressure, psia			
0 F	37.8	11.5	7.3
70 F	124.3	45.0	31.3
100 F	188.7	71.8	51.6
100 F (ASTM[10]), psig max	210		70
130 F	274.5	109.5	80.8

* Properties are for commercial products and vary with composition.

PROPERTIES

Properties of LP-gases are summarized in Tables 5-4, 5-5, 5-17, 2-17, 2-18, and 2-19. Figure 1-3c is the pressure-enthalpy diagram for propane.

Table 5-5 Combustion Data* for LP-Gases[6]

(all values at 60 F and 14.696 psia unless otherwise stated)

	Propane	*iso*-Butane	Butane
Flash temperature, °F (calculated)	−156	−117	−101
Ignition temperature, °F	932	950	896
Maximum flame temperature in air, °F			
Observed	3497	3452	3443
Calculated	3573	3583	3583
Flammability limits, % gas in air			
Lower	2.37	1.80	1.86
Higher	9.50	8.44	8.41
Maximum rate flame propagation in 1 in. tube			
Inches per second	32	33	33
Percentage gas in air	4.6–4.8	3.6–3.8	3.6–3.8
Required for complete combustion (ideal gas)			
Air, cu ft per cu ft gas	23.9	31.1	31.1
lb per lb gas	15.7	15.5	15.5
Oxygen, cu ft per cu ft gas	5.0	6.5	6.5
lb per lb gas	3.63	3.58	3.58
Products of combustion (ideal gas)			
Carbon dioxide, cu ft per cu ft gas	3.0	4.0	4.0
lb per lb gas	2.99	3.03	3.03
Water vapor, cu ft per cu ft gas	4.0	5.0	5.0
lb per lb gas	1.63	1.55	1.55
Nitrogen, cu ft per cu ft gas	18.9	24.6	24.6
lb per lb gas	12.0	11.8	11.8

* Properties are for commercial products and vary with composition.

ANALYSIS

Analysis of LP-gases for various hydrocarbon constituents is usually made by low temperature fractional distillation methods. Analyses may also be made by mass spectrometer, infra-red devices, and chromatography.

Unsaturated hydrocarbons in LP-gases may be determined by the foregoing methods or by the ASTM silver-mercuric nitrate method.[7] Presence of unsaturated hydrocarbons is determined by selective absorption in silver-mercuric nitrate reagent, in a conventional Orsat-type gas analysis apparatus.

VAPOR PRESSURE

The NGAA[1] method determines vapor pressure in psig at 100 F. It is identical with ASTM D1267-55.[4]

Apparatus (Fig. 5-11). Two chambers, with capacities in ratio of four to one, are connected by straight-thru valve with bleeder valve assembly for purging. Assembled chambers are certified by manufacturer to withstand 1000

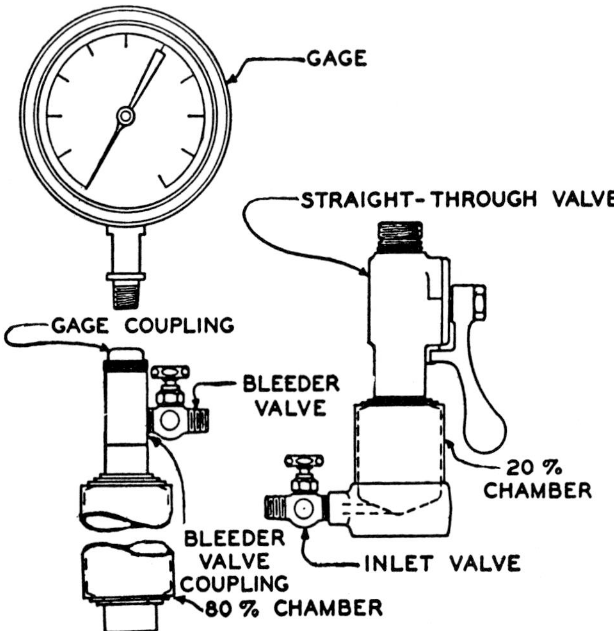

Fig. 5-11 Typical LP-gas vapor pressure apparatus.[1]

psig hydrostatic pressure without permanent deformation. Entire apparatus except bleeder valve assembly meets requirements for ASTM Method D323 Test for Vapor Pressure of Petroleum Products (Reid Method).[8]

Procedure. With assembled apparatus thoroughly clean, inlet valve is connected to sample source and entire apparatus is purged thru slightly opened bleeder valve until liquid escapes. Without disconnection, first bleeder and then inlet valves are closed, apparatus is quickly inverted, and then it is held in that position with bleeder valve opened, blowing out all liquid and allowing vapor to escape until pressure is approximately atmospheric.

Bleeder valve is then closed, apparatus is returned to original position, and inlet valve is again opened. Chilling of chamber exteriors may be necessary to permit them to become liquid full, as determined by appearance of liquid immediately upon opening bleeder slightly. If gas escapes, sample is expelled by inverting apparatus and again blowing all liquid from it. Sampling operation is repeated as before until both chambers are liquid full.

Bleeder and inlet valves are then closed in order and apparatus disconnected. Valve between containers is then closed and inlet valve opened with apparatus upright. Inlet valve is then closed as soon as no more liquid escapes, and communicating valve is immediately opened. Apparatus now contains test sample with 20 per cent outage for safety purposes.

Apparatus is next inverted, shaken vigorously, righted, and placed in 100 F water bath completely covering top of bleeder valve coupling. After 5 min, apparatus is withdrawn from bath, inverted, shaken, and replaced in bath in normal position. Operation is repeated at 2-min intervals until constant gage readings are maintained before removal from bath. To insure equilibrium, 20 to 30 min are normally required. Constant reading thus obtained represents uncorrected vapor pressure. Gage connection, if any, is applied to obtain corrected vapor pressure.

Corrected vapor pressure is converted to standard barometric pressure of 29.92 in. Hg as follows:

$$\text{LP-gas vapor pressure} = \text{corrected vapor pressure} - (29.92 - P)(0.49)$$

where P = observed barometric pressure, in. Hg

SPECIFIC GRAVITY

NGPA Hydrometer Method*

This method, which determines specific gravity of liquefied petroleum products at 60/60 F, should not be used with liquids of higher than 200 psig vapor pressure at this temperature.

Apparatus (Fig. 5-12). This special hydrometer is enclosed in a *Lucite* pressure cylinder. Cylinder ends are sealed with plates and gaskets. Liquid inlet and outlet valves are connected to cylinder bottom and vapor outlet valve to top. A water bath large enough to permit complete immersion of this cylinder is required.

Procedure. After purging, pressure cylinder is filled with liquid sample to level at which enclosed hydrometer floats freely. Cylinder is then disconnected from sample source and placed in water bath which is maintained at about 60 F. When temperature of cylinder contents reaches 60 F ± 0.5°F, apparatus is removed from bath and specific gravity is observed immediately. Hydrometer is read to nearest 0.001 while free of contact at any point with cylinder walls.

Observed specific gravity is recorded at 60/60 F.

To convert specific gravity to degrees API, see Table 1-46 and its accompanying formula.

Pressure Pycnometer Method

This method,[1,2] which also determines specific gravity of liquefied petroleum products at 60/60 F, consists of weighing a known volume of LP-gas as liquid in a pressure pycnometer (Fig. 5-13) previously calibrated against water.

* Also see ASTM D1657-59T.[10]

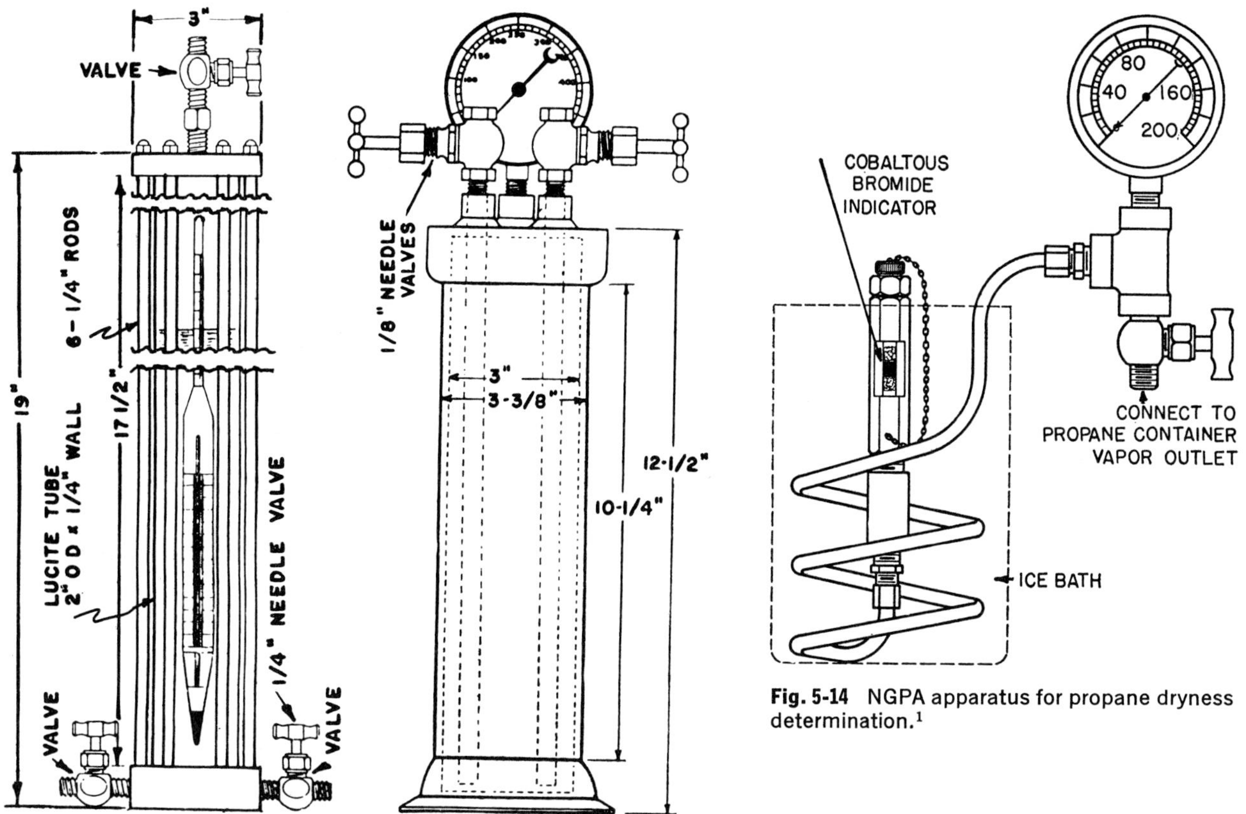

Fig. 5-14 NGPA apparatus for propane dryness determination.[1]

Fig. 5-12 (left) Combined hydrometer and pressure cylinder for determining liquid specific gravity.[1]
Fig. 5-13 (right) Pressure pycnometer for determining liquid specific gravity.[1]

DRYNESS

Dryness tests guard against stoppages due to freezing. Bottled gas systems are susceptible to moisture, which tends to concentrate in the vapor phase and to freeze out at regulator orifices on pressure reduction, resulting in shut off of gas flow.

Both NGPA[1] and CNGA[2] employ a cobaltous bromide test for determining dryness of LP-gas. A dryness test is required for all grades of LP-gas by CNGA; NGPA requires such a test for commercial propane only.

The NGPA method (Fig. 5-14) consists of passing propane vapor under 50 psig thru a specially prepared indicator impregnated with cobaltous bromide. The vapor is held at a temperature of 32 to 35 F by first passing it thru a cooling coil immersed in an ice bath. If the indicator color remains blue during test exposure, the sample is considered "dry." If a change to lavender or pink occurs, it is considered "wet."

Dryness of commercial propane may also be determined from its dew point, obtained by using Bureau of Mines Dew Point Apparatus (Fig. 6-21). Propane having a dew point of −15 F or lower obtained by this procedure is considered free of moisture.

CORROSIVITY

Tests for corrosivity and sulfur content protect brass or copper gas-carrying parts. Formation of deposits and possible stoppages may result from presence of corrosive constituents in the fuel. Excessive amounts of sulfur compounds in flue gases may also corrode appliance flueways and vent connections.

The NGPA **Copper Strip Method**[1] detects corrosive compounds in LP-gas. It consists of immersing a polished copper strip, of specified size and quality, in a sample of approximately 100 ml of the LP-gas liquid, which is in a special corrosion test cylinder (Fig. 5-15). About 1 ml of distilled water is added just ahead of sample to assist in developing latent corrosivity. Cylinder is then immersed in a water bath

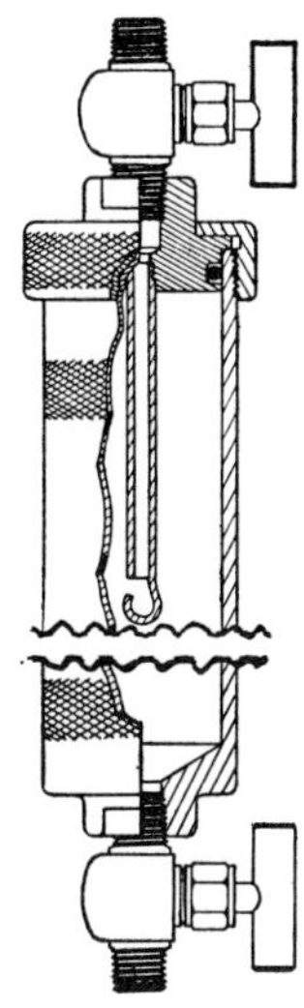

Fig. 5-15 Corrosivity test cylinder.[1]

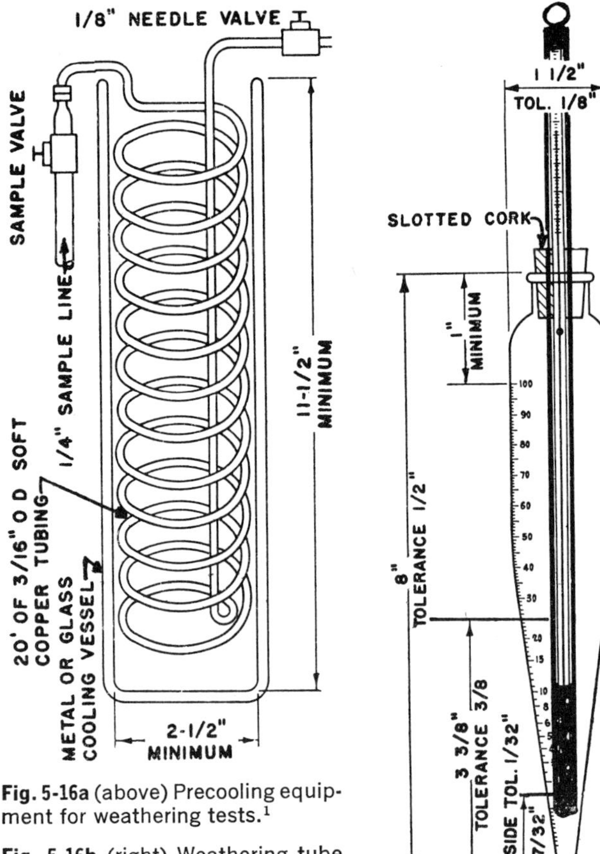

Fig. 5-16a (above) Precooling equipment for weathering tests.[1]

Fig. 5-16b (right) Weathering tube for LP-gases.[1]

Graduation Tolerances

Range, ml	Scale division, ml	Limit of error, ml
0.0 - 0.1	0.05	0.02
0.1 - 0.3	0.05	0.03
0.3 - 0.5	0.05	0.05
0.5 - 1.0	0.1	0.05
1.0 - 3.0	0.1	0.1
3.0 - 5.0	0.5	0.2
5.0 - 25.0	1.0	0.5
25.0 - 100.0	1.0	1.0

The shape of the lower tip of the tube is especially important. The taper shall be uniform and the bottom shall be rounded as shown in the drawing. Tubes must be made of Pyrex glass and thoroughly annealed; wall thickness shall comply with ASTM centrifuge tube requirements.

held at 100 F. After 1 hr immersion, copper strip is removed and immediately compared with a freshly polished strip.

The sample is reported as not meeting the test when the exposed strip, as compared with one freshly polished, shows more than a very slight discoloration, equivalent to corrosion copper strip No. 1 in ASTM Method D130-56. Isolated brown spots on the copper strip are frequently caused by the 1 ml distilled water. Presence of such spots should be disregarded in evaluating discoloration.

It is reported that this method is capable of detecting H_2S or elementary sulfur in concentration as low as 1 ppm and methyl mercaptan in concentration of 20 ppm.

TOTAL SULFUR

ASTM[5] and NGPA[1] describe similar methods for determination of total sulfur in LP-gas in the *liquid phase.*

These methods consist of burning a weighed sample of LP-gas in a special burner in an atmosphere of carbon dioxide and oxygen. Resulting oxides of sulfur are absorbed in hydrogen peroxide and oxidized to sulfuric acid, which is then titrated with standard alkali solution. Sulfur content in per cent by weight is then calculated.

Grains of total sulfur per 100 cu ft of gas (15 permitted[11]) at 60 F and 29.92 in. Hg may be calculated using the following equations:

For propane: $R = 814S$

For butane: $R = 1072S$

For butane–propane mixtures: $R = S\ [3374(G - 0.5077) + 814]$

where: R = grains of total sulfur per 100 cu ft of gas
S = per cent by weight of sulfur in LP-gas
G = specific gravity of mixture at 60/60 F

Apparatus for determination of sulfur in LP-gas in the gaseous state is shown in Fig. 6-20.

WEATHERING

Weathering tests limit LP-gas composition both on light and heavy ends. Presence of light ends may produce excessive vapor pressure and combustion difficulties. Heavy ends may cause combustion trouble and insufficient pressure when the container approaches exhaustion. They may also cause deposits which would interfere with proper metering or pressure regulation and produce stoppages in appliances.

Thermometer Method[1] (for commercial propane) determines per cent hydrocarbon residue after weathering commercial propane to the melting point of mercury (−38.0 F). Its purpose is to limit inclusion of butane and higher hydrocarbons in commercial propane.

Test procedure consists of passing a liquid sample thru precooling equipment shown in Fig. 5-16a and collecting a 100 ml sample in a Pyrex graduate (Fig. 5-16b). The sample is then allowed to weather, and its temperature is read when a 5 ml residue remains, i.e., at the 95 per cent boiling point. A corrected temperature under −38.5 F indicates an *iso*-butane content less than 2.0 per cent by liquid volume. ASTM[11] states that the 95 per cent evaporated temperatures are −37 F and −36 F for commercial propane and butane, respectively.

Test for Butane and Butane–Propane Mixtures[1]

Indication of higher boiling hydrocarbon constituents in commercial butane and butane–propane mixtures is given by this test:

A Pyrex centrifuge tube as shown in Fig. 5-16b forms the weathering vessel. A 100 ml sample cooled by passage thru equipment illustrated in Fig. 5-16a is collected in the weathering tube. The sample is then allowed to weather, with a thermometer in place at the bottom of the tube. The temperature is read when the sample falls to a hydrocarbon residue of 5 ml. Observed temperature is then corrected to barometric pressure of 760 mm Hg, as well as for thermometer error.

CNGA[2] Test Methods

The choice between either of two weathering tests depends upon the composition of the LP-gases. One method applies to mixtures which are predominantly propane, *iso*-butane, and normal butane. The other covers only samples predominantly propane.

REFERENCES

1. Natural Gas Processors Assn. "N.G.P.A. Liquefied Petroleum Gas Specifications and Test Methods." (Pub. 2140-62) *Plant Operations Test Manual.* Tulsa, Natural Gasoline Assn. of Am.
2. Calif. Natural Gasoline Assn. *Tentative Specifications and Tentative Standard Methods of Test for Liquefied Petroleum Gases.* (Bul. TS-441) Los Angeles, 1953.
3. A.S.T.M. "Standard Method of Sampling Liquefied Petroleum (LP) Gases." (Standard D1265-55) *ASTM Standards* 8: 1531–4. Philadelphia, 1961.
4. ——. "Standard Method of Test for Vapor Pressure of Liquefied Petroleum (LP) Gases." (Standard D1267-55) *ASTM Standards* 8: 1553–7. Philadelphia, 1961.
5. ——. "Tentative Method of Test for Sulfur in Petroleum Products and Liquefied Petroleum (LP) Gases (Lamp Method)." (Standard D1266-59T) *ASTM Standards* 8: 1535–52. Philadelphia, 1961.
6. ——. "Proposed Method of Test for Residues in Liquefied Petroleum (LP) Gases (End Point Index Method)." *ASTM Standards on Petroleum Products and Lubricants* 1: 1096–1100. Philadelphia, 1961.
7. ——. "Standard Method of Test for Unsaturated Light Hydrocarbons." (Standard D1268-55) *ASTM Standards* 7: 673–8. Philadelphia, 1961.
8. ——. "Standard Method of Test for Vapor Pressure of Petroleum Products (Reid Method)." (Standard D323-58) *ASTM Standards* 7: 160–8. Philadelphia, 1961.
9. Clifford, E. A. *Practical Guide to LP-Gas Utilization*, rev. ed. New York, Moore, 1957.
10. A.S.T.M. "Tentative Method of Test for Specific Gravity of Light Hydrocarbons by Pressure Hydrometer." (Standard D1657-59T) *ASTM Standards* 7: 857–9. Philadelphia, 1961.
11. ——. "Tentative Specifications for Liquefied Petroleum (LP) Gases." (Standard D1835-61T) *ASTM Standards* 7: 979–83. Philadelphia, 1961.

Chapter 4

Transportation, Storage, Distribution, and Measurement of (Non-Utility) LP-Gases

by E. Martin Anderson, Walter H. Johnson, W. R. Fraser, C. George Segeler, F. E. Vandaveer, and A. H. Withrow

TRANSPORTATION

Portable Bottled Gas Cylinders

For shipment across state lines, portable cylinders must be constructed and marked according to the specifications of the Interstate Commerce Commission (e.g., ICC 4B240 and 4BA240). Capacities of ICC cylinders are limited to 120 gal or 1000 lb of water capacity. They are used primarily for domestic installations, but commercial and small industrial loads are sometimes handled by multiple cylinder installations.

The portable cylinders are used primarily for propane; hence, their sizes are usually designated by the propane capacity in pounds. A 100 lb cylinder, for example, is one which will contain 100 lb of propane when properly filled. The 20 lb cylinder is popular for cash-and-carry transportation by small domestic users, for trailers, temporary installations, appliance demonstrations, and other applications where portability is essential.

Markings. ICC cylinders must be marked with the following (Fig. 5-17):

1. *ICC code number.* This indicates the code under which the cylinder was built. Code 4B240, the most common, shows that the container is of the Type 4B, designed for a working pressure of 240 psig.
2. *Water capacity.* This is simply the number of pounds of water that the full container will hold. The 100 lb cylinder is marked W.C. 238 or W.C. 239, i.e., the container would hold up to 239 lb of water but only 100 lb of propane.
3. *Date of manufacture and subsequent test dates.* ICC requirements call for reinspection every 5 years. Visual inspection is permitted when performed carefully by an experienced person.
4. *Ownership.* This must be shown on the container and must be registered with the Bureau of Explosives, 63 Vesey St., New York, N. Y.
5. *Serial number.*
6. *Tare weight.*

Miniature cylinders containing 2 lb or less of propane are widely used for hand torches, lights, and camping cook stoves.

Typical cylinder sizes are listed in Table 5-6.

Table 5-6 Typical Portable Cylinder Capacities and Dimensions

Capacity, lb			Container dimensions, in.	
Propane	Butane	Water	ID	Length
20	24.2	47.6	12	15.25
25	30.3	59.5	12	18.38
40	48.5	95.2	9.9	38.88
60	72.8	143.0	12	39.50
100	121.4	238.1	14.5	44.75
150	182.1	357.2	18.5	42
200	242.8	476.2	18.5	54
300	364.2	715.0	24	50
420	510.0	1000.0	29	47.75

Stress Calculations in Cylinders. The calculation of wall stress may be made by the formula:

$$S = \frac{P(1.3D^2 + 0.4d^2)}{D^2 - d^2}$$

where: S = wall stress, psi
P = container pressure,* psig
D = outside diameter, in.
d = inside diameter, in.

Allowable Wall Stresses and Pressures for ICC Cylinders

Specification	Allowable wall stress, psi	Pressure
3B	24,000	Test pressure or 450 psig, whichever is greater
4	18,000	Test pressure
4A (with longitudinal seam)	18,000	Same as for 3B
4A (without longitudinal seam)	24,000	Same as for 3B
4B (with longitudinal seam)	18,000	Same as for 3B
4B (without longitudinal seam)	24,000	Same as for 3B
4B (with copper brazed seam)	22,800	Same as for 3B

* Minimum test pressure prescribed by water-jacket test.

SPECIFICATIONS

SPECIFICATIONS

Mfgr.'s and owner's names & addresses
Total water capacity and tare (incl. valves), lb
ICC Spec. No. and cylinder No.
Owner's and inspector's marks
Dates of manufacture and retests

Fig. 5-17 Identification markings required on ICC LP-gas cylinders.

Railway Tank Cars

In 1961 it was estimated that there were 20,000 cars in such service; 3.0 billion gal of LP-gases were shipped by rail in 1960. The approximate cost of transportation from Texas to the Middle Atlantic States is between 8 and 11 cents per gal. Table 5-7 shows the ICC classification for railroad tank cars for LP-gas service. All tank cars were required to be insulated to a thermal conductance of not more than 0.075 Btu per hr-sq ft-°F. In 1956 ICC granted approval to noninsulated cars using steel of higher tensile strength. These cars were painted with light-colored or white reflective paints. Size for size, the newer cars hold about 650 more gallons of propane than the older insulated cars.

Table 5-7 Classes of Railroad Tank Cars for LP-Gas Service

Class	Test pressure of tank, psig	Safety valve setting, psig	Designed for
104A	100	75	Butanes
104A W	100	75	Butanes
105A 300	300	225	Propane
105A 300 W	300	225	Propane
105A 400	400	300	Propane
105A 400 W	400	300	Propane
105A 500	500	375	Propane
105A 500 W	500	375	Propane
105A 600	600	450	Ethane–propane
105A 600 W	600	450	Ethane–propane
112A 400 W	400	300	Propane–butane

The water capacity of the typical older tank car is about 11,600 gal. Consequently, such cars would hold about 9500 gal of propane, with the exact amount depending upon the maximum permitted filling density. In 1961, much larger, uninsulated tank cars became available—water capacity 30,398 gal. Maximum filling densities permitted by ICC regulations are shown in Table 5-8. Figure 5-18 compares LP-gas shipments by various transportation methods.

Table 5-8 Filling Density for Transporting LP-Gas in Lagged Tank Cars

(per cent)

Sp gr*	Filling density	Sp gr*	Filling density	Sp gr*	Filling density
0.500	45.500	0.550	51.500	0.600	56.900
.510	46.750	.560	52.750	.610	57.800
.520	48.125	.570	53.800	.620	58.700
.530	49.250	.580	54.900	.630	59.600
0.540	50.400	0.590	56.000	0.635	60.050

* At 60 F.

Note: A car capable of holding 30,000 gal H_2O could be filled with 30,000 × 8.33 × 0.455 = 114,000 lb of 0.5 sp gr LP-gas.

Tank Trucks

Four types of motor transport for LP-gases are in general use. These are built to the ASME code for 250 psig design pressure and, if intended for interstate use, must meet ICC Specification MC330. They should also meet NFPA Std. 58:[1]

1. *Large transports.* Tank truck and trailer with carrying capacities up to 11,000 gal.
2. *Semitrailers.* Generally used for 6500 to 11,000 gal requirements.
3. *Small transports.* Consumer delivery trucks or "bobtails"—generally 1500 to 3300 gal capacity for local distributions.
4. *Skid tanks.* Carried on flat-rack trucks—generally 300 to 2000 gal capacity.

Many, though not all, of the motor transports in the first three groups are equipped with pumps for unloading and liquid meters for billing. Some transports also carry compressors.

Interstate transportation by tank truck must comply with Motor Carrier Tariff No. 13, published by the ICC, which is made available thru the Bureau of Explosives, 63 Vesey St., New York, N. Y. Also available are equipment design standards for tank truck transportation, including the design pressure for cargo trucks, the required valves, piping and other accessories, as well as details for mounting and operating the equipment. Requirements for parking and garaging of LP-gas truck vehicles are available, too.[1]

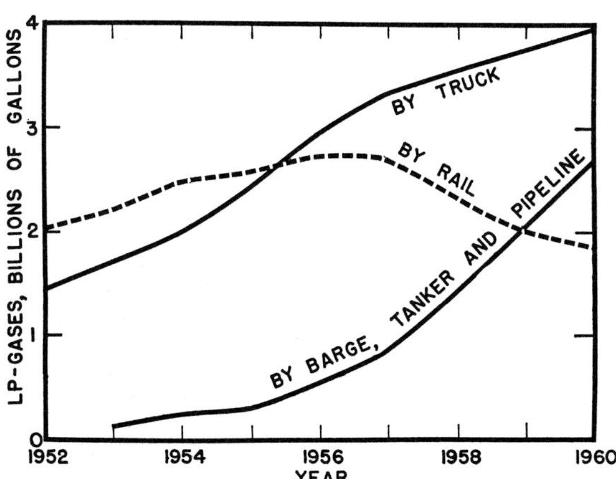

Fig. 5-18 Comparison of LP-gas equipment by various transportation methods.

Not counting local distribution, the number of tank truck operations for shipment of LP-gases has grown steadily, and the total now exceeds that shipped by rail.

Pipelines

In 1960, 22 companies shipped 1.4 billion gal of LP-gases thru pipelines (Fig. 5-18). LP-gas is transported by pipelines in four ways: as finished products (propane or butane), in "raw" streams, in crude oil, and in natural gas. The last three categories require processing to separate LP-gases, which may require still further treatment before marketing. The practice of transporting crude oil–LP-gas mixtures by pipeline provides two advantages: (1) Reduction of the oil viscosity, thus making it easier to pump, and, (2) Transportation of LP-gas to the major consuming areas.

The transportation of LP-gas with natural gas depends primarily upon the marketing economies. Several large petrochemical plants extract LP-gas from natural gas.

Boats, Tankers, and Barges

Transportation of LP-gas by water has increased steadily, but it still constitutes only a small fraction of the U. S. shipment. Outside this country, however, the world-wide interest in transporting LP-gas by tanker is expanding rapidly as new sources of LP-gas in out of the way producing areas are developed. Much larger seagoing vessels have been constructed. Consideration is being given to refrigerated as well as pressurized containers. In mid-1957, 17 barges suitable for LP-gas or anhydrous ammonia shipments were in domestic operations. Five of these were seagoing units, and the rest were for inland water use. The seagoing barges carried more than 500,000 gal and the trend was for larger barges. Inland barges were somewhat smaller—360,000 gal. U. S. regulations for water transportion of LP-gas are prescribed by the Bureau of Marine Inspection and Navigation, U. S. Dept. of Commerce, and the U. S. Coast Guard.

STORAGE

Fluctuations in demand and economic considerations necessitate LP-gas storage at production, consumption, or intermediate points. In the U. S., LP-gas sales are approximately 50 per cent greater in winter than in summer, necessitating substantial storage at the point of use.

Reference may be made to the following extracts from NFPA No. 58[1] regarding bulk installations using ASME type containers:

SYSTEMS UTILIZING CONTAINERS OTHER THAN ICC

2.2. Container Valves and Accessories, Filler Pipes and Discharge Pipes.

(a) The filling pipe inlet terminal shall not be located inside a building. For containers with a water capacity of 125 gal or more, such terminals shall be located not less than 10 ft from any building, see B.6(b), and preferably not less than 5 ft from any driveway, and shall be located in a protective housing built for the purpose.

(b) The filling connection* shall be fitted with one of the following:

1. Combination back-pressure check valve and excess flow valve,
2. One double or 2 single back-pressure check valves,
3. A positive shutoff valve, in conjunction with either:
   (a) An internal back-pressure valve, or
   (b) An internal excess flow valve.

(c) All openings in a container shall be equipped with approved automatic excess flow valves except in the following: filling connections as provided in 2.2(b); safety relief connections, liquid level gaging devices as provided in B.7(d), B.18(d) and B.18(i); pressure gage connections as provided in B.7(e), as provided in 2.2(d) and (f) and (g).

(d) No excess flow valve is required in the withdrawal service line providing the following are complied with:

1. Such systems' total water capacity does not exceed 2000 U. S. gallons.
2. The discharge from the service outlet is controlled by a suitable manually operated shutoff valve:
   (a) threaded directly into the service outlet of the container; or
   (b) is an integral part of a substantial fitting threaded into or on the service outlet of the container; or
   (c) threaded directly into a substantial fitting threaded into or on the service outlet of the container.
3. The shutoff valve is equipped with an attached handwheel or the equivalent.
4. The controlling orifice between the contents of the container and the outlet of the shutoff valve does not exceed 5/16 in. in diameter for vapor withdrawal systems and 1/8 in. in diameter for liquid withdrawal systems.
5. An approved pressure-reducing regulator is directly attached to the outlet of the shutoff valve and is rigidly supported, or that an approved pressure-reducing regulator is attached to the outlet of the shutoff valve by means of a suitable flexible connection, provided the regulator is adequately supported and properly protected on or at the tank (see 2.8).

(e) All inlet and outlet connections except safety relief valves, liquid level gaging devices and pressure gages on containers of 2000 gal water capacity, or more, and on any container used to supply fuel directly to an internal combustion engine, shall be labeled to designate whether they communicate with vapor or liquid space. Labels may be on valves.

(f) In lieu of an excess flow valve; openings may be fitted with a quick-closing internal valve which except during operating periods shall remain closed. The internal mechanism for such valves may be provided with a secondary control which shall be equipped with a fusible plug (not over 220 F melting point) which will cause the internal valve to close automatically in case of fire.

(g) Not more than two plugged openings shall be permitted on a container of 2000 gal or less water capacity.

(h) Containers of 125 gal water capacity or more manufactured after July 1, 1961, shall be provided with an approved device for liquid evacuation, the size of which shall be 3/4 in. National Pipe Thread minimum. A plugged opening will not satisfy this requirement.

2.3. Safety Devices.

(a) General: All safety devices shall comply with the following:

1. All container safety relief devices shall be located on the containers and shall have direct communication with the vapor space of the container.
2. In industrial and gas manufacturing plants, discharge pipe from safety relief valves on pipe lines within a building shall discharge vertically upward and shall be piped to a point outside a building.
3. Safety relief device discharge terminals shall be so located as to provide protection against physical damage and such discharge pipes shall be fitted with loose raincaps. Return bends and restrictive pipe fittings shall not be permitted.
4. If desired, discharge lines from two or more safety relief devices located on the same unit, or similar lines from two or more different units, may be run into a common discharge header, provided that the cross-sectional area of such header be at least equal to the sum of the cross-sectional area of the individual discharge lines, and that the setting of safety relief valves are the same.
5. Each storage container of over 2000 gal water capacity shall be provided with a suitable pressure gage.
6. When the delivery pressure from the final stage regulator is not more than 5 psig, the low pressure side shall be equipped with a relief valve, set to start to discharge at not less than two times, and not more than three times the delivery pressure, but

---

* B.7(h) Containers of 2000 gal water capacity or less, filled on a volumetric basis, and manufactured after December 1, 1963, shall be equipped for filling into the vapor space.

not more than 5 psi in excess of the delivery pressure. When the delivery pressure is more than 5 psi, the relief valve shall be set to not less than 1¼ times and not more than two times the delivery pressure. This requirement may be waived on liquid feed systems utilizing tubing specified in B.8(b). When a regulator or pressure relief valve is installed inside a building, the relief valve and the space above the regulator and relief valve diaphragms shall be vented to the outside air with the discharge outlet located not less than 5 ft horizontally away from any opening into the building which is below such discharge. (These provisions do not apply to individual appliance regulators when protection is otherwise provided. In buildings devoted exclusively to gas distribution purposes, the space above the diaphragm need not be vented to the outside.)

(b) ABOVEGROUND CONTAINERS: Safety devices for aboveground containers shall be provided as follows:

1. Containers of 1200 gal water capacity or less which may contain liquid fuel when installed aboveground shall have the rate of discharge required by Appendix A provided by spring loaded relief valve or valves. In addition to the required spring loaded relief valve(s), suitable fuse plug(s) may be used provided the total discharge area of the fuse plug(s) for each container does not exceed 0.25 sq in.

2. The fusible metal of the fuse plugs shall have a yield temperature of 208 F minimum and 220 F maximum. Relief valves and fuse plugs shall have direct communication with the vapor space of the container.

3. On a container having a water capacity greater than 125 gal, but not over 2000 gal, the discharge from the safety relief valves shall be vented away from the container vertically upward and unobstructed to the open air in such a manner as to prevent any impingement of escaping gas upon the container; loose fitting rain caps shall be used. Suitable provision shall be made for draining condensate which may accumulate in the relief valve or its discharge pipe [see B.10(i)].

4. On containers of 125 gal water capacity or less, the discharge from safety relief devices shall be located not less than 5 ft horizontally away from any opening into the building below the level of such discharge.

5. On a container having a water capacity greater than 2000 gal, the discharge from the safety relief valves shall be vented away from the container vertically upward to a point at least 7 ft above the container, and unobstructed to the open air in such a manner as to prevent any impingement of escaping gas upon the container; loose fitting rain caps shall be used. Suitable provision shall be made so that any liquid or condensate that may accumulate inside of the safety relief valve or its discharge pipe will not render the valve inoperative. If a drain is used, a means shall be provided to protect the container, adjacent containers, piping or equipment against impingement of flame resulting from ignition of product escaping from the drain [see B.10(i)].

(c) UNDERGROUND CONTAINERS: On all containers, which are installed underground and which contain no liquid fuel until buried and covered, the rate of discharge of spring loaded relief valve installed thereon may be reduced to minimum of 30 per cent of the specified rate of discharge in Appendix A. Containers so protected shall not be uncovered after installation until the liquid fuel has been removed therefrom. Containers which may contain liquid fuel before being installed underground and before being completely covered with earth are to be considered aboveground containers when determining the rate of discharge requirement of the relief valves.

(d) On underground containers of more than 2000 gal water capacity, the discharge from safety relief devices shall be piped vertically and directly upward to a point at least 7 ft above the ground.

1. Where there is a probability of the manhole or housing becoming flooded, the discharge from regulator vent lines shall be above the highest probable water level. All manholes or housings shall be provided with ventilated louvers of their equivalent, the area of such openings equaling or exceeding the combined discharge areas of the safety relief valves and other vent lines which discharge their content into the manhole housing.

(e) VAPORIZERS: Safety devices for vaporizers shall be provided as follows:

1. Vaporizers of less than 1 qt total capacity, heated by the ground or the surrounding air, need not be equipped with safety relief valves provided that adequate tests certified by any of the authorities listed in B.2, demonstrate that the assembly is safe without safety relief valves.

2. No vaporizer shall be equipped with fusible plugs.

3. In industrial and gas manufacturing plants, safety relief valves on vaporizers within a building shall be piped to a point outside the building and be discharged upward.

2.4. Reinstallation of Containers.

Containers may be reinstalled if they do not show any evidence of harmful external corrosion or other damage. Where containers are reinstalled underground, the corrosion resistant coating shall be put in good condition; see Par. 2.6(f). Where containers are reinstalled aboveground the requirements for safety devices and gaging devices shall comply with Sec. 2.3 and B.18, respectively, for aboveground containers.

2.5. Capacity of Liquid Containers.

No liquid storage container shall exceed 90,000 standard U. S. gallons water capacity.

2.6. Installation of Storage Containers.

(a) Containers installed aboveground except as provided in Par. 2.6(g) shall be provided with substantial masonry or noncombustible structural supports on firm masonry foundation.

(b) Aboveground containers shall be supported as follows:

1. Horizontal containers shall be mounted on saddles in such a manner as to permit expansion and contraction. Structural metal supports may be employed when they are protected against fire in an approved manner. Suitable means of preventing corrosion shall be provided on that portion of the container in contact with the foundations or saddles.

2. Containers of 2000 gal water capacity or less may be installed with nonfireproofed ferrous metal supports if mounted on concrete pads or footings, and if the distance from the outside bottom of the container shell to the ground does not exceed 12 in.

(c) Any container may be installed with non-fireproofed ferrous metal supports if mounted on concrete pads or footings, and if the distance from the outside bottom of the container to the ground does not exceed 5 ft, provided the container is in an isolated location and such installation is approved by the authority having jurisdiction.

(d) Containers may be partially buried providing the following requirements are met:

1. The portion of the container below the surface and for a vertical distance not less than 3 in. above the surface of the ground is protected to resist corrosion, and the container is protected against settling and corrosion as required for fully buried containers [see 2.6(f)].

2. Spacing requirements shall be as specified for underground tanks in B.6(b).

3. Relief valve capacity shall be as required for aboveground containers.

4. Container is located so as not to be subject to vehicular damage, or is adequately protected against such damage.

5. Filling densities shall be as required for aboveground containers.

(e) Containers buried underground shall be placed so that the top of the container is not less than 6 in. below grade. Where an underground container might be subject to abrasive action or physical damage due to vehicular traffic or other causes, then it shall be:

1. placed not less than 2 ft below grade, or
2. otherwise protected against such physical damage.

It will not be necessary to cover the portion of the container to which manhole and other connections are affixed; however, where necessary, protection shall be provided against vehicular damage. When necessary to prevent floating, containers shall be securely anchored or weighted.

(f) UNDERGROUND CONTAINERS.

1. Containers shall be given a protective coating before being placed underground. This coating shall be equivalent to hot dip galvanizing or to two coatings of red lead followed by a heavy coating of coal tar or asphalt. In lowering the container into place, care shall be exercised to prevent damage to the coating. Any damage to the coating shall be repaired before backfilling.
2. Containers shall be set on a firm foundation (firm earth may be used) and surrounded with earth or sand firmly tamped in place. Backfill should be free of rocks or other abrasive materials.

(g) Containers with foundations attached (portable or semi-portable containers with suitable steel "runners" or "skids" and popularly known in the industry as "skid tanks") shall be designed, installed and used in accordance with these rules subject to the following provisions: (See also Par. 3.19.)

1. If they are to be used at a given general location for a temporary period not to exceed 6 months they need not have fire-resisting foundations or saddles but shall have adequate ferrous metal supports.
2. They shall not be located with the outside bottom of the container shell more than 5 ft above the surface of the ground unless fire-resisting supports are provided.
3. The bottom of the skids shall not be less than 2 in. or more than 12 in. below the outside bottom of the container shell.
4. Flanges, nozzles, valves, fittings, and the like, having communication with the interior of the container shall be protected against physical damage.

   Note: It is recommended that such containers should have outlets only in the heads.
5. When not permanently located on fire-resisting foundations, piping connections shall be sufficiently flexible to minimize possibility of breakage or leakage of connections if container settles, moves, or is otherwise displaced.
6. Skids, or lugs for attachment of skids, shall be secured to container in accordance with the code or rules under which the container is designed and built (with a minimum factor of safety of four) to withstand loading in any direction equal to four times the weight of the container and attachments when filled to the maximum permissible loaded weight.

(h) Field welding where necessary shall be made only on saddle plates or brackets which were applied by manufacturer of tank.

(i) For aboveground containers secure anchorage or adequate pier height shall be provided against possible container flotation wherever sufficiently high flood water might occur.

(j) When permanently installed containers are interconnected, provision shall be made to compensate for expansion, contraction, vibration and settling of containers and interconnecting piping. Where flexible connections are used, they shall be of an approved type and shall be designed for a bursting pressure of not less than five times the vapor pressure of the product at 100 F. The use of nonmetallic hose is prohibited for permanently interconnecting such containers.

(k) Container assemblies listed for interchangeable installation aboveground or underground shall conform to the requirements for aboveground installations with respect to safety relief capacity and filling density. For installation aboveground all other requirements for aboveground installations shall apply. For installation underground all other requirements for underground installations shall apply.

2.7. Dikes and Embankments.

(a) Because of the pronounced volatility of liquefied petroleum gases, dikes are not normally necessary, hence their general requirement is not justified as in the case of gasoline or similar flammable liquids. It should be borne in mind that the heavy construction of the storage containers makes failure unlikely.

2.8. Protection of Container Accessories.

(a) Valves, regulating, gaging, and other container accessory equipment shall be protected against tampering and physical damage. Such accessories shall also be so protected during the transit of containers intended for installation underground.

Note: The use of other than frangible shank type locks is not desirable because it prevents access to gas controls in case of emergency.

(b) All connections to underground containers shall be located within a substantial dome, housing or manhole and with access thereto protected by a substantial cover.

2.9. Drips for Condensed Gas.

Where vaporized gas on low-pressure side of system may condense to a liquid at normal operating temperatures and pressures, suitable means shall be provided for revaporization of the condensate.

2.10. Damage from Vehicles.

When damage to LP-gas systems from vehicular traffic is a possibility, precautions against such damage shall be taken.

2.11. Pits and Drains.

Every effort should be made to avoid the use of pits, except pits fitted with automatic flammable vapor detecting devices. No drains or blow-off lines shall be directed into or in proximity to sewer systems used for other purposes.

2.12. General Provisions Applicable to Systems in Industrial Plants (of 2000 gal water capacity and more) and to Bulk Filling Plants.

(a) When standard watch service is provided, it shall be extended to the LP-gas installation and personnel properly trained.

(b) If loading and unloading are normally done during other than daylight hours, adequate lights shall be provided to illuminate storage containers, control valves and other equipment.

(c) Suitable roadways or means of access for extinguishing equipment such as wheeled extinguishers or fire department apparatus shall be provided.

(d) To minimize trespassing or tampering the area which includes container appurtenances, pumping equipment, loading and unloading facilities and cylinder filling facilities shall be enclosed with at least a 6-ft high industrial type fence unless otherwise adequately protected. There shall be at least two means of emergency access.

2.13. Container Charging Plants.

(a) The container charging room shall be located not less than:

1. 10 ft from bulk storage containers.
2. 25 ft from line of adjoining property which may be built upon.

(b) Tank truck filling station outlets shall be located not less than:

1. 25 ft from line of adjoining property which may be built upon.

2. 10 ft from pumps and compressors if housed in one or more separate buildings.

(c) The pumps or compressors may be located in the container charging room or building, in a separate building, or outside of buildings. When housed in a separate building, such building (a small noncombustible weather cover is not to be construed as a building) shall be located not less than:

1. 10 ft from bulk storage tanks.
2. 25 ft from line of adjoining property which may be built upon.
3. 25 ft from sources of ignition.

(d) Where a part of the container charging building is to be used for a boiler room or where open flames or similar sources of ignition exist or are employed, the space to be so occupied shall be separated from container charging room by a partition wall or walls of fire resistant construction continuous from floor to roof or ceiling. Such separation walls shall be without openings and shall be joined to the floor, other walls and ceiling or roof in a manner to effect a permanent gas tight joint.

2.14. Fire Protection.

(a) Each bulk plant shall be provided with at least one approved portable fire extinguisher having a minimum rating of 12-B,C. Ratings shall be in accordance with the Standard for Installation, Maintenance, and Use of Portable Fire Extinguishers, NFPA No. 10.

(b) In industrial installations involving containers of 150,000 gal aggregate water capacity or more, provision shall be made for an adequate supply of water at the container site for fire protection in the container area, unless other adequate means for fire control are provided. Water hydrants shall be readily accessible and so spaced as to provide water protection for all containers. Sufficient lengths of fire hose shall be provided at each hydrant location on a hose cart, or other means provided to facilitate easy movement of the hose in the container area. It is desirable to equip the outlet of each hose line with a combination fog nozzle. A shelter shall be provided to protect the hose and its conveyor from the weather.

(c) If in the opinion of the authority having jurisdiction, the use of fixed water spray nozzles will better serve to protect the containers and area, these may be specified. The method of release of water spray and alarm facilities shall be at the discretion of the authority having jurisdiction.

2.15. Painting.

(a) Aboveground containers shall be kept properly painted.

2.16. Lighting.

(a) At the discretion of the authority having jurisdiction, industrial installations shall be illuminated.

Reference may be made to the following extract from NFPA No. 58[1] regarding the location of containers and regulating equipment, except at gas utility company plants, refineries, and certain other specified locations:

B.6. Location of Containers and Regulating Equipment.

[Fig. 5-19 shows container spacing.—EDITOR]

(a) Containers and first stage regulating equipment shall be located outside of buildings other than buildings especially provided for this purpose, except containers and regulating equipment may be used indoors under the following conditions:

1. If temporarily used for demonstration purposes and the container has a maximum water capacity of 12 lb.
2. If used with a completely self-contained gas hand torch or similar equipment, and the container has a maximum water capacity of 2½ lb.
3. As provided in 1.7 and in Divisions IV and V.

(b) Each individual container shall be located with respect to the nearest important building or group of buildings or line of

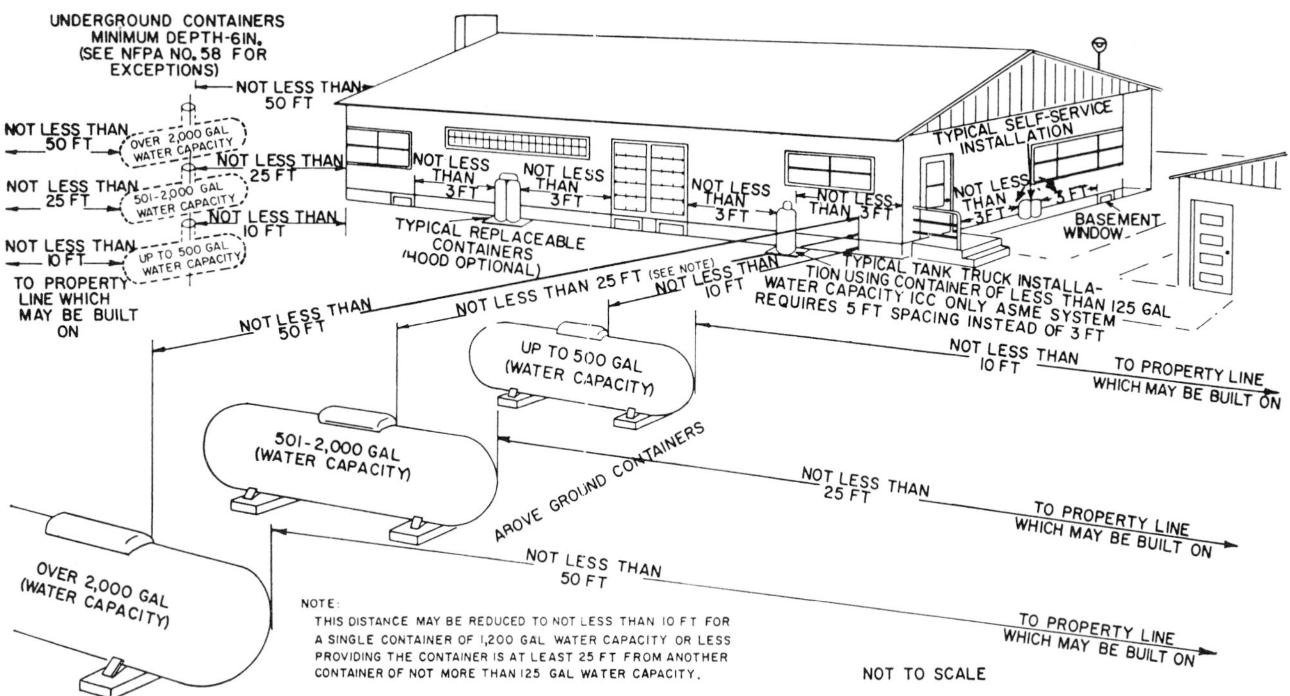

Fig. 5-19 Container spacing.[1] Above 30,000 gal water capacities, see the table in B.6(b) of NFPA No. 58 extracted in this chapter.

adjoining property which may be built on in accordance with the following table:

Water capacity per container, gal	Minimum distances, ft		
	Containers		Bet. aboveground containers
	Underground	Aboveground	
Less than 125	10	None	None
125 to 500	10	10	3
501 to 2000	25*	25*	3
2001 to 30,000	50	50	5
30,001 to 70,000	50	75	¼ of sum of diam of adjacent containers
70,001 to 90,000	50	100	

* Note: The above distance requirements may be reduced to not less than 10 ft for a single container of 1200 gal water capacity or less, providing such a container is at least 25 ft from any other LP-gas container of more than 125 gal water capacity.

(c) No containers while installed for use shall be stacked one above the other.

(d) In cases of bulk storage in heavily populated or congested areas, the authority having jurisdiction shall determine restrictions of individual tank capacity, total storage, and distance to line of adjoining property which may be built on and other reasonable protective methods.

(e) In industrial installations involving containers of 180,000 gal aggregate water capacity or more, where serious mutual exposures between the container and adjacent properties prevail, the authority having jurisdiction may require fire walls or other means of special protection designed and constructed in accordance with good engineering practices.

(f) In the case of buildings devoted exclusively to gas manufacturing and distributing operations the above distances may be reduced provided that in no case shall containers of water capacity exceeding 500 gal be located closer than 10 ft to such gas manufacturing and distributing buildings.

(g) Any container used in domestic or commercial service, where transfer of liquid is made from such containers into portable containers such as on tractors, skid tanks, or similar applications shall be located not less than 50 ft from nearest important building except as provided in Par. 8.5(a)1. Special attention shall be given to maintaining the above distances on such transferring in trailer camps with respect to any trailer, except as provided in Par. 6.13.

(h) Readily ignitable material such as weeds and long dry grass shall be removed within ten feet of any container.

(i) The minimum separation between liquefied petroleum gas containers and flammable liquid tanks shall be 20 ft, and the minimum separation between a container and the center line of the dike shall be 10 ft. The foregoing provision shall not apply when LP-gas containers of 125 gal or less capacity are installed adjacent to Class III flammable liquid tanks of 275 gal or less capacity.

(j) Suitable means shall be taken to prevent the accumulation of flammable liquids under adjacent liquefied petroleum gas containers, such as by diking, diversion curbs or grading.

(k) When dikes are used with flammable liquid tanks, no liquefied petroleum gas containers shall be located within the diked area.

Reference may be made to the following extract from NFPA No. 58[1] for systems utilizing containers constructed in accordance with the Interstate Commerce Commission Specifications:

CYLINDER SYSTEMS

(Sometimes called Bottled Gas)

1.2. Description of a Division I System.

(a) A Division I system shall include the container base or bracket, containers, container valves, connectors, manifold valve assembly, regulators, and relief valves.

1.3. Location of Containers and Regulating Equipment.

(a) Containers shall not be buried below ground. However, this shall not prohibit the installation in a compartment or recess below grade level, such as a niche in a slope or terrace wall which is used for no other purpose, providing that the container and regulating equipment are not in contact with the ground and the compartment or recess is drained and ventilated horizontally to the outside air from its lowest level, with the outlet at least 3 ft away from any building opening which is below the level of such outlet. Except as provided in B.10(k), the discharge from safety relief devices shall be located not less than 3 ft horizontally away from any building opening which is below the level of such discharge and shall not terminate beneath any building unless such space is well ventilated to the outside and is not enclosed on more than two sides.

(b) Containers shall be set upon firm foundation or otherwise firmly secured; the possible effect on the outlet piping of settling shall be guarded against by a flexible connection or special fitting.

1.7. Use of Gas for Industrial Applications Where Oxygen Is Not Required.

(a) Where portability of containers is necessary making their location outside the building or structure impracticable, containers may be located for use but not for storage inside the building or structure, only: where gas is to be used for industrial processing or repair work in an industrial building or structure being employed for industrial purposes; or where it is to be used in the construction, repair or improvement of buildings or structures and their fixtures and equipment. Such installations are subject to the following additional rules:

1. The regulator employed may be connected directly to the cylinder valve or located on a manifold which is connected to the cylinder valve. The regulator must be of a type suitable for use with liquefied petroleum gas.
2. The aggregate capacity of the containers connected to each portable manifold shall not exceed 300 lb of gas by weight, and not more than one such manifold with containers may be located in the same room unless separated by at least 50 ft.
3. Manifolds and fittings connecting containers to the pressure reducing regulator inlets shall be designed to withstand without rupture at least 500 psig.
4. Containers, regulating equipment, and manifolds, shall be located where they are not subjected to excessive rise in temperature, physical damage or tampering by unauthorized persons.

Refineries, Natural Gasoline Plants, Terminals, and Tank Farms

A standard for the construction of facilities in these locations is available.[2] This Standard covers installations which are specifically excluded from Standards No. 58[1] and No. 59.[3]

The large-scale storage requirements for LP-gas operations have been met both by aboveground containers and by underground caverns.

Refrigerated storage, particularly for butanes and butenes, has been used at refineries.

The earliest LP-gas containers were cylindrical tanks of riveted construction erected either horizontally or vertically.

Individual tanks containing 150,000 gal were in use by 1935 in refinery service.

Spheroids intended to operate under pressures of about 20 psig have been built to sizes up to 5 million gal. By the use of refrigeration, vapor pressures can be maintained below the design working pressure of the container.

When higher pressures are used, spherical containers provide the most desirable shape. Such containers, built in accordance with the ASME Code designed for 50 psig, have been constructed in sizes up to 1250 million gal.

Other relevant codes include Sec. VIII of the ASME Unfired Pressure Vessel Code for storage containers designed for working pressures above 15 psig, and the API Low Pressure Storage Tank Code 620 for containers between 0.5 and 15 psig.

A range of comparative costs (1955) of LP-gas aboveground storage systems is shown in Table 5-9.

Table 5-9 Comparative Costs of LP-Gas Aboveground Storage[4]

Design working press., psig	Aboveground storage	Container cost per bbl, $
250	30 M gal horizontal	24.0–31.0
125	30 M gal horizontal	20.3–27.5
125	31 M gal horizontal	15.3–21.0
125	4 M & 6 M bbl spheres	18.9–26.0
75	45 M gal horizontal	19.0–24.9
75	63 M gal horizontal	14.7-19.5
75	6 M bbl spheres	13.8–19.9

Underground LP-Gas Storage

Storage of LP-gas in cavities in salt and in mined caverns has found wide economical use. Some attempts have also been made to store LP-gas products in depleted oil reservoirs. However, this system has not been widely used because of technical factors and limited availability of suitable structures. In 1960, underground storage capacity was 2.5 billion gal (92 per cent dissolved salt type and 8 per cent mined cavern type).[5]

Use of mined caverns is based upon the availability of structurally strong, unfractured, relatively homogeneous, impervious rock at appropriate depths. To provide the necessary strength, depths of 200 ft or more are required for propane and 100 ft for butane. A standard for underground storage of LP-gas which provides some basic rules and general information is available.[6]

Cost of caverns involves substantial fixed amounts for the evaluation and testing needed to select a suitable site for the sinking and lining of the shaft (commonly 42 in. diam), and for drilling the ventilation shafts used during construction and later used for the handling of the product, together with pumps, piping, valves, meters, and miscellaneous equipment. Approximately $400,000 (1959) will be needed, regardless of the cavern size. Mining costs in general average $3.75 per barrel; thus total cost may run from $7.75 for a 100,000 bbl cavern to $4.55 for the 500,000 bbl cavern. These figures do not include other costs for land, terminal facilities, and any necessary product treating equipment.

Other Storage Systems

In addition to steel pressure containers and steel containers for refrigerated LP-gas products, limited use has been made of prestressed concrete pipe containers installed underground. It is planned that future installations of this type be built in the form of a torus to eliminate the problem of end bracing for the heads. Proposals also have been made to store propane[7] and refrigerated propane in appropriate prestressed concrete containers.

Utility Plants. LP-gas storage at utility plants should conform to NFPA No. 59.[3]

DISTRIBUTION OF LP-GAS

Most producer storage of LP-gases is at the point of production, although it is becoming more common to construct and operate large volume storage nearer major consuming areas. Most of this secondary storage is located on pipelines or at water terminals. However, there are several major storage terminals which receive their product by tank car during the "off" season.

Storage by the marketer is done mostly in the horizontal cylindrical containers of ASME construction for 250 psig design pressure. Butane storage may utilize tanks having a design pressure of 125 psig (the minimum set forth in NFPA No. 58[1]). Tanks containing 12,000, 18,000, and 30,000 gal are the most common for bulk storage plants. A few marketers have developed their own underground storage where their operations are large enough to justify the investment and where their marketing area is close to suitable geological formations.

Consumer storage may include all types and sizes mentioned above for producers and marketers. The size and type will vary with the load demand by the domestic, commercial, industrial, petrochemical, or utility user.

Large Industrial Plants. Propane storage systems at large industrial plants are covered by NFPA No. 58.[1] Several major points of difference exist between NFPA No. 58 and NFPA No. 59.[3] For example, Par. 2.5 in NFPA No. 58 limits liquid containers to 90,000 U. S. gallons water capacity, while NFPA No. 59 allows over a million gallons. Space does not permit discussion of the technical differences. However, the pertinent standard should be used without change.

To facilitate liquid transfer into the storage tanks, a vapor equalization line may be used. When this line is used and the vapor quantities involved are substantial, it may be desirable to account for vapor flow.

Domestic Consumer Premises. For domestic cooking, refrigeration, incineration, water heating, and auxiliary space heating, a 100 lb (water capacity) cylinder or, more often, two 100 lb cylinders are adequate for vaporization capacities and for at least a week's supply. Tank truck service is widely used for space heating in addition to cooking and water heating loads. Consumers using LP-gas for central heating should have larger tanks adequate for providing sufficient fuel for the coldest month. The preferred size is twice the minimum needed to insure adequate storage and convenient fuel delivery. Common tank sizes for such purposes range from 350 to 1500 gal capacity.

Unloading LP-Gas

LP-gases are unloaded from tank cars and tank trucks into storage tanks thru a closed system under pressure or by means of a liquid pump, a gas compressor, or gas pressure. In the first two methods of unloading, connections are made thru flexible hose or pipe (fitted with special swivel joints) from the liquid outlet on the tank car or tank truck to the liquid inlet of the storage system. Another flexible hose or flexible piping system connects the vapor line of the storage system to the vapor space of the tank car or truck. If a pump is used, liquid is pumped from the transport tank into the storage system, and the pressures in the storage container and the transport tank are equalized thru the vapor connections.

If a compressor is used (Fig. 5-20), vapor is taken from the storage container and discharged into the vapor space of the tank car or truck, creating a pressure differential between the two, which forces liquid into the storage container. A compressor may have its connections reversed, to remove most of the vapors remaining after the liquid is removed from tank cars. This procedure is usually not economical with tank truck deliveries because of the additional time required to remove the vapors.

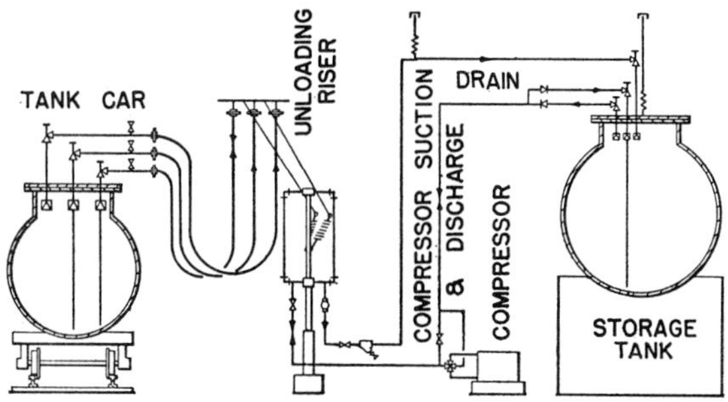

Fig. 5-20 Tank car unloading by pressure differential (compressor).

Gas under pressure may be used to increase the tank car pressure above that of the storage tank, thereby forcing liquid to flow to the storage tank. LP-gases are normally loaded into tank cars and tank trucks from bulk storage by means of pumps.

All flexible hose connections should be protected with either excess-flow valves or back-flow check valves installed in the piping, to prevent the escape of LP-gas in event of hose failure. The hose should be resistant to the action of liquid LP-gas. During tank car loading and unloading operations, a "Stop—Tank Car Connected" sign should be prominently displayed, and the wheels at both ends of the car should be blocked on the rails. The rails on which tank cars may be loaded or unloaded should be electrically grounded to the transfer piping in accordance with railroad or Association of American Railroads (AAR) requirements.

All electrical equipment used in connection with transferring LP-gases should meet the Class 1, Group D requirements of the *National Electrical Code.*

Bulk delivery of LP-gas by tank truck involves the following steps:

1. Position the truck reasonably near the customer's tank in such a manner that there will be a minimum of interference with others using the service areas. Set brakes and chock rear wheels before connecting hose.
2. Gage the contents of the receiving tank, or determine in other ways that there is sufficient room to accommodate the delivery.
3. Caution people not to smoke in the vicinity of the truck. Check delivery area for open flames (as in construction markers), and for other sources of ignition. Do not unload until the hazard has been eliminated.
4. Slowly open the proper valves on the truck and receiving tank, and commence the delivery of the product.
5. Immediately check all connections for leaks, and if any are evident, stop the unloading operation and make necessary repairs.
6. Be alert and in attendance during the delivery. Remain in a position to observe the hose connections at container being filled, and to have quick accessibility to the unloading shutoff valves.
7. If customer's tank is accidentally overfilled, evacuate excess to truck tank.
8. Should a leak occur, take the following steps:
    a. Stop flow of product, if possible.
    b. Warn everyone away from area.
    c. If flow of product continues, call fire department where possible and stand by with fire extinguishers.
    d. Do not start vehicle engines—turn off any that are running.
    e. Do not leave scene until area has been made safe.
9. After completing delivery, close valves, disconnect the hose from the receiving tank, and replace cover. Rewind hose on reel or coil in hose tub.
10. Walk around vehicle before departing, check all truck valves to make sure they are closed and that hoses are disconnected and properly secured, and pick up wheel chocks. Note any improper installation or other unsafe conditions—correct if possible (report to supervisor).
11. Delay in unloading, because of hazardous condition or damage to vehicle, should be reported to the supervisor.

Cylinder deliveries of LP-gas involve the following steps:

1. Position the truck near the customer's cylinder installation in such a manner that there will be a minimum of interference with others using the service area and so that the distance for cylinder transporting will be shortened. Set brakes properly.
2. Determine if cylinder is actually empty or if out-of-gas call was due to some other difficulty.
3. Caution people not to smoke in the vicinity of the truck. Check delivery area for open flames or other sources of ignition. Do not unload cylinders until hazards have been eliminated.
4. Lower cylinder carefully from truck bed to the ground.
5. Use hand truck to wheel cylinder to installation locations. Do not remove cap.
6. Close valve on used cylinder, disconnect it, and place cap on cylinder until cylinder is in place.
7. Connect full cylinder and:
    a. For single cylinder installation or a two cylinder installation where cylinders are empty, see "out-of-gas" procedures.

b. For two cylinder installation with manual cylinder changeover regulator, leave cylinder valve closed.
c. For two cylinder manual installation without a changeover regulator, do not open cylinder valve, but tag it to notify customer that cylinder was changed.
d. For two cylinder automatic changeover regulator installations, open cylinder valve.

8. Place used cylinder carefully on truck bed.
9. Should a leak occur, take the following steps as far as possible:

a. Stop flow of product or remove leaking cylinder to a safe area.
b. Warn everyone away from area and remove sources of ignition.
c. If flow of product continues, call fire department where possible, and stand by with fire extinguisher.
d. No vehicle engines should remain running or be started.
e. Do not leave scene until area has been made safe.

"Out-of-Gas Call" Procedures

1. After arriving at the customer's location, *make sure* that the tank or cylinder and all appliance valves are closed.
2. Fill the customer's tank or replace the cylinders with filled ones.
3. Open the tank or cylinder valve and then put all appliances back into service, making sure that all pilots are properly relighted.
4. If no one is home, leave the supply valve closed. Then hang tags on all doors and the appropriate tag on the supply valve. These tags read as follows:

IMPORTANT NOTICE

On (date) we refilled your gas supply tank with (trade name) gas. Because you were not at home, the gas supply valve on the tank has been left closed. Before opening the gas supply valve (open slowly) be sure that gas control valves on *all* appliances are closed.

IMPORTANT NOTICE

On (date) we delivered and connected a full cylinder of (trade name) gas. Because you were not at home, the gas supply valve on the cylinder has been left closed. Before opening the gas supply valve (open slowly) be sure that gas control valves on *all* appliances are closed.

NOTICE

Gas supply to appliances has been shut off by closing this cylinder valve. See that burners and pilot lights on all appliances are turned off before turning on gas at cylinder valve. Remove this tag before turning on gas.

Piping and Fittings

Reference may be made to the following extract from NFPA No. 58:[1]

B.8. Piping, Tubing and Fittings.

(**a**) Piping, except as provided in Division III, Par. 3.3(c), Division IV, Par. 4.5(a), Division VI, Par. 6.9(c) and Division VII, Par. 7.8(c), shall be wrought iron or steel (black or galvanized), brass, copper, or aluminum alloy pipe; or seamless copper, brass, steel or aluminum alloy tubing.

1. Vapor piping and tubing with operating pressures not exceeding 125 psig shall be suitable for a working pressure of at least 125 psig. Piping shall be at least Schedule 40 (ASTM A-53, Grade B Electric Resistance Welded and Electric Flash Welded Pipe or equal).
2. Vapor piping and tubing with operating pressures over 125 psig and all liquid piping and tubing shall be suitable for a working pressure of at least 250 psig. Pipe shall be at least Schedule 80 if joints are threaded, or threaded and back welded. At least Schedule 40 (ASTM A-53, Grade B Electric Resistance Welded and Electric Flash Welded Pipe or equal) shall be used if joints are welded, or welded and flanged.

(**b**) Aluminum alloy pipe shall be in accordance with specification ASTM B-241 except that the use of alloy 5456 is prohibited. Copper tubing may be of the standard grade K or L, or equivalent. Aluminum alloy tubing shall be of standard type A or B, or equivalent as covered in specification ASTM B-318. Aluminum alloy pipe and tubing shall be suitably marked every 18 in. indicating compliance with appropriate ASTM specifications. Copper and aluminum alloy tubing shall have a minimum wall thickness of 0.032 in. Aluminum alloy tubing and piping shall be protected against external corrosion when (a) it is in contact with dissimilar metals other than galvanized steel, (b) its location is subject to repeated wetting by such liquids as water (except rain water), detergents, sewage or leakage from other piping, (c) it passes thru flooring, plaster, masonry, or insulation. Galvanized sheet steel or pipe galvanized inside and out may be considered suitable protection. Aluminum alloy pipe and tubing shall be limited to ⅝ in. nominal tubing size or ¾ in. nominal pipe size and shall not be used for pressure in excess of 20 psig. Aluminum alloy pipe or tubing shall not be installed within 6 in. of the ground.

(**c**) In systems where the gas in liquid form without pressure reduction enters the building [see B.13(a)] only heavy walled seamless brass or copper tubing with an internal diameter not greater than 3/32 in., and a wall thickness of not less than 3/64 in. shall be used. This requirement shall not apply to research and experimental laboratories, buildings or separate fire divisions of buildings used exclusively for housing internal combustion engines, and to commercial gas plants or bulk stations where containers are charged, nor to industrial vaporizer buildings.

(**d**) Pipe joints may be screwed, flanged, welded, soldered, or brazed with a material having a melting point exceeding 1000 F. Joints on seamless copper, brass, steel or aluminum alloy gas tubing shall be made by means of approved gas tubing fittings, or soldered or brazed with a material having a melting point exceeding 1000 F.

(**e**) For operating pressures of 125 psig or less, fittings shall be designed for a pressure of at least 125 psig, except for tank truck requirements, see Par. 3.3(b). For operating pressures above 125 psig, fittings shall be designed for a minimum of 250 psig.

(**f**) The use of threaded cast iron pipe fittings such as ells, tees, crosses, couplings, and unions is prohibited. Aluminum alloy fittings shall be used with aluminum alloy pipe and tubing. Insulated fittings shall be used where aluminum alloy pipe or tubing connects with a dissimilar metal.

(**g**) Strainers, regulators, meters, compressors, pumps, etc., are not to be considered as pipe fittings. This does not prohibit the use of malleable, nodular or higher strength gray iron for such equipment.*

(**h**) All materials such as valve seats, packing, gaskets, diaphragms, etc., shall be of such quality as to be resistant to the action of liquefied petroleum gas under the service conditions to which they are subjected.

(**i**) All piping, tubing, or hose shall be tested after assembly and proved free from leaks at not less than normal operating pressures. After installation, piping and tubing of all domestic and commercial systems shall be tested and proved free of leaks using a manometer or equivalent device that will indicate a drop in pressure. Test shall not be made with a flame.

(**j**) Provision shall be made to compensate for expansion, contraction, jarring and vibration, and for settling. This may be accomplished by flexible connections.

---

* See ASTM Stds.: A47-52, A395-56T, A126-42, Class B or C.

(k) Piping outside buildings may be buried, aboveground, or both, but shall be well supported and protected against physical damage. Where soil conditions warrant, all piping shall be protected against corrosion. Where condensation may occur, the piping shall be pitched back to the container, or suitable means shall be provided for revaporization of the condensate.

Vaporization

Heat to vaporize the liquefied petroleum must come thru walls of the container from the atmosphere or ground surrounding it, or from heat exchangers or vaporizers on large installations. If adequate container surface is not provided, the liquid will become too cold and will not vaporize fast enough for supply demands. Latent heat of vaporization of propane is 183.3 Btu per lb at −44 F, and 151 Btu per lb at 65 F.

For **aboveground tanks** a coefficient of heat transfer of 2 Btu per hr-sq ft of container surface and °F temperature difference is a low figure. For **underground containers** a coefficient of 0.5 Btu per hr-sq ft of container surface and °F temperature difference may be used. Frost formation on container walls acts as an insulating layer, slowing up heat flow. Appearance of frost is a sign that the container is too small. Temperature of frost formation on LP-gas containers is shown in Table 5-10. Table 5-11 shows the number of 100 lb cylinders required for various withdrawal rates and temperatures.

Selection of an ASME tank may be made from Table 5-12. This table is based upon the standard sizes used in the industry. An alternate "rule of the thumb" method follows.

Sizing ASME Tanks. Vaporization rate, H, in Btu, for tank ⅓ full is given by:

$$H = DLC$$

where: D = tank diameter, in.
L = tank length, in.
C = vaporization factor (Table 5-13)

Table 5-10 Temperature of Frost Formation (°F) on LP-Gas Containers[8]

Atmospheric temp, °F	Relative humidity, per cent				
	30	40	60	80	100
32	7	13	21	27	32
30	5	11	19	25	30
20	−3½	2	9½	15½	20
10	−12½	−7½	0	6	10
0	−21½	−16½	−9½	−4	0
−10	−31	−26	−19	−14	−10
−20	−40	−35½	−28½	−23	−20
−30	...	...	−38	−33	−30

Table 5-11 Number of 100 lb LP-Gas Cylinders Required for Various Demands and Ambient Temperatures[9]

Withdrawal rate*		Lowest outdoor temperature, °F (avg for 24 hr period)						
cfh	MBtu/hr	32	20	10	0	−10	−20	−30
10	25	1	1	1	1	1	1	2
25	62.5	1	1	1	2	2	3	4
50	125	2	2	3	3	4	5	9
100	250	4	4	5	6	7	10	20

* Average rate for 8 hr period (not absolute maximum).

Table 5-12 Sizing LP-Gas Customer Aboveground Tanks for Various Demands and Ambient Temperatures[9]

(in gallons water capacity)

Withdrawal rate*		Lowest outdoor temperature, °F (avg for 24 hr period)						
cfh	MBtu/hr	32	20	10	0	−10	−20	−30
50	125	115	115	115	250	250	400	600
100	250	250	250	250	400	500	1000	1500
150	375	300	400	500	500	1000	1500	2500
200	500	400	500	750	1000	1200	2000	3500
300	750	750	1000	1500	2000	2500	4000	5000

* Average rate for 8 hr period (not absolute maximum).

Table 5-13 Vaporization Factor, C, for ASME Aboveground Tanks[9]

Temp, °F	C	Temp, °F	C
70	235	10	110
60	214	0	90
50	193	−10	70
40	172	−20	48
30	152	−30	28
20	131		

Note: Dimensions, tank shape, and various other factors affect vaporization rate. For example, a sphere has the minimum surface area per unit volume, and consequently the poorest vaporization rate.

Table 5-14 Vaporization Capacities of 100 lb Propane Cylinders

(New England Area)[8]

No. of cylinders required	Rate of withdrawal per cylinder, lb per hr					
	0.5	1	2	3	5	6
	Vaporization capacity, MBtu per hr					
1	11	22	43	65	109	130
2	22	43	87	130	217	260
4	43	87	174	260	434	520
6	65	130	260	391	651	781
8	87	174	347	520	868	1041
12	130	260	520	782	1302	1562
16	174	347	695	1042	1737	2083
20	217	434	868	1302	2169	2604

For underground tanks, a ground temperature of 50 F has been assumed with a minimum liquid temperature of 35 F. To take full advantage of the 15° differential, propane content of the fuel stored must be sufficient to assure a satisfactory operating pressure at 35 F.

Table 5-14 shows vaporization capacity of 100 lb propane cylinders. For straight butane vaporization, capacities will be from a third to a half of those shown in Table 5-14. It is assumed that the tank or cylinder size will be adequate to meet the most severe conditions of one-third full capacity at extremely cold atmospheric conditions. The allowable temperature differential in any given area will depend upon the relative atmospheric humidity, which should be checked against temperatures in Table 5-10.

Tank sizes in Table 5-15 are the minimum required for adequate vaporization. For economical operation, larger sizes should be recommended so that no more than five or six deliveries per year will be needed. In an installation designed for a large demand, it may be necessary to use either a number

Table 5-15 Tank Sizing Table (New England Area)[8]
(for a 4:1 length to diameter ratio)

Tank size, gal water capacity	Load factors*						
	0.1	0.2	0.3	0.4	0.6	0.8	1.0
	Vaporization capacity, MBtu per hr						
100	120	68	50	42	39	38	37
150	178	98	71	60	51	50	49
200	235	127	92	76	64	60	59
300	335	181	130	106	88	80	78
400	440	233	167	135	108	98	94
600	625	330	236	191	147	131	124
800	800	420	300	242	182	160	150
1,000	970	508	360	290	215	187	175
1,500	1360	700	496	400	290	248	228
2,000	1720	880	620	498	356	300	277
2,500	2040	1050	740	588	418	350	322
3,000	2350	1210	855	672	475	395	362
4,000	2930	1510	1070	832	582	480	440
5,000	3460	1800	1275	972	682	556	510
6,000	3920	2070	1465	1100	776	627	573
7,000	4320	2330	1645	1225	864	695	636
8,000	4700	2560	1825	1350	950	760	697
9,000	5100	2770	2000	1470	1035	820	752
10,000	5400	2950	2150	1575	1110	880	806

* The **load factor** for an installation or an appliance is the ratio of the actual fuel consumption, for a given period of time, to the maximum possible consumption with all burners operating continuously at full capacity. One month is the usual base period.

of small tanks or a vaporizer (with larger tanks) to insure sufficient vaporization of LP-gas. This is an economic balance that must be determined from an analysis of the specific installations. Vaporizer equipment is listed by the Underwriters' Laboratories, Inc.

Reference may be made to the following extract from NFPA No. 58[1] for regulations on vaporizers and housing:

B.11. Vaporizer and Housing.

Note: B.11 does not apply to motor fuel vaporizers (see Division IV, Sec. 4.7), nor to integral vaporizer-burners used for such purposes as weed burners or tar kettles.

(a) Indirect fired vaporizers utilizing steam, water or other heated medium shall be constructed and installed as follows:

1. Vaporizers shall be constructed in accordance with the requirements of the ASME Unfired Pressure Vessel Code or API-ASME Unfired Pressure Vessel Code and shall be permanently marked as follows:

With the code marking signifying the specifications to which vaporizer is constructed.

With the allowable working pressure and temperature for which the vaporizer is designed.

With the sum of the outside surface area and the inside heat exchange surface area expressed in square feet. (See Appendix B.)

With the name or symbol of the manufacturer.

2. Vaporizers having an inside diameter of 6 in. or less exempted by the ASME Unfired Pressure Vessel Code shall have a design pressure not less than 250 psig and need not be permanently marked.

3. Heating or cooling coils shall not be installed inside a storage container.

4. Vaporizers may be installed in buildings, rooms, sheds, or lean-tos used exclusively for gas manufacturing or distribution, or in other structures of light, noncombustible construction or equivalent, well ventilated near the floor line and roof.

5. Vaporizers shall have at or near the discharge, a safety relief valve providing an effective rate of discharge in accordance with Appendix B, except as provided in Sec. 2.3(e)(1).

6. The heating medium lines into and leaving the vaporizer shall be provided with suitable means for preventing the flow of gas into the heat systems in the event of tube rupture in the vaporizer. Vaporizers shall be provided with suitable automatic means to prevent liquid passing through the vaporizers to the gas discharge piping.

7. The device that supplies the necessary heat for producing steam, hot water, or other heating medium may be installed in a building, compartment, room or lean-to which shall be ventilated near the floor line and roof to the outside. This device location shall be separated from all compartments or rooms containing liquefied petroleum gas vaporizers, pumps and central gas mixing devices by a wall of substantially fire resistant material and vapor tight construction. This requirement does not apply to the domestic water heaters which may supply heat for a vaporizer in a domestic system.

8. Gas-fired heating systems supplying heat exclusively for vaporization purposes shall be equipped with automatic safety devices to shut off the flow of gas to main burners, if pilot light should fail.

9. Vaporizers may be an integral part of a fuel storage container directly connected to the liquid section or gas section or both.

10. Vaporizers shall not be equipped with fusible plugs.

11. Vaporizer houses shall not have unprotected drains to sewers or sump pits.

(b) Atmospheric vaporizers employing heat from the ground or surrounding air shall be installed as follows:

1. Buried underground, or

2 Located inside building close to a point at which pipe enters the building provided capacity of unit does not exceed 1 qt.

3. Vaporizers of less than 1 qt capacity heated by the ground or surrounding air, need not be equipped with safety relief valves provided that adequate tests certified by any of the authorities listed in B.2 demonstrate that the assembly is safe without safety relief valves.

4. Vaporizers designed primarily for domestic service shall be protected against tampering and physical damage.

(c) Direct gas-fired vaporizers shall be constructed, marked, and installed as follows:

1. (a) With the requirements of the ASME Code that are applicable to the maximum working conditions for which the vaporizer is designed. (See Sec. B.3.)

(b) With the name of the manufacturer; rated Btu input to the burner; the area of the heat exchange surface in square feet; the outside surface of the vaporizer in square feet; and the maximum vaporizing capacity in gallons per hour.

2. (a) Vaporizers may be connected to the liquid section or the gas section of the storage container, or both; but in any case there shall be at the container a manually-operated valve in each connection to permit completely shutting off when desired, of all flow of gas or liquid from container to vaporizer.

(b) Vaporizers with capacity not exceeding 35 gal per hour shall be located at least 5 ft from container shutoff valves. Vaporizers having capacity of more than 35 gal but not exceeding 100 gal per hour shall be located at least 10 ft from the container shutoff valves. Vaporizers having a capacity greater than 100 gal per hour shall be located at least 15 ft from container shutoff valves.

3. Vaporizers may be installed in buildings, rooms, housings, sheds, or lean-tos used exclusively for vaporizing or mixing of liquefied petroleum gas. Vaporizing housing structures shall be of noncombustible construction, well ventilated near the floor line and the highest point of the roof. See Sec. B.10(a) for venting of relief valves.

4. Vaporizers shall have at or near the discharge, a safety relief valve providing an effective rate of discharge in accord-

ance with Appendix B. Relief valve shall be so located as not to be subjected to temperatures in excess of 140 F.

5. Vaporizers shall be provided with suitable automatic means to prevent liquid passing from the vaporizer to the gas discharge piping of the vaporizer.

6. Vaporizers shall be provided with means for manually turning off the gas to the main burner and pilot.

7. Vaporizers shall be equipped with automatic safety devices to shut off the flow of gas to main burners if pilot light should fail. When flow through pilot exceeds 2000 Btu per hour, the pilot also shall be equipped with automatic safety device to shut off the flow of gas to the pilot should the pilot flame be extinguished.

8. Pressure regulating and pressure reducing equipment if located within 10 ft of a direct fired vaporizer shall be separated from the open flame by a substantially airtight noncombustible partition or partitions.

9. Except as provided in Sec. B.11(c)3 the following minimum distances shall be maintained between direct fired vaporizers and nearest important building or group of buildings or line of adjoining property which may be built upon:

10 ft for vaporizers having a capacity of 15 gal per hour or less vaporizing capacity.

25 ft for vaporizers having a vaporizing capacity of 16 to 100 gal per hour.

50 ft for vaporizers having a vaporizing capacity exceeding 100 gal per hour.

10. Direct fired vaporizers shall not raise the product pressure above the design pressure of the vaporizer equipment nor shall they raise the product pressure within the storage container above the pressure shown in the second column of the table in Sec. 2.1(a).

11. The minimum capacity of the storage container feeding the vaporizer shall not be less than ten times the hourly capacity of the vaporizer in gallons.

12. Vaporizers shall not be provided with fusible plugs.

13. Vaporizers shall not have unprotected drains to sewers or sump pits.

(d) Direct gas-fired tank heaters shall be constructed and installed as follows:

1. Direct gas-fired tank heaters, and tanks to which they are applied, shall only be installed aboveground.

2. Tank heaters shall be permanently marked with the name of the manufacturer, the rated Btu input to the burner, and the maximum vaporizing capacity in gallons per hour.

3. Tank heaters may be an integral part of a fuel storage container directly connected to the container liquid section, or vapor section, or both.

4. Tank heaters shall be provided with a means for manually turning off the gas to the main burner and pilot.

5. Tank heaters shall be equipped with an automatic safety device to shut off the flow of gas to main burners, if pilot light should fail. When flow thru pilot exceeds 2000 Btu per hour, the pilot also shall be equipped with automatic safety device to shut off the flow of gas to the pilot should the pilot flame be extinguished.

6. Pressure regulating and pressure reducing equipment if located within 10 ft of a direct fired tank heater shall be separated from the open flame by a substantially airtight noncombustible partition.

7. The following minimum distances shall be maintained between a storage tank heated by a direct fired tank heater and nearest important building or group of buildings or line of adjoining property which may be built upon:

10 ft for storage containers of less than 500 gal water capacity.

25 ft for storage containers of 500 to 1200 gal water capacity.

50 ft for storage containers of over 1200 gal water capacity.

8. No direct fired tank heater shall raise the product pressure within the storage container over 75 per cent of the pressure set out in the second column of the table of Sec. 2.1(a).

9. The minimum capacity of a storage container being heated by a direct fired tank heater shall be not less than ten times the hourly vaporizing capacity of the tank heater in gallons.

(e) The vaporizer section of vaporizer-burners used for dehydrators or dryers shall be located outside of buildings. They shall be constructed and installed as follows:

1. Vaporizer-burners shall have a minimum design pressure of 250 psig with a factor of safety of five.

2. Manually operated positive shutoff valves shall be located at the containers to shut off all flow to the vaporizer-burners.

3. Minimum distances between storage containers and vaporizer-burners shall be as follows:

Water capacity per container, gal	Minimum distances, ft
Less than 501	10
501 to 2000	25
Over 2000	50

4. The vaporizer section of vaporizer-burners shall be protected by a hydrostatic relief valve. (See B.10(j).) The relief valve shall be located so as not to be subjected to temperatures in excess of 140 F. The discharge of the relief valve shall be directed away from component parts of the equipment and operating personnel.

5. Vaporizer-burners shall be provided with means for manually turning off the gas to the main burner and pilot.

6. Vaporizer-burners shall be equipped with automatic safety devices to shut off the flow of gas to the main burner and pilot in the event the pilot is extinguished. (See NFPA No. 93.)

7. Pressure regulating and control equipment shall be located so that the temperatures surrounding this equipment shall not exceed 140 F. If necessary, the equipment shall be separated from the open flame by a noncombustible partition.

8. Pressure regulating and control equipment when located downstream of the vaporizer shall be designed to withstand the maximum discharge temperature of the vapor.

9. The vaporizer section of vaporizer-burners shall not be provided with fusible plugs.

10. Vaporizer coils or jackets shall be made of ferrous metal or high temperature alloys.

11. Equipment utilizing vaporizer-burners shall be equipped with automatic shutoff devices upstream and downstream of the vaporizer section connected so as to operate in the event of excessive temperature, flame failure and, if applicable, insufficient air flow. (See Standard for Dehydrators and Dryers for Agricultural Products, NFPA No. 93.)

PRESSURE REGULATION

The A.G.A. Laboratories test LP-gas appliances under ASA standards. **Test pressures** for domestic appliances are 8 in. w.c. min and 13 in. w.c. max. Industrial appliances may operate at these pressures or up to several pounds, depending upon burner design.

Since vapor pressure in the tank may be as high as 188.7 psig for propane at 100 F and 51.6 psig for butane at 100 F, delivery pressure must be reduced to and held at controlled appliance pressures. This is done with either a single regulator or two regulators in series. Pressure regulators for installation on LP-gas systems are tested and listed by Underwriters' Laboratories, Inc.; appliance regulators are tested and listed either by A.G.A. Laboratories or Underwriters' Laboratories, Inc.

An LP-gas pressure regulator is similar to the *pounds-to-pounds* or *pounds-to-inches* regulators used in distribution of utility company gases.

Relief valves are incorporated in pressure regulators to relieve excess pressures at not less than two and not more than three times discharge pressures. Normal setting of relief valves is about 22 to 27 in. w.c. Relief setting may depend upon height of a mercury column or, more frequently, upon spring loading.

If pressure reduction is accomplished in two stages, there is less refrigerating effect and less chance of freeze-up. Protection of the regulator from rain and snow is essential. Its vent pipe should be pointed downward and protected from the entrance of insects and other foreign matter.

LP-GAS CONDITIONING

Dirt and iron scale in LP-gas tanks may cause trouble in regulators, other controls, and appliances. Thus, tanks should be cleaned before they are put into service. Impurities may be filtered out with liquid strainers or filters. Such strainers are used as regular equipment on trucks (but not on tank cars). Ice and hydrate prevention are two other major considerations; the supplier should keep the dew point low and/or remove any water from the product to avoid these formations. Some precautions follow.

Ice Prevention. Moisture remaining in new tanks or inadvertently accumulated in LP-gas or tanks during handling may cause freeze-ups in regulators or pipelines. Addition of 1 pt of methyl alcohol for each 100 gal of fuel, where conditions warrant, has been successful in preventing freeze-ups. Some operators add 5 gal of methyl alcohol to each tank car received during winter months.

Two-stage pressure reduction decreases refrigeration effects. Moisture absorbents like activated alumina or silica gel have been used in containers placed in the gas line, but this method is not considered the best practice.

Addition of heat via heat exchangers in large industrial installations has also been used to prevent freeze-ups.

Hydrate Prevention. When water is present, hydrates of propane and butane may form under certain temperature and pressure conditions and be mistaken for ice. The same prevention methods are used as those for ice formation. Hydrate or butane may form at atmospheric pressure and up to its vapor pressure at temperatures below 32 F. It cannot exist above 32 F. Hydrate of propane may form at from 6 to 500 psig, at temperatures below 42.5 F. It cannot exist above 42.5 F.

MEASUREMENT

Three methods of measurement are used in sale of LP-gas to the consumer.

By Pound

Where propane is handled in reusable ICC cylinders, it is normally sold by the pound. A tank is weighed before and after filling, and the difference is the number of pounds of propane added. Measurement of deliveries can be made by subtracting unloaded from loaded weight of truck.

By Gallon

When delivery and sale are made by tank truck to the user's storage tank, liquid LP-gas is measured by the gallon. Liquid meters or the gage on the customer's tank may be used. A system for LP-gas usually consists of a pump, meter, differential valve (diaphragm type, back pressure regulator valve, with tank truck pressure applied above diaphragm causing condensation of vapors—only liquids remain for metering), and other equipment.

By Cubic Foot of Vapor

Many customers find it easier to pay for a month's fuel after use rather than in advance. The meters used pass LP-gas vapor in the same way as natural gas. They may be geared to register cubic feet, pounds, gallons, therms, or decitherms. Equivalents of these terms are listed in Table 5-16.

Gaging

Since the sampling of receipts of LP-gas is made on arrival of a tank car, gaging of the car may be a part of the testing procedure.[10,11]

1. Remove plug from thermometer well, lower thermometer to bottom of well, and allow it to remain there for at least 10 min. Raise thermometer quickly and read it at once. Return thermometer to well for a check reading. Record temperature.

Table 5-16 Table of Equivalents of Typical Commercial LP-Gases

(at 60 F and 30 in. Hg)

Unit	Btu	Pound	Therm	Decitherm	Cu ft	Gallon
Propane* (specific gravity = 1.52)						
Pound	21,690	1	0.2169	2.169	8.604	0.2376
Therm	100,000	4.614	1	10	39.67	1.095
Decitherm	10,000	0.4614	0.1	1	3.967	0.1095
Cu ft	2,521	0.1162	.0252	0.2521	1	0.02761
Gallon	91,300	4.23	0.913	9.13	36.45	1
Butane (specific gravity = 2.00)						
Pound	21,340	1	0.2134	2.134	6.532	0.2072
Therm	100,000	4.686	1	10	30.61	.971
Decitherm	10,000	0.4686	0.1	1	3.061	.0971
Cu ft	3,267	0.1531	0.0327	0.3267	1	0.0317
Gallon	103,000	4.86	1.03	10.3	31.79	1
Butane 70%, Propane 30% (specific gravity = 18.6)						
Pound	21,445	1	0.2145	2.145	7.047	0.2155
Therm	100,000	4.663	1	10	32.86	1.005
Decitherm	10,000	0.4663	0.1	1	3.286	0.1005
Cu ft	3,043	0.1418	.0304	0.3043	1	0.0306
Gallon	99,490	4.639	0.9949	9.949	32.69	1
Butane 50%, Propane 50% (specific gravity = 1.76)						
Pound	21,515	1	0.2152	2.1515	7.434	0.2215
Therm	100,000	4.648	1	10	34.55	1.029
Decitherm	10,000	0.4648	0.1	1	3.455	0.1029
Cu ft	2,894	0.1347	0.0289	0.2894	1	0.0298
Gallon	97,150	4.545	0.9715	9.715	33.57	1

* As with natural gases, exact values will vary somewhat with the source. Thus many vapor meters have been built using the following usual propane values[9] at 30 in. Hg and 60 F:

Cu ft per lb	8.50	Cu ft per gal	36.18
Btu per cu ft	2537	Btu per lb	21,570
Sp gr	1.53	Btu per gal	91,800

Table 5-17 Volume* and Specific Gravity† Correction Factors for LP-Gas Liquids at Various Temperatures and Specific Gravities (Liquid)

Temp, °F	Specific gravities at 60 F/60 F													
	0.500	Propane 0.5079	0.510	0.520	0.530	0.540	0.550	0.560	*iso*-Butane 0.5631	0.570	0.580	*n*-Butane 0.5844	0.590	0.600
−50	1.160	1.155	1.153	1.146	1.140	1.133	1.127	1.122	1.120	1.116	1.111	1.108	1.106	1.102
−40	1.147	1.142	1.140	1.134	1.128	1.122	1.117	1.111	1.110	1.106	1.101	1.099	1.097	1.093
−30	1.134	1.129	1.128	1.122	1.116	1.111	1.106	1.101	1.100	1.096	1.092	1.090	1.088	1.084
−20	1.120	1.115	1.114	1.109	1.104	1.099	1.095	1.090	1.089	1.086	1.082	1.080	1.079	1.076
−10	1.105	1.102	1.100	1.095	1.091	1.087	1.083	1.079	1.078	1.075	1.072	1.071	1.069	1.066
0	1.092	1.088	1.088	1.084	1.080	1.076	1.073	1.069	1.068	1.066	1.063	1.062	1.061	1.058
4	1.086	1.083	1.082	1.079	1.075	1.071	1.068	1.065	1.064	1.062	1.059	1.058	1.057	1.054
8	1.081	1.078	1.077	1.074	1.070	1.066	1.063	1.060	1.059	1.057	1.055	1.053	1.052	1.050
12	1.075	1.072	1.071	1.068	1.064	1.061	1.059	1.056	1.055	1.053	1.051	1.049	1.048	1.046
16	1.070	1.067	1.066	1.063	1.060	1.056	1.054	1.051	1.050	1.048	1.046	1.045	1.044	1.043
20	1.064	1.062	1.061	1.058	1.054	1.051	1.049	1.046	1.046	1.044	1.042	1.041	1.040	1.039
24	1.058	1.056	1.055	1.052	1.049	1.046	1.044	1.042	1.042	1.040	1.038	1.037	1.036	1.034
28	1.052	1.050	1.049	1.047	1.044	1.041	1.039	1.037	1.037	1.035	1.034	1.034	1.032	1.031
32	1.046	1.044	1.043	1.041	1.038	1.036	1.035	1.033	1.033	1.031	1.030	1.030	1.028	1.027
36	1.039	1.038	1.037	1.035	1.033	1.031	1.030	1.028	1.028	1.027	1.025	1.025	1.024	1.023
40	1.033	1.032	1.031	1.029	1.028	1.026	1.025	1.024	1.023	1.023	1.021	1.021	1.020	1.019
44	1.027	1.026	1.025	1.023	1.022	1.021	1.020	1.019	1.019	1.018	1.017	1.017	1.016	1.016
48	1.020	1.019	1.019	1.018	1.017	1.016	1.015	1.014	1.014	1.013	1.013	1.013	1.012	1.012
52	1.014	1.013	1.012	1.012	1.011	1.010	1.010	1.009	1.009	1.009	1.009	1.009	1.008	1.008
56	1.007	1.007	1.006	1.006	1.005	1.005	1.005	1.005	1.005	1.005	1.004	1.004	1.004	1.004
60	1.000	1.000	1.000	1.000	1.000	1.000	1.000	1.000	1.000	1.000	1.000	1.000	1.000	1.000
64	0.993	0.993	0.994	0.994	0.994	0.994	0.995	0.995	0.995	0.995	0.996	0.996	0.996	0.996
68	.986	.986	.987	.987	.988	.989	.990	.990	.990	.990	.991	.991	.991	.992
72	.979	.980	.981	.981	.982	.983	.984	.985	.986	.986	.987	.987	.987	.988
76	.972	.973	.974	.975	.977	.978	.979	.980	.981	.981	.982	.982	.983	.984
80	.965	.967	.967	.969	.971	.972	.974	.975	.976	.977	.978	.978	.979	.980
84	.957	.959	.960	.962	.965	.966	.968	.970	.971	.972	.974	.974	.975	.976
88	.950	.952	.953	.955	.958	.961	.963	.965	.966	.967	.969	.969	.970	.971
92	.942	.945	.946	.949	.952	.955	.957	.959	.960	.962	.964	.965	.966	.967
96	.935	.938	.939	.942	.946	.949	.952	.954	.955	.957	.959	.960	.961	.963
100	.927	.930	.932	.936	.940	.943	.946	.949	.950	.952	.954	.955	.957	.959
110	.907	.911	.913	.918	.923	.927	.932	.936	.937	.939	.943	.944	.946	.949
120	.887	.892	.894	.900	.907	.912	.918	.923	.924	.927	.931	.932	.934	.938
130	.865	.871	.873	.880	.888	.895	.901	.908	.909	.913	.918	.921	.923	.927
140	0.842	0.850	0.852	0.861	0.870	0.879	0.886	0.893	0.895	0.900	0.905	0.907	0.910	0.915

* A volume of 1000 gal of propane measured at −50 F corresponds to 1000 × 1.155 = 1155 gal at 60 F.
† Propane at −50 F has a specific gravity of 0.5079 × 1.155 = 0.587.

2. Release slip tube lock and raise slip tube until 6 in. mark is opposite index pointer. Slip tube may be forcibly ejected by container pressure, therefore, operator should stand clear of tube. Open valve slightly to prevent excess flow valve from closing. Gas should now flow from valve. If liquid appears, raise tube until gas flows. Then raise tube ¼ in. at a time until gas reappears. Read tube marking opposite pointer and record reading. To check reading, lower tube ¼ in., at which point liquid should flow. If it does not, reading was in error and operation must be repeated until a reading is found that can be checked.

From the tank outage thus determined and the temperature of the liquid, the contents received may be found from the outage tables supplied by the vendor.

Gaging and Filling Containers

LP-gases in their liquid state increase considerably in volume with temperature rise. Therefore, a container should never be filled completely at ordinary temperatures. If room is not allowed for expansion of liquid with increasing tem-

Table 5-18 Maximum Permitted Filling Density[1,3]

Specific gravity at 60 F	Aboveground containers: 0 to 1200 gal total water cap., per cent	Aboveground containers: Over 1200 gal total water cap., per cent	Underground containers, all capacities, per cent
0.496–0.503	41	44	45
.504– .510*	42	45	46
.511– .519	43	46	47
.520– .527	44	47	48
.528– .536	45	48	49
.537– .544	46	49	50
.545– .552	47	50	51
.553– .560	48	51	52
.561– .568	49	52	53
.569– .576	50	53	54
.577– .584†	51	54	55
.585– .592	52	55	56
0.593–0.600	53	56	57

* 0.508 for propane.
† 0.584 for butane.

Table 5-19 Capacities of Storage Tanks at 60 F

Tank capacity		Maximum aboveground capacity for LP-gas liquids							
		Propane		Butane		Butane 70%, propane 30%		Butane 50%, propane 50%	
Gallons	(Water), lb	lb	gal	lb	gal	lb	gal	lb	gal
100	833	350	83	425	87	408	87	392	86
150	1,250	525	124	637	131	612	131	587	129
200	1,666	700	166	849	175	816	175	783	172
250	2,082	875	207	1,061	218	1,021	218	978	215
300	2,500	1,050	248	1,274	262	1,227	262	1,173	258
400	3,330	1,400	331	1,699	349	1,632	350	1,566	344
500	4,165	1,750	414	2,123	436	2,040	437	1,958	431
600	5,000	2,100	496	2,547	524	2,447	525	2,345	516
800	6,660	2,800	662	3,400	699	3,264	699	3,132	689
1,000	8,330	3,499	827	4,248	874	4,082	874	3,915	861
2,000	16,660	7,500	1772	9,000	1850	8,660	1854	8,330	1832
3,000	24,990	11,250	2660	13,500	2776	13,000	2780	12,500	2747
4,000	33,320	15,000	3543	18,000	3700	17,330	3708	16,680	3664
5,000	41,650	18,750	4432	22,500	4626	21,660	4635	20,830	4579
6,000	49,980	22,500	5320	27,000	5552	26,000	5562	25,000	5494
8,000	66,640	30,000	7086	36,000	7400	34,660	7416	33,380	7328
10,000	83,300	37,485	8860	44,982	9250	43,316	8740	41,650	9160
Multipliers for intermediate tank sizes up to 1200 gal									
1	8.33	3.4986	0.827	4.2483	0.874	4.0817	0.874	3.9151	0.861
Multipliers for intermediate tank sizes over 1200 gal									
1	8.33	3.7485	0.886	4.4982	0.925	4.3316	0.927	4.1650	0.916

To find the capacity of a container for underground installation, divide aboveground capacity by the filling density on which it is based and multiply by the filling density for underground container.

perature, its volume may become greater than the container capacity, with the extra volume being discharged thru the pressure relief valve. Table 5-17 gives volumetric and specific gravity correction factors.

A "strapping stick" is sometimes used to gage liquid volume. This stick is graduated and marked for a given size tank. When held beside the slip tube on the tank, with its point corresponding with the height to which the tube is raised, the tank contents are indicated in gallons or per cent.

Magnetic float, rotary, and fixed liquid level gages are also used to determine liquid tank content. Maximum weight of fuel permitted in various containers is the *product* of the container capacity *in pounds of water* (2nd column of Table 5-19) and filling density (Table 5-18). Table 5-19 also tabulates the results of this multiplication for butane, propane, and two mixtures of these for aboveground containers.

In general:[3]

$$V = DC/100GF$$

where:

V = max liquid LP-gas volume, at any temperature, T, which can be placed in container, gal

C = container water capacity, gal

D = filling density (Table 5-18), per cent

G = specific gravity of liquid LP-gas *at 60 F* placed in container

F = correction factor, used when liquids are at a temperature, T, other than 60 F (Table 5-17)

Example: An LP-gas liquid of 0.55 sp gr at 60 F is to be loaded into an aboveground tank of 10,000 gal water capacity. How many gallons of liquid at 82 F should be placed in the container?

$$V = \frac{DC}{100GF} = \frac{50 \times 10{,}000}{100 \times 0.55 \times 0.971} = 9350 \text{ gal}$$

REFERENCES

1. Natl. Fire Protection Assn. *Standard for the Storage and Handling of Liquefied Petroleum Gases.* (NFPA No. 58) Boston, 1963.
2. Am. Pet. Inst. *Design and Construction of Liquefied-Petroleum-Gas Installations at Marine and Pipeline Terminals, Natural-Gasoline Plants, Refineries, and Tank Farms.* (Std 2510) New York, 1957.
3. Natl. Fire Protection Assn. *Standard for the Storage and Handling of Liquefied Petroleum Gases at Utility Gas Plants.* (NFPA No. 59) Boston, 1963.
4. Kramer, W. H. and others. "Recent Developments in LP-Gas Storage" (CEP-55-14) *A.G.A. Proc.* 1955: 900–914.
5. Dyer, A. F. and Hale, J. W. "Underground Storage of Liquefied Petroleum Gas." *NFPA Quarterly* 55: 5–16, July 1961.
6. Natural Gasoline Assn. of Am. *NGAA Tentative Standards for the Underground Storage of Liquefied Petroleum Gas,* amended. Tulsa, 1954.
7. Underwriters' Labs. *Fact-Finding Report on Prestressed Concrete-Embedded Cylinder Pipe for the Underground Storage of Liquefied Petroleum Gas.* (Miscell. Hazard 6544) Chic., 1957.
8. Clifford, E. A. *Practical Guide to LP-Gas Utilization,* rev. ed. New York, Moore, 1957.
9. Texas Univ. Ind. Ed. Dept. and Liquefied Pet. Gas Assn. *LP-Gas Service Training Course.* Austin, 1958–60.
10. A.S.T.M. "Tentative Method of Gaging Petroleum and Petroleum Products." (Standard D1085-57T) *ASTM Standards* 8: 1007–59. Philadelphia, 1961.
11. Natural Gas Processors Assn. "N.G.P.A. Liquefied Petroleum Gas Specifications and Test Methods." (Pub. 2140–62) *Plant Operations Test Manual.* Tulsa, Natural Gasoline Assn. of Am., 1962.

Chapter 5

Design of LP-Gas Plants for Utility Companies

by A. E. Wastie and H. Emerson Thomas

Utility companies install LP-gas plants for complete conversion from city gas, new gas supply, substitute for gas or oil, enrichment of low Btu-gas, emergency standby, peak load shaving, or catalytic cracking. The plant size is usually 30,000 gal water capacity or more, stored in one or more tanks. Underground storage, as in mined caverns and salt domes, is also used. Design for LP-gas plants is basically the same, regardless of ultimate use, and certain items of equipment are required in all cases. Modifications may be made as needed to suit individual conditions. Major equipment items include the storage system, unloading facilities, vaporizer, regulator, mixing equipment, piping and fittings, and electrical equipment.

NFPA Standard No. 59[1] covers the storage and handling of LP-gases at utility plants; NFPA Standard No. 58[3] applies up to aggregate water capacity of 2000 gal.

As a prerequisite to proper selection of equipment, the following information is necessary: maximum hourly demand, heating value, pressures, physical space requirements, proposed plant layout, availability of railroad siding or trucking facilities, allowable soil pressures for foundation support, and type of soil.

STORAGE CONTAINERS

Both nonrefrigerated (Fig. 5-21) and refrigerated containers are in use. General considerations follow for selecting pressure containers for LP-gas storage in both of these categories:

1. Design and construction of steel containers should conform to the ASME unfired pressure vessel code,[2] and also to government regulations which apply at installation location.
2. Container design pressure should be determined by the product to be stored. Working pressure is determined under NFPA standards on the basis of vapor pressure at 100 F of the particular LP-gas to be stored. Normally, Type 200 (propane) containers are installed to provide flexibility in the product stored. These standards provide for safety valves set at specified working pressure.
3. Government regulation, if any, for initial and periodic inspection should be determined.
4. If permits to operate are required, their provisions should be checked to make certain that compliance is obtained.
5. Data sheets should be required from the manufacturer certifying that container is constructed of allowable material, that welders are properly qualified, and that vessel has been inspected and tested by a certified inspector. This sheet also provides pertinent information on requirements for securing an operating permit when it is required by law. Even if no such permit is required, inspection of container by a properly qualified inspector is advisable.

Foundation Design. The major factor is the bearing capacity of the soil. Terzaghi[4,5] derived two sets of equations: (1) for *general shear*, (2) for *local shear*. General and local shear are distinguished by the load settlement curve.[4] Local shear is presumed to occur in loose or highly compressible soils, and general shear in dense or slightly compressible soils. For a two

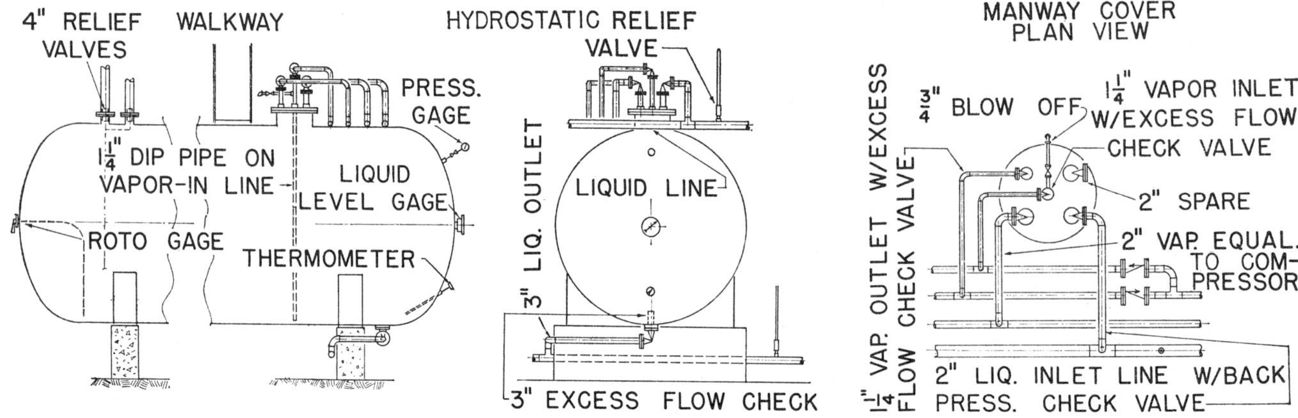

Fig. 5-21 A 30,000 gal nonrefrigerated propane storage tank for utility plants. (Drake & Townsend, Inc.) See Fig. 5-24 for relief valves. The 2 in. vapor equalizing line is also used for liquid return pump relief and tank pressurizing.

dimensional, shallow, square footing in ideal cohesive soil—i.e., soil with zero friction angle in general or local shear (for loose granular material *angle of friction* and angle of repose are the same; in dense soils angle of friction exceeds angle of repose).

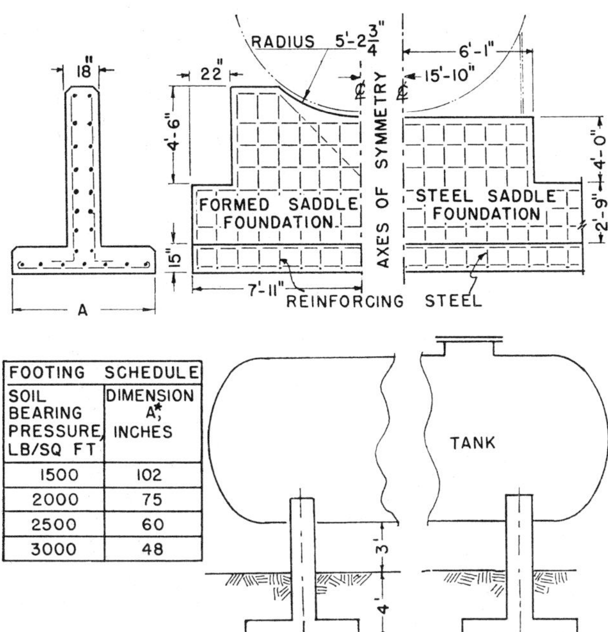

FOOTING SCHEDULE	
SOIL BEARING PRESSURE, LB/SQ FT	DIMENSION A*, INCHES
1500	102
2000	75
2500	60
3000	48

Fig. 5-22 Two types of reinforced concrete foundations for a nonrefrigerated LP-gas tank. (Drake & Townsend, Inc.) Foundation and piers for these containers are designed to carry the weight of the storage tank filled with water. Preferably, containers should rest on two supports; more than two may cause excessive stresses because of unequal foundation settlement. Footing bases must be below possible frost line.

Table 5-20 Allowable Bearing Pressure for Footings

(to be used as a guide only)*

Soil description	Soil condition	Safe allowable pressure, psf
Fine-grained soils:		
Clays, silts, very fine sands, or mixtures of these containing few coarse particles of sand or gravel	Soft, unconsolidated, high-moisture content (mud)	1,000
	Stiff, partly consolidated, medium moisture content	4,000
	Hard, well-consolidated, low-moisture content (slightly damp to dry)	8,000
Sands and well-graded sandy soils containing some silt and clay	Loose, not confined	3,000
	Loose, confined	5,000
	Compact	10,000
Gravel and well-graded gravelly soils containing some sand, silt, and clay	Loose, not confined	4,000
	Loose, confined	6,000
	Compact	12,000
	Cemented sand and gravel	16,000
Rock	Poor quality, soft and fractured; also hardpan	10,000
	Good quality; hard and solid	20,000†

* For dynamic loads only a fraction of these soil bearing pressures can be used—depending on frequency of forces acting.[7]

† Minimum.

The simplified equation is:

$$q = 7.4C + WZ$$

where:

q = bearing pressure, psf

C = average unconfined compressive strength of soil, from test, psf

W = weight of soil, lb per cu ft

Z = depth of footing, ft

Table 5-20 gives bearing pressure for some common soils and Fig. 5-22 shows a typical reinforced cement concrete foundation for a 30,000 gal tank.

Nonrefrigerated Containers

Reference may be made to the following extracts from NFPA No. 59:[1]

22. Design Pressure and Classification of Nonrefrigerated Containers

220. Shop fabricated storage containers for nonrefrigerated storage shall be designed and classified as follows:

Container type	For gases with vapor press. not to exceed lb per sq in. gage at 100 F (37.8 C)	Minimum design pressure of container, psig: 1949 and earlier editions of ASME Code (Par. U-68, U-69)	Minimum design pressure of container, psig: 1949 edition of ASME Code (Par. U-200, U-201); 1950, 1952, 1956 and 1959 editions of ASME Code; all editions of API-ASME Code‡
80*	80*	80*	100*
100	100	100	125
125	125	125	156
150	150	150	187
175	175	175	219
200†	215	200	250

* New storage containers of the 80 type have not been authorized since Dec. 31, 1947.

† Container type may be increased by increments of 25. The minimum design pressure of containers shall be 100% of the container type designation when constructed under 1949 or earlier editions of the ASME Code (Par. U-68 and U-69). The minimum design pressure of containers shall be 125% of the container type designation when constructed under: (1) the 1949 ASME Code (Par. U-200 and U-201), (2) 1950, 1952, 1956 and 1959 editions of the ASME Code, and (3) all editions of the API-ASME Code.

‡ Construction of containers under the API-ASME Code is not authorized after July 1, 1962.

Note: Because of low soil temperature usually encountered, and the insulating effect of the earth, the average vapor pressure of products stored in underground containers will be materially lower than when stored aboveground. This reduction in actual operating pressure therefore provides a substantial corrosion allowance for these containers when installed underground.

221. Field-erected nonrefrigerated containers shall be built in accordance with applicable provisions of the 1959 edition of the ASME Boiler and Pressure Vessel Code Section VIII, Unfired Pressure Vessels, except that construction using joint efficiencies in Tables UW12, Column C is not permitted.

222. Field-erected containers for nonrefrigerated storage shall be designed for a pressure not less than 125% of the maximum vapor pressure of the product at 100 F to be stored in the containers, but in no case shall the container be designed for a pressure of 25 psig or less.

223. The shell or head thickness of any nonrefrigerated container shall not be less than 3⁄16 in.

24. Location of Nonrefrigerated Containers

240. Nonrefrigerated Aboveground Containers

(a) Containers shall be located outside of buildings.

(b) Containers shall be located in accordance with the following table:

Water capacity of each container, gal	Minimum distances: Between containers,* ft	Minimum distances: From container to nearest important building or group of buildings, or a property line which may be built upon, ft
2001 to 30,000	5	50
30,001 to 70,000	¼ of sum of diameters of adjacent containers	75
70,001 to 125,000	"	100
125,001 to 200,000	"	200
200,001 to 1,000,000	"	300
1,000,001 or more	"	400

* The minimum distance requirement for spacing between containers when the water capacity of a container is 180,000 gal or more shall be at least 25 ft. The minimum distance requirement for spacing between groups of containers when a group of two or more containers has an aggregate water capacity of 180,000 gal or more shall be at least 25 ft.

(c) A container or containers with an aggregate water capacity in excess of 180,000 gal, and their loading stations should be located 100 ft or more from buildings occupied for generation, compression or purification of manufactured gas, or from natural gas compressor buildings, or from outdoor installations essential to the maintenance of operation in such buildings. Such container or containers and their loading stations should be 100 ft or more from aboveground storage of flammable liquids and from any buildings of such construction or occupancy which constitutes a material hazard of exposure to the containers in the event of fire or explosion in said buildings. If the container or containers are located closer than 50 ft to any such buildings or installations, then the latter shall be protected by walls adjacent to such storage containers or by other appropriate means against the entry of escaped liquefied petroleum gas, or of drainage from the storage container area and its loading points—all in such a manner as may be required and approved by the authority having jurisdiction.

(d) Nonrefrigerated liquefied petroleum gas containers shall not be located within dikes enclosing flammable liquid tanks, and shall not be located within dikes enclosing refrigerated liquefied petroleum gas tanks.

241. Nonrefrigerated Underground Containers

(a) Underground containers shall include both buried and partially buried (or mounded) containers.

(b) Containers shall be located outside of any buildings. Buildings or roadways shall not be constructed over any underground containers. Sides of adjacent containers shall be separated by not less than 3 ft.

(c) When containers are installed parallel with ends in line, any number of containers may be in one group. When more than one row is installed, the adjacent ends of the tanks in each row shall be separated by not less than 10 ft.

(d) Containers and their loading stations shall be located not less than 50 ft from the nearest important building or group of buildings or line of adjacent property which may be built upon.

(e) The containers and their loading stations should be located not less than 50 ft from buildings occupied for generation, compression or purification of gas, or from outdoor installations essential to the maintenance of operation in such buildings. They should be located not less than 50 ft from aboveground storage of flammable liquids and from any buildings of such construction or occupancy which constitutes a severe exposure to any aboveground appurtenances of the underground installation in the event of fire or explosion in said buildings. If the underground installations by necessity are located closer than 50 ft to any such buildings or installations, then the latter shall be protected against the entry of escaping liquefied petroleum gas, in such a manner as may be required and approved by the authority having jurisdiction.

25. Installation of Nonrefrigerated Storage Containers

250. Nonrefrigerated Aboveground Containers

(a) Every container shall be supported to prevent the concentration of excessive loads on the supporting portion of the shell or heads.

(b) Supports for containers shall be of solid masonry, concrete or steel. Structural metal supports may be employed when they are protected against fire in an approved manner. Steel supports shall be protected against fire with a material having a fire resistance rating of at least two hours. Steel skirts having only one opening shall be protected as above but fireproofing need only be applied to the outside of the skirt.

(c) Horizontal containers shall be mounted on saddles in such a manner as to permit expansion and contraction, not only of the container but also of the connected piping. Only two saddles shall be used.

(d) Suitable means to prevent corrosion shall be provided on that portion of the container in contact with the foundations or saddles.

(e) Containers should be kept properly painted or otherwise protected from the elements.

251. Nonrefrigerated Underground Containers

(a) Buried containers shall be placed so that the top of the container is not less than 6 in. below the grade of the surrounding area. Partially buried (or mounded) containers shall have not less than 12 in. of cover, sufficient to provide surface drainage without erosion or other deterioration.

(b) The container manway shall not be covered with the backfill or mounding material. Under conditions where the container manway cover is below the ground level, a manway providing sufficient access shall be installed. No other part of the container shall be exposed.

(c) The containers shall be set on a firm foundation or firm undisturbed earth and surrounded with soft earth or sand well tamped into place. Provision shall be made to take care of settling and rotation.

(d) Containers shall be adequately protected against corrosion.

(e) Bottom connections to the container shall be prohibited. All connections shall be in the container manway or at openings along the top length of the container.

252. Field welding where necessary shall be made only on saddle plates or brackets which were applied by manufacturer of container, except as provided by the code under which the container was fabricated.

253. Secure anchorage or adequate pier height shall be provided to protect against container flotation wherever sufficiently high water might occur.

27. Gaskets

270. The gaskets for use on storage containers shall be resistant to the action of liquefied petroleum gas in the liquid phase. Gaskets shall be made of metal having a melting point of over 1500 F or shall be confined within an assembly having a melting point of over 1500 F. Aluminum "O" rings and spiral wound metal gaskets are also acceptable. When a flange is opened, the gasket shall be replaced.

28. Filling Densities

280. The "filling density" is defined as the per cent ratio of the weight of the gas in a container to the weight of water at 60 F that the container will hold. [See Tables 5-17, 5-18, and 5-19—EDITOR]

282. For individual underground nonrefrigerated installations, the authority having jurisdiction may authorize the use of increased filling densities where the maximum ground temperatures do not exceed 60 F. These filling densities shall be based upon sound engineering practices for the operating conditions involved.

Refrigerated Containers

Reference may be made to the following extracts from NFPA No. 59:[1]

31. Requirements for Construction, Design and Original Test of Refrigerated Containers

310. Refrigerated containers shall be built in accordance with applicable provisions of one of the following codes as appropriate for conditions of maximum allowable working pressure, design temperature, and hydrostatic testing:

(a) For pressures of 15 psig or more use the 1959 edition of the ASME Boiler and Pressure Vessel Code Section VIII, Unfired Pressure Vessels, except that construction using joint efficiencies in Table UW12, Column C is not permitted.

(b) For pressures below 15 psig use API Standard 620, Recommended Rules for the Design and Construction of Large, Welded, Low Pressure Storage Tanks.* (Tentative) First Edition 1956 and Addenda 1958.

311. Field-erected containers for refrigerated storage shall be designed as an integral part of the storage system including tank insulation, compressors, condensers, controls, and piping. Proper allowance shall be made for the service temperature limits of the particular process and the products to be stored when determining material specifications and the design pressure. Welded construction shall be used. All main shell seams shall receive complete acceptable radiographic examination.

312. Materials having ductility and impact resistance at the design temperature equal to or superior to those listed in Appendix F shall be used in the fabrication of containers for refrigerated storage of liquefied petroleum gas.

* The thickness of the container may be determined using the maximum expected density of the product to be stored instead of the density used in API Standard 620 but, in no case during test, should any portion of the container be stressed more than 80% of the minimum specified yield strength of any of the material.

* * * *

APPENDIX F

Minimum Material Requirements for Shells and Bottoms of Refrigerated Storage Tanks for Various Temperatures and Thicknesses Constructed of Ferritic Steels

Design Tempera-ture	Thickness	Material Spec.	Qualifications to be Added to the Basic Specification
65 F to 25 F incl.	Up to ½ in., incl.	Any Approved Steel with specified min. T.S.* not exceeding 60,000 psi (See Note 1)	None
	Over ½ in to 1 in., incl.	A-131 B (or C) A-442 A-201 A & B	None None FGP (Fine grain practice)
	Over 1 in. to 1⅜ in., incl.	A-131 C A-442 A-201 A & B	None None FGP
	Over 1⅜ in	A-131 C A-442 A-201 A & B	Normalized Normalized FGP, Normalized
Below 25 F to —5 F incl.	Up to ½ in., incl.	A-442 A-201 A & B	None None
	Over ½ in. to 1⅜ in., incl.	A-131 C A-442 A-201 A & B	FGP FGP FGP, High Mang. (See Note 2)
	Over 1⅜ in.	A-131 C A-442 A-201 A & B	Normalized Normalized FGP, High Mang., Normalized
Below —5 F to —20 F	Up to ½ in., incl.	A-442 A-201 A & B	FGP FGP, High Mang.
Below —5 F to —20 F	Over ½ in. to 1⅜ in., incl.	A-131 C A-442 A-201 A & B	Normalized FGP, Normalized FGP, High Mang., Normalized
Below —20 F	All thicknesses	A-300	Test temperature to be 20°F lower than design temperature
	Over 1⅜ in.	A-300 Class 1	A-201 A & B only

Note 1: For this thickness-temperature category approved steels include all those listed in API 620. Materials for ASME Code Vessels must comply with requirements of that Code, and any additional requirements of this table; A-131 steel is not approved by ASME; some Code cases have not been approved by some local jurisdictions. All specific materials listed in table are satisfactory for all designs based on API 620.

Note 2: Manganese content of 0.85% to 1.20% is preferred in lieu of usual content of 0.80% maximum.

* T.S. = tensile strength.

* * * *

313. When austenitic steels or nonferrous materials are used, the ASME Code shall be used as a guide in the selection of materials for use at the design temperature.

314. Materials for nozzles, attached flanges, structural members which are in tension, and other such critical elements shall be selected for the design temperature based on impact test requirements. Materials which are certified by impact testing shall absorb 15 ft-lb Charpy V-notch tests at the lowest expected service temperature.

315. The design temperature shall be the lower of the following:

(a) The minimum temperature to which the tank contents will be refrigerated.

(b) The minimum anticipated tank shell temperature due to atmospheric temperature considering the effectiveness of the insulation in keeping shell temperature above expected minimum atmospheric temperature where atmospheric temperature below the refrigerated temperature may be expected.

316. The provisions of 310 shall not be construed as prohibiting the continued use of reinstallation of containers constructed and maintained in accordance with the Code in effect at the time of fabrication.

33. Location of Refrigerated Containers

330. Refrigerated Aboveground Containers

(a) Containers shall be located outside of buildings.

(b) Containers shall be located in accordance with the following table:

	Minimum distances	
Water capacity of each container, gal	Between containers, ft	From container to nearest important building or group of buildings, or a property line which may be built upon, ft
200,001 to 1,000,000	¼ of sum of diameters of adjacent containers	300
1,000,001 or more		400

(c) A container or containers with an aggregate water capacity in excess of 180,000 gal, and their loading stations should be located 100 ft or more from buildings occupied for generation, compression or purification of manufactured gas, or from natural gas compressor buildings, or from outdoor installations essential to the maintenance of operation in such buildings. Such container or containers and their loading stations should be 100 ft or more from aboveground storage of flammable liquids and from any buildings of such construction or occupancy which constitute a material hazard of exposure to the containers in the event of fire or explosion in said buildings. If the container or containers are located closer than 50 ft to any such buildings or installations, then the latter shall be protected by walls adjacent to such storage containers or by other appropriate means against the entry of escaped liquefied petroleum gas, or of drainage from the storage container area and its loading points—all in such a manner as may be required and approved by the authority having jurisdiction.

(d) Refrigerated liquefied petroleum gas containers shall not be located within dikes enclosing flammable liquid tanks or within dikes enclosing nonrefrigerated liquefied petroleum gas tanks.

331. Refrigerated containers shall not be installed one above the other.

332. The ground within 25 ft of any aboveground refrigerated container and all ground within a diked area shall be kept clear of readily ignitible material such as weeds and long dry grass.

333. In cases where refrigerated containers are to be installed in heavily populated or congested areas, the authority having jurisdiction shall determine restrictions of individual tank capacity, total storage, distance to line of adjoining property which may be built on or other reasonable protective methods.

34. Installation of Refrigerated Containers

340. Refrigerated aboveground containers shall be installed on the ground, or on foundations or supports of concrete, masonry piling, or steel. Foundations and supports shall be protected to have a fire-resistance rating of not less than 2 hr.

341. Containers for product storage at less than 30 F shall be supported in such a way or heat supplied to prevent the effects of freezing and consequent frost heaving.

342. Any insulation used shall be noncombustible and shall resist dislodgment by fire hose streams.

343. Refrigerated storage containers shall be provided with a means for containment having a volumetric capacity of 150% of the container or containers within the area. Except where protection is provided by natural topography, dikes or retaining walls shall be required and shall be of earth, concrete, or solid masonry designed to be liquidtight and to withstand a full hydraulic head, and so constructed as to provide the required protection. Earthen dikes shall have a flat section at the top not less than 2 ft wide. The slope shall be consistent with the angle of repose of the material of which the dikes are constructed. The walls of the dikes shall be as low as practicable but not less than 5 ft in height. When provision is made for draining rain water from diked areas, such drains shall be kept closed and shall be operated so that when in use they will not permit tank contents to enter natural water courses, public sewers, or public drains. When pumps control drainage from the diked area, they shall not be self-starting.

344. Field welding on container where necessary shall be made only on saddle plates or brackets which were applied by manufacturer of container, except as provided by the code under which the container was fabricated.

345. Secure anchorage or adequate pier height shall be provided to protect against container flotation wherever sufficiently high water might occur.

346. When flammable liquid storage tanks are in the same general area as liquefied petroleum gas containers, the flammable liquid storage tanks shall be diked or diversion curbs or grading used to prevent accidentally escaping flammable liquids from flowing into liquefied petroleum gas container areas.

347. The container storage area shall be fenced or otherwise protected where necessary and at least two points of access thru the fencing, if used, shall be provided.

36. Gaskets

360. The gaskets for use on storage containers shall be resistant to the action of liquefied petroleum gas in the liquid phase. Gaskets shall be made of metal having a melting point of over 1500 F or shall be confined within an assembly having a melting point of over 1500 F. Aluminum "O" rings and spiral-wound metal gaskets are also acceptable. When a flange is opened, the gasket shall be replaced.

37. Filling Densities

370. The filling limits for refrigerated storage containers shall be based upon sound engineering practice for the individual design and operating conditions involved. Since negligible expansion of the liquid can take place within the possible range of operating pressure and temperature of a refrigerated tank, the maximum liquid volume in per cent of the total container capacity may be greater for a refrigerated tank than normally employed for a nonrefrigerated tank.

The largest refrigerated LP-gas storage facility[6] to date has been installed in Minneapolis. The tank, which is 106 ft OD, double wall (30 in. of insulation) by 100 ft high, will hold

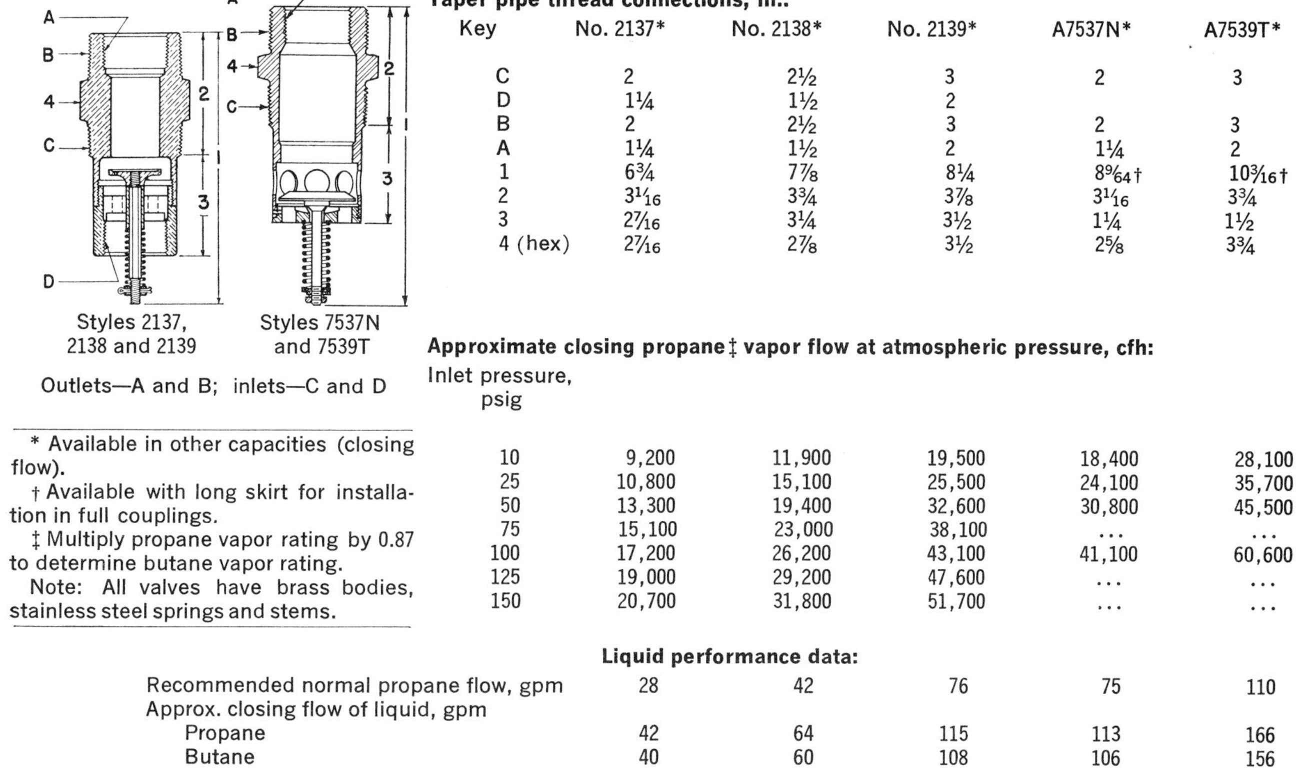

Taper pipe thread connections, in.:

Key	No. 2137*	No. 2138*	No. 2139*	A7537N*	A7539T*
C	2	2½	3	2	3
D	1¼	1½	2		
B	2	2½	3	2	3
A	1¼	1½	2	1¼	2
1	6¾	7⅞	8¼	8 9/64†	10 3/16†
2	3 1/16	3¾	3⅞	3 1/16	3¾
3	2 7/16	3¼	3½	1¼	1½
4 (hex)	2 7/16	2⅞	3½	2⅝	3¾

Approximate closing propane‡ vapor flow at atmospheric pressure, cfh:

Inlet pressure, psig	No. 2137*	No. 2138*	No. 2139*	A7537N*	A7539T*
10	9,200	11,900	19,500	18,400	28,100
25	10,800	15,100	25,500	24,100	35,700
50	13,300	19,400	32,600	30,800	45,500
75	15,100	23,000	38,100	...	...
100	17,200	26,200	43,100	41,100	60,600
125	19,000	29,200	47,600	...	...
150	20,700	31,800	51,700	...	...

* Available in other capacities (closing flow).
† Available with long skirt for installation in full couplings.
‡ Multiply propane vapor rating by 0.87 to determine butane vapor rating.
Note: All valves have brass bodies, stainless steel springs and stems.

Liquid performance data:

	No. 2137*	No. 2138*	No. 2139*	A7537N*	A7539T*
Recommended normal propane flow, gpm	28	42	76	75	110
Approx. closing flow of liquid, gpm					
Propane	42	64	115	113	166
Butane	40	60	108	106	156

Fig. 5-23 RegO excess flow valves. (Bastian Blessing Co.)

120,000 bbl of propane at −50 F. The refrigeration system accompanying it is designed to accommodate the receipt of 120,000 gal of propane at 90 F.

CONTAINER VALVES AND ACCESSORIES

All liquid containers should be equipped with automatic excess flow valves and, where necessary, with remote-controlled shutoffs or fusible links to shut off liquid flow in case of fire. Figure 5-23 shows one type of excess flow valve.

Every container should be provided with one or more safety relief devices (Fig. 5-24), each of which is properly marked to indicate pressure at which it is set to start to discharge, discharge rate at full open position, and type of container on which it is designed to be installed. A vent stack with loose fitting rain cap should be installed to discharge at a safe distance above the container.

Typical steel storage tank (30,000 to 80,000 gal water capacity) openings and connections follow:

1. Two or three in. liquid fill connections complete with back pressure check valve and approved type shutoff valve.
2. Two or more 2 in. vapor outlet connections complete with excess flow valve and approved type shutoff valve (one of which shall be used as an equalizer, pump return, or repressuring line).
3. One 1¼ in. compressor suction connection complete with excess flow valve and approved type shutoff valve.
4. One 1¼ in. compressor discharge connection normally equipped with dip pipe to within ½ in. from bottom of container, excess flow valve and approved type shutoff valve.
5. One or more 2 or 3 in. bottom connections for liquid withdrawal from container complete with excess flow valve and approved type shutoff valve.
6. One or more relief valve openings of adequate size (NFPA No. 59).
7. One ¾ in. connection for pressure gage complete with excess flow, ball check, or No. 54 orifice valve, and approved shutoff valve.
8. One 1 in. opening and one 2½ in. opening in center of each head of container for liquid level gages (one rotary and one magnetic).
9. One thermometer well connection near bottom edge of head for obtaining liquid temperatures in the container.
10. It is considered good practice to install a maximum liquid level gage.

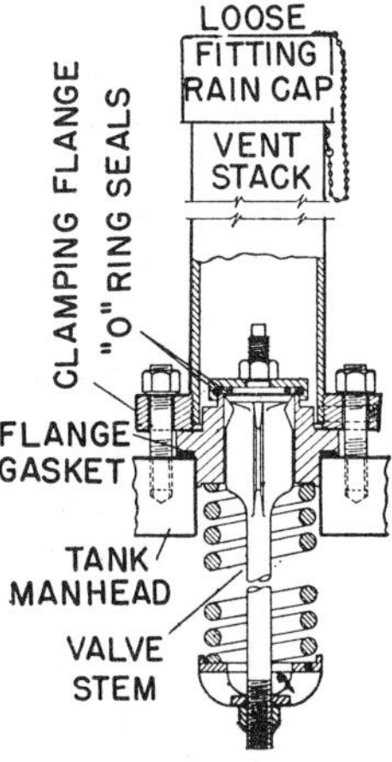

Fig. 5-24 Relief valve with stack and rain cap. A vent hole in the flange of the valve (not shown) drains any water in the vent stack to the atmosphere.

11. *Hydrostatic relief valves* should be placed on all liquid lines between shutoff or blocking valves. Relief is thus provided for any build-up in pressure, such as from external heat in event of closure of such valves.

The need for valve connections and pipeline sizes varies with the particular installations. Typical connections for a propane storage tank are shown in Fig. 5-22.

For additional information reference may be made to the following extract from NFPA No. 59:[1]

42. Container Valves and Accessories

420. All shutoff valves and accessory equipment (liquid or gas) shall be suitable for use with liquefied petroleum gas, and designed for not less than the maximum extreme pressure and temperature to which they may be subjected. Valves for use with nonrefrigerated containers which may be subjected to container pressure shall have a rated working pressure of at least 250 psig. Cast iron valves, piping, and fittings shall be prohibited on liquefied petroleum gas containers and their connections. This does not prohibit the use of container valves or fittings made of malleable or nodular iron.*

421. All connections to containers, except safety relief connections, liquid level gaging devices, and plugged openings, shall have shutoff valves located as close to the container as practicable.

422. Excess flow valves where required by this standard shall close automatically at those rated flows of vapor or liquid as specified by the manufacturer. The connections or line including valves, fittings, etc., downstream of an excess flow valve shall have a greater capacity than the rated flow of the excess flow valve.

423. Except as provided in 424 and 443,† all liquid and vapor connections on containers except safety relief connections shall be equipped with approved automatic excess flow valves, or with back pressure check valves, or a remotely controlled automatic quick-closing valve which shall remain closed except during operating periods. The mechanism for remotely controlled, quick-closing valves shall be provided with a secondary control equipped with a fusible release (not over 220 F melting point) which will cause the quick-closing valve to close automatically in case of fire.

424. Openings from a container or thru fittings attached directly on the container to which pressure gage connection is made, need not be equipped with an excess flow valve if such openings are not larger than No. 54 drill size.

425. Excess flow and back pressure check valves where required by this standard shall be located inside of the container or at a point outside where the line enters the container; in the latter case, installation shall be made in such a manner that any undue stress beyond the excess flow or back pressure check valve will not cause breakage between the container and such valve.

426. Excess flow valves shall be designed with a by-pass, not to exceed a No. 60 drill size opening to allow equalization of pressures.

427. All inlet and outlet connections except safety valves, liquid level gaging devices and pressure gages on any container shall be labeled or color coded to designate whether they are connected to vapor or liquid space. Labels may be on valves.

428. Each storage container shall be provided with a suitable pressure gage.

43. Filler and Discharge Pipes, Manifolds

430. Piping connections between container and manifold should be designed to provide adequate allowances for contraction, expansion, vibration, and settlement. Compression type couplings shall not be considered suitable for this purpose.

431. It is desirable that liquid manifold connections be located at nonadjacent ends of parallel rows of containers.

432. The use of nonmetallic hose is prohibited for interconnecting stationary containers.

433. A good test for determination of piping stresses consists of unbolting piping at a flange and noting whether the flange remains in proper alignment.

434. The filling pipe inlet terminal shall not be located inside a building. Such terminals shall be located not less than 10 ft from any building, and preferably not less than 5 ft from any driveway, and shall be properly supported and protected from physical damage.

435. A shutoff valve shall be provided in liquid piping for each section of pipe containing 500 gal capacity when the pipe is within 300 ft of storage containers or other important aboveground structures.

436. When the liquid line manifold connecting containers in a group has a volumetric capacity of more than 100 gal, such container manifolds shall be located not less than 100 ft from the nearest adjacent property owned by others which may be built upon. The manifold piping terminates at the first line valve which may be used to isolate the manifolded containers from any other part of the liquid line system.

437. If more than three storage containers discharge liquid into a manifold whose nominal diameter is greater than 2 in. and if the flow capacity of such manifold is less than the total discharge capacity of the discharge lines from the containers, one of the following for each container shall be provided:

(a) A remotely controlled external shutoff valve in combination with an excess flow valve.

(b) A remotely controlled quick-closing valve which shall remain closed except during operating periods. The mechanism for such valves may be provided with a secondary control equipped with a fusible release (not over 220 F melting point) which will cause the quick-closing valve to close automatically in case of fire.

45. Hose Specifications

450. Hose shall be fabricated of materials that are resistant to the action of liquefied petroleum gas.

451. Hose subject to container pressure shall be designed for a bursting pressure of not less than five times the pressure for which the container was designed. Hose connections when made shall be capable of withstanding a test pressure of twice the pressure for which the container is designed.

452. Hose and hose connections located on the low pressure side of regulators or reducing valves shall be designed for a bursting pressure of not less than 125 psi but not less than five times the pressure setting of the safety relief devices protecting that portion of the system. There shall be no leakage from assembled hose connections.

47. Pumps and Compressors

470. Each pump and compressor shall be suitable for the liquefied petroleum gas service intended. Each pump and compressor shall be marked with its maximum working pressure.

471. Refrigerated storage systems shall be provided with sufficient capacity to maintain all containers at a pressure not in excess of the operating pressure under summer weather conditions and shall be provided with additional capacity for filling or stand-by service. Unless facilities are provided to safely dispose of vented vapors while the refrigeration system is inoperative, at least two compressors shall be installed where compressors and condensers are used. Compressor capacity provided for stand-by service shall be capable of handling the volume of vapors necessary to be evolved to maintain operating pressure. Auxiliary equipment, such as fans, circulating water pumps, and instrument air compressors, shall be provided with spare or stand-by facilities sufficient to insure that prolonged failure of refrigeration may be prevented.

---

* For information as to the suitability of malleable or nodular iron for this use, refer to Standards of the American Society for Testing and Materials (A47–52 or A339–51T).

† Gaging devices shall have a maximum of a No. 54 drill size bleed valve unless provided with an excess flow valve.

472. Adequate means shall be available for operating equipment in event of failure of normal facilities.

SECTION 6. RELIEF DEVICES

61. General

610. Relief devices on containers shall be so arranged that the possibility of tampering will be minimized; if the pressure setting or adjustment is external, the relief devices shall be provided with an approved means for sealing the adjustment.

611. Each nonrefrigerated shop fabricated container relief device shall be plainly and permanently marked with the "Container Type," of the pressure vessel on which the device is designed to be installed, with the pressure in pounds per square inch gage at which the device is set to start to discharge, with the actual rate of discharge of the device at its full open position in cubic feet per minute of air at 60 F and atmospheric pressure, and with the manufacturer's name and catalog number; for example, T-200—250–15,000 AIR—indicating that the device is suitable for use on a Type 200 container, that it is set to start to discharge at 250 psi gage, and that its rate of discharge at full open position is 15,000 cfm of air. Each field erected nonrefrigerated and refrigerated container relief device shall be similarly marked except "Container Type" indication is not required.

612. The rate of discharge of container relief valves shall be in accordance with the provisions of Appendix A for nonrefrigerated containers and Appendix E for refrigerated containers.

* * * *

APPENDIX A

Minimum Required Rate of Discharge in Cubic Feet per Minute of Air at 120% of the Maximum Permitted Start to Discharge Pressure for Safety Relief Devices to be Used on Nonrefrigerated Containers Other Than Those Constructed in Accordance with Interstate Commerce Commission Specifications

Surface area, sq ft	Flow rate, cfm air	Surface area, sq ft	Flow rate, cfm air	Surface area, sq ft	Flow rate, cfm air
20, or less	626	170	3620	600	10,170
25	751	175	3700	650	10,860
30	872	180	3790	700	11,550
35	990	185	3880	750	12,220
40	1100	190	3960	800	12,880
45	1220	195	4050	850	13,540
50	1330	200	4130	900	14,190
55	1430	210	4300	950	14,830
60	1540	220	4470	1000	15,470
65	1640	230	4630	1050	16,100
70	1750	240	4800	1100	16,720
75	1850	250	4960	1150	17,350
80	1950	260	5130	1200	17,960
85	2050	270	5290	1250	18,570
90	2150	280	5450	1300	19,180
95	2240	290	5610	1350	19,780
100	2340	300	5760	1400	20,380
105	2440	310	5920	1450	20,980
110	2530	320	6080	1500	21,570
115	2630	330	6230	1550	22,160
120	2720	340	6390	1600	22,740
125	2810	350	6540	1650	23,320
130	2900	360	6690	1700	23,900
135	2990	370	6840	1750	24,470
140	3080	380	7000	1800	25,050
145	3170	390	7150	1850	25,620
150	3260	400	7300	1900	26,180
155	3350	450	8040	1950	26,750
160	3440	500	8760	2000	27,310
165	3530	550	9470		

Surface Area = Total outside surface area of container in square feet.

When the surface area is not stamped on the name plate or when the marking is not legible, the area can be calculated by using one of the following formulas:

(1) Cylindrical container with hemispherical heads
Area = Overall length × outside diameter × 3.1416

(2) Cylindrical container with semi-ellipsoidal heads
Area = (Overall length + 0.3 outside diameter) × outside diameter × 3.1416

(3) Spherical container
Area = Outside diameter squared × 3.1416

Flow Rate, cfm air = Required flow capacity in cubic feet per minute of air at standard conditions, 60 F and atmospheric pressure (14.7 psia).

The rate of discharge may be interpolated for intermediate values of surface area. For containers with total outside surface area greater than 2000 sq ft, the required flow rate can be calculated using the formula, Flow Rate, cfm air = 53.632 $A^{0.82}$.

where:

A = total outside surface area of the container in square feet.

Valves not marked "Air" have flow rate marking in cubic feet per minute of liquefied petroleum gas. These can be converted to ratings in cubic feet per minute of air by multiplying the liquefied petroleum gas ratings by the factors listed below. Air flow ratings can be converted to ratings in cubic feet per minute of liquefied petroleum gas by dividing the air ratings by the factors listed below.

Air Conversion Factors

Container Type	100	125	150	175	200
Air Conversion Factor	1.162	1.142	1.113	1.078	1.010

* * * *

APPENDIX E

Note: The safety relief valve capacity in addition to preventing excessive pressure in the event of fire exposure also protects the container from excessive pressure in event the refrigeration system does not function.

Minimum required rate of discharge in cubic feet per minute of air at 120 per cent of the maximum permissible start-to-discharge pressure as specified in Paragraph 632 for safety relief devices to be used on refrigerated containers shall be computed by the following formula:

$$Qa = \frac{633{,}000\, FA^{0.82}}{LC}\sqrt{\frac{ZT}{M}}$$

where:

Qa = Minimum required flow capacity of air, in cubic feet per minute, at 60 F and 14.7 psia.

F = A composite environmental factor, as tabulated in Table E-1.

A = Total exposed wetted surface, in the case of spheres or spheroids, to the elevation of maximum horizontal diameter of the tank, in sq ft.

L = Latent heat of gas at *flowing* conditions in Btu/pound

Z = Compressibility factor at *flowing* conditions.

T = Absolute temperature at *flowing* conditions.

M = Molecular weight of gas.

C = Constant for gas which is a function of the ratio of specific heats at standard conditions. While not strictly applicable to flows at pressures under 15 psig, its use will produce conservative results.

k = C_p/C_v, value for C is then taken from graph of k vs. C, or the table on page 5/40.

k	Constant C	k	Constant C	k	Constant C
1.00	315	1.26	343	1.52	366
1.02	318	1.28	345	1.54	368
1.04	320	1.30	347	1.56	369
1.06	322	1.32	349	1.58	371
1.08	324	1.34	351	1.60	372
1.10	327	1.36	352	1.62	374
1.12	329	1.38	354	1.64	376
1.14	331	1.40	356	1.66	377
1.16	333	1.42	358	1.68	379
1.18	335	1.44	359	1.70	380
1.20	337	1.46	361	2.00	400
1.22	339	1.48	363	2.20	412
1.24	341	1.50	364		

where:

$$C = 520 \sqrt{k \left(\frac{2}{k+1}\right)^{\frac{k+1}{k-1}}}$$

Table E-1—Environmental Factors

Environment	Factor F
1. Bare vessel	1.0
2. Insulated vessels, with the following typical conductance values, in Btu per hr-sq ft-°F, based on 1600 degrees Fahrenheit temperature difference:	
a. 4.0	0.3
b. 2.0	0.15
c. 1.0	0.075

* * * *

613. Connections to which relief devices are attached, such as couplings, flanges, nozzles, and discharge lines for venting, shall have internal dimensions that will not restrict the net relief area.

614. The size of the relief device outlet connection shall not be smaller in diameter than the nominal size of the relief outlet connection and shall not appreciably restrict flow thru the relief.

615. All container relief devices shall be located on the containers and shall be connected with the vapor space of the container.

616. No shutoff valve shall be installed between the relief device and the container, equipment, or piping to which the relief device is connected except that a shutoff valve may be used where the arrangement of this valve is such that full required capacity flow thru the relief device is always afforded.

Note: The above exception is made to cover such cases as a three-way valve installed under two relief devices, each of which has the required rate of discharge. The installation will allow either of the reliefs to be closed but does not allow both reliefs to be closed at the same time. Another exception to this may be where two separate reliefs are installed with individual shutoff valves. In this case the two shutoff valve stems shall be mechanically interconnected in a manner which will allow full required flow of one relief at all times.

617. Relief device discharge vents shall be installed in a manner which will provide protection against physical damage and such discharge pipes shall be fitted with loose fitting rain caps. Return bends and restrictive pipe fittings shall not be permitted.

618. If desired, discharge lines from two or more relief devices located on the same unit, or similar lines from two or more different units, except those located on storage containers, may be run into a common discharge header provided the header is designed with a flow capacity sufficient to limit the maximum back pressure to (a) not exceeding 10 per cent of the lowest start-to-discharge pressure setting for conventional relief valves, and (b) not exceeding 50 per cent of the lowest start-to-discharge pressure setting for balanced valves. Header design shall assume that all valves connected to the header are discharging at the same time.

619. Discharge from a relief device shall not terminate in any building, beneath any building, or in any other kind of confined area. The discharge from all relief devices, except those installed between shutoff valves, shall be piped to a point not less than 3 ft above the highest point of any building within 50 ft.

62. Testing Relief Devices

620. Frequent testing of relief devices, as would be required where there is a probable increase or decrease in the releasing pressure of the device due to clogging, sticking, corrosion or exposure to elevated temperatures, is not necessary for such devices on liquefied petroleum gas containers for the following reasons:

(a) The gases are so-called "sweet gases," i.e , they have no corrosive effect on metals; the devices are constructed of materials not readily subject to corrosion and are protected against the weather when installed in pressure vessels. Further, the temperature variations are not sufficient to bring about any permanent set of spring mechanisms.

(b) Therefore the testing and inspecting of relief devices to check relief pressure settings is required only at about five-year intervals.

63. On Aboveground Containers

630. Every container shall be provided with spring loaded relief valves or their equivalent.

631. The discharge from the relief devices shall be vented away from the container, and unobstructed to the open air in a manner to prevent any impingement of escaping gas upon the container, adjacent containers, piping and other equipment. The vents shall be fitted with loose fitting rain caps. Suitable provision shall be made to prevent any liquid or condensate that may accumulate inside the relief device or its vent from rendering the relief device inoperative. If a bottom drain is used, a means shall be provided to protect the container, adjacent containers, piping of equipment against impingement of flame resulting from ignition of product escaping from the drain. The vent piping shall extend upward at least 7 ft above the top of the container.

632. Container relief devices shall be set to start to discharge as follows with relation to the design pressure or maximum allowable working pressure of the container as appropriate for the applicable code:

Containers	Minimum	Maximum*
ASME Code; Par. U-68, U-69—1949 and earlier editions	110%	125%
ASME Code; Par. U-200, U-201—1949 edition	88	100
ASME Code—1950, 1952, 1956 and 1959 editions	88	100
API-ASME Code—All editions	88	100
API Standard 620 (First Edition)		100
API Standard 650 (First Edition)		100

* Note: A plus tolerance of 10% is permitted.

633. Relief devices on containers shall be constructed to discharge at not less than the rates shown in Appendix A or E, before the pressure is in excess of 120 per cent of the maximum permitted start to discharge pressure setting of the devices.

634. In certain locations sufficiently sustained sun temperatures prevail which will require the use of a lower vapor pressure product to be stored or the use of a higher designed pressure vessel in order to prevent the container relief device from opening as a result of these temperatures. As an alternative the containers may be protected by cooling devices such as water sprays, by shading, or other effective means.

635. For refrigerated storage, consideration shall be given to making proper provisions for vacuum conditions.

64. On Underground Containers

640. Relief devices shall meet all the conditions outlined for Aboveground Containers except the rate of discharge for relief devices installed thereon may be reduced to a minimum of 30

per cent of the specified rate of discharge shown in Appendix A. The discharge pipe from safety relief devices shall extend directly, vertically upward at least 7 ft above the ground. If liquid product is placed in containers while they are not buried, these containers should be considered as aboveground containers.

641. Where there is a probability of the manhole or housing becoming flooded, the discharge from regulator vent lines should be above such water level. All manholes or housings shall be provided with ventilated louvers or their equivalent.

65. On Vaporizers

650. Each vaporizer shall be provided with a relief device providing an effective rate of discharge in accordance with Appendix B.

* * * *

APPENDIX B

Minimum Required Rate of Discharge for Safety Relief Valves for Liquefied Petroleum Gas Vaporizers (Steam Heated, Water Heated, and Direct Fired)

The minimum required rate of discharge for relief valves shall be determined as follows:

1. Obtain the total surface area by adding the surface area of vaporizer shell in square feet directly in contact with liquefied petroleum gas and the heat exchange surface area in square feet directly in contact with liquefied petroleum gas.
2. Obtain the minimum required rate of discharge in cubic feet of air per minute, at 60 F and 14.7 psia from Appendix A for this total surface area. [Appendix A follows Paragraph 612, which appears in this chapter.—Editor]

* * * *

651. Relief valves on direct fired vaporizers shall be located so that they shall not be subjected to temperatures in excess of 140 F. (See 61 for other requirements on relief devices.)

66. Between Shutoff Valves

660. A relief device shall be installed between each pair of shutoff valves on liquefied petroleum gas liquid piping so as to relieve into a safe atmosphere. It is recommended that the start to discharge pressure of such relief devices be not in excess of 500 psig.

67. At Discharge of Final Stage Regulators.

670. When the discharge pressure from the final stage regulator is not more than 5 lb, the low pressure side shall be equipped with a relief device, set to relieve at not less than two times, and not more than three times the discharge pressure but not more than 5 lb in excess of the discharge pressure. When the discharge pressure is more than 5 lb, the relief shall be set to not less than 1¼ times and not more than two times the discharging pressure. Regulator breather vents shall be piped outside the building and equipped with insect-proof terminal screens.

VAPORIZERS

Liquefied petroleum gases are used in gaseous form. A vaporizer is required when the heat transferred to the liquid is inadequate to vaporize sufficient gas for maximum demand. Size of the vaporizer, i.e., heat exchanger required, depends upon such factors as maximum gas demand, size and location of storage tank, minimum amount of gas carried in storage tank, climatic conditions, and gas pressure to be supplied by plant. A steam, hot water, or direct fired type vaporizer may be used.

A brief explanation of what occurs within the container follows. When saturated vapor is drawn off from the space above the liquid, the state of equilibrium that exists between liquid and vapor is disturbed. As a result, the pressure drops, and the liquid begins to boil. This promotes distillation (*weathering*), and the percentages of the different constituents in the vapor change continuously from the full to empty tank—the most volatile constituents evaporate first. A vaporizer minimizes the weathering phenomenon.

A vaporizer should be equipped with an automatic means of preventing liquid passing from vaporizer to gas discharge piping. Normally this is done by a liquid level controller and positive shutoff of liquid inlet line or by a temperature control unit for shutting off liquid line at low temperature conditions within vaporizer.

Some installations operate on "flash vaporization," whereby the liquid is converted to a gas as soon as it enters the vaporizer, while others maintain a liquid level in the vaporizer. The process, demand, and auxiliary controls determine the best practical method of operation.

Reference may be made to the following extract from NFPA No. 59:[1]

51. General

510. Liquefied petroleum gas storage containers shall not be directly heated with open flames.

511. Heating or cooling coils shall not be installed inside of a storage container.

512. Vaporizers shall not be equipped with fusible plugs for pressure relief.

513. Vaporizer houses shall not have drains to sewers or sump pits.

52. Vaporizers Not Directly Heated With Open Flames

520. Vaporizers constructed in accordance with the requirements of the ASME Unfired Pressure Vessel Code shall be permanently marked as follows:

(a) With the code marking signifying the specifications to which vaporizer is constructed.

(b) With the allowable working pressure and temperature for which the vaporizer is designed.

(c) With the sum of the outside surface area and the inside heat exchange surface area expressed in square feet.

(d) With the name or symbol of the manufacturer, date of manufacture, and serial number.

521. Vaporizers having an inside diameter of 6 in. or less exempted by the ASME Unfired Pressure Vessel Code shall have a design working pressure not less than 250 psig and need not be permanently marked.

522. Vaporizers shall not be installed in the same room with units furnishing air other than for a liquefied petroleum gas mixing device. Vaporizers may be installed in buildings, rooms, sheds, or lean-tos, other than those in which open flames or fires may exist. Such structures shall be of light fire resistive construction or equivalent, well ventilated near the floor line and at the highest point in the roof.

523. A shutoff valve shall be installed on the liquid line to the liquefied petroleum gas vaporizer unit at least 50 ft away from the vaporizer building.

524. The heating medium lines into and leaving the vaporizer shall be provided with suitable means for preventing the flow of gas into the heat systems in the event of tube rupture in the vaporizer. Vaporizers shall be provided with suitable automatic means to prevent liquid passing from the vaporizers to the gas discharge piping.

525. The device that supplies the necessary heat for producing steam, hot water, or other heating medium shall be separated

from all compartments or rooms containing liquefied petroleum gas vaporizers, pumps, and central gas mixing devices by a wall of substantially fire resistive material and vaportight construction.

53. Direct Fired Vaporizers

530. Each vaporizer shall be marked to show the name of the manufacturer; rated British Thermal Unit input to burners; the area of the heat exchange surface in square feet; and the maximum vaporizing capacity in gallons per hour, and date and serial number.

531. No direct fired vaporizers shall be located closer than 50 ft to line of adjoining property upon which structures may be built. They shall also be located a minimum distance of 50 ft away from any liquefied petroleum gas storage container.

532. No direct fired vaporizer shall be connected to a container that has a storage capacity in gallons, less than 10 times the hourly capacity of the vaporizer in gallons. Vaporizers may be connected to the liquid section or the gas section of the storage container, or both; but in any case there shall be at the container a manually operated valve in each connection to permit complete shutting off, when desired, all flow of gas or liquid from container to vaporizer.

533. Vaporizers may be installed in buildings, rooms, housings, sheds, or lean-tos used exclusively for vaporizing or mixing of liquefied petroleum gas. All vaporizer housing structures shall be of light fire resistive construction, well ventilated near the floor line and the highest point of the roof.

534. When vaporizers and mixing equipment are installed in structures that house other facilities, the vaporizers and mixing equipment room shall be separated from the other parts of the building with fire resistive, vaportight walls.

535. Vaporizers shall be provided with suitable automatic means to prevent liquid passing from the vaporizer to the gas discharge piping of the vaporizer.

536. Vaporizers shall be provided with a means for turning off the gas to the main burner and pilot from a remote location.

537. Vaporizers shall be equipped with automatic safety devices to shut off the flow of fuel to main burners and pilot, if the ignition device should fail.

538. Pressure control equipment which is a pertinent part of the vaporizer, if located within 10 ft of the vaporizer, shall be separated from the open flame by a substantial vaportight, fire resistive partition or partitions.

539. No direct fired vaporizer shall raise the product pressure over the designed working pressure of the vaporizer equipment.

Indirect Fired Vaporizers

This type is commonly used in larger LP-gas installations. Its arrangement involves heating a water bath by direct firing. This bath in turn heats coils which contain LP-gas.

Steam and Hot Water Vaporizers

Propane vaporizers consume about one pound of steam per gallon of propane vaporized. They are of the vertical tubular type for capacities up to about 2500 gal per hr. For larger capacities the horizontal type is generally used. A typical vertical type steam vaporizer is shown in Fig. 5-25. Piping details are presented in Fig. 5-26. A horizontal type vaporizer with controls is shown in Fig. 5-27.

Hot water and steam vaporizers have many common design features. The capacity of a hot water unit is lower than a steam unit of equal size.

Fig. 5-25 Capacities and dimensions (in inches) of Paracoil LP-gas vaporizers.

Unit size	Cap., gph	Shell A	Shell B	Piping connections D	E	F	G	H	J	L min	M	Approx. wt, lb
F-48E	600	8⅝	0.277	2	2½	2½	¾	2	2	12	76	465
H-48E	960	10¾	.307	2	2½	2½	¾	2	2	12	76⅝	700
K-48E	1400	12¾	0.330	2½	3	3	1	2	3	12	78	850
L-48E	1725	14	5/16	2½	3	3	1	2	3	12	78¼	1145
N-48E	2300	16	⅜	3	4	4	1½	2	3	19½	79½	1340
P-48E	3000	18	⅜	3	4	4	1½	2	3	19½	81	1960
R-48E	4000	20	⅜	4	6	4	1½	2	4	19½	84½	2130
T-48E	5000	22	⅜	4	6	6	2	2	4	24½	86¼	2850
V-48E	6000	24	½	4	6	6	2	2	4	24½	87¼	3425

Capacities based on propane boiling at 32 F, steam at 212 F (0 psig).

	Design Pressures, psig		
	Working at 250 F	Hydrostatic test	Hammer test
Shell	250 (liq & vapor)	375	375
Tubes	100 (steam & condenser)	150	150

CODE:

- A—Shell diameter
- B—Shell thickness
- D—Liquid inlet
- E—Vapor outlet
- F—Steam inlet
- G—Condensate outlet
- H—Bottom float connection
- J—Top cage and safety valve connection
- K—Tube sheet
- N—Tubes—Outer vaporizing—⅞ in. OD
  Inner steam feeder—½ in. OD
- T—Auxiliary drain (plugged)
- V—¾ in. P. T. pressure regulator connection
- Y—P.T. shell drain
- Z—¼ in. air vent (plugged)
- AA—¾ in. P.T. vacuum breaker connection
- AB—½ in. P.T. drain (plugged)

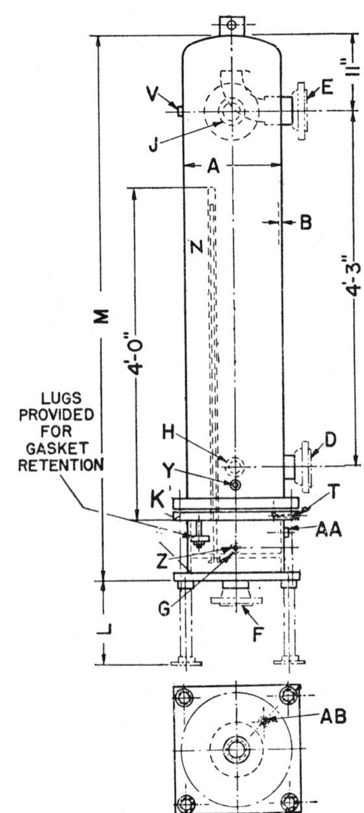

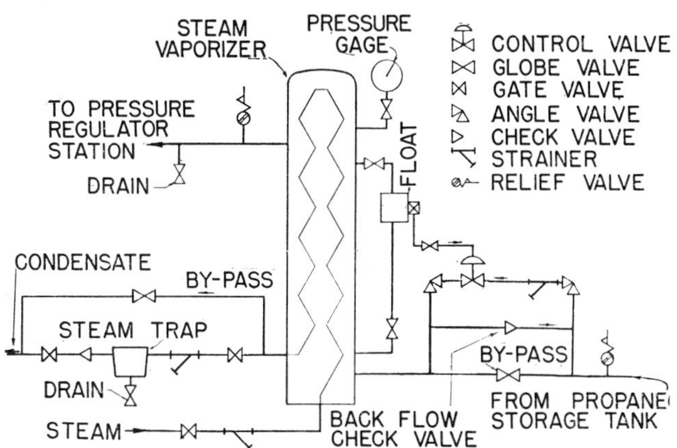

Fig. 5-26 Vertical type steam vaporizer installation. (H. Emerson Thomas and Assoc., Inc.)

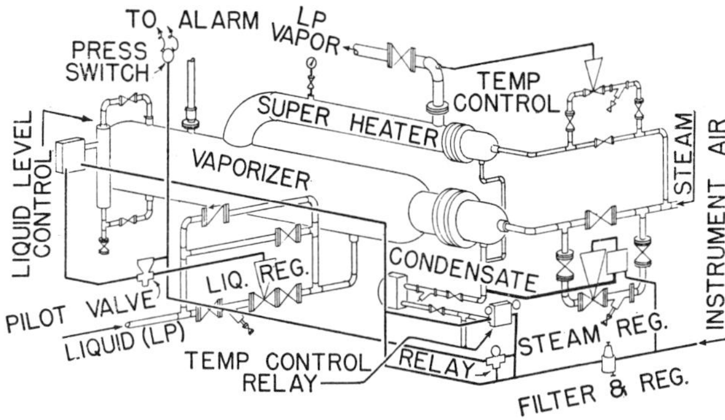

Fig. 5-27 Horizontal type steam vaporizer. (Drake & Townsend, Inc.)

Direct Fired Vaporizers

Vaporization is obtained by direct heating of the vaporizer chamber. Figure 5-28 shows the general design of one direct-fired unit and a diagram of its connection to a storage tank.

MIXING EQUIPMENT

This equipment combines various fuel gases or gas–air mixtures in required proportions, e.g., for a specified Btu sendout.

Ratio Controller

In this device (Fig. 5-29), LP-gas and air, for example, at atmospheric pressure, are aspirated thru the mixer by a motor-driven pump. Proportion control is very accurate at rated flow. The governor (regulator) shown reduces the LP-gas pressure to atmospheric.

Venturi or Jet Mixers

There are both high and low pressure units which use the pressure energy of the constituents. In a low pressure unit, the kinetic energy of the LP-gas inspirates atmospheric air, while in a high pressure one the LP-gas inspirates compressed air at approximately the required discharge pressure.

These mixers are usually arranged in parallel (Fig. 5-30a) to cover a large capacity range in, say, equal increments. For example, four jets with capacity ratios of 1:2:4:8 will give 15 equal rates of flow. If the ratio of the absolute pressures of the LP-gas to air exceeds two, phenomena associated with critical velocity will occur, resulting in an energy loss. However, high LP-gas to air pressures are necessary to achieve high downsteam pressure or to change the gas–air ratio.

On practically all venturi jet installations the LP-gas is supplied at approximately 30 psig to inspirate atmospheric air; at a 1 to 1 ratio of the air–gas mixture (1300 Btu), the downstream pressure approximates 4.5 psig. Equation **1** (Sec. 12, Chap. 13) and Table 12-92a relate the factors involved.

In Fig. 5-30a, the pilot pressure difference regulator controls the butterfly valve in the air suction header to hold a constant pressure difference between this suction header and the mixed gas header. The pressure difference regulator has two diaphragms which have the differential pressure on the outside of them and the pilot control valve for the diaphragm motor between them. The force produced by the pressure difference is balanced by loading springs, one of which is adjustable from the Btu control motor. Thus, the pilot valve position will be determined by the pressure difference on the diaphragms and the force exerted by the loading springs. Hence, the pressure on the top of the diaphragm-motor operator will be controlled in response to the pressure difference,

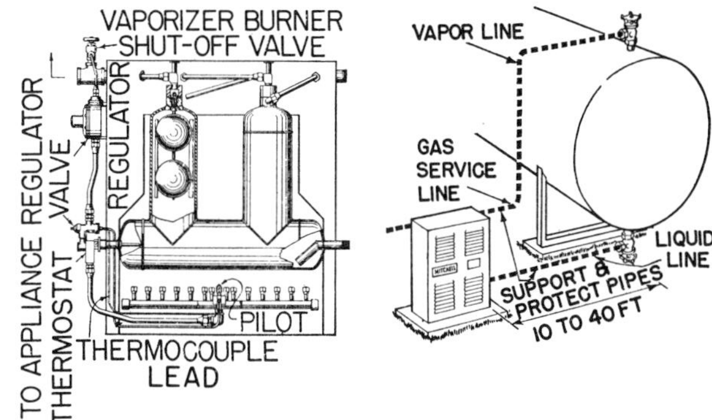

Fig. 5-28 (left) Mitchell direct-fired vaporizer and (right) its connection to an LP-gas storage tank. (H. Emerson Thomas and Assoc., Inc.)

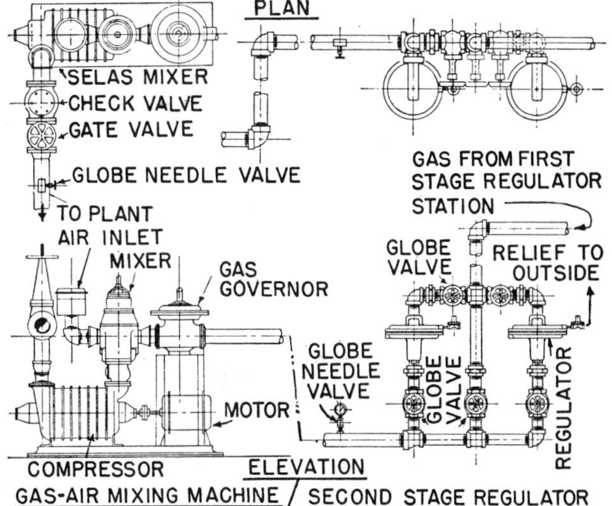

Fig. 5-29 Selas gas–air mixing machine with piping connections. (H. Emerson Thomas and Assoc., Inc.)

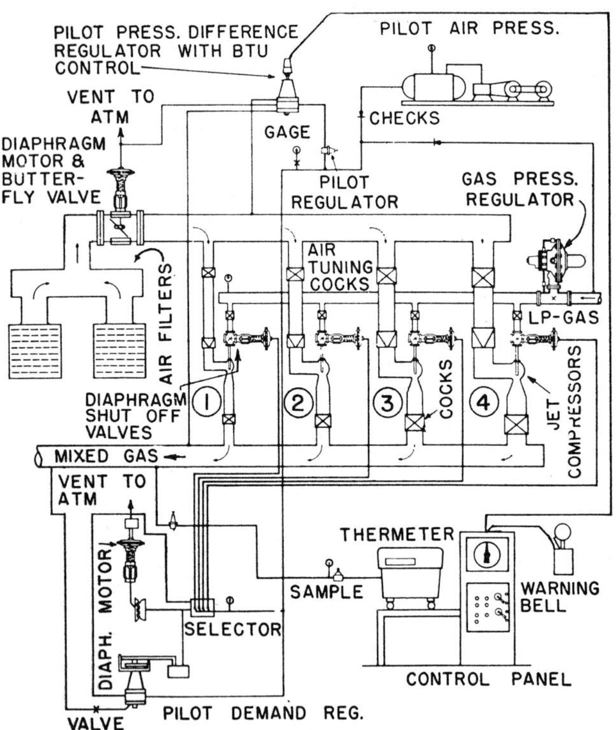

Fig. 5-30a Jet compressor mixing of LP-gas and air. (Cutler-Hammer, Inc.)

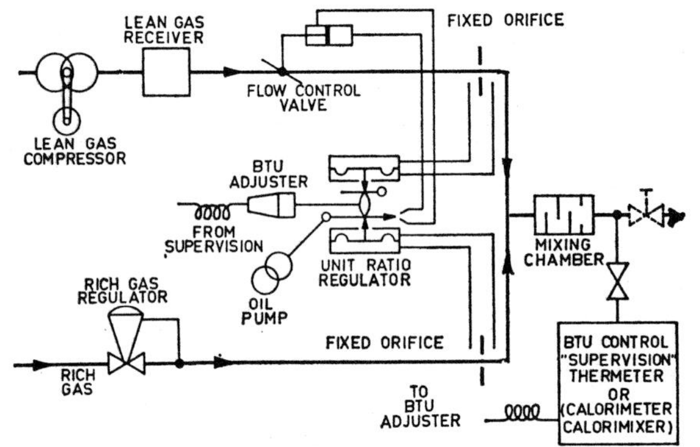

Fig. 5-30b Fixed orifice type mixer plant. (Cutler-Hammer, Inc.)

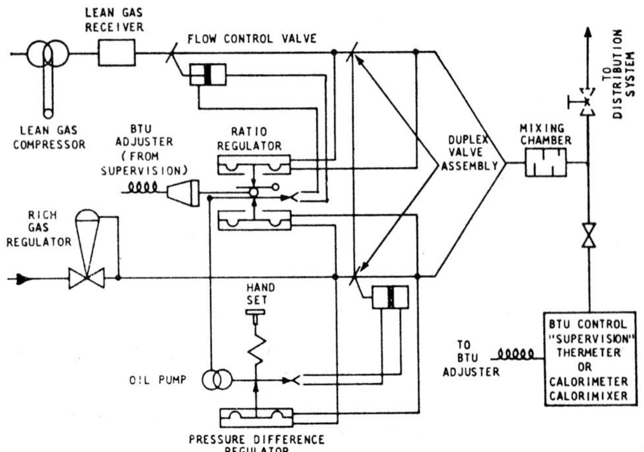

Fig. 5-30c Adjustable orifice mixer plant. (Cutler-Hammer, Inc.)

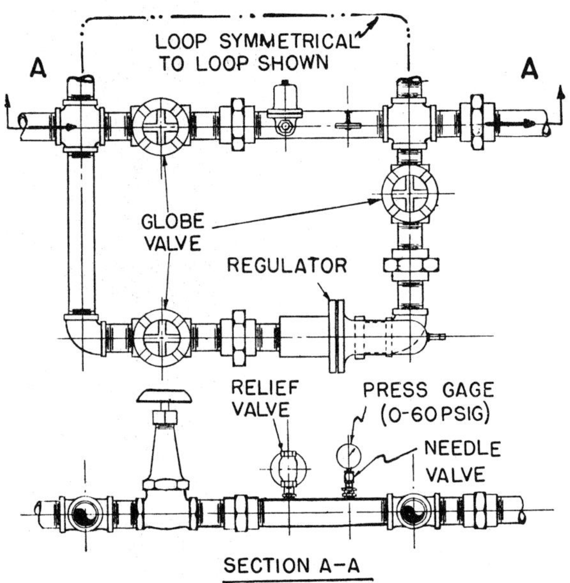

Fig. 5-31 Standard first stage regulator station. (H. Emerson Thomas and Assoc., Inc.)

and the butterfly valve position will be adjusted to produce the pressure difference required to satisfy the loading of the regulator.

Assume that jet compressor No. 1 was meeting the demand, and that suddenly the demand goes up to the capacity of jet No. 2. The selector switch would automatically start jet No. 2 and shut off jet No. 1. If demand then increases, jet No. 1 will also operate. The other paralleling units will also be automatically activated and shut down as required. Automatic Btu sendout control is initiated by a calorimeter. Regulators are air controlled. Alarm for low LP-gas or natural gas pressures and low or high Btu of mixed gas are also provided.

Orifice Mixers

Fixed Orifice (Fig. 5-30b). Orifices meter the flow of each constituent. Upstream of the orifices, a manually adjustable regulator controls the higher pressure (and higher Btu) gas while a ratio controller automatically adjusts the flow rate of the lean gas to compensate for any changes in the rich gas regulator setting or Btu value of the constituents. Capacities as high as 600 MSCFH at 50 psig are available; minimum output—25 per cent of rated capacity.

Adjustable Orifice. Like the fixed orifice setup, this system permits a much larger range in capacity. This greater flexibility is accomplished by means of a hand setting of a duplex valving arrangement. These valves or adjustable orifices pass a controlled flow of gas at as low as 3 per cent of rated flow. The mixed gas delivery pressure is largely determined by the setting of the rich gas regulator.

The duplex valving arrangement in Fig. 5-30c is more automatic. Hand setting of the pressure difference regulator constitutes the initial coarse duplex valve setting. Thereafter the pressure difference regulator controls the duplex valve.

REGULATOR STATIONS

Suitable regulator stations are essential for proper control of operation of LP-gas systems. In design it is advisable to provide regulator stations in duplicate, so arranged that one regulator may be taken out of service for repairs when necessary without shutting down entire plant. Relief valves should be installed in the piping on the downstream side.

Maximum flow rates and pressure, both inlet and outlet, are factors governing selection of regulators. Capacities given in sizing charts and manufacturer's data are usually based on air (1.00 sp gr) or gas of 0.60 specific gravity.

Arrangement of a standard regulator station, including necessary valves and piping, is shown in Fig. 5-31.

PIPING AND FITTINGS

Steel pipe and fittings should be designed for container pressure. Piping over 2 in. nominal diameter preferably should be of welded construction and with flanged joints. Pipe joint compounds, gaskets, packing, and valve seats should be resistant to LP-gas.

Tank connections should be of the swing joint type to provide flexibility against stresses due to temperature, pressure, and movement. Expansion joints should be provided when welded pipe is used. Overhead piping should be properly supported with necessary allowances for expansion. Where cars or trucks pass below piping, adequate head room should be provided. Underground piping should be protected against mechanical damage where it passes under roads, paved areas, or trackage, and against corrosion as local conditions may require.

REFERENCES

1. N.F.P.A. *Standard for the Storage and Handling of Liquefied Petroleum Gases at Utility Gas Plants.* (NFPA 59) Boston, 1963.
2. A.S.M.E. *ASME Boiler and Pressure Vessel Code.* (Rules for Construction of Unfired Pressure Vessels, Sec. VIII) New York, 1962.
3. N.F.P.A. *Standard for the Storage and Handling of Liquefied Petroleum Gases.* (NFPA 58) Boston, 1963.
4. Karol, R. H. *Soils and Soil Engineering.* New York, Prentice-Hall, 1960.
5. Terzaghi, K. *Theoretical Soil Mechanics.* New York, Wiley, 1943.
6. "World's Largest Refrigerated Storage Facility to be Installed in Minneapolis." *Gas Age* 127: 57, Jan. 19, 1961.
7. Newcomb, W. K. "Principles of Foundation Design for Engines and Compressors." *A.S.M.E. Trans.* 73: 307–318, 1951.

Chapter 6

Operation and Safety of Utility LP-Gas Plants

by John L. Turnan and F. E. Vandaveer

The operation of an LP-gas plant depends upon its function, size, variation in demand, and storage system (i.e., underground or aboveground, refrigerated or nonrefrigerated). Plant investment varies from \$7.50[1] to \$50.87[2] per MCF. The amount of LP-gas mixed with air or other gases depends upon the heating value (Fig. 5-32) required—outside the flammability or dew point ranges. Figure 5-33 shows dew point of LP-gas–air mixtures. *Cost of LP-gas–air* varies from \$1.40 to \$3.50 per MCF for a 1000 Btu gas, depending upon the cost of LP-gas and the amount of mixture produced. Ease of shutdown and starting, as well as minimum maintenance, are the major desirable factors for stand-by plants. Starting time for an LP-gas plant is usually less than an hour.

GENERAL LP-GAS HANDLING PRECAUTIONS

1. Smoking and open flames shall be prohibited in area of process plant, storage tanks, and unloading stations.
2. Test for leaks with soap suds, liquid detergents, or similar leak indicating material.
3. Ground tank cars and truck trailers before any unloading or loading operation is started, to prevent electrical shock to personnel.
4. Wear rubber (neoprene) gloves when gaging tank cars, truck trailers, and storage tanks.
5. Keep vapor or liquid off skin and clothing. Liquid in contact with skin produces same injury as a burn.
6. Require thorough ventilation in all process building, especially at floor level.
7. Where flammable gas–air mixtures exist, prohibit work which may cause sparking from tools.
8. For artificial illumination, use a flashlight or an approved safety lamp.
9. Open all valves slowly. If excess flow valves close, they may be opened by closing line discharge valve for a few minutes; this permits pressure above and below excess flow valve to equalize thru small hole provided for this purpose, thus enabling valve to open.
10. Fill storage tanks to proper levels.
11. Do not confine liquid in any isolated section of pipe, except where a pressure relief valve is provided, since the coefficient of thermal expansion of liquefied gas is high.
12. Do not pressurize tank cars above safety valve setting. If faulty car equipment prevents normal unloading, do not loosen or remove car valves under any circumstances. Fill in "Bad Order Tag" and attach to defective car part. Inform proper authorities of defective car, and await further instructions on its disposition.
13. Bleed vapor and liquid unloading and loading hoses before uncoupling from tank cars or truck trailers.
14. Bleed pressure from vapor compressor crankcase before removing oil filler plug.
15. Before undertaking repairs, isolate container or section of pipe to be repaired, by closing valves or by physical disconnection from communicating lines as situation warrants, followed by thorough purging with CO_2 or other inert gas.
16. Maintain hoses and fittings in good order. Damaged or unsafe hose must be removed from unloading area and must not be used for any other purpose.
17. Never hold head directly over tank car gaging device after releasing hold-down latch, because high pressure may force tube out rapidly.
18. Remove plugs and caps on manhole and tank connections slowly. Verify that shutoff valves are tightly closed and not leaking before removing plug-in valves. Check for leakage by slowly loosening plug.
19. Never place head over relief valves.
20. Be sure that operating personnel know the physical properties of LP-gases, particularly their flammable limits, vapor pressures at various temperatures, and that these gases are heavier than air.
21. Avoid accumulations of gas in rooms or pits. Preferably, avoid pits and ventilate rooms drawing air from floor level.
22. Some companies have a monitoring system for automatic continuous sampling of air at various hazardous locations. An alarm is rung when escaping gas reaches 20 per cent of lower explosive limit in air, and plant is shut down thru hydraulic system if gas concentration reaches 70 per cent of lower explosive limit.
23. Make positive identification of piping system content by lettered legend and color bands giving the name of the content in full or abbreviated form. Arrow may be used to indicate direction of flow.[3]

National Standards

Reference may be made to the following extract from NFPA No. 59:[4]

71. Transfer of Liquids Within a Utility Plant

710. Liquefied petroleum gas in liquid form may be transferred from tank cars, or tank trucks, or storage within a utility plant either by liquid pump or by pressure differential.

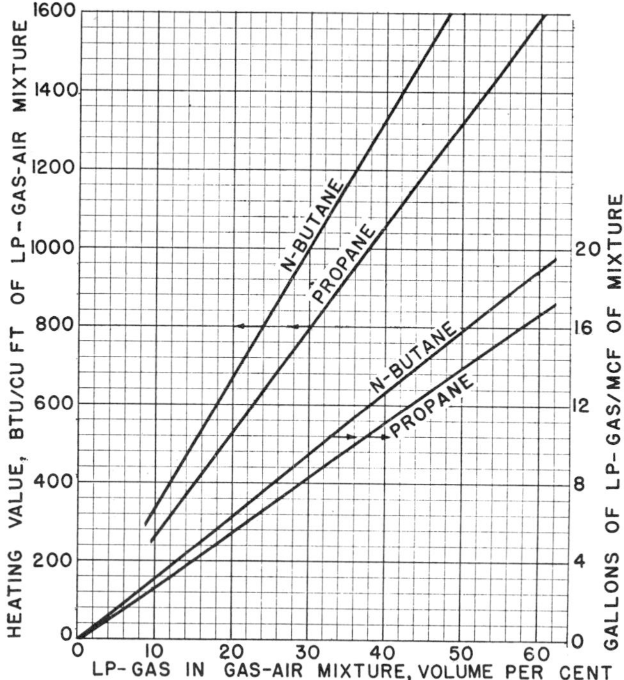

Fig. 5-32 Heating values* of average LP-gas–air mixtures and gallons of LP-gas per MCF of mixture.

* Curves start at upper flammability limits.

(a) Pumps and compressors used for transferring liquefied petroleum gas shall be designed for the product handled.

(b) Pressure differential for transferring liquid should be developed by a vapor compressor which takes suction from the vapor space of the liquefied petroleum gas container being filled and discharges into the vapor space of the container being emptied.

711. Under certain conditions, it may be necessary to create a pressure differential by using fuel gas, air, or inert gas, which is at a pressure higher than the pressure of the liquefied petroleum gas in the container being filled. This may be done under the following conditions:

(a) Adequate precautions must be taken to prevent liquefied petroleum gas from flowing back into the fuel gas, air, or inert gas line or system by installing two back flow check valves in series in these lines at the point where they connect into the liquefied petroleum gas system. In addition, a manually operated positive shutoff valve shall be installed at this point.

(b) Any fuel gas, air, or inert gas used to obtain a pressure differential to move liquid liquefied petroleum gas shall be noncorrosive and dried to avoid stoppage by freezing.

(c) If a fuel gas, air, or inert gas is used to obtain a pressure differential to move liquid liquefied petroleum gas, consideration should be given, after the operation is discontinued, to removing the fuel gas, air, or inert gas from the container into which it was placed, such as by venting. This should be done only if the vented gas can be conducted to a proper vent, preferably a distance from the plant and then properly disposed of.

(d) Before any fuel gas, air, or inert gas is placed in a tank car for unloading liquefied petroleum gas by pressure differential, permission should be obtained from the vendor of the liquefied petroleum gas to introduce such vapors into the tank car or a tank truck.

EXAMPLE:

Read down from the dew point temperature scale to the proper reference line then right to a point above the per cent LP in the mixture finding the highest practical storage or distribution pressure by interpolating between the lines of gauge pressures.

Thus to find the dew point gauge pressure of a 750 BTU N-Butane gas air mixture of 0 F read down from 0 F to the N-Butane reference line and from there read right to a point directly above 22% finding the dew point pressure is 20 PSIG.

NOTE:

Propane and Butane are shown in liquid volume per cent.

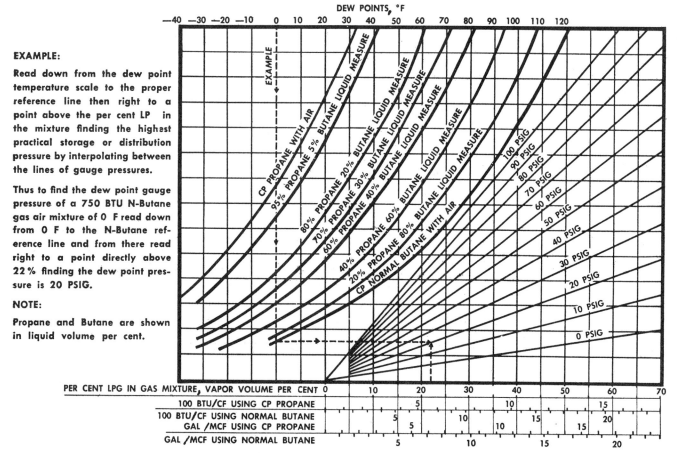

Fig. 5-33 Dew points of LP-gas–air mixtures, °F at various pressures.

712. At least one attendant shall remain close to the transfer connection from the time the connections are first made until they are finally disconnected, during the transfer of product.

713. The maximum vapor pressure of the product at 100 F which may be transferred into a container shall be in accordance with 220 or 221 and 222. [See Chap. 5.—EDITOR]

714. Where needed unloading piping or hoses shall be provided with suitable bleeder valves for relieving pressure before disconnection.

715. Precaution shall be exercised to assure that only those gases for which the system is designed, examined, and listed, are employed in its operation, particularly with regard to pressures.

72. Tank Car Loading and Unloading Point

720. The track of tank car siding shall be relatively level.

721. A TANK CAR CONNECTED sign, as covered by ICC (Interstate Commerce Commission) rules, shall be installed at the active end or ends of the siding while the tank car is connected for unloading.

722. While cars are on side-track for unloading, the wheels at both ends shall be blocked on the rail.

723. A man shall be in attendance at all times while the tank car or cars are being unloaded.

724. The pipe line to which the tank car unloading hoses are connected shall be equipped with a back flow check valve to prevent discharge of the liquefied petroleum gas from the receiving container and line in case of rupture of line hose or fittings.

725. The tank car unloading point should be located with due safety consideration to the following:

(a) Proximity to railroad and highway traffic.
(b) The distance of such unloading point from adjacent property.
(c) With respect to buildings on installer's property.
(d) Nature of occupancy.
(e) Topography.
(f) Type of construction of buildings.
(g) Number of tank cars that may be safely unloaded at one time.
(h) Frequency of unloading.

726. Where practical, the distance of the tank car unloading point should conform to the distance in 240 [See Chap. 5.—EDITOR] except that lesser distances may be used, keeping in mind the above items and upon approval of the authority having jurisdiction.

Reference may be made to the following extract from ASA B31.8–1963:[5]

862 LIQUEFIED PETROLEUM GAS [LPG] SYSTEMS

862.1 Liquefied petroleum gases, generally, include butane and propane, and mixtures of them that can be stored as liquids under moderate pressures (approximately 80 to 250 psig) at ambient temperatures.

862.2 This code is concerned only with certain safety aspects of liquefied petroleum gases when they are vaporized and used as gaseous fuels.

862.3 All the requirements of Standards No. 58 and No. 59 of the National Board of Fire Underwriters' and the National Fire Protection Association and of this code, concerning design, construction, and operation and maintenance of piping facilities shall apply to piping systems handling butane, propane, or mixtures of these gases.

862.4 Special Safety Requirements for LPG Systems

862.41 *Odorization.* Liquefied petroleum gases are usually nontoxic, but for safety when distributed for consumer use, or used as fuel in a place of employment, they shall be odorized.

862.42 *Ventilation.*

(a) All liquefied petroleum gases are heavier than air, hence structures above ground for housing regulators, meters, etc., shall have open vents near the floor level. Such equipment shall not be installed in pits or in underground vaults, except in cases where suitable provisions for forced ventilation are made.

(b) Special care is required in the location of relief valve discharge vents releasing LPG to the atmosphere, to prevent accumulation of the heavy gases at or below ground level. Likewise, special precautions are necessary for adequate ventilation where excavations are made for the repair of leaks in an underground LPG distribution system.

UNLOADING LP-GAS TANK CARS

The extent to which a tank car or other container of liquid LP-gas should be depressurized (i.e., its LP-gas vapor content reduced), after unloading its liquid contents, is basically an economic consideration.

Reference may be made to the foregoing Par. **72** extracted from NFPA No. 59.[4]

Using Natural Gas Under Pressure and a Vapor Compressor[6]

Use of natural gas in tank car unloading (if permitted by LP-gas vendor) serves a dual purpose:

1. Mixing natural gas with LP-gas in tank car lowers gravity of gases transferred from cars thru depressuring lines into storage holder. Resultant mixed gas disperses readily into atmosphere in case of line leakage.

2. The small volume of gas thus left in a tank car is composed largely of natural gas, thereby insuring a high percentage of recovery of LP-gas from tank cars.

Procedure for Propane.

1. Attach ground wire to car. Block car wheels. Close roadway gates. Place "*Stop—Tank Car Connected*" signs ahead of first car on unloading siding.

2. Determine temperature and gage contents of car. Record readings in LP-Gas Plant Log Book.

3. Connect unloading fitting to car vapor valve and couple vapor hose from unloading rack to fitting. Open car vapor valve slowly.

4. Connect unloading fittings to car liquid eduction valves and couple hoses from unloading rack to fittings. Open car liquid eduction valves slowly and fully. (These valves must be fully open during car unloading, to insure functioning of excess flow valves in liquid eduction lines in case of a line break.)

5. Open valve of storage tank filling line slowly, taking care not to open it too wide if tank car pressure is considerably higher than storage tank pressure. (Gage on filling line indicates storage tank pressure; gage on vapor line at rack indicates car pressure.) Car will unload while car pressure remains 5 to 10 psi higher than storage tank pressure.

6. When pressures in tank car and storage tanks approach equalization:

(a) Open vapor line from unloading rack to car.
(b) Start vapor compressor, putting control on "automatic." Adjust differential pressure switch to desired minimum and maximum pressures to be maintained.
(c) Withdraw vapor from storage tanks being filled by admitting to inlet manifold of vapor compressor.
(d) Impose compressor discharge pressure on car thru vapor line connection. Car will unload.

7. Tank car is empty when a flow of vapor, instead of liquid, is observed thru sight flow indicator. Another indication is rapid equalization of tank car and storage tank pressures.

8. When tank car is empty, close storage tank filling line valve at unloading rack.

9. Shut off vapor compressor. Close inlet and outlet valves on vapor compressor manifold.

10. Open depressuring line valve at unloading rack.* When tank car pressure reaches 70 psig, close vapor line valve at unloading rack and close car vapor valve. Bleed vapor hose to atmosphere thru ½ in. bleed valve on car unloading fitting.

11. Disconnect vapor hose from vapor line at unloading rack, and connect to natural gas line (at about 100 psig) at unloading rack.

12. Open car vapor valve and partially open natural gas line at unloading rack to admit natural gas (not in excess of 200 psig) into tank car for purpose of mixing natural gas with residual vapors in car. The LP-gas–natural gas mixture is passed into a gas holder (where economics warrant).

13. Continue depressuring car until a level of 10 to 15 psig is reached, and then shut off natural gas to car. Discharge goes into a low pressure holder.

14. Close depressuring line valve at unloading rack.

15. Close car vapor line valve. Close car liquid eduction line valve.

16. Bleed natural gas and liquid transfer hoses to atmosphere thru ½ in. bleed valves on car unloading fittings.

17. Disconnect hose couplings at fittings and place hose on racks.

18. Remove unloading fittings from car and replace plugs in car valves.

19. Secure car dome cover in place with iron pin provided.

20. Reverse steps taken in item 1.

21. Record car depressuring data in LP-Gas Log Book.

22. Report car empty to office for return billing.

Procedure for Butane.

1. thru 5. Follow instructions 1 thru 5 for unloading propane.

6. When pressures in tank car and storage tanks approach equalization: partially open natural gas line valve at unloading rack to impress natural gas on car contents. Maintain from 5 to 15 psi differential between car and storage tanks by adjusting natural gas admission valve. Car will unload. *Caution:* Do not repressure tank car in excess of car relief valve setting stenciled on side of every LP-gas tank car.

7. When car is unloaded, close natural gas admission valve.

8. Close storage tank filling line valve at unloading rack.

9. Open depressuring line valve at unloading rack and depressure car to 10 to 15 psig.

10. thru 17. Follow instructions 15 thru 22 for unloading propane.

Using Typical Unloading Compressor[7]

1. See that car is properly blocked and that handbrakes are set. Place **"Tank Car Connected"** sign on track facing switching end. Ground tank car dome with grounding connection provided.

2. Note condition of car seal and report if broken. Note car number and initials.

3. Place thermometer in thermometer well (G in Fig. 5-34) and allow it to remain 10 min before taking temperature of liquid. Withdraw thermometer quickly, read immediately to minimize atmospheric temperature effects, and record temperature.

4. Remove cover from gaging device (D) and proceed as follows: Depress tube and unlock from hold-down device. If equipped with gage tube brake, depress brake to raise tube. One hand should be on gage tube at all times since car pressure may force it upward rapidly. Raise gage tube until figure "5" appears opposite or above top surface of gaging pointer. Open valve at top of tube to permit gas vapor to escape. If liquid appears raise tube another inch or so. With tube valve open, lower gage tube into car with a rotary motion, in steps, until liquid appears at gage tube valve. Shut valve and read outage indicated on gage tube opposite top surface of pointer. Check reading by raising gage tube 1 in., opening valve until vapor appears, and then lowering tube as before until liquid appears. Shut off valve and take reading. Lower tube, replace locking device and cover. Make note of gage tube reading.

5. Connect liquid unloading hoses (J) to liquid outlet valves (A) on tank car, making sure that locking clips on sides of couplings are parallel with hose. Connect vapor hose (J) to vapor outlet valve on car (B).

6. Make sure that shutoff valves (N) at top of tower are closed and also that bleed valves (M) are closed; slowly open tank car valves (A) to allow pressure to build up in unloading hose to full open position.

* Vapors are discharged into a large low-pressure holder in which the pressure is not materially changed by the pressure in the tank car or the quantity of LP-gas vapors.

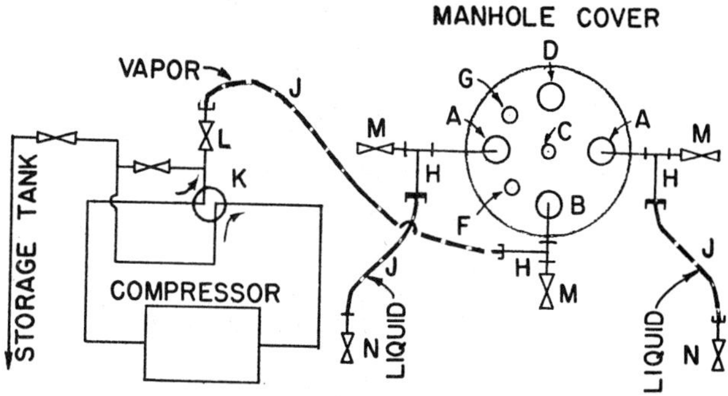

Fig. 5-34 Unloading connections for LP-gas tank cars. Note: 4-way valve is shown in position for pressurizing tank car to unload liquid.

LEGEND

A—liquid outlet valve	H—tank car connectors
B—vapor valve	J—flexible hose
C—relief valve	K—4-way valve
D—gaging device	L—vapor line valve at tower
F—sample valve	M—bleed valves
G—thermometer well	N—liq. line valves at tower

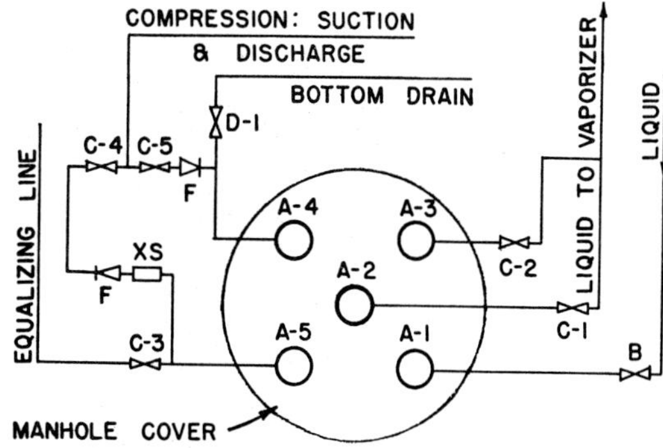

Fig. 5-35 Connections at manholes of LP-gas storage tanks.

LEGEND

A-1	Angle valve—liquid inlet
A-2	Angle valve—liquid outlet to vaporizer
A-3	Angle valve—liquid outlet to vaporizer
A-4	Vapor inlet valve and bottom drain
A-5	Vapor outlet valve
B	Liquid inlet shut-off valve
C-1-2-3-4-5	Auxiliary shut-off valves
D-1	Bottom drain valve
XS	Excess flow valve
F	Check valves

7. Make sure valve in vapor line at top of tower (L) is closed and slowly open vapor valve on car (B) to allow pressure to build up in hose to full open position.

8. On storage tank to be filled (Fig. 5-35), open wide angle valve (A-1) and plug cock (B) on liquid inlet line. Open wide angle valve (A-5) on vapor-out line and slowly open shutoff valve (C-4) to allow pressure to build up in vapor line to tank car.

9. At unloading compressor (Fig. 5-34), place 4-way valve (K) in proper position for pressurizing tank car. Slowly open all valves (except by-pass) in vapor line from storage tank and from tank car, including vapor valve (L) at top of tower.

10. Pressure gages on compressor should now indicate storage tank pressure and tank car pressure.

11. If storage tank and tank car pressures are equal at beginning, or if tank car pressure is less than storage tank pressure, proceed with item 14 immediately.

12. If the tank car pressure is 5 psi or more higher than storage tank pressure, open valves (N) in liquid line at top of tower until flapper in sight-flow glass is almost horizontal.

13. Permit liquid to flow from tank car to storage tank, opening valves (N) as required to maintain flapper near horizontal position.

14. When storage tank pressure and tank car pressure have equalized, start compressor. After several minutes' operation, tank car pressure should be from 25 to 50 psi higher than storage tank pressure.

15. If, during course of operations, flapper in sight-flow glass drops to a vertical position, it is an indication that excess flow valves in tank car have closed. If storage tank pressure, as indicated by gage at compressor, shows a steady rate of decline, it is an indication that either excess flow valves at storage tank have closed (there are two of them) or that some valve in suction line has not been opened. With compressor in operation, all liquid and vapor valves should be wide open, since its capacity is designed to maintain proper liquid flow.

16. Proceed with liquid transfer as above until car is empty, as indicated by: (a) flapper in sight-flow glass dropping to an almost vertical position, and (b) vapor appearing upon opening valve (F).

17. Shut off compressor, turn 4-way valve (K) 90 degrees, shut liquid valves (N) at top of tower.

18. At storage tank, shut off both valves in liquid inlet line (A-1 and B) and both valves in vapor outlet line (A-5 and C-4). Open angle valve (A-4) and valve (C-5) in vapor inlet line.

19. Start compressor and withdraw vapor from tank car, reducing pressure in tank car to approximately 50 psig in summer and 25 psig in winter.

20. When tank car pressure has been reduced, shut off compressor, turn 4-way valve (K) 90 degrees, shut off all liquid and vapor valves on tank car and also vapor valve (L) at top of tower.

21. Open three bleed valves (M) and bleed all pressure from hoses. Disconnect hoses, leaving bleed valves open during operation. Remove grounding clamp.

22. Remove pipe connections on tank car, close dome cover, and remove or reverse placards on four sides of car.

23. Remove "Tank Car Connected" sign from track.

24. Close angle valve (A-4) and valve (C-5).

UNLOADING LP-GAS TRUCKS AND TRAILERS[6]

1. Do not operate engines where flammable gas–air mixtures exist.

2. Remain close to transfer connection and shut off truck engines while unloading and depressuring carriers.

3. Ground carrier to loading rack. Block trailer wheels.

4. Determine temperature and gage contents of carrier (truck or trailer).

5. Connect vapor line hose at unloading rack to carrier vapor line. Open carrier vapor line valve.

6. Connect liquid transfer hoses from unloading rack to carrier liquid discharge manifold. Open carrier liquid discharge line valves slowly and fully.

7. When pressure on carrier contents exceeds storage tank pressure, leave vapor line valve at unloading rack in closed position. Open storage tank filling line valve slowly. Carrier will unload.

8. When carrier and storage tank pressures approach equalization:

(a) Open vapor line valve at unloading rack.
(b) Start vapor compressor, putting control on "automatic." Set differential pressure switch to desired minimum and maximum pressures to be maintained on carrier.
(c) Withdraw vapor from battery of storage tanks being filled by admitting to inlet of vapor compressor.
(d) Impose compressor discharge pressure on carrier thru vapor line connection. Carrier will unload.

9. Carrier is empty when a flow of vapor, instead of liquid, is observed thru sight-flow indicator.

10. When carrier is empty, close storage tank filling line valve at unloading rack.

11. Shut off vapor compressor. Close inlet and outlet valves on compressor manifold.

12. Open depressuring line valve at unloading rack. Depressure carrier to 10 to 15 psig (where economics warrant or shipment invoice specifies); 50 psig is a more common value since most transport deliveries give credit for vapor left in container.

13. Close depressuring line valve and vapor line valve at unloading rack.

14. Operator of truck-trailer unit shall close valves on carrier.

15. Bleed liquid transfer hoses to atmosphere thru ½ in. bleed valve on unloading rack. Bleed vapor transfer hose thru bleed valve on carrier manifold.

16. Disconnect hoses from carrier and return them to unloading rack.

17. Remove ground wire from carrier and return to unloading rack.

18. Make complete circuit of carrier with truck operator, removing wheel blocks and closing carrier hose connection chambers.

19. Sign carrier operator's invoice and deliver one copy to office.

20. Record following data in LP-Gas Plant Log Book:

(a) Carrier "*Time In*" and "*Time Out.*"
(b) Shipment invoice number.
(c) Initial and final readings of depressured gas meter.

LOADING LP-GAS TRUCKS AND TRAILERS

1. Do not operate engines where flammable gas–air mixtures exist.

2. Remain close to transfer connection.

3. Ground carrier to loading rack. Block trailer wheels.

4. Connect vapor line hose from loading rack to carrier. Open carrier vapor line valve.

5. Connect liquid transfer hoses from "filling line" on rack to carrier manifold. Open liquid admission valves on carrier manifold.

6. Open filling line valve on rack slowly. If storage tank pressure exceeds carrier pressure, liquid will flow into carrier.

7. When carrier pressure and storage tank pressure approach equalization:

(a) Open vapor transfer line valve at rack.
(b) Start vapor compressor. Withdraw vapor from carrier by admitting to inlet of compressor. Impose compressor discharge pressure on storage tanks from which liquid is being expelled. Operate vapor compressor on *"manual control"* as required to maintain sufficient differential pressure between storage tanks and carrier to complete loading of carrier.

8. Operator of carrier unit shall determine when his unit is sufficiently filled with liquid.

9. When carrier is filled, close "filling line" valve at rack.

10. Shut off vapor compressor, closing inlet and outlet manifold valves.

11. Close vapor transfer line valve at rack.

12. Operator of carrier unit shall close carrier valves.

13. Depressure liquid transfer hoses to holder pressure thru depressuring line. Bleed gas remaining in hoses to atmosphere thru ½ in. bleed valve on loading rack.

14. Bleed vapor transfer hose to atmosphere thru bleed valve on carrier manifold.

15. Disconnect hoses and return them to loading rack.

16. Remove ground wire from carrier and return to loading rack.

17. Make complete circuit of carrier with truck operator, removing wheel blocks and closing carrier hose connection chambers.

18. Verify carrier operator's gaging and temperature determination of carrier contents. Have carrier operator sign shipment card. Deliver signed shipment card to office.

19. Record following data in LP-Gas Log Book:

(a) Carrier tractor and trailer number.
(b) Carrier *"Time In"* and *"Time Out."*

CONTAINER LIQUID LEVEL

Each storage container is required to be equipped with a liquid level gaging device.[4] The two basic gage types are mechanical and magnetic. The latter may be operated by a float, a set of bevel gears, and a permanent magnet. No stuffing box is used. As container liquid changes level, the float, thru its bevel gears, rotates permanent magnet in back of gage base. A pointer on front of gage face follows permanent magnet. Gage pointer indicates percentage of total tank volume. If gage reads 53, as shown in Example 1 of Fig. 5-36, liquid content is 53 per cent of tank capacity. To determine maximum filling height of a tank, the specific gravity of the liquid must be known. It can be secured from shipping papers for particular car being loaded. If not available, use 0.500 for propane. Temperature of tank liquid must also be known. This may be secured from either tank car temperature reading or that of thermometer at bottom of storage tank. If these temperatures differ, the higher should be used. With specific gravity 0.525 and temperature 0 F, maximum filling height is indicated when pointer is directly over intersection of these two lines as shown in Example 2 of Fig. 5-36. Pointer reading is approximately 82.5 per cent. Example 3 on same sketch shows maximum filling height for a liquid with a specific gravity of 0.525 and at a temperature of 60 F. Reading is approximately 93 per cent.

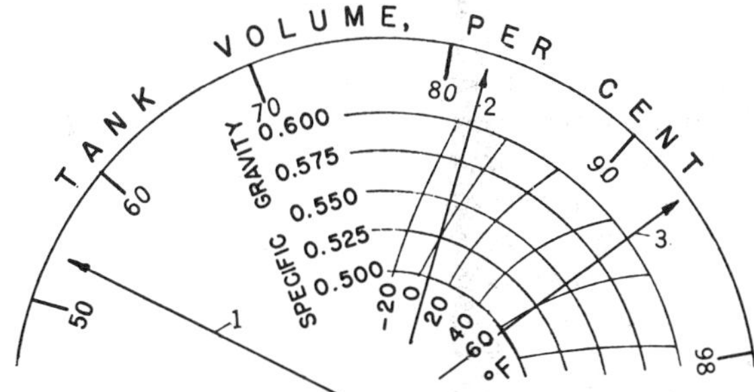

Fig. 5-36 Liquid level gage dial. (Not to scale)

OPERATION OF LP-GAS–AIR PLANT

Specific operating procedures are needed for each plant because of variation in steps required in starting up and shutting down. They depend upon the make and method of control—whether manual, semiautomatic, or automatic; whether jet compressors or other types are used; whether a high or low pressure gas system is supplied.

Suggested general procedures for starting up and shutting down used by one company follow:[7]

Starting Plant

It is assumed that all plant units have been properly serviced and are ready for operation. All valves—especially the plug cock on top of the mixer in the vaporizer room, and the glove valve and by-pass on the propane–air inlet to the natural gas mixer—should be in the shutdown position. All ventilators, windows, and doors should be open. Sequence of steps for start-up follows:

1. Start *Thermeter* on natural gas.

2. Start steam boiler.

3. Verify that chain operated plug cock in liquid line between storage tanks and vaporizer is in open position. (This is an emergency shutoff valve.)

4. At storage tank, open angle valves A-2 and A-3 (Fig. 5-35) to wide open position on each of the two liquid outlet lines. Open globe valve C-1, slowly at first, to allow liquid pressure to build up in line to vaporizer. Then open fully both globe valves, C-1 and C-2. If these valves are opened too fast, excess flow checks may close. If this occurs, close both globe valves and start again. Repeat steps for additional tanks are required. If the maximum safe liquid flow thru an excess flow valve on one tank is reached, liquid from second tank must be turned on. Open equalizing line (valves C-3 and A-5) on each tank to give equal pressure in all tanks.

5. Select vaporizer to be used and close inlet and outlet valves on steam trap; open condensate by-pass valve; open steam valve. Leave by-pass valve open until tubes and steam chest are warmed up and long enough to insure that system is cleared of condensate, dirt, oil, sludge, and rust. Then, open both blocking valves at trap and close by-pass valve, testing trap action at this time to assure satisfactory operation. Check operation of trap at frequent intervals while plant is in operation.

6. Where two vaporizers are installed and only one is to be used, turn immersion thermostat by-pass switch to *"on"* position for vaporizer not used, and turn similar by-pass switch to the *"off"* position on unit used.

7. Select compressor to be used, and place in operation.

8. Turn on water to aftercooler.

9. On panel board in vaporizer room, set *Fisher Wizard* controller for vaporizer being used to zero position; set *Foxboro* vapor controller for automatic operation and zero pressure.

10. Turn on compressed air to instruments.
11. Vapor and liquid control valves should now go to "closed" position.
12. On *Cutler-Hammer* control panel:

(a) Turn oil pump and mixing control disconnect switches to "on" position.
(b) Turn bell and Btu control selector switches to "off" position.
(c) Turn oil pump selector switch to "on" position.

13. Set loading plunger on *Askania* air pressure differential regulator to give approximately 1440 Btu propane-air mixture.
14. Move air flow control valve to "closed" position by opening by-pass valve in oil lines to cylinder and by moving valve manually. Close by-pass valve tight. Close throttle valve.
15. Move duplex valve (coupled valves) to positions on scale, as previously determined by opening by-pass valve on oil lines to cylinder and moving valve manually. Close by-pass valve.
16. Light pilot on flare.
17. At vaporizer, slowly open liquid regulator inlet and outlet plug cocks, keeping by-pass closed.
18. Slowly open plug cock (if any) at vaporizer outlet.
19. Raise pressure in vaporizer by gradually raising setting on *Fisher Wizard* controller to predetermined setting.
20. At vapor regulator on outlet of vaporizer, slowly open inlet and outlet plug cocks, keeping by-pass closed.
21. At this point, alarm bell should stop ringing, and solenoid toggle on air valve should "pull in."
22. Raise vapor pressure by raising setting on *Foxboro* controller slowly and in steps to predetermined pressures. If regulator does not hold pressure down when set at 5 psig, go immediately to next step.
23. Start vapor flow to flare by opening plug cock and globe valve in line to flare. If vapor regulator does not "hold tight," this operation must be completed quickly to prevent vapor pressure build-up and opening of safety valve in vapor line.
24. Check operation of vapor and liquid control valves.
25. If vapor regulator controller is set at 5 psig, raise to pressure selected in step 22; recheck operation of vapor and liquid control valves.
26. Open all impulse shutoff valves in gage lines and *Askania* diaphragm lines just above cylinder-operated butterfly valves. Recheck and make sure by-pass valves in oil lines to cylinders are closed.
27. On gage board on vapor differential "U" gage, open by-pass valve and then inlet and outlet valves. Close by-pass valve slowly; gage should show a low differential.
28. Set *Askania* vapor pressure differential regulator *"set index"* to selected pressure. Open inlet and outlet valves on impulse lines to diaphragm. Close by-pass valve.
29. Start oil pump by closing switch on wall. Duplex valves should move toward closed position.
30. Set flow of vapor to flare to give a reading of about 3.0 to 3.25 on duplex valve scale, by adjusting globe valve. Differential as shown on vapor "U" gage should increase to setting of *Askania* hand set unit.
31. On air "U" gage located on gage board:

(a) Open by-pass valve.
(b) Open inlet and outlet valves to "U" gage.
(c) Close by-pass valve slowly. "U" gage reading should be practically zero.

32. Open inlet and outlet lines to diaphragm on *Askania* air pressure differential regulator and close by-pass.
33. At mixing chamber, open gate valve in air line, slowly at first, and then to full open position.
34. At cylinder-operated air flow butterfly valve, open throttle valve in oil line slowly. At this point, butterfly valve should open, and air "U" gage should show a differential approximately equal to setting of plunger on *Askania* air pressure differential regulator. If surging starts, increase flow rate to flare.
35. Open all valves in sample line from mixer to *Thermeter* and close natural gas valve at *Thermeter*, thereby transferring *Thermeter* from operation on natural gas to operation on propane–air mix. Start *Ranarex* specific gravity recorder.
36. Allow *Thermeter* recorder to indicate heating value of propane–air mix. If Btu reading tends to level out at more than 50 Btu above or below Btu setting, use "raise" or "lower" buttons to bring reading within above range. Buttons should only be pressed momentarily, and sufficient time should be allowed after pressing to allow recorder to indicate change made.
37. When recorder is indicating a Btu value within 50 Btu above or below *Thermeter* setting, turn bell and Btu control selector switches to "on" position.
38. In vaporizer room, slowly open worm-gear-operated plug cock on top of mixer until a small flow of gas is established. Stop at this point, allow pressure in line to natural gas building to build up, and then open plug cock wide.
39. Turn off electric ignition switch to flare pilot and, if flare is not required, gas to pilot may be turned off.
40. Turn to "off" position motor starter by-pass switch (if used) and engine ignition by-pass switch (if used).
41. Push up lever on solenoid-operated air valve, and engage latch. Open inlet and outlet air valves and close by-pass valve. Plant is now on automatic operation for Btu control, volume control, and safety shutdown.
42. Start *Foxboro* flow ratio controller by setting dial at approximate ratio desired.
43. Place transfer switch in "M" (manual) position.
44. Check to see that both flow meters are turned on. Recheck again to make sure globe valve at inlet to LP-gas–air–natural gas mixer is closed.
45. Turn on supply gas to controller and back off on pressure control knob until supply pressure reads zero. Butterfly control valve should now be in closed position. Check operation of butterfly control valve by slowly turning pressure regulator knob in a clockwise direction, increasing supply and output pressure until butterfly valve opens. Close butterfly valve by slowly turning pressure regulator knob in a counterclockwise direction until supply pressure reads zero. Be sure butterfly valve is in closed position before proceeding to next step.
46. When propane–air mixture is at desired Btu level and plant is functioning smoothly, slowly open valve at inlet to LP-gas–air–natural gas mixer.
47. Very slowly "screw in" on pressure regulator knob in controller. As supply and output pressures rise to about 3 psig, flow control valve should start to open. Keep increasing pressure very slowly until LP-gas–air flow, as shown by blue pen on chart, reaches blue pointer.
48. When blue pen and pointer are aligned, throw transfer switch to "A" (automatic) position and quickly screw in on regulator knob to bring supply pressure to 17 psig.
49. Gradually reduce volume of gas going to flare by closing plug cock until it is completely shut off, provided there is sufficient gas flowing into system to keep proportioning equipment from surging. If surging occurs, it will be necessary to relight flare and burn sufficient gas to stop surging.
50. Where more than one vaporizer is installed, recheck and see that immersion by-pass switch is "off" for vaporizer in use and "on" for vaporizer not in use.
51. Open by-pass valves on both "U" gages and close inlet and outlet valves if desired.
52. Check operation of all compressors, boilers, instruments, and other units. Repeat frequently.
53. Observe position of cylinder-operated air flow control butterfly valve while plant is in operation. When it reaches 70 degrees open, or if air pressure starts to drop, whichever comes first, and if there is a very small flow of air from air regulator bleed valve, it indicates that if load on plant continues to increase, an additional compressor will be required for continued operation.

Shutting Down Plant

1. Turn on pilot light for flare.
2. Turn on flare burners by opening plug cock only to full position. (Globe valve has been previously set for minimum plant load.)

3. If "U" gages are in use, open by-pass valves and close inlet and outlet valves.
4. Reduce LP-gas–air gas flow at *Foxboro* flow ratio controller.
5. Close globe valve on inlet line to natural gas mixer.
6. Shut off supply gas to controller.
7. Turn off *Ranarex* gravity recorder.
8. Slowly and completely close plug cock on propane–air mixer.
9. With engine-driven compressor in use, turn ignition by-pass switch to "on" position. With motor-driven compressor in use, turn motor starter by-pass switch to "on" position.
10. Open by-pass valve on solenoid-operated air valve and close inlet and outlet valves.
11. On *Cutler-Hammer* panel, turn bell and Btu selector switches to "off" position.
12. Transfer *Thermeter* to natural gas:
    (a) Turn on natural gas valve in sample line.
    (b) Turn off propane–air valve in sample line and purge burner.
13. Stop oil pump by opening electric switch in vaporizer room.
14. On both *Askania* pressure differential regulators:
    (a) Open by-pass valves.
    (b) Close both impulse lines to diaphragms.
15. Close gate valve in air line to mixer in vaporizer room, slowly as valve approaches closed position.
16. At vaporizer close both plug cocks on liquid inlet line.
17. On gage board set *Fisher Wizard* controller to zero setting.
18. Allow vapor pressure to be reduced to approximately 10 psig, as shown on pressure gage on vapor line to mixer. As pressure in vaporizer drops, bell should ring and solenoid air valve should be de-energized.
19. Shut off flare at this point by closing both plug cock and globe valve.
20. Shut off pilot light.
21. Close plug cock, if any, at vaporizer outlet.
22. At vapor regulator, close inlet and outlet plug cocks.
23. On gage board turn *Foxboro* controller setting to zero.
24. Close by-pass valve on solenoid-operated air valve. Close valves in air supply lines to various instruments.
25. On *Cutler-Hammer* panel, turn to "off" position and disconnect following switches: oil pump and mixing control. (Bell should stop ringing at this point.) Turn recorder switch to "off" position.
26. It is recommended that natural gas to *Thermeter* be left on and that *Thermeter* motor be left operating.
27. Turn off water to aftercooler.
28. Shut down steam boiler unless required for building heating purposes.
29. Shut down air compressor.
30. Just before leaving plant:
    (a) Close all valves in liquid outlet lines at storage tank.
    (b) Close steam valve at vaporizer.
    (c) Open condensate trap by-pass valve.
    (d) Close condensate trap valves.

Refrigerated Propane Storage

The first refrigerated propane storage tank[8] used by a utility company was completed in 1957. It is 70 ft in diameter and 52 ft high, with 8 in. foam glass insulation and an aluminum coating. Capacity is 1.29 MM gal of liquid propane (at −46 F and 0.5 psig).

A continuous refrigeration cycle (Fig. 5-37) is employed. Operation of system is as follows:

1. Liquid propane from the transfer line enters a flash drum thru a flow control valve which reduces the 100 psig line pressure to approximately 50 psig. Flashing in this tank reduces the temperature of the liquid to 20 F. The vapor produced in flashing is released to the interstage piping of the filling compressor and continues thru the refrigeration cycle. The liquid is released into the storage tank at approximately 0.5 psig thru a control valve where additional flashing and cooling take place, resulting in temperature reduction to −46 F.

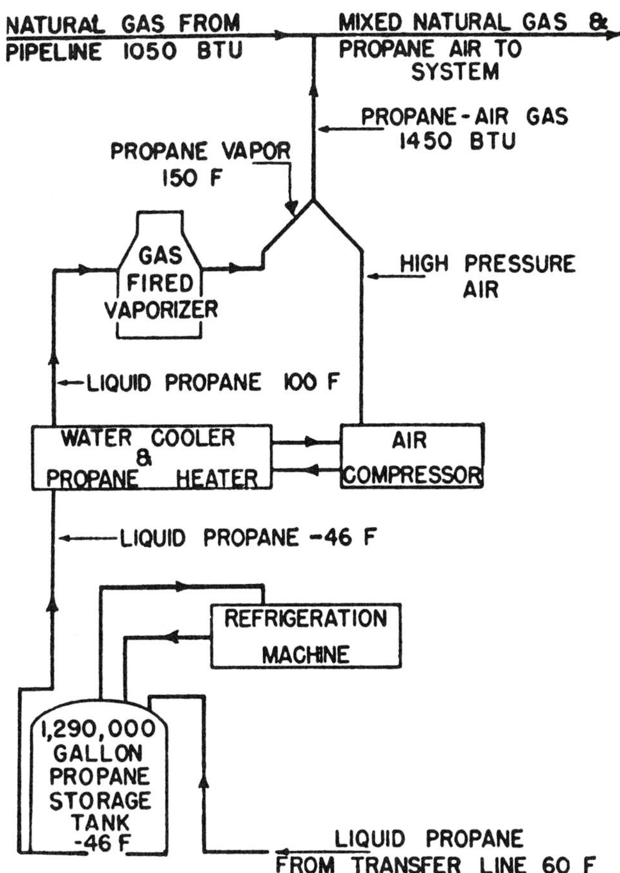

Fig. 5-37 Flow diagram of refrigerated propane–air peak shaving plant. (Atlanta Gas Light Co.)

2. Propane vapor is withdrawn from the top of the tank and compressed to 275 psig. The resultant heat of compression and a portion of the heat in the vapor before compression are removed by air-cooled condensers. The propane then goes to a separator where noncondensables are separated for release to the purger. The liquid is discharged into the 50 psig flash drum, and from there back to the tank.
3. When propane–air is needed to supplement natural gas supply, liquid propane is pumped from tank and passed thru a heat exchanger which warms liquid to about 100 F. Heat exchanger also cools compressors.
4. Warmed liquid propane then enters a gas-fired vaporizer where it becomes vapor at 150 F. The LP-gas is then mixed with metered natural gas and passed into the distribution system.

Automatic Remote-Controlled Stations

One, fully automatic, remote-controlled, unmanned, high pressure propane–air mixing plant operates in the following manner. It is controlled by telemetering from a point about 16 miles away and is remotely started, operated, and shut down; it can be operated independently of any power failures.

Basically, the mixing plant has two orifice runs, with two pressure regulators for maintaining constant pressures on each run. Instead of two orifice plates, two *Hancock Flow Control* valves are used to meter the air and propane for the Btu required. The valves are calibrated to accommodate any required change in Btu.

Once the dispatcher starts the plant, he may both control and record the flow from the station, as well as the flow from the natural gas station. From this he knows the make-up of his sendout at all times.

This station is used both for pressure control on peak hours and for peak shaving, since the dispatcher can start and stop it as he pleases. This type of station can be made to come on automatically in case of a pressure drop or any interruption of gas supply; for example, one due to a failure in the natural gas lateral line. This type of station is also used in large industrial plants like glass plants.

APCO Portable Propane–Air Plants

These fully automatic, factory built, completely packaged plants are available either as fully portable or as semiportable units. When designed as a fully portable unit, it is contained within a standard (e.g., 28 ft long by 8 ft wide by 8 ft high) highway trailer for ready attachment to any tractor unit. The semiportable unit (weighing 24,000 lb) has skids in lieu of wheels and is designed to be placed on a simple concrete pier foundation.

The only field connection made is to a source of liquid propane. No water or electrical services are required. The entire unit is designed for automatic operation after manual start-up and is equipped with a full complement of safety controls.

The building is of galvanized ribbed sheet metal construction on a structural steel frame. The partition walls separating the engine room, mixer room, vaporizer area, and office are vapor-tight. A safety glass window in the office wall and in the wall to the engine room enable visual observation of all of the equipment. A chemical toilet is also provided.

To heat the office as well as to prevent equipment freeze-up, a finned pipe radiator system is provided throughout the plant. This is fed by a forced circulation pump, using water from the vaporizer.

The gas engine-driven radial air compressor used incorporates a special jack shaft to drive the air aftercooler. The compressor speed is varied with demand, and an air relief valve is provided (no air receiver). The engine is equipped with an exhaust muffler, and an air inlet silencer is included for the compressor. A propane liquid booster pump driven off the compressor shaft can also be furnished. The vaporizer is an insulated water bath unit equipped with all firing controls, liquid level controls, etc.

The electric generator provided for control current and building lights is gasoline-driven so that it can be used to furnish lighting during hook-up of the plant.

The mixer is a standard APCO balanced flow unit containing cross balanced regulators, safety shutoff valves, orifice valves, mixing chamber, check valves, and manual cocks and a flow control valve in the sendout line. The sendout line from the mixer is piped thru an orifice meter run and then into a natural gas/propane–air mixing chamber. One plant was designed for an output of 9.6 to 48 MCFH of 1400 Btu propane–air mix (range 1000 to 1500 Btu) at up to 100 psig. Such plants may be used to supply new housing developments or large industrial plants until mains can be extended to them, and to delay the relaying of larger mains by using them to peak shave on days of high sendouts. They may also be used in a *Hortonsphere,* or as a replacement for peak shaving in outlying sections or for pressure control. These plants may be moved from one location to another as needed.

PLANT INSPECTION AND MAINTENANCE

To assure satisfactory condition and performance of LP-gas plants, the routine steps required are:

1. A periodic tank farm inspection that includes observation of grounds, storage tanks, piping, valves, fire extinguishers, and fire control buildings.
2. Good housekeeping—no paper, rags, or other debris shall be permitted in tank farm areas, car unloading areas, and process buildings. If combustible material is used in maintenance or construction work in LP-gas areas, it must be removed immediately upon completion of job. Weeds and grass must not be allowed to grow in tank farm areas.
3. Plug type valves in tank farm, unloading station, compressor building, and process building shall be "worked" frequently. When necessary, small quantities of lubricant shall be used. Do not attempt to stop leakage in plug valves with lubricant. Tightening valve packing gland will prevent leakage.
4. Tank farm leak survey, with soapsuds, shall be made twice a month or at other selected intervals.
5. An unwarranted increase or decrease in volume of liquid in a storage tank or battery shall be investigated immediately.
6. Storage tank water blow-off shall be made twice a year, preferably in May and November.
7. Screens in process supply line and tank filling line shall be examined twice a year.
8. Residue in vaporizers shall be blown out as operations require.
9. Process air filters shall be examined and cleaned as operations require.
10. Stacks extending from tank relief valves should be removed annually so that relief valve gasket may be changed.
11. Ground connections on tank batteries should be checked for resistance between tank and earth. If the resistance exceeds two ohms, regrounding may be advisable.
12. Provide regular service as specified by manufacturers of all instruments and control equipment.
13. Many companies have weekly and monthly inspection and service with check lists to be filled out by inspectors.

FIRE PROTECTION

LP-gas can be safely handled, stored, and processed for utilization, if precautions essential to handling any fuel gas are observed.

If LP-gas from a leak is ignited, the resulting fire should be either extinguished or controlled. It can be extinguished quickly if leakage can be stopped by shutting off source of gas supply. If gas leakage cannot be readily stopped, it may be desirable to control the fire without attempting to extinguish it until the gas supply is entirely consumed. A fire can be confined, while the uncontrolled release of combustible gases may cause an explosion.

When a fire occurs near a storage tank, every effort should be made to shut off leaking gas. If fire is large enough to heat metal of tank vapor space, water should be applied to top of tank. Do not attempt to extinguish a fire at outlet of tank pressure relief risers. Effort should be directed to controlling and extinguishing fire around tank and reducing tank pressure sufficiently to close pressure relief valves. *Dry powder fire extinguishers* are quickly useful in controlling and extinguishing small fires. For large fires, water spray is most effective for controlling fire and keeping surrounding equipment cool.

For fire protection purposes, a complete water deluge system may be installed in tank farm, providing sufficient water so that all tank surfaces and piping will be constantly wetted. This system is activated automatically by: (a) excessive temperature rise at each tank location, (b) excessive pressure rise in vapor lines, (c) manual trips from various locations, or (d) loss of pressure for any reason in pneumatic actuating system.

Reference may also be made to the following extract from NFPA No. 59:[4]

81. Fire Protection

810. The wide range in the size, design, and location of utility plant liquefied petroleum gas installations makes the recommendations of any specific kind or method of fire protection impractical. The planning of effective fire protection should initially be coordinated with the protection practices followed in other sections of the particular utility company and should give due consideration to the requirements of the authority having jurisdiction.

811. Gas fires should not be extinguished until the source of the burning gas can be shut off. Remotely operated or remotely located pipe line valves may be advantageously used for fire control under many circumstances (see 435, 437 and 523). [See Chapter 5.—EDITOR]

812. Hand or wheeled fire extinguishers designed for gas fires, preferably of the dry chemical type, should be available at each strategic location within a liquefied petroleum gas plant.

813. Supplies of water may be utilized thru hose lines preferably equipped with combination (spray and straight stream) nozzles to permit widest adaptability in fire control. If sufficient quantities of water can be made available, complete water spray protection can be given consideration. The water is used for the sole purpose of cooling equipment, foundations, and piping. It shall not be relied upon for extinguishing gas fires.

814. Fire-resistive insulation may be utilized for protecting metal against heat. Care in selecting and applying such insulation is necessary since effectiveness is dependent on its ability to stay in place on a container during a fire.

815. Where standard watchman service is provided it shall be extended to the liquefied petroleum gas installation; such personnel shall be properly trained.

816. Suitable roadways or means of access shall be provided for extinguishing equipment such as wheeled extinguishers or other fire department apparatus.

817. Routine fire drills and inspections should be scheduled and operating personnel thoroughly trained in the use of available fire-fighting equipment and the location and use of all gas and liquid piping and valves.

REFERENCES

1. Parker, P. S. "Cost and Problems in Storing Btu's for Peak Loads." *A.G.A. Proc.* 1949: 545.
2. Lauderbaugh, A. B. "Propane." *A.G.A Nat. Gas Proc.* 1949: 98.
3. Am. Standards Assn. *Scheme for the Identification of Piping Systems.* (A13.1–1956) New York, A.S.M.E., 1956.
4. Natl. Fire Protection Assn. *Standard for the Storage and Handling of Liquefied Petroleum Gases at Utility Gas Plants.* (NFPA No. 59) Boston, 1962.
5. Am. Standards Assn. *Gas Transmission and Distribution Piping Systems* (B31.8–1963) New York, A.S.M.E., 1963.
6. Michigan Consolidated Gas Co. *Operating Procedures for Liquefied Petroleum Gas Plant at River Rouge Station.* Detroit.
7. Jones, H. C. *Suggested Procedure for the Operation of LP Gas–Air Plants, New England Electric System, Gas Division.* Malden, Mass.
8. "A First: Refrigerated Propane Storage." *Gas Age* 119: 13, Feb. 21, 1957.

SECTION 6

GAS TESTING

C. W. Wilson, *Section Chairman,* and Chapters 1, 2, 5–7, and 9, Baltimore Gas & Electric Co., Baltimore, Md.
Hugh E. Ferguson (deceased), Chapters 10 and 11, The Peoples Gas Light and Coke Co., Chicago, Ill.
D. V. Kniebes, Chapters 3 and 4, Institute of Gas Technology, Chicago, Ill.
D. M. Mason, Chapter 3, Institute of Gas Technology, Chicago, Ill.
F. E. Vandaveer, Chapters 2, 10, and 11, Con-Gas Service Corp., Cleveland, Ohio
Donald L. White, Chapter 8, Washington Gas Light Co., Washington, D. C.

CONTENTS

Chapter 7—Determination of Suspended Particles

Chapter 8—Gas Calorimetry

Chapter 9—Gas Density and Specific Gravity

Chapter 10—Pressure Measurement

Chapter 11—Temperature Measurement

Tables

Figures

Chapter 1
Sampling

by C. W. Wilson

Selection of sampling procedures and suitable analytical and testing techniques for gas analyses will be governed by the purpose for which the analysis or test is required. Analyses are required most frequently: (1) to see that gas meets requirements of public utility commissions or other regulatory bodies; (2) for process or product quality control, to insure economic and efficient operation of plant processing equipment; (3) for adjustment and control of efficiency and operation of consumers' equipment, both domestic and industrial.

For a sample to be representative of the whole, consideration must be given to possible changes in composition of the material sampled during and after the sampling process. Some items to be considered in choosing a sampling procedure are: (1) size or volume of sample required for the selected analytical procedure, (2) period of time to be covered by the sample, and (3) chemical reactivity of the components of the sample.

For more complete information on sampling natural, manufactured, and liquefied petroleum gases see References 1, 2, 3, and 4.

SIZE AND TYPE OF SAMPLE

The minimum sample required for extended chemical volumetric analysis of a gas by Orsat-type equipment is usually about 125 ml. Some flue gas or combustion product analyses are made with samples of approximately 75 ml. A sample for mass-spectrometer or gas chromatographic analysis need be only a few milliliters, while the amount for a low temperature distillation should be from two to five cubic feet or more.

Samples, regardless of volume, may be classified from the point of view of time. A sample may be a "snap," "grab," or "spot," collected with a minimum expenditure of time and representing only that material passing the sample point at the moment it was drawn. If the gas composition varies only negligibly or very slowly with time, such a spot sample may be fairly representative of a much larger quantity of material. On the other hand, it cannot be expected that a spot sample of gas from a water-gas machine, for example, or from any rapidly fluctuating stream will be representative of the entire product of that process.

A series of spot samples taken consecutively over a period of time may be considered as an average sample. How representative the average is, however, may be uncertain if unusual fluctuations occur in either the rate of flow or composition of the stream from which the sample is taken.

Continuous analyses of gas streams, made for example thru calorific value by means of a continuous recording calorimeter, or thru specific gravity by means of a recording gravitometer, may be considered as using automatically collected time-average samples. Depending upon sensitivity of the instruments, along with the lag or delay inherent in measuring instruments and sample lines, individual points on the recorder trace may be considered as representative of "instantaneous" or spot values for the property being measured.

STANDARD CONDITIONS FOR SAMPLING

There are no universal sampling standards[5,6] for temperature and pressure. Whenever the saturated condition is specified, avoid the partial expansion of gas in the sampling container in the absence of water. Table 1-38 (or its formula) may be used to correct the sample to 60 F and 30 in. Hg at 32 F.

SAMPLE CONTAINERS

All sample containers, for collecting and storing the sample and delivering it to the point of analysis, should be gas-tight and easily handled. Containers should be made of material which will not react with the sample. Glass and stainless steel are probably least affected and most frequently used. Other metals or alloys may be employed, as conditions permit. For example, ordinary steel or brass should be satisfactory if there are no sulfur compounds in the gas sampled.

Sample containers may be of a number of types and designs. Bottles, cans, tubes, cylinders, or tanks may be employed. Sealing may be done by means of stopcocks, valves, corks, or rubber clips, as required by conditions of sampling and use. If samples are to be shipped under pressure, containers should comply with the requirements of the Interstate Commerce Commission and be approved by it.

Size of container may vary from about 125 cu cm to several cubic feet, depending upon type of sample and purpose for which it is required.

Lubricants for container stopcocks should be carefully selected and used sparingly with samples containing hydrocarbons. Metaphosphoric acid, graphite, or starch-glycerine silicones should be considered for special purposes where contamination or loss from hydrocarbons is important. Packless valves may be used to advantage with metallic containers.

Several types of sample containers are shown in Fig. 6-1.

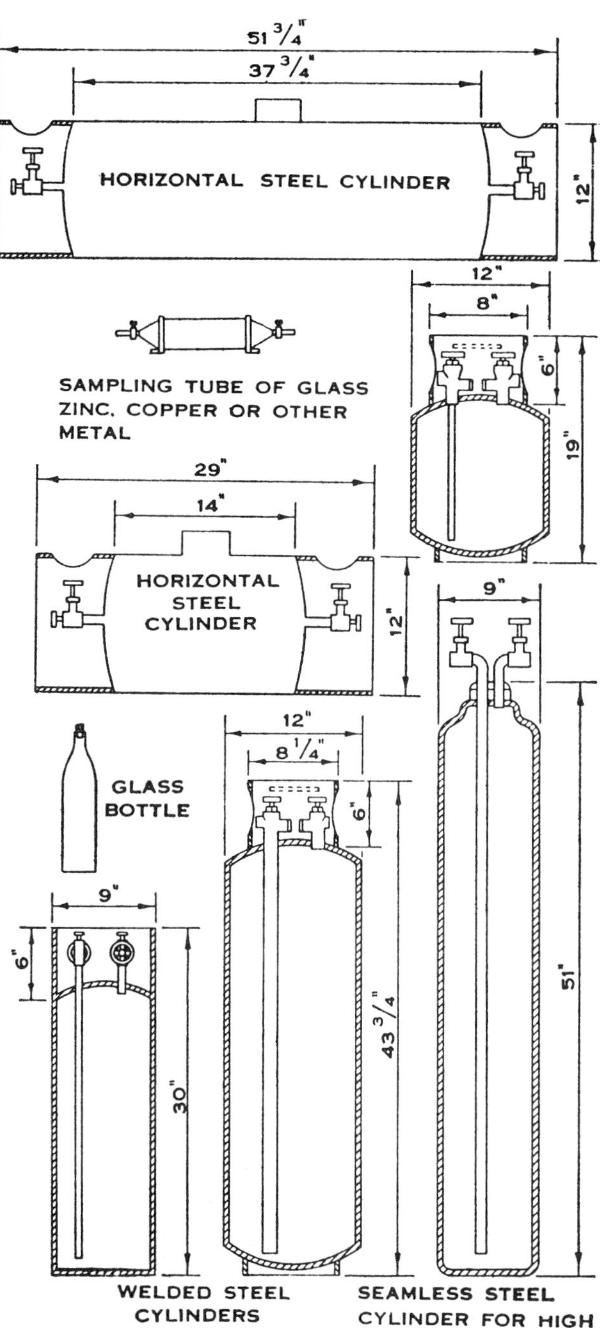

Fig. 6-1 Gas sample containers.[1]

SAMPLE LINES AND CONNECTIONS

Samples for analysis must be withdrawn from pipes, containers, or other equipment thru sampling tubes and lines. Continuous sample lines are preferable, but permanent installations are frequently constructed from piping. All joints should be tight to prevent sample loss or contamination. Semirigid or flexible tubing is convenient for temporary use.

Sample lines must be of comparatively small diameter, with total volume small in proportion to volume of sample passing. Lines must be purged of unwanted constituents before sample collection is begun. Where large sample lines are installed permanently, as for continuous sampling by calorimeters or specific gravity recorders, it is common practice to purge continuously at such a rate of flow that the residence time of gas within the line is small. Excess gas is discharged into a holder, burned, or pumped back to its source.

Materials for sample lines are selected according to the same principles that govern selection of containers. Material must not react with the gas sample passing. It must not be permeable to or absorb any constituent. Materials frequently employed are glass, porcelain, rubber, neoprene, plastics, copper, tin, and aluminum. Iron or steel pipe is suitable for certain permanent lines.

Rubber tubing connections should be kept to a minimum. Rubber is somewhat permeable to hydrogen, helium, and some other components. It absorbs hydrocarbons (especially aromatics) and sulfur compounds. Vulcanized rubber may introduce sulfur or sulfur compounds into the gas. Using glass, plastic, and semirigid metal tubing can overcome some of these difficulties. If rubber connections are used with glass tubing, for example, they should be short with glass tube ends butting together to minimize exposure of rubber to gas.

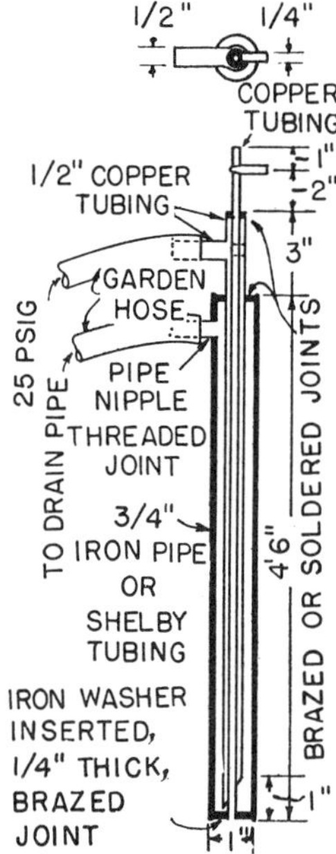

Fig. 6-2 Water-cooled sampling tube.[7] Used by the U. S. Bureau of Mines for sampling hot furnace gases and blast-furnace gases. Inner tube conveys gas sample; two outer tubes constitute water jacket. Such a sampler made of steel or copper tubing with brazed joints is light, convenient, and free from water leakage.

If sampling conditions indicate that condensation of some constituent may occur within sample tubes or lines at ambient temperature, companion steam lines or other heater jackets may be required. Conversely, withdrawing samples of combustion products from furnaces or ovens calls for a water-cooled sample probe to minimize changes in composition during the sampling process. A typical water-cooled probe is illustrated in Fig. 6-2.

CONFINING FLUIDS

Chemical reactivity of components of the gas stream must be recognized in determining sampling procedure. Samples may be collected by displacement of some liquid confining medium. Mercury, water, or aqueous solutions are the most common. Means must be provided for preventing absorption of reactive components by the confining medium. Mercury is a most satisfactory liquid for confining or displacing samples, unless the sample contains constituents like hydrogen sulfide or organic sulfur compounds, with which mercury reacts chemically. Since its cost and weight usually prohibit its use in large quantities, mercury is most frequently used to confine small samples in glass containers.

Water alone is sometimes used as a displacement fluid. However, great care must be taken to insure that: (1) sample is not contaminated with gases already dissolved in the water, or (2) sample composition is not changed by solution of some components. Table 1-25 shows solubilities of some common gases in water (*note the caution at the bottom of this table*). Unsaturated hydrocarbons, such as ethylene, propylene, butylenes, acetylene, and aromatic hydrocarbons, along with the higher paraffins, are also rather soluble.

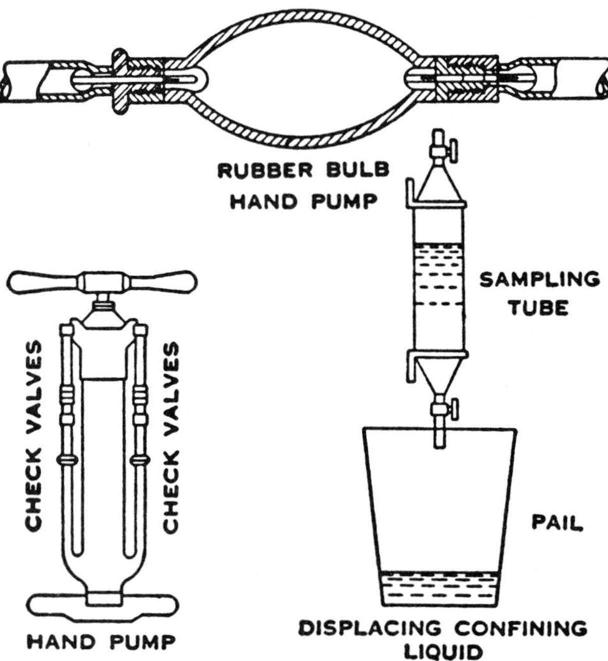

Fig. 6-3 Apparatus for inducing or forcing gas flow.[1]

Losses thru solubility are sometimes minimized by using nearly saturated solutions of sodium chloride or magnesium chloride, in either case slightly acidified with hydrochloric acid. These solutions should not be completely saturated since a slight lowering of ambient temperature might cause some of the salt to crystallize out in the equipment. Aqueous confining fluids should be brought into solubility equilibrium with the gas to be sampled, by bubbling it thru the solution before the sample is taken. However, where *trace* components are to be determined, samples taken with confining fluids are often unreliable.

Use of evacuated containers or displacement of air, by purging a container with the gas before confining the sample, is a means of avoiding sample changes due to use of a confining fluid. If the container initially contains air or other gas, it may be adequately purged by passing thru it a volume not less than ten times that of the container, provided sample line connections are so arranged that the purging fluid does not "short-circuit" or "by-pass" the volume. This procedure is particularly useful when samples are to be collected under pressure.

METERING OF SAMPLES

Determinations of trace components, such as organic sulfur compounds, cyanogen, hydrocyanic acid, and oxides of nitrogen, usually require metering the gas sample before analysis. Wet test meters can rarely be used in such cases because even a small loss of these components by solution in the meter water causes a large percentage error in the results; saturating the water in the meter with a preliminary quantity of the gas

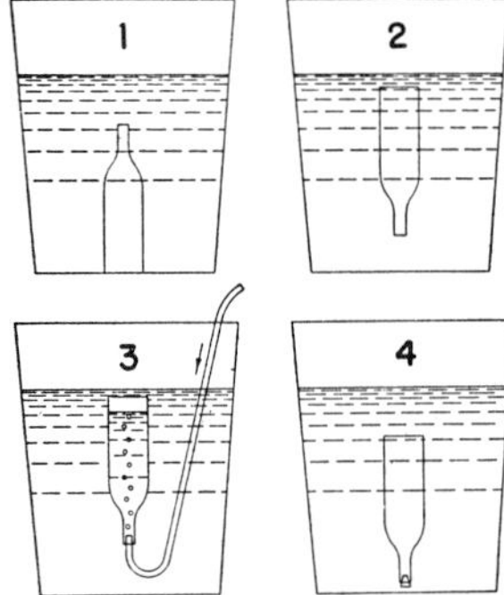

Fig. 6-4 Steps in taking a sample by water displacement from a glass bottle:[1] (1) completely fill glass quart bottle with water; (2) invert bottle under water—ascertain absence of air bubbles; (3) bubble gas into bottle—one minute to displace water plus two minutes to increase sample pressure; (4) cork bottle as tubing is withdrawn.

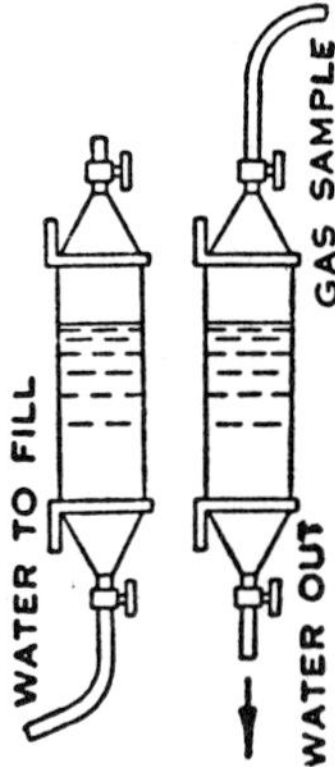

Fig. 6-5 Steps in taking a sample by water displacement in a sample container.[1]

to be analyzed is a questionable expedient. Capillary flow meters and orifice meters are superior within limits of their precision of volumetric measurement. Dry positive displacement meters are usually satisfactory for those cases in which the meter dressing will not be an absorbing or contaminating factor, and in which the accuracy limits of the meter are acceptable.

COLLECTION OF SAMPLES

Fuel gases are usually under sufficient pressure to purge and fill sample containers. When this is not the case, means shown in Fig. 6-3 may be used.

Methods of collecting various types of samples of fuel gases of ordinary composition are outlined in the following paragraphs. When gases to be sampled contain hydrogen sulfide, organic sulfur or other sulfur contaminants, carbon dioxide in excess of 0.5 per cent, gasoline, or condensables, special procedures may be needed.[1,2]

Small Low Pressure Samples

In collecting small samples in low pressure containers for analysis of chemical constituents, glass bottles or containers with stopcocks at each end may be used. Figures 6-4 and 6-5 illustrate methods of sample collection.

Large High Pressure Samples

In collecting relatively large samples in high pressure sample containers, as for analysis and heating value or specific gravity determinations, steel containers may be employed. Figure 6-6 shows a possible arrangement of equipment.

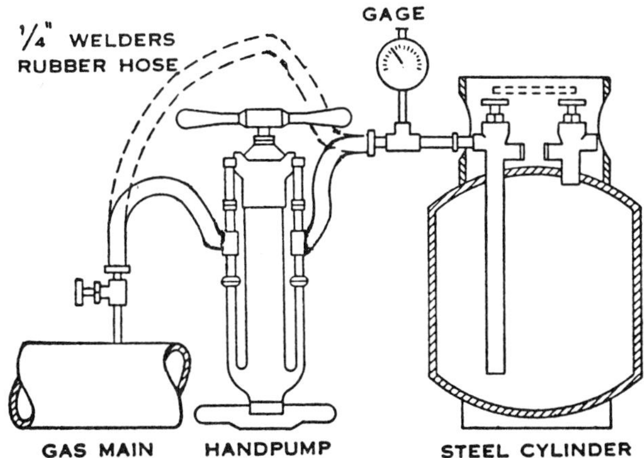

Fig. 6-6 Apparatus for collecting a sample in a steel container under pressure, either directly from gas main or using a hand pump.[1]

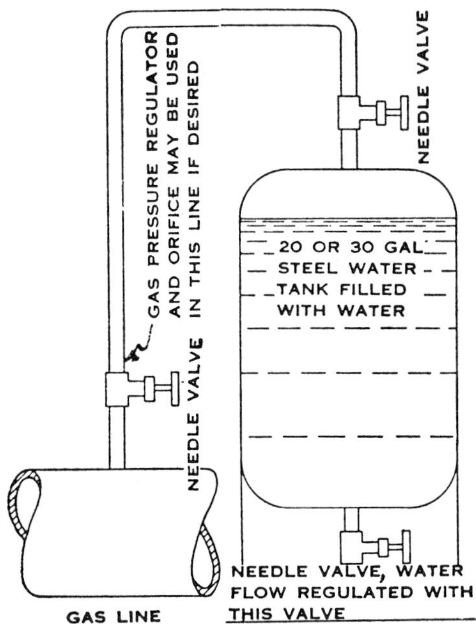

Fig. 6-7 Collecting an average sample by water displacement.[1]

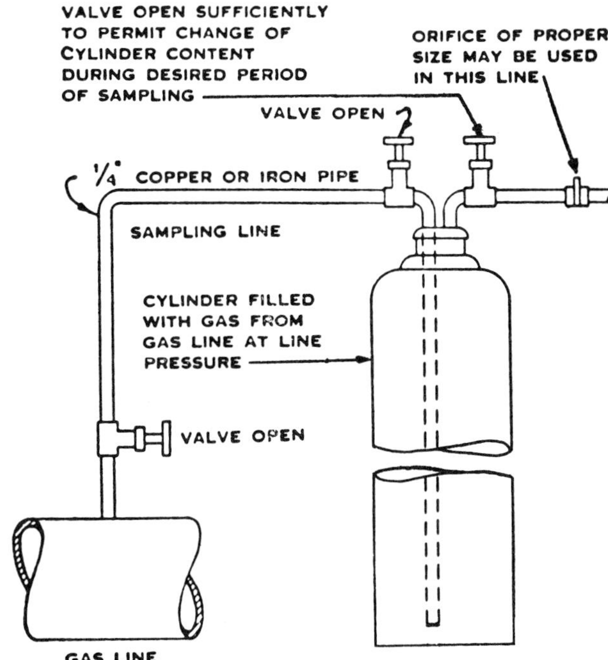

Fig. 6-8 Arrangement of apparatus for collecting an average dry sample under pressure.[1]

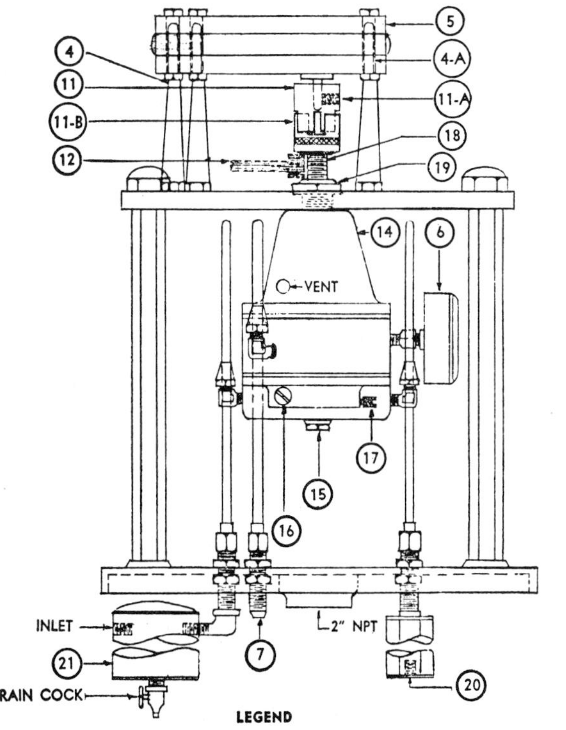

Fig. 6-9 ARCCO-Anubis continuous gas sampler.[9]

If large samples are required, as in fractionation analysis, several high pressure containers may be needed. These may be filled from the pressure source using a manifold arrangement which permits filling all containers simultaneously.

Representative Average Samples

In collecting average samples in portable containers, the arrangement shown in Fig. 6-7 may be employed.

When an average dry sample is required, the method illustrated in Fig. 6-8 may be used.

The ARCCO-Anubis continuous gas sampler (Fig. 6-9) is used quite extensively in obtaining representative gas samples at a uniform rate over a selected period. It operates on the principle that a gas sample can be collected in a container at a uniform flow rate by maintaining a uniform rate of increase in gas pressure on the line supplying the container. This uniform increase in gas pressure is accomplished by using a specially designed pressure regulation system with a clock mechanism. It is protected by a weatherproof metal housing, easily removable for inspection and adjustment.

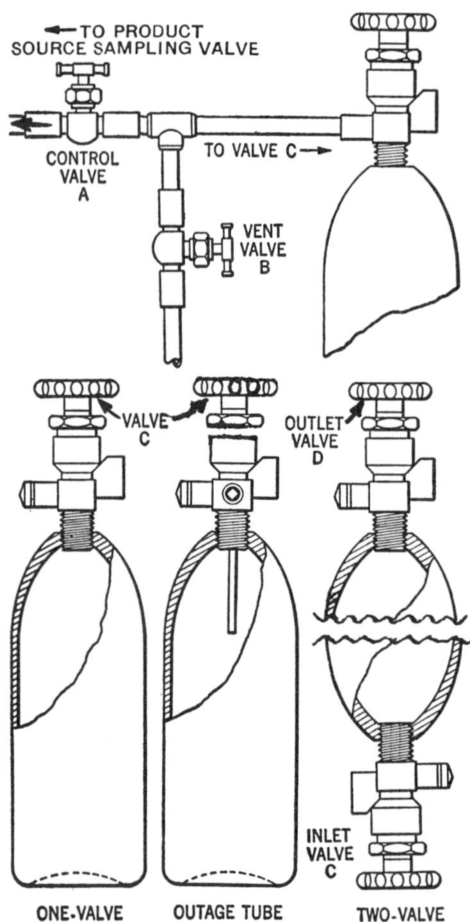

Fig. 6-10 LP-gas sample containers and transfer line.[3]

Variable Pressures.[8] The Edwards bottle is an arrangement for collecting samples at a constant rate under variable pressure.

Liquefied Petroleum Gas Sampling

A liquid sample of propane, butane, or mixtures of the two, is transferred from the source into a sample container by purging the container and filling it with liquid to 80 per cent of capacity. Obtaining a representative sample may require considerable effort, especially if the material being sampled is a mixture of liquefied petroleum gases. The following procedures must be followed:

1. Obtain samples of the liquid phase only.
2. When it is definitely known that the material being sampled is predominantly composed of only one liquefied petroleum gas, a liquid sample may be taken from any part of the vessel.
3. When the material being sampled has been agitated until uniformity is assured, a liquid sample may be taken from any part of the vessel.

Typical liquefied petroleum gas containers and sample transfer line are shown in Fig. 6-10.

NGPA[10] and CNGA[11] standards for LP-gas specifications and test methods are available.

REFERENCES

1. A.S.T.M. "Standard Method of Sampling Natural Gas." (Standard D1145-53*) *ASTM Standards,* 1961, pt. 8. Philadelphia, 1962.
2. ——. "Standard Method of Sampling Manufactured Gas." (Standard D1247-54*) *ASTM Standards,* 1961, pt. 8. Philadelphia, 1962.
3. ——. "Standard Method of Sampling Liquefied Petroleum (LP) Gases." (Standard 1265-55*) *ASTM Standards,* 1961, pt. 8. Philadelphia, 1962.
4. California Natural Gasoline Assn. *Tentative Standard Procedure for Sampling Natural Gasoline, Liquefied Petroleum Gas, Natural Gas, Rich and Lean Absorption Oils, and Crude Oils.* (Bull. TS-342) Los Angeles, 1956.
5. Am. Pet. Inst. *Recommended Practice for Measuring, Sampling and Testing Natural Gas.* (RP50A) Dallas, 1957.
6. A.S.T.M. "Standard Methods for Measurement of Gaseous Fuel Samples." (Standard D1071-55*) *ASTM Standards,* 1961, pt. 8. Philadelphia, 1962.
7. Altieri, V. J. *Gas Analysis & Testing of Gaseous Materials.* New York, A.G.A., 1945.
8. British Standards Institution. *Methods for the Sampling and Analysis of Fuel Gases.* (Standard 3156: 1959; UDC 662.76: 543) London, 1959.
9. ARCCO-Instrument Co. *Operating Instructions for ARCCO-Anubis Continuous Gas Samples.* Los Angeles.
10. Natural Gas Processors Assn. "N.G.P.A. Liquefied Petroleum Gas Specifications and Test Methods." (Pub. 2140-62) *Plant Operations Test Manual.* Tulsa, Natural Gasoline Assn. of Am.
11. California Natural Gasoline Assn. *Tentative Specifications and Tentative Standard Methods of Test for Liquefied Petroleum Gases.* (Bull. TS-441, rev.) Los Angeles, 1953.

* Available separately from A.S.T.M.

Chapter 2

Volumetric Analysis of Gas by Chemical Methods

by C. W. Wilson and F. E. Vandaveer

For many years, volumetric chemical absorption and combustion methods have been the accepted procedures in the industry determining composition of gaseous mixtures. Before the development of mass spectrometers, infra-red instruments, thermal conductivity instruments, fractional distillation, and gas chromatography, they were the only methods available for gas analysis. In recent years, apparatus, procedures, and reagents have undergone much development and refinement, but the fundamental process remains the same, i.e., measuring decrease in volume of the gas sample after preferential absorption of one constituent in an appropriate reagent solution, or combustion by spark, platinum coil, copper oxide, or catalysts. Many varieties of apparatus (see References, particularly 1) are available from which suitable selection may be made, depending upon kind of gas, nature of analysis, accuracy and precision desired, and other governing factors.

THEORY OF GAS ANALYSIS

The theory of constant-pressure gas analysis is simple. A measured volume of gas is subjected to successive treatments which are suitable for removing its various components. After removal of a component, the volume is again measured under the original pressure and the volume decrease is recorded. Results are reported on a volume percentage basis.

The methods most commonly used for removal of various gases are listed below.

ANALYTICAL APPARATUS

Various gas analysis apparatus[1,2] which have been and may still be used in this country include Hempel, Barnhart and Randall, Gockel, Elliott, Morehead, Illinois, Burrell–Orsat U. S. Bureau of Mines Orsat, U. S. Steel Orsat, Shepherd

Methods for Removal of Various Gases

Gas	*Procedure(s)*
Carbon dioxide	1. Absorb in concentrated KOH solution. 2. NaOH may be used, but it deposits carbonate more readily.
Illuminants (unsaturated hydrocarbons)	1. Absorb in fuming H_2SO_4 containing 15–20 per cent SO_3. 2. Absorb by bromine water to form solid bromine products.
Oxygen	1. Absorb in alkaline pyrogallol. 2. Absorb in chromous chloride solution.
Carbon monoxide	1. Absorb in acid cuprous chloride solution. For accurate work, two absorption pipettes should be used, one for bulk and one for traces. 2. Absorb in a suspension of cuprous sulfate and beta-naphthol in sulfuric acid. 3. Use two pipettes, one containing acid cuprous chloride, followed by one containing cuprous sulfate beta-naphthol suspension. 4. Oxidize to carbon dioxide and absorb in KOH. Oxidation may be accomplished by: (a) passing over copper oxide at 572 F; (b) slow combustion in oxygen in presence of a glowing platinum coil.
Hydrogen	1. Oxidize to water by a method for oxidizing CO. Volume contraction is a measure of hydrogen present, since volume of water formed is negligible.
Methane and ethane	1. Oxidize to carbon dioxide and water by slow combustion in oxygen in presence of a glowing platinum coil. Contraction in volume is measured, and carbon dioxide is then absorbed as above.
Combination of combustibles (H_2, CO, CH_4, and C_2H_6)	1. Slow combustion may be used for any one or any two of the four. When two are present, volume contraction due to combustion is determined, followed by a measurement of contraction due to carbon dioxide absorption. 2. If hydrogen, carbon monoxide, and one of the hydrocarbons are present, an additional measurement is necessary. Oxygen consumed by combustion must be determined. This procedure is not particularly recommended, since any errors of measurement are greatly magnified. 3. Hydrogen or carbon monoxide may be removed by either of previous procedures, and hydrocarbons can then be determined by slow combustion.
Nitrogen	1. Residue after removing reactive components is assumed to be nitrogen.

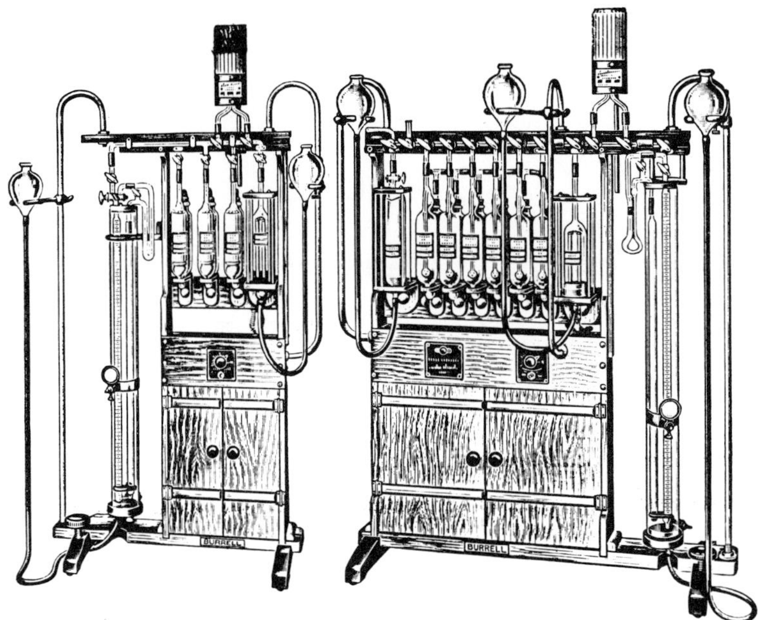

Fig. 6-11a (left) Burrell–U. S. Bureau of Mines Orsat type gas analysis apparatus.
Fig. 6-11b (right) Burrell Premier–U. S. Steel Corporation Orsat type gas analysis apparatus.

Burrell–Haldane, Dennis, Orsat–Muencke, Orsat–Lunge, Williams, Hays–Orsat, Ellison, and Institute of Gas Technology Orsat.[3] There are also variations, modifications, and additions to this equipment to suit the purpose and convenience of the analyst.

In 1874, Orsat[4] patented the device which has been applied to gas volumetric apparatus, where the pipettes and measuring burette are permanently connected, as contrasted to the Hempel method where they are temporarily connected for each operation. This Orsat device has emerged as the most widely used volumetric analysis apparatus.[5] It consists of a group of specially designed pipettes, each containing a different absorption reagent, all connected thru stopcocks to a manifold, to which a measuring burette is also connected. With this arrangement the gas sample can be passed successively thru the reagent solutions, and the residual volume can be measured in the burette before and after each constituent is absorbed. In modern laboratory equipment, precise adjustment of gas pressure, before each volume measurement, is provided by a mercury manometer. Ambient pressure and temperature changes are compensated for by a closed bulb connected to the opposite end of the manometer.

In nonportable laboratory apparatus, constituents like hydrogen and saturated hydrocarbons not amenable to absorption by chemical reagents are determined by combustion followed by measurement of contraction due to burning to water and absorption of the CO_2 thus formed. Carbon monoxide may also be determined by combustion methods. In portable equipment, combustion steps may not be included. Such apparatus is therefore used primarily for flue gas analysis for CO_2, O_2 and CO, and where small residual quantities of unburned combustibles, if present, are not of great importance. When CO is found, it is probable that an equal amount of H_2 is present.

In the nonportable laboratory gas analysis apparatus, three Orsat modifications are usually featured in scientific apparatus catalogues.[5] Other equipment previously listed also can be obtained:

1. **Bureau of Mines Model**[6] (Fig. 6-11a): Contains three contact pipettes for CO_2, O_2, and illuminants; a slow combustion pipette for CH_4 and C_2H_6; a copper oxide tube and heater for CO and H_2; and a nonflushing manifold.

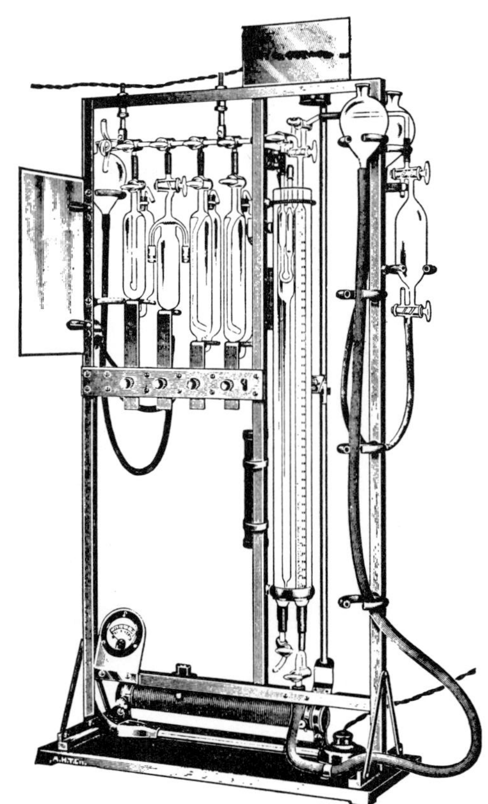

Fig. 6-11c Shepherd gas analysis apparatus.

2. **U. S. Steel Corporation Model**[7] (Fig. 6-11b): Contains six Francis auto-bubbler absorption pipettes, one each for CO_2, O_2, and illuminants; two for CO; one extra for storage purposes; slow combustion pipette for CH_4 and C_2H_6; copper oxide tube and heater for H_2.

3. **Shepherd Model**[8] (Fig. 6-11c): Contains absorption pipettes of the bubbling type, slow combustion pipette, and storage pipette. It was designed for greatest precision and simplicity of manipulation. H_2, CO, and CH_4 are burned together with O_2 on a hot platinum filament. Subsequent absorption of CO_2 and excess of unconsumed O_2 is required to complete the determination.

A fourth unit, called the **Precision-Shell gas analysis apparatus,**[9] extends this type of volumetric analysis into gases encountered in the petroleum industry. It features a mercury lift, ball and socket glass joints, a constant pressure reservoir, and special types of contact and bubbler pipettes. It provides for determination by absorption of:

Carbon dioxide	in 50 per cent potassium hydroxide,
Acetylene	in potassium iodomercurate,
Isobutylene	in 65 per cent sulfuric acid,
Propylene	in 87 per cent sulfuric acid,
Ethylene	in acid mercuric sulfate,
Oxygen	in chromous chloride, and
Carbon monoxide	in cuprous chloride;

for determination by combustion of:

Hydrogen	over copper oxide, and
Saturated hydrocarbons	over copper oxide with one per cent iron oxide at 1112 F.

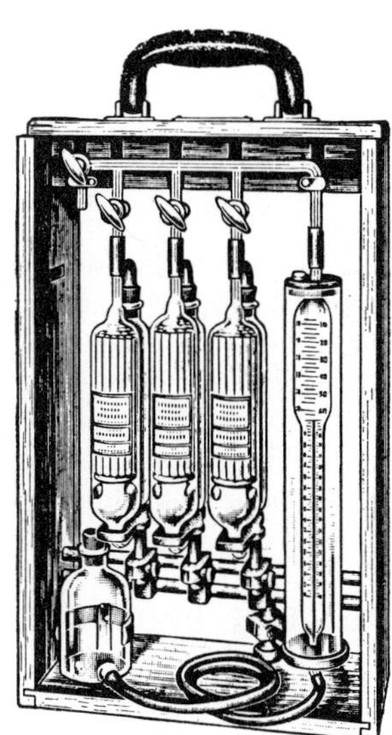

Fig. 6-11d Portable type Orsat gas analysis apparatus.

Portable Orsat models are made in a number of shapes and sizes which are essentially shorter, smaller models of the laboratory sizes described. Figure 6-11d shows a typical Orsat apparatus with a shortened 100 cu cm burette and three pipettes for carbon dioxide, oxygen, and carbon monoxide. The hydrogen produced by incomplete combustion is roughly equal to the amount of CO present, but no provision is generally made for analytical determination, nor is the amount of H_2 reported. Longer models, including other pipettes, copper oxide heaters, and slow combustion pipettes, are also available.

ACCURACY AND PRECISION

It should be recognized that accuracy and precision of volumetric absorption analysis are limited. In usual procedures, unsaturated hydrocarbons of different molecular weight are not separated or distinguished. Accuracy of the combined unsaturate volume is influenced by the number of times the sample is passed thru the absorbent. The choice of a suitable number of passes represents a compromise between complete absorption of unsaturates and partial reaction of the reagent with any higher saturated hydrocarbons present. The combustion procedure will not permit determination of more than two saturated hydrocarbons, and errors caused by presence of other hydrocarbons depend upon their identity and amounts. For example, if propane is present, it will be calculated as ethane and methane. Any error in saturated hydrocarbon determination will be reflected in the concentration of nitrogen determined by the difference.

Table 6-1 gives estimates of the probable precision and accuracy which should be attainable in determining some components, as revealed by cooperative studies sponsored by the American Society for Testing Materials and conducted thru the National Bureau of Standards.[10–12]

Table 6-1 Probable Precision and Accuracy Attainable in Analyses for Various Gases by Volumetric Chemical Methods

Constituent	Probable accuracy, per cent	Reproducibility, per cent: Different laboratories and apparatus	Reproducibility, per cent: One laboratory and apparatus
Carbon dioxide	0.05	0.05	0.02
Oxygen	0.1 to 0.2	0.1	0.03
Unsaturated hydrocarbons as a group	...	...	0.01
Nitrogen	0.6	0.6	0.1
Methane	...	1.0	0.2
Ethane	...	1.0	0.2

ABSORBENT SOLUTIONS FOR GASEOUS COMPONENTS

Some latitude is permitted in concentration of reagents in the various absorbing solutions, as indicated by directions for their preparation found in various sources. The following may be considered representative of good practice.

Carbon Dioxide

Potassium hydroxide solution is almost universally used for absorption of CO_2. Sodium hydroxide can also be used, although its capacity is limited by the solubility of solid sodium carbonate which will stop up gas passageways upon precipitation.

Potassium hydroxide sticks or pellets (ACS grade) are dissolved in distilled water until an excess of solid KOH remains. The solution is cooled to about 3°C below the lowest temperature that will exist in the pipette during analysis.

Oxygen

Alkaline pyrogallol solution has proved successful for oxygen absorption, though chromous chloride or yellow phosphorous has been used for special purposes.

Seventeen grams of reagent grade pyrogallol crystals are added to each 100 ml of the above potassium hydroxide solution. Before use, the solution must be protected from air in a glass-stoppered bottle. It may be placed in a refrigerator to keep cool until the crystals dissolve. This solution will generate no significant amount of CO when in use.[10]

Chromous chloride absorbs oxygen rapidly and cleanly, but it is limited in capacity. In a solution of 10 ml concentrated hydrochloric acid and 190 ml distilled water, 75 grams of $CrCl_2$ are dissolved. The resulting solution is reduced in a Jones reductor by amalgamated zinc and is admitted directly to the gas analysis pipette. This solution will absorb all oxygen from air in two to four passes of the sample thru the pipette. It will give off no hydrogen during use[7].

Oxsorbent is supplied by Burrell for oxygen absorption.

Unsaturated Hydrocarbons

These constituents (absent in natural gas), often termed "illuminants," are usually determined as a group, since the analytical reactions which take place are those of an unsaturated double bond.

Fuming sulfuric acid is the most commonly used reagent. ACS grade, containing at least 15 per cent free SO_3, is suitable as purchased.

Bromine water has been used, though it does not absorb aromatic compounds like benzene as rapidly as fuming sulfuric acid does. A suitable solution for olefins is prepared by adding bromine to 100 ml of distilled water. This solution is saturated by shaking with an excess of bromine, and is stored in a glass-stoppered bottle on a layer of liquid bromine.

In separating individual members of the olefin group during special purpose analyses, sulfuric acid in different concentrations may be used. Addition of catalysts, such as silver sulfate, has also been recommended.[13]

Lubsorbent is supplied by Burrell for illuminants absorption.

Carbon Monoxide

Acid Cuprous Chloride. In precision laboratory equipment, carbon monoxide is determined by oxidation with copper oxide, not by absorption. However, in the interests of portability, convenience, and speed, acid cuprous chloride is used in portable equipment for CO determination. Its use is ordinarily limited to mixtures containing relatively small quantities of CO, such as flue gases or furnace atmospheres produced by burning fuel with deficiency of air.

Although a wide variation in composition of cuprous chloride is possible, the following may be considered typical.[1] Chemically pure (cp) cuprous chloride is dissolved in concentrated hydrochloric acid, 90 grams of the salt in one liter of the cp acid, 1.18 sp gr. Oxygen is rapidly absorbed by acid cuprous chloride, and the reagent is appreciably oxidized during preparation. To the solution, in a glass-stoppered bottle, some copper turnings, sheet, wire, or gauze is added. Several days may be required to reduce the greenish-black solution to a straw-yellow color. After reduction, the remaining copper will keep it reduced if the bottle remains closed.

Cuprous Sulfate–Beta Naphthol. This solution has been recommended for absorption of CO, but it is not as frequently used as the more easily prepared acid cuprous chloride solution. It offers the advantage of leaving no residual acid vapors in the gas sample, which must be removed after each absorption; however, at times sediment is found in the pipette, which interferes with good operation. It is probably most satisfactory when obtained already made from supply houses, sometimes under proprietary names such as *Cosorbent* supplied by Burrell.

COMBUSTION METHODS FOR GASEOUS COMPONENTS

Fractional Combustion Over Copper Oxide

Hydrogen is oxidized to water by copper oxide at 482 F. **Carbon monoxide** is oxidized to CO_2 by copper oxide at 572 F. **Methane** and **ethane** are oxidized to CO_2 and water at 1112 F. Usually, CO and H_2 are oxidized together by maintaining the copper oxide at about 572 F. Hydrogen is determined by the contraction in volume after passing the gas mixture thru copper oxide several times. Carbon dioxide formed is absorbed in potassium hydroxide, contraction giving amount of CO present. Methane and C_2H_6 do not oxidize at 572 F.

Slow Combustion

Slow combustion of saturated hydrocarbons, CO, H_2, or a combination of the three, may be done over a hot platinum coil at nearly white heat, or in a catalyst tube at 932 F when sufficient oxygen is present to burn the gases. When C_2H_6 is present, oxygen is added to the pipette to about four times the volume of that portion of sample taken for analysis. Gases burn to CO_2 and water. Calculations are made as indicated later for the gas constituents.

Explosion

Oxygen or air is added to the gas sample in a proportion that will make this mixture explosive. After thorough mixing of O_2 and gas, the mixture is kept in the explosion pipette and an electric spark discharged thru it, with combustion to CO_2 and water taking place. Measurement of the contraction, CO_2, and remaining O_2 permits calculation of combustibles present. This method is hazardous and permits errors not involved in other combustion methods.

ANALYTICAL PROCEDURES

Manipulative procedures for volumetric gas analyses vary in detail, depending upon constituents sought and particular apparatus used. Generally, applicable steps are outlined here. For details see manufacturers' instruction manuals.

Systematic determination of the sample composition may be made as follows:

1. A sample quantity is taken into the burette. Its initial volume is measured at the reference pressure and temperature, to which the manometer and compensator are adjusted in the laboratory type apparatus, or at atmospheric pressure and ambient temperature in the portable apparatus. The latter is accomplished by equalizing the confining liquid levels in the burette and leveling bulb.

2. Carbon dioxide is absorbed by passing the sample thru the KOH solution four or five times by alternately raising and

lowering the confining fluid in the burette by means of the leveling bulb. Residual volume is measured by returning it to the burette, adjusting its pressure and temperature as done for the initial volume, and noting the burette reading. Difference between initial volume, V_1, and that after absorption of CO_2, V_2, represents volume of CO_2 in the sample, or:

$$\text{per cent } CO_2 = 100 \frac{V_1 - V_2}{V_1}$$

3. The second determination is the unsaturated hydrocarbons group (*illuminants*), if present. In a manner similar to that described for absorption of CO_2, the sample is contacted with fuming sulfuric acid in the second pipette. After four such passes, the residual sample is returned to the burette and again passed several times thru KOH solution to remove acid vapors introduced by the fuming sulfuric acid. Residual volume, V_3, is measured at the reference pressure and temperature as in steps 1 and 2. Then:

$$\text{per cent illuminants} = 100 \frac{V_2 - V_3}{V_1}$$

4. Oxygen is determined by passing the residual sample thru alkaline pyrogallol solution in the third pipette. Number of passes required for complete absorption of O_2 depends upon its concentration in the sample, residual capacity of absorbing solution, and efficiency of the pipette. Complete absorption is attained when several successive passes produce no decrease in volume. After each series of passes, volume is measured at the reference temperature and pressure, as previously. Final volume, V_4, is subtracted from V_3 (or V_2 if no determination of illuminants was made). Then:

$$\text{per cent } O_2 = 100 \frac{V_3 - V_4}{V_1}$$

5. At this point different procedures may be used, depending upon analytical requirements and apparatus employed:

(a) With portable apparatus or with laboratory equipment in which CO is estimated by absorption, the residual sample is passed thru acid cuprous chloride or cuprous sulfate–beta naphthol, until CO is completely absorbed. If acid cuprous chloride is used, HCl vapors must be removed from the residual sample before measuring its volume. If volume after absorption of CO is V_5, then:

$$\text{per cent CO} = 100 \frac{V_4 - V_5}{V_1}$$

When CO is determined by absorption in laboratory apparatus, the remaining combustibles, hydrogen, and paraffin hydrocarbons are usually determined by slow combustion over a hot platinum filament in presence of excess air. The residual sample is stored in an auxiliary pipette over mercury or KOH solution. A suitable quantity of air or O_2 is taken into the burette, and its volume, V_6, is measured at reference temperature and pressure. It is then transferred to the slow combustion pipette and confined at very nearly atmospheric pressure by adjustment of mercury levels in the pipette and leveling bulb. A portion of the residual gas sample is again drawn into the burette, and its volume, V_7, is measured. Volumes of air or O_2 and gas taken must be such that an excess of O_2 remains after combustion, and total volume of products is less than the burette volume. The platinum filament in the combustion pipette is heated electrically to a temperature high enough to ignite the gas, and the sample is admitted very slowly. Several successive passes back and forth from pipette to burette over the hot filament (but not allowing the mercury to contact same) are usually required to complete the combustion.

The volume of products, V_8, remaining after combustion is measured in the burette at reference temperature and pressure. Then the CO_2 produced by combustion of the hydrocarbons is determined by absorption in KOH solution as in step 2. The residual volume after this absorption is V_9.

Paraffin hydrocarbons in the original sample are calculated as methane, CH_4, by assuming that the combustion is represented by the equation:

$$CH_4 + 2O_2 \rightarrow CO_2 + 2H_2O$$

Then:

$$\text{per cent } CH_4 = 100 \frac{V_1}{V_7}(V_8 - V_9)$$

A contraction in volume occurs upon combustion because water, which is formed by burning the free H_2 present plus the H_2 present in the hydrocarbons, condenses to liquid. Thus, both H_2 and O_2 disappear and are replaced by only a negligible volume of liquid:

$$H_2 + 0.5O_2 = H_2O$$

The free H_2 present in the sample burned is equal to:

$$\tfrac{2}{3}\,[\text{contraction} - 2(CO_2 \text{ moles formed})], \text{ or}$$

$$\text{per cent } H_2 = \frac{200}{3} \times \frac{V_5}{V_7}\left[\frac{V_6 + V_7 + 2V_9 - 3V_8}{V_1}\right]$$

(b) In laboratory apparatus equipped with fractional combustion copper oxide pipettes, H_2 and CO are determined in the residual volume of sample, V_4, after absorption of O_2. The copper oxide is maintained at a temperature of 554 to 590 F. The sample is passed slowly thru the fractional combustion tube by transfer from the burette to the mercury-filled slow combustion pipette, with substantially atmospheric pressure maintained. After returning the sample thru the copper oxide to the burette, its volume is determined. Contraction results from the reaction of hydrogen and copper oxide, which forms water. Then the sample is passed thru KOH in order to absorb the CO_2 formed by the reaction of CO with copper oxide. Repeated passes of the sample over copper oxide—noting *contractions* followed by passages thru KOH and noting the *reductions* in volume—are continued until no further changes are evident. The separate summations of contractions and reductions are V_{10} and V_{11}, respectively. Then:

$$\text{per cent } H_2 = 100 \left(\frac{V_{10}}{V_1}\right)$$

$$\text{per cent CO} = 100 \left(\frac{V_{11}}{V_1}\right)$$

The residual volume, V_{12}, may then be burned by slow combustion with air or O_2 over the heated platinum wire, as described in (a).

Since two measurements are made after slow combustion, taking account of H_2O and CO_2 (and sometimes a third measurement of residual oxygen), the residual paraffin hydrocarbons can be calculated as ethane and methane, as follows:

$$\text{Volume of } C_2H_6 = \tfrac{2}{3}(2CO_2 - \text{contraction})$$

$$\text{Volume of } CH_4 = CO_2 - 2C_2H_6$$

The per cent of each constituent in the initial gas sample is then found by multiplying by the ratio representing the proportion of the sample taken for combustion.

$$\text{per cent } C_2H_6 = 100 \left(\frac{V_{12}}{V_{13}}\right)\left(\frac{\text{Volume of } C_2H_6}{V_1}\right)$$

$$\text{per cent } CH_4 = 100 \left(\frac{V_{12}}{V_{13}}\right)\left(\frac{\text{Volume of } CH_4}{V_1}\right)$$

Where V_{13} is the volume of that portion of V_{12} actually burned.

6. The nitrogen content of the original sample of gas is ordinarily determined by difference. In certain procedures it is suggested that, when oxygen is used for slow combustion of hydrocarbons, residual nitrogen be measured after absorption of unconsumed oxygen. It appears more practical to employ the former, or:

$$\% N_2 = 100 - (\% CO_2 + \% O_2 + \% \text{illuminants} + \% H_2 + \% CO + \% \text{paraffins})$$

CORRECTIONS FOR DEVIATION OBSERVED FROM THEORETICAL MOLECULAR VOLUMES

The assumption of ideal molecular volume (22.412 liters at 0 C and 760 mm Hg pressure) is not always accurate enough for analysis, since ordinary chemical equations are on a weight basis and do not exactly represent the volume relation.

For example, the equation $CH_4 + 2O_2 = CO_2 + 2H_2O$ expresses the chemical reaction from the standpoint of gas; that is, one *molecular weight* of methane reacts with two molecular weights of oxygen to form one of CO_2 and two of water. Usually this is also taken to indicate that one *volume* of CH_4 reacts with two volumes of O_2 to form one volume of CO_2 and two volumes of water vapor, most of which condenses to liquid water. Methane, however, deviates slightly from an ideal gas, and CO_2 deviates markedly, so that instead of the volume relation given above, the equation is $0.998\ CH_4 + 2.000\ O_2 = 0.994\ CO_2 + 2.000\ H_2O$. Thus, 100 cu cm of methane will produce only about 99.5 cu cm of CO_2 instead of 100, as indicated by the usual equation.

For precise analysis of natural gas (or other gas mixtures in which hydrocarbons predominate), it is advisable to use a large volume of sample and to apply corrections for deviation from the gas laws.[15] CO_2, CH_4, C_2H_6, and other hydrocarbons deviate from the gas laws, but CH_4 deviates only slightly. CO behaves like an ideal gas.

In combustion analysis, necessary dilution with oxygen or air reduces partial pressures and thus decreases deviation errors. However, it is not always desirable to resort to excessive dilution in order to minimize deviation errors because so little of the original gas is then involved that manipulation errors would be unduly magnified thru calculation to a percentage basis. For manufactured fuel gases (or other multicomponent gases with low partial pressures), the quantity of sample should be such that CO_2 doesn't exceed 30 per cent of the total gas measured after combustion.[14]

For reference purposes, Table 6-2 gives the relative molecular volumes (perfect gas = 1) of CO_2 and C_2H_6—two constituents likely to require the use of deviation correction. The alternative would be the method for calculating heating value and specific gravity of a gas mixture from its composition by taking account of compressibility (at 30 F and 30 in. Hg) by means of the **summation factors** given in Table 2-63.

Table 6-2 Molecular Volumes (Perfect Gas = 1) of Carbon Dioxide (at 20 C) and Ethane (at 0 C) at Various Partial Pressures

Pressure, mm Hg	Molecular volume	
	CO_2	C_2H_6
100	0.999	0.999
200	.999	.998
300	.998	.997
400	.997	.996
500	.996	.994
600	.996	.993
700	.995	.992
760	0.994	0.992

SOAP FILM METHOD[16]

Gas at a constant temperature and pressure is passed thru a series of meters and scrubbers to measure each constituent. The components, in order of removal, are CO_2, O_2, unsaturated hydrocarbons, H_2, CO, and saturated hydrocarbons. Temperature and pressure corrections are unnecessary; mercury is not used in the system. A two liter sample size is required.

REFERENCES

1. Altieri, V. J. *Gas Analysis & Testing of Gaseous Materials.* New York, A.G.A., 1945.
2. Dennis, L. M. and Nichols, M. L. *Gas Analysis*, rev. ed. New York, Macmillan, 1929.
3. Hakewill, H. and Miyoji, M. C. "Apparatus and Methods Employed for Volumetric Gas Analysis at the Institute of Gas Technology." A.G.A. Proc. 1951: 470–5.
4. Orsat, L. H. "Improved Apparatus for Analyzing Gas." (Patent no. 1853) *Chemical News* 29: 177, Ap. 17, 1874.
5. Scientific apparatus catalogs of Burrell Corp., Pittsburgh; Arthur H. Thomas Co., Philadelphia; Central Scientific Co., Chicago; E. H. Sargent & Co., Chicago; Chemical Rubber Co., Cleveland; Harshaw Chemical Co., Cleveland; Fisher, Eimer and Armend, Pittsburgh and New York; Precision Scientific Co., Chicago.
6. Fieldner, A. C. and others. *Bureau of Mines Orsat Apparatus for Gas Analysis.* (Bur. of Mines T. P. 320) Washington, G.P.O., 1925.
7. U. S. Steel Corp. *Methods of the Chemists of the United States Steel Corporation for the Sampling and Analysis of Gases*, 3rd ed. Pittsburgh, Carnegie Steel Co., 1927.
8. Shepherd, M. "Improved Apparatus and Method for the Analysis of Gas Mixtures by Combustion and Absorption." (RP 266). *Natl. Bur. Std. J. Res.* 6: 121–67, Jan. 1931. (Also his "Some Common Errors of Gas Analysis and Their Remedies." *Am. Gas Jo.* 134: 49–52, April; 67–73, May 1931.)
9. Brooks, F. R. and others. "Analysis of Gases by Absorption and Combustion." *Anal. Chem.* 21: 1105–16, Sept. 1949.
10. A.S.T.M. "Standard Method for Analysis of Natural Gases by the Volumetric-Chemical Method." (D 1136-53) *A.S.T.M. Standards*, 1961, pt. 8. Philadelphia, 1962. (Available as pamphlet from A.S.T.M.)
11. Shepherd, M. "Analysis of a Standard Sample of Natural Gas by Laboratories Cooperating with the American Society for Testing Materials." (RP 1759) *Natl. Bur. Std. J. Res.* 38: 19–51, Jan. 1947.
12. ——. "Analysis of a Standard Sample of the Carburetted Water-Gas Type by Laboratories Cooperating with the American Society for Testing Materials." (RP 1704) *Natl. Bur. Std. J. Res.* 36: 313–49, Mar. 1946.
13. Frey, F. E. and Yant, W. P. "Fractionation Analysis of Several Fuel Gases with Special Reference to Illuminants." *Ind. Eng. Chem.* 19: 1358–61, Dec. 1927.
14. Haber, F. *Thermodynamics of Technical Gas—Reactions.* New York, Longmans, 1908.
15. Burrell, G. A. and Seibert, F. M. *Errors in Gas Analysis Due to Assuming that the Molecular Volumes of All Gases Are Alike.* (Bur. of Mines TP 54) Washington, G.P.O., 1913.
16. British Standards Institution. *Methods for the Sampling and Analysis of Fuel Gas.* (Standard 3156: 1959; UDC 662.76: 543) London, 1959.

Chapter 3

Low-Temperature Fractional Distillation Analysis

by D. M. Mason and D. V. Kniebes

INTRODUCTION

Low-temperature fractional distillation is a method of separating mixtures of two or more volatile substances into their individual components by distillation procedures; it utilizes differences in boiling points of the substances. This method is principally applied to analyses of liquefied petroleum gases and natural gases, and in oil refinery operations and butadiene manufacture. The process is essentially one of separating constituents by distillation and then identifying and measuring them. The sample is liquefied by cooling with liquid nitrogen, liquid air, or other suitable refrigerant. Individual substances are identified by their boiling points as they distill, and each is measured by collecting it as vapor and determining its pressure in an evacuated bottle. Boiling points of various gases or vapors found in manufactured gases, natural gases, and liquefied petroleum gases are given in Table 2-74.

The methods of analyzing gases thru differences in vapor pressure were probably first developed by Travers[1] in his work on separation of atmospheric gases. These methods were based upon repeated distillation of a sample from one bulb, maintained at a definite low temperature, to another maintained at a lower temperature. Application was later made to hydrocarbon gases.[2–4]

Analyses of natural and manufactured gases, separation of natural gases, and separation of illuminants in manufactured gases into individual hydrocarbon fractions were made by this original apparatus.[2] However, the procedures are long and tedious, requiring a large number of bulb-to-bulb distillations with recombination of fractions; thus, they are seldom used today.

Development of laboratory fractionating columns using reflux was necessary to provide a rapid and practical method of analysis for general use. With temperatures as low as the boiling point of methane (−259 F), this was only possible after development of the silvered vacuum jacket by which heat leakage could be reduced. Various workers[4–7] made contributions in this field. Methods and apparatus are described in several publications.[8–11]

FRACTIONATING APPARATUS

Figure 6-12 shows a diagram of a manual type fractionating apparatus. The two distillation columns shown differ only in the size of the still pot. The column itself is a glass tube about 3 ft long and 3 to 5 mm ID, with wire spiral as packing, and a separable double-wall evacuated jacket with silvering or a metal reflector inside. At the upper end, an annular metal cooling vessel surrounds the column and is itself surrounded by an enlarged section of the vacuum jacket. Vaporization of liquid nitrogen or liquid air in the vessel cools the column for production of liquid reflux. Sealed into the top of the column are a capillary metal tube for delivery of gaseous distillate, and a thermocouple for indicating the boiling point of the reflux at its uppermost boundary. The delivery tube leads to a manifold with connections to a manometer which indicates the column pressure, to fraction-collecting burettes or evacuated bottles, to a manometer which indicates the pressure, and thereby the amount, of distillate collected in the bottle, and to a vacuum pump. Other apparatus includes Dewar vessels for holding liquid nitrogen, a Dewar vessel to shield the still pot, a millivoltmeter or potentiometer to measure emf of the thermocouple, a variable transformer or other means of controlling heat input to the still pot, and a scrubbing train to remove carbon dioxide and moisture from the gas before it enters the column.

Details of a wire spiral column and of other parts of a manual type apparatus for low temperature fractional distillation are available.[8,10] An automatic distillation apparatus with columns of greatly improved efficiency has also been developed.[9,12] Devices incorporating these so-called *Super-Cool* columns with *Heli-Grid* packing, may be obtained in units of various degrees of automation, from manual control to a unit in which column pressure, rate of distillation, and reflux cooling are automatically controlled, and in which the distillate boiling point is automatically plotted against the amount of distillate. Some of these are designated as *Semi-Robot* and *Hyd-Robot;* the latter is a fully automatic recording model.

Operation

Before an analysis, the apparatus, including column, manifold, manometers, and receiving bottles, is evacuated and tested for leaks. The top of the column is cooled by liquid nitrogen or liquid air forced into the cooling vessel to a temperature slightly below the boiling point of the lowest-boiling component, or, if noncondensables are present, to the minimum obtainable temperature. The still pot is also cooled to a temperature dependent upon the volatility of the sample. The drying train is purged with sample; then, with the column connected to its manometer but closed to the remainder of

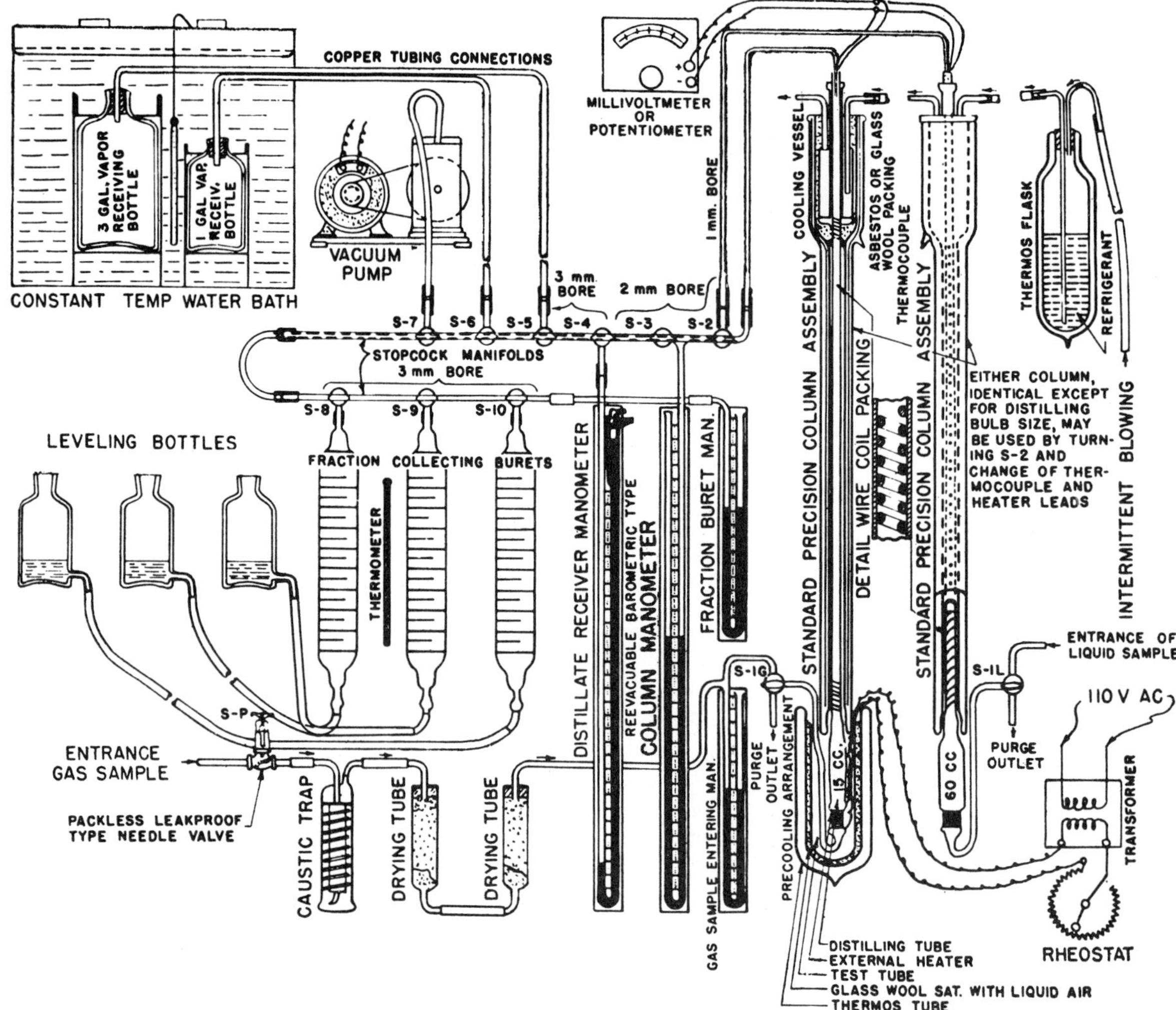

Fig. 6-12 Low temperature fractional distillation apparatus (manual type).[9]

the manifold and the receivers, sample gas is allowed to enter. Column pressure is allowed to approach atmospheric and, provided the sample contains little or no noncondensables, may be maintained there by cooling at top and bottom while the sample enters and condenses. Presence of noncondensables (nitrogen, oxygen, etc.) is indicated if column pressure builds up and the reflux temperature, as indicated by the thermocouple, falls below the boiling point of methane. Such gases must be bled off to the receiver during sample entry.

After sample admission, the still pot is heated to produce vapors rising thru the column, and its top is cooled to condense reflux down the column along its wire spiral packing. The lowest-boiling constituent of the sample mixture concentrates at the top of the column and is obtained substantially pure in the case of lower-boiling members of the paraffin series. When the reflux temperature becomes constant, indicating that the column is at equilibrium, distillation is started at a controlled rate by allowing vapor distillate to flow thru the partially opened distillation rate stopcock to the evacuated receiver. Distillation rate, still pot heat, and reflux cooling are so balanced throughout distillation that they maintain the proper reflux and column pressure.

To distill C_5 (pentane) and higher-boiling components which are liquid at room temperature and atmospheric pressure, the column pressure must be reduced to maintain distillate in vapor form. C_5 and C_6 may be distilled in this manner, though at a reduced rate. Distillation of C_7 and higher hydrocarbons is generally not considered practical.

By plotting reflux temperature against cumulative pressure rise on the distillate receiver, a distillation curve is obtained. Figure 6-13 shows such a curve from distillation of a dry natural gas sample. The horizontal sections, indicating take-off of pure methane and ethane, are termed "plateaus." The abrupt rise of a boiling point from one separable component to the next is termed a "break." The "break point" indicates quantitatively the transition from one component to the next, and it is usually located at a definite temperature. The distance on the curve between break points (or number of mm Hg pressure rise on the receiver) measures the quantity of each component. The natural gas sample as shown in Fig.

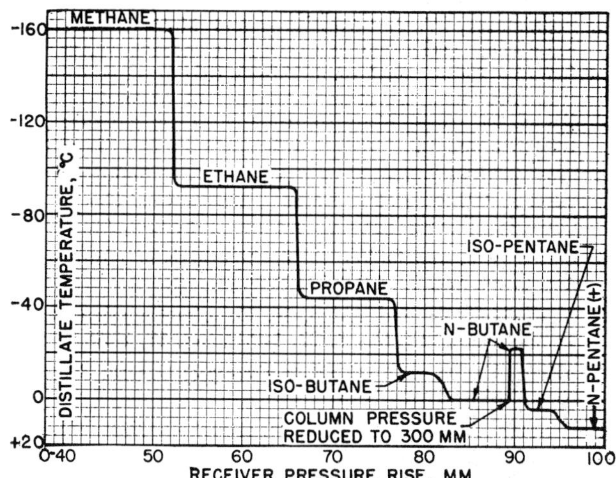

Fig. 6-13 Low temperature fractional distillation curve for a dry natural gas sample. (Note: pressure rises measured in mm Hg.)

6-13 contained, in per cent: Methane 52, Ethane 14, Propane 11, *iso*-Butane 6, *n*-Butane 8, *iso*-Pentane 5, and *n*-Pentane+ 4.

When one component is not completely separated from the next, for example, propylene from propane, or *iso*-butane from *n*-butane, a gradual rise in boiling point is obtained. The incompletely separated hydrocarbon fractions are then treated as a group (C_3, C_4, etc.) Such fractions may be segregated and subjected to further analysis if desired.

The method of finishing the distillation depends upon the type of sample. Residues from gaseous samples, consisting of C_5+, C_6+, or C_7+ hydrocarbons, may be vaporized at low pressure and measured by the pressure rise produced on the receiver. Those from gasoline or other samples containing high-boiling constituents must be removed from the kettle as liquid, and measured by weight or volume.

APPLICATION OF FRACTIONATING APPARATUS

Thru low-temperature fractionation it is possible to characterize hydrocarbon constituents of fuel gases to a greater degree than thru volumetric chemical methods. Natural gas, for example, can be separated into methane plus noncondensables, ethane, propane, and C_4, C_5, and C_6 fractions with the simplest apparatus (manual type with wire spiral packing). Separation of *iso*-butane and *n*-butane can also be accomplished with this apparatus, but only with some difficulty and greater expenditure of time.

With the more efficient *Heli-Grid* type of column, this separation and that of *iso*-pentane and *n*-pentane are easily accomplished, thus affording analysis for all individual hydrocarbons in natural gas, up to and including C_5 fractions. When both olefins and paraffins are present, as in high-Btu oil gas and refinery oil gas, samples may be separated into fractions containing hydrocarbons of the same number of carbon atoms. They may then be segregated and analyzed for unsaturates by Orsat to yield individual hydrocarbons up to and including C_3, C_4, and C_5 saturates and unsaturates. With the *Heli-Grid* type of column, *iso*-butane and *n*-butane can be separated in the presence of olefins,[11] and determined by difference thru auxiliary Orsat analysis for unsaturates. *Iso*-butane and *iso*-butylene can be separated from *n*-butane and the 2-butenes with somewhat greater difficulty. A procedure combining distillation and chemical methods for determining individual hydrocarbons in the paraffinic-olefinic C_5 fraction has been described.[13]

Accuracy and Precision

Several light hydrocarbon samples containing C_3, C_4, and C_5 components in known concentration were prepared and analyzed in part by low temperature fractional distillation[11] during a test program conducted by the Rubber Reserve Corp. This program was part of its work on evaluating and standardizing analytical methods used in specification of butadiene and butadiene plant feed stocks. The Podbielniak Hyd-Robot distillation apparatus was used in all cases.

Test samples contained both saturated and olefinic hydrocarbons. Several variations in procedure were tested. In some, the C_4 fraction as a whole was segregated from C_3 and C_5 for subsequent chemical or spectrometric analysis. Low results for C_3, and high results for C_4, were no doubt due to holdup of vapor and/or liquid reflux between the thermocouple and distillate measuring system. Also tested were methods by which *iso*-butane and *n*-butane were segregated into separate fractions thru distillation, and the amount of each was determined by subsequent chemical analysis for unsaturates.

Conclusions from these analyses were:

> "Individual laboratories were able to check their own analyses with excellent precision, by any of the standard methods employed. On the average, a given laboratory can be expected to make determinations for most light hydrocarbon components of a gas within a probable error of ±0.2 to 0.3 mole per cent. Furthermore, the extent to which laboratories check each other has been found to be within values two to three times the probable error for a given laboratory.
>
> "Distillation and mass spectrometer methods show about the same accuracy and precision for the analysis of total C_3, total C_4, and total C_5. The distillation method is a necessity when it is desired to separate a complex mixture so that the infra-red method can be applied to the analysis of the individual butenes."

Many commercial transactions in the natural gasoline industry involve low-temperature fractional distillation analysis. NGAA accordingly conducted a test program to evaluate errors prevailing in application of the method and to promote adoption of correct technique.[14]

Seventy-seven industrial and commercial laboratories participated and submitted a total of 510 results. In this program it was concluded, that:

> "The mean deviation from the true value, averaged for all components, ranged from 0.55 to 1.04 mole per cent for the various types of samples analyzed.
>
> "Greatest deviation occurs on separations of isomers.
>
> "The application of holdup corrections has reduced but not eliminated the tendency towards low analytical values for the lowest boiling component.
>
> "Accuracy has improved since the earlier gas analysis program and again since the first phase of the liquid analysis program, but still further improvement is needed.

"Major errors are not due to basic procedures or apparatus and are attributed to human element factors, such as improper application, lack of understanding, inadequate training, miscalibrations, etc."

A Low-Temperature Fractional Analysis School was subsequently set up by the NGAA to train analysts in the industry, with a view to minimizing human-element errors. Methods for determining vapor holdup, calibrating receivers, applying compressibility corrections, and calculating results were taught, and in addition a method was devised for eliminating liquid holdup. With these improvements, the modified Podbielniak automatic apparatus was reported to yield analyses with an average deviation from the synthesis composition of less than 0.2 per component on an LP-gas sample.

REFERENCES

1. Travers, M. W. *Experimental Study of Gases.* London, Macmillan, 1901.
2. Burrell, G. A. and Robertson, I. W. "Rapid Method of Fractionating Gases at Low Temperatures." *Ind. Eng. Chem.* 7: 210–11, Mar. 1915. Also: "Separation of Gases by Fractional Distillation in a Vacuum at Low Temperatures." *Ind. Eng. Chem.* 7: 209–10, Mar. 1915. Also: *Analysis of Natural Gas and Illuminating Gas by Fractional Distillation at Low Temperatures and Pressures.* (Bur. of Mines TP 104) Washington, G.P.O., 1915.
3. Shepherd, M. and Porter, F. "Improved Method for the Separation of Gas Mixtures." *Ind. Eng. Chem.* 15: 1143–46, 1923.
4. Podbielniak, W. J. "Apparatus and Methods for Precise Fractional Distillation Analysis: New Method of Analysis." *Ind. Eng. Chem., Anal. ed.* 3: 177–88, 1931; 5: 119–42, 172–78, 1933. Also his: "Fractional Analyses of Natural Gas." *Oil Gas J.* 27: 30+, May 16, 1929. Also his patents: U. S. 1,909,315 and 1,917,272; Brit. 380,220; French 744,353.
5. Frey, F. E. and Yant, W. P. "Separation of Individual Saturated and Unsaturated Hydrocarbons in Coal Gas by Fractional Distillation." *Ind. Eng. Chem.* 19: 492–3, Ap. 1927.
6. Lucas, H. J. and Dillon, R. T. "Synthesis of 1-Butene." *J. Am. Chem. Soc.* 50: 1460–64, 1928.
7. Oberfell, G. G. and Alden, R. C. "Fractionation at Low Temperature." *Oil Gas J.* 27: 142+, Oct. 18, 1928.
8. Altieri, V. J. *Gas Analysis & Testing of Gaseous Materials.* New York, A.G.A., 1945.
9. Rohman, A. and Krappe, J. M. *Handbook, Butane-Propane Gases,* 3rd ed. Los Angeles, Western Business Papers, 1942.
10. Calif. Nat. Gasoline Assn. *Tentative Standard Method for Analysis of Natural Gas and Gasoline by Fractional Distillation.* (Bull. TS-411) Los Angeles, 1941.
11. Burke, O. W. and others. *Light Hydrocarbon Analysis.* New York, Reinhold, 1951.
12. Podbielniak, W. J. "Apparatus and Methods for Precise Fractional Distillation Analysis." *Ind. Eng. Chem., Anal. ed.* 13: 639, 1941.
13. Robey, R. F. and Wiese, H. K. "Chemical Analysis of Refinery C_5 Hydrocarbon Fractions." *Anal. Chem.* 20: 926–33, Oct. 1948.
14. Miller, A. J. "NGAA Fractional Analysis Program." *Petrol. Engr.* 24:C31+, Aug. 1952.

Chapter 4
Other Analytical Methods

by D. V. Kniebes

MASS SPECTROMETER

The mass spectrometer is an electronic instrument capable of making rapid analyses of complex mixtures of gases or light liquids. Gas molecules are ionized at a very low pressure (about 10^{-5} mm of mercury),* and the resulting charged molecules and molecular fragments are then separated into their respective masses by means of crossed electric and magnetic fields. Finally, the abundance of each mass is measured.

Operation

As shown in Fig. 6-14, gas molecules enter the tube in the ionizing chamber where they are bombarded with a stream of electrons. Resulting positive particles, or ions, are then accelerated thru a slit system by an electrostatic field. A magnetic field is also applied normal to the ion path, causing the ions to deviate into a circular path. Light ions like hydrogen travel in short-radius circles, while heavy ions like butane take larger-radius paths. Only one of these separated ion beams reaches the ion collector. The others strike the tube walls where they discharge themselves and are pumped from the system.

Ions reaching the ion collector also discharge themselves, but their charge is measured. Resulting voltage is amplified and used to drive a galvanometer, meter, or recording potentiometer, thus obtaining a quantitative measurement of the number of ions in the beam. Ions of other weights can be caused to strike the collector simply by varying accelerating voltage or magnetic field intensity, thereby changing the radius of curvature of their paths to permit them to enter the slit and collector. In this manner, each mass group of ions produced from a gas can be measured quantitatively.

The instrument produces a mass spectrum, in the form of a strip chart, of the gas or mixture being analyzed. The chart has a horizontal base line with sharply defined peaks at intervals corresponding to the mass-to-charge ratio of the molecules and molecular fragments produced by the ionization process. Peak heights are proportional to numbers of these ions of each type formed. As a result, compounds of different molecular structures will produce unique mass spectra and will therefore permit qualitative and quantitative analyses of mixtures of compounds. Detailed description of the instrument's operation and application is available.[1–3]

* Instruments operating at 10^{-9} mm Hg and up to 842 F are available.

The mass spectrometer must be calibrated periodically with pure samples of each component that occurs in the samples to be analyzed. Although valuable qualitative information can be obtained from the mass spectrum of an unknown material, an accurate quantitative analysis can be made only when all of the components of a sample have been identified.

Samples as small as 0.1 ml of gas and 0.01 ml of liquid can be analyzed by this instrument; its threshold of detection is 0.001 mole per cent. Since the sample must enter the ionization chamber in gaseous form, all its components should have a vapor pressure of fifty microns or more at room temperature. All gases and hydrocarbon liquids to about the C_{10} group are, therefore, included. Special heated inlet systems can be used for introducing much heavier hydrocarbons, such as C_{44} waxes.

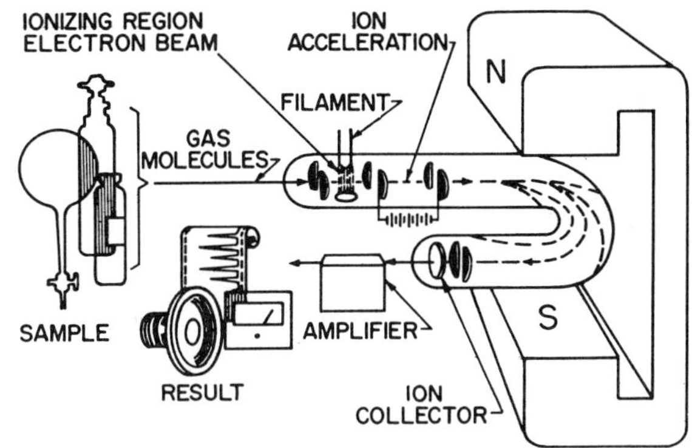

Fig. 6-14 Operating diagram of mass spectrometer.

Composition of a sample analyzed by a mass spectrometer is in mole per cent, not volume per cent at one atmosphere as in most other methods. The sorting and counting processes are conducted according to molecular weight or mass, thus providing a direct approach to the determination of composition independent of other properties of the molecule. A compressibility correction, therefore, is necessary in making a mass spectrometer analysis comparable to one by volumetric methods. This correction is normally not significant, except for mixtures with components having widely different compressibilities such as hydrogen-butane mixtures.

Time required to complete an analysis varies from 20 min to about 2 hr, depending upon sample complexity and computing equipment available. Only 10 to 20 min are required to obtain the mass spectrum of a sample; remaining time is used to calculate the mole per cent composition. It is not uncommon for a control laboratory to make 20 or more analyses per 8-hr shift with a single operator; however, from one to six technicians would be required for computation of these analyses. High speed automatic computing facilities reduce these personnel requirements.

A mass spectrometer, including necessary calibrating materials and computing equipment, represents a large initial investment. After the instrument is placed in service, it is operated continuously and is not fully shut down at any time except for major maintenance.

Accuracy

The accuracy of a mass spectrometer analysis depends somewhat upon sample constituents; however, an accuracy equivalent to or better than low temperature distillation and volumetric gas analyzer methods can be expected. Table 6-3 shows comparative analyses of a synthetic oil gas made by mass spectrometer and by Orsat methods. It illustrates the accuracy and complete breakdown of all constituents which can be obtained in a mass spectrometer analysis. In some instances as little as 0.005 mole per cent of a component can be determined.

Table 6-3 Comparison of Analyses of a Synthetic Oil Gas by Mass Spectrometer and Orsat Methods

Composition	Known	M.S.*	Orsat
Condensable at −40 C			1.5
Benzene	0.7	0.4	
Toluene	0.1	Trace	
Pentenes	0.3	0.4	
Pentanes	0.5	0.7	
Carbon dioxide	5.0	4.8	5.1
Ethene (ethylene)	16.7	16.5	16.5
Propane and C_4 olefins			
Propene	8.1	8.1	7.8
C_4 olefins			3.2
Butenes	1.6	1.4	
Butadienes	1.6	1.5	
Oxygen	0.7	0.7	0.8
Hydrogen	13.0	13.3	12.9
Carbon monoxide	2.0	1.9	2.3
Paraffins			48.4
Methane	39.2	39.1	
Ethane	8.0	7.8	
Propane	1.1	1.2	
Butanes	0.7	0.8	
Nitrogen	0.7	1.4	1.5
	100.0	100.0	100.0

* M.S.—Mass Spectrometer; Consolidated Engineering Corp. Type 21-103.

Applications

This relatively expensive equipment requires highly trained personnel and is best suited for an organization where it can be put to continuous use. Accurate and complete analysis of all utility fuel gases, as well as light oils and organic sulfur compounds used as odorants, may be made by the mass spectrometer. Complex mixtures such as high-Btu oil gas with numerous components can be analyzed by the instrument. Table 6-3 illustrates its ability to separate all components of illuminant and paraffin fractions of an oil gas. This contrasts with results of volumetric analysis, which give total paraffins and partial separation of illuminants. With complete analysis available as given by the mass spectrometer, heating value and specific gravity can be computed, thus permitting a check on the analysis.

The mass spectrometer has been found useful in identifying gaseous components from air-diluted samples of undetermined origin, such as in leakage cases. Results are then matched against possible sources.

INFRA-RED ANALYZERS

Infra-red spectroscopy is based on the differences in infrared absorption characteristics of various compounds, which depend upon differences in molecular structure. Such quite similar gases as ethane, propane, and butane have sufficiently different spectra to permit accurate analysis of their mixtures. Early studies demonstrated that each gas has a unique absorption spectrum which can be utilized for its identification in a gaseous mixture. Also, by calibrating the attenuation of an infra-red beam directed thru a volume of gas, concentration of each absorbing component can be measured.

Spectrometers and nondispersion instruments constitute two general classes of infra-red analyzers.[4]

Spectrometers

These can be used for qualitative and quantitative analysis by determining infra-red spectra. They are best adapted to applications where complete analysis of many different types of mixtures for numerous constituents is required, such as in laboratory and pilot plant operations.

A spectrometer consists basically of an infra-red source, a sample cell, a dispersing system, and a detecting system. Sample absorption at each frequency of infra-red light is determined, and sample transmission is plotted as a function of wave length. Modern instruments are capable of automatically plotting sample spectra in less than 15 min. Compounds can be identified from absorption spectra and compositions of mixtures calculated.

Nondispersion Instruments

These instruments are adapted for continuous analysis of product flow or for analysis by intermittent sampling—for example, in analysis of natural gas for methane, ethane, or butane, and of manufactured gas for carbon dioxide, carbon monoxide, methane, or illuminants. Fuel gas water vapor content may be determined from dew points of 100 F to −100 F. Analysis of combustion products in industrial operations or in appliance testing may be rapidly and accurately made.

A nondispersion instrument incorporates an infra-red source, a sample cell, a filter cell, and a detection system. The instrument can be sensitized to a specific compound by appropriate detector design and selection of filter cells. Concentrations of this compound in mixtures supplied to the sample cell may be determined from the proportional energy absorption in the spectral region for which the instrument has been sensitized.

Nondispersion instruments employ either selective or nonselective detectors. Figure 6-15a is a diagram of a nonselective detector instrument. Radiation from sources S_1 and S_2 is directed to detectors D_1 and D_2. These are usually bolometers or thermopiles, connected in opposition electrically to indicate any energy difference in the two beams after traversing the gas cells.

The four gas cells used are a sample cell, two sensitizing cells, X and Y, and a filter cell, F. With the sensitizing cells filled with nonabsorbing gases, there is no difference in the beams regardless of the mixture in the sample cell. If cell X is filled with an absorbing gas, A, this gas will absorb all energy that it is capable of absorbing, and detector D_1 behind that cell will receive less energy than D_2, with detector outputs thus unbalanced. If a mixture containing a nonabsorbing gas and the absorbing gas then is placed in the sample cell, energy reaching D_2 will be reduced, while that reaching D_1 remains unchanged. Output of D_2 thus approaches that of D_1 and is restored to balance at a lower level. Degree of balance restoration indicates concentration in the sample.

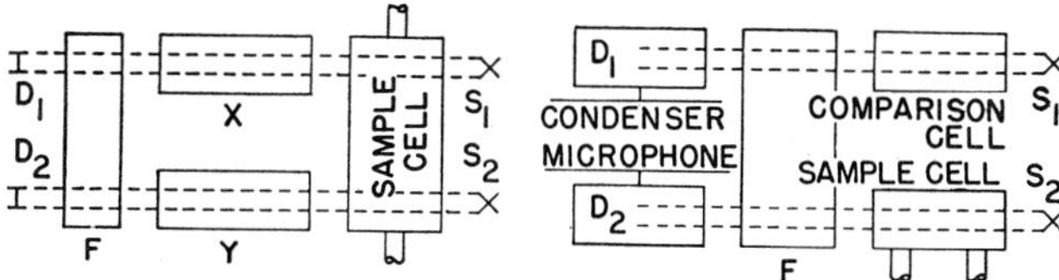

Fig. 6-15a (left) Nonselective detector instrument diagram.
Fig. 6-15b (right) Selective detector instrument diagram.

If the sample contains a second gas, B, with absorption characteristics similar to A, it will also tend to restore the detector output balance. By filling filter cell F with this interfering gas, the analyzer loses most of its sensitivity to the spectrum of that gas but gains selectivity to the first gas.

To increase desired selectivity, the second sensitizing cell, Y, may be filled with a mixture of the interfering gas and a nonabsorbing gas, in such proportions that both beams are equally affected when the interfering gas is placed in the sample cell.

Figure 6-15b is a diagram of a selective detector instrument. Radiation from sources S_1 and S_2 is directed to detector chambers D_1 and D_2, both of which are filled with an absorbing gas. Absorption of radiation by the gas increases its temperature and pressure. One wall of each detector is a membrane, mechanically connected to a condenser microphone whose output measures any energy difference in the two beams after traversing the gas cells.

Three gas cells are used: a sample cell, a comparison cell, and a filter cell, F. In contrast to the nonselective detector instrument, this type analyzer is sensitized at the detection point by selecting a specific gas to fill the detector chambers. If a mixture containing this gas is placed in the sample cell, energy reaching detector D_2 is reduced, pressures for the two chambers are unbalanced, and the microphone produces an output signal. Effects of an interfering gas can be compensated for by filling filter cell F with this gas.

Commercially available selective detector type analyzers, e.g., the *Lira*, use interrupted radiation. In this basic arrangement, both beams are simultaneously blocked and passed, producing a pulsating output from the condenser microphone tuned to the interrupting frequency. Ambient temperature changes do not cause errors. *Lira* recorders may be used for CO traces.

Applications. Both types of nondispersion instruments embody many variations in material and construction. Each analyzer, while basically the same, requires its own sensitizing technique for its particular application. In theory, both types are capable of the same selectivity and sensitivity. Practically, selective detectors are best utilized where highest sensitivity and selectivity are demanded and where proper installation may be provided with skilled maintenance available. Nonselective detectors may be employed where maximum sensitivity and selectivity are not essential and where conditions require rugged equipment free from effects of vibration and stray currents.

GAS CHROMATOGRAPH

The gas chromatograph is an instrument with which virtually all gas mixtures and mixtures of liquids, with components boiling to about 300 C, can be analyzed. [5-7] The procedure employs a chromatographic column, typically a 3 to 30 ft length of ¼ in. OD copper tubing filled with a selected stationary phase packing material thru which a steady flow of carrier gas such as helium or air is passed. A small sample of the mixture to be analyzed is injected into the carrier gas stream and passes thru the column, where components of the mixture are selectively retarded according to their individual affinities, to the stationary phase. Each component will, therefore, elute from the column at a characteristic time. Detectors, such as a thermal conductivity sensing element or a catalytic combustion cell, are placed in the carrier gas stream as it emerges from the column, and the presence of each component is detected as it is eluted with the carrier gas. The composition of the mixture is represented by a chromatogram (Figs. 6-17a, 6-17b, 6-17c) which is a strip chart record of the detector output. The time of appearance of each peak is characteristic of a particular component, and each peak's height and area are proportional to the quantity of that component in the mixture.

Apparatus

Most gas chromatographs operate according to the simplified flow diagram of Fig. 6-16. The carrier gas flows in sequence thru the preheater coil, the reference side of the thermal conductivity cell, the chromatographic column, the sample side of the thermal conductivity cell, and a flow meter, where it is

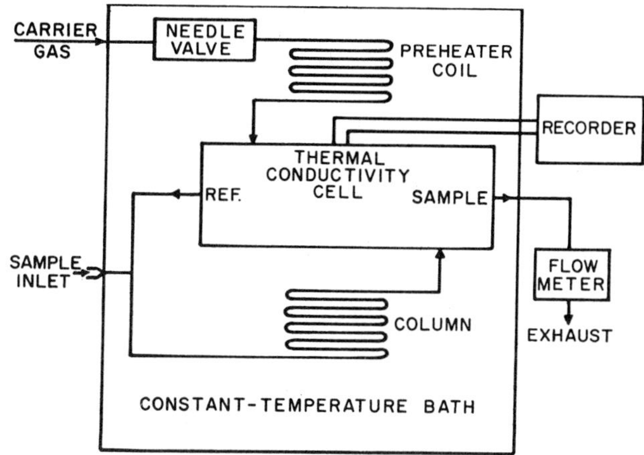

Fig. 6-16 Flow diagram of gas chromatograph.

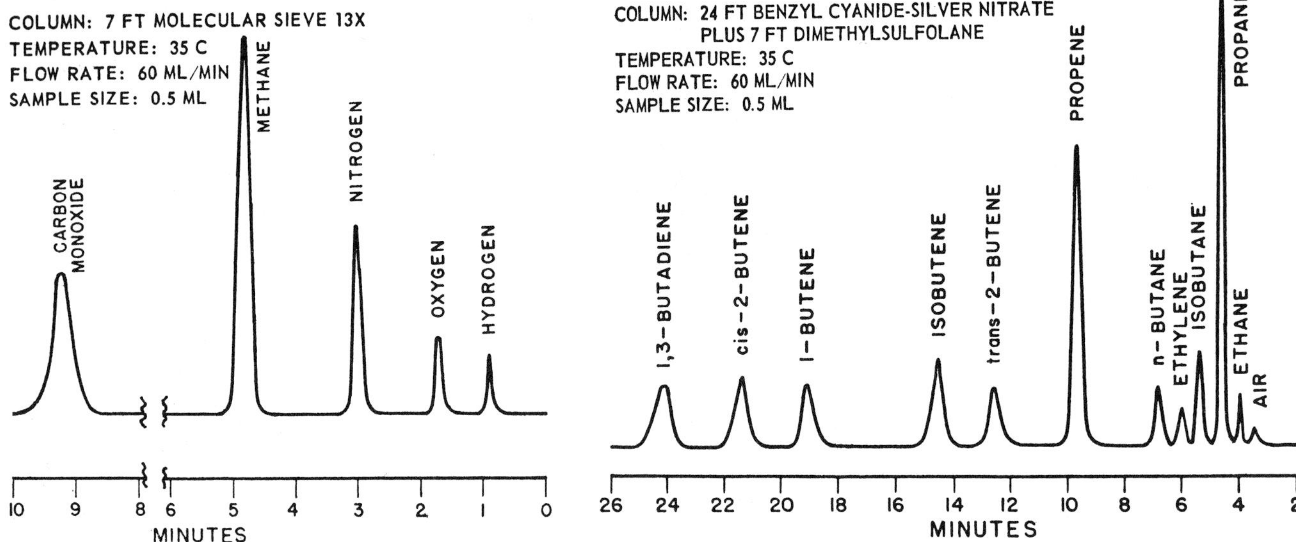

Fig. 6-17a (left) Chromatogram of reformed gas on molecular sieve column.

Fig. 6-17b (right) Chromatogram of a synthetic LP-gas sample.

finally exhausted. Flow rate of the carrier gas is controlled in the range of 30–100 ml per minute by a needle valve as the gas enters the constant temperature bath in which the principal elements of the instrument are enclosed. A measured sample is injected at a point just ahead of the chromatographic column. Both liquid and gas samples can be injected with a hypodermic syringe which pierces a rubber septum. Gas samples are best injected, however, by means of a gas sampling valve. These valves permit reproducible gas samples of 0.25 ml and greater to be injected.

Analytical columns used for separation of mixture components are of two types: the solid adsorbent column and the liquid partition column. Solid adsorbents are used primarily for separation of low molecular weight components in gas mixtures. For example, as shown in Fig. 6-17a, the components oxygen, nitrogen, methane, and carbon monoxide separate distinctly on a molecular sieve 5A or 13X column. Other useful solid adsorbents are silica gel, alumina, and activated charcoal.

A liquid partition column which can be used to separate light paraffins and olefins has been reported.[8] This column is packed with benzyl cyanide-silver nitrate on *Chromosorb* and is used in the analysis of liquefied petroleum gases. The performance of a column of this type, slightly modified by the addition of a 7 ft section of dimethylsulfolane, is shown in Fig. 6-17b. It was also reported that propane–propylene separation was effected by silica gel at 200 F.

Liquid partition columns are packed with a support, such as diatomaceous earth, which has been impregnated by a selected high-boiling liquid, normally 10 to 30 per cent by weight. The liquid selected for this stationary phase will depend upon the type of separation and operating temperature required for the analysis. A liquid partition column which is useful for separating natural gas hydrocarbons consists of a series-connected 16-ft dimethylsulfolane and 8-ft diisodecylphthalate column. Performance of this column is shown in the chromatogram of Fig. 6-17c.

Other columns shown to be satisfactory for this separation are hexamethylphosphoramide (20 ft) and silicone 200/500 (30 ft). *Chromosorb* pink is used as the solid support for each of these four columns, and the liquids constitute about 30 per cent by weight of the final column packing.

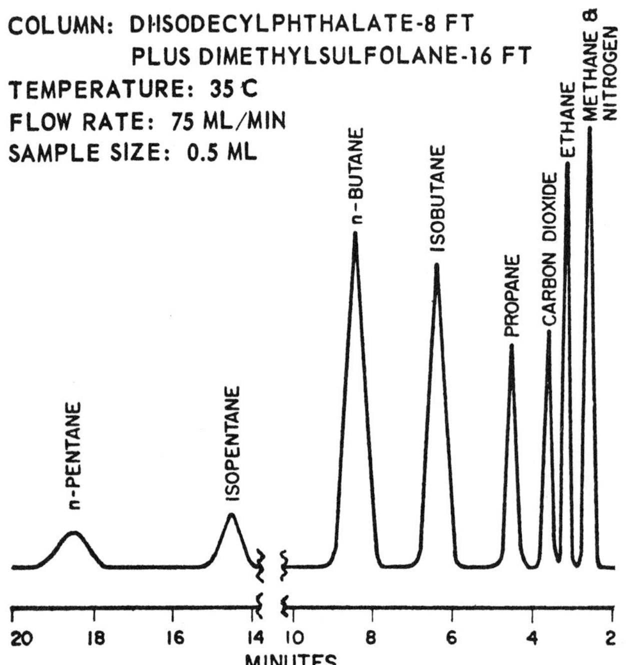

Fig. 6-17c Chromatogram of natural gas.

Applications

Since the cost of a gas chromatograph is low compared to that of a mass spectrometer, the chromatograph can be economically applied to many analytical problems in the laboratory, plant, and field. **Analytical procedures** have been developed for analysis of several types of fuel gases,[8–10] including natural gas, reformed gas, and LP-gas. Trace component determinations can be made when the instrument is equipped with high-sensitivity detectors such as the flame ionization and argon diode detectors; alternatively, sensitivity can be increased by concentrating the sample. Thus, odorants can be detected in less than part per million concentrations, and traces of methane and other hydrocarbons can be determined in samples containing mostly air. Specialized gas chromatographs, which can be used in the field to detect ethane and methane in air, have been developed for leak survey and gas source identification purposes.

Process stream analyzers based on the principles of gas chromatography are also used. These instruments monitor, on a short cycle basis, the concentration of one or more components of a gas stream and provide appropriate signal output for plant control purposes.

REFERENCES

1. Consolidated Engineering Corp. *Operation Manual, Vol. I, Model 21-103 Mass Spectrometer*. Pasadena.
2. A.S.T.M. "Standard Method for Analysis of Natural Gases and Related Types of Gaseous Mixtures by the Mass Spectrometer." (Standard D 1137-53) *ASTM Standards*, 1961, pt. 8. Philadelphia, 1962. (Standard available as separate pamphlet.)
3. Berl, W. G., ed. *Physical Methods in Chemical Analysis*. New York, Academic Press, 1956–62. 4 V.
4. Waters, J. L. and Hartz, N. W. "Improved Luft-Type Infrared Gas and Liquid Analyzer." *Instr. Soc. Am. Proc.* 6: 79–83, 1951.
5. Keulemans, A. I. M. *Gas Chromatography*, 2d ed. New York, Reinhold, 1959.
6. Pecsok, R. L., ed. *Principles and Practice of Gas Chromatography*. New York, Wiley, 1959.
7. Scott, R. P. W., ed. *Gas Chromatography, 1960*. Washington, Butterworth, 1960.
8. A.S.T.M. "Tentative Method for Analysis of Certain Light Hydrocarbons by Gas Chromatography." (Standard D 1717-61T) *ASTM Standards*, 1961, pt. 7, Philadelphia, 1961. (Standard available as separate pamphlet.)
9. Kniebes, D. V. *Utility Gas Analysis by Gas Chromatography*. (I.G.T. Tech. Rept. 4) New York, A.G.A., 1962.
10. Natural Gasoline Assn. of Am. *N.G.A.A. Tentative Method for Natural Gas Analysis by Gas Chromatography*. (Pub. 2261-61) Tulsa, 1961.

Chapter 5
Determination of Miscellaneous Constituents

by C. W. Wilson

HYDROGEN SULFIDE

Hydrogen sulfide, H_2S, may be present in utility fuel gases in varying amounts. It occurs in manufactured gases as a product of pyrolysis of sulfur-bearing coal or oil, and is found in natural gases from certain fields in amounts varying from very small to relatively large concentrations in *sour* gas. Liquefied petroleum gases from oil refineries may contain small amounts of H_2S, but the salable product of the refinery is substantially free of it. Presence of H_2S is undesirable because it corrodes equipment with which it comes in contact. It also has deleterious effects on catalysts in certain processes, and along with its combustion products, has toxic effects on people. For these reasons, every effort is made to remove H_2S from fuel gases distributed by utilities, and quantitative estimation of its concentration in both raw and sendout gases is essential. Analytical procedures may be selected, depending upon concentration present and desired accuracy of test.

In 1963, the Chemical and Engineering Committee of the A. G. A. Operating Section suggested that the *methylene blue test* (a colorimetric method) be used to determine H_2S in natural gas. See page 6/32.1.

Methods

U. S. Steel Chemists.[1,2] This method is a typical general procedure for H_2S determination. A measured gas sample is passed thru a potassium or sodium hydroxide solution, or thru an ammoniacal solution of cadmium chloride or zinc sulfate. After acidifying with hydrochloric acid, liberated H_2S is titrated with standard iodine or iodate solution.

Apparatus: A 500 ml flask, carrying gas inlet bubbling tube and gas outlet tube; wet test meter or four liter aspirating bottle; one liter graduated cylinder; sulfur-free rubber tubing; and miscellaneous glassware.

Reagents:

1. Starch Indicator. Shake 6 g of starch with 100 ml cold water; add to one liter of boiling water; boil 5 min; cool. Add 6 g zinc chloride dissolved in 50 ml cold water; mix well; let stand 24 hr with occasional shaking. Decant the supernatant liquid into a glass-stoppered bottle; add 3 g potassium iodide and dissolve.

2. Ammoniacal Cadmium Chloride Solution. Dissolve 5 g cadmium chloride in 375 ml water, and add 625 ml concentrated ammonium hydroxide.

3. Ammoniacal Zinc Sulfate Solution. Add 10 g zinc sulfate to 50 ml concentrated ammonium hydroxide in a 1000 ml calibrated flask; stir; add water to make one liter.

4. Sodium or Potassium Hydroxide Solution. Dissolve 20 g of sodium or potassium hydroxide in one liter water.

5. Standard $0.1N$ Potassium Iodate Solution. To a 1000 ml graduated flask add 3.57 g potassium iodate, 13.8 g potassium iodide, and 300 ml water; shake well; add 1 g potassium hydroxide; shake until dissolved; add water to make one liter.

6. Standard $0.1N$ Iodine Solution. Transfer 12.7 g iodine and 18 g potassium iodide to a one liter graduated flask; add abut 50 ml cold water; shake well; dilute to one liter; let stand 24 hr before standardizing.

Procedure: Place 20 ml of one of the absorbent solutions in 500 ml flask; add about 200 ml water and mix well. Purge gas sampling line and connect with gas inlet tube. Connect gas outlet tube with measuring device, and pass about 0.1 cu ft of gas thru absorbent. Record temperature and pressure of metered sample and barometer reading.

Disconnect flask. Wash bubbling tube with about 100 ml water, containing 1–2 ml of 1:1 hydrochloric acid; add about 5 ml starch indicator and excess of 1:1 hydrochloric acid. Immediately titrate liberated sulfide with standard $N/10$ iodine or iodate solution, adding standard solution rapidly at first, with little shaking, to avoid loss of H_2S. H_2S content may then be calculated as follows:

$$V_1 = \frac{100V_2C + 0.0042D}{100 - J}$$

$$G = 2.63D/V_1$$

$$P = 0.0003757D/V_1$$

$$H = G/636.4$$

where:

V_1 = volume of gas entering apparatus, cu ft at 60 F, 30 in. Hg, sat.
V_2 = volume of treated gas as metered, cu ft
C = correction factor converting V_2 to cu ft of gas at 60 F, 30 in. Hg, sat.
J = per cent CO_2 in entering gas, absorbed simultaneously with H_2S
D = ml of titrating standard used
G = grains H_2S per 100 cu ft
P = lb H_2S per 100 cu ft
H = per cent H_2S in entering gas

This procedure can be adapted to determine the average concentration of H_2S in a cumulative gas sample over a period such as 24 hr, by suitably reducing gas flow rate so that absorptive capacity of the solutions will not be exceeded. Rate will depend upon estimated average H_2S concentration of sample.

Tutwiler.[2] When an instantaneous sample is desired and H_2S concentration is ten grains per 100 cu ft or more, a 100 ml Tutwiler burette is used. For concentrations less than ten grains, a 500 ml Tutwiler burette and more dilute solutions are used. In principle, this method consists of titrating hydrogen sulfide in a gas sample directly with a standard solution of iodine.

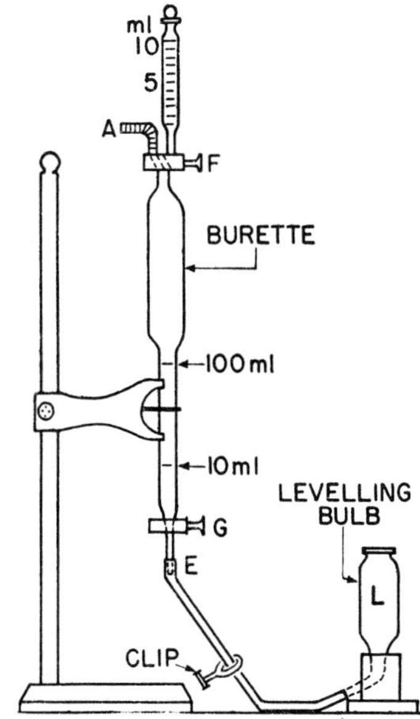

Fig. 6-18 Tutwiler burette. (Lettered items mentioned in text.)

Apparatus: (See Fig. 6-18.) A 100 or 500 ml capacity Tutwiler burette, with two-way glass stopcock at bottom and three-way stopcock at top which connect either with inlet tubulature or glass-stoppered cylinder, 10 ml capacity, graduated in 0.1 ml subdivision; rubber tubing connecting burette with leveling bottle.

Reagents:

1. Iodine Stock Solution, 0.1N. Weigh 12.7 g iodine, and 20 to 25 g cp potassium iodide for each liter of solution. Dissolve KI in as little water as necessary; dissolve iodine in concentrated KI solution, make up to proper volume, and store in glass-stoppered brown glass bottle.

2. Standard Iodine Solution, 1 ml = 0.00171 g I. Transfer 33.7 ml of above 0.1N stock solution into a 250 ml volumetric flask; add water to mark and mix well. Then, for 100 ml sample of gas, 1 ml of standard iodine solution is equivalent to 100 grains H_2S per 100 cu ft of gas.

3. Starch Solution. Rub into a thin paste about one teaspoonful of wheat starch with a little water; pour into about a pint of boiling water; stir; let cool and decant off clear starch solution. Make fresh solution every few days.

Procedure: Fill leveling bulb with starch solution. Raise (L). open cock (G), open (F) to (A), and close (F) when solution starts to run out of gas inlet. Close (G). Purge gas sampling line and connect with (A). Lower (L) and open (F) and (G). When liquid level is several ml past the 100 ml mark, close (G) and (F), and disconnect sampling tube. Open (G) and bring starch solution to 100 ml mark by raising (L); then close (G). Open (F) momentarily, to bring gas in burette to atmospheric pressure, and close (F). Open (G), bring liquid level down to 10 ml mark by lowering (L). Close (G), clamp rubber tubing near (E) and disconnect it from burette. Rinse graduated cylinder with a standard iodine solution (0.00171 g I per ml); fill cylinder and record reading. Introduce successive small amounts of iodine thru (F); shake well after each addition; continue until a faint permanent blue color is obtained. Record reading; subtract from previous reading, and call difference D.

With every fresh stock of starch solution perform a blank test as follows: Introduce fresh starch solution into burette up to 100 ml mark. Close (F) and (G). Lower (L) and open (G). When liquid level reaches the 10 ml mark, close (G). With air in burette, titrate as during a test and up to same end point. Call ml of iodine used C. Then,

$$\text{Grains } H_2S \text{ per 100 cu ft of gas} = 100\,(D - C)$$

Greater sensitivity can be attained if a 500 ml capacity Tutwiler burette is used with a more dilute (0.001N) iodine solution. Concentrations less than 1.0 grains per 100 cu ft can be determined in this way. Usually, the starch–iodine end point is much less distinct, and a blank determination of end point, with H_2S-free gas or air, is required.

Lead Acetate Paper.[3] This is a good qualitative test to determine presence or absence of H_2S in gas. Sometimes it is used to determine compliance with regulatory requirements which specify that gas distributed shall be "free from hydrogen sulfide." A quick method is to hold the wetted lead acetate paper in the gas stream and note speed of discoloration. A more refined method follows:

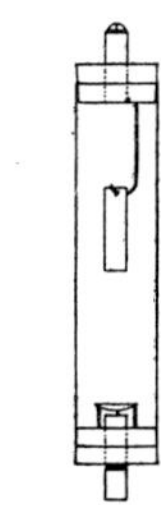

Fig. 6-19 Apparatus for lead acetate test.

Apparatus: (See Fig. 6-19.) Glass cylinder about 1.75 in. diam by 8 in. long; bottom stopper, fitted with gas inlet tube and glass baffle about 1 in. diam; top stopper, fitted with burner tip, passing gas at rate of about 5 cu ft per hour at normal pressure; glass suspension hook.

Procedure: Dip strip of white filter paper 1 in. wide by 4 in. long in five per cent lead acetate solution; press strip between clean blotters to remove excess solution and immediately suspend midway between baffle and upper stopper. Lead acetate papers which need only to be dipped into distilled water to be prepared for this test may be purchased from chemical supply houses. Pass gas at 4.5 to 5.5 cu ft per hr for 1 min. Immediately compare exposed strip with another moistened with same solution but not exposed to gas. If exposed strip is not distinctly darker, gas shall be considered free from hydrogen sulfide for distribution purposes.

Sensitivity: Limit of detection is 0.3 to 0.4 grains H_2S per 100 cu ft of gas with 1 min exposure. If other conditions remain the same, limit is about 0.45 and 0.2 grains per 100 cu ft, for tests of 30 sec and 30 min duration, respectively. Factors influencing reproducibility and sensitivity are paper surface, opacity, moisture content, and method of preparation; strength of lead acetate solution; time of contact, gas flow rate and humidity; and form and size of apparatus.

Shaw.[4] For accurate determinations of small concentrations of H_2S, a more elaborate procedure is necessary. This technique, which requires a special absorption bottle and careful manipulative technique, provides adequate precautions for accurate determination of H_2S concentrations ranging from a small fraction of one grain to a few grains per 100 cu ft. Cadmium sulfide is precipitated by H_2S from an absorbent of mixed cadmium chloride and sodium carbonate solutions. The absorption flask is designed to avoid errors which would be introduced by loss of H_2S or iodine during determinations. Precipitated cadmium sulfide is subsequently decomposed by an acid iodine solution, with excess back-titrated with thiosulfate. Mercaptans, if present, are determined by a procedure modification giving an immediate correction to the apparent H_2S concentration. The method is applicable to gases from any source or of any composition.

Colorimetric. For special purposes, determination of extremely minute H_2S concentrations in gas may be necessary. These methods are probably most suitable when concentrations are on the order of a few parts per million or less (0.1 grain of H_2S per 100 cu ft is equivalent to 1.5 ppm by volume). Probably the most reliable of these are the methylene blue method[5] and the bismuth sulfide method.[6] When procedures are followed carefully as described, either method appears capable of determining sulfide sulfur in the absorbent with a precision of ± 10 per cent for a concentration of about 1 ppm. Corresponding concentration in the gas will depend upon sample size.

A hydrogen sulfide detector,[7] made by Mine Safety Appliances Co., draws gas thru a glass tube containing granules of activated alumina impregnated with silver cyanide, by means of an aspirator bulb. Granules turn dark gray if H_2S is present, and length of discoloration is a measure of its amount. Detector tubes and calibration scale are available in two ranges, from 0 to 50 ppm (0 to 3.2 grains per 100 cu ft) and 0 to 0.04 per cent (0 to 26 grains per 100 cu ft).

ORGANIC AND TOTAL SULFUR

A permissible limit of about 30 grains of sulfur per 100 cu ft of fuel gas is specified in some state or municipal utility codes.

Organic sulfur is that sulfur in gas which is combined in molecules also containing carbon. Organic sulfur compounds are probably found in greatest variety in coke–oven gas and in products of pyrolysis of oils. Prominent among the sulfur-bearing molecules are carbon disulfide, CS_2; carbon oxysulfide (carbonyl sulfide), COS; mercaptans, *RSH*,* sulfides, *R-S-R;* disulfide, *R-S-S-R;* and thiophene.

* *R* stands for organic, or carbon-containing radical.

† See also page 6/32.1.

Methods

The following procedures† will give estimates of organic sulfur alone, if any residual H_2S is first removed from the sample by a procedure which will not at the same time influence organic sulfur compounds. Carefully conditioned iron-oxide purifier bottles have been used for this purpose, but their use makes subsequent analytical results rather uncertain, since iron oxide and wood shavings, with which it is usually mixed, absorb organic sulfur compounds and may release them with a change in ambient conditions. Removal of H_2S may be suitably accomplished by a solution of potassium ferricyanide, made by dissolving 150 g of $K_3Fe(CN)_6$ and 185 g Na_2CO_3 in 1 liter of water. Any appropriate gas-washing bottle or absorber, containing enough solution to purify the sample, may be inserted in the sample supply train.

Referee. In absence of other accepted standard tests for total or organic sulfur in gaseous fuels, this test has long been used by gas manufacturing companies. It possesses a degree of authority with most regulatory bodies for that reason, although its precision and accuracy leave much to be desired for present-day use. Its precision may be estimated as about ± 1 grain per 100 cu ft.

Apparatus: (See Fig. 6-20.) Special burner with steatite top and stand above it; heavy glass trumpet tube; heavy glass condensing tower which has tubulature near bottom with constricted space above, packed with glass balls; bent glass condensing tube; glass drip tube.

Apparatus as illustrated can usually be purchased as a complete unit from chemical apparatus supply sources. A small pressure regulator and $\frac{1}{10}$ cu ft wet test meter are also required.

Reagents:

1. Barium Chloride. Dissolve 100 g barium chloride in one liter distilled water.
2. Bromine Water. Cover bottom of a 250 ml glass bottle with bromine to about $\frac{1}{4}$ in. depth. Fill with distilled water and shake gently until water becomes saturated.
3. Silver Nitrate. Dissolve 50 g silver nitrate in one liter distilled water. Keep in dark brown bottle.
4. Hydrochloric Acid. Concentrated acid of 1.10 sp gr (1*N* acid).
5. Methyl Orange. Dissolve 0.1 g methyl orange in 100 ml distilled water.
6. Ammonium Carbonate. Chemically pure ammonium carbonate lumps.

Procedure: Rinse glass parts of apparatus thoroughly with distilled water. Dry glass trumpet tube, allowing other parts to remain wet. Attach meter to gas source, followed by Referee apparatus, using glass-to-glass connections joined by short rubber sections. Adjust gas burner to rate of 0.3 to 0.6 cu ft per hr with clean-cut blue flame, and allow to burn in air for several minutes. (Depending upon gas under test, burner input should be approximately 300 Btu per hr. If hydrogen content of test gas is less than 10 per cent, a slow water drip should be introduced at condensing tower top.) Place several layers of fresh ammonium carbonate lumps around stem of burner. With all parts of apparatus in place, set trumpet tube over burner and quickly connect with condenser, recording meter reading at instant trumpet tube is put in place.

Two to five cubic feet of gas should be burned so that sufficient barium sulfate may be produced upon precipitation for determination by usual gravimetric methods. If more than 3 cu ft are burned, ammonium carbonate should be replaced with a fresh supply.

When sufficient gas has been burned, read meter, turn off, and allow apparatus to cool. Wash out trumpet tube and condensing tube once each and condensing tower four times, each time with 50 ml wash water. Transfer condensed liquid in Erlenmeyer flask and all washings to 600 ml beaker for further treatment.

Add two or three drops of methyl-orange to beaker contents and neutralize with concentrated hydrochloric acid drop-by-drop, finally adding 1 ml of 1*N* acid in excess. Introduce 10 ml of saturated bromine water and boil solution until all bromine fumes are expelled. Then add 10 ml hot barium chloride solution and boil for 5 min. Allow to stand on a warm plate until all precipitate of barium sulfate has settled. Filter precipitate thru a quantitative filter paper and wash with hot water until a few drops of washings show no cloudy appearance on addition of silver nitrate solution. Place filter paper in weighed porcelain crucible, ignite, cool, and weigh. Increase in weight represents barium sulfate.

Sulfur concentration is calculated from observed data from expression:

Grains total sulfur per 100 cu ft of gas =

$$\frac{\text{Grams } BaSO_4 \times 212}{\text{Cubic feet of gas burned (60 F, 30 in. Hg, sat.)}}$$

Where laboratory air is contaminated with sulfur compounds, burner base should be surrounded with suitable housing into which purified air is introduced. Purification may be accomplished by passing air under slight pressure thru soda lime and activated carbon, for example a gas mask canister.

Sulfur Lamp. Total sulfur, including organic sulfur, may be determined by methods in which combustion products are passed thru absorbent solutions to absorb oxides of sulfur that form when sulfur compounds in the fuel are burned. Two modifications of this general procedure have been accepted by the American Society for Testing Materials.[8] Both appear adequate and afford a sensitivity of approximately 0.1 grain of sulfur per 100 cu ft of gas.

In one procedure, D1072-56, oxides of sulfur are absorbed in dilute Na_2CO_3 solution, after burning gas with air. Sulfate in absorbent solution is subsequently determined by titration with standard $BaCl_2$ solution, using tetrahydroxyquinone as indicator.

In D1266-57T, gas is burned in an atmosphere composed of a mixture of carbon dioxide and oxygen. Oxides of sulfur are absorbed in a neutral solution of three per cent hydrogen peroxide and converted to sulfuric acid. The latter is then titrated with standard alkaline solution. A CO_2–O_2 atmosphere is provided to avoid formation of oxides of nitrogen during combustion, which, when absorbed in peroxide, would introduce an indeterminate error in titration of sulfuric acid.

D1266-57T describes handling of liquefied petroleum gas samples, while in D1072-56 samples are assumed to be gaseous and at nearly atmospheric pressure. Either method is adaptable to both types of gases. Sulfate in either type absorbent solution may be determined gravimetrically as $BaSO_4$, or turbidimetrically at low concentrations.

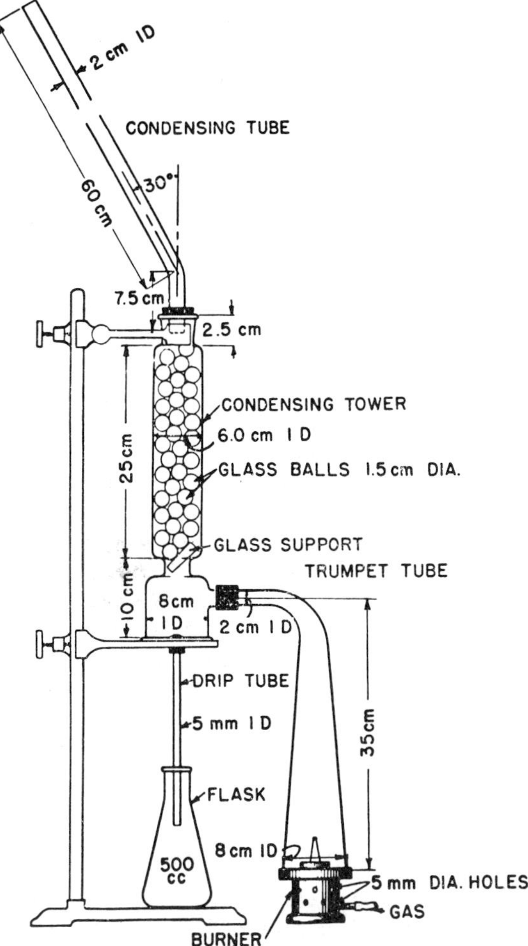

Fig. 6-20 Referee apparatus for determination of total sulfur in gas.

Reduction. Several analytical methods have been described in which organic sulfur compounds present in gases are converted at high temperatures to H_2S. These have been found quite useful with manufactured gases containing substantial hydrogen concentrations. When applied to hydrocarbon gases like natural gas, it is necessary to dilute with hydrogen, usually by adding moisture, to provide the necessary reducing atmosphere and at the same time to minimize cracking of hydrocarbons and carbon formation.

Catalysts are employed to promote the reduction of sulfur to H_2S; for example, a reduction tube containing an electrically heated spiral of platinum wire,[9] or 14–20 mesh alumina, Al_2O_3, at 900 C.[6,10] A silica gel catalyst has been used by the Southern California Gas Co. More difficulties, such as cracking and absorption of sulfur compounds, are encountered with granulated or pelletized catalysts than with platinum wire. They are thus less readily reconditioned for subsequent determinations.

Hydrogen sulfide, the product of reduction of organic sulfur compounds in gas, may be determined by any convenient procedure. The **Tutwiler** method will give almost instantaneous or running values of organic sulfur concentration. Iodimetric procedures are adaptable to a cumulative or extended test to give a time-average concentration.

Titrilog. This chemical-electronic instrument is an automatic recording titrator[11] for determining trace quantities of oxidizable sulfur in gas or in an atmosphere. It has been widely used for sulfur determinations, particularly for gas odorant concentration, atmospheric pollutants, and various chemical manufacturing processes. Measurement of sulfur concentration is accomplished by titration with bromine, generated electrolytically in solution in which sulfur compounds are absorbed from the gas stream. Course of reaction appears to be thru hypobromous acid, and it is given for oxidation or organic sulfides as: $RSR + Br_2 + H_2O = R_2SO + 2H + 2Br_2$.

Sulfur compounds, such as SO_2, H_2S, mercaptans, thioethers, thiophenes, and organic sulfides are titrated. Carbonyl sulfide and carbon disulfide do not titrate. Olefins and diolefins titrate at two to three per cent the efficiency of the sulfur compounds. Thus, their interference is negligible, unless they are present in large concentrations, as in manufactured gas. Limit of threshold sensitivity is about 0.1 ppm or 0.005 grains per 100 cu ft. Instrument can be adjusted and calibrated for a range of operation in five steps from 0 to 100 grains of sulfur per 100 cu ft with sensitivity varying for these amounts from 0.005 to 1.0 grains per 100 cu ft. By routing the sample thru selective fluid absorbers, H_2S, mercaptans, and total sulfur compounds may be determined automatically.

Individual Organic Sulfur Compounds

Determination of individual organic sulfur compounds or types of compounds containing sulfur in gas is not a usual procedure. It is known that in coke-oven gas, carbureted water gas, and oil gas, such organic sulfur compounds as COS, CS_2, and thiophene predominate. In refinery gases, mercaptans, sulfides, and disulfides result from processing of oil. In natural gas, mercaptans and organic sulfides may be found. Currently available odorizing agents contain predominantly mercaptans, cyclic sulfides, or straight chain hydrocarbon sulfides. It has usually been found sufficient to know the total concentration of sulfur contributed by these compounds.

By a modification of the Shaw method[4] for the determination of H_2S in gas, concentration of mercaptan sulfur can also be found. A method has been proposed for determining concentration of the respective classes of compounds.[12] It makes use of selective absorbents, combining determinations of the sulfur compounds absorbed with data on residual sulfur in gas after passing thru absorbents, to estimate H_2S, COS, CS_2, mercaptans, and thiophenes. The differential solubility of sulfur compounds in an absorbent oil was used as a basis for their classification and determination.[6]

For detecting H_2S, elemental sulfur, and mercaptans in liquid hydrocarbons associated with natural gas, the **Doctor test,**[2] sodium plumbite, may be used. Systematic analyses for groups of sulfur compounds in liquid fractions of gas were reported[13] and reproduced.[2] A qualitative field test for presence of H_2S and mercaptans in a new natural gas well is to pass gas directly from the well thru a metallic condenser surrounded by dry ice, collect condensate in a glass flask, and add cadmium chloride to one part of liquid and silver nitrate to another. A precipitate containing cadmium chloride indicates presence of H_2S; one with silver nitrate indicates presence of mercaptans.

WATER VAPOR

Determination of concentration of water vapor in gas is necessary for control of hydrate formation in high pressure transmission lines, operation of dehumidification equipment, and control of rehydration or humidification where it is practiced in low pressure distribution systems. Concentrations encountered are of varying magnitude for these different applications, and procedures must be chosen accordingly.

Methods

Direct Gravimetric. This tedious procedure is seldom used for control purposes, although the method is basic and adaptable to a fairly wide range of concentration. It consists of passing gas containing water vapor thru a desiccant and determining weight of water removed from a given volume.

A general procedure may be followed and adapted to specific conditions. For example, since gas pressure in apparatus (glass) should be substantially atmospheric, one or more stages of pressure reduction will be required between a high pressure gas supply and the desiccant tubes. Or, since gases in a manufactured gas plant may contain entrained droplets of liquid water, gas samples are often preheated to vaporize these droplets and thus ensure collection of all water present.

Three tubes containing desiccant are connected so that the sample passes thru them in series. They may be glass U-tubes of the customary form. The first two tubes are weighed on an analytical balance at beginning and end of test. The third tube protects the train outlet from back-diffusion of moisture from the meter atmosphere. Sample volume is measured at the third drying tube outlet (i.e., downstream) by an accurate test meter.

Suitable desiccants are calcium chloride or *Dehydrite.* Silica gel and activated alumina are unsuitable because they absorb hydrocarbons. Use of perchlorate desiccants in presence of combustible gases is not considered advisable.

With the desiccant tubes filled and connected in series, a test is begun by purging the train with gas under test. When the train is purged and pressure within the tubes has been adjusted to atmospheric, the first two tubes are carefully weighed. After reconnecting, gas is passed at a suitably slow rate. Meter reading, temperature, and barometric pressure are recorded. After an appropriate period, depending upon gas flow rate and expected moisture content, drying tubes are disconnected, adjusted to atmospheric pressure, and weighed again. Meter reading, temperature, and barometric pressure are recorded at end of test.

Concentration of water vapor in the gas is obtained from test data by the expression:

$$C_w = 1543W/V$$

where: C_w = concentration of water vapor in the gas, grains per 100 cu ft of dry gas
W = gain in weight, grams, of the two desiccant tubes
V = sample volume, SCF, dry

Concentration found can be converted to other units such as per cent by volume, relative humidity, or dew point, with aid of psychrometric tables or charts.

If gas sample is measured by wet test meter, corrected volume is obtained by the expression:

$$V = V_o \left(\frac{520}{460 + T}\right)\left(\frac{P - p_w}{30.0}\right)$$

where: V_o = observed volume, cu ft
T = meter temperature, °F
P = barometric pressure, in. Hg
p_w = saturated vapor pressure of water at meter temperature T, in. Hg

Dew Point. Water vapor in a gas sample may be determined by measuring its dew point. The dew point is that temperature at which saturation vapor pressure of liquid water equals partial pressure of water vapor in the gas. It may also be defined as the temperature at which water just begins to condense from gas to liquid phase.

The dew point is usually determined by cooling a highly polished metallic or mirror surface exposed to the gas sample, noting temperature at which surface becomes hazy from condensation. Subsequent cooling is discontinued, and the metal is allowed to warm up slowly, with gas continuously passed over it. Temperature is noted when condensed moisture disappears. Average of such successive observations is taken as the dew point.

Laboratory forms of dew point apparatus, Waidner and Mueller and U. S. Steel Chemists' types, were described and illustrated.[2] In their use it is good practice to cool and warm the metal surface slowly, so that thermometer lag will not influence observed temperature. A difference of ±1°F between temperatures at which moisture haze appears and disappears is usually satisfactory. This represents an uncertainty in water vapor concentration of about ±5 per cent. If dew point is very low, additional precautions should be taken to ensure that gas flow rate past the cooled surface is large enough and quantity of water condensed is only a small fraction of that present in the sample. This will be attained if the observed dew point does not change when gas flow rate is varied.

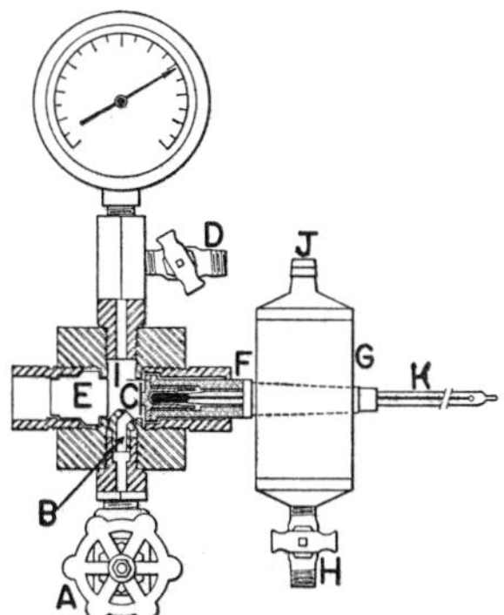

Fig. 6-21 Bureau of Mines dew point apparatus.

The standard method of test for water vapor content of gaseous fuels by measurement of dew point temperature, included in ASTM D1142-58, is shown in Fig. 6-21. This method covers a range of water vapor contents from 0 to 100 per cent. However, its use is arbitrarily limited to conditions where: (1) dew point temperature at test pressure is at least 3°F lower than both ambient temperature of testing equipment and that of gas line to apparatus; (2) determined dew point temperature is not lower than 0 F; and (3) ice crystals do not form. Condensation of vapors other than water, such as gasoline or glycols, may interfere with measurement if their dew point temperature is higher than that of water vapor.

The dew point apparatus[14] shown in Fig. 6-21 consists of a metal chamber, into and out of which test gas is permitted to flow thru control valves (A) and (D). Gas entering thru valve (A) is deflected by nozzle (B) toward the cold portion of apparatus (C). Gas flows across face of (C) and out thru valve (D). The highly polished stainless steel "target mirror" (C) is cooled by means of copper cooling rod (F). The mirror is silver-soldered to a nib on the copper thermometer well fitting (I), which is soft-soldered to cooling rod (F). Thermometer well is integral with fitting (I). Cooling of rod (F) is accomplished by vaporizing a refrigerant such as liquid butane, propane, carbon dioxide, or some other liquefied gas in chiller (G). Refrigerant is throttled into chiller thru valve (H) and passes out at (J). Temperature of target mirror (C) is indicated by a calibrated mercury-in-glass thermometer (K) whose bulb fits snugly into thermometer well. Observation of dew deposit is made thru pressure-resisting transparent window (E).

Tests permit a determination with a precision (reproducibility) of ±0.2°F and with an accuracy of ±0.2°F when dew points range from room temperature to 32 F.

The *Dewtector* (UGC Instruments, Div. of United Gas Corp.), an optical device that helps to distingush dew points and to identify their nature, attaches to the conventional Bureau of Mines Type Dew Point Instrument. The device employs a 16× magnification of an illuminated mirror (by means of a self-contained light source). Identification and separation of the dew points of amine, glycol, hydrocarbon, water, and ice during a single run are possible with this attachment.

Automatic instruments made by the Surface Combustion Corp. and General Electric Co. use photoelectric tubes to detect the point of fog formation on a chilled mirror surface. These instruments have wide ranges (−90 F to +100 F) and are quite accurate if no substances are present with dew points higher than that of water vapor.

Stationary Psychrometer or Hygrometer. These tests depend upon the comparison of the temperatures registered by a dry bulb thermometer and a wet bulb thermometer in the gas stream (preferably moving at about 600 fpm) and, thereby, obtaining the relative humidity of the gas from Table 1-29. From the per cent saturation so obtained and the temperature, the water vapor content in gallons or per cent by volume can be computed using Tables 4-33, 1-27, or steam or psychrometric tables or charts.

It was reported[22] that reservoir-fed wicks result in erroneous wet bulb readings; it is believed that the wick furnishes a means of transferring heat from the water to the thermometer sensing element. Reductions in wet bulb depressions of as much as five or more degrees were found in testing air streams. Intermittent wet bulb wetting appeared to overcome this problem. The relative accuracy of humidity measuring instruments was also included.

Electrical Conductivity. A sudden increase in conductivity will take place when moisture begins to condense on a cooled nonmetallic surface like glass, and a sudden conductivity decrease will occur when it evaporates again. Warming and cooling cycles can be controlled automatically and the dew point temperature recorded if a thermocouple is attached to the cooled surface. An automatic dew point recorder using this principle is manufactured by the General Electric Co.

Instruments Using Hygroscopic Salts. Certain salts which absorb moisture, like lithium chloride, are used[15] to indicate, by electrical measurement, concentration of water vapor in gas to which they are exposed. Refrigeration systems are not required by these instruments, since the element temperature is at or above the gas temperature and the effects of other condensable gases or vapors are only slight. Changes in electrical conductivity of the lithium chloride coating, when it is exposed to varying water vapor concentrations, are measured in micro-amperes; the water vapor content must then be derived from data correlating conductivity and relative vapor saturation. Portable indicating models and recording instruments are available.

A recording instrument, developed by the Bureau of Mines, has been used to monitor water vapor concentration in high pressure natural gas transmission lines and in dehydration processing.[16] The Bureau of Mines hygrometer[15] unit consists of two fine platinum wires wound in parallel around a $\frac{3}{8}$ in. plastic tube. The tube is coated with a water-sensitive plastic film of water polyvinyl alcohol solution containing a small percentage of lithium chloride. A small alternating current passes thru the lithium chloride film on the element, varying with the amount of water vapor present. The electical measurement is translated by suitable means into a recorder reading. The instrument has a range from −20 F dew point to ambient temperature and from atmospheric to 2500 psig. A companion instrument that gives a continuous record of dew point under changing conditions in transmission lines up to 1500 psig is available.[23]

Another water vapor detector consists of a pair of platinum electrodes separated by an insulator, the surface of which is coated with a film of sulfuric acid and phosphoric acid mixture.[17]

The *American Instrument Co.* recorder covers a relative saturation range of 7 to 100 per cent with eight separate sensing elements (records in micro-amperes). The *Brown Instrument Co.* recorder covers a range of 35 to 93 per cent relative saturation with a single sensing element (records relative saturation directly). The *Foxboro Co.* recorder covers a range of 12 to 100 per cent relative saturation with a single element (records dew points). Each of these instruments can be converted into a controller. The American Instrument Co. controller provides for on-and-off switching of an electrical circuit of 15 amp at 115 v. The Brown Instrument Co. combines its recorder with its *Air-O-Line* control unit, using compressed air to operate a throttling type of control valve. The Foxboro Co. uses its *Stabilog* air controller to actuate a throttling type automatic valve. The Consolidated Electrodynamics Corp., Pasadena, Calif., produces a moisture monitor for determining water vapor concentration in the 0 to 100 ppm range. Moisture is quantitatively absorbed in a cell by phosphorus pentoxide. The current required to electrolize this moisture gives a measure of its quantity. This instrument will indicate, record, and telemeter continuously.

Direct Chemical Method.[24] Here, the absorption of water vapor by ethylene glycol is followed by titration of the absorbed water with Karl Fischer reagent (a methanol solution of iodine, pyridine, and sulfur dioxide). It was developed for the purpose of referee use where conventional methods based on dew point fail to give definite and reproducible results.

Refinery Supply Co. Moisture Tester. Water vapor content of the gas is determined by the color of cobalt bromide contained in the indicator. A description of the instrument and procedure is available.[25]

NITROGEN COMPOUNDS

Ammonia, cyanogen or hydrocyanic acid, and various oxides of nitrogen represent nitrogen-containing compounds most commonly encountered in manufactured fuel gases.

Determination of ammonia and cyanogen or hydrocyanic acid is primarily of interest in coke-oven gas purification practice. Oxides of nitrogen are found in carbureted water gas and in oil gas when blast products are included in the make gas.

Ammonia

Direct titration after absorption in a standard acid is the simplest procedure for determining ammonia.

Apparatus: Two gas-washing bottles, 250 to 500 ml capacity; $\frac{1}{10}$ cu ft wet test meter.

Reagents:

1. Standard Acid (1 ml = 0.006 g NH_3). Add 1.25 to 1.5 ml concentrated sulfuric acid to two liters of distilled water. Standardize by precipitation of sulfate as $BaSO_4$.
2. Standard Alkali. Dissolve about 1.8 g sodium hydroxide in two liters distilled water. Stir until thoroughly mixed. Standardize by titration with standard acid, with methyl orange or sodium alizarin sulfonate as indicator. Mix one gram sodium alizarin sulfonate with 100 ml water and filter. At end point, color changes from greenish yellow to light brown. Beyond end point, color changes to red.

Procedure: By means of burette, measure 15.0 ml of standard acid into each of two gas scrubbing bottles; add enough water to obtain good scrubbing and a few drops of indicator. Connect meter to outlet of bottles and their inlet to gas supply. Record meter reading. Pass gas thru train at rate of 0.5 to 0.6 cu ft per hr, for 2 to 5 hr; turn off gas and again record meter reading. Transfer solutions to 400 ml beaker and titrate with standard acid or alkali to end point.

Ammonia content of gas is obtained from test data by the expression:

$$X = 26.18(a - b)/C$$

where:

X = grains ammonia per 100 cu ft of gas at 60 F, 30 in. Hg, sat.
a = millequivalents of standard acid used
b = millequivalents of standard alkali used
C = volume gas passed, cu ft at 60 F, 30 in. Hg, sat.

Cyanogen and Hydrogen Cyanide

Hydrogen cyanide, HCN, and cyanogen, $(CN)_2$, are closely related compounds and their chemical reactions are similar. Therefore, they are often determined together and reported as HCN. Many determination methods have been devised to meet different requirements for speed and accuracy of results. Volumetric, gravimetric, and colorimetric methods are available.

Procedure: Assemble gas-purifying train to pass stream of gas thru three gas-washing bottles or test tubes and a meter. Place 25 ml of 10 per cent H_2SO_4 solution in bottle No. 1 to remove free NH_3. In each of the other two bottles, place 25 ml of concentrated H_2SO_4. Purge sample line and pass representative stream of gas at rate of about 1.0 cu ft per hr until measurable amount of HCN is absorbed; then disconnect train. Observing safe practices, transfer concentrated acid from last two bottles to a one liter Kjeldahl flask, containing about 250 ml of water. Rinse bottles and add washings to flask. Install flask in ammonia distillation apparatus; add excess of strong sodium hydroxide solution, distill and collect the ammonia in measured excess of $0.1N$ H_2SO_4. Titrate with $0.1N$ NaOH using methyl red as indicator.

When HCN reacts with strong sulfuric acid, ammonium acid sulfate is formed according to the equation:

$$HCN + H_2SO_4 + H_2O \rightarrow NH_4 \cdot HSO_4 + CO$$

HCN content of gas is obtained from test data by the expression:

$$G = 4.17A/V$$

where:
G = grains HCN per 100 cu ft of gas
A = ml of standard acid consumed
V = volume of gas sample, cu ft at 60 F, 30 in. Hg, sat.

This procedure is not recommended for carbureted water gas, since concentrated sulfuric acid reacts with unsaturated hydrocarbons present. It is applicable to blue gas. A number of other methods are available for HCN determination.[2]

Nitric Oxide

Presence of either nitrogen dioxide, NO_2, or nitric oxide, NO, in very small amounts, as low as 0.003 grains per 100 cu ft, in manufactured gases containing unsaturated hydrocarbons and oxygen, may result in formation of vapor phase gum particles. These minute particles may remain in suspension until consumers' appliances are reached, causing service difficulties.

Typical procedures are available[2] for determining nitric oxide in various manufactured gases (or in combustion products, etc., when the concentration is at about the same level). One setup is illustrated in Fig. 6-22.

Apparatus: Small pump (1) passes a stream of gas thru pressure regulator (2), maintaining 20 in. of water pressure; KOH solution scrubber (3) removes H_2S, CO_2, and much of water vapor; solid KOH tower (4) is for further purification; maleic anhydride-xylene sintered glass-type scrubber (5) removes interfering gum-forming compounds; and cock (6) maintains constant rate of flow of 0.4 cu ft per hr. Commercial butylene, containing butadiene catalyst from

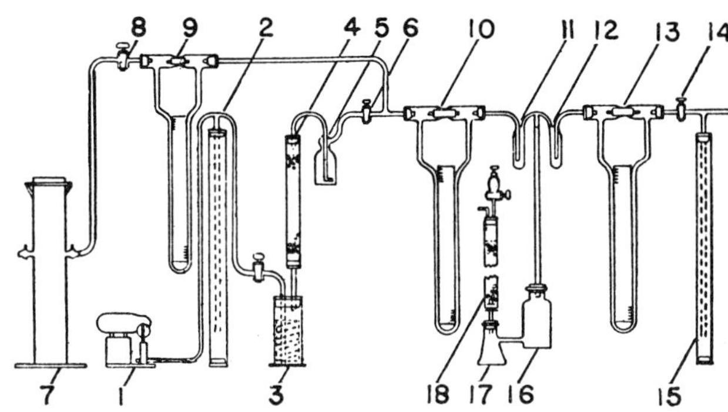

Fig. 6-22 Fulweiler's or U.G.I. spot tester for nitric oxide.[18]

vessel (7), passing thru cock (8), maintaining constant rate of 0.01 cu ft per hr as indicated by flowmeter (9), joins gas leaving cock (6). Mixture passes thru flowmeter (10) and trap (11) into reaction bottle (16). Oxygen from vessel (15) passes thru cock (14) maintaining constant flow of 0.4 cu ft per hr as indicated by flowmeter (13), thru trap (12) and into reaction bottle (16). Volume of (16) is selected to allow optimum reaction time, namely, 1.5 min. Gas-oxygen mixture from reaction bottle passes thru vessel (17) and glass-bead-packed absorption tower (18), where a measured amount of Griess reagent flowing downward scrubs ascending gas stream.

Reagents:

1. Griess Reagent. Mix 10 ml of sulfanilic acid solution with 10 ml of alpha-naphthylamine solution, and dilute with water to 100 ml.
2. Sulfanilic Acid Solution. Dissolve four grams of sulfanilic acid in 600 ml of boiling water; cool; add 200 ml of glacial acetic acid, and dilute with water to one liter.
3. Alpha-Naphthylamine Solution. Dissolve 2.51 g of alpha-naphthylamine in 200 ml of cold glacial acetic acid, dilute to one liter, and store in an amber colored bottle.
4. Sodium Nitrite Solution. Dissolve 0.027 g of $NaNO_2$ in 200 to 300 ml of water, and dilute to one liter.

Procedure: When a suitable color has developed in Griess reagent used, proceed as follows: Disconnect suction flask (17); wash tower (18) twice with 30 ml of water. Transfer to Nessler tube; add water to mark, let stand 10 min, and then compare with nitrite standards. A nitric oxide recorder based on this method is available.[19] The Guyer and Weber method described by Hollings[20] uses a scrubbing train of ammoniacal zinc sulfate, ten per cent sulfuric acid, flow gage or meter, potassium permanganate to oxidize NO to NO_2, and the Griess reagent.

Shaw-Schuftan.[21] This method (Fig. 6-23), employed for determination of nitric oxide in coke-oven gas, uses metaphenylenediamine (*mpd*) instead of the Griess reagent. The procedure is used to determine low concentrations of nitric oxide in coal gas and, with adequate modifications, the nitric oxide content of other gases.

Reagents: Sulfuric acid (25 ml of 95 per cent H_2SO_4 added to 75 ml of water); 30 per cent aqueous solution of KOH; large KOH pellets; butylene; metaphenylenediamine (1–3 diaminobenzene); sodium nitrite; activated charcoal (*Nuchar* powder); oxygen; and glacial acetic acid.

1. Metaphenylenediamine Solution. Dissolve 5 g *mpd* in hot water; add 25 ml glacial acetic acid and dilute to one liter.

This solution will keep indefinitely. Portions must be clarified with charcoal before use, since light and air slowly cause its darkening. To clarify, put 300 to 500 ml in a one liter beaker. Add about one heaping teaspoon of *Nuchar*. Heat to about 60 C, but avoid boiling. Stir to wet charcoal thoroughly. Filter thru free-flowing paper, using considerable portion at once to form filter mat. When filtrate has started to come thru clear, return it to original beaker and refilter whole, catching filtrate in clean glass-stoppered bottle. Solution should be water-white. Keep away from strong light.

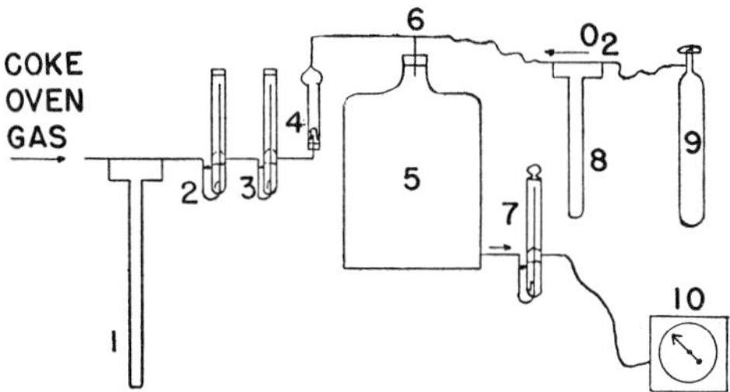

Fig. 6-23 Shaw apparatus for nitric oxide in coke-oven gas.[2] Legend: (1) flowmeter—regulates coke-oven gas; (2) absorber—10 per cent by volume H_2SO_4; (3) absorber—30 per cent KOH; (4) drying tube—KOH pellets (large size); (5) delay bottle—2 liters; (6) glass T; (7) absorber—*m*-phenylenediamine reagent; (8) flowmeter—oxygen; (9) lecture bottle—oxygen; and (10) wet test meter.

2. Standard Sodium Nitrite Solution. Stock solution of 0.300 g $NaNO_2$ dissolved in one liter of nitrite-free water.

3. Standard Solution. Exactly 10 ml of stock solution diluted to one liter with nitrite-free distilled water. A drop of chloroform acts as a preservative.

There is uncertainty concerning the absolute values of NO concentrations in different gases, determined by different procedures. All procedures appear to require empirical correction factors, but it seems established that a given method should successfully indicate NO values proportional to their actual concentration. Relations may also be established between NO concentrations in a given gas as indicated by different procedures.

REFERENCES

1. U. S. Steel Corp. *Methods of the Chemists of the United States Steel Corporation for the Sampling and Analysis of Gases*, 3rd ed. Pittsburgh, Carnegie Steel Co., 1927.
2. Altieri, V. J. *Gas Analysis & Testing of Gaseous Materials.* New York, A.G.A., 1945.
3. McBride, R. S. and Edwards, J. D. *Lead Acetate Test for Hydrogen Sulphide in Gas.* (Bur. of Standards, T41) Washington, G.P.O., 1914.
4. Shaw, J. A. "Rapid Determination of Hydrogen Sulfide and Mercaptan Sulfur in Gases and in Aqueous Solutions." *Ind. Eng. Chem., Anal. ed.* 12: 668–71, Nov. 15, 1940.
5. Sands, A. E. and others. *Determination of Low Concentrations of Hydrogen Sulfide in Gas by the Methylene Blue Method.* (Bur. of Mines RI 4547) Washington, G.P.O., 1949.
6. Field, E. and Oldach, C. S. "Determination of Hydrogen Sulfide in Gases." *Ind. Eng. Chem., Anal. ed.* 18: 665–7, Nov. 1946.
7. Littlefield, J. B. and others. *Detector for Quantitative Estimation of Low Concentrations of Hydrogen Sulfide.* (Bur. of Mines RI 3276) Washington, G.P.O., 1935.
8. A.S.T.M. "Standard Method of Test for Total Sulfur in Fuel Gases" (Standard D 1072-56) and "Tentative Method of Test for Sulfur in Petroleum Products Including Liquefied Petroleum Gas by Lamp Combustion" (Standard D 1266-57T) *ASTM Standards*, 1961, pt. 8. Philadelphia, 1962.
9. Lusby, O. W. "Quantitative Determination of Organic Sulfur." *A.G.A. Proc.* 1936: 752–6.
10. Hakewill, H. and others. *Determination of Total Sulfur in Natural Gas.* (Progress Rept. on A.G.A. Proj. CPR-25) Chicago, I.G.T., 1950.
11. Austin, R. R. "The Automatic Recording Titrator and Its Application to the Continuous Measurement of the Concentration of Organic Sulfur Compounds in Gas Streams." *A.G.A. Proc.* 1949: 505–15. Consolidated Electrodynamics Corp. *Titrilog—Operation and Maintenance Manual.* Pasadena, 1953.
12. Hakewill, H. and Rueck, E. M. "Tentative Procedures for Determining Individual Organic Sulfur Compounds in Gas." *A.G.A. Proc.* 1946: 529–38.
13. Ball, J. S. *Determination of Types of Sulfur Compounds in Petroleum Distillates.* (Bur. of Mines RI 3591) Washington, G.P.O., 1941.
14. Deaton, W. M. and Frost, E. M., Jr. *Bureau of Mines Apparatus for Determining the Dew Point of Gases Under Pressure.* (Bur. of Mines RI 3399) Washington, G.P.O., 1938.
15. Dunmore, F. W. "An Improved Electric Hygrometer for Measurement of Upper-Air Humidities." (Bur. of Standards RP 1265) *Natl. Bur. Std. J. Res.* 23: 701–14, Dec. 1939.
16. Deaton, W. M. "Instrument for the Measurement of Water Vapor in Natural Gases." *A.G.A. Proc.* 1953: 1062–66.
17. Weaver, E. R. "Determination of Water in Natural Gas by the Measurement of Electrolytic Conductivity." *A.G.A. Proc.* 1956: 476–82.
18 Fulweiler, W. H. "The Gum Problem—Recent Developments." *A.G.A. Proc.* 1933: 829–46.
19. Stackhouse, W. E. *Recording Apparatus for Determining Trace Concentration of Certain Impurities in Industrial Gas.* New York, A.G.A. Production and Chemical Conference, 1942.
20. Hollings, H. "Formation of Nitrogenous Gum During the Storage and Distribution of Gas." (I.G.E. Communication 147) *Inst. Gas Engrs. Trans.* 86: 501–77, 1936–37.
21. Shaw, J. A. "Determination of the Nitric Acid in Coke-Oven Gas." *Ind. Eng. Chem., Anal. ed.* 8: 162–7, 1936.
22. Flanigan, F. M. "Comparison of the Accuracy of Humidity Measuring Instruments." *ASHRAE J.* 2: 56–9, Dec. 1960.
23. Refinery Supply Co., 625 E. 4th, Tulsa, Okla.
24. Brickell, W. F. "Determination of Water Vapor in Natural Gas by a Direct Chemical Method." (PC-52-17) *A.G.A. Proc.* 1952: 753–60.
25. Natural Gas Processors Assn. "N.G.P.A. Liquefied Petroleum Gas Specifications and Test Methods." (Pub. 2140-62) *Plant Operations Test Manual.* Tulsa, Natural Gasoline Assn. of Am., 1962.

Note: Addendum covering tests for H_2S and total sulfur starts on the following page.

ADDENDUM

TESTS FOR H_2S AND TOTAL SULFUR*

Methylene Blue Test for H_2S in Natural Gas

The hydrogen sulfide content in natural gas as distributed is generally less than 1 grain per 100 SCF. Iodine titration, e.g., the U. S. Steel chemists' method, and other techniques are applicable when H_2S exceeds 0.3 grain per 100 SCF. The methylene blue colorimetric test (proposed for ASTM adoption) determines H_2S when it is present in concentrations not exceeding 0.3 grain per 100 SCF.

Experience at I.G.T. indicates that with a gas sample of 0.1 cu ft an accuracy of 2 per cent or better will be obtained when H_2S is present in the range from 0.3 down to 0.1 grain per 100 SCF. An accuracy of 0.002 grain per 100 SCF can be obtained below that level.

The proposed I.G.T. method of testing for H_2S in natural gas (methylene blue method) follows:

Scope. This method is intended for the determination of H_2S in natural gas containing not more than 0.3 grain H_2S per 100 SCF.

Summary of Method. A measured sample of the gas is bubbled thru zinc acetate solution. N,N-dimethyl-*p*-phenylenediamine in acid solution and ferric chloride are added to the zinc acetate solution and react with zinc sulfide to form the dye methylene blue. The methylene blue is determined by measurement of optical transmittance of the solution at a specified wave length or wave length range.

Apparatus. The apparatus consists of the following items:

(*1*) *Absorption flask* (shown at the right in Fig. 6-23.1) constructed of a 50-ml volumetric flask with enlarged neck (Kohlrausch flask).

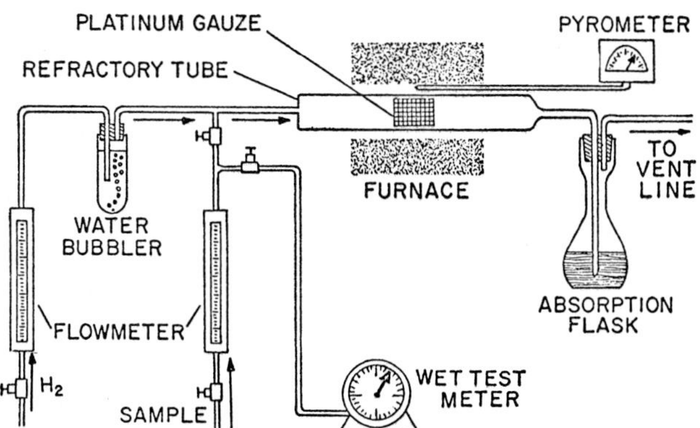

Fig. 6-23.1 Apparatus for the hydrogenation method for total sulfur determination.

(*2*) *Photoelectric photometer or spectrophotometer.* Any instrument designed for the requirements of practical absorption photometry and suitable for measurements at approximately 745 mμ may be used. A glass color filter of the low-wave length cutoff type having less than 37 per cent transmission below 567 mμ and more than 80 per cent transmission from 608 to 750 mμ, or an interference filter with the peak of its band pass at any wave length between 620 and 750 mμ and having less than 0.5 per cent transmission below 570 mμ, is used with instruments in which the spectral range is isolated by filters. The photometer should be equipped with matched absorption cells: two or more with a 20- to 25-mm path length, and two or more with a 40- to 50-mm path length.

(*3*) *Wet test meter*, 0.05 cu ft per revolution, or *calibrated aspirator bottle* of approximately 0.1 cu ft capacity.

Reagents. The following reagents are required:

(*1*) *Arsenious acid, standard solution (0.01N).* Dissolve 0.4945 g of As_2O_3 (Bureau of Standards) in 15 ml of 1N solution of NaOH. Transfer to 1-liter volumetric flask, neutralize, and make slightly acid to litmus paper with 1N solution of HCl. Add water to volume.

(*2*) *Ferric chloride solution.* Dissolve 2.7 g of ferric chloride ($FeCl_3 \cdot 6H_2O$) in 50 ml HCl (sp gr 1.19) and dilute to 100 ml with water.

(*3*) *Hydrogen sulfide.* Cylinder gas, 99.5 per cent minimum purity.

(*4*) *Iodine, standard solution (0.01N).* Dissolve 1.27 g of iodine (I_2) and 4 g potassium iodide (KI) in water and dilute to 1 liter. Store in a cool place in a dark-colored, glass-stoppered bottle. Standardize as follows: Pipet 25 ml of standard As_2O_3 solution into a 250-ml Erlenmeyer or iodine flask, add 25 to 50 ml of water, 1 g of $NaHCO_3$, and 5 ml of starch solution. Titrate with iodine solution until the appearance of the first blue color. The endpoint will be sharper if the solution is cooled to 5 C or lower.

(*5*) *Methylene blue* ($C_{16}H_{18}N_3SCl$). Certified biological staining reagent.

(*6*) *N,N-dimethyl-*p*-phenylenediamine sulfate, purified (diamine solution).* Dissolve 0.11 g in 100 ml of 2:1 H_2SO_4 in a brown bottle. Make a fresh solution every two weeks.

(*7*) *Sodium thiosulfate, standard solution (0.01N).* Dissolve 2.5 g of sodium thiosulfate ($Na_2S_2O_3 \cdot 5H_2O$) in 1 liter of freshly boiled and cooled water in a sterile glass bottle. If sulfur precipitates during preparation or upon subsequent use, discard the solution and prepare a new one. Standardize as follows: Pipet 25 ml of standard iodine solution into a 250-ml Erlenmeyer or iodine flask, add 25 to 50 ml of water, and 5 ml of HCl. Titrate with $Na_2S_2O_3$ solution until the iodine solution fades to a pale yellow. Add 5 ml of starch solution and titrate to faint blue or colorless. The endpoint will be sharper if the solution is cooled to 5 C or lower.

(*8*) *Starch solution.* Make a paste of 2 g of soluble starch and 10 mg of mercuric iodide (HgI_2), used as a preservative, with a little cold water. Slowly add to 1 liter of boiling water. Cool and transfer to a glass-stoppered bottle.

(*9*) *Sulfuric acid (2:1).* Place one volume of distilled water in an Erlenmeyer flask and add, in small portions and with cooling, two volumes of H_2SO_4 (sp gr 1.84). Store in a tightly stoppered bottle to prevent absorption of moisture and change of concentration.

(*10*) *Zinc acetate solution (2 per cent).* Dissolve 23.9 g of zinc acetate [$Zn(C_2H_3O_2)_2 \cdot 2H_2O$] in water to make 1 liter, and add approximately three drops of glacial acetic acid.

---

* Mason, D. M. "Better H_2S Analysis in Natural Gas." *Hydrocarbon Process Petrol Refiner.* 43 (No. 10): 145–150, Oct. 1964.

Calibration. Calibrate the photometer by either of the following methods:

(*1*) *Methylene Blue Reagent Method.*

(a) Pipet 200 ml of 2:1 H_2SO_4 and 40 ml of $FeCl_3$ solution into a 2-liter volumetric flask and dilute to the mark with water while cooling.

(*b*) Weigh out 0.25 ± 0.01 g of methylene blue to the nearest 0.5 mg, and dissolve in the H_2SO_4–$FeCl_3$ solution. Transfer the solution to a 500-ml volumetric flask and bring up to volume with the H_2SO_4–$FeCl_3$ solution. Pipet 5 ml into a 500-ml volumetric flask and bring up to volume with the H_2SO_4–$FeCl_3$ solution. This solution contains approximately 0.005 g methylene blue per liter.

(c) Prepare calibration solutions by diluting 50, 35, 25, 20, 15, 10, and 5 ml of 0.005 g per liter methylene blue solution to 100 ml in volumetric flasks with the H_2SO_4–$FeCl_3$ solution.

(d) If a spectrophotometer is used, determine the wave length setting for peak absorbance near 745 mμ.

(e) Read the absorbency of the calibration solutions against the H_2SO_4–$FeCl_3$ solution in the reference cell with each set of cells used in the analysis. Read at the peak wave length setting if a spectrophotometer is used or with the specified color filter if a filter photometer is used.

(f) Calculate the concentration of methylene blue in each calibration solution according to the assay value of the methylene blue given on the label.

(g) Calculate the equivalent H_2S content for each methylene blue calibration solution according to the equation:

$$H_2S \text{ concentration, grains per 50 ml} = (0.157)(\text{methylene blue concentration, g per liter})$$

(h) On rectangular coordinate paper plot the equivalent H_2S content in grains H_2S per 50 ml against absorbency.

(i) Prepare a new calibration curve whenever a new lot of 2:1 H_2SO_4 is prepared and used.*

(*2*) *Hydrogen Sulfide Method.*

(a) Place about 500 ml water in a 750-ml glass-stoppered Erlenmeyer flask. Dissolve 15 ml of H_2S, measured in a syringe, length of glass tubing, or gas burette, in the water.

(b) Pipet 25 ml of standard 0.01N I_2 into each of two iodine flasks, add 2 ml concentrated HCl, and cool in an ice bath. Fill a 500-ml volumetric flask to below the neck with zinc acetate solution. Pipet 10 ml of the H_2S solution into the 500-ml flask, mix, and pipet 50 ml of the H_2S solution into each of the iodine flasks and shake.

(c) Back-titrate each of the mixtures in the iodine flasks with standard 0.01N $Na_2S_2O_3$ to a pale yellow, add 5 ml starch solution, and continue the titration until the first appearance of the blue color.

(d) Bring the 500-ml volumetric flask to volume with zinc acetate reagent, and mix. Pipet aliquots of 40, 25, 10, 5, 2 and 1 ml of the resulting ZnS solution into six long-necked 50-ml volumetric flasks and bring the last five to about 10 ml below the neck with zinc acetate solution. Cool these standard solutions to 5 C and develop the methylene blue color by pipetting 5 ml of diamine solution into each, mixing gently, then adding 1 ml of $FeCl_3$ solution and again mixing gently. Remove the solutions from the ice bath and let stand 15 to 30 min, then bring up to volume with zinc acetate solution.

(e) Prepare a reference solution in a 50-ml volumetric flask with 5 ml diamine solution, 1 ml $FeCl_3$ solution, and the remainder zinc acetate solution. Check the reference by reading its per cent transmittance in a 20- or 25-mm cell, with water as the reference. The transmittance of the blank must be 98 per cent or better.

(f) Place the reference solution in one of a pair of matched cells and the sample solution in the other. Obtain the absorbency of the sample at the peak wave length if a spectrophotometer is used. Obtain the absorbency with the proper color filter if a filter photometer is used.

(g) Calculate the average net milliequivalents of iodine, using the formula:

$$I_{Calib} = (\text{ml } I_2)N - (\text{ml } Na_2S_2O_3)n$$

where: N = normality of the iodine solution
n = normality of the $Na_2S_2O_3$

(h) Calculate the concentration of the H_2S in the six standard solutions by the formula:

$$H_2S \text{ concentration, grains } H_2S \text{ per 50 ml} = (0.2630/50D)I_{Calib}$$

where: D = dilution factor = 50/(vol ZnS aliquot)

(i) On rectangular coordinate paper plot H_2S content in grains H_2S per 50 ml against absorbency.

Procedure. Proceed as follows:

(1) Place 30 ml of zinc acetate solution in the 50-ml absorption bottle. Insert the stopper with a bubbler tube and connect the inlet to the sample container with aluminum or polyvinylidene chloride tubing and a minimum of rubber or polyvinyl chloride tubing. Connect the outlet of the absorption bottle to a wet test meter or to a calibrated aspirator bottle.

(2) Note the meter reading, then pass 0.1 cu ft of sample gas† thru the absorber at a rate not exceeding 1 cu ft per hr. Read the meter and obtain an average meter temperature to the nearest 1 F and a barometer reading to the nearest 0.1 in. for the period of the absorption.

(3) Disconnect the absorber, add 10 ml of zinc acetate solution, and place in an ice bath to cool. When it has cooled to 5 C, pipet 5 ml of diamine solution into the absorber and stir gently. Add 1 ml of ferric chloride solution and again stir gently. Draw the solution in and out of the bubbler tube. Remove the absorber from the ice bath and allow to stand for

* The spectrum of methylene blue changes with acid concentration. Error from this source should be less than 1 per cent if the concentration of the 2:1 H_2SO_4 is within ±1.5 per cent of that of the acid used in the calibration.

† A larger sample may be used to obtain greater accuracy when the H_2S concentration is low. However, to avoid absorbency readings over 0.8, the product of H_2S concentration in grains per 100 SCF and sample volume in cubic feet should not exceed 0.03.

15 to 30 min. Remove the bubbler tube, rinse it with zinc acetate solution, bring the volume in the flask up to the mark with zinc acetate solution, and mix.

(4) Prepare a reference solution containing the same amount and kind of reagents as the sample solution.

(5) Place the reference solution in one of a pair of matched cells and the sample solution in the other. Obtain the absorbency of the sample at the peak wave length if a spectrophotometer is used. Obtain the absorbency with the proper color filter if a filter photometer is used. If the absorbency is above 0.8 when read with 40- to 50-mm cells, redetermine the absorbency in cells of shorter path. If the absorbency is below 0.3 when read with cells of 20- to 25-mm path length, redetermine the absorbency in cells of longer path.

Calculation. Proceed as follows:

(1) Convert the absorbency reading for the sample solution to grains H_2S per 50 ml by means of the calibration curve.

(2) Calculate the concentration of H_2S in the sample as follows:

$$H_2S \text{ content, grains per 100 SCF} = 100H/VF$$

where:

H = grains H_2S per 50 ml
V = observed volume of gas sample, cu ft
F = meter factor to correct gas volume to standard conditions from average barometric pressure and average meter (or aspirator bottle) temperature, from Table 1-38

Interferences. A number of materials that may be present in natural gas only in trace amounts are nevertheless possible sources of interference with the methylene blue colorimetric test. Although their concentrations may be low, these materials may be present in amounts as great as, or considerably greater than, that of H_2S. Accordingly, tests were made with a number of such compounds to determine whether they would cause spuriously high or low results for H_2S. The following compounds were tested, each at one concentration within the range 0.5 to 5 grains of S per 100 SCF:

Methyl mercaptan	*n*-Amyl mercaptan
Ethyl mercaptan	Methyl sulfide
n-Propyl mercaptan	Thiophane
n-Butyl mercaptan	Methyl isopropyl sulfide
Isobutyl mercaptan	Methyl disulfide
sec-Butyl mercaptan	*tert*-Butyl disulfide
tert-Butyl mercaptan	

A mixture of monoethanolamine, diethanolamine, methanol, benzene, toluene, and ethylene glycol was also tested. The effect of any of these compounds was less than 0.001 grain per 100 SCF, whether or not H_2S was present. It was concluded that interference from constituents of natural gas was negligible.

Total Sulfur Determination in Natural Gas

The present ASTM combustion method for total sulfur determination is suitable only for concentrations above 1 grain sulfur per 100 SCF. A combustion method used at I.G.T. has an accuracy (average deviation) of 0.1 grain per 100 SCF, obtained either with gravimetric finish or a faster turbidimetric finish.* It is likely that better accuracy could be obtained with some of the newer titration methods and micro- or semimicrotechnique.

Other methods are based on the conversion of organic sulfur to H_2S by hydrogenation. Sulfur in synthesis gas has been determined at I.G.T. by passing the gas with the added hydrogen over magnesium oxide at 900 C, and determining the H_2S by iodine titration or, in the low-sulfur range, by the methylene blue procedure.* The Bureau of Mines has used the Lusby platinum spiral method† for the conversion of organic sulfur in synthesis gas to H_2S; this is followed by the methylene blue analysis. Both the platinum spiral and the hot magnesium oxide methods have been tested briefly on natural gas at I.G.T., with inconclusive results. In France, however, a platinum catalyst in a small tube furnace is reported to be used in a hydrogenation method for total sulfur in natural gas.‡ In addition to the advantage of high sensitivity (and, hence, small sample size), total sulfur determination with the methylene blue procedure will be particularly attractive when it is also adopted for determination of H_2S per se.

An investigation of a hydrogenation method with methylene blue finish was conducted at I.G.T. Platinum gauze, heated in a small tube furnace, was selected as a catalyst likely to operate indefinitely at 900 C without the deterioration characteristic of finely divided forms of platinum. In an analysis, the sample flowing at a measured rate is mixed with four parts of humidified hydrogen and passed first over the catalyst and then thru a zinc acetate solution (Fig. 6-23.1). The sulfide solution is analyzed by the methylene blue procedure, as in the H_2S method.

Results of analyses on three samples prepared in the laboratory are given in Table 6-3.1. To obtain reasonably accurate analysis of these samples by combustion, they were prepared with sulfur contents of about 1 grain per 100 SCF. Since this content is higher than the hydrogenation method is designed to handle, it was necessary to increase the amount of absorbing solution and decrease the length of the run.

Table 6-3.1 Reproducibility and Accuracy of Total Sulfur Determination

Sulfur compounds	S content, by combustion, grains S/100 SCF	Hydrogenation: S content, avg grains S/100 SCF	Hydrogenation: No. of analyses	Hydrogenation: Standard deviation
Thiophane	1.05	0.99	7	0.03
Mercaptans	0.95, 0.96	1.02	4	0.003
Sulfides	0.85, 0.95	1.02	6	0.05

* Mason, D. McA., and Hakewill, H., Jr. "Identification and Determination of Organic Sulfur in Utility Gases." *Inst. Gas Technol., Res. Bull. 5.* Jan. 1959.

† Sands, A. E., and others. *Organic Sulfur in Synthesis Gas: Occurrence, Determination and Removal.* (Bur. of Mines RI 4669) Washington, G.P.O., 1950.

‡ Duchemin, M., Direction des etudes et techniques nouvelles, Gaz de France. Private communication. May 8, 1961.

The reproducibility of the analyses, as shown by a pooled standard deviation of 0.04 grain sulfur per 100 SCF, or about 4 per cent, is considered satisfactory. The maximum difference between combustion analysis and the average of the hydrogenation analyses is 0.2 grain per 100 SCF, or about 12 per cent. In view of the good reproducibility of the hydrogenation method and the difficulties in obtaining sufficient accuracy by the combustion method, it appears that the differences may not be highly significant as a measure of the accuracy of the hydrogenation method. Further work is required to establish the accuracy of this method more precisely.

The presence of higher hydrocarbons in quantities greater than normally found in dry natural gas may cause difficulty in the analysis, since they may deposit carbon, which is known to retain sulfur at high temperatures. Moisture in the hydrogen during the analysis period tends to prevent carbon deposition and should remove carbon deposits during the purge period. Since the natural gas used in the tests described contained little, if any, of the higher hydrocarbons, additional tests were conducted. For these tests a sulfur-free natural gas sample containing 0.5 per cent *n*-pentane, 0.2 per cent benzene, and 0.05 per cent each of toluene, *n*-heptane, and *iso*-octane was prepared. Two analyses were made on samples consisting of nine parts of this high-hydrocarbon sample with one part of the thiophane sample listed in Table 6-3.1. Results of the two analyses agreed within 2 per cent with the amount found in the thiophane sample by direct test. In these and other tests with the higher hydrocarbon samples, no H_2S was formed when, after the test, humidified hydrogen was passed thru the tube. Thus, it appears that no interference should occur with the amounts of higher hydrocarbons found in dry natural gas.

Chapter 6

Determination of Carbon Monoxide in Minute Concentrations

by C. W. Wilson

INTRODUCTION

Carbon monoxide is a colorless, almost odorless, highly toxic, nonirritating gas. Although a normal constituent (1 to 25 per cent) of many manufactured gases, CO does *not* occur in natural gas. It may be formed by incomplete combustion of any fuel containing carbon in the solid, liquid, or gaseous state, in which case it would exist with other gases in the products of combustion. Some gases, such as aldehydes and alcohols, which often accompany CO formation, are very odorous. If combustion is complete, no trace of CO will be present. However, very small CO quantities are permitted under the American Standard Approval Requirements for most gas appliances without being judged "incomplete" combustion. Instruments and methods of test used for detection of CO must be capable of making analysis accurate to hundredths or thousandths of one per cent.

Maximum Allowable Concentration of Carbon Monoxide in Air Breathed

Carbon monoxide can enter the body only thru the respiratory system. It acts as an asphyxiant[1,2] by combining with the hemoglobin of the blood to exclude oxygen. This combination is more stable than that of hemoglobin with oxygen, and affinity of CO for hemoglobin is much greater than that of oxygen. It is estimated that one part of CO in the blood stream requires 210 parts of oxygen to replace it.

If a person is exposed to carbon monoxide, the amount absorbed in his blood will depend upon several factors. These include the CO concentration in the air, time of exposure, number and length of periods of breathing fresh air between exposures, subject's degree of physical activity, and his physical health.

Table 6-4 lists effects of various percentages of CO in air on human beings. Figure 6-24 shows the effects of exposure time on per cent of blood saturation and on resulting physical impairment.

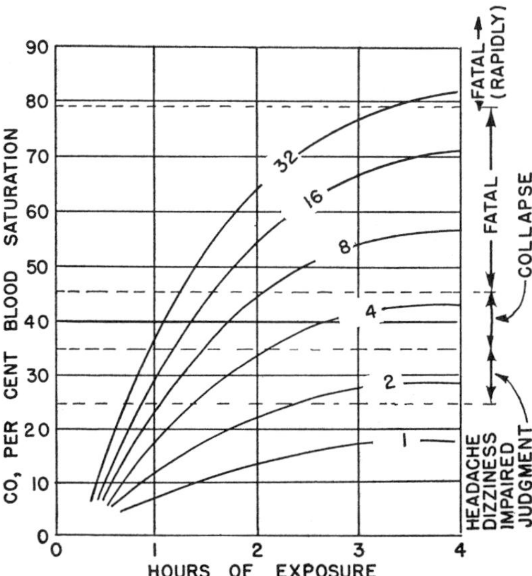

Fig. 6-24 Carbon monoxide absorption by human blood.[3] Curves indicate parts of CO in 10,000 parts of air, i.e., hundredths of a per cent.

Table 6-4 Physiological Effect of Carbon Monoxide[2]

CO concentration in air		
Parts per million	Per cent	Effects
100	0.01	Concentration allowable for several hours exposure. Dept. of Labor, New York State Health Bulletin, 1930, states that this concentration can be tolerated indefinitely. (See * footnote to Table 1-30.)
400 to 500	0.04 to 0.05	Concentration which can be inhaled for one hour without appreciable effect.
600 to 700	0.06 to 0.07	Concentration causing barely appreciable effect after 1 hr exposure.
1000 to 1200	0.10 to 0.12	Concentration causing unpleasant but not dangerous symptoms after 1 hr exposure.
4000 and over	0.40 and over	Fatal concentration in exposures less than 1 hr.

All gas appliances tested for compliance with *American Standard Approval Requirements* and bearing the Approval Seal of the American Gas Association have met very rigid tests for combustion, together with many other safety tests. For example, for domestic gas ranges[4] the following requirement applies:

"A gas range shall produce no carbon monoxide. This requirement shall be deemed met when a concentration not in excess of 0.01 per cent is produced in a room of 1000 cu ft content with four air changes occurring during the combustion of an amount of gas liberating 60,000 Btu."

This maximum permissible concentration is within the limits of tolerance for an 8 hr exposure with no perceptible effects. *American Standards* for other types of domestic gas appliances contain provisions limiting carbon monoxide during test to less than one-half the amount listed above. Test details are given in standards sponsored and published by the American Gas Association and approved by the American Standards Association. To date no solid or liquid fuel-burning domestic appliances or combustion engines have been subjected to such limitations on completeness of combustion.

METHODS FOR TRACE DETERMINATION

In volumetric analysis of fuel gases for CO, an accuracy of approximately 0.2 per cent is considered acceptable. Since the limit of CO concentration in air which will produce no perceptible effect is 0.01 per cent, analytical equipment accurate to more than 0.01 per cent must be used. Instruments sensitive and accurate to 0.001 per cent are sometimes required. Most of the instruments described in this chapter are capable of detecting or determining the presence of CO in air or flue gases in hundredths of one per cent or less.

Stationary Laboratory Apparatus

Iodine Pentoxide Apparatus. This apparatus, as perfected from the original Bureau of Standards model, was used at the A.G.A. Laboratories in Cleveland, Ohio, for over 25 years (superseded by the M.S.A. *Lira*) to determine carbon monoxide in air and in various gaseous mixtures (Fig. 6-25).[5,6] It is based on oxidation of CO to CO_2 in a gaseous sample by heated iodine pentoxide which liberates iodine vapor; this liberated iodine is then determined by titration with sodium thiosulfate, using starch as indicator. A precision of 0.002 per cent may be obtained, and a determination may be made in 20 to 40 min. Variations of the iodine pentoxide method for analysis of different mixtures of gases, using special purification materials such as liquid air, dry ice, or absorbing chemicals ahead of the iodine pentoxide, have been reported.[7,8]

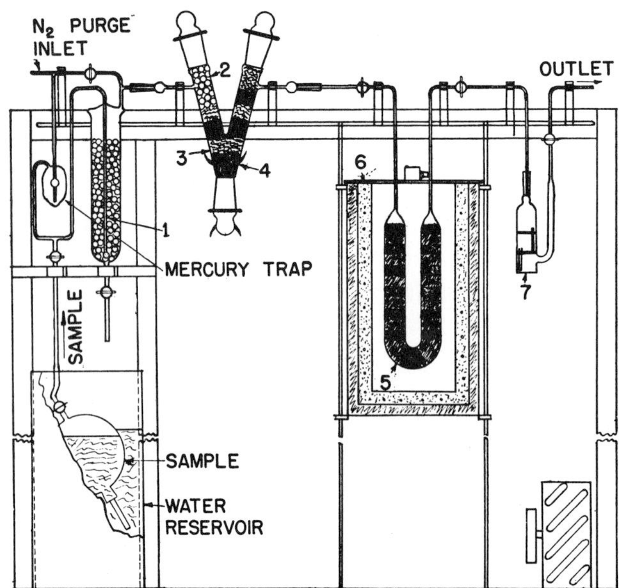

Fig. 6-25 Iodine pentoxide apparatus for analysis of CO in flue gas. (Numbered items mentioned in text.)

Apparatus: (See Fig. 6-25.) Glass reagent train consisting of chromic acid scrubbing tower (1); Y-tube containing potassium hydroxide (2) to absorb acid, phosphorus pentoxide to remove all moisture (3), and layers of glass wool (4); U-tube partially filled with I_2O_5 and glass wool, for oxidation of CO in sample (5), enclosed in electric heater (6) which maintains I_2O_5 at 150 C, and Gomberg bulb containing potassium iodide solution for retention of liberated iodine (7). Connections made by glass tubing with tapered joints are ground stopcocks.

Reagents:

1. Iodine Pentoxide. I_2O_5 of high purity, known as iodic anhydride, is used. After heating U-tube and pentoxide at temperature of 205–215 C for two days, with air or nitrogen purge, temperature is reduced to 150 C, and purging is continued two more days. Addition of starch to KI solution in Gomberg bulb should show no trace of I_2 after 2 to 3 hr purging with nitrogen. I_2O_5 so conditioned should be usable for six months to one year of daily use with samples containing reasonably low CO percentages.
2. Chromic Acid. Concentrated sulfuric acid (cp) is saturated with potassium dichromate.
3. Potassium Iodide Solution. Solution is made of ten per cent by weight of KI and 90 per cent boiled distilled water. Solution decomposes in sunlight and must be kept in dark. Fresh solution should show no trace of free iodine and with proper care should remain satisfactory for use for several weeks. Preparation of fresh solution every few days has been found preferable.

Starch Indicator: Dissolve 2 to 3 g of potato starch in a few cu cm of cold water, add to 200 cu cm hot distilled water just after boiling ceases, then cool to room temperature. Solution should preferably be prepared fresh daily. If not, distinct change from blue to white will not be obtained when titrating iodine. Best results are obtained when titrating solutions of high iodine concentrations by delaying addition of indicator until most of yellow color has been discharged with thiosulfate.

Sodium Thiosulfate: Approximately 0.001 normal solution should be used, 0.2482 g per l. Before standardizing, it should be allowed to stand about four weeks to attain equilibrium. Strength will then remain constant several months. A satisfactory procedure is to make up a quantity of 0.1N solution and periodically to dilute a portion as necessary to approximately 0.002N. After storing for two to three weeks, solution has weakened to about 0.001N. It is standardized just before using, and then checked for normality at intervals.

Standardization is made against pure iodine solution, using starch as indicator.

Pure, doubly sublimed iodine is weighed accurately to give approximately 0.001N solution. Because of its hygroscopic and volatile nature, special precautions are needed. About 2 g potassium iodide and a few drops of water may be placed in weighing bottle which is then weighed. About 0.2 g of iodine is then added and weighing repeated, with difference in weight giving exact weight of iodine. Solution is then made up to one liter.

Typical calculation to convert thiosulfate into carbon monoxide terms follows:

If 0.0974 g I_2 is dissolved per liter, and titration of 10 cu cm I_2 = 8.37 cu cm $Na_2S_2O_3$, then:

$$\begin{aligned} 1 \text{ cu cm } Na_2S_2O_3 &= 1.195 \text{ cu cm } I_2 \\ &= 1.195 \times 0.0000974 \text{ g } I_2 \\ &= 0.0001164 \text{ g } I_2 \end{aligned}$$

By appropriate factors* determine the CO (in cu cm) equivalent of $Na_2S_2O_3$:

$$\begin{aligned} 1 \text{ cu cm } Na_2S_2O_3 &= \frac{30{,}701{,}720}{64{,}674.59} \times 0.0001164 \text{ cu cm CO} \\ &= 0.0552 \text{ cu cm CO at 60 F and 760 mm Hg} \end{aligned}$$

Procedure: Apparatus is thoroughly purged with inert CO-free nitrogen. Sample for analysis is drawn by water displacement into glass bottle of accurately known volume of approximately 500 cu cm. Bottle is then immersed in water container. After reaching equilibrium temperature, sample is drawn into apparatus by vacuum. Purging with nitrogen is continued until sample is partially withdrawn and is then shut off until remainder of sample is withdrawn. Purge is then resumed.

Note text under *Apparatus.* Drawing sample thru apparatus at proper rate requires about 7 min. Purging with nitrogen for another 23 min is usually sufficient. If sample contains 0.1 per cent CO or more, a longer purge may be needed.

At end of purge period, stopcocks at inlet and outlet of Gomberg bulb are closed and bulb is removed. Contents are poured into small Erlenmeyer flask, bulb washed with distilled water, and washings added to flask.

Starch solution is added, giving immediate blue color if iodine is present. Solution is then titrated with standardized thiosulfate until blue first disappears. Volume of CO in sample is calculated from volume of thiosulfate added.

M.S.A. Carbon Monoxide Recorder.[7,9] This recorder was originally developed by the Bureau of Mines for detecting and controlling CO to less than four parts per 10,000 in vehicular tunnels. It is manufactured by the Mine Safety Appliances Co., Pittsburgh, Pa., and has found wide use in tunnels and garages, in testing flue gases from gas appliances, in manufacturing processes where CO is involved (as in hydrogenation and ammonia synthesis), and in the general testing of gases.

The recorder (Fig. 6-26) employs a motor-driven exhaust pump which draws a continuous test sample. A measured (orifice meter) part of the sample flows thru a charcoal scrubbing tower and a CO detector cell containing *Hopcalite.* Heat liberated in this cell during oxidation of CO to CO_2, measured by a differential thermopile, is indicated by a recording potentiometer. The potentiometer may be equipped to operate warning signals and ventilation controls.

Hopcalite is a mixture consisting mainly of copper oxide and manganese dioxide. Certain mixtures will oxidize CO completely at room temperature, while others require a constant bath temperature as high as 212 F. By controlling temperature and mixture, *Hopcalite* will preferentially oxidize CO without oxidizing hydrocarbons, hydrogen, or other combustible gases which might be present.

The instrument can be constructed to give any range of sensitivity desired. It may be calibrated to cover a range of 0 to 200 ppm (0.02 per cent), permitting a sensitivity of 1 ppm. For a wider range, the instrument can cover a range of 0 to 0.1 per cent or greater, as desired.

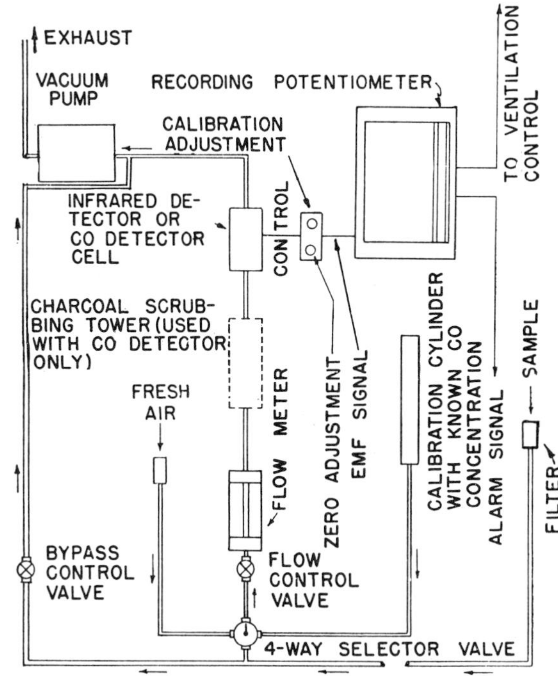

Fig. 6-26 Schematic of a CO recording and alarm system. (Mine Safety Appliances Co.)

Lira Carbon Monoxide Recorder.[10] This instrument (Fig. 6-15b) is an infra-red gas analyzer designed for determination of CO in flue gases in very small quantities—0 to 0.1 per cent. It may be used as a portable instrument if a source of electricity is provided.

Lira stands for Luft-Type Infra-Red Analyzer. It is sensitive to 0.001 per cent CO and will determine and record CO concentration in a flue gas sample in about 1 min. Because of its accuracy, ease of operation and servicing, and speed of analysis, one instrument has replaced eight iodine pentoxide instruments at the A.G.A. Laboratories, where as many as 200 CO determinations per day are made. After many check tests against the iodine pentoxide apparatus, on known and unknown samples, the conclusion was reached that the *Lira* CO recorder was just as accurate and probably more dependable. Periodically, it requires checking with known samples of CO, about twice per day. National Bureau of Standards certifies CO samples to limits of ±0.0003 per cent.

Precautions: Solids in the sample, e.g. smoke, absorb infra-red radiation better than gases. In this regard, smoke accumulation on the lenses tends to change the calibration. Fog, on the other hand, generally evaporates at the *Lira* operating temperature of 120 F. Since the *Lira* measures the partial pressure of a component of a gaseous mixture, calibra-

* For derivation of factor 30,701,720/64,674.59 see Ref. 5.

tions must be checked whenever a marked change in barometric pressure occurs (particularly with the CO_2 *Lira*).

Portable Laboratory or Field Apparatus

Colorimetric Carbon Monoxide Tester or Indicator. Two companies[11] have developed such devices, based on use of a detector indicator tube developed by the National Bureau of Standards. They are fast (1 to 2 min for a test), simple to operate, weigh less than ½ lb, and indicate CO in amounts from 0.001 to 0.1 per cent.

These instruments are based on color change of a chemically treated yellow silica gel contained in a glass tube thru which a sample under test is aspirated in measured volume (bulb size) and at a set rate (orifice size). CO concentration is estimated by comparing resulting gel color with a standard color chart. Figure 6-27 shows one make of instrument. Another has a separate color scale card; its glass color tube is merely inserted in a rubber tube attached to an orifice and aspirator bulb.

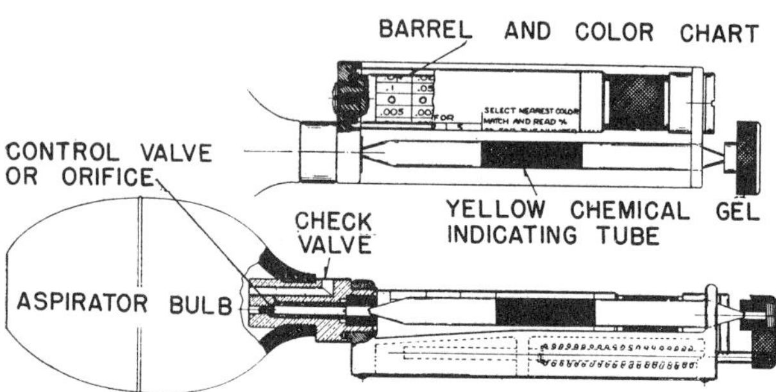

Fig. 6-27 Two views of a colorimetric CO tester. (Mine Safety Appliances Co.)

Indicating gel, light yellow in color, is made by impregnating purified silica gel with a sulfuric acid solution of ammonium molybdate and palladous sulfate, to form palladous silicomolybdate. This compound is reduced by CO and is changed to an intense molybdenum blue. A layer of indicating gel is placed in a glass tube between layers of white guard gel at its ends. These guard layers serve to protect the indicating gel against moisture and partially against action of other reducing gases. After sealing, tubes remain sensitive for a year or more if protected against strong light.

To use, tips of the glass tube are broken off and tube is inserted in the instrument. Sample is then aspirated thru the indicator tube by means of a 60 ml rubber bulb. Resulting color change of indicating gel from yellow to shades of green depends upon *concentration of CO present and number of bulb squeezes;* i.e., intensity of test color is a function of the *product* of CO concentration and time, at a definite rate of sample aspiration. Thus, percentage concentration is estimated by comparing the test color with a series of color standards for a particular number of aspirated sample volumes.

A scrubber attachment is available for removing certain interfering gases when determining low CO concentrations in some flue gases.

Supersensitive indicator tubes may be prepared for laboratory use.[7] They are sensitive to ±0.002 per cent CO in air, in the range of 0 to 0.01. In contrast to sealed tubes for field use, these supersensitive tubes must be prepared and used the same day.

M.S.A. Carbon Monoxide Detector and Detector Ampoule. This detector is based on color change of a chemical (activated iodine pentoxide) from gray-white to a shade of green, varying in intensity with the amount of CO present. It requires 10 sec to operate, with ten squeezes of the aspirator bulb, and has a range of CO indication from 0.1 to 1.0 per cent. Squeezing the bulb forces the sample thru a cylinder of charcoal, which cleanses the sample before it enters the detector tube.

The detector ampoule is not as sensitive or as fast as the CO tester or indicator, but it will detect CO concentrations of 0.06 per cent or more. Each ampoule consists of a chemical solution in a sealed glass vial surrounded by absorbent cotton. The ampoule is crushed and suspended in the atmosphere for at least 10 min. The saturated cotton turns gray in presence of CO. Degree of concentration is determined by comparison with a color chart card.

M.S.A. Carbon Monoxide Indicator. This instrument may be run with a hand pump, a 110 v a-c pump, or a 6 v d-c battery pump. Its principle of operation, oxidation of CO by *Hopcalite* and electrical measurement of the heat thus produced, is similar to that of the M.S.A. CO recorder.

As shown in Fig. 6-28, a small pump draws the sample continuously into the instrument. It first passes thru a flow meter, consisting of an orifice and differential pressure gage, which is used to maintain a constant flow. This is accomplished by sustaining a constant differential on the flow meter thru adjustment of the volume control valve. Sample then passes thru a dehydrating canister which removes any moisture. From the canister it enters the cell containing *Hopcalite* and then passes out the exhaust valve of the pump.

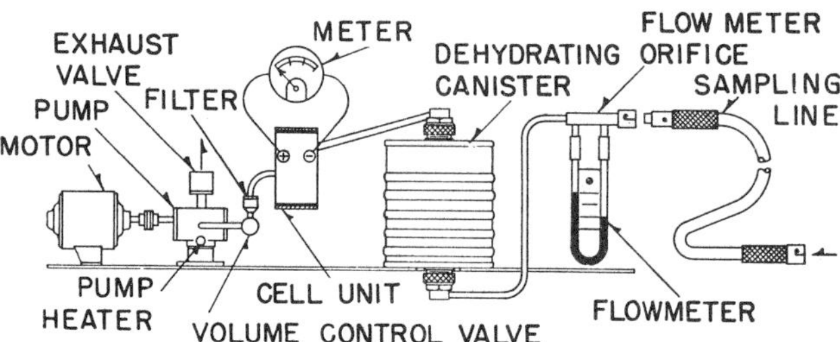

Fig. 6-28 Flow diagram of CO indicator. (Mine Safety Appliances Co.)

In the cell, any CO is oxidized to CO_2 by catalytic action. Heat liberated by this oxidation is directly proportional to amount of CO present and is measured by a series of thermocouples in series with the indicating meter. The meter is calibrated to read directly in per cent CO, with its scale ranging from 0 to 0.15 per cent. It can be read directly to 0.005 per cent and estimated to 0.001 per cent (one part per 100,000).

Monoxor Carbon Monoxide Indicator. This instrument, produced by the Bacharach Instrument Co., indicates directly the per cent CO in an air sample. A brownish stain is imparted by CO to a chemical contained in an indicating tube. The length of this stain, measured on the instrument scale, is directly proportional to the CO content of the sample.

Instrument readings are from 0.001 to 0.2 per cent CO. They may be readily taken over a wide range of lighting conditions.

Carbon Monoxide Alarms

The M.S.A. carbon monoxide recorder, previously described, can operate an alarm, a light, and ventilating equipment, when a certain limit of CO is reached.

Another compact unit is the M.S.A. CO alarm. This continuously operating instrument gives both audible and visible warning when CO increases to a predetermined setting—between 0.01 and 0.04 per cent. A small motor-driven blower draws a continuous air sample and forces it thru an electric heating unit and thru a cell containing thermocouples and active *Hopcalite* which converts CO to CO_2. Heat thus liberated is measured by a thermocouple and indicated on a contacting meter dial in terms of CO concentration. When the needle reaches the predetermined setting, the alarm circuit closes. For continuous use, the *Hopcalite* must be replaced about every two weeks.

Carbon Monoxide Saturation in Blood

M.S.A. CO Poisoning Test Kit. The degree of CO poisoning of both conscious and unconscious people can be checked in less than 1 and 5 min, respectively. Conscious subjects inflate a balloon by exhaling (alveolar air) into this device (Fig. 6-29—left). The mouthpiece tube is then replaced by a CO indicator tube (similar to that shown in Fig. 6-27) and the air is withdrawn thru the tube. The degree of color change of the indicating chemical measures the per cent CO in blood.

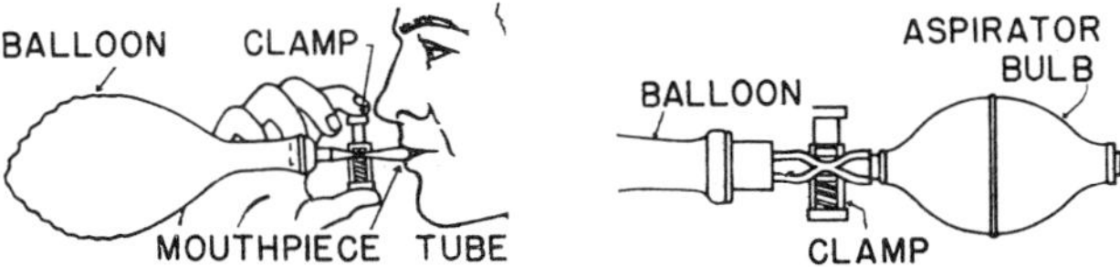

Fig. 6-29 CO poisoning test kit; (left) inflating balloon with breath sample; (right) setup for adding air to blood-plus-reagent. (Mine Safety Appliances Co.)

When a patient is unable to supply an air sample, competent medical personnel are required to withdraw a 1 cu cm blood sample. This blood is mixed with a reagent (to release CO) in a balloon, and a fixed amount of air is then pumped into the balloon by means of an aspirator bulb (Fig. 6-29—right). The air in the balloon is withdrawn thru an indicator tube as with a breath sample.

Van Slyke Manometric Blood Gas Apparatus. This apparatus[12] is used by medical technicians for measuring blood gases. Measurement is based upon direct observation of the amount of substance obtained. Gases so collected can be analyzed for various constituents, including carbon monoxide.

REFERENCES

1. Drinker, C. K. *Carbon Monoxide Asphyxia.* New York, Oxford U. Press, 1938.
2. Henderson, Y. and Haggard, H. W. *Noxious Gases and the Principles of Respiration Influencing Their Action,* 2nd ed. New York, Reinhold, 1943.
3. Shnidman, L. *Gaseous Fuels,* 2d ed. New York, A.G.A., 1954.
4. A.G.A. *American Standard Approval Requirements for Domestic Gas Ranges,* effective Jan. 1, 1962. New York, 1961.
5. Vandaveer, F. E. and Gregg, R. C. "Simplified Iodine Pentoxide Apparatus for Determination of Carbon Monoxide in Flue Gas." *A.G.A. Monthly* 11: 469–74+, Aug. 1929.
6. Vandaveer, F. E. "Improved Iodine Pentoxide Apparatus." *Gas* 18: 24–29, Oct. 1942.
7. Beatty, R. L. *Methods for Detecting and Determining Carbon Monoxide.* (Bur. of Mines Bull. 557) Washington, G.P.O., 1955.
8. Altieri, V. J. *Gas Analysis & Testing of Gaseous Materials.* New York, A.G.A., 1945.
9. Mine Safety Appliances Co. *Continuous Carbon Monoxide Recorder.* (Bull. DB-2) Pittsburgh.
10. Water, J. L. and Hartz, N. W. "Improved Luft-Type Infrared Gas and Liquid Analyzer." *Instr. Soc. Am. Proc.* 6: 79–83, 1951.
11. Mine Safety Appliances Co., Pittsburgh, and United States Safety Service Co., Kansas City.
12. Arthur H. Thomas Co. *Catalogue.* Philadelphia.

Chapter 7
Determination of Suspended Particles

by C. W. Wilson

The terms *suspended particles* or *suspensoids*, as they are used in gas utility practice, refer to minute particles of solid, liquid, or semisolid material, suspended in and carried along with gas during transmission and distribution. These materials may be iron oxide, dust, salt, tar, oil fog, gum, water vapor, naphthalene, light oil, lampblack, scrubber oil, gasoline, or silica. Their particles are colloidal in size (less than five microns) and are kept in suspension by Brownian movement, gravity, thermal gradients, electrical forces, acoustical forces, and centrifugal forces. They may properly be called aerosols or hydrosols.

Suspended particles can cause distribution and service difficulties, such as regulator troubles, flow reduction, pilot stoppage, erosion of component parts, and valve sticking. On the other hand, liquid particles in the form of oil fog or water vapor may be introduced into a distribution system in order to prevent drying out of jute-packed cast iron pipe joints, to reduce odor fading or to wet dust deposits in mains. Studies[1] sponsored by the American Gas Association have established procedures for the collection, monitoring, and identification of particles in gas distribution systems. Other developments in determining suspended particles in air have been reported.[2,3]

In some cases, determination of gas-carried material can be made with relative ease. This is particularly true with particles of large size. Here, mass of material, rather than number of particles or size distribution, is important, and larger particles may be retained by ordinary filters. Examples are dusts picked up and carried along at high velocities in transmission lines and tar at inlets of manufactured gas tar extractors.

The technical worker in the gas industry normally desires information about weight or mass concentration of particles per cubic foot of gas, size distribution, and chemical composition of the suspended particles.

SAMPLING[1,4]

Proper sampling technique is necessary to determine a valid representation of the actual concentration of particles in the flowing gas stream. *Iso*-kinetic sampling is required. The main precautionary measures of this technique follow (the reference contains a schematic diagram of a recommended sampling system):[4]

1. Sampling point should be located along a straight line of pipe where the gas flows undisturbed.
2. Sampling nozzle should face directly upstream with its axis parallel to pipe axis. Internal diameter of probe should be not less than 0.25 in., and preferably greater than 0.40 in., with constant internal diameter from probe to particle detector.
3. The leading edge of the probe nozzle should be feathered in order to minimize turbulent effects.
4. The probe should be short, to prevent deposition of material, and should contain a minimum number of bends.
5. To prevent thermal deposition, sampling line should be maintained at or slightly above gas temperature in pipe when volatile droplets are collected.

DUST DETERMINATIONS

U. S. Steel Corporation Method. Dust from a measured amount of gas is filtered and weighed. It is necessary to withdraw the gas sample without changing its velocity; otherwise, its dust content will change on account of greater inertia of the solid particles.

Apparatus: Short section of 1¼ in. standard pipe, threaded along its full length for screwing into main, encases sample tube and two pressure tubes for connection to draft gage. Brady filter holding 33 × 75 mm extraction shell to serve as filter is screwed into sampling tube outlet. Cylindrical electric heater, for vaporizing any moisture present and keeping filter dry, encases filter. Outlet from filter provides meter and thermometer connections.

Procedure: Dry, tared, filter shell is placed in cylinder; test meter and pressure gage are connected; and current is turned on to extraction shell. When shell has been heated to desired temperature (usually slightly above 100 C), connection to meter is shut off and sample tube is inserted as quickly as possible into gas line with its point at right angles to gas flow. With blast furnace and other manufactured gases, great care should be taken against inhaling escaping gas. To avoid this danger, operation should not be performed in an enclosure, and a gas mask should be worn.

As soon as tube has been inserted, it is gradually turned toward gas flow. At same time, tube is opened to permit gas flow to filter, and suction is adjusted to maintain zero pressure differential. Gas is cooled to dew point or to atmospheric temperature before passing thru meter. Amount passed depends upon quantity of dust present and is limited by operator's ability to maintain proper flow thru apparatus. Aspirator is started as soon as necessary to keep proper flow rate, as indicated by pressure gage. When this rate can no longer be maintained, test is stopped and sample tube removed. Extraction shell is removed, dried, and weighed. Method of drying depends upon nature of matter filtered from gas. If it is all mineral matter, filter is dried at 100 to 105 C. If it contains tar or other volatile substances, drying is done over concentrated sulfuric acid.

Dust is reported in grains per cubic foot of dry gas at standard conditions.

Blaw-Knox Method.[5] A weighed ceramic filter is mounted in a high pressure shell, gas is passed thru the thimble-shaped filter at the desired rate, and the thimble is weighed again to determine weight increase. Oil can be burned off the filter in a gas furnace or dissolved with proper solvents. Laboratory tests using various known sizes of dust particles showed that the equipment was accurate to 0.005 lb of dust per MMCF of gas, or 0.000035 grain per cubic foot. As applied to liquid entrainment, the test apparatus was accurate to 0.002 gal per MMCF.

Absolute Filter.[1] This is the most direct method for determining weight loading of dust in gas. The gas is passed thru a filter fine enough to catch suspended particles and is then metered. The catch on the filter is weighed accurately. The filtering media is an all-glass filter paper composed of half-micron diameter fibers supported on three-micron fibers and bound together by an acrylic resin. The resin is necessary only for packaging and can be removed either by heating the paper to red heat or by washing it with solvent. The all-glass filter paper is inert.

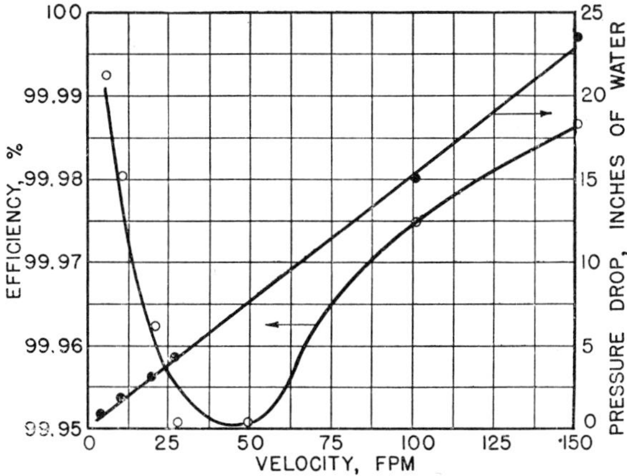

Fig. 6-30 Penetration and pressure drop vs. velocity for A.D.L. Absolute filter media, using test particles 0.3 microns diameter.[1]

In a specially designed filter holder, the paper is placed on a porous fritted glass support which has an area of 0.01 sq ft. Flow rate of a gas thru this filter for a pressure drop of 15 in. w.c. is 100 fpm. Efficiency of removal is more than 99.9 per cent of 0.3 micron test particulates, as shown in Fig. 6-30.

The M.S.A. ultra-efficient filter media,[3] constructed of glass fibers of micron size, will also filter out over 99.9 per cent of particles as small as 0.3 micron.

Millipore Filter.[1] This porous membrane of cellulose ester, with pore diameter of approximately 0.5 micron, is suitable for filtering solid material for direct microscopic observation. (Made by Millipore Filter Corp., Watertown, Mass.) This filter *cannot* be used for liquid droplets. By adding oil of suitable index of refraction to the filter paper, it becomes transparent in the microscope field. Therefore, the particles can be examined with transmitted light and oil immersion optics.

The general method for using the Millipore filter for gas main suspensoids is to pass a measured volume of gas thru the filter. By subsequent examination under a microscope, the total number and size distribution of the collected particles can be determined. The concentration per unit volume can then be calculated, since the volume from which the sample was drawn is known.

A.G.A. Thermal Precipitator.[1] This device employs a hot wire adjacent to a cool collecting surface. The thermal gradient causes suspensoids in a gas stream, passed slowly between wire and surface, to precipitate in a line on a removable glass slide. This device is a good collector for small particles, and for oil fog. Droplets of oil are not agglomerated as in other devices. Microscopic examination of the deposit reveals its qualitative composition. Particle size distribution can be estimated.

Four other instruments of this type are described in Reference 1.

A difficulty encountered in using this type instrument is its necessarily slow rate of sample flow, resulting either in small samples or long sampling periods. In addition, larger suspensoids may be lost thru settling out before they reach the instrument. Isokinetic sampling with the thermal precipitator would be difficult for gas industry applications.

A.G.A. Cascade Impactor.[1] This instrument comprises a series of narrow slots, usually three or four, set at right angles to adhesive coated plates. The gas sample, drawn at high velocity thru this arrangement, deposits its particles on the plates, since the particles cannot follow the lines of flow of the gas stream. Because each successive slot is narrower than the preceding, gas linear velocity increases. This allows a classification of particle sizes, due to inertial forces, into about four size or weight brackets, ranging from 20 microns down to approximately 0.1 micron diameter.

References 1 and 2 illustrate and describe instruments of this type.

Advantages of cascade impactors are: obtaining sample for analysis in a short time; classifying particles according to size; and collecting particles as small as 0.1 micron. Disadvantages include: shortness of collection time—a few seconds—which does not allow precise measurement; settling of larger particles in the sampling line; exceptional cleanliness needed in sampling lines; and inability to collect liquid particles.

No standardized procedures have been proposed for analyzing suspensoid deposits. Determination of composition can be made by usual macro- or micro-techniques of chemistry, which depend in turn upon circumstances indicating the material expected. Separation into organic compounds (e.g., oils or gums) and inorganic substances (e.g., dirt, mill-scale, or corrosion products), using suitable solvents, is often made first followed by more or less extensive quantitative estimates.

Electrostatic Precipitator. An electrostatic device can be used for collecting aerosol samples, since electrostatic separating forces can be made as much as 100 times greater than inertial forces. It would seem to offer advantages in collecting large samples for composition or particle size analysis. Limitations have been encountered in collecting the sample deposit on a small area for observation. Either liquid or solid particles should be handled.

For sampling particulates in industrial atmospheres, the M.S.A. electrostatic sampler[3] is available. It will sample at a rate of 3 cfm and will collect up to 200 mg of sample. The sample can then be analyzed for weight, size of particles, and composition.

Dust Counting

Particle size distribution or population is important in designing and operating gas cleaning equipment and in studying dispersal of oil fog in a distribution system. At present, microscopic particle counting appears to be the best available technique. Some progress is being made in devising instruments for automatic counting and monitoring. Discussion of use of microscopes, their necessary equipment and attachments, how to count particles, and calibration of reticule grid with stage micrometer is available.[1] M.S.A.[3] supplies a dust-counting microscope and a *dust-vue* microprojector which projects an enlarged image of the microscope's field on a ruled screen.

Another dust-counting instrument[6] combines both the necessary air sampling device and a dark field microscope viewing and counting system. It is mounted on a circular base and incorporates an illuminating apparatus. The air-sampling mechanism consists of a moistening chamber thru which air is drawn by means of an accurately calibrated hand pump of $\frac{1}{1000}$ cu ft capacity, and an impinging device with a rectangular slit which deposits suspended dust particles on an enclosed circular glass plate. Dust deposit is in the form of a ribbon. The slide holder has an externally visible numbered index which is used to identify the sample. By rotating the knob, the sample is brought under the dust-counting microscope. Twelve samples may be collected on one slide, viewed, and counted at once without removal. They may also be preserved for future reference by sealing a cover glass to the slide. When desired, the sample may be examined on the stage of a regular laboratory-type compound microscope.

The viewing and counting apparatus consists of a built-in compound microscope of 200× magnification with a dark field illuminating system. It is fitted with a hyperplane eyepiece and has a micrometer disk ruled in two rows of squares with sides of 30 microns. These squares are used for dust counting. An extra line is ruled alongside the squares to permit approximate measurement of particles for size classification. The apparatus is so calibrated that, by multiplying the number of dust particles appearing in these square fields by 100,000, the total dust count per cubic foot is obtained.

The attachable illuminator is supported in a fixed position under the dark field condenser. This construction insures accurate and positive instrument location in the best position for dark field illumination. Therefore, counting conditions are standardized and all readings are comparable. This is an essential feature for running tests over extended periods. The illuminator is detachable from the condenser mount.

Monitoring Instruments[1]

These give a continuous time record of a characteristic of the suspended particles. The A.G.A. Light-Scattering Photometer, for example, measures the intensity of light scattered by the suspended particles. This intensity is related to particle concentration. Such instruments have a fast response time and give a continuous record. For systems of reasonably constant properties, the scattered light is proportional to the amount of suspended matter in the gas. Concentrations as low as 0.005 gram per MMCF can be detected; concentrations as low as 0.01 gram per MMCF have been measured.

Another recording device,[7] developed principally for measuring the concentration of lampblack in manufactured gas, is also adaptable for indicating presence of other solids in gas. Essentially, this unit allows gas to make contact with a filter paper which collects such substances on its surface. The paper is then subjected to a photoelectric test where the amount of light passed (or not passed) thru it indicates the quantity of the collected substance. The intensity of light thus passed is converted into electrical energy which drives an indicating instrument.

TAR DETERMINATIONS

Tar Extraction Shell. This method is used for determining tar fog in gas. The sample tube is the same as the U. S. Steel Method employed for determining dust, and method of use is similar.

Procedure: Extraction shell is first dried to constant weight in desiccator over sulfuric acid. During test, shell is heated to temperature necessary to keep it dry enough to permit free gas passage. Excessive heat should be avoided to prevent loss of some tar oils or naphthalene. When meter registers desired volume of gas, which varies according to amount of tar and solids present, shell is removed and dried over concentrated sulfuric acid to constant weight, as was done before test. Tube is rinsed with chemically pure benzene into tared porcelain dish, benzene is evaporated at atmospheric temperature in current of dry air, and residual tar is weighed. If gas contains no coal dust or other mineral dust, this weight is added to gain in weight of shell to find total weight of tar in metered gas sample.

If gas contains coal and mineral dust, as does producer gas for example, tar on shell is extracted with chemically pure benzene. Total dust is found by drying shell in current of dry air and reweighing it. Mineral dust is then found by burning shell, weighing ash, and subtracting from this weight the average weight of shell ash, determined by burning several shells separately.

Results are expressed in grams or grains per cubic foot at standard conditions.

Tar Camera. A metered gas sample is drawn thru a filter paper—for example, No. 1 Whatman Paper (4.25 cm diam) or equivalent—supported in a device known as a tar camera[7,8] connected directly to supply source. Tar adhering to the filter paper is determined either by weighing or by comparing paper discoloration with a standard colorimetric chart.

This method can be used to determine tar only, but it is not suitable for gas containing large amounts of tar. Experience has shown that to obtain reasonably accurate results, gas must be under pressure.

Procedure: Preliminary run is made to ascertain gas flow rate that may be used without tearing filter paper, since paper strength is affected by gas moisture content. Three 11-cm filter papers are placed in camera, gas connection is made, and meter and flow are adjusted until allowable rate has been established.

Three new filter papers then are placed in camera, and 1 cu ft of gas is allowed to pass at the predetermined rate. After shutting off gas, filter papers are removed. If colorimetric method is used, tar is determined by matching their color against chart supplied with instrument. With gravimetric method, papers are dried in vacuum and weighed between watch glasses both before and after test. Increase in weight is taken as weight of tar per cubic foot of raw wet gas.

REFERENCES

1. Doyle, A. W. and others. *Collection, Monitoring and Identification of Particles in Gas Distribution Systems.* (Final Report, Project PF-15)(46/OR) New York, A.G.A., 1959.
2. Manufacturing Chemists' Assn. *Air Pollution Abatement Manual,* ch. 6 and 7. Washington, 1952.
3. Mine Safety Appliances Co. *Catalog of Industrial Safety Equipment.* Pittsburgh, Pa.
4. Little, Arthur D., Inc. *Investigation of Sampling Procedure Requirements.* (Project NFX-12; PF-15)(11/PR) New York, A.G.A., 1957.
5. Blaw-Knox Co. *Gas Cleaner Catalogue.* Pittsburgh, Pa.
6. Bausch & Lomb Co., Rochester, N. Y.
7. Gilkinson, R. W. "Determining Suspended Solids in Gas." *Gas Age* 114: 31+, Sept. 9, 1954; also in (inc. bibliog.) *A.G.A. Proc.* 1954: 606–611.
8. Altieri, V. J. *Gas Analysis & Testing of Gaseous Materials.* New York, A.G.A., 1945.

Chapter 8
Gas Calorimetry

by Donald L. White

The heating value of gas (calorific value) is an important factor in its purchase and sale. Heat content of the fuel is the principal measure of its ability to perform a heating service (except in certain applications where flame intensity is the principal criterion, for example, in lamp manufacture). In many locations gas is purchased and sold on a therm basis, which necessitates an accurate handling of its calorific value. Most state and municipal regulatory bodies[1-3] include calorific value limits in their regulations. Suitable equipment and test procedures for determining calorific value are therefore essential. Standards for the two widely recognized types of equipment are available—water flow calorimeters,[4] commonly known as Junkers calorimeters, and Cutler-Hammer Recording Calorimeters.[5]

DEFINITIONS

These definitions apply directly to water flow calorimetry,[4] but they are also applicable, where appropriate, to calorimetric determinations using other apparatus.

British Thermal Unit (Btu): quantity of heat that must be added to one avoirdupois pound of pure water to raise its temperature from 58.5 F to 59.5 F under standard pressure.

Combustion Air: air passing into combustion space of calorimeter (theoretical air plus excess air).

Excess Air: quantity of air passing thru combustion space in excess of theoretical air.

Flue Gases: products of combustion remaining in gaseous state together with excess air.

Net Calorific Value: number of Btu evolved by the complete combustion at constant pressure of one standard cubic foot of gas with air; temperature of gas, air, and products of combustion being 60 F, all water formed by combustion reaction remaining in the *vapor* state.*

Observed Calorific Value: number of Btu obtained by multiplying the pounds of water heated in a calorific-value test by its corrected temperature rise in degrees F, and dividing by number of standard cubic feet of gas burned.

Products of Combustion: sum of all substances resulting from burning of gas with its theoretical air, including inert constituents of gas and theoretical air, but excluding excess air.

* Net calorific value is total calorific value minus latent heat of vaporization at standard temperature of water formed by combustion reaction.

Standard Cubic Foot of Gas (SCF): quantity of any gas that at standard temperature and pressure will fill a space of 1 cu ft when in equilibrium with liquid water. (Partial pressure of gas = 30.000 — 0.522 = 29.478 in. Hg.)

Standard Pressure: absolute pressure of a column of pure mercury 30 in. high, at 32 F, under standard gravity (32.174 ft per sec-sec).

Standard Rate of Combustion: equivalent to 3000 Btu per hour of total calorific value (for any gas with a heating value of 300 to 3500 Btu per SCF in the water flow calorimeter).†

Standard Temperature: 60 F, based on the international temperature scale.

Theoretical Air: volume of air that contains the quantity of oxygen, in addition to that in gas itself, which is consumed in complete combustion of a given quantity of gas.

Total Calorific Value: number of Btu evolved by the complete combustion at constant pressure of one standard cubic foot of gas with air; temperature of gas, air, and products of combustion being 60 F, all water formed by combustion reaction being condensed to *liquid* state.

Calorific Value of Gas as Purchased or Sold: seldom measured at standard conditions as defined above. It may be practically dry, at temperatures above or below 60 F, and at pressures considerably above 30 in. Hg (or below, at elevations significantly above sea level). Conditions of temperature, pressure, and water vapor content are accordingly agreed upon by the contracting parties, and calorific value as determined or calculated at standard conditions is converted to contract conditions. For complete information on conversion, see Reference 4. Two examples follow:

Example 1: Given the heating value of a gas as determined by a calorimeter at standard conditions (30 in. Hg, 60 F, saturated). Convert this value to a different set of pressure P, temperature t, and moisture content conditions.

Multiply the calorimeter value by C from Eq. 1.

$$C = \frac{(P - D)(60 + 460)}{(30 - 0.522)(t + 460)} = 17.64 \frac{P - D}{t + 460} \tag{1}$$

† When the heating value of a gas is as low as, say, 100 Btu per SCF, the required combustion rate of 3000 Btu per hr will result in an excessively high (and thus inaccurate) rotation rate of the 0.1 ft meter.

where: P = total pressure (barometric plus gage), in. Hg
t = temperature, °F
D = vapor pressure of water corresponding to the dew point of the gas, in. Hg

Example 2: Given that the heating value of a gas as determined by a calorimeter equals 1020 Btu per cu ft (at 30 in. Hg, 60 F, saturated). Find the heating value of this gas at metered conditions of 14.65 psia, 60 F, dry. Btu per cu ft at metered conditions = 1020 × 1.0118 = 1032; where 1.0118 is taken from Table 6-5.

Table 6-5 Factors to Convert Heating Values at Standard Conditions to Other Base Conditions[6]

(Btu per cu ft at 30 in. Hg, 60 F, saturated)

Special base conditions			
Pressure, abs	Temp, °F	Moisture	Factor
14.4 psi	60	Dry	0.9945
14.65 psi	60	Dry	1.0118
14.70 psi	60	Sat.	0.9976
14.73 psi	60	Dry	1.0173
14.73 psi	60	Sat.	0.9997
14.90 psi	60	Dry	1.0291
30 in. Hg	32	Sat.	1.0693
30 in. Hg	60	Dry	1.0177

Note: This table was calculated from Eq. 1 using the following conversions: 30 in. Hg (at 32 F) = 14.735 psi, saturated vapor pressure at 32 F = 0.1804 in. Hg = 0.089 psi, saturated vapor pressure at 60 F = 0.522 in. Hg = 0.256 psi.

WATER-FLOW CALORIMETER

The water-flow calorimeter has been widely employed for gas calorific value determinations and has been accepted as a standard apparatus for calorific value measurement by many regulatory bodies and utility companies. It is manually operated, of rugged construction, requires little space for installation, and, if properly installed and operated, gives accurate results.

An American Standard Method of Test[4] for calorific value of gaseous fuels is based on this instrument. In its use, the heat evolved by complete combustion of a measured amount of gas is absorbed by a measured quantity of water flowing thru the instrument. Measurements of the amount of gas burned, weight of water heated, and its temperature rise provide the primary data for calculating the calorific value of the gas.

Apparatus

Figure 6-31 shows a section of a typical water-flow calorimeter,[4,7,8,9] available from several sources. For detailed description of a water-flow calorimeter and accompanying equipment, including its assembly, see Reference 4.

Full operating instructions for calorific value determinations by the water-flow calorimeter are also given in Reference 4. They cover both the humidity correction and the humidity control procedures, together with necessary test record forms. Instructions for calculation of total and net calorific values per SCF are also included, along with their conversion to other bases of field measurement.

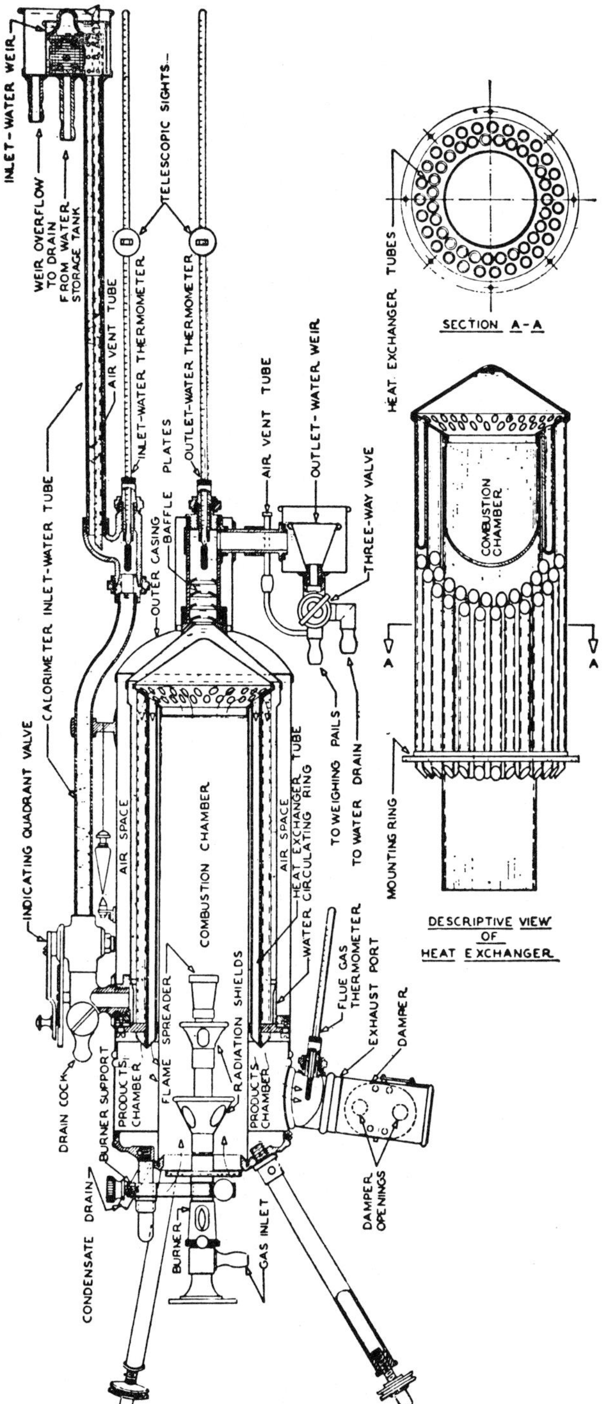

Fig. 6-31 Sectional view of a water-flow gas calorimeter.[4]

Reproducibility of Results[4]

A.S.T.M. Standard D900-55 states:[4] "For manufactured gases of about 540 Btu, natural gases of about 1050 Btu, and liquefied petroleum gases such as propane and butane of about 2400 to 3200 Btu, the maximum permissible difference between the results of duplicate determination of total calorific value of a given sample of gas by the same observer using a single set of apparatus is 0.3 per cent. The maximum permissible difference between results of determinations of calorific value of a given sample of gas by different observers using different apparatus is 0.5 per cent.

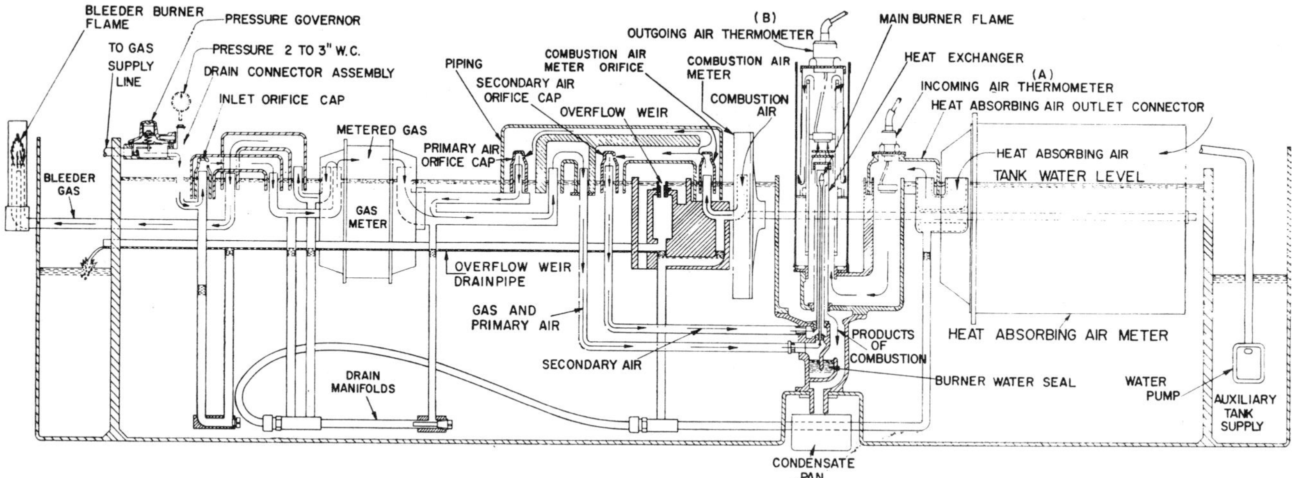

Fig. 6-32 Tank unit schematic flow diagram, Cutler-Hammer recording calorimeter.

"Caution—In order to obtain the indicated reproducibility it is necessary to follow very carefully the procedures specified in this method. This statement applies particularly to gases of high calorific values, for example those consisting largely of propane or butane or mixtures thereof."

CUTLER-HAMMER RECORDING CALORIMETER

The operating principle[10,11] involves burning of gas at a constant rate and absorption of heat developed by an air stream. Flow rates of gas, combustion air, and heat-absorbing air are regulated thru electrically driven metering devices which are geared together so that a constant gas-to-air ratio is maintained. Products of combustion from a specially designed burner are kept separate from heat-absorbing air and are cooled almost to initial tank temperature, thus condensing the water formed in combustion. Rise in temperature of heat-absorbing air is measured by two resistance thermometers forming part of a Wheatstone bridge circuit, and it is automatically recorded in terms of Btu per standard cubic foot. Accuracy: within 0.5 per cent from half to full scale with room temperature between 60–85 F.

Use of air as the heat-absorbing medium eliminates need of corrections for room temperature and pressure variations since heat-imparting gas and absorbing air follow the same physical laws governing behavior with varying temperature and pressure.

Construction

Figure 6-32 is a flow diagram of the calorimeter tank unit. All moving parts are supported and aligned by a heavy base. Meters for gas, combustion air, and heat-absorbing air are suspended on lower side of base; motor, gear reduction, and water pump are mounted on upper side. Entire base is pivoted so that it can be swung back for inspection and repair to give access to parts mounted on lower side. All underwater parts are brass, which eliminates corrosion or electrolytic action between dissimilar metals. Design permits draining all liquids from combustion products into a small condensate pan, eliminating a major source of tank water contamination.

Operation

Gas enters inlet of the gas meter (Fig. 6-32) thru a pressure-reducing orifice. About 200 Btu per hr are metered. Balance of the gas passes to a bleeder burner and is burned. By using bleeder burner, sufficient gas flow is obtained to purge sample line rapidly, and gas in the meter is regulated exactly to atmospheric pressure.

Primary air joins metered gas at meter outlet, and mixture passes to main burner. Metered combustion air is divided by suitable orifices into primary, which mixes with gas, and secondary, which flows around outside of burner tube to support combustion. Chamber surrounding burner is closed at top and open at bottom, so that combustion products must flow downward, releasing heat of combustion thru chamber walls to a series of fins.

Stream of metered air, delivered by heat-absorbing air meter, passes over entrance thermometer and around combustion chamber and fluted tube where it absorbs heat of combustion. Heated air passes over exit thermometer, thru series of baffle tubes, then discharges outside burner assembly.

Water pump automatically dips into tank storage compartment, raises small quantities of water, and discharges into main tank to maintain an exact water level, excess water overflowing a weir.

Entrance and exit thermometers form two legs of a Wheatstone bridge for a resistance-type recorder. A calibrated slide wire rheostat is automatically moved together with a pen, to balance bridge circuit. Movement required to balance circuit measures temperature difference between two thermometers. This is registered by the recorder and shown directly in Btu per cubic foot.

Gas and air, both fully saturated, are measured at atmospheric pressure at tank water temperature. Accordingly, calorimeter readings entirely independent of pressure, temperature, or humidity changes are assured, with no corrections for these variables necessary.

Table 6-6 shows various chart scale ranges available. Changes in inlet gas orifice, primary and secondary air orifices, and burners are required for the various ranges.

Table 6-6 Operating Ranges for Cutler-Hammer Recording Calorimeters

(All values in Btu per standard cubic foot)

75–150	400–600	800–1200
100–200	350–700	750–1500
150–300	450–900	1000–1500
300–600	600–900	900–1800
	600–1200	1800–3600

Changes are required in the standard instrument to adapt it to gases which are fully combustible in themselves, such as natural gas-air premix gas. Principal modifications are in both main and bleeder burners and in combustion air meter when sufficient air for combustion is present in the gas. When used on blast furnace gas, oxygen has been found preferable to air for supporting combustion.

Installation

Manufacturer's recommended specifications for housing a single Cutler-Hammer recording calorimeter follow:

1. Cubic Space: 1000 cu ft min—2000 cu ft max
2. Ceiling Height: 10 ft min
3. Side Wall Widths: 10 ft min—13 ft max
4. Window: one—North side preferred in northern hemisphere.
5. Door: (a) east side preferred, (b) width 3 ft min, (c) provided with door check.
6. Location of Calorimeter Tank Unit: side of room opposite door preferred.
7. Heating and/or Cooling: (a) temperature to be controlled in range from 60 to 85 F; (b) thermal insulation where ambient conditions cause large or rapid changes in room temperature.
8. Floor, Foundation: (a) 3000 lb load to produce no deflection; (b) tank unit feet to be on bare floor or foundations; (c) no deflection in floor due to variable floor loading adjacent to tank unit.
9. Lighting: (a) no direct sunlight on calorimeter tank unit; (b) ample lighting for making observations; (c) electrical plug receptacle within 10 ft of tank unit.
10. Condition of Air: (a) free from dust, (b) free from combustible gas.
11. Vibrations: free from vibrations or shock.
12. Water: pure, clean water to be obtainable for filling tank and replenishing; desirable to have water supply and drain in room.
13. Power Supply: (a) 110 v, 1 phase, 60 cycle a-c is standard for calorimeter; (b) 1000 w for tank unit motor and recorder; (c) lighting and air conditioning requirements in addition.
14. Gas Supply: (a) sample line to be capable of delivering about 4 cfh; (b) sample line to have minimum capacity to reduce time lag; (c) desirable to reduce pressure of sample line within room or building to 1 psig or less; (d) supply pressure at calorimeter 6 to 30 in. w.c., sulfur-free.
15. Tank Unit: shielded from any hot radiating surfaces.

Note: Fresh paint, floor wax, or volatile cleaning materials can affect accuracy.

Master recorder may be located in a room separated from the tank unit—maximum distance 200 ft. Distant recorders may also be installed in number desired at other locations, as required.

Accuracy and Calibration

The heating value of the gases to be measured influences the choice of calibrating gas; i.e., hydrogen (319.3 Btu per cu ft) is normally used as the standardizing gas for all calorimeter ranges, except for those with 1200 Btu full scale, for which cylinders of methane, certified as to heating value and/or specific gravity by the Institute of Gas Technology, are normally used. (The National Bureau of Standards certifies the primary standard gases used by IGT.)

A number of companies prefer to use natural gas for calibration. In some instances, large underground tanks were filled with natural gas, and its heating value was determined accurately on the calorimeter by hydrogen calibration, or by calculation from mass spectrometer analysis.[12] This gas would be usable over a long period. In some instances, a tank of high purity methane gas, which had been tested on an accurate calorimeter, was obtained and used for calibration. Results of a cooperative research project by the National Bureau of Standards and the Gas Operations Research Committee of the American Gas Association on accuracy of the Cutler-Hammer Recording Calorimeter were reported.[13] Gas, composed almost entirely of methane, was obtained from a California source, purified to 99.93 per cent methane and 0.07 per cent nitrogen, and its heating value accurately determined to within ±0.1 Btu per cu ft by mass spectrometer. In the course of this study[13] it was found that:

1. Variations between 60 and 80 F in room temperature cause only small changes in accuracy, but outside this range the amount of inaccuracy becomes progressively greater.
2. If the tank water temperature changes greatly from that at which the calibration was made, inaccuracies up to 0.75 per cent can be common, and greater are possible.
3. Although a calibrating gas like standard methane is preferred, other sources of calibrating gas—hydrogen from *Hydrone*, hydrogen from tanks, and standard methane—are suitable if proper precautions are taken.
4. No extension should be added in any way to the bleeder burner during calibration tests.
5. The inlet gas pressure to the instrument must be maintained constant.
6. The instrument may be used satisfactorily to determine heating value of gases as high as propane (2529 Btu per cu ft).
7. With proper maintenance, an average operating accuracy of 0.25 per cent or better can be maintained on any normally distributed gases.

Customarily, calibration is in Btu per standard cubic foot (30 in. Hg, 60 F, saturated), but it may be made on any basis desired.

Maintenance

Information on care and operation of the instrument is given in Reference 10. Periodic inspection and maintenance are essential. Certain maintenance is required on a weekly, monthly, bi-monthly, and every four months' basis. Further discussion of installation and maintenance will be found in References 3 and 14.

OTHER RECORDING CALORIMETERS

The **Cutler-Hammer Recording Thermeter**[15] is a small recording calorimeter consisting of a tank unit 21⅝ in. high × 23$\frac{9}{16}$ in. wide × 20$\frac{1}{16}$ in. deep with a circular recorder. It operates like the recording unit previously discussed and, with regular inspection and servicing, is capable of a high degree of accuracy. It is not used in gas purchase and sale. As a smaller, less expensive unit, it is used in industrial operations, and where great accuracy is not required.

The **Sigma Recording Calorimeter—Mark 2,** made by Sigma Instrument Co., Ltd.,[16] Letchworth Hertfordshire, England, measures the differential expansion of two concentrically mounted steel tubes thru a frictionless mechanical magnifying mechanism. Tubes are expanded by heat obtained when burning a controlled gas flow. Very few installations have been made in the United States to date.

REFERENCES

1. Nat. Assn. of Railroad and Utilities Commissioners. "Suggested NARUC Regulations Governing Service Supplied by Gas Companies." *Report of Committee on Engineering,* 1961, Appendix B. Washington, 1961.
2. U. S. Nat. Bur. of Standards. *Standards for Gas Service.* (Circ. 405) Washington, G.P.O., 1934.
3. Calif. Public Utilities Com. *Standards of Calorimetry for Gaseous Fuels.* (General Order 58-B) Sacramento, 1956.
4. A.S.T.M. "Standard Method of Test for Calorific Value of Gaseous Fuels by the Water-Flow Calorimeter." (Standard D900-55) *ASTM Standards,* 1961, pt. 8. Philadelphia, 1962. (Available also as a separate pamphlet)
5. A.S.T.M. "Tentative Method of Test for Calorific Value of Gases in Natural Gas Range by Continuous Recording Calorimeter." (Standard D1826-61T) *ASTM Standards,* 1961, pt. 8. Philadelphia, 1962. (Available also as a separate pamphlet)
6. Warner, C. W. "Fundamentals of Gas Calorimetry." (CEP-57-9) *A.G.A. Proc.* 1957.
7. McGlashen, W. A. "Convenient and Accurate Method of Making Calorimetric Determinations." *P.C.G.A. Proc.* 29: 105–110, 1938.
8. "Fuel Gas Calorimetry, Water-Flow Method." *P.C.G.A. Proc.* 32 suppl.: 1–64, 1941.
9. Jessup, R. S. and Weaver, E. R. *Gas Calorimeter Tables.* (Bur. of Standards Circ. 464) Washington, G.P.O., 1948.
10. Cutler-Hammer, Inc. *Instructions for the Care and Operation of the Cutler-Hammer Recording Calorimeter.* (Pub. 8293 and 70–252 for "Type AB" and electronic models respectively) Milwaukee.
11. Jessup, R. S. "The Thomas Recording Gas Calorimeter." (Bur. of Standards RP 519) *Natl. Bur. Std. J. Res.* 10: 99–121, Jan. 1933.
12. Brewer, A. K. and Dibeler, V. H. *Mass Spectrometric Analyses of Hydrocarbon and Gas Mixtures.* (Bur. of Standards RP 1664) Washington, G.P.O., 1945.
13. Eiseman, J. H. and Potter, E. A. "Accuracy of the Recording Gas Calorimeter When Used with Gases of High Btu Content." (CEP-55-13) *A.G.A. Proc.* 1955: 885–900.
14. Gilbert, T. M. and others. "Installation and Maintenance of Cutler-Hammer Inc. Recording Calorimeter for Accuracy and Reliability." *Appalachian Gas Measurement Short Course, Proc.* 13: 289–303, 1955.
15. Cutler-Hammer, Inc. *Instructions for the Care and Operation of the Thermeter.* (Pub. 6651 and 70-2913 for "Type AB" and later model, respectively) Milwaukee.
16. Sigma Instrument Co. *Sigma Recording Calorimeter—Mark 2.* Letchworth, Eng. (Also pub. by Cosa Corp., New York).

Chapter 9

Gas Density and Specific Gravity

by C. W. Wilson

DENSITY

The density of a gas is the mass per unit volume under specified conditions of temperature and pressure. Since weight is proportional to mass, density of a gas is ordinarily expressed in pounds per cubic foot at 60 F, 30 in. Hg pressure, and saturation. In accordance with Dalton's law, the saturation criterion is equivalent to stating that the partial pressure of the gas is 30 *less* 0.522, or 29.478 in. Hg.

Density of a complex gas mixture may be determined directly or by reference to specific gravity determinations, or it may be calculated from the gas analysis. The density of such a mixture is the sum of densities of individual constituents and their partial pressures, mole fractions, or percentages in the mixture. It may be obtained from the expression:

$$d_m = d_1p_1 + d_2p_2 + d_3p_3 + \dots + d_np_n \qquad \textbf{(1a)}$$

where:

d = density

p = partial pressure, mole fraction, or percentage of a mixture component

m = subscript representing the mixture

1,2, 3,...,n = subscripts representing individual components

Values derived from gas analyses may be used for engineering calculations if all components are determined. If some components—for example, those of illuminants in manufactured gas—are not included in its analysis, density should be obtained by direct determination.[1]

Wet vs. Dry Basis

Dry gas density is the only true density of a gas. *Wet* gas density is the density of a mixture of the gas and water vapor, and is of considerable practical importance. It is related to dry gas density as follows:

$$d_{gw} = \frac{d_{gd}\,(p - p_w) + d_u p_w}{p} \qquad \textbf{(1b)}$$

where: d_{gw} = density of *wet* gas

d_{gd} = density of *dry* gas

d_w = density of water vapor

p = total absolute pressure of gas and water vapor

p_w = partial pressure of water vapor in the wet gas

Densities must be expressed in consistent units and pressures. When gas has been brought into temperature and pressure equilibrium conditions, p_w may be taken as vapor pressure of water at existing temperature. Unless such equilbrium is attained, water vapor partial pressure must be determined experimentally.[1]

Applications

Density is a property of major importance where transport, measurement, or mixing of gases is involved. It is an important factor in many more complex processes such as heat transfer and combustion. Work required to move a volume of gas from one place to another depends in part upon its mass. Static pressures registered by a distribution system are determined by weights of corresponding columns of gas. Transfer of momenta, as in certain mixing and combustion operations, involves mass as an important element of momentum. Locating a gas manufacturing plant at a distribution system low point takes advantage of gas buoyancy because most manufactured gases are lighter than air. Hot flue gases flow thru appliances and vent thru chimneys because of their lowered density.

Dynamic methods for measuring flow of gases, such as orifice meters, venturi meters, and pitot tubes, are affected by density since differential pressures registered by the meters are determined by density of the gas stream. Consequently, accurate calculations of gas volume obtained by such meter readings for a gas undergoing changes in density can be made only if such changes are recorded as they occur.

Maintenance of good operating conditions in a gas distribution system involves certain limitations on gas density variations. Density is important in this regard because of its effect on pressure and on primary air entrainment in Bunsen type gas burners.

Determination

Direct determination of density may be made by carefully weighing the gas in a globe of known volume under known conditions of temperature and pressure. A volume of from 200 to 500 ml is ordinarily large enough. To eliminate errors due to possible changes in buoyancy of a globe, another globe with the same external volume as the first is used as a counterweight.

It is usually simpler to determine the specific gravity of a gas experimentally, from which its density can then be computed. Discussions of density and specific gravity determination are available.[1-3]

SPECIFIC GRAVITY

The specific gravity of a gas is the ratio of its density, under the observed conditions of temperature and pressure, to the density of dry air of normal carbon dioxide content, at the same temperature and pressure.[3] This definition assumes that pressure is approximately one atmosphere.

Wet vs. Dry Basis

Specific gravity of a *dry* gas:

$$s_{gd} = \frac{d_{gd}}{d_{ad}} \quad \text{(2a)}$$

where:

s_{gd} = specific gravity of *dry* gas

d_{gd}, d_{ad} = density, or weight per unit volume, of *dry* gas and *dry* air, respectively, at same temperature and pressure

If gas specific gravity is about 0.6, correction for water vapor is ordinarily negligible. If gas is saturated or contains determined amounts of water vapor, its specific gravity on a *dry* basis, s_{gd}, is calculated from:

$$s_{gd} = \left[\frac{\text{density of gas and water} - d_w p_w}{\text{density of air and water} - d_w p_w}\right] \times \left[\frac{\text{pressure of air and water} - p_w}{\text{pressure of gas and water} - p_w}\right] \quad \text{(2b)}$$

where: d_w = density of water vapor

p_w = partial pressure of water vapor in the wet gas

Specific gravity of a gas *containing water vapor:*

$$s_{gs} = \frac{s(p - p_w) + s_w p_w}{p} \quad \text{(3)}$$

where:

s_{gs} = specific gravity on a *wet* basis

s_{gd} = specific gravity on a *dry* basis

p = total absolute pressure of gas and water vapor

p_w = partial pressure of water vapor as found by determination or, if saturated, from water vapor pressure table corresponding to temperature

s_w = specific gravity of water vapor, generally taken as 0.622

Specific gravity on a *dry* basis may be converted to a *wet* basis by the expression:

$$s_{gs} = \frac{s_{gd} + k}{1 + k} \quad \text{(4)}$$

where: $k = d_w p_w / d_{ad}(p - p_w)$

d_{ad} = density of air

d_w/d_{ad} = 0.622 (usual value)

Applications

Specific gravity is widely employed in gas engineering practice for calculating gas flow thru orifice meters, pitot tubes, and appliance orifices, as well as for gas interchangeability evaluations. Illustrations discussed under density are also pertinent to specific gravity. An error of two per cent in specific gravity determination affects by one per cent measurement of gas flow by an orifice meter, pitot tube, oı venturi meter. Because a gas measurement error of one per cent cannot be tolerated when large volumes are involved, specific gravity determination must be made with as great an accuracy as possible and preferably to less than 0.003 sp gr for orifice meter calculations. This reduces measurement calculation error from this source to less than one fourth of one per cent. It also requires a continuous recording gravitometer which will follow closely any changes in gas specific gravity.

Specific gravity apparatus[3] may be expected to give results with an accuracy of plus or minus two per cent. For a 0.60 sp gr gas, a two per cent error would be 0.0120 sp gr and, as stated above, accuracy should be held to about 0.003 sp gr. These methods are based on extensive tests on 11 instruments using 15 test gases of known specific gravity.[2]

Accuracy required has been accomplished by: (1) incorporation in the best instruments of many suggestions[2] for design improvement; (2) selection of best instruments available for purpose intended; (3) regular and frequent instrument servicing and calibration; (4) attention to installation details to secure good operating conditions, (5) use on recorders of charts with full scale range just within limits of specific gravity variations; (6) periodic checks of calibrating instruments on gases of known specific gravity.

Manual or Indicating Specific Gravity Instruments

Direct Weighing. This method,[3] which is seldom used except for referee or calibration purposes, is not practical for field or routine laboratory tests.

Involved here is the differential weighing, on a very accurate chemical balance, of two suitable containers of as nearly as possible equal weight and volume, each filled successively with gas and air under similar conditions of temperature and pressure. Pressure is always approximately atmospheric. The samples are reversed and reweighed, thus doubling weight difference and canceling errors from several sources. If this method is properly applied, specific gravity determinations can be made with great accuracy.

Apparatus:

1. Balance. Analytical type with capacity of at least 250 g, sensitivity of 0.1 mg, and room for suspension of a one liter round flask on each balance arm.
2. Flasks. Two glass flasks with well ground stopcocks and connecting tubes with standard type male ends; flasks closely matched in form, weight, and volume and marked for identification.
3. Bath. Container with thermostatic control of temperature (constant to 0.2°F), with space for two flasks and facilities for their connection, support, and manipulation.
4. Manifold. Facilities and connections in a permanent manifold for evacuating flasks, filling with gas sample,

filling with purified air, and connecting to outer air without danger of contamination.

5. Barometer. Accurate mercurial type.

Procedure: With flasks in bath slightly above room temperature, fill one with gas to be tested and other with purified outdoor air, after filling and evacuating each three times. Allow flask to stand in bath 30 min. After manifold tubes have been purged with sample gas and air, respectively, open valves of flasks for a few seconds to equalize pressure. Leave flasks closed about 5 min, open valves to air for 1 sec and then close.

Remove flasks from bath, clean carefully, hang on balance, and let stand 30 min to attain balance temperature. Bring balance to equilibrium by adding weights as required.

Repeat above procedure exactly as before, except that flask formerly filled with purified gas is now filled with air, and vice versa. Weigh same flask on same balance pan, regardless of contents.

Accuracy of controls and manipulation should be checked by determining specific gravity of air in place of gas. Specific gravity of sample can then be calculated from expression:

$$S = \frac{D_g}{D_a} = \frac{P_2V_1 + P_1V_2}{P_1V_1 + P_2V_2} + \left(\frac{760}{D_a}\right) \frac{(W_r - W_l)_1 + (W_l - W_r)_2}{P_1V_1 + P_2V_2}$$

where:*

S = specific gravity of sample gas

D_g = gas density at 760 mm Hg at bath temperature, g per l

D_a = air density at 760 mm Hg at bath temperature, g per l

$(W_r)_1$, $(W_l)_1$ = weights added to right and left pan, respectively, to balance in first weighing, g

$(W_r)_2$, $(W_l)_2$ = weights added to pans to balance after interchanging air and gas in flask, g

P_1, P_2 = barometric pressure during first and second weighings, respectively, mm Hg

V_1, V_2 = volume of gas in flasks (1) and (2), respectively, at 760 mm Hg and bath temperature, l

Improved Ac-Me Gas Gravity Balance and Junior Gas Gravity Balance. These instruments of the pressure balance type are used for laboratory or field determinations. A beam carrying a bulb and counterweight is brought to balance successively in air and in gas, by adjusting the balance case pressure. Absolute pressures are determined by a barometer and by a mercury filled manometer. Specific gravity is computed from ratio of absolute pressures. According to Boyle's law, gas density is proportional to its pressure and buoyant force is proportional to its density.

Figure 6-33 shows the Improved Ac-Me Gas Gravity Balance. Its description, together with operating instructions, is given[3] in ASTM D1070-52, and by the manufacturer.[4]

* A consistent system of units given; other consistent unit systems may be employed.

Apparatus: The *Senior* instrument (Fig. 6-33) is cylindrical in shape, about 6 in. O.D. and 20 in. long, mounted on cast aluminum base with leveling screws for table or tripod support. Table height is about 11 in. Mounted on accompanying tripod stand with accessories, height is about 64 in., weight about 76 lb.

The *Junior* balance is 4 in. diam and 13 in. long, with a capacity of 125 cu in. Essentially, it is a metal rod balance beam with large bulb at one end, large and small adjustable counterweights at other end, and cross arm at center. Beam is

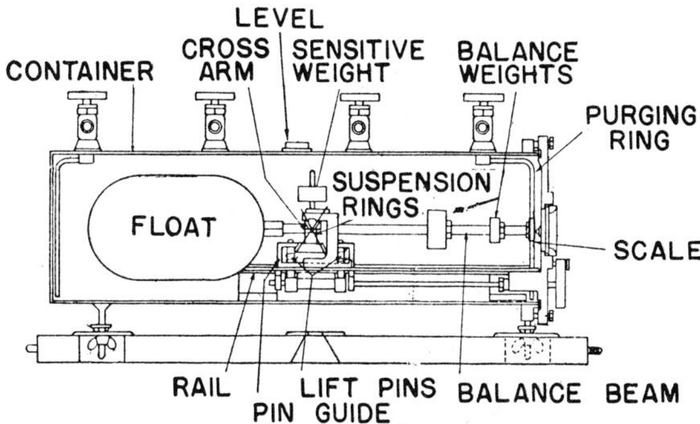

Fig. 6-33 Improved Ac-Me gas gravity balance (four-spring type).

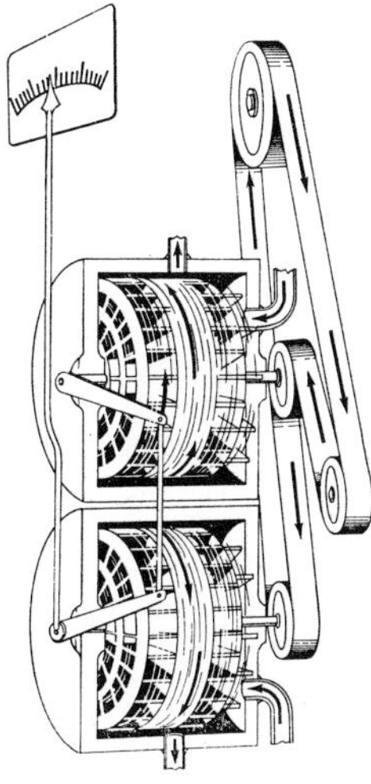

Fig. 6-34 Schematic of Ranarex gravitometers (differential torque type). (The Permutit Co.)

suspended from frame attached to face plate, with frame in turn sliding into cylinder, to open end of which plate is attached by cap screws. Construction provides for raising balance beam and securely locking it in place.

Accessories consist of a reciprocating, plunger-type hand pump with check valve for obtaining vacuum or pressure, and air dryer of indicating silica gel in a transparent plastic tubing. Connection from balance case to pump is made by needle valves and heavy rubber tubing with metal union fittings.

Procedure: In determining gas specific gravity, the following data should be recorded: (1) average barometer reading, (2) air reading, (3) gas reading, and (4) air check reading.

All readings should be made without change in adjustment or position of balance. Temperature should be recorded for each of the air, gas, and air check readings, with level checked before and after each, particularly when tripod mounting is used.

Reports are available on the following manually-operated specific gravity instruments.[2,5,6] A description of the **effusion method** is also available.[7]

Edwards Gas Density Balance (pressure balance).

ARCCO-Anubis Portable Gas Balance (balance between air and gas made successively at atmospheric pressure).

Metric Indicating Gravitometer (differential pressure type, gas and air pressure created by centrifugal blowers on a common shaft registered by manometers).

Ranarex Portable Specific Gravity Indicator (differential torque type)—Fig. 6-34.

Schilling Specific Gravity Apparatus (effusion type).

National Bureau of Standards Specific Gravity Apparatus (effusion type).

Recording Gravitometers

Ac-Me Recording Gravitometer. This instrument operates on the principle of weighing a volume of gas in comparison with an equal volume of air. As shown in Fig. 6-35a, it employs a float of $1\frac{1}{4}$ cu ft volume, thru which gas flows at a selected rate of 4 to 8 cfh. The float, along with balancing rod and weights, is suspended from one end of a balance bar; a calibrating rod, a pen arm linkage, and weights are on the other end. An automatic compensator corrects for changing temperature and barometric pressure. If properly installed and serviced regularly (say once or twice a month), this recorder will give accurate specific gravity readings within 0.003 sp gr. It requires no electricity and can be operated satisfactorily in unheated buildings. Operating characteristics were described.[2,4,8]

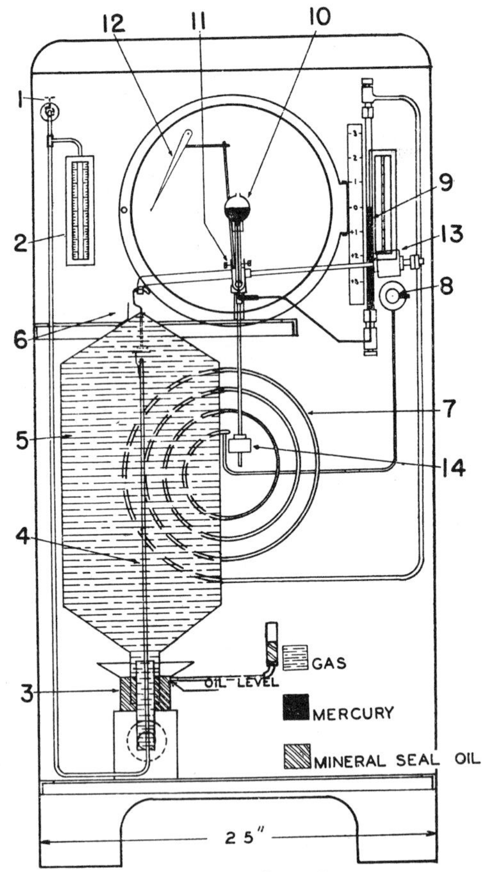

Fig. 6-35a Ac-Me recording gravitometer.

1—Inlet valve
2—Manometer
3—Oil seal
4—Inlet tube
5—Float
6—Pin for test weights
7—Compensator air tube
8—Compensator valve
9—Mercury gage
10—Mercury container
11—Adjusting screws
12—Pen
13—Balancing weight
14—Calibrating weight

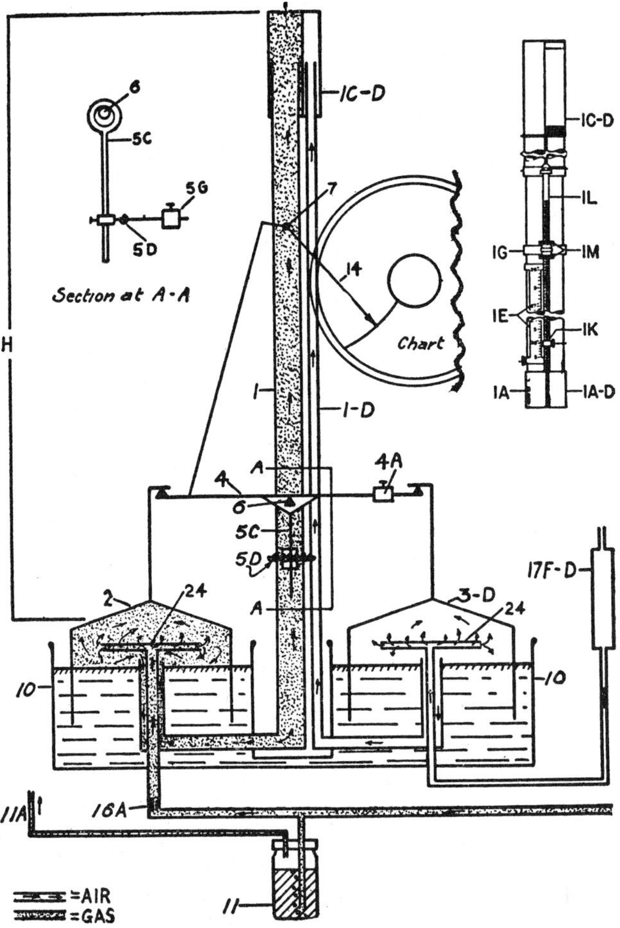

Fig. 6-35b Flow diagram for double column ARCCO-Anubis recording gas gravitometer.

1—Base gas column
1A—Removable gas column
1E—Barometric scale
1K—Index pointer
1L—Column adjusting screw
1-D—Base dry air column
1A-D—Removable air column
1C-D—Air column dryer
2—Working bell
3-D—Dry air bell
4—Beam
4A—Vernier zero weight
5C—Pendulum rod
5D—Thermostatic coil
5G—Pendulum weight
6—Front fulcrum link
7—Pen shaft
10—Tanks
11—Pressure regulator
11A—Regulator vent
16A—Orifice
17F-D—Case dryer
24—Diffusers

ARCCO-Anubis Recording Gas Gravitometer. This instrument, shown in Fig. 6-35b, is housed in a case about 25 in. high, $18\frac{3}{4}$ in. wide, and 11 in. deep. It consists of a balance beam with arms of equal length, at the ends of which are suspended two bells of the same diameter, wall thickness, and height. Their open ends are sealed by immersion in a sealing liquid which is contained in two interconnected tanks. The space above this liquid, in the working (gas) bell and in a connected vertical gas column, is supplied with about 0.7 cu ft of gas per hour. Similar space, in the balancing (dry air) bell and in a connected vertical air column, receives dry air thru an air dryer in the column top.

Gas at nearly atmospheric pressure passes thru the working bell into the gas column, and from there out its top to atmosphere. This gas column exerts upward pressure on the working bell, while a dry air column of equal height exerts upward pressure on the balancing bell. Thus, the actuating force of the instrument is the difference in pressure between the gas and dry air columns of the same height acting on the bell. To compensate for changes in buoyant forces, a pendulum weight is suspended below the beam fulcrum.

The compensated motion of the balance beam is transmitted to a pen mechanism which records the specific gravity of the gas on a clock-driven chart.

This gravitometer, as described, embodies several important improvements recommended by testing agencies and its users. Information based on previous models is available.[2,3]

Ranarex Specific Gravity Recorder (and Portable Indicator). The stationary indicator and recorder is about 27 in. high, 11 in. wide, 11 in. deep, and weighs about 90 lb. The portable indicator is about 17 in. × 8 in. × 8 in., and weighs about 32 lb. Both instruments operate on 110 v a-c (the portable unit also runs on 6 or 12 v d-c).

As shown in Fig. 6-34, a rotating motion is imparted to the gas by a motor-driven impeller running in a cylindrical chamber. This fan drives the gas against the blades of a similar fan or impulse wheel facing (but not touching) the first, in the same chamber. A torque is thus produced on the impulse wheel shaft. To eliminate influence of changes in fan speed, temperature, humidity, and atmospheric pressure, a torque (in the opposite direction) is produced in the second chamber by impinging air on another impulse wheel by means of a second fan driven by the same motor and belt. The two impulse wheels' shafts are coupled together by two lever arms and a connecting link, thus preventing complete rotation of the impulse wheels. Differences between the two opposing torques causes a limited movement of the system that is transmitted to a pointer. The pointer indicates the ratio of gas torque to air torque. These torques are proportional to the gas and air densities, respectively, i.e., the specific gravity. Scales reading in selected ranges from 0.07 to 3.0 sp gr are available.

The manufacturer claims 0.5 per cent accuracy in some applications and 1.5 per cent accuracy over total range in other applications. Full response within 20 sec was also claimed; inaccurate leveling, shock, and vibration were also claimed to be insignificant. The instrument may be furnished with means to saturate both the gas and air to any desired degree before whirling.

Calculation from Analysis

The **specific gravity** of a gas mixture can be readily calculated if an accurate determination of its composition has been made. The mass spectrometer and gas chromatograph yield analyses of gas mixtures (at 60 F and 1 atm) that are more accurate and complete than those obtainable by most other methods. Calculated **heating values** of equal reliability can also be obtained.

For a great many purposes, it is sufficient to estimate *ideal* values for specific gravity and heat of combustion. The equations used are:

$$G = x_1G_1 + x_2G_2 + x_3G_3 + \ldots + x_nG_n \quad \textbf{(5a)}$$

$$H = x_1H_1 + x_2H_2 + x_3H_3 + \ldots + x_nH_n \quad \textbf{(5b)}$$

where:

G = *ideal* specific gravity of gas mixture

$G_1, G_2, G_3, \ldots, G_n$ = *ideal* specific gravities of component gases

$x_1, x_2, x_3, \ldots, x_n$ = mole fractions of component gases

H = *ideal* heating value of gas mixture, Btu per SCF

$H_1, H_2, H_3, \ldots, H_n$ = *ideal* heating values of component gases, Btu per SCF (from Table 2-59)

If, however, greater precision in the calculated values is desired, by taking into account deviations from ideal gas behavior, the *ideal* values can be converted to the *real* basis by applying a calculated compressibility factor:[9]

$$G_r = G/z \quad \textbf{(6a)}$$

$$H_r = H/z \quad \textbf{(6b)}$$

$$z = 1 - (x_1b_1^{0.5} + x_2b_2^{0.5} + x_3b_3^{0.5} + \ldots + x_nb_n^{0.5})^2 + (2x_H - x_H^2)(0.0005) \quad \textbf{(7)}$$

where:

G_r = *real* specific gravity of gas mixture

H_r = *real* heating value of gas mixture, Btu per SCF

z = compressibility factor at 60 F and 1 atm

$b_1^{0.5}, b_2^{0.5}, b_3^{0.5}, \ldots, b_n^{0.5}$ = summation factors of component gases (from Table 2-63)

x_H = mole fraction of hydrogen in mixture

The magnitude of the compressibility correction is usually very small and may not be justified if the accuracy of the mixture analysis is not of the highest order.

REFERENCES

1. Shnidman, Louis, ed. *Gaseous Fuels,* 2nd ed. New York, A.G.A., 1954.
2. Smith, F. A. and others. *Tests of Instruments for the Determination, Indication, or Recording of the Specific Gravities of Gases.* Washington, G.P.O., 1947.
3. A.S.T.M. "Standard Methods of Test for Specific Gravity of Gaseous Fuels." (Standard D 1070-52) *ASTM Standards,* 1961, pt. 8. Philadelphia, 1962. (Available as a separate pamphlet.)
4. Refinery Supply Co., 621 East 4th St., Tulsa, Okla.
5. Altieri, V. J. *Gas Analysis & Testing of Gaseous Materials.* New York, A.G.A., 1945.
6. Edwards, Arthur H. Thomas, Philadelphia; Anubis, Am. Recording Chart Co., Los Angeles; Metric, Am. Meter Co., Erie; Ranarex, Permutit Co., New York; Schilling, Fisher Scientific Co., Pittsburgh.
7. Wilson, C. W. "Determination of the Specific Gravity of Gas by the Effusion Method." *A.G.A. Proc.* 1937: 777–82.
8. Chandler, A. W. "Ac-Me Gravitometer." *Appalachian Gas Measurement Short Course* 14: 273–77, 1954.
9. Mason, D. McA. and Eakin, B. E. "Proposed Standard Method for Calculating Heating Value and Specific Gravity from Gas Composition." *A.G.A. Proc.* 1961: CEP-61-11.

Chapter 10
Pressure Measurement

by Hugh E. Ferguson and F. E. Vandaveer

Pressure is the force borne by a unit of area, measured in such units as pounds per square inch or dynes per square centimeter. It may be expressed as fractions or multiples of standard atmospheric pressure; for example, 2 atm would be 29.4 psia. It may also be expressed in terms of height of a balancing column of liquid. Thus, normal atmospheric pressure (at sea level and latitude 45°) is balanced by a mercury column (in a barometer) 29.9213 in. or 760 mm high. Conversion factors for units of pressure are given in Table 1-39.

Measurement of gas and barometric pressures is probably the most common laboratory and field determination in the gas industry. Gas pressures are measured, indicated, or recorded at all compressor, regulator, and measurement stations along cross-country transmission lines; also at city border stations, at various locations in city distribution systems, and at numerous points in gas manufacturing plants. Pressure observations of domestic and industrial gas appliances are made to determine adequacy of gas supply and proper orifice sizing. Orifice meter measurements involve determining differential pressure across an orifice plate. Drafts in flues and chimneys are measured as pressures. Air pressure, steam pressure, oil pressure, and water pressure measurements are also frequently required.

Various kinds of pressure measurements are: (a) static, (b) velocity, (c) total, (d) differential, (e) pressure drop, and (f) vacuum.

Boyle's law (for ideal gases) is approximately applicable to real gases at low pressures. Atmospheric pressure is so low that error introduced by the assumption of ideal gas laws is less then one per cent. Actual error varies with each gas. For example, a change of 4.0 in. water column pressure causes a change of approximately one per cent in gas volume.

Normal or standard atmospheric pressure in most scientific work is 14.696 psia, 29.92 in. Hg, or 760 mm Hg. In gas industry testing, 30 in. Hg (at 45° latitude) is generally used as standard atmospheric pressure. The standard pressure base for natural gas measurement may be another value, defined by purchase contract or by regulatory authorities. It is usually expressed in lb per sq in.

Absolute atmospheric pressure can be measured by a barometer. Absolute or total gas pressure is the sum of barometric pressure plus line gas pressure (gage). Effects of elevation and latitude on barometric pressure are shown in Tables 6-7 and 6-8, respectively.

In 1963, A.G.A. endorsed the adoption of 14.73 psia and 60 F as the standard base conditions of pressure and temperature to be used in gas volume measurements, wherever this measurement is at such a pressure (greater than normal utilization pressure) that a correction is desired. Density of mercury varies with temperature; thus, height of 29.9213 in. (normal atmospheric —Table 1-39) applies exactly only at 32 F. Barometer scales also vary in length as temperature changes. Therefore, accurate barometer readings require corrections for both of these factors (Table 6-9).

Table 6-7 Elevation Effects on Pressure and Temperature[1]

(at about 40° latitude in North America, assuming temperature at zero altitude = 59°F, decreasing at the rate of 3.566°F per 1000 ft of elevation)

Height,* ft	Press., in. Hg	Height,* ft	Press., in. Hg
−500	30.466	6,000	23.978
0	29.921	6,500	23.530
500	29.385	7,000	23.088
1000	28.856	7,500	22.653
1500	28.335	8,000	22.225
2000	27.821	8,500	21.803
2500	27.315	9,000	21.388
3000	26.817	9,500	20.979
3500	26.326	10,000	20.577
4000	25.842	10,500	20.181
4500	25.365	11,000	19.791
5000	24.896	11,500	19.407
5500	24.434	12,000	19.029

* See Fig. 8-40 for data (in psia) up to 15,000 ft elevation.

Table 6-8 Barometer Corrections for Latitude[2]

	Observed barometer readings, in. Hg					
Latitude	25	26	27	28	29	30
25°	−0.043	−0.045	−0.047	−0.049	−0.050	−0.052
30°	−0.034	−0.035	−0.037	−0.038	−0.040	−0.041
35°	−0.024	−0.025	−0.026	−0.027	−0.027	−0.028
40°	−0.013	−0.013	−0.014	−0.014	−0.015	−0.015
45°	0.000	0.000	0.000	0.000	0.000	0.000
50°	+0.010	+0.011	+0.011	+0.012	+0.012	+0.012
55°	+0.021	+0.022	+0.023	+0.024	+0.025	+0.026
60°	+0.032	+0.033	+0.034	+0.036	+0.037	+0.038
65°	+0.041	+0.043	+0.045	+0.046	+0.048	+0.050

Table 6-9 Correction of Brass Scale Mercury Barometer for Temperature

°F	Observed barometer readings, in. Hg							
	24	25	26	27	28	29	30	31
0	+0.063	+0.065	+0.068	+0.070	+0.073	+0.076	+0.078	+0.081
10	+0.041	+0.042	+0.044	+0.046	+0.047	+0.049	+0.051	+0.053
20	+0.019	+0.020	+0.020	+0.021	+0.022	+0.023	+0.024	+0.024
30	−0.003	−0.003	−0.003	−0.003	−0.003	−0.004	−0.004	−0.004
40	−0.025	−0.026	−0.027	−0.028	−0.029	−0.030	−0.031	−0.032
50	−0.046	−0.048	−0.050	−0.052	−0.054	−0.056	−0.058	−0.060
60	−0.068	−0.071	−0.074	−0.077	−0.080	−0.082	−0.085	−0.088
70	−0.090	−0.094	−0.097	−0.101	−0.105	−0.109	−0.112	−0.116
80	−0.111	−0.116	−0.121	−0.125	−0.130	−0.135	−0.139	−0.144
90	−0.133	−0.138	−0.144	−0.150	−0.155	−0.161	−0.166	−0.172
100	−0.154	−0.161	−0.167	−0.174	−0.180	−0.187	−0.193	−0.200

Note: Zero correction to barometer at 28.5 F (a composite of a zero mercury correction at 32 F and a zero brass scale correction at 62 F). Example: For a 29 in. reading, at 70 F, about 0.11 in. Hg should be deducted from the observed value.

CONNECTION OF PRESSURE GAGES

For accurate pressure measurement, care must be taken in location and size of connection, tapping on pipe or vessel, and piping between it and the pressure gage.

Static pressure tappings, as normally used for measuring line pressure, should not be located where the venturi action of flowing gas thru a valve, control, or orifice causes a reduction in pressure; nor should it be placed where gas flow impinging on point of measurement, for example at top of an elbow or bend, causes the addition of velocity pressure to static pressure. Size of connection should be adequate to transmit changes in pressure rapidly. For low pressures, as in gas distribution systems, a 1/8 or 1/4 in. connection is adequate. For high pressure and for permanent installations, 1/4 in. or larger connections are used. Inside surfaces of tappings or fittings should be reasonably smooth. For transmitting pressure to gages at a considerable distance, tubing or piping of 1/4 to 3/4 in. diam is used, with the larger sizes for longer distances and for line strength and rigidity. Height h in Fig. 6-36a illustrates this concept.

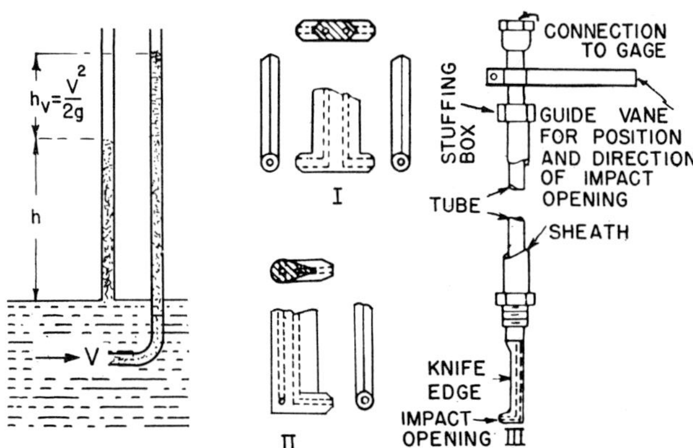

Fig. 6-36a (left) Velocity pressure, h_v, and static pressure connections.[4] A Pitot tube (Fig. 6-36b) incorporates both connections.

h = height of column of fluid created by static pressure, ft
$h_v = V^2/2g$ = height of column of fluid created by velocity pressure, ft
V = velocity of fluid flowing, ft per sec
g = acceleration constant of gravity, ft per sec-sec

Fig. 6-36b (right) Pitot tubes for insertion in pipes.[4] Flow in Sketch I can be from either direction; Sketch III shows a commercial version of Sketch II.

Velocity pressure connections are made with a tube, e.g., Pitot tube, with open end facing gas flow and usually located at pipe center (Fig. 6-36a). Note that Fig. 6-36a is directly applicable to flowing liquids. While the principle is the same for gases, the density difference between the flowing gas and the indicating fluid (commonly converted to feet of water) necessitates a different arrangement—see Fig. 6-37a.

PITOT TUBES[1]

Pitot tubes for measuring either *static* or *velocity* pressure or *sum* of both are illustrated in Fig. 6-36b. Representative

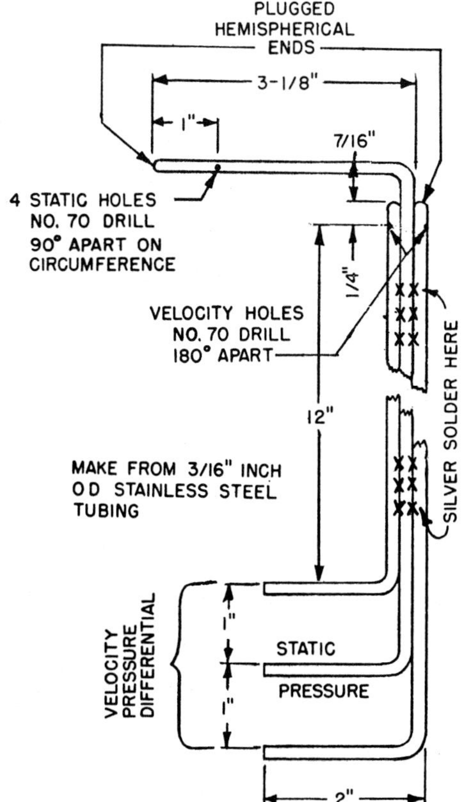

Fig. 6-36c Double acting special Pitot tube for indicating velocity pressure differential and static pressure in vent systems.[5] (Velocity Pressure Differential Multiplication Factor = 1.4; i.e., pressures obtained using this special tube must be divided by 1.4 to obtain the true local pressure.)

Pitot tubes of a compact type are illustrated in I and II of Fig. 6-36b. Such tubes have coefficients less than 1. For tubes like those in the I category, $V = 0.84\sqrt{2gh_v}$, i.e., the indicated water-column difference, h_v, is about $(1/0.84)^2$, or 42 per cent greater than the h_v of Fig. 6-36a. (Terms are defined under Fig. 6-36a.) Very slight changes in shape of a compact-type side-pressure-opening tube, like that in II, or in location of pressure openings, cause a change in tube coefficient. Here, $V = 0.84$ to $0.88\ \sqrt{2gh_v}$, depending upon location of side openings and shape of tube. For precise work such tubes should be calibrated. A single-opening tube in connection with a "wall piezometer" (i.e., either a single hole, a pair of diametrically opposite holes, or a ring piezometer) is the simplest arrangement and least liable to errors resulting from construction imperfections. The coefficient is unity ($V = \sqrt{2gh_v}$) for a tube having its impact opening projecting a little upstream from its body (III of Fig. 6-36b). Static pressure holes in pipe should be about 1 in. upstream from impact point, so that flow past them will not be disturbed by presence of tube.

A special Pitot tube for indicating velocity pressure differential and static pressure in flue pipe and venting systems is illustrated in Fig. 6-36c.

For orifice meter measurement of pressure, very definite specifications apply for pressure tap holes and for pipe lengths preceding and following an orifice.[3]

LIQUID COLUMN MANOMETERS

Manometers are U-shaped tubes, either vertical or inclined, filled with a liquid of known density. Difference in pressure between manometer ends is determined from displacement of liquid. A manometer with one end open to atmosphere gives a direct reading of equivalent gage pressure or vacuum. A U-tube manometer (Fig. 6-37a) may be used in either upright or inverted position. Air or indicating liquid at top of an inverted U-tube must be trapped before inversion, or it may be pumped in against existing pressure. Correction must be made for the difference in heights of columns of indicating liquid above the manometric fluid in each leg. Even with air above such fluid, correction should be made at high pressures, e.g., where density of air at 100 psig is about 0.01 that of water.

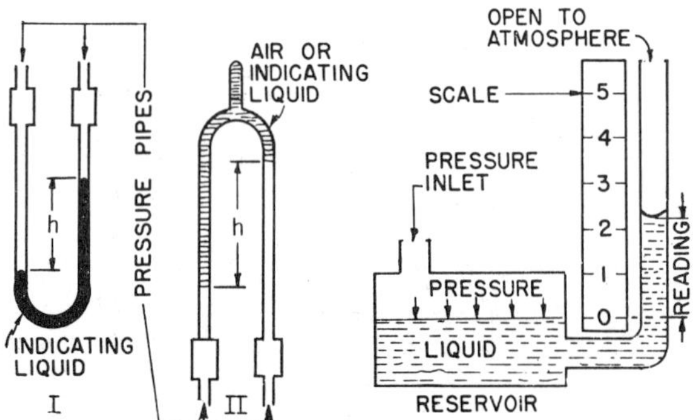

Fig. 6-37a (left) Glass U-tube manometers; h indicates pressure difference in units of the measuring fluid.

I Upright: measured fluid lighter than indicating fluid.

II Inverted: measured fluid heavier than indicating fluid.

Fig. 6-37b (right) Well, reservoir, or cistern-type manometer.

A manometer, with a large chamber or cistern as one U-tube leg, is shown in Fig. 6-37b. Scale may be calibrated to take account of the small changes in well level. Closing the tube marked "open to atmosphere" and evacuating volume above "reading" convert the device shown to a barometer.

The most commonly used manometer liquids are mercury, water, oil, and alcohol. Table 6-10 gives specific gravities of these liquids at 32 and 70 F and conversion factors for changing manometer readings to pressure in psi.

Tables 6-11 and 6-12 give equivalents in psi of water and mercury columns, respectively, at various temperatures.

Table 6-10 Specific Gravities and Conversion Factors for Manometer Liquids

Manometer liquids	Specific gravity relative to water at 39 F*		To convert inches of liquid to psi, multiply by	
	Manometer temp, 70 F	Manometer temp, 32 F	Manometer temp, 70 F	Manometer temp, 32 F
Mercury	13.543	13.595	0.489	0.491
Water	0.9980	...	.0361	...
Kerosene (44–46 deg API)	.799	0.812	.0289	.0293
Red draft gage oil (approx.)	.820	.835	.0296	.0301
Methyl (wood) alcohol	.790	.810	.0285	.0293
Ethyl (grain) alcohol	.789	.805	.0285	.0291
Gasoline (58–60 deg API)	0.739	0.752	0.0267	0.0270

* Density of water at 39 F = 62.43 lb per cu ft.

MULTIPLYING GAGES[6]

To increase precision in measuring small pressure differences by liquid-column manometers, it is often necessary to devise means of magnifying readings. With open manometers, a low density indicating fluid will give some magnification. For major magnification, means described below may be used. The first may give tenfold multiplication; the second, as much as thirtyfold.

Draft Gage

An inclined manometer or slope gage (Fig. 6-38) is commonly used for measuring low gas heads. One leg is a reservoir of much larger bore than that of the inclined leg. Variations

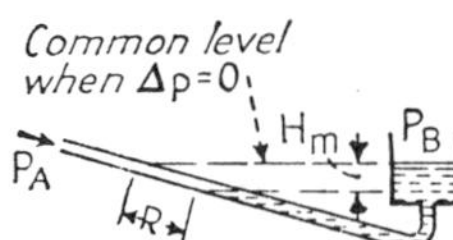

Fig. 6-38 Draft gage for low gas heads.

P_A and P_B are the pressures at A and B, respectively, in the same units; Δp or H_m is the difference in pressure between P_A and P_B (equals zero when $P_A = P_B$); R readings are magnified representations of Δp.

in inclined tube level therefore produce very little change in reservoir level. Although H_m may be readily computed in terms of the reading, R, and tube dimensions, gage calibration is preferable because changes in reservoir level often are not negligible. Moreover, variations in tube diameter may intro-

Table 6-11 Equivalents, in psi, of Water Columns in Inches at Various Temperatures

Water, in.	Water temperature, °F								
	40	50	60	70	80	90	100	150	200
10.0	0.3613	0.3612	0.3609	0.3605	0.3600	0.3595	0.3588	0.3541	0.3479
20.	0.7225	0.7223	0.7218	0.7211	0.7201	0.7189	0.7175	0.7083	0.6958
30.	1.0838	1.0835	1.0828	1.0816	1.0802	1.0784	1.0763	1.0624	1.0438
40.	1.4451	1.4447	1.4437	1.4422	1.4402	1.4378	1.4351	1.4166	1.3917
50.	1.8063	1.8058	1.8046	1.8027	1.8002	1.7973	1.7938	1.7707	1.7396
60.	2.1676	2.1670	2.1655	2.1632	2.1603	2.1567	2.1526	2.1248	2.0875
70.	2.5289	2.5282	2.5264	2.5238	2.5204	2.5162	2.5113	2.4790	2.4354
80.	2.8901	2.8893	2.8874	2.8843	2.8804	2.8756	2.8701	2.8331	2.7834
90.	3.2514	3.2505	3.2483	3.2449	3.2404	3.2351	3.2289	3.1873	3.1313
100.0	3.6126	3.6117	3.6092	3.6054	3.6005	3.5945	3.5876	3.5414	3.4792

Example: The psi equivalent of 86.95 in. w.c. at 95 F may be computed as the sum of the psi equivalents of 80.0 plus 6.0 plus 0.9 plus 0.05 in. w.c. = (2.8728 + 0.2155 + 0.0323 + 0.0018) = 3.1224 psi.
(Note that the submultiples of the tabulated data were obtained by suitably moving the decimal points in the data given.)

Table 6-12 Equivalents, in psi, of Mercury Columns in Inches at Various Temperatures

Mercury, in.	Mercury temperature, °F								
	20	30	40	50	60	70	80	90	100
10.0	4.9175	4.9126	4.9076	4.9026	4.8977	4.8928	4.8879	4.8830	4.8780
20.	9.8351	9.8251	9.8152	9.8053	9.7955	9.7856	9.7757	9.7659	9.7561
30.	14.753	14.738	14.723	14.708	14.693	14.678	14.664	14.649	14.634
40.	19.670	19.650	19.630	19.611	19.591	19.571	19.551	19.532	19.512
50.	24.588	24.563	24.538	24.513	24.489	24.464	24.439	24.415	24.390
60.	29.505	29.475	29.446	29.416	29.386	29.357	29.327	29.298	29.268
70.	34.423	34.388	34.353	34.319	34.284	34.249	34.215	34.181	34.146
80.	39.340	39.300	39.261	39.221	39.182	39.142	39.103	39.064	39.024
90.	44.258	44.213	44.168	44.124	44.080	44.035	43.991	43.947	43.902
100.0	49.175	49.126	49.076	49.026	48.977	48.928	48.879	48.830	48.780

Example: The psi equivalent of 48.36 in. Hg at 85 F may be computed as the sum of the psi equivalents of 40.0 plus 8.0 plus 0.3 plus 0.06 in. Hg = (19.542 + 3.908 + 0.147 + 0.029) = 23.626 psi.
(Note that the submultiples of the tabulated data were obtained by suitably moving the decimal points in the data given.)

duce serious error. Commercial gages are often provided with a scale giving H_m directly in inches of water, if a *particular* indicating fluid is used. Failure to appreciate that scale is incorrect, unless gage is filled with specified liquid, is a frequent source of error. If scale reads correctly when density of gage liquid is d_o, then reading must be multiplied by d/d_o if density of fluid in actual use is d.

Two-Fluid U-Tube

These tubes (Fig. 6-39) are very sensitive instruments for measuring small gas heads. Let A be the cross-sectional area of each reservoir, and a that of U-tube; let d_1 be density of lighter fluid, d_2 that of heavier, and d_w density of water—all in pounds per cubic foot. Then for Type I (Fig. 6-39), $P_A - P_B = (R - R_o)[(ad_1/A) + d_2 - d_1]/d_w$ in. w.c., where the

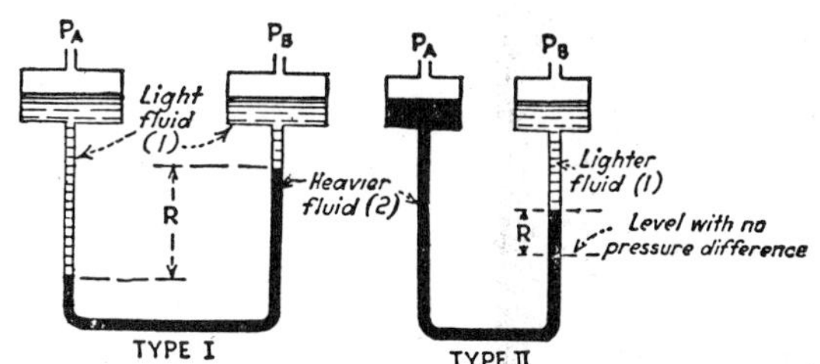

Fig. 6-39 Two-fluid U-tube.[6]

reading R is in inches and R_o is its value* with zero pressure difference. For Type II, if the reading R (in.) is taken as shown in Fig. 6-39,

$$P_A - P_B = \frac{R}{d_w}\left[d_2 - d_1 + \frac{a}{A}(d_2 + d_1)\right] \text{ in. w.c.}$$

When a/A is sufficiently small, terms $(a/A)d_1$ and $(a/A)(d_2 + d_1)$ of foregoing formulas become negligible in comparison with difference $(d_2 - d_1)$. *However, these terms should not be omitted without due consideration.* Since reading magnification varies inversely with difference in densities, the tendency is to choose a liquid pair for which that difference is very small. Consequently, to account for preceding omissions, reservoirs required may be of excessive diameter. In applying the formulas, densities of gage liquids may not be taken from tables without introducing serious error, because each will dissolve appreciable quantities of the other. Before filling gage, the liquids should be shaken together, and densities of the two layers should be determined for temperature at which gage is to be used. When high magnification is sought, it may be necessary to enclose the U-tube in a constant temperature bath, so that $(d_2 - d_1)$ may be accurately known. In general, if accuracy is desired, the gage should be calibrated.

* Assuming that the same end of the heavier fluid is always lower than the other end.

MICROMANOMETERS[6]

Micromanometers, based on the liquid-column principle, of extreme precision and sensitivity (0.001 – 0.00001 in. liquid), have been developed for measuring minute pressure differences and for calibrating low-range gages. Although greatest accuracy can be obtained only under laboratory conditions, these gages have been used successfully in Pitot-tube tests at low gas velocities and for similar purposes, when the pressure has been reasonably steady. In use, they should be placed on a firm support with all parts at a uniform temperature and glass portions clean. These micromanometers are free from errors due to capillarity, and, aside from checking micrometer scale, they require no calibration.

The simplest micromanometers are merely upright U-tubes or vertical tubes provided with a delicate means, like a **hook gage** or microscope, for determining manometric liquid levels. Legs of such a gage are at least 2 or 3 in. in diameter, to avoid errors due to capillarity.

Tilting Manometer

In the Chattock and Chattock-Fry instruments, a U-tube, with widely separated legs of large diameter connected by a smaller tube, is mounted in a vertical plane on a rigid beam that can be raised at one end by a micrometer screw (Fig. 6-40a). A slight excess of pressure at opening II, slightly depresses right arm fluid level, causing bubble of immiscible liquid at **A** to move to some position **B**. The greater the reservoir diameter is, compared with connected tube diameter, the greater will be the distance $\overline{\mathbf{AB}}$ for a given change in level. Tilting gage by turning micrometer screw returns bubble to its zero position at **A.** In practice, a microscope mounted on the instrument beam is used to observe position of bubble. From instrument dimensions and micrometer readings, pressure difference can be calculated. Capillary errors are avoided, because at time of reading all interfaces are always at same position relative to U-tube. An instrument of this type with a sensitivity of 0.00001 in. of manometric liquid has been devised.

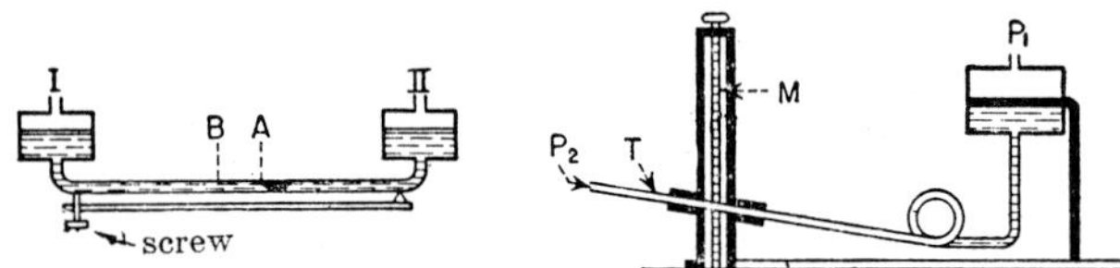

Fig. 6-40a (left) Tilting manometer.[6]
Fig. 6-40b (right) Micromanometer.[6]

U-Tube with Flexible Inclined Leg

Illustrated in Fig. 6-40b is a U-tube with one flexible leg. A rigidly held reservoir is connected to an inclined glass tube, T, by a rubber tube. Tube, T, is carried by a sliding block that may be raised or lowered by micrometer screw, M. Readings are made by adjusting micrometer until the meniscus is at a scratch previously placed on tube. Difference between zero reading of micrometer and reading under pressure gives pressure difference directly. Differential heads of less than 0.001 in. of manometric fluid can be determined. Sensitivity may be varied by changing inclination of tube, T.

Wahlen Gage

This two-fluid inclined tube micromanometer uses fluids of very nearly equal density.[7]

GAGES FOR LOCATING LIQUID SURFACE[6]

Several instruments are used for precisely locating position of a free liquid surface (for, say, a manometer liquid).

Hook Gage

Figure 6-41a shows a hook gage. It utilizes the very distinct optical effect that is produced when a sharp point pierces a liquid surface from beneath. Upturned point should be a sharp cone of large vertex angle (45, 90, or even 120 deg).

Point Gage

A point gage resembles a hook gage, but point is directed downward. In use, point is lowered until a "bubble" is formed on liquid surface. A plumb bob suspended by a metal tape measure makes a serviceable point gage.

Float Gage

A float gage (Fig. 6-41b), for exact measurements, is usually a hollow metal float attached to a slender vertical spindle which moves in guides along a scale. The float may be made to operate a pointer on a dial or pen of a recording device. When precision is desired, these additions are unwise because of possibility of lost motion in, or inertia of, mechanism. When using a float gage, bounding walls of liquid surface should be at least 1 in. from float at all points, to avoid errors due to capillarity.

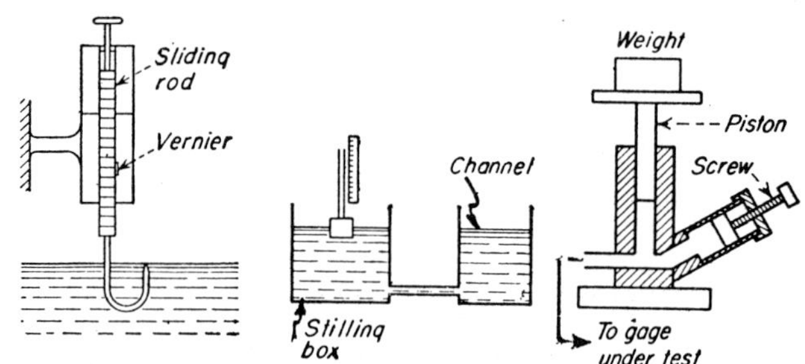

Fig. 6-41a (left) Hook gage.[6]
Fig. 6-41b (center) Float gage.[6]
Fig. 6-41c (right) Dead-weight tester.[6]

If liquid surface is substantially stationary, little difference appears between hook and float gages, so far as accuracy is concerned. Both can be made to give correct readings within less than 0.01 in. If level fluctuates even slightly, however, it is difficult to obtain satisfactory hook-gage readings. The float gage, with the advantage of being an indicating gage, is then preferable. When head of a fluid flowing in an open channel is to be measured accurately, all these gages must be used in a stilling box which has free communication with flowing fluid by an opening flush with channel wall.

MECHANICAL PRESSURE GAGES

Bourdon Tube

The most common of the spring-type gages, the Bourdon tube is elliptical in cross section, bent into a circular arc. When internal pressure is applied to such a tube, it tends to straighten out. Single-spring and double-spring arrangements are used in pressure gages. A variety of Bourdon gages is available, from the nominally priced common production-type gage to the precision gage, with special adjustments including a micro zero setting. Such precision gages are accurate to about 0.25 per cent of full-scale reading. They are made in large sizes, usually 10 to 16 in. For accurate work, a Bourdon gage should be calibrated frequently on a dead-weight gage tester. In use, it should be protected from vibration, from excessive temperatures, and from corrosive liquids or gases. Gages of special materials are available, in case corrosion is a problem.

Diaphragm and Bellows

These gages are used mainly for low pressures, though metallic-bellows gages are available for pressures as high as 200 psi. Diaphragm or bellows gages provide larger forces to actuate their indicating or recording mechanism than bent-tube gages. Consequently, they are especially suitable for measuring pressures in the manometer or low Bourdon-type ranges. Slack-diaphragm gages use soft elastic diaphragms of leather, treated cloth, or rubber, externally spring-loaded. Several commercial makes of draft gages operate on this principle. They have the advantage of not involving any liquid or requiring accurate leveling.

GAGE CALIBRATION

Simple liquid-column manometers do not require calibration, if their construction minimizes capillarity errors. If their measuring scales have been checked against a standard, their accuracy depends solely upon precision of determining the position of their liquid surfaces. It follows that liquid-column manometers are used to standardize other gages.

For high pressures and, with commercial mechanical gages, even for quite moderate pressures, a **dead-weight gage** (Fig. 6-41c) is often used as the primary standard, because it is safer and more convenient than a manometer. A piston moves in a cylinder of defined cross-sectional area, usually 1.0 or 0.5 or 0.25 sq in. Instrument to be examined is connected to tester by a small metal tubing, filled with oil and provided with a screw-pump for slowly and simultaneously increasing pressure in both tester and instrument. A known weight is placed on the piston tray of the tester, and oil pressure increased until it supports tray and weight. Known weight of tray, its burden, and area of piston indicate pressure actually imposed on instrument, in psi. Instrument reading may then be adjusted to this true value. In calibrating and adjusting a pressure instrument, a series of at least five tests is desirable, at points approximating 10, 30, 50, 70, and 90 per cent of its normal range.

When manometers are used as high-pressure standards, an inordinately high mercury column may be avoided by connecting a number of U-tubes in series. Multiplying gages are standardized by comparing them with any of the micromanometers. Procedure in calibrating a gage merely consists of connecting it in parallel with a standard instrument to a reservoir in which any desired constant pressure may be maintained. Readings of the unknown gage are then made for various reservoir pressures, as determined by the standard.

Table 6-13 Characteristics of Gages for Measurement of Low Absolute Gas Pressures[8]

	Recommended range,* mm Hg	Operating principles	Effect of gas composition	Contamination of gas by gage	Contamination of gage by gas	Temperature independence	Utility at low temp	Utility at high temp	Remarks
Bourdon tube	10 to >1000	Mech. deflect. & indic.	A†	A	A	A	A	A	High pressure only
Aneroid bellows	0.1 to 1000	Mech. deflect. & indic.	A	A	A	A	B	B	...
Diaphragm gage	$>10^{-3}$ to 10	Mech. deflect. & elect. indic.	A	A	A	C	D	D	Rather delicate
Liquid manometer	>0.1 to >1000	Mech. deflect. & indic.	A	C	C	A	B	B	Vap. press. gas solubility
McLeod	$>10^{-5}$ to >10	Compress.: Boyle's law	B	C	C	A	B	C	Noncondensable gases only
Knudsen	10^{-6} to $>10^{-4}$	Molecular rebound from heated surface	B	A	A	C	B	B	Delicate, temp affects calibration
Pirani & thermocouple	$>10^{-3}$ to <1	Thermal conductivity	C	A	C	C	C	C	Temp affects calibration
Ionization gage	10^{-7} to 10^{-3}	Ionization by low-energy electrons	C	C	C	A	B	B	Hot filament may injure gas
Philips gage	$<10^{-5}$ to $>10^{-2}$	Electrical discharge in magnetic field	C	C	C	A	A	A	Some gettering by discharge
Alphatron ionization gage	$>10^{-4}$ to 1000	Ionization by high-energy alphas	C	A	B	A	A	A	No gettering

* A single instrument will not necessarily cover the entire given range.
† Guide to usage: A—recommended; B—slight trouble; C—moderate trouble; D—severe trouble.

VACUUM OR LOW ABSOLUTE GAS PRESSURE GAGES[8]

Vacuum here refers to pressures less than atmospheric, measured either from the base of *zero* pressure, i.e., 0 in. Hg *absolute;* or from the base of atmospheric pressure, i.e., 0 in. Hg *vacuum* = 1.0 atm. Vacuum pressures are commonly expressed in inches or in millimeters of mercury. Data on pressure gages for the range 10^3 to 10^{-7} mm Hg are given in Table 6-13.

The standard for high vacuum measurement is the McLeod gage, from which all other types of vacuum gages are commonly calibrated. This device involves direct measurement of vacuum in terms of absolute pressure of mercury. One type of McLeod gage (Stokes, Fig. 6-42) operates on a swivel from charging position **a** to measurement position **b**, and cuts off a definite volume of rarefied gas at the unknown pressure of the vacuum system. Trapped gas is compressed to a smaller volume at a higher pressure, equal to difference in level between mercury in center tube, C, and the compensating capillary. From Boyle's law, a suitable scale can be calibrated to represent pressure value in the vacuum system. This unit can be used for recording by means of a resistance-type electrical measuring system which indicates height of mercury in measuring tube by changing resistance in accordance with changes in mercury level. Resistance value, in effect, measures pressure in vacuum system. This type of gage can measure only at intervals. It has a wide range from a fraction of a micron* to 5000 μ. Since Boyle's law applies only to non-condensable gases, where water vapor or other condensable gases are present, a false reading results in indicating a higher vacuum than actually exists. This source of error may be eliminated by a chemical trap to absorb condensable vapor, or by a freezing trap utilizing dry ice or liquid air.

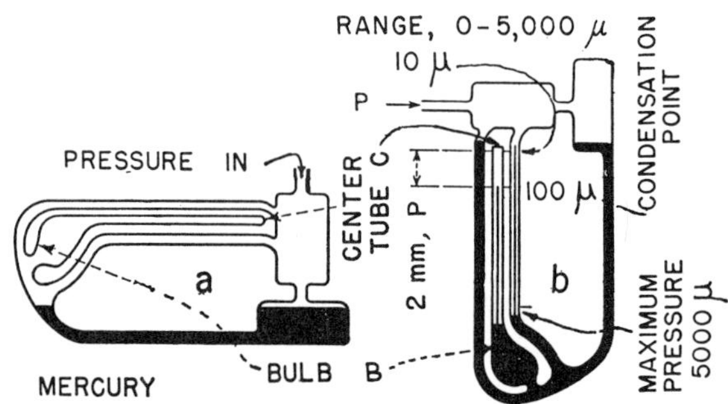

Fig. 6-42 McLeod gage (Stokes).[6]

LABORATORY DETERMINATION OF GAS APPLIANCE PRESSURE DROP

For determining pressure drop thru a manifold or thru individual gas controls, the American Gas Association has established the following test method[9] for use with natural, manufactured, and mixed gases:

* 1 micron = μ = 10^{-3} mm.

Complete manifold assembly shall be connected to a gas or air line beyond outlet of a meter of adequate capacity. A suitable gage which may be read directly to 0.005 in. water column pressure shall be connected as close to manifold as practicable.

When separate controls are being tested, length of straight run of pipe before and after them shall be not less than ten pipe diameters (ID), and connections to gage shall be located five pipe diameters (ID) from inlet and outlet connections.

Discharge end of valves shall be open to the atmosphere (spud or hood removed). If outlet end of valve is tapped, that portion which actually holds spud shall be sawed off and burrs removed.

All controls (except certain temperature limiting controls) shall be in fully opened position, which, in case of electrically operated controls, is equivalent to position assumed in normal (not manual) operation. Gas shall be burned as far from point of discharge as practical.

Gas rate shall be adjusted to give an indication on gage approximately equal to 1.0 or 1.5 in. pressure drop (depending on design) and necessary observations made and recorded. Pressure drop at manufacturer's hourly Btu input rating shall be calculated from these data on equivalent of: (1) a 500 Btu per cu ft, 0.6 sp gr gas for controls to be used with gases having a total heating value of less than 800 Btu per cu ft, (2) an 800 Btu per cu ft, 0.7 sp gr gas for controls to be used with mixed gases having a total heating value of 800 to 950 Btu per cu ft, (3) a 1000 Btu per cu ft, 0.64 sp gr gas for controls to be used with natural gases having a total heating value of 950 Btu per cu ft or more, or (4) a 2500 Btu per cu ft, 1.53 sp gr gas for controls to be used with liquefied petroleum gases. The following formula will be applied:[9]

$$pd_2 = \frac{pd_1 \times T \times Q_2^2 \times (sp\ gr)_s}{P_t \times sp\ gr \times Q_1^2} \times 0.0577$$

where:

pd_2 = pressure drop, in. w.c., thru manifold and controls at manufacturer's hourly Btu input rating, in terms of: (1), (2), (3), or (4) in foregoing text.

P_t = absolute pressure of test gas (or air) as metered, in. Hg

$sp\ gr$ = specific gravity of test gas (or air) determined as follows:

$$\frac{sp\ gr_1(P_t - at)}{P_t} + \frac{at \times sp\ gr_2}{P_t}$$

$sp\ gr_1$ = specific gravity of dry test gas (or air) referred to dry air as 1.0 (at standard temperature and pressure)

$sp\ gr_2$ = 0.62 = specific gravity of water vapor referred to dry air as 1.0

$sp\ gr_s$ = 0.60 when calculations are based on a 500 Btu gas
= 0.70 when calculations are based on an 800 Btu gas
= 0.64 when calculations are based on a 1000 Btu gas
= 1.53 when calculations are based on a 2500 Btu gas

at = aqueous tension of water vapor in test gas (or air), in. Hg

T = temperature of test gas (or air) as metered, °F abs

Q_1 = quantity of test gas (or air) as metered, cfh

Q_2 = manufacturer's rating for appliance in terms of: (1), (2), (3), or (4) in foregoing text, cfh

pd_1 = observed pressure drop corrected for difference in velocity head, if any, due to change of area at points tappings are taken, in. w.c.

In event areas of inlet and outlet tappings are different:

$$pd_1 = pd_o + h_{v1} - h_{v2}$$

where: Velocity head, in. w.c., at inlet tapping (h_{v1}) or outlet tapping (h_{v2}) is found by following formula:

$$h_v = \frac{1.0335 \times 10^{-5} \times Q_1^2 \times P \times sp\ gr}{D^4 \times T}$$

and:

pd_o = pressure drop (may be negative) between inlet and outlet pressure tappings on manifold as observed, in. w.c.

D = inside diameter of pipe at inlet (for h_{v1}) or outlet (for h_{v2}) pressure tapping, in.

P = absolute pressure of test gas (or air) at inlet (for h_{v1}) or outlet (for h_{v2}) pressure tapping, in. Hg

REFERENCES

1. U. S. Nat. Advisory Committee for Aeronautics. *Standard Atmosphere, Tables and Data for Altitudes to 65,800 Feet.* (Report 1235) Washington, G.P.O., 1956.
2. Lange, N. A. *Handbook of Chemistry*, 9th ed., p. 1687. Sandusky, Ohio, Handbook Publishers, 1956.
3. A.G.A. *Orifice Metering of Natural Gas.* (Gas Measurement Com. Rept. 3) New York, 1955.
4. Marks, L. S. *Mechanical Engineers' Handbook*, 6th ed., New York, McGraw-Hill, 1958.
5. Reed, H. L. *Field Survey of Gas Appliance Venting Conditions—Part I*, p. 23. (Research Rept. 1243) Cleveland, A.G.A. Laboratories, 1956.
6. Perry, J. H. *Chemical Engineers' Handbook*, 3rd ed. New York, McGraw-Hill, 1950.
7. Willard, A. C. and others. *Investigation of Warm-Air Furnaces and Heating Systems.* (Univ. of Illinois Experiment Station Bul. 120) Urbana, U. of Ill., 1921.
8. Lawrence, R. B. "Survey of Gauges for Measurement of Low Absolute Gas Pressures." *Chem. Eng. Progr.* 50: 155–60, Mar. 1954.
9. A.G.A. Laboratories. *American Standard Approval Requirements for Central Heating Gas Appliances, Vol. II: Gravity and Forced Air Central Furnaces.* (Z21.13.2-1961) Cleveland, 1961.

Chapter 11
Temperature Measurement

by Hugh E. Ferguson and F. E. Vandaveer

Measurement of temperature enters into many phases of the gas business. The temperatures measured range from approximately −258 F for liquefied natural gas to about 3700 F for some flame temperatures. Most common measurements are for gas temperatures in pipelines, air temperature, flue gas temperature of appliances, oven and furnace temperatures, and operating temperatures in gas manufacturing plants. Temperature measurement also frequently becomes a part of accurate volume measurement (**Charles' law**), since there is approximately one per cent volume change per 5°F temperature change. Table 1-44 shows relationships among the various temperature scales.

Thermocouples are the most important of the elements which depend upon changes in electrical characteristics to indicate temperature change. Units in this group, with the exception of **resistance thermometers,** are one class of **pyrometers.**

For correct temperature measurement the thermal measuring element must be properly located and, if necessary, shielded from hotter or colder radiation sources. Thermometers should be immersed to the point at which they were calibrated or a correction must be applied. ASME Codes recommend attaching a second thermometer with its bulb near the middle of exposed stem portion, with correction K made by the formula:

$$K = 0.00009D(t_1 - t_2)$$

where: D = extent of exposed mercury filament, °F
t_1 = reading of main thermometer, °F
t_2 = reading of attached thermometer, °F

When stem is cooler than bulb, correction K is added.

When measuring temperature of rapidly moving gases, an axial thermometer location is desirable. This arrangement will not give average temperature indication because of temperature gradient between bulb and walls. If container is well insulated externally, or if a thin metal shield is interposed between its walls and thermometer bulb to absorb radiation from surrounding walls, good temperature results will be obtained. A small diameter wire thermocouple in such a location would indicate approximately correct temperature. Where the cross section permits, the traverse method of Table 7-6 and Fig. 7-12 should be used.

External pressure corrections should be applied to glass-stem thermometers when bulbs are exposed to high pressures. Correction: subtract about 0.01°F per psi in case of a high grade thermometer with bulb wall thickness of 0.5 to 0.7 mm.

Proper selection of measuring instruments is essential for accuracy.[1–7] For example, base-metal thermocouples are recommended for determinations of temperatures of gases containing combustibles, since noble metal thermocouples were found to act as catalytic agents.[7a]

FLUE GAS TEMPERATURE MEASUREMENT

To obtain a good representation of the temperature, readings should be taken at various locations in a given cross section to obtain a traverse; radiation errors are inherent. If the measurement device can "see" (or is enclosed by) surfaces hotter than the gases, the readings will be high; if the device can "see" (or is enclosed by) surfaces colder than the gases, the reading will be low.

There are many ways to approach true gas temperature measurement; reference 2 describes 26 methods, of which the four simplest and most often used follow (in order of increasing accuracy):

1. **Bare bead thermocouple.** A.G.A. studies have shown that an error of ±50°F may result from the location of the measurement for 525 F flue gas temperatures. Generally at about 5 ft downstream of the draft hood of an appliance, a bare bead thermocouple traverse is considered accurate enough for determining flue loss from standard charts, using CO_2 concentration at the plane of temperature measurement.

2. Reduction of the emissivity of the bare bead reduces radiation errors.[8] **Aluminum foil** pressed over the bead is a good reduction means.

3. **Thermocouples of different sizes.**[8] By using, say, three couples (of known bead diameter) of decreasing size and by extrapolating a plot of these measurements to zero diameter, fairly good accuracy can be obtained.

4. A simple **high velocity thermocouple** will give gas temperatures to within ±5 to 10°F of the true gas temperature at temperatures around 525 F.[9] A traverse of temperatures is still needed with this device.

High-Temperature Illustration. Data for a laboratory-type gas-fired furnace are given in Table 6-14.

Table 6-14 Readings of Different Types of Thermocouples[1a]
(in a 4-in. square duct*)

	Reading, °F
Estimated true gas temperature	2105
Multiple-shield, high-velocity thermocouple	2103
Single-shield, high-velocity thermocouple	1963
Optical pyrometer "hot wire"	1607

* Errors would decrease with increasing duct size.

LIQUID-IN-GLASS THERMOMETERS

Mercury-filled glass thermometers, the most common in this class, are mainly used for the range −35 to +750 F. For temperatures below freezing point of mercury, liquids like toluol, alcohol, or pentane are used. Glass thermometers are calibrated for either total or partial immersion.

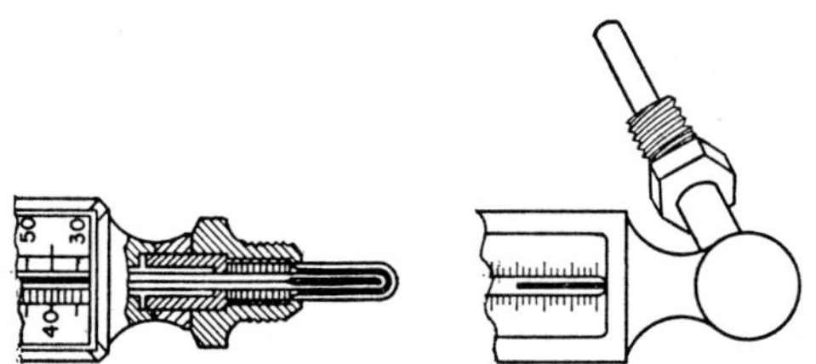

Fig. 6-43 Industrial thermometers[10]—partial immersion straight (left) and angle (right) types.

Figure 6-43 shows two forms of industrial thermometers. These are partial-immersion types, usually employing mercury, protected with a glass-covered metal scale and a metal tube over the bulb. They are available in various lengths with scale faces set at various angles. Temperature lag is reduced by filling annular space between bulb and protecting tube with a heat-transfer medium, such as oil, mercury, copper powder, graphite, or low-melting metal. For measuring temperature of gases, protecting tubes are perforated to decrease temperature lag.

LIQUID-EXPANSION THERMOMETERS

Liquid-expansion thermometers are usually filled with mercury (Class V).* Organic fluids, like hydrocarbons, may also be used; nonvolatile liquids other than metals fall in Class I.* Liquid expansion results in an approximately linear relationship between temperature and movement of receiving element. **Mercury-filled** thermometers are satisfactory over a range of −40 to +1000 F. **Organic fillings** may be available in a range of −170 to +500 F.

Volume changes of capillary fluid, due to ambient temperature variations along the tube, are a major error source. Uncompensated mercury-pressure thermometers are usually limited to a maximum capillary length of 25 ft. Compensation for ambient temperature changes (Class V-A is fully compensated via a bimetal strip) is generally necessary if capillary is longer than a few feet, particularly with organic fillings. Compensated mercury-filled thermometers may have capillary lengths up to about 200 ft. A fully compensated liquid-filled system (other than a metal) is in Class I-A;† case-compensated (via bimetal) liquid-filled systems are in Class I-B.†

* Designated by the Scientific Apparatus Makers Association (SAMA).

LIQUID-PRESSURE THERMOMETERS

Confined liquids exert an increased pressure when heated. These pressures, which are proportional to temperature, actuate devices like that in Fig. 6-44. It consists of a bulb,

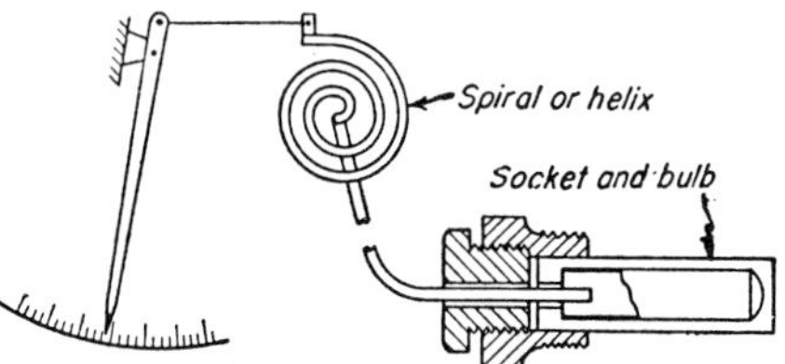

Fig. 6-44 Pressure thermometer with spiral element.[10]

often enclosed in a protecting socket, connected by capillary tubing to an element which is capable of expanding or otherwise altering its dimensions under increasing pressure.

GAS- AND VAPOR-PRESSURE THERMOMETERS

Gas-pressure instruments (Class III)† are primarily used to calibrate other types of thermometers. They depend upon the pressure-volume-temperature perfect gas relationship of **Boyle's** and **Charles' laws,** which at low pressures is approximately true for real gases. Nitrogen, helium, hydrogen, and oxygen are commonly used. The gas selection is influenced by the indicating fluid used, e.g., O_2 would not be used in conjunction with Hg. Instrument types include: *constant pressure, constant volume,* and composites of these.

Gas-pressure thermometers are generally satisfactory over ranges between −125 and +800 F. Gas-filled thermometers are affected by ambient conditions along capillary tubes and can be compensated in the same way as mercury thermometers. However, gas-filled thermometers are often fully compensated (Class III-A)† if the large bulb volume is less than 30 times the capillary volume. Partial compensation via a bimetal strip is designated Class III-B.† Large bulbs are usually satisfactory, since this type of thermometer is often used for averaging temperatures. Maximum capillary length is about 200 ft.

Vapor-pressure thermometers (Class II)† are made in ranges from about −40 to +600 F, depending upon filling medium. Bulb systems are partly filled with any one of a wide variety of volatile liquids, including methyl chloride, ether, sulfur dioxide, benzol, toluol, butane, propane, and hexane. Element is actuated by liquid vapor pressure, which varies with temperature according to thermodynamic laws. Moderately small bulbs and high responsiveness are characteristic of vapor-pressure thermometers. Their maximum tubing length is usually 250 ft.

The following subclasses of the SAMA classification apply to the measured temperature, *m*, as related to the rest of the system temperature, *s:* II-A for *m* greater than *s;* II-B for *m*

† SAMA designation.

less than s; II-C for m either less than or greater than s; and II-D for m less than, equal to, or greater than s (using liquid, volatile and nonvolatile).

BIMETALLIC THERMOMETERS

These depend upon differential expansion of strips of dissimilar metals of relatively high and low coefficients of expansion, such as brass, monel, or steel, and iron nickel-iron, or invar; they are laminated by welding, brazing, soldering, or even riveting. The material may also be coiled into spirals or helixes. As bimetal temperature changes, the system deflects. The motion may operate only an indicating device, or it may be used to actuate a control device. This type of unit is used in many thermostats.

THERMOCOUPLES[11]

An electric current is generated and flows in a closed circuit made up of wires of two dissimilar metals, if the temperature of one junction exceeds that of the other (**Seebeck effect**). The corresponding emf is usually measured by a potentiometer or millivoltmeter. The junction held at a constant temperature is known as the *cold* or *reference junction*, and the other as the *hot* or *measuring junction*.

The iron-constantan couple is available with an iron element in the form of a tube surrounding a constantan wire element. This "pipe" or "pencil" type device, commonly used without a protecting tube, has high response speed in small sizes. Another wire-within-tube arrangement, the **needle thermocouple,**[12] consists of an insulated thermocouple wire within a small diameter tube; the assembly is run thru a die, thereby bonding the tube to the wire.

Table 6-15 shows temperature-emf relations (entirely empirical) for various thermocouples. Cold junctions at other than 32 F may be compensated for. For example, when the cold junction temperature is 75 F and a copper-constantan couple is used, a millivoltmeter would read 3.967 *less* 0.945, or 3.022 mv for a 200 F hot junction.

The cold junctions of potentiometers can usually be compensated by varying the resistance in their bridge

Table 6-15 Temperature-Millivolt Relations for Various Thermocouples[11,13]
(Reference junction at 32 F)

Deg F	Copper-constantan (Millivolts)	Iron-constantan	Chromel-alumel	Platinum/10% rhodium vs. platinum	Deg F	Iron-constantan (Millivolts)	Chromel-alumel	Platinum/10% rhodium vs. platinum
50	0.389	0.50	0.40	0.056	700	20.26	15.18	2.977
75	0.945	1.22	0.95	0.137	800	23.32	17.53	3.506
100	1.517	1.94	1.52	0.221	900	26.40	19.89	4.046
150	2.711	3.41	2.66	0.401	1000	29.52	22.26	4.596
200	3.967	4.91	3.82	0.595	1200	36.01	26.98	5.726
250	5.280	6.42	4.97	0.800	1400	42.96	31.65	6.897
300	6.647	7.94	6.09	1.017	1600	50.05	36.19	8.110
350	8.064	9.48	7.20	1.242	1800	57.07	40.62	9.365
400	9.525	11.03	8.31	1.474	2000	64.11	44.91	10.662
450	11.030	12.57	9.43	1.712	2500	...	...	13.991
500	12.575	14.12	10.57	1.956	3000	...	...	17.292
600	15.773	17.18	12.86	2.458				

Table 6-16 Recommended Temperature Limits, Atmospheres, and Wire Sizes* for Various Protected Thermocouples†

Thermocouple	Lower limit, °F	Upper limit, °F, for various wire sizes (AWG)‡: 8 ga	14 ga	20 ga	24 ga	28 ga	Suitable atmospheres
Copper-constantan§	—300	...	700	500	400	400	Oxidizing and reducing
Iron-constantan	0	1400	1100	900	700	700	Reducing <500 F; oxidizing or reducing, 500–1200 F
Chromel-alumel**	0	2300	2000	1800	1600	1600	Oxidizing
Platinum/rhodium vs. platinum	0	...	...	...	2700	...	Oxidizing
Rhenium-tungsten††	...			4000			Reducing, inert, vacuum
Rhenium-molybdenum††	...			3200			Reducing, inert, vacuum
Iridium vs. iridium/rhodium††	...			3600			Oxidizing, inert, vacuum, sulfur-free

* AWG (American Wire Gage), see Table 1-18.

† In closed-end protecting tubes.

‡ Instrument Soc. of Am. *Recommended Practice—Thermocouples and Thermocouple Extension Wires.* (RP 1.1-7) Pittsburgh, 1959.

§ Constantan composition: 60 per cent Cu and 40 per cent Ni.

** Trademark—Hoskins Corp.; Chromel comp.: 90% Ni, 10% Cr; Alumel comp.: 94% Ni,2% Al, 3% Mn, 1% Si.

†† Pearse, D. J. "High Temperature Measurement Methods, Part II—Thermocouples." *Nonmetallic Minerals Processing* 2: 23–26, May 1961

Table 6-17 Symbols and Error Limits (for Standard Wire Sizes) of Thermocouples and Extension Wires[11]

	Copper(+) vs. constantan(−)	Iron(+) vs. constantan(−)	Chromel(+) vs. alumel(−)	Platinum/10% rhodium(+) vs. platinum(−)
Couple type or symbol	T	J†	K	S‡
Error limits*—std (temp range, °F)	±2% (−150 to −75) ±1.5°F (−75 to +200) ±0.75% (200 to 700)	±4°F (0 to 530) ±0.75% (530 to 1400)	±4°F (0 to 530) ±0.75% (530 to 2300)‡‡	±5°F (0 to 1000) ±0.5% (1000 to 2700)
Extension wire symbol§	TX	JX	KX, WX, VX	SX
Error limits—std †† (temp range, °F)	±1.5°F (−75 to +200)	±4°F (0 to 400)	±4°F** (0 to 400)	±12°F (75 to 400)

* Where in per cent—apply to reading, e.g., Type J limits for 1000 F = 1000 ± 7.5°F.
† Most widely used of the two calibrations, other known as Type Y.[15]
‡ Another combination, Type R, contains 13 per cent rhodium in the (+) wire.
§ Of same material as couple except for WX, VX, and SX, which are iron vs. alloy, copper vs. constantan, and copper vs. alloy, respectively.
** For Type KX only; ±6°F (75° to 400°) for Type WX and ±6°F (75 ° to 200°) for Type VX.
†† For 75 F reference junction only—extensions WX, VX, and SX.
‡‡ Do not exceed 1400 F if accurate measurements below 1000 F are also required.

circuits. A **thermopile,** a group of thermocouple junctions in series, is used when the emf per degree change in temperature is too small to assure accurate temperature measurements with a single couple, or where emf amplification is required. *Paralleling* of thermocouples on, say, a traverse, requires equal resistance in each circuit.

Suction pyrometers[1] measure the temperature of still or *moving* gases. These instruments are also called *aspirated, velocity,* and *high-velocity* pyrometers; *high-velocity thermocouple* is another common term (since a thermocouple is usually the sensing element). These instruments, by means of multiple (no more than five) shields, baffles, and ambient temperature compensating devices, minimize the temperature effects of surrounding surfaces and phenomena, such as flame radiation. Aspiration velocities of 500 to 700 ft per sec have been recommended (available equipment limited to 100 to 200 ft per sec).[23]

Selection

Table 6-16 outlines primary thermocouple usage parameters. In general, selection depends upon: (1) emf generated per degree temperature change (high ratio and linear relationship preferred); (2) temperature range; (3) resistance to corrosion by moisture, oxidation, or media in which couple is used; (4) tolerance on reproducibility; and (5) sensitivity. The emf generated is relatively independent of both conductor size and resistance.

Extension wires may be used to join thermocouples to indicating instruments. Advantages of these may include greater economy of the more expensive thermocouple materials and the greater strength of extension wires, compared to thermocouple leads. Sizes recommended: 14, 16, and 20 AWG—the latter and smaller may be used when bundled and reinforced for pulling. See Table 6-17 for recommended materials.

High Temperature Measurement. Above, say, 1500 F, the following combinations are common:

Chromel-alumel (Table 6-16) was reported satisfactory to 2100 F for continuous use and to about 2400 F for intermittent use.[14] Nickel content of both these alloys imparts a high resistance to corrosion and oxidation, since the oxides form a protective skin.

Use in a neutral or a reducing atmosphere (especially with alternating oxidation cycles) ultimately results in metal segregation, which in the presence of thermal gradients causes a loss of accuracy. However, the use of protection tubes or sheathed, ceramic-insulated couples permits selective application in reducing atmospheres and the higher temperature ranges.

Platinum/rhodium vs. platinum (Table 6-16). Both Type S and Type R (Table 6-17) have the same temperature range, need nonmetallic protection tubes, and are used in oxidizing (free oxygen present) atmosphere. The 10 per cent alloy combination is more refractory—consequently more stable—and has a lower thermal-emf output. Platinum has a marked affinity for metallic vapors, is susceptible to contamination from phosphorus, silicon, hydrogen, and carbon at the higher temperatures, and can combine chemically with the silicon in a refractory protection tube; it is, therefore, advisable to use procelain type tubes where protection is required.

Rhenium-tungsten and rhenium-molybdenum (Table 6-16) have excellent stability, calibration accuracy, and high emf per degree temperature change.

Iridium vs. iridium/rhodium (Table 6-16). Applications involving excessive vibration or shock should be avoided because temperature cycling causes brittleness. Oxides of alumina, zirconia, and magnesium are used for insulation below 2000 F. For temperature measurements between 2000–4000 F beryllium oxide is used for insulation and structural support.

In all *high temperature* media, it is also necessary to provide contamination protection and radiation shielding.

Precautions. The following practices will help to increase accuracy and reproducibility of results:

1. Thermocouple must have same calibration as the instrument with which it is to be used when indicator scale is in temperature unit.

2. Purchase both thermocouple and extension wiring from same manufacturer at same time.

3. If emf is measured by a millivoltmeter, resistance of extension wires should be taken into account.

4. Keep all wiring away from live electrical conductors.

5. Use radiation shield when measuring surface temperature.

6. Static gas temperature measurements may be made using either shielded[16,17] or bare thermocouples. However, electrically heated radiation compensating shields are best.

7. Details of experiments involving measurements of stagnation temperatures were reported.[1,18,19]

8. Small-diameter thermocouple wires and tubes should be used on applications where it is essential to detect small temperature changes rapidly, if such light wire can be expected to give reasonable life.

Thermocouple Fabrication. Techniques for welding and soldering methods appropriate to different thermocouples were reported.[11] Butted joints were preferred for gas temperature measurement. A special welding technique joins wires as small as 0.0005 diam.[20]

Error Limits and Calibration. Table 6-17 suggests limits for some *standard* couples and extensions. Thus, calibration and periodic checking are necessary for accurate measurements.[11,21] Check extension wire circuitry for grounds with a magneto set (first disconnecting instrument and thermocouple).

Protecting Tubes

Practically all thermocouples for industrial temperature measurement require protecting tubes or *wells*. Desirable properties of the tube include mechanical strength to withstand thermal stresses, pressures, shocks, sagging, and erosion. Vertical installations are preferred, especially at high temperatures, and low porosity is also recommended when tubes are used in injurious atmospheres.

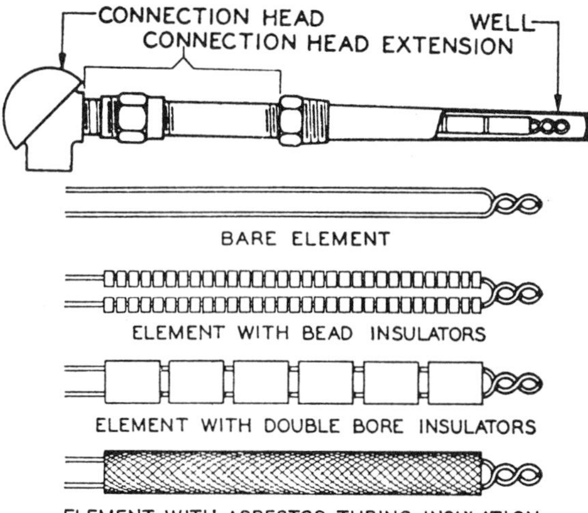

Fig. 6-45 A thermocouple in well assembly and four types of thermocouples.[11]

Insulation, required for most thermocouples, is provided by porcelain, lava, or asbestos tubes, thru which wires are threaded (Fig. 6-45). Tube is slipped over element and joined to a connection head which isolates the element terminal block. Figure 6-45 shows a well arrangement. Connection head temperature should not exceed 250 F. Materials for this protection service depend upon specific conditions of temperature and corrosive medium. They include glass, porcelain, fused silica, silicon carbide, graphite, clay, nickel, steel, iron, and other materials and alloys.

Immersion.[11] Ten times the outside diameter of the protecting tube is minimum immersion necessary to minimize heat transfer along tube length. With flowing liquids six diameters may be used, if both pipe and external portion of protecting tube are well insulated.

RESISTANCE THERMOMETERS[22]

These depend upon the principle that electrical resistance of a metallic conductor changes with temperature and that this resistance-temperature relationship is sufficiently linear to permit accurate temperature measurement—nominally to 1800 F, although use to 3000 F has been reported. The temperature-sensitive element is usually a coil of wire* with high temperature coefficient of resistance, usually nickel or platinum, installed as one arm of a Wheatstone-bridge circuit, as shown in Fig. 6-46. Fixed resistances, R_1, R_2, R_3, consist of a metal with low temperature coefficient of resistance, such as *manganin*. These three resistances and the bulb usually have the same resistance at room temperature.

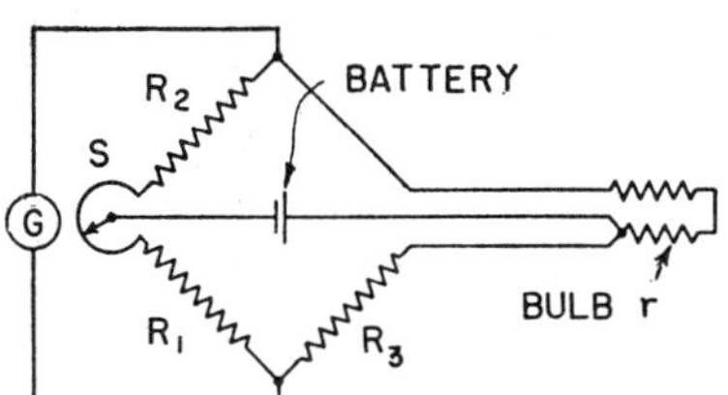

Fig. 6-46 Three-lead resistance thermometer circuit.[10]

In operation, any unbalance of galvanometer, **G**, due to variation in resistance r with temperature change, can be compensated by moving slide wire, **S**, which is calibrated in temperature. Temperature can also be read directly by deflecting a millivoltmeter substituted for **G**, without using a slide wire. The first, or *null* method, using a self-balancing instrument for adjusting slide wire to a zero galvanometer deflection is most common. Ratio of coil to lead resistance should be high, to reduce error due to change in lead resistance with temperature. Three-lead connection (Fig. 6-46), with all leads having equal resistance, is preferred over two lead setup, since the former automatically compensates for variations in lead resistance and thus permits use of longer leads.

Over a range from −150 to +300 F, nickel resistance bulbs are commonly used, while platinum bulbs, with accuracy better than 0.005°F, are used for standard work. Industrial units encased in stainless steel tubes have ±1 per cent full scale accuracy between −400 F and +1800 F with ±0.2 per cent reproducibility.

* Doped germanium crystals, encapsulated and hermetically sealed, have been used up to 100 K.

Thermistor Thermometers

A semiconductor replaces the metallic bulb used for temperature sensing in this type of resistance thermometer. This substitution converts the sensing element from a low to a high electrical resistance, thus minimizing effects of changes in lead and contact resistance whenever a single recording or indicating bridge unit is used with more than one sensing element. The sensitivity is also greatly improved. At ordinary room temperatures, a one-degree change in temperature, which results in about 0.3 per cent change in the electrical resistance of a metallic-resistance sensing element, causes about 3.0 per cent change in the electrical resistance of a thermistor sensing element.

RADIATION PYROMETERS

These devices may be used, from 100 F to 7000 F or more, for measurements of moving items or those which cannot otherwise be physically contacted, and in cases where there is excessive shock, vibration, or prohibitive corrosive atmosphere, or where average area temperatures are desired. These instruments (e.g., Fig. 6-47) measure the radiant energy absorbed by various elements—thermocouples, thermistors, thermometers, **bolometers** (sensitive resistance thermometers using foil element or thermistors); a peep hole is usually provided for aiming at the object. For example, a **thermopile** can be calibrated to read flux density in Btu per hr-sq ft. This device, used in conjunction with a potassium bromide window, is recommended for the radiation measurement of gas-fired sources.

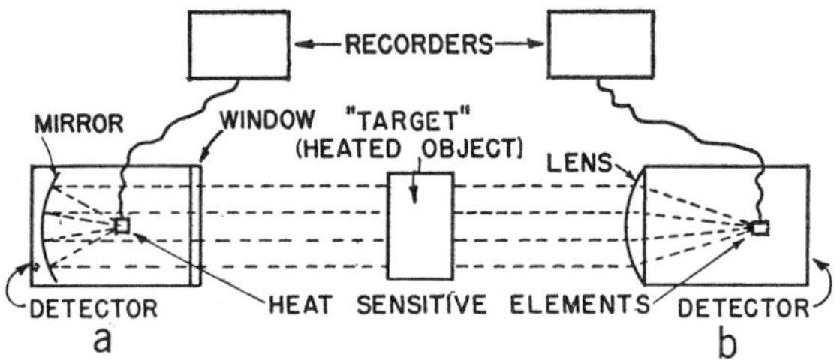

Fig. 6-47 Mirror, a, and lens, b, focusing radiation pyrometer systems. Distance from target to detector ranges from 20 to 30 times the minimum target diameter.[25]

Radiation pyrometers have an advantage over **optical pyrometers** (except photoelectric units) in not requiring visual comparison. These instruments are of two general types, fixed-focus and adjustable-focus. Radiation from a particular area, a function of both temperature and emissivity, is concentrated on, say, the hot junction of a thermocouple; thermocouple potential indicates intensity of all radiation reaching it. The spectral distribution of the radiation received can be determined with a series of bandwidth filters.

Non-black-body conditions can be readily compensated for, and it is common practice to use a closed tube of refractory material in which the instrument is sighted. This corrects for non-black-body errors, although it also slows instrument response. When thermocouples are employed, it is necessary to provide cold-junction compensation. Air jets may be used to keep sighting tube free of smoke or flame. Although radiation-type units offer advantage of having no physical contact with object of temperature measurement, they have attendant disadvantage of large errors when radiation from object must pass thru intervening atmospheres, smoke, or dust, which may absorb or otherwise interfere with transmission of radiation. **Calibration** of both optical and radiation pyrometers is accomplished by sighting on a furnace of known temperature under black-body conditions.

Most **lens and window materials** transmit radiation of wave lengths of 0.3 micron and larger. *Pyrex* is favored from 1400 to 3400 F; fused silica from 1000 to 7000 F; calcium fluoride has the widest transmission band, but is preferred for low temperature ranges. Optical glass is a poor transmitter of infra-red and should never be used below 800 F.

Globe Thermometer

This 8 in. OD $\times$ 0.022 in. wall copper sphere coated with flat black paint (0.94 emissivity) was recommended[24] for the measurement of **mean radiant temperature*** (MRT). The temperature, T_g, is measured at the center of the globe by means of a mercury thermometer (calibrated from 0 to 120 F) or a thermocouple (say, No. 34 wire). The following equation applies:

$$MRT = 100\ [1.014V^{0.5}\ (T_g - T_a) + (0.01T_g)^4]^{0.25}$$

where:

MRT = mean radiant temperature, °F abs
V = air velocity, fpm
T_g = globe temperature, °F abs
T_a = ambient air temperature, °F abs—by means of an aspirated thermometer located 12 in. from the side of the globe. If $T_a - T_g < 10$°F, T_a may be measured with a thermocouple (No. 30 or smaller wires) located 2 in. from the sphere.

Note: The constant, 1.014, for an 8 in. OD globe may be corrected for different globe sizes as follows:

Globe size, in.	8	6	4	3	2
Multiplier	1	1.03	1.10	1.15	1.23

OPTICAL PYROMETERS

A radiation pyrometer utilizing only the narrow band of visible spectrum is an optical pyrometer. Generally used in the 1400 to 5200 F range (7600 F max—Leeds and Northrup Co.), they depend upon a comparison of black-body temperatures of two radiating bodies. This is done by comparing intensity of light of a given wave length given off by each. A common method is by using the disappearing-filament pyrometer, which requires visual comparison. The Leeds and Northrup type (Fig. 6-48) requires an incandescent filament for comparison. Light from the radiating body (0.003 in. diam min) enters thru the lens in the objective and is viewed thru the eyepiece. A telephoto lens permits the temperature measurement of a ¼ in. diam object at a distance of 10 ft. Temperature of

* The temperature of a uniform black enclosure in which a solid body or occupant would exchange the same amount of radiant heat as in the existing non-uniform environment. It is very close to an effective room temperature.

filament in lamp is altered by means of an adjustable resistance (not shown), until filament image merges with that of the object being measured and "disappears." At this point filament current, measured by balancing the potential drop across a calibrated slide wire against the emf of a standard cell, is a measure of black-body temperature. A red-glass filter interposed between eyepiece

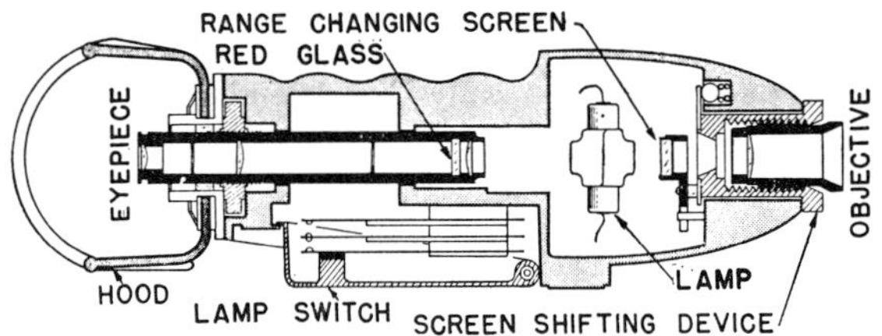

Fig. 6-48 Optical pyrometer telescope—disappearing-filament type. (Leeds and Northrup Co.)

and filament gives *monochromatic* light, while the screen shown is rotated into line with the objective for measurements above 2250 F. This radiation absorption screen reduces object brightness to within the calibrated range of filament brightness. One disadvantage of the optical pyrometer is that it depends upon emissivity of the heated object. This is eliminated in **two-color** or **ratio**[26] pyrometers.

A **photographic technique**[27] for measuring the flame temperature of jet engine exhaust was reported accurate within 18°F between 1832 and 3632 F. Involved is the photographing of four standard comparison lamps, whose brightnesses span those of the flame under study. Account is taken of the fact that the flame does not emit black-body radiation. Heterogeneous flames, flames of varying temperature, and luminous particle density are among the factors the authors point out as requiring further study.

ELECTRICAL TEMPERATURE-MEASURING INSTRUMENTS

Two types of instruments, **millivoltmeters** and **potentiometers,** are commonly used to measure small voltages or changes in potential generated by thermoelectric primary elements, along with other electric variables. Millivoltmeters are simpler and less expensive. Most millivoltmeters for this service operate on the d'Arsonval galvanometer principle.

Inherently a more versatile and accurate instrument than the millivoltmeter, the potentiometer in its various forms is widely used in industry to record temperature, pH, conductivity, and other variables that can be measured electrically. Its principal advantage over the millivoltmeter is that its measurements are not affected by variations in external resistance, as in lead wires or connections. Most potentiometer circuits are fully balanced (null); some are partly balanced (deflection).

The *Pyromaster* circuit (Bristol) employs a contacting galvanometer to balance. Another system of potentiometer balancing is found in the *Celectray* (Tagliabue). This instrument employs a mirror galvanometer which reflects a beam of light on or off a photocell to control the balancing motor. No galvanometer is employed in the *Speedomax* (Leeds & Northrup).

A simplified version of the continuous-balance potentiometer is made by Brown Instrument Co. The *Dynalog* (Foxboro) is still another electronic potentiometer. Continuous balancing is provided by a simple rotating variable air capacitor, run by a double-solenoid-type drive which moves only when instrument is rebalancing.

Most record type recorders have automatic standard cell balancing at about 20 min intervals. Expected accuracy of electronic indicators and recorders is ¼ per cent of full scale—after 20 min warmup; i.e., readings accurate to 2°F on a 0 to 800 F scale.

The **Weston standard cell,** used to calibrate most **potentiometers,** has an exact voltage of 1.0186 v at 20 C (the international standard of the absolute volt). A rectifier and a nonmetallic (**Zener**) diode have been used instead of a standard cell to obtain a constant voltage.

PYROMETRIC CONES, CRAYONS, PELLETS, LACQUER

Pyrometric cones are a simple and inexpensive form of fusion pyrometer. They are small pyramids, about 2 in. high. Each one is prepared from mixtures of oxides and glass, to give a definite melting point. A series of cone melting points 20° to 70° apart covers the range from about 1100 to 3600 F as shown in Table 6-18. Accuracy of measurement is about the same as interval between successive cones, although it may not be as close because behavior of cones depends somewhat upon rate of heating and upon furnace atmosphere. Pyrometric cones are used mainly in ceramic industries.

Table 6-18 Melting Points of Seger Cones

Cone No.	Deg C	Deg F	Cone No.	Deg C	Deg F	Cone No.	Deg C	Deg F
022	600	1110	7	1270	2320	30	1670	3040
016	750	1380	15	1435	2615	35	1770	3220
010	950	1740	20	1530	2790	39	1880	3420
02	1110	2030	26	1580	2880	42	~2000	3600

Pyrometric crayons, pellets, and lacquer[28] which melt at specific temperatures are available for surface temperature measurements and for places that are difficult to reach with other instruments. Crayons and lacquer are available for temperatures from 113 F to 2000 F, in intervals of 12 to 50 F; pellets from 113 F to 2500 F, in 12 to 50 F intervals.

Paper thermometers are available[29] to determine temperature in the range 115–490 F. Response time is under one second; accuracy is about one per cent. They change color permanently from white to black at specific temperature.

CALIBRATION OF TEMPERATURE INSTRUMENTS

Calibration of temperature instruments is usually accomplished by one of three methods:

1. Comparison with a standard instrument of suitable type. This should be done in a well-stirred fluid, with sensitive elements of both instruments fastened together.

2. Readings of instrument at melting or solidifying point of pure solid materials. Points generally used are ice, tin, lead, zinc, aluminum, copper, and sodium chloride. See Table 1-1 for element melting and boiling points and Table 6-19 for selected compound data. Cooling curves (temperature vs. time) should be plotted and point should be determined from flat portion of curve.

3. Readings of instrument at boiling points or saturation temperatures of pure liquids. Water is generally used and temperatures taken from steam tables. Pressures must be measured very accurately, as by dead-weight apparatus. Atmospheric boiling points of naphthalene, 425 F, and of sulfur, 832 F, are often used.

The National Bureau of Standards, Washington, D. C., calibrates all types of thermometers and pyrometers. A fee schedule can be obtained on request. A certificate of calibration is provided by the Bureau for each instrument.

Table 6-19 Fixed Points for Thermometer and Pyrometer Standardization

(compounds only—see Table 1-1 for elements)

Substance	Phase change	Temp, °F
Toluene	Melts	−139.0
Carbon dioxide	Sublimes	−109.3
Glauber's salt	Melts	90.3
Ethyl alcohol	Boils	172.9
Toluene	Boils	230.0
Naphthol (α)	Boils	532.4
Potassium dichromate	Melts	747.5
Manganous sulfate	Melts	1292.0
Sodium chloride	Melts	1472.7
Sodium carbonate	Melts	1565.6
Sodium sulfate	Melts	1624.5
Stannic oxide	Melts	2061
Lithium silicate	Melts	2194
Alumina	Melts	3632

SURFACE TEMPERATURES

Surface temperatures are best measured with thermocouples, but may also be measured with thermometers or with radiation instruments. A thermometer pressed firmly against a hot surface and sealed with plastic material will read low. Error varies from about 5°F when true temperature is 50°F above ambient, to 20°F when temperature is 200°F above ambient. More accurate surface temperature measurements may be made with thermocouples, either attached to surface with adhesive or imbedded in cement in a shallow groove. Fine wires (24 to 40 AWG) should be used, with insulated leads in contact with surface for some distance from couple. For high temperatures or for quick readings from inaccessible surfaces, a radiation thermopile or a radiation pyrometer may be used.

The Alnor *Pyrocon*[30] is made for surface temperature measurements in ranges below 800 F and for temperatures from 800 to 1200 F.

BTU METERS

Energy in the form of flowing hot or chilled water may be metered. The measurement devices register the heat content of a stream by integrating the product of a constant or varying water flow rate and the difference in temperature between inlet and outlet water. The water meter component is generally installed in the water line of most nearly constant temperature. Displacement- and orifice-type meters are installed in horizontal lines; propeller-type meters may be installed in lines that run at any angle.

Some meters do not require any electrical connections. Others use 110-volt, 60-cps power. The latter instruments compensate for any fluctuation in supply voltage.

Table 6-20 Accuracy of Btu Meter Components

(American Meter Co., 1964)

Component	Tolerance*	Average accumulated error†
Water meter, displacement type	1 per cent	½ per cent
Water meter, propeller type	2 per cent	1 per cent
Btu computer	0.6 per cent	0.3 per cent
Thermometers for refrigeration—		
25°F full-scale differential	¼°F	Under ¼°F
Thermometers for heating		
50°F full-scale differential	½°F	Under ½°F
100°F full-scale differential	1°F	Under 1°F
200°F full-scale differential	2°F	Under 2°F

* Tolerances are manufacturers' guarantees, computed over the recommended range of flow rates or temperature differentials. Figures on displacement meters cover cold water operation.

† Based on average operating conditions, at flows varying throughout the recommended ranges.

Table 6-20 gives the accuracy of one manufacturer's meter components. These data may be used to determine the *maximum* possible error (all errors in the same direction) in a metering system.

Example: A Btu meter with a propeller-type water meter (2 per cent tolerance) passes one pound of chilled water at an inlet temperature of 39 F; the outlet temperature is 64 F (0.25°F tolerance*). The tolerances given, and the tolerance for the Btu computer component (0.6 per cent tolerance), are as shown in Table 6-20. Find the *maximum* possible overall meter error.

$$Error = 100\,\frac{25 - [(1 - 0.020)(25 - 0.25)(1 - 0.006)]}{1 \times 25} = 3.6\%$$

The actual overall error would probably be considerably less, since the component errors would partially cancel.

Temperature and Pressure Ratings

Application of the following data for hot water systems precludes inaccuracies in metering that result from water flashing into vapor:

Energy system temperature range, °F	Water vapor pressure, psig	Recommended† meter rating, psig
125–300	55 at 300 F	150
300–350	123 at 350 F	200
350–400	232 at 400 F	400

* 64 − 39 = 25°F, which corresponds to the full-scale differential.

† By American Meter Co.

REFERENCES

1. Baker, H. D. and others. *Temperature Measurement in Engineering.* 2 vols. New York, Wiley, 1953–61.

1a. Mullikin, H. F. and Osborn, W. J. "Accuracy Tests of the High-Velocity Thermocouple." (In: Am. Inst. of Physics. *Temperature, Its Measurement and Control,* p. 805–29. New York, Reinhold, 1941.)

2. Am. Institute of Physics. *Temperature; Its Measurement and Control in Science and Industry.* New York, Reinhold, 1941.

3. Severinghaus, W. L. "Reducing Radiation Errors in Gas-Temperature Measurement." *Mech. Eng.* 59: 334+, May 1937.

4. "How to Measure Accurately the Temperature of a Flowing Gas." *Glass Ind.* 19: 423–4, Nov. 1938.

5. Fishenden, M. "Use of Thermocouples for Temperature Measurement." *Inst. of Heating Ventilating Engrs. J.* 7: 15–35, Mar. 1939.

6. Foote, P. D. and others. *Pyrometric Practice.* (Nat. Bur. of Standards T170) Washington, G.P.O., 1921.

7. Weil, S. A. and others. *Fundamentals of Combustion of Gaseous Fuels.* (Research Bul. 15) Chicago, I.G.T., 1957.

7a. Stanforth, C. M. *Catalytic Effects of Thermocouple Materials.* (SAE Paper 524G) New York, Soc. of Automotive Engineers, 1962.

8. Kreisinger, H. and Barkley, J. F. *Measuring the Temperature of Gases in Boiler Settings.* (Bur. of Mines Bul. 145) Washington, G.P.O., 1918.

9. Haslam, R. T. and Chappell, E. L. "Measurement of the Temperature of a Flowing Gas." *Ind. Eng. Chem.* 17: 402–8, April 1925.

10. Perry, J. H. *Chemical Engineers' Handbook,* 3rd ed., p. 1269–70. New York, McGraw-Hill, 1950.

11. Instrument Soc. of Am. *Recommended Practice—Thermocouples and Thermocouple Extension Wires.* (RP 1.1-7) Pittsburgh, 1959.

12. Rauch, W. G. "Design and Construction of Needle Thermocouples." *Metal Progr.* 65: 71–4, Mar. 1954.

13. Shenker, H. and others. *Reference Tables for Thermocouples.* (Nat. Bur. of Standards C 561) Washington, G.P.O., 1955.

14. Pearse, D. J. "High Temperature Measurement Methods, Part II—Thermocouples." *Nonmetallic Minerals Processing* 2:23-26, May 1961.

15. Roeser, W. F. and Dahl, A. I. *Reference Tables for Iron-Constantan and Copper-Constantan Thermocouples.* (Nat. Bur. of Standards RP 1080) Washington, G.P.O., 1938.

16. Rohsenow, W. M. and Hunsaker, J. P. "Determination of the Thermal Correction for a Single-Shielded Thermocouple." *A.S.M.E. Trans.* 69:699–704, 1947.

17. Dahl, A. I. and Fiock, E. F. "Shielded Thermocouples for Gas Turbines." *A.S.M.E. Trans.* 71: 153–61, 1949.

18. Hottel, H. C. and Kalitinsky, A. "Temperature Measurements in High-Velocity Air Streams." *A.S.M.E. Trans.* 67: A-25—A-32, 1945.

19. Babcock and Wilcox Co. "Temperature Measurement." (In: *Steam, Its Generation and Use,* 37th ed., chap. 6. New York, 1960.)

20. Carbon, M. W. and others. "Response of Thermocouples to Rapid Gas-Temperature Changes." *A.S.M.E. Trans.* 72: 655–57, 1950.

21. Roeser, W. F. and Lonberger, S. T. *Methods of Testing Thermocouples and Thermocouple Materials.* (Nat. Bur. of Standards C590) Washington, G.P.O., 1958.

22. Pearse, D. J. "High Temperature Measurement Methods, Part I—Resistance Thermometer." *Nonmetallic Minerals Processing* 2: 131–34, April 1961.

23. Land, T. and Barber, R. "Suction Pyrometers in Theory and Practice." *J. Iron & Steel Inst.* 184, pt. 3: 269–73, Nov. 1956.

24. Nelson, D. W. and others. "Measurement of the Physical Properties of the Thermal Environment." *Heating, Piping Air Conditioning* 14: 382–85, June 1942.

25. Pearse, D. J. "High Temperature Measurement Methods, Part III—Total Radiation Pyrometers." *Nonmetallic Minerals Processing* 2: 31–34, June 1961.

26. ——. "——, Part IV—Optical Pyrometers." *Ibid:* 327, Jl. 1961.

27. "Film Tells Flame Temperature." *Chem. Eng.* 65: 64, 66, Jan. 27, 1958.

28. Tempil Corp., 132 W. 22nd St., New York City.

29. Paper Thermometer Co., 10 Stagg Dr., Natick, Mass.

30. Illinois Testing Laboratories, Inc., 420 N. La Salle St., Chicago 10.

SECTION 7

GAS MEASUREMENT

John E. Overbeck (retired), *Section Chairman*, Columbia Gas System Service Corp., Columbus, Ohio
H. S. Bean, Consultant, Kensington, Md. (formerly with National Bureau of Standards, Washington, D. C.)
S. R. Beitler, Ohio State University, Columbus, Ohio
A. V. Brashear, Michigan Consolidated Gas Co., Detroit, Mich.
R. W. Davis, Columbia Gas System Service Corp., Columbus, Ohio
P. H. Miller, Texas Eastern Transmission Corp., Shreveport, La.
F. M. Partridge, Pacific Northwest Pipeline Co., Salt Lake City, Utah
H. C. Schroeder (retired), The East Ohio Gas Co., Cleveland, Ohio
E. E. Stovall (retired), Lone Star Gas Co., Dallas, Texas

The authors of this Section chose to be identified with the Section as a whole rather than with individual chapters.

CONTENTS

Tables

Figures

Chapter 1

General

UNIT OF GAS MEASUREMENT

Normally, gas is measured by volume which is expressed in cubic feet at some specified reference pressure and temperature, as measured by a meter or computed from a meter record. In 1963, A.G.A. endorsed the adoption of 14.73 psia and 60 F as the standard base conditions of pressure and temperature to be used in gas volume measurements, wherever this measurement is at such a pressure (greater than normal utilization pressure) that a correction is desired.

Standard Cubic Foot

As generally used in laboratory tests, the standard cubic foot of gas is that quantity of gas saturated with water vapor, which at a pressure of 30 in. of mercury (temperature of mercury = 32 F, density = 13.595 grams per cubic centimeter, acceleration due to gravity = 32.174 ft/sec-sec) and at a temperature of 60 F, occupies one cubic foot.

For natural gas measurement, both the "dry" and "as measured" cubic foot are normally used. Generally, the two are practically the same, since most gas contracts stipulate a very low water vapor content, i.e., about 7 lb per MMSCF.

A common practice is to express the quantity of natural gas measured thru orifice meters in standard cubic feet, at an absolute pressure of 14.73 pounds per square inch,* a temperature of 60 F, and without adjustment for water vapor content.

Low Pressure—Metered Cubic Foot

The quantity of gas recorded in cubic feet under the pressure (normally up to 0.5 psig) and temperature conditions existing at the consumer's meter is used for billing purposes, without correction.

High Pressure—Metered Cubic Foot

Gas measurements made at pressures above the usual low pressure levels require some definite agreement upon a pressure base and a temperature base, besides other factors which may affect delivery computations. Under such conditions, it is the usual practice to stipulate the "pressure base" and "temperature base" in a contract, agreement, or tariff; as well as to specify other factors affecting delivery computations, such as specific gravity, supercompressibility, and possibly, moisture content.

Atmospheric Pressure

The metered gas volume also depends upon the absolute static pressure of the gas at the point of measurement. This absolute pressure is the sum of the gage pressure and the atmospheric pressure. (The latter in turn depends upon the altitude and barometric conditions. Because the barometric pressure varies, it is customary to establish an agreed-upon local average value.)

Pressure Base

The pressure base (14.73 psia endorsed by A.G.A.) is the absolute pressure for which the cubic foot is the unit of measurement, according to the contract, agreement, or tariff.

Temperature Base

The temperature base is the temperature for which the cubic foot is the unit of measurement, according to the contract, agreement, or tariff. A frequently used value is 60 F.

SAMPLE GAS MEASUREMENT SPECIFICATIONS FOR A CONTRACT

The following specifications might be included in a contract between a seller and a large volume buyer. This type of contract is generally not given to industrial users by gas utilities, but it may be given on a direct sale from a pipeline to a customer. The following material is for illustrative purposes; it does not indicate that the numerical values given are recommended or that the specific clauses are standard.

1. The *unit of measurement and metering base* shall be one cubic foot of gas at a pressure base of 14.73 psia, a temperature base of 60 F (approximately 520 R), and without adjustment for water vapor content.
2. The *average atmospheric pressure* shall be assumed to be 14.4 psia, regardless of actual elevation or location of the delivery point above sea level or variations in actual barometric pressure from time to time. (The average absolute atmospheric pressure usually takes into account the location of the buyer.)
3. The *temperature* of the gas flowing thru the meter or meters shall be the arithmetic average of the hourly temperature record; or shall be read from established tables of monthly averages for the location involved.

* A 1960 A.G.A. survey[1] showed that the 14.73 psia base was specified in nine states; 24 states, however, had no definition for this base; the remaining states specify a variety of pressure bases. Base temperature, where specified, is 60 F (except Ohio, 62 F). The advantages of standardization are also covered.

4. The *specific gravity* of the gas shall be determined by test, using a gas gravity balance, or other approved instrument, at the beginning of deliveries hereunder and as often thereafter as deemed necessary; or if mutually agreed upon, by the use of a specific gravity recorder, periodically checked with an approved instrument, or by other accepted methods.

5. The *deviation* from Boyle's law shall be determined by one of the following ways:

a. Deviation factors shall be computed by approved methods, or read from standard tables; computations or selection of factors from tables are based upon the gas composition and conditions at the measurement point. Factors used shall be checked by tests of the gas and determined with such frequency as may be found necessary.

b. The deviation factors shall be determined from tests of the gas. They shall be determined with such frequency as may be found necessary.

6. The *measuring equipment at the points of delivery,* unless otherwise agreed upon, shall be properly fitted with either displacement or orifice meters or both and with any other necessary measuring equipment by which the volumes of gas delivered hereunder may be determined. Orifice meters shall be installed and operated and gas volumes computed according to specifications recommended in Gas Measurement Committee Report No. 3.[16] Volumes calculated from displacement meter dial readings shall be corrected to contract base temperature and pressure.

7. The *accuracy of measuring equipment* shall be verified at reasonable intervals and, if requested, in the presence of representatives of the other party; but neither seller nor buyer shall be required to verify the accuracy of such equipment more frequently than once in any 30 day period. If either party at any time desires a special test of any measuring equipment, or if either party at any time believes there is an error in any such measuring equipment, he will promptly notify the other party. Both parties shall then cooperate to secure a prompt verification of the accuracy of such equipment.

If, upon test, measuring equipment is found to have an error of less than two per cent, previous recordings of such equipment shall be considered accurate in computing deliveries hereunder; but such equipment shall be adjusted at once to record correctly. If, upon test, measuring equipment is found in error by more than two per cent, at a recording corresponding to the average hourly rate of gas flow for the period since the last preceding test, previous recordings of such equipment shall be corrected to zero error for the period the error is suspected or known to have existed, or for a period agreed upon. If the period is not known definitely or agreed upon, correction shall be for a period extending over one-half the time elapsed since the date of the last test, but not exceeding 16 days.

In the event that any measuring equipment is out of service, or is found registering inaccurately and the error is not determinable by test, previous recordings or deliveries through such equipment shall be estimated: (a) by using the registration of a check meter or meters, if installed and accurately registering; or if (a) is not feasible, (b) by correcting the error, if the percentage of error is ascertainable by calibration, special test, or mathematical calculation; or if both (a) and (b) are not feasible, (c) by estimating the quantity of delivery on the basis of deliveries during periods under similar conditions when the meter was registering accurately. (Other bases and values, as agreed upon, may be substituted for those shown without changing the basic intent and purpose.)

Some contracts state that the gas moisture content shall not exceed that corresponding to saturation at the gas temperature and pressure in the pipeline, at a point approximately 50 ft ahead of the meter inlet header at or near any delivery point; and that no water shall be present in liquid phase.

To determine the volume of gas delivered, factors for pressure, temperature, specific gravity, and deviation from Boyle's law shall be applied. A factor for specific gravity is not applicable to measurement by displacement type meters.

GAS LAWS AS APPLIED TO GAS MEASUREMENT

Barometric Pressure

The absolute metering pressure is equal to the static gage pressure of the flowing gas plus the barometric pressure which is based upon some average condition. It is necessary to consider the barometric pressure when computing the delivery of gas thru a meter at elevated pressures. Variation in barometric pressure with changing elevation is shown in Fig. 8-40 and Table 6-7.

For deliveries made with *displacement type meters,* where no computations or adjustments are made for the static pressure of the flowing gas (such as domestic size meters on city distribution pressure), local differences in altitude may not appreciably affect measurement for billing purposes.

Dalton's Law

This law states that each gas in a mixture exerts a partial pressure which the same mass of the gas would exert if it were present alone in the given space at the same temperature. Therefore, when measuring gas containing water vapor, the total pressure indicated is the sum of the water vapor pressure and the gas pressure.

At constant temperature, the change in water vapor pressure, dp, with change in the gas pressure, dP, is equal to the ratio of the molal volume of the water, V_l, to that of its vapor V_g, under its own vapor pressure, p. Thus:

$$\frac{dp}{dP} = \frac{V_l}{V_g}$$

Since V_l/V_g is a very small fraction, it follows that, for all small values of dP, the variation in vapor pressure with total pressure is generally negligible. However, at the high pressures now often used in gas transmission, variations in water vapor pressure will be manifest. Thus, under a pressure of 300 psia (at 60 F), the vapor pressure of water increases by over 10 per cent (compared to the pressure at one atmosphere).

Since the foregoing equation cannot be conveniently used, it may be integrated to give:

$$\log_e \frac{p_2}{p_1} = \frac{V_l}{R(t + 460)} (P_2 - P_1) \tag{1}$$

where:

p_2 and

p_1 = the vapor pressures under total pressures P_2 and P_1, respectively (all pressures are expressed in psia)

t = temperature, °F

V_l = molal volume of the liquid water, cu ft per lb-mole

R = universal gas constant for the ideal gas, 10.73 [(psia)(cu ft)/(°F abs)(lb-mole)]

The foregoing equation can be used with any consistent set of units.

Effect of Water Vapor on Gas Measurement. Natural gas as measured in the majority of large volume gas meters is only partially saturated with water vapor at base conditions, because of its previous compression to a much higher pressure during transmission or because of its intentional dehydration. Water vapor content is not used in ordinary measurement computations of natural gas volumes where volumes "as measured" are used. In case it is desirable or necessary to compute the water vapor content of measured volumes or to convert measured volumes to a saturated basis, the degree of saturation must be determined.

The actual water vapor content of saturated compressed gases is larger than that calculated by using the laws of a perfect gas and vapor pressure alone. It was reported that water content of air at 600 psi and 60 F shows an increase of about 24 per cent over calculated values.[2] The saturated water content of the natural gases studied was 30 to 35 per cent greater than calculated values at this pressure and temperature. These investigations have indicated that the variation in water carrying capacity of two widely different natural gases is probably not much greater than the present precision of experimental data. On this basis, a correlation of these data has been accomplished.[3] Table 4-33 is sufficiently accurate for engineering work with natural gases of widely varying composition.

For commercial use, the following equation may be used to compute *water vapor factor*, F_{wv}, for use in converting gas measurements to a saturated basis:

$$F_{wv} = \frac{P_b(P_f - aR_h)}{(P_b - A)P_f} \quad (2)$$

where:

P_b = pressure base used in the measurement, abs, usually psia

A = partial pressure (in same units as P_b) due to saturated water vapor at temperature base used in measurement

P_f = average metering pressure, abs

a = partial pressure (in same units as P_f) due to saturated water vapor at average metering temperature

R_h = relative gas humidity at average metering absolute pressure and temperature

A and a may be read from any standard water vapor pressure tables. R_h may be determined from the equation:

$$R_h = W_1/W_2 \quad (3)$$

where:

W_1 = actual weight of water vapor in a given volume of gas divided by the quantity of gas *at base conditions*, lb per MMCF

W_2 = weight of water vapor in a given volume of *saturated* gas divided by the quantity of gas *at base conditions*, lb per MMCF

A **dew point apparatus** (e.g., Fig. 6-21) may also be used to determine the flowing temperature of the gas. The actual weight of the water vapor, W_1, may then be read directly from Table 4-33. W_1 may also be determined by estimating the line pressure and temperature at which the gas was last at its dew point.

Example. Assume that the dew point of a particular gas at 100 psia is 20 F. Also assume an average metering pressure of 100 psia and a temperature of 50 F. Then:

$$W_1 = 27.4 \text{ (from Table 4-33)}$$
$$W_2 = 89.0 \text{ (from Table 4-33)}$$
$$R_h = W_1/W_2 = 27.4/89.0 = 0.308$$

Since the factor, $P_b/(P_b - A)$, of Eq. **2** is always a constant for any particular base condition, and the factor, $(P_f - aR_h)/P_f$, does not greatly vary from unity for the average natural gas, this factor may be applied with considerable accuracy, without much complication.

Where water vapor factors are applied to **orifice meter** measurements, specific gravity should be taken into account. However, since the specific gravity of water vapor differs little from that of the average natural gas, any resulting error would be negligible, particularly when the specific gravity of gas is determined without drying.

The water vapor factor, F_{wv}, may be applied to orifice meter deliveries by including it in the **orifice flow constant,** or as a multiplier of the total cubic feet measured. This factor may be applied to **displacement meter** deliveries by including it along with the other factors used, such as pressure and temperature.

Specific Gravity

The specific gravity of the gas being measured is generally considered as the ratio of its specific weight under the observed conditions of temperature, pressure, and water vapor content, to the specific weight of *dry* air of normal carbon dioxide content at the same temperature and pressure. It is generally determined by: (1) periodic determination at meter location or from samples taken for laboratory test; and (2) specific gravity recorders.

The following procedures have been found practical:

For meters where the specific gravity is reported periodically by determination at meter location, or by laboratory test of spot samples taken for the purpose, the volume of gas passing may be calculated by using an assumed specific gravity or that given by the latest test. The *total volume or coefficient* should be corrected at the end of the billing period, if the sample is taken monthly, but at the beginning of the subsequent billing period, if the sample is taken at intervals greater than one month.

For meter stations with a recording gravitometer, the average recorded gravity covering the same period as the meter charts should be used for computing gas volume.

For meters in a local area with a common source of gas, the gas passing may be calculated by using a specific gravity which represents the continuous, thirty-day average reported specific gravity from a continuous sampling recorder at the common source. However, where daily deliveries must be furnished for purposes like demand billing, the reported specific gravity from the recorder should be used for computing gas volumes on all meters having a common source of gas; and the specific gravity is applied chart for chart, as if the recorder were at the meter.

For orifice meters in an area with multiple sources of gas, the gas passing may be calculated by using an assumed specific gravity. The total volume at the end of the billing period may then be corrected for the actual specific gravity, which is obtained by weighting the various reported specific gravities for the area. The weighting should recognize such factors as pipeline network, source volume, demand, and season. Weighted values, when used, should be periodically checked, either by a sufficient number of spot samples or by periodic use of a specific gravity recorder. A specific gravity error of two per cent affects flow measurement by one per cent.

The total volume calculated for an *assumed* specific gravity, G_a, may be corrected for the *actual weighted* specific gravity, G_w, using the factor, $(G_w/G_a)^{0.5}$, as a multiplier.

SUPERCOMPRESSIBILITY

According to Boyle's law, the specific weight of a gas is directly proportional to the absolute pressure (at constant temperature). All gases deviate from this law by varying amounts. This deviation is called "supercompressibility." Within the normal range of conditions in the natural gas industry, the actual specific weight under the higher pressure is usually *greater than* the theoretical.

A factor for taking supercompressibility into account is frequently necessary in gas measurement, particularly at high line pressure. This factor may be expressed by the following equation:

$$F_{pv} = (RT/PV)^{0.5} \tag{4a}$$

where:

F_{pv} = supercompressibility factor
R = universal gas constant
T = gas temperature, abs
P = gas pressure, abs
V = molar gas volume

In terms of **compressibility** of a particular gas, its supercompressibility factor may be expressed as:

$$F_{pv} = (1/Z_f)^{0.5} \tag{4b}$$

where:

Z_f = compressibility at flowing gas pressure, psia

The compressibility factor depends upon pressure, specific gravity, temperature, and gas composition. In **orifice meter** measurement, flowing conditions like pressure, specific gravity and temperature are expressed under the radical in the flow equation; the term $1/Z_f$ also depends on these variables and, therefore, must also be under the radical. Typical F_{pv} factors for orifice meter practice appear in **Table 16-A** of Chap. 2, Sec. 7.

To compensate for supercompressibility in **displacement meter** practice, the volume as measured may be multiplied by the factor, $(F_{pv})^2$. Normally, most displacement metering is done at pressures only slightly above atmospheric; thus, the correction is small and seldom used. However, some conditions exist where displacement meters receive gas at higher pressures and all measurements are corrected for supercompressibility.

Determination of Supercompressibility Factors

The deviation of any gas from the perfect gas laws is a function of its composition. To know this deviation accurately for any gas, it is necessary either to determine it experimentally for that gas, if a representative sample is available, or to know the gas composition and the variations of deviation with composition. In many cases it is not practical to determine the deviations and supercompressibility factors by tests. This is especially true if the composition of the gas measured does not remain constant because the factors change with the composition.

A continuing study has been conducted to develop a method which will give a reasonably accurate value of the supercompressibility factor from properties which are normally recorded. Several methods for determining supercompressibility factors of natural gas mixtures are available.[4–7] Giving consideration to overlapping of the various methods, the one proposed by Dunkle[7] appears to be the most promising, not only because predicted results and test data apparently agree, but primarily because it is based on methane as the only hydrocarbon. Values computed from Reference 4 are very close to the true value, if operating pressures are less than 500 psig. Those computed from References 5 and 6 are very nearly correct for higher pressures if the specific gravity of the gas is high.

For pressures greater than 500 psig and gravities less than 0.9, the method which seems to give best results is based on the "law of corresponding states"; this applies the concept of pseudo-critical pressure and temperature and the generally accepted rule that critical pressure and temperature of a paraffin hydrocarbon mixture may be defined by its specific gravity. The method also assumes that for basic calculations methane is the most important hydrocarbon in natural gas; therefore, it can be used to represent the consistent properties of natural gas mixtures. The supercompressibility for methane, expressed in terms of reduced temperature and reduced gage pressure, is used to evaluate supercompressibility factors of mixtures of hydrocarbon fractions, nitrogen, and carbon dioxide.

Table 7-1a gives supercompressibility factors of a hydrocarbon gas over a range of reduced pressures (from 0 to 3.0) and reduced temperatures (from 1.16 to 1.80). To use this table in determining the supercompressibility factor for a mixture, it is necessary only to determine the reduced pressure and temperature of the mixture and then read its supercompressibility factor by interpolation.

Table 7-1a Supercompressibility Factors, F_{pv}, for a Hydrocarbon Gas (0.6 sp gr)*

Reduced temp, T_R	Reduced pressure, P_R, dimensionless														
	0.2	0.4	0.6	0.8	1.0	1.2	1.4	1.6	1.8	2.0	2.2	2.4	2.6	2.8	3.0
1.16	1.0254	1.0512	1.0795	1.1122	1.1525	1.2006									
1.18	1.0237	1.0481	1.0743	1.1037	1.1393	1.1802									
1.20	1.0223	1.0452	1.0696	1.0964	1.1277	1.1631									
1.22	1.0210	1.0425	1.0652	1.0896	1.1176	1.1489	1.1835								
1.24	1.0198	1.0400	1.0612	1.0839	1.1092	1.1317	1.1667	1.1972							
1.26	1.0186	1.0375	1.0574	1.0786	1.1017	1.1265	1.1527	1.1804							
1.28	1.0176	1.0354	1.0540	1.0737	1.0948	1.1172	1.1411	1.1657	1.1926						
1.30	1.0167	1.0334	1.0509	1.0693	1.0888	1.1093	1.1310	1.1532	1.1766	1.1970					
1.32	1.0158	1.0315	1.0480	1.0651	1.0831	1.1021	1.1219	1.1418	1.1622	1.1807	1.1961				
1.34	1.0150	1.0298	1.0453	1.0613	1.0780	1.0957	1.1139	1.1317	1.1496	1.1662	1.1811	1.1944			
1.36	1.0142	1.0282	1.0429	1.0577	1.0734	1.0896	1.1063	1.1226	1.1385	1.1533	1.1675	1.1798	1.1907	1.1993	
1.38	1.0134	1.0267	1.0405	1.0545	1.0690	1.0840	1.0994	1.1142	1.1286	1.1423	1.1552	1.1665	1.1765	1.1847	1.1905
1.40	1.0125	1.0253	1.0384	1.0513	1.0650	1.0788	1.0930	1.1067	1.1199	1.1323	1.1438	1.1545	1.1636	1.1712	1.1765
1.42	1.0119	1.0240	1.0362	1.0484	1.0612	1.0740	1.0870	1.0997	1.1119	1.1234	1.1340	1.1438	1.1517	1.1588	1.1639
1.44	1.0112	1.0226	1.0342	1.0458	1.0573	1.0696	1.0816	1.0932	1.1045	1.1153	1.1250	1.1340	1.1414	1.1475	1.1521
1.46	1.0106	1.0214	1.0324	1.0433	1.0544	1.0656	1.0768	1.0874	1.0977	1.1078	1.1168	1.1250	1.1321	1.1376	1.1412
1.48	1.0100	1.0202	1.0306	1.0410	1.0515	1.0618	1.0723	1.0820	1.0915	1.1010	1.1094	1.1172	1.1236	1.1288	1.1322
1.50	1.0095	1.0191	1.0289	1.0388	1.0487	1.0583	1.0680	1.0771	1.0859	1.0947	1.1025	1.1099	1.1158	1.1206	1.1238
1.52	1.0090	1.0181	1.0275	1.0367	1.0461	1.0550	1.0640	1.0725	1.0808	1.0890	1.0960	1.1032	1.1088	1.1132	1.1164
1.54	1.0085	1.0171	1.0260	1.0348	1.0437	1.0520	1.0605	1.0684	1.0710	1.0836	1.0901	1.0969	1.1020	1.1061	1.1092
1.56	1.0082	1.0163	1.0245	1.0330	1.0414	1.0492	1.0570	1.0644	1.0716	1.0786	1.0846	1.0908	1.0957	1.0996	1.1024
1.58	1.0078	1.0155	1.0233	1.0312	1.0392	1.0465	1.0539	1.0607	1.0674	1.0740	1.0795	1.0850	1.0898	1.0935	1.0961
1.60	1.0074	1.0148	1.0221	1.0295	1.0371	1.0441	1.0510	1.0574	1.0635	1.0695	1.0748	1.0798	1.0842	1.0877	1.0901
1.62	1.0070	1.0140	1.0210	1.0280	1.0351	1.0419	1.0482	1.0543	1.0600	1.0655	1.0705	1.0750	1.0791	1.0825	1.0844
1.64	1.0065	1.0133	1.0200	1.0266	1.0333	1.0396	1.0456	1.0514	1.0569	1.0618	1.0665	1.0705	1.0742	1.0773	1.0795
1.66	1.0063	1.0125	1.0190	1.0253	1.0315	1.0375	1.0432	1.0486	1.0537	1.0583	1.0626	1.0664	1.0699	1.0727	1.0750
1.68	1.0060	1.0119	1.0180	1.0240	1.0299	1.0355	1.0409	1.0459	1.0508	1.0551	1.0591	1.0625	1.0657	1.0684	1.0705
1.70	1.0056	1.0114	1.0171	1.0227	1.0284	1.0335	1.0386	1.0434	1.0481	1.0521	1.0558	1.0589	1.0618	1.0643	1.0663
1.72	1.0054	1.0108	1.0163	1.0210	1.0269	1.0318	1.0365	1.0411	1.0455	1.0492	1.0526	1.0556	1.0581	1.0605	1.0625
1.74	1.0050	1.0102	1.0154	1.0205	1.0255	1.0301	1.0346	1.0388	1.0429	1.0465	1.0496	1.0524	1.0547	1.0568	1.0586
1.76	1.0049	1.0098	1.0146	1.0194	1.0241	1.0285	1.0327	1.0365	1.0404	1.0439	1.0469	1.0494	1.0515	1.0534	1.0549
1.78	1.0046	1.0093	1.0138	1.0185	1.0229	1.0270	1.0310	1.0345	1.0380	1.0413	1.0440	1.0465	1.0484	1.0500	1.0513
1.80	1.0044	1.0089	1.0130	1.0174	1.0216	1.0254	1.0291	1.0324	1.0356	1.0387	1.0412	1.0435	1.0451	1.0466	1.0478

* See Table 7-1b.

For gaseous mixtures of hydrocarbons, i.e., those containing no diluents, the pseudo-critical temperatures and reduced temperatures can be determined from the following formulas:

$$T_c = 157.5 + 336.6G \tag{5}$$

$$T_R = \frac{t + 460}{T_c} \tag{6}$$

where:

T_c = pseudo-critical temperature of the hydrocarbon mixture, °F abs
G = specific gravity of the hydrocarbon mixture (use below 0.7 sp gr only[8])
T_R = reduced temperature, dimensionless
t = temperature of the gas at which the supercompressibility factor is to be computed, °F

For the same mixtures the pseudo-critical pressures and reduced pressures can be computed similarly by the following formulas:

$$P_c = 690 - 31G \tag{7}$$

$$P_R = \frac{P_g}{P_c} \tag{8}$$

where:

P_c = pseudo-critical pressure, psia
G = specific gravity of the hydrocarbon mixture (use below 0.7 sp gr only[8])
P_g = pressure at which the supercompressibility factor is to be determined, psia
P_R = reduced pressure, dimensionless

If the gas contains N_2 and/or CO_2, the specific gravity no longer defines the critical pressure and temperature. The mole fractions of N_2 and/or CO_2 are required, to permit computing the supercompressibility factor. If only nitrogen is present, the heating value may be used in place of the mole fraction of nitrogen. If CO_2 is present, the mole fraction of gas which it represents must be known from analysis. The pseudo-critical pressure and temperature for gases containing hydrocarbon gases, N_2 and/or CO_2, can be calculated by using the following formulas:

$$T_c = 127.0 + 92.10G + 243.2X_c + 0.1642H \tag{9}$$

$$P_c = 670.8 - 184.8G + 681.9X_c + 0.1033H \tag{10}$$

where:

G = specific gravity of the mixture
X_c = mole per cent (per cent by volume) of CO_2 in the mixture
H = gross heating value of the mixture on dry basis, Btu per cubic foot at 60 F and 14.73 psia

Other terms are defined under Eq. **8**.

The values of critical temperature and pressure can be substituted in Eqs. **6** and **8** to determine the reduced temperatures and pressures, from which the supercompressibility factors can be determined from Table 7-1a.

In summary, it is believed that for 0 to about 3000 psia, 0 to 180 F, 0.55 to roughly 0.75 sp gr, and up to about 12 per cent diluents, the supercompressibility factor, F_{pv}, can be predicted with an *average* error less than 0.1 per cent.

A.G.A. Method. A study[8] completed in 1962 both extended (Table 7-1b) the range of earlier[14] compressibility studies (including A.G.A. Gas Measurement Committee Report No. 3)[16] and provided a basic equation for calculating these supercompressibility factors, F_{pv}. A digital computer may be used to solve this equation. The F_{pv} factor was basically determined from pseudo-critical pressure and temperature data. None of the data developed in this study are presented here since they are very extensive and do not lend themselves to usable summaries.

F_{pv} factors are tabulated over the full ranges of temperature and pressure indicated in Table 7-1b. However, when gases with specific gravities above 0.750 are involved, calculations of certain adjustments are necessary. These adjustments as well as the orifice flow constant (Eq. 2, Chap. 2, Sec. 7) may be calculated on a digital computer.

Table 7-1b Range of Applicability of A.G.A. Supercompressibility Factors[8]

Pressure, psig	0 to 5000
Temperature, °F	−40 to 240
Specific gravity	0.554 to 1.000
CO_2, mole %	0 to 15
N_2, mole %	0 to 15

National Bureau of Standards Apparatus. The essential features of the apparatus[9,10] and their connections are shown schematically in Fig. 7-1. The volume of the mercury reservoir should be about 10 per cent greater than that of the measuring bulb **B.** The zero of the scale **F** should be level with mark **9** on the lower neck of **B.** Valve **2,** which is the main sample control valve used to make a determination, should be a needle type that will permit easy and careful control of gas flow and will close tightly. It is necessary to know the volume of the small gas sample cylinder **A** including that of the tubing to the control valve **2** and the trap mark **8,** and the volume of **B** from the mark **9** to the core of the 3-way cock **3.**

If determinations of F_{pv}, the supercompressibility factor, are to be made on samples at initial pressures much in excess of 500 psig, the oil (or mercury-oil) trap **D** should be replaced by a suitable diaphragm type pressure balance indicator. Also, it will be convenient to have the volume of **B** between two and four times that of **A.** Valve **7** is closed at all times except when filling **D** with oil or mercury and oil.

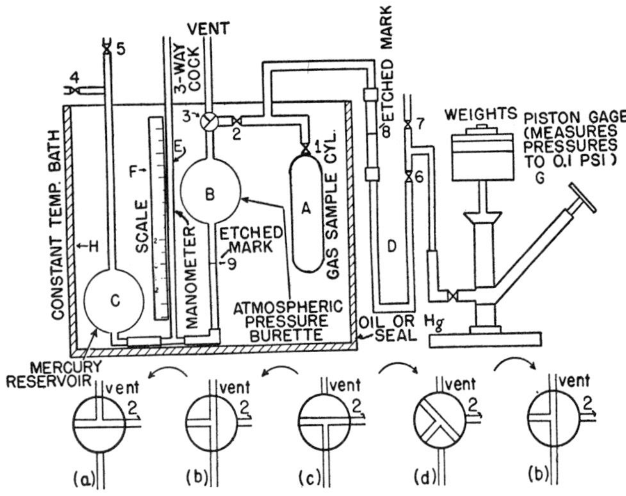

Fig. 7-1 Apparatus for the equal step method of supercompressibility factor determination. Needle valves are used at points 1, 2, 4, 5, 6, and 7.

Test Procedure: (See Fig. 7-1.) Starting with the sample charged and in place, barometric pressure and apparatus temperature readings should be taken periodically throughout the test, and each one averaged at its conclusion.

1. Purge tubing; turn cock **3** to vent pressure line, position **a.** With valves **2** and **6** closed, open **1** slightly until pressure in tubing is approximately equal to cylinder pressure. Close **1** and open **2.** When pressure in tubing drops to atmospheric, close **2,** turn **3** in order to vent **B,** and close off line to **2,** position **b.**
2. With a low pressure (e.g., 10 psig) air supply connected at **4,** close **5** and open **4** carefully, causing mercury to flow into **B.** Just as the mercury in **B** reaches the edge of the core of cock **3,** valve **4** is closed and **3** is turned to connect **B** with the line to **2,** position **c.**
3. With valves **2** and **6** still closed, valve **1** is opened slowly until pressure in tubing is equal to that in cylinder, then it is opened one or two turns. Place weights on piston gage to give a pressure equal to that of gas in cylinder, and open **6** cautiously, readjusting the weights so that with the plunger floating freely, the top of the fluid in trap **D** is in the plane of mark **8.** Note and record the pressure equivalent of the weights on the piston gage, referred to as W in the computations.
4. Open valve **5** slightly and carefully crack control valve **2,** thus letting gas from **A** to **B** as mercury flows back to **C.** Open valve **5** fully. As the surface of the mercury leaving **B** nears the lower neck, start closing valve **2,** and when the mercury surface is just below line **9,** close valve **2** tight.
5. Turn cock **3** so that all outlets are closed, position **d.** Close valve **5** and by very carefully cracking valve **4** the surface of the mercury in the lower neck of **B** is brought up to the place of line **9.** Read on scale **F** the position of the mercury surface in **E.** This is the gage pressure of the gas in **B,** referred to as p_b in the computations.
6. Return cock **3** to position **b** and repeat step 2. Remove weights from piston gage **G** and obtain the next value of W as in step 3. Repeat steps 4 and 5, and so on.
7. When it is estimated that sufficient gas remains in **A** to fill **B** three to five times, divide the last value of W by the average change in W. If the quotient is nearly a whole number, continue with the same quantities of withdrawals. If the quotient is not a whole number, increase or decrease the quantities withdrawn so that the final pressure in **A** and **B** will be within the range of manometer **E.** (Note that it may be necessary to apply suction to **C** at **4** or **5** in order to bring the surface of the mercury in **B** down to line **9.**)

8. After the last filling of **B** with gas, cock **3** and valve **2** are open so that the passage between **A** and **B** is open. The piston gage plunger is resting on its lower stop and by manipulation of the oil pump on the piston gage the surface of the liquid in trap **D** is brought to line **8**. By application of pressure or suction to **C**, the mercury surface in **B** is adjusted to line **9** and the pressure is read by manometer **E**. This is the last value of p_b and is also the final or residual pressure in **A**. It will be designated $(p_b)_n$ in the computations.

9. Take final readings of the barometer and of the apparatus temperature.

Computations. Observed quantities and constants applying to the apparatus are:

W = pressure equivalent of weights on piston, psig
p_b = readings of manometer **E**, in. Hg
t = average temperature of apparatus during a test, °F
B = average barometer reading during test corrected to 32 F, in. Hg
n = number of cycles of operation in test
K = the constant for trap **D**, psig. It is the pressure (or suction) that may have to be applied at the cylinder connection to **D** in order to depress the trap liquid to line **8** (or to bring the diaphragm seal into neutral position), with the piston floating freely and all weights removed.
V_a = volume of gas cylinder **A**, including tubing to valve **2** and trap reference line **8**
V_b = volume of glass bulb **B** from line **9** to surface of plug of cock **3**. (V_a and V_b must be measured in same units of volume.)

1. For each value of W compute:

$$P_a \text{ (in psia)} = W + K + 0.491B$$

2. For each value of p_o compute:

$$P_b \text{ (in psia)} = 0.491(B + p_b)$$

3. Starting with the nth or last value of $(P_b)_n$ corresponding to $0.491[B + (p_b)_n]$, add to it the preceding value of P_b; add to this sum the $(n - 2)$ value; and so on up to the first. Call these sums $\bar{P}_b$.
4. Multiply each value of $\bar{P}_b$ by V_b/V_a.
5. To each value of $\bar{P}_b(V_b/V_a)$ add $(P_b)_n$.
6. Divide each value of $[\bar{P}_b(V_b/V_a) + (P_b)_n]$ by the corresponding values of P_a. The quotients are the values of $(F_{pv})^2$ corresponding to the several values of P_a.

U. S. Bureau of Mines-Burnett Compressibility Apparatus. This apparatus[10,11] is unique in that it permits determining isothermal compressibility factors of a gas mixture (including multiphase mixtures[15]) from pressure measurements only, without knowing any gas volumes or quantities. It is based upon variation from the constancy associated with ideal gas behavior of the pressure ratios accompanying expansions of a gas thru a constant volume ratio. Figure 7-2 shows a section thru a heavy-walled type of this apparatus, designed by P. V. Mullins* for 4000 psi max working pressure, along with a diaphragm cell and a sensitive rotating piston precision pressure gage. The diaphragm cell, interposed in the pressure connection to the gage, greatly facilitates operative manipulation and pressure measurements. It serves the triple purpose of separating gas under test from oil transmitting pressure to the gage, of acting as an automatic and positive dual check-valve against any difference between gas and oil pressures, and of indicating electrically a sensitive balance of such pressures.

* P. V. Mullins, General Manager, Helium Operations, U. S. Bureau of Mines, Amarillo, Texas.

Isothermal Expansion Method: After evacuating the two chambers V′ and V″, a sample of monophase gas at high pressure, P_0, is isolated in V′, at temperature, T_s, then expanded into evacuated V″ to a pressure, P_1, observed after temperature equilibrium is attained. Next, the expansion valve is closed, gas in V″ is discarded, and V″ is re-evacuated; then the gas remaining in V′ is expanded into V″, equalizing as P_2. Consecutive expansions as above are continued to as low a pressure, P_n, as may be feasible or accurately measurable.

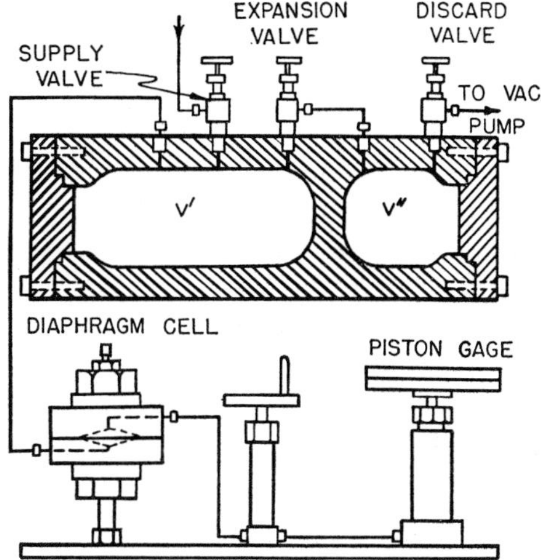

Fig. 7-2 Schematic of Burnett compressibility apparatus.

Treatment of Data: The observed pressure ratios, P_0/P_1, P_1/P_2, . . ., P_{n-1}/P_n, are not equal to the volume ratio $(V' + V'')/V' = N$, a constant, as would be the case with ideal gas expansions, but they approach that value as the number of expansions is increased indefinitely, i.e., as the pressures approach zero. Hence, a plot of the pressure ratios (whether consecutive or random) as ordinates, vs. either their higher or their lower pressures as abscissas, extended graphically to the ordinate axis accurately evaluates N as that intercept.

The isothermal compressibility factors, Z, are related to the pressure ratios and to the volume ratio, N, by the expressions:

$$P_0/P_1 = NZ_0/Z_1,\ P_1/P_2 = NZ_1/Z_2,\ \ldots,\ P_{n-1}/P_n = NZ_{n-1}/Z_n$$

From which, by combination and writing Z' for Z:

$$Z_1'/Z_0' = NP_1/P_0,\ Z_2'/Z_0' = N^2P_2/P_0,\ \ldots,\ Z_n'/Z_0' = N^nP_n/P_0$$

Assume Z_0' = unity at $P = P_0$; calculate Z_1', Z_2', . . ., Z_n' and plot them as a function of their corresponding pressures, P_1, P_2, . . ., P_n. Note the value, Z_b', where a curve thru these Z' points intercepts the pressure ordinate at $P = P_b$, the standard or base reference pressure. Then, at any pressure, P, along the curve, the corresponding Z is equal to the Z' at that pressure divided by Z_b' or $Z = Z'/Z_b'$.

Isochoric Method: Assuming that the function Z_s of P_s at T_s is known for a mixture of composition X in its monophase condition, follow the steps on page 7/10.

1. Fill both evacuated containers to pressure P_s at T_s and close their interconnecting valve.

2. Cool both containers to a minimum temperature, T_m, recording concomitant temperatures and pressures in V′ during the cooling. (Extract phase samples from V″ at T_m if desired for analyses.)

3. With interconnecting valve still closed, warm both containers to T_s, recording associated temperatures and pressures in V′ during this process to check those observed during the cooling. (More phase samples may be taken from V″ during this step.)

4. At T_s discard the contents of V″ only, and re-evacuate it.

5. Expand into V″ the charge remaining in V′ and record the resulting reduced pressure after temperature equalization at T_s.

6. Close the interconnecting valve between V′ and V″. Then repeat the above procedure from step 2 on.

Along these isochors, Z may be computed at associated T and P observed, by the relation $(ZT/P = Z_sT_s/P_s)_{X,V}$. Also from pressure ratio data obtained at T_s, Z_s may be determined by the isothermal expansion method as the assumed function of P_s in this isochoric method; the latter automatically handles multiphase conditions for constant composition fluid mixtures.

For high accuracy, particularly at high pressures, the containers should be supplied with appropriate pressure jackets.

Beckman Design Apparatus. An apparatus, based upon the same principle as the National Bureau of Standards apparatus (with methods of both calculation and calibration likewise similar), but simpler and more rugged because of its all-metal construction, was developed by the Pacific Gas and Electric Co.[12] (Fig. 7-3). The volume ratio of this apparatus is about four.

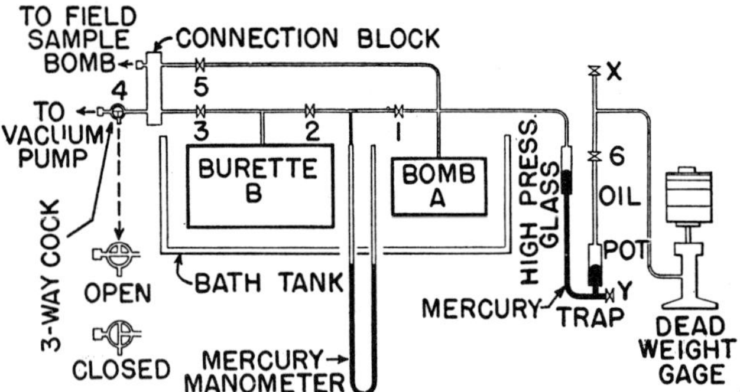

Fig. 7-3 Beckman supercompressibility determination apparatus.

Principle: Gas at high pressure in steel bomb **A** is expanded to atmospheric pressure thru the control valve **1** into steel burette **B,** which has been previously evacuated. After several evacuations and fillings, the pressure in **A** is reduced to one atmosphere. The gas will then have been metered at atmospheric pressure. With pressure and capacity of **A** known, an atmospheric volume may be computed by Boyle's law. The actual metered volume divided by this computed volume gives the deviation factor.

Operation: The bomb is not removed from the apparatus for receiving a sample, but it is filled at the connection block with valves **1** and **6** closed. After closing valve **5,** valve **6** is opened and a pressure reading taken. With valves **2** and **3** open, the burette is evacuated; then valve **3** is closed and the manometer is read. Valve **1** is then cracked and the burette filled, after which **1** is closed and a manometer reading taken. This procedure is repeated until the pressure in the bomb has been reduced to about one atmosphere. Valve **2** is used only for checking the tightness of valve **1**.

Calibration: The ratio of burette to bomb volumes is sufficient for both pieces of apparatus, but it must be known to an accuracy of 0.05 per cent. Actual measurement of each volume has been proved unsatisfactory and the ratio is usually found directly. This is accomplished by conducting a test on dry air free of CO_2; the deviation of air is known and the volume ratio may be computed.[12,13]

REFERENCES

1. Goodholm, P. R. "How Much Gas in a Cubic Foot?" *A.G.A Monthly* 44: 33–5, Ap. 1962.
2. Deaton, W. M. and Frost, E. M., Jr. "Water Content of Compressed Gases." *A.G.A. Nat. Gas Dept. Proc.* 1941: 143–53.
3. Hammerschmidt, E. G. "Equilibrium Water Vapor Content of Compressed Gases." *Gas Age* 97: 24+, June 27, 1946. Also in: *Am. Gas J.* 165: 21–3, Aug. 1946.
4. Calif. Natural Gasoline Assoc. *Tentative Standards for the Determination of Superexpansibility Factors in High Pressure Gas Measurement.* (Bull. TS-354) Los Angeles, 1936.
5. ——. *Tentative Standard Procedure for the Determination of Superexpansibility and Manometer Factors Used in Measurement of Natural Gas by Orifice Meter at Pressures in Excess of 500 Psig.* (Bull. TS-461) Los Angeles, 1947.
6. Natural Gasoline Assoc. of Am. *Tentative Standard Method for Calculation of High Pressure Gas Measurement.* (Standard 4142) Tulsa, 1942.
7. Dunkle, R. V. "Pseudo-Critical Method for Evaluating Deviation of Natural Gas from Boyle's Law." *Gas* 20: 41–52, Oct. 1944.
8. A.G.A. *Manual for the Determination of Supercompressibility Factors for Natural Gas.* New York, 1963.
9. Bean, H. S. "Apparatus and Method for Determining the Compressibility of a Gas and the Correction for Supercompressibility." *Bur. Std. J. Res.* 4: 645–61, May 1930. Also in: *Oil Gas J.* 28: 42+, Jan. 16, 1930.
10. Bloomer, O. T. *Measurement of Gas Law Deviations with Bean and Burnett Apparatus.* (I.G.T. Research Bull. 13) Chicago, 1952.
11. Burnett, E. S. "Compressibility Determinations Without Volume Measurements." *J. Appl. Mech., A.S.M.E. Trans.* 3: A-136-40, Dec. 1936.
12. Beckman, P. E. "Simplified Apparatus for Determining the Deviation of Gas from Boyle's Law." *Am. Gas J.* 136: 20+, Mar. 1932.
13. ——. "Deviation of Natural Gas from Boyle's Law." *P.C.G.A. Proc.* 23: 329–44, 1932.
14. A.G.A. *Supercompressibility Factors for Natural Gas.* Temperature range 0–180 F; specific gravity range 0.554–0.750. Vols. I–VI: Pressure range 0–3000 psig in 500 degree increments. Vol. VII: Tables for the determination of supercompressibility factors for natural gas containing nitrogen and/or carbon dioxide, pressure range 0–3000 psig. New York, 1955.
15. Burnett, E. S. *Application of the Burnett Method of Compressibility Determinations to Multiphase Fluid Mixtures.* (Bur. of Mines R.I. 6267) Washington, 1963.
16. A.G.A. *Orifice Metering of Natural Gas.* (Gas Measurement Com. Rept. 3) New York, 1955.

Chapter 2
Rate-of-Flow or Inferential Meters

The types of measuring equipment discussed here are those which have proved from actual industry experience to be accurate and practical. However, it should not be inferred that other types are not equally accurate.

The **differential pressure group**[8] not discussed in this Handbook includes: eccentric and segmental orifices, flow nozzles, critical flow nozzles (also called critical flow orifices), venturi tubes, gate valve meters, elbow meters, and linear resistance meters. **Nondifferential**[8] type meters not covered in this Handbook are variable area meters and turbine meters. There are four other nondifferential methods described in Reference 8: (1) *method of mixtures*—addition of small concentrations of a substance at a steady known rate and testing for the proportion of the foreign substance downstream; (2) *tracer method*—similar to preceding except that presence, not proportion, is noted downstream; (3) *sonic or sound velocity methods*—depend upon comparing the velocity of the stream to the velocity of sound in the stream; (4) *thermal meters*—these depend upon the measurement of the temperature change resulting from the addition or removal of heat from the stream.

The term rate-of-flow or inferential pressure meter is applied to certain meters thru which the fluid passes in a continuous stream rather than in isolated (separately counted) quantities. The movement of this fluid stream thru the "primary element" is directly or indirectly utilized to actuate the "secondary element."

The *primary element* of an orifice meter is a straight piece of pipe containing a restriction. On each side of this restriction a pressure tap hole is placed to measure pressure differential. The primary element for metering the gas is usually complemented by orifice fittings, straightening vanes, and thermometer wells.

The *secondary element* is the device(s) which measure the differential pressure across the plate and other factors required for determining the flow rate. These devices include differential pressure gages, static pressure gages, thermometers, and gravitometer.

The orifice meter is the rate-of-flow meter most commonly accepted by the gas industry for large volume measurement in both purchase and sale of gas.

ORIFICE METERS

General data on orifice meter station design are available.[1,8] In its simplest and most familiar form, the orifice fitting is a circular hole in a thin flat plate located between flanges at a joint in the pipeline, with its plane normal to the axis of the pipe and the hole concentric with the pipe. (Other shapes and nonconcentric openings have been used.) Pressure connections for attaching a differential gage and a separate static gage are made at side holes in the pipe wall on both sides of the plate or integral with the flanges, fixed so that the difference in the fluid pressure between the sides of the orifice can be measured. This differential pressure is utilized to determine the rate of flow.

If a series of static-pressure holes were made in the pipe along its axis on each side of the orifice plate, and if each hole were connected to a manometer, the manometer readings would illustrate variations in static pressure.

As the gas approaches the inlet side of the orifice, the static pressure in the pipe increases slightly and reaches its maximum value at the orifice entrance. The pressure of the fluid drops abruptly as it flows thru the orifice and it reaches a minimum value a short distance beyond. The pressure then increases again, slowly at first, rapidly for a short distance, then again more slowly until its second maximum is reached several pipe diameters beyond the orifice plate. Since no guiding of the stream occurs on either the inlet or outlet of an orifice, the acceleration and deceleration of the fluid stream, which the pressure gradient manifests, are accompanied by considerable turbulence and consequent dissipation of energy (pressure), especially on the outlet side.

Selection of pressure tap locations with respect to the plane of the inlet side of an orifice plate is usually governed by one or more of the following considerations:

1. The pressure taps should be located in the same positions relative to the plane of the orifice for all sizes of pipe.
2. Locations should be such that either the maximum or minimum pressure value will be observed, because slight errors in the tap locations will then have less effect on the observed values.
3. Locations should be selected to facilitate the tap installation in either shop or field and to minimize the possibility of locating taps incorrectly.

The pressure tap locations generally used in commercial gas measurement work are known as "flange taps" and "pipe taps."

Design and Installation of Primary Elements

The following Sections, **II—Scope** and **III—Construction & Installation Specifications,** have been extracted from A.G.A. Gas Measurement Committee Report No. 3.[2]

II—SCOPE

(10) Report No. 3 and its specifications and recommendations relate to and should be understood to be limited to the following two types of orifice meters:

a. *Orifice meters with circular orifices placed concentrically in the pipe line, and having upstream and downstream pressure taps as specified for Flange Taps in Section III.*
b. *Orifice meters with circular orifices located concentrically in the pipe line, and having upstream and downstream pressure taps as specified for Pipe Taps in Section III.*

(11) This report is concerned only with the primary element, that is, the orifice plate, the orifice flanges or fitting, and adjacent pipe sections, and does not cover the equipment used in the determination of the pressures, temperatures, and other values which must be known for the accurate measurement of gas.

III—CONSTRUCTION AND INSTALLATION SPECIFICATIONS

A. ORIFICE PLATES

(12) The thickness of the orifice plate for 4 in. nominal diameter pipe and smaller shall be at least 0.060 in. and not over 0.130 in. For pipes of 6 in. nominal diameter the thickness shall be at least 0.100 in. and not over 0.255 in. For use in pipes larger than 6 in. nominal diameter, the thickness shall be at least 0.100 in. and not over 1/30 of the inside diameter of the pipe, but in no case to exceed 0.505 in.

(13) The upstream face of the orifice plate shall be as flat as can be obtained commercially and shall be perpendicular to the axis of the pipe, when in position between the orifice flanges, or in the orifice fitting. Since perfect flatness is difficult to obtain, any plate which does not depart from flatness along any diameter by more than 0.010 in. per inch of the dam height, $(D - d)/2$, shall be considered flat. The upstream face of the orifice plate shall be smooth and with a finish at least equivalent to that obtained in commercial cold-finished sheet stock of the non-rusting alloy stainless steels or monel metal.

(14) The upstream edge of an orifice shall be square and sharp so that it will not appreciably reflect a beam of light when viewed without magnification. It shall be maintained in this condition at all times. Moreover, the plate is to be kept clean at all times and free from accumulations of dirt, ice, and other extraneous material.

(15) The thickness of the orifice plate at the orifice edge shall not exceed:

a. 1/30 of the pipe diameter, D,
b. 1/8 of the orifice diameter, d,
c. 1/4 of the dam height, $(D - d)/2$,

the minimum of these requirements governing in all cases. It is recommended that wherever practicable the requirement "a" be reduced to 1/50 of the pipe diameter D.

(16) In some cases the thickness of the orifice plate will be greater than permitted by the limitations in par. 15, in such cases the downstream edge shall be cut away (beveled or recessed) at an angle of 45° or less to the face of the plate, leaving the thickness of the orifice edge within the requirements as set forth in par. 15. All orifice plates which are beveled should have the square-edged side (i.e., the side opposite the beveling) stamped INLET or the beveled side stamped OUTLET.

(17) In centering orifice plates, the orifice must be concentric with the inside of the meter tube or orifice fitting. The concentricity shall be maintained within 3% of the inside diameter of the tube or fitting along all diameters.

(18) The measured orifice diameter shall be as close as practical to that used in computing the Basic Orifice Factor. At least three different diameters shall be measured. No measured diameter shall differ from the diameters used in computing the Basic Orifice Factor, nor from any other measured diameter, by an amount greater than the tolerances shown in Table 1. Where the tolerance is less than .001 in., plug gages of all types may be used. For those cases in which the tolerance is greater than .001 in., only plug gages of the types that will permit checking for out-of-roundness of the orifice may be used.

TABLE 1 PRACTICAL TOLERANCES FOR ORIFICE DIAMETERS

Orifice size, in.	Tolerance plus or minus, in.	Orifice size, in.	Tolerance plus or minus, in.
0.2500	0.0003	1.2500	0.0014
.3750	.0005	1.5000	.0017
.5000	.0006	1.7500	.0020
.6250	.0008	2.0000 to 5.0000	.0025
.7500	.0009	over 5.0000	0.0005 per inch of diameter
0.8750	.0010		
1.0000	0.0012		

(19) It is recommended that for the commercial measurement of gases the orifice to pipe diameter ratio β, d/D, be limited as follows:

a. With meters using flange taps, β shall be between 0.15 and 0.70.
b. With meters using pipe taps, β shall be between 0.20 and 0.67.

(20) With either type of pressure tap, diameter ratios as low as 0.10 may be used, while ratios as high as 0.75 may be used with flange taps and as high as 0.70 may be used with pipe taps. The flow constants for these extreme values of β are subject to higher tolerances, and the use of these extreme ratios is to be avoided wherever practicable.

B. METER TUBES

(21) The term "meter tube" shall mean the upstream pipe (dimension A and A′ on installation sketches, including the straightening vanes) between the orifice and the nearest fitting and the downstream pipe (dimension B on installation sketches) between the orifice and the nearest fitting.

(22) The sections of pipe to which the orifice flanges are attached or the sections adjacent to the orifice flange or fitting shall comply to the following:

a. Inside pipe walls shall not be polished, but should be as smooth as is commercially practicable where pipe walls are not machined or ground. When pipe walls are machined or ground, the finish should simulate that of new smooth pipe. Seamless pipe or cold drawn seamless tubing may be used.

b. Grooves, scoring, pits, raised ridges resulting from seams, distortion caused by welding, offsets, etc. (regardless of the size of such irregularities) which affect the inside diameter at such points by more than the tolerance given in Figure 1, shall not be permitted. When these measurements are exceeded, the roughness may be corrected by filling in or grinding or filing off, so as to obtain smoothness within the tolerance.

(23) In order to use the Tables of Basic Orifice Factors given in Section V of this report,* the *measured inside diameter* of the upstream section of the meter tube should be as nearly as possible the same as the *published inside diameter* given in Table 2. These published inside diameters were used in calculating the constants and factors of Section V. If the measured (i.e., the *actual*) inside diameter of the upstream meter tube differs from the published inside diameter by an amount greater than the tolerance recommended in Figure 1, then the measured inside diameter of the upstream section should be used in computing the orifice to pipe diameter ratio β and also the coefficient of discharge and other factors should be calculated for this exact value of β as explained in Appendix B.† Although the diameter ratio β

* [Table 3 for flange tap installations given here; Table 8 for pipe tap installations, see original.—EDITOR]

† [Not given in this Handbook; see original.—EDITOR]

is based on the diameter of the upstream section of the meter tube, it is suggested that the diameter of the downstream section be held to twice the tolerance allowed by Figure 1.

TABLE 2 DIMENSIONS OF PIPE

Nominal size, inches	Published inside diameter, inches			
2		1.689	1.939	2.067
3	2.300	2.626	2.900	3.068
4	3.152	3.438	3.826	4.026
6	4.897	5.189	5.761	6.065
8	7.625	7.981	8.071	
10	9.564	10.020	10.136	
12	11.376	11.938	12.090	
16	14.688	15.000	15.250	
20	18.814	19.000	19.250	
24	22.626	23.000	23.250	
30	28.628	29.000	29.250	

(24) The actual inside diameter of the meter tubes should be determined as follows:

a. Measurements are to be made on at least three diameters one inch from the face of the orifice plate.

b. A corresponding number of check measurements of the diameter are to be made at two additional cross sections. The actual location of the check measurements of the diameter, circumferentially and axially along the tube, are not specified (except for the one inch location) in that the checks should be taken at points which will indicate the maximum and minimum of diameter that exist, covering at least one pipe diameter from the face of the orifice plate.

(25) The tolerances for the measurements of the meter tubes are:

a. The average of all measurements shall agree with the *published inside diameters* given in Table 2 within the tolerance allowed by Figure 1.

b. The difference between the maximum and the minimum of all measured diameters shall not exceed the tolerance allowed by Figure 1 referred to the *published inside diameters* in Table 2.

c. Under all circumstances, abrupt changes in diameter (shoulders, offsets, ridges, etc.) should not exist in meter tubes in those sections of the tube following the straightening vanes or in the designated minimum lengths of straight pipe required when straightening vanes are not used. This applies to both upstream and downstream sections.

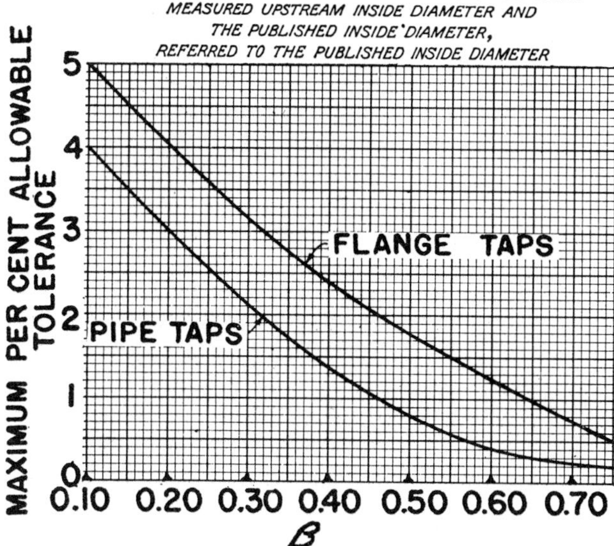

Note: It is recommended that for new installations in which the orifice to pipe diameter ratio is liable to be changed, the tolerance permitted for variations in pipe size, as shown by Figure 1, be the same as that given for the maximum orifice to pipe diameter ratio which may be used.

C. LENGTH OF PIPE PRECEDING AND FOLLOWING AN ORIFICE

(26) The orifice flow constants and other factors given in this report are based on the results of many experiments. In all cases, normal flow conditions were obtained by the use of long straight lengths of pipe, both upstream and downstream from the orifice, or by the use of straightening vanes. To obtain satisfactory measurement of flow, these conditions should be approximated in practice.

(27) Any serious distortion of the flow will produce errors, and there are many locations at which the orifice will not produce satisfactory results. Certain of the more common types of installations have been studied as to their effect on metering accuracy, and the results of these studies are shown in the following sketches. These sketches show the minimum lengths of straight pipe required, both preceding and following an orifice, if the errors caused by such disturbances are to be eliminated or kept within the tolerance limits stated in this report. For those installations not explicitly covered in the installation sketches, it is recommended that installation sketch Figure 6 be followed.

FIGURE 2

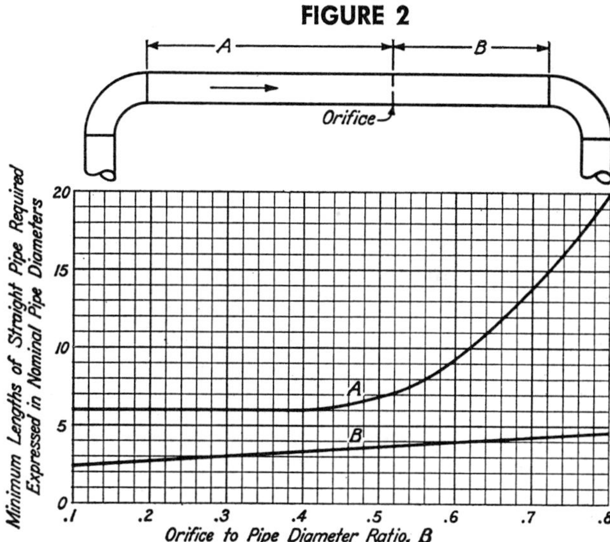

Note 1: When pipe taps are used, A should be increased by 2 pipe diameters and B by 8 pipe diameters.

Note 2: When the diameter of the orifice may require changing to meet different conditions, the lengths of straight pipe should be those required for the maximum orifice to pipe diameter ratio that may be used.

(28) The graphs accompanying the installation sketches, Figures 2 through 6, indicate that the minimum length of a straight pipe required varies with the orifice to pipe diameter ratio, longer lengths of straight pipe being required for the higher ratios. When the diameter of the orifice may require changing to meet different conditions, the length of straight pipe installed should be the same as that given for the maximum orifice to pipe diameter ratio that may be used. Meter tubes longer than those indicated by the graphs are desirable when circumstances permit. When straightening vanes are used with longer meter tubes the

dimension indicated by C should not be less than that indicated by the graphs.

(29) In determining whether or not straightening vanes are required, it may not always be the first fitting on the inlet end of the upstream meter tube that governs. Each individual station design may have a different set of conditions. Therefore, it would be impractical to set up specifications that would suit all conditions. The main consideration should be that there is no disturbance at the orifice plate caused by any fittings upstream. This may be accomplished by the installation of straight pipe of sufficient length (in the meter tube) for the fitting having the worst disturbance, or having sufficient length following each type fitting as though it were a meter tube having an orifice to pipe diameter ratio the same as in the meter tube, either using or not using straightening vanes.

1. Installation with One Ell Preceding the Meter Tube

(30) The installation sketch, Figure 2, shows one ell preceding the straight run of pipe in front of the orifice, a very common type of installation.

2. Installation with Two Ells or Bends Preceding the Meter Tube (Bends in Same Plane)

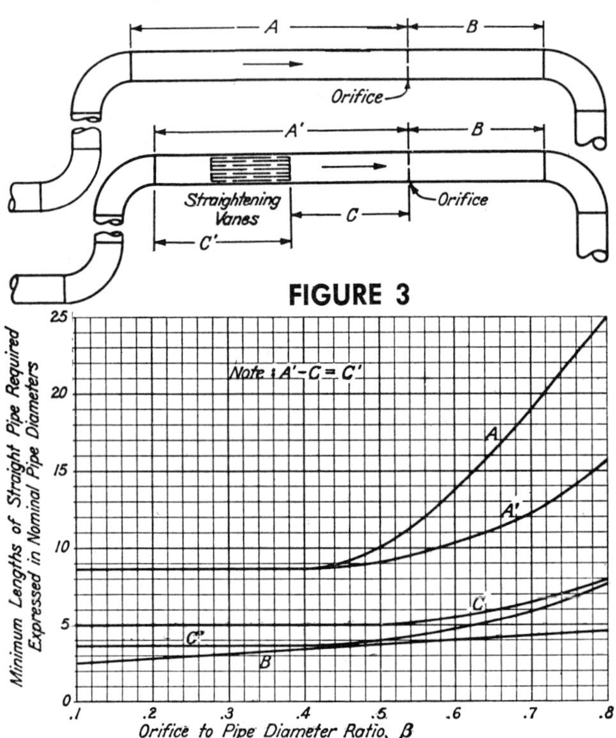

Note 1: When pipe taps are used, A, A', and C should be increased by 2 pipe diameters, and B by 8 pipe diameters.

Note 2: When the diameter of the orifice may require changing to meet different conditions, the lengths of straight pipe should be those required for the maximum orifice to pipe diameter ratio that may be used.

(31) The installation sketch, Figure 3, shows two ells or two bends preceding the straight run of pipe upstream of the orifice. There is more of a distortion of the flow lines in passing through the two fittings than through one as shown in Figure 2 even though they are in the same plane.

3. Installation with Two Ells or Bends Preceding the Meter Tube (Bends Not in the Same Plane)

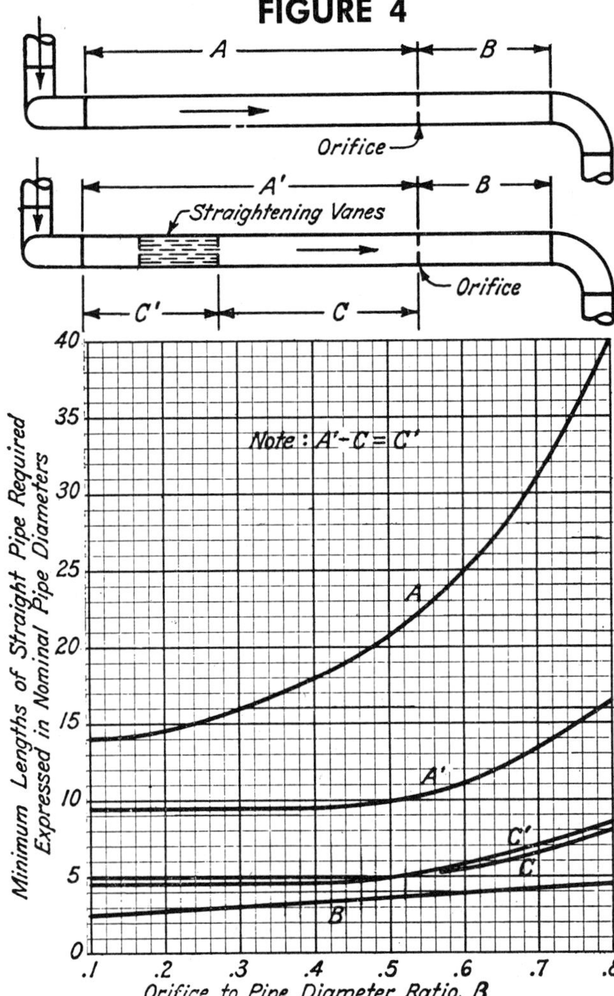

Note 1: When pipe taps are used, A, A', and C should be increased by 2 pipe diameters and B by 8 pipe diameters.

Note 2: When the 2 ells shown in the above sketches are closely preceded by a third which is not in the same plane as the middle or second ell, the piping requirements shown by A should be doubled.

Note 3: When the diameter of the orifice may require changing to meet different conditions, the lengths of straight pipe should be those required for the maximum orifice to pipe diameter ratio that may be used.

(32) The installation sketch, Figure 4, shows two ells, or two bends, at right angles to each other (not in the same plane). This type of bend produces considerable disturbance that is ironed out only by long runs of straight pipe, or by straightening vanes, or both, on the inlet side of the orifice.

4. Installation with Reducer Preceding the Meter Tube

(33) The installation sketch, Figure 5, shows the use of a reducer preceding the run of straight pipe. This type of installation makes a rather easy installation insofar as obtaining accuracy is concerned inasmuch as the minimum lengths of the straight pipe required between the reducer and the orifice are quite moderate in amount.

5. Installation with Valve or Regulator Preceding the Meter Tube

(34) The installation sketch, Figure 6, shows a valve which restricts the flow of the gas, such as a regulator or a partially closed gate valve, globe valve, or plug valve. However, if the gate valve, globe valve, or plug valve is wide open, it may be considered as not creating any serious disturbance and shall be located according to the requirements of the fitting immediately preceding it, but in no case to be located closer to an orifice than permitted by curves A, A′, or B in the appropriate figure of installation sketches, Figures 3, 4, or 5. Where there are no other fittings, installation sketch Figure 3 shall apply in locating the above mentioned wide open valves.

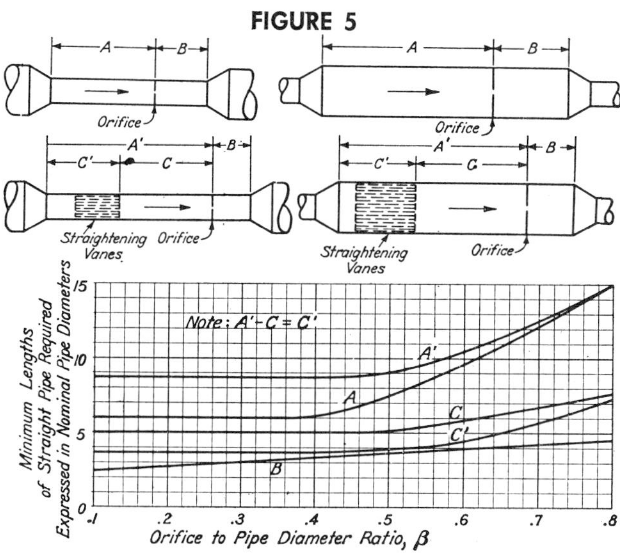

Note 1: When pipe taps are used, length A, A', and C should be increased by 2 pipe diameters, and B by 8 pipe diameters.

Note 2: Straightening vanes will not reduce lengths of straight pipe A. Straightening vanes are not required because of the reducers, they are required because of other fittings which precede the reducer. Length A is to be increased by an amount equal to the length of the straightening vanes whenever they are used (see bottom sketches).

Note 3: When the diameter of the orifice may require changing to meet different conditions, the lengths of straight pipe should be those required for the maximum orifice to pipe diameter ratio that may be used.

D. STRAIGHTENING VANES

(35) The installation of straightening vanes as indicated in the installation sketches will reduce considerably the amount of straight pipe required preceding an orifice. The purpose of the vanes is to eliminate swirls and crosscurrents set up by the pipe fittings and valves preceding the meter tube. While the specifications which follow apply particularly to the type of vanes shown in Figure 7, vanes of other designs must meet these specifications.

(36) In the construction of vanes the maximum transverse dimension, a, Figure 7, of any passage through the vanes shall not exceed one-fourth ($\frac{1}{4}$) the inside diameter, D, of the pipe. Also, the cross-sectional area "A" of any passage within the assembled vanes shall not exceed one-sixteenth ($\frac{1}{16}$) of the cross-sectional area of the containing pipe. The length, L, of the vanes shall be at least ten (10) times the inside dimension, a.

(37) The vanes, Figure 7, may be built of standard weight pipe, or thin walled tubing, either welded together and properly and securely attached into the meter tubes, or mounted into two end rings small enough to slip into the pipe. The amount of passage blockage caused by the end rings should be kept as small as practical. All tubes should be reamed as thin as practical at both ends of each tube.

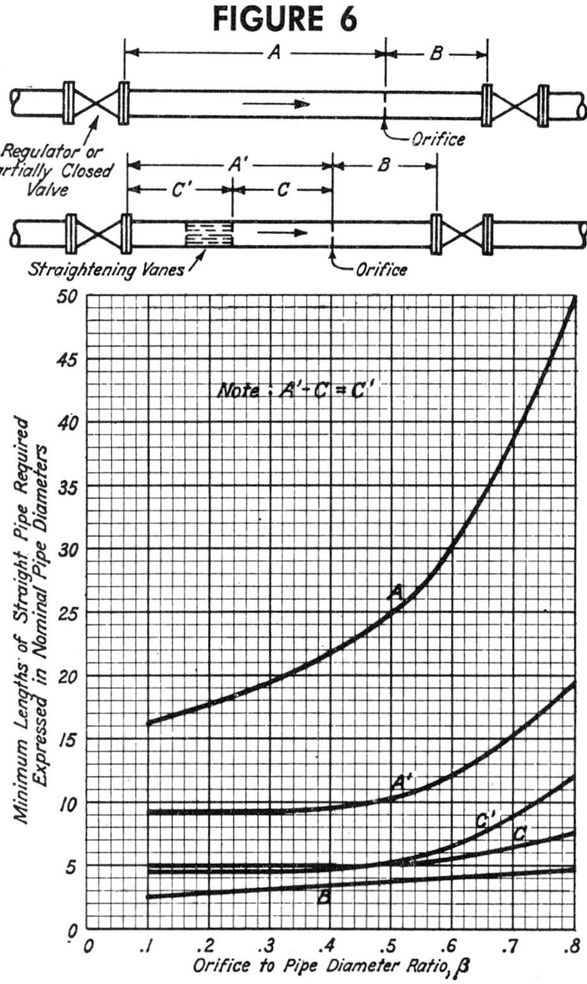

Note 1: When pipe taps are used, length A, A', and C should be increased by 2 pipe diameters, and B by 8 pipe diameters.

Note 2: Line A, A', C, and C' apply to regulators or gate valves, globe valves, and plug valves which are used for throttling the flow (partially closed).

Note 3: When the diameter of orifice may require changing to meet different conditions, the lengths of straight pipe should be those required for the maximum orifice to pipe diameter ratio that may be used.

(38) Square, hexagonal, or other shaped tubing may be used in making the vanes. It is not necessary that all the vane passages be of the same size, but their arrangement should be symmetrical.

(39) Straightening vanes must be firmly constructed. After being inserted in the pipe, they are to be so securely fastened in place as to prevent their being dislodged and pushed down against the orifice plate.

E. ORIFICE FLANGES

(40) Orifice flanges for new or remodeled orifice meter installations should be so constructed and attached to the pipe that the inner surface of the passage extends (or is extended) through the orifice flange so that there is no recess greater than $\frac{1}{4}$ in. at the orifice plate as measured parallel to the axis of the tube. The section of pipe to which the flanges are attached should be of such length that it extends upstream at least to the straightening vanes where vanes are used, and where vanes are not used, the section of pipe should be equal to the minimum length of straight pipe required. The downstream length of straight pipe should also equal the minimum length of straight pipe required.

(41) If in existing installations there is a recess preceding the orifice plate, either in the orifice flange or between the end of the pipe and the plate, the length of which, measured parallel to the axis of the pipe, is greater than ¼ in., the orifice flow constants given in this report may be used only if the orifice to pipe diameter ratio β does not exceed 0.3 with a 2-in. line, 0.4 with 3-in. and 4-in. lines, and 0.5 with all line sizes greater than 4 in. If this ratio is to be exceeded, then the upstream orifice flange shall be replaced, or the recess filled up with poured lead or other material which will not be attacked by the fluid in the pipe, so as to make the inner surface smooth and continuous up to the orifice plate. The downstream orifice flange may be treated likewise.

FIGURE 7

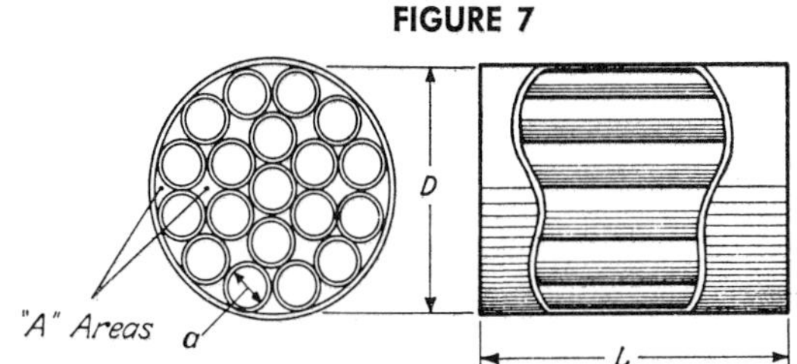

(42) With the availability of many different weights of flanges and the accompanying variation in bolt circle diameter, the method of centering orifice plates in meter tubes is left to the personal judgment of the designer. Sizing the outside diameter of the orifice plate to the bolt circle of the flange, the use of spacer rings on the flange bolts, and drilling the flanges for dowel pins are commonly used methods of centering plates. All methods of centering plates must adhere to the tolerance specified for concentricity in paragraph 17.

F. ORIFICE FITTINGS

(43) The extensive use of orifice fittings of the various types now available, and the probable development of still other fittings, obviates the necessity of full consideration of the requirements of their construction and use. The tolerances that are to be applied with regard to tap hole location, diameter dimensions upstream and downstream from the orifice plate, and centering of the plate in the fitting shall be the same as with orifice flanges. These tolerances are to be adhered to if the Tables of Basic Orifice Factors given in this report* are to be used and the results are to fall within the allowances of the report.

(44) When an orifice fitting is used, the average inside diameter of the section of pipe connected to the inlet side of the fitting should agree with the inside diameter of the fitting within the tolerance given in Figure 1. When installing the fitting, the inlet side should be connected to the upstream section of pipe first and carefully centered, after which any sharp edges at this junction are to be smoothed off. In order to prevent future slipping at this joint, when a welded connection has not been made, it is suggested that two diametrically opposite bolt holes be reamed and snug fitting bolts installed, or dowel pins used, or the same result obtained in some other manner.

G. PRESSURE TAP HOLES

(45) For orifice meters using flange taps, the center of the upstream pressure tap hole shall be placed 1 in. from the upstream face of the orifice plate. The center of the downstream pressure hole shall be 1 in. from the downstream face of the orifice plate. If the pressure tap holes are located by measuring from the bearing face of the flange, allowance must be made for the thickness of the gasket that will be used. The tap hole shall be located at the 1 in. dimensions within plus or minus the tolerances specified in Figure 8.

* [Table 3 for flange tap installations given here; Table 8 for pipe tap installations, see original.—Editor]

FIGURE 8
ALLOWABLE VARIATION IN PRESSURE TAP HOLE LOCATION

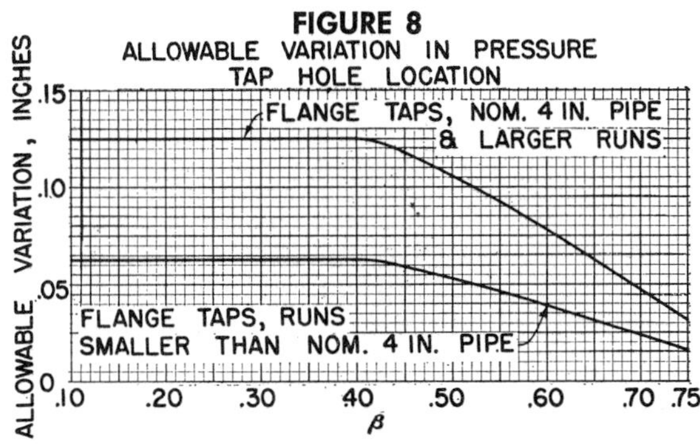

Note: For pipe taps, use plus or minus ten times the tolerances given.

(46) For orifice meters using pipe taps the upstream tap shall be placed two and one half times the published inside diameter from the upstream face of the orifice plate. The downstream tap shall be placed 8 times the published inside diameter from the downstream face of the orifice plate. The tap holes shall be located at the specified distances within plus or minus ten times the tolerances given in Figure 8.

(47) Pressure tap holes shall be drilled radial to the meter tube, that is, the centerline of the tap hole shall approximately intersect and form a right angle with the axis of the meter tube.

(48) The diameter of the pressure tap holes at the inner surface of the pipe shall never be less than ¼ in. This diameter shall not exceed ⅛ of the inside diameter of the pipe for pipes of from 2 in. to 4 in. inside diameter nor exceed ½ in. for pipes of 4 in. inside diameter and larger.

(49) The edges of the pressure tap holes on the inner surface of the pipe shall be free from burrs and slightly rounded.

(50) With orifice meters using flange taps, the outer ends of the pressure holes in the flanges or fittings may be drilled out and threaded to receive the desired size of pressure piping. With orifice meters using pipe taps, the location for each pressure hole shall be spotted on the pipe with a centering punch and a hole of the proper size then drilled through the pipe. This hole shall not be threaded for connecting the pressure piping, but a fitting should be fastened to the pipe at this point. When a fitting is fastened onto the pipe, great care should be taken to make sure that the inside of the pipe is not distorted in any way. If the fitting is welded to the pipe, the tap hole should not be drilled until after the welding is done.

H. THERMOMETER WELLS

(51) Thermometer wells should be placed on the downstream side of the orifice and not closer to the plate than dimension B on the installation sketches included in this report. However, if straightening vanes are used the thermometer well may be located 1 pipe diameter or more upstream from the vanes.

Design and Installation of Secondary Elements

These elements are generally regarded as including devices for recording or integrating the differential pressure between the two pressure taps, and also the static pressure, at either the upstream or downstream pressure tap or some intermediate tap, together with their piping connections.

Devices like temperature and specific gravity recorders are sometimes considered as secondary elements. Here they are not so regarded, but are included under other headings.

Gas Piping. Orifice meter gage piping is generally considered as including small connections between the pressure tap holes and the orifice meter gages. Many different types of arrangements are in industry use. Generally, any arrangement will serve the same purpose so long as the true pressure at the pressure tap is transmitted to the orifice meter gage. A by-pass arrangement between the upstream and downstream connections is necessary for test purposes, but it is essential that no flow passes between the two connections when the gages are in operation. This is usually accomplished by using a bleed valve between double by-pass valves or a 3-way valve.

Where intermediate static pressure connections are installed in the gage piping, it is essential that the small gas flow thru the connection (usually located near the pressure taps) does not affect the correct pressures to the differential recorder within the readable limits. Separate taps are sometimes required.

Intermediate Pressure Connection Device. As described elsewhere, the use of the mean static pressure between the upstream and downstream pressures may be advantageous, particularly with flange taps.

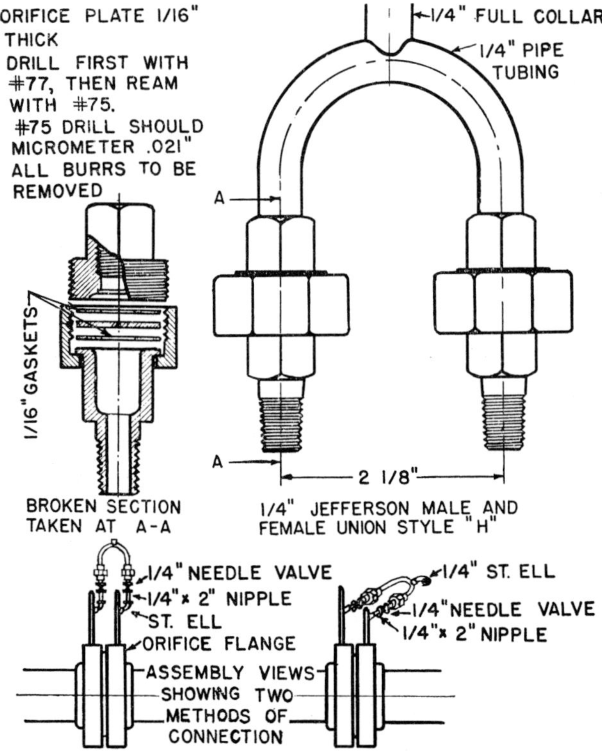

Fig. 7-4 Intermediate pressure connection device.

Figure 7-4 shows a type using small orifices. Figure 7-5 shows a cross section of the *Metric* intermediate pressure connection. The pressure obtained at the ¼ in. vertical tap at the center is the mean of the upstream and downstream static pressures, because with the tap midway between two devices of equal resistance, pure viscous flow is created from the upstream to the downstream side of the orifice.

Gas entering this device first passes thru an oiled horsehair filter and then thru a capillary tubing to the connection to the pressure tube. Then, before entering the main lines it must pass thru a secondary capillary tubing which has the same resistance as the first capillary. The quantity of gas passing is a negligible percentage of that passing thru the orifice for ordinary commercial orifice meter settings.

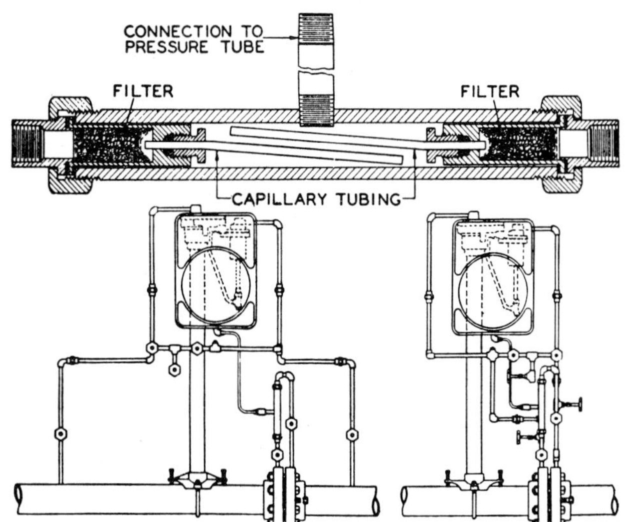

Fig. 7-5 Metric intermediate pressure connection and two methods for its installation. Pipe tap arrangement (left), upstream tap 2½ diameters (ID) from orifice, downstream tap eight diameters (ID) from orifice. Flange tap arrangement (right).

This device is made entirely of steel except for the brass capillary tubing. The filters can be readily removed for cleaning or replacing their elements.

Figure 7-5 also shows this device connected to a line at regular flange taps. The upstream connection may also be made anywhere from the upstream face of the orifice to five pipe diameters upstream, and the downstream connection can be made anywhere from the downstream face of the orifice to ¾ pipe diameter downstream. Both cuts show the device in a vertical riser from the line. It may also be installed in a horizontal position between two vertical risers.

Effect of Pulsation

It is impossible to calculate the pulsation error produced on any given meter, even if the exact pulsating wave form is known. Thus, although the use of a **mechanical pulsometer** is not theoretically valid, it does indicate that some corrective means must be designed into the facilities. Restriction plates or pulsation dampeners are normally required when the pulsation effect on the pressure differential exceeds the error limit of Fig. 7-7.

Pulsation Measuring Instrument. The mechanical-type pulsometer shown in Fig. 7-6 consists of two cylindrical volumes, "A" and "B," and it is connected to the inlet- and outlet-pressure connections of the meter, as close to the primary element as possible. "A" and "B" are separated by a sylphon bellows to prevent leakage between them. "A" is connected to the high-pressure tap and contains a coil spring, the tension of which can be changed by turning an adjusting wheel outside the instrument. A scale with each division representing 0.1 in. indicates the spring elongation. Electrical connections are arranged in "B," which is connected to the low-pressure tap, so that any bellows motion caused by differential pressure variation will be indicated by an exterior lamp or flashlight.

In operation, the tension of the spring is gradually increased until the light goes out, indicating that the differential

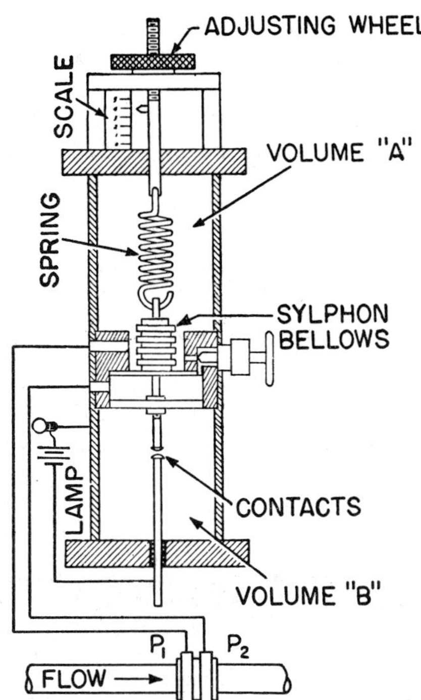

Fig. 7-6 Mechanical pulsometer. (Refinery Supply Co., Tulsa, Oklahoma.) The by-pass valve shown may be used to deactivate the instrument (joins volume "A" to volume "B"). P_1 and P_2 are the upstream and downstream pressure taps, respectively.

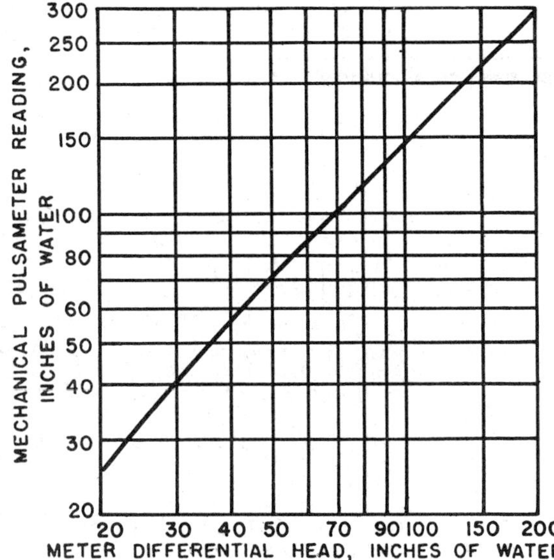

Fig. 7-7 Preliminary pulsation error-limit chart for flange and pipe taps.

Method of using chart: (1) Locate pulsometer reading on left margin. (2) Trace to right to perpendicular line representing meter differential on bottom scale. (3) If this point is below curve, pulsation effect is within tolerance of one per cent. (4) If this point is above curve, pulsation effect is greater than one per cent.

pressure has been equalized by the spring tension. The tension of the spring should then represent the differential pressure due to the flow plus the maximum amplitude of the differential pulsation pressures.

The instrument spring can be calibrated by comparing its scale readings with any differential gage when no pulsations are present. When measuring pulsation amplitude, the instrument is almost totally unaffected by inertia, since the sylphon diaphragm is practically at rest when the determination is made. For practical use, all pulsometers must have the same general dimensions as the one used to determine the effect of pulsation, since their shape and dimensions will undoubtedly have some effect on the instrument reading.

Determination of Existence of Pulsation Error. Figure 7-7, which applies to both flange and pipe taps, can be used to determine whether a meter measuring pulsating flow is accurate. It is necessary only to attach the pulsometer to the pressure taps, take its reading at the same time that the differential pressure is determined, and apply Fig. 7-7.

The curve of Fig. 7-7 was determined by experiment. To show its relationship to some more or less rational curves, Fig. 7-8 was plotted. Its dotted line represents the curve of Fig. 7-7 and Curve 1 gives the values of pulsometer and differential-head readings for one per cent error, computed with the assistance of several assumptions.

These assumptions are: (1) that the reason for the error is that the flow calculated from the meter readings is the square root of the average differential, (2) that the pulsometer reading represented the true maximum value of the differential pressure, and (3) that the differential pressure plotted against time would be a sine wave. Using these assumptions, the true rate of flow calculated from the average differential and the maximum pulsation wave would be:

$$Q_a = K \frac{1}{2\pi} \int_0^{2\pi} (h + p \sin \theta)^{0.5} d\theta$$

where:

Q_a = quantity flowing in proper units
K = meter constant
h = average differential pressure as read by secondary device on meter
p = maximum height of pulsation differential pressure, measured above average differential pressure (pulsometer reading *minus* reading of differential gage)
θ = phase angle of differential pulsation wave

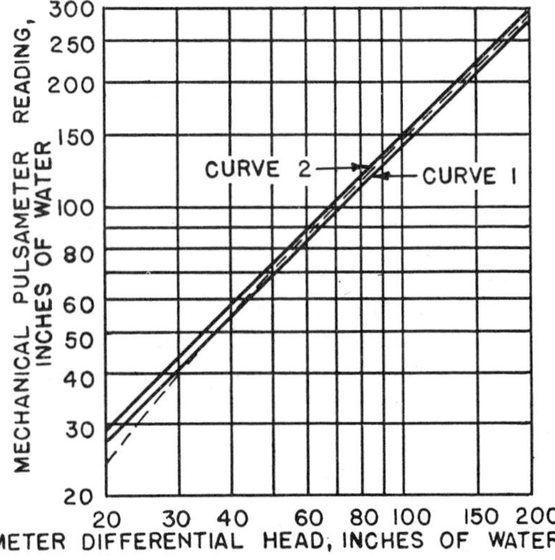

Fig. 7-8 Preliminary pulsation error-limit chart.

This equation may be integrated by a series approximation. The result for the first five terms of the series may be expressed as:

$$Q_a = Kh^{0.5}\left[1 - \frac{p^2}{16h^2} - \frac{15p^4}{1024h^4}\right]$$

Terms are defined under previous equation.

The values for the fifth term in the binomial expansion become insignificant for the limits stated.

Since the flow indicated by the meter secondary device is

$$Q_m = Kh^{0.5}$$

the error for any values of h and p can be computed by dividing $(Q_m - Q_a)$ by Q_a. There will be some definite relation between h and p for any percentage error which will not change with the absolute values of h and p.

Curve 1 of Fig. 7-8 represents one per cent error with these assumptions, and Curve 2 represents 1.5 per cent error. The empirical curve lies between the two calculated curves for all differential heads greater than 32 in. of water. This indicates that the assumptions concerning the effect of differential pulsation are reasonably correct. For lower differential pressures, the experimental curve is intentionally drawn below the one per cent rational curve, since operating experience shows that measurements made at low differential with pulsating flow are very sensitive to changes in pulsation. Thus, it was agreed that the line should be lower than normal to allow for the effect of these variations. Because of erratic results the empirical curve was not drawn for differentials below 20 in. of water. If pulsations are present, it is believed better not to attempt to meter the flow where the differentials are so low. These curves are plotted on logarithmic paper so that points with the same percentage of error would be approximately the same distance from the curve for any value of the differential head.[3]

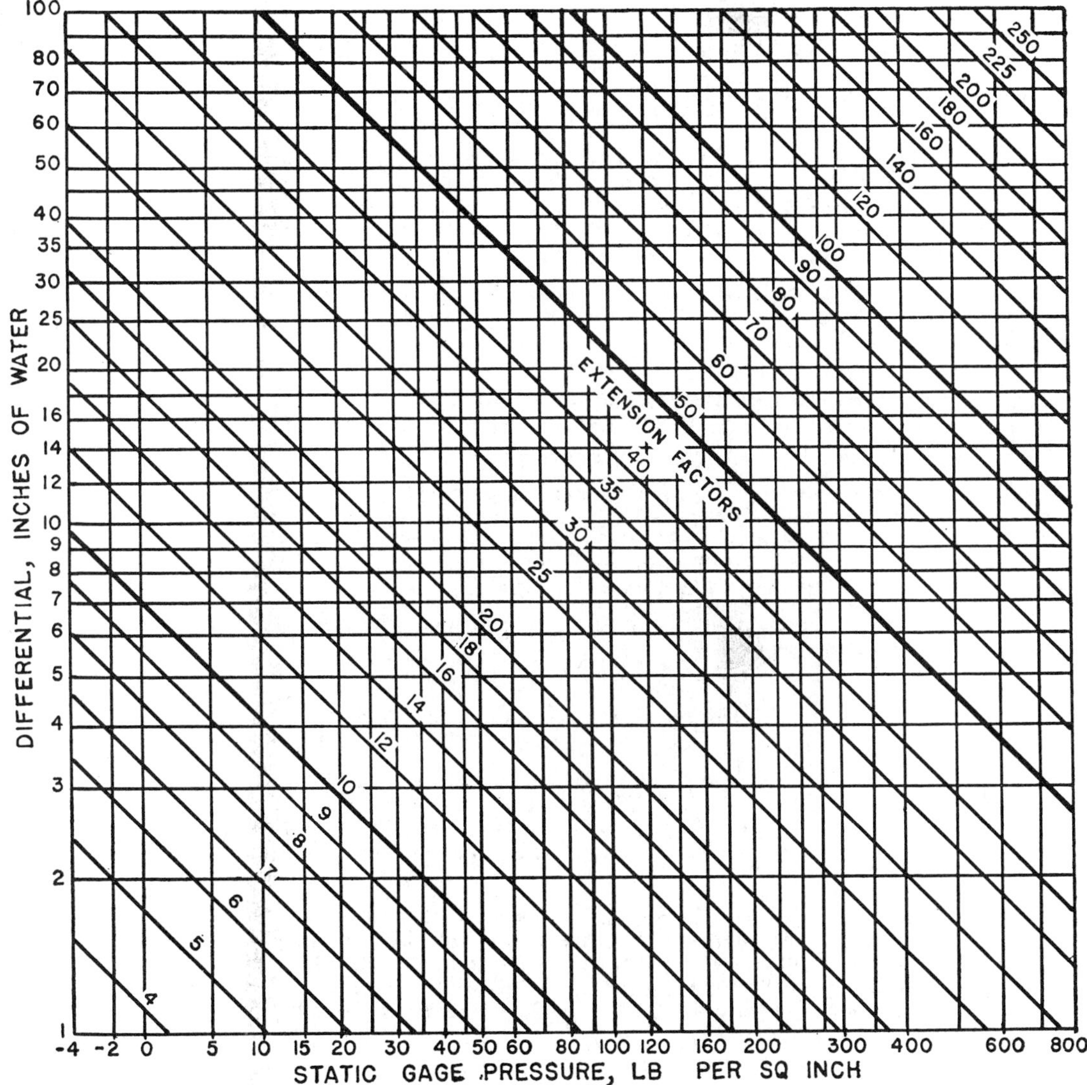

Fig. 7-9 Method of determining approximate hourly flow thru an orifice meter.

Example: Assume an orifice meter with a 2 in. orifice in a 4 in. pipe run, having a differential of 40 in. w.c. and a static pressure of 200 psig, using flange taps.

Extension = 93 (from above chart); Orifice flow constant = 1070 (from Table 7-2*); Capacity = 93 × 1070 = 100,000 cfh.

* For pipe taps use Table 7-3.

Table 7-2 Approximate Hourly Orifice Flow Constants —Flange Taps

(Based on 0.65 specific gravity gas at average atmospheric pressure and temperature.)
Note: These values should not be used in commercial gas measurement computations.

Orifice diam, in.	Standard pipe sizes (ID), in.						
	2 (2.067)	3 (3.068)	4 (4.026)	6 (6.065)	8 (8.071)	10 (10.136)	12 (12.090)
0.25	16	16	16				
.50	64	64	64	64			
.75	146	144	143	143			
1.00	267	258	256	255	254		
1.25	438	409	403	399	398	397	397
1.50	688	600	585	577	574	573	572
1.75	...	842	805	788	783	781	780
2.00	...	1150	1070	1030	1,030	1,020	1,020
2.25	...	1550	1380	1320	1,300	1,300	1,290
2.50	...	...	1760	1630	1,610	1,600	1,600
2.75	...	...	2220	2000	1,960	1,940	1,940
3.00	...	...	2790	2400	2,340	2,320	2,310
3.25	...	...	...	2660	2,760	2,730	2,710
3.50	...	...	...	3370	3,220	3,170	3,150
3.75	...	...	...	3950	3,720	3,650	3,630
4.00	...	...	...	4600	4,270	4,170	4,130
4.50	...	...	...	6220	5,520	5,330	5,260
5.00	...	...	...	...	7,020	6,660	6,540
5.50	...	...	...	...	8,820	8,190	7,980
6.00	...	...	...	...	11,000	9,960	9,600
6.50	...	...	...	...	...	12,000	11,400
7.00	...	...	...	...	...	14,400	13,500
7.50	...	...	...	...	...	17,200	15,800
8.00	...	...	...	...	...	...	18,400
8.50	...	...	...	...	...	...	21,400
9.00	...	...	...	...	...	...	24,800

Table 7-3 Approximate Hourly Orifice Flow Constants —Pipe Taps

(Based on 0.65 specific gravity gas at average atmospheric pressure and temperature.)
Note: These values should not be used in commercial gas measurement computations.

Orifice diam, in.	Standard pipe sizes (ID), in.						
	2 (2.067)	3 (3.068)	4 (4.026)	6 (6.065)	8 (8.071)	10 (10.136)	12 (12.090)
0.25	16	16	16				
.50	67	65	65	64			
.75	159	150	147	145			
1.00	309	276	266	259	257		
1.25	556	453	428	410	404	402	400
1.50	...	698	637	598	586	581	578
1.75	...	1040	906	829	805	795	789
2.00	...	1530	1250	1110	1,060	1,050	1,040
2.25	...	...	1690	1430	1,360	1,330	1,320
2.50	...	...	2270	1820	1,710	1,660	1,640
2.75	...	...	3050	2270	2,100	2,030	2,000
3.00	...	...	...	2810	2,550	2,440	2,390
3.25	...	...	...	3440	3,050	2,900	2,830
3.50	...	...	...	4190	3,620	3,410	3,320
3.75	...	...	...	5090	4,260	3,980	3,840
4.00	...	...	...	6190	4,990	4,600	4,420
4.50	...	...	...	...	6,760	6,030	5,730
5.00	...	...	...	...	9,070	7,780	7,280
5.50	...	...	...	...	12,200	9,920	9,100
6.00	...	...	...	...	...	12,600	11,200
6.50	...	...	...	...	...	15,900	13,800
7.00	...	...	...	...	...	20,000	16,800
7.50	...	...	...	...	...	...	20,400
8.00	...	...	...	...	...	...	24,800

Table 7-4 Estimated Uncertainties in the Measurement or Evaluation of Items Involved in Fluid Metering[4]

Item or factor	Uncertainty range* "Good" laboratories	"Field practice"
Throat diameter of orifice or flow nozzle: <5 in.	±0.0001 in.	to ±0.002 in.
>5 in.	±0.0002 in.	to ±0.005 in.
Pipe diameter: 1-in. pipe to 30-in. OD pipe	±0.001 in.	to ±0.10 in.
Static pressure: dead weight gage, single reading	±0.05 lb	to ±1.0 lb
average 10 readings	±0.01 lb	to ±0.2 lb
test gage, single reading	±0.1 lb	to ±2.0 lb
average 10 readings	±0.02 lb	to ±0.5 lb
indicating gage, single reading	±0.5 lb	to ±10. lb
average 10 readings	±0.1 lb	to ±2. lb
recording chart, regardless of chart scale	±0.01 in.	to ±0.05 in.
Differential pressure: mercury or water manometer, single reading	±0.001 in.	to ±0.25 in.
average 10 readings	±0.0002 in.	to ±0.05 in.
recording chart, regardless of chart scale	±0.01 in.	to ±0.05 in.
Temperature: mercury-in-glass, thermocouple, resistance thermo.	±0.01°F	to ±5.°F
temperature recorder, regardless of chart scale	±0.01 in.	to ±0.05 in.
Specific weight (or density): water at <140 F	±1 in 60,000	to ±1 in 6000
at >140 F	±1 in 12,000	to ±1 in 3000
other liquid, direct determination	±1 in 50,000	to ±1 in 200
air	±1 in 5,000	to ±1 in 200
steam, superheated	±1 in 2,000	to ±1 in 200
wet	±1 in 400	to ±1 in 50
other gases, direct determination	±1 in 2,000	to ±1 in 50
Specific gravity: liquids	±1 in 5,000	to ±1 in 50
gases	±1 in 2,000	to ±1 in 50
Weighings, commercial scales	±1 in 10,000	to ±1 in 100
Coefficient of discharge or of flow: by calibration	±0.05%	to ±1.0%
from table or curve	±0.5%	to ±5.0%
Compressibility or supercompressibility of gases	±0.1%	to ±5.0%
Expansion factor, compressible fluids $\left[x = \frac{p_1 - p_2}{p_1} 100\right]$	±0.5% of x	to ±2% of x

* Signifies an estimate of the uncertainty; i.e., difference to be expected between observed or determined values of the measurement and the time value, if such were known or could be determined exactly.

Orifice Meter Capacity

The capacity of an orifice meter for measuring gas depends upon the differential pressure, static pressure, type of pressure taps, orifice size, and pipe run diameter.

Figure 7-9, used with Table 7-2 for *flange* type pressure connections and with Table 7-3 for *pipe* type pressure connections, constitutes a convenient method for computing orifice meter capacities.

For a given orifice and pipe run diameter, **capacity in** cfh = **extension factor × orifice flow constant.** See **Example** under Fig. 7-9. Note that the extension factor = $(h_w p_f)^{0.5}$ (and may be read from Fig. 7-9).

where: h_w = differential pressure, inches of water
p_f = static pressure, psia

Orifice flow constant is obtained from Tables 7-2 or 7-3, depending upon type of pressure taps.

Orifice Meter Selection

Preliminary to proper meter selection, it is necessary to know the following about the characteristics and conditions of the flow to be metered: maximum peak hourly rate; duration of maximum peak or uniformity of flow; minimum hourly rate; duration of minimum rate period or its uniformity; metering gage pressure required and available; and permissible pressure variations.

Range. The quantity of gas flowing thru an orifice at constant pressure varies as the square root of the differential pressure. Accordingly, for half of a given rate of flow, the differential pressure will be one-fourth of that for the given rate. Because of mechanical and installation limitations, it has been considered impractical to construct a differential gage that will continuously record pressures with acceptable accuracy below about one-sixteenth of its maximum range. Therefore, the working range of one orifice plate and one differential gage is from maximum capacity to about one-fourth of maximum. The maximum capacity may be changed by changing the orifice size; however, this changes only the maximum and minimum capacities but not the ratio between them.

Under some operating conditions, the ratio between maximum and minimum capacities at a single orifice meter may be increased by using two differential gages connected to the same pressure taps. The lower range gage is used for the lower flow rates and the higher range gage for the higher rates.

Computation of Flow thru Orifice Meters

Table 7-4 may be used to estimate the reliability of measurement data. The methods given in the following extracts[2] from A.G.A. Gas Measurement Committee Report No. 3 are representative of practical means which may be used in making calculations of this sort. Methods other than those included may be equally practical. Appendix B of Report No. 3 gives the derivation of formulas and method for calculating the tables involved.

IV—INSTRUCTIONS FOR COMPUTING THE FLOW OF GAS THRU ORIFICE METERS

(56) The following recommendations of the committee concerning the calculation of the flow thru the orifice meter and the computation of the necessary constants for use in these calculations are confined strictly to orifice meters of the type specified in Section II* of this report, and installed and operated according to the provisions outlined in Section III.*

A. GENERAL EQUATION

(57) In the measurement of most gases, and especially natural gas, it is almost the universal practice (in the United States) to express the flow in cubic feet per hour referred to some specified reference or base condition of pressure and temperature. For the calculation of the quantity of gas the committee recommends the continued use of the formula:

$$Q_h = C'\sqrt{h_w p_f} \qquad (1)$$

in which

Q_h = quantity rate of flow at base conditions, cu ft/hr
C' = orifice flow constant
h_w = differential pressure in inches of water at 60 F
p_f = absolute static pressure in psi

B. ORIFICE FLOW CONSTANT

(58) The orifice flow constant C′ may be defined as the rate of flow in cubic feet per hour, at base conditions, when the extension $\sqrt{h_w p_f}$ equals one. It was formerly known as the "flow coefficient." It is here called "orifice flow constant," and should not be confused with the former coefficient or with the coefficient of discharge mentioned in the appendix of this report. It is to be calculated by the equation:

$$C' = F_b F_r Y F_{pb} F_{tb} F_{tf} F_g F_{pv} F_m \qquad (2)$$

in which

F_b = basic orifice factor
F_r = Reynolds number factor
Y = expansion factor
F_{pb} = pressure base factor
F_{tb} = temperature base factor
F_{tf} = flowing temperature factor
F_g = specific gravity factor
F_{pv} = supercompressibility factor
F_m = manometer factor

(59) The values of all the factors F_b, F_r, etc., are obtained from the tables in Section V of this report.† The following method is for a meter with flange taps. The method for a meter with pipe taps is exactly the same, except that the tables for pipe taps are used.

C. BASIC ORIFICE FACTOR

(60) The basic orifice factor F_b is taken directly from Table 3 or 8‡ for the correct orifice and pipe size. For sizes not listed in this table, the value of F_b may be found by calculation as explained in Appendix B‡ of this report. (Interpolation should not be relied upon.) The pipe diameter of the meter tubes, it is important to point out, should be within the limits specified in Figure 1. If the value of D is outside of those limits the flow factors of Table 3 or 8‡ cannot be relied upon. Then, the exact value of F_b should be calculated for the particular value of β based on the actual value of D by using the equations given in the Appendix.

D. REYNOLDS NUMBER FACTOR

(61) The Reynolds number factor F_r is obtained by using Table 4 or 9.‡ For its determination the average extension at which the meter operates must be known in addition to the orifice

* [See foregoing Sections, SCOPE, and CONSTRUCTION AND INSTALLATION SPECIFICATIONS, respectively.—EDITOR]

† [See following Tables 3 thru 7 and 12 thru 17.—EDITOR]

‡ [Not given in this Handbook; see original.—EDITOR]

and pipe size. The value of $\sqrt{h_w p_f}$ (extension) used in calculating the F_r factor from Table 4 or 9* may be based upon the meter record, or estimated from a knowledge of the average static pressure and the average differential at which the meter may operate. This extension, it should be noted, is an index with which a factor is selected, and does not enter directly into the computation of the gas volume. An extension value selected as suggested will probably be sufficiently close to the average operating conditions of the meter for selecting the proper value of F_r, especially since the variations in F_r corresponding to values of $\sqrt{h_w p_f}$ above or below the selected average will be compensating (plus or minus) over any appreciable length of time.

(62) Tables 4 and 9,* it should be noted, have been calculated by using average values of viscosity, 0.0000069 lb/ft-sec, of temperature, 60° F, and of specific gravity, 0.65, applying particularly to natural gas. If the gas being metered has a viscosity, temperature, or specific gravity quite different from these, the value of F_r in Table 4 or 9* may not be applicable. However, for variations in viscosity of from .0000059 to .0000079, in temperature of from 30° to 90° F or in specific gravity of from .55 to .75, the variations in the factor F_r would be well within the tolerances given for this report.

E. EXPANSION FACTOR

(63) The expansion factor Y is to be taken from Table 5, 6, 7, 10*, or 11* depending upon the tap location from which the static pressure is taken. Here the ratio h_w/p_f is an index with which the value of Y is selected, and may be based upon the meter record, or estimated from the average static pressure and the average differential at which the meter may operate. The effects of operating variations from an "average" h_w/p_f will ordinarily be compensating.

(64) For orifice meters with flange taps, the variations to be expected in meter accuracy from the average will be greater or less depending on whether upstream, downstream, or intermediate pressure connections are used. As the h/p ratio is increased, it will be observed, the use of the intermediate static pressure gives the minimum departure from the mean condition used in the selection of the expansion factor.

(65) With orifice meters using pipe taps, the variation from the mean estimated operating value of h/p used in the determination of the expansion factor Y result in the greatest variation in the value of Y when upstream static pressures are used. The use of intermediate static pressures in connection with pipe taps has little advantage, since it only slightly reduces the variations in Y resulting from departures from the mean. The use of downstream static pressures for this type of connection results in the least change in the values of Y due to changes in the values of the h/p ratio from the mean operating value.

Note: In use it may be found that a group of meters in the same locality, or operating under the same conditions, will have the same values for F_r and Y. If this is true the flow constants in this group will be the same for all plates of the same size and ratio. Likewise, for individual stations the values for F_r and Y may be based on estimated average static and differential pressures.

F. PRESSURE BASE FACTOR

(66) The pressure base factor F_{pb} is taken from Table 12.

G. TEMPERATURE BASE FACTOR

(67) The temperature base factor F_{tb} is taken from Table 13.

H. FLOWING TEMPERATURE FACTOR

(68) The flowing temperature factor F_{tf} is taken from Table 14 and should be based on the actual flowing temperature of the gas.

* [Not given in this Handbook; see original.—EDITOR]

I. SPECIFIC GRAVITY FACTOR

(69) The specific gravity factor F_g is taken from Table 15 and should be based on the actual specific gravity of the gas as determined by test.

J. SUPERCOMPRESSIBILITY FACTOR [See Table 7-1b. —EDITOR]

(70) The development of the general hydraulic flow equation involves the actual specific weight of the fluid at the point of measurement. In the measurement of gas this depends upon the flowing pressure and temperature. To translate the calculated volume at the flowing pressure and temperature to base pressure and temperature, it is necessary to apply the law for an ideal gas. All gases deviate from this ideal gas law to a greater or lesser extent. This deviation has been termed "Supercompressibility." A factor to take account of this supercompressibility is necessary in the measurement of some gases. This factor is particularly appreciable at high line pressures.

(71) If not obtained from actual determinations, the supercompressibility factor F_{pv} may be obtained from Tables 16-A, 16-B, 16-C, 16-D, and 16-E. This is an empirical method of evaluating supercompressibility factors for normal natural gas mixtures. The accuracy of determining the factors from this method will be within the allowances of this report if a specific gravity of 0.75 and diluent contents of 12 mole per cent nitrogen and/or 5 mole per cent carbon dioxide are not exceeded. In the ranges of pressure and temperature covered in these tables, certain pseudo values may be indicated which become necessary in order to determine required temperature and pressure adjustment indexes. Supercompressibility tests with approved apparatus may be used to establish the suitability of using the tables for various gas mixtures.

(72) Two methods are available for applying the tables to supercompressibility factor determination. The first method, which will be referred to as the "specific gravity method," requires knowledge of the specific gravity and of the mole fraction nitrogen, X_n, and mole fraction carbon dioxide, X_c, contents, if any is present. The second method, which will be referred to as the "heating value method," requires knowledge of the heating value, the specific gravity and of the mole fraction carbon dioxide content, if any is present. Tables 16-A, 16-B, and 16-C are for use with the specific gravity method and Tables 16-A, 16-D, and 16-E are for use with the heating value method.

(73) Slight differences will be found between the supercompressibility factors evaluated by the two methods. These differences are due to normal uncertainties in evaluating the specific gravity, heating value, and diluent contents of the gas combined with unavoidable uncertainties incident to developing empirical relations between these several gas properties. The comparative accuracy of both methods is essentially the same. While the determination of supercompressibility factors using these basic Tables (No. 16) is complex, they are intended to be used as a basis for developing convenient operating tables from which factors can be selected directly and rapidly. Such tables, developed from the basic tables in this report, are available from the American Gas Association.

K. MANOMETER FACTOR

(74) The manometer factor F_m is taken from Table 17.

(75) The weight of the gas column over the mercury reservoir of orifice meter gauges introduces an error in determining the differential pressure across the orifice unless some adjustment is made. This error is consistently in one direction and becomes increasingly important with increasing pressure. The application of the manometer factor may be used to compensate for this error where conditions warrant.

L. APPLICATION AND EXAMPLES

(76) In the application of these factors, it will be noted that F_r, Y, F_{pb}, F_{tf}, F_g, and F_{pv} are multipliers, and may be applied

either to the basic orifice factor F_b or to the calculated quantity rate of flow, as preferred.

(77) An example of the calculation of the quantity rate of flow using the tables in this report follows:

Given:

Meter equipped with flange taps, with static pressure from downstream tap.

D = line size = 8.071 in. actual inside diameter
d = orifice size = 1.000 in.
flowing temperature = 65 F
ambient temperature = 70 F
P_b = contract pressure base = 14.65 psia
temperature base = 50 F = 510° absolute
G = specific gravity = .570
H_w = total heating value = 999.1 Btu per cu ft
X_n = mole fraction nitrogen content = .011 = mole % ÷ 100
X_c = mole fraction carbon dioxide content = .000 = mole % ÷ 100
h_w = average differential head = 50 in. water
P_{f2} = average downstream gage static pressure = 370 psig
p_{f2} = average downstream absolute static pressure = P_{f2} + average barometric pressure = 370 + 14.4 = 384.4 psia

Required: The orifice flow constant and the quantity rate of flow for one hour at base conditions.

Solution: $\beta = d/D = .1239$

Published inside diameter of pipe = 8.071 in.

Ratio actual inside diameter to published inside diameter = 1.000

Per cent difference = 0.0; therefore, the tables can be used

$$\text{Average } \sqrt{h_w p_{f2}} = 138.64$$
$$\text{Average } h_w/p_{f2} = .1301$$

From Table 3 for 1 in. plate in 8.071 in. ID line:

$$F_b = 200.38$$

From Table 4 for 1 in. plate in 8.071 in. ID line:

$$b = .0680$$

$$F_r = 1 + \frac{b}{\sqrt{h_w p_{f2}}} = 1 + 0.00049 = 1.0005$$

From Table 6 (Y_2 for downstream static pressure):

Interpolating for $h_w/p_{f2} = .1301$ and $\beta = .1239$, $Y_2 = 1.0008$
From Table 12, for $P_b = 14.65$ psia, $F_{pb} = 1.0055$
From Table 13, for temperature base = 50 F, $F_{tb} = .9808$
From Table 14, for flowing temperature = 65 F, $F_{tf} = 0.9952$
From Table 15, for specific gravity = 0.570, $F_g = 1.3245$

* * *

Using the "Specific Gravity Method" of Supercompressibility Factor Evaluation

Referring to Table 16-B, Pressure Adjustment Index =

$$f_{pg} = G - 13.84X_c + 5.420X_n$$
$$= .570 - 13.84(0) + 5.420(.011) = 0.630$$

Interpolating for P_{f2} = 370.0 psig and f_{pg} = .630,

Pressure Adjustment = ΔP = 0.52 psi, therefore
Adjusted Pressure = $P_f = P_{f2} + \Delta P$ = 370.0 + .52 = 370.5 psig

Referring to Table 16-C, Temperature Adjustment Index =

$$f_{tg} = G - .472X_c - .793X_n$$
$$= .570 - .472(0) - .793(.011) = 0.561$$

Interpolating for flowing temperature = 65 F and f_{tg} = 0.561,

Temperature Adjustment = ΔT = 19.90°F, therefore
Adjusted Temperature = Flowing Temperature + ΔT = 65 + 19.90 = 84.9 F

Referring to Table 16-A:

Interpolating for P_f = 370.5 psig and temperature = 84.9 F, F_{pv} = 1.0254

* * *

Using the "Heating Value Method" of Supercompressibility Factor Evaluation

Referring to Table 16-D, Pressure Adjustment Index =

$$f_{ph} = G - .5688\frac{H_w}{1000} - 3.690X_c$$
$$= .570 - .5688(.9991) - 3.690(0) = 0.0017$$

Interpolating for P_{f2} = 370.0 psig and f_{ph} = 0.0017,

Pressure Adjustment = ΔP = 0.52 psi, therefore

Adjusted Pressure = $P_f = P_{f2} + \Delta P$ = 370.0 + .52 = 370.5 psig

Referring to Table 16-E, Temperature Adjustment Index =

$$f_{th} = G + 1.814\frac{H_w}{1000} + 2.641X_c$$
$$= .570 + 1.814\,(.9991) + 2.641(0) = 2.382$$

Interpolating for flowing temperature = 65 F and f_{th} = 2.382,

Temperature Adjustment = ΔT = 19.92°F
Adjusted Temperature = Flowing Temperature + ΔT = 65 + 19.92 = 84.9 F

Referring to Table 16-A:

Interpolating for P_f = 370.5 psig and temperature = 84.9 F, F_{pv} = 1.0254

* * *

From Table 17:

Interpolating for P_{f2} = 370 psig, ambient temperature = 70 F, and G = .570, F_m = 0.9993

Then the Orifice Flow Constant is:

$$C' = 200.38 \times 1.0005 \times 1.0008 \times 1.0055 \times .9808 \times .9952 \times 1.3245 \times 1.0254 \times .9993 = 267.25$$

and the rate of flow for one hour at base conditions is:

$$Q_h = 267.25\sqrt{50 \times 384.4} = 37052$$

Note: When certain factors such as flowing temperature, specific gravity, or supercompressibility are applied directly to the gas quantities or to extensions, the factor of 1.0000 should be substituted for these factors in Equation 2, paragraph (58).

V — ORIFICE METER TABLES FOR NATURAL GAS

The tolerances necessary in the use of any orifice meter do not warrant taking the values in these tables to be accurate beyond one in 500. Four figures are given in all cases solely to enable different computers to agree within 1 or 2 in the fourth significant figure regardless of whether it is on the right or left of the decimal.

In some of the tables values of the constants for a few of the smaller orifices are marked with an asterisk, these orifices have diameter ratios lower than the minimum value for which the formulas used were derived and this size of plate should not be used unless it is understood that the accuracy of measurement may be relatively low.

A. TABLES APPLYING TO ORIFICE FLOW CONSTANTS FOR FLANGE TAP INSTALLATIONS.

C. TABLES APPLYING TO ORIFICE FLOW CONSTANTS FOR FLANGE TAP AND PIPE TAP INSTALLATIONS.

TABLE 3 F_b, BASIC ORIFICE FACTORS—FLANGE TAPS

Base temperature = 60 F Flowing temperature = 60 F $\sqrt{h_w p_f} = \infty$
Base pressure = 14.73 psia Specific gravity = 1.0 $h_w/p_f = 0$

Pipe sizes—nominal and published inside diameters, inches

Orifice diam, in.	2			3				4	
	1.689	1.939	2.067	2.300	2.626	2.900	3.068	3.152	3.438
.250	12.695	12.707	12.711	12.714	12.712*	12.708*	12.705*	12.703*	12.697*
.375	28.474	28.439	28.428	28.411	28.393	28.382	28.376	28.373	28.364
.500	50.777	50.587	50.521	50.435	50.356	50.313	50.292	50.284	50.258
.625	80.090	79.509	79.311	79.052	78.818	78.686	78.625	78.598	78.523
.750	117.09	115.62	115.14	114.52	113.99	113.70	113.56	113.50	113.33
.875	162.95	159.56	158.47	157.12	156.00	155.41	155.14	155.03	154.71
1.000	219.77	212.47	210.22	207.44	205.18	204.04	203.54	203.33	202.75
1.125	290.99	276.20	271.70	266.35	262.06	259.95	259.04	258.65	257.63
1.250	385.78	353.58	345.13	335.12	327.39	323.63	322.03	321.37	319.61
1.375		448.57	433.50	415.75	402.18	395.80	393.09	391.97	389.03
1.500			542.26	510.86	487.98	477.36	472.96	471.14	466.39
1.625				623.91	586.82	569.65	562.58	559.72	552.31
1.750					701.27	674.44	663.42	658.96	647.54
1.875					834.88	793.88	777.18	770.44	753.17
2.000						930.65	906.01	896.06	870.59
2.125						1091.2	1052.5	1038.1	1001.4
2.250							1223.2	1199.9	1147.7
2.375									1311.7
2.500									1498.4

Orifice diam, in.	4		6				8		
	3.826	4.026	4.897	5.189	5.761	6.065	7.625	7.981	8.071
.250	12.687*	12.683*							
.375	28.353*	28.348*							
.500	50.234	50.224	50.197	50.191*	50.182*	50.178*			
.625	78.450	78.421	78.338	78.321	78.296	78.287			
.750	113.15	113.08	112.87	112.82	112.75	112.72			
.875	154.40	154.27	153.88	153.78	153.63	153.56	153.34	153.31	153.31
1.000	202.20	201.99	201.34	201.19	200.96	200.85	200.46	200.39	200.38
1.125	256.69	256.33	255.31	255.08	254.72	254.56	253.99	253.89	253.87
1.250	318.03	317.45	315.83	315.48	314.95	314.72	313.91	313.78	313.74
1.375	386.45	385.51	382.99	382.47	381.70	381.37	380.25	380.96	380.02
1.500	462.27	460.79	456.93	456.16	455.03	454.57	453.02	452.78	452.72
1.625	545.89	543.61	537.77	536.64	535.03	534.38	532.27	531.95	531.87
1.750	637.84	634.39	625.73	624.09	621.79	620.88	618.02	617.60	617.50
1.875	738.75	733.68	721.03	718.69	715.44	714.19	710.32	709.77	709.64
2.000	849.41	842.12	823.99	820.68	816.13	814.41	809.22	808.50	808.34

* [See Note in Section V.—EDITOR]

TABLE 3 (Continued)

Orifice diam, in.	4		6				8		
	3.826	4.026	4.897	5.189	5.761	6.065	7.625	7.981	8.071
2.125	970.95	960.48	934.97	930.35	924.07	921.71	914.79	913.86	913.64
2.250	1104.7	1089.9	1054.4	1048.1	1039.5	1036.3	1027.1	1025.9	1025.6
2.375	1252.1	1231.7	1182.9	1174.2	1162.6	1158.3	1146.2	1144.7	1144.3
2.500	1415.0	1387.2	1320.9	1309.3	1293.8	1288.2	1272.3	1270.3	1269.8
2.625	1595.6	1558.2	1469.2	1453.9	1433.5	1426.0	1405.4	1402.9	1402.3
2.750	1797.1	1746.7	1628.9	1608.7	1582.1	1572.3	1545.7	1542.5	1541.8
2.875		1955.5	1801.0	1774.5	1740.0	1727.5	1693.4	1689.3	1688.4
3.000		2194.9	1986.6	1952.4	1907.8	1891.9	1848.6	1843.5	1842.3
3.125			2187.2	2143.4	2086.4	2066.1	2011.6	2005.2	2003.8
3.250			2404.2	2348.8	2276.5	2250.8	2182.6	2174.6	2172.9
3.375			2639.5	2569.8	2479.1	2446.8	2361.8	2352.0	2349.9
3.500			2895.5	2808.1	2695.1	2654.9	2549.7	2537.7	2535.0
3.625			3180.8	3065.3	2925.7	2876.0	2746.5	2731.8	2728.6
3.750				3345.5	3172.1	3111.2	2952.6	2934.8	2930.8
3.875				3657.7	3435.7	3361.5	3168.3	3146.9	3142.1
4.000					3718.2	3628.2	3394.3	3368.5	3362.9
4.250					4354.8	4216.6	3879.4	3842.3	3834.2
4.500						4900.9	4412.8	4360.5	4349.0
4.750							5000.7	4928.1	4912.2
5.000							5650.0	5551.1	5529.5
5.250							6369.3	6236.4	6207.3
5.500							7170.9	6992.0	6953.6
5.750								7830.0	7777.8
6.000									8706.9

Orifice diam, in.	10			12			16		
	9.564	10.020	10.136	11.376	11.938	12.090	14.688	15.000	15.250
1.000	200.20								
1.125	253.55	253.48	253.47						
1.250	313.31	313.20	313.18	312.94	312.85	312.83			
1.375	379.44	379.29	379.26	378.94	378.82	378.79			
1.500	451.95	451.76	451.72	451.30	451.14	451.10	450.53	450.48	
1.625	530.87	530.63	530.57	530.04	529.83	529.78	529.06	528.99	528.94
1.750	616.21	615.90	615.83	615.16	614.90	614.84	613.94	613.85	613.78
1.875	707.99	707.61	707.51	706.68	706.36	706.28	705.18	705.07	704.99
2.000	806.23	805.76	805.65	804.61	804.23	804.13	802.78	802.65	802.55
2.125	910.97	910.38	910.24	908.98	908.51	908.39	906.77	906.61	906.49
2.250	1,022.2	1,021.5	1,021.3	1,019.8	1,019.2	1,019.1	1,017.1	1,017.0	1,016.8
2.375	1,140.1	1,139.2	1,139.0	1,137.1	1,136.4	1,136.2	1,133.9	1,133.7	1,133.5
2.500	1,264.5	1,263.4	1,263.1	1,260.8	1,260.0	1,259.8	1,257.1	1,256.8	1,256.6
2.625	1,395.6	1,394.2	1,393.9	1,391.1	1,390.1	1,389.9	1,386.7	1,386.4	1,386.1
2.750	1,533.4	1,531.7	1,531.3	1,528.0	1,526.8	1,526.5	1,522.7	1,522.4	1,522.1
2.875	1,678.0	1,675.9	1,675.4	1,671.4	1,670.0	1,669.6	1,665.2	1,664.8	1,664.5
3.000	1,829.4	1,826.9	1,826.3	1,821.4	1,819.7	1,819.3	1,814.1	1,813.7	1,813.3
3.125	1,987.8	1,984.7	1,984.0	1,978.1	1,976.1	1,975.6	1,969.6	1,969.0	1,968.6
3.250	2,153.2	2,149.5	2,148.6	2,141.5	2,139.2	2,138.6	2,131.5	2,130.9	2,130.4
3.375	2,325.7	2,321.2	2,320.2	2,311.7	2,308.9	2,308.2	2,299.9	2,299.2	2,293.7
3.500	2,505.6	2,500.1	2,498.9	2,488.7	2,485.4	2,484.6	2,474.9	2,474.1	2,473.5
3.625	2,692.8	2,686.2	2,684.7	2,672.6	2,668.7	2,667.7	2,656.4	2,655.5	2,654.8
3.750	2,887.6	2,879.7	2,877.9	2,863.5	2,858.8	2,857.7	2,844.6	2,843.5	2,842.7
3.875	3,090.1	3,080.7	3,078.5	3,061.4	3,055.9	3,054.6	3,039.4	3,038.1	3,037.2
4.000	3,300.6	3,289.3	3,286.8	3,266.4	3,260.0	3,258.5	3,240.8	3,239.4	3,238.3
4.250	3,746.1	3,730.2	3,726.7	3,698.4	3,689.6	3,687.5	3,663.8	3,661.9	3,660.5
4.500	4,226.0	4,204.1	4,199.2	4,160.4	4,148.4	4,145.5	4,113.9	4,111.5	4,109.7
4.750	4,742.7	4,712.8	4,706.2	4,653.4	4,637.2	4,633.4	4,591.5	4,588.4	4,586.0
5.000	5,298.6	5,258.5	5,249.6	5,179.0	5,157.4	5,152.3	5,097.2	5,093.1	5,090.1
5.250	5,897.4	5,843.6	5,831.8	5,738.5	5,710.0	5,703.3	5,631.4	5,626.1	5,622.3
5.500	6,543.1	6,471.9	6,456.3	6,333.8	6,296.6	6,287.9	6,194.8	6,188.1	6,183.1
5.750	7,240.0	7,146.9	7,126.5	6,966.9	6,919.0	6,907.8	6,788.1	6,779.6	6,773.3
6.000	7,993.3	7,873.0	7,846.6	7,640.4	7,579.0	7,564.7	7,412.3	7,401.5	7,393.6
6.250	8,808.9	8,654.8	8,621.1	8,357.3	8,278.9	8,260.7	8,068.4	8,054.8	8,044.8
6.500	9,693.3	9,498.1	9,455.3	9,121.0	9,021.7	8,998.7	8,757.3	8,740.3	8,727.9
6.750	10,654	10,409	10,355	9,935.2	9,810.5	9,781.6	9,480.4	9,459.4	9,444.0
7.000	11,711	11,394	11,327	10,804	10,649	10,613	10,239	10,213	10,194

TABLE 3 F_b, BASIC ORIFICE FACTORS—FLANGE TAPS (Continued)

Orifice diam, in.	10			12			16		
	9.564	10.020	10.136	11.376	11.938	12.090	14.688	15.000	15.250
7.250		12,467	12,381	11,732	11,540	11,496	11,035	11,003	10,980
7.500		13,656	13,541	12,725	12,489	12,434	11,869	11,831	11,803
7.750				13,787	13,500	13,433	12,745	12,698	12,664
8.000				14,927	14,578	14,498	13,664	13,607	13,566
8.250				16,158	15,730	15,633	14,628	14,560	14,511
8.500				17,505	16,962	16,845	15,642	15,560	15,501
8.750					18,296	18,148	16,706	16,609	16,539
9.000						19,565	17,826	17,711	17,628
9.250							19,004	18,868	18,770
9.500							20,245	20,085	19,969
9.750							21,552	21,365	21,230
10.000							22,930	22,712	22,555
10.250							24,385	24,132	23,948
10.500							25,924	25,628	25,416
10.750							27,567	27,210	26,962
11.000							29,331	28,899	28,600
11.250								30,710	30,348

Orifice diam, in.	20			24			30		
	18.814	19.000	19.250	22.626	23.000	23.250	28.628	29.000	29.250
2.000	801.40	801.35	801.29						
2.125	905.11	905.06	904.98						
2.250	1,015.2	1,015.1	1,015.0						
2.375	1,131.6	1,131.5	1,131.4	1,130.2	1,130.1	1,130.0			
2.500	1,254.4	1,254.3	1,254.2	1,252.8	1,252.6	1,252.6			
2.625	1,383.6	1,383.5	1,383.3	1,381.7	1,381.5	1,381.4			
2.750	1,519.1	1,519.0	1,518.8	1,517.0	1,516.8	1,516.7			
2.875	1,661.0	1,660.9	1,660.7	1,658.6	1,658.4	1,658.3	1,656.0		
3.000	1,809.4	1,809.2	1,809.0	1,806.6	1,806.4	1,806.2	1,803.7	1,803.5	1,803.4
3.125	1,964.1	1,963.9	1,963.7	1,961.0	1,960.7	1,960.6	1,957.7	1,957.5	1,957.4
3.250	2,125.3	2,125.1	2,124.8	2,121.7	2,121.5	2,121.3	2,118.0	2,117.9	2,117.7
3.375	2,292.9	2,292.6	2,292.3	2,288.9	2,288.6	2,288.4	2,284.8	2,284.5	2,284.4
3.500	2,466.9	2,466.6	2,466.3	2,462.4	2,462.1	2,461.8	2,457.8	2,457.6	2,457.5
3.625	2,647.3	2,647.0	2,646.6	2,642.4	2,642.0	2,641.7	2,637.3	2,637.0	2,636.8
3.750	2,834.2	2,833.9	2,833.5	2,828.7	2,828.3	2,828.0	2,823.1	2,822.8	2,822.6
3.875	3,027.6	3,027.3	3,026.8	3,021.5	3,021.0	3,020.7	3,015.2	3,014.9	3,014.7
4.000	3,227.5	3,227.1	3,226.5	3,220.6	3,220.1	3,219.8	3,213.8	3,213.5	3,213.2
4.250	3,646.7	3,646.2	3,645.6	3,638.3	3,637.7	3,637.2	3,630.1	3,629.7	3,629.4
4.500	4,092.1	4,091.5	4,090.6	4,081.8	4,081.0	4,080.5	4,071.9	4,071.4	4,071.1
4.750	4,563.7	4,562.9	4,551.1	4,551.1	4,550.1	4,549.5	4,539.4	4,538.8	4,538.4
5.000	5,061.8	5,060.8	5,059.6	5,046.4	5,045.2	5,044.5	5,032.5	5,031.8	5,031.4
5.250	5,586.6	5,585.4	5,583.8	5,567.7	5,566.4	5,565.5	5,551.3	5,550.5	5,550.0
5.500	6,138.2	6,136.7	6,134.8	6,115.3	6,113.6	6,112.6	6,095.8	6,094.9	6,094.4
5.570	6,717.0	6,715.2	6,712.8	6,689.1	6,687.2	6,685.9	6,666.2	6,665.2	6,664.5
6.000	7,323.4	7,321.1	7,318.2	7,289.4	7,287.1	7,285.6	7,262.5	7,261.3	7,260.5
6.250	7,957.5	7,954.7	7,951.2	7,916.4	7,913.6	7,911.9	7,884.7	7,883.4	7,882.5
6.500	8,620.0	8,616.5	8,612.2	8,570.2	8,566.9	8,564.8	8,533.0	8,531.4	8,530.4
6.750	9,311.1	9,306.9	9,301.6	9,251.1	9,247.2	9,244.7	9,207.4	9,205.6	9,204.4
7.000	10,031	10,026	10,020	9,959.3	9,954.6	9,951.7	9,908.0	9,905.9	9,904.6
7.250	10,782	10,776	10,768	10,695	10,689	10,686	10,635	10,633	10,631
7.500	11,562	11,555	11,546	11,459	11,452	11,448	11,388	11,386	11,384
7.750	12,374	12,365	12,354	12,250	12,243	12,238	12,168	12,165	12,163
8.000	13,218	13,207	13,194	13,071	13,062	13,056	12,975	12,971	12,969
8.250	14,095	14,082	14,066	13,920	13,910	13,903	13,809	13,805	13,802
8.500	15,005	14,990	14,971	14,799	14,787	14,779	14,669	14,665	14,661
8.750	15,950	15,933	15,911	15,708	15,693	15,684	15,557	15,552	15,548
9.000	16,932	16,911	16,885	16,648	16,630	16,620	16,473	16,466	16,462

TABLE 3 (Continued)

Orifice diam, in.	20			24			30		
	18.814	19.000	19.250	22.626	23.000	23.250	28.628	29.000	29.250
9.250	17,950	17,926	17,895	17,618	17,598	17,585	17,416	17,409	17,404
9.500	19,007	18,979	18,943	18,620	18,597	18,582	18,387	18,379	18,373
9.750	20,104	20,071	20,030	19,655	19,628	19,611	19,386	19,377	19,371
10.000	21,243	21,205	21,157	20,723	20,692	20,672	20,414	20,403	20,396
10.250	22,426	22,382	22,326	21,825	21,789	21,767	21,471	21,458	21,450
10.500	23,654	23,603	23,538	22,962	22,921	22,895	22,556	22,542	22,533
10.750	24,931	24,872	24,797	24,134	24,087	24,058	23,672	23,656	23,646
11.000	26,257	26,190	26,104	25,344	25,290	25,257	24,817	24,799	24,787
11.250	27,636	27,559	27,460	26,592	26,531	26,492	25,992	25,972	25,959
11.500	29,070	28,982	28,870	27,878	27,809	27,766	27,199	27,176	27,161
11.750	30,562	30,462	30,334	29,205	29,126	29,077	28,437	28,411	28,394
12.000	32,116	32,001	31,856	30,574	30,485	30,429	29,706	29,677	29,659
12.500	35,417	35,270	35,084	33,444	33,330	33,259	32,343	32,306	32,283
13.000	39,003	38,817	38,581	36,502	36,357	36,267	35,114	35,068	35,039
13.500	42,913	42,673	42,375	39,762	39,581	39,467	38,025	37,968	37,932
14.000	47,244	46,921	46,523	43,241	43,015	42,874	41,082	41,012	40,968
14.500				46,958	46,679	46,505	44,291	44,206	44,151
15.000				50,934	50,591	50,378	47,662	47,557	47,490
15.500				55,192	54,774	54,513	51,202	51,075	50,993
16.000				59,759	59,251	58,935	54,923	54,769	54,671
16.500				64,701	64,060	63,670	58,835	58,649	58,531
17.000					69,288	68,792	62,950	62,728	62,586
17.500							67,282	67,017	66,848
18.000							71,844	71,530	71,330
18.500							76,653	76,282	76,046
19.000							81,725	81,289	81,012
19.500							87,079	86,568	86,244
20.000							92,734	92,140	91,761
20.500							98,728	98,025	97,584
21.000							105,130	104,280	103,750
21.500								110,980	110,340

TABLE 4 "b" VALUES FOR REYNOLDS NUMBER FACTOR F_r DETERMINATION—FLANGE TAPS

$$F_r = 1 + \frac{b}{\sqrt{h_w p_f}}$$

Pipe sizes—nominal and published inside diameters, inches

Orifice diam, in.	2			3				4	
	1.689	1.939	2.067	2.300	2.626	2.900	3.068	3.152	3.433
.250	.0879	.0911	.0926	.0950	.0979	.0999	.1010	.1014	.1030
.375	.0677	.0709	.0726	.0755	.0792	.0820	.0836	.0844	.0867
.500	.0562	.0576	.0588	.0612	.0648	.0677	.0695	.0703	.0728
.625	.0520	.0505	.0506	.0516	.0541	.0566	.0583	.0591	.0618
.750	.0536	.0485	.0471	.0462	.0470	.0486	.0498	.0504	.0528
.875	.0595	.0506	.0478	.0445	.0429	.0433	.0438	.0442	.0460
1.000	.0677	.0559	.0515	.0458	.0416	.0403	.0402	.0403	.0411
1.125	.0762	.0630	.0574	.0495	.0427	.0396	.0386	.0383	.0380
1.250	.0824	.0707	.0646	.0550	.0456	.0408	.0388	.0381	.0365
1.375		.0772	.0715	.0614	.0501	.0435	.0406	.0394	.0365
1.500			.0773	.0679	.0554	.0474	.0436	.0420	.0378
1.625				.0735	.0613	.0522	.0477	.0457	.0402
1.750					.0669	.0575	.0524	.0500	.0434
1.875					.0717	.0628	.0574	.0549	.0473
2.000						.0676	.0624	.0598	.0517
2.125						.0715	.0669	.0642	.0563
2.250							.0706	.0685	.0607
2.375									.0648
2.500									.0683

TABLE 4 "b" VALUES FOR REYNOLDS NUMBER FACTOR F_r DETERMINATION—FLANGE TAPS (Continued)

Orifice diam, in.	4		6				8		
	3.826	4.026	4.897	5.189	5.761	6.065	7.625	7.981	8.071
.250	.1047	.1054							
.375	.0894	.0907							
.500	.0763	.0779	.0836	.0852	.0880	.0892			
.625	.0653	.0670	.0734	.0753	.0785	.0801			
.750	.0561	.0578	.0645	.0665	.0701	.0718			
.875	.0487	.0502	.0567	.0587	.0625	.0643	.0723	.0738	.0742
1.000	.0430	.0442	.0500	.0520	.0557	.0576	.0660	.0676	.0680
1.125	.0388	.0396	.0444	.0462	.0498	.0517	.0602	.0619	.0623
1.250	.0361	.0364	.0399	.0414	.0447	.0464	.0549	.0566	.0571
1.375	.0347	.0344	.0363	.0375	.0403	.0419	.0501	.0518	.0523
1.500	.0345	.0336	.0336	.0344	.0367	.0381	.0457	.0474	.0479
1.625	.0354	.0338	.0318	.0322	.0337	.0348	.0418	.0435	.0439
1.750	.0372	.0350	.0307	.0306	.0314	.0322	.0383	.0399	.0403
1.875	.0398	.0370	.0305	.0298	.0298	.0303	.0353	.0366	.0371
2.000	.0430	.0395	.0308	.0296	.0287	.0288	.0327	.0340	.0343
2.125	.0467	.0427	.0318	.0300	.0281	.0278	.0304	.0315	.0318
2.250	.0507	.0462	.0334	.0310	.0281	.0274	.0286	.0295	.0297
2.375	.0548	.0501	.0354	.0324	.0286	.0274	.0271	.0278	.0280
2.500	.0589	.0540	.0378	.0342	.0295	.0279	.0259	.0264	.0265
2.625	.0626	.0579	.0406	.0365	.0308	.0287	.0251	.0253	.0254
2.750	.0659	.0615	.0436	.0391	.0324	.0300	.0246	.0245	.0245
2.875		.0647	.0468	.0418	.0343	.0314	.0244	.0240	.0240
3.000		.0673	.0500	.0448	.0366	.0332	.0245	.0238	.0237
3.125			.0533	.0479	.0389	.0353	.0248	.0239	.0237
3.250			.0564	.0510	.0416	.0375	.0254	.0242	.0240
3.375			.0594	.0541	.0443	.0400	.0263	.0248	.0244
3.500			.0620	.0569	.0472	.0426	0273	.0255	.0251
3.625			.0643	.0597	.0500	.0452	.0286	.0265	.0260
3.750				.0621	.0527	.0479	.0300	.0274	.0271
3.875				.0640	.0553	.0505	.0316	.0289	.0283
4.000					.0578	.0531	.0334	.0304	.0297
4.250					.0620	.0579	.0372	.0338	.0330
4.500						.0618	.0414	.0386	.0366
4.750							.0457	.0416	.0405
5.000							.0500	.0457	.0446
5.250							.0539	.0497	.0487
5.500							.0574	.0535	.0524
5.750								.0569	.0559
6.000									.0588

Orifice diam, in.	10			12			16		
	9.564	10.020	10.136	11.376	11.938	12.090	14.688	15.000	15.250
1.000	.0738								
1.125	.0685	.0701	.0705						
1.250	.0635	.0652	.0656	.0698	.0714	.0718			
1.375	.0588	.0606	.0610	.0654	.0671	.0676			
1.500	.0545	.0563	.0568	.0612	.0631	.0635	.0706	.0713	
1.625	.0504	.0523	.0527	.0573	.0592	.0597	.0670	.0678	.0684
1.750	.0467	.0485	.0490	.0536	.0555	.0560	.0636	.0644	.0650
1.875	.0433	.0451	.0455	.0501	.0521	.0526	.0604	.0612	.0618
2.000	.0401	.0419	.0414	.0469	.0488	.0492	.0572	.0581	.0587
2.125	.0372	.0389	.0383	.0438	.0458	.0463	.0542	.0551	.0558
2.250	.0346	.0362	.0356	.0410	.0429	.0434	.0514	.0523	.0529
2.375	.0322	.0337	.0330	.0383	.0402	.0407	.0487	.0496	.0502
2.500	.0302	.0315	.0308	.0359	.0377	.0382	.0461	.0470	.0476
2.625	.0283	.0296	.0287	.0336	.0354	.0358	.0436	.0445	.0452
2.750	.0267	.0278	.0269	.0316	.0332	.0336	.0413	.0422	.0428
2.875	.0254	.0263	.0253	.0297	.0312	.0317	.0391	.0399	.0406
3.000	.0243	.0250	.0252	.0278	.0294	.0298	.0370	.0378	.0385

TABLE 4 (Continued)

Orifice diam, in.	10			12			16		
	9.564	10.020	10.136	11.376	11.938	12.090	14.688	15.000	15.250
3.125	.0234	.0239	.0241	.0264	.0278	.0282	.0350	.0358	.0365
3.250	.0226	.0230	.0231	.0251	.0263	.0266	.0331	.0339	.0346
3.375	.0221	.0223	.0224	.0239	.0250	.0253	.0314	.0321	.0328
3.500	.0219	.0218	.0218	.0229	.0238	.0241	.0298	.0305	.0311
3.625	.0218	.0214	.0214	.0221	.0228	.0230	.0282	.0290	.0295
3.750	.0218	.0213	.0212	.0214	.0219	.0221	.0268	.0275	.0281
3.875	.0221	.0213	.0211	.0208	.0212	.0213	.0255	.0262	.0267
4.000	.0225	.0214	.0212	.0204	.0206	.0207	.0243	.0249	.0254
4.250	.0238	.0222	.0219	.0200	.0198	.0198	.0223	.0228	.0232
4.500	.0256	.0236	.0231	.0201	.0195	.0194	.0206	.0210	.0213
4.750	.0279	.0254	.0249	.0207	.0196	.0194	.0193	.0196	.0198
5.000	.0307	.0277	.0270	.0217	.0202	.0199	.0184	.0185	.0187
5.250	.0337	.0303	.0295	.0231	.0212	.0208	.0178	.0178	.0179
5.500	.0370	.0332	.0323	.0249	.0226	.0221	.0176	.0174	.0174
5.750	.0404	.0363	.0354	.0270	.0243	.0237	.0176	.0174	.0172
6.000	.0438	.0396	.0386	.0294	.0263	.0255	.0180	.0176	.0173
6.250	.0473	.0437	.0418	.0320	.0285	.0277	.0186	.0180	.0177
6.500	.0505	.0462	.0451	.0347	.0309	.0300	.0195	.0188	.0183
6.750	.0536	.0493	.0483	.0376	.0335	.0325	.0206	.0198	.0192
7.000	.0562	.0523	.0513	.0406	.0362	.0351	.0220	.0210	.0202
7.250		.0550	.0540	.0435	.0390	.0379	.0235	.0224	.0216
7.500		.0572	.0564	.0463	.0418	.0407	.0252	.0240	.0230
7.750				.0491	.0446	.0434	.0271	.0257	.0246
8.000				.0517	.0473	.0461	.0291	.0276	.0264
8.250				.0540	.0498	.0487	.0312	.0296	.0283
8.500				.0560	.0522	.0511	.0334	.0317	.0303
8.750					.0543	.0534	.0357	.0338	.0324
9.000						.0553	.0380	.0361	.0346
9.250							.0402	.0383	.0368
9.500							.0425	.0406	.0390
9.750							.0447	.0428	.0412
10.000							.0469	.0449	.0434
10.250							.0489	.0470	.0455
10.500							.0508	.0490	.0475
10.750							.0526	.0509	.0495
11.000							.0541	.0526	.0513
11.250								.0541	.0528

Orifice diam, in.	20			24			30		
	18.814	19.000	19.250	22.626	23.000	23.250	28.628	29.000	29.250
2.000	.0667	.0671	.0676						
2.125	.0640	.0644	.0649						
2.250	.0614	.0618	.0622						
2.375	.0588	.0592	.0597	.0659	.0665	.0669			
2.500	.0563	.0568	.0573	.0636	.0642	.0646			
2.625	.0540	.0544	.0549	.0614	.0620	.0624			
2.750	.0517	.0521	.0526	.0592	.0599	.0603			
2.875	.0494	.0499	.0504	.0571	.0578	.0582	.0662		
3.000	.0473	.0477	.0483	.0551	.0557	.0562	.0644	.0649	.0652
3.125	.0452	.0457	.0462	.0531	.0538	.0542	.0626	.0631	.0634
3.250	.0433	.0437	.0442	.0511	.0520	.0523	.0608	.0613	.0616
3.375	.0414	.0418	.0423	.0493	.0500	.0504	.0590	.0596	.0599
3.500	.0395	.0399	.0405	.0474	.0481	.0486	.0574	.0579	.0582
3.625	.0378	.0382	.0387	.0457	.0464	.0468	.0557	.0562	.0566
3.750	.0361	.0365	.0370	.0440	.0447	.0451	.0541	.0546	.0550
3.875	.0345	.0349	.0354	.0423	.0430	.0435	.0525	.0530	.0534
4.000	.0329	.0333	.0339	.0407	.0414	.0419	.0509	.0515	.0518
4.250	.0301	.0304	.0310	.0376	.0384	.0388	.0479	.0485	.0488
4.500	.0275	.0279	.0283	.0348	.0355	.0360	.0450	.0456	.0460
4.750	.0252	.0256	.0260	.0322	.0328	.0333	.0423	.0429	.0433
5.000	.0232	.0235	.0239	.0297	.0304	.0308	.0397	.0403	.0407

TABLE 4 "b" VALUES FOR REYNOLDS NUMBER FACTOR F_r DETERMINATION—FLANGE TAPS (Continued)

Orifice diam, in.	20			24			30		
	18.814	19.000	19.250	22.626	23.000	23.250	28.628	29.000	29.250
5.250	.0214	.0217	.0220	.0275	.0281	.0285	.0373	.0378	.0382
5.500	.0199	.0201	.0204	.0254	.0260	.0264	.0349	.0355	.0359
5.750	.0186	.0188	.0191	.0236	.0241	.0245	.0327	.0333	.0337
6.000	.0176	.0177	.0179	.0219	.0224	.0228	.0306	.0312	.0316
6.250	.0167	.0168	.0170	.0204	.0208	.0212	.0287	.0292	.0296
6.500	.0161	.0162	.0163	.0191	.0195	.0198	.0269	.0274	.0277
6.750	.0157	.0157	.0157	.0179	.0183	.0185	.0252	.0257	.0260
7.000	.0155	.0155	.0154	.0169	.0172	.0174	.0236	.0240	.0244
7.250	.0155	.0154	.0153	.0161	.0163	.0165	.0221	.0226	.0229
7.500	.0157	.0155	.0154	.0154	.0156	.0157	.0208	.0212	.0215
7.750	.0160	.0158	.0156	.0148	.0150	.0151	.0195	.0199	.0202
8.000	.0166	.0163	.0160	.0144	.0145	.0146	.0184	.0187	.0190
8.250	.0172	.0169	.0165	.0142	.0142	.0142	.0174	.0177	.0179
8.500	.0180	.0177	.0172	.0141	.0140	.0140	.0164	.0168	.0170
8.750	.0190	.0186	.0180	.0141	.0140	.0139	.0156	.0159	.0161
9.000	.0201	.0196	.0190	.0143	.0141	.0140	.0149	.0152	.0153
9.250	.0213	.0208	.0201	.0146	.0143	.0141	.0143	.0145	.0146
9.500	.0226	.0220	.0213	.0150	.0146	.0144	.0138	.0139	.0141
9.750	.0240	.0234	.0226	.0155	.0150	.0147	.0133	.0135	.0136
10.000	.0256	.0249	.0240	.0161	.0155	.0152	.0130	.0131	.0132
10.250	.0271	.0264	.0255	.0168	.0162	.0158	.0128	.0128	.0128
10.500	.0288	.0280	.0270	.0176	.0169	.0164	.0126	.0126	.0126
10.750	.0305	.0297	.0286	.0185	.0176	.0172	.0125	.0125	.0125
11.000	.0322	.0314	.0303	.0194	.0186	.0181	.0125	.0124	.0124
11.250	.0340	.0332	.0320	.0205	.0196	.0190	.0126	.0125	.0124
11.500	.0358	.0349	.0338	.0216	.0207	.0200	.0128	.0126	.0125
11.750	.0376	.0367	.0355	.0228	.0218	.0211	.0130	.0128	.0127
12.000	.0394	.0385	.0373	.0241	.0230	.0223	.0134	.0131	.0129
12.500	.0429	.0420	.0408	.0267	.0255	.0248	.0142	.0138	.0136
13.000	.0463	.0454	.0442	.0296	.0282	.0274	.0153	.0148	.0145
13.500	.0494	.0485	.0474	.0326	.0311	.0302	.0166	.0160	.0157
14.000	.0520	.0512	.0502	.0356	.0341	.0331	.0182	.0175	.0171
14.500				.0386	.0370	.0360	.0199	.0192	.0187
15.000				.0415	.0400	.0390	.0218	.0209	.0204
15.500				.0443	.0428	.0418	.0239	.0230	.0224
16.000				.0470	.0455	.0446	.0260	.0250	.0244
16.500				.0494	.0480	.0471	.0283	.0273	.0266
17.000					.0503	.0494	.0307	.0296	.0288
17.500							.0331	.0319	.0312
18.000							.0355	.0343	.0335
18.500							.0379	.0366	.0358
19.000							.0402	.0390	.0382
19.500							.0424	.0412	.0404
20.000							.0446	.0434	.0426
20.500							.0466	.0455	.0448
21.000							.0485	.0475	.0467
21.500								.0492	.0485

TABLE 5 Y_1, EXPANSION FACTORS—FLANGE TAPS*

Static pressure taken from upstream taps

$\frac{h_w}{p_{f_1}}$ Ratio	$\beta = \frac{d}{D}$ Ratio																
	.1	.2	.3	.4	.45	.50	.52	.56	.60	.62	.64	.66	.68	.70	.72	.74	.75
0.0	1.0000	1.0000	1.0000	1.0000	1.0000	1.0000	1.0000	1.0000	1.0000	1.0000	1.0000	1.0000	1.0000	1.0000	1.0000	1.0000	1.0000
0.2	.9977	.9977	.9977	.9977	.9976	.9976	.9976	.9975	.9975	.9974	.9974	.9974	.9973	.9973	.9972	.9971	.9971
0.4	.9954	.9954	.9954	.9953	.9953	.9952	.9952	.9951	.9949	.9949	.9948	.9947	.9946	.9945	.9944	.9943	.9942
0.6	.9932	.9932	.9931	.9930	.9929	.9928	.9927	.9926	.9924	.9923	.9922	.9921	.9919	.9918	.9916	.9914	.9913
0.8	.9909	.9909	.9908	.9907	.9906	.9904	.9903	.9901	.9899	.9897	.9896	.9894	.9892	.9890	.9888	.9886	.9884
1.0	.9886	.9886	.9885	.9884	.9882	.9880	.9879	.9877	.9874	.9872	.9870	.9868	.9865	.9863	.9860	.9857	.9855
1.2	.9863	.9863	.9862	.9860	.9859	.9856	.9855	.9852	.9848	.9846	.9844	.9841	.9838	.9835	.9832	.9828	.9826
1.4	.9841	.9840	.9840	.9837	.9835	.9832	.9831	.9827	.9823	.9821	.9818	.9815	.9812	.9808	.9804	.9800	.9798
1.6	.9818	.9818	.9817	.9814	.9811	.9808	.9806	.9803	.9798	.9795	.9792	.9788	.9785	.9781	.9776	.9771	.9769
1.8	.9795	.9795	.9794	.9791	.9788	.9784	.9782	.9778	.9772	.9769	.9766	.9762	.9758	.9753	.9748	.9743	.9740
2.0	.9772	.9772	.9771	.9767	.9764	.9760	.9758	.9753	.9747	.9744	.9740	.9735	.9731	.9726	.9720	.9714	.9711
2.2	.9750	.9749	.9748	.9744	.9741	.9736	.9734	.9729	.9722	.9718	.9714	.9709	.9704	.9698	.9692	.9685	.9682
2.4	.9727	.9726	.9725	.9721	.9717	.9712	.9710	.9704	.9697	.9692	.9688	.9683	.9677	.9671	.9664	.9657	.9653
2.6	.9704	.9704	.9702	.9698	.9694	.9688	.9686	.9679	.9671	.9667	.9662	.9656	.9650	.9643	.9636	.9628	.9624
2.8	.9681	.9681	.9679	.9674	.9670	.9664	.9661	.9654	.9646	.9641	.9636	.9630	.9623	.9616	.9608	.9600	.9595
3.0	.9658	.9658	.9656	.9651	.9647	.9640	.9637	.9630	.9621	.9615	.9610	.9603	.9596	.9588	.9580	.9571	.9566
3.2	.9636	.9635	.9633	.9628	.9623	.9616	.9613	.9605	.9595	.9590	.9584	.9577	.9569	.9561	.9552	.9542	.9537
3.4	.9613	.9612	.9610	.9604	.9599	.9592	.9589	.9580	.9570	.9564	.9558	.9550	.9542	.9534	.9524	.9514	.9508
3.6	.9590	.9590	.9587	.9581	.9576	.9568	.9565	.9556	.9545	.9538	.9532	.9524	.9515	.9506	.9496	.9485	.9480
3.8	.9567	.9567	.9564	.9558	.9552	.9544	.9540	.9531	.9520	.9513	.9505	.9497	.9488	.9479	.9468	.9457	.9451
4.0	.9545	.9544	.9542	.9535	.9529	.9520	.9516	.9506	.9494	.9487	.9479	.9471	.9462	.9451	.9440	.9428	.9422

* [Condensed from original.—EDITOR]

TABLE 6 Y_2, EXPANSION FACTORS—FLANGE TAPS*

Static pressure taken from downstream taps

$\frac{h_w}{p_{f_2}}$ Ratio	$\beta = \frac{d}{D}$ Ratio																
	.1	.2	.3	.4	.45	.50	.52	.56	.60	.62	.64	.66	.68	.70	.72	.74	.75
0.0	1.0000	1.0000	1.0000	1.0000	1.0000	1.0000	1.0000	1.0000	1.0000	1.0000	1.0000	1.0000	1.0000	1.0000	1.0000	1.0000	1.0000
0.2	1.0013	1.0013	1.0013	1.0013	1.0012	1.0012	1.0012	1.0011	1.0011	1.0010	1.0010	1.0010	1.0009	1.0009	1.0008	1.0008	1.0007
0.4	1.0027	1.0027	1.0026	1.0026	1.0025	1.0024	1.0024	1.0023	1.0022	1.0021	1.0020	1.0019	1.0018	1.0017	1.0016	1.0015	1.0014
0.6	1.0040	1.0040	1.0040	1.0039	1.0038	1.0036	1.0036	1.0034	1.0033	1.0032	1.0030	1.0029	1.0028	1.0026	1.0025	1.0023	1.0022
0.8	1.0054	1.0053	1.0053	1.0052	1.0050	1.0049	1.0048	1.0046	1.0044	1.0042	1.0041	1.0039	1.0037	1.0035	1.0033	1.0030	1.0029
1.0	1.0067	1.0067	1.0066	1.0065	1.0063	1.0061	1.0060	1.0058	1.0055	1.0053	1.0051	1.0049	1.0047	1.0044	1.0041	1.0038	1.0037
1.2	1.0080	1.0080	1.0080	1.0078	1.0076	1.0073	1.0072	1.0069	1.0066	1.0064	1.0061	1.0059	1.0056	1.0053	1.0050	1.0046	1.0044
1.4	1.0094	1.0094	1.0093	1.0091	1.0089	1.0086	1.0084	1.0081	1.0077	1.0074	1.0072	1.0069	1.0066	1.0062	1.0058	1.0054	1.0052
1.6	1.0108	1.0107	1.0106	1.0104	1.0101	1.0098	1.0096	1.0093	1.0088	1.0085	1.0082	1.0079	1.0075	1.0071	1.0067	1.0062	1.0060
1.8	1.0121	1.0121	1.0120	1.0117	1.0114	1.0111	1.0109	1.0104	1.0099	1.0096	1.0093	1.0089	1.0085	1.0080	1.0076	1.0070	1.0068
2.0	1.0135	1.0134	1.0133	1.0130	1.0127	1.0123	1.0121	1.0116	1.0110	1.0107	1.0103	1.0099	1.0095	1.0090	1.0084	1.0078	1.0075
2.2	1.0148	1.0148	1.0147	1.0143	1.0140	1.0136	1.0133	1.0128	1.0122	1.0118	1.0114	1.0109	1.0104	1.0099	1.0093	1.0087	1.0083
2.4	1.0162	1.0162	1.0160	1.0156	1.0153	1.0148	1.0146	1.0140	1.0133	1.0129	1.0124	1.0120	1.0114	1.0108	1.0102	1.0095	1.0091
2.6	1.0176	1.0175	1.0174	1.0170	1.0166	1.0161	1.0158	1.0152	1.0144	1.0140	1.0135	1.0130	1.0124	1.0118	1.0111	1.0103	1.0099
2.8	1.0189	1.0189	1.0187	1.0183	1.0179	1.0173	1.0170	1.0164	1.0156	1.0151	1.0146	1.0140	1.0134	1.0127	1.0120	1.0112	1.0107
3.0	1.0203	1.0203	1.0201	1.0196	1.0192	1.0186	1.0183	1.0176	1.0167	1.0162	1.0157	1.0150	1.0144	1.0137	1.0129	1.0120	1.0116
3.2	1.0217	1.0216	1.0214	1.0209	1.0205	1.0198	1.0195	1.0188	1.0179	1.0173	1.0167	1.0161	1.0154	1.0146	1.0138	1.0128	1.0124
3.4	1.0230	1.0230	1.0228	1.0223	1.0218	1.0211	1.0208	1.0200	1.0190	1.0184	1.0178	1.0171	1.0164	1.0156	1.0147	1.0137	1.0132
3.6	1.0244	1.0244	1.0242	1.0236	1.0231	1.0224	1.0220	1.0212	1.0202	1.0196	1.0189	1.0182	1.0174	1.0165	1.0165	1.0146	1.0140
3.8	1.0258	1.0258	1.0255	1.0249	1.0244	1.0236	1.0233	1.0224	1.0213	1.0207	1.0200	1.0192	1.0184	1.0175	1.0165	1.0154	1.0148
4.0	1.0272	1.0271	1.0269	1.0263	1.0257	1.0249	1.0245	1.0236	1.0225	1.0218	1.0211	1.0203	1.0194	1.0185	1.0174	1.0163	1.0157

* [Condensed from original.—EDITOR]

TABLE 7 Y_m, EXPANSION FACTORS—FLANGE TAPS*

Static pressure mean of upstream and downstream

$\frac{h_w}{p_{fm}}$ Ratio	$\beta = \frac{d}{D}$ Ratio																
	.1	.2	.3	.4	.45	.50	.52	.56	.60	.62	.64	.66	.68	.70	.72	.74	.75
0.0	1.0000	1.0000	1.0000	1.0000	1.0000	1.0000	1.0000	1.0000	1.0000	1.0000	1.0000	1.0000	1.0000	1.0000	1.0000	1.0000	1.0000
0.2	.9995	.9995	.9995	.9995	.9994	.9994	.9994	.9993	.9993	.9992	.9992	.9992	.9991	.9991	.9990	.9989	.9989
0.4	.9991	.9990	.9990	.9990	.9989	.9988	.9988	.9987	.9986	.9985	.9984	.9983	.9982	.9981	.9980	.9979	.9978
0.6	.9986	.9986	.9986	.9984	.9984	.9982	.9982	.9980	.9978	.9977	.9976	.9975	.9974	.9972	.9970	.9969	.9968
0.8	.9982	.9981	.9981	.9980	.9978	.9977	.9976	.9974	.9971	.9970	.9968	.9967	.9965	.9963	.9961	.9958	.9957
1.0	.9977	.9977	.9976	.9974	.9973	.9971	.9970	.9968	.9964	.9963	.9961	.9959	.9956	.9954	.9951	.9948	.9946
1.2	.9972	.9972	.9972	.9970	.9968	.9965	.9964	.9961	.9958	.9955	.9953	.9951	.9948	.9945	.9942	.9938	.9936
1.4	.9968	.9968	.9967	.9965	.9963	.9960	.9958	.9955	.9951	.9948	.9946	.9943	.9939	.9936	.9932	.9928	.9926
1.6	.9964	.9964	.9962	.9960	.9957	.9954	.9952	.9949	.9944	.9941	.9938	.9935	.9931	.9927	.9922	.9918	.9915
1.8	.9959	.9959	.9958	.9955	.9952	.9949	.9947	.9942	.9937	.9934	.9930	.9927	.9923	.9918	.9913	.9908	.9905
2.0	.9955	.9955	.9954	.9950	.9947	.9943	.9941	.9936	.9930	.9927	.9923	.9919	.9914	.9909	.9904	.9898	.9895
2.2	.9951	.9951	.9949	.9946	.9942	.9938	.9936	.9930	.9924	.9920	.9916	.9911	.9906	.9901	.9895	.9888	.9885
2.4	.9947	.9946	.9945	.9941	.9937	.9932	.9930	.9924	.9917	.9913	.9908	.9903	.9898	.9892	.9885	.9878	.9874
2.6	.9943	.9942	.9941	.9936	.9932	.9927	.9924	.9918	.9911	.9906	.9901	.9896	.9890	.9883	.9876	.9868	.9864
2.8	.9938	.9938	.9936	.9932	.9928	.9922	.9919	.9912	.9904	.9899	.9894	.9888	.9882	.9875	.9867	.9859	.9854
3.0	.9934	.9934	.9932	.9927	.9923	.9917	.9914	.9906	.9898	.9892	.9887	.9881	.9874	.9866	.9858	.9849	.9845
3.2	.9930	.9930	.9928	.9923	.9918	.9912	.9908	.9901	.9891	.9886	.9880	.9873	.9866	.9858	.9849	.9840	.9835
3.4	.9926	.9926	.9924	.9918	.9913	.9906	.9903	.9895	.9885	.9879	.9873	.9866	.9858	.9850	.9840	.9830	.9825
3.6	.9922	.9922	.9920	.9914	.9909	.9901	.9898	.9889	.9879	.9872	.9866	.9858	.9850	.9841	.9831	.9821	.9815
3.8	.9919	.9918	.9916	.9910	.9904	.9896	.9893	.9884	.9872	.9866	.9859	.9851	.9842	.9833	.9823	.9812	.9806
4.0	.9915	.9914	.9912	.9905	.9899	.9891	.9887	.9878	.9866	.9859	.9852	.9844	.9835	.9825	.9814	.9802	.9796

* [Condensed from original.—EDITOR]

TABLE 12 F_{pb}, FACTORS TO CHANGE FROM A PRESSURE BASE OF 14.73 PSIA TO OTHER PRESSURE BASES

Pressure base, psia	F_{pb}
14.4	1.0229
14.525	1.0141
14.65	1.0055
14.70	1.0020
14.73	1.0000
14.775	0.9970
14.90	.9886
15.025	.9804
15.15	.9723
15.225	.9675
15.275	.9643
15.325	.9612
15.40	.9565
15.525	.9488
15.65	.9412
15.775	.9338
15.90	.9264
16.025	.9192
16.15	.9121
16.275	.9051
16.40	.8982
16.70	0.8820

TABLE 13 F_{tb}, FACTORS TO CHANGE FROM A TEMPERATURE BASE OF 60 F TO OTHER TEMPERATURE BASES

Temp, °F	F_{tb}	Temp, °F	F_{tb}
40	0.9615	65	1.0096
41	.9635	66	1.0115
42	.9654	67	1.0135
43	.9673	68	1.0154
44	.9692	69	1.0173
45	.9712	70	1.0192
46	.9731	71	1.0212
47	.9750	72	1.0231
48	.9769	73	1.0250
49	.9788	74	1.0269
50	.9808	75	1.0288
51	.9827	76	1.0308
52	.9846	77	1.0327
53	.9865	78	1.0346
54	.9885	79	1.0365
55	.9904	80	1.0385
56	.9923	81	1.0404
57	.9942	82	1.0423
58	.9962	83	1.0442
59	0.9981	84	1.0462
60	1.0000	85	1.0481
61	1.0019	86	1.0500
62	1.0038	87	1.0519
63	1.0058	88	1.0538
64	1.0077	89	1.0558
		90	1.0577

TABLE 14 F_{tf}, FACTORS TO CHANGE FROM FLOWING TEMPERATURE OF 60 F TO ACTUAL FLOWING TEMPERATURE

Temp, °F	F_{tf}	Temp, °F	F_{tf}	Temp, °F	F_{tf}
1	1.0621	51	1.0088	101	0.9628
2	1.0609	52	1.0078	102	.9619
3	1.0598	53	1.0068	103	.9610
4	1.0586	54	1.0058	104	.9602
5	1.0575	55	1.0048	105	.9594
6	1.0564	56	1.0039	106	.9585
7	1.0552	57	1.0029	107	.9577
8	1.0541	58	1.0019	108	.9568
9	1.0530	59	1.0010	109	.9560
10	1.0518	60	1.0000	110	.9551
11	1.0507	61	0.9990	111	.9543
12	1.0496	62	.9981	112	.9535
13	1.0485	63	.9971	113	.9526
14	1.0474	64	.9962	114	.9518
15	1.0463	65	.9952	115	.9510
16	1.0452	66	.9943	116	.9501
17	1.0441	67	.9933	117	.9493
18	1.0430	68	.9924	118	.9485
19	1.0419	69	.9915	119	.9477
20	1.0408	70	.9905	120	.9469
21	1.0398	71	.9896	121	.9460
22	1.0387	72	.9887	122	.9452
23	1.0376	73	.9877	123	.9444
24	1.0365	74	.9868	124	.9436
25	1.0355	75	.9859	125	.9428
26	1.0344	76	.9850	126	.9420
27	1.0333	77	.9840	127	.9412
28	1.0323	78	.9831	128	.9404
29	1.0312	79	.9822	129	.9396
30	1.0302	80	.9813	130	.9388
31	1.0291	81	.9804	131	.9380
32	1.0281	82	.9795	132	.9372
33	1.0270	83	.9786	133	.9364
34	1.0260	84	.9777	134	.9356
35	1.0249	85	.9768	135	.9349
36	1.0239	86	.9759	136	.9341
37	1.0229	87	.9750	137	.9333
38	1.0218	88	.9741	138	.9325
39	1.0208	89	.9732	139	.9317
40	1.0198	90	.9723	140	.9309
41	1.0188	91	.9715	141	.9302
42	1.0178	92	.9706	142	.9294
43	1.0168	93	.9697	143	.9286
44	1.0158	94	.9688	144	.9279
45	1.0147	95	.9680	145	.9271
46	1.0137	96	.9671	146	.9263
47	1.0127	97	.9662	147	.9256
48	1.0117	98	.9653	148	.9248
49	1.0108	99	.9645	149	.9240
50	1.0098	100	0.9636	150	0.9233

[For temperature range 150 thru 249 F, see Table 4-12.—EDITOR

TABLE 15 F_g, FACTORS TO ADJUST FOR SPECIFIC GRAVITY

Specific gravity, G	0.000	0.002	0.004	0.006	0.008	Specific gravity, G	0.000	0.002	0.004	0.006	0.008
0.550	1.3484	1.3460	1.3435	1.3411	1.3387	.780	1.1323	1.1308	1.1294	1.1279	1.1265
.560	1.3363	1.3339	1.3316	1.3292	1.3269	.790	1.1251	1.1237	1.1222	1.1208	1.1194
.570	1.3245	1.3222	1.3199	1.3176	1.3153	.800	1.1180	1.1166	1.1152	1.1139	1.1125
.580	1.3131	1.3108	1.3086	1.3063	1.3041	.810	1.1111	1.1097	1.1084	1.1070	1.1057
.590	1.3019	1.2997	1.2975	1.2953	1.2932	.820	1.1043	1.1030	1.1016	1.1003	1.0990
.600	1.2910	1.2888	1.2867	1.2846	1.2825	.830	1.0976	1.0963	1.0950	1.0937	1.0924
.610	1.2804	1.2783	1.2762	1.2741	1.2720	.840	1.0911	1.0898	1.0885	1.0872	1.0859
.620	1.2700	1.2680	1.2659	1.2639	1.2619	.850	1.0846	1.0834	1.0821	1.0808	1.0796
.630	1.2599	1.2579	1.2559	1.2539	1.2520	.860	1.0783	1.0771	1.0758	1.0746	1.0733
.640	1.2500	1.2480	1.2461	1.2442	1.2423	.870	1.0721	1.0709	1.0696	1.0684	1.0672
.650	1.2403	1.2384	1.2365	1.2347	1.2328	.880	1.0660	1.0648	1.0636	1.0624	1.0612
.660	1.2309	1.2290	1.2272	1.2254	1.2235	.890	1.0600	1.0588	1.0576	1.0564	1.0553
.670	1.2217	1.2199	1.2181	1.2163	1.2145	.900	1.0541	1.0529	1.0518	1.0506	1.0494
.680	1.2127	1.2109	1.2091	1.2074	1.2056	.910	1.0483	1.0471	1.0460	1.0448	1.0437
.690	1.2039	1.2021	1.2004	1.1986	1.1969	.920	1.0426	1.0414	1.0403	1.0392	1.0381
.700	1.1952	1.1935	1.1918	1.1901	1.1884	.930	1.0370	1.0358	1.0347	1.0336	1.0325
.710	1.1868	1.1851	1.1834	1.1818	1.1802	.940	1.0314	1.0303	1.0292	1.0281	1.0270
.720	1.1785	1.1769	1.1752	1.1736	1.1720	.950	1.0260	1.0249	1.0238	1.0228	1.0217
.730	1.1704	1.1688	1.1672	1.1656	1.1640	.960	1.0206	1.0196	1.0185	1.0174	1.0164
.740	1.1625	1.1609	1.1593	1.1578	1.1562	.970	1.0153	1.0143	1.0132	1.0122	1.0112
.750	1.1547	1.1532	1.1516	1.1501	1.1486	.980	1.0102	1.0091	1.0081	1.0071	1.0060
.760	1.1471	1.1456	1.1441	1.1426	1.1411	0.990	1.0050	1.0040	1.0030	1.0020	1.0010
.770	1.1396	1.1381	1.1366	1.1352	1.1337	1.000	1.0000				

[The increments of specific gravity were increased from 0.001 to 0.002 in this condensed version of the referenced[2] table.—EDITOR]

TABLE 16-A F_{pv}, SUPERCOMPRESSIBILITY FACTORS

Base data—0.6 specific gravity hydrocarbon gas

P_f, psig	Temperature, °F							
	−40	−35	−30	−25	−20	−15	−10	−5
0	1.0000	1.0000	1.0000	1.0000	1.0000	1.0000	1.0000	1.0000
40	1.0062	1.0060	1.0059	1.0057	1.0055	1.0053	1.0051	1.0049
80	1.0125	1.0122	1.0119	1.0115	1.0111	1.0107	1.0103	1.0099
120	1.0192	1.0186	1.0178	1.0175	1.0168	1.0162	1.0156	1.0151
160	1.0262	1.0251	1.0241	1.0235	1.0228	1.0218	1.0210	1.0202
200	1.0333	1.0319	1.0307	1.0296	1.0288	1.0275	1.0265	1.0255
240	1.0406	1.0388	1.0373	1.0360	1.0347	1.0334	1.0321	1.0309
280	1.0482	1.0461	1.0442	1.0425	1.0408	1.0394	1.0379	1.0365
320	1.0562	1.0537	1.0514	1.0494	1.0474	1.0456	1.0439	1.0422
360	1.0642	1.0614	1.0589	1.0564	1.0541	1.0520	1.0500	1.0480
400	1.0727	1.0695	1.0666	1.0638	1.0611	1.0586	1.0563	1.0540
440	1.0816	1.0779	1.0746	1.0713	1.0682	1.0654	1.0627	1.0601
480	1.0909	1.0866	1.0828	1.0791	1.0756	1.0723	1.0693	1.0664
520	1.1004	1.0956	1.0911	1.0869	1.0830	1.0794	1.0761	1.0728
560	1.1106	1.1051	1.1000	1.0952	1.0908	1.0868	1.0830	1.0793
600	1.1213	1.1149	1.1091	1.1038	1.0989	1.0944	1.0901	1.0860
640	1.1323	1.1252	1.1186	1.1127	1.1072	1.1021	1.0973	1.0928
680	1.1439	1.1359	1.1286	1.1218	1.1156	1.1099	1.1047	1.0998
720	1.1562	1.1469	1.1388	1.1313	1.1245	1.1181	1.1123	1.1069
760	1.1692	1.1587	1.1496	1.1413	1.1337	1.1267	1.1202	1.1143
800	1.1826	1.1708	1.1607	1.1516	1.1432	1.1355	1.1283	1.1217
840	1.1967	1.1835	1.1723	1.1622	1.1528	1.1443	1.1365	1.1293
880	1.2116	1.1968	1.1843	1.1731	1.1627	1.1533	1.1449	1.1373
920	1.2269	1.2103	1.1965	1.1842	1.1728	1.1625	1.1534	1.1450
960	1.2427	1.2245	1.2093	1.1956	1.1832	1.1721	1.1620	1.1530
1000	1.2591	1.2391	1.2221	1.2072	1.1936	1.1815	1.1706	1.1610
1040	1.2756	1.2537	1.2351	1.2190	1.2044	1.1913	1.1796	1.1690
1080	1.2922	1.2685	1.2485	1.2310	1.2152	1.2011	1.1886	1.1772
1120	1.3091	1.2834	1.2619	1.2431	1.2260	1.2109	1.1978	1.1854
1160	1.3259	1.2985	1.2753	1.2552	1.2370	1.2209	1.2068	1.1939

Factors for intermediate values of pressure and temperature should be interpolated.

[The increments of the pressure argument were increased from 20 psi to 40 psi in this condensed version of the referenced[2] table; interpolation without loss of essential accuracy is still possible. Values at smaller increments are also available.[9]—EDITOR]

TABLE 16-A F_{pv}, SUPERCOMPRESSIBILITY FACTORS (Continued)

Base data—0.6 specific gravity hydrocarbon gas

P_f, psig	Temperature, °F							
	−40	−35	−30	−25	−20	−15	−10	−5
1200	1.3412	1.3127	1.2883	1.2669	1.2477	1.2305	1.2154	1.2018
1240	1.3559	1.3264	1.3009	1.2783	1.2580	1.2399	1.2240	1.2098
1280	1.3692	1.3390	1.3128	1.2894	1.2682	1.2493	1.2326	1.2176
1320	1.3812	1.3505	1.3240	1.3000	1.2782	1.2586	1.2411	1.2252
1360	1.3917	1.3611	1.3344	1.3101	1.2878	1.2675	1.2491	1.2326
1400	1.4002	1.3699	1.3432	1.3186	1.2960	1.2754	1.2568	1.2398
1440	1.4069	1.3774	1.3508	1.3264	1.3038	1.2830	1.2640	1.2466
1480	1.4118	1.3833	1.3571	1.3331	1.3105	1.2894	1.2703	1.2530
1520	1.4152	1.3878	1.3621	1.3384	1.3161	1.2954	1.2763	1.2586
1560	1.4172	1.3910	1.3661	1.3428	1.3207	1.3004	1.2813	1.2638
1600	1.4179	1.3930	1.3690	1.3462	1.3247	1.3047	1.2860	1.2683
1640	1.4176	1.3938	1.3708	1.3488	1.3278	1.3079	1.2895	1.2720
1680	1.4162	1.3936	1.3716	1.3504	1.3300	1.3108	1.2928	1.2755
1720	1.4139	1.3926	1.3715	1.3513	1.3317	1.3130	1.2951	1.2782
1760	1.4111	1.3909	1.3707	1.3513	1.3321	1.3143	1.2970	1.2804
1800	1.4075	1.3884	1.3693	1.3507	1.3323	1.3151	1.2983	1.2819
1840	1.4035	1.3855	1.3673	1.3496	1.3321	1.3153	1.2990	1.2831
1880	1.3989	1.3818	1.3647	1.3478	1.3312	1.3150	1.2992	1.2838
1920	1.3940	1.3779	1.3617	1.3457	1.3298	1.3142	1.2989	1.2840
1960	1.3888	1.3737	1.3584	1.3431	1.3279	1.3129	1.2982	1.2839
2000	1.3834	1.3691	1.3547	1.3400	1.3254	1.3110	1.2971	1.2833
2040	1.3778	1.3642	1.3506	1.3368	1.3228	1.3089	1.2955	1.2823
2080	1.3720	1.3591	1.3462	1.3332	1.3196	1.3065	1.2937	1.2809
2120	1.3660	1.3539	1.3416	1.3292	1.3164	1.3039	1.2915	1.2793
2160	1.3600	1.3486	1.3367	1.3250	1.3129	1.3010	1.2891	1.2774
2200	1.3538	1.3431	1.3318	1.3206	1.3091	1.2978	1.2864	1.2753
2240	1.3476	1.3373	1.3268	1.3162	1.3051	1.2943	1.2835	1.2729
2280	1.3412	1.3315	1.3217	1.3116	1.3011	1.2907	1.2804	1.2702
2320	1.3349	1.3257	1.3164	1.3068	1.2969	1.2870	1.2772	1.2674
2360	1.3285	1.3199	1.3110	1.3019	1.2925	1.2831	1.2737	1.2643
2400	1.3223	1.3141	1.3056	1.2969	1.2880	1.2790	1.2700	1.2611
2440	1.3159	1.3082	1.3002	1.2919	1.2834	1.2748	1.2663	1.2577
2480	1.3096	1.3022	1.2948	1.2869	1.2788	1.2706	1.2624	1.2542
2520	1.3033	1.2963	1.2893	1.2817	1.2741	1.2663	1.2585	1.2506
2560	1.2970	1.2904	1.2835	1.2766	1.2693	1.2620	1.2546	1.2470
2600	1.2909	1.2846	1.2780	1.2714	1.2645	1.2575	1.2505	1.2433
2640	1.2847	1.2787	1.2725	1.2661	1.2596	1.2530	1.2462	1.2394
2680	1.2785	1.2729	1.2670	1.2609	1.2547	1.2484	1.2420	1.2356
2720	1.2723	1.2670	1.2614	1.2557	1.2498	1.2438	1.2377	1.2315
2760	1.2663	1.2612	1.2559	1.2505	1.2448	1.2391	1.2334	1.2275
2800	1.2603	1.2555	1.2504	1.2454	1.2400	1.2345	1.2290	1.2234
2840	1.2543	1.2497	1.2448	1.2401	1.2349	1.2298	1.2246	1.2193
2880	1.2483	1.2441	1.2394	1.2349	1.2300	1.2251	1.2202	1.2152
2920	1.2424	1.2384	1.2341	1.2298	1.2252	1.2205	1.2158	1.2110
2960	1.2366	1.2328	1.2287	1.2246	1.2202	1.2157	1.2112	1.2067
3000	1.2309	1.2273	1.2234	1.2195	1.2153	1.2111	1.2069	1.2027

P_f, psig	Temperature, °F											
	0	5	10	15	20	25	30	35	40	45	50	55
0	1.0000	1.0000	1.0000	1.0000	1.0000	1.0000	1.0000	1.0000	1.0000	1.0000	1.0000	1.0000
40	1.0048	1.0047	1.0045	1.0044	1.0042	1.0041	1.0040	1.0038	1.0037	1.0036	1.0034	1.0033
80	1.0096	1.0093	1.0090	1.0087	1.0084	1.0081	1.0078	1.0076	1.0073	1.0070	1.0068	1.0066
120	1.0146	1.0141	1.0136	1.0131	1.0127	1.0122	1.0118	1.0114	1.0110	1.0106	1.0103	1.0100
160	1.0195	1.0188	1.0182	1.0176	1.0169	1.0163	1.0158	1.0152	1.0147	1.0142	1.0138	1.0133
200	1.0245	1.0237	1.0229	1.0220	1.0213	1.0206	1.0198	1.0192	1.0185	1.0179	1.0173	1.0167
240	1.0298	1.0288	1.0277	1.0267	1.0257	1.0248	1.0239	1.0231	1.0223	1.0215	1.0208	1.0201
280	1.0351	1.0339	1.0327	1.0315	1.0303	1.0292	1.0281	1.0271	1.0261	1.0252	1.0244	1.0236
320	1.0406	1.0391	1.0377	1.0363	1.0349	1.0336	1.0324	1.0312	1.0300	1.0290	1.0280	1.0270
360	1.0462	1.0444	1.0427	1.0411	1.0395	1.0380	1.0366	1.0353	1.0340	1.0328	1.0316	1.0305

Factors for intermediate values of pressure and temperature should be interpolated.

[The increments of the pressure argument were increased from 20 psi to 40 psi in this condensed version of the referenced[2] table; interpolation without loss of essential accuracy is still possible. Values at smaller increments are also available.[9]—EDITOR]

TABLE 16-A (Continued)

P_f, psig	Temperature, °F											
	0	5	10	15	20	25	30	35	40	45	50	55
400	1.0519	1.0498	1.0479	1.0461	1.0444	1.0427	1.0410	1.0395	1.0381	1.0366	1.0352	1.0340
440	1.0577	1.0553	1.0531	1.0511	1.0492	1.0472	1.0453	1.0437	1.0421	1.0405	1.0389	1.0375
480	1.0636	1.0609	1.0585	1.0562	1.0540	1.0519	1.0498	1.0479	1.0461	1.0444	1.0427	1.0411
520	1.0697	1.0667	1.0639	1.0613	1.0588	1.0565	1.0543	1.0522	1.0503	1.0484	1.0465	1.0447
560	1.0759	1.0726	1.0695	1.0666	1.0639	1.0612	1.0587	1.0565	1.0544	1.0523	1.0502	1.0483
600	1.0822	1.0787	1.0753	1.0721	1.0691	1.0661	1.0634	1.0609	1.0586	1.0562	1.0540	1.0519
640	1.0886	1.0848	1.0811	1.0775	1.0742	1.0710	1.0680	1.0653	1.0628	1.0602	1.0578	1.0556
680	1.0953	1.0910	1.0869	1.0830	1.0793	1.0760	1.0728	1.0698	1.0670	1.0643	1.0617	1.0593
720	1.1020	1.0973	1.0928	1.0885	1.0847	1.0810	1.0775	1.0742	1.0712	1.0684	1.0656	1.0630
760	1.1089	1.1038	1.0989	1.0943	1.0900	1.0860	1.0822	1.0788	1.0756	1.0725	1.0694	1.0667
800	1.1159	1.1103	1.1050	1.1000	1.0954	1.0911	1.0870	1.0833	1.0798	1.0765	1.0733	1.0704
840	1.1229	1.1169	1.1112	1.1057	1.1008	1.0962	1.0919	1.0879	1.0841	1.0805	1.0771	1.0740
880	1.1301	1.1236	1.1175	1.1117	1.1064	1.1015	1.0968	1.0925	1.0885	1.0847	1.0811	1.0778
920	1.1373	1.1303	1.1237	1.1175	1.1118	1.1066	1.1016	1.0970	1.0928	1.0887	1.0849	1.0813
960	1.1448	1.1372	1.1301	1.1234	1.1175	1.1119	1.1065	1.1016	1.0971	1.0928	1.0887	1.0850
1000	1.1520	1.1440	1.1365	1.1294	1.1230	1.1170	1.1114	1.1062	1.1013	1.0968	1.0925	1.0885
1040	1.1595	1.1509	1.1428	1.1353	1.1285	1.1222	1.1163	1.1107	1.1057	1.1008	1.0964	1.0922
1080	1.1669	1.1578	1.1492	1.1411	1.1340	1.1273	1.1211	1.1153	1.1099	1.1048	1.1001	1.0957
1120	1.1744	1.1647	1.1555	1.1471	1.1395	1.1325	1.1259	1.1198	1.1141	1.1088	1.1038	1.0993
1160	1.1819	1.1716	1.1619	1.1531	1.1451	1.1377	1.1307	1.1243	1.1184	1.1128	1.1075	1.1028
1200	1.1895	1.1784	1.1682	1.1588	1.1505	1.1427	1.1354	1.1287	1.1225	1.1167	1.1113	1.1063
1240	1.1968	1.1852	1.1745	1.1646	1.1558	1.1477	1.1401	1.1331	1.1266	1.1206	1.1149	1.1097
1280	1.2040	1.1918	1.1805	1.1703	1.1611	1.1526	1.1446	1.1374	1.1307	1.1244	1.1184	1.1130
1320	1.2109	1.1983	1.1867	1.1758	1.1663	1.1574	1.1492	1.1417	1.1347	1.1281	1.1219	1.1163
1360	1.2178	1.2048	1.1926	1.1814	1.1714	1.1622	1.1536	1.1458	1.1386	1.1317	1.1253	1.1195
1400	1.2244	1.2108	1.1983	1.1866	1.1763	1.1667	1.1577	1.1496	1.1422	1.1352	1.1287	1.1226
1440	1.2307	1.2166	1.2037	1.1918	1.1810	1.1712	1.1620	1.1536	1.1459	1.1386	1.1318	1.1256
1480	1.2365	1.2220	1.2088	1.1966	1.1856	1.1754	1.1658	1.1572	1.1493	1.1418	1.1349	1.1285
1520	1.2421	1.2273	1.2137	1.2012	1.1900	1.1795	1.1697	1.1608	1.1526	1.1450	1.1378	1.1313
1560	1.2469	1.2320	1.2182	1.2054	1.1940	1.1834	1.1733	1.1642	1.1559	1.1480	1.1407	1.1340
1600	1.2514	1.2365	1.2225	1.2095	1.1979	1.1871	1.1769	1.1676	1.1590	1.1510	1.1435	1.1366
1640	1.2555	1.2406	1.2265	1.2132	1.2015	1.1905	1.1802	1.1707	1.1620	1.1537	1.1461	1.1390
1680	1.2591	1.2441	1.2299	1.2166	1.2049	1.1938	1.1832	1.1736	1.1647	1.1563	1.1485	1.1413
1720	1.2620	1.2471	1.2331	1.2198	1.2079	1.1967	1.1861	1.1764	1.1674	1.1587	1.1508	1.1435
1760	1.2645	1.2497	1.2357	1.2227	1.2106	1.1993	1.1887	1.1787	1.1697	1.1610	1.1529	1.1455
1800	1.2665	1.2519	1.2381	1.2251	1.2130	1.2017	1.1910	1.1810	1.1718	1.1631	1.1549	1.1473
1840	1.2680	1.2538	1.2401	1.2272	1.2151	1.2038	1.1930	1.1830	1.1738	1.1649	1.1566	1.1490
1880	1.2690	1.2551	1.2417	1.2289	1.2169	1.2056	1.1948	1.1848	1.1755	1.1667	1.1583	1.1506
1920	1.2697	1.2561	1.2429	1.2303	1.2184	1.2072	1.1964	1.1864	1.1770	1.1682	1.1598	1.1521
1960	1.2700	1.2566	1.2438	1.2314	1.2197	1.2085	1.1978	1.1877	1.1784	1.1696	1.1612	1.1534
2000	1.2699	1.2569	1.2443	1.2321	1.2207	1.2097	1.1990	1.1890	1.1796	1.1707	1.1623	1.1545
2040	1.2695	1.2569	1.2446	1.2326	1.2213	1.2104	1.1998	1.1898	1.1805	1.1716	1.1632	1.1554
2080	1.2686	1.2565	1.2445	1.2328	1.2216	1.2109	1.2005	1.1906	1.1813	1.1725	1.1640	1.1563
2120	1.2674	1.2556	1.2440	1.2327	1.2217	1.2111	1.2009	1.1912	1.1819	1.1730	1.1646	1.1569
2160	1.2658	1.2545	1.2433	1.2322	1.2216	1.2112	1.2011	1.1915	1.1823	1.1735	1.1651	1.1574
2200	1.2640	1.2531	1.2423	1.2315	1.2212	1.2110	1.2011	1.1916	1.1825	1.1737	1.1654	1.1577
2240	1.2621	1.2514	1.2410	1.2307	1.2206	1.2106	1.2009	1.1916	1.1825	1.1738	1.1655	1.1579
2280	1.2600	1.2495	1.2394	1.2296	1.2197	1.2100	1.2004	1.1913	1.1824	1.1738	1.1656	1.1579
2320	1.2576	1.2475	1.2378	1.2282	1.2186	1.2092	1.1998	1.1909	1.1821	1.1736	1.1655	1.1579
2360	1.2549	1.2454	1.2360	1.2267	1.2173	1.2081	1.1991	1.1903	1.1816	1.1732	1.1652	1.1576
2400	1.2521	1.2430	1.2339	1.2249	1.2158	1.2070	1.1983	1.1895	1.1810	1.1727	1.1648	1.1572
2440	1.2491	1.2405	1.2318	1.2231	1.2142	1.2056	1.1971	1.1886	1.1802	1.1720	1.1643	1.1567
2480	1.2459	1.2377	1.2294	1.2210	1.2125	1.2041	1.1958	1.1875	1.1793	1.1712	1.1636	1.1562
2520	1.2427	1.2349	1.2269	1.2188	1.2106	1.2025	1.1944	1.1863	1.1782	1.1703	1.1626	1.1555
2560	1.2395	1.2320	1.2242	1.2164	1.2086	1.2007	1.1928	1.1849	1.1770	1.1693	1.1617	1.1546
2600	1.2361	1.2287	1.2214	1.2140	1.2064	1.1987	1.1910	1.1834	1.1757	1.1681	1.1606	1.1537
2640	1.2325	1.2256	1.2185	1.2113	1.2040	1.1966	1.1892	1.1817	1.1742	1.1668	1.1596	1.1527
2680	1.2290	1.2223	1.2155	1.2086	1.2015	1.1944	1.1872	1.1798	1.1725	1.1653	1.1584	1.1516
2720	1.2253	1.2189	1.2124	1.2058	1.1990	1.1922	1.1852	1.1780	1.1709	1.1638	1.1569	1.1503
2760	1.2216	1.2155	1.2092	1.2029	1.1964	1.1898	1.1830	1.1761	1.1691	1.1622	1.1554	1.1490
2800	1.2178	1.2120	1.2060	1.1999	1.1936	1.1872	1.1806	1.1740	1.1672	1.1605	1.1539	1.1476
2840	1.2140	1.2084	1.2027	1.1968	1.1908	1.1846	1.1782	1.1718	1.1652	1.1586	1.1522	1.1461
2880	1.2100	1.2048	1.1993	1.1937	1.1878	1.1818	1.1757	1.1696	1.1632	1.1568	1.1504	1.1445
2920	1.2062	1.2011	1.1959	1.1905	1.1848	1.1790	1.1731	1.1673	1.1610	1.1547	1.1486	1.1428
2960	1.2023	1.1974	1.1924	1.1872	1.1817	1.1761	1.1705	1.1648	1.1587	1.1527	1.1468	1.1411
3000	1.1984	1.1937	1.1889	1.1839	1.1786	1.1733	1.1678	1.1622	1.1564	1.1505	1.1448	1.1393

TABLE 16-A F_{pv}, SUPERCOMPRESSIBILITY FACTORS (Continued)

Base data—0.6 specific gravity hydrocarbon gas

P_f, psig	Temperature, °F												
	60	65	70	75	80	85	90	95	100	105	110	115	120
0	1.0000	1.0000	1.0000	1.0000	1.0000	1.0000	1.0000	1.0000	1.0000	1.0000	1.0000	1.0000	1.0000
40	1.0032	1.0031	1.0030	1.0029	1.0028	1.0027	1.0027	1.0026	1.0025	1.0024	1.0023	1.0022	1.0022
80	1.0064	1.0062	1.0061	1.0058	1.0056	1.0054	1.0052	1.0051	1.0049	1.0047	1.0046	1.0044	1.0043
120	1.0097	1.0094	1.0091	1.0088	1.0085	1.0082	1.0079	1.0076	1.0073	1.0071	1.0069	1.0067	1.0065
160	1.0129	1.0125	1.0121	1.0117	1.0112	1.0108	1.0105	1.0101	1.0098	1.0095	1.0092	1.0089	1.0087
200	1.0162	1.0156	1.0151	1.0146	1.0140	1.0135	1.0131	1.0127	1.0123	1.0119	1.0115	1.0111	1.0108
240	1.0194	1.0188	1.0181	1.0175	1.0168	1.0163	1.0158	1.0153	1.0148	1.0143	1.0138	1.0133	1.0129
280	1.0228	1.0220	1.0212	1.0205	1.0197	1.0191	1.0185	1.0178	1.0173	1.0167	1.0162	1.0155	1.0150
320	1.0261	1.0252	1.0243	1.0235	1.0227	1.0219	1.0212	1.0205	1.0198	1.0191	1.0185	1.0178	1.0173
360	1.0294	1.0284	1.0273	1.0264	1.0256	1.0247	1.0238	1.0230	1.0222	1.0215	1.0207	1.0200	1.0194
400	1.0328	1.0317	1.0305	1.0294	1.0285	1.0275	1.0265	1.0256	1.0246	1.0238	1.0230	1.0223	1.0215
440	1.0361	1.0349	1.0336	1.0324	1.0313	1.0302	1.0292	1.0281	1.0272	1.0262	1.0253	1.0244	1.0236
480	1.0395	1.0381	1.0367	1.0354	1.0341	1.0329	1.0318	1.0307	1.0297	1.0287	1.0276	1.0267	1.0258
520	1.0430	1.0414	1.0399	1.0385	1.0371	1.0357	1.0345	1.0333	1.0321	1.0310	1.0299	1.0289	1.0279
560	1.0465	1.0448	1.0432	1.0416	1.0400	1.0385	1.0372	1.0359	1.0346	1.0334	1.0322	1.0311	1.0300
600	1.0499	1.0481	1.0463	1.0446	1.0430	1.0414	1.0399	1.0384	1.0370	1.0358	1.0345	1.0333	1.0321
640	1.0534	1.0514	1.0495	1.0476	1.0460	1.0442	1.0426	1.0410	1.0396	1.0381	1.0368	1.0355	1.0341
680	1.0570	1.0547	1.0527	1.0507	1.0488	1.0470	1.0453	1.0436	1.0420	1.0405	1.0390	1.0377	1.0363
720	1.0605	1.0580	1.0559	1.0537	1.0517	1.0497	1.0479	1.0461	1.0444	1.0428	1.0412	1.0398	1.0383
760	1.0640	1.0614	1.0591	1.0568	1.0546	1.0524	1.0505	1.0487	1.0468	1.0451	1.0435	1.0419	1.0403
800	1.0676	1.0648	1.0623	1.0598	1.0575	1.0552	1.0532	1.0513	1.0492	1.0474	1.0456	1.0440	1.0424
840	1.0711	1.0681	1.0654	1.0628	1.0603	1.0580	1.0558	1.0536	1.0517	1.0497	1.0478	1.0460	1.0443
880	1.0745	1.0714	1.0686	1.0658	1.0631	1.0607	1.0584	1.0562	1.0540	1.0519	1.0500	1.0481	1.0463
920	1.0779	1.0746	1.0716	1.0688	1.0660	1.0634	1.0610	1.0586	1.0563	1.0541	1.0520	1.0501	1.0482
960	1.0814	1.0779	1.0748	1.0718	1.0689	1.0662	1.0636	1.0610	1.0586	1.0563	1.0541	1.0521	1.0501
1000	1.0847	1.0811	1.0778	1.0746	1.0717	1.0687	1.0660	1.0634	1.0608	1.0585	1.0562	1.0539	1.0519
1040	1.0882	1.0843	1.0809	1.0775	1.0744	1.0714	1.0685	1.0658	1.0631	1.0606	1.0582	1.0559	1.0538
1080	1.0916	1.0875	1.0839	1.0804	1.0771	1.0740	1.0709	1.0681	1.0654	1.0628	1.0602	1.0578	1.0556
1120	1.0950	1.0908	1.0870	1.0834	1.0800	1.0766	1.0734	1.0703	1.0676	1.0649	1.0623	1.0598	1.0574
1160	1.0983	1.0939	1.0899	1.0862	1.0826	1.0791	1.0758	1.0727	1.0698	1.0669	1.0643	1.0616	1.0592
1200	1.1016	1.0970	1.0928	1.0889	1.0851	1.0816	1.0782	1.0750	1.0718	1.0689	1.0661	1.0634	1.0610
1240	1.1048	1.1001	1.0957	1.0916	1.0876	1.0840	1.0805	1.0771	1.0739	1.0709	1.0681	1.0652	1.0626
1280	1.1079	1.1030	1.0985	1.0942	1.0901	1.0863	1.0827	1.0791	1.0758	1.0728	1.0699	1.0670	1.0643
1320	1.1110	1.1059	1.1012	1.0968	1.0925	1.0886	1.0849	1.0812	1.0778	1.0746	1.0716	1.0686	1.0659
1360	1.1140	1.1087	1.1039	1.0993	1.0949	1.0909	1.0870	1.0833	1.0797	1.0764	1.0733	1.0703	1.0675
1400	1.1168	1.1114	1.1065	1.1017	1.0973	1.0931	1.0891	1.0853	1.0816	1.0782	1.0750	1.0719	1.0690
1440	1.1197	1.1141	1.1090	1.1042	1.0995	1.0952	1.0912	1.0873	1.0834	1.0800	1.0767	1.0735	1.0705
1480	1.1225	1.1167	1.1115	1.1064	1.1016	1.0973	1.0931	1.0891	1.0852	1.0816	1.0783	1.0750	1.0719
1520	1.1251	1.1191	1.1138	1.1087	1.1038	1.0993	1.0950	1.0909	1.0870	1.0833	1.0799	1.0766	1.0734
1560	1.1276	1.1215	1.1161	1.1108	1.1059	1.1012	1.0969	1.0927	1.0887	1.0850	1.0815	1.0780	1.0748
1600	1.1301	1.1238	1.1183	1.1129	1.1078	1.1031	1.0987	1.0944	1.0904	1.0866	1.0830	1.0795	1.0762
1640	1.1323	1.1260	1.1203	1.1149	1.1097	1.1049	1.1004	1.0960	1.0920	1.0881	1.0844	1.0809	1.0775
1680	1.1345	1.1281	1.1223	1.1167	1.1115	1.1066	1.1020	1.0976	1.0934	1.0895	1.0858	1.0822	1.0787
1720	1.1366	1.1300	1.1241	1.1185	1.1132	1.1082	1.1036	1.0992	1.0950	1.0910	1.0872	1.0835	1.0799
1760	1.1385	1.1318	1.1258	1.1201	1.1147	1.1097	1.1051	1.1006	1.0964	1.0923	1.0884	1.0847	1.0811
1800	1.1402	1.1334	1.1273	1.1216	1.1161	1.1111	1.1064	1.1019	1.0976	1.0935	1.0896	1.0858	1.0821
1840	1.1418	1.1349	1.1288	1.1230	1.1175	1.1124	1.1077	1.1031	1.0988	1.0947	1.0907	1.0868	1.0831
1880	1.1433	1.1364	1.1302	1.1243	1.1187	1.1137	1.1089	1.1043	1.0999	1.0957	1.0916	1.0877	1.0840
1920	1.1447	1.1378	1.1315	1.1255	1.1199	1.1148	1.1099	1.1053	1.1009	1.0967	1.0925	1.0886	1.0848
1960	1.1460	1.1389	1.1326	1.1266	1.1209	1.1158	1.1109	1.1063	1.1019	1.0976	1.0934	1.0894	1.0856
2000	1.1470	1.1399	1.1336	1.1276	1.1219	1.1168	1.1119	1.1073	1.1027	1.0984	1.0942	1.0902	1.0864
2040	1.1480	1.1408	1.1344	1.1284	1.1227	1.1176	1.1127	1.1081	1.1034	1.0992	1.0950	1.0909	1.0870
2080	1.1488	1.1416	1.1353	1.1291	1.1234	1.1184	1.1134	1.1088	1.1042	1.0999	1.0956	1.0915	1.0876
2120	1.1494	1.1422	1.1358	1.1297	1.1240	1.1189	1.1140	1.1094	1.1048	1.1004	1.0961	1.0919	1.0880
2160	1.1499	1.1427	1.1363	1.1302	1.1245	1.1194	1.1145	1.1098	1.1052	1.1008	1.0965	1.0923	1.0884
2200	1.1503	1.1431	1.1367	1.1306	1.1250	1.1198	1.1149	1.1102	1.1056	1.1012	1.0969	1.0927	1.0888
2240	1.1505	1.1433	1.1369	1.1310	1.1254	1.1201	1.1152	1.1105	1.1059	1.1015	1.0972	1.0930	1.0891
2280	1.1505	1.1434	1.1371	1.1312	1.1257	1.1204	1.1154	1.1108	1.1061	1.1017	1.0974	1.0932	1.0893
2320	1.1504	1.1434	1.1371	1.1312	1.1258	1.1205	1.1156	1.1110	1.1063	1.1019	1.0976	1.0934	1.0895
2360	1.1502	1.1432	1.1370	1.1312	1.1258	1.1205	1.1156	1.1110	1.1063	1.1020	1.0978	1.0936	1.0897

Factors for intermediate values of pressure and temperature should be interpolated.

[The increments of the pressure argument were increased from 20 psi to 40 psi in this condensed version of the referenced[2] table; interpolation without loss of essential accuracy is still possible. Values at smaller increments are also available.[9]—EDITOR]

TABLE 16-A (Continued).

P_f, psig	Temperature, °F												
	60	65	70	75	80	85	90	95	100	105	110	115	120
2400	1.1499	1.1429	1.1368	1.1311	1.1256	1.1205	1.1156	1.1110	1.1063	1.1020	1.0978	1.0937	1.0897
2440	1.1495	1.1426	1.1366	1.1309	1.1255	1.1204	1.1155	1.1109	1.1063	1.1020	1.0978	1.0937	1.0897
2480	1.1491	1.1422	1.1363	1.1306	1.1253	1.1201	1.1153	1.1107	1.1061	1.1018	1.0976	1.0935	1.0895
2520	1.1485	1.1417	1.1358	1.1302	1.1249	1.1198	1.1151	1.1105	1.1059	1.1016	1.0974	1.0933	1.0893
2560	1.1478	1.1411	1.1352	1.1297	1.1244	1.1194	1.1147	1.1101	1.1055	1.1013	1.0972	1.0931	1.0891
2600	1.1470	1.1404	1.1345	1.1290	1.1239	1.1189	1.1142	1.1097	1.1051	1.1010	1.0968	1.0929	1.0889
2640	1.1461	1.1396	1.1337	1.1283	1.1232	1.1183	1.1136	1.1091	1.1047	1.1006	1.0965	1.0925	1.0885
2680	1.1450	1.1387	1.1329	1.1275	1.1225	1.1177	1.1130	1.1085	1.1042	1.1001	1.0960	1.0921	1.0881
2720	1.1440	1.1377	1.1320	1.1266	1.1217	1.1170	1.1123	1.1079	1.1036	1.0995	1.0955	1.0917	1.0876
2760	1.1428	1.1366	1.1310	1.1257	1.1208	1.1162	1.1116	1.1072	1.1030	1.0989	1.0949	1.0911	1.0872
2800	1.1414	1.1354	1.1299	1.1247	1.1199	1.1153	1.1108	1.1065	1.1024	1.0983	1.0943	1.0904	1.0866
2840	1.1401	1.1343	1.1288	1.1237	1.1188	1.1144	1.1099	1.1057	1.1016	1.0975	1.0936	1.0898	1.0860
2880	1.1387	1.1330	1.1276	1.1225	1.1177	1.1134	1.1090	1.1047	1.1008	1.0968	1.0929	1.0892	1.0854
2920	1.1371	1.1316	1.1263	1.1213	1.1166	1.1123	1.1079	1.1037	1.0998	1.0959	1.0921	1.0884	1.0847
2960	1.1355	1.1302	1.1249	1.1201	1.1155	1.1111	1.1069	1.1027	1.0988	1.0950	1.0913	1.0876	1.0840
3000	1.1339	1.1288	1.1235	1.1187	1.1142	1.1099	1.1058	1.1017	1.0978	1.0941	1.0904	1.0867	1.0831

P_f, psig	Temperature, °F												
	125	130	135	140	145	150	155	160	165	170	175	180	185
0	1.0000	1.0000	1.0000	1.0000	1.0000	1.0000	1.0000	1.0000	1.0000	1.0000	1.0000	1.0000	1.0000
40	1.0022	1.0020	1.0020	1.0020	1.0019	1.0018	1.0018	1.0017	1.0016	1.0016	1.0016	1.0015	1.0014
80	1.0042	1.0040	1.0039	1.0039	1.0038	1.0036	1.0035	1.0034	1.0032	1.0031	1.0030	1.0029	1.0028
120	1.0063	1.0061	1.0059	1.0057	1.0056	1.0054	1.0052	1.0050	1.0048	1.0047	1.0045	1.0044	1.0042
160	1.0084	1.0081	1.0078	1.0076	1.0074	1.0072	1.0069	1.0067	1.0064	1.0063	1.0061	1.0058	1.0056
200	1.0104	1.0101	1.0097	1.0094	1.0092	1.0089	1.0086	1.0083	1.0080	1.0078	1.0075	1.0073	1.0070
240	1.0125	1.0121	1.0117	1.0114	1.0110	1.0107	1.0103	1.0100	1.0096	1.0094	1.0090	1.0087	1.0084
280	1.0146	1.0142	1.0137	1.0132	1.0128	1.0125	1.0121	1.0117	1.0112	1.0109	1.0105	1.0102	1.0098
320	1.0167	1.0161	1.0156	1.0151	1.0146	1.0142	1.0138	1.0133	1.0129	1.0124	1.0119	1.0116	1.0112
360	1.0187	1.0181	1.0175	1.0169	1.0164	1.0159	1.0154	1.0149	1.0144	1.0139	1.0134	1.0129	1.0125
400	1.0208	1.0201	1.0195	1.0189	1.0182	1.0177	1.0171	1.0165	1.0160	1.0154	1.0149	1.0143	1.0138
440	1.0228	1.0220	1.0213	1.0207	1.0200	1.0193	1.0187	1.0181	1.0175	1.0168	1.0162	1.0156	1.0151
480	1.0248	1.0239	1.0232	1.0225	1.0218	1.0211	1.0204	1.0197	1.0190	1.0183	1.0176	1.0169	1.0163
520	1.0269	1.0258	1.0251	1.0243	1.0235	1.0228	1.0220	1.0212	1.0204	1.0197	1.0190	1.0183	1.0177
560	1.0289	1.0278	1.0269	1.0261	1.0252	1.0245	1.0236	1.0227	1.0219	1.0211	1.0204	1.0196	1.0189
600	1.0309	1.0298	1.0288	1.0279	1.0270	1.0261	1.0251	1.0242	1.0233	1.0225	1.0217	1.0209	1.0201
640	1.0329	1.0317	1.0307	1.0296	1.0287	1.0277	1.0267	1.0257	1.0248	1.0239	1.0230	1.0222	1.0214
680	1.0350	1.0337	1.0325	1.0314	1.0304	1.0293	1.0282	1.0272	1.0262	1.0253	1.0244	1.0235	1.0226
720	1.0369	1.0355	1.0343	1.0331	1.0320	1.0309	1.0298	1.0287	1.0276	1.0266	1.0257	1.0247	1.0237
760	1.0388	1.0374	1.0361	1.0349	1.0336	1.0324	1.0313	1.0301	1.0290	1.0280	1.0269	1.0259	1.0249
800	1.0408	1.0393	1.0380	1.0366	1.0353	1.0340	1.0327	1.0315	1.0303	1.0292	1.0281	1.0271	1.0260
840	1.0427	1.0412	1.0396	1.0382	1.0368	1.0355	1.0342	1.0329	1.0317	1.0306	1.0294	1.0283	1.0272
880	1.0446	1.0430	1.0414	1.0399	1.0384	1.0370	1.0356	1.0343	1.0330	1.0318	1.0306	1.0294	1.0282
920	1.0464	1.0448	1.0431	1.0415	1.0400	1.0385	1.0371	1.0357	1.0343	1.0330	1.0317	1.0305	1.0293
960	1.0483	1.0465	1.0448	1.0431	1.0415	1.0400	1.0385	1.0370	1.0356	1.0342	1.0329	1.0315	1.0303
1000	1.0501	1.0481	1.0463	1.0446	1.0429	1.0413	1.0398	1.0383	1.0369	1.0354	1.0340	1.0326	1.0313
1040	1.0518	1.0498	1.0480	1.0461	1.0444	1.0428	1.0412	1.0396	1.0381	1.0365	1.0350	1.0336	1.0323
1080	1.0535	1.0514	1.0495	1.0476	1.0459	1.0442	1.0425	1.0409	1.0393	1.0377	1.0361	1.0346	1.0333
1120	1.0552	1.0530	1.0510	1.0491	1.0472	1.0454	1.0437	1.0420	1.0404	1.0388	1.0372	1.0357	1.0343
1160	1.0569	1.0546	1.0525	1.0505	1.0485	1.0467	1.0450	1.0432	1.0415	1.0399	1.0383	1.0367	1.0352
1200	1.0585	1.0562	1.0540	1.0519	1.0499	1.0479	1.0461	1.0443	1.0426	1.0410	1.0393	1.0377	1.0361
1240	1.0601	1.0577	1.0554	1.0532	1.0512	1.0492	1.0473	1.0455	1.0437	1.0420	1.0403	1.0386	1.0370
1280	1.0618	1.0592	1.0568	1.0546	1.0524	1.0504	1.0485	1.0466	1.0448	1.0430	1.0412	1.0395	1.0379
1320	1.0633	1.0607	1.0583	1.0559	1.0536	1.0515	1.0496	1.0477	1.0457	1.0439	1.0421	1.0404	1.0386
1360	1.0647	1.0621	1.0597	1.0572	1.0548	1.0527	1.0506	1.0487	1.0467	1.0449	1.0431	1.0412	1.0394
1400	1.0661	1.0635	1.0610	1.0585	1.0560	1.0537	1.0516	1.0497	1.0477	1.0458	1.0439	1.0420	1.0402
1440	1.0676	1.0648	1.0622	1.0596	1.0571	1.0548	1.0527	1.0506	1.0486	1.0466	1.0447	1.0428	1.0410
1480	1.0690	1.0662	1.0635	1.0609	1.0583	1.0560	1.0538	1.0515	1.0494	1.0474	1.0454	1.0435	1.0417
1520	1.0704	1.0675	1.0647	1.0620	1.0594	1.0570	1.0547	1.0525	1.0503	1.0482	1.0462	1.0443	1.0425
1560	1.0717	1.0687	1.0658	1.0631	1.0605	1.0579	1.0556	1.0534	1.0511	1.0490	1.0470	1.0450	1.0432

TABLE 16-A F_{pv} SUPERCOMPRESSIBILITY FACTORS (Continued)

Base data—0.6 specific gravity hydrocarbon gas

P_f, psig	Temperature, °F												
	125	130	135	140	145	150	155	160	165	170	175	180	185
1600	1.0730	1.0699	1.0670	1.0642	1.0615	1.0589	1.0566	1.0543	1.0520	1.0498	1.0477	1.0457	1.0439
1640	1.0743	1.0711	1.0681	1.0652	1.0625	1.0597	1.0574	1.0551	1.0528	1.0505	1.0484	1.0464	1.0445
1680	1.0754	1.0721	1.0691	1.0662	1.0634	1.0606	1.0582	1.0558	1.0535	1.0512	1.0491	1.0470	1.0450
1720	1.0765	1.0732	1.0701	1.0672	1.0643	1.0615	1.0591	1.0567	1.0542	1.0519	1.0497	1.0476	1.0456
1760	1.0776	1.0742	1.0711	1.0681	1.0652	1.0624	1.0598	1.0574	1.0549	1.0525	1.0503	1.0482	1.0462
1800	1.0786	1.0752	1.0720	1.0690	1.0659	1.0631	1.0605	1.0580	1.0556	1.0531	1.0508	1.0487	1.0466
1840	1.0796	1.0761	1.0729	1.0698	1.0667	1.0639	1.0612	1.0586	1.0562	1.0537	1.0514	1.0492	1.0471
1880	1.0805	1.0769	1.0737	1.0706	1.0675	1.0647	1.0618	1.0592	1.0567	1.0542	1.0519	1.0497	1.0475
1920	1.0813	1.0777	1.0745	1.0713	1.0682	1.0653	1.0625	1.0598	1.0572	1.0546	1.0523	1.0501	1.0479
1960	1.0820	1.0784	1.0752	1.0719	1.0688	1.0659	1.0630	1.0602	1.0576	1.0550	1.0526	1.0505	1.0482
2000	1.0826	1.0790	1.0757	1.0724	1.0693	1.0664	1.0635	1.0607	1.0580	1.0554	1.0530	1.0508	1.0485
2040	1.0833	1.0796	1.0762	1.0729	1.0698	1.0669	1.0640	1.0611	1.0584	1.0558	1.0534	1.0511	1.0488
2080	1.0839	1.0801	1.0767	1.0734	1.0703	1.0673	1.0644	1.0615	1.0588	1.0562	1.0537	1.0514	1.0490
2120	1.0843	1.0807	1.0772	1.0738	1.0707	1.0676	1.0647	1.0619	1.0590	1.0565	1.0540	1.0516	1.0492
2160	1.0847	1.0811	1.0776	1.0742	1.0711	1.0679	1.0650	1.0621	1.0592	1.0567	1.0542	1.0518	1.0493
2200	1.0851	1.0814	1.0780	1.0746	1.0713	1.0681	1.0652	1.0623	1.0594	1.0569	1.0544	1.0520	1.0495
2240	1.0854	1.0817	1.0782	1.0748	1.0715	1.0683	1.0654	1.0625	1.0596	1.0570	1.0545	1.0521	1.0495
2280	1.0856	1.0819	1.0784	1.0750	1.0717	1.0685	1.0655	1.0626	1.0597	1.0571	1.0546	1.0521	1.0496
2320	1.0857	1.0820	1.0785	1.0751	1.0719	1.0686	1.0655	1.0626	1.0597	1.0571	1.0546	1.0522	1.0496
2360	1.0858	1.0821	1.0786	1.0752	1.0719	1.0687	1.0656	1.0627	1.0598	1.0572	1.0547	1.0522	1.0497
2400	1.0859	1.0822	1.0787	1.0752	1.0719	1.0687	1.0657	1.0628	1.0599	1.0572	1.0546	1.0521	1.0496
2440	1.0859	1.0822	1.0787	1.0752	1.0719	1.0687	1.0657	1.0628	1.0599	1.0572	1.0546	1.0521	1.0496
2480	1.0858	1.0822	1.0786	1.0751	1.0718	1.0687	1.0657	1.0627	1.0598	1.0571	1.0545	1.0520	1.0495
2520	1.0856	1.0820	1.0784	1.0749	1.0716	1.0685	1.0656	1.0627	1.0598	1.0571	1.0544	1.0519	1.0493
2560	1.0854	1.0818	1.0782	1.0747	1.0715	1.0683	1.0654	1.0625	1.0596	1.0569	1.0542	1.0517	1.0492
2600	1.0852	1.0814	1.0779	1.0745	1.0713	1.0681	1.0651	1.0622	1.0594	1.0566	1.0540	1.0515	1.0490
2640	1.0848	1.0811	1.0776	1.0742	1.0710	1.0679	1.0648	1.0619	1.0591	1.0563	1.0537	1.0513	1.0488
2680	1.0844	1.0807	1.0772	1.0739	1.0707	1.0676	1.0645	1.0616	1.0588	1.0560	1.0534	1.0511	1.0485
2720	1.0840	1.0803	1.0768	1.0734	1.0703	1.0673	1.0643	1.0614	1.0586	1.0558	1.0532	1.0507	1.0482
2760	1.0835	1.0799	1.0764	1.0730	1.0699	1.0669	1.0639	1.0610	1.0582	1.0555	1.0528	1.0503	1.0478
2800	1.0830	1.0793	1.0759	1.0726	1.0695	1.0664	1.0635	1.0606	1.0577	1.0551	1.0525	1.0499	1.0474
2840	1.0824	1.0788	1.0754	1.0721	1.0690	1.0659	1.0631	1.0602	1.0573	1.0547	1.0521	1.0495	1.0470
2880	1.0819	1.0783	1.0749	1.0716	1.0685	1.0653	1.0626	1.0598	1.0569	1.0542	1.0516	1.0491	1.0466
2920	1.0812	1.0777	1.0743	1.0710	1.0679	1.0648	1.0620	1.0592	1.0564	1.0537	1.0511	1.0486	1.0461
2960	1.0805	1.0770	1.0737	1.0704	1.0673	1.0643	1.0614	1.0586	1.0558	1.0531	1.0506	1.0481	1.0456
3000	1.0797	1.0764	1.0731	1.0698	1.0667	1.0637	1.0608	1.0580	1.0554	1.0527	1.0501	1.0476	1.0451

Factors for intermediate values of pressure and temperature should be interpolated.

[The increments of the pressure argument were increased from 20 psi to 40 psi in this condensed version of the referenced[2] table; interpolation without loss of essential accuracy is still possible. Values at smaller increments are also available.[9]—EDITOR]

TABLE 16-B SUPERCOMPRESSIBILITY PRESSURE ADJUSTMENTS, ΔP

Based on: carbon dioxide and nitrogen contents and specific gravity
Pressure Adjustment Index = $f_{pg} = G - 13.84\ X_c + 5.420\ X_n$

Pressure Adjustment Index, f_{pg}	Pressure, psig (adjustment at 0 psig = 0)							
	200	600	1000	1400	1800	2200	2600	3000
−0.7	−11.32	−33.97	−56.62	−79.27	−101.92	−124.56	−147.21	−169.86
−0.6	−10.50	−31.49	−52.49	−73.49	−94.48	−115.48	−136.47	−157.47
−0.5	−9.67	−29.00	−48.33	−67.66	−86.99	−106.33	−125.66	−144.99
−0.4	−8.83	−26.48	−44.13	−61.78	−79.43	−97.09	−114.74	−132.39
−0.3	−7.98	−23.93	−39.89	−55.85	−71.80	−87.76	−103.71	−119.67
−0.2	−7.12	−21.37	−35.62	−49.87	−64.12	−78.36	−92.61	−106.86
−0.1	−6.26	−18.78	−31.30	−43.82	−56.34	−68.86	−81.38	−93.90
0	−5.39	−16.17	−26.95	−37.73	−48.51	−59.29	−70.07	−80.85
+0.1	−4.51	−13.54	−22.56	−31.58	−40.61	−49.63	−58.66	−67.68
+0.2	−3.63	−10.88	−18.13	−25.38	−32.63	−39.89	−47.14	−54.39
+0.3	−2.73	−8.20	−13.66	−19.12	−24.59	−30.05	−35.52	−40.98
+0.4	−1.83	−5.49	−9.15	−12.81	−16.47	−20.13	−23.79	−27.45
+0.5	−0.92	−2.76	−4.60	−6.43	−8.27	−10.11	−11.95	−13.79
+0.6	0	0	0	0	0	0	0	0
+0.7	0.93	2.78	4.64	6.49	8.35	10.20	12.06	13.91
+0.8	1.86	5.59	9.32	13.05	16.78	20.50	24.23	27.96
+0.9	2.81	8.42	14.04	19.66	25.27	30.89	36.50	42.12
+1.0	3.76	11.29	18.81	26.33	33.86	41.38	48.91	56.43
+1.1	4.73	14.18	23.63	33.08	42.53	51.99	61.44	70.89
+1.2	5.70	17.09	28.49	39.89	51.28	62.68	74.07	85.47
+1.3	6.68	20.04	33.40	46.76	60.12	73.48	86.84	100.20
+1.4	7.67	23.01	38.35	53.69	69.03	84.37	99.71	115.05
+1.5	8.67	26.01	43.35	60.69	78.03	95.37	112.71	130.05
+1.6	9.68	29.04	48.40	67.76	87.12	106.48	125.84	145.20
+1.7	10.70	32.10	53.50	74.90	96.30	117.70	139.10	160.50
+1.8	11.73	35.19	58.65	82.11	105.57	129.03	152.49	175.95
+1.9	12.77	38.31	63.85	89.39	114.93	140.47	166.01	191.55
+2.0	13.82	41.46	69.10	96.74	124.38	152.02	179.66	207.30

Factors for intermediate values of pressure adjustment index and pressure should be interpolated. [Referenced[2] table gives data at 200 psi intervals.—EDITOR]

TABLE 16-C SUPERCOMPRESSIBILITY TEMPERATURE ADJUSTMENTS, ΔT

Based on: carbon dioxide and nitrogen contents and specific gravity
Temperature Adjustment Index = $f_{tg} = G - 0.472\ X_c - 0.793\ X_n$

Temperature Adjustment Index, f_{tg}	Temperature, °F						Temperature Adjustment Index, f_{tg}	Temperature, °F					
	0	40	80	120	160	200		0	40	80	120	160	200
0.45	75.16	81.70	88.24	94.77	101.31	107.84	0.60	0	0	0	0	0	0
0.46	69.41	75.45	81.49	87.52	93.56	99.59	0.61	−4.27	−4.64	−5.01	−5.38	−5.75	−6.12
0.47	63.76	69.30	74.84	80.39	85.93	91.48	0.62	−8.45	−9.19	−9.93	−10.66	−11.40	−12.13
0.48	58.24	63.30	68.36	73.43	78.49	83.56	0.63	−12.57	−13.66	−14.75	−15.85	−16.94	−18.03
0.49	52.81	57.40	61.99	66.58	71.18	75.77	0.64	−16.61	−18.05	−19.49	−20.94	−22.38	−23.83
0.50	47.52	51.65	55.78	59.91	64.05	68.18	0.65	−20.57	−22.36	−24.15	−25.94	−27.73	−29.52
0.51	42.33	46.01	49.69	53.37	57.05	60.73	0.66	−24.47	−26.60	−28.72	−30.85	−32.98	−35.11
0.52	37.25	40.48	43.72	46.96	50.20	53.44	0.67	−28.29	−30.76	−33.22	−35.68	−38.14	−40.60
0.53	32.26	35.07	37.88	40.68	43.49	46.29	0.68	−32.06	−34.84	−37.63	−40.42	−43.21	−46.00
0.54	27.38	29.76	32.14	34.52	36.90	39.28	0.69	−35.75	−38.86	−41.97	−45.08	−48.19	−51.30
0.55	22.60	24.56	26.52	28.49	30.45	32.42	0.70	−39.38	−42.81	−46.23	−49.66	−53.08	−56.51
0.56	17.90	19.46	21.01	22.57	24.12	25.68	0.71	−42.95	−46.69	−50.42	−54.16	−57.90	−61.63
0.57	13.29	14.45	15.61	16.76	17.92	19.07	0.72	−46.46	−50.50	−54.54	−58.58	−62.62	−66.66
0.58	8.78	9.54	10.30	11.07	11.83	12.59	0.73	−49.91	−54.25	−58.59	−62.93	−67.27	−71.61
0.59	4.35	4.73	5.10	5.48	5.86	6.24	0.74	−53.31	−57.95	−62.59	−67.22	−71.86	−76.49
							0.75	−56.67	−61.60	−66.53	−71.46	−76.38	−81.31

Factors for intermediate values of temperature adjustment index and temperature should be interpolated. [Referenced[2] table gives data at 20°F intervals.—EDITOR]

7/40

TABLE 16-D SUPERCOMPRESSIBILITY PRESSURE ADJUSTMENTS, ΔP

Based on: carbon dioxide content, heating value and specific gravity

$$\text{Pressure Adjustment Index} = f_{ph} = G - 0.5688\,\frac{H_w}{1000} - 3.690\,X_c$$

Pressure Adjustment Index, f_{ph}	Pressure, psig (adjustment at 0 psig = 0)							
	200	600	1000	1400	1800	2200	2600	3000
−0.22	−11.25	−33.76	−56.27	−78.78	−101.29	−123.79	−146.30	−168.81
−0.20	−10.27	−30.80	−51.34	−71.88	−92.41	−112.95	−133.48	−154.02
−0.18	−9.27	−27.82	−46.36	−64.90	−83.45	−101.99	−120.54	−139.08
−0.16	−8.27	−24.80	−41.33	−57.86	−74.39	−90.93	−107.46	−123.99
−0.14	−7.25	−21.74	−36.24	−50.74	−65.23	−79.73	−94.22	−108.72
−0.12	−6.22	−18.66	−31.10	−43.54	−55.98	−68.42	−80.86	−93.30
−0.10	−5.18	−15.55	−25.91	−36.27	−46.64	−57.00	−67.37	−77.73
−0.08	−4.13	−12.40	−20.66	−28.92	−37.19	−45.45	−53.72	−61.98
−0.06	−3.07	−9.21	−15.35	−21.49	−27.63	−33.77	−39.91	−46.05
−0.04	−2.00	−5.99	−9.98	−13.98	−17.97	−21.96	−25.96	−29.95
−0.02	−0.91	−2.74	−4.56	−6.38	−8.21	−10.03	−11.85	−13.68
0.00	0.18	0.56	0.92	1.30	1.67	2.04	2.41	2.78
+0.02	1.29	3.88	6.47	9.06	11.65	14.24	16.82	19.41
+0.04	2.42	7.25	12.08	16.91	21.74	26.58	31.41	36.24
+0.06	3.55	10.65	17.75	24.85	31.95	39.05	46.15	53.25
+0.08	4.70	14.09	23.48	32.87	42.26	51.66	61.05	70.44
+0.10	5.86	17.57	29.28	40.99	52.70	64.42	76.13	87.84
+0.12	7.03	21.08	35.14	49.20	63.25	77.31	91.36	105.42
+0.14	8.22	24.65	41.08	57.51	73.94	90.38	106.81	123.24
+0.16	9.42	28.25	47.08	65.91	84.74	103.58	122.41	141.24
+0.18	10.63	31.89	53.15	74.41	95.67	116.93	138.19	159.45
+0.20	11.86	35.57	59.29	83.01	106.72	130.44	154.15	177.87
+0.22	13.10	39.30	65.50	91.70	117.90	144.10	170.30	196.50

Factors for intermediate values of pressure adjustment index and pressure should be interpolated. [Referenced[2] table gives data at 200 psi intervals.—EDITOR]

TABLE 16-E SUPERCOMPRESSIBILITY TEMPERATURE ADJUSTMENTS, ΔT

Based on: carbon dioxide content, heating values and specific gravity

$$\text{Temperature Adjustment Index} = f_{th} = G + 1.814\,\frac{H_w}{1000} + 2.641\,X_c$$

Temperature Adjustment Index, f_{th}	Temperature, °F						Temperature Adjustment Index, f_{th}	Temperature, °F					
	0	40	80	120	160	200		0	40	80	120	160	200
2.10	56.03	60.90	65.77	70.64	75.52	80.39	2.66	−15.49	−16.84	−18.18	−19.53	−20.88	−22.22
2.12	53.13	57.75	62.37	66.99	71.61	76.23	2.68	−17.68	−19.22	−20.75	−22.29	−23.83	−25.36
2.14	50.19	54.55	58.91	63.28	67.64	72.01	2.70	−19.85	−21.58	−23.30	−25.03	−26.75	−28.48
2.16	47.29	51.40	55.51	59.62	63.74	67.85	2.72	−22.00	−23.91	−25.82	−27.74	−29.65	−31.56
2.18	44.46	48.33	52.20	56.06	59.93	63.80	2.74	−24.12	−26.22	−28.32	−30.42	−32.51	−34.61
2.20	41.64	45.26	48.89	52.51	56.13	59.75	2.76	−26.23	−28.51	−30.79	−33.07	−35.35	−37.63
2.22	38.86	42.24	45.61	48.99	52.37	55.75	2.78	−28.31	−30.78	−33.24	−35.70	−38.16	−40.62
2.24	36.10	39.24	42.37	45.51	48.65	51.79	2.80	−30.38	−33.02	−35.66	−38.30	−40.94	−43.59
2.26	33.37	36.28	39.18	42.08	44.98	47.88	2.82	−32.42	−35.24	−38.06	−40.88	−43.70	−46.52
2.28	30.67	33.34	36.01	38.67	41.34	44.01	2.84	−34.45	−37.45	−40.45	−43.44	−46.44	−49.43
2.30	28.01	30.44	32.88	35.32	37.75	40.19	2.86	−36.46	−39.63	−42.80	−45.97	−49.14	−52.31
2.32	25.37	27.58	29.78	31.99	34.19	36.40	2.88	−38.45	−41.80	−45.14	−48.48	−51.82	−55.17
2.34	22.76	24.74	26.72	28.70	30.68	32.66	2.90	−40.42	−43.94	−47.45	−50.96	−54.48	−57.99
2.36	20.18	21.93	23.68	25.44	27.19	28.95	2.92	−42.37	−46.06	−49.74	−53.42	−57.11	−60.79
2.38	17.62	19.16	20.69	22.22	23.75	25.28	2.94	−44.31	−48.16	−52.01	−55.86	−59.72	−63.57
2.40	15.09	16.40	17.72	19.03	20.34	21.65	2.96	−46.23	−50.25	−54.27	−58.29	−62.31	−66.33
2.42	12.59	13.69	14.78	15.88	16.98	18.07	2.98	−48.12	−52.30	−56.48	−60.67	−64.85	−69.04
2.44	10.12	11.00	11.88	12.76	13.64	14.52	3.00	−50.00	−54.35	−58.70	−63.05	−67.39	−71.74
2.46	7.67	8.34	9.00	9.67	10.34	11.00	3.02	−51.89	−56.40	−60.91	−65.42	−69.94	−74.45
2.48	5.24	5.70	6.16	6.61	7.07	7.52	3.04	−53.73	−58.40	−63.07	−67.74	−72.42	−77.09
2.50	2.85	3.10	3.34	3.59	3.84	4.08	3.06	−55.57	−60.40	−65.23	−70.06	−74.90	−79.73
2.52	0.47	0.51	0.55	0.60	0.64	0.68	3.08	−57.36	−62.35	−67.34	−72.33	−77.31	−82.30
2.54	−1.88	−2.04	−2.20	−2.37	−2.53	−2.69	3.10	−59.16	−64.30	−69.44	−74.59	−79.73	−84.88
2.56	−4.20	−4.57	−4.94	−5.30	−5.67	−6.03	3.12	−60.95	−66.25	−71.55	−76.85	−82.15	−87.45
2.58	−6.50	−7.07	−7.64	−8.20	−8.77	−9.33	3.14	−62.70	−68.15	−73.60	−79.05	−84.51	−89.96
2.60	−8.79	−9.55	−10.31	−11.08	−11.84	−12.61	3.16	−64.45	−70.05	−75.65	−81.26	−86.86	−92.47
2.62	−11.04	−12.00	−12.96	−13.92	−14.89	−15.85	3.18	−66.19	−71.95	−77.71	−83.46	−89.22	−94.97
2.64	−13.28	−14.43	−15.58	−16.74	−17.89	−19.05	3.20	−67.90	−73.80	−79.70	−85.61	−91.51	−97.42

Factors for intermediate values of temperature adjustment index and temperature should be interpolated. [Referenced[2] table gives data at 20°F intervals.—EDITOR

TABLE 17 F_m, MANOMETER FACTORS

(Factor = 1.0 at 0 psig for all values of G)

Specific Gravity, G	Flowing pressure, psig					
	500	1000	1500	2000	2500	3000
	Ambient temperature = 0 °F					
.55	0.9989	0.9976	0.9960	0.9943	0.9930	0.9921
.60	.9988	.9972	.9952	.9932	.9919	.9910
.65	.9987	.9967	.9941	.9920	.9908	.9900
.70	.9985	.9961	.9927	.9907	.9896	.9890
	Ambient temperature = 40 °F					
.55	.9990	.9979	.9967	.9954	.9942	.9932
.60	.9989	.9976	.9962	.9946	.9933	.9923
.65	.9988	.9973	.9955	.9937	.9923	.9913
.70	.9987	.9970	.9947	.9926	.9912	.9903
.75	.9986	.9965	.9937	.9915	.9902	.9893
	Ambient temperature = 80 °F					
.55	.9991	.9981	.9971	.9960	.9950	.9941
.60	.9990	.9979	.9967	.9955	.9943	.9933
.65	.9989	.9977	.9963	.9948	.9935	.9925
.70	.9988	.9974	.9958	.9940	.9926	.9915
.75	.9987	.9971	.9951	.9931	.9916	.9906
	Ambient temperature = 120 °F					
.55	.9992	.9983	.9974	.9965	.9956	.9948
.60	.9991	.9981	.9971	.9960	.9950	.9941
.65	.9990	.9979	.9967	.9955	.9944	.9934
.70	.9989	.9977	.9963	.9950	.9937	.9926
.75	0.9988	0.9975	0.9959	0.9943	0.9929	0.9918

Factors for intermediate values of pressure, temperature, and specific gravity should be interpolated.

Note: This table is for use with mercury manometer type recording gages that have gas in contact with the mercury surface.

Computing Orifice Flow Meter Charts

Orifice meter gages measuring gas flow have two pens, one for recording differential, the other for static pressure. Twenty-four hour chart rotations are commonly used. Where extreme fluctuating flows are encountered, 2, 3, 4, 6, 8, and 12 hour chart rotations are often desirable; strip charts may also be used. Where flow conditions are steady enough to give readable static pressure and differential lines, chart rotations up to 31 days may be used.

Charts may be computed by the following three methods:

Manual or Observation Method. On charts graduated in inches of water differential and pounds of static pressure, average values of h_w and p_f are used for each hour, and are written in each hourly spacing or equivalent time period, depending upon chart rotation. Hourly extensions should be inserted on the chart, to correspond to the values of pressure and differential.

For each hour, the extension equals $(h_w p_f)^{0.5}$ (where: h_w = differential pressure, in. of water; p_f = static pressure psia). Extension table books are published by several meter manufacturers, with a separate page for each even pound of pressure and with differentials from zero to 100 in. After extensions are written on a chart for each pressure and differential reading, they are totaled, and the result is multiplied by a combined coefficient with various connections for temperature, specific gravity, and orifice coefficient.

Where pressures or differentials show wide fluctuation it is common practice to use 15-min extensions with 15-min coefficients. This procedure will be closer to the average square root than to the square root of the average.

When computing **square root charts,** the same basic procedure is used to get the total volume. For example, if the pressure pen reads *6* and the differential pen *7*, the extension for any hour or time interval would be their product, or *42*. No extension tables are required for computing square root charts.

Values on the square root chart are equal to the square root of the equivalent values on a regular chart for installations where the differential range is 100 in. w.c. and the pressure range is 100 psia.

It is apparent that, without changing the meter setting, the value $(h_w)^{0.5}$ may be read directly. However, the square root of the *absolute* static pressure must be known. On meters using a direct reading chart, the static pressure pen is set so that zero *gage* pressure is zero on the chart. Thus, to obtain the *absolute* static pressure, the gage pressure reading on the direct reading chart is added to the atmospheric pressure at point of measurement. When using the square root chart on any given scale, the static pressure pen is set so that *absolute* zero is zero on the chart. This is accomplished by setting the pen at zero *gage* pressure at a point on the scale which corresponds to the square root of the *absolute* pressure—*with no pressure on the gage.* Thus, if complete vacuum could be applied to the pressure element, the pen would read zero on the chart. For example, on a 100 psia range meter, to make the gage read absolute pressure, where atmospheric pressure reads 14.40 psia at the point of measurement, the pen is set so that it reads $(14.40)^{0.5}$ on the chart when zero gage pressure is applied to pressure element. The actual deflection of the pen per pound pressure change is not affected, only the initial zero setting. The reading that corresponds to 14.4 psia (zero gage pressure) on a direct reading chart becomes $(14.40)^{0.5}$ = 3.795 on the square root chart. Therefore, to arrive at the reading on the square root chart, using a 100 psia element, it is necessary only to set the pen for zero gage pressure at 3.795, and then read the chart directly.

Thus far we have considered only installations where the differential range is 100 in. and the static pressure range 100 psia. For other pressure and differential ranges, the same 0–10 square root chart may be used. However, a chart constant must be multiplied into the coefficient. For example, if the differential range of the meter is 50 in. w.c., a reading of 50 in. would be, for calculating purposes, its square root, or 7.071. Since the square root chart is reading 10, its factor would be 0.7071. Thus, the chart reading must be multiplied by 0.7071. Likewise, for a static pressure range of 50 psia, the factor used in reading the static would be 0.7071. Considering the differential reading on the square root chart as h_{ws} and the static reading as p_{fs}, the computation of volume would be given by:

$$Q_h = C_s \times 0.7071 h_{ws} \times 0.7071 p_{fs} = 0.5000 C_s h_{ws} p_{fs}$$

Here, the figure 0.5000 is really a factor to be combined with the chart coefficient. The proper factor should be incorporated with the chart coefficient regardless of what method is used in reading the charts. Table 7-5 shows square root chart coefficients for various static and differential pressure ranges of meters, and pressure pen settings for atmospheric pressures of 14.4 psia.

Table 7-5 Chart Coefficients and Static Settings for Square Root Charts

(base: 100 in. w.c. differential and 100 psia static)

Diff. range, inches water	Static range, psia	Chart coefficient, C_s*	Chart values for setting static pen at atmospheric pressure† at 14.4	at 14.7
5	No static	0.2236	—	—
10	No static	.3162	—	—
20	30 in. vacuum to 35 psia	.3162	—	—
20	No static	.4472	—	—
20	50	.3162	5.367	5.422
20	100	.4472	3.795	3.834
20	250	.7071	2.400	2.425
50	50	.5000	5.367	5.422
50	100	0.7071	3.795	3.834
50	250	1.1180	2.400	2.425
50	500	1.5811	1.697	1.715
100	50	0.7071	5.367	5.422
100	100	1.0000	3.795	3.834
100	250	1.5811	2.400	2.425
100	500	2.2361	1.697	1.715

* Chart coefficient, $C_s = 0.01\,(h_r p_{ar})^{0.5}$
where: h_r = differential range of chart, in. w.c.
p_{ar} = static range of chart, psia

† Chart values for setting static pen Atmospheric setting = $10(A_p/R_p)^{0.5}$
where: A_p = atmospheric pressure, psia
R_p = pressure range of spring, lb

The equation for calculating gas flow, measured by an orifice meter using a square root chart, follows:

$$Q = C'C_s p_{fs} h_{ws}$$

where:

Q = flow rate at pressure base and temperature base used in determining coefficient C'
C' = coefficient
C_s = chart factor = $0.01\,(h_r p_{ar})^{0.5}$

$$p_{fs} = 10\left(\frac{p_f' + A_p}{p_{ar}}\right)^{0.5} = 10\left(\frac{p_f}{p_{ar}}\right)^{0.5}$$

$$h_{ws} = 10\left(\frac{h_w}{h_r}\right)^{0.5}$$

p_{fs} = pressure reading on the square root chart
p_f' = static pressure, psig
A_p = atmospheric pressure, psia
p_{ar} = static pressure range of meter or chart, psia
p_f = static pressure, psia
h_{ws} = differential reading on the square root chart
h_w = differential pressure, in. w.c.
h_r = differential range of meter or chart, in. w.c.

Planimeter Method. The planimeter is a mechanical device equipped with a scale type wheel. By tracing differential pressure in one operation and then tracing static pressure, the average values can be obtained. This simple radial planimeter can be used, if the static pressure and differential pressure values do not vary appreciably during the recorded time.

The **square root planimeter** is an improved type, operated in the same manner as the simple radial planimeter. It gives the average of the square root values of the differential and static pressures accurately, providing the static pressure does not vary greatly. In practice, if either the static pressure or the differential pressure record remains reasonably constant, the square root planimeter provides a fairly accurate, convenient, and reliable means of interpreting chart records.

The planimeter is faster than the hourly observation method of reading charts and has taken its place for obtaining operating information. Figure 7-10 shows a square root planimeter which is often used for computing flow from orifice meter charts when the number of charts does not warrant purchase of an integrating instrument.

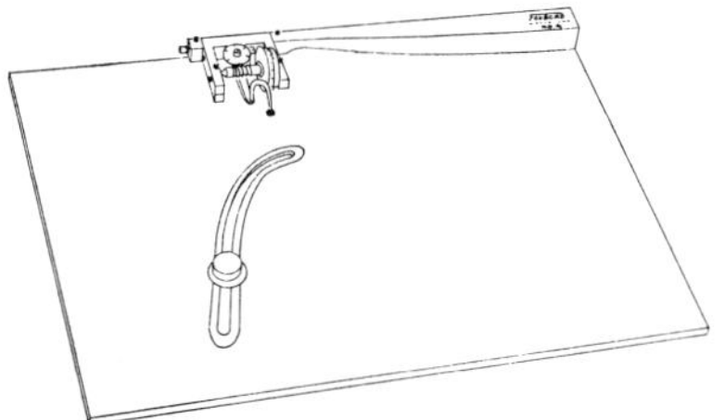

Fig. 7-10 Square root planimeter. (Foxboro Co.)

Integrator Methods. These are motor-operated, mechanical devices which employ the square root cam principle. Rotating plates drive a dial counter. It accumulates the average of the infinite number of instantaneous square roots of products of absolute static pressures and differential pressures for an exact time period.

The operator traces over the lines of static pressure and differential pressure recorded with the integrator pens. The speed of the chart plate is governed by a rheostat type foot control.

The fountain pens on the instrument superimpose ink lines of contrasting shades over the original chart records, thus leaving a permanent check. The continuous type dial is read at the start of a tracing and again at the end. The difference between the two readings is the summation of the average instantaneous square root extensions of static pressure and differential pressure by time, or a mathematically perfect reading of the recorded values by the meter, regardless of how much the pressure and differential may vary.

Time Lag. The time lag between the static pressure and differential pressure pen on the orifice meter is a very important factor in all methods of chart calculation. When charts are read by observation or planimeter method, the lag should be held comparable to where the two pens pass each other on the meter.

The differential pressure pen follows the arc line of the instrument chart. On a 24-hr chart, the static pressure pen is set 15 min in advance of the differential pen, at the zero position. This gives the same linear distance between the two pens throughout the chart radius. The integrated pens should be set in a like manner, to assure synchronization of the pressure and differential pens when integrating charts.

Some integrators are designed with a machine constant of one (1) on a 500 psia-100 in. w.c. chart. In this case, the subtracted difference in counter readings is the total extension on a 24-hr chart. Machine constants for various pressure and differential ranges may be multiplied by the counter difference reading to arrive at the total extension, or they may be incorporated with the orifice coefficient when computing the total volume.

MAX d/D ____
MAX. PLATE SIZE ____ "

STATION NAME ____
STATION No. ____ METER No. ____
DATE ____ TIME, HR ____ MIN ____ A. M. / P. M. ____
TYPE OF SETTING SKETCH 1 ☐ SKETCH 2 ☐ SKETCH 3 ☐

TUBE DIAMETER MICROMETER READINGS

TYPE OF FLANGE OR FITTING ____ SERIAL No. ____ PROPER ALIGNMENT YES ☐ NO* ☐

UPSTREAM					DOWNSTREAM				
			MAX d/D ____					MAX d/D ____	
POSITION	A	B	C	D	POSITION	E	F	G	H
|					|				
\					\				
/					/				
AVERAGE					AVERAGE				

UPSTREAM	AVERAGE MICROMETER READINGS	DOWNSTREAM
"	Average of All Inside Diameters	"
"	Published Diameter Schedule ____ Pipe	"
"	Difference Between Ave. & Published I D	"
%	Difference Between Ave. & Published I D	%
β	Maximum Allowable d/D Ratio (AGA Fig. 1)	β
"	Maximum Allowable Orifice Bore Diameter	"
"	Largest Micrometer Reading Obtained	"
"	Smallest Micrometer Reading Obtained	"
"	Out-of-Round Difference (Largest - Smallest)	"
%	Out-of-Round Difference " ÷ Published I D	%
β	Maximum Allowable d/D Ratio (AGA Fig. 1)	β
"	Maximum Allowable Orifice Bore Diameter	"

LEFT	RIGHT	FITTING PRESSURE TAPS	LEFT	RIGHT
"	"	Distance From Face of Orifice Pl. To ℄ Tap	"	"
"	"	Difference Between Distance & One Inch	"	"
β	β	Maximum Allowable d/D Ratio (AGA Fig. 8)	β	β
"	"	Maximum Allowable Orifice Bore Diameter	"	"

Left & Right Pressure Taps to be Determined by Facing Direction of Flow

METER TUBE LENGTHS

LENGTH	ACTUAL	.75 β REQUIREMENT
A	"	"
A′	"	"
B	"	"
C	"	"
C′		

MAX. d/D ____

STRAIGHTENING VANES

D
a
L

Number of Vanes ____
Length of Vanes (L) ____ "
10 x "a" Dimension ____ "
Within Tolerance Yes ☐ No* ☐
Diameter of Largest Vane (a) I D ____ "
¼ Meter Tube I D ____ "
Within Tolerance Yes ☐ No* ☐
MAX d/D ____

ORIFICE PLATE

Serial No. ____ Size ____ x ____ Edges Sharp: Yes ☐ No* ☐
Orifice Diameter ____ " ____ " ____ " Average ____ "
Thickness ____ O D ____ Buckle Within Tol. Yes ☐ No* ☐
Plate Should Be Replaced Yes ☐ No ☐

MISCELLANEOUS

Location of Thermometer Well:
Condition of Meter Tubes:
* (Explain)

Witnessed By ____ Test By ____

Fig. 7-11a Orifice meter tube inspection report (primary elements).

ORIFICE METER TEST REPORT

STATION NAME ______ MAX d/D ______
STATION No. ______ METER No. ______
DATE ______ TIME, HR ______ MIN ______ A. M. / P. M. ______
DATE LAST TEST: MAN. ______ PLATE ______ METER TUBE ______
GAUGE LINE LEAKAGE: DIFF. CHANGED ______ "/MIN STATIC CHANGED ______ #/MIN

DIFFERENTIAL ELEMENT

Make ______ Type ______
Chart Rotation: 24 Hr ☐ 7 Day ☐ 31 Day ☐
Gauge Range ______ Oper. Diff. ______
Pen Arc ______ Pen Lag. ______
Zero: Under Press. ______ Atm Press. ______
Friction: Under Press. ______ Atm Press. ______
Manometer Fluid: Water ☐ Mercury ☐

UP TEST		DOWN TEST	
Man.	Gauge	Man.	Gauge

PULSATION: Yes ☐ No ☐ No Test ☐

STATIC ELEMENT

Press. Conn.: P1 ☐ P2 ☐ Pm ☐
Gauge Range ______ Oper. Press. ______
Deadweight ☐ Spring Gauge ☐
Mercury Man. ☐ Test Gauge ☐

Test Gauge	Recorder	Inter. Press. Conn.
0		Tested Yes ☐ No ☐
		Location Correct
		Yes ☐ No* ☐

TEMPERATURE

Gauge Range ______ °F

Test	Recorder

SPECIFIC GRAVITY

Gauge Range ______

Test	Recorder

Sample Date

ORIFICE PLATE

No. ______ Size ______ " x ______ "
I D ______ ______ ______ Avg. ______ "
Plate Thickness ______ " O D ______ "
Edges Sharp Yes ☐ No* ☐
Plate Clean Yes ☐ No* ☐
Buckle Within Tol. Yes ☐ No* ☐

METER TUBES

Foreign Material in Tubes Yes* ☐ No ☐ Tap Holes Clean Yes ☐ No* ☐
Straightening Vanes Clean Yes ☐ No* ☐ Correct Gasket Yes ☐ No* ☐
Auto Valve Test ☐ Leaks Yes ☐ No* ☐ By Pass Valve Seal ☐ Lub. ☐ Oper. ☐

REMARKS

* (Explain)

Witnessed By ______ Test By ______

Fig. 7-11b Orifice meter test report (secondary elements).

Integrators are designed to read Foxboro, Metric, and Rockwell orifice meter charts. Separate pen arm assemblies are designed to follow the arc and the pen radius of these three makes of orifice meters. Static pen arm assemblies are designed with different set-back positions for various pressure ranges of charts and combinations of atmospheric pressures.

Special groups of cams can be incorporated on the instrument to correct for supercompressibility for a given average specific gravity and flowing temperature of gases, thereby taking the F_{pv} factor into account simultaneously with the assimilation of $(h_w)^{0.5} \times (p_f)^{0.5}$ values on the chart.

Integrators are approximately eight times faster than the hourly observation method. Claimed accuracy, when used by skilled personnel, is 0.1 per cent. To check accuracy on charts recording fluctuating flow it is necessary to use the 15-min observation method, thereby approaching the instantaneous reading given by the machine.

Electroscanner. This UGC Instruments Co. device is a high-capacity electronic flow meter chart integrator which employs a solid-state computer, coupled to an optical scanning system. A series of lenses on the periphery of a disk turns above the chart being processed. The relative positions of the disc and chart are such that the paths followed by the lenses across the face of the chart are the same as the paths followed by the meter pen arms in making the record on the chart. As each lens moves across the face of the chart, the portion of the chart that it views is focused on a photomultiplier tube. The chart rotates, and the timing relationship between the chart and the optical scan disk is such that, as one lens completes a scan across the chart, the rotation of the chart will present a different reading path to the following lens. Thus, whenever the chart has completed one revolution, the scanning disk has moved the optical system across the face of the chart 800 times and has taken 800 observations of the chart record. The photomultiplier tube senses the location of the differential pressure and static pressure records on the chart. The cumulative value of all 800 scans is totaled, and the resulting number is the chart extension.

The tube selected has a color sensitivity that does not register the grid lines. The scanner does not have to discriminate between the static and differential lines on the chart (it is necessary that the static pens on each meter be set to absolute pressure). It makes no difference which line is read first, since all readings are made in percentage of full-scale. The charts should be reasonably clear, since the optical system reads what it sees. The lines can cross, and no scanning or computing difficulty is experienced. The scanning and computing process produces an output that is in terms of per cent of chart scale, and the actual scaling factors are introduced in subsequent computation. The extension can be read on the front of the scanning cabinet, printed on the back of the chart after it is removed from the scan table, or reproduced on punched cards, paper tape, or magnetic tape.

No special operating skill, training, or judgment is involved in determining the readings; production rate is 250 flow charts per hour. Special calibration disks are provided that are standardized to ±0.19 per cent of full scale readings. The *Electroscanner* will reproduce repeatable readings on these standards within ±0.5 per cent of full scale.

Orifice Meter Testing Forms and Methods

Figure 7-11a shows a sample orifice meter tube inspection report form. It includes those dimensions which would be required only once for each station, unless changes in equipment were made. Certain other tests or inspections should be made periodically (Fig. 7-11b); the manufacturer's recommendations should be followed in testing these secondary elements.

PITOT TUBES

The velocity of a gas stream at the mouth of an impact tube is given by:[8]

$$U = \frac{11.13E}{F_{pv}}\left[\frac{hT}{pG}\right]^{0.5}$$

where:

U = velocity of fluid at tube, fps
E = instrument constant; for carefully constructed tubes, properly installed, $E = 1.00$
F_{pv} = supercompressibility factor
h = velocity head, in. w.c.
T = gas temperature, °R
p = static pressure, psia
G = specific gravity of gas, (air = 1.0)

The velocity of a fluid flowing in a pipe varies from a minimum at its surface to a maximum at its center. Consequently, to determine the quantity of fluid flowing, an average velocity must be obtained. To do so, the Pitot tube may be located at the pipe center, where the maximum velocity exists. For approximate work, involving the measurement of gas or liquid flow where turbulent flow is known to exist, the ratio of the mean to the maximum velocity may be assumed to be 0.82.

Another method is to place the Pitot tube at a point in the pipe where the actual velocity is equal to the mean velocity for the entire cross section. This point is usually assumed to be on the circumference of a circle of a radius 0.74 and concentric with a pipe of radius unity.

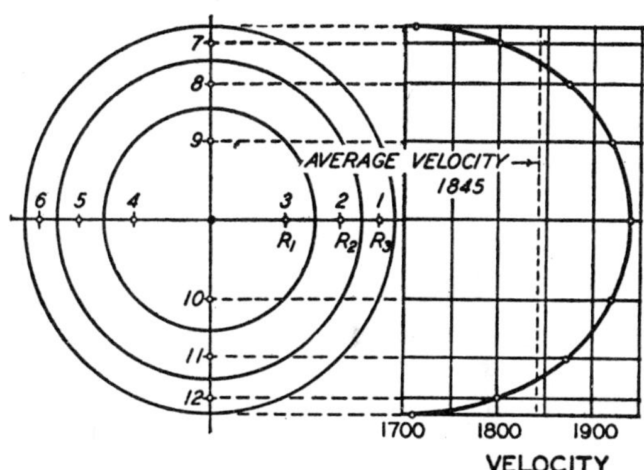

Fig. 7-12 Finding average velocity thru a round duct by means of a traverse. The average of the velocities in each area is shown plotted, and a curve is drawn thru these points to show the relative distribution of velocity across the duct.

For more accurate work, it is better to make a traverse of the pipe and either determine the coefficient for the case in question or take the average of all readings. A rectangular duct may be divided into a number of small squares or rectangles, and a reading may be taken in the center of each. The average of all velocities corresponding to these pressures will give the true velocity in the duct. A round pipe may be divided into a number of concentric zones or rings of equal area, and four readings may be taken in each area, horizontally and vertically across it. The points at which readings are taken are on a circle dividing the ring into equal areas (see Fig. 7-12). However, two readings taken in each equal area are usually found to provide sufficient accuracy for intercompany gas operation measurements.

When traversing a section of pipe, it is most convenient to know the location of the traverse points in relation to their distance from the inside of the pipe. This may be computed from the following formulas. Each formula yields two values, because division into three areas requires six traverse points.

$$Y_1 = D\left(\frac{1 \pm \sqrt{1 - \frac{1}{2x}}}{2}\right)$$

$$Y_2 = D\left(\frac{1 \pm \sqrt{1 - \frac{3}{2x}}}{2}\right)$$

$$Y_3 = D\left(\frac{1 \pm \sqrt{1 - \frac{5}{2x}}}{2}\right)$$

$$Y_n = D\left(\frac{1 \pm \sqrt{1 - \left(\frac{2n-1}{2x}\right)}}{2}\right)$$

where:

Y = distance from inside of pipe to traverse point
D = inside diameter of pipe
x = number of equal areas in which traverse readings are taken

The two values received for each computation will be the dimension to each of the two points to be traversed in each ring, one being diametrically opposite the other.

Table 7-6 is convenient for computing the points at which traverse readings are to be taken. For example, assume that a traverse is to be made thru five equal areas of a 24-in. OD pipe, which has a wall thickness of $^5/_{16}$ in. The internal diameter would be 24.000 − 2(0.3125) = 23.375.

The first reading is taken at 23.375 × 0.0257 = 0.601 in. from inside edge; the second reading is taken at 23.375 × 0.0817 = 1.910 in. from inside edge; the third reading is taken at 23.375 × 0.1464 = 3.422 in. from inside edge; repeat for each of the ten readings to be taken.

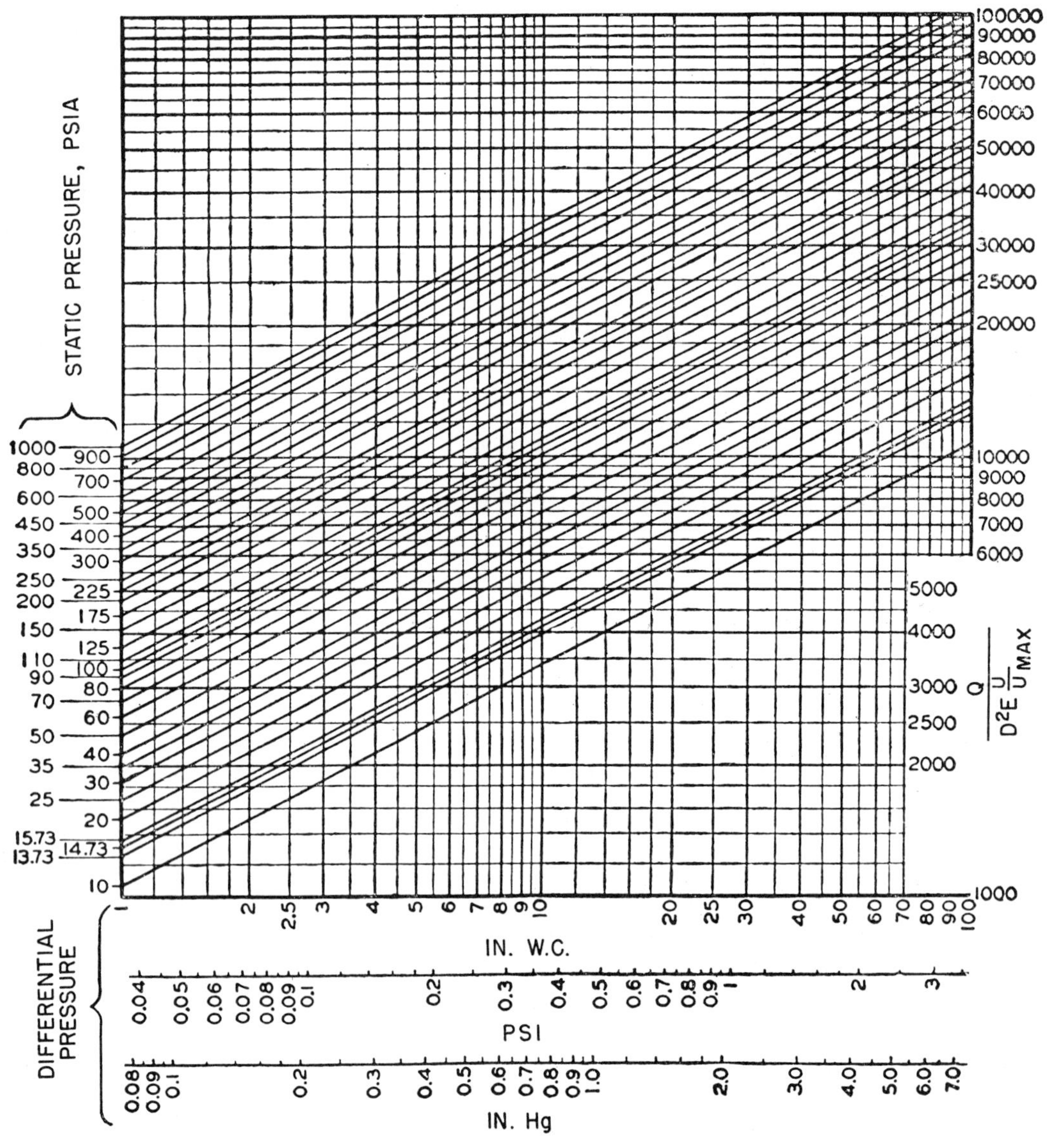

Fig. 7-13 Pitot-tube chart for gases.

Special Formulas Applicable to Gas and Air Measurement

Using units generally employed in gas measurement, the following equations are applicable to the flow of gases. The Pitot-tube chart, shown in Fig. 7-13, is based on these equations.

$$Q = 40{,}600ED^2 \frac{u}{u_{\max}} \sqrt{\frac{(P_1 - P_2)P_a}{G_aT_a}}$$

$$Q = 28{,}440ED^2 \frac{u}{u_{\max}} \sqrt{\frac{H_gP_a}{G_aT_a}}$$

$$Q = 7713ED^2 \frac{u}{u_{\max}} \sqrt{\frac{H_wP_a}{G_aT_a}}$$

where:

Q = standard cubic feet of gas per hour (60 F, 14.73 psia)

E = calibration constant for Pitot-tube installation. For well-constructed Pitot tubes, properly installed, $E = 1.00$

$(P_1 - P_2)$ = differential pressure, psi

T_a = temperature of flowing gas, °F abs

G_a = gravity of flowing gas (air = 1)

P_a = static pressure, psia

D = diameter of pipe, in.

$u/u_{\max}$ = ratio of mean velocity to $u_{℄}$, the velocity at center of the pipe

H_g = differential pressure, in. Hg at 32 F

H_w = differential pressure, in. H_2O at 60 F

The chart is plotted for cases where gas temperature is 60 F and gravity is 1.00. If the actual temperature and gravity differ from these values, the Q read from the chart should be multiplied by temperature and gravity correction factors. The range of this chart can be extended by multiplying the abscissas by 100 and the ordinates by 10, or by dividing the abscissas by 100 and the ordinates by 10. In measuring gas with Pitot tubes, the use of a calculation sheet similar to Fig. 7-14 is suggested.

Date ________, Location ________, Sta. No. ________, Pitot tube No. ________,
Description ________, Pipe size: OD ________, Wall thickness ________,
ID ________, Flow temp. ________ °F, Gravity ________, Avg press. ________ psig

FLOW CALCULATIONS

Quantity, $Q = C \times F_f =$ ________ Mcfh × 24 = ________ Mcfd
Constant, $C = 436.67^* \times E \times D^2 \times u/u_{℄} =$ ________
Variable, $F_f = (hp)^{0.5} \times F_g \times F_{tf} \times F_{pv} =$ ________
E ________ F_g ________ h ________ D^2 ________
F_{tf} ________ p ________ $u/u_{℄}$ ________ F_{pv} ________ $(hp)^{0.5}$ ________
$(hp)^{0.5}$, F_g, F_{tf}, & F_{pv} same as in "Computation of Flow Thru Orifice Meters."

***E* DETERMINATION**

$$E = \frac{\text{Known Volume}}{\text{Pitot Measurement}} = \text{\_\_\_\_\_\_\_\_} = \text{\_\_\_\_\_\_\_\_}$$

u/u_{max} or $u/u_{℄}$ DETERMINATION

Traverse point: Per cent ID	Inches	h	"u" ~ (h)$^{0.5}$‡
2.57			
8.17			
14.64			
22.61			
34.19			
50.00	℄	†	℄
65.81			
77.38			
85.35			
91.83			
97.43			

$$\frac{\text{Mean "u"}}{\text{"u"}_{℄}} = \frac{u}{u_{℄}} = \text{\_\_\_\_\_\_\_\_}$$

* Assumes pressure base = 14.93 psia, temp base = 60 F, and sp gr = 0.60.

† Do not include in average.

‡ This form may be used to calculate the desired ratio; a precise relationship is given in the text.

Fig. 7-14 Gas measurement Pitot-tube calculation sheet.

Table 7-6 Pipe Traverse Points for Pitot Tube Calibration Distance from Inside of Pipe to Point of Reading in Per Cent of Internal Diameter

Traverse point No.	Number of equal areas in traverse					
	3	4	5	6	7	8
1	4.36	3.23	2.57	2.13	1.82	1.59
2	14.64	10.47	8.17	6.70	5.68	4.93
3	29.58	19.38	14.64	11.80	9.91	8.54
4	70.41	32.32	22.61	17.72	14.64	12.50
5	85.35	67.67	34.19	25.00	20.14	16.93
6	95.64	80.61	65.81	35.57	26.85	22.05
7	...	89.52	77.38	64.43	36.64	28.35
8	...	96.77	85.35	75.00	63.36	37.50
9	...	...	91.83	82.27	73.14	62.50
10	...	...	97.43	88.20	79.87	71.65
11	...	...	...	93.30	85.35	77.95
12	...	...	...	97.87	90.09	83.07
13	...	...	...	...	94.31	87.50
14	...	...	...	...	98.18	91.45
15	...	...	...	...	...	95.06
16	...	...	...	...	...	98.41

It will be noted that this gas calculation sheet is set up for placing the Pitot tube impact orifice at the pipe center as the most practical procedure. This practice is normally followed in gas operations because of the relatively increased magnitude of the impact readings obtained at the pipe center, and the relatively small errors introduced by slight deviations in locating the opening, as compared to those which would result from similar deviations when attempting to locate the impact opening at the point of average velocity

Regardless of the point for placement of the impact orifice or of the Pitot tube design, it is very important that the impact orifice faces directly upstream and is kept clean and entirely free of any obstruction.

In Fig. 7-15, the static pressure connection consists of a simple tap thru the wall of the pipe. With this type of static connection it is necessary that the opening thru the pipe wall be round and smooth, with its inside edges slightly rounded off, absolutely free from burrs, and clean at all times. Since this connection serves the low side artery to the differential gage as well as to the static pressure element, extreme care must be exercised to avoid leakage. The slightest leak in any part of this system will cause serious error.

This typical installation sketch shows a combination static and differential recording gage as used in practical applications on natural gas transmission systems. In such applications exceedingly sensitive high pressure and low range differential gages are required.

Factors Affecting Accuracy of Pitot Tubes.[5]

1. The velocity on the diameters where the Pitot-tube readings are being taken may be constantly changing during the time necessary to obtain the readings.

2. The gas flow can only approach, never reach, the ideal condition of parallel flow, and the Pitot tube is correct, theoretically, only when the tube is exactly parallel to the current of gas.

3. The Pitot tube as a means of measuring gases is reliable within approximately one per cent when the static pressure is correctly obtained and when all readings are taken with a sufficient degree of refinement. To obtain this degree of accuracy, the Pitot tube should be preceded by a straight run of pipe 20 to 38 times the pipe diameter, in order to make the flow of gas as nearly uniform across the section of pipe as possible.

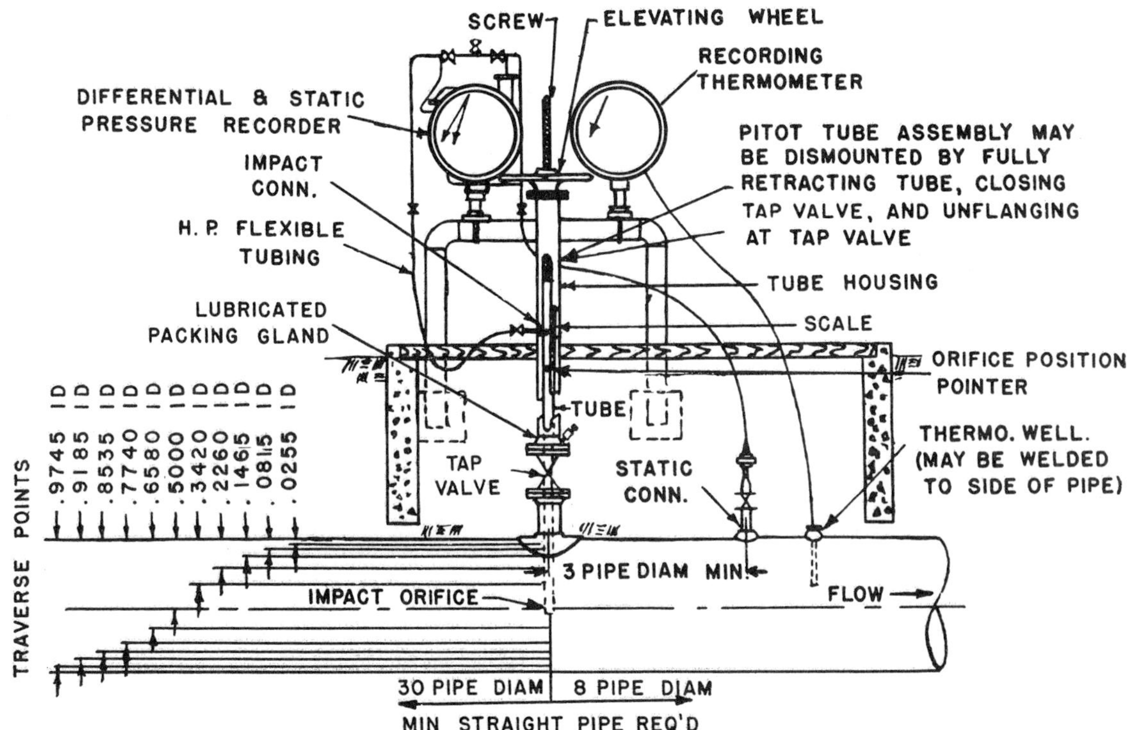

Fig. 7-15 Typical Pitot measuring installation used in natural gas operations.

4. All the methods of obtaining the dynamic head give accurate results.

5. The most reliable and accurate means of obtaining the static pressure is the piezometer or its equivalent. The results showed, beyond any doubt, that the static pressure is constant across any section of pipe in which gas is flowing at a uniform rate.

6. In obtaining the static pressure by the Pitot tube itself (i.e., a Pitot static tube), the best means is a very small hole in a very smooth surface.

7. The results obtained when slots are used for static pressure openings may be in error from 3.5 to 10 per cent. Neither the length of the slot nor the thickness of the tube appears to affect these results.

MASS FLOW MEASUREMENT

Considerable attention has recently been devoted to gas measurement by weight. Some advantages of this measurement method over conventional volumetric methods have been outlined,[6] and different ways of accomplishing it have been described. Mass measurement eliminates the necessity of defining standard conditions on which volumetric measurements are normally based. The heating value of any given gas on a pound basis is the same, regardless of delivery conditions. In addition, mass measurement makes determining gas specific gravity and supercompressibility unnecessary.

Three methods of mass measurement have been tested under field conditions. These methods are described in the following:

Mass Flow with Orifice Meter. Substitution of a density measurement device in place of the static pressure element permits weight measurement with the orifice meter. This device may be a hollow spinner rotated at constant speed in the gas measured. Centrifugal force creates a low pressure at its center. The difference between this pressure and the line pressure, as shown on a standard differential pressure recorder, is directly proportional to the specific weight of the gas at flowing conditions. The recorder actually consists of two different recorders, one for specific weight and one for the orifice differential, and their readings are integrated by a standard instrument.

The specific weight of the gas in pounds per cubic foot is equal to the differential reading multiplied by a cell factor. The following equation may be used for flow computations:[2]

$$W = 1.0618 F_b F_r Y (h_w \gamma)^{0.5}$$

where:

W = flow rate, lb per hr
F_b = basic orifice factor (Table 3 for flange taps)
F_r = Reynolds number factor (Table 4 for flange taps)
Y = expansion factor (Tables 5, 6, or 7 for flange taps)
h_w = differential pressure, in. w.c.
γ = specific weight of gas at flowing conditions, lb per cu ft

Vortex-Velocity Mass Flow Meter. Basically, this meter is a volumetric device which measures gas flow in cubic feet at flowing conditions. A section of tubing has a cylindrical chamber welded normal to one side. Gas flowing thru the tubing generates a vortex in this chamber. A rotor or vortex cage mounted in the chamber revolves within the vortex at a speed proportional to the gas flow velocity. Volume rate is the product of rotor speed and an area factor. The volume is totaled on a counter geared to the vortex cage.

Weight flow is determined by the product of volume flow in cubic feet at flow conditions and the specific weight in pounds per cubic foot.

Velocity-Change Mass Flow Meter.[7] Two essentially identical elements are mounted independently in series on an axis colinear with the pipe axis. Each is composed of two thin-walled concentric cylinders, with the annular space between them divided into narrow longitudinal passages. The flowing fluid is directed thru these passages. The first element, the impeller, is rotated at constant speed, and imparts to the gas a constant angular component of velocity normal to its linear velocity. This component is removed from the gas stream by the second element, the turbine, which is restrained from rotating. The force produced by the torque required to restrain that rotation is the meter output. It is directly proportional to the weight flow rate.

A specially designed gyroscope is caused to precess around its major axis when the torque from the turbine is applied to its minor axis. The speed of this precession is directly proportional to the weight-flow rate. Thru a geared counter operated by this precession, the cumulative mass flow is recorded. Black, Sivalls & Bryson, Inc. reports an accuracy within one per cent of any reading (and a repeatability of ¼ per cent) for their meters.

REFERENCES

1. West Virginia University, School of Mines. *Designing and Installing Measuring and Regulating Stations.* (Practical Methods Rept. 3) Morgantown, 1956.
2. A.G.A. *Orifice Metering of Natural Gas.* (Gas Measurement Com. Rept. 3) New York, 1955.
3. Beitler, S. R. and others. "Developments in the Measuring of Pulsating Flows with Inferential-Head Meters." *A.S.M.E. Trans.* 65: 353, 1943.
4. Bean, H. S. "Comments on How Good Are Good Fluid Flow Measurements." *Gas* 37: 70–2, Nov. 1961.
5. Rowse, W. C. "Pitot Tubes for Gas Measurement." *A.S.M.E. Trans.* 35: 633, 1913.
6. Bean, H. P. "Mass Flow Measurement." (In: Texas College of Arts and Industries. *Proc. of 14th Annual Short Course in Gas Technology,* p. 27–37. Kingsville, 1959).
7. Halsell, C. M. and others. "Fluid Mass Flow." *Pipe Line News* 32: 23–9, July 1960.
8. Terrell, C. E., ed. *A.G.A. Gas Measurement Manual.* New York, A.G.A., 1963.
9. A.G.A. *Supercompressibility Factors for Natural Gas.* Temperature range 0–180 F; specific gravity range 0.554–0.750. Vols. I–VI: Pressure range 0–3000 psig in 500 degree increments. Vol. VII: Tables for the determination of supercompressibility factors for natural gas containing nitrogen and/or carbon dioxide, pressure range 0–3000 psig. New York, 1955.

Chapter 3
Quantity or Displacement Meters

In quantity or displacement meters the fluid passes thru the primary element in successive and more or less completely isolated quantities of either weight or volume, by alternately filling and emptying containers of known or fixed capacities. They have sometimes been called "positive" meters but the term has led to confusion and is not generally employed. Reference 1 gives a good survey of displacement meters. A glossary of terms for diaphragm meters and metering is available.[2]

The secondary element of a quantity meter consists of a counter with suitably graduated dials for registering the total quantity that has passed thru the meter.

LARGE CAPACITY DIAPHRAGM TYPE

Large capacity diaphragm type displacement meters are of the same general construction and design as domestic size tinned steel case, cast iron, and aluminum case type displacement (diaphragm) meters, except that their cases are heavier, in most instances to provide more structural strength and in some instances to withstand higher working pressures. Diaphragm type meters should not be overloaded. Also, it is impractical to operate a displacement type meter continuously at maximum capacity, since excessive wear of its parts and, hence, more deviation in metering accuracy may result. It is generally recommended that displacement meters be operated at their *badged* capacity only for peak loads, rather than for continuous operation. Care should also be taken not to select meters which are too large. It is not good practice to operate meters at a low percentage of their capacity, unless peak demands require large sizes. Large capacity displacement meters can be installed in manifold with orifice meters where extreme ratios of maximum to minimum rates of flow exist.

The **maximum capacity** of a large capacity displacement meter may be considered as the capacity at which the differential pressure across it is 2 in. w.c.* when operating at an inlet pressure slightly above atmospheric. This capacity may be increased for higher inlet pressures. To obtain approximate maximum capacity with various inlet pressures (psig), the maximum capacity in standard cubic feet at atmospheric pressure may be multiplied by the factor [(meter pressure +14.4)/14.73]. One large utility uses the square root of this factor to determine maximum capacity at higher pressures, in order to achieve longer meter life and more dependable accuracy. Depending upon the particular meters, other factors may be recommended by their manufacturers.

* Some gas companies rate their meters at a 0.5 or a 1.0 in. w.c. differential.

Meter Gages

Since displacement meters measure the volumetric displacement of the gas at the meter pressure, it is usually necessary, for high pressure measurements, to record the pressure or to provide some other means of adjusting the meter displacement for the pressure different from the base pressure. The following four types of gages have been found practical:

Pressure-Time. (P.T.) The pressure chart is driven by a time clock. No method is provided for weighting the average pressure against gas volumes. This type is suitable only when either the pressure or delivery rate is relatively constant.

Pressure-Volume-Time. The pressure chart is driven by a time clock. A second pen attached to the meter index indicates when each unit of volume passes. Data are thus provided from which the weighted average pressure may be computed.

Pressure-Volume. The pressure chart is driven by the index of the meter. With this method, the average recorded pressure represents the weighted average pressure and is suitable only when the time of the gas demand is not required.

Pressure-Volume-Temperature-Time. This type records the pressure and temperature, with the chart driven by the index. A separate pen records time on the chart. With this type, the weighted average temperature and pressure may be obtained, along with the time that gas was delivered.

Testing Methods

Test Rates of Flow. It is recommended that large capacity displacement meters be tested at three different rates of flow: maximum, 50 per cent of maximum, and 10 per cent of maximum rate.

The maximum rate test should be made at a differential pressure across the meter at least equal to the maximum differential the meter encounters in service. This may be accomplished by testing the meter under operating pressure at its maximum operating index rate. Where testing is impractical under operating pressure, the meter may be tested at any other (lower) pressure and (higher) operating flow rate which causes the same differential pressure, providing its designed maximum rate of flow is not exceeded. See Table 7-7 for Factor A for use in determining the designed maximum index rate. Also, where it is known that accurate tests will be obtained at higher differentials produced by over-speeding

Table 7-7 Factors for Obtaining Test Flow Rates from Badged Capacity for Various Test Pressures

Testing pressure, psig	Factor A	Factor B	Testing pressure, psig	Factor A	Factor B
0	1.00	1.00	110	0.56	0.34
15	0.84	0.70	120	.55	.33
20	.80	.65	130	.53	.32
25	.77	.61	140	.51	.31
30	.75	.57	150	.50	.30
35	.73	.54	160	.49	.29
40	.71	.52	180	.47	.27
45	.70	.50	200	.45	.26
50	.68	.48	225	.43	.25
60	.65	.45	250	.41	.24
70	.63	.42	300	.38	.22
80	.61	.39	350	.35	.20
90	.60	.37	400	.33	.19
100	0.58	0.36	500	0.30	0.17

the meter rather than by increasing density or pressure of the testing medium, this method may be used, if it does not conflict with regulatory requirements.

The index rate of flow at *operating pressure* may be multiplied by the operating pressure in psia and divided by testing pressure in psia, to obtain the index rate of flow at *testing pressure* which will cause the same differential pressure across the meter.

A study of differential pressures across displacement meters will reveal that it is sometimes impossible to obtain a differential pressure at low pressure (approximately atmospheric) equal to that existing at high operating pressure, without exceeding designed rates of flow at atmospheric gas pressure. That is to say, high pressure meters operating at differential pressures above that at which they are designed to operate (usually 2 in. w.c.) cannot be tested at low pressure without exceeding their badged capacity.

It is quite difficult to make high pressure tests on displacement meters by using either a low pressure flow prover or a test meter, since it is necessary to regulate the pressure between the test meter and the equipment. This adds materially to the complications. Also, the size or capacity of necessary testing equipment makes it more difficult to handle. However, accurate tests can be made using the high pressure method if instructions furnished by the meter manufacturer are followed. In testing meters operating at differentials above those at which the meter is designed to operate at low pressure, the use of a **critical flow prover** may be more practical.

Where meters operating on high pressure are limited in their maximum operating dial rate, to keep the differential pressure equal to or below the differential pressure at which the meter is designed to operate on low pressure, they may be tested at low pressure. Low pressure tests may be made up to the maximum differential existing with the meter in service, without exceeding the badged capacity.

Table 7-7 gives factors for various testing pressures by which the badged meter capacity at 2 in. w.c. differential may be multiplied to obtain:

1. The designed maximum index rate of flow. Using Factor A gives approximate permissible dial speeds.

2. The corresponding index rate of flow at the same differential (2.0 in. w.c.) pressure if atmospheric test pressure is used. Factor B is used.

Example: Assume a meter with a badged capacity of 5000 cu ft of gas per hour (at 2 in. w.c. differential) is to be tested at 100 psig.

a. To test at the designed maximum index rate of flow, the index rate of flow is 5000 × 0.58 (Factor A from Table 7-7), or 2900 cfh.

b. To test at the index rate of flow with a differential corresponding to that at capacity using atmospheric pressure, the required index rate of flow is 5000 × 0.36 (Factor B from Table 7-7), or 1800 cfh.

Bell Type Prover Test. Most accurate testing of large capacity meters (units badged at 2500 cfh and over) can usually be done with a bell-type prover of a volume of 50 to 100 cu ft. This prover is also useful for checking the accuracy of test meters or flow provers in the field. However, it is difficult and cumbersome to obtain satisfactory tests with a 10 cu ft bell-type prover, even if it is equipped with connections large enough to permit making tests at a high rate of flow.

Low Pressure Flow Prover. The low pressure flow prover (Fig. 7-16) has been used for many years to test field meters on location. The apparatus consists of two pieces of aluminum pipe, a set of straightening vanes in the upstream section of pipe, special flanges equipped with hook bolts and thumbscrews, orifice disks for various flow rates, a differential pressure gage, a stop watch, and a thermometer. An aneroid barometer is also desirable.

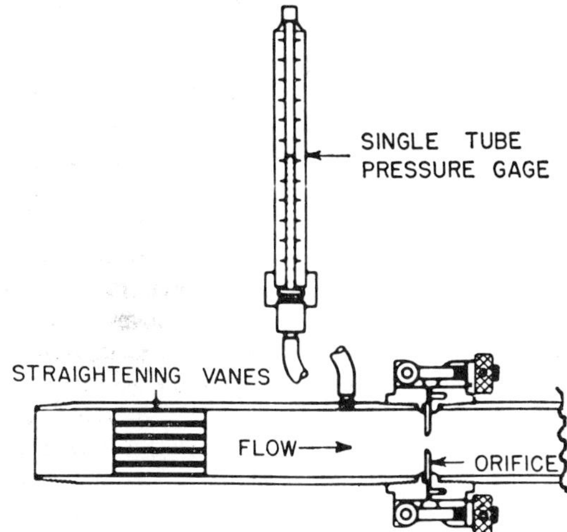

Fig. 7-16 Low pressure flow prover. (Bulletin E-4, American Meter Co.)

Each orifice is marked on the outlet side with its standard time. This standard time usually is the interval in seconds required for 1 cu ft of air to pass (at the upstream pressure) when the absolute pressure at eight pipe diameters downstream from the orifice is 29.0 in. Hg, the differential across the orifice is 8.00 in. w.c., and the temperature is 60 F.

The orifices are placed in the flow prover with a gasket on the inlet side. To indicate the differential across the orifice, it is advisable to use a single column well-type differential gage. The stop watch gives the time required for the passage

Table 7-8 Pressure Factors

Differential pressure, in. w.c.	Pressure at eight diameters downstream, in. Hg, abs					
	28.0	28.4	28.8	29.2	29.6	30.0
5.0	1.236	1.245	1.253	1.262	1.270	1.279
5.1	1.224	1.233	1.241	1.250	1.258	1.267
5.2	1.213	1.221	1.230	1.238	1.246	1.255
5.3	1.202	1.210	1.218	1.227	1.235	1.243
5.4	1.191	1.199	1.207	1.215	1.223	1.232
5.5	1.180	1.188	1.196	1.204	1.212	1.221
5.6	1.170	1.178	1.186	1.194	1.202	1.210
5.7	1.160	1.168	1.176	1.184	1.191	1.199
5.8	1.150	1.158	1.166	1.174	1.181	1.189
5.9	1.140	1.148	1.156	1.164	1.172	1.179
6.0	1.131	1.139	1.147	1.154	1.162	1.170
6.1	1.122	1.130	1.137	1.145	1.153	1.160
6.2	1.113	1.121	1.128	1.136	1.144	1.151
6.3	1.104	1.112	1.120	1.127	1.135	1.142
6.4	1.096	1.104	1.111	1.119	1.126	1.133
6.5	1.088	1.095	1.103	1.110	1.118	1.125
6.6	1.080	1.087	1.095	1.102	1.109	1.117
6.7	1.072	1.079	1.087	1.094	1.101	1.108
6.8	1.064	1.071	1.079	1.086	1.093	1.100
6.9	1.057	1.064	1.071	1.078	1.085	1.093
7.0	1.049	1.056	1.064	1.071	1.078	1.085
7.1	1.042	1.049	1.056	1.063	1.070	1.078
7.2	1.035	1.042	1.049	1.056	1.063	1.070
7.3	1.028	1.035	1.042	1.049	1.056	1.063
7.4	1.021	1.028	1.035	1.042	1.049	1.056
7.5	1.015	1.022	1.029	1.036	1.042	1.049
7.6	1.008	1.015	1.022	1.029	1.036	1.043
7.7	1.002	1.009	1.016	1.022	1.029	1.036
7.8	0.996	1.002	1.009	1.016	1.023	1.029
7.9	.990	0.996	1.003	1.010	1.016	1.023
8.0	.984	.990	0.997	1.004	1.010	1.017
8.1	.978	.984	.991	0.998	1.004	1.011
8.2	.972	.978	.985	.992	0.998	1.005
8.3	.966	.973	.979	.986	.992	0.999
8.4	.961	.967	.974	.980	.987	.993
8.5	.955	.962	.968	.975	.981	.987
8.6	.950	.956	.963	.969	.976	.982
8.7	.944	.951	.957	.964	.970	.976
8.8	.939	.946	.952	.958	.965	.971
8.9	.934	.941	.947	.953	.959	.966
9.0	.929	.936	.942	.948	.954	.961
9.1	.924	.931	.937	.943	.949	.956
9.2	.919	.926	.932	.938	.944	.950
9.3	.915	.921	.927	.933	.939	.946
9.4	.910	.916	.922	.928	.935	.941
9.5	.905	.911	.918	.924	.930	.938
9.6	.901	.907	.913	.919	.925	.931
9.7	.896	.902	.908	.915	.921	.927
9.8	.892	.898	.904	.910	.916	.922
9.9	.888	.894	.900	.906	.911	.917
10.0	0.883	0.889	0.895	0.901	0.907	0.913

of gas for one or more complete revolutions of the index proving hand. The test time should not be less than 100 sec. Table 7-10 indicates the approximate flow rates in cubic feet per hour for orifices of various diameters and approximate times for the passage of 1 cu ft of air at standard conditions, as stated above. Before using this table, it is advisable to check the manufacturer's data for the prover involved, since those tables may be based on other conditions.

The prover is attached to the meter or to a connection close to its outlet between the meter and a gate valve. A small orifice is placed between the flanges for the check rate, and one column of the differential gage is attached to a connection on the upstream side of the orifice. The valve on the upstream side of the field meter is adjusted so that the reading of the differential gage stays constant at between 5 and 10 in. of water. Readings recorded include: differential gage, time required for the increase in index reading, temperature at the prover outlet, barometric pressure, and, if necessary, the static pressure at a point eight diameters downstream from the orifice.

When the gas cannot be discharged directly to the atmosphere, a special downstream section should be used, about 12 diam in length, discharging into a vent line. The differential pressure is then obtained between a tap 1 diam upstream and another 8 diam downstream. Usually, the gas is discharged directly into the atmosphere, using the standard downstream section, 8 diam in length. The downstream static pressure connection is then unnecessary; the upstream pressure is the differential, since the downstream pressure is atmospheric.

Table 7-9 Specific Gravity Factors

Factor = $(\text{sp gr})^{0.5}$

Sp gr	Factor	Sp gr	Factor	Sp gr	Factor	Sp gr	Factor	Sp gr	Factor
0.40	0.632	0.60	0.775	0.80	0.894	1.00	1.000	1.20	1.095
.41	.640	.61	.781	.81	.900	1.01	1.005	1.22	1.105
.42	.648	.62	.787	.82	.906	1.02	1.010	1.24	1.114
.43	.656	.63	.794	.83	.911	1.03	1.015	1.26	1.123
.44	.663	.64	.800	.84	.917	1.04	1.020	1.28	1.131
.45	.671	.65	.806	.85	.922	1.05	1.025	1.30	1.140
.46	.678	.66	.812	.86	.927	1.06	1.030	1.32	1.149
.47	.686	.67	.819	.87	.933	1.07	1.034	1.34	1.158
.48	.693	.68	.825	.88	.938	1.08	1.039	1.36	1.166
.49	.700	.69	.831	.89	.943	1.09	1.044	1.38	1.175
.50	.707	.70	.837	.90	.949	1.10	1.049	1.40	1.183
.51	.714	.71	.843	.91	.954	1.11	1.054	1.42	1.192
.52	.721	.72	.849	.92	.959	1.12	1.058	1.44	1.200
.53	.728	.73	.854	.93	.964	1.13	1.063	1.46	1.208
.54	.735	.74	.860	.94	.970	1.14	1.068	1.48	1.217
.55	.742	.75	.866	.95	.975	1.15	1.072	1.50	1.225
.56	.748	.76	.872	.96	.980	1.16	1.077	1.52	1.233
.57	.755	.77	.877	.97	.985	1.17	1.082	1.54	1.241
.58	.762	.78	.883	.98	.990	1.18	1.086	1.56	1.249
0.59	0.768	0.79	0.889	0.99	0.995	1.19	1.091	1.58	1.257

The meter is tested at the maximum rate by using a larger orifice between the flanges and by obtaining the same data as for the check rate.

To conduct a proof determination, the following data must be obtained: (1) differential, inches of water; (2) pressure at 8 diam downstream from the orifice, inches of mercury absolute; (3) temperature at prover outlet (at a point 2 in. upstream); (4) specific gravity of gas; (5) standard time for 1 cu ft; (6) time required for meter index to indicate that a definite number of cubic feet of gas has passed thru the meter; and (7) difference between pressures at field meter inlet and at upstream side of flow prover orifice.

The pressure factor from Table 7-8, the temperature factor from Table 14 in Chap. 2 (up to 150 F) or Table 4-12 (above 150 F), and the specific gravity factor from Table 7-9 are multiplied together to obtain the combined factor. This factor is multiplied by the increase in the index read-

Table 7-10 Approximate Capacities and Air Passage Times Under Standard Conditions for 2 in., 3 in., and 4 in. Low Pressure Flow Provers Using Various Orifices

Orifice diam, in.	Capacity, cfh	Time for 1 cu ft, sec
	Two-inch prover	
0.21	100	36.0
0.42	400	9.0
0.62	900	4.0
0.80	1,600	2.25
0.98	2,500	1.44
1.13	3,600	1.00
	Three-inch prover	
0.30	200	18.0
0.83	1,600	2.25
1.02	2,500	1.44
1.20	3,600	1.00
1.49	6,000	0.60
1.81	10,000	0.36
	Four-inch prover	
0.30	200	18.0
0.84	1,600	2.25
1.04	2,500	1.44
1.23	3,600	1.00
1.56	6,000	0.60
1.94	10,000	0.36

ing and by the standard time of the orifice, to obtain the *correct* gas time. The **proof** is equal to the test time divided by the *correct* time.

If the pressure drops appreciably between the field meter inlet and that of the prover, or if the temperature changes unavoidably between the meter and the flow prover, the **proof** determination should be adjusted to correct for the volume change, corresponding to the pressure drop or temperature change.

Critical Flow Orifice Prover. When the downstream absolute pressure of a stream of gas flowing thru an orifice is about one-half the upstream absolute pressure, a condition of *critical flow* exists. For such flow, the velocity in the throat equals the velocity of sound in the gas at the existing pressure and temperature. While critical flow prevails, the volume measured upstream of the orifice does not change with any

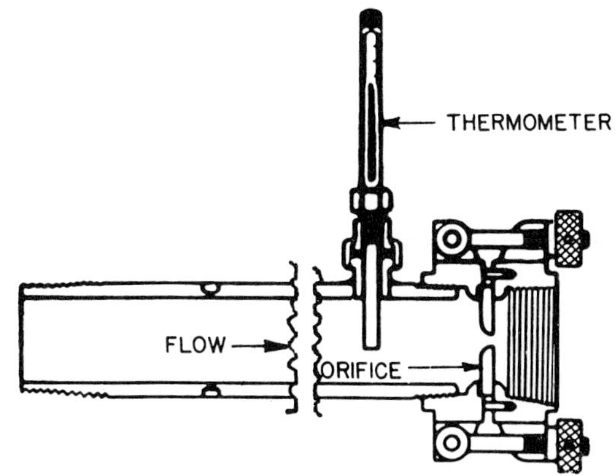

Fig. 7-17 Critical flow prover. (Bulletin E-4, American Meter Co.)

change in upstream pressure. Consequently, if an orifice is attached to the outlet of a displacement meter and gas is passed thru the two in series at constant temperature and various pressures, with each sufficient to give critical flow in the orifice (Fig. 7-17), the rate of index registration will remain constant.*

By equating the expression for velocity of sound in a gaseous medium and the fundamental adiabatic equation, an exact relation—prerequisite of *critical flow*—may be derived be-

* Note that constant flow in this context, as indicated by constant index registration, refers only to volume; the *weight* flow of gas passing thru a critical flow prover under critical flow conditions increases with increasing upstream pressures.

Table 7-11 F Factors for Temperature and Ratio of Specific Heats
(for Equation 1)

Temp, °F	Value of k, ratio of specific heats										
	1.26	1.27	1.28	1.29	1.30	1.32	1.34	1.36	1.38	1.40	Air
10	1.094	1.091	1.088	1.085	1.082	1.076	1.071	1.065	1.060	1.054	1.052
15	1.088	1.085	1.082	1.079	1.076	1.071	1.065	1.059	1.054	1.049	1.046
20	1.083	1.079	1.076	1.074	1.071	1.065	1.059	1.054	1.049	1.043	1.041
25	1.077	1.074	1.071	1.068	1.065	1.060	1.054	1.049	1.043	1.038	1.035
30	1.072	1.069	1.066	1.063	1.060	1.054	1.049	1.043	1.038	1.033	1.030
35	1.066	1.063	1.060	1.057	1.054	1.049	1.043	1.038	1.033	1.027	1.025
40	1.061	1.058	1.055	1.052	1.049	1.043	1.038	1.033	1.027	1.022	1.020
44	1.057	1.053	1.051	1.048	1.045	1.039	1.034	1.028	1.023	1.018	1.016
48	1.052	1.049	1.046	1.044	1.041	1.035	1.030	1.024	1.019	1.014	1.012
52	1.048	1.045	1.042	1.040	1.037	1.031	1.026	1.020	1.015	1.010	1.008
56	1.044	1.041	1.038	1.036	1.033	1.027	1.022	1.016	1.011	1.006	1.004
60	1.040	1.037	1.034	1.032	1.029	1.023	1.018	1.013	1.008	1.002	1.000
64	1.036	1.033	1.030	1.028	1.025	1.019	1.014	1.009	1.004	0.999	0.996
68	1.032	1.029	1.026	1.024	1.021	1.015	1.010	1.005	1.000	.995	.992
72	1.028	1.025	1.023	1.020	1.017	1.012	1.006	1.001	0.996	.991	.989
76	1.025	1.022	1.019	1.016	1.013	1.008	1.003	0.997	.992	.987	.985
80	1.021	1.018	1.015	1.012	1.009	1.004	0.999	.994	.989	.984	.981
84	1.017	1.014	1.011	1.008	1.006	1.000	.995	.990	.985	.980	.978
88	1.013	1.010	1.008	1.005	1.002	0.997	.991	.986	.981	.976	.974
92	1.010	1.007	1.004	1.001	0.998	.993	.988	.983	.978	.973	.971
96	1.006	1.003	1.000	0.998	.995	.990	.984	.979	.974	.969	.967
100	1.002	0.999	0.997	0.994	0.991	0.986	0.981	0.976	0.971	0.966	0.964

tween upstream and downstream pressures. From this, the expression for the flowing volume is evolved:

$$q_1 = ca\sqrt{gk\frac{RT_1}{m}\left(\frac{2}{k+1}\right)^{\frac{k+1}{k-1}}}$$

where:

q_1 = flowing volume at upstream pressure and temperature, cfs
c = orifice coefficient of discharge
a = orifice area, sq ft
R = universal gas constant, ft-lb per (lb-mole)(°F)
T_1 = upstream temperature, °F abs
m = molecular weight of the gas, lb per lb-mole
g = acceleration of gravity, ft per sec-sec
k = ratio of specific heats, C_p/C_v or c_p/c_v, from Table 2-74

The following simplified expression may be used in the field:

$$t_g = Nt_aF\sqrt{G} \qquad (1)$$

where:

t_g = time to pass N cu ft (index registration), sec
t_a = air time to register 1 cu ft as stamped on orifice, sec
$\sqrt{G}$ = square root of the specific gravity of the gas, from Table 7-9

$$F = \sqrt{\frac{245}{(460 + °F)k}\left(\frac{k+1}{2}\right)^{\frac{k+1}{k-1}}}, \text{ from Table 7-11}$$

The ratio of specific heats, k, for a gas mixture may be determined by adding the *products* of k (from Table 2-74) for the various gases and their respective fractions in the mixture.

To test a meter under operating conditions, the prover may be attached to a connection close to the meter outlet.

Approximate index rates in cubic feet per hour at the upstream pressure for various orifices and approximate times required for the passage of one cubic foot of air follow. The orifice size for use for proof tests may thus be readily determined.

Orifice diam, in.	Air time for 1.0 cu ft index reading, sec	Index readings, cfh	
		Air	Gas ($G = 0.64$)
0.125	18.00	200	250
.250	4.50	800	1000
.375	2.00	1800	2250
.500	1.13	3200	4000
0.625	0.75	4800	6000

Because of the small size of critical flow prover orifices, it is impractical to calculate the standard air time by the measurement of orifice diameter. The calibration, stamped on the outlet side of the disk, is always determined by test.

In field testing of meters special precaution should be used, because the rate of gas flow thru the meter is considerably higher than in ordinary testing. Consequently, if there is any explosion hazard, the gas should be conveyed from the prover to the outside of the building housing the meter to avoid collection of any explosive mixture.

Orifice disks should be handled very carefully to avoid any possible damage with change of calibration. They should be examined after each test, to be sure that no foreign matter has collected in the orifice.

The factor for temperature and ratio of specific heats from Table 7-11 is multiplied by the factor for specific gravity from Table 7-9, to obtain the combined factor. This combined factor is multiplied by the increase in index reading and by the standard time on the orifice, to obtain the correct time. The proof is equal to the test time divided by the correct time.

The critical flow prover should be installed as close as possible to the outlet of the field meter to avoid any change in volume due to any pressure or temperature change between meter and prover. If there is an appreciable pressure drop between the meter inlet and that of the prover, or if an unavoidable temperature change occurs between the meter and prover, the proof determination should be adjusted to correct for the corresponding change in volume.

Test Meter. Typical test meters used for testing large capacity meters are similar in construction to standard commercial gas meters, but they are usually equipped with a large dial for accurate reading of the small quantities measured.

After accurate adjustment in the shop on a bell type prover, a test meter is taken to location and connected by a large gas-tight hose to the outlet of the meter to be tested. It is usually good practice to equip both meters with individual pressure gages and thermometers for accurate gas pressure and temperature measurements at each. The test is made by allowing the same gas to pass thru both meters. Any errors in the meter under test are then determined by comparing the measured quantities.

The rates of flow at which the test is to be run may be determined by using Table 7-7. Gas is then allowed to flow at these rates thru the meter under test and the test meter. The test meter should not be run at rates above its capacity and should not be used with gas at rates higher than 125 per cent of the highest rate at which it was tested with air on a bell type meter prover. The accuracy of the meter under test is then determined by comparing the "registered quantity" shown by it, using only complete revolutions of its test hand, with the "actual quantity" registered by the test meter.

The per cent error in the meter under test, using a test meter, may be computed from the expression:

$$\text{Per cent Error} = \frac{(\text{Difference in volumes by two meters}) \times 100}{\text{Volume by test meter}}$$

Where the connections between the two meters are of such size that the pressure drop between them may be appreciable (2 in. w.c. or more), a compensating factor should be applied to the test results. For practical purposes, each 4 in. of water pressure difference may be considered equivalent to one per cent. In the interests of convenience, such pressures should be determined at the inlets to both meters. Since the pressure on the meter under test is always greater, the effect of such pressure difference is always in the same direction. Therefore,

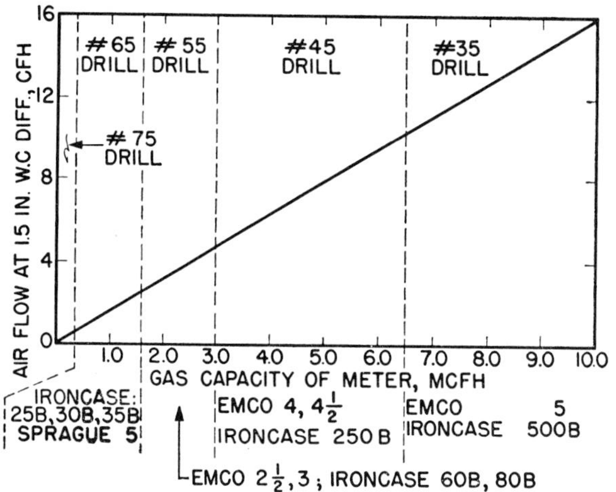

Fig. 7-18 Recommended orifice sizes for use with various size meters operating at 1.5 in. w.c. differential.

plus one per cent may be added algebraically to the test results for each 4 in. of pressure difference. For example, assume that a meter under test is found one per cent slow, with a pressure difference of four inches of water. This difference is the equivalent of plus one per cent, which added algebraically to minus one per cent, the amount by which the meter appeared slow, equals zero. Thus, the meter is correct.

Table 7-12 Air Passed by Various Meters per Tangent Revolution

Make of meter	Cu ft per one rev. of meter tangent	Make of meter	Cu ft per one rev. of meter tangent
Emco		Iron Case	
No. 2½	0.8	25 B	0.2
3 gear type	1.1	30 B	.5
4 gear type	2.0	35 B	.5
4½	3.3	60 B	.8
5	5.6	80 B	0.8
Sprague 5	1.2	250 B	1.7
		500 B	4.0

Slow motion test procedures for shop and field tests are as follows:

Shop Use: Recommended orifice sizes for use with various size meters at a pressure differential of 1.5 in. w.c. are shown in Fig. 7-18. The orifice sizes recommended on this graph have gas capacities equal to approximately 0.2 per cent of the meter's badged capacity at 2 in. w.c. differential pressure.

To pass the slow motion test, a meter should register an amount of air equal to one complete revolution of the meter tangent. Table 7-12 shows the number of cubic feet of air passed by various meters in making one revolution.

Field Use: Recommended orifice sizes for use with various size meters and at different pressures are shown on Fig. 7-19. For example, assume a No. 5 *Emcometer* operating at 3 psig. The recommended orifice size is shown by intersection of the 3 psig pressure curve with the horizontal line designated "No. 5 *Emco.*" The closest orifice size is No. 56. Orifice sizes recommended on Fig. 7-19 have gas capacities equal to approximately 0.2 per cent of the meter's badged capacity at 2.0 in. w.c. differential.

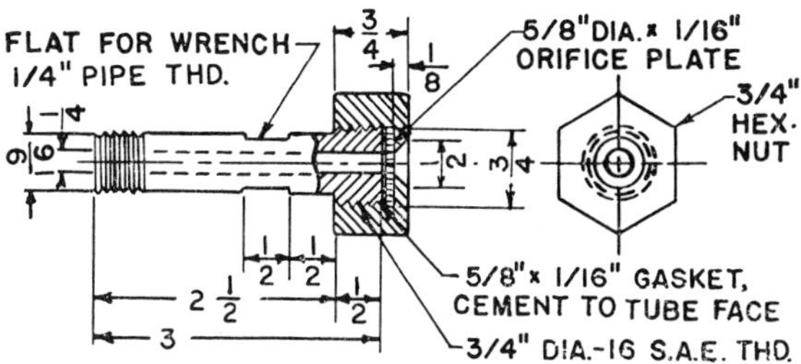

Fig. 7-20 Dimensions of fixture used in calibrating small orifices.

Note: Furnish orifice plates (one each) drilled in the following twist drill sizes: 80, 75, 70, 65, 60, 56, 55, 52, and 50. Finished diameter or orifice opening to be as near actual drill size as practicable.

Figure 7-20 shows the dimensions of the fixture employed both to calibrate and to hold these small orifices in use.

To pass the slow motion test, a meter should register an amount of gas equal to at least one complete revolution of the meter tangent.

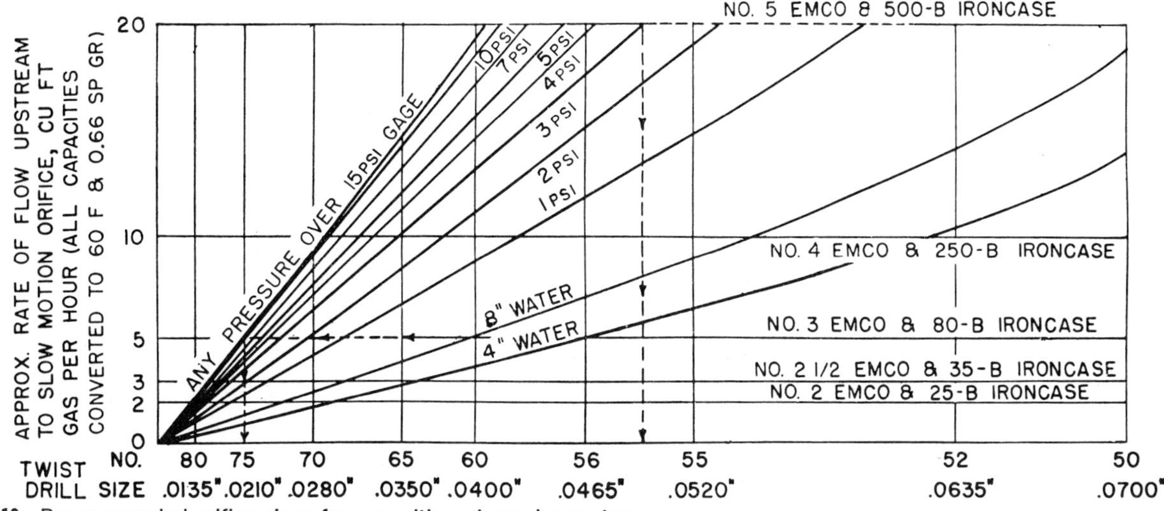

Fig. 7.19 Recommended orifice sizes for use with various size meters.

Definitions of Meter Capacities

Rated (or Badged) Gas Capacity of Meters. The capacity shown on a badge on the front gallery plate of a meter shall be 1.25 times the air capacity of the meter. This definition was adopted in 1922 by the Consumer's Meters Committee of the United Gas Association. The factor 1.25 derives from the relationship $(1/0.64)^{0.5}$. Although the specific gravity 0.64 may still be used in the rating of tinned steel case meters, by 1960 a specific gravity of 0.60 had come into use for the rating of large capacity meters. Thus, in terms of the air capacity of such meters, the gas capacity would be 1.29 per cent times air capacity. The air capacity is the cubic feet of air which a correctly adjusted meter will register in one hour, with a differential pressure equal to 0.5 in. w.c. Under the 1922 definitions, the gas capacity would also have been on the basis of a pressure drop of 0.5 in. w.c., but since large capacity meters may be operated under a 2 in. w.c. differential, manufacturers' capacity tables show the capacities at both 0.5 and 2.0 in. w.c.

ROTARY TYPE (GEARED IMPELLER)

Specifications and capacities are given in Table 7-13. Figure 7-21 illustrates the principle upon which this type of rotary positive displacement meter operates. Two figure-eight shaped impellers revolve in a fixed relationship to each other within a cylindrical housing. Measurement is accomplished in compartments formed by one side of one impeller, the wall of the housing, and the flat headplates at either end.

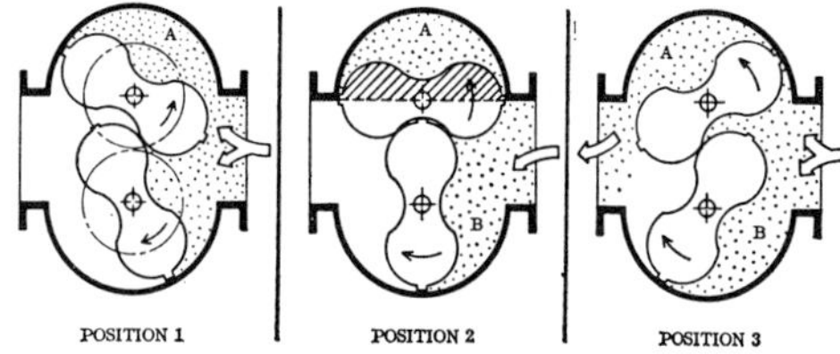

Fig. 7-21 Impeller rotation and gas flow thru a vertical rotary meter. (Roots-Connersville Div. of Dresser Ind., Inc.)

As the upper impeller rotates counterclockwise to the horizontal position, gas enters the space between it and the cylinder. At the horizontal position, a definite volume of gas "A" is contained in the measuring compartment. As the impeller continues to turn, due to a slight pressure differential between the inlet and outlet, the measured volume of gas is moved around and discharged from the meter. The lower impeller, rotating in the opposite direction, has meanwhile closed to the horizontal position confining another known volume of gas "B." This same action takes place four times per revolution of the impellers, i.e., 2A + 2B are delivered. Rotary or impeller type displacement meters measure the volumetric displacement of the gas *at the pressure within the meter*. Thus, when used for metering gas at pressures above those of normal base pressures, the meter measurement must be corrected.

The general equation for converting meter readings at high pressure to a base pressure and temperature is:

$$Q_s = Q_d P_m T_m (F_{pv})^2 \tag{2}$$

where:

Q_s = Quantity of gas at the contract base pressure and temperature, cu ft

Q_d = Actual (displaced) gas passed at existing metering conditions, cu ft

P_m = Pressure multiplier =

$$\frac{\text{Weighted average existing gage pressure} + \text{barometric pressure}}{\text{absolute pressure base}}$$

T_m = Temperature multiplier = $\dfrac{\text{Temperature base} + 460}{\text{Average flowing gas temperature} + 460}$

F_{pv} = Supercompressibility factor (**Table 16-A** in Chap. 2)

It is important that the pressure multiplier, P_m, be determined from the weighted average metering pressure (i.e., the pressure being weighted according to the metered volume), particularly where the pressure is found to vary widely with rate of flow. Meter gages may be used.

Table 7-13 Characteristics, Dimensions, and Capacities of Rootsmeters

Model*	Maximum capacity,§ MSCFH	Diff., in. w.c.‡	Overall dimensions,† L × W × H	Flanged connect., in.	Approx. shipping wt, lb
3M125	28.3	1.0	14.4 × 6.8 × 7.1	2	55
7M125	66.2	1.0	17.9 × 9.5 × 9.4	3	125
16M125	151.4	1.6	32.1 × 14 × 14.4	4	300
23M125	217.6	1.8	35.4 × 16 × 16.4	6	475
38M125	359.5	1.9	40.0 × 18 × 19.3	6	750
56M125	529.9	2.0	43.7 × 21 × 22.4	8	1100
102M125	965.2	1.9	52.2 × 28 × 29.6	10	2150
7M600	291.9	1.0	30.6 × 15.5 × 25	3	450
16M600	667.3	1.6	34.9 × 20.5 × 28.1	4	600
23M600	959.3	1.8	38.1 × 20.8 × 28.6	6	800
38M600	1584.0	1.9	37.8 × 26.3 × 30.3	6	1200
3M1200	247.3	1.0	17.6 × 7.7 × 11.4	2	125
7M1200	577.0	1.0	21.8 × 10.7 × 13.4	3	280

* The number preceding "M" is the continuous rating for maximum displaced volume at 14.73 psia and 60 F, MSCFH; the number following "M" is the pressure rating, psig. Previous meter lines (not shown) cover 25, 60, and 125 psia and also carry *one hour peak overload ratings*; these are 1.5 *times* the continuous ratings.

† To nearest 0.1 in.; length, L, is measured perpendicular to direction of flow; width, W, is measured in the direction of flow.

‡ Pressure drop on air (1.0 sp gr) at atmospheric pressure for the maximum displaced volume (see *).

§ Capacities vary directly with absolute pressures; thus, values in this column and the pressure ratings given in the preceding one correspond to each other. Intermediate values are found by using a simple equation; e.g., to find the capacity of 7M600 at 400 psig:

$$\text{Capacity} = \frac{(400 + 14.73)}{600 + 14.73} \times 291.9 = 196.9 \text{ MSCFH}$$

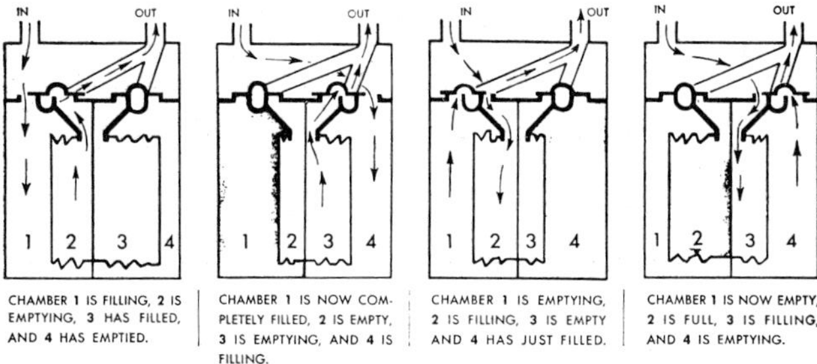

IN
OUT
1
2
3
4
CHAMBER 1 IS FILLING, 2 IS EMPTYING, 3 HAS FILLED, AND 4 HAS EMPTIED.
CHAMBER 1 IS NOW COMPLETELY FILLED, 2 IS EMPTY, 3 IS EMPTYING, AND 4 IS FILLING.
CHAMBER 1 IS EMPTYING, 2 IS FILLING, 3 IS EMPTY AND 4 HAS JUST FILLED.
CHAMBER 1 IS NOW EMPTY, 2 IS FULL, 3 IS FILLING, AND 4 IS EMPTYING.

Fig. 7-22 Operation of diaphragm type gas meters.

In many instances a pressure and volume recording gage may be eliminated by using a special index which is constructed to mechanically convert meter readings at existing conditions to equivalent readings at base conditions. The supercompressibility factor, F_{pv}, may be applied by including it in the same tables with the pressure multipliers, P_m. To do so, it must be based on some predetermined average gas composition and average flowing temperature. Using these combined factors simplifies calculations considerably. When no recording thermometers are installed, the volume is at whatever the existing temperature may be (standard practice for low pressure measurement).

SMALL DIAPHRAGM TYPE

The domestic diaphragm meter is the most common type of displacement unit (Fig. 7-22).

Standards[3] were initiated thru the combined efforts of meter manufacturers and gas utilities to minimize nonessential differences in meters and to ensure maximum interchangeability.

Class. This denotes the designed capacity in cubic feet per hour of 0.64 sp gr gas (air = 1.0) at 60 F and 30 in. Hg that the meter will pass with not more than a 0.5 in. w.c. pressure drop. There are three class designations: 50, 175, and 250.

A comprehensive standard for the use of **differential gages** to measure meter differentials was proposed separately.[14] This reference also showed that only 20 per cent of the true fluctuation of the differential reading is indicated by an inclined fluid type differential gage (generally used in meter shops) when a domestic meter is tested at 20 cycles per minute.

Pressure Rating. A normal working pressure of ½ psig max (with a design safety factor of 10) was specified.

Dimensions. Center to center for above three classes: iron and aluminum—6 in.; tin—11¼ in. (13½ in. is also standard for Class 250).

Other Requirements. Performance—accuracy, pressure drop, and pressure drop fluctuation at partial loads; pertinent dimensions; exterior finish; and general specifications. The proof setting has not been standardized because of the various legal limits established by state regulatory bodies; a proof spread of ± one per cent, for *all* gas utility company meters, falls within the requirements of 44 states (including states without any regulations).[12]

Diaphragms in new meters are generally made of synthetic materials but many older meters employ specially selected and processed leather diaphragms. Meter cases are most commonly tinned sheet steel, although many hard case meters are used; new hard case domestic meters generally have aluminum cases, and steel and cast iron are also available. Meter sizing practices vary[12] with different companies;

Table 7-14 Ranges in Capacities and Dimensions of Various Diaphragm Type Meters

Working pressure	Capacity, cfh at 0.5 in. w.c.	Dimensions,‡ Height	Width	Depth	Pipe connections, in. Size	On center‡	Shipping weight, lb
American Meter Co. Soldered Tinned Steel Case Meters,* 28 models§							
5 psig	45–300	11–15	7–13	8–11	½–¾	6–11	9–18
20 in. w.c.	150–4000	16–43	12–33	10–30	½–3	11–32	18–285
Iron Case and Aluminum Meters,† 24 models							
5–15 psig	110– 650	12–22	9–16	8–16	¾–1½	6–11	38–165
20-75 psig	175–4800	13–45	9–30	9–30	1-4	6–26	45–1040
100–1000 psig	400–5000	18–46	13–31	14–31	1¼–4	8–26	54–1420
Rockwell Mfg. Co. Tinned Steel Case Meters,† 7 models							
5 psig	150– 500	16–18	12–15	10–11	½–1	11–13	16–23
Iron and Aluminum Meters, 24 models							
5–15 psig	130– 750	10–20	8–16	7–13	½–1½	6–11	11–52
50–75 psig	175–5000	13–51	10–40	9–20	1-4	6–37	48-917
100–1000 psig	800–5000	20–40	16–39	14–27	1½–4	11–32	70–976
Cleveland Gas Meter Co. Tinned Steel Case Meters,* 10 models							
...	150–1800	16–33	13–31	10-21	½–1½	11–25	14–128
Sprague Meter Co. Aluminum and Cast Iron Meters,† 7 models							
5–10 psig	175– 400	14–17	10–12	8–10	¾–1½	6– 7	12-52
25 psig	675–1000	23–27	16–18	14	1¼–2	10–11	62–148
50 psig	400**	17	12	10	1¼–1½	7	55
Superior Tinned Steel Case Meters,* 26 models							
8 in. w.c.	80–6000	12–45	9–33	7-37	½–4	8–37	9–410
Aluminum Meters, 2 models							
8 in. w.c.	175– 340	12 & 14	9 & 11	9 & 10	¾–1¼	6	7–16

* Capacities based on 0.64 sp gr gas (air = 1.0) at 60 F and 30 in. Hg. † Capacities based on 0.6 sp gr gas. ‡ Nearest inch. § Welded-type meters and open- and closed-top units. ** Capacity based on 0.6 sp gr gas at 4 oz.

e.g., 100 per cent of the badged rating of automatic equipment plus 70 per cent of the nonautomatic equipment.

Capacities, Design Features, and Principal Dimensions

Table 7-14 shows capacities and principal dimensions of diaphragm type meters produced by several leading manufacturers. Additional details may be obtained by referring to their literature. Meters with built-in bimetallic compensation for temperature changes are available.

Meter Performance Records and Repair Procedures

Meter Records. Unsatisfactory meter performance such as high percentages of fast, slow, nonregister, or stopped meters, can be detected by meter test records. The same records can also indicate overloaded units.[4] Meter records should fulfill requirements of regulatory commissions. Suggested requirements for meter records are available.[5]

Classification of Repairs. A practical system of classifying meters for repair is essential. It should take into consideration: the age of the meter, its time in service, proof, previous repairs, and condition at the time of intesting. The initial classification for repair will generally place a meter in one of four classes established by the extent of the repair required: namely, *O.K.*, *Readjust, Partial*, or *Condemned.* Subsequent tests may require that a meter be reclassified for more extensive repair in the General or New Diaphragm classes.

1. *O.K. Repair.* A meter with proof test within prescribed limits and with a period of service not greater than the average life of diaphragms; or a meter previously given partial, general, or diaphragm repairs within a period determined reasonable by local experience.

2. *Readjust Repair.* A meter not testing within prescribed limits, but meeting other requirements of an *O.K. Repair.*

3. *Partial Repair.* A meter not testing within prescribed limits, and conditions indicating top repairs necessary, but diaphragm inspection unnecessary.

4. *General Repair.* A meter not testing within prescribed limits, and conditions indicating diaphragm inspection necessary; or a meter in service long enough to warrant diaphragm inspection irrespective of proof test; or a meter requiring repairs below the "table," i.e., in the diaphragm section.

5. *New Diaphragm.* A meter with diaphragm age greater than the average useful life, or with defective diaphragms.

Repair Operations.

Class 1—O.K. Repair. Repair steps and tests include the following: examine index and proving circle, check-rate, and rated capacity tests; conduct immersion test for outside leaks; make pilot burner test or slow motion test; clean and paint; make final check-rate and rated capacity tests; and place in stock.

Class 2—Readjust Repair. The following steps and tests are required in addition to the foregoing: remove top, tin top, and flange; repack stuffing-boxes if necessary; test division of flag arms; replace worn or damaged king posts, tangent arms, or indexes; test for binds; conduct four-point test or slow motion test; readjust; retop meter; and test top.

Class 3—Partial Repair. The following steps and tests are required in addition to the foregoing: remove valve-box cover and tin; oil diaphragms; test diaphragms; test crank arms for lost motion; remove valves, grind if necessary, replace, and test for division; replace valve-box cover; and replace fittings as required.

Class 4—General Repair. The following steps and tests are required, in addition to the foregoing: remove front and back; examine diaphragms, and repair leaks where satisfactory repairs can be made; tin case and parts; and replace back and front.

Class 5—New Diaphragms. The following steps and tests are required, in addition to the foregoing: inspect interior of meter; remove diaphragms; boil out meter and parts as required; inspect case after boiling; test channels; install new diaphragms and disks; attach flag to disk, and replace guide wire (if used).

The interrelation and/or combination of the above steps depends upon the size of the shop, the number of employees, and the number of meters repaired per year.

The procedure for repairing iron, aluminum, and steel case meters is like that given above for tinned case meters, substituting the dismantling and assembly process appropriate to the particular type of meter.

Periodic Testing. Analyzing tests of meters removed from service is an important function in the operation of all metering organizations. Such analyses, properly made, furnish the basis for determining the effectiveness of previous repairs, types of meters best suited for the particular service, and many other important items. These include proper parts replacements, selection of diaphragm oils, valve grinding methods, and stuffing box packing materials and methods. Such analyses should also disclose the most appropriate period for which meters may be left in service between tests or repairs.

Generally, the gas industry has agreed to separate meters returned from service into the following six classifications:

1. *Commercially Correct Meters.* Meters testing within a certain limit of O.K., usually between plus or minus two per cent.

2. *Fast Meters.* Meters faster than commercially correct meters.

3. *Slow Meters.* Meters slower than commercially correct meters.

4. *Don't Pass Gas (D.P.G.) Meters.* Meters not passing air when connected to a bell prover balance to deliver air at a pressure of 20 in. w.c.

5. *Don't Register (D.R.) Meters.* Meters connected to a bell prover but not registering this flow.

6. *No Test (N.T.) Meters.* Meters not capable of test due to damage from fire, etc.

Besides segregating meters into these six classifications, it is general practice to compute the "meter weighted mean error" for commercially correct, fast and slow meters, and to eliminate the *D.P.G.*, *D.R.*, and *N.T.* meters from this analysis.

Diaphragm type meters are generally intested at 20 per cent of their rated capacity, and the proof at this rate of flow is accepted as the proof of the meter while it was in service. When air is used as the test medium, the rate of flow may be established to approximate that of gas at 0.64 specific gravity.

Table 7-15 Example of Annual Test Report for Tinned Steel Case Meters

Years in service	Stuck and don't register				Slow meters					Accurate		Fast meters					Total tested
	D.P.G.	D.R.	Total	%	Over 10%	10–5.1%	5–2.1%	Total	%	Total	%	2.1–5%	5.1–10%	Over 10%	Total	%	
0	28	17	45	2.8		3	115	118	7.4	1,388	87.5	33	2		35	2.3	1,586
1	41	48	89	1.7		2	239	241	4.7	4,641	90.9	133	1		134	2.7	5,105
2	46	31	77	1.2	1	11	303	315	5.0	5,623	89.3	274	4	1	279	4.5	6,294
3	31	12	43	0.7		10	302	312	5.0	5,628	90.3	245	4		249	4.0	6,232
4	19	5	24	.4		4	174	178	2.9	5,563	90.2	395	4	1	400	6.5	6,165
5	36	15	51	.8		4	161	165	2.7	5,421	88.3	498	6		504	8.2	6,141
6	17	8	25	.5		3	159	162	3.2	4,412	88.3	393	4		397	8.0	4,996
7	26	9	35	.6		8	203	211	3.6	4,993	85.4	600	8		608	10.4	5,847
8	22	8	30	.6		6	158	164	3.1	4,571	85.1	596	8	1	605	11.2	5,370
9	21	8	29	.5		5	152	157	2.7	4,915	83.5	772	12		784	13.3	5,885
10	37	10	47	.2		7	447	454	1.9	20,034	82.8	3,598	55	2	3,655	15.1	24,190
Over 10	31	22	53	.3	2	29	460	491	2.3	17,368	83.0	2,974	37		3,011	14.4	20,923
Total	355	193	548	0.6	3	92	2873	2968	3.0	84,557	85.6	10,511	145	5	10,661	10.8	98,734

This will create a higher differential pressure, but will not adversely affect the proof of a meter.

Some companies record meter intest results according to the six classifications described, without regard to the length of service or last repair, and compute the percentage in each classification. Other companies also relate these results to the length of service and type of previous repairs.

This kind of breakdown can be made for each make and type of meter and type of service, as well as for a total of all types of meters returned from service. Thus, information may be obtained concerning the type or manufacture of meter that performs best on any particular kind of service.

Further refinement of this method of analysis is advisable in obtaining data on changes in meter accuracy versus years of service since the meter was tested in the shop. The tabulation of the meter intest data is made in the same manner, except that the meters are further segregated by years in service since last test or repair.

Table 7-15 shows a tabulation of meter intest results, classified by years in service, by a company where this method of analyzing is used. It has been found advisable to accumulate this information for consecutive years for the purpose of studying trends.

Proof Points. A statistical treatment of meter test data is used to find the *mean error*. First, *proof points* are obtained by multiplying the number of meters tested in each percentage category by the proof. The algebraic total *divided* by the number of meters tested gives the mean error. For example:

Category	*No. of meters*	× *Proof*	= *Proof points*
1% slow	150	−1	−150
2% fast	250	+2	+500
Totals	400		350

$$\text{Mean error} = \frac{+350}{400} = +0.9 \text{ per cent}$$

Figure 7-23 shows a typical analysis of tests of cast iron meters for a five year period compared to a one year period. A chart developed on this basis for the particular meters involved, together with other local information, should furnish the basis for selecting the most appropriate period

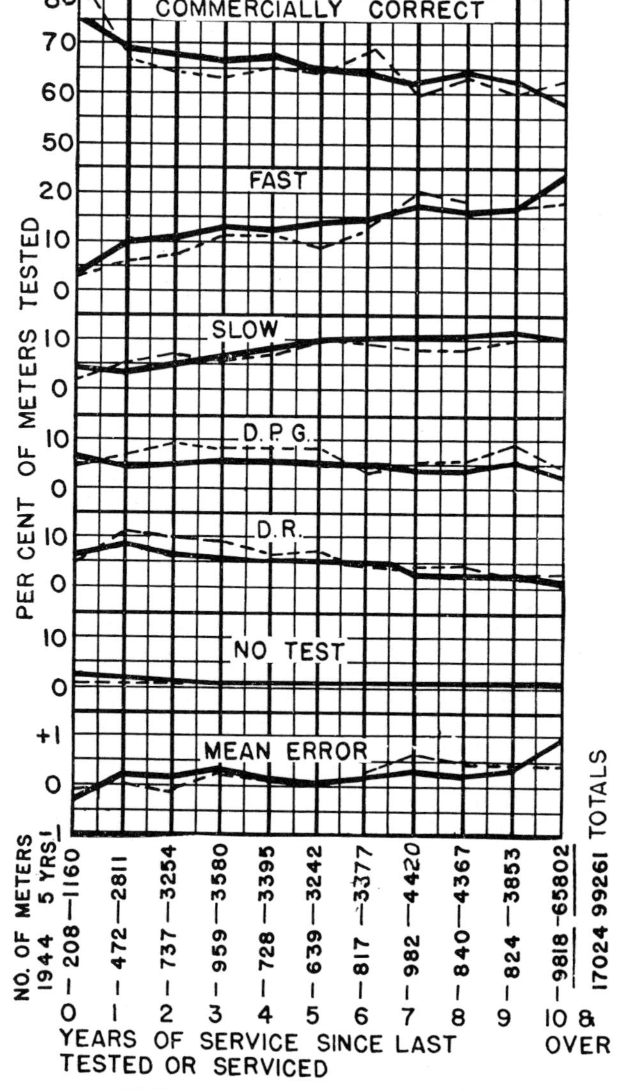

Fig. 7-23 Intest results of cast iron meters returned from service, classified by years in service since last test or repair.

during which meters may be left in service between tests or repairs.

A discussion of the application of one of the sampling plans —MIL STD 105B[6]—is available.[7]

Percentage Error

In all tables and charts in this section the percentage error in a gas measurement is defined by the expression:

$$\text{Percentage error} = \frac{\text{Meter reading} - \text{Prover reading}}{\text{Prover reading}} \times 100$$

This definition is usually used in physical measurement work. It expresses the error in terms of per cent of the prover reading, and has been adopted for this section because it is the one more generally used. However, it is by no means universal, and for several practical reasons the following is more convenient:

$$\text{Correction factor} = \frac{\text{True volume}}{\text{Indicated volume}}$$

The correction factor can be applied directly to a meter reading to obtain the true volume. Percentage errors calculated by the first method must first be converted to per cent of the meter reading before application to the meter reading. The second definition is used in data published by some meter manufacturers and is recommended by the Bureau of Standards.[8]

The **proof** of a meter is the ratio, expressed as a percentage, of the prover scale reading to the meter proving dial reading.

Meter Testing Equipment

Bell Type Meter Prover. Gas meter provers of two, five, and ten cubic feet capacity are usually calibrated by comparison with a cubic foot bottle or standard. The procedure consists of measuring air out of or into the prover by means of the standard, one cubic foot at a time, noting the reading of the prover scale at the start and finish of each transfer.

Provers should be located in a well-lighted room which is provided with temperature regulation adequate to maintain the temperature within ½°F of the desired average temperature.[3] Location should be free of outside influences like steam lines, radiators, and windows. The prover tank should be raised from the floor by legs or blocks, since this not only reduces the lag between the prover and room temperatures, but decreases the accumulation of moisture on the underside of the tank. The sealing fluid used in provers (and in cubic foot bottles and standards) should be a low viscosity, low vapor pressure oil. Suggested specification follows:[3]

Viscosity, Saybolt-sec = 65 to 75 at 100 F; vapor pressure, mm Hg = 0.6 at 200 F, 11.0 at 300 F, 88.0 at 400 F; sp gr = 0.848 to 0.858 (water = 1) at 60 F; pour point = 25 F; flash point = 310 F; fire point $\geq$ 310 F.

The use of oil (at ambient temperature) as a sealing fluid will decrease the cooling effect due to evaporation, when the prover bell is raised from the tank. It will also retard any tendency of the bell to corrode.

Before starting a calibration, the bell should be examined to see that it is tight, clean, and free of dents. It should move freely throughout its entire travel with neither binding nor excessive play within its guides at any position. The wheels in the prover should only momentarily touch the guide rod during its movement and should never contact the rods to the extent that rolling would occur, especially at the extreme position of the bell. To facilitate reading the prover scale to one decimal place beyond that normally used in testing meters, the regular scale pointer may be replaced with a vernier or optical measuring device.

Meter Prover Calibration.[1] While it is possible to measure air out of a prover into an immersion bottle under the usual prover pressure, it is difficult to avoid losing some air as the lower neck of the bottle is raised close to the surface of the sealing fluid in its tank. Therefore, it is advisable to make the test at atmospheric pressure, in which case the counterweight must be increased until it just balances the bell. This adjustment is also necessary if, conversely, air is to be measured into a prover from an immersion bottle.

To measure air *into the bottle*, start with the prover bell raised and the connection between prover and bottle open; adjust the position of the prover bell to zero scale reading. Raise the bottle so that the air will be drawn into it from the prover. As the lower neck of the bottle reaches the surface of the sealing fluid, care should be taken to stop just short of breaking the seal. Close the valve between prover and bottle and record the scale reading. Vent the air from the bottle as it is again lowered into the tank. Open the valve between prover and bottle, adjust prover bell to scale reading of 1.00, and repeat the process of removing another cubic foot of air from the prover.

In measuring the air *into the prover*, the above procedure is reversed. In this case, adjust the prover bell to a scale reading at one of the even foot marks and hold it there while lowering the bottle until the bottom of the lower neck just meets the sealing fluid surface. Then release the prover bell and measure one cubic foot of air into it by lowering the bottle.

With either a moving-tank type of bottle or a Stillman-type portable cubic foot standard, the calibration may be carried out under the usual prover pressure. This requires, when using a moving-tank type of bottle, that the valves in the bottle and prover connections be open while adjusting the quantity of water in the tank and the positions of the stops, so that the water will come to rest in the planes of the gage marks about the upper and lower necks of the bottle. Since transfer of air to or from the prover takes place within a completely closed system, there is no possibility of losing a small amount of air at one end of the transfer, as with an immersion type bottle.

The procedure followed with either type of standard is very simple. After the connections have been checked for leaks, and with the valves between prover and standard open, line up an even foot mark on the prover scale with the index zero. Transfer one cubic foot of air to the standard and record the prover scale reading. Then discharge the air in the standard from the system and repeat the cycle.

If desired, several transfers may be made for the same 1 cu ft interval of the prover scale before proceeding to the next interval. In so doing, the prover scale reading should be readjusted to the even foot mark before a transfer in either direction is started, keeping the connection between prover and standard open so that both are under full prover pressure.

The method to be used in calibrating provers of over 10 cu ft capacity depends upon the capacity, design, and mode of operation of the gasometer. If it is not too large (100 cu ft or less), it may be most convenient to use a cubic foot standard or a five to ten cubic foot prover that has been calibrated. For other gasometers, it will probably be necessary to determine the capacity from measurement of the dimensions. The usual procedure is to measure the outside circumference of the prover bell at several sections. From these measurements and from the metal thickness, the average inside cross-sectional area and capacity per unit height are computed. In the calculations, it may be necessary to take account of the changes of the sealing fluid height produced by raising and lowering of bell.

COMPUTATION OF FLOW THRU DISPLACEMENT METERS

Displacement type meters measure gas in cubic feet under the conditions of pressure and temperature at the point of measurement. Therefore, where gas is measured at other than normal distribution pressure, meter readings must be converted to the base pressure and temperature conditions.

DEMAND METERING

Industry experience has indicated the desirability of means for measuring gas flow over a given period of time, usually 15, 30, or 60 min intervals. This kind of measurement is termed demand metering. It is useful for several purposes. As outlined by Lovretin,[9] these include determination of (1) diversity factor and load characteristics of a particular customer, class of customer, or appliance; (2) rate of return on investments for various class loads; (3) customer's cost determinants for various classes of service (rate studies); (4) meter and regulator sizing; (5) maximum gas consumption for establishment of billing rates; and (6) flow rates in leaking problems.

For satisfactory application and use, gas demand metering requires a device which, within the specified time interval, can record the metered gas. Such a device must be suitable for use with all types and sizes of meters. Also, it must be dependable, accurate at all flow rates, inexpensive, and easily installed; and it must provide data which can be tabulated readily.

Much interest was taken in demand metering during the 1940 decade. In 1949 Griffin[10] summarized replies received to inquiries made to operating utilities about their experiences. These replies showed that several types of demand devices were in use. The simplest among them mechanically registered meter flow. Another type employed dry cells to energize a solenoid, which operated a chart pen. Other devices printed metered consumption at stated intervals. The last, used successfully by several utilities, included the Volco Recorder and the Niroc Demand Recorder.

According to Griffin, many positive displacement meters may be equipped by their manufacturer with recording attachments of different designs. Pickford and Hoag have outlined methods employed by a large eastern utility to adapt tinned case meters in common sizes for use with General Electric type demand units.[11] Means utilized consist of three general components: (1) meter contactor device and impulse wiring; (2) relaying equipment; and (3) demand meter, either General Electric Type G-9 or M-16.

Recent experiences of a large eastern utility with two types of recorders have been described.[9] One is a numerical printing device which represents a further development of the Niroc Demand Recorder previously mentioned. The other utilizes magnetic tape to record simultaneously time interval and load impulses. Resulting data are analyzed by a digital computer.

AUTOMATIC STATIONS[13]

Manifolding one or more diaphragm meters (the primary line) with either a rotary meter or an orifice meter (the secondary line) facilitates measurement where there are large variations in flow. Differential limit controllers automatically activate meters as required. Rotary meters are generally used in the secondary line to measure fluctuating loads of automatic equipment, while orifice meters are preferred for relatively constant loads.

REMOTE RESIDENTIAL METER READING[15]

A 16-month test program investigated the feasibility of remotely and automatically reading gas and electric meters. Approximately 1100 readings were coded, programmed into telephone lines, and sent to a central telephone office. Reportedly, the system, as tested, showed "its average accuracy of reading to be within 2 per cent of the actual quantity of energy consumed and its operation to be at least 98 per cent reliable."

REFERENCES

1. Am. Meter Co. *Displacement Gas Meters.* (Handbook E-4) Erie, Pa., 1952.
2. Peters, F. "Glossary of Meter and Metering Terminology." *A.G.A. Op. Sect. Proc.* 1960: DMC-60-17.
3. "Standard Purchase Specifications for Small Gas Meters of the Diaphragm Type." *A.G.A. Op. Sect. Proc.* 1958: OP-58-2.
4. Larche, T. F. "Sizing of Domestic Types of Meters to Known Gas Loads at Low Pressure." (DMC-59-15) *A.G.A. Op. Sect. Proc.* 1959: D-79-82.
5. Milliron, A. R. "Essential Meter Records." (DMC-58-29). *A.G.A. Op. Sect. Proc.* 1958: D-223-6.
6. U. S. Dept. of Defense. *Sampling Procedures and Tables for Inspection by Attributes.* (MIL-STD-105B) Washington, G.P.O., 1958.
7. Morey, C. V. "Selective Sampling Reduces Metering Costs." *A.G.A. Op. Sect. Proc.* 1960: DMC-60-16.
8. U. S. Nat. Bur. of Standards. *Gas Measuring Instruments.* (Circ. 309) Washington, 1926.
9. Lovretin, A., Jr. "Recent Developments in Demand Metering." *A.G.A. Op. Sect. Proc.* 1960: DMC-60-31.
10. Griffin, G. E. "Demand Metering." *A.G.A. Proc.* 1949: 596–607.
11. Pickford, J. M. and Hoag, R. J. "The Demand Meter Situation." *A.G.A. Proc.* 1948: 395–403.
12. Lovretin, A., Jr. "Challenges in Metering." *A.G.A. Op. Sect. Proc.* 1962: DMC-62-7.
13. Terrell, C. E., ed. *A.G.A. Gas Measurement Manual.* New York, A.G.A., 1963.
14. Price, R. A. and Kemp, L. J. "Standardization of Meter Differential Pressure Determination." *A.G.A. Op. Sect. Proc.* 1962: DMC-62-12.
15. Consumers Power Co., and others. *A Report: Automatic Remote Reading of Residential Meters.* Jackson, Mich., 1964.

Note: Addendum covering international standard units for fuel gas industry will be found on page 7/62.

ADDENDUM

Table 7-16 International Standard Units for the Fuel Gas Industry Recommended by the International Gas Union Subcommittee on Units

1 Item No.	2 Name of item or quantity	3 Symbol	4 Units in SI* system	5 Name of SI unit or units	6 Common U. S. engineering units	7 Conversion factor—to convert col. 4 units to col. 6 units, *divide* by:
Length						
1	Pipeline length	L	m	meter	ft	0.3048
			km	kilometer	miles	1.6093
2	Diameter of pipe	D	mm	millimeter	inch	25.4000
Volume						
3	Gas sales (volume basis)	V	m^3	cubic meter	ft^3	0.028317
4	Standard volume	V_s	m_s^3	cubic meter at 15 C, 1013 mbar, dry	ft^3 at 60 F, 14.73 psia, dry	0.028334
5	Swept volume of a gas meter, compressor, gas engine (displacement)	V (V_d)	dm^3	cubic decimeter = $(0.1m)^3$	$in.^3$	.016387
6	Geometric holder volume		m^3	cubic meter	ft^3	.028317
			dam^3	cubic decameter = $(10m)^3$	Mft^3	.028317
7	Holder capacity		dam_s^3, m_s^3		Mft_s^3, ft_s^3	.028334
8	Underground storage		hm^3, m^3	cubic hectometer = $(100m)^3$ = $10^6 m^3$	$MMft^3 (= 10^6 ft^3)$	0.028317
9	Gas field reserve		km^3	cubic kilometer = $(1000m)^3$ = $10^9 m^3$	$MMft^3$	28.317
Volume rate of flow						
10	Consumption of an appliance, volume basis	Q	dm^3/s	cubic decimeter per sec	ft^3/sec	28.317
			m^3/h	cubic meter per hour	ft^3/hr	0.028317
11	Capacity of a gas meter governor, booster, main		dm_s^3/s	$(0.1m)^3$/sec at std cond	scfs	28.3346
			m_s^3/s	m^3/sec at std cond	scfh	0.28334
Mass						
12	(Absolute) density†	q, (ρ)	g/cm^3	grams/cubic centimeter	lb_m/ft^3	0.01602
			g/ml	grams/milliliter	lb_m/ft^3	16.018
			kg/m_s^3	kilograms/cubic meter	lb_m/ft^3	16.018
13	Specific gravity Relative density†	d, (G)	ratio			
Speed of rotation and frequency						
14	Rate of revolutions (of meters, compressors, engines)	ns^{-1}	n/s	number/sec	rpm	1/60
15	Frequency	Hz	cycles/s	Hertz	cps	1
Pressure						
16	Atmospheric pressure	B	mbar	millibar	in. Hg at 32 F	33.864
17	Gage pressure	p or p_e	bar	bar	psig	0.06895
			mbar	millibar	in. Hg at 60 F	33.769
18	Absolute pressure	p	bar		psia	0.06895
19	Differential pressure	Δp	mbar	millibar	in. H_2O at 60 F	2.488
					in. Hg at 60 F	33.769
			N/m^2	Newton per sq meter (= 10 dynes/sq cm)	in. H_2O at 60 F	248.87
					in. Hg at 60 F	3376.9
Calorific (heating) value and Wobbe index						
20	Gross calorific value	H or H_s	MJ/m_s^3	megajoule/std cubic meter	Btu/scf	0.03722
21	Net calorific value	H_i	MJ/m_s^3	megajoule/std cubic meter	Btu/scf	.03722
22	Relative Wobbe index	W or W_r	MJ/m_s^3	megajoule/std cubic meter	Btu/scf	.03722
	(The) Wobbe index		$(MJ/m_s^3)/\sqrt{sp\ gr}$			
23	(Absolute) Wobbe index	W_a	$(MJ/m_s^3)/\sqrt{kg/m_s^3}$ = $MJ/\sqrt{kg \times m_s^3}$	$(MJ/scm)/\sqrt{std\ density}$ = $MJ/\sqrt{kg \times (std\ cubic\ meter)}$	$(Btu/scf)/\sqrt{lb_m/scf}$	0.009303
Quantity of heat						
24	Gas sales (heat basis)		MJ	megajoule (= 10^6 joule)	therm (= 10^5 Btu)	105.47
			kwh	kilowatt-hour		29.302
Power, heat flow rate						
25	Input, output of an appliance (heat basis)	Q (q)	kw	kilowatt	Btu/hr	0.000293
			w	watt	Btu/min	17.58

* International Standard Units.

† In French, "density" is used to mean relative density, i.e., specific gravity; in England, Germany, and the U. S., density is understood to be "mass per unit volume."

SECTION 8

TRANSMISSION OF GAS

Grove Lawrence, *Section Chairman,* Pacific Natural Gas Exploration Co., Los Angeles, Calif.
Guy Corfield, Editorial Reviewer, Southern California Gas Co., Los Angeles, Calif.
M. J. Binckley, Chapter 1, Southern California Gas Co., Los Angeles, Calif.
M. V. Burlingame, Chapter 4, Natural Gas Pipeline Co. of America, Chicago, Ill.
C. H. Burnham, Chapter 2, Natural Gas Consultant, Merriam, Kan.
Rex V. Campbell, Chapter 3, King Tool Co., Ltd., Longview, Tex.
H. L. Fruechtenicht, Chapter 2, Consumers Power Co., Jackson, Mich.
George W. McKinley, Chapter 4, Con-Gas Service Corp., Pittsburgh, Pa.
D. R. Pflug, Chapters 2 and 5, Houston, Tex.
H. A. Rhodes, Chapter 5, Transcontinental Gas Pipe Line Corp., Houston, Tex.
C. A. Sweningsen, Chapter 2, Southern Counties Gas Co., Los Angeles, Calif.
F. E. Vandaveer (retired), Chapters 3 and 4, Con-Gas Service Corp., Cleveland, Ohio
Carl J. Veit, Chapter 3, The East Ohio Gas Co., Cleveland, Ohio

CONTENTS

Tables

Figures

Chapter 1
Location of Transmission Lines

by M. J. Binckley

INTRODUCTION

Authorization is required from government agencies to construct, operate, and maintain a gas transmission pipeline. Where gas is transported across state lines, the economic feasibility and public necessity of the project must be demonstrated to the Federal Power Commission and authorization obtained from it. Where the gas to be transported is not and has never been in interstate transit, the state utilities commission exercises control, and its authorization must be obtained.

Development of a major gas transmission pipeline project consists of two principal phases. The first is preliminary: to establish economic feasibility of the project and, if gas will move across state lines, to support required applications for governmental authority for it. The second comprises its detailed design, construction, and operation when completed.

Proper preliminary development requires establishment of the source and quantity of gas reserves and evaluation of present and future gas markets. Generally, the gas reserves must be adequate to supply the anticipated market for at least 20 years. When the source of gas and its market areas have been determined, the general area traversed by the pipeline will also be known. Pipe size, maximum operating pressure, compressor station type, size, and spacing must all be determined as a basis for preliminary cost estimates. A reasonably exact route for the pipeline must be selected, in order to discover the existence of any unfavorable conditions which might materially raise the project cost or cause abnormally high operating or maintenance expenses.

Any attempts to base estimates upon assumptions instead of more accurate data from detailed studies should be avoided, particularly where inaccurate estimates could lead to a material increase in pipeline length. It may be expected that several possible routes and various combinations of pipe sizes, compressor horsepowers, and operating pressures will be found worthy of further investigation. Each of these combinations should be subjected to a complete economic study, and the one which best meets present and anticipated future requirements should be selected. Future market growth should be kept fully in mind. Possibilities of later increases in system capacity thru installation of additional compressor stations or loop lines should be evaluated in the preliminary development phase.

Design, construction, operation, and maintenance of gas transmission systems are covered by American Standard Gas Transmission and Distribution Piping Systems (ASA B31.8–1963—Section 8 of American Standard Code for Pressure Piping). Pertinent extracts from this Code are quoted in Chapters 2 and 3. They should be consulted in the preliminary and detailed development of any gas transmission project. There were 194,970 miles of transmission lines in the U. S. at the end of 1962 plus 58,680 miles of field and gathering lines.[1] In addition, there were about 1400 miles of mixed gas transmission lines.

LOCATION PRELIMINARIES

Location study of a gas transmission pipeline project starts with examination of general maps of the area between the selected gas supply source and anticipated markets. This study continues thru the preliminary development phase of the project into that of construction, until the line has been completed. The required degree of precision in establishing location increases as work progresses.

A preliminary location is necessary as a basis for proper economic study of the project. Since pipe and construction costs constitute such a large part of the total, and since system capacity decreases as length increases, choice of the preliminary location should be based upon the most complete study possible within the available time. Major considerations include the economic problems of the project, rights of way acquisition, pipeline and compressor station design, construction costs and methods, and operation and maintenance procedures.

Preliminary studies will ascertain existence of natural and other obstacles like impassable mountains, rivers, swamps, restricted areas, and large communities. Such obstructions, together with availability of favorable compressor station sites and location of present and future market centers, may indicate several possible routes from which the most favorable can be selected.

Many considerations influence the pipeline location. These include the shortest possible route; the function of the line; the degree of security which must be provided against natural or other hazards such as landslides, washout areas, and very corrosive areas like coal or slag piles; contour of terrain; and accessibility for operation and maintenance functions. The problem of deviation from a direct route for any reason becomes increasingly important as the line grows longer. Increased pipe costs and reduction in overall system capacity, resulting from added length required to avoid construction obstacles, often prove unjustified when compared to the construction costs added by following the more direct route. Large diameter transmission lines should not deviate from the most direct route merely to make a closer approach to less important market centers.

Accessibility of a pipeline for operation, maintenance, and repair must be consistent with its function. Where the line is the sole source of gas supply and any interruption would be disastrous to the market served, accessibility for men and equipment to the entire line must be provided at all times, regardless of weather or surrounding conditions.

Reconnaissance of Area

General topographic contour maps like those available from government sources (scale: 1 inch = 0.9864 mile), covering the area between gas source and market, are suitable for reconnaissance purposes. They are used in selection of preliminary locations, in determination of the approximate line length, and in locating important design and construction features such as rivers, mountains, communities, congested valleys, and intensively cultivated areas. Aerial photographs, if available, are quite useful as supplements to maps.

Examinations from the air and ground of the area to be traversed by the proposed line are essential for proper evaluation of maps and photographs, as well as for more detailed study of any location where construction difficulties are indicated.

The primary purpose of reconnaissance should be to establish the practicability of several possible routes and to develop only the details required to form the basis for cost estimates to be used in the economic study. Secondary purposes are to permit development of policies covering franchises and private rights of way and, after their adoption, to permit immediate progress in obtaining the rights required.

Where possible routes are not widely separated and are limited in number, aerial photographs should be taken. They will aid immediately in selecting the most favorable general route and will later be useful during the detailed phase of the project. These photographs, on some convenient scale such as 1 inch = 500 ft or 1000 ft, can later serve as bases for construction drawings and permanent mapping of the system.

Reconnaissance of the general area should be directed particularly to disclosure of crossings of major rivers or streams, since the availability of suitable stream crossings will determine the route of the line. Crossings are of two general types, underwater and overhead. The characteristics of an area suitable for an underwater crossing are generally entirely unsuitable for an overhead crossing; the reverse is also true. A reasonably accurate evaluation of the possibilities of each site must be made, to permit best selection. Since permission to cross navigable streams at the desired points may not always be obtainable, a general route should be plotted with reasonable assurance that the crossing points will be approved.

Photogrammetry, the application of photographic principles to the science of mapping, uses photographs from special aerial cameras, either directly to form a picture, or indirectly to make a scale map. Accurate or controlled mosaics require establishment of known ground points. A discussion of the techniques pertinent to pipeline routing is available.[2]

Field Surveys

As a rule, field surveys should be conducted only as far and as accurately as required to meet immediate needs. Much time will be lost and essential activities delayed by attempting to make a final survey during the preliminary development phase of the project.

During location preliminaries, field surveys will be required for ground control of aerial photography and for an occasional approximate ground tie to correlate aerial photographs with terrain features or property lines.

For acquisition of rights of way, the extent of detailed surveys will depend upon the policy adopted for such acquisition. Under a policy of obtaining options, field surveys will be limited to determining the approximate point of entrance and exit to each land parcel. Where landowners refuse to grant options or where the policy is to obtain initially a precisely defined easement, a detailed and accurate field survey will be necessary. Where the character of the terrain is such that a serious error might result from basing the length of the line upon distances scaled from a map, more exact measurements may be required to determine the necessary amount of pipe.

Before requesting bids from contractors, a field survey must be made to plan the project fully and to give the bidders as complete a description as possible of the work to be performed. Most attention should be directed to those matters which require more time for investigation than can be allowed the bidders for that purpose. Any elements which will be apparent to the bidders in their examination of the location can be omitted. For their benefit, the proposed line should be flagged to indicate general or exact location, as the local situation requires. Field surveys will be necessary immediately preceding actual construction and throughout the job, until each easement is located exactly and all physical features of the pipeline are determined.

RIGHTS TO CONSTRUCT, OPERATE, AND MAINTAIN

General types of rights to construct, operate, and maintain gas transmission pipelines are: (1) franchises for use of public roads, and (2) rights covering use of other lands.

Franchises are rights to construct, operate, and maintain pipelines in public roadways, granted by appropriate state, county, or municipal governments. Granting of such rights may be subject to a tax imposed by a government agency. Pipeline companies have to relocate their facilities when so required by a government agency.

Rights to use other lands are obtained by grants from the landowners, who may include the several government agencies and departments; federal, state, and municipal corporations; private companies; and individuals. Such rights are granted by outright sale of the property, by perpetual or limited term easements, or by revocable licenses. The perpetual easement is the most common form of right obtained by pipeline companies.

While pipeline companies are usually recompensed when they have to give up water rights of way for public projects, they may or may not receive compensation for surrendering their franchises for public rights of way.

The right to use certain classifications of government-owned properties, particularly federal, is often difficult or impossible to obtain. This is the case especially with national parks, national monuments, Defense Department-owned or controlled areas, and navigable rivers. The right to use Tribal Indian lands can be obtained only by a grant from the Tribal Council. Such lands cannot be condemned.

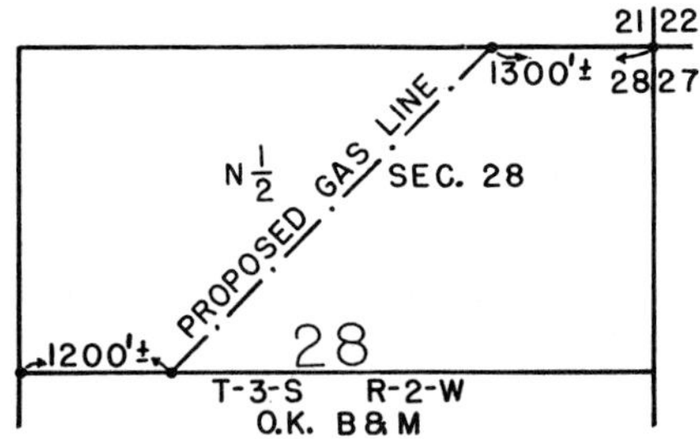

Fig. 8-1 General option covering entire parcel; distances approximate as scaled from aerial photo.

General Policies Covering Rights[3]

Policies must be established concerning the use of rights obtained by franchise, contrasted with those obtained from property owners. The total franchise tax payments over the life of the project must be related to the possibly higher initial cost of private rights of way and accumulated charges.

The policy adopted for acquisition of rights of way will be influenced by the terrain, size of the parcels, state of their development, and time available for acquisition before start of construction. In some cases, where it can be foreseen that the need of future increased system capacity will be met by the addition of paralleling lines, outright purchase of right of way may prove economical. Limited term easements or revocable licenses should be avoided, even if a slightly less desirable location must be used.

Sites for installations requiring full use of the land surface, such as compressor stations, metering stations, or employee housing, are generally purchased outright or leased for a long term.

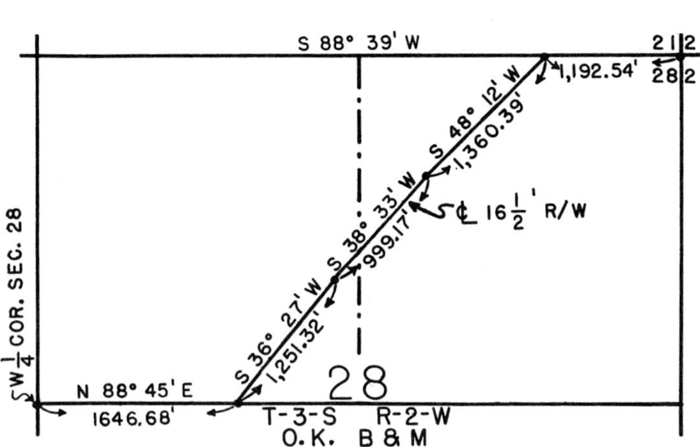

Fig. 8-2 Final drawing for a permanent $16\frac{1}{2}$ ft wide easement with pipeline at center. Surveyed after line was constructed under option obtained (Fig. 8-1).

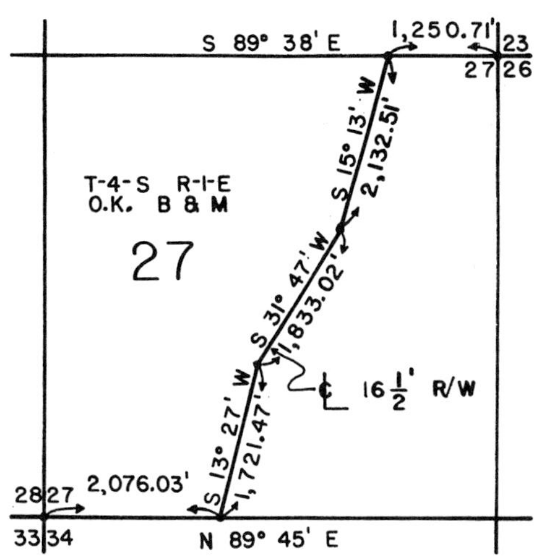

Fig. 8-4 Final location drawing of a pipeline at center of $16\frac{1}{2}$ ft easement. Surveyed after line was constructed under option obtained (Fig. 8-3).

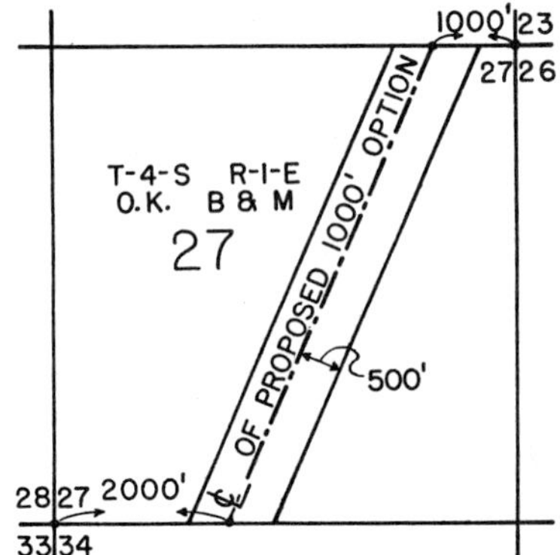

Fig. 8-3 Option drawing restricted to a 1000 ft band across a parcel; (entry and exit points established by scaling maps, photos, or approximation on ground).

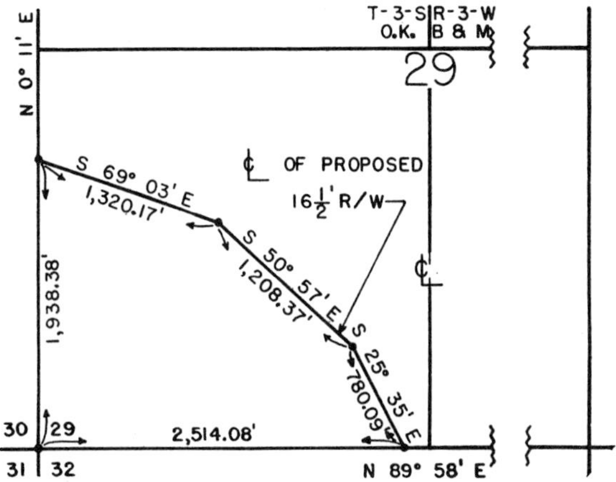

Fig. 8-5 Typical detailed survey where owner granted no option; work must be completed before an easement is granted.

Most gas transmission pipeline companies are public utilities and have the right of eminent domain thru which it is possible for them to acquire necessary rights of way by condemnation proceedings. This method is generally time-consuming and invariably creates adverse public reaction.

Difficulties in Securing Rights of Way. Overcoming difficulties encountered in obtaining authorities, rights, and permits may require all types of business negotiations. These difficulties include many situations requiring time-consuming procedures to obtain approval or decisions from government agencies and courts. Situations involving estates, missing owners, minors, incompetents, or family differences, are frequently encountered. Corporations often put up resistance where the pipeline project will create competition for them. Opposition based on differences over public vs. private ownership may be met in franchise negotiations.

Personnel. The kind of organization required to locate and acquire rights of way for transmission lines naturally varies with size of the project, length of the line, and time available for its completion. Two groups are usually needed: (1) an engineering force to select the desired location and perform the necessary mapping work, and (2) a legal staff to negotiate for and acquire the necessary rights for construction and operation of the line. Management policies with respect to right of way personnel show a variety of duties.[4]

Acquisition of Easements. Both perpetual and limited term easements in final form generally require a precise description of the area based upon an accurate field survey. This is a time-consuming procedure and any part of it that can be deferred will expedite the overall task of obtaining construction rights.

Some companies first obtain from the property owners an option granting as broad rights of location as possible on the property; after construction they determine the precise location of the easement for inclusion in the permanent document. The option method permits taking initial action toward obtaining the right to proceed as soon as it becomes known that a parcel of property is to be crossed. If the option covers the entire parcel, no survey is required, since the property description is sufficient. Where the option is to cover only a portion of the property, such as a wide strip, that strip can be described by using distances and bearings scaled from maps or photographs or determined by crude survey methods on the ground.

Options allow the location engineer the maximum freedom in making detailed changes to meet requirements of adjoining landowners. Also, during construction, they permit him the best use of the terrain to ensure security, less costly construction, and a minimum of property damage. Selection of the final detailed location may also be deferred until it becomes necessary to clear and grade the working strip.

Where several possible routes are being considered and time is limited, the option method, with only a partial payment made on granting, permits investigation of the several routes, pending a final decision. Advantages gained from the time thus saved usually more than offset the loss of the partial payments made on the rejected routes.

Figures 8-1 thru 8-5 show various forms of drawings involved in securing easements.

Compensation to Property Owners. Since the easement is only a *part* of the ownership, compensation paid to the property owner is usually less than the market value per square foot or acre of the land acquired for right of way purposes. Each case, however, is subject to negotiation. Costs may vary widely, depending upon location of the land, its current or possible future use, and other factors which normally influence real estate transactions.

PUBLIC RELATIONS

Several elements of pipeline location are very important from a public relations standpoint. Good judgment, therefore, must be exercised in selecting line locations that will be generally acceptable to the public.

After the route for a line has been selected and it becomes necessary to enter private lands for the purpose of mapping and staking out locations, it is good practice to inform owners and residents before undertaking the work. Once construction work is under way, adequate supervision should be exercised over the construction crew or the independent contractor, to make sure that work causes the least possible inconvenience and expense to property owners and the public. Some companies have found that claims for damages due to construction work can best be adjusted by representatives of the right of way department. Having originally negotiated with the property owners, they fully understand the commitments and arrangements made in advance of construction.

LOCATING BUILDINGS NEAR PIPELINES

Existing pipelines have fully demonstrated that they are safe structures.[5,6] Sufficient clearance to permit adequate maintenance and repair work on the pipeline is one accepted procedure for determining the amount of land that is left clear. The pipeline owner is in the best position to specify the space requirements for maintenance of the particular line in question and negotiate for an acceptable solution should a problem exist.

In addition, building location should satisfy two criteria:

1. Local codes with respect to the building lines on the lot.
2. Building restrictions, if any, included in the right of way of the gas line owner.

REFERENCES

1. A. G. A. *Gas Facts.* New York, annual.
2. Barry, B. A. "Pipeline Engineering by Photogrammetry." *A.G.A. Proc.* 1961: GSTS-61-4.
3. Dunshee, B. K. and Putnam, H. I. "On Making Friends and Getting Rights-of-Way." *A.G.A. Nat. Gas Dept. Proc.* 1952: 17–21.
4. Spangler, W. S. "Present Management Concepts of Right of Way Acquisition." *American Right of Way Assn. Seventh Annual National Seminar Proc.* 1961. Also in: *Right of Way* 8: 9–14, Aug. 1961.
5. Owens, S. "Pipeline Operations Found Safe." *A.G.A. Monthly.* 44: 39–40, Jan. 1962.
6. ——. "Study Confirms Pipeline Safety." *A.G.A. Monthly.* 40: 2+, Dec. 1958.

Chapter 2
Transmission Pipelines

by C. H Burnham, H. L. Fruechtenicht, D. R. Pflug, and C. A. Sweningsen

PIPELINE FLOW FORMULAS

For isothermal gas flow in horizontal lines, the general steady flow equation derived[1] from an energy balance* over a pipeline may be expressed as:

$$Q = 38.77 \frac{T_b}{P_b}\left(\frac{1}{f}\right)^{0.5}\left[\frac{P_1^2 - P_2^2}{GTLZ}\right]^{0.5} D^{2.5} \qquad (1)$$

where:

Q = flow rate, cu ft/day at P_b and T_b; when Q is expressed in cfh the constant becomes 1.616; when Q is expressed in cfh and L is expressed in feet, the constant is 117.4.
P_b = base pressure, psia
T_b = base temperature, °F abs, °R
$(1/f)^{0.5}$ = transmission factor (Table 8-1), dimensionless
f = friction coefficient, dimensionless
P_1 = inlet pressure, psia
P_2 = outlet pressure, psia
G = gas specific gravity (air = 1.0)

* Flow is at sufficiently high velocity to neglect change in kinetic energy; kinetic energy may be considered in short sections.

T = average flowing gas temperature, °F abs, °R
L = length of pipe, miles
D = inside diameter of pipe, in.
Z = compressibility factor at *average* conditions, dimensionless; may be determined at T and at *average* pressure, P_m, where:

$$P_m = 0.6667(P_1^3 - P_2^3)/(P_1^2 - P_2^2)$$

(Z may be omitted for pressures under 100 psig[2]; see Eq. **4** when pressure is unknown.)

Some remarks as to the applicability of various transmission flow formulas are given in Table 8-1. It is suggested that the flow formula selected be "assembled" by inserting the factor from Table 8-1 (Table 9-29 for distribution formulas) into Eq. **1**; this procedure assures that all of the variables involved are accounted for. For ready reference, however, the following three formulas (in the same units as Eq. **1** and applying efficiency factors from Table 8-1) are given:[4]

Weymouth:

$$Q = 433.5 \frac{T_b}{P_b}\left[\frac{P_1^2 - P_2^2}{GTLZ}\right]^{0.5} D^{2.667} \qquad \textbf{(1a)}$$

Table 8-1 Transmission Factors for Flow Formulas

(for use in Eq. **1**; terms defined at end of table)

Flow formula	Transmission factor, $(1/f)^{0.5}$	Remarks
Smooth-pipe[1] (laminar)	$4 \log_{10}(f^{0.5}R) - 0.6$	Seldom applicable to large diameter natural gas transmission lines.[3] Table 8-2 gives a number of solutions to this implicit function. Approximated[3] by the factor for the "improved flow equation," $(1/f)^{0.5} = 5.18R^{0.0909}$ with a deviation of ±1 per cent for Reynolds numbers from 60,000 to approximately 7,000,000.
Rough-pipe (fully turbulent)	$4 \log_{10}(3.7D/k)$	Characterizes most natural gas transmission operating conditions.[3] Table 8-3 gives a number of solutions.
Weymouth[4]	$1.10^* \times 11.2D^{0.167}$	Reasonably good approximation of preceding (rough-pipe) formula for $D = 10$ in. and $k = 0.002$. Constant is for pipe over 24 in. ID.
Panhandle "A"[4]	$0.92^* \times 6.87R^{0.073}$	Large diameter transmission piping where R varies from 5 million to 20 million.
New Panhandle[4]	$0.90^* \times 16.5R^{0.0196}$	

* Average steel pipeline efficiency factor; aluminum varies from 0.92 to 0.96.
R = Reynolds number (see Eq. **2**).
f = friction coefficient = reciprocal of transmission factor squared.
D = inside diameter of pipe, in.
k = effective roughness (function of height of surface irregularities), in.

Panhandle "A":

$$Q = 435.7 \left(\frac{T_b}{P_b}\right)^{1.0788} \left[\frac{P_1^2 - P_2^2}{G^{0.8539}TLZ}\right]^{0.5392} D^{2.6182} \quad \textbf{(1b)}$$

New Panhandle:

$$Q = 737 \left(\frac{T_b}{P_b}\right)^{1.02} \left[\frac{P_1^2 - P_2^2}{G^{0.961}TLZ}\right]^{0.51} D^{2.53} \quad \textbf{(1c)}$$

Table 8-2 Selected Smooth Pipe Transmission Factors[1]

$(1/f)^{0.5} = 4 \log_{10}(f^{0.5}R) - 0.6$

Reynolds No., R, (in millions)	Factor, $(1/f)^{0.5}$
0.0327	13.0
.0625	14.0
.119	15.0
.226	16.0
.427	17.0
0.804	18.0
1.51	19.0
2.83	20.0
5.28	21.0
9.83	22.0
18.3	23.0
33.9	24.0

Transmission Factors

Of the many variables in Eq. **1**, the transmission factor, $(1/f)^{0.5}$, has long been the most difficult to evaluate. Thus, the literature contains many different empirical transmission factors which have been used to meet the needs of pipeline engineers. This chapter does not contain all of the many variations which have been used to date. Instead, the authors have pre-

Table 8-3 Selected Rough Pipe Transmission Factors[1]

$(1/f)^{0.5} = 4 \log_{10} (3.7D/k)$ where $k = 0.0007$ in.*

D, pipe ID, in.	$(1/f)^{0.5}$
10.00	18.9
13.50	19.4
15.44	19.7
19.38	20.0
23.25	20.4
25.31	20.5
29.25	20.8

* Average effective roughness for clean steel pipe in transmission service. Calculated values of effective roughness of commerical gas pipe ranged from 0.000468 to 0.00209 and seemed to be independent of pipe diameter for diameters from 2 to 36 in.[1,3]

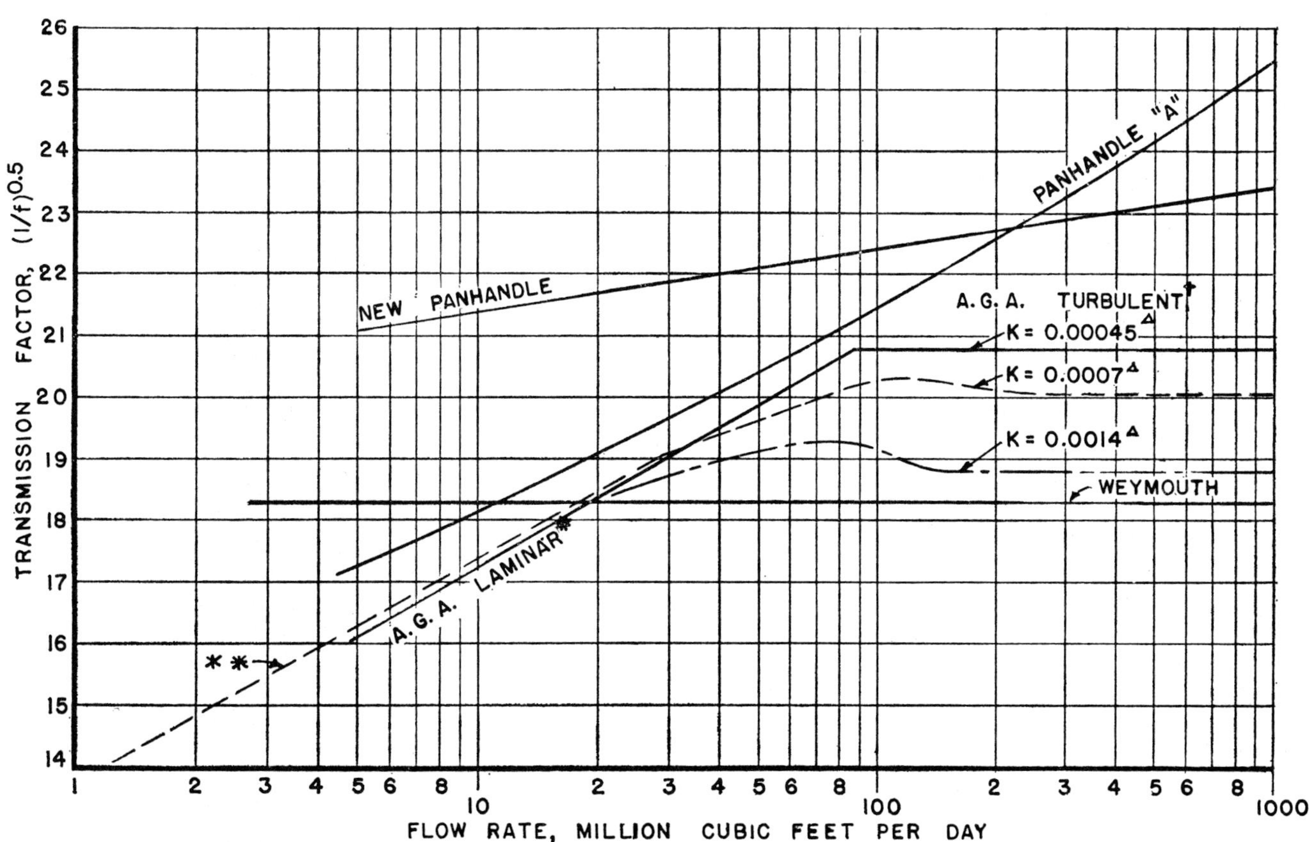

Fig. 8-6 Transmission factor vs. flow thru 20 in. (19.25 in. ID) pipeline using various flow formulas. Curves shown dotted are based on experimental test data from Bureau of Mines Monograph 9.

* $(1/f)^{0.5} = 5.18R^{0.0909}$, "the improved flow equation," which is an approximation of the smooth-pipe formula.
† $(1/f)^{0.5} = 16.21D^{0.084}$, an approximation of the rough-pipe formula.
△ $(1/f)^{0.5} = 4 \log_{10}(3.7D/K)$, the rough-pipe formula.
** $(1/f)^{0.5} = 4 \log_{10} (f^{0.5}R) - 0.6$, the smooth-pipe formula.

sented those transmission factors (Table 8-1) which they believe to be the most significant and have either best stood the tests of usage or have strong foundations in basic flow theories. The factors given in Table 8-1 were plotted on Fig. 8-6 for 20 in. pipe. Generally, similar curves apply to other sizes.[1]

Regarding the latter, comparison of the first two expressions in Table 8-1 shows that for laminar flow (low Reynolds number, R) the transmission factor is a function of R, while for turbulent flow (high Reynolds number) the transmission factor is a function of pipe roughness, k, and diameter, D. Between these two stable ranges lies the *partially turbulent* or transition range.

Reynolds Number

Equation **2** expresses the dimensionless Reynolds number, R, in natural gas engineering terms:

$$R = 477.5 \times 10^{-6} \frac{GQP_b}{VDT_b} \tag{2}$$

where:

V = absolute gas viscosity, lb-mass per sec-ft; average value of 7.4×10^{-6} may be used. Note Fig. 1-5 (V in lb-mass per sec-ft = micropoises/14.882).

Other terms are defined under Eq. **1**.

Compensation for Elevation

Account should be taken of substantial pipeline elevation changes.[1,5] Equation **3**, a rearrangement and modification of Eq. **1**, highlights the additional terms which must be evaluated:

$$Q = A \left[\frac{P_1^2 - e^S P_2^2}{L_e}\right]^{0.5} \tag{3}$$

where: $A = 38.77 \dfrac{T_b}{P_b}\left(\dfrac{1}{f}\right)^{0.5}\left[\dfrac{1}{GTLZ}\right]^{0.5} D^{2.5}$

e = base of natural logarithms = 2.718

$S = 0.0375GH/TZ$

H = outlet elevation minus inlet elevation, ft

For a uniform slope:

$$L_e = (e^S - 1)L/S$$

For a non-uniform slope (where elevation change cannot be simplified to a single section of constant gradient, an approach in steps to any number of sections, n, will yield:

$$L_e = (e^{S_1} - 1)L_1/S_1 + e^{S_1}(e^{S_2} - 1)L_2/S_2 + e^{S_1 + S_2}(e^{S_3} - 1)L_3/S_3 + \ldots + e^{\Sigma_1^n S_{n-1}}(e^{S_n} - 1)L_n/S_n$$

where: $S_1 = 0.0375GH_1/TZ$,

$S_2 = 0.0375GH_2/TZ$, . . .

$S_n = 0.0375GH_n/TZ$

Other terms are defined under Eq. **1**.

Note: $S = S_1 + S_2 + S_3 + \ldots + S_n$. The numerical subscripts refer to the individual sections of the overall line which is operating under pressure differential, $P_1 - P_2$; for example, if the line is divided into four sections, n would equal 4 and subscript 2 would pertain to the properties of the second section. A method of facilitating the solution of L_e is available.[19]

Solution When Pressure Is Unknown

Since compressibility Z in Eq. **1** was derived in terms of pressures, the following determination[6,7] for Z may be used when inlet or outlet pressure is unknown:

$$\frac{1}{Z} = 1 + JP \tag{4}$$

where:

Z = compressibility factor, dimensionless

J = a function of gas composition and temperature, which may be found from compressibility correlations; i.e., reduced pressures and temperatures or test data. J is often referred to as compressibility per pound

P = pressure, absolute

As an alternative Eq. **1** may be modified as follows:

$$Q = 38.77 \frac{T_b}{P_b}\left(\frac{1}{f}\right)^{0.5} \left\{\frac{P_1^2[1 + (2/3)JP_1] - P_2^2[1 + (2/3)JP_2]}{GTL}\right\}^{0.5} D^{2.5} \tag{5}$$

Terms are defined under Eqs. **1** and **4**.

Values for the general term $P^2[1 + (2/3)JP]$ of Eq. **5** are given in Table 8-4 for a 1020 Btu per cu ft (gross)dry natural gas, 0.670 sp gr, and negligible CO_2. Thus trial-and-error solutions of Eq. **5** in conjunction with Table 8-4 do not require knowledge of the compressibility factor, Z.

Table 8-4 $P^2\,[1 + (2/3)\,JP]$ **vs.** P **for a Natural Gas**[7]
(used in Eq. 5)

P^2 [1 + (2/3) JP]	P, psia
10,000	100
40,000	200
90,000	300
168,000	400
261,000	500
380,000	600
525,000	700
695,000	800
880,000	900
1,105,000	1000

Equivalent Lengths

When a pipeline consists of several different pipe sizes or of looped or of parallel line sections, it may be conveniently reduced to an equivalent length of a selected pipe size, D_s, and handled as a single line of that diameter.

The *equivalent length,** which will pass gas at the same rate and with the same total pressure drop as pipe with any L and D dimensions, may be calculated from:

$$L_E = \left(\frac{D_s}{D}\right)^A L \tag{6}$$

* Different meanings are also assigned to this term.

where: L_E = equivalent length of selected size (e.g., Table 4-28)

D_s = ID of selected size
L = length of actual pipe of size D
D = ID of actual pipe section
A = exponent from Table 8-5

Table 8-5 Exponents for Equations 6, 7 and 8

	A	B	C
For Weymouth Formula[8]	5.333	0.50	2.667
For Panhandle "A"[9]	4.854	0.5394	2.618
For New Panhandle[7]	4.961	0.51	2.53

Series and Parallel Circuits. The equivalent length of different sizes of pipe connected in series is the sum of the individual equivalent lengths. When lines are connected in parallel their combined equivalent length is derived from:

$$L_{EP} = \left[\frac{1}{(1/L_{E1})^B + (1/L_{E2})^B + (1/L_{E3})^B + \ldots}\right]^{1/B} \quad \textbf{(7)}$$

Terms are defined under Eq. **6** and exponents in Table 8-5.

If each of the parallel lines is of a single pipe size and of equal length, L, Eq. **7**, may be expressed as:

$$L_{EP} = \left[\frac{1}{D_1{}^C + D_2{}^C + D_3{}^C + \ldots}\right]^{1/B} D_S{}^A L \quad \textbf{(8)}$$

Terms are defined under Eq. **6** and exponents given in Table 8-5.

TWO-PHASE FLOW

When gas and condensate are brought to a central treating plant thru common large diameter trunk gathering lines, they can be considered as more main transmission lines.

Drops in transmission pressure in a horizontal two-phase gas-condensate line are greater than in a dry gas line. Even relatively small amounts of liquid increase such pressure drops considerably. Contributing factors are (1) energy lost in accelerating and transporting the liquid; (2) increased roughness caused by wetting the pipe wall and by waves on the surface of the liquid; and (3) reduction in available flow area due to introduction of a liquid phase. In hilly terrain, even more energy is lost in lifting the liquid over individual rises. This additional pressure drop appears to depend largely upon the gas velocity in the pipeline.

Six main types of flow patterns, as shown in Fig. 8-8, have been recognized in two-phase flow. Assuming a horizontal line initially running full of flowing liquid, the following flow types may be anticipated due to effects of adding increasing volumes of gas:

1. *Bubble Flow.* Separate gas bubbles travel along the upper part of the pipe at about the same velocity as the liquid.
2. *Stratified Flow.* Liquid flows along the bottom of the pipe and gas flows above it with a relatively smooth interface.
3. *Wave Flow.* This is similar to stratified flow, except that the gas moves at higher velocity and the interface is disturbed by a wave action.
4. *Slugging Flow.* The interface level rises and falls, with frothy slugs forming periodically which pass along the pipe at higher velocities than the average liquid velocity.
5. *Annular Flow.* Liquid flows in a uniform film on the pipe wall, while gas travels at high velocity thru the core.
6. *Fog or Dispersed Flow.* Most of the liquid is entrained as fog or spray at extremely high gas velocity.

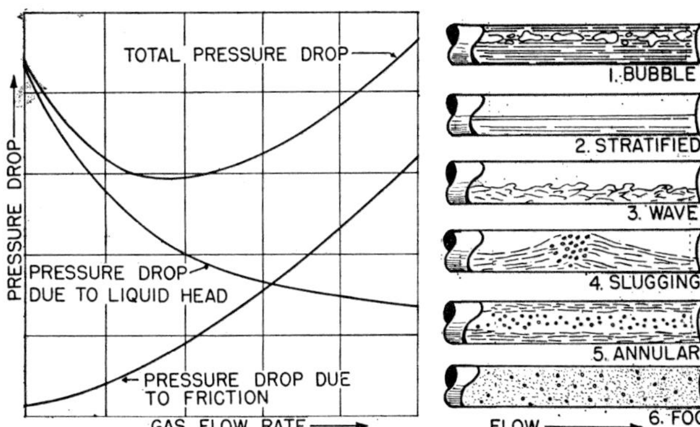

Fig. 8-7 (left) Pressure drops in two-phase flow thru hilly terrain.

Fig. 8-8 (right) Types of flow patterns in a horizontal two-phase pipeline.

Figure 8-7 illustrates the effect of pressure drops due to friction and liquid head on hills. The total pressure drop is the sum of these two major components. Locating an optimum region of minimum total pressure drop is an important part of designing any system in hilly terrain.

When a two-phase system is expanded to where it requires a large-diameter trunk line extension connecting several fields many miles apart, possibilities other than stabilized two-phase flow should be considered. For example, lowest transportation costs of both liquid and gas may sometimes be achieved by separate gas and liquid lines rather than with the additional horsepower required on the two-phase system. In other cases, routine on-stream pigging of wet lines may reduce overall pressure drop. The design situation may be rather complex, involving viscosity, density, and many other physical and mechanical variables.

Field test work and fundamental research are now being conducted on this problem. A joint A.G.A.–API committee was established in 1959 for a Multiphase Flow Study, designated Project NX-28 and scheduled to be completed in 1964. Its goal is to determine horizontal, vertical, and inclined flow characteristics of multiphase fluids in pipelines of two inches or larger, and to set up formulas for line sizing.

PIPELINE DESIGN

Piping System Symbols for Mapping[10]

Figure 8-9 shows a group of easily interpreted symbols for mapping of pipeline systems. They have been assembled from the suggestions of more than 60 gas distribution and transmission companies to simplify map-making and reading.

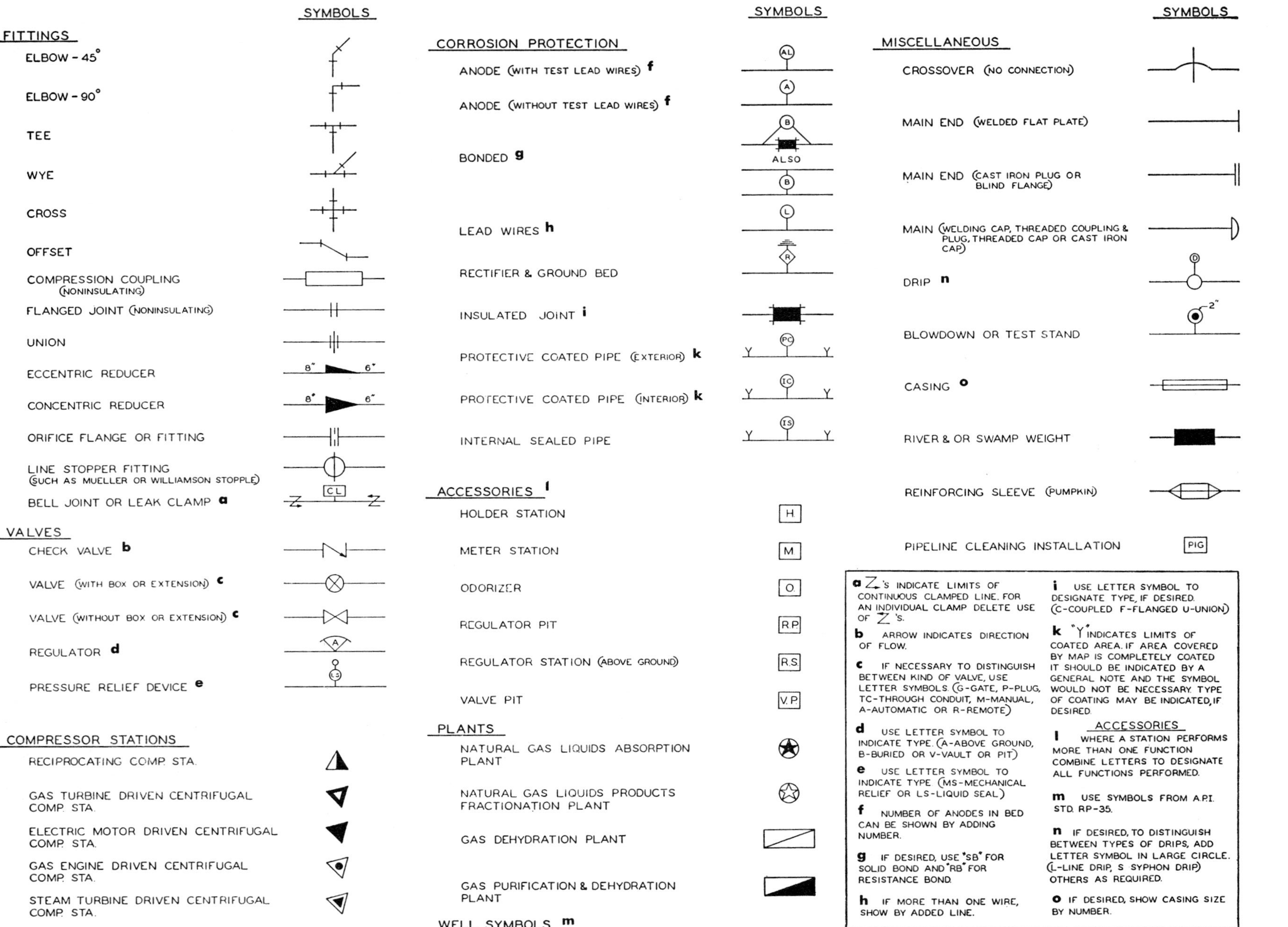

Fig. 8-9 Symbols for mapping piping systems.[10] (Also see Graphic Symbols for Pipe Fittings, Valves, and Piping, ASA Z32.2.3-1949, rev. 1953.)

Table 8-6 Abbreviations for Mapping Pipeline Systems[10]

	Symbols		Symbols
Mains:*		Type joint:	
Existing	———	Bell & spigot	B.S.
Proposed	-—-—--	Compression	C.
Pipe material:		Mechanical	M.J.
Steel	S.	Threaded	T.
Cast iron	C.I.	Welded	W.
Ductile iron	D.I.	Pressure:†	
Wrought iron	W.I.	High	H.P.
Copper	CU.	Medium	M.P.
Plastic	PL.	Low	L.P.

* It is suggested that lines representing gas mains be very heavy on the drawing so that they show up distinctly. It is also recommended that line size, operating pressure, material and type of joint be designated by letters and figures. Year of installation may also be shown. For example, 2″ S.-M.P.-T. written above ——— would mean two-inch existing pipe, steel, medium-pressure, threaded.

† Pressure limits for each category to be established by individual companies. Other similar designations may be adopted by individual companies to meet varying requirements.

General Provisions

For general information on pipeline design, reference may be made to the following extracts from ASA B31.8-1963:

811 QUALIFICATIONS OF MATERIALS AND EQUIPMENT.

811.1 Materials and equipment fall into five categories insofar as methods of qualification for use under this code are concerned:

(a) Items which conform to standards or specifications listed in this code.

(b) Items that are important from a safety standpoint, of a type for which standards or specifications are listed in this code, but the specific item in question does not conform to a listed standard. Example: Pipe manufactured to a specification not listed in the code.

(c) Items of a type for which standards or specifications are listed in this code but which do not conform to the standards and are relatively unimportant from a safety standpoint because of their small size or because of the conditions under which they are to be used.

(d) Items of a type for which no standard or specification is listed in this code: Example: Gas compressor.

(e) Unidentified or used pipe.

811.252 Used steel pipe, unidentified new steel pipe, and steel pipe purchased under Specification ASTM A 120 may be used for low-stress level service (hoop stress less than 6000 psi) where no close coiling or bending is to be done, provided careful visual examination indicates that it is in good condition, free from split seams or other defects that would cause leakage, and provided further that, if the pipe is to be welded and is of unknown specification or ASTM Specification A 120, it shall satisfactorily pass weldability tests prescribed in 811.253 E.

811.253 Used steel pipe, unidentified new steel pipe, and steel pipe purchased under Specification ASTM A 120 may be qualified for use at stress levels above 6000 psi or for service involving close coiling or bending by the procedures and within the limits outlined in the table below.

The letters in the table refer to the corresponding paragraphs following:

	New or used pipe unknown or ASTM A-120 Specification	Used pipe known Specification (ASTM A-120 excluded)
Inspection	A	A
Bending Properties	B	
Thickness	C	C
Joint Efficiency	D	D
Weldability	E	
Defects	F	F
Yield Strength	G	
S Value (841.1)	H	
Test	I	I

A. *Inspection.* All pipe shall be cleaned inside and outside, if necessary, to permit good inspection, and shall be visually inspected to insure that it is reasonably round and straight, and to discover any defects which might impair its strength or tightness.

B. *Bending Properties.* For pipe 2 inches and under in nominal diameter, a sufficient length of pipe shall be bent cold through 90° around a cylindrical mandrel, the diameter of which is twelve times the nominal diameter of the pipe, without developing cracks at any portion and without opening the weld.

For pipe larger than 2 inches in diameter, flattening tests as prescribed in ASTM A 53 shall be made. The pipe shall meet the requirements in this test except that the number of tests required to determine flattening properties shall be the same as required in Paragraph G below to determine yield strength.

C. *Determination of Wall Thickness.* Unless the nominal wall thickness is known with certainty, it shall be determined by measuring the thickness at quarter points on one end of each piece of pipe. If the lot of pipe is known to be of uniform grade, size and nominal thickness measurement shall be made on not less than 10% of the individual lengths, but not less than 10 lengths; thickness of the other lengths may be verified by applying a gage set to the minimum thickness. Following such measurement, the nominal wall thickness shall be taken as the next commercial wall thickness below the average of all the measurements taken, but in no case greater than 1.14 times the least measured thickness for all pipe under 20-inches OD, and no greater than 1.11 times the least measured thickness for all pipe 20-inches OD and larger.

D. *Joint Efficiency.* If the type of longitudinal joint can be determined with certainty, the corresponding Longitudinal Joint Factor "E" (Table 841.12) may be used. Otherwise, the factor "E" shall be taken as 0.60 for pipe 4 inches and smaller, or 0.80 for pipe over 4 inches.

E. *Weldability* shall be determined as follows: A qualified welder shall make a girth weld in the pipe. The weld shall then be tested in accordance with requirements of API Standard 1104. The qualifying weld shall be made under the most severe conditions under which welding will be permitted in the field and using the same procedure as to be used in the field. The pipe shall be considered weldable if the requirements set forth in API Standard 1104 are met. At least one such test weld shall be made for each 100 lengths of pipe in sizes over 4 inches in diameter. On sizes 4 inches and under one test will be required for each 400 lengths of pipe. If, in testing the weld, the requirements of API Standard 1104 cannot be met, the weldability may be established by making chemical tests for carbon and manganese (see 824.23), and proceeding in accordance with the provisions of ASA B31.1, Section 6, Chapter IV. The number of chemical tests shall be the same as required for circumferential weld tests, stated above.

F. *Surface Defects.* All pipe shall be examined for gouges, grooves and dents, and shall be qualified in accordance with the provisions of 841.24.

G. *Determination of Yield Strength.* When the manufacturer's specified minimum yield strength, tensile strength or elongation for the pipe is unknown, and no physical tests are made, the minimum yield strength for purposes of design shall be taken as not more than 24,000 psi. Alternately, the tensile properties may be established as follows:

Perform all tensile tests prescribed by API Standard 5LX, except that the number of such tests shall be as follows:

Number of Tensile Tests—All Sizes

Lot of	
10 lengths or less	1 Set of tests from each length
11 to 100 lengths	1 Set of tests for each 5 lengths, but not less than 10
Over 100 lengths	1 Set of tests for each 10 lengths, but not less than 20

All test specimens shall be selected at random.

If the yield-tensile ratio exceeds .85, the pipe shall not be used except as provided in 811.252.

H. *S Value.* For pipe of unknown specification, the yield strength, to be used as *S* in the formula of 841.1, in lieu of the specified minimum yield strength shall be 24,000 psi, or determined as follows:

Determine the average value of all yield strength tests for a uniform lot. The value of *S* shall then be taken as the lesser of the following:

(1) 80% of the average value of the yield strength tests.

(2) The minimum value of any yield strength tests, provided, however, that in no case shall *S* be taken as greater than 52,000 psi.

I. *Hydrostatic Test.* New or used pipe of unknown specification and all used pipe the strength of which is impaired by corrosion or other deterioration, shall be re-tested hydrostatically either length by length in a mill type test or in the field after installation before being placed in service, and the test pressure used shall establish the maximum allowable operating pressure subject to limitations described in 841.14 (a) and (b).

Pipe purchase specifications do not provide a means of controlling fracture toughness. A research report on this and other properties is available.[17]

PIPING SYSTEM COMPONENTS

Reference may be made to the following extracts from ASA B31.8-1963:

831.2 Flanges.

831.21 *Flange Types and Facings.*

(a) The dimensions and drilling for all line or end flanges shall conform to the following standards:

ASA B16 Series—(For Iron and Steel)
MSS SP-44—Steel Pipe Line Flanges
Appendix H—Light-Weight Steel Flanges
ASA B16.24—Brass or Bronze Flanges and Flanged Fittings.

Flanges cast or forged integral with pipe, fittings or valves will be permitted in sizes and for the maximum service rating covered by the Standards listed above, subject to the facing, bolting and gasketing requirements of this paragraph and 831.22 and 831.23.

(b) Screwed companion flanges which comply with the B16 group of American Standards will be permitted in sizes and for maximum service ratings covered by these standards.

(c) Lapped flanges will be permitted in sizes and pressure standards established in the American Standard ASA B16.5.

(d) Slip-on welding flanges will be permitted in sizes and pressure standards established in American Standard ASA B16.5 and MSS SP-44. Slip-on flanges of rectangular section may be substituted for hubbed slip-on flanges provided the thickness is increased as required to produce equivalent strength as determined by calculations made in accordance with Section VIII, Unfired Pressure Vessels, of the ASME Boiler and Pressure Vessel Code.

(e) Welding neck flanges will be permitted in sizes and pressure standards established in American Standard ASA B16.5 and MSS SP-44. The bore of the flange should correspond to the inside diameter of the pipe used. For permissible welding end treatment see Fig. 823-B.

(f) Cast iron and steel flanges shall have contact faces finished in accordance with MSS Standard Finishes for Contact Faces of Connecting-End Flanges of Ferrous Valves and Fittings, SP-6.

(g) Nonferrous flanges shall have contact faces finished in accordance with American Standard ASA B16.24.

(h) 25 psi and Class 125 cast iron integral or screwed companion flanges may be used with a full-face gasket or with a flat ring gasket extending to the inner edge of the bolt holes. When using a full-face gasket, the bolting may be of alloy steel (ASTM A-193). When using a ring gasket, the bolting shall be of carbon steel equivalent to ASTM A-307, Grade B, without heat treatment other than stress relief.

(i) When bolting together two Class 250 integral or screwed companion cast iron flanges, having 1/16 inch raised faces, the bolting shall be of carbon steel equivalent to ASTM A-307 Grade B, without heat treatment other than stress relief.

(j) 150 psi steel flanges may be bolted to Class 125 cast iron flanges. When such construction is used, the 1/16 inch raised face on the steel flange shall be removed. When bolting such flanges together using a flat ring gasket extending to the inner edge of the bolt holes, the bolting shall be of carbon steel equivalent to ASTM A-307, Grade B, without heat treatment other than stress relief. When bolting such flanges together using a full-face gasket, the bolting may be alloy steel (ASTM A-193).

(k) 300 psi steel flanges may be bolted to Class 250 cast iron flanges. Where such construction is used, the bolting shall be of carbon steel, equivalent to ASTM A-307 Grade B, without heat treatment other than stress relief. Good practice indicates that the raised face on the steel flange should be removed, but also in this case, bolting shall be of carbon-steel equivalent to ASTM A-307 Grade B.

(l) Forged steel welding neck flanges having an outside diameter and drilling the same as American Standard ASA B16.1, but with modified flange thicknesses, hub dimensions, and special facing details, may be used to bolt against flat faced cast iron flanges and operate at the pressure-temperature ratings given in American Standard ASA B16.1 Class 125 Cast Iron Pipe Flanges, provided:

a. The minimum flange thickness "T" is not less than that specified in Appendix H for sizes 6 inch and larger.

b. Flanges are used with non-metallic full face gaskets extending to the periphery of the flange.

c. The design has been proved by test to be suitable for the ratings.

(m) Flanges made of nodular cast iron shall conform to material specifications and dimensional standards listed in Paragraph 831.11 (b) and shall be subject to all service restrictions as outlined for valves in that paragraph. The bolting requirements for nodular cast iron flanges shall be the same as for carbon and low-alloy steel flanges as listed in Paragraph 831.22

831.22 *Bolting.*

(a) For all flange joints, the bolts or stud-bolts used shall extend completely through the nuts.

(b) For all flange joints other than described in Paragraph 831.21 (h), (i), (j), and (k), the bolting shall be made of alloy steel conforming to ASTM Specifications A 193, A 320, or A 354, except that bolting for Class 150 and Class 300 flanges at temperatures between minus 20 F and 450 F may be made of Grade B of ASTM A 307.

(c) Alloy-steel bolting material conforming to ASTM A 193 or ASTM A 354 shall be used for insulating flanges if such bolting is made 1/8 inch undersized.

(d) The materials used for nuts shall conform with ASTM Specifications A 194 and A 307. A 307 nuts may be used only with A 307 bolting.

(e) All carbon- and alloy-steel bolts, stud-bolts, and their nuts shall be threaded in accordance with the following thread series and dimension class as required by American Standard ASA B1.1.

Carbon Steel—All carbon-steel bolts and stud-bolts shall have coarse threads, Class 2A dimensions and their nuts, Class 2B dimensions.

Alloy Steel—All alloy-steel bolts and stud-bolts of 1 in. and smaller nominal diameter shall be of the coarse-thread series; nominal diameters 1⅛ in. and larger shall be of the 8-thread series. Bolts and stud-bolts shall have a Class 2A dimension, and their nuts shall have Class 2B dimension.

(f) Bolts shall have American Standard regular square heads or heavy hexagonal heads and shall have American Standard heavy hexagonal nuts conforming to the dimensions of American Standard ASA B18.2.

(g) Nuts cut from bar stock in such a manner that the axis will be parallel to the direction of rolling of the bar may be used in all sizes for joints in which one or both flanges are cast iron and for joints with steel flanges where the pressure does not exceed 250 psig. Such nuts shall not be used for joints in which both flanges are steel and the pressure exceeds 250 psig except that, for nut sizes ½ in. and smaller, these limitations do not apply.

831.23 *Gaskets.*

(a) Material for gaskets shall be capable of withstanding the maximum pressure and of maintaining its physical and chemical properties, at any temperature to which it might reasonably be subjected in service.

(b) Gaskets used under pressure and at temperatures above 250 F shall be of non-combuslible material. Metallic gaskets shall not be used with 150 lb standard or lighter flanges.

(c) Asbestos composition gaskets may be used as permitted in the American Standard for Steel Pipe Flanges and Flanged Fittings (ASA B16.5). This type of gasket may be used with any of the various flange facings except small male and female, or small tongue and groove.

(d) The use of metal or metal-jacketed asbestos gaskets (either plain or corrugated) is not limited as to pressure provided that the gasket material is suitable for the service temperature. These types of gaskets are recommended for use with the small male and female or the small tongue and groove facings. They may also be used with steel flanges with any of the following facings: lapped, large male and female, large tongue and groove, or raised face.

(e) Full-face gaskets shall be used with all bronze flanges, and may be used with 25 psi or Class 125 cast iron flanges. Flat ring gaskets with an outside diameter extending to the inside of the bolt holes may be used with cast iron flanges, with raised face steel flanges, or with lapped steel flanges.

(f) In order to secure higher unit compression on the gasket, metallic gaskets of a width less than the full male face of the flange may be used with raised face, lapped or large male and female facings. Width of gasket for small male and female or for tongue and groove joints shall be equal to the width of the male face or tongue.

(g) Rings for ring joints shall be of dimensions established in ASA B16.20. The material for these rings shall be suitable for the service conditions encountered and shall be softer than the flanges.

(h) The insulating material shall be suitable for the temperature, moisture and other conditions where it will be used.

831.3 Fittings Other Than Valves and Flanges.

831.31 *Standard Fittings.*

(a) The minimum metal thickness of flanged or screwed fittings shall not be less than specified for the pressures and temperatures in the applicable American Standards or the MSS Standard Practice.

(b) Steel butt-welding fittings (not flanged) shall comply with the American Standard for Steel Butt-Welding Fittings (ASA B16.9) and shall have pressure and temperature ratings based on stresses for pipe of the same or equivalent material. To insure adequacy of fitting design, the actual bursting strength of fittings shall at least equal the computed bursting strength of pipe of the designated material and wall thickness. Mill hydrostatic testing of factory made steel butt-welding fittings is not required, but all such fittings shall be capable of withstanding a field test pressure equal to the test pressure established by the manufacturer, without failure or leakage, and without impairment of their serviceability.

(c) Steel socket-welding fittings shall comply with American Standard for Steel Socket-Welding Fittings (ASA B16.11).

831.32 *Special Fittings.* When special cast, forged, wrought, or welded fittings are required to dimensions differing from those of regular shapes specified in the applicable ASA and MSS Standards, the provisions of 807 shall apply.

831.33 *Branch Connections.*

(a) Welded branch connections on steel pipe must meet the design requirements of 831.4 and 831.5.

(b) Threaded taps in cast iron pipe are permitted, without reinforcement, to a size not more than 25% of the nominal diameter of the pipe, except that 1¼ inch taps are permitted in 4 inch pipe. Larger taps shall be covered by a reinforcing sleeve.

(c) Mechanical fittings may be used for making hot taps on pipelines and mains; provided they are designed for the operating pressure of the pipeline or main, and are suitable for the purpose.

831.34 *Special Components Fabricated by Welding.*

(a) This section covers piping system components other than assemblies consisting of standard pipe and fittings joined by circumferential welds.

(b) All welding shall be performed using procedures and operators that are qualified in accordance with the requirements of 824.

(c) Branch connections shall meet the design requirements of 831.4 and 831.5.

(d) The design of other components shall be in accordance with recognized engineering practice and applicable requirements of this code. When the strength of such components cannot be computed or determined with reasonable accuracy under the provisions of this code, the allowable working pressure shall be established as prescribed by Paragraph UG-101, Section VIII, ASME Boiler and Pressure Vessel Code.

(e) Prefabricated units, other than regularly manufactured butt-welding fittings, which employ plate and longitudinal seams as contrasted with pipe that has been produced and tested under one of the specifications listed in this code, shall be designed, constructed and tested under requirements of the ASME Boiler and Pressure Vessel Code. It is not intended to apply ASME Code requirements to such partial assemblies as split rings or collars or other field welded details.

(f) Orange-peel bull plugs and orange-peel swages are prohibited on systems operating at stress levels of 20% or more of the specified minimum yield strength of the pipe material. Fish tails and flat closures are permitted for 3 inch diameter pipe and smaller, operating at less than 100 psi. Fish tails on pipe larger than 3 inch diameter are prohibited. Flat closures larger than 3 inch diameter shall be designed according to Section VIII, Unfired Pressure Vessels, of the ASME Boiler and Pressure Vessel Code.

(g) Every prefabricated unit produced under this section of the code shall successfully withstand a pressure test without failure, leakage, distress or distortion other than elastic distortion, at a pressure equal to the test pressure of the system in which it is installed, either before installation or during the system test. When such units are to be installed in existing systems, they shall be pressure tested before installation, if feasible; otherwise, they shall withstand a leak test at the operating pressure of the line.

831.4 Reinforcement of Welded Branch Connections.

831.41 *General Requirements.* All welded branch connections shall meet the following requirements:

(a) When branch connections are made to pipe in the form of a single connection or in a header or manifold as a series of connections, the design must be adequate to control the stress levels in the pipe within safe limits. The construction shall take cogni-

FIGURE 831-A
REINFORCEMENT OF BRANCH CONNECTIONS

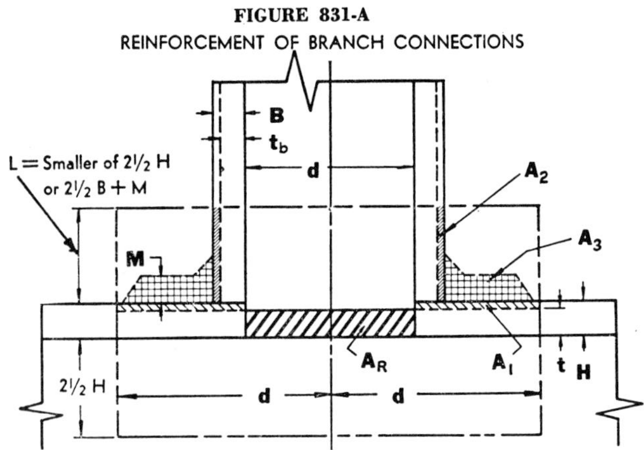

"Area of reinforcement" enclosed by — — — — lines
Reinforcement area required $A_R = dt$
Area available as reinforcement $= A_1 + A_2 + A_3$

$A_1 = (H - t)d$
$A_2 = 2(B - t_b)L$
$A_3 =$ Summation of area of all added reinforcement, including weld areas which lie within the "Area of reinforcement"
$A_1 + A_2 + A_3$ must be equal to or greater than A_R

where:
H = Nominal wall thickness of header
B = Nominal wall thickness of branch
t_b = Required nominal wall thickness of the branch (Under the appropriate section of the Code)
t = Required nominal wall thickness of the header (under the appropriate section of the Code)
d = The length of the finished opening in the header wall (measured parallel to the axis of the header)
M = Actual (by measurement) or nominal thickness of added reinforcement

zance of the stresses in the remaining pipe wall due to the opening in the pipe or header,* the shear stresses produced by the pressure acting on the area of the branch opening, and any external loadings due to thermal movement, weight, vibration, etc. The following paragraphs provide design rules for the usual combinations of the above loads, except excessive external loads.

(b) The reinforcement required in the crotch section of a welded branch connection shall be determined by the rule that the metal area available for reinforcement shall be equal to or greater than the required area as defined in this paragraph and in Figure 831-A in the Appendix.

(c) The required cross-sectional area A_R is defined as the product of d times t:

$$A_R = d \times t$$

where, d = the length of the finished opening in the header wall measured parallel to the axis of the run.

t = the nominal header wall thickness required by Section 841.1 of this code for the design pressure and temperature.

When the pipe wall thickness includes an allowance for corrosion or erosion, all dimensions used shall be those that will result after the anticipated corrosion or erosion has taken place.

(d) The area available for reinforcement shall be the sum of:

(1) The cross-sectional area resulting from any excess thickness available in the header thickness [over the minimum required for the header as defined in 831.41 (c) above] and which lies within the reinforcement area as defined in 831.41 (e) below.

(2) The cross-sectional area resulting from any excess thickness available in the branch wall thickness over the minimum thickness required for the branch and which lies within the reinforcement area as defined in 831.41 (e) below.

(3) The cross-sectional area of all added reinforcing metal which lies within the reinforcement area, as defined in 831.41 (e) below, including that of solid weld metal which is conventionally attached to the header and/or branch.

(e) The area of reinforcement is shown in Figure 831-A and is defined as a rectangle whose length shall extend a distance "d" on each side of the transverse centerline of the finished opening and whose width shall extend a distance of 2½ times the header wall thickness on each side of the surface of the header wall, except that in no case shall it extend more than 2½ times the thickness of the branch wall from the outside surface of the header or of the reinforcement, if any.

[* The lowest stress level (**stress intensification factor**) that can reasonably be expected in fabricated *branch connections* is about 1.6 times the nominal pressure stress in the header pipe. Results of a research program, which can be applied directly to branch connection problems, are available.[18]—EDITOR]

(f) The material of any added reinforcement shall have an allowable working stress at least equal to that of the header wall, except that material of lower allowable stress may be used if the area is increased in direct ratio of the allowable stresses for header and reinforcement material respectively.

(g) The material used for ring or saddle reinforcement may be of specifications differing from those of the pipe, provided the cross-sectional area is made in correct proportion to the relative strength of the pipe and reinforcement materials at the operating temperatures and provided it has welding qualities comparable to those of the pipe. No credit shall be taken for the additional strength of material having a higher strength than that of the part to be reinforced.

(h) When rings or saddles are used which cover the weld between branch and header, a vent hole shall be provided in the ring or saddle to reveal leakage in the weld between branch and header and to provide venting during welding and heat treating operations. Vent holes should be plugged during service to prevent crevice corrosion between pipe and reinforcing member, but no plugging material should be used that would be capable of sustaining pressure within the crevice.

(i) The use of ribs or gussets shall not be considered as contributing to reinforcement of the branch connection. This does not prohibit the use of ribs or gussets for purposes other than reinforcement, such as stiffening.

FIGURE 831-B
WELDING DETAILS FOR OPENINGS WITHOUT REINFORCEMENT OTHER THAN THAT IN HEADER AND BRANCH WALLS

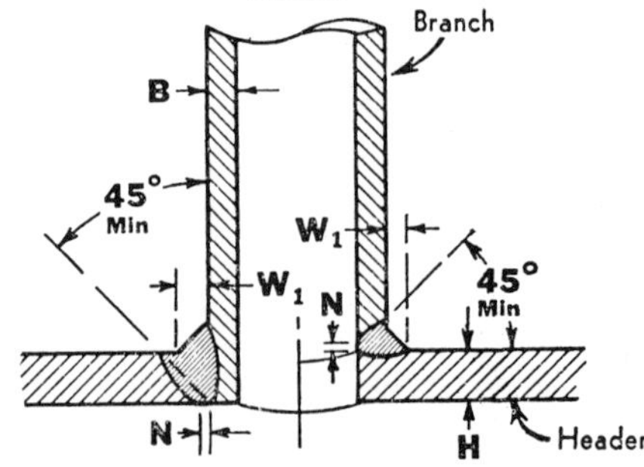

When a welding saddle is used it shall be inserted over this type of connection

$W_1 = \frac{3}{8}$ B but not less than ¼ in.

N = 1/16 in. min, ⅛ in. max, (Unless back welded or backing strip is used)

FIGURE 831-C

WELDING DETAILS FOR OPENINGS WITH LOCALIZED TYPE REINFORCEMENT

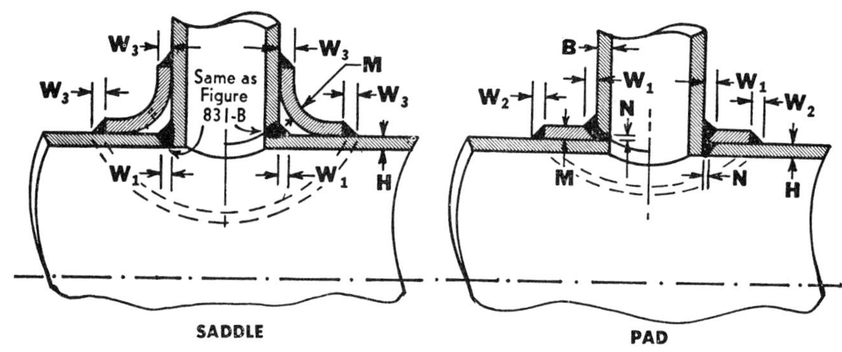

W_1 min = 3/8 **B** but not less than 1/4 in.
W_2 min = 1/2 **M** but not less than 1/4 in.
W_3 min = **M** but not greater than **H**

N = 1/16 in. min, 1/8 in. max (Unless back welded or backing strip is used)

All welds to have equal leg dimensions and a minimum throat = 0.707 × leg dimension.

Note 1: If **M** is thicker than **H** the reinforcing member shall be tapered down to the header wall thickness.

Note 2: Provide hole in reinforcement to reveal leakage in buried welds and to provide venting during welding and heat treatment. (Par. 831.41h)

FIGURE 831-D

WELDING DETAILS FOR OPENINGS WITH COMPLETE ENCIRCLEMENT TYPES OF REINFORCEMENT

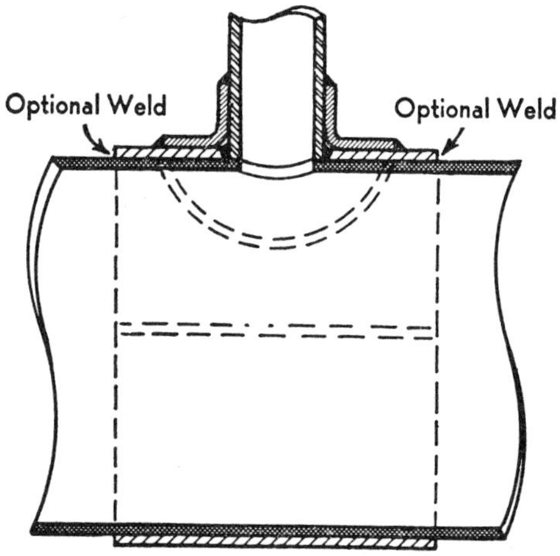

Optional Weld
Optional Weld

SADDLE AND SLEEVE TYPE

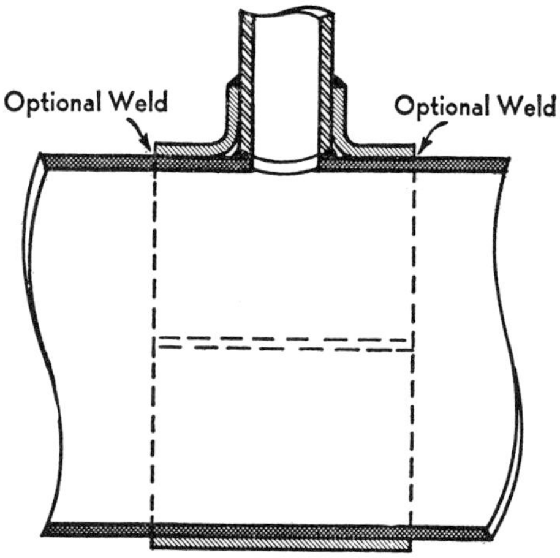

Optional Weld
Optional Weld

SADDLE TYPE

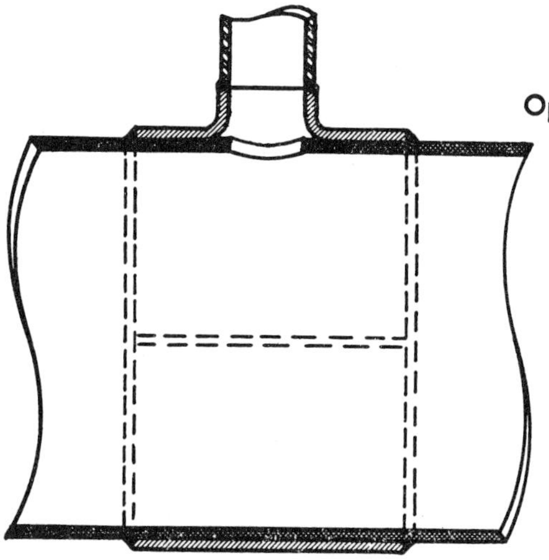

TEE TYPE

Note: Since fluid pressure is exerted on both sides of pipe metal under tee, the pipe metal does not provide reinforcement.

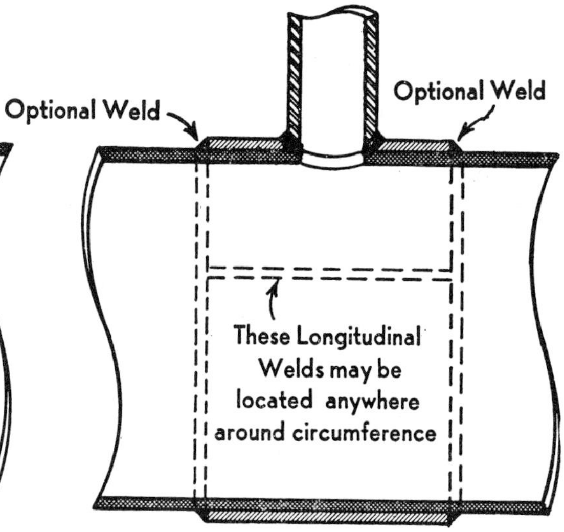

Optional Weld
Optional Weld
These Longitudinal Welds may be located anywhere around circumference

SLEEVE TYPE

Note: Provide hole in reinforcement to reveal leakage in buried welds and to provide venting during welding and heat treatment. (Par. 831.41h) Not required for Tee Type.

(j) The branch shall be attached by a weld for the full thickness of the branch or header wall plus a fillet weld "W_1," as shown in Figures 831-B and 831-C. The use of concave fillet welds is to be preferred to further minimize corner stress concentration. Ring or saddle reinforcement (Figures 831-C) shall be attached as shown by the figure. When a full fillet is not used it is recommended that the edge of the reinforcement be relieved or chamfered at approximately 45° to merge with the edge of the fillet.

(k) Reinforcement rings and saddles shall be accurately fitted to the parts to which they are attached. Figures 831-C and 831-D illustrate some acceptable forms of reinforcement.

(l) Branch connections attached at an angle less than 85° to the run become progressively weaker as the angle becomes less. Any such design must be given individual study and sufficient reinforcement must be provided to compensate for the inherent weakness of such construction. The use of encircling ribs to support the flat or re-entering surfaces is permissible, and may be included in the strength calculations. The designer is cautioned that stress concentrations near the ends of partial ribs, straps or gussets may defeat their reinforcing value.

831.42 *Special Requirements.* In addition to the requirements of 831.41, branch connections must meet the special requirements given in the following Table 831.421.

A. Smoothly contoured wrought steel tees of proven design are preferred. When tees cannot be used, the reinforcing member shall extend around the circumference of the header. Pads, partial saddles, or other types of localized reinforcement are prohibited.

B. Smoothly contoured tees of proven design are preferred. When tees are not used, the reinforcing member should be of the complete encirclement type but may be of the pad type, saddle type, or a welding outlet fitting.

C. The reinforcement member may be of the complete encirclement type, pad type, saddle type, or welding outlet fitting type. The edges of reinforcement members should be tapered to the header thickness. It is recommended that legs of fillet welds joining the reinforcing member and header do not exceed the thickness of the header.

D. Reinforcement calculations are not required for openings 2 inch and smaller in diameter; however, care should be taken to provide suitable protection against vibrations and other external forces to which these small openings are frequently subjected.

E. All welds joining the header, branch and reinforcing member shall be equivalent to those shown in Figures 831-B and 831-C.

TABLE 831.421 Reinforcement of Welded Branch Connections Special Requirements

Ratio of design hoop stress to minimum specified yield strength in the header	Ratio of nominal branch diameter to nominal header diameter		
	Less than 25%	25 to 50%	50% and more
Less than 20%	G	G	H
20 to 50%	I D	I	I H
50% and more	C D E	B E	A E F

F. The inside edges of the finished opening shall, whenever possible, be rounded to a ⅛ inch radius. If the encircling member is thicker than the header and is welded to the header, the ends shall be tapered down to the header thickness and continuous fillet welds made.

G. Reinforcement of openings is not mandatory; however, reinforcement may be required for special cases involving pressures over 100 psi, thin wall pipe or severe external loads.

H. If a reinforcement member is required, and the branch diameter is such that a localized type of reinforcement member would extend around more than half the circumference of the header, then a complete encirclement type of reinforcement member shall be used, regardless of the design hoop stress; or a smoothly contoured wrought steel tee of proven design may be used.

I. The reinforcement may be of any type meeting the requirements of 831.41.

831.5 Reinforcement of Multiple Openings.

831.51 When two or more adjacent branches are spaced at less than two times their average diameter (so that their effective areas of reinforcement overlap) the group of openings shall be reinforced in accordance with 831.4. The reinforcing metal shall be added as a combined reinforcement, the strength of which shall equal the combined strengths of the reinforcements that would be required for the separate openings. In no case shall any portion of a cross section be considered to apply to more than one opening, or be evaluated more than once in a combined area.

831.52 When more than two adjacent openings are to be provided with a combined reinforcement, the minimum distance between centers of any two of these openings shall preferably be at least 1½ times their average diameter, and the area of reinforcement between them shall be at least equal to 50% of the total required for these two openings on the cross section being considered.

831.53 When two adjacent openings as considered under 831.52 have the distance between centers less than 1⅓ times their average diameter, no credit for reinforcement shall be given for any of the metal between these two openings.

831.54 Any number of closely spaced adjacent openings, in any arrangement may be reinforced as if the group were treated as one assumed opening of a diameter enclosing all such openings.

EXAMPLE ILLUSTRATING THE APPLICATION OF THE RULES FOR REINFORCEMENT OF WELDED BRANCH CONNECTIONS

An 8 in. outlet is welded into a 24 in. header. The header material is API 5LX 46 with 5⁄16 in. wall. The outlet is API 5L Grade B (Seamless) Sched. 40 with 0.322 in. wall. The working pressure is 650 psi. The construction is Type B, used in Location Class 1, in accordance with 841.01. The joint efficiency is 1.00. The temperature is 100 F. Design Factors (841.1), $F = 0.60$, $E = 1.00$, $T = 1.00$. For dimensions see "Figure for Example."

Header

Nominal wall thickness:

$$t = \frac{PD}{2SFET} = \frac{650 \times 24}{2 \times 46{,}000 \times .60 \times 1.00 \times 1.00} = 0.283 \text{ in.}$$

Excess thickness in header wall:

$$(H - t) = .312 - .283 = 0.029 \text{ in.}$$

Outlet

Nominal wall thickness:

$$t_b = \frac{650 \times 8.625}{2 \times 35{,}000 \times .60 \times 1.00 \times 1.00} = 0.133 \text{ in.}$$

Excess thickness in outlet wall:

$$(B - t_b) = .322 - .133 = 0.189 \text{ in.}$$

$$d = \text{diameter of opening} = 8.625 - (2 \times .322) = 7.981 \text{ in.}$$

Reinforcement required

$$A_R = d \times t = 7.981 \times .283 = 2.26 \text{ sq in.}$$

Reinforcement provided

$$A_1 = (H - t)d = .029 \times 7.981 = 0.23 \text{ sq in.}$$

Effective area in outlet

Height: $L = 2\tfrac{1}{2}\,B + M$ (assume ¼ inch pad) $= (2\tfrac{1}{2} \times .322) + 0.25 = 1.05$ in.

or

$$L = 2\tfrac{1}{2}\,H = 2.5 \times .312 = 0.78 \text{ in.} \quad \text{Use 0.78 in.}$$

$$A_2 = 2(B - t_b)\,L = 2 \times .189 \times .78 = 0.295 \text{ sq in.}$$

This must be multiplied by 35,000/46,000. [831.41(f)]

Effective $A_2 = 0.295 \times 35{,}000/46{,}000 = 0.22$ sq in.

Required area $A_3 = A_R - A_1 - A_2 = 2.26 - 0.23 - 0.22 = 1.81$ sq in.

Use reinf. pl. ¼ inch thick (minimum practicable) × 15.5 in. diam

Area $(15.50 - 8.62) \times 0.25 = 1.72$ sq in.

Fillet welds (assuming two ¼ inch welds each side)
$.25 \times .25 \times .50 \times 2 \times 2 = 0.12$ sq in.
Total A_3 provided $= 1.84$ sq in.

FIGURE FOR EXAMPLE

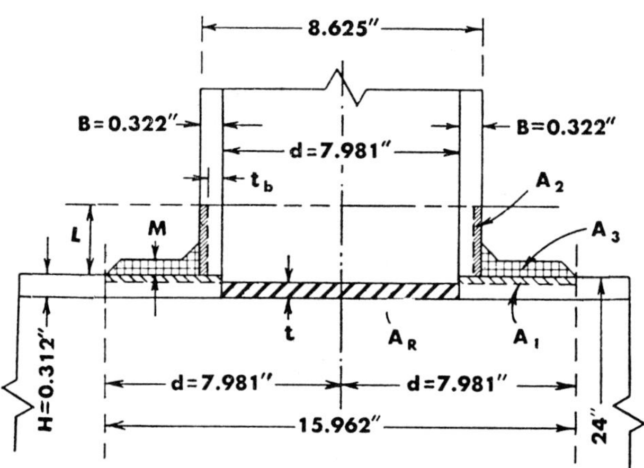

832 EXPANSION AND FLEXIBILITY.

832.1 This section is applicable to above ground piping only and covers all classes of materials permitted by this code up to temperatures no greater than 450 F.

832.2 Amount of Expansion. The thermal expansion of the more common materials used for piping shall be determined from Table 832.21. The expansion to be considered is the difference between the expansion for the maximum expected operating temperature and that for the expected average erection temperature. For materials not included in this table, or for precise calculations, reference shall be made to authoritative source data, such as publications of the National Bureau of Standards.

TABLE 832.21 Thermal Expansion of Piping Materials

(Carbon and low alloy high tensile steel and wrought iron)

Temperature, Degree F	Total expansion in inches per 100 feet above 32 F
32	0.
60	0.2
100	0.5
125	0.7
150	0.9
175	1.1
200	1.3
225	1.5
250	1.7
300	2.2
350	2.6
400	3.0
450	3.5

832.3 Flexibility Requirements.

832.31 Piping systems shall be designed to have sufficient flexibility to prevent thermal expansion or contraction from causing excessive stresses in the piping material, excessive bending or unusual loads at joints, or undesirable forces or moments at points of connection to equipment or at anchorage or guide points. Formal calculations shall be required only where reasonable doubt exists as to the adequate flexibility of the system.

832.32 Flexibility shall be provided by the use of bends, loops, or offsets; or provision shall be made to absorb thermal changes by the use of expansion joints or couplings of the slip joint type or expansion joints of the bellows type. If expansion joints are used, anchors or ties of sufficient strength and rigidity shall be installed to provide for end forces due to fluid pressure and other causes.

832.33 In calculating the flexibility of a piping system the system shall be treated as a whole. The significance of all parts of the line and all restraints, such as solid supports or guides, shall be considered.

832.34 Calculations shall take into account stress intensification factors found to exist in components other than plain straight pipe. Credit may be taken for the extra flexibility of such components. In the absence of more directly applicable data, the flexibility factors and stress intensification factors shown in Table 1, Appendix D, may be used.

832.35 Properties of pipe and fittings for these calculations shall be based on nominal dimensions, and the joint factor E (841.12) shall be taken as 1.00.

832.36 The total range in temperature shall be used in all expansion calculations, whether piping is cold-sprung or not. In addition to the expansion of the line itself, the linear and angular movements of the equipment to which it is attached shall be considered.

832.37 *Cold-springing.* In order to modify the effect of expansion and contraction, runs of pipe may be cold sprung. Cold spring may be taken into account in the calculations of the reactions as shown in 833.5 provided an effective method of obtaining the designed cold spring is specified and used.

832.38 Flexibility calculations shall be based on the modulus of elasticity E_c at ambient temperature.

833 COMBINED STRESS CALCULATIONS.

833.1 Using the above assumptions, the stresses and reactions due to expansion shall be investigated at all significant points.

833.2 The expansion stresses shall be combined in accordance with the following formula:

$$S_E = \sqrt{S_b^2 + 4S_t^2}$$

where:

$S_b = iM_b/Z$ = Resultant bending stress, psi.
$S_t = M_t/2Z$ = Torsional stress, psi
M_b = Resultant bending moment, lb-in.
M_t = Torsional moment, lb-in.
Z = Section modulus of pipe, cu in.
i = Stress intensification factor.

833.3 The maximum computed expansion stress range, S_E, shall not exceed $0.72S$, where S is the specified minimum yield strength, psi; subject to the further limitation of 833.4.

APPENDIX D

TABLE 1 Flexibility Factors k and Stress Intensification Factors i

Description	Flexibility factor, k	Stress intens. factor	Description	Flexibility factor, k*	Stress intens. factor, i*	Flexibility characteristic, h
Butt welded joint, reducer, or welding neck flange	1	1.0	Welding elbow, or pipe bend† (Sketch I)	$\frac{1.65}{h}$	$\frac{0.9}{h^{2/3}}$	$\frac{tR}{r^2}$
Double-welded slip-on or socket welding flange	1	1.2	Mitre bend†‡ with close spacing: $s < r\,(1 + \tan a)$ (Sketch II)	$\frac{1.52}{h^{5/6}}$	$\frac{0.9}{h^{2/3}}$	$\frac{\cot a}{2}\,\frac{ts}{r^2}$
Fillet welded joint, or single-welded socket welding flange	1	1.3	Mitre bend‡ with wide spacing:§ $s \geqq r\,(1 + \tan a)$ (Sketch III)	$\frac{1.52}{h^{5/6}}$	$\frac{0.9}{h^{2/3}}$	$\frac{1 + \cot a}{2}\,\frac{t}{r}$
Lap joint flange (with ASA B16.9 lap joint stub)	1	1.6	Welding tee** per ASA B16.9	1	$\frac{0.9}{h^{2/3}}$	$4.4\frac{t}{r}$
Screwed pipe joint, or screwed flange	1	2.3	Reinforced fabricated tee,** with pad or saddle	1	$\frac{0.9}{h^{2/3}}$	$\frac{(t + {}^1\!/_2\,T)^{5/2}}{t^{3/2}r}$
Corrugated pipe, straight or curved, or creased bend	5	2.5	Unreinforced fabricated tee**	1	$\frac{0.9}{h^{2/3}}$	$\frac{t}{r}$

* The flexibility factors k and stress intensification factors i in the Table apply to fittings of the same nominal weight or schedule as the pipe used in the system, and shall in no case be taken as less than unity. They apply over the effective arc length (shown by dash-dot lines in the sketches) for curved and mitre elbows, and to the intersection point for tees.

† Where flanges are attached to one or both ends the values of k and i in the Table shall be multiplied by the following factors:

One end flanged: $h^{1/6}$; Both ends flanged: $h^{1/3}$

‡ Subject to limitations of 841.236.

§ Also includes single-mitre joint.

** where: t = thickness of tee

r = 0.5 (outside radius + inside radius)

T = thickness of pad or saddle

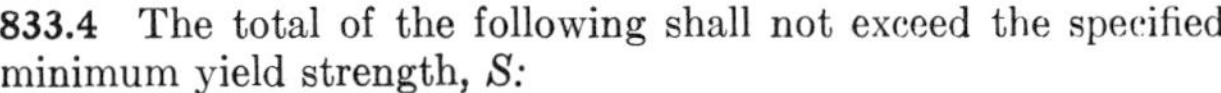

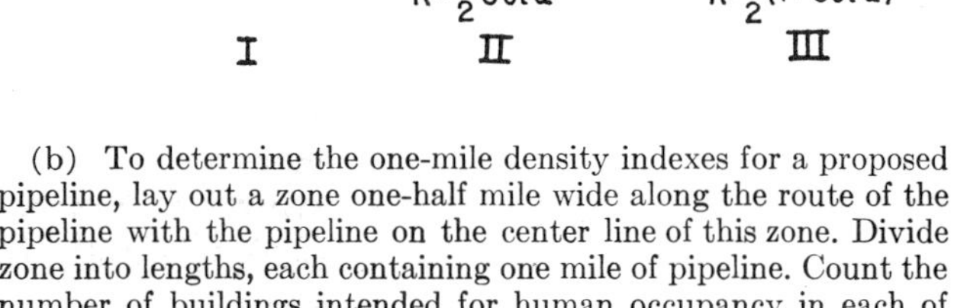

833.4 The total of the following shall not exceed the specified minimum yield strength, S:

(a) The combined stress due to expansion, S_E

(b) The longitudinal pressure stress

(c) The longitudinal bending stress due to external loads, such as weight of pipe and contents, wind, etc.

The sum of (b) and (c) shall not exceed 75% of the allowable stress in the hot condition ($S \times F \times T$, Section 84.1).

833.5 The reactions R' shall be obtained as follows from the reactions R derived from the flexibility calculations:

$R' = (1 - \frac{2}{3}\,C_sR)$ When C_s is less than 0.6

$R' = C_sR$, when C_s is between 0.6 and 1.0

where:

C_s = The cold spring factor varying from zero for no cold spring to one for 100 percent cold spring.

R = Range of reactions corresponding to the full expansion range based on E_c.

E_c = The modulus of elasticity in the cold condition.

R' is maximum reaction for the line after cold-springing. The reactions so computed shall not exceed limits which the attached equipment or anchorage is desired to sustain.

841 STEEL PIPE.

841.001 *Population Density Indexes.*

(a) Two population density indexes, determined at the time of initial construction, are used to classify locations for design and testing purposes: (1) the one-mile density index, which applies to any specific mile of pipeline; and (2) the ten-mile density index, which applies to any specific ten-mile length of pipeline.

(b) To determine the one-mile density indexes for a proposed pipeline, lay out a zone one-half mile wide along the route of the pipeline with the pipeline on the center line of this zone. Divide zone into lengths, each containing one mile of pipeline. Count the number of buildings intended for human occupancy in each of these lengths. These numbers are the one-mile indexes for the pipeline.

(c) To determine the ten-mile density indexes for any given ten-mile length of pipeline, proceed as follows: Add the one-mile density indexes for the ten-mile section. In case a one-mile index equals or exceeds 20, it is to be included in the sum as 20. Divide the sum thus obtained by 10. The quotient is the ten-mile density index for the section.

841.01 *Classification of Locations.*

841.011 *Class 1 Locations:* Class 1 locations include waste lands, deserts, rugged mountains, grazing land, and farm land, and combinations of these; provided, however, that:

(a) The ten-mile density index for any section of the line is 12 or less.

(b) The one-mile density index for any one mile of line is 20 or less.*

841.012 *Class 2 Locations:* Class 2 locations include areas where the degree of development is intermediate between Class 1 locations and Class 3 locations. Fringe areas around cities and towns, and farm or industrial areas where the one-mile density index exceeds 20 or the ten-mile density index exceeds 12 fall within this location class.

* It is not intended here that a full mile of lower-stress-level pipeline shall be installed if there are physical barriers or other factors that will limit the further expansion of the more densely populated area to a total distance of less than 1 mile. It is intended, however, that where no such barriers exist, ample allowance shall be made in determining the limits of the lower-stress design to provide for probable further development in the area.

841.013 *Class 3 Locations:* Class 3 locations include areas subdivided for residential or commercial purposes where, at the time of construction of the pipeline or piping system, 10% or more of the lots abutting on the street or right-of-way in which the pipe is to be located are built upon, and a Class 4 classification is not called for. This permits classifying as Class 3, areas completely occupied by commercial or residential buildings with the prevalent height of three stories or less.

841.014 *Class 4 Locations*: Class 4 locations include areas where multistory* buildings are prevalent, and where traffic is heavy or dense and where there may be numerous other utilities underground.

841.015 It should be emphasized that *Location Class* (1, 2, 3 or 4), as described in the foregoing paragraphs, is defined as the general description of a geographic area having certain characteristics as a basis for prescribing the types of construction and methods of testing to be used in those locations or in areas that are respectively comparable. A numbered Location Class refers only to the geography of that location or a similar area, and does not necessarily indicate that a correspondingly numbered Construction Type will suffice for all construction in that particular location or area. Example: In Location Class 1, all aerial crossings require Type B construction. (See 841.143)

* Multistory means 4 or more "floors" above ground including the first or ground floor. The depth of basements or number of basement floors is immaterial.

841.016 When classifying locations for the purpose of determining the type of pipeline construction and testing that should be prescribed, due consideration shall be given to the possibility of future development of the area. If at the time of planning a new pipeline this future development appears likely to be sufficient to change the location class, this should be taken into consideration in the design and testing of the proposed pipeline.

It is also anticipated that some increase in population density will occur in all areas after a line is constructed, and this possibility has been taken into account in establishing the design, construction, and testing procedures for each location class.

841.02 Classification of Steel Pipe Construction*

Four types of steel pipe construction are prescribed in this code. The distinguishing characteristics of each type and the location in which each type shall be used are as follows:

A. Characteristics	Type A Construction	Type B Construction	Type C Construction	Type D Construction
1. Design factor F (See 841.11)	.72	.60	.50	.40
B. Location where type of construction shall be used	(a) On private rights of way in Class 1 locations. (b) Parallel encroachments on: Privately owned roads in Class 1 locations. Unimproved roads in Class 1 locations. (c) Crossings without casings of privately owned roads in Class 1 locations. (d) Crossings in casings of unimproved public roads, hard-surfaced roads, highways or public streets and railroads in Class 1 locations.	(a) On private rights of way in Class 2 locations. (b) Parallel encroachments on: Privately owned roads in Class 2 locations. Unimproved public roads in Class 2 locations. Hard surfaced roads, highways or public streets and railroads in Class 1 and Class 2 locations. (c) Crossings without casings of: Privately owned roads in Class 2 locations. Unimproved public roads in Class 2 locations. Hard surfaced roads, highways or public streets and railroads in Class 1 locations. (d) Crossings in casings of: Hard surfaced roads, highways or public streets and railroads in Class 2 locations. (e) On bridges in Class 1 and Class 2 locations. (See 841.143) (f) Fabricated assemblies in pipelines in location Classes 1 and 2. (See 841.142)	(a) On private rights of way in Class 3 locations. (b) Parallel enroachmentments on: Privately owned roads in Class 3 locations. Unimproved public roads in Class 3 locations. Hard surfaced roads, highways or public streets and railroads in Class 3 locations. (c) Crossings without casings of: Privately owned roads in Class 3 locations. Unimproved public roads in Class 3 locations. Hard surfaced roads, highways or public Class 2 and 3 locations. (d) Compressor station piping.	(a) In all locations in location Class 4.

* It is necessary to distinguish between construction types, as defined by Section A of this table, and location classes, as defined in 841.01, to avoid confusion. If pipelines or mains are located in private rights of way, the code prescribes that Type A construction be used in Class 1 locations, Type B construction in Class 2 locations, Type C construction in Class 3 locations, and Type D construction in Class 4 locations. There are many exceptions to this association of Class 1 with Type A, etc., however, as Table 841.02 shows, most of which are cases where pipelines or mains are located in highways or on bridges, etc.

841.03 Construction Types Required for Parallel Encroachments of Pipelines and Mains on Roads and Railroads

Kind of thoroughfare	Construction type required			
	Location Class 1	Location Class 2	Location Class 3	Location Class 4
(a) Privately owned roads	Type A	Type B	Type C	Type D
(b) Unimproved public roads	Type A	Type B	Type C	Type D
(c) Hard surface roads, highways or public streets and railroads	Type B	Type B	Type C	Type D

841.04 Construction Types Required for Pipelines and Mains Crossing Roads and Railroads

Kind of thoroughfare	Construction type required			
	Location Class 1	Location Class 2	Location Class 3	Location Class 4
(a) Privately owned roads	Type A without casing	Type B without casing	Type C without casing	Type D without casing
(b) Unimproved public roads	Type A with casing Type B without casing	Type B without casing	Type C without casing	Type D without casing
(c) Hard surface roads, highways or public streets and railroads	Type A with casing Type B without casing	Type B with casing Type C without casing	Type C without casing	Type D without casing

841.1 Steel Pipe Design Formula. The design pressure for steel gas piping systems or the nominal wall thickness for a given design pressure shall be determined by the following formula:

$$P = \frac{2St}{D} \times F \times E \times T$$

where (for exceptions see 841.4):

P = Design pressure, psig.
S = Specified minimum yield strength, psi, stipulated in the specifications under which the pipe was purchased from the manufacturer or determined in accordance with 811.253H. The specified minimum yield strengths of some of the more commonly used piping steels, whose specifications are incorporated by reference herein, are tabulated for convenience in Appendix C. For special limitation on S see 841.14 (e) and (f).
D = Nominal outside diameter of pipe, inches.
t = Nominal wall thickness, inches.
F = Construction type design factor obtained from 841.11.
E = Longitudinal joint factor obtained from 841.12.
T = Temperature derating factor obtained from Table 841.13.

TABLE 841.11 Values of Design Factor F

Construction type (See 841.02)	Design factor F
Type A	0.72
Type B	0.60
Type C	0.50
Type D	0.40

APPENDIX C

Specified minimum yield strength (See 841.1) for steel and iron pipe commonly used in piping systems. Note: This table is not complete.* For the minimum specified yield strength of other grades and grades in other approved specifications, refer to the particular specification.

Specification	Specified minimum yield strength, psi
API 5L Grade A seamless or electric welded	30,000
API 5L Grade B seamless or electric welded	35,000
API 5L Butt-welded Class I open hearth	25,000
API 5L Butt-welded Class II open hearth	28,000
API 5L Butt-welded Bessemer	30,000
API 5L Butt-welded open-hearth iron or wrought iron	24,000
API 5LX Grade X42	42,000
API 5LX Grade X46	46,000
API 5LX Grade X52	52,000
ASTM A53 Grade A	30,000
ASTM A53 Grade B	35,000
ASTM A53 Butt-welded open hearth or Electric Furnace	25,000
ASTM A53 Butt-welded Bessemer steel	30,000
ASTM A72	24,000
ASTM A106 Grade A	30,000
ASTM A106 Grade B	35,000
ASTM A135 Grade A	30,000
ASTM A135 Grade B	35,000
ASTM A139 Grade A	30,000
ASTM A139 Grade B	35,000
ASTM A381 Class Y-35	35,000
ASTM A381 Class Y-42	42,000
ASTM A381 Class Y-46	46,000
ASTM A381 Class Y-48	48,000

[* Considerable API 5LX Grade X 60 was installed in 1963.—EDITOR]

TABLE 841.12 Longitudinal Joint Factor E

Spec. number	Pipe class	E factor
ASTM A53	Seamless	1.00
	Electric resistance welded	1.00
	Furnace butt welded	.60
ASTM A106	Seamless	1.00
ASTM A134	Electric fusion arc welded	.80
ASTM A135	Electric resistance welded	1.00
ASTM A139	Electric fusion welded	.80
ASTM A155	Electric fusion arc welded	1.00
ASTM A211	Spiral welded steel pipe	.80
ASTM A381	Double submerged arc welded	1.00
API 5L	Seamless	1.00
	Electric resistance welded	1.00
	Electric flash welded	1.00
	Furnace butt welded	.60
API 5LX	Seamless	1.00
	Electric resistance welded	1.00
	Electric flash welded	1.00
	Submerged arc welded	1.00

Note: Definitions for the various classes of welded pipe are given in Paragraph 805.51.

TABLE 841.13 Temperature Derating Factor T For Steel Pipe

Temp., °F	Temp. derating factor T
250 F or less	1.000
300 F	0.967
350 F	0.933
400 F	0.900
450 F	0.867

Note: For intermediate temperatures interpolate for derating factor.

841.14 *Limitations of Pipe Design Values.*

(a) P for furnace butt-welded pipe shall not exceed the restrictions of 841.1 or 60% of the mill test pressure, whichever is the lesser.

(b) P shall not exceed 85% of the mill test pressure for all other pipes; provided, however, that pipe, mill tested to a pressure less than 85% of the pressure required to produce a stress equal to the specified minimum yield, may be retested with a mill type hydrostatic test or tested in place after installation. In the event the pipe is retested to a pressure in excess of the mill test pressure, then P shall not exceed 85% of the retest pressure rather than the initial mill test pressure. It is mandatory to use a liquid as the test medium in all tests in place after installation where the test pressure exceeds the mill test pressure. This paragraph is not to be construed to allow an operating pressure or design pressure in excess of that provided for by 841.1.

(c) Transportation, installation or repair of pipe shall not reduce the wall thickness at any point to a thickness less than 90% of the nominal wall thickness as determined by 841.1 for the design pressure to which the pipe is to be subjected.

(d) "t" shall not be less than shown in Table 841.141.

(e) When pipe that has been cold worked for the purpose of meeting the specified minimum yield strength is heated to 600 F or higher (welding excepted), the maximum allowable pressure at which it can be used shall not exceed 75% of the value obtained by use of the steel pipe design formula given in 841.1.

(f) In no case where the code refers to the specified minimum value of a physical property can the actual value of the property be substituted in design calculations, unless the actual is less than the specified minimum.

841.142 *Fabricated Assemblies.* When fabricated assemblies, such as connections for separators, main line valve assemblies, cross-connections, river crossing headers, etc., are to be installed in areas defined as location Class 1, Type B construction is required throughout the assembly, and for a distance of 5 pipe diameters in each direction beyond the last fittings. Transition pieces at the end of an assembly and elbows used in place of pipe bends are not considered as fittings under the requirements of this paragraph. See also 830.

841.143 Pipelines or mains supported by railroad, vehicular, pedestrian, or pipeline bridges shall be in accordance with the construction type prescribed for the area in which the bridge is located, except that in Class 1 locations Type B construction shall be used.

841.15 *Protection of Pipelines and Mains from Hazards.* When pipelines and mains must be installed where they will be subjected to natural hazards, such as washouts, floods, unstable soil, land slides, or other conditions which may cause serious movement of, or abnormal loads on the pipeline, reasonable precaution shall be taken to protect the pipeline, such as increasing the wall thickness, constructing revetments, erosion prevention, installing anchors, etc. Where pipelines and mains are exposed, such as at spans, trestles, and bridge crossings, the pipelines and mains shall be reasonably protected by distance or barricades from accidental damage by vehicular traffic or other causes.

TABLE 841.141 Least Nominal Wall Thicknesses, t

(all dimensions in inches)

	Nominal pipe size	Outside diameter	Location classes* 1	2	3 & 4	Compressor stations
Threaded or plain end	⅛	0.405	0.068	0.068	0.068	0.095
	¼	0.540	0.088	0.088	0.088	0.119
	⅜	0.675	0.091	0.091	0.091	0.126
	½	0.840	0.109	0.109	0.109	0.147
	¾	1.050	0.113	0.113	0.113	0.154
	1	1.315	0.133	0.133	0.133	0.179
	1¼	1.660	0.140	0.140	0.140	0.191
	1½	1.900	0.145	0.145	0.145	0.200
	2	2.375	0.154	0.154	0.154	0.218
Plain end only	2½	2.875	0.103	†0.125	†0.125	0.203
	3	3.500	0.104	†0.125	†0.125	0.216
	3½	4.000	0.104	†0.125	†0.125	0.226
	4	4.500	0.104	†0.125	†0.125	0.237
	5	5.563	0.104	†0.125	†0.125	0.250
	6	6.625	0.104	0.134	0.156	0.250
	8	8.625	0.104	0.134	0.172	0.250
	10	10.75	0.104	0.164	0.188	0.250
	12	12.75	0.104	0.164	0.203	0.250
	14	14.0	0.134	0.164	0.210	0.250
	16	16.0	0.134	0.164	0.219	0.250
	18	18.0	0.134	0.188	0.250	0.250
	20	20.0	0.134	0.188	0.250	0.250
	22, 24, 26	22, 24, 26	0.164	0.188	0.250	0.250
	28, 30	28, 30	0.164	0.250	0.281	0.281
	32, 34, 36	32, 34, 36	0.164	0.250	0.312	0.312

* If threaded pipe is to be used in those sizes for which least nominal wall thicknesses are given for "plain end pipe only," those sizes marked by † shall be increased as follows: 2½ inch to 0.203, 3 inch to 0.216, 3½ inch to 0.226, 4 inch to 0.237, 5 inch to 0.258, and add 0.100 inch to all other wall thicknesses given in Table 841.141.

841.16 *Cover and Casing Requirements Under Railroads, Roads, Streets or Highways.*

(a) All buried pipelines, mains, and casings when used, shall be installed with a minimum cover of 24 inches unless otherwise provided herein.

(b) Buried pipelines and mains operating at hoop stresses of less than 20% of the specified minimum yield strength and located within private rights-of-way, private thoroughfares, sidewalks or parkways may be installed with less than the minimum cover of 24 inches if it appears that external damage to the pipe will not be likely to result.

(c) Abandoned pipe having a cover less than 24 inches may be used as a casing or conduit for pipelines and mains operating at hoop stresses less than 20% of the specified minimum yield strength.

(d) Buried pipelines and mains installed in areas where farming or other operations might result in deep plowing, or in thoroughfares or other locations where grading is done, or where the area is subject to erosion, should be provided with more cover than the minimum otherwise required.

(e) Where it is impractical to comply with the provisions of 841.16(a) and it is necessary to prevent damage from external loads, the pipe shall be cased or bridged.

(f) Casings shall be designed to withstand the superimposed loads. Where there is a possibility of water entering the casing, the ends of the casing shall be sealed. If the end sealing is of a type that will retail the full pressure of the pipe, the casing shall be designed for the same pressure as the pipe but according to Type A construction requirements. Venting of sealed casings is not mandatory; however, if vents are installed they should be protected from the weather to prevent water from entering the casing.

841.161 *Clearance Between Pipelines or Mains and Other Underground Structures.* There should be at least 2 inches clearance wherever possible between any gas main or pipeline and any other underground structure not used in conjunction with the pipeline or main. When this clearance cannot be attained, other suitable precautions to protect the pipe shall be taken, such as the installation of insulating material, installation of casing, etc.

846 VALVES.

846.1 Required Spacing of Valves.

846.11 *Transmission Lines.* Sectionalizing block valves on transmission lines shall be installed at a spacing not to exceed 20 miles within areas conforming to location Class 1, 15 miles within areas conforming to location Class 2, 8 miles within areas conforming to location Class 3, and 5 miles within areas conforming to location Class 4.

846.2 Location of Valves.

846.21 *Transmission Valves.*

(a) Sectionalizing block valves shall be accessible and protected from damage and tampering. If a blowdown valve is involved it shall be located where the gas can be blown to the atmosphere without undue hazard.

(b) Sectionalizing valves may be installed above ground, in a vault, or buried. In all installations an operating device to open or close the valve shall be readily accessible to authorized persons. All valves shall be suitably supported to prevent settlement, or movement of the attached piping.

(c) Blow-down valves shall be provided so that each section of pipeline between main line valves can be blown down. The sizes and capacity of the connections for blowing down the line shall be such that under emergency conditions the section of line can be blown down as rapidly as is practicable.

(d) This code does not require the use of automatic valves, nor does the code imply that the use of automatic valves presently developed will provide full protection to a piping system. Their use and installation shall be at the discretion of the operating company.

Pipe Data

Use Table 8-7 and Fig. 8-10 to facilitate the calculation of allowable pressures in a given pipe. The table was calculated for selected pipe sizes and wall thicknesses on the basis of pipe having a minimum yield strength of 35,000 psi. Figure 8-10, used in conjunction with the table, furnishes the multiplier to determine the yield point of any given grade of pipe and for any selected percentage of minimum yield, in accordance with preceding extract 841.02 from ASA B31.8-1963. Optimum wall thickness is a function of economics and safety. Choice is made only after consideration of all available wall thicknesses and grades of pipe. Pressure values in Table 8-7 are based on *Barlow's formula:*

$$P = \frac{2St}{D}$$

where: P = internal pressure, psig
S = minimum yield stress, (35,000 psi for Grade B)
t = wall thickness, in.
D = OD of pipe, in: (Both a simplifying and safe assumption, ID is theoretically proper.)

Example:

For 30 in. OD × 3/8 in. wall pipe of 52,000 psi min yield, determine the internal pressure based on 80 per cent of a mill test of 90 per cent of its minimum yield.

From Table 8-7, internal pressure at 100 per cent min yield for 30 in. OD × 3/8 in. wall pipe is 870 psig. Percentage of minimum yield desired = 80 per cent × 90 per cent = 72. Referring to Fig. 8-10, intersection of vertical dotted line from 72 per cent of minimum yield with line for 52,000 psig minimum yield establishes multiplication factor of 1.07, as read on the ordinate scale.

Internal pressure for conditions stated is thus 1.07 × 870 = 930 psig.

In relating internal pressures determined in this manner to test, design and working pressures, it may be necessary to consider other factors, as required by applicable codes. (See 841.1 of ASA B31.8-1963.)

Table 8-7 Selected Pipe Data

OD,‡ in.	Wall thick., in.	Weight,§ ton/mile	Internal displace., cu ft/mile	Internal press. for 35,000 psi pipe† (based on Barlow's formula), psig
*12.75	0.219	77.30	4,370	1200
12.75	.250	88.12	4,320	1370
12.75	.281	98.87	4,280	1540
12.75	.312	109.59	4,230	1710
12.75	.330	115.55	4,210	1810
12.75	.344	120.25	4,190	1890
12.75	.375	130.84	4,150	2060
12.75	.438	151.88	4,060	2400
12.75	.500	172.71	3,970	2740
12.75	.562	193.5	3,900	3090
12.75	.687	234.0	3,720	3770
*14	.219	85.01	5,300	1090
*14	.250	96.91	5,240	1250
*14	.281	108.79	5,200	1400
14	.312	120.60	5,150	1560
14	.344	132.37	5,100	1720
14	.375	144.06	5,050	1870
14	.438	167.30	4,960	2190

Table 8-7 (continued)

OD,‡ in.	Wall thick., in.	Weight,§ ton/mile	Internal displace., cu ft/mile	Internal press. for 35,000 psi pipe† (based on Barlow's formula), psig
14	.500	190.32	4,860	2500
14	.593	224.40	4,720	2960
14	.750	280.20	4,500	3750
*16	.219	97.34	6,980	960
*16	.250	111.01	6,920	1090
*16	.281	124.66	6,860	1230
16	.312	138.23	6,800	1360
16	.344	151.75	6,750	1500
16	.375	165.21	6,700	1640
16	.438	191.98	6,590	1920
16	.500	218.51	6,470	2190
16	.656	283.8	6,200	2870
16	.843	360.2	5,900	3690
*18	.250	125.11	8,810	970
18	.312	155.84	8,690	1210
18	.344	171.12	8,630	1340
18	.375	186.36	8,560	1460
18	.438	216.64	8,440	1700
18	.500	246.71	8,320	1940
18	.625	277	8,200	2430
18	.750	364.80	7,840	2910
18	.937	451.40	7,480	3640
*20	.250	139.21	10,950	870
20	.312	173.47	10,800	1090
20	.344	190.50	10,740	1200
20	.375	207.50	10,670	1310
20	.438	241.32	10,520	1530
20	.500	274.90	10,390	1750
20	.593	324.50	10,130	2070
20	0.812	439.30	9,610	2840
20	1.031	551.40	9,260	3620
*22	0.250	153.30	13,310	790
22	.312	191.08	13,150	990
22	.344	209.91	13,080	1090
22	.375	228.65	13,000	1190
22	.438	265.98	12,840	1390
22	.500	303.10	12,690	1590
22	.687	413.20	12,610	2180
22	0.812	485.80	11,960	2580
22	1.0312	609.80	11,430	3280
*24	0.25	167.40	15,890	730
24	.312	208.72	15,720	910
24	.344	229.28	15,650	1000
24	.375	249.80	15,560	1090
24	.438	290.66	15,390	1280
24	.500	331.29	15,220	1460
24	.562	371.70	15,050	1640
24	.687	451.90	14,720	2000
24	0.937	609.60	14,100	2730
24	1.281	821.00	13,220	3740
*26	0.250	181.50	18,710	670
26	.312	226.33	18,530	840
26	.375	270.94	18,350	1010
26	.438	315.32	18,170	1080
26	.500	359.49	17,990	1340
26	0.687	491.00	17,450	1850
26	1.281	892.30	15,800	3440
*28	0.250	195.60	21,780	620
*28	.281	219.81	21,680	700
28	.312	243.96	21,570	780
28	.344	268.04	21,410	860
28	.375	292.09	21,370	930
28	.438	340.01	21,190	1090
28	.500	387.68	20,980	1250
28	0.687	528.00	20,400	1710
28	1.281	966.20	18,630	3200
*30	0.25	209.70	25,060	580
*30	.281	235.67	24,960	650
30	.312	261.57	24,840	720
30	.344	287.44	24,730	800
30	.375	313.24	24,620	870
30	.438	364.66	24,430	1020
30	.500	415.88	24,200	1160
30	0.687	567.60	23,600	1600
30	1.281	1037.50	21,700	2980
30	1.531	1227.60	20,850	3670
*32	0.25	223.79	28,580	540
*32	.281	251.54	28,460	610
32	.312	279.21	28,350	680
32	.344	306.82	28,240	750
32	.375	334.38	28,120	820
32	.438	389.35	27,900	950
32	.500	444.07	27,670	1090
32	.625	552	27,250	1360
32	.750	662	26,800	1640
32	0.875	763	26,400	1910
32	1.000	875	26,000	2190
*34	0.250	237.89	32,320	510
*34	.281	267.38	32,200	570
34	.312	296.82	32,080	640
34	.344	326.20	31,960	700
34	.375	355.53	31,840	770
34	.438	414.00	31,600	900
34	.500	472.27	31,360	1030
34	.625	590	30,900	1280
34	.750	705	30,450	1540
34	0.875	818	30,000	1800
34	1.000	932	29,450	2060
*36	0.250	251.99	36,300	480
*36	.281	283.25	36,170	540
36	.312	314.45	36,040	600
36	.344	345.58	35,910	660
36	.375	376.68	35,780	720
36	.438	438.69	35,530	850
36	.500	500.47	35,280	970
36	.625	625	34,700	1210
36	.750	746	34,300	1460
36	0.875	865	33,800	1700
36	1.000	960	33,300	1940
38	0.312	332.06	40,230	570
38	.344	364.98	40,090	630
38	.375	397.82	39,960	690
38	.438	463.35	39,960	800
38	.500	528.66	39,430	920
40	.312	349.14	44,650	540
40	.344	384.65	44,510	600
40	.375	418.97	44,360	650
40	.438	488.58	44,080	760
40	.500	556.86	43,800	870
42	.312	366.72	49,300	520
42	.344	404.03	49,150	570
42	.375	440.11	49,000	620
42	.438	513.27	48,700	730
42	.500	585.05	48,410	830
42	.625	730	47,900	1040
42	.750	873	47,200	1250
42	0.875	1020	46,600	1450
42	1.000	1160	46,100	1600

* These are special lightweight sizes. All other sizes are regular weight.
† Pressure values are based on minimum yield strength of 35,000 psi, and nominal dimensions specified by API.
‡ Outside diameter and wall thicknesses are subject to tolerances as given by Table 5 of API Std 5LX March 1958.
§ Based on steel weight of 0.2833 lb per cu in.

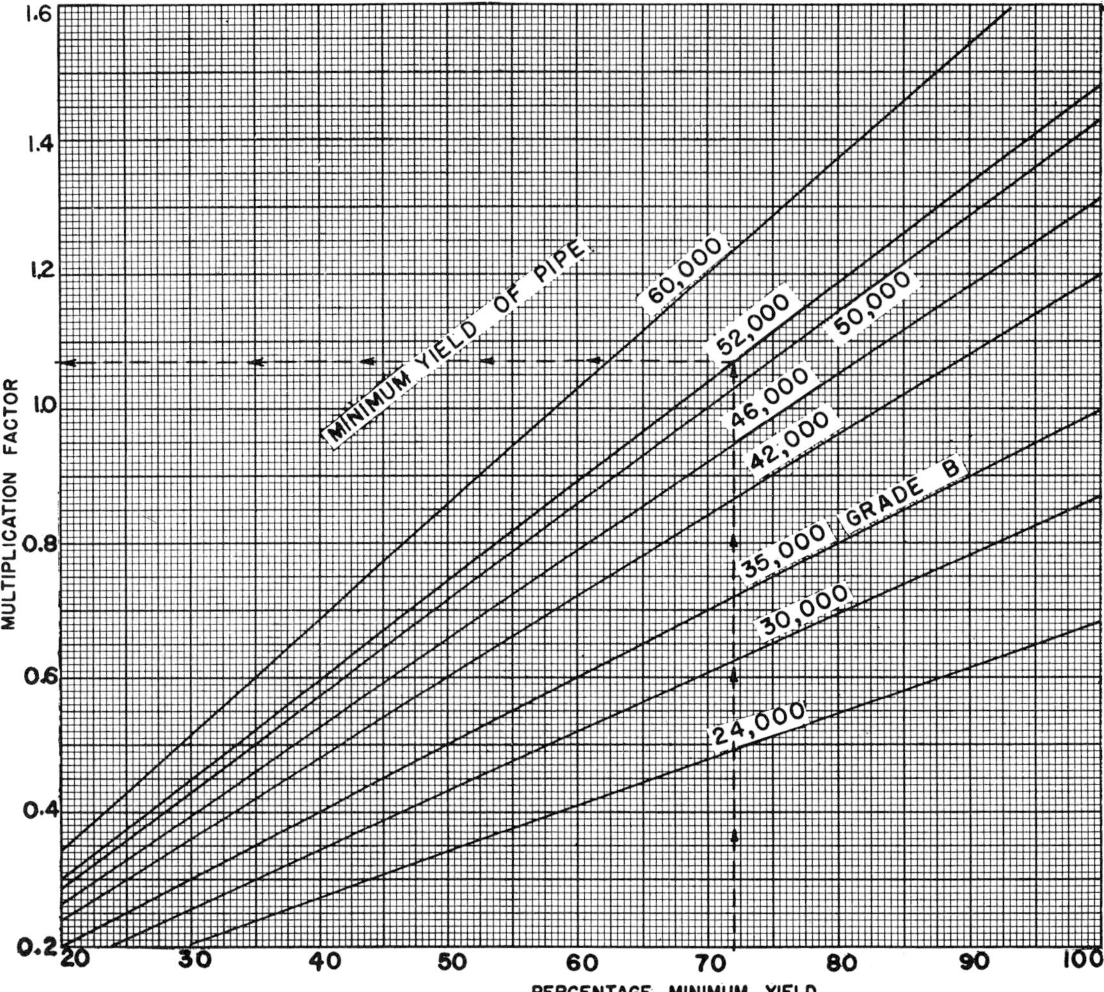

Fig. 8-10 Pressure multipliers for API Specification pipe (based on 35,000 psi minimum yield pipe, e.g., API 5L Grade B).

Valves

Many companies use full opening *gate valves* in main transmission lines to permit passage of scrapers. *Plug valves* may be used for blowoffs.

Main line valves should be located where operation and maintenance convenience may require. Valve spacing depends upon the area traversed by the pipeline; 5 to 20 miles apart is the order of magnitude for spacing block valves on large diameter lines.

Valves may be of the weld-end, flange-end, or weld by flange-end type, according to their location and use. Weld-end type valves are generally preferred.

Concrete pads (size depending upon valve weight) may be placed under main line valves to provide adequate support.

Main line gate valves should be equipped with a by-pass and blowoff arrangement to equalize pressure differentials and relieve excessive stresses required to open the unit with unequal pressure across its disk. Figure 8-11 shows a typical assembly. On large diameter valves with weld ends, it is considered good practice to have forged steel transition pieces shop-welded in place. They provide proper cross-sectional area and are made of a material easily welded under severe field conditions.

Operators. Large diameter valves are preferably equipped with some type of motor operator. Figure 8-12 shows one of several types available. A motor type operator using gas as a power supply may be used for these valves in main lines. This unit may be directly connected either to the main line or to a bottle located beside it for automatic operation.

Drips

It is common practice to provide drips on all pipelines except those which may be cleaned with pigs or scrapers. On such lines, drips may not be necessary because fluids can be pushed ahead of the pig and thereby removed from the line. Instances have occurred where a pig had to be run thru a line several times before a slug of liquid was removed. Dehydration of natural gas, LP-gas removal, and gasoline absorption from natural gas, where practiced, lessen the need for drips since the amount of fluid collecting in the line is materially reduced. On lines carrying only dry gas, drips are not required.

Figure 8-13 shows one type of drip, with an enlarged cross-sectional area in the pipe. The resulting decrease in gas velocity improves fluid removal. Baffles in the pipe stop fluids from moving along its bottom with the gas stream.

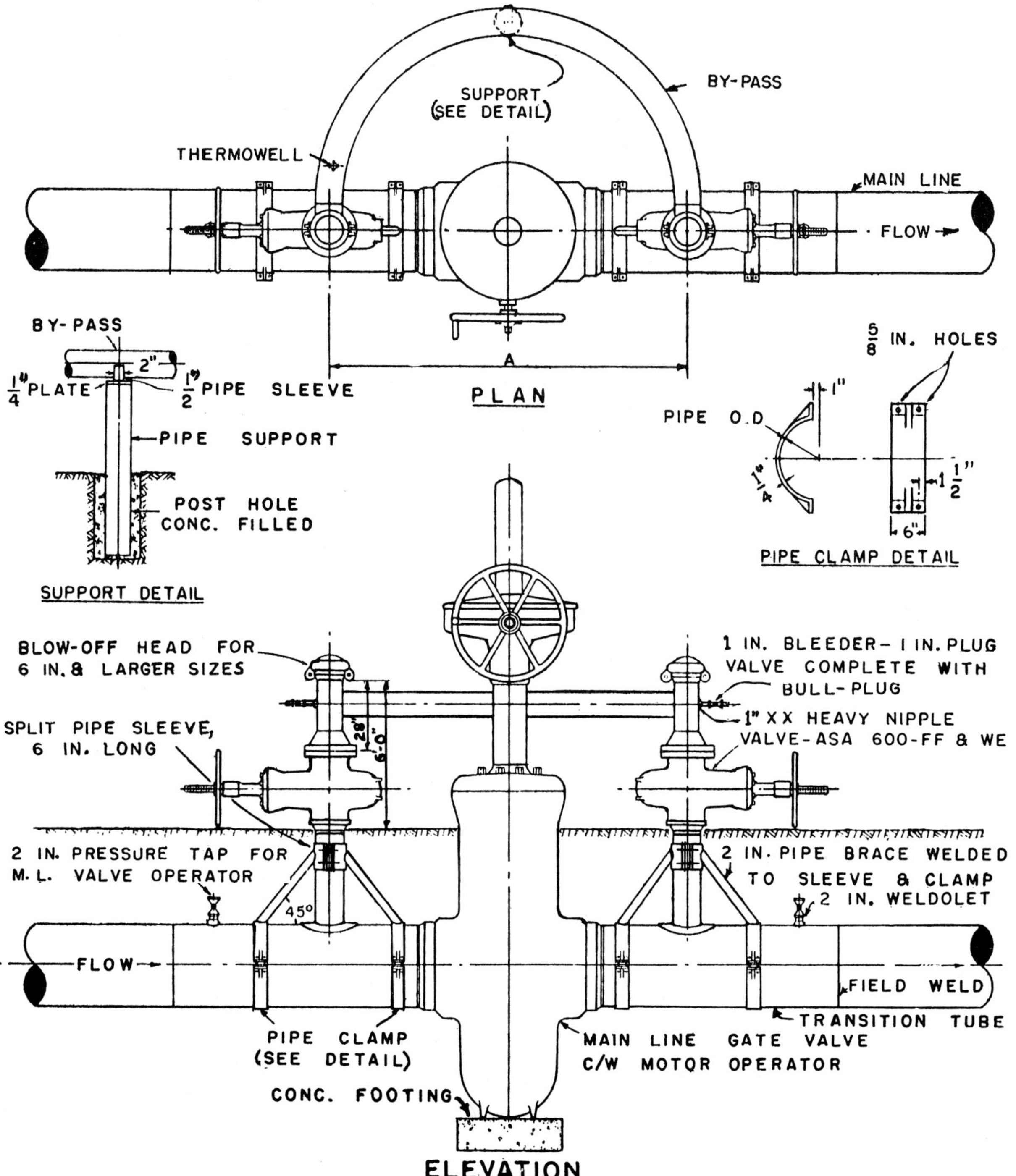

Note: Main line sizes 10 in. and smaller, by-pass to be omitted. Where unequal blowoffs occur, by-pass shall be size of smallest valve.

Main line	Blowoff and by-pass	Concrete block	Dim "A"
10	4	12 × 30 × 30	108
12	4	12 × 32 × 32	108
14	6	12 × 34 × 34	108
16	6	15 × 36 × 36	120
18	6	15 × 38 × 38	120
20	6	15 × 40 × 40	120
22	8	18 × 42 × 42	126
24	8	18 × 44 × 44	127
26	10	18 × 46 × 46	127
30	10	18 × 50 × 50	132

(All values in inches)

Fig. 8-11 Main line gate valve equipped with by-pass and blow-off arrangement.

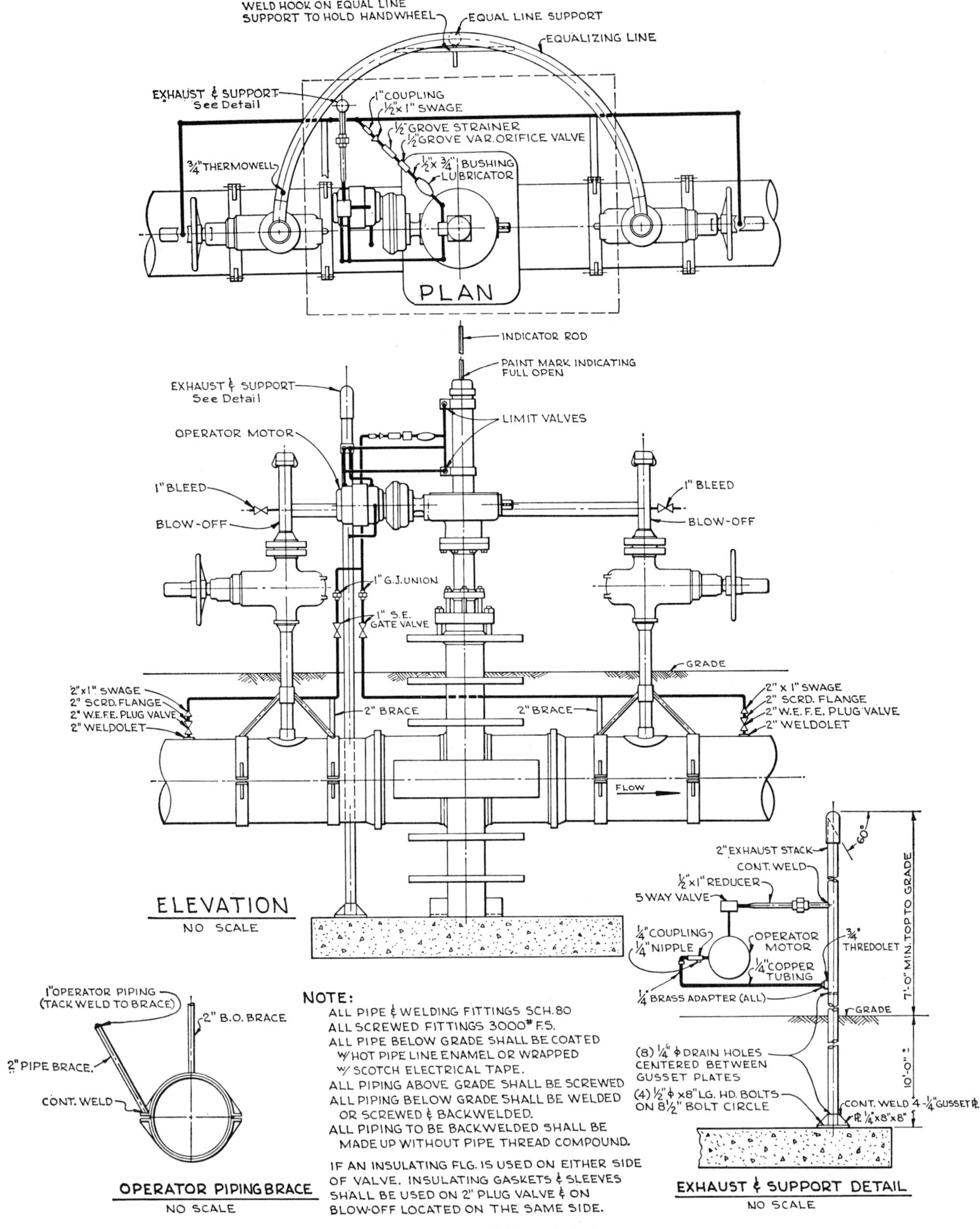
WELD HOOK ON EQUAL LINE SUPPORT TO HOLD HANDWHEEL
EQUAL LINE SUPPORT
EQUALIZING LINE
EXHAUST & SUPPORT See Detail
1" COUPLING
1/2"x 1" SWAGE
1/2" GROVE STRAINER
1/2" GROVE VAR. ORIFICE VALVE
3/4" THERMOWELL
1/2"x 3/4" BUSHING
LUBRICATOR
PLAN
INDICATOR ROD
PAINT MARK INDICATING FULL OPEN
EXHAUST & SUPPORT See Detail
LIMIT VALVES
OPERATOR MOTOR
1" BLEED
BLOW-OFF
1" BLEED
BLOW-OFF
1" G.J. UNION
1" S.E. GATE VALVE
GRADE
2"x1" SWAGE
2" SCRD. FLANGE
2" W.E.F.E. PLUG VALVE
2" WELDOLET
2" BRACE
2" BRACE
2" x 1" SWAGE
2" SCRD. FLANGE
2" W.E.F.E. PLUG VALVE
2" WELDOLET
FLOW
ELEVATION
NO SCALE
2" EXHAUST STACK
60°
CONT. WELD
1/2"x1" REDUCER
5 WAY VALVE
1/4" COUPLING
1/4" NIPPLE
OPERATOR MOTOR
3/4" THREDOLET
1/4" COPPER TUBING
1/4" BRASS ADAPTER (ALL)
GRADE
7'-0" MIN. TOP TO GRADE
10'-0" ±
(8) 1/4" φ DRAIN HOLES CENTERED BETWEEN GUSSET PLATES
(4) 1/2" φ x8" LG. HD. BOLTS ON 8 1/2" BOLT CIRCLE
CONT. WELD
4-1/4" GUSSET PL
PL 1/4"x8"x8"
EXHAUST & SUPPORT DETAIL
NO SCALE
1" OPERATOR PIPING (TACK WELD TO BRACE)
2" B.O. BRACE
2" PIPE BRACE
CONT. WELD
OPERATOR PIPING BRACE
NO SCALE
NOTE:
ALL PIPE & WELDING FITTINGS SCH. 80
ALL SCREWED FITTINGS 3000# F.S.
ALL PIPE BELOW GRADE SHALL BE COATED W/ HOT PIPE LINE ENAMEL OR WRAPPED W/ SCOTCH ELECTRICAL TAPE.
ALL PIPING ABOVE GRADE SHALL BE SCREWED
ALL PIPING BELOW GRADE SHALL BE WELDED OR SCREWED & BACKWELDED.
ALL PIPING TO BE BACKWELDED SHALL BE MADE UP WITHOUT PIPE THREAD COMPOUND.
IF AN INSULATING FLG. IS USED ON EITHER SIDE OF VALVE. INSULATING GASKETS & SLEEVES SHALL BE USED ON 2" PLUG VALVE & ON BLOW-OFF LOCATED ON THE SAME SIDE.

Fig. 8-12 Typical motor-operated system for underground valves. (Walworth Co.)

Another type of drip is illustrated in Fig. 8-14. It is more commonly used on field gathering lines where considerable fluid collection is likely. A vertical baffle is provided to deflect the gas and fluid downward and prevent carry-through.

Scraper Barrels

Incoming and outgoing scraper barrels may be provided as means for clearing a pipeline. Their use requires equipping the main line with full-opening valves. Scraper barrels are preferably located not more than 80 miles apart and, if possible, at compressor stations, since the pressure difference between station suction and discharge favors this operation.

Barrel location should be adequate for maneuvering loading equipment without property damage. To facilitate loading and unloading of the scraper, barrels should be located above ground. An elevation of two to three feet is usually satisfactory. Barrel length is determined by length of the scraper to be used. Inside diameter of the barrel should be 2 in. larger than main line diameter.

Incoming scraper barrels should be tapped for an equalizer line at 90° from the vertical, or vertically at a point near their end closure. This equalizer line then connects to a point downstream from the station suction line valve. The outgoing barrel should be equipped with a kicker line connected from a point 90° from the vertical, near the barrel end closure, to one upstream of the station discharge line valve. This barrel should also be equipped with a 1 in. equalizer line connected to its upstream end, to prevent the scraper from being pushed back into the oversize portion of the barrel when the scraper gate is opened. Both scraper barrels should also be provided with blowoffs. That on the incoming valve should be the larger since its purpose is to relieve pressure ahead of the scraper traveling into the barrel area. See Fig. 8–15.

Crossings Under Highways and Railroads

Studies were initiated in 1959 to obtain fundamental research data for designing casings and uncased carrier pipe crossings of railroads and highways. This work also included basic research pertaining to the evaluation of the modulus of passive resistance of soils. Participants in the program include AAR, A.G.A., API, AREA, ASCE, and U. S. Bureau of Public Roads. In 1964 this group submitted an interim specification on crossings to the various industries involved.

A survey of the design practices of 28 companies is available.[21]

Construction. An opening for the installation may be made by using the open cut method or by boring, jacking, or tunneling. Approval of the method to be used should be agreed upon by the authorized agent of the facility being crossed and the representative of the pipeline company installing the crossing. Care should be exercised to eliminate the formation of voids and waterways around the casing or pipeline.

Casings.[20] These should be considered a tool or a means of construction and the need for their use should be carefully evaluated. They can be used to avoid interference with traffic on major highways or railroads, or to satisfy other construction and engineering situations. A casing may interfere with cathodic protection of the carrier pipe; this factor should be considered when evaluating casing usage.

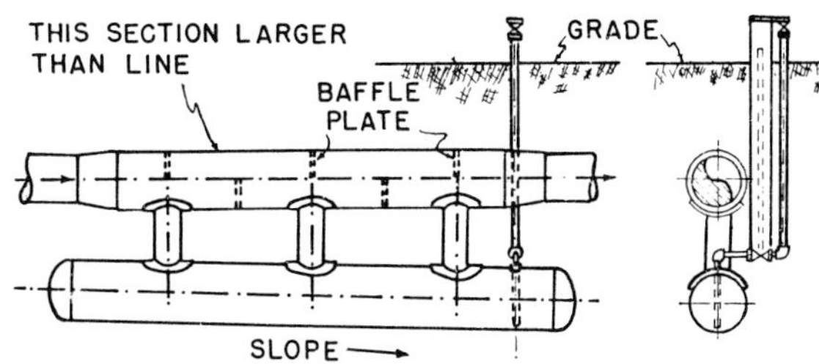

Fig. 8-13 Transmission line drip.

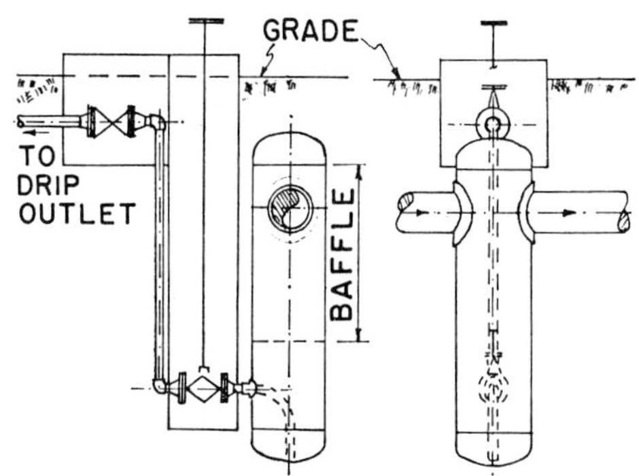

Fig. 8-14 Drip for gathering lines.

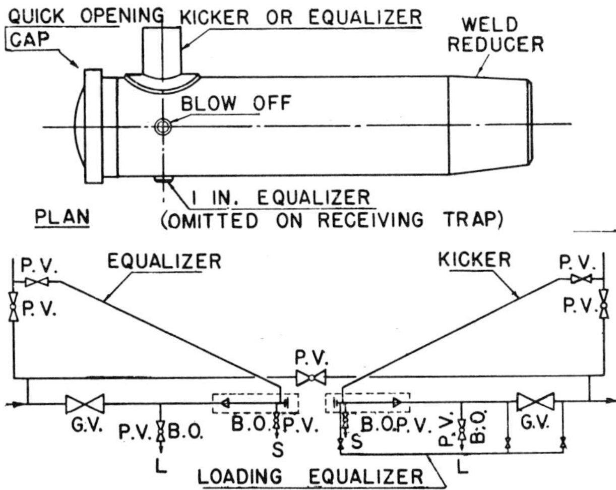

Fig. 8-15 Typical scraper barrels and connections. P.V. = plug valve; G.V. = gate valve; B.O. = blowoff; L = line; S = scraper barrel.

Main line OD	Scraper barrel OD	Equalizer line		Kicker line	Blowoffs			
					Incoming trap		Outgoing trap	
		In-coming	Out-going		L	S	L	S
$10^3/_4$	$12^3/_4$	4	1	6	4	2	4	2
$12^3/_4$	14	4	1	6	4	2	4	2
14	16	6	1	6	6	4	6	2
16	18	6	1	6	6	4	6	2
18	20	6	1	6	6	4	6	2
20	22	8	1	8	8	4	6	2
22	24	8	1	8	8	4	6	2
24	26	8	1	8	8	4	6	2
26	28	8	1	8	8	4	6	2
30	32	10	1	10	10	4	8	2

(All values in inches.)

The diameter of casing pipe should be at least four inches greater (nominal pipe size) than that of the carrier pipe. The wall thickness of the casing should be adequate to withstand the dead and live loads to which it may be subjected. Casings are installed at depths prescribed by the regulatory body or railroad usually at a minimum depth of three feet below grade at the lowest point in the right of way.

Venting. It is no longer a universal practice to vent casings. However, when a vent is used, the vent pipe is installed from the top of the casing to near the right of way line and extended three feet above the ground. The top of the vent should be equipped with a screened 180° bend, or with a relief valve.

Carrier Pipe. The pipe should be supported in its casing by insulated spacers, the locations of which depend upon the size of the pipeline and the type of spacer used. The top of the carrier pipe or spacer should not have contact with the top of the casing at any time. Generally, spacers are located within 12 in. of the ends of the casing.

Immediately after installation is completed, a check should be made to ensure complete electrical insulation between the carrier pipe and the casing pipe.

Submerged Stream Crossings

Pipelines under rivers may be subjected to loads due to bank recession, stream bed scour and fill, buoyancy, drag, movement of debris and sand, and temperature change. In navigable streams, effects of river traffic and of possible future improvements for navigation and flood control should be considered. Exposure of the pipe to forces from these sources should be eliminated as far as possible. Low maintenance costs and freedom from service interruptions may fully justify additional expense of conservative crossing design.

For proper selection of a suitable crossing location, available data should include a complete historical survey detailing movements of banks and thalweg, recorded depths of scour, and the hydrograph of the stream. These data may be obtained from field investigation and from government hydrographic and topographic surveys. They should cover a considerable distance on each side of the proposed crossing location. Soil borings along the proposed location and detailed inspection of them are also helpful.

Submerged pipeline crossings should be at right angles to the stream and located at a level below river scour. Multiple lines should be laid when sizable streams are crossed.

Valves should be provided at each side of the stream. They should preferably be located on high ground and, if possible, on the land side of levees.

Trenching should be extended far enough behind the permanent bank line to insure adequate pipe cover. Where bank is eroding, length of entrenchment should be based on a study of the erosion rate. Soil borings from banks of alluvial streams are useful in determining slope of banks and of crossing approaches.

Flexibility is a principal requirement for crossing pipe, both for laying and for resistance to subsequent forces. It is proportional to material of the pipe and to the fourth power of its diameter. For adequate flexibility, crossing pipe should be made of heavier wall mild steel with properly designed transition, because: (a) Greater mechanical strength is needed to resist unknown forces present in a submerged crossing; and (b) Part of the negative buoyancy is obtained thru additional metal weight.

Figure 8-16 illustrates a typical submerged river crossing.

Weights. A submerged line should be sufficiently weighted to resist buoyancy and to insure that it will remain in position in its trench. Either metal or concrete clamps or a concrete coating may be used. Clamp design and data are presented in Fig. 8-17. Table 8-8 gives cost data for concrete coating.

It is impractical to provide sufficient weight to prevent fluttering of a large pipeline if it is exposed to turbulent action in a swift stream. Cast iron or steel clamps provide maximum weight per unit of volume, with less area exposed to river current during laying. Continuous coatings of concrete permit ease of handling without coating damage, and if fluttering occurs, stress concentrations do not result. Additional weight and less bulk may be obtained by using coatings containing heavy aggregates. Weighting should be carried up and over any overbends and to all overflow areas.

Buoyancy Calculations. Weight requirements should be based upon a desired 20 per cent negative buoyancy in appropriate medium, say 72 lb per cu ft fluid. Number and spacing of weights depends upon their size. A weight of 490 lb per cu ft

Table 8-8 Typical Costs for Concrete Coating of Pipe

(1960–61 avg in Louisiana—Gulf Coast area,* based on 40 ft joints)
All costs in $ per ft

Pipe size & wall thick., in.	Concrete thickness, in., & density, lb/cu ft	Handling†	Wrap coat 3/32 in. asphalt or tar w/glass wrap	Concrete coating 2M–10M ft	10M–40M ft	Over 40M ft
14 × 0.312	2½—140	0.13	0.50	2.26	2.15	2.03
14 × .500	1 —140	.13	.50	0.92	0.87	0.83
16 × .312	2⅞—140	.16	.57	3.00	2.85	2.70
16 × .500	1⅝—140	.17	.57	1.58	1.50	1.42
20 × .312	4 —140	.25	.72	4.44	4.22	4.00
20 × .500	2⅞—140	.41	0.72	3.04	2.89	2.75
24 × .312	3¾—165	.31	1.15	6.76	6.42	6.10
24 × .500	4 —140	.37	1.15	4.83	4.59	4.35
30 × .312	4 —190	.43	1.42	10.04	9.54	9.00
30 × 0.500	3⅜—190	0.48	1.42	8.45	8.03	7.50

* In Northeastern United States add 20 per cent to costs for 20 in. pipe and smaller and add 10 per cent to pipe larger than 20 in.
† Sum of unloading bare pipe and re-loading coated pipe.

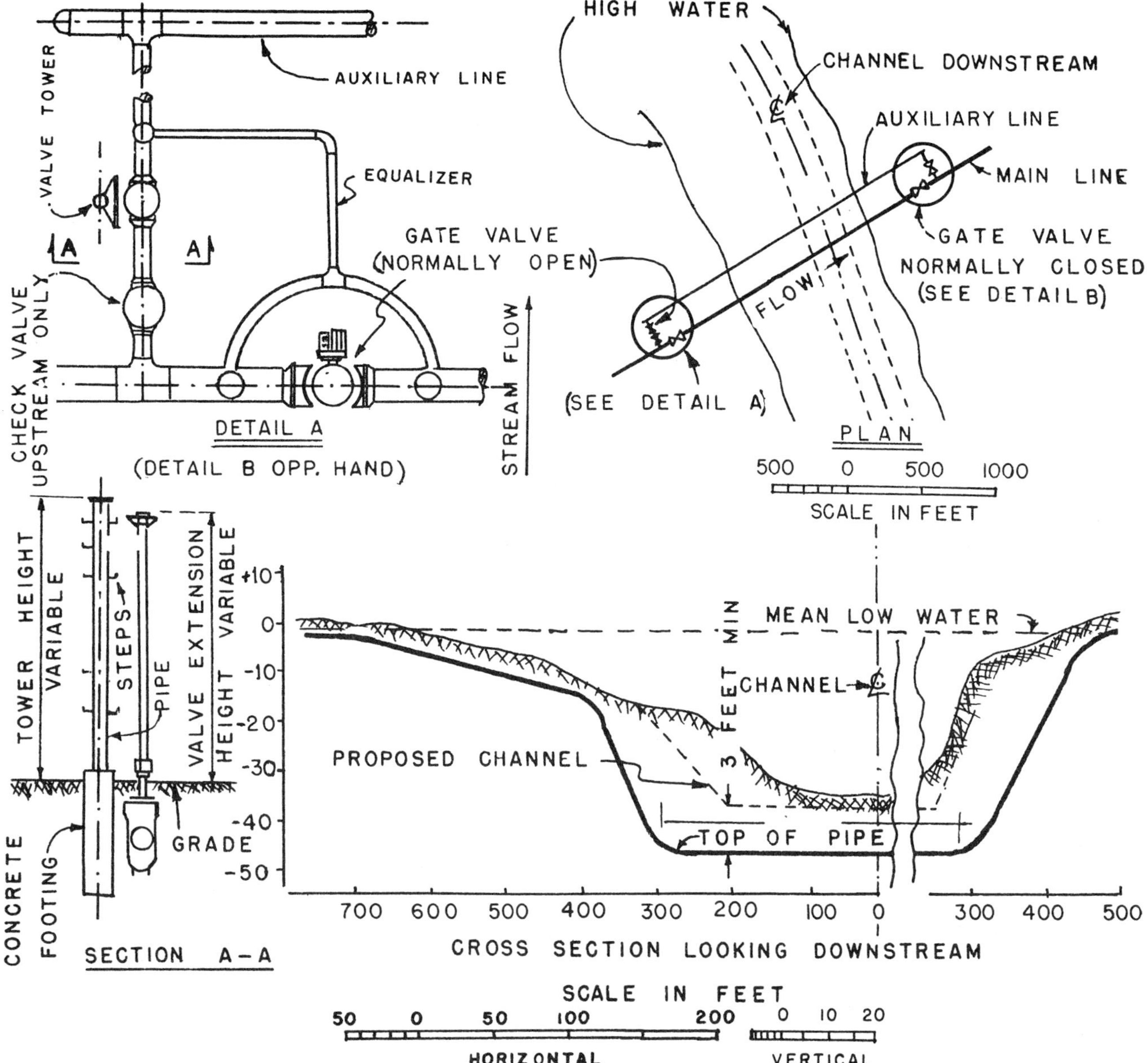

Fig. 8-16 Typical submerged pipeline crossing.
Note 1: Crossovers both sides far enough back into bank to furnish protection.
Note 2: Mainline valve and crossover valve operators piped so that both valves may be operated from the same valve tower.

is 85 per cent efficient. Heavy concrete may be considered as weighing 205 lb per cu ft, and thickness required may be calculated on this basis. Actual samples of the fluid gathered along the line of lay may also be used to determine pipeline weighting. The specific gravity of soil *solids* normally averages 2.6 to 2.8.

The equation for static equilibrium in terms of *pounds per foot of pipe* is:

(steel *plus* corrosion control coating and wrapper *plus* concrete) = overall cross sectional area per foot *times* environmental density *times* negative buoyancy factor. The factor for 20 per cent negative buoyancy, i.e., 1.2 times as dense as the environment = 1.20

Overhead Crossings

In general, location of overhead river crossings should be determined in much the same manner as submerged crossings.

A pipeline bridge should be strong enough to accommodate the weight of the line filled with water. Adequate provision should be made for any stresses imposed by passage of pipe scrapers. It may also be necessary to meet requirements of local regulatory authorities.

Bridge towers should be located well back from the bank line on soil considered safe from erosion. Pilings with a concrete pier foundation should be used to provide maximum protection against future erosion or failure of supporting soil. Adequate provision should be made for expansion of the line.[11]

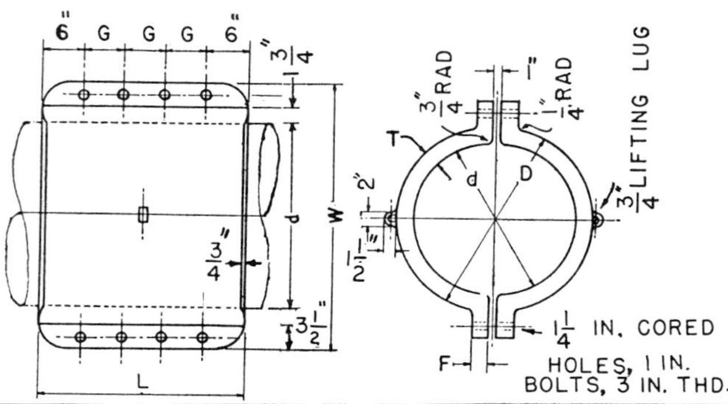

Fig. 8-17 River clamp data.

NOTE. CLAMPS WILL HAVE EXTRA SHEATH TO MAKE FIT TIGHT TO PIPE. BURRS ON INSIDE WALL TO BE GROUNDED OFF TO BASE.

$1\frac{1}{4}$ IN. CORED HOLES, 1 IN. BOLTS, 3 IN. THD.

* Coating thickness, $t_c = 0.25$ in.

Pipe, inches		Pounds per lineal foot							Clamp dimensions, inches							Bolts	Clamp weights, pounds			Clamp spacing, feet and inches	
					Pipe and coating		Buoyancy														
OD	Wall	Pipe	Coat*	Total	In 72 fluid	In 82 fluid	In 72 fluid	In 82 fluid	D	d	T	F	W	G	L	No.-size	In air	In 72 fluid	In 82 fluid	In 72 fluid	In 82 fluid
30	0.500	157.53	15	172.5	362.8	414.0	190.3	241.5	37	31	3	2.5	44	12	48	8-1 × 8	4293	3623	3512	19′ ½″	14′ 6½″
30	.375	118.65	15	133.7	362.8	414.0	229.1	280.3	37	31	3	2.5	44	12	48	8-1 × 8	4293	3623	3512	15′ 9¾″	12′ 6¼″
30	.3125	99.08	15	114.1	362.8	414.0	248.7	299.9	37	31	3	2.5	44	12	48	8-1 × 8	4293	3623	3512	14′ 6¾″	11′ 8½″
28	.500	146.85	14	160.9	316.8	361.5	155.9	200.6	35	29	3	2.5	42	11	45	8-1 × 8	3804	3211	3112	20′ 7″	15′ 6″
28	.375	110.64	14	124.6	316.8	361.5	192.2	236.9	35	29	3	2.5	42	11	45	8-1 × 8	3804	3211	3112	16′ 8½″	13′ 1¾″
28	.3125	92.41	14	106.4	316.8	361.5	210.4	255.1	35	29	3	2.5	42	11	45	8-1 × 8	3804	3211	3112	15′ 3″	12′ 2½″
26	.500	136.17	13	149.2	273.9	312.5	124.7	163.3	33	27	3	2.5	40	10	42	8-1 × 8	3345	2824	2736	22′ 7¾″	16′ 9″
26	.375	102.63	13	115.6	273.9	312.5	158.3	196.9	33	27	3	2.5	40	10	42	8-1 × 8	3345	2824	2736	18′ 2½″	13′ 11″
26	.3125	85.73	13	98.7	273.9	312.5	175.2	213.8	33	27	3	2.5	40	10	42	8-1 × 8	3345	2824	2736	16′ 1½″	12′ 9½″
24	.500	125.49	12	137.5	234.1	267.1	96.6	129.6	31	25	3	2.5	38	10	42	8-1 × 8	3115	2629	2548	27′ 2½″	19′ 8″
24	.375	94.62	12	106.6	234.1	267.1	127.5	160.5	31	25	3	2.5	38	10	42	8-1 × 8	3115	2629	2548	20′ 7½″	15′ 10½″
24	.3125	79.06	12	91.1	234.1	267.1	143.0	176.0	31	25	3	2.5	38	10	42	8-1 × 8	3115	2629	2548	18′ 4½″	14′ 6″
22	.500	114.81	11	125.8	197.4	225.3	71.6	99.5	28	23	2.5	2.5	35	10	42	8-1 × 8	2454	2071	2007	28′ 11″	20′ 2″
22	.375	86.61	11	97.6	197.4	225.3	99.8	127.7	28	23	2.5	2.5	35	10	42	8-1 × 8	2454	2071	2007	20′ 9″	15′ 8½″
22	.3125	72.38	11	83.4	197.4	225.3	114.0	141.9	28	23	2.5	2.5	35	10	42	8-1 × 8	2454	2071	2007	18′ 2″	14′ 1¾″
20	.375	78.60	10	88.6	163.9	187.0	75.3	98.4	26	21	2.5	2.5	33	10	42	8-1 × 8	2282	1926	1867	25′ 7″	19′ 0″
20	.3125	65.71	10	75.7	163.9	187.0	88.2	111.3	26	21	2.5	2.5	33	10	42	8-1 × 8	2282	1926	1867	21′ 10″	16′ 9″
20	.250	52.73	10	62.7	163.9	187.0	101.2	124.3	26	21	2.5	2.5	33	10	42	8-1 × 8	2282	1926	1867	19′ 0″	15′ 0″
18	.375	70.59	9	79.6	133.5	152.3	53.9	72.7	23	19	2	2	30	10	42	8-1 × 7	1641	1385	1342	25′ 8½″	18′ 5½″
18	.250	47.39	9	56.4	133.5	152.3	77.1	95.9	23	19	2	2	30	10	42	8-1 × 7	1641	1385	1342	18′ 0″	14′ 0″
16	.375	62.58	8	70.6	106.2	121.2	35.6	50.6	21	17	2	2	28	8	36	8-1 × 7	1290	1088	1053	30′ 6½″	20′ 9½″
16	.250	42.05	8	50.1	106.2	121.2	56.1	71.1	21	17	2	2	28	8	36	8-1 × 7	1290	1088	1053	19′ 4¾″	14′ 9½″
14	.375	54.57	7	61.6	82.0	93.6	20.4	32.0	19	15	2	2	26	8	36	8-1 × 7	1171	988	957	48′ 5¼″	29′ 11″
14	.250	36.71	7	43.7	82.0	93.6	38.3	49.9	19	15	2	2	26	8	36	8-1 × 7	1171	988	957	25′ 9½″	19′ 2¼″
12.75	.375	49.56	6.5	56.1	68.5	78.1	12.4	22.0	16.5	13.5	1.5	1.5	23.5	12	36	6-1 × 6	774	653	633	52′ 11″	28′ 11″
12.75	.250	33.38	6.5	39.9	68.5	78.1	28.6	38.2	16.5	13.5	1.5	1.5	23.5	12	36	6-1 × 6	774	653	633	22′ 11¼″	16′ 7¾″
10.75	.375	41.55	5.5	47.1	49.4	56.3	2.3	9.2	14.5	11.5	1.5	1.5	21.5	12	36	6-1 × 6	686	580	561	253′ ½″	61′ 3½″
10.75	.250	28.04	5.5	33.5	49.4	56.3	15.9	32.8	14.5	11.5	1.5	1.5	21.5	12	36	6-1 × 6	686	580	561	36′ 7½″	24′ 9″
8.625	.375	33.04	4.5	37.5	32.5	37.1	−5.0	−0.4	12.5	9.5	1.5	1.5	19.5	9	30	6-1 × 6	500	422	409		
8.625	.250	22.36	4.5	26.9	32.5	37.1	5.6	10.2	12.5	9.5	1.5	1.5	19.5	9	30	6-1 × 6	500	422	409	75′ 4¼″	40′ 1¼″
6.625	.375	25.03	3.5	28.5	19.8	22.6	−8.7	−5.9	10.5	7.5	1.5	1.5	17.5	9	30	6-1 × 6	427	360	349		
6.625	0.250	17.02	3.5	20.5	19.8	22.6	−0.7	2.1	10.5	7.5	1.5	1.5	17.5	9	30	6-1 × 6	427	360	349		166 2½″

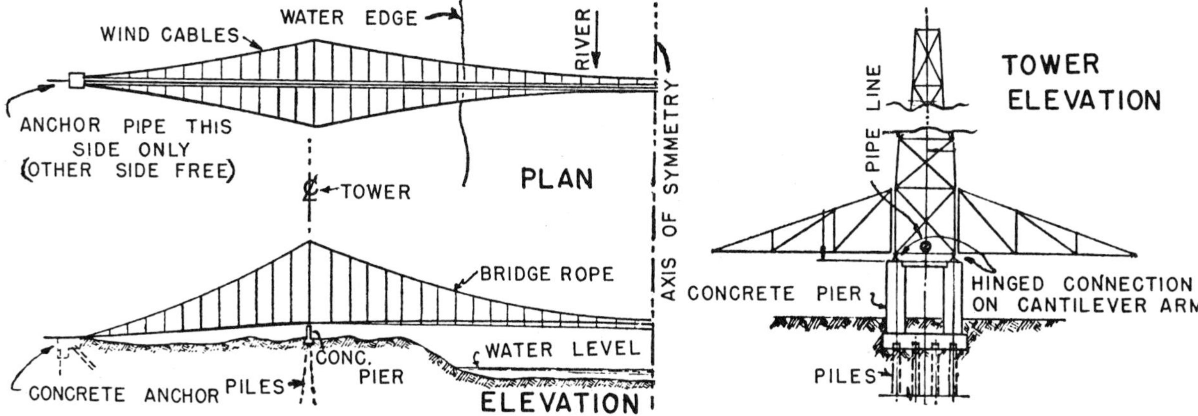

Fig. 8-18 Typical overhead pipeline crossing.

One of the following means is commonly used: (1) large ground loop, either buried or exposed on bank at one end of bridge; (2) U-shaped loop supported on outrigger arm; or (3) side 90° bend on bank side of tower at bridge approach with leg designed to required length and sloped to ground.

Main supporting cables should be prestressed. Galvanize all cables, fittings, and, where practicable, structural steel.

Towers should be constructed with either pivoted or rigid base. Pivoted towers are considered preferable because they eliminate undue stresses due to improper cable placement and structural misalignments.

Pipe for bridge crossings may be of the same nominal diameter and specification as for the main line. For high yields, a greater wall thickness may be employed to give additional mechanical strength. Figure 8-18 shows a typical bridge crossing.

See Ref. 21 for design practices.

Unsupported Spans

Relatively short unsupported spans may be used for crossing gullies, canyons, and creeks where banks provide firm and undisturbed support. Use of such spans is convenient, economical and structurally safe. In general, these spans are subject to internal pressure stresses and thermal stresses similar to those affecting buried pipelines. Wider temperature fluctuation, of course, may be expected. Additional tensile and compressive stresses result from loading the unsupported pipe as a beam. Loading calculations may be based upon the weight of the pipe itself plus the weight of water filling its internal volume.[12]

Figure 8-19 shows span length plotted against bending stress for 30 in. diameter pipe of several wall thicknesses. A formula relating the different variables is also included.

A self-supporting pipe arch (approx. 150 ft radius) spanning 150 ft, rising 20 ft vertically, was constructed of two 16 in. OD × 0.375 wall pipes.[13]

Industrial Area Crossings

Buried pipelines in industrial areas should be properly protected against corrosive action of any chemicals which may be present in the soil as determined by a preliminary survey. A suitable resistant coating should be selected in such cases. Cathodic protection may be advisable and a thorough investigation should be made both before and after installation of the line. Isolation of sections of the line passing thru corrosive areas, by use of insulating flanges for example, may be desirable.

Aboveground piping may be used in industrial areas where investigation shows it would be difficult to protect and maintain buried pipe. A minimum elevation of 1 ft above grade is preferable for such piping. Supporting piers should be spaced so that permissible stresses from bending and internal pressures combined are not exceeded. Piers may be made of concrete or masonry, topped with a steel shoe or wearing plate which permits movement due to temperature changes without abrasive damage. Confining straps should be placed over the pipe at each support to hold it in place on the support, while permitting slight movements of expansion and contraction. Exposed piping should be protected against attack by airborne chemicals.

Pipeline Blowdowns

Each section of a pipeline system usually contains two blowdown valves for exhausting gas from the line when required by repairs or other conditions. One such valve is placed on each side of and within about 25 ft of the main sectionalizing valve.

Proper blowdown design is of major importance, because the issuing gas jet produces a violent reaction thrust. To withstand this thrust an adequately designed foundation and a line reinforcement below the blowdown riser are required. Angle supports for the blowdown riser may be necessary. Figure 8-20 shows a typical blowdown installation. The section to be exhausted is isolated by closing the sectionalizing valves controlling it.

Blowdown valve size may be selected by using a chart like that given in Fig. 8-21, covering the particular type of valve considered. With line volume known, blowdown time for emptying the line may be read from the ordinate scale for different line pressures and valve sizes of the type covered. The size having the blowdown time closest to that desired may then be selected.

If both blowdown valves are to be operated simultaneously, the chart may still be employed. Since each valve then empties half of the sysem, its size is selected using half of the total line volume.

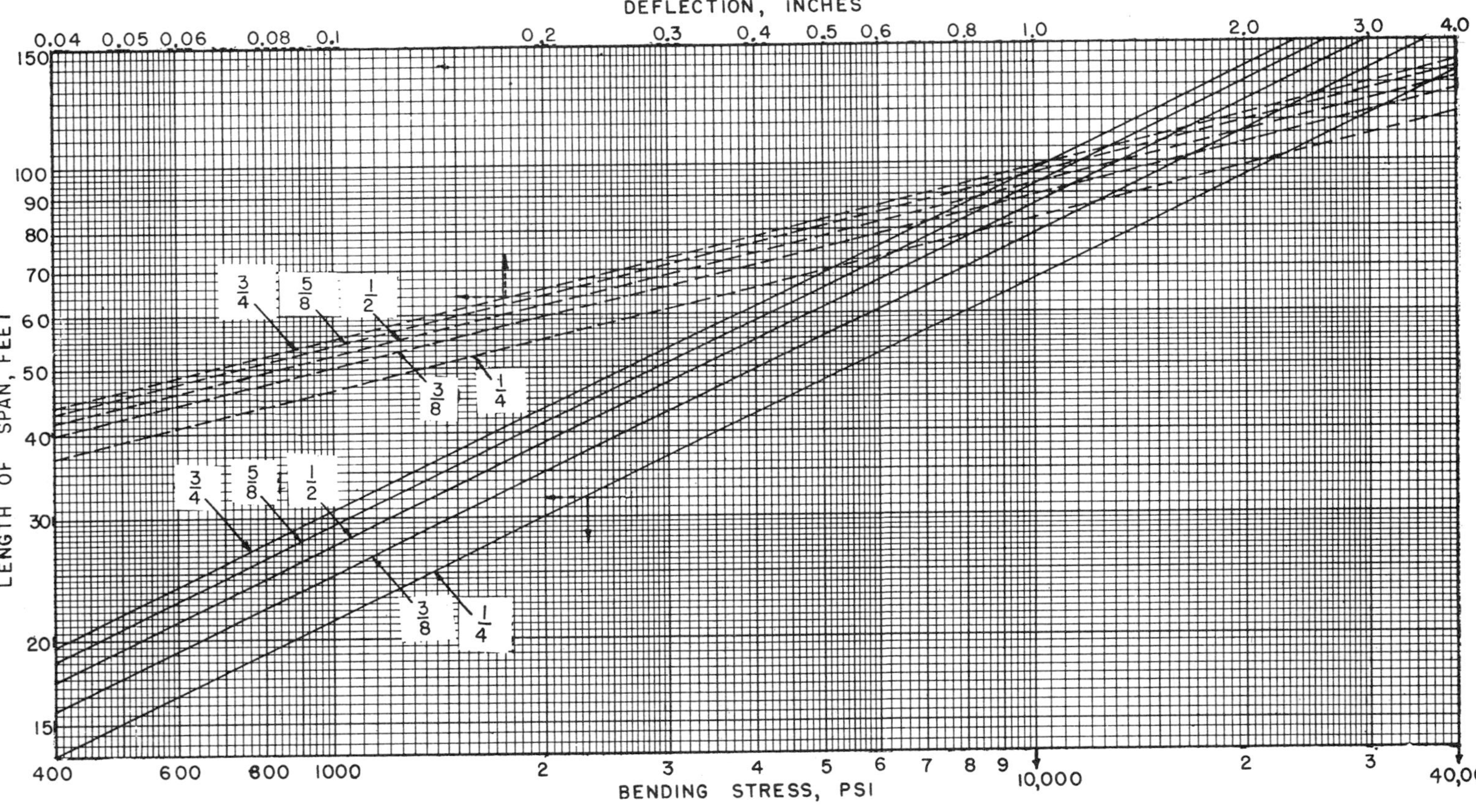

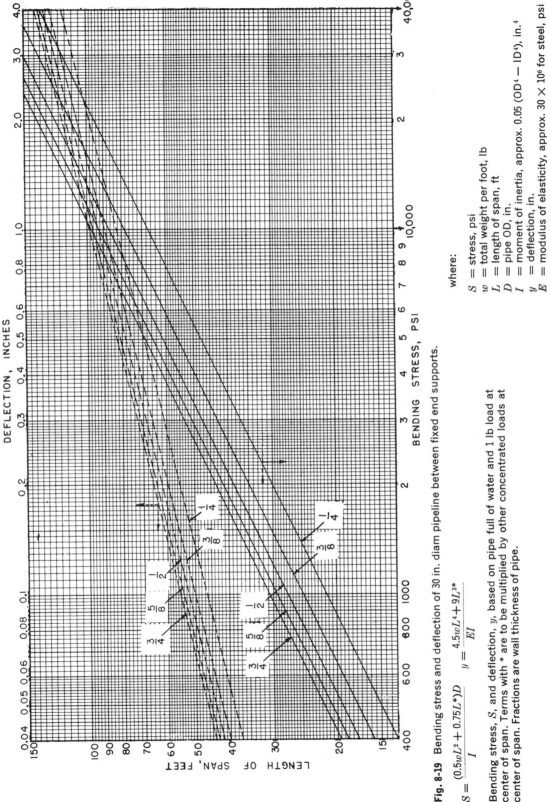

Fig. 8-19 Bending stress and deflection of 30 in. diam pipeline between fixed end supports.

$$S = \frac{(0.5wL^2 + 0.75L^*)D}{I} \qquad y = \frac{4.5wL^4 + 9L^{3*}}{EI}$$

Bending stress, S, and deflection, y, based on pipe full of water and 1 lb load at center of span. Terms with * are to be multiplied by other concentrated loads at center of span. Fractions are wall thickness of pipe.

where:

S = stress, psi
w = total weight per foot, lb
L = length of span, ft
D = pipe OD, in.
I = moment of inertia, approx. 0.05 (OD^4 — ID^4), in.^{4}
y = deflection, in.
E = modulus of elasticity, approx. 30 × 10^6 for steel, psi

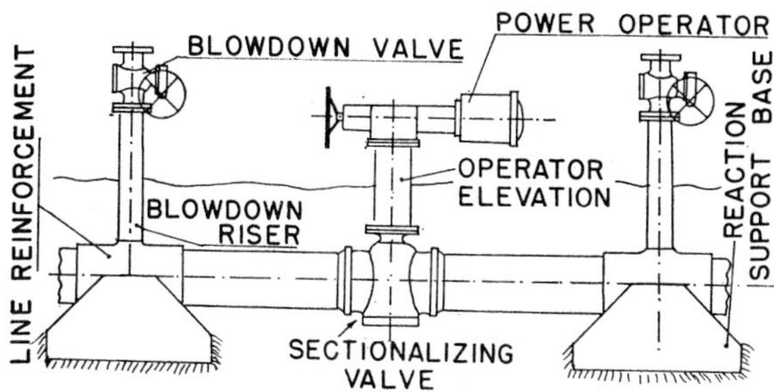

Fig. 8-20 Typical blowdown and sectionalizing valve installation on gas pipeline.

for payment on any work items not definitely covered by unit prices. Such work items, commonly termed "extra work," are usually paid on the basis of actual cost plus a fixed percentage. In determining the cost items included under extra work, general practice is to agree with the contractor on labor and equipment charge schedules and to incorporate them into the contract.

Most pipeline construction contracts are awarded as a result of competitive bidding. Before issuing "Invitations to Bidders," those work items for which unit prices are to be bid should be carefully studied.

Methods of formulating Invitations to Bidders and final

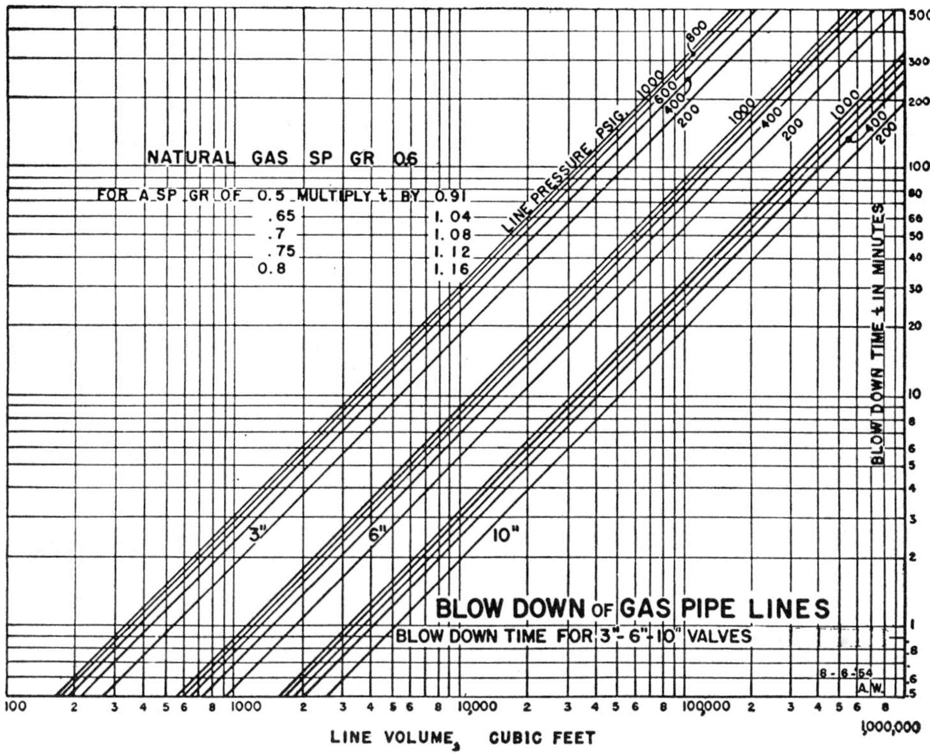

Fig. 8-21 Typical chart for selection of blowdown valve size.

PIPELINE CONSTRUCTION

Construction Contracts

Particularly for larger transmission lines (16 in. and over), actual construction is usually done under contract by a firm specializing in such work. In this way a utility or pipeline owner may avoid purchasing or leasing expensive special equipment and organizing and maintaining skilled personnel required for the operation.

The contractor may be paid on a cost plus fee basis or at a fixed rate. Under the first contract, actual cost plus fee, the fee may be a stipulated percentage of the actual cost of a fixed amount. The fee may also be based on a percentage of the cost but limited to a set maximum amount, thus providing an incentive for the contractor to complete his work as rapidly and economically as possible.

The second contract, a firm bid, usually calls for the contractor to perform specified work items in estimated quantities at agreed unit prices. Specific provision should also be included

contracts vary considerably. Various state laws and differing opinions of pipeline owners have so far made standardization of such forms impractical.

Contract Items. In addition to construction details, the following items may be included in preparing invitations to bidders and in final contracts:

Insurance requirements with liability limits.
Provision for contract modification.
Furnishing of bond and acceptance of responsibility by contractor.
Procurement of permits and licenses.
Status of subcontractors if employed.
Suspension of work if contract is violated.
Time and order of completion.
Hindrances, delays, and losses.
Quality and quantity of materials furnished by contractor and owner.
Extra work or materials.

Compliance by contractor with laws and regulations.
Protection by contractor of persons and property.
Type of contractor's personnel.
Safe use of explosives.
Night work provisions, including inspection.
Infringement of patents.
Liens, notification, and protection of owner's rights.
Lands for construction purposes.
Inspection and right of access by owner.
Progress estimates and payments.
Right to require replacement of defective work not waived by provisional acceptance of portions of work.
Final payment.

General Provisions

For general information on welding procedures and other details of pipeline construction reference may be made to the following extracts from ASA B31.8-1963:

820 WELDING.

821 GENERAL.

821.1 This chapter* concerns the arc and gas welding of pipe joints in both wrought and cast steel materials, and more specifically covers butt joints in pipe, valves, flanges and fittings, and fillet welded joints in pipe branches, slip-on flanges, socket weld fittings, etc., as applied in pipelines and connections to apparatus or equipment. When valves or equipment are furnished with welding ends suitable for welding directly into a pipeline, the design, composition, welding, and stress relief procedures must be such that no significant damage will be likely to result from the welding or stress relieving operation. This chapter* does not apply to the welding of longitudinal joints in the manufacture of pipe.

821.2 These standards apply to manual shielded metal arc and gas welding, automatic submerged arc welding, and are recommended for other manual and automatic welding where applicable.

821.3 These standards are based on the principle that a welding procedure has been established and qualified for sound and ductile welds. In applying these standards, the welder is required to qualify under the procedure employed. These standards establish the groupings of materials that can be welded under a procedure which has been qualified with any one of the materials included in the group. The changes in material, filler metal, process or procedure that require requalification of either welding procedure or welder are set out in 824.21 (a), (b), and (c).

821.4 The standards of quality for pipelines and mains to operate at 20% or more of the specified minimum yield strength are established under section 829, "Standards of Acceptability" and the methods of non-destructive and destructive examination are set out.

821.5 All the welding done under the standards of this code shall be performed under a specification which at least embodies the requirements of this code. Example of such specification is API Standard 1104, "A Standard for Field Welding of Pipe Lines."

821.6 Welding Terms. Definitions pertaining to welding as used in this code conform to the standard definitions established by the American Welding Society and contained in AWS publication, "Standard Welding Terms and Their Definitions"—AWS A3.0.

[* Refers to Chapter II in ASA B31.8-1963.—Editor]

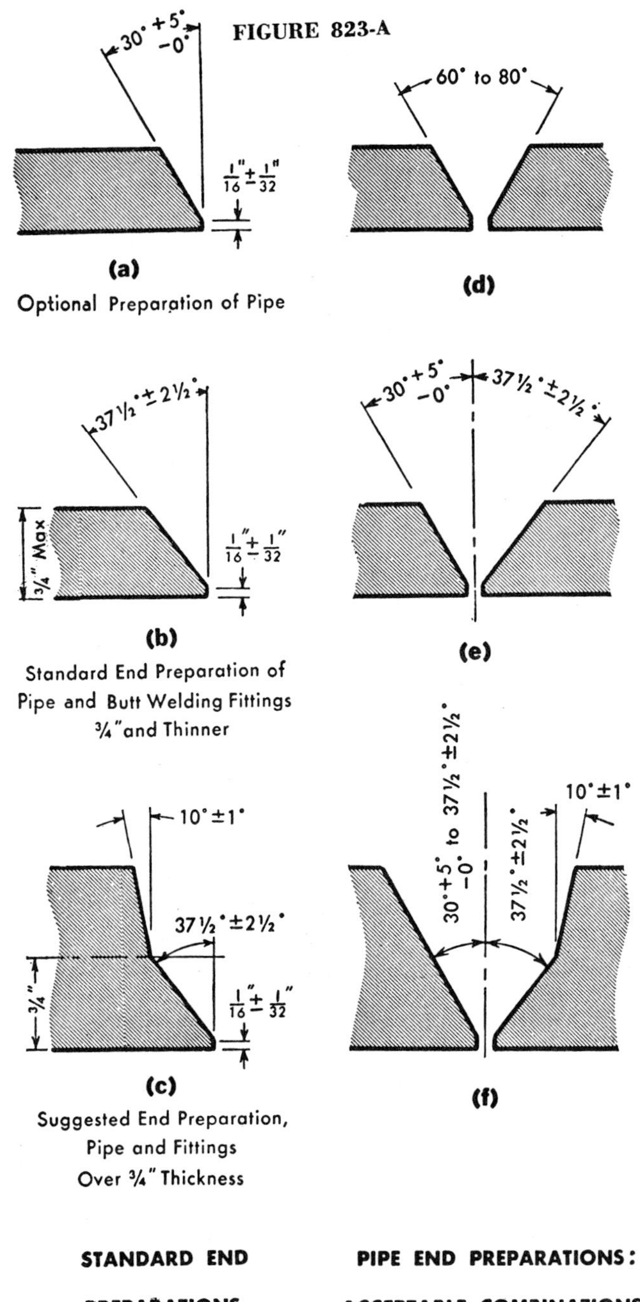

822 TYPES OF WELDS.

822.1 Butt Welds. Butt joints may be of the single Vee, double Vee, or other suitable type of groove. Joint designs shown in Figure 823-A or applicable combinations of these joint design details are recommended. See Figure 823-B for acceptable preparation for butt welding of pieces of unequal thickness.

822.2 Fillet Welds. Fillet welds may be concave to slightly convex. The size of a fillet weld is stated as the leg length of the largest inscribed right isosceles triangle as shown in Figure 823-C.

822.3 Seal Welds. Seal welding shall be done by qualified welders. Seal welding of threaded joints is permitted but the seal welds shall not be considered as contributing to the strength of joints.

FIGURE 823-B

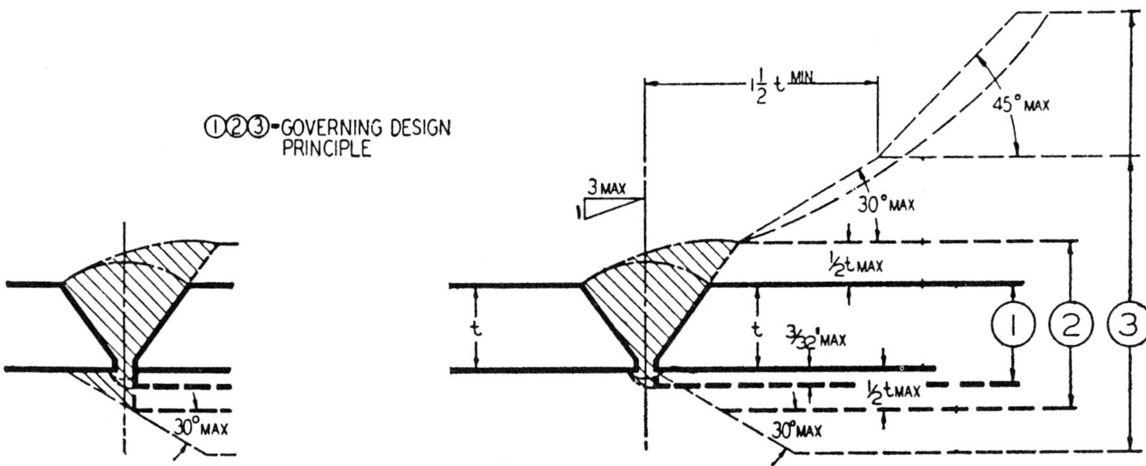

Note: This internal preparation optional if ID of pipe affords excess.
t = nominal thickness of thinner section

Internal and/or external diameters unequal.

Explanatory Notes:

The sketch in Figure 823-B is designed to illustrate some acceptable preparations of ends having unequal thickness providing the following basic precepts are complied with:

1. Where materials of unequal unit strengths (specified minimum yield) are joined together, and the full strength of the higher unit strength material is required, design conditions require that the thickness of the end having the lower physical property be such that its strength be at least equal to that of the adjoining part.
2. The transition between ends of unequal thickness may be accomplished by taper or welding as illustrated or by means of a prefabricated transition ring.
3. The transition weld shall have a slope not greater than 1:3. (Approx. 18°). Excess metal thickness of the heavier section may be tapered for a smooth transition by an angle not exceeding 30° with reference to the pipe surface.
4. Physical properties of the deposited weld metal shall be at least equal to those of the higher strength pipe material.

The design principles governing the preparation of ends having unequal thickness should comply with the following:

1. If the nominal wall thickness of adjoining ends are equal in thickness or do not vary more than 3/32 in. no special treatment is necessary provided full penetration and bond is accomplished in welding.
2. Where the nominal internal or external offset does not exceed 1/2 the thinner wall section, the transition may be made by welding or taper, provided full penetration and bond is accomplished and the basic precepts are adhered to.
3. Where the nominal wall section of valves, fittings, etc., is greater than required for the design strength of the joint, such additional metal may be tapered to the accepted re-entrant angle as illustrated.
4. For piping to operate at hoop stresses of less than 20 per cent of specified minimum yield strength, if the nominal wall thicknesses do not vary more than 1/8 in. no special treatment is necessary provided adequate penetration and bond is accomplished in welding.

823 PREPARATION FOR WELDING.

823.1 Butt Welds.

(a) The welding surfaces shall be clean, and free of material that may be detrimental to the weld.

(b) *Welding Groove Details.* End preparation as given in Figure 823-A represents present acceptable practice.

(c) *Pipe Alignment.* The ends of pipe to pipe, or pipe to fitting joints shall be aligned as accurately as practicable giving consideration to existing commercial tolerances on pipe diameters, pipe wall thickness and out of roundness. Alignment shall provide the most favorable condition for the deposition of the root bead and shall be preserved during welding of the root bead.

(d) Root opening of the joint shall be as given in the Procedure Specification employed.

823.2 Fillet Welds. Minimum dimensions for fillet welds used in the attachment of slip-on flanges, for socket welded joints, are shown in Figure 823-C. Similar minimum dimensions for fillet welds used in branch connections are shown in Figures 831-B and 831-C.

824 QUALIFICATION OF PROCEDURES AND WELDERS.

824.1 Requirements for Qualification of Procedures and Welders on Piping Systems Operating at Hoop Stresses of Less Than 20% of the Specified Minimum Yield Strength.

824.11 Welders whose work is limited to the application of the oxy-acetylene or manual arc welding processes on piping operating at hoop stresses of less than 20% of the specified minimum yield strength, shall be qualified under any of the references given in 824.21 or in accordance with Appendix F of this code.

FIGURE 823-C

RECOMMENDED ATTACHMENT DETAILS OF FLANGES

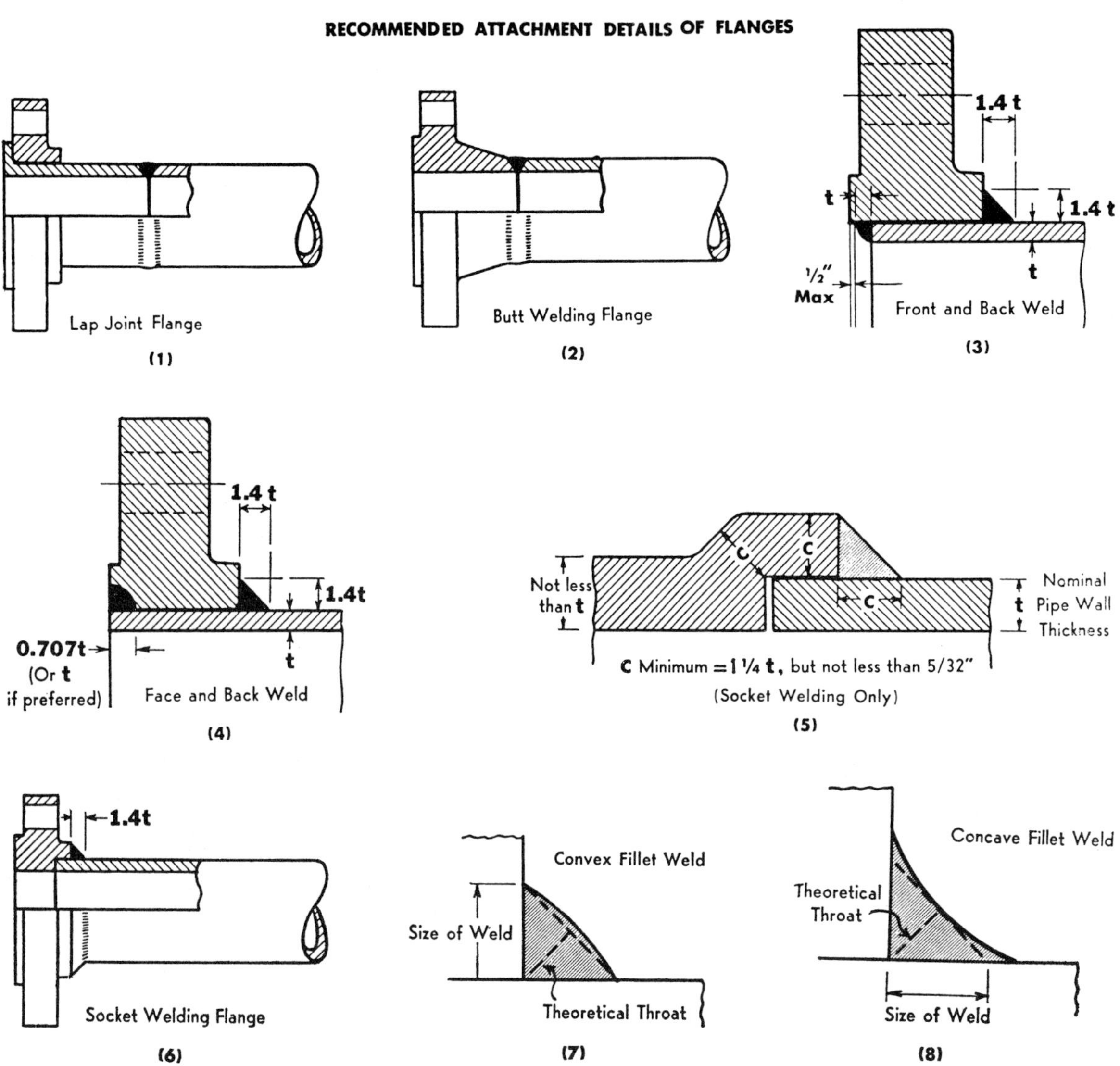

* * * *

APPENDIX F

(Referred to in 824.11)

TESTS OF WELDERS WHO ARE LIMITED TO WORK ON LINES OPERATING AT HOOP STRESSES OF LESS THAN 20% OF THE SPECIFIED MINIMUM YIELD STRENGTH:

(1) An initial test shall qualify a man for work and thereafter his work shall be checked either by requalification at one year intervals or by cutting out and testing production work at least every six months.

(2) The test may be made on pipe of any diameter 12 inches or smaller. The test weld shall be made with the pipe in a horizontal fixed position so that the test weld includes at least one section of overhead position welding.

(3) The beveling, root opening and other details must conform to the procedure specification under which the welder is qualified.

(4) The test weld shall be cut into four coupons and subjected to the root bend test. If, as a result of this test, a crack develops in the weld material or between the weld and base metal more than 1/8 inch long in any direction, this shall be cause for rejection. Cracks occurring on the corner of the specimen during testing shall not be considered. If no more than one coupon is rejected, the weld is to be considered as acceptable.

(5) Welders who are to make welded service connections to mains should be required to satisfactorily pass the following tests:

(a) Weld a service connection fitting to a pipe section having the same diameter as a typical main. This weld should be made in the same position as this type of weld is made in the field.

The weld should be rejected if it shows a serious undercutting or if it has rolled edges.

(b) The weld should be tested by attempting to break the fitting off the run pipe by any available means (knocking off).

A sample shall be rejected if the broken weld at the junction of the fitting and run pipe shows incomplete fusion, overlap, or poor penetration.

(6) For the periodic checking of welders who work on small services only (2 inches or smaller in diameter), the following special field test may be employed. This test should not be used as a substitute for the original qualifying test.

(a) Two sample welds made by the welder under test should be taken from steel service pipe. Each sample should be cut 8 inches long with the weld located approximately in the center. One sample shall have the ends flattened and the entire joint subjected to the tensile strength test. Failure must be in the parent metal and not adjacent to or in the weld metal to be acceptable. The second sample shall be centered in the guided bend testing machine and bent to the contour of the die for a distance of 2 inches on each side of the weld. The sample to be acceptable must show no breaks or cracks after removal from the bending machine.

When a tensile strength testing machine is not available, two bend test samples will be acceptable in lieu of one tension and one bending test.

(7) *Tests for Copper Joints.* Personnel who are to work on copper piping should satisfactorily pass the following tests:

(a) A brazed copper bell joint should be made on any size of copper pipe used, with the axis of the pipe stationary in the horizontal position. The joint so welded is to be sawed open, longitudinally at the top of the pipe. (The top being the uppermost point on the circumference at time joint is brazed.) The joint should be spread apart for examination. The bell end of the joint must be completely bonded. The spigot end of the joint must give evidence that the brazing alloy has reached at least 75% of the total area of the telescoped surfaces. At least 50% of the length at the top of the joint must be joined.

(8) Records shall be kept of the original tests and all subsequent tests conducted on the work of each welder.

* * * *

824.2 Requirements for Qualification of Procedures and Welders on Pipelines to Operate at Hoop Stresses of 20% or more of the Specified Minimum Yield Strength.

824.21 Welding procedures and welders performing work under this classification must be qualified under one of the following standards:

(a) ASA B31.1, Code for Pressure Piping, Section 6, Chapter IV.

(b) ASME Boiler and Pressure Vessel Code, Section IX.

(c) API Standard 1104, "Standard for Field Welding of Pipe Lines."

824.22 When welders qualified under API Standard 1104 are employed on compressor station piping, their qualifying test shall have been based upon the guided bend test.

824.23 *Variables Requiring Separate Qualification of Welding Procedure and Welders.* The references given in 824.21 (a), (b) and (c) contain sections entitled "Essential Variables" applicable to welding procedures and also to welders. These shall be followed except that for the purposes of this code all carbon steels which have a carbon content not exceeding 0.32% by ladle analysis, and a carbon equivalent (C + ¼ Mn) not exceeding 0.65% by ladle analysis, are considered to come under material grouping P1. Alloy steels having weldability characteristics demonstrated to be similar to these carbon steels shall be welded, preheated and stress relieved as prescribed herein for such carbon steels. Other alloy steels shall be welded, preheated, and stress relieved as prescribed in ASA B31.1, Section 6, Chapter IV.

824.24 *Welder Requalification Requirements.* Welder requalification tests shall be required if there is some specific reason to question a welder's ability or the welder is not engaged in a given process of welding (i.e., arc or gas) for a period of six months or more.

824.25 *Qualification Records.* Records of the tests that establish the qualification of a welding procedure shall be maintained as long as that procedure is in use. The operating company or contractor shall, during the construction involved, maintain a record of the welders qualified, showing the date and results of tests.

825 WELDING PROCEDURE—GENERAL.

825.1 The welding procedure followed during the qualifying tests shall be recorded in detail, and shall be adhered to during subsequent construction.

825.2 Welding shall not be done when the quality of the completed weld would be likely to be impaired by the prevailing weather conditions including, but not limited to, airborne moisture, blowing sand, or high wind. Wind shields may be used when practicable.

826 PREHEATING.

826.1 Carbon steels having a carbon content in excess of 0.32% (ladle analysis) or a carbon equivalent (C + ¼ Mn) in excess of 0.65% (ladle analysis) shall be preheated as prescribed in ASA B31.1, Section 6, Chapter IV. Preheating may also be advisable for steels having lower carbon or carbon equivalent, when conditions exist that either limit the welding technique that can be used, or that tend to adversely affect the quality of the weld.

826.2 When welding dissimilar materials having different preheating requirements, the material requiring the higher preheat shall govern.

826.3 Preheating may be accomplished by any suitable method, provided that it is uniform and that the temperature does not fall below the prescribed minimum during the actual welding operations.

826.4 The preheating temperature shall be checked by the use of temperature indicating crayons, thermocouple pyrometers or other suitable method to assure that the required preheat temperature is obtained prior to and maintained during the welding operation.

827 STRESS RELIEVING.

827.1 Carbon steels having a carbon content in excess of 0.32% (ladle analysis) or a carbon equivalent (C + ¼ Mn) in excess of 0.65% (ladle analysis) shall be stress relieved as prescribed in ASA B31.1, Section 6, Chapter IV. Stress relieving may also be advisable for steels having lower carbon or carbon equivalent when adverse conditions exist which too rapidly cool the weld.

827.2 Welds in all carbon steels shall be stress relieved when the wall thickness exceeds 1¼ inches.

827.3 When the welded joint connects parts that are of different thicknesses but of similar materials, the thickness to be used in applying the rules in 827.1 and 827.2 shall be:

(a) The thicker of the two pipes joined.

(b) The thickness of the pipe run or header in case of branch connections, slip-on flanges or socket weld fittings.

827.4 In welds between dissimilar materials, if either material requires stress relieving, the joint shall require stress relieving.

827.5 All welding of connections and attachments shall be stress relieved when the pipe is required to be stress relieved by the rules of 827.3 with the following exceptions:

(a) Fillet and groove welds not over ½ inch in size (leg) that attach connections not over 2-inch pipe size.

(b) Fillet and groove welds not over ⅜ inch in groove size which attach supporting members or other non-pressure attachments.

827.6 Stress Relieving Temperature.

(a) Stress relieving shall be performed at a temperature of 1100 F or over for carbon steels, and 1200 F or over for ferritic alloy steels. The exact temperature range shall be stated in the procedure specification.

(b) When stress relieving a joint between dissimilar metals having different stress relieving requirements, the material requiring the higher stress relieving temperature shall govern.

(c) The parts heated shall be brought slowly to the required temperature and held at that temperature for a period of time proportioned on the basis of at least 1 hour per inch of pipe wall thickness, but in no case less than ½ hour, and shall be allowed to cool slowly and uniformly.

827.7 Methods of Stress Relieving.

(a) Heating the complete structure as a unit.

(b) Heating a complete section containing the weld or welds to be stress relieved before attachment to other sections of work.

(c) Heating a part of the work by heating slowly a circumferential band containing the weld at the center. The width of the band which is heated to the required temperature shall be at least 2 inches greater than the width of the weld reinforcement. Care should be used to obtain a uniform temperature around the entire circumference of the pipe. The temperature shall diminish gradually outward from the ends of this band.

(d) Branches, or other welded attachments for which stress relief is required, may be locally stress relieved by heating a circumferential band around the pipe on which the branch or attachment is welded with the attachment at the middle of the band. The width of the band shall be at least 2 inches greater than the diameter of the weld joining the branch or attachment to the header. The entire band shall be brought up to the required temperature and held for the time specified.

827.8 Equipment for Local Stress Relieving.

(a) Stress relieving may be accomplished by: electric induction, electric resistance, fuel-fired ring burners, fuel-fired torch or other suitable means of heating provided that a uniform temperature is obtained and maintained during the stress relieving.

(b) The stress relieving temperature shall be checked by the use of thermocouple pyrometers or other suitable equipment to be assured that the proper stress relieving cycle has been accomplished.

828 WELDING INSPECTION AND TESTS.

828.1 Inspection of Welds on Piping Systems Intended to Operate at Less than 20% of the Specified Minimum Yield Strength. The quality of welding should be checked visually on a sampling basis, and if there is any reason to believe that the weld is defective, it shall be removed from the line and tested in accordance with the specification or it may be subject to a non-destructive test as outlined in 828.2.

828.2 Inspection and Tests of Welds on Piping Systems Intended to Operate at 20% or More of the Specified Minimum Yield Strength.

(a) The quality of welding should be checked by non-destructive tests or by removing completed welds as selected and designated by the operating company. Non-destructive testing may consist of radiographic examination, magnetic particle testing, or other acceptable methods. The trepanning method of non-destructive testing is prohibited.

(b) When radiographic examination is employed, the procedure set forth in API Standard 1104 (Standard for Field Welding of Pipe Lines) shall be followed; and the number and location of welds examined shall be at the discretion of the operating company.

(c) To be acceptable, completed welds which have been removed for inspection shall successfully meet the testing requirements outlined under the welder qualification procedure, and in addition, shall meet the standards of acceptability contained in 829.

829 STANDARDS OF ACCEPTABILITY OF WELDS ON PIPELINES INTENDED TO OPERATE AT 20% OR MORE OF THE SPECIFIED MINIMUM YIELD STRENGTH.

829.1 Inadequate Penetration and Incomplete Fusion. Any individual inadequate penetration or incomplete fusion shall not exceed 1 inch in length. In any 12-inch length of weld, the total length of inadequate penetration or incomplete fusion shall not exceed 1 inch. The total length of the inadequate penetration or incomplete fusion in any two succeeding 12-inch lengths shall not exceed 2 inches and individual defects shall be separated by at least 6 inches of sound weld metal.

829.2 Burn-Through Areas. Any individual burn-through area shall not exceed ½ inch in length. In any 12-inch length of weld, the total length of burn-through area shall not exceed 1 inch. The total length of burn-through area in any two succeeding 12-inch lengths shall not exceed 2 inches, and individual defects shall be separated by at least 6 inches of sound weld metal.

829.3 Elongated Slag Inclusions. Any elongated slag inclusion shall not exceed 2 inches in length or 1/16 inch in width. In any 12-inch length of weld, the total length of elongated slag inclusions shall not exceed 2 inches. The total length of elongated slag inclusions in any two succeeding 12-inch lengths, shall not exceed 4 inches, and individual defects shall be separated by at least 6 inches of sound weld metal. Parallel slag lines shall be considered as individual defects if they are wider than 1/32 inch.

829.4 Isolated Slag Inclusions. The maximum width of any isolated slag inclusion shall not exceed ⅛ inch. In any 12-inch length of weld, the total length of isolated slag inclusions shall not exceed ½ inch, nor shall there be more than four isolated slag inclusions of the maximum width of ⅛ inch in this length. Any two such inclusions shall be separated by 2 inches of sound weld metal. In any 24-inch length of weld, the total length of isolated slag inclusions shall not exceed one inch.

829.5 Gas Pockets. The maximum dimensions of any individual gas pocket shall not exceed 1/16 inch. Maximum distribution of gas pockets shall not exceed that shown in Figs. 7 and 8 of API Standard 1104.

829.6 Cracks. No welds containing cracks, regardless of size or location shall be acceptable until such welds have been repaired in conformance with 829.9.

829.7 Accumulation of Discontinuities. Any accumulation of discontinuities having a total length of more than 2 inches in a weld length of 12 inches is unacceptable. Any accumulation of discontinuities which total more than 10% of the weld length of a joint is unacceptable.

829.8 Undercutting. Undercutting adjacent to the cover bead on the outside of the pipe shall not exceed 1/32 inch in depth and 2 inches in length. Undercutting adjacent to the root bead on the inside of the pipe shall not exceed 2 inches in length.

829.9 Repair of Defects.

(a) Except as provided in (g) below, defective welds shall be repaired or removed from the pipeline at the request of the company representative. The company may authorize repairs of defects in the root and filler beads, but any weld that shows evidence of repair work having been done without authorization by the company may be rejected.

(b) Minor cracks in the surface and filler beads may be repaired when so authorized by the company, but any crack penetrating the root bead or the second bead shall be cause for complete rejection of the weld. The entire weld shall then be cut from the pipe line and replaced. Minor cracks shall be defined as cracks visible in the surface bead and not over 2 inches in length.

(c) Before repairs are made, injurious defects shall be removed by chipping, grinding, or oxygen gouging to clean metal. All slag and scale shall be removed by wire brushing.

(d) It is recommended that all such areas be preheated before the repair weld is started.

(e) Repaired areas shall be carefully inspected and radiographed when considered necessary.

(f) No further repairs shall be allowed in repaired areas.

(g) Repairs may be made to pin holes and undercuts in the final bead without authorization, but must meet with the approval of the company.

841.2 Installation of Steel Pipelines and Mains.

841.21 *Construction Specifications.* All construction work performed on piping systems in accordance with the requirements of this code shall be done under construction specifications. Preferably the construction specifications shall cover all phases of the work and should be in sufficient detail to cover the requirements of this code.

841.22 *Inspection Provisions.*

841.221 The operating company shall make provision for suitable inspection. Inspectors shall be qualified by either experience or training.

841.222 The installation inspection provisions for pipelines and other facilities to operate at hoop stresses of 20% or more of the specified minimum yield strength should be adequate either to make possible the following inspections at sufficiently frequent intervals or to do other things that will assure good quality of workmanship.

(a) Inspect the surface of the pipe for serious surface defects just prior to the coating operation. See 841.242 (a).

(b) Inspect the surface of the coated pipe as it is lowered into the ditch to find coating lacerations that indicate the pipe might have been damaged after being coated. Damage during the lowering-in process should be found during this inspection.

(c) Inspect the fit-up of the joints before the weld is made.

(d) Visually inspect the stringer beads before subsequent beads are applied.

(e) Inspect the completed welds before they are covered with coating.

(f) Inspect the condition of the ditch bottom just before the pipe is lowered in.

(g) Inspect the fit of the pipe to the ditch before backfilling.

(h) Inspect all repairs, replacements or changes ordered before they are covered up.

(i) Perform such special tests and inspections as are required by the specifications, such as the radiographing of a portion of the welds and the electrical testing of the protective coating.

841.223 The inspector shall have authority to order the removal and replacement of any section that fails to meet the standards of this code.

Welding Fittings. Several shapes of welding fittings are available or can be fabricated for use with steel mains. These include turns, ells, reducers, tees, drips, and insulating joints. Minor changes of direction can be obtained thru the flexibility of the pipe itself. More abrupt changes in direction can be achieved by hot or cold wrinkle bending, by miter joints, or by prefabricated elbows. Likewise, reductions in diameter can be accomplished either with field fabricated or prefabricated reducers. The decision whether to field fabricate or use manufactured fittings is basically one of economics. Although prefabricated fittings are quite expensive, their use frequently avoids delays in the *field*. Wrinkle bending and miter joints are subject to limitations stated in the following excerpt from ASA B31.8-1963:

841.23 *Bends, Elbows, and Miters in Steel Pipelines and Mains.* Changes in direction may be made by the use of bends, elbows, or miters under the following limitations:

841.231 The bends shall be free from buckling, cracks or other evidence of mechanical damage. For field cold bends on sizes 12 inch and larger, the longitudinal axis of the pipe shall not be deflected more than 1½ degrees in any length along the pipe axis equal to the diameter of the pipe. All bends other than wrinkle bends shall not have a difference between the maximum and minimum diameters in excess of 2.5% of the nominal diameter.

841.232 When a circumferential weld occurs in a bend section, it shall be subjected to X-ray examination after bending.

841.233 Hot bends made on cold worked or heat treated pipe shall be designed for lower stress levels in accordance with 841.14 (e).

841.234 Wrinkle bends shall be permitted only on systems operating at less than 30% of the specified minimum yield strength. When wrinkle bends are made in welded pipe, the longitudinal weld shall be located as nearly to 90° with the top of the wrinkle as conditions will permit. Wrinkle bends with sharp kinks shall not be permitted. Wrinkles shall have a spacing not less than the distance equal to the diameter of the pipe measured along the crotch. On pipe 16 inch and larger, the wrinkle shall not produce an angle of more than 1½ degrees per wrinkle.

841.235 The longitudinal weld of the pipe shall preferably be near the neutral axis of the bend.

841.236 Mitered bends are permitted subject to the following limitations:

(a) In systems intended to operate at 40% or more of the specified minimum yield strength, mitered bends are not permitted. Deflections caused by misalignment up to 3° are not considered as miters.

(b) In systems intended to operate at 10% but less than 40% of the specified minimum yield strength, the total deflection angle at each miter shall not exceed 12½°.

(c) In systems intended to operate at less than 10% of the specified minimum yield strength, the total deflection angle at each miter shall not exceed 90°.

(d) In systems intended to operate at 10% or more of the specified minimum yield strength, the minimum distance between miters measured at the crotch shall not be less than one pipe diameter.

(e) Care shall be taken in making mitered joints to provide proper spacing and alignment and full penetration.

841.237 Factory-made wrought-steel welding elbows or transverse segments cut therefrom may be used for changes in direction provided that the arc length measured along the crotch is at least 1 inch on pipe sizes 2 inches and larger.

841.24 *Pipe Surface Requirements Applicable to Pipelines and Mains to Operate at a Hoop Stress of 20% or More of the Specified Minimum Yield Strength.* Gouges, grooves, and notches have been found to be a very important cause of pipeline failures and all harmful defects of this nature must be prevented or eliminated. Precautions shall be taken during manufacture, hauling, and installation to prevent the gouging or grooving of pipe.

841.241 *Detection of Gouges and Grooves.*

(a) The field inspection provided on each job shall be suitable to reduce to an acceptable minimum the chances that gouged or grooved pipe will get into the finished pipeline or main. Inspection for this purpose just ahead of the coating operation and during the lowering-in and backfill operation is recommended.

(b) When pipe is coated, inspection shall be made to determine that the coating machine does not cause harmful gouges or grooves.

(c) Lacerations of the protective coating should be carefully examined to see if the pipe surface has been damaged.

841.242 *Field Repair of Gouges and Grooves.*

(a) Injurious gouges or grooves shall be removed.

(b) They may be removed by grinding, provided that the resulting wall thickness is not less than the minimum prescribed by this code for the conditions of usage. See 841.14 (c).

(c) When the conditions outlined in 841.242 (b) cannot be met, the damaged portion of pipe shall be cut out as a cylinder and replaced with a good piece. Insert patching is prohibited.

841.243 *Dents*

(a) A dent may be defined as a depression which produces a gross disturbance in the curvature of the pipe wall (as opposed to a scratch or gouge which reduces the pipe wall thickness). The depth of a dent shall be measured as the gap between the lowest point of the dent and a prolongation of the original contour of the pipe in any direction.

(b) A dent, as defined in 841.23(a), which contains a stress concentrator such as a scratch, gouge, groove, or arc burn shall be removed by cutting out the damaged portion of the pipe as a cylinder.

(c) All dents which affect the curvature of the pipe at the longitudinal weld or any circumferential weld shall be removed. All dents which exceed a maximum depth of ¼ inch in pipe 12¾ inch OD and smaller or 2% of the nominal pipe diameter in all pipe greater than 12¾ inch OD, shall not be permitted in pipelines or mains intended to operate at 40% or more of the specified minimum yield strength. When dents are removed the damaged portion of the pipe shall be cut out as a cylinder. Insert patching and pounding out of the dents are prohibited.

841.244. *Arc Burns.* Arc burns have been found to cause serious stress concentration in pipelines of grade API 5LX or equal and shall be prevented or eliminated in all lines corresponding to these specifications intended to operate at 40% or more of the specified minimum yield strength.

841.245 *Elimination of Arc Burns.* The metallurgical notch caused by arc burns shall be removed by grinding, provided the grinding does not reduce the remaining wall thickness to less than the minimum prescribed by this code for the conditions of use*. In all other cases repair is prohibited and the portion of pipe containing the arc burn must be cut out as a cylinder and replaced with a good piece. Insert-patching is prohibited.

841.25 *Application and Inspection of Protective Coatings for Underground Piping.* (Also see 841.17.)

(a) Protective coatings for underground piping shall be applied in accordance with either the coating manufacturer's recommendations or the Company's coating specifications for the particular conditions encountered. These recommendations or specifications shall also cover the patching of damaged spots, the coating of joints, short lengths of pipe and fittings coated in the field.

(b) Crews that apply protective coatings shall be suitably instructed and provided with all of the equipment necessary to accomplish their work in a satisfactory manner.

(c) It is recommended that the protective coating be inspected and tested either completely or on a sampling basis using a recognized "flaw detector" before or after backfilling.

841.26 *Electrical Test Leads for Corrosion Control or Electrolysis Testing on Pipelines or Mains to Operate at 20% or More of the Specified Minimum Yield Strength.*

841.261 When electrical test leads for corrosion control or electrolysis testing are required, care should be exercised in their installation, particularly on pipelines that are stressed to near the maximum stress levels permitted by this code, to avoid stress concentration.

841.262 Electrical test leads may be attached directly on to the pipe by the Thermit welding process using aluminum powder and copper oxide provided the charge is limited to #15 (15 gram) cartridges and the size of electrical conductor restricted to #6 AWG or smaller. Where the application involves the attachment of a larger wire, use a multi-strand conductor and rearrange the strands into groups no larger than #6 AWG and attach each group to the pipe separately, using a #15 (15 gram) cartridge of powder. Attaching electrical test leads directly onto the pipe by other methods of brazing is prohibited.

841.263 All test lead connections and all bare leads shall be protected by coating and/or wrapping.

841.27 *Miscellaneous Operations Involved in the Installation of Steel Pipelines and Mains.*

841.271 *Handling, Hauling and Stringing.* Care should be taken in the selection of the handling equipment and in handling, hauling, unloading, and placing the pipe so as not to damage the pipe.

* Complete removal of the metallurgical notch created by an arc burn can be determined as follows: After visible evidence of the arc burn has been removed by grinding, swab the ground area with a 20% solution of ammonium persulfate. A blackened spot is evidence of a metallurigical notch and indicates that additional grinding is necessary.

841.272 *Installation of Pipe in the Ditch.* On pipelines operating at stresses of 20% or more of the specified minimum yield strength, it is very important that stresses induced into the pipeline by construction be minimized. This includes grading the ditch so that the pipe has a firm substantially continuous bearing on the bottom of the ditch. The pipe shall fit the ditch without the use of external force to hold it in place until the backfill is completed. When long sections of pipe that have been welded alongside the ditch are lowered in, care shall be exercised so as not to jerk the pipe or impose any strains that may kink or put a permanent bend in the pipe. Slack loops are not prohibited by this paragraph where laying conditions render their use advisable.

841.273 *Backfilling.*

(a) Backfilling should be performed in a manner to provide firm support under the pipe.

(b) If there are large rocks in the material to be used for backfill, care should be used to prevent damage to the coating, by such means as the use of rock shield material, or by making the initial fill with rock free material to a sufficient depth over the pipe to prevent rock damage.

(c) Where flooding of the trench is done to consolidate the backfill, care shall be exercised to see that the pipe is not floated from its firm bearing on the trench bottom.

841.274 *Hot Taps.* All hot taps shall be installed by trained and experienced crews.†

841.28 *Precautions to Avoid Explosions of Gas-Air Mixtures or Uncontrolled Fires during Construction Operations.*

841.281 Operations such as gas or electric welding and cutting with cutting torches can be safely performed on pipelines and mains and auxiliary equipment, provided that they are completely full of gas, or air that is free from combustible material. Steps shall be taken to prevent a mixture of gas and air at all points where such operations are to be performed.

841.282 When a pipeline or main can be kept full of gas during a welding or cutting operation, the following procedures are recommended:

(a) Keep a slight flow of gas moving toward the point where cutting or welding is being done.

(b) The gas pressure at the site of the work shall be controlled by suitable means.

(c) Close all slots or open ends immediately after they are cut, with tape, and/or tightly fitted canvas or other suitable material.

(d) Do not permit two openings to remain uncovered at the same time. This is doubly important if the two openings are at different elevations.

841.283 No welding or acetylene cutting should be done on a pipeline, main or auxiliary apparatus that contains air if it is connected to a source of gas, unless a suitable means has been provided to prevent the leakage of gas into the pipeline or main.

841.284 In situations where welding or cutting must be done on facilities which are filled with air and connected to a source of gas and the precautions recommended above cannot be taken, one or more of the following precautions, depending upon circumstances of the job, are suggested:

(a) Purging of the pipe or equipment upon which welding or cutting is to be done, with combustible gas or inert gas.

(b) Testing of the atmosphere in the vicinity of the zone to be heated before the work is started and at intervals as the work progresses, with a combustible gas indicator or by other suitable means.

(c) Careful verification before the work starts that the valves that isolate the work from a source of gas do not leak.

[† It is generally agreed that specific instructions covering the work and the equipment should be available. Competent supervision is necessary at all times. Examples of the detailed instructions desirable for such work are the procedures used by the Southern California Gas Co. Manufacturers of hot tapping equipment can best advise on the application of their products.—EDITOR]

841.285 *Purging of Pipelines and Mains.*

(a) When a pipeline or main full of air is placed in service, the air in it can be safely displaced with gas provided that a moderately rapid and continuous flow of gas is introduced at one end of the line and the air is vented out the other end. The gas flow should be continued without interruption until the vented gas is free from air. The vent should then be closed.

(b) In cases where gas in a pipeline or main is to be displaced with air and the rate at which air can be supplied to the line is too small to make a procedure similar to, but the reverse of that described in 841.285 (a) feasible, a slug of inert gas should be introduced to prevent the formation of an explosive mixture at the interface between gas and air. Nitrogen or carbon dioxide can be used for this purpose.

(c) If a pipeline or main containing gas is to be removed, the operation may be carried out in accordance with 841.282 or the line may be first disconnected from all sources of gas and then thoroughly purged with air, water or with inert gas before any further cutting or welding is done.

(d) If a gas pipeline or main or auxiliary equipment is to be filled with air after having been in service and there is a reasonable possibility that the inside surfaces of the facility are wetted with a volatile inflammable liquid, or if such liquids might have accumulated in low places, purging procedures designed to meet this situation shall be used. Steaming of the facility until all combustible liquids have been evaporated and swept out is recommended. Filling of the facility with an inert gas and keeping it full of such gas during the progress of any work that might ignite an explosive mixture in the facility is an alternative recommendation. The possibility of striking static sparks within the facility must not be overlooked as a possible source of ignition.

841.286 Whenever the accidental ignition in the open air of a gas-air mixture might be likely to cause personal injury or property damage, precautions shall be taken as, for example:

(a) Prohibit smoking and open flames in the area, and

(b) Install a metallic bond around the location of cuts in gas pipes to be made by other means than cutting torches, and

(c) Take precautions to prevent static electricity sparks, and

(d) Provide a fire extinguisher of appropriate size and type in accordance with NBFU No. 10, Standard for Portable Fire Extinguishers.

841.3 Testing After Construction.

841.31 *General Provisions.* All pipelines, mains and services shall be tested after construction, except as follows:

Tie-ins. Because it is sometimes necessary to divide a pipeline or main into test sections and install test heads, connecting piping, and other necessary appurtenances for testing, it is not required that the tie-in sections of pipe be tested.

841.4 Test Requirements.

841.41 *Test Required to Prove Strength of Pipelines and Mains to Operate at Hoop Stresses of 30% or More of the Specified Minimum Yield Strength of the Pipe.*

841.411 All pipelines and mains to be operated at a hoop stress of 30% or more of the specified minimum yield strength of the pipe shall be given a field test to prove strength after construction and before being placed in operation.

841.412 (a) Pipelines and mains located in Location Class 1 shall be tested either with air or gas to 1.1 times the maximum operating pressure or hydrostatically to at least 1.1 times the maximum operating pressure. See 841.5.

(b) Pipelines or mains located in Location Class 2 shall be tested either with air to 1.25 times the maximum operating pressure or hydrostatically to at least 1.25 times the maximum operating pressure. See 841.5.

(c) Pipelines and mains in Location Classes 3 and 4 shall be tested hydrostatically to a pressure not less than 1.4 times the maximum operating pressure.

(d) The test requirements given in 841.412(a), (b) and (c) above are summarized in Table 841.412(d).

TABLE 841.412(d) Test Requirements for Pipelines and Mains to Operate at Hoop Stresses of 30% or More of the Specified Minimum Yield Strength of the Pipe

1	2	3	4	5
Location class	Permissible test fluid	Prescribed test press. Min	Prescribed test press. Max	Max allowable operating press., the lesser of
1	Water	1.1 m.o.p.	None	t.p. ÷ 1.1 or d.p.
	Air	1.1 m.o.p.	1.1 d.p.	
	Gas	1.1 m.o.p.	1.1 d.p.	
2	Water	1.25 m.o.p.	None	t.p. ÷ 1.25 or d.p.
	Air	1.25 m.o.p.	1.25 d.p.	
3	Water	1.40 m.o.p.	None	t.p. ÷ 1.40 or d.p.
4	Water	1.40 m.o.p.	None	t.p. ÷ 1.40 or d.p.

m.o.p. = maximum operating pressure (not necessarily the maximum allowable operating pressure)
d.p. = design pressure
t.p. = test pressure

Note: This table brings out the relationship between test pressures and maximum allowable operating pressures subsequent to the test. If an operating company decides that the maximum operating pressure will be less than the design pressure a corresponding reduction in prescribed test pressure may be made as indicated in Column 3. However, if this reduced test pressure is used the maximum operating pressure cannot later be raised to the design pressure without retesting the line to the test pressure prescribed in Column 4. See 805.14, 845.22 and 845.23.

841.413 Requirements of 841.412 (c) for hydrostatic testing of mains and pipelines in Location Classes 3 and 4 do not apply if at the time the pipeline or main is first ready for test, one or both of the following conditions exist:

(a) The ground temperature at pipe depth is 32 F or less, or might fall to that temperature before the hydrostatic test could be completed, or

(b) Water of satisfactory quality is not available in sufficient quantity.

(c) In such cases an air test to 1.1 times the maximum operating pressure shall be made and the limitations on operating pressure imposed by 841.412(d) above do not apply.

841.414 Other provisions of this code notwithstanding, pipelines and mains crossing highways and railroads may be tested in each case in the same manner and to the same pressure as the pipeline on each side of the crossing.

841.415 Other provisions of this code notwithstanding, fabricated assemblies, including mainline valve assemblies, cross connections, river crossing headers, etc., installed in pipelines in Class 1 locations and designed in accordance with Type B construction, as required in 841.142, may be tested as required for Class 1 locations.

841.416 Notwithstanding the limitations on air testing imposed in 841.412(c), air testing may be used in Location Classes 3 and 4, provided that all of the following conditions apply:

(a) The maximum hoop stress during test is less than 50% of the specified minimum yield strength in Class 3 locations, and less than 40% of the specified minimum yield strength in Class 4 locations.

(b) The maximum pressure at which the pipeline or main is to be operated does not exceed 80% of the maximum field test pressure used.

(c) The pipe involved is new pipe having a longitudinal joint factor E in Table 841.12 of 1.00.

841.417 *Records.* The operating company shall maintain in its file for the useful life of each pipeline and main, records showing the type of fluid used for test and the test pressure.

841.42 *Tests required to prove strength for pipelines and mains to operate at less than 30% of the specified minimum yield strength of the pipe, but in excess of 100 psi.* Steel piping that is to operate at stresses less than 30% of the specified minimum yield strength in location Class 1 shall at least be tested in accordance with 841.43. In location Classes 2, 3, and 4, such piping shall be tested in accordance with Table 841.412 (d), except that gas or air may be used as the test medium within the maximum limits set in Table 841.421.

TABLE 841.421 Maximum Hoop Stress Permissible During Test
(Per cent of specified minimum yield strength)

Test medium	Location class		
	2	3	4
Air	75	50	40
Gas	30	30	30

841.43 *Leak Tests for Pipelines or Mains to Operate at 100 psi or More.*

841.431 Each pipeline and main shall be tested after construction and before being placed in operation to demonstrate that it does not leak. If the test indicates that a leak exists, the leak or leaks shall be located and eliminated, unless it can be determined that no undue hazard to public safety exists.

841.432 The test procedure used shall be capable of disclosing all leaks in the section being tested and shall be selected after giving due consideration to the volumetric content of the section and to its location.

841.433 In all cases where a line is to be stressed in a strength-proof test to 20% or more of the specified minimum yield strength of the pipe, and gas or air is the test medium, a leak test shall be made at a pressure in the range of 100 psi to that required to produce a hoop stress of 20% of the minimum specified yield, or the line shall be walked while the hoop stress is held at approximately 20% of the specified minimum yield.

841.44 *Leak Test for Pipelines and Mains to Operate at Less Than 100 psi.*

841.441 At the time of or prior to placing in operation distribution mains and related equipment to operate at less than 100 psi, they shall be tested to determine that they are gas-tight.

841.442 Gas may be used as the test medium at the maximum pressure available in the distribution system at the time of the test. In this case the soap bubble test may be used to locate leaks if all joints are accessible during the test.

841.443 Testing at available distribution system pressures as provided for above in 841.442 may not be adequate if substantial protective coatings are used that would seal a split pipe seam. If such coatings are used, the leak test pressure shall be 100 psi.

841.5 Safety During Tests. All testing of pipelines and mains after construction shall be done with due regard for the safety of employees and the public during the test. When air or gas is used, suitable steps shall be taken to keep persons not working on the testing operations out of the testing area during the period in which the hoop stress is first raised from 50% of the specified minimum yield to the maximum test stress, and until the pressure is reduced to the maximum operating pressure.

Personnel, Equipment, and Operations

Tables 8-9 and 8-10 show personnel and equipment requirements respectively for typical construction groups or *spreads*, used in normal installation of transmission pipelines of 8 in. to 30 in. diameter inclusive. Table 8-11 based on one company's experience, shows progress which may be expected for a typical construction group or spread under conditions shown.

Clearing and Grading. Preparation of the right of way is usually the first step in actual pipeline construction. Permission must be obtained from individual landowners for use of access roads over their property. Fences on the right of way must be cut and temporary gates erected where necessary. In general, the right of way must be cleared and graded sufficiently to permit passage of stringing trucks, ditching machines, and other equipment. Roots and stumps along the proposed trench must be removed to facilitate ditching machine operations.

Pipe Stringing. As soon as the right of way has been prepared, the pipe is hauled and strung. This operation is often performed by pipe stringing contractors, particularly when pipe arrives in large shipments and prompt unloading is necessary to avoid demurrage charges.

Since ditching machines normally throw the spoil from the pipeline trench to their left, pipe should be strung on the right, so that remaining operations may be conducted without interference from the spoil bank. Special care should be taken to see that the right amount of pipe is strung, with the different sizes and wall thicknesses all being in proper locations. Some additional pipe should be available to allow for any damaged lengths, for bends, or for possible rejects. Joints should be laid down in staggered positions so that ends will be exposed for the crews following.

Ditching. Excavation of the trench is usually done by wheel-type ditching machines, with back hoes and drag lines often used where the ground is too soft to support a heavier machine. A rooter plow is sometimes run ahead of the ditching machine to cut roots and loosen the top strata of earth, thus making excavating easier. In rocky areas, use of rock drills, air hammers, and dynamite may also be necessary.

Where the subsoil is rock, caliche, or some other undesirable material, *double ditching* may be needed. This consists of separating the topsoil from the remainder so that it can be replaced in its proper position during backfilling.

Welding. Electric arc welding is generally employed for transmission line joints. Oxy-acetylene welding may be used, particularly in detail work, cutting, and bending.

The usual steps in pipeline welding are:

1. Clean, square, and buff the beveled ends.
2. Align the joints, using external lineup clamps on small pipe and internal lineup clamps on large pipe.
3. Tack or spot weld at several spots to hold the joint to be welded.
4. Run the stringer bead or first complete circumferential weld. This bead may be followed at once by a "hot pass" which deposits less metal than other welds but prevents the stringer bead from cooling too rapidly. It also fills any pin holes present.
5. Run the filler welds, followed by the cap or outside weld. (Step 6 given on page 8/47.)

Table 8-9 Typical Construction Group for Normal Transmission Line Construction
(1954, except 36 in. data—1962)
(Excluding special features like highway crossings or heavy clearing)

Jobs and skills	Pipe sizes, in.								
	8	12	14	16	18	20	24	30	36
Fences and Light R/W Clearing									
Foreman	1	1	1	1	1	1	1	1	1
Operator, tow	—	—	1	1	1	1	1	1	2
Powder man	1	1	1	1	1	1	1	1	1
Power saw operator	2	2	2	2	2	2	2	2	2
Light truck driver	1	1	1	1	1	1	1	1	1
Helper	1	1	1	1	1	1	1	1	3
Fence man	1	1	1	1	1	1	1	1	1
Saw filer	—	—	—	—	1	1	1	1	1
Common labor	15	18	18	20	20	20	20	20	20
Grub and Grade									
Operator, dozer	2	2	2	2	3	3	3	3	4
Assistant foreman	—	—	—	—	1	1	1	1	1
Light truck driver	—	—	—	—	—	—	—	—	2
Swamper	2	2	2	2	3	3	3	3	4
Ditching									
Foreman	1	1	1	1	1	2	2	2	2
Operator, dozer	—	—	—	—	—	—	—	—	1
Operator, ditcher	1	1	2	2	2	2	2	2	3
Operator, dragline	1	1	1	1	2	2	2	2	4
Assistant foreman	—	—	—	—	1	1	1	1	1
Light truck driver	—	1	1	1	1	1	1	1	2
Oiler	1	1	2	2	3	4	4	4	6
Swamper	1	1	2	2	2	2	2	2	5
Helper	1	—	—	—	—	—	—	—	—
Common labor	14	15	16	16	16	17	17	18	20
Watchman	—	1	1	1	1	1	1	1	1
Bending									
Foreman	—	—	1	1	1	1	1	1	1
Engineer	—	—	—	—	—	—	1	1	1
Operator, sideboom	—	1	1	1	1	1	1	1	2
Operator, bending machine	—	—	1	1	1	1	1	1	1
Swamper	—	1	1	1	1	1	1	1	2
Common labor	—	—	1	1	1	1	1	1	2
Line and Tack									
Foreman	1	1	1	1	1	1	1	1	1
Welder (stringer)	1	2	2	2	2	2	3	3	6
Stabber-spacer	1	2	2	2	3	3	6	7	7
Operator, sideboom	2	2	2	2	2	2	2	2	4
Operator, tow	—	—	1	1	1	1	1	1	1
Clam man	—	—	1	1	1	1	2	2	2
Light truck driver	2	2	2	2	2	2	3	3	3
Buffing machine operator	1	1	1	1	1	1	2	2	2
Swamper	2	2	2	2	2	2	2	2	4
Helper	1	2	2	2	2	2	3	3	6
Skid man	1	2	2	2	2	2	6	6	6
Semiskilled labor	—	—	3	3	4	4	2	3	2
Common labor	8	10	10	10	10	14	15	16	16
Watchman	—	—	—	1	—	—	1	1	1
Welding									
Foreman	1	1	1	1	1	1	1	1	1
Welder (line)	3	4	7	8	12	13	14	15	19
Operator, tow	1	1	1	1	1	1	1	1	2
Light truck driver	1	1	1	1	1	1	1	1	1
Swamper	1	2	2	2	2	2	2	2	2
Helper	3	4	7	8	10	10	14	15	20
Semiskilled labor	—	—	2	2	4	4	2	2	3
Watchman	—	—	—	—	—	1	1	1	1
Clean and Prime									
Foreman	—	—	—	1	1	1	1	1	1
Operator, sideboom	1	1	1	1	1	1	1	1	2
Operator, cleaning machine	1	1	1	1	1	1	1	1	1
Light truck driver	—	1	1	1	1	1	1	1	1
Swamper	1	2	2	2	2	2	2	2	4
Skid man	—	—	—	—	2	2	2	2	3
Semiskilled labor	3	3	3	3	2	2	2	2	3
Coat and Wrap									
Foreman	1	1	1	1	1	1	1	1	1
Operator, sideboom	1	1	1	1	1	2	2	2	3
Operator, coat machine	1	1	1	1	1	1	1	1	1
Operator, tow	1	1	1	1	1	1	1	1	1
Heavy truck driver	1	1	1	1	1	1	1	1	1
Assistant foreman	—	—	1	1	1	1	1	1	1
Dope pot foreman	1	1	3	3	3	4	4	5	5
Light truck driver	2	2	2	2	2	2	2	2	3
Swamper	2	3	3	4	4	4	4	5	5
Dope chopper	2	2	3	3	3	3	3	4	5
Semiskilled labor	—	—	4	4	5	6	6	7	7
Common labor	11	14	14	14	15	17	18	19	20
Watchman	1	—	1	—	1	1	1	1	1
Lower-In									
Foreman	1	1	1	1	1	1	2	2	2
Welder (line)	1	1	1	1	1	1	1	2	4
Operator, sideboom	2	4	3	3	4	4	5	6	8
Operator, tow	—	—	—	—	—	—	—	—	2
Operator, dragline	1	1	1	1	2	2	2	2	2
Assistant foreman	—	—	—	—	1	1	1	1	1
Dope pot foreman	1	1	1	1	1	1	1	1	1
Light truck driver	1	1	1	1	2	2	2	2	2
Oiler	1	1	1	1	2	2	2	2	2
Swamper	2	3	3	3	4	4	4	4	8
Helper	1	1	1	1	1	2	1	1	1
Dope chopper	—	—	1	1	1	1	1	1	1
Semiskilled labor	—	—	—	—	—	2	2	2	2
Common labor	10	13	15	15	15	17	18	18	20
Watchman	—	—	—	—	—	—	1	1	1
Backfill and Cleanup									
Foreman	1	1	1	1	1	1	1	1	1
Operator, dozer	2	2	2	2	2	2	4	4	4
Operator, backfiller	—	—	—	—	1	1	2	2	2
Assistant foreman	—	1	1	—	1	2	2	2	2
Light truck driver	2	2	2	2	2	2	2	2	2
Oiler	—	—	—	—	—	—	—	—	1
Swamper	2	2	2	2	4	4	4	4	4
Helper	—	—	—	—	—	3	—	—	—
Fence man	1	1	1	1	1	1	1	1	1
Common labor	15	23	25	25	25	25	27	30	30
Watchman	—	—	—	1	1	1	1	1	1
Utility–General									
Foreman	—	—	—	—	—	—	—	—	1
Welder (stringer)	1	1	1	1	1	1	1	1	4
Welder (line)	—	—	—	—	1	1	1	1	1
Mechanic	1	1	2	2	2	2	3	3	5
Heavy truck driver	1	2	3	3	3	3	4	4	6
Fuel man	—	1	1	1	1	1	1	1	3
Grease man	1	1	1	1	1	1	1	1	3
Light truck driver	—	—	1	1	1	1	1	1	2
Swamper	—	—	—	—	2	2	2	—	2
Helper	2	2	—	—	3	3	3	3	6
Semiskilled labor	—	—	—	—	—	—	—	—	3
Watchman	1	1	1	1	1	1	1	1	1
Administration and Supervision									
Job superintendent	1	1	1	1	1	1	1	1	1
Assistant job superintendent	—	—	—	1	1	1	1	1	1
Paymaster	1	1	1	1	1	1	1	1	1
Timekeeper	1	1	1	1	1	2	2	2	2
Timechecker	—	—	—	—	2	3	3	3	3
Material man	—	—	—	—	1	1	1	1	2
Warehouseman	1	1	1	1	1	1	2	2	2
Helper	—	—	—	—	—	—	—	—	2

Table 8-10 Equipment Required by Typical Construction Group for Normal Transmission Line Construction

(1954, except 36 in. data—1962)

(Excluding special features like highway crossings or heavy clearing)

Equipment	Pipe sizes, in.								
	8	12	14	16	18	20	24	30	36
Fences and Light R/W Clearing									
Tow tractor, 65–70 hp	—	—	1	1	1	1	1	—	—
Tow tractor, 80–90 hp	—	—	—	—	—	—	—	1	2
Pickup, ½ ton	1	1	1	1	1	1	1	1	2
Winch truck, 1½ ton	1	1	1	1	1	1	1	1	1
Power saw	2	2	2	2	2	2	2	2	2
Grub and Grade									
Dozer, angle/bull 65–70 hp	1	1	1	—	—	—	—	—	—
Dozer, angle/bull 80–90 hp	1	1	1	2	3	3	1	1	1
Dozer, angle/bull 109–130 hp	—	—	—	—	—	—	2	—	—
Dozer, angle/bull 148–175 hp	—	—	—	—	—	—	—	2	3
Ditching									
Dozer, angle/bull 148–175 hp	—	—	—	—	—	—	—	—	1
Ditching machine, medium	1	1	1	1	1	1	—	—	—
Ditching machine, large	—	—	1	1	1	1	2	2	3
Dragline-backhoe-clam, ½ yard	1	1	1	1	—	—	—	—	—
Dragline-backhoe-clam, ¾ yard	—	—	—	—	2	2	2	2	4
Pickup, ½ ton	1	1	1	1	1	2	2	2	4
Winch truck, ½ ton	—	1	1	1	1	1	1	1	1
Light plant	—	—	—	—	1	1	1	1	2
Bending									
Sideboom, 60–70 hp	—	—	1	1	1	—	—	—	—
Sideboom, 80–90 hp	—	1	—	—	—	1	1	—	—
Sideboom, 109–130 hp	—	—	—	—	—	—	—	1	—
Sideboom, 140–175 hp	—	—	—	—	—	—	—	—	2
Bending machine	—	—	1	1	1	1	1	1	1
Pickup, ½ ton	—	—	1	1	1	1	1	1	2
Line and Tack									
Sideboom, 40–43 hp	1	—	—	—	—	—	—	—	—
Sideboom, 65–70 hp	1	2	2	1	1	—	—	—	—
Sideboom, 80–90 hp	—	—	—	1	1	2	1	—	—
Sideboom, 109–130 hp	—	—	—	—	—	—	1	1	—
Sideboom, 148–175 hp	—	—	—	—	—	—	—	1	4
Tow tractor, 40–43 hp	—	—	1	—	—	—	—	—	—
Tow tractor, 65–70 hp	—	—	—	1	—	—	—	—	—
Tow tractor, 80–90 hp	—	—	—	—	1	1	1	1	1
Welding machine, 300 amp	1	2	2	2	2	2	3	3	6
Pickup, ½ ton	1	1	1	1	1	1	1	1	1
Winch truck, 1½ ton	2	2	2	2	2	2	3	3	2
Internal line clamp	—	—	1	1	1	1	2	2	2
Beveling machine	1	1	1	1	1	1	1	1	2
Oxyacetylene rig	1	1	1	1	1	1	1	1	1
Light plant	1	1	1	1	1	1	1	1	1
Radio equipment	—	—	—	—	—	1	1	1	1
Welding									
Tow tractor, 40–43 hp	1	1	1	—	—	—	—	—	—
Tow tractor, 65–70 hp	—	—	—	1	—	—	—	—	—
Tow tractor, 80–90 hp	—	—	—	—	1	1	1	1	2
Welding machine, 200 amp	3	4	7	8	12	13	14	15	19
Pickup, ½ ton	1	1	1	1	1	1	1	1	1
Winch truck, 1½ ton	1	1	1	1	1	1	1	1	1
Tensile testing machine	1	1	1	1	1	1	1	1	1
Clean and Prime									
Sideboom, 65–70 hp	1	—	—	—	—	—	—	—	—
Sideboom, 80–90 hp	—	1	1	—	—	—	—	—	—
Sideboom, 109–130 hp	—	—	—	1	1	1	1	—	—
Sideboom,148–175 hp	—	—	—	—	—	—	—	1	2
Tow tractor	—	—	—	—	—	—	—	—	1
Clean and prime machine	1	1	1	1	1	1	1	1	1
Pickup, ½ ton	—	—	—	1	1	1	1	1	2
Winch truck, 1½ ton	—	1	1	1	1	1	1	1	1
Float truck	—	—	—	—	—	—	—	—	1
Coat and Wrap									
Sideboom, 65–70 hp	1	—	—	—	—	—	—	—	—
Sideboom, 80–90 hp	—	1	1	—	—	1	1	1	—
Sideboom, 109–130 hp	—	—	—	1	1	1	1	—	1
Sideboom, 148–175 hp	—	—	—	—	—	—	—	1	2
Tow tractor, 40–43 hp	1	1	—	—	—	—	—	—	—
Tow tractor, 65–70 hp	—	—	1	1	—	—	—	—	—
Tow tractor, 80–90 hp	—	—	—	—	1	1	1	1	1
Coat and wrap machine	1	1	1	1	1	1	1	1	1
Dope pot, 10 bbl	3	—	—	—	—	—	—	—	—
Dope pot, 25 bbl	—	2	2	2	2	3	1	2	2
Dope pot, 30 bbl	—	—	—	—	—	—	2	2	3
Pickup, ½ ton	1	1	1	1	1	1	1	1	2
Winch truck, 1½ ton	2	2	2	2	2	2	2	2	2
Winch truck, 2½ ton	1	1	1	1	1	1	1	1	2
Lower-In									
Sideboom, 40–43 hp	—	1	—	—	—	—	—	—	—
Sideboom, 65–70 hp	2	1	—	—	—	—	—	—	—
Sideboom, 80–90 hp	—	2	3	2	2	—	1	1	—
Sideboom, 109–130 hp	—	—	—	1	2	4	4	1	3
Sideboom, 148–175 hp	—	—	—	—	—	—	—	4	5
Tow tractor, 65–70 hp	—	—	—	—	—	—	—	—	2
Dragline-backhoe-clam, ½ yard	1	1	1	1	1	1	1	—	—
Dragline-backhoe-clam, ¾ yard	—	—	—	—	1	1	1	2	2
Welding machine, 200 amp	1	1	1	1	1	1	1	2	3
Welding machine, 300 amp	—	—	—	—	—	—	—	—	1
Dope pot, 10 bbl	1	1	1	1	1	1	1	1	—
Dope pot, 25 bbl	—	—	—	—	—	—	1	1	2
Pickup, ½ ton	1	1	1	1	1	1	2	2	3
Winch truck, 1½ ton	1	1	1	1	2	2	2	2	—
Winch truck, 2½ ton	—	—	—	—	—	—	—	—	2
Beveling machine	1	1	1	1	1	1	1	2	2
Oxyacetylene rig	1	1	1	1	1	1	1	2	2
Backfill and Cleanup									
Dozer, angle/bull 65–70 hp	2	1	1	1	1	1	—	—	—
Dozer, angle/bull 80–90 hp	—	1	1	1	1	1	2	3	1
Dozer, angle/bull 109–130 hp	—	—	—	—	—	—	2	—	2
Dozer, angle/bull 148–175 hp	—	—	—	—	—	—	—	1	1
Backfiller, 65–70 hp	—	—	—	—	1	1	2	2	2
Dragline-backhoe-clam, ¾ yard	—	—	—	—	—	—	—	—	1
Pickup, ½ ton	1	1	1	1	1	1	1	1	2
Winch truck, 1½ ton	2	2	2	2	2	2	2	2	2
Utility–General									
Welding machine, 200 amp	1	1	—	—	1	1	1	1	2
Welding machine, 300 amp	—	—	1	1	1	1	1	1	2

Table 8-10 (Continued)

Equipment	Pipe sizes, in.								
	8	12	14	16	18	20	24	30	36
Pickup, ½ ton	—	—	—	—	—	—	—	—	2
Winch truck, 2½ ton	—	—	1	1	1	1	2	2	2
Lowboy truck	—	1	1	1	1	1	1	1	2
Float truck	1	1	1	1	1	1	1	1	2
Beveling machine	—	—	—	—	1	1	1	1	2
Oxyacetylene rig	—	—	—	—	1	1	1	1	2
Light plant	—	—	1	1	1	1	1	2	2
Tensile testing machine	—	—	—	—	—	—	1	1	1
Radio equipment	—	2	2	2	2	2	2	2	2
Miscellaneous small tools lot	1	1	1	1	1	1	1	1	1
Administration and Supervision									
Automobile	3	3	3	4	5	7	8	8	8
Pickup, ½ ton	—	1	1	1	3	3	3	3	4
Winch truck, 1½ ton	—	—	1	1	1	1	1	1	1
Radio equipment	—	2	2	2	2	3	3	3	3

6. Inspect welds. This may be done with X-rays, gamma rays, or visual observation. As a further check, one or more welds may be cut out daily and coupons cut from them. These coupons are then tested for tensile strength, ductility, and soundness.

Pipe Laying. Placing the pipe in position in the trench, or *laying*, is closely associated with welding operations. It is usually done by either the *stove pipe* or the *firing line method.*

With the stove pipe method, the joints are welded successively to the line. With the firing line method, from 10 to as many as 40 pipe lengths may be welded together as a unit, with several welders working simultaneously, and the unit formed is thus tied into the finished line.

Variations of the stove pipe method consist of welding two or three pipe lengths together in a central yard or on the right of way ahead of the main welding crew.

Pipe Coating. It is usually advisable to apply a coating (e.g., *Extrucoat* which is an extruded plastic coating) and cathodic protection to the line to protect it from underground corrosion. The coating may be applied to the pipe before stringing, at the factory or yard for instance, or just before laying. Cathodic protection in the form of sacrificial anodes is usually provided during construction; if supplied by rectifiers this protection may be installed about a year later.

In the case of precoated pipe, all field joints must be coated manually after welding is completed. When coating is done in the field, the pipe must be thoroughly cleaned first. Machines equipped with moving brushes are generally employed, and sand blasting may be used on mill coating. Machine application of priming then supplies a bonding agent between pipe and coating. Finally, coating and wrapping are applied mechanically

Great care should be taken to protect the coating from damage at all times before backfilling. Inspection of the coating may be made by visual examination during application and by use of electronic holiday detectors. These are run over the coated line to detect flaws, which can then be patched. Selection of the proper coating depends upon local conditions which must be met.

Interior Pipe Coatings. A survey of oil and gas industry experience, including methods used and suitability of coating materials, is available.[14] Coating properties considered include adhesion, abrasion resistance, and surface roughness—all within the limits of 15 to 130 F. Cyclic decompression from 1000 psig is also considered. It has been found that the cost of applying coatings in the field ranges up to 70 per cent of the total cost of internal coating.

Valves and Fittings. Main line valves, particularly the larger sizes, are customarily installed after the line has been laid. It is difficult to properly position a heavy valve welded or bolted to the line. By removing a short section of the line, after the line has been given time to settle, and then installing the valve, possible exterior stresses on the valve body may be eliminated. Also, if plug or venturi gate type main line valves are used, it may be necessary to run cleaning scrapers before valves are set. *Pressure recovery tube* inserted downstream of lubricated plug valves permits recovery of 90 per cent of velocity pressure (instead of 50 per cent).[15]

Meter installations, drips, side taps, cross-overs between parallel lines, foreign lines, and similar connections also are customarily installed after the main line has been laid.

Backfilling and Cleanup. When backfilling is completed, excess soil, rock, and similar materials should be removed from the right of way together with accumulated construction debris. In cultivated and pasture lands the ground should be replaced as close as possible to its former level. Drainage ditches, terraces, field roads, and fence should be restored to their former conditions. Pipeline

Table 8-11 Expected Performance of a Pipeline Spread*
(Miles per day)

Line size, in.	Level to rolling terrain		Hilly terrain		Low wet terrain	
	Progress per working day	Progress per calendar day at 15% lost time	Progress per working day	Progress per calendar day at 15% lost time	Progress per working day	Progress per calendar day at 15% lost time
8	1.85	1.57	1.57	1.34	0.94	0.80
12	1.45	1.23	1.38	1.18	.81	.69
16	1.32	1.12	1.25	1.06	.76	.65
20	1.18	1.00	1.11	0.95	.66	.56
24	1.12	0.95	1.06	.90	.62	.53
30	0.98	.83	0.94	.80	.56	.48
36	0.95	0.81	0.91	0.77	0.50	0.43

* Figures shown do not include heavy clearing, cased crossings, main valve installation, major stream crossing, weight or anchor installation, rock excavation, or special right-of-way cleanup.

markers, mile posts, and warning signs should be erected. All surplus construction materials should be satisfactorily disposed of. Purging, testing, and filling (not necessarily in this order) follow.

Cleaning. It is common practice to employ internal scrapers to remove debris, scale, water or other foreign material which may have accumulated in the line during construction. Natural gas under pressure is the usual driving force for scrapers. Compressed air may be used if gas is unavailable. Before scrapers are used, care should be taken to see that no design features such as short radius bends, plug valves, or changes in line size exist or that no other factors are present which might cause a scraper to become lodged or would otherwise prevent its use.

PIPELINE OPERATION, MAINTENANCE, REPAIR

Operation

Ability of a pipeline to perform properly the functions for which it was designed and constructed depends upon its safe, economical and efficient operation. This, in turn, requires adequate maintenance and, when necessary, prompt and effective repairs. The line must be capable of delivering a supply of gas to the markets served, according to contract provisions. In view of the importance of the service provided, all possible precautions should be taken to guard against its interruption.

Pipeline Flow Efficiency. Flow efficiency of a gas pipeline may be defined as the ratio of measured flow rate under a given set of conditions to calculated flow rate under the same conditions. In a flow test, accurate measurements of the following are necessary: flow rate, pipe diameter and length, upstream and downstream pressures, barometer, gas temperature, gas specific gravity, and elevation change between ends of pipe.

The calculated rate may be obtained by using pipeline flow formulas. Errors in measurement may be expected to produce corresponding errors in efficiency as follows:

	Measurement error	Efficiency error
Flow rate	1%	1%
Pressure		
for 20 psi difference	0.2% at 700 psia	1.0%
for 100 psi difference		0.2%
Temperature	2°F	0.2%
Specific gravity	0.002	0.2%

The standard industry method for *gas measurement* in transmission lines is by *orifice meter*. With properly maintained equipment, errors involved should be less than one per cent.

Ammonia Slug Method.[16] This technique can produce results comparable in accuracy to the orifice meter. It consists of injecting a slug of ammonia into the line at an upstream point and measuring the time needed for its detection at a downstream point. Volumetric flow is determined from the travel time and the intermediate pipe volume. This assumes no change in relative placement of the leading edge of the ammonia slug during travel, since diffusion effects appear negligible. The use of ammonia is limited because of corrosion, erratic results in the presence of large quantities of water, and, when CO_2 is present, because of the formation of ammonium carbonate.

Depending upon the size of the transmission line, from one-half to about four pounds of ammonia may be used. It may be injected into the middle of the gas stream thru a ¼-in. pipe inserted thru a blowoff, using a source of gas from 200 to 300 psi above line pressure and of sufficient volume to insure quick displacement into the line.

Detection of ammonia at the downstream point is done by a phenolphthalein solution which changes to pink on contact with ammonia as gas is bubbled thru it. A high pressure vent is placed just before a needle valve controlling gas flow to the indicator. Sufficient gas must be vented so that gas from the middle of the line will reach the indicator without appreciable lag.

The test pipe run should be a continuous single line from 5 to 15 miles long, fairly level, and with no take-offs. The line must be in steady flow when the test is made. Pressures are read at the injection and detection points at the start, finish, and during passage of the slug. Flowing gas temperature should be nearly constant.

The measured flow rate is calculated as follows:

$$Q_m = \frac{1440 V P_m T_o}{t T_m Z P_o} \tag{9}$$

where:

Q_m = measured flow at P_o and T_o, SCF per day
V = pipeline volume between injection and detection points, cu ft
t = travel time of slug, min
P_m = mean presure, psia
T_m = mean temperature, °F abs
Z = average compressibility evaluated at P_m and T_m
P_o = standard pressure, psia
T_o = standard temperature, °F abs

$$P_m = \frac{2}{3}\left[\frac{P_1^3 - P_2^3}{P_1^2 - P_2^2}\right]$$

If the pipeline is not level, a simple approximate correction for change in elevation often makes it sufficiently accurate. One method compensates for the static gas head caused by the difference in elevation, by modifying the downstream pressure as follows:

$$P_2 = P_2' + C \tag{10}$$

where:

P_2' = observed downstream pressure, psia
C = elevation correction = $P_m G h / 53.3 T_m Z$, psi

where:

h = change in elevation, ft (negative if elevation at P_2 is less than at P_1)
G = specific gravity of gas
Other terms are defined under Eq. **9**.

This simplified correction method should be used only where C is less than ten per cent of $P_1 - P_2$.

Maintenance

Principal objectives of maintenance and repair of gas transmission lines are to keep the facilities in proper operating condition, to provide lowest cost of service, to insure continuity of service, and to maintain transmission at design capacity.

Proper maintenance includes preservation of the ground surface over and near the line, to prevent stresses due to earth displacement and metal loss from rust or corrosion. Surface maintenance includes weed and brush control, right of way cultivation of cover crops, and care of access and patrol roads, together with gates, fences, and markers. Area checks made by airplane patrol or by line walkers are the usual methods of surface inspection. They serve to keep the pipeline company informed of conditions along its right of way. Reports from adjoining landowners are also helpful. Discoloration of vegetation along the right of way or emission of dust at specific points provides evidence of leakage.

Detailed *inspection of suspected areas* may be made by use of bar holes at regular intervals over the line, with leakage detected by a portable gas analyzer. Bell-holes may be used periodically to expose the line for complete check of the condition of its coating and for any corrosive effects of the surrounding soil on the pipe. Periodic checks should be conducted at points of concentration of stress, such as at branch connections, at river-crossing manifolds, or where change of direction occurs. If surface conditions at such points indicate movement, the line should be uncovered for further examination.

Periodic inspection should be made of all valve operators, blowoffs, and other exposed controls; of river, bridge, and highway crossings; and of casing and exposed branch connections. Moving parts should be checked for wear and, if practical, run thru a complete cycle.

A permanent record should be kept of all inspections performed and any irregularities noted. When further work is found necessary, upon its completion a separate report should be prepared stating what was done and the results achieved.

Pipeline leaks should be reported and classified thru routine procedure, with their locations recorded on alignment sheets.

Necessary maintenance records include maps showing location details and specifications of pipe, valves, fittings, and coatings. Description of special construction featu:es should also be available. The original soil survey and details of the cathodic protection system, if used, are essential for the proper evaluation of the record of each pipe perforation by external corrosion and of the condition of pipe and coating whenever the line is exposed.

Maintenance crews and supervisory personnel should be aware at all times of other activities in their area, such as land improvement, deep plowing, highway construction, drainage work, and installation of other underground structures.

Repairs

Transmission pipeline failures requiring repair operations are of the following four general classes:

1. Those which can be repaired without complete interruption of service, such as leaks due to exterior corrosion, leaks in gaskets or welds, failure of defective material, and accidental line puncture. In such cases, repairs can be made by installing repair sleeves under full or reduced line pressure, with a minimum of excavation and equipment.

2. Failures too extensive for sleeve repair and requiring line shutdown for several hours, with complete interruption of service for blowdown and installation of new pipe or fittings in the line.

3. Extensive line damage, like that caused by earthquakes or floods, with complete interruption of service. Immediate repairs in these cases require temporary installation of minimum facilities for restoration of service.

4. Repairs involving extensive reconditioning or replacement. These are frequently performed under contract.

Trained personnel and necessary material and equipment should be available at strategic points along the transmission system so that any emergency repairs which may be expected on the line can be handled promptly and effectively. Handling facilities should require a minimum of manpower and time.

REFERENCES

1. Smith, R. V. and others. *Flow of Natural Gas through Experimental Pipelines and Transmission Lines.* (Bur. of Mines Mono. 9) New York, A.G.A., 1956.
2. Wilson, G. G. and Ellington, R. T. "Selection of Flow Equations for Use in Distribution System Network Calculations." (DMC-58-31) *A.G.A Proc.* 1958: D-231-42.
3. Staats, W. R. and Ellington, R. T. "Flow Phenomena in Large-Diameter Natural Gas Pipelines." (GSTS-58-10) *A.G.A. Proc.* 1958: T-61-8.
4. Bukacek, R. F. "Flow Equations for Natural Gas Pipelines." *Am. Soc. of Civil Eng. Proc.* 84: PL2, pt. 1, paper 1667, June 1958.
5. Ferguson, J. W. "Mathematical Analysis of the Effect of Differences in Elevation on Flow Formulae for Gas Pipe Lines." *Gas Age Record* 78: 39+, July 11; 72, July 18, 1936.
6. Beitler, S. R. and Zimmerman, R. H. "Determination of Supercompressibility Factor." *Proceedings of the Southwestern Gas Measurement Short Course.* Univ. of Okla., 1953.
7. Schomaker, J. F. and Hanna, L. E. "How to Work Revised Panhandle Formula." *Pet. Eng.* 28: D25-31, May; D47-9, June 1956. See also: *Oil & Gas J.* 54: 192+, July 30, 1956.
8. Weymouth, T. R. "Problems in Natural Gas Engineering." *A.S.M.E. Trans.* 34: 185–234, 1912.
9. Grizzle, B. F. "Simplification of Gas Flow Calculation by Means of a New Special Slide Rule." *Pet. Eng.* 16: 154+, Sept. 1945.
10. Wortman, H. R. "Report of Map Piping Symbols Task Group." *A.G.A. Proc.* 1962: DMC-62-4.
11. Watkins, E. G. "Design of Gas Piping on Bridges." (DMC-52-21) *A.G.A. Proc.* 1952: 502-16.
12. Reynolds, P. E. "Designing Pipeline Spans." *Gas* 28: 102–11.
13. Donnelly, W. H. and McMullin, E. V. "Crossing of Coldwater Creek." *Am. Gas Jo.* 188: 22–7, Aug. 1961.
14. Wilson, G.G. *Interior Surface Coating of Pipe for Natural Gas Service.* (I.G.T. Tech. Rept. 3) New York, A.G.A., 1960.
15. Van Deventer, F. M. "Technical Aspects of Valve Selection for Gas Transmission and Distribution Systems." (DMC-58-32) *A.G.A. Proc.* 1958: D-243-8.
16. Wright, J. C. "New Displacement Method of Measuring Gas Flow through Pipe Lines." *Pet. Eng.* 22: D45+, Mar. 1950. See also: *Oil & Gas J.* 48: 99–103, Mar. 16, 1950; *Gas* 26: 116+, May 1950.
17. McClure, G. M. and others. *Summary Report, Research on the Properties of Line Pipe.* New York, A.G.A., 1962.
18. Atterbury, T. J. and others. *Branch Connections, Data for Design.* (Project NG-11) New York, A.G.A, 1962.
19. Parker, R. F. "The Effect of Elevation Changes on Gas Flow Calculations." *Pipe Line News* 35: 9+, August 1963.
20. Trouard, S. E. "The Fallacy of Casing." *A. G. A. Proc.* 1963: DMC-63-29.
21. Janssen, J. S. and others. "Design Practices for Gas Lines Crossing Bridges, Railroads and Highways." *A. G. A. Operating Sect. Proc.* 1964: 64-D-151.

Chapter 3
Compressor Stations

by Rex V. Campbell, F. E. Vandaveer, and Carl J. Veit

INTRODUCTION

In 1962 there were 557 compressor stations on 104,140 miles of pipe of 28 major transmission lines (systems with more than 500 miles of transmission mains and $5,000,000 operating revenues). The stations had a total of 6,303,147 hp, averaging 11,316 hp per section, spaced about 187 miles apart. Combined capacity on a peak day was 41,695 MMCFD of natural gas, amounting to movement of approximately 953,000 tons of gas.[1]

For every pressure increase of about 14.7 psig, one volume of a gas is roughly compressed into another so that at 1000 psig approximately 68 cu ft are compressed into 1 cu ft. Also, gases compress more at higher pressures and occupy a smaller volume than indicated by Boyle's law. This *compressibility* characteristic differs for each constituent of a gas mixture. Natural gases may deviate as much as 21.9 per cent from Boyle's law at 900 psig.

Adiabatic compression causes an increase in temperature, usually between 0.2 and 3°F per 1.0 psi increase in pressure, depending upon kind of gas and the pressures involved. Adiabatic compression volume is about 0.2 to 0.6 per cent greater than the **isothermal** volume for same pressure range.

Types of Compressor Stations

Whenever a gas has insufficient potential energy for its required movement, a compressor station must be used. The following five types of compressor stations are in general use:

1. *Field or gathering stations* gather gas from wells in which pressure is not sufficient to produce a desired rate of flow into a transmission or distribution system. Such stations may handle suction pressures from below atmospheric pressure to 750 psig, and volumes from a few thousand to many million cubic feet per day.
2. *Relay or main line stations* boost pressures in transmission lines. They are generally of large volume and operate with low compression ratios, usually less than two. Their pressure range is usually between 200 and 1000 psig, sometimes as high as 1300 psig.
3. *Repressuring or recycling stations* are an integral part of a processing or secondary recovery project, generally not involving transportation of natural gas to market. They may discharge at pressures above 6000 psig.
4. *Storage field stations* compress trunk line gas for injection into storage wells. These stations may discharge at pressures up to 4000 psig, employing compression ratios as high as four. Designs of some storage stations permit withdrawal of gas from storage and forcing it into high pressure lines. Storage field stations require precision design engineering because of their wide range of pressure-volume operating conditions.
5. *Distribution plant stations* ordinarily pump gas from holder supply to medium or high pressure distribution lines at about 20 to 100 psig, or pump into bottle storage up to 2500 psig.

In addition to suitable housing, auxiliary equipment at compressor stations may include: cleaners and filters for inlet and outlet gas (wet or dry types), water cooling systems for engines, compressed gas cooling system, water treatment facilities, fuel gas regulatory system, continuous duty or standby generators, compressed air system, lubricating oil purifiers, combustion air filters, steam heating system, safety devices, instrumentation, emergency shutdown system, repair shop, plant for introducing odorant, dehydration plant, gas measurement equipment, and remote control devices.

Types of Compressors and Prime Movers

Gas industry *compressors* fall into the following three distinct types:

1. *Jet.*
2. *Rotary.*
   a. Rotary blowers, impellers rotating in opposite directions.
   b. Centrifugal (radial flow type).
3. *Reciprocating.*
   a. Single acting.
   b. Double acting.

Other compressors, like the vane type and axial flow type, are also used.

All types of prime movers have been applied as compressor drives. Choice is dictated largely by economic factors.

POWER EXTRACTION PUMPING

Jet compressors[2,3,4a] are used in gas and oil fields where motive and suction gas pressures do not vary appreciably. One example is on dually completed gas wells, where both high and low pressure gas is available and where there is an intermediate pipeline pressure.

Rotary blowers are employed primarily in distribution systems where the pressure differential between suction and discharge is not over 15 psi.

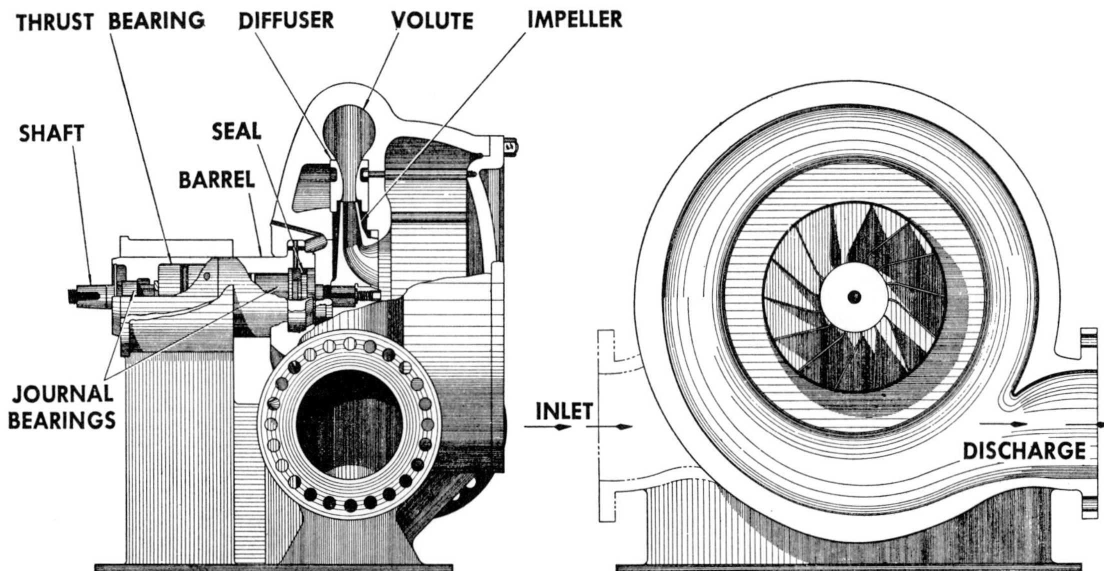

Fig. 8-22 Basic construction of centrifugal compressor. (Nordberg Mfg. Co.)[8]

Turbine expansion of natural gas may be applied to drive a compressor in order to increase pressure in a second pipeline; the availability of a low temperature heat sink (the expanded gas) would be a by-product.[4b]

CENTRIFUGAL COMPRESSORS

Historical data on the applications of centrifugal compressors are available.[5–7]

Characteristics

In a centrifugal compressor[8] work is done on the gas by an impeller. Gas is then discharged at a high velocity into a diffuser, where its velocity is reduced and its kinetic energy converted to static pressure, *without* confinement and physical *squeezing*, which is done by positive displacement machines.

Essentially, the centrifugal compressor (Fig. 8-22) consists of a housing with flow passages, a rotating shaft upon which the impeller is mounted, bearings, and seals to prevent gas from escaping along the shaft. Units differ in shape of volute, impeller, and diffuser.

Features governing centrifugal compressor performance include:

1. Few moving parts, since only the impeller and shaft rotate. Thus, lubricating oil consumption and maintenance costs should be low.
2. Inability to attain as high a compression ratio as reciprocating units (due to absence of positive displacement).
3. Continuous delivery without cyclic variations.
4. Cooling water normally not required because of lower compression ratio and lower friction loss (multistage units for process compression may require some form of cooling).

Centrifugal compressors with relatively unrestricted passages and continuous flow are inherently high-capacity, low-pressure-ratio machines that lend themselves easily to *series* arrangements within a station. In this way, each compressor is required to develop only part of the station compression ratio.

Figure 8-23 shows typical efficiency curves for centrifugal and reciprocating compressors. These curves indicate relative efficiencies over the usual range of operation as compression ratio changes.

Overall efficiency of a centrifugal compressor is based on 99 per cent mechanical efficiency and its manufacturer's adiabatic efficiencies. A mechanical efficiency of 95 per cent is assumed for reciprocating compressors. However, losses in suction and discharge valves and in passageway restrictions vary greatly between different cylinder and valve designs. An assumption of 17.5 per cent for a 1.2 compression ratio varying down to 9.5 per cent for a 1.7 ratio may be made as currently representative of the more efficient cylinders. Adiabatic efficiency for a reciprocating compressor is taken as 100 per cent.

Engineering Data

Curves of Fig. 8-23, in a limited way, show the range of economical operation of each type of unit. They illustrate higher compression efficiency of the reciprocating unit at

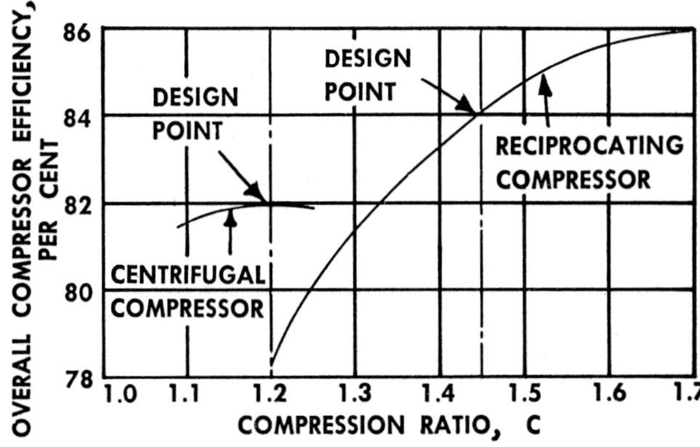

Fig. 8-23 A comparison of centrifugal and reciprocating compressor efficiencies. (Nordberg Mfg. Co.)[8]

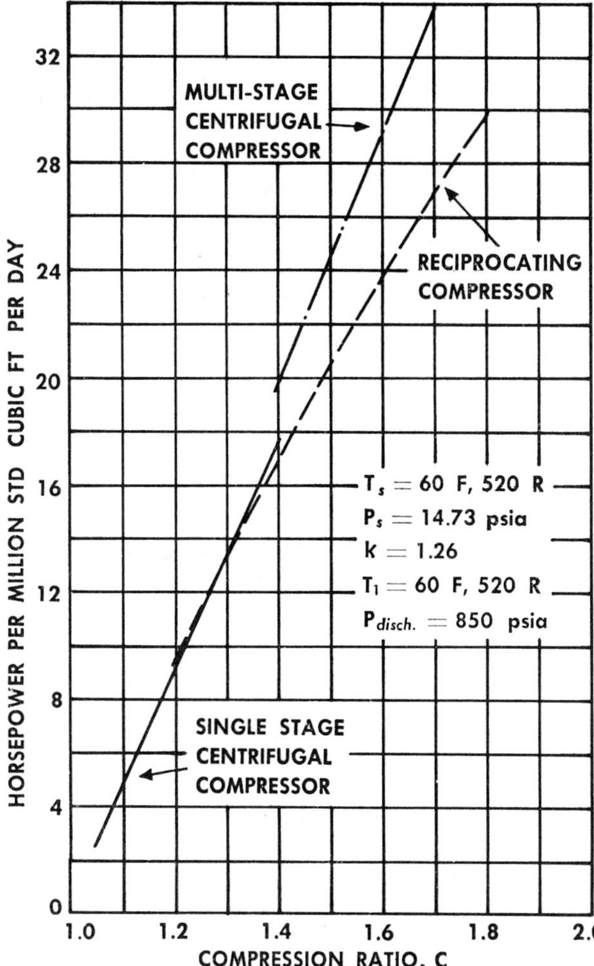

Fig. 8-24 Compressor horsepower requirements at various compression ratios. (Nordberg Mfg. Co.)[8]

pressure ratios within its normal range and its steadily increasing efficiency advantage as the ratio increases.

For pipeline calculations, a more useful comparison of efficiencies may be made on the basis of horsepower per MMSCFD of gas compressed at design conditions. Figure 8-24 shows the power requirements at each ratio using a compressor economically designed *for that particular ratio.* These curves were calculated using the formula:

$$hp = \frac{3.0325}{10^6} Q_s \frac{P_s T_1 Z}{T_s e} \left(\frac{k}{k-1}\right)\left(C^{(k-1)/k} - 1\right) \quad (1)$$

where:

Q_s = flow, SCFD
P_s = std pressure, 14.73 psia
T_s = std temperature, 520 R
T_1 = inlet temperature, 520 R
Z = compressibility factor
e = compressor efficiency, isentropic horsepower/actual horsepower
k = ratio of specific heats = 1.26
C = pressure ratio, abs discharge pressure/abs inlet pressure

Values of Z were taken from the curve in Fig. 8-25 using inlet pressures at each pressure ratio corresponding to 850 psia discharge pressure. Many other empirical formulas[9] may also be used.

The reciprocating compressor curve in Fig. 8-24 extends from 1.2 to 1.8 pressure ratio. This covers the useful range for pipeline relay or main line compressor operation, with the lower limit imposed by physical size of the compressor cylinders and valve losses, and the upper limit by economical pipeline design. The compression ratio for a single stage centrifugal compressor extends from 1.05 to 1.4. Single stage centrifugals are limited to a maximum of approximately 1.4

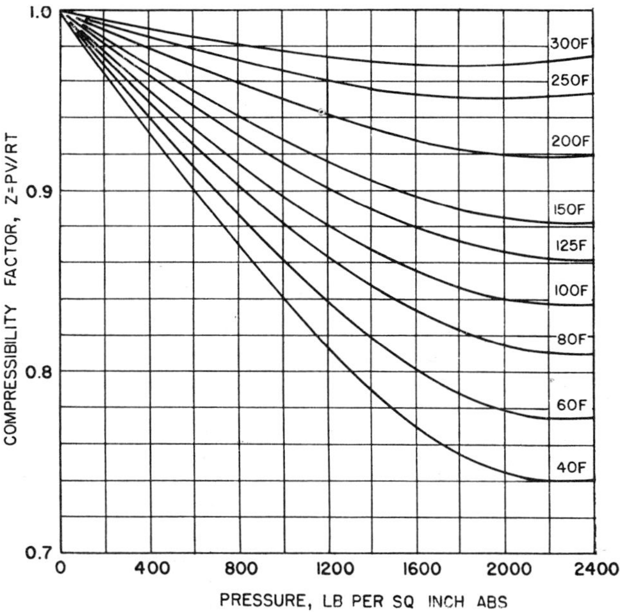

Fig. 8-25 Compressibility factors for typical natural gas at various inlet temperatures. Composition, per cent: CH_4—94.75, C_2H_6—3.36, C_3H_8—0.74, N_2—0.43, CO_2—0.34, remainder C_4+.

pressure ratio because of inherent compressor characteristics. The 1.05 ratio is the approximate practical minimum for pipeline service, because at lower ratios either the distance between stations becomes too short to be economical or the number of units in series within a station becomes too great. At approximately 1.26 ratio, the curves cross for reciprocating and single stage centrifugal compressors.

When compressor systems are compared on the basis of overall station horsepower requirements, the losses in station piping, valves, or headers must be considered along with compressor horsepower requirements.

Example: Compare a typical station having three centrifugal units arranged in series and one having reciprocating units in parallel arrangement.

Given: Suction pressure to station = 586 psia
Discharge pressure from station = 850 psia
Scrubber and suction piping loss for station = 3 psi
Discharge piping loss for station = 2 psi
Loss between each series unit = 0.5 per cent
Loss in each parallel unit: suction = 1 psi
discharge = 1 psi
Loss in pulsation dampeners for reciprocating units = 3 psi

For centrifugal compressors: Pressure ratio or compression ratio, C, required for each centrifugal unit—

$$C = \sqrt[n]{C_o \times 1.005^{n-1}} = \sqrt[3]{1.46 \times 1.005^2} = 1.14$$

where:

n = number of compressors in series = 3

$$C_o = \frac{\text{station discharge pressure} + \text{discharge losses}}{\text{station suction pressure} - \text{suction losses}} = \frac{850 + 2}{586 - 3} = 1.46$$

From Fig. 8-24, hp per MMSCFD = 6.4
Total hp per MMSCFD = 6.4 × 3 units = 19.2

For reciprocating compressors: Pressure ratio required for each reciprocating unit—

$$C = C_o = \frac{\text{station discharge pressure} + \text{discharge losses}}{\text{station suction pressure} - \text{suction losses}}$$

$$C = \frac{850 + 2 + 1 + 3}{586 - 3 - 1} = 1.47$$

From Fig. 8-24, hp per MMSCFD = 19.5 Total

The curve in Fig. 8-24 for a **multistage centrifugal** compressor illustrates its characteristics (not in use in 1958 in gas pipelines). It is evident that the advantage in horsepower requirement of a reciprocating compressor steadily increases with compression ratio and is substantial in the range of the multistage unit.

The centrifugal compressor can be adapted to varying conditions by changing speed or by using inlet guide vanes. Figure 8-26 illustrates its typical operation at different ratios, speeds, and horsepowers. A smaller horsepower variation occurs for a given change in compression ratio in a reciprocating machine, as compared to a centrifugal one. However, since centrifugal compressors operate in series, greater variations in ratio can be made by adding or subtracting units. From Fig. 8-26, a station with three centrifugal units, with a load variation of only 5 per cent and a speed variation from 95 to 105 per cent, would have a range of ratios from 1.19 to 1.295 per machine. A total station ratio range of 1.19 to 2.15, from the ratio formula given previously, would be possible.

The **stability limit** indicated in Fig. 8-26 represents the maximum pressure ratio at a given speed, beyond which the compressor becomes unstable and surging may result. In design and selection of equipment, placing the stability limit[10] beyond the operating range will avoid instability and surging during normal operation.

Where constant speed drivers are used with centrifugal compressors, additional range of operation over the design head and volume curve can be obtained by employing *inlet guide vanes.* These are adjustable stationary blades located on the suction side of the impeller, which impart a swirl to the entering gas either in the direction of the impeller's rotation or against it. The range of ratios made available by this means is adequate for most pipeline operations; however, using guide vanes does sacrifice efficiency, compared to that afforded by variable speed operation.

Centrifugal units, like reciprocating compressors, must be selected with operating characteristics that allow for future expansion. The initial range can usually be selected to include

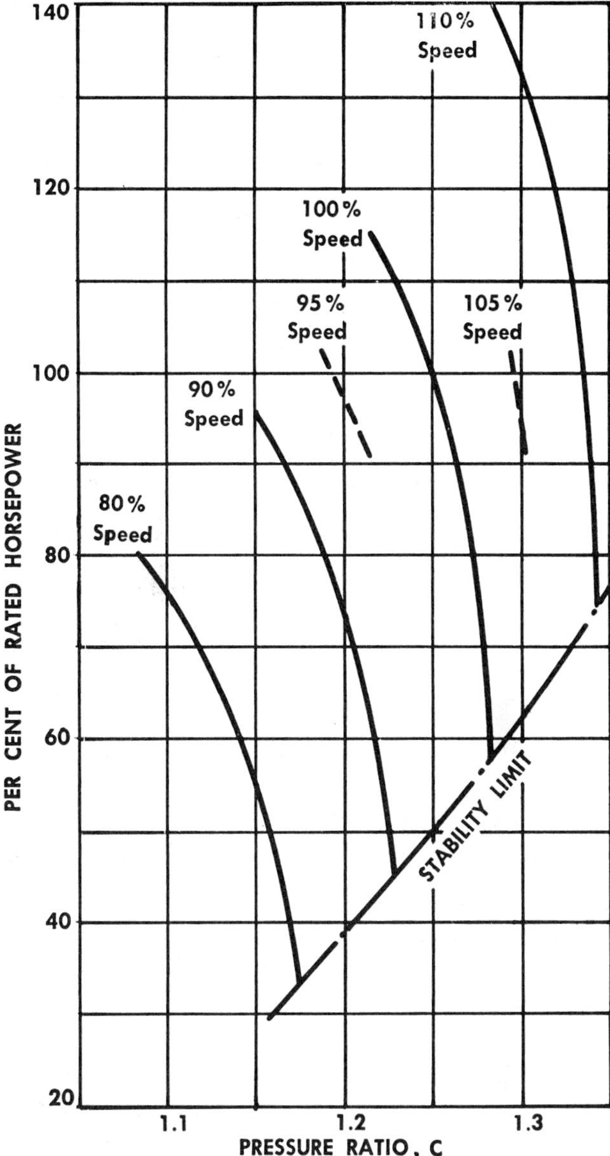

Fig. 8-26 Performance of centrifugal compressor at various speeds. (Nordberg Mfg. Co.)[8]

some expansion, with a change of impellers allowing for further expansion. In extreme cases, a complete new unit may be installed on the existing prime mover, possibly with a change in rated compressor speed.

Non-Overloading Characteristics. Impellers with backward bending vanes have a pressure-volume characteristic that, at constant speed, causes the discharge pressure to decrease gradually with increasing volume (Fig. 9-1). Thus, at design inlet temperature and pressure, it is not possible to overload a properly selected prime mover since both the discharge pressure and brake horsepower will decrease appreciably as the suction volume increases above 120 per cent of its normal level.

Stability. Percentage change in volume between normal capacity and the surge point at rated discharge pressure is a measure of the stability of centrifugal compressors. This change varies from approximately 70 per cent for compressors developing very low pressure ratios to as low as

30 per cent for those developing very high ratios. In initial design, provision can be made for high stability at slightly reduced efficiency if partial loads of long duration will be likely. When the design load is to be normally sustained, efficiency can be improved by reducing stability.

Speed Change. Discharge pressure differs greatly with small changes in speed. Centrifugal compressor prime movers are generally designed for operation at between 95 and 105 per cent of normal speed.

Design and Selection

The three major items usually necessary in centrifugal compressor design for a given discharge pressure and capacity are: shaft horsepower, operating speed, and discharge temperature.

Determination of horsepower and speed is predicated on calculation of the *head* required for compression. Head, which actually represents work done *per pound of fluid handled,* may be expressed in terms of *feet* (i.e., ft lb per lb) as for a liquid pump, as follows:

$$H = 144 \int V dP \tag{2}$$

where: H = head, ft
V = specific volume, cu ft per lb
P = pressure, psia

When the specific volume or density is constant, Eq. **2** is readily integrated to:

$$H = 144V(P_2 - P_1) \tag{3}$$

where P_2 and P_1 = outlet and inlet pressure, respectively, psia
Other terms are defined under Eq. **2**.

For a centrifugal compressor, where the specific volume is a variable, a somewhat more complex relation is obtained. Assuming that the compression is *polytropic,* it may be represented by the equation:

$$PV^n = \text{constant}$$

where P and V are the absolute pressure and volume, respectively.

Equation **2** may be integrated and rearranged as follows:

$$H = \frac{144nP_1V_1}{n-1}\left[\left(\frac{P_2}{P_1}\right)^{\frac{n-1}{n}} - 1\right] \tag{4}$$

where: n = polytropic exponent of compression
V_1 = initial volume, cu ft
Other terms are defined under Eqs. **2** and **3**.

Equation **4** may be alternately expressed in the form:

$$H = \frac{nZR'T_1}{n-1}\left[\left(\frac{P_2}{P_1}\right)^{\frac{n-1}{n}} - 1\right] \tag{5}$$

where:
R' = specific gas constant $\cong$ 1545/mole wt (see Table 2-71a)
T_1 = inlet temperature, °R
Z = compressibility factor at inlet
Other terms are defined under Eqs. **2**, **3**, and **4**.

For ease in calculation, Eqs. **4** and **5** may be expressed in the form:

$$H = 144P_1V_1\beta = ZR'T_1\beta \tag{6}$$

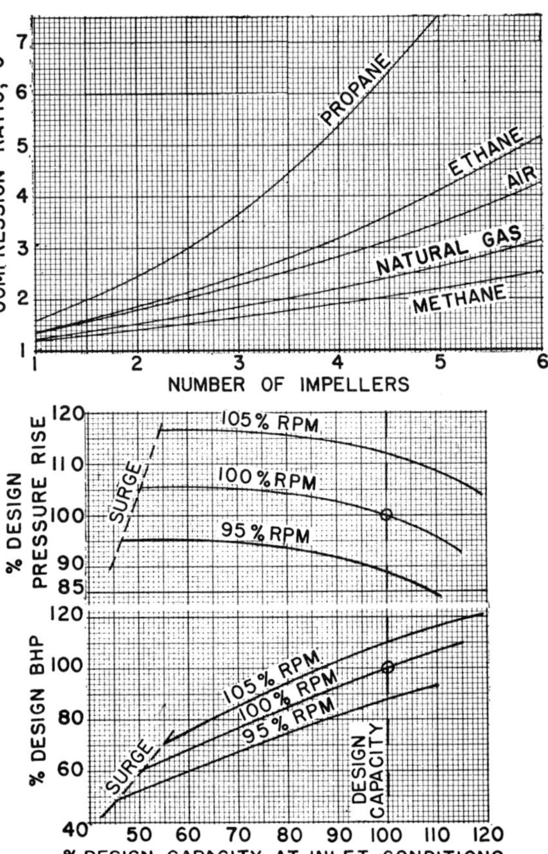

Fig. 8-27a (top) Compression ratio vs. number of impellers. Uncooled centrifugal compressor; suction at 14.7 psia and 100 F; avg μ = 0.55; tip speed = 750 fps; polytropic eff. = 75 per cent. (Clark Bros. Co.)[11]
Fig. 8-27b (bottom) Typical centrifugal performance characteristics. Four impeller uncooled air compressor; μ is the pressure coefficient. (Clark Bros. Co.)[11]

where: β = $[(P_2/P_1)^M - 1]/M$, plotted in Figs. 8-31a and 8-31b.
M = $(r - 1)/n$, plotted in Fig. 8-28
Other terms are defined under Eqs. **2**, **3**, and **4**.

Head, and hence the horsepower, for a given compression ratio varies directly with the absolute inlet temperature and inversely with the molecular weight of the gas being handled (Eqs. **5** and **6**). Since the amount of head which a single impeller will develop is limited, it follows that gases having a high molecular weight will require fewer impellers (i.e., stages) than gases having a low molecular weight when both are compressed thru the same pressure range. This is indicated on Figs. 8-27a and 8-27b for several common gases.

For a perfect gas the compressibility factor is unity, whereas for a real gas this factor deviates from unity. Where this deviation is not large, i.e., between 0.95 and 1.02, or where it remains fairly constant over the range of compression, an average value of the compressibility factor may be used in Eq. **5**, with negligible error. In other instances, where the compressibility factor is subject to greater variation over the range of compression, the head may be approximated from Eq. **7**.

$$H = 165.7(P_1V_1 + P_2V_2)\log_{10}\frac{P_2}{P_1} \qquad (7)$$

Terms defined under Eqs. **2** and **3**.

Equation **7** is not strictly correct, since it is based on the assumption that the log mean value of PV equals the arithmetic mean. This results in an error of 1 to 2 per cent for high compression ratios, and is based only on the assumption that the compression is polytropic, and can be represented by a single exponent, n.

To be correct, the following formula should be used:

$$H = 144\left[\frac{P_2V_2 - P_1V_1}{\log_{10}\left(\frac{P_2V_2}{P_1V_1}\right)}\right]\log_{10}(P_2/P_1) \qquad (7a)$$

Terms defined under Eqs. **2** and **3**.

Equation **7** is particularly useful for hydrocarbon gases at moderate or high pressures and/or low temperatures.

It should be noted that the successful use of Eqs. **4** and **5** depends upon the determination of the polytropic exponent. This exponent may be readily determined by using the following equations, which follow from the definition of hydraulic or polytropic efficiency:

$$\eta = \frac{144\int V dP}{\Delta h} \qquad (8)$$

or,

$$\eta = \frac{n(k-1)}{k(n-1)} \qquad (9)$$

where: η = polytropic efficiency
Δh = change in enthalpy, ft lb per lb
k = isentropic exponent, c_p/c_v

Other terms are defined under Eqs. **3** and **4**.

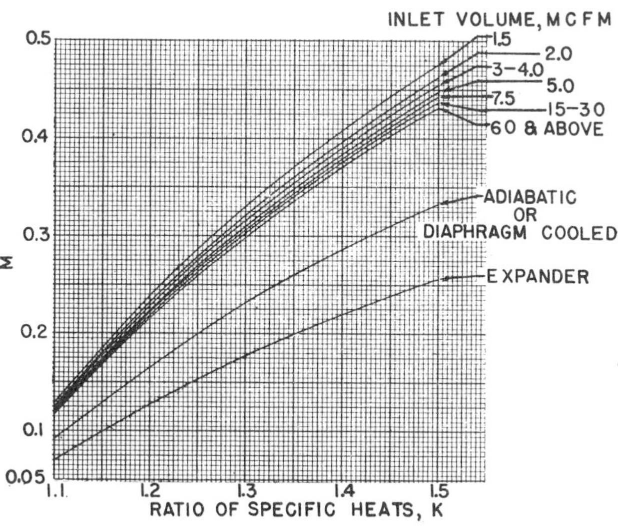

Fig. 8-28 M (for centrifugal compressors) vs. ratio of specific heats, k, where $k = c_p/c_v$ (from Table 2-74) and $M = (n-1)/n$. The upper six curves give $(n-1)/n = [(k-1)/k]$/hydraulic efficiency, where the efficiency varies with inlet volume (from 71 to 75 per cent). Expander values (lowest curve) are 75 per cent of adiabatic values. (Clark Bros. Co.)[11]

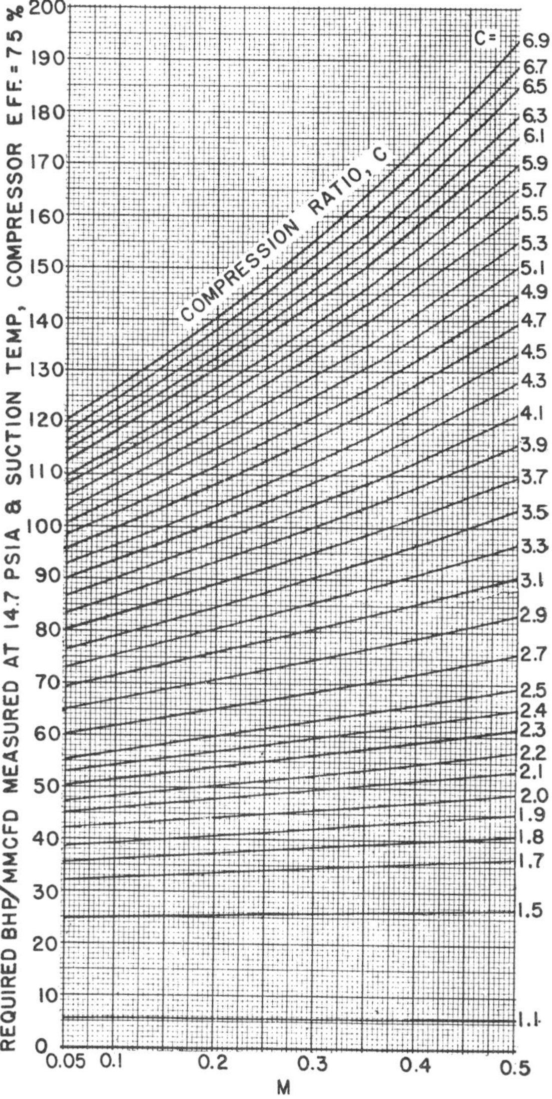

Fig. 8-29 Brake horsepower per MMCFD for centrifugal compressors vs. M for various compression ratios, where $M = (n-1)/n$ (see Fig. 8-28) and $C = P_2/P_1$. These curves incorporate a centrifugal compressor efficiency of 75 per cent. (Clark Bros. Co.)[11]

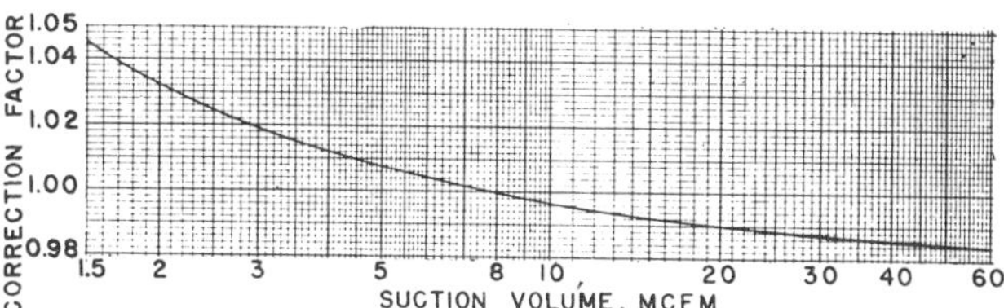

Fig. 8-30 Correction factor for centrifugal compressor **BHP/MMCFD** at 14.7 psia and suction temperature vs. suction volume. (Clark Bros. Co.)[11]

The polytropic efficiency is established by the manufacturers' tests, and is generally a function of the capacity at inlet conditions to the compressor.

The head in feet, H, which a centrifugal compressor stage consisting of an impeller and diffuser will develop, may be related to the peripheral velocity by Eq. **10**.

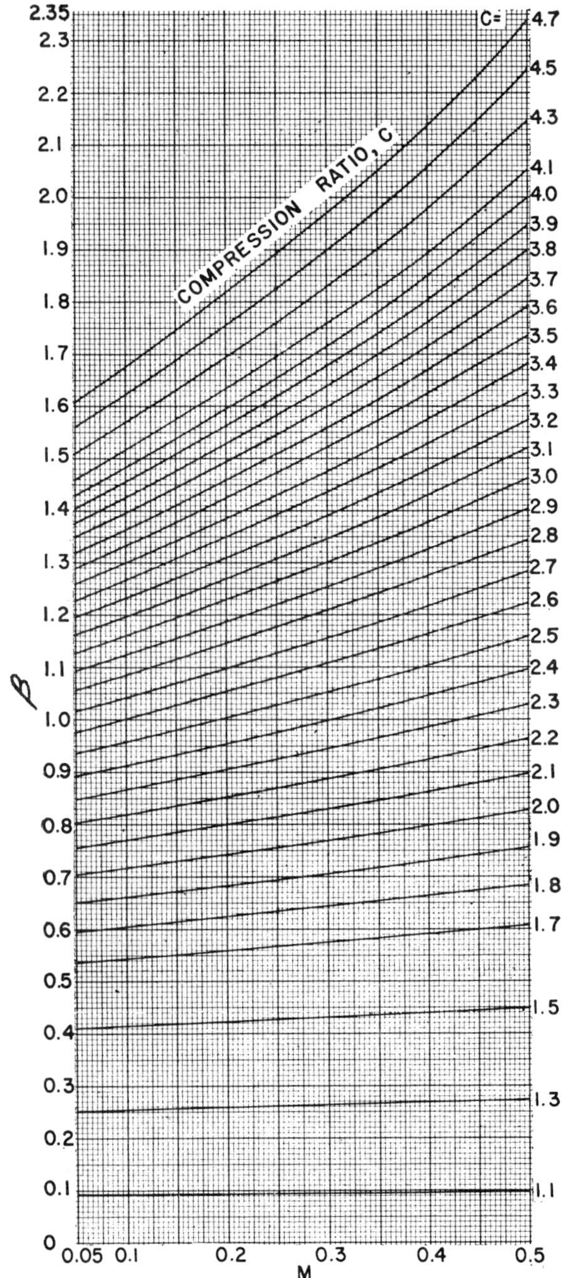

Fig. 8-31a β vs. M for 1.1 to 4.7 compression ratios, where $\beta = (C^M - 1)/M$ and $M = (n - 1)/n$. (Clark Bros. Co.)[11]

$$H = \mu \frac{u^2}{g}, \text{ plotted in Fig. 8-32 (upper part)} \tag{10}$$

where:

μ = pressure coefficient
u = peripheral velocity, $\pi DN/720$, fps
where D = impeller diameter in inches and N = rotative speed in rpm
g = gravitational constant, 32.2 ft per sec-sec

The value of the pressure coefficient μ is a characteristic of the stage design. An average value for one stage of a Clark multistage centrifugal compressor is 0.55. If a normal peripheral velocity of 770 ft per sec is assumed, it can be seen

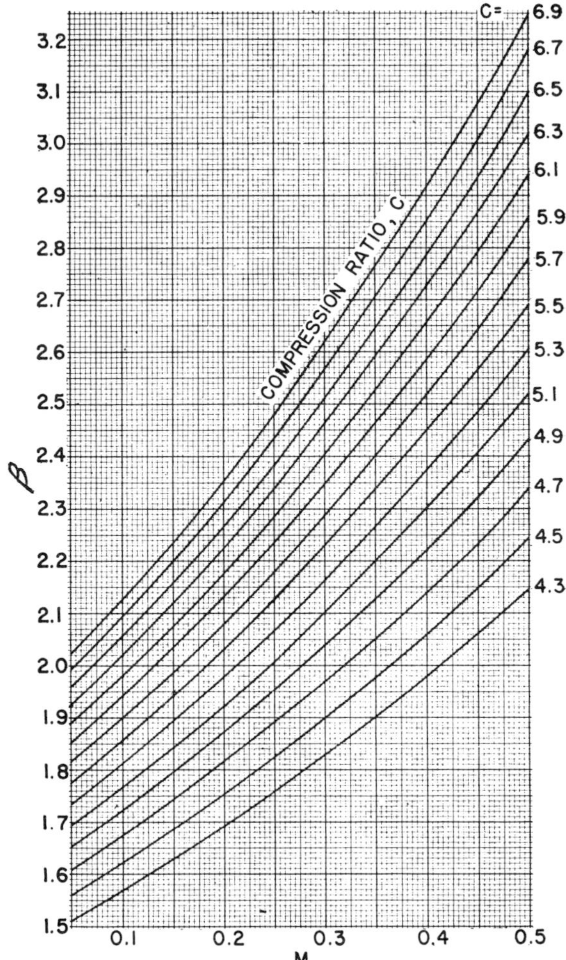

Fig. 8-31b β vs. M for 4.3 to 6.9 compression ratios, where $\beta = (C^M - 1)/M$ and $M = (n - 1)/n$. (Clark Bros. Co.)[11]

that the normal head per stage is approximately 10,000 ft. This assumption permits ready approximation of the number of stages required to develop the head corresponding to the particular compression process.

The power required for the compression of a gas may be calculated from the following:

$$ghp = \frac{W \Delta h}{33{,}000} \tag{11}$$

where: ghp = gas horsepower
W = gas flow, lb per min
Δh = change in enthalpy, ft lb per lb

Therefore:

$$ghp = \frac{WH}{33{,}000\ \eta} \tag{11a}$$

Terms are defined under Eqs. **2**, **8**, and **11**.

To obtain actual horsepower required, the following may be used:

$$\text{hp} = \frac{WH}{33{,}000\, e_0} \tag{12}$$

where: e_0 = actual overall efficiency.

Other terms are defined under Eqs. **2** and **11.**

The compressor **shaft horsepower** equals the gas horsepower divided by the mechanical efficiency. For most centrifugal compressor applications, mechanical losses are relatively small, and an average mechanical efficiency of 99 per cent may be used for estimating purposes.

The **rotative speed** of a centrifugal compressor (Eq. **13**) is fixed by the peripheral velocity and diameter of the impellers. As shown in Eq. **10,** the peripheral velocity is related to the head to be developed; the impeller diameter is determined by the capacity to be handled, as measured at inlet conditions. Thus:

$$N = \frac{1300}{D}\sqrt{\frac{H'}{\mu}} \qquad \textbf{(13)}$$

where: N = rotative speed, rpm
D = impeller diameter, inches
H' = head per stage, ft
μ = pressure coefficient

Discharge temperature, for an uncooled compression, may be calculated from the fundamental relation:*

$$T_2 = T_1\left(\frac{P_2}{P_1}\right)^M \qquad \textbf{(14)}$$

where:

T_2 and T_1 are the outlet and inlet temperatures, respectively, °R

P_2 and P_1 are the outlet and inlet pressures, respectively, abs

$M = (k - 1)/k$, where k, the isentropic exponent, $= c_p/c_v$

In applications where the discharge temperatures for an uncooled compression would be prohibitive (i.e., would exceed 450 F), internal diaphragm cooling may be used. In such cases, the average exponent of compression is approximated by the isentropic exponent, and the head may be estimated on that basis.

Example:[11]

Fluid handled	atmospheric air
Inlet: capacity, cfm	10,000
pressure, psia	14.7
temperature, °R	85 F + 460 = 545
Relative humidity, per cent	70
Discharge pressure, psia	39.7
Isentropic exponent, $k = c_p/c_v$	1.40
Compression ratio, C	2.70

Find number of impellers, compressor size, and discharge temperature.

1. From a psychrometric chart: inlet specific volume = 14.1 cu ft per lb
2. Weight flow = 10,000/14.1 = 710 lb per min
3. From Fig. 8-28: $M = (n - 1)/n = 0.38$
4. From Fig. 8-29:
   BHP/MMCFD = 71 (at M = 0.38 and C = 2.7)
   MMCFD = 10,000 cfm × 1440/10⁶ × 14.7/14.7 = 14.4
   bhp (uncorrected) = 14.4 × 71 = 1020

* Factor $(P_2/P_1)^M$ can be read directly from Fig. 8-42.

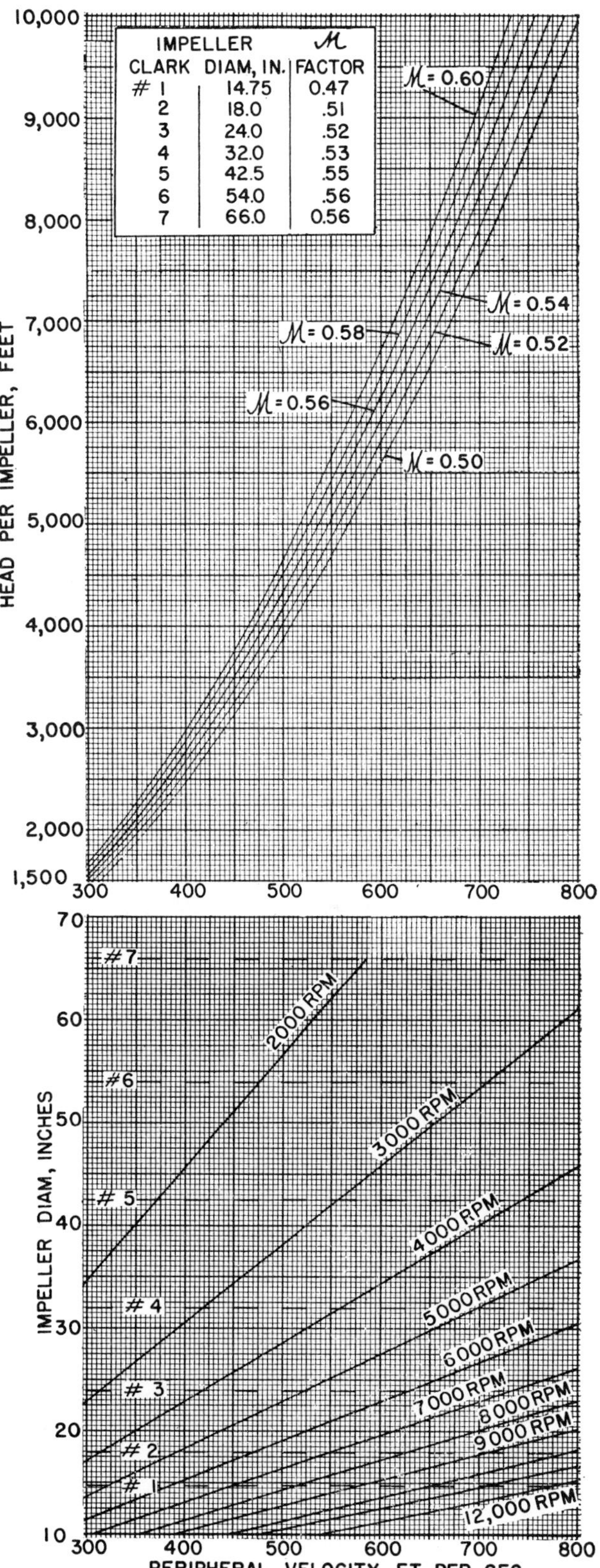

Fig. 8-32 Head per impeller and impeller diameter vs. peripheral velocity for various pressure coefficients, μ, and rotative speeds, respectively. (Clark Bros. Co.)[11]

5. From Fig. 8-30:
   Correction factor to **BHP/MMCFD** = 0.996 (at 10 MCFM)
   bhp (corrected) = 1020 × 0.996 = 1020
6. From Fig. 8-31a:
   $\beta = 1.2$ (at $M = 0.38$ and $C = 2.7$)
   Applying Eq. **6**:
   $$H = 144\,P_1V_1\beta = 144 \times 14.7 \times 14.1 \times 1.2 = 35{,}800 \text{ ft}$$

Thus, the *number of impellers* required, based upon a maximum head per impeller of 10,000 ft, is 35,800/10,000 = 4.

Actual head per impeller = 35,800/4 = 8950 ft

From manufacturer's tabulation (Clark Bros. Co.) a Size 3 compressor is required for an inlet capacity of 10,000 cfm.

From Fig. 8-32:
For a Size 3 unit, $\mu = 0.560$
u = Peripheral velocity = 720 fps (at the 8950 ft head)

From Fig. 8-32: N = speed = 6900 rpm

7. From Fig. 8-33:
   C^M = Temperature factor = 1.45 (at $M = 0.38$ and $C = 2.7$)

Substituting into Eq. **14**:

Discharge temp = $T_2 = T_1(P_2/P_1)^M$ = 545 × 1.45 = 790 R = 330 F

Note 1: For those applications where the required head exceeds that which can be developed by a single case, with the maximum number of impellers available, two or more cases in series must be used. Where intercooling is employed, estimates may be based upon equal compression ratio for each case. The overall compression ratio should be corrected to compensate for intercooler pressure drop by means of the following:

$$\text{Actual overall ratio} = \frac{\text{required compression ratio}}{0.965^{S-1}}$$

where: S = cases in series

Note 2: For those applications where the required head can be developed within a single case, but where the discharge temperature for an uncooled compression would exceed 450 F, internal diaphragm cooling may be used. In such instances, determine the M value from Fig. 8-28.

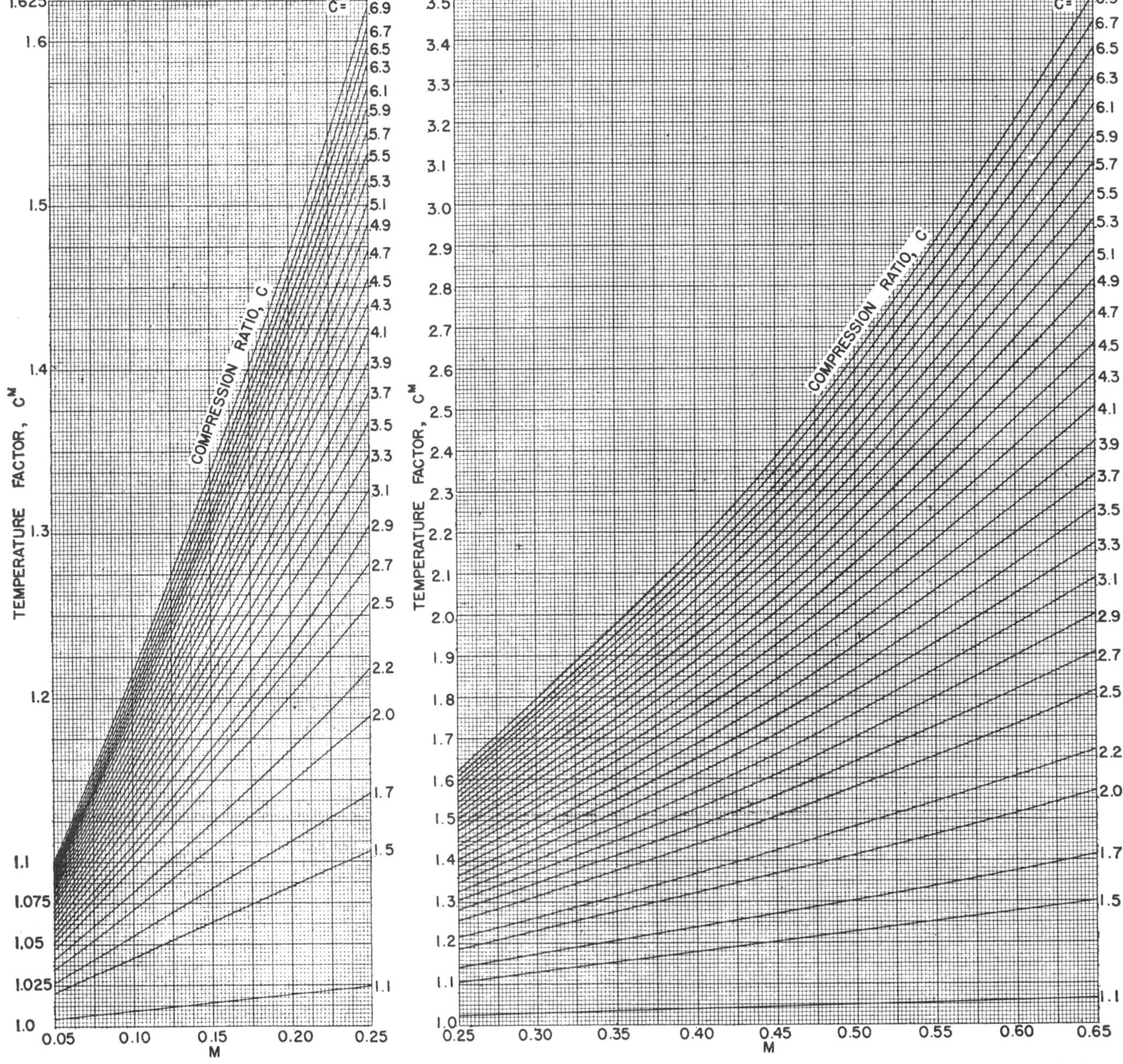

Fig. 8-33 Temperature factor, C^M, vs. M for various compression ratios, where $C = P_2/P_1$ and $M = (n-1)/n$. (Clark Bros. Co.)[11]

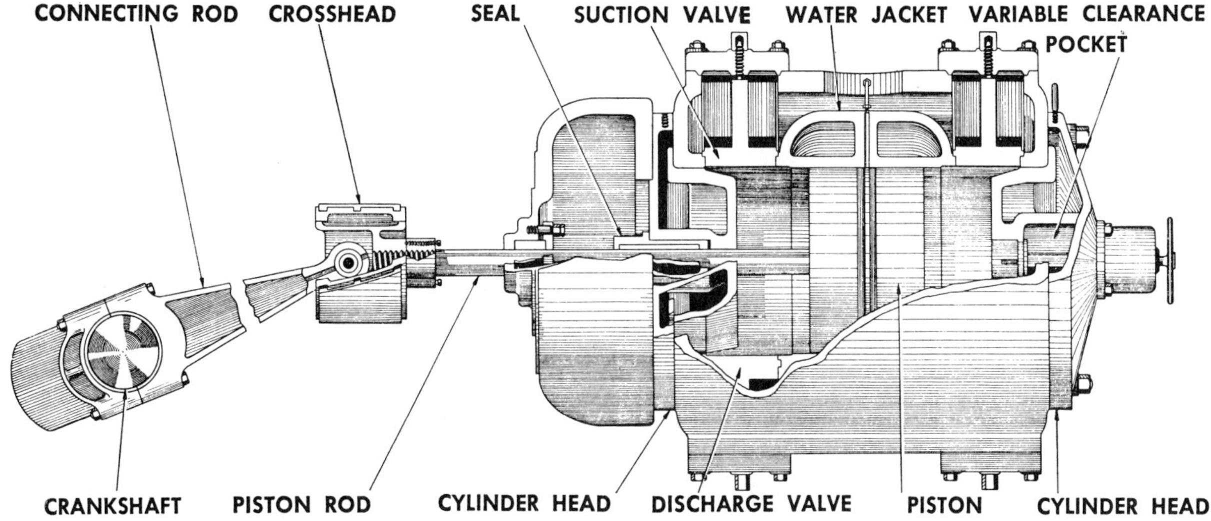

Fig. 8-34 Basic construction of reciprocating compressor. (Nordberg Mfg. Co.)[8]

Operating Problems

The following operating problems have been reported with centrifugal compressors:[7]

1. *Surge or Pumping.* There is a minimum capacity for each centrifugal compressor at every speed, below which its operation becomes unstable. This instability is accompanied by a characteristic noise known as *pumping* or *surging.* The pumping limit is determined largely by the impeller discharge angle. For the average centrifugal compressor, it is about 50 per cent of capacity at the point of greatest efficiency. The primary cause of surging lies in the shape of the head-capacity curve which, after reaching a maximum at about half rated capacity, begins to *droop* toward the zero-capacity point. When capacity is reduced quickly below this point, pressure in the discharge pipe exceeds that produced by the blower and the flow tends to reverse momentarily. As soon as flow is further reduced, discharge pipe pressure drops and the blower begins to discharge again. Response of the compressible gas in the discharge system magnifies such pulsations in pressure and capacity.[12]

The severity of surge seems to vary according to the brand of centrifugal compressor. It may vary from a howling noise during surge to a severe pounding, with enough energy to shake and damage the compressor casing, piping, equipment, and foundation.

2. *High speed operation,* 4000 to 6000 rpm, requires good dynamic balance.

3. *Operating personnel*[13] must be trained in centrifugal compressor operation. The line cannot be packed as it is with reciprocating compressors.

4. *Oil leakage* must be watched at high pressure and at gas seal. Normal usage is 5 to 10 qt of oil per 100 hr of operation. Some units use as much as 100 qt per 100 hr.

RECIPROCATING COMPRESSORS

Reciprocating compressors (Fig. 8-34) have more moving parts and, therefore, lower mechanical efficiencies than centrifugal machines. Both size limitations of compressor cylinders and efficiency drop-off at low compression ratios may preclude series arrangement of reciprocating machines. In parallel operation, each cylinder may handle a small part of the total flow while developing the whole station pressure ratio. Two or more stages of compression may be accomplished on one machine when high compression ratios are encountered and multicylinder compressors are used.

Characteristics

Each cylinder assembly of a reciprocating compressor consists of a piston, cylinder, cylinder heads, suction and discharge valves, and the parts necessary to convert rotary motion to reciprocating motion (connecting rod, crosshead, wrist pin, and piston rod).

A reciprocating machine is designed for a certain range of compression ratios thru the selection of proper piston displacement and clearance volume within the cylinder. This clearance volume can either be fixed or variable, depending upon the extent of the operating range and the per cent of load variation desired. Figure 8-35 illustrates the ranges of

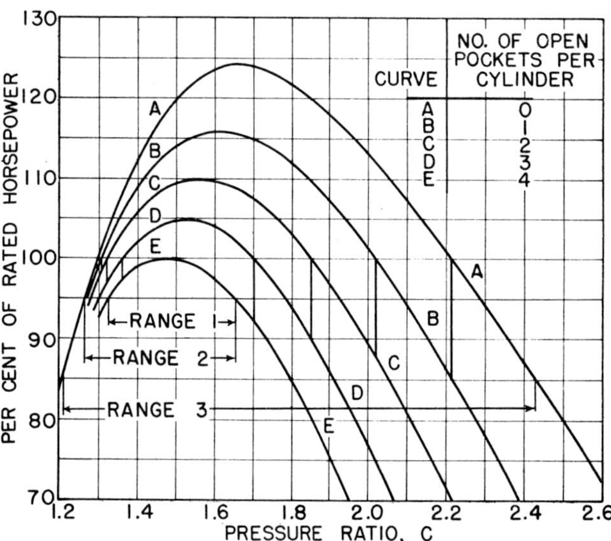

Fig. 8-35 Performance of reciprocating compressor with various clearance volumes. (Nordberg Mfg. Co.)[8]

operation obtainable with a compressor having four double acting cylinders. If operation were desired in Range 1, ratios from 1.33 to 1.66, with load kept between 95 and 100 per cent, a fixed clearance volume of 90 per cent (Curve E) of piston displacement would work satisfactorily. To extend this range to a lower ratio, it would be necessary to use a smaller fixed clearance and pockets which could be opened or closed, depending upon operating conditions. In Range 2, ratios of 1.26 to 1.66 would be available and the horsepower loading would be kept between 95 and 100 per cent by using cylinders of 59 per cent fixed clearance, with one clearance pocket of 62 per cent in each of the four cylinders. Thus, operation would proceed along the fixed clearance line, Curve A from 1.26 to 1.3, after which it would be necessary, in order to prevent overloading, to open one or more pockets. Operation over Range 3 would be possible if a load variation of 85 to 100 per cent were permitted, using the same four clearance pockets. At higher ratios, the pockets are progressively closed to prevent loads under 85 per cent.

There are usually several reciprocating units in a station; thus large capacity variations may be accomplished by either adding or subtracting units. Smaller variations are obtained by varying operating speeds, although care must be taken to avoid critical torsional vibrations which may occur at some speeds.

Valve lifting devices may also be used to control compressor capacities. These devices either lift the entire valve or depress the valve plate, making that end of the compressor cylinder inoperative and thereby reducing effective displacement, decreasing capacity, and unloading the unit.

For future expansion, a suitable range of ratios may be selected for meeting all probable operating conditions. Sometimes it is best to purchase new cylinders at the time of expansion, for more efficient operation or greater flexibility. In some cases different size cylinder liners and pistons can be purchased for radical changes in the capacity of compressor cylinders.

Engineering Data

Figure 8-36 shows an ideal pressure-volume diagram for a compressor cylinder with corresponding piston locations.

Piston displacement is the actual volume displaced by the piston as it travels the length of its stroke from position 1, bottom dead center, to position 3, top dead center. It is normally expressed as the volume in cubic feet displaced per minute, cfm.

For a double acting cylinder, the head end displacement exceeds crank end displacement by the volume of the piston rod. Piston displacement for a single acting cylinder may be calculated from:

$$PD = \frac{A_{HE}LN}{1728} \quad (15)$$

where: PD = piston displacement, cfm
A_{HE} = area of head end of piston, sq in.
L = length of stroke, in.
N = speed, rpm

and for a double acting cylinder:

$$PD_{DA} = \frac{(2A_{HE} - A_R)LN}{1728} \quad (16)$$

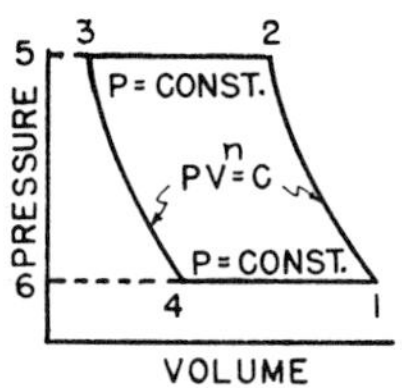

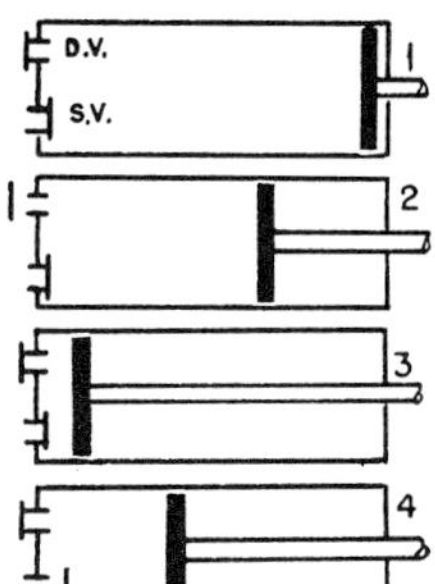

Position 1. Compression stroke starts, suction and discharge valves closed.

Position 2. Discharge valve opens.

Position 3. Compression stroke ends, suction stroke starts. Discharge valve closes.

Position 4. Suction valve opens.

Fig. 8-36 Ideal pressure-volume cylinder diagram for single acting compressor, showing corresponding piston positions.

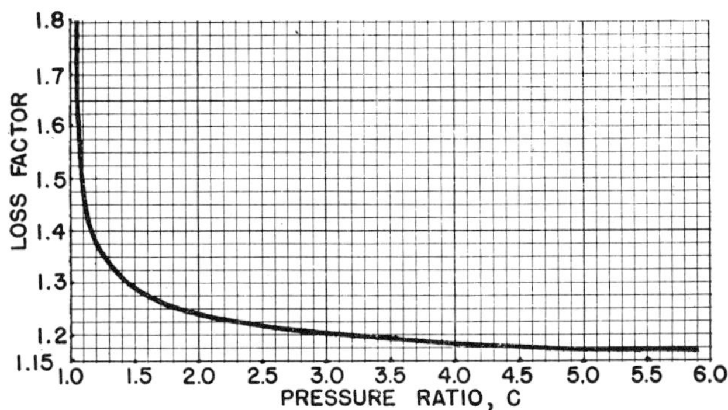

Fig. 8-37 Loss factor curve for reciprocating compressors. (Cooper-Bessemer Corp.)

where: PD_{DA} = piston displacement—double acting, cfm
A_R = area of rod, sq in.
Other terms are defined under Eq. **15**.

The compression ratio is defined as the ratio of the absolute discharge pressure to the absolute suction pressure.

Clearance volume is the volume remaining in the compressor cylinder at the end of the discharge stroke, or at position 3 in the ideal *PV* diagram (Fig. 8-36). It includes the space between the end of the piston and the cylinder head, and the volumes in the valve ports, in the suction valve guards, and in the discharge valve seats.

Per cent clearance or clearance volume is usually expressed as a per cent of piston displacement.

Volumetric efficiency, e, is the ratio of actual cylinder capacity to piston displacement, stated as a percentage. If there were no clearance volume to re-expand and delay the opening of the suction valve, the cylinder could deliver its entire piston displacement as capacity. Effects of clearance volume depend upon the compression ratio and gas characteristics—in this case, its k value. Increases in valve areas reduce valve losses and improve volumetric efficiency.[14]

Actual volumetric efficiency in per cent, e, may be calculated from Eq. 17. The $-C$ term covers losses.

$$e = 100 - C - V_c(C^{1/k} - 1) \quad (17)$$

where: C = compression ratio, P_2/P_1, absolute pressures
V_c = per cent clearance volume, based on piston displacement
k = ratios of specific heats, c_p/c_v

Theoretical horsepower may be calculated from the following formula.*

$$\text{Theoretical hp} = \frac{P_1V_1}{229} \times \frac{k}{k-1}\left[\left(\frac{P_2}{P_1}\right)^{\frac{k-1}{k}} - 1\right] \quad \textbf{(18)}$$

while: Actual hp = theoretical hp × loss factor **(19)**

where: k = ratio of specific heats, c_p/c_v
P_1 = suction pressure, psia
P_2 = discharge pressure, psia
V_1 = suction volume, cfm
loss factor = losses due to pressure drop thru valves and manifold, and to friction of piston rings and rod packing (Fig. 8-37)

Design and Selection

In selecting a reciprocating unit for a given set of conditions, the horsepower and the cylinder volumes are usually determined first. Then the actual capacity and the corresponding actual horsepower load are ascertained.

Horsepower Requirements.

$$\text{bhp} = \textbf{BHP/MMCFD} \times \text{capacity} \times 10^{-6} \quad \textbf{(20)}$$

where:
bhp = brake horsepower required
BHP/MMCFD = brake horsepower required to compress one million cubic feet of gas per day, measured at 14.4 psia and suction temperature. Figures 8-38a thru 8-38c give curves for **BHP/MMCFD** if compression ratios, C, and ratio of specific heats, k, of the gas are known.

Capacity is at the base measuring conditions of 14.4 psia and suction temperature, cu ft per day.

Cylinder Volume Required.

$$e(PD) = \frac{\text{bhp} \times 10^4}{\textbf{BHP/MMCFD} \times P_1} \quad \textbf{(21)}\dagger$$

where:
$e(PD)$ = volume handled, measured at suction pressure and temperature to the cylinder, cfm
PD = compressor cylinder piston displacement, cfm. One manufacturer's data are given in Table 8-12. These values may also be computed directly, using the physical dimensions of the cylinder and the gas speed, as previously explained.
e = volumetric efficiency, per cent/100. This can be found in Figs. 8-39a thru 8-39c for selected k values. The curves were plotted from Eq. **17**. Manufacturers' catalogues tabulate per cent clearances for various classes and diameters of compressor cylinders, along with piston displacements.

* Assumes adiabatic compression.
† Compare with Eq. **47** in Sec. 9, Chap. 1.

P_1 = suction pressure, psia; note qualifications for Eq. **24**.
Other terms are defined under Eq. **20**.

Actual Capacity Handled (combining Eqs. **20** and **21**).

$$\text{Capacity} = 100\, eP_1(PD) \quad \text{(22)}$$

Terms are defined under Eqs. **20** and **21**.

Table 8-12 Nine-Inch Stroke Compressor Cylinders for GMSH Units at 400 rpm

Diam, in.	*PD*, piston displ., cfm S.A.	D.A.	v_c, clearance, per cent	
3	14.72	...	10.5	C6BS*
4	26.18	...	6.8	
5	40.91	78.16	16.1	CED
6	58.89	114.1	11.3	
8	104.7	205.8	10.7	CFD
9	132.5	261.4	8.6	
10	163.6	323.6	16.9	CGD
12	235.6	467.6	7.5	

(Cooper-Bessemer Corp.)
* Values with one inch end clearance. End clearance may vary from 1/16 in. min to 2 1/16 in. max.
S.A. means single acting, D.A. means double acting.
All double-acting values for 1.5 in. diam piston rod.

Actual Horsepower Load (substituting capacity, Eq. **22**, into Eq. **20** for bhp).

$$\text{bhp} = eP_1(PD) \times \textbf{BHP/MMCFD} \times 10^{-4} \quad \text{(23)}$$

Terms are defined under Eqs. **20** and **21**.

Example:‡

Capacity = 10,500,000 SCFD (measured at 60 F + 460 or 520 R and 14.7 psia)
Pressure conditions, psig: suction = 5
discharge = 40
Suction temperature, °R = 90 F + 460 = 550
Isentropic exponent, $k = c_p/c_v$ = 1.25
Altitude, ft = 3000

Find (**1**) *horsepower of engine(s) required;* (**2**) *size, class, and number of compressor cylinders for engine(s);* (**3**) *standard reference capacity.*

1. *Horsepower of engine(s)*

(a) Find compression ratio and discharge temperature. At 3000 ft, atmospheric pressure = 13.14 psia, from Fig. 8-40.

$$P_1 = 5 + 13.1 = 18.1 \text{ psia}$$

$$P_2 = 40 + 13.1 = 53.1 \text{ psia}$$

$$C = 53.1/18.1 = 2.93$$

Check discharge temperature (Eq. **14**) to see that limit of, say, 350 F has not been exceeded. Assuming adiabatic compression:

$$T_2 = T_1\left(\frac{P_2}{P_1}\right)^M = 550 \times 2.93^{\frac{1.25-1}{1.25}} = 683 \text{ R} = 223 \text{ F}$$

Factor $(P_2/P_1)^M$ can be read directly from Fig. 8-42.

(b) From Figs. 8-38a thru 8-38c:
BHP/MMCFD = 64.9 (at C = 2.93 and k = 1.25 in Fig. 8-38b)

(c) Convert *capacity* measured at 14.7 psia and 520 R to the base pressure and temperature of the **BHP/MMCFD** curves (14.4

‡ More examples are given under **Reciprocating Compressors**, Sec. 9, Chap. 1.

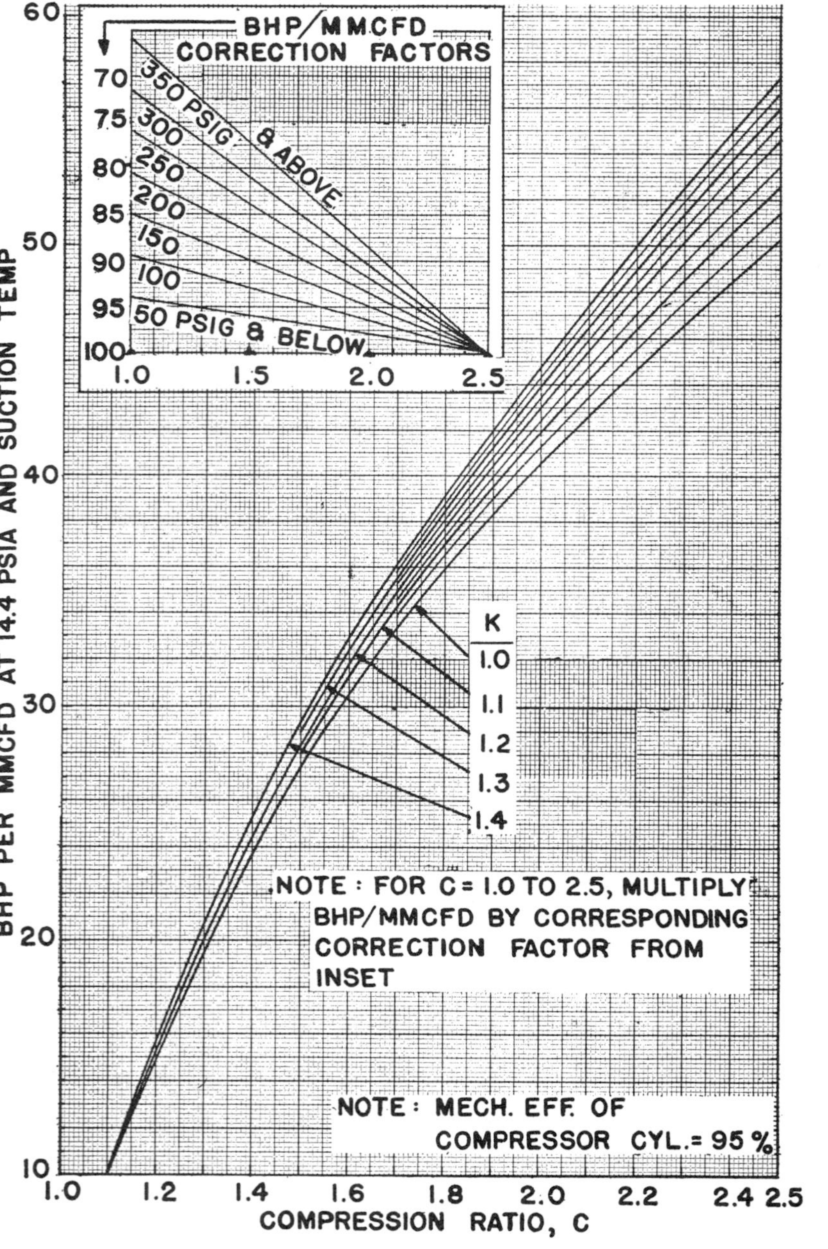

Fig. 8-38a Brake horsepower required to deliver one MMCFD in a reciprocating compressor; correction factor is in terms of inlet pressure. (Cooper-Bessemer Corp.)

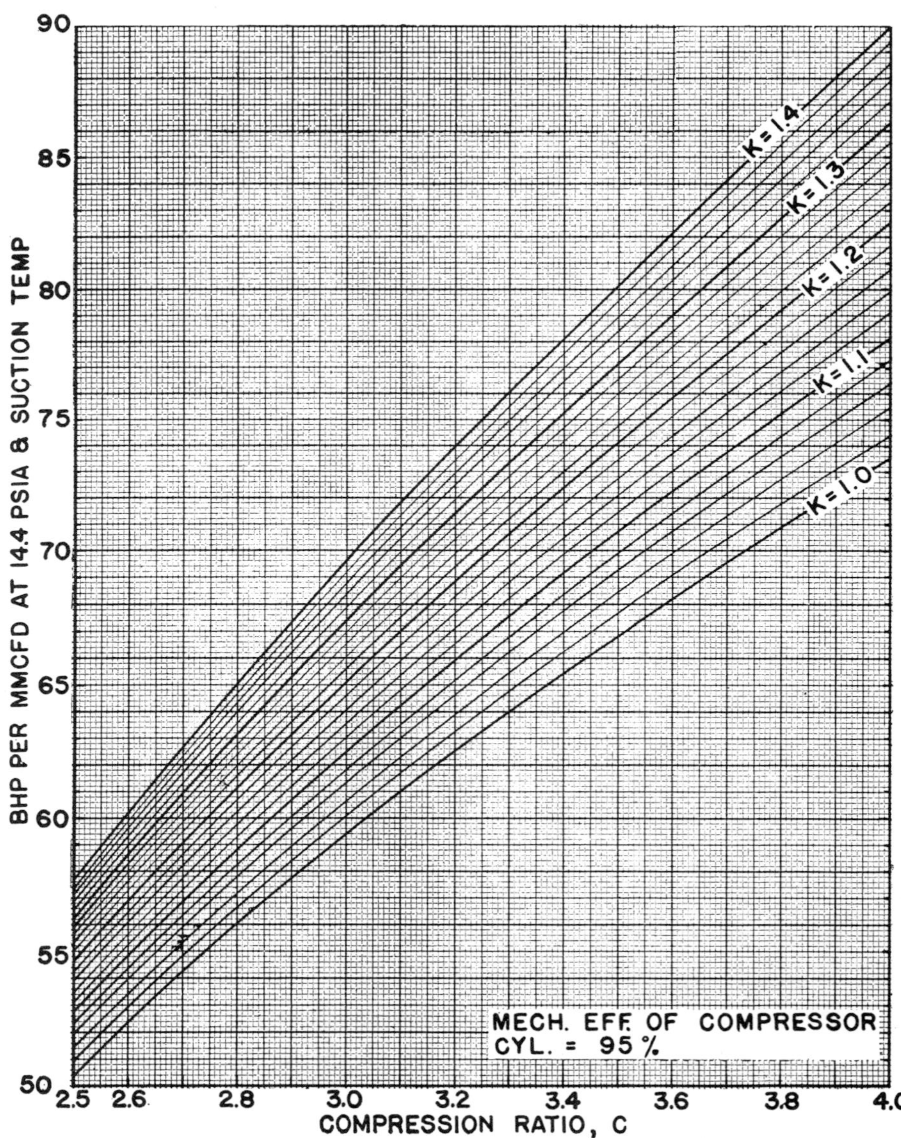

Fig. 8-38b Brake horsepower required to deliver one MMCFD in a reciprocating compressor. (Cooper-Bessemer Corp.)

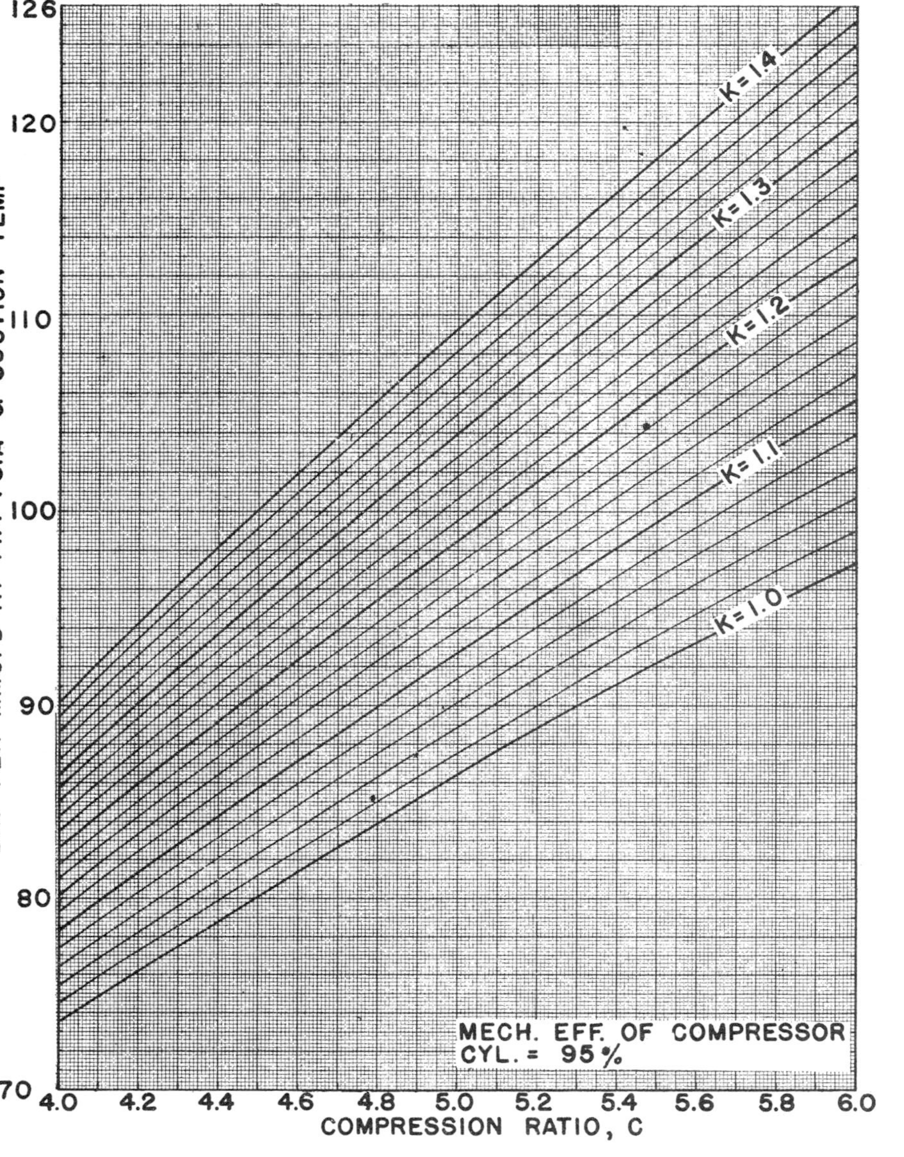

Fig. 8-38c Brake horsepower required to deliver one MMCFD in a reciprocating compressor. (Cooper-Bessemer Corp.)

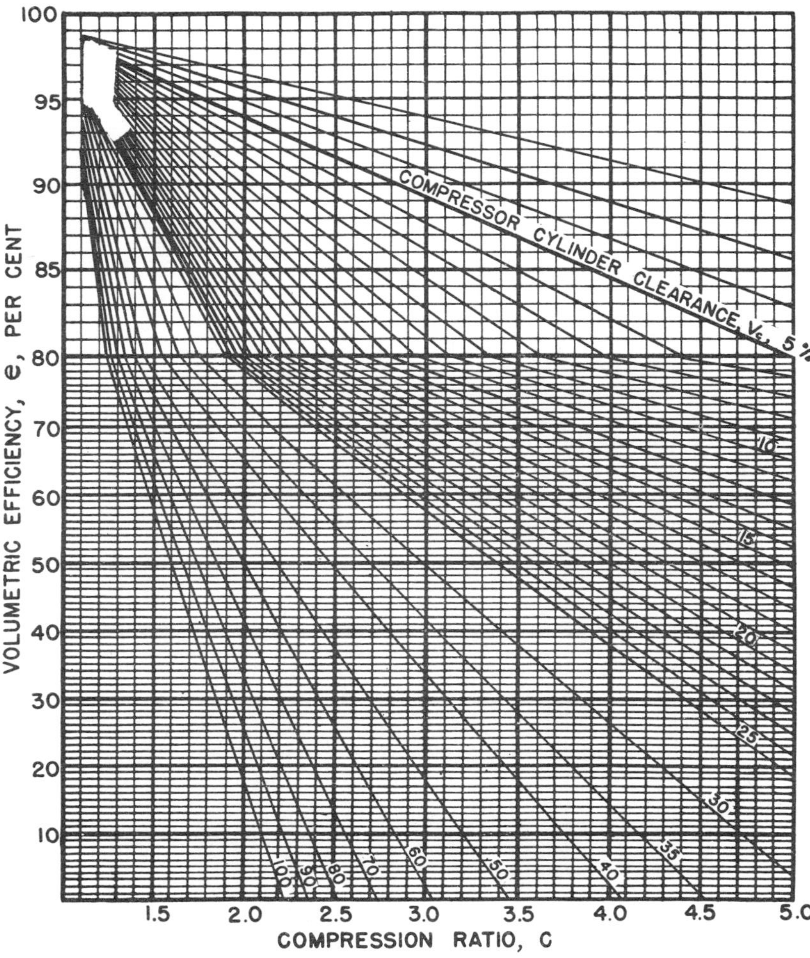

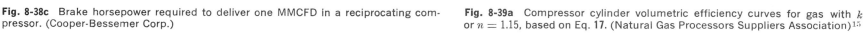

Fig. 8-39a Compressor cylinder volumetric efficiency curves for gas with k or $n = 1.15$, based on Eq. 17. (Natural Gas Processors Suppliers Association)[15]

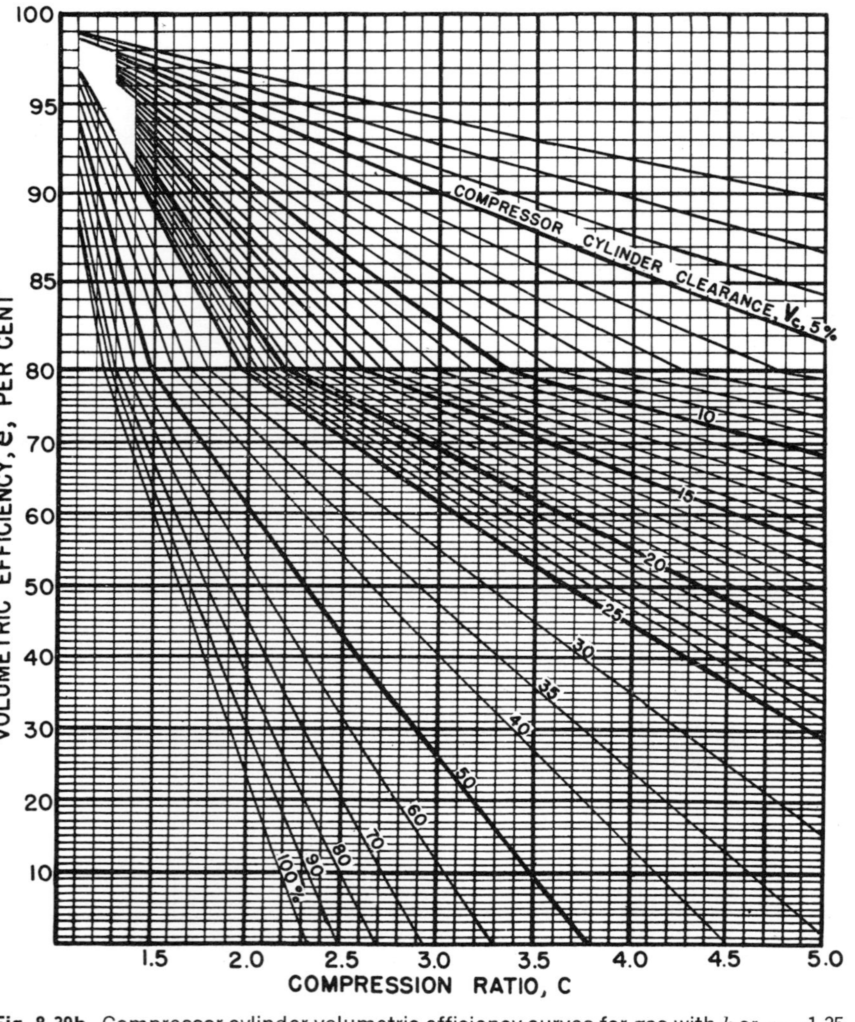

Fig. 8-39b Compressor cylinder volumetric efficiency curves for gas with k or $n = 1.25$.

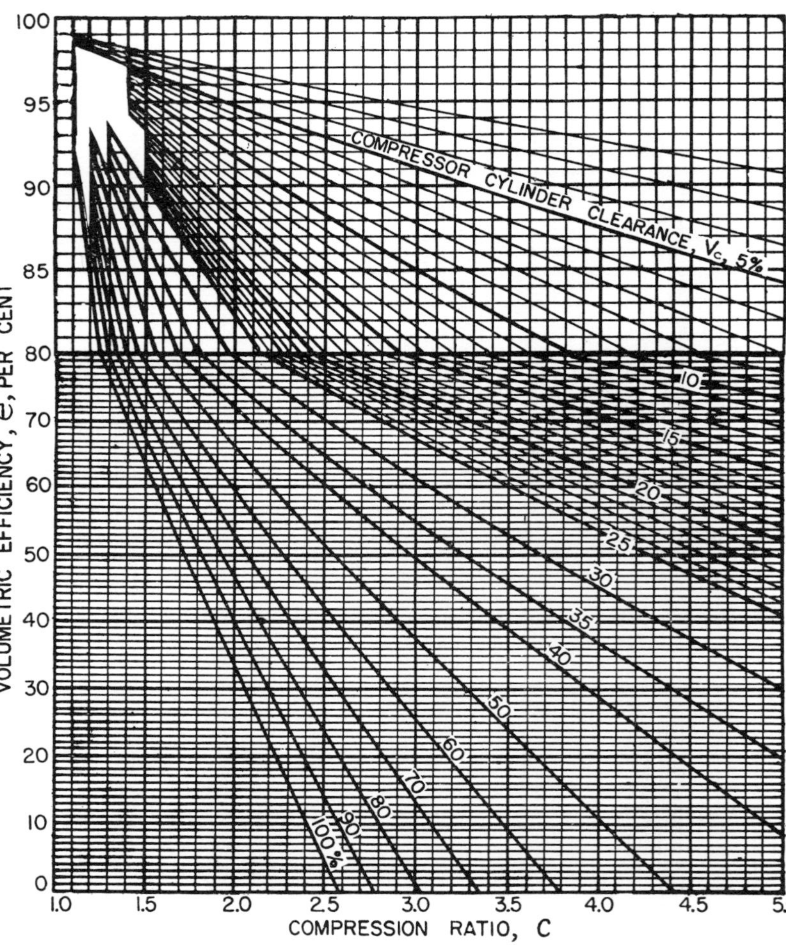

Fig. 8-39c Compressor cylinder volumetric efficiency curves for gas with k or $n = 1.40$.

(Both figures are based on Eq. 17)
(Natural Gas Processors Suppliers Association)[15]

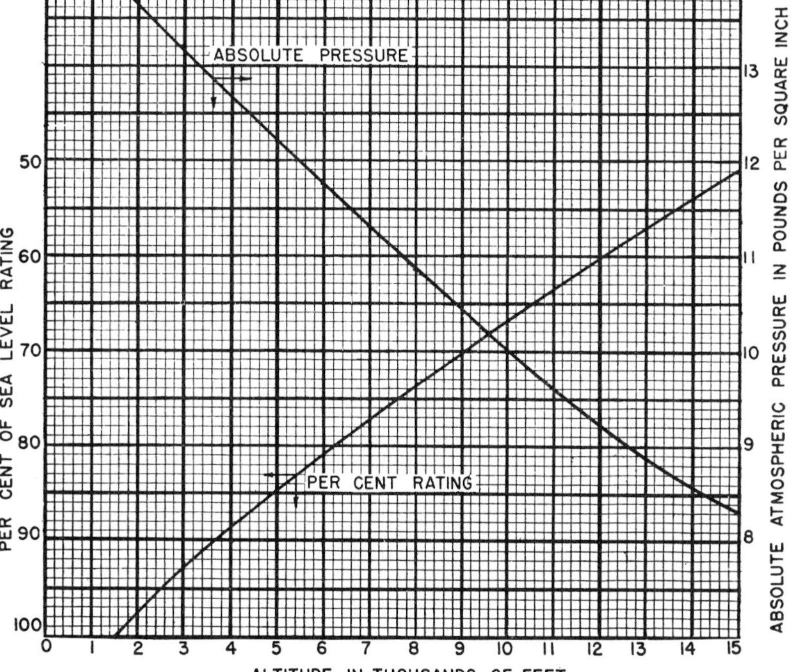

Fig. 8-40 Absolute pressure and gas engine derating vs. altitude. (Gas engine naturally aspirated.)

psia and suction temperature, 550 R in this case) by using Boyle's and Charles' laws:

$$10{,}500{,}000 \times \frac{14.7}{14.4} \times \frac{550}{520} = 11{,}330{,}000 \text{ cfd at}$$

14.4 psia and suction temperature

(d) From Eq. **20**:
Required bhp = **BHP/MMCFD** × *capacity* × 10^{-6} = 64.9 × 11.33 = 735

(e) Thus, required horsepower, accounting for deration,

$$735/0.934^* = 787 \text{ hp}$$

2. *Compressor cylinders*

(a) Find cylinder volume handled, $e(PD)$, that a cylinder or cylinders must provide, by substituting into Eq. **21**. For suction pressure to a stage less than or equal to 10 psig, use Eq. **24**.

$$e(PD) = \frac{\text{bhp} \times 10^4}{\textbf{BHP/MMCFD} \times (P_1 - 0.5)} \tag{24}$$

Terms are defined under Eqs. **20** and **21**.

Thus, $$e(PD) = \frac{735 \times 10^4}{64.9(18.1 - 0.5)} = 6470 \text{ cfm}$$

(b) Select compressor cylinders. For suction pressure to a stage less than or equal to 10 psig *only*, the volumetric efficiency should be determined from the *e* curves, Figs. 8-39a, b, and c, by using a compression ratio, C', calculated from the following equation:

$$C' = \frac{P_2}{P_1 - 0.5} \tag{25}$$

This higher compression ratio, C', is to be used only in the step indicated above, determining volumetric efficiency for cylinders with inlet pressures below 10 psig. The reason is that when gas is drawn into a compressor cylinder, flow losses occur as the gas passes thru the valve passages and thru the valves into the cylinder. These losses cause a pressure drop. With higher suction pressures, this drop is so small that its effect is negligible. With suction pressures below 10 psig, however, capacity may be reduced. The above procedure, therefore, must be followed to insure an adequate cylinder capacity.

* Engine deration factor at 3000 ft, from Curve A of Fig. 8-41.

(c) Assuming that the engine selected has a 14 in. stroke and operates at 300 rpm, the manufacturer's piston displacement tables will indicate which compressor cylinders have suitable diameters and per cent clearances.

If it is further assumed that the largest available cylinder piston displacement, PD, is not sufficient for the required $e(PD)$ = 6470,† two compressor cylinders must be used, providing 6470/2 = 3240 cfm per cylinder. Now a cylinder must be found that will provide the desired $e(PD)$. The approximate size can be determined by noting the per cent clearances for the general piston displacement range in which the cylinder will fall and by checking *e* curves of Figs. 8-39a, b, and c for $C' = 53.1/(18.1 - 0.5) = 3.02$ (from Eq. **25**) and $k = 1.25$ to ascertain the approximate *e*. It can now be determined whether this approximate $e(PD)$ is close to what is desired, bearing in mind that the working pressure required will also determine the cylinder class selected.

(d) Check to determine:

(1) *Maximum allowable working pressure.* For this problem the discharge pressure is well below the allowable pressure of 150 psig for the cylinder selected.

(2) *Whether cylinders selected may be used on adjacent throws.* Reference to manufacturers' Cylinder Combination Charts shows that two cylinders will not fit on adjacent throws. It will, therefore, be necessary to place these cylinders on alternate throws.

(3) *Compressor piston rod loads.* Allowable rod loads for the engine selected are:

Tension = 52,500 psi
Compression = 76,000 psi
Tension load = $T = (P_2 - P_1)A_{HE} - P_2A_R$
Compression load = $C = (P_2 - P_1)A_{HE} + P_1A_R$

where: A_{HE} = head end area, sq in.
A_R = rod area, sq in.
P_2 and P_1 are in psig

Example: Check rod loading for a 32 in. diameter cylinder (A_{HE} = 804 sq in.) with a 3 in. rod (area = 7.07 sq in.).

† Calculated in preceding step 2(a).

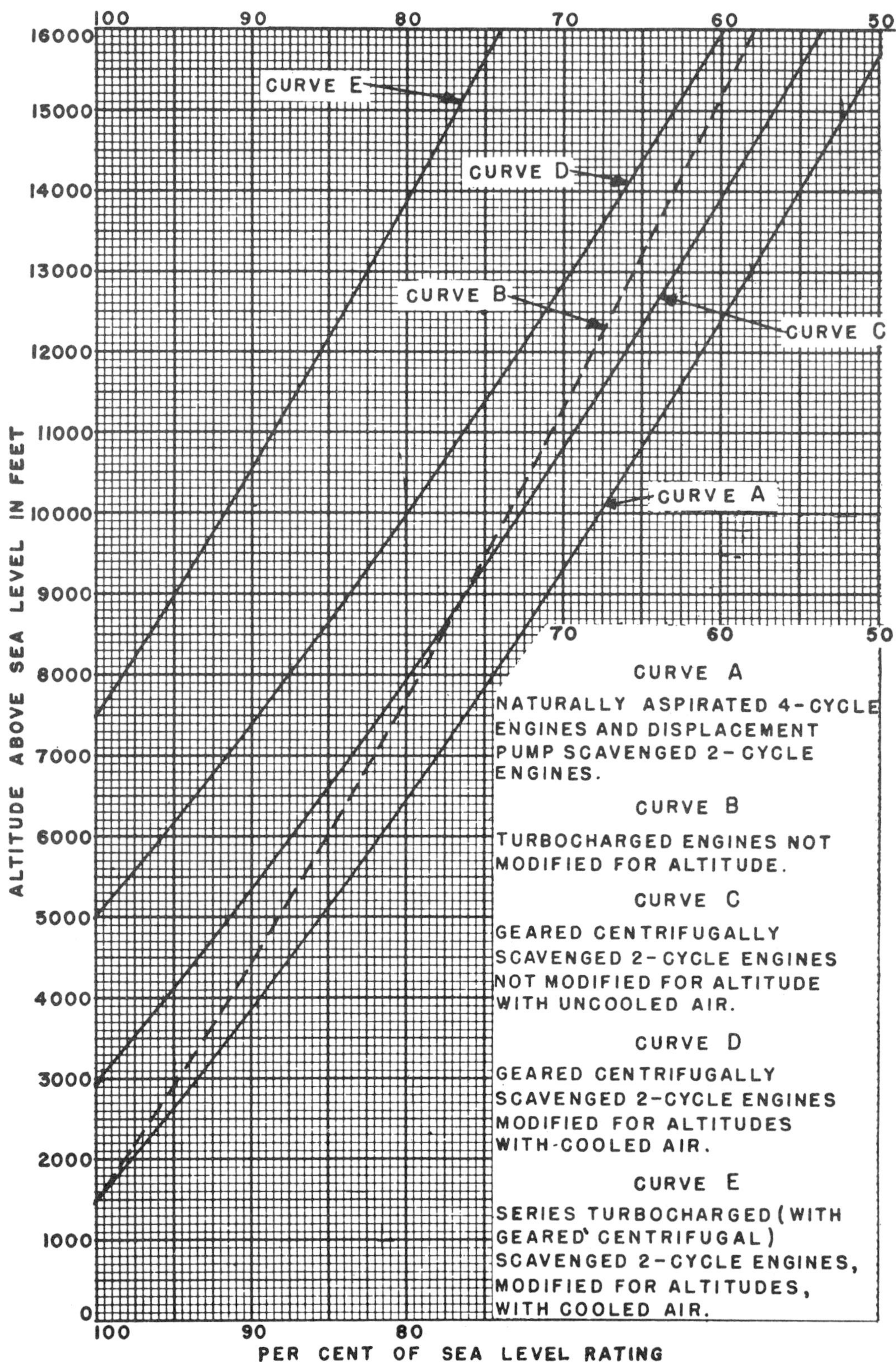

Fig. 8-41 Altitude deration factors for horsepower of 2-cycle compressor engines. (Cooper-Bessemer Corp.)

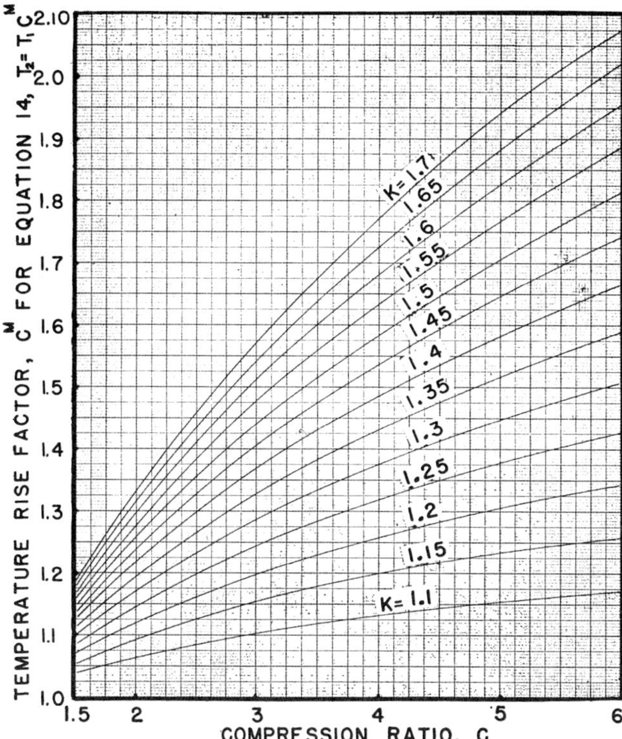

Fig. 8-42 Temperature rise factor curves for calculating adiabatic temperature rise, where $C = P_2/P_1$ and $M = (k - 1)/k$. (Cooper-Bessemer Corp.)

Pressure conditions, psig: $P_2 = 40$
$P_1 = 5$

$T = (35 \times 804) - (40 \times 7.07) = 27{,}800$ psi
Acceptable, since $27{,}800 < 52{,}500$

$C = (35 \times 804) + (5 \times 7.07) = 28{,}100$ psi
Acceptable, since $28{,}100 < 76{,}000$

(4) *From Eq.* **24**, *actual horsepower with the cylinders selected.*

bhp = $e(PD) \times$ **BHP/MMCFD** $\times 10^{-4} \times (P_1 - 0.5) \times$ number of cylinders

bhp = $0.846 \times 3980 \times 64.9 \times 10^{-4} \times (18.1 - 0.5) \times 2 =$ 769 at 2.9 per cent overload

Therefore, the unit required to meet the given specifications will be an 800 hp engine with two 32 in. diam $\times$ 4 in. stroke compressor cylinders.

3. *Standard reference capacity* is found by solving Eq. **20** for *capacity* and is based on rated horsepower of unit at the altitude at which it is to operate.

Standard reference capacity =

$$\frac{\text{hp} \times 10^6}{\textbf{BHP/MMCFD}} = \frac{787^* \times 10^6}{64.9} =$$

12.1 MMCFD at 14.4 psia and suction temperature (90 F)

Convert to 14.7 psia and 60 F.

Standard reference capacity =

$$12.1 \times \frac{14.4}{14.7} \times \frac{520}{550} = 11.2 \text{ MMCFD}$$

Use of Mollier Diagram. A rapid way to solve a compressor problem is by the use of a Mollier (CH_4: Fig. 8-43, N_2: Fig. 8-44) or pressure-enthalpy diagram (CH_4: Fig. 1-3a). All compressor problems for which suitable Mollier charts exist should be worked in this manner.

Solution. As in other compressor problems, the horsepower requirements must first be determined so that the size and type of engine can be chosen. Horsepower required can be found from the basic relationship:

$$\text{Work done} = h_2 - h_1$$

where: h_2 and h_1 are the gas enthalpies at discharge and suction, respectively, Btu per lb

Convert this to horsepower based on pounds per minute flow and apply loss factor curve (Fig. 8-37), which makes allowance for losses due to pressure drop thru the compressor valves and manifold, and for losses due to friction of piston rings and rod packing. Thus, the equation becomes:

$$\text{Required bhp} = \frac{(\text{lb/min})(h_2 - h_1)(\text{Loss Factor})}{42.4} \quad \textbf{(26)}$$

If the desired capacity is given in cfm, this cfm must be converted from the measuring conditions to the suction conditions for the cylinder being sized. This will give the cfm at suction, or the $e(PD)$, that the cylinder(s) must provide.

Quite often the capacity will be given in pounds per minute. In this case, multiply the flow by the specific volume, v_s, in cu ft per lb that the gas occupies at *suction* conditions to the cylinder, i.e., cfm = lb per min $\times v_s$. Thus,

$$e(PD) = \text{lb per min} \times v_s \quad \textbf{(27)}$$

where:

e = volumetric efficiency, per cent/100, found by substituting v_s/v_d for $C^{1/k}$ in Eq. **17**.
PD = compressor cylinder piston displacement, cfm
v_s = specific volume of gas at suction, cu ft per lb
v_d = specific volume of gas at discharge, cu ft per lb

A cylinder is then selected that will provide the required $e(PD)$ and checked to see if the selection is satisfactory (see previous example).

Unloading Problems

Every compressor unit with a given set of compressor cylinders has a characteristic horsepower curve. Over a wide range of pressure conditions, this curve will rise, peak, and then fall. It is thus necessary, when varying pressure conditions exist, to minimize engine overloading and still take full advantage of the available horsepower.

Note in Eq. **23** that there are three variables, e, P_1, and **BHP/MMCFD,** affecting the bhp curve. The volumetric efficiency, e, is the one most capable of being controlled. This is accomplished by adding or subtracting per cent clearance volume, V_c, in Eq. **17**. The amount of clearance can be adjusted in any compressor cylinder by using unloading devices consisting of head end unloader pockets or bottles, variable piston type head end unloaders, and, when used in conjunction with double deck valves, valve cap unloaders. In the case of single-acting cylinders, end clearance can usually be adjusted by screwing the piston rod further into the crosshead. Suction valve lifters, which make entire cylinder ends ineffective, may also be used. They do not change clearance but unload by reducing effective piston displacement.

---
* From step 1(e).

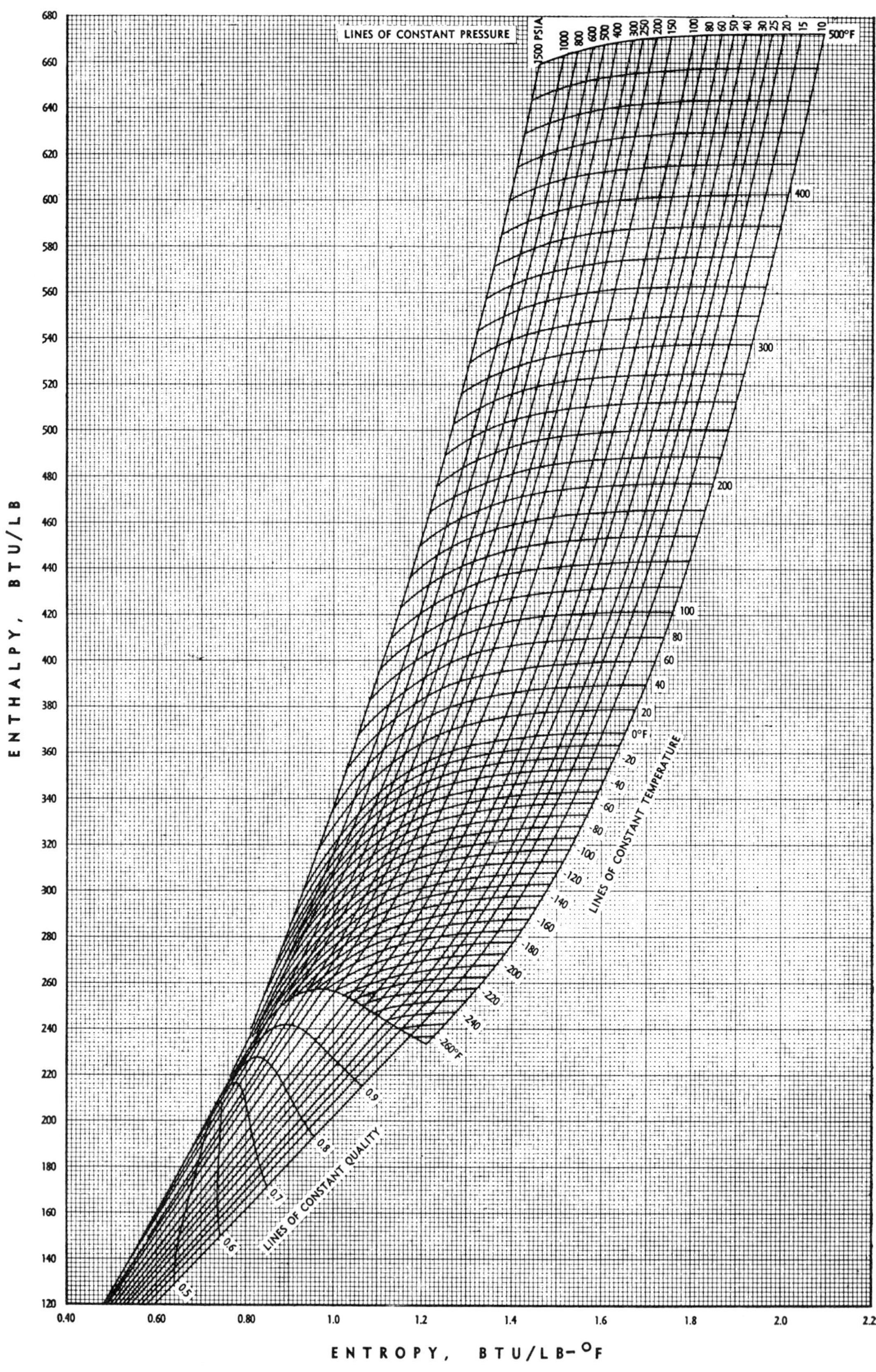

Fig. 8-43 Mollier chart for methane.[16]

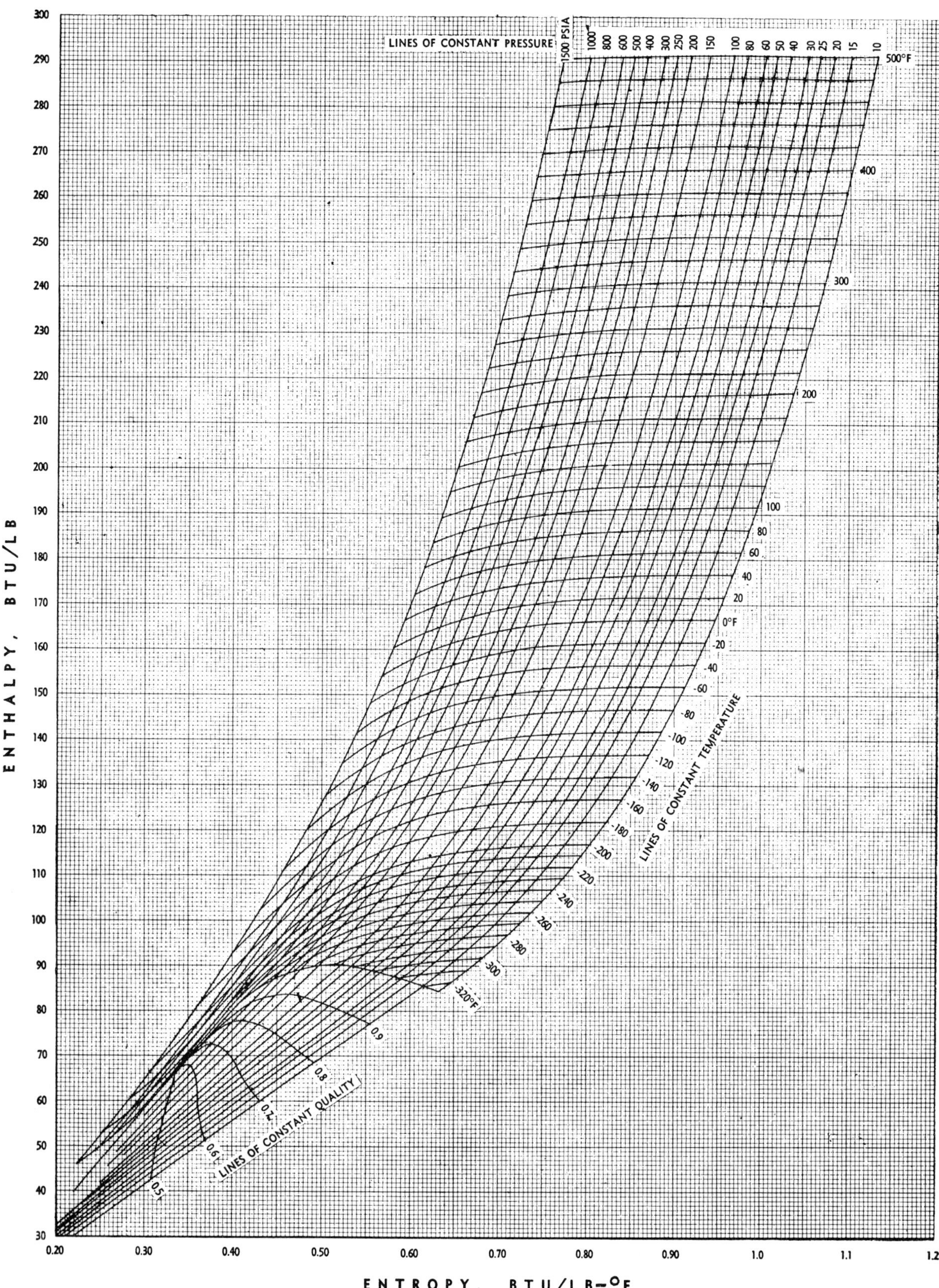

Fig. 8-44 Mollier chart for nitrogen.[16]

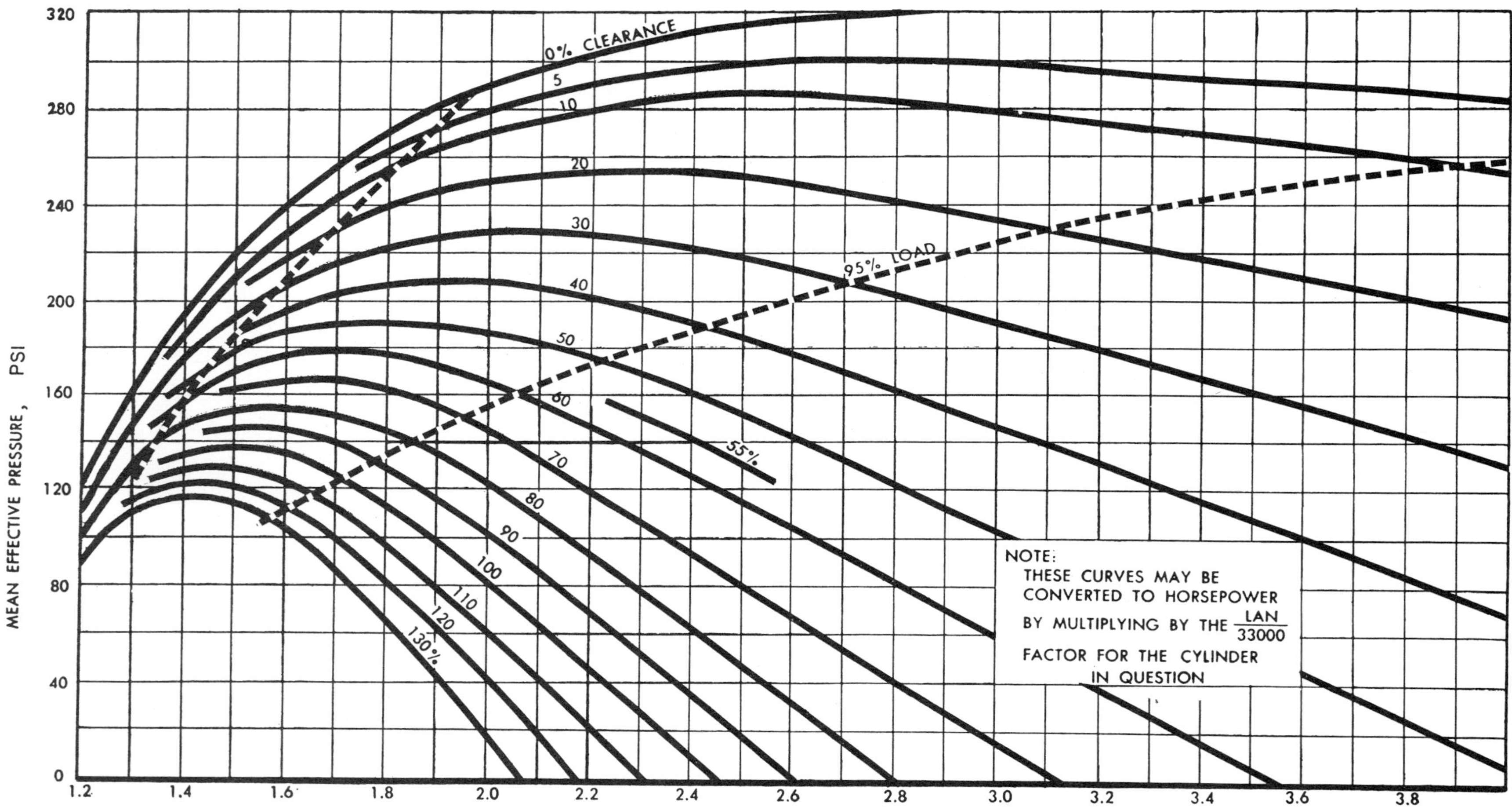

Fig. 8-45a Mean effective pressure characteristic curves for a reciprocating compressor with variable suction and constant discharge pressure.[17] (760 psig discharge, 0.6 sp gr, $n = 1.28$)

$$\text{net ihp} = \left(\frac{LAN}{33{,}000}\right)\left(\frac{n}{n-1}\right)P_s\left[C^{(n-1)/n} - 1 - M(1 + C - C^{1/n} - C^{(n-1)/n})\right]$$

(L, A, and N defined in Eq. 28)

Note: The unlabeled dotted line in the figure represents the 105% load.

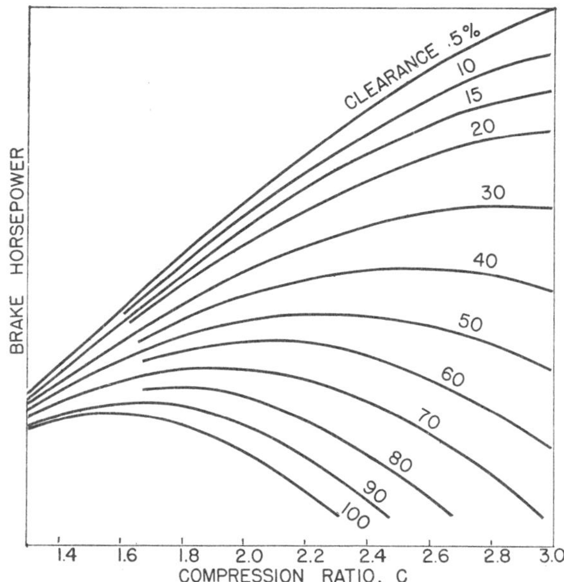

Fig. 8-45b Horsepower characteristic curves for variable discharge pressure and constant suction pressure. Note: $k = 1.3$; at lower compression ratios, mean correction factor at 200 psia suction pressure is used, see Fig. 8-38a. (Cooper-Bessemer Corp.)

Figure 8-45a shows characteristic mean effective pressure curves for different per cent clearances when discharge pressure is constant and suction pressure varies. For a particular piston, engine speed, and stroke, horsepower depends only upon mean effective pressure (Eq. 28). Figure 8-45b covers constant suction and varying discharge pressures.

These curves are useful in solving compressor problems. When varying suction or discharge pressures are first encountered, selecting the cylinder sizing point becomes a problem. The point selected must be one at which the unit will load up rather than become underloaded as pressure conditions change. Knowing the normal per cent clearance for most cylinders, the characteristic horsepower, or the mean effective pressure vs. per cent clearance curves for the ratios involved, yields an indication of the slope of the horsepower curve. Thus, the sizing point can be selected.

Note that for each per cent clearance there is a range of ratios thru which the curve is comparatively flat. In many cases where the desired compression ratios fall into one of these ranges, a fixed per cent clearance can be selected, providing a relatively flat horsepower curve thru these ratios. Thus, a minimum amount of unloading would be required, often none.

Frequently, it is not possible to operate entirely in the flat region of a horsepower curve. For example, conditions may fall in a range where all the characteristic curves slope in the same direction, or the range of the conditions may be of such a wide extent that a single curve would not span them without excessive overloading or underloading. In other instances, the required amount of fixed clearance would be prohibitive.

In these cases the horsepower or mean effective pressure curve must be adjusted and kept between the usual operating limits of 3 per cent overload and 5 per cent underload. This can be done by changing the per cent clearance in the cylinder, thru opening or closing unloading devices. For example, note the 95 and 105 per cent load lines in Fig. 8-45a.

Occasionally only one unloading step is necessary. Often a series of steps is required to keep the horsepower within desirable operating limits. Most unloading problems fall into one of these categories.

PRIME MOVERS[8]

Since centrifugal compressors are fundamentally high speed machines and reciprocating compressors are low speed machines, certain prime movers are more suited to one than to the other. Certain types, like reciprocating steam engines, are rarely used for the newer gas compression stations, and are not covered here. Comparisons are limited to 1960 designs of prime movers.

Electric Motors

The two characteristics limiting the general application of electric motors are their limited maximum speed and the high cost of variable speed drive, if that feature is desired. However, the governing considerations in using electric motors for pipeline service are almost always the cost and availability of electric power.

Initial investment is usually lower than for other types of drivers, if it is not necessary to build a long electric transmission line to provide station power. Motors lend themselves readily to remote or automatic control and require little or no attendance during operation, thus reducing labor costs to a minimum. Maintenance costs are also low.

Expense of variable speed drive is a decided disadvantage of electric motors from the standpoint of flexibility of centrifugal compressor operation. Adjustable guide vanes in the compressor inlet overcome this deficiency in part, but they do not give the broad range of efficient operation afforded by variable speed. Because the maximum speed of motors on 60 cycle current is 3600 rpm, a further limitation is imposed unless speed increasing gears are used. The majority of recently built centrifugal compressors operate at higher speeds.

Motor limitations of maximum speed and high cost of variable speed drives pose no particular problem for reciprocating compressors built to operate at a constant speed in the 250 to 400 rpm range. Electric motor drive for reciprocating compressors in the larger sizes is not widely used, however, because of competition from gas engine driven units.

Steam Turbines

Steam turbines dominate the field as prime movers in applications requiring 10,000 hp units and over. They are used so frequently because their boilers can burn any type of fuel—solid, liquid, or gaseous—utilizing the one that is economical and plentiful in a certain area. At present engines and gas turbines are restricted to liquid or gas fuels. For gas pipeline service, the ability to burn solid fuels is no longer an advantage, since natural gas is the most economical fuel available in almost all cases.

Any comparison between the utility of steam turbines and that of other prime movers for gas pipeline service must be made on the basis of the size of unit. Steam turbines are built in sizes up to 250,000 hp and larger and plant efficiencies vary over a wide range. To provide adequate flexibility and reliability in compression service on larger pipelines, unit prime mover size should be limited to approximately 5000 hp.

Consideration of steam turbines for gas pipeline service should include the following:

1. High rotary speed, which may be varied according to demands, makes steam turbines very well suited for direct connection to centrifugal compressors.
2. Fuel consumption of units in sizes suitable for gas pipelines is relatively high compared with gas engines and about the same as for gas turbines. Specific fuel consumption at partial load is reasonably good compared with full load.
3. Although steam turbines themselves are essentially simple and reliable, the large amount of auxiliary equipment necessary makes their installation, maintenance, and operating labor costs approximately equal to those of gas engine stations.
4. A plentiful supply of water for make-up and cooling is necessary.

Gas Turbines

Gas turbines lend themselves readily to gas pipeline centrifugal compressors for the following reasons:

1. Units of the two-shaft design have an adequate speed range to meet varying requirements.
2. Operating speeds make direct connection to centrifugal compressors practicable.
3. During peak gas demands[18,19] in winter, substantial overloads can be carried because ambient temperatures are low.
4. Cooling water requirements are low.
5. Operating labor costs and expected maintenance costs are moderate.
6. Lubricating oil consumption is negligible.
7. There are few auxiliaries; hence, little auxiliary power is required.

Disadvantages of the gas turbine as a prime mover for gas pipeline compressors include its high rate of fuel consumption at rated load and rather steep fuel consumption curve (very poor efficiencies at partial loads). Full load thermal efficiency of simple cycle units was approximately 18 per cent and fuel consumption was 14,150 Btu per bhp-hr. For regenerative cycle units, thermal efficiency is 27.5 per cent max at the optimum rating point, and fuel consumption is 9250 Btu per bhp-hr.

Although the inverse relationship between turbine capacity and ambient temperature is advantageous at low temperatures, at higher temperatures this characteristic decreases load carrying ability approximately 5 to 7 per cent for each 10 F above the rated temperature of 80 F. This deficiency can be offset in part by precooling the intake air, but such cooling requires considerable water evaporation and additional equipment.

Jet Engine.[20] A 10,500 hp Cooper-Bessemer Corp. RT-248 gas turbine (incorporating a Pratt & Whitney Aircraft J-57 jet engine for its energy source) was installed on a Columbia Gulf Transmission Co. line near Clementsville, Ky., in 1960. Installation data: pumping 600,000 MMSCFD from 750 to 950 psig, weighing 85 tons, jet compression ratio—10, no regeneration, 4000 lb jet engine section requiring less than 4 hr combined removal and replacement time between overhauls at estimated 8000 hr intervals (equivalent to a year's continuous operation). The following are performance data* at 1000 ft altitude and 80 F inlet temperature covering the first 6375 operational hours:[21]

Rated load, per cent	Thermal eff, %	Fuel consumption, MBtu per bhp-hr
100	25	10.2
80	22.8	11.2
60	19.5	13.1
40	16.1 (est.)	15.8 (est.)
30	14 (est.)	18.1 (est.)

Gas Engines

The reciprocating internal combustion gas engine is the leading prime mover for gas pipeline service, and for the past 30 years it has remained the most efficient independent driver for compressors. The most popular designs are composite gas engine reciprocating compressor units, like the angle machines.

Size and general design of recent gas engines are significantly different from those of early horizontal units. More important, their horsepower range has been increased and efficiency greatly improved. Gas engines are available (1961) from very small sizes (below 100 hp) to approximately 5500 hp. Guaranteed thermal efficiencies of some gas engines exceed 40 per cent.

Gas engine designs vary to such an extent that it is not possible to consider them as a group in comparing them with other prime movers. Their range of sizes, speed characteristics, and efficiencies is such that they may be used to drive either reciprocating or centrifugal compressors, while gas or steam turbines definitely lend themselves best to centrifugal types.

Although all gas engines are basically reciprocating machines with rotary crankshafts, they have several major differences. Perhaps the most fundamental is their cycle of operation—*either two or four strokes*. Neither type has any clear-cut overall advantage and years of competition among engine builders have failed to eliminate either. However, the following generalizations can be made:

1. *Specific fuel consumption* is about the same for the two basic types except that some four-cycle engines have been successfully built with higher expansion ratios and correspondingly lower fuel consumption.
2. *Space requirement* for a unit of a given horsepower, in general, depends upon ratio of horsepower to piston displacement. The two-cycle engine has twice as many power strokes and thus, half the brake mean effective pressure, bmep, of a four-cycle engine running at the same speed, in order to be equivalent in horsepower-to-displacement ratio. Since the application of turbochargers to engines, bmep ratings of the two engine types have increased appreciably; bmep approaches 100 and 200 for 2- and 4-cycle engines, respectively. Therefore, the long-held advantage of the two-cycle engine in

* Based on the lower heating value of gas fuel—920 Btu per cu ft (higher heating value = 1020 Btu per cu ft).

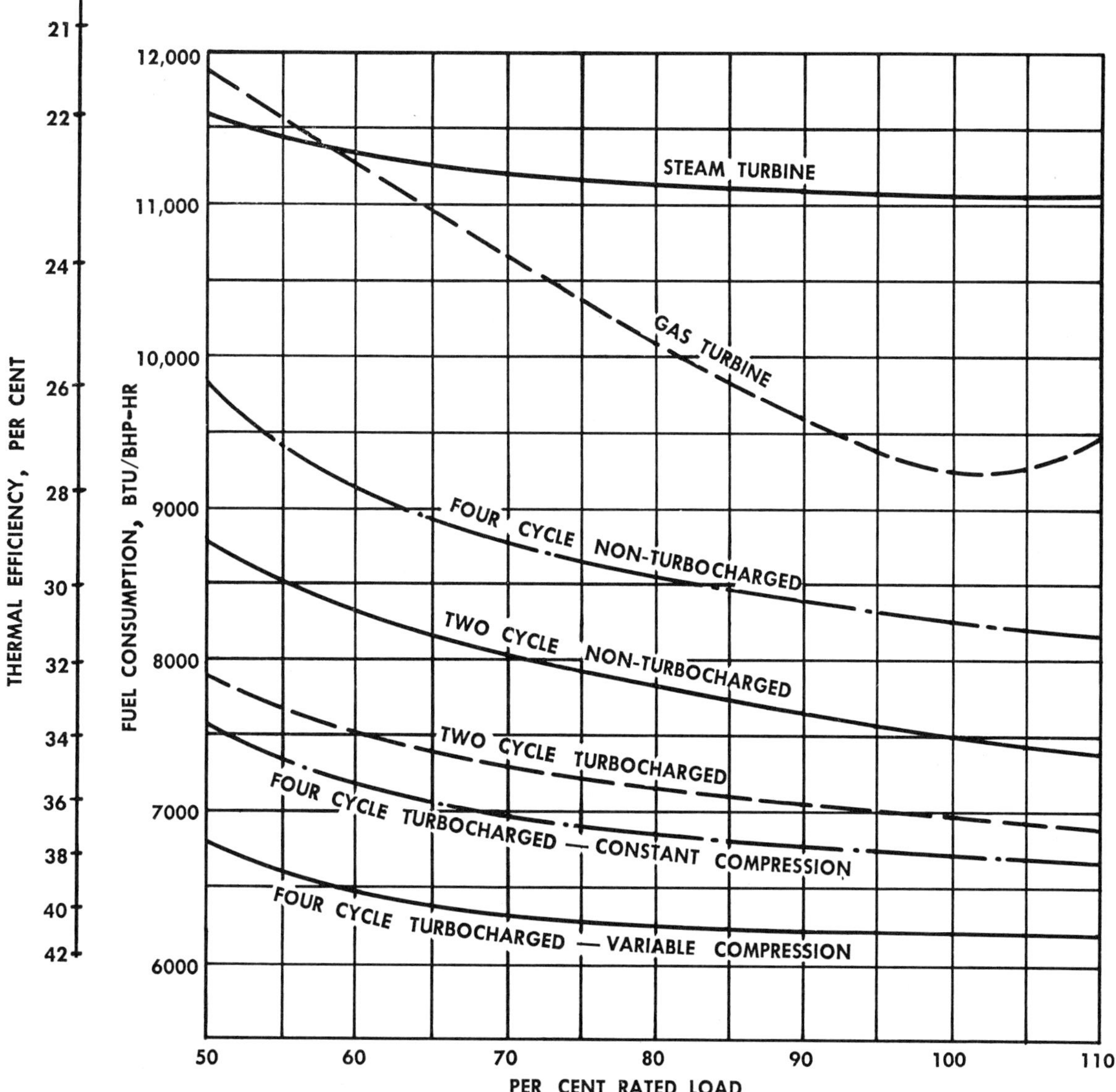

Fig. 8-46 Fuel consumption and thermal efficiency of prime movers. (Nordberg Mfg. Co.)[8]

compactness and low space requirements has largely been nullified.

3. *Lubricating oil consumption* for two-cycle engines is approximately 1 gal per 4000 to 8000 bhp hr. For four-cycle engines, it is about 1 gal per 6000 to as high as 20,000 bhp hr. Considerable evidence indicates that lubricating oil consumption remains practically constant for a four-cycle engine regardless of its bmep rating. Therefore, assuming good piston ring design and normal temperatures, bhp hours per gallon increase as the bmep rating is raised. Lubricating oil consumption is greater in two-cycle engines because of location of exhaust and intake ports and because of hotter cylinder surfaces.

4. *Maintenance*—the two-cycle engine with port scavenging has fewer moving parts because it does not require inlet and exhaust valves. On the other hand, certain major parts of a two-cycle port scavenged engine are subjected to more severe operating conditions.

Horsepower ratings of engines depend mainly upon the amount of fuel which can be burned. Because a rather definite air-fuel ratio is required, a very important design factor is the method of supplying air for combustion. In naturally aspirated four-cycle engines, the pressure created by the piston forces out the burned exhaust gases. Weight of air drawn in is limited because of burned exhaust gases in the clearance volume. Supercharged four-cycle engines have either a mechanically driven blower or, more generally, an exhaust driven turbocharger. They supply air at pressures up to 30 psig and thus increase the weight of air in a given cylinder and, with valve overlap, exhaust the burned gases normally left in the clearance volume. Two-cycle engines invariably have pumps or blowers for scavenging air because the exhaust and intake strokes are combined. Recently, exhaust turbochargers have also been applied to two-cycle engines to increase the amount of combustion air and improve efficiency. Capacity gains of as much as 20 per cent were anticipated.[14]

Fuel consumption and thermal efficiency data for prime movers are given in Fig. 8-46.

Maintenance and operating labor costs for a gas engine approximate those for a steam turbine installation, but exceed those for a gas turbine. These costs, along with relatively high lubricating oil consumption and substantial cooling requirements, comprise the major disadvantages of the gas engine. Its limitation in maximum speed is readily overcome by use of speed-increasing gears when driving a centrifugal compressor.

ENGINE AND COMPRESSOR TESTS

Use of Indicators[22]

An engine indicator records the pressure in a cylinder as a function of piston travel, time, or some other variable. It is a means of measuring work done in the cylinder, of gaining information on the timing of various events in the engine or compressor cycle, and of discovering faulty adjustments and malfunctional design.

The several available types of indicators include:

1. MIT high speed, balanced pressure type (American Instrument Co., Inc., Silver Spring, Md.) is the most accurate instrument available for determining horsepower.[42]
2. *Maihak* standard, of the low speed (under 250 rpm) pencil and drum type.
3. *Maihak* high speed (over 250 rpm), with lightweight pencil motion and stiffer springs.
4. Sampling valve, for high speed work.
5. Electronic, with a catenary diaphragm pressure pickup or cathode ray oscilloscope.

Selection. Consider three items:

1. *Ultimate result desired.*
   a. Are cards for calculation purposes?
   b. Is the entire cycle or only a portion to be covered?
   c. Are data for discovering faulty operation or for engine balancing?
2. *Is the speed of the equipment to be tested slow, medium, or high?*
3. *Is the cycle to be studied recurring or non-recurring?*

Balancing Engine Cylinders. One of the widest uses of an indicator card, particularly in the field, is to balance the engine load among the cylinders. For many years this load balancing was done by means of exhaust temperature. This method is still used in some instances to provide a good criterion of performance.

Prevalent methods for balancing the load among cylinders on an engine are: (1) by means of peak pressures, (2) by measuring time-averaged cylinder pressures, and (3) by determining indicated horsepower (measured in terms of mean effective engine pressure). The first two methods are faster and require less effort but more care in both reading and interpretation than the third method.

Two types of indicators may be used: (1) the pencil drum type, either low speed or high speed, and (2) the peak pressure indicator, giving the pressure reading on a dial or gage.

Record peak pressures of all cylinders consecutively on one card by attaching the indicator to each cylinder in succession, pressing its pencil lightly on the paper-covered drum for a few strokes, and moving the drum slightly by pulling its cord so that the cards are not drawn one on the other. Either before or after taking the peaks, with the indicator *not* on the engine, draw the atmospheric line. These cards may then resemble Fig. 8-47a. Now take compression pressures of each

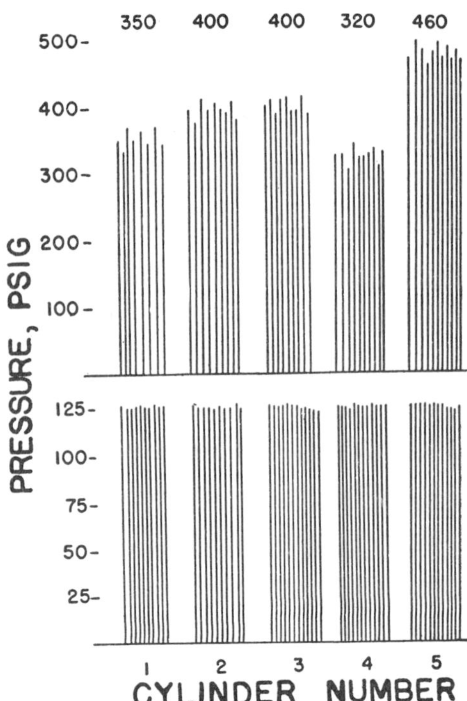

Fig. 8-47a (top) Firing pressure pull cards.

Fig. 8-47b (bottom) Compression pressure pull cards.[23] (Cooper-Bessemer Corp.)

cylinder to aid in balancing the engine. To take these pressures, the load on the engine should be reduced, so that when one cylinder is cut out the remainder will carry the load at running speed. The compression card is taken in the same manner as the firing card, except either the fuel or the ignition is cut off from the cylinder while the pressures are being checked. In this way, a card like that shown in Fig. 8-47b is obtained.

Interpretation of Figs. 8-47a and 8-47b follows: On the compression card (Fig. 8-47b) compression pressures are the same; therefore compression has no bearing on the firing balance. Figure 8-47a shows that cylinders 2 and 3 are about normal, while cylinders 1 and 4 are low (not carrying their share of the load), and 5 is quite high (overloaded).

If the card (Fig. 8-47a) were taken on a gas engine, further interpretation might show faults in any of the following:

1. Individual gas adjustment
2. Ignition timing
3. Fuel injection valve timing
4. Valve timing

After making necessary corrections a second card should be taken to see that the peak pressures are now more nearly even.

This method of balancing cylinder loads on a multiple cylinder engine is widely used. However, it does not always provide sure proof that the engine is in balance, because all the peak pressures may be equal, while loads on each individual cylinder may vary.

The *second method* of balancing cylinders on a multiple cylinder engine is by taking complete indicator cards and calculating the mean effective pressure, mep, and indicated horsepower, ihp, of each cylinder, then making adjustments until the ihp of the cylinders is the same. This procedure requires considerably more equipment and labor, but the results show the load distribution and also give necessary information for diagnosing what should be done to put the engine in balance.

A *peak pressure indicator* may be used in a somewhat similar manner, to obtain cylinder balance. With this device, the operator observes and records both peak firing pressures and peak compression pressures for all cylinders. After making a comparison, the proper adjustments or repairs are made, to bring each of these pressures to a uniform value.

Measuring Mean Effective Engine Pressure. Figure 8-48 is a card taken with a balanced pressure indicator on a two-cycle gas engine of moderate speed. From it, peak pressure, terminal pressure at exhaust port opening, pressure when scavenging ports open, and indicated horsepower can be measured.

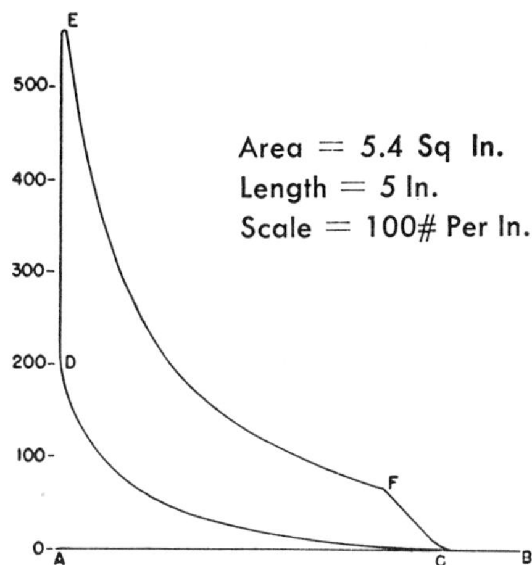

Fig. 8-48 Two-cycle gas engine indicator card.[23] (Cooper-Bessemer Corp.)

To calculate indicated horsepower, the mean effective pressure, mep, in the cylinder is found by measuring the indicator card area with a planimeter, then dividing it by the length of the card and multiplying the quotient by the pressure scale. This resulting mep is the constant pressure acting throughout the stroke, which would produce the same work as the variable pressure shown by the indicator card. The following equation is used for determining the indicated horsepower, ihp, developed in the cylinder of single acting engines and compressors, or in each cylinder end of double acting engines or compressors.

$$\text{ihp} = \frac{PLAN}{33{,}000} \tag{28}$$

where:

P or mep = mean effective pressure, psig
L = piston stroke length, feet
A = piston area, square inches
N = number of power strokes per minute. For a two-cycle engine this is the same as the engine rpm. For a four-cycle engine it is rpm/2.

The total indicated horsepower of the double acting cylinder is the sum of the horsepower developed in the two ends. That of a multicylinder engine is the sum of the horsepower developed in all cylinders.

Interpreting Indicator Cards. For diagnosis of gas engine trouble Fig. 8-49 shows indicator card shapes on a four-cycle engine.

No. 1 of Fig. 8-49 shows what is called a perfect card. Others should be like it if design and settings are correct. On No. 3, timing was late and the spark did not occur until after top dead center. No. 7 shows a premature burning caused by the gas mixture being ignited too early, possibly because of hot spots in the cylinder chamber or mistiming of the spark, the latter occurring very early. Nos. 6, 8 and 9 show toe cards as they would appear with correct, early, and late exhaust valve timing, respectively.

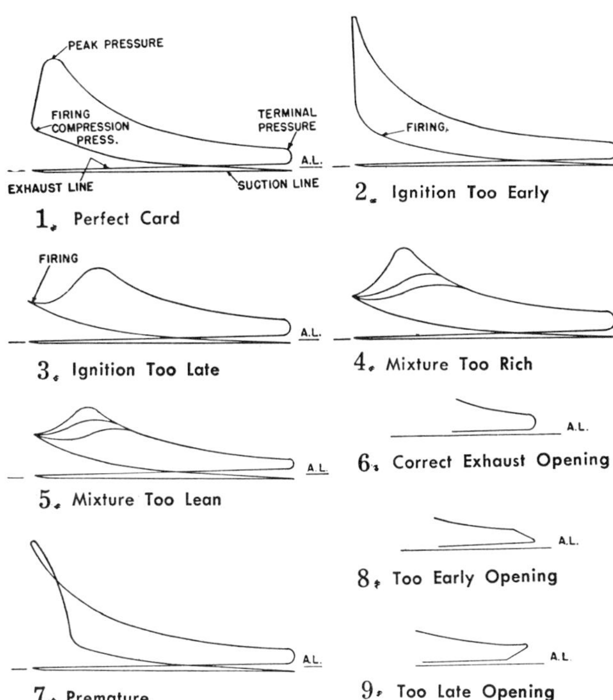

Fig. 8-49 Malfunctioning of four-cycle gas engine. (Cooper-Bessemer Corp.)[22]

Ported engines. The timing of the intake and exhaust are fixed by design and have no adjustment. The operator can, by means of an indicator card, determine if they are functioning properly, because any effects of back pressure, exhaust pulsation, or clogged ports will be shown. For a valved uniflow cylinder, the indicator card may be used to determine proper valve timing similar to the four-cycle engine.

In newer types of two-cycle gas engines, the fuel gas is injected directly into the combustion space, and air and gas mixing takes place in the cylinder. The mixture for lighter loads then becomes very lean and missing often results. On some engines the scavenging air is throttled on light loads to aid the mixture, and this eliminates missing to a certain extent. Because of these characteristics, the indicator can seldom be employed to determine malfunctioning of the burning cycle, except by using the 90° or offset card to determine spark timing and rate of pressure rise.

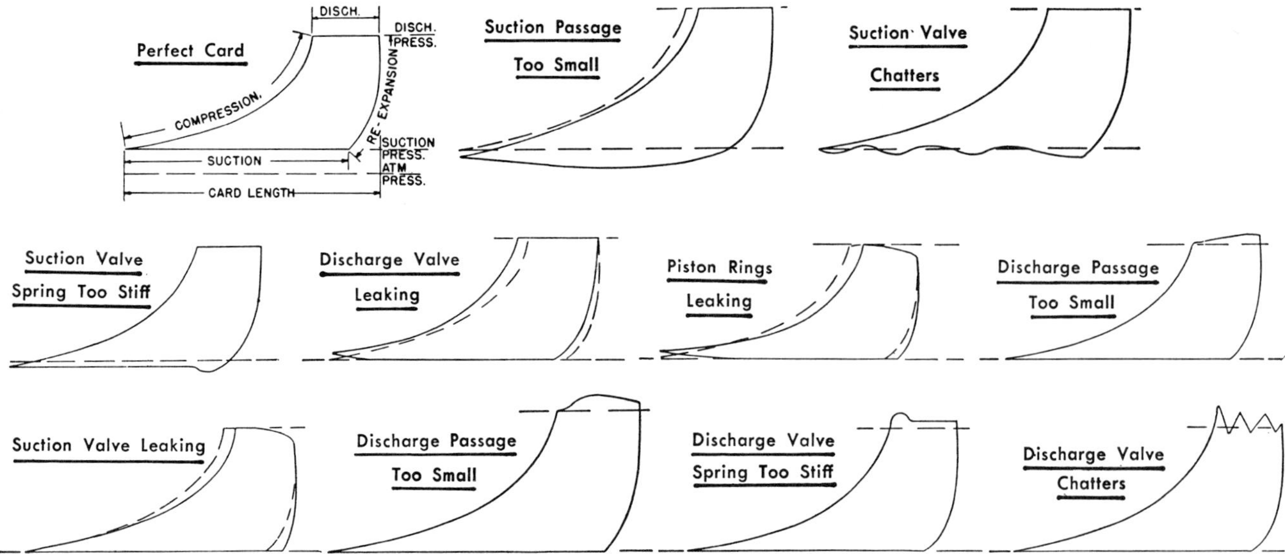

Fig. 8-50 Compressor cards showing various faults. (Cooper-Bessemer Corp.)[22]

Checking Compressors. Figure 8-50 shows compressor faults by means of cards made to show the defect. In some cases the perfect card shape is shown dotted to illustrate more clearly effects of faulty performance. Indicators and their drives can easily cause distortions and irregular cards; thus, considerable care must be exercised to make sure that all equipment is in good condition.

Indicated Compressor Horsepower. Use formula for indicated engine horsepower (see Eq. **28**). When working with compressor indicator cards, it must always be remembered that they show horsepower *absorbed* instead of work delivered. To determine the horsepower necessary to drive the compressor, the indicated compressor horsepower must be divided by the compressor mechanical efficiency.

Efficiency. The *mechanical efficiency*, e_m, of an engine compressor unit is the work done in the compressor cylinders divided by the work done in the engine cylinders, or: ihp compressor/ihp engine $= e_m$.

Volumetric efficiency of a compressor may be determined by means of an indicator card. This efficiency is defined as the ratio of the actual volume of air or gas drawn in at intake temperature and pressure, to the compressor displacement. From the indicator diagram, the measurement of air or gas being pumped by the compressor may be determined. A reciprocating gas or air compressor is, in effect, a displacement type gas meter.

Miscellaneous Performance Tests

Kiene Indicator. This peak pressure indicator, developed for use on Diesel engines, can also be used for quick checks of compression and firing pressures on gas engines. It consists of a maximum reading pressure gage and release with tapered plug and wing nut connector. In use, this indicator is attached directly to the indicator cock on each cylinder. By opening the cock, the peak pressure is indicated on the hydraulic gage.

Pi Meter. This device is used in much the same way as the Kiene indicator to balance engines rapidly and accurately. It consists of a pendulum-dampened pressure indicating gage, actuated by a spring loaded piston. This gage gives the time-averaged cylinder pressure (related to the brake mean effective pressure). The device is attached to each cylinder in turn by the indicator cocks; adjustments are made until all cylinder pressures are approximately equal.

Exhaust Gas Analysis. A minimum of excess air and CO and a maximum of CO_2 are obviously desirable in engine exhaust gases. Analysis of exhaust for these gases permits setting of a proper gas-air mixture. While such analyses can be made on four-cycle engines, they are difficult on two-cycle engines because of the air scavenging sequence.

Instruments capable of fast, accurate analysis such as by chromatography, infra-red spectroscope indicators, and the Pauling oxygen indicator, are preferable to the Orsat type apparatus. Special instruments for exhaust gas analysis are available as power provers and testers. Proper sampling technique is essential.

Measuring Exhaust Temperature. For periodic checking of exhaust gas temperatures of a multicylinder gas engine, a thermocouple-pyrometer may be used. It is installed as standard equipment on most modern engines. For many years it was the major method of balancing engine load.

Spectrographic Analysis of Lubricating Oil.[23] This analysis for compressor engines is a valuable, although highly sensitive, method. After a number of oil analyses have determined a normal contaminant level for a particular engine, a significant and consistent departure in one or more of the metal concentrations in the oil would indicate excessive engine wear. Such warnings of engine trouble permit scheduling major maintenance when required, rather than at arbitrary intervals.

Emission spectrographic analyses for eight elements are useful: the normal contaminant levels shown were not exceeded by 30 of the 54 engines tested:

Silicon, a measure of dirt in oil, normally indicates malfunctioning air cleaner, 4 ppm.

Iron is a measure of wear in cylinders and liners, and of general engine corrosion, 10 ppm.

Chromium indicates wear on chromium-plated rings and liners (and chromate-treated water in crankcase), 1 ppm.

Copper, lead, and *tin* indicate wear on bearing surfaces, 2 ppm.

Aluminum indicates engine wear with aluminum power cylinders,* 2 ppm.

Boron indicates borax-treated water in crankcase.

Error sources and variables which must be evaluated include lead-containing sealing compounds, molybdenum disulfide and other detergents, inefficient oil filter systems, air-borne contamination with open oil sumps, upset of equilibrium between oil and engine of from 200 to 1000 hr following oil change, load fluctuations, and operation and maintenance history (e.g., wearing in new parts).

COMPRESSOR STATION DESIGN

For general information, reference may be made to texts,[24] equipment manufacturers' literature, and the following extract from American Standard Code for Gas Transmission and Distribution Piping Systems, ASA B31.8-1963:

843 COMPRESSOR STATIONS.

843.1 Compressor Station Design.

843.11 *Location of Compressor Building.* The main compressor building for gas compressor stations should be located at such clear distances from adjacent property not under control of the company as to minimize the hazard of communication of fire to the compressor building from structures on adjacent property. Sufficient open space should be provided around the building to permit the free movement of fire-fighting equipment.

843.12 *Building Construction.* All compressor station buildings which house gas piping in sizes larger than 2 inches in diameter, or equipment handling gas (except equipment for domestic purposes) shall be constructed of noncombustible materials as defined by the National Board of Fire Underwriters (Special Interest Bulletin No. 294: Definition of Noncombustible Building Construction Material and National Building Code).

843.13 *Exits.* A minimum of two exits shall be provided for each operating floor of a main compressor building and basements and any elevated walkway or platform 10 feet or more above ground or floor level. Individual engine catwalks shall not require two exits. These exits may be fixed ladders, stairways, etc. of each such building. The maximum distance from any point on an operating floor to an exit shall not exceed 75 feet measured along the centerline of aisles or walkways. Said exits shall be unobstructed doorways so located as to provide a convenient possibility of escape and shall provide unobstructed passage to a place of safety. Door latches shall be of a type which can be readily opened from the inside, without a key. All swinging doors located in an exterior wall shall swing outward.

843.14 *Fenced Areas.* Any fence which may hamper or prevent escape of persons from the vicinity of a compressor station in an emergency shall be provided with a minimum of two gates. These gates shall be so located as to provide a convenient opportunity for escape to a place of safety. Any such gates located within 200 feet of any compressor plant building shall open outward and shall be unlocked (or openable from the inside without a key) when the area within the enclosure is occupied. Alternatively, other facilities affording a similarly convenient exit from the area may be provided.

* Measure of aluminum bearing wear not established.

843.2 Electrical Facilities. All electrical equipment and wiring installed in gas transmission and distribution compressor stations shall conform to the requirements of the National Electrical Code, ASA C1, insofar as the equipment commercially available permits.

843.3 Corrosion Control. Suitable investigation shall be made and if it indicates that corrosion protection is needed, gas piping within compressor stations shall be protected by any recognized method or combination of methods including coating with protective material, the application of cathodic current, or electrical isolation by sections. After installation of piping, periodic inspections or tests of the piping shall be conducted to determine whether or not the pipe metal is adequately protected.

843.4 Compressor Station Equipment.

843.41 *Gas Treating Facilities.*

843.411 *Liquid Removal.* When condensable vapors are present in the gas stream in sufficient quantity to liquefy under the anticipated pressure and temperature conditions, the suction stream to each stage of compression (or to each unit, for centrifugal compressors) shall be protected against the introduction of dangerous quantities of entrained liquids into the compressor. Every liquid separator used for this purpose shall be provided with manually operated facilities for removal of liquids therefrom. In addition, automatic liquid removal facilities or an automatic compressor-shutdown device or a high liquid level alarm shall be used where slugs of liquid might be carried into the compressors.

843.412 *Liquid Removal Equipment.* Liquid separators, unless constructed of pipe and fittings and no internal welding is used, shall be manufactured in accordance with Section VIII, Unfired Pressure Vessels, of the ASME Boiler and Pressure Vessel Code. Liquid separators when constructed of pipe and fittings without internal welding, shall be in accordance with Type D construction requirements.

843.42 *Fire Protection.* Fire protection facilities should be provided in accordance with the National Fire Protection Association and National Board of Fire Underwriters' recommendations. If fire pumps are a part of such facilities, their operation shall not be affected by emergency shut-down facilities.

843.43 *Safety Devices.*

843.431 *Emergency Shutdown Facilities.*

(a) Each transmission compressor station shall be provided with an emergency shutdown system by means of which all gas compressing equipment, all gas fires, and all electrical facilities in the vicinity of gas headers and in the compressor building can be shut down and the gas can be blocked out of the station and the station gas piping blown down. The emergency shutdown system shall be operable from any one of at least two locations outside the gas area of the station, preferably near exit gates in the station fence, but not more than 500 feet from the limits of the station. Blowdown piping shall extend to a location where the discharge of gas is not likely to create a hazard to the compressor station or surrounding area. Unattended field compressor stations of 1000 horsepower and less are excluded from the provisions of this paragraph.

(b) Each compressor station supplying gas directly to a distribution system shall be provided with emergency shutdown facilities located outside of the compressor station buildings by means of which all gas can be blocked out of the station provided there is another adequate source of gas for the distribution system. These shutdown facilities can be either automatic or manually operated as local conditions designate. When no other gas source is available, then no shutdown facilities shall be installed that might function at the wrong time and cause an outage on the distribution system.

843.432 *Engine Overspeed Stops.* Every compressor prime mover except electrical induction or synchronous motors shall be provided with an automatic device which is designed to shut down the unit before the speed of the prime mover or of the driven unit exceeds the maximum safe speed of either, as established by the respective manufacturers.

843.44 *Pressure Limiting Requirements in Compressor Stations.*

843.441 Pressure relief or other suitable protective devices of sufficient capacity and sensitivity shall be installed and maintained to assure that the maximum allowable operating pressure of the station piping and equipment is not exceeded by more than 10%.

843.442 A pressure relief valve shall be installed in the discharge line of each positive-displacement transmission compressor between the gas compressor and the first discharge block valve. The relieving capacity shall be equal to or greater than the capacity of the compressor. If the relief valves on the compressor do not prevent the possibility of overpressuring the pipeline, as specified in 845,* a relieving device shall be installed on the pipeline to prevent it from being over-pressured.

843.443 An acceptable relief device, in accordance with 845,* or automatic compressor shutdown device shall be installed in the discharge of each positive displacement distribution compressor between the gas compressor and the first discharge block valve. The relieving device shall be installed and maintained to prevent the maximum allowable operating pressure of the compressor and discharge piping from being exceeded by more than 10%.

843.444 Vent lines provided to exhaust the gas from the pressure relief valves to atmosphere shall be extended to a location where the gas may be discharged without undue hazard. Vent lines shall have sufficient capacity so that they will not interfere with the performance of the relief valve.

843.45 *Fuel Gas Control.* An automatic device shall be provided on each gas engine operating with pressure gas injection, which is designed to shut off the fuel gas when the engine stops. The engine distribution manifold shall be simultaneously automatically vented.

843.46. *Cooling and Lubrication Failures.* All gas compressor units shall be equipped with shutdown or alarm devices to operate in the event of inadequate cooling or lubrication of the units.

843.47 *Explosion Prevention.*

843.471 *Mufflers.* The external shell of mufflers for engines using gas as fuel shall be designed in accordance with good engineering practice and shall be constructed of ductile materials. It is recommended that all compartments of the muffler shall be manufactured with vent slots or holes in the baffles to prevent gas from being trapped in the muffler.

843.472 *Building Ventilation.* Ventilation shall be sufficient so that employees are not endangered, under normal operating conditions (or such abnormal conditions as a blown gasket, packing gland, etc.), by accumulations of hazardous concentrations of flammable or noxious vapors or gases in rooms, sumps, attics, pits, or similarly enclosed places, or in any portion thereof.

843.5 Compressor Station Piping.

843.51 *Gas Piping* (general provisions applicable to all gas piping).

843.511 *Specifications for Gas Piping.* All compressor station gas piping, other than instrument, control and sample piping, to and including connections to the main pipeline, shall be of steel and shall be Type C construction. Nodular cast iron may not be used in the gas piping components in this location.

843.512 *Installation of Gas Piping.* The provisions of 841.2,† "Installation of Steel Pipelines and Mains," shall apply where appropriate to gas piping in compressor stations.

843.513 *Testing of Gas Piping.* All gas piping within a compressor station shall be tested after installation in accordance with the provisions of Paragraph 841.4† for pipelines and mains in location Class 3, except that small additions to operating stations need not be tested where operating conditions make it impractical to test.

[* Given in Sec. 9, Chap. 2—Editor]
[† Given in Chap. 2 of this Section—Editor]

843.514 *Identification of Valves and Piping.* All emergency valves and controls shall be identified by signs. All important gas pressure piping shall be identified by signs or color codes as to their function.

843.52 *Fuel Gas Piping* (specific provisions applicable to fuel gas piping only).

843.521 All fuel gas lines within a compressor station, serving the various buildings and residential area, shall be provided with master shutoff valves located outside of any building or residential area.

843.522 The pressure regulating facilities for the fuel gas system for a compressor station shall be provided with pressure limiting devices to prevent the normal operating pressure of the system from being exceeded by more than 25%, or the maximum allowable operating pressure by more than 10%.

843.523 Suitable provision shall be made to prevent fuel gas from entering the power cylinders of an engine and actuating moving parts while work is in progress on the engine or on equipment driven by the engine.

843.524 All fuel gas used for domestic purposes at a compressor station, which has an insufficient odor of its own to serve as a warning in the event of its escape, shall be odorized as prescribed in 861.‡

843.53 *Air Piping System.*

843.531 All air piping within gas compressing stations shall be constructed in accordance with Division 1 under Section 2 of the ASA B31.1 Code for Pressure Piping.

843.532 The starting air pressure, storage volume, and size of connecting piping shall be adequate to rotate the engine at the cranking speed and for the number of revolutions necessary to purge the fuel gas from the power cylinder and muffler. The recommendations of the engine manufacturer may be used as a guide in determining these factors. Consideration should be given to the number of engines installed and to the possibility of having to start several of these engines within a short period of time.

843.533 A check valve shall be installed in the starting air line near each engine to prevent backflow from the engine into the air piping system. A check valve shall also be placed in the main air line on the immediate outlet side of the air tank or tanks. It is recommended that equipment for cooling the air and removing the moisture and entrained oil be installed between the starting air compressor and the air storage tanks.

843.534 Suitable provision shall be made to prevent starting air from entering the power cylinders of an engine and actuating moving parts while work is in progress on the engine or on equipment driven by the engines. Acceptable means of accomplishing this are installation of a blind flange, removal of a portion of the air supply piping or locking closed a stop valve and locking open a vent downstream from it.

843.535 *Air Receivers.* Air receivers or air storage bottles, for use in compressor stations, shall be constructed and equipped in accordance with Section VIII, Unfired Pressure Vessels, of the ASME Boiler and Pressure Vessel Code.

843.54 *Lubricating Oil Piping.* All lubricating oil piping within gas compressing stations shall be constructed in accordance with Division A under Section 3 of ASA B31.1 Code for Pressure Piping.

843.55 *Water Piping.* All water piping within gas compressing stations shall be constructed in accordance with Section 1 of ASA B31.1 Code for Pressure Piping.

843.56 *Steam Piping.* All steam piping within gas compressing stations shall be constructed in accordance with Section 1 of ASA B31.1 Code for Pressure Piping.

843.57 *Hydraulic Piping.* All hydraulic power piping within gas compressing stations shall be constructed in accordance with Division A under Section 3 of ASA B31.1 Code for Pressure Piping.

[‡ Given in Sec. 9, Chap. 11—Editor]

853 COMPRESSOR STATION MAINTENANCE.

853.1 Compressors and Prime Movers. The starting, operating and shutdown procedures for all gas compressor units shall be established by the operating company and the operating company shall take appropriate steps to see that the approved practices are followed.

853.2 Inspection and Testing of Relief Valves. All pressure relieving devices in compressor stations except rupture disks shall be inspected and/or tested in accordance with 855* and shall be operated periodically to determine that they open at the correct set pressure. Any defective or inadequate equipment found shall be promptly repaired or replaced. All remote control shutdown devices shall be inspected and tested periodically to determine that they function properly.

853.3 Inspection for Corrosion. In existing plants where corrosive or potentially corrosive situations exist, procedures shall be set up for periodic inspections at sufficiently frequent intervals to enable the discovery of corrosion before serious impairment of the strength of the piping or equipment has occurred. Prompt repairs or replacements shall be made when needed.

853.4 Isolation of Equipment for Maintenance or Alterations. The operating company shall establish procedures for isolation of units or sections of piping for maintenance, and for purging prior to returning units to service, and shall follow these established procedures in all cases.

[* Given in Sec. 9, Chap. 5—Editor]

853.5 Storage of Combustible Materials. All flammable or combustible materials in quantities beyond those required for everyday use or other than those normally used in compressor buildings, shall be stored in a separate structure built of non-combustible material located a suitable distance from the compressor building. All above ground oil or gasoline storage tanks shall be protected in accordance with the National Fire Protection Association and the National Board of Fire Underwriters Standard No. 30.

853.6 No Smoking Signs. Smoking shall be prohibited in all areas of a compressor station in which the possible leakage or presence of gas constitutes a hazard of fire or explosion. Suitable signs shall be posted to serve as warnings of these areas.

Buildings

Rigid frame steel structures, with corrugated metal or transite siding and roofing, have been used for compressor buildings. Rigid frame steel buildings with suitable overhead clearances for equipment facilitate the use of a crane in maintenance operations. The steel structure (rigid frame), asbestos or sheet metal clad type, with provisions for travelling cranes and jib hoists, is generally accepted as best within the economical range for large compressor stations. All buildings around compressor stations should be properly ventilated and, to a certain extent, insulated.

Since the advent of centrifugal compressors, small brick structures with commercial sash, a flat roof, and a structural steel frame to support the crane rails, have been used ad-

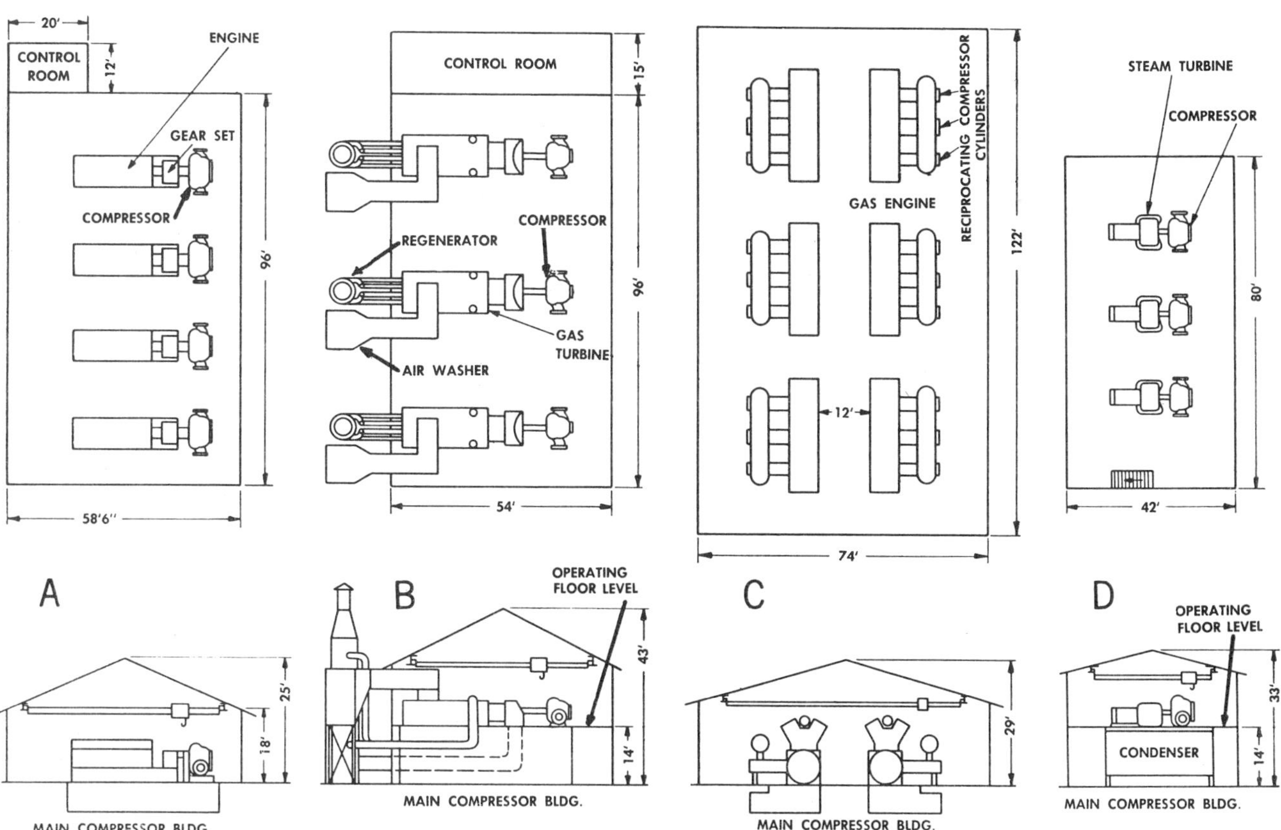

Fig. 8-51 Typical compressor stations. Dimensions for elevations given from grade line. (Nordberg Mfg. Co.)[8]

A. Gas engine centrifugal (plus auxiliary bldg. 28′ w. × 32′ d. × 18′ h.)
B. Gas turbine centrifugal (plus auxiliary bldg. 25′ w. × 25′ d. × 18′ h.)
C. Angle reciprocating (plus auxiliary bldg. 32′ w. × 37′ d. × 18′ h.)
D. Steam turbine centrifugal (plus auxiliary bldg. 40′ w. × 70′ d. × 27′ h. and one boiler and stack per compressor).

vantageously and at a reasonable cost. Such buildings are easy to insulate and heat, and can be constructed with materials desirable from a maintenance standpoint. Because compressors reject heat, proper building ventilation is essential to remove heat in summer and to retain it in winter. Building insulation and tightness should be a prime concern to provide needed comfort to operating personnel. In areas where there are severe dust conditions, it may be necessary to provide washed air and to maintain the interior pressure slightly above atmospheric, to control dust entry.

Figure 8-51 shows sketches of main and auxiliary buildings for stations of some 14,000 to 15,000 hp, powered by each of the four main types of prime movers. Outlines of the main items of machinery are sketched. One-floor structures are sufficient for the centrifugal engines and angle units. Gas and steam turbine stations require two floor levels. The extra space is taken up by the complicated duct work and piping for regenerative cycle gas turbines. Steam turbines require an elevated operating floor since it is convenient to mount the compressor unit on top of the condenser. The lower floor may then be used for certain smaller auxiliaries and the condensers.

Foundations

For comparison purposes, approximate foundation requirements for four stations using different prime movers are given in Table 8-13. Equipment manufacturers can supply requirements for mass and bearing needed for their units.

The following data indicate safe bearing capacities of soils. They should not be substituted for on-the-site test values.[25] One method is given by ASTM.[26]

	Bearing capacity, tons/sq ft
Sand, clean and dry (natural beds)	2–4
Earth, dry, solid (natural beds)	4
Clay, dry	3–4
Gravel, dry, coarse	6
Sandstone	20
Limestone	25
Granite	30

Table 8-13 Approximate Foundation Requirements for Compressor Stations

(cubic yards of concrete)

	Engine centrifugal, 4 units 14,200 hp	Gas turbine centrifugal, 3 units 15,000 hp	Angle reciprocating, 6 units 15,000 hp	Steam turbine centrifugal, 3 units 15,000 hp
Main compressor unit	480	900	750	60*
Auxiliary generating sets	60	40	90	40
Accessories including regenerators and boilers where required	15	70	20	240

* Mounting on its condenser practically eliminates need of concrete foundation for main units.

In designing foundations for station facilities, consideration should be given to the potential of vibration, the frost line, and depth of saturation from excessive water during seasonal changes, as in thawing and rainy seasons. Some soils in their dry state furnish adequate support, but when water in large quantities is added, they become plastic and unsuitable for supporting certain types of foundations. Thus, care should be taken to investigate the complete seasonal stability of the soils and footings should be designed accordingly.

Where several pieces of equipment, such as compressors, are to be placed in the same building, it is common practice to pour a reinforced concrete mat on which the individual unit foundations are placed and secured in position by "dowelling" to it. This method provides uniform support for all foundations and maintains alignment. Foundations for walls and piers are usually set on footings which provide a lower unit load on the subsoil and at the same time help to provide lateral stability. Preliminary soil resistance surveys and seismic tests are used; also bore holes, impact tests, and induced oscillation tests.

Station Piping and Surge Tanks

For formulas of maximum allowable operating pressure and nominal wall thickness of piping, and data on its expansion and flexibility, see extracts from ASA B31.8-1963, quoted in this chapter and in Chap. 2.

Efficiency of compressor installations may be reduced and drive units may be unnecessarily overloaded by using improper compressor pipe sizes and failing to provide adequate surge volumes for the compressor cylinders. Some methods

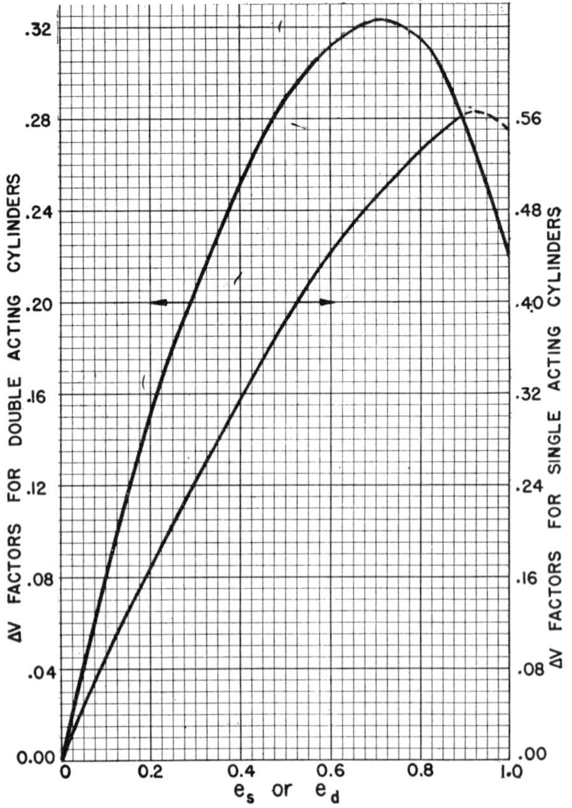

Fig. 8-52 ΔV factor curve for single and double acting cylinders for Eq. **29.** (Cooper-Bessemer Corp.)

often used in selecting pipe sizes fail to consider the actual manner of behavior of the gas flowing to and from the cylinders. Compressor units are selected and cylinders sized on the assumption that stated operating pressures will exist at their suction and discharge flanges. For satisfactory operation, it is imperative that compressor surge drum and pipe sizes should be adequate. Adequately sized cylinder by-pass piping is necessary to insure complete unloading during start-up.

Pipe and surge drum sizes recommended by one compressor manufacturer are selected by the following methods:

Calculation of Surge Tank (Receiver) Size. Equation **29** is for a single compressor cylinder. For a drum to take care of a number of cylinders of the same size, multiply the volume required for one of them by the number of cylinders. Note sample calculations in Table 8-15.

$$V = \frac{(PD/\text{stroke}) \times \Delta V}{(P'/P)^{1/k} - 1} \qquad (29)$$

where:

V = volume of surge drum, cu ft
PD/stroke = $PD/(2 \times \text{rpm})$ for double acting cylinders, cu ft
PD/stroke = PD/rpm for single acting cylinders, cu ft
PD = compressor cylinder piston displacement, cfm
ΔV = volume rate of change factor determined from Fig. 8-52 for single and double acting cylinders. ΔV_s used for a *suction* drum is determined from these curves, using the suction volumetric efficiency, e_s. ΔV_d used for *discharge* drums is determined from the same curves, using the discharge volumetric efficiency, e_d.

$$e_d = \frac{e_s}{C^{1/k}} \qquad (30)$$

where:

C = compression ratio, P_2/P_1, absolute pressures
k = ratio of specific heats, c_p/c_v
P'/P = the allowable pressure fluctuation ratio within the drum. It has been found that a P'/P of 1.05 will give the most satisfactory results. Values for $[(P'/P)^{1/k} - 1]$ are given in Table 8-14.

Table 8-14 Values for $[(P'/P)^{1/k} - 1]$ When $P'/P = 1.05$

k	$(1.05)^{1/k} - 1$
1.40	0.0355
1.35	.0368
1.30	.0383
1.26	.0395
1.25	.0398
1.20	.0415
1.15	0.0434

The surge tank diameter D to length L ratio should be about 1:2, or

$$D = 10.32V^{1/3} \qquad (31a)$$

where: D = min ID of drum, = $0.5L$, in.
V = volume in drum, cu ft

A rough sketch should be made when sizing surge tanks showing the center distance between adjacent compressor cylinders. Cooper-Bessemer sizing data follow:

Engine	Center distance between cylinders	Frame	Center distance between cylinders
GMXA	36 in.	EM	...
GMVA	48 in.	FM	36 in.
GMWA	40 in.	JM	60 in.

Table 8-15 Compressor Piping Data

(see Eqs. **29** and **30** for definitions of terms)

Compressor Specification: Requirement for 440 bhp at 300 rpm to be satisfied by three Model GMV-4-TF, double acting, one cylinder, 15 in. bore by 14 in. stroke, class C5B. $PD = 842$ cfm, suction: 36.2 psia and 60 F, discharge: 144.8 psia and 313 F, 9.2 per cent clearance, $k = 1.4$, $C = 4.0$, capacity = 2.45 MCFD at 14.4 psia and 60 F, 220 bhp, working pressure = 525 psig max, $e_s = 0.804$, $e_d = 0.298$, $\Delta V_s = 0.314$, $\Delta V_d = 0.205$, $P'/P = 1.05$, $PD/\text{stroke} = 1.40$ cu ft, 8 in. suction and discharge nozzles, 30 ASA series, air service.

	Columns:‡	A	B	C	D	E	F
Rows	Suction and discharge piping	Rate factor	Flow rate (avg), cfm	Area (min), sq ft	Diam (min), in.	Pipe size, in.	
1	Station suction header	2.41	2030	1.015	13.65	14	
2	Header to suction surge drum	0.804	676	0.338	7.86	8	
3	Suction surge drum $V = 1.40 \times 0.314/0.0355 = 14.4$ cu ft, use 24 in. OD × 4 ft long drum						
4	Suction surge drum to cylinder	1.140	960	0.480	9.40	8*	
5	Cylinder to discharge surge drum	0.812	684	0.342	7.92	8*	
6	Discharge surge drum $V = 1.402 \times 0.205/0.0355 = 8.1$ cu ft, use 22 in. OD × 42 in. long drum						
7	Discharge surge drum to station header	0.298	251	0.1255	4.8	5†	
8	Station discharge header	0.894	751	0.376	8.3	10	

(Cooper-Bessemer Corp.)

Notes: $(P'/P)^{1/k} - 1 = 0.0355$, $PD/\text{stroke} = 842/(2 \times 300) = 1.40$ cu ft, $e_d = e_s/C^{1/k} = 0.298$.

* This pipe size may be kept the same as the cylinder flange size, providing the drum is kept within a distance of twice the cylinder flange diameter and the calculated diameter does not exceed the cylinder flange diameter by more than two standard pipe diameter increments.

† In installations where 5 in. diam pipe is not commonly used, 6 in. or larger pipe may be used instead.

‡ Derivation of values detailed in text.

This offers a simple way of spotting interference between tanks on adjacent cylinders and finding whether the length determined by Eq. **31** is too great to be practical. If it is too great or causes interference, select a practical length, L, and calculate the new diameter, D, which will give the necessary volume, or:

$$D = 46.9(V/L)^{0.5} \tag{31b}$$

where: D = new diameter, in.
V = drum volume, cu ft
L = selected length, in.

Calculation of Compressor Piping Sizes. One manufacturer has prepared a form (Table 8-15) for pipe size and surge drum calculations. Minimum pipe diameters may be found from the following relation:

$$D = 13.54\sqrt{\frac{PD \times \text{rate factor}}{2000}} \tag{32}$$

where: D = pipe diameter, in.
PD = compressor cylinder piston displacement, cfm

This form (Table 8-15) presents a simple sequence of operations for solving this equation for the sections of pipe which it is usually necessary to size.

Rows 3 and 6 are for surge drum selections, calculated as previously described. Where single or independently operating units are involved it will not be necessary to consider Rows 1 and 8.

The vertical columns of Table 8-15 list the sequence of operations necessary to solve for the required minimum diameter.

Column A—Rate Factor. It is necessary to know the actual *average momentary flow rate* for the periods during which gas enters and leaves the compressor cylinders to determine the pipe diameter. This rate is the average of the instantaneous rates at which gas enters and leaves the compressor cylinder. The time during which the suction or discharge valve is open and the corresponding piston speed during that portion of the stroke when the valve is open determine the average flow rate; Fig. 8-53 gives these rate factors in terms of volumetric efficiencies, e_s and e_d. The rate factor multiplied by the piston displacement then gives the resultant average flow rate.

The piping to the *suction* surge drum and from the *discharge* surge drum is based on a rate factor incorporating the actual e_s and e_d, because these sections are away from the pulsations caused by the compressor cylinder, and the average flow rate thru them is considered the actual capacity or $e(PD)$ of the cylinders (see Eq. **21**). The PD is applied in Column B of Table 8-15.

Tests have indicated that pipe and surge drums selected in the manner described in Table 8-15 and accompanying text operate satisfactorily. Compressor installations can be expected to operate without losses in efficiency and overloading of drive units. Resonant lengths of pipe and the resulting vibration problems (discussed later) are not considered in this procedure.

Commercial pulsation dampening equipment is also available. In many instances, use of these specially designed dampeners will result in smaller units which may prove more economical than those indicated by common methods of computation. Consideration always should be given to providing pulsation dampening equipment whenever a reciprocating compressor is installed.

Determination of Data Given in Table 8-15.

Rate Factors (Column A)
For single acting cylinders:

*Row**
1 e_s × No. of cylinders × No. of units
2 e_s × No. of cylinders
4 At e_s enter Fig. 8-53, read Curve A for CE and Curve C for HE
5 At e_d enter Fig. 8-53, read Curve A for HE and Curve C for CE
7 e_d × No. of cylinders
8 e_d × No. of cylinders × No. of units

For double acting cylinders:

*Row**
1 e_s × No. of cylinders × No. of units
2 e_s × No. of cylinders
4 Read Curve B of Fig. 8-53 at e_s
5 Read Curve B of Fig. 8-53 at e_d
7 e_d × No. of cylinders
8 e_d × No. of cylinders × No. of units

Average Flow Rate, cfm (Column B)
For single acting cylinders:
Column B value = Column A value × $2PD$†

For double acting cylinders:
Column B value = Column A value × PD

Minimum Area, sq ft (Column C)
For both double and single acting cylinders:
Column C value = Column B value/allowable velocity, say 2000 fpm

Minimum Diameter, in. (Column D)
Column D value = (Column C value)$^{0.5}$ × 13.54, or

$$D = 13.54\sqrt{A} \tag{33}$$

where: D = diam, in.
A = cross sectional area, sq ft

Pipe Size (Column E)
Round off Column D to the next largest standard pipe size (see asterisk footnote to Table 8-15).

Schedule Number (Column F)
Select pipe specification and thickness based upon operating pressure and temperature.

Piping Expansion and Flexibility. Allowance for expansion, contraction, and flexibility should be considered in the design of compressor station piping. Tempera-

* Piping identified in Table 8-15. e_s and e_d are the volumetric efficiencies at suction and discharge, respectively, HE and CE.

† In a double acting cylinder the total PD is the sum of the crank end and head end piston displacement. However, for a single acting cylinder all the PD is in one end and the flow is handled in magnitudes of twice that of a single end of a double acting cylinder of the same piston displacement. Therefore, it is necessary to use the factor 2 × PD.

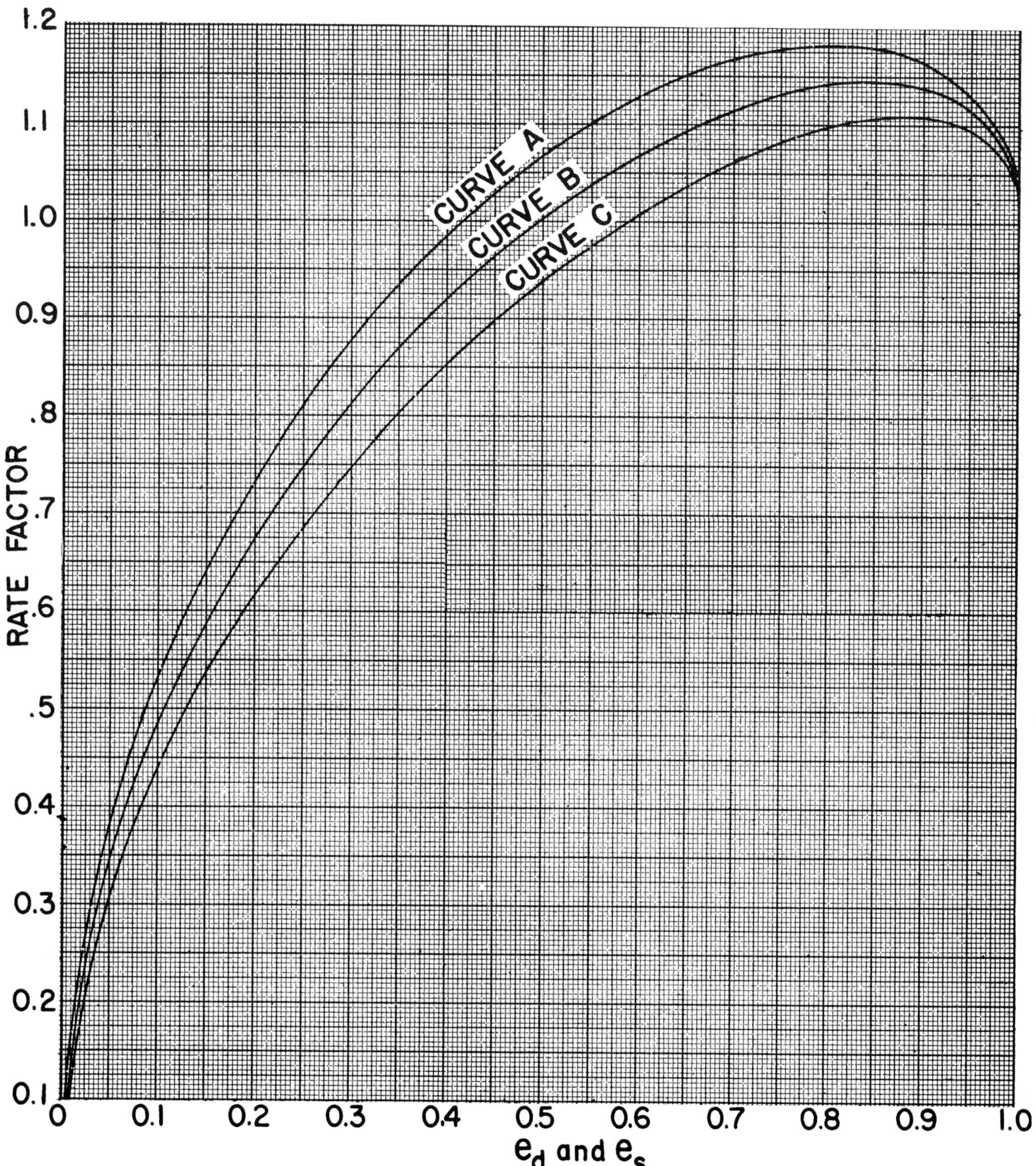

Fig. 8-53. Rate factor curves. For D.A. cyls., use Curve B; for S.A. cyls., use Curve A with HE e_d and CE e_s, use Curve C with HE e_s and CE e_d (Cooper-Bessemer Corp.)

Notes: D.A. = double acting, S.A. = single acting; e_d and e_s equal discharge and suction volumetric efficiency, respectively.

ture differentials are the primary consideration at centrifugal compressor stations. At reciprocating stations not only must the temperature differential be absorbed, but also provision must be made for preventing vibration amplification.

See also Paras. 832 and 833 of ASA B31.8-1963 quoted in Chap. 2.

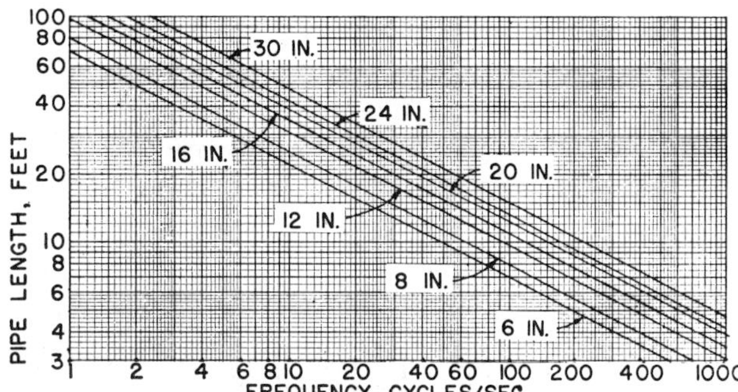

Fig. 8-54 Resonant lengths of pipe, supported at both ends. All pipe Schedule 40 (dimensions given in Table 1-9).

Pulsation and Vibration

Pulsation in a piping system can be described as a complex pressure wave having frequency components that may increase or decrease in magnitude from one point in the system to another. In the case of normal transmission systems the maximum peak-to-peak amplitudes are frequently more than 10 per cent of the static pressure.[27]

Causes of gas pulsation lie within the system as a whole and are not governed by any individual component to the exclusion of others. Among major factors[28a] causing pulsation are compressor-cylinder configuration, including clearances, number of cylinders on a given unit and their crank-angle relationship, and components like nozzles, volume bottles, lateral lines, and yard piping.

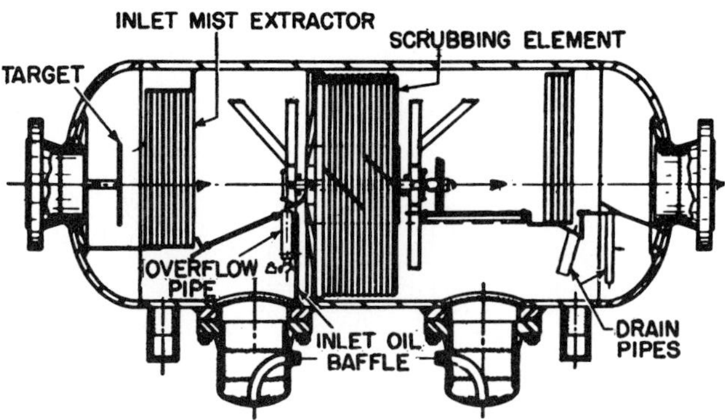

Fig. 8-55a Horizontal oil bath gas cleaner. (King Tool Co., Ltd.)

Probably the best known and most easily recognized effect of gas pulsation is piping vibration which leads to piping and support failures. To combat support failures, hold-down straps have sometimes been increased in size, but this increase often results in breaking foundation bolts. If bolts are then strengthened, cracks or failure of supporting concrete piers or foundations may result. Gas pulsation is often the cause of vibration in compression cylinders, too, which leads to maintenance problems for both equipment and foundations.

Most serious from the standpoint of economics and system design are the effects of gas pulsations on compressor loading, pumping capacity, and overall efficiency. Waste horsepower of 4 to 20 per cent, caused by gas pulsation, has been encountered.

The Southern Gas Association's pulsation research project at Southwest Research Institute (1952-1958) developed the following countermeasures:

1. An electrical analog of reciprocating compressor installations.
2. Refinement of both new and known principles in acoustics, compressor, and piping design thru use of the analog.
3. Effective utilization of developments in correcting problems at existing stations.

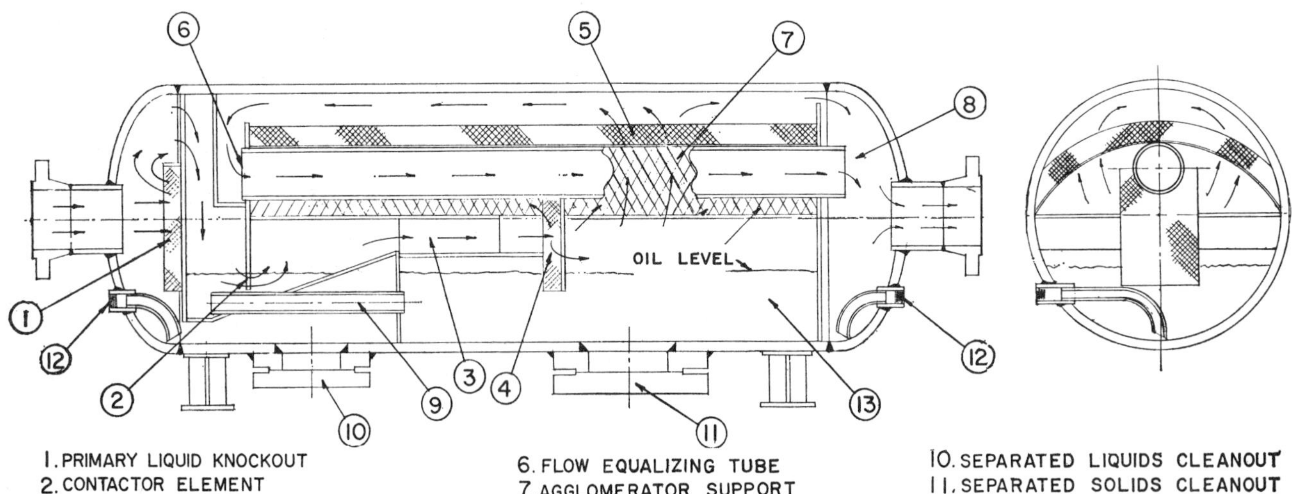

Fig 8-55b Horizontal oil bath gas cleaner. (Taylor Forge & Pipe Works Inc.)

4. Effective utilization of developments in design of piping and equipment for new stations to insure that:

(a) Compressor operates at rated capacity and maximum efficiency.

(b) Pulsations are effectively dampened or controlled with minimum pressure loss.

(c) Forces resulting from pulsations are effectively controlled to prevent damaging vibration.

The test equipment for another method of correcting pulsation and vibration consists of a pressure transducer, amplifiers, and oscilloscope. In one case a variation in compressor speed was caused by periodic overloads due to surge resonance in the piping. Pulsation snubbers placed between the compressor and the piping (on both suction and discharge) eliminated the difficulty.[28b]

A piping system may be designed to be comparatively free of vibration. This may be accomplished by avoiding resonant lengths of pipe (Fig. 8-54), by providing proper passages for the gas moved to and from the compressor, and by securely anchoring all piping. It is good practice to consult with experts on pulsation dampening during design. If vibrations are experienced after installation and a period of operation thru the complete loading range, specialized technicians should be consulted to determine the best corrective means. In evaluating equipment proposed for dampening vibration, effects on the volume of gas handled thru the station should be considered.

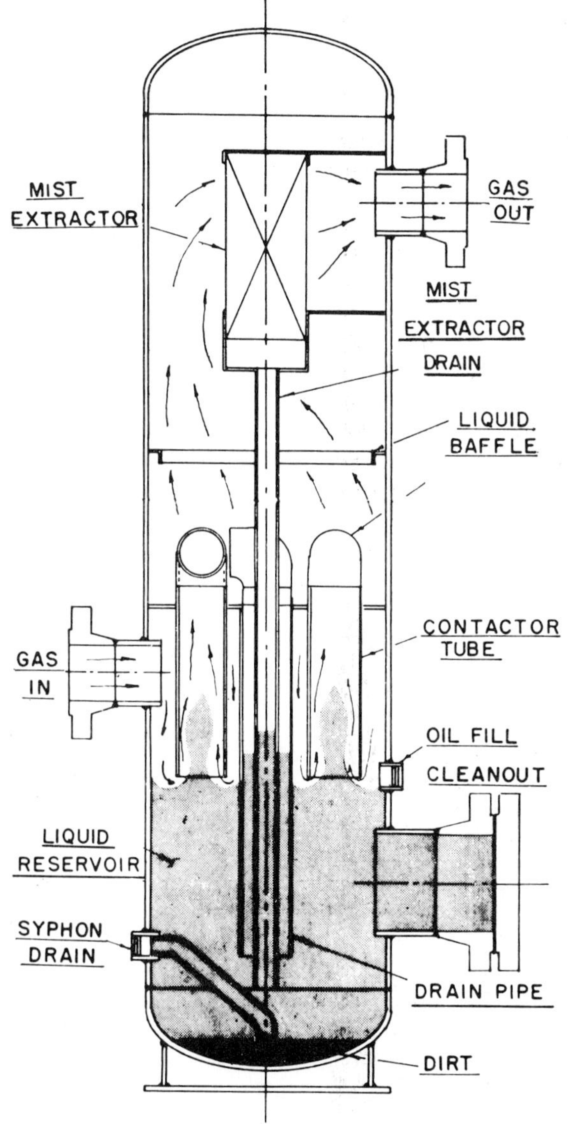

Fig. 8-56 Vertical oil bath gas cleaner. (Peerless Mfg. Co.)

Gas Cleaners

Gas cleaners, usually installed in the suction piping to compressor stations, remove liquids and solid particles which might damage the compressors. Separators should be used after discharge coolers to prevent liquid from entering the line.

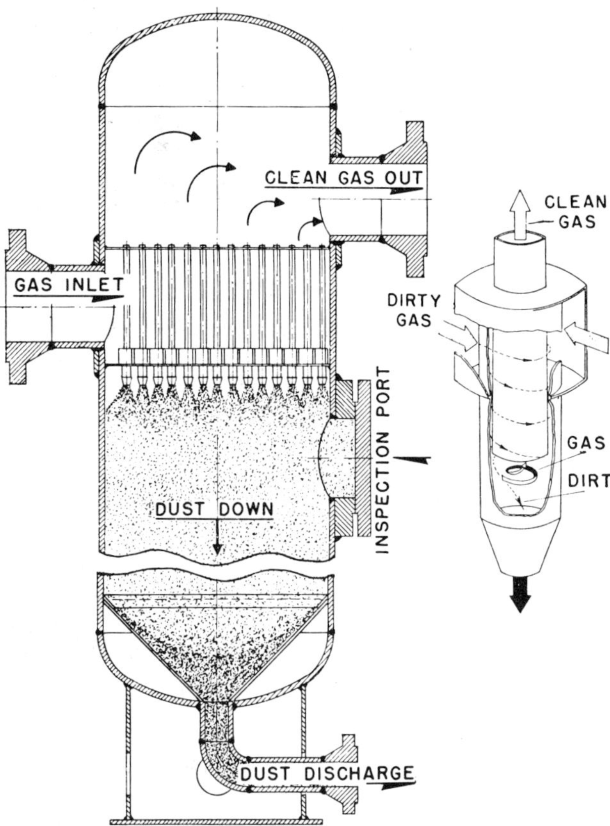

Fig. 8-57 Centrifugal gas scrubber employing a battery of parallel centrifugal units; inset at right shows one of these centrifugal tube-within-tube units. (Aerotec Industries Inc.)

Capacity for 0.6 sp gr gas at 2 psi pressure drop, MMSCFD

Scrubber diam, in.	No. of tubes	Operating pressure, psig 150	300	600	900
6	1	0.64	0.89	1.24	1.51
8	2	1.28	1.78	2.48	3.02
10	4	2.56	3.56	4.96	6.04
12	9	5.76	8.0	11.16	13.60
18	15	9.6	13.3	18.6	22.6
24	28	17.9	24.9	34.8	42.2
30	45	28.8	40.0	56.0	68.0
36	71	45.4	63.0	88.0	107
42	102	65.2	90.5	127	154
48	135	86.3	120	167	204
54	172	110	153	212	260
60	212	135	187	262	320
66	262	167	232	325	395

The four general types of cleaners used at compressor stations are:

1. Oil contacting type, horizontal (Figs. 8-55a and 8-55b) or vertical (Fig. 8-56). Particles too light and small (below 5 microns) to be gravitationally separated even when subjected to 50 g are increased in weight by wetting them with oil. Such treatment also increases agglomeration of smaller particles.
2. Centrifugal or inertial separators (Fig. 8-57). Such equipment is recommended for particles above 5 microns; however, a substantial percentage of smaller particles tend to agglomerate.
3. Dry filters, e.g., glass fiber or cloth.
4. Strainer.

Gas, Oil, and Water Cooling

Whether or not cooling is employed for various compressor station operations depends upon type of equipment and the compression ratio, since each situation requires special engineering and cooling methods themselves vary with location, water supply, and type of station.

When gas is compressed for transmission at ratios of 1.4 or less, with outlet temperature below 150 F, no cooling may be needed. For ratios above 3.5 at storage stations and other such locations and where temperatures of 250 F and higher are reached, cooling may be advisable.

Engine jacket water temperature is kept at approximately 160 F with a maximum of 180 F. Return water, after cooling, is usually 5 to 10°F lower. It is considered good practice to maintain as low a differential as possible both for jacket water and lubricating oil. A study of re-entrainment of liquids from inertial separators, due to pickup of the liquid from separator surfaces by the gas feed stream, is available.[29] Experiments at velocities and liquid loads approximated conditions encountered in pipelines, except that atmospheric pressure was employed.

Engine lubricating oil outlet temperature is usually maintained at approximately 150 F, with a maximum of 170 F. There is some tendency toward oils which permit a higher operating temperature. The oil is cooled 5 to 10°F before return. A 5°F differential is advocated by engine manufacturers, but 10°F is satisfactory. Centrifugal seal oil is maintained at 150 to 160 F or slightly lower. A temperature of 185 to 190 F has caused trouble with bearing seals.[7]

There are four sources of coolants: suction or inlet gas, dry air, water, and the earth. Several variations may be used for accomplishing the heat transfer.

Suction Gas as Coolant.[30] Inlet gas to the station may be used as a coolant for some of the smaller loads, like oil cooling, centrifugal seal oil cooling, small desulfurization plants at wellhead, and glycol dehydration plants on the discharge site of a compressor station.

The heat exchange surface may consist of a number of copper tubes thru which the coolant gas flows, and a containing vessel with the oil flowing around the tubes inside it.

Horsepower requirements increase slightly with increasing suction temperatures. For a main line transmission station with 60 to 70 F inlet gas temperature and 1.4 compression ratio, increase in power due to a gas oil cooler is estimated at about 0.7 per cent.

Air as Coolant. Use of air for cooling compressed gas, engine jacket water, lubricating oil, and centrifugal seal oil is increasing in the newer stations. In what is known generally as the fin-fan system (Fig. 8-58), the fluid to be cooled flows thru finned copper tubes in numerous shapes and designs over which air is blown. The tubes, in banks or rows, may be placed vertically, horizontally, or at an angle. The fan usually blows air over the tubes but it may also induce a draft thru them.

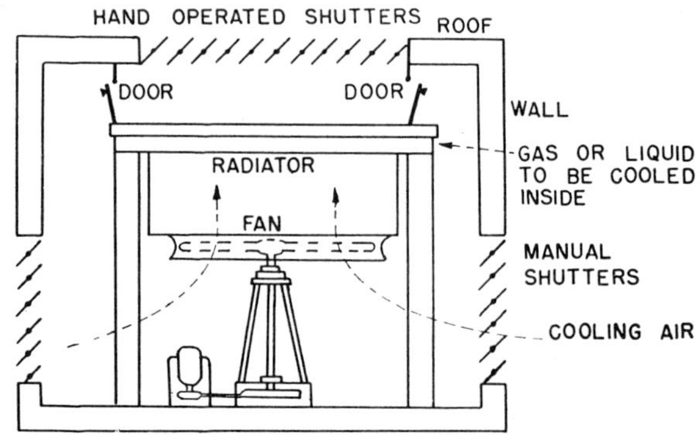

Fig. 8-58 Typical air-ccoled heat exchanger.

Louvers and shutters control air flow over tubes and thereby control the temperature drop of the fluid cooled. Adjustable fan speeds and fans with adjustable blade pitch may be used to conserve power. Lubricating oil may be cooled directly by passing it thru the cooling tubes, or indirectly by interchange of heat as by circulation of water thru the tubes immersed in the oil collecting tank.

Advantages of this system include:

1. Ability to locate station on most favorable site without regard to raw water supply, thus being unaffected by floods or dry weather.
2. Possibility of good water temperature control.
3. Low maintenance costs with no chemical treatment needed.
4. Operation of air fans directly from the main engine.

Disadvantages include:

1. Slightly higher initial costs than for a cooling water tower.
2. More complicated water temperature control.
3. Provisions against freezing of circulating water during very cold weather.
4. Higher operating costs due to fan horsepower required.

Manufacturers' heat exchange data and curves vary with each design of finned tube and with fan speed and design.

Basic heat exchange formulas are:

a. Heat load, Btu per hr of liquid being cooled = gal per min × 500 × sp gr of liquid × sp ht × fluid temperature difference

b. Heat load, Btu per hr for coil or tubes = No. of rows × coil face area in sq ft × overall heat transfer coefficient in Btu per hr-sq ft × mean temperature difference in °F

c. Heat load, Btu per hr of air coolant = standard coil face velocity in fpm (measured at 70 F and 0 ft altitude) × coil face area (sq ft) × 1.085 × air temp rise (diff. in inlet and outlet cooling air) in °F

Water as Coolant. Evaporative cooling is the most widely used cooling method for compressor stations with adequate water supply. It involves both a latent heat transfer, due to change of state of a small portion of the water from liquid to vapor, and a sensible heat transfer, due to difference in temperature of water and air. Approximately 1000 Btu (amount of heat lost in cooling 100 lb of water 10°F) are required to evaporate 1 lb of water. This is equivalent to 1 per cent loss of water due to evaporation. There are also drift losses of 0.5 per cent for mechanical towers; 1 per cent for atmospheric towers; 3 per cent for spray towers; and 10 per cent for spray ponds, depending upon size.

Cooling Ponds. Cooling action of ponds is independent of water depth, varying directly with exposed surface, water temperature, relative humidity, and air temperature and velocity.

Under average conditions, 1 sq ft is sufficient to cool 4 to 6 lb of water per hour from 100 to 70 F. Another method is to allow 3.5 Btu per hr-sq ft (of pond surface)-°F difference between wet-bulb temperature of the air and heated water temperature. The surface required is so great that cooling ponds are seldom used.

Spray Ponds. Since space requirements often make cooling ponds impractical, cooling sprays are used to accomplish the same purpose. Nozzles must be designed to break up emerging water jets into fine sprays under pressure of about 6 psig, with a capacity of about 50 gpm per nozzle. A rough rule for the spray pond surface is to allow 130 Btu per sq ft-°F difference between air and heated water. To prevent spray from being carried from the pond, the outer nozzles should be 24 ft from its border. If less space is available, a fence should be erected with sufficient openings for air passage.

Cooling Towers. These are generally classified as natural-draft or mechanical-draft. Natural-draft towers are further subdivided into: (1) the atmospheric type in which prevailing winds provide all ventilation, and (2) the chimney type which is becoming obsolete. Draft towers have a decided advantage over cooling and spray ponds. They are, however, more costly and require more maintenance and higher pumping head. The static lift is 35 to 40 ft. They must also be located so that they are unobstructed and broadside to the prevailing wind. From 0.7 to 1.0 sq ft of ground area is required per gpm of water circulated.

Mechanical-draft towers are divided into forced-draft, in which a fan at the base pushes air upward, and induced-draft types, where a fan at the top draws air upward. The induced-draft type is often favored over the forced-draft because it eliminates recirculation of the heated air. The static lift of water, with a cooling range of 25° to 35°F and for 15° to 20°F approach to the wet-bulb temperature, is 15 to 20 ft; for 8° to 15°F approach, 25 to 30 ft; for 4° to 8°F approach, 35 to 40 ft. The cooling range, as used here, is equivalent to the temperature rise in the condenser.

Mechanical-draft towers have many advantages over other types of water-cooling equipment. These are: (1) area required is about half that of the atmospheric type, viz., 0.3 to 0.5 sq ft per gpm, (2) independence of wind, which in turn allows (3) freedom in choice of location, (4) minimum of draft nuisance, which in turn results in (5) small make-up water requirement. Disadvantages include higher initial, operating, and maintenance costs.

Table 8-16 illustrates reduction of tower size and fan horsepower with elevation of wet-bulb temperature. It is also apparent that reduction of cooling range increases the size and horsepower of a tower for a given wet-bulb cold-water temperature and heat load.

Water Treatment. Cooling water is commonly filtered, deaerated, softened, or neutralized. Corrosion inhibitors, like chromates, are added to decrease corrosion and fouling of heat transfer surfaces. Corrosion of aboveground equipment is inhibited by painting. Equipment below ground can be protected cathodically. Spray from cooling water towers is often largely responsible for corrosion of equipment exposed above ground.

Certain agents, like chlorine or copper sulfate, should be added to prevent deposition of organic slime in cooling towers and in equipment. Three types of organisms, algae, bacteria, and fungi, are sources of trouble. Chlorine seems to be the only universally potent additive. The type of controlling agent is determined by the character of the organism. Additives not harmful to the wood of the tower should be selected.

Wood Deterioration.[31] Three kinds of wood deterioration in cooling towers are surface chemical, surface biological, and internal biological. All can be minimized by correct chemical treatment of the water to hold its pH value between 7.0 and 7.5.

Earth as Coolant. When a gas or liquid flows thru buried pipe, its temperature will approach that of the ground. Gas flowing between compressor stations may actually attain ground temperature if the stations are far

Table 8-16 Typical Performance of a Mechanical Cooling Tower
(based on cooling 1000 gpm)

	Temperature, °F						
Wet-bulb	Hot water	Cold water	Range	Flow, gpm per sq ft	Area, sq ft	Fan hp	Capacity, 10^6 Btu per hr
60	95	70	25	1.9	526	22	12.5
	105	75	30	2.3	435	18	15
65	100	75	25	2.0	500	21	12.5
	110	80	30	2.5	400	17	15
70	105	80	25	2.4	416	17	12.5
	115	85	30	2.7	370	16	15
75	110	85	25	2.5	400	17	12.5
	120	90	30	3.0	333	14	15
80	115	90	25	2.8	358	15	12.5
	124	94	30	3.0	333	14	15

enough apart. This is not a practical method of cooling where temperature must be dropped rapidly and controlled to a selected degree. Such cooling does, however, occur in almost all transmission lines and should be considered in station design calculations.

Compressor Station Noise

Noise at compressor stations may come from five distinct sources:[32,33] (1) engine exhaust, (2) engine intake, (3) engine room noise from engines and compressors, (4) heat exchangers (cooling tower and fin-fan), and (5) blowdown.

These noises may be objectionable to the plant operators or to the surrounding residential areas. It is practically impossible to converse when the noise level is 100 decibels or over.[34] Temporary hearing impairment results from prolonged exposure to noise levels in the region of 110 db. A few people who are very susceptible to hearing damage require noise levels under 90 db.[35] Permanent damage to hearing results when the level exceeds 115 db.

A survey of regulator and meter stations showed that their noise frequency spectrum was between 600 and 10,000 cycles per second with peaks between 2000 and 4000 cycles per second.[35] The detrimental effects of ultrasonic frequencies were reported.[36]

Generation and Effective Abatement of Noise. Practically all sound is a result of vibration, the frequencies depending upon the frequencies of the sources. *Vibration suppression*[35,37] can be accomplished by adding to or subtracting from the mass of the system. *Decoupling devices* which prevent transmission of vibration from one part to another may also be used to decrease noise; these devices consist of materials which absorb shocks. Rigid materials, double walls, and preformed glass-wool insulation are good *sound barriers.*[37] *Mufflers* are commonly used to dissipate exhaust noise thru absorption and/or successive reflections.

Measures taken by one company[32] to reduce noise are reported as an example of the problems involved and extreme measures taken to deaden the sound.

1. The compressor building was brick and had large entrance doors with interior surfaces covered with sound absorbing material and windows of glass brick. Roof was double ridged to provide emplacement for mufflers to deflect upward any noise passing them. Building interior was lined with sound absorbing material. Ventilation was done by blowers pulling air thru ventilators in the roof and blowing downward into the engine room.
2. All yard piping was located in brick walled trenches.
3. Residential protection from pipe noise was gained by a wall and shrubbery.
4. A blowdown silencer to permit blowdown of 5 MMSCFH from 1500 psig to 0 was provided with a 6½ ft diam stack, 24 ft high.
5. Expansion bottles, exhaust lines, and intake lines were lagged.
6. Residential type mufflers[33] were used for low frequency attenuation.
7. Cost of extras for noise control, including extra expenditure for brick building, amounted to about 3 per cent of the plant investment.
8. Total sound level in engine room was reduced to about 95 db. Noise radiated from plant was virtually unnoticeable.

Three methods of noise suppression in engine rooms are:[34]

1. Use of half walls in the building to let noise escape. This is possible in favorable climates and in sparsely populated areas.
2. Sound barriers around each engine-compressor with provision for forced ventilation air and barrier removal for engine repair.
3. Use of sound absorbing double cones (10–20 in. diam) suspended from ceiling.

Emergency Shutdown Facilities

ASA B31.8-1963, Sec. 843.431 (extracted in this chapter), makes it mandatory that each transmission compressor station be equipped with an emergency shutdown system, and lists the facilities required.

The system shown in Fig. 8-59a depends upon pressure differentials across orifice restrictions. This differential occurs with a release of pressure, since the system is constantly under pressure. The large valve operators are balanced torque units operating on oil and powered by gas pressure. Power gas is supplied thru an arrangement of poppet valves connected thru a lever to a differentially activated piston. The level may also be manually controlled for individual valve operation.

For a remote control emergency shutdown system see Fig. 8-59b. It employs a manually operated, gas energized control system and gas operators. In an emergency, operation of one or both of two valves will automatically close the station intake and discharge valves, interrupt ignition and thus shut down main engines, and open blowdown valves to vent gas in the station piping to the atmosphere.

Valves indicated (closed) confine the high pressure gas to the energizing portion of the system. The gas supply line may be tied into any high pressure line in the station. The check valve, **A**, between the gas supply and valves **1** and **2** confines gas to the energizing system, in case a failure of high pressure piping should take place within the station. A reservoir of sufficient capacity to operate all emergency equipment is installed in the energizing portion of the system.

Opening either valve **1** or **2** allows gas from the energizing portion of the system to flow to the valve operators and main engine ignition interrupter. However, a more acceptable way of shutting down an engine is to close the fuel gas valve automatically and vent the fuel gas piping between this safety valve and the engine. Then the ignition system may be interrupted or the magnetos grounded.

The valve operators are of the piston cylinder type. One end of the cylinder is connected to the control portion of the shutdown system and the other end is open to the atmosphere. Gas from the control piping enters the cylinder at one end and builds up pressure, thereby creating a pressure differential across the piston, which causes it to move. By a mechanical arrangement, the piston movement closes the valve, and a pressure trip electrical switch is used to interrupt the main engine ignition.

To summarize:

Open valve **1** or **2** to close header valves, open blowdown valves, and break main engine ignition circuit.

Close valve **3** beforehand, to prevent opening blowdown valves.

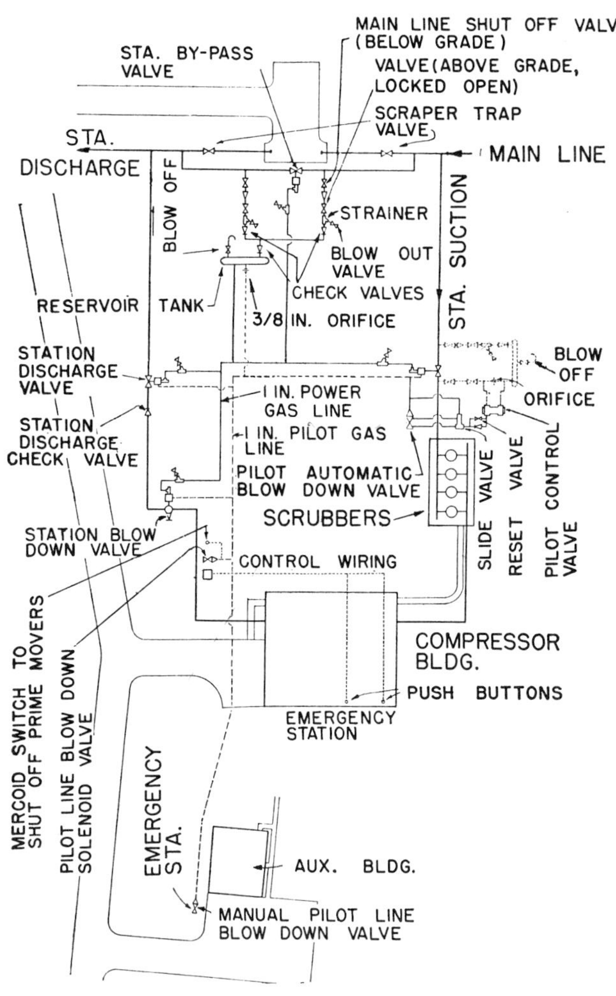

Fig. 8-59a Typical compressor station emergency shutdown system.

Close valve **4** beforehand, to prevent breaking main engine ignition circuit.

Close valve **5** to prevent filling reservoir.

Close valve **6** only, to keep pressure in valve cylinders without wasting gas.

In case of fire, a piping break within the plant, or any other serious emergency in which escaping high pressure gas will increase the hazard, the station is to be shut down immediately by opening valve **1** or **2**. This will close the header valves, open the blowdown valves, and stop the main engines by breaking their ignition circuit. The dispatcher should be notified as quickly as possible.

If a line break occurs on the discharge side of the station, the station is to be shut down immediately by closing valve **3** and opening valve **1** or **2**. This will close the header valves and stop the main engines. The dispatcher should be notified promptly.

If a break occurs on the intake side of the station, the dispatcher must be notified at once. Header valves should not be closed until the dispatcher gives instructions to do so. The station should be continued in operation until ordered shut down, or until intake pressure becomes so low that operation is no longer feasible because of overloading or compressor

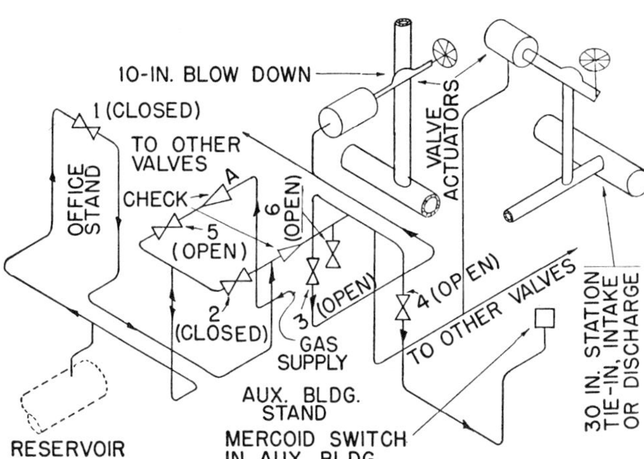

Fig. 8-59b Diagrammatic sketch of remote control for valves at compressor stations. Normal position of valves indicated as (open) or (closed).

heating. When these pressures are reported, the dispatcher may order that certain remote control valves be operated manually.

Remote Controlled Push-Button Compressor Stations

Such stations have become a reality in recent years, and in the future the trend will probably be toward an always greater use of them. One example may be cited of transmission line service using gas engine driven centrifugal compressors[38,39] and a second of underground gas storage and removal at 1000 psig using gas engine driven reciprocating compressors.[40] Both stations are unattended, except for maintenance men and watchmen.

The operation of five remote controlled stations was reported[38] on the Gulf Interstate Line. Each station is unattended, and all are operated from one control center. Each has four 3500 hp supercharged gas engines driving centrifugal compressors thru speed increasers. Digital indicators are used to show the flow rates, static pressures, differential pressures, and number of metering tubes in operation at two locations. On the one-unit compressor station there are 52 safety devices covering vibration of various equipment, gas surges, breaks in pipeline, operating pressures, temperature, power failure, and communication failure.

At the Hope Natural Gas Co. station,[40] designed for a maximum flow of 250 MMCFD at a peak pressure of 1000 psig, four 1350 hp single-stage engines, each with compressors, are operated from a distant location. With its own power, light and water facilities, gas scrubbers, and coolers, it is almost completely automatic. Compressed air is used to start the engines. Push-button controls start, load, and stop compressor engines and yard piping valves. An instantaneous blowdown safety system can be operated from one of four *panic stations*.

STATION OPERATION AND REPAIR

Compressor station operation is fundamentally the control of power to keep gas flowing continuously. Station personnel requirements vary with size, scope, and the degree of automatic operation. Thus, no typical example covers adequately the various policies regarding station crews. Basically (for a 40-hr week), a nonautomatic station requires one man in charge of and responsible for the plant. Main line station manpower requirements might consist of superintendent, clerk, repair foreman, four stationary engineers, four auxiliary engineers, one oiler for each two or three horizontal engines of 1000 hp or over, or one oiler for each three or four vertical engines of 1600 hp or less, plus two to four laborers or repairmen for maintenance work.

An electrician may be assigned to plants of more than 12,000 hp, or all electrical work may be done by division crews. A division electrician may be responsible for looking after three to eight plants, providing the system and stations are not too large.

Operating to Minimize the Need for Maintenance

Among the many practices used to minimize the need for machinery maintenance, the following may be considered:

Proper Loading. Departure from the established load and speed limits for each engine should be approved by the compressor department superintendent. The loading curves used to guide him should be prepared with the most accurate instruments available.

Engine Balance. Appropriate steps should be taken to balance engines, usually on a weekly basis but oftener if necessary.

Calibration of Instruments. In order that the operator may have confidence in the instrument readings, it is essential that all such equipment be calibrated on a routine basis.

Water Treatment. No detailed recommendations can be given because of the many differences in water supply. However, water troubles can be the cause of extensive down time. Water testing and water treating are essential to proper maintenance.

Lubricating Oil. The principal problem is the possibility of carbon buildup. Thus, quantity and quality control of lubricating oil should be the essential part of the program. One company, Transcontinental Gas Pipe Line Corp.,[41] uses straight mineral oil in its two-cycle engines, with a Fuller's earth sidestream filter. The guides for filter change used by this company are: (1) neutralization number 2.0 on the ASTM D664 acid test, and (2) maximum sediment concentration 0.5 per cent.

Routine Inspections. All parts of the compressor station should be checked regularly. Suggested major items include:[41]

Annual (show month completed)

1. High pressure gas—pressure drop test the main line to headers
2. Station shutdown with emergency shutdown system (complete cycle)
3. Visit to main line block valves by plant personnel
4. Relief valves test, temperature and pressure devices test
5. Domestic water bacteriological test (by laboratory)
6. Company housing inspection
7. Electrical switchboard inspection (when possible)
8. External corrosion inspection (engineering dept.)
9. Fuel gas meter internal inspection (measurement dept.)
10. Internal scrubber inspection

Semi-Annual (show month completed)

1. Review of "Outline of Emergency Action"
2. Emergency shutdown system check and test
3. Flushing of fire hydrants, and fire hoses and nozzles test
4. Service and operation of high pressure gas valves
5. Megging of electric motor circuits

Quarterly (show month completed)

1. Revision of emergency call list
2. Test of all unit shutdown devices
3. Fire extinguishers inspection
4. Shipping of oil samples to vendor for analysis

Monthly

1. Emergency shutdown system inspection
2. Fuel gas meter test and inspection (measurement dept.)
3. Safety meeting

Repairs may be set up on a continuing or a periodic basis. On a continuing basis, equipment is run until it needs attention and then it is repaired. On a periodic basis, equipment is taken out of service at definite periods, thoroughly checked and, if necessary, repaired. Both systems have their advantages, but the key to the success of either is in complete and up-to-date operating and repair records.

Use of air cleaners for combustion air; full flow oil filters on lubricating oil; gas scrubbers on fuel gas; correct fuel mixtures; and closed system large volume low temperature rise cooling systems for power cylinders help to insure long life with a minimum of repairs for prime movers. Removal of compressor lubricating oil and other undesirable foreign matter from the discharge gas stream is desirable. Good ignition is essential to efficiency. Record should be kept of point and plug life, and no overrunning should be permitted.

Economically, a compressor station should pump the most gas for the least cost. The first step necessary to attain this goal is to have an adequate and accurate means of measuring the gas pumped and fuel gas used, along with accurate engine tachometers and operating pressure gages. With data gathered from these instruments properly correlated, any deviation from maximum efficiency can be readily determined. A check should be made at least once a day, but preferably for every shift. A slight drop in indicated efficiency may lead to the detection of a bad compressor valve, a bad set of rings, or prime mover troubles long before they would become apparent in ordinary observation.

REFERENCES

1. A.G.A. *Gas Facts*. New York, annual.
2. "Gas Jet Compressor Lifts Line Pressure." *A.G.A. Mo.* 39: 15, Mar. 1957.
3. Fletcher, G. R. "Gas Jet Compressors." *Power and Fluids* (Worthington Corp.) 2: 12–6, Fall 1954.

4a. Warner, C. W. "Gas Mixing with Jet Compressors." *Gas* 30: 50–6, Sept. 1954.

4b. Miaskiewicz, R. F. "Use of Natural Gas Pressure Reduction to Operate Compressors." *Gas Age* 127: 25–27, June 22, 1961.

5. Salls, D. M. "Design and Application of Centrifugal Compressors." *Gas* 32: 105–7, Dec. 1956.
6. Reed, P. "Texas Eastern Operates First High-Pressure Gas-Transmission Centrifugal Compressors." *Oil & Gas J.* 46: 60+, Jan. 22, 1948.
7. Walsh, E. A. and Carameros, A. H. "What El Paso Natural Learned about Centrifugal Compressors for Natural-Gas Pipelines." *Oil & Gas J.* 55: 148+, Mar. 11, 1957.

8. Nordberg Manufacturing Co. *Selection of Compressing Equipment for Gas Pipe Lines* (Bulletin 226) Milwaukee, 1954.
9. Katz, D. L. and others. *Handbook of Natural Gas Engineering*, p. 643. New York, McGraw-Hill, 1959.
10. Fulleman and Parris. *The Centrifugal Compressor*. Mt. Vernon, Ohio, Cooper-Bessemer Corp., Sept. 12, 1958.
11. Clark Bros. Co., Inc. Bulletin 150. Olean, N. Y.
12. Stepanoff, A. J. *Turboblowers*, p. 25. New York, Wiley, 1955.
13. Lafferty, H. B. "Gas Dispatching Problems Encountered in the Operation of Gas Turbines and Centrifugal Compressors." (GSTS-55-20) *A.G.A. Proc.* 1955: 1176–82.
14. Murray, E. S. and Mertz, R. V. "Improved Station Design and Operation." *Oil & Gas J.* 53: 101–3, May 10, 1954.
15. Natural Gasoline Supply Men's Assn. *Engineering Data Book*, 7th ed. Tulsa, 1957.
16. Bloomer, O. T. and others. *Thermodynamic Properties of Methane-Nitrogen Mixtures*. (Research Bul. 21) Chicago, I.G.T., 1955.
17. Mast, B. T. "Compression Cylinder Design for Gas Transmission Pipe Line Systems." *Petrol. Engr.* 22: D14+, June 1950.
18. White, A. O. and others. "Design Concepts of Low Cost Gas Turbine Generating Plant for Peak Load Service." *Combustion* 30: 56–60, May 1959.
19. Vaughan, A. L. "Centrifugal Compressors Find Their Place." *Oil & Gas J.* 54: 91–3, Oct. 22, 1956.
20. "Jet Engine Starts Moving Gas." *Oil & Gas J.* 58: 147, Nov. 21, 1960.
21. Boyer, R. L. *A New 10,500 H.P. Gas Turbine*. A.S.M.E., Gas Turbine Power Conf., 1960.
22. Hines, J. D. *The Engine Indicator for Performance Evaluation*. Mt. Vernon, Ohio, Cooper-Bessemer Corp., 1952.
23. Kniebes, D. V. *Wear in Pipeline Engines and Compressors: Methods of Measurement*. (I.G.T. Tech. Rept. 1) New York, A.G.A., 1959.
24. Villbrandt, F. C. and Dryden, C. E. *Chemical Engineering Plant Design*, 4th ed. New York, McGraw-Hill, 1959.
25. Novosad, T. L. "Needed: More Knowledge on How to Design Stable Compressor Foundations." *Oil & Gas J.* 55: 129–32, Mar. 11, 1957.
26. A.S.T.M. *Procedures for Testing Soils*, 3d ed. Philadelphia, 1958.
27. Bain, R. C. "Pulsation in the Compressor Station." *A.G.A. Proc.* 1957: GSTS-57-7.
28a. Henderson, E. N. "Gas Pulsations." *Oil & Gas J.* 56: 115–22, May 12, 1958.
28b. Paddock, S. G. "Snubbers Stop Serious Vibrations." *Gas Age* 116: 38–9, Dec. 15, 1955.
29. Semran, K. T. *Investigation of Entrainment Separation*. (37/PR) New York, A.G.A., 1961.
30. Young, F. S. "Aspects of Gas Heat Exchangers as Applied to Transmission System Problems." *Gas Age* 105: 40+, Feb. 16, 1950. See also *Petrol. Engr.* 22: D11+, Aug. 1950; *Oil & Gas J.* 48: 84–8, Feb. 9, 1950.
31. "Cooling Tower Wood Deterioration." *Petrol. Process.* 9: 211–5, Feb. 1954. See also *Oil & Gas J.* 52: 78+, Jan. 11, 1954; *Power* 98: 113–7, Feb. 1954.
32. Baird, R. C. "Control of Compressor Room Noise." *Gas Age* 117: 29–33, Ap. 19, 1956. See also *Oil & Gas J.* 54: 98–102, Ap. 9, 1956; *Gas* 32: 142–6, April 1956.
33. Baird, R. C. "Compressor Plant Acoustics." *P.C.G.A.* 46: 78–83, 1955.
34. Culver, C. A. "Reducing Noise in Gas Compressor Stations." *Gas* 29: 141+, Oct. 1953.
35. Lascoe, S. "Noise Abatement in Gas Pipeline Operations." (GSTS-58-4) *A.G.A. Proc.* 1958: T21–3.
36. Sharp, J. M. and others. *Noise Abatement at Gas Pipeline Installations, Vol. I: Physiological, Psychological and Legal Aspects of Noise*, p. 1. New York, A.G.A., 1959.
37. *Ibid., Vol. II: Noise Suppression at Pressure Regulating and Metering Stations*. New York, A.G.A., 1960.
38. Orlofsky, S. "First Push Button Gas Pipeline Operation Successful." *Oil & Gas J.* 56: 114–9, July 14, 1958.
39. Lochiano, R. "How to Build an Automatic Compressor Station." *Gas Age* 118: 22–26, Oct. 18, 1956. Also in: *Gas* 32:109+, Dec. 1956; *Oil & Gas J.* 54: 94–6, Nov. 5, 1956; *Pipe Line Ind.* 5: 34–8, Nov. 1956.
40. "Hope Natural's Push-Button Compressor Station." *Gas Age* 121: 37–42, Apr. 3, 1958. Also in: *A.G.A. Monthly* 40: 24–26, April 1958.
41. Boehm, John C. "Compressor Maintenance—Cause and Effect." *Gas Age* 127: 34, 46, June 22, 1961.
42. Dixon, R. R. and House, S. A. "Performance Acceptance Test." *A.G.A. Op. Sec. Proc.* 1963: (GSTS-63-4).

Chapter 4
Economics of Gas Transmission

by M. V. Burlingame, George W. McKinley, and F. E. Vandaveer

GENERAL FACTORS AFFECTING ECONOMICAL TRANSMISSION SYSTEM DESIGN

The initial steps in designing a gas transmission system are fixing the origin and terminus of the lines, plotting present and future market requirements,[1] ensuring an adequate supply of gas and determining availability and economics of gas storage fields near the market. After that, pipe costs can be balanced against those of compressor horsepower required, and yearly system operating and maintenance expenses may be added to the annual fixed charges on the capital investment. There are studies available which break down pipeline and station construction and operating costs for the different types of prime movers.[2–4] The cost data and comparisons in this Chapter are only valid for the dates noted.

At the outset, the field pressure, the pressure required at the market end of the line, and the volume of gas to be handled must be known. Field pressure will govern the location of the first compressor station. In medium or low pressure areas (about 450 psig or under) this station should be built in the field or immediately adjacent to it. If field pressures are high, it may be practical to locate the first station 100 miles or farther away. But field pressure will ultimately decline, which will necessitate a station in the field. Therefore, the station nearest the field should be located so that there will be a balanced system when field power is installed. With this in mind, calculations can be made for various sizes of line, compression ratios, and station spacings to find the best combination.

Three formulas are needed for the general design calculation: namely, (1) flow—Eq. 1 in Chap. 2, Sec. 8, (2) horsepower, and (3) Barlow's formula for pipe wall thickness. Horsepower calculations can be aided by using curves relating compression ratio to horsepower per MMCFD of gas compressed (e.g., Figs. 8-38a thru c).

After deciding on the general design, the route between the producing area and markets must be established.

Investment Cost Estimates

The volume of gas which a pipeline may carry depends upon the maximum allowable operating pressure (assuming other conditions fixed). This pressure is limited by the physical and chemical properties of the available pipeline steel. Line flow may be calculated for several sizes and grades of pipe at a maximum operating pressure, as prescribed by the American Standard Code for Pressure Piping, B31.8-1963. Distances between compressor stations should also be considered. As such distance increases, inlet pressures decrease, and when the maximum compressor outlet pressure is constant, the compression ratio and power at each station increase. The converse is also true.

Applying comparative construction labor and materials costs to these calculated flows will enable a determination of the minimum investment required to transport the desired maximum capacity. The annual operating and maintenance expenses should also be estimated for each calculated system, together with its annual fixed charges on the capital investment. Weighing the capital investment with the annual operating expenses will determine which systems can be installed most economically. The following factors should be investigated in estimating investment cost.

Mainline (block) valves may be either manual or automatic. Cost will be affected by the number and type to be installed, which in turn will be governed by the spacing desired.

Depending upon the construction area, it may sometimes be necessary to use large quantities of specially fabricated pipe which materially increase construction costs. The construction area itself also affects the type of coating and wrapping material, since tough terrain and adverse soil conditions require more expensive materials. Whether they are shop- or field-fabricated also affects the estimated cost.

Labor costs play a major role in estimating pipeline construction costs, since they vary considerably with the terrain and proximity of the proposed pipeline route to railroads and highways. Rough terrain greatly increases not only hauling costs, but also those of stringing pipe, ditching, and other work along the right of way.

To the base costs of pipe and miscellaneous material must be added freight charges to the job site.

The degree of inspection, both at the mill and in the field, should also be considered in computing pipeline costs. If inspection includes a large labor force and extensive investigation of mill fabrication, welding, and other construction operations, inspection costs will necessarily be higher.

Right of way and associated damages should also be included in estimating capital investment. These are highest in farmlands and metropolitan areas. The number of purchase and sales meter stations required should also be considered. These stations are acquired either in fee simple or under a permanent easement.

Compressor station design depends upon the total line length and pipe size from the supply source to the market areas. For a smaller line, pressure drop per mile increases; therefore, distance between stations will decrease, requiring a greater number. For a given capacity, one combination of line size and compressor stations will result in the lowest capital investment, but not necessarily in the lowest annual operating expense. If the same number of stations is maintained and the line size is decreased, with resulting increase in pressure drop, the compression ratio at each station will increase. Thus, additional horsepower will be required to handle the maximum capacity. As the volume handled increases or decreases, greater or less horsepower will be required.

Station spacing and location will be influenced by such factors as availability of a water supply (when required*), adaptability of the proposed site, and its cost. The proximity of the site to existing communities will, to a small degree, limit the need for company housing of permanent station personnel.

Freight charges for necessary construction materials must also be considered in the total station cost. Under certain state laws governing trucking on highways, complete engines, because of their weight, cannot be hauled from the railroad siding to the compressor station location. In that case, the engine must either be shipped before assembly or disassembled on arrival and transported in parts to the station site.

In addition to the pipeline and compressor station construction costs, an estimate of capital investment should include engineering, general, and administrative expense, and other such unavoidable expenses as office overhead, company transportation, and legal fees.

DESIGN OF A LONG TRANSMISSION LINE (1000 MILES)

Design Factors[5]

A proper balance between investment in pipe and stations and total system operating costs must be reached. In comparing compressing equipment this balance is upset by the substitution of stations with different types of equipment, which vary in initial investment and operating cost. Therefore, minimum total transportation cost for a certain type of equipment may involve the use of different pipe and greater station horsepower and spacing than is required by another type, even though the gas volume transported and the distance are the same.

Variables involved in pipeline design are gas volume, length of line, pipe diameter, operating pressure, and station compression ratio. From them, station spacing and horsepower can be established, along with pipe cost, station cost, and total operating cost. Curves of Figs. 8-60a, 8-60b and 8-60c illustrate trends of these variables with regard to total pipeline horsepower for various capacities. A length of 1000 miles was arbitrarily selected for illustration. However, the curves would have the same trend, regardless of length.

Pipe selection affects gas transportation costs more than any other single factor. The horsepower reduction gained from using larger diameter pipe with thicker walls (higher operating pressure) must be weighed against the additional investment in pipe tonnage. Figures 8-60a and 8-60b show the effects of pipe diameter and operating pressure on total pipeline horsepower.

* Many stations collect and store rain water for make-up purposes; air in place of water for cooling is also used to a considerable extent.

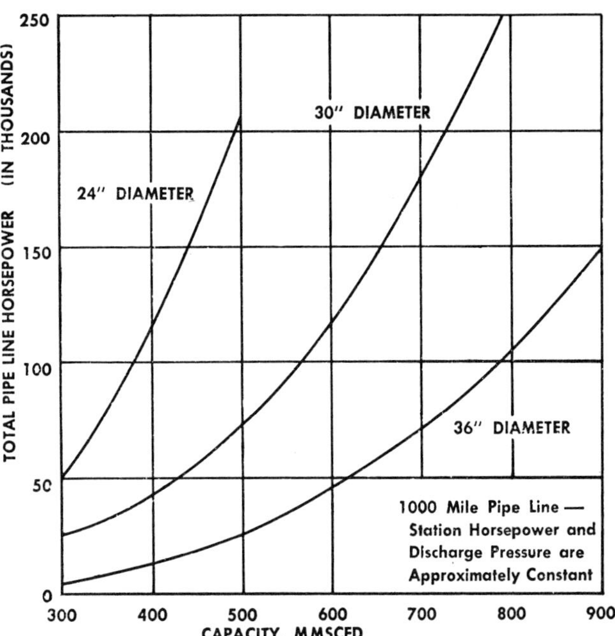

Fig. 8-60a Effect of pipe diameter on total line horsepower. (Nordberg Mfg. Co.)[5]

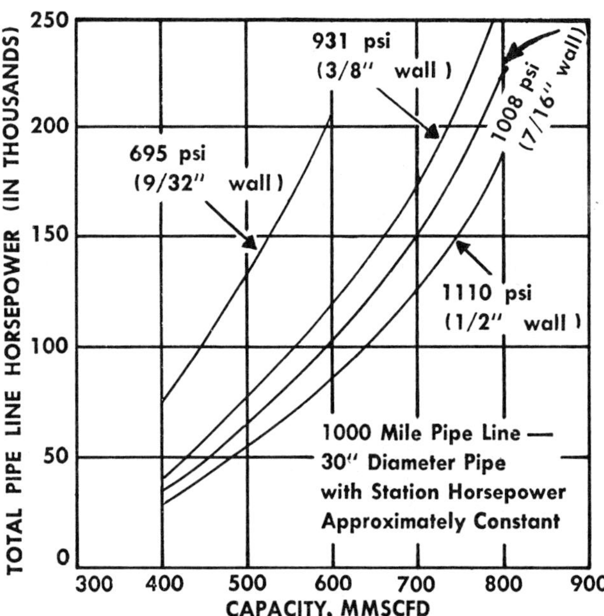

Fig. 8-60b Effect of station discharge pressure on total line horsepower. (Nordberg Mfg. Co.)[5]

Telescoping. If pipe of constant wall thickness is used throughout a line, only the portion near the discharge side of each station is stressed up to the allowable limit. Therefore, capital investment may be reduced by telescoping, that is, by installing pipe of progressively thinner walls as the distance downstream from the station increases and as the falling gas pressure permits. Thus, pipe costs can be greatly reduced, if the distance and pressure drops between stations are relatively large. Telescoping makes possible the use of

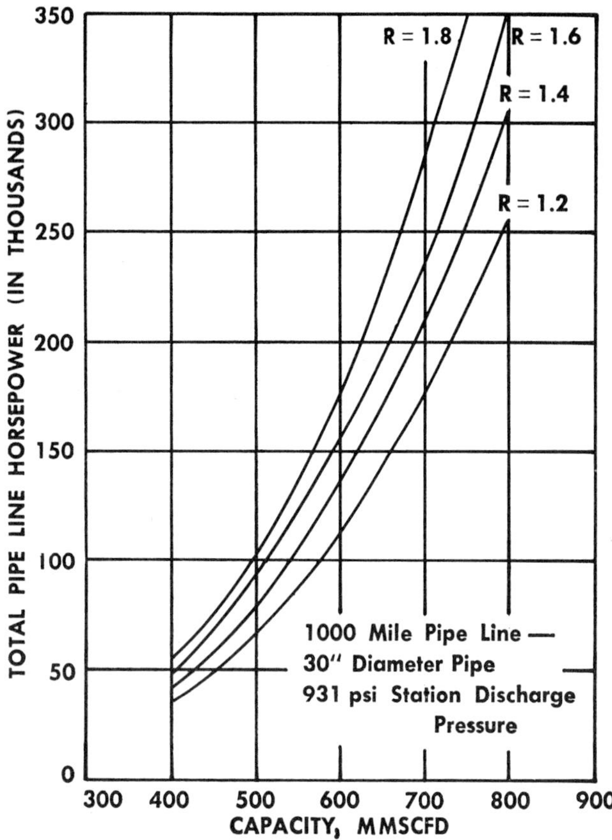

Fig. 8-60c Effect of station compression ratio on total line horsepower. (Nordberg Mfg. Co.)[5]

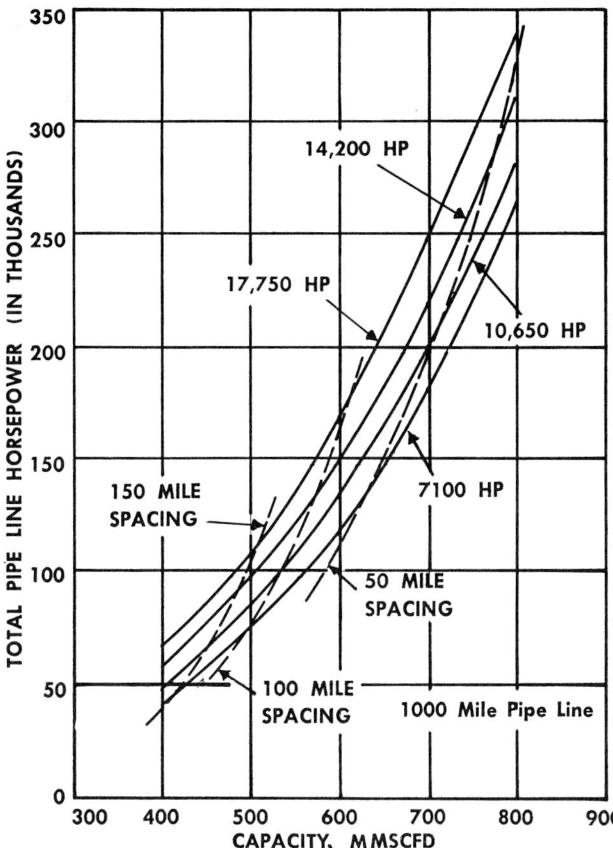

Fig. 8-60d Effect of station horsepower and station spacing on total line horsepower. (Nordberg Mfg. Co.)[5]

larger horsepower stations, spaced further apart than would otherwise be economical.

Although the effect of reduced pipe cost may lower transportation cost, definite disadvantages in the operation and expansion of a telescoped line are:

1. Operation at reduced capacities which lower pressure drop between stations and necessitate cutting back discharge pressure correspondingly, in order to prevent overpressuring pipe with thinner walls. This increases horsepower requirements and may seriously limit capacity, if the flow reduction was caused by compressor units being out of service.

2. Dispatching is complicated, and chance of pipe failure is somewhat increased—unless adequate relief devices are provided.

3. Line packing is limited.

4. Expansion of a system by means of intermediate stations or looping must be planned before initial line construction. Provisions to do this reduce actual savings in investment.

Station Compression Ratio. Use of stations of low horsepower, spaced relatively close together, and thus operating at low compression ratios, brings a considerable saving in pipeline horsepower, as illustrated in Fig. 8-60c. The increasing efficiency of centrifugal compressors at low ratios further contributes to this saving, while the decreasing efficiency of the reciprocating compressor reduces it. Although the curves indicate that even further reductions in compression ratio would be advantageous, a practical limit is reached when the total number of stations becomes too large and the horsepower of each station too small to be economically possible. In small stations, initial investment per horsepower and operating cost per horsepower become so high that it is not advantageous to decrease compression ratio beyond a certain point.

Figure 8-60d shows somewhat similar curves, except that station size was held constant instead of compression ratio. This is a more practical approach, since station compression ratio is usually adjusted to load up a station consisting of an integral number of standard units. Curves of constant station spacing are also shown.

Flexibility. In designing a pipeline and selecting compressing equipment, a flexible arrangement must be provided to operate efficiently over a range of capacities and under all conditions likely to be encountered. For example, Fig. 8-60d shows that overall horsepower requirements are lowered by using stations of low horsepower, spaced closely. However, low horsepower stations have less operational flexibility; e.g., one unit out of service in a two-unit station reduces line capacity more than one out of service in a station of higher horsepower. Similarly, normal fluctuations in capacity in larger stations more easily allow removal of a unit from service for maintenance work.

Expansion. It is usually not desirable to design a pipeline for a specific set of conditions without allowing for possible future growth in capacity, in addition to normal operating fluctuations. Such expansion is always facilitated if initial plans have appropriately provided for it, with the most flexible arrangement possible consistent with economy.

An increase in capacity is usually made by one or more of the following methods:

1. Increasing operating pressure by adding horsepower to existing stations or by constructing new intermediate stations.
2. Looping line, i.e., parallel existing line on the same right of way.

Planned expansion normally goes in steps, with capacity added as required.

The first steps usually involve adding units to existing stations; this addition affords small increases in capacity with relatively little increase in investment and operating cost. However, Fig. 8-60d shows that adding horsepower to a constant station spacing increases total pipeline horsepower requirements sharply for a large expansion. Thus, a point is reached at which it is economically advantageous to install intermediate stations or to add pipe.

Expansion by looping, because of the large investment required, is usually considered only for large increments of expansion or as a provision for further growth later.

Selection of a method of expansion in a pipeline system, as in designing a new line, involves detailed economic study of each case.

Economic Relationships in Selecting Compressing Equipment

Gas pipelines in the U. S. are regulated either by the Federal Power Commission (if interstate) or by State Commissions (if intrastate lines). Since these commissions generally limit the annual return on the total investment from 5¼ to 6¾ per cent (1961), the most satisfactorily designed line is one on which this return can be realized while cost of transporting gas is kept at a minimum.

Cost data in this chapter are based on 1954 prices, unless otherwise specified. Price levels for 1962 were about 30 per cent higher than the 1954 quotations. Annual indices of pipeline equipment costs are available.[10]

Table 8-17 Representative Pipe Cost Data[11] (1962)

Outside diam, in. 1	Wall thick., in. 2	Allowable operating pressure, psi 3	Weight, tons/mile 4	Pipe cost, \$1000/mile 5	Laying cost, \$1000/mile 6	Total cost, \$1000/mile 7
36	½	1067	498	116	50	166
36	7/16	932	440	105	50	155
36	⅜	795	377	88	50	138
30	½	1290	416	97	40	137
30	7/16	1120	365	85	40	125
30	⅜	960	313	73	40	113
24	⅜	1210	250	58	30	88
24	5/16	1005	209	49	30	79

(Nordberg Mfg. Co.)

Explanatory Notes

Yield strength = 52,000 psi.

Column 3: Allowable operating pressure—based on 72% of yield strength as average allowable stress.

Column 5: Pipe cost—\$140 per ton (\$143 per ton for wall thicknesses under 11/32 in.) plus \$22.60 per ton freight and 15% general overhead.

Column 6: Laying cost—includes right of way and all other costs pertaining to placing pipe in ground ready for operation, plus 15% general overhead.

To determine actual transportation cost, the total investment must be determined first. This consists of pipe costs, including freight, laying, right of way, ditching, wrapping, and welding; of station costs, including equipment and buildings; and of the fixed or overhead costs on each.

Pipe Cost and Laying Cost. For practically any pipeline, it is estimated that 70 to 90 per cent of the total capital investment represents the cost of the pipe in the ground. Average cost figures cannot be applied to all lines. Laying costs may vary from \$5.00 to \$20.00 per foot, depending upon the terrain and right of way cost (1954). While pipe cost per ton remains approximately the same, freight costs vary. However, these variations do not materially affect the selection of compressing equipment. Table 8-17 gives some representative pipe costs, adequate for comparing compressing equipment although not necessarily for estimating any specific line. A range of pipe sizes is shown, best suited to capacities of 300 MMSCFD and up.

Initial Station Costs. An important factor in comparing compressing equipment is the total cost of the particular type of station. Typical items considered are the number of employee houses, which should be proportional to the labor force necessary to operate the station; building requirements for compressors; and gas piping, which differs for centrifugal and reciprocating compressors.

Table 8-18 lists cost estimates for compressor stations of 14,000 to 15,000 hp. Different companies may show variations in costs because of differences in station design, local conditions, or estimating methods. However, the various costs have been treated alike for all competing types of equipment, and their relative position should remain unchanged.

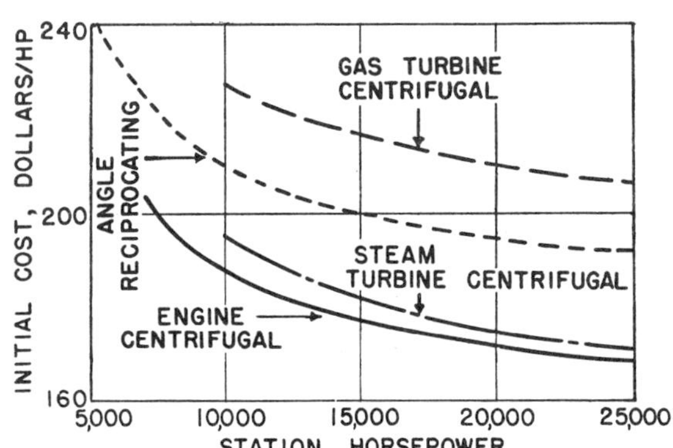

Fig. 8-61 Initial station investment for various prime movers (1954). (Nordberg Mfg. Co.)[5]

Figure 8-61 shows a comparison of initial costs of stations from 5000 to 25,000 hp. The curves were not extended below the size of two-unit stations, because those of one unit were considered impractical for general use. Costs shown in Table 8-18 and curves in Fig. 8-61 were based on 2500 hp angle units, 5000 hp gas turbines and steam turbines, and 3550 hp engine centrifugal units. Figure 8-61 shows highest initial investment for gas turbine centrifugals and lowest for engine centrifugals.

Operating Costs. Since pipeline cost factors are significant only if correlated to the amount of gas transported, it becomes necessary to add up all expenses entering into the cost of service. These expenses which determine the sales price of the gas include transportation cost, field cost, allowable return on investment, depreciation, and total taxes.

Analysis of fixed costs is very significant, in view of the relatively large capital costs inherent in pipeline operation. Fixed costs, as defined by the Federal Power Commission, include "... those which relate entirely or predominately to the capital outlay necessary to provide the system capacity and also those operating expenses which do not vary materially with the quantity of gas transported thru the pipeline system."[6]

Table 8-19 summarizes station operating costs, as calculated by Nordberg.[2] All costs previously listed are included, along with operating overhead and fixed costs on investment.

The curves in Fig. 8-62 were plotted using the total yearly operating costs in dollars per horsepower from this summary and values for stations of other sizes. They illustrate the inverse relationship between cost per horsepower and station size. As stated previously, however, use of stations of high horsepower spaced farther apart increases total horsepower requirements and their cost; thus, total operating costs are not decreased. These curves establish the relative economic position of various types of equipment for pipeline service and include all economic factors relating to selection of compressor equipment for any given size station. Pipeline design also has some bearing on station size and kind of equipment. Where the design is fixed, these curves become the criterion for selecting compressing equipment.

Table 8-18 Estimated Costs for Compressor Stations of 14,000 to 15,000 HP
(Based on 1954 equipment prices)[5]

	Engine centrifugal, 4 units— 14,200 hp*	Gas turbine centrifugal, 3 units— 15,000 hp*	Angle reciprocating, 6 units— 15,000 hp*	Steam turbine centrifugal, 3 units— 15,000 hp*
Group I—Mechanical & Electrical Equipment (prime movers, compressors, accessories and auxiliaries, auxiliary generating units, switchgear, instrumentation and controls).				
Group cost	$1,398,000	$1,889,000	$1,569,500	$1,276,400
Group II—Installation of Mechanical & Electrical Equipment (counting $25 per day for common and skilled, and $60 per day for supervisory labor).				
Group cost	$ 119,000	$ 177,400	$ 173,600	$ 386,200
Group III—Equipment Foundations.				
For compressor units, cu yd	480	900	750	60
For auxiliary generating units, cu yd	60	40	90	40
For accessories incl. regenerators & boilers where required, cu yd	15	70	20	240
Group cost	$ 27,700	$ 50,500	$ 43,000	$ 17,000
Group IV—Buildings (main compressor, auxiliary, office and storage at 80 cents per cu ft, houses for personnel at $15,000 ea.—see Fig. 8-51 for sizes).				
Group cost	$ 242,000	$ 309,800	$ 366,500	$ 275,400
Group V—Land and Improvements (all 25 acres at $300 per acre).				
Group cost	$ 37,500	$ 37,500	$ 37,500	$ 47,500
Group VI—Main Gas Piping (pipe, valves, fittings, gas scrubbers, cathodic protection, installation labor).				
Group cost	$ 352,000	$ 329,500	$ 379,000	$ 329,500
Group VII—Miscellaneous (freight at $1.50 per cwt on applicable items, general overhead 15 per cent of total cost).				
Group cost	$ 352,200	$ 463,400	$ 427,500	$ 390,500
GRAND TOTAL	$2,528,400	$3,257,100	$2,996,600	$2,722,500
Dollars/horsepower	$ 178	$ 217	$ 200	$ 182

(Nordberg Mfg. Co.)
* Total

Table 8-19 Total Yearly Operating Costs for 14,000 to 15,000 HP Station[5] (1954)

	Engine centrifugal, 4 units— 14,200 hp*		Gas turbine centrifugal, 3 units— 15,000 hp*		Angle reciprocating, 6 units— 15,000 hp*		Steam turbine centrifugal, 3 units— 15,000 hp*	
Item	$/year	$/hp-yr	$/year	$/hp-yr	$/year	$/hp-yr	$/year	$/hp-yr
1. Labor	64,800	4.60	55,400	3.70	108,600	7.20	80,800	5.40
2. Maintenance and supplies	58,100	4.10	41,000	2.70	69,200	4.60	48,300	3.20
3. Lubricating oil	4,600	0.30	1,800	0.10	12,500	0.80	1,700	0.10
4. Fuel	120,300	8.50	184,700	12.30	144,600	9.60	238,800	15.90
5. Operating overhead 40% of items 1, 2, 3, 4	99,100	7.00	113,000	7.60	134,000	9.00	147,900	9.90
5. Fixed costs 12% of investment	303,500	21.40	391,000	26.10	360,000	24.00	327,500	21.90
Total	$650,400	$45.90	$786,900	$52.50	$828,900	$55.20	$845,000	$56.40

(Nordberg Mfg. Co.)
* Total.

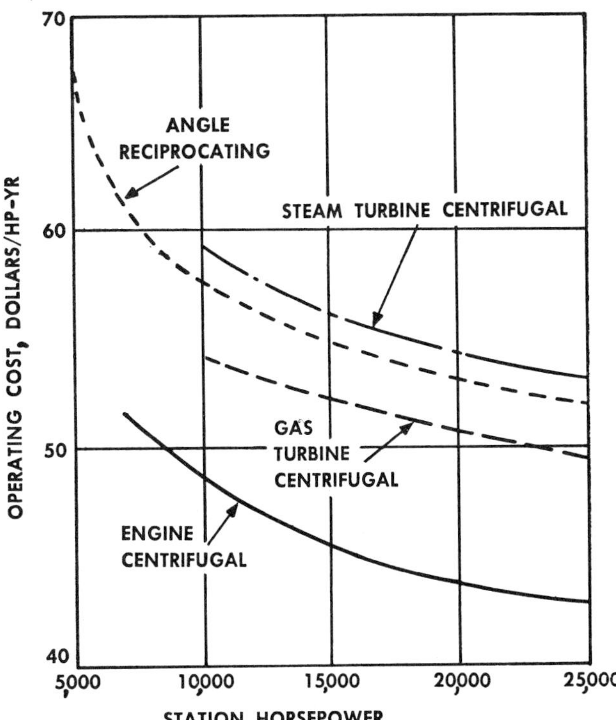

Fig. 8-62 Total annual station operating costs for various prime movers (1954). (Nordberg Mfg. Co.)[5]

Minimum Transportation Cost

In a general comparative study of compressing equipment, it is convenient to base a hypothetical pipeline design on minimum gas transportation cost. Equipment can then be compared for a complete pipeline project. The following procedure determines proper pipe diameter, operating pressure, station spacing, and station size for each type of compressing equipment, to arrive at minimum gas transportation cost for each sales volume.

A minimum of two compressing units per station was set to provide adequate flexibility. No stand-by units were included and no provisions were made to facilitate future expansion or to limit capital investment. But, because fixed costs on investment represent a very large portion of overall transportation cost, the capital investment for a pipeline design giving minimum transportation cost is only slightly higher than the absolute minimum capital investment for any given sales volume.

A hypothetical 1000 mile pipeline was chosen for purposes of illustration. For each of the four types of compressing equipment, pipe sizes of 24 in., 30 in., and 36 in. were chosen for a range of 300 MMSCFD to 900 MMSCFD capacity. Several suitable pipe wall thicknesses were used to investigate operating pressures of 700 to 1000 psi, with allowable design stress of 70 per cent of yield strength. Telescoping was not considered. Station compression ratios below 1.2 were eliminated for reciprocating compressors, since this was held to be their practical limit. The sales volume delivered at the end of the line was considered constant at full design capacity. Within these limits, the minimum transportation cost at each sales volume was established for the various types of units.

Figure 8-63 shows curves of minimum transportation cost and total capital investment. These curves were derived as follows:

Using pipe diameters, operating pressures, and capacities as described with suitable pressure ratios, curves of station horsepower vs. station spacing were drawn, using the *Panhandle "A"* flow formula and the compressor horsepower formula. From them, the required number of stations of each size and their operating horsepowers were established. To obtain total investment, station investment was added to pipe investment. Transportation cost was then computed by adding a fixed percentage of total investment to operating costs plus fuel costs and by dividing this sum by actual sales volume delivered.

Fuel costs for the main compressor units were determined for the entire pipeline, rather than for each station individually. To evaluate fuel at each station, it is necessary to add the field cost and pumping cost to the consumption point. A simple method, used in this study, was to base pipeline design on a volume representing the average of the amounts purchased and delivered. Since fuel is a part of the total gas purchased, it was evaluated only at field cost. Sales volume was based on total gas purchased less fuel consumed. A field cost of 10 cents per MCF was assumed* and the following full load fuel consumption rates (*net* Btu per bhp-hr)* were used for each type of equipment:

Engine centrifugal...............	6,250
Gas turbine......................	9,250
Steam turbine....................	11,100
Angle reciprocating..............	7,000

In this way, the line was designed to carry the extra volume represented by the fuel, while the unit transportation cost was based on the actual volume available for sale. In practice, the initial stations pump practically all the fuel and would require more horsepower or closer spacing than those downstream. This method gives average station horsepower and total number of stations.

The curves of Fig. 8-63 for each nominal pipe size represent the minimum transportation cost for that size. For each sales volume, the optimum station size and pipe wall thickness were established by the foregoing procedure for each possible combination, and the minimum transportation cost was selected.

The curves (dashed lines) tangent to the actual pipe curves (smooth solid curves) give theoretical minimum transportation costs for an infinite number of intermediate pipe sizes. The smooth curves (dashed line), in the upper portions of the figure for each compressor type, represent the capital investment for this theoretical minimum transportation cost. Curves of 40 in. pipe were also plotted (dot-dash lines); this size was not readily available. Multiple pipes of smaller diameter are often used to provide added reliability of service.

Minimum transportation cost at all points for engine centrifugals, gas turbine centrifugals, and steam turbine centrifugals was provided by two unit stations. The angle stations range from two units each at the lower capacities to four each at 900 MMSCFD, with three unit stations providing

* These assumptions may be modified to adjust to specific local circumstances and the curves replotted accordingly.

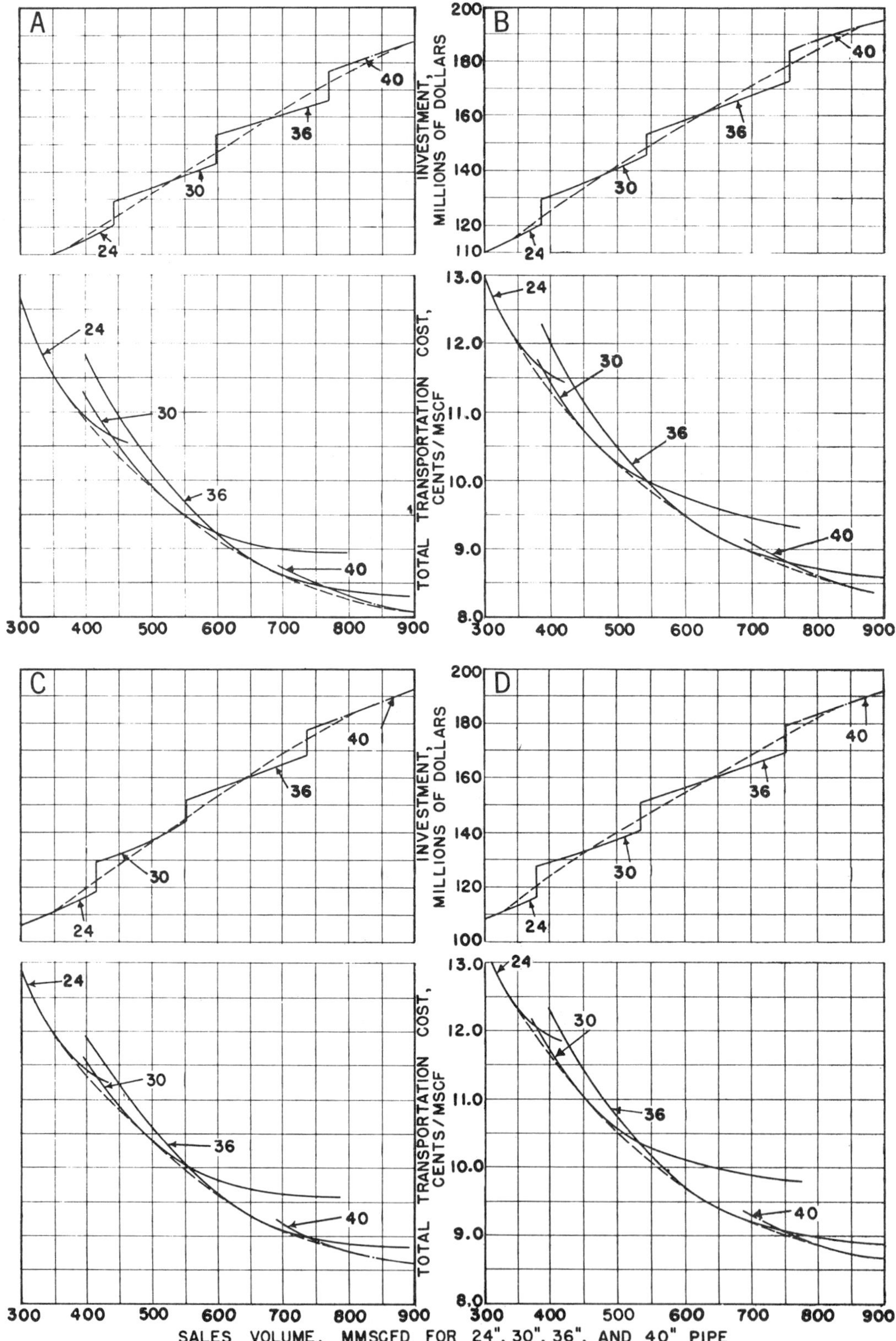

Fig. 8-63 Total capital investment and minimum transportation costs for various prime movers (1954). (Nordberg Mfg. Co.)[5] A—engine centrifugal compressors; B—gas turbine centrifugal compressors; C—angle reciprocating compressors; D—steam turbine centrifugal compressors.

minimum cost at most points. Sizes of units used were: 3550 hp engine centrifugal, 5000 hp gas turbine, 5000 hp steam turbine, and 2500 hp angle compressors.

Table 8-20 shows variation in transportation costs for stations with different numbers of units.

Theoretical minimum transportation costs using each of the four types of equipment are plotted in Fig. 8-64. It illustrates graphically the relative position of each type, with all economic factors considered. All its curves are roughly parallel.

A 1962 study of engine centrifugal compressor systems is available.[11]

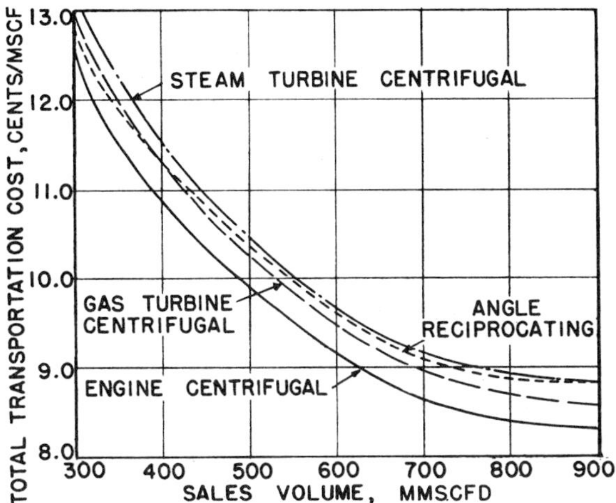

Fig. 8-64 Theoretical minimum transportation costs for various prime movers (1954). (Nordberg Mfg. Co.)[5]

Transportation costs indicated by these curves are, in general, considerably lower than for earlier lines of this length (prior to 1954), for the following reasons:

1. No provisions were made for future expansion.
2. No limitations were placed on capital investment.
3. Only certain pipe sizes are available at commercial (i.e., competitive) prices. Thus, at some capacities, actual costs would be higher.
4. Because of improvements in pipe and development of compressing equipment for lower compression ratios, these curves are based on slightly higher operating pressures and lower compression ratios than are found on most earlier lines (prior to 1954).
5. Because efficient types of compressing equipment were used, these curves reflect cost reductions afforded by design improvements.
6. These transportation costs are based on continuous operation at full design capacity. In practice, this would not be possible because of operational limitations or demand and supply fluctuations.
7. No stand-by compressor units were included.

Table 8-21 Comparison of Minimum Transportation Costs with Fixed Cost on Installed Pipe Deducted[5]

Type of unit	Total unit transportation cost ¢/MSCF	% Difference	Fixed cost of pipeline, ¢/MSCF	Total station fixed and operating costs plus overhead ¢/MSCF	% Difference
Engine centrifugal	9.95	0	7.80	2.15	0
Gas turbine	10.33	+3.8	7.80	2.53	+17.7
Angle reciprocating	10.40	+4.5	7.80	2.60	+20.9
Steam turbine	10.57	+6.2	7.80	2.77	+28.8

(Nordberg Mfg. Co.)

Differences in transportation costs are relatively small among the various types of compressing equipment, because of the huge fixed costs of the pipeline itself. Nevertheless, proper equipment selection is a principal way to lower operating costs. In this regard, Table 8-21 gives the minimum transportation costs from Table 8-20 and deducts the portion for the fixed costs of the installed pipe itself. Note that while the total unit transportation cost spread is only 6.2 per cent, the differences in costs attributable to stations range from 17.7 to 28.8 per cent.

Generally, for a given gas flow rate, a particular combination of pipe size, wall thickness, and horsepower will show minimum initial and transportation costs. However, the combination showing minimum operating cost will seldom be the one giving minimum transportation cost, as the following discussion on short lines will indicate.

Table 8-20 Transportation Costs for Various Sizes of Stations[5] (1954)

Sales volume = 500 MMSCFD; 30 in. pipe, 3/8 in. wall

Type of unit	No. of units per station	No. of stations on 1000 mile pipeline	Total capital investment, $MM	Station operating hp	Actual sales volume delivered, MMSCFD	Total daily transportation cost, $	Unit transportation cost, ¢/MSCF
Engine	2	11	134.4	7,000	494.2	49,190	9.95
centrifugal	3	9	136.3	9,300	493.7	50,340	10.20
Gas turbine	2	9	139.0	9,300	490.7	50,670	10.33
centrifugal	3	7	141.3	14,400	488.8	52,220	10.68
Angle	2	15	136.8	4,600	494.2	51,780	10.48
reciprocating	3	11	136.8	6,800	493.7	51,360	10.40
	4	9	137.0	9,000	493.2	51,870	10.52
Steam turbine	2	9	136.1	9,300	488.9	51,670	10.57
centrifugal	3	7	137.6	14,400	486.6	53,020	10.90

(Nordberg Mfg. Co.).

DESIGN OF A SHORT TRANSMISSION LINE (100 MILES, 1957)

There are two general classifications of transmission pipelines. One is the short line, not exceeding 150 miles long (often much shorter), whose outlet pressure is an important, in fact the deciding, consideration. The second is the long line, over 150 miles long, in which outlet pressure becomes less important as length increases.

A method of planning a short line and some general conclusions about expected results are presented here. Actual cost of gas or investment for any particular installation or project is not covered.

Assume a pipeline 100 miles long, 250 psig outlet pressure, with a supply pressure of 100 psig for one set of examples and 500 psig supply pressure for a second. Volumes of gas flowing are set at 50, 100, 200, and 250 MMCFD. This wide range is used to show the influence of volume on ultimate system design.

Pipe material was assumed to be API 5LX-52 (minimum specified yield point of 52,000 psi). Using a design factor of 72 per cent, allowable stress in the pipe equals 37,400 psi. Minimum pipe wall thickness was set at 0.250 in. as the minimum for practical use in field operations. No allowance was made for corrosion.

The outlet pressure was set at 250 psig, because this seemed adequate for a town border station. Outlet pressure can be varied to a degree without too great an effect on the inlet pressure.

The supply pressure was set at 100 psig for the first set of examples, to represent the suction pressure of a compressor station located in a production area. The supply pressure for the other set of examples was set at 500 psig to represent the pressure at which gas might be received from a long cross-country transmission line.

The *Grizzle Gas Pipeline Slide Rule*[7] was used to calculate gas flow thru various sizes of pipe. This rule is based on the *Panhandle "A"* formula. No allowance was made for deviation from Boyle's law, since its effects here would be very slight. No efficiency or experience factor was applied, other than that included in the formula and slide rule.

Pipe wall thickness calculations were made using *Barlow's formula* (also see Table 8-7 and Fig. 8-10), while weight of pipe in tons per mile was calculated by the following formula:

$$W = 28.2T(D - T)$$

where: D = pipe OD, in.

T = pipe wall thickness, in.

In determining the first cost of a pipeline, a number of factors must be considered. After studying cost figures (1957) from various companies, the following values can be taken for representative costs per mile (keeping in mind that these rule-of-thumb figures are not reliable for detailed analyses):

Engineering and survey	$300 per mile
Cost of pipe	$150 per ton
Freight	$ 15 per ton
Laying costs (clearing, ditching, laying, welding, and backfilling; see Fig. 8-68)	$528 per in. diam-mile + $7500 per mile
Valves, flanges, bolts, welding rod	$ 20 per in. diam-mile
Inspection and supervision	$ 40 per in. diam-mile + $400 per mile
Painting and wrapping	$200 per in. diam-mile + $800 per mile
Rights of way and damages	$160 per in. diam-mile + $1000 per mile

Combining these figures:

First cost per mile = $165 per ton *plus* $948 per in. diam *plus* $10,000 per mile.

Cost per required brake horsepower ($275) was taken to be a representative value for the first cost of compressor stations. In all examples, compressor station cost was based on required horsepower and no attempt was made to size units or allow for spare equipment.

The annual total operating expense consists of two items: (1) annual fixed expense, and (2) annual variable expense or direct expense.

The *fixed expense* is a direct function of the first cost or investment. It consists of return on investment, depreciation, and certain taxes and insurance. Return was set at 6.5 per cent per year and depreciation at 3.0 per cent (expected life of 33⅓ years). A charge of 1.5 per cent for taxes and insurance was made.

Annual *variable costs* were determined from operating cost records. An average or representative value was chosen for each charge, as follows:

1. Pipe lines
   (a) Direct operating costs—$185 per mile-yr
   (b) Direct maintenance costs—$115 per mile-yr
2. Compressor stations
   (a) Direct operating labor—$13 per hp-yr
   (b) Fuel and other operating costs—$25 per hp-yr
3. Payroll taxes
   (a) Pipelines—$10 per mile-yr
   (b) Compressor stations—1.73% per year on labor only
4. Administrative and general expense
   (a) Pipelines—$229 per mile-yr
   (b) Compressor stations—67% per yr on labor only

Adding all operating costs as outlined gives the equation:

Total annual operating costs = $539 per mile-yr + $46 per hp-yr + 11 per cent of total first cost.

Data given in Table 8-22 and Figs. 8-65a thru 8-65g were calculated by using the assumed gas flow, outlet pressure, supply pressure, and length of line. Required inlet pressures were calculated for various pipe diameters by using the *Grizzle Slide Rule*.[7] Knowing the required inlet pressure, the pipe wall thickness was calculated for different pipe diameters. Weight of pipe in tons per mile was next determined, and the first cost of the pipe in the ground was calculated.

From the compression ratio, the total required horsepower was determined. The first cost of the compressor stations was then readily obtained. This cost plus the first cost of the pipe in the ground gave the total first cost of the system.

The annual operating costs, both variable and fixed, were determined by using the foregoing equation. Papers giving further discussion of these methods are available.[8,9]

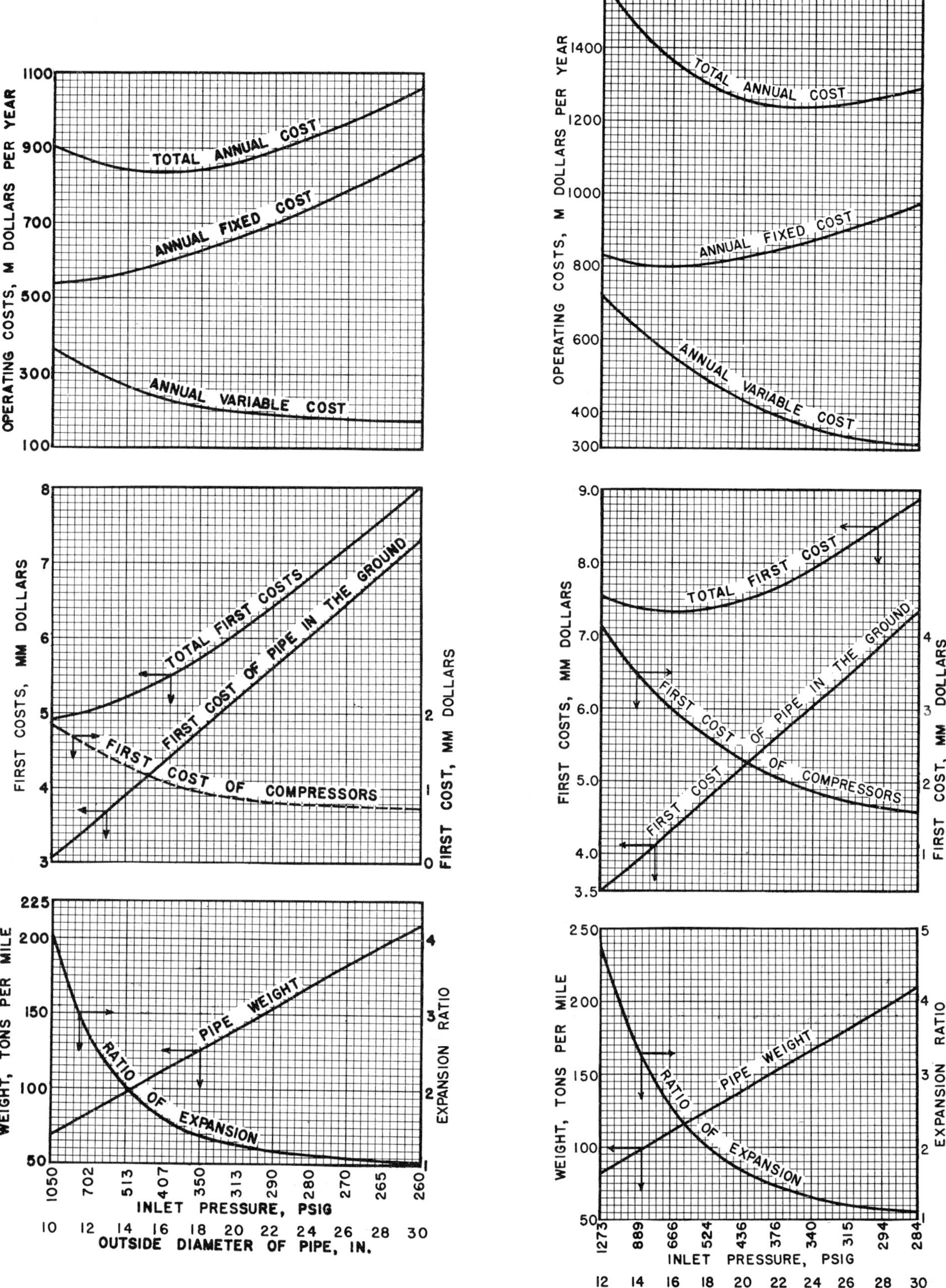

Fig. 8-65a Cost curves for 100 mile line, flow rate 50 MMSCFD, outlet pressure 250 psig, supply pressure 100 psig.

Fig. 8-65b Cost curves for 100 mile line, flow rate 100 MMSCFD, outlet pressure 250 psig, supply pressure 100 psig. See Table 8-22 for supplementary data.

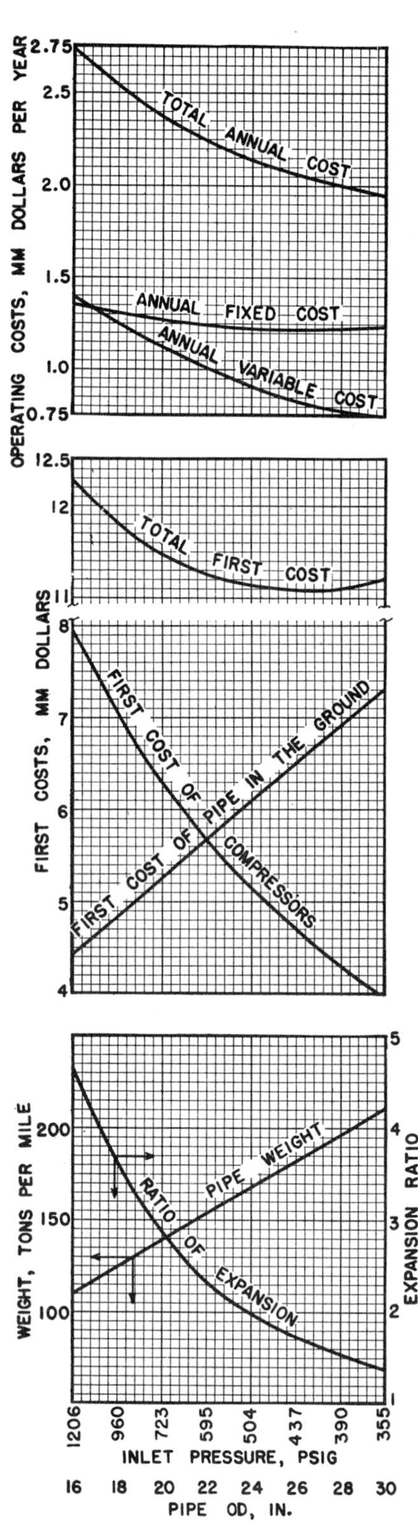

Fig. 8-65c Cost curves for 100 mile line, flow rate 200 MMSCFD, outlet pressure 250 psig, supply pressure 100 psig.

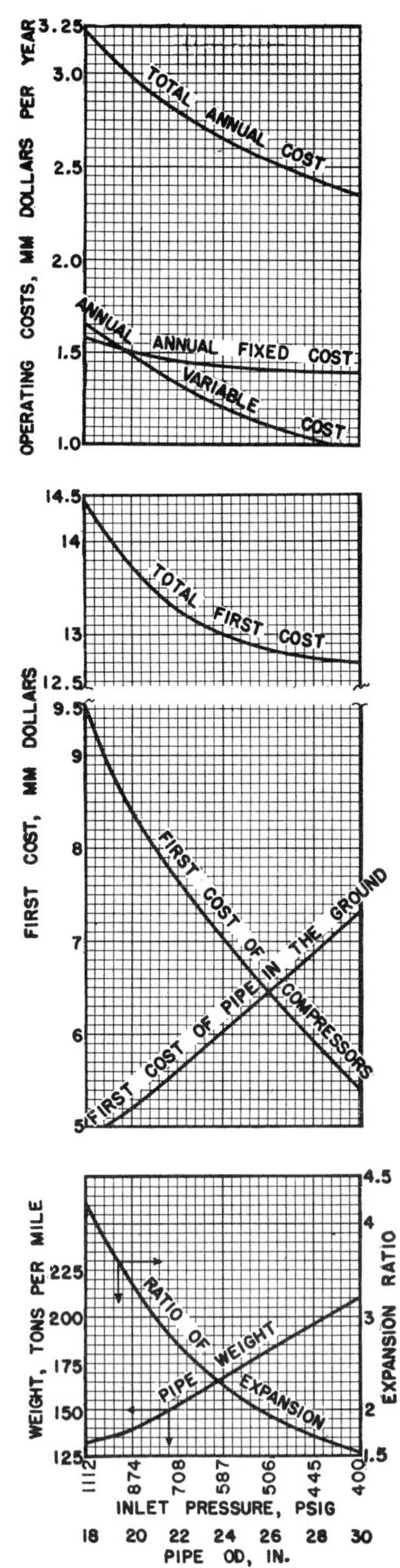

Fig. 8-65d Cost curves for 100 mile line, flow rate 250 MMSCFD, outlet pressure 250 psig, supply pressure 100 psig.

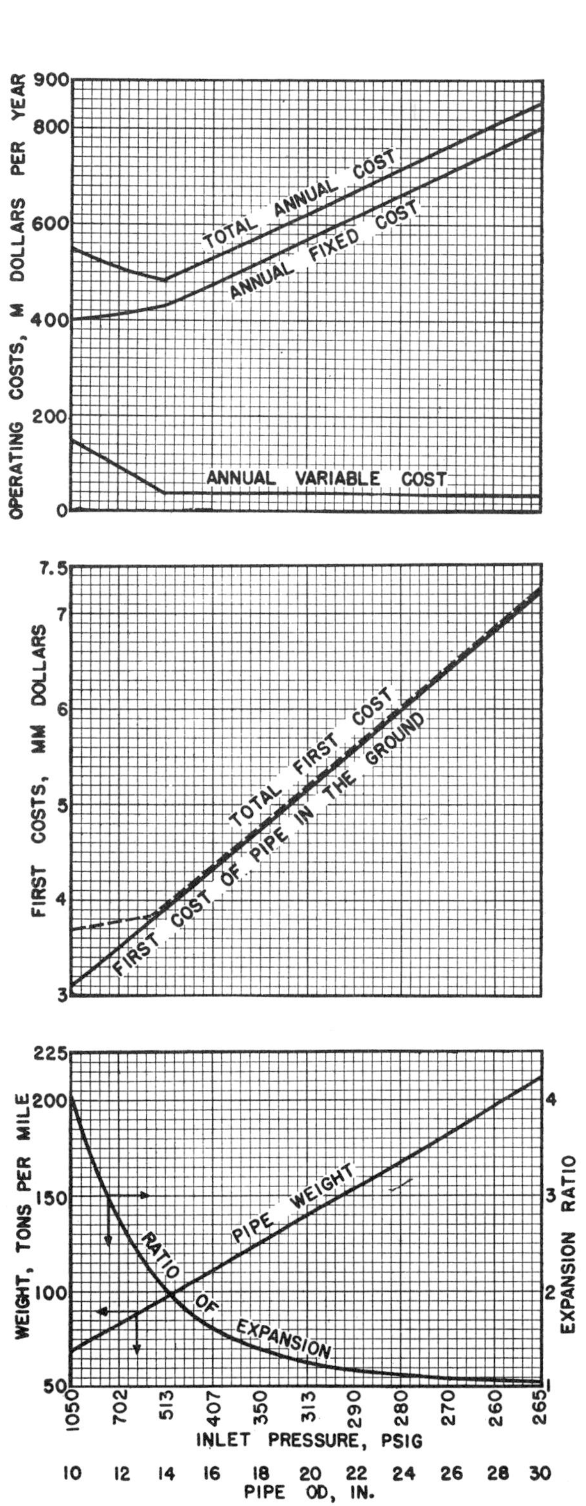

Fig. 8-65e Cost curves for 100 mile line, flow rate 50 MMSCFD, outlet pressure 250 psig, supply pressure 500 psig.

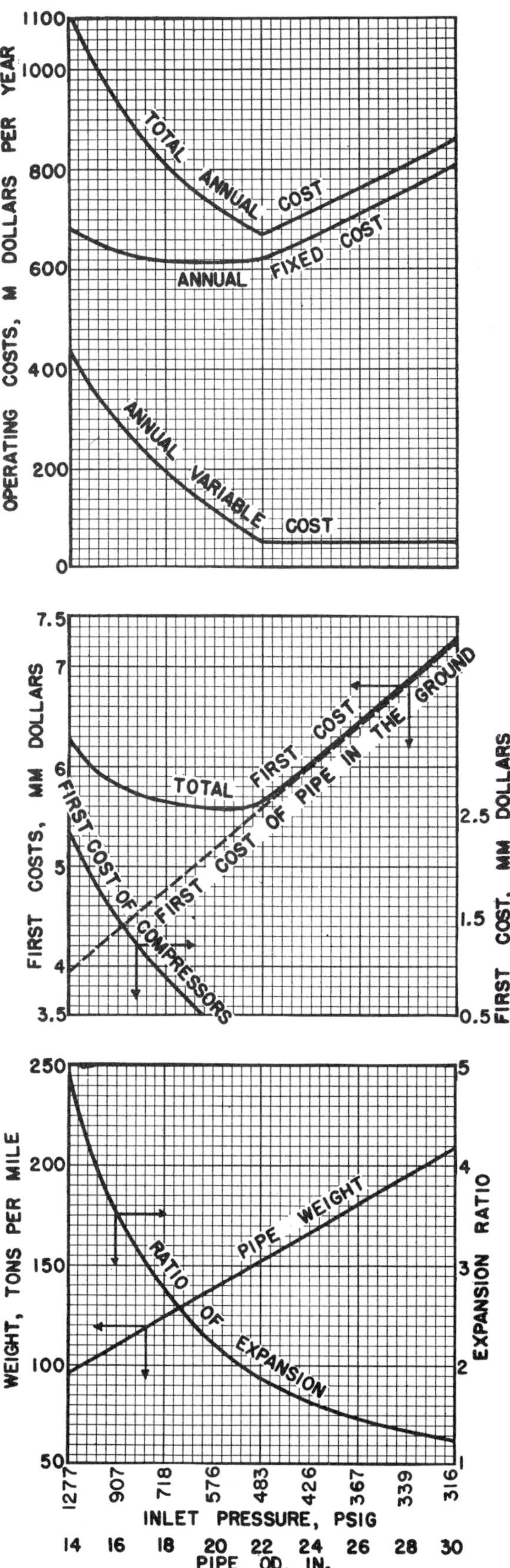

Fig. 8-65f Cost curves for 100 mile line, flow rate 150 MMSCFD, outlet pressure 250 psig, supply pressure 500 psig.

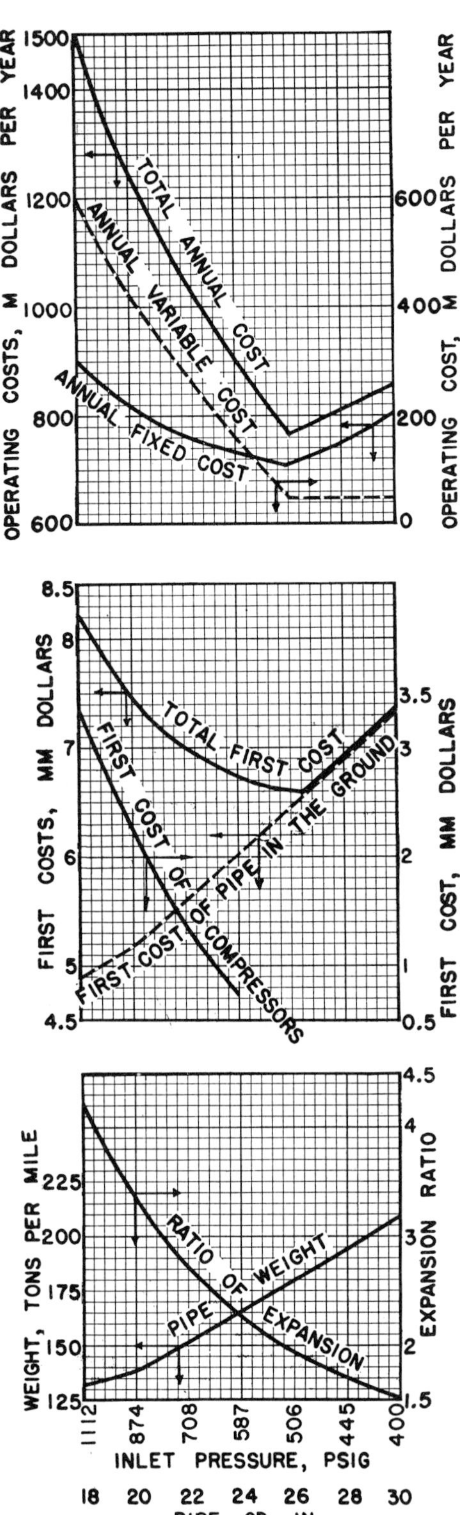

Fig. 8-65g Cost curves for 100 mile line, flow rate 250 MMSCFD, outlet pressure 250 psig, supply pressure 500 psig.

Curves for each pipe diameter were plotted to show the influence of various trends and factors. A different inlet pressure applies for each pipe diameter (placed above the pipe diameters to show the pressure variations). Other pertinent data not plotted are tabulated in Table 8-22 (related to Fig. 8-65b).

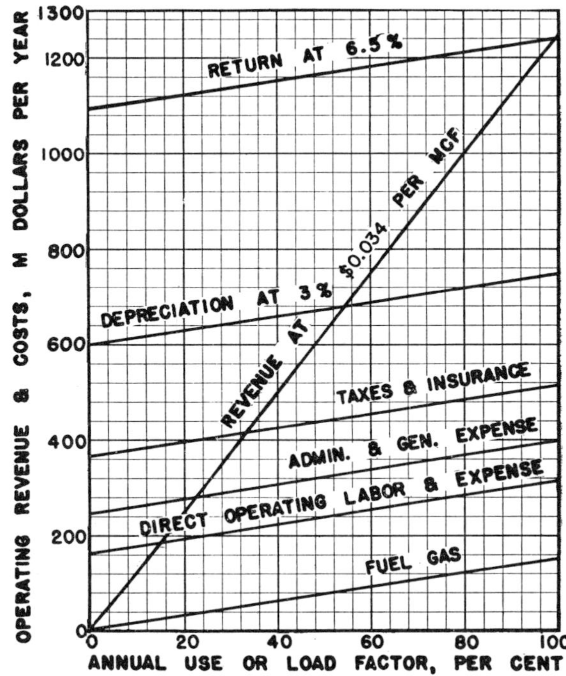

Fig. 8-66 Minimum annual operating costs: gas flowing 100 MMSCFD, 22 in. pipeline, first cost $7,661,000, depreciation 3 per cent per year.

Table 8-22 Supplement to Data Plotted in Figure 8-65b (1957)

Gas flowing—100 MMSCFD
Supply pressure—100 psig
Outlet pressure—250 psig
Length of line—100 miles
API 5LX—52 pipe, design factor 72%
Limit minimum wall thickness to 0.25 inch

Nom. pipe size, in.	Total horse-power	Annual operating cost, $M: Variable operating costs: Pipelines	Stations	Total	Total operating cost*
30	5,800	53.9	267.0	320.9	1,298.9
28	6,000	53.9	276.5	330.4	1,267.4
26	6,380	53.9	294.0	347.9	1,251.9
24	6,860	53.9	316.0	369.9	1,241.9
22	7,460	53.9	344.0	397.9	1,240.9
20	8,420	53.9	388.0	441.9	1,268.9
18	9,520	53.9	438.0	491.9	1,303.9
16	10,900	53.9	503.0	556.9	1,363.9
14	12,560	53.9	578.0	631.9	1,443.9
12	14,800	53.9	682.0	735.9	1,569.9

* Total operating cost equals total variable cost plus fixed cost; the latter is 11 per cent of total first cost.

Two break-even charts for the 100 MMCF example, supply pressure of 100 psig, have been included. Figure 8-66 shows conditions where the pipeline has been designed for *minimum operating cost,* while Fig. 8-67 shows the same conditions when the pipeline is designed for *minimum total first costs.* These curves show the importance of the annual use or load factor in pipeline design problems. No return on the investment is realized when the use factor drops to about 55 per cent. If the expected life is shorter and a higher depreciation rate is figured, the return vanishes at an even higher load factor.

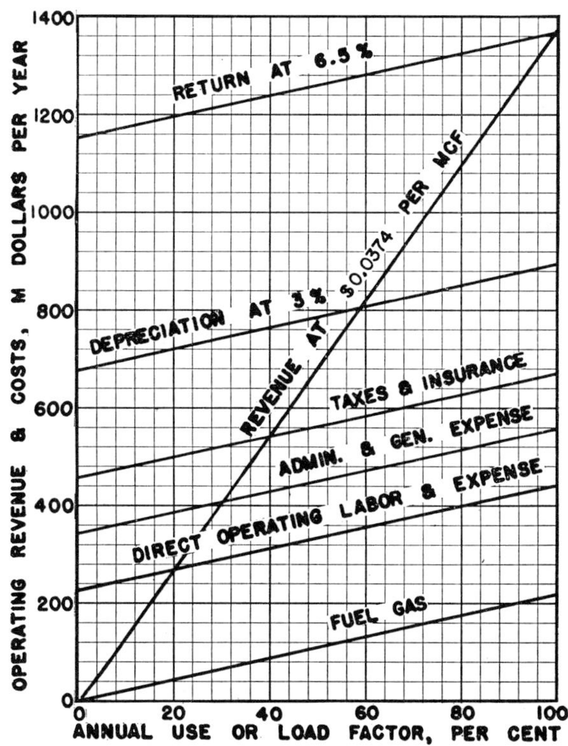

Fig. 8-67 Minimum total first costs: gas flowing 100 MMSCFD, 16 in. pipeline, first cost $7,349,000, depreciation 3 per cent per year.

For short lines the following general statements can be made:

1. No single best pressuıe condition exists for all gas transmission problems.

2. A pipeline system can be designed for minimum first cost or minimum annual operating cost, but not for both.

3. In designing for *minimum first cost*, small volumes will be transported at high pressures and large volumes at lower pressures.

4. In designing for *minimum annual operating cost*, all volumes will be transported at pressures lower than they would be for first cost design.

5. Outlet pressure is an important factor in determining best design pressure.

6. Annual use or load factor is an important consideration.

7. All cost factors must be evaluated correctly and realistically when beginning a design problem.

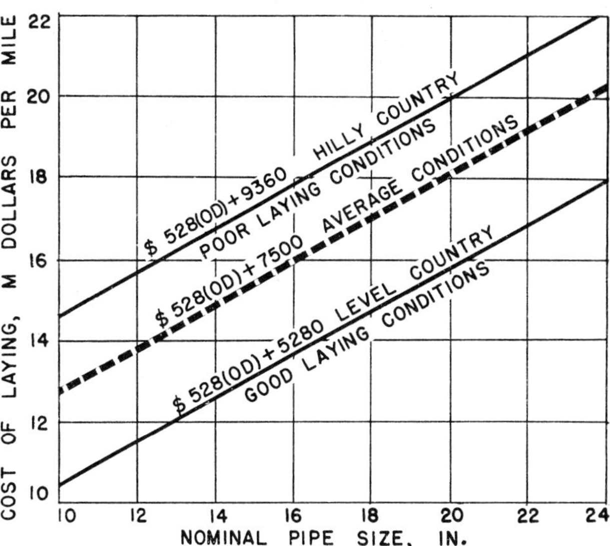

Fig. 8-68 Pipeline laying costs. (Cleaning, ditching, laying, back filling, and welding.)

REFERENCES

1. Poor, W. B. "What Factors Affect Pipe Line Design?" *Gas Age* 114: 29–30, Nov. 18, 1954.
2. Kinney, G. "What It Costs to Build and Operate Gas Pipelines." *Oil & Gas J.* 58: 99, June 27, 1960.
3. Reed, P. and Kinney, G. "Pipeline Contractors' Costs." *Oil & Gas J.* 56: 105, Jan. 20, 1958.
4. ——— ———. "Operating Costs of Gas Lines Take Bigger Bite Out of Revenue." *Oil & Gas J.* 57: 93, July 13, 1959.
5. Nordberg Manufacturing Co. *Selection of Compressing Equipment for Gas Pipe Lines.* (Bul. 226) Milwaukee, 1954.
6. A.G.A. Rate Committee. *Gas Rate Fundamentals*, p. 217. New York, 1960.
7. Grizzle, B. F. "Simplification of Gas Flow Calculation by Means of a New Special Slide Rule." *Pet. Eng.* 16: 154+, Sept. 1945.
8. Kepler, W. R. "Gas Pipe Line Factors Effecting a Minimum Cost." *A.G.A. Proc.* 1930: 797–819.
9. Laverty, F. W. and Huntington, R. L. *High Pressure Pipeline Research.* A.G.A. Transmission Conf., Nat. Gas Sect., 1942.
10. O'Donnell, J. P. "Pipeline Installation and Equipment Costs." *Oil & Gas J.* 61: 105, July 8, 1963.
11. Nordberg Manufacturing Co. *Nordberg Engine Centrifugal Compressor Units for Gas Pipe Lines.* (Bul. 306) Milwaukee, 1962.

Chapter 5
Communications and Dispatching

by D. R. Pflug and H. A. Rhodes

INTRODUCTION

Communications are an essential part of any gas pipeline system. In early gas transmission, few commercial communication circuits were available in areas thru which pipelines passed. Consequently, pole lines were constructed along the right of way. The first circuits were built in the early 1920's at costs of about $400 per circuit mile; they satisfied all communication requirements then necessary to operate pipeline systems. Such requirements were not great, since pressures were then maintained at relatively low levels, with constant contact between control points and field desirable, though not absolutely necessary. An ordinary magneto-type telephone or a telegraph system maintained contact between control points and field.

As gas demand grew, throughput was increased either by installing intermediate compressor stations or by raising operating pressures, and communication systems were expanded to provide the additional communication channels needed. In 1961 it was not unusual to find a private line equipped with from two to five and, in some instances, even 12 carrier channels, plus several telemetering or control channels. These larger circuits became necessary as pipeline operation became more complex, in order to ensure good service to customers and optimum throughput. Along with expansion of channel capacity came the need to maintain contact with all field forces from central control points. Such contact was provided first by mobile radio and immediately afterward by microwave, which made it possible to provide all channel requirements necessary for pipeline operation.

In 1958 pipeline companies owned and operated more than 400,000 channel-miles of communication circuits excluding their mobile units.[1]

METHODS OF COMMUNICATION

Telephone. This is the most frequently used method of communication in the gas industry. Most operating information and orders are transmitted by telephone. It is fast and relatively accurate, but has the disadvantage of providing no record of the information exchanged. Telephones are either of the magneto type, with code ringing facilities, or dial type, with selective ringing devices.

Telegraph. Within the past few years the telephone has largely supplanted the telegraph for communication within the gas industry. On many gas lines dispatching was originally handled exclusively by telegraph operators. This means of communication had the inherent advantage of providing a record of all orders transmitted, but it was slow and required the services of skilled operators at all communication points. Also, in many instances today orders are issued only after an exchange of pertinent information by telephone.

Teletype. The teletype has replaced the telegraph, since it is faster, does not require specially trained operators, and provides a written record of all communication. Teletype equipment, with terminal and line facilities, is more costly than its telegraph counterparts. Both teletype and telegraph use are limited within a gas company to dispatching and orders. Most routine information on operating problems must still be handled by telephone; consequently, both telegraph and teletype must be considered as adjuncts to a primary system.

Mobile Radio. This tool has wide and varied use in the gas industry, but it plays its principal role in emergencies, where constant contact is required between local operating headquarters and field forces which are either moving or capable of being moved.

Mail. Letters, despite their obvious slowness, are still one of the most important communication methods, because they furnish a record of principal operations and transactions, many of which are confirmations of previous information exchanges. Airmail had expedited this means of communication.

Telemetering. Despite its present limited use, this method is expanding as a communication tool. Any intelligence capable of conversion to an electrical impulse can be telemetered, giving a constant record and useful check of remote operations. In addition, instructions in the form of preselected functions or operations can be initiated from a remote point, with the results recorded locally. Many operations, events, or conditions can thus be observed and controlled from a central location. The use of telemetering, however, is limited to preselected operations.

COMMUNICATION SYSTEMS

Company-Owned Pole Line along Right of Way. Many right-of-way agreements incorporate this feature. The older pipeline systems used pole lines for most communication needs, since pipelines usually followed sparsely inhabited routes. Repair crews and line walkers also used these facilities to maintain contact with their headquarters before the advent of mobile radio.

Pole line construction differs considerably in various sections. Pipeline companies in the North and in recognized sleet belts normally use a more rugged type of construction with shorter span length than companies in the South. Costs for the reliable communication systems required in 1960 were about $1000 per mile or higher. Pole spacings range roughly from 150 ft to a little over 200 ft where Nos. 8, 9, and 12 copper wire are used; local conditions are an important factor. Where longer spans are desired, Copperweld or some other high tensile strength conductor is used. To keep transmission losses down, copper, Copperweld, or copper alloys are ordinarily used on long line circuits. If a single talking circuit is adequate, brackets are employed for attaching wires to poles, with four to ten pin crossarms used to accommodate additional circuits. Use of a single wire with a ground return for providing a talking circuit has generally been abandoned, because of the high longitudinal ground and stray current flows, which make this type of line practically useless except for telegraph or teletype service. Originally, when more than one talking circuit was required, four wires were installed, transposed to eliminate objectionable cross talk, and connected into a phantom group to provide three phone lines and one d-c telegraph circuit, as indicated on the left of Fig. 8-69.

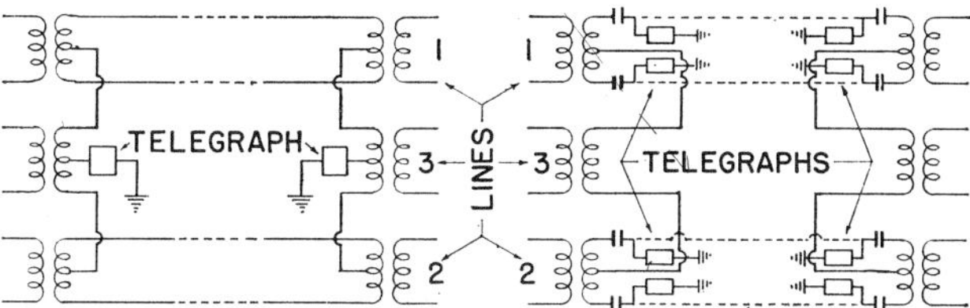

Fig. 8-69 Phantom circuits, showing single telegraph on the left and four composite telegraph circuits on the right.

If additional d-c telegraph channels were required, duplex operation could be employed on the telegraph or the lines could be "composited" to derive four telegraph channels, all of which could be duplexed if necessary. This is accomplished as indicated on the right in Fig. 8-69.

Most private systems employ frequency division carriers for deriving additional talking or telegraph (teletype) circuits. These carriers are relatively cheap and will provide an additional circuit for a fraction of the initial plant investment. The carrier-derived channel, voice or telegraph, is generally superior to the physical circuit, and it can be repeated, that is, it can be amplified for long transmission, more easily than physical circuits. For voice circuits, carriers are generally single side band and amplitude-modulated; telegraph or teletype channels are either amplitude- or frequency-modulated. Voice channel carriers can be added either as single units or in groups of three at the lower frequencies. At the higher frequencies up to 12 channel groups can be obtained.

Telegraph channels can usually be added in groups as required. In some instances, a 3000-cycle carrier or physical voice circuit is used for multichannel telegraph, teletype, or telemetering requirements. The total number derived depends upon transmission speeds and band width requirements, with 15 to 18 narrow band circuits not uncommon. To avoid objectionable cross talk with other systems on the same pole lines, output levels and a uniform transmission direction should be agreed upon and all installations should conform to these. For long carrier circuits, telephone, telegraph, or some form of automatic gain control is desirable for maintaining transmission levels at reasonably constant values.

Carrier transmission distance is a function of line loss, transmitter output, and receiver gain. Line loss varies inversely with conductor size and spacing between conductors, and directly with wire resistance, leakage between conductors, and frequency. Amplifiers make up for line loss.

Company-Owned Wire on Foreign Poles. Where space is available on foreign poles, it is usually desirable from an economic standpoint to lease crossarm space and string wires as required. An annual rental is paid for this privilege, but the lessee is relieved of pole maintenance responsibilities. The gas company lines are more open to troubles on heavily loaded pole lines than they are when constructed along company-owned rights of way. Generally, however, these more heavily loaded pole lines are more accessible for maintenance.

Since existing pole lines usually do not follow pipeline routes, many taps and loops must be constructed to join pipeline installations to pole line rights of way. Carrier installations similar to those made on pole lines along a pipeline right of way can be made on these wire facilities. However, since most heavily loaded pole lines already have many circuits with carrier installations, the pipeline company facilities must be carefully transposed to avoid undesirable pickup from these other lines. These transportation installations are relatively costly and in many instances can be partially avoided on systems built along a pipeline.

Company-Owned Cable along Right of Way. Companies have never used cable installations along the pipeline rights of way to any great degree for their communications, mainly because of hazards to cables when pipeline outages are repaired, and because of the high initial cost of lead covered cables. Recent developments in cable insulations, however, have reduced cable costs to a point at which they are competitive with wire lines. This is especially true where there is a need for multiconductors, since the costs per circuit decrease rapidly as the number of conductors per cable increases. Cables can be utilized for all communication circuits needed in pipeline use. Automatic gain controls are not as important on cable as they are on open wire circuits, because leakage caused by rain and other adverse weather is not a factor.

Transmission loss is much higher with cables than with open wire; consequently, boosters are required at much more frequent intervals. Inserting loading coils at regular intervals materially increases the communication range of cables. Loading also has the advantage of raising cable impedances, thus improving line to equipment matching.

Maintenance costs of cable installations should compare quite favorably with those of pole lines.

Leased Telephone Facilities. Commercial communication system facilities may be leased for providing telephone, telegraph, teletype, telemetering, and some radio service at published rates, usually carrying a 25 per cent tax. Telephone costs are on a per unit basis, except on long circuits where a sliding scale applies as distance increases. The sliding scale does not apply if intermediate stations requiring services are added. This service is generally reliable and, where circuit requirements are not too great, financially attractive. On long systems reliability of communication is generally backed up by alternate routing possibilities. On taps or way stations at which alternate routes are not available, reliability suffers.

Leased Teletype Facilities. Rates for the line portion of teletype facilities leased from commercial communication systems are about one-half the charges for leased telephone lines, with additional costs depending upon the terminal equipment requirements. Generally, a pipeline company's teletype channel requirements are not extensive. Consequently, leased facilities are usually financially attractive despite their limited application. Also available is a teletype service similar to ordinary long distance telephone service, with cost based on time used. Since it is limited to communication exchanges between subscribers, it has had restricted application.

Company-Owned Mobile Radio. Radio is employed intermittently, principally for mobile service. Microwave communication is principally point to point, with continuous carrier radiation for telephone service. Generally, gas pipeline companies are eligible for mobile radio usage in the petroleum service, and gas distribution companies hold eligibility in the utility service. Some eligibility latitude is given by federal regulatory bodies when a need develops for pipeline companies to use frequencies allocated to utility service and vice versa.

Primarily, mobile service in either of the above categories is confined to frequencies between 25 and 160 megacycles; the bands below 50 megacycles being the most used. Additional frequencies are available in other bands. Complete details are covered in the FCC (Federal Communications Commission) rules and regulations.

Transmission distance depends upon local noise conditions at the receiver, frequency, height of the radiating element, effective radiated power, and receiver gain capabilities. Line of sight transmission is approached as mobile frequencies increase. However, most mobile frequencies have a range considerably beyond line of sight. Line of sight in *miles* is approximately $1.25\ H^{0.5}$, where H is the height of the radiating element in feet.

Conservatively estimated curves to indicate range in miles from base stations to mobile units at two frequencies with various antenna heights and signal inputs to receivers are given in Fig. 8–70.

Company-Owned Microwave System. Microwave is a term applied to communication systems using very short radio waves. Many pipeline systems employ microwave for multichannel communication facilities. Frequencies for this service begin at approximately 800 megacycles and extend beyond the present usable portion of the frequency spectrum.

Attractive aspects of microwave are its multichannels, with the possibilities for utilizing them for many operational functions beyond normal telephone, telegraph, and teletype service; its low initial and maintenance costs; and its relative freedom from the effects of adverse weather, which hurt other communication facilities.

It should be noted that propagation vagaries exist at microwave frequencies, particularly the tendency of signals to "fade" during the night and early morning hours, when there is an absence of air turbulence. Signals also tend to "overshoot" during such periods and interfere with the desired signal at a station further along the line. Adequate margins or other safeguards must be provided against fades, and either station staggering or multiple frequencies should be used to eliminate overshooting effects.

Various manufacturers employ radically different methods in making microwave communication systems. Some systems utilize crystal-controlled transmitters and others employ temperature-compensated cavities to maintain frequency stability. Again, some systems employ frequency division carriers which frequency-modulate the RF carrier and others use time division multiplexing to derive the required channels.

Where the earth's surface between the antennas is a good reflecting medium such as water, the reflected signal must also be considered. The phase relationship between the direct and reflected signal may be such that the resultant signal is greater or less than the direct signal alone. This phase relationship depends upon the height of the antennas, so that, as they are raised on a tower, the received signal will pass thru maximum and nulls. When the effective earth curvature changes, because of abnormal dielectric gradients, it will also change this phase relationship and produce variations in signal strength. Generally, the reflection coefficient of the earth is great enough to allow the reflected waves to produce an appreciable change in received signal strength.

Although complex factors are involved in microwave propagation, most microwave systems are designed around a set of empirical and experimentally derived curves and tables. These may be obtained from equipment manufacturers.

GAS DISPATCHING

Centralized control of gas in a gas pipeline system is necessary. To accomplish this, dispatchers are placed at strategic points throughout the system to control flow, pressure, field withdrawal rates, and deliveries to customers. A pipeline system design must include adequate facilities for furnishing dispatchers with the information necessary to carry out these duties.

There are three distinct types of dispatching operations; namely,[2] long line, network, and distribution.

Long line dispatching consists of collecting large quantities of gas from many sources and directing its flow thru hundreds of miles of pipelines to network or distribution systems.

Network dispatching encompasses control of the movement of gas to a small community from a single source, as well as the movement of gas from long line companies' local producing wells, storage, and stand-by facilities thru thousands of miles of transmission line to customers in several states.

Distribution dispatching ranges from serving one city with gas from one source to directing the deliveries of a complicated combination of natural and manufactured gases throughout a metropolitan area.

The basic information that the dispatcher receives and uses to control pipeline flow is normally recorded each hour or each quarter-hour on a dispatcher's sheet or daily pressure and delivery report. It includes gas pressures, rates of flow, dew points, and gas temperatures. Depending upon specific requirements, water vapor content, specific gravity, heating

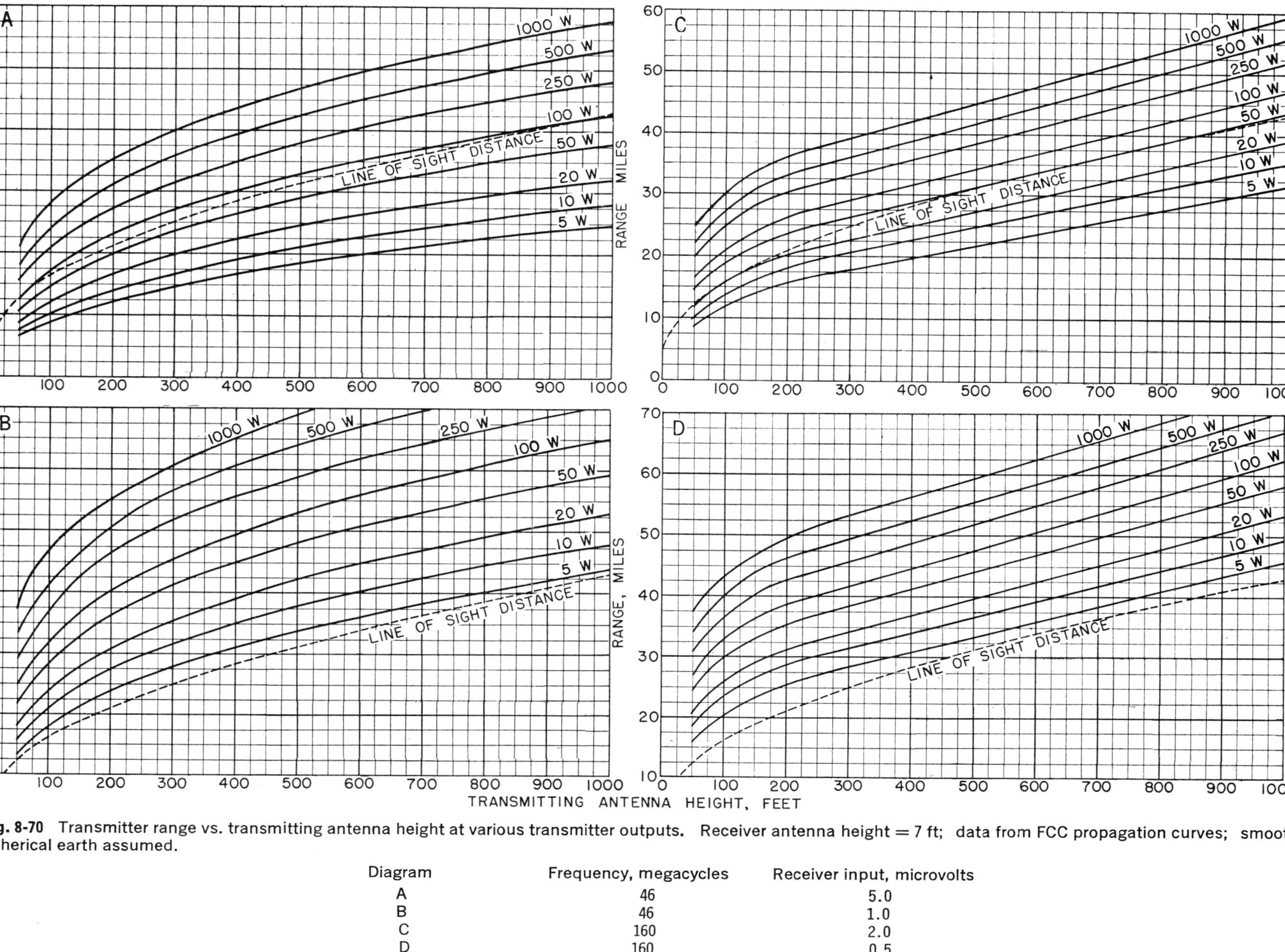

Fig. 8-70 Transmitter range vs. transmitting antenna height at various transmitter outputs. Receiver antenna height = 7 ft; data from FCC propagation curves; smooth spherical earth assumed.

Diagram	Frequency, megacycles	Receiver input, microvolts
A	46	5.0
B	46	1.0
C	160	2.0
D	160	0.5

value, presence of foreign matter such as sulfur or dust, and weather information at key locations from supply source to delivery points may also be included in the report.

As a ready reference, printed forms for posting this information usually carry the maximum working pressure of the pipelines at each location from which a pressure is received, along with the number of engines and their horsepower at each compressor station. The daily average pressure; monthly minimum, maximum, and average pressures; daily total deliveries; monthly total deliveries; and daily load factors of the major delivery points are tabulated at the end of each 24-hr period. Such information as requests for increases and decreases in the production fields, changes in the number of engines in operation at compressor stations, and major changes in loads at delivery points is also recorded.

If communications are an essential part of any gas pipeline system, they are particularly so to the dispatcher, who must have contact with all parts of the system at all times and especially at stated intervals, for example, hourly. To handle gas from the field efficiently, dispatchers should have the following information available at all times: location of wells, proration as outlined in gas purchase contracts, annual minimum quantity and suppliers' obligations to deliver, the number and location of delivery points, delivery pressures, and contract terms.

Some states have regulations controlling use and preventing waste of natural gas. The dispatcher must be familiar with their provisions and regulate withdrawals accordingly.[3] When covered by Federal Power Commission regulations, a pipeline company must control the quantity of gas in accordance with the allocation. The dispatcher must keep an accurate record of daily deliveries. Since several days may be required for final computation of the gas meter charts, the dispatcher must devise methods to keep his figures up to date. Deliveries are approximated from spot readings received from field personnel. Computed deliveries (modified by correction factors) should check closely with the actual values obtained later.

Weather information is important to the dispatcher. It is obtained from the U. S. Weather Bureau and from the CAA teletype service. Some companies even employ private weather forecasting services. Dispatchers should receive information on the weather over the pipeline system as changes occur from hour to hour, and they should be able to interpret it in terms of potential load requirements. Because it usually takes many hours to transport gas from supply to market, it is necessary for the dispatcher to plan its movement thru the pipelines in advance for the times of peak requirements. Wind velocity, atmospheric conditions, and temperature are all equally important in calculating load requirements and in anticipating loads for the following day. A temperature change of even one degree during the winter heating season generally causes gas requirements to change appreciably. Accurate forecasting of gas requirements over a monthly period is also necessary to tie in with field operations.

Dispatchers maintain complete pipeline test records, along with permanent operating records of shutdowns for pipe replacements or repairs. They confirm whether alternate supplies of gas are available to customers during normal supply shutdowns because of construction or emergencies. They also advise customers of any impending interruption of service.

Many long line systems and some network systems are interconnected for emergency purposes. There have been instances in which a break in one long line or a loss of supply in one area has been met without loss of service by obtaining a supply of gas promptly from one or more sources thru these interconnections. The dispatcher plays a major role in these emergencies.

Gas storage and stand-by LP-gas and manufacturing facilities furnish the dispatcher with great quantities of readily available gas on short notice. Dispatching gas to or withdrawing it from storage has added to the year-round responsibilities of dispatching personnel. The chief dispatcher's recommendation for starting up stand-by facilities for peak loads governs the operation of such plants. He also participates in decisions to curtail industrial gas when necessary.

Recent developments in the automation of compressor stations on long line systems and at storage fields permit greater control over the start-up and shutdown of pumping facilities. Pressure can thus be regulated directly from the dispatching office. This has increased dispatching responsibility, at the same time permitting better control of gas flow.

An outline of a training program for gas dispatchers[4] and details for the preparation and maintenance of a dispatcher's wall map[5] are available.

REFERENCES

1. Moon, C. L. "What's New, and What Are the Trends in Pipeline Communication." *Oil Gas J.* 56: 103, July 14, 1958.
2. Chadwell, S. A. "Gas Dispatching in Network Systems." *A.G.A. Proc.* 1953: 423.
3. Kelly, T. B. "Load Dispatching." *A.G.A. Proc.* 1953: 121.
4. A.G.A. Operating Section. *Manual on Job Training for Gas Dispatchers.* (OP-60-2) New York, 1960.
5. ——. *Wall Maps for Gas Dispatching Departments.* (OP-58-3) New York, 1958.

SECTION 9

DISTRIBUTION OF GAS

R. J. Ott (deceased), *Section Chairman,* and Chapters 2, 5, and 6, Philadelphia Gas Works Div. of U.G.I. Co., Philadelphia, Pa.
G. P. Binder (retired), Chapter 6, Con-Gas Service Corp., Pittsburgh, Pa.
Guy Corfield, Chapters 3 and 6, Southern California Gas Co., Los Angeles, Calif.
J. K. Dawson, Chapter 1, The Peoples Gas Light & Coke Co., Chicago, Ill.
Harold Emerson, Chapter 7, Public Service Electric & Gas Co., Newark, N. J.
H. E. Ferguson (deceased), Chapter 14, The Peoples Gas Light & Coke Co., Chicago, Ill.
B. E. Hunt, Chapter 6, Illinois Power Co., Decatur, Ill.
Lester B. Inglis, Jr., Chapter 3, American Gas Association, New York, N. Y.
C. F. Kleck (deceased), Chapter 1, The Peoples Gas Light & Coke Co., Chicago, Ill.
J. A. Lane, Chapter 9, Public Service Electric & Gas Co., Newark, N. J.
T. J. Noonan, Chapter 9, The East Ohio Gas Co., Cleveland, Ohio
J. M. Pickford, Chapters 5, 10–14, Northern Indiana Public Service Co., Hammond, Ind.
J. S. Powell, Chapter 11, Southern California Gas Co., Los Angeles, Calif.
H. C. Roemmele, Chapters 4, 7, and 8, Public Service Electric & Gas Co., Newark, N. J.
C. George Segeler, Chapter 2, American Gas Association, New York, N. Y.
F. E. Vandaveer (retired), Chapters 4, 5, 6, 10–15, Con-Gas Service Corp., Cleveland, Ohio

CONTENTS

Chapter 5—Pressure Regulation in Distribution

Chapter 6—Distribution Design for Increased Demand

Chapter 7—Main Installation

Chapter 8—Services

Chapter 9—Maintenance of Mains and Services

Chapter 10—Distribution System Deposits

Chapter 11—Gas Odorization

Chapter 12—Unaccounted-for Gas

Chapter 13—Oil Fogging of Gas

Chapter 14—Humidification

Chapter 15—Dust Filters and Gas Cleaners

Tables

Figures

Chapter 1
Distribution Pumping

by J. K. Dawson and C. F. Kleck

Distribution pumping is employed in those systems in which the gas supply is received at the inlet of the system at a pressure below that needed to maintain good pressure on the distribution system or below that needed to transmit gas thru the lines to areas at distant points. It is confined primarily to manufactured or mixed (manufactured and natural) gas systems (including peak shaving) in which the manufactured gas is produced and stored at pressures only slightly above atmospheric. Natural gas is usually received at the city border station at a pressure sufficiently high to require reduction rather than boosting. In special instances natural gas may be compressed to a higher pressure in a distribution system to fill a high-pressure holder or to boost the pressure when gas has been stored in low-pressure holders.

Distribution of gas may be at high* pressure direct to the consumer, or thru a medium* pressure transmission system feeding a low* pressure distribution system which supplies the consumer. When manufactured and natural gas mixtures represent the system sendout, pumping thru some automatically controlled mixer will be necessary. Pressures considered in this chapter are limited to about 125 psi.

Boosting Pressure on Consumer's Premises. The gas supply for prime movers and other commercial and industrial applications may have to be at a pressure above that available from a distribution system.

TYPES OF PUMPING EQUIPMENT

Various types of equipment are available for pumping and for combined pumping and mixing. They are classified according to the duty to be performed and to their pressure–volume characteristics. Figures 8–23 and 8–24 compare, in a limited way, efficiencies and power requirements of centrifugal and reciprocating compressors.

A **centrifugal machine** (Fig. 8-22) compresses gas or air by means of centrifugal force. For large volume distribution pumping, it is generally used for a maximum discharge pressure of approximately 30 psig. The discharge pressure is dependent upon the inlet conditions of the gas.

A **positive displacement unit** compresses successive volumes of gas or air by confining them within a closed space wherein the pressure is increased as the volume of the closed space is decreased. Such a compressor may be of either the rotary or the reciprocating type. A *rotary vane type* of compressor is generally used to compress relatively small volumes of gas against a maximum discharge pressure of approximately 50 psig, while a *rotary impeller type* of positive pump is generally used for relatively low discharge pressures. The Lysholm compressor and the liquid (water) ring compressor are two rotary types which may also be used. A *reciprocating compressor* is generally used for high discharge pressures.

A *power extraction unit* is a combined compressor and mixer which utilizes the available energy of high-pressure gas to compress and mix with a low-pressure gas. Dynamic mixers, expander engine compressors, and jet compressors are available for pumping and mixing installations.

Selecting the Unit. Consideration must be given not only to the pressure–volume characteristics, but also to the type of driver. Small rotary compressors (vane or impeller type) are generally driven by electric motors. Large volume positive compressors operate at lower speeds and are usually driven by steam or gas engines. They may be driven thru reduction gearing by means of a steam turbine or an electric motor. Reciprocating compressors are designed in varying patterns to employ steam, electric, or gas power for the driver. The most widely used reciprocating compressor in the gas industry is the conventional gas engine-driven *angle-type* compressor unit.

Centrifugal-type compressors are high-speed machines, usually driven by steam turbines or electric motors. Because of the large load fluctuation, gas turbines are usually not suitable as drivers for centrifugal units designed for gas distribution pumping. They may be considered when a centrifugal unit is to be operated for base load purposes.

THERMODYNAMIC EQUATIONS

Basic thermodynamic equations most generally used in gas compressor problems are:

Adiabatic† *(Isentropic)*‡ *Compression*

$$p_1V_1^k = p_2V_2^k = \text{constant} \tag{1}$$

where:

p_1 = initial pressure, psfa
p_2 = final pressure, psfa
V_1 = initial specific volume, cu ft per lb
V_2 = final specific volume, cu ft per lb
k = c_p/c_v = ratio of specific heats

* Defined in Chapter 2 of this Section.

† No heat transferred to or from the working substance.
‡ An ideal or reversible process.

where: c_p = specific heat at constant pressure, Btu per lb-°F
c_v = specific heat at constant volume, Btu per lb-°F

Pressure Characteristic, β

$$\beta = \frac{k}{k-1}\left[\left(\frac{P_2}{P_1}\right)^{(k-1)/k} - 1\right] \qquad (2)$$

where: P_1 = initial pressure, psia
P_2 = final pressure, psia
k = ratio of specific heats
Equation **2** is plotted in Figs. 8–31a and b, where $n = k$ and $C = P_2/P_1$.

Mean Effective Pressure, mep, psia

$$\text{mep} = P_1\beta \qquad (3)$$

Temperature Variation

$$T_2 = T_1\left(\frac{P_2}{P_1}\right)^{(k-1)/k} \qquad (4)$$

where: T_1 = initial temperature, °R
T_2 = final temperature, °R
Other terms are defined under Eqs. **1** and **2**.
The term $(P_2/P_1)^{(k-1)/k}$ can be read in Fig. 8–42.

Adiabatic Horsepower

$$\text{ahp} = p_1Q_1\beta/33{,}000 \qquad (5)$$

$$\text{ahp} = wR'T_1\beta/33{,}000 \qquad (6)$$

$$\text{ahp} = P_1Q_1\beta/229 \qquad (7)$$

where:

ahp = adiabatic horsepower, ft-lb per min
R' = specific gas constant* = $1544/M$ (M = molecular weight of gas), ft per °F
Q_1 = inlet capacity at inlet conditions, cfm
w = weight of flow, lb per min
P_1 = initial pressure, psia
p_1 = initial pressure, psfa
P_2 = final pressure, psia
p_2 = final pressure, psfa
For other terms, see Eqs. **2** and **4**.

Adiabatic Efficiency (overall), e_a

$$e_a = \text{ahp}/\text{bhp} \qquad (8)$$

where: ahp = adiabatic horsepower
bhp = brake horsepower

Isothermal Horsepower, ihp

$$\text{ihp} = \frac{p_1Q_1}{33{,}000}\log_e\frac{p_2}{p_1} \qquad (9)$$

where p_1 and p_2 are in pounds per square foot absolute
Terms are defined under Eqs. **1** and **7**.

$$\text{ihp} = \frac{wR'T_1}{33{,}000}\log_e\frac{P_2}{P_1} \qquad (10)$$

$$\text{ihp} = \frac{P_1Q_1}{229}\log_e\frac{P_2}{P_1} \qquad (11)$$

where P_1 and P_2 are in pounds per square inch absolute
Terms are defined under Eqs. **4** and **7**.

Isothermal Efficiency, e_i

$$e_i = \text{ihp}/\text{bhp} \qquad (12)$$

where: ihp = isothermal horsepower
bhp = brake horsepower

Actual compression curves seldom follow either the adiabatic or isothermal processes. They are generally of the form:

$$p_1\bar{V}^n = \text{constant} \qquad (13)$$

where: n = any constant number
Other terms defined under Eq. **1**.

Usually **centrifugal compressors** are uncooled and are calculated on an adiabatic basis. **Reciprocating compressors** are cooled; consequently, they are calculated on an isothermal basis.

Frequently, for centrifugal blower and jet compressor design, it is necessary to calculate thermodynamic properties of gases being compressed and gases used for motor power. This involves the following equations for **relationships of specific heats:**

$$1544/M = J(c_p - c_v) \qquad (14)$$

$$c_p = c_v + (1.987/M) \qquad (15)$$

$$c_p = c_v + (R'/J) \qquad (16)$$

$$c_v = [R'/(k-1)J] \qquad (17)$$

$$k = c_p/c_v \qquad (18)$$

where: J = Joule's constant, 778 ft-lb per Btu
c_p = specific heat at constant pressure, Btu per lb-°F
c_v = specific heat of constant volume, Btu per lb-°F
Other terms defined under Eq. **7**.

CALCULATION OF PROPERTIES OF GAS MIXTURES

Table 9-1 shows the calculation for the specific heat at constant pressure, c_p, for a natural gas of the given analysis.

Since the specific gravity of air is unity and its molecular weight is 28.95, the specific gravity of the gas and its ratio of specific heats, k, may be determined as follows:

$$\text{sp gr} = \frac{\text{mole weight of gas}}{28.95} = \frac{19.55^\dagger}{28.95} = 0.675$$

Substituting in Eq. **15**:

$$c_v = 0.458^\dagger - \frac{1.987}{19.55} = 0.357$$

Substituting in Eq. **18**:

$$k = 0.458/0.357 = 1.28$$

* See Table 2-71a for more accurate values.

† From Table 9-1.

Table 9-1 Calculation of the Specific Heat of a Gas Mixture (illustrative problem)

Gas	Composition by vol, V	Mole wt, M	VM	$\frac{VM}{\Sigma VM}$	c_p, Btu/lb-°F	$\frac{VM}{\Sigma VM}c_p$
CO_2	0.001	44.01	0.044	0.00225	0.199	0.0004
CH_4	.789	16.04	12.655	.64721	.526	.3404
C_2H_6	.060	30.07	1.804	.09226	.409	.0377
C_3H_8	.034	44.09	1.499	.07666	.388	.0304
C_4H_{10}	.010	58.12	0.581	.02972	.397	.0118
N_2	0.106	28.02	2.970	0.15190	0.248	0.0377
	1.000	$\Sigma VM = 19.55$*			Overall c_p =	0.458

* Composite mole weight M.

Volume of gas to be compressed is frequently stated at standard conditions (60 F dry, and 30 in. Hg), whereas the gas to be pumped by the compressor will be saturated with water vapor. Volume must be corrected for actual gas inlet temperature, pressure, and saturation as shown by the following example:

With inlet gas volume of 20,000 cfm measured at 60 F dry, and 30 in. Hg, find corrected volume when saturated at 100 F and 14.0 psia at compressor inlet.

Partial pressure of steam at 100 F: 0.95 psia
Pressure correction: 14.73/14.0 = 1.052
Temperature correction: 560° abs/520° abs = 1.077
Saturation correction:

$$\frac{\text{total pressure}}{\text{total pressure} - \text{partial pressure}} =$$

$$\frac{14.0}{14.0 - 0.95} = 1.073$$

Corrected inlet volume to compressor:
20,000 × 1.052 × 1.077 × 1.073 = 24,320 cfm

The compressor must therefore be designed to pump 24,320 cfm at inlet conditions in order to compress the required 20,000 cfm measured at standard conditions.

CENTRIFUGAL MACHINES

These are generally classified as fans, blowers, or compressors.

Fans are used for compressing large volumes at low discharge pressures up to about 1 psig.

Centrifugal **blowers** are used for compression to discharge pressures not over 35 psig. Such blowers operating at inlet pressures *below* atmospheric are termed centrifugal *exhausters*. With inlet pressures *above* atmospheric, they are termed centrifugal *boosters*.

Centrifugal **compressors** (Fig. 8–22) are machines compressing to discharge pressure above 35 psig.

Blower Head and Pressure Coefficient

For centrifugal boosters or exhausters, *where no cooling is employed*, the pressure head and overall efficiency are usually based on *adiabatic* compression. Since the velocity head is small, it is neglected. The total adiabatic head is generally based upon the static pressure developed and may be expressed as follows:

$$H = R'T_1\beta \qquad (19)^*$$

where:

H = adiabatic head, feet of the fluid flowing
T_1 = inlet temperature, °R
R' = specific gas constant† = $1544/M$ (M = molecular weight of gas), ft per °F
β = pressure characteristic (see Eq. **2** or Figs. 8–31a and 8–31b)

Another expression for the head developed per stage is

$$H = f(u^2/g) \qquad \textbf{(20)}$$

plotted in Fig. 8–32 (where $f = \mu$)

where: f = overall pressure coefficient (see Eq. **22**)
g = acceleration of gravity, 32.2 ft per sec-sec
u = tip speed of impeller wheel, ft per sec

This coefficient, f, is frequently determined from test and is expressed as follows:

$$f = R'T_1g\beta/u^2n = 1720T_1\beta/u^2Gn \qquad (21)$$

$$f = 1720T_1\beta/\Sigma u^2G \qquad (22)$$

where:

n = number of stages of compression
G = specific gravity of gas
Σu^2 = summation of tip speeds squared (for machines of varying wheel size)

Other terms defined under Eqs. **2**, **19**, and **20**.

Wheel tip speed is:

$$u = dN/229 \qquad (23)$$

where: u = tip speed, ft per sec
d = impeller wheel, OD, in.
N = speed, rpm

Vane Angle

The overall pressure coefficient, f, is dependent upon the blade design, since the pressure–volume characteristic varies with the exit angle of the blades. Figure 9–1 illustrates the volume–head–horsepower characteristics for three values of exit vane angle. The theoretical or ideal head curves AB are straight lines and vary as a function of the tangent of the exit vane angle. The friction losses, which are similar for all cases, are represented by the area between curves AB and AC. The circulatory and shock losses are similar in all cases and are represented by the area between the friction curve AC and the resultant head-capacity curve DE. The design point, M, or point of maximum efficiency, occurs when the friction, circulatory, and shock losses are a minimum. The stable operating range for all units consists of all points to the right of the maximum pressure point K on all curves.

Figure 9-1 also shows that the *backward curved vanes* provide the greatest operating range because of the drooping pressure–volume characteristic. In addition, the horsepower

* Comparable to Eqs. **4**, **5**, and **6** of Sec. 8, Chap. 3.
† See Table 2-71a for more accurate values.

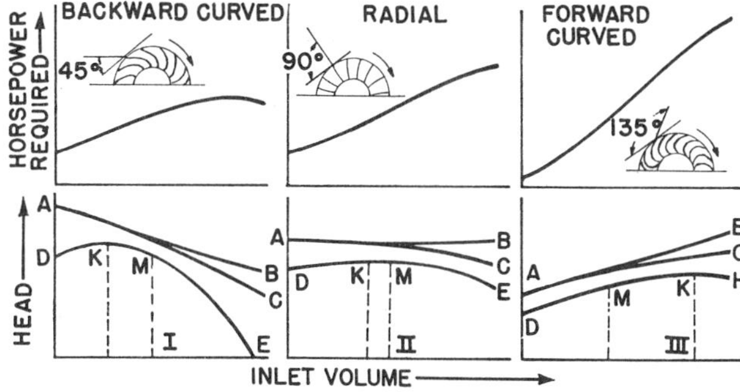

Fig. 9-1 Effect of vane angle on performance.

characteristic of this type of blading is self-limiting, and hence the unit cannot overload its driver. *Radial-type* impellers have a flat pressure–volume characteristic with a reasonable operating range. They are usually used when a higher pressure is required than can be provided with backward curved vanes. *Forward curved vanes* result in a high pressure for an impeller of a given size. However, the operating range is beyond the maximum efficiency point (refer to points K and M, Fig. 9-1). Consequently, the use of forward curved vanes is generally limited to fans.

The value of the pressure coefficient, f, varies from approximately 0.40 for backward curved vanes to 0.65 for radial-type vanes. When estimating the performance of a backward curved impeller at the design point, the value of f is generally taken to be 0.50.

Blower Performance Curves

Flow of gas in cfm at constant inlet conditions of pressure and temperature is customarily plotted against the discharge pressure together with the horsepower requirement. Typical blower performance curves are shown in Fig. 9-2. Such performance curves vary widely with the characteristics of the gas and the inlet conditions. Thus, it is more desirable for analysis purposes to plot the overall pressure coefficient, f, and the overall adiabatic efficiency, e_a, against the inlet volume divided by the speed, Q_1/N. The curves so plotted are independent of operating speed and gas inlet conditions. Typical characteristic curves for the same blower are shown in Fig. 9-3.

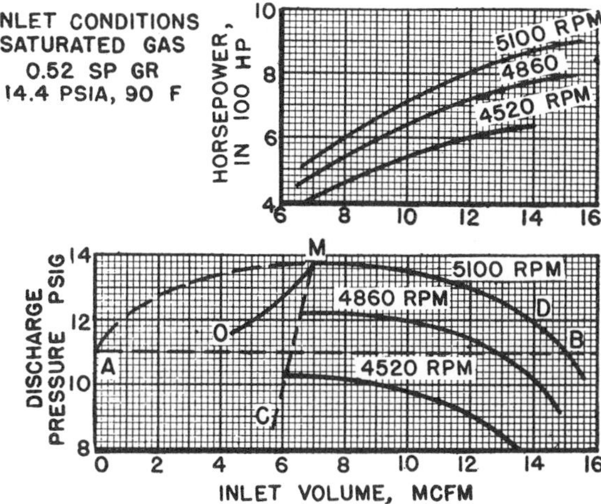

Fig. 9-2 Blower performance curves.

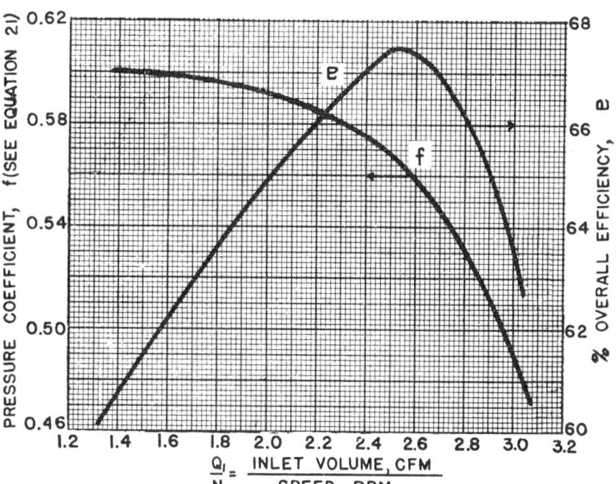

Fig. 9-3 Blower characteristics (30-in. impeller, five-stage, 5100 rpm, 0.52 sp gr, 90 F inlet temp, 14.4 psia inlet pressure, steam turbine drive).

Figure 9-2 illustrates volume–pressure–horsepower characteristics of a blower for the indicated operating speeds and the specified constant inlet conditions. The design point of the unit is indicated as D (at 14,000 cfm, 12 psig, and 5100 rpm).

Dashed curve MC in Fig. 9-2 indicates the *minimum* pumping limit of the unit. This usually occurs at from 40 to 50 per cent of the design volume and decreases directly with a decrease in speed. If the unit were pumping 14,000 cfm against 12 psig pressure and the distribution system pressure decreased, the volume flow from the machine would increase along the performance curve. With a gradually increasing distribution system pressure, the delivered volume from the unit would decrease until the pumping limit was reached. A further increase in the system pressure would be greater than the 13.7 psig delivery pressure (since the latter theoretically tends to drop off along the dotted curve from M to A) and gas would tend to flow back thru the blower unit, the operating point dropping off to point A, the shutoff condition. Since gas is continually being removed from the gas distribution system, the required pressure will drop below the 11 psig pressure of point A and the machine will rapidly deliver the volume (15,000 cfm) corresponding to the system discharge pressure (11 psig) at point B. With this increased delivery, the pressure will increase rapidly to the maximum point M of the curve until delivery again ceases; when the system pressure decreases sufficiently below that of point A, the unit again begins pumping. This cycle, repeated rapidly, causes pulsation. Figure 8-26 indicates the stability limit of centrifugal compressors at various speeds.

Pulsation

Repetition of the cycle just described can bring about damaging pulsation in a resonant piping system. Pulsation will continue until conditions causing it are remedied or the volume flow required by the system is greater than the flow (7000 cfm) at the maximum pressure point M on the performance curve shown in Fig. 9-2.

Table 9-2 Blower Performance for Various Inlet Conditions

(illustrative problem)

1	2	3	4	5	6	7	8	9	10	11	12
Q_1/N	N	Q_1	f	β	C	P_1, psia	P_2, psia	P_2, psig	ahp	e_a	bhp
1.4	5460	7,644	0.600	1.371	3.225	14.4	46.44	32.05	659	0.608	1085
1.8	5460	9,828	0.597	1.364	3.210	14.4	46.22	31.80	843	.636	1326
2.2	5460	12,012	0.585	1.337	3.145	14.4	45.29	30.90	1009	.661	1528
2.4	5460	13,104	0.576	1.316	3.090	14.4	44.50	30.10	1084	.671	1617
2.6	5460	14,196	0.559	1.277	3.005	14.4	43.27	28.90	1139	.674	1691
2.8	5460	15,288	0.530	1.211	2.860	14.4	41.18	26.80	1164	.662	1760
3.0	5460	16,380	0.489	1.117	2.660	14.4	38.30	23.90	1150	0.636	1808

One method of preventing pulsation is to throttle the gas on the inlet side of the blower, causing the characteristic pressure–volume curve to drop off more rapidly, as indicated by curve MO on Fig. 9-2. Another method is to install a by-pass line from the discharge piping back to the storage holder. The by-pass line could be either manually operated or automatically controlled by means of a volume flow control regulator opening the by-pass valve when the minimum flow at point M is reached. The unit would then operate at some point to the right of the pulsating zone, as indicated in Fig. 9-2. If gas is by-passed to the suction side of the unit, care must be taken to prevent the creation of high blower inlet temperature from recirculation of compressed gas.

If adjustable inlet guide vanes are installed to give the gas entering the first stage of the unit a whirling flow in the direction of rotation, the centrifugal unit will develop lower discharge pressure and require less horsepower; thus, it will tend to move the pulsation point to the left. The pulsation point of a blower equipped with guide vanes ranges between 35 and 40 per cent of the design volume.

For the throttling of constant speed machines, some manufacturers incorporate a reaction turbine power wheel at the inlet of the first stage of the unit. Gas is expanded thru adjustable inlet guide vanes, impinges on the reaction blades of the wheel, and expands, thus producing a greater pressure drop. The torque developed is transmitted to the shaft and reduces the power requirement of the driver by an equivalent amount.

Blower Performance under Changed Inlet Conditions

Although Figs. 9-2 and 9-3 apply to a specific installation, these curves may be used in determining blower performance under other inlet conditions, as shown by the following example:

Find the pressure–volume characteristic for this blower (Figs. 9-2 and 9-3) when compressing an 0.80 sp gr gas saturated at 60 F and 14.4 psia with the unit operating at a maximum speed of 5460 rpm. Calculated ratio of specific heats, k, is 1.36.

The overall pressure coefficient, f, and the efficiency, e, remain constant for the same ratio of inlet volume to speed. They are obtained from Fig. 9-3.

Tip speed of the five 30-in. diam impeller wheels operating at 5460 rpm is, by Eq. **23**:

$$u = \frac{dN}{229} = \frac{30 \times 5460}{229} = 715 \text{ ft per sec}$$

Solve Eq. **21** for β:

$$\beta = \frac{u^2Gnf}{1720T_1} = \frac{511{,}000 \times 0.80 \times 5 \times f}{1720 \times 520} = 2.285\, f$$

Since $k = 1.36$, $k/(k-1) = 3.78$. From Eq. 2,

$$\beta = 3.78 \quad [C^{1/3.78} - 1]$$

where C = pressure ratio = P_2/P_1

Solving Eq. **2** for C yields:*

$$C = \frac{P_2}{P_1} = \left(\frac{\beta}{3.78} + 1\right)^{3.78}$$

Also, the adiabatic horsepower is, by Eq. **7**:*

$$\text{ahp} = 14.4Q_1\beta/229 = Q_1\beta/15.91$$

and the brake horsepower is (Eq. **8**):*

$$\text{bhp} = \text{ahp}/e_a$$

Calculated performance characteristics are given in Table 9-2 and Fig. 9-3.

Figure 9-4 shows the inlet volume of column 3 (Table 9-2) plotted against the outlet pressure from column 9 and the brake horsepower from column 12. This permits comparison with operation at other conditions so that the overall effect of the variables involved may be determined.

The following relationships may also be used in calculating blower performance:

1. Flow is directly proportional to the speed, head and mean effective pressure are directly proportional to the *square* of the speed ratio, and horsepower is directly proportional to the *cube* of the speed ratio, efficiency remaining constant.
2. Horsepower, head, and mean effective pressure vary *directly* with the ratio of specific gravities.
3. Horsepower, head, and mean effective pressure vary *inversely* with the ratio of the absolute inlet temperatures.
4. Flow varies directly as the *cube* of the ratio of impeller outside diameters, head and mean effective pressure vary as the *square* of the ratio of diameters, and horsepower varies as the *fifth* power of the ratio of diameters.
5. With the suction to the blower *throttled*, flow is directly proportional to the ratio of the inlet pressures. As speed is unchanged, head and pressure ratio remain constant; thus,

* See Table 9-2.

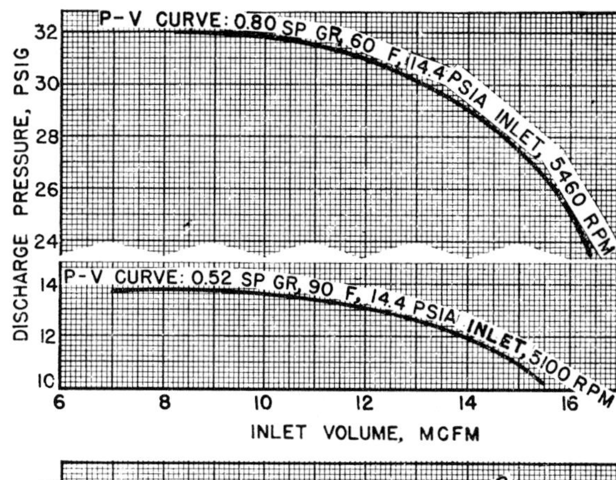

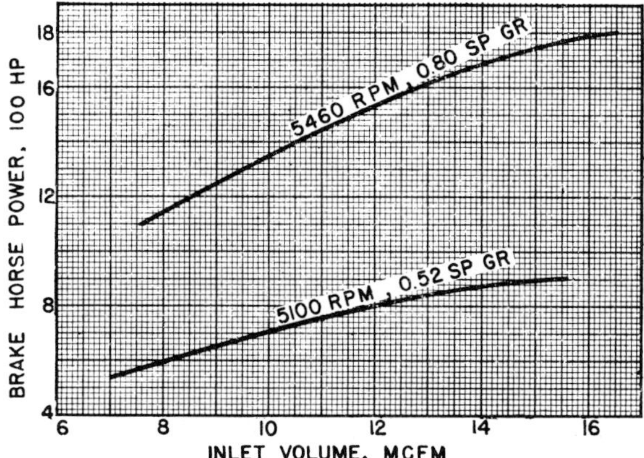

Fig. 9-4 Characteristics of example blower (see Table 9-2).

the delivery pressure will be less. With only inlet volume changed by throttling, brake horsepower will vary directly with the ratio of inlet pressures.

These relationships may be summarized by the following equations (subscript o refers to the original condition and subscript r to the required condition; temperatures and pressures are at the blower inlet):

$$Q_r = Q_o \frac{N_r}{N_o}\left(\frac{D_r}{D_o}\right)^3 \tag{24}$$

$$w_r = w_o \frac{N_r}{N_o}\frac{P_r}{P_o}\frac{T_o}{T_r}\left(\frac{D_r}{D_o}\right)^3 \tag{25}$$

$$H_r = H_o\left(\frac{N_r}{N_o}\right)^2\left(\frac{D_r}{D_o}\right)^2\frac{G_r}{G_o}\frac{T_o}{T_r} \tag{26}$$

$$\beta_r = \beta_o\left(\frac{N_r}{N_o}\right)^2\frac{G_r}{G_o}\frac{T_o}{T_r}\left(\frac{D_r}{D_o}\right)^2 \tag{27}$$

$$e_r = e_o \tag{27a}$$

$$\text{bhp}_r = \text{bhp}_o\left(\frac{N_r}{N_o}\right)^3\frac{P_r}{P_o}\frac{G_r}{G_o}\frac{T_o}{T_r}\left(\frac{D_r}{D_o}\right)^5 \tag{28}$$

$$\text{bhp}_r = \text{bhp}_o \frac{P_r}{P_o}\frac{Q_r}{Q_o}\frac{\beta_r}{\beta_o} \tag{28a}$$

With suction throttled:

$$Q_r = Q_o \frac{P_r}{P_o}\frac{G_r}{G_o} \tag{29}$$

$$H_r = H_o \tag{29a}$$

$$\beta_r = \beta_o \tag{29b}$$

$$\text{bhp}_r = \text{bhp}_o\left(\frac{P_r}{P_o}\right) \tag{30}$$

where: Q = volume flow
w = weight flow
N = speed
H = head
G = specific gravity
P = inlet pressure, abs
T = inlet temperature, abs
D = outside diameter of impeller
β = pressure characteristic (see Eq. 2)
e = efficiency

The previous example may be checked by using the preceding group of equations.

The new volume at the design point (Eq. 24) is:

$$Q_r = Q_o \frac{N_r}{N_o} = 14{,}000 \times \frac{5460}{5100} = 14{,}990 \text{ cfm}$$

Solving for the new pressure characteristic:
By Eq. 2,

$$\beta_o = 3.78\left[\left(\frac{26.4}{14.4}\right)^{1/3.78} - 1\right] = 0.6577$$

Then:

$$\beta_r = \beta_o\left(\frac{N_r}{N_o}\right)^2\frac{G_r}{G_o}\frac{T_o}{T_r} = 0.6577\left(\frac{5460}{5100}\right)^2 \times \frac{0.80}{0.52}\times\frac{550}{520} = 1.226$$

$$C = \left[\frac{\beta_r}{3.78}+1\right]^{3.78} = \left[\frac{1.226}{3.78}+1\right]^{3.78} = 2.89$$

$$P_2 = P_1C = 14.4 \times 2.89 = 41.6 \text{ psia}$$

The pressure at the design point = 41.6 − 14.4 = 27.2 psig.

The horsepower requirement is:

$$\text{bhp}_r = \text{bhp}_o\left(\frac{N_r}{N_o}\right)^3\frac{G_r}{G_o}\frac{T_o}{T_r} = 890 \times \left(\frac{5460}{5100}\right)^3 \times \frac{0.80}{0.52}\times\frac{550}{520} = 1750 \text{ hp}$$

The new design point checks with the curve of Fig. 9-4, which shows that 14,960 cfm of 0.80 sp gr gas saturated at 60 F and 14.4 psia inlet may be discharged against 27.5 psig and requires 1750 hp.

Critical Speeds

As the shaft speed of a centrifugal machine is increased, it can be observed that at certain speeds the shaft may vibrate excessively, whereas at both higher and lower speeds, the shaft rotates relatively quietly. The speeds at which the shaft vibrates excessively are generally called the *critical speeds* of the machine. These phenomena are due to small inaccuracies in balancing, i.e., the center of gravity and the center of rotation do not coincide. Although this difference is slight, rotation of the shaft causes a centrifugal force of the mass center, which produces a force on the shaft. At the critical speed, the shaft center will rotate about the center of mass, and the resisting force of the shaft will tend to build up large amplitudes of vibration. If the calculated critical speed of a centrifugal unit occurs above its design speed, the shaft is referred to as a *rigid shaft.* If the critical speed occurs below its design speed, the shaft is referred to as a *flexible shaft.*

If a centrifugal machine were permitted to operate at the critical speed, the resultant vibration would produce large stresses or rubbing of parts. Usually, the critical speed of a flexible shaft machine is at 60 per cent of the design speed. Safe operation of the unit would permit an operating speed range from 20 per cent above the critical speed to the design speed. Thus, if the design speed of a flexible shaft machine were 5000 rpm, the safe operating range of the unit would be from 3600 to 5000 rpm.

Tip Speed

Since the stresses induced by centrifugal force are extremely high in disks rotating at high speeds, one of the most important factors in the selection of a centrifugal pumping unit is the tip speed of the impellers. Table 9-3 indicates the maximum safe limit for tip speeds in feet per second for various types of impeller wheels.

Table 9-3 Tip Speeds of Impellers

Type of wheel	Safe tip speed, fps
Aluminum	525
Cast steel	560
Forged and riveted steel	800
Forged milled radial steel	1000

Pumping Flexibility

To obtain pumping flexibility when daily sendout varies widely, it is advisable to divide the sendout among several parallel pumping units. Such parallel operation generally employs units in which the *pressure–volume characteristic* contains an appreciable amount of droop or slope. Figure 9-5 illustrates the pressure–volume characteristics for two centrifugal units. The one represented by curve *A* would be preferred for parallel operation. Units represented by the flat curve *B* are more difficult to parallel because one unit tends to deliver more gas than the other, thus causing one of the units to start pulsating. When parallel operation is considered, it is advisable to equip centrifugal pumping units with a variable speed driver so that a speed control can be used to operate each unit at a desired operating point.

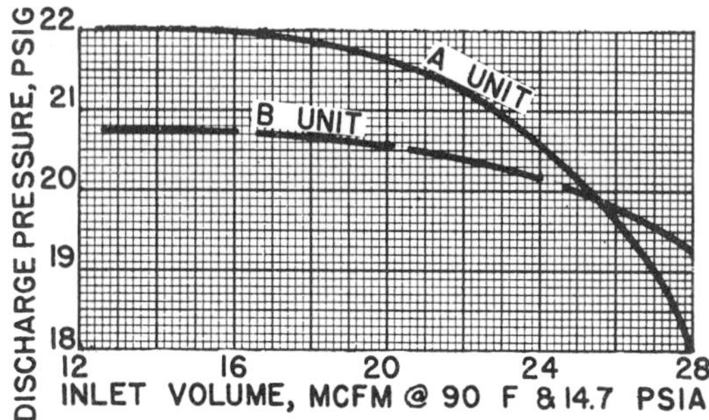

Fig. 9-5 Pressure–volume characteristics of centrifugal blowers.

POSITIVE DISPLACEMENT UNITS

Such units may be of either the rotary or the reciprocating type. Either type of machine is not affected seriously by changes in gas specific gravity and is, therefore, ideal for pumping over wide ranges of volume and pressure within its design capacities.

Rotary Compressors

Rotary compressors may be of either the sliding vane or the two-impeller positive type. A rotary sliding vane type of compressor is shown by Fig. 9-6. These compressors are available in sizes up to 5300 cfm and 50 psig for single-stage compression at speeds varying from 3600 rpm for the small sizes to 450 rpm for the large units. They are usually driven by an electric motor, steam turbine, gas, or gasoline engine.

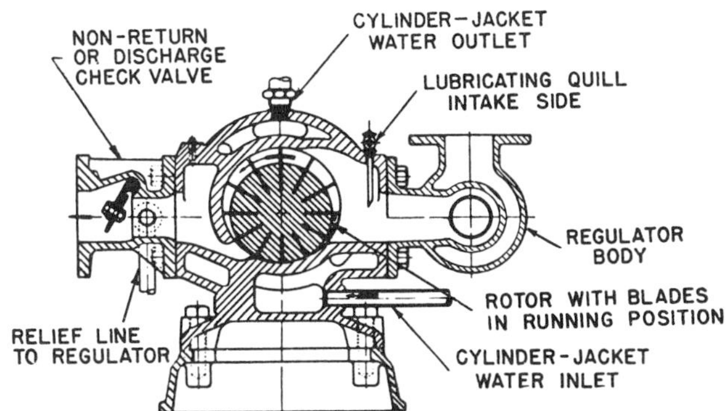

Fig. 9-6 Rotary sliding vane positive-type compressor.

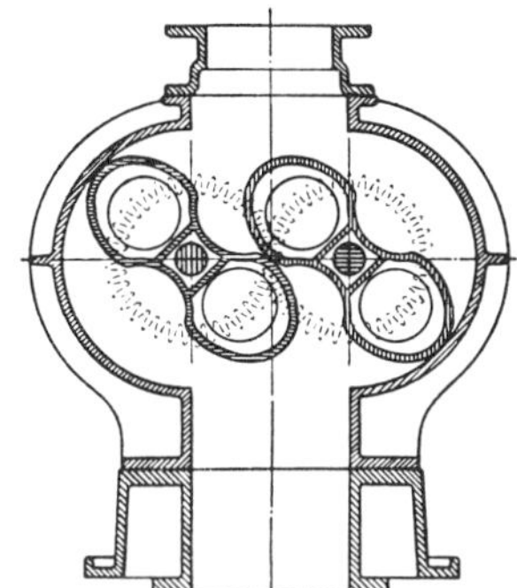

Fig. 9-7 Rotary two-impeller positive-type blower.

A rotary two-impeller positive-type blower is shown in Fig. 9-7. These units are built for capacities of from 5 to 50,000 cfm and for discharge pressures of up to 15 psig for single-stage compression. Individual units may be compounded to produce higher pressures. Figure 9-7 shows the symmetrically formed lobed impellers driven by lubricated gears located in the external housing. The clearance between impellers is accurately set to prevent rubbing, thereby establishing a predetermined air slippage at any given pressure. No seal or internal lubrication is required. Operating speeds vary from 1750 rpm for the smaller units to 125 rpm for the larger.

Rotary Compressor Performance. Figure 9-8 shows typical performance curves of a 26 in. × 22 in. Roots-Connersville rotary-type gas pump operating at a speed of 400 rpm when compressing air, with a 0.40 sp gr gas. Effects of speed variation on the performance of this compressor are shown by Fig. 9-9. A family of curves for various operating discharge pressures results from plotting inlet capacity and horsepower requirement against operating speed. This curve also illustrates variations in performance between air and a 0.40 sp gr gas for a discharge pressure of 10 psig.

The following general relationships govern performance of rotary impeller-type compressors:

1. Flow varies *directly* with the relative speed (operating speed minus slip).
2. Slip is constant at all speeds, but varies as the *square root* of the differential pressure, temperature, and specific gravity.
3. Horsepower varies *directly* with the speed and pressure differential across the unit.

Equations for calculating rotary impeller-type compressor performance follow:

$$\text{bhp} = DN\Delta P/200 \tag{31}$$

$$Q_1 = D(N - S) \tag{32}$$

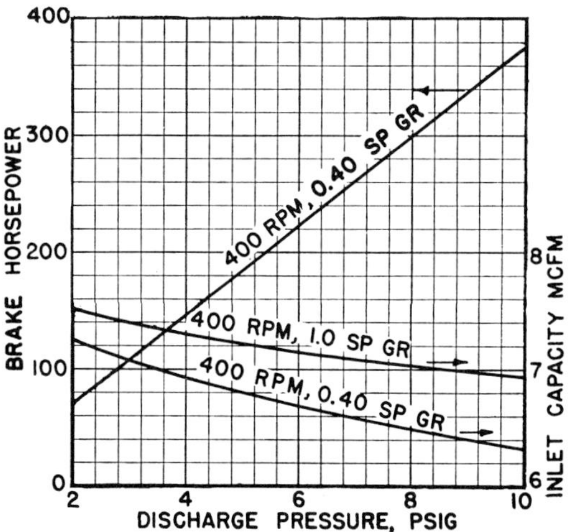

Fig. 9-8 Typical performance curves of a rotary positive gas pump.

Pump data: 26 in. × 22 in. Roots-Connersville rotary type, 20 cu ft displacement per revolution, 53 rpm air slippage at 10 psig discharge and 60 F, 400 rpm operating speed with 375 bhp driver.

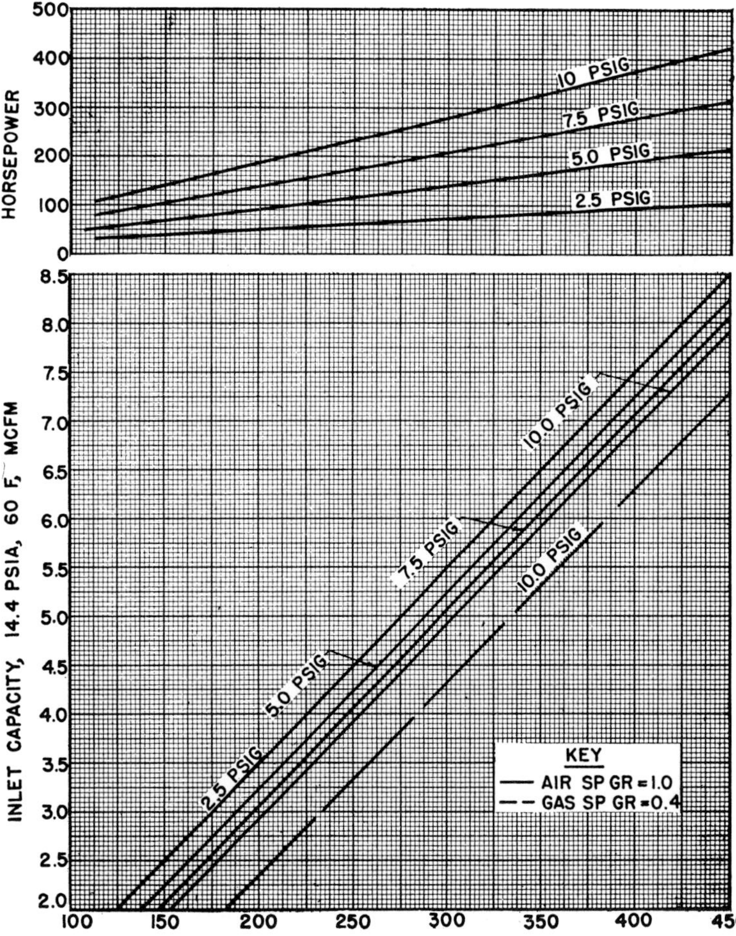

Fig. 9-9 Performance curves for a typical Roots-Connersville rotary positive gas pump.

Note: Same pump as in Fig. 9-8.

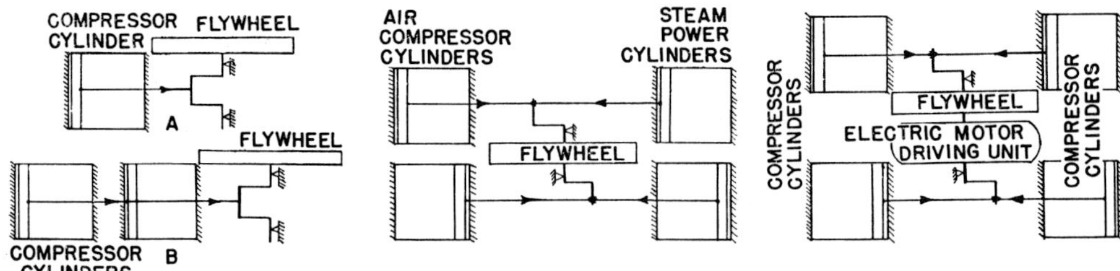

Fig. 9-10 (left) A, Single-frame double-acting compressor; and B, single-frame tandem double-acting compressor.
Fig. 9-11 (center) Duplex four-cornered steam-driven compressor.
Fig. 9-12 (right) Four-cornered motor-driven compressor.

where:

bhp = brake horsepower
D = actual displacement, cu ft per revolution
N = operating speed, rpm
ΔP = pressure differential, psi
200 = empirical constant, assuming efficiency of 87.5 per cent
Q_1 = inlet capacity, cfm
S = slip, rpm

When **slip**, S, is known at one condition, it may be calculated for a required condition from Eq. **33**, where subscript o refers to original condition and r to the required condition.

$$S_r = S_o \sqrt{\frac{\Delta P_r T_r G_o}{\Delta P_o T_o G_r}} \tag{33}$$

where: ΔP = pressure differential, psi
T = absolute temperature, °R
G = specific gravity

Calculation of performance under different operating conditions may be shown by an *example:*

Find the capacity and horsepower requirement when the unit (Fig. 9-8) is operating at 400 rpm and compressing a 0.60 specific gravity gas at 60 F from atmospheric pressure to 8 psig discharge pressure.

Horsepower requirement (Eq. **31**):

$$\text{bhp} = \frac{20 \times 400 \times 8}{200} = 320 \text{ hp}$$

Gas slippage at 8 psi differential and 0.60 sp gr (Eq. **33**):

$$S_r = S_o \sqrt{\frac{\Delta P_r G_o}{\Delta P_o G_r}} = 53 \sqrt{\frac{8 \times 1.0}{10 \times 0.4}} = 75 \text{ rpm}$$

Capacity (Eq. **32**):

$$Q_1 = 20(400 - 61) = 6780 \text{ cfm}$$

Reciprocating Compressors

Reciprocating compression of gas or air is by means of a piston within a cylinder. A *single-acting* unit effects compression at only one end of its cylinders; a *double-acting* unit effects compression at both ends of the cylinder.

Compressor cylinders are arranged in varying patterns to meet space requirements and permit single or multistage compression. *Single-frame straightline* compressors may be either horizontal or vertical, with double-acting cylinders in line with a single frame having one crank throw and one connecting rod and crosshead (Figs. 9-10,A and 8-34). A unit with two compressor cylinders mounted on a single frame with one crank throw and one connecting rod and crosshead is termed a single-frame tandem double-acting compressor (Fig. 9-10,B). Duplex compressors are machines with cylinders mounted on two parallel frames connected on a common crankshaft. A duplex four-cornered steam driven compressor is shown in Fig. 9-11. It consists of parallel frames with one or more compressor cylinders on one end of each frame and one or more steam power cylinders on the opposite end of each frame. Figure 9-12 shows a four-cornered motor driven compressor. It consists of a driving motor mounted between two parallel frames with one or more compressor cylinders on each end of each frame. A gas engine *angle-type* compressor consists of a vertical in-line or V-type gas engine with the compressor cylinders mounted horizontally and connected to a common crankshaft (Fig. 9–13).

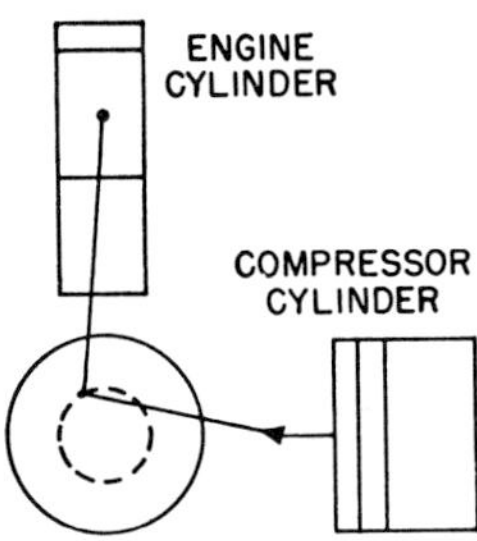

Fig. 9-13 Angle-type engine-driven compressor.

Reciprocating Compressor Performance. Problems of main importance to the pumping engineer are first, horsepower requirements and second, compressor cylinder requirements.

The horsepower required in a gas engine angle-type compressor is normally computed from charts furnished by the manufacturers. Typical charts, Figs. 8-38a thru 8-38c, indicate the brake horsepower required per million cubic feet of gas per day referred to a suction pressure of 14.4 psia plotted against compression ratio for a number of ratios of specific heats, k. The required brake horsepower is based upon a mechanical efficiency of 95 per cent and a compression efficiency of approximately 83.5 per cent. The latter is an assumed value and varies with the compression ratio, piston speed, and compressor valve design. Quantity of gas is measured at 14.4 psia and actual intake temperature.

The brake horsepower required for compression may be expressed as:

$$\text{bhp} = \frac{VP_bT_1}{14.4T_b} \times (\text{BHP/MMCFD}) \quad \textbf{(34a)}$$

or

$$\text{bhp} = \frac{QP_bT_1}{T_b(10)^4} \times (\text{BHP/MMCFD}) \quad \textbf{(34b)}$$

where:

V = inlet capacity of compressor, million cu ft per 24 hr day
Q = inlet capacity of compressor, cfm
P_b = pressure base at which the volume is measured, psia
T_b = temperature base at which the volume is measured, °R
14.4 = pressure base of charts, psia
T_1 = inlet temperature of compressor, °R
(BHP/MMCFD)
= factor determined from Figs. 8-38a thru 8-38c

The compression ratio, C, is as follows:

$$C = \frac{\text{discharge pressure, abs}}{\text{inlet pressure, abs.}} = \frac{P_2}{P_1} \quad \textbf{(35)}$$

Example:*

Find the brake horsepower required to compress 14,000 cfm of gas measured at 14.7 psia and 60 F from 25 psig to 75 psig with an inlet temperature of 90 F. The barometric pressure is 14.4 psia and ratio of specific heats is 1.35.

The compression ratio is (Eq. **35**):

$$C = \frac{75 + 14.4}{25 + 14.4} = \frac{89.4}{39.4} = 2.27$$

Therefore, from Fig. 8-38a:

BHP/MMCFD = 51.4 (at k = 1.35 and C = 2.27)

The horsepower requirement is (Eq. **34b**):

$$\text{bhp} = \frac{14{,}000 \times 14.7 \times 550}{520 \times 10^4} \times 51.4 = 1120 \text{ hp}$$

When compression ratios are *greater than* 5, use two-stage compression to minimize valve failure. Horsepower losses encountered in high compression ratios are avoided with two-stage compression.

The horsepower requirement for multistage compression is decreased by intercooling between stages. If intercooling is not employed, the heat of compression in the first stage increases the volume to be compressed in the second stage with resultant increase in horsepower required. The horsepower required by the second stage is in direct proportion to the absolute inlet temperature of the first stage. Hence, the second-stage horsepower requirement will be equal to the first-stage requirement multiplied by the ratio of temperature increase across the first stage.

* Another example is given under Reciprocating Compressors, Sec. 8, Chap. 3.

This temperature ratio may be expressed by (see Eq. **4**):

$$c^{(k-1)/k}$$

where: c = compression ratio per stage
k = ratio of specific heats

The second-stage horsepower may be determined from the following formula, in which subscript 1 refers to the first stage and subscript 2 to the second stage.

$$(\text{BHP/MMCFD})_2 = (\text{BHP/MMCFD})_1 c_1^{(k-1)/k} \quad \textbf{(36)}$$

A good rule of thumb for compression ratio *per stage* is to extract the *square root* of the overall compression ratio for two-stage compression and the *cube root* of the overall ratio for three-stage compression problems. Hence:

$$c = C^{0.5} \quad \text{for two-stage compression} \quad \textbf{(37)}$$

$$c = C^{0.333} \quad \text{for three-stage compression} \quad \textbf{(38)}$$

where: C = overall compression ratio
c = compression ratio *per stage*

If intercooling is used, the total horsepower required is roughly equal to the horsepower requirement of the first stage multiplied by the number of stages.† The compression ratio per stage must be corrected for the pressure drop resulting from the intercooler. This pressure drop is generally taken as four per cent of the first-stage discharge pressure. Horsepower requirements with and without intercooling may be calculated as shown by the following example:

Find horsepower required with and without intercooling when compressing 16,000 cfm of natural gas, k = 1.28, measured at 60 F and 14.7 psia, from atmospheric pressure of 14.4 psia to 125 psig. Inlet temperature is 70 F.

Without intercooling (see Eq. **35**):

$$C = \frac{P_2}{P_1} = \frac{125 + 14.4}{14.4} = 9.68$$

and from Eq. **37**:

$$c = 9.68^{0.5} = 3.11$$

From Fig. 8-38b:

BHP/MMCFD, *1st stage* (at k = 1.28 and c = 3.11) = 69.1

From Eq. **36**:

BHP/MMCFD, *2nd stage* = $69.1(3.11)^{0.2185}$ = 88.5

Total **BHP/MMCFD** = 157.6

The corrected horesepower required (Eq. **34b**) is:

$$\text{bhp} = \frac{16{,}000 \times 14.7 \times 530}{520 \times 10^4} \times 157.6 = 3780 \text{ hp}$$

With intercooling:

Allowing 4 per cent of first-stage discharge pressure or approximately 1.0 psi per stage as a pressure drop between stages, the *corrected* compression ratio, c, per stage is:

$$c = 3.11 + \frac{1.0}{14.4} = 3.18$$

† Assuming the same inlet temperature for all stages and the use of the aforementioned rule of thumb.

From Fig. 8-38b:

BHP/MMCFD per stage (at $k = 1.28$ and $c = 3.18$) = 70.6

Total **BHP/MMCFD** = 70.6 × 2 = 141.2

The corrected horsepower requirement is (Eq. **34b**):

$$\text{bhp} = \frac{16{,}000 \times 14.7 \times 530}{520 \times 10^4} \times 141.2 = 3385 \text{ hp}$$

Intercooling shows a saving of approximately 395 horsepower.

Conversely, Figs. 8-38a thru 8-38c may be used to determine the pumping capacity of any given size engine. However, to determine the capacity of a given engine operating as a multistage unit, allowance must be made for the pressure losses between stages. It is customary to assume full horsepower rating for a single-stage unit, 98 per cent of the horsepower rating for a two-stage unit, and 96 per cent of the horsepower rating for a three-stage unit.

The following formulas may be used to determine the pumping capacity of a given size unit:

For Single-Stage Compression:

$$V = \frac{\text{Rated bhp} \times 14.4 \times T_b}{(\textbf{BHP/MMCFD})P_bT_1} \qquad (39)$$

$$Q = \frac{\text{Rated bhp} \times T_b \times (10)^4}{(\textbf{BHP/MMCFD})T_1P_b} \qquad (40)$$

For Two-Stage Compression:

$$V = \frac{\text{Rated bhp} \times 0.98 \times 14.4 \times T_b}{(\textbf{BHP/MMCFD})nT_1P_b} \qquad (41)$$

$$Q = \frac{\text{Rated bhp} \times 0.98 \times T_b \times (10)^4}{(\textbf{BHP/MMCFD})nT_1P_b} \qquad (42)$$

For Three-Stage Compression:

$$V = \frac{\text{Rated bhp} \times 0.96 \times 14.4 \times T_b}{(\textbf{BHP/MMCFD})nT_1P_b} \qquad (43)$$

$$Q = \frac{\text{Rated bhp} \times 0.96 \times T_b \times (10)^4}{(\textbf{BHP/MMCFD})nT_1P_b} \qquad (44)$$

where:

BHP/MMCFD is determined by using the compression ratio per stage
V = inlet capacity of compressor, MMCFD
Q = inlet capacity of compressor, cfm
14.4 = pressure base of charts
P_b = pressure base of V or Q, psia
T_b = temperature base of V or Q, °R
T_1 = inlet temperature, °R
Rated bhp = manufacturer's rating of engine
n = number of stages

An alternate method of determining the capacity of a given size engine unit would be to allow a reasonable pressure drop between stages and correct the pressure ratio per stage to compensate for this loss, as in the previous example. The capacity may be determined by the following formula:

$$V = \frac{\text{Rated bhp} \times 14.4 \times T_b}{n(\textbf{BHP/MMCFD})P_bT_1} \qquad (45)$$

$$Q = \frac{\text{Rated bhp} \times T_b \times (10)^4}{n(\textbf{BHP/MMCFD})P_bT_1} \qquad (46)$$

where **BHP/MMCFD** is determined from Figs. 8-38a thru 8-38c, using c as the corrected compression ratio per stage.

Example: Find the capacity at standard conditions (60 F and 14.7 psia) for an 1100-hp two-stage gas engine compressor unit compressing natural gas, $k = 1.28$, from atmospheric pressure 14.4 psia at 90 F inlet condition to a discharge pressure of 126 psig. Check by the alternate method.

From Eq. **35**:

$$C = \frac{P_2}{P_1} = \frac{125 + 14.4}{14.4} = 9.68$$

From Eq. **37**:

$$c = \sqrt{9.68} = 3.11 \textit{ per stage}$$

From Fig. 8-38b:

BHP/MMCFD (at $c = 3.11$ and $k = 1.28$) = 69.1

For two-stage compression, Eq. **42**:

$$Q = \frac{1100 \times 0.98 \times 520 \times (10)^4}{69.1 \times 2 \times 550 \times 14.7} = 5000 \text{ cfm}$$

Using the alternate method gives us $C = 9.68$ and $c = 3.11$ from the foregoing.

Allowing 4 per cent of the first-stage discharge pressure or approximately 1.0 psi per stage, we get the *corrected* compression ratio per stage:

$$c = 3.11 + \frac{1.0}{14.4} = 3.18$$

From Fig. 8-38b:

BHP/MMCFD (at $k = 1.28$ and $c = 3.18$) = 70.6

From Eq. **46**:

$$Q = \frac{1100 \times 520 \times (10)^4}{2 \times 70.6 \times 14.7 \times 550} = 5010 \text{ cfm}$$

Engine capacity performance curves may be developed from Figs. 8-38a thru 8-38c. It is customary to plot the inlet capacity in cubic feet per minute or per day at 14.7 or 14.4 psia and at suction temperature against the discharge pressure to obtain the performance curve of a given size engine compressor. The following example will illustrate the development of an engine compressor capacity curve.

Example: Develop the engine capacity curve for an assumed 1320-hp gas engine, compressing a mixture of natural gas and manufactured gas from atmospheric pressure of 14.4 psia at temperatures varying between 40 and 90 F to varying discharge pressures between 15 and 40 psig. The pressure base is 14.7 psia and $k = 1.34$.

Since both the charts and the capacity curve are based on the *actual* inlet temperature, no temperature correction need be applied in the capacity formula. Therefore, substituting in Eq. **46**:

$$Q = \frac{\text{Rated bhp} \times (10)^4}{(\textbf{BHP/MMCFD})P_b} = \frac{1320 \times 10^4}{(\textbf{BHP/MMCFD}) \times 14.7} = \frac{8.98 \times 10^5}{\textbf{BHP/MMCFD}}$$

Table 9-4 Calculated Capacity of a 1320-hp Gas Engine

Discharge pressure, P_2				Capacity, Q, cfm		
psig	psia	Compr. ratio, C	BHP/MMCFD	Full load	90% load	80% load
40.0	54.4	3.775	83.60	10,750	9,675	8,600
37.5	51.9	3.605	80.40	11,170	10,050	8,940
35.0	49.4	3.430	77.00	11,660	10,500	9,330
32.5	46.9	3.255	73.60	12,200	10,980	9,760
30.0	44.4	3.085	70.10	12,810	11,530	10,250
27.5	41.9	2.910	66.25	13,550	12,200	10,840
25.0	39.4	2.735	62.40	14,390	12,950	11,510
22.5	36.9	2.560	58.15	15,440	13,900	12,350
20.0	34.4	2.390	54.25	16,550	14,900	13,240
17.5	31.9	2.218	49.95	17,980	16,180	14,380
15.0	29.4	2.040	45.45	19,760	17,780	15,810
12.5	26.9	1.868	40.30	22,280	20,050	17,825
10.0	24.4	1.694	35.50	25,290	22,760	20,235

The brake horsepower per million cubic feet per day is determined from Figs. 8-38a thru 8-38c for each of the corresponding compression ratios using a k value of 1.34. Capacities so calculated are shown in Table 9-4. Figure 9-14 is constructed by plotting these capacities against corresponding discharge pressure.

When gas engine compressors are to be used at altitudes over 1500 ft, the manufacturer's sea level rating must be derated. Figure 8-40 may be used to determine the per cent of sea level rating and the atmospheric pressure for various elevations above sea level.

If the assumed 1320-hp engine were to operate at a 4000-ft elevation, then, from Fig. 8-40, the rated 1320 bhp would be reduced to $0.89 \times 1320 = 1175$ hp and the corresponding atmospheric pressure would be 12.65 psia.

Compressor Cylinder Calculations. The performance of a gas engine-driven compressor is greatly influenced by the maximum available piston displacement. It is customary for the compressor manufacturers to calculate compressor cylinder capacities in cubic feet either per minute or per day, using a pressure base of either 14.4 or 14.7 psia and suction temperature. These capacities may be converted to conform to the "horsepower per million cubic feet" charts which are based on 14.4 psia base pressure and suction temperature. The piston displacement of a compressor cylinder can be computed easily from the cylinder bore, piston stroke, diameter of connecting rod, and speed.

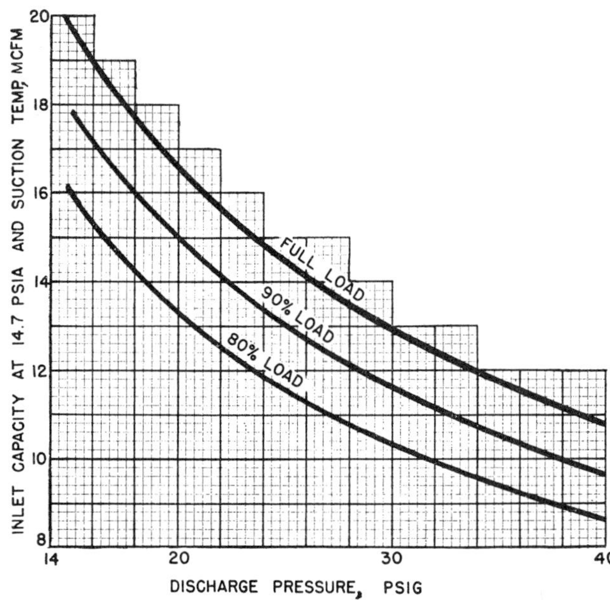

Fig. 9-14 Typical engine capacity performance curves.

To design a compressor cylinder, it is necessary to provide a clearance at each end of the piston stroke. The delivery of the cylinder is dependent upon the amount of clearance volume. Gas trapped in the clearance space at the end of the stroke must expand, after the piston changes direction, to a pressure below that of the suction header before the suction valves open and filling begins. Thus, the full stroke of the piston is not utilized on the suction cycle and the *volumetric or filling efficiency* is less than unity.

The capacity of a cylinder may be calculated from the following formula:

$$Q = (PD) \frac{e}{100} \frac{P_1}{P_b} \tag{47}^*$$

where:

Q = capacity at pressure base P_b and suction temperature, cfm

PD† = piston displacement, cfm

e = volumetric or filling efficiency based on the piston displacement, per cent

P_1 = inlet pressure to cylinder, psia

P_b = pressure base, psia

For all practical purposes, cylinder volumetric efficiency may be determined from Figs. 8-39a thru 8-39c. These charts are constructed by plotting per cent volumetric efficiency against compression ratio for numerous values of cylinder clearance. They are based on Eq. 48. The $-C$ term covers losses.

$$e = 100 - C - V_c(C^{1/k} - 1) \tag{48}$$

where: e = volumetric efficiency, per cent

C = compression ratio

V_c = clearance volume based on the piston displacement, per cent

k = ratio of specific heats

Some manufacturers prefer to use the following empirical equation when calculating volumetric efficiency:

$$e = F - V_c(C^{1/k} - 1) \tag{49}$$

where: F = loss of leakage factor

Other terms defined under Eq. **48**.

F usually varies from 95 to 98 and is dependent upon length of stroke and type of gas compressed. Generally, either equation will give sufficiently accurate results for preliminary engineering problems.

Compressor cylinder performance can be calculated from the foregoing equations and combined with horsepower capacity curves to give complete performance curves for the

* Compare with Eq. **21** of Sec. 8, Chap 3.

† See Eqs. **15** and **16** of Sec. 8, Chap. 3.

Table 9-5 Calculated Compressor Cylinder Performance*

Disch. press., psig	Comp. ratio, C	$C^{1/k} - 1$	$V_c(C^{1/k} - 1)$	e, per cent	Capacity per cylinder, cfm: Double acting	Capacity per cylinder, cfm: Single acting	Total double-acting capacity 4 cyl, cfm
5.0	1.347	0.249	2.366	96.3	4730	2380	18,900
17.5	2.215	0.810	7.695	90.1	4430	2230	17,700
40.0	3.777	1.690	16.103	80.1	3940	1980	15,750

* Terms defined under Eq. **48**.

unit. Compressor cylinders used in gas distribution compressors are usually designed with clearance volumes ranging from 6 to 15 per cent of the piston displacement.

Example: Calculate the performance of an assumed 1320-hp gas engine compressor operating at 300 rpm, equipped with four 33-in. × 17-in. compressor cylinders with assumed clearance volume of 9.5 per cent and 3.25-in. diam connecting rods, compressing gas from atmospheric pressure of 14.4 psia to discharge pressures ranging between 5 and 40 psig. The base pressure is 14.7 psia and $k = 1.34$.

The engine capacity has been resolved already, and is shown by Fig. 9-14.

Therefore, substituting in Eq. **15** of Sec. 8, Chap. 3:

$$PD = \frac{(33)^2 \pi \times 17 \times 300}{4 \times 1728} = 2524 \text{ cfm } \textit{per cylinder}$$

The crank end displacement per cylinder is:

$$\frac{(33^2 - 3.25^2) \pi \times 17 \times 300}{4 \times 1728} = 2500 \text{ cfm } \textit{per cylinder}$$

$$\text{Total displacement} = 5024 \text{ cfm per cylinder}$$

The compressor cylinder performance calculated from Eqs. **47** and **48** is given in Table 9-5.

Figure 9-15 shows discharge pressure plotted against engine capacity from Table 9-4 and cylinder capacity from Table 9-5. Curve ABC is the full load capacity curve of the 1320-hp gas engine. Curve DBE indicates the performance of the four 33-in. × 17-in. compressor cylinders. Operation along the curve DBE at any point to the left of point B will be at partial load. Operation at any point to the right of point B will cause an overload on the engine. Therefore, maximum performance of the unit must be along the curve DBC.

Operation along the full load curve BC may be accomplished by adding either fixed volume or variable volume clearance pockets to the compressor cylinders. This will increase the normal clearance, decreasing the volumetric efficiency and the capacity of the cylinders. Another method of unloading the engine would be to equip the compressor cylinders with suction valve lifters, thereby converting the cylinders from double-acting to single-acting compressor cylinders. Valve lifters may be installed on the head or crank end of one or more cylinders.

The number and size of fixed volume clearance pockets are dependent upon the compressor cylinder design. Since this compressor unit is hypothetical, assume that each compressor cylinder is to be equipped with one fixed volume clearance pocket. Then the volumetric efficiency must be decreased in order to decrease the capacity of 15,750 cfm at 40 psig (point E on Fig. 9-15) to 10,750 cfm (point C). Care must be taken to avoid installing too large a clearance pocket, for that would result in a negative volumetric efficiency. The required clearance and volumetric efficiency may be computed from Eqs. **47** and **48**.

Decrease in capacity required per pocket:

$$\frac{15{,}750 - 10{,}750}{4} = 1250 \text{ cfm}$$

Thus, at 40 psig discharge pressure:

$$Q = 3940 - 1250 = 2690 \text{ cfm}$$

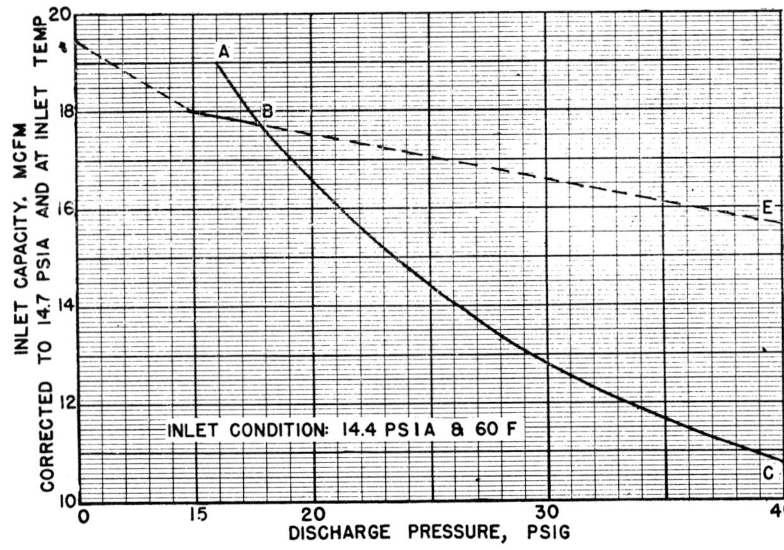

Fig. 9-15 Typical engine and compressor cylinder performance curve. Point D at 19.4 MCFM and 0 psig; note scale change at 15 psig.

Solving Eq. **47** for *e:*

$$e = 100 \frac{QP_b}{PDP_1} = 100 \times \frac{2690}{5024} \times \frac{14.7}{14.4} = 54.6 \text{ per cent}$$

Substituting the required *e* in Eq. **48** and solving for the required clearance:

$$V_c = \frac{100 - C - e}{C^{1/k} - 1} = \frac{100 - 3.777 - 54.6}{1.69} = 24.6 \text{ per cent}$$

Table 9-6 gives cylinder capacities based on the required clearance (24.6 per cent) for various discharge pressures.

Table 9-6 Calculated Cylinder Performance with Clearance Pocket*

Disch. press., psig	Comp. ratio, C	$(C^{1/k} - 1)$	$V_c(C^{1/k} - 1)$	e, per cent	Cylinder capacity, cfm
5.0	1.347	0.249	6.125	92.53	4550
17.5	2.215	0.810	19.926	77.86	3830
40.0	3.777	1.690	41.697	54.53	2690

* Terms defined under Eq. **48**.

When the capacity of a compressor cylinder operating with a clearance pocket open and the suction valves lifted on the crank end is computed, adjustment must be made for the value of the percentage of clearance. The clearance pocket affects the delivery from the *head end* of the piston. The required 24.6 per cent clearance is based on the piston displacement and incorporates both the normal clearance and the pocket clearance.

The adjusted clearance percentage may be determined as follows:

Piston displacement in cubic inches:

$$pd = \frac{PD \times 1728}{N} \tag{50}$$

where: pd = piston displacement, cu in.
PD = piston displacement, cfm
N = speed, rpm

Therefore, from Eq. **50**:

$$pd = \frac{5024 \times 1728}{300} = 28{,}900 \text{ cu in.}$$

Normal cylinder clearance:

$$28{,}900 \times 0.095 = 2745 \text{ cu in.}$$

Volume of the clearance pocket:

$$28{,}900\ (0.246 - 0.095) = 4360 \text{ cu in.}$$

Swept volume with the crank end suction valve lifted:

$$pd = \frac{2524 \times 1728}{300} = 14{,}540 \text{ cu in.}$$

Adjusted clearance of the head end:

$$\frac{(14{,}540 \times 0.095) + 4360}{14{,}540} \times 100 = 39.4 \text{ per cent of displacement}$$

Since the pressure base is 14.7, the head end piston displacement is:

$$PD = 2524 \times \frac{14.4}{14.7} = 2472 \text{ cfm}$$

The capacity of the cylinder can then be computed, with the help of Eqs. **47** and **48**.

Table 9-7 gives calculated capacities of one 33-in. × 17-in. cylinder operating at 300 rpm with a clearance pocket open and crank end suction valves lifted.

The capacities of the four cylinders under various operating conditions, computed from the data in Tables 9-5 thru 9-7, are listed in Table 9-8.

Table 9-7 Cylinder Capacity with Pocket Open and Valves Lifted*

Disch. press., psig	Comp. ratio, C	$C^{1/k} - 1$	$V_c(C^{1/k} - 1)$	e, per cent	PD, cfm	Cylinder capacity, $e(PD)$, cfm
5.0	1.347	0.249	9.786	88.87	2472	2200
17.5	2.215	0.810	31.833	65.95	2472	1630
40.0	3.777	1.695	66.417	30.81	2472	760

* Terms defined under Eqs. **47** and **48**.

Table 9-8 Capacity of Four 33-in. × 17-in. Cylinders Equipped with Four Clearance Pockets and One Suction Valve Lifter

Operation	Capacity at discharge pressure, cfm — Suction valve not lifted		Suction valve lifted	
	5 psig	40 psig	5 psig	40 psig
All pockets closed	18,900	15,750	16,550	13,780
One pocket open	18,740	14,510	16,370	12,530
Two pockets open	18,560	13,260	16,190	11,280
Three pockets open	18,380	12,010	16,010	10,030
Four pockets open	18,200	10,760	15,850	8,830

Figure 9-16a illustrates the performance of the hypothetical angle type of gas engine pumping unit at full speed, 300 rpm. Curves of engine capacities in Table 9-4 and cylinder capacities in Table 9-8 are plotted against discharge pressures. By means of the four clearance pockets, the compressor cylinders can be unloaded to maintain the engine horsepower at approximately full load between points R and L (18 to 40 psig discharge pressure) along the curve KRL. Operation with four pockets open on curve E will cause a slight overload condition between 30 and 40 psig discharge pressure. The family of *dashed* curves on Fig. 9-16a illustrates the performance of the unit with the crank end suction valves of one cylinder lifted, together with various combinations of clearance pockets for operation of the unit at full speed. A further reduction in capacity can be accomplished by equipping another cylinder with a suction valve unloader.

A reduction of speed will further reduce capacity, since in positive displacement units the capacity varies *directly* with speed. The minimum operating speed is approximately **60** per cent of the design speed for two-cycle gas engines and **50** per cent of the design speed for four-cycle gas engines.

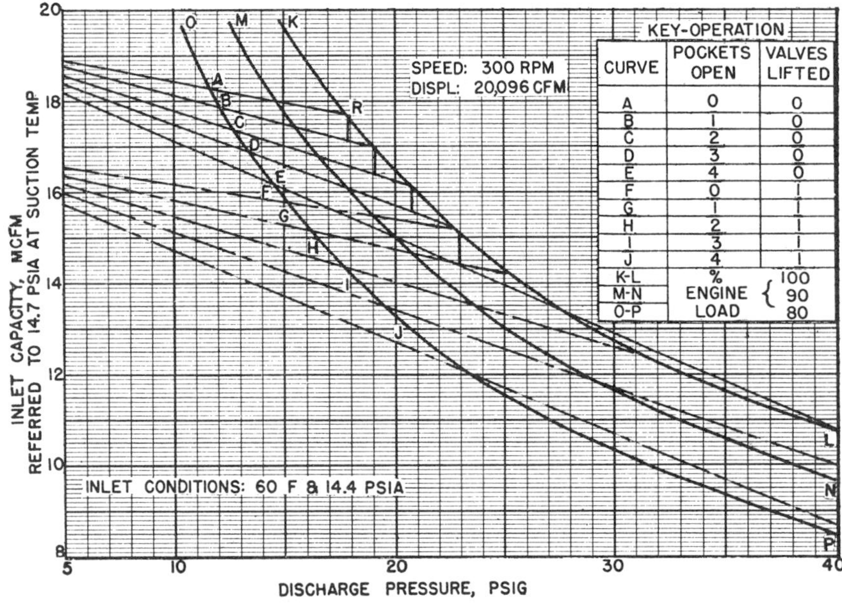

Fig. 9-16a Expected performance curve for typical 1320-hp angle-type compressor unit.

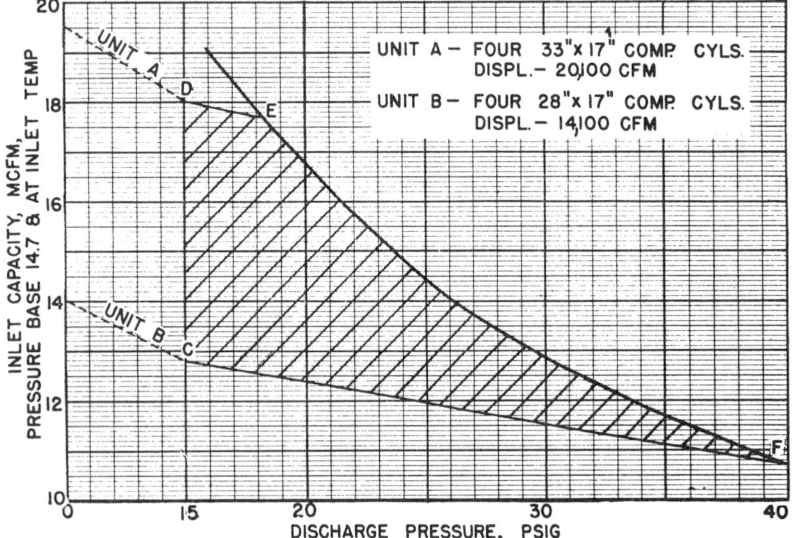

Fig. 9-16b Typical maximum capacity performance curves for 1320-hp angle-type compressor units.

Pulsation. Positive displacement units pumping into a pipe network frequently create pulsations of such frequency and amplitude that connected controls and metering are seriously affected. This is particularly true in reciprocating units operating at moderate speeds, but can also become a factor with the lobe or vane type of unit.

Complete elimination of pulsations would be extremely difficult, if not impossible, but it is possible to dampen the impulses in such a manner that the interference with controls and metering would be of a minor character.

When positive displacement pumping units are installed, careful consideration must be given to the design of the piping network to ensure that there are proper dampening devices between the compressors and the piping network on both the suction and discharge sides.

Pumping Flexibility. To obtain pumping flexibility when sendout varies widely, it is advisable to divide the sendout among several pumping units. Major consideration should be given to the selection of reciprocating pumping units for parallel operation. Figure 9-16b illustrates the pressure–volume characteristics of two 1320-hp hypothetical angle-type gas engines. Unit A is assumed to be equipped with four 33-in. × 17-in. compressor cylinders having a displacement of approximately 20,100 cfm. Its performance is indicated by the curve DEF. Unit B is assumed to be equipped with four 28-in. × 17-in. compressor cylinders having a displacement of 14,100 cfm. Its performance is indicated by the curve CF. The distance between curves CF and DEF shown cross-hatched on Fig. 9-16b represents the greater capacity of unit A over unit B for system pressures lower than the maximum 40 psig discharge pressure. For a system in which the pressures vary widely with seasonal changes, it is advantageous to select units with the maximum compressor cylinder displacement, in order to utilize the greater capacity at low discharge pressures.

POWER EXTRACTION PUMPING

A power extraction pumping unit is usually applied as a combined compressor and mixer, utilizing the available energy in a high-pressure gas to compress a low-pressure gas to some intermediate pressure. After or during compression the two gases are usually mixed in some common outlet or header.

Several types of equipment are designed for this purpose. All require a natural gas transmission line terminal pressure which is relatively high compared with the distribution system pressure, and all are applicable for utilities having a mixed natural and manufactured gas sendout. The high-pressure natural gas is used to compress and enrich the manufactured gas to a heating value somewhat below that of the sendout gas. Final calorific control of the sendout is accomplished by adding a controlled amount of natural gas on a by-pass of the pumping unit.

Dynamic Mixer

The dynamic mixer is essentially a multistage centrifugal blower and a single-stage reaction turbine on a common shaft and integrally cased (Fig. 9-17). High-pressure natural gas is expanded thru the turbine wheel and exhausted at the pressure required in the distribution system. The power extracted in this manner drives the blower compressing the manufactured gas. Blower discharge and turbine exhaust mix in a common outlet.

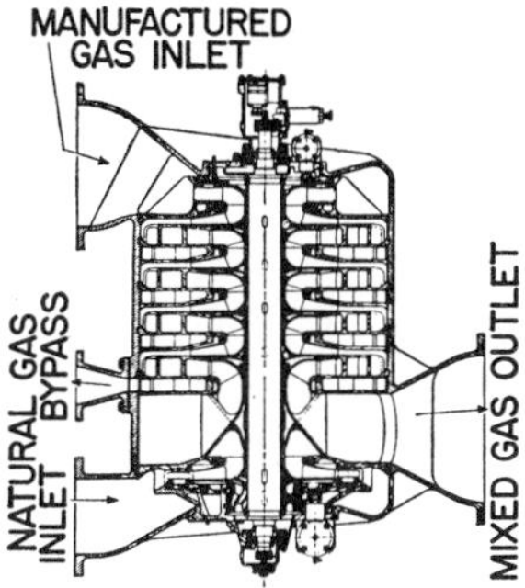

Fig. 9-17 Dynamic mixer.

The dynamic mixer is very sensitive to variations in the properties of either the motive gas or the manufactured gas. Preheating the natural gas serves two purposes: (1) it counters excessively low temperatures resulting from the expansion of the gas in the turbine nozzle, and (2) it makes a greater amount of energy available to the driver, thereby reducing both the quantity of natural gas thru the driver and the heat content of the effluent gas. (In some instances reduction of the heat content may be required for the calorific control of the sendout.)

Performance can be calculated from the fundamental thermodynamic equations and assumed efficiencies applying to a multistage blower and a reaction turbine.

Expander Engine Compressor

The expander engine compressor is a basic four-cornered duplex steam engine as shown in Fig. 9–11, except that it is designed to use natural gas in place of steam as the motive power. Natural gas is expanded in a power cylinder while manufactured gas is compressed in the opposing cylinder. Discharge from all cylinders is at the pressure required for distribution and is mixed in a common header.

This type of unit requires more installation space than other power extracting types. It is independent of specific gravity changes and can, therefore, pump a wider range of manufactured gases. As in the dynamic mixer, preheating is advantageous to prevent icing of the exhaust valves from the natural gas expansion and to assure Btu control.

Jet Compressor

The jet compressor operates on the principle of the steam jet air ejector, natural gas replacing steam as the motive power (Fig. 9-18). High-pressure natural gas expands thru a single divergent nozzle, resulting in a high-velocity, low-pressure jet stream. Manufactured gas flows into the low-pressure suction chamber surrounding the nozzle and is entrained in the high-velocity jet. The mixture enters the throat of the venturi-type diffuser, where the kinetic energy of the high-velocity jet is converted into the pressure required for distribution.

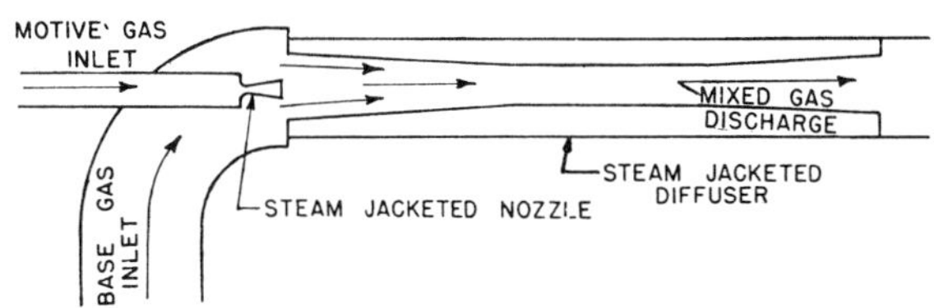

Fig. 9-18 Schematic jet compressor.

The jet compressor is essentially *a constant volume machine.* Design of the nozzle and diffuser is critical and fixed for given operating conditions, resulting in a lack of flexibility. Discharge volume changes are effected by having several jets of different capacity in a battery and cutting them in and out in the proper combination to get the desired total capacity.

Steam jacketing or other means of heating the nozzle and diffuser barrel is required to prevent the formation of ice and naphthalene crystals.

Efficiency of the jet compressor is low compared with that of the dynamic mixer and the expander engine. Despite low

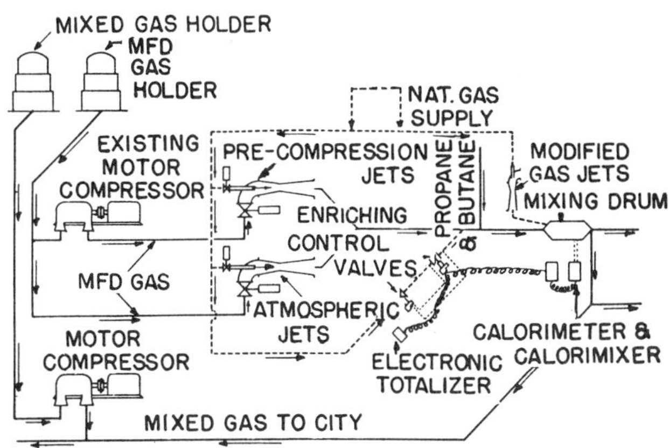

Fig. 9-19 Typical flow diagram for jet compressor installation.

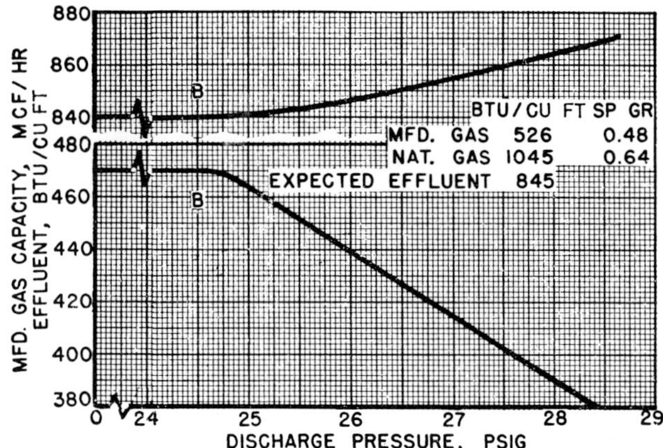

Fig. 9-20a Typical jet compressor performance curves with constant suction pressure. Upper curve, effluent Btu; lower curve, manufactured gas capacity.

efficiency and lack of flexibility, the jet compressor offers the following principal advantages:

1. Low installation cost—outdoor installation possible, eliminating building construction.
2. No auxiliary equipment needed except source of heat.
3. No moving parts—lubrication unnecessary.
4. Low maintenance.
5. Simplicity of operation—with motorized valves, remote operation is possible.

A typical installation is shown in Fig. 9-19 (see also Fig. 5-30a). The atmospheric jet represents a battery of six jets sized to give a manufactured gas capacity from 200 MCFH to 1000 MCFH in steps of 200 MCFH, with a motive gas pressure of 115 psig and a discharge pressure of up to 14.5 psig. Similarly, the precompression jet battery has the same capacity, with an inlet manufactured gas pressure of 10 psig and a discharge pressure of 25 psig.

Discharge of each of these batteries has a calorific value of approximately 850 Btu per cu ft. This is brought up to the sendout 900 Btu by adding natural gas thru the enriching control valves. These valves are controlled by a Calorimixer, which takes a continuous sample of the mixed gas and is, in turn, monitored by a calorimeter on the same sample line. This same calorific control system can be applied to a dynamic mixer or expander engine installation. With the jet batteries, an electronic totalizer is sometimes used on the enriching valve control to handle the sudden changes when going from one combination of jets to another.

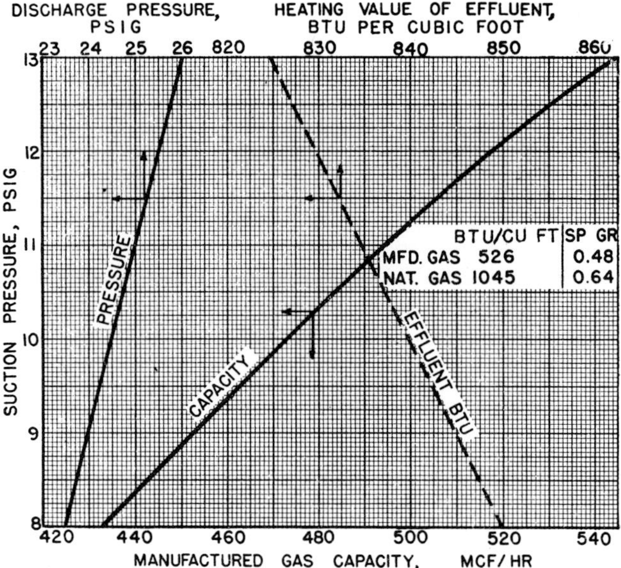

Fig. 9-20b Typical jet compressor performance curves with variable suction pressure.

Some operating characteristics of the jet compressor are shown in Figs. 9-20a and 9-20b. Curves in Fig. 9-20a for manufactured gas capacity and heating value vs. discharge pressure illustrate the constant volume performance up to the break point B, where a further increase in discharge pressure results in lowered capacity and higher effluent heating value.

Relationships of the break point capacity, the effluent Btu, and the discharge pressure to a variable suction pressure on the precompression jets are shown by Fig. 9-20b.

BIBLIOGRAPHY

Church, A. H. *Centrifugal Pumps and Blowers.* New York, Wiley, 1944.

Compressed Air and Gas Institute. *Compressed Air Handbook,* 2nd ed. New York, McGraw-Hill, 1954.

Faires, V. M. *Applied Thermodynamics,* rev. ed. New York, Macmillan, 1947.

Kearton, W. J. *Turbo-Blowers and Compressors.* New York, Pitman, 1926.

Marks, L. S. *Mechanical Engineers' Handbook,* 6th ed. New York, McGraw-Hill, 1958.

Natural Gasoline Supply Men's Assoc. *Engineering Data Book,* 7th ed. Tulsa, Okla., 1957.

Chapter 2
Distribution Pressures

by R. J. Ott and C. George Segeler

Pressures used in the distribution of gas are divided into two categories in ASA B31.8–1963:

Low Pressure or Distribution Pressure. Gases within this pressure range may be piped to domestic appliances without the use of a service regulator. Detailed discussion of this range follows later.

High Pressure or Transmission Pressure. These terms are used to describe pressures above the low-pressure range. There is no definite top limit but at the present time few *distribution* systems carry pressures in excess of 150 psig. Natural gas pipeline companies, however, use pressures up to 1200 psig because high pressures are economical for long range transportation.

Many gas distribution companies use the terms **medium pressure** (the intermediate pressure level in networks having three pressure levels) and **high pressure** to define certain pressure levels adopted for local convenience. There are no recognized definitions for such pressures.

PRESSURES IN THE LOW-PRESSURE MAINS

The prime consideration is to maintain pressure within satisfactory limits at the appliances; what is "satisfactory" depends upon: (a) the kind of gas used; (b) the pressure for which the appliances are adjusted; and (c) the kind of appliances employed.

In order to pass the pertinent *American Standard Approval Requirements,* an appliance is adjusted properly at the "normal test pressure" for the given type of gas and must then operate satisfactorily, without additional adjustment, within specified limits of appliance inlet pressure. **Test pressures** for a number of gases are given in Table 9-9. At 1.5 normal pressure, the appliance would be burning 22.5 per cent more gas than at normal pressure.* Conversely, at 0.5 normal pressure, the appliance would be burning 41.4 per cent less gas than at normal pressure.

Table 9-9 Test Pressures at Appliance Inlet[1]

Test gas (Btu/cu ft)	Test pressures,* in. w.c.		
	Reduced	Normal	Increased
A. Natural (1075)	3.5	7.0	10.5
B. Manufactured (535)	1.75	3.5	5.25
C. Mixed (800)	3.0	6.0	9.0
D. Butane (3175)	8.0	11.0	13.0
E. Propane (2500)	8.0	11.0	13.0
F. Butane–air (525)	3.0	6.0	9.0
G. Butane–air (1400)	3.0	6.0	9.0

* There are some variations for the different appliances.

The range of test pressures stipulated in the *American Standard Approval Requirements* is broad enough to ensure satisfactory appliance flexibility in operation under any pressure condition likely to be encountered. The need for such broad test requirements is shown in Table 9-10, which summarizes the results of a typical gas industry survey. Table 9-11 shows the range of pressures delivered to customers' services by state.

Table 9-10 Average Pressures Maintained by Gas Companies in the United States[2]

Main normal press., oz	Thousands of meters reporting pressures shown				
	Natural	Mfd.	Mixed	LP-gas	Total
2.0	3.8	...	825.5	1.4	830.7
2.5	694.8	1.2	967.9	5.9	1669.8
3.0	3411.0	12.6	120.9	15.8	3560.3
3.5	5616.2	139.0	528.1	40.4	6323.7
4.0	6997.7	50.3	73.2	35.6	7156.8
4.5	2644.4	24.7	...	4.5	2673.6
5.0	175.9	...	...	4.9	180.8
5.5	0.7	...	...	...	0.7
6.0	1299.8	...	...	0.5	1300.3
6.5	3.4	...	...	2.9	6.3
7.0	5.6	5.5	...	2.7	13.8
8.0*	857.1	...	...	†	857.1
10.0	1.2	...	...	...	1.2
11.0	0.4	...	...	...	0.4
12.0	0.8	...	...	...	0.8
Totals	21712.8	233.3	2515.6	114.6	24576.3

* Upper pressure limit in scope of ASA Z21.30—1959.
† Less than 50 customers.

Practices employed by 19 companies (serving about 40 per cent of U. S. gas meters) in the sizing of house piping are as follows:[3] Nine companies reported a pressure drop in house piping of less than 0.5 in. w.c.; five companies reported a pressure drop of 0.5 to 1.0 in. w.c.; and five companies, 1.0 to 3.0 in. w.c. The viewpoints of the responding companies on pressure variations and their effects are given in Table 9-12.[3]

Pressure regulators are mandatory on certain appliances and optional on others. The test pressures are applied up-

* Flow is proportional to the square root of pressure drop.

Table 9-11 Range of Gas Pressures Delivered to Customers' Services by State[2]

(all types of gases)

State	Pressure, oz per sq in.*		State	Pressure, oz per sq in.*	
	Minimum†	Maximum†		Minimum†	Maximum†
Alabama	3.5	8.1	Montana	4.0	6.0
Arizona	4.0	4.6	Nebraska	3.5	6.4
Arkansas	4.0	8.1	Nevada	1.2	5.8
California	3.5	7.0	New Hampshire	1.7	4.0
Colorado	3.5	6.4	New Jersey	2.0	6.4
Connecticut	2.9	4.6	New Mexico	3.5	8.1
Delaware	3.5	3.5	New York	2.3	6.9
Dist. of Col.	2.9	2.9	North Carolina	2.3	6.9
Florida	2.9	4.6	North Dakota	3.5	5.8
Georgia	3.5	5.0	Ohio	2.0	10.0
Idaho	4.0	4.0	Oklahoma	3.0	12.0
Illinois	1.7	6.0	Oregon	2.6	6.0
Indiana	2.3	10.0	Pennsylvania	1.2	8.1
Iowa	2.9	4.6	Rhode Island	2.9	4.6
Kansas	2.9	12.0	South Carolina	3.5	4.0
Kentucky	4.0	8.0	South Dakota	3.5	5.0
Louisiana	2.9	8.0	Tennessee	3.5	10.0
Maine	3.5	4.9	Texas	2.3	12.0
Maryland	1.7	8.7	Utah	2.9	4.0
Massachusetts	2.3	4.6	Vermont	2.3	3.5
Michigan	2.3	8.7	Virginia	2.3	6.4
Minnesota	3.0	4.6	Washington	2.6	4.0
Mississippi	3.5	4.3	West Virginia	2.0	8.7
Missouri	2.3	8.1	Wisconsin	2.3	4.6
			Wyoming	3.5	10.0

* Multiply values given by 1.73 to convert to inches of water column.

† The minimum and maximum shown may not apply to a single company within a state.

Table 9-12 Reported Pressure Variations* for Satisfactory and Safe Operation[3]

Appliance	Pressure	Variation, per cent 25	30	35	50	100	No reply
Ranges	Above normal	12	1	2	3	0	1
	Below normal	8	2	1	7	0	1
Water heaters	Above normal	7	1	2	6	1	2
	Below normal	6	1	1	9	0	2

* Tabulated are the number of responding companies (29) that tolerate the per cent variance shown; e.g., eight companies of the 29 tolerate a pressure 25 per cent below normal for ranges.

Table 9-13 Pressure Drops Suitable for Low-Pressure Systems

(in inches of water column)

Minimum pressure at appliance inlet	3.5
Pressure drops:	
House piping	0.3
Meter	.5
Meter to head of service	.3
Service pipe	0.2
Minimum pressure at main	4.8

stream of all controls, including regulators. Regulators are adjusted for an outlet pressure of 4.0 in. w.c. when testing is done with gases A, B, C, F, and G (Table 9-9). If an appliance is tested with a regulator in use, an outlet pressure setting of 10.0 in. w.c. is used with gases D and E.

In certain states, the allowable variation in pressure is included in commission orders. However, one rule used by many gas companies is that for *any one customer* the minimum pressure should *not be less* than one-half the maximum. A more liberal approach[4] used by some other companies permits variation in delivered pressure, maximum to minimum, as *twice* to *one-half* the normal pressure. Regulatory commissions usually specify that pressures are to be measured at the meter outlet; sometimes they specify the outlet of the service pipe.

As indicated in Table 9-13, when a gas company normally maintains main pressures at relatively low levels, pressure drops are naturally held to a minimum in sizing services and house piping systems.

Gas appliances are so flexible in design that they give satisfactory operation under all average pressures (Tables 9-10 and 9-11) listed. Many appliances are equipped with pressure regulators and are therefore nearly independent of gas line pressure variations. It cannot be assumed, however, that in appliances without regulators satisfactory operation will be obtained at high and low extremities if fluctuations from the average adjustment pressure are as much as ±50 per cent. Operation will be satisfactory as long as pressure is maintained close to that at which the appliance was adjusted. However, this adjustment pressure should be within the wide range of 4 in. to 14 in. w.c. on natural gas. It is the variation from the adjustment pressure that may cause trouble. At pressures more than 50 per cent below adjustment, the tendencies are insufficient heat, flashback, and poor ignition. At high pressures, possibly more than 50 per cent above adjustment, the tendencies are incomplete combustion, flames blowing off burner, and carbon deposition.

Reference may be made to the extract from ASA B31.8-1963[5] given on page 9/25.

845 CONTROL AND LIMITING OF GAS PRESSURE

845.1 Basic Requirement for Protection against Accidental Overpressuring. Every pipeline, main, distribution system, customer's meter and connected facilities, compressor station, pipe-type holder, bottle-type holder, container fabricated from pipe and fittings, and all special equipment, if connected to a compressor or to a gas source where the failure of pressure control or other causes might result in a pressure which would exceed the maximum allowable operating pressure of the facility (refer to 805.14), shall be equipped with suitable pressure relieving or pressure limiting devices. Special provisions for service regulators are set forth under paragraph 845.5.

845.2 Control and Limiting of Gas Pressure in Holders, Pipelines, and All Facilities that Might at Times Be Bottle Tight.

845.21 Suitable types of protective devices to prevent over-pressuring of such facilities include:

(a) Spring-loaded relief valves of types meeting the provisions of the ASME Unfired Pressure Vessel Code.

(b) Pilot-loaded back-pressure regulators used as relief valves, so designed that failure of the pilot system or control lines will cause the regulator to open.

845.22 *Maximum Allowable Operating Pressure for Steel Pipelines or Mains.* This pressure is by definition the maximum operating pressure to which the pipeline or main may be subjected in accordance with the requirements of this Code. For a pipeline or main in good operating condition, the maximum allowable operating pressure is the lesser of the two pressures described in (a) and (b) below.

(a) The design pressure (defined in 805.11) of the weakest element of the pipeline or main. Assuming that all fittings, valves and other accessories in the line have an adequate pressure rating, the maximum allowable operating pressure of a steel pipeline or main shall be the design pressure determined in accordance with 841.1.

(b) The pressure obtained by dividing the pressure to which the pipeline or main is tested after construction by the appropriate factor for the location class involved, as follows:

Class No. location	Pressure
1	Test pressure/1.10
2	Test pressure/1.25
3	Test pressure*/1.40*
4	Test pressure*/1.40*

*Other factors than 1.4 should be used if the line was tested under the special conditions described in 841.413, 841.416 and 841.42. In such cases use factors that are consistent with the applicable requirements of these seetions.

(c) In some cases the operating company will consider that the maximum operating pressure to which a pipeline or main should be subjected is less than the pressure determined by either (a) or (b) above. Pipelines that are known to be seriously corroded or that have other defects seriously affecting their strength and which have been operated for years at lower pressures fall into this category. In such cases the operating company shall decide the maximum pressure it considers safe, and shall install over-pressure protective devices designed to prevent accidentally exceeding this maximum pressure, if there is a reasonable possibility that the pressure will be exceeded.

(d) If services are connected to the pipeline or main, there are additional considerations that might in some cases limit the maximum allowable operating pressure of the facility. See 845.33.

845.3 Control and Limiting of Gas Pressure in High-Pressure Steel or Cast Iron Distribution Systems.

845.31 Each high-pressure distribution system or main, supplied from a source of gas which is at a higher pressure than the maximum allowable operating pressure for the system, shall be equipped with pressure regulating devices of adequate capacity, and designed to meet the pressure, load and other service conditions under which they will operate or to which they may be subjected.

845.32 In addition to the pressure-regulating devices prescribed in 845.31, a suitable method shall be provided to prevent accidental over-pressuring of a high-pressure distribution system.

Suitable types of protective devices to prevent over-pressuring of high pressure distribution systems include:

(a) Relief valves as prescribed in 845.21(a) and (b).

(b) Weight-loaded relief valves.

(c) A monitoring regulator installed in series with the primary pressure regulator.

(d) A series regulator installed upstream from the primary regulator, and set to continuously limit the pressure on the inlet of the primary regulator to the maximum allowable operating pressure of the distribution system or less.

(e) An automatic shut-off device installed in series with the primary pressure regulator, and set to shut off when the pressure on the distribution system reaches the maximum allowable operating pressure, or less. This device must remain closed until manually reset. It should not be used where it might cause an interruption in service to a large number of customers.

845.33 *Maximum Allowable Operating Pressure for High-Pressure Distribution Systems.* This pressure shall be the maximum pressure to which the system can be subjected in accordance with requirements of this Code. It shall not exceed:

(a) The design pressure of the weakest element of the system as defined in 805.11.

(b) 60 psig if the services in the system are not equipped with series regulators or other pressure limiting devices as prescribed in 845.53.

(c) 25 psig in cast iron systems having unreinforced bell and spigot joints as prescribed in 842.15(a).

(d) The pressure limits to which the joint could be subjected without possibility of parting.

(e) 2 psig in high-pressure distribution systems equipped with service regulators not meeting the requirements of 845.51 and which do not have an over-pressure protective device as required in 845.52.

In some cases the operating company will consider the maximum pressure to which a system should be subjected is less than the pressure obtained by applying the applicable limits in 845.33 (a), (b), (c), (d) or (e). Systems that are known to be corroded and that have been operated for years at lower pressures than these limits fall into this category. In such cases the operating company shall decide the maximum pressure it considers safe, and shall install over-pressure protective devices to prevent accidentally exceeding this maximum pressure if there is a reasonable possibility that the pressure will be exceeded.

845.34 *Qualifying a High-Pressure Steel Distribution System for a New and Higher Maximum Allowable Operating Pressure.* Note: This paragraph applies to high-pressure distribution mains and to pipelines where the new and higher maximum allowable operating pressure is less than that required to produce a hoop stress of 30% of the specified minimum yield strength of the pipe. When the new and higher maximum allowable operating pressure is more than this value the provisions of 845.23 shall apply.

(a) Before increasing the maximum allowable operating pressure of a high-pressure distribution system that has been operating at less than the applicable maximum pressure stated in 845.33 to a new maximum allowable operating pressure equal to or less than the maximum applicable pressure in 845.33, it is recommended that the following factors be taken in consideration:

(1) The design of the system including kinds of material and equipment used.

(2) Past maintenance records including results of any previous leakage surveys.

(b) Before increasing the pressure the following steps should be taken:

(1) Make a leakage survey, if past maintenance records indicate that such a survey is advisable, and repair leaks found.

(2) Repair or replace parts of the system found to be inadequate for the higher operating pressure.

(3) Install suitable devices on the services to regulate and limit the pressure of the gas in accordance with 845.53 if the new maximum allowable operating pressure is to be over 60 psig.

(4) Adequately reinforce or anchor offsets, bends and dead ends in coupled pipe to avoid movement of the pipe should the offset, bend or dead end be exposed in an excavation.

(c) The rate of pressure increase to the new maximum allowable operating pressure should be gradual so as to allow sufficient time for periodic observations of the system.

845.35 *Qualifying a Cast Iron High Pressure Main or System for a New and Higher Maximum Allowable Operating Pressure.*

(a) The maximum allowable operating pressure of a cast iron main or system shall not be increased to a pressure in excess of that permitted in 845.33 or 842.11, whichever is the lesser. Where records are not complete enough to permit the direct application of 842.11, the following procedures shall be used:

(1) *Laying Condition:* Where the original laying condition cannot be ascertained, it shall be assumed that Condition D exists (i.e., pipe supported on blocks, backfill tamped).

(2) *Cover:* Unless the actual maximum cover depth is known with certainty, it shall be determined by exposing the main or system at three or more points and making actual measurements. The main or system shall be exposed in those areas where the cover depth is most likely to be greatest. The greatest measured cover depth shall be used for computations.

(3) *Nominal Wall Thickness:* Unless the nominal thickness is known with certainty, it shall be determined by cutting coupons from three or more separate pipe lengths and making actual wall thickness measurements. The coupons shall be cut from pipe lengths located in those areas where the cover depth is most likely to be greatest. The average of all measurements taken shall be increased by the allowance indicated in the following table:

Pipe size, in.	Allowance, in. Pit cast pipe	Allowance, in. Centrifugally cast pipe
3–8	0.075	0.065
10–12	.08	.07
14–24	.08	.08
30–48	.09	0.09
54–60	0.09	...

The nominal wall thickness shall be that standard thickness listed in Table 10 of ASA A21.1 nearest the value thus obtained, or Table 11, whichever is applicable.

(4) *Manufacturing Process:* Unless the pipe manufacturing process is known with certainty, it shall be assumed to be pit cast pipe having a bursting tensile strength (S) of 11,000 psi and a modulus of rupture (R) of 31,000 psi.

(b) Before increasing the maximum allowable operating pressure, the following measures shall be taken:

(1) Review the design of the main or system, the materials and equipment used, and previous tests.

(2) Determine the condition of the main or system from past maintenance records including results of previous leakage surveys.

(3) Make a leakage survey or a gas detector survey.

(4) Repair, reinforce or replace any part of the main or system found to be inadequate for the higher operating pressure.

(5) Adequately reinforce or anchor offsets, bends and dead ends in coupled or bell and spigot pipe to avoid movement of the pipe should the offset, bend or dead end be exposed by excavation.

(6) Install suitable devices on the services to regulate and limit the pressure of the gas in accordance with 845.53 if the new and higher maximum allowable operating pressure is to be over 60 psig.

(c) If after compliance with 845.35 (b) above, it is established that the main or system is capable of safely withstanding the proposed new and higher maximum allowable operating pressure, the pressure in the main or system shall be increased gradually, in steps, to the new maximum allowable operating pressure. After each incremental pressure increase, the main or system shall be checked before making the next increase to determine the effect of the previous pressure increase.

845.4 Control and Limiting of Gas Pressure in Low-Pressure Distribution Systems.

845.41 Each low-pressure distribution system or low-pressure main supplied from a gas source which is at a higher pressure than the maximum allowable operating pressure for the low-pressure system shall be equipped with pressure-regulating devices of adequate capacity, designed to meet the pressure, load and other service conditions under which they will have to operate.

845.42 In addition to the pressure-regulating devices prescribed in 845.41, a suitable device shall be provided to prevent accidental over-pressuring. Suitable types of protective devices to prevent over-pressuring of low-pressure distribution systems include:

(a) A liquid seal relief device that can be set to open accurately and consistently at the desired pressure.

(b) Weight loaded relief valves.

(c) An automatic shut-off device as described in 845.32(e).

(d) A pilot loaded back-pressure regulator as described in 845.21(b).

(e) A monitoring regulator as described in 845.32(c).

(f) A series regulator as described in 845.32(d).

845.43 *Maximum Allowable Operating Pressure for Low-Pressure Distribution Systems.* The maximum allowable operating pressure for a low-pressure distribution system shall not exceed either (a) or (b) below.

(a) A pressure which would cause the unsafe operation of any connected and properly adjusted low-pressure gas burning equipment, or

(b) A pressure of 2 psig.

845.44 *Conversion of Low-Pressure Distribution Systems to High-Pressure Distribution Systems.*

(a) Before converting a low-pressure distribution system to a high-pressure distribution system, it is recommended that the following factors be taken into consideration:

(1) The design of the system including kinds of material and equipment used.

(2) Past maintenance records including results of any previous leakage surveys.

(b) Before increasing the pressure the following steps (not necessarily in sequence shown) should be taken:

(1) Make a leakage survey, if past maintenance records indicate that such a survey is advisable, and repair leaks found.

(2) Reinforce or replace parts of the system found to be inadequate for the higher operating pressures.

(3) Install a service regulator on each service, and test each regulator to determine that it is functioning. In some cases it may be necessary to raise the pressure slightly to permit proper operation of the service regulator.

(4) Isolate the system from adjacent low-pressure systems.

(5) At bends or offsets in coupled or bell and spigot pipe, reinforce or replace anchorages determined to be inadequate for the higher pressures.

(c) The pressure in the system being converted should be increased by steps, with a period to check the effect of the previous increase before making the next increase. The desirable magnitude of each increase and the length of the check period will vary depending upon conditions. The objective of this procedure is to afford an opportunity to discover before excessive pressures are reached any unknown open and unregulated connections to adjacent low-pressure systems or to individual customers.

845.5 Control and Limiting of the Pressure of Gas Delivered to Domestic, Small Commercial and Small Industrial Customers from High Pressure Distribution Systems.

Note: When the pressure of the gas and the demand by the customer are greater than that which is applicable under the provisions of 845.5, the requirements for control and limiting of the pressure of gas delivered are included in 845.1.

845.51 If the maximum actual operating pressure of the distribution system is between 2 psig and 60 psig and a service regulator having the characteristics listed below is used, no other pressure limiting device is required:

(a) A pressure regulator capable of reducing distribution line pressure (pounds per square inch) to pressures recommended for household appliances (inches of water column).

(b) Single port valve with orifice diameter no greater than that recommended by the manufacturer for the maximum gas pressure at the regulator inlet.

(c) The valve seat shall be made of resilient material designed to withstand abrasion of the gas, impurities in gas and cutting by the valve, and to resist permanent deformation when it is pressed against the valve port.

(d) Pipe connections to the regulator shall not exceed 2 inches in diameter.

(e) The regulator must be of a type that is capable under normal operating conditions of regulating the downstream pressure within the necessary limits of accuracy and of limiting the build-up of pressure under no-flow conditions to 50% or less of the discharge pressure maintained under flow conditions.

(f) A self-contained service regulator with no external static or control lines.

845.52 If the maximum actual operating pressure of the distribution system is between 2 psig and 60 psig and a service regulator not having all of the characteristics listed in 845.51 is used, or if the gas contains materials that seriously interfere with the operation of service regulators, suitable protective devices shall be installed to prevent unsafe over-pressuring of the customer's appliances should the service regulator fail. Some of the suitable types of protective devices to prevent over-pressuring of customer's appliances are:

(a) A monitoring regulator. (b) A relief valve.

(c) An automatic shut-off device.

These devices may be installed as an integral part of the service regulator or as a separate unit.

845.53 If the maximum actual operating pressure of the distribution system exceeds 60 psig, suitable methods shall be used to regulate and limit, to the maximum safe value, the pressure of gas delivered to the customer, such as the following:

(a) A service regulator having the characteristics listed in 845.51 above and a secondary regulator located upstream from the service regulator. The secondary regulator in no case shall be set to maintain a pressure higher than 60 psi. A device shall be installed between the secondary regulator and the service regulator to limit the pressure on the inlet of the service regulator to 60 psi or less in case the secondary regulator fails to function properly. This device may be either a relief valve or an automatic shut-off that shuts, if the pressure on the inlet of the service regulator exceeds the set pressure (60 psi or less), and remains closed until manually reset.

(b) A service regulator and a monitoring regulator set to limit to a maximum safe value the pressure of the gas delivered to the customer.

(c) A service regulator with a relief valve vented to the outside atmosphere, with the relief valve set to open so that the pressure of gas going to the customer shall not exceed a maximum safe value. The relief valve may either be built into the service regulator or it may be a separate unit installed downstream from the service regulator. This combination may be used alone only in those cases where the inlet pressure on the service regulator does not exceed the manufacturer's safe working pressure rating of the service regulator, and is not recommended for use where the inlet pressure on the service regulator exceeds 125 psi. For higher inlet pressures, method (a) or (b) above should be used.

EFFECTS OF ELEVATION ON PRESSURE

The pressure in a pipeline changes with elevation. Pressures of gases *heavier* than air, such as propane and butane, *decrease* with elevation. Conversely, pressures in gases *lighter* than air, for example, natural gases, *increase* with elevation. The following equation may be used to calculate the pressure at various elevations in a stagnant vertical column of low-pressure gas:

$$P_2 = P_1 + 0.0146(1 - G)\Delta H + P_1(3.61 \times 10^{-5}G\Delta H) \quad (1)$$

where:

P_2 and P_1 = gage pressures at an upper and a lower point in the column, respectively, in w.c.

G = specific gravity of the gas

ΔH = difference in elevation between points 2 and 1, ft

The equation was derived for a value of P_1 = 8 in. w.c. The third term is usually insignificant for utility gases, since $G \cong 1$.

Examples: Find P_2 for a natural gas (G = 0.60), propane (G = 1.56), and butane (G = 2.07); P_1 = 7.0 in w.c. and P_2 is measured 100 ft above P_1. Substituting in the foregoing equation:

Natural Gas

$$P_2 = 7.0 + 0.0146(1 - 0.60)100 + 7.0(3.61 \times 10^{-5} \times 0.60 \times 100)$$

$$P_2 = 7.0 + 0.58 + 0.01 = 7.59 \text{ in. w.c.}$$

Propane

$$P_2 = 7.0 - 0.82 + 0.04 = 6.22 \text{ in. w.c.}$$

Butane

$$P_2 = 7.0 - 1.56 + 0.05 = 5.49 \text{ in. w.c.}$$

In sizing low-pressure natural gas piping systems, for example, those used in high-rise buildings, the pressure gain due to elevation may be taken into account to reduce pipe sizes. However, if this is done the variation in pressure at the higher elevations may vary excessively when the loads are at a minimum. Consequently, it is usually considered better practice to disregard the pressure gain due to elevation.

The preferred procedure is to size piping on the basis of acceptable pressure loss. When this is done, loads at higher elevation will receive gas at higher pressure. The appliances must be adjusted for this higher pressure (or a suitable pressure regulator must be used to reduce the pressure to the normal level). With this arrangement there is a minimum pressure variation due to load variations.

SURGE TANKS

Surge tanks may be used to neutralize low-frequency pulsations such as those caused by intermittent operation of industrial appliances.

During the high-flow phase of the cycle, when more gas is drawn from the tank than is supplied to it, the tank must be of sufficient size to maintain pressure above the minimum allowable. Conversely, during that phase of the cycle when less gas is withdrawn from the tank than is supplied to it, the tank must be of sufficient size to receive the excess by means of an increase in pressure to the maximum pressure allowed.

Example: A 14-sec cycle consists of 200 MCFH (or 55.5 cu ft per sec) of gas for eight seconds and then 160 MCFH (44.4 cu ft per sec) of gas for six seconds. Calculate tank size for pressure limits of 18 to 22 psig.

Volume of gas delivered *per cycle* = $(55.6 \times 8) + (44.4 \times 6) = 711$ cu ft

Average flow rate during cycle = $711/14 = 50.8$ cu ft per sec

During the first part of the cycle *the tank supplies* gas at average flow rate = $55.6 - 50.8 = 4.8$ cu ft per sec.

During the second part of the cycle *the tank receives* gas at average flow rate = $50.8 - 44.4 = 6.4$ cu ft per sec.

The quantity of gas supplied by the tank during the first part of the cycle is $(4.8 \times 8) = 38.4$ cu ft which, of course, is the same as the quantity received by the tank during the second part of the cycle, $(6.4 \times 6) = 38.4$ cu ft.

Now:

$$V_t = \frac{p}{\Delta p_t} V_s$$

where: V_t = required tank volume, cu ft
p = pressure at STP, 14.73 psia
Δp_t = pressure range, minimum to maximum, psi
V_s = volume at STP, cu ft

Therefore:

$$V_t = \frac{14.73}{(22 - 18)} \times 38.4 = 142 \text{ cu ft}$$

REFERENCES

1. A.G.A. Laboratories. *American Standard Approval Requirements for Central Heating Gas Appliances, Vol. I: Steam and Hot Water Boilers; Addenda, effective Jan. 1, 1962.* (ASA Z21.13.1a, 1962) Cleveland, Ohio, 1961.
2. A.G.A. Bur. of Statistics. *Survey of Residential Gas Service by County.* New York, 1958.
3. Miller, C. F. "House Piping—House Pressures and Appliance Regulators." (DMC-56-8) *A.G.A. Proc.* 1956: 239–45.
4. Natl. Assoc. of Railroad and Utilities Commissioners. "Proposed Uniform Rules for Gas Utilities: Report of the Committee on Engineering." *Proc.* 1951: 371.
5. ASME. *Gas Transmission and Distribution Piping Systems.* (ASA B31.8-1963) New York, 1963.

Chapter 3
Materials for Mains and Services

by Guy Corfield and Lester B. Inglis, Jr.

In the selection of the material (steel, wrought iron, cast iron, ductile iron, copper, aluminum, or plastic) and the installation method to be employed for a given main system, the following factors should be considered: (1) cost; (2) location; (3) soil; (4) depth of cover; (5) external loads; (6) gas pressure; (7) internal (e.g., that due to gas components) and external (e.g., burrowing animals) corrosion.

STEEL PIPE

Steel pipe for mains is made of open-hearth, electric furnace, or Bessemer steel, or of open-hearth iron or wrought iron. The pipe is formed from plate, ingots, or billets by the butt-weld, electric-weld, or seamless processes. Applicable welding processes are defined in the following extract from ASA B31.8-1963:

805.51 *Welding Nomenclature.* Types of welds and names of welded joints are used herein according to their common usage as defined in the American Welding Society Publication "Standard Welding Terms and Their Definitions" (AWS A3.0), or as specifically defined as follows:

A. *Electric-Resistance-Welded Pipe:* Pipe produced in individual lengths, or in continuous lengths from coiled skelp and subsequently cut into individual lengths, having a longitudinal butt joint wherein coalescence is produced by the heat obtained from resistance of the pipe to the flow of electric current in a circuit of which the pipe is a part, and by the application of pressure.
Typical specifications: ASTM A 53
ASTM A 135
API 5L
API 5 LX

B. *Furnace Butt-Welded Pipe:*

(1) *Bell-Welded:* Furnace-welded pipe produced in individual lengths from cut-length skelp, having its longitudinal butt joint forge welded by the mechanical pressure developed in drawing the furnace-heated skelp through a cone-shaped die (commonly known as a "welding bell") which serves as a combined forming and welding die.
Typical specifications: ASTM A 53
API 5L

(2) *Continuous-Welded:* Furnace-welded pipe produced in continuous lengths from coiled skelp and subsequently cut into individual lengths, having its longitudinal butt joint forge welded by the mechanical pressure developed in rolling the hot-formed skelp through a set of round pass welding rolls.
Typical specifications: ASTM A 53
API 5L

C. *Electric-Fusion-Welded Pipe:* Pipe having a longitudinal butt joint wherein coalescence is produced in the preformed tube by manual or automatic electric-arc welding. The weld may be single or double and may be made with or without the use of filler metal.
Typical specifications:

ASTM A 134 } Single or double weld is permitted
ASTM A 139 } with or without the use of filler metal
ASTM A 155—Requires both inside and outside welds and the use of filler metal

Spiral-welded pipe is also made by the electric-fusion-welded process with either a butt joint, a lap joint or a lock-seam joint.
Typical specifications:

ASTM A 134 } Butt joint
ASTM A 139 }
ASTM A 211—Butt joint, lap joint, or lock-seam joint

D. *Electric-Flash-Welded Pipe:* Pipe having a longitudinal butt joint wherein coalescence is produced, simultaneously over the entire area of abutting surfaces, by the heat obtained from resistance to the flow of electric current between the two surfaces, and by the application of pressure after heating is substantially completed. Flashing and upsetting are accompanied by expulsion of metal from the joint.
Typical specifications: API 5L
API 5LX

E. *Double Submerged-Arc-Welded Pipe:* Pipe having a longitudinal butt joint produced by at least two passes, one of which is on the inside of the pipe. Coalescence is produced by heating with an electric arc or arcs between the bare metal electrode or electrodes and the work. The welding is shielded by a blanket of granular, fusible material on the work. Pressure is not used and filler metal for the inside and outside welds is obtained from the electrode or electrodes.
Typical specifications: ASTM A 381
API 5LX

Typical piping specifications derived from the foregoing welding techniques are referenced in the table, Appendix C, of ASA B31.8–1963, shown in Sec. 8, Chap. 2.

Thin Wall Tubing Services. Use of about 250,000 ft of ⅝ in. OD electric weld tubing made of 20 gage low carbon strip was reported.[12] The material has a tensile strength of 45,000 psi min and a yield strength of 35,000 psi min. Manufacturing steps include (1) nondestructive testing by eddy current methods as it comes from the welder, (2) coiling, (3) pressure testing with air, (4) plastic coating, and (5) paper wrapping. The coating (X-Tru-Coat process) consists of 10 mils of a hot-applied adhesive plus a 25 mil thickness of a high density copolymer polyethylene.

Dimension Standards. The three common methods of designating steel pipe dimensions are: (1) ASA steel pipe schedules; (2) ASA steel pipe nominal wall thickness lesignations (S, standard; XS, extra strong; XXS, double extra strong); and (3) lightweight stainless steel pipe schedules. Of the three designations, the first is becoming most popular because its flexibility permits more standard sizes (see Table 1-9).

Testing and Inspection. Steel pipe is subjected to hydrostatic tests and visual surface inspection prior to leaving the manufacturer. The hydrostatic test pressure for API 5L and API 5LX pipe is computed from the Barlow formula. (Also, see Par. 841.1 in ASA B31.8-1963, extracted in Chap. 2 of Sec. 8.) Many pipe manufacturers are turning to the use of electronic nondestructive testing of pipe. Many manufacturers utilize magnetic, induced eddy currents and ultrasonic inspection techniques to detect irregularities in the critical weld area of the pipe.

A.G.A. Research Project NQ-33, "Nondestructive Inspection of Steel Pipe," led to the development of a successful prototype testing device for transmission by the end of 1962. The program was under way in 1964 to extend the range of application to include distribution piping. The inspection system uses ultrasonics, magnetics, and mechanical gaging to resolve defects by position, type, and depth of penetration simultaneously. Development of similar equipment by others is also under way.

API and ASTM specifications for line pipe state that a representative of the purchaser "shall have free entry at all times while work on contract of the purchaser is being performed . . . to the manufacturer's works which will concern the manufacture of the pipe ordered." An additional inspection after shipping will help to ensure against placing any unsuitable pipe.

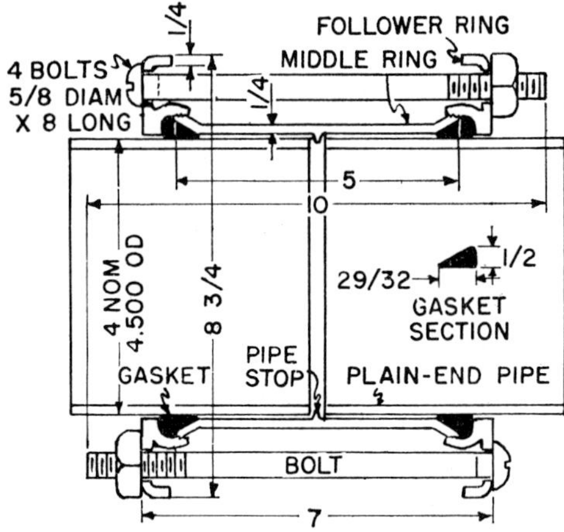

Fig. 9-21 A typical 4-in. mechanical coupling (all dimensions in inches).

Joints and Fittings

The common methods of joining lengths of steel pipe are welding,* threaded joints (couplings), and mechanical fittings. **Welding** of pipe for distribution mains is the most common and satisfactory joining method.

Threaded joints. These are not commonly used in modern distribution systems. They are structurally weaker than welded joints due to stress concentration in the thread area. Joints in small diameter house service lines are sometimes threaded; however, Par. 849.21d of the ASA B31.8 Code states, "where practical, welded joints or compression type fittings should be used in all underground steel services."

Mechanical Fittings. Plain end steel pipe may be readily connected by such couplings (Fig. 9-21). Since mechanical joints are particularly susceptible to parting (slipping out) if the line is subjected to longitudinal tension, adherence to Par. 835 of ASA B31.8 is vital (extracted in Chap. 6 of this Section).

The prefabrication of headers and bends has resulted in cost savings. Extruded headers containing multiple openings are economical because of the elimination of casting fittings (ells, tees, and caps) and circumferential welds. Recently, extensive tests were made on a 4-in. diam Grade B 0.188-in. wall pipe that had been cold bent to 45° on a 16-in. bending radius. Diameter distortion was below the 2.5 per cent maximum set by the ASA B31.8 Code. Thinning and elongations of the unit were within required limits.

* See "WELDING" in Par. 820 of ASA B31.8-1963, extracted in Chap. 2 of Sec. 8.

CAST IRON PIPE

Cast iron shows excellent resistance to corrosion. Cast iron pipe for gas distribution is usually manufactured by either the pit cast or centrifugal casting method. Bell and spigot, plain end, and mechanical joint pipe may be made by either method.

Pit cast pipe is formed vertically and is not as homogeneous as centrifugally cast pipe (cast horizontally). In the latter process, the mold is spun about its axis and filled at a particular speed. Centrifugal action forces the molten metal to the wall of the mold. Spinning is continued until the metal cools.

Specification and Design

Reference may be made to the following extract from ASA B31.8-1963:

842.1 Cast Iron Pipe Design.

842.11 *Basic Equation to Determine Required Wall Thickness.* Cast iron pipe shall be designed in accordance with the methods set forth in the ASA A21.1 "American Standard Practice Manual for the Computation of Strength and Thickness of Cast Iron Pipe."

842.12 *Maximum Allowable Values of S and R.* The values of bursting tensile strength, S, and modulus of rupture, R, to be used in the equations given in ASA A21.1 are:

Specification	Type of pipe	Bursting tensile strength, S	Modulus of rupture, R
ASA A21.3	Pit cast	11,000 psi	31,000 psi
ASA A21.7	Centrifugal (Metal mold)	18,000 psi	40,000 psi
ASA A21.9	Centrifugal (Sand-lined mold)	18,000 psi	40,000 psi

842.13 *Allowable Thicknesses for Cast Iron Pipe.* The least cast iron pipe thicknesses permitted are the lightest standard classes for each nominal pipe size as shown in American Standards A21.3, A21.7 and A21.9.

TABLE 842.142

STANDARD THICKNESS OF CAST IRON GAS PIPE CENTRIFUGALLY CAST IN METAL MOLDS OR SAND-LINED MOLDS

Thickness in Inches. Working Pressure in Pounds per Square Inch.
Thicknesses Include Allowances for Foundry Practice and Corrosion.

Laying Condition A—Flat Bottom Trench, Without Blocks, Untamped Backfill
Laying Condition B—Flat Bottom Trench, Without Blocks, Tamped Backfill
Laying Condition C—Pipe Laid on Blocks, Untamped Backfill
Laying Condition D—Pipe Laid on Blocks, Tamped Backfill

Size, in.	Working pressure	Laying condition for 3½ Ft of cover A	B	C	D	5 Ft of cover A	B	C	D	8 Ft of cover A	B	C	D
4	10	0.35[1]	0.35	0.35	0.35	0.35	0.35	0.35	0.35	0.35	0.35	0.41	0.35
		.38[2]	.38	.38	.38	.38	.38	.38	.38	.38	.38	.41	.38
	50	.35[1]	.35	.35	.35	.35	.35	.35	.35	.35	.35	.41	.35
		.38[2]	.38	.38	.38	.38	.38	.38	.38	.38	.38	.41	.38
	100	.35[1]	.35	.35	.35	.35	.35	.35	.35	.35	.35	.41	.35
		.38[2]	.38	.38	.38	.38	.38	.38	.38	.38	.38	.41	.38
	150	.35[1]	.35	.35	.35	.35	.35	.38	.35	.35	.35	.41	.35
		.38[2]	.38	.38	.38	.38	.38	.38	.38	.38	.38	.41	.38
6	10	.38[1]	.38	.41	.38	.38	.38	.41	.38	.38	.38	.48	.38
		.41[2]	.41	.41	.41	.41	.41	.41	.41	.41	.41	.48	.41
	50	.38[1]	.38	.41	.38	.38	.38	.41	.38	.38	.38	.48	.38
		.41[2]	.41	.41	.41	.41	.41	.41	.41	.41	.41	.48	.41
	100	.38[1]	.38	.41	.38	.38	.38	.44	.38	.38	.38	.48	.38
		.41[2]	.41	.41	.41	.41	.41	.44	.41	.41	.41	.48	.41
	150	.38[1]	.38	.41	.38	.38	.38	.44	.38	.38	.38	.48	.38
		.41[2]	.41	.41	.41	.41	.41	.44	.41	.41	.41	.48	.41
8	10	.41	.41	.44	.41	.41	.41	.48	.41	.41	.41	.52	.41
	50	.41	.41	.44	.41	.41	.41	.48	.41	.41	.41	.52	.41
	100	.41	.41	.48	.41	.41	.41	.48	.41	.41	.41	.56	.41
	150	.41	.41	.48	.41	.41	.41	.48	.41	.41	.41	.56	.41
10	10	.44	.44	.48	.44	.44	.44	.52	.44	.44	.44	.60	.44
	50	.44	.44	.48	.44	.44	.44	.52	.44	.44	.44	.60	.44
	100	.44	.44	.52	.44	.44	.44	.52	.44	.44	.44	.60	.48
	150	.44	.44	.52	.44	.44	.44	.56	.44	.48	.44	.60	.48
12	10	.48	.48	.52	.48	.48	.48	.56	.48	.48	.48	.60	.52
	50	.48	.48	.52	.48	.48	.48	.56	.48	.48	.48	.60	.52
	100	.48	.48	.56	.48	.48	.48	.56	.48	.52	.48	.65	.52
	150	.48	.48	.56	.48	.48	.48	.56	.48	.52	.48	.65	.52
16	10	.54	.50	.58	.54	.54	.50	.63	.58	.58	.54	.73	.63
	50	.54	.50	.63	.54	.54	.50	.63	.58	.63	.58	.73	.63
	100	.54	.54	.63	.58	.58	.54	.68	.58	.63	.58	.73	.68
20	10	.62	.57	.67	.62	.62	.57	.72	.67	.67	.62	.78	.72
	50	.62	.57	.72	.62	.67	.57	.72	.67	.72	.62	.78	.72
	100	.62	.57	.72	.67	.67	.62	.78	.67	.72	.67	.84	.78
24	10	.68	.63	.73	.68	.73	.63	.79	.73	.79	.68	.85	.79
	50	.68	.63	.79	.68	.73	.63	.79	.73	.79	.73	.85	.79
	100	.73	.63	.79	.73	.73	.68	.85	.79	.79	.73	.92	.85
30	10	.79	.73	.85	.79	.85	.73	.92	.85	.92	.79	.99	.92
	50	.85	.73	.85	.85	.85	.79	0.92	.85	0.92	.85	0.99	0.92
36	10	.87	.81	0.94	.87	0.94	.81	1.02	.94	1.02	.87	1.10	1.02
	50	0.94	.81	1.02	.94	1.02	.87	1.10	0.94	1.10	.94	1.19	1.02
42	10	1.05	.90	1.05	0.97	1.05	.90	1.13	1.05	1.13	0.97	1.22	1.13
	50	1.05	.90	1.13	1.05	1.13	.97	1.13	1.05	1.22	1.05	1.32	1.13
48	10	1.14	.98	1.14	1.06	1.14	0.98	1.23	1.14	1.33	1.06	1.33	1.23
	50	1.14	0.98	1.23	1.14	1.23	1.06	1.33	1.14	1.33	1.14	1.44	1.33

[1] Class 22 Thickness.

[2] Class 23 Thickness offers increased factor of safety and is recommended for use in areas of dense population and heavy traffic.

Note.—This table is taken from ASA A21.7 and A21.9.

Table 9-14 Thicknesses, Diameters, and Weights of Centrifugally Cast Pipe for Gas[1]

(dimensions in inches, weights in pounds)

							Wt based on 16 ft laying length	
Nom. pipe size	Thickness class*	Thickness	OD	ID	Bbl wt per foot	Bell & bead wt	Per length†	Average per ft‡
4	22	0.35	4.80	4.10	15.3	16	260	16.4
	23	.38	4.80	4.04	16.5	16	280	17.6
	24	.41	4.80	3.98	17.6	16	300	18.6
6	22	.38	6.90	6.14	24.3	22	410	25.7
	23	.41	6.90	6.08	26.1	22	440	27.6
	24	.44	6.90	6.02	27.9	22	470	29.3
	25	.48	6.90	5.94	30.2	22	505	31.6
8	22	.41	9.05	8.23	34.7	30	585	36.7
	23	.44	9.05	8.17	37.1	30	625	39.0
	24	.48	9.05	8.09	40.3	30	675	42.2
	25	.52	9.05	8.01	43.5	30	725	45.4
	26	.56	9.05	7.93	46.6	30	775	48.5
10	22	.44	11.10	10.22	46.0	40	775	48.5
	23	.48	11.10	10.14	50.0	40	840	52.5
	24	.52	11.10	10.06	53.9	40	900	56.4
	25	.56	11.10	9.98	57.9	40	965	60.4
	26	.60	11.10	9.90	61.8	40	1,030	64.3
12	22	.48	13.20	12.24	59.8	50	1,005	62.9
	23	.52	13.20	12.16	64.6	50	1,085	67.7
	24	.56	13.20	12.08	69.4	50	1,160	72.5
	25	.60	13.20	12.00	74.1	50	1,235	77.2
	26	.65	13.20	11.90	80.0	50	1,330	83.1
16	21	.50	17.40	16.40	82.8	95	1,420	88.7
	22	.54	17.40	16.32	89.2	95	1,520	95.0
	23	.58	17.40	16.24	95.6	95	1,625	101.5
	24	.63	17.40	16.14	103.6	95	1,755	109.6
	25	.68	17.40	16.04	111.4	95	1,875	117.3
	26	.73	17.40	15.94	119.3	95	2,005	125.2
20	21	.57	21.60	20.46	117.5	134	2,015	125.9
	22	.62	21.60	20.36	127.5	134	2,175	135.9
	23	.67	21.60	20.26	137.5	134	2,335	145.9
	24	.72	21.60	20.16	147.4	134	2,490	155.7
	25	.78	21.60	20.04	159.2	134	2,680	167.6
	26	.84	21.60	19.92	170.9	134	2,870	179.3
24	21	.63	25.80	24.54	155.4	177	2,665	166.5
	22	.68	25.80	24.44	167.4	177	2,855	178.5
	23	.73	25.80	24.34	179.4	177	3,045	190.4
	24	.79	25.80	24.22	193.7	177	3,275	204.8
	25	.85	25.80	24.10	207.9	177	3,505	219.0
	26	.92	25.80	23.96	224.4	177	3,765	235.4
30	21	.73	32.00	30.54	223.7	285	3,865	241.5
	22	.79	32.00	30.42	241.7	285	4,150	259.5
	23	.85	32.00	30.30	259.5	285	4,435	277.3
	24	.92	32.00	30.16	280.3	285	4,770	298.1
	25	0.99	32.00	30.02	300.9	285	5,100	318.7
	26	1.07	32.00	29.86	324.4	285	5,475	342.2
36	21	0.81	38.30	36.68	297.7	395	5,160	322.4
	22	.87	38.30	36.56	319.2	395	5,500	343.9
	23	0.94	38.30	36.42	344.2	395	5,900	368.9
	24	1.02	38.30	36.26	372.7	395	6,360	397.4
	25	1.10	38.30	36.10	401.1	395	6,815	425.8
	26	1.19	38.30	35.92	432.9	395	7,320	457.6
42	21	0.90	44.50	42.70	384.6	510	6,665	416.5
	22	0.97	44.50	42.56	413.9	510	7,130	445.7
	23	1.05	44.50	42.40	447.2	510	7,665	479.1
	24	1.13	44.50	42.24	480.4	510	8,195	512.3
	25	1.22	44.50	42.06	517.6	510	8,790	549.5
	26	1.32	44.50	41.86	558.7	510	9,450	590.6
48	21	0.98	50.80	48.84	478.6	645	8,305	519.0
	22	1.06	50.80	48.68	516.8	645	8,915	557.1
	23	1.14	50.80	48.52	554.9	645	9,525	595.2
	24	1.23	50.80	48.34	597.6	645	10,205	637.9
	25	1.33	50.80	48.14	644.9	645	10,965	685.2
	26	1.44	50.80	47.92	696.7	645	11,790	737.0

* Heavier thickness classes can be furnished when specified.

† Including bell. Calculated weight of pipe rounded off to nearest 5 lb; 18 and 20 ft lengths are also available.

‡ Including bell. Average weight per foot based on calculated weight of pipe before rounding.

Table 9-15 Dimensions and Weights of Ends of Centrifugally Cast Pipe for Gas[2],*

(dimensions in inches, weights in pounds)

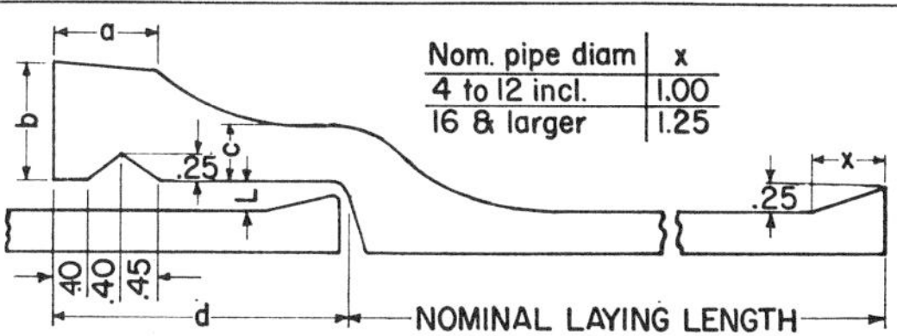

Nom. diam	Pipe thickness From	Pipe thickness To	Pipe, OD	Dimensions of bells: Socket diam	Joint thickness, L	Socket depth, d	a	b	c	Bell and spigot bead, wt
4	0.35	0.38	4.80	5.80	0.50	3.50	1.25	1.20	0.45	18
	.41	...	4.80	5.80	.50	3.50	1.25	1.20	.60	20
6	.38	.44	6.90	7.90	.50	3.50	1.38	1.25	.54	27
	.48	...	6.90	7.90	.50	3.50	1.38	1.25	.66	30
8	.41	.48	9.05	10.05	.50	4.00	1.38	1.35	.58	40
	.52	.56	9.05	10.05	.50	4.00	1.38	1.35	.72	45
10	.44	.52	11.10	12.10	.50	4.00	1.50	1.45	.63	52
	.56	.60	11.10	12.10	.50	4.00	1.50	1.45	.78	59
12	.48	.56	13.20	14.20	.50	4.00	1.50	1.50	.69	67
	.60	.65	13.20	14.20	.50	4.00	1.50	1.50	.85	75
16	.50	.63	17.40	18.66	.63	4.00	1.75	1.62	.76	101
	.68	.73	17.40	18.66	.63	4.00	1.75	1.62	.93	113
20	.57	.72	21.60	22.86	.63	4.00	1.75	1.77	0.88	140
	.78	.84	21.60	22.86	.63	4.00	1.75	1.77	1.08	159
24	.63	.79	25.80	27.06	.63	4.00	2.00	1.92	0.91	180
	.85	.92	25.80	27.06	.63	4.00	2.00	1.92	1.16	207
30	.73	0.92	32.00	33.26	.63	4.50	2.00	2.12	1.07	264
	.99	...	32.00	33.26	.63	4.50	2.00	2.12	1.36	302
36	0.81	1.02	38.30	39.56	.63	4.50	2.00	2.32	1.22	371
	1.10	1.19	38.30	39.56	.63	4.50	2.00	2.32	1.54	426
42	0.90	1.13	44.50	45.76	.63	5.00	2.00	2.52	1.35	486
	1.22	1.32	44.50	45.76	.63	5.00	2.00	2.52	1.70	561
48	0.98	1.23	50.80	52.06	.63	5.00	2.00	2.72	1.48	607
	1.33	1.44	50.80	52.06	0.63	5.00	2.00	2.72	1.86	708

* The 1962 version of the ASA *American Standard for Cast-Iron Pipe Centrifugally Cast in Sand-Lined Molds for Gas* does not include bell and spigot joints, claiming they are "no longer used for gas service."[1]

842.14 *Standard Thickness for Cast Iron Pipe.* The wall thickness, diameter, and maximum working pressure permitted under ASA A21.1 for the type and sizes of cast iron pipe most commonly used for gas piping are shown in Tables 842.141* and 842.142. For pipe sizes, pressure, thicknesses, or laying conditions not shown in these tables, reference should be made to ASA A21.1 for the method of calculation.

842.15 *Cast Iron Pipe Joints.*

(a) *Caulked Bell and Spigot Joints.* Dimensions for caulked bell and spigot joints shall conform to American Standards A21.3, A21.7, A21.9 and A21.10. This type of joint shall not be used for pressures in excess of 25 psig, unless reinforced with mechanical clamps.

(b) *Mechanical Joints.* Mechanical joints shall utilize gaskets made of a resilient material as their sealing medium. The material selected for gaskets shall be of a type not adversely affected by the gas or condensates in the main. The gaskets shall be suitably confined and retained under compression by a separate gland or follower ring. A joint of this type is shown in ASA A21.11.

(c) *Threaded Joints.* The use of threaded joints to couple lengths of cast iron pipe is not recommended.

(d) *Flanged Joints.* The dimensions and drilling for flanges shall conform to the ASA B16 series of the American Standard Cast Iron Pipe Flanges and Flanged Fittings. Flanges shall be cast integrally with fittings or valves.

(e) *Special joints.* Special joints are not prohibited provided they are properly qualified and utilized in accordance with appropriate provisions of this Code.

* [Not reproduced here, since pit-cast pipes are no longer specified for gas.—Editor]

Tables 9-14 thru 9-17b cover cast iron pipe, applicable mechanical joints, and two classes of flanges and fittings.

Ductile Cast Iron Pipe

The crystalline structure of gray cast iron contains random graphite flakes which break up the continuity of the metal. The addition of a small amount of magnesium causes the graphite to take a spheroidal shape, which reduces the surface area of the cleavage planes and discontinuities, resulting in a strong ductile cast iron. Compared with gray iron, it has twice the bursting strength, three to four times the deflection, and eight to ten times the impact load; compared with steel, it has better machinability and castability. The corrosion rate of ductile iron is the same as that of gray cast iron.

Different combinations of strength, ductility, and toughness are available in ductile cast iron pipe. For example, a type

Table 9-16 Mechanical Joint Dimensions of Cast Iron Pipe[3]
(dimensions in inches, weights in pounds)

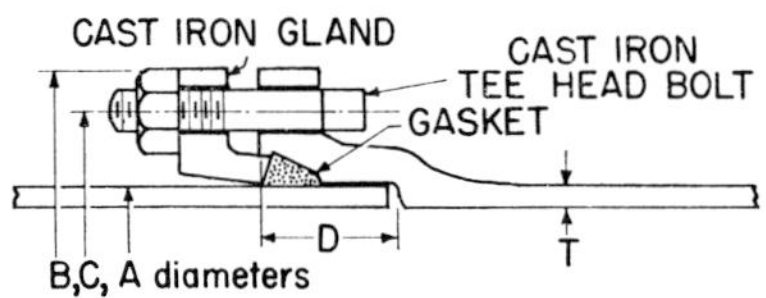

Nom. pipe size	A	B	C	D	Thickness, T A.G.A.	Class 150 WWP-421	Class 250 WWP-421	Class† 22 ASA	Bolts No. × Size × Length	Gland, bolts, and gaskets, wt
3	3.96	7.50	6.19	2.50	0.39	0.33*	0.36*	0.32	4 × ⅝ × 3	6
4	4.80	8.88	7.50	2.50	.40	.34	.38	.35	4 × ¾ × 3½	11
6	6.90	10.88	9.50	2.50	.43	.37	.43	.38	6 × ¾ × 3½	16
8	9.05	13.13	11.75	2.50	.45	.42	.50	.41	6 × ¾ × 4	22
10	11.10	15.50	14.00	2.50	.49	.47	.57	.44	8 × ¾ × 4	28
12	13.20	17.75	16.25	2.50	.54	.50	0.62	.48	8 × ¾ × 4	35
14	15.30	20.25	18.75	3.50	.57	.55	...	.51	10 × ¾ × 4	47
16	17.40	22.50	21.00	3.50	.62	.60	...	.54	12 × ¾ × 4½	60
18	19.50	24.75	23.25	3.50	.64	.65	...	.58	12 × ¾ × 4½	73
20	21.60	27.00	25.50	3.50	.68	.68	...	.62	14 × ¾ × 4½	83
24	25.80	31.50	30.00	3.50	.76	.76	...	.68	16 × ¾ × 5	112
30	32.00	38.75	36.50	4.00	.85	0.89*	...	.79	20 × 1 × 5½	207
36	38.30	45.50	43.25	4.00	0.95	1.01*	...	.87	24 × 1 × 5½	276
42	44.50	52.50	50.00	4.00	1.07	1.11*	...	0.97	28 × 1¼ × 6	377
48	50.80	59.00	56.50	4.00	1.26	1.22*	...	1.06	32 × 1¼ × 6	431

* Thickness computed by Fairchild formula.
† Thickness computed by ASA Method of Design. Pipe in steps of 8 per cent increase in thickness over Class 22 can be furnished.

used for gas piping, 60–45–10 ductile iron, has 60,000 psi minimum tensile strength, 45,000 psi minimum yield strength, and 10 per cent minimum elongation.[6] Standards for nodular ductile iron castings were adopted by ASTM in 1955.

However, as of 1960, there were no ASA specifications for ductile cast iron pipe. Pipe manufacturers in the United States have been using a combination of the ASA specifications for cast iron pipe centrifugally cast (A21.9) and the ASTM specifications for nodular iron castings (A339).

Numerous gas utilities are using ductile cast iron pipe for distribution and feeder piping in low, intermediate, and high pressure services; a number of companies are making trial installations.

Ductile cast iron pipe is installed, using standardized mechanical joints which have been used for gray cast iron gas pipes for many years. It can be drilled and tapped for service connections with the same equipment used for gray cast iron pipe.

One utility has standardized on the use of ductile cast iron pipe in its distribution system, installing over 220 miles from 1955 thru 1962 for both medium and low pressure service in sizes 6 to 36 in. This utility has encountered no failures in ductile cast iron pipe installed since the beginning of its use in 1955. It therefore expects to continue using it in the future.

COPPER PIPE AND TUBING

Copper has not been used extensively in distribution mains because of the cost. It has been widely used for high-pressure services for house services (Table 9-37), especially for insertion thru corroded steel services. Listing of applicable ASTM specifications for seamless copper pipe and tubing follows: B 42, pipe; B 75 and B 88,* tubing; B 251, copper and copper alloy pipe and tubing. The material used is basically at least 99.40 per cent pure copper with nominal O_2 and 0.04 per cent sulfur maximum. The three common grades of copper tubing, in order of increasing strength and brittleness, are annealed (or "soft"), light-drawn, and hard-drawn. Standard lengths are: straight tubing, 20 ft; coils of annealed tubing, 60 ft (100 ft for up to 1 in. diam).

Reference may be made to the following extract from ASA B31.8-1963:

849.4 Copper Services (and Mains).

849.41 *Copper Pipe Design Requirements.* The following requirements shall apply to copper pipe or tubing, when used for gas mains or services:

(a) Copper pipe or tubing shall not be used for services or mains where the pressure exceeds 100 psig.

(b) Copper pipe or tubing shall not be used for services or mains where the gas carried contains more than an average of 0.3 grains of hydrogen sulfide per 100 standard cubic feet of gas. This is equivalent to a trace as determined by the lead-acetate test.

(c) Copper pipe or tubing shall not be used for services or mains where the piping strain or external loading may be excessive.

(d) Copper services may be installed within buildings, provided that the service is not concealed and is suitably protected against external damage.

(e) Copper tubing or pipe for mains shall have a minimum wall thickness of 0.065 inches and shall be hard drawn.

(f) The minimum wall thickness for copper pipe or tubing used for gas services shall be not less than Type "L" as specified in ASTM Specifications for Copper Water Tube, designation B 88.

(g) An underground copper service installed through the outer foundation wall of a building shall be either encased in a sleeve

* Wall thickness Types K and L only.

Table 9-17a Cast Iron Flanges and Fittings, Class 125[4]

(dimensions in inches)

Nom. pipe size	Flange diam	Flange thickness,† min	Bolt circle diam	No.‡ of bolts	Bolt diam	Bolt hole diam‡	Bolt length§ Machine	Stud
1	4¼	7/16	3⅛	4	½	⅝	1¾	...
1¼	4⅝	½	3½	4	½	⅝	2	...
1½	5	9/16	3⅞	4	½	⅝	2	...
2	6	⅝	4¾	4	⅝	¾	2¼	...
2½	7	11/16	5½	4	⅝	¾	2½	...
3	7½	¾	6	4	⅝	¾	2½	...
3½	8½	13/16	7	8	⅝	¾	2¾	...
4	9	15/16	7½	8	⅝	¾	3	...
5	10	15/16	8½	8	¾	⅞	3	...
6	11	1	9½	8	¾	⅞	3¼	...
8	13½	1⅛	11¾	8	¾	⅞	3½	...
10	16	1 3/16	14¼	12	⅞	1	3¾	...
12	19	1¼	17	12	⅞	1	3¾	...
14	21	1⅜	18¾	12	1	1⅛	4¼	...
16	23½	1 7/16	21¼	16	1	1⅛	4½	...
18	25	1 9/16	22¾	16	1⅛	1¼	4¾	...
20	27½	1 11/16	25	20	1⅛	1¼	5	...
24	32	1⅞	29½	20	1¼	1⅜	5½	...
30	38¾	2⅛	36	28	1¼	1⅜	6¼	...
36	46	2⅜	42¾	32	1½	1⅝	7	...
42	53	2⅝	49½	36	1½	1⅝	7½	...
48	59½	2¾	56	44	1½	1⅝	7¾	...
54*	66¼	3	62¾	44	1¾	2	8½	10½
60*	73	3⅛	69¼	52	1¾	2	8¾	10¾
72*	86½	3½	82½	60	1¾	2	9½	11½
84*	99¾	3⅞	95½	64	2	2¼	10½	12¾
96*	113¼	4¼	108½	68	2¼	2½	11½	14

* These sizes are included for convenience and do not carry a definite rating.

† Facing: plain faced, finished in accordance with MSS SP-6-1951.

‡ Flange bolt holes: multiples of four, on center line, to permit rotation to any quarter.

§ Spotfacing:

Size, in.	*Faced flange, if oversized*	*Spotface to minimum thickness*	
Up to 12	⅛ in.	−0, + ⅛ in.	flanges
14 to 24	3/16 in.	−0, +3/16 in.	flanges
Over 30	¼ in.	−0, +3/16 in.	flanges
Below 18	Same as noted above		fittings
18 to 24	Same as noted above	−0, +3/16 in.	fittings
30 to 48	Same as noted above	−0, +3/16 in.	fittings

Where spotfacing is required, it shall be done in accordance with MSS SP-9-1955.

Bolts and nuts per ASA B18-2-1952.

Gaskets per ASA B16.21-1951.

or otherwise protected against corrosion. The service pipe or tubing and/or sleeve shall be sealed at the foundation wall to prevent entry of gas or water.

(h) A copper service installed underground under buildings shall be encased in a conduit designed to prevent gas leaking from the service and getting into the building. When joints are used, they shall be of the brazed or soldered type in accordance with 849.44.

849.42 *Valves in Copper Lines.* Valves installed in copper lines may be made of any suitable material permitted by this Code, except that ferrous valves installed on underground copper services shall be protected from contact with the soil and/or insulated from the copper pipe.

849.43 *Fittings in Copper Lines.* It is recommended that fittings in a copper line and exposed to the soil, such as service tees, pressure control fittings, etc., be made of bronze, copper or brass. If iron or steel fittings are used, they shall be protected as specified above for valves.

849.44 *Joints in Copper Pipe and Tubing.* Copper pipe shall be joined by using either a compression-type coupling or a brazed or soldered lap joint. The filler material used for brazing shall be a copper-phosphorus alloy or silver base alloy. Butt welds are not permissible for joining copper pipe or tubing. Copper tubing shall not be threaded but copper pipe with wall thickness equivalent to the comparable size of Schedule 40 steel pipe may be threaded and used for connecting screw fittings or valves.

849.45 *Protection against Galvanic Action Caused by Copper.* Provision shall be made to prevent harmful galvanic action where copper is connected underground to steel. This can be accomplished in most cases by using one or the other of the following methods:

(a) Install an insulating-type coupling or an insulating flange between the copper and the steel, or

Table 9-17b Cast Iron Flanges and Fittings, Class 250[5]

(dimensions in inches)

Nom. pipe size	Flange diam	Flange thickness,* min	Raised face diam	Bolt circle diam	Bolt hole diam†	No.† of bolts	Bolt diam	Bolt length‡ Machine	Bolt length‡ Stud
1	4⅞	11/16	2 11/16	3½	¾	4	⅝	2½	...
1¼	5¼	¾	3 1/16	3⅞	¾	4	⅝	2½	...
1½	6⅛	13/16	3 9/16	4½	⅞	4	¾	2¾	...
2	6½	⅞	4 3/16	5	¾	8	⅝	2¾	...
2½	7½	1	4 15/16	5⅞	⅞	8	¾	3¼	...
3	8¼	1⅛	5 11/16	6⅝	⅞	8	¾	3½	...
3½	9	1 3/16	6 5/16	7¼	⅞	8	¾	3½	...
4	10	1¼	6 15/16	7⅞	⅞	8	¾	3¾	...
5	11	1⅜	8 5/16	9¼	⅞	8	¾	4	...
6	12½	1 7/16	9 11/16	10⅝	⅞	12	¾	4	...
8	15	1⅝	11 15/16	13	1	12	⅞	4½	...
10	17½	1⅞	14 1/16	15¼	1⅛	16	1	5¼	...
12	20½	2	16 7/16	17¾	1¼	16	1⅛	5½	...
14	23	2⅛	18 15/16	20¼	1¼	20	1⅛	6	...
16	25½	2¼	21 1/16	22½	1⅜	20	1¼	6¼	...
18	28	2⅜	23 5/16	24¾	1⅜	24	1¼	6½	...
20	30½	2½	25 9/16	27	1⅜	24	1¼	6¾	...
24	36	2¾	30¼	32	1⅝	24	1½	7¾	9½
30	43	3	37 3/16	39¼	2	28	1¾	8½	10½
36*	50	3⅜	43 11/16	46	2¼	32	2	9½	11¾
42*	57	3 11/16	50 7/16	52¾	2¼	36	2	10¼	12½
48*	65	4	58 7/16	60¾	2¼	40	2	10¾	13

* Fittings in these sizes are not produced and used in sufficient quantities to warrant standardization. However, the flange dimensions are included for convenience where special fittings larger than 30 in. are required. When these fittings are made, the body structure should be designed to be the equivalent of the flanges in service pressure ratings.

† Flange bolt holes: multiples of four, on center line, to permit rotation to any quarter.

‡ See Table 9-17a, footnote §, for spotfacing, bolts, nuts, and gaskets.

(b) Protect the copper and steel for a distance of two feet or more in all directions from the junction with insulating pipe corrosion protection material.

849.46 *Service Connections to Copper Mains.*

(a) Connections using a copper or cast bronze service tee or extension fitting sweat-brazed to the copper main are recommended for copper mains.

(b) Butt welds are not permitted.

(c) Fillet-brazed joints are not recommended.

(d) The requirements of 849.44 shall apply to:

(1) Joints not specifically mentioned above, and

(2) All brazing material.

ALUMINUM PIPE

As of 1961, approximately 500 miles of aluminum pipe had been laid in the United States, some experimental, and some to meet particular corrosion problems. The extensive use of aluminum was prohibitive in 1962 because of cost. Aluminum is practical where the corrosive properties of the soil and the gas transported require frequent replacement of steel or cast iron lines.

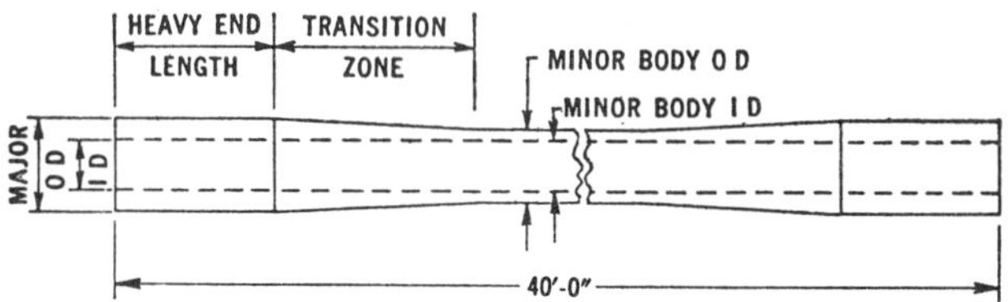

Fig. 9-22 Configuration of a heavy-end aluminum pipe length. (Reynolds Metals Co.)

Aluminum pipe is extruded to close tolerances, which results in cost savings otherwise impossible, using stocked wall thicknesses. Where the narrow, heat-affected zones adjacent to circumferential welds require local extra pipe thickness (not otherwise required by proposed service), manufacturers provide a heavy-end pipe (Fig. 9-22). This manufacturing method can provide savings in material of up to 37 per cent. Other attributes of aluminum pipe are that it (1) is lightweight; (2) has a smooth internal bore for high flow efficiency; (3) is nonsparking under most service conditions. Service conditions will dictate what, if any, corrosion protection is required.

The computation of pressure drop and flow thru an aluminum pipe is comparable to that of a steel or cast iron line—refer to Chap. 2, Sec. 8. Pipeline efficiency factors for aluminum pipe[7] vary from 0.92 to 0.96 (see Table 8-1).

PLASTIC PIPE AND TUBING

Plastic pipe first made its appearance during World War II as a substitute for metal pipe. Since 1952, *Gas*[8] has kept abreast of gas industry use of plastic pipe thru periodic user surveys. As of the beginning of 1964, there were about 6000 miles of such pipe in the ground.

The ASA B31.8-1963 Code does not include the use of plastic pipe and fittings. Since this Code does not specifically prohibit its use, Code Par. 811.24 is of interest:

"Items of a type for which no standards or specifications are listed in this code . . . may be qualified by the user by investigation and tests (if needed) that demonstrate that the item of material or equipment is suitable and safe for the proposed service and provided further that the item is recommended for that service from the standpoint of safety by the manufacturer. . ."

Such piping would have to: (1) withstand the appropriate test and operating pressures, and (2) maintain tight joints—whether lengths are joined to themselves or to existing metal piping.

Reliable methods for detecting the possibility of failures by brittle fracture are needed, as is information on failures by weeping or pinholing. These and other factors were recognized by the ASA B31.8 committee, which appointed a task force in 1963 to prepare a formal proposal covering plastic piping for the main committee's consideration.

Of the 1200 miles of plastic pipe laid in 1963, approximately 45 per cent was used as replacement for gas services by inserting it in the existing metal service, 21 per cent was used for new services, 32 per cent for new mains, and 2.2 per cent for main insertions. Lengths available are: rigid, 20 ft; and flexible, 100 ft coils (400 ft for small diameters). Reports by an A.G.A.-sponsored Subcommittee on Plastic Pipe which investigated the available materials and made recommendations on their use for transmitting gas are available.[9] Thermoplastic materials and design fiber stresses recommended for gas piping are given in Table 9-18.

Table 9-18 Tentative Design Stresses* for Thermoplastic Pressure Pipe at 100 F for Gas Service[9]

Type of thermoplastic pipe	Design stress, psi	Based on data obtained with pipe made of
ABS Type I Grade 2 (ABS 1210)	500	Kralastic W1601; Cycolac LL4001
CAB† (CAB MHO8)	400	Tenite Butyrate 205MH, 293MH, 426MH, and 466MH
PE Type III (PE 3306)	315	AC Polyethylene; Marlex TR212
PVC Type I (PVC-1120, PVC-1220)	1000	Geon 8750; DACO 52701
PVC Type II (PVC-2110)	500	Geon 8700A; Geon 8714; DACO 52702
Acetal resin‡	630	Delrin

* The values for ABS, PE, and PVC pipes are based on 73.4 F hydrostatic design stress data. All values have been reduced by 25 per cent to allow for temperature up to 100 F and another 25 per cent for gas service environment. Environmental factors, such as impact resistance and aging, and piping system factors, such as joint efficiency, have not been specifically covered by these reductions.

† There is no proposed Commercial Standard hydrostatic stress value for pipe made from this material. The value given is for MH flow only.

‡ Tentative approval only; stress value listed is based on data supplied by the material supplier and pipe producer, using the same factors as used for ABS, PE, and PVC pipes.

The maximum working pressure of plastic pipe in distribution systems should be limited to 60 psig and computed by the basic Barlow formula. Maximum operating temperature recommended is 100 F. The use of plastic pipe in a distribution system that may contain manufactured gas (such as in peak shaving operations) is not recommended. It is believed that the aromatic hydrocarbons contained therein attack plastic molecules. Plastic pipe should be solvent- or cement-welded with the material supplied by the manufacturer specifically for the pipe used (some products require heat for fusion). Threaded joints are not recommended.

Experiments were under way[10] to test a gas distribution development using plastic piping suspended from steel cables attached to existing utility poles for support. The proposed method of gas distribution appeared to reduce the cost of new systems materially. It would be most useful in bringing gas service to small communities being added to the system.

FLEXIBLE ARMORED PIPELINE[11]

Installation of 1400 ft of 2-in. diam flexible tubing in less than 24 hr under turbulent tidal waters was reported. The submarine pipeline, designed for up to 1000 psi gas service, has a tensile strength of nearly 67 tons. The pipline, made by the Simplex Wire and Cable Co. of Cambridge, Mass., consists of polyethylene tubing covered successively with spirally wound layers of jute- and asphalt-coated steel strands.

REFERENCES

1. Am. Standards Assoc. *American Standard for Cast-Iron Pipe Centrifugally Cast in Sand-Lined Molds for Gas.* (ASA A21.9-1962) New York, 1962.
2. *Ibid.* (ASA A21.9-1953) New York, 1953.
3. Cast Iron Pipe Research Assoc. *Manual of Mechanical Joint Cast Iron Pipe and Fittings for the Gas Industry.* Chicago, Ill., 1947.
4. ASME. *Cast-Iron Pipe Flanges and Flanged Fittings, Class 125.* (ASA B16.1-1960) New York, 1960.
5. ——. *Cast-Iron Pipe Flanges and Flanged Fittings, Class 250.* (ASA B16.2-1960) New York, 1960.
6. Clay, R. "Ductile Iron Pipe." *Gas* 36: 54–8, June 1960.
7. Reynolds Metals Co. *Aluminum Pipelines: Design and Construction, Welding, Applications.* Richmond, Va., 1961.
8. Clay, R. "1960 Plastic Pipe Survey Shows: the Infant Is Becoming a Child." *Gas* 36: 78–81, Mar. 1960.
9. Webster, I. S. "Report of Subcommittee on Plastic Pipe Standards." *A.G.A. Operating Sec. Proc.* 1963: DMC-63-14.
10. Denham, D. W. "Aerial System Put to Test." *Gas Age* 130: 34–36, Dec. 1963.
11. "Flexible Armored Pipeline: New Hampshire Gas Utility Makes First Installation." *Gas Ind.* 6: 2, Oct. 1962.
12. Martin, J. A. "Thin Wall Steel Tubing for Services—X-Trube." *A.G.A. Operating Sec. Proc.* 1964: 64-D-201.

Chapter 4

Valves

by H. C. Roemmele and F. E. Vandaveer

The valves used in a gas distribution system include shutoffs for mains, services, and meters, electrically and pneumatically operated automatic valves, check valves, and pressure relief valves. Table 9-19 gives a valve schedule used by one company for 15 and 60 psig; for 120 psig, in all valve sizes, flanged ends are used—plug valves are specified (gate valves may be used for construction purposes).[1]

Table 9-19 Specifying Valves for 15 and 60 psig[1]

	Sizes for types of ends*		
Valve type	Threaded	Compression	Flanged
Plug	2 in.	2 and 3 in.	3 in.†
Gate	...	3 in.	3 to 20 in.†

* Valves with mechanical joint ends may be installed in mechanical joint cast iron mains.

† 24 in. and larger, vertical gate valves; except that horizontal gate or plug valves may be used when cover requires.

Reference may be made to the following extract from ASA B31.8-1963:

831.1 Valves.

831.11 Valves shall conform to American Standards governing minimum wall thickness and dimensions and shall be used only in accordance with the service recommendation of the manufacturer.

(a) Valves manufactured in accordance with standards listed in this paragraph may be used in accordance with pressure–temperature ratings contained in these standards:

ASA B16.5—Steel Pipe Flanges and Flanged Fittings
API 6D —Specification for Steel Gate, Plug and Check Valves for Pipeline Service
MSS SP-44 —Steel Pipe Line Flanges
MSS SP-52 —Cast Iron Pipe Line Valves
ASA B16.24—Brass or Bronze Flanges and Flanged Fittings.

(b) Valves having parts made of nodular cast iron in compliance with ASTM Specifications A 395 or A 445 and having dimensions conforming to ASA B16.1, B16.2, B16.5, or API 6D may be used at pressures not exceeding 80 per cent of the pressure ratings for comparable carbon steel valves at their listed temperature provided the adjusted service pressure does not exceed 1000 psi and welding is not employed either in fabrication of the valves or in their assembly as a part of the piping system. Valves having parts constructed of nodular cast iron may not be used in the gas piping components of compressor stations.

831.12 Screw-end valves shall be threaded according to the American Standard Pipe Threads (ASA B2.1) or API Specification for Line Pipe (5L) or API Specification for Threads in Valves, Fittings and Flanges (6A).

A comprehensive report is available covering characteristics, construction, application, costs, and comparisons among the various valves used in hydrocarbon service.[2]

Globe valves[2] are superior for throttling applications—valve seats wear relatively little. While these valves give tight shutoff, they are generally more costly than gate valves, and are not commonly used above the 1½ in. size. Angle and Y-types of globe valves overcome some of the large pressure drop inherent in conventional models. Angle valves, however, are seldom used because of stress considerations. **Ball valves[2]** using synthetic seats are recommended by the manufacturers for "bubbletight" service. These economical valves are quick-opening, long-wearing, and relatively low in resistance to flow.

MAIN SHUTOFFS

Location in Low Pressure Mains. In general, valves are not required here because of the ease of obtaining a shutoff with **bags** and **stoppers.** Special locations where valves may be installed for convenience or safety are: (1) at major bridge, stream, or railroad crossings; (2) at the inlet and outlet of district regulator stations, meters, and compressors, to permit servicing and inspection; and (3) at various points in a network to sectionalize the system.

Location in Intermediate and High Pressure Mains. Valves may be installed for operating and emergency purposes at the following locations:

1. At specified intervals in a feeder main (spacing determined by conditions and company practice).
2. Where by-pass or shutoff is necessary to permit servicing of equipment, such as pressure regulators, relief valves, meters, and compressors.
3. At points of connection to existing or planned branch lines.
4. At points of connection to mains of lower or higher pressure, for emergency use, e.g., as additional gas supply.
5. At connections at which main taps are made under pressure (hot taps).
6. At blowoff connections.
7. At major bridge, submarine, or railroad crossings.
8. At points in a distribution network to limit the number of customers affected by a shutdown for repairs.

One company uses cast iron body valves for up to 120 psig in up to 8 in. mains; all valves 12 in. and larger subject to 120 psig are of steel body construction.

Reference may be made to the following extract from ASA B31.8-1963:

846.12 Valves on distribution mains, whether for operating or emergency purposes, shall be spaced as follows:

(a) *High Pressure Distribution Systems.* Valves shall be installed in high pressure distribution systems in accessible locations in order to reduce the time to shut down a section of main in an emergency. In determining the spacing of the valves consideration should be given to the operating pressure and size of the mains and local physical conditions as well as the number and type of consumers that might be affected by a shutdown.

(b) *Low Pressure Distribution Systems.* Valves may be used on low pressure distribution systems, but are not required except as specified in 846.22 (a).

846.22 *Distribution System Valves.*

(a) A valve shall be installed on the inlet piping of each regulator station controlling the flow or pressure of gas in a distribution system. The distance between the valve and the regulator or regulators shall be sufficient to permit the operation of the valve during an emergency, such as a large gas leak or a fire in the station.

(b) Valves on distribution mains, whether for operating or emergency purposes, shall be located in a manner that will provide ready access and facilitate their operation during an emergency. Where a valve is installed in a buried box or enclosure, only ready access to the operating stem or mechanism is implied. The box or enclosure shall be installed in a manner to avoid transmitting external loads to the main.

Gate Valves

These valves are suitable for the majority of applications that require full open or closed positioning—with infrequent operation. "Bubbletight" service, however, cannot be achieved, since sealing depends upon deformation of mating surfaces. Temperature changes and external pipe stresses interfere with such sealing.[2]

Gate valves are available in sizes up to 36-in. diam pipe (in some cases, up to 48-in. diam pipe) with cast iron, cast steel, or bronze bodies; nonrising or rising stems; brass seat rings and gates; and connections for flanges (American Standards: cast iron valves, 125 lb; steel valves, 150 lb), compression joints, threads, or welding. Two examples are shown in Fig. 9-23. Gate valves on underground mains with a valve box over them are usually of double parallel gate construction, with inside, nonrising stem of bronze or stainless steel, bodies of cast iron or cast steel, and brass seat rings in the body and on the gates. Gate valves with iron and cast steel bodies are also available with a duct system so that a lubricant-seal may be applied to the valve seating surfaces. Gate valves used aboveground, in vaults, or in buildings, are usually of the rising stem type. They are installed in vertical lines where there is any possibility of sediment depositing in the bonnet. Operation of gate valves may be accomplished by manually

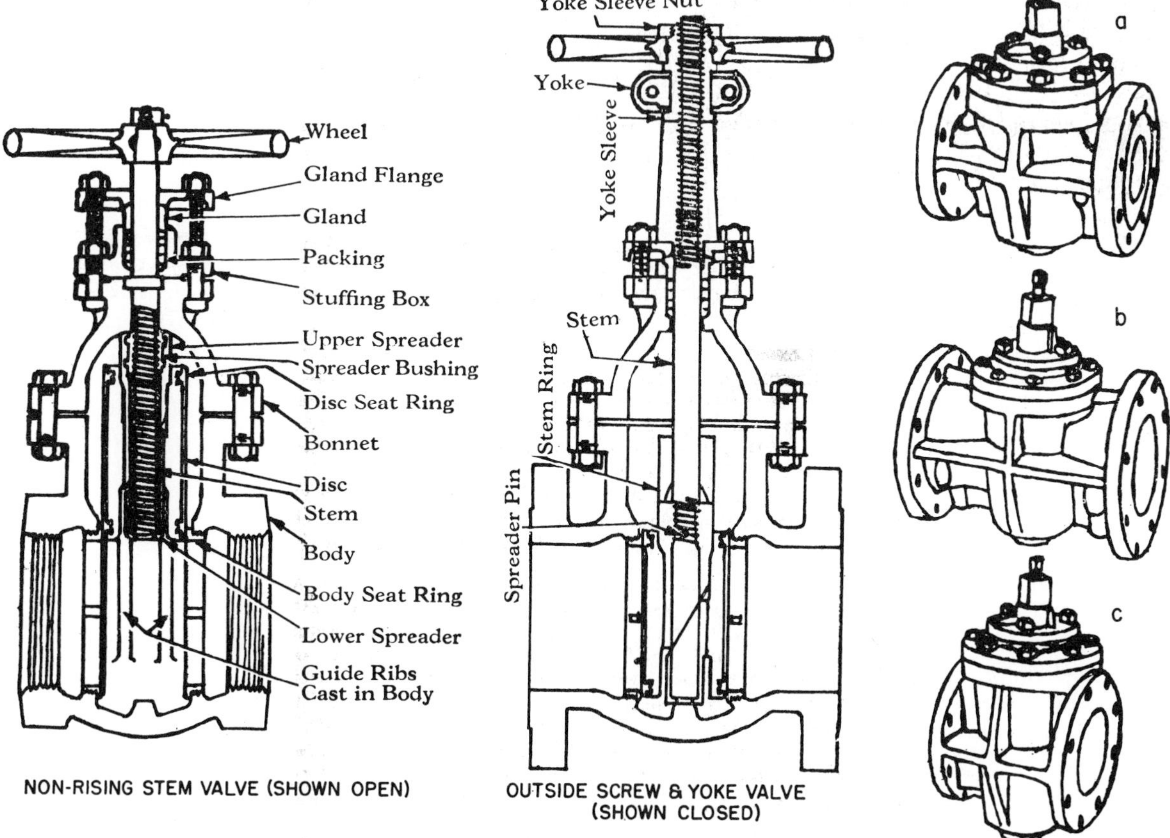

Fig. 9-23 (left and center) Two types of iron body double disk gate valves.

Fig. 9-24 (right) Three types of lubricated plug valve bodies: a, regular; b, venturi; c, gate valve replacement (short pattern).

turning a wheel or key, by an electric motor, or by a pneumatic operator.

Inspection and Maintenance. Periodic inspection and maintenance of gate valves is necessary to ensure good operation. Gate valves left in the open position for a long period of time may become stuck or difficult to operate. Dry packing in the stuffing box, tar, gum, or other deposits on the thread of the stem or on the gates and seats, may cause sticking. Among the ways to free the stem are loosening the gland, saturating the packing with penetrating oil, applying heat in the form of steam, or applying a solvent to the valve interior. Gastight closure of the valve may be hindered by a collection of dust, rust, scale, or tar in the bottom of the valve. These deposits may sometimes be removed by partially closing the valve (thereby increasing the velocity of gas flow over the deposits), and by the repeated opening and closing of the valve.

The introduction of steam thru an *oiler* or *bleed* pipe connected to the bonnet, and the addition of a solvent such as hot drip oil, will frequently soften tarry deposits so that the valve may be closed reasonably tight.

Lubricated Plug Valves

These have the basic advantage of "bubbletight" closure. They are made in a variety of patterns, each suitable for a field of use as described by the manufacturer. They may have a full round opening thru the plug, a rectangular venturi opening, or a multiport opening. However, they all retain the fundamental V or plug shutoff mechanism. Typical plug valves are shown in Fig. 9–24. They are available with screwed, flanged, butt-welding, or plain ends for mechanical coupling. Valves may be wrench- or gear-operated* (between travel stops), and a wide variety of arrangements is available. Plug valves are used for throttling[3] for ratios of maximum to minimum flow up to 100 to 1.

Valve equipment may include electric motor operated units consisting of motor, reduction gears, declutching device, limit switches and contacts for controlling position, and indicating lights for remote or local control. Plug valves, equipped with controls and operators, are used as large volume regulators.

For emergency pipeline protection, automatic cylinder operators for valves are available. Valves so equipped can be adjusted to close following a line break, but not to respond to changes occurring in normal operation. Power for emergency operation is provided by gas stored in a cylinder attached to the line at all times.

A thin film of lubricant between the plug and barrel permits easy turning (Fig. 9-25). A stuck tapered plug valve may be *freed* by the continued introduction of lubricant (production of a hydraulic jacking effect to separate plug from barrel). Valves with cylindrical plugs are also available.

Lubricants determine the ease with which lubricated plug valves will operate. They are special compounds and are not to be confused with ordinary petroleum greases. Such lubricants are available in bulk as well as in the form of plastic sticks. A softer consistency is used in *grease guns*. The chief functions of a plug lubricant are to provide a plastic seal between the seating surfaces while resisting dissolution by the line fluid, and to be thermally resistant at the operating temperature of the valve. The lubricant must also provide satisfactory characteristics for hydraulically jacking the valve plug from its seat. Lubricants for various gases, temperatures, and pressures are available. Occasionally some lubricants deposit downstream of the valves.

* One company specifies ¼ turn to open in up to 6 in. size; 8 in. and larger, gear operated.

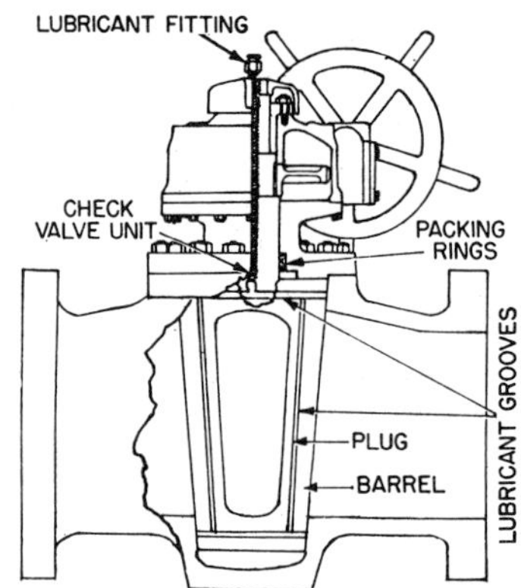

Fig. 9-25 One type of lubricated plug valve showing lubricant sealing grooves, lubricant chamber, lubricant check valve, and lubricant screw. The lubricant fitting is usually extended via piping to an accessible location in the valve box or vault.

Lubricated plug valves (4 in., 12 in., and 24 in.) were tested with gas flowing at 125, 55, and 15 psig. The pressure drop caused by the lubricated plug valves ranged from 0.25 to about 5 per cent of that in the piping grid, while the reduction in deliverability ranged from 0.13 to 2 per cent.[4] Lubricated plug valves must be lubricated in fully opened or closed position and partially opened and closed periodically to ensure quick operation in times of emergency.

Installation

Main shutoffs are generally installed vertically in valve boxes or vaults set to grade so as to be readily available for operation. One company specifies a minimum of 8 in. clearance between the top of the stem and the finished grade.[1] Two test or pressure pipes (usually ¾ in. diam) are generally connected to the mains, one on each side of the valve, and brought into the valve box or vault to permit the attachment of gages for regulation of pressures when the valve is operated. On gate valves, a third pipe, sometimes called a bleed or oiler pipe, is usually connected to the bonnet of the valve. This pipe

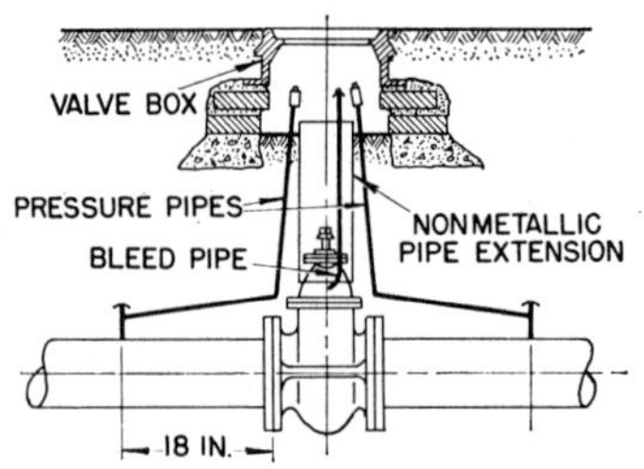

Fig. 9-26 Gate and plug valve installations, 4 in. and larger.

Valve	*Main*	*Remarks*
C.I.	Steel	Add compression fitting (flange to be 3 to 6 ft from flange shown) in welded mains.
Both	C.I.	Add C.I. adapters to both sides of valve.

permits testing for tight closure of the gates, and can also be used to introduce steam, solvents, or a sealing medium between the gates. Figure 9-26 shows a gate valve installation. See also Figs. 8-11 and 8-12.

SERVICE SHUTOFFS

Service shutoffs, service stops, and curb cocks (Fig. 9-27a) are terms used interchangeably for plug valves for services. There are a number of designs of iron body and brass, bronze plug, or brass throughout. They may or may not have spring-loaded plugs. Threaded, union, or mechanical joint connections may be used. They always have a rectangular or square head for turning with a curb cock wrench or key handle. A curb cock is usually protected with a curb box (Fig. 9-27b).

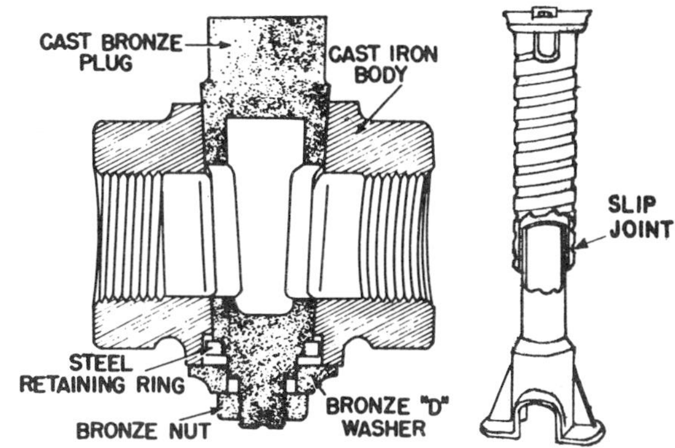

Fig. 9-27a (left) One type of curb cock.
Fig. 9-27b (right) One type of curb box.

Reference may be made to Par. 849.12 in ASA B31.8-1963 (see Chap. 8 of Sec. 9).

Location of Service Shutoffs

In general, each meter set assembly is provided with a shutoff (which can be locked in shut position) to facilitate meter removal or exchange and to furnish means of turning off gas to the premises. In addition, a second shutoff on services is installed under certain circumstances as a safety control in case of fire or other damage to the premises. In some areas at least one shutoff outside the building is always required. This valve, generally located near the curb or property line, is protected by a curb box (Fig. 9-27b). Outside meter installations may have just one shutoff.

Reference may be made to Par. 849.13 in ASA B31.8-1963 (see Chap. 8 of Sec. 9).

METER SHUTOFFS

Meter stops or meter cocks (Fig. 9-28) are available in a wide variety of designs, always embodying plug cocks. Variations are in the body (either iron or brass), in the plug (usually brass), in the key head (lubricated* or nonlubricated plug), in methods for making the valve tamper-resistant, and in provisions for holding the plug in place. Meter stops are usually separate cocks, but may be incorporated as part of a rigid meter bar. Specifications are available for meter stops (½ to 2 in., 60 psig max).[10]

CHECK VALVES AND AUTOMATIC GAS SHUTOFF VALVES

Some utilization equipment may produce back pressure in the gas supply system or permit entrance of air or oxygen. Protection against such conditions is essential to avoid possible damage to meters and regulators or interruption of service in the adjacent area. Equipment to accomplish this objective is installed downstream from the meter.

Check Valves

Check valves (Fig. 9-29) for prevention of back pressure or reverse flow in gas lines are available in various sizes and combinations of materials. They may be of the lift check, swing check, or diaphragm types. The latter are not recommended where back pressures above 50 psig may be encountered.

Reference may be made to the following extract from ASA B31.8-1963:

848.32 A suitable protective device such as a back-pressure regulator, or a check valve, shall be installed downstream of the meter if and as required under the following conditions:

(a) If the nature of the utilization equipment is such that it may induce a vacuum at the meter, install a back-pressure regulator downstream from the meter.

(b) Install a check valve or equivalent if

(1) The utilization equipment might induce a back-pressure.

(2) The gas utilization equipment is connected to a source of oxygen or compressed air.

(3) Liquefied petroleum gas or other supplementary gas is used as standby and might flow back into the meter. A three-way valve installed to admit the standby supply and at the same time shut off the regular supply can be substituted for a check valve if desired.

* The use of the lubricated plug valve will frequently result in grease in the meter unless there is a filter between lubricated plug valve and meter.

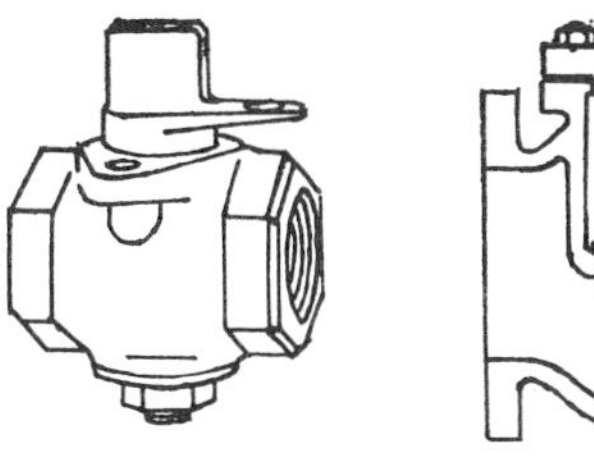

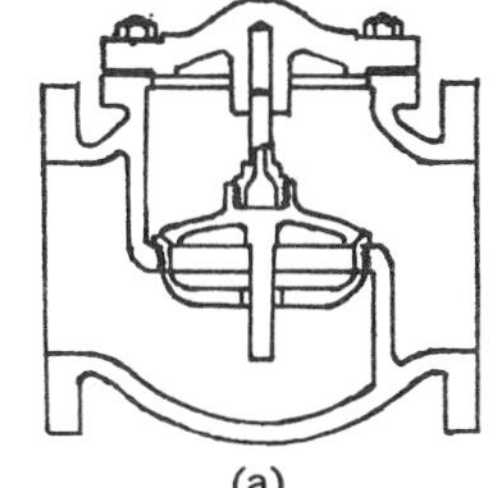

(a)

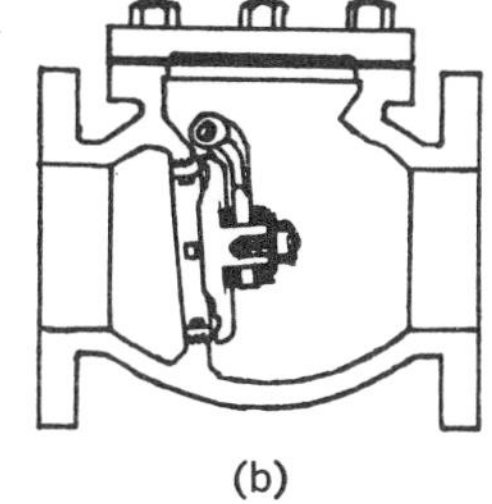

(b)

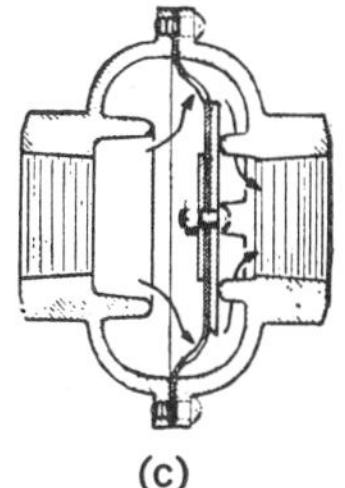

(c)

Fig. 9-28 (left) One type of meter stop with iron body and lock wing.
Fig. 9-29 (right) Three types of check valves: (a) lift type; (b) swing type; (c) diaphragm type.

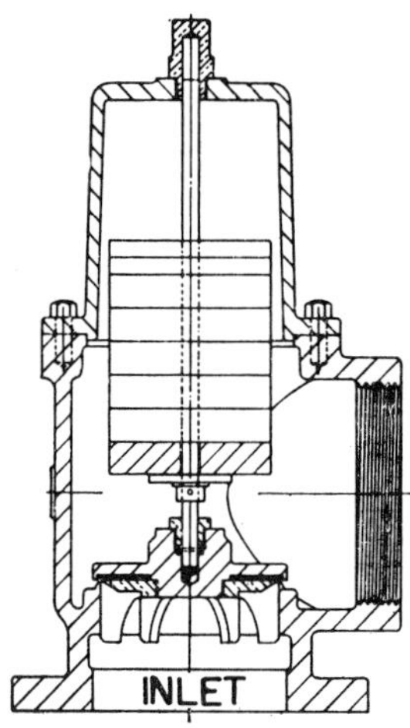

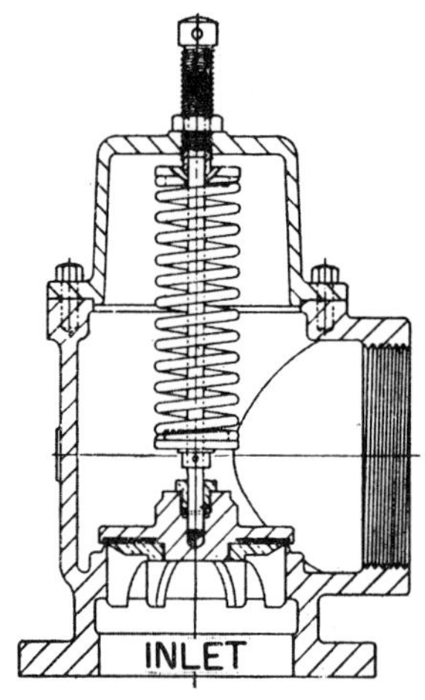

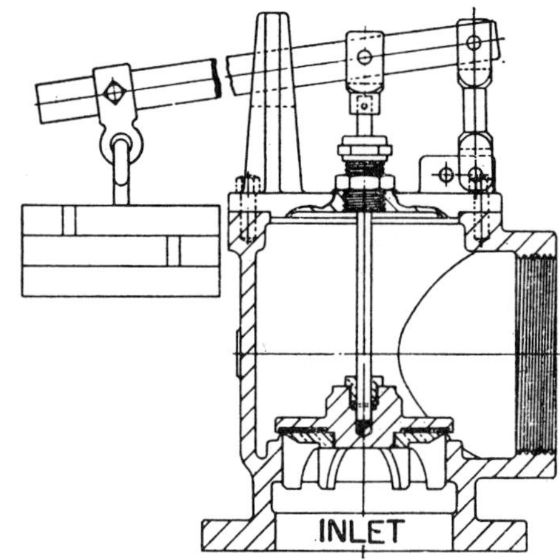

Fig. 9-30 Three types of relief valves: (left) weight-loaded; (center) spring-loaded; (right) lever and weight-loaded.

Automatic Shutoff Valves

These valves shut off the gas when abnormal pressures are indicated, e.g., in the event of failure in a controlling regulator (Fig. 9-42b). The valve must be reset manually. One type is built in as part of a service regulator. Automatic shutoff valves are available for turning off gas in case of fire. Such valves are *not* specified in recognized codes.[5-7]

PRESSURE RELIEF VALVES AND OTHER RELIEF DEVICES

The term *relief valves*, as used herein, applies to valves which automatically relieve to the atmosphere any unwanted build-up of gas pressure in mains or services. Other means are also used to prevent excessive gas pressure: automatic shutoff valves; relieving into a lower pressure gas line; or a monitoring regulator in series with the controlling regulator.

Location

Common locations for relief valves, when used, are: (1) downstream from the pressure regulator separating the high-pressure feeder line from the intermediate- or low-pressure gas line; (2) as an integral part of service regulator on an intermediate- or high-pressure system, or as a separate unit; and (3) on farm taps to high-pressure lines after each stage of pressure reduction, or as a part of each pressure regulator.

Types and Capacities

Weight-loaded valves, spring-loaded valves, or a combination of the two, and the liquid seals may be obtained either built into a service regulator or as separate units to be connected to the gas line as desired. Examples of the various types of relief valves are shown in Figs. 9-30 thru 9-34.

Examples of relief capacities of mercury seal relief valves are given in Figs. 9-35a and 9-35b. An example of relief capacity of three sizes of **dead weight** relief valves is given in Fig. 9-36 and of an **internal relief** valve in a service regulator in Fig. 9-37. As an indication of the range of relieving capacities of various types of relief devices, Table 9-20 is presented.

Table 9-20 Typical Relieving Capacity Ranges of Various Types of Relief Valves and Devices

Type	Normal relief pressure range	Valve size, in.	Flow to atmosphere, cfh of 0.6 sp gr gas
Weight loaded, small	7 in. w.c. to 2 psig	⅜–1	1,100–2,500 (Fig. 9-36)
large	14 in. w.c. to 7.5 psig	2–6	2,920–123,000 (2 to 6 in. valves)
Spring loaded, small	8.5 in. w.c. to 150 psig	⅜–1	*
large	1 to 25 psig	2–6	2,760–295,000 (2 to 6 in. valves)
Combination weight and spring loaded	1 to 15 psig	2–6	*
Lever and weight loaded	5 to 125 psig	¾–6	14,900–447,000 (2 to 6 in. valves)
Diaphragm-operated spring loaded or weight loaded	7 in. w.c. to 1 psig	¾–3	Fig. 9-37*
Diaphragm-operated spring loaded, high capacity	11 in. w.c. to 2 psig	4–8	158,000–1,260,000 (4 to 8 in. valves)
Mercury seals	9 to 21 in. w.c.	¾–2	Figs. 9-35a and 9-35b*
Pop valve	100 to 250 psig	1	81,000–165,000
Liquid seal tanks†	9 to 30 in. w.c. or more	...	...
4 in. inlet and outlet pipe	18 in. w.c.	...	60,000 (oil retained), 120,000 (no oil retained)
6 in. inlet and outlet pipe	18 in. w.c.	...	120,000
Expansible tube	‡	2	1,000,000 at 1000 psig§

* See manufacturers' catalogs.
† 12-in. diam × 36 in., 15-in. oil seal, single tank, no baffles.
‡ Intermediate and high.
§ Fox, G. M. "Relief Valves on Customer Meter Sets." *Gas*, p. 57–60, June 1954. Figure 9-48 shows a regulator using an expansible tube.

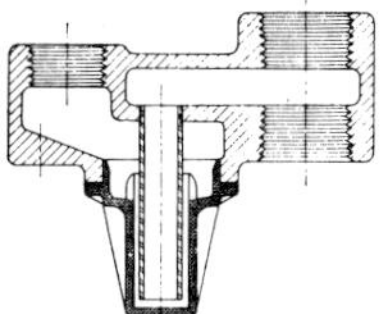

Fig. 9-31 One type of mercury seal. Mercury is placed in the annular space between the cross hatched cup and the shaded tube. Relief is thru mercury and to vent at left.

Testing. A code[8] is available for testing gas relief valves to determine one or more of the following characteristics: (1) relieving pressure; (2) relieving capacity at the relieving pressure; (3) theoretical relieving capacity; (4) coefficient of discharge; (5) start-to-leak pressure; (6) opening pressure; (7) closing pressure; (8) seal-off pressure; (9) blowdown; (10) reproducibility of valve performance; and (11) mechanical characteristics of the valve as determined by seeing, feeling, or hearing, such as: (a) ability to reseat satisfactorily, (b) tightness before and after relieving, and (c) absence of chatter, flutter, sticking, and vibration. If vent pipes are used in actual installation, the pressure drop in such vent pipes must be taken into account.

Operation of relief valve shown in Fig. 9-32a:

In normal operation, spring (1) holds valve (3) on its seat (2). If, because of dirt on the main regulator valve (B) or a nicked main valve seat (A), the main valve fails to close tightly, pressure will build up under the main diaphragm (4) in excess of the normal outlet pressure. When this pressure increase, multiplied by the area of the large diaphragm (4), is sufficient to overcome spring (1), valve (3) is raised from its seat and the excess pressure is relieved into the top diaphragm chamber. From here the gas flows thru vent connection (5)

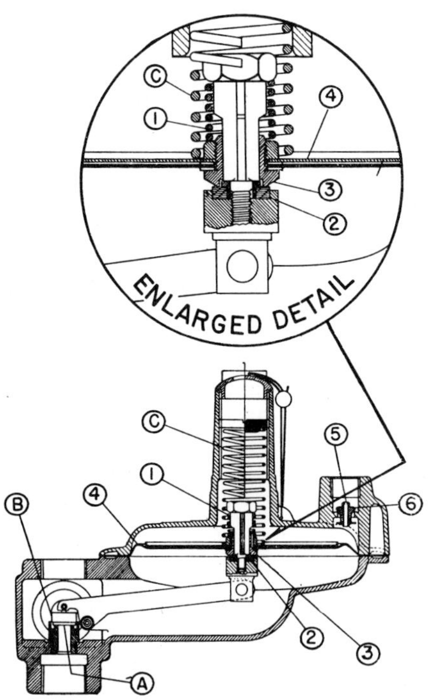

Fig. 9-32a Diaphragm-operated, spring-loaded relief valve, small size, built into a service regulator.

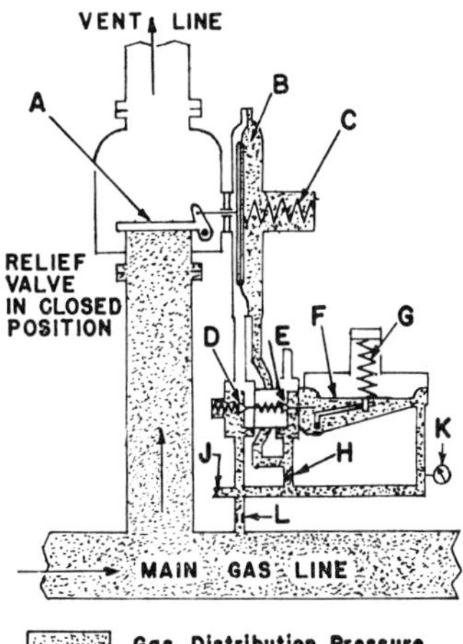

Fig. 9-32b Diaphragm-operated, spring-loaded, high-capacity relief valve.

by raising breather disk (6). This vent must be piped to a safe place for discharge of unburned gas if regulator is installed inside or under a building. Upon relieving the overpressure, spring (1) will again close valve (3). It will be noted that the relief valve spring is selected to operate at a definite rise in pressure above the normal outlet pressure. Hence, the relieving pressure will vary with the outlet pressure, but it will always be approximately four ounces above the outlet pressure. The relief valve spring in no way influences operation or adjustment of the main regulator control spring (C).

The relief valve (3) is made of brass with a specially formed seating lip. The relief valve seat is made either of synthetic material or of treated leather, depending on the service.

The breather valve (6) incorporated in the vent outlet prevents regulator pulsations, yet does not interfere with the free discharge of gas from the internal relief valve. When flow through the vent ceases due to the opening of the relief valve, the breather valve falls back to its seat. During normal regulator operation, a small orifice in the center of the valve serves as the breather hole.

Operation of relief valve shown in Fig. 9-32b:

In normal operation, main valve A is held closed by main spring C and the loading pressure in chamber B.

If the upstream district regulator fails to open, system pressure would build up above the value for which spring G is set, thus permitting upward movement of auxiliary diaphragm assembly F, which thru suitable linkages opens switching valves D and E. The pressure in chamber B bleeds out exhaust valve E faster than it can build up thru restriction H. Distribution pressure passes thru switching valve D and builds up the pressure on the left side of the main diaphragm, causing the diagragm to move from left to right, compressing the main spring C and opening valve A, thus allowing excess pressure to flow straight thru. Intermediate pressures underneath diaphragm F result in intermediate positions of main valve A. This valve has a throttling range of slightly less than 4 in. w.c.

Operation of relief valve shown in Fig. 9-33:

In normal operation, the plunger assembly is held on its seat by the greater force from the larger area above the

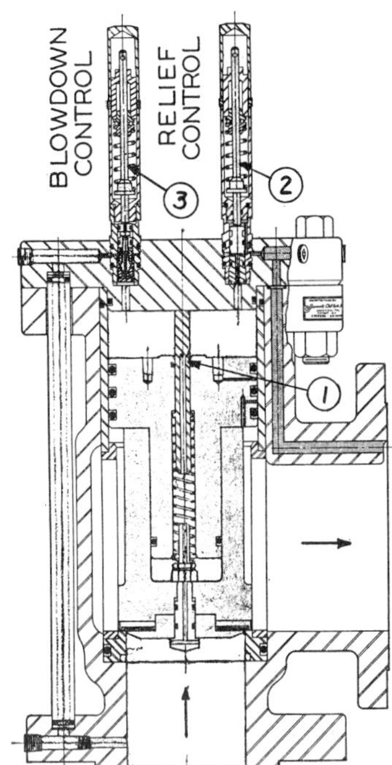

Fig. 9-33 Relief valve with plunger slide valve assembly.

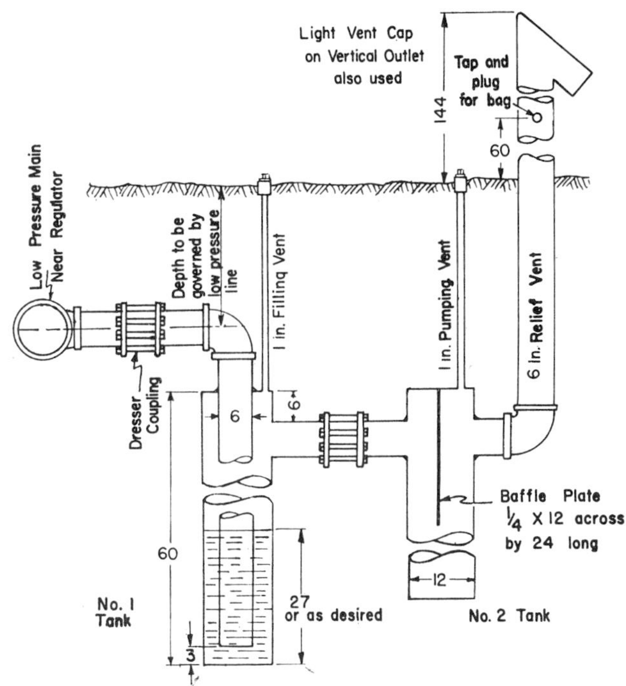

Fig. 9-34 Typical liquid seal relief valve assembly (all dimensions in inches).

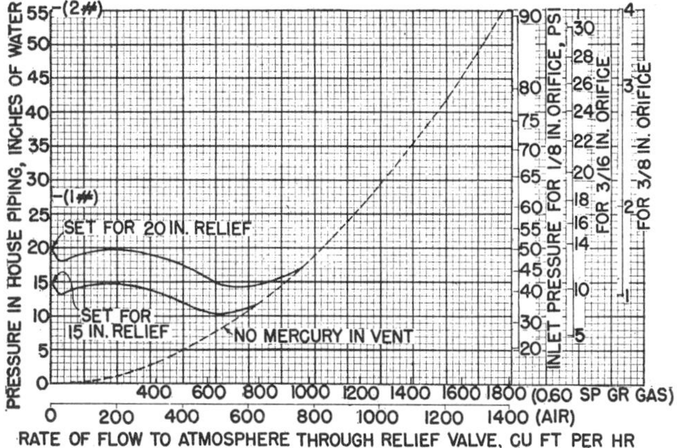

Fig. 9-35a Relief capacity of a mercury seal in one type of service regulator of low capacity; ¾ in. valve (no vent line), valve full open, with ⅛, 3/16 and ⅜ in. orifice in regulator.

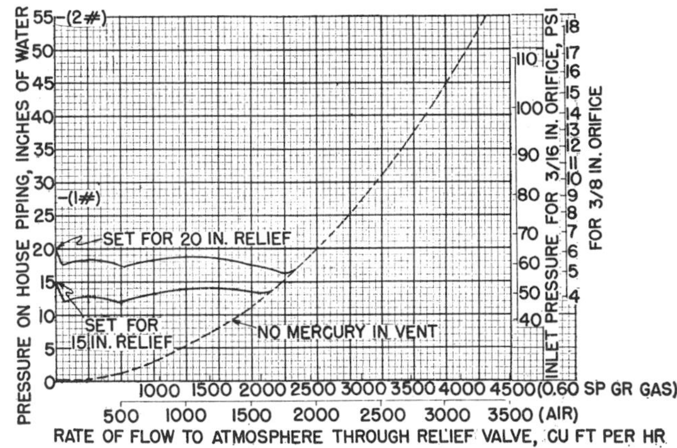

Fig. 9-35b Relief capacity of a mercury seal in one type of service regulator of full capacity; ¾ in. vent (no vent line), valve full open.

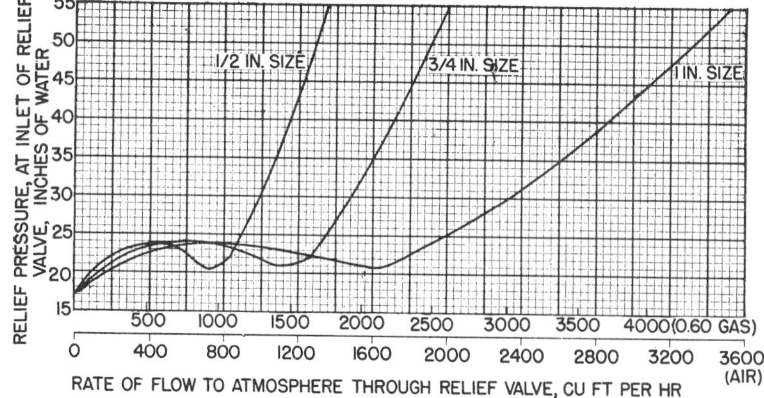

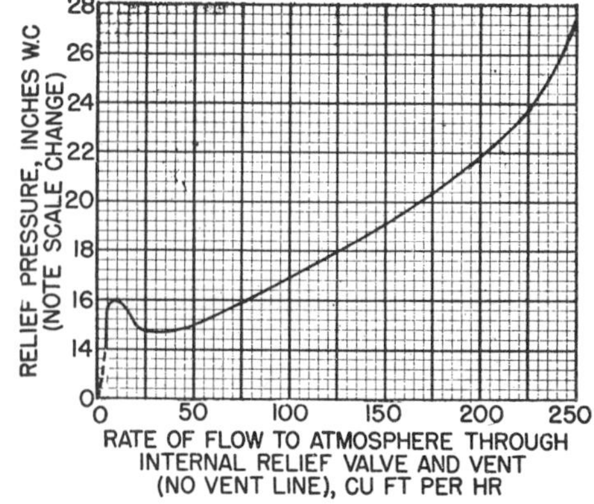

Fig. 9-36 (left) Relief capacity of three sizes of dead weight relief valves of one manufacturer (no vent line attached).

Fig. 9-37 (right) Relief capacity of one type of diaphragm-operated, spring-loaded, internal relief valve. Note change in pressure scale.

plunger. This differential in force keeps the valve securely seated.

The port in the plunger slide valve (1) connects directly thru a filter screen to the pressure at the valve inlet. When the pressure on the system increases beyond normal, it is transmitted thru the plunger slide valve to the chamber above the plunger. When the set pressure of the relief valve is reached, relief control (2) opens, releasing the pressure above the plunger to the downstream side of the valve. The line pressure then moves the plunger assembly to the top of the cylinder. This movement closes the plunger slide valve. The relief control closes when the plunger reaches the top of the stroke.

The relief valve plunger remains open until the pressure has relieved or "blown down" to normal or to any preset pressure. At this point, blowdown control (3) opens, feeding line pressure thru a filter screen into the cylinder above the plunger assembly. The plunger assembly moves rapidly toward its seat. Approximately 1/4 in. off the seat, the plunger slide valve reopens and the flow thru the plunger slide valve, added to that thru the blowdown control, accelerates the plunger assembly to its seat.

When desired, the relief valve can be supplied with a manually operated control valve to permit blowing down a system at any time.

Liquid seals (Fig. 9-34) are commonly used to prevent overpressuring of low-pressure systems since they are virtually foolproof (no moving parts), very reliable, inexpensive, and require little maintenance. Main gas pressure in excess of the sealing fluid *head* will permit gas to pass thru the seal and escape to the atmosphere by way of a relief vent stack* terminating at least 10 ft (usually 20 ft) above the ground. Maximum recommended venting capacity for liquid seals is about 150 MCFH.[9] Sealing liquids include neutral meter oil, SAE 20 oil, glycol–water mixture, straw oil, and mercury.

Vent Lines

In many cases, relief valve discharge must be carried to a safe point by adding a vent line to the assembly. It is essential that the vent line pressure drop does not significantly reduce the system discharge capacity. The vent line capacity can be found by assuming a reasonable pressure drop. Then a check should be made to establish that the available pressure at the end of the vent is adequate to discharge the desired volume thru the vent termination.

Reference may be made to the following extract from ASA B31.8-1963.

845.6 Requirements for Design of All Pressure Relief and Pressure Limiting Installations.

845.61 All pressure relief or pressure limiting devices shall:

(a) Be constructed of materials such that the operation of the device will not normally be impaired by corrosion of external parts by the atmosphere, or of internal parts by gas.

(b) Have valves and valve seats that are designed not to stick in a position that will make the device inoperative and result in failure of the device to perform in the manner for which it was intended.

(c) Be designed and installed so that they can be readily operated to determine if the valve is free; and can be tested to determine the pressure at which they will operate; and can be tested for leakage when in the closed position.

845.62 The discharge stacks, vents or outlet ports of all pressure relief devices shall be located where gas can be discharged into the atmosphere without undue hazard. Consideration should be given to all exposures in the immediate vicinity. Where required to protect devices, the discharge stacks, or vents, shall be protected with rain caps to preclude the entry of water.

845.63 The size of the openings, pipe and fittings located between the system to be protected and the pressure relieving device, and the vent line, shall be of adequate size to prevent hammering of the valve and to prevent impairment of relief capacity.

845.64 Precautions shall be taken to prevent unauthorized operation of any stop valve which will make a pressure relief valve inoperative. This provision shall not apply to valves, the operation of which will isolate the system under protection from its source of pressure. Acceptable methods for complying with this provision are:

(a) Lock the stop valve in the open position. Instruct authorized personnel of the importance of not inadvertently leaving the stop valve closed and of being present during the entire period that the stop valve is closed so that they can lock it in the open position before they leave the location.

(b) Install duplicate relief valves, each having adequate capacity by itself to protect the system, and arrange the isolating valves or 3-way valve so that mechanically it is possible to render only one safety device inoperative at a time.

845.65 Precautions shall be taken to prevent unauthorized operation of any valve which will make pressure limiting devices inoperative. This provision applies to isolating valves, by-pass valves, and valves on control or float lines which are located between the pressure limiting device and the system which the device protects. A method similar to 845.64 (a) shall be considered acceptable in complying with this provision.

845.66 (a) When a monitoring regulator, series regulator, system relief or system shut-off is installed at a district regulator station to protect a piping system from overpressuring, the installation shall be designed and installed to prevent any single incident such as an explosion in a vault or damage by a vehicle from affecting the operation of both the overpressure protective device and the district regulator. (See 846 and 847.)

(b) Special attention shall be given to control lines. All control lines shall be protected from falling objects, excavations by others, or other foreseeable causes of damage and shall be designed and installed to prevent damage to any one control line from making both the district regulator and the overpressure protective device inoperative.

845.7 Required Capacity of Pressure Relieving and Pressure Limiting Stations.

845.71. Each pressure relief station or pressure limiting station or group of such stations installed to protect a piping system or pressure vessel shall have sufficient capacity and shall be set to operate to prevent the pressure from exceeding the maximum allowable operating pressure plus 10%, or the pressure which produces a hoop stress of 75% of the specified minimum yield strength, whichever is the lower, or in a low pressure distribution system, a pressure which would cause the unsafe operation of any connected and properly adjusted gas burning equipment.

845.72 When more than one pressure regulating or compressor station feeds into a pipeline or distribution system and pressure relief devices are installed at such stations, the relieving capacity at the remote stations may be taken into account in sizing the relief devices at each station. However, in doing this the assumed remote relieving capacity must be limited to the capacity of the

---

* Such stacks of small diameter or containing numerous bends, if long, will seriously reduce relief vent capacity.

piping system to transmit gas to the remote location or to the capacity of the remote relief device, whichever is less.

845.8 Proof of Adequate Capacity and Satisfactory Performance of Pressure Limiting and Pressure Relief Devices.

845.81 Where the safety device consists of an additional regulator which is associated with or functions in combination with one or more regulators in a series arrangement to control or limit the pressure in a piping system, suitable checks shall be made to determine that the equipment will operate in a satisfactory manner to prevent any pressure in excess of the established maximum allowable operating pressure of the system should any one of the associated regulators malfunction or remain in the wide open position.

845.82 Suitable checks shall be made periodically to insure that the combined capacity of the relief devices on a piping system or facility is adequate to limit the gas pressure at all times to values prescribed by this Code. This check should be based on the operating conditions that create the maximum probable requirement for relief capacity in each case, even though such operating conditions actually occur infrequently and/or for only short periods of time.

REFERENCES

1. Public Service Electric and Gas Co. Gas Distribution Dept. *Construction Standards.* Newark, N. J., 1958 + (loose-leaf).
2. Evans, F. L., Jr. "Valves for Today's Hydrocarbon Processing Plant." *Hydrocarbon Process. Petrol. Refiner.* 40: 121–36, Oct. 1961.
3. Standard, J. H., Jr. "Use of the Plug Valve as a Control Valve." *A.G.A. Proc.* 1957: GSTS-57-2.
4. Van Deventer, F. M. "Technical Aspects of Valve Selection for Gas Transmission and Distribution Systems." (DMC-58-32) *A.G.A. Proc.* 1958:D243-8.
5. Natl. Fire Protection Assoc. *Standard for the Installation of Gas Appliances and Gas Piping.* (NFPA 54) Boston, Mass., 1959.
6. A.G.A. Laboratories. *American Standard Installation of Gas Piping and Gas Appliances in Buildings.* (ASA Z21.30-1959) Cleveland, Ohio, 1959.
7. ASME. *Gas Transmission and Distribution Piping Systems.* (ASA B31.8-1963) New York, 1963.
8. ASME. *Safety and Relief Valves.* (Power Test Code PTC 25-1958) New York, 1958.
9. Anuskiewicz, M., Jr. "Methods to Prevent Overpressure in Distribution Systems—A Task Group Report." (DMC-55-3) *A.G.A. Proc.* 1955: 381–416.
10. A.G.A. Operating Sect. *Specifications for Manually Operated, Non-Pressure Lubricated Iron Body, Brass Plug, Gas Stops One-Half Inch to Two-Inch* (X 50664). New York, 1964.

Chapter 5
Pressure* Regulation in Distribution

by R. J. Ott, J. M. Pickford, and F. E. Vandaveer

Pressure gages throughout the system monitor the gas supply. Pressure data are reported to a central office at regular intervals, by telephone, telemeter, distant recorders, or radio. Pressures are then increased or decreased as the demand varies in different parts of the system, via dispatching orders to gas manufacturing plants, storage fields, producing fields, pipeline companies, border regulator stations, and district regulator crews.

Distribution Piping. The *utility transmission main* (60–250 psig) picks up gas at the *city gate station*, where the gas is measured and reduced in pressure, and carries it to the distribution system *trunk main*. Gas then flows into *feeder mains* and from there thru *distributor mains* (6 in. w.c., 100 psig) and services.

Table 9-21 shows types of regulators used in distribution systems. Pressure conditions, as well as the methods of handling pressures, vary greatly within these systems. Any one method, as well as combinations of several methods, may be used. Regardless of the system, low pressure is required for residential and some commercial and industrial usage, while intermediate and high pressures are required by other commercial and industrial customers. The "order of magnitude" among the data in Tables 9-22 and 9-23 is representative of the types of regulators described. A specific regulator may have somewhat different specifications.

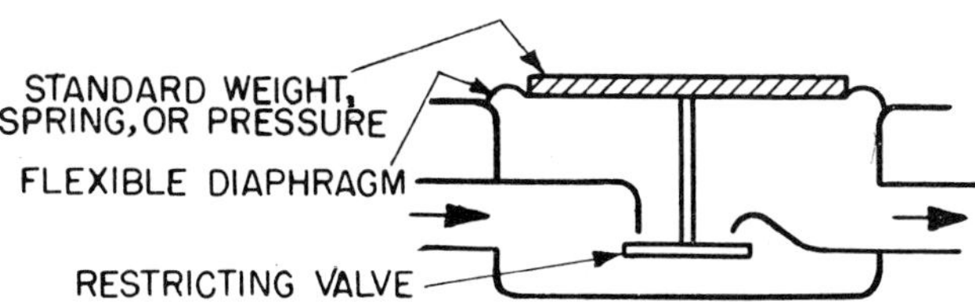

Fig. 9-38a The three essential parts of a pressure regulator.

PRINCIPLES OF GAS PRESSURE REGULATORS[1,2]

A gas pressure regulator automatically varies the rate of gas flow thru a pipeline to maintain a preset outlet pressure. The three essential parts of a regulator (Fig. 9-38a) are:

1. A sensing element (diaphragm, Bourdon tube, etc.) subject to the controlled pressure. Any pressure change on one side of this diaphragm will move the restricting valve.

2. A standard to which the controlled pressure is referred. This may be a dead weight, a spring, or a gas pressure acting on the diaphragm, its force opposing that exerted by the pressure to be controlled. The valve-opening force is called "loading."

3. A variable restriction in the flowing gas stream, positioned by the sensing element or diaphragm. This restriction may be a poppet valve, slide valve, piston, butterfly valve, plug valve, or similar device.

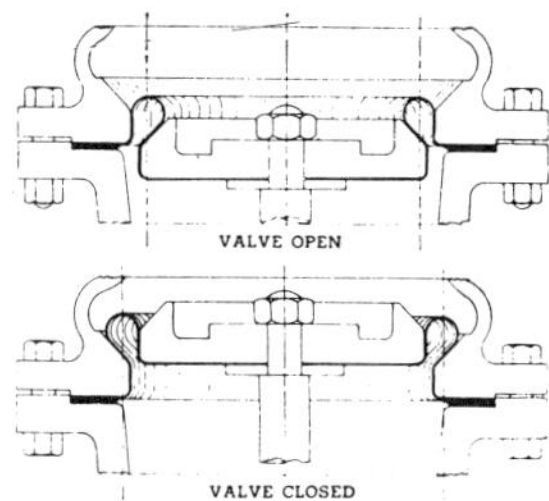

Fig. 9-38b The "roll-out" diaphragm (used in spring loaded regulators). The increase in effective area (outer center lines) as valve moves from the OPEN to the CLOSED position compensates for the increase in spring compressive force as the valve closes. (Chaplin-Fulton Mfg. Co.)

Ideally, a regulator would give a constant outlet pressure at all flows from zero up to the capacity of the valve. However, there are many factors which may prevent such ideal pressure characteristics. Such factors include: (1) variation of spring constant over the length of stroke of the valve; (2) inlet pressure variation; (3) minimum flow; (4) valve body design; and (5) diaphragm area change. Regulators and their systems include features that compensate (Fig. 9-38b) for deviations from ideal performance. For example, the "roll-out" diaphragm compensates for diaphragm area change.

Spring Loading

Figure 9-39a shows a self-operated, spring loaded regulator. A change in downstream pressure, transmitted thru the control line, allows the spring to move the diaphragm and connected valves. Service (house) regulators are mainly of the spring loaded type. Spring loaded regulators are available in sizes of one-quarter inch and larger with various restricting valve and leverage arrangements. These regulators are used for: (1) house service regulators—pounds-to-inches service; (2) farm tap regulators and field regulators—pounds-to-

* This chapter uses 14.4 psia as the pressure base because of the historical use of this level in regulator work. A.G.A. recommends 14.73, which is used elsewhere in this handbook.

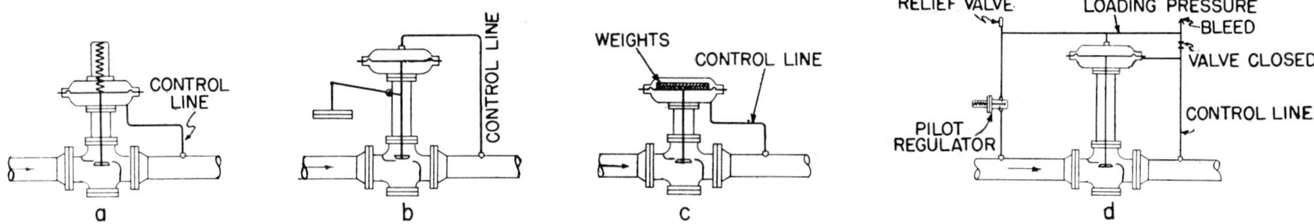

Fig. 9-39 Schematics of four types of pressure regulators: a, spring loaded; b, lever and weight loaded; c, dead weight loaded; and d, pressure loaded—self-operated. Internal control lines are used on small regulators, that is, most service regulators are self-contained. Double valves are available in 2 in. regulators, and balanced (double) reduced size inner valves for them are made as small as ¾ in.

Table 9-21 General Characteristics of Regulators*

Type of regulator	Connection sizes, in.	Pressure range, psig† Inlet	Outlet	Capacity range, 0.6 sp gr, cu ft per hr
Service	¾–2	0.5–125	3–15 in. w.c., 10 psig	0.5–8,000
"Industrial"	2	0.5–125	3 in. w.c., 50 psig	5,000–50,000
District,‡ low pressure, lb-to-in.	2–20§	0.5–100	2 in. w.c., 5 psig	1,000–5,000,000
Auxiliary pilot controlled	2–16§	0.5–125	2 in. w.c., 3 psig	1,000–5,000,000
Power pilot controlled	2–12§	0.5–125	3 in. w.c., 10 psig	1,000–5,000,000
District and city border station, high pressure, lb-to-lb				
Lever and weight	2–12	50–1200**	2–500	5,250–33,000,000
Spring	2–6	50–125	2–100	5,250–4,000,000
Roll diaphragm	2–6	50–675	2–75	132,000–23,000,000
Power pilot controlled	2–12††	50–1200	2–300	5,250–16,000,000
Instrument controlled	2–12	50–1200	2–1000	5,250–33,000,000
Piston operated	2–24‡‡	50–1440	2–1000	150,000–218,000,000
Field and farm taps	1–2	10–3000	5–400	490–21,440

* Conventional flat disk-type synthetic valves are widely used where lockup does not exceed 200 psi. Molded polyurethane valves (soft seated) are available where tight lockup to about 400 psi differential is needed. For still higher differentials, inner valves incorporating O-ring construction are available.

† Data given are representative of extreme ranges for the types listed; e.g.; cast iron bodies for low pressures and steel bodies for high pressures.

‡ Also industrial and commercial customers.

§ Larger sizes are available in both globe and butterfly designs.

** Lower value for cast iron, upper value for steel.

†† Piston operated valves with power pilot controllers are also available (1000 psig max inlet pressure and 100–450 psig outlet).[3]

‡‡ 30 in. available as special.

Note: Valve seats are made of numerous materials, depending on the valve's use, for example, steel for gas containing abrasive material, Teflon for soft seat and tight shutoff requirements. Manufacturers' information bulletins list all available materials.

pounds; (3) commercial and small industrial regulators—pounds-to-pounds, pounds-to-inches, and inches-to-inches; and (4) pilot loading regulators—pressures as required.

Weight Loading

Figures 9-39b and c show two basic forms of the weight loaded regulator. A change in downstream pressure transmitted thru the control line allows the weight to move the diaphragm and connected valves. The weight-and-lever system is common to reducing or back pressure (control of upstream pressure) regulators in which pressure cuts are not excessive and the variation in controlled pressure is allowable. Dead weight loading is sometimes used in large low-pressure regulators, resulting in an excellent outlet pressure. The diaphragm of the weight loaded regulator must be relatively heavy in order to resist the controlled pressure; thus, the thick diaphragm and the inertia of the weights are factors limiting its application. Generally, unit capacity is based on a 20 per cent drop (or droop) in outlet pressure.

Pressure Loading

Pressure loading consists of applying a constant pressure to a regulator diaphragm assembly to oppose or balance the downstream pressure applied to the opposite side of the diaphragm. Pressure loading is particularly adaptable to high outlet pressures in either self-operated or relay-operated regulators. The force exerted by the loading pressure may be opposed by downstream pressure, springs, weights, or combinations thereof.

A pressure loaded regulator with downstream pressure in opposition is shown in Fig. 9-39d. A small spring loaded regulator (called a pilot regulator when so used) imposes a controlled pressure on top of the main regulator diaphragm. The weight of the main regulator parts also acts in the same direction as the loading pressure. Therefore, the downstream pressure will be equal to the loading pressure plus the pressure equivalent of the weight of the regulator stem, valve, diaphragm, etc.

A bleed or vent to atmosphere is necessary to relieve excess pressure above the main diaphragm when downstream pressure requires the diaphragm to move upward.

AUXILIARY CONTROLS FOR IMPROVED PERFORMANCE AND RELIABILITY

Basically, regulators are either self- or relay-operated. In the former, the downstream pressure may act on one side

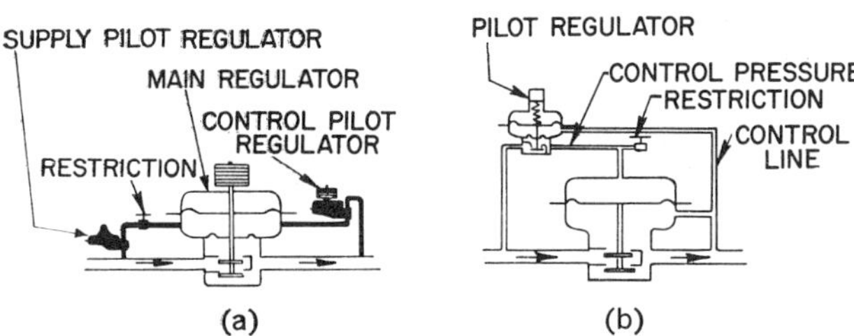

Fig. 9-40 (a) Auxiliary pilot control system on large regulators at low pressure for better shutoff and more positive positioning of valves on dirty gas. (b) Power pilot control system effected by controlling pilot loading regulator from main line outlet pressure.

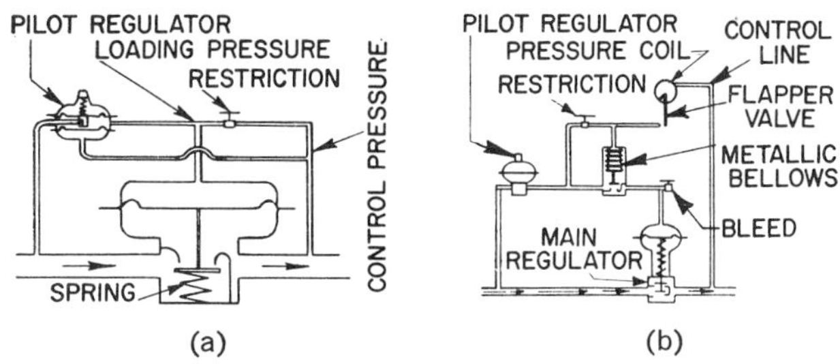

Fig. 9-41 (a) Power pilot control system using two diaphragms of equal size in the pilot regulator. (b) Instrument control for pressure loading using a pressure element to position a pilot valve.

of the diaphragm (Fig. 9-39a, b, and c). On the other hand, in auxiliary-controlled regulators, the downstream pressure is *either supplemented or amplified*, commonly by the upstream pressure. Thus, auxiliary-controlled regulators are used where restricting valves must be large and heavy, to ensure large flows and precise control and to overcome large frictional forces.

Auxiliary Pilot Control Systems

A *supply pilot regulator*, set for a pressure greater than the desired outlet pressure, feeds gas thru a *restriction* into the *main regulator* with sufficient pressure to close the main regulator—Fig. 9-40(a). However, if the controlled pressure is below the desired value, the main regulator will remain open, since the *control pilot regulator* will bleed off gas from under the main valve diaphragm faster than it can pass thru the *restriction*.

The controlled pressure may be as low as a few inches of water column. A change of a fraction of an inch of water column in the downstream pressure operates the control pilot. The pilot supply regulator is usually set at twice the output pressure setting of the controlled regulator. Thus, a small change in the downstream pressure causes a relatively large change in the pressure under the main diaphragm which positions the main restricting valve. With this multiplication of power the regulator has a very good lockup characteristic and ample power to overcome a sluggish condition due to dirt accumulation. However, this sensitivity means that the slightest upset in the downstream pressure system tends to start a cycle of hunting.

An auxiliary pilot control system is not very suitable for high-pressure control because the gas pressure is only on one side of the diaphragm. This situation requires a thick diaphragm, which detracts from the regulator sensitivity.

Power Pilot Control Systems

Pressure loading and pilot control can be combined by controlling the outlet pressure of the pilot loading regulator by means of the outlet pressure of the main line, as shown in Fig. 9-40(b). If the outlet pressure in the main line decreases, the pilot regulator opens up, increasing the pressure on the main diaphragm. This forces the main valves to open wider, increasing the main outlet pressure. This type of system supplies additional force for operating the regulator, since the loading pressure does not remain constant, but changes according to the demand for more or less pressure in the discharge line. A thin and flexible main diaphragm may be used, since very little differential pressure is needed across it.

Quite often this system of pressure loading is accomplished by using two diaphragms, usually of equal size, in the pilot regulator. The loading pressure is between the two diaphragms and is connected to the main regulator diaphragm case, with the control pressure acting beneath the lower pilot regulator diaphragm—Fig. 9-41(a).

When a variable pressure is used for loading a regulator controlled by the outlet pressure, the unbalanced forces across the main valves have no effect on the set outlet pressure. Therefore, with a pilot control system or a combination of pilot loading and pilot control, a single valve construction may be used. For high pressures, however, a double valve design is necessary in three-inch and larger sizes to avoid unbalanced forces.

Instrument Control for Pressure Loading Systems

Instrument control, which is almost identical to pilot control, but more flexible, employs a Bourdon tube rather than a diaphragm as a means of measuring downstream pressure. Instead of a diaphragm type of pilot regulator, the pressure element is used to position a pilot valve—Fig. 9-41(b). This pressure element is operated by the outlet pressure in the main line. As pressure decreases, the coil moves the flapper valve, shutting off the nozzle flow and increasing the pressure in the line. This increases the pressure in the bellows, which will move to close the valve in the supply line, thereby reducing the pressure on the control head of the main valve. This valve will then open to maintain downstream pressure.

Monitors[4,5]

Installations (Fig. 9-42a) consisting of two regulators (not necessarily of equal capacity) installed in series, with the control of point for both regulators downstream of both regulators, are called monitored regulator settings. The monitoring regulator is set slightly above (in some cases as little as 1.0 in w.c. over district pressure) the downstream pressure; thus, it remains open and does not come into operation except upon failure of the main regulator.

A pressure loaded regulator with an auxiliary spring or lever and weight control may also be used, instead of two regulators in series. The auxiliary spring or lever and weights are set for an outlet pressure which is slightly lower than the

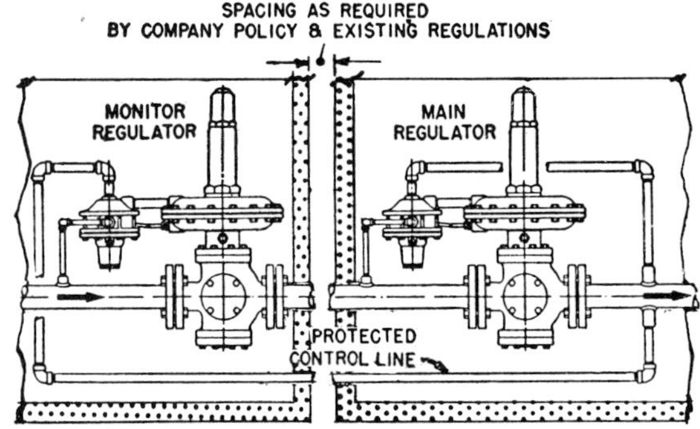

Fig. 9-42a Overpressure protection system with monitoring regulator. A distance of 20 to 25 ft between vaults is considered adequate for low-pressure systems.[5]

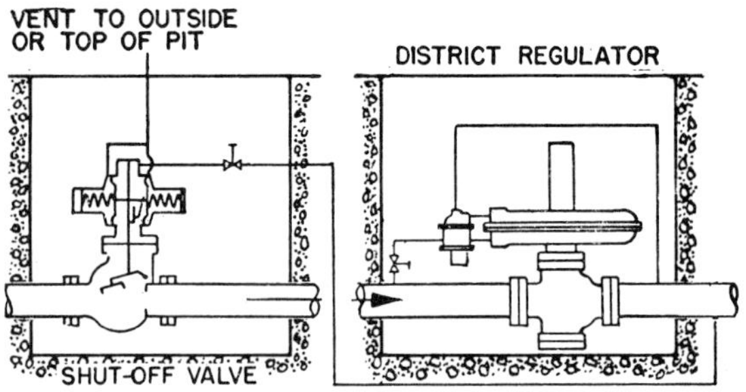

Fig. 9-42b Automatic shutoff valve system for complete gas shutoff on either increasing or decreasing line pressure.

desired outlet pressure, while the pressure loading system is set for the desired outlet pressure. The auxiliary controls remain inoperative until such time as the pressure loading system fails, then take over at the lower pressure. Another system (Fig. 9-42b) completely shuts off gas flow if the line pressure becomes either excessive or inadequate.*

The **advantages of upstream monitors**[6] are that (1) relatively higher flowing gas temperature minimizes freezing difficulties at the monitor; (2) greater gas density minimizes pressure drop in the monitor; (3) lower gas velocity results in less turbulence thru valve, which in turn results in quieter valve operation.

The **advantages of downstream monitors**[6] are that (1) primary unit protects monitor against foreign objects; (2) the turbulent gas stream helps to keep the monitor relatively free of foreign matter; (3) the pressure surge resulting from the abrupt failure of the primary unit would undoubtedly be less.

Maximum Capacity. Two equal-capacity units in series have a combined capacity of 70 per cent of *one* of them; a rule of thumb for finding the combined capacity of two units of unequal size is to take 70 per cent of the *mean*. It is necessary to take into account the pressure drops across both regulators while they are operating at maximum capacity. The following example illustrates the calculations involved.

Example: Find the maximum capacity of the monitored regulator setting for an inlet pressure of 700 psig (to first regulator) and an outlet pressure of 600 psig (from second regulator); note that both regulators are wide open at this maximum flow.

1. Assume that 0.4 (or 40 psi) of the total pressure drop occurs across the first regulator and 0.6 (or 60 psi) of the total pressure drop occurs across the second regulator.

2. Compute maximum flow for each regulator based on the above assumed pressure drops. First regulator: inlet pressure = 700 psig, outlet pressure = 660 psig; flow (from manufacturer's capacity tables) = 2600 MCFH. Second regulator: inlet pressure = 660 psig, outlet pressure = 600 psig; flow (from manufacturer's capacity tables) = 3000 MCFH.

3. If flows computed in step 2 are not equal, repeat steps 1 and 2 for different assumed pressure drops until the computed flows are approximately equal.

Boosters

Boosters maintain a constant pressure A at a remote point by creating a variable pressure B at the point of regulation so that $B = A + C$, where C equals the pressure drop between points A and B. This type of operation may be desirable in distribution systems in which system leakage (and frequently pumping cost) can be materially reduced by operating at the lowest possible pressures. Various devices may be used to operate booster regulators. These may be actuated by time, ambient temperature, pressure at a remote point, or existing flow conditions.

TYPES OF REGULATORS

Table 9-21 lists the operating ranges and capacities of various types of distribution regulators.

Service Regulators

Service regulators are installed primarily on individual house lines or lines to small commercial establishments. When there are freezing temperatures and a high degree of moisture in the gas, indoor installation may be required. A service regulator at each house permits main pressure variations without noticeable variation in gas pressure at appliances.

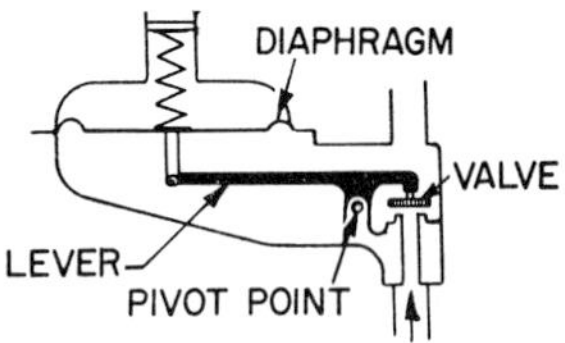

Fig. 9-43 Large diaphragm and leverage system as used on a service regulator. In many recent designs the lever arms are at right angles to each other.

An example of a service regulator is shown in Fig. 9-43. Other common designs differ mainly in the lever arrangement. Available features include: internal or external relief valves; boosters or ejectors to produce outlet pressure increasing proportionately with flow; vertical positioning of diaphragm to permit drainage of condensate from vent; lockup pressures as low as 0.5 in. w.c.; automatic gas cutoff on low and high pressures; and access for repair and maintenance.

Standards[7] were prepared for two-inch and smaller pounds-to-inches types of spring loaded regulators thru the combined

---

* See Par. 845.32(e) of ASA B31.8-1963 (Chap. 2, Sec. 9).

Table 9-22 Capacity of Service Regulators of Type Similar* to That Shown in Fig. 9-43
(in cu ft per hr of 0.60 sp gr gas at 4 oz above 14.4 psi atmospheric pressure—connections: 3/4, 1, 1 1/4 in.)

Orifice diam, in.	Outlet press. range, in. w.c.	Inlet pressure, psig 2	5	10	20	50
3/8	3–14†	500	900	1000	...	...
3/8	3–14‡	...	600	1000	1200	...
3/16	3–14	...	...	650†	700‡	1200‡
1/2	3–6.75	900	1100	1200	1200	...
1/2	3–14	900	1100	1200	1200	...
3/16	3–6.75	...	...	...	1000	1200
3/16	5.5–8.5	...	...	...	1100	1250
3/16	3–14	...	...	...	900	1200
3/16	6–28	...	...	...	800	1100
3/8	3–14	700	1000	1100	1200	...
3/8	6–28	300	600	900	1200	...
3/8	3–14	850	1100	1200	...	...

* Short lever and an additional fulcrum are used in lower group, in lieu of long lever.
† High boost.
‡ Low boost.

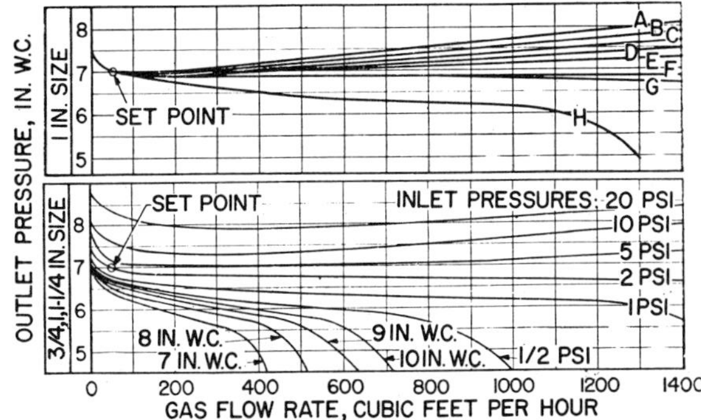

Fig. 9-44 Performance curves for typical booster-type service regulators (0.6 sp gr gas, base: 14.7 psia, 60 F).

Curve	A	B	C	D	E	F	G	H
Inlet pressure, psig	150	35	20	75	5	10	2	1
Orifice diam, in.	0.187	0.250	0.375	0.187	0.500	0.437	0.500	0.500

efforts of regulator manufacturers and gas utilities. Objectives: specifications for materials, strengths, operating characteristics, and testing *by manufacturer*.

Materials and Strengths. Valve bodies: cast iron, 125 psig nom. (175 psig max at 150 F); aluminum and other materials for which standards are available, 125 psig nom. Melting point of body materials, 1000 F min (of diaphragm case material, 700 F). Outlet pressure of 2 psig withstood without leakage or permanent deformation.

Features and Operating Characteristics. Service regulators are self-contained (no external static or control lines). Relief, overpressure, and underpressure features are optional; internal relief, where specified, is from 7 in. w.c. to 1 psig *above outlet setting*. Body connections are made of 3/4, 1 1/4, 1, 1 1/2, and 2 in. ASA female pipe thread. Testing provisions intended for use by the manufacturer are also included in Ref. 7.

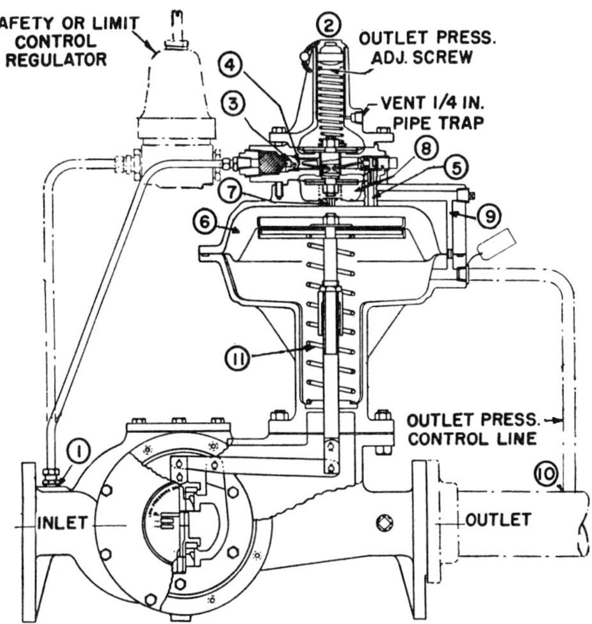

Fig. 9-45 Power pilot loading of regulator. (Rockwell Mfg. Co.) See Tables 9-22 and 9-23 for capacities and constants, respectively. A tap (1) at the upstream side of the main regulator allows a small quantity of gas to flow into the pilot regulator (2). The pressure of this gas is controlled by the valve orifice (3) and valve (4). From this point the gas flows thru channel (5) to the main loading chamber (6). The loading pressure in chamber (6) is discharged thru a fixed orifice (7) into the control chamber (8), then thru the control line (9) to the outlet piping on the main regulator (10). For the loading pressure to be discharged into the main outlet line, the pressure in the loading chamber (6) must be higher than the pressure in the main regulator outlet piping. To accomplish this, a spring (11), which is initially loaded to overcome the weight of the moving parts and to effect a lockup of the main balance valves, is placed under the main diaphragm.

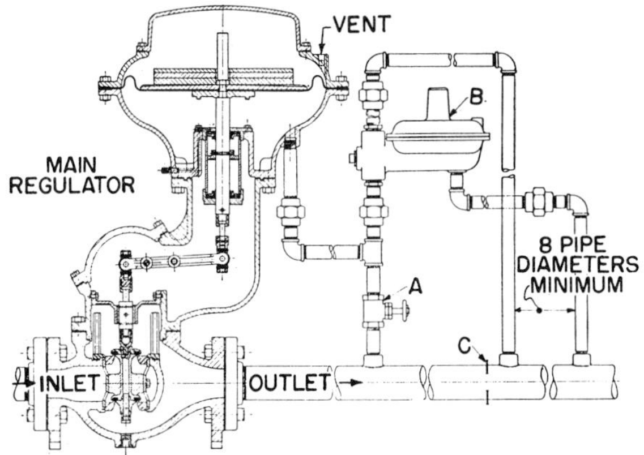

Fig. 9-46 Pressure boosting installation—low-pressure type. (Reliance Regulator Co.)

Capacity. A service regulator must be able to regulate the pressure to one appliance pilot (approximately 0.2 cu ft per hr—mini-pilots use as little as 0.05 cu ft per hr) as well as to all gas appliance burners running simultaneously (possibly 250 cu ft per hr in a residence; more in commercial and industrial installations). Regulator outlet pressures should be suitable to the appliances served. If, for example, a nominal 7

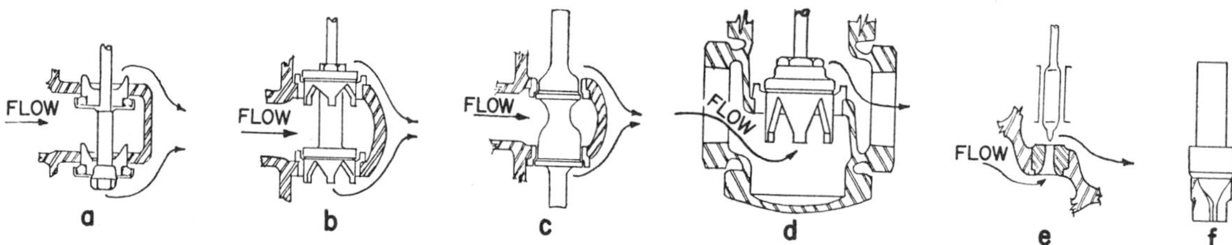

Fig. 9-47 Various types of inner valves for regulators: a, disk type; b, skirt guided V-port; c, throttle plug; d, single V-port; e, microflute pup with globe valve seat; f, microflute pup. Valve seat(s) and body are shown shaded; the inner valve(s), unshaded.

in. w.c. pressure is desired at the appliances, a pressure tolerance of ±0.7 in. w.c. would result in a Btu delivery within five per cent of the nameplate rating. A capacity table and performance curves for typical service regulators from one manufacturer are given in Table 9-22 and Fig. 9-44.

District Regulators, Pounds-to-Inches

Figures 9-45 and 9-46 illustrate some widely used regulators for district and city gate stations where pressure reduction is from pounds to inches.

Provisions for changing valve size easily if load conditions change, for renewing valve seats, and for adjusting balanced valves are important. Valve shape may be varied to obtain desired operating characteristics; commonly used shapes are illustrated in Fig. 9-47. The quick opening valve gives maximum capacity. The other two types give greater range of flow (rangeability) at any specified pressure condition.

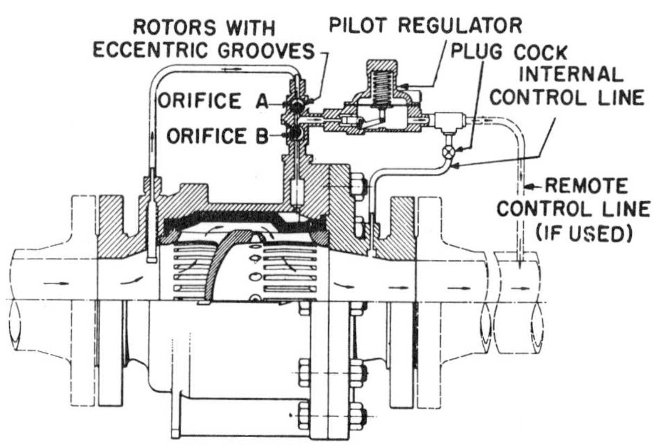

Fig. 9-48a Regulator using an expansible tube as both diaphragm and valve; pounds-to-pounds or pounds-to-inches. (Grove Valve and Regulator Co.) Maximum inlet pressure is 230 psig (up to 1500 psig on some models); outlet pressure range, ounces to 120 psi and up; sizes, 1 in. to 12 in.; differential pressure required, 10 psi minimum; capacity range, 13 to 10,500 MCFH—0.6 sp gr gas.

The stroke or lift of the valve in a regulator can be a major factor in control accuracy. For most simple valve designs, this stroke should not be any greater than one-fourth of the valve diameter.[8] Some increase in stroke can be allowed for V-port, skirted, or other special valves.

A pressure regulator that uses an expansible tube (known as an "unloading pilot") as both diaphragm and valve is shown in Fig. 9-48a. The device closes when inlet line pressure is applied to the "jacket" space around the outside of the expansible tube. It opens to throttle if gas is bled from the jacket space to reduce this pressure. Auxiliary control or self-operation is possible. The regulator, depending on the auxiliary system used, may be a reducing valve, a back pressure valve, a diaphragm motor valve, or a constant flow valve. The self-operated regulator can be employed only as a back pressure regulator. A somewhat similar device uses a metal cylinder slide valve (Fig. 9-48b) in place of the expansible tube.

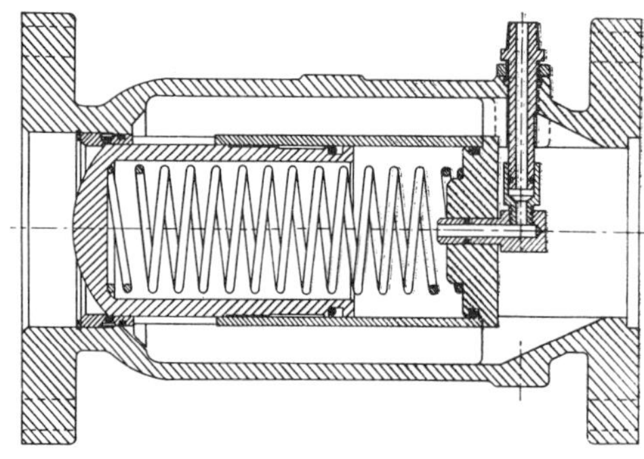

Fig. 9-48b Slide valve-type regulator. The projection connects to a pilot regulator. (Rockwell Mfg. Co.)

This type of pilot operation is particularly suited for tight shutoff service and *pounds-to-inches* reduction, using a suitable low-pressure regulator as a pilot. It is also used for *pounds-to-pounds* reduction, with a small reducing regulator of proper pressure range as a pilot. A similar type of regulator uses an expansive element in conjunction with a grid.[3]

An example of a *pressure differential-operated* boosting arrangement in a low pressure system is shown in Fig. 9-46. The main regulator is loaded for the average required downstream pressure (within the range of 4 to 14 in. w.c.). The setting in pilot regulator *B* determines the *maximum* pressure increase (boosting) in the main regulator. The orifice at *C* is used to indicate the pressure differential. There is an adjustable restriction at *A* which is set for a flow less than the capacity at *B*.

Capacity. Manufacturers' *K* factors for the illustrated typical low-pressure district regulators are shown in Table 9-23.

District and City Gate Station Regulators, Pounds-to-Pounds

General characteristics of high-pressure *pounds-to-pounds* regulators suitable for city gate stations or transmission lines (also large industrial customers) are shown in Table 9-21. Many low-pressure regulators can be adapted by appropriate construction changes to *pounds-to-pounds* control.

Table 9-23 Examples of Constants for Various Types of Regulators*

Type	See Fig.	Use Eq:										
Balanced valve	9–45	**1 or 2**	Regulator size:	⅞ in.†	1 in.†	2 in.	3 in.	4 in.	6 in.	8 in.	10 in.	12 in.
			K factor:	540	990	2600	6000	10,000	23,000	39,000	65,000	91,000
	. . .	**1a, 1b,**	K_l factor:	2200	4000	10,000	23,000	39,000	89,000	150,000	250,000	350,000
		or **2a**	(for valve size)	(⅞ in.)	(1 in.)	(1½ in.)	(2¼ in.)	(2⅞ in.)	(4¼ in.)	(5½ in.)	(7¼ in.)	(8½ in.)
Pressure boosting	9-46	**1 or 2**	Regulator size:	2 in.	3 in.	3 in.	4 in.	6 in.	8 in.	8 in.	10 in.	10 in.
			(valve size)	(1¼ in.)	(1⅝ in.)	(1⅞ in.)	(2⅞ in.)	(4¼ in.)	(5 in.)	(5¾ in.)	(6 in.)	(7¼ in.)
			K factor:	2000	4100	5900	15,000	33,000	47,000	63,000	68,000	100,000
Lever and weight loaded	9-39 (b)	**1 or 2**	Regulator size:	2 in.	3 in.	4 in.	6 in.	8 in.	10 in.	12 in.		
			(valve size)	(1¾ in.)	(2⅛ in.)	(3 in.)	(4¼ in.)	(5¾ in.)	(7⅛ in.)	(8½ in.)		
			K factor:	4800	7100	14,000	28,000	60,000	92,000	130,000		
Farm tap	. . .	**1 or 2**	(valve size)‡	(⅛ in.)	(¼ in.)	(⅜ in.)	(½ in.)					
			K factor:	30	110	210	360					
High-pressure balanced valve, lever and weight, power pilot	. . .	**1 or 2**	Regulator size:	¾ in.†	1 in.†	2 in.	3 in.	4 in.	6 in.	8 in.	10 in.	12 in.
			(valve size)	(¾ in.)	(1 in.)	(1½ in.)	(2 in.)	(2⅜ in.)	(3½ in.)	(5 in.)	(5¾ in.)	(7 in.)
			K factor:	430	1400	2900	5400	9300	18,000	31,000	45,000	66,000

* In installations in which the regulator is equipped with the inner valve of a smaller size regulator, the constants capacity will be that of the inner valve. Thus, in an 8-in. regulator in which the inner valve of a 4-in. regulator is installed, the capacity will be that of a 4-in. regulator.

† Inner valve size for 2-in. regulator bowl.

‡ All in 1-in. regulators.

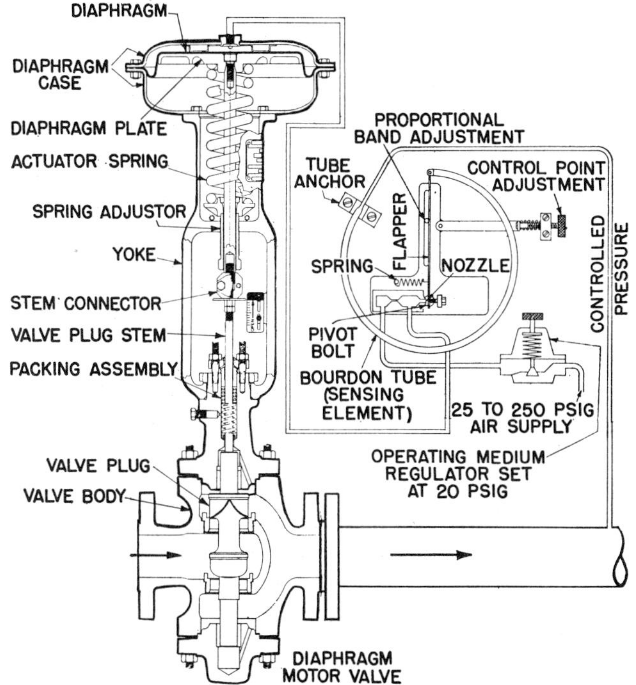

Fig. 9-49a Diaphragm motor valve operated by a relay pilot instrument. (Valve portion from Fisher Governor Co.; instrument portion from Ref. 10.)

Capacity. Data for typical regulators for high-pressure *pounds-to-pounds* operation are shown in Table 9-23.

Diaphragm Motor Valves. These throttle, open, or close by taking positioning energy indirectly from a source *other than* the controlled pressure (Fig. 9-49a). This pressure is controlled by a primary pilot valve which responds to changes in the controlled system.

Cylinder Actuated Instrument Control (Positioner and Controller) Lubricated Plug-Type Valves. Figure 9-49b shows one automatic system.

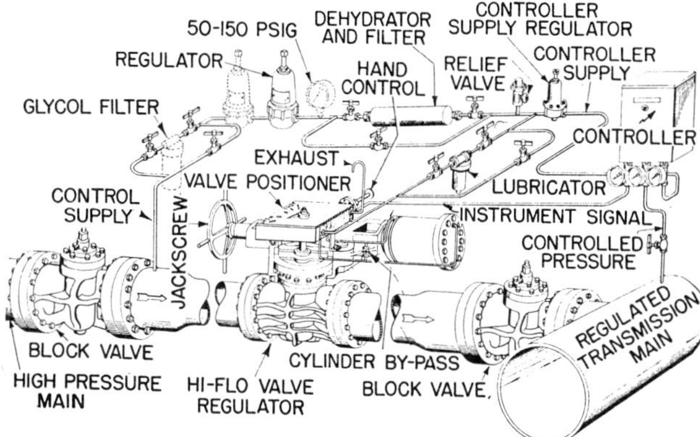

Fig. 9-49b Plug-type valve, cam positioned. (Rockwell Mfg. Co.)

Boosting. Use of boosting devices on regulators increases the possibility of "hunting."

Field and Farm Tap Regulators

These regulators (Fig. 9-50a) are designed to serve individual farms or small industrial customers directly from high-pressures lines. They must operate at high inlet pressure and reduce to pressures at which downstream regulators of both the *pounds-to-pounds* and *pounds-to-inches* classes may func-

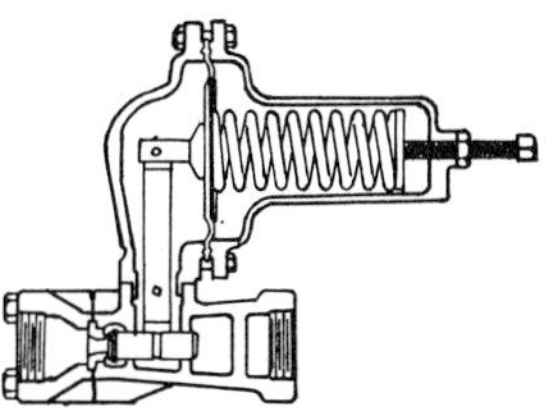

Fig. 9-50a High-pressure field regulator. (Fisher Governor Co.) Inlet pressure = 1500 psig max; outlet pressure = 3 to 275 psig; high-tensile iron body with 1 or 2 in. screwed connections.

tion satisfactorily. Figure 9-50b and Table 9-23 give capacities and K factors, respectively, of typical field and farm tap regulators.

CAPACITY CALCULATIONS

Table 9-23 contains K factors for various types of regulators. Basically, the capacity or flow rate thru a regulator is a function of the pressure drop as well as the pressure ratio across this device. When the ratio of outlet-to-inlet pressure, absolute values, exceeds 0.55:*

$$Q = K(\Delta P \times P_2)^{0.5} \quad (1)$$

$$= K_l(\Delta P)^{0.5} \quad (1a)$$

$$= 0.19K_l(\Delta P_w)^{0.5} \quad (1b)$$

where:

Q = capacity at outlet pressure, cfh
K = regulator constants
ΔP = pressure drop across valve, $P_1 - P_2$, psi
P_1 = inlet pressure, psia min
P_2 = outlet pressure, psia max
K_l = regulator constants at an outlet pressure of 4 oz above 14.4 psi atmospheric pressure; see Table 9-23
ΔP_w = pressure drop in regulator, in. w.c.

When the ratio of outlet-to-inlet pressure is less than 0.55, Eq. 1 becomes:

$$Q = 0.5KP_1 \quad (2)$$

In a similar manner, Eq. 1a becomes:

$$Q = 0.13\,KP_1 = K_cP_1 \quad (2a)$$

where K_c = constant for the particular valve for flow above critical velocity.†

Example: Find the K for a lever and weight loaded regulator to pass 92.4 MCFH when the inlet and outlet pressures are 25 and 10 psig, respectively, and the pressure base is 0.25 psi above 14.4 psi atmospheric pressure:

1. Outlet-to-inlet pressure ratio $= (10 + 14.4)/(25 + 14.4) = 0.62$ (which is greater than 0.55; therefore, Eq. 1 is applicable).
2. $K = Q/(\Delta P \times P_2)^{0.5} = 92{,}400/[(25 - 10) \times (10 + 14.4 + 0.25)]^{0.5} = 4810$.

Correction Factors for Specific Gravity and Temperature

Most capacity tables are based on a 0.60 specific gravity gas. For a gas of another specific gravity, G, *multiply* capacity by $(0.60/G)^{0.5}$ or *divide* by factors in Table 7-9.

A correction factor for *temperature* should be applied if there is a wide departure from 60 F—the basis of most capacity tables. For the majority of installations this factor is small enough to be neglected. The flow thru a regulator varies *inversely* as the square root of the gas density; in turn, the density varies *inversely* as the first power of the absolute temperature. Therefore, for gas at a temperature, t, multiply the capacity by the correction factor $(460 + t)/520^{0.5}$ or *divide* by the factors in Table 14 extracted in Chap. 2, Sec. 7.

* $0.55 = (2/k + 1)^{k/(k-1)}$ where $k = c_p/c_v = 1.3$ (Ref. 9).

† This relates to the acoustic velocity (speed of sound), which determines flow characteristics of orifices, valves, etc.

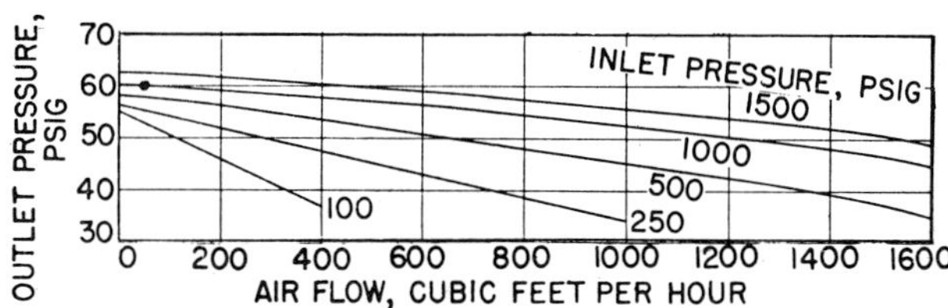

Fig. 9-50b Capacities of typical field and farm tap regulators. Initial set point: 1000 psig inlet and 60 psig outlet; 50 cu ft per hr flow.

FACTORS IN REGULATOR SELECTION

Flow Rate. Regulator should be sized for both minimum flow rate at maximum inlet pressure and maximum flow rate at minimum inlet pressure, respectively. Regulators may be designed in terms of allowable fall-off in the outlet pressure over the working range. Data on this are available for balanced valve units.[8] An oversized regulator which operates with its inner valves in an almost closed position at minimum flows will often chatter, pulsate, or otherwise erratically control flow. Ideally, a regulator should not have to operate at flows that are less than 10 per cent of its maximum capacity. Actually, service regulators must operate down to almost zero flow and lock up tight. Continuous operation at low flows, however, may result in valve seat cutting, and with sensitive regulators, there is a greater chance of vibration.

Type of Gas. With most natural gases it is desirable to use synthetic rubber material for soft seat valves. Leather may also be used. Special diaphragm treatment or synthetic diaphragms must be used with LP-gases, since these gases displace the oil used to treat leather.

Inlet and Outlet Pressures. Inlet pressures affect the selection of the regulator type and size. The equipment must withstand the pressure and minimize any abrasive action due to high gas velocity. The desired range of outlet pressure determines the size of diaphragm case in self-operated spring-type and weight-type regulators using flat diaphragms. The larger the diaphragm the more responsive it will be to pressure changes. A single size diaphragm case may be used for pilot loading regulators.

The pressure drop across a regulator will generally dictate whether series regulation is necessary. For inlet pressures up to 1200 psig, not more than two stages may be needed unless reduction is to ounces.[10] The second stage regulator should be capable of taking full line pressure in case the first stage fails.

Accuracy of Control. On large regulators many types of loading diaphragms are available. The simplest and most rugged design giving the necessary accuracy should be used. For example, the weight loaded high-pressure balanced valve regulator is a simple type and should be used where the flow is fairly constant or the outlet pressure can vary about 5 to 10 per cent. If better control is needed, the power pilot regulator can be used to load this regulator and obtain constant outlet pressure over a wide range of flow. The **proportional band** is the range thru which the pressure being controlled must vary to cause the regulator to move from the closed to open (or open to closed) position.[10]

It may be expressed in percentage of the scale (in Fig. 9-49, for example, scale is the rating of the Bourdon tube). In general, the wider the proportional band of a regulator, the more stable it is. When the proportional band is made wider in an instrument-controlled regulator, regulator "hunting" where metering is done can be alleviated by a reset feature used to correct the pressure set-point (instrument-type regulators) and thus facilitate pressure stability thru the orifice meter.

Type of Valve Seat. Shutoff and lockup are functions of pressure differential. When a complete shutoff of flow is required, a soft seat valve should be used. On high pressures, hard seats are a more practical choice since they can resist the erosion action of high-speed gas stream particles better.

Sizes of Connections. Regulators may be furnished with standard threaded or flanged (including ring joint) connections, or weld ends. The size of the regulator is generally given as the nominal pipe size of its connection; i.e., a two-inch valve comes with two-inch standard pipe conections.

INSTALLATION OF REGULATORS

Regulators should be accessible for maintenance. Simple piping connections minimize pressure loss. Whenever possible, gas lines should be cleaned before the regulator is installed, since line dirt may cut the valve seat and prevent valve closure. Where continual dirt troubles are encountered, a dust scrubber or filter should be installed upstream of the regulator. Various arrangements for meter and regulator stations are available.[10]

In locations at which continuous gas service must be maintained, a regulator by-pass or dual regulation in parallel should be installed for use during inspection or repair.

Large seasonal flow variations may be accommodated by two regulators (of fail-open type) in parallel rather than one large regulator. When the load is small, as in summer, only one regulator would be in use. During peak loads both regulators may be in use, but set for slightly different outlet pressures, so that they will not buck each other. The control line should be run from the lower diaphragm case of the regulator to a point of smooth flow in the main outlet line.

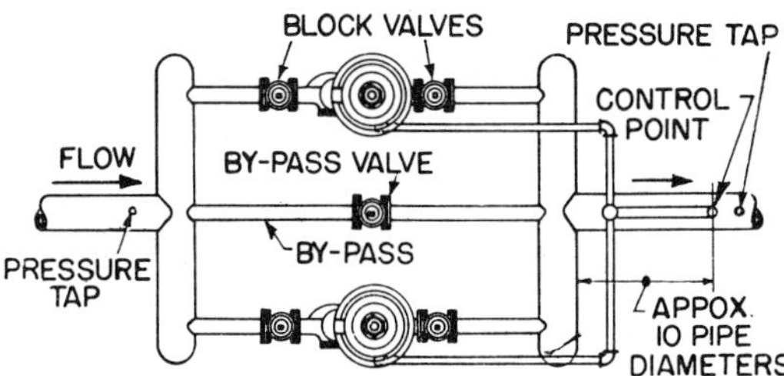

Fig. 9-51 Parallel regulator installation.

As an alternative, seasonal load variations may be accommodated by changing the inner valve and seat sizes, especially in pilot regulators. Thus, one regulator can cover the entire capacity requirement independently, while the paralleling unit is available as a by-pass.

Figure 9-51 is a typical installation of parallel regulators. When the **pilot control** system is used, enough valves should be installed to allow the control regulators to be shut off and the regulator to be operated by direct loading. Note the pressure gage taps on Fig. 9-51.

To avoid regulator freeze-up, gas may be dehydrated before it enters the system to a point at which internal freezing or hydrate formation will not occur. If exterior icing inhibits free movement of parts, heat may be added, usually by steam or water heat exchangers upstream of the regulator. Bleed or vent holes should be watched for formation of ice which would prevent free passage of gas.

Two or more regulators may be installed in series to ensure safer operation and minimize trouble from freezing.* It is common practice to use one regulator when the maximum inlet pressure does not exceed 60 psig and two regulators for inlet pressures between 60 and 200 psig.†

Initiate flow thru regulators by first opening the downstream block valve and then slowly opening the upstream block valve. These valves and regulators should be separated by a *minimum* of six pipe diameters.[10]

Pressure Control Line.[10] This line runs from the outlet side of the regulator to the sensing element of the regulator installation. Installation requirements include: (1) sloping line into straight run of downstream pipe; (2) connecting to top (or side) of piping—never to the bottom; (3) overcoming turbulent flow at connection by means of straightening vanes or a lock shield needle valve; (4) sizing line—¼ or ⅜ in. for 8 to 15 ft runs and ½ in. for 20 to 30 ft runs. A break in a downstream control line will open the regulator. With a back pressure regulator (upstream control line), a broken line will close the valve.

Location of Stations

A survey[11] of 80 companies in 35 states revealed the following distribution of regulator station sites: parkway areas, 40 per cent; streets, 24 per cent; sidewalk space, 18 per cent; private property, 18 per cent. Over 80 per cent of the stations were installed *below* ground. The majority of those above-ground were located at company plants.

Reference may be made to the following extract from ASA B31.8-1963:

848.1 Location for Customers' Meter and Regulator Installations.

(a) Customers' meters and regulators may be located either inside or outside of buildings, depending upon local conditions, except that on services requiring series regulation, in accordance with 845.53 (a),‡ the upstream regulator shall be located outside of the building.

(b) When installed within a building, the service regulator shall be in a readily accessible location near the point of gas service entrance and, whenever practical, the meters shall be installed at the same location. Meters shall not be installed in bedrooms, closets, bathrooms, under combustible stairways or in unventilated or inaccessible places, nor closer than three feet to sources of ignition, including furnaces and water heaters. On services supplying large industrial customers or installations where gas is utilized at higher than standard service pressure, the regulators may be installed at other readily accessible locations.

(c) When located outside of buildings, meters and service regulators shall be installed in readily accessible locations where they will be reasonably protected from damage.

* When sufficient heat transfer to gas is possible between regulators.

† See ASA B31.8-1963, Par. 845.53 (extracted in Chap. 2, Sec. 9).

‡ [Extracted in Chap. 2, Sec. 9.—EDITOR]

(d) Regulators requiring vents for their proper and effective operation shall be vented to the outside atmosphere in accordance with the provisions of 848.33.

848.31 Meters and service regulators shall not be installed where rapid deterioration from corrosion or other causes is likely to occur.

848.33 All service regulator vents, and relief vents where required, shall terminate in the outside air in rain and insect resistant fittings. The open end of the vent shall be located where, if a regulator failure resulting in the release of gas occurs, the gas can escape freely into the atmosphere and away from any openings into the buildings. At locations where service regulators might be submerged during floods, either a special antiflood type breather vent fitting shall be installed, or the vent line shall be extended above the height of the expected flood waters.

848.34 Pits and vaults, housing customers' meters and regulators, shall be designed to support vehicular traffic when installed in the following locations:
(a) Travelled portions of alleys, streets and highways.
(b) Driveways.

848.4 Installation of Meters and Regulators.
All meters and regulators shall be installed in such a manner as to prevent undue stresses upon the connecting piping and/or the meter. Lead connections, or other connections made of material which can be easily damaged, shall not be used. The use of standard weight close nipples is prohibited.

847 VAULTS.

847.1 Structural Design Requirements.
Underground vaults or pits for valves, pressure relieving, pressure limiting or pressure regulating stations, etc., shall be designed and constructed in accordance with the following provisions:

(a) Vaults and pits shall be designed and constructed in accordance with good structural engineering practice to meet the loads which may be imposed upon them.

(b) Sufficient working space shall be provided so that all of the equipment required in the vault can be properly installed, operated and maintained.

(c) In the design of vaults and pits for pressure limiting, pressure relieving and pressure regulating equipment, consideration shall be given to the protection of the equipment installed from damage, such as that resulting from an explosion within the vault or pit, which may cause portions of the roof or cover to fall into the vault.

(d) Pipe entering, and within, regulator vaults or pits shall be steel for sizes 10 inches and less except that control and gage piping may be copper. Where piping extends through the vault or pit structure, provision shall be made to prevent the passage of gases or liquids through the opening and to avert strains in the piping. Equipment and piping shall be suitably sustained by metal, masonry, or concrete supports. The control piping shall be placed and supported in the vault or pit so that its exposure to injury or damage is reduced to a minimum.

(e) Vault or pit openings shall be located so as to minimize the hazards of tools or other objects falling upon the regulator, piping, or other equipment. The control piping and the operating parts of the equipment installed shall not be located under a vault or pit opening where workmen can step on them when entering or leaving the vault or pit, unless such parts are suitably protected.

(f) Whenever a vault or pit opening is to be located above equipment which could be damaged by a falling cover, a circular cover should be installed or other suitable precautions taken.

847.2 Accessibility
Consideration shall be given, in selecting a site for a vault, to its accessibility. Some of the important factors to consider in selecting the location of a vault are as follows:

(a) *Exposure to traffic.* The location of vaults in street intersections or at points where traffic is heavy or dense should be avoided.

(b) *Exposure to flooding.* Vaults should not be located at points of minimum elevation, near catch basins, or where the access cover will be in the course of surface waters.

(c) *Exposure to adjacent subsurface hazards.* Vaults should be located as far as is practical from water, electric, steam, or other facilities.

847.3 Vault Ventilation.
Underground vaults and closed top pits composing either a pressure regulating or reducing station, or a pressure limiting or relieving station, shall be ventilated as follows:

(a) When the internal volume exceeds 200 cubic feet, such vaults or pits shall be ventilated with two ducts each having at least the ventilating effect of a pipe 4 inches in diameter.

(b) The ventilation provided shall be sufficient to minimize the possible formation of a combustible atmosphere in the vault or pit.

(c) The ducts shall extend to a height above grade adequate to disperse any gas–air mixtures that might be discharged. The outside end of the ducts shall be equipped with a suitable weatherproof fitting or vent-head designed to prevent foreign matter from entering or obstructing the duct. The effective area of the openings in such fittings or vent-heads shall be at least equal to the cross-sectional area of a 4-inch duct. The horizontal section of the ducts shall be as short as practical and shall be pitched to prevent the accumulation of liquids in the line. The number of bends and offsets shall be reduced to a minimum and provisions shall be incorporated to facilitate the periodic cleaning of the ducts.

(d) Such vaults or pits having an internal volume between 75 cubic feet and 200 cubic feet may be either tightly closed or ventilated. If not ventilated, all openings shall be equipped with tight fitting covers without open holes through which an explosive mixture might be ignited. Means shall be provided for testing the internal atmosphere before removing the cover.

(e) If vaults or pits referred to in (d) above are ventilated by means of openings in the covers or gratings and the ratio of the internal volume, in cubic feet, to the effective ventilating area of the cover or grating, in square feet, is less than 20 to 1, no additional ventilation is required.

(f) Such vaults or pits having an internal volume less than 75 cubic feet may be ventilated or not at the option of the operating company.

847.4 Drainage and Waterproofing.

(a) Provisions shall be made to minimize the entrance of water into vaults, and vault equipment shall always be designed to operate safely, if submerged.

(b) No vault containing gas piping shall be connected by means of a drain connection to any other substructure, such as a sewer.

(c) Electrical equipment in vaults shall conform to the requirements of Class 1, Group D, of the National Electrical Code, ASA C1.

Service Regulators

Service regulators designed for use in vertical pipelines must be so installed; units for horizontal installation are also available. If gas is supplied to more than one meter, as in an apartment house, the header between the regulator and meters should be sized large enough to avoid the possibility of pulsation.

When the service regulator is installed outside the building, a suitable housing may be used or special vent caps may be supplied to prevent closure of vents due to freezing, other stoppage, or loading due to rain water accumulating on the upper diaphragm.

District Regulators, Low Pressure, Pounds-to-Inches, Balanced Valve

Installation should be in a regulator vault or house, properly ventilated and free from possible flooding conditions. If the regulator must be placed in an underground vault, where possibility of flooding exists, controls should be in a watertight enclosure and the diaphragm vent piped above any possible water line.

Noise. Regulator noise, particularly at high flow rates, may be very objectionable to nearby residents and to workmen at the station. Methods for reducing noise at regulator stations include:[12,13]

1. Cover piping and regulator body with at least a four-inch layer of sand held in place by a box or an asphalt outer layer.

2. House the regulator and as much of the adjoining piping as practical in a concrete or brick house. Omission of windows or double glass windows with 1 in. air space further reduces noise transmission. Solid wood doors or double doors with an 8 in. dead space reduce noise further. In extreme cases, a building with double walls, on separate foundations, and lined with acoustical insulation (total wall thickness of 22 in.) has been used.

3. Cover station piping with two inches of acoustical Fiberglas faced on the outside with aluminum foil. Below the regulator, add a double wrap of Fiberglas. Box in outlet header with ¾ in. plywood to allow for the addition of a 1 in. layer of sound-absorbing board inside and of a 6 in. layer of sand at bottom. Fill space around the pipe with glass wool. Insulate balance with alternate layers of deadening felt and hair felt. Cover straight runs with layers of 45 lb roofing felt and 26 gage aluminum.

Substitution of a different type of inner valve in a regulator sometimes reduces noise level. Noise from district regulators transmitted to house meters may be reduced by using iron case meters in place of tin case meters. Vibration is another consideration.

City Gate Station Regulators, High Pressure, Pounds-to-Pounds, Balanced Valve

In general, information on low-pressure balanced valve regulators also applies to high-pressure regulators.

It is best to have regulators some distance apart when making two pressure cuts, in order to minimize chances of one starting to work against the other and creating surging. In most cases, however, there is only a short section of pipe between the two regulators. It is usually better not to have a surge tank between two regulators, particularly when using pilot loading. There is less chance of surging if the first stage regulator is of the spring, weight, or weight and lever type (for lower pressures), with the second regulator using pilot loading for accurate control. To conserve space it is possible to install high-pressure balanced valve regulators back to back (Fig. 9-52) by using an *opposite-hand* (left-hand) body along with the standard body.

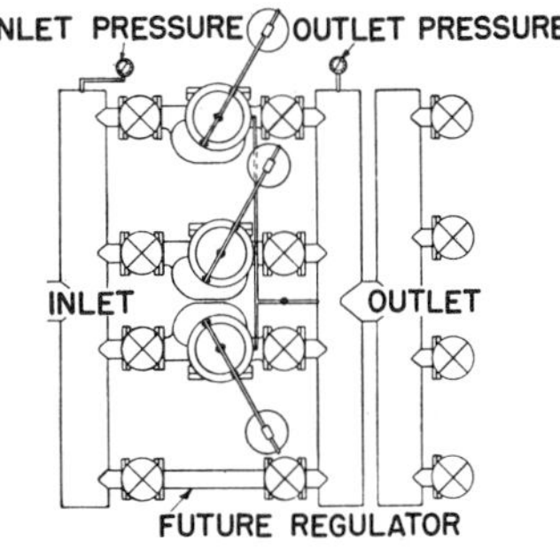

Fig. 9-52 High-pressure regulators installed back to back to conserve space.

Field and Farm Tap Regulators

Protection from weather and accessibility for service are of major importance. Where regulators are not housed, a vent cap must be used which will assure that the vent line is always kept open. Special caps are available to help to prevent freezing or closure by insects. It is usually best to enclose the regulator completely.

Generally, field regulators are installed in horizontal lines with the spring housing vertical. Design of the regulator often permits variations in installation position. Often the primary regulator is installed underground.

Prevention of Freezing. Regulators for farm tap service from a very high pressure line may freeze and stick even at atmospheric temperatures considerably above 32 F. Even with relatively dry gas (2 to 8 lb of water vapor per MMCF), hydrate-forming temperature may be high (Fig. 4-44). Methyl alcohol added to natural gas, using a drop tank and wicking, will aid in preventing regulator freeze-up. Published data indicate that 0.002 to 0.04 gal per MCF will prevent freezing down to −5 F.

AUTOMATIC AND REMOTE PRESSURE CONTROL SYSTEMS

Automatic remote control of pressure regulators in districts and at city gate stations has been gradually increasing in recent years. Advantages of such control include lowering operating costs, maintenance of more uniform distribution pressure, and quick response. Several systems are now in use.

"Round-the-Block" Control

In such systems, the regulator sensing line goes to the low-pressure point in the network. Auxiliary boosting may be used. Figure 9-53 shows an application of an early automatic station control using opposing orifices (a type of peak load booster). The system consists of a group of mains arranged to form a small district with pressure drops from station to control line similar to those from the station to the low-pressure point

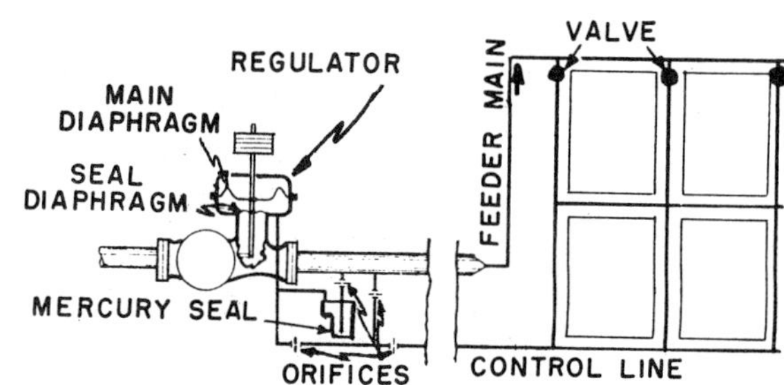

Fig. 9-53 "Round-the-block" control system.

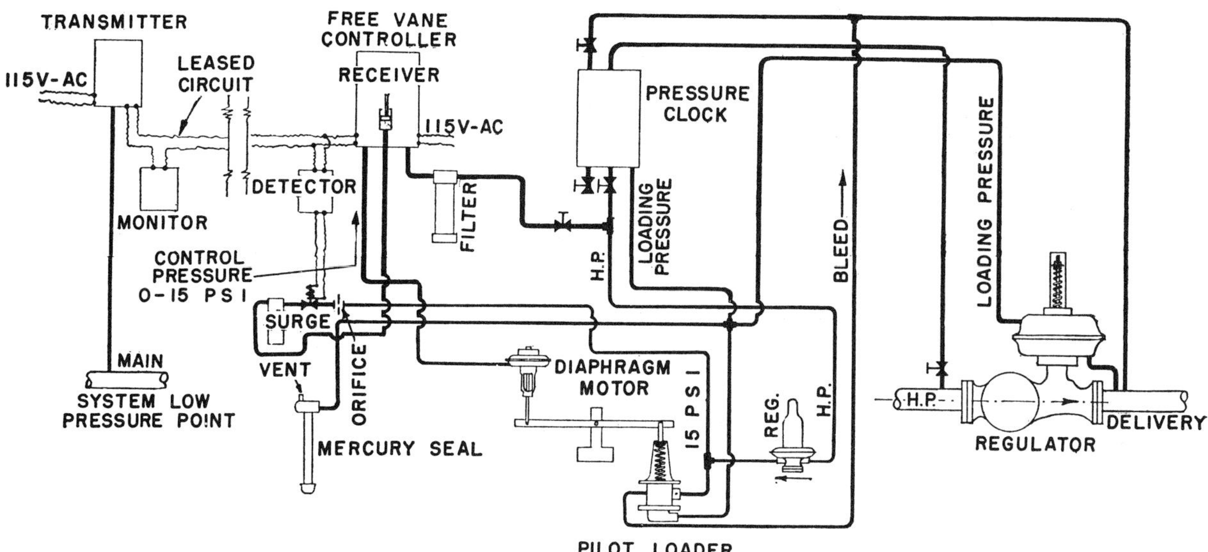

Fig. 9-54a Free vane controller system (telemeter type). H.P. = high pressure.

in the district served. The regulator control line is connected into the chamber between the main diaphragm and the seal diaphragm.

In some cases this system falls short of producing the results desired; that is, the desired *peaking* of delivery pressures during heavy loads requires too much *droop* in control line pressures, because the interconnecting orifices installed to stabilize control tend to oppose the building up of large differences between control line and delivery pressures. Adjustments made to reduce drooping of control line pressures often result in overcontrol and violent hunting action in which the regulator valves cycle from open to fully closed. Pressure loading thru a pilot loader instead of by weights was found highly desirable, since this avoids excessive pressures on the diaphragm. Difficulties involved in adaptation of pilot loaded regulators to *round-the-block* control fluctuations indicated additional need for a different control mechanism.

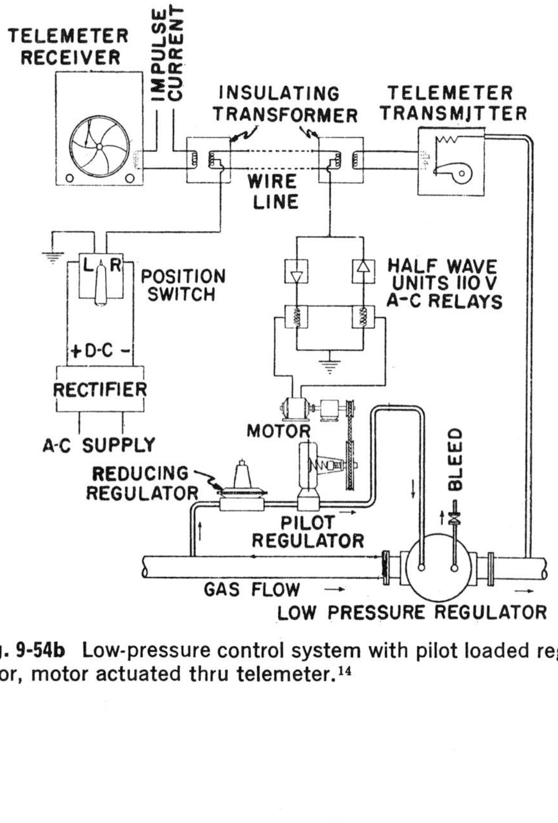

Fig. 9-54b Low-pressure control system with pilot loaded regulator, motor actuated thru telemeter.[14]

Free Vane Control

The free vane controller consists of a pressure gage element or telemeter receiver which indicates or records the pressure maintained. Pressure fluctuations cause the measuring element to move a metal vane between two very small opposing nozzles. A high-pressure regulator feeds the nozzles, which constantly bleed a small amount to the atmosphere. As the vane moves between the nozzles, back pressure is created in the nozzle supply system, causing gas to be delivered at pressures ranging from 0 to 15 psig. This outlet pressure loads the station regulator, so that when pressure in the measuring element tends to fall or rise, the instrument increases or decreases the pressure delivered from the station. An application of the free vane controller to automatic operation of a district station by telemetering is illustrated in Fig. 9–54a.

Pilot Loaded Regulator, Motor Actuated thru Telemeter

A low-pressure control system with a pilot loaded regulator, motor actuated thru a telemeter, is shown in Fig. 9-54b.[14]

Free Vane Control with Rotating Cam and Diaphragm Operator

A free vane control system for a regulator with rotating cam and a diaphragm operator is shown in Fig. 9-54c.[15]

Pilot Loaded Regulator Control[16,17]

Figure 9-54d shows a pilot loaded regulator control system. Control information may be transmitted to the master station thru the use of a tone or combination of tones over a single pair of telephone wires. This allows a saving in operating thru the reduction in the number of communication circuits required.

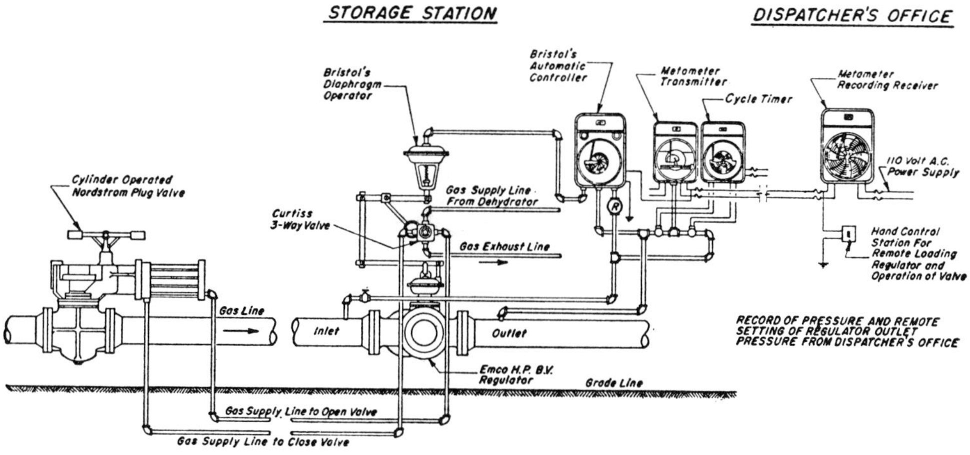

Fig. 9-54c Free vane control system for regulator, using rotating cam and diaphragm operator.[15]

REGULATOR MAINTENANCE

A periodic inspection and repair program is the best way to ensure safe and continuous regulator operation. Regulator size and load characteristics determine the amount of inspection and care needed. **Service regulators** should be checked in the field when the meter is changed. Maintenance of large regulators of the **district** and **border station** type is continuous thru inspection of pressure charts taken throughout the system and by visual inspection of the units.

Low-pressure balanced valve regulators are not subject to as hard usage as high-pressure regulators. Usually, a complete check once a year is sufficient to ensure their good operation. Where dirt or gum troubles exist, the regulators should be checked more often. The main checks on a large regulator consist of testing the diaphragm for leakage by noting any escape of gas thru the vent, seeing that the external control line piping is in good condition, checking diaphragm and valve linkage for wear, and determining if the valves are closing tightly so that outlet pressure cannot build up during off-peak loads.

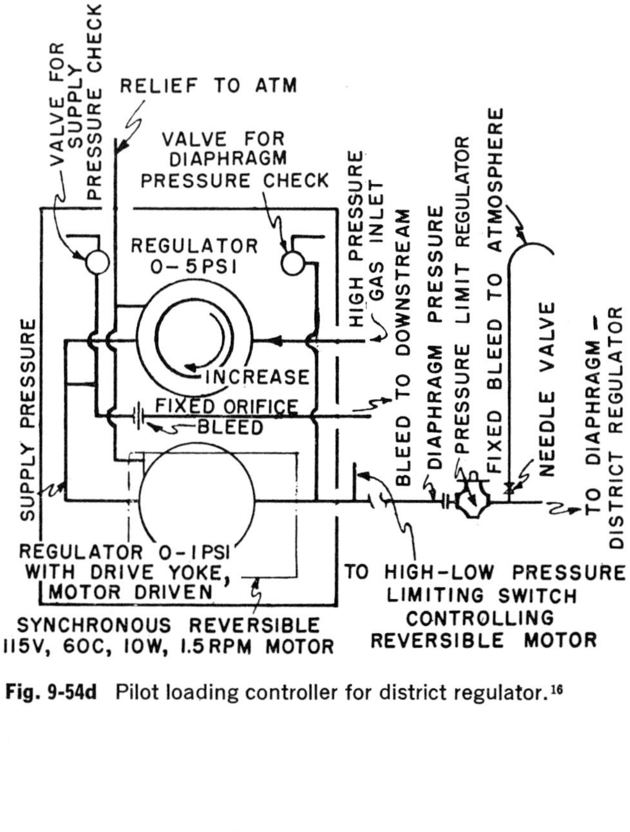

Fig. 9-54d Pilot loading controller for district regulator.[16]

The frequency of inspection of high-pressure balanced valve regulators depends on the pressure drop, volume of gas passed, condition of the gas, and the importance of the particular regulator in the system.

Reference may be made to the following extract from ASA B31.8-1963:

855 MAINTENANCE OF PRESSURE LIMITING AND PRESSURE REGULATING STATIONS.

855.1 All pressure limiting stations, relief devices, and pressure regulating stations and equipment shall be subjected to systematic periodic inspections and/or tests to determine that they are:

(a) In good mechanical condition.
(b) Adequate from the standpoint of capacity and reliability of operation for the service in which they are employed.
(c) Set to function at the correct pressure.
(d) Properly installed and protected from dirt, liquids, or other conditions that might prevent proper operation.

855.2 (a) Every distribution system supplied by more than one district pressure regulating station shall be equipped with telemetering or recording pressure gages to indicate the gas pressure in the district.

(b) On distribution systems supplied by a single district pressure regulating station the operating company shall determine the necessity of installing such gages in the district. In making this determination the operating company shall take into consideration the operating conditions such as the number of customers supplied, the operating pressures, and the capacity of the installation, etc.

(c) If there are indications of abnormal high or low pressure the regulator and the auxiliary equipment shall be inspected and the necessary measures shall be employed to rectify any unsatisfactory operating conditions. Suitable periodic inspections of single district pressure regulation stations not equipped with telemetering or recording gages shall be made to determine that the pressure regulating equipment is functioning properly.

855.3 Whenever it is practicable to do so, pressure relief valves should be tested in place to determine that they have sufficient capacity to limit the pressure on the facilities to which they are connected to the desired maximum pressure. If such tests are not feasible, periodic review and calculation of the required capacity of the relieving equipment at each station should be made and these required capacities compared with the rated or experimentally determined relieving capacity of the installed equipment for the operating conditions under which it works. If it is determined that the relieving equipment is of insufficient capacity, steps shall be taken to install new or additional equipment to provide capacity.

857 VAULT MAINTENANCE.

Regularly scheduled inspections shall be made of each vault housing pressure regulating and pressure limiting equipment and having a volumetric internal content of 200 cubic feet or more to determine if it is in good physical condition and adequately vented. This inspection shall include the testing of the atmosphere in the vault for combustible gas. If gas is found in the vault atmosphere, the equipment in the vault shall be inspected for leaks and leaks found shall be repaired. The ventilating equipment shall also be inspected to determine if it is functioning properly. If the ventilating ducts are obstructed, they shall be cleared. The condition of the vault covers shall be carefully examined to see that they do not present a hazard to public safety.

Excessive Outlet Pressure Variation

Dirt or gum accumulation on the valves or around the diaphragm stem where it enters the lower diaphragm case will cause erratic regulation. This effect is most noticeable when the load drops suddenly. Cleaning the valves, seats, pivot bearings, and all connections between the valves and regulator will usually correct this condition.

The use of filters should be weighed against the possibility that they may clog and cause an outage. Systems which preclude this possibility are recommended.

The **partial closure of the vent or control piping** will also make the regulator perform as described above. These conditions preclude quick changes in flow rate without a temporary deviation of the outlet pressure. However, if the flow rate changes very slowly, performance may be satisfactory.

Improper control line installation (for example, if the control point is taken ahead of line restrictions, such as pipe elbows) may allow the regulator to perform satisfactorily but cause the main line pressure to drop excessively at high flows. This condition is easily corrected by connecting the control line downstream of line restrictions.

An **undersized regulator** permits drop-off in outlet pressure during high rates of flow. On a *weight type of regulator* it is easy to notice whether or not the valves are wide open. If they are, the regulator is operating beyond its capacity.

Inlet pressure variations do not change outlet pressure appreciably on a *balanced valve regulator*, unless accompanied by a large change in flow rate. The effect of inlet pressure changes on *pilot loaded regulators* is significant at very low outlet pressures. An extra pilot regulator, placed ahead of the main pilot regulator, will minimize this effect and maintain a constant pressure to the pilot loading system.

Sudden large changes in load may change the line pressure faster than the regulator can respond. The less line capacity between the regulator and the load, the more pronounced the pressure fluctuation will be.

Diaphragm stiffness or dryness may cause poor response to pressure variations in the diaphragm case. Leather diaphragms should be re-oiled according to manufacturer's instructions or replaced periodically. Diaphragms made of synthetic fabric eliminate the need for oiling. Many older regulators that used excessively thick diaphragms can now be equipped with thinner and more flexible material of greater strength.

Incorrect valve adjustment (i.e., when the two valves do not seat at the same time) prevents shutoff. A soft seat valve may have been clamped too tight, deforming the seating surface. Misalignment of valves and valve seats permits leakage and poor response. A bent valve stem is a common cause of this condition.

The **spring** under the main diaphragm of a *power pilot loaded regulator* is a critical component. Improper replacements may prevent complete shutoff. The problem appears mainly in connection with controls that bleed back to the outlet line rather than with those bleeding to the atmosphere.

Hunting

Hunting or pulsation in a regulator is often difficult to overcome since it may be caused by the overall distribution system rather than just the regulator characteristics. Hunting must not be confused with vibration, since entirely different methods must be used to correct it. When hunting, the valve stem actually travels up and down, opening and closing the valves, within from 1 to 30 sec. This type of pulsation will definitely be recorded on a pressure gage.

A regulator operated by pilot loading or pilot control is more likely to hunt than a spring or weight type of regulator. The weight of the moving parts also has an effect. Another type of regulator which may often hunt is one with a boosting effect causing the outlet pressure to increase with flow.

In general, a regulator is hunting because it does not respond quickly enough to flow changes, due to slight binds either in the regulator itself or in the volume of the line fed by it. A regulator too large for the load is more likely to hunt than one properly sized.

With a simple *direct spring or weight loaded regulator*, an increase in diaphragm vent size should result in faster diaphragm movement. Any mechanical binds should be removed. A more flexible diaphragm will also help. Connection between the lower diaphragm case and outlet line should be as short as possible and unrestricted. However, a restriction may be placed between the regulator control point and an oversized line. The restriction acts as a buffer which may break up the hunting action.

On a *power pilot loaded regulator* mechanical friction in the regulator may cause the control pressure to overcorrect and start hunting. An extra heavy spring in the pilot regulator will slightly decrease regulator sensitivity and thus help prevent hunting. The toggles in the pilot regulator can be adjusted so that the pilot valve will be closed with very little movement of the pilot regulator diaphragm.

When using the power pilot regulator on *double valve regulators*, it is possible to decrease sensitivity by reducing the lower diaphragm area of the pilot regulator slightly. Sometimes just the relocation of the control point may stop pulsation.

Some pilot control regulators have sensitivity adjustment features to increase regulator stability.

Two regulators installed in series may "pump" against each other. Both should be examined and all friction removed.

It often helps to have the first regulator of the weight and lever type, using the downstream regulator only to obtain accurate outlet pressure control.

On instrument loading, a *sensitivity* adjustment can be made to increase regulator stability. Unless the instrument is furnished with an automatic reset, any change in sensitivity means that an increase in flow rate will result in decidedly more reduction in outlet pressure. Instrument loaded regulators will pulsate if excessive binding exists in the stuffing box or the motor head.

Vibration

Vibration is caused by a very rapid movement of the valve and definitely produces a noisy regulator. It is entirely different from hunting and requires other means for correction. Vibration is harmful since it can loosen regulator parts and cause fatigue with resulting breakdown. Usually, a vibrating regulator does *not* cause variations in outlet pressure.

The most usual cause of vibration is too much freedom of movement of the valves without diaphragm movement. This slack is the result of wear in the coupling block between diaphragm stem and valve stem. Excess wear on the valve guides will allow side motion. The first step in eliminating vibration is to replace or repair all worn parts at which lost motion occurs.

On small regulators, *reducing* the vent hole size will often eliminate vibration. Reduction of volume above the diaphragm, as by raising the diaphragm position, also helps.

Vibration can often be stopped by added friction around the diaphragm stem. Repacking or tightening a stuffing box will eliminate vibration temporarily, but the result is usually not permanent. A *dash pot* or constant friction device is a more effective means of correction. Since dampening stops vibration, a heavier diaphragm in the regulator is often satisfactory.

Restriction in the control line between the lower diaphragm case and the outlet line can eliminate vibration, but the required very small orifice may also cause pulsation.

A regulator is much more likely to vibrate on small flows when the valve operates very close to its seat. Use of a smaller size regulator will often provide correction. If size cannot be reduced because of a large flow range, a V-port valve guide on parabolic valves will help. On *high- and low-pressure balanced valve regulators*, plates may be installed between the valves and valve guides to make the valves travel further away from the seats. Plate thickness can vary from ⅛ to ¼ in., depending on regulator size.

On pilot loaded valves, changing the regulator spring to a heavier one, calibrated to operate over a pilot output range of 3 to 30 psi instead of the normal 3 to 15 psi range, will often eliminate valve vibration.

REFERENCES

1. Rockwell Mfg. Co. *Gas Regulator Handbook.* (Bull. 1090) Pittsburgh, Pa., 1953.
2. List, R. C. "Fundamental Principles of Regulators." *Southwestern Gas Measurement Short Course* 35: 241–6, 1960.
3. Krupp, W. J. "What's New in Distribution Regulators." *A.G.A. Operating Sec. Proc.* 1961: DMC-61-8.
4. Anuskiewicz, M., Jr., "Development of an Overpressure Protection Program for a Distribution System." *A.G.A. Proc.* 1954: 394–424.
5. ——. "Methods to Prevent Overpressure in Distribution Systems." (DMC-55-3) *A.G.A. Proc.* 1955: 381–416.
6. Escolas, E. J. "Monitored Regulator Characteristics." *A.G.A. Operating Sec. Proc.* 1960: DMC-60-28.
7. A.G.A. *A.G.A.-GAMA Service Type Regulator Specifications.* (OP-63-2) New York, 1963.
8. MacLean, A. D. "Capacity of Balanced Valve Regulators." *Gas* 25: 34–41, Oct. 1949.
9. A.G.A. *Orifice Metering of Natural Gas*, p. 86. (Gas Measurement Comm. Rept. 3) New York, 1956.
10. A.G.A. *Gas Measurement Manual.* New York, 1963.
11. "Gas Handbook/Catalog Issue." *Am. Gas J.*, Aug. 15, 1962.
12. Partridge, F. M. "Reducing Noise at Regulator Stations." *Gas* 27: 91–8, Dec. 1951:
13. Serafino, C. A. "Experiments in Sound Abatement." *Gas* 30: 49–50, Dec. 1954.
14. LaViolette, W. A. "Remote Control of Regulator Stations." (DMC-55-5) *A.G.A. Proc.* 1955: 432–54.
15. Glamser, J. H. "Remote Control of an Underground Gas Storage." *A.G.A. Proc.* 1955: 455–6.
16. Steger, S. J. "Remote Control for a Low Pressure District Regulator." *A.G.A. Proc.* 1955: 457–62.
17. Humphreys, T. G., Jr. "Operating Experience With Remote Supervisory Control and Telemetering." *Southwestern Gas Measurement Short Course Proc.* 1953: 327.

Chapter 6

Distribution Design for Increased Demand

by Guy Corfield, B. E. Hunt, R. J. Ott, G. P. Binder, and F. E. Vandaveer

Designing a distribution system to meet increased gas demand may involve two major considerations: (1) reinforcement of an existing system to meet increased heating and industrial loads, and (2) extension of mains into new areas or towns. Both problems are basically similar, requiring provisions for adequate gas supply and adequate pressure to customers at all times. Networks of existing mains influence design conditions. In new areas, a completely new system may be installed involving different pressures and main and service sizes. In some companies the distribution system ends at the curb or a few feet from the customer's property line; i.e., the customer owns the service.

Adequacy of any system is measured by its ability to meet peak hour customer demands on a peak day with the pressure available at system input points. Analyses of distribution networks are made from time to time to determine need for system reinforcement; effect of valve and regulator operation; pumping pressure schedules; production or supply requirements for interconnected networks; economics of fringe areas; expected system pressures under various conditions; existence of system faults or stoppages; effect of new loads; and effect of equipment or line failure.

Reference may be made to the following extract from ASA B31.8-1963:

831 PIPING SYSTEM COMPONENTS.

All components of piping systems, including valves, flanges, fittings, headers, special assemblies, etc., shall be designed to withstand operating pressures, and other specified loadings, with unit stresses not in excess of those permitted for comparable material in pipe in the same location and type of service. Components shall be selected that are designed to withstand the field test pressure to which they will be subjected, without failure or leakage, and without impairment of their serviceability.

840 DESIGN, INSTALLATION, AND TESTING.

840.1 General Provisions. The design requirements of this Code are intended to be adequate for public safety under all conditions usually encountered in the gas industry. However, special conditions that may cause additional stress in any part of a line or its appurtenances shall be provided for, using good engineering practice. Examples of such special conditions include: long self-supported spans, unstable ground, mechanical or sonic vibrations, weight of special attachments, and thermal forces other than seasonal.

845.23 *Qualifying a Steel Pipeline or Main for a New and Higher Maximum Allowable Operating Pressure.* Note: This paragraph applies to pipelines or mains where the new and higher maximum allowable operating pressure will produce a hoop stress of 30% or more of the specified minimum yield strength of the pipe. When the new and higher maximum allowable operating pressure is equal to or less than this value the provisions of 845.34 shall apply.

Before increasing the maximum allowable operating pressure of a pipeline or main that has been operating for a period of several years or more at a pressure less than that determined by 845.22-(a)*, it is required that:

(a) The following investigative and corrective measures be taken:

(1) The design and previous testing of the pipeline and the materials and equipment in it be reviewed to determine that the proposed increase in allowable operating pressure is safe and in general agreement with the requirements of this Code.

(2) The condition of the line be determined by field inspections, examination of maintenance records, or other suitable means.

(3) Repairs, replacements or alterations in the pipeline disclosed to be necessary by steps (1) and (2) be made.

(b) The maximum allowable operating pressure may be increased after compliance with (a) above and one of the following provisions:

(1) If the physical condition of the line as determined by (a) above indicates that the line is capable of withstanding the desired increased operating pressure in accordance with the design requirements of this Code and the line has previously been tested to a pressure equal to or greater than that required by this Code for a new line for the proposed new maximum allowable operating pressure, the line may be operated at the increased maximum allowable operating pressure.

(2) If the physical condition of the line as determined by (a) above indicates that the ability of the line to withstand the increased maximum operating pressure has not been satisfactorily verified or the line has not been previously tested to the levels required by this edition of the Code for a new line for the proposed new maximum allowable operating pressure, the line may be operated at the increased maximum allowable operating pressure if the line shall successfully withstand the test required by this edition of the Code for a new line to operate under the same conditions.

(3) If, under the foregoing provisions of (b) above, it is necessary to test a pipeline or main before it can be up-rated to a new maximum allowable operating pressure, and if it is not practical to test the line either because of the expense or difficulties created by taking it out of service, or because of other operating conditions, a new and higher maximum allowable operating pressure may be established as follows:

3.1 Perform the requirements of (a) above.

3.2 Select a new maximum allowable operating pressure consistent with the condition of the line and the design requirements of this Code; provided, however, that,

[* Extracted in Chap. 2, Sec. 9.—EDITOR]

3.3 In no such case shall the new maximum allowable operating pressure exceed 80% of that permitted for a new line of the same design in the same location.

(c) In no case shall the maximum allowable operating pressure of a pipeline be raised to a value higher than would be permitted by this Code for a new line constructed of the same materials and in the same locations.

The rate of pressure increase to the new maximum allowable operating pressure should be gradual so as to allow sufficient time for periodic observations of the pipeline.

DESIGN CONSIDERATIONS

The three major *classes of gas service* are **residential, industrial,** and **commercial** (Fig. 2-2). The following discussion presents data relevant to the forecasting of these loads. These data are strictly applicable only to the studies which evolved them.

Customer demand per day and per hour can be calculated *theoretically* from input* ratings of gas appliances, and *actually* by metering input to equipment and checking total sendouts. However, design calculations require reliable estimates of usage during *peak periods* so that the **maximum coincident demand** for the three classes of service may be supplied. **Demand metering** may be used to measure and record gas flow over short periods.

Basically, the classes of service mentioned above can be divided into space heating and nonspace heating applications. The space heating requirements are primarily influenced by outdoor temperatures, wind velocities, and thermostat settings (or the extent of manual thermostat turndown). The gas requirements for nonspace heating are usually estimated by applying appropriate diversity data to the estimated total connected load. One report[1] showed that facilities designed for 50 per cent of the total connected load would have been adequate to take care of residential and commercial total loads. Industrial loads may require special consideration.

Field experience on daily and hourly demands varies widely, thus making it desirable for each company to collect its own data. A detailed discussion of the method of selecting customers for load characteristic testing is available.[2] The number and type of applications, the climate, and the kind of gas supplied are all major influences on the load. The peak load of most gas systems occurs on the coldest day because of the space heating load.

* Neglecting modulation of appliance inputs by thermostat action or burner turndown as well as diversity of use.

Table 9-24 Typical Inputs of Domestic Gas Appliances
(all burners operating at full rate)

Appliance		MBtu/hr
Boilers		45–400*
Central furnaces		50–400*
Warm air per room		20
Steam or hot water per room		30
Clothes dryers		20–40
Conversion burners		45–400*
Floor furnaces		30–90
Hot plates and laundry stoves, per burner		9
Incinerators		2–40
Light		3
Ranges:	Single oven and broiler, 4 top units	65
	Separate oven and broiler, 4 top units	85
	Six top units and two ovens	105
Refrigerators		3–4
Room heaters:	Radiant type, per radiant	2
	Wall heaters	4–25
	Circulators, unvented	5–50
	Circulators, vented	5–125
Vented wall heaters		3–75
Water heaters:	Automatic storage, 30–40 gal	45
	Automatic instantaneous (2–4 gal/min)	150–300
	Domestic, circulating or side-arm	35

* Most homes need less than 150 cfh.

Residential Customer Demand

The averages of annual estimated gas usages, in therms, for residential appliances reported to the A.G.A. Committee on Comparison of Competitive Services in 1962 were: range, 106; refrigerator, 136; clothes dryer, 43 (electric pilot), 93 (gas pilot); incinerator, 150; water heater, 255;† yard light, 167.

† Usage for a gas dishwasher was reported elsewhere at 60 to 90 therms per year.

Table 9-25 Peak Hour Average Demand per Customer Expressed in Btu/MBtu of Rated Heating Equipment Input*

	Company conducting study†								
Temp, °F	1	2	3	4	5	6	7	8	9
60	38	33	...	95	66	95	206	88	148
50	185	147	333	254	148	202	310	176	304
40	250	267	370	400	238	368	420	279	445
30	325	338	466	495	330	440	528	371	572
20	388	444	564	565	420	496	563	456	603
10	425	567	600	635	505	565	592	522	778
0	490	...	...	705	594	...	...	...	...
Input,‡ Btu/hr	80,000	75,000	30,000§	78,000	121,000	123,000	163,000	136,000	67,200
No. of houses in study	86	134	30	94	99	79	80	86	59

* Data for columns 1 thru 7 from Reference 5; data for columns 8 and 9 from Reference 8. Values calculated by *dividing* the recorded peak hour (7 to 8 A.M., except column 8, which was 8 to 9 A.M.) average gas heating demand per customer.

† The numbers refer to: 1, Worcester Gas Light Co.; 2, Baltimore Gas and Electric Co.; 3, Public Service Electric & Gas Co. of N. J. (1952); 4, Long Island Lighting Co.; 5, New York State Electric & Gas Corp.; 6, Philadelphia Electric Co. (1951–2); 7, Philadelphia Electric Co. (1946–7); 8, Public Service Electric & Gas Co. of N. J. (1958); 9, Alabama Gas Corp. (1956–7).

‡ Average rated input of heating equipment in study. Data for input and number of houses in study for 1 thru 3 are from Reference 9; for 4, 5, and 7, from Reference 1; for 6, from Reference 10; and for 8 and 9, from Reference 8.

§ Kitchen heating.

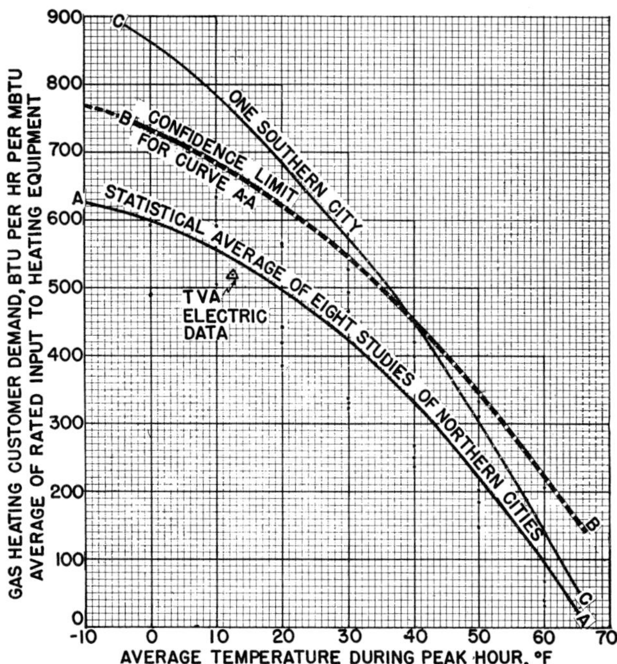

Fig. 9-55a Coincident peak hour average demand in Btu per hr for each gas heating customer and for each 1000 Btu of average rated input of heating equipment. Data for Curves A-A and C-C are shown in Table 9-25 (columns 1 thru 8 and column 9, respectively); Curve B-B is the upper confidence limit of Curve A-A—a minimum of 90 per cent of the observations can be expected to fall below this limit. Point in triangle is electric coincident demand expressed in Btu per hr per MBtu of electric heating equipment input from TVA data[7] on 400 low-rent houses.

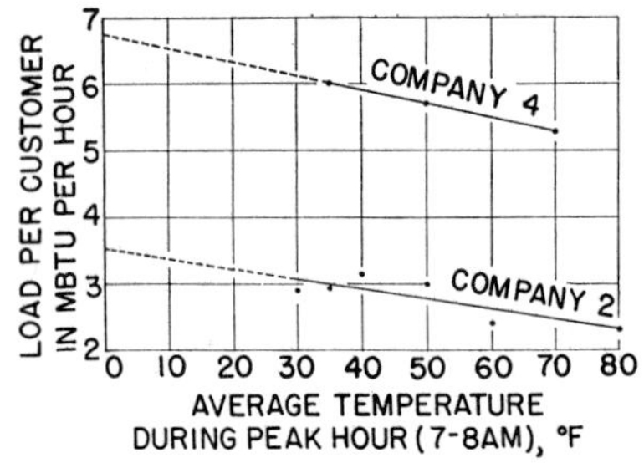

Fig. 9-55b Nonheating load coincident with peak hour heating demand[5] for customers using gas for heating as well as for those using gas for other purposes only.

Load Factors.[3] These indices relate the average gas usage rate over a long period to the usage that would have occurred in the same interval if the maximum short-term usage rate had prevailed. See Fig. 9-57a and Table 9-24 for some typical load factors.

Input ratings to both heating and other domestic appliances are given* in Table 9-24. Many gas utility companies develop load data for forecasting their requirements.[2,4,8,10–12]

Examples of load factors:

Daily annual for system, per cent =

$$\frac{100 \times \text{total annual sendout, MCF}}{365 \times \text{maximum daily sendout, MCF}}$$

* Use actual values where known.

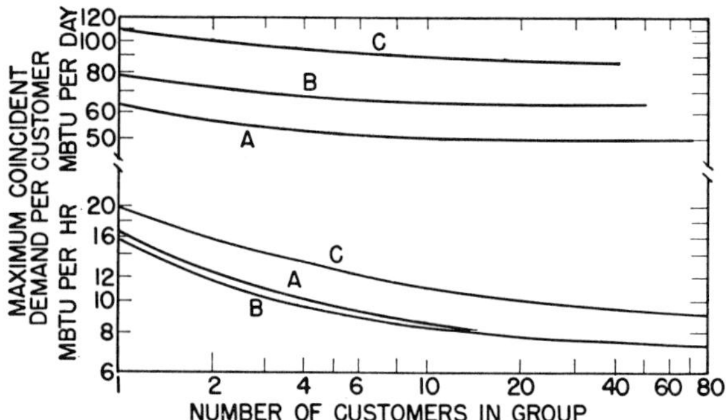

Fig. 9-56 Maximum coincident daily and hourly demands for combinations of appliances in one family houses, and corresponding coincidence factors. Daily data based on daily use over a weekly test period, hourly data based on Sunday group half-hour use. A, cooking only; B, cooking and refrigeration; and C, cooking and year-round water heating.[6]

Corresponding average coincidence factors, per cent:

No. of customers	1	2	4	10	20	40	80
Daily factor	100	90	85	82	80	80	80
Hourly factor	100	77	64	54	50	47	46

The composite daily load curve was obtained by taking the actual daily load curves derived from tests and weighting each curve by the proportion of customers with each appliance combination within each annual consumption. It was found that the maximum half-hour occurred at 12:30 P.M. on Sunday.

Hourly annual† for avg heating customer, per cent =

$$\frac{\text{total annual use of avg space heating customer, MCF}}{87.6 \times \text{max hourly use of avg space heating customer, MCF}}$$

Heating Load. The effect of gas house-heating saturation can generally‡ be expected to establish peak conditions for domestic customer demand. Therefore, in the absence of local metered demand data, the first step in evaluating the probable maximum *hourly* demand in any area should be to determine the saturation of gas heating, i.e., the number of gas heating customers, N. Second, the estimate of the *average rated input* in MBtu, I, per unit, of these N heating units is prepared. Third, a value for the coincident peak hour average demand for heating customers[5] may be selected from Fig. 9–55a. The *product* of these three factors is the estimated peak hour average demand in Btu per hr; i.e., $N \times I \times$ (value from Fig. 9–55a).

Peak hour occurrence depends on a number of factors. The most significant of these are the customers' rising habits, the types of heating controls, and the types of heating systems. Note that the curves in Fig. 9-55a (plot of Table 9-25) intersect the horizontal axis at approximately 65 F, which is the usual base for the degree day heating theory.

Limited data[5] are available (two companies) showing the relationship between the peak hour *maximum* demand and the

† For example, northern U. S. space heating load factors were reported as 17 per cent for hourly annual and 26 per cent for daily annual.

‡ In areas in which house heating either is not needed or is used only slightly, some other peak condition must be determined, such as, perhaps, water heating.

Table 9-26a Monthly Consumption of Apartment House Customers[12]

Monthly use,* therms/customer	Number of customers: Cooking only	Cooking and refrigeration	Total
0– 2.24	177	2	179
2.28– 4.48	548	17	565
4.52– 6.71	513	26	539
6.76– 8.95	196	60	256
9.00–11.19	61	176	237
11.23–13.43	12	215	227
13.47–15.66	7	136	143
15.71–17.90	6	64	70
17.94–20.14	0	32	32
20.18–22.38	0	8	8
22.42–24.61	0	3	3
Total	1520	739	2259

* Intervals are result of conversion of units from MCF of 537 Btu per cu ft gas.

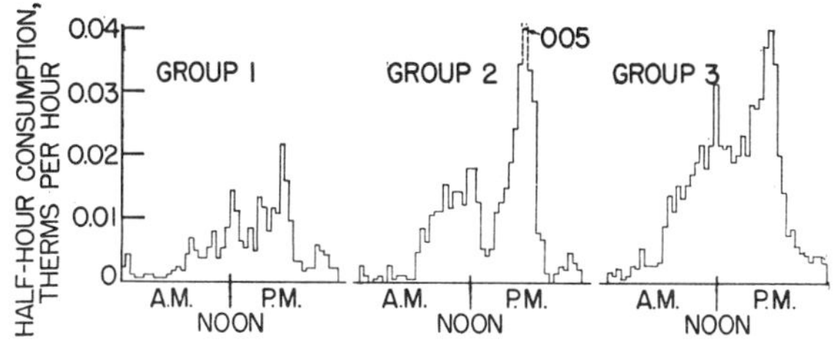

Fig. 9-57a Apartment house customer load characteristic study for small, medium, and large cooking-only customers on the 24-hr peak day, Friday.[12]

	Group 1, Small	2, Medium	3, Large
Avg monthly consumption, therms per customer	2.91	4.83	7.88
Avg 24-hr use, therms per customer	0.130	0.231	0.333
Avg peak half-hr use, therms per customer	0.011	0.252	0.020
Hourly load factor, per cent	26.8	19.1	34.7

peak hour *average* demand. The value of the ratio of *maximum* to *average* varied from about 2.4 to 1.0. Load research studies on the characteristics of gas house heating[6] may be of value in preparing such reports.

Nonheating Load. Variations depend on the variety and number of appliances involved. Data may not be as consistent as those for house heating. In general, the hour of peak nonheating load does not coincide with the hour

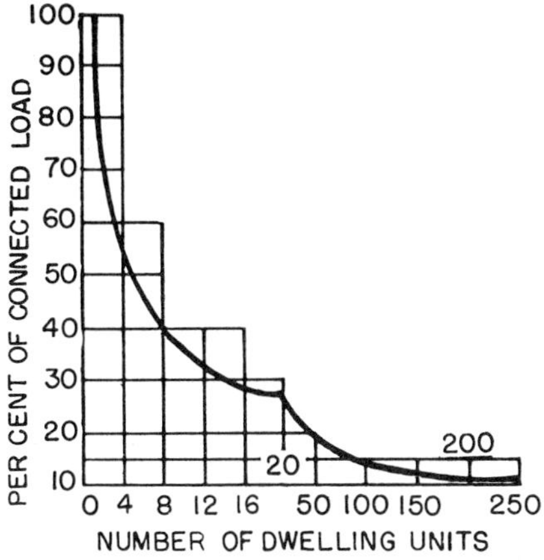

Fig. 9-57b Design load of piping as a per cent of rated connected load for ranges and refrigerators in multiple dwellings.[13] Note change in scale.

of peak house heating load. Thus, when house heating is the important factor, the coincident nonheating load should be taken into account. Figure 9-55b shows the results of two field studies of nonheating load *coincident* with the heating load data in Fig. 9-55a. Confirming data (somewhat lower values) from other companies are available.[1,8,9] Therefore, the application of the upper curve should be conservative. The saturation of appliances in the area involved (in per cent) was: gas ranges, 88.5; water heaters, 100; and clothes dryers, 12.7.

One Family Residences (nonheating loads). The **maximum diversity** among customers is substantially reached when 20 or more customers are treated as a group.[6] The **maximum coincident** hourly and daily demands for groups of up to 80 customers are given in Fig. 9-56. The corresponding coincidence factors (for groups of 20 or more) ran about 48 and 82 per cent, respectively.

The maximum hourly nonheating load gas consumption for a group of 19 Indiana cities (30,417 meters) ranged from 4440 to 9500 Btu per hr per meter (6870 average); temperatures ranged from 35 to 74 F during this *maximum* hour.[11] This reference suggests using a value of 8000 Btu per hr per meter in the absence of definite data.

Apartment Houses. Table 9-26a shows the monthly gas consumption per customer for a 9.8 per cent sample of 230,000 such customers. The same study[12] also reported both the 24-hr consumption for each day of the week and the maximum half-hour consumption for apartment house customers (Table 9-26b). In light of the Friday 24-hr peak and the little variation in daily maximum half-hours, the Friday maximum half-hour was chosen for analysis. Figure 9-57a shows this analy-

Table 9-26b Daily and Peak Half-Hour Consumption for Apartment House Customers[12]

Day	Cooking only, therms per customer: in 24 hr	in max half-hr	Cooking and refrigeration, therms per customer: in 24 hr	in max half-hr
Monday	0.181	0.020 (6:00 P.M.)	0.443	0.018 (6:00 P.M.)
Tuesday	.170	.017 (6:30 P.M.)	.460	.020 (5:30 P.M.)
Wednesday	.176	.019 (6:00 P.M.)	.433	.019 (6:00 P.M.)
Thursday	.181	.016 (6:00 P.M.)	.456	.019 (6:30 P.M.)
Friday	.224	.018 (6:00 P.M.)	.502	.022 (6:00 P.M.)
Saturday	.158	.009 (7:00 P.M.)	.448	.018 (6:30 P.M.)
Sunday	0.190	0.011 (12:00 Noon)	0.474	0.017 (11:30 A.M.)

Table 9-27a Commercial Customer Load Characteristics Data[12]

	Restaurants	Sidewalk bakeries	Bars, grills, ice cream parlors	Tailors	Grocers and butchers	Misc. retail stores	Rooming houses	Apt. house misc.*	Misc.†
Number of customers	40	5	11	27	5	5	6	9	9
Average annual consumption per customer, therms	3522	1699	1314	1504	197	1146	1403	1815	1449
Group maximum coincident demands:									
Half-hour, therms per hr-customer	0.763	0.773	0.404	0.666	0.178	0.619	0.395	0.430	0.610
Half-hour ending	10:30 A.M.	9:00 A.M.	4:00 P.M.	2:00 P.M.	3:30 P.M.	8:30 A.M.	7:30 P.M.	6:00 P.M.	11:00 A.M.
Day of week	Wed.	Fri.	Sat.	Fri.	Sat.	Mon.	Fri.	Tues.	Tues.
Daily, therms per day-customer	11.53	5.53	5.16	6.91	0.95	4.97	3.80	4.78	5.67
Day of week	Fri.	Wed.	Sun.	Thurs.	Sat.	Mon.	Fri.	Mon.	Tues.
Load factors on annual basis:									
Half-hour, per cent	52.7	25.1	37.1	25.8	12.7	21.2	40.5	48.3	27.1
Daily, per cent	83.7	84.2	69.8	59.6	57.0	63.2	101.1	104.0	70.6
Group noncoincident maximum:									
Half-hour demand, therms per hr-customer	1.10	1.01	0.56	0.85	0.23	0.82	0.47	0.66	0.79
Hourly coincidence factor, per cent	69.4	76.6	72.7	78.7	79.0	75.4	84.0	64.8	77.2

* Primarily water heating, laundry dryers, and oil burner pilots.
† Primarily service, supply, and repair agencies.

Table 9-27b Coincident Demand for Trailer Parks[14]

No. of sites	Demand per site, MBtu/hr	No. of sites	Demand per site, MBtu/hr
1	125	11 to 20	66
3	104	21 to 30	62
5	92	31 to 40	58
8	81	41 to 60	55
10	77	Over 60	50

sis for Friday in terms of the cooking-only load. Note that Group 3, consisting of the largest consumers, has the best load factor.

Coincident demand data may also be presented in terms of *per cent of rated connected load,* as shown in Fig. 9-57b for gas ranges and refrigerators in multiple dwellings.

Commercial Customer Demand

This class of service may be studied by dividing it into several groups, such as restaurants, tailor shops, etc. One company reported (Table 9-27a) that a selection of 117 out of 47,000 commercial customers (as few as five customers were found representative of any one group) yielded meaningful data. Although a wide diversity of peak hours was noted among the groups, the overall load factor was high. Trailer park demands are given in Table 9-27b.

Another approach[15] to forecasting commercial loads relates commercial (including space heat) usage to the residential nonheating and residential heating loads, respectively.

Table 9-28 Industrial Customer Load Characteristics Data[12]

	Ceramic and glass products	Bakeries	Other food products	Metal products	Clothing and millinery
Number of customers in test group	11	8	10	11	8
Average consumption per customer for month of maximum industrial sales* of company, therms	7256	14,265	8048	16,874	742
Group maximum coincident demands:					
Half-hour, therms per hr-customer	23.1	36.5	33.3	53.2	4.3
Half-hour ending	9:00 A.M.	9:00 A.M.	9:00 A.M.	9:00 A.M.	9:00 A.M., 3:00 P.M.
Daily, therms per day-customer	317	623	387	713	41
Day of week	Thurs.	Thurs.	Tues.	Wed.	Thurs.
Load factors on basis of month of maximum industrial sales:					
Half-hour, per cent	44.2	54.4	33.9	44.4	25.0
Daily, per cent	76.3	76.3	69.3	78.9	60.5
Group maximum noncoincident demands:					
Half-hour, therms per hr-customer	25.8	39.2	36.0	61.8	5.4
Daily, therms per day	335	630	405	774	43
Coincidence factors:					
Half-hour, per cent	89.6	93.2	92.5	86.1	80.0
Daily, per cent	94.7	98.9	95.5	92.0	93.8

* Following World War II, when industrial activity was high.

Industrial Customer Demand

This class of service is difficult to analyze in terms of load characteristics because of the wide variations in both size and activity among seemingly similar enterprises. Since industrial customers frequently vary production rates by varying the hours worked, rather than by varying the use of equipment during a particular hour, development of load curves is difficult. One suggested approach is to analyze approximately ten of the largest customers in each of five of the largest subclasses of industry in the area.[16] Table 9-28 gives an early study made along these lines. Note that the maximum coincident half-hour for each subclass occurred at 9 A.M.

Diversity and Coincidence Factors

Diversity factor is defined as the ratio of the sum of the noncoincident maximum demands of two or more loads to their coincident maximum demand for the same period.[16] As defined, the diversity factor can never be less than unity. The coincidence factor, which is the reciprocal of the diversity factor, may also be defined as the ratio of a group's actual sustained usage rate divided by the rated connected load. When field data such as those already discussed are available, the diversity factor is included as a part of such data, and no additional factor is used.

Design Temperature for Heating

Conservatively, a winter temperature equal to or less than the lowest temperature ever recorded in the area may be selected for distribution system design. However, the recommendations of the American Society of Heating, Refrigeration and Air Conditioning Engineers might be considered;[17] namely, that the design temperature for heating systems be such that a lower average daily temperature will only have a probability of once in thirteen years. Such temperatures for principal U. S. cities are available.[17] Periods of very high winds with rather low temperatures may cause system peak demands.

Another approach to the selection of a design temperature is based on observed trends in the weather, i.e., winters are not as severe in many areas of the country as those experienced in former years.[18] These trends are expressed in equation form for various cities. To use such data to predict temperatures in future years, demonstrated changes in the *rate* and even the *direction* of some of the trends should be considered as well as the dispersion of the data used in computing these equations.

It is customary to define a *design day* as one in which weather conditions produce the maximum 24-hr sendout. For such a day, sendout is tabulated opposite temperature for each hour. This design day would not necessarily have its lowest temperature at the hour of maximum sendout.

For example, the minimum temperature might occur at 6 A.M., yet the thermostat may not "call for" heat until 7 A.M. The average of the 24 hourly temperatures might be several degrees higher than the lowest temperature.

PREVENTING AND SOLVING LOW-PRESSURE AREA PROBLEMS

Usually, low-pressure conditions develop gradually, and correlating reports of difficulties causes a pattern to emerge which indicates a need for correction. Locating and solving pressure problems may include taking the following steps:

1. Channel all applications for house heating, commercial, and industrial installations thru a central engineering office so that calculations may be made of adequacy of gas supply before installation is made.
2. Keep records of locations of low pressure during cold weather.
3. Make pressure tests, both static and flowing, on customer's premises where low pressure is indicated.
4. Provide gages to record pressure fluctuations in trouble areas.
5. Divide the system into areas according to source of supply, designate these areas on a map, and show for each the present and estimated future customers, appliances, and peak loads.

METHODS FOR REINFORCING A DISTRIBUTION SYSTEM

Increasing Main Pressure Without Use of Service Regulators and/or Appliance Regulators. Permissible increase in distribution pressure is limited by its effect on appliance performance, increase in system leakage, and allowable operating pressure. A.G.A.-approved appliances are tested for satisfactory operation at pressures outlined in Table 9-9.

The performance of equipment not tested for conformity with ASA requirements may limit a proposed pressure increase, since the burners involved may not have the flexibility to burn the resulting higher gas input properly. If pressure is to be increased more than 25 per cent, such appliances should be checked and, where necessary, an appropriate reduction in input should be made. Certain classes of burners, such as those in older model commercial coffee urns, side arm water heater burners, and old unvented space heaters, may not even tolerate a 25 per cent increase.

Increasing gas distribution pressure during peak load periods by boosting pressure at the source is limited in low-pressure systems to a 2 to 3 in. w.c. increment.

Increasing Main Pressure with Use of Individual Service Regulators. Main pressure in such systems can be increased, usually from 0.5 to 15 psig, thereby increasing system capacity without affecting appliance performance. Pressures that may be carried in an older system without extensive repair will depend upon its gas tightness, pipe joints involved, and corrosion protection afforded. At least one large city has made this type of conversion successfully and many others are using it for conversions or for system extensions.[19–21]

Increasing Main Pressure with Use of Appliance Pressure Regulators. Regulators are required parts in a number of appliances, such as central heating units, unit heaters, incinerators, and clothes dryers. Addition of regulators to gas ranges, water heaters, and other appliances permits main pressure to be increased above a normal pressure of about 7 in. w.c. without affecting appliance performance. This has been done by at least one large company.

Increasing the Number of High-Pressure Regulator Stations. Where a low-pressure system is fed by a high-pressure network, setting additional regulators will bolster pressures in areas in which they have fallen below the level necessary for good service, or in which such undesirably low pressures may be expected. In an emergency, temporary regulators can be installed in protected locations and converted to permanent installations later. One very large system met increases from 30 to 85 per cent in house heating saturation, accompanied by other large load increases, by this general method.[22]

Increasing Heating Value. A natural gas of 1000 Btu per cu ft will in effect *double* the distributive capacity of a system supplying manufactured gas of 500 Btu per cu ft. However, heating value, or any factor affecting **interchangeability,** cannot be varied except within limits of good appliance performance without a complete readjustment of all appliances.

Changes in the carbon–hydrogen ratio resulting from the Btu increase under discussion will affect the performance of special atmosphere generators such as the ones used in heat treating operations. The endothermic type of generator is affected most because the catalyst may become covered with soot. Control instruments can be added to such generators to make compensating adjustments automatically.

Renewal, Enlargement, or Paralleling (and Tie-In) of Key Feeder Mains. These relatively costly alternatives are considered when the expedients mentioned above are not possible. Renewals and enlargements are made preferably at the same time that replacement becomes necessary because of corrosion, road improvement, or other reasons. Renewals by running new steel pipe (to be operated at high pressures) thru old low-pressure cast iron mains were reported;[23] the steel was protected with an epoxy resin coating and casting insulators.

Many utilities tie together, wherever possible, all ends of distribution lines to form loops or networks. Such ties are useful because (1) they ensure more uniform distribution pressure, since gas may feed from two directions; (2) repairs may be made without interruption in service, since the affected section can be blocked off completely; and (3) delivery to the area can be increased.

MAIN SIZING

It is customary for each utility to standardize a few pipe sizes to facilitate inventory and repairs. These sizes are based on experience, system pressure, kind of gas, and pipe material. For a low-pressure system in urban areas, some companies use a schedule of standard diameters—4 or 6 in. min. For intermediate or regulated pressure, 2 and 3 in. diam mains are often used with feeder mains at higher pressure in larger sizes, as needed. Some companies use 2 in. distributor mains[21] for three-block spans between larger and higher pressure feeder mains for urban and other areas.

The following are some of the factors involved in main sizing:

1. **Calculated size** which is based on the volume of gas the main will be required to handle at present (e.g., as determined from load studies) plus an allowance for anticipated future loads.

2. **Pipe material** (cast iron, steel, ductile iron, plastic). For cost reasons, cast iron mains tend to be sized closer to calculated size than steel mains.

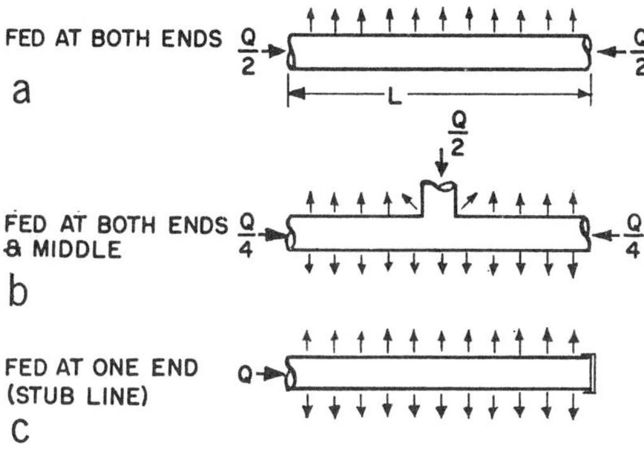

Fig. 9-58 Pressure drops for three input and delivery arrangements for low-pressure mains: (a) $AQ^2L/23$; (b) $AQ^2L/185$; (c) AQ^2L/x, $x = 3$ for 40 or more equally spaced and equally discharging services on the main. For lesser numbers of services, Y, use the following schedule:

Y	2	4	6	8	10	15	20	30
X	1.59	2.13	2.38	2.50	2.59	2.72	2.79	2.86

Application explained in text.

3. **Joints,** whether mechanical or welded. Any size pipe may be mechanically coupled. Welding procedure becomes more difficult on 2 in. diam or smaller pipe and standard wall thicknesses.

4. **Pipe wall thickness** increases with increase in pipe diameter for standard weight pipe up to approximately 5⁄16 in. wall thickness in the larger sizes. Thicker wall pipe costs more per foot, but affords more resistance to corrosion. When coated and cathodically protected, thin wall pipe may be adequate.

5. **Pipe strength** is not a major factor in selecting steel pipe to withstand normal distribution pressures, but it is a factor with regard to its structural strength.

6. **Cost of installation,** pipe, and fittings is a major factor in main sizing. On a first cost basis, the smallest pipe in the smallest trench yields lowest installation costs. However, if there is any possibility of increasing load and the next larger size of pipe is only slightly more costly, it may be advisable to use it. In certain instances[24] it is contended that "if a main is 20 per cent loaded in five years it is a good investment if there is anticipated continued growth."

Inputs and Deliveries. The pressure drop for three arrangements of supplying and delivering gas may be approximated by the factors in Fig. 9-58. For example, the pressure drop in a **stub main** of length L, feeding equal volumes of gas Q/Y at Y equally spaced points, may be approximated by $AQ^2L/3$, where Q is the input of the gas, Y is the number of points, and A is the resistance factor of the diameter of the stub line per unit length of pipe (A is further defined under Eq. **2a** and is tabulated in Table 9-30). This factor is based on equal deliveries to 40 equally spaced services (accurate up to 10,000 services on the main).

Stub Mains. These mains, which are not part of a closed loop of piping, feed gas only to the services connected to them. Thus, the gas flow rate in such mains decreases at each service take-off. The pressure drop may be calculated as the sum of the series of drops in the short pipe sections between services, or the loads can be treated as if they were concentrated (see Fig. 9-58c).

"Mini-Main" System.[25,26] Elements of one economic 60 psig distribution system follow: (1) designed for coincident peak demand of 100 cfh per residential customer; (2) outside metering (hard case) of dry natural gas with full-relief regulators; (3) 3/8 in. OD copper tubing services;* (4) pressure at most remote service tap on the 1 in. mains is 15 psig min; (5) 45 to 60 psig inputs to the 1 in. mains from a 6 in. steel header; supplies 160 MCFH and takes care of 1600 customers per idealized square mile; (6) header, with a 60 psig input, is 1 in. oversize to provide some "cushion"; (7) unexpectedly large loads are handled by looping.

Adequacy of Existing Mains.[27] The relationship between pressure drop and capacity in a medium- or high-pressure main is nonlinear. When the loads increase to the point where $P_{abs,\ downstream}/P_{abs,\ upstream} = 0.60$, only 20 per cent reserve main capacity remains. Consideration should then be given to increasing the main capacity or reducing the gas load. The effective main capacities in terms of the foregoing pressure ratio are:

$P_{abs,\ downstream}/P_{abs,\ upstream}$	0.95	.90	.80	.70	.60	.40	0.20
Main capacity, %	30	43	60	71	80	91	97

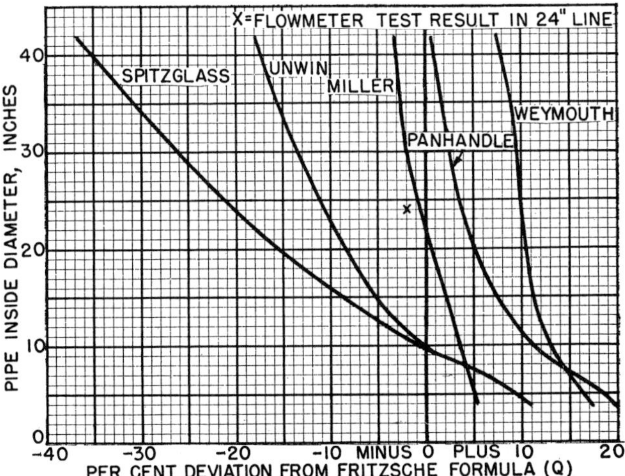

Fig. 9-59 Variations of flow formulas from Fritzsche equation at various pipe diameters[11] where $P_1 = 7.0$ psig, $P_2 = 6.2$ psig, sp gr $= 0.65$, and $L = 0.75$ miles.
Note: The Fritzsche equation was selected as the base for convenience only.

GAS FLOW FORMULAS

The **general formula for flow of gas** thru pipelines is given in Eq. **1** of Chap. 2, Sec. 8. Many flow equations have been developed and used in the gas industry. Most of these are based on experimental results over different ranges of flow conditions for piping with varying roughness. Thus, as shown in Fig. 9-59, different equations do not give the same results.

The Spitzglass formulas appear to be favored for low-pressure distribution studies and the Weymouth and Panhandle formulas for high-pressure calculations. Each company must determine the formula most applicable to its situation. Formulas may be modified to fit local conditions, such as the eventual reduction of the carrying capacity of piping because of the accumulation of deposits which are not readily removable in distribution lines; the use of distribution networks, which have more fittings per mile than their transmission counterparts; and anticipated increases in demand. Negligible "line pack" or storage capacity in low-pressure systems precludes designing for less than the instantaneous maximum demand.

The most important difference among the many formulas for natural gas flow lies in the evaluation of the coefficient of friction, f, or the transmission factor, $(1/f)^{0.5}$. The classic determination of the coefficient is thru charts such as Fig. 9-60. The gas distribution engineer, however, will find the collection of factors in Table 9-29 more applicable. They are all consistent with Eq. **1** of Chap. 2, Sec. 8. The following Eqs. **1** thru **3** are flow formulas which incorporate their respective transmission factors. It is suggested that the reader "assemble" flow formulas by inserting the appropriate factor from Table 9-29 or Table 8-1 into Eq. **1** of Chap. 2, Sec. 8. Using this procedure ensures that all the variables involved are taken into account. Factors for Eqs. **2** and **2b** are available in tabulations and charts.[20]

Table 9-29 Transmission Factors, $(1/f)^{0.5}$, for Distribution Flow Formulas[29]

(clean pipe factors—used in Eq. **1** in Chap. 2, Sec. 8; see also Tables 8-1 and 8-3)

Flow formula	Transmission factor
Constant Transmission Factors	
Rix	14.72
Pole	
for 3/4, 1 in. ID of pipe	9.56
for 1 1/4, 1 1/2 in. ID of pipe	10.51
for 2 in. ID of pipe	11.47
for 3 in. ID of pipe	12.43
for 4+ in. ID of pipe	12.90
Transmission Factors Expressed as a Function of Diameter	
Spitzglass (both high- and low-pressure versions)	$\left[\frac{354}{1 + (3.6/D) + 0.03D}\right]^{0.5}$
Unwin	$\left[\frac{227}{1 + (1.714/D)}\right]^{0.5}$
Oliphant	$13.0 + 0.433D^{0.5}$
Transmission Factor Expressed as a Function of Re*	
Blasius†	$3.56\ \mathbf{Re}^{0.125}$
Mueller†	$3.35\ \mathbf{Re}^{0.130}$
"Improved" flow equation	$5.18\ \mathbf{Re}^{0.09091}$
Lees	$\left(\frac{\mathbf{Re}^{0.35}}{0.0018\ \mathbf{Re}^{0.35} + 0.153}\right)^{0.5}$
Transmission Factor Expressed as a Function of Re and D	
IGT distribution	$4.169\ \mathbf{Re}^{0.100}$
Fritzsche	$5.145\ (\mathbf{Re}D)^{0.071}$

* **Re** = Reynolds number (see Eq. **2** in Chap. 2, Sec. 8; note that Q is in cu ft per day in this equation).
† Accurate in low-pressure systems for up to 8 in. pipe; in medium-pressure systems, for up to 2 in. pipe.[29]

The formulas given on page 9/71 have been used for low pressures (under 1 psig)

* Safety aspects: It was shown that a spade will kink the tubing and shut off the gas without cutting the tubing; a leak from a broken "high-pressure" 3/8 in. copper service is easier to stop than a 10 psig leak from a 1 in. steel service; and less gas escapes per unit time from a broken 3/8 in. tube than from a broken low-pressure 1 1/4 in. line (or intermediate-pressure 3/4 in. line).

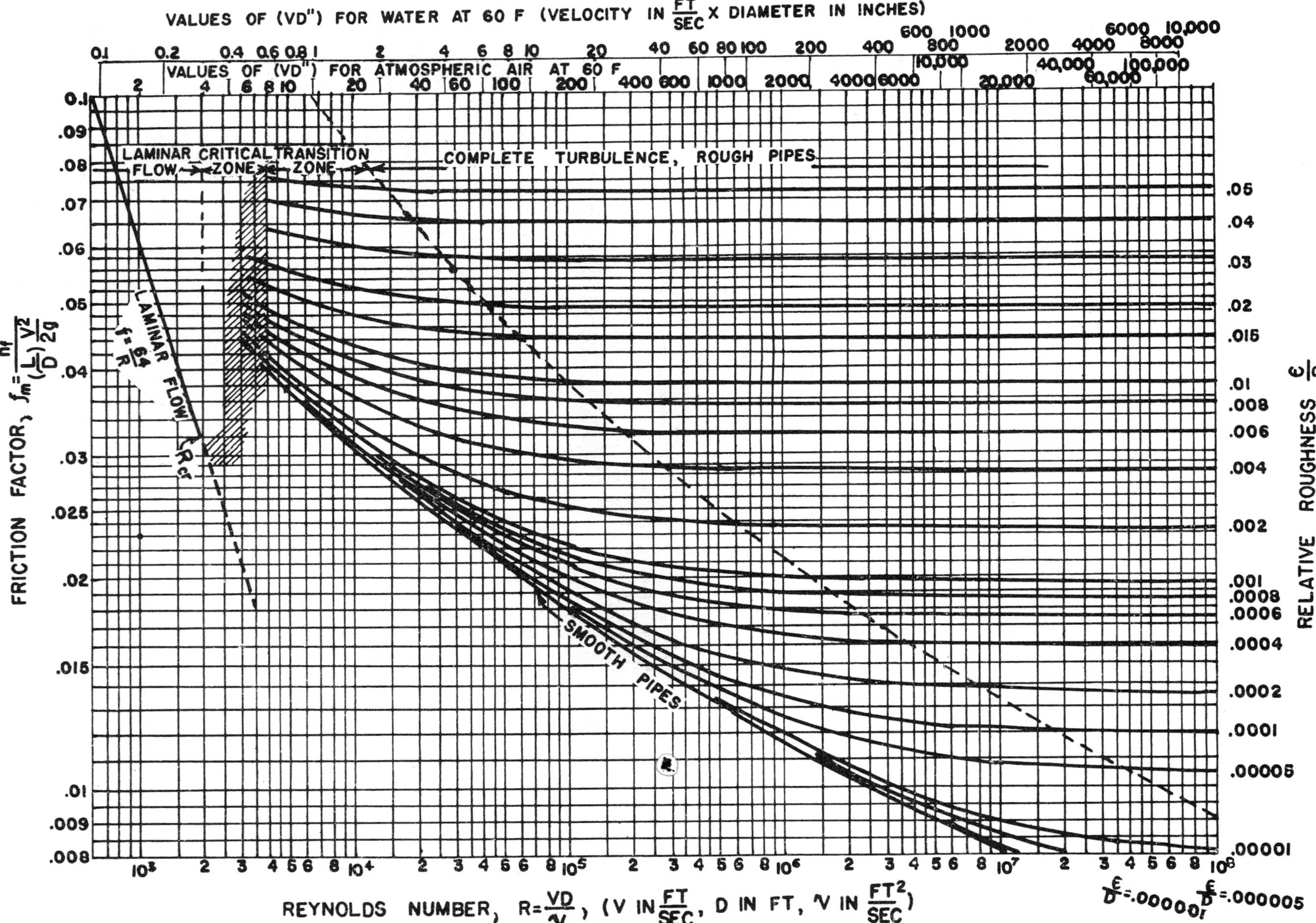

Fig. 9-60 Friction factor, f_m, for pipe flow.[39] The head loss, h_f, in feet may be taken as equal to $\Delta p/\rho$, where Δp = pressure drop in lb per sq ft and ρ = fluid density in lb per cu ft; g = acceleration due to gravity in ft per sec-sec. The friction factor is given as four times the value used in the Fanning equation, $\Delta p = 4f\,(L)(\rho V^2)/(D)(2g)$.

Pole low-pressure formula (widely used for 4 to 10 in. w.c. pressures in 2 to 4 in. pipe*[,29]):

$$Q_h = 876\left[\frac{h}{GL_f}\right]^{0.5} D^{2.5} \quad \text{(from Ref. 28)} \qquad (1)$$

where:

Q_h = flow rate at 60 F, cfh
h = pressure drop, in w.c., approximately equal to $(P_1^2 - P_2^2)/1.062$, where P_1 and P_2 are the inlet and outlet pressures, respectively, in psia
G = gas specific gravity (air = 1.0)
L_f = length of pipe, ft
D = inside diameter of pipe, in.

The preceding old-style formula is based on a constant friction for all pipe diameters. Objections to this may be overcome by substitution of the following schedule of constants† (pipe diameter, in.): 1800 (3/4 to 1); 1980 (1 1/4 to 1 1/2); 2160 (2); 2340 (3); 2420 (4).

Spitzglass low-pressure formula (widely used for under 1 psig):

$$Q_h = 3550\dagger\left[\frac{D^5}{1 + (3.6/D) + 0.03D}\right]^{0.5}\left[\frac{h}{GL_f}\right]^{0.5} \qquad (2)$$

All terms are defined under Eq. **1**.

Equation **2a**, suitable for Hardy Cross calculations, is an alternate form of Eq. **2**:

$$h = AL_fQ_h^2 \qquad \textbf{(2a)}$$

where:

$$A = \frac{G}{126 \times 10^5}\left[\frac{1 + (3.6/D) + 0.03D}{D^5}\right]$$

(or is obtained from Table 9-30)

Other terms are defined under Eq. **1**.

The following formulas have been used for higher distribution pressures[31,32] (above 1 psig):

Spitzglass high-pressure formula (widely used in up to 12 in. pipe*[,29]):

$$Q_h = 3415\dagger\left[\frac{D^5}{1 + (3.6/D) + 0.03D}\right]^{0.5}\left[\frac{P_1^2 - P_2^2}{GL_f}\right]^{0.5} \qquad \textbf{(2b)}$$

All terms are defined under Eq. **1**.

Fritzsche formula‡ (widely used for from 2 to 15 psig, in 8 in. or smaller pipes[28]).

Weymouth formula (widely used for from 3 to 20 psig.[30]) See Eq. **1a** of Chap. 2, Sec. 8.§

IGT distribution‡ (widely used for 10 in. w.c. to 60 psig pressures in 2 in. and larger pipe[29]).

Gustafson formula (widely used for from 15 to 60 psig[30]):

$$Q = 4.214\frac{T_b}{P_b}\left[\frac{(P_1^2 - P_2^2)^{0.5} D^{2.4445} \log 3.7\ D/\epsilon}{G^{0.47355}T^{0.500}L^{0.500}ZV^{0.02645}}\right]^{1.0272} \qquad (3)$$

where:

ϵ = effective roughness of steel pipe, in. (use 0.0007 in. if exact roughness is unknown)
V = absolute gas viscosity, lb-mass per sec-ft (0.00000712 at 60 F)

Other terms are defined under Eq. **1** in Chap. 2, Sec. 8.

Panhandle formula. See Eq. **1b** in Chap. 2, Sec. 8.§

Unwin formula‡ (widely used in up to 12 in. pipe*[,29]).

NETWORK ANALYSIS

Practices in network analysis vary widely, especially for other than computer solutions, since individual companies simplify calculations by using approximations applicable to their individual situations (see, for example, Ref. 33).

The trend is toward *computer* solutions, since the great number of calculations involved in complex distribution analyses are rapidly performed, checked, and conveniently stored within a computer. Simpler network problems may be solved with manual calculators. Organizations that cannot justify investment in computers may use computer service facilities or participate in a group facility.[34] The magazine *Datamation* (Thompson Publications, Inc., New York) presents annually in its November issue a comprehensive tabulation of most of the commercially available computers.

Route Book Billing Data (Plus Calculations and Maps) Method[35–37]

1. Estimate peak hour load demand for the town. This procedure cannot be usefully condensed. It generally involves preparing two charts, one showing use per average firm gas customer per day vs. *mean daily temperature*, and the other showing growth trends for a stated period in terms of cubic feet per customer per day at the *lowest mean daily temperature* encountered and the increase in number of customers for the same period. Peak firm demand per day is then calculated from these charts. Peak hour sendout is determined as a per cent of total daily sendout from a curve showing hourly sendout. This value *times* the peak firm daily demand gives the peak hour load demand.

2. Obtain a cross section of demand in each area of town by compiling firm gas billing totals from meter route books for a peak winter month. Routes are combined to give relatively large areas, 20 or more base areas per town, depending upon its size. Outline the base areas on a town map. Tabulate for each base area the number of customers, total sales per day, total per customer per day, total peak hour demand, and total peak hour demand per customer.

3. Convert peak hour load in base area into line load by means of three charts: (1) regulator site locations in a high-pressure area plus a *load concentration point* from preceding data; (2) base area pattern outline for same area; and (3)

* With effective roughness up to ten times that of clean, commercial steel pipe (see Tables 8-1 and 8-3).

† The constants were derived for a base pressure, P_b, of 14.73 psia, with the base temperature, T_b, and flowing gas temperature, T, both at 520 F abs. The appropriate transmission factor(s) for use in Eq. 1 of Chap. 2, Sec. 8, is (are) given in Table 9-29.

‡ Insert the appropriate transmission factor (Table 9-29) into Eq. 1 of Chap. 2, Sec. 8; flow equations may be compared on the basis of their respective transmission factors alone.[30]

§ Or insert the appropriate transmission factor (Table 8-1) into Eq. 1 of Chap. 2, Sec. 8.

base area boundaries superimposed on site and concentration areas.

4. Using these demand data as distributed along the high-pressure arteries, calculate ability of existing system to deliver the estimated gas volumes. Results of calculations should be recorded on an area trunk line map to obtain a composite network picture under peak load.

Temperature–Pressure Drop Method[38]

Step 1. Install recording pressure gages at customers' meters in a number of selected locations and obtain data for 24 or more hours.

Step 2. Record average hourly temperatures for each 24 hr period.

Step 3. Plot a curve of gas usage per customer in cubic feet per day against average temperature. Usage data are based on the monthly firm gas sendout vs. the degree days for the month. To obtain values in the lower temperature ranges, use sendout on several peak winter days vs. degree days.

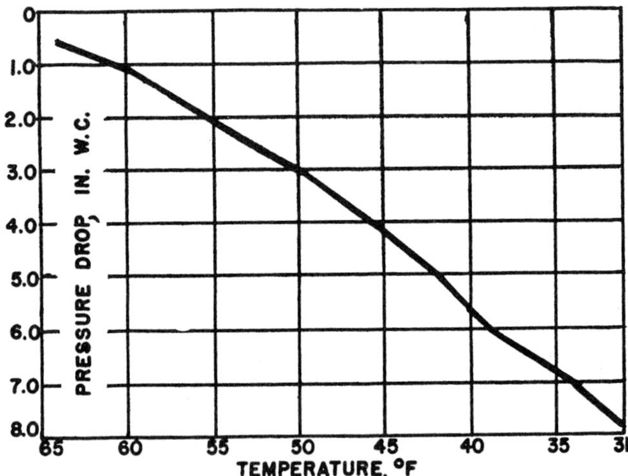

Fig. 9-61 Example of calculated pressure drop vs. temperature for 2876 degree day area.

Step 4. Construct a curve of pressure drop vs. temperature, as shown in Fig. 9-61. Pressure drop data are obtained by subtracting the *minimum* recorded pressure from the *maximum* recorded pressure on charts from customers' meters (step 1).

Calculate other pressure drops using the equation:

$$h_2 = h_1(Q_2/Q_1)^2$$

where:

h_1 = max allowable pressure drop on a peak day
h_2 = calculated pressure drop at Q_2 volume and temperature
Q_1 = gas volume from chart in step 3 on same peak day
Q_2 = gas volume from step 3 at desired temperature

Step 5. Establish from pressure chart at customer's meter the drop from maximum indicated pressure to minimum indicated pressure. Plot this drop against the mean temperature of that day (similar to Fig. 9-61). If point falls below the curve, low pressure will be experienced on a cold peak day.

Pressure Surveys on Intermediate-Pressure System

The following method[21] involves continuous pressure surveys during the heating season for a system with individual service regulators, bolstered by higher pressure feeder mains:

1. Set recording pressure gages in a selected area on several customers' services in front of regulators and take simultaneous recordings. Low pressures can be detected readily.
2. Record additions to heating load, as noted in city building permits, on card files and maps.
3. Plot average annual system pressures for each month of the year. Include average maximum delivery pressure, average delivery pressure on average day, average end pressure on average day, and mean system pressure on average day.

Pressure Calculations, No-Flow or Equal Pressure Method

The following is one trial-and-error method[40] for calculating pressures:

1. Show all district regulators and approximate the areas they serve on a system map. District boundaries may be estimated to pass halfway between adjoining regulators.
2. Prepare and keep up to date a district record load map showing each house and its rated heating unit input (checked by meter). Nonhouse heating customers are figured at 8 cfh each.
3. Calculate total load per block and place on another map showing mains only. Peak hour requirement for the house heating portion of the load is taken at 0.8 *times* rated input.
4. Using information on load map, you can calculate pressures which may be expected during a peak hour at various area points; i.e., by arbitrarily "cutting" the network at estimated points of no-flow or equal pressure, you can make trial-and-error determinations of pressures and district boundaries.
5. The no-flow method is illustrated in Fig. 9-62a. Length, in feet, of each block is designated above the street lines, and gas load, in cfh, per block is listed below the lines. At the end of each block, calculated pressure by Spitzglass' formula, in inches of water column, is indicated by a diagonal line. The points of no-flow or boundary district between regulator (5) and regulator (6) are shown by double lines across three streets.

Hardy Cross Method

The following principles apply to this method for calculating network pressure drops and flow rates:

1. Total gas flow arriving at a junction equals total leaving it.
2. Sum of pressure drops along any closed path must be zero for the network to be in balance.
3. Flow formula is expressed in the form $h = AL_fQ^n$, where h is pressure drop (see Eq. **2**) or $\Delta(P^2)$, *difference in squares* of initial and final absolute pressure (see Eq. **2b**), and A is resistance to flow per unit length.
4. Resistance to change in gas flow in any pipe equals approximately $nAL_fQ^{(n-1)}$.

Data required to solve a network problem include:

1. Diameter and length of mains, locations of main connections, and locations of supply sources.

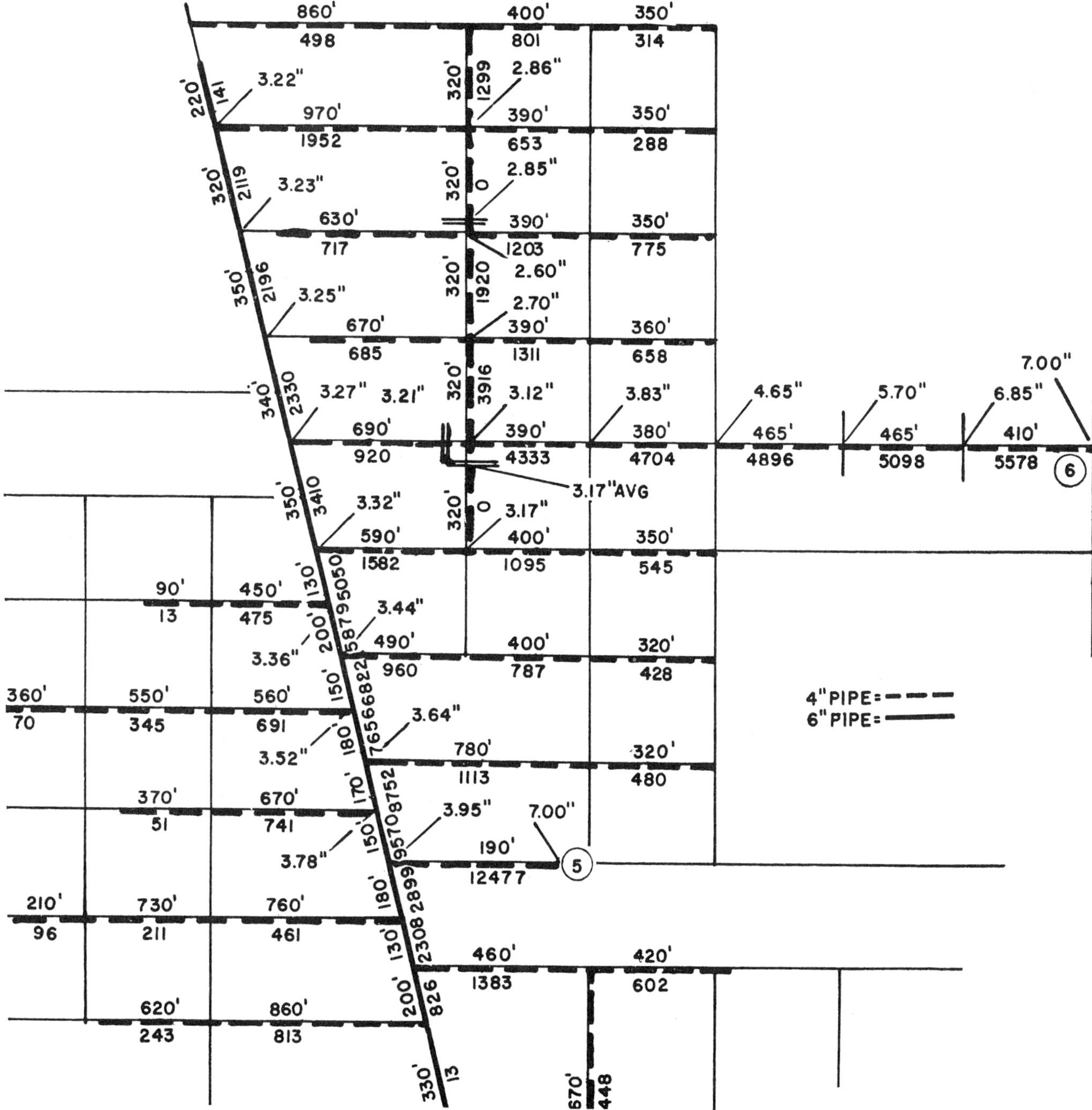

Fig. 9-62a Illustration of no-flow method of calculating network pressures.

2. Volume of gas flowing into network at each source of supply, and estimated load distribution, i.e., gas taken at all points of delivery from network.

3. Maximum allowable pressure drop in network, and minimum or maximum allowable pressure.

4. A gas flow formula best fitting the type of distribution system.

Main Layout. The first step in solving a problem is to make a network map showing main sizes and lengths, connections between mains (junctions), and sources of supply.

For convenience in locating mains, assign each loop and each main a code number. Note that mains on the network periphery are common to one loop and those in the network interior are common to two loops. Special cases may occur in which two mains cross each other but are not connected, resulting in certain mains being common to three or more loops. The distribution network then becomes "three-dimensional" rather than two-dimensional.

Figure 9-62b is an example of a three-dimensional network because main 15 is not connected to main 6. Mains 1 to 14 form a two-dimensional network of four loops. Loop 5 consists of mains 15, 9, 10, 11, and 12. Mains 9, 10, and 11 are each common to two loops (5 and 3, 5 and 3, and 5 and 2, respectively) and main 12 is common to three loops (2, 4, and 5). Loop 5 could have been chosen along several other

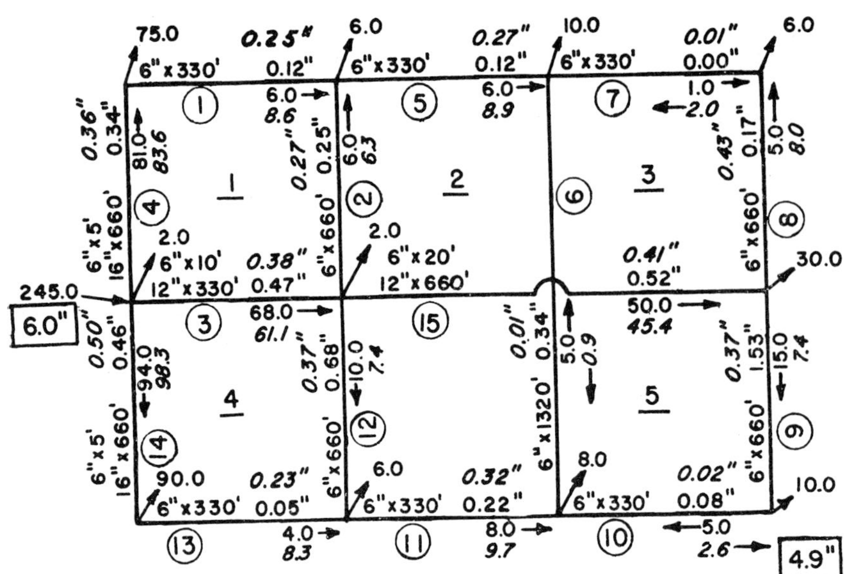

Fig. 9-62b Low-pressure gas distribution network analyzed by Hardy Cross method.

Table 9-30 Constant A for Spitzglass Low-Pressure Formula and Factors for Computing Equivalent Lengths of Pipe

(0.60 sp gr gas)

Nominal size of pipe, in.	A*	Equivalents† 6 in. pipe	16 in. pipe
4	9.036×10^{-11}	8.78	1169.
6	1.029×10^{-11}	*1.00*	133.
8	2.482×10^{-12}	2.41×10^{-1}	32.1
10	7.818×10^{-13}	7.59×10^{-2}	10.1
12	3.170×10^{-13}	3.08×10^{-2}	4.10
16	7.728×10^{-14}	7.51×10^{-3}	*1.00*
20	2.651×10^{-14}	2.57×10^{-3}	3.43×10^{-1}
24	1.116×10^{-14}	1.08×10^{-3}	1.44×10^{-1}
30	3.954×10^{-15}	3.84×10^{-4}	5.11×10^{-2}
36	1.718×10^{-15}	1.67×10^{-4}	2.23×10^{-2}

Note: When gas flow is in MCFH, A must be multiplied by 10^6.
* See Eq. **2a.**
† Values tabulated correspond to A/A_e, where A_e is the desired equivalent diameter. Equivalents for sizes other than 6 and 16 in. may be determined from A data given, e.g., one foot of 30 in. pipe in terms of 10 in. pipe is: $A_{30\text{ in.}}/A_{10\text{ in.}} = 3.954 \times 10^{-15}/7.818 \times 10^{-13} = 0.005$ ft. See also Eq. **6** in Chap. 2, Sec. 8.

paths in the two-dimensional network; for example, by starting at the right end of main 15 via main 8 and returning to the left end of main 15 via main 2 or 3.

The next step is to write the gas load distribution on the map. A convenient method[41] of load distribution is to locate at each end of the main a concentrated load equal to one-half the distributed load on it.* The resulting load distribution places points of delivery from the network at main junctions. This will produce practically the same pressure drop as the distributed load if the distributed load is less than one-half the total flow entering the main. Computed pressure drops are always less than actual drops—a negligible error since the load data are estimated.

* The magnitude of the error introduced into network flow calculations by this assumption is:

Distributed load/total load	0.3	0.4	0.6	0.8	1.0
Error, per cent	1.0	2.0	6.0	13.0	25.0

Equivalent Length of Pipe. In using the Hardy Cross method, it is desirable to calculate the equivalent length of each main in terms of a single reference diameter. Factors for determining equivalent lengths of 6 in. and 16 in. diam pipe are shown in Table 9-30. The equivalent length is obtained by multiplying the actual length of pipe by the appropriate factor in one of the last two columns of this table. For example, 330 ft of 12 in. main is the equivalent of 330 $(3.08 \times 10^{-2}) = 10$ ft of 6 in. main.

Slide Rule Calculation of Sample Problem.[42] A low-pressure gas distribution system, composed of fifteen mains (Fig. 9-62b), has been analyzed by the Hardy Cross method† to determine the individual main flow rates and pressure drops. Numbers underlined refer to the loops; those in circles refer to the mains. Gas flow into the network from a *source* on the left side is 245 MCFH. Points of delivery are at junctions of mains, with the arrows pointing to volumes delivered (summation of these deliveries equals 245 MCFH). Assumed gas flow and its direction, also indicated by an arrow, appear either to the right or below the main for the first trial, and the corresponding pressure drop is to the left or above the main. The diameter, length of main, and *equivalent* length of 6 in. main for the large diameter mains are shown next to the pressure drop. When equivalent lengths of main are used, the pipe resistance factor, A, becomes the same for all mains in the network.

After the network map with its main and loop numbers and delivery and supply data has been prepared, the next step is to assume a flow pattern in the network. This may be done by starting at sources‡ with volumes of gas delivered into the system, and distributing these volumes thru the mains until they have been allocated to the various delivery points. The flows thus assumed are entered next to their respective

† Operations described here are similar to those used in Kirchhoff's method for solving electrical circuits. However, the resistances involved are not constants. Thus, the trial-and-error method presented here is more complex. For example, although the algebraic sum of the flows in all mains that meet at a point is zero, the sum of all the pressure drops acting around a loop usually approaches zero only in later trials.

‡ A single source in this problem.

Table 9-31 Hardy Cross Analysis for the Network in Fig. 9-62b

$A \pm 1.03 \times 10^{-5}$*			Trial 1						Trial 2						Trial 3						Trial 4		
						Correction						Correction						Correction					
Loop	Main	Length	Q	ALQ	ALQ^2	1	2	3	Q	ALQ	ALQ^2	1	2	3	Q	ALQ	ALQ^2	1	2	3	Q	ALQ	ALQ^2
1	1	330	+6.0	0.0204	+0.12	−1.8	...	...	+7.8	0.0265	+0.21	−0.5	...	...	+8.3	0.0282	+0.23	−0.3	...	...	+8.6	0.0292	+0.25
1	2	660	−6.0	0.0408	−0.25	−1.8	±1.9	...	−6.1	0.0415	−0.25	−0.5	±0.5	...	−6.1	0.0415	−0.25	−0.3	±0.5	...	−6.3	0.0428	−0.27
1	3	10	−68.0	0.0070	−0.47	−1.8	‡3.4	...	−62.8	0.0065	−0.40	−0.5	‡0.6	...	−61.7	0.0064	−0.39	−0.3	‡0.3	...	−61.1	0.0063	−0.38
1	4	5	+81.0	0.0042	+0.34	−1.8	...	...	+82.8	0.0043	+0.36	−0.5	...	...	+83.3	0.0043	+0.36	−0.3	...	...	+83.6	0.0043	+0.36
				0.0724	−0.26					0.0788	−0.08					0.0804	−0.05						−0.04
				Δ = −0.26/2(0.0724) = −1.8						Δ = −0.08/2(0.0788) = −0.5						Δ = −0.05/2(0.0804) = −0.3							
2	5	330	+6.0	0.0204	+0.12	−1.9	...	...	+7.9	0.0269	+0.21	−0.5	...	...	+8.4	0.0286	+0.24	−0.5	...	...	+8.9	0.0296	+0.27
2	6	1320	−5.0	0.0680	−0.34	−1.9	‡3.6	...	+0.5	0.0068	+0.00	−0.5	=0.2	...	+0.8	0.0109	+0.01	−0.5	=0.4	...	+0.9	0.0122	+0.01
2	11	330	−8.0	0.0272	−0.22	−1.9	∓2.8	...	−8.9	0.0302	−0.27	−0.5	∓1.0	...	−9.4	0.0320	−0.30	−0.5	∓0.8	...	−9.7	0.0330	−0.32
2	12	660	−10.0	0.0680	−0.68	−1.9	‡3.4	∓2.8	−7.5	0.0510	−0.38	−0.5	‡0.6	∓1.0	−7.4	0.0503	−0.37	−0.5	‡0.3	∓0.8	−7.4	0.0503	−0.37
2	2	660	+6.0	0.0408	+0.25	−1.9	=1.8	...	+6.1	0.0415	+0.25	−0.5	=0.5	...	+6.1	0.0415	+0.25	−0.5	=0.3	...	+6.3	0.0415	+0.27
				0.2244	−0.87					0.1564	−0.19					0.1633	−0.17						−0.14
				Δ = −0.87/2(0.2244) = −1.9						Δ = −0.19/2(0.1564) = −0.5						Δ = −0.17/2(0.1633) = −0.5							
3	7	330	+1.0	0.0034	+0.00	+3.6	...	...	−2.6	0.0089	−0.02	−0.2	...	...	−2.4	0.0082	−0.02	−0.4	...	...	−2.0	0.0068	−0.01
3	8	660	−5.0	0.0340	−0.17	+3.6	...	...	−8.6	0.0585	−0.50	−0.2	...	...	−8.4	0.0570	−0.48	−0.4	...	...	−8.0	0.0544	−0.43
3	9	660	+15.0	0.1020	+1.53	+3.6	‡2.8	...	+8.6	0.0585	+0.50	−0.2	‡1.0	...	+7.8	0.0530	+0.41	−0.4	‡0.8	...	+7.4	0.0503	+0.37
3	10	330	+5.0	0.0170	+0.08	+3.6	‡2.8	...	−1.4	0.0048	−0.01	−0.2	∓1.0	...	−2.2	0.0075	−0.02	−0.4	∓0.8	...	−2.6	0.0081	−0.02
3	6	1320	+5.0	0.0680	+0.34	+3.6	=1.9	...	−0.5	0.0068	−0.00	−0.2	±0.5	...	−0.8	0.0109	−0.01	−0.4	±0.5	...	−0.9	0.0122	−0.01
				0.2244	+1.78					0.1375	−0.03					0.1366	−0.12						−0.11
				Δ = +1.78/2(0.2244) = +3.6						Δ = −0.03/2(0.1375) = −0.2						Δ = −0.12/2(0.1366) = −0.4							
4	3	10	+68.0	0.0070	+0.47	+3.4	=1.8	...	+62.8	0.0065	+0.40	+0.6	=0.5	...	+61.7	0.0064	+0.39	+0.3	=0.3	...	+61.1	0.0063	+0.38
4	12	660	+10.0	0.0680	+0.68	+3.4	=1.9	∓2.8	+7.5	0.0510	+0.38	+0.6	=0.5	∓1.0	+7.4	0.0503	+0.37	+0.3	=0.5	∓0.8	+7.4	0.0503	+0.37
4	13	330	−4.0	0.0136	−0.05	+3.4	...	...	−7.4	0.0252	−0.19	+0.6	...	...	−8.0	0.0271	−0.22	+0.3	...	...	−8.3	0.0282	−0.23
4	14	5	−94.0	0.0048	−0.46	+3.4	...	...	−97.4	0.0050	−0.49	+0.6	...	...	−98.0	0.0051	−0.49	+0.3	...	...	−98.3	0.0051	−0.50
				0.0934	+0.64					0.0877	+0.10					0.0889	+0.05						+0.02
				Δ = +0.64/2(0.0934) = +3.4						Δ = +0.10/2(0.0877) = +0.6						Δ = +0.05/2(0.0889) = +0.3							
5	15	20	+50.0	0.0103	+0.52	+2.8	...	...	+47.2	0.0099	+0.46	+1.0	...	...	+46.2	0.0095	+0.44	+0.8	...	...	+45.4	0.0091	+0.41
5	9	660	+15.0	0.1020	+1.53	+2.8	‡3.6	...	+8.6	0.0585	+0.50	+1.0	±0.2	...	+7.8	0.0530	+0.41	+0.8	±0.4	...	+7.4	0.0503	+0.37
5	10	330	+5.0	0.0170	+0.08	+2.8	‡3.6	...	−1.4	0.0048	−0.01	+1.0	=0.2	...	−2.2	0.0075	−0.02	+0.8	=0.4	...	−2.6	0.0081	−0.02
5	11	330	−8.0	0.0272	−0.22	+2.8	=1.9	...	−8.9	0.0302	−0.27	+1.0	=0.5	...	−9.4	0.0320	−0.30	+0.8	=0.5	...	−9.7	0.0030	−0.32
5	12	660	−10.0	0.0680	−0.68	+2.8	=1.9	‡3.4	−7.5	0.0510	−0.38	+1.0	=0.5	‡0.6	−7.4	0.0503	−0.37	+0.8	=0.5	‡0.3	−7.4	0.0503	−0.37
				0.2245	+1.23					0.1544	+0.30					0.1023	+0.16						+0.09
				Δ = +1.23/2(0.2245) = +2.8						Δ = +0.30/2(0.1544) = +1.0						Δ = +0.16/2(0.1023) = +0.8							

* From Table 9-30 (for 6 in. pipe), when gas flow is in MCFH. Note that factor A would not affect corrections if it were eliminated from products ALQ and ALQ^2.

mains, with arrows to indicate direction. *The total gas flow arriving at a junction must equal the total gas flow leaving it.* The assumed flow pattern will approximate the correct flow pattern if consideration is given to the relative flow capacity of various network mains.

Four trials are presented in Table 9-31. Results of the fourth trial, after the network has been brought into approximate balance, are also shown in Fig. 9-62b. With an assumed source pressure of 6 in. w.c., the lowest pressure in the network will be approximately 4.9 in. w.c. at the junction of mains 9 and 10.

Loop and main numbers are listed in the first and second columns of the table, respectively. Main length, expressed as equivalent length of 6 in. main, is listed in the third column of Table 9-31. The assumed gas flow in each main for Trial 1 is listed in the fourth column. The plus or minus preceding the flow, Q, indicates the direction of the main flow for the *particular loop*. A *plus sign* denotes *clockwise* flow in the main within the loop; a *minus sign, counterclockwise.*

The first computation—resistance to change in gas flow in each main, AL_fQ—is listed in Table 9-31. The coefficient n, which in this case equals 2, has been factored out of this product, and is applied later. The pressure drop in each main, AL_fQ^2, is listed, and carries the same sign as the gas flow. Column AL_fQ is added *arithmetically* for each loop. Column AL_fQ^2 is added *algebraically* for each loop. A flow correction,* Δ, equal to $(\Sigma + AL_fQ^2)/2\Sigma AL_fQ$ when $n = 2$, is computed for each loop. This correction must be *subtracted algebraically* from the assumed gas flow. A main common to two loops receives two corrections, and a main common to three or more loops receives three or more corrections.

Corrections listed for Trial 1 are for the gas flows in each main. Correction 1 is from the particular loop under consideration. Corrections 2 and 3 are from the second and third loops to which a main belongs. The *upper plus or minus sign* shown indicates direction† of flow in that main in these two loops and is obtained from Q for Trial 1. The upper sign is the same as the sign in front of Q if the flow direction in each loop coincides with the assumed flow direction in the particular loop under consideration, and opposite if it does not.

The flow, Q, and corrections are totaled across to obtain the Q listed under Trial 2 according to the following rules:

1. The algebraic operation for Correction 1 should be the *opposite* of its sign; i.e., *add when the sign is minus.*

2. The algebraic operation for Corrections 2 and 3 should be the opposite of their lower signs when their upper signs are the same as the sign in front of Q, and as indicated by their lower signs when their upper signs are opposite to the sign in front of Q.

Computation according to these rules is by an *algebraic subtraction* of the flow correction terms, and is shown by the following examples for Trial 1:

Loop 1, main 1: $+6.0 + 1.8 = +7.8$
Loop 1, main 2: $-6.0 + 1.8 - 1.9 = -6.1$
Loop 2, main 12: $-10.0 + 1.9 + 3.4 - 2.8 = -7.5$
Loop 4, main 12: $+10.0 - 3.4 - 1.9 + 2.8 = +7.5$
Loop 5, main 12: $-10.0 - 2.8 + 1.9 + 3.4 = -7.5$

* $(\Sigma \pm ALQ^n)/n\Sigma ALQ^{n-1}$.
† Plus sign is clockwise; minus sign, counterclockwise.

The calculation procedure is repeated for Trial 2 and for successive trials until the net pressure drop around each loop, $\Sigma \pm AL_fQ^2$, is as close to zero as the degree of precision desired demands.

The network is then in approximate balance. The gas flows, with their directions and pressure drops as calculated in the final trial, are inserted on the network map next to the corresponding mains, as shown in Fig. 9-62b. An examination of the map will disclose the junction of lowest pressure.

Pressure drops in mains along a path from the junction of lowest pressure to the supply source are summed to obtain the total pressure drop in the network. When the gas flow in a main is in the same direction as the path taken between two junctions, there is a pressure loss. When the flow is opposite to the direction of the path, there is a pressure gain. Since the network is in approximate balance, the total pressure drop should be computed along several paths and averaged to obtain a better value.

Punched Card Calculating Machine

A transition from slide rule to punched card machine calculation of the flows and pressure drops in a network may be completed in the following four steps:[42] (1) prepare a deck of punched cards consisting of main cards and *trailer* cards containing main length, assumed gas flow, pipe resistance factor, and codes; (2) calculate pressure drops and flow corrections in an electronic calculating punch machine; (3) punch flow corrections into pertinent cards in a reproducing punch machine; and (4) print an iteration or results-of-trial report, compute new gas flow, and punch summary into a new deck of cards by means of an accounting machine and reproducing punch machine.

Analog Computer Methods

Gas distribution networks are not easily duplicated by models using fluids. However, since many similarities exist between the flow of fluids and the flow of electricity, it is possible, for example, to solve many gas distribution problems via electrical analogs.

The upper diagram of Fig. 9-63 shows a simple pipeline network supplied with gas at locations A and G, and dis-

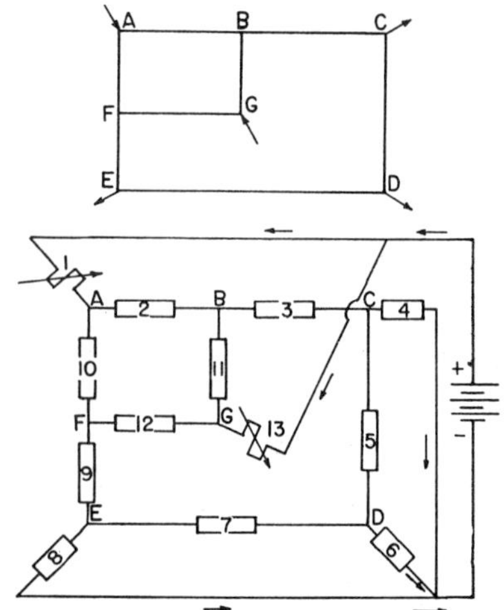

Fig. 9-63 A distribution network and its electrical analog.

charging at C, D, and E. The lower diagram represents an electric circuit analogous to this network. Source rheostats 1 and 13 and the load resistors (or constant current devices) 4, 6, and 8 are adjusted so that currents flowing thru them are proportional to desired gas flows. At all times, current thru and voltage drop across any resistor are, respectively, analogous, but not necessarily proportional, to gas flow and pressure drop h (or $P_1^2 - P_2^2$, in high-pressure networks) in the corresponding pipeline. Current in conductors at circuit terminals and voltages around circuit loops, however, inherently satisfy Kirchhoff's requirements* that flow rates balance at pipeline junctions and that pressure drops balance around pipeline loops. The source of current, represented as a battery in Fig. 9-63, may be any constant potential source of direct current of adequate capacity. Table 9-32 lists comparable quantities in the two networks.

Table 9-32 A Gas Main Network and Its Electrical Analog (supplements Fig. 9-63)

Gas main system	Electrical analog
Pressure at source, or square of absolute pressure*	Voltage at source
Pressure drop due to friction, or $(P_1^2 - P_2^2)$*	Voltage difference due to resistance
Gas flow rate:	Current:
Thru pipe	Thru resistor
At supply points A and G	At sources A and G
At load centers C, D, and E	At load centers

* For high-pressure flow.

To provide "complete" proportionality between corresponding quantities in the pipeline network and in the electric circuit, the relation between the voltage across and the current thru each resistor must be similar to that between the pressure drop along and the flow rate thru the pipeline the resistor represents. Since the pipeline relationships are nonlinear, an ordinary linear resistor cannot provide a complete analogy to pipeline performance—a successive approximation procedure must be employed. The approximation procedures are considerably improved by means of nonlinear resistors with current–voltage characteristics approximating those of pressure drop vs. flow in pipelines.

Pressure Drop vs. Flow Relation. Gas flow rate calculations are commonly based on the general flow equation (Eq. **1** in Chap. 2, Sec. 8). In fluid flow systems, resistance depends on flow rate, Q.† Flow rate (the analog of current) is nonlinearly related to pressure drop (the analog of voltage). According to Eq. **2a**, for example, the pressure drop varies as the *square* of the flow rate. There is also a difference in the manner in which flow rate is related to diameter in various flow equations, but this is of no greater importance, since either a single diameter or networks in terms of an equivalent diameter are involved. In general,

$$h = R_p Q^n \tag{4}$$

where R_p takes account of specific gravity, length, diameter, etc.

* The basis of the Hardy Cross analysis.

† As reflected in the Reynolds number; the relationship does not hold[43] in the transition region between smooth and turbulent flow for Reynolds numbers above 10,800,000.

The foregoing nonlinear relationships are of no concern in simple systems, as instruments with nonlinear scales can be made. However, difficulty is realized in interconnected systems. For example, referring to the lower diagram in Fig. 9-63, all current, I, entering at A divides into two streams: $I = I_{AB} + I_{AF}$. Similarly, in the upper diagram, $Q = Q_{AB} + Q_{AF}$. If linear resistors are used in the calculator, I cannot be used to represent Q, since in the gas flow formula h varies as Q^n, where n is between 1.75 and 2. On the other hand, if Q^n is represented by I, an equation inconsistent with the foregoing one would result.

Calculators Using Linear Resistors. Consider a single resistance in an electrical network, which will simulate a particular section of pipe in the gas distribution system. Let voltage be proportional to pressure drop:

$$B = E/h \qquad \text{(volts per unit pressure drop)} \tag{5}$$

Similarly, let current be proportional to flow:

$$G = I/Q \qquad \text{(amperes per unit flow rate)} \tag{6}$$

and let electrical resistance R_e be proportional to pipe resistance R_p:

$$C = R_e/R_p \tag{7}$$

It is possible to prescribe initially the value of any two of the proportionality constants B, G, and C, and to evaluate the third by the flow equation as follows.

Substituting Eqs. **5** and **6** into Eq. **4**:

$$E = R_p B I^n / G^n \tag{8}$$

Accordingly, in the electrical network, since $E = IR_e$, where R_e = electrical resistance:

$$R_e = R_p B I^{n-1}/G^n \tag{9}$$

Several procedures are available for evaluating the G term.[44–49] Basically, all involve measuring the current in each branch and adjusting each branch resistance, R_e, as determined by Eq. **9**, using the measured current value. This procedure is continued until successive values of currents are in agreement. Replacing h by $(P_1^2 - P_2^2)$ adapts the procedure to high-pressure systems. (See Ref. 49 for use of an oscilloscope to obviate need for repetitive calculations.)

One version of a network analyzer consists of linear rheostats arranged in rows on panels. Each rheostat assembly consists of three elements: the rheostat, a metering jack for measuring current and resistance, and jacks for interconnecting the various rheostat assemblies into a configuration similar to the distribution system studied. The panels also provide network connections to the battery at supply points.

Located on other panels are electronic load circuits, which require currents proportional to gas rates withdrawn from corresponding points in the gas system. These circuits, once set to withdraw a certain current from the network, will maintain this current regardless of network changes. Each load circuit consists of an electron tube, a circuit jack for interconnecting the load unit with the proper network point, a measuring jack for measuring current withdrawn by the tube, and a potentiometer for adjusting the amount of current carried by the load unit.

Metering equipment consists of a voltmeter and a combination ammeter-ohmmeter for measuring electrical values in the network. For more precise voltage measurements a potentiometer circuit is used.

Network calculators of this type have been used for many years by the electrical industry. Some calculators employ calibrated rheostats; others use different methods for constructing the network to be studied. Most of these use a d-c power supply, although a-c network calculators have been used by some investigators.[46,47]

Calculators Using Nonlinear Resistors. The number of successive approximations necessary to prepare an analog of a gas flow network is reduced by using nonlinear resistors. For example, the McIlroy Pipeline-Network Analyzer, manufactured by the Standard Electric Time Co., Springfield, Mass.,[41,50–54] employs nonlinear resistors known as *fluistors*. Each fluistor performs over its normal operating range; for example, from 1 to 4.6 volts according to the relation $E = kI^{1.85}$ or $k = E/I^{1.85} = R_p(B/G^{1.85})$* where k is a constant called the coefficient of the fluistor. Fluistors are manufactured with a large number of coefficients, varying by increments of approximately 5 per cent, to represent the ranges of G, L, and D involved in network studies.

Digital Computer Methods[55]

A modified Hardy Cross method will be followed in discussing these computer techniques. Data as indicated in the first four columns of the sample slide rule calculation of the Hardy Cross method given in Table 9-31 are coded onto tapes and fed into the memory unit of the computer. These tapes form a permanent record of the problem. Note that although these tapes are easily changed to reflect system modifications and revised loadings, the computer time required to achieve subsequent solutions will not be significantly reduced. The program or outline of calculations to be performed is also fed into the computer. The computer then begins a series of trial-and-error network solutions. The major difference between computer trials and those shown in Table 9-31 is that in computer trials each loop flow correction is applied to the flow rates in the pipe sections comprising the loop *before* the flow correction for the succeeding loop is calculated.

Means of following the progress of the solution include a *print-out* (1) of the loop flow correction as soon as it is determined; (2) following each trial of the number of loops for which either no flow correction was required or the correction made was less than an assigned value; (3) of the arithmetic sum of the absolute values of loop corrections; and (4) of the algebraic sum of loop pressure losses. These print-outs enable the computer technician to make adjustments that will lessen the number of trials necessary to reach the desired degree of *convergence* as well as to follow the progress of the solution.

A convergence plot, the summation of the *absolute* value of ΔQ (indicated as Δ in Table 9-31) vs. the trial or iteration number, does not necessarily show a decreased absolute ΔQ for each successive iteration. In general, an absolute ΔQ equal to or less than 0.05 MCFH is considered acceptable for an "average" distribution system. Higher values of ΔQ could be set for intermediate- and high-pressure feeder networks.

Methods for increasing the rate of convergence and decreasing the time required per trial include treating every possible low-resistance path between feed points as a closed loop and minimizing, if possible, the loops having combinations of high-flow low-resistance pipe sections and low-flow high-resistance pipe. It was found in regard to the latter that the elimination of low-flow high-resistance sections seldom had any significant effect on flow in the remainder of the problem.

Pressure regulators are accounted for by considering the paths between them as loops (the number of these paths equals one less than the number of regulators).

* By substitutions from Eq. **8**.

SUPPORT AND ANCHORAGE OF MAINS

Reference may be made to the following extract from ASA B31.8-1963:

834 SUPPORTS AND ANCHORAGE FOR EXPOSED PIPING.

834.1 General. Piping and equipment shall be supported in a substantial and workmanlike manner, so as to prevent or damp out excessive vibration, and shall be anchored sufficiently to prevent undue strains on connected equipment.

834.2 Provision for Expansion. Supports, hangers and anchors should be so installed as not to interfere with the free expansion and contraction of the piping between anchors. Suitable spring hangers, sway bracing, etc., shall be provided where necessary.

834.3 Materials, Design and Installation. All permanent hangers, supports, and anchors shall be fabricated from durable incombustible materials, and designed and installed in accordance with the good engineering practice for the service conditions involved. All parts of the supporting equipment shall be designed and installed so that they will not be disengaged by movement of the supported piping.

834.4 Forces on Pipe Joints.

(a) All exposed pipe joints shall be able to sustain the maximum end force due to the internal pressure, i.e., the design pressure (psi) times the internal area of the pipe (sq in.); as well as any additional forces due to temperature expansion or contraction, or to the weight of pipe and contents.

(b) If compression or sleeve-type couplings are used in exposed piping, provision shall be made to sustain the longitudinal forces noted in 834.4 (a). If such provision is not made in the manufacture of the coupling, suitable bracing or strapping shall be provided; but such design must not interfere with the normal performance of the coupling nor with its proper maintenance. Attachments must meet the requirements of 834.5.

834.5 Attachment of Supports or Anchors.

(a) If the pipe is designed to operate at a hoop stress of less than 50% of the specified minimum yield strength, structural supports or anchors may be welded directly to the pipe. Proportioning and welding strength requirements of such attachments shall conform to standard structural practice.

(b) If the pipe is designed to operate at a hoop stress of 50% or more of the specified minimum yield strength, support of the pipe shall be furnished by a member which completely encircles it. Where it is necessary to provide positive attachment, as at an anchor, the pipe may be welded to the encircling member only; the support shall be attached to the encircling member, and not to the pipe. The connection of the pipe to the encircling member shall be by continuous, rather than intermittent, welds.

835 ANCHORAGE FOR BURIED PIPING.

835.1 General. Bends or offsets in buried pipe cause longitudinal forces, which must be resisted by anchorage at the bend, by restraint due to friction of the soil, or by longitudinal stresses in the pipe.

835.2 Anchorage at Bends. If the pipe is anchored by bearing at the bend, care shall be taken to distribute the load on the soil so that the bearing pressure is within safe limits for the soil involved.

835.3 Restraint Due to Soil Friction. Where there is doubt as to the adequacy of anchorage by soil friction, calculations should be made.

835.4 Forces on Pipe Joints. If anchorage is not provided at the bend (835.2), pipe joints which are close to the points of thrust origin shall be designed to sustain the longitudinal pullout force. If such provision is not made in the manufacture of the joint, suitable bracing or strapping shall be provided, unless calculations show the joint to be safe.

835.5 Supports for Buried Piping. In pipelines, especially those which are highly stressed from internal pressure, uniform and adequate support of the pipe in the trench is essential. Unequal settlements may produce added bending stresses in the pipe. Lateral thrusts at branch connections may greatly increase the stresses in the branch connection itself, unless the fill is thoroughly consolidated or other provisions made to resist the thrust.

835.51 When openings are made in a consolidated backfill to connect new branches to an existing line, care must be taken to provide firm foundation for both the header and the branch, to prevent both vertical and lateral movements.

835.6 Interconnection of Underground Lines. Underground lines are subjected to longitudinal stresses due to changes in pressure and temperature. For long lines, the friction of the earth will prevent changes in length from these stresses, except for several hundred feet adjacent to bends or ends. At these locations the movement, if unrestrained, may be of considerable magnitude. If connections are made at such a location to a relatively unyielding line, or other fixed object, it is essential that the interconnection shall have ample flexibility to care for possible movement, or that the line shall be provided with an anchor sufficient to develop the forces necessary to limit the movement.

Anchorage of Mains*,[56]

Movements seldom occur in a piping system since its components are interlocked and restrained by the surrounding soil, but in most piping systems there are thrusts which tend to pull joints apart. Thus, additional restrains are occasionally required, especially where compression couplings or joints are extensively used.

When to Reinforce. Multiply the maximum operating pressure, in psig, by the pipe *internal* diameter, in inches, for all joints adjacent to bends greater than 12° as well as for dead ends. Reinforcement is usually necessary if this *product* is greater than 200; products under 200 may also call for reinforcement.

Safe Embedment Distance. For buried piping, the major restraints tending to prevent compression joint slippage are: (1) the weight of the earth cover above and (2) the soil pressure from the sides, both acting upon the pipe between the component subject to thrust and the nearest unreinforced compression joint. The length of structurally continuous† buried pipe required to resist a given thrust force and thereby prevent the pull-out of adjacent compression joints is known as the safe embedment distance.

In an *all-welded* piping system, safe embedment distance is practically inherent since longitudinal pipe stresses are transferred from joint to joint until the restraint due to earth equals the thrust.

Calculating Effect of Earth Cover. The effect of the earth cover on buried pipe can be calculated in the same way as loads on ditch conduits.[57] The values obtained by such formulas should be *doubled,* however, in view of the fact that both the top and bottom pipe surfaces tend to restrain pipe movement.

Lateral Soil Pressure. Active, rather than *passive,* lateral soil pressure should be considered when calculating safe embedment distance. Active lateral soil pressure is the maximum pressure exerted by an earth mass on a stationary body; e.g., on a retaining wall.

Other Restraints. In addition to resistance due to earth, there are, of course, other restraints which tend to prevent joint separations. Some of these are anchors, pull-out resistance of compression joints, resistance of certain types of coupling reinforcements, and pipe weight. The resistance capacity of the foregoing varies, of course, both within and among systems. Generally speaking, however, these restraints sometimes materially reduce the safe embedment distance.

Weight of Pipe and Anchors. These are usually classified as mere *dead weights.* Thrust- or bearing-type anchors (i.e., anchors which distribute horizontal thrust over a rather large expanse of soil) should be considered only if they bear against sufficient earth (there is seldom any assurance that the earth adjacent to the anchors will not be disturbed in urban areas). Mechanical anchors (e.g., the *screw type*) are so dependent on design and installation method and conditions that manufacturer's design data are necessary.

Pull-Out Resistance of Compression Joints. Variables include the condition of the pipe and coupling and the care used in joint assembly; a value of *50 pounds per linear inch of circumference* is conservative under normal conditions.

Coupling Reinforcement. These sleeves or *pumpkins* provide compression couplings with additional bearing surface. The reinforcements may be considered the equivalent of *bearing-type* anchors. Note, however, the precautionary remark on bearing-type anchors.

Conditions and Factors Affecting Pipe Movement. Although safe embedment distance and the effect of anchors, compression joint pull-out resistance, etc., can be calculated, there are many factors affecting pipe movement which are as yet not too well understood. For example:

1. *Coated pipe.* Some coatings become plastic under certain conditions and permit the piping to creep or slip. These phenomena seem to be confined to desert areas and piping near compressor stations. For the usual distribution or transmission system, no distinction between coated and bare piping is generally made.

2. *New systems.* Backfill does not develop the calculated pressures until after it has settled over a period of time. On the other hand, there have been reports that, perhaps as a result of thermal contractions and expansions due to seasonal temperature variations, voids form at the pipe–soil interface. Derating factors for these phenomena are not in general use.

3. *Vertical thrusts.* At overbends and similar points at which upward vertical thrusts can occur, the weight of the backfilled earth generally acts as the primary resistance to the

* By Thaddeus J. Miller.

† Longitudinal stresses developed at the point of thrust origin must be transferred to the adjacent pipe; i.e., pipe joints within that distance are either welded or, in the case of compression joints, reinforced to transmit the stresses parallel to the axis.

Table 9-33 Values for Terms Used in Embedment Formulas*

Pipe diam, nom., in.	R_j,† lb	A_r, sq ft
6	1000	0.73
8	1350	0.88
10	1700	1.08
12	2000	1.23
14	2200	1.33
16	2500	1.48
18	2800	1.64
20	3100	1.80
22	3500	1.94
24	3800	2.10
26	4100	2.25
30	4700	2.56

Soil type	F_c,‡ psf
Muck, peat, etc.	0
Soft clay	500
Sand	1000
Sand & gravel	1500
Sand & gravel w/clay	2000
Shale	5000

Ditch fill mat'l	w,‡·§ lb/cu ft
Sand, dry	90–110
moist	100–115
wet	110–120
Earth (loam or silt), dry	80–100
moist	90–110
wet	100–120
Gravel (round to angular)	100–135
Gravel, sand, & clay	100–120
Clay	120–130

Materials	u_1§
Masonry on:	
Wet clay	0.33
Dry clay	.50
Sand	.40
Gravel	0.60

Materials	u_2**
Pipe on:	
Wet clay & sat. soil	0.20
Sandy silt, top soil, & clay	.33
Dry clay, dense gravel, sand, etc.	0.45

Ditch fill mat'l	C‡ at following H/B ratios (fill height, H, to breadth, B—both above pipe) 0.5	1.0	1.5	2.0	2.5	3.0	3.5	4.0	4.5	5.0	K_a††‡‡
Gran. w/o tamp or set	0.455	0.830	1.140	1.395	1.606	1.780	1.923	2.041	2.136	2.219	0.320
Sand, general	.461	.852	1.183	1.464	1.702	1.904	2.075	2.221	2.344	2.448	.155
Top soil, saturated	.464	.864	1.208	1.504	1.764	1.978	2.167	2.329	2.469	2.590	.151
Clay, general	.469	.881	1.242	1.560	1.838	2.083	2.298	2.487	2.650	2.798	.079
Clay, saturated	0.474	0.898	1.278	1.618	1.923	2.196	2.441	2.660	2.856	3.032	0.055

* T. J. Miller. "When Should Compression Joints Be Reinforced?" (DMC-59-25) *A.G.A. Operating Sec. Proc.* 1959: D145–56.

† Based on data in: Consolidated Edision Co. *Steel Gas Mains—Anchorage Requirements and Determination of Embedment Length.* (Spec. EO-8079) New York, 1953

‡ S. Crocker. *Piping Handbook*, 4th ed. New York, McGraw-Hill, 1945

§ L. C. Urquhart, ed. *Civil Engineering Handbook*, 3rd ed., p. 697. New York, McGraw-Hill, 1950.

** L. L. Elder and L. B. Inglis. *Discussion on a Method of Determining a Safe Imbedment Distance for Gas Pressure Pipe with a Thrust* (unpublished report), Gas. Eng. Dept., Columbia Gas System Service Corp., Columbus, Ohio, Oct. 7, 1955.

†† D. P. Krynine. *Soil Mechanics, Its Principles and Structural Applications*, p. 135. New York, McGraw-Hill, 1947.

‡‡ M. G. Spangler, *Supporting Strength of Rigid Pipe Culverts*, (Bull. 112) Iowa Eng. Expt. Sta., Iowa State College, Ames, Iowa, Feb. 8, 1933; and "Stresses in Pressure Pipelines and Protective Casing Pipes," *Proc. Am. Soc. Civil Engrs.*, 82(ST5): Sept., 1956.

vertical component of the thrust. Hence, calculations should be made to ascertain whether additional anchorage may be required.

4. *Unstable soils.* In areas of unstable oils, or in areas which are subject to flooding or erosion, conditions may occur in which soil friction will be nonexistent. In such areas, piles, tie rods attached to anchors, ballast, and similar devices should be considered.

Embedment Formulas and Charts. Equations **10** thru **14** plus Table 9-33 constitute *Method A* for determining embedment distance for steel pipe. Alternate *Method B* is given in Fig. 9-64.

Method A:

$$L = \frac{1.5\,[T - (R_r + R_a)]}{R_{cp}} \qquad \textbf{(10)}$$

where:

L = safe embedment distance from thrust origin for buried pipe without size, pressure, or material* limitations; for systems using compression couplings (with and without welded joints) in stable soils (not subject to erosion or flooding) for both horizontal and vertical downward thrust; ft

T = thrust, lb (see Eqs. **11a** thru **11d**)

R_r = bearing resistance of coupling reinforcement, lb (see Eqs. **12a** and **12b**)

R_a = anchor resistance,† lb (see Eq. **13**)

R_{cp} = resistance due to cover and pipe weight, lb per linear ft (see Eq. **14**)

$$T = (PA - R_j) \sin 0.5\theta \qquad \textbf{(11a)}$$

for bend inside compression coupling.‡

$$T = PA\,(1 - \cos\theta) - R_j \qquad \textbf{(11b)}$$

* Except cast iron and coated and wrapped piping near compressor stations, in deserts or other high-temperature surroundings.

† The bearing-type anchor, installed where possible disturbance to it is remote, may be valued at design rating.

‡ Thrust is *twice* indicated value. The thrust, however, is resisted by restraints acting on both legs of the bend. Therefore, for the purpose of determining the safe embedment distance along each leg of the bend, only *half* of the actual thrust need be considered.

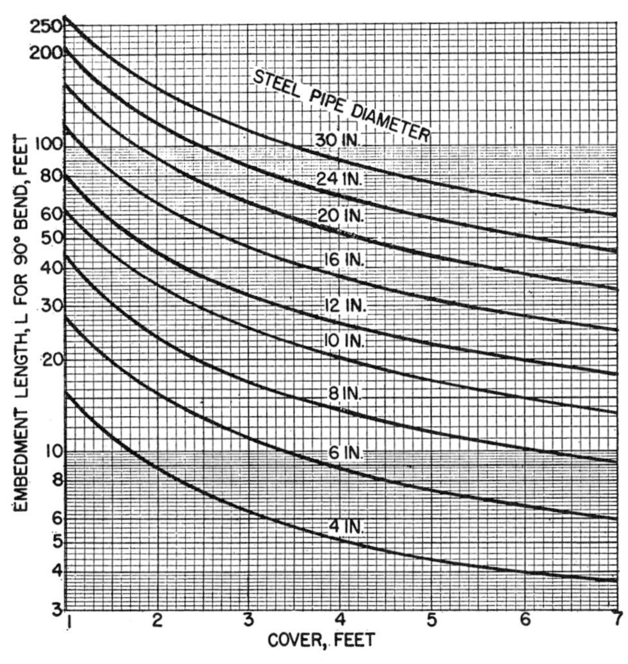

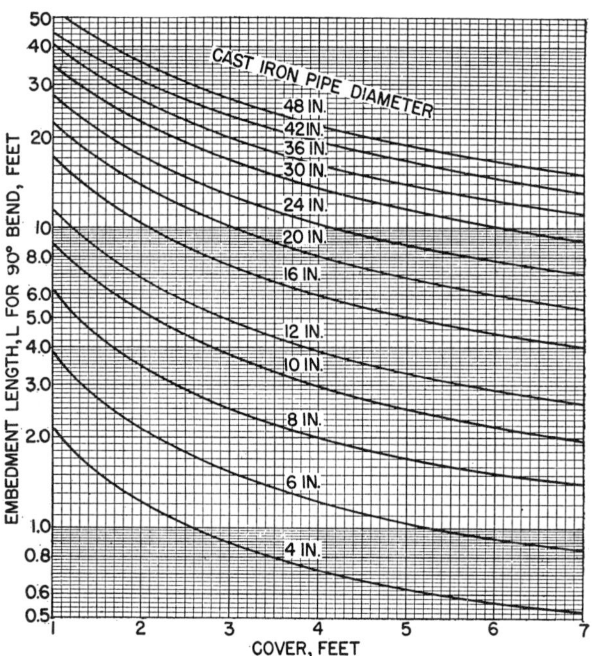

Fig. 9-64 Embedment lengths: (left) 4 to 30 in. steel mains at 100 psig; (right) 4 to 48 in. cast iron mains at 15 psig.[58]

for miter bend or bend made with fittings,* e.g., butt-welding ell, Dresser end ell, etc.

$$T = PA - R_j \tag{11c}$$

for dead end; e.g., plug, cap, tee, valve, etc.

$$T = P(A_1 - A_2) - R_j \tag{11d}$$

for reducer, where

P = maximum system operating pressure, psig
A = cross-sectional flow area of pipe, sq in.
A_1 = cross-sectional flow area of *large* end of reducer, sq in.
A_2 = cross-sectional flow area of *small* end of reducer, sq in.
R_j = resistance of compression joints, lb; see Table 9-33
θ = angle of bend, degrees

$$R_r = nA_rF_c \sin 0.5\theta \tag{12a}$$

for bend inside compression coupling.

$$R_r = nA_rF_c \tag{12b}$$

for miter bend or bend made with fittings and for dead ends, reducers, tees, etc., where

n = No. of compression couplings in distance L when calculated with $R_r = 0$ (first calculation of L determines n)
A_r = cross-sectional reinforcement area resisted by the soil, sq ft; see Table 9-33 for maximum values for "pumpkins" or sleeve-type reinforcements

* Thrust acts parallel to the pipe axis. Due consideration should be given, however, to the normal component of the thrust force. This normal component has a value equal to (PA) (tan 0.5 θ) (cos θ). An investigation should be made in each case to determine if this normal component is adequately resisted.

F_c = bearing capacity of earth, lb per sq ft; see Table 9-33 for horizontal thrusts with 2 ft min pipe cover; $F_c = 0$ for less than 2 ft cover and vertical upward thrusts; bearing for vertical *downward* thrusts *twice* tabulated value

$$R_a = u_1W_aF_L \tag{13}$$

where:

u_1 = friction coefficient between anchor and earth, dimensionless; see Table 9-33
W_a = anchor weight,† lb
F_L = location factor = 0.5 at bends, 1.0 elsewhere

$$R_{cp} = u_2[2CwB(B + DK_a) + W_p] \tag{14}$$

where:

u_2 = friction coefficient between pipe and earth, dimensionless; see Table 9-33
C = ditch coefficient, dimensionless; see Table 9-33
w = ditch filling material weight, lb per cu ft; see Table 9-33
B = trench width at top of pipe, ft
D = pipe OD, ft
K_a = backfill coefficient, dimensionless; see Table 9-33
W_p = pipe weight, lb per linear ft

Method B:[58]

When reinforcement is necessary, use Fig. 9-64 to determine embedment length L. The correction factor for *pressures* other than noted is the actual pressure in psig/noted pressure. The denominators for steel and cast iron mains are 100 and 15, respectively.

† Density usually 150 lb per cu ft.

Correction factor for bend angles other than 90°:

Bend angle, deg	12	22½	45	60
Factor	0.02	.08	.30	0.50

Anchorage Design. Angle iron or U-shaped braces may be welded to steel pipes on both sides of a compression joint and tied together with steel rods. An insulating bushing and washers are used at one end of each rod to effect the insulation of the individual lengths of pipe. Table 9-34 shows an applicable tie rod schedule.

Table 9-34 Tie Rods for Compression Coupling Anchorages on Steel Pipe[58]

Pressure, psig	Pipe size, in.	No. of rods	Rod diam, in.
100	2	1	½
100	4	2	⅝
100	6–12	2	¾
100	16–24	4	1
250	4–6	2	¾
250	8–12	4	1
350	6–8	2	1
350	10–12	2	1
350	16	4	1
350	20	4	1
350	24	4	1
350	30	6	1

REFERENCES

1. A.G.A. Bur. of Statistics. *Load Characteristics of Gas Heating Customers: Rept. 2.* New York, 1956.
2. Quinby, R., and others. "Report of Subcommittee on Customer Load Characteristics." *A.G.A. Proc.* 1949: 77–129.
3. Wilson, G. G. *Distribution Pipe Sizing Manual.* Chicago, Ill., I.G.T. To be published.
4. Morgan, G. A. "Development of Load Data for Distribution System Design." (DMC-53-27). *A.G.A. Proc.* 1953: 650–4.
5. Segeler, C. G. "Footnote for Load Characteristics." *Gas Age* 118: 24–6, Dec. 27, 1956.
6. "Report of Rate Committee." *A.G.A. Proc.* 1949: 60–135.
7. New, U. R. "7 Things TVA Has Learned About Its House Heating Load." *Elec. World* 144: 129+, Nov. 14, 1955.
8. A.G.A. Bur. of Statistics. *Load Characteristics of Gas Heating Customers: Rept. 3.* New York, 1958.
9. ——: *Rept. 1.* New York, 1955.
10. Bary, C. "Characteristics of the Gas House Heating Load." *A.G.A. Monthly* 31: 12+, Mar. 1949.
11. "Distribution System Capacities." *Gas* 23: 51–8, May 1947.
12. Gardner, J. R., and others. "Report of Subcommittee on Customer Load Characteristics." (In: "Report of Rate Committee," Appendix C) *A.G.A. Proc.* 1950: 117–43.
13. *A.G.A. Reference Manual of Modern Gas Service.* New York, 1946–7.
14. A.G.A. Laboratories. *Installation of Gas Appliances and Gas Piping.* (ASA Z21.30-1964) Cleveland, Ohio. 1964.
15. Hunt, B. E. "A Method Used in Forecasting Loads and Making Load Studies." *A.G.A. Proc.* 1953: 602–5.
16. A.G.A. Rate Comm. *Gas Rate Fundamentals,* New York, 1960.
17. Am. Soc. Heating, Refrig. Air-cond. Engrs. *ASHRAE Guide and Data Book.* New York, 1963.
18. A.G.A. Bur. of Statistics. *Trends in Annual and Peak Day Heating Degree Days, 1906–1907—1955–1956.* New York, 1958.
19. Taylor, J. C. *How Detroit Increased Its Distribution Capacity to Supply an Additional 100,000 Spaceheating Customers.* (DMC-50-4) New York, A.G.A., 1950.
20. Partridge, F. M. and Pomeroy, R. D. "Some Problems in Metropolitan Main Replacement." *P.C.G.A. Proc.* 40: 145, 1949.
21. Miller, D. J. "Denver Distribution Design for House-Heating Load." *A.G.A. Proc.* 1948: 657–64.
22. East Ohio Gas Co., Cleveland, Ohio. (Private memorandum.)
23. Kulman, F. E. "Operation Dig Less." *Am. Gas J.* 190: 24–7, Jan. 1963.
24. Cooper, G. A. S. "Large Mains Are Cheaper than Small Mains If You Expect Your Business to Expand." *A.G.A. Proc.* 1949: 634–40.
25. Patterson, M. K. "How Well Do Minnegasco's 'Mini-Mains' Work?" *Gas* 39: 53–7, Feb. 1963.
26. ——. "Investment Per Customer: How Can It Be Reduced?" *Am. Gas J.* 188: 25–8, Dec. 1961.
27. McCafferty, D. W., and Kroeger, C. V. "Gas Main Capacity." *Am. Gas J.* 182: 14–5, Sept. 1955.
28. "Gas Handbook Issue." *Am. Gas J.* 189: 34–37, Aug. 15, 1962.
29. Wilson, G. G., and Ellington, R. T. "Selection of Flow Equations for Use in Distribution System Network Calculations." (DMC-58-31) *A.G.A. Operating Sec. Proc.* 1958: D231–42.
30. Gustafson, R. J. "Flow Conditions Peculiar to Medium High Pressure Distribution Systems." *A.G.A. Operating Sec. Proc.* 1960: DMC-60-23. (Abstract of thesis for M.S. in Chem. Eng., Univ. of Rochester, 1959).
31. Johnson, T. W., and Berwald, W. B. *Flow of Natural Gas through High-Pressure Transmission Lines.* (Bur. of Mines Mono. 6) New York, A.G.A., 1935.
32. Beggs, C. W. "Distribution Analysis." *Gas Age* 117: 32–7, Feb. 23, 1956.
33. Peters, W. C. "A Practical Method of Utilizing Gas Load Data in Distribution System Design." (DMC-55-18) *A.G.A. Proc.* 1955: 567–80.
34. Griffith, M. P. "A Joint Participating Group Computer Facility." (DMC-58-8) *A.G.A. Operating Sec. Proc.* 1958: D53-7.
35. Beckman, P. E., and others. "Forecasting Winter-Day Load in a Natural Gas Territory." *P.C.G.A. Proc.* 32: 65–75, 1941.
36. Pugh, J. J. "A Method of Testing the Adequacy of Gas Distribution Systems." *A.G.A. Proc.* 1947: 517–28.
37. ——. "Method of Forecasting the Adequacy of Natural Gas Distribution Networks to Supply Potential Heating Load." *P.C.G.A. Proc.* 45: 74–8, 1954.
38. Smith, D. K. "Determination of Distribution System Adequacy by Temperature–Pressure Drop Method." *P.C.G.A. Proc.* 41: 118–20, 1950.
39. Moody, L. F. "Friction Factors for Pipe Flow." *Trans. ASME* 66: 671, 1944.
40. Stringer, R. B. "A Method of Studying Small Gas Utility Distribution Systems to Determine Best Means of Reinforcement to Carry Heavier Loads." *A.G.A. Proc.* 1954: 340–51.
41. Hunt, B. E. "The Uses and Applications of the Distribution System Calculator." *A.G.A. Proc.* 1952: 516–30.
42. Sickafoose, R. D. "A Punched Card Program for Calculating Gas Distribution Network Pressures by the Hardy Cross Method." (DMC-55-1) *A. G. A. Proc.* 1955: 345–66.
43. Smith, R. V., and others. *Flow of Natural Gas through Experimental Pipe Lines and Transmission Lines.* (Bur. of Mines Mono. 9) New York, A.G.A., 1956.
44. Camp, T. R., and Hazen, H. L. "Hydraulic Analysis of Water Distribution Systems by Means of an Electric Network Analyzer." *New Engl. Water Works Assoc. J.* 48: 383–404, Dec. 1934.
45. Perry, H. A., Jr., and others. "Network-Flow Analysis Speeded by Modified Electrical Analogy." *Eng. News Record* 143: 19–23, Sept. 22, 1949.
46. Haupt, L. M. *Solution of Hydraulic Flow Distribution Problems through the Use of the Network Calculator.* (Research Rept. 18) College Station, Texas, Texas Agr. Mech. College, Eng. Expt. Sta., 1950.
47. Clennon, J. P., and Dawson, J. K. "Gas Distribution Problems Solved by Electric Network Calculators." (DMC-51-14) *A.G.A. Proc.* 1951: 389–414.
48. Hunt, B. E. *Development of a Distribution System Calculator.* (I.G.T. Rept. No. 1) Chicago, Ill., 1951.

49. Stephenson, R. E., and others. "Simplify Analyzer Adaptation." *A.G.A. Monthly*. 35: 27+, Feb. 1953.
50. McIlroy, M. S. "Direct-Reading Electric Analyzer for Pipeline Networks." *Am. Water Works Assoc. J*. 42: 347–66, April 1950.
51. ——. "Water-Distribution Systems Studied by Complete Electrical Analogy." *New Engl. Water Works Assoc. J*. 65: 299–318, Dec. 1951.
52. Barker, C. L. "Computing Flow Conditions in Gas Lines With the McIlroy Analyzer." *Pipe Line News* 24: 21–22, Aug. 1952.
53. McIlroy, M. S. "Gas Pipe Networks Analyzed by Direct-Reading Electric Analogue Computor." (DMC-52-10) *A.G.A. Proc*. 1952: 433–47.
54. Trunk, E. F., and others. "Uses of the Hardy Cross Analysis for Calculating Distribution Network Pressures." *A.G.A. Proc*. 1953: 596–602.
55. Wilson, G. G., and others. "Practical Aspects of Distribution System Analysis with an Electronic Digital Computer." *A.G.A. Operating Sec. Proc*. 1957: DMC-57-19.
56. Miller, T. J. "When Should Compression Joints Be Reinforced?" (DMC-59-25) *A.G.A. Operating Sec. Proc*. 1959: D145–56.
57. Spangler, M. G. *Supporting Strength of Rigid Pipe Culverts*. (Bull. 112) Ames, Iowa State College, Eng. Expt. Sta., 1933.
58. Consolidated Edison Co. *Steel Gas Mains—Anchorage Requirements and Determination of Embedment Length*. (Spec. EO-8079) New York, 1953.

Chapter 7
Main Installation

by Harold Emerson and H. C. Roemmele

This text was prepared with utility company operations in mind. However, mains and services installed, owned, or operated by others may be handled in the same way.

PRELIMINARY PLANNING

Main installation may be done by the utility company or by a contractor, the choice depending in part on the availability of specialized equipment. Factors of cost, quality of work, public and employee relations, complexities of the project, and availability of utility personnel and equipment must be evaluated. Repaving, for example, is usually done by a contractor, since it requires special equipment not commonly owned by utility companies.

Contractors' equipment may also be employed on a rental basis. Sometimes this arrangement may be used advantageously when the utility makes part of the installation and desires full control over the progress of the project.

When main laying is done by contract, adequate utility inspection of the work as it progresses should ensure quality of the finished installation. Terms relating to workmanship and material should be clearly defined in the contract.

A published survey[1] reports on the equipment of contractors serving gas distribution companies.

Location

Data on the location, size, and type of existing and proposed underground structures should be obtained from available records and plotted. Then a field survey should be made to locate other possible obstacles before a main location is finally agreed upon. Authorities having jurisdiction over the use of streets by public utilities generally must approve and record the location selected.

Mains under unpaved streets or streets with unpaved or treated shoulders are usually laid in the roadway outside of the main line of traffic.

Where the entire roadway is paved, a sidewalk location will have the advantages of lower installation costs, less interference with traffic for installation or repair, and greater safety for workmen. Provision must be made for the installation of utility poles and light standards. Where the building line adjoins the sidewalk, avoid placing mains close to basement walls.

In built-up areas and on streets of unusual width, parallel mains under the sidewalks on each side may be more economical than a single main supplying services to both sides. Higher initial cost of a two-main system may be offset by lower maintenance costs, as mains and services are accessible with out breaking up expensive paving. System comparisons should also include an estimate of main lengths and number of services eventually required in the area and should reveal whether jacked or bored services are possible without interference to other utilities. Cathodic protection requirements should also be ascertained.

Many highway authorities request or require that mains not be located in roadways of heavy traffic arteries. Preferred locations, where main installation and maintenance will cause minimum interference with traffic, include sidewalks and center islands of divided highways.

Cover

Distance from top of main to ground surface normally depends on size of main, type of pipe, traffic conditions, depth of frost, geographic location, corrosiveness of soil at various depths, and characteristics of gas distributed. Where low winter temperatures occur, mains are usually placed below the frost line to reduce stresses due to excessive temperature changes and ground movement caused by freezing and thawing. When the gas contains moisture or other condensables, cover must be sufficient to prevent freezing and to permit grading the services back toward the main.

Protection of the main from traffic shocks and vibrations is another factor in determining cover, particularly for small cast iron mains. Cover of 30 to 36 in. of soil may be considered normal in most areas. Where very low temperatures occur, a greater depth may be needed. In mild climates, a lesser depth (24 in. and less) may be satisfactory, as provided in Par. 841.16 of ASA B31.8-1963 (extracted in Chap. 2 of Sec. 8).

Adequate cover for mains of average size crossing other underground structures is not normally a problem. Problems may arise with large mains. Placing the gas main below the other structure, such as a sewer, may make it too inaccessible. The alternative is to run it above the other structure with shallow (less than 18 in.) cover using a specially fabricated steel main section, a manifold of small pipes, or a steel pipe with a welded offset. Protection can also be provided by a steel or concrete casing, or by flat or curved steel plates[2] placed in the backfill about one foot above the top of the main. A two-inch clearance between structures is a desirable minimum.

INSTALLATION

Main and service installation and maintenance, and care of regulatörs, valves, and drips are generally the responsibility of the street division of the distribution department. Size and personnel of the main laying group depend on the nature of the individual project; i.e., pipe size and length, kind of joints, and extent to which mechanical equipment is employed dictate crew make-up. If joints are welded, qualified welders are required. Supervision is customarily provided by a foreman responsible to the general foreman of the street division; field engineering and surveying are frequeutly employed. (Films and other visual aids which emphasize safety in various distribution operations are available from A.G.A. on either a sale or a no-charge loan basis.)

Stringing of Pipe

In uncongested areas where it is possible to do so, pipe should be strung (blocked to prevent rolling) along the route in advance of trenching so as to be immediately available for laying. It should be placed as close as practical to the proposed main location so that it can be moved into the trench without crossing a traveled highway. Damage to the pipe in handling should be avoided, especially if coated pipe is involved. In congested areas it may be necessary to lay pipe off the truck and not string it along the ditch. When possible, double random lengths of steel pipe are used to reduce the number of welds required.

Trenching

Open trenching usually affords the most economical method for installing mains. Where special conditions, such as expensive pavements, heavy traffic, railroad crossings, or concrete driveways prevail, tunneling or boring may be resorted to for installation of sections of limited length. For small steel mains laid in paved streets, alternate trenching and boring may be preferable. Horizontal earth (and rock) boring machines are available with capacities up to 36 in. in diameter and over 200 ft. in depth.[3]

If underground structures such as water mains, sewers, and electric or telephone conduits are present and their location is indefinite, hand digging rather than trenching equipment is used. Support should be provided where necessary for any existing underground facilities in or near the construction area during both trenching and backfilling operations. Some companies hand-dig all utility crossings prior to machine trenching. Trench width should be kept to a minimum consistent with proper installation. One utility's minimums for basic (pipe joints and fittings require larger minimums) trench widths follow:[2]

	Pipe Size	
	Up to 4 in.	*Over 4 in.*
Cast iron	18 in.	4 in. each side
Steel	12 in.	4 in. each side

Sheeting. Trenching in very soft or unstable ground requires restraints to protect workmen and maintain trench shape. Sheeting is the most commonly used method. The usual procedure is to drive sharpened wooden planks into the ground with their flat sides against the trench walls. The surface formed by the planks joined* at the edges and securely braced prevents collapse of the walls. Planks are added as trenching proceeds. Steel plating may also be used for this purpose.

Water in the trench is usually removed by gasoline sump pumps. In sandy soil with a high water table, on the other hand, well-point tipped piping is driven into the ground around the area to be excavated and manifolded to a pump to withdraw water and entrained air. By this method entry of water and free-flowing soil into the trench is avoided.

Pipe Laying

Steel Pipe. When welded joints are used, the pipe is generally placed on supports alongside or above the trench. Each length is connected to the preceding one by a line-up clamp. Adjoining ends are either tack welded or continuously welded (*stringer bead* only) around their circumference to maintain position.

The line-up clamp can then be removed and the joint completely welded. Either arc or gas welding may be specified for the particular job. The length of pipe assembled outside the trench depends on the length of unobstructed trench and on the ability of available handling equipment to lower pipe into the trench. When the pipe must be laid under an obstruction crossing the trench, adjoining pipe sections must be joined by a *bell hole weld* after being finally positioned. Steel pipe in smaller sizes joined by mechanical couplings may be assembled adjacent to the trench and lowered into position in the same way as welded pipe. Larger sized pipe joined by mechanical couplings is best assembled in the trench to avoid undue strain on the joints.

When dry gas is to be distributed and exact grade need not be maintained, the pipe may rest either on the bottom of the trench as excavated or upon a layer of sand to avoid damage to pipe coating. See Par. 841.2 of ASA B31.8-1963 (extracted in Chap. 2 of Sec. 8).

Cast Iron Pipe. Mechanical joint cast iron pipe in smaller sizes may be assembled along the trench in sections like steel pipe, unless a joint employing lead in addition to a rubber (neoprene for manufactured gases) gasket is used. In such cases assembly should be made in the trench to avoid deformation of the lead due to handling.

Joints in larger pipe should be made with the pipe finally positioned in the ditch. This requires digging of bell holes to provide room for making joints. Each length of pipe should be tested for soundness before being placed in the ditch.

The inside of pipes should be thoroughly cleaned before make-up. When gas contains moisture or other condensables the main should be graded to low points by means of suitable earth supports at various levels.

When a significant change is anticipated in the settlement rates of adjoining subgrades, the denser section should be overexcavated to the pipe joint nearest to the junction of the hard and soft subgrades.[2]

Reference may also be made to the extracts from ASA B31.8–1963 given on page 9/86.

---

* Skeleton or noncontinuous sheeting is sometimes used.

842.2 Installation of Cast Iron Pipe.

842.21 Underground cast iron pipe shall be laid in accordance with the applicable field conditions described in the ASA "American Standard Practice Manual for the Computation of Strength and Thickness of Cast Iron Pipe" A21.1.

842.22 Underground cast iron pipe shall be installed with a minimum cover of 24 inches unless prevented by other underground structures.

842.23 Where sufficient cover cannot be provided to protect the pipe from external loads or damage and the pipe is not designed to withstand such external loads, the pipe shall be cased or bridged to protect the pipe.

842.24 Cast iron pipe installed in unstable soils shall be provided with suitable supports.

842.25 Suitable harnessing or buttressing shall be provided at points where the main deviates from a straight line and the thrust if not restrained would part the joints.

842.26 *Making and Testing of Cast Iron Field Joints.*

(a) Cast iron pipe joints shall conform to 842.15, and shall be assembled according to recognized AWWA, or ASA Specifications, or in accordance with the manufacturer's written recommendations.

(b) Cast iron pipe joints shall be leak tested in accordance with 841.44 of this code.

Backfilling and Tamping

Backfilling and tamping procedure will depend on such factors as location, kind of soil, type of pipe laid, and traffic conditions. Generally, a primary backfill consisting of good soil should be placed and hand-tamped around the pipe to a depth of about one foot above the pipe. Organic soil or soil containing ashes or cinders should not be used as backfill. The remainder of the trench may then be backfilled using mechanical backfilling and tamping equipment if possible. Backfills in sandy soil can be compacted by puddling. Before the puddling is done, the trench may be filled to within about a foot of its top. The top layer of fill is then added and tamped. Clay soil does not compact by puddling. Backfills should not contain large rocks which may work thru the primary backfill and damage the pipe.

Heavily traveled streets require thorough compacting; sand may be specified for backfill in such locations. It is sometimes necessary to replace large quantities of unstable soils or even to construct underground bridges supported by piles to avoid significant pipe settlements. Inadequate compaction in freezing weather results in settlement when soil thaws. In general, the material removed to expose a pipe for repairs should be backfilled in its entirety.

Repaving

Safe and durable construction, in compliance with local requirements, is the principal objective of repaving. Paving construction, a specialized activity, is usually best performed by a qualified contractor. Acceptance of finished work should be subject to satisfactory inspection.

TESTING NEW MAINS

Except for short extensions of small size pipe, completed sections of new mains should be tested for leakage before they are placed in service. Short extensions are often isolated by means of end caps and tested under available gas pressure, each joint being checked with soapsuds unless the main is so substantially protected by coatings that they would hide a split seam. In that case a 100 psig pressure test should be used.

The procedure and test pressure employed for determining tightness of a new line depend on location, kind of pipe, and operating pressure. It is preferable to use the maximum operating pressure plus a factor of safety as the test pressure and to make the test over a 24 or 48 hr. period with a recording pressure gage. Bends should be braced when testing precedes backfilling.

Leakage tests may be made by filling the new line with air and noting if any appreciable drop in test pressure occurs. One method of locating leaks is to introduce an appropriate odorant into the line and drive bar holes near the joints; leaks may be detected by smell. A second method employs a gas indicator at bar holes. See Pars. 841.285 and 841.3 of ASA B31.8-1963 (extracted in Chap. 2 of Sec. 8).

REFERENCES

1. Hale, D. "Construction Equipment Survey." *Am. Gas J.* 188: 22–7, Mar. 1961.
2. Moran, Proctor, Mueser & Rutledge. *Manual on Pipe Trenching and Backfill.* New York, Public Service Elec. and Gas Co. 1962.
3. "New Method: Horizontal Earth Boring." *Gas Age.* 118: 19–21, Sept. 20, 1956.

Chapter 8

Services

by H. C. Roemmele

A gas service is the pipe between a distribution main or pipeline and the customer's meter. Usually it supplies a single building housing one or more customers. The service piping between the pipe entrance to the building and a more distant meter location in the same building is termed a **service extension** (or building service).

Two or more meters located in different buildings or rooms may be supplied by a single service if utility policy permits. Such a service, known as a **branch service,** may consist of underground or exposed piping. Some companies do not permit branch services; others limit their use to special conditions because of safety considerations when service work is being done on equipment connected to one branch only.

Steel pipe with welded or mechanical joints is commonly employed for gas services. Copper pipe or copper tubing is extensively used by some utilities where installation and operating conditions make it desirable. Plastic pipe and tubing of various compositions (Table 9-18) are used to some extent by more than 100 gas companies for new services, service renewals, and mains.

SERVICE SIZING

Local utility experience normally establishes the size of services installed. For low-pressure distribution systems, 1¼ in. IPS (or NS) is the minimum employed by most companies. For high- and intermediate-pressure systems with individual service regulators, ¾ in. IPS is the most commonly accepted minimum. The two minimum sizes given for low and high pressure, respectively, are normally ample for single family residences. They provide adequate allowance for future increases in load or restrictions to flow due to the accumulation of deposits in the line. In cases in which large demand appliances are connected or anticipated or services are of more than average length, some increase above the minimum size may be recommended.

Replies to a 1963 poll[20] of a limited number of gas companies regarding high-pressure service sizing practices showed: (1) half are either using copper or studying its use; (2) several use ½ in. copper or steel services.

Sizing Factors

Service sizing considerations include total connected load, length of service, allowable pressure loss, specific gravity of gas, and diversity factor.

For single family residences, the service is designed to handle the total connected load; that is, no diversity factor is used. On the other hand, when a service supplies multiple customers, as in an apartment house, a diversity factor should be applied.

Length of service for calculation purposes should include the actual length of service pipe plus the equivalent lengths of service tap, valves, and other fittings in the line.

In low-pressure main systems, most gas companies permit the allowable pressure drop between the main and the meter to be either 0.3 or 0.5 in. w.c. (Table 9-13). For systems operating between 1 and 15 psig, the allowance main-to-meter drop is generally limited to 0.5 psig. For high-pressure systems the allowable drop is from 0.5 to 3.0 psig;[1] 10 psi and larger drops are practical.[20]

Sizing Calculations

Flow Characteristics of Low-Pressure Services.[2] Experiments in 1960 showed that the widely used **Spitzglass** equation for low-pressure gas flow required new resistance constants. It was shown that flow is a function of inside diameter rather than of surface smoothness. Thus, tubing will not have the capacity of the larger size steel pipe thru which it may be drawn for replacement purposes. The flow equation for copper or steel services from ¾ in. CTS copper to 1½ in. NS steel was found to follow the form (applicable to velocities of 5 to 16 fps, a laminar range appropriate to domestic utilization situations):

$$Q = \left(\frac{0.6h}{K_p s(L + L_{ef})}\right)^{0.54} \qquad (1)$$

where: Q = flow rate, cfh
h = pressure drop in service, in. w.c.
K_p = pipe resistance constant (Table 9-35)
s = specific gravity of gas
L = length of service, ft
L_{ef} = equivalent length of fittings (Table 9-36)

Table 9-37 presents capacities for various lengths of a number of types of services used in low-pressure systems. The tabulation is for a 0.5 in. w.c. pressure drop; for other drops, multiply the values given by

$$\left(\frac{\text{desired drop, in. w.c.}}{0.5}\right)^{0.54}$$

The first column of Table 9-37 gives the size and material of a number of services. Note that these are listed separately for 4- and 6-in. mains. The second column gives various inlet connections.

Table 9-35 Resistance Constants, K_p, for Selected Tubes and Pipes[2]

(used in Eq. 1)

Service size, in.	K_p
¾ CTS copper*	1.622×10^{-6}
1 in. ID plastic	0.279×10^{-6}
1 in. CTS copper*	$.383 \times 10^{-6}$
1¼ CTS copper*	$.124 \times 10^{-6}$
1¼ NS steel	$.080 \times 10^{-6}$
1½ NS steel	0.037×10^{-6}

* CTS, standard copper tube size, corresponds to the wall thickness of Type K tubing. See Table 9-37.

Table 9-36 Equivalent Length of Various Fittings[2]

(tabulated data give length of straight pipe with equivalent flow resistance, ft)

Fitting	Copper tubing		Steel pipe	
	1 in.	1¼ in.	1¼ in.	1½ in.
Street tee*†	5.5*	9.5‡	10.5	15.0‡
Street elbow‡	5.0*	4.0§	7.5	7.5
Curb cock	3.5	3.5	13.5	12.0
Outlet fittings**	6.0††	4.5	8.0	22.0

* 1¼ in. fittings.
† Equivalent lengths for other street tees: 1¼ × 1 × 1¼ in 1¼ steel, 23 ft; 1½ × 1¼ × 1½ in 1½ steel, 19 ft.
‡ 1½ in. tee attached to sleeve around main; 1¼ in. hole in main.
§ 1½ in. elbow instead of 1¼ in. elbow.
** Tee and service cock.
†† Also applies to 1 in. plastic fitting; 2.0 for ¾ in. copper.

Table 9-37 Recommended Allowable Gas Flow thru Low-Pressure Services[2]

(in cu ft per hr at main pressure = 6.0 in. w.c., total pressure drop = 0.5 in. w.c., 0.6 sp gr gas)

Size of service, in.*	Inlet connection†	Length of service, ft										
		20	30	40	50	60	70	80	90	100	110	120
From 4 in. main												
¾ CTS‡	P	155	130	115	105	95	85	80	75	70	70	65
1 plastic§	P	365	315	280	255	235	215	205	195	185	175	170
1 CTS**	P††	305	265	235	215	195	185	175	165	155	145	140
	S††	285	250	225	205	190	180	170	160	150	145	140
1¼ CTS copper (1¼ ID,	P	535	470	425	390	360	335	315	295	280	270	260
1⅜ OD)	P with C‡‡, S	515	455	410	375	350	330	310	295	280	270	260
	S with C	490	435	395	365	345	325	305	285	275	265	255
1¼ NS steel (1.38 ID, 1.66 OD)	P	530	485	450	420	400	380	360	340	330	320	310
	P with C	470	440	410	390	370	350	335	325	315	305	295
	S	500	460	430	405	385	365	345	330	320	310	300
	S with C	450	420	400	380	360	340	330	320	310	300	290
1½ NS steel (1.61 ID, 1.90 OD)	P	815	740	685	645	605	575	545	525	505	485	465
	P with C	730	680	635	600	570	540	520	500	480	460	450
	S	760	700	655	615	585	555	530	510	490	470	455
	S with C	690	645	610	580	550	525	505	485	465	455	445
From 6 in. main												
¾ CTS‡	P	155	130	115	105	95	85	80	75	70	70	65
1 plastic§	P	365	315	280	255	235	215	205	195	185	175	170
1 CTS**	P††	330	280	250	225	205	190	180	170	160	150	145
	S	300	260	235	215	195	185	175	165	155	145	140
1¼ CTS copper (1¼ ID,	P	545	475	425	380	360	335	315	295	280	270	260
1⅜ OD)	P with C	520	460	415	380	350	330	310	290	280	270	260
	S	515	455	410	375	350	330	310	290	280	270	260
	S with C	490	435	395	365	345	325	305	285	275	265	255
1¼ NS steel (1.38 ID, 1.66 OD)	P	600	540	495	455	425	400	380	360	345	335	325
	P with C	520	480	445	415	395	375	355	340	330	320	310
	S	555	505	465	435	410	390	370	350	335	325	315
	S with C	495	455	425	405	385	365	345	330	320	310	300
1½ NS steel (1.61 ID, 1.90 OD)	P	780	720	670	630	595	565	535	515	495	475	460
	P with C	710	660	620	590	560	535	515	495	475	455	445
	S	740	685	640	605	575	545	520	500	480	460	450
	S with C	680	635	600	570	540	520	500	480	460	450	440

* CTS, standard copper tube size, corresponds to the wall thickness of Type K tubing.
† P = plain end connection; S = swing joint connection; C = curb cock.
‡ ¾ in. CTS copper (¾ ID, ⅞ OD).
§ 1 in. plastic (1 ID, 1¼ OD).
** 1 in. CTS copper (1 ID, 1⅛ OD).
†† With or without curb cock.
‡‡ Curb cock with plain inlet only.

INSTALLATION

New service installation is closely associated with main laying and can often be performed to best advantage in whole or part when the main is laid. Particularly when mains are located in roadways already paved or soon to be paved, it is important that all underground work be completed as part of the initial operation so that future street openings may be avoided.

Reference may be made to the following extract from ASA B31.8-1963:

849 GAS SERVICES.

849.1 General Provisions Applicable to Both Steel and Copper Services.

849.11 *Installation of Services.*

(a) Services shall be installed at a depth which will protect them from excessive external loadings, and local activities, such as gardening. It is recommended that a minimum depth of 12 inches in private property and a minimum depth of 18 inches in streets and roads be maintained. Where this cannot be done, due to existing substructures, etc., less cover is permitted provided however that where such services are subject to excessive superimposed loads, those portions of the service shall be cased or bridged to avoid harmful additional loads on the pipe, or strengthened to resist them.

(b) Service piping shall be properly supported at all points on undisturbed or well compacted soil, so that the pipe will not be subject to excessive external loading by the backfill. The material used for the backfill shall be free of rocks, building materials, etc., that might cause damage to the pipe or the protective coating.

(c) Where there is evidence of condensate in the gas in sufficient quantities to cause interruptions in the gas supply to the customer, the service shall be graded so as to drain into the main or to drips at the low points in the service.

849.12 *Types of Valves Suitable for Service Shut-Offs.*

(a) Valves or cocks used as service shut-offs shall meet the applicable requirements of 810 and 831.1.*

(b) The use of soft seat shut-off valves or cocks is not recommended.

(c) A valve incorporated in a meter bar which permits the meter to be by-passed does not qualify under this Code as a service shut-off.

(d) Service shut-offs on high pressure services, installed either inside of buildings or in confined locations outside of buildings where the blowing of gas would be hazardous, shall be designed and constructed to minimize the possibility of the removal of the core of the valve or cock accidentally or willfully with ordinary household tools.

(e) The operating company shall make certain that the shut-off valves or cocks installed on high pressure services are suitable for this use either by making their own tests or by reviewing the tests made by the manufacturers.

(f) On services designed to operate at pressures in excess of 60 psig the service shut-off valve or cock shall be the equivalent of a pressure lubricated cock or a needle type valve. Other types of valves or cocks may be used where tests by the manufacturer or by the user indicate that they are suitable for this kind of service.

849.13 *Location of Service Shut-Offs*

(a) Service shut-offs shall be installed on all new services (including replacements) in a readily accessible location.

(b) Shut-offs shall be located upstream of the meter if there is no regulator, or upstream of the regulator, if there is one.

(c) All gas services operating at a pressure greater than 10 psig, and all services 2 inches in diameter or larger, shall be equipped with a shut-off located on the service line outside of the building, except that whenever gas is supplied to a theatre, church, school, factory or other building where large numbers of persons assemble, an outside shut-off in such case will be required regardless of the size of the service or of the service pressure.

(d) Underground shut-offs shall be located in a covered durable curb box or standpipe, which is designed to permit ready operation of the valve. The curb box or standpipe shall be supported independently of the gas service line.

849.14 *Location of Service Connections to Main Piping.* It is recommended that services be connected to either the top or the side of the main. The connection to the top of the main is preferred, in order to minimize the possibility of dust and moisture being carried from the main into the service.

849.15 *Testing of Services After Construction.* Each service shall be tested after construction and before being placed in service to demonstrate that it does not leak.

Services to operate at a pressure between 1 psig and 40 psig, shall be given a stand-up air or gas pressure test at not less than 50 psig for at least five minutes before being placed in service.

Services to operate at pressures in excess of 40 psig, but stressed less than 20% of the specified minimum yield, shall be tested to the maximum operating pressure or 100 psig, whichever is the the lesser. Services stressed to 20% or more of the specified minimum yield shall be tested in accordance with the requirements for mains.

The service connection to the main need not be included in these pressure tests if it is not feasible to do so.

849.2 Steel Services.

849.21 *Design of Steel Services.*

(a) Steel pipe, when used ror gas services, shall conform to the applicable requirements of Chapter I.†

(b) Underground steel services, when installed below grade through the outer foundation wall of a building, shall be either encased in a sleeve or otherwise protected against corrosion. The service pipe and/or sleeve shall be sealed at the foundation wall to prevent entry of gas or water.

(c) Steel services, where installed underground under buildings, shall be encased in a gas tight conduit. When such a service supplies the building it subtends, the conduit shall extend into a normally usable and asscessible portion of the building and, at the point where the conduit terminates, the space between the conduit and the service pipe shall be sealed to prevent the possible entrance of any gas leakage.

(d) Where practical, welded joints or compression type fittings should be used in all underground steel services.

(e) Consideration shall be given to insulating, near or within the building, those services which are connected through the house piping to water services, electrical ground, etc., so as to eliminate possible galvanic corrosion. This is especially important in areas where stray current electrolysis is prevalent, or where copper or lead water services are used.

849.22 *Installation of Steel Services in Bores.* When coated steel pipe is to be installed as a service pipe in a bore, care should be exercised to prevent damage to the coating during installation. For all installations to be made by boring, driving or similar methods or in a rocky type soil, the following practices or their equivalents are recommended:

(a) When a service is to be installed by boring or driving and a coated steel pipe is to be used for the service, the coated pipe should not be used as the bore pipe or drive pipe and left in the ground as part of the service. It is preferable to make such installations by first making an oversize bore, removing the pipe used for boring and then inserting the coated pipe.

(b) Coated steel pipe preferably should not be inserted through a bore in exceptionally rocky soil where there is a likelihood of damage to the coating resulting from the insertion.

* [Par. 831.1 is the first extract in Chap. 4, Sec. 9.—Editor]

† [Refers to chapter in ASA B31.8-1963, many paragraphs of which are given in Chap. 2, Sec. 8.—Editor]

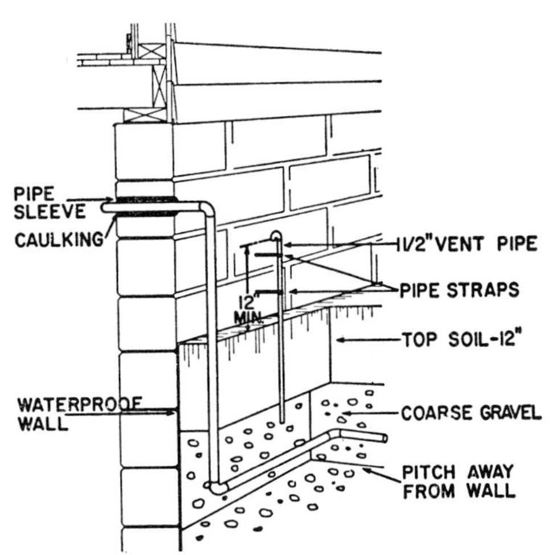

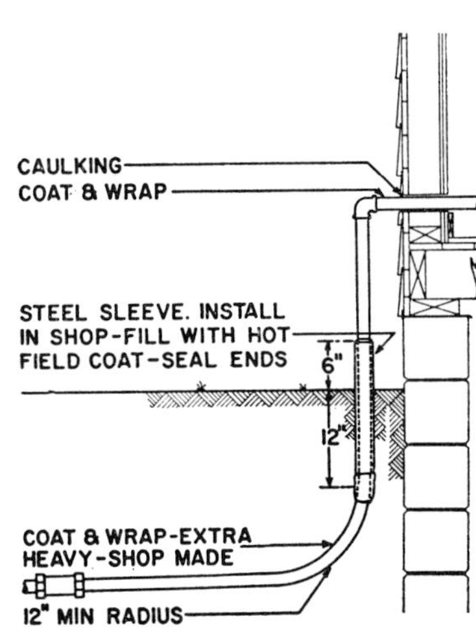

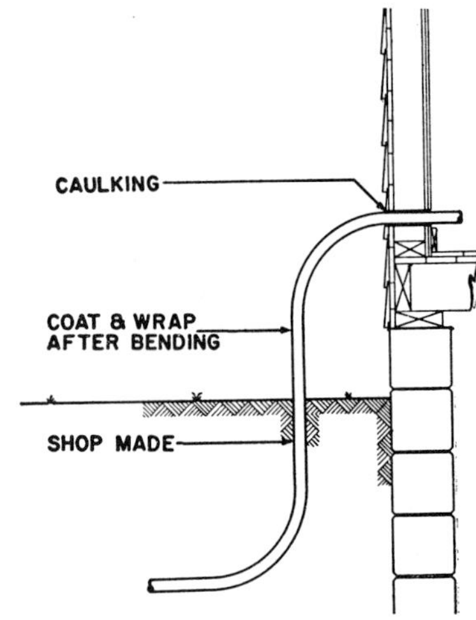

Fig. 9-65a (left) Above-grade service entrance with vented service trench.[4]

Fig. 9-65b (center and right) Above-grade service entrances.[1]

849.23 The recommendations in (a) and (b) above do not apply when bare steel pipe is used as the service pipe, or where coated pipe is installed under conditions where the coating is not likely to be damaged, such as in sandy soil.

849.24 *Service Connections to Steel Mains.* Services may be connected to steel mains by:

(a) Welding a service tee or similar device to the main.

(b) Using a service clamp or saddle.

(c) Compression fittings using rubber or rubberlike gaskets or welded connections may be used to connect service pipe to the main connection fitting. Gaskets used in a manufactured gas system shall be of a type that resists effectively that type of gas.

849.3 Cast Iron Services.

849.31 *Use of Cast Iron Services.* When used for gas services, cast iron pipe shall meet the applicable requirements of 842.* The use of cast iron pipe less than 6 inches in diameter for gas services is prohibited. Cast iron pipe 6 inches or larger in diameter, may be used for gas services except for that portion of the service which extends through the building wall. The latter portion shall be of steel pipe. Cast iron services shall not be installed in unstable soils or under buildings.

849.32 *Service Connections to Cast Iron Mains.* Services may be connected to cast iron mains by:

(a) Drilling and tapping the main; provided, however, that the diameter of the tapped hole shall not exceed the limitations imposed by 831.33 (b).†

(b) Using a reinforcing sleeve.

849.321 Service connections shall not be brazed directly to cast iron mains.

849.322 Compression fittings using rubber or rubberlike gaskets or welded connections may be used to connect the service pipe to the main connection fitting. Gaskets used in a manufactured gas system shall be of a type that resists effectively that type of gas.

Service Design

Apart from the requirements of ASA B31.8, special stipulations for gas service installations into certain classes of buildings will be encountered from time to time. For example, Public Housing Administration Bulletin No. LR-7, Part V, recommends that local housing authorities specify that above-grade service entries (Fig. 9-65a) be installed in low-rent housing projects. Service entrances to project buildings without basements (slab on ground or concealed crawl-space constructions) should be made above the floor line.

For buildings insured by the Associated Factory Mutual Fire Insurance group, reference should be made to the recommendations of the Factory Mutual Engineering Division.[3] Above-grade service entries are preferred. When below-grade service entry is essential, it must be encased.

The Rochester Gas & Electric Co. has developed flanged gasketed sleeves[21] for improved long life service entry. These provide for sealing against entry of water or gas by means of tight fitting gaskets placed against both the interior and exterior faces of the wall. Wedge gaskets seal against the gas service pipe.

Figure 9-65b shows two types of above-grade entrances. If the gas contains moisture and atmospheric temperatures are low, the exposed portions of above-grade services must be insulated to prevent freezing.

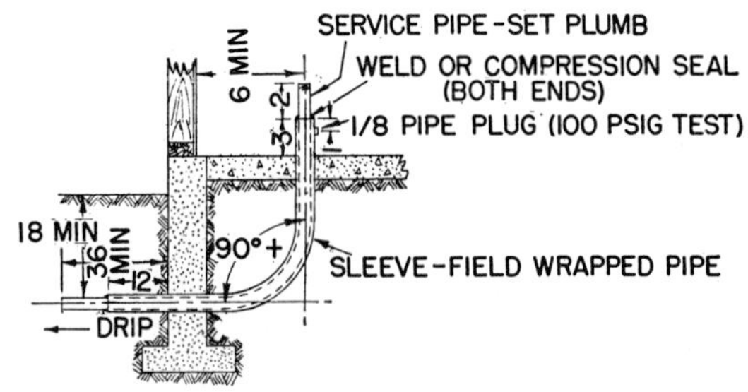

Fig. 9-65c Typical encased service entrance for slab on ground buildings. All dimensions in inches except as noted. Plain end of service (at left) is extended by welding or a compression fitting. Casing schedule follows (all dimensions in inches):

Service	¾	1¼–1½	2	3	4	6
Type of bend	8 R	12R	MW	MW	MW	MW
Sleeve	1¼	2	3	4	6	8

where R = cold bend, radius of bend given in multiples of service size, and MW = 45° miter weld for service and sleeve.

* [The material under Para. 842 is given in Chap. 3, Sec. 9.—Editor]

† [Par. 831.33 is given in Chap. 2, Sec. 8.—Editor]

Figure 9-65c shows a type of encased service entrance used for slab on ground buildings; the same design is suitable for crawl-space construction.

The integrity of basement walls at and near gas and other service lines should be ensured by the building owner's using suitable means to prevent the entry of ground water or gas. Gas could be present in case of failure of external underground gas piping.

Contractor Damage to Buried Services.[22] Good communication between contractors and utility companies is a basic protection requirement. Additionally, some gas companies use a compression coupling, installed anywhere from ten feet from the service entrance to as close to the main as possible, to provide a weak point in the event that a service is inadvertently struck.

Excess Flow Valves. Many gas utility companies have been studying the use of excess flow valves as a means of reducing leakage from broken services. As shown in the following, the shutoff flows for excess flow valves are far below the calculated gas flows that occur with broken services:

Pressure before valve closes, psig	Fuel flow in service, cfh		Shutoff flow, cfh
	¾ in. diam	1 in. diam	
10	2,900	5,400	330
50	9,100	17,000	580
100	17,000	32,000	890

Other tests indicated that leakage thru a closed excess flow valve is in the range of 5–30 cfh at pressures up to 75 psig.

Some companies feel that limiting the size of service taps and properly sizing services provide sufficient protection against excessive gas loss from broken services without the need for excess flow valves.

Location and Cover

Where possible, a service should be run in a straight line from the main to as close as possible to the meter or meters to be supplied. It should be laid in an individual trench, properly supported on undisturbed or solidly compacted earth. Locations near lateral sewer ditches and cesspool overflows should be avoided most. Services which cannot avoid proximity to cesspools and sewer laterals should be coated and wrapped. Offsets in services should be avoided whenever possible. If obstructions are present and the service must be offset, the use of 90 degree ells makes it easier to find buried services.

Basement Installations. The service should enter thru the foundation wall as close as practical to the meter location. A sleeve previously provided in the foundation wall for the entrance of the service is desirable and should be furnished in all large buildings as part of the original construction.

Outdoor Installations. The service should be run underground close to the location selected for the meter, with a vertical riser extending above the ground. Cover of the service is necessarily limited by that of the main. Where low temperatures are encountered, it is desirable to keep the service below frost line. If the gas contains moisture or other condensables, the service should be graded back to the main. This may result in insufficient cover near the building end, especially with long services, unless care is taken in installing.

Steel Services

Methods of connecting steel services to mains vary, depending on the material of the main. Since these connections are often under paving, selection and proper installation of the type of joint most suitable for the conditions involved are of major importance.

For connection to a steel main, a service tee or similar fitting is usually welded to the main. The service pipe may then be connected to the fitting by welding or by use of a compression fitting. Alternate procedures include attaching a service clamp or saddle to the main or tapping the main for a fitting to which the service pipe is connected (Fig. 9-66). A 1962 development is a device called a "punch-it" tee, which eliminates the use of special tapping tools.

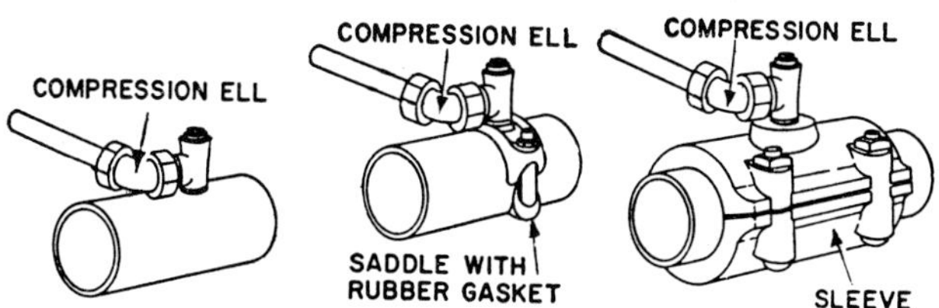

Fig. 9-66 Three types of main connection.[10] Malleable iron fittings are commonly used for steel services from cast iron mains; steel tees with malleable plugs are used with steel mains; brass fittings are used with copper services. The combination of a tee and an elbow, called a swing joint, is widely used with cast iron mains (and sometimes with steel mains) to reduce the stress at the main tap produced by relative movement between the main and service.

Cast iron mains are customarily tapped to provide service connections. The maximum tap size is limited by the main diameter, requisite number of threads which should be engaged, and weakening effect of the tap. Generally, the maximum tap size permitted in a cast iron main is limited to 25 per cent of the main's internal diameter, except that a 1¼ in. tap is permissible in a 4 in. main.

When the size of the service requires a tap larger than is permissible, a service clamp, saddle, or sleeve is used, and when the size of the service approaches the size of the main, a tee is usually inserted. On high-pressure cast iron mains a service clamp or saddle is generally used with a smaller size hole made in the main. Corporation cocks or special street tees which can be screwed into tapped holes in mains under pressure without the escape of gas ("no-blow") are available.

Figure 9-66 illustrates three types of service connection to cast iron mains. Table 9-38 gives a sizing schedule used by one company.

Connections are available which may be joined to "live" mains by means of pressure-tight installation equipment. Such fittings may incorporate a service shutoff valve which may be operated without the use of any special equipment.

The Lowell Gas Co. of Massachusetts designed the *Service Plug Extracting Pouch* which makes possible the replacement of the plug in a service tee at the head of a live low-pressure service. The gas-tight pouch permits access to the fittings while confining the resultant gas leakage.

Severe corrosion in the service is most likely to occur where the service enters the building underground; on threaded joints; on defective areas in the pipe; on contacts between pipe and low-resistance conducting materials; at or near connections to dissimilar metals such as cast iron or copper; and at defects in coatings.

Table 9-38 Schedule of Service Connection for Cast Iron Mains*

Service size, in.	Size of taps, in., for following main sizes: 16 in.	12 in.	8 in.	6 in.	4 in.
1¼	1¼	1¼	1¼	1¼	1
1½	1½	1½	1½	1¼	1
2	2	2	1½	1¼ tap or 2 S.S.†	2 S.S.
3	3	2	3 S.S.	3 S.S.	Tee
4	3	4 F.S.S.‡ or tee	tee	Tee	Tee

* Public Service Electric & Gas Co., Newark, N. J.
† 2 S.S. = split sleeve tapped with 2 in. hole.
‡ 4 F.S.S. = flanged split sleeve with 4 in. hole.
Note: When a service sleeve is used, the main tapping may be one size smaller than that of the sleeve. When not used for service or other pipe connection, the following schedule may be used:

Main, in.	3	4	6
Tap, in.	1	1¼	1½

Methods for protecting services against corrosion depend upon local conditions and severity of corrosion involved. They include:

1. Use of bare steel pipe and fittings in relatively noncorrosive situations. This practice is declining.
2. Use of bare extra heavy steel pipe and corresponding wall thickness on fittings; increased wall thickness greatly increases pipe life. However, this is not a desirable substitute for other protective measures.
3. Use of insulating couplings at street end of services which insulate services from mains.
4. Use of insulating unions, meter bars, swivels, or other insulating fittings at meter outlets (or inlets). This isolates the service from electrical contact with various current sources in the house and the galvanic effect of the dissimilar metals, e.g., the water service (usually copper or lead).
5. Cathodic protection of bare steel pipe by magnesium or zinc anodes with insulating fittings at each end of the service or insulation of the meter end with cathodic protection of the service aligned with that of the main distribution system.
6. Coating of standard weight services with coal tar enamel, asphalt, wax mixture, plastic, or other suitable coatings. This is accompanied by cathodic protection, using magnesium or zinc anodes or a tie-in with the protection applied to street mains. Field coating of joints and fittings may be done with the foregoing materials applied hot, with cold applied coatings, or with plastic or coal tar tapes. Use of coatings has increased because of the protection provided against most conditions of corrosion and the long service life it ensures.

Copper Services

A considerable amount of copper tubing is used, both for replacement by insertion thru former steel services and for new installations. Experience with such services has been satisfactory. (Capacities of copper tubing are included in Table 9-37). The economics of sizing service renewals for current or for ultimate loads should be carefully investigated. Utilities serving large cities in widely separated parts of the United States have installed a large number of copper services and have reported details of their installation practice.[5-11]

Material used is usually soft drawn or hard drawn Type K copper tubing. In some cases "bending temper" tubing, with characteristics between those of standard soft drawn and standard hard drawn tubing, is employed. Soft drawn tubing is supplied in rolls; medium and hard drawn tubing, in lengths of about 20 ft. Hard solder joints with or without sleeves or brass neoprene compression fittings are applied. A few companies have encountered internal corrosion of copper services by copper sulfide formed thru the reaction of the copper with H_2S in the gas. However, tin-lined or aluminum-lined tubing which keeps H_2S out of contact with the copper is available.

Copper tubing services may be a minimum of ½ in. or ¾ in. CTS for intermediate-pressure services, depending on local practice. For low pressure, 1 or 1¼ in. CTS for insertion thru 1¼ or 1½ in. steel pipe and 1¼ in. CTS for new services are commonly employed.

Connection to cast iron mains (Fig. 9-66) is usually made by a brass service tee and a service ell that has a compression outlet. In some cases, a service tee carrying a compression outlet is used. For steel mains, a steel service tee is welded to the main, with copper reducing coupling brazed to the side outlet.

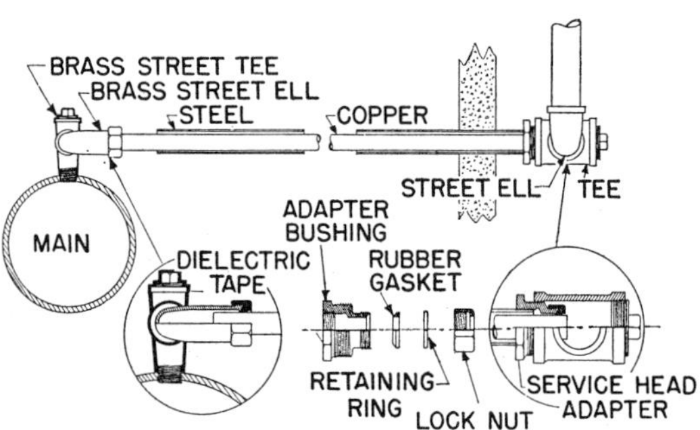

Fig. 9-67a Service renewal by copper tubing insert.

When using copper tubing for service renewal, it is common practice to employ the old steel service as a protective sleeve (see Fig. 9-67a). After the necessary cleaning of the old service interior, the tubing is pulled thru it. Renewal can often be accomplished by the use of a single street opening over the main where gas connection is made. All necessary operations are carried out from this opening and from the basement of the building supplied.

Plastic Services

Installation practices[12-14] with plastic tubing are generally similar to those used with copper, except for joints. These are made by use of special plastic fittings into which the tubing is cemented or by use of compression fittings. With compression fittings, an inner steel sleeve is generally inserted into the tubing to provide reinforcement. Figure 9-67b shows a method of service renewal with plastic tubing.

Table 9-39 Plastic Pipe Dimensions and Weight[15]

(allowable working pressure = 60 psig)

Pipe size, nom., in.	Schedule	OD, in.	Wall, in.	Lb/ft
½	CTS	0.625	0.062	0.051
½	IPS	0.840	.094	.099
1¼	IPS	1.660	.094	.208
2	IPS	2.375	.094	.304
3	IPS	3.500	.131	0.625
4	IPS	4.500	0.163	1.001

A digest of standards for plastic service installation designed by the major American user in 1962 follows:[15]

(1) Plastic pipe dimensions as given in Table 9-39; (2) black ABS plastic used for both pipe and fittings; see Table 9-18; (3) extreme care in storage and handling, since plastic pipe is highly sensitive to impact damage; (4) methyl ethyl ketone with dissolved ABS used as joining cement; cure 10 to 15 min before moving, 24 hr for complete cure; (5) plastic-to-steel connections made with compression fittings, with an internal metal reinforcing sleeve for the plastic end; (6) a No. 18 gage plastic coated copper wire brazed to the steel riser pipe portion of the service and extended alongside the plastic pipe run, acting as a tracer for finding the buried service; and (7) replacement of the damaged section of plastic pipe if an emergency necessitates shutting off gas flow by squeezing the pipe (tools shown in Ref. 15). Reference 15 also covers connecting services to "live" and inactive mains, repairing breaks or holes in plastic pipe, and tools used in installing plastic piping.

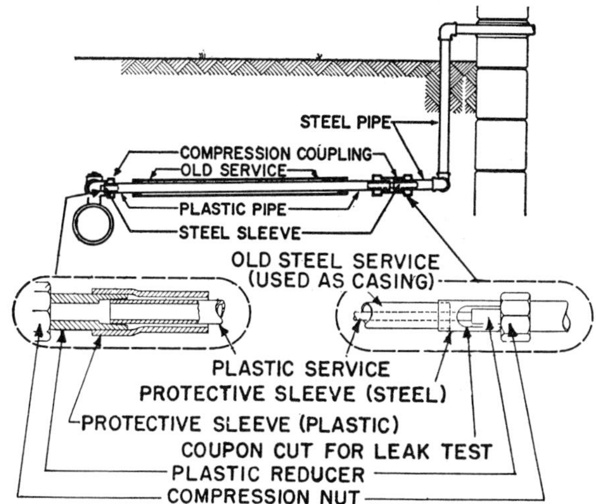

Fig. 9-67b Service renewal by plastic tubing insert.[4]

Meters

Meters should be installed where they can be read easily and where their connections are readily accessible for servicing. The location and dimensions of the space as well as the type of installation should be acceptable to the utility company.[16] Inside meter locations should be as near as practicable to the point at which the service enters the building. Outside mounted registry devices for inside meter installations are available.[17] Exterior meter locations are common where gas is "dry" or where favorable temperatures and other climatic conditions permit. Meters may be changed without interrupting service, by means of portable cylinders of fuel gas.[18]

A 1960 survey[19] of residential gas meter installations, covering 152 gas utilities (80 per cent of U. S. gas meters), concluded: (1) metered gas temperature and ambient gas temperature are practically the same; (2) uncompensated outside meters show losses—an Illinois study showed 3.7 per cent less registration annually for outside sets compared with inside sets (greater variance during nonheating season); (3) on all test locations at which temperature-compensated meters were set in outdoor-indoor series, a reading variance of under 1 per cent was reported; (4) outdoor vs. indoor meter reading data were as shown in Table 9-40; and (5) 18 companies reported using compensated meters.

The same survey also reported the various installation practices (and the reasons for them) and innovations in reading equipment.

Table 9-40 Indoor vs. Outdoor Meter Reading Data*[,19]

Per cent of reported meters:	set inside	97.58	...
	set outside	...	96.6
No. of meters read, average/man-day		226	354
Per cent of skip reads, avg		8.72	0.76
Annual meter reading cost, avg, $		1.20	0.93

* Sample data from two of the nine U. S. regions in the reference. The differences between these two columns of data should not be attributed solely to the indoor vs. outdoor location. One major factor, for example, is the distance between meters.

House Trailer Service

A trailer service is defined as the piping extending from the meter outlet (or service regulator outlet when a meter is not provided) to the terminal of the gas riser at each trailer site. These risers, ¾ in. min (for other than undiluted LP-gases), should be placed in the rear third of each site, at least 18 in. from the roadside wall of the trailer. Demand factors are given in Table 9-27b.

The break in a gas service which would occur if a trailer or mobile home were inadvertently driven off without the gas service having been closed off and disconnected can be avoided.[20] The arrangement involves the use of a plastic nipple upstream of an automatic shutoff valve. Any appreciable external load on the service would break the nipple and close the automatic valve, which would respond to the temporary drop in pressure.

Supplementary Shutoff Devices

Discontinuance of gas supply to a meter is normally performed by closing the service shutoff on single meter installations or the meter shutoff on multiple meter installation. Service shutoffs equipped with "lock wings" can be locked in the closed position by inserting a locking stud, the removal of which requires a special key. Since not all shutoffs are equipped with lock wings and some customers will open shutoffs to make unauthorized use of gas, supplementary shutoff devices are used by some companies to prevent unauthorized use and to prevent possible leakage.

One common device is a blind disk of metal or plastic which is inserted at the meter inlet connection whenever a meter is shut off and made inactive. This device prevents gas leakage if the service shutoff does not hold tight and the house line is open at a disconnected appliance.

Another device, used under extreme conditions when customers remove blind disks or locking devices, is a rubber plug inserted in the piping or service ahead of the meter. The plug is inserted and expanded by a special key which is then disengaged and removed. Normally, such plugs are inserted only in low-pressure services or in the piping downstream from the service regulator.

Where tampering is a serious problem, it may be necessary to remove the meter and inlet piping and to cap the service with a special cap requiring a key for removal.

REFERENCES

1. Hunt, B. E., and others. "Gas Service Design." *Am. Gas J.* 183: 12–17, July 1956.
2. Menegakis, D., and Luntey, E. H. "Experimental Investigation of Flow Characteristics of Low Pressure Services." *A.G.A. Operating Sec. Proc.* 1961: DMC-61-15.
3. Associated Factory Mutual Fire Ins. Cos., Factory Mutual Eng. Div. *Handbook of Industrial Loss Prevention.* New York, McGraw-Hill, 1959.
4. Hunt, B. E., and others. "Gas Service Design—Final Report of Task Group." (DMC-56-14) *A.G.A. Proc.* 1956: 272–87.
5. Thorson, L. H. "Copper Service Pipe Installations." *A.G.A. Proc.* 1949: 581–91.
6. Alexander, R. W. "Copper Services, Materials and Methods Used in Cincinnati." *A.G.A. Proc.* 1949: 592–5.
7. Bradfield, S. A. "Experience in the Use of Copper Tubing for Mains and Services." *A.G.A. Proc.* 1949: 613–22.
8. Hall, F. J. "Renewing Services with Copper Tubing." *A.G.A. Proc.* 1949: 674–8.
9. Webster, I. S. "Equipment and Procedure for Service Renewals with Soft Copper Tube." *A.G.A. Proc.* 1951: 361–78.
10. Catell, R. B. "Service Renewals with Copper Tubing." *Am. Gas J.* 188: 27–9, June 1961.
11. McCarthy, T. F. "Copper Tubing for Service Renewals." *Gas Age* 126: 36–7, Sept. 15, 1960.
12. Dye, G. G. "Seven Years Experience with Plastic Services." *Gas* 28: 45–8, Apr. 1952.
13. Fugazzi, J. F. "Plastic Tubing Use Found Practical at Public Service Co. of Colorado." *Gas* 28: 49–52, Oct. 1952.
14. Clark, W. W. "Installation and Experience Data Pile Up as More Utilities Turn toward Plastic Pipe." *Gas* 31: 58–61, Mar. 1955.
15. "Industry's Largest User Sets Up Its Own Standards." *Gas* 38: 84–91, Mar. 1962.
16. A.G.A. Laboratories. *American Standard Installation of Gas Appliances and Gas Piping.* (ASA Z21.30–periodically revised) Cleveland, Ohio.
17. Clay, R. "Realistically Priced Remote Registry Ready for Meter Readers." *Gas* 38: 50–5, Aug. 1962.
18. Petersen, A. D. "Changing Meters without Interrupting Service." *A.G.A. Operating Sec. Proc.* 1962: DMC-62-29.
19. Gallagher, C. A. "Residential Meter Location." *Gas Age* 125: 21–42, Apr. 28, 1960.
20. Hendrickson, C. P. "What's New in Service Design?" *A.G.A. Proc.* 1963: DMC-63-13.
21. Kleinberg, J. S. "New Ideas in Construction and Maintenance—A Panel Service Entry Sleeve." *A.G.A. Operating Sec. Proc.* 1963: DMC-63-40.
22. Skibinski, E. A. "Design Against Contractor Damage." *A.G.A. Operating Sec. Proc.* 1964: 64-D-166.

Chapter 9
Maintenance of Mains and Services

by J. A. Lane and T. J. Noonan

The investment in the network of mains and services and their accessories is such a large part of the total system investment that their adequate maintenance is essential. Other important considerations are the public safety aspects of distribution maintenance programs, the conservation of gas itself, and the capacity of the distribution system. In 1961, distribution system maintenance costs for U. S. straight natural gas distribution companies averaged 14.9 per cent of their total operating expense exclusive of purchased gas cost.[1]

Primary maintenance objectives (not necessarily in order of importance) are: (1) leak prevention—clamping and sealing of cast iron mains, installing equipment to mitigate corrosion, replacing corroded mains and services, and preventing damage during construction operations; (2) leak detection, followed by repair or replacement; (3) prevention and removal of stoppages, e.g., water, dusts, gums; and (4) prevention of overpressure conditions.

A number of guides are available for use in setting up a maintenance program to accomplish these objectives.[2−9]

Maintenance is usually the job of a division (e.g., the street division) of a utility company. Factors such as size and nature of the territory, characteristics of the distribution system, and company policies determine the make-up of the maintenance organization. Maintenance operations include the location and repair of street leaks, clearance of service stoppages, cutting off and abandonment of services, renewal of services, installation of anodes, operation and lubrication of valves, and general overhauling of mains and services.

A watch over the distribution system is maintained by one or more of the following:

1. The company obtains information from city or state authorities as to all street excavation permits granted. For example, one state requires that utility companies give reasonable advance notice of excavations to known owners of underground facilities in or near the construction.[10] Another law requires 72 hr advance notice to a gas company prior to excavation or blasting in or near streets.[11] A company inspector may then be assigned to visit these sites, advise on the location of gas lines, and protect company property.
2. District foremen keep in touch with all street or service excavation projects, as they progress and are carried out.
3. Leakage survey crews patrol congested areas and test for gas leaks at regular intervals.

In some of the larger cities with a multiplicity of underground structures, gas company inspectors patrol the distributing system. These men investigate street openings made by other utility companies or contractors whereby mains or services may be exposed. They arrange for the protection or repair of the gas facilities as required.

Legal Responsibility for Leakage.[12] Action against a gas utility in the matter of alleged damages resulting from leakage would probably be a tort action related to negligence. Elements of such an action are: (1) a legal duty to conform to a standard of conduct for the protection of others against unreasonable risks; (2) a failure to conform to the standard; (3) any connection between the standard of conduct and the resulting injury; and (4) actual loss or damage resulting to the interests of another.

The gas utility can easily become a "three-time" loser when excessive gas leakage is involved. First there is the cost of gas lost thru system leakage. When leakage results in damages to others, the company pays again. It must pay a third time in increased insurance costs and loss of future revenues because of damaged public relations.

Operating Pressure vs. Leakage Rate.[8] The perforations resulting from corrosion are irregular orifices. The flow rate thru the orifices may be expressed as: $Q = K(Ph)^{0.5}$, where Q is the flow rate; K, the discharge coefficient, depends on the number and size of leaks, the specific gravity, and the gas temperature; P is the operating pressure, abs; and h is the pressure drop across the leak(s). For pipelines buried in tightly packed soils, the P required for critical flow (where Q would be a linear function of P) is much greater than 1.9 times the atmospheric pressure because of the pressure drop thru the soil adjacent to the leak.

The leakage rate thru either soil or the porous packing in a *bell-and-spigot joint* can be taken as proportional to the gage pressure in the pipe for gas pressures from 1.7 to 10.2 in. w.c. Similarly, the *leakage rate* from steel lines varies linearly with gage pressure for pressures over 30 psig. No recommendations were given for lower operating pressures.

COMMUNICATIONS

Many maintenance operations, particularly the location and repair of leaks and damages to any part of the distribution system, represent emergencies which must be handled promptly for reasons of safety. Other cases, such as gas supply interruptions, require prompt attention in the interest of good service. Notices of complaints received by the utility at any of its locations are usually transmitted to a central dispatcher. Thus, good communications between the dispatcher and maintenance crews are essential.

Transmittal means include two-way radio, telephone, and teletype. When radio equipment is not available, maintenance crews telephone the dispatcher at definite times. Employees on call at home for emergency service are also used.

GENERAL EQUIPMENT

Distribution maintenance equipment is generally similar to that employed for main and service installations. Transportation of maintenance crews is usually by truck equipped with a self-contained air compressor to operate pneumatic tools such as paving breakers, jack hammers, and diggers. Each maintenance truck is provided with welding equipment (if appropriate), hand tools, gas stopping devices, fittings, and leak repair clamps or sleeves (in the smaller sizes). Instruments for detection of gas leakage and for location of underground piping and other structures (such as valve and curb boxes which may become covered with soil or other material) should also be carried. Safety equipment such as hand fire extinguishers and first aid kits must also be on hand.

Specially equipped mobile units are used for such purposes as clearing of main and service stoppages.[13] These may provide for heating, pumping, and storing the flushing liquid. Other special units may be equipped for copper service installation, anode installation, or corrosion protection.

LEAKAGE SURVEYS

ASA Code B31.8-1963 (Par. 852.22) provides for surveys and recommends *minimum* survey frequencies—once a year in business districts and once in five years outside business areas. These surveys may be made by one or more of the methods described below or by other effective procedures. The primary types used in the United States in 1959 were manhole and bar hole surveys with combustible gas indicators, vegetation surveys, and mobile infra-red surveys.[8] Prompt investigation and repair should be made of all leaks found, and complete records should be kept of all operations.

Gas Detector Surveys. Tests are made with combustible gas indicators in every available opening in the street, including valve and curb boxes, manholes, catch basins, pavement cracks, and bar holes. Frequency of these tests can be established after a trial period during which a higher than "normal" testing frequency might be followed. Periodic review of company practices on this matter is desirable.

Bar Test Surveys. Bar holes are driven at regularly spaced intervals over mains and services and tested with a gas indicator. In street or sidewalk pavements, such bar holes are sometimes fitted with sealing devices so that they will remain clear and accessible for periodic reuse. After the pavement has been drilled, bar holes are driven, usually with hand tools, although a small motor-driven vehicle equipped to drive a probe hydraulically is available.

All street opening tests for leakage should reflect the possibility that other combustible gases or vapors, such as gasoline vapor, marsh gas, or sewer gas, may be present, alone or in combination with utility gas. Filtering devices, special instruments, or laboratory analysis of gas samples may be necessary to identify gases or vapors originating from sources other than the gas distribution system. Other users of street substructures, such as electric, telephone, telegraph, sewer, water, and oil companies, may also make combustible gas tests. Arrangements have been set up in many communities to make these findings mutually available when positive indications are found.

Vegetation Surveys. These surveys are often made by plant pathologists who can recognize whether damage to vegetation is the result of manufactured gas leakage, natural gas soil dehydration, or pathological conditions. Such surveys must be conducted during the growing seasons, when the indicative dry and withered foliage will be evident. Suspected locations are tested with the use of bar holes and combustible gas indicators.

Pressure Drop Surveys. Isolation valves are installed for an area containing 100 to 200 gas meters.[14] This group of meters and their isolating valves are shut off and gas, routed into the area thru a by-pass, is measured by a displacement meter for about 2 hr. Aboveground piping is tested with soapsuds. When area leakage exceeds 27 cfh per mile of main, underground leaks are searched for and repaired, and all services are pressure tested. Continuous pressure records have great value in giving first indications of major leakage within a system. An unusual pressure drop entry warrants immediate investigation.

Soapsuds Testing of Exposed Pipe and Fittings. Components exposed outside or inside buildings are coated with soap solution or a similar liquid and inspected for leakage, which is indicated by the formation of bubbles.

Public Building Surveys. Buildings in which many people may congregate, such as schools, hospitals, and theaters, are inspected at regular intervals, depending on their location and on utility policy. Generally included are visual inspection of exposed portions of the service, meter, regulator, and accessible house piping and testing with a combustible gas indicator in the vicinity of the gas service and other utility lines entering the building. The service shutoff is checked for accessibility and operating condition. Combustible gas indicator tests in bar holes over the service may also be made.

Greenhouse Surveys. Although natural gas is not toxic to plants, elements sometimes present in manufactured or peak shaving gases may have harmful effects. When the latter gases are present, periodic tests with combustible gas indicators are made in the vicinity of gas piping within the structures and in bar holes over the service and main to detect leaks.

Pressure Tests. House piping and services are tested by some utilities on complaint of odor or fumes, on meter replacements, and on renewed or reconnected gas services.

Increase in Gas Odorization. An appreciable increase in odorization for a short time in a town or area may facilitate smelling leaks from gas mains, services, and house lines. This usually results in an increase in leaks reported by customers. Many of these leaks turn out to be of trivial size or are not bona fide (such as an odor during ignition which the customer may notice because of the higher odorant level). In one case[15] in which two or more pipes operated by different companies were in close proximity, determination of which pipe was leaking was accomplished by the use of different odorants.

Line Walking or Patrolling. Patrolling gas lines in open country is done regularly by some companies. Gas leaks may be detected by color of vegetation, by gas bubbling thru the ground, by smell, or by sound.

Meter Reader Inspection. Inspectors trained to locate gas leaks and hazardous conditions on consumer premises may cover a town periodically. Such inspectors take about one-third of the regular meter reader's route per day.

Mobile Leak Detection Surveys. Some companies use mobile leak detection surveys to supplement other types of surveys. A vehicle equipped with an infra-red detector (operating principle is similar to that described for Fig. 6-15b) and sampling tubes positioned close to the street surface travels slowly over the main or near the curb. Samples are continually aspirated thru the detector, which is extremely sensitive to the gas for which it is calibrated (5 ppm for methane[25]). The presence of gas is indicated on a recording chart. Bar hole and combustible gas indicator tests are made to verify the location of a leak indicated on the chart. In areas of heavy traffic and continuous parking, the mobile detector may be mounted on a small hand truck which is pushed along the sidewalk.

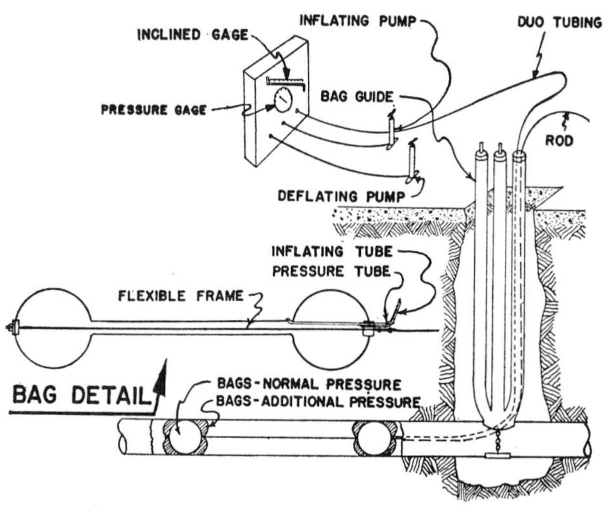

Fig. 9-68 Chicago leak detector.[16]

Leak Pinpointing. This refers to determining the location on the ground surface that is directly over a leak in a buried pipe. Reasonable accuracy is necessary to avoid needless excavations, particularly where expensive heavy pavements are involved. Many bar holes must be driven and allowed to vent before gas indicator tests can be taken which, by their relative intensity, will indicate the location of the leak.

Observation of withered grass or other ground cover near the leak will assist in locating the leak, since the area affected will be within a circle drawn with the leak at its center. When only a portion of the circle is covered with grass, the center of rotation of the arc between the damaged and undamaged grass indicates the leak.

One company[16] uses a Chicago leak detector (Fig. 9-68) to pinpoint leaks in low-pressure mains where it is impractical to bar (e.g., at intersections, car tracks, or areas where subsurface structures exist). This device, consisting of two rubber bags about 9 ft apart on a flexible frame, is placed in the main and moved along with a steel tape (150 to 200 ft in each direction from entry). The bags are inflated (about 3 to 6 psig) simultaneously, thus trapping a volume of gas between the bags. A drop in the gas pressure of the trapped volume indicates a leak in the main between the two bags. This technique is used on *live* mains where the main tested is supplied with gas from both directions.

Instruments which detect the sound of gas leaking from a buried pipe have been tried for pinpointing purposes. These range from nonelectrical devices which are adaptations of the medical stethoscope to electronic devices which amplify particular bands of sound frequencies picked up by a microphone in contact with the gas in the pipe. One sonic leak detector[17] places the microphone in the gas main and moves it along with a crawling device. The microphone diaphragm of another sonic instrument[18] contacts the gas stream by means of an attachment to a service or other pipe connected to the main.

An A.G.A. research project[19] (still in process in 1964) investigated methods and instruments to pinpoint the location of leaks by introducing sound vibrations of known frequency into the gas stream in low-pressure mains. At a point of leakage the sound waves can be detected with electronic sound detection instruments.

GAS DETECTION INSTRUMENTS

Gas detection instruments detect and measure gas concentrations in air.

Combustible and Conductivity Gas Indicators

Although suitable for use in confined spaces, these instruments are not sufficiently sensitive to detect leakage over an exterior gas pipe route.[8]

In the **combustible type,** an aspirated sample of combustible gas is oxidized by a hot-wire, catalytic combustion element. Normal sensitivity is about 0.1 per cent natural gas in air (0.01 per cent is also claimed) to the lower explosive limit (i.e., about 4.5 per cent natural gas in air). The combustible gas indicator uses the change in resistance of a hot platinum wire which is part of a Wheatstone bridge to measure gas concentrations. The range of this instrument may be extended by attaching a device to air-dilute samples proportionally. The instrument contains a flame arresting feature.

Conductivity-type instruments, based on the fact that the thermal conductivity of natural gas differs appreciably from that of air, are not as sensitive as the instruments of the type described above. However, they do indicate 0 to 100 per cent combustible gas in air.

Combination instruments (Fig. 9-69), which use a catalytic combustion element for the lower gas concentrations (below the L.E.L.) and a conductivity cell for higher concentrations, are available.

Any combustible gas or vapor (e.g., gasoline or combustible cleaning fluid) will register on a gas indicator of the combustion filament type. The thermal conductivity type of instrument will only give reliable readings on the gas on which it has been calibrated, and it *cannot* be used on vapors.

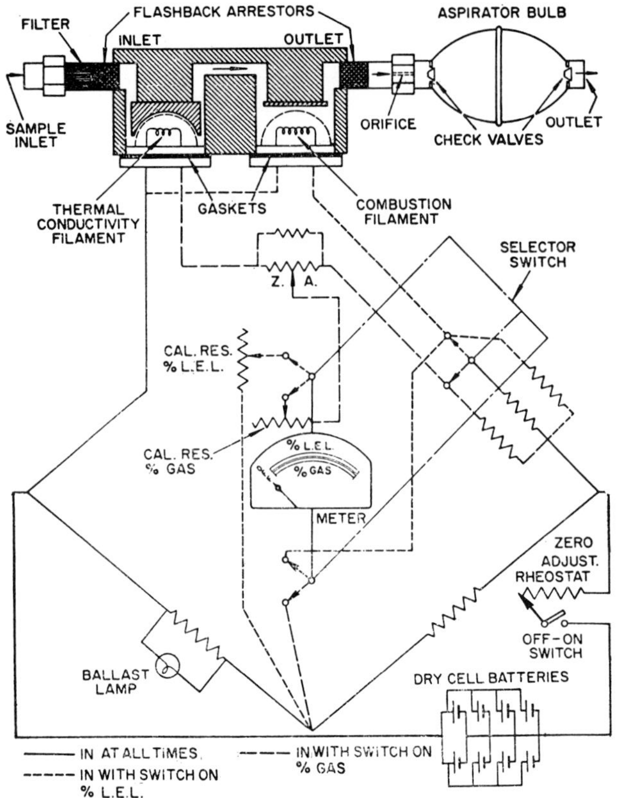

Fig. 9-69 Circuits for leakage testing instrument containing a combustible gas indicator and a thermal conductivity cell.

As leaks from service stations will be, in most instances, gasoline containing tetraethyl lead, care should be taken to select a leakage indicator whose accuracy is not affected by vapors containing this additive. Underwriters' Laboratories lists separately instruments not affected by vapors containing tetraethyl lead and instruments whose filaments are poisoned by this compound. The Laboratories of the Factory Mutual Insurance Companies also have two such separate approvals.

To distinguish between natural gas and all other hydrocarbons heavier than propane, including gasoline, butanes, alcohols, acetone, and benzene, filters such as activated carbon or silica gel may be used. These condense out the higher hydrocarbons and pass the lower ones. Thus, an indication of condensation without the filter and no indication of condensation with the filter in place means that the flammable material is a higher rather than a lower hydrocarbon. It follows that detection instruments do not distinguish between natural gas and sewage gas (or marsh gas). The latter differs from natural gas in that it does not contain ethane, but does contain large amounts of CO_2. Some instruments make use of differences in ignition temperatures of petroleum vapors and natural gas.

Oxygen or Air Indicators

Testing the air in work areas such as manholes and tanks with combustible gas indicators gives no indication of oxygen deficiency, since as little as one per cent oxygen may be enough for an accurate combustible gas indicator reading. Where it is suspected that oxygen has been displaced by noncombustible gases, tests for oxygen may be made with a portable oxygen indicator or a safety lamp.[24] Serious oxygen deficiency will extinguish the lamp. For safe operation with all gases, the lamp should not be placed in the atmosphere under test, but at a remote location with the gas sample aspirated thru the lamp by means of a sampling tube. Positive and continuous ventilation of confined spaces with an air blower before and during occupancy will eliminate the possibility of oxygen deficiency.

Carbon Monoxide Indicators

Manufactured gas may usually be identified by the presence of its component, CO. Portable CO indicators may also be used to detect incomplete combustion in gas appliances. Since these instruments react only to CO, they are not generally used for leak detection work, which can involve methane and other gases.

PIPE, VALVE, AND CURB BOX LOCATORS

Instruments for locating pipe and other metallic structures underground are widely used to minimize digging.[26,27] They may also be used to: (1) pinpoint pipe location; (2) approximate pipe depth; (3) locate laterals connected to a main; (4) locate short stubs; and (5) locate valves, curb boxes, meter boxes, and manhole covers.

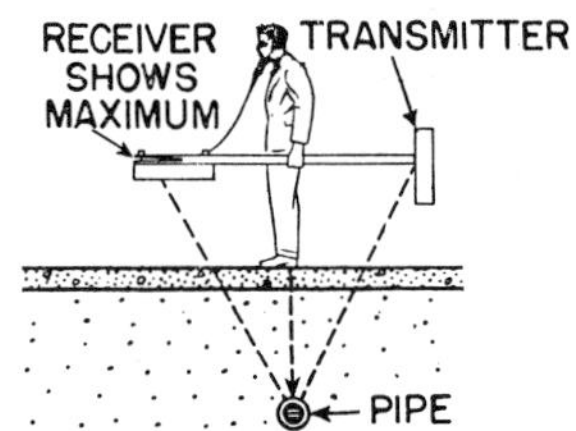

Fig. 9-70 Pipe locator in use.

The two-box electronic locator consists of directional transmitting and receiving units, earphones, and a connecting handle (Fig. 9-70). In use, the transmitter induces electromagnetic waves into the earth (or thru the pipe, if attached to same). Any pipe in the wave field acts as a path for return current, which is, in turn, picked up by the receiver. To determine exact locations, the transmitting and receiving units are detached from the handle and kept apart while the pipe is traced. Trained operators are thus able to find the exact centerline of the pipe, its course, and its depth.

For locating curb boxes, valves, drip boxes, and other structures which are just under the ground surface, electronic box locators and magnetic dip needles are used.

LEAK REPAIR

Steel Mains

Gas leakage from steel mains can be stopped with clamps of various kinds, depending on the nature of the leak. A few of the more common types of leaks and the applicable repair fittings are shown in Fig. 9-71. Various sizes of clamps are available for both high- and low-pressure lines.

A suitable length of a band, saddle, or split repair clamp or a combination of several such clamps may be used to re-

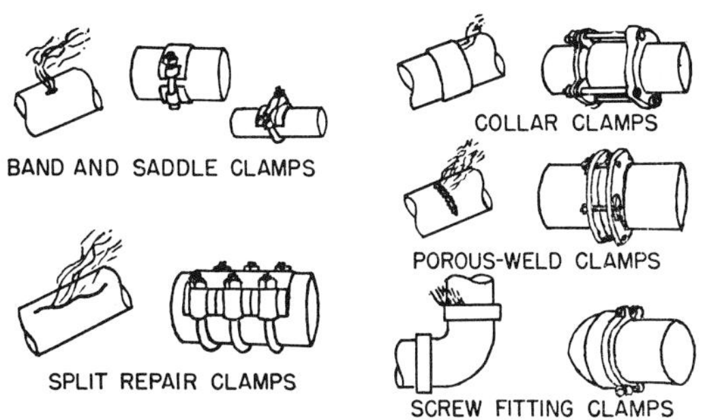

Fig. 9-71 Typical leak repair clamps for steel pipe. Rubber gaskets, neoprene gaskets, or buttons are secured over the leak (not shown).

pair multiple main holes. Bands may be of stainless or band steel of about the same thickness as the pipe. Neoprene or rubber gaskets, cone-shaped rubber buttons, or various thicknesses of sponge rubber sheets may be used as the sealing medium under the steel band. The clamps are designed to last as long as the pipe on which they are placed. Some utilities install a magnesium anode at each leak repair location to protect the pipe from further corrosion.

Cast Iron Mains

Cast iron mains with bell-and-spigot joints packed with jute and lead or cement are used by a number of utilities. A study of 60 four- to eight-inch bell joints in gas systems from 30 to 103 years old showed that the controlling factors in the occurrence of joint leakage are the method and type of backing used (lead or cement) to make up the joint. There was no evidence of any relationship between joint leakage and age, condition of packing, or the practices of **oil fogging** and **humidification.**

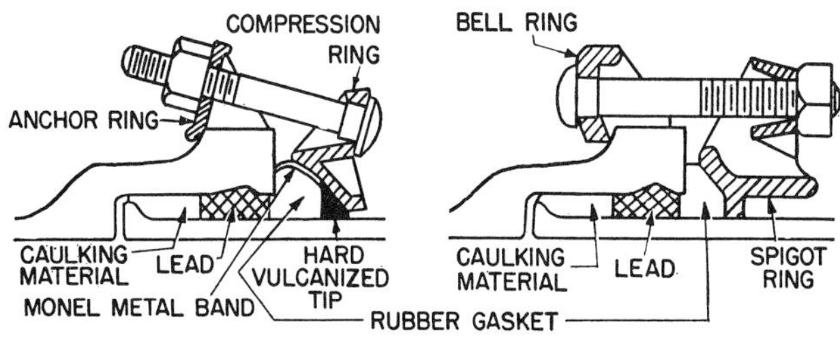

Fig. 9-72 Two designs for bell joint clamps.

Bell Joint Clamps. Various clamp designs (see, for example, Fig. 9-72) are used to stop joint leaks which may result from ground movement, pipe movement caused by temperature changes, settling, traffic, and higher pressures. Satisfactory performance in service is dependent largely on proper clamp installation.

The clamp design should provide:[28] (1) ample gasket cross section—width should substantially exceed the joint space; (2) sufficiently rigid clamp ring segments with secure and substantial means of assembling them into an essentially solid ring; (3) accommodation of clamp to varying dimensions of bell; and (4) close fitting of clamp to pipe and face of bell to prevent undue motion and possible cold flow of the gasket.

Preparation of joints to receive bell joint clamps includes: (1) thorough cleaning of the joint surface and adjacent portions of the bell and spigot; (2) recaulking and refacing of lead joints to bring surface as near flush with face of bell as possible (if recessed more than 1/8 in., refacing with lead, wool, or plaster of Paris is necessary); and (3) cutting out cement joints 1/2 to 3/4 in. and refacing with neat cement, flush with face of bell. A variety of pneumatic hand tools, such as scalers, chipping hammers, wire brushes, and grinders, are used to clean the joint surfaces. Abrasive blasting is an effective final cleaning operation.

Some companies recaulk joints to stop leakage before installing the clamp. Others prefer to install the clamp on the leaking joint to permit an immediate test for tightness.

Gaskets. These may be either rubber or synthetic and either plain, or armored to prevent cold flow of the gasket. Before the spigot ring is installed, the gasket should be fitted to the joint according to the manufacturer's instructions. Precautions should be taken to prevent foreign matter from adhering to the face of the bell or the spigot before the gasket is put in place. A liberal coating of a soap solution on the gasket and joining surfaces will facilitate proper seating. Gasket compression is obtained by tightening the clamp bolts with a torque recommended by the manufacturer. The bolts may be coated with mastic compound for greater resistance to corrosion.

Sleeves. Circumferential breaks or longitudinal cracks in cast iron pipe may be repaired by applying split cast iron sleeves or steel band clamps with rubber or synthetic gaskets. The breaks may be covered with tape first to stop the gas flow and thus to permit the cleaning of the adjacent areas of the pipe prior to the application of the sleeve or clamp. Where a marked angular deflection or offset exists at the break, a split sleeve with end gaskets is more likely to seal than a band clamp with a full circle cylindrical gasket. Special split sleeves enclose an entire bell-and-spigot or mechanical type of joint in order to repair a split at the bell or a circumferential break in the pipe too close to the bell to permit the use of a standard split sleeve.

Other Repair Methods. Although the use of external clamps or sleeves for leak repairs is widespread, high cost and the excavation accompanying each repair make other methods attractive. These methods include treatments applied externally or internally, such as the use of liquids to rejuvenate the jute caulking and sealants which solidify in joints and other cracks. Joint-treating liquids which are applied while the main is in service have given good results in pipe up to 8 in. in diameter.[8]

Results of sealing tests on packing using various sealants, data on the chemical nature of jute fiber and on packing deterioration, and methods of measurement of physical properties of sealants are available.[9]

Carboseal. Carboseal is a liquid of low volatility, essentially nonflammable, having a specific gravity of 1.017 at 60 F and weighing 9.19 lb. per gal. It has been used successfully on 8-in. and smaller diameter pipes. Carboseal treatments, which do not interrupt service, are based on saturating jute packing of cast iron mains with a glycol; it swells jute fibers about 44 per cent. However, Carboseal has no effect on cracks or other breaks in the piping, nor is it intended for use in steel lines, welded or mechanically coupled. The treatment is ineffective on soaped, cement-grouted, or heavily tarred jute, poor joint packing, and packing extensively dried out or reduced to dust.

Methods of application:

1. Gravity flow. Generally used to apply Carboseal to low-pressure mains, it involves successive injections of the liquid at high points in the system, from which it flows by gravity to drips, where excess material is recovered. Capillary action carries the liquid around the joint.

2. Hose application. The spray nozzle (80 to 100 psig) is inserted into and drawn thru intermediate-pressure mains at 100 to 200 ft intervals. Large diameter pipe joints can be treated by drilling and tapping a hole into the joint; Carboseal is then applied directly to the jute.

3. Low input application. The liquid is injected by means of a small battery-operated positive displacement pump of the oscillating cylinder type, capable of delivering 1½ to 2 gal in 24 hr. The pump delivers Carboseal from a small tank to the high point of the main, from which it flows by gravity. Up to two months may be required to complete the treatment.

Favorable test results[29] have been reported by a number of cities which have used Carboseal.

"Never Leak" Method.[30] In this internal sealing method, the main is sectionalized, all services are shut off, and the isolated main section is completely filled with the sealing liquid. A pressure of 65 psig is then applied to the liquid (for mains up to 8 in. in diam) and maintained for 1.5 hr. The liquid contains approximately 40 per cent rubberlike solids which upon drying form an elastic seal. The liquid is forced into the passage of gas leakage, penetrating the jute and coating the surfaces. After treatment the liquid is pumped out, filtered (it is not necessary to clean the main before treatment), and collected for reuse. Service is restored during that same day.

Equipment used includes a 2000 gal tank truck equipped with a pump for filling and draining the gas main. A 300 gal tank on a separate truck is used in conjunction with inert gas cylinders to pressurize the liquid in the main. When a medium-pressure main, 2 to 15 psig, is treated by this method, a maximum pressure of 20 in. w.c. is advised for a six to eight week period to permit the sealant to come up to full strength.

Tests over a period of two years have indicated "absolutely no loss of gas." These tests are being continued. Furthermore, infra-red leak detection surveys over all treated areas showed favorable results.

Thiokol. Bell joint clamps cannot be applied properly to some bell-and-spigot joints because of angularity of the pipe or lack of sufficient clearance to adjoining structures. In one alternate repair method[31] for low-pressure systems, a sealer is applied to the outside of a clean joint. This sealer, unpolymerized Thiokol, and an accelerator polymerize when mixed to form a gel in 3 to 4 hr; the gel cures to tough rubber in two to three days. It bonds well to metal and resists solvents, including aromatic hydrocarbons. As the sealer is ineffective until cured, a Thiokol putty should first be worked into the joint as a preliminary seal. Then the Thiokol sealer should be applied by trowel or brush to form a substantial fillet around the joint.

The repaired joint should be left uncovered overnight so that any leakage may be detected and repaired. Do not test joint with soapsuds immediately following repair, as soap impairs bonding between the sealer and metal. For protection during backfilling, wrapping of the repaired joint is advisable.

Thiokol sealer was initially applied internally to joints of large pipe, 30 in. and over, by a workman who first cleaned the interior surface at the joint. Forced ventilation and lighting were necessary. For mains too small for a man to enter, machines designed by the C. W. Fuelling Co., Inc., of Decatur, Ind., have been used to apply Thiokol and similar sealants in 8- to 24-in. diam mains. Access to the main is by removal of an 8-ft section. The joints in both directions are located by an electronic locator and cleaned by an air-driven machine; the sealant is handled by a machine which mixes the material and applies it with trowel-like paddles.

The Consolidated Edison Co. developed the internal spring band method[32] for internal sealant application. Here, the sealant is applied to a spring steel band whose diameter is slightly less than the internal diameter of the pipe. Premature curing of the coating is prevented by keeping the device under refrigeration until it is used. The band is folded and wired so that it can be carried into the pipe. When it is positioned on the joint, the wires are cut, the band thus assuming a circular shape and forcing the sealant into the joint. It is left in place for 24 hr to allow the sealant to harden. Aluminum foil placed on the band before the sealant is applied allows it to be separated from the sealant, and refolding and rewiring permits its removal from the pipe for reuse. In another internal method, a one-piece gasket was secured to the joints of a 42-in. diam cast iron main, 2.7 miles long, by the use of rings which were assembled within the pipe.[33]

Gutentite.[34] Gutentite is a plastic in colloidal suspension which forms a rubberlike material. It is used to impregnate jute in joints. Before it is applied, the main is sectionalized and all services are shut off. The isolated main section is filled with water, drained, and then filled with Gutentite. A pressure of 25 psig is maintained on the liquid for about 1 hr. The liquid is then removed, and air pressure is used to force the deposited rubberlike material into the joints.

Epoxy Resins.[35,36] Both internal and external applications of epoxy resins are used to seal holes, cracks, and leaking joints. Cured epoxy resins are strong, adhesive, resistant to chemical attack, and relatively unchanged by time, heat, or pressure.

Epoxy resins cannot be applied to a wet or oily surface. The metal should be thoroughly cleaned, preferably by sandblasting. Either leaks must be stopped before application of the resin or a vented channel must be provided to allow time for the resin to harden. The vented channel, formed from sheet

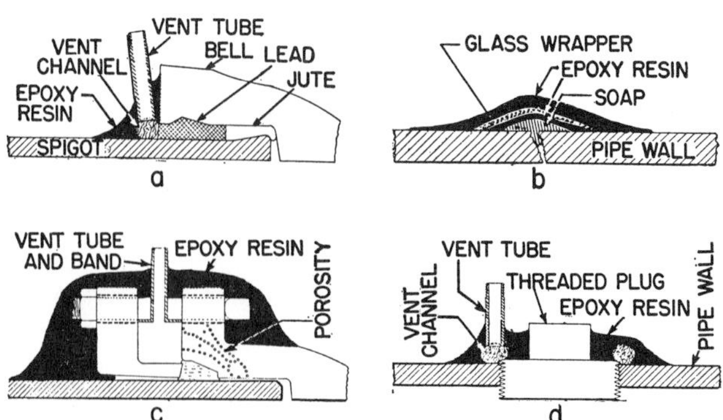

Fig. 9-73 Typical external leak repair methods, using epoxy resins: a, leak in bell and spigot joint; b, break in main; c, porosity in mechanical joint; d, leak in threaded connection. Vents are capped after epoxy sets.

metal, a coil spring, rope, or porous cotton cord, is capped later. Curing time varies inversely with both the ambient and surface temperatures of the pipe.

It is recommended that the surface to be coated first be primed with epoxy resin. This will force the material into all the tiny craters of the metal, increasing the total contact area. The mass of epoxy should be applied with a trowel immediately afterward. It is claimed that as much as 40 per cent increased adhesion is obtained when the preliminary brush-in technique is used. Figure 9-73 shows some typical repairs.

OVERHAULING

Periodic overhauling of mains and services is usually associated with street repaving, a leak reduction program, or system operating changes. Many cast iron mains, particularly, have had long service, and joint leakage may exist to an extent that requires correction. Accumulations of deposits may also require attention. Overhauling is primarily confined to corrosive areas and to the older companies with long manufactured gas experience.

Main Overhauling Procedures

The choice of the overhauling procedure depends on knowledge of the condition of the main. Applicable data include leak surveys, records of leak clamps installed, and inspections of the pipe at various excavations. When repaving is to be done, information should be secured in advance on grade or other changes which may necessitate main or service relocation such as lowering to maintain adequate cover.

Repairing leaks is the major item in main overhauling. Consideration should also be given to **anchorage of mains,** especially when the capacity of the system is to be increased. Valve boxes and pressure pipes, as well as drip boxes and risers, should be inspected during overhauling and renewed if necessary. It may be necessary to raise or lower them to accommodate grade changes.

Lowering Mains. This may be done by excavating along and below the main to the new depth, leaving earthen pipe supports at intervals. After wooden blocks are placed below the main, the remaining earth is dug out, leaving the main resting on the blocks. The main is then lowered to the position desired by removing the blocks one at a time. Another method is to suspend the main by ropes from stringers across the trench. After excavation to the required depth, the main is gradually lowered with the help of these ropes. All joints should be examined after lowering and any needed repairs made.

Renewing Mains. Running high-pressure steel mains thru larger cast iron mains has proved to be an economical procedure; in some cases about 80 per cent less excavation was required.[37] Insulating collars, placed at intervals on epoxy-coated pipe, are coated with grease to facilitate sliding the new pipe thru the old main; this can be done at the rate of 20 fpm with pulling devices or 1.5 fpm with pushing jacks.

Deposits. Excessive accumulations of deposits (Table 9-41a) in mains are often removed during overhauls.

Abrasive Blasting. This is an efficient way of cleaning many irregularly shaped surfaces, particularly prior to applying sealants of any kind. It may also be used prior to painting exterior piping and to prepare joints for mechanical clamps (wire brushing usually suffices). The blasting grit may be common sand (inexpensive and readily available), steel grit, or a smelting and refining industry by-product grit (a recent product). Some of the substitute grit contains extremely little free silica, eliminating the silicosis hazard (at least on outside jobs).

Protection.[41] Air-fed abrasive-blasting respirators (Bureau of Mines approved) are mandatory when sand is used and the operation is not isolated from personnel. Filter-type respirators may be used for small dust hazards.

Service Overhauling Procedures

Any distribution line overhauling program should include the services. A review of service complaints helps to indicate how much overhauling may be advisable. At least all the old low-pressure services in sizes up to 1 in. usually need renewal.

When services under paving must be renewed, it is desirable to minimize openings in both size aud number. One procedure is to run copper or plastic tubing (Figs. 9-67a and 9-67b, respectively) thru the old pipe. If the steel services were originally sized for manufactured gas, the smaller size tubing used for renewal is generally adequate for natural gas. Another technique is to drive out the existing service and introduce the new one into the opening which remains, using pneumatic driving tools, pipe pushers, or jacks. Work is performed thru a small street opening at the main and in the basement of the building supplied.

Service Rehabilitation Programs. In many cities in which all the gas services in a particular area were installed at the same time, renewal of all services over a certain age (say, over 50 years old) may be made. Sometimes, all services in a block are renewed after a specified number have developed leaks. Programmed mass renewals can be done more economically than scattered single renewals. In general, an actuarial approach may be used to determine the life and failure probabilities of services.[38] A discussion of survivor curves (the curves that show the per cent of an original group still remaining in service at a given age) is available.[39] Plots of annual leakage rate and degree of corrosion vs. soil resistivity are also used in service maintenance programs.[40]

RECORDS

Complete and permanent records of mains, services, and related auxiliaries should be maintained by the utility company at a central location. These records should provide all data needed for locating all utility company structures. As an initial step in recording new construction, dimensioned sketches may be prepared on the job; these data are then transferred to permanent records. When street openings are made for maintenance, existing records may be checked and supplemented. Records covering the underground work by others in the proximity of gas lines may also be maintained.

Main records should detail tie-ins with connecting mains and clearances from adjacent structures such as water mains, sewers, ducts, and manholes. Mains should be located with respect to permanent reference points such as curb or property lines in established communities. In open areas, the most suitable reference points may be poles, catch basins, manholes, or roadway centerlines. Main data should include size, material, and type of main, and type of joint and cover. All street names, curb-to-curb street widths, and kind of paving should be reported. Before any street work that requires digging or barring is undertaken, a copy of all pertinent data should be obtained from the permanent records file.

Main valve records should cover each valve in the distribution system. They should include specific location and number, size, type, make, direction to open, number of turns required, and line identification. Results of periodic valve inspections should be recorded.

Drip records (where condensation is a factor) should give location, size and type of main, kind and direction of drip, capacity, and high and low points. Separate records should be kept of the date of each removal of condensate and its amount. Periodic review of such records helps to determine the necessary pumping frequency and the causes and effects of condensate in the system.

Service records should give address, customer's name, type of building, curb cock (if any), service size and length, size of main, and type of service connection. Location of service should be given in terms of distance from the property line or intersecting street. Any unusual condition, such as a run parallel to the curb, an offset, or an extension, should be noted. If the service is a so-called "stub" service or an extension of a stub service, the record should so indicate. These data are usually gathered by the service crew chief at the time of installation or renewal. Permanent service cards are then prepared from the information.

ECONOMICS OF REPAIR VS. REPLACEMENT

Before making a major main repair or overhaul, consider the entire situation, including the possibility of laying a new main of the same or larger size or prolonging the life of the present line with protection against corrosion at the same time the repair is made. Factors to be studied include the size of the main, its age and general condition, present and future street paving, and kind and condition of joints. The character and the potential development of the area supplied by the main should be considered. The possibility that a system can provide additional gas by an increase in distribution pressure and the effect this would have on leakage from the repaired main should be evaluated. Probable rising costs of labor and material which would affect later construction costs should also be kept in mind. Installation costs of a replacement main will entail no expense for paving where no street paving exists yet or where repaving is to be done for other reasons.

REFERENCES

1. A.G.A. Bur. of Statistics. *Gas Facts.* New York, annual.
2. ASME. *Gas Transmission and Distribution Piping Systems.* (ASA B31.8-1963) New York, 1963.
3. Wisconsin Public Service Comm. "Interim Gas Safety Rules and Regulations." *Wisconsin Administrative Code,* Chap. PSC 135. Madison, Wisc. 1963.
4. Assoc. of Casualty and Surety Companies. *Protection of Utilities from Damage during Construction Operations.* (Spec. Hazards Bull. Z-99) New York, 1956.
5. Natl. Fire Protection Assoc. *Flammable Liquids and Gases in Manholes, Sewers.* (NFPA 328M) Boston, Mass., 1956.
6. Alexander, R. W. "Report of Task Committee on the Prevention of Sub-Structure Damage." *A.G.A. Operating Sec. Proc.* 1961: DMC-61-26.
7. Los Angeles, Calif. *Manual on Surface Traffic Interference Problem as Caused by Substructure Construction and Operation.* 1952.
8. Wilson, G. G., and others. "Evaluation of Current Practices in the Detection, Repair and Prevention of Gas Leaks." (DMC-59-32) *A.G.A. Operating Sec. Proc.* 1959: D225-41.
9. —— *Bell Joint Leakage and Test Methods for Packing-Space Sealants.* (I.G.T. Tech. Rept. 5) New York, A.G.A., 1962.
10. Illinois Commerce Comm. *In the Matter of Adapting Rules Relating to Underground Public Utility* (Genl. Order 185, 1st Suppl. Order) Urbana, Ill., Jan. 16, 1962.
11. New York State. "Construction or Blasting Near Pipes Conveying Combustible Gas." *Laws of the State.* (Chap. 731, Sec. 1918) 2: 1610–1, 1953.
12. Kroeger, C. V. "Distribution Maintenance." *Gas* 37: 45–50, Dec. 1961.
13. Heuser, W. L. "New Equipment for Cleaning Drips and Services." *Am. Gas J.* 170: 24+, Feb. 1949.
14. Hough, F. A., "Economics of Unaccounted-for Gas." *P.C. G.A. Proc.* 27: 86–92, 1936.
15. George, E. S. "How Houston Natural Approaches Distribution System Maintenance." *Gas* 31: 64–6, Apr. 1955.
16. Campbell, E. G. "Why Planned Program of Leak Detection Pays Continuing Dividends." *Gas Age* 108: 26+, Dec. 20, 1951.
17. McElwee, L. A., and Scott, T. W. "Sonic Leak Detector." *Am. Gas J.* 184: 14–17, Aug. 1957.
18. "Novel Device Determines and Locates Gas Leaks by Sound." *Gas Age* 124: 21–2, Sept. 17, 1959.
19 Reid, J. M., and others. "New Approach to Pinpointing Gas Leaks with Sonics." *A.G.A. Operating Sec. Proc.* 1961: DMC-61-22.
20. Tomkins, S. S. "Gas Detection Instruments." *A.G.A. Proc.* 1932: 810–25.
21. Yeaw, J. S. "New Developments in Instruments." *A.G.A. Proc.* 1947: 547–90.
22. Mueller, F. P. "Gas Leak Detection Practices." (PC-50-14) *A.G.A. Proc.* 1950: 639–46.
23. Hartz, N. W. "New Approach to Gas Leak Detection." *Gas* 30: 38–40, July 1954.
24. Miller, R. W. "How Four Utilities Cooperate to Reduce Leakage." *Gas* 27: 46–8, Apr. 1951.
25. Thomas, E. R. "Finding Leaks with a Mobile Unit." *Gas Age* 117: 31+, May 31, 1956. Also (DMC-56-7) in *A.G.A. Proc.* 1956: 232-8
26. Eggleston, H., Jr. "Operating Principles of Pipe Locator." *Gas* 27: 40–4, Sept. 1951.
27. Cassingham, J. L. "How to Find Buried Lines with Pipe Locators." *Gas* 25: 42–3, Jan. 1949.
28. Knapp, K. R. "Progress Report of Pipe Joint Research." *A.G.A. Proc.* 1932: 714–22.
29. Carbide and Carbon Chem. Co. *Stop Gas Main Leakage.* (Bull. F4506-B) New York, 1951.
30. Xenis, C. P. "The 'Never Leak' Method for Internal Sealing of Gas Mains." *Am. Gas J.* 186: 18–21, June 1959.
31. Wilby, F. V. "External Sealing of Bell and Spigot Joints on Cast Iron Mains." *Gas* 28: 69–72, Apr. 1952.
32. Xenis, C. P., and Hale, D. "Sealing Pipe Joints by the Internal Spring Band Method." *Am. Gas J.* 185: 17–9, Dec. 1958.
33. Kooke, C. A. "Ingenious Internal Clamps Upgrade 43-Year-Old Baltimore 42-In. Main." *Am. Gas J.* 190: 20–5, Feb. 1963.
34. Ungethuem, E. "Gutentite." *A.G.A. Operating Sec. Proc.* 1957: DMC-57-9.
35. Detlefsen, R. J. "Chicago: External Repairs Made with Epoxy Resins." *Am. Gas J.* 184: 12–4, July 1957.
36. Blain, H. M. "New Orleans: Joints Coated from Interior." *Am. Gas J.* 184: 17–9, July 1957.
37. Kulman, F. E. "Operation Dig Less." *Am. Gas J.* 190: 24–7, Jan. 1963.
38. Park, W. R. "How to Insure Effective Service Renewal." *Gas Age* 129: 24–7, Dec. 1962.
39. Marston, A., and others. *Engineering Valuation and Depreciation,* 2nd ed. New York, McGraw-Hill, 1953.
40. Shepard, R. M. "Top Management Will Look at Corrosion Control." *A.G.A. Operating Sec. Proc.* 1960: DMC-60-20. Also in *Am. Gas J.* 187: 36–41, Sept. 1960.
41. Natl. Safety Council. *Abrasive Blasting.* (Data Sheet D-433) Chicago, Ill., 1956.

Chapter 10
Distribution System Deposits

by J. M. Pickford and F. E. Vandaveer

ORIGIN OF DEPOSITS

Liquid and solid deposits, which may be found in manufactured and natural gas distribution systems in varying amounts and locations, are listed in Table 9-41a. Note that some deposits are *not* common to both systems. The amount and nature of deposits and the extent of stoppages of mains and service lines will vary for each company and each system.

Factors contributing to the formation of deposits include gas type and composition, manufacturing process or source of gas; purification or cleaning processes employed and how satisfactorily they are operated; gas temperature and pressure; atmospheric temperature; whether filters, dehydration, oil fogging, or humidification are used; and tightness of the gas mains.

Deposits may be classified as to origin (Table 9-41a) within each of the five groups:[1]

1. *Suspended matter in gas entering system.* This may include iron oxide,* dirt,* sand,* salt,† tar,‡ and lampblack.‡

2. *Material deposited by condensation.* This may include water,* ice,* scrubber oil,† hydrates,†[2] drip oil,‡ indene–styrene gum (liquid phase),‡ naphthalene,‡ deliquescent calcium–magnesium salts,† and hydrocarbon liquids.†

3. *Material formed by chemical reactions in gas.* Vapor phase gum formed in manufactured gases by interaction of nitric oxide, oxygen, and unsaturated hydrocarbons is an example of chemical reaction in the gas phase. Ammonium carbonate would also be formed if ammonia were present in a gas containing carbon dioxide.

4. *Material picked up from one part of system and deposited in another.* Gas traveling at high velocity (in the high-pressure feeder systems, for instance) may pick up iron oxide, dirt, sand, construction debris, salt, calcium carbonate, and other solids or liquids, and carry them into the low-pressure system and house lines.

5. *Chemical reaction between gas components and pipe.* Iron oxide formed by reaction of oxygen and water in gas with iron pipe is the major constituent formed by this process. Other solids so formed include iron sulfide (iron pipe and hydrogen sulfide or mercaptans), iron sulfate (iron pipe and sulfur dioxide or sulfur trioxide), copper sulfide (copper tubing and sulfides), and ferrocyanide (iron pipe and cyanogen). These reactions proceed slowly at ground temperatures, but they are more rapid at elevated temperatures and pressures.

* In both natural and manufactured gas.
† In natural gas only.
‡ In manufactured gas only.

DETERMINATION OF DEPOSIT NATURE

Rub deposit in hand to distinguish metallic substances. Ignite a small portion on the blade of a knife. **Iron oxide** will give a bright red flame.

If the substance forms a ball when rubbed, it can be classified as tar or gum. **Tar** will dissolve in benzene, whereas gum will not. However, **gum** will dissolve in ethyl alcohol.

To test for **naphthalene,** take a small portion, add glacial acetic acid, shake well, and add picric acid. Evaporate. If naphthalene is present, white naphthalene picrate will form.

To test for **cyanides,** take a portion of the sample in a test tube and add 10 cu cm of hydrochloric acid and 30 cu cm of water. Shake well. A deep blue color (Prussian blue) indicates ferric ferrocyanide.

A tar camera[3] may be used to determine the kind and quantity of material carried by the gas. The method involves passing a known volume of gas thru a filter paper. Methods for testing manufactured gas for deposits are also available.

METHODS OF REDUCING DEPOSITS[4]

A brief listing of methods to reduce, dissolve, or remove deposits from a distribution system is given in Table 9-41a. Table 9-41b shows the effectiveness of selected solvents in the removal of deposits from manufactured gas (variations of water gas and oil gas) mains. Reduction of deposits starts with the cleaning of internal rust and construction debris from a new line, followed by thorough cleaning of the gas entering the system, careful control of the manufacturing process, use of good coal or oil, purification of corrosive constituents from the gas, filtering of small solid and liquid particles from the gas at selected locations, and use of main and service materials nonreactive with the gas.

MAIN AND SERVICE STOPPAGES

Stoppages in mains or services in *areas supplied with natural gas,* although rare, are usually caused by water condensed out of a saturated gas or forced into a low-pressure line thru a hole or leaky joint by water pressure greater than the gas pressure; by scrubber oil siphoned out of

Table 9-41a Nature, Origin, and Methods of Removal of Deposits in Distribution Lines

Nature of deposits in: Manufactured gas	Natural gas	Origin	Methods for reducing, dissolving, or removing deposit
		A. Liquid Deposits	
Water	Water	Leakage into lines, manufacturing process or wells, humidification.	Pump or blow drips, pump low place in line, install dehydrators, force pig thru line. Cool gas to condense out liquids in manufactured gas plant.
Light oil	Scrubber oil	Manufacturing process or dust scrubbers.	Pump or blow drips, pump low place in line, install filters, improve light oil removal at plant.
Fog oil	Fog oil	Dry gas conditioning.	Same as above.
...	Lubricating oil	Equipment such as compressors.	Same as above.
Drip oil	Crude oil, hydrocarbon liquids	Manufacturing process, drips, or wet natural gas.	Same as above. For crude oil, install oil separator at wells.
Tar	...	Manufacturing process.	Solvents such as gasoline, benzol, acetone, kerosene and naphtha, heat and steam.
Gum (liquid phase)	...	Manufacturing process.	Solvents: benzol, acetone. Improve manufacturing processes by reducing organic sulfur, oxygen, condensable hydrocarbons, and indene and styrene. Pump drips often.
		B. Solid Deposits	
Iron oxides	Iron oxides	Rust inside pipe before laying; rusting due to O_2, H_2O, S, etc. in gas. Purifier carry-over in manufactured gas.	Clean pipe before laying. Blow or pig after installation. Install dust traps or filters. Reduce O_2 and H_2O in gas. Reduce gas velocity. Avoid reversal of gas flow.
Dirt, dust	Dirt, dust	Construction, gas plant, gas wells. Dusting of jute in cast iron pipe joints.	Clean pipe by blowing or pigging after laying. Install dust filters. Oil wet pipe interior. Reduce velocity of gas; avoid reversal of gas flow.
Naphthalene	...	All manufactured gas processes.	Remove from gas by condensing and scrubbing equipment. Solvents include kerosene, naphtha, hot oil, steam.
Tar	...	Manufactured gas processes.	Remove from gas by electrical precipitation, impingement, absorption, scrubbing. Solvents include gasoline, kerosene, naphtha, benzol, acetone. Heat and steam.
Lampblack	...	Manufactured gas processes.	Same as above.
Cyanides	...	Manufactured gas processes.	Remove hydrogen cyanide by oxide or liquid purification. Soda and ammonium polysulfide dissolve HCN. Alcohols and acids may dissolve solid cyanides.
Sand	Sand	Manufactured gas operations. Gas wells, construction.	Clean pipe by blowing or pigging. Install filters.
...	Salt (NaCl)	Gas wells.	Soluble in water.
...	Calcium chloride	Gas wells.	Fairly soluble in cold water. More soluble in hot water.
Iron sulfate	Iron sulfate	Reaction of sulfur compounds in gas with iron pipe. Manufactured gas operations.	Blow or pig line. Filters. Slightly soluble cold water; decomposes in hot water.
Gum	...	Liquid phase gum oxidized to a solid. Manufactured gas processes.	Solvents: benzol, acetone, pyridine, Cellosolve. Pumping drips often improves manufactured gas processes and purification.
Iron sulfide	Iron sulfide	Reaction of iron pipe with sulfides or mercaptans in gas.	Keep hydrogen sulfide, mercaptans, and oxygen in gas at minimum. Filter. Soluble weak acids.
Copper sulfide	Copper sulfide	Reaction of copper tubing with sulfides or mercaptans in gas.	Keep hydrogen sulfide, mercaptans, and oxygen in gas at a minimum. Filter. Soluble weak nitric acid.
Calcium carbonate	Calcium carbonate	Manufactured gas production, gas wells, construction.	Filter. Soluble in acids and ammonium chloride.
Ice	Ice	Freezing of water in the gas or leakage into gas lines.	Dissolve with methyl or ethyl alcohol. Apply heat and pressure. Dehydrate the gas.
...	Hydrates	Solidification of hydrocarbons and water.	Seldom found in distribution lines because of low pressure. Soluble in methyl alcohol and ammonia. Reduce pressure or increase temperature to discharge hydrates.

Table 9-41b Effectiveness of Solvents on Manufactured Gas Deposits[11]

Solvent	Deposit dissolved in 24 hr, %
Dimethyl formamide	83
Dimethyl formamide + 5% water	80
Dimethyl formamide + 10% water	80
Concentrated nitric acid	79
Glacial acetic acid	51
1:1:1 mixture acetone: methanol: benzene	49
Dimethyl formamide + 20% water	26
Cities Service solvent 62	25
1:1 mixture methanol: benzene	25

oil scrubbers; by ice; or by iron oxide and dirt in the line. For water or oil, pumping the drips may clear the line. If it doesn't, it may be necessary to locate the low spot, tap the line, and pump out the liquid. Ice in that part of the service line or outside meter connections aboveground can usually be located and cleared readily. If the ice is below ground, introduction of methyl alcohol forced to the obstruction or heat applied to the affected line may clear it. When the obstruction is iron oxide or dirt, it will be identified easily in meters and connections and proper steps to clean the line and filter the gas can be taken. Iron oxide (Fe_3O_4) is magnetic and is readily soluble in hydrochloric acid; $Fe_2O_3 \cdot nH_2O$, however, may be indicated as present by a chemical test.

Main and service stoppages in *areas recently converted to natural gas* or in *manufactured gas areas* are more frequently encountered. Clearing of these mains and services requires different treatment.

Mains

Locate the stoppage by pressure readings on successive services, drips, or other sources. Starting from a clear point, uncover the main at 100 ft spacing.[5] Cut out a section of main at both excavations and cap live ends. Pass a tape (100 ft × 1 in. × 1/8 in.) thru the plugged section (mains 3 in. and smaller), insert a second time, and leave it in the pipe as air pressure is applied gradually at one end of the section. Meters on service lines are disconnected and *Y* connections are screwed on service tees. As air comes thru the line, gradually withdraw tape, thus preventing plugging and facilitating discharge of the deposit. Blow main, clear services by air pressure thru the *Y* connections to the main, and blow main again. If stoppage extends beyond a 100 ft section repeat the cutting, rodding, and blowing at 100 ft intervals.

On 4 in. and larger mains the same procedure is used except that a chain is pulled thru the main after the tape. When a sizable hole has been made thru the deposit the chain is pulled to the uphill end and a water wash is poured into the main until it comes out at the other end. Water is blown out of services and mains by air pressure.

For **tar stoppages** the water wash is heated to 150 F. This is followed by a kerosene wash at 150 F and working of the chain back and forth to secure agitation. The kerosene is followed by another hot water wash and, finally, the section is air-blown.

Naphthalene stoppages can be treated in the same manner as tar stoppages, by use of kerosene, naphtha, hot oil, or steam as solvents.

Frost accumulations yield to hot calcium chloride or alcohol solutions. The latter is preferred, since calcium chloride is corrosive.

Ice deposits that will not yield to alcohol can be cleared with a lance and hot solutions. The main is tapped (1¼ in. tap) on the downgrade side of the stoppage. A ⅜ in. copper tube lance, not over 60 ft long, is inserted and connected by hose to a pump of a drip truck containing a hot calcium chloride solution. The solution is then fed into the main until the stoppage is reached and melted. If there are no drips nearby to catch the hot solution, a *bag* can be placed in the main on the downstream side of the hole so that the hot liquid can run out of the tap hole. In emergencies, electric thawing machines are used successfully. Currents from 100 to 500 amp to 10 to 12 v have been used for this purpose, particularly by water companies to thaw water lines.

Another method[6] of cleaning mains involves sections of 200 to 300 ft length. The line is bagged at street openings, a 3 to 4 ft section of pipe is removed, and the live ends are capped. Meter cocks are shut off at each house on the line to be cleaned. A sewer rod of 5⁄16 in. Flexicrome is inserted. When it emerges at the opposite end, a small auger and a cable are attached. The rod, auger, and cable are pulled thru the main, a compressor hose is inserted, and air is blown thru the main. Larger augers, followed by brushes, are pulled thru the main until a full sized brush can be pulled thru it easily. The services are then blown toward the main and a final pass of the brush and air thru the main clears it, making it ready to be put back into service. Total cost of cleaning by this method is 30 per cent of the estimated cost of replacement of the pipe.

Another method used in Indianapolis[7] to clean 3 and 4 in. cast iron mains, is to pump the lines up to 100 psig with air at one end and open a valve quickly at the other end, discharging dirt and air into a canvas bag 4 ft in diameter × 12 ft long. All meters on that section of line are disconnected and the work done on one block at a time, excavations being at street intersections. Services are blown out from basements and turned off at the curb during the main cleaning operation.

A similar procedure[8] for blowing dust out of a utility transmission line and catching the dust in a canvas bag used natural gas at both 5 and 20 psig at flows of 150 and 350 MCF, respectively, for 1/2 hr.

Pigs and scrapers forced through lines by gas or air pressure have also been used to clean pipe lines successfully.[9]

Services

A serviceman answering a low-pressure or no-gas complaint may proceed as follows, unless there is visual evidence* of the trouble:

1. Light burners and observe flames A gradual decrease of pressure as successive burners are turned on would indicate partial stoppage in the service or house piping, particularly if no other complaints have been received in this area. A slow, regular rising and falling of the flame indicates an overloaded or a sticky meter. A sharp, irregular jumping of the flame indicates liquid† matter in the service meter or piping.

* Including undersized meter or piping.

† Water and liquid gum are manufactured gas rather than natural gas problems.

2. Where it is possible to take the necessary precautions, the cap may be removed from the end of a live service to clear stoppages (by blow out or snake). These precautions include the use of a two-man team, shutdown of electrical devices that produce arcing, adequate ventilation, and use of a stopper to minimize the escape of gas

3. A solid service stoppage may be attributed to naphthalene, ice, or rust. Pouring or forcing alcohol into the service, followed by blowing, will usually free the service of ice. Kerosene or gasoline will usually free a service of naphthalene. Sometimes a service or house piping must be cleaned with a rod. For safety, the service is often cut either at the main or outside the building wall. There must be good ventilation and no open flame nearby.

For frost stoppages,[5] alcohol is poured into the service at the house end. After the alcohol has worked its way thru the frost to give partial clearance, the service is flushed with alcohol. Sometimes a slight pressure is applied (with a hand pump) to inject the alcohol.

For stoppages of dirt, rust, and ice,[5] a motor-driven rodding device operated under a 3 in. Hg vacuum is used to remove the deposit from the service. For very difficult situations, an air jet exhauster in conjunction with an air compressor is used, giving a 26 in. Hg vacuum which is applied in successive jerks. If the stoppage does not yield to any of the above methods, the service must be dug up and replaced.

Pilot stoppage or outage on gas appliances may be a major source of customer complaint. Only occasionally are pressures so low or stoppages so great in certain areas as to cause complete outage of all appliances on a service. Pilot outage, excluding drafts and dust from the house, may be caused by dust storms in pipelines in an area; oil; gum; copper sulfide flaking off the inside of a copper pilot or service line; carbon from high-temperature cracking of the gas, especially in the presence of nickel or nickel steel; or excessively high or low gas pressure. Dust storms or excessive oil droplets will usually cause outages in the areas affected. If the outages continue, installation of pilot or house line filters, or a filter at a district regulator station, may be necessary. Gums, identified by a sticky feeling, by a magnifying glass, or by dissolving in alcohol, can be filtered out with pilot gum filters. Outage caused by copper sulfide can be stopped by changing the pilot tube to aluminum or plain iron. Copper sulfide can be identified by its shiny black, flaky appearance, or by dropping weak hydrochloric acid on a small sample and noting hydrogen sulfide odor. Carbon, on the other hand, is a dull black, will not generate H_2S when acid is dropped on it, and will burn. Pressure fluctuations may be determined by pressure gages. It may be necessary to leave a sensitive recording pressure gage on the line for a day or so to catch the momentary pressure surges which may cause pilot outage.

REFERENCES

1. Shnidman, L. "Deposits—Distribution Systems and Appliances." In Schnidman, L. *Gaseous Fuels*, 2nd ed., Chap. XII. New York, A.G.A., 1954.
2. Deaton, W. M., and Frost, E. M., Jr. *Gas Hydrates and Their Relation to the Operation of Natural-Gas Pipe Lines* (Bur. of Mines Mono. 8) New York, A.G.A., 1946.
3. Altieri, V. J. *Gas Analysis and Testing of Gaseous Materials*, p. 284. New York, A.G.A., 1945.
4. A.G.A. Library. *Bibliography on Distribution System Deposits, 1932–1952*. New York, 1952.
5. Peters, W. C. "Cleaning Mains and Services of Foreign Matter." *A.G.A. Proc.* 1948: 652–6.
6. Tolford, J. K., and Wigglesworth, G. L. "Cleaning Mains." *Gas* 26: 56–7, Oct. 1950.
7. "Vacuum Cleaning Mains." *Gas* 27: 56, Sept. 1951.
8. Senatoroff, N. K., and Niederer, E., Jr. "How Southern Counties Developed Method for Blowing Dust from Lines." *Gas Age* 112: 23+, July 16, 1953.
9. "This Pig Has a Potent Poke." *Gas Age* 110: 23–4, Sept. 25, 1952.
10. Symnoski, S. C., and others. "Undesirable Deposits in a Utility Distribution System—Their Detection and Control." (CEP-56-5) *A.G.A. Proc.* 1956: 486–504.
11. "How Citizens Removes Deposits from Gas Mains." *Gas* 38: 79–2, Sept. 1962.

Chapter 11

Gas Odorization

by J. S. Powell and F. E. Vandaveer

Odorization of natural gas was suggested as early as 1885 as a means of facilitating leak detection.[1] Addition of odorants to city gases in this country became a commercial procedure with the use of Pintsch gas condensates[2] to odorize water gas in 1905 and of stenches as a warning in mines[3] in 1919. Odorant use in blue water gas and producer gas[4] was required by law in certain German cities as early as 1918. Twenty-four stenches for detecting leakage of blue water gas and natural gas[5] were studied in 1920. An investigation of warning agents for fuel gases by the Bureau of Mines in cooperation with A.G.A. resulted in the publication of Monograph No. 4 in 1931.[6] Some 89 substances were evaluated and odor or irritating properties measured for 57.

The American Gas Association sponsored research on odorants at Arthur D. Little, Inc. for a number of years thru 1959. Although more than 300 materials were screened without leading to an ideal material, the program stimulated manufacturers to develop at least eight better odorants. An odorant test room and odor testing equipment were designed and used in the evaluation of commercial odorants thru the applica-

Table 9-42 Odor Characteristics of Pure Hydrocarbons Which May Be Present in Natural Gas and Natural Gasoline

(dispersed in air in 500 cu ft room; six observers)

Hydrocarbon	Hydrocarbon in air, %	Drops of hydrocarbon*	Calc. odorant, lb/MMCF of air	Odor description† and human reaction
Methane	100‡	...	...	Sweet, pleasant, distinct
	Up to 4	...	...	Almost odorless
	Above 4	...	...	Weak, gas smell
Ethane	100‡	...	...	Pleasant, weak, gas smell
	Up to 4	...	...	Pleasant, very weak gas smell
	Above 4	...	...	Weak but recognizable gas smell
Propane	100‡	...	...	Distinct gas smell
	0.2	...	230	Distinct gas smell
	0.4	...	460	Distinct, rather offensive; slight watering of eyes
n-Butane	100‡	...	...	Strong, unpleasant, rather nauseating
	0.6	...	920	Distinct, penetrating
iso-Butane	100‡	...	...	Strong, penetrating
n-Pentane	100‡	...	...	Weak, sweet, penetrating
	...	20	0.78	Distinct, sweet, not identifiable as gas
iso-Pentane	100‡	...	...	Weak, musty, less distinct than *n*-pentane
	...	16	0.58	Faint, sweet, sickening
n-Hexane	100‡	...	...	Sweet, pleasant, weak
	...	15	0.61	Faint, sweet, persistent
n-Heptane	100‡	...	...	Sweet, weak, gasoline-like
	...	15	0.63	Faint
n-Octane	100‡	...	...	Weak, sweet, distinct
	...	15	0.65	Faint, distinct, sweet, persisting
n-Nonane	100‡	...	...	Rather weak, sweet
	...	15	0.74	Faint, sweet, disappearing
Sweetened gasoline, field A	...	3	0.39	Strong, gas smell
Unsweetened gasoline, field A	...	10	1.1	Definite gas smell
Unsweetened gasoline, field B	...	10	1.1	Strong, gas smell, objectionable

* Measured using eye dropper; approximately three drops per cubic centimeter.

† The term "gas smell" used herein refers to the natural gas then distributed by the East Ohio Gas Co.

‡ 100% tests were made directly from a container. The gas was discharged from a cylinder, with a man's nose very close to the outlet. The man, of course, inhaled air as well as gas.

tion of an odor profile method. This method depended on the quantitative summation of independent observers' evaluation of five odor characteristics of each material.

Sufficient odorant may be added to an otherwise odorless gas so that the gas can be detected by sense of smell long before the concentration of gas in the area becomes hazardous. *In order to build up an explosive concentration of gas in a dwelling, the leakage rate has to be at least as great as the rate at which the gas would be dissipated at the explosive concentration by the air change within the dwelling which occurs thru openings to the outside.* Experimentally, it has been found that at a rate of gas leakage equal to 1.5 per cent of the volume of a room per hour, the concentration of gas within the room will not reach the lower explosive limit of natural gas which is about five per cent by volume.[7]

Effect on Plants. A three-year investigation[8] showed lack of toxicity in natural gas or natural gas containing odorants toward shade trees, tomatoes, and rye grass. A tree exposed over a period of eight years to natural gas remained unharmed. Manufactured gases do have harmful effects on trees.

ODOR INTENSITY OF DISTRIBUTED NATURAL GAS

Natural gas as it leaves the well usually contains enough hydrocarbons higher than ethane (principally the liquefied petroleum gases and other components of casinghead gasoline) or sour (sulfur bearing) constituents such as hydrogen sulfide or mercaptans to be detected easily by smell. High methane gases obtained from some wells are more difficult to detect by smell. To avoid trouble from condensation, a major portion of the odor-bearing constituents, along with most of the condensable hydrocarbons, are removed from natural gas in dehumidification and other purification processes.

Odor characteristics of hydrocarbons present in natural gas and natural gasoline, as determined by test[9] using pure research grade materials, are listed in Table 9–42. The major constituents of natural gas—methane and ethane—are almost odorless up to four per cent in air, but they are easily identified in very high percentages in air. The strongest smelling hydrocarbons in natural gas are propane, butane, and gasolines, either sweetened or unsweetened.

An investigation was made[10] of the possibility of using natural gasoline as an odorant. It was found that, to achieve an odor profile similar to that of an unscrubbed natural gas, 60 lb (10.9 gal) of *sweet* gasoline or 15 lb (2.7 gal) of *sour* gasoline per MMCF would be required.

A study[11] of the sensitivity of human olfactory nerves to changes in the concentration of *thiophane* in natural gas revealed that a 150 per cent change in concentration is necessary to establish a definite change in odor level; with an equal change in odorant concentration, **mercaptans** affect the odor level to a greater degree than the **cyclic sulfide**. Mercaptans also have a lower threshold odor level than *thiophane*.[12]

ODOR INTENSITY OF MANUFACTURED GAS

Intensity of odor varies with type of gas and method of manufacture. Several constituents of manufactured gases, such as CO_2, O_2, CO, H_2, CH_4, and N_2 are almost odorless.

Table 9-43 Odor Intensity of Various Percentages of Manufactured Gases in Air[6]

Type of gas	Very faint	Faint to moderate	Easily noticeable	Strong
42% coke oven, 58% carbureted water gas	0.033	0.04	0.07	0.20
Coal gas	.07	.14	.27	.55
Carbureted water gas	.05	.12	.27	.55
Refinery oil gas	.05	.10	.15	.27
Mixture of carbureted water gas, coal gas, and oil refinery gas	0.05	0.11	0.20	0.37

Producer gas, blue water gas, and reformed natural gas are difficult to smell. Strong odor intensity gases are illuminants (unsaturated hydrocarbons) and residual trace components of hydrogen sulfide, mercaptans, nitric oxide, oil vapors, tar, and naphthalene.

Odor levels of some manufactured gases[6] are shown in Table 9-43. It will be noted that all gases shown give a strong odor when much less than one per cent gas is present in the air. Carbureted water gas contains from 100 to 300 gal of oil vapor per MMCF, which contributes to the odor.

STATUS OF GAS ODORIZATION

Odorization of gas is usually done for one or more of the following three reasons:

1. **Customer Protection.** If customers can smell and identify gas they can locate and stop a leak or obtain assistance. It should be recognized that odorization is only one phase of customer protection and other practices for minimizing and locating gas leaks are usually maintained.

2. **Gas Leak Detection.** Leaks may be detected not only on customers' premises, but also in service lines, street mains, regulators, stations, and everywhere that gas is handled.

3. **State Regulation.** More than half the states require that gas have an odor (Table 9-44). Reference may be made to the following extract from ASA B31.8-1963 (adherence to this Code is required to varying degrees in 24 states):

861 ODORIZATION. Any gas, distributed to customers through gas mains or gas services or used for domestic purposes in compressor plants, which does not naturally possess a distinctive odor to the extent that its presence in the atmosphere is readily detectable at all gas concentrations of one-fifth of the lower explosive limit and above, shall have an odorant added to it to make it so detectable. Odorization is not necessary, however, for such gas as is delivered for further processing or use where the odorant would serve no useful purpose as a warning agent.

Most states have adopted NFPA Standards Nos. 58 and 59 for the storage and handling of LP-gases. In these standards (NFPA No. 58, Para. B.1 and NFPA No. 59, Para. 140) odorization is required "provided, however, that the odorization is not required if harmful in the use of further processing of the LP-gas, or if odorization will serve no useful purpose as a warning agent in such use or further processing." The ICC has no requirement for the odorization in tank cars but does require odorization in tank trucks. The ICC require-

Table 9-44 Summary of State Commission Regulations on Odorization of Natural Gas

State	Reference	Date	Min detectable gas in air	Requirements
Arkansas	Rules & Regulations Governing Utility Service	3/8/56	1%	Distinctive odor readily perceptible to normal person coming from fresh air into an area where gas is present in concentrations of not more than one part gas to ninety-nine parts air.
California	General Order No. 112	7/1/61	Not over 1/5 the lower limit of combustibility	Distinctive odor of sufficient intensity....
Colorado	Statutes and Rules Governing Public Utilities	1961	None specified	Detectable and recognizable odor.
Connecticut	Docket No. 8950	5/11/54	None specified	Distinctive odor to act as indicator to its presence.
	Docket No. 8612	2/27/52	0.5%	Distinctive odor of sufficient intensity . . . readily perceptible to normal person coming from fresh air into closed room containing one part of the gas in 199 parts of air.
District of Columbia	Order No. 3156	3/31/47	None specified	Distinctive odor to serve as warning agent.
Florida	Docket No. 5563–Rule, Order No. 2793	8/15/59	None specified	Shall have a distinctive and readily detectable odor to act as an indication of its presence.
	Senate Bill No. 321	6/15/59	None specified	Distinctive odor to the extent that its presence in the atmosphere is readily detectable at concentrations well below that required to produce an explosive mixture.
Idaho	General Order No. 98	8/1/55	Over 20% of concentration at the lower limit of combustibility	Possess a distinctive odor to the extent that its presence in the atmosphere is readily detectable.
Illinois	Order No. 159	8/1/48	None specified	Gas not having a natural odor shall be artificially odorized in a manner satisfactory to the commission.
Kansas	Docket 34,856-U	1/16/61	1/5 of the lower explosive limit and above	Distinctive odor.
Kentucky	Rules Governing Gas Utilities	11/28/59	1% (approximately 20% of the lower explosive limit)	Distinctive odor readily detectable to normal person coming from fresh ungasified air into a closed room, or by appropriate instruments.
Louisiana	Act No. 409	10/1/39	1%	Indicate by distinctive odor the presence of gas and should be readily perceptible by normal person on entering from fresh air when gas is present not more than one part to ninety-nine parts air.
Maryland	Case No. 5374, Order No. 51157	1/1/55	1%	Distinctive odor to act as an indication of its presence.
Massachusetts	D.P.U. 11725-B	6/11/62	0.15%	Distinctive odor readily perceptible to normal person coming from fresh air into closed room containing one part of the gas in 666 parts of air.
	D.P.U. 9734-B	1/23/63	0.2%	Distinctive odor readily perceptible to normal person coming from fresh air into closed room containing one part of gas in 332 parts of air.
Michigan	Order No. 1982	1/1/44	None specified	Add plainly detectable odorant approved by commission.
Mississippi	Rules and Regulations Governing Public Utility Service	9/2/57	1%	Distinctive odor readily perceptible to normal person coming from fresh air into an area where gas is present in concentrations of not more than one part gas to ninety-nine parts air.
Missouri	General Order No. 20	3/2/53	None specified	Recommended that gas possess a strong and distinctive odor. If the cost of introducing an odor is excessive, a suitable odorant shall be introduced during the early part of the heating season and once during the nonheating season of each year.

Table 9-44 Summary of State Commission Regulations on Odorization of Natural Gas (Continued)

State	Reference	Date	Min detectable gas in air	Requirements
New Hampshire	Supplemental Order No. 7435	1/11/60	1%	Distinctive odor readily perceptible to normal person entering from fresh air into a closed room containing one part gas in ninety-nine parts air.
New Jersey	Regulations	10/1/60	20% of the lower explosive limit	All gas transmitted and distributed which does not naturally possess a distinctive odor to serve as a warning agent in the event of the escape of unburnt gas shall be odorized with a suitable odorant.
New Mexico	General Order No. 6	11/17/41	1%	Distinctive and readily detectable odor which can easily be detected by at least four persons who are confined in or enter a closed room.
New York	Case 15686	11/7/52	None specified	Readily detectable.
North Carolina	Docket G100, Sub 3	11/12/62	1/5 of the lower explosive limit and above	Distinctive odor . . . readily detectable.
Rhode Island	Rules & Regulations for High Pressure Gas Transmission Pipelines	11/1/55	0.5%	Distinctive non-nauseating gassy odor readily perceptible to normal person coming from uncontaminated air into a closed room containing one part of gas in 199 parts of air.
Texas	Gas Utilities Act (Art. 6053, Sec. 2)	1939	None specified	. . . Indicate by a distinctive odor the presence of gas.
Utah	General Order No. 70	8/12/57	None specified	Distinctive odor to act as an indication of its presence.
Washington	First Supplemental General Order No. U-8336	8/18/52	1/5 the lower level of combustibility	Indicate positively by a distinctive odor.
West Virginia	General Order No. 127-A	3/1/56	1%	Indicate the presence of gas by distinctive odor readily perceptible to normal person coming from ungasified air when gas is present in concentrations of not more than one part to ninety-nine parts air.
Wisconsin	Administrative Code—Chap. PSC 135 of Departmental Rules for Gas Utilities	9/5/52	20% of the lower limit	Detectable and recognizable odor at the most remote customer's utilization equipment.

ment contains the same provision for the elimination of odorization given in NFPA Standards Nos. 58 and 59. In these standards, odorization is covered under Basic Rule B.1 (extracted in Sec. 5, Chap. 1). LP-gases are generally odorized by the producer, although they may be odorized by the shipper or marketer.

PROPERTIES OF ODORANTS

Present day commercial odorants contain one or more of the mercaptans, aliphatic sulfides, or cyclic carbon sulfur ring compounds. The mercaptans, typified by an SH group, normally consist of ethyl (e.g., C_2H_5SH), propyl, isopropyl, butyl, tertiary butyl, or amyl mercaptans, either as relatively pure compounds or as selected mixtures. The aliphatic sulfides, sometimes called thioethers, and typified by a sulfur molecule attached to a straight carbon chain, may consist of dimethyl, ethyl, diethyl, propyl, and/or isopropyl sulfides, either as relatively pure compounds or as a mixture of sulfides, and various mercaptans. The cyclic sulfides typified by thiophane are carbon sulfur ring compounds and may be supplied either as a relatively pure chemical or mixed with aliphatic sulfides and mercaptans.

Most commercial odorants have at least an 0.8 sp gr, 18 psia Reid vapor pressure, and a −50 F freezing point (some freeze at 10 F). The suggested odorant rate varies from 0.25 to 1.5 lb per MMCF. The compositions of different batches of a commercial odorant may vary considerably; the trace constituents in an odorant often have important effects on odorant properties. Available commercial odorants are periodically described in A.G.A.[13] and other[14] publications.

Desirable characteristics for a gas odorant vary considerably, depending upon utility objectives and needs, geographic location, and whether natural or manufactured gas has recently been distributed in the system. A list of odorant considerations might include:

1. Odor. The odor should be unpleasant, distinctive of manufactured gas in a conversion area and of natural gas in an old natural gas area. It should be readily identifiable as gas and dissimilar to other household odors or to odors prevailing in area. It should not fatigue the olfactory senses unduly.

2. Volatility. Odorant should not condense out of the gas at pressures, temperatures, and odorizing rates employed.

3. Inertness. Odorant should be inert enough not to polymerize, decompose, or react with other constituents of the gas or with materials in the distribution system or appliances.

4. Adsorption by soil. Gas passing thru the soil should retain sufficient odor to remain detectable.

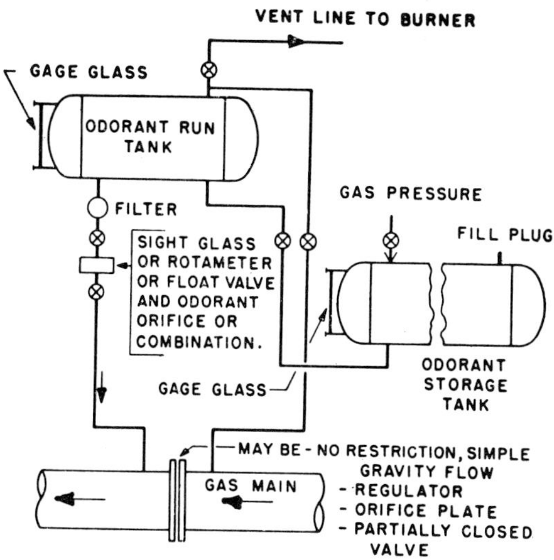

Fig. 9-74 Drip-type odorizer with various means both for regulating odorant flow and for applying pressure on odorant.

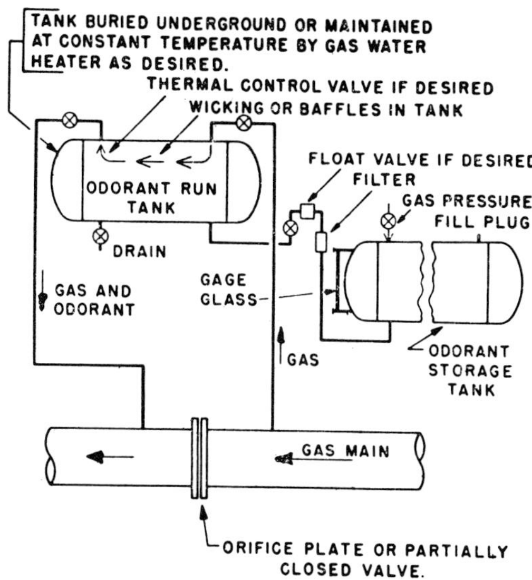

Fig. 9-75 By-pass absorption-type odorizer with various refinements.

5. Corrosion. Odorant should be noncorrosive under all conditions encountered in transmission, distribution, and utilization. This calls for both a lower sulfur content and a low reactive type of sulfur bonding in the molecule.

6. Combustion Products. Odorant should burn completely in the gas flame to form products which are not corrosive, irritating, or toxic.

7. Toxicity. No harmful effects should result from inhalation of or contact with odorant vapors. Odorant should also be relatively nontoxic in the liquid state.

8. Potency. Odorant should be strong smelling at normal odorizing rates of 0.5 to 1.5 lb per MMCF of gas.

9. Economy. Delivered cost should be low, preferably below $1.00 per MMCF of gas odorized.

10. Boiling Range. A narrow boiling range is desirable for certain types of odorizers. It should also provide for uniform evaporation without leaving deposits of heavy ends. However, the boiling point should not be so low as to cause difficulty in field handling.

11. Freezing Point. Odorant must not freeze at distribution temperatures.

The conclusions and recommendations of a 1961 literature survey of odorant compound chemistry[15] indicate that: (1) odor fading can be minimized or prevented by high-concentration dosing;[16] (2) the causes of odor fading are unknown; (3) catalytic oxidation (in oxygen-bearing gas) and catalytic reformation of mercaptans are probable; (4) decomposition of mercaptans to olefins is improbable; (5) thiophane oxidizes less readily than mercaptans; and (6) chemical transformation of odorants can be reduced by removing catalysts (e.g., iron oxide) and by introducing inhibitors.

ODORIZING EQUIPMENT

Criteria for selecting or designing odorizing equipment include maximum operating pressure, kind of odorant to be used, maximum and minimum gas flow rate, injection accuracy required, maximum and minimum ambient temperature, and time interval desired between odorant storage tank fillings (generally 30 to 90 days). A 1959 survey[17] indicated that odorant capacity requirements had a minimum range of 0 to 0.8 gph and a maximum range of 0.02 to 4.8 gph. Initial costs, installation considerations, and in-service problems of odorizing facilities are discussed in Reference 18.

Fractionation. Fractionation may occur in by-pass units with odorants containing a mixture of compounds. The composition of the evaporated odorant may be ascertained by partial pressure relationships.[41]

Drip Type

Figure 9-74 indicates various means both for regulating odorant flow and for applying pressure on the odorant. The simplest type has manual adjustable valves and sight glass for visual field adjustment. A rotameter added as shown enables the operator to measure the odorant flow. A liquid level float valve minimizes the effects of variable liquid head on the odorizer tank and permits more uniform flow of odorant. An orifice in the odorant line may improve regulation of odorant flow.

By-Pass Absorption Types

Horizontal Type. Figure 9-75 shows a horizontal by-pass absorption type of odorizer with various refinements. An orifice plate or partially closed valve in the line forces a portion of the gas thru a tank of odorant provided with baffles or wicking. Normally, the tank is so sized that this gas becomes saturated with odorant; the gas is then returned to the main gas line. A thermocontrol valve can be provided to vary the by-pass gas flow rate *inversely* with increase in the odorant's volatility caused by its temperature rise. The odorant tank can be buried for more uniform temperature control or heated to a selected temperature by an automatic gas water heater. There are exceptions, but generally these units are used on gas loads of up to 500 MCFH.[19] One manufacturer's units are rated at 300 and 500 psig.[20]

Vertical Type. The flow of the by-passed gas is directed on the surface of the odorant by means of a float arrange-

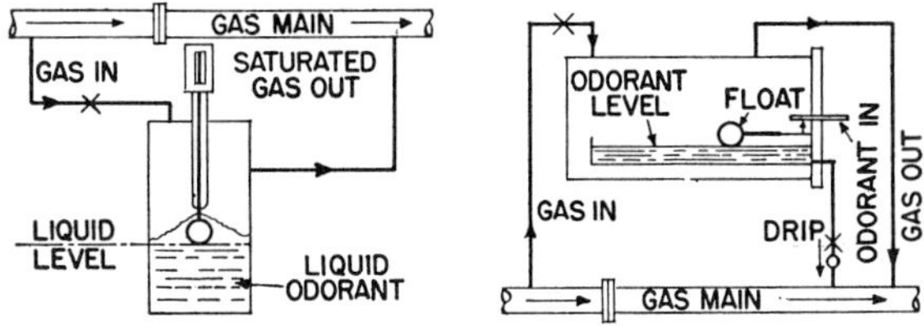

Fig. 9-76a (left) Vertical by-pass odorizer.

Fig. 9-76b (right) Combination by-pass and drip odorizer.

ment to assure good gas odorant contact at all gas flow rates. These units are available in sizes up to 1 MMCFH (Fig. 9-76a).

Combination Drip and By-Pass[19]

The drip portion of this unit is set for the *minimum gas load* with the by-pass portion handling the variations in flow. There is no theoretical limitation to the capacity of these units, although the available storage tanks generally limit their use to flows not greater than 2 MMCFH (Fig. 9-76b). One advantage claimed for this type is that the dripping feature eliminates the accumulation of the *heavy ends* of the odorant.

Meter-Type By-Pass and Dipper Pump

Figure 9-77 shows a meter type of by-pass odorizer. By-pass gas at pressures of up to a *maximum* of 100 in. w.c. differential, controlled by a precision valve, is forced thru a gas meter by an orifice or valve restriction of the main line. The meter drives a dipper type of pump by means of a chain and sprocket drive, the odorant dipping into the outlet gas line. A safety tank is provided to catch overflow odorant if the float valve fails. The setting of the precision valve must be changed with changes in pipeline pressure.

Meter units are available to handle gas loads of almost any size. With *dilute* odorant, they will efficiently odorize flows of 10 MCFH, and the larger units with concentrated odorant can easily odorize flows of 10 MMCFH.[19] However, in recent years the larger flows have generally been handled by pump installations. Units are made for various design pressures; one manufacturer's line includes 200, 550, and 1000 psig equipment.[20]

Pump Type

Variations are shown in Figs. 9-78 and 9-79. In Fig. 9-78 the main line gas flow is measured by an orifice or pitot venturi in the gas line. The flow meter sets the Thy-mo-trol (electronic speed control, made by General Electric Co.) that regulates the speed of the pump drive motor. A high degree of control can thus be obtained over the full range of flow. Adjustment of length of pump stroke permits any desired ratio of odorant to gas flow within limitations of the equipment.

In Fig. 9-79 the main line gas flow is measured by a pitot venturi tube or orifice plate and controller transmitter. The transmitter sets the controller that regulates the piston stroke length and therefore the amount of odorant pumped into the line. This system is actuated by about 20 psig gas pressure. Control is pneumatic. The motor runs at constant speed. Pump stroke length can be adjusted manually or automatically for any desired ratio of odorant to gas flow within limitations of the selected equipment size. One manufacturer's line[17] covers a capacity range of from 0.097 to 6.0 gph (based on a 1000 psig discharge pressure*). *Maximum* discharge pressure is 2500 psig; motor size, 1/6 hp.

* As a rule of thumb, capacity is *decreased* approximately one per cent for each 100 psi increase in discharge pressure above 1000 psig.

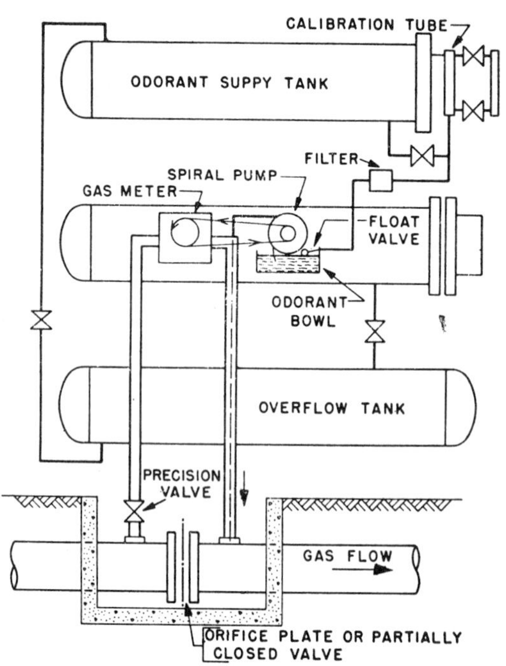

Fig. 9-77 Meter type of by-pass and dipper pump odorizer.

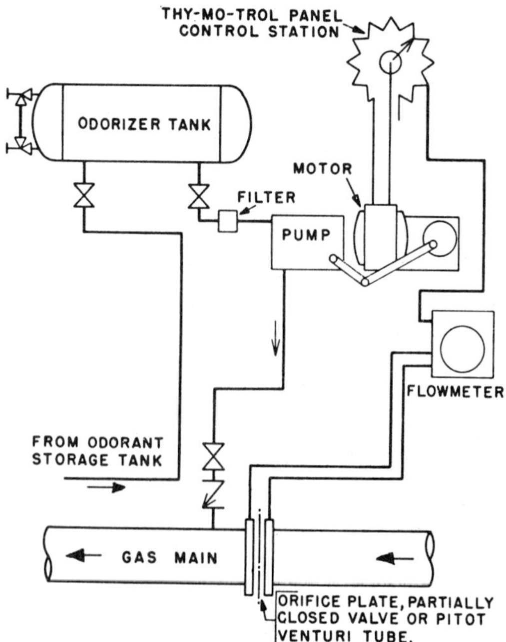

Fig. 9-78 Pump-type (Milton Roy) odorizer with automatic electronic control of odorant flow.

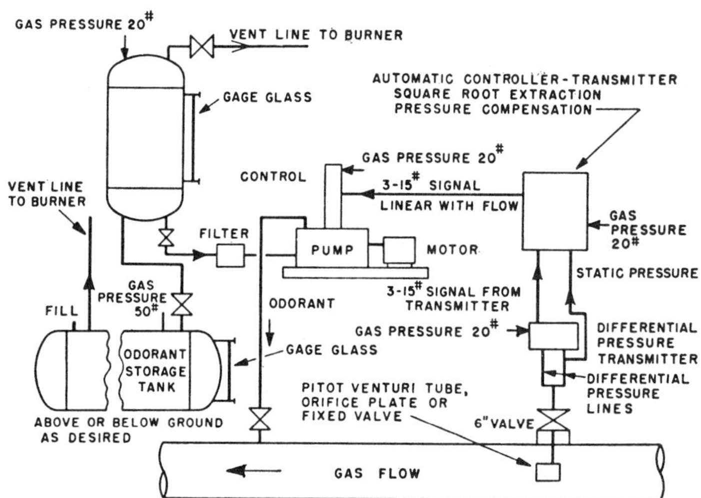

Fig. 9-79 Pump-type odorizer automatic control system, gas pressure operated for Lapp Pulsafeeder diaphragm pump with Conoflow control or piston pump with Hagan control.

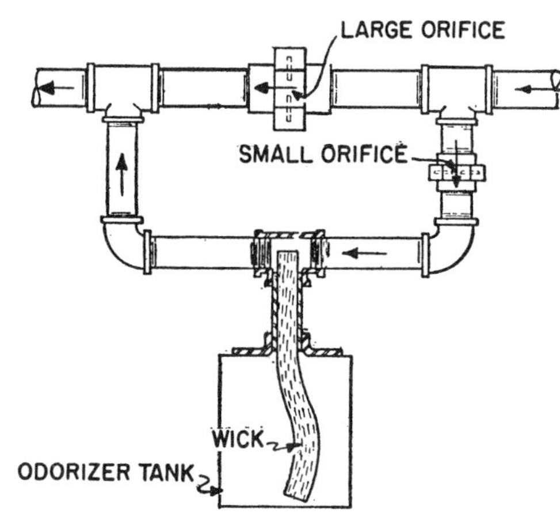

Fig. 9-80 Wick type of by-pass odorizer for service lines.

Wick-Type By-Pass for Individual Services

Small wick-type by-pass odorizers for individual service lines (e.g., schools, camps, small towns, farm taps) are shown in Fig. 9-80. Accuracy of odorant introduction depends upon proper sizing of by-pass restriction, temperature, wick action, and other factors. One manufacturer's units have a storage capacity of 0.66 gal and 3 gal.[20]

Bubble Type[21]

By-pass gas is bubbled thru liquid odorant. The pressure differential across the line constriction should not exceed 1.5 psig at *maximum* flow. Control is by valve adjustments on the by-pass. Capacity is 20 to 200 MCFH; 200 and 500 psig models are available.

Installation

Piping and Materials. Steel, stainless steel, copper-free steel alloys, and aluminum are used. Copper should be avoided, since it reacts with mercaptans. Seamless steel tubing is commonly used (preferably stainless steel type 304) for process lines. The following data pertain mainly to high-pressure installations. Forged welding fittings and flanged construction are recommended in sizes above 1 in.[22] Where possible, pipe joints should be welded. For threaded fittings, pipe dope resistant to the odorant should be applied after cleaning oil from threads. Packless valves and mechanical seals are preferable. For valves with packing, Resistoflex No. B-1001, Teflon, and Koroseal black sheet packing in formed gasket rings have proved satisfactory. For piston pumps, braided Teflon piston packing is preferable; pistons should be ¼ in. diam or larger. One manufacturer suggests Belmont 754 and ACE-O-Pax 90 for steel and aluminum pumps, respectively.[22] Because of the difficulty of preventing odorant leakage around the piston, the trend is toward use of diaphragm-type pumps.

Ordinary steels, costing less than nonreacting metals, may be used for odorant piping and equipment. Often, certain precautions are taken, including pretreatment or *pickling* the steel with mercaptan or H_2S and maintaining the ferrous sulfide coating thus formed in a dry and inert atmosphere. Untreated steel reacts with odorants, reducing their concentration in the gas until a coating of *ferrous* sulfide has been formed. In the presence of air, ferrous sulfide scale oxidizes rapidly to *ferric* sulfide. If the heat evolved in this reaction is not dissipated, oxide particles may reach red heat, which could be a source of ignition for organic materials. Nevertheless, untreated mild steel is often used, since it reacts with mercaptans relatively slowly and the resulting products can be filtered out of the system.

Tanks and Vessels. Fabricate in accordance with the *ASME Unfired Pressure Vessel Code.* Steel tanks and equipment are either conditioned with moist mercaptan or H_2S, or lined with Amercoat 77 (baked phenolic resins may be satisfactory) to prevent color change in the stored product. Type 316 or 304 low carbon stainless steel or Inconel is preferred where steel corrosion is high and product contamination must be minimized.

Maintenance of Equipment.[22] Steam cleaning and venting of resultant vapors thru a dilute solution of hypochloride destroys mercaptan. Scale from iron equipment in mercaptan service is disposed of in metal containers; the surface of this oxide is kept below water.

Tank Car Handling.[22] Cars for ethyl and methyl mercaptan are of the ICC-106A-500 type, insulated, designed for chlorine service, and rated at 500 psig. They are unloaded only from the top. *Dry* nitrogen and natural gas are recommended pressurizing media for unloading. Cars are connected to unloading lines with flexible stainless steel or neoprene hose. The higher mercaptans (propyl, butyl, and amyl) are shipped in ICC-103W cars, rated at 25 psig.

Tank Truck Handling.[22] Two-compartment ICC-MC-330 approved vehicles with a total capacity of 3000 gal of odorant have been used.

LP-Gas Odorizers

Some LP-gas producers odorize a tank car of product by pouring in the contents of an individual container of odorant.

Equipment is available to prevent the excessive escape of vapors during this operation. For example, an odorizer may be inserted as a by-pass in the line leading to the LP-gas storage or shipping container.[23] Line pressure is used to drive a pointed bar thru the tops of individual containers stacked in this odorizer. The liquefied gas is then passed thru the device, where it picks up the charge of odorant. After the product run is completed, the by-pass valves are closed, the device is opened, and the plunger and empty odorant containers are removed.

Housing. Maintenance of odorant and odorizing equipment at a reasonable temperature, avoiding extremes of cold and heat, will increase the accuracy of the odorant introduction rate. At times of repair and servicing and of occasional leakage, odorant will escape from the equipment. For these and other reasons many odorizers are housed in buildings.[24] Such buildings should be as tight as possible and provided with blowers that pass all air leaving the building through a bed of activated charcoal which removes odors. This construction reduces odor complaints in the area and provides better working conditions for service men.

Disposal of Gas Vented from Odorizer Tanks. When it is uneconomical to deodorize vented gases by activated charcoal, silica gel, or bubbling thru caustic solutions, flaring may be used. The elements of one system[25] to depressurize a 250 gal odorant supply tank at 250 psig included: (1) regulation of the vented gas pressure to 30 psig to a flare burner located 40 ft from the odorant tank; (2) combustion of the vapor (required 600 cfh of 1000 Btu gas for 15 min); (3) electrical ignition of constant pilot; (4) combustion-draft arrangement consisting of a 22 in. diam pipe within a 30 in. pipe, both 10 ft high.

TEST METHODS FOR DETERMINING ODORIZATION ADEQUACY

Several test methods and instruments are available for determining adequacy of odorization. A study[26] was made of the characteristics of four instruments that can be used to analyze natural gas for sulfur compounds in trace concentrations. Three of these instruments, the Titrilog, the Austin I BR 313 gas titrator, and the Beckman mercaptan analyzer, depend on chemical reactions to measure sulfur content. They cannot determine individual sulfur compounds. The fourth instrument is a conventional *gas chromatograph* equipped with an Argon ionization diode detector. This instrument does appear capable of measuring individual sulfur compound concentrations at typical odorant levels.

The A.G.A. Committee on Odorization established some odor level terms in 1955. These terms are applicable to the sensations detected by a healthy adult coming from the outside air upon entering a room containing some escaped gas.

No.	*Term*	*Description*
0	Nil	No change from ambient odor
1	Threshold	First change from ambient odor
2	Barely detectable	First recognition of odor of gas
3	Readily detectable	Adequate recognizable odor of gas
4	Strong	More than adequate gas odor

General Methods for Odor Estimation

1. One or more observers may smell the gas from an opened valve. This is not accurate, as the amount of gas in air cannot be evaluated, although a relative odor evaluation can be made.
2. Chart gas leak complaints from customers on a per 1000 meter per month basis. This operating information will indicate need for changes in odorant injection rate. The proper rate is affected by the season, other odors in the atmosphere, neighborhood odors, sewers odors, smog, etc.
3. Increase odorant rate for short periods and evaluate the resultant increase in leak complaints received.
4. Record volume of odorant used per day or per week and calculate the pounds of odorant used per MMCF of gas sold.

Colorimeter Methods (for Amyl Mercaptan Only)[27]

Pass 2.5 cu ft of gas thru copper oleate treated paper and expose paper to sun or other light for 15 min. Comparison of resultant color (shade of brown) with a Pentalarm color standard chart permits determining amount of odorant present.

Another method involves passing 3 cu ft of gas thru a mixture of copper butyl phthalate and *n*-butanol in an absorber. The solution is then transferred to a colorimeter tube, exposed to ultraviolet light, and compared in a colorimeter with a standard solution. Intensity of color gives results in terms of pounds of odorant per MMCF.

The Austin Gas Titrator

This titrator (Micro-Path Inc.) is a lightweight, compact battery operated instrument for making analyses of sulfur odorant compounds in gas streams.[28] It is used in the field by nontechnical personnel, as well as for research and control problems. Various classes of sulfur compounds are separated by bubbling gas thru selective absorption solutions prior to titration. A complete analysis of odorant compounds in a gas stream is made in a matter of a few minutes at the sample site. The Titrilog is another commonly used titrator.

Odorometers

Companies which have used odorometers, with varying degrees of success, include Oronite Chemical Co., San Francisco, Calif.; Southern California Gas Co., Los Angeles, Calif.; Consolidated Edison Co.,[29] New York, N. Y.; Industrial Chemical Division of Pennsalt Chemicals Corp.,[30] Philadelphia, Pa.; and Rochester Gas & Electric Co.,[31] Rochester, N. Y. The principle of the odorometer is based on supplying accurate mixtures of up to two per cent gas in air at an outlet at which observers may smell the mixture.

Test Rooms

Odor level impressions of a panel of six people entering an odorized gas filled test room of 500 or 1000 cu ft are generally used for an odorant *evaluation* test (*not* for establishment of odorization rates or for field tests). Gas is usually admitted to the room until 0.5 or 1.0 per cent gas is present. Observers report the odor impressions at various gas levels.

Table 9-45 Odor Levels of Typical Commercial Odorants[12]

Odorant	Mfr. recommended odorizing rate, lb/MMCF	Lb of odorant per MMCF of gas diluted to 1% gas in air			Remarks
		Threshold*	Recognition†	Warning‡	
B P Captan	0.25–1.0	0.0014	0.0050	0.050	Rotten meat & cabbage—sulfide
Captan	.25–1.0	.0014	.0050	.010	Sour, meaty, and onion—mercaptans
Alert 80	.50–1.0	.0035	.0100	.010	Eggy, sour, and putrid
Calodorant "C"	0.50	.0035	.0070	.014	Minty, garlic breath, eye and nose irritation
Calodorant "C" Spec.	1.0	.0040	.0200	.080	Garlic breath, gassy, sour
Spotleak 1008	0.25–1.0	0.0050	0.0140	0.050	Sulfide coal gas, metallic

* First change from inherent room odor.
† The point at which an odor becomes recognizable.
‡ Impact sufficient to arrest attention.

A recent type of odor test room[12] is lined with 0.0025 in. aluminum foil sealed to Transite with sodium silicate. An air flushing system takes air from inside the building thru a battery of activated charcoal canisters. A Gilmont Ultra-microburet is used to introduce odorant into the room.

By another method,[7] with the observers present in the room, odorized gas is admitted at such a rate that a maximum concentration of 1 per cent is realized when the entering rate equals the rate lost by normal air exchange. The required gas rate is 0.33 per cent of the room volume per hr.* For the gas to be considered adequately odorized, all observers should detect the odor of gas soon after it is injected.

ODORANT CONCENTRATION FOR DETECTION

Tests by trained observers in a closed room have demonstrated that the *warning level* of various commercial odorants is reached with from 1/5 to 1/125 of the amounts recommended by the manufacturers under ideal test conditions (see Table 9-45). *Recognition level* may be from 1/10 or more of the amount for warning, and the *threshold* of smell is from 1/5 to 1/2 of that for recognition.

The amount of odorant actually used in a distribution system usually falls within the manufacturer's recommended range. However, in some instances the amount used will range from 1/4 of the minimum recommended to double the maximum. Reasons for these variations follow:

1. Condition and age of mains are a factor. Rusty, dirty mains, as well as new mains, may consume considerable odorant.

2. Some odorants are more reactive with iron oxide and other main deposits than others. For example, a high reactivity of mercaptans and iron oxide was suggested.[15] Oil (from oil fogging) tends to be preferentially adsorbed; thus, the odorant is unaffected (or at least is less affected).

3. Some operating policies call for a highly odorized gas sufficient to make consumers report gas leaks even of nuisance size. In other cases just enough odor is preferred for identification of gas leaks.

4. Conversion from manufactured gas to natural gas may make it desirable to maintain the high odor level of the former gas.

* The atmosphere in the special test room thus contaminated (containing two doors and windows, with one of the doors and the windows in two outside walls, and with wind outside the room varying from no wind to a slight wind) will be at approximately 1/5 the lower flammability limit of natural gas.

A common way of deciding on the optimum odorant rate is to total the number of monthly leak complaints which are due only to inadvertent release of gas, e.g., gas escape preceding ignition. Tolerances among company practices appear to be as many as 3 to 45 leak complaints per 1000 meters per month, with the majority between 3 and 13, and the average about 7 complaints per 1000 meters per month.

ODORIZATION PROBLEMS

Fading of Odor

A gas adequately odorized at the station may lose some of its odor under certain conditions before it reaches all points in the system. Two factors may reduce the odor level. First, the odorant may be chemically unstable, and thus prone to break down or to react with materials in the pipeline. Second, the odorant may be physically adsorbed or absorbed by the pipe wall or materials thereon.

Mercaptans[32] are oxidized by hydrated ferric oxide (usually present in pipelines) to disulfides—compounds only about one-eighth as odorous as the parent mercaptans. If oxygen is present, iron oxide acts as a catalyst and the mercaptan is oxidized by the oxygen to a disulfide. Because mercaptan odorants are used in low concentrations, only low concentrations of oxygen are required for complete conversion. A given amount of iron oxide will convert ten times as much mercaptan to disulfides with oxygen present as without.

In odorizing *for the first time*, extra odorant should be added to compensate for odorant sorption by pipeline. Odor conditioning of new mains[16] has been done with small by-pass odorizers. Odorant has also been added in one slug or from a drip pot that might feed for several days. Reduction of gas oxygen content (when air is added to control heating value), use of a cyclic carbon–sulfur ring odorant, and oil wetting permit early odorant transmission thru new mains.

Fading of odor within piping has reportedly been overcome by a company which used oil fogging[33] and local injection of oil. Oil wet mains appear to maintain odorant better than dry mains.

Sorption by Soils. Odorants are absorbed in varying degrees during passage thru different soils.[34] Low molecular weight, insolubility in water, and low oxidation rate are indications of low soil sorption.

"Any gas with a strong odor leaking from underground pipelines will retain its odor to a great degree under ordinary conditions. The extent of adsorption depends upon several factors, some of which are velocity and volume of gas, porosity of soil, moisture content, type and concentration of odorant, depth of gas line, and time of exposure. Numerous crevices, cracks, and voids usually exist, particularly in soil and backfill surrounding gas lines. This condition permits leaking gas to usually escape with very little restriction or filtration. Leaks from underground lines are regularly being discovered by smell of odorized gas expelled from the earth."[35]

It was concluded from a study[36] of the resistance of odorants to soil adsorption that odorant selection should be based on general needs without undue emphasis on resistance to soil adsorption. Factors evaluated in this *laboratory* study of *pure* soil compounds include space velocity, moisture in the gas and soil, oxygen and fogging oil in the gas, and iron oxide in the soil.

Corrosivity of Odorants

Copper and iron readily form sulfides, but they are only slightly reactive with odorants at normal atmospheric temperatures. A trace of hydrogen sulfide, however, may permit reactions at low temperature. Black copper sulfide has been found in copper services after a few years' operation by using mercaptan odorant in the presence of traces of hydrogen sulfide. However, at room temperature, common odorants are not generally corrosive. At elevated temperatures, such as may occur in pilot lines, burner boxes, or combustion chambers, odorants may attack copper and iron under suitable conditions. A thin film of copper sulfide forms on copper after many years of service. Aluminum, however, is not affected. The degree of corrosion *decreases* in the following order of types of odorants:[7] mercaptan; disulfide; thioether; carbon–sulfur ring. Internally tinned copper tubing, if kept well below the melting point of tin (449 F), would be resistant to sulfur attack. Plain copper pilot tubing may be affected by stoppages caused by flaking oxide.

Mercaptans undergo thermal decomposition to form hydrogen sulfide at certain active points on the copper surface with consequent formation of copper sulfide. At room temperature, hydrogen sulfide corrosion will form some copper sulfide even at concentrations below 0.25 grains per 100 cu ft. At elevated temperatures, mercaptans actively attack copper and brass.[37]

Polysulfides, which are corrosive to copper, may be formed if the gas contains traces of hydrogen sulfide, oxygen, and mercaptans. Removal of one or more of the reactants will prevent formation of the polysulfides. Amines work as catalysts in these reactions and increase the reaction rates.

A study[38] of the effect of sulfur in flue gases on corrosion of gas appliance heating elements and flues indicated that the corrosion of mild steel for both cyclic and continuous condensation is about equal. Under continuous heating at temperatures above the dew point, little or no effect was noted on the corrosiveness of flue gas by the addition of sulfur.

DEODORIZING OR COUNTERACTING ODORANTS

Eliminating Odor from Vent Gases

Vent gases from odorant storage tanks or run tanks, discharged when filling, can be burned with a gas burner or passed thru a container of activated charcoal. The burner must be designed so that all vent gases pass thru the flame and are oxidized to SO_2, CO_2, and H_2O.[39] The odor threshold of SO_2 (about 4 ppm) is so much higher than that of a mercaptan or organic sulfide (about 0.0002 ppm) that the latter odors are relatively insignificant.[40] Activated charcoal will retain about 25 per cent of its weight in odorant before it reaches saturation. From buildings housing odorizing equipment vents, gases can be discharged thru activated charcoal to remove odors.[24]

Counteracting Spills of Odorant

Potassium permanganate solution (one teaspoon per gallon of water) is an effective oxidizer of mercaptan odorants; Chlorox, Purex, Dreft, Sharples Destenching Compound (70% calcium hypochlorite), or sodium hypochlorite solutions (about 16 parts of water to one part of bleach) may also be used. However, both types of solutions are ineffective on tertiary butyl mercaptan in ratios as great as 200 parts of permanganate or hypochlorite per part of odorant. Never use these materials in powder form, since the heat of reaction with the odorant or other organic materials will cause violent combustion. About forty quarts of bleach solution are required per quart of mercaptan.

Activated charcoal is effective; the average heat of absorption is 250 Btu released (up to 1000 Btu per lb initially) per lb of odorant absorbed. Cover the spillage with a 3/8 in. layer of charcoal on concrete (1 in. on soil), and allow to remain for several hours. Remove charcoal and burn or bury it. Repeat the operation if necessary.

Neutroleum gamma (Fritsche Bros., New York City), Captan Wafto, and Airkem (Airkem, Inc., New York City) are aerosol sprays of highly effective odor neutralizer. They do not react with odorant vapors, but change the odor to a nonobjectionable type.

Lime, iron oxide, vinegar, and acetic acid have also been used as counteragents (reduce rather than modify or mask odor) for odorants. Masking agents—essential oils with pleasant odors—may be applied in dilute kerosene or water solutions.

Treatment of Empty Barrels

Residual odorant in barrels may be removed by adding 3 lb of Columbia "G" activated carbon (or its equivalent) to each drum. The bung is replaced for three or more days. If the odor still persists, additional carbon may be added and further sorption time allowed. The used carbon should be buried or burned. The drum can be used for other purposes after treatment.

Use of potassium permanganate, sodium hypochlorite,[41] or other bleaches to eliminate odorants has also been reported. However, handling these chemicals and disposing of the odoriferous reaction products involved is cumbersome. Steaming of drums and burning of the vent gases may be practical where appropriate facilities are available. Storage tanks large enough to permit truck or tank car shipments eliminate the barrel disposal problem.

Removal of Odorant from Skin

Use soap and water to remove odorant from the skin. For further treatment, very weak potassium permanganate (strawberry soda color) may be used and then washed off. The aerosol sprays mentioned above are also effective.

When a vapor-filled atmosphere must be entered, a suitable respiratory device should be worn. The Mine Safety Appliance Co. Tank Gauger's Mask (GMC-SS-1 Connector) has been recommended.

REFERENCES

1. Hilt, L. "Chronology of the Natural Gas Industry." *Am. Gas J.* 172: 29–36, May 1950.
2. Senatoroff, N. K. "Odorization Practices." (PC-50-20) *A.G.A. Proc.* 1950: 684–702.
3. Katz, S. H., and others. *Use of Stenches as a Warning in Mines.* (Bur. of Mines T.P. 244) Washington, D. C., 1920.
4. Palmer, C. S. "Producer Gas, Its Manufacture and Use." *Engineers' Soc. of Western Penna. Proc.* 34: 357, 1918–19.
5. Katz, S. H., and Allison, V. C. *Stenches for Detecting Leakage of Blue Water Gas and Natural Gas.* (Bur. of Mines T.P. 267) Washington, D. C., 1920.
6. Fieldner, A. C., and others. *Warning Agents for Fuel Gases.* (Bur. of Mines Mono. 4) New York, A.G.A., 1931.
7. Powell, J. S. "Selection of an Odorant for Natural Gas Odorization." *P.C.G.A. Proc.* 41: 134–8, 1950.
8. Pirone, P. P. "Natural Gas and Odorants Do Not Harm Shade Trees." *Gas Age* 125: 31+, Mar. 17, 1960.
9. East Ohio Gas Co., Cleveland, Ohio. Unpublished.
10. Kendall, D. A., and McKinley, R. W. *Development of Unique Odorants for Natural Gas: Final Rept.* (Research Proj. PF-7) New York, A.G.A., 1953.
11. McClure, J. S. "Odorant Concentration as Compared to Odor Intensity." *A.G.A. Operating Sec. Proc.* 1960: CEP-60-10.
12. McKinley, R. W., and Larratt, A. E. *Study of Commercial Odorants for Natural Gas: Final Rept.* (Research Proj. PF-7) New York, A.G.A., 1954. Also 1955 and 1957 Supplements.
13. Van der Pyl, L. M. "Bibliography on Odorization of Gas." (CEP-59-11) *A.G.A. Operating Sec. Proc.* 1959: P109-10; Suppl., 1961: CEP-61-1.
14. Cable, C. R., and others. "Properties of Natural Gas Odorants." *Am. Gas J.* 177: 12+, Aug. 1952.
15. Miller, S. A., and others. *Survey of the Chemistry of Odorant Compounds.* (Gas Operations Research Proj. PM-34) New York, A.G.A., 1961.
16. Rader, A. M. "Odor Conditioning of New Gas Mains." (CEP-55-7) *A.G.A. Proc.* 1955: 836–9.
17. Hills-McCanna Co. *Catalog.* Chicago, Ill.
18. Dormer, G. G. "Installing an Odorizer." *Gas* 30: 60-6, Sept. 1954.
19. Natural Gas Odorizing, Inc., *Catalog.* Houston, Tex.
20. Peerless Manufacturing Co. *Catalog.* Dallas, Tex.
21. ToPaz Inc. *Catalog.* Houston, Tex.
22. Pennsalt Chemicals Corp. *Pennsalt Gas Odorants.* Philadelphia, Pa.
23. Humphrey-Wilkinson Inc. *Catalog.* North Haven, Conn.
24. Dunkley, W. A. "New Gas Odorizer Buildings at Memphis." *Gas Age* 106: 13+, Sept. 14, 1950.
25. Manfred, N. A. "Waste Gas Burner Disposes of Gas Vented from Odorizer Tanks." *Gas* 29: 56, June 1953.
26. Tarman, P. B., and others. "Comparison of Instrumental Methods of Analysis for Odorants and Other Sulfur Compounds in Natural Gas." *A.G.A. Operating Sec. Proc.* 1961: CEP-61-17.
27. White, D. L., and Reichardt, P. E. "Colorimetric Tests for Determining Mercaptan Odorants in Natural Gas." *Gas* 25: 38–9, June 1949.
28. Austin, R. R. "Automatic Recording Titrator and Its Application to the Continuous Measurement of the Concentration of Organic Sulfur Compounds in Gas Streams." *A.G.A. Proc.* 1949: 505–15.
29. Coryell, R. L. "Gas Odor Test Equipment." *A.G.A. Proc.* 1954: 714–20.
30. Nevers, A. D. "How Odorants Are Evaluated by Use of New Apparatus." *Am. Gas J.* 182: 20–3, Feb. 1955.
31. Gilkinson, R. W. "Rochester Gas Develops a Portable, Lightweight, Battery-Operated Odorometer." *Gas* 36: 67–9, Dec. 1960.
32. Powell, J. S. "Experiences with Odor and Odorization of Midcontinent Natural Gas." *P.C.G.A. Proc.* 40: 191–4, 1949.
33. Joachim, J. L. "Gas Odorization Experiences." *A.G.A. Proc.* 1951: 491–3. Also in *Gas Age* 108: 19+, July 5, 1951.
34. Johnson, E. E. "New Developments in Odorants." (PC-50-21) *A.G.A. Proc.* 1950: 703–5.
35. Henderson, E. L. "Odorization of Gas." (OS-52-1) *A.G.A. Proc.* 1952: 294–303.
36. Tarman, P. B., and Linden, H. R. *Soil Adsorption of Odorant Compounds.* (Research Bull. 33) Chicago, I.G.T., 1961.
37. Kruger, H. O., and Robinson, M. L. "Corrosion of Copper by Trace Constituents of Natural Gas." *P.C.G.A. Proc.* 41: 141–5, 1950.
38. Pray, H. A., and others. *Corrosion of Mild Steel by the Products of Combustion of Gaseous Fuels.* (Proj. DGR-4-CH, Rept. 2) Columbus, Ohio, Battelle Memorial Inst., 1949.
39. Manfred, N. A. "Factors Involving Design of Odorizer Installations." *A.G.A. Proc.* 1953: 711–8.
40. Turk, A. "Odor Control in Gas Odorization." (CEP-55-4) *A.G.A. Proc.* 1955: 822–8.
41. Olund, S. A. "Odorant Fractionation in By-Pass Absorption Odorizers," *Am. Gas J.* 190: 36–42, Oct. 1963.

Chapter 12
Unaccounted-for Gas

by J. M. Pickford and F. E. Vandaveer

The term unaccounted-for gas is applied to the discrepancy between the total gas available from all sources and the total gas accounted for in sales, net interchange, company use, production of other gases, storage, and other uses. It is usually calculated on an annual basis but may be established monthly or for any other designated period. For example, unaccounted-for gas in the United States averaged 2.6 per cent for 1958–1960 and 4.0 per cent for 1946–1950.[1] An annual compilation of unaccounted-for gas for many gas utility companies is available.[2]

Because of the growth of gas sales, percentage figures of unaccounted-for gas do not serve as truly valid reference points[3]—they tend to minimize the problem when sales are expanding. On the other hand, calculations on the basis of cubic feet of unaccounted-for gas per mile of 3-in. equivalent main may make system loss appear worse than it really is, since there are great differences in sales per mile of main due to appliance saturation, climate, economic level, amount of industry, and similar factors.

Therefore, considerations of unaccounted-for gas should include a clear understanding of the immediate objective. For example, if leakage control is being studied, a direct approach to this problem by leakage test methods would be preferable to the use of inferences drawn from unaccounted-for gas.

REPORTING UNACCOUNTED-FOR LOSSES

There is no recognized standard basis for calculating and reporting unaccounted-for gas losses. Various means of expressing such losses include: (1) percentage of total gas available lost; (2) volume lost; (3) therms lost; (4) volume lost per mile of main per year; (5) volume lost per mile of main and service per year; (6) volume lost per mile of 3-in. equivalent main per year; and (7) dollars lost, based on cost of gas.

FACTORS IN UNACCOUNTED-FOR LOSSES

Many factors and sources are involved in unaccounted-for losses and no one item can be singled out as the major cause. In the design of a program to minimize any or all of these factors, all of them should be given full recognition at all times. Conditions will vary for each company. In 1959, an A.G.A. Operating Section Task Group presented data on factors affecting unaccounted-for gas. Included were simplified methods of estimating the contribution of each of these more significant factors:[4] (1) variation in customer's metering temperatures and pressures; (2) meter error, both slow and fast; (3) main leakage by classes: Class A, small leaks; Class B, large leaks; and Class C, leaks with a relatively high hazard potential regardless of size; and (4) service leakage.

Leakage from Mains and Services

The A.G.A. report offered suggestions on how to develop an index of leakage for a particular distribution system which could be used to monitor the effectiveness of corrosion control and maintenance activities from year to year. An example of using only significant factors in approaching the question of leakage is the exclusion of steel pipe which has been coated, wrapped, and cathodically protected from leakage calculations. This will tend to furnish more realistic data, since leakage from protected systems may be expected to be very low.

Volume Corrections

Gas purchased and gas sold must be corrected to a common basis of temperature, pressure, moisture, and gravity. A review of a number of pipeline tariffs shows them all to be corrected to 60 F, but the pressures vary from 14.65 (other sources show 14.4) to 15.025 psia.[4,*]

A gas *temperature* difference of 5°F will cause a change of approximately 1 per cent in gas volume. Ground temperature data should be used to correct gas temperature; summer and winter averages or a 12-month average may be used. Uncompensated outdoor meter sets require additional temperature corrections. The insertion of a thermometer in a gas stream may give valid data for continuous flows at small temperature variations. Intermittent flows at variable rates, however, require more responsive temperature measuring instruments.

A *pressure* difference of 4 in. w.c. causes a change in volume of approximately 1 per cent. An elevation difference of 1000 ft causes a change in volume of about 3 per cent; therefore, negative unaccounted-for gas is a possibility.[4] A calibrated barometer provides correction factors for fluctuations of atmospheric pressure as well as of elevation.

---

* A 1950 A.G.A. survey[1] showed that the 14.73 psia base was specified in nine states; 24 states, however, had no definition for this base; the remaining states specify a variety of pressure bases. Base temperature, where specified, is 60 F (except Ohio, 62 F). The advantages of standardization are also covered.

A specific *gravity* difference of about 0.003 for a 0.600 specific gravity gas causes a change in volume calculations for orifice meters of about 0.25 per cent. A change in gas water *vapor* content from saturated to dry will *decrease* the volume 1.74 per cent; oil fog condensation will also decrease gas volume.

Because gas sales are greater in winter than in summer and there is a lag in meter reading dates, some companies calculate their yearly unaccounted-for gas at the end of a summer month rather than at the end of the year. This decreases the error of matching gas sendout volumes with gas sales at any given time.

General Meter Problems

Gas meters are accurate if properly installed and maintained. They must be designed, sized, proved, and tested (Table 7-15) for the service conditions involved. Regular attention to orifice meters, proper plate size, and proper design of orifice runs are essential, as is periodic replacement of displacement meters to avoid excessive wear of parts and diaphragm trouble. Other metering considerations include lost meters, wrong meter indexes, and unmetered company use. Some large meters are incapable of registering small flows, such as those used for pilots.

Normal Operating Losses

These include such items as unavoidable loss of gas by blowing drips; purging pipelines, holders, and other equipment; replacing and repairing gas mains; and testing service lines during meter replacement. Changes in billing dates and meter reading dates may also affect unaccounted-for gas data. Wherever possible, an estimate of such losses should be made and reported.

Theft and Line Breakage

These items are self-evident. Line breakage may be caused by mechanical failure or accidental breakage from various sources.

Lack of Adequate Measurement Facilities

Some companies cannot determine unaccounted-for losses accurately because of insufficient gas measurement facilities. Meters may not have been put on the feed lines to a town. Gas put into storage or taken out of storage may not be measured. In other instances, gas from certain company owned wells may not be measured. In these cases, the amount of gas purchased, produced, or manufactured cannot be determined except by estimating.

REFERENCES

1. A.G.A. *Gas Facts.* New York, annual.
2. *Brown's Directory of American Gas Companies.* New York, Moore Pub. Co., annual.
3. Thomas, H. E., and Peacock, P. E., Jr. "True Pucture of Unaccounted-for Gas Shown by Line Loss Studies." *Am. Gas J.* 171: 20+, Oct. 1949.
4. Rohret, L. C. "Final Report of Task Group on Factors Affecting Unaccounted-for Gas." (DMC-59-14) *A.G.A. Proc.* 1959: D-69-78.

Chapter 13
Oil Fogging of Gas

by J. M. Pickford and F. E. Vandaveer

The object of oil fogging is to produce an oil-wetted condition inside the piping to prevent the gas from carrying iron rust and dirt, thereby reducing pilot outages and other control troubles as well as drying out joint packing in cast iron mains. In the last regard, note that a study[1] could not establish any relationship between joint leakage and oil fogging.

Oil particles of 0.00004 in. diam (approximately 1 micron) or smaller are required to produce a persistent fog.[2] One cubic foot of oil will produce 30 trillion oil droplets of the above size. Dust particles as large as 0.000145 in. diameter (3.7 microns) will be floated and carried along by a gas stream moving at a pilot rate of 0.15 cu ft per hr thru a ¼ in. pipe.[3] Thus, oil fog could clog pilot orifices.

Oil fog* rather than oil vapor is required, since the former can be carried in larger quantities by the gas stream. Another disadvantage of vapor is that it requires a drop in temperature before it will form condensate. Fog can be deposited from the gas by impingement on pipe walls. If a persistent fog is produced (more readily accomplished with hot foggers than with cold ones), the mains will gradually be coated with oil by this impingement. A nonpersistent fog will be largely eliminated by settling in a relatively short distance from the point of introduction. One utility reports evidence of satisfactory oil coating at least ten miles from the point of hot oil fogging. When pressure is sharply reduced in an oil-fogged gas, some dropout (drained thru drips) will occur.

Extent of Use. Of 59 companies replying to a questionnaire[4] in 1957, 41 did *not* oil fog. Of the 18 that did, nine used hot foggers, four used cold ones, and five used both. While half the companies that do oil fog reported practically no trouble due to fogging, others reported oil accumulations in district regulators, service regulators, and meters as well as pilot and service outages.

Gas Cleaners of the Oil Scrubbing Type. Although these are not regarded as oil foggers, they will introduce oil into the gas lines at rates of from 0.03 to 0.07 gal per MMCF of gas. At such low oil injection rates, a long time is needed to coat the interior of mains in the distribution system. One company's experience showed that over a period of 15 to 25 years, during which only oil contact gas cleaners and no other oil fogging equipment had been used, oil had saturated the system to the extent that small amounts were removed from drips throughout the system, and as far as 17 miles from the oil cleaners. Iron rust removed by cloth-type filters at district regulator stations more than 10 miles from the cleaners contained as much as 15 per cent oil by weight.

Oil from Compressors. Lubricating† oil from gas compressors on transmission lines is discharged into the gas at a rate of approximately 0.22 gal per MMCF for relatively new compressors and at higher rates for older compressors.

* Fogging involves mechanical entrainment, which means that vapor pressures cannot measure the phenomenon.

HOT OIL FOGGERS

These either bubble a portion of the gas thru heated oil or counterflow heated gas thru a downflow of oil. The hot gas saturated with oil vapor mixes with the main gas stream, where it is chilled and produces a fine fog. A separator removes the larger oil particles.

Metal Products Division of Koppers Co. GH- and GM-type foggers are the same, except that the GM-type is furnished with a gas booster to provide the required differential across a governor or other device. The maximum oil capacity of the unit is 50 gal per day. Reductions in primary sweep gas rate, pressure differential, and operating temperature reduce capacity.

Blaw-Knox. The Electroil Fogger[5] has an oil fogging capacity of up to 15 gal per hr. It is automatic, requiring only infrequent attendance. The *portable fogger* was designed for use at district regulator stations and at city gate stations at which the load is not too great. It is electrically heated and explosion-proof.

Baltimore Type.[6,7] Three hot oil foggers have been designed and used in Baltimore. One is for an outlying station at which oil requirements are limited to a few gallons per day on low-pressure gas (Fig. 9-81). The second is for high-pressure stations to fog up to 10 gal per day with sweep gas of 5000 to 6000 cfh. The third is of high capacity to fog up to 100 gal of oil per day and 30 MCFH of steam heated gas. Temperatures above 500 F are to be avoided to prevent cracking of the fog oil.

FOGGING OIL

Specifications for fogging oil are given in Table 9-46. Gas oils have been used, but it is doubtful whether this is good practice.

† Similar to cleaner oil.

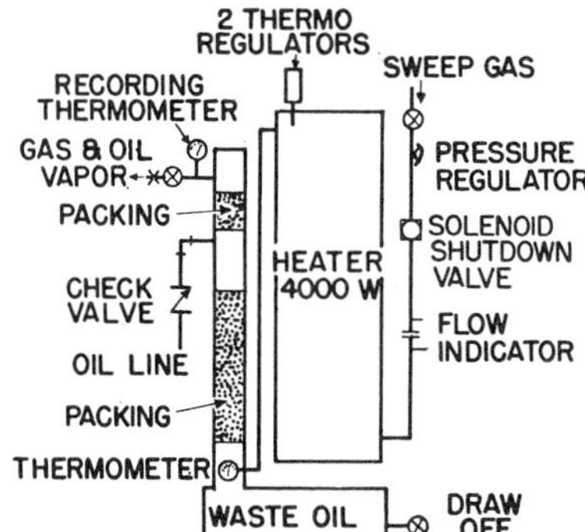

Fig. 9-81 Baltimore hot oil fogger for low-pressure stations.[6] Of all steel construction, this type is designed to fog 7.5 gal of oil per day, using 800–1000 cu ft of sweep gas per day.

Usage Rates

Usage varies from ¾ to 5 gal per MMCF, with normal usage at 1½ to 2 gal per MMCF. Rates as high as 40 gal per MMCF have been reported, although pilot stoppages have occurred at the high rates. It was claimed by one source that rates up to 8 to 10 gal have been used with little difficulty In Baltimore, with a nearly saturated gas, a rate of 1½ gal was used initially and 1 gal per MMCF after a few weeks when the system became oil-wet. In this case, evidence of fogging oil was found 8.8 miles from the fogger. Another company, using a dry natural gas, found that 0.05 gal per MMCF was adequate once the system was oil-wet.

The following oil rates have been suggested[2] for gas with varying moisture contents:

Saturation of gas with water, %	Gal oil per MMCF gas originally containing no oil vapor
90–70	2
70–40	3
40–5	5

The amount of oil theoretically required to saturate natural gas has been calculated in terms of the equilibrium concentrations of pure hydrocarbon compounds of various molecular weights:[8]

Mole wt	Lb per MMCF at 40 F	Lb per MMCF at 80 F
140	150	800
200	0.4	6.5
270	0.00008	0.01

Test Methods for Determining Oil Fog

Impingement on Glass Slide and Measurement of Refraction Index. Impingement tests[9] involve the deposit of oil, if present, on a glass slide. The oil on the slide is transferred to a thin Hycar rubber disk approximately ⅞ in. in diameter. The refractive index and thus the identification of the oil may be made with any standard refractometer by swinging the auxiliary prism out of the way and holding the disk hard against the refractometer prism. The sample is illuminated thru the back window of the refractometer prism housing. In this way, the refractive index is measured by reflected light rather than by transmitted light, which would be the usual procedure. The refractive index of a small drop of oil may be measured in this way. By comparing the index found with known indices of fractions of the oil used for fogging, the presence of fogging oil may be identified.

Absolute Filter Paper.[10] This consists of glass fibers of about three-micron diameter (80 per cent by weight) and the remainder of ½-micron diameter fiber together with a binder. This paper will remove most particles of oil or dirt from a gas stream flowing thru it. The paper is held in a union fitting and backed by a glass frit for support. The filter paper can be weighed before and after a large volume of measured gas has been passed thru it for quantitative determination of the oil present in the gas.

Thermal Precipitator.[10] If a dust or oil particle suspended in gas passes between a hot and cold surface, it will be accelerated toward, and be precipitated on, the cold surface. The precipitating forces are so gentle that they eliminate shattering or splashing of particles. The Thermopositor is made by Roy A. Martin Co., Atlanta, Ga.

Table 9-46 Two Sample Specifications for Fogging Oil

Initial boiling point, °F	475 min
End point, °F	650 max
Gravity, deg API	36 to 42
Viscosity, SSU at 100 F	35 to 45
Flash, °F	250 min
Pour	To meet local conditions
A paraffin base oil, acid washed, and distilled	
1 per cent over at	587 F
50 per cent over at	610 F
99.3 per cent over at	668 F
Viscosity, SSU at 100 F	50

REFERENCES

1. Wilson, G. G., and others. "Bell-Joint Leakage and Test Methods for Packing-Space Sealants." (*I.G.T. Tech. Rept.* 5) New York, A.G.A., 1962.
2. Garrison, C. W., and Shively, W. L. "Gas Conditioning." *A.G.A. Proc.* 1930: 1630–44.
3. Wills, F. "Removal of Liquids and Dust from Natural Gas." *P.C.G.A. Proc.* 35: 128–39, 1944.
4. Laudani, H. "Survey of Current Gas Conditioning Practices." (CEP-59-7) *A.G.A. Operating Sec. Proc.* 1959: P35-38.
5. "Brooklyn Union Installs Permanent Oil Foggers." *Gas* 28: 35, Dec. 1952. Also: Blaw-Knox Co. *Blaw-Knox Gas Equipment.* Blawnox, Pa., 1954.
6. Lusby, O. W. "Design and Operation of Oil Foggers at Baltimore." (PC-52-3) *A.G.A. Proc.* 1952: 543–9.
7. Kooke, C. A. "Effects of Gas Conditioning on the Distribution System." (OS-52-10) *A.G.A. Proc.* 1952: 280–4.
8. Wilson, G. G., and others. "Detection, Repair and Prevention of Leaks in Gas Distribution Systems." (*I.G.T. Tech. Rept.* 2) New York, A.G.A., 1959.
9. Wilson, C. W. "Method for the Determination of the Quantity of Suspended Material in Gas." *A.G.A. Monthly* 24: 325–7, Sept. 1942.
10. Moore, C. B., and others. "Development of Analytical Techniques for Suspensoids in Fuel Gas." (CEP-55-25) *A.G.A. Proc.* 1955: 972–1000.

Chapter 14
Humidification

by J. M. Pickford, H. E. Ferguson, and F. E. Vandaveer

Neither manufactured nor natural gas is humidified *except* when the latter is distributed in *some* systems formerly used for manufactured gas. Humidification may have been used[1,2] in those conversions in which jute fibers were used to seal joints, and it was believed desirable to use water to wet down dust until the piping could be cleaned. Some of the meter packing and diaphragm materials used with manufactured gas may need moisture to prevent their drying.

Most natural gas transmitted at high pressure to city border stations has been thoroughly dehydrated (2 to 12 lb of water vapor per MMCF). A 1957 survey[3] indicated that 72 per cent of the responding utilities were not humidifying their gas (70 per cent in 1949[2]). The following were some of the reasons dry gas is preferred:

1. Combustion characteristics of the dry gas are excellent.
2. Distribution systems are predominantly of steel pipe and remain gastight on dry gas. The small number of cast iron joints still in service are allowed to dry out. Leaking joints are clamped of otherwise sealed, or the line is replaced.
3. Water vapor cannot hold dust in place in all parts of a distribution system, as moisture content varies greatly with temperature and pressure. Gas can be too dry in some parts of the system and too wet in others at the same time.
4. Water vapor causes increased internal corrosion of pipelines, with formation of additional iron rust, particularly where wetting and drying occur alternately.
5. A dryer gas causes less trouble from freezing in meters, regulators, and piping. It lessens or eliminates drip pumping and stoppages of gas due to accumulation of water in pipes.
6. Low-pressure saturated gas at 60 F contains 1.74 per cent by volume water vapor, thus decreasing heating value and increasing gas transportation costs.
7. Leather meter diaphragms do not dry out when operated continuously on dry natural gas, providing the leather is well oiled. During repairs diaphragm oil can be added if necessary. Synthetic diaphragms do not need water vapor or oil to remain pliable.
8. If water is sprayed into a pipeline for humidification, sediment is deposited. Steam does not introduce sediment.

UNDERGROUND TEMPERATURE EFFECTS ON GAS IN MAINS

Figure 9-82a presents a five-year average of gas temperature observations in large mains in Chicago at 3 and 6 ft depths compared with ground surface and atmospheric temperature observations.

In most northern localities, atmospheric and ground surface temperatures are higher than underground or main temperatures from April to September. During this time gas in holders or in aboveground plant equipment will be warmer and can contain more water vapor than that retained by gas in underground mains under prevailing pressures and temperatures. From September to April, however, atmospheric and surface temperatures may be lower than the underground temperature, and gas entering the mains will tend to be less saturated than the gas underground. Depending on main pressures. dehydration or reduction of water vapor content may be indicated during spring and summer, and humidification during fall and winter.

Gas attains the ground temperature prevailing at the main depth within a relatively short distance from its entry point, possibly one to three miles, depending on velocity. Thus, regardless of its original water vapor content, any moisture in excess of that required to saturate the gas at main pressure and temperature will be condensed out within a few miles of its entry point. Figure 9-82b shows concentrations of water vapor required to saturate a gas at various pressures under prevailing seasonal conditions in Chicago mains. The solid lines show the *maximum* concentrations for this location if excessive condensation, rusting, and leaching within three miles of the sendout point are to be avoided. These concentrations are also the *minimums* necessary to avoid drying out at points more than three miles away. Increase in distribution pressure decreases permissible concentration of water vapor, as indicated by the lower lines in the figure.

In Figure 9-82b the curve marked "distribution mains—5 psi" averages dew point determinations of the water vapor content of gas in Chicago mains at a point 3.5 miles from sendout over a five-year period. Simultaneous tests ten miles downstream yielded practically identical results.

When, for example, gas pressure is reduced from 25 psig to between 5 and 20 in. w.c., the concentration of water vapor needed for saturation is increased (see Fig. 9-82b). Thus, condensation does not necessarily result from a pressure decrease. However, if the temperature of the saturated gas in high-pressure lines is more than about 10°F above the temperature of the saturated gas in downstream low-pressure mains, condensation may occur after pressure change. Under these conditions appreciable condensation of water vapor can occur during the spring and summer months in those low-

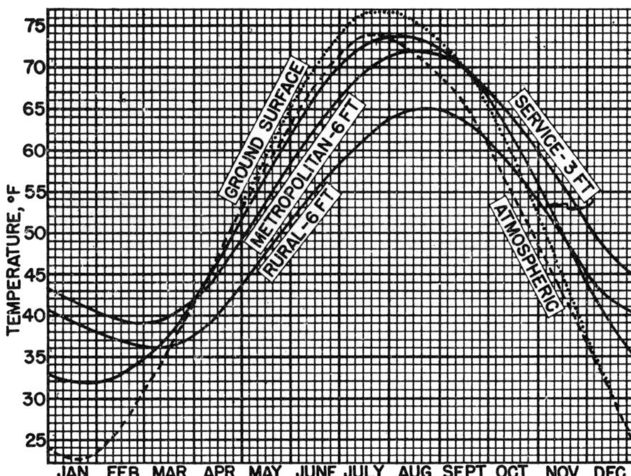

Fig. 9-82a Comparison of gas temperatures in mains with ground surface and atmosphere, five-year average in Chicago.

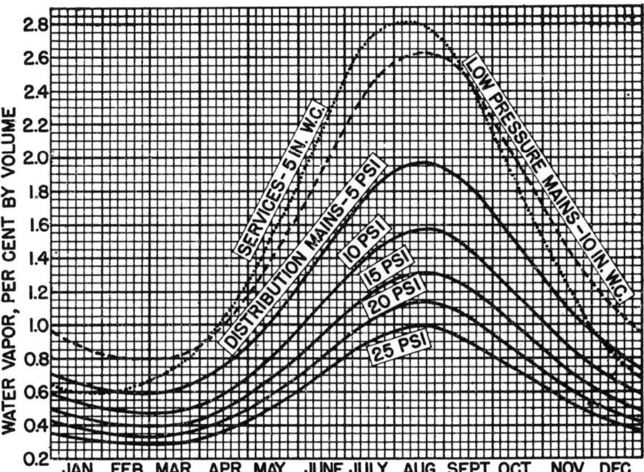

Fig. 9-82b Water vapor content of saturated gas in Chicago mains at various temperatures and pressures.

pressure mains and services which are supplied thru regulators located within one to three miles of the sendout point.

Services are generally located just below normal frost penetration, seldom more than 3 ft below ground surface. Thus, they are subject to larger temperature variations than the more deeply buried mains. It will be noted in Fig. 9-82b that during the winter months the saturation content curve for services at 3-ft depths passes below that for 5-psi mains at 6-ft depths. Thus, even where conditions are such that there is little or no condensation in low-pressure mains, there can be condensation and corrosion in services supplied by these mains.

Many distribution systems have short sections of main with less earth cover than the remainder of the system; e.g., where the main is raised to bridge underground obstacles; where it cannot be lowered to conform to surface depressions; or at exposed locations such as bridges, railroads, etc. Such sections of main tend to have exceptional deposition of water in winter and severe drying out during summer.

HUMIDIFYING EQUIPMENT

Most gas humidifying has been done by injecting steam. One such system[4] (Fig. 9-83) consists of (1) pipe with an adjustable disk damper to create a pressure drop and cause a portion of the gas to flow thru the humidifier; (2) by-pass pipe and condensate separator; (3) low-pressure steam admission to the by-pass piping, regulated according to temperature in the steam separator thru a simple air-actuated controller; and (4) pressure and temperature indicators and controllers installed at appropriate locations. The system illustrated is based on by-passing a portion of the gas to be humidified (say 10 per cent), raising the temperature of that portion by

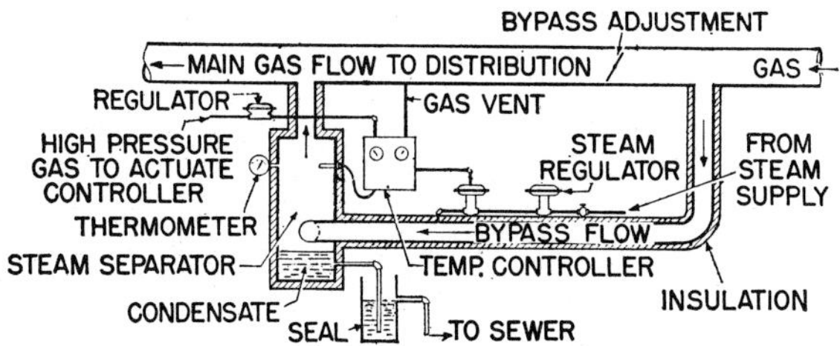

Fig. 9-83 A humidification system using steam.[3]

injecting sufficient steam so that when the 10 per cent by-passed portion rejoins the main stream, the entire flow will be humidified to the desired dew point.

Example: The water required to saturate gas at 50 F (from Table 4-33) = 580 lb per MMCF. Therefore, the water content required in the 10 per cent by-pass gas = 5800 lb per MMCF. Interpolating from Table 4-33, the temperature of 10 per cent by-pass gas to contain 5800 lb of water per MMCF = 122 F.

After the desired dew point is determined, the humidifier is put into operation. Adjustments are made to obtain the correct amount of by-pass gas or the correct temperature setting of the controller. Once set, the equipment will maintain a constant dew point for wide ranges of flow and gas temperatures. The dew point setting is changed periodically to conform with changes in ground temperatures.

In another system,[1] steam and *water spray* were introduced into the gas stream approximately 6 ft ahead of the point at which oil fog was introduced. A 90° change in the direction of gas flow between these two points facilitated removal of heavy water droplets. Moisture in the gas was controlled so that the amount of condensate removed from drip pots in the system was held to approximately the same amount as that removed from them before conversion to natural gas.

Automatic control of humidification is desirable in all cases. The amount of moisture introduced must be changed throughout the year, depending upon main temperatures. Otherwise, difficulties will be experienced with freeze-ups, excessive drip pumping, and condensation in meters and lines. If the humidity is controlled so that it does not exceed 85 per cent of saturation at the lowest main temperature in the system, such difficulties will be avoided.

REFERENCES

1. Kooke, C. A. "Effects of Gas Conditioning on the Distribution System." (OS-52-10) *A.G.A. Proc.* 1952: 280–4.
2. Anuskiewicz, M., Jr. "Gas Conditioning." *A.G.A. Proc.* 1949: 622–33.
3. Laudani, H. "Survey of Current Gas Conditioning Practices." (CEP-59-7) *A.G.A. Operating Sec. Proc.* 1959: P35-8.
4. Tuttle, L. W. "Humidification System." *A.G.A. Proc.* 1941: 543–4.

Chapter 15

Dust Filters and Gas Cleaners

by F. E. Vandaveer

INTRODUCTION

Dust, a good portion of which is iron oxide, originates from several sources. Deposits of dust associated with manufactured gas may be found in a mixed gas system or in a system converted from manufactured to natural gas. These deposits often remain dormant. Sometimes they are removed.

Gas Cleaning Practices

A number of companies install gas cleaners, usually of the **oil scrubbing type,** at city border stations on distribution system inlets. The trend has been to install **dry cloth-type** dust filters in addition to border station scrubbers at district regulator stations. Use of filters has also increased on intermediate-pressure service lines. Various kinds of filters for pilots are employed in both natural and manufactured gas areas.

Intensified studies of air pollution[1,2] and disposal of atomic energy wastes have developed improved materials for filtering out very small particles from a gas stream. Information on aerosols reported in those studies is directly applicable to filtering dust from gas.

Minimizing Dust Problems and Pilot Outages

Several suggestions for minimizing **dust storms** and pilot outages follow. The relative usefulness of each of these varies for each distribution system.

1. Prevent corrosion of inner pipe wall by minimizing water vapor, oxygen, sulfides, and carbon dioxide in the gas.

2. Protect stored pipe from corrosion. Internal coating prevents or minimizes internal corrosion during both storage and service.[3]

3. Blow and pig transmission lines and distribution feeder mains wherever possible. Sometimes mains are cleaned by rodding and suction techniques.

4. Keep the gas flow in the same direction as much as possible. Fluctuations in network demand may reverse flow in some lines.

5. Maintain the lowest possible gas flow velocity. This would entail adequately sized lines, particularly in high- and intermediate-pressure systems, where velocity is much higher than in low-pressure systems. Variations in velocity also stir up dust.

6. Filter or oil scrub gas at border stations. Use filters at

Table 9-47a Actual Chemical Analyses of Pipeline Dust in Natural Gas Lines, Per Cent

Constituent	Sample[5] M-3	Sample[5] P-1	Sample[6] 1	Sample[7] 2	Sample* on ¼-in. screen	Sample* on centrifugal separators	Sample* 1 on cloth filter	Sample† 2 on cloth filter
H_2O	1.7	2.8	3.65	3.10	...	...	...	...
Fe_2O_3	41.5	...	50.19	76.71	...	...	...	...
$Fe_2O_3 \cdot H_2O$	44.4	64.9	...	...	...	...	...	...
$2Fe_2O_3 \cdot 3H_2O$	...	...	...	...	97.5	99.0	99.5	84.63
Fe_3O_4	...	...	26.95	...	...	...	...	...
SiO_2	4.6	1.2	1.67	3.42	2.5	1.0	0.5	...
Silicon	...	...	0.22	...	...	...	...	...
$FeSO_4$	4.3‡	14.9‡	...	0.16‡	...	...	...	...
Combustible matter	...	...	13.42	16.61	...	...	...	15.32 (scrubber oil)
Hydrogen	0.7	0.4	...	...	...	...	...	...
Carbon	3.6§	3.7§	...	...	...	...	...	...
NaCl	...	12.5**	...	...	...	...	...	...
$CaCl_2$	...	1.2	...	...	...	...	...	...
$CaCO_3$	...	...	...	...	...	...	...	...
$MgCl_2$	...	1.4	...	...	...	...	...	...

* Data taken from an old high-pressure 6⅝ in. diameter gas line by the East Ohio Gas Co.
† Data taken from an old cloth filter at a district regulator by the East Ohio Gas Co.
‡ The result of computing all sulfur as $FeSO_4$.
§ Believed to have been carried into pipeline as oily and/or waxy products from gas well deposits.
** Materials such as sulfur, carbon, magnesium salts, and silica carried from gas at well.

Table 9-47b Actual Screen Analyses of Pipeline Dust in Natural Gas Lines, Per Cent

Screen,* on mesh No.	Sample[5] P-1	Sample[6] 1	Sample[7] 2	Sample† on ¼ in. screen	Sample† on centrifugal separators	Sample† 1 on cloth filter	Sample‡ 2 on cloth filter
10	0.22	1.33	...	37.70	3.78	0.41	31.5
20	0.52	1.87	...	19.51	7.04	2.03	10.0
40	7.40	...	...	11.40	8.51	13.33	17.1
50	...	7.42	...	7.50	11.19	31.10	15.7
60	16.58	...	...	4.63	6.60	8.14	5.0
80	14.11	8.35	...	6.31	18.51	35.73	6.4
100	9.90	3.59	...	3.21	8.18	4.88	2.9
150	...	11.05	...	3.80	11.19	0.89	2.9
170	25.18	...	...	...	...	3.33	2.8
200	7.67	18.15	...	3.41	16.86	...	...
270	10.68	14.43	...	...	...	...	...
325	4.18	13.85	...	...	...	...	...
On pan	3.56	19.88	97.0 (<5 μ)	3.21	8.14	0.16	5.7

* Sizes are given in U. S. sieve series numbers. See Table 9-48.
† Data taken from an old high-pressure 6⅝ in. diameter gas line by the East Ohio Gas Co.
‡ Data taken from an old cloth filter at a district regulator by the East Ohio Gas Co.

Table 9-48 Standard Screen Sizes

U. S. and ASTM Std. sieve No.	Actual opening In.	Actual opening μ	U. S. and ASTM Std. sieve No.	Actual opening In.	Actual opening μ
10	0.0787	2000	100	0.0059	149
12	.0661	1680	120	.0049	125
14	.0555	1410	140	.0041	105
16	.0469	1190	170	.0035	88
18	.0394	1000	200	.0029	74
20	.0331	840	230	.0024	62
25	.0280	710	270	.0021	53
30	.0232	590	325	.0017	44
35	.0197	500	400*	.00142	36
40	.0165	420	550*	.00099	25
45	.0138	350	625*	.00079	20
50	.0117	297	1,250*	.000394	10
60	.0098	250	1,750*	.000315	8
70	.0083	210	2,500*	.000197	5
80	0.0070	177	5,000*	.000099	2.5
			12,000*	0.0000394	1

* Estimated.

Table 9-49 Chemical Analyses of Pipeline Dust in Manufactured Gas Lines, Per Cent

Constituent	Oil gas operation[8]			Carbureted water gas[9]
H_2O	...	...	...	2.93 and oil
Fe_2O_3	77.00	38.03	62.70	73.69
SiO_2	2.00	1.38	13.50	0.47
Silicon	6.59	...	15.00	0.93
$FeSO_4$	...	4.01	1.00	...
Combustible matter	13.90 (oil)	15.53 (tar, gum, naphthalene)	8.75 (oil)	12.60
Carbon	0.01	...	...	...
NaCl	0.50	...	...	...
Sulfocyanide	...	...	...	0.20
Prussian blue, $Fe_4[Fe(CN)_6]_3$	...	35.86	...	3.19
Unknown	...	...	...	5.99

district regulator stations, on individual service lines, and in pilot lines.

7. Oil fog to reduce dust movement.
8. Use pilots with fixed orifices*[4] or use pilot filters.

Dust Particle Composition and Size

Table 9-47a gives several analyses of pipeline dust in natural gas lines, measured in *gas* samples taken under varying conditions and by several media (sample numbers shown originated with the sources referenced or footnoted). Ignoring the scrubber oil content, it is apparent that forms of iron oxide generally constitute over 90 per cent of the dust mixture. These oxides were reported[5] to consist of:

1. *Mill scale,* chemically: ferrous ferrite (Fe_3O_4), also called *magnetite;* black in color.
2. *Ferric oxide* (Fe_2O_3), known as *hematite;* color varies from reddish brown to black.
3. *Goethite,* which is monohydrated ferric oxide ($Fe_2O_3 \cdot H_2O$). Variation in the amount of *water of crystallization* found with ferric oxide yields this substance, which is yellow, red, or brown. *Limonite,* yellow brown in color and known as trihydrated diferric oxide ($2Fe_2O_3 \cdot 3H_2O$), is also formed by this process.

* Larger dust particles will pass thru a circular orifice than thru an annular ring of equal area.

Sieve or screen analyses (Table 9-47b) vary considerably, depending upon the sampling location and the conditions causing the dust. Particle sizes vary from large oxide scale masses to 1 or 2 μ average diameter. These analyses show that pipeline dust consists mostly of extremely small particles.

Analyses of **deposits in manufactured gas** lines are given in Table 9-49. The major constituent is iron oxide or an iron compound. Tar, oil, and naphthalene contents are high; silica and sulfur contents are also fairly high.

Both the very small sizes of particles involved and the sizes of particles which can be carried in a gas stream were studied[10] (see Table 9-50; also see Figs. 1-6a and b for general dust loading values). Dust in natural gas lines may be of the order of 0.7 grains per MCF. In a laboratory gas line, 3,400,000 dust particles per cubic foot were noted, with a maximum particle size of 15 μ. A gas sample direct from a well contained 3,900,000 particles per cubic foot, with a maximum particle size of 5 μ.

Table 9-50 Thickness of Laminar Layer and Particle Size Carried in Natural Gas[10]

System	Diam, in.	Avg press., atm	Pressure drop	Thickness of laminar layer,* μ	Particle size carried†
Transmission	24	36	30 psi/mile	5.17	Large as pipe
	12	36	30	7.3	Large as pipe
	6	36	30	10.3	Large as pipe
	24	16	2.5	27.0	6.7 in.
	12	16	2.5	38.2	2.7
	6	16	2.5	54.0	1.1
High-pressure distribution	24	2	2.5	76.2	6.4
	12	2	2.5	108.0	1.22
	6	2	2.5	152.0	0.83
Low-pressure distribution	24	1	1½ in. w.c./1000 ft	318	2350μ
	12	1	1½	450	400
	6	1	1½	637	104
Low-pressure service	1¼	1	0.5 in. w.c./100 ft	765	113
	1	1	.5	855	85
	¾	1	0.5	986	60
Pilot	5⁄16	1	At 0.15 cfh	Viscous flow	14.3
	5⁄16	1	.50	Viscous flow	47.7
	¼	1	.15	Viscous flow	3.7
	¼	1	0.50	Viscous flow	12.3

* *Laminar layer* is defined as the layer of gas in viscous motion along the pipe wall with the velocity gradients well defined. Calculations in this column are based on data prepared by R. V. Dunkle and reported by Wills.[15] These data are the starting point for derivations of fundamental equations for transport of dust particles in mains. For further information see References 16 and 17.

† Specific gravity at 4.0 (water = 1); these are *calculated* values based on Reference 15.

It was also shown[11,12] that the dust particles causing pilot stoppages were mainly from 20 to 0.5 μ in size (at a pilot flow of 0.5 cfh). At a 7 in. w.c. pressure and 0.5 cfh natural gas, the orifices or pilot valve openings (Fig. 9-84) for equal area ranged in size as follows:

1. In a needle valve (No. 47) the width of the annular orifice was 0.0005 in.
2. In a circular orifice (No. 80 drill) the diameter was 0.0135 in.
3. In a pilot cock orifice the opening was 0.0050 in. by 0.0410 in.

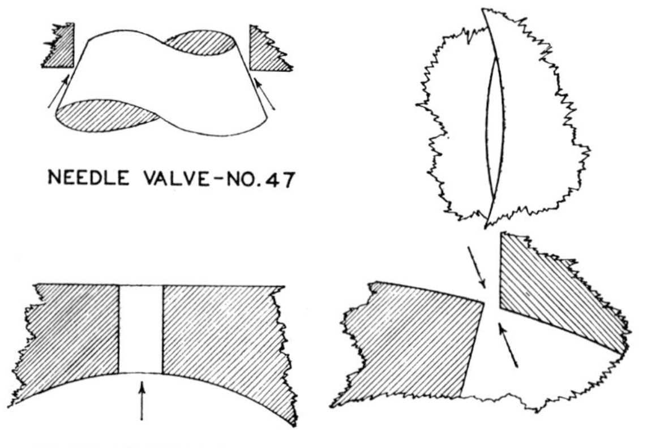

Fig. 9-84 Three types of pilot orifices.[10] Both the needle valve and pilot cock types were adversely affected by dust particles as small as 0.5μ.

GAS CLEANERS AND DUST FILTERS FOR STATION INSTALLATION

Oil Contacting Cleaners (Scrubbers)

In these cleaners oil (or other liquid, if preferred) is brought into contact with the dust particles in the gas, which will then separate the oil and oil-coated dust from the outgoing gas stream. The ideal cleaner of this type should accomplish the following:

1. Facilitate intimate oil-to-dust particle contact to assure subsequent removal of dust.
2. Divide oil into five particles or mist, to assure wetting of dust at the micron size level.
3. Provide baffles to trap, agglomerate, and thus remove oil-dust particles from the main gas stream.

The criteria for **selecting an oil contacting cleaner** follow:

1. Type: Horizontal (Figs. 8-55a and 8-55b) or vertical (Fig. 8-56). Efficiencies and capacities of oil scrubber installations have been reported.[13,14]

2. Capacity: The parameters are gas volume, cleaner diameter, and inlet pressure. See Fig. 9-85 for capacity ratings of typical cleaners. Sizing is ordinarily based on the hourly load at *minimum* pressure.

3. Pressure Drop: The difference in gas pressure across the cleaner. The inner configuration of the cleaner affects gas velocity and pressure.

4. Oil Carry-Over (Entrainment): The oil lost by transfer to the gas. The average carry-over of an oil contacting filter is between 0.01 and 0.07 gal per MMCF of gas. Any substantial excess of carry-over beyond 0.07 gal will have the effect of oil fogging.

Several **border station cleaners** (60 in. diam) at the outlet of transmission lines over 100 miles long remove about

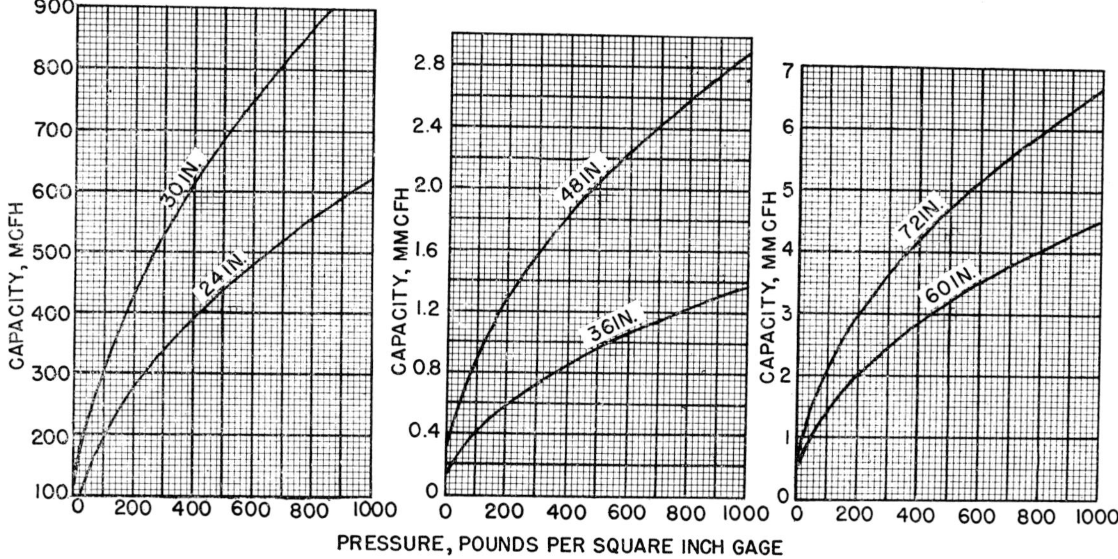

Fig. 9-85 Capacities of various sizes of typical oil contacting gas cleaners made by the Blaw-Knox Co., Blawnox, Pa.

125 lb of iron oxide, sand, and adhering oil mixture per year. Of this amount about 100 lb is iron oxide; the remainder is mostly oil. In 18.9 per cent of this mixture passed thru a 200 mesh screen the largest particles were 4 μ; the smallest, about 0.5 μ; and the average, 1 to 2 μ.

At rated capacity, scrubbers discharge from 0.03 to 0.07 gal of oil per MMCF of gas. Oil carry-over increases greatly at more than 10 per cent above or less than 25 per cent below capacity. Sudden changes in pressure or flow rate can result in the discharge of the entire tank of oil into the line. Use of a downstream *mist extractor* will minimize oil carry-over in situation in which it is not desirable to *oil-fog* the gas stream (oil foggers add as much as 1.5 to 2 gal of oil per MMCF of gas).

Under normal conditions, dust separators are cleaned before they have completely filled up. However, the weight of accumulated dust may have to be taken into account in designing the structural supports. The weight of the dust is believed to range from 50 to 125 lb per cu ft.

Dry Filters

Dry filters from 2 to 12 in. pipe size are frequently used upstream of district regulators. The filtering medium is normally a 1/8 in. thick felt cloth formed over metal supports. Gas filters and filter elements of this type are shown in Figs. 9-86 and 9-87. These cloth filters remove dust particles down to 2 μ. As they become laden with dust they remove smaller particles more efficiently. No cases are reported of dust storms in pipelines downstream from such filters Capacities of a typical cloth filter are listed in Table 9-51.

Table 9-51 Capacities of a Typical Cloth Filter*
(free air at 80 psig with a 7 in. w.c. drop)

Pipe size, in.	Dimensions, in. Diam	Length	Capacity, MCFH	Approx. cloth filter area, sq in.	Cu ft per hr-sq in. cloth area
2	10¼	20¼	40	...	...
4	19½	35	96	2700	35.5
6	26½	43½	168	...	...
8	30½	44	288	4612	62.3

* Data from Dollinger Corp., Rochester, N. Y.

Reported test data[18] on cloth filters show that:

1. Dust collection efficiency remains high (99.4 per cent) where the gas flow rate varies from 38 to 53 cu ft per hr per sq in. of cloth filter area.
2. The 1/8 in. cloth and 1/4 in. wool felt cloth filter materials are equally efficient in collecting dust (99.4 per cent).
3. Dust collection efficient drops slightly as gas flow rates drop.
4. A dust cake on the filter cloth increases the dust collecting efficiency.

Conclusions from these studies are that cloth filters may be used ahead of district regulators without an unduly rapid increase in pressure drop and necessity for frequent servicing, providing the gas flow per unit area of filter is kept sufficiently low. The gas capacity of a given filter depends upon operating

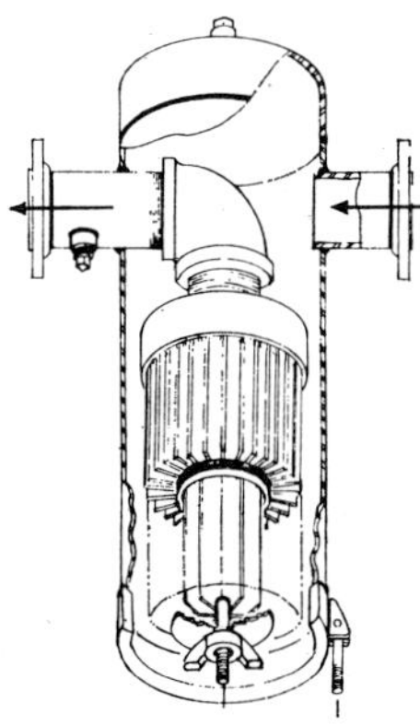

Fig. 9-86 A commercial cloth-type gas filter widely used in 2 to 12 in. pipe sizes. This 8 in. size has two cloth filters. Flanged bottom and second filter element omitted for clarity.

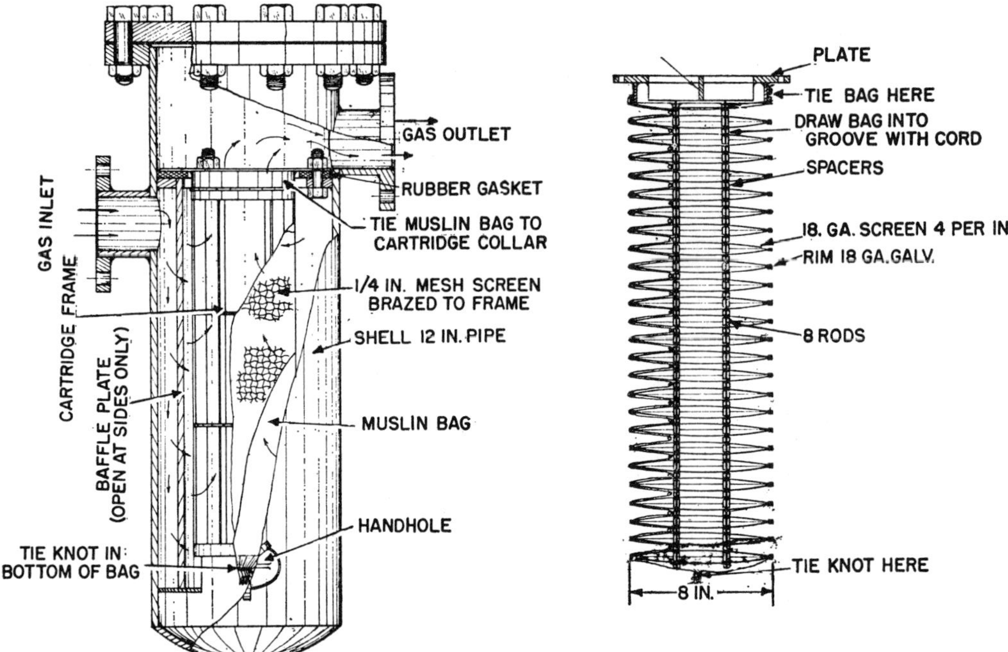

Fig. 9-87 Typical cloth-type gas filter with disk-type filter element shown at right.[18]

pressures and permissible pressure drop. Dust collection efficiency of cloth filters is about 99 per cent (practically independent of gas flow per unit area of filter and amount of dust collected on it). It does depend on particle size when dust is very fine.

The condition of operating filters may be determined[19] to some extent by (1) observing pressure drop at any known gas flow; (2) determining corresponding pressure drop if that flow were increased to peak hour flow; and (3) comparing this pressure drop with the maximum selected for that filter at peak hour flow.

Another type of dry filter, using a number of seamless tubes made of resin impregnated glass fiber, is illustrated in Fig. 9-88a. This filter is designed to pass up to 30 MMSCFH at pressures up to 3000 psia; it is claimed that it traps 99 per cent of all solids and liquid droplets ranging in size from 1 μ

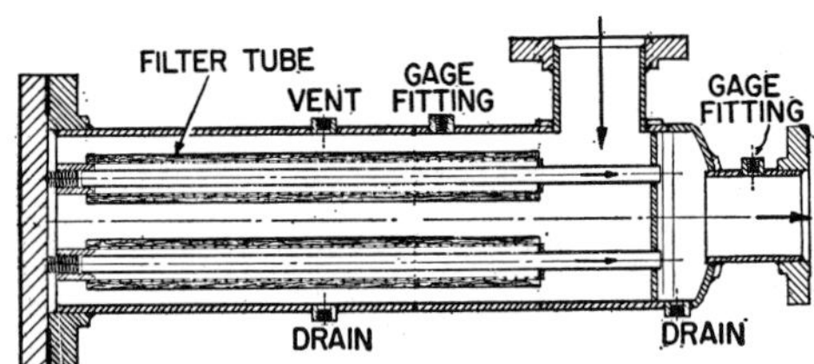

Fig. 9-88a Glass fiber gas filter. (Perry Equipment Corp., Mineral Wells, Texas.)

up, with an efficiency of 99.95 per cent removal by volume (assuming normal distribution of particle sizes) when operated within 10 to 100 per cent of design capacity. Units are rated at a 2 psi pressure drop; this loss varies as the square of the flow rate.

One successful composite filter consisted of two layers of 1 in. thick dust-stopping glass fiber mat wrapped around an 18 gage, 3/4 in. mesh Monel screen, a layer of 1/4 in. hardware cloth, and two layers of 1/4 in. glass wool flotation matting.[20,21] The 4 in. pipe size unit is shown in Fig. 9-88b. Such a filter will remove particles as small as 1 μ, including liquid particles.

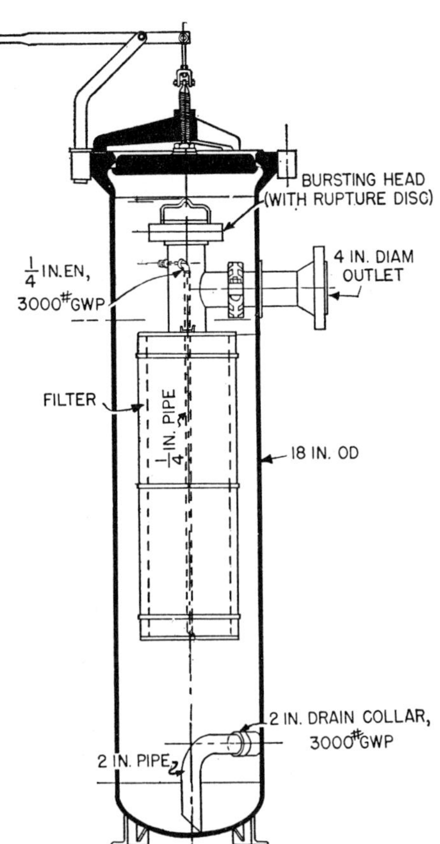

Fig. 9-88b Glass fiber gas filter (mat type). (Lone Star Gas Co., Dallas, Texas.) Filter consists of window screen (next to metal sleeve) and Fiberglas flotation mat 1 in. thick, covered with 1/4 in. hardware cloth.

Dry Filter Developments. Rodebush[22] made the following comments about filters:

> Small particles whose diameter is in the neighborhood of 1 micron will not be precipitated by centrifugal action since the inertia of the particle is not sufficient to overcome the resistance of the air. Air flows around the fibers of a filter in stream lines and the particles are carried around with the stream lines. There is a range of particle sizes for which a high velocity will improve the operation of the filter since the inertial effects will carry the particles across the stream lines into collision with the fibers of the filter. For particles smaller than 1 micron diameter no inertial effects exist, but the kinetic diffusion becomes of greater importance in the smaller particle. Very small particles (0.01 micron) are precipitated very rapidly by diffusion. The process is analogous to the condensation of a vapor on a cold surface. The particles most difficult to remove by filtration are those in the range 0.1 to 1.0 micron, i.e., smokes. In order to obtain efficient filtration without excessive resistance the filter must contain fibers of small diameter approaching that of the particles themselves.

It was reported that the electrostatic behavior of filter fabrics and their rates of charge dissipation exert important effects of filter operation.[23] Fine glass wool, having fibers of about 1 μ diam, was used successfully to filter over four billion cubic feet of gas.[21] Glass wool wrapped around a felt filter pad was used to remove oil fog from gas upstream of compressor stations.[24] This glass wool contained fibers of 1 to 3 μ diam. The use[25] of glass webs of high physical strength fabricated in 10 mil (0.010 in.) thickness as a filter material was reported.

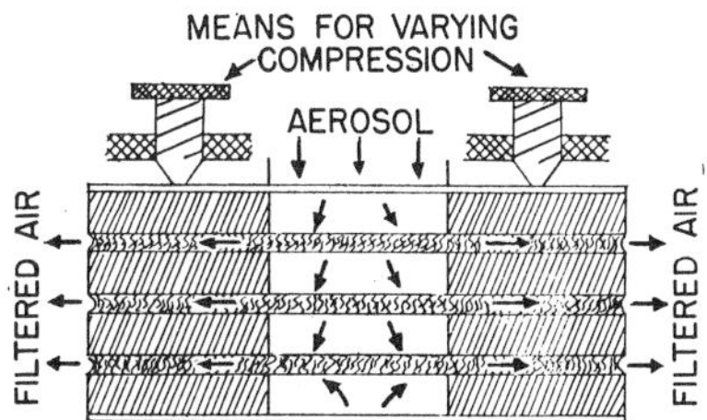

Fig. 9-89 Variable compression gas filter.[25]

At a velocity of 5 fpm, they have an efficiency of 99.98 per cent, or penetration of 0.02 per cent on dioctyl phthalate smoke which has a mean size of 0.3 μ. One manufacturer has used aluminum silicate in the fabrication of glass webbing which consists of extremely fine fibers of 1 μ and less in size. Another has developed a resin bonded glass wool with 1 to 3 μ glass fibers. A variable compression filter made of this material (Fig. 9-89) permits adjustment of the filter resistance. The configuration is one of a compression bed of fibers arranged to filter radially. As resistance increases the compression is relaxed, allowing the maintenance of a constant resistance to gas flow.

Centrifugal or Inertial Separators

Separation of dust particles from the gas in such devices is based on the particle terminal velocity, which is a function of size, size distribution, shape, surface, and specific gravity. It is also somewhat dependent on gas velocity and density. One widely used centrifugal separator contains a battery of centrifugal units (Fig. 8-57). The number of these tube-within-tube units used depends on volume of gas handled. Gas travels between the two tubes to the bottom of the inner tube, where centrifugal force prevents the dust from entering the inner tube.

These centrifugal separators are made for vertical or horizontal transmission lines. The greatest efficiency in dust removal occurs with a pressure drop of 2 to 5 psi across the tubes. Data show that such units will remove all particles larger than 8.0 μ and approximately 50 per cent of the particles in the 2.0 μ range, depending upon their specific gravity.[26] Results of a test made by the East Ohio Gas Co., Cleveland, Ohio, using a centrifugal separator on a 0.7 mile section of 6⅝ in. pipe known to contain an appreciable quantity of dust, are shown in Tables 9-47a and b. Line gas pressure was built up to 35 psig, the outlet valve was opened wide, and the line was blown down to 0 psig. This was repeated several times. Pipe velocity varied from 456 to 13,390 fpm. About 55.9 per cent of the dust was collected in the first ¼ in. mesh section and 21.4 per cent in the centrifugal section, while the remaining 22.7 per cent was collected by a cloth filter.

Electrostatic Precipitators

Giuricich[27] makes the following comments about electrostatic precipitators:

> The principle of electrostatic precipitation depends on the electrical charging of particles entrained in a carrier gas and subjecting these particles to a high voltage unidirectional electric field where under the force of the field the charged particles are attracted to flat plates. Basically all precipitators are divided into gas passages made up of flat plates or tubes in which are suspended thin wires or rods. The thin wires or emitting electrodes are given a negative high voltage, direct current potential, and the flat plates or collecting electrodes are grounded. When the voltage is sufficiently high a "corona discharge" occurs about the wire electrodes and under the influence of the field negative ions are formed which migrate to the grounded plates. Entrained particles carried by a gas stream through a gas passage will be charged by the negative ions and consequently will be carried to the collecting electrode plates from which they must be removed. The ability of a precipitator to operate properly depends on the nature of the carrier gas, velocity of the gas stream, the entrained material, the moisture content of the gas stream and numerous other factors.

The two general classes of electrical precipitators are:[28]

1. *Single-stage*, in which ionization and collection are combined. They[29] are most generally used for dust and mist collection from industrial process gases.

2. *Two-stage*, in which ionization is achieved in one portion of the equipment, followed by collection in another. This type of precipitator* is also used to a great extent in air conditioning.

Table 9-52 gives some data on charge and motion of spherical particles in an electrical field.

Electrical precipitators are generally designed and maintained for a **collection efficiency** in the range of from 90 to 99.9 per cent. **Electrical power consumption** is generally 0.2 to 0.6 kw per 1000 cfm of gas handled. Pressure drops across the units are usually less than 0.5 in. w.c. Applied potentials range from 30,000 to 100,000 v. Gas velocities generally range from 3 to 10 fps and retention times from 1 to

* Precipitron Electronic Air Cleaner, Westinghouse Electric Corp., Sturtevant Div., Boston, Mass.

6 sec. Chief disadvantages are high initial costs and, in some cases, high maintenance costs.

An electrical precipitation unit was applied to high-pressure natural gas. Operation was at 500 psig with a capacity of 7 MMCF of gas per day. All suspended particles in the gas were collected.[30] Note that the velocity of gas in contact with liquid surfaces must be kept low at this high pressure to prevent reentrainment of liquid.

Table 9-52 Charge and Motion of Spherical Particles in an Electrical Field

Particle diam, μ	Number of elementary electrical charges*	Particle migration velocity, ft/sec
0.10	10	0.27
.25	25	.15
0.50	50	.12
1.00	105	.11
2.50	655	.26
5.00	2,620	.50
10.00	10,470	0.98
25.00	65,500	2.40

* Dimensionless.

Magnetic Attraction and Ultrasonic Agglomerators

As stated previously, a large portion of dust in gas pipelines (70 to 99 per cent) is iron oxide, part of which—Fe_3O_4 (magnetite)—is magnetic. This has suggested the possibility of using large magnets with baffles designed to direct the gas flow against the magnets for dust deposition. One manufacturer, the Eriez Manufacturing Co., Erie, Pa., made a magnetized device in sizes up to 20 in. pipe diameter for removing tramp iron. The goals in designing this type of dust remover are: (1) to get all particles within the magnetic field; (2) to provide for cleaning the magnet; (3) to hold initial equipment cost within reason; and (4) to keep magnets permanently magnetized.[31]

Ultrasonic agglomerators have not been used in the gas industry to date for collecting dust particles in pipelines. High-intensity acoustic vibrations cause collisions among particles and tend to flocculate fumes and mists, so that they can be collected in conventional apparatus. The U. S. Bureau of Mines[32] has done considerable work on this subject, and the Ultrasonics Corp., Cambridge, Mass., has developed a gas-siren generator, with a frequency range of from 1,000 to 200,000 cycles per second. Power consumption for **aerosol agglomeration** is normally in the range of from 2 to 5 kw per MCFM of gas handled. Sonic precipitation is limited to a particle concentration greater than one grain *per cubic foot* (since natural gas in pipelines seldom contains more than 0.7 grain per MCF). Conceivably, the method could be used for the collection of extremely heavy dust concentrations in gas lines.

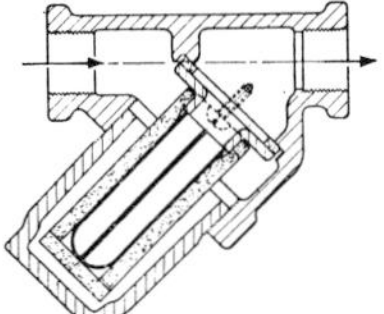

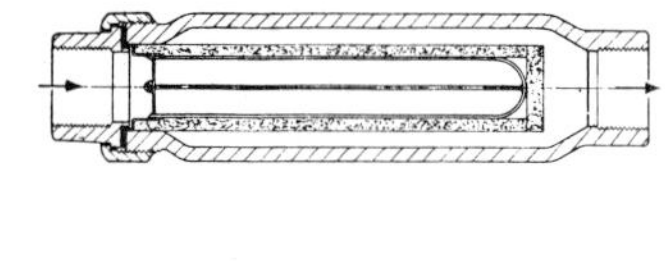

Fig. 9-90 Removable wool felt pad filters. (left) Nondisconnect type. (right) Disconnect type. (Reliance Regulator Corp., Alhambra, Calif.)

DUST FILTERS FOR SERVICE AND HOUSE LINES

Dust filters for service and house lines serve mainly to protect appliances. The following data are representative of many of the different units of this type.

Removable Wool Felt Pad Filters

Filters having replaceable wool felt pads are shown in Fig. 9-90. It is claimed that these filters remove dust particles as small as 3 μ. They are available for pressures up to 1000 psig (depending on casing material). The standard filter element is of wool felt whose seams are cemented with a waterproof compound. Supporting frames for the elements make them

Table 9-53 Capacities of Removable Wool Felt Pad Filters

Size, in.	Inlet press., psig	Differential, in. w.c. 0.1	0.5	7	14
¾ NPT	0.25	50	160	...	...
¾ NPT	25.0	...	...	1,750	2,625
1 NPT	0.25	100	240	...	...
1 NPT	25.0	...	...	2,000	3,250
1½ NPT	5.0	...	...	7,000	10,250
1½ NPT	25.0	...	...	10,100	14,350
2 NPT	5.0	...	...	9,750	13,800
2 NPT	25.0	...	...	14,600	20,000

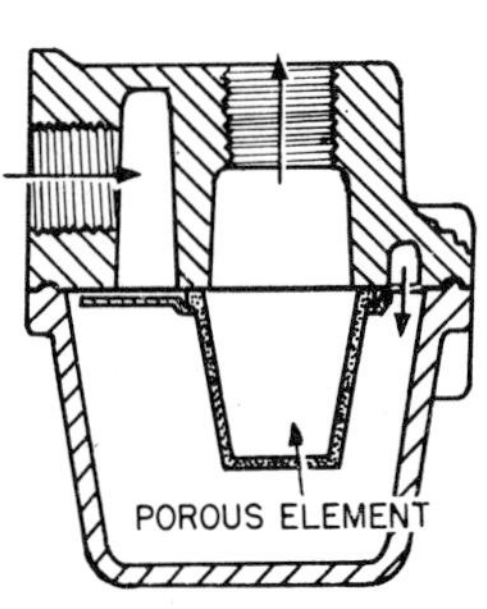

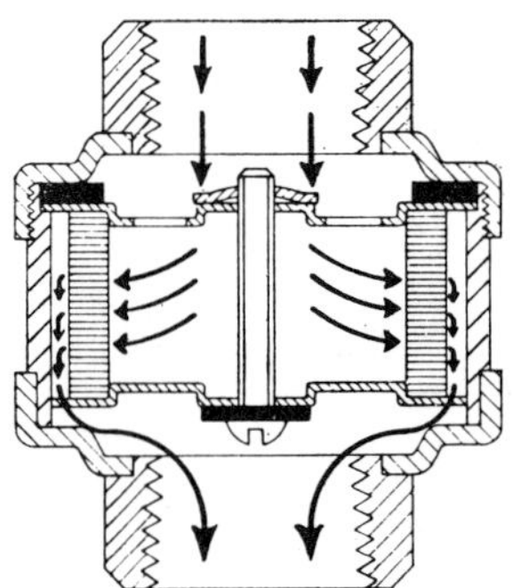

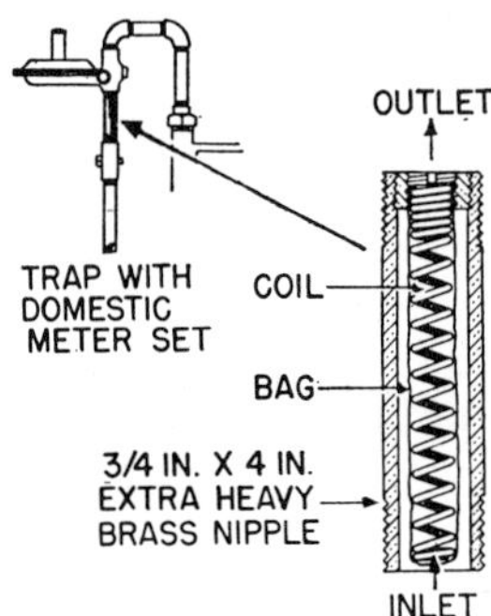

Fig. 9-91 (left) Metal trap dust filter. (Sprague Meter Co., Bridgeport, Conn.)

Fig. 9-92 (center) Large volume pressed paper filter. (Patrol Valve Co., Cleveland, Ohio.)

Fig. 9-93 (right) Filter bag dust trap.

collapseproof. The unit shown at the left in Fig. 9-90 (as opposed to that at the right) may be inspected and serviced with a new element, without disconnecting piping. Capacities (SCFH of 0.64 sp gr gas) of these filters, supplied by the manufacturer, are shown in Table 9-53.

Metal Trap Dust Filters

An all-metal trap[33] for collecting dust (other than fine dust) in intermediate- and high-pressure service lines is shown in Fig. 9-91. In addition to filtering, centrifugal force is utilized to throw the larger and heavier particles into the bowl. The element can be inspected and replaced without any pipe disconnection. It can be cleaned and reused.

Large Volume Pressed Paper Filters

A typical dust filter of this type is shown in Fig. 9-92. A cylinder of laminated and pressed paper forms the filter element. It filters out dust particles as small as 40 μ. The filter element can be cleaned or replaced. The low-pressure units are made in ½, ¾, and 1 in. sizes, with capacities of from 75 to 1200 cfh at 0.5 in. w.c. pressure drop. It may be installed in low-pressure lines (for example, just upstream of meters) or at appliances.

The high-pressure 2, 3, and 6 in. sizes have rated capacities ranging from 0.75 to 30 MCFH of air at 0.5 in. w.c. drop and 150 psig max. They may be used by utility and industrial companies.

Filter Bag Dust Traps

Dust traps installed upstream of domestic regulators were used in San Diego, Calif.[34] These consisted of a brass nipple enclosing a filter bag of a double thickness of Piquot sheeting held in shape by a wire coil (Fig. 9-93). This trap will remove dust down to 10 μ. The filter bag may be cleaned several times by brushing before replacement is necessary.

REFERENCES

1. U. S. Tech. Conf. on Air Pollution. *Air Pollution: Proceedings, 1950.* New York, McGraw-Hill, 1952.
2. Friedlander, S. K., and others. *Handbook on Air Cleaning; Particulate Removal.* Washington, D.C., Harvard Univ. School of Public Health and U. S. Atomic Energy Comm., 1952.
3. Johnson, R. M. "Internal Coating of Gas Transmission Lines." *Gas* 34:19, May 1958.
4. Consterdine, C. "Fixed Orifices for Prevention of Pilot Valve Stoppages." *Gas* 28:43–6, July 1952.
5. Bissey, L. T. "Natural Gas Pipeline Deposits." (PC-52-1) *A.G.A. Proc.* 1952: 531–6.
6. Blaw-Knox Catalog, form 1867, Blawnox, Pa.
7. Senatoroff, N. K., and Niederer, E., Jr. "How Southern Counties Developed Method for Blowing Dust from Lines." *Gas Age* 112: 23+, July 16, 1953.
8. Tomlinson, K. C. "Dust and Moisture Control." *P.C.G.A. Proc.* 18: 339–55, 1927.
9. Shnidman, L. *Gaseous Fuels,* 2nd ed., p. 321. New York, A.G.A., 1954.
10. Wills, F. "Removing Liquids and Dust from Gas." *Gas* 21: 21–8, Feb. 1945.
11. Corfield, G. "Dust Stoppage." *P.C.G.A. Proc.* 25: 146+, 1934.
12. Wallace, T. G., and Corfield, G. "Gas Main Dust." *Western Gas* 11: 8+, Jan. 1935.
13. McDonald, P. W., and Woods, E. M. "Performance Tests on Oil Contact Gas Scrubbers." *Gas* 26: 48–55, Sept. 1950.
14. ——. "Oil Contact Scrubbers." *Gas* 26: 43–7, Aug. 1950.
15. Wills, F. "Removal of Liquids and Dust from Natural Gas." *P.C.G.A. Proc.* 35: 128–39, 1944.
16. Holman, R. G., and Hoff, N. L. "Laboratory Experiments on the Transport of Gas Main Dust." *P.C.G.A. Proc.* 30: 77–9, 1939.
17. Torrance, E. M. "Dust Filters." *P.C.G.A. Proc.* 25: 149, 1934.
18. Oberseider, J. L., and others. "Dust Control Equipment—Cloth Filters and Centrifugal Separators." *P.C.G.A. Proc.* 30: 69–76, 1939.
19. Emmerichs, L. O. "Selecting, Sizing, and Servicing Cloth Type Dust Filters." *P.C.G.A. Proc.* 31: 79–86, 1940.
20. Bacon, T. S. "Gas Cleaners—Today and Tomorrow." *Gas Age* 113: 35+, Apr. 22, 1954.
21. "Filter Unit Screens Minute Particles from Natural Gas." *Gas* 31: 120–1, Feb 1955.
22. Rodebush, W. H. "General Properties of Aerosols." In: U. S. Atomic Energy Comm. *Handbook on Aerosols,* Chap. 4. Washington, D. C., 1950.
23. Frederick, E. R. "How Dust Filter Selection Depends on Electrostatics." *Chem. Eng.* 68: 107–14, June 26, 1961.
24. Lurvey, D. T. *Removal of Particulates from Compressor Natural Gases: Investigation of Filtering.* (Fritch Lab. Special Rept.) Chicago, Natural Gas Pipeline Co., 1952.
25. Silverman, L. "New Developments in Air Cleaning." *Am. Ind. Hyg. Assoc. Quart.* 51: 183, Sept. 1954.
26. Dennis, R., and others. *Particle Size Efficiency Studies on a Design Two Aerotec Tube.* (NYO-1583) Washington, D. C. Harvard Univ. School of Public Health and U. S. Atomic Energy Comm., 1952.
27. Giuricich, N. L. "Buell Gas Cleaner for Gas Industry." *A.G.A. Proc.* 1954: 518.
28. Perry, J. H. "Electrical Precipitators." *Chemical Engineers' Handbook,* 4th ed., p. 20–82. New York, McGraw-Hill, 1963.
29. Sproull, W. T., and Nakada, Y. "Operation of Cottrell Precipitators." *Ind. Eng. Chem.* 43: 1350–70, June 1951.
30. Anderson, E. "Electrical Precipitation in the Gas Industry." *P.C.G.A. Proc.* 30: 142–4, 1939.
31. Sloan, C. D. "How Permanent Is a Permanent Magnet?" *Inco Nickel Topics* 6(7): 4, 1953.
32. St. Clair, H. W. "Sonic Flocculator as a Fume Settler: Theory and Practice." In: U. S. Bureau of Mines. *Progress Reports—Metallurgical Division.* (RI 3400) Washington, D. C., 1938. Also: St. Clair, H. W., and others. *Flocculation of Aerosols by Intense High-Frequency Sound.* (Bureau of Mines R.I. 4218) Washington, D. C., 1948.
33. Stewart, C. W. "Customer Service Dust Traps . . . Why Use Them?" *Gas* 31: 53–4, Aug. 1955.
34. Capwell, C. W. "Elimination of Dust." *P.C.G.A. Proc.* 29: 98–102, 1938.

SECTION 10

STORAGE OF GAS

H. K. Thomas (deceased), *Section Chairman*, Chapter 1, Philadelphia Gas Works Div., The U.G.I. Co., Philadelphia, Pa.

Wm. S. Bunnell, Chapter 1, Reading Gas Div., The U.G.I. Co., Reading, Pa.

J. R. Lamplugh (deceased), Chapter 1, Philadelphia Gas Works Div., The U.G.I. Co., Philadelphia, Pa.

Marvin D. Ringler, Chapter 4, American Gas Association, New York, N. Y.

E. E. Roth, Chapter 3, Columbia Gas System Service Corp., New York, N. Y.

B. C. White, Chapter 2, Pacific Gulf Oil, Ltd., Tokyo, Japan

CONTENTS

Tables

Figures

Chapter 1

Low Pressure Holders

by H. K. Thomas, Wm. S. Bunnell, and J. R. Lamplugh

Low pressure holders store gas either in an inverted cylindrical bell which is sealed in water and free to move within a steel "guide frame," or beneath a piston in a cylinder. The former are the familiar **water-sealed holders,** the latter the piston type—frequently called **waterless** or dry holders. The pressure within the holder is produced by the weight of the lifts (the bell) or the piston, and it is virtually uniform in piston type holders and single-lift water-sealed holders; multilift water-sealed holders vary their pressure according to the number of lifts engaged, usually about 2.0 to 2.5 in. w.c. for each lift, in addition to that obtained from the inner lift. Figure 10-3 shows a three-lift holder; note that the inner section collapses into the middle section, and that both in turn collapse into the outer section.

The estimated construction costs for both dry and water-sealed holders—including foundations, painting, alarms, and piping—are shown in Fig. 10-1, while Fig. 10-2 shows their maintenance costs including water heating and hose replacement for water-sealed holders, painting at six year intervals, seal maintenance, and general repairs.

WATER-SEALED HOLDERS

Dimensions and Nomenclature

Table 10-1 gives the weights and dimensions of riveted gas holders. Table 10-2 gives the dimensions and painting surfaces of holders with one to five lifts.

Figures 10-3 and 10-4 identify most of the major components of water-sealed holders. The list on page 10/4 is the key to these figures.

Table 10-1 Weights and Dimensions of Riveted Water-Sealed Gas Holders

					Crown pressure, in. w.c.		Foundation loads				Load per lineal ft on tank bottom curb		Foundation	
							Rest blocks			Load on, tank bottom, psf				
Capacity, Mcf	No. of lifts	Tank dimensions, diam × height, ft	No. of stds.	Wt of holder, M lb	Max	Min	No.	Length, ft	Load on each, lb		Steel weight, lb	Wind at 20 psf proj. area	diam, ft	vol, cu yd
25	1	42.3 × 20.8	6	98	..	4.3	12	1.0	2,320	1330	380	570	44.4	61
50	1	53.3 × 25.5	8	164	..	4.9	12	1.0	3,900	1620	475	650	55.3	91
75	1	62.3 × 27.7	10	226	..	4.6	20	2.0	2,850	1750	590	680	64.3	120
75	2	51.5 × 22.3	9	210	9.8	5.8	18	2.0	4,670	1420	560	820	53.6	86
100	1	66.3 × 32.5	9	294	..	4.8	18	1.3	4,160	2060	800	870	68.4	160
100	2	62.3 × 20.0	10	251	7.8	4.7	20	2.0	5,170	1270	520	760	64.3	120
150	1	84.0 × 30.0	10	404	..	4.4	20	1.3	4,840	1900	790	940	86.3	245
150	2	68.7 × 24.2	10	324	8.0	4.8	20	2.0	6,390	1540	635	1020	71.3	145
200	2	76.3 × 25.6	10	399	7.9	4.8	20	2.0	7,760	1620	710	1020	78.3	173
250	2	82.3 × 27.3	10	470	7.7	4.7	20	2.0	7,680	1730	800	1090	84.4	198
300	2	84.0 × 31.4	10	553	8.3	5.1	20	2.0	8,810	1990	970	1400	86.7	250
500	2	103.7 × 33.5	12	840	7.9	5.0	24	2.0	10,060	2120	1250	1290	106.0	363
500	2	109.0 × 30.5	12	833	7.1	4.5	24	2.0	10,400	1930	1120	1020	111.7	402
500	3	96.5 × 27.3	10	810	11.0	5.3	20	3.0	15,740	1730	1090	1670	99.3	320
750	3	108.0 × 32.0	12	1085	11.1	5.5	24	3.5	16,920	2020	1380	1960	110.7	395
1,000	3	124.5 × 31.7	12	1376	10.9	5.3	24	3.5	21,880	2000	1480	1670	127.5	517
1,500	3	143.7 × 36.5	15	1957	10.4	5.1	30	5.0	22,170	2260	1940	1830	146.5	725
1,500	4	134.0 × 32.0	14	1908	14.3	5.7	28	5.0	29,030	2220	1840	2400	136.7	639
2,000	4	143.7 × 36.5	15	2375	14.2	5.7	30	5.3	30,960	2310	2260	2950	146.5	730
3,000	4	171.8 × 37.5	18	3258	13.0	5.4	36	5.3	33,860	2370	2650	2600	174.8	1018
5,000	5	200.0 × 37.5	21	4832	14.9	5.8	42	7.0	45,760	2370	3290	3180	203.3	1368
6,000	5	218.3 × 37.7	22	5466	13.7	5.5	44	7.5	47,760	2370	3660	2930	221.7	1617
10,000	5	264.7 × 42.0	28	8800	13.5	5.5	84	7.5	38,000	2660	5000	3000	267.0	2281

(Cruse-Kemper Co., Ambler, Pa.)

Key to Figs. 10-3 and 10-4

1. Foundation
2. Inlet and outlet pipes (cast iron)
3. 1 in. dry grout (riveted holders only)
4. Tank bottom
5. Outer course tank bottom
6. Lowering screw plates (riveted bottoms only)
7. Tank bottom curb
8. Tank shell
9. Standard supports
10. Standard support angle
11. Tank top curb
12. Deck (or balcony)
13. Deck curb or outer top curb
14. Deck braces
15. Tank rest blocks
16. Tank guide rails
17. Inlet and outlet riser pipes (steel)
18. Overflow swing joint
19. Overflow shield
20. Overflow discharge
21. Steel center pier
22. Permanent wood frame (crown support)
23. Oil skimmer
24. Inner section shell
25. 2nd or middle section shell
26. Outer section shell
27. Bottom rollers
28. Outer section bottom curb
29. Cup channels
30. Cup rest blocks
31. Cup sheets
32. Inner section top curb
33. Grip channels
34. Grip rest blocks
35. Grip sheets
36. Dam sheets
37. Inner section legs or vertical stiffeners
38. Inner section leg bracket
39. Legs or vertical stiffeners (mid. & out. sect.)
40. Crown outer course
41. Crown secondary curb
42. Inner section carriage
43. Inner section carriage brace rod
44. Middle section carriage
45. Outer section carriage
46. Radial rollers
47. Tangential rollers
48. Livesey seal
49. Livesey cover
50. Livesey seal dipping sheet
51. Livesey seal by-pass connections
52. Hand railings
53. Guide frame standards
54. Guide frame girders
55. Guide frame diagonals
56. Guide frame wind ties
57. Guide frame wind tie cantilever braces
58. Tank stairway
59. Guide frame stairway
60. Guide frame stair platform on standard
61. Guide frame stair platform on girders
62. Guide frame ladder
63. Crown landing platform
64. Middle section landing platform
65. Tank heating piping (deep discharge)
66. Tank heating piping (surface discharge)
67. Heating pipe riser ladder
68. Riser platform
69. Piping: siphon & hose (cup heating system)
70. Drip pot
71. Drip pump
72. Test cock
73. Gas main disconnect piece

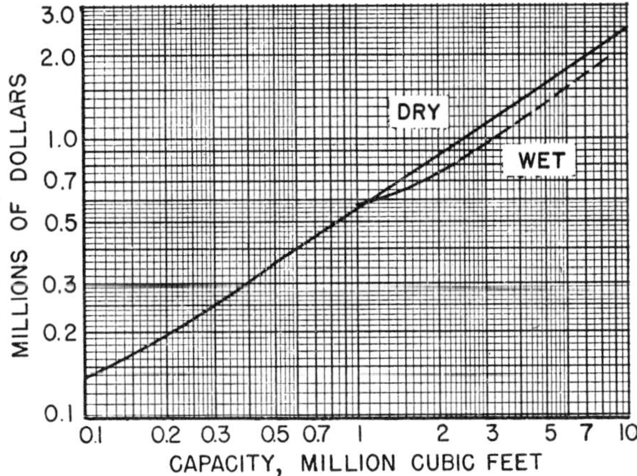

Fig. 10-1 Construction costs of low pressure gas holders.[1] Based on Construction Cost Index of 323.6 (1926 = 100).

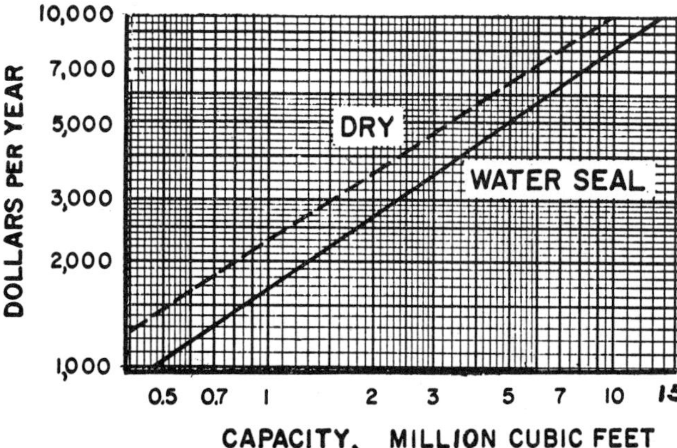

Fig. 10-2 Annual maintenance costs of low pressure gas holders.[1] Based on $2.50 labor rate (includes 20 per cent indirect labor), 1.5 cents per kwh, 3 cents per therm.

Livesey Seal. This arrangement, Fig. 10-5 (also item 48 in Figs. 10-3 and 10-4), was intended to provide a method of inspecting for corrosion of the inlet and outlet pipes of a water-sealed gas holder.[2] The holder was lowered until the skirt of the attached seal was below the water line, thus permitting removal of the seal cover without escape of the gas between the water line and the crown. This procedure probably would not be undertaken today.

The more recent purpose of the seal was to prevent the holder from landing inadvertently. However, where a pusher or exhauster draws gas from a holder, it is possible for the suction from the exhauster or pusher to pull the water over the top of the outlet pipe, thus flooding the gas mains. Livesey seals are also fitted with an external by-pass which is necessary for the initial inflation and final deflation of a holder.

Heating

It is necessary to supply heat to holders in cold weather to prevent the freezing of the holder tank water.

Heating with Steam. Steam is obtained either from the plant steam system or, for isolated holders, by specially provided boilers. The steam is usually supplied to the holder thru encircling loop piping from which vertical risers, attached to the guide frame, ascend to the required height. At suitable intervals on the risers there are takeoffs for flexible

Table 10-2 Dimensions and Painting Surfaces for Water-Sealed Gas Holders[1]

	One lift	Two lift								Three lift							
Capacity, Mcf	100	100	150	200	250	300	350	400	500	300	400	500	600	750	1,000	1,500	2,000
General Dimensions, ft:																	
1st section diameter	69.5	54.0	62.3	67.7	76.0	79.0	85.5	90.0	96.8	71.5	74.0	88.0	88.0	104.0	114.5	136.5	152.0
Height	26.5	22.0	24.5	28.0	28.0	30.5	31.0	32.0	34.0	24.6	31.0	27.5	32.7	30.0	32.0	34.2	36.5
2nd section diameter	..	56.0	64.5	69.7	78.0	81.0	87.5	92.0	99.5	73.5	76.0	90.0	90.0	106.0	116.9	139.0	154.5
Height	..	22.0	24.5	28.0	28.0	30.5	31.0	32.0	34.0	24.6	31.0	27.5	32.7	30.0	32.0	34.2	36.5
3rd section diameter	..	..	..	..	..	..	..	..	..	75.5	78.0	92.0	92.0	108.0	119.3	141.5	157.0
Height	..	..	..	..	..	..	..	..	..	24.6	31.0	27.5	32.7	30.0	32.0	34.2	36.5
Steel tank diameter	72.3	58.5	67.0	72.3	80.5	83.5	90.0	94.5	102.0	78.0	80.5	94.5	94.5	110.5	122.0	144.0	160.0
Height	27.7	23.0	25.5	29.0	29.0	31.5	32.0	33.0	35.0	25.7	32.0	28.5	33.7	31.0	33.0	35.3	37.5
No. columns	8	9	9	10	10	10	10	10	12	10	10	10	10	12	12	14	16
Areas, sq ft:																	
Steel tank bottom	4,125	2,688	3,527	4,100	5,217	5,480	6,360	6,947	8,220	4,778	5,090	6,950	6,947	9,600	11,690	16,286	20,110
Steel tank shell	6,280	4,229	5,367	6,582	7,335	8,265	9,050	9,797	11,200	6,288	8,083	8,465	10,019	10,760	12,650	15,985	18,350
Tank walk	1,290	1,570	1,960	2,036	2,050	2,065	2,280	2,300	2,480	2,000	2,050	2,300	2,300	2,900	3,430	5,100	8,400
Tank shell 4 ft inside	904	735	840	904	1,000	1,050	1,135	1,184	1,280	980	1,000	1,185	1,184	1,390	1,535	1,810	2,010
Guide frame	2,530	4,800	4,910	6,242	6,580	7,320	8,700	9,200	10,000	12,000	13,300	10,000	13,600	19,030	19,376	20,000	38,000
Stairway and ladder	340	560	125	130	432	520	1,490	1,500	1,700	1,600	1,700	1,890	2,000	2,810	2,000	3,330	3,600
Crown	3,795	2,410	3,166	3,716	4,626	4,995	5,845	6,383	7,520	4,100	4,410	6,190	6,190	8,800	10,355	14,495	18,900
1st section	5,786	4,290	5,455	6,650	7,460	8,365	9,190	10,020	11,550	6,374	7,990	8,490	9,610	10,850	12,655	15,475	19,000
2nd section	..	4,218	5,760	7,210	8,070	9,010	9,890	10,769	12,200	7,270	8,743	10,080	10,850	12,700	14,605	17,487	21,800
3rd section	..	..	..	..	..	..	..	..	..	7,230	8,624	9,370	10,800	11,850	14,216	17,302	21,000
Carriages	180	350	350	400	400	600	600	800	1,000	800	1,000	1,000	1,200	1,200	1,200	1,450	1,700
Total area, sq ft:	25,230	25,850	31,460	37,970	43,170	47,670	54,540	58,900	67,150	53,420	61,990	65,920	74,700	91,890	103,710	128,720	172,870

	Four lift						Five lift					
Capacity, Mcf	1,000	1,500	2,000	2,500	3,000	4,000	3,000	4,000	5,000	6,000	8,000	10,000
General Dimensions, ft:												
1st section diameter	102.0	122.5	133.5	150.0	167.1	177.7	147.5	164.8	189.7	198.5	225.7	244.8
Height	30.0	31.0	34.8	35.0	34.0	39.7	34.7	36.7	34.7	38.3	39.5	42.0
2nd section diameter	104.0	125.2	136.2	152.5	169.7	180.3	150.0	167.3	192.6	201.0	228.7	247.8
Height	30.0	31.0	34.8	35.0	34.0	39.7	34.7	36.7	34.7	38.3	39.5	42.0
3rd section diameter	106.0	127.8	138.8	155.0	172.4	183.0	152.5	169.7	195.5	203.5	231.6	250.9
Height	30.0	31.0	34.8	35.0	34.0	39.7	34.7	36.7	34.7	38.3	39.5	42.0
4th section diameter	108.0	130.5	141.5	157.5	175.0	185.0	155.0	172.0	198.4	206.0	234.5	253.9
Height	30.0	31.0	34.8	35.0	34.0	39.7	34.7	36.7	34.7	38.3	39.5	42.0
5th section diameter	..	..	..	..	..	..	157.5	174.4	201.3	208.5	237.4	257.0
Height	..	..	..	..	..	..	34.7	36.7	34.7	38.3	39.5	42.0
Steel tank diameter	110.5	133.0	144.0	160.0	177.5	188.5	160.0	177.5	204.2	211.3	241.0	260.0
Height	31.0	32.0	36.0	36.0	35.5	41.1	36.0	38.0	36.0	39.7	41.0	43.5
No. columns	12	14	14	16	18	20	16	18	22	22	24	26
Areas, sq ft:												
Steel tank bottom	9,600	13,895	16,286	20,110	24,745	27,907	20,110	24,900	32,600	35,100	45,300	53,000
Steel tank shell	10,760	13,371	16,286	18,095	19,796	24,328	18,095	21,400	23,000	26,400	30,500	35,500
Tank walk	2,700	4,500	5,100	8,400	5,450	5,800	8,400	9,560	11,500	13,000	9,600	10,750
Tank shell 4 ft inside	1,390	1,671	1,810	2,011	2,224	2,369	2,011	2,260	2,560	2,660	3,030	3,270
Guide frame	25,650	27,000	36,180	40,851	46,900	67,000	57,600	75,400	88,000	96,000	124,000	148,000
Stairway and ladder	2,810	3,000	3,330	4,220	4,100	4,800	4,490	4,600	4,800	4,600	5,100	6,400
Crown	8,180	11,757	14,126	18,420	22,780	25,116	17,082	22,850	29,500	32,200	50,800	64,000
1st section	10,630	12,735	15,486	18,155	18,990	23,400	17,248	21,400	22,450	26,400	32,640	37,340
2nd section	11,850	14,535	17,394	21,350	21,318	25,855	20,030	25,850	26,150	31,930	40,000	45,610
3rd section	12,090	14,845	17,760	21,657	21,605	26,287	20,518	26,500	26,500	32,710	40,490	46,160
4th section	11,740	14,671	17,612	21,371	21,332	25,898	20,864	27,100	27,000	33,450	40,920	46,730
5th section	..	..	..	..	..	..	19,910	26,900	27,500	33,200	37,890	43,530
Carriages	1,500	1,800	1,800	1,800	2,500	3,500	3,000	3,420	4,000	4,180	5,300	5,300
Total area, sq ft	108,900	133,780	163,170	196,440	211,740	262,260	229,360	292,140	325,560	371,830	465,570	545,590

(Stacey Manufacturing Co.)

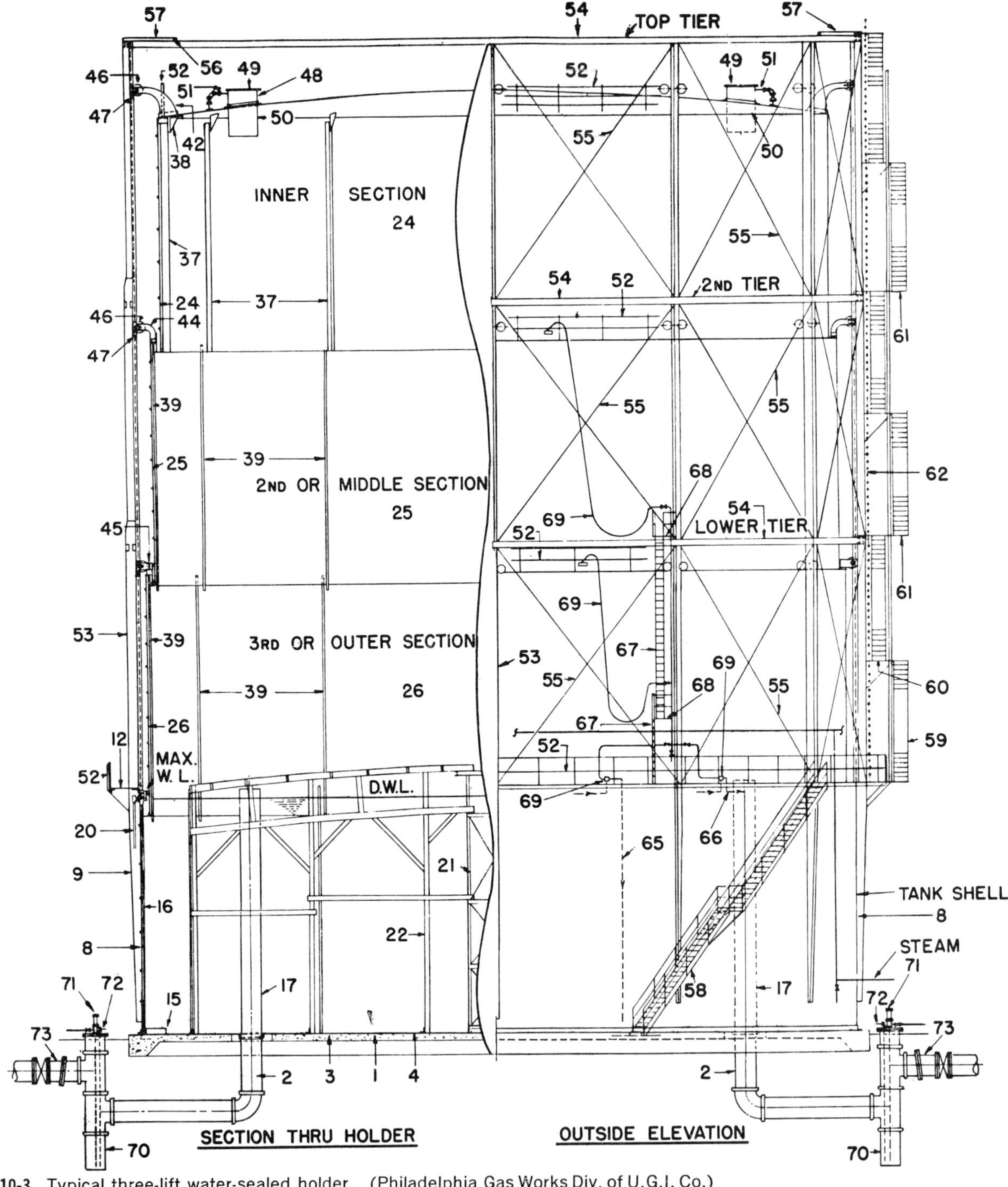

Fig. 10-3 Typical three-lift water-sealed holder. (Philadelphia Gas Works Div. of U.G.I. Co.)

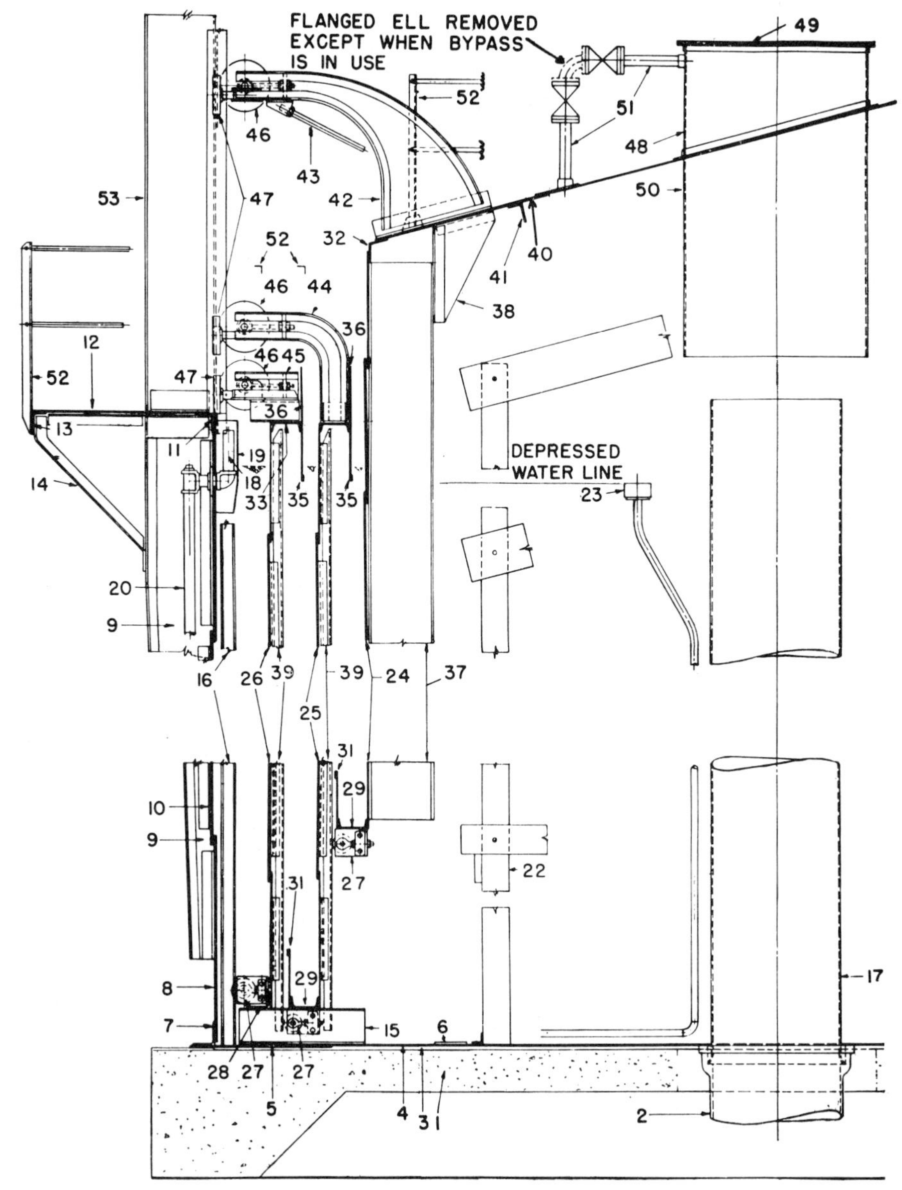

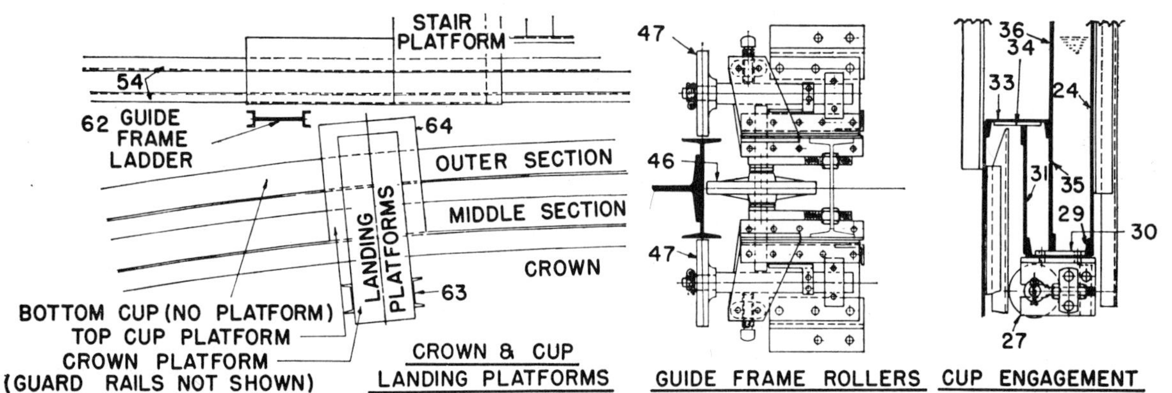

Fig. 10-4 Details of typical three-lift water-sealed holder of riveted construction. (Philadelphia Gas Works Div. of U.G.I. Co.)

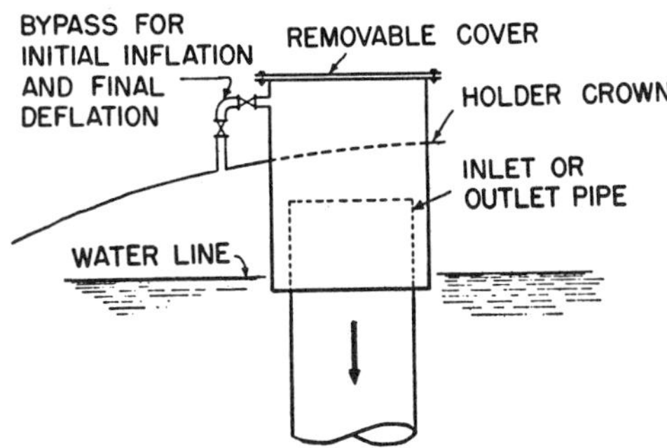

Fig. 10-5 Livesey seal and by-pass.

hoses which lead to the cups; the takeoff points should be about midway along the path traveled by the cup for which they provide steam. The steam can be introduced into the cup water by a nozzle, but since this is accompanied by an objectionable amount of noise, a steam siphon is usually used. The jets or siphons should all discharge in the same direction (e.g., clockwise) so that a continuous circumferential current will be produced.

Tank siphons are larger or more numerous than those used in each cup. There are various arrangements of the tank siphons: in some cases half of them have their suction and discharge pipes at the surface, while the other half have a suction pipe near the bottom of the tank and the discharge on the surface. A reverse arrangement, with the suction on the surface and the discharge at the bottom, seems more positive in its action. Usually only the surface jets are used in mild weather; it is necessary to start the deep water jets when the water temperature at the bottom of the tank approaches 40 F.

A failure of the steam supply to a siphon may cause vacuum sufficient to drain the water from the cups. One way of preventing this siphoning action is to fix the discharge pipes no lower than the safe water level in the cups.

Heating with Circulated Water. In these systems, water is withdrawn from the holder tank, preferably near the bottom, pumped thru heat exchangers, and returned to the holder cups and tank. The tank water in a large holder stores considerable heat for the cup heating during mild weather. The value of this heat sink probably depends upon the rapidity of the transition from summer to winter weather. In climates where there is a long, cool fall season, much of the heat of the tank water is lost by radiation. The most favored custom appears to be to start the pumps when falling air temperature reaches 35 F, and to stop them when rising air temperature returns to 35 F. For adding heat to holders in the plants, steam—either live or exhaust—or some source of waste heat may be used, such as condenser water.[3] For outlying holders, gas-fired boilers are preferred. After the temperature of the tank water (a function of location, design, and dependability) is lowered to a safe minimum—about 40 F—the water becomes merely a conveyor of heat from the boilers to the holders, and for this reason the economy of circulating a large volume of water is questionable. Circulating water systems have been designed to operate under automatic control in unattended holder stations.[4–6] In water systems, as in steam systems, the design of the system must include safeguards against siphoning water from the cups into the tank.

Compressor Cooling with Circulating Holder Water. Heat may be transferred from gas compressors to holder water (by means of a reservoir). A stand-by source of additional heat for the holder may be necessary during cold months. Conversely, a cooling tower for the compressor may be required for the times when the holder water temperature is too high.

In one system of this type there are two 3 MMCF holders and four gas engine-driven compressors, with a total of 3845 brake horsepower; in another arrangement, one 5 MMCF holder and three compressors, with a total of 3200 brake horsepower. In both systems, the water for holder heating purposes is drawn from the bottom center of the tank, and the water for cooling the compressors is drawn from the top of the outer annulus, pumped thru the engine and compressor jackets, and discharged into the suction pipe of the holder heating system.

It is important to note that this amount of horsepower can be provided for in these two systems only because their operation is confined to winter loads; if they were to operate all year, the tank water would become overheated.

Corrosion

The water-sealed holder is exposed to many oxidation sources, including the atmosphere, the contained gas, the sealing water, and the ground upon which the holder is situated

Atmospheric. Breaks in the protective coating allow metal to combine with the active gases and vapors present in the atmosphere. Edges of plates are especially prone to these breaks. Similarly, rivet head corrosion is accelerated[7a] by greater exposure, oxide inclusions, overheating during construction, loose surface scale, and metal heterogeneity. The overall effect of these may tend to render the rivet heads very anodic to the rest of the surface.

Corrosion occurs where rain water collects and does not evaporate or drain away. Such points are found on composite guideframe structures where the rivet spacing permits crevices. These openings expand as the exposed metal rusts.

Exposure to the Contained Gas. Certain surfaces, notably the underside of the crown and the interior of the cups and grips, are corroded by the water vapor content of the contained gas, as well as by H_2S and cyanogen when these are present. Also, these surfaces cannot be painted since they are not accessible.

Sealing Water. The corrosive action of the sealing water has several causes:

1. *Water Line Corrosion*—Water with pH less than 6.8 tends to produce pitting and the sometimes large eroded area at or just below a water line which has an air exposure. Such locations are in the water tank and the cup structures. Hydrogen sulfide tends to intensify this effect.[7b]

Corrosion just below the air-water-metal contact results from differential aeration. Surfaces wetted by "creep" at and above the water line are very well aerated and consequently cathodic to the surfaces which are in contact with less well aerated water below the water line.

With some waters and perhaps some paint coatings, there is a kind of water line corrosion[7c] occurring exactly at the water line. Here the metal is highly anodic, and the penetration is usually deep and rapid. These crannies retain the corrosion products formed between a broken film of paint pulled

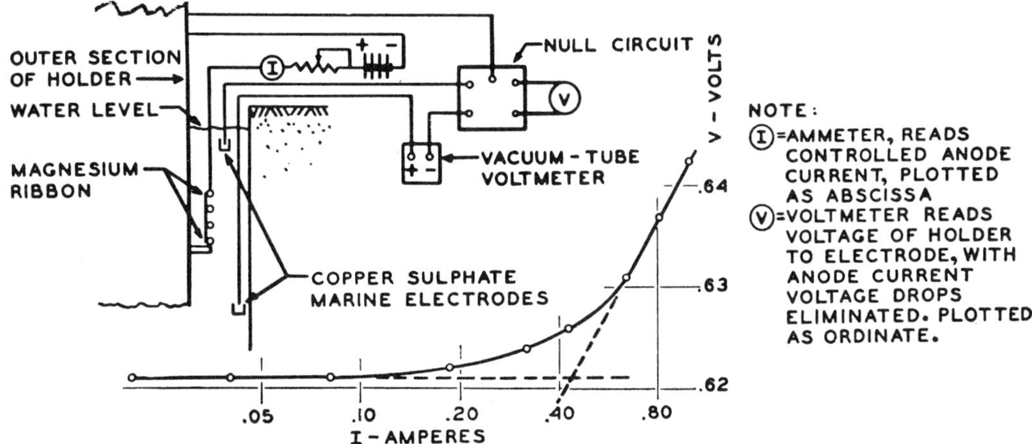

Fig. 10-6 Engineering data to determine cathodic currents for polarization. (Philadelphia Gas Works Div. of U.G.I. Co.)

out by the surface tension of the water meniscus and the metal.

2. *Corrosion on Lifts*—Surfaces alternately exposed to tank water and air show effects similar to (but more pronounced than) those observed on surfaces exposed to weather. This corrosion often appears as tubercles,[8a] concealing a pit in the metal which sets up a closely coupled oxygen concentration cell. This condition results from a small anodic area of bare iron and a surrounding cathodic area of permeable paint over sound mill-scale. Since rivet heads tend to hold water, paint on them fails before the paint on adjacent flat surfaces. Good surface preparation and a water resistant paint should practically eliminate this form of corrosion.

3. *Deep Water Corrosion*—Corrosion rarely occurs in the deeply submerged portions of a holder, except when cyanogen or hydrogen sulfide is present. Corrosion, however, may build up between the metal plates and the inside stiffeners of the lifts, which have been forced apart between the attaching rivets.[8b]

Because of its very low concentration in most manufactured gases, cyanogen causes little corrosion. Hydrogen sulfide, however, because of its activity in the presence of iron, causes heavy losses in holder structures, particularly in relief holders containing unpurified gas. Hydrogen sulfide is a contaminant to some degree in all manufactured gas, in natural gases from some fields (before purification), and in much of the refinery gas. In the presence of moisture, hydrogen sulfide combines directly with the iron, probably by removing the iron ions from solution as insoluble sulfides.

Biological Activity. Water in pit holders may accumulate more debris than in tank holders, and the resulting putrefaction generates H_2S which fouls the stored gas.

During most of the year when the water is quiescent, biological activity is taking place to some degree. The hydrogen sulfide, formed in the debris where the oxygen content is low or absent, diffuses thru the water and is partly used up by the submerged metal. The rest saturates the cold deeper body of water and hardly contaminates the gas[9] until the water begins to circulate. In the fall, the sharp temperature drop at night cools the surface layers below the temperature of the deeper body of water and an inversion* of the water occurs; this inversion frees the accumulated hydrogen sulfide which diffuses into and fouls the stored gas; the water often looks milky and has an odor strongly suggesting sewage. In a short time the water turns black because of a suspension of iron sulfide. Ordinary putrefactive organisms may be active in hard water and where organic matter or sulfates are present. The above agents are very common in many soils.[8c,10,11] "Iron" bacteria[9] also form tubercles on iron, under which the anaerobic bacteria may continue the work of pitting and destruction.

Ground Conditions. Serious corrosion occurs on large underside areas of holder bottoms. This attack is often by anaerobic bacteria (sulfate reducing), especially under the following conditions:

1. Frequent entry of ground water between the steel tank bottom and its base. Note that ground water level varies.
2. Water-borne sulfates or sulfites, or both, and organic matter in contact with the steel.
3. Water is depleted of oxygen, making it possible for the sulfate reducing bacteria to form hydrogen sulfide. The oxygen from the aerated ground waters combines with the polarizing hydrogen on the steel. This latter action also furthers the loss of metal.

Oxidation of Sulfides.[12] Oxygen from ground water and sulfides in the presence of certain organisms yield sulfates. These organisms, called thiobacilli,[12] create an acidic condition which furthers corrosion.

Control of Biological Activity. Chemical treatment of holder water has been advanced as the most effective means of controlling biological activity in holder tanks.

Electrolysis. Holders near d-c power sources, or supplied by gas lines running across such sources, may be affected by stray current electrolysis.

Mitigation techniques include: (1) bonding of gas lines to negative returns of electric system; (2) applying cathodic protection to the holder shell; and (3) insulating joints in gas lines.

Figure 10-6 shows the data obtained to determine the cathodic currents required to establish polarization on a holder. Note that a definite break occurs at 0.43 amp, indicating the point of transition at which the steel surface is covered with a hydrogen film. Consequently, it is impossible for metallic ions to go into solution.

* Confined to pit holders, since convection currents are common in other holders.

Discoloration of Holder Paint and Its Prevention

The reddish-brown stain which discolors the paint on the lifts of water-sealed gas holders has plagued gas operators for many years, particularly with regard to good public relations. A method of preventing discoloration follows.

The Sodium Bicarbonate–Sodium Dichromate Method. J. R. Skeen[13] has reported that holder waters contain micro-organisms similar to "iron bacteria," which discolor paint. Briefly, the organisms absorb dissolved ferrous iron and oxidize it to the insoluble red ferric hydrate. The presence of organic matter in the water accelerates the reaction. The organisms then become embrittled and form sediments, while the iron-laden capsules form floating scums which adhere to the moving lifts.

Skeen has reported that a combination of two salts, sodium bicarbonate (0.03 to 0.05 per cent by weight) and sodium dichromate (0.0024 to 0.005 per cent by weight) in dilute solution in the tank water, suppresses the growth of the bacteria.

When CO_2 is present, soda ash is used since it reacts with the carbon dioxide dissolved in the water to form sodium bicarbonate. The water that results from the addition of these salts is nearly neutral (pH 7.0 $\pm$ 0.1). The tank waters are sampled twice each year at three points: near the surface, midway, and near the bottom. Salt addition (to maximum concentration) was found necessary usually only once a year. During a ten-year period, the average salt requirement per year for a 3 MMCF holder was 655 lb of sodium dichromate and 1200 lb of soda ash, and for a 5 MMCF holder, 980 lb of sodium dichromate and 1670 lb of soda ash. The addition of the sodium dichromate presents no problem because it is easily soluble and will not attack the paint; but the soda ash must be added slowly, about 200 lb each week, and conducted inside the lifts to avoid harm to the paint. In the examples discussed here, this is accomplished by the use of a permanent oil skimmer. After dilution by rain or by steaming, the water in the cups is replaced with water from the tank thru a pump provided for this purpose. The use of these two salts in waters containing cyanogen in excess of 10 ppm is not recommended.

Paints containing lead pigments must not be used for a finish coat, since lead reacts with sodium dichromate. However, one finish coat of a nonlead paint has been found sufficient protection for a lead prime coat. While the protection has for its primary purpose the prevention of discoloration of the paint, it also helps to reduce corrosion.

Disposal of Holder Tank Water*

Dilution with fresh water and the addition of a masking agent are usually adequate treatment for excess holder water. Larger quantities require aeration towers, chemical treatment, or both.

Safety dictates that the concentration of explosive gas, in the sewer lines charged with holder effluent, be monitored to ensure that only a negligible fraction of the lower explosive limit is reached. For example, a tower scrubber, 66 in. diam by 22 ft high was used to reduce the dissolved gases in 350 gpm of holder water discharge; about 525 cfm of air was required. The sewer atmosphere was at 55 per cent and 4 per cent of the lower explosive limit, before and after aeration, respectively, and the dissolved O_2 was conversely increased from under 2 to about 8 ppm. A surface tension depressant, 2 to 5 ppm, was used to control foaming.

Disposal activities should be coordinated with pertinent regulatory agencies. For example, authorities required a reduction of the sodium dichromate† concentration from 50 to 5 ppm to assure the function of a sewage disposal plant. Ferrous sulfate (copperas) and hydrated lime were used to reduce the chromate. The lime acted to increase the pH and helped to settle the precipitate. It was noted, however, that a better plan would have been to reduce the dichromate by attrition. On the other hand, the presence of 500 ppm of sodium bicarbonate was not considered objectionable to either streams or disposal plants.

* This material was prepared primarily for manufactured gas storage.

† An oxidizing agent such as sodium hypochlorite was sometimes used.

Welded Construction

The chief advantages of this construction method are that fabrication and erection are facilitated and plate warpage is decreased. Corrosion is also reduced thru continuous welding of vertical stiffeners to the inside of the lift plates. This is also true of the guide frame if continuous welding is used.

The soundest engineering approach to the field welding of such structures is in the use of quality materials and construction, including x-raying and stress-relieving. The additional cost should be compensated for by the difference in weight of steel used and in the smaller volume of weld metal used between the plates. The latter is important because the amount of weld metal increases nearly as the square of plate thickness.

PISTON TYPE HOLDERS

Pressures available in these holders may vary up to a practical limit of about 20 in. of water. The pressure produced by the dead weight of the piston may be increased by adding weights, usually concrete blocks, on the piston.

The additional weight necessary to produce a desired increase in pressure may be calculated from the equation:

$$W = 5.204PA \tag{1}$$

where: P = pressure increase, in. w.c.
A = piston area, sq ft
W = weight, lb

These holders require comparatively light foundations; the interior of the shell may be readily inspected without removal from service.

Maintenance of the seal ensures that the space above the piston and within the tank shell does not accumulate enough gas for an explosive mixture; a detection device may be used to indicate the hazard.

Koppers Co. Inc., B-H Waterless Holder (M.A.N. Type)

This waterless holder has a polygonal shell and utilizes a rubbing bar in conjunction with a liquid tar to seal (Fig. 10-7) the gas contained below the piston of the holder. Table 10-3 gives dimensional data. The depth of the tar in the holder cup

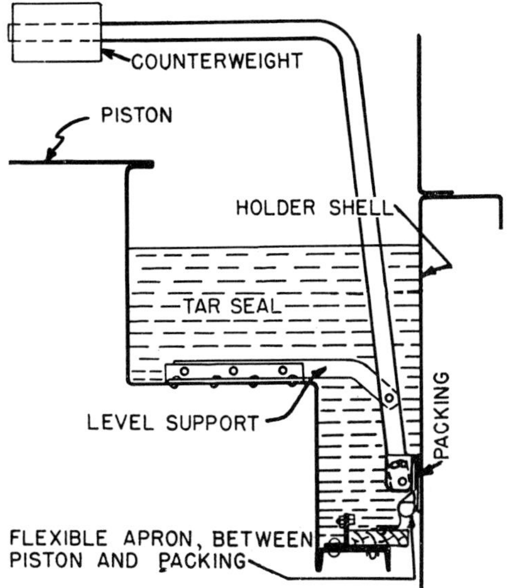

Fig. 10-7 Seal mechanism of B-H Koppers M.A.N. waterless holder.

Table 10-3 Dimensions and Capacities of Koppers Company, Inc., B-H Waterless Holders

Capacity, MMCF	Diam, ft	Lift, ft	Total height, ft	No. of sides	Chord, feet	Area, sq ft
0.1	45	70.0	78	8	17.2	1,428
.2	56	88.0	100	10	17.2	2,276
.3	63	96.0	108	10	20.2	3,140
.5	82	102.0	115	10	25.2	4,886
.6	82	124.0	137	10	25.2	4,886
.75	98	105.5	138	10	30.28	7,056
.8	98	112.6	145	10	30.28	7,056
0.9	98	126.8	160	10	30.28	7,056
1.0	98	142.2	174.7	10	30.28	7,056
1.5	114	156.2	185.5	12	29.51	9,747
2.0	124	172.5	206.3	14	27.59	11,674
3.0	140	200.3	334.3	16	27.31	15,001
4.0	153.8	221.0	258	16	30.00	18,100
5.0	172	222.2	263.8	18	29.87	22,766
6.0	188	222.3	265.2	20	29.41	27,305
7.0	188	259.0	301.8	20	29.41	27,305
8.0	200	261.9	306.3	20	31.29	30,902
9.0	200	292.0	337	20	31.29	30,902
10.0	218.1	272.9	320	22	31.05	36,868
11.0	218.1	298.0	346	22	31.05	36,868
12.0	218.1	326.9	375.1	22	31.05	36,868
15.0	254.3	299.2	349	28	28.47	50,370
17.0	254.3	339.0	389	28	28.47	50,370
20.0	280.3	329.5	415.7	28	31.38	61,199
22.0	280.3	360.0	445.7	28	31.38	61,199

is maintained by circulation. Pumps transfer the tar from the dam at the bottom of the holder to the cup. An indicator records each time these pumps operate; seal trouble is thus noted as a variation in the number of pump runs.

Viscosity of the tar varies with seasonal temperatures and is increased if some of the lighter oils are absorbed by the gas. These viscosity variations are controlled by the addition of thinners. Thinners should be miscible in all proportions with oil, have a high boiling point (above 400 F), be completely fluid at the lowest temperatures ever experienced in the area, and be free from naphthalene crystals and residue. During the warmer months, it is good practice to keep the tar level in the cup as much as 2 in. lower by regulating the amount of tar in the system. This prevents excess splashing of the then more fluid tar on the piston.

Moisture, condensed on the holder shell, will drop into the tar dam. Occasionally, this water will float on the tar surface, but it usually mixes intimately with the tar to form an emulsion. After stopping the pump and allowing the mixture to separate, the water can sometimes be removed with skimmers. However, if this fails, it is necessary to dehydrate the mixture externally or to replace it with dry tar. Tar losses are reduced if the mixture is placed in a separate tank and given adequate time for separation.

In some areas where dry natural gas is distributed, a petroleum base oil heavier than water is used instead of tar, thus preventing water-rot of the canvas trough.

Maintenance consists of regular inspection, replacement of parts when necessary, and periodic lubrication. Procedures to be followed in case of electrical failure should be prepared. Emergency gas engine-driven electrical power could be provided.

Stacey-Klonne Holder

This equipment consists of a piston (free to turn) within a vertical cylindrical shell. It is sealed by a dilatable steel ring attached to the piston. The packing, composed of several layers of heavy cotton fabric and synthetic rubber material, is fastened to the ring. A series of weights acting thru levers expand the ring outward, forcing the packing against the shell. (See Fig. 10-8.) The space between the dilatable ring

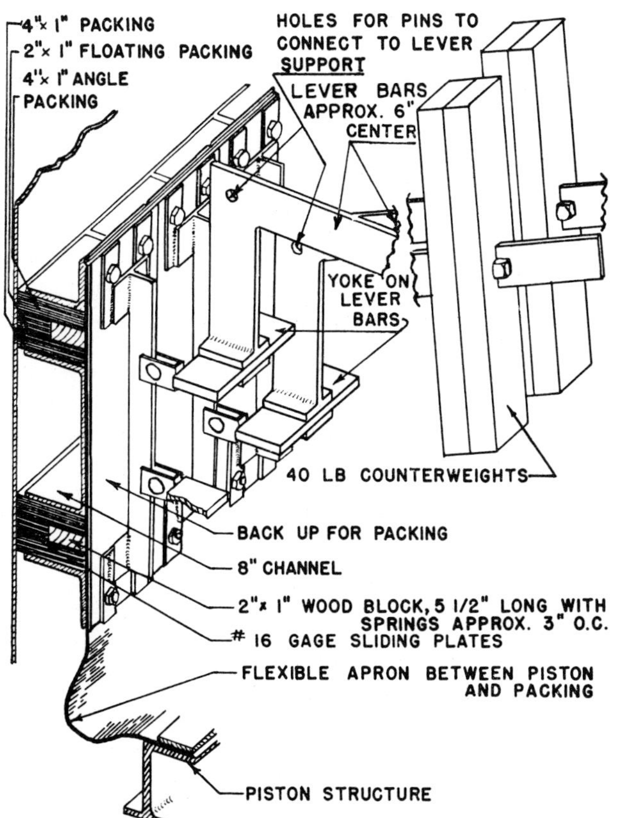

Fig. 10-8 Seal assembly with lever bars and counterweights for the improved Stacey-Klonne dry seal holder. The "back up for packing" is a dilatable fluted steel ring.

Table 10-4 General Dimension and Foundation Data for Stacey-Klonne Dry Seal Gas Holders

Holder capacity, MCF	Number of columns	Holder dimensions Diam, ft	Holder dimensions Height, ft	Radius outside foundation, ft	Max load on bottom curb, lb per ft of circum.	Max uplift per column, lb	Concrete in holder foundation, cu yd
25	6	31.4	45.0	16.0	3,000	18,000	50
30	6	31.4	51.8	16.0	3,540	19,500	50
50	7	36.6	61.5	18.8	3,800	26,500	75
100	9	47.0	73.0	24.0	4,260	30,200	90
150	10	52.3	86.8	26.6	5,150	37,300	120
200	11	57.5	95.5	29.2	5,850	40,600	140
300	12	62.7	117.9	32.1	7,440	53,300	190
500	14	73.2	140.3	37.3	7,530	54,700	230
750	16	83.6	159.5	42.6	9,880	55,300	280
1,000	20	104.5	142.3	52.3	9,550	41,600	320
1,500	22	115.0	170.6	58.2	11,500	52,500	394
2,000	24	125.4	189.4	62.5	11,400	66,000	440
3,000	26	135.9	241.1	68.9	13,000	89,300	670
4,000	30	156.8	239.4	79.5	14,500	80,700	685
5,000	32	167.2	261.0	83.6	1,500	80,100	770
6,000	34	177.7	276.7	89.8	17,050	93,100	897
7,000	36	188.1	287.3	95.1	16,900	93,500	1040
8,000	36	188.1	326.3	95.1	20,000	108,500	1140
10,000	40	209.0	330.5	105.5	20,800	107,000	1220
12,500	42	219.5	373.0	110.8	23,000	100,000	1300
15,000	44	229.9	405.0	116.0	27,600	107,700	1395
20,000	48	250.8	451.0	126.4	29,700	113,550	1475

and the piston is sealed by a flexible apron which is a composition of fabric and synthetic rubber. The packing is lubricated with a special grease.

Vents, equally spaced around the shell, are provided near the top of the holder to release gas if the piston should be raised to the maximum height. The vents are closed normally with hinged plates which permit the seal to pass over the vent openings without loss of lubricant. However, if the seal should pass above the vents, the vents are automatically opened by the difference in pressure between the stored gas and the atmosphere.

Automatic lubrication may be by means of the *Farval System*, in which the lubricant is stored on top of the holder and fed to the top of the packing ring by means of pipe and a flexible hose mounted on a reel. In addition to the automatic lubricating system, an automatic de-icing system is available which sprays alcohol or glycerine lightly on the shell just below the packing ring. Such a system is required only when storing a highly saturated gas.

Flooding connections can be provided to offset damage from buoyancy resulting if high water surrounds the holder.

Details of dimensions, capacity, and foundation are shown in Table 10-4.

Wiggins Holder

This holder differs from other waterless holders in using a seal which consists of fabric impregnated and coated with synthetic rubber to seal the annular space between piston and shell. The seal is distended with an upward loop when gas is confined beneath the piston.

The area above the piston is constantly ventilated and accessible by means of entry ports thru the gas shell at any height above the piston. Clearances and tolerances for this holder are so large that even a shifting of foundation would not impede operating action.

The range of available capacities is smaller than those of the Stacey-Klonne and Koppers B-H holders. Table 10-5 gives dimensional data.

The world's largest Wiggins holder was started up in 1960 at Long Beach, Calif. The holder capacity of 5 MMCF of gas provides for fluctuations in demand.

Table 10-5 Basic Wiggins Gas Holder Dimensions

Holder capacity, MCF	Diam, ft	Height, ft
50	44.0	41.4
150	66.6	58.0
200	79.8	51.3
300	79.8	74.7
500	94.0	88.6
750	106.5	103.7
1000	119.7	109.7
1500	131.8	123.0
2000	152.8	135.0
3000	179.4	146.8
4000	192.8	171.0
5000	211.0	168.0

INSPECTION AND MAINTENANCE

Scheduling inspections is a matter of either individual company policy or Public Utilities Commission Codes;[14] therefore, the following discussion is indicative only of inspection and maintenance procedures.

Basic inspection and maintenance are performed by the holder attendant. Additional inspections are made periodically by inspectors and holder experts. The following is a general breakdown of inspections in terms of operation, measurement, and visual checks:

Operation. Elevator systems (including alarms), heaters (in cold weather), holder volume alarm, hand line, starting switches for fan pump and skimmer, and the overflow and skimmer devices must be run to ascertain their operability.

Measurement. Determine piston tilt by taking simultaneous readings of piston level at 180 degree intervals. Check volume indicator reading by comparing with measured internal piston height. Tar depth in seals, temperature of tar sealant, and the depth of sediment in cups are also measured.

Visual. Alignment and operation of the moving portions of the holder involve checking the following: guide roller lubrication, action, and tension; clearances between guide block toes and holder shell; rubbing bar counterweights for freedom of motion and clearance; leveling or dump pipe in each tar cup segment for clearance; and the guide frame.

Corrosion prevention primarily involves checking the condition of the paint surface. In this regard, the tank (especially at the water line), the guide frame, lifts, cups, and stairway are checked. The condition of the surfaces normally under water, along with the bottom rollers of the outer lifts, should be checked when the holder is in its uppermost position. Poor drainage causes corrosion; checkpoints include bottom curb angle of tank, periphery of holder bottom, tank deck, and stairway platforms. The amount of tank sediment should be reported; the surface of the tank water can be skimmed clean.

The specific items that should be inspected for gas leakage include pipe tunnels, holder connection valves, tank sides, seal canvas around the piston, and the tank bottom. The latter is generally indicated by water seepage between the foundation and the bottom curb.

Tar leaks may be indicated on the outside of the holder shell and on tar lines. The condition of the tar dam floats which start the tar pumps should also be checked.

Accident prevention measures include good housekeeping and fence maintenance.

Checks are made on miscellaneous items, including electric lines and equipment, concrete foundation spalling, external lights, and indicator cable and clamps.

Welding on the active holder proper or on its components is permissible as long as the possibility of air combining with the holder gas is extremely remote. Any such holder maintenance activities should be supervised by a competent authority.

REFERENCES

1. Stone and Webster Eng. Corp. *Gas Storage at the Point of Use.* New York, A.G.A., 1957.
2. Meade, A. *Modern Gasworks Practice,* p. 453. New York Van Nostrand, 1916.
3. Rogers, P. C. "Dependable Gas Holder." *Am. Gas J.* 125: 365, Oct. 12, 1926.
4. Putnam, A. A. "District Holder Heating Plant." *Am. Gas J.* 165: 29, July 1946.
5. Campbell, D. R. "Automatic Heating of a 1500 m., 3 Lift Gas Holder." *Am. Gas J.* 165: 29–30, July 1946.
6. Facey, P. G. "Unattended Holder Station at East Hampton, Mass." *Am. Gas J.* 165: 37–60, July 1946. Also in *Gas Age* 98: 30+, July 11, 1946.
7. Evans, U. R. *Metallic Corrosion, Passivity and Protection.* New York, Longmans, Green, 1946.
   a. "Honeycombing," p. 511.
   b. "Water Seals of Gas Holders," p. 334.
   c. "Box," p. 552.
8. Speller, F. N. *Corrosion, Causes and Prevention,* 3d ed. New York, McGraw-Hill, 1951.
   a. "Tuberculation," p. 218.
   b. p. 645(h).
   c. "Anaerobic Microorganisms," p. 208.
9. "Generation of Hydrogen Sulphide in Water-Sealed Gasholders." *Gas World* 103: 446, Nov. 9, 1935.
10. Hadley, R. F. "Studies in Microbiological Anaerobic Corrosion." *A.G.A. Proc.* 1940: 764–88.
11. "Corrosion of Iron Pipes." *Science* (*Science News Suppl.*) 96: 10-12, Oct. 2, 1942.
12. Beckwith, T. D. "Corrosion of Iron by Biological Oxidation of Sulfur and Sulfides." *Gas* 21: 47–8, Dec. 1945.
13. Skeen, J. R. "Control of Holder Discoloration." *Pa. Gas Assn. Proc.* 1938: 86–103.
14. Calif. Public Utilities Commission. "Rules Governing the Design, Construction, Operation, Maintenance and Inspection of Gas Holders and Liquid Hydrocarbon Vessels." (General Order 94-A) (In its *General Orders,* p. 133–53. San Francisco, 1952).

Chapter 2
High Pressure Holders

by B. C. White

SPHERES AND CYLINDRICAL TANKS

Large tanks have been used for high pressure storage for many years. They are called "bullets" in many companies and are usually operated at pressures between 50 and 60 psi, although design operating pressures of 100 psi are practical.

The ASME Unfired Pressure Vessel Code applies to construction of high pressure storage holders. This code requires stress-relieving after welding for carbon steel plates over 1.25 in. thick, and for high strength materials 5/8 in. or over. Design is thus limited by material thickness, since it is impractical to do stress-relieving in the field.

Capacities of horizontal and vertical high-pressure cylinders are shown in Tables 10-6 and 10-7, respectively. Spherical tanks (Table 10-8) require shell thicknesses less than those needed for a cylindrical tank, but this saving is somewhat offset by higher construction costs.

Construction costs* and maintenance costs† for both spheres and cylinders are shown on Figs. 10-9 and 10-10, respectively.

Table 10-6 Capacity of Horizontal High Pressure Gas Holders at Various Pressures

Diam × length, ft	Volume, MCF	At 30 psig, MCF	At 40 psig, MCF	At 50 psig, MCF	At 60 psig, MCF
18.0 × 45.3	10	20.4	27.2	34	40.8
24.0 × 63.1	25	51	68	85	102
24.0 × 118.5	50	102	136	170	204
27.0 × 140.0	75	153	204	255	306
32.0 × 135.0	100	204	272	340	408
32.0 × 197.2	150	306	408	510	612

(Stacey Manufacturing Co.)

BOTTLE- AND PIPE-TYPE UNDERGROUND HOLDERS

High pressure bottle- or pipe-type holders for natural gas have been used, particularly for small to moderate requirements. Installations have varied in capacity from 1.5 to 120 MMCF.

This type of high pressure gas storage has lower initial costs than either low pressure storage in holders or storage in bullets or spheres. Such a high pressure storage installation can discharge at any desired pressure, with a minimum of mechanical equipment, and can operate even without an outside power source.

A number of bottle-type underground high pressure gas storage holders were built to the API specification for grade N-80 casing for oil wells. ASA B31.8-1963, Paragraph 844 covers these types of holders. The salient requirements of the paragraph include a fenced-in storage site; expression of holder design as a function of site location and the clearance between the holders and the site fence; and pipe location classes that are the same as those used for transmission systems. Table 10-9 shows the design factors.

Table 10-7 Capacity of Vertical High Pressure Gas Holders at Various Pressures

Diam and height* above foundation	Volume, MCF	At 30 psig, MCF	At 40 psig, MCF	At 50 psig, MCF	At 60 psig, MCF
20.0 × 72.3	20	40.8	54.4	68	81.6
24.0 × 65.1	25	51	68	85	102
30.0 × 68.7	40	81.6	108.8	136	163.2
30.0 × 82.8	50	102	136	170	204
38.0 × 80.8	75	153	204	255	306
38.0 × 102.8	100	204	272	340	408

(Stacey Manufacturing Co.)
* Height includes 2 ft between bottom of tank and foundation.

Table 10-8 Spherical Tank Capacity for High Pressure Gas Storage (Welded Construction)

Diam, ft	Working pressure, psig	Thickness of steel, in.	Capacity,* MCF	Painting area, sq ft
	30	0.30	68	
40	50	.50	114	5,025
	75	.75	171	
	30	.37	133	
50	50	.62	222.5	7,855
	75	.94	333.7	
	30	.45	230	
60	50	0.75	384	11,310
	75	1.13	575	
70	30	0.53	366	15,395
	50	.88	611	
80	30	0.60	547	20,105
	60	1.20	1,093	

(Chicago Bridge and Iron Co.)
* Amount of gas liberated from working pressure to 0 psig.

* Complete including foundations, painting, piping-in; as of December 1, 1955, Construction Cost Index 323.6 (1926 = 100).

†Includes painting every six years, testing, and inspection; $2.50 labor rate (includes 20 per cent indirect labor).

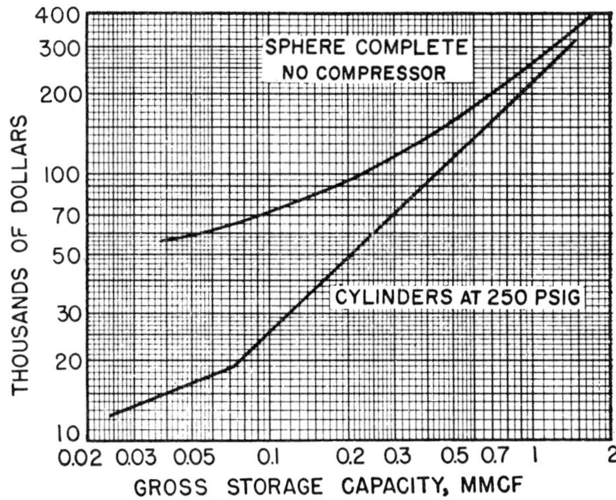

Fig. 10-9 Construction costs for sphere and cylinder type pressure vessel holders.[1]

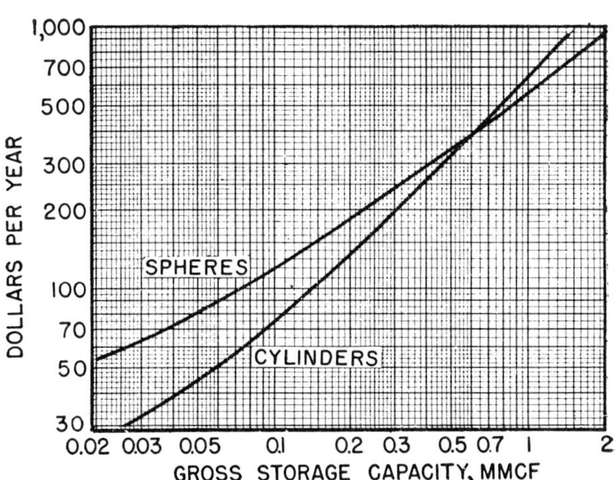

Fig. 10-10 Annual maintenance costs for sphere and cylinder type pressure vessel holders.[1]

The required gross storage capacity may be determined by Eq. **1**:

$$C_g = \frac{C_n P'}{P' - 2P_s} \tag{1}$$

where: C_g = gross storage capacity, cu ft
C_n = usable storage capacity, cu ft
P' = design storage pressure, psia
P_s = system pressure, psia

The wall thickness of the container is obtained by Eq. **2**:

$$t = \frac{PD}{2SFE} \tag{2}$$

where:
t = normal wall thickness, in.
P = design storage pressure, psig
D = nominal pipe OD, in.
S = specified minimum yield strength, psi, stipulated in the specifications under which the material was purchased
F = design factor obtained from Table 10-9
E = longitudinal joint factor obtained from Table 841.12 of ASA B31.8-1963 (extracted in Sec. 8, Chap. 2) which shows the factors for commonly used piping steel

There are special provisions applicable to bottle type holders only. These may be manufactured from steel which is not weldable under field conditions, subject to certain limitations: (a) alloy steels used shall meet the requirements given in API Standard 5A or ASTM Standard A372 (tentatively the E factor may be taken as 1.00); (b) in no case shall the ratio of actual yield strength to actual tensile strength exceed 0.85; (c) welding shall not be performed after bottles have been heat-treated or stress-relieved, or both, except that it is permissible to attach small copper wires to the small diameter portion of the bottle closure for cathodic protection using localized Thermit welding.

Table 10-9 Design Factor, F, for Pipe- and Bottle-Type Holders[6] on Company Sites

Holder site location* class	Minimum clearance between containers and fenced boundaries of site 25 to 100 ft	Minimum clearance between containers and fenced boundaries of site 100 ft or over
1	0.72	0.72
2	.60	.72
3	.60	.60
4	0.40	0.40

* Location classes are defined under paragraph 841.01 of ASA B31.8-1963 (extracted in Sec. 8, Chap. 2).

Minimum clearance between pipe containers or bottles is determined by Eq. **3**:

$$C = \frac{3DPF}{1000} \tag{3}$$

where:
C = minimum clearance between containers, in.
D = OD of bottle or pipe container, in.
P = max allowable operating pressure, psig. This operating pressure is obtained in the same manner as for a transmission line. See Table 841.412(d) of ASA B31.8-1963 (extracted in Sec. 8, Chap. 2)
F = design factor, see Table 10-9

Figure 10-11 shows the arrangement frequently used in this type of storage system. In this instance the units are approximately four feet below ground and eight feet between ends.

The minimum clearance between containers and the fenced boundaries of the site is fixed by the maximum operating pressure of the holder as follows:

Maximum operating pressure	*Minimum clearance*
Less than 1000 psig	25 ft
1000 psig or more	100 ft

In a typical system, like that shown in Fig. 10-11, high carbon alloy steel units were used. As many as 12 units have been placed in one string (eight units in a string was more common); it was general practice to connect approximately five of these strings to each manifold. Two inch manifolds and shutoff valves are located at one end of each group. To protect the public, storage units were kept approximately 150 ft away from property lines. Spacing should be determined by Eq. **3**.

A typical group of 160 units holds approximately 67 MMCF of gas at 2240 psig, of which 57 MMCF may be withdrawn at desired outlet pressure. The actual quantity of gas stored will be influenced by the compressibility factor given in Fig. 8-25. Gross storage capacity of such units has varied from 1.5 to 120 MMCF.

The sendout facilities for the installation shown in Fig. 10-11 include one or more high pressure gas heaters followed by a single-step pressure reducing valve of the balanced, single-seated, angle type. These gas heaters counteract the refrigeration effect of expanding gas, thus minimizing hydrate formation in the gas stream and ice outside the pipe, and reducing pipe stresses by reducing contraction. These gas heaters raise the high pressure gas stream temperature to approximately 140 F, but they are also designed to heat the gas at lower pressures to not less than 80 F. The specification of these units requires that adequate heating of the gas occur at both maximum and minimum outlet pressures and that the gas pressure drops thru the heat exchanger at these pressures do not become excessive, which would limit the withdrawal rate.

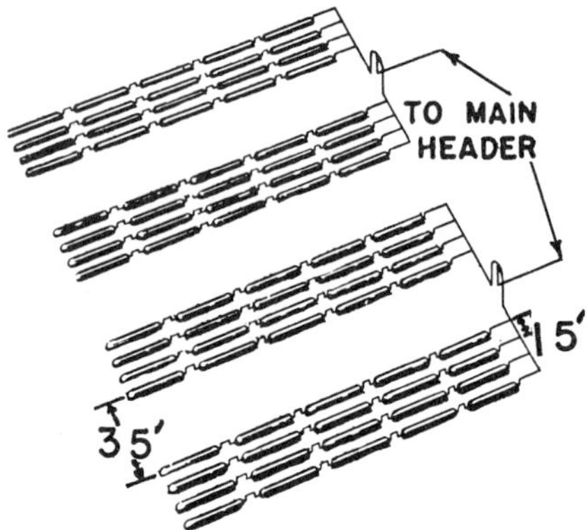

Fig. 10-11 Arrangement of 80 bottles for underground high pressure storage. Note Eq. 3.

The equipment required to place the gas in storage includes dehydrators,* which reduce the water content of the gas to 0.5 lb† per MMCF (at 60 F, 30 in. Hg), gas compressors, and gas intercoolers and aftercoolers. In some cases, depending upon the rate of gas delivery to storage and the compression ratio, it was possible to omit the aftercoolers (serving to minimize softening of internal coatings and reduce thermal stresses). Generally, the compressors were so sized that the total gas storage was filled in from 8 to 21 days, depending upon the requirements of the gas system. Eight to ten days was the more usual figure; three days was minimum.

The gas sendout facilities may be combined with propane-air blending facilities so that a mixture of propane, air, and natural gas is available for emergency periods. Similarly, it is possible to combine the duties of the air compressors and gas compressors, placing cylinders for each on the same machine and selecting the cylinders so that the total horsepower available from the crankshaft can be absorbed by either the gas or the air compression cylinders.

An installation of this type utilizes approximately 1.25 acres per MMCF of gas storage capacity. In view of this, the

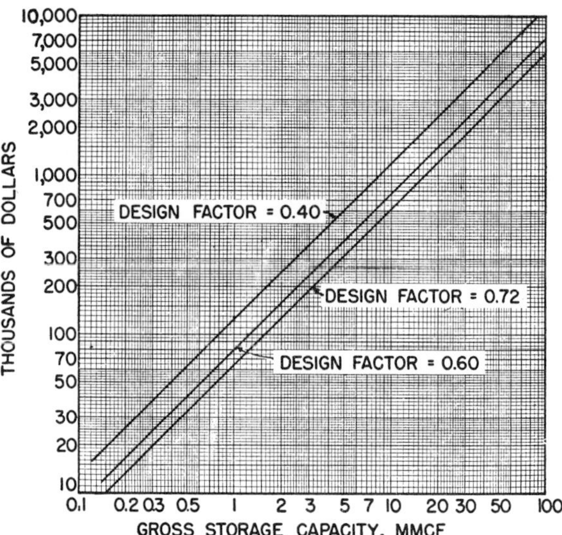

Fig. 10-12 Construction costs for bottle (1 to 100 MMCF) and pipe battery (0.1 to 100 MMCF) type pressure vessel holders.[1] (See Table 10-9 for correlation of Design Factors.)

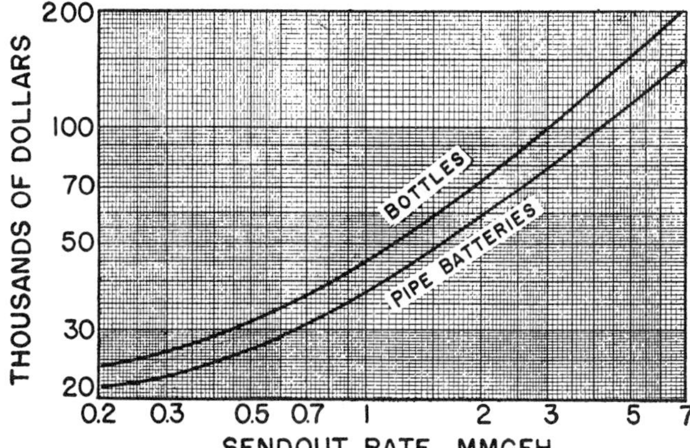

Fig. 10-13 Construction costs of a sendout plant for bottle and pipe battery type pressure vessel holders.[1]

facilities are suitable only for rural installation. Few operators are required; in fact, some installations have been designed as stand-by stations to which no regular operators are assigned.

Costs

Estimated construction costs‡ for bottle holders are shown in Fig. 10-12. These costs include all storage facilities with the exception of sendout (see Fig. 10-13) and compression equipment§ (see Fig. 10-14).

Construction costs‡ for pipe-type holders are also shown in Fig. 10-12 (based on using 30 in. diam API 5LX52). The corresponding construction costs for sendout and compressor§ plants are shown in Figs. 10-13 and 10-14, respectively. Usually, large pipe diameters are favored because the saving in compressor horsepower more than balances the cost of an incremental increase of pipe diameter, assuming equal storage capacities.

* A dew point under 0 F at storage pressure is the common practice.

† Allowable H_2O content in high pressure transmission lines is 2 to 5 lb per MMCF.

‡ Based on Construction Cost Index of 323.6 (1926 = 100).

§ Note that corrections must be made for inlet pressures other than 200 psi.

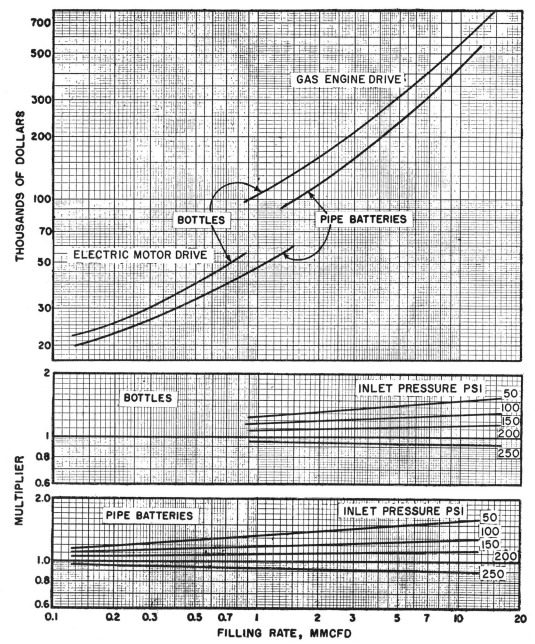

Fig. 10-14 Construction costs of compressor plants for bottle and pipe battery type pressure vessel holders.[1] Break in curves represents change from electric motor to gas engine drive at 150 bhp.

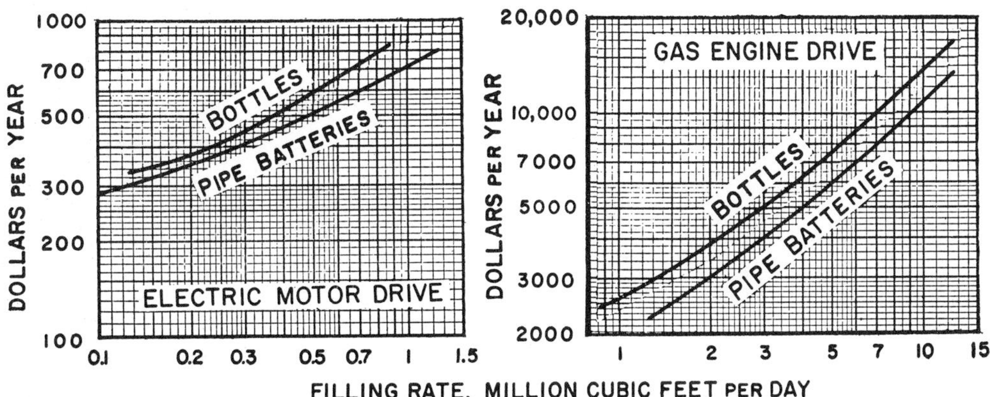

Fig. 10-15 Annual maintenance costs for pressure vessel holders: electric motor drive (left); gas engine drive (right).

The annual maintenance costs (based on $2.50 labor rate which includes 20 per cent indirect labor) for pipe- and bottle-type holders are shown on Fig. 10-15. The annual labor costs for these facilities is about $920 per year for part-time attendance for filling rates under one MMCF per day. At or above one MMCF per day filling rates, full time attendance was common. Costs ranged from about $21,500 for up to two MMCF, $33,000 for up to ten MMCF, and finally $44,000 at a filling rate of ten MMCF per day.

AN EXAMPLE OF TRANSMISSION LINE TYPE STORAGE[2]

Instead of installing a high-pressure system which requires expensive containers, high compression, and extra dehydration, some companies extend high pressure transmission lines to increase their distribution and storage capabilities simultaneously. Since line type storage design involves non-steady flow computations of a complex nature, this discussion is presented as a "case" study rather than in conventional handbook form.

In designing the Texas-Pacific pipeline, additional storage in the form of line pack was built into the California section of the line. This feature permitted varying terminal delivery rates from hour to hour to meet changing load conditions while maintaining a constant rate of input into the initial section of line. The storage feature did not impair the 24-hr daily deliverability of the line. Computations indicated that line pack storage could be built into the pipeline system at a fraction of the cost of an equivalent number of conventional aboveground low pressure or high pressure tank holders of the type which had previously been used for daily load balance.

Most natural gas companies with more than one transmission line used line pack for storage regularly. Take, for example, a 30 in. transmission line 100 miles long normally operating at 300 psig, with the source of supply at 500 psig or more. An increase in line pressure of 100 psi would increase the gas storage 6.8 *times* the cubic capacity of the line (6.8 × 2,445 MCF = 16 MMCF). The mechanics of storage in natural gas transmission pipelines have been presented by Beckman.[3]

Design and Operation

The California portion of the line which comprises the storage section runs from the Colorado River at Blythe to Santa Fe Springs near Los Angeles and consists of approximately 214 miles of 30 in. OD pipe, with a booster station at Blythe. The original design capacity of the California section was 305 MMCF per day.

The storage feature was obtained by increasing the pipe wall thickness and providing more compressor horsepower at Blythe than that required for a similar line operating with uniform terminal rates. This enables packing gas into the California section by increasing the operating pressure. The only costs chargeable to the storage thus made available are the incremental construction costs for the additional compressor horsepower at Blythe and the increased pipe wall thickness.

Although the input into the California section at Blythe is maintained at a substantially uniform rate of approximately 305 MMCF per day, deliveries from the line during peak load hours often exceed a rate of 500 MMCF per day, and during light load periods at night deliveries from the line may fall to a rate of 100 MMCF per day.

The volume of gas available from storage in line pack is from 52 to 75 MMCF. The exact amount depends upon the manner in which the pipeline is operated, but 52 MMCF can be delivered under all practical conditions. Under favorable and carefully controlled operating conditions, a considerably greater volume of equivalent holder capacity can be realized.

Figure 10-16 shows actual flow data collected from operating logs. Curve 1 shows the uniform deliveries into the line at Blythe; Curve 2 shows varying deliveries from the line terminus. The area between Curves 1 and 2 represents the amount of gas being packed or withdrawn.

Design Computations. It was relatively easy to select a design that would provide "pack" in the pipeline. This

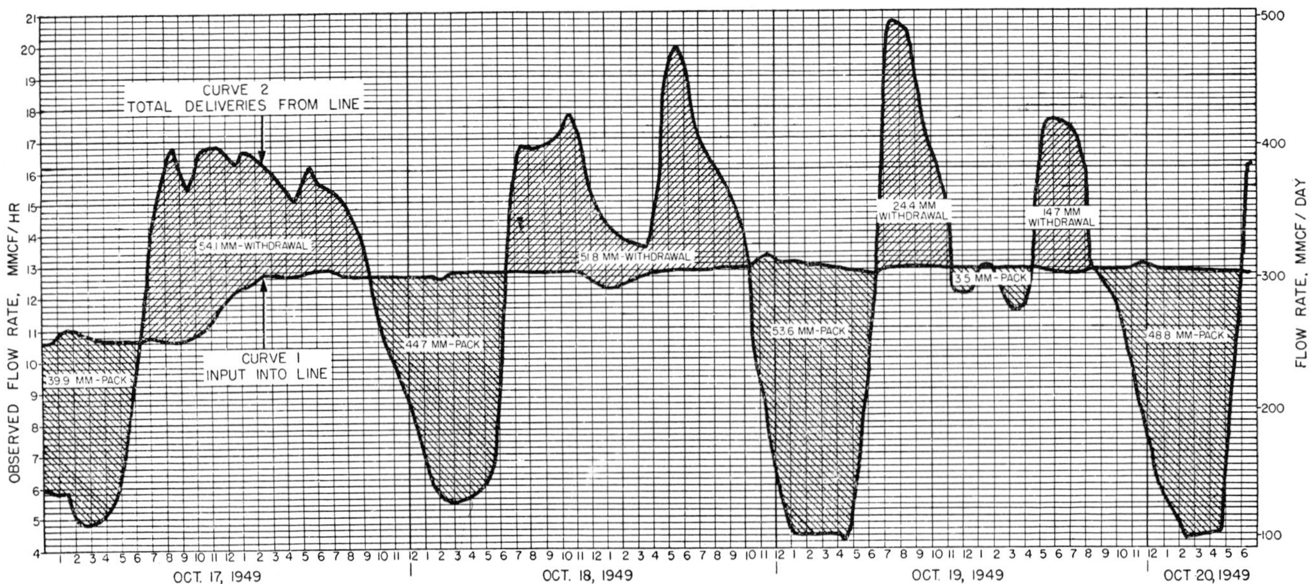

Fig. 10-16 Input and withdrawal from pipeline employing line pack storage.

Table 10-10 Estimate of Cost of Storage Built into California Section of Texas-Pacific Pipeline

(Prices and Piping Design Standards as of 1950)

SPECIFICATIONS

Design A—Line Operating Under Uniform Flow Conditions without Storage (305 MMCF Daily Delivery)
Computations indicated most favorable design consists of 214 miles of 30 in. OD line.* Maximum operating pressure at Blythe—675 psig. Terminal pressure at Santa Fe Springs—250 psig. Installed horsepower at Blythe—7410 (490 psig compressor inlet pressure).

Design B—52 MMCF† of Storage Built into the Pipeline (305 MMCF Daily Delivery)
Optimum design—214 mile, 30 in. OD line. Maximum working and operating pressure at Blythe—807 psig, with terminal delivery pressures at Santa Fe Springs varying from 250 psig to approximately 500 psig. Blythe compressor horsepower required—approximately 10,220 (490 psig compressor inlet pressure).

Note: In each Design, **A** and **B** above, the line is divided into four sections with the pipe wall thickness of the sections progressively reduced from the initial section at Blythe to the terminal section at Santa Fe Springs. Each section is protected by a pressure-limiting station that prevents build-up over the design working pressure during low rate deliveries.

Comparison of Weight of Pipeline

	Design **A** without storage, total wt, tons	Design **B** with storage, total wt, tons
Colorado River—Blythe		
0.13 miles	47	47
2.75 miles	791	791
3.53 miles	1,107	1,107
Blythe—Shaver		
65.74 miles	17,213	18,913
Shaver—Garnet		
52.6 miles	11,041	13,772
0.23 miles	54	66
0.14 miles	47	58
Garnet—Olinda		
51.11 miles	10,728	12,056
1.38 miles	433	574
0.37 miles	78	97
22.62 miles	5,336	6,508
0.02 miles	5	6
Olinda—Santa Fe Springs		
12.04 miles	2,527	3,464
Totals	49,407	57,459

The difference of 8,052 tons is the additional weight of pipe required to provide storage.

ESTIMATED CAPITAL COST OF STORAGE

Pipe—8,052 tons at $135/ton	$1,087,000
Incremental pipe installation cost—8,052 tons at $30/ton	242,000
Incremental Blythe compressor station cost—2810 hp (10,220 − 7410) at $210/hp	590,000
Total Incremental Capital Cost	$1,919,000
Capital Cost/MCF of Storage†	$37/MCF

ESTIMATED ANNUAL COST OF STORAGE

Fixed charges on incremental investment—$1,919,000 × .14‡	$ 268,000
Additional pipeline operation and maintenance	....
Incremental Blythe compressor station operation and maintenance cost for 24,600,000 horsepower hours/year (2810 × 24 × 365) 24,600,000 at 4.38 mills§	108,000
Total Incremental Annual Cost	$ 376,000
Annual Cost/MCF of Storage†	$7.24/yr

* Estimates indicate that a 26 in. line, while having slightly lower capital cost, would involve greater annual cost.

† The effective storage capacity available from the line is equivalent to a holder capacity of from 52 to 75 MMCF, depending upon the manner in which the line is operated. The estimated unit cost of storage is based upon an available storage of 52 MMCF.

‡ Fixed charges, 14% of installed cost based upon 30-year life, include 6% return; and ad valorem, federal income and franchise taxes, and sinking fund depreciation.

§ Includes 2.78 mills/bhp hour for fuel, lubricant, and maintenance cost; and 1.60 mills/bhp hour incremental operating costs.

Table 10-11 Comparison of Capital Cost and Annual Cost of Transmission Line Storage with Other Types of Storage

(1960 Prices)

	Dollars per MCF	
	Capital cost	Annual cost
Transmission Line		
Cost range	74 to 200	...
Cost of 52 MMCF of storage built into California section of Texas-Pacific line	74	14.50
High Pressure Cylindrical Tank Type Holders with Compressors		
Discharge pressure, psig		
5	652	95
25	685	99
Water-Seal Low Pressure Holders		
Discharge pressure, psig		
5	473	71
25	560	87
50	610	93

Diam, in.	Transmission pressure, psig	Type C*	Type A*	Type C*	Type A*
Pipe Holders					
24	50	415	346	76	76
	250	304	243	54	54
	500	266	192	44	43
30	550	...	115	...	23
34	445	...	111	...	22
with compressor					
30	550	...	232	...	47
34	445	...	204	...	41
Bottle Holders					
24	50	400	382	84.20	75.20
	250	299	286	54.20	54.20
	500	251	233	43.60	43
30	550	...	...	...	63
	1265	...	316	...	...
34	445	...	...	...	63
	2150	...	317	...	...
with compressor					
30	550	...	...	...	106
	1265	...	531	...	...
34	445	...	...	...	158
	2150	...	790	...	...

(Pacific Gas and Electric Co.)

* Defined in Table 841.02 of ASA B31.8-1963 (extracted in Sec. 8, Chap. 2).

could be done by: (a) increasing the operating pressure of the line; (b) increasing the diameter of the line; (c) paralleling a portion of the line with additional pipe; or (d) a combination of these means. This does not establish that the "pack" could be withdrawn from the terminus within the period of seven to fifteen hours each day in which the additional gas is needed. Gas flow friction during high terminal deliveries would create a large pressure gradient which might preclude withdrawing the gas that has been stored in the upstream portion of the line within the time available. Furthermore, when the line is being repacked during the light load periods at night, the pressure at the initial end at Blythe cannot be permitted to increase to the extent that input delivery at that point would be reduced. If this should occur, the desired daily deliverability of 305 MMCF would be reduced. An analytical-graphical method[4] has been developed for computing the effective storage capacity that can be procured from a pipeline operating under line pack conditions.

This computation method provides an accurate and reliable means of computing and designing transmission line type storage, but the work is laborious and time-consuming and it can be performed only by persons who possess considerable mathematical skill and knowledge.

A simplified method of computation has been devised by the Texas-Pacific Pipe Line Co., which, while mathematically not so rigorous, furnishes fairly close approximations.

Costs. Pipeline costs with and without storage are compared in Table 10-10; Table 10-11 shows the estimated capital and annual cost of the storage built in the Texas-Pacific pipeline along with the cost estimates of storage in conventional holders, pipe holders, and bottle holders.[5]

Application of Line Type Storage

The main disadvantage of transmission line type storage is that if a line break should occur, not only would the transmission capacity to the area be lost but also a major portion, or perhaps all, of the gas stored in the pipeline. Thus, it appears unwise to build too great a portion of a utility's storage in a single pipeline. Another answer is to use valves that close automatically in case of a line break. Where an area is supplied from two or more transmission lines, it would be desirable to put a part of the storage in each of the lines, thus providing diversity and greater reliability. A combination of transmission line type storage and conventional holders, either bottle or aboveground, might be attractive. The transmission line type storage, which involves relatively small capital and operating expense, could be used daily throughout the year for load balance. The bottle holders or conventional holders, involving greater capital cost and considerably greater operating cost, could be reserved for extreme peak day use and for stand-by. In such a combination storage system, the principal advantages and economies of each type of storage would be realized.

REFERENCES

1. Stone and Webster Engineering Corp. *Gas Storage at the Point of Use.* New York, A.G.A., 1957.
2. Mosteller, W. C. "Transmission Line Type Storage—Capacity Calculations and Costs." *A.G.A. Nat. Gas Dept. Proc.* 1950: 34.
3. Beckman, P. E. "Mechanics of Storage in Natural Gas Transmission Pipe Lines." *P.C.G.A. Trans.* 29: 163, 1938.
4. Olds, R. H. and Sage, B. H. "Transient Flow in Gas Transmission Lines." (Pet. Trans., A.I.M.E. 192:217–22) *Jo. of Pet. Technology* III, July 1951.
5. Hough, F. A. and others. "Study of the Comparative Costs of Storing Natural Gas in Conventional Holders and in Pipe Holders and Bottle Holders." *A.G.A. Nat. Gas Dept. Proc.* 1949: 64–97.
6. A.S.M.E. *Gas Transmission and Distribution Piping Systems.* (ASA B31.8—1963) New York, 1963.

Chapter 3

Underground Storage of Natural Gas

by E. E. Roth

Natural gas is stored underground when it can be injected into natural rock or sand reservoirs which have suitable connected pore spaces, and it is retained there for future use. Such storage sites are usually depleted oil and gas fields. Aquifers are also being put to use by displacing water with the gas to be stored.

As of December 31, 1962 there were 258 underground storage pools with 10,521 wells and 184 compressor stations having 571,535 hp, in 21 states.[1] Maximum gas in storage was 2216 billion cu ft; ultimate storage capacity was 1954 billion cu ft, with a maximum day output of 15,705 MMCF. West Virginia, Pennsylvania, Ohio, and Michigan are the leaders in all phases of underground storage to date.

DEFINITIONS[2]

Current Gas—The total volume of extraneous gas injected into a storage reservoir in excess of the total cushion gas. This is the total maximum volume of gas available for delivery during any input-output cycle.

Cushion Gas—The total volume of unrecoverable and economically recoverable native or foreign gas which exerts a pressure, within the storage reservoir, from 0 psig to the pressure in a storage reservoir which will maintain a required minimum rate of delivery during any output cycle.

Cushion Gas, Capitalized or Foreign—That part of the volume of injected cushion gas which is extraneous to the storage reservoir.

Cushion Gas, Native—That part of the volume of cushion gas which is indigenous to the storage reservoir.

Deliverability—The output of gas from a storage reservoir, expressed as a rate in MCF per 24 hr, at a given total volume of gas in storage with a corresponding reservoir pressure and at a given flowing pressure at the wellhead.

Foreign Gas—The volume of extraneous gas injected into the storage reservoir which exerts, within the storage reservoir, a gage pressure above that gage pressure (psi) at which storage was started.

Formation, Storage (**or Storage Zone**)—The drillers' name of that stratum of the earth's crust within which the storage reservoir is located.

Injectability—The input of gas to a storage reservoir, expressed as a rate in MCF per 24 hr at a given total volume of gas in storage with a corresponding reservoir pressure and at a given flowing pressure at the wellhead.

Input, Total—Volume of foreign or extraneous gas injected into a storage reservoir during a given period of time.

Maximum Gas in Storage—The total highest volumetric balance of total input over total output of gas in storage.

Native Gas—The volume of gas indigenous to the storage reservoir. This should include the total volume of unrecoverable gas and economically recoverable gas within the storage reservoir, which exerts a pressure from 0 psig to the gage pressure (psi) at which gas storage is started.

Output, Total—Volume of gas withdrawn from a storage reservoir during a given period of time.

Pressure, Ultimate Reservoir—The maximum gage pressure—psi—(either wellhead or bottom hole, as designated) exerted by the volume of gas at the ultimate reservoir capacity.

Reservoir, Storage—That part of a storage formation or zone having a defined limit of porosity which can effectively be used to retain gas at a given ultimate reservoir pressure.

Ultimate Reservoir Capacity—The total volume of gas within a reservoir which exerts a pressure from 0 psig to the maximum or ultimate reservoir gage pressure (psi). This should include all native gas (recoverable and unrecoverable), cushion gas, and current gas.

Well, Gas Storage—A bore hole extending from the surface into the storage reservoir, which is used primarily for either the input or output of gas or for observation of pressure, or the extraction or injection of fluids in connection with a gas storage project.

PURPOSE OF UNDERGROUND GAS STORAGE

Primarily, underground storage provides an economical way to supply large volumes of gas for space heating consumption. Storage improves the transmission line load factor by providing a choice of delivering gas either to the users or to the underground storage reservoir. Another use is in the transfer of gas from a highly competitive field to a field wholly controlled by one company. Under this arrangement, the gas can be withdrawn as needed and used to best economic advantage. Also, the storage field can be used advantageously to store gas from low pressure wells, usually the smaller wells, during the off-peak season. Thus, low pressure gas can be compressed and concentrated into a small area, and the storage wells in that small area will have a much greater deliverability available for the peak season. In the case of long transmission lines, underground storage near the consuming centers also acts as a safeguard or reservoir in case of pipeline failures. Also, it may be possible to gain a price advantage by obtaining off-peak gas during the summer and storing it.

A gas storage field must: (1) be capable of delivering, over normal deliverability, the excess peak day gas at any time during the winter, while meeting the near-peak days during this period, and (2) be able to deliver the seasonal volume necessary to supplement the supply of regular pipeline gas, in order to meet the normal winter demand. To handle gas in this manner, the storage field must be located reasonably near markets, near compressing stations and, generally, near the main transmission lines. Only rarely do fields meet these ideal requirements. However, many storage projects now in use are still quite satisfactory.

STORAGE FIELD RESERVOIR CONSIDERATIONS

A reservoir should be able to deliver both the daily rate and the total winter output without excessive compression or too much "cushion volume." The terms "excessive" and "too much" are defined by the overall economics of the system. Cushion gas is of little value in supplying either peaks or seasonal load directly, but it is essential in building a foundation for the turnover of usable storage gas. If for any reason storage operations should be ended in a particular reservoir, the cushion volume would be recovered, just as remaining reserves in any well or pool are recovered.

Essential reservoir features include: (1) an impermeable reservoir cap rock to prevent leakage and pressure loss; (2) high porosity and permeability in the reservoir rock; (3) reservoir formation depth sufficient to allow for a safe pressure; (4) either an absence of water or an easily controllable water condition in the reservoir; (5) a thick vertical reservoir formation, rather than a thin horizontal formation; (6) an oil-free formation, for although exhausted oil-producing formations have been successfully used,* it is usually more satisfactory to use one which does not contain any liquid that might interfere with storage operations. In a few instances, water in the formation has been used to advantage in storage operations. In such cases the water not only acts as a seal retaining the stored gas, but by slowly advancing as the gas is withdrawn it has the effect of a piston, maintaining the pressure to some extent.

Many gas fields are too large for effective storage operations, at least in the matter of total volume. Thus, they require too great a volume of cushion gas to increase their reservoir pressures to satisfactory levels.

Excessive injection pressures might force gas into less porous sections of the formations, from which recovery would be relatively slow. In several storage fields, however, the pressures have been raised somewhat higher than the original reservoir pressure without noticeable migration or loss of gas. Furthermore, many fields have original formation pressures much higher than could be calculated on the basis of depth and hydrostatic head, a factor commonly used as a guide in determining what is often referred to as "reservoir pressure." In some water flooding operations in oil reservoirs, pressures exerted on the formations are very much higher than would be considered normal for the reservoir, considering its depth.

There is the rule of thumb that compact rock, like that found in a storage reservoir, will not be fractured until the pressure exerted is greater than one psi for each foot of depth, which is over twice the hydrostatic head. The "hydrafrac" method of fracturing the producing formation to obtain increased production is based to a considerable extent upon this principle.

The general matter of maximum pressure is not formalized, and additional experience and research will be necessary before a definite rule can be followed. In all probability, it will be found that each field must be treated on its own merits.

ACQUISITION OF AN UNDERGROUND STORAGE FIELD

The factors involved in determining the suitability of a given location are: (1) general geological features including the reservoir area, formation thickness, and geological structure; (2) the relationship of gas, oil, and water to each other, and the position of their contacts; (3) the areal extent or perimeter of the reservior as ascertained by the above studies, related surface ownership, and the exact location of all drilling operations; (4) a preliminary study of the amount of gas the reservoir will hold, based on past production and reservoir pressures, or volumetric computations; and (5) the approximate rate of deliverability, also based on the records of past production.

Following these determinations, legal rights must be obtained to convert a gas field into a storage field. Practically all useful acreage was originally leased for the production of oil and gas, and not for gas storage. The ordinary oil or gas lease grants to the lessee the right to produce and remove those original substances underlying the lessor's property, but it does not grant rights to any subsurface rock as a reservoir into which gas may be injected, stored, and later removed. Therefore new storage contracts become necessary.

Various methods and legal forms may have to be used in acquiring the desired control. Such forms may include: (1) combination oil and gas and gas storage leases; (2) oil and gas leases, or assignments thereof, and supplemental gas storage leases; (3) gas storage deeds; (4) mineral (oil and gas) deeds; (5) royalty conveyances and assignments; (6) ordinary deeds; (7) easements; and (8) special forms or special provisions written into ordinary forms. These agreements, particularly on tracts not having wells, should be made to cover as long a period of time as possible, because storage operations may extend for many years.

Experience has shown that a wide variety of terms must be considered for a storage agreement; therefore, not all of the following would normally be included in any given contract. After the usual introduction and description of both parties, the agreement generally states the proposal, i.e., the establishment and operation of an underground storage reservior. A complete description of the land involved and the payment for the lease are also included. The contract then deals with the general rights and privileges of the lessee or grantee to use for gas storage any and all sands and strata in and under the previously described lands.

Exceptions to these rights are then noted. Sands and strata which supply water wells may be specifically excluded. In

* E.g., Lone Star Gas Company's New York City Field; the value of the liquids recovered more than offsets their storage costs.

an argument for storage alone, sands containing oil or indigenous gas capable of being produced in paying quantities may be excepted from the contract.

Specific items included with the storage rights may deal with migratory gas from other reservoirs and evacuation and disposal of water, both residual and migratory, from the reservoir strata. Permission may also be granted for the establishment and expiration of all necessary equipment, housing, and accesses to and from the reservoir and surface facilities.

More detailed points are then considered. In such details, the grantee agrees to pay for drilling wells, to be penalized for failure to perform the contract in a stipulated time, and to the method and time of these payments. The grantee also reserves the rights to all personal property he uses and to determine the size of the reservoir at any time. The grantor agrees to meet all taxes, liens, and mortgages against the property, to keep the title clear, and to notify the grantee of any change in ownership. The activities of the grantee may be restricted regarding location of wells and depth of pipe, and he shall be responsible for damages to growing crops or damages resulting from actionable negligence. In return, the grantor may not do anything which will harm the reservoir or the grantee's equipment. Arrangements are also made for joint ownership of the lands and extension of coverage of the agreement.

Laws pertaining to the principle of eminent domain for natural gas storage have been adopted (up to 1955) in Kentucky, Michigan, Illinois, Kansas, Oklahoma, West Virginia and Pennsylvania. In the other states, acquiring contracts is often a matter of patient and lengthy negotiation.

ACTIVATING A STORAGE FIELD

Preliminary steps include a detailed geological and engineering survey of the reservoir, since the preliminary study mentioned previously was sufficient only to ascertain the advisability of acquiring the field for storage purposes. All structure maps should then be rechecked and brought up to date. If at all possible, formation thickness or isopach maps should be constructed, and the general thickness of the reservoir should be determined. The positions of the "pay" sections in the reservoir rock should be located. The relative merits of permeability, local capacities, and well performance can often be worked out from these studies. Several methods are available for estimating well capacities and gas reserves.

Reconditioning

Steps should be taken to establish that the reservior is free from physical defects thru which gas might be lost. For active wells, it is often advisable to pull the tubing or casing, clean the well thoroughly, and shoot, acidize, or hydraulically fracture[3] the formation in order to increase its deliverability. New tubing and casing should then be installed and packed or cemented so that the wells will withstand design pressure. Abandoned wells inadequately plugged for storage operations should be either cleaned out for use as storage wells, or replugged to prevent leakage.

It is important to know the exact position of the effective porous parts of the formations constituting the reservoir section. Therefore, it is advisable to obtain electric logs—in old fields preferably a gamma ray log—and also a temperature survey.

PRESSURE AND VOLUME OBSERVATION

In a closed reservoir there is usually a fairly consistent relationship between a unit of gas pumped in and the corresponding rise in the reservoir pressure throughout the range of storage operation. However, this relationship does not always remain constant, because it is affected by such variables as

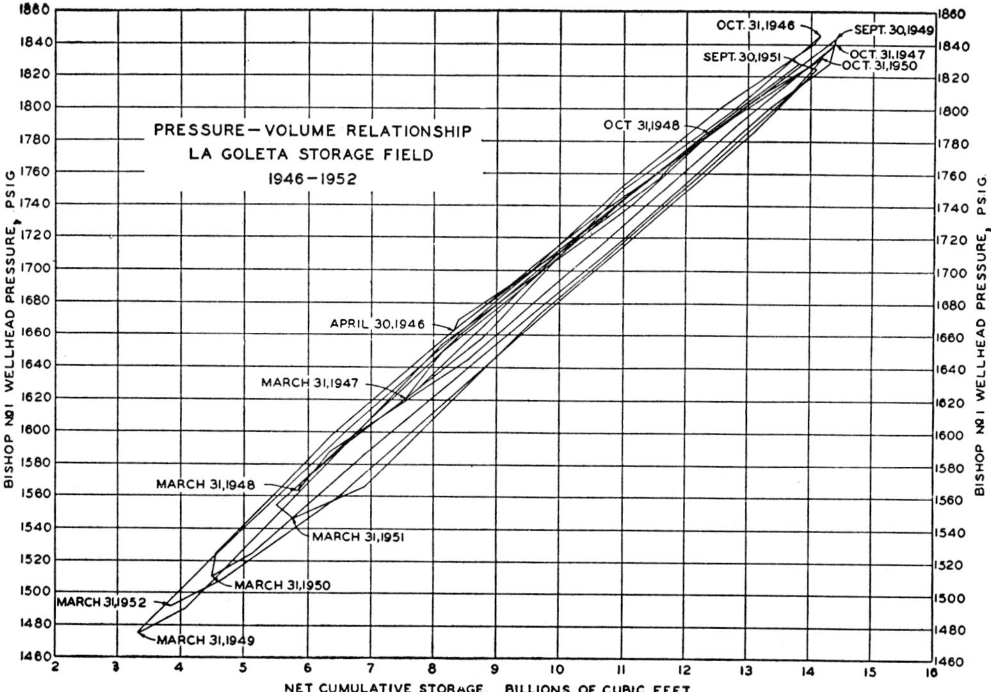

Fig. 10-17 Pressure-volume relationships for the La Goleta gas storage field, 1946-1954.[5]

permeability, porosity, presence of connate water, and formation thickness. Also, since each well in a field is usually equipped with a pressure gage instead of an individual meter, pressure measurements are averaged and used to obtain total volume.

Figure 10-17 shows the pressure-volume relationship for the La Goleta (Calif.) gas storage field. Both yearly variations and variations between injection and withdrawal are clearly seen. "Capillary control" may be the factor which causes the variation that results in a curved line in such plotting.[4] This theory states that these curves are portions of true parabolas.

VOLUME IN STORAGE AND DELIVERABILITY

With a given volume of gas in storage, the deliverability from a field varies with the size and number of wells in it and with the pressure against which the wells must feed. When the field has been completely activated with all the wells, lines, stations, and other equipment installed, it is possible to determine how much daily delivery can be expected from the field when it contains various volumes. Several schemes are used to show the relationship between volume and deliverability, but one of the simpler methods is to plot deliverability against the volume in storage (Fig. 10-18).

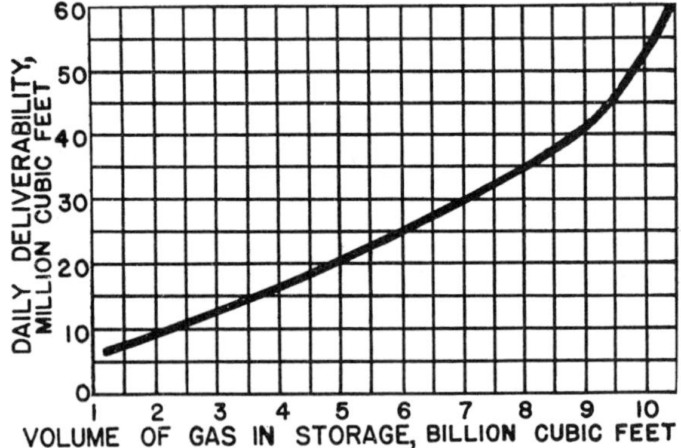

Fig. 10-18 Example of daily deliverability estimate for a gas storage field.

Figure 10-18 assumes fairly uniform operating conditions in a rather low permeability reservoir. For example, if the wells were feeding against a consistently high line pressure and if it then became necessary to lower this line pressure, the position of the curve on the graph would change to reflect this new operating condition. It is also assumed in the graph that the field will have a fairly continuous usage. If it should be decided later to use the field only under peak conditions, then because of the "heading up" of wells and for other reasons, the field would produce at a higher rate—with a given volume of gas in the reservoir—than is reflected by the graph. It is therefore essential to know the specific operating expectations before reasonable accuracy can be obtained in determining deliverability with a given volume in storage.

FLOW OF STORAGE GAS THRU THE STORAGE AREA

Flow of storage gas from injection wells to other outlying producing or dormant wells in the storage areas can be determined by various methods.[6]

These methods include:

a. Recording changes in gas pressure on the wells before and during injection.

b. Adding helium[7] or other tracers to the injection gas and testing for these tracers at outlying wells.

c. Analyzing the gas at outlying wells for heating value and specific gravity[6] at regular intervals during injection, provided the injection gas is different from the native gas in these characteristics.

During nine years' experience with this last method, a company that determined whether migration was occurring to over 400 wells pointed out that an increase in reservoir pressure may not indicate migration, and that migration may occur with no increase in reservoir pressure. Usually, with an injection pressure of 1400 psi, gas will not flow more than one mile underground, but instances of gas flowing as far as 8360 ft have been found.

STORAGE OF NATURAL GAS IN AQUIFERS, SALT CAVITIES, AND MINED CAVERNS

A typical water **aquifer** used to store natural gas by displacement of water is shown in Fig. 10-19; Herscher Dome south of Chicago in the Galesville sand is about 100 ft thick, about 1750 ft below ground surface, 18.5 per cent porous, covers about 6000 acres, and is capable of holding 150 billion cu ft of gas. Eighteen billion cubic feet were stored between

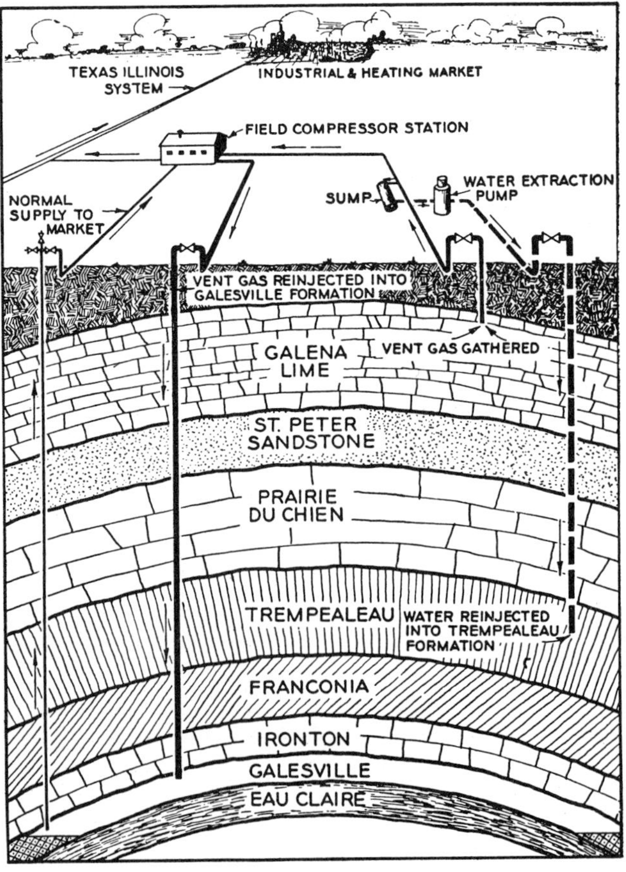

Fig. 10-19 A water aquifer storage area in the Galesville sand near Chicago, and the vent gas gathering system.

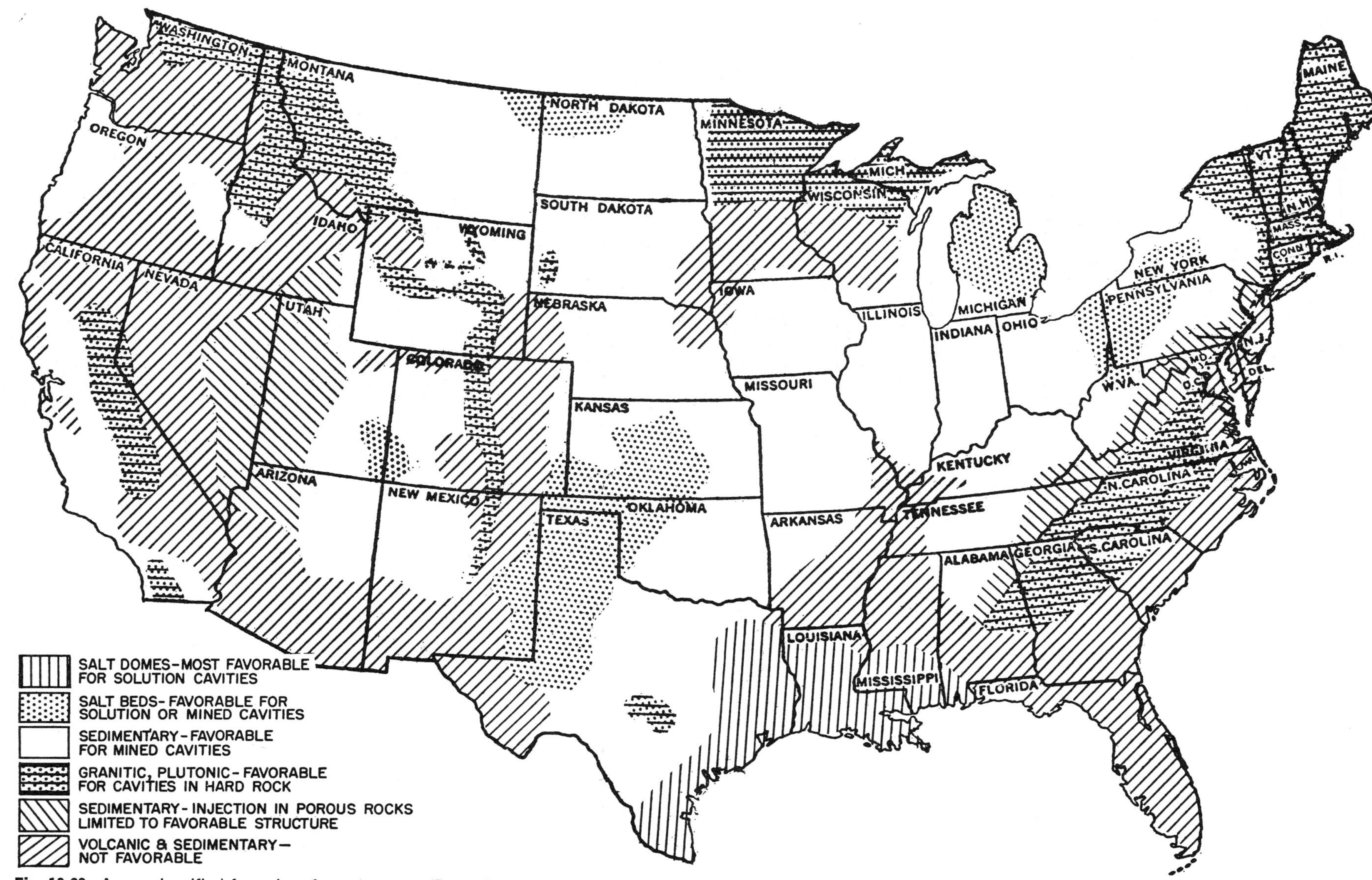

Fig. 10-20 Areas classified for subsurface storage. (From the National Petroleum Council Report of the Committee on Underground Storage for Petroleum, April 22, 1952; updated in Oct. 1962 by C. T. Brandt, Underground Storage and Mining Consultant, Bartlesville, Okla.)

1953 and 1956. Gas leaking from the storage sand to a shallower formation is collected in a vent gas gathering system, purified of sulfur, and returned to storage, as illustrated. Other aquifer locations[8] are at Doe Run, Ky.; East Hastings, Tex.; Redfield, Iowa (ultimate capacity—140 billion cu ft, working capacity—40 billion cu ft); Florissant, Mo. (2 billion cu ft withdrawn during 1959–60 heating season); and Troy Grove, Ill.

Storage of gas in cavities dissolved out of **salt beds** is a possibility, but it has not been used to date (1960). The interesting features of this method have been described,[6] and the locations where this type of storage would be possible are shown in Fig. 10-20.[9] An abandoned salt cavern has been successfully applied to natural gas storage—capacity 410 MMSCF at 1200 psig. Formations that can withstand a pressure of 500 psig appear to be suitable for such storage.[15]

The possibilities of storing gas underground in **mined** or **natural caverns** have also been described,[10] and the possible favorable locations are shown in Fig. 10-20. In 1960, progress was reported[11] on using an abandoned coal mine for gas storage. Approximately 30 per cent of two separated portions of the mine were unwatered. The pressures in these sections, 82 and 104 psig, respectively, permitted storage of a total of 280 MMCF. It was anticipated that ultimately (about 1964) 3 billion cu ft would be stored there at 300 psig.

Equations **16** and **17** in Chap. 4 of Sec. 4 may be used to estimate gas reserves in pore space volumes. A study is available on the effects of the cyclic nature of storage operations on underground water drive occurring in gas storage reservoirs.[14]

UNDERGROUND STORAGE OF MANUFACTURED GAS, PROPANE-AIR, AND HELIUM

Billions of cubic feet of coke oven gas,[12] millions of cubic feet of propane-air gas,[13] and large volumes of helium[6] have been stored underground successfully. Oxygen and hydrogen sulfide in the gas are reduced considerably during storage. Nitric oxide is removed from coke oven gas to prevent gum formation.

REFERENCES

1. A.G.A. Com. on Underground Storage. *Underground Storage of Gas in the United States: Statistics, 1962.* (OP-63-1) New York, 1963.
2. ——.——, 1953; p. 11: Appendix A. New York, 1954.
3. Waters, A. B. and Ayres, H. J. "Hydraulic Fracturing as Applied to Underground Gas Storage Reservoirs." (GSTS-58-51) *A.G.A. Proc.* 1958: T-105–116.
4. Herold, S. C. *Analytical Principles of the Production of Oil, Gas and Water from Wells.* Stanford U. Press, 1928.
5. Todd, R. W. "Operation of La Goleta Gas Storage Field." *A.G.A. Nat. Gas Dept. Proc.* 1952: 76–83.
6. Vandaveer, F. E. and Schmidt, J. J. "Underground Storage and Migration of Natural Gas." (PC-50-24) *A.G.A. Proc.* 1950: 768–77. Abridged in *A.G.A. Monthly* 32: 13+, Sept. 1950, and *Gas* 26: 121–8, Oct. 1950.
7. Frost, E. M., Jr. *Helium Tracer Studies in the Elk Hills, Calif., Field.* (RI 3897) Washington, Bur. of Mines, 1946.
8. Kornfeld, J. A. "Aquifers for Underground Natural Gas Storage." *Am. Gas J.* 187: 45–8, Ap. 1960.
9. Kramer, W. H. and others. "Recent Developments in LP-Gas Storage." (CEP-55-14) *A.G.A. Proc.* 1955: 900–14.
10. Raymond, L. C. "Profit Possibilities in Cavern Storage of Natural Gas Near Major Markets." *Gas Age* 112: 21–4, Oct. 8, 1953.
11. Personal communication from Public Service Co. of Colorado.
12. Bircher, J. R. "Storage of Coke Oven Gas." *Gas* 23: 48–51, Dec. 1947.
13. Fruechtenicht, H. L. and Simpson, J. B. "Underground Storage." *A.G.A. Monthly* 31: 18+, June 1949. Also in *Gas Age* 103: 27+, June 23, 1949.
14. Katz, D. L. and others. *Movement of Underground Water in Contact With Natural Gas.* New York, A.G.A., 1963.
15. Gentry, H. L. "Storage of High Pressure Natural Gas in underground Salt or Rock Caverns." *A.G.A. Op. Sect. Proc.* 1963: GSTS-63-5.

A bibliography consisting of 560 technical articles which have appeared in trade and scientific journals, as well as unpublished papers or brochures which have any bearing on storing gas underground was prepared as a project of the Committee on Underground Storage of the Operating Section of A.G.A.:
Grow, George C., (ed.): Bibliography of Underground Storage, OP-59-4, A.G.A., New York, July 1, 1959 (the break-off point was Jan. 31, 1959).

Semi-annual supplements to this bibliography have subsequently been published.

Chapter 4

Liquefied Natural Gas and Nonconventional Natural Gas Storage

by Marvin D. Ringler

Of the storage systems considered in this chapter, only liquefaction and absorption appeared at all attractive in the U. S. in 1962. Liquefaction should cause fewer operational difficulties and require much smaller storage volumes than the other systems, and for these reasons it is examined in greater detail.

Liquefaction is a means of preparing natural gas for either shipment or storage. One cubic foot (or 0.177 bbl) of liquid *methane,* at −260 F and 1 atm, is the equivalent of approximately 630 SCF of methane gas. Higher temperatures could be used if the liquid were stored under pressure. Some of the engineering and economics involved in storing liquefied natural gas at pressures of about 325 psig are examined on page 10/36. The highest temperature at which methane could be liquefied is −116.3 F (**critical temperature**); the corresponding critical pressure is 673 psia. Commercial storage at or approaching this condition is not considered practical.

The bubble points for natural gases which contain 98 per cent methane and 72 per cent methane are approximately −259 F and −250 F, respectively. A gallon of gas liquid at −263 F weighs 3.46 lb (0.42 sp gr) and has a heating value of approximately 86,000 Btu. Gas density (in lb per cu ft) varies appreciably at low temperatures; e.g., 0.0448 at 32 F, 0.109 at −259 F (density, in lb per MSCF = 42.4).

The removal of about 16 Btu converts an SCF of gas at 40 F to liquid at −260 F. The heat of vaporization of liquefied methane at one atmosphere is approximately 10 Btu per SCF (see Fig. 1-3a); i.e., the addition of 10 Btu to liquefied methane at 1 atm will cause the formation of a quantity of gaseous methane *at −260 F* that is equivalent to 1 cu ft at standard conditions (60 F and 14.7 psia). It takes 6575 Btu to vaporize a cubic foot of liquid CH_4. About 20 lb of steam (or 20 MBtu) are required to vaporize enough liquid methane at 1 atm to form an MSCF of gas at standard conditions.

Early notable events in the history of natural gas liquefaction include the development of *helium separation* in Texas in 1924,[1] and the building of the first *pilot plant,* Hope Natural Gas, W. Va., 1940 (300 MCF per day, one MMSCF storage).[2] The *first commercial plant* was built by The East Ohio Gas Co., Cleveland, Ohio (Feb. 1941 to Oct. 1944, 4 MMSCF per day liquefaction, 240 MMSCF storage, 72 MMSCF regasification).[3,4] The ammonia–ethylene cascade system used gas engine-driven compressors (3300 hp) and gas-fired boilers (300 hp). This equipment consumed 333 cu ft of natural gas per MCF natural gas (at 30 psig) liquefied. Holder failure and subsequent fire destroyed part of the plant and the remainder was dismantled. A safety standard for liquefaction facilities is available;[5] it notes that clearances for liquefied natural gas storage need not be more stringent than that for gasoline.[6]

The *first foreign installation:*[7,8] A U. S.-built plant has been in operation in a suburb of Moscow since 1954. This cascade cycle uses ammonia, ethylene, and natural gas as refrigerants. Liquefaction: 320 MCF per hr; storage: 7.25 MMCF at 9 to 12 psig. Liquid methane distribution by specially designed tank trucks serves Moscow suburbs beyond regular pipelines. It was claimed that this practice is technically and economically feasible up to a distance of 60 miles.

TREATMENT PRECEDING LIQUEFACTION

Before liquefaction, CO_2, H_2S, and water vapor must be removed, since their presence would plug the liquefaction unit. Purification or sweetening, e.g, by **scrubbing** with *MEA* (monoethanolamine), removes both H_2S (plus other sulfur compounds) and CO_2. The resultant scrubber solution is regenerated by means of heat.

Dehydration

Process gas leaving the scrubber is water-saturated and must be dried to avoid icing in the cold unit. Parallel solid-desiccant type adsorber units are alternately operated and then regenerated with a high temperature gas stream. The regenerator gas is cooled, separated from the resulting water, boosted in pressure to account for system pressure drop, heated, and recycled.

LIQUEFACTION CYCLES

Some cycles cool feed gas, initially above 400 psia, to below its liquefaction temperature, and thence flash the gas in successive stages to approximately atmospheric (−260 F) storage. Factors influencing choice of cycle include: capacity, initial and operating costs, and weight and space limitations. Location, i.e., shore or shipboard,* is also a factor. Refrigerant selections, efficiency aside, are also influenced by ease of replacement (for example, use of refrigerants which could be derived from the feedstock).

The choice[9] of liquefaction cycle for lowest overall cost seems to be between the Heylandt expansion cycle (with

* A Conch Methane Services, Ltd. barge-mounted liquefaction unit of 8 MMSCF per day capacity was started up in 1956.

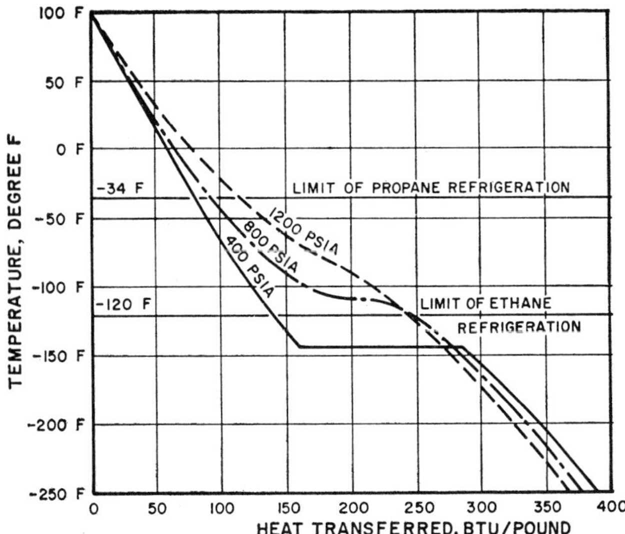

Fig. 10-21a Cooling curves for methane at several pressures.[12]

precooling to avoid high pressure in the expansion engine) and cascade cycles. The latter are more complex but they operate at lower pressures and require less heat transfer area.

The two basic types of liquefiers, according to the method of cooling, are: (1) those which depend mainly upon the compression and expansion of the gas to be liquefied—either primarily by the Joule-Thomson effect or thru an expansion engine generating external work;* and (2) cascade systems, which depend mainly upon auxiliary refrigerants—either via a vapor recompression cycle or an expansion engine.

Joule-Thomson Effect. The expansion (throttling) of a *real* gas, from a point on or within the temperature-pressure plot of the inversion curve—locus of the maxima of isenthalpic curves plotted on a temperature pressure diagram[10] —of the gas results in a cooler gas. Specifically, expanding compressed methane from pressures as high as 200 atm and temperatures not too far above critical yields a cooling effect. This refrigeration over a range in temperature is used to cool a gas to the liquid state. The extent of the Joule-Thomson effect increases when there are high boiling point impurities in the product stream.

Vapor Recompression Principle. The influence of pressure on the boiling point of a liquid is used to obtain cooling at a constant temperature.

Small or mobile (e.g, shipboard) units generally employ an expander cycle because of its lower capital investment. Assume the liquefaction of one MMSCF per day of stripped natural gas received at 700 psig. A well-designed *cascade* unit thus rated requires about 460 bhp and uses about 12 per cent of the input stream. On the other hand, a comparable *expander* plant requires 830 bhp and uses about 20 per cent of the input stream;[11] i.e., the process output is 100 bhp, net input is 730 bhp.

Figure 10-21a shows how much heat must be removed from methane to cool it to a temperature as low as −250 F at feed gas pressures ranging from 400 to 1200 psia. Note that the discontinuity in the cooling curve disappears well above the critical pressure; i.e., the specific heat becomes approximately constant at high pressures. This figure also shows the design limit temperature levels of propane and ethane refrigeration chosen to preclude a vacuum at the refrigerant compressor. Note the significance of feed gas pressure, i.e., approximately 243 Btu per lb are removed from 800 and 1200 psia methane feed gas at temperatures above the −120 F ethane refrigerant design limit, while at a feed gas pressure *below the critical pressure,* e.g., at 400 psia, methane refrigeration at −145 F (7.25 bhp per ton of refrigeration) is required to remove *latent heat* (the horizontal portion of 400 psia curve in Fig. 10-21a). Refrigeration horsepower requirements vary *inversely* to feed gas pressure, since the shape of the **cooling curve** dictates the proportion of heat that can be removed at various levels. Estimates of brake horsepower as a function of refrigerant evaporating temperature are available.[13,14]

Cooling curve construction for a given natural gas requires: dew point, bubble point, and proportion and composition of the vapor and liquid present at various temperatures. A consistent set of vapor-liquid equilibrium data should be used.[13,15] The assignment of consistent enthalpy values involves employment of *partial* enthalpies since some components of the feed gas will be in the vapor phase when their properties as a pure material indicate they should be liquid. Other components will be in the liquid phase above the critical point. Available enthalpy data[16–19] contain many inconsistencies.

Venting of gaseous impurities like air, nitrogen, oxygen, helium, and argon is necessary because these do not liquefy at or below the liquefaction temperature for methane (−259 F and one atmosphere).

Draining of liquid impurities, like aromatics, is necessary because they are insoluble in the main body of liquid at approximately −200 F.

Saturated hydrocarbons can remain in true solution in liquid methane. They *raise* the boiling point (thereby *reducing* the amount of energy required for liquefaction), calorific value, and density of the liquid.

Expander Cycles

Natural gas at high pressure (say, 1500 psia) is liquefied by a progressive gas-to-gas heat exchange against approximately twice its volume of cold recycle natural gas or methane, also at high pressure. The recycle gas (refrigerant) is then expanded† *isentropically* to approximately −190 F (150 psia) before heat exchanging against the liquefying feedstock gas, which is cooled below its *critical* temperature. The refrigerant is exhausted as saturated vapor at, say, −225 to −250 F.

Flashing a portion of the feedstock into the storage tank (at one atmosphere) cools the remainder to its final temperature; boil-off gas from storage is also heat-exchanged against the feedstock. Nitrogen is stripped from the liquid feedstock.

Cascade Systems

These systems use several refrigerants in series, each employed over the temperature range in which it is most

* If a system used 1300 hp for a compressor and recovered 300 hp in an expansion engine, a net of 1000 hp would be consumed.

† Causing a temperature drop thru the Joule-Thomson effect and the removal of work in expander engines.

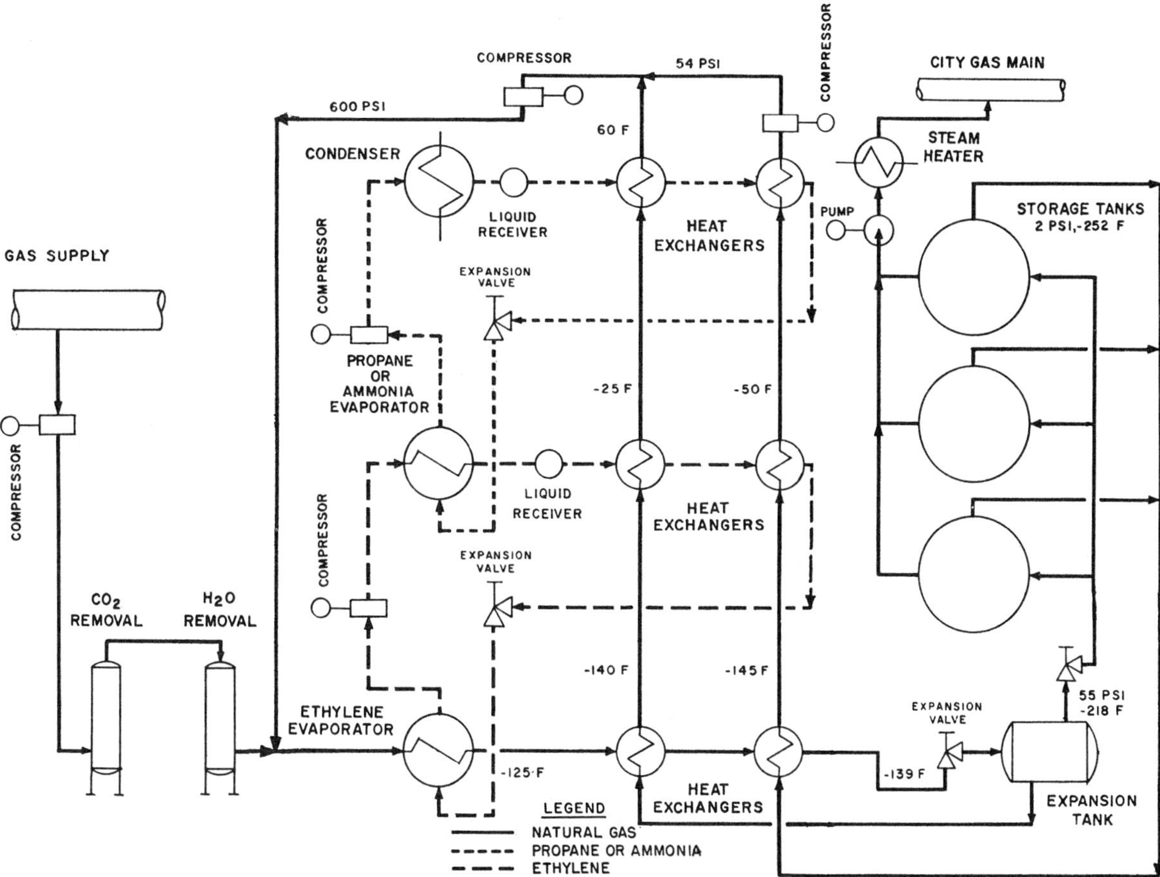

Fig. 10-21b Flow diagram of a cascade cycle for liquefying natural gas.[20]

efficient. For example, in Fig. 10-21b, water or air condenses propane; propane, in turn, cools natural gas to about −34 F and also condenses ethane refrigerant. The ethane cools and condenses the natural gas to −116 F and condenses methane refrigerant. Methane refrigerant can then chill the liquid gas to the storage temperature, approximately −250 F. Combinations of refrigerants other than those shown in Fig. 10-21b can be used. The cascade system is thermodynamically more efficient than the expander system, because it is more nearly reversible.

Recycle gas consists of methane gathered from the storage boil-off vapors (resulting from heat leakage to storage), compressed and combined with, say, 50 psia flash vapors (corresponds to −228 F). This stream is further compressed and combined with 150 psia expanded gas. This total recycle stream is then compressed thru two more stages to a final pressure of 1500 psia and cooled by heat exchange with ammonia or propane refrigeration cycles. The nitrogen stripped from the feedstream also serves as an additional refrigerant.

Isentropic expansion of the recycle gas powers the NH_3 or C_3H_8 refrigerant compressors.

STORING LIQUEFIED NATURAL GAS

Cost estimates for the major liquefied natural gas storage techniques were made by firms with long experience in the required operations (based on a billion SCF total storage, i.e., 285,000 bbl).[21]

The estimates were:

	$/MCF storage
Aboveground double-wall tanks	2.85 to 6.50
Mined and lined cavern	1.55 to 2.12
Quarried limestone pit, lined and covered	0.71 to 0.77
Buried prestressed concrete tank, lined	1.25 to 1.50

Another report suggested that natural gas could be liquefied, stored, and revaporized for $1.10 to $1.50 per MSCF for a three billion cubic foot facility. (This assumes 200 days of valley gas available for filling storage and one complete turnover of the storage each year.)[22]

A storage vessel is designed for an evaporation rate that reflects the reason for the liquefaction; i.e., a peak shaving unit should have a low evaporation rate, say, 0.1 to 0.2 per cent per day. A heat transfer rate under 2 Btu per hr-sq ft of outer tank surface is a good rule of thumb.

Safety Considerations.[5] Besides the several modes of underground storage, liquefied natural gas can also be stored safely in suitably designed aboveground tanks surrounded by earthen dikes, in much the same manner as gasoline.[6] Experiments[5] evaluated the potential hazards associated with liquefied natural gas compared to other common fuels.

Some interesting factors may be considered: (1) Vaporization—massive spillage and evaporation of the liquefied gas may sufficiently cool surrounding surfaces so that ground level

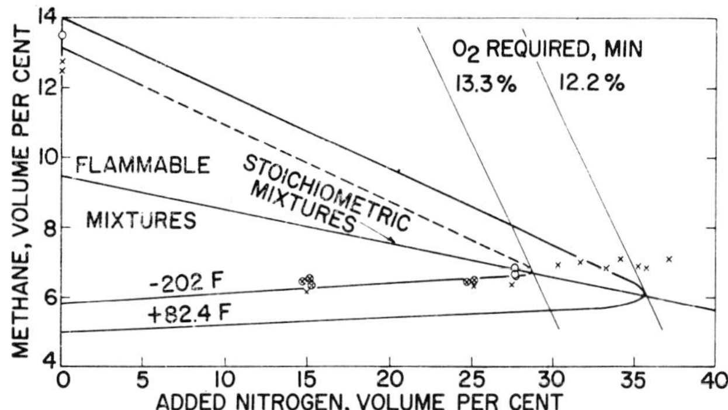

Fig. 10-22 Limits of flammability by volume of methane in air combined with various amounts of nitrogen at −202 F and +82.4 F.[5]

layering of gas may result. Since this gas is cold, it would be heavier than ambient air. (2) Assuming that this evolved gas ignited following* the initial 20 to 30 sec of initial film boiling, a steady rate of combustion would be established over the liquid pool—roughly 23 per cent (percentages as high as 34 were reported) of the heat generated would radiate into the liquid pool under windless conditions. (3) A pit, 20 ft square by 21 in. deep, filled within 5 in. of the rim, burned more than 0.5 hr without frothing or burnover—the same results are anticipated for any size liquefied natural gas fire. (4) Although the radiation from a natural gas fire exceeded that from a comparable gasoline fire, the flame size was roughly the same and the former was more easily extinguished than the latter (with finely powdered sodium bicarbonate). (5) Both sodium bicarbonate and potassium bicarbonate can be used to extinguish fires quickly, applying about 0.14 lb of powder per sq ft-sec over the entire fire surface; longer times are required if less than the entire area is attacked at once.

Flammability Limits at Low Temperature.[5] The following equation may be used to determine the lower flammability limit of methane-air mixtures at atmospheric pressure and at temperatures between −247 and +77 F:

$$L_L = 5.0 + 0.0029\ (77 - t)$$

where: L_L = lower flammability limit of methane, per cent by volume

t = temperature, °F

In general, a decrease in temperature narrows the flammability range of a gas (Table 2-78). Figure 10-22 compares ranges for methane in air with various percentages of nitrogen diluent at −202 F and +82.4 F.

Metal Storage Tanks

Insulated tanks store liquefied gas at or slightly above atmospheric pressure (oxygen usually at −300 F). Aluminum (e.g., alloys 5083 and 5456), a 9 per cent nickel-steel alloy,† and certain stainless steels‡ may be used as an inner liner, while carbon steel makes an effective outer covering; resistance to stresses caused by large temperature changes and resistance to low temperature embrittlement are important material properties.

Thermal stresses on inner walls when a tank is filled are due to the temperature differential of over 100°F which may exist from liquid level to tank top, with a particularly sharp gradient immediately above the liquid. This condition is aggravated in a moving transport.

The **tank bottom** may consist of an inner liner on top of layers of perlite concrete and insulation (plus concrete, sand, and crushed stone in on-earth installations). Keys or anchors secure the tank bottom to the perlite, to preclude buckling the bottom liner because of stresses due to temperature differentials. Heaters may be used to prevent the earth under tank from freezing and heaving; mounting the tank on pilings is an alternative solution.

Insulation consisting of three feet of granular perlite (expanded volcanic lava) is typical. Daily heat leakage losses of 0.2 per cent of maximum tank volume were reported for the Conch Methane Services, Ltd. tank (67 ft diam, 56 ft high, dome-roofed) at Lake Charles, La.[11] One 46 ft diam spherical shell is insulated with 5 ft of perlite within an annular space evacuated to 10μ Hg.[24]

The insulated annulus between the outer and inner walls may be pressurized with a dry gas to prevent the entry of moisture into the insulation. Either the tank design or the insulation itself (e.g., a section of resilient insulation) should take account of dimensional changes in the annulus which might otherwise compress the insulation sufficiently during temperature transitions (fillings and emptyings) to resist tank movements. Of course, the inner tank should be protected by relief devices against pressure buildups beyond design rating.

Balsa wood, although permeable, was used as an insulator. However, any liquid which tends to penetrate the insulation reaches a point where it boils. The resultant gas blocks further liquid penetration.

Costs.[25] Component costs of a 60 Mbbl, flat bottom, double shell, insulated tank (as percentages of overall tank cost) follow: outer carbon steel shell of ASTM A-283 Grade C, 17.5; inner vertical shell plus stiffeners, both of nine per cent Ni steel, 41 + 9 ; inner tank roof and supports of nine per cent Ni steel, 9; inner tank bottom, 5.5; insulation: between shells (6.2), in double roof space (1.8), inner tank bottom (10). Below 45 Mbbl, a suitable aluminum alloy would be the economical material for the inner tank.

Diameter-to-Height Ratio.[25] Minumum costs for the 20 M to 130 Mbbl capacity range is for a 1.5 ratio; above 130 Mbbl, the ratio approaches 2.5.

Relief Valve Sizing. The LNG Task Group (1963)—Chemical and Engineering Committee of the A.G.A. Operating Section, compared the heat transfer aspects of liquefied natural gas storage with that for LP-gas. It was noted that while relief valve sizing for LP-gas is based on heat transfer during a fire (Appendix E of NFPA Standard No. 59—extracted in Sec. 5, Chap. 5), relief sizing for liquefied

* Earlier ignition (within one minute of spillage) produces a large momentary flash, but no evidence of overpressure or splashing of liquid.

† Following suitable heat treatment (quenched and tempered), recommended for service as low as −320 F ("Operation Cryogenics," Fairless Works of U. S. Steel, October 1960). Code[23] permits use of this steel up to 1¼ in. thick without post-fabrication stress relief (up to 2 in. thick, if relieved).

‡ Several 18-8 stainless steels are suitable, e.g., austenitic type 304.

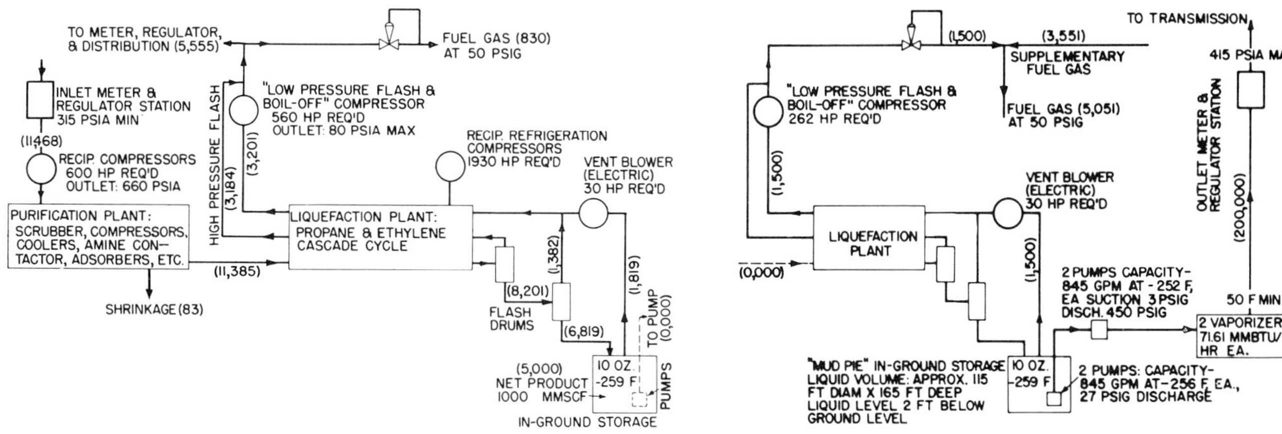

Fig. 10-23 Liquefied natural gas pit storage (Mudpie) facility flow diagrams: (left) injection of 5 MMSCFD; (right) withdrawal of 200 MMSCFD. Numbers in parentheses are gas flow rates in MSCFD at 14.7 psia.[27]
Fuel Gas Requirements:
Injection— 830 MSCFD for inlet gas, "low pressure flash and boil-off" and liquefaction plant compressors, and purification plant.
Withdrawal—5051 MSCFD for inlet gas and "low pressure flash and boil-off" compressors and vaporizers.
Notes: Flash gas is gas that was in the liquid state but returned to the gaseous state when its pressure was reduced during the liquefaction process. Boil-off gas is gas vaporizing in the storage container due to heat transfer into the container. (Transcontinental Gas Pipe Line Corp. facility, to be completed in 1965–6, in Bergen County, New Jersey)

natural gas should take account of heat gain during both (1) fire—heat gain $\sim \Delta t_1 A^{x<1}$* and (2) refrigeration failure—heat gain $\sim \Delta t_2 A^{x=1}$. Note that while the temperature differential, $\Delta t_1 > \Delta t_2$, the area term, A^x is smaller for fire considerations than for refrigeration failure. Therefore, at some value of A, the heat transfer attributable to a fire equals the heat transfer attributable to usual ambient conditions. Should refrigeration failure occur, even ambient heat transfer would require means of relief. In general, the total net heat influx to the above grade portion of a tank exposed to a fire includes: (1) gain due to fire, (2) gain due to ambient conditions, and (3) increased gain resulting if insulation deteriorates or is dislodged.

Pit Storage (Mudpie)[26]

This process involves freezing the ground that surrounds the area to be excavated, digging the hole, and installing a vapor-tight sealed aluminum roof over the hole before filling it with liquefied natural gas. A hole 20 ft in diam by 20 ft deep, containing 850 bbl (3 MMSCF) liquefied gas, showed no leaks or ground swelling problems in a 30 day heat. The rate of boil-off decreased with time to about four per cent of the contents of the pit per day. Figure 10-23 shows the first commercial application of this technique.

Concrete Tanks[28,29]

The largest (16,000 bbl) concrete tank for cryogenic storage in use (1962) is located in East Chicago, Ill. A 1961 study suggested the following tank design features: (1) single shell prestressed concrete structure consisting of a roof, wall, base ring, and floor—all as separate members; (2) either poured-in-place or precast wall and roof sections; (3) roof, domed or flat (supported by pillars), made from concrete or nine per cent nickel-steel; (4) insulation, sufficient to restrict boil-off to desired value—0.1 to 0.2 per cent of tank capacity daily considered reasonable for a peak load facility; (5) continuous metal barrier, stainless steel or nine per cent nickel steel, located on the inside at the floor and outside of the concrete wall and roof (suggested with reservations since its need was not fully established and its cost may approach 40 per cent of the complete structure); (6) backfill of coarse sand and gravel around and below the tank: (7) heating coils below and around the tank to prevent frost heaving—32 F isotherm held within select backfill.

Design Study.[28] A general design developed for a 285,000 bbl tank included: (1) experimental determinations of low-temperature properties of materials—moist concrete compressive strength about 17,000 psi at −260 F and 5000 psi at 75 F; thermal conductivity of concrete at 2.0 Btu-ft per hr-sq ft-°F at −140 F and 1.8 at 80 F; various aspects of concrete reinforcement, e.g., tensile strength of reinforcing wire at −250 F exceeded the room temperature value by up to 30 per cent; (2) heat transfer calculations to determine insulation requirements and related boil-off rates, and thermal stresses in cool-down and operation; (3) low-temperature effects on soil as related to installation design.

Cavern Storage[30]

Where suitable porous structures or caverns are feasible (Fig. 10-20) in proximity to the market, this type of liquefied natural gas storage may be used.[31] Advantages include increased safety (even at several atmospheres),[31] lower initial and maintenance costs and savings in both space and materials.

Heat transfer from the surrounding rock depends upon the cavern surface area and the type of surrounding rock. The slope of the temperature gradient thru the rock is also a function of the length of time the storage facility has been in service. For example, at the end of 10 hours the temperature of a 50 ft diam uninsulated cavern in average rock, 1

* The exponent x accounts for the fact that the surface of a large tank is not as severely exposed to an open fire as that of a small tank.

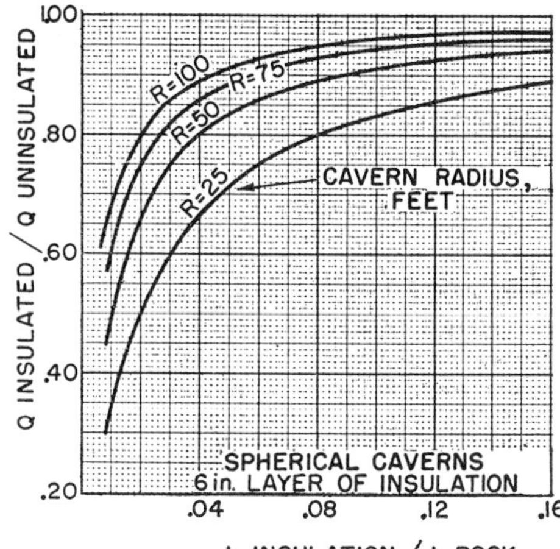

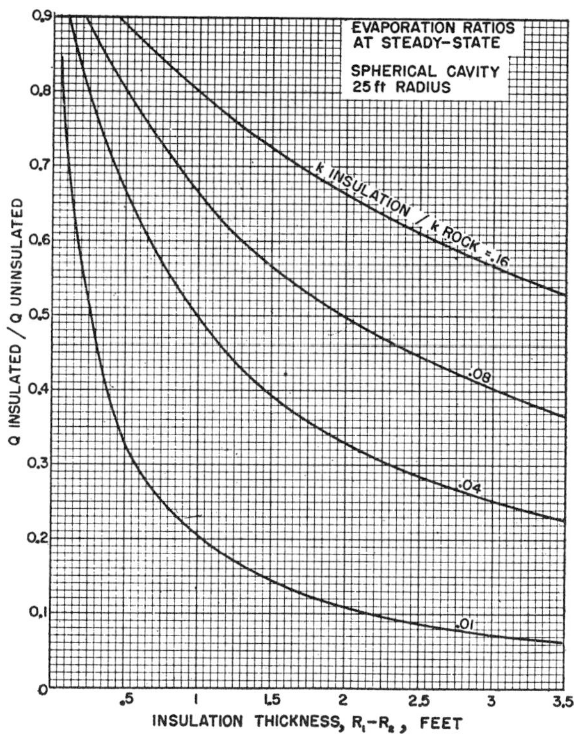

Fig. 10-24a (above) Effect of cavern size and insulation on methane evaporation rate (constant: 6 in. layer of insulation).[30]
Fig. 10-24b (right) Effect of insulation on methane evaporation rate (constant: 25 ft cavity radius).[30]

Note: Uninsulated caverns are not desirable since it may take an unreasonable length of time to attain steady-state conditions.

in., 6 in., and 12 in. from the methane-rock interface, would measure −240, −105, −15 F, respectively. The steep temperature gradient indicates the thermal insulation properties of rock.* Thus, the possibility is relatively small that the liquefied gas would become contaminated from gases in the surrounding structure; i.e., H_2O, CO_2, H_2S, NH_3, and other gases would solidify before reaching the storage cavern.

The following formula may be used to determine the heat loss from spherical caverns (see Table 10–12 for representative values):

$$Q = 4\pi k R \Delta T\left[1 + \frac{R}{\sqrt{\pi \alpha t}}\right]$$

where: Q = heat loss, Btu per hr
k = thermal conductivity, Btu-ft per hr-sq ft-°F
R = cavern radius, ft
ΔT = temperature difference, cavern to earth, °F
α = thermal diffusivity of rock, $k/\rho c$, sq ft per hr
where ρ = rock density, lb per cu ft
c = specific heat of rock, Btu per lb-°F
t = elapsed time, hr

Insulation of cavern walls reduces liquid evaporation; the ratio of thermal conductivities, k (insulation) to k (rock), is a measure of insulation effectiveness. Figures 10-24a and 10-24b show insulation effectiveness for various cavern sizes and thicknesses of insulation, respectively.

Feasibility Study.[31] A report of theoretical studies, laboratory investigations, and cavern experiments noted the following:

1. A complete mathematical solution for the soil temperature distribution surrounding a cavern was feasible only for *spherical* reservoirs.
2. Surface-to-cavern top depth of 100 ft precludes surface frost.
3. Very high lateral stresses with maximum occurring in continuous service (stress analysis for a spherical cavity assuming constant entropy and constant rock expansion). Thus, it is unnecessary to program the cooling of the cavern walls. Insulation substantially reduces these stresses.
4. The differences between hypothetical and actual caverns should be considered; i.e., samples cannot ascertain cavern shape, non-homogeneity of soil (e.g., horizontal stratification), physical properties of rock,erratic behavior of water in rock, pre-existing pressure in the soil, and properties of rock *in situ*.
5. While *non-porous* concrete linings showed satisfactory endurance, *porous* concretes intended as heat insulation had a high water content and failed as insulators.
6. Frost formation caused numerous difficulties, particularly by fouling mechanical equipment in the excavation.
7. A procedure placing the reservoir in service.

Table 10-12 Thermal Properties of Various Materials[30]
(in air at ordinary temperatures except as noted)

Material	k*	c	ρ	α
Granite	1.6	0.19	168	0.050
Limestone	1.2	0.22	168	0.032
Marble	1.3	0.21	168	0.037
Sandstone	1.5	0.21	162	0.044
Average rock	1.4	0.21	165	0.04
Concrete, average stone	0.54	0.20	144	0.019
Calcareous earth, 43 per cent water	0.41	0.53	104	0.007
Quartz sand, medium fine, dry	0.15	0.19	103	0.008
Ice	1.28	0.49	57	0.046
Asbestos	0.09	0.25	36	0.010

* For water-saturated rock at low temperatures, measured conductivity = conductivity of dried specimen × 100/(100 − rock porosity).[31]

* In one month: 1 in. at −250 F, 1 ft at −210 F, 5 ft at −70 F, 10 ft at +25 F.

8. Only liquids should be withdrawn (not gases) to avoid increasing the concentration of heavy liquid products and, possibly, sedimentation.

9. Vapor formed by the heat gain should be allowed to increase the pressure in the reservoir to the allowable installation limit.

Auxiliary Equipment

Aluminum and some of its alloys, austenitic stainless steel, and copper and some of its alloys, have proved to be suitable materials for the low temperature service. Dimensional changes due to the temperature differentials involved are a major design criterion. Flanged pipeline joints, for example, require components that maintain closure when suddenly chilled.

Transfer.[32] Minimization of heat gain, with its accompanying gasification of a portion of the liquid stream, is the primary objective. Analyses of two-phase flow with boiling heat transfer were not reported. Treatment of single-phase transfer of liquefied He, H_2, Ne, N_2, and O_2 is available. Although methane was not included in these, the techniques involved are similar.

Basically, the four losses minimized are: (1) *pumping*—pressure drop in single flow* very much less than in two-phase; (2) *flashing*—heat leak in transfer line plus heat added to liquid by pump; (3) *cool-down*—heat gain to system (from ambient to operating temperature); and (4) *trapped liquid*—volume of transfer system. The latter two losses occur in the start-up and shutdown of the transfer system, respectively.

Pipe Insulation. Use of high vacuum (under 10^{-5} mm Hg) and evacuated powder (10^{-2} mm Hg), where justified from the economic viewpoint, results in very minor heat losses compared to conventional insulations such as glass wool or foam glass. A basic transfer line arrangement may consist of two concentric casings; the annular space thus formed confines the high vacuum or evacuated powder. A more efficient adaptation of this arrangement adds two additional concentric casings. The inner of the annular rings thus formed conveys a shielding liquid (easily refrigerated), while the outer ring confines a second insulating annulus.

A comparison of heat losses for a 4 in. diam transfer line shows:

System	Heat loss, Btu/hr-ft	
	With liq. N_2 shield	Without N_2 shield
1. High vacuum, 0.04 wall emissivity	0.0167	3.82
2. —80-mesh perlite at 10^{-2} mm Hg, 5 to 1 ratio—insulation to flow diameter	0.117	1.43

Valves. The gate type with extended covers permits the gland location outside the cold region. Vertical orientation of the valve spindles permits an insulating gas pocket to form under the gland. Relief valves are located wherever the possibility of trapping liquid methane exists.

Pumps. Provision for a positive suction head prevents cavitation which might otherwise occur under mixed phase flow. Thermal contraction of moving parts, bearing lubrication, and seal closure are the major design difficulties.

TRANSPORTING LIQUEFIED NATURAL GAS

The tanker, *Methane Pioneer,* first crossed the Atlantic in 1960 (from Lake Charles, La., to Canvey Island near London, England) with a cargo of 32,000 bbl (112 MMSCF) of liquefied methane. The gas was liquefied at a Lake Charles facility of six to eight MMSCFD capacity.

This voyage indicated *technical* feasibility of ocean transportation of liquefied natural gas. The *economic* attractiveness of this activity depends mainly upon the price and reliability of the source of supply. A transport capacity of 500 MMSCF was considered an economic necessity.[22] Preparations were under way in 1962 to pipe Sahara gas to Port Arzew (near Oran), Algeria for liquefaction.[34] Two tankers, each with a capacity of 12,000 tons of liquid methane, as well as an expanded receiving and storage facility (Canvey Island, England) were also under construction. The French have started construction of a receiving facility at Le Havre which will have a capacity of 15.5 billion cu ft per year starting in late 1964.[38]

Shipboard Installations[33]

Design features include double hull construction and free-standing tanks insulated from the inner hull. Deep hulls are practical, since cargo specific gravity is only 0.42 (water = 1.0).

The welded seams employed throughout the aluminum tanks of the *Methane Pioneer* were X-rayed in accordance with the low porosity standard set by API Tentative Standard 12G (1957). There were no connections to tank bottoms, and all piping, vents, and other auxiliary equipment were directed from the top of the tanks thru the weather deck of the ship.

The space between the cargo tanks and the insulation is maintained at a small positive pressure by means of dry N_2, to prevent the ingress of moist air into this cold temperature region, as well as to provide an inert atmosphere in the insulation space. Continuous *infra-red analysis* of this atmosphere checks for ingress of methane.† An additional *temperature detection* system monitors inner hull temperatures to disclose cargo leakage. Heaters are provided to return inner hull temperatures to normal is such cargo leaks should occur. Severe leaks may necessitate jettisoning the contents of the tank involved.

It was concluded from both tests and properties that liquid methane was potentially less hazardous than gasoline. Pertinent considerations include the narrow range of flammability (approximately 5 to 15 per cent), the high ignition temperature (approximately 1200 F), the low flame velocity (approximately 1.25 fps), and the maintenance of a gas pressure above the liquid, which prevents the ingress of air. These design features also preclude the development of major problems from minor collisions.

* Liquid transfer rate of 120,000 gph thru a well-insulated, 10 in. diam., 2300 ft pipe with a 6 psi drop was reported.[33]

† The evaporation of methane can be avoided by the addition of sufficient inert liquefied gas (with a boiling point below CH_4), say N_2, to absorb the heat transferred thru the container walls.[35]

REGASIFICATION

The basic elements in a vaporization system are: (1) a pump and piping to deliver the liquid methane at, say, 100 psig to (2) a heat exchanger (tube and shell type and tube in strip type have been used), and (3) a heat transfer medium (both river water and propane heated by 50 psig steam have been used).

The liquid natural gas (at −260 F) is heated in a vaporizer where the liquid is first vaporized; the resulting vapor is then superheated to the range of 30 to 60 F. Theoretically, 15,800 Btu per MSCF (actually 20 lb of steam) are required to vaporize to 60 F and 400 psia. The sendout should be warm enough both to be compatible with pipeline materials and to avoid problems in mixing with existing pipeline contents, e.g, hydrates. Vaporizers may be direct-fired or heated with steam. To avoid the possibility of freezing the steam condensate, a binary cycle may be used (e.g., a steam-to-propane exchange, followed by a propane-to-methane exchange). Shell-tube arrangements are recommended—natural gas in stainless steel tubes—otherwise carbon steel may be used throughout.

High pressure gas can be obtained from the vaporizing equipment due to the high vapor pressure of methane. If CH_4 at the critical point (−116.3 F and 673 psia) is superheated at constant volume, a pressure of 1020 psia can be developed at 60 F; thus, gas recompression plants are unnecessary.

A portion of the energy used in liquefying the gas can be recovered; possibilities include direct mechanical power or generation of electricity thru a reverse cascade system. Of course, the most effective use of such energy would be refrigeration at a temperature level slightly above the boiling point of methane (−260 at 1 atm). Such applications include steps in liquefying O_2 and solidifying CO_2.

Heating Boil-Off Vapor. When the liquefaction plant is in operation and the tanks are being filled, this vapor would normally be heat-exchanged against the incoming gas stream to reclaim refrigeration, then it would be recompressed and reliquefied. However, under conditions of static storage, it may be more economical to superheat the vapors from −260 F to +60 F and use them for fuel or sendout. This operation represents a minor duty of the vaporizers (approximately 29,000 Btu per hr to heat 100 MCFD of boil-off).

ECONOMICS

Estimated costs for constructing liquefied natural gas storage and sendout facilities are available, as well as annual costs of operation.[20]

Ocean-going tanker costs vary widely, but $700 to $850 per ton of carrying capacity seems a reasonable estimate. Corresponding prices for vessels carrying LP-gas were 30 per cent less.

Comparisons of pipeline transmission of natural gas with liquefaction and ocean transportation of same have indicated a break-even point at 3000 miles for moving large volumes (500 MMSCFD). Pipeline transmission distances of 1300 miles were indicated as equivalent (investment costs) to 4000 miles of ocean transport of 50 MMSCFD liquefied gas.[36]

NONCONVENTIONAL NATURAL GAS STORAGE[21]

(For the most part, the following is quoted directly from the reference.)

Absorption. Natural gas is slightly soluble in many materials and substantially more soluble in other hydrocarbons. Possibly the best known sorbent is propane, for which the solubility of natural gas (assumed to be pure methane) is presented in Fig. 10-25. The maximum quantity of natural gas that can be dissolved in propane at a given temperature is that which will raise the vapor pressure of the liquid mixture to the storage pressure. For example, the operating pressure of existing caverns is governed by their depth, with a pressure gradient between 0.44 (hydrostatic) and 1.0 psi per ft depth, or, based on the depth of present caverns, a probable maximum pressure of about 600 psia.

From Fig. 10–25 ~~it~~ can be seen that at 600 psia the volume of natural gas dissolved in propane per cu ft of liquid mixture would be: 39 SCF at 104 F (compare with the fact that 40 SCF of methane would occupy one cubic foot when com-

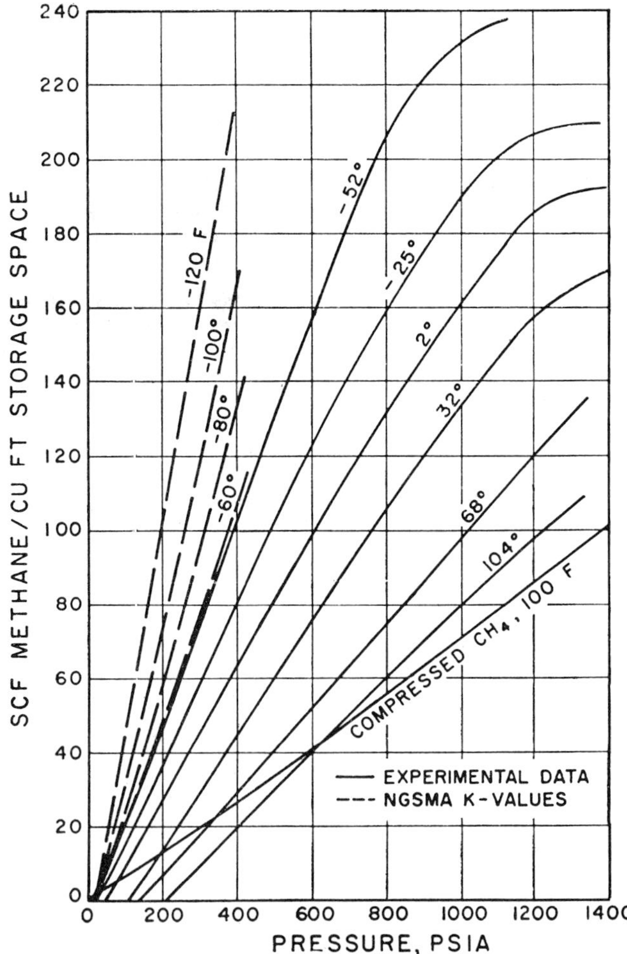

Fig. 10-25 Solubility of methane in liquid propane at various temperatures and pressures.[21] (K is the vapor-liquid equilibrium ratio. It may be defined as the mole fraction of the component in the vapor phase, divided by the mole fraction of the component in the liquid phase, when both phases are in equilibrium. NGSMA is the Natural Gasoline Supply Men's Association.)

pressed to 600 psia), 52 SCF at 68 F, and 155 SCF at −52 F (refrigeration would therefore appear essential).

The disadvantages of liquid absorbents are similar to those of the solid adsorbents. For large volume storage of natural gas, enormous volumes of LP-gas would be required. In order to receive any benefit from absorption, the system should be refrigerated and operated at elevated pressure. An additional problem would be that natural gas vaporized directly from the LP-gas sorbent would contain large quantities of LP-gas vapor, with the LP-gas concentration increasing as more of the methane is removed. One distinct advantage is that the heat of solution of methane in liquid LP-gas is apparently much less than the heat of condensation.

Adsorption. Natural gas tends to form an adsorbed layer or film on most solid surfaces, with the quantity depending upon temperature, pressure, and the activity of the solid surface. Approximately 16 SCF of methane can be adsorbed on one cubic foot of activated carbon at 70 F and 40 psia, which is six times as much as could be stored as compressed gas at the same conditions. The ratio decreases rapidly with increasing pressure, however, limiting the usefulness of this process to low pressures only.

The use of solid adsorbents has two very distinct disadvantages. First, the initial cost of the sorbent is high, and the volume required is large. Second, it is quite difficult to obtain rapid release rates of the adsorbed natural gas.

Reversible Chemical Combination. Natural gas will form reversible chemical bonds only with relatively few other materials. The most familiar combination is with water, to form **hydrates.** Each of the four lightest hydrocarbons, plus CO_2, and H_2S, will form hydrates in the presence of water under certain conditions of pressure and temperature.

The use of water makes for a cheap combining chemical, and it is possible to store about 170 SCF of methane per cubic foot of hydrate, at 35 F and 450 psia. A stable hydrate can be maintained at atmospheric pressure, if refrigerated to about −75 F.

Storage of natural gas by formation of the solid hydrate has several disadvantages. The heat of formation, which must be removed at 35 F or below, is very large for hydrates—74,000 Btu per MCF of methane. This heat must be resupplied when releasing the methane from the hydrate, and it is difficult to obtain enough heat transfer to the solid to give rapid release rates. It would also be necessary to handle large volumes of refrigerated slurry or solid at elevated pressures, and the storage would have to be refrigerated. The principal advantages are a cheap combining chemical (water), and refrigeration requirements at a much less severe temperature than for liquefaction.

A mathematical analysis of a proposed facility to store *manufactured* gas in underground sand strata is available.[37]

REFERENCES

1. Ormston, R. H. "Liquefaction—the Answer to Storage?" *Gas Age* 113: 27–33, Feb. 25, 1954.
2. Clark, J. A. and Miller, R. W. "Liquefaction, Storage, and Regasification of Natural Gas." *A.G.A. Proc.* 1940: 192–8.
3. Robinson, J. F. "Storage of Natural Gas." *Gas Age* 93: 23+, Feb. 10, 1944.
4. Turner, C. F. "Liquefying and Storing Natural Gas for Peak Loads." *A.G.A. Monthly* 26: 243–6, June 1944.
5. Burgess, D. and Zabetakis, M. G. *Fire and Explosion Hazards Associated With Liquefied Natural Gas.* (RI 6099) Washington, Bur. of Mines, 1962.
6. Nat. Fire Protection Assn. *Flammable Liquids Code.* (NFPA 30) Boston, 1962.
7. Parker, P. S. and Ormston, R. H. "Liquefaction Behind the Iron Curtain." *Gas Age* 113: 29–32, Mar. 11, 1954.
8. Newman, L. L. "The Gas Industry in the U.S.S.R." *Gas* 36: 59–65, Nov. 1960.
9. Barber, N. R. and Haselden, G. G. "Liquefaction of Naturally Occurring Methane." *Inst. Chem. Eng. Trans.* 35: 77–86, Ap. 1957.
10. Zemansky, M. W. *Heat and Thermodynamics*, 4th ed. New York, McGraw-Hill, 1957.
11. DeLury, J. "Liquefaction: New Gas Market." *Chem. Eng.* 66: 165–8, Dec. 14, 1959.
12. Young, A. R. "Economic Significance of Liquefied Natural Gas: the Supplier's Viewpoint." *Am. Gas J.* 187: 27–30, May 1960.
13. Natural Gasoline Supply Men's Assn. *Engineering Data Book*, 7th ed. Tulsa, 1957.
14. Lederman, P. B. and Williams, B. "Commercial Liquefaction of Natural Gas. *Oil & Gas J.* 55: 97–102, Sept. 30, 1957.
15. DePriester, C. L. "Light Hydrocarbon Vapor Liquid Distribution Coefficients." *Chem. Eng. Prog. Symposium Ser.* 49: 1, 1953.
16. Canjar, L. N. and Edmister, W. C. "Hydrocarbon Partial Enthalpies—Values for Methane, Ethene, Ethane, Propene, Propane, and n-Butane." *Chem. Eng. Prog. Symposium Ser.* 49: 73, 1953.
17. Canjar, L. N. and Peterka, V. J. "Enthalpy of Hydrocarbon Mixtures." *AIChE Jo.* 2: 343–7, Sept. 1956.
18. Papadopoulos, A. and others. "Partial Molal Enthalpies of Lighter Hydrocarbons in Solution." *Chem. Eng. Prog. Symposium Ser.* 49: 119, 1953.
19. Peters, H. F. "Partial Enthalpies of Light Hydrocarbons." *Petrol. Ref.* 28: 109–16, May 1949.
20. Stone and Webster Eng. Corp. *Gas Storage at the Point of Use.* New York, A.G.A., 1957.
21. Eakin, B. E. "Nonconventional Storage of Natural Gas." *A.G.A. Proc.* 1960: CEP-60-5.
22. Mellen, A. W. and others. "Storage and Transportation of Low Temperature Fuels." *A.G.A. Proc.* 1961: CEP-61-19.
23. "9 Per Cent Nickel Steel, SA-353 Modified." (Code Case 1308, ASME Boiler and Pressure Vessel Code) *Mech Eng.* 84: 92, Ap. 1962.
24. Clapp, M. B. and Wissmiller, I. L. "Liquefied Natural Gas Storage in Above Ground Tanks." *A.G.A. Proc.* 1962: CEP-62-9.
25. Stuchell, R. M. and Werninck, L. R. "New Steel Design Concepts for Liquefied Natural Gas Storage." *A.G.A. Proc.* 1962: CEP-62-15.
26. "Operation Mudpie." *Gas Age* 128: 43-8, Nov. 9, 1961.
27. "A Famous First: Frozen Earth LNG Storage." *Gas Age* 130: 22–8, Ap. 1963.
28. Eakin, B. E. and others. *Belowground Storage of Liquefied Natural Gas in Prestressed Concrete Tanks.* (I.G.T. Tech. Rept. 8) New York, A.G.A., 1963.
29. Morse, W. F. "LNG Storage in Buried Prestressed Concrete Tanks." *Am. Gas J.* 189: 32–4, Dec. 1962.
30. Flanagan, D. A. and Crawford, P. B. "How Feasible Is Underground Storage of Liquefied Methane?" *Am. Gas J.* 187: 34–7, Oct. 1960.
31. Bresson, H. "Cavern Storage of Liquefied Natural Gas." *A.G.A. Proc.* 1962: CEP-62-12.
32. Jacobs, R. B. *Single-Phase Transfer of Liquefied Gases.* (Bur. of Standards C 596) Washington, G.P.O., 1958.
33. Clark, L. J. *Sea Transport of Liquid Methane.* (World Power Conf., Sectional Meeting, III A/4) Madrid, 1960.
34. Ritter, C. L. "How Conch Will Carry Liquefied Natural Gas from Algeria to England." *Gas Age* 129: 28+, Oct. 1962.
35. "Latent-Heat Method Decreases Evaporation of Liquefied Natural Gas in Transport." *Gas Age* 127: 6, Ap. 13, 1961.

36. Szeszich, L. von. "Ocean Transportation of Liquefied Natural and Petroleum Gases." *Oil & Gas J.* 57: 76–7, Jan. 12, 1959.
37. Densham, A. B., and others. "Physical and Chemical Aspects of Underground Storage." *Gas Times* 97: 44–47, May 1963.
38. Delsol, R. and Verret, P. "Problèmes et Perspectives du Transport Maritime du Gaz Natural par Liquéfaction." *J. des Industries du Gaz* 87: 51–60, Feb. 1963.

A 27-page bibliography on liquefied natural gas which includes summaries of the listed articles has been published by the American Gas Association.

Roess, A. C., "Bibliography on Liquefied Natural Gas, 1937–1963." *A.G.A. Operating Sect. Proc.* 1964:64–P–1.

ADDENDUM

Liquefied Natural Gas Storage Under Pressure*

The critical temperature (−116.3 F) and pressure (673 psia) for natural gas indicate that it is possible to liquefy and store it under conditions other than approximately −260 F and 1 atm pressure. For example, at −155 F and 325 psig, equipment available from commercial refrigeration suppliers, as well as from cryogenic specialists, could be used. In ordinary usage, −155 F is not considered to be in the cryogenic range, taking the liquefaction point of ethylene at atmospheric pressure as a criterion.

One engine horsepower applied at various temperatures yields various refrigeration capacities; Fig. 10-26 shows these relationships.

The presence of CO_2 within the expected range will not require separate removal when storage at approximately −155 F is employed.

Because 325 psig is likely to be close to the utility company gas pressure involved, a concomitant advantage of storage at that pressure is the reduction in the required pump capacity for supplying the vaporized gas to a distribution system.

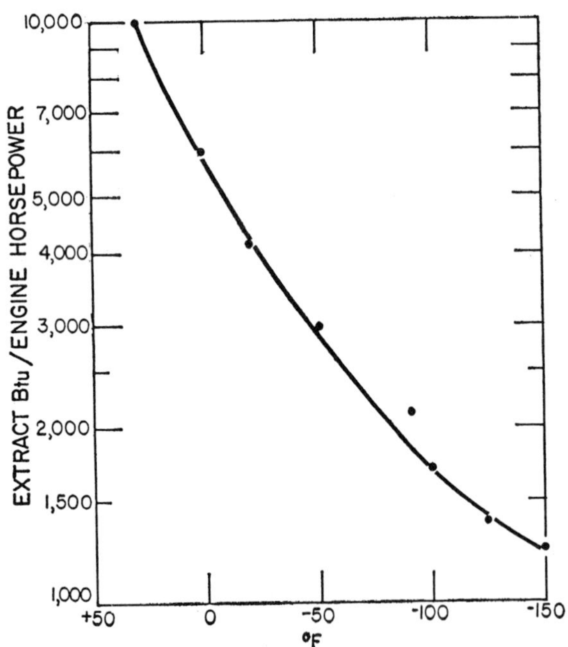

Fig. 10-26 Extract Btu per engine horsepower vs. temperature. The data apply to the liquefaction of natural gas in the temperature range appropriate to commercial refrigeration equipment. The curve should not be extrapolated, since the more sophisticated equipment used in cryogenic processes achieves a different order of extract Btu per engine horsepower.

Storage is not complex. For the small plant, shop-fabricated containers of low-nickel steels would be used. These containers would have to be constructed in accordance with the appropriate ASME Code.

When storage requirements indicate and geologic conditions are appropriate, an insulated mine cavern would be a logical consideration. These caverns must be insulated for use with liquefied natural gas (LNG). The economics of construction may be considered from two mining viewpoints: conventional vertical shaft procedures; and mining by a sloping tunnel.

Insulation of the mined cavern may be accomplished by (1) direct application of insulation to the cavern wall or (2) by construction of an insulated container of high-density urethane, polyvinyl chloride foam, or other appropriate insulation within the cavern. In the first case, the gradual cooling of the rock would be likely to introduce thermal stresses in the rock that would have to be taken into account. In the second case, the suitably lined or sealed insulated container would not touch the walls, floor, or roof of the cavern. The rock wall would be kept at a uniform temperature (in the range from −20 to 0 F) by diverting boil-off gas into the annular spaces outside the LNG container. This gas temperature could be altered by the addition of warm gas when necessary. Leakage of water thru the rock could be readily controlled by freezing.

A point to consider in selecting the insulation is that it must withstand the gas pressure plus the hydraulic pressure (rather small) on the inside and the gas pressure on the outside. In other words, it must have adequate crushing resistance.

Advantages and Limitations of Pressure Storage

1. LNG storage under pressure should require lower capital investment costs. Complete capital costs in 1965 for pressure storage of LNG were estimated at $2.00 to $5.00 per MSCF for a system having one billion SCF storage capacity. The capital cost of LNG storage at atmospheric pressure was estimated at $6.00 to $12.00 per MSCF. The capital cost of storing gas in an aquifer was estimated at $0.50 to $1.00 per MSCF.
2. Such storage should require lower operating costs.
3. The system can be designed so that an adequate differential pressure between storage and delivery can be obtained without the use of a pump.
4. The ratio of liquefied to gaseous standard cubic feet at −155 F and 325 psig is 450, as contrasted with 620 for storage at −260 F and 1 atm.
5. In the event that cavern leaks or other factors require lower storage pressure, liquid discharge pumps may be required.

* Prepared by D. W. Linscott, Liquefied Natural Gas Consultant, Allenhurst, New Jersey.

SECTION 11

CORROSION

J. L. Adkins, *Section Chairman,* The Peoples Gas Light & Coke Co., Chicago, Ill.
J. V. Adkin, Chapter 5, Rochester Gas & Electric Corp., Rochester, N. Y.
C. W. Beggs, Chapter 10, Public Service Electric & Gas Co., Newark, N. J.
J. C. Berringer, Chapter 7, Panhandle Eastern Pipe Line Co., Kansas City, Mo.
R. C. Buchan, Chapter 8, Humble Oil & Refining Co., Houston, Tex.
Guy Corfield, Chapter 2, Southern California Gas Co., Los Angeles, Calif.
I. A. Denison (retired), Chapter 1, National Bureau of Standards, Washington, D. C.
C. G. Deuber, Chapter 1, Deuber Laboratories, New York, N. Y.
H. W. Dieck, Chapter 5, Long Island Lighting Corp., Hicksville, N. Y.
W. P. Koster, Chapter 9, Metcut Research Associates, Inc., Cincinnati, Ohio
W. J. Kretschmer, Chapter 5, Columbia Gas System Service Corp., Columbus, Ohio
F. E. Kulman (deceased), Chapters 1 and 4, Consolidated Edison Co., New York, N. Y.
J. O. Mandley, Chapter 4, Michigan Consolidated Gas Co., Detroit, Mich.
P. F. Marx (retired), Chapter 5, Consultant, Bradford, Pa.
M. C. Miller, Chapter 6, M. C. Miller Co., Upper Saddle River, N. J.
N. P. Peifer, Chapters 3, 4, and 5, Manufacturers Light, Heat & Power Co., Pittsburgh, Pa.
T. L. Powers, Chapter 10, Public Service Electric and Gas Co., Newark, N. J.
W. J. Schreiner, Chapter 9, Cincinnati Gas & Electric Co., Cincinnati, Ohio
Peter P. Skule, Chapters 3 and 5, The East Ohio Gas Co., Cleveland, Ohio
A. L. Stegner, Chapter 5, Tenneco Chemical Co., Houston, Texas
S. E. Trouard, Chapter 1, New Orleans Public Service Inc., New Orleans, La.
F. E. Vandaveer (retired), Chapters 2 and 3, Con-Gas Service Corp., Cleveland, Ohio
C. L. Woody, Chapter 5, United Gas Corp., Shreveport, La.

CONTENTS

Chapter 5—Cathodic Protection

Chapter 6—Instrumentation for Corrosion Control

Chapter 7—Compressor Station Corrosion

Chapter 8—Corrosion of Gas Wells

Chapter 9—Aboveground Corrosion

Chapter 10—Safety in Corrosion Mitigation

Tables

Figures

Chapter 1

Causes of Underground Corrosion

by I. A. Denison, C. G. Deuber, F. E. Kulman, and S. E. Trouard

GENERAL THEORY

Corrosion of construction materials underground, as in other natural environments, is associated with flow of electric current between areas differing in potential. This difference in potential may be caused by variance in chemical composition of the soil, or it may result from the use of different or nonhomogeneous metals in construction. Table 11-1 lists potentials of certain metals and alloys. From these it is possible to predict which combinations of metals will produce galvanic currents when coupled together and exposed to natural waters or to soils. Because this current flows from the element of higher potential thru an electrolyte to the one of lower potential, one element of a couple will be corroded (anode), and the other will be immune to corrosion (cathode). Comparison of potential differences of various bimetallic couples gives no indication of the relative magnitude of currents produced by them, because metal potentials are affected differently by the current flow. For example, an aluminum alloy coupled to 18-8 stainless steel produces less current in sodium chloride solution than the same aluminum alloy coupled to copper, despite the fact that the electromotive force of the aluminum alloy-stainless steel couple is greater than that of the other couple.

Table 11-1 Potentials of Metals and Alloys in Sodium Chloride Solution

(measured against a saturated calomel half-cell reference electrode)

Metal or alloy	Potential, volts
Magnesium	−1.73
Magnesium containing 4% Al	−1.68
Zinc	−1.10
Aluminum containing 4% Zn	−1.02
Aluminum containing 4% Mg	−0.87
Aluminum	− .84
Cadmium	− .82
Aluminum containing 4% Cu	− .69
Iron	− .63
Lead	− .55
Iron containing 5% Cr	− .50
Tin	− .49
Iron containing 12% Cr	− .27
Copper containing 30% Zn (70-30 brass)	− .25
Copper	− .20
Stainless steel (18% Cr + 8% Ni)	− .15
Copper containing 10% Al	− .15
Monel	− .10
Copper containing 5% Sn	− .08
Silver	− .08
Nickel	−0.07

The galvanic current is related to the rate of corrosion by the equation:

$$W = itK \tag{1}$$

where:

W = weight loss of metal corroded, grams
i = current, amperes
t = time, seconds
K = electrochemical equivalent (weight of metal in grams liberated by one coulomb or an ampere-second)

When two metals differing in potential are connected externally and immersed in an electrolyte of low resistivity, the potentials of the anode and cathode tend to approach a value intermediate between the open-circuit potentials of the metals. If the anode potential is relatively unchanged by current flow, as is usually the case with ferrous metals in corrosive soils, corrosion rate is controlled almost entirely by the reaction at the cathode, and the observed potential of the entire surface approaches the anode open circuit potential. Conversely, if only the anode potential is changed by the galvanic current, as when slightly soluble corrosion products are precipitated in immediate contact with the anode, the corrosion rate would be anodically controlled and the observed potential would approach the cathode open circuit potential.

Because the potential of a corroding surface may be affected by the type of control of corrosion rate and also by the relative areas of the anode and cathode, measurements of potential alone have little meaning with respect to corrosion rate or severity of corrosion. A metallic surface, the measured potential of which is close to its open circuit potential, may corrode at a negligible rate, if at all. For example, although the measured potential of zinc underground may correspond to that given in Table 11-1, the corrosion rate may be negligible because of the marked polarization of the local cathodes.

Current-potential relations of couples of dissimilar metals also apply to electrolytic cells produced by differences in potential on the surfaces of two electrodes made of a single metal. Because oxygen causes a steel surface to assume a more cathodic potential than a steel surface from which oxygen has been excluded, the aerated electrode becomes, in effect, the cathode, and the unaerated one the anode of a galvanic couple. Consequently, whether potential differences on buried metal structures arise from interconnection of different metals or from dissimilarities in soil environment, the same basic electrical relations apply. The rate of corrosion, current required for cathodic protection, anode and cathode resistances of the corrosion circuits, and type of control of the corrosion reaction can be obtained from measurements of the potential of a corroding surface as external currents are anodically and cathodically applied. The corrosion rate or corrosion potential is also affected by the relative areas of the anode and cathode.

Specific Causes

Electrolytic cells chiefly responsible for underground corrosion may be broadly classified as: (1) dissimilar electrode cells; (2) dissimilar electrolyte cells; and (3) cells resulting from externally applied potential (stray-current electrolysis).

Dissimilar Electrode Cells. These cells may be formed, for example, by different metals or different phases of the same metal, as in a binary alloy. If a galvanized pipe is installed in an underground system of bare steel or cast iron, its zinc coating will be rapidly removed because zinc is more anodic than steel or cast iron. Although copper is highly resistant to corrosion in soils, the good service provided by copper in gas distribution systems may be due in part to the cathodic protection provided by steel services or mains.

Cells produced by differences in potential associated with the condition of the metal surface might be considered a special case of dissimilar electrode cells. Development of a rust coating causes iron and steel pipe to assume a potential more cathodic than that of bright steel. Consequently, if a section of new pipe is installed in an old line or if a new line is interconnected with an adjacent one long in service, relatively faster corrosion of new pipe may result from action of the galvanic cell formed by the same metal at different stages of corrosion.

Dissimilar Electrolyte Cells. Electrolytic cells produced by differences in concentration of oxygen or of other reacting constituents are a common cause of corrosion of underground piping. Areas inaccessible to oxygen are anodic to areas to which oxygen is accessible; corrosion tends to occur where the oxygen supply is deficient. Pipelines ordinarily discharge current in swampy areas and in depressed areas of heavy clay soils, because sections exposed to these poorly aerated soils are anodic to those in adjacent well aerated soils. Even within the same soil, local aeration differences may cause severe corrosion.

Differential aeration circuits originate not only from mechanical shielding of parts of a metal surface from oxygen, but also from consumption of oxygen by biochemical processes. Depletion of oxygen at or near a metal surface by decomposition of organic matter causes the affected area to assume a potential anodic to that of normally aerated areas. Corrosion of stainless steels in sea water resulting from attachment of barnacles illustrates the development of anodic areas not only by mechanical shielding from oxygen but also by consumption of oxygen incident to decomposition of the barnacle.

Although corrosion by differential aeration may occur in soils which are moderately to well aerated, it is most common when the buried metal is inaccessible to oxygen. Corrosion under such conditions is characterized by the presence of sulfides in the corrosion products. Anaerobic bacteria reduce sulfates present in the soil to sulfides, with an equivalent amount of iron converted to ferrous sulfide.

Stray Current Electrolysis. Electric current supplied for operation of electric railways is delivered by overhead conductors and is frequently returned thru the rails. If an underground pipeline is adjacent, a fraction of the current may leak electrolytically from the rails into the earth, and from there onto one section of the pipeline and discharge from another section of the line in returning to the rails. The sections of a steel pipeline on which the stray currents collect are cathodically protected from corrosion, while at the areas of current discharge severe corrosion may occur.

Miscellaneous Causes

Chemical. Such corrosion may cause pipeline metal to go into solution directly, without any electric current flow.

High Pipeline Temperature. Hot gases, say 150–240 F, on the outlet of a compressor cause a rapid oxidation of underground steel pipe; the metal flakes off as rust. Leaking steam near pipelines often produces the same type of corrosion.

Mechanical Wear. Probably the best example of so-called mechanical wear is that which occurs when water is continually flowing over the surface of a line, as in a creek crossing where the top of the pipe is level with the flow line of the creek. Actually, chemical corrosion may also assist in the process by bringing oxygen directly to the pipe surface.

GALVANIC CELL CORROSION

The potential difference between two metals in neutral solution or soil can be predicted from Table 11-1. If two of the listed metals are connected electrically, the one closer to the bottom of the list will be the cathode, and the other will be the anode. Potential difference between them is a measure of the driving emf, tending to cause corrosion currents to flow, but the true measure of galvanic corrosion rate is the magnitude of the current. Galvanic current causes anode corrosion at a rate conforming to Faraday's law, and tends to reduce that normally occurring at the cathode.

Current magnitude for a given potential difference between two electrodes depends upon soil resistivity, chemical constituents of the soil electrolyte, geometric separation between anode and cathode, anode and cathode polarization, and relative areas of anode and cathode surfaces.

Soil Resistivity. Temperature, moisture content, and concentrations of ionized salts present strongly influence soil resistivity and corrosion current flow. Galvanic corrosion is most likely to be severe in low resistivity soils, and, conversely, it may be comparatively rare in very high resistivity soils.

Chemical Constituents of Soil. Nature of corrosion products depends upon the type of salts present in the soil. Deposits of insoluble and poorly conducting products on metal retard corrosion current. For example, slightly soluble ferric oxide may be formed on iron in well aerated soils. In the absence of sufficient oxygen in the soil, the corrosion product may be ferrous hydroxide, which is less protective, and thus the corrosion current will be higher.

Geometric Separation between Anode and Cathode. Corrosion is likely to occur where anode and cathode metals are close together, i.e., where the current path in the soil is relatively short. Other conditions being equal, doubling the separation between two dissimilar metals should reduce corrosion current intensity.

Anode and Cathode Polarization. With the passage of current in the soil, protective films formed at the anode or cathode surfaces may be characterized by a back emf or an added resistance or both. Films of gaseous hydrogen formed at cathode surface of buried metal tend to insulate it from the soil and thus reduce corrosion current magnitude. This hydrogen can be removed by oxidation with oxygen in the soil, or it may be consumed by sulfate-reducing bacteria. Thus, the rate at which cathodic depolarization occurs influences the corrosion rate at the anode. Polarization by hydrogen gas is an important limiting factor when cathode and anode areas are about equal. Where the ratio of cathode to anode area is very large, cathodic polarization becomes of less practical significance and anode polarization may be influential in determining the rate of galvanic corrosion.

Relative Surface Area of Anode and Cathode. For a given magnitude of corrosion current, the rate of anode penetration (depth of corrosion) will be inversely proportional to anode area. This follows from the application of Faraday's law of electrolysis. Theoretical rates of corrosion for a current intensity of one milliampere per square inch at the anode are:

Anode metal	*Penetration, in. per yr*
Magnesium	0.139
Zinc	.091
Aluminum	.065
Mild steel	.071
Lead	.182
Copper	0.142

If two dissimilar metals are to be used and cannot be insulated, good design favors coupling a large anode to a small cathode, rather than the reverse.

Galvanic Corrosion of Gas Distribution Systems. Several galvanic cells or couples follow:

Anode	*Cathode*
Steel main	Cast iron main
Steel main	Corroded steel main
Steel main	Copper service
Steel main	Concrete coated pipe
Steel main	Cinders
Steel service	Cast iron main
Steel service	Copper service
Steel service	Brass valve
Galvanized steel	Black steel

To protect against galvanic corrosion, insulating joints are widely used to couple pipes of dissimilar metals. Locations of these joints should be specified by the corrosion engineer.

Where coatings are applied only to a limited degree, preference should be given to coating the cathode, rather than the anode, to avoid accelerated corrosion at accidental breaks in the coating.

MICROBIOLOGICAL CORROSION[1-4]

Bacteria in soil or water may influence the corrosion process by direct participation in its electro-biochemical reactions or by physically or chemically creating a more corrosive environment at the metal surface. Microorganisms chiefly concerned are sulfur and iron bacteria. The scope of such corrosion appears to be enlarging to include certain aspects of the activities of fouling organisms such as iron bacteria and associations of bacteria and various sessile organisms.

For microbiological corrosion, moisture must be present in the immediate vicinity of the metal structures for a considerable portion of the year. Fresh, salt, or variously contaminated waters are involved as well as many soils. In most water and soil environments, the bacteria identified with microbiological corrosion are present and can adjust to various levels of nutrient supply and a wide temperature range. In dry sites these bacteria are either absent or inactive. A normal oxygen supply favors development of aerobic bacteria, some of which are concerned with aerobic microbiological corrosion. Limited or no oxygen permits anaerobes to develop, some of which are involved with anaerobic microbiological corrosion. The metabolism of many bacteria is adjusted to either an aerobic or to an anaerobic environment, while others can partially adjust to either state.

Principal Mechanisms Involved

1. Direct participation of one phase of the metabolic processes of the microorganism in the corrosion reactions. This occurs under anaerobic conditions and involves removal of polarizing hydrogen from the cathode.
2. Formation of differential aeration cells by growth on limited regions of the metal surface. Both aerobic and anaerobic microorganisms, or associations of both types, may be involved in increasing difference in potential between metal surfaces under the bacterial colonies or under the deposits resulting from their growth and metal not so enclosed.
3. Creation of a more corrosive environment by products, usually acids, produced by the metabolism of microorganisms.

Anaerobic Microbiological Corrosion

Anaerobic sulfate-reducing bacteria are generally found in low lying, wet, poorly aerated, and poorly drained soils. Thru the production of sulfide, they are generally responsible for the black discoloration of the soil. Corrosion products of pipelines laid thru these areas are characterized by high sulfide contents.

The theory advanced is that sulfate-reducing bacteria utilize the hydrogen formed at the cathode surface, causing depolarization and thereby continuing the corrosion process, with ferrous sulfide as one of its major products. With steel, the characteristic effect is pitting. With cast iron, the metal is removed and the carbon remains in the form of graphite. This condition is termed *graphitization.*

It is difficult to determine by visual observation of corrosion products whether they were caused by sulfate-reducing bacteria. The detection of sulfide is made by applying hydrochloric acid to the corrosion products. If an odor of hydrogen sulfide results, the sulfate-reducing bacteria are causing the corrosion.

Various strains of sulfur bacteria (sulfate-reducing) have been isolated. They account for anaerobic corrosion experienced in the brackish water of gas holders, well casings, salt water installations, tanks containing oil emulsions, and at both *low* and *normal* temperatures as well as in *hot* mineral springs. The most favorable condition for active development of these bacteria is near neutrality, but they will grow in a range of pH 5.5 to 8.5. Presence of such organic matter as cellulose and sewage stimulates their growth.

A soil probe[1] for the measurement of pH and oxidation-reduction intensity, E_h, of the soil in place, is valuable in determining these factors at pipe depth. From numerous measurements with this probe, the following scale was arranged for estimating the degree of bacterial corrosiveness from *Redox* potential data.[5,6]

Range of soil E_h, mv	Degree of corrosiveness
Below 100	Severe
100 to 200	Moderate
200 to 400	Slight
Above 400	Noncorrosive by bacteria

Influence of Seasonal Variations.[7] Bacterial populations in soils undergo seasonal changes in number, dominant type, and activity. They are also influenced by precipitation thru its effect on water table height. See Table 11-2.

Table 11-2 Average Values of Site Factors and Bacterial Population at a New York Location

	March	May	July	October
Soil temperature, °C	8.0	13.0	20.0	17.0
Moisture in soil, per cent	19.2	21.0	16.9	14.9
Soil reaction, pH	6.75	5.4	5.95	6.25
Organic matter present, per cent	7.2	6.3	5.5	5.2
Reducing bacteria, No. per ml	1890	6250	577	14
Oxidizing bacteria, No. per ml	3390	3170	3140	2090

Saprophytic bacteria, under favorable conditions while decomposing organic residues at or near a pipe surface, deplete the oxygen supply and reduce the oxide film normally present on it. Corrosion by differential aeration results.

Aerobic Microbiological Corrosion[8,9]

The principal microorganisms concerned with aerobic corrosion are sulfur bacteria and iron bacteria. Chief interest is in the genus Thiobacillus and in the species T. thiooxidans. These bacteria are notable for living in more acid media and for producing more acid end products. They will thrive in media with a pH of 2.0 or less. Sulfuric acid is a product of their metabolism. Their presence in soil or water increases corrosion by increasing the acidity of the soil.

Iron bacteria secure energy for growth by oxidation of ferrous iron. This may be available in the environment or may be secured from the surface of ferrous metal, in which case the metal is pitted.

Prevention of Microbiological Corrosion[10,11]

To control bacterial corrosion on a bare line, it is necessary to raise the pH at the metal interface to 9.0 or higher. Under some conditions where certain salts are present in the soil, this can be done by applying cathodic protection. This will cause the pipe to become the cathode, with the formation of a film or membrane whose pH may exceed 9.0.

On new pipeline installations the pipe should be coated with a good protective coating. The application of cathodic protection to the coated pipe will cause a film formation at any faults, the pH of which generally exceeds the value required to inhibit the action of bacteria.

STRAY CURRENT ELECTROLYSIS[12-18]

Such corrosion on metallic underground structures is caused by stray direct current and occurs where stray currents discharge from underground structures to earth while returning to their source.

Usual sources of stray currents are electric railway systems, cathodic protection systems, electroplating systems, Edison three-wire d-c underground power distribution systems, and central generator systems for charging batteries at telephone branch offices, with only one metallic path between generator and batteries. Other sources are shipyards utilizing large banks of d-c welding machines with only one insulated conductor and the earth as the other path.

Corrosion on underground structures may be caused by self-generated or galvanic currents. These currents are in turn caused by certain dissimilarities either in the structures themselves or in their environment. Stray currents may either increase or retard corrosion caused by galvanic currents, depending upon direction of flow. *Electrolysis caused by alternating current* is usually negligible, being about one per cent of the damage produced by direct current of equal magnitude.

According to Faraday's law, 96,500 coulombs or 26.8 amp-hr removes one gram equivalent of the metal. The ratio of actual corrosion loss to theoretical loss, expressed as a percentage, is called the corrosion efficiency. For ferrous metals it may vary from 20 to 140 per cent. For iron, the commonly accepted weight loss is 20 lb per amp-yr.

Street Railway Electrolysis Problems and Mitigative Methods

Figure 11-1 shows stray current conditions on a system where no electrolysis mitigative measures have been employed. Direct current, generated at a railway substation usually at 500–700 v, is fed thru positive feeders and trolley wires into streetcars, and thence into the rails. If the rails are light, inadequately bonded at joints, switches, or crossings, and in contact with the earth, relatively little current returns to the substation thru the rails. A voltage drop builds up in them because of high linear resistance between the end of the line and the point of drainage of the current near the substation. Thus, the rails are at a higher potential than the earth in areas distant from the substation, and much of the current discharges from them to earth which acts as a parallel return path. Some of this current in turn collects on underground structures which serve as additional parallel return paths. In the area near the substation current discharges from underground structures to earth and from there back to the rails, causing corrosion of underground structures.

Figure 11-2 shows electrical relationships (for Fig. 11-1) among rails, earth, and pipes from a point near the substation to one near the end of the line.

Sectionalized Three-Wire System. This system, employing two 600 v generators in series, with their midpoint grounded to the rails, is divided into sections in which the trolley wires are insulated from one another. Alternate sections are supplied by feeders tied to the positive side of one generator and to the negative side of the other. Since only current caused by unbalanced loading returns on the track, overall rail voltage drops, and stray currents are thereby minimized.

Double Trolley System. Here, rails are not used as electrical conductors, since insulated positive and negative feeders and trolley wires compose the entire circuit.

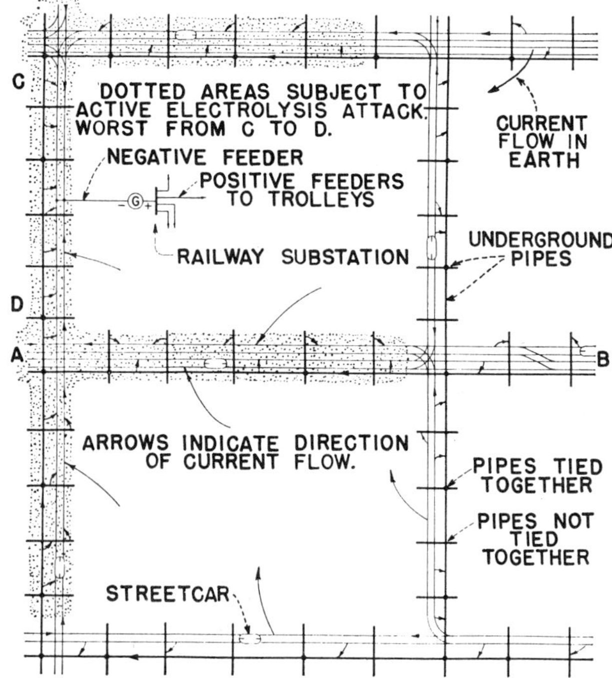

Fig. 11-1 System with poor electrolysis conditions. Lightweight rails, no joint bonding, no cross bonding, no special work bonding, single rail drain point, no electrolysis drainage, underground pipes not all tied together. Majority of current does not follow rail path because of poor conductivity, but leaks into earth and onto pipes to return to station, leaving pipes and earth as shown in dotted areas, causing corrosion. Despite poor electrolysis conditions, majority of pipes are collecting current, thereby receiving some cathodic protection.

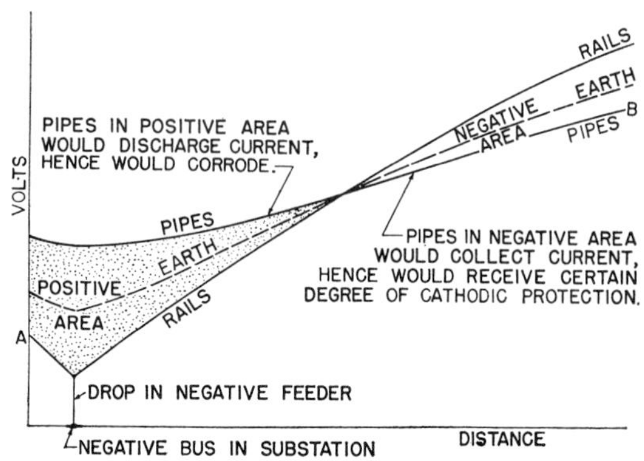

Fig. 11-2 Potential profile along line A-B of system with poor electrolysis conditions shown in Fig. 11-1.

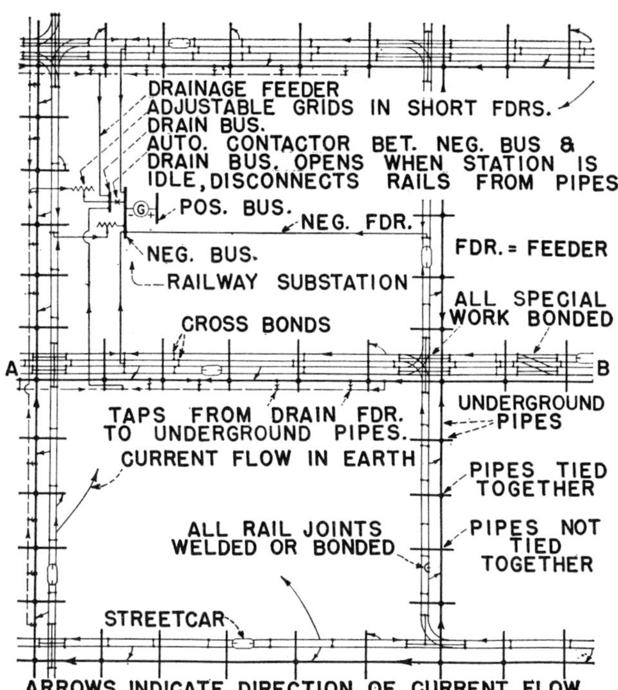

Fig. 11-3 System with improved electrolysis conditions. Heavy rail weight, joints bonded, rails well cross-bonded, special work bonded, multiple rail feed points, insulated negative feeders, adjustable resistors in short feeders, adequate drainage feeders, numerous drainage taps, adjustable resistor in short drainage feeder, automatic contactor between drain bus and negative bus, underground structures all cross-bonded in area near substation. Less rail leakage because of improved rail conductivity. Moderate rail leakage into earth with part collecting on pipes. Current on pipes near substation returns thru copper drainage feeders rather than discharging to earth, and from there to rails. Since stray current collects on all pipes, all receive some cathodic protection.

Reversed Polarity System. Connecting the trolley wire to the negative side of the generator and the rails to the positive side eliminates acute electrolysis problems adjacent to the substation, but spreads them out over a large remote area. The practical impossibility of providing electrical drainage in remote areas makes this method objectionable. A modification consists of periodic reversal of trolley polarity.

Insulation of Rails and Underground Structures. In this method, efforts are made to insulate the rails from the earth insofar as is feasible, and to locate underground structures as far as practicable from them. Linear insulated joints are installed to break up underground structures into isolated sections and minimize current flow on them. The complexity of maintaining such insulation, joints, and isolation for all systems involved limit use of this method, especially in cities and where the water table is high and soil resistances are low. However, some gas utilities consider insulating joints very desirable for controlling stray and long-line current.

Electrical Drainage and Insulated Negative Feeders. Figure 11-3 shows how the poor electrolysis conditions of Fig. 11-1 were improved by employing electrical drainage and insulated negative feeders. Adequate rail bonding and reasonable rail weight keep overall rail voltage drops lower than those shown in Fig. 11-2. Current flowing on continuous metallic pipes does not discharge to earth near the substation, but returns thru separate low resistance feeders called drainage feeders. These are connected to a separate bus in the substation called a drainage bus and are tied to the underground structures, which in turn are all cross-bonded in the area near the substation. Stray current removal from pipelines near substations is increased by inserting a resistance between negative return wires and the negative terminal of the generator. Drainage or current removal cables from the pipeline are connected to this negative terminal also. This procedure tends to raise the potential of the rails with respect to the pipeline, eliminating a positive area on the pipeline and increasing the current drained from it.

All rail current is not taken off at one point but from several points, by insulated negative feeders. These are insulated from each other, from earth, and from drainage feeders, except at the substation. Grid resistors are inserted in short negative feeders, so that voltage drops in all negative feeders are nearly equal. Thus, all rail feed points are nearly equipotential, with the result that steep rail drops are minimized and rail potential gradients are flattened out, particularly in the area near the substation.

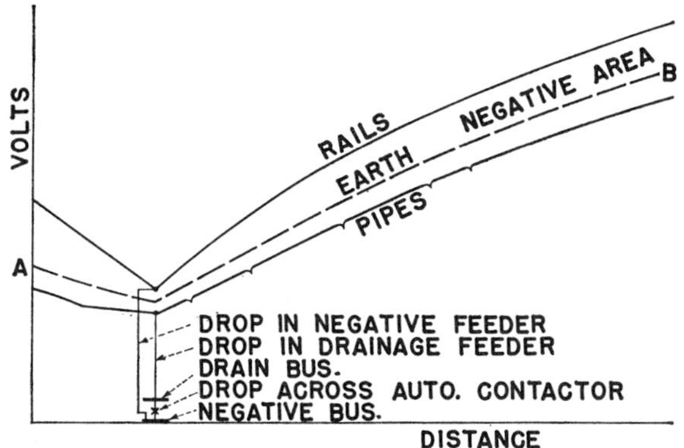

Fig. 11-4 Potential profile along line A-B of system with improved electrolysis conditions shown in Fig. 11-3. Pipes are at all places negative to rails and to earth, so that no corrosion occurs from stray currents. Collection of current gives some cathodic protection to pipes.

Figure 11-4 shows electrical relationships (for Fig. 11-3) among rails, earth, and pipes from a point near the substation to one near the end of the line. Successful operation of this electrolysis control method depends largely upon providing enough pipe drainage taps, adequate cross-bonding of underground structures, and occasional electrolysis surveys to assure good rail bonding and to cope with physical system changes.

In some systems, a modification of the above method consists of utilizing fewer drainage feeders and connecting underground structures to rails in positive area fringes thru unidirectional switches, or selenium or copper oxide rectifier stacks. These devices are designed to permit current flow thru them from pipes to rails, but not the reverse. Under certain conditions, value of rectifier stacks is doubtful because high voltage drops across them may cause some current to return to the rails thru the earth as a shunt path.

Trolley Coach Systems. Design and maintain negative feeders free from accidental grounds (particularly to system neutrals and metal conduits on poles) to prevent flow of large currents in underground structures and thus avoid serious electrolysis conditions. Also, switches should be provided to permit sectionalizing and testing for such grounds. In systems which utilize both streetcars and trolley coaches with substations operating in parallel, it is usually wise to keep negative feeders normally ungrounded at substations used entirely for trolley coach operation.

Electrolysis Testing

Current in electrolysis surveys is usually measured on rails, on negative and drainage feeders, on drainage taps, and on underground structures. Normally, millivolt drops are taken across chosen lengths of structures of known resistance. Voltages are also taken between various underground structures, between underground structures and a copper sulfate electrode in the earth, using a voltmeter of at least 50,000 ohms per volt resistance, between different points in the earth, between various parts of the same underground structure, and between rails and underground structures. Using recording meters greatly facilitates data analysis and interpretation.

Complexity of electrolysis problems affecting structures owned by various utilities operating in an area has resulted, in many cities, in formation of electrolysis committees, for joint study and solution of mutual electrolysis problems.

REFERENCES

1. Starkey, R. L. and Wight, K. M. "Anaerobic Corrosion of Iron in Soil." *A.G.A. Proc.* 1945: 307–412.
2. Hadley, R. F. "Microbiological Anaerobic Corrosion of Steel Pipe Lines." *Oil & Gas J.* 38: 92+, Sept. 21, 1939.
3. ——. "Corrosion by Micro-Organisms in Aqueous and Soil Environments." (In: Uhlig, H. H. *Corrosion Handbook*, p. 466–81. New York, Wiley, 1948.)
4. Speller, F. N. *Corrosion, Causes and Prevention*, 3d ed. New York, McGraw-Hill, 1951.
5. Deuber, C. G. and Deuber, G. B. *Development of the Redox Probe.* (Project PM-20) (6/OR) New York, A.G.A., 1956.
6. Costanzo, F. E. and McVey, R. E. "Development of Redox Probe Field Technique." *Corrosion* 14: 26–30, June 1958.
7. Kulman, F. E. "Clues to Bacterial Corrosion." *A.G.A. Monthly* 32: 26, 30, Oct. 1950.
8. Starkey, R. L. "Physiology of Thiobacillus Thioöxidans, and Autotrophic Bacterium Oxidating Sulfur under Acid Conditions." *J. Bacteriol.* 10: 135–63, 1925.
9. ——. "Transformations of Iron by Bacteria in Water." *Am. Water Works Assn. J.* 37: 963–84, Oct. 1945.
10. Hunter, J. B. and others. "Control of Anaerobic Bacterial Corrosion." *Corrosion* 4: 567–81, Dec. 1948. See also: *Oil & Gas J.* 47: 249–50, Nov. 11, 1948.
11. Rogers, T. H. "Inhibition of Sulphate-Reducing Bacteria by Dyestuffs." *Soc. Chem. Ind. J.* 59: 34–9, Feb. 1940.
12. McCollum, B. and Ahlborn, G. H. *Influence of Frequency of Alternating or Infrequently Reversed Current on Electrolytic Corrosion.* (Tech. Paper 72) Washington, Bur. of Standards, 1916. Summarized in Circular 401, 1933.
13. Morgan, P. D. and Double, E. W. W. *Corrosion of Lead by Alternating Current.* (F/T73) London, Brit. Elec. & Allied Industries Res. Assn., 1934. Communication to 1937 Conf. on Underground Corrosion.
14. Greve, L. F. and Levine, D. L. "Prevention of Corrosion of Lead Cable Sheaths." *4th Soil Corrosion Conf.* Washington, Bur. of Standards, 1937.
15. Shepard, E. R. "Electrolytic Corrosion of Lead by Continuous and Periodic Currents." *Am. Electrochemical Soc. Trans.* 39: 239–52, 1921.

16. Ewing, S. P. "Corrosion by Stray Current." (In: Uhlig, H. H. *Corrosion Handbook*, p. 601–6. New York, Wiley, 1948).
17. Kuhn, R. J. "Methods of Controlling Electrolysis in New Orleans." *Internat. Assn. of Municipal Electricians Convention*, 1928.
18. Am. Committee on Electrolysis. *Report, 1921*. New York, 1921.

BIBLIOGRAPHY

Bunker, H. J. "Anaerobic Soil Corrosion: the Function of the Sulfate-Reducing Bacteria." *Iron & Steel Inst., Corrosion Com. Fifth Report*, Sec. F, Part 3: 431–4, 1938.

Bunker, H. J. "Microbiological Anaerobic Corrosion." *Chem. & Ind.* 59: 412–14, June 15, 1940.

Bunker, H. J. "Microbiological Aspect of Anaerobic Corrosion." Congrès Mondial du Pétrole. *Rapport*, p. 3. Paris, 1937.

Bunker, H. J. "Micro-Biological Experiments in Anaerobic Corrosion." *Soc. Chem. Ind. J.* 58: 93–100, Mar. 1939.

Butlin, K. R. "Bacteria That Destroy Concrete and Steel." *Discovery* 9, May 1948.

Clark, C. L. and Nungester, J. W. "Corrosion of Water Pipes in a Steel Mill." *Am. Soc. Metals Trans. Preprint* 26, 1942.

Denison, I. A. "Chemical Aspects of Underground Corrosion and Corrosion Prevention." *A.G.A. Proc.* 1948: 517–33.

Denison, I. A. and Romanoff, M. "Soil Corrosion Studies, 1946: Ferrous Metals and Alloys." (RP 2057) *Bur. Standards J. Res.* 44: 47–76, Jan. 1950.

Denison, I. A. and Romanoff, M. "Soil Corrosion Studies, 1946 and 1948: Copper Alloys, Lead, and Zinc." (RP 2077) *Bur. Standards J. Res.* 44: 259–89, Mar. 1950.

Deuber, C. G. "Present Status of Bacterial Corrosion Investigations in the United States." *Corrosion* 9: 95–9, Mar. 1953.

Doig, K. and Wachter, A. "Bacterial Casing Corrosion in the Ventura Field." *Corrosion* 7: 212–24, July 1951.

Ganser, P. "Pipe-Line Corrosion Caused by Anaerobic Bacteria." *Gas Age* 86: 25+, July 18, 1940. Also in: *Gas J.* 231: 371–3, Aug. 28, 1940.

Gt. Brit. Admiralty. Corrosion Com. Hull Subcom. *Corrosion of Iron and Steel Associated With the Presence of Sulphate-Reducing Bacteria.* (Rept. acc./H105.1/46) London, 1946. Abstract: Nat. Res. Council, *Deterioration* 5, Met. 214 B0334, 1948.

Hadley, R. F. "Studies in the Microbiological Anaerobic Corrosion of Natural Gas Pipe Lines." *Gas* 16: 36, July 1940. Full text in *A.G.A. Proc.* 1940: 764–88.

Huddleston, W. E. "Results Obtained from Five Years of Cathodic Protection on 24-Inch Gas Line Rapidly Deteriorating from Bacterial Corrosion." *Corrosion* 3: 1–7, Jan. 1947.

Kulman, F. E. *Corrosion Control on High Pressure Gas Distribution System.* A.G.A. Conf. on Distribution, Motor Vehicles & Corrosion, 1949.

Kulman, F. E. "Microbiological Corrosion of Buried Steel Pipe." *Corrosion* 9: 11–18, Jan. 1953.

Liberthson, L. "Effect of Sulphur Bacteria on Corrosion." *Iron & Steel Eng.* 24: 69–73, June 1947.

Miller, L. P. "Rapid Formation of High Concentrations of Hydrogen Sulfide by Sulfate Reducing Bacteria." *Boyce Thompson Inst. Contrib.* 15: 437–65, 1949.

Minchin, L. T. "Bacterial Corrosion of Underground Pipes." *Coke & Gas* 22: 392, Sept. 1960.

Pomeroy, R. "Corrosion of Iron by Sulfides." *Water Works & Sewerage* 92: 133–38, April 1945.

Romanoff, M. *Underground Corrosion.* (Circ. 579) Washington, Bur. of Standards, 1957.

Thomas, A. H. "Role of Bacteria in Corrosion." *Water Works & Sewerage* 89: 367–72, Sept. 1942.

Vernon, W. H. J. "Some Recent Contributions of a British Corrosion Research Group." *Pittsburgh Internat. Conf. on Surface Reactions*, p. 141. Pittsburgh, Corrosion Pub. Co., 1948.

Von Wolzogen Kuhr, C. A. H. and Van der Vlught. "Graphitization of Cast Iron As an Electro-Biochemical Process in Anaerobic Soils." *Water* 18: 147–65, 1934.

Chapter 2

Corrosion by Soils

by Guy Corfield and F. E. Vandaveer

Where coated pipelines with supplementary cathodic protection for holidays are not used in corrosive soils, corrosion of metallic structures installed underground or underwater is a serious and costly problem. The magnitude of the problem varies with the electrical conductivity of the soil. Beyond conductivity, other properties or conditions of soil and underground environment may influence the deterioration of protective coatings applied to metal. One such property is soil moisture. It renders soil conductive and corrodes metals thru electrolysis or galvanic currents, it may absorb oils and plasticizers from protective pipe coatings (making them porous and conductive), and it is necessary for bacteria culture. In addition, soil may adhere to a protective coating, expanding and contracting with variable moisture, temperature, and pressure. The resultant stress may crack the coating and ultimately tear it off. A modification of this same effect would be the movement of the pipe and soil from settlement and other causes. Another influence is the presence of rocks and clods, which can penetrate the coating.

Soil corrosion involves factors not encountered in atmospheric corrosion. Procedures, practices, and materials for successful control of atmospheric corrosion are not necessarily applicable to underground metallic structures. In addition to an understanding of corrosion processes, a general knowledge of soil properties and an examination of the soils in which metallic structures are to be installed are essential.

SOIL COMPOSITION AND STRUCTURE[1–6]

Soil can be classified by consistency, geology, grain size, origin, structure, or texture. Table 11-3 outlines a field method of identification by texture and Fig. 11-5 shows grain size classification. Acidity and moisture saturation of soil are given in Tables 11-4 and 11-5, respectively.

Except for hard gravels and clean sands, most soils are widely varying conglomerations of mineral, vegetable, and animal matter. Some of the materials are in the colloidal state. Such chemical salts as chlorides, sulfates, and carbonates may be present, as well as nitrogen, carbon dioxide, and oxygen. Microscopic plants and animals, including certain bacteria which cause severe corrosion, may also be found. Soil properties may also change considerably with depth.

Table 11-6 gives soil characteristics and constituents in terms of degree of corrosivity. There is no distinct line of demarcation between these groupings; they may be modified by such factors as dryness and aeration.

SOIL EXAMINATION AND TESTING

Part of a soil survey is to examine and record surface topography, noting hills and valleys, creeks and rivers, mining areas, swamps, pastures, and plowed grounds. It should include gathering of information on rainfall, probable future surface development of an agricultural, residential, or industrial nature, and previous experience with buried structures.

Examination of surface topography can be made when the line is staked out or at the time of the soil survey. It can indicate which areas are dry and which are or may be wet. Low-lying or flat land may be of a clay type which may shrink or crack in the dry season, thereby damaging the pipe coating. Such soil can then be corrosive in the wet season and promote

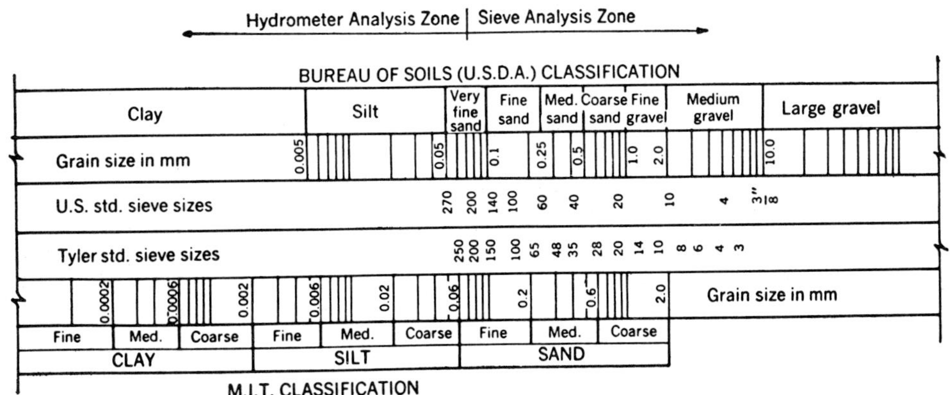

Fig. 11-5 Grain size classification of soil.[6]

Table 11-3 Field Method for Identification of Soil by Texture[6]

	Visual detection of particle size and general appearance of the soil	Squeezed in hand and pressure released: When air dry	Squeezed in hand and pressure released: When moist	Soil ribboned between thumb and finger when moist
Sand	Soil has a granular appearance in which the individual grain sizes can be detected. It is free-flowing when in a dry condition.	Will not form a cast and will fall apart when pressure is released.	Forms a cast which will crumble when lightly touched.	Cannot be ribboned.
Sandy loam	Essentially a granular soil with sufficient silt and clay to make it somewhat cohesive. Sand characteristics predominate.	Forms a cast which readily falls apart when lightly touched.	Forms a cast which will bear careful handling without breaking.	Cannot be ribboned.
Loam	A uniform mixture of sand, silt, and clay. Grading of sand fraction quite uniform from coarse to fine. It is mellow, has somewhat gritty feel, yet is fairly smooth and slightly plastic.	Forms a cast which will bear careful handling without breaking.	Forms a cast which can be handled freely without breaking.	Cannot be ribboned.
Silt loam	Contains a moderate amount of the finer grades of sand and only a small amount of clay; over half of the particles are silt. When dry it may appear quite cloddy; can be broken readily and pulverized to a powder.	Forms a cast which can be freely handled. Pulverized it has a soft flour-like feel.	Forms a cast which can be freely handled. When wet, soil runs together and puddles.	Will not ribbon but has a broken appearance, feels smooth, and may be slightly plastic.
Silt	Contains over 80 per cent of silt particles with very little fine sand and clay. When dry, it may be cloddy; readily pulverizes to powder with a soft flour-like feel.	Forms a cast which can be handled without breaking.	Forms a cast which can freely be handled. When wet, it readily puddles.	It has a tendency to ribbon with a broken appearance, feels smooth.
Clay loam	Fine textured soil breaks into hard lumps when dry. Contains more clay than silt loam. Resembles clay in a dry condition. Identification is made on physical behavior of moist soil.	Forms a cast which can be freely handled without breaking.	Forms a cast which can be handled freely without breaking. It can be worked into a dense mass.	Forms a thin ribbon which readily breaks, barely sustaining its own weight.
Clay	Fine textured soil breaks into very hard lumps when dry. Difficult to pulverize into a soft flour-like powder when dry. Identification based on cohesive properties of the moist soil.	Forms a cast which can be freely handled without breaking.	Forms a cast which can be handled freely without breaking.	Forms long thin flexible ribbons. Can be worked into a dense compact mass. Considerable plasticity.
Organic soils	Identification based on the high organic content. Muck consists of thoroughly decomposed organic material with considerable amount of mineral soil finely divided with some fibrous remains. When considerable fibrous material is present, it may be classified as peat. The plant remains or sometimes the woody structure can easily be recognized. Soil color ranges from brown to black. They occur in lowlands, in swamps, or swale. They have high shrinkage upon drying.			

Table 11-4 Acidity of Soils[4]

Degree of acidity	pH*
Extremely acid	Below 4.5
Very strongly acid	4.5–5.0
Strongly acid	5.1–5.5
Medium acid	5.6–6.0
Slightly acid	6.1–6.5
Neutral	6.6–7.3
Mildly alkaline	7.4–7.8
Moderately alkaline	7.9–8.4
Strongly alkaline	8.5–9.0
Very strongly alkaline	Over 9.0

* pH is measure of degree of existing hydrogen ion concentration, and not of the total quantity of ionizable hydrogen in the soil, which is referred to as total soil acidity.

Table 11-5 Moisture in Soil Classification[6]

Condition	Degree of saturation, %
Dry	0
Humid	1 to 25
Damp	25 to 50
Moist	50 to 75
Wet	75 to 99
Saturated	100

Table 11-6 Soil Characteristics and Constituents Effecting Corrosion of Underground Steel Pipe (Ohio Data)

Indicators	Very corrosive*	Moderately corrosive*	Relatively noncorrosive
1. Resistivity, ohm-cm	<3000	3000 to 10,000	>10,000
2. pH	<4.0	4.0 to 6.5	6.5 to 8.0
3. Sulfides in soil	Strong H_2S smell, porous stone containing sulfides	Weak H_2S smell when HCl is dropped on soil sample	None
4. Coal	†	Light showing of coal blossom	...
5. Miscellaneous	Cinders, blue-gray clay, swampy muck, fertilizer, manure, decaying vegetation	Brown and black shale	Sand, gravel, rock, sandy loam, very dry soil the year round

* One or more characteristics determine the degree of corrosion.
† Coal, coal blossom, coal gob piles, water drainage from coal mining operations.

deterioration of the metal at the points of damage. Creeks, rivers, and swamps represent areas continuously wet. Dry washes, gullies, and valleys can be dangerous at times of flood because the pipe may be undermined or exposed, and its coating may be damaged by moving rocks and debris. Plateau, hillside, cultivated land, or other well drained areas may be high or low in salt and organic matter, but if rocky, they will provide a poor environment because the pipe coating may be damaged by resting on rocks or by impact during ditch backfill. Lush vegetation indicates presence of water. Alkaline areas are often shown by certain types of coarse grass. The typical white alkali scum or efflorescence may also be visible. It should be remembered, when looking at the surface, that the pipeline will be about 2 or 3 ft underground so that excavations or borings are necessary to make sure that the surface is not misleading. For example, a thin layer of sand often overlies pebbles or rocks, and sandy loam may overlie a corrosive blue-gray clay or coal blossom. Borings by soil auger are often made at the same time as the topographical survey.

Consideration of present and probable future surface developments is important. Cultivation means irrigation or watering, plowing, and use of fertilizers, which may be harmful to buried metal. Industrial plants, coal mining, or other mining and drilling may release highly corrosive water.

Corrosivity Tests

The ferrous metals which constitute the bulk of underground metal structures become corroded to some extent in practically all soils. Physical characteristics of some soils also tend to deteriorate protective coatings. Therefore, to provide adequate protection, it is important when planning a pipeline to examine and test the soil along its route, even when it has been decided to coat and cathodically protect the pipe. In some locations where severe corrosion is suspected and it is doubtful that it can be adequately controlled, pipe of extra-heavy wall thickness may be installed or extra-thick or special coating may be applied. In other locations it may be necessary to protect the coating with a rock shield or to grind up the rock for backfill or haul in good dirt or sand. Sometimes it may even be advisable, where very severe corrosion and coating deterioration may be expected, to select another route even if it is less advantageous in other respects.

Numerous tests and test instruments have been developed to estimate soil corrosivity. *Soil corrosivity maps* of a pipeline route or an entire area are plotted to indicate the expected degree of corrosion and, therefore, the amount or type of protection that should be applied. Tests usually include the electrical conductivity (or its reciprocal, resistivity) and chemical characteristics of the soil. Based on the test results, the soils are usually grouped into such broad classifications as mildly, moderately, and severely corrosive. Each group may be assigned a color such as green, yellow, and red, respectively; the map, then, can consist of colored dots, bands, or areas.

An area map as described above usually has the appearance of a patchwork quilt, although some general differentiation generally appears. For example, Los Angeles is predominantly slightly corrosive in the foothill area of Altadena and Pasadena, severely corrosive along the low flat plain to the south, and moderately corrosive in the Hollywood and westerly area. Soil maps are used not only to decide on the degree of protection that should be given an underground metal structure when installed, but also to indicate the frequency of subsequent inspection and the degree of care required in maintenance and repairs.

Soil Sampling. Samples should be taken at selected regular spacings along the exact route of the line at pipe depth and possibly at some other depth above its top. Variation of a few feet horizontally or a few inches vertically from the pipe location may give erroneous results since soil strata may change very abruptly. Spacing usually selected varies from 300 to 1000 ft in regular terrain, with additional tests where the topography changes, at all creek and river crossings, at all valleys and swamps, at both sides of road crossings, and at any other major variation in terrain. Samples are taken with a soil auger (a solid ½ in. diam or larger metal rod with tee handle and cutting end and auger) by boring down to selected depth and withdrawing the auger with the sample held in its grooves. The sample should be tested on the spot or put in a tight container for transmission to the laboratory. Air exposure and drying will alter the characteristic of some soils, so tests at the site are preferable whenever possible.

Field Tests. At selected spacing along the pipeline route one or more of the following would be accomplished:

1. Remarks on location, color and type of soil, levelness of ground, creek, and swamp.
2. Determination of electrical resistivity with Shepard rods[6,7] (or other selected rods) at 6 in. depth and at pipe depth.
3. Determination of pH of wet soil, creek, drainage, and river water using Hydrion test papers of 0.5 pH range.
4. Sulfide test of all suspected soil samples from 6 in. and pipe depth using 10 per cent HCl, dropping it on sample of soil on a watchglass. Strong H_2S smell indicates very corrosive sulfides; weak smell indicates moderate corrosiveness.

Open Trench Inspection and Tests. Performed before pipe laying, these disclose much reliable information on soil corrosivity, while before trenching, soil tests may not be made at proper depth or exact horizontal location. Visual inspection usually is sufficient to locate such very corrosive

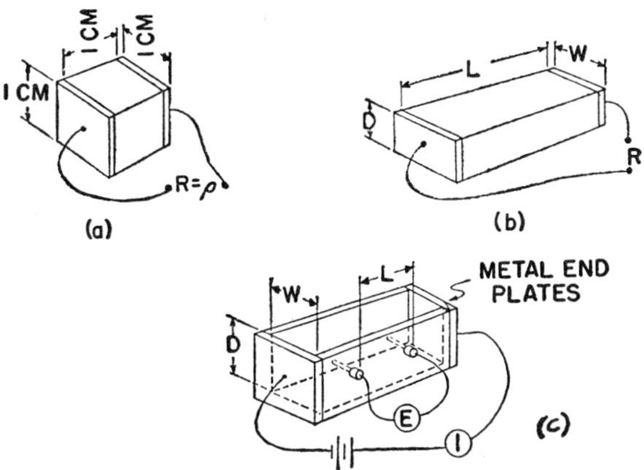

Fig. 11-6 Resistivity, resistance, and soil box relationships.[8] (a) Resistivity, ρ, in ohm-cm is numerically equal to resistance, R, in ohms between opposite faces of a cube one centimeter on the side. (b) Resistance of a rectangular solid, $R = \rho L/WD$. (c) Soil box, in which ρ is obtained by measuring the resistance between the planes of the potential pins, or $\rho = RWD/L$, where $R = E/I$.

areas as coal, blue-gray clay, or vegetation (Table 11-6). At other questionable locations one or more of the preceding soil survey tests can be made in the trench. Protection which may be needed can be specified and installed, based on the information revealed.

Laboratory Tests. Electrical and/or chemical tests may be applied to field soil samples. The most direct electrical test consists of determining soil resistance by means of a suitable apparatus with a box to hold the soil, electrodes on opposite sides, and a resistance-measuring instrument (Fig. 11-6). Tests on such disturbed samples are not as meaningful as field tests. Another test consists of immersing a weighed metal coupon in the soil and passing current from it to the soil by impressing a controlled voltage between the coupon and the metallic soil container. If all assembly dimensions, voltage and time are kept constant, loss of weight of the coupon becomes an index of corrosivity of the soil.

A number of chemical tests may also be applied. One measures soil acidity or alkalinity, either by chemical titration of a water extract, or by determining the hydrogen ion concentration or pH. Care must be exercised in interpreting such tests. A very useful test for one type of soil may not be so dependable for another type. If a soil area is known to be acid or alkaline, it should be suspected of being corrosive, and would be expected to be most corrosive where the acidity or alkalinity is highest. However, a neutral soil can also be very corrosive, as evidenced by the corrosivity encountered in tide flats impregnated with sea water, or sodium chloride, which is a neutral salt but very corrosive. The actual soluble chemical content of the soil sample can also be determined by regular analytical procedures. It is often important to know the presence and concentration of such salts as sodium chloride, sulfate, and bicarbonate.

REFERENCES

1. Marbut, C. F. "Soils of the United States." (In: U. S. Dept. of Agri. *Atlas of American Agriculture.* Washington, G.P.O., 1936.)
2. Logan, K. H. "Corrosion by Soils." (Uhlig, H. H. *Corrosion Handbook,* p. 446–66. New York, Wiley, 1948.)
3. Ewing, S. P. *Soil Corrosion and Pipe Line Protection.* New York, A.G.A., 1938.
4. Romanoff, M. *Underground Corrosion.* (Nat. Bur. of Standards C 579) Washington, G.P.O., 1957.
5. Speller, F. N. "Prevention of Corrosion Underground." *Corrosion, Causes and Prevention,* 3rd ed., Chap. 13. New York, McGraw-Hill, 1951.
6. Karol, R. H. *Soils and Soil Engineering.* New York, Prentice-Hall, 1960.
7. Shepard, E. R. "Pipe-Line Currents and Soil Resistivity as Indicators of Local Corrosive Soil Areas." (RP 298) *Nat. Bur. Standards, J. Res.* 6: 683–708, April 1931.
8. Parker, M. E. *Pipe Line Corrosion and Cathodic Protection.* Houston, Gulf Pub. Co., 1954.

Chapter 3

Pipe Corrosion Surveys

by N. P. Peifer, Peter P. Skule, and F. E. Vandaveer

OBJECTIVES

Corrosion surveys are a method for determining the physical condition of underground pipelines, to ascertain that they can contain, without loss, the liquids or gases transported thru them at existing or higher pressures. Such surveys may also be conducted before installing a pipeline, to determine what type of protection should be applied and where.

Specific objectives of corrosion surveys may be: (1) to avoid gas leakage, (2) to specify and install means for reducing or eliminating corrosion, (3) to determine need for partial or complete replacement, (4) to observe effectiveness of an existing protection system, (5) to inspect condition of existing pipe coatings, (6) to obtain knowledge of corrosion causes for establishing control policy, (7) to evaluate existing lines for rate structure or investment purposes, (8) to appraise ability of a line to carry increased gas pressures, (9) to determine where protection should be added, and (10) to avoid service interruptions from line failure because of corrosion.

Classification of Corrosion

The appearance of a corroded surface usually falls in one or more of the following forms:

Pitting. This is perhaps the most common form of pipe corrosion encountered. It may be isolated or general. The pits are usually counted and measured across and for depth with a depth gage.

Local or General Corrosion. In dry soil areas oxygen may reach the pipe surface and form a thick rust coating, completely encasing the entire line. In wet areas, where little or no oxygen reaches the pipe surface (assumes O_2 is not dissolved in H_2O) little or no scale may develop. On the other hand, on lines operated at relatively high temperatures, general rusting occurs in dry areas (less in wet areas).

Graphitization. This form of corrosion occurs on cast iron pipe, due to the removal of iron from the pipe wall. Appearance of the pipe is hardly changed because the graphite remaining, which originally was well distributed throughout the metal, maintains the dimensional properties of the pipe wall.

Dezincification. Occurring in brass, where the zinc is removed from the metal by electrochemical corrosion, the phenomenon results in a porous copper remnant of the original brass mixture.

Slabbing. This occurs when the line is laid in a heavy wet clay soil, where oxygen is practically excluded and where microbiological action finally causes disintegration of the pipe wall over a large area into a thick loose scale.

Bright or Dull. A bright shiny steel color of a pipeline usually indicates that corrosion is active either in pit form or in general area corrosion by microbes or by acid attack, while a dull steel color may indicate little active corrosion.

METHODS

Leak Record Review

Areas of previous pipeline leakage caused by corrosion may be located by a review of operating records of leak repairs. These records would undoubtedly show the date the pipe was installed and the date, type (mechanical or corrosion), and location of each leak. This information has proved valuable when it is used in conjunction with a potential survey and current measurement described below.

Exposure of Pipeline

This method is often used to establish the pipe condition for investment data. At locations picked for special reasons or at random, the pipeline is exposed for a distance up to 10 ft in length (completely around the pipe circumference). Pipe is cleaned by wire brush or sand blast which removes rust and scale, and a physical examination of its condition is made. At the most corroded location, the pits in a one square foot area are counted and their depth is measured by a micrometer, or a dial or *Morlane* pit depth gage. The pH of the soil is tested with pH paper placed on the soil near the pipe. The soil resistivity of the earth adjacent to the pipe may be measured, and a *McCollum Earth Current* test may be made; pipe-to-soil potential is also recorded. The following form is usable for permanent record.

Pipe Condition Record

District division__________ Street location; Farm________ Township________
Borough________ City________ County________ State________
Line No.____ Size____ Pressure____ Construction____ Bare or coated____
No. pits per sq ft____ Pit depth (deepest pit)____ Soil type____
Soil: Wet____ Damp____ Dry____ pH____ Resistance________
Pipe-to-soil potential__________ Earth current reading + −
Purpose of exposure____________________________
Inspector______________ Remarks____________________
______________________________________ Date________

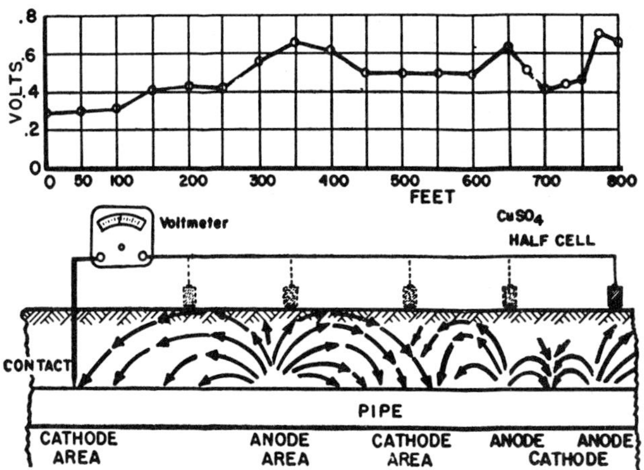

Fig. 11-7 A corrosion survey profile.[3]

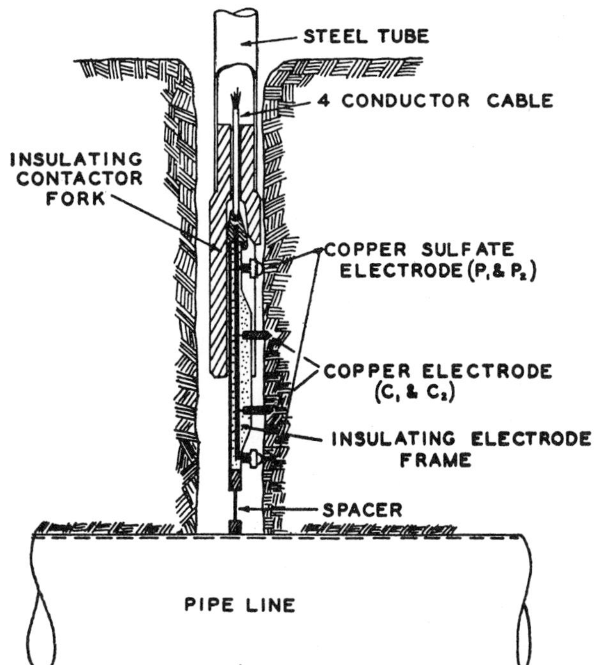

Fig. 11-8 A cantilever electrode.[3]

Potential Survey

By connecting the (+) terminal of a high resistance, sensitive voltmeter to the pipe and its other terminal to a copper-copper sulfate electrode, and then placing the electrode on the surface of the earth directly over or closely adjacent to the pipe, a change in potential will be noted when the electrode is moved along the line over an active galvanic or corrosion cell area.[1] A peak potential indicates an anodic area and a valley or lower potential indicates a cathodic area. Experience has shown that the maximum distance between test electrodes should not exceed 50 ft and under certain conditions may be as small as 10 ft (or less in critical areas). Figure 11-7 shows a curve of readings at 50 ft intervals.

To eliminate the necessity of contacting the pipe at each 50-ft test it is possible to contact the pipe at one location by using a steel bar connected to one terminal of the voltmeter, and then to connect the other terminal to a reel containing from 500 to 1500 ft of wire. The copper-copper sulfate elec-

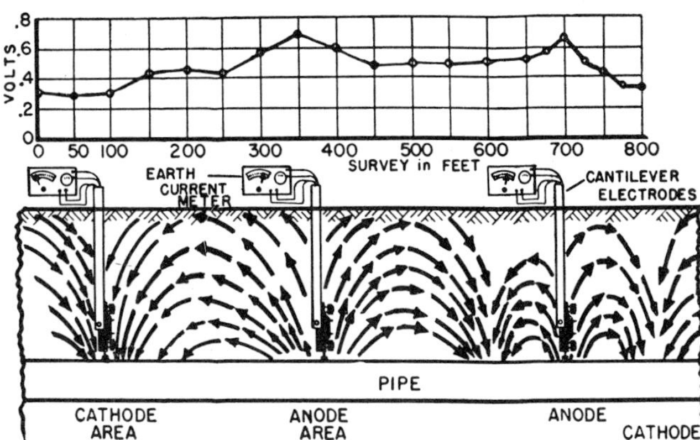

Fig. 11-9 Current measurements of anodes and cathodes.[3]

trode is connected to the loose end of the wire on the reel and the wire is pulled off the reel for tests at the desired intervals. By this method, it is possible to test 1.5 miles or more of pipe per day in open ground and pinpoint most areas requiring protection.

Use of Pipe-to-Soil Potential Survey.[1] It is necessary to evaluate the condition of pipe in a distribution system for safety and economic reasons. Use of potential surveys for transmission lines has long been practiced, but it is also most useful for distribution lines.

Preliminary Considerations. Data are collected where mains are exposed for maintenance, and soil conditions and details on the progress of corrosion are noted, i.e., a report is made on findings obtained with a pit gage. The data are recorded on appropriate maps which are then used to indicate areas to be studied by a pipe-to-soil potential survey.

Procedure. Gas mains and services and other adjacent metal structures must be located. As long as other underground structures are at least three feet away from the gas main, reliable survey results can be obtained. Direct contact with water mains and other structures may occur thru the house piping, and this as well as close proximity to such structures would interfere with obtaining good results. Systems using insulation are better suited for these procedures.

First, the steel main is accurately located with a pipe locater. If it is satisfactorily situated, i.e., not under pavement and separated from other pipe, a connection is made between the main and the negative terminal of a high resistance voltmeter. The positive terminal is connected to 1500 feet of #16 stranded wire. The end of the wire is connected to a copper-copper sulfate reference cell. The survey is then run by contact with moist ground over the main every 10 ft. As a reference base, a pipe to remote soil reading (100 ft perpendicular to the main) is taken about every 1000 ft. These data are plotted as straight lines. Readings above the datum line are anodic and below the line, cathodic. Peak readings are selected as locations for test holes, and if the condition of the pipe is judged serviceable at such spots, it should generally be serviceable throughout.

One of three recommendations is made: (1) renew a section of the main; (2) install magnesium anodes; (3) take no action.

Measurement of Current Flowing between Pipe and Soil

This type of survey was developed by McCollum[2] for measuring current flowing to or from the earth as a result of

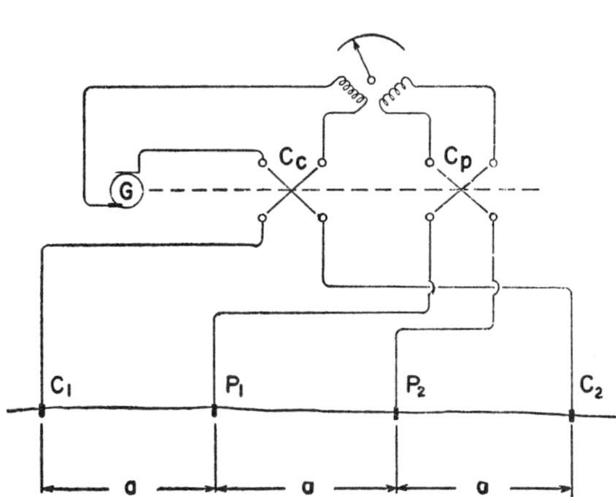

Fig. 11-10 Soil resistivity Megger.[4] Direct current from hand-cranked generator G is converted to a-c power by the commutator C_c, and enters the soil thru C_1 and C_2, after passing thru the current coil; the a-c potential picked up between P_1 and P_2 is converted to d-c by commutator C_p, and is fed to the potential coil. The needle comes to rest, under the opposing forces from the two coils, at a point which indicates the resistance R. Resistivity ρ is then obtained from $\rho = 2\pi aR$.

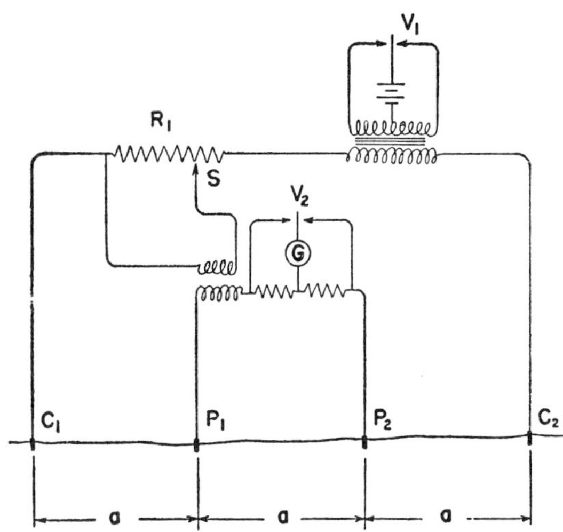

Fig. 11-11 Measurement of soil resistivity via Vibroground.[4] Vibrators V_1 and V_2 are synchronized; V_1 converts the d-c battery power into a-c; the slidewire S is adjusted until the voltage drop across R_1 just bucks out the current in the potential circuit, as indicated by a zero deflection on the galvanometer G; resistance R is then read directly on the slidewire, and resistivity ρ is $\rho = 2\pi aR$.

stray currents from railway operations. It is also useful in measuring galvanic current. The principle involved is the measurement of the resistance of the soil between two points on a four point electrode in contact with the soil (Figs. 11-8 and 11-9).

The resistance is determined by measuring a drop in potential caused by a flow of battery current thru the earth. The battery current is then shut off, and the drop in potential caused by current flowing between the earth and the buried pipe is measured.

Calculations yield the milliamperes per square foot of pipe surface flowing radially to the pipe surface or from the pipe surface thru the earth.[2] When the current being discharged from the pipe is known, it is possible by Faraday's law to determine the quantity of metal lost by corrosion (error is about 10 and 20 per cent for fast and slow corrosion rates, respectively). This is the only electrical test method of determining the corrosion rate of the pipe; however, as excavations are required, it is infrequently used. Combined with any or all of the previous methods it will give a good indication of the condition of the buried pipe.

Coated Pipeline Survey

Corrosion surveys on coated pipelines differ somewhat from those on bare pipelines. In constructing a pipeline with a protective coating, it is customary to install test stations at each road crossing or at intervals of about one mile or more, to provide contacts to the buried pipe. These stations are installed because it is not good practice to make bar contacts thru protective coating.

It is difficult to determine the points (mainly at coating discontinuities) at which corrosion is taking place on the coated pipeline. For this reason, it has been established that an adequately coated pipeline should have direct current applied to it until the potential reaches a minimum value of −0.85 v, as measured by a copper-copper sulfate electrode contacting the earth and the negative terminal of a zero center high resistance sensitive voltmeter. The other meter terminal is connected to the pipe wire at the test station. Another criterion[4] for adequate protection calls for a potential change of at least 0.3 v from the previous static level (before installing cathodic protection).

The procedure in making a current requirement test is to supply d-c power to the pipe thru a current interrupter. The positive terminal of the power source is connected to ground rods driven or placed in the earth. The negative terminal of the power source is connected to the pipe to be tested. The purpose of the interrupter is to open and close the circuit at regulated intervals. It should be adjusted so that the current is applied for a short period and then disconnected for a longer period. This is done to prevent test polarization of the pipe and to distinguish current on and off positions.

Test Procedure. Each test wire station along the pipeline is tested for the potential of pipe to earth by a copper sulfate electrode with the d-c power applied and then disconnected by an interrupter. The object of this test is to determine the current required to raise the potential of the pipe to 0.85 v if static potential* is below 0.80 v. On some lines the potential of the pipe to earth without the application of current will be found to be already 0.85 v or more. Under this condition it is necessary to apply sufficient current to raise the potential of the pipe to soil 0.2 v higher than the existing potential.

At the same time that the pipe-to-soil measurements are made, perform a current flow test using the potential drop method. The purpose of the latter test is to determine the direction of the current flow and the amount of current picked

* Potential of the pipe to earth when no d-c power is applied.

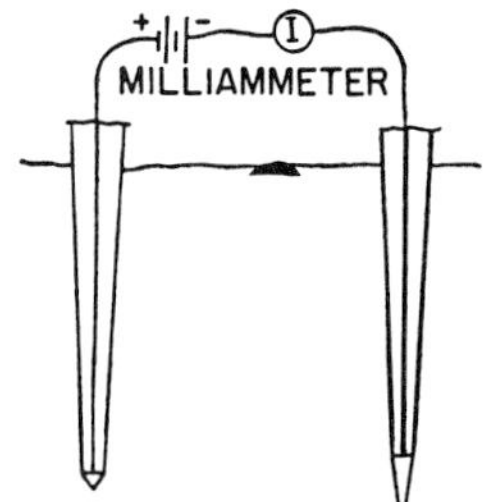

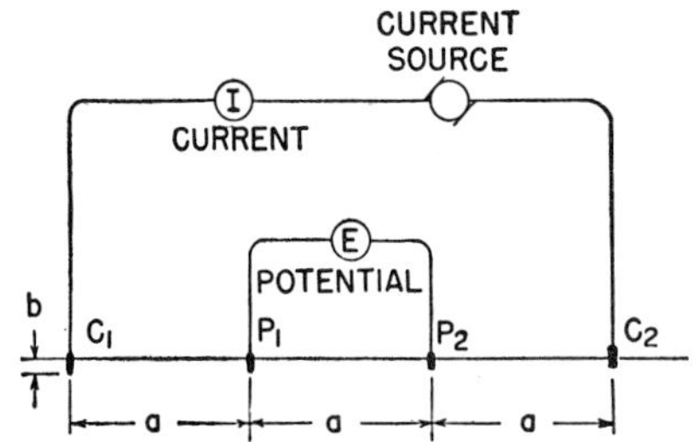

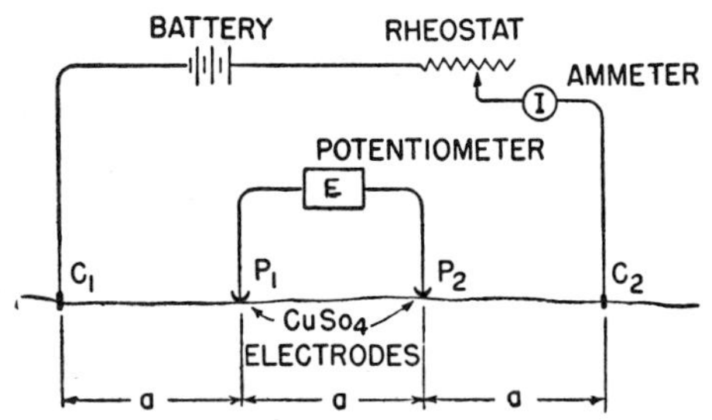

Fig. 11-12 (left) Shepard Canes for soil resistivity measurement.[4] Current from a three-volt battery (two flashlight cells) is passed thru the soil between two iron electrodes mounted on insulating rods. Current flow is measured by a double range milliammeter (0–25 and 0–100) graduated to read directly in ohm-cm (10,000–400 and 500–100). Cathode is made larger to avoid polarization; accuracy is within six per cent when tips are separated eight inches or more. Meter, batteries, and switch are mounted on anode rod.
Fig. 11-13 (center) Four-terminal (Wenner) measurement of soil resistivity.[4] Depth of electrode b must be small compared to a. The basic formula is $\rho = 2\pi aE/I$ where resistivity ρ is "averaged" to a depth approximately equal to the electrode spacing a.
Fig. 11-14 (right) Ammeter-voltmeter measurement of resistivity.[4] A storage battery or dry cells may be used for current. Copper sulfate electrodes will help avoid polarization errors. The voltmeter used must be high resistance, or a potentiometer may be employed. If the distance a is made 62.5 inches, then the formula simplifies to $\rho = 1000E/I$. Direction of current should be reversed, and readings averaged, to balance out extraneous currents or potentials in the soil.

up from holidays in the coating in each section of line between the test stations.

Soil Survey Before Pipe Installation

Surveys of this nature are made when it is not desired to apply a protective coating over the entire line, or when it is to be protected by the use of expendable anodes like magnesium or zinc. Such surveys include soil resistivity measurements, pH determinations, and *Columbia Soil Rod* galvanic cell tests.

Soil Resistivity Tests. Resistivity ρ is measured in ohm-cm as explained in Fig. 11-6. These tests are made by using the *Megger* (Fig. 11-10), *Vibroground* (Fig. 11-11), *McCollum Earth Current Meter*, and soil box (Fig. 11-6). Tests conducted by using the first three are made in the field on undisturbed soil. The *Wenner Method* (Fig. 11-13), the ammeter-voltmeter method (Fig. 11-14), and *Shepard Canes* (Fig. 11-12) are also used. The latter and similar methods[5–8] are valid only for soil in the immediate neighborhood of the electrodes. The value of soil resistivity tests is questionable, since they cannot differentiate between acid soils which are detrimental to ferrous metal and alkaline soils which are not. Soil resistivity test results can be improved by taking pH and total acidity of the soil at each test location. Tests should be made at 50-ft intervals along the right of way and more frequently in critical areas.

Calibrated induction type pipe locators may be used to evaluate soil resistivity with reasonable accuracy, particularly in scouting low resistivity areas.[4]

The *Columbia Soil Rod* is an instrument which consists of a steel tip and a copper sleeve insulated from each other and mounted on a steel rod. Wires from the tip and the sleeve are connected to the terminals of a microammeter. A hole is bored nearly to the pipe depth, and the rod is inserted into the hole, so that both the steel tip and the copper sleeve contact the soil. A galvanic cell is developed, the steel tip becomes the anode and the copper sleeve the cathode, and the current developed is measured by the microammeter connected across the external circuit. Readings of the maximum galvanic cell current are observed and recorded every three seconds until the current becomes stabilized, indicating that the copper cathode is polarized. Readings taken over a period of years yield a curve showing the corrosivity of the soil.

REFERENCES

1. Saunders, R. H. "Saving Dollars With Corrosion Surveys." *Gas Age* 128: 32–5, Dec. 7, 1961.
2. McCollum, B. and Logan, K. H. "Practical Applications of the Earth Current Meter." (T351) *Bur. of Standards Tech. Papers* 21: 683, 1927.
3. Columbia Gas System Service Corp. *A Manual on Underground Corrosion.* New York, 1952.
4. Parker, M. E. *Pipe Line Corrosion and Cathodic Protection,* 2d ed. Houston, Gulf Pub. Co., 1962.
5. Shepard, E. R. "Pipe-Line Currents and Soil Resistivity as Indicators of Local Corrosive Soil Areas." (RP 298) *Bur. Standards, J. Res.* 6: 683–708, April 1931.
6. Ewing, S. "Electrical Methods for Estimating the Corrosiveness of Soils." *A.G.A. Monthly* 14: 356–61, Aug. 1932.
7. Kindsvater, E. F. "Soil Resistivity Rods Using Alternating Current." *4th Soil Corrosion Conf.* Washington, Bur. of Standards, 1937.
8. Gish, O. H. and Rooney, W. J. "Measurement of Resistivity of Large Masses of Industrial Earth." *Terrestrial Magnetism & Atmospheric Elec.* 30: 161–88, Dec. 1925.

Chapter 4

Corrosion-Resistant Pipe Metals and Corrosion Preventive Means

by F. E. Kulman, J. O. Mandley, and N. P. Peifer

The standard method for preventing corrosion on transmission lines has been to protect the steel piping with a nonmetallic coating and to supplement the coating with cathodic protection, if necessary. However, in distribution systems, particularly with small diameter services, corrosion protection can be obtained by using metallic coatings which sacrificially protect the underlying steel, or by using pipe metals more resistant to corrosion than plain steel or iron. The Bureau of Standards[1] has conducted soil burial tests on a wide variety of metals since 1932. Much of the information available on underground corrosion of nonferrous metals and coatings is based on these tests.

CORROSION-RESISTANT PIPE METALS

Ferrous Alloys

Although atmospheric corrosion of steel is reduced by small additions of copper as an alloying element, this alloying gives no benefits in buried steel. Moderate to substantial proportions of chromium, nickel, and molybdenum are needed to improve underground corrosion resistance of steels. Beneficial results from small additions of these alloys in atmospheric and aerated exposures are attributed to formation of a dense, adherent rust film or layer of corrosion products. However, in underground exposure, especially in soils with relatively high moisture content and poor aeration, the iron oxide films formed on low-alloy steels are usually too porous to provide corrosion protection, and little advantage is gained.

Soil burial tests of the Bureau of Standards showed that although the corrosion resistance of copper-bearing steels was not appreciably better than mild steel, copper-nickel steel (2.5 per cent Ni and 1.0 per cent Cu) was superior to mild steel in the 15 test sites.

Chromium contents of four to six per cent reduced weight losses appreciably, in some soils by as much as 50 per cent, and reduced pitting depth in all environments except in the organic, reducing types of soil. Chromium additions beyond six per cent tended to cause very localized corrosion in heavy, poorly drained clay soils. Deeper pitting than on ordinary steel resulted in one case. The tendency of high chromium contents to accelerate pitting in steels is neutralized by alloying with nickel. Steels with chromium and nickel contents higher than 18 and 8 per cent, respectively, showed very little corrosion in all soils including Docas clay—a poorly aerated soil containing two per cent sodium chloride. These test results were obtained on small specimens and should be applied cautiously in predicting the life of actual pipe installations in widely varying soils.

Nonferrous Metals and Alloys

As with other metals, corrosion of **copper and red brass** usually occurs in poorly aerated soils.[1] Specifically, copper has been found to corrode: (1) in poorly aerated organic soils highly acid in reaction, (2) in neutral or alkaline organic soils rich in soluble salts, (3) in a clay containing two per cent sodium chloride, and (4) in cinders. Corrosion rates, however, tend to decrease with exposure time in most soils, especially when they are aerated. Thus, copper tubing has been found satisfactory in original installations of services. The electrochemical potential of copper is noble with respect to that of iron. Therefore, in replacing steel services by inserting a copper tube thru the existing pipe, the copper tube will be protected sacrificially by the iron.

Dezincification of **yellow brasses** (high-zinc) may occur in corrosive soils. However, this is not necessarily associated with the normal corrosion process of zinc, in which loss of weight and pitting occur. Despite the tendency for high-zinc brasses to dezincify, such corrosion can be prevented by adding as little as 0.08 per cent arsenic. **Red brasses** (low-zinc) show little or no tendency to dezincify.

Aluminum has not been installed underground to any great extent. In 1950 the first aluminum gas line, consisting of 1.8 miles of extruded type 63S-T6 aluminum alloy pipe 8⅝-in. OD was installed in Alabama.[2] Since then, increasing quantities have been installed.[3] It has been reported that aluminum can be buried unprotected in soils with resistivities greater than 1500 ohm-cm.[4]

Lead, used extensively for water services and as sheathing for underground electric cables, corrodes in soils which have a deficiency of oxygen and in those which have organic acidity. However, soil salts like sulfates tend to inhibit corrosion.

METALLIC COATINGS

Metallic coatings on buried pipes are subject to slow corrosion and therefore have found only limited application underground; nonmetallic coatings are usually superior.

Zinc is the only metallic coating that is used to any appreciable degree on underground pipes. It adheres well to iron and may be applied by hot-dip galvanizing, electroplating, or spraying. Zinc applied by hot-dipping is often used for protecting small pipes (up to 4 in.), especially in gas and water services. Pitting of zinc is less than that of iron in most soils, but it is excessive in highly acid soils (pH 2.6).

The protection given to iron by zinc probably results in part from the anodic potential of zinc, and possibly of intermediate layers of zinc-iron alloy, at normal temperatures, with respect to iron. This causes a protective current to flow from the zinc thru the soil electrolyte to the iron. The zinc-iron layers are probably influential in prolonging zinc life underground. Eventually, however, the zinc is completely dissolved and its protective effect ceases.

Tests on uncoupled galvanized pipe show that the protective value of a zinc coating of three ounces per square foot may disappear in less than ten years for most soils. Other tests show that a similar weight of coating completely prevented pitting over a ten-year period. If galvanized services are installed they should be insulated from the main and from the service equipment. Otherwise, the galvanizing life may be shortened by galvanic action between the zinc and adjacent uncoated steel pipes.

Calorized coatings were developed to prevent oxidation and scaling of the underlying steel at high temperatures. These coatings consist of an alloy of aluminum with base metal and may be applied by powder or dip processes. To determine their merits as an underground coating of aluminum oxide, developed on surfaces of calorized metals, the Bureau of Standards placed specimens of pipe, calorized by each process, in seven soils. All calorized specimens showed less weight loss and, with one exception, less pitting than unprotected steel. Although calorizing increased pipe resistance to soil corrosion, pitting was not prevented in any soil. Highest weight losses and rates of pitting occurred in poorly drained soils.

When steel pipe is **lead-coated,** the lead may be sufficiently corroded in spots to cause coating penetration, exposing the underlying metal. This tends to accelerate corrosion of steel since it is anodic to lead in most soils. Tests by the Bureau of Standards on lead-coated steel pipe indicate that the performance of lead coatings can be considered satisfactory in only a limited number of corrosive soils, chiefly those which contain sulfates.

NONMETALLIC COATINGS

A good nonmetallic pipe coating is one which maintains intimate unbroken contact with the pipe surface, has low moisture transmission, is insoluble in soil electrolytes, and is inert to soil chemicals and microorganisms. Ability of a given coating to meet these requirements can be determined by tests on the coating itself and on the coated pipe. Other properties usually desired in a satisfactory pipe coating follow:

1. Resistance to soil pressure and deformation.
2. Resistance to impact during transportation and installation.
3. Sustained high electrical insulation resistance.
4. Ease of application and repair.
5. Low cost.

Bituminous Materials

These are usually processed **coal tars** from coke ovens or **asphalts** from oil refineries. As such, the properties of the raw materials cover a wide range.[5] Coating manufacturers blend these materials and fillers to produce products to meet pipe-coating requirements.[6]

The American Water Works Association has adopted detailed specifications for coating and lining steel water pipe. These are based on underground use of coal tar coatings. The gas industry has no standard coating. Bituminous coatings of coal tar enamels, asphalt enamels, and asphalt mastics are most frequently employed for corrosion protection.

Coal tar coatings consist of either coal tar pitch and mineral fillers, or processed coal tar pitches of crushed coal, coal tar pitch, and heavy coal tar oil, combined with mineral fillers. The component materials are mixed by the manufacturer in required proportions to give satisfactory performance under specified conditions of use.

Primers. Hot-applied enamels used as a pipeline coating will not bond to a steel pipe unless a primer has been previously applied. The primer for this purpose may be one of the following types:

Bituminous Primers. These are a solution of the enamel used as a coating material in a solvent. Care should be exercised to select a primer designed for use with the particular hot enamel to be applied.

The bond between primer and enamel is the result of the combination of the solution or vehicle in the primer with the applied enamel. If the applied enamel is not at too low a temperature, this combination will take place. If the soluble material is completely evaporated from the primer, no combination will take place. The latter condition will be observed when the primer is brittle and flakes off when scratched.

The drying time of this type of primer depends primarily upon atmospheric conditions and pipe temperature, and it may vary from 10 min to several hours. Damage to the enamel in the form of gas bubbles will occur if it is applied immediately after priming. Good bond will result if the primer is slightly tacky or dry but not brittle. If it becomes dead and brittle, the pipe is again cleaned and primed.

Synthetic Primers. These consist of a synthetic resin dissolved in a suitable vehicle. Unlike primers made from coal tar enamels, synthetic primers may be used without modification with all coal tar enamels. Drying time is fast, and the pipe is suitable to receive enamel either immediately or months after its priming. Apparently, the bond between metal and primer is a polar one, and that between primer and hot enamel is the result of a polar and a fusion bond.

Coal Tar Pitch. Pitch is the still residue left from partial distillation of raw coal tar. It has a dark color and a viscous to solid consistency. It is comparatively nonvolatile and fusible, of variable composition, and sometimes associated with carbonaceous matter, with its noncarbonaceous constituents largely soluble in carbon disulfide. In its original form pitch is not suitable for use as a protective coating because of impurities. These are eliminated by further processing. The melting or softening point of both coal tar and petroleum pitches is used as a gage of the consistency or hardness of the pitches and certain compositions made from them.

Table 11-7 Specifications for Five Types of Pipeline Enamels[7]

	Narrow range		Medium range		Wide range		Hot line		AWWA*	
	Min	Max	Min	Max	Min	Max	Min	Max	Min	Max
Specific gravity at 77 F (ASTM D71-52)	1.40	1.55	1.4	1.50	1.4	1.5	1.4	1.5	1.40	1.60
Softening point, °F (ASTM D36-26)	190	200	195	205	220		240		220	
Penetration, tenths of a mm (ASTM D5-52)										
Load: 100 g for 5 sec at 77 F	0	2	2	6	7	11	2	6	10	20
50 g for 5 sec at 115 F	0	8	10	20	14	28	5	15	15	55
Filler ash, per cent (ASTM D271-58)	20	25	20	25	20	25	25	30	25	35
Sag test—2/32 in. to 3/32 in. on 12 in. square plate, 5 hr at 130 F		1/32 in.		1/32 in.		1/32 in.		1/16 in.	1/16 in. at 160 F for 24 hr	
Cold test—above plate 5 hr at 20 F, breaking and disbonding	None		None		None		None		No cracking at −20 F	
Stripping or peeling (ASTM D1664-59T)	None		None		None		None (10 to 180 F)		None (80 to 160 F)	
Dielectric strength test—break thru 1/16 in. coating at 10,000 v and low amperage	None		None		None		None			
Handling temp (min or range), °F†	45	110	32	140	25	150	40	180	45	
Disbonding temperature, °F	20		0		20		0			

* American Water Works Assoc. Std. for Coal-Tar Enamel Protective Coatings for Steel Water Pipe, C 203-62, 1962.
† Coating on pipe would be brittle or would cold-flow, below and above the range, respectively.

Filler and Ash Content. The mineral content of most pipeline enamels results from the addition of minerals of various kinds during manufacture. The ash or filler content in itself is of little significance until correlated with other data. Basic considerations follow:

1. High melting point bases and low filler content offer maximum resistance to soil stress or pressure deformation.

2. Low melting point bases and high filler content are less resistant to slide or sag.

3. Generally, bases having a high filler content are of lower melting point and not highly resistant to soil stress or pressure deformation.

4. In a given melting point range, lower filler content compounds can be applied at lower temperatures. They are not subject to large evaporation losses. In contrast, higher filler content compounds may suffer greater evaporation losses.

Enamels. Hot-applied pipeline enamel specifications are given in Table 11-7. Quantities of primer and enamel required for over-the-ditch coating application are given in Table 11-8.

Narrow Range Enamels. These are mineral-filled coal tar pitch—an improved version of enamels used in the early days.

Medium Range Enamels. Consisting of mineral-filled coal tar pitch with powdered bituminous coal digested into the compound, these enamels are quick-setting and flexible.

Wide Range Enamels. These consist of a fully plasticized, mineral-filled coal tar pitch. The plasticizing agent is pulverized coal digested into the mix. They are not quick setting when pipe temperatures range from 90 F to 135 F. Although they are very soft at 160 F, no sagging occurs. Faults occur when pipe is placed on skids. The ditch bottom must be of soft earth when pipe is installed directly after coating during the summer.

Hot Line Enamels. These consist of a coal digestion pitch with a mineral content slightly higher than wide range enamel. Developed chiefly for application to pipe operated at elevated temperature, they are also used under ordinary conditions. There is no sliding at 180 F.

Table 11-8 Primer and Enamel Requirements per 1000 Ft of Pipe
(over-the-ditch coating application in gallons)

Pipe diam, in.	Primer	Narrow range enamels	Medium range enamels	Primer	Wide range enamels	Hot line enamels
4 ID	1.96	1,249	1,293	2.14	1,331	1,390
6 ID	2.88	1,838	1,904	3.15	1,959	2,046
8 ID	3.76	2,396	2,479	4.11	2,551	2,664
10 ID	4.69	2,985	3,090	5.12	3,180	3,320
12 ID	5.56	3,542	3,665	6.07	3,772	3,939
16 OD	6.98	4,445	4,599	7.62	4,733	4,943
20 OD	8.72	5,556	5,749	9.52	5,917	6,178
24 OD	10.47	6,667	6,899	11.42	7,100	7,414
28 OD	12.21	7,779	8,048	13.33	8,283	8,649
32 OD	13.96	8,890	9,199	15.23	9,467	9,886
36 OD	15.54	10,001	10,349	17.14	10,650	11,215

AWWA Enamels. These enamels are quite similar to the wide range type but have a higher filler content and a higher penetration range. They are generally used as an internal lining. They have been mixed with wide range enamels to increase the penetration at lower application temperatures.

Asphalts. These are a species of bitumen of dark color and variable hardness. Comparatively nonvolatile and composed of hydrocarbons, they are substantially free of oxygenated bodies, contain little to no crystallizable paraffins, and are sometimes associated with mineral matter; the nonmineral constituents are fusible, and largely soluble in carbon disulfide. This definition is applicable to natural or native asphalts and petroleum or pyrogenous asphalts obtained from distillation and blowing of petroleum.

Asphalt Emulsions. A stable asphalt emulsion can be formed by adding water, soap or dust. A chromate corrosion inhibitor may also be incorporated. The pipe to which the emulsion is to be applied should be primed with an asphalt paint to prevent rust formation from the water in the emulsion. Otherwise, the rust prevents a bond to the pipe.

Asphalt emulsion is applied cold in a coating 1/8 in. thick by brush or spray. Drying is by evaporation of the water, reducing the coating thickness to 1/16 in. Drying time is from 12

Table 11-9 Composition of Somastic Coatings
(an asphalt mastic)

Primer—0.35 gal per 100 sq ft min (produced from manufactured asphalt, natural asphalt, and appropriate petroleum thinner).

Asphalt—(12 to 14 per cent by weight)

Grade	I	II	III	IV
Operating temperature, °F	80	125	160	210
Softening point (Ring and Ball), °F	150–175	175–200	210–220	250–265
Penetration at 77 F, 100 g, 5 sec	21–26	15–17	7–11	5–8
Flash point (Cleveland Open Cup), °F	450+	450+	450+	450+
Loss on heating at 325 F, 5 hr, %	0.5−	0.5−	0.5−	0.5−
Ductility at 77 F, cm	3.5+	2.5+	1.0+	0+
Per cent soluble in CCl_4	99.0+	99.0+	99.0+	99.0+

Aggregate—(85 to 87 per cent by weight)
Clean, nonmicaceous and graded to maximum density. (a) Sand: type No. 1 (for pipe sizes up to and including 4 in. nom diam), 100 per cent passing 8 mesh U. S. standard screen; type No. 2 (for pipe sizes 5⁹⁄₁₆ in. nom diam and over), 100 per cent passing 6 mesh U. S. standard screen. (b) Mineral filler: crushed stone or equivalent added until 20 to 26 per cent of the total aggregate passes 200 mesh U. S. standard screen.

Fiber—(one per cent by weight)
To be long, free from lumps or foreign matter; 100 per cent passing ⅜ in. screen opening; 60 per cent max passing 10 mesh; 20 per cent max passing 40 mesh.

to 24 hr or longer. As drying progresses, the dielectric strength of the coating becomes greater, reaching a maximum in from 72 to 144 hr.

Dry Portland cement may be dusted on the emulsion to expedite moisture removal. To be effective, it must be applied to a thickness to absorb all water present. Pipelines primed with a good paint and with emulsion applied over it have remained underground for as long as 20 years in fair condition. The difficulty encountered in using emulsions is their extended drying time.

Asphalt Mastics. These contain a large proportion of sand and other inert materials for reinforcing the asphalt.[8] Table 11-9 shows typical composition of one type of asphalt mastic. Table 11-10 gives thicknesses and weights. The mixture is heated to a viscous state and is extruded onto the cleaned and primed pipe with a thickness of approximately ⁵⁄₁₆ to ⅝ in. (depending upon pipe size). This type of coating showed the best performance in the API Pipe Coating Tests.[9] Ingredient proportions are chosen to produce a mastic of maximum tensile strength and minimum water absorption. Since the asphalt absorbs water slowly, the insulation resistance of the coating gradually decreases. Asphalt mastics appear to be inert to soil chemicals and microorganisms. Because of the mastic thickness, the coating may add considerably to the pipe weight, possibly an advantage in swampy areas.

Asphalt Enamels.[10] Fillers are added to the asphalt bitumen to form an asphalt enamel. The enamels are applied hot to cleaned and primed pipe, and together with reinforcing wrappers and an outer wrap constitute the coating. Two grades of coating are available, corresponding to minimum softening points of 210 F and 240 F for the asphalt enamel.

Nonbituminous Materials

Nonbituminous coatings for prevention of underground corrosion include those based on petroleum waxes and greases, Portland cement, vitreous enamels, and synthetic resins. Some of these coatings have given good performance in transmission lines as well as in distribution systems. In distribution work, the trend appears to be toward cold-applied coatings on small fittings in service pipes.

Petroleum Wax. These coatings, easily applied at the mill or yard over a prime coat,[11] appear to have met with considerable approval in some parts of the country. Their principal constituents are usually hydrocarbons derived from petroleum. These microcrystalline waxes are inert, not subject to deterioration, resistant to moisture penetration, and of high electrical resistance.[12] Their melting point is relatively low, approximately 170 F. Chemical inhibitors are usually incorporated for added corrosion protection.

Table 11-11 gives specifications for physical properties of a representative wax, blended from several waxes, each with its own particular melting point. Such a blend provides the high temperature strength and low temperature ductility required for pipeline application.

In constructing the coating, a minimum thickness of ¹⁄₃₂ in. of wax is flowed onto the revolving pipe at approximately 270 F. No objectionable fumes are emitted. To shield against soil pressures, a special wrapping of laminated tape is spirally applied over the wax. It usually consists of an asbestos sheet, wax-impregnated and placed between cotton cloth and an outer layer of impervious plastic sheet. For protection under the most severe conditions, a hard outside wax coating with good mechanical properties (e.g., abrasion resistance) is then

Table 11-10 Thicknesses and Weights of Somastic Coatings
(all sizes not shown)

	Nominal pipe diameter, in.								
	1¼	2	4	6	8	10	12	18	24
Nominal coating thick., in.	⁵⁄₁₆	⅜	⅜	½	½	½	½	⅝	⅝
Minimum coating thick., in.	¼	⁵⁄₁₆	⁵⁄₁₆	⅜	⅜	⅜	⅜	½	½
Approx. coating wt, lb/lin ft	2	3	5½	10½	13½	16¾	19½	34½	45½
Nominal thick. of coating for field joint, in.	⁹⁄₁₆	⅝	⅝	¹¹⁄₁₆	¾	¾	¾	⅞	⅞
Nominal length of field joint mold, in.	18	18	20	20	20	20	20	20	20
Minimum end overlap of field joints, in.	¾	¾	¾	¾	¾	¾	¾	¾	¾
Approx. weight of field joint, lb	6	9	18	30	41	50	59	93	123
Maximum tiers in nested stockpile on soft level earth	11	9	7	6	5	5	4	3	2

Table 11-11 Specifications for a Typical Wax for Pipe Coating

Melting point (ASTM D 127-60)	165–175 F
Viscosity, SUS at 210 F (ASTM D 88-56)	75–100
Penetration, tenths of a mm, 100 g for 5 sec at 77 F (ASTM D 5-52)	28–32
Ash, per cent (ASTM D 482-59)	0.1–0.3
Specific gravity at 32 F	0.947
(ASTM D 70-52) at 77 F	0.928
(ASTM D 287-55) at 170 F	0.816
at 250 F	0.789
Flash point (ASTM D 92-57)	525 F min
Fire point (ASTM D 92-57)	550 F min
Inhibitors	Selected
Wetting agents	Selected

applied. Coatings and wrappings are usually applied by coating machines.

Petroleum Grease. These coatings consist mainly of petrolatum which is obtained from steam distillation of paraffin-base crudes. A chromate inhibitor may be added for corrosion prevention. Consistency of the grease is approximately that of soft butter and, since it repels water, it can be applied to wet pipe. Its mobility tends to heal any openings in the coating. The grease coating is protected by a wrapper to prevent its absorption into the soil. For pipes larger than six inches, the wrapper may be asbestos sheet, impregnated with wax and backed by a plastic sheet. For smaller sizes the wrapper is usually a plastic sheet with a flexible cloth backing impregnated with wax.

Low temperatures apparently have no effect on the petrolatums as corrosion preventive coatings. Adverse effects may result from flow at higher temperatures. Rubbed-on petrolatum coatings appear to bond well to the underlying metal and are difficult to remove by scraping. The coating thickness should be about $\frac{1}{32}$ in.

The petrolatums have found favor as a field coating of small fittings since they can be applied by hand without heating. The grease has no apparent adverse affect on bituminous pipe coatings which may be contacted.

Portland Cement. For 3 in. sizes and smaller, the coating is usually $\frac{1}{4}$ in. thick, consisting of about 1:2 of cement-sand mortar. The thickness is gradually increased to $\frac{1}{2}$ in. on 8-in. or larger sizes. Under severe corrosive conditions in swampy lands, a shield of reinforced concrete $2\frac{1}{2}$ in. thick has been found valuable for corrosion protection, and also as an anchor to keep the pipes from floating.

The hardness of cement coatings provides necessary resistance against soil stresses. The weight of concrete coatings is advantageous in river crossings and swamp lands to overcome the buoyancy of soft mud. Since cement is alkaline, the iron surface may be protected locally by any moisture reaching it from the shield. An alkaline environment causes the iron surface to become more noble in its potential than in a neutral electrolyte. However, galvanic action may occur between portions of pipe covered with concrete and other portions either uncoated or covered with bituminous or other coatings. Another limitation is the porosity of mortars and concretes. Concrete has been found to deteriorate readily in waters and soils containing sulfur compounds. Therefore, these coatings should not be considered waterproof barriers to soil moisture but mainly as mechanical shields to the coating underneath. A concrete shield over a bituminous or wax coating affords adequate protection in the most corrosive soils.

Paints. These mechanical mixtures of pigments and vehicles (oils, varnishes, thinners and driers) have proved ineffective in preventing underground pipe corrosion. Failures have been traced to abrasion of the paint during pipe installation and local deterioration by soil microorganisms. Thicker coats help to alleviate these problems.

Vitreous Enamels. The enamel consists of a mixture of grit, clay, borax, feldspar, and water, applied to the pickled pipe at the mill by spraying or dipping. Baking at 1500 to 1600 F produces fusion and gives a glazed surface. Two coating and firing operations are generally necessary to avoid holidays.

The bond between vitreous enamel and the metal probably exceeds that of any coating flowed or painted on. Enamel absorbs little water, has high temperature resistance, is generally immune to aging, and has good resistance to soil stress. However, vitreous enamel is inherently brittle and may require extra care in handling and installation. Field coating of joints and fittings, and field repairs of the vitreous pipe coating are necessarily made with nonvitreous materials.

Synthetic Resins. These coatings can be produced with properties designed to meet specific corrosion protection requirements. One present disadvantage is their higher cost compared to bituminous and other materials, although the uniformity of the products permits use of a thinner coating. Synthetic materials can be extruded, sprayed, spread from solution, or taped on the pipe.

Cellulose acetate and cellulose acetate butyrate are among the oldest thermoplastic resins. Acetate butyrates are considerably more stable at higher temperatures and humidities than straight acetates. These materials possess high dielectric strength. As pipe coatings they may be extruded onto the pipe, spread, or wrapped spirally as a tape. In the form of wrappers they may be used for reinforcing and shielding asphalt and wax type coatings.

Wrappers for wax coatings commonly consist of a wax-saturated asbestos felt laminated to a film of polyethylene or cellulose acetate plastic. An unsupported plastic film of polyvinylidene chloride or a cellulose acetate-glass laminate wrapper may be wrapped over wax as another alternative. The plastic is wrapped over hot applied microcrystalline wax. Heavy kraft paper is commonly used as an overwrap.

Polyvinyl chloride has excellent resistance to chemicals, water, electricity, and aging. It is applied to pipes as an extruded coating, spread-coated from solutions, and sprayed on. It may also be formed into pressure-sensitive tapes with an adhesive facing. The latter form has been used for field coating of small services and simple fittings.

Polyethylenes are chemically inert, unusually resistant to moisture penetration, and have excellent electrical properties. The material can be extruded onto the pipe or made into pressure-sensitive tapes that may be used for field coating of pipes and simple fittings. Polyethylene coatings can also be applied from hot solutions of the resin, from dispersions in organic solvents or by flame-spraying.

Neoprene is synthesized by polymerizing chloroprene under controlled conditions. The unvulcanized product is a thermoplastic with considerable plasticity, like unvulcanized rubber. It may be vulcanized by heat alone or by use of metallic oxides. Its physical qualities may be modified over a wide range by proper choice of pigments, accelerators, and antioxidants. Vulcanized neoprene is more resistant to oils, heat, and sunlight than natural rubber, while its physical properties are comparable to rubber. Neoprene sheets have been applied over field welds and mechanical couplings and have been bonded to the pipe with adhesive. One such method employs a solid sleeve which is rolled onto the pipe during installation and unrolled over the fitting to be protected. A neoprene paint is applied between neoprene and pipe to obtain a good bond.

PIPE WRAPPERS

The coating of underground pipe generally includes some type of wrapper. Wrappers are used to reinforce and strengthen coating and to protect it from abrasion, and may be grouped according to their materials, the coatings to be used with them, and purposes of use. They are fabricated from cotton, glass, plastic, and asbestos, and are made for all types of coatings for cold and hot application.

Cotton cloth of open weave and saturated with tar or asphalt cutback is used with cold-applied asphalt and tar paints, mastics, and emulsions. After application of coating, the wrapper is wound spirally over it while wet, and the coating is thus forced thru the open weave in the wrapper and appears on its surface. Another coating application is made over the wrapper and the combination is permitted to dry. Coating thickness can be increased by additional applications of coating material and wrapper. Cotton is deteriorated by moisture-borne fungi and bacteria often present in soil.

Glass in coarse and fine filaments performs two protective services. Single (coarse) filaments laid down in random form and bonded together with one of many available materials constitute one type of wrapper. Generally used with hot-applied coatings, it acts as a reinforcing wrapper by adding its strength to that of the coating. Wrapper construction is such that it breaks up gas and air bubbles when pulled into the enamel. Care must be taken in application to assure that sufficient tension is applied to the wrapper to pull it into, but not completely thru the enamel; i.e., the wrapper should not be exposed above the enamel surface, since this would make it possible for water to follow the glass filaments to the metal surface by capillary action.

When these same glass filaments are formed into a thicker mat and saturated with a low melting point tar and asphalt mixture, the wrapper thus formed is used as an outer covering to protect the enamel from abrasion and damage in backfill operations. To prevent the layers from sticking together when formed into rolls, the material is usually dusted with mica or slate dust which may have to be removed before application, since it may prevent a strong bond between the wrapper and the hot enamel.

Glass in the form of very fine filaments is woven into an open mesh fabric called glass fab or fabric. When saturated with asphalt, tar cutback, or other bonding material, this fabric is used with cold coatings in the same manner as cotton fabric. Unlike cotton, it is unaffected by soil bacteria or fungi.

Asbestos wrappers, more commonly known as asbestos felts, consist of a mixture of asbestos and cellulose fibers, held together with a binder and saturated with either an asphalt, tar, or wax cutback. The cellulose content may vary from 10 to 20 per cent. The asbestos may be either of long or short fiber. This wrapper tends to pick up moisture if not properly stored or handled, which becomes apparent if a vapor forms during application when the felt contacts hot enamel or wax. Such vapors trapped under the wrapper may form bubbles in the coating, leaving a weak spot where failure will eventually occur.

The wrapper is applied directly over the hot coating as it is flooded or sprayed on the pipe, thus becoming tightly bonded to the hot coating surface. Tension should be adjusted so that the wrapper does not pull thru the enamel or wax and contact the metal surface. This condition can be checked by cutting out a section of the coating where the wrapper laps and examining it. It is often indicated by the size of the enamel bead formed at the lap. A bead of considerable size may indicate that the felt has been pulled in too tightly, squeezing out enamel at the lap. Excessive tearing of the felt may indicate too much tension. It may also be caused by evaporation of volatile liquids from the cutback used for felt saturation.

Formation of white crystals on the felt surface indicates that naphthalene has evaporated from the cutback and formed naphthalene crystals. When the wrapper becomes heated (from the hot enamel) these crystals will evaporate. If the wrapper is a reinforcing agent, voids in the enamel coating may result from vapor trapped under it, with later failure at this point. If used as an abrasion or perforated type wrapper, the gas or vapor may escape thru the holes. It is believed that when the crystals evaporate naturally, they cause the saturant to become stiff, leaving a wrapper that may be easily torn because of stiffness imparted by the saturant.

Fifteen pound felt was the standard weight for many years. Recently an 8 lb felt, reinforced with glass threads, has been introduced. Both are used in abrasion resistant wrappers.

Care should be taken to secure the right wrapper for the hot applied coating used. Thus, if a coal tar enamel is used, the felt should be saturated with a coal tar cutback. Similarly, an asphalt cutback should be used when the coating is asphalt, and wax with a wax type coating. These wrappers are also used as an abrasion resistant wrap for plastic tapes. When so used, they may be saturated with either tar or asphalt cutbacks.

Asbestos felt wrappers not only act as an abrasion resistant covering when applied over hot coatings, but also resist coating deformation by cold flow and lessen penetration dangers.

Wax and grease wrappers of asbestos felt saturated with wax have found considerable use in both hot wax coatings and cold grease applications. Wrappers for such use include asbestos felt with a fabric termed tobacco cloth bonded to it, asbestos felt with tobacco cloth on one side and an acetate film on the other, and asbestos felt reinforced with glass threads.

One specification for coating application calls for $\frac{3}{32}$-in. thickness of enamel, followed by an outer wrap of kraft paper. This specification would seem suitable for pipelines and lower in cost than one calling for asbestos felt. Experience has shown, however, that under some conditions, such a protective coating may fail. The paper is applied directly over and bonded to hot enamel or wax. When alternately wet by rain, and then permitted to dry, the paper tends to shrink. Since it is bonded to the enamel, it pulls the enamel or waxes apart at laps in the wrapper.

Paper is used as an outer wrap over mill coated pipe. In some specifications it is applied over and bonded to the enamel; in others, it is applied over asbestos felt or other type of outer wrap. The paper prevents coated pipe from sticking to racks in the mill, freight cars, or haulage trucks. It also serves to reduce pipe temperature while stored or strung by deflecting heat from the sun. It does not increase the dielectric strength of the coating, nor does it act as a water vapor barrier.

In certain localities protection offered by abrasion resistant wrappers is not sufficient to prevent coating penetration by rocks on ditch bottoms or in backfills. Under such conditions an additional wrapper is generally applied over the coated pipe. Materials for this purpose include 90 lb roofing paper, sometimes called tar paper, and a specially fabricated material known as rock shield, manufactured of roofing material scraps, mixed with asphalt and rolled into a film 1/8 to 1/4 in. thick. This film is backed on both sides with an asbestos felt or a heavy cardboard for warm weather use and with a light paper or tissue for cold weather use. The cardboard-backed rock shield can be used in cold weather by keeping it warm to conform with the shape of the pipe. Paper or tissue-backed rock shield is not suitable for hot weather use because it becomes very soft and can be penetrated by rocks.

Wood slats 1/2 to 1 in. thick, 2 in. wide, and from 3 to 6 ft long, are used as a shield on lines thru river beds, where pipe is pushed thru casing or conduit, and in areas where much large, sharp rock is encountered. Slats are laid parallel to the pipe and spaced from 1 to 2 in. apart around its circumference. Their purpose is to prevent damage from sharp rocks which might penetrate the rock shield, to eliminate danger of scraping off coating when pipe is pushed thru casing or conduit at road and railroad crossings, and to eliminate damage from large rocks falling uncontrolled over the pipe. When used thru casings, they distribute the pipe weight over a larger area, reducing possibilities of coating penetration.

A simple method of installation consists of placing two springs, such as screen door springs, around the pipe circumference, fastened together to form a continuous ring. Spacing between springs is determined by length of slats used. With springs in position, the slats are slipped under them at the desired spacing and then locked into position by steel banding. Bands 0.015 in., by 3/4 in., are generally satisfactory.

Wood slats should not be used on bare pipe. It is general procedure to use a layer of rock shield under slats on coated pipe. Coated pipe thus protected has been lifted by wire rope, dragged over the ground for several hundred feet, and even rolled without coating damage.

Plastic Tapes

Plastic tapes (Table 11-12), both polyvinyl chloride and polyethylene, occupy a unique position in the protective coating field. They possess the ease of application of paints and cold-applied coatings as well as the high dielectric and insulating resistance of hot-applied coatings many times their thickness.[13] Polyethylene and polyvinyl plastic tapes show excellent resistance to sunlight, acids, alkalies, corrosive salts, salt water, plain water, and aging. They show good resistance to alcohols, oils, and greases, and poor resistance to aromatic hydrocarbons.

Table 11-12 Physical Properties of Plastic Tapes

	Polyethylene		Polyvinyl chloride	
Thickness overall, mils	12	20	10	20
Backing, mils	8	15	8.5	18
Adhesive, mils	4	5	1.5	2
Type of adhesive	Synthetic		Synthetic	
Adhesion to backing, oz/in. width	48	48	25	25
Adhesion to steel, oz/in. width	60	60	25	25
Tensile strength, psi	25	32	25	45
Elongation at break, per cent	100	120	200	200
Tear strength, Elmendorf				
Machine direction	...	1200	...	...
Transverse direction	...	1800	...	...
Water vapor transmission rate, g per 100 sq in.-24 hr	0.20	0.10	1.55	1.21
Water absorption, per cent	0.02	0.02	0.19	0.19
Dielectric strength, volts	14,000	20,000	10,000	14,000
Insulating resistance, megohms	1,000,000		1,000,000	
Application temperature, °F				
Low	0	0	40	40
High	200	200	200	200
Application requirements	Clean pipe		Clean pipe	
Primer	Advantageous		Advantageous	
Type primer	Any kind		Rubber base	
Outer wrap*	Asbestos felt		Asbestos felt	

* Use of an outer wrap is recommended for both types of plastic tape to reduce damage when rocks are present in backfill. In sandy or loam type soil, wrappers are not required.

Tape materials are calendered into films coated on one side with a pressure-sensitive adhesive. To secure a strong bond between pipe surface and tape, it is necessary that the pressure-sensitive adhesive contact the entire metal surface. This s composed of innumerable small craters, many having a depth of several mils. When crater depth is greater than adhesive thickness, the metal at the crater bottom will not be in contact with the film. Where pipe has been factory-coated with a pipeline enamel primer or lacquer, tape can be applied directly to its surface, provided the dust has been wiped off and a tape with a four mil adhesive is used. For a tape with a two mil adhesive, it is recommended that a special rubber base primer be applied over the mill-applied primer.

All protective coatings are used for the single purpose of preventing contact between a metal surface and its environment. Since moisture is present in all soils, coatings must be as nearly impervious to water as possible, and they should have a low water vapor transmission rate. Table 11-12 gives water absorption and water vapor transmission rates of polyethylene and polyvinyl chloride tapes. Both these rates are the same as or less than that for coal tar enamels and much less than that for asphalt enamels.

Although plastic tapes are not subject to cold flow, they are affected by pressure deformation and are subject to damage from backfill. To avoid such damage, it is advisable to apply over the tape an abrasion resistant wrapper such as eight or fifteen pound asbestos felt, saturated with either tar or asphalt cutback to resist moisture absorption. To keep the felt from unwrapping, it is customary to apply at intervals of 100 to 200 ft a wrapping of paper, nylon tape, or glass filament tape with a pressure-sensitive adhesive on one side.

Plastic tapes can be applied by hand, by hand-operated wrapping machine, or by a power-operated wrapping machine. In addition, they can be added to pipelines at low temperatures, without warming of tape or pipe. This makes them useful during winter months when coal tar, asphalt, and wax hot-applied coatings become brittle and crack.

COATING TESTS[14,15]

Several tests, which not only determine uniformity of coating materials but also suitability of application, have been developed for laboratory use. They serve to evaluate various coating materials and to screen new ones.

Salt Crock Tests

These are used primarily to evaluate the continuity of pipe coatings and wraps and are also a means of measuring the resistance of pipe coatings to current flow and of establishing their effectiveness.

The tests consist of placing a coated specimen in a solution of chemical compounds frequently present in ground waters. A potential is impressed between the pipe and a carbon electrode, both immersed in the solution. The function of the chemical compounds is merely to make the water conduct current more readily. When the water penetrates the coating, current begins to flow and this flow is measured. Current magnitude is considered as a measure of the extent of penetration of the coating by the water.

Three salt crock tests are used for testing pipe coatings. Two are water penetration tests, one with the pipe as the anode and the other with it as the cathode. The third is a "creep" test which measures ability of a coating to localize a fault.

Resistance to water penetration is determined by immersing a specimen nipple in a salt solution, causing it to be either electrically positive or negative, and measuring resulting current. The test duration is 60 days. The specimen is coated with the material under study and placed vertically in the salt crock. The immersed end must be sealed to prevent electrical contact between bare metal and solution.

For the anodic (pipe is positive) water penetration test, a 60 day failure criterion of 40 micromhos per sq ft of coating surface has been suggested. The specimen is removed when this value is exceeded, and the failure site is determined. Coatings, the conductivity of which does not exceed 40 micromhos per sq ft during the 60 day test period, are considered satisfactory. No failure criterion has been established for the cathodic water penetration test or for the creep test.

The creep test is identical with that for anodic water penetration except that a small hole is cut in the coating. Current readings are taken so that the test may be stopped when the pipe wall is pierced by corrosion. When the test is completed, the coating near the hole is removed, and the extent of penetration under it is observed and measured.

Clay Burial Test

This test provides information on behavior of exterior pipe coatings and wrappings in clay soils subject to alternate wetting and drying and on their relative resistance to damage by rocks in backfill.

The clay burial test utilizes a wooden box lined with galvanized sheeting. A 2-in. layer of gravel is placed in its bottom, then a 2-in. layer of clay. Specimen nipples are coated with the material under study, their ends closed with tight-fitting wooden plugs, and then sealed. Specimens are placed horizontally on the clay mat, and a clay slurry (33 per cent water by weight) is added to provide a cover of 2 in. over them. The clay should be of a type which is not exceptionally severe in expansion and contraction. Its corrosion index (Corfield test) should be in the range of moderate corrosivity. After the slurry has set for two days, a bank of infra-red lamps is turned on for a 12 day period to accelerate drying by uniform heating of the clay surface. During this drying period, air is circulated by an electric fan over the surface of the clay, its surface temperature ranging between about 83 and 112 F. Top of pipe temperature may range from 82 to 107 F and that at bottom from 77 to 94 F.

The clay is thoroughly dry and extensively cracked after the drying period. It is broken up by light tapping with a hammer and lifted away in segments from the test specimens, which are then cleaned for examination by soaking in water and scrubbing.

The test specimens are carefully examined and the condition of the coating is recorded. A spark test may be used to determine failure. The test may be repeated and results reported as the number of cycles required to cause failure.

INSULATING FITTINGS AND SIMILAR DEVICES

The term "insulating materials" in the pipeline industry applies to materials for pipe coatings used essentially to keep one type of pipe from contacting another, and not to those employed as a protection against metal contact with the soil or with waters in the soil. One example of their use is the separation of distribution systems from transmission systems, and another is the sectionalization of both transmission and distribution systems into smaller units to facilitate cathodic protection, with less chance of damage to other structures.[16] Insulating qualities of materials used in all types of insulating fittings are far in excess of that required solely for corrosion protection, e.g., some protection against lightning. Practically all these materials will stand a potential of over 500 v before breakdown.

Insulating Meter Swivels

In appearance, an insulating meter swivel is much the same as a regulator or noninsulating swivel. A piece of molded nylon or other insulating material prevents the swivel shank and lip from making electrical contact with the union nut. The washer used as a gas seal is shaped to prevent contact between the meter inlet or outlet connection and the bottom of the swivel.

Union nuts are of Bakelite, Formica, or Transite, reinforced with a steel shell. Probably the most widely used insulating swivel employs a molded nylon ring and a special rubber gasket. Where Bakelite or Formica union nuts are used, meter weight may pull the meter away from the swivel at temperatures over 350 F. This condition will not occur with a Transite swivel reinforced with a steel shell. However, the mechanical strength of the Transite threads will not withstand the wrench pressure applied to tighten the union.

Insulating Screwed Fittings and Compression Couplings

Where *reducing bushings* are permitted, their use should be limited to outdoors where there is no possibility of temperatures of 300 F or over. *Insulating couplings* for joining screwed pipe are available in two types. One is limited to low pressures and is similar in appearance to a regular coupling but is constructed of Bakelite-type material. A similar coupling fitted into a steel shell is rated by its manufacturer to stand pressures up to 1000 psig. Insulating couplings are subject both to temperature and to longitudinal or angular strain. *Insulating nipples* are of *Bakelite, Formica,* or *Synthane* in various lengths. These materials are affected by temperatures above 300 F and do not possess great mechanical strength. Nipples carry a male IPS thread at each end. Available wall thickness range is from ⅛ to ¼ in. An *insulating union* looks very much like a regular union, except that a sleeve of insulating material extends thru the nut on its female section.

Insulating bolt or boltless type compression couplings differ from standard couplings in that one end of the center ring is slightly larger in diameter to accommodate an insulating rubber or nylon sleeve covering the pipe wall and pipe end. The inside diameter of the follower rings is also turned to a slightly larger diameter to accommodate the insulating sleeve extending beyond them. An insulating cap is applied over one end of the pipes to be coupled to prevent their contact.

Insulating Flanges

These are standard flanges assembled to prevent electrical contact of the pipe on one side of the flange with the one on the other side. This is accomplished by an insulating gasket between the flange faces, an insulating sleeve over each bolt to prevent metal contact in the bolt holes, and an insulating washer under each nut to prevent flange contact. A steel washer is placed between each nut and insulating washer to prevent breakage of the insulating washer.

The standard drilling of the bolt holes is sufficient to allow insulated sleeves to fit. The regular thickness of insulated sleeves is $\frac{1}{32}$ in. Sometimes sleeves of greater thickness are used; this requires reduction of bolt diameter and use of material of greater tensile strength. It is general practice to use stud bolts on insulating flanges since these are accurately machined and thus permit ready passage of a sleeve without cracking or damage.

One method of assuring good workmanship is to assemble the flange aboveground and weld a pup joint to each side. This assembled unit is then welded into the line. By this procedure, it is possible to make sure that the bolt holes are in proper alignment and that the flange is insulated before its installation.

The flange should also be insulated from earth contact. There are many methods of doing this, such as coating the flange, installing a shield around the flange and then waterproofing the shield, or by installation in a concrete pit. Considerable difficulty may be encountered if the flange is encased in hot tar unless the tar temperature is kept below the flow temperature of the insulated sleeves.

Use of Insulating Devices

Except for bolt type insulating couplings in larger sizes, the insulating devices discussed are most often used to separate electrically cast-iron water mains and copper services from steel gas mains and services. Some companies install the device at the end of their facility, separating the customer's service from the utility portion of the service. Other companies install insulation at the meter, separating both the utility and the customer's service from the water service.

Bolt type insulating couplings are used to insulate distribution mains and services. In many localities it has been found advantageous to install an insulating coupling at each cross street on distribution mains. This segregates them into 600 to 800 ft lengths that can be more easily protected by magnesium anodes spaced along their length. This is of advantage whether the lines are bare or coated.

There is no set rule for spacing insulating flanges on long transmission lines. In some instances they are spaced up to 20 miles, in others from seven to ten miles. With many foreign pipelines along the right of way, it has been found that if insulating flanges are spaced as close as seven miles on a good coated line, such a low current will be required for its cathodic protection that little, if any, cathodic interference will be encountered. Where there are few, if any, foreign lines in the area, spacings may be greatly increased because the increased current for cathodic protection will not cause cathodic interference.

Use of insulating material in gas lines can be of great value in corrosion protection and, for greatest effectiveness, the location of insulation should be specified by the corrosion engineer. This is particularly true in stray current areas where considerable amounts of stray current may be flowing in the pipe.

Where insulating coupling or flanges are installed in a coated line, it is advantageous to install lightning arresters or spark gaps to prevent insulation from being damaged by current surges, caused by lightning or any faults on high tension overhead electric power lines.[16]

Testing Insulating Joints

Insulating joints are tested for electrical resistance by impressing a direct current on the pipe and measuring the current I flowing thru the joint and the voltage V across the joint. The resistance R equals V/I. This is the most accurate method of determining whether the insulating joint has insulating properties or not. Usually a resistance of 15 ohms or more is considered satisfactory.

When an insulating joint is reinforced by insulated tie rods (as some large diameter high pressure mains are), the resistance measured is the composite of that of the insulating joint and that of the tie rods. If a low resistance is obtained, tests will be required to determine whether the short circuit is in the joint or in the tie rods.

In the case of an insulated *flanged* joint which has a low resistance, the specific bolt which conducts current thru the joint can be determined by an ohmmeter instrument if the bolts are insulated from both flanges. If only one end of each bolt is insulated, the ohmmeter measurement is without value since all the bolts are in electrical connections with one of the flanges. In this case, removal, inspection and replacement of each bolt in turn will be required. In general, determining the quality of an insulating joint is less difficult if the insulating joint is detached from the piping than if connected.

Considerable expense may be involved in excavating mains for the application and measurement of direct current. Hence, the following qualitative methods may be used to indicate whether the insulating joint is conductive or not:

1. The measurement of a voltage drop across an insulating joint in a main or service, especially if the voltage drop is within the range of expected magnitude and polarity.

2. Buzzer and audio tones impressed on the buried pipe and picked up on the ground surface by instruments sensitive to alternating magnetic fields. The method requires experience on the part of the user, since a portion of the tone may be conducted thru the insulation of a well-insulated joint, especially at the higher frequencies.

3. Exposed insulating joints in service pipes may be checked with a test kit.[17] In this method, an alternating current is applied to the piping on each side of the insulating joint. A pick-up coil is placed over the pipe at the insulating joint. The small current induced in the coil is amplified by a transistorized amplifier and measured by an indicating instrument. If the insulator is good, no indication is observed; if the insulator is not good, an indication is obtained.

REFERENCES

1. Romanoff, M. *Underground Corrosion.* (Circ. 579) Washington, Bur. of Standards, 1957.
2. "Underground Aluminum Gas Line." *Ind. & Welding* 23: 30–1, Mar. 1950.
3. Dalrymple, R. S. *Aluminum Pipe Line Case History Records —A Report of NACE Technical Unit Com. T-2M.* Houston, Nat. Assn. of Corrosion Engineers, 1962.
4. Whiting, J. F. and Wright, T. E. "General Corrosion Considerations in Laying Aluminum Pipe." *NACE Conf.* 1962.
5. Seymour, R. B. *Hot Organic Coatings.* New York, Reinhold, 1959.
6. Shideler, N. T. "Hot Applied Coal Tar Coatings." *Pet. Eng.* 23: D5–8, Feb.; D18+, Mar. 1951.
7. "Tentative Recommended Specifications and Practices for Coal Tar Coatings for Underground Use." *Corrosion* 12: 75–6, Jan. 1956.
8. "Tentative Recommended Specifications for Asphalt Type Protective Coatings for Underground Pipe Lines—Mastic Systems." *Corrosion* 13: 77–80, May 1957.
9. Logan, K. H. "API Pipe-Coating Tests—Final Report." *A.P.I. Proc.* 21, Sect. IV: 32–69, 1940.
10. "Tentative Recommended Specifications for Asphalt Type Protective Coatings for Underground Pipe Lines—Wrapped Systems." *Corrosion* 13: 75–7, Apr. 1957.
11. Cramer, A. H. "Pipe Corrosion Mitigation Practices." *Am. Gas J.* 168: 23+, June 1948. Also in *World Oil* 128: 218–22, Nov. 1948.
12. Brouwer, A. A. "Microcrystalline Wax-Polyvinylidene Chloride Plastic Film Pipeline Coating and Wrapper Virtually Unchanged After 10 Years Underground." *Materials Protection* 1: 16–21, June 1962.
13. "Plastic Pressure-Sensitive Tapes as Pipe Coating Materials." *Am. Gas J.* 187: 26–35, Mar. 1960.
14. Gally, S. K. "Preliminary Laboratory Evaluation of Pipe Coatings by the Salt Crock Test." *P.C.G.A. Proc.* 40: 218–20, 1949.
15. Williamson, K. F. "Improvements in Salt Crock Testing Techniques." *P.C.G.A. Proc.* 42: 125–9, 1951.
16. Hamilton, H. L. "Insulating Joints—Engineering Aspects." *A.G.A. Proc.* 1954: 246–51.
17. Snedden, T. "Checking Insulators in Gas Distribution System." *Materials Protection* 1: 50–5, July 1962.

Chapter 5
Cathodic Protection

by J. V. Adkin, H. W. Dieck, W. J. Kretschmer, P. F. Marx, N. P. Peifer, P. P. Skule, A. L. Stegner, and C. L. Woody

GENERAL THEORY

Cathodic protection is a method of reducing or preventing corrosion of underground metallic structures by impressing potentials on them which make them cathodic throughout with respect to surrounding soil (electrolyte). The impressed current source may be a rectifier, a direct current generator, or dissimilar metals.

Figure 11-15 shows local current flow from a pit (anodic section) in corroding metal to the adjacent cathodic area. The corrosion rate, a function of this current, is proportional to the potential difference between the anode and the cathodes, and inversely proportional to the resistance of these elements with respect to the electrolyte. Counteraction of this corrosion by cathodic protection is illustrated in Fig. 11-16. The total field current will be the vector sum of the galvanic currents and any impressed currents. Note that there may be *local* current flow, even though *average* current is zero. It is, therefore, necessary for the applied voltage to be sufficient for current to flow into every part of the protected metal surface. Current in excess of the minimum necessary for complete protection is wasteful and injurious to the overall protection system. This procedure transfers corrosion from a protected structure to a ground bed or auxiliary anode, which is comparatively economical to replace.

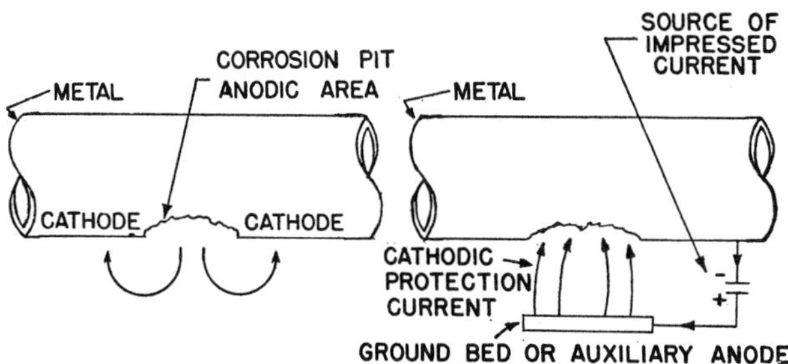

Fig. 11-15 (left) Galvanic corrosion.
Fig. 11-16 (right) Counteraction of corrosion by cathodic protection.

GROUND BEDS

The term ground bed, as used here, applies to a buried structure used to introduce protection currents into the earth from an external emf source. In designing an adequate ground bed, such items as anode materials, construction and design methods, effects of soil conditions, and anode bed configuration play an important part.

Annual costs involve investment charges, power consumption, anode replacement cost, and other maintenance costs. Once the site has been located, the type and number of anodes can be selected in keeping with the soil resistivity. Because this may not be uniform, the design is likely to require modification while the installation is being made.

Parker[1] suggests connecting the pipeline to the anodes by temporary means. The total loop resistance between the pipe and anodes should be measured, with the anodes connected one at a time. A fully charged storage battery (6 v) may be used; first one, then two, and then all three cells. The current values measured by an ammeter are plotted against the impressed voltage. The voltage where this plot intersects the zero current point is the galvanic potential between the anode and the steel pipe.

From these data, changes in the number of anodes may be determined. Alternatively, a rectifier could be used to produce the voltage needed to deliver the desired current thru the anode bed as already constructed.

Materials

Iron, Carbon and Graphite. Formerly, the most widely used material for ground beds was scrap steel rails, steel or cast iron pipe, automobile engine blocks, and manhole covers. Within recent years much success has been obtained with preformed carbon and graphite rods with or without a coal-coke breeze (carbonaceous) backfill. Where the ground bed must be installed in salt water, specially treated graphite anodes are available. A recent material is a high silicon (14.5 per cent) cast iron.

Aluminum and Magnesium. Light metals like aluminum and magnesium should also be considered in designing ground beds. In applying cathodic protection to internal surfaces of large water storage tanks, corrosion products, either as oxides or gases, must not color or make the water less potable or tend to increase the corrosion rate. For these reasons, aluminum or magnesium is often selected as the electrode or ground bed material. It is also possible to use them for ground beds installed to protect buried metallic structures.

An important factor in using these metals is the possibility of segregation (discontinuity) after a period of service. If this occurs, all metal separated from the electrode is useless, thereby reducing ground bed service life. The cause of segregation is uneven corrosion of the metal, possibly due to unequal resistance of the various contacting soils. Danger of segregation is greatly reduced if electrode wires are connected at several places on the metal surface. Their points of attachment, as well as a one inch "jumper" joining them, should be well insulated with a coating material.

Backfills. It has become general practice to surround the electrode with a prepared backfill of uniform conductance, which will hold moisture. The backfill composition also supplies the excess of anions necessary for corrosion of the sacrificial anode. The composition varies with type of metal used.

Calcium sulfate ($CaSO_4$ or gypsum) is the chief, sometimes only, ingredient for backfills for magnesium and zinc anodes. Volcanic ash or bentonite is sometimes mixed in the proportion of one part of three parts gypsum. In some locations wood fiber plaster, available at most building supply stores, gives excellent results. A mixture of one part sodium sulfate to ten of calcium sulfate may also be used. This may reduce backfill resistance (see Table 11-16).

A backfill for use with aluminum can be prepared by mixing 15 lb of lime with 15 lb of salt and adding enough water to make 1 cu ft. Painting the aluminum anode with calomel before installation has given good results. Magnesium oxychloride, a mixture of magnesia and magnesium chloride, in cement form, has also been successfully used as a backfill for aluminum. Examination of magnesium and aluminum in use as ground beds at two installations show them to be performing satisfactorily (efficiency of 90 per cent or better).

Coke breeze is used as a backfill with carbon, graphite, or high silicon cast iron rods. Special backfills are available for certain cathodic protection applications that use these materials. Examination of ground beds so constructed have shown excellent results after years of service.

Construction and Design

The corrosion rate and ground bed life depend upon the current discharged by the bed and its material. The theoretical weight loss for iron or steel is approximately 20 lb per amp-yr. For a graphite or carbon ground bed, theoretical weight loss is about 2 lb per amp-yr, and for a high silicon cast iron ground bed less than 1 lb. These weight losses for graphite or high silicon cast iron ground beds are based upon a definite limit of current density at the anode surface. A limit of 0.25 amp per sq ft of anode surface is recommended with earth backfill, or 1.0 amp if the anode is suspended in salt water or located in a carbonaceous backfill.

Consideration must be given also to installation of feeder cables between the rectifier and anode bed. Feeder cables and leads between anodes are usually buried, and therefore subject to a current discharge and eventual corrosion of any point of leakage in the insulation. Thus anode connections and feeder cable splices must be made with great care and must be well insulated from earth.

Soil Characteristics

A most important characteristic of an anode or ground bed is resistance to current flow. Most of this resistance appears in the earth near the anode. Soil resistivity determines the size of the ground bed necessary to pass a given current. Required resistance of a proposed ground bed is usually known, and the problem is to locate a site where soil resistivity is low enough to permit construction of a bed of economical size.

Relations between the required size of bed and soil resistivity are best determined by using Dwight's[2] expressions for resistance of electrodes buried in uniform earth. Curves based on these equations at a given soil resistivity are helpful; electrode resistances at other soil resistivity values are then proportional. Since determination of soil resistivity by four-terminal method takes account of depth, it may be used to locate anodes in the lowest resistivity stratum. Areas where the resistance rises appreciably with depth are to be avoided.

Drainage is another important soil characteristic, since the ground bed must not dry out because of seasonal changes. Fortunately, locations of lowest resistivity are usually poorly drained areas which make superior ground bed sites.

RECTIFIERS

Composite Conduction Rectifiers. The copper oxide and selenium types are most generally used for cathodic protection because they are constructed of solid metals with no moving parts.

Two commonly used types of rectifier circuits are shown in Figs. 11-17 and 11-18. In the half-wave circuit (Fig. 11-17), current flows only during the positive half-cycle of the a-c supply, giving a pulsating d-c output with a large ripple. The full-wave bridge circuit (Fig. 11-18) takes advantage of both directions of a-c flow and produces a pulsating d-c flow with less than half the ripple of the half-wave circuit. Figure

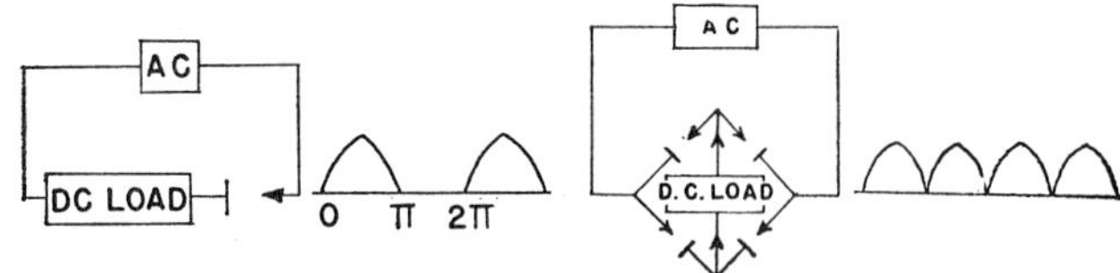

Fig. 11-17 (left) Half-wave rectification circuit and d-c output voltage wave.
Fig. 11-18 (right) Full-wave rectification circuit and d-c output voltage wave (bridge).

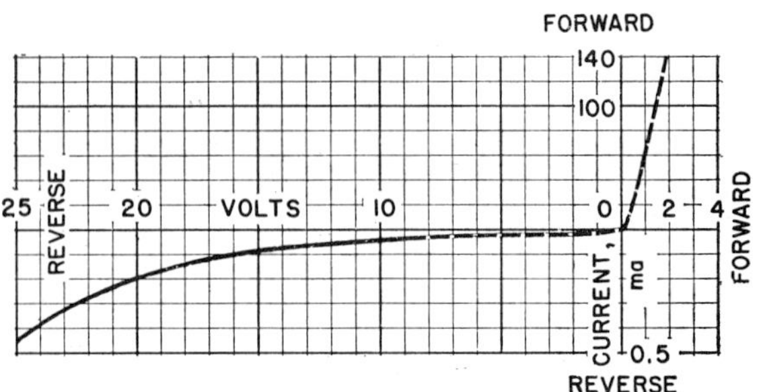

Fig. 11-19 Current-voltage relations for single rectifier plate. Reverse current scale is 200 times the forward current scale.

11-19 is a plot of current flow vs. applied voltage in both forward and reverse directions for a single rectifier plate. Note that the reverse current scale is 200 *times* the forward current scale.

Practically all single phase rectifiers for cathodic protection use the full-wave bridge circuit with a tap changing transformer (Fig. 11-20). They give a pulsating d-c voltage of about 75 per cent of the transformer secondary voltage. A polyphase rectifier has a d-c output with very little ripple.

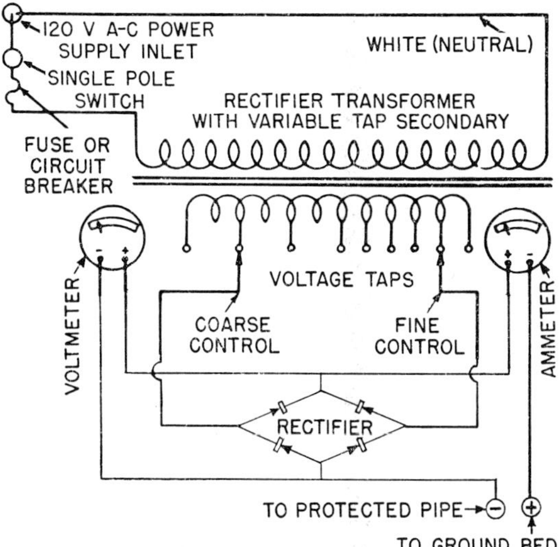

Fig. 11-20 Full-wave bridge circuit with tap changing transformer, single phase rectifier.

Commercial rectifiers are either air-cooled or oil-immersed, depending upon local requirements and preference. Oil-immersed units are usually higher in initial cost but are more adaptable to temperature extremes and corrosive surroundings. When fully enclosed, dust, rain, and insects are completely excluded. Air-cooled units are generally less expensive, lighter in weight and smaller for a given capacity. Both types will give years of satisfactory service when properly located and maintained.

Rectifier manufacturers supply weatherproof cases and hangers for pole mounting when desired. Large oil-immersed units are generally best mounted on concrete pedestals.

Characteristics

Rectifier efficiency ranges from 50 per cent in single phase to 75 per cent in three phase units. It is highest at or near full capacity and very low at fractional capacity. Figure 11-21 shows efficiency curves of the rectifier plates alone (transformer losses for an operating unit not included).

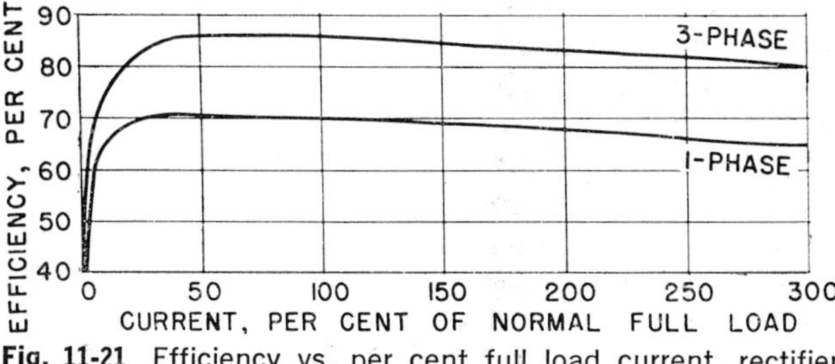

Fig. 11-21 Efficiency vs. per cent full load current, rectifier plant alone.

Copper oxide (below, left)—Where 24 hr average ambient temperature of operation exceeds 35 C, it is necessary to derate the output voltage rating in proportion to the increase in temperature above 35 C. Diagram at left shows a curve for derating to proper values for higher ambient temperature.

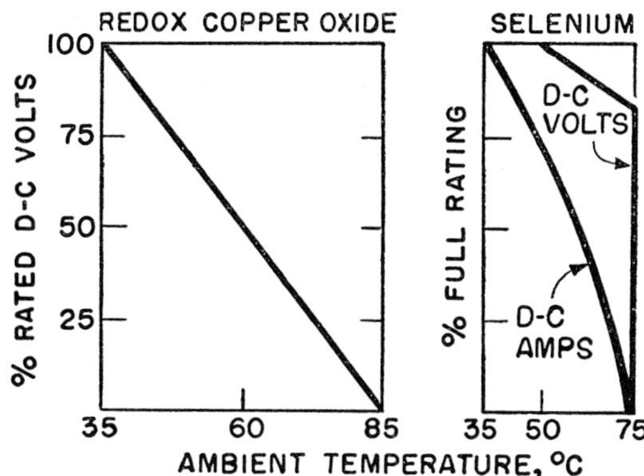

Selenium (above, right)—Voltage ratings are derated above ambient temperatures of 50 C (122 F). Current ratings are reduced for ambient temperatures above 35 C (95 F). For any 24 hr average ambient temperature, the proper derating factors are obtained from curves shown at right. Where rectifiers are to be operated in ambient temperatures beyond this range, refer to the factory.

Fig. 11-22 Derating output of selenium and copper oxide rectifiers for high (24 hr avg) ambient temperatures.

Both selenium and copper oxide rectifiers are rated at an ambient temperature of 35 C (95 F). They must be operated below their rated capacity at higher temperatures. Percentages of full load current and voltage which can be used above 35 C are shown in Fig. 11-22.

Application

Rectifiers are rated by their d-c output in volts and amperes. They are commercially available in any combination from a few volts and amperes up to several hundred. Table 11-13 gives the more common voltages with corresponding maximum current ratings.

Table 11-13 Common Output Voltages of Rectifiers and Corresponding Maximum Current Ratings

d-c voltage	d-c current, ma
6	500
10	250
15	200
20	150
25	150
30	120
40	120
60	120

In general, single-phase air-cooled units are adapted to low output requirements, and polyphase oil-immersed units are adapted to extra heavy current demands. Choice between power-fed and galvanic anodes is one of economics. Rectifiers are usually more economical when current requirements are high or when soil resistivity is above 5000 ohm-meters.

Operation

Rectifiers are closely designed and should never be operated at outputs above their design ratings. Installations should be provided with accurate recording means because the current output varies seasonally and with consumption of the ground bed. Sample record and report cards are shown in Figs. 11-23 and 11-24. The test meter reading is usually the potential difference between the protected line and an unprotected one. Other readings are self-explanatory.

Rectifiers should have adequate grounding for lightning protection at the distribution transformer serving the rectifier. Distribution system neutrals should be tied in to the rectifier case, with grounding resistance never exceeding five ohms. Contact or low resistance must be avoided between both terminals of a rectifier and the neutral of a secondary distribution system or the common neutral of electrical transmission lines. In new construction, guys and anchors are commonly bonded to the neutral, to short-circuit the ground bed if contact is made.

TOWN *Anytown* RECTIFIER NO. *5* LOCATION *Aspenway St.* YEAR *1950*

DATE OF INST. *4-10-49* KWH METER NO. *43681* TOTAL MILES OF 3" EQUIV PROTECTED *1.54*

DATE OF CHECK	CHECKED BY	KWH METER READING	KWH CONSUMPTION SINCE LAST READING	RECTIFIER METER READINGS		SOIL CONDITION (WET, MOIST, OR DRY)	METER READING TEST POINT VOLTS
				VOLTS	AMPS		
1.3	BSB	3028	72	11.5	8.7	MOIST	0.28
2.2	BSB	3100	75	12.0	8.1	DRY	0.30
3.1	BSB	3175	70	12.2	8.0	DRY	0.29
4.3	CID	3245	73	12.1	8.4	MOIST	0.28
5.4	CID	3316	71	11.2	8.7	WET	0.27
6.2	BSB	3385	69	11.3	8.5	WET	0.28
7.1	CID	3460	75	12.0	8.7	DRY	0.31
8.3	CID	3536	76	12.2	8.7	DRY	0.32
9.4			72	11.4	8.8	WET	0.26
				13.0	8.2	DRY	

GENERAL NOTES

Fig. 11-23 Cathodic protection—rectifier record form.

TOWN *Anytown* **DATE** *July 1*, **19**

MONTHLY REPORT ON RECTIFIER AND GROUND BED INSTALLATION

A. UNIT NUMBER *5* STREET *Aspenway St.*

B. METER NUMBER *43681* INDEX *3460*

C. DC AMMETER READING *8.7 amperes*

D. DC VOLTMETER READING *13.0 volts*

E. CONDITION OF SOIL (WET) (MOIST) (DRY)

F. TEST METER READING *0.31 volt*

G. GENERAL REMARKS

By *John Doe*

Fig. 11-24 Report form for field readings on rectifier and ground bed.

During installation of a rectifier, it should be assured by test that the positive terminal is properly connected to the ground bed and the negative one to the pipe. Current directional flow tests should be made and pipe-to-soil potential measurements taken.

The a-c power supply of a rectifier unit is "hot" to ground and can cause dangerous shock. All power should be cut off when servicing.

SACRIFICIAL ANODES

The more negative metals in the electrochemical series have long been considered as electrical energy sources for protecting more positive metals in a corrosive electrolyte. Galvanic anodes are, in general, the most convenient and economical current source for requirements of a few tenths of an ampere when installed in groups, or of a few amperes when distributed. Economics of galvanic anodes vs. rectifiers depends upon many variables, including power source and cost, labor, soil resistivity, and locations of other underground structures. Figure 11-25 shows relations between cost per ampere-year and ampere-years for one operating company. These curves may shift either horizontally or vertically, depending upon conditions at the desired anode location. Practical limitations are mainly confined to soil resistivity values and current requirement per unit length of pipe.

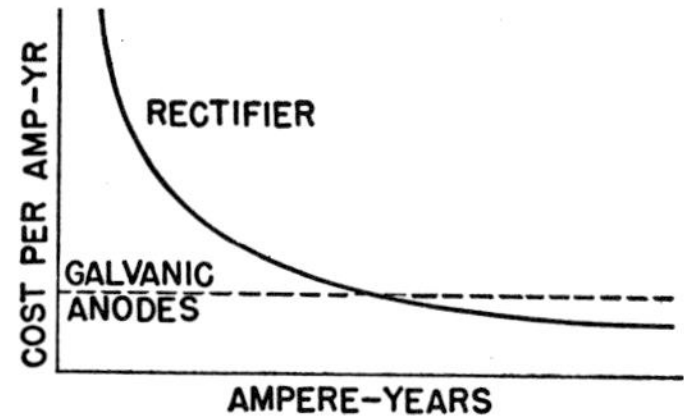

Fig. 11-25 Cost per year vs. ampere-years for cathodic protection current sources. Scale values are determined by considering all factors affecting the cathodic protection installation. The curves are based on a minimum power service charge and on average commercial power rates.

Pertinent data for three metals used commercially for galvanic anodes are given in Table 11-14.

Table 11-14 Pertinent Characteristics of Metals Used as Galvanic Anodes in Soil (Average Values)

	Zinc	Magnesium Cell	Magnesium Dow-metal H	Aluminum + 5% zinc
Specific gravity	7	1.73	1.94	2.92
Density, lb per cu ft	440	108	121	182
Theoretical amp-hr per lb	372	1000	1000	1300
Theoretical lb per amp-yr	23.5	8.7	8.7	6.5
Current efficiency, per cent	90	49	55	39
Actual amp-hr per lb	335	490	550	500
Actual lb per amp-yr	26	18	16	17
Solution potential, $CuSO_4$ reference, v	1.1	1.7	1.55	1.1
Driving potential, cathode volts minus 0.85 volt to $CuSO_4$ electrode, v	0.25	0.85	0.70	0.25
Backfill	Gypsum plus clay			Lime salt-calomel

Zinc Anodes

An early reported installation of zinc for underground pipe protection was made in 1935 in Colorado on a 6-in. and 8-in. welded gas line installed in 1929. High-purity zinc rods 1⅜-in. diam × 4 ft, weighing 18 lb, were installed along it. Tests indicated that substantial currents were delivered to the pipe in areas of favorable moisture and soil conditions and that zinc anodes in soil containing gypsum mixtures performed better than where gypsum was absent. Later tests led to use of artificial backfill to increase efficiency of these anodes. Extensive installations in the Gulf Coast States have indicated that zinc has a relatively long service life.

Commercial zinc anodes are available of high-purity 99.9 per cent zinc in plate or cast bars and of special high-purity 99.99 per cent zinc normally in cast bars only.

ASTM Designation B6-49 allows the following impurities in slab zinc:

Zinc grade	*Per cent maximum impurities* Lead	Iron	Cadmium	Aluminum	*Per cent total*
Special high	0.006	0.005	0.004	None	0.01
High	0.07	0.02	0.07	None	0.10

Cast and rolled anodes are used interchangeably. Results indicate about equal performance in similar gypsum backfill mixtures. The special high grade may perform better in backfill or electrolyte containing chlorides. The electrical properties of zinc make it a very suitable anode material when small currents are required for a considerable period as in well coated distribution piping systems. The open circuit voltage of zinc in a gypsum-bentonite backfill (1.1 v) is too low to overcome an appreciable circuit resistance—limiting zinc to low current applications in low resistance mediums.

Circuit resistance is of primary importance in designing zinc applications because of its relatively low driving voltage. Test installations have indicated that zinc anodes, ½ × 3 × 36 in., installed in a gypsum-bentonite backfill in 1000 ohm-m soil, will give a current output of from 10 to 30 ma with a useful life of approximately 335 amp-hr per lb. Useful life will be determined mostly by segregation and by the drop in current output below a useful value because of reduction in anode size or passivation of the zinc.

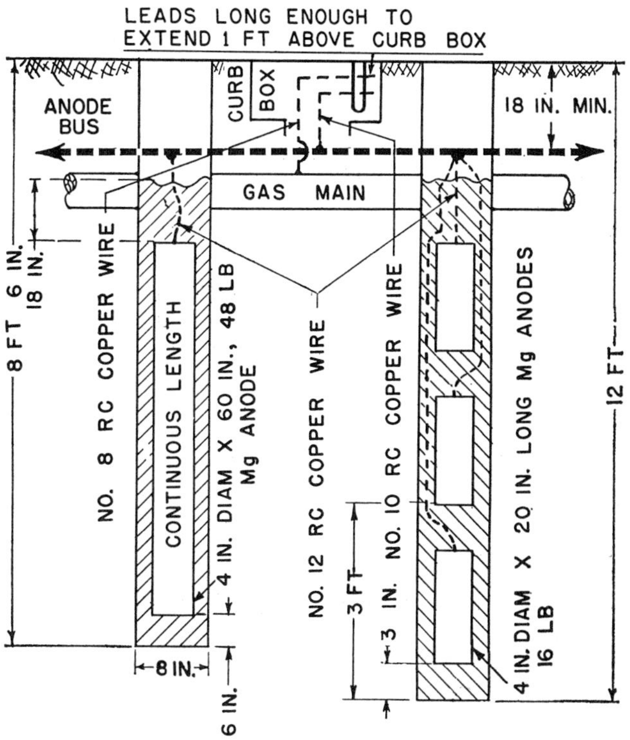

Fig. 11-26 Vertical installation of magnesium anodes in continuous (left) and broken lengths (right).

Notes:

1. Space magnesium anodes at least ten feet apart except where limited by space requirements or where it is desired to reduce current output per anode, as in the case of low resistance soils.
2. Where bare pipe is to be protected, anodes should be at least ten feet from pipe.
3. Backfill per anode:

	lb	gal
Continuous (left)	115	13
Broken lengths (right)	60	7

4. Solder wire to gas main.
5. Solder and tape all electrical connections to anode bus with rubber tape and friction tape.
6. In curb box, solder connections or use brass or bronze split bolt connector covered with heavy grease.

Magnesium Anodes

It was reported (1958) that the use of magnesium for anodes has far exceeded that of any other metal since 1944. Cell magnesium, *Dowmetal H* and secondary magnesium (Table 11-15) are grades used in cathodic protection. Cell or commercially pure magnesium was used in some of the first installations. *Dowmetal H*, a structural alloy, was modified for anode use by limiting certain impurities like nickel and iron. Secondary magnesium consists of remelted magnesium-base alloy scrap similar to *Dowmetal H*.

Table 11-15 Composition of Magnesium Anode Metals

Metal	Cell magnesium, %	Dowmetal H, %
Magnesium (by diff.)	99.878	90
Aluminum	0.003	5.3 to 6.7
Zinc	...	2.5 to 3.5
Manganese	.08	0.15 min
Silicon	.005	.30 max
Copper	.003	.05 max
Nickel	.001	.003
Iron	0.03	.003
Other impurities	...	0.3 max
Total	100.000	100.0

Effects of numerous impurities on the self-corrosion rate of magnesium in salt water were determined. With a highly purified magnesium, additions of small amounts of nickel or iron were found particularly objectionable. An increased *iron* content of otherwise pure magnesium from 0.01 to 0.03 per cent increased the corrosion rate in 3 per cent salt solution about 500 times. *Nickel* contents exceeding 0.001 per cent affected the corrosion rate greatly. Tolerances for various impurities changed with alloying elements of aluminum, zinc, and manganese present. Extensive experiments showed that higher current efficiencies were realized with a Mg–Al–Zn–Mn alloy than with cell magnesium, particularly in anodes operating at low current densities. While conclusive data are not available, general indications are that impurity tolerances are much less critical in a gypsum-clay backfill than in sea water applications.

Sizes of magnesium anodes range from the extruded type of magnesium ribbon or rods to cast anodes weighing hundreds of pounds. Readily available commercial sizes are either round, rectangular, or D-shape, weighing approximately 3, 6, 9, 15, 30, 50, 60, and 100 lb. Physical dimensions other than the weight-area ratio are believed to have little significance in actual use. Many sizes are available precast or packed in backfill in an organic bag with leads attached. Such packaging simplifies handling and installation materially. Figures 11-26 thru 11-29 show different magnesium anode installations and theoretical performance.

Its practical electrical properties make magnesium attractive for galvanic anodes (Table 11-14). It is primarily used where greater current is required, such as for large lines, for poorly coated lines, and in high soil resistivity areas. Test installations have indicated that a 60 lb magnesium anode, 4 in. sq × 60 in., will give approximately 250 ma for ten years in a clay-gypsum backfill. A steel core is used in all commercial magnesium anodes to prevent segregation of anode materials and also to provide means of attaching lead wires.

Introducing high potential magnesium anodes has permitted their use in higher soil resistivity areas where a larger driving potential is necessary. These anodes have an open circuit potential of from 1.7 to 1.9 v.

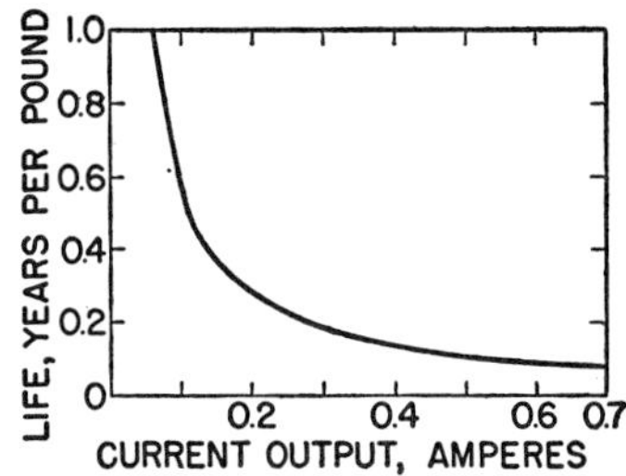

Fig. 11-27 Theoretical life of magnesium anodes. Fifty per cent current efficiency assumed. Curve derived from life (in years per pound) = 0.058/current output (in amperes).

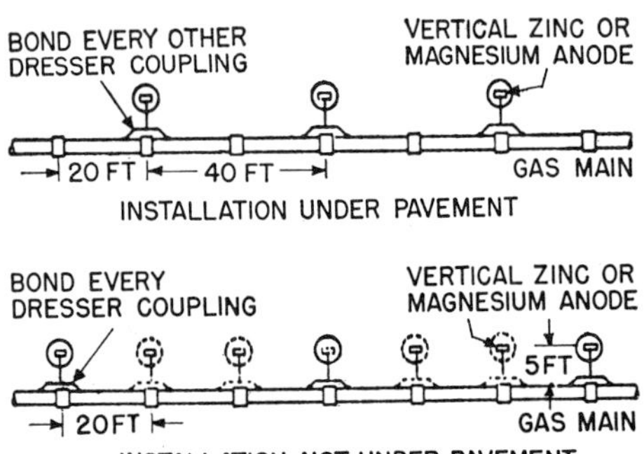

Fig. 11-28 Use of distributed anodes to protect Dresser-coupled bare pipeline. Depth of shafts, spacing, and number of anodes per shaft will depend on current requirements and soil resistivity.

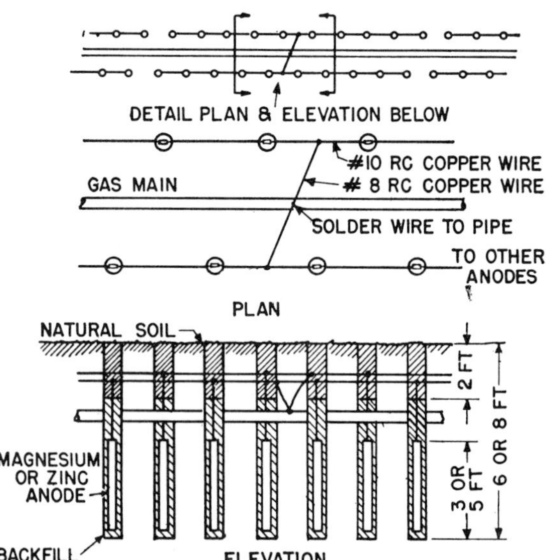

Fig. 11-29 Use of distributed anodes to protect uninsulated sections of bare pipe. Depth of shafts, spacing, and number of anodes per shaft will depend on current requirements and soil resistivity. Solder and tape all electrical connections with rubber tape and friction tape.

Aluminum Anodes

Although aluminum has high theoretical ampere-hours per pound, it is not considered a general purpose anode on account of its behavior in practical backfills. It was reported that aluminum containing approximately five per cent zinc will give better results than pure aluminum.

Extensive field tests with aluminum for galvanic anodes have been in progress for several years. The National Association of Corrosion Engineers Galvanic Anode Committee, in its first interim report, claimed some success with aluminum using specially prepared cast backfill.

Backfill

Backfills must be considered when installing commercial galvanic anodes. The principal problem is to provide an environment capable of maintaining the anodes in active condition over a long period. Many metals build up high resistance films and reduce solution potentials in some soils.

Early zinc anodes installed in soil containing gypsum crystals showed less tendency to become passive than where gypsum was absent. In later installations gypsum was tamped around the anodes for improved performance, or gypsum plaster was mixed with clay to form a slurry. Zinc was found to perform better when installed in gypsum-clay backfills than in natural soils.

Gypsum-clay backfills were later used experimentally with extruded rods of cell magnesium. Results were much better than expected and indicated that the improved behavior of magnesium in a gypsum backfill is more pronounced than when zinc is used. Further field and laboratory investigations indicated that gypsum backfills were the most practical for both zinc and magnesium.

Important practical advantages of gypsum as a backfill material are: (1) availability in practically every locality; (2) relative cheapness; (3) low solubility (about 1 to 400), which permits design of installations for long life. From the standpoints of convenience, simplicity and low cost, use of pure gypsum molding plaster, installed dry and then flooded with water, is advantageous, particularly for small installations. Molding plaster alone sets up hard when not mixed with clay or bentonite. If the soil has sufficient moisture, the setting does not introduce adverse reactions, but the resistance to earth of the anode is initially somewhat higher where moisture is inadequate. Gypsum readily absorbs moisture and tends to hold it better than most soils.

In localities of limited soil moisture or in well drained soils, use of gypsum-clay or gypsum-bentonite muds may result in voids around the anode. This increases the resistance to earth with low or erratic current outputs. One cause of voids is shrinkage as moisture is absorbed by the soil. This difficulty is particularly pronounced when gypsum is used in hydrated form. With gypsum molding plaster, much excess water is absorbed as water of crystallization with reduced shrinkage. The mixing sequence of gypsum and bentonite is also important.

Another cause for voids between mud backfills and magnesium anodes is accumulation of gas, probably hydrogen, which may insulate a large portion of the anode surface with resulting low current output. This effect apparently does not occur with dry backfills when moisture penetrates to the anode. Evolution of hydrogen would be expected to be particularly pronounced in acid soils or backfills. Data indicate that the addition of lime reduces gas blocking to a point where it is unimportant. Adding magnesium hydrate in acid soils has been recommended.

Table 11-16 Composition of Several Commercial Backfills
(per cent by weight)

Backfill No.	Gypsum: Hydrated	Gypsum: Mold plaster	Bentonite: Wyoming	Bentonite: Southern	Sodium sulfate "salt cake"	Molecular mixture: soda ash & lime
1	50	..	50	..	..	...
2	25	..	50	..	25	...
5	..	50	..	50	..	...
6	75	..	20	..	5	...
9	..	75	20	..	5	...
10	..	75	17	..	..	8

Several commercially available backfills may be obtained prepacked with the anodes or in bulk quantities. Table 11-16 shows their composition. Usually nonsoluble materials (No. 5) are preferred in low resistance areas of high soil water content. In dry areas of relatively high resistance, a backfill of high soluble sodium sulfate content (No. 2) may be preferred. Some are considered general purpose backfills (Nos. 6 and 10). Backfill No. 10 is not considered satisfactory for zinc, but it is used extensively with magnesium.

CRITERIA OF PROTECTION

Criteria are desirable for determining that current flow from earth to the structure to be protected is sufficient to counteract any flowing from the structure to earth.

Both the anode and cathode areas of a buried structure possess potentials. The combined effect is a net flow of current into the electrolyte from the anodes and an equal net flow from the electrolyte into the cathodes. If the buried structure is connected to a source of protective current to polarize it cathodically, the local anodes are "canceled out" by superposition. A portion, X, of the polarizing current goes to the anode areas, and the remainder, $1 - X$, reaches the cathode areas. For corroding steel the resistance per unit area of the cathodes can be as much as ten times that of the anodes, so that the polarizing current flows mainly (as desired) to the anodes (see Figs. 11-15 and 11-16). Consequently, if the total polarizing current is made large enough, the fixed fraction flowing to the anodes will eventually cancel the anode discharge and stop the anode reaction.

In the case of a corroding structure, the potential of a local circuit (Fig. 11-15) changes little until a reversal of current occurs at the anodes. With steel, this reversal results in discharge of hydrogen rather than plating out of ferrous ions. The Null method has been used to measure the polarization potential of the buried structures.

Another criterion of protection is based on equalization of surface potentials. It is accomplished by polarizing the cathodic areas of the metal until their potentials equal the open circuit potential of the anodic areas. Current leaving the anodic areas thus becomes zero and corrosion is stopped.

Since corrosion is a result of current flowing from the surface of the buried structure to earth, currents leaving the pipe in the anodic area and flowing to the pipe in the cathodic areas are measured. An instrument for such measurements is the **McCollum Earth Current Meter.**

This instrument is operated by using a special contactor constructed of a Formica rod carrying four electrodes. The two inner electrodes are copper pins, a measured distance apart, and the two outer ones are copper sulfate half cells at a measured distance from the pins. Current is caused to flow between the copper pins in contact with an electrolyte or earth and the resulting potential drop is measured by the copper sulfate electrodes. The test current is continuously reversed by a built-in commutator.

Current is observed on a milliammeter and the potential drop in the soil on a megohm per volt voltmeter. With current from the instrument shut off, the voltmeter reading is the IR drop in the soil resulting from the current flowing to or from the pipe surface. If the reading is on the negative side of the meter, current is flowing from earth to the metal surface. If on the positive side, current is flowing from the metal surface to earth, indicating corrosion.

Current flow may be calculated from the following equation:

$$M = \frac{E_e I_c}{E_c} K \qquad (1)$$

where:

M = current, ma per sq ft
I_c = milliammeter reading when battery current is flowing
E_c = voltmeter reading when battery current is flowing
E_e = voltmeter reading when the battery current is shut off, i.e., potential drop in the earth because of current flowing to or from the pipe
K = electrode constant (depends on diameter of pipe)

Where cathodic protection is applied to a coated pipeline, ordinary test methods for bare lines are sometimes inadequate. Coating faults (holidays) may be closely associated or a considerable distance apart. Under these conditions, it may be time-consuming and costly to locate and test each fault.

CURRENT REQUIREMENT SURVEYS

Coated Pipelines

The current necessary for cathodic protection is only that required to protect the metal at its coating faults. However, the potential difference between widely spaced faults may be greater than the potentials occurring on bare pipe. The quality of coating materials and methods used may reduce the number of faults so that application of current at one point will protect several miles of line. Location of this point can be determined by the location of low resistance earth and effects of cathodic interference on other buried structures.

A temporary location for a ground bed is determined by measuring the soil resistivity at several points and by selecting one of fairly low resistance in an area where there are few if any other pipelines. Foreign pipelines will collect current if they are in the potential gradient of the ground bed. This current will leave the pipe at some point causing corrosion. Such a condition is termed **cathodic interference.**

Field tests to determine the change in potential caused by the application of direct current are required at the cathodic power installation and at the ends of the line. They may be made as follows, assuming a new welded line, electrically isolated from others by insulating couplings or flanges.

After locating a temporary ground bed, ground rods are driven to a depth of several feet and connected to the positive terminal of a d-c power source. With power on and an interrupter in operation, the current is set at a value estimated to cause a change in potential at the ends of the line. This may vary from 2 to 10 amp, controlled by the power source potential and the resistance between the pipe and ground bed.

In reading the potential between the pipe and earth at the power source, the copper sulfate electrode must be located some distance from the pipe on the side away from the ground bed. To determine the proper location, readings may be taken normal to the pipe every 25 ft until three fairly constant readings are obtained.

The positive terminal of a zero or zero-center voltmeter is connected to the pipeline by a separate test wire, and its negative terminal is attached to the copper sulfate electrode contacting earth (reverse leads with a left-handed zero meter). The voltmeter reading with power off is recorded as E and the one with power on as E'. The difference between them is the change in potential caused by the current read on the ammeter connected to the power source.

Similar readings are made at the ends of the line and recorded. These readings indicate the change in potential caused by the current delivered or applied, with the location of the power source remaining constant.

Example: A new pipeline five miles long was coated with a good hot-applied coating and electrically insulated by insulating flanges.

A ground bed was located at the center of the line because no other pipelines were closer than two miles. The resistance between the ground bed and the coated line measured 6.4 ohms, indicating a required potential of more than 16 v to secure a current over 2 amp.

The power source was an engine-driven generator delivering 110 a-c v. It was connected to a small portable rectifier which had an output of 10 amp at 24 v. Taps permitted variation of output voltage from 2 to 24 v in 2-v steps. An interrupter with a time cycle of 7 sec off and on was connected in the negative lead from the rectifier to the pipe. The positive lead was connected to the ground bed thru an ammeter.

The generator was turned on and the rectifier set to deliver 2.5 amp at about 16 v. The copper sulfate electrode was set 100 ft from the pipe (determined by tests previously described). The observed readings here were 1.1 v off, E, and 1.4 v on, E'. At the north end of the line the observed readings were $E = 0.8$ v and $E' = 1.2$ v. Change in potential caused by 2.5 amp was 0.4 v, or 0.16 v per amp. At the south end of the line $E = 0.78$ v and $E' = 1.18$ v, giving the same change in potential. The d-c power required was 40 w.

When the permanent installation was made, an abandoned line, paralleling the new one, was used for the ground bed. The resistance between the new and old lines measured only 1.6 ohms. A potential of 3.0 v was sufficient to cause 2.5 amp to flow to the coated line. Power consumption on the d-c side of the rectifier was only 7.5 w, reducing operating power cost.

Since the abandoned pipeline was continuous and paralleled the line to be protected, no cathodic interference was experienced on the pipelines crossing it. If a standard ground bed had been installed near these lines, possibly 25 per cent or more cathodic protection would have required removal from the foreign line.

The high potential of the coated line to earth at the rectifier location was the result of stray currents. Therefore, the potential was raised above the stray current potential, effectively blocking stray current flow to the line. Care must always be exercised to insure that the line potential does not reach the hydrogen overvoltage potential of steel.

Bare Pipelines

Current requirement surveys for bare pipelines differ from those for coated lines. What are called long line currents may exist on long continuous bare pipelines. The cause of these currents has not been fully explained. They may reach two or more amperes and neutralize currents flowing in the opposite direction from galvanic cells without affecting currents from such cells flowing in the same direction. Changes in chemical composition of the soil, as in oxygen or salt content, are generally recognized as the major causes of galvanic cell formation.

Since galvanic cells of varying potential and current are formed continuously along a pipeline, protection can be achieved only when the cells of maximum current output are nullified. To locate them, **potential surveys** are made; readings are taken at intervals of 50 ft or less, using the same equipment as for coated lines. The buried pipe is connected to the positive side of a zero voltmeter (to negative side of a left-handed zero meter). The negative side is connected to one end of the test wire on the reel, the other end of which is connected to the copper sulfate half cell electrodes.

On plotting the readings (Fig. 11-7), defined peaks appear at the points of high potential and valleys at those of lower potential. Since current always flows from high to low potential, the points of high potential are the anodic areas, and those of low potential are the cathodic areas. Corrosion takes place at the high potential points and the pipe is cathodically protected by galvanic cell currents at the low potential points. These tests show the points where corrosion is taking place and where it is not, but do not indicate either by potential value or in any other manner the corrosion rate. This rate can be determined only if the conductance or resistance of the earth between the anode and the cathode is known. This resistance *cannot* be measured by four point ground tests because effects of the pipe as a conductor are included.

Current discharged from the anode on the pipe surface to earth may be measured as well as that discharged from earth to the pipe at the cathode. The same method may be employed as the one that is used to determine the current flowing on a conductor of unknown resistance. First, the potential drop over a measured length of the wire is measured, then the resistance of the conductor, using a rapidly reversing battery or alternating current. With the potential drop and the resistance known, current flowing on the wire is given by Ohm's law and its direction by the voltmeter deflection.

The same method can also be used to determine current flowing from the anode to the cathode on a buried bare pipe except that a calibrated electrode is used. Since the earth is a conductor, it reflects a potential drop if a measurement is made over a measured distance. This measurement can be made by using an Earth Current Meter, previously described.

In field practice, the potential survey is plotted and an earth current reading is made at the point of highest potential in each 500 or 1000-ft section of pipe surveyed. The current required for protection at this point is then used as that required for each square foot of pipe surface in the section.

Magnesium anodes have proved a most successful source of cathodic protection on bare lines because they are distributed along the pipe length and are not affected by the large potential drops experienced when rectifiers are used. Excess current from the anode to the pipe at point of closest approach is reduced to a minimum. The spread of protective current can be calculated and the calculations verified after installation of the anodes.

The magnesium anode is installed at the point of highest potential as shown by the survey. The current flowing from the anode to the pipe is measured with a milliammeter. The earth current calculation has shown the current required per square foot of pipe surface. The current output from the anode divided by the square feet of pipe surface per running foot gives the number of running feet protected, i.e., the required anode spacing.

To prove the calculations, a 2½ in. hole may be drilled to the buried pipe, the earth current electrode inserted and the potential drop measured. With the anode disconnected the meter reading will be positive. When the anode is connected, the reading will change to negative, indicating that current from the anode has overcome that from the pipe to earth and is now flowing from earth to the pipe giving it cathodic protection. Currents may be 0.1 ma or lower at this location.

COORDINATION TESTS

Cathodic Protection Interference

When cathodic protection is applied to underground structures, current is caused to flow in the earth. Part may be intercepted by nearby "foreign" structures, and the current picked up may cause their corrosion as it discharges to the earth. Interception of cathodic protection currents by such a structure is called *interference,* the current on the structure *interference current,* and the point of discharge to the earth *exposure.*

Theory. When a *point anode* is used in a homogeneous earth and no other structure is in the electrical field caused by it, the equipotential lines form concentric spheres. If a structure such as a pipeline traverses this field, it crosses many equipotential lines. This gradient causes a current to flow on the pipe away from the anode in both directions toward the outer equipotential circles as indicated by the arrows (Fig. 11-30, part A). It is assumed that the cathode is remote and does not affect the anode field. If the pipeline crossed the equipotential lines at a point considerably removed from the anode (part B), it would traverse only a slight potential gradient, and current flow on it would be small compared to the line in part A.

The field of a cathode, if a point, would be exactly the same as that of a point anode except for a volcanic "dent" in the ground potential. However, if the cathode were a pipeline "extending to infinity," equipotential lines would not be concentric circles, but elongated ellipses (Fig. 11-30 part C). Excluding anode effects, if a foreign pipeline were installed adjacent and parallel to the cathodic pipeline, as indicated by the broken line, the foreign line would cross a series of equipotential lines. Current flow on it would be from its extremities toward the center where the cathode attachment was made.

If a foreign line is near a cathodically protected pipe and the anode is near both structures, the anode potential field may predominate. Thus, the current flow on foreign structure may be away from the anode location, while on the protected line toward the cathode attachment.

If the pipeline arrangement remains the same, and the anode is moved out normal to the line to a remote point, the anode field would have little effect and the cathode equipotential lines would predominate. Current on the protected structure would then flow toward its center and to the point of cathode attachment. Current flow on the foreign structure would also be toward its center and point opposite the anode, a reverse of the flow with the anode close to both structures. Thus, at some point on the line normal to the parallel pipelines, the anode will have a minimum effect on the foreign structure because the anode and cathode fields will be more nearly balanced there. Since the two fields of the point anode and the line cathode are not symmetrical, an exact balance point is never realized.

Point anode effects on foreign structures which have nonparallel geometric relations to a protected pipeline are shown in Fig. 11-30, parts D thru G. The E arrangement produces the least interference on foreign lines. Part G shows a common city network situation.

Part H shows the potential of an anode some distance from

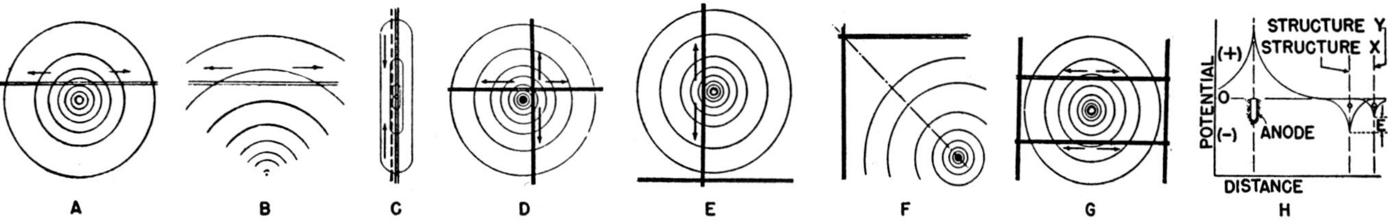

Fig. 11-30 Indicated current flow for various relative locations of point anode, cathodically protected pipeline, and foreign structures.

a cathodically protected pipeline **X** (connected to the anode) and a foreign line **Y** paralleling line **X**. The profile for the negative dent at **X** is made up of the anode effect plus the potential due to the cathode effect, while pipeline **Y** merely has the cathode effect. Thus, a potential difference as indicated by Δ exists between pipelines **X** and **Y**.

Factors Affecting Cathodic Interference

The extent of interference current problems depends in general upon the three following factors. The first is largely under the control of the system designer. The second and third are inherent in the structures and their environment, but an understanding of them will help to estimate probable stray current and necessary corrective measures.

Influence of Protective Installation. *Influence* is the tendency of a cathodic protection installation to produce interference currents. It depends upon amount of current used, location of anodes or ground bed and their configuration and resistance to earth, and type of structure (bare, poorly coated, or well-coated) protected. For any fixed anode arrangement, type of structure, and soil resistivity, influence is directly proportional to current.

Susceptiveness of Exposed Structure. *Susceptiveness* is the ability of a structure to pick up interference current and discharge it in a manner likely to cause corrosion. The most susceptive structure is an uncoated one, the least, one with a highly insulative coating. A pipeline with a poor coating which remains fairly uniform is much less susceptive than a bare structure. The longitudinal conductivity, that is, ability of the exposed structure to carry electrical current, is also a factor. Such a line is less likely to transmit interference currents when broken up by insulating joints.

Coupling between Structures. *Coupling* is the change in potential of an unprotected structure when protective current is applied to one nearby. Assuming a constant influence and susceptiveness, the tendency of a cathodically protected structure and its anode to cause interference current flow in a nearby structure depends upon their physical separation and earth resistivity. The less the separation, the greater, in general, will be the current picked up and discharged by an unprotected structure, particularly where lines cross near an anode, or where laterals of two parallel lines cross. For a given configuration of lines, coupling increases with earth resistivity. In soils with relatively high resistivities (more than 10,000 ohm-cm) and resultant high coupling, certain favorable conditions result. A relatively small current will establish protection of the one line, and the interference is likely to leave the unprotected line over a large area, thereby minimizing its corrosive effect.

Where earth resistivity is low and confined to relatively small areas, the resultant coupling is also low. However, large currents are likely to be picked up and discharged by the unprotected line within short distances. It may be advantageous to select a right of way that would minimize soil resistivity effects. For instance, if two structures must cross, their crossing in soils of high resistivity is advisable. If two lines are parallel in low resistance soil, the best relative location for the unprotected structure is far enough from the protected one so that coupling is at a maximum.

Methods for Determining and Mitigating Cathodic Interference

Various methods are used to determine necessity for a drainage bond and amount of current that should be drained from the unprotected line. They are often combined to solve a bonding problem.

Line-Current Method. This, one of the oldest methods, is still relied on to verify results of newer ones. It often determines need for bonds where other methods apparently fail. It is based on the theory that no electrolysis can be caused on another line by a protection system if this system does not cause a current loss from the unprotected line. Line currents are calculated from potential drops measured along the line and from its known or measured resistance. Measurements obtained before and after protection are compared. If they do not vary, no bond is needed. However, if any current loss to soil due to the protection system is detected, a bond is installed and adjusted to drain sufficient current from the unprotected line to eliminate such current loss. Accurate measurements and considerable work are involved. The indicated bond adjustment may vary appreciably without noticeable effect in current gain or loss.

Pipe-to-Soil Potential Method. The theory is that no electrolysis will be caused by a protection system if the potential gradient along an adjacent unprotected line is the same before and after protection. It requires measurement of enough pipe-to-soil potentials both before and after protection to determine potential gradients. If the protection appreciably increases the slope of the potential gradient, a bond should be installed. It is adjusted until enough current is drained from the unprotected to the protected line to reestablish the natural potential gradient. Considerable field work is required, and it is often difficult to restore the potential gradient when the gradient is irregular, as on bare lines. As with the line-current method, before and after protection surveys should be made within a short period.

On-and-Off Method. This method is so named because the protection unit must be turned on and off intermittently to determine if interference exists and also to adjust any bond if one is indicated. Theoretically, no current loss will occur because of the protection system if the "on-and-off" causes no potential change from the unprotected line to earth at the point of greatest exposure. Assuming a case of two lines crossing at right angles, an electrode is placed in undisturbed soil between the lines near the unprotected one (Fig. 11-31). If no electrolysis exists because of the unit, the potential measured between the unprotected line and the electrode will not change when the protection unit is turned on and off. If it should change, however, a bond is installed and adjusted until sufficient current is drained from the unprotected to the protected line to maintain the same potential either on or off. This method has the advantage of requiring no "before protection" survey. It gives a definite indication and is usually effective with a minimum of work.

Zero-Potential Method. This newer method is based on the theory that a tendency for electrolysis is present if the protection system causes a potential to exist from the soil near the unprotected line to that near the pro-

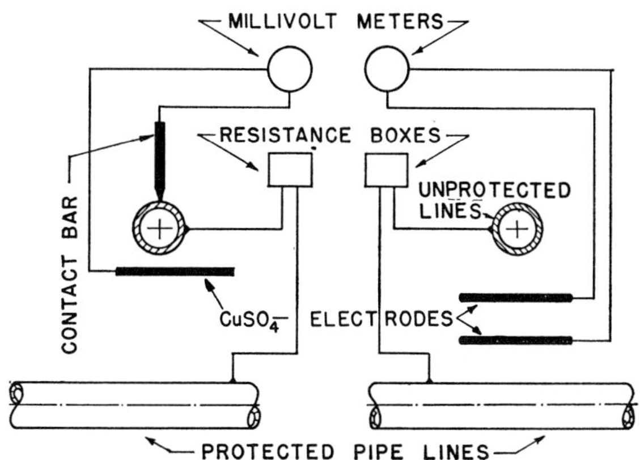

Fig. 11-31 (left) On-and-off method of determining cathodic interference.

Fig. 11-32 (right) Zero-potential method of determining cathodic interference.

Note: For illustration purposes, soil is not shown.

tected one along the plane of nearest proximity. Assuming that the two lines cross, a hole is dug at the crossing without disturbing the soil between the lines (Fig. 11-32). Two electrodes are placed between the lines, one half-way between them, the other near the unprotected line. If the protection system creates a potential of several millivolts from the unprotected to the protected line, a bond is indicated. The bond is adjusted until enough current is drained from the unprotected line to bring the potential between electrodes to zero.

This method is simple, gives a definite indication, and requires less time and work than those previously discussed. However, it has not been used enough to permit definite conclusions on its effectiveness.

CALCULATION OF CATHODIC INTERFERENCE ON PARALLEL STRUCTURES

Causes of such interference follow:

1. Leaky coating on pipes
2. Variation in soil resistivity
3. Change in longitudinal conductivity of structures
4. Change in diameter of structures
5. Change in distance to anode
6. Change in separation of structures

Computations are made in terms of : (1) maximum interference current; (2) maximum rate of current loss; and (3) length of corrosion exposures.

For illustration, the following examples are considered:

Case 1: Parallel 4-in. bare welded steel pipes, 10 meters apart, buried at a depth of one meter in soil of 100 ohm-m resistivity, with the anode 100 meters from the drained (protected) structure. Longitudinal resistance of each pipe is 0.0575 ohm per km.

Case 2: Same as in *Case 1*, except for uniformly leaky coatings on both structures with a leakage resistance of 1000 ohms per meter of pipe length. Leaky coating assumed because coatings of high insulation would practically eliminate interaction between structures.

Case 3A: Same as *Case 1* except for soil resistivity of 200 ohm-meters.

Case 3B: Same as *Case 1* but longitudinal resistance of structures halved.

Case 4: Same as *Case 1* but pipe diameter doubled.

Case 5: Same as *Case 1* but distance to anode doubled.

Case 6: Same as *Case 1* but separation of structures is doubled.

Since no metallic connections to the interference structure exist, its current, which results from alteration of earth potential, must be picked up from and returned to earth. The drained structure and its anode are the two sources of this alteration. They may be considered separately as *structure effect* and *anode effect*. The *net effect* may be obtained by combining these. In investigating the structure effect the anode is assumed to be so extremely remote that its effect on either structure is negligible. To investigate the anode effect, current is assumed to be transferred from a remote earth electrode to another electrode at the assumed anode location. Because separation of the structures is small compared with the distance to the anode, there is essentially no difference in the anode effect on each structure.

It was reported[3] that the cathodic protection system for over 60 miles of large uncoated mains in congested city streets required only *one* interference bond to a foreign structure. This system was installed in lieu of a main replacement program.

Structure Characteristics

Both anode effect and structure effect depend upon the attenuation constant of the structure. This in turn depends upon its longitudinal resistance and leakage resistance as indicated by the following equation:

$$\alpha = \frac{R}{R_L} \quad (2)$$

where α = attenuation constant, per meter
R = longitudinal resistance of structure, ohms per m
R_L = leakage resistance of structure, ohms

As R_L includes resistance of the soil path and that of any coating:

$$R_L = R_c + R_E \quad (3)$$

where: R_c = resistance of coating, ohms
R_E = resistance of soil path, ohms

R_E can be computed from the equation:

$$R_E = \frac{\rho}{\pi} \log_e \frac{1.12}{\alpha a_1} \quad (4)$$

where: $\log_e = 2.303 \log_{10}$
ρ = soil resistivity, ohm-meters
a_1 = equivalent electrical radius of structure, meters

For a single structure:

$$a_1 = (2ad)^{0.5} \quad (5)$$

where: a = outside radius of structure, meters
d = depth of structure center below surface, meters

For two close similar parallel structures:

$$a_2 = (a_1 y_1)^{0.5} \tag{6}$$

where:

a_2 = equivalent electrical radius of second structure, meters
y_1 = separation of structures, meters

By combining the above, the following equation is developed for computing α, the attenuation constant.

For one structure (α_1):

$$\alpha_1^2\left(R_c + \frac{\rho}{\pi}\log_e\frac{1.12}{\alpha_1 a_1}\right) = R \tag{7}$$

For two structures (α_2):

$$\alpha_2^2\ \frac{R_c}{2} + \frac{\rho}{\pi}\log_e\frac{1.12}{\alpha_2 a_2} = \frac{R}{2} \tag{8}$$

The values of α per meter and other characteristics are summarized for the seven cases in Table 11-17.

For uncoated pipes, $R_c = 0$ and the above equations become, *for one structure:*

$$\frac{\rho}{\pi}\left(\log_e\frac{1.12}{\alpha_1 a_1}\right)\alpha_1^2 = R \tag{9}$$

and for two structures:

$$\frac{\rho}{\pi}\log_e\frac{1.12}{\alpha_2 a_2}\alpha_2^2 = \frac{R}{2} \tag{10}$$

These equations show that on bare structures an increase in soil resistivity has the same effect on the value of α_1 and α_2 as a proportional decrease in longitudinal structure resistance.

Anode Effect

Figure 11-33 illustrates conditions considered in the anode effect. When separation of the structures, y_1, is small compared with the anode distance, y, currents in each of the two similar structures at any point at distance x are essentially equal. Since the anode affects both structures, their joint characteristics must be used in calculations. For the current in either structure due to anode current, I_a, the following equation for a uniformly conducting earth applies:

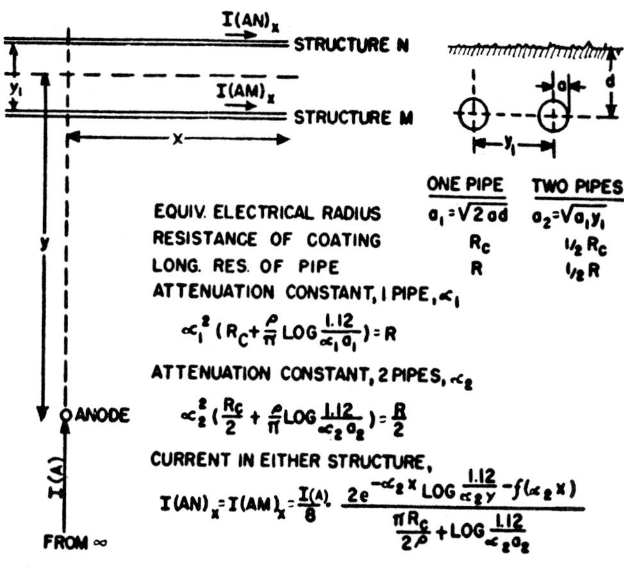

Fig. 11-33 Nearby anode effect on forced drainage interference.

$$I_{nx} = I_{mx} = \frac{I_a}{8}\ \frac{2e^{-\alpha_2 x}\log_e\dfrac{1.12}{\alpha_2 y} - f(\alpha_2 x)}{\dfrac{\pi R_c}{2\rho} + \log_e\dfrac{1.12}{\alpha_2 a_2}} \tag{11}$$

where:

I_a = anode current
I_{mx} = current in drained structure at distance x from drainage point, meters
I_{nx} = current in other structure due to anode effect at distance x from drainage point, meters
y = distance from anode to structure (if two structures—to their mid-point), meters

Values of $f(\alpha_2 x)$ are given in Table 11-18. Equation **11** applies when x is greater than y and $\alpha_2 y$ is less than 1.

For uncoated pipes ($R_c = 0$) their characteristics are reflected in the values of α_2, a_2, and y which are identical for *Cases 3A* and *3B*, similar results thus being obtained (Table 11-17).

The anode may, depending upon earth resistivity, affect earth potential at points quite remote from it. Such effects can be predicted only when the assumed soil resistivity value

Table 11-17 Sample Cases of Cathodic Interference

Description	*1* Bare	*2* Coated	*3A* Double soil resist.	*3B* Half longit. resist.	*4* Double radius	*5* Double anode distance	*6* Double separation
Depth of pipe, d, meters	1	1	1	1	1	1	1
Separation of pipes, y_1, meters	10	10	10	10	10	10	20
Radius of pipe, a, meters	0.057	0.057	0.057	0.057	0.114	0.057	0.057
Equiv. elect. radius of 1 pipe, a_1, meters	0.338	0.338	0.338	0.338	0.477	0.338	0.338
Equiv. elect. radius of 2 pipes, a_2, meters	1.838	1.838	1.838	1.838	2.185	1.838	2.60
Longit. res., R, microhm per meter	57.5	57.5	57.5	28.75	57.5	57.5	57.5
Resistance of coating, R_c, ohm per meter	0	1000	0	0	0	0	0
Soil resistivity, ρ, ohm-meters	100	100	200	100	100	100	100
Anode distance, y, meters	100	100	100	100	100	200	100
Attenuation const. 1 pipe, α_1, per kilometer	0.451	0.21	0.311	0.311	0.460	0.451	0.451
Attenuation const. 2 pipes, α_2, per kilometer	0.35	0.195	0.237	0.237	0.352	0.35	0.358

holds for considerable depth. A surface condition may be misleading if subsurface resistivity differs materially. This is illustrated in Fig. 11-34. When the anode distance is large compared to the upper layer depth, the anode effect is controlled largely by the deeper resistivity. In most cases the variations in soil resistivity are not so large as in Fig. 11-34 and a reasonable average value can be used for both effects.

Figure 11-35 shows calculated results of anode effects for *Case 1*. It compares the current in two structures with that which would have been found in one structure with the other absent. Current at a point opposite the anode is zero. Moving along the structures, a current away from the anode is created, making them cathodic up to about 1000 meters away where maximum current is attained. Beyond this point the structures lose current and are anodic.

Table 11-18 Values of $f(\alpha_2 x)$

$\alpha_2 x$	$f(\alpha_2 x)$	$\alpha_2 x$	$f(\alpha_2 x)$
0.01	8.056	0.45	0.796
.02	6.670	.50	.647
.03	5.860	.60	.404
.04	5.291	.70	.224
.05	4.846	.80	.087
.06	4.486	0.90	− .021
.07	4.182	1.0	− .102
.08	3.918	1.1	− .163
.09	3.688	1.2	− .210
.10	3.483	1.3	− .244
.15	2.704	1.4	− .273
.20	2.166	1.5	− .288
.25	1.763	1.6	− .302
.30	1.448	1.7	− .303
.35	1.190	1.8	− .308
0.40	0.977	1.9	− .315
		2.0	−0.351

Figure 11-36 shows calculated results for all seven cases. It is seen that doubling pipe diameters (without changing longitudinal resistance), or doubling their separation, increases anode effect current only slightly. Also, that doubling soil resistivity or pipe conductance increases maximum current about 10 per cent. With the anode twice as far away, maximum current is reduced about 42 per cent. Leaky coating reduces maximum anode effect current about 60 per cent.

Structure Effect

For determining current in the adjacent structure, a series of curves are derived giving interference currents at distances from the drainage point in terms of the ratio, μ, of mutual leakage resistance between the structures to the leakage resistance of each. This ratio is approximated by Eq. **12** on page 11/42.

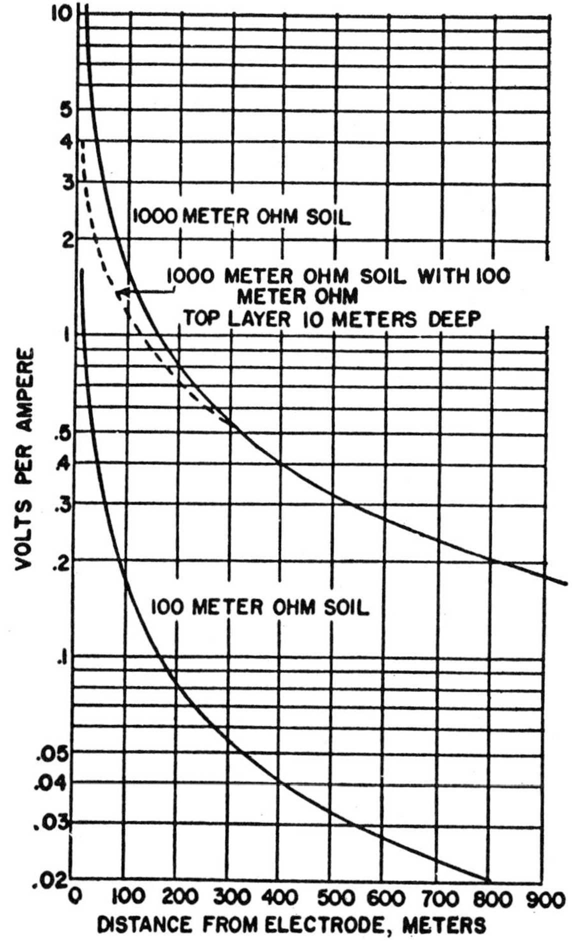

Fig. 11-34 Effect of two-layer soil on earth potential around a point electrode.

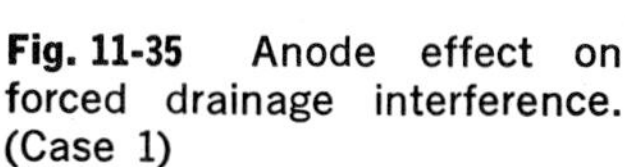

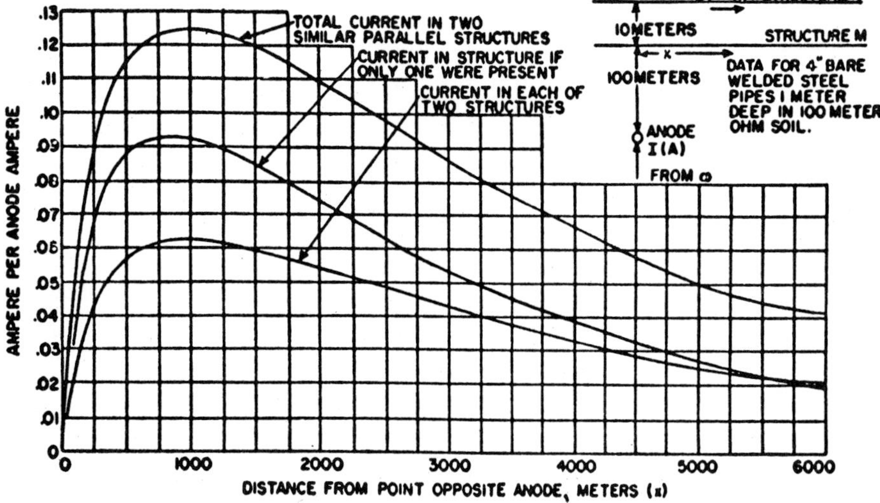

Fig. 11-35 Anode effect on forced drainage interference. (Case 1)

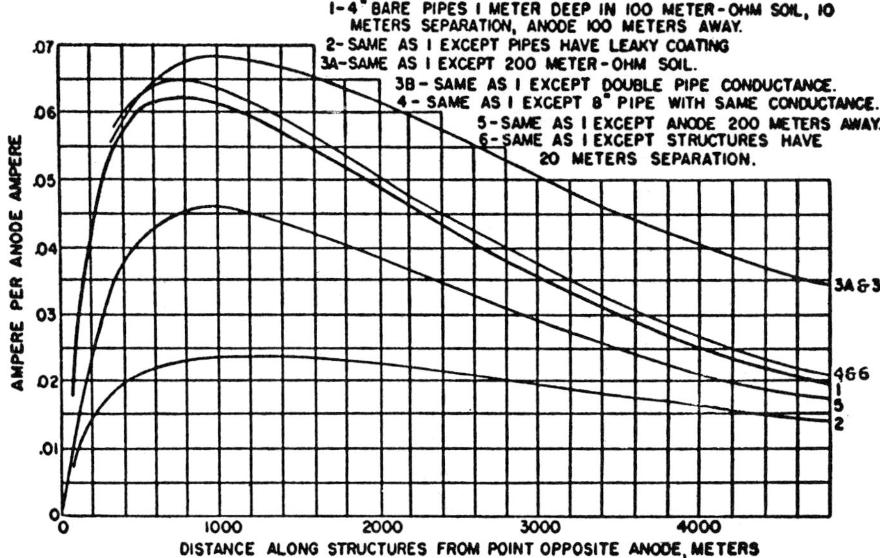

Fig. 11-36 Anode effect on forced drainage interference. (All 7 cases)

$$\mu \cong \frac{\log_e \dfrac{1.12}{\alpha_1 y_1}}{\log_e \dfrac{1.12}{\alpha_1 a_1} + \dfrac{\pi R_c}{\rho}} \quad (12)$$

Terms are defined under Eqs. **3**, **4**, **6**, and **7**.

Figure 11-37 shows curves for interference current as a function of $\alpha_1 x$ for various values of μ.

For uncoated pipes ($R_c = 0$) structure characteristics are reflected in the values of α_1, a_1 and y_1 so that *Cases 3A* and *3B* are again identical for structure effect as well as anode effect.

Figure 11-38 shows structure effects for six cases. Structure effect produces no current in the interference structure at a point on it opposite the drainage point. However, other structure points have current toward the drainage point instead of from it, as caused by the anode effect. Thus, current is lost from the adjacent structure near the drainage point. Since current is produced in the opposite direction from that caused by anode effect, values in Fig. 11-38 are plotted as negative. For *Case 1*, the value of current in the adjacent structure increases up to about 1700 m from the drainage point, then decreases, making the structure cathodic at the more remote locations.

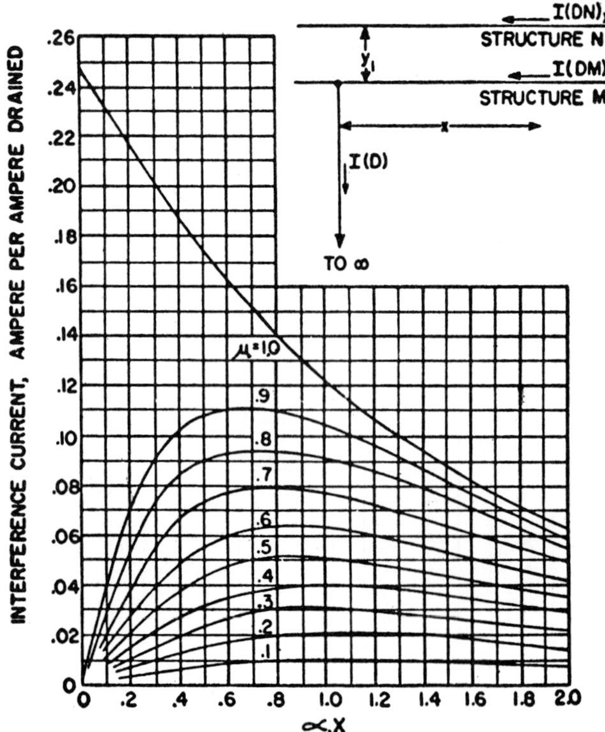

Fig. 11-37 Structure effect on forced drainage interference caused by drainage on a structure (see Eq. 12).

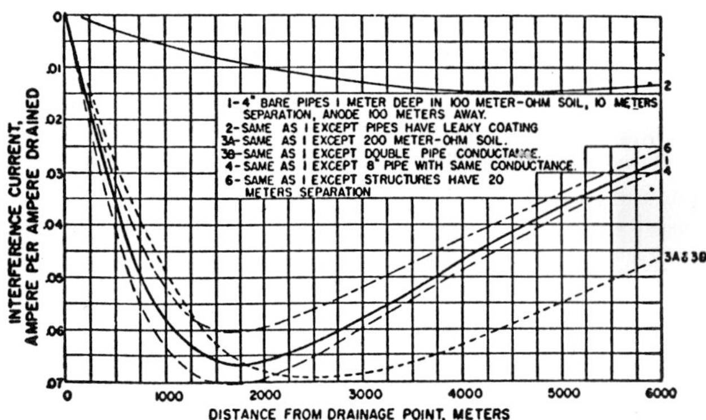

Fig. 11-38 Structure effect on forced drainage interference.

For the assumed conditions, doubling soil resistivity or pipe conductance increases maximum current slightly and moves its point of occurrence farther away. Also, doubling structure separation reduces maximum current about 10 per cent and a uniform leaky coating reduces it about 70 per cent.

Net Drainage Effect

Figure 11-39 shows anode and structure effect currents for bare pipes (*Case 1*). These effects tend to neutralize each other, but not completely, since they are not symmetrical about the zero current axis. Graphic combination of these curves produces one of net interference current. Progressing along the adjacent structure from a point opposite the drainage point, current increases (current pickup) to a maximum of about 0.025 amp per amp drained at 325 m.

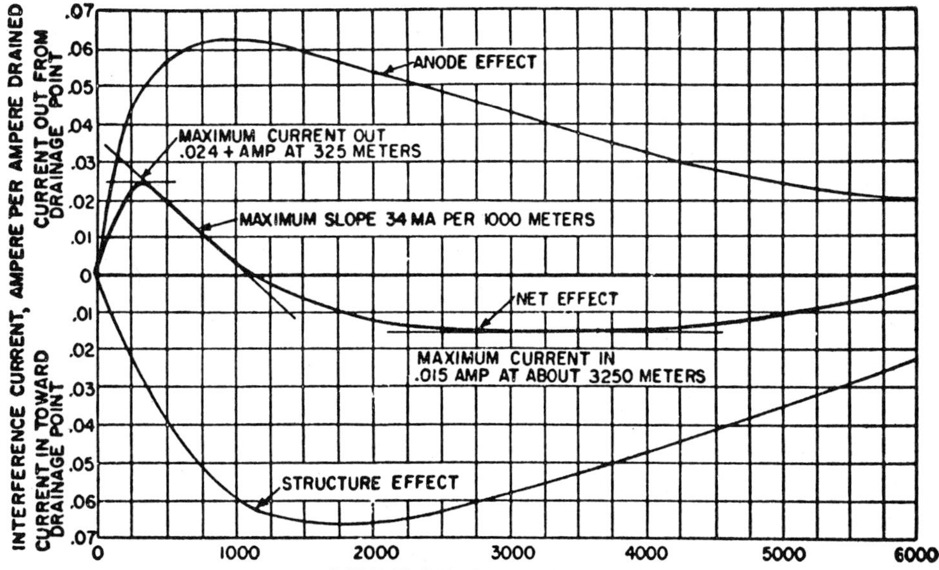

Fig. 11-39 Combined anode and structure effects on forced drainage interference. (Case 1)

Then it decreases (representing current loss) to zero at about 1150 m, beyond which it increases toward the drainage point (also indicating current loss) to a maximum of about 0.015 amp per amp drained around 3250 m. Total loss is about 40 ma per amp drained over a 3000 m length, much of which represents an insignificant loss rate with no practical effect on structure life. Assuming a current loss rate below 10 ma per 1000 meters per amp drained as insignificant, the length of exposure created by the interference is about 1600 m. More important, however, is the maximum rate of current loss, measured by the maximum negative slope of the net effect curve. It is shown here as about 34 ma per 1000 meters per amp drained at a distance of 500 m.

Figure 11-40 shows the net effect curves for all seven cases. Except for the coating, the variations represented have little effect on the corrosion exposure created by the interference currents.

Effects of moving the anode away were investigated further, to locate the point of minimum interference referred to as the *conjugate location.* First computations were made for *Case 1* with the anode 200, 250, and 300 meters away. At the 250 meter distance, no current transfer and, therefore, zero potential, was indicated. Since the zero potential change did not prove a real criterion of corrosion exposure, additional calculations were made for *Case 1* for closer and more distant anodes, and the maximum current loss rate was determined. Figure 11-41 shows results with maximum rate of current loss plotted against distance to the anode. For *Case 1*, an anode too

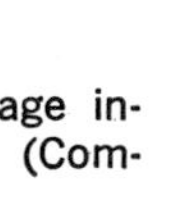

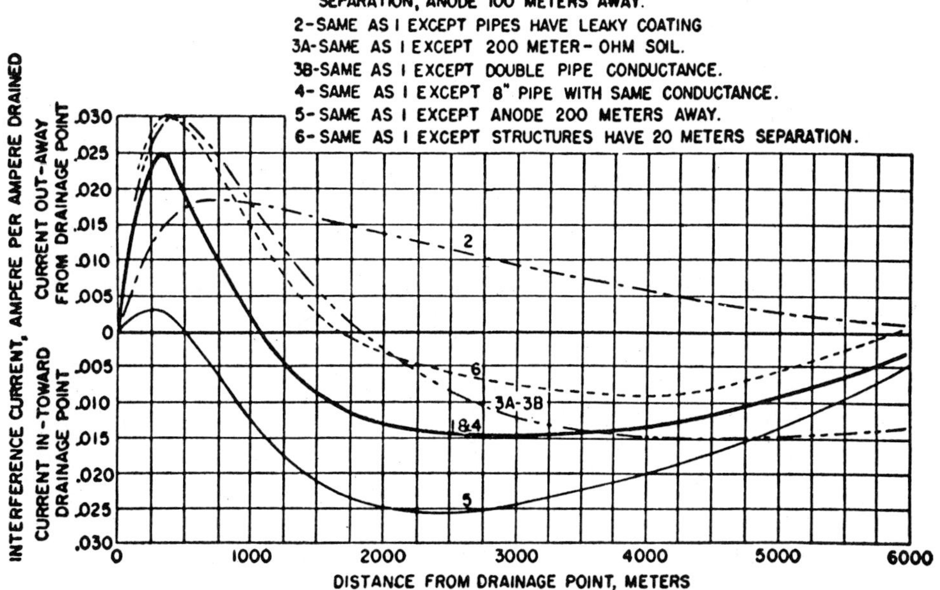

Fig. 11-40 Forced drainage interference—net effect. (Comparison of 7 cases)

remote can be worse than a close one from the standpoint of maximum current loss, minimum exposure being between 150 and 375 meters.

When interference effects are to be neutralized and the anode is very remote, it is usually good practice to place resistance bonds between the structures at the drainage point.

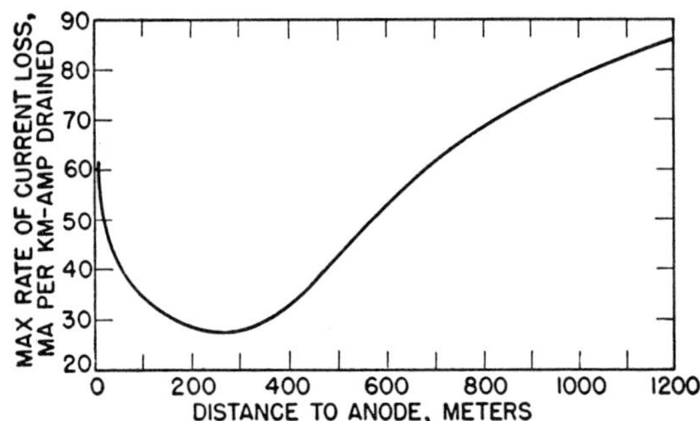

Fig. 11-41 Effects of anode position on forced drainage interference. Illustration for 4 in. bare welded steel pipes, 1.0 meter deep in 10 m-ohm soil, pipes separated one meter.

With a less remote anode, better neutralization may be obtained by using two resistance bonds a short distance on each side of the drainage point where the maximum positive structure-to-earth potential is created on the adjacent structure. The most economical arrangement frequently indicates an anode location resulting in minimum interference. Where soil conditions or other considerations complicate such placement of the anode, another location, either more remote or closer, can be used with proper bonding.

Bond Resistance

An electrical bond installed between the cathodically protected pipe and the foreign structure must produce a current equal to, or greater than, the loss of interference current as measured between the zero-potential points. Terminals may therefore be selected (Fig. 11-42) between the two structures and increments of current produced in these terminals with a test battery. From measurement of these

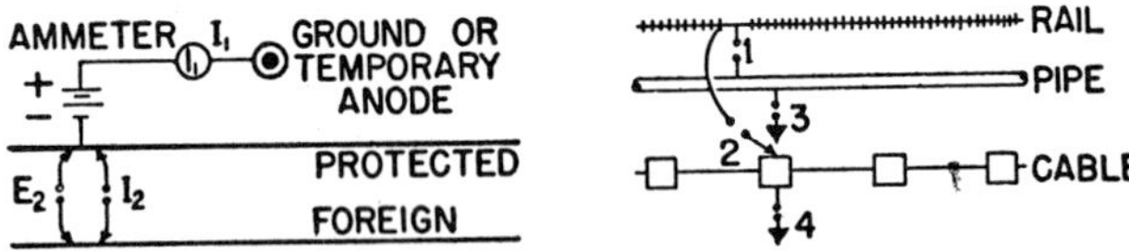

Fig. 11-42 (left) Arrangement for measuring bond resistance. **Fig. 11-43** (right) Rail, pipe, and cable layout for stray current drainage.

current increments and corresponding potentials at the terminals along the foreign structure, the amount of current which must be delivered between the two structures to eliminate loss of interference current can be determined. Likewise, the effects of these current increments can also be determined for the cathodically protected line.

If the internal resistance between the two structures is known, the electrical bond to remove the interference current can be designed. To make the description of the design of this bond simpler, no extraneous potentials are assumed to exist between the two pipelines.

Using current from a test battery (not shown) at I_2, with anode battery circuit open, the internal resistance between the two pipelines (Fig. 11-42) is measured at I_2 and the corresponding voltage drop is read at E_2; separate circuits are used to avoid errors produced by voltage drops in commonly connected leads. The internal resistance $R_{2,2}$ is then equal to $\Delta E_2/\Delta I_2$. With the voltmeter still connected at E_2 (with I_2 circuit open) and the anode battery circuit closed, the anode I_1 is recorded. The coupling resistance $R_{2,1}$ is then equal to $\Delta E_2/\Delta I_1$.

When the amount of anode current has been decided, its value *times* $R_{2,1}$ gives the voltage difference produced between the two lines, E_2 (Fig. 11-42). Thus, with this voltage, a bond between the two structures must deliver at least the current between the zero-potential points. Resistance of this bond, B_2, is determined from the following current flow equation:

$$I_2 = \frac{E_G - E_2}{R_{2,2} + B_2}$$

Solving for B_2,

$$B_2 = \frac{E_G - E_2}{I_2} - R_{2,2} \tag{13}$$

where: B_2 = bond resistance, ohms
I_2 = bond current, amp
E_2 = open circuit potential produced by rectifier = $I_1R_{2,1}$, v
E_G = galvanic potential between structures, v
$R_{2,2}$ = internal resistance between structures, ohms

Installation between the two structures of a bond wire which has a total resistance equal to or less than B_2 will eliminate effects of cathodic interference current on the foreign structure.

Interference Coordination in Stray Current Drainage

Existing stray direct currents on a pipeline or cable may be removed or drained to control corrosion. Establishing a bond between a pipe and a rail at a point will remove stray current from the pipe at all points under the influence of rail leakage current. Potential and current relations of a nearby cable system with respect to earth, pipe, and rail are thus altered. Existence of an electrolysis drainage system on one structure will probably cause interference currents on a nearby foreign structure or portions of it. Therefore, mitigation of increased corrosion exposure to a foreign structure due to the deliberate draining of another becomes a problem in coordination. In many situations, more satisfactory coordination between drainage requirements on separate underground structures can be obtained by designing the drainages at the same time rather than independently.

A reasonably precise method of designing coordinated bonds follows:

Figure 11-43 shows a cable and piping system discharging current to a traction rail system. Assume that preliminary tests show that points **3** (on pipe) and **4** (on cable) are best for making drainage connections. *Trial* drainage wires are connected as shown, **1** being the pipe drainage terminals and **2** those for cable drainage. Terminals **3** and **4** are the pipe-to-earth and cable-to-earth terminals, respectively.

Drainage effects on the pipe are first determined by introducing current in **1** and measuring coupling resistances $R_{1,1}$, $R_{2,1}$,* $R_{3,1}$, and $R_{4,1}$. Similarly, resistances $R_{2,2}$, $R_{1,2}$, $R_{4,2}$, and $R_{3,2}$, are determined by introducing current in terminals **2**. It should be noted that current toward the rails in **1** will make potentials E_1, E_2, and E_3 *less* positive but will make E_4 *more* positive. Also, current toward the rails in **2** will make potentials E_2, E_1, and E_4 *less* positive but will make E_3 *more* positive. $R_{2,1}$, of course, is equal to $R_{1,2}$ by reciprocity and need be determined only once.

In addition to the couplings, it is also necessary to determine relations among the four open circuit voltages and to express them as equations:

$$E_2 = a_2 + B_2E_1$$

$$E_3 = a_3 + B_3E_1$$

$$E_4 = a_4 + B_4E_1$$

The drainage system may be designed as follows:

(A) If terminals **1** are closed thru a bond having a resistance of J_1 ohms, the current in **1** is:

$$I_1 = \frac{E_1}{R_{1,1} + J_1}$$

(B) However, if terminals **2** are also closed and pass I_2 amperes, E_1 is reduced by $I_2R_{2,1}$, so that, under this condition:

$$I_1 = \frac{E_1 - I_2R_{2,1}}{R_{1,1} + J_1} \qquad \textbf{(14)}$$

Similarly,

$$I_2 = \frac{E_2 - I_1R_{2,1}}{R_{2,2} + J_2} \qquad \textbf{(15a)}$$

or, substituting for E_2:

$$I_2 = \frac{a_2 + B_2E_1 - I_1R_{2,1}}{R_{2,2} + J_2} \qquad \textbf{(15b)}$$

(C) The potentials to earth (V_g) of points **3** and **4** with *both* bonds in place are:

$$V'_{g3} = a_3 + B_3E_1 - I_1R_{3,1} + I_2R_{3,2} \qquad \textbf{(16a)}$$

$$V'_{g4} = a_4 + B_4E_1 - I_2R_{4,2} + I_1R_{4,1} \qquad \textbf{(16b)}$$

By assigning desired values to V'_{g3} and V'_{g4} (terms of E_1) Eqs. **16a** and **16b** can be solved for I_1 and I_2 in terms of E_1. The values of I_1 and I_2 can then be substituted in Eqs. **14** and **15b**, and J_1 and J_2 may be found by simultaneous solution.

This method of designing coordinated bonds applies to cases where two structures are to be drained. The same method can be used in designing separate drainage wires for any number of structures. However, data required for more than three are prohibitive and the accuracy of solution becomes poorer because it depends upon differences between small quantities. Where more than two structures are to be drained, it is usually desirable to consider bonding some of them together with resistance as near zero as possible. Those so bonded can then be considered as a unit, thus reducing the number of drainage wires. Near a proposed drainage point, all iron structures (gas and water pipes) might be bonded together, as might all lead structures (power and communication cables). Two drainage wires could then take care of all structures. Usually, it is not desirable to bond cables and pipes directly because: (1) galvanic currents operating to the disadvantage of one might develop; and (2) the pipes' low resistance to earth would make it difficult to drain the cables properly.

After the designed drainage wires have been installed check tests should be made as before on both structures to see that expected performance is obtained. The resistances of either or both bonds may require adjustment, depending upon the accuracy of the original data.

* Read notation as: resistance as measured by the change in voltage per ampere at terminal 2 due to a change in anode current I_1.

TEST LEADS

Electrical measurements relative to buried pipe are facilitated by attaching permanent low electrical resistance metallic contacts. One end of a lead is integrally attached to the pipe, and the other end is located for easy access. In the absence of test leads, electrical contact to a buried pipe must be obtained by driving a pointed metal bar thru the earth until it contacts the pipe or by exposing the pipe and contacting it with an electric conductor. Such manual means are inconvenient (e.g., under paving) and result in errors due to contact resistance, especially for current flow measurements.

Test leads usually consist of a No. 12 solid copper wire conductor (Type TW), plastic coated, suitable for underground use. They are often attached to the pipe by Thermit welding† or to a steel coupon electrically welded to the pipe.

Test leads on transmission lines are usually placed one to five miles apart and at road crossings, foreing lines, railroad and bridge crossings, at casings, and at any other locations where current interchange can take place with metallic structures.

Single Wire Leads

Perhaps the most commonly used test lead is the single wire type (Fig. 11-44, part a), employed in conjunction with a high resistance (zero-center) voltmeter, and an earth electrode (usually copper sulfate) half-cell in the circuit, to measure data on pipe-to-soil d-c potentials. The ground electrode is placed either over the pipe or remote and at right angles to it. Normally, a negative reading will result and it is often recorded, for example, as +GAS PIPE TO $CuSO_4$ ELECTRODE = −0.65 VOLTS over pipe, or remote, as the case may be.

Measurements with this simple circuit indicate the potential of either cathodically or noncathodically protected lines referenced to ground. These data may be utilized to deter-

† See paragraph 841.262 of ASA B31.8-1963 (given in Chap. 2 of Sec. 8).

mine either the effectiveness or requirement for such protection. Galvanic voltage between the half-cell and the steel pipe is included in the recorded pipe-to-soil voltage. For some investigations an average galvanic voltage is subtracted from the recorded data but, in most cases, the recorded voltage is used as read.

This same circuit can be utilized to obtain the soil potential gradients surrounding a line, by referencing the soil potentials to the line itself as a datum line. By moving the electrode along and across the line, an electrical topographical map can be produced, which is useful in locating anodic and cathodic

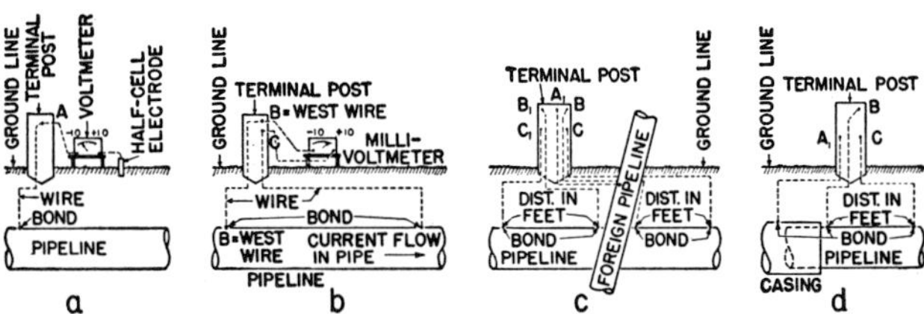

Fig. 11-44 Test lead connections: (a) single wire; (b) dual wire; (c) foreign line crossing; (d) encased line.

currents of noncathodically protected lines or defects in coatings on protected lines. Sometimes it is used to detect current interchange between two separate pipelines crossing each other. The location and soil gradient of anodes, either sacrificial or power fed, can also be traced and plotted.

Single test leads also can be used to measure the potential difference between the line at hand and any other metallic underground structure, reading directly from the voltmeter. This method is useful in investigating interchange of current between two or more lines or between a line and its casing.

Dual Wire Leads

Two identified test leads may be attached a suitable distance apart, to continuous pipelines, unbroken by couplings (Fig. 11-44, part b). These leads, termed *pipe gradient* leads, *span* leads, *pipe potential drop* leads, or *pipe current* leads, are used in detecting the direction, intensity, and persistence of a longitudinal current flow along a pipeline. Any stray a-c or d-c flow which originates from foreign structures thru earth or direct contact may be detected by current leads. Table 11-19 gives resistance data for pipe.

These leads are used in corrosion surveys on unprotected as well as on cathodically protected and coated lines to determine the pipeline current flow. Likewise, at foreign line crossings, checks for current interchange can be made (Fig. 11-44, part c). Where low longitudinal current flow is likely on large diameter pipelines, the spacing between span leads may be as much as 500 ft. Readings may be in the order of a fraction of a millivolt. Other combinations of test wire leads may be used, such as in Fig. 11-44, part d, where a combined pipe potential lead and span wires are placed on one terminal post.

Some companies have well designed wire terminal posts with polarized receptacles for pipe potential readings and also for span tests which automatically polarize the voltmeter with the test wires. Conduits and condulets are often used in addition to combination line markers and terminal posts.

EFFECTIVENESS OF INSTALLATIONS

Cathodic protection has been used for several decades. Most systems in use in 1960 were installed since 1940. Operating results show that properly applied, controlled, and maintained cathodic protection is a practical solution to pipeline corrosion problems.

Most cathodic protection failures have been due to lack of understanding of the principles involved, faulty system analysis, and inadequate protection. The latter may be due to physical conditions like locations of foreign underground structures, which may make complete cathodic protection uneconomical. In such cases, when protection is applied only at the more corrosive locations, it may be difficult to determine effectiveness of protection. Successful application cannot be made by a rule-of-thumb method. Installations should be checked to determine not only the extent of protection to the desired structure but also to make sure that no adverse effect is produced on other underground metallic structures.

When judging the results of cathodic protection on pipelines with long leak records, it must be remembered that leaks from earlier mechanical failure may exist even though corrosion has been completely arrested.

If cathodic protection is designed into a new underground structure, no addition has to be made to pipe wall thickness to take account of corrosion. Well coated pipelines have been protected to increase line life indefinitely with installation of a cathodic protection system at a cost of less than one per cent of that of the structure.

Many companies report very favorable results in reduction of new leaks by applying cathodic protection. One large city has reported that such protection eliminated need of replacing 24,000 ft of mains in its business section with an estimated operating expense saving of about 70 per cent of previous yearly piping cost. Another large city reported that its coated steel gas mains, cathodically protected, are buried in soils, generally considered as among the most corrosive in the nation, without a single corrosion failure or leak in the first 20 years of operation.

Many companies have reported reduction in operating costs thru cathodic protection. Yearly returns from expenditures for such protection have been reported as high as 30 to 50 per cent thru reduction of corrosion losses.

REFERENCES

1. Parker, M. E. *Pipe Line Corrosion and Cathodic Protection*, 2nd ed. Houston, Gulf Pub. Co., 1962.
2. Dwight, H. B. "Calculation of Resistances to Ground." *Elec. Eng.* 55: 1319–28, Dec. 1936. Also in: Uhlig, H. H., ed. *Corrosion Handbook*, p. 938–9. New York, Wiley, 1948.
3. Ganser, P. "Cathodic Protection for an Uncoated Distribution System." *A.G.A. Op. Sec. Proc.* 1963: DMC-63-16.

BIBLIOGRAPHY

Am. Committee on Electrolysis. *Report.* New York, 1921.

Am. Tel. and Tel. Co. *Design of Electrolysis Drainage Wires.* Section AB67.410-1, Issue 1, Feb. 1947. (Standard, Bell System Practices, Transmission Engineering and Data, Electrolysis Practices) New York, 1947.

——. "Methods of Measurement." Section AB67.410.2.

Anderson, H. H. "Cathodic Protection of Buried Metallic Structures Against Corrosion." *Am. Water Works Assn. J.* 40: 485–8, May 1948.

Assn. of Am. Railroads. "Possibilities of Improper Signal Operations Due to Cathodic Protection." *Proc.* XL: 153A, 1942.

Bonner, W. F. "Selenium Rectifiers for Cathodic Protection." *Corrosion* 2: 249–60, Nov. 1946. Also in: *Nat. Petrol. News* 38: R450-2, June 5, 1946, sec. 2.

Bridge, A. F. "A Review of Cathodic Protection of Pipe Line." *A.G.A. Proc.* 1934: 180–9.

Burton, L. W. and Hamann, C. E. "Construction and Ratings of Copper-Oxide Rectifiers for Cathodic Protection of Pipe Lines." *Corrosion* 3: 75–95, Feb. 1947.

Columbia Gas System Service Corp. *Manual on Underground Corrosion.* New York, 1952.

Corfield, G. "Relations between Protective Coatings and Cathodic Protection." *Cathodic Protection Symposium, Electrochemical Soc. and NACE.* Washington, 1949.

Denison, I. A. and Romanoff, W. "Behavior of Experimental Zinc-Iron Couples Underground." *Cathodic Protection Symposium, Electrochemical Soc. and NACE.* Washington, 1949.

Doremus, E. P. and others. "Engineering Aspects of Cathodic Protection as Applied to Pipe Lines." *Corrosion* 5: 273–81, Sept. 1949.

Dwight, H. B. "Calculation of Resistances to Ground." *Elec. Eng.* 55: 1319–28, Dec. 1936.

Ebasco Services Inc. *Zinc as a Galvanic Anode: A Report Prepared for the American Zinc Institute.* New York, 1951.

Ewing, S. "Cathodic Protection of Pipe Lines from Soil Corrosion." *Gas Age-Record* 75: 179+, Mar. 2, 9, 16, 23, 1935.

Glass, D. C. "Economics of Rectifier Installation for Cathodic Protection of a Bare Pipe Line." *Corrosion* 7: 322–5, Oct. 1951.

Good, D. B. "Location and Selection of Anode Systems for Cathodic Protection Units." *Corrosion* 3: 539–48, Nov. 1947; 632–5, Dec. 1947.

Gorman, L. J. "Electrolysis Surveys on Underground Cables." *Corrosion* 1: 163–77, Dec. 1945.

Hanawalt, J. D. and others. "Corrosion Studies of Magnesium and its Alloys." (Am. Inst. Mining & Met. Eng. Tech. Pub. 1353) *Metals Tech.* Sept. 1941.

Handbook of Chemistry and Physics, 43rd ed. Cleveland, Chemical Rubber Co., 1963.

Hart, P. and others. "Development of Magnesium for Cathodic Protection." *Corrosion* 1: 59, June 1945.

Holler, H. D. "Role of Current Distribution in Cathodic Protection." (N.B.S. Research Paper 2220) *Bur. of Standards, J. of Res.* 47: 1–6, July 1951.

Holloway, J. A. "Installation and Economics of Placing Magnesium Anodes at Leaks Repaired on a Pipe Line." *Corrosion* 6: 157–61, May 1950.

Holsteyn, D. "Practical Design and Economics of Cathodic Units Applied in a Refinery." *Petrol. Refiner* 22: 179–82, June 1943.

——. "Cathodic Protection in the Refinery." *Petrol. Engr.* 17: 154–8, May 1946.

Hoxeng, R. B. and others. "Galvanic Aluminum Anodes for Cathodic Protection." *Corrosion* 3: 263–74, June 1947.

Knox, E. S. and Keeling, H. J. "Pipe-Line Maintenance Costs Cut by Cathodic Protection." *Elec. World* 103: 392–5, Mar. 17, 1934.

Logan, K. H. and others. "Determination of the Current Required for Cathodic Protection." *Petrol. Engr.* 14: 168–80, July annual issue, 1943.

Marx, P. F. "Practical Application of Electrolysis Prevention on Pipe Lines and in Compressor Stations." *Petrol. Engr.* 11: 48+, Jan.; 51–6, Feb. 1940.

——. "Interference on Telephone Lines Caused by Cathodic Units." *Petrol. Engr.* 15: 124+, Nov.; 86+, Dec. 1943.

May, T. P. "Anodic Behavior of Zinc and Aluminum-Zinc Alloys in Sea Water." *Cathodic Protection Symposium, Electrochemical Soc. and NACE.* Washington, 1949.

McCollum, B. and Logan, K. H. "Practical Applications of the Earth Current Meter." *Bur. of Standards, Tech. Papers* 21: 683, 1927.

—— ——. *Electrolysis Testing.* (T355) Washington, Bur. of Standards, 1927.

McGraw, E. C. "Why We Incorporate a Corrosion Control Program in the Early Stages of Construction in a Modern Gas Transmission System." *A.G.A. Proc.* 1951: 464–6.

McRaven, C. H. "Resistance Bonding in Connection with Cathodic Protection." *Petrol. Ind. Elec. Assn. Elec. News* 11: 23, June 1941.

——. "Cathodic Protection." *Corrosion* 2: 320–9, Dec. 1946.

Mears, R. B. and Brown, R. H. "A Theory of Cathodic Protection." *Trans. Electrochem. Soc.* 74: 519–31, 1938.

Mears, R. B. and Brown, C. D. "Light Metals for the Cathodic Protection of Steel Structures." *Corrosion* 1: 113–8, Sept. 1945.

Miller, M. C. "Galvanic Couples and Cathodic Protection." *Petrol. Engr.* 17: 55–8, May 1946.

——. "Hidden Corrosion Costs." *A.G.A. Proc.* 1948: 461–2.

——. "Characteristics and Field Use of Electrical Instruments for Corrosion Investigations and Cathodic Protection." *Cathodic Protection Symposium, Electrochemical Soc. and NACE.* Washington, 1949.

Morgan, C. L. "Zinc Anodes for Preventing Corrosion of Distribution Mains." *Petrol. Engr.* 16: 196+, Sept. 1945. Also in: *A.G.A. Op. Sec. Distrib. Conf.*, 1946.

Mudd, O. C. "Experiences with Zinc Anodes." *Petrol. Ind. Elec. Assn. Elec. News* 13: 11, May 1943.

——. *Control of Pipe Line Corrosion.* Houston, Nat. Assn. of Corrosion Eng., 1946. Also in: *Corrosion* 1: 192–218, Dec. 1945; 2: 25–58, Mar. 1946.

Nat. Assn. of Corrosion Engineers. *Cathodic Protection: Report of Correlating Committee.* Houston, 1951.

——. *First Interim Report on Ground Anode Tests of T.P.C. No. 3—Anodes for Impressed Currents.* (Pub. 50-1) Houston, 1950.

——. *First Interim Report on Galvanic Anode Tests of T.P.C. No. 2* (Pub. 50-2) Houston, 1950. Condensed in: *Corrosion* 6: 274–5, Aug. 1950.

Nelson, L. B. "Coordination of Cathodic Protective Installations To Avoid Interference with Adjacent Structures." *Cathodic Protection Symposium, Electrochemical Soc. and NACE.* Washington, 1949.

Nelson, C. E. "Secondary Magnesium." (Am. Inst. Mining & Met. Engrs. Tech. Pub. 1642, p. 77–84) *Metals Tech.* Oct. 1943.

Olson, G. R. "Cathodic Protection Co-ordination." *Gas* 21: 43–4, Jan. 1945.

Olson, G. R. and McRaven, C. H. "Control of Stray Current from Cathodic Protection Installations." *Petrol. Ind. Elec. Assn. Elec. News*, Feb. 24, 1943.

Peabody, A. W. "Performance of Impressed Current Ground Beds." *A.G.A. Proc.* 1951: 448–55.

Peabody, A. W. and Woody, C. L. "Experience and Economic Benefits from Cathodic Protection on Gas Distribution Systems." *Corrosion* 5: 369–76, Nov. 1949.

Pearson, J. M. "Concepts and Methods of Cathodic Protection." *Petrol. Engr.* 15: 216+, Mar.; 199+, Apr.; 219+, May 1944.

——. "Electrical Instruments and Measurements in Cathodic Protection." *Corrosion* 3: 549–66, Nov. 1947.

Peifer, N. P. "Current Required for Cathodic Protection." *Cathodic Protection Symposium, Electrochemical Soc. and NACE.* Washington, 1949.

Pope, R. "Interference from Forced Drainage." *Corrosion* 6: 201–7, June 1950.

——. "Attenuation of Forced Drainage Effects on Long Uniform Structures." *Corrosion* 2: 307–19, Dec. 1946.

Reid, K. K. and others. "Elementary Mechanism of Cathodic Protection." *A.G.A. Proc.* 1951: 351–60.

Rhodes, G. I. "Cathodic Protection or Electric Drainage of Bare Pipe Lines." *A.G.A. Nat. Gas Dept. Proc.* 1935: 240–64.

Robinson, H. A. "Magnesium Anodes for the Cathodic Protection of Underground Structures." *Corrosion* 2: 199–218, Oct. 1946. Also in: *Nat. Petrol. News* 38: R452, June 5, 1946 (sec. 2).

Rogers, W. F. "Calculating Current and Potentials Distribution in Cathodic Protection Systems," *Petrol. Engr.* 12: 66+, Dec. 1940.

Rohrman, F. A. "Criteria for Cathodic Protection." *World Oil* 131: 177–90, July 1, 1950.

Roush, W. L. and Wood, E. E. "Cathodic Protection Rectifiers." *Corrosion* 3: 169–72, Apr. 1947.

Scherer, L. F. "Cooperative Problems Involved in Cathodic Protection." *Oil & Gas J.* 38: 179+, Nov. 17, 1939.

Schneider, W. R. "Improvement in Electrical Pipe Protection Equipment." *Gas* 12: 22+, June 1936.

——. "Economics of Cathodic Pipe Protection." *Western Gas* 10: 12+, Jan. 1934.

Simpson, A. D., Jr. "Mitigation of Corrosion on City Gas Distribution Systems." *Corrosion* 5: 56–9, Feb. 1949.

——. "Practical Cathodic Protection for Gas Distribution Systems." *A.G.A. Proc.* 1949: 641–68.

Smith, A. V. "Theory and Use of Cathodic Protection." *A.G.A. Proc.* 1937: 816–21.

——. "Cathodic Protection Interference." *A.G.A. Monthly* 25: 421–7, Oct. 1943. Also in *Gas Age* 92: 21+, Aug. 26, 1943; *A.G.A. Proc.* 1943: 379–89.

Smith, W. T. and Marshall, T. C. "Zinc for Cathodic Protection of Pipe." *Gas Age* 84: 15+, Aug. 17, 1939.

Speller, F. N. *Corrosion, Causes and Prevention,* 3rd ed. New York, McGraw-Hill, 1951.

Standring, J. M. "Attenuation of Drainage Effects on a Long Uniform Structure with Distributed Drainage." *Corrosion* 3: 301–9, June 1947.

Stegner, A. L. "Rectifiers and Galvanic Anodes for Cathodic Protection." *A.G.A. Proc.* 1949: 669–73.

Sudrabin, L. P. and MacDonald, F. P. "Principles of Cathodic Protection Design." *Am. Water Works Assn. J.* 40: 489–94, May 1948.

Sunde, E. D. "Currents and Potentials along Leaky Ground Return Conductors." *Elec. Eng.* 55: 1338–45, Dec. 1958.

——. *Earth Conduction Effects in Transmission Systems.* New York, Van Nostrand, 1949.

Trouard, S. E. *Practical Application of Cathodic Protection to Gas Distribution System in New Orleans.* Galveston, Southern Gas Assn., 1948. Digest in: *Gas* 24: 46–9, June 1948.

Uhlig, H. H. *Corrosion Handbook.* New York, Wiley, 1948.

Wahlquist, H. W. "Use of Zinc for Cathodic Protection." *Corrosion* 1: 119–47, Sept. 1945.

Wahlquist, H. W. and Fanett, H. M. "Practical Use of Galvanic Anodes." *Cathodic Protection Symposium, Electrochemical Soc. and NACE,* p. 114–43. Washington, 1949.

Wainwright, R. M. "Cathodic Protection on Distribution Systems." *Petrol. Ind. Elec. Assn. Elec. News* 12: 33, May 1942.

Wight, K. M. and Hadley, R. F. "The Anodic Behavior of Sacrificial Metals in Specific Environments, Interim Report," *A.G.A. Proc.* 1947: 771–85.

Chapter 6
Instrumentation for Corrosion Control

by M. C. Miller

INTRODUCTION

The extent and location of corrosion on underground piping, and the amount of current and potential required to stop corrosion by cathodic protection are determined by making electrical measurements of potentials of pipe to earth, and of current flow in pipes, in the earth to and from pipes, and from rectifiers, batteries, galvanic anodes, and ground beds. Measurements of soil resistivity are also necessary.

Factors Influencing Instrument Choice

Extent of Testing. Simple-to-use, rugged, low-cost instruments may be suitable for routine measurements of potentials and current output of rectifiers and anodes, potentials of structures to earth, and *IR* drops. Where corrosion investigations, cathodic protection tests, resistance measurements, soil-resistivity tests, and similar testing procedures are to be carried on, higher grade, more versatile instruments are preferred.

Testing Speed. Many instruments are designed specifically for corrosion, electrolysis and cathodic protection testing. These have scales, ranges, polarity-reversing switches or center-zero scales, damping characteristics, and other features suitable for the potential and current values and conditions encountered in everyday field testing. Multipurpose meters and meters with multiple ranges and circuitry save time in making connections, especially for such tests as resistance, soil-resistivity measurements, and cathodic protection tests. These multipurpose instruments usually cost much less than the many separate instruments which would be required to accomplish the same extent of testing.

FACTORS IN MEASURING POTENTIALS

In measuring potentials, factors which must be considered to avoid large errors include: resistance of coatings and of the structure to earth, internal resistance of instruments used, internal resistance and resistance to earth of reference electrodes, polarization of structures, and polarization of reference electrodes.

Practically all the factors causing errors result in indicated readings which are lower than the correct ones.

INSTRUMENTS

Data on instruments used in corrosion testing may be found in *Reference Book on Instruments for Electrolysis, Corrosion and Cathodic Protection Testing* (X52164) prepared by the Instrumentation Task Group of the A.G.A. Corrosion Committee. A complete set of corrosion and cathodic protection instruments should include at least one from each of the following groups:

For measurement of potentials to earth:

1. High-resistance voltmeter
2. Potentiometer or potentiometer/voltmeter
3. Vacuum-tube voltmeter, or equivalent
4. Combinations of two or more of above

For measurement of IR drops and rectifier, battery, and structure to structure potentials:

1. Low-resistance millivoltmeter
2. Low-reading potentiometer or potentiometer/voltmeter
3. Combination of both

For current measurements:

1. Milliammeter; ammeter
2. Low-resistance, high-sensitivity millivoltmeter (for measuring *IR* drop)
3. Combination of both

For soil resistivity measurements:

1. Megger, vibroground, multicombination meter or equivalent
2. Resistivity rod or equivalent
3. Soil pins, soil box, leads, etc.
4. Multiconductor soil resistivity set

Reference electrodes:

1. Copper sulfate reference electrodes (two or more required)

Test leads:

1. Short and long test leads with suitable test clips, color-coded to reduce errors
2. Test lead reels for long leads

Pipe locators:

1. Induction or contact
2. Interrupted tone or constant tone

Recording voltmeters:

Recorders should be suitable for field use; i.e., weatherproof, self-contained, battery-operated motor, or spring-wound drive. Strip-chart type with sensitized paper is desirable.

1. Recording high-resistance voltmeter with suitable ranges. Strip-chart type with sensitive paper preferred. Usually must be self-contained with battery-operated motor drive or spring-wound drive.
2. Recording low-resistance millivolt/voltmeter and milliammeter/ammeter with suitable ranges, and external shunts for large current.

Portable shunts:

Ranges and millivolt drop should match instruments selected.

For engineers investigating corrosion, designing cathodic protection installations, and making installation and final tests, cost of a complete set of suitable instruments, excluding recording units, may vary from $1000 to $2000.

High-sensitivity, high-accuracy instruments are necessarily delicate and easily damaged if misused. However, with reasonable care and use, these instruments have given long reliable service in the field, even under extremely adverse operating and climatic conditions.

FIELD MEASUREMENTS

Pipe-To-Soil Potentials

Suitable lower cost instruments are available for use by junior corrosion engineers and operating men to make periodic inspections and measurements of pipe-to-soil potentials and anode output current. These tests may require potentials-to-earth instruments and a reference electrode.

If only a voltmeter is used, each reading should be cross-checked (read on two ranges) to determine if any error is caused by a high electrode-to-earth resistance. If a potentiometer with an insensitive galvanometer is used, an error may occur in its balancing because of polarization in the structure or the reference electrode.

A low-cost potentiometer-type voltmeter with a high-resistance voltmeter as a galvanometer is a good solution. With care and with each reading cross-checked for electrode resistance, most readings may be taken with good accuracy by using the high-resistance voltmeter; the potentiometer/voltmeter could then be used only where the cross-check shows it is needed. Thus, a minimum of testing time is required.

If the sensitivity of the high-resistance voltmeter/galvanometer is low, better results can be obtained by using an electrode with a large clean wood plug, and possibly by wetting the earth around it to lower its resistance to earth.

Very few corrosion engineers use a vacuum-tube voltmeter for measuring pipe-to-soil potentials because this instrument with zero stability and overall accuracy suitable for field testing is large and cumbersome.

Current Flow

The four general methods of determining current flow are:

1. Connecting an ammeter or milliammeter directly in series in the circuit.

2. Connecting a shunt of known resistance in series in the circuit and measuring the millivolts drop across it. It may be necessary to calibrate the shunt for use with a specific meter and set of leads.

3. Measuring the *IR* drop along the structure between measured points along it. Magnitude and direction of direct current flowing on a wire, pipe, or other metallic structure can be so determined. It is necessary to know or determine the structure resistance between the points contacted (Fig. 11-45). Usually, to obtain accurate measurements of current flow, meter readings must be corrected for resistance of the leads. Good low-resistance contacts to the pipe are essential.

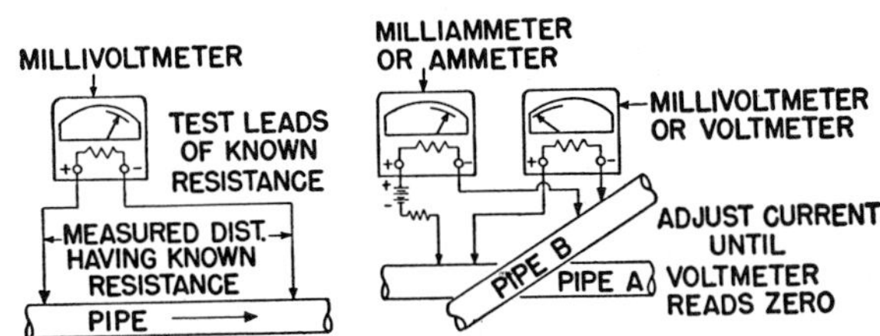

Fig. 11-45 (left) Determining current flow in a pipe by measuring the voltage drop in a known pipe length (of known resistance).

Fig. 11-46 (right) Current measurement by zero-resistance type ammeter or equivalent.

4. Using a "zero-resistance-type ammeter" or equivalent (Fig. 11-46). This may be desirable for measuring current which would flow between two low-resistance structures if they were in solid contact. Current is "pumped" thru ammeter and rheostat by battery until millivoltmeter reads zero. Ammeter shows magnitude of current which would flow (neglecting polarization) if pipes A and B were in solid contact. (A zero-resistance-type milliammeter/ammeter is available in all models of multicombination meters.)

REFERENCE ELECTRODES

The copper sulfate reference electrode is commonly used in investigating corrosion on pipelines and for cathodic protection testing on pipelines. It is almost ideal for this purpose since it is available in forms practical for field use with suitable accuracy if ordinary precautions are observed, and special sizes and shapes can be made of readily available materials. The small diameter tube type can be set on the ground which can be wetted down if necessary.

The wood plugs are of close-grain pine, bois-de-arc, oak, or birch, each of which has advantages and disadvantages. Transparent plastic holders are supplied by one manufacturer to permit easy and frequent observation of amount of copper and $CuSO_4$ solution and their condition, thus eliminating accidental use of a "dry" electrode. Some electrodes are supplied with a rubber cap to slip over the wood plug to keep it from drying out when not in use. This eliminates the

Table 11-19 Size, Weight, and Resistance to Current Flow of Steel Pipe* and Wrought Iron Pipe†

	Standard steel		Extra strong steel		Wrought iron			
Nominal ID, in.	Weight, lb/ft	Resistance, microhm/ft	Weight, lb/ft	Resistance, microhm/ft	Weight, lb/ft	Resistance, microhm/ft	Weight, lb/ft	Resistance, microhm/ft
0.125	0.24	900.0	0.31	700.0	0.24	870.0	0.29	720.0
.250	.42	510.0	.54	400.0	.42	498.0	.54	387.0
.375	.57	379.0	0.74	292.0	.56	374.0	0.74	283.0
.500	0.85	254.0	1.09	198.0	0.84	249.0	1.09	192.0
0.750	1.13	191.0	1.47	147.0	1.12	187.0	1.39	150.0
1.000	1.68	129.0	2.17	100.0	1.67	125.0	2.17	96.0
1.250	2.27	95.0	3.00	72.0	2.25	93.0	3.00	70.0
1.500	2.72	79.0	3.63	60.0	2.69	78.0	3.63	58.0
2.000	3.65	59.0	5.02	43.0	3.66	57.0	5.02	41.7
2.500	5.79	37.3	7.66	28.2	5.77	36.3	7.67	27.3
3.000	7.58	28.5	10.25	21.1	7.54	27.8	10.25	20.4
3.500	9.11	23.7	21.51	17.3	9.05	23.1	12.47	16.8
4.000	10.79	20.0	14.98	14.4	10.72	19.5	14.97	14.0
4.500	...	...	...	...	12.49	16.8	18.22	11.5
5.000	14.62	14.8	20.78	10.4	14.56	14.4	20.54	10.2
6.0	18.97	11.40	28.57	7.60	18.76	11.2	28.58	7.3
7.0	23.54	9.20	38.05	5.70	23.41	8.90	37.67	5.60
8.0	24.70	8.70	43.39	4.98	25.00	8.40	43.00	4.87
8.0	28.55	7.60	...	...	28.34	7.40	...	...
9.0	33.91	6.40	48.73	4.43	33.70	6.20	48.73	4.29
10.0	31.20	6.90	54.74	3.94	32.00	6.50	...	...
10.0	34.24	6.30	...	...	35.00	6.00	...	...
10.0	40.48	5.30	...	...	40.00	5.20	54.74	3.82
11.0	45.56	4.74	60.08	3.59	65.00	6.70	60.08	3.68
12.0	43.77	4.93	65.42	3.30	45.00	4.70	...	...
12.0	49.56	4.36	...	...	49.00	4.27	65.42	3.20
13.0	54.57	3.96	72.09	3.00				
14.0	58.57	3.69	77.43	2.79				
15.0	62.58	3.45	82.77	2.61				
16.0	57.47	3.75	62.57	3.44				
20.0	72.16	2.99	78.59	2.74				
22.0	72.38	2.98	86.61	2.49				
24.0	79.06	2.73	94.70	2.29				
26.0	85.73	2.51	102.62	2.10				
28.0	92.40	2.33	110.62	1.96				
30.0	99.00	2.18	119.00	1.81				

* National Tube Co. Resistivity = 215.8 microhms-ft. † Bayers tables of weights. Resistivity = 209.3 microhms-ft.

Table 11-20 Copper Wire and Cable Characteristics

B & S gage	Circular mils	No. wires in strand	Diam of each wire, in.	Diam of strand, in.	Resistance per 1000 ft at 68 F	Amp per mv/ft	Diam of solid wire, in.
	2,000,000	91	0.1482	1.6302	0.00530	188.7	
	1,000,000	61	.1280	1.1520	.01060	94.3	
	750,000	61	.1109	0.9981	.01413	70.7	
	500,000	37	.1162	.8134	.02116	47.4	
	250,000	19	.1147	.5738	.04233	23.60	0.5000
0000	211,600	19	.1055	.5277	.04997	20.05	.4600
000	167,772	19	.0940	.4700	.06293	15.90	.4096
00	133,079	7	.1380	.4134	.07935	12.62	.3648
0	105,625	7	.1228	.3684	.10007	10.0	.3248
1	83,694	7	.1093	.3279	.12617	7.93	.2893
2	66,358	7	.0973	.2919	.15725	6.36	.2576
3	52,624	7	.0867	.2601	.19827	5.05	.2294
4	41,738	7	.0772	.2316	.25000	4.0	.2043
6	26,244	7	.0612	.1836	.39767	2.52	.1620
8	16,512	7	.0486	.1458	0.62686	1.60	.1285
10	10,384	7	.0385	.1155	1.00848	0.997	.1019
12	6,528	7	.0305	.0915	1.59716	.625	.0808
14	4,108	7	0.0242	0.0726	2.54192	0.387	.0640
16	2,583				4.009		.05082
18	1,624				6.374		.04030
20	1,022				10.140		.03196
22	642				16.120		0.02535

need for carrying the electrode in a container of water or copper sulfate solution.

High-purity copper-sulfate crystals are recommended. However, "blue stone" available in any drug store is often used, probably with only minor error. Distilled water is recommended, although tap water is frequently used. A surplus of copper-sulfate crystals in the electrode is recommended, so that water can be added when needed in the field. The copper rod, plate, or tube should be removed periodically, thoroughly cleaned, and filled with a fresh supply of copper sulfate and distilled water.

The "half-cell" potential of the copper sulfate cell is affected slightly by temperature changes, but these may be considered negligible for purposes of practical field measurements. The copper sulfate solution freezes at about 29 F (electrode cannot be used when frozen). If freezing does not crack or otherwise harm the electrode case or plug, no permanent harm results and it can be used after thawing.

To avoid polarization, do not pass a heavy current thru the electrode. The safe magnitude depends upon the area of copper in the solution. If care is taken, polarization will not occur with a high-resistance voltmeter or a potentiometer or potentiometer/voltmeter having a high-sensitivity indicating meter or galvanometer. It is believed that currents thru the electrode in excess of 1.0 ma may polarize some types of electrodes. In general, the larger size electrodes should be used with low-sensitivity meters to reduce errors in readings as well as polarization.

Location of the reference electrode with respect to the structure to be tested is important. Conditions which dictate the location include: (1) whether pipeline is bare or coated (and effectiveness of the coating); (2) whether a line is "spot" protected (i.e., at a road or creek crossing) or protected over many miles of line; (3) area being protected, e.g., pipeline or a distribution system; (4) whether line is electrically insulated from other structures, e.g., tap lines, services; (5) method of providing protection, i.e., whether galvanic anodes are installed close to the line, or protection is provided by rectifiers with remote ground or with distributed ground bed anodes close to the line.

The reference electrode placed directly over the pipeline being protected is probably the "safest" location and is probably used by the majority of companies. However, under some conditions, the pipe-to-soil potentials as measured to a reference electrode "remote" from the line can show that full protection is achieved with considerably less protection current than might be indicated as necessary with the reference electrode directly over the pipeline. There is no "one" correct location. Pipe-to-soil potential readings reflect the surface potential of the pipe, polarization of the pipe, and *IR* drops in the soil. The reference electrode should not be placed over, or relatively close to, the galvanic anode or the ground bed anode when measuring pipe-to-soil potentials to determine the extent of cathodic protection. The minimum distance from an anode or ground bed will be different for every installation.

INTERNAL CORROSION SURVEYS

The United Gas Corporation corrosion detector measures and records internal pipe corrosion. A series of the instrument's free-hinged arms press against the internal surface of the pipe and enter any cavities present. Depth and location of these and other irregularities are recorded on a cylindrical chart. A distance-marking wheel keeps the chart in continuous motion as the instrument is pulled thru pipe (4 to 36 in. diam range). Calibration permits corrosion readings from 0.005 in. depth to pipe thickness.

ELECTRICAL RESISTANCE DATA

The following tables present data helpful in corrosion testing of pipelines. Various characteristics of steel and wrought iron pipes are given in Table 11-19, and those for copper wire and cables comprise Table 11-20. Data for cable sheath flow computations are found in Table 11-21.

Table 11-21 Data For Computing Flow on Cable Sheaths

Circumference, in.	Diam, in.	Current, amp per mv/ft	Circumference, in.	Diam, in.	Current, amp per mv/ft
1.0	0.318	0.57	5.0	1.59	4.88
1.1	.350	.65	5.1	1.62	5.07
1.2	.382	.72	5.2	1.65	5.25
1.3	.414	.81	5.3	1.69	5.41
1.4	.446	0.90	5.4	1.72	5.57
1.5	.447	1.04	5.5	1.75	5.71
1.6	.509	1.11	5.6	1.78	5.85
1.7	.541	1.21	5.7	1.81	6.0
1.8	.573	1.32	5.8	1.85	6.13
1.9	.604	1.42	5.9	1.88	6.29
2.0	.636	1.51	6.0	1.91	6.48
2.1	.668	1.6	6.1	1.94	6.60
2.2	.700	1.69	6.2	1.97	6.74
2.3	.732	1.74	6.3	2.02	6.89
2.4	.764	1.81	6.4	2.04	7.0
2.5	.795	1.90	6.5	2.07	7.12
2.6	.828	1.98	6.6	2.10	7.24
2.7	.860	2.06	6.7	2.13	7.38
2.8	.892	2.14	6.8	2.16	7.51
2.9	.923	2.22	6.9	2.20	7.63
3.0	.954	2.31	7.0	2.23	7.80
3.1	0.986	2.41	7.1	2.26	7.93
3.2	1.020	2.52	7.2	2.29	8.08
3.3	1.050	2.62	7.3	2.32	8.19
3.4	1.080	2.75	7.4	2.36	8.30
3.5	1.110	2.89	7.5	2.39	8.42
3.6	1.140	3.01	7.6	2.42	8.55
3.7	1.180	3.15	7.7	2.45	8.70
3.8	1.210	3.28	7.8	2.48	8.90
3.9	1.240	3.40	7.9	2.52	9.01
4.0	1.270	3.50	8.0	2.54	9.16
4.1	1.310	3.61	8.1	2.57	9.38
4.2	1.330	3.72	8.2	2.61	9.51
4.3	1.370	3.85	8.3	2.64	9.72
4.4	1.400	3.98			
4.5	1.430	4.09			
4.6	1.460	4.22			
4.7	1.490	4.36			
4.8	1.530	4.53			
4.9	1.560	4.70			

Chapter 7
Compressor Station Corrosion

by J. C. Berringer

COMPRESSOR ENGINES AND APPURTENANT EQUIPMENT

Experience has indicated that many of the corrosion problems in compressor station engines and appurtenant equipment can be remedied or minimized by cathodic protection, corrosion inhibitors, and water treatment; proper selection of metals is a major consideration. Different types of corrosion present special problems, each influenced by its particular environment and by other individual factors. Therefore, it is essential that the cause and type of corrosion or metal loss be accurately identified before applying mitigation measures

Corrosion from Cooling Water

Cathodic protection of metal surfaces exposed to circulating water in power pistons and jackets is a major problem with equipment cooled by an open water system. Use of magnesium anodes in various shapes is required for equipment with such complex physical shape that it prohibits other types of protection. It is essential that sufficient space be provided between the placed anode and adjacent structure so that interference to fluid flow will not impair operation of the equipment. The main factor to be considered in designing such installations is the type of water encountered.

Various shapes of magnesium anodes allow their installation close to the surface requiring protection. Their suitability for such application is a major factor in numerous installations of this type of cathodic protection for compressor station equipment. Use of cathodic protection with impressed currents is limited because of area available for anode material.

Corrosion in the water jacket of power cylinders has been reduced or eliminated by mounting magnesium slabs on insulating pads attached to the jacket wall. Current output of the anodes should be limited to that which will permit anode replacement at the time of scheduled engine overhaul. Such current output can be limited by using resistor washers installed under the stud bolt attaching the magnesium to the protected surface. The stud bolt head and resistor washer must be protected by filling the anode recess with sealing compound.

Cavitation and Impingement Corrosion

Magnesium ribbon has been used in power piston rods to reduce metal loss, although such installations are difficult to maintain. Metal loss at the water discharge end of the rod has been in many cases the result of cavitation erosion, a particular type of metal loss observed at numerous other locations. It occurs when the liquid pressure falls below the vapor pressure and vapor pockets or bubbles form which collapse as they are carried to the area of high pressure. Their collapse then causes erosion. These highly localized pressure pulses result in pitting and claw-like corrosion.

This type of corrosion has been reduced by replacing the worn water orifice in the center of the piston (during overhaul). In use, this orifice becomes enlarged, and the resulting increased water flow increases cavitation erosion near the end of the power piston rod.

Cavitation can also be reduced or eliminated by increasing water pressure, by improving design to reduce liquid pressure differential, and also by coating exposed surfaces with a hard alloy.

Impingement attack or metal loss associated with turbulent liquid flow results in smooth, rounded pits that are often undercut on the downstream side. This type of metal loss can also be reduced or eliminated by improved design which reduces turbulence, or by coating the metal surfaces with a hard alloy.

Corrosion of Aftercooler Coils

To mitigate corrosion of coils and tanks of compressed air aftercoolers at compressor stations is a difficult problem, since a protective coating applied to the coils would reduce heat transfer and effectiveness of their operation and replacement is also very difficult.

Immediately after the air is compressed, it is passed thru aftercooling coils. Because of the elevated temperature and corrosive conditions present, cathodic protection is required to maintain this equipment in service. Cathodic installations in aftercoolers in tanks approximately 4 ft long, 1 ft wide, and 2 ft deep consist of three extruded 1.5 in. diam magnesium rods, suspended in the tanks on 1 ft centers between coils. They extend within 2 in. of the tank bottom and are connected by suitable conductors to the aftercooling coils and tanks.

These magnesium installations have provided the necessary cathodic protection and eliminated coil failure due to corrosion. Average anode life is approximately nine months.

Corrosion of Water Tanks

Elevated water tanks at compressor stations are subject to corrosion and are very difficult to repair. In one installation before the application of cathodic protection, the usual procedure was to clean and repaint tank interiors yearly. In addition to labor and material costs for this work, station operation was impaired while the tank was out of service.

Adequate cathodic protection can be applied to such tanks by installing an anode system. To protect tanks of 75,000-gal capacity, five metal rods are suspended from the roof and insulated from the tank. They are arranged in a circular design approximately 4 ft from the tank wall and extending to within 2 ft of its bottom. If junk iron rods are used for anodes, each ampere of current discharged requires 20 lb of material. It is difficult to install a quantity of junk iron sufficient for more than one year's service. Aluminum anodes or anode rods of other material may be substituted, however, thus extending anode life to three or four years.

Junk iron for anode material is now being replaced by *Duriron* in elevated tanks. *Duriron* is a 14.5 per cent silicon cast iron. It is very dense, hard, and resistant to electrochemical attack. It is very difficult to machine and is therefore cast in various diameters up to 5 ft lengths. If *Duriron* rods are to be joined together to provide greater anode length or if electrical connections are required, inserts can be cast in them or they can be brazed. The manufacturer reports that the low metal loss per ampere-year permits replacing five iron anodes by one *Duriron* anode 1 in. diam $\times$ 10 ft long. To provide a suitable current distribution in the tank, however, installation of three anodes in each tank may be required. Data indicate that at a current density of 1 amp per sq ft of anode material, metal loss is negligible in waters of 7 to 10 pH.

Average station operation prevents ice formation in the tank, which would damage anodes as water elevation fluctuates. Therefore, installation of anodes that provide more than one year's service is feasible.

A rectifier unit for the cathodic protection of these tanks should provide 14 v with a d-c output of 10 amp. It may be mounted on one leg of the tank. Current requirements vary from four to eight amperes for a 75,000-gal tank, depending upon water analysis. A magnesium anode may be placed at the riser base on an insulating pad and the magnesium lead connected to it, to provide the riser pipe with the additional protection it may require. The quantity of magnesium may be increased to ensure that replacement will not be required until inspection is made under the regular maintenance program.

Hydrogen Sulfide Corrosion

Hydrogen sulfide in fuel gas generally corrodes copper valve seals and valve cover gaskets in compressor cylinders.[1] On the other hand, cylinder walls, pistons, piston rings and rods, rod packing, and discharge valves are usually well coated with lubricating oil so that corrosion of these parts is seldom encountered in compressor cylinders. However, if the H_2S content of the compressed gas is very high, it may be advisable to use more lubricating oil than is otherwise considered good practice.

The gasket problem is easily solved by using soft iron or aluminum instead of copper or its alloys. Protection of the intake valves is easily accomplished by injecting lubricating oil or gas oil into the gas just before it reaches the compressor. The oil coats the intake valves and protects them from the corrosive gas. When flushing oil is so used, a heavier lubricating oil may be used for compressor cylinders.

Corrosion from H_2S in the gas appears to vary directly with pressure. Appreciable H_2S will prevent satisfactory engine operation. An acid sludge is formed in the crankcase, and acid attack is experienced on the entire combustion area of the engine. Vapor-phase cooling has been used as a remedy.

EXTERNAL CORROSION OF BURIED STRUCTURES IN COMPRESSOR STATIONS

Corrosion control within a compressor station presents many problems not encountered in the cathodic protection of cross-country pipelines. Station piping comprises many different piping systems, all interlaced and/or located in the same ditches. They may vary in size from gage lines under one inch diameter to main transmission lines 24 in. diam or larger. Not all these lines may be of the same material. Steels of different compositions may be buried in the same immediate area and interconnected to offer the possibility of galvanic cells. In addition, galvanized electrical conduit or pipe may be buried nearby and connected to steel pipe, with resulting damage to the zinc galvanizing thru electrolytic action.

Protective Coatings

Selection and application of protective coatings for buried pipes and other facilities within a compressor station present many problems. Cross-country lines are usually designed to carry gas at a relatively constant temperature. Thus, the entire line between compressor stations may be coated with material of the same grade and specification. A compressor station embraces many systems of piping that carry fluids over a wide temperature range. Inlet piping to the compressors and outlet piping beyond the gas cooling stage will have approximately the same temperature as the main lines. However, the jacket water system and the gas piping between the compressors and the cooling stage will carry fluids at 130 F or higher. Different coating materials for the various systems may thus prove advisable.

Special problems will be encountered in applying whatever type of protective coating is selected. On cross-country lines coating and wrapping may be applied uniformly and efficiently by machines. In compressor stations, however, straight runs long enough to justify use of machines are rare.

Special attention should be given to points where piping leaves the ground. If the coating terminates below the surface, an area of bare pipe is exposed. It is impractical to rely on cathodic protection to prevent corrosion on such areas. All coating and wrapping should be extended well above the ground line and sealed in order to prevent entrance of moisture beneath it.

Cathodic Protection

On cross-country lines with one or more running parallel to each other in a reasonably narrow right of way, no particular difficulty is encountered in impressing the required protective current on all surfaces. The many buried piping systems in a compressor station, however, are generally arranged to form a series of squares and/or squares within squares when viewed in plan. If it is considered necessary to maintain all steel pipe surfaces at a potential of −0.85 v referred to an adjacent copper sulfate plug, protection may prove difficult. This problem is a result of shielding, i.e., when the ground bed is concentrated, most of the current may be absorbed by the pipes forming the outer squares. Even though the potential of the entire structure, referred to a remotely situated copper sulfate plug, may be maintained at the selected value of −0.85 v, potentials of the inner squares to local plugs may be less.

Low negative potentials on inner shielded surfaces may be brought to the selected value by greatly increasing the current output of the concentrated ground bed. Potentials of the outer squares will, of course, be increased proportionately, representing current waste thru overprotection.

Bonding of Buried Structures. When it has been decided to apply cathodic protection, it must be made absolutely certain that all buried metallic structures are bonded together and made a part of the system to be protected. Any pipes or other buried structures in the protective current field but not tied into the protected system will be subject to damage from stray currents picked up by such structures and discharged back to the protected system thru the soil.

There are several methods of checking the various systems to determine whether they are bonded satisfactorily. If cathodic protection is already in operation or a temporary ground bed and source of current are available, pipe-to-soil potentials may be used. Some means should be provided for interruption of the applied current while pipe-to-soil potentials are taken on all station piping. A variation in pipe-to-soil potential with application and removal of the protective current indicates a properly bonded system. If the pipe-to-soil potentials are taken to a truly remote copper sulfate plug, this variation will be approximately the same for all points throughout the station. If local potentials are used, these variations will be less on some portions of the system.

Another method of checking for proper bonding is structure-to-structure resistance readings; a source of protective current is not required. By employing a megger or similar instrument for direct measurement of low resistances, those between the various structures are determined. However, low readings may be found between two structures even if *not* mechanically bonded, because of their extent, proximity, and large numbers of coating holidays. These readings represent the resistance of the earth path between the two structures. The resistance of the leads used must also be considered.

Sales or purchase connections are often present at a compressor station. When such lines are not owned or operated by the company operating the station, it is often desirable to insulate them electrically from the protected plant. This may be accomplished by means of a flange insulating set. These sets are preferably installed in a flange, either above ground or in a dry-valve box. Consideration should be given to using resistance bonds across points of insulation to eliminate possibilities of damage to such foreign-owned lines by stray or interference currents.

Application of Cathodic Protection. Several methods may be used. The first choice is between galvanic anodes and impressed current anodes. The second choice is between a concentrated or a distributed ground bed. *Concentrated* means that all individual anodes are located in one immediate area spaced only far enough apart as required to achieve the desired efficiency. *Distributed* means that the individual anodes comprising the ground bed are situated throughout the area containing the buried pipelines and other structures.

When the complex piping forms a series of squares, these individual anodes are located within these squares and spaced so that all buried structures are, within reasonable limits, the same distance from an anode. Since piping is often grouped in a single ditch or alley, one of two squares of the same physical dimension may contain many more square feet of buried pipe surface than the other. Thus, it may be necessary to put more than one anode of the selected size, or a larger anode, in some squares. It will, of course, be impossible to establish exactly equal spacing or to regulate the anode surface to meet exactly the current requirements of each square; it may prove desirable to provide a means of regulating the amount of current taken by each of the anodes. Thus, by a combination of varying the anode surface in each square and regulating the individual anode current, reasonably uniform current distribution to all buried surfaces may be achieved.

Another type of impressed current anode system for station protection is the deep ground bed which may extend from 300 to 500 ft vertically into the earth. This anode system consists of special type rods and backfill which should give a service life of 10 to 15 years, and when installed in a suitable soil strata, should also give a reasonably uniform current distribution to the various station piping.

A compressor station includes much more buried metal surface than is apparent from a cursory examination. Foundations for all buildings, main compressor engine support blocks, scrubbers, mufflers, and catwalk support piers, very often contain steel reinforcing physically bonded to the piping system to be protected, and such bonds are most often impossible to eliminate. Also, when auxiliary engines furnish electrical power, some system of grounding is provided, and it may prove impractical to separate this ground from the piping system to be protected. Both these conditions result in the unavoidable addition of many square feet of bare metal to the buried system. The protective current thus required may be high even with excellent coating and wrapping of the pipe.

Reduction of current requirements for electrical grounding systems at compressor stations may be achieved by using zinc and other types of anodes which provide suitable low resistance to ground and require less current than a copper grid.

REFERENCE

1. Hegeman, A. K. "Frequent Causes of Corrosion in Gas Engines and Compressors." *Gas* 29: 95–98, Jan. 1952.

Chapter 8
Corrosion of Gas Wells

by R. C. Buchan

GENERAL

Internal corrosion in gas and condensate wells is an increasingly important problem, due principally to development of deeper reservoirs with higher pressures and temperatures. Gas composition, water content, and type of water produced are important factors. Corrosion of gas well casings is similar to that of any well, but the major corrosion problem is in deep condensate wells. Most gas wells encounter some corrosion if and when they produce salt water. Corrosion of steel and its alloys in wells takes the form of intercrystalline attack resulting in pitting and ultimate perforation of tubular goods and fittings. Erosion has very little effect, except that it usually removes corrosion products and exposes fresh metal to chemical attack.

CAUSES OF WELL CORROSION

Chemical

Most internal corrosion in gas wells is due to acidic constituents of the gas dissolved in water, and is, therefore, a chemical phenomenon. When salt water is present internally or externally, electrolytic corrosion due to currents flowing must also be considered. External casing corrosion may also be affected by chemical conditions, or it may be induced or sustained by bacterial action.

Produced gases contain various amounts of such acidic materials as carbon dioxide, organic acids, and hydrogen sulfide, which will cause corrosion in the presence of water. The pH of water produced from wells has been found to be 4.0 or lower under well conditions. In wells where tubing is not set on a packer, H_2S causes corrosion in the upper part of the casing-tubing annulus. CO_2 does not seem to have this effect. Water vapor in the annulus is thought to condense near the surface and circulate vertically to keep parts of the casing and the outside of the tubing water-wetted. H_2S then causes corrosion and liberates hydrogen. This process results in blistering and cracking of steels under tension.

In all gas and condensate wells, gas leaving the formation is usually saturated with water vapor. This vapor condenses and deposits on the steel as the gas proceeds toward the surface. With a sufficient concentration of acidic materials, a corrosive environment exists. Where gas is not saturated with water, either naturally or by dehydration, corrosion cannot occur.

Electrical

The flow of electric currents in and into wells is common. Long line currents and those generated by the corrosion of flow lines can flow either toward the well as in most cases, or away from it. These currents are discharged from the outside of the casing into the shallowest low resistivity formation, usually the first salt water sand penetrated by the well. In addition, currents usually flow from the various formations penetrated, and these flows are either up or down, depending upon conditions. There is some evidence that static currents are generated by gas flow in a well.

LOCATING CORROSION

Precautions should be taken to prevent corrosion at the time wells are completed; however, it is good practice to determine periodically the location, extent, and rate of corrosion. Under some conditions such as in shallow, low-pressure, dry-gas wells, corrosion may not be of economic importance. Where shut-in tubing pressures exceed about 2000 psig, corrosion should always be suspected. One of the simplest and best means of detecting internal corrosion is by the iron content of the produced water. After a gas well is completely cleaned of drilling fluids, this iron content is a direct measure of the loss of iron from the tubing. Fresh water produced from high-pressure corrosive condensate wells usually has an iron content of from 150 to 400 ppm. If removed uniformly from the tubing, this amount would not constitute an economic problem; however, since it is removed from pits, tubing perforation results in a few years. Where chemical means are used to prevent corrosion, control is frequently considered effective when iron content is reduced to 50 ppm or less. In many cases this treatment reduces iron content to less than 5 ppm. When major changes are made in producing rates, the corrosion problem also changes. Corrosion does not ordinarily occur in a shut-in sweet gas well.

In questionable cases, wellhead assembly interiors should be inspected. Frequently, the first corrosion noted is on choke parts. The inexperienced observer frequently calls corrosion of choke parts "erosion," but close inspection usually reveals sharp-edged pits.

Corrosion rate and severity can be measured with mild steel corrosion coupons inserted in the flow line near the well. Most operators use the well tubing as the coupon and make periodical tubing caliper surveys to determine the exact location, extent, and severity of pitting. Caliper surveys, water analyses, and coupon tests are useful tools for determining the effectiveness of corrosion prevention measures.

To determine electric current flow into wells, the voltage drop in ten or twenty feet of flow line connected to the well may be measured with a potentiometer. Knowing the physical dimensions of the pipe and its electric resistance, current flow can then be calculated.

CASING CORROSION

Casing must be protected both externally and internally. Except for flow of electric currents into the well, external casing corrosion can be prevented by suitable mud and cement programs before and during the casing installation. Cement provides excellent protection. Drilling muds not treated during drilling of the well to increase their pH content should be treated just before setting casing, so that the mud left behind has a pH of from ten to twelve. Sulfate-reducing bacteria cannot multiply at these high pH values, and neutralizing effects of alkalies in the mud on any acidic materials in the wellbore are usually adequate to prevent attack during life of the well.

Most operators consider it good practice to insulate electrically flow lines from gas wells with insulating flanges. These have proved effective even in wells producing several hundred barrels of salt water daily. In some extreme cases, it may be desirable to protect the casing exterior cathodically. This has been reported to be feasible to a depth of 10,000 ft. Galvanic metal anodes or rectified currents are suitable for such installations.

"Sand-cutting" or erosion occurs in gas condensate wells. Although sand-cutting sometimes occurs as a purely mechanical action, the damage is usually a combination of corrosion and erosion. This type of damage is found near restrictions where velocities and turbulence are relatively high.[1]

TUBING CORROSION

Failure of tubing in high-pressure wells is the most important corrosion problem in gas production. In wells which have a shut-in tubing pressure above 2000 psig, experience has shown that 85 to 90 per cent of such wells are corrosive, and with no preventive measures taken, average tubing life is about three years. Failure has been experienced as early as six months after completion. Even where tubing life might be eight or ten years, it is normally economical to use preventive measures. A limited number of high-pressure wells produce gas containing H_2S, and failures from corrosion and embrittlement of steel tubing have been rapid in such cases.

In average wells, corrosion starts in the upper third or half of the tubing string, since water leaving the formation as vapor is condensed in this portion. The pitting follows no uniform pattern; it is the result of intercrystalline attack on the steel and is affected by variations in its grain structure, both locally and throughout each length. Where threaded tubing connections have not been carefully made and mechanical leaks exist, corrosive gas escaping thru them also corrodes the threads.

Prevention

Means of prevention which may be used to increase tubing life follow

Alloys. Nine per cent nickel steel, costing three to four times that of API steels is generally satisfactory except in wells producing H_2S. Economics of the higher investment should be compared to the cost of using chemicals in individual fields or wells. This nickel steel is not necessarily effective where wells produce salt water.

Plastic Coatings. Baked-on phenolic resins are satisfactory coatings when six to ten layers are applied, to obtain an impermeable coat. These coatings can be damaged by wire line tools run into the tubing string. Their use is economical in some fields because the coated pipe costs only about twice as much as plain steel, and if the coating fails, the pipe can still be protected with inhibitors.

Inhibitors. The method of injecting chemical inhibitors depends upon their nature and upon whether the tubing is set on a packer. With no packer, it is practical to pump chemicals or their solutions into the casing so that they can descend and enter the tubing at its bottom. Where tubing is set on a packer, it is necessary to dump chemicals periodically into the tubing string and keep it shut in long enough for the inhibitor to reach the well bottom.

Neutralizing agents like soda ash must be fed rather continuously, or at least batch-fed daily. Soda ash solutions, fed continuously into a well casing, can control corrosion rather economically. They should not be used in wells producing salt water because of formation of calcium carbonate and other scales. Sodium chromates with or without caustic soda are sometimes used, but they require careful control to prevent deposition of thick scales. They may be objectionable from a water disposal standpoint.

Organic chemicals are the most widely used. Their advantages are low ultimate corrosion control cost, in comparison with alloys and coated pipe, and flexibility not inherent in inorganic chemicals. With no packer, these chemicals can be fed either continuously or in batches; the chemical is adsorbed on the steel surface, and resulting films appear to last for weeks or months. Where tubing is set on a packer and the chemical is dumped into the tubing string periodically, films appear effective for two weeks or longer.

Organic chemicals used should be nontoxic, safe, and easy to handle and pump at operating temperatures. They should not cause emulsification of liquid hydrocarbons and water, and preferably should tend to break such emulsions. They should be internally stable, i.e., films on steel should not polymerize under subsurface conditions to form increasingly thick deposits.

Amounts of chemical used vary considerably. The average well requires about ¼ lb of soda ash per MMCF of gas. Most operators, when using organic adsorbing-type inhibitors, start by using approximately one quart per MMCF of produced gas for corrosion control. After this is accomplished, the amount is gradually reduced to an optimum quantity. Where chemical is dumped periodically into the tubing, the amount used is that necessary to treat the gas until the next treatment. With batch treatments, the well must be left shut in long enough for the chemicals to reach the bottom. Diluted solutions of any chemical tend to require shorter shut-in times than undiluted chemicals, which are more viscous. Shut-in times ordinarily vary from 1 to 6 hr. Most corrosive gas wells with tubing set on a packer are treated weekly to once every three weeks, with 1 pt to 1 qt per MMCF of gas that will be produced until the next treatment. The well is left shut in at least 6 hr wherever feasible.

Liquid chemicals can be dumped or fed with volume-type batch lubricators. Suitable chemicals are also available in solid forms. One popular shape is a 1.5 in. diam bar or "stick" 18 in. long. These sticks can be dropped effectively thru a "lubricator" above the wellhead. Although their chemical cost is higher than for liquids, it is in part offset by eliminating pumps. Stick chemicals are frequently used for treating one well or a few isolated wells.

Each field (and often each well) constitutes a separate problem. Controlling internal corrosion by chemical means is effective and economical. Best results are generally obtained only after experimenting with the actual wells, trying several chemicals and varying application techniques.

CORROSION OF WELLHEAD ASSEMBLIES

Valves and fittings of wellhead assemblies are subject to corrosion except where they are protected by inhibitors. It is good practice to use corrosion-resistant materials for tubing, master valves, flange gaskets, choke parts on all condensate wells, and on new high-pressure wells where it is not known whether corrosion will develop. In extremely corrosive service, all wellhead assembly parts in contact with the well fluids are of corrosion-resistant materials. Those found to be resistant in this service include 11 to 13 per cent chrome iron, the most widely used material for valves and choke parts, *Stellite*, *Hastelloy*, *Monel*, and 18-8 stainless steel. Gasket rings of 18-8 stainless steel or of reinforced phenolic plastic are ordinarily used.

CORROSION OF FLOW LINES AND CONDENSATE TANKS

Flow lines and gathering systems serving high-pressure gas wells are subject to both internal and external corrosion. It is generally economical, however, to use externally coated and wrapped flow lines, with or without cathodic protection, to prevent external corrosion. Internal corrosion can be prevented by injecting inhibitors or neutralizing agents at the well. Where inhibitors are used to prevent subsurface corrosion, they automatically protect the flow lines and gathering systems against internal corrosion. Where the gas is dehydrated or where its dew point is always lower than its temperature, internal corrosion does not occur.

Interior corrosion of condensate tanks can be fairly rapid. The deck and the upper part of the sides pit as a result of attack by condensate containing CO_2 and/or H_2S. Stagnant water in the tank bottom is also corrosive because of its low pH. Tank corrosion can be prevented by vinyl resin coatings. Ammonia in the upper part of uncoated tanks and alkaline materials like soda ash dissolved in the water in the tank bottom, also effectively prevent corrosion.

WELL COMPLETIONS

Some recent developments in completing and equipping wells differ from what might be called "conventional well settings." Dual- and triple-completed wells are becoming more numerous, and "permanent" completions are sometimes used.

Inhibitors that effectively prevent corrosion of tubing are also effective in preventing corrosion of gas wells producing thru the casing, provided the chemical is applied in the same manner. Since more steel is exposed in the casing well annulus in a dually completed well than in the tubing string, it is good practice to use at least twice as much chemical per volume of gas produced from the annulus well as for the tubing well, until corrosion is definitely under control, as indicated by the iron content of the produced water. It is also advisable to make a casing caliper survey whenever tubing is out of the well, to verify the state of internal casing corrosion.

Wells are permanently completed because it is then possible to perform workover jobs at lower costs by using small tools run inside the tubing string. In most instances where the tubing packer is set more than a short distance off bottom, with the bottom of the tubing string above the producing interval, it is good practice to install a tail pipe to extend the tubing bottom to or below the producing interval. The tail pipe serves to keep water from accumulating in the casing below the packer, thus preventing casing corrosion by a stagnant pool of low pH and/or saline water.

REFERENCE

1. Nat. Assn. of Corrosion Engineers and Am. Pet. Inst. *Corrosion of Oil- and Gas-Well Equipment.* Dallas, A. P. I., 1958.

Chapter 9
Aboveground Corrosion

by W. J. Schreiner and W. P. Koster

Places where aboveground corrosion occurs include: (1) metallic structures and equipment; (2) water systems; and (3) generating and mechanical apparatus.

METALLIC STRUCTURES AND EQUIPMENT

No single material is suitable for every corrosive environment, and many different types of corrosion-resistant materials are used. The variables most responsible for differences in corrosion rates are air moisture content and amount and nature of atmospheric contaminants; SO_2 and NaCl are the most common and most potent.

Iron, Steel, and Ferrous Alloys

Environment and composition are the two main factors influencing corrosion of unprotected iron and steel. Corrosion tests of short duration (one year) show that sulfur dioxide, which is found in many industrial locations, will cause much more corrosion in that time than sodium chloride will at some marine locations. Extended tests, however, indicate a gradual decrease of corrosion in a sulfur dioxide atmosphere, because a film of corrosion product builds up and becomes somewhat protective.

Iron and steel structures are also corroded to a considerable extent by other industrial contaminants, like carbon dioxide, chlorine, ammonia, acid fumes, and soot. Moisture is also necessary; atmospheres with high moisture contents generally increase the corrosion rate. Frequently, dust and soot depositing on materials will absorb moisture and maintain a relatively wet surface which is conducive to corrosion. For this reason, many industrial stacks and flues, operating at rather high temperatures and not ordinarily susceptible to corrosion, deteriorate rapidly when idle.

There are two common methods of decreasing the atmospheric attack on ferrous metals. One is adding alloying elements which form a very dense and protective type of corrosion product; this in turn greatly reduces future corrosion. The other is to cover the metal with a protective coating which is corrosion-resistant for the particular application.

Added in small quantities, **copper** has been found effective in controlling the corrosion rate of low alloy steels. Ferrous materials containing 0.1 to 0.3 per cent copper form a more dense and protective rust than without such addition. **Phosphorus** (0.01 to 0.3 per cent) has a very marked beneficial effect on the atmospheric corrosion resistance of steel when copper is also present. Adding phosphorus alone is impractical, since the quantities necessary would have an adverse effect on cold-forming properties of the steel.

Stainless steels, containing substantial amounts of **chromium,** stand alone in their ability to resist atmospheric corrosion. The chromium present forms a tough and very impermeable surface oxide which inhibits subsequent corrosion. Marine atmospheres are sometimes injurious because of presence of chloride ions, and sulfurous atmospheres may discolor stainless steel to an important degree. Less corrosion will occur on surfaces exposed to rain than on those exposed only to atmospheric moisture and dust. Rain keeps the surface relatively free of material which would tend to retain a high moisture level on it.

Aluminum and Aluminum-Base Alloys

Aluminum base alloys as a group are highly resistant to normal atmospheric exposure. Pure aluminum is the most resistant. Its only alloys noticeably lacking in this property are those which contain more than one per cent copper. Gases found in ordinary industrial atmospheres have little effect in accelerating corrosion of aluminum alloys, although settled particles of carbon may accelerate it by galvanic action. Therefore, a rain wash is not harmful, but actually beneficial. Sulfur compounds have no important effect on aluminum, except when water containing dissolved sulfur dioxide or sulfur trioxide comes continuously in contact with the surface. Whenever the aluminum is in contact with moist, porous materials, like wood, cloth, and some types of insulation, an accelerated attack is likely. Use under very humid conditions, even with appreciable condensation, should be satisfactory if these porous materials are avoided.

Aluminum-base alloys are anodic to most commonly used structural alloys. If they are exposed in moist locations while in electrical contact with dissimilar metals, galvanic attack of the aluminum will probably result. Contact with copper alloys is likely to be the most serious. Steel does not present a serious problem except in marine atmospheres. Contact with stainless steel is usually less harmful than with ordinary structural steel. Since cadmium has about the same potential as aluminum, no mutual corrosion will result. Zinc and magnesium are anodic to aluminum and will therefore be corroded in its presence. These anodic metals may be used sacrificially in some cases to protect aluminum cathodically.

For mechanical reasons, it is often necessary to employ one of the stronger aluminum alloys. *Alclad* products, consisting of a section of a desired alloy with a thin sheet of pure aluminum rolled onto its surfaces, have helped greatly in making aluminum alloys satisfactory for many applications. *Alclad* materials are now available as sheet, plate, wire, rod, and tubing. Corrosion resistance above that of pure aluminum can be obtained by applying various nonmetallic coatings or thru using sprayed or electro-deposited metallic coatings.

Protective Coatings

These prevent contact between the atmosphere and the surface protected.

Metallic Coatings. Zinc coatings by various means are widely used. In hot dipping, parts are cleaned and dipped into a bath of molten zinc in the presence of appropriate fluxes. Electrolytic coating results in ductility and excellent brightness of the pure zinc deposited. In *sheradizing*, parts are placed in a drum with zinc dust, heated to about 600 F, and "rumbled" or actively mixed with zinc, to form a coating which is particularly suitable for small parts.

In the *Schoop* metal-spraying process, a stream of very small hot metal particles (perhaps molten), formed by atomizing (by gas pressure) molten coating material, is played over the surface to be coated. Both metallic and nonmetallic materials have been so coated. By using the proper gas and temperature, lead, zinc, tin, cadmium, copper, and nickel can be sprayed. These coatings tend toward porosity, but may be entirely satisfactory if produced by applying several thin coatings rather than a single thick one. This process is used chiefly to coat assembled structures such as bridges, ship sections, railway car frames, and storage tanks. Its obvious advantage is that all joints are sealed, in addition to the protection given to the major metal sections. This feature is important since joints and seams are often locations of major corrosion problems.

Cadmium behaves about like zinc as an iron and steel coating. It offers one advantage in that thin coatings may be used (except in industrial atmospheres). Cadmium is usually applied by electro-deposition or metal spraying.

Oxide and phosphate coatings are also used. Various processes give a surface coating of magnetic iron oxide, Fe_3O_4, which is highly resistant to all normal corrosive environments. Phosphate coatings depend upon formation of a rust-resisting deposit of basic ferrous phosphate. This coating is only mildly protective, but oiling greatly increases its effectiveness. Phosphating (e.g., *Bonderizing*) a steel surface is an excellent method of priming before painting or lacquering. Phosphate treatment may also be applied to zinc coated material before painting.

Nonmetallic Coatings. Paints and varnishes are protective coatings classified as air drying. They dry or set by oxidation, as contrasted to those which dry by evaporation of a solvent or by simple hardening. Paint and varnish films usually display excellent water resistance but permit enough water vapor and gases to penetrate to a metallic surface to be damaging. The type of paint pigment influences mechanical properties of the film and also neutralizes effects of damaging gases. To provide real protection, priming coats must be secure and the paint itself must offer resistance to sun, water, and abrasion.

Adequate surface preparation is essential for a sound, protective coating. All extraneous matter must be removed by chip hammering, sand blasting, air blasting, pickling, flame cleaning, and wire brushing. Surfaces must be completely dry before painting.

The best rust-inhibiting pigments are the reactive type, chiefly compounds of lead and zinc. Neutral pigments like iron oxide, magnesium silicate, and silica have also demonstrated definite advantages when used with reactive pigments. These include film strength, improved adhesion, less reactivity with the vehicle, and a "tooth," making for better bonding between coats. A priming paint for structural steel should contain thoroughly inhibitive and some inert pigments. Included are basic chromates of lead and zinc, basic sulfates of lead, and oxides of lead, zinc, and iron. Where severity of corrosive attack is great, it may be beneficial to add small amounts of powdered lead and/or zinc to the primer.

The many natural and synthetic organic materials now available for manufacture of lacquers and enamels allow an almost infinite variety of such coatings. Each type of resin has properties which fit it for a special service; none combines all desirable properties. Metal lacquers and varnishes, rubber paints, bituminous coatings, and shellacs are a few of the common varieties.

The protection afforded structural steel encased in concrete should not be overlooked. The alkaline nature of cement assures great permanence to structural steel so protected. Some degree of fire protection also is provided. Members should be insulated from excessive ground current and other stray current leaks to avoid deterioration by direct electrolysis.

WATER SYSTEMS

Dissolved O_2, CO_2, as well as pH and temperature, affect corrosion rates. Increased velocity accelerates corrosion. Pitting can cause failure in a pipe when only five per cent of its weight has been lost. Therefore, pitting or localized corrosion, caused by contact of dissimilar metals or by variations in solution composition or concentration, is worse than a uniform corrosion attack.

Tuberculation is closely associated with interior pitting of iron and steel pipe. Tubercles consist of mounds of corrosion products which cover and shield areas of pitting. They accelerate its rate at first, and later retard it as they build up and become more impermeable. Main objections to tuberculation are the resulting frictional increase and corresponding flow reduction which may completely clog small pipe.

Tuberculation effects corrosion by setting up a series of oxygen concentration cells. Within a tubercle, the amount of oxygen in solution will be lower than in the water passing, because some oxygen has combined with iron in the corrosion process at the metal-liquid interface. This differential in oxygen content causes the area within the tubercle to become anodic, thus promoting local corrosion. Even without oxygen, iron can corrode rapidly if sulfate-reducing bacteria are present, and by their metabolic processes reduce dissolved sulfates to sulfides and depolarize cathodic areas of iron.

Corrosion proceeds as rapidly as bacterial action permits. These bacteria do not thrive in chlorinated water.

Even when corrosion is restricted to attack by water under conditions of total immersion, deterioration of metal can take place thru any one or a combination of several mechanisms. However, it can be controlled in nearly every case by: (1) changing chemical or physical properties of the water; (2) selecting a suitable metal or alloy for the particular application; (3) applying a protective coating to the metal surface; or (4) using cathodic protection. The last procedure is similar to the steps taken to protect water tanks at compressor stations.

Water Treatment

For economic reasons, water treatment for very large distribution systems is limited to pH control and intentional deposition of calcium carbonate scale. Subsequent local treatment is usually warranted when domestic water is used for boiler feed and hot water supply in large buildings. Treatment of potable water is also limited to that which will not render it objectionable or injurious for drinking and cooking.

One method of treatment involves adding calcium carbonate in amounts large enough to cause deposition of a continuous protective layer on the water main interior. This layer hinders diffusion of oxygen to the metal. Deposits are more likely in cathodic areas because of their greater alkalinity.

Sodium silicate has been proven effective in protecting iron, lead, and brass water pipe. A solution of $Na_2O \cdot 3SiO_2$ is generally recommended. For local treatment of hot water systems, sodium silicate, equivalent to 12 or 16 ppm silica, should be fed to the water for the first month, reducing the amount to about 8 ppm afterward. In copper hot water systems, using this silica mixture to give water pH of about eight generally is satisfactory.

Sodium hexametaphosphate has been used to prevent scale deposits from excess calcium and magnesium salts and to prevent iron discoloration. It affords increased protection against corrosion when it is added simultaneously with sodium silicate. Corrosion protection by converting carbonates to sulfates has been reported.[1]

Cooling water recirculated in contact with air can generally be inhibited to prevent serious corrosion of steel by adding about 300 ppm of sodium chromate. However, chromate may cause pitting if its concentration falls below a certain minimum. Likewise, pitting may be due to unequal concentrations of sodium chromate if previous rust tubercles and loose scale are present. If a system contains such rust and scale deposits, chromate concentrations as high as 500 ppm may be desirable, with the solution pH maintained at about 8.5.

Water in closed recirculating coolant systems, as in internal combustion engines, can also be efficiently treated by adding sodium chromate. Organic anti-freeze mixtures, however, necessitate special organic inhibitors.

Treatment of nonrecirculating cooling water is relatively expensive. One method is adding enough lime or caustic soda to raise the solution pH to about 11. This is possible only when no aluminum or other metals attacked by high alkalinity water are present. Mechanical deaeration has been applied to such systems.

Dissolved oxygen is usually the controlling factor in steel corrosion in water. Free carbon dioxide can also be of prime importance under certain conditions. The various treatments described above are used to prevent diffusion of either oxygen or hydrogen, thus controlling corrosion. In some cases, particularly at elevated temperatures, more adequate steps to eliminate oxygen effects are necessary. Maximum amounts of dissolved oxygen ordinarily permissible without serious corrosion are:

Cold water systems	0.30 ppm
Hot water systems (160 F)	0.10 ppm
Low-pressure steam boiler	0.03 ppm
High-pressure steam boiler	0.005 ppm

Mechanical deaeration is useful for removing the largest portion of dissolved gases from water. Cold water is sprayed over trays in a chamber under high vacuum and the noncondensable gases which separate are removed by steam ejectors. In this cold water process, the oxygen content can be lowered to 0.33 ppm. Deaerators operating in conjunction with a hot water system often spray water into an atmosphere of steam under pressure as low as 1.5 psig. This method will reduce water oxygen content to practically zero.

For specialized applications, like steam generating plants, further reduction of oxygen may be necessary. Deactivation is then used. One method is to allow hot water to stand in contact with perforated iron sheet. Exposure for 30 min at 160 F is sufficient to exhaust all oxygen of a typical water sample. A much better method for oxygen removal is by inoculation with sodium sulfite. This is economically feasible if most of the oxygen has been previously removed mechanically. A slight excess of sulfite is desirable in the boiler to insure complete oxygen removal.

Resistant Metals or Alloys

Common pipe materials for water are ferrous alloys, copper, and brass. Numerous ferrous materials are available, but experience has shown that moderate variations in alloy content make no material difference in corrosion rate. Condition of the water contacting the metal is much more significant than metal composition.

Galvanized iron pipe is frequently used. It resists corrosion best in waters of moderate carbonate and bicarbonate content; copper is better in soft waters, but the carbon dioxide content must not exceed 2 ppm. Brass pipes low in zinc (15 per cent max) are also suitable, but when zinc content is about 35 per cent, dezincification may occur (even in moderately corrosive waters). Cupro-nickel (70-30) pipe and tubing give the best service in sea water systems, but care should be taken to avoid galvanic effects because of high conductivity of sea water.

Protective Coatings

Interior protection of cast iron and steel pipe is possible thru hot application of bituminous enamel. Air-drying paints lack the durability afforded by enamel. Paints based on phenolic, chlorinated rubber, and other synthetic resins have been developed for underwater service at normal temperatures. Portland cement coatings are satisfactory linings for large water mains and for old mains after cleaning.

GENERATING AND MECHANICAL EQUIPMENT

Corrosion of these facilities is, in some respects, peculiar to the particular type of equipment involved.

Boilers

During the normal operation of a boiler, an oxide of the boiler material is formed. Corrosion control consists largely of maintaining this film on the boiler surface. Type and composition of the steel used in boiler construction are relatively unimportant from the corrosion viewpoint.

Combinations of chemicals for maintaining this protective film will vary greatly with temperature, pressure, and heat transfer rate. In general, the chemicals keep the boiler water alkaline and free of dissolved oxygen, and maintain clean heating surfaces. Oxygen can be removed chemically by sulfites, ferrous hydroxide, and some organic reducers. For high temperature operation, sulfites will probably be most effective. Oxygen may also be removed by mechanical deaeration, but this is economical only in large, high-pressure installations.

Caustic soda provides maximum protection of steel at pH values between 11 and 12, with a pH of about 11 commonly used. Very high pressure systems are commonly treated with solutions of a slightly lower pH value due to greater possibility of metal attack. Caustic soda concentration of about 100 ppm affords maximum protection; higher concentrations are dangerous. Areas of poor circulation are particularly susceptible to caustic attack because of accumulation of concentrated solution. Caustic attack may also be caused by overheating of the steel, which occurs when heat input is high and localized steam blanketing and film boiling take place. Each bubble of vapor that forms has a film of highly concentrated liquid surrounding it. Thus, the boiler tube surface may be continuously subjected to a liquid much more concentrated than the average. One answer is to use a neutral salt in place of part of the caustic, thus materially reducing caustic attack and still maintaining a slightly alkaline solution.

Localized pitting is probably the most serious type of corrosion. It is very significant in both boilers and feed water heaters. Segregated impurities and residual stresses both form areas anodic to the balance of the steel. Dissolved oxygen, an active depolarizer, is probably the greatest promoter of this type of corrosion. Also, concentration cells can easily be originated under scale and sludge, where a stagnant condition exists. Dissolved carbon dioxide, existing in solution as carbonic acid, also promotes pitting. Both oxygen and carbon dioxide can be removed from solution chemically and by mechanical deaeration.

Idle Boilers. These must be maintained completely dry or completely filled with water containing an inhibitor like a sulfite or chromate. During cleaning, the protective iron oxide layer will be removed in several places and may not be replaced satisfactorily during subsequent operation. It appears wise to subject boilers to an alkaline boiling-out procedure immediately after cleaning, as is done with new boilers before they are put into service.

Embrittlement in steam boilers is a type of intercrystalline cracking which takes place in areas of high stress and is produced by solutions which attack the grain boundaries. Riveted joints are particularly susceptible. This phenomenon is the combined result of caustic soda solutions and of high stresses near riveted joints, where very slight leakage occurs. These salts build up in the seams and other crevices to the necessary concentration, which is many times that found in boiler waters. Welded seam construction is an excellent remedy. Riveted boilers in use can be protected by adding sodium nitrate to the boiler water or by replacing caustic soda with another type of inhibitor.

Steam Apparatus

Superheater tubes, steam lines, and turbine blades are subjected to reaction with high-temperature, superheated steam. Pressure has been found to have little influence on corrosion rate, its rapid acceleration being caused by increased temperature. The addition of chromium seems to protect steel effectively against high-temperature steam; required chromium content increases with temperature.

Corrosion of inner surfaces of superheater tubes may also be due to steam carry-over of boiler water salts which deposit on tube walls and cause direct chemical attack or blistering. Small amounts of gases, such as oxygen and carbon dioxide thus carried over, are not generally believed to cause any serious corrosion in passing thru the superheater. Steam lines and turbine blades are corroded by the same mechanisms except for a few minor exceptions. Turbine blades also suffer from cavitation.

Condensers

Corrosion within condensing units can occur from two sources: (1) the coolant, and (2) either the condensing steam or the condensate.

Surfaces Exposed to Coolant. Corrosion from water at ambient temperature is usually not severe except in marine service, or when salt or brackish water is used. In inland installations it may occur when cooling water is polluted by industrial wastes or acid mine water. Condenser parts attacked include the shell, water boxes, and miscellaneous plates and baffles, as well as the condenser tubes. Exterior and interior tube corrosion is usually by far the most critical. It may be caused by an excessive amount of entrained air in the circulating water, foreign material lodged or deposited in the tubes, and improper tube material for corrosive water conditions.

Most condenser tubes are fabricated of copper or copper-base alloys, primarily because these materials exhibit excellent corrosion-resistant properties under various service conditions and have high thermal conductivity. Admiralty metal in various forms is excellent for withstanding fresh, salt, and mine water as well as corrosive agents in the condensate. Aluminum, brass, and several copper-nickel alloys have also been found quite satisfactory for condenser tubes.

During initial service, condenser tubes usually undergo slight surface corrosion. The resulting coating or film, if insoluble, impervious, continuous, and adherent, protects the metal against further corrosion. Impingement of rapidly moving water may break down such films with subsequent localized attack of condenser surfaces. Abrasive solid matter, carried with rapidly circulating cooling water, will also have a damaging effect.

Desirable condenser design features include streamlining

water boxes, injection nozzles, and piping, and avoiding abrupt changes in low-pressure pockets, direction of flow, and any other feature causing local turbulence. Treatment of cooling water is also possible, but except with a recirculating system, the cost normally would be prohibitive. Where economically practicable, sodium chromate or caustic soda may be added.

Surfaces Exposed to Steam or Condensate. Occasionally, condensed water droplets absorb small amounts of corrosive materials like hydrochloric acid, ammonium hydroxide, or sulfur dioxide. This may lead to the formation of deep corrosion grooves in the condenser tubes where water droplets form repeatedly at the same spot and roll down the same cold surface. Considerable longitudinal grooving or pitting may occur on the tube bottom where droplets of corrosive liquid may hang for some time. Usually, considerably more thinning of the tube wall occurs at the bottom than at the top.

Serious corrosion may sometimes result from concentration cells at or adjacent to baffles and tube sheets, either because of crevices which exist at these locations or because of accumulation of local corrosion products or foreign material. This attack may also be severe beneath nonprotective film coatings like corrosion products of sulfides.

Within condensate lines, dissolved oxygen and carbon dioxide account for practically all corrosion. Slight attacks can occasionally be traced to sulfur compounds. Corrosion accelerated by oxygen is typified by presence of reaction products at its site or farther downstream, which frequently leads to pitting. Attack caused by dissolved carbon dioxide is characterized by clean, bright surfaces, and is generally not serious at condenser temperatures. Previous feed water treatment will usually remove these impurities, but some may be carried thru the system. Whenever condensate contains oxygen and carbon dioxide in appreciable quantities, they can be removed either chemically or by venting to the atmosphere.

Condensate containing oil, as from a reciprocating engine, is practically noncorrosive. However, inadequate oil content actually accelerates corrosion, as is typical of other anodic inhibitors. It has been claimed that sodium silicate, ammonia, and one of several organic inhibitors also serve to curb corrosive activity in condensate lines. The type and extent of treatment of condensate will depend largely upon its subsequent use.

Uneven Residual Stresses. These stresses in the assembled condenser also promote corrosion, since highly strained areas tend to become anodic to the remainder. While largely a matter of proper manufacture, such difficulties can also be partly overcome by suitable design.

Selecting suitable materials is usually the best solution to problems of condenser corrosion. A balance must be struck between installation costs and the costs of continual water treatment.

REFERENCE

1. Russell, J. T., "How Panhandle Eastern Combated Five Common Cooling-Water Problems." *Oil & Gas J.* 58: 101-8, Feb. 1, 1960.

BIBLIOGRAPHY

Am. Soc. for Metals. "Corrosion-Resistant Steel Castings." *Metals Handbook*, 8th ed, vol. 1, 432–38. Novelty, Ohio, 1961.

Speller, F. N. *Corrosion, Causes and Prevention*, 3rd ed. New York, McGraw-Hill, 1951.

Uhlig, H. H. *Corrosion Handbook.* New York, Wiley, 1948.

Zapffe, C. A. *Stainless Steels.* Cleveland, Am. Soc. for Metals, 1949.

Chapter 10

Safety in Corrosion Mitigation

by C. W. Beggs and T. L. Powers

Corrosion mitigation practices involve some operations which require attention to safety. Chief among these are protection of operating personnel and safe working practices.

EMPLOYEE PROTECTION

Work Area Protection.[1–7] Personnel working on a public street or highway must provide for protecting both themselves and the public. This protection is furnished by using properly applied standard safety equipment, such as signs, lights, flags, signal-men, barricades, high-level warning, and traffic cones. Companies doing road work should set up standards and instructions governing this work.

Work in Confined Spaces.[8–10] A major hazard to an employee is entry into a manhole or vault. Important precautions to be observed include:

1. Do not open a sealed vault without permission.
2. Do not enter a vault or manhole belonging to a company which does not permit entry except in the presence of its own employees.
3. Do not enter a vault or manhole containing unfamiliar apparatus.
4. Do not enter a vault unless it has first been tested for toxic, explosive, or suffocating gas (oxygen deficient). *Only a gas-tested manhole is a safe manhole.*[11,13]
5. Do not enter a vault or manhole unless a second man is on the surface.
6. Do not make electrical connections in a vault if the atmosphere might be combustible.
7. Where the atmosphere might be combustible, ventilate a vault continuously with a portable blower or a sail.
8. Wear a safety belt or harness. Attach the hook of the safety rope to the harness *before* entering the vault or manhole. Ladders (wooden only) are used wherever feasible. The extra rope should be coiled up on the surface of the road away from traffic and should be readily available to the surface man.
9. To prevent slipping on the ladder or manhole frame, see that boots are free of oil.
10. Be sure that the men on the job are qualified to administer artificial respiration. A fresh air mask and associated equipment to provide air should be available at the job site.
11. Come to the surface for fresh air occasionally. Check ault or manhole from time to time for gas. The surface man must do this, using an instrument, because the man in the vault will probably not notice changes in his atmosphere. If the vault man feels a headache or dizziness he should come up until he is sure it is not caused by gas or lack of oxygen.
12. Watch for foreign vapors, e.g., seepage from gasoline stations, dry cleaners, chemical plants.

Pipe coating personnel should wear proper apparel. Neoprene-coated canvas gloves with tight wrists, an oil cloth or leather apron, high shoes, goggles, and a cartridge respirator are recommended. Sleeves should be kept down and pants kept out of boots or shoe tops.

SAFE WORKING PRACTICES

Pipe Coating

When pipe coating is used, every effort must be made to see that it survives construction hazards and is sound when backfilled. A poorly applied coating will permit corrosion to proceed at an accelerated rate at each holiday unless the line is cathodically protected.

Coated pipe should be handled carefully at all times. Unless proper inspection is made during construction, the coating cost may be wasted and pipe life decreased. When coating is exposed above ground, it may be damaged by the weather. Rough handling from the freight car to the construction site will cause stone bruises, skid marks, sling marks, and gouges of other kinds. Unless it is protected by rock shields or clean fine earth during backfilling operations, the coating may be damaged by stony material dropped or forced around it.

In operating the melting pot, the coating material should be drawn from the spigot into dry pouring pots which are in good condition. They should not be filled more than two-thirds full. Coating material should be kept dry when it is broken up and loaded into the kettle. Chunks should be charged carefully, to avoid splashing. Wet stock should not be charged. Pots should be kept level and free of carbon deposit. Lids should be closed while moving equipment. Proper temperature should be maintained as shown by a thermometer. Employees should stand clear when lighting the kettle. Production of yellow fumes should be avoided. Coating residues and spillages should be cleaned up promptly to prevent injury to livestock. Employees carrying pitch pots should walk, not run. When pouring, the spout should be kept close to the pipe.

Rectifiers[12]

Installing a rectifier involves several safety factors. It should be safe from tampering, yet easily reached for test and maintenance. The rectifier must be adequately ventilated, yet protected from the elements. It should never be installed where it could start a fire, even by overheating. It should not be installed in a gaseous atmosphere or in a location that might become gaseous. Wiring and fusing should be done in accordance with the National Electrical Code, or with local regulations where applicable. Where the wiring passes underground thru conduit, the conduit should be sealed where it enters the rectifier box, to prevent gas infiltration from an underground leak.

One large utility company recommends lightning arresters between the phase leads of the service and the negative rectifier lead (connected to the gas main). Also, if over 100 ft of aerial wire are used between the rectifier and anode, a lightning arrester should be placed between this wire and the negative side of the rectifier. The arrester on the anode lead will prevent a high lightning charge and extreme earth gradients over the anode bed. Lightning strikes to installations without lightning arresters have been known to cause loss of both pole and rectifier.

Polarity of rectifier connections should be checked upon installation by a corrosion engineer to avoid a reversed connection which would cause accelerated corrosion and rapid failure of the protected structure.

Ground Beds

For a cathodic protection system, select ground bed site that will permit a minimum of stray current to other buried structures. Existence of other structures in the affected area should be determined beforehand. An offer should be made to run cathodic protection coordination tests with the owners of these other structures after the ground bed or anode is installed, and to make any changes found necessary.[12]

In designing a ground bed or anode in a location where anode resistance is high and current requirements call for a high voltage rectifier, a high potential gradient over the anode is possible. To prevent injury to persons or livestock, the gradient should be checked from the anode center at 30 in. intervals outward. Voltage drop should not be more than ten volts across each 30 in. circle, or injury may result. Tests should be made with the soil saturated with water, for instance after a heavy rain.

Insulating Joints

When the use of an insulating joint in a pipeline is desired, great care should be taken in selecting its type, location, and method of installation. Insulating joints have a definite place in corrosion work, but their use should be indicated only after a careful check by a competent corrosion engineer or by reference to design standards. With a new line, it will be found economical to install these joints rather freely during construction, rather than later. When so installed, the insulated joints should be bonded with the bond leads carried to a roadway box for joining and testing. They should remain bonded (closed circuit) until disposition of the joints is determined by test. A No. 12 or larger wire should be used for bonding.

Generally, insulating joints are used to isolate branches from a protected main line, or to sectionalize a main line into elements requiring different amounts of protection. When used to break up stray current flow, careful study is necessary to determine the location and method of installation. Otherwise, serious damage may result from currents by-passing the joint thru earth and corroding the end of the pipe near the joint with the higher potential.

Several types of insulating joints are available. To be effective, they must be wrapped with pipe coating, enclosed in a waterproof envelope, or boxed in coating material. The coating and protection should extend to a reasonable distance, varying from one to 20 ft on either side of the joint, depending upon conditions and accepted standards.

Some large transmission companies require that the joint be installed above ground, away from buildings and machinery. They do not permit its installation in a compressor house or other location where its failure, or arcing caused by tools laid across it or by a lightning stroke, could cause an accident.

An insulating joint may be by-passed by such structures as walkways, stairways, tracks, fences, old pipes, the water system, and electric ducts. Even though voltage is low, the extremely low resistance to ground of extensive metallic structures may permit a current of several amperes to flow thru an accidental short circuit, such as from contact with a wrench or other tool. On interrupting this circuit, an arc may develop which is capable of igniting flammable vapors. For these reasons, care must be used in selecting a location for an insulating joint.

An insulating joint installed above ground out-of-doors offers the following inherent advantages:

1. Ease of maintenance, observation, and checking.
2. Freedom from explosion hazard.
3. Elimination of expensive coating material box.
4. More effective insulating value.
5. Preclusion of leakage into a building resulting from insulating material failure.

Lightning Arresters[14]

Lightning may short-circuit insulating joints and permit a surge of power current. In such failures, the insulating gasket may burn thru inside the pipe at the bottom. A stream of metal flowing thru this notch may weld to a pool of metal on each side. The removal of the solid connection requires an expensive repair job. Occasionally, bolt thimbles or insulating washers are burned across or thru.

From the study of such an incident, it was concluded that the lightning across a nearby power line insulator triggered a follow-up power surge which continued for 3 to 20 cycles (up to $\frac{1}{3}$ sec) until the breaker tripped out. This power follow-up was believed to have caused the failure of the joint. Emphasis was laid on ease and desirability of placing separate lightning arresters on each side of an insulating joint when installed. One method utilizes a ground wire lug bolted to each lead wire lug from the joint, with a perforated sheet of mica inserted between the lugs. Note that the holding bolt must be insulated from the lugs to prevent grounding of the lead under normal conditions. Arresters must be inspected periodically because they are sometimes fused by a power discharge. Recently, *Thyrite* discharge resistors have been used to arrest lightning.

Where a steel line is well insulated and carries a good coating, extreme care must be used in disconnecting lightning arresters or bonds. Lethal currents may flow in such buried lines, because of stray current, lightning, or other sources.

One large company has reported that fault currents from power cable failure have induced currents in paralleling gas mains which caused joints to arc and catch fire.

Cross Bonds and Jumpers[15–17]

The danger of spark ignition is always present in a hazardous atmosphere. A spark may come from an open flame, hot chips, hot metal, stray currents, falling objects, tools and equipment, and static electricity. A person may acquire a 10,000 v potential, if properly insulated from ground (equal to a charge of 0.015 joule).

The possibility always exists that a pipe will serve as a conductor for static electricity, for stray currents from power lines, or for soil currents, so that a spark may result when the line is severed or reconnected. A bonding wire should be securely fastened to the pipe on both sides of a proposed cut before the line is separated or reconnected. This eliminates one source of ignition.

In electric railway areas or where currents from cathodic protection or other stray currents are involved, care should be taken to see that no accidental contacts occur between pipe being repaired and other exposed structures in the same area. Sufficient potential differences may be present to cause a small spark.

If there is a 0.3 v potential difference between two parallel or intersecting pipelines, or if transmission lines cross grids of a distribution system, permanent cross bonds or other equivalent means are desirable to prevent damaging crossflow of current. The function of cross bonds or their equivalents between gas and water distribution systems is performed by their interconnection thru the house services, provided that there are no insulating bushings on the services. Sometimes current interchange thru house services is quite large because of high resistance joints in the main, with resultant arcing when meter connections are broken. Insulating bushings, flanges, or unions in house services or meter installations provide protection to piping upstream of the insulators from the effects of potentials which may be applied to gas piping within a building. They should be shunted by a bond while the pipe is disconnected or the meter is removed.

If the voltage is sufficient to cause an electric shock or to create a spark while making meter installations, tests should be made to see whether lighting system wires have become grounded to the house piping. There is also a possibility that an unbalanced electric neutral current will cause a spark hazard.

When bonding in a hazardous atmosphere, a No. 10 or No. 12 wire should be attached to each pipe. Individual wires should be carried away from the hazardous area and connected at a safe distance. Connections should be solid and should remain until the lines are reconnected.

Miscellaneous

No matter how much protection is designed and installed in buried pipelines, it may be nullified by hazards resulting from carelessness, slovenly methods, or willful disregard for the property of others. Chemicals may be spilled on the ground or dumped into streams where they attack buried lines. Excavators may slash the coating of a protected structure. Hand diggers may pick holes in coating, and drilling machines of various kinds may strike and damage coated structures. Stray current from railways or industry using direct current may spread over large areas, causing damage where they leave buried structures. The improper installation of cathodic protection drainage systems can cause damaging interference currents in other structures in the area.

To aid in solution and control of mutual problems, corrosion and electrolysis engineers often form cooperative corrosion committees. These bodies serve as a clearing house for corrosion problems, solutions and mitigation plans.[12]

Legal Considerations[18,19]

It is beyond the scope of this chapter to cover all the laws, regulations, and ordinances covering safety of employees and safe practices. It is, however, incumbent upon the reader to consider and follow all possible laws and rules which might be applicable in his locality.

REFERENCES

1. National Safety Council. *Barricades and Warning Devices for Highway Construction Work.* (Data Sheet D-239). Chicago.
2. ——. *Trench Excavation.* (Data Sheet D-254) Chicago.
3. Consolidated Edison System Cos. *Field Manual, Planned Work Area Protection.* New York.
4. Public Service El. & Gas Co. *Work Area Protection Manual.* Newark, N. J.
5. Northern Ill. Gas Co. *Work Area Protection: a Manual.* Aurora, Ill.
6. Pacific Gas & El. Co. *Planned Work Area Protection: a Manual.* San Francisco.
7. A.G.A. *Suggested Safe Practices for Gas Distribution Men,* 2d ed. New York, 1957.
8. National Safety Council. *Black Widow Spiders.* Rev. ed. (Data Sheet D-258) Chicago.
9. ——. *Atmospheric Conditions in Underground Structures.* (Data Seet D-250) Chic., 1953.
10. Public Service El. & Gas Co. *Manhole Manual.* Newark, N. J.
11. "A Gas-Tested Manhole Is a Safe Manhole." *Plant Eng.* 6: 132, Jan. 1952.
12. Nat. Assn. of Corrosion Engineers. *Cathodic Protection.* Houston, 1951. Includes 56-item bibliography.
13. Nat. Safety Council. *Static Electricity.* (Safe Practices Pam. No. 52) Chicago, 1942.
14. Kleinheksel, S. "Lightning Arresters for Protection of Insulated Joints in Buried Pipe Lines." *Petrol. Engr.* 22: D87-90, Mar. 1950.
15. Huff, W. J. *Sources of Gas Ignition.* Presented before A.G.A. by permission of Bur. of Mines.
16. Factory Mutual Ins. Co. *Static Electricity.* (Loss Prev. Bul. 12.21) Boston, 1950.
17. Klinkenburg, A. and Van der Minne, J. L. *Electrostatics in the Petroleum Industry.* New York, Am. Elsevier, 1957.
18. A.S.M.E. *Gas Transmission and Distribution Piping Systems.* (A.S.A. B31.8-1963) New York, 1963.
19. N. J. Dept. of Labor & Ind. *Safety Regulation No. 10 Relating to Work in Confined Spaces.* Trenton.

SECTION 12
UTILIZATION OF GAS

C. George Segeler, *Section Chairman,* Chapters 2, 3, 5, and 17–20, American Gas Association, New York, N. Y.
J. C. Agarwal, Chapter 21, United States Steel Corp., Monroeville, Pa.
D. A. Campbell (deceased), Chapter 13, Eclipse Fuel Engineering Co., Rockford, Ill.
A. H. Cramer, Chapter 2, Michigan Consolidated Gas Co., Detroit, Mich.
H. J. Cullinane, Chapter 5, American Gas Association, New York, N. Y.
Keith T. Davis, Chapter 5, Bryant Manufacturing Co., Indianapolis, Ind.
J. W. Farren (retired), Chapter 6, Rheem Mfg. Co., Kalamazoo, Mich.
E. J. Horton, Chapter 4, Southern California Gas Co., Los Angeles, Calif.
E. A. Jahn, Chapters 6, 7, 9, 15, 16, and 22, American Gas Association, New York, N. Y.
Evelyn M. Kafka, Chapter 7, American Gas Association, New York, N. Y.
M. Khan, Chapter 10, American Gas Association, New York, N. Y.
K. R. Knapp (retired), Chapters 2, 6, 7, and 19, American Gas Association Laboratories, Cleveland, Ohio
A. H. Koch (deceased), Chapter 19, Surface Combustion Div., Midland-Ross Corp., Toledo, Ohio
Robert C. LeMay, Chapter 19, Selas Corp. of America, Dresher, Pa.
C. J. Mathieson, Chapters 5 and 16, The East Ohio Gas Co., Cleveland, Ohio
C. S. O'Neil (deceased), Chapter 9, Hamilton Mfg. Co., Two Rivers, Wisc.
B. A. Phillips, Chapter 8, Whirlpool Corporation, St. Joseph, Mich.
Marvin D. Ringler, Chapters 23 and 24, American Gas Association, New York, N. Y.
W. Roger Sarno, Chapter 25, American Gas Association, New York, N. Y.
E. H. Smith, Chapter 9, Lovell Mfg. Co., Erie, Pa.
Richard L. Stone, Chapter 3, American Gas Association Laboratories, Cleveland, Ohio. Present address: Metalbestos Div., William-Wallace, Co., Belmont, Calif.
F. E. Vandaveer (retired), Chapters 4, 5, 10–12, 16–18, 22, and 24, Con Gas Service Corp., Cleveland, Ohio
E. J. Weber, Chapters 12 and 14, American Gas Association Laboratories, Cleveland, Ohio
H. C. Weller (deceased), Chapter 19, Surface Combustion Div., Midland-Ross Corp., Toledo, Ohio
D. D. Williams, Chapter 6, A. O. Smith Corp., Kankakee, Ill.
W. Wirt Young, Chapter 19, New Haven Heat Treating Company, Inc., New Haven, Conn.
James Yund, Chapter 5, Whirlpool Corporation, Benton Harbor, Mich.

CONTENTS

Chapter 4—Appliance Controls and Accessories

Chapter 5—Gas Comfort Heating and Cooling

Chapter 6—Gas Water Heating

Chapter 7—Domestic Gas Cooking Appliances

Chapter 8—Domestic Gas Refrigerators

Chapter 9—Direct-Fired Gas Clothes Dryers

Chapter 10—Incinerators

Chapter 11—Gas Illumination

Chapter 12—Gas Burner Design

Chapter 13—Industrial Combustion Systems

Chapter 14—Interchangeability of Fuel Gases

Chapter 15—Commercial Kitchen Gas Appliances

Chapter 16—Other Commercial Gas Appliances

Chapter 17—Prepared Atmospheres

Chapter 18—Industrial Gas Ovens and Dryers

Chapter 19—Industrial Furnaces

Chapter 20—Kilns for Lime, Cement, and Other Products

Chapter 21—Blast Furnace Fuel Injection

Chapter 22—Fuel Comparisons

Chapter 23—Gas Total Energy Systems

Chapter 24—Gas Internal Combustion Engines (Reciprocating)

Chapter 25—Gas Turbines

Tables

Figures

Chapter 1

Gas Appliance and Equipment Installation

DOMESTIC AND COMMERCIAL APPLIANCES

The basic requirements for installing domestic and commercial appliances supplied at gas pressures of 0.5 psi or less are covered in ASA Z21.30–1964. This standard is also applicable to such appliances when installed in office or cafeteria spaces of industrial occupancies. Extracts covering gas piping appear in Chapter 2 and those dealing with venting in Chapter 3. Pertinent sections of the Standard relating to clearance, mounting, and location of specific appliances—both listed and unlisted—follow.

Part 3—Appliance Installation

3.1 GENERAL

3.1.1 Appliances, Accessories and Equipment to be "Approved."

Gas appliances, accessories, and equipment shall be "Approved." "Approved" shall mean "acceptable to the authority having jurisdiction."

Note: In determining acceptability, the authority having jurisdiction would normally base acceptance on compliance with NFPA, ASA or other appropriate standards. In the absence of such standards, said authority would normally require evidence of proper installation, procedure or use. The authority having jurisdiction would normally refer to the listings* or labeling† practices of nationally recognized testing laboratories,‡ *i.e.*, laboratories qualified and equipped to conduct the necessary tests, in a position to determine compliance with appropriate standards for the current production of listed items, and the satisfactory performance of such equipment or materials in actual usage.

* LISTED. Equipment or materials included in a list published by a nationally recognized testing laboratory that maintains periodic inspection of production of listed equipment or materials and whose listing states either that the equipment or material meets nationally recognized standards or has been tested and found suitable for use in a specified manner.

† LABELED. Equipment or materials to which has been attached a label of a nationally recognized testing laboratory that maintains periodic inspection of production of labeled equipment or materials and by whose labeling is indicated compliance with nationally recognized standards or the conduct of tests to determine suitable usage in a specified manner.

‡ Among the laboratories nationally recognized by the authorities having jurisdiction in the United States and Canada from whom listings are available are the Underwriters' Laboratories, Inc., the Factory Mutual Engineering Division, the American Gas Association Laboratories, the Underwriters' Laboratories of Canada, the Canadian Standards Association Testing Laboratories, and the Canadian Gas Association Approvals Division.

The National Fire Protection Association and the American Standards Association, Inc., do not approve, inspect or certify any installations, procedures, equipment or materials, nor do they approve or evaluate testing laboratories.

3.1.2 Type of Gas:

It shall be determined whether the appliance has been designed for use with the gas to which it will be connected. No attempt shall be made to convert the appliance from the gas specified on the rating plate for use with a different gas without consulting the serving gas supplier or the appliance manufacturer for complete instructions.

3.1.3 Automatic Pilots for LP-Gas Appliances:

Manually controlled water heaters and automatically controlled appliances, except domestic ranges and commercial cooking equipment having pilot input ratings of 500 Btu per hour or less, for use with undiluted liquefied petroleum gases, shall be equipped with automatic pilots of the complete shutoff type.

3.1.4 Use of Air or Oxygen under Pressure:

When air or oxygen under pressure is used in connection with the gas supply, effective means such as a back pressure regulator and relief valve shall be provided to prevent air or oxygen from passing back into the gas piping. The serving gas supplier shall be consulted for details. When oxygen is used, see the Standard for Installation and Operation of Oxygen-Fuel Gas Systems for Welding and Cutting, NFPA No. 51–1964.§

3.1.5 Flammable Vapors:

Gas appliances shall not be installed in any location where flammable vapors are likely to be present, unless the design, operation and installation are such as to eliminate the possible ignition of the flammable vapors.

3.1.6 Installation in Residential Garages:

(a) Gas appliances may be installed on the floor of a residential garage provided a door of the garage opens to an adjacent ground or driveway level that is at or below the level of the garage floor. When this condition does not exist, appliances shall be installed so that the burners and pilots are at least 18 inches above the floor.

(b) Gas appliances shall be located, or reasonably protected, so that they are not subject to physical damage by a moving vehicle.

3.1.7 Installation in Commercial Garages:

(a) Floor mounted heaters in commercial garages for more than 3 motor vehicles shall be installed as follows:

1. Heaters may be located in a room separated from other parts of the garage by construction having at least a one hour fire-resistance rating. This room shall not be used for combustible storage and shall have no direct access from the garage

§ Available from the National Fire Protection Association, 60 Batterymarch Street, Boston, Mass. 02110

storage or repair areas. All air for combustion purposes entering such a room shall be from outside of the building, or

2. Floor mounted heaters may be located in the garage if they are installed so that the bottom of the combustion chamber is at least 18 inches above the floor and outside grade level. Such heaters shall be protected from physical damage by vehicles.

(b) Overhead heaters shall be installed at least 8 feet above the floor.

(c) Sealed combustion system heaters may be located within a garage. When necessary, they shall be protected against physical damage.

3.1.8 Installation in Aircraft Hangars:

Heaters in aircraft hangars shall be installed in accordance with NFPA No. 409–1962, Standard on Aircraft Hangars.*

3.1.9 Venting of Flue Gases:

Appliances shall be vented in accordance with the provisions of Part 5, Venting of Appliances.†

3.1.10 Extra Device or Attachment:

No device or attachment shall be installed on any appliance which may in any way impair the combustion of gas.

3.1.11 Adequate Capacity of Piping:

When connecting additional appliances to a gas piping system, the existing piping shall be checked to determine if it has adequate capacity (see 2.4‡). If inadequate, the existing system shall be enlarged as required or separate gas piping of adequate capacity shall be run from the meter or from the service regulator when a meter is not provided, to the appliance.

3.1.12 Avoid Strain on Gas Piping:

Gas appliances shall be adequately supported and so connected to the piping as not to exert undue strain on the connections.

3.1.13 Venting of Pressure Regulators:

(a) Gas appliance pressure regulators requiring access to the atmosphere for successful operation shall be equipped with vent piping leading to the outer air or into the combustion chamber adjacent to a constantly burning pilot, unless constructed or equipped with a vent limiting means to limit the escape of gas from the vent opening in the event of diaphragm failure.

(b) Vent limiting means on appliance pressure regulators for use with natural, manufactured or mixed gases or LP-gas–air mixtures shall limit the escape of gas in the event of diaphragm rupture to not more than 1.0 cubic foot per hour of 0.6 specific gravity gas at 7.0 inches water column pressure.

(c) Vent limiting means on appliance pressure regulators for use with undiluted liquefied petroleum gases shall limit the escape of gas in the event of diaphragm rupture to not more than 0.5 cubic foot per hour of 1.53 specific gravity gas at 11.0 inches water column pressure. (Appliance pressure regulators complying with the Addenda to the American Standard Listing Requirements for Domestic Gas Appliance Pressure Regulators, Z21.18a–1960, are required to be equipped with a device that will comply with this limitation.)

(d) In the case of vents leading to the outer air, means shall be employed to prevent water from entering this piping and also to prevent stoppage of it by insects and foreign matter.

(e) In the case of vents entering the combustion chamber, the vent shall be located so that the escaping gas will be readily ignited from the pilot flame and the heat liberated will not adversely affect the operation of the thermal element of the automatic pilot. The terminus of the vent shall be securely held in a fixed position relative to the pilot flame. For manufactured gas, a flame arrester in the vent piping may also be necessary.

3.1.14 Combination of Appliances:

Any combination of appliances, attachments, or devices used together in any manner shall comply with the standards which apply to the individual appliances.

3.1.15 Installation Instructions:

The installing agency shall conform with the appliance manufacturer's specific recommendations in completing an installation to assure satisfactory performance and serviceability. The installing agency shall also leave the manufacturer's installation, operating and maintenance instructions in a location on the premises where they will be readily available for reference and guidance of the authority having jurisdiction, servicemen and the owner or operator.

3.1.16 Protection of Outdoor Appliances:

Appliances not listed for outdoor installation but installed outdoors shall be provided with protection to the degree that the environment requires and be accessible for service. (See 3.3.1.)

* * *

3.3 ACCESSIBILITY AND CLEARANCE

3.3.1 Accessibility for Service:

(a) Every gas appliance shall be located with respect to building construction and other equipment so as to permit access to the appliance. Sufficient clearance shall be maintained to permit cleaning of heating surfaces; the replacement of filters, blowers, motors, burners, controls and vent connections; the lubrication of moving parts where required; and the adjustment and cleaning of burners and pilots. For attic installation the passageway and servicing area adjacent to the appliance shall be floored.

(b) Appliances listed for outdoor installation may be installed without protection in accordance with the provisions of their listing and shall be accessible for servicing.

3.3.2 Permissible Temperatures on Combustible Materials:

(a) All gas appliances and their vent connectors shall be installed so that continued or intermittent operation will not create a hazard to persons or property. They shall not, during operation, raise the temperature of unprotected combustible walls, partitions, floors, or ceilings more than 90°F above normal room temperature when measured with mercury thermometers or conventional bead type thermocouples. (When wall and partition temperatures are measured with disc type thermocouples as specified in American Standard Approval Requirements for the types of appliances involved, an indicated temperature rise of 120° F will correspond to the 90° F rise measured with thermometers or conventional bead type thermocouples.)

(b) Minimum clearances between combustible walls and the back and sides of various conventional types of appliances and their vent connectors are specified in Parts 4 and 5.†

3.4 AIR FOR COMBUSTION AND VENTILATION

3.4.1 General

(a) The provisions of 3.4 are intended to apply to appliances that are installed in buildings and which require air for combustion, ventilation and draft hood dilution from within the building. They are not intended to apply to (1) sealed combustion system appliances which are constructed and installed so that all air for combustion is derived from the outside atmosphere and all flue gases are discharged to the outside atmosphere, or (2) enclosed furnaces which incorporate an integral total enclosure and use only outside air for combustion and draft hood dilution.

(b) Appliances shall be installed in a location in which the facilities for ventilation permit satisfactory combustion of gas, proper venting and the maintenance of ambient temperature at

---

* Available from the National Fire Protection Association, 60 Batterymarch Street, Boston, Mass. 02110. [Covered in Chap. 5 of this Section, under "Installation of Unit Heaters."—EDITOR]

[† For pertinent portions of Part 5 see Chap. 3 of this Section.—EDITOR]

[‡ Not given in this Handbook but subject is covered in Chap. 2 of this Section.—EDITOR]

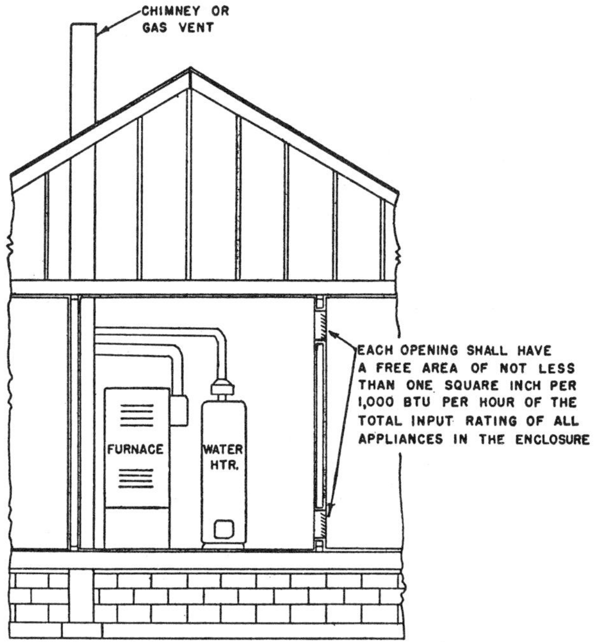

Fig. 5. Appliances Located in Confined Spaces. All Air from Inside the Building. (See 3.4.3-a.)

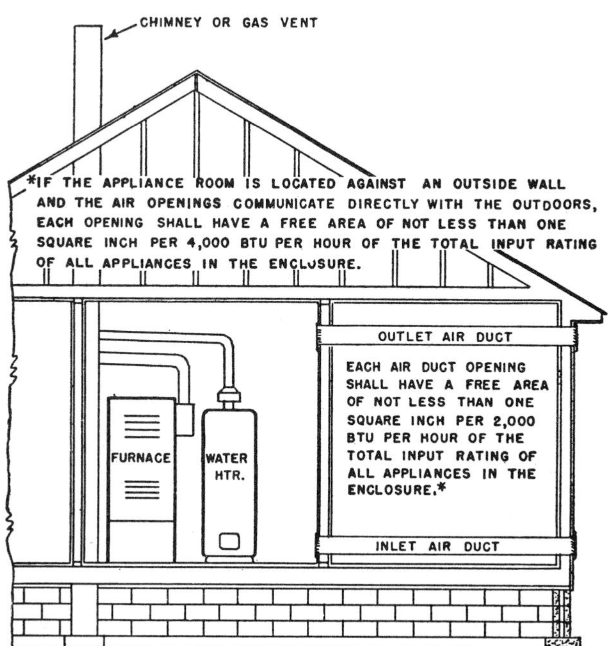

Fig. 6. Appliances Located in Confined Spaces. All Air from Outdoors. (See 3.4.3-b.)

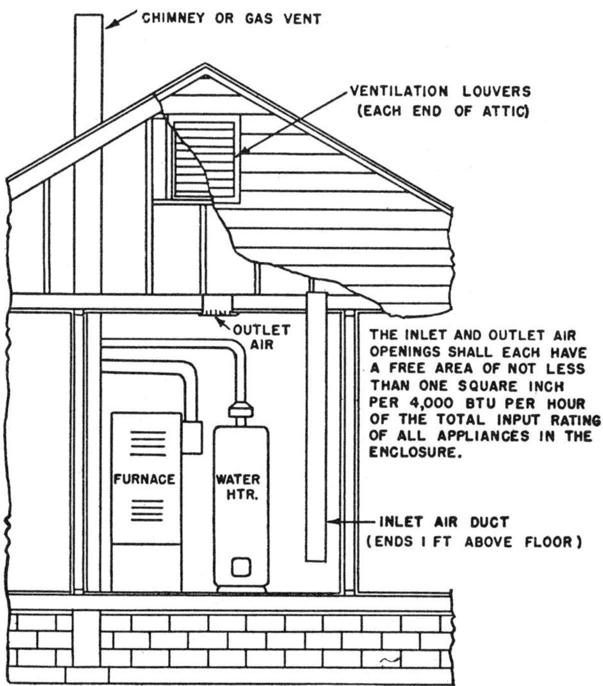

Fig. 7. Appliances Located in Confined Spaces. All Air from Outdoors Through Ventilated Attic. (See 3.4.3-b.)

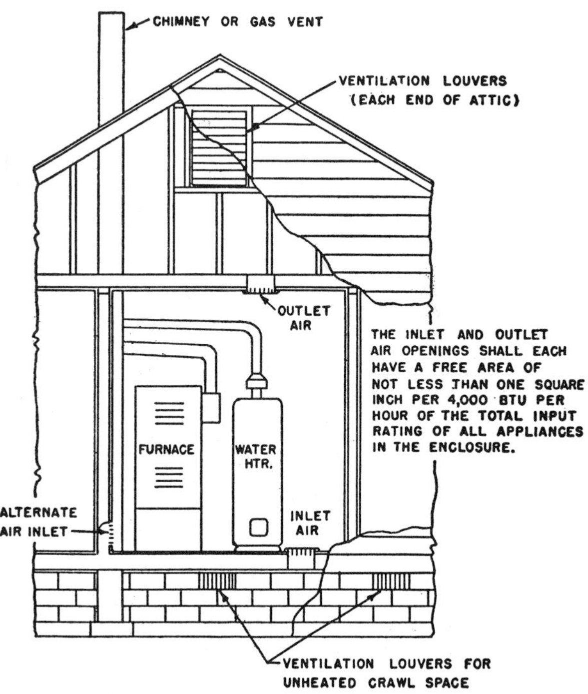

Fig. 8. Appliances Located in Confined Spaces. All Air from Outdoors—Inlet Air from Ventilated Crawl Space and Outlet Air to Ventilated Attic. (See 3.4.3-b.)

safe limits under normal conditions of use. Appliances shall be located so as not to interfere with proper circulation of air within the confined space. When buildings are so tight that normal infiltration does not meet air requirements, outside air shall be introduced.

(c) While all forms of building construction cannot be covered in detail, air for combustion, ventilation and draft hood dilution for gas appliances vented by natural draft normally may be obtained by application of one of the methods covered in 3.4.2, 3.4.3 and 3.4.6.

3.4.2 Appliances Located in Unconfined Spaces:

(a) In unconfined spaces in buildings of conventional frame, brick or stone construction, infiltration normally is adequate to provide air for combustion, ventilation, and draft hood dilution.

(b) If the unconfined space is within a building of unusually tight construction, air for combustion, ventilation, and draft hood dilution shall be obtained from outdoors or from spaces freely communicating with the outdoors. Under these conditions a permanent opening or openings having a total free area of not less than one square inch per 5,000 Btu per hour of total input rating of all appliances shall be provided. Ducts used to convey make-up air from the outdoors shall be of the same cross-sectional area as the free area of the openings to which they connect. Such ducts connected to the outside air only may be connected to the cold air return of the heating system. The minimum dimension of rectangular air ducts shall be not less than 3 inches.

3.4.3 Appliances Located in Confined Spaces:

(a) *All Air from Inside Building:*

The confined space shall be provided with two permanent openings, one near the top of the enclosure and one near the bottom. Each opening shall have a free area of not less than one square inch per 1,000 Btu per hour of the total input rating of all appliances in the enclosure, freely communicating with interior areas having in turn adequate infiltration from the outside. (See Fig. 5.)

(b) *All Air from Outdoors:*

The confined space shall be provided with two permanent openings, one in or near the top of the enclosure and one in or near the bottom. The openings shall communicate directly, or by means of ducts, with outdoors or to such spaces (crawl or attic), that freely communicate with outdoors. (See Figs. 6, 7, and 8.)

When directly communicating with outdoors or by means of vertical ducts, each opening shall have a free area of not less than one square inch per 4,000 Btu per hour of total input rating of all appliances in the enclosure. If horizontal ducts are used, each opening shall have a free area of not less than one square inch per 2,000 Btu per hour of total input of all appliances in the enclosure.

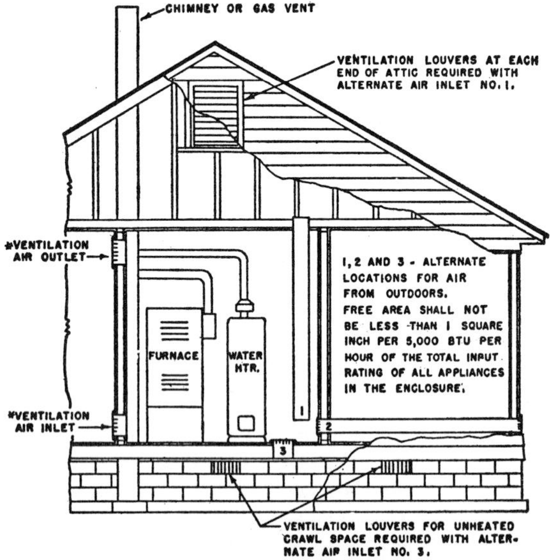

Fig. 9. Appliances Located in Confined Spaces. Ventilation Air from Inside Building – Combustion and Draft Hood Dilution Air from Outside, Ventilated Attic or Ventilated Crawl Space. (See 3.4.3-c.)

Ducts shall be of the same cross-sectional area as the free area of the openings to which they connect. The minimum dimension of rectangular air ducts shall be not less than 3 inches.

(c) *Ventilation Air from Inside Building—Combustion and Draft Hood Dilution Air from Outdoors:*

The enclosure shall be provided with two openings for ventilation, located and sized as described in 3.4.3(a). In addition, there shall be one opening directly communicating with outdoors or to such spaces (crawl or attic) that freely communicate with outdoors. This opening shall have a free area of not less than one square inch per 5,000 Btu per hour of total input of all appliances in the enclosure. A duct used to convey make-up air shall be of the same cross-sectional area as the free area of the opening required. Such ducts connected directly to outdoor air only may be connected to the cold air return of the heating system. The minimum dimension of rectangular air ducts shall be not less than 3 inches. (See Fig. 9.)

3.4.4 Louvers and Grilles:

In calculating free area in 3.4.2 and 3.4.3, consideration shall be given to the blocking effect of louvers, grilles or screens protecting openings. Screens used shall not be smaller than ¼ inch mesh. If the free area through a design of louver or grille is known, it should be used in calculating the size opening required to provide the free area specified. If the design and free area is not known, it may be assumed that wood louvers will have 20–25 per cent free area and metal louvers and grilles will have 60–75 per cent free area.

3.4.5 Special Conditions Created by Mechanical Exhausting or Fireplaces:

Operation of exhaust fans, kitchen ventilation systems, clothes dryers, or fireplaces may create conditions requiring special attention to avoid unsatisfactory operation of installed gas appliances.

3.4.6 Specially Engineered Installations:

The size of combustion air openings specified in 3.4.2 and 3.4.3 shall not necessarily govern when special engineering assures an adequate supply of air for combustion, ventilation, and draft hood dilution.

* * *

3.6 ELECTRICAL CONNECTIONS

3.6.1 Electrical Connections:

Electrical connections between gas appliances and the building wiring shall conform to the National Electrical Code, ASACl–1962 (NFPA No. 70)*

3.6.2 Electric Ignition and Control Devices:

No devices employing or depending upon an electrical current shall be used to control or ignite a gas supply if of such a character that failure of the electrical current could result in the escape of unburned gas or in failure to reduce the supply of gas under conditions which would normally result in its reduction unless other means are provided to prevent the development of dangerous temperatures, pressures or the escape of gas.

3.6.3 Electrical Circuit:

The electrical circuit employed for operating the automatic main gas-control valve, automatic pilot, room temperature thermostat, limit control or other electrical devices used with the gas appliance shall be in accordance with the wiring diagrams supplied with the appliance.

3.6.4 Continuous Power:

All gas appliances using electrical controls shall have the controls connected into a permanently live electric circuit, i.e., one

* Available from the National Fire Protection Association, 60 Batterymarch St., Boston, Mass. 02110 in pamphlet form and in the National Fire Codes, Volume 5. Also available from the American Standards Association, Inc., 10 East 40th St., New York, N. Y. 10016.

that is not controlled by a light switch. It is recommended that central heating gas appliances for domestic use be provided with a separate electrical circuit.

3.6.5 Transformers:

It is recommended that any separately mounted transformer necessary for the operation of the gas appliance be mounted on a junction box, and a switch with "On" and "Off" markings installed in the hot wire side of the transformer primary.

3.6.6 Wire Size:

It is recommended that multiple conductor cable, not lighter than No. 18 American Wire Gage, having type "T" (formerly type SN) insulation or equivalent be used on control circuits. Multiple conductor cables should be color coded to assist in correct wiring and to aid in tracing low-voltage circuits.

3.7 ROOM TEMPERATURE THERMOSTATS

3.7.1 Locations:

Room temperature thermostats should be located in the natural circulating path of room air. The device should not be placed so that it is exposed to cold air infiltration, drafts from outside openings such as windows and doors, air current from warm or cold air registers, or so that the natural circulation of the air is cut off such as behind doors, in shelves, or in corners.

Thermostats controlling floor furnaces shall not be located in a room or space which can be separated from the room or space in which the register of the floor furnace is located.

3.7.2 Exposure:

A room temperature thermostat should not be exposed to heat from nearby radiators, fireplaces, radios, television sets, lamps, rays of the sun, or mounted on a wall containing pipes or warm air ducts, or a chimney or gas vent, which would affect its operation and prevent it from properly controlling the room temperature.

3.7.3 Drafts:

Any hole in the plaster or panel through which the wires pass from the thermostat to the appliance being controlled shall be adequately sealed with suitable material to prevent drafts from affecting the thermostat.

Part 4—Installation Requirements for Specific Appliances

4.1 GENERAL

A listed appliance or accessory may be installed in accordance with its listing, or as elsewhere required in Part 4.

4.2 DOMESTIC RANGES

4.2.1 Clearance from Combustible Material:

(a) Listed domestic ranges, except as noted in 4.2.1(b) and 4.2.1(c), when installed on combustible floors shall be set on their own bases or legs and shall be installed with clearances of not less than that shown on the marking plate and the manufacturer's instructions. In the absence of clearance information on the marking plate, the range shall be installed with clearances of not less than that shown in Table 8. The clearance shall not interfere with requirements for combustion air, accessibility for operation, or servicing.

(b) Listed domestic ranges with listed gas room heater sections shall be installed so that the warm air discharge side shall have a minimum clearance of 18 inches between it and adjacent combustible material. A minimum clearance of 36 inches shall be provided between the top of the heater section and the bottom of cabinets. The minimum clearance between the back of the heater section and combustible material shall be in accordance with Table 10, Minimum Clearances for Listed Room Heaters.

Table 8
Minimum Clearances for Listed Domestic Ranges, Unless Otherwise Marked

		Distance from combustible material, inches			
		Sides		Rear	
Type of range	Spacing of center line of top burners from side of range	Wall not extending above cooking top	Wall extending above cooking top	Body of range	Projecting flue box
Insulated	Less than 10 in.	½	4½	1	1
Insulated	10 in. or more	½	½	1	1
Flush to wall	Less than 10 in.	Flush	4½	Flush	...
Flush to wall	10 in. or more	Flush	Flush	Flush	...

(c) Domestic ranges which include a solid or liquid fuel burning section shall be spaced from combustible material and otherwise installed in accordance with the standards applying to the supplementary fuel section of the range.

(d) Unlisted domestic ranges shall be installed with at least a 6-inch clearance at the back and sides to combustible material. Combustible floors under unlisted appliances shall be protected in an approved manner.*

4.2.2 Vertical Clearance above Cooking Top:

Domestic ranges shall have a vertical clearance above the cooking top of not less than 30 inches to combustible material or metal cabinets except the clearance may be reduced to not less than 24 inches as follows:

(a) The underside of the combustible material or metal cabinet above the cooking top is protected with asbestos millboard at least ¼-inch thick covered with sheet metal not lighter than No. 28 manufacturer's standard gage, or,

(b) A metal ventilating hood of not lighter than No. 28 manufacturer's standard gage sheet metal is installed above the cooking top with a clearance of not less than ¼ inch between the hood and the underside of the combustible material or metal cabinet and the hood is at least as wide as the range is and is centered over the range.

4.2.3 Install Level:

Ranges shall be installed so that the cooking top or oven racks are level.

4.3 BUILT-IN DOMESTIC COOKING UNITS

4.3.1 Installation:

Listed built-in domestic cooking units shall be installed in accordance with their listing and the manufacturer's instructions. Listed built-in domestic cooking units may be installed in combustible material unless otherwise marked.

The installation shall not interfere with the requirements for combustion air and accessibility for operation and servicing.

Unlisted built-in domestic cooking units shall not be installed in, or adjacent to, combustible material.

4.3.2 Vertical Clearance above Top Cooking Unit:

Built-in domestic top (or surface) cooking units shall have a vertical clearance above the cooking top of not less than 30 inches to combustible material or metal cabinets except the clearance may be reduced to not less than 24 inches as follows:

---

* For details of protection, refer to NBFU Code for the Installation of Heat Producing Appliances, available from the National Board of Fire Underwriters, 85 John St., New York, N. Y. 10038.

(a) The underside of the combustible material or metal cabinet above the cooking top is protected with asbestos millboard at least ¼-inch thick covered with sheet metal not lighter than No. 28 manufacturer's standard gage, or,

(b) A metal ventilating hood of not lighter than No. 28 manufacturer's standard gage sheet metal is installed above the cooking top with a clearance of not less than ¼ inch between the hood and the underside of the combustible material or metal cabinet and the hood is at least as wide as the unit is and is centered over the unit.

4.3.3 Horizontal Clearance of Listed Top Cooking Units from Walls Extending above Top Panel:

The minimum horizontal distance from the center of the burner head(s) of a top (or surface) cooking unit to vertical combustible walls extending above the top panel shall be not less than that distance specified by the permanent marking on the unit.

4.3.4 Install Level:

Built-in cooking units shall be installed so that the cooking top, broiler pan, or oven racks are level.

4.4 OPEN TOP BROILER UNITS

4.4.1 Listed Units:

Listed open top broiler units shall be installed in accordance with their listing and the manufacturer's instructions.

4.4.2 Unlisted Units:

Unlisted open top broiler units shall be installed in accordance with the manufacturer's instructions, but shall not be installed in combustible material.

4.4.3 Protection above Domestic Units:

Domestic open top broiler units shall be provided with a metal ventilating hood of not lighter than No. 28 manufacturer's standard gage with a clearance of not less than ¼ inch between the hood and the underside of combustible material or metal cabinets. A minimum clearance of 24 inches shall be maintained between the cooking top and the combustible material or metal cabinet and the hood shall be at least as wide as the open top broiler unit is and be centered over the unit.

4.4.4 Commercial Units:

Commercial open top broiler units shall be provided with ventilation in accordance with "Ventilation of Restaurant Cooking Equipment," NFPA No. 96-1964.*

4.5 WATER HEATERS

4.5.1 Prohibited Installations:

Water heaters, with the exception of those having sealed combustion systems, shall not be installed in bathrooms, bedrooms, or any occupied rooms normally kept closed.

Single-faucet automatic instantaneous water heaters, as permitted under 5.1.2† in addition to the above, shall not be installed in kitchen sections of light housekeeping rooms or rooms used by transients.

4.5.2 Location:

Water heaters shall be located as close as practicable to the chimney or gas vent. They should be located so as to provide short runs of piping to fixtures.

4.5.3 Clearance:

(a) Listed water heaters shall be installed in accordance with their listing and the manufacturer's instructions. In no case shall the clearances be such as to interfere with the requirements for combustion air, draft hood clearance and relief, and accessibility for servicing. (See Table 9.)

* Available from the National Fire Protection Association, 60 Batterymarch Street, Boston, Mass. 02110.

[† Given in Chap. 3 of this Section.—EDITOR]

Table 9
Minimum Clearances for Listed Water Heaters

Type of heater	Distance from combustible material, inches: Nearest part of jacket	Flat side
Type A	6	...
Type B	2	...
Type C	...	Flush
Counter type unit	In accordance with manufacturer's instructions.	

Type A—Miscellaneous (including circulating tank, instantaneous, uninsulated, underfired).

Type B—Underfired, insulated automatic storage heaters.

Type C—Type B units with one or more flat sides and tested for installation flush to wall.

Counter type—Type B units specifically designed for installation in or beneath a counter.

(b) Unlisted water heaters shall be installed with a clearance of 12 inches on all sides and rear. Combustible floors under unlisted water heaters shall be protected in an approved manner.‡

4.5.4 Connections:

Water heaters shall be connected in a manner to permit observation, maintenance, and servicing.

4.5.5 Closed Systems:

No water heater shall be installed in a closed system of water piping unless an approved water pressure relief valve is provided.

4.5.6 Temperature Limiting Devices:

An automatic storage type water heater or a hot water storage vessel shall be installed with an automatic gas shutoff system, or a temperature relief valve, or a combination temperature and pressure relief valve.

4.5.7 Temperature, Pressure and Vacuum Relief Valves:

The installation and adjustment of temperature, pressure, and vacuum relief valves or combinations thereof, and automatic gas shutoff valves or devices shall be in accordance with the requirements of the authority having jurisdiction, or, with the manufacturer's instructions accompanying such devices.

4.5.8 Automatic (Instantaneous or Storage) Types:

(a) *Independent Gas Piping:* Gas piping shall be separate and direct from the meter, or service regulator when a meter is not provided, to the appliance unless gas piping of ample capacity exists or is installed (see 2.4§).

4.5.9 Automatic Instantaneous Type:

(a) *Cold Water Supply:* The water supply to any automatic instantaneous water heater shall be such as to provide sufficient pressure to properly operate the water actuated control valve, when drawing hot water from a faucet on the top floor.

4.5.10 Circulating or Tank Types:

(a) *Connection to Boiler or Tank:* The method of connecting the circulating water heater to the tank shall assure proper circulation of water through the heater, and permit a safe and useful temperature of water to be drawn from the tank [see Fig. 12-77].

‡ For details of protection refer to the NBFU Code for the Installation of Heat Producing Appliances, available from the National Board of Fire Underwriters, 85 John Street, New York, N. Y. 10038.

[§ Not given in this Handbook but subject is covered in Chap. 2 of this Section.—EDITOR]

(b) *Size of Water Circulating Piping:* The size of the water circulating piping, in general, shall conform with the size of the water connections of the heater.

(c) *Sediment Drain:* A suitable water valve or cock, through which sediment may be drawn off or the tank emptied, shall be installed at the bottom of the tank.

(d) *Anti-Siphoning Devices:* Means acceptable to the authority having jurisdiction shall be provided to prevent siphoning in any boiler or tank to which any circulating water heater is attached. A cold water tube with a hole near the top is commonly accepted for this purpose [see Fig. 12-77].

4.6 ROOM HEATERS

4.6.1 Installations in Sleeping Quarters:

Room heaters installed in sleeping quarters for use of transients, as in hotels, motels and auto courts, shall be of the vented type and shall be connected to an effective chimney or gas vent and equipped with an automatic pilot. It is recommended that room heaters installed in all sleeping quarters or rooms generally kept closed be of the vented type and be connected to an effective chimney or gas vent and equipped with an automatic pilot.

4.6.2 Installations in Institutions:

Room heaters installed at any location in institutions such as Homes for the Aged, Sanitariums, Convalescent Homes, Orphanages, etc., shall be of the vented type and shall be connected to an effective chimney or gas vent and equipped with an automatic pilot.

Table 10
Minimum Clearances for Listed Room Heaters

Types of appliance	Distance from combustible material, inches	
	Jacket, sides and rear	Projecting flue box or draft hood
Warm air circulators	6	2
Radiant heaters	6	2
Wall heaters	Flush	...

4.6.3. Clearance:

A room heater shall be placed so as not to cause a hazard to walls, floors, curtains, furniture, doors when open, etc., and to the free movements of persons within the room. Appliances designed and marked "For use in noncombustible fire-resistive fireplace only," shall not be installed elsewhere. Listed room heaters shall be installed with clearances not less than specified in Table 10, except that appliances listed for installation at lesser clearances may be installed in accordance with their listings. In no case shall the clearances be such as to interfere with the requirements of combustion air and accessibility. (See 3.3.1 and 3.4.)

Table 11
Clearances to Combustible Material for Furnaces and Boilers Installed in Rooms Which Are Large in Comparison with Size of Appliance, except As Provided in 4.7.3 (a) (See Note 9)

	Minimum clearance, inches				
	Above and sides of bonnet or plenum	Jacket sides and rear	Front (see Note 1)	Projecting flue box or draft hood	Vent connector (see Note 2)
I. Listed automatically fired, forced air or gravity system, with 250 F temperature limit control.	2 (see Notes 3 and 4)	6	18	6	6
II. Unlisted automatically fired, forced air or gravity system, equipped with temperature limit control which cannot be set higher than 250 F.	6 (see Note 5)	6	18	18 (see Note 6)	18 (see Note 6)
III. Listed automatically fired heating boilers—steam boilers operating at not over 15 psi gage pressure and hot water boilers operating at not in excess of 250 F.	6 (see Note 7)	6	18	6	6
IV. Unlisted automatically fired heating boilers—steam boilers operating at not over 15 psi gage pressure and hot water boilers operating at not in excess of 250 F.	6 (see Note 7)	6	18	18 (see Note 6)	18 (see Note 6)
V. Central heating boilers and furnaces, other than above.	18 (see Note 8)	18	18	18 (see Note 6)	18 (see Note 6)

Notes Applicable to Table 11

1. Front clearance shall be sufficient for servicing the burner and furnace or boiler.
2. The vent connector clearance does not apply to listed Type B gas vents.
3. This clearance may be reduced to 1 inch for a listed forced air or gravity furnace equipped with:
   a. A limit control that cannot be set higher than 200 F, or
   b. A marking to indicate that the outlet air temperature cannot exceed 200 F.
4. Clearance from supply ducts within 3 feet of the plenum shall not be less than that specified from the bonnet or plenum. No clearance is required beyond this distance.
5. Clearance from supply ducts within 6 feet of the plenum shall not be less than 6 inches. No clearance is required beyond this distance.
6. For unlisted gas appliances equipped with an approved draft hood, this clearance may be reduced to 9 inches.
7. This clearance is above top of boiler.
8. Clearance from supply ducts shall not be less than 18 inches out to 3 feet from the bonnet or plenum, not less than 6 inches from 3 feet to 6 feet, and not less than 1 inch beyond 6 feet.
9. Rooms which are large in comparison with the size of the appliance are rooms having a volume equal to at least 12 times the total volume of a furnace and at least 16 times the total volume of a boiler. Total volume of furnace or boiler is determined from exterior dimensions and is to include fan compartments and burner vestibules, when used. When the actual ceiling height of a room is greater than 8 feet, the volume of a room shall be figured on the basis of a ceiling height of 8 feet.

Table 12
Clearances, Inches, with Specified Forms of Protection*

Type of protection Applied to the combustible material unless otherwise specified and covering all surfaces within the distance specified as the required clearance with no protection. (See Fig. 11.) Thicknesses are minimum.	Where the required clearance with no protection is:											
	36 inches			18 inches			12 inches		9 inches	6 inches		
	Above	Sides & rear	Vent connector	Above	Sides & rear	Vent connector	Above	Sides & rear	Vent connector	Above	Sides & rear	Vent connector
(a) ¼ in. asbestos millboard spaced out 1″†	30	18	30	15	9	12	9	6	6	3	2	3
(b) 28 gage sheet metal on ¼″ asbestos millboard	24	18	24	12	9	12	9	6	4	3	2	2
(c) 28 gage sheet metal spaced out 1″†....................	18	12	18	9	6	9	6	4	4	2	2	2
(d) 28 gage sheet metal on ⅛″ asbestos millboard spaced out 1″†....................	18	12	18	9	6	9	6	4	4	2	2	2
(e) 1½″ asbestos cement covering on heating appliance....	18	12	36	9	6	18	6	4	9	2	1	6
(f) ¼″ asbestos millboard on 1″ mineral wool bats reinforced with wire mesh or equivalent.................	18	12	18	6	6	6	4	4	4	2	2	2
(g) 22 gage sheet metal on 1″ mineral wool bats reinforced with wire or equivalent........................	18	12	12	4	3	3	2	2	2	2	2	2
(h) ¼″ asbestos cement board or ¼″ asbestos millboard....	36	36	36	18	18	18	12	12	9	4	4	4
(i) ¼″ cellular asbestos........	36	36	36	18	18	18	12	12	9	3	3	3

* Except for the protection described in (e), all clearances shall be measured from the outer surface of the appliance to the combustible material, disregarding any intervening protection applied to the combustible material.

† Spacers shall be of noncombustible material.

Unlisted room heaters shall be installed with clearances from combustible material not less than the following:

(a) *Circulating Type.* Room heaters having an outer jacket surrounding the combustion chamber, arranged with openings at top and bottom so that air circulates between the inner and outer jacket, and without openings in the outer jacket to permit direct radiation, shall have clearance at sides and rear of not less than 12 inches.

(b) *Radiating Type.* Room heaters other than those described above as of circulating type shall have clearance at sides and rear of not less than 18 inches; except that heaters which make use of metal, asbestos or ceramic material to direct radiation to the front of the appliance shall have a clearance of 36 inches in front, and if constructed with a double back of metal or ceramic may be installed with a clearance of 18 inches at sides and 12 inches at rear. Combustible floors under unlisted room heaters shall be protected in an approved manner.*

4.6.4 Wall Type Room Heaters:

Wall type room heaters shall not be installed in or attached to walls of combustible material unless listed for such installation.

4.6.5 Connection:

The provisions of 3.5, Appliance Connections to Building Piping, shall be observed. [See page 12/35.]

4.7 CENTRAL HEATING BOILERS AND FURNACES

4.7.1 Independent Gas Piping:

Gas piping shall be separate and direct from the meter, or service regulator when a meter is not provided, to the boiler or furnace, unless gas piping of ample capacity exists or is installed (see 2.4†).

4.7.2 Manual Main Shutoff Valves:

When a complete shutoff type automatic pilot system is not utilized, a manual main shutoff valve shall be provided ahead of all controls except the manual pilot gas valve.

When a complete shutoff type automatic pilot system is utilized, a manual main shutoff valve shall be provided ahead of all controls.

A union connection shall be provided downstream from the manual main shutoff valve to permit removal of the controls.

4.7.3 Clearance:

(a) Central heating boilers and furnaces installed in rooms which are large in comparison with the size of the appliance, shall be installed with clearances not less than specified in Table 11 except as provided in 4.7.3(a) 1, 2, and 3.

1. Central heating furnaces and boilers listed for installation at lesser clearances than specified in Table 11 may be installed in accordance with their listing and the manufacturer's instructions.

2. Central heating furnaces and boilers listed for installation at greater clearances than specified in Table 11, shall be installed in accordance with their listing and the manufacturer's instructions unless protected as specified in 4.7.3(a) 3.

3. Central heating furnaces and boilers may be installed in rooms, but not in confined spaces such as alcoves and closets, with reduced clearances to combustible material provided the combustible material or the appliance is protected as described in Table 12.

(b) Central heating furnaces and boilers shall not be installed in confined spaces such as alcoves and closets unless they have

* For details of protection, refer to the NBFU Code for the Installation of Heat Producing Appliances, available from the National Board of Fire Underwriters, 85 John St., New York, N. Y. 10038.

[† Not given in this Handbook but subject is covered in Chap. 2 of this Section.—EDITOR]

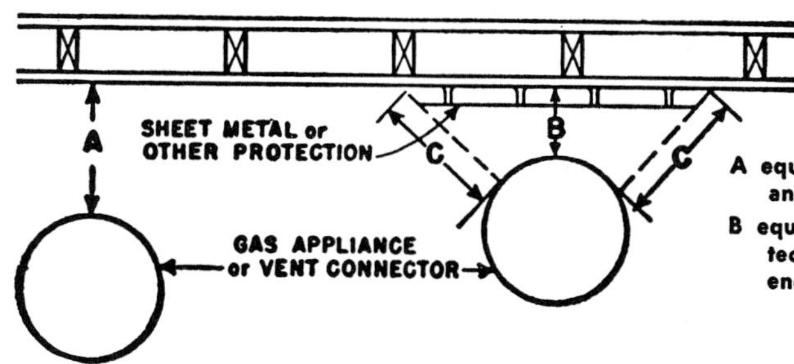

CONSTRUCTION USING COMBUSTIBLE MATERIAL, PLASTERED OR UNPLASTERED

A equals the required clearance with no protection specified in Tables 11 and 13 and in the sections applying to various types of appliances.

B equals the reduced clearance permitted in accordance with Table 12. The protection applied to the construction using combustible material shall extend far enough in each direction to make C equal to A.

Fig. 11. Extent of Protection Required to Reduce Clearances From Gas Appliances or Vent Connectors.

[Table 13 given in Chap. 3 of this Section.—Editor]

been specifically listed for such installation and are installed in accordance with their listing. The installation clearances for furnaces and boilers in confined spaces shall not be reduced by the protection methods described in Table 12.

When the plenum is adjacent to plaster on metal lath or noncombustible material attached to combustible material the clearance shall be measured to the surface of the plaster or other noncombustible finish when the clearance specified is 2 inches or less.

The clearance to these appliances shall not interfere with the requirements for combustion air, draft hood clearance and relief, and accessibility for servicing. (See 3.3.1 and 3.4.)

4.7.4 Erection and Mounting:

A central heating boiler or furnace shall be erected in accordance with the manufacturer's instructions and shall be installed on a floor of fire-resistive construction with noncombustible flooring and surface finish and with no combustible material against the underside thereof or on fire-resistive slabs or arches having no combustible material against the underside thereof unless listed for installation on a combustible floor, or the floor is protected in an approved manner.*

4.7.5 Connection of Flow and Return Piping:

The method of connecting the flow and return piping on steam and hot water boilers shall be in accordance with the manufacturer's recommendations to facilitate a positive, balanced and unobstructed flow of water or steam through the boiler. The direction of flow through the boiler shall be established by use of normal return and flow connections.†

4.7.6 Feed Water and Drain Connections:

Steam and hot water boilers shall be provided with means of introducing feed or make-up water from a water supply through an individual control valve and connection to the boiler piping system. A drain valve shall also be provided and connected with the lowest water space practicable for the purpose of draining or flushing the boiler.

4.7.7 Temperature or Pressure Limiting Devices:

Steam and hot water boilers respectively shall be provided with approved automatic limiting devices for shutting down the burner(s) to prevent boiler steam pressure or boiler water temperature from exceeding the maximum allowable working pressure or temperature.

4.7.8 Low Water Cutoff:

Steam boilers shall be provided with an automatic low water fuel cutoff for shutting down the burner in the event that the boiler water level drops to the lowest safe water line.

4.7.9 Steam Safety and Pressure Relief Valves:

[Steam boilers shall be equipped with listed steam safety valves and hot water boilers shall be equipped with listed pressure relief valves] of appropriate discharge capacity and conforming with ASME requirements.‡ Steam safety valves and pressure relief valves shall be set to discharge at a pressure not to exceed the maximum allowable working pressure of the boiler.

4.7.10 Plenum Chambers and Air Ducts:

(a) A plenum chamber supplied as a part of a furnace shall be installed in accordance with the manufacturer's instructions.

(b) When a plenum chamber is not supplied with the furnace, any fabrication and installation instructions provided by the manufacturer shall be followed. The method of connecting supply and return ducts shall facilitate proper circulation of air.§

(c) When the furnace is installed within a confined space, the air circulated by the furnace shall be handled by ducts which are sealed to the furnace casing and are entirely separate from the means provided for supplying combustion and ventilation air.

4.7.11 Refrigeration Coils:

(a) A refrigeration coil shall not be installed in conjunction with a forced air furnace when circulation of cooled air is provided by the furnace blower unless the blower has sufficient capacity to overcome the external static resistance imposed by the duct system and cooling coil at the air throughput required for heating or cooling, whichever is greater.

(b) Furnaces shall not be located upstream from cooling units unless the cooling unit is designed or equipped so as not to develop excessive temperature or pressure.

(c) Refrigeration coils shall be installed in parallel with or on the downstream side of central furnaces to avoid condensation in the heating element unless the furnace has been specifically listed for downstream installation. With a parallel flow arrangement, the dampers or other means used to control flow of air shall

* For details of protection refer to the NBFU Code for the Installation of Heat Producing Appliances, available from the National Board of Fire Underwriters, 85 John St., New York, N. Y. 10038.

† For common piping systems reference may be made to the American Society of Heating, Refrigerating and Air Conditioning Engineers Guide, available from The American Society of Heating, Refrigerating and Air Conditioning Engineers, 345 East 47th Street, New York, N. Y. 10017, and the Institute of Boiler and Radiator Manufacturers Installation Guides, available from the Institute of Boiler and Radiator Manufacturers, 608 Fifth Avenue, New York, N. Y. 10020.

‡ For details of requirements on low pressure heating boiler safety devices refer to ASME Boiler and Pressure Vessel Code Low Pressure Heating Boilers, Section IV, available from The American Society of Mechanical Engineers, United Engineering Center, 345 East 47th Street, New York, N. Y. 10017.

§ Reference may be made to the Standard for the Installation of Air Conditioning and Ventilating Systems of Other than Residence Type, NFPA No. 90A, Standard for the Installation of Residence Type Warm Air Heating and Air Conditioning Systems, NFPA No. 90B, available from the National Fire Protection Association, 60 Batterymarch St., Boston, Mass. 02110, and to the Design and Installation Manuals of the National Warm Air Heating and Air Conditioning Association, available from the National Warm Air Heating and Air Conditioning Association, 640 Engineers Bldg., Cleveland, Ohio 44114.

be sufficiently tight to prevent any circulation of cooled air through the furnace.

(d) Adequate means shall be provided for disposal of condensate and to prevent dripping of condensate on the heating element.

4.7.12 Cooling Units Used with Heating Boilers:

(a) Boilers, when used in conjunction with refrigeration systems, shall be installed so that the chilled medium is piped in parallel with the heating boiler with appropriate valves to prevent the chilled medium from entering the heating boiler.

(b) When hot water heating boilers are connected to heating coils located in air handling units where they may be exposed to refrigerated air circulation, such boiler piping systems shall be equipped with flow control valves or other automatic means to prevent gravity circulation of the boiler water during the cooling cycle.

4.8 WALL FURNACES

4.8.1 Installation:

(a) Listed wall furnaces shall be installed in accordance with their listing and the manufacturer's instructions. They may be installed in or attached to combustible material.

(b) Unlisted wall furnaces shall not be installed in or attached to combustible material.

(c) Vented wall furnaces connected to a Type BW gas vent system listed only for single story shall be installed only in single story buildings or the top story of multistory buildings. Vented wall furnaces connected to a Type BW gas vent system listed for installation in multistory buildings may be installed in single story or multistory buildings. Type BW gas vents shall be attached directly to a solid header plate which may be an integral part of the vented wall furnace, and which serves as a fire stop at that point. The stud space in which the vented wall furnace is installed shall be ventilated at the first ceiling level by installation of the ceiling plate spacers furnished with the gas vent. Fire stop spacers shall be installed at each subsequent ceiling or floor level penetrated by the vent. (See Fig. 12 for Type BW gas vent installation requirements.)

(d) Sealed combustion system wall furnaces shall be installed with the vent-air intake terminal in the outside atmosphere. The thickness of the walls on which the appliance is mounted shall be within the range of wall thickness marked on the appliance and covered in the manufacturer's installation instructions.

(e) Panels, grilles and access doors which must be removed for normal servicing operations shall not be attached to the building.

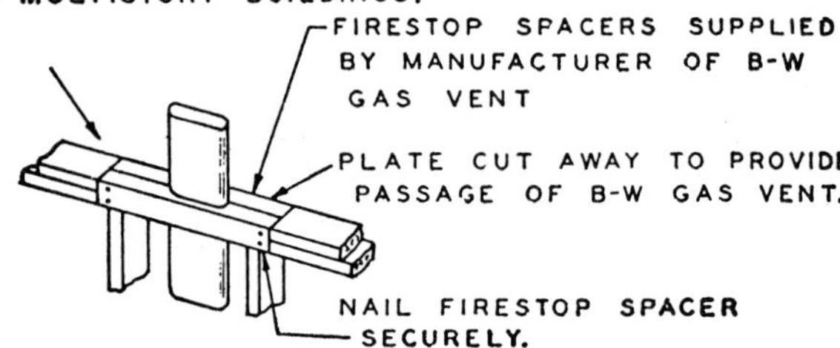
INSTALLATION OF B-W GAS VENT FOR EACH SUBSEQUENT CEILING OR FLOOR LEVEL OF MULTISTORY BUILDINGS.
FIRESTOP SPACERS SUPPLIED BY MANUFACTURER OF B-W GAS VENT
PLATE CUT AWAY TO PROVIDE PASSAGE OF B-W GAS VENT.
NAIL FIRESTOP SPACER SECURELY.

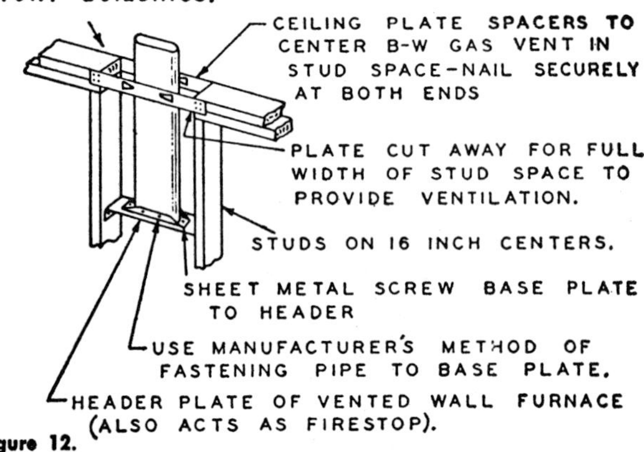
INSTALLATION OF B-W GAS VENT FOR ONE STORY BUILDINGS OR FOR FIRST FLOOR OF MULTI-STORY BUILDINGS.
CEILING PLATE SPACERS TO CENTER B-W GAS VENT IN STUD SPACE-NAIL SECURELY AT BOTH ENDS
PLATE CUT AWAY FOR FULL WIDTH OF STUD SPACE TO PROVIDE VENTILATION.
STUDS ON 16 INCH CENTERS.
SHEET METAL SCREW BASE PLATE TO HEADER
USE MANUFACTURER'S METHOD OF FASTENING PIPE TO BASE PLATE.
HEADER PLATE OF VENTED WALL FURNACE (ALSO ACTS AS FIRESTOP).

Figure 12. **Installation of Type B-W Gas Vents for Vented Wall Furnaces.**

4.8.2 Location:

Wall furnaces shall be located so as not to cause a hazard to walls, floors, curtains, furniture, or doors. Wall furnaces installed between bathrooms and adjoining rooms shall not circulate air from bathrooms to other parts of the building.

4.8.3 Manual Main Shutoff Valve:

A manual main shutoff valve shall be installed ahead of all controls including the pilot gas valve.

4.8.4 Combustion and Circulating Air:

Adequate combustion and circulating air shall be provided (see 3.4).

4.9 FLOOR FURNACES

4.9.1 Installation:

Listed floor furnaces may be installed in combustible floors. Unlisted floor furnaces shall not be installed in combustible floors.

4.9.2 Manual Main Shutoff Valve:

A separate manual main shutoff valve shall be provided ahead of all controls and a union connection shall be provided downstream from this valve to permit removal of the controls or the floor furnace.

4.9.3 Combustion and Circulating Air:

Adequate combustion and circulating air shall be provided (see 3.4).

4.9.4 Placement:

The following requirements apply to furnaces to serve one story.

(a) Floor furnaces shall not be installed in the floor of any aisle or passageway of any auditorium, public hall, or place of assembly, or in an exitway from any such room or space.

(b) *Walls and Corners.* The grille of a floor furnace with a horizontal warm air outlet shall not be placed closer than 6 inches to the nearest wall. A distance of at least 15 inches from two adjoining sides of the floor grille to walls shall be provided to eliminate the necessity of occupants walking over the warm air discharge from grilles. Wall-register models shall not be placed closer than 6 inches to a corner.

(c) *Draperies.* The furnace shall be placed so that a door, drapery, or similar object cannot be nearer than 12 inches to any portion of the register of the furnace.

(d) *Central Location.* The furnace should be installed in a central location favoring slightly the sides exposed to the prevailing winter winds.

4.9.5 Bracing:

The space provided for the furnace shall be framed with doubled joists and with headers not lighter than the joists.

4.9.6 Support:

Means shall be provided to support the furnace when the floor grille is removed.

4.9.7 Clearance:

The lowest portion of the floor furnace shall have at least a 6-inch clearance from the general ground level, except that when the lower 6-inch portion of the floor furnace is sealed by the manufacturer to prevent entrance of water, the clearance may be reduced to not less than 2 inches. When these clearances are not present, the ground below and to the sides shall be excavated to form a "basin-like" pit under the furnace so that the required clearance is provided beneath the lowest portion of the furnace. A 12-inch clearance shall be provided on all sides except the control side, which shall have an 18-inch clearance.

4.9.8 Access:

The space in which any floor furnace is installed shall be accessible by an opening in the foundation not less than 24 by 18 inches or a trap door, not less than 24 by 24 inches in any cross section thereof, and a passageway not less than 24 by 18 inches in any cross section thereof. The serving gas supplier should be consulted with reference to the access facilities for servicing when it provides service.

4.9.9 Seepage Pan:

When the excavation exceeds 12 inches in depth or water seepage is likely to collect, a watertight copper pan, concrete pit, or other suitable material shall be used, unless adequate drainage is provided or the equipment is sealed by the manufacturer to meet this condition. A copper pan shall be made of not less than 16-ounce-per-square-foot sheet copper. The pan shall be anchored in place, so as to prevent floating, and the walls shall extend at least 4 inches above the ground level, with at least 6 inches clearance on all sides except the control side, which shall have at least 18 inches clearance.

4.9.10 Wind Protection:

Floor furnaces shall be protected, where necessary, against severe wind conditions.

4.9.11 Upper Floor Installations:

Listed floor furnaces may be installed in an upper floor provided the furnace assembly projects below into a utility room, closet, garage, or similar nonhabitable space. In such installations, the floor furnace shall be enclosed completely (entirely separated from the nonhabitable space) with means for air intake to meet the provisions of 3.4, with access for servicing, with minimum furnace clearances of 6 inches to all sides and bottom, and with the enclosure constructed of portland cement plaster on metal lath or material of equal fire resistance.

4.9.12 First Floor Installation:

Listed floor furnaces installed in the first or ground floors of buildings need not be enclosed unless the basements of these buildings have been converted to apartments or sleeping quarters, in which case the floor furnace shall be enclosed as specified for upper floor installations and shall project into a nonhabitable space.

4.10 DUCT FURNACES

4.10.1 Clearance:

(a) Listed duct furnaces shall be installed with clearances of at least 6 inches between adjacent walls, ceilings and floors of combustible material and the appliance projecting flue box or draft hood, except that duct furnaces listed for installation at lesser clearances may be installed in accordance with their listings. In no case shall the clearance be such as to interfere with the requirements for combustion air and accessibility. (See 3.3.1 and 3.4.)

(b) Unlisted duct furnaces shall be installed with clearances to combustible material in accordance with the requirements for unlisted furnaces and boilers, Table 11. Combustible floors under unlisted duct furnaces shall be protected in an approved manner.*

4.10.2 Erection of Appliance:

Duct furnaces shall be erected and firmly supported in accordance with the manufacturer's instructions.

* For details of protection refer to the NBFU Code for the Installation of Heat Producing Appliances, available from the National Board of Fire Underwriters, 85 John St., New York, N. Y. 10038.

4.10.3 Access Panels:

The ducts connected to duct furnaces shall have removable access panels on both the upstream and downstream sides of the furnace.

4.10.4 Location of Draft Hood and Controls:

The controls, combustion air inlet, and draft hoods for duct furnaces shall be located outside the ducts. The draft hood shall be located in the same enclosure from which combustion air is taken.

4.10.5 Circulating Air:

When a duct furnace is installed in a confined space, the air circulated by the furnace shall be handled by ducts which are sealed to the furnace casing and which separate the circulating air from the combustion and ventilation air.

4.10.6 Duct Furnaces Used with Refrigeration Systems:

(a) A duct furnace shall not be installed in conjunction with a refrigeration coil when circulation of cooled air is provided by the blower unless the blower has sufficient capacity to overcome the external static resistance imposed by the duct system, furnace and the cooling coil at the air throughput required for heating or cooling whichever is greater.

(b) To avoid condensation within heating elements, duct furnaces used in conjunction with cooling equipment shall be installed in parallel with or on the upstream side of cooling coils unless the duct furnace has been specifically listed for downstream installation. With a parallel flow arrangement, the dampers or other means used to control the flow of air shall be sufficiently tight to prevent any circulation of cooled air through the unit.

(c) When duct furnaces are to be located upstream from cooling units, the cooling unit shall be so designed or equipped as to not develop excessive temperatures or pressures.

(d) Duct furnaces may be installed downstream from evaporative coolers or air washers if the heating element is made of corrosion-resistant material. Stainless steel, ceramic-coated steel, or an aluminum-coated steel in which the bond between the steel and the aluminum is an iron-aluminum alloy, are considered to be corrosion-resistant. Air washers operating with chilled water which delivers air below the dew point of the ambient air at the appliance are considered as refrigeration systems.

4.10.7 Installation in Commercial Garages and Aircraft Hangars:

Duct furnaces installed in garages for more than three motor vehicles or in aircraft hangars shall be of a listed type and shall be installed in accordance with 3.1.7 and 3.1.8.

4.11 CONVERSION BURNERS

Installation of conversion burners shall conform to American Standard Installation of Domestic Gas Conversion Burners, Z21.8–1958.†

4.12 CONVERSION BURNERS FOR DOMESTIC RANGES

Installation of conversion burners in ranges originally designed to utilize solid or liquid fuels shall conform to American Standard Requirements for Installation of Gas Conversion Burners in Domestic Ranges, Z21.38–1957.†

4.13 UNIT HEATERS

4.13.1 Support:

Suspended type unit heaters shall be safely and adequately supported with due consideration given to their weight and vibration characteristics. Hangars and brackets shall be of noncombustible material.

† Available from American Gas Association, Inc., 605 Third Ave., New York, N. Y. 10016.

4.13.2 Clearance:

(a) *Suspended Type Unit Heaters.*

1. Listed unit heaters shall be installed with clearance from combustible material of not less than 18 inches at the sides, 12 inches at the bottom and 6 inches above the top when the unit heater has an internal draft hood or 1 inch above the top of the sloping side of a vertical draft hood.

2. Unit heaters listed for reduced clearances may be installed in accordance with the clearance marked on the unit which will require not less than 6 inches from the draft hood relief opening and 6 inches above an elbow attached directly to the draft hood outlet.

3. Unlisted unit heaters shall be installed with clearance to combustible material of not less than 18 inches.

4. Additional clearances required for servicing shall be in accordance with the manufacturer's recommendations contained in the installation instructions.

(b) *Floor Mounted Type Unit Heaters.*

1. Listed unit heaters shall be installed with clearance from combustible material at the back and one side only of not less than 6 inches. When the flue gases are vented horizontally, the 6 inch clearance shall be measured from the draft hood or vent instead of the rear wall of the unit heater.

2. Unit heaters listed for reduced clearances may be installed at the clearances marked on the unit from the back, two side walls and ceiling. Walls and ceiling will be required to have at least 6 inches clearance from the draft hood relief openings and the nearest point of the draft hood exterior to the unit.

3. Floor mounted type unit heaters may be installed on combustible floors if listed for such installation.

4. Combustible floors under unlisted floor mounted unit heaters shall be protected in an approved manner.*

5. Additional clearances required for servicing shall be in accordance with the manufacturer's recommendations contained in the installation instructions.

4.13.3 Combustion and Circulating Air:

Adequate combustion and circulating air shall be provided (see 3.4).

4.13.4 Ductwork:

A unit heater shall not be attached to a warm air duct system unless listed and marked for such installation.

4.13.5 Installation in Commercial Garages and Aircraft Hangars:

Unit heaters installed in garages for more than 3 motor vehicles or in aircraft hangars shall be of a listed type and shall be installed in accordance with 3.1.7 and 3.1.8.

4.14 INFRARED RADIANT HEATERS

4.14.1 Support:

Suspended type infrared radiant heaters shall be safely and adequately fixed in position independent of gas and electric supply lines. Hangers and brackets shall be of noncombustible material.

4.14.2 Clearance.

(a) Listed heaters shall be installed with clearances from combustible material of not less than shown on the marking plate and in the manufacturer's instructions.

(b) Unlisted heaters shall be installed in accordance with clearances from combustible material acceptable to the authority having jurisdiction.

---

* For details of protection refer to the NBFU Code for the Installation of Heat Producing Appliances, available from the National Board of Fire Underwriters, 85 John Street, New York, N. Y. 10038.

4.14.3 Combustion and Ventilating Air:

Adequate combustion and ventilating air shall be provided.

4.14.4 Installation in Commercial Garages and Aircraft Hangars:

Overhead heaters installed in garages for more than 3 motor vehicles or in aircraft hangars shall be of a listed type and shall be installed in accordance with 3.1.7 and 3.1.8.

4.15 CLOTHES DRYERS

4.15.1 Clearance:

(a) Listed Type 1 clothes dryers shall be installed with minimum clearance of 6 inches from adjacent combustible material except that clothes dryers listed for installation at lesser clearances may be installed in accordance with their listing.

(b) Listed Type 2 clothes dryers shall be installed with clearances of not less than shown on the marking plate and in the manufacturer's instructions. Type 2 clothes dryers designed and marked "For use only in fire-resistive locations" shall not be installed elsewhere.

(c) Unlisted clothes dryers shall be installed with clearances to combustible material of not less than 18 inches. Combustible floors under unlisted clothes dryers shall be protected in an approved manner.*

4.15.2 Exhausting to the Outside Air:

(a) Type 1 clothes dryers should not be installed in bathrooms or bedrooms unless exhausted to the outside air.

(b) Type 2 clothes dryers shall be exhausted to the outside air

4.15.3 Provisions for Make-up Air:

(a) When a Type 1 clothes dryer is exhausted to the outside, consideration shall be given to provision for make-up air. (See 3.4.5.)

(b) Provision for make-up air shall be provided for Type 2 clothes dryers, with a minimum free area (see 3.4.4) of one square inch for each 1,000 Btu per hour total input rating of the dryer(s) installed.

4.15.4 Exhaust Ducts for Type 1 Clothes Dryers:

(a) A clothes dryer exhaust duct shall not be connected into any vent connector, gas vent or chimney.

(b) Ducts for exhausting clothes dryers shall not be put together with sheet metal screws or other fastening means which extend into the duct and which would catch lint and reduce the efficiency of the exhaust system.

4.15.5 Exhaust Ducts for Type 2 Clothes Dryers:

(a) Exhaust ducts for Type 2 clothes dryers shall comply with 4.15.4.

(b) Exhaust ducts for Type 2 clothes dryers shall be constructed of sheet metal or other noncombustible material. Such ducts shall be of adequate strength to meet the conditions of service with minimum thicknesses equivalent to No. 22 galvanized sheet gage.

(c) Exhaust ducts for Type 2 clothes dryers shall have a clearance of at least 6 inches to combustible material except as provided in 4.15.5 (d).

(d) Exhaust ducts for Type 2 clothes dryers may be installed with reduced clearances to combustible material provided the combustible material is protected as described in Table 12.

(e) When ducts pass through walls, floors or partitions, the space around the duct shall be sealed with noncombustible material.

(f) Multiple installations of Type 2 clothes dryers shall be made in a manner to prevent adverse operation due to back pressures that might be created in the exhaust systems.

4.15.6 Multiple Family or Public Use:

Clothes dryers installed for multiple family or public use shall be equipped with approved automatic pilots.

4.16 INCINERATORS

4.16.1 Clearance:

(a) Listed incinerators shall be installed in accordance with their listing and the manufacturer's instructions, provided that in any case the clearance shall be sufficient to afford ready accessibility for firing, clean-out and necessary servicing.

(b) The clearances above a charging door to combustible material shall be not less than 48 inches. The clearance may be reduced to 24 inches provided the combustible material is protected with sheet metal not less than No. 28 manufacturer's standard gage spaced out 1 inch on noncombustible spacers, or equivalent protection. Such protection shall extend 18 inches beyond all sides of the charging door opening. Listed incinerators designed to retain the flame during loading need not comply with this paragraph.

(c) Unlisted incinerators shall be installed with clearances to combustible material of not less than 36 inches at the sides and top and not less than 48 inches at the front, but in no case shall the clearance above a charging door be less than 48 inches. Unlisted wall mounted incinerators shall be installed on a noncombustible wall communicating directly with a chimney.

(d) Domestic type incinerators may be installed with reduced clearances to combustible material in rooms, provided the combustible material is protected as described in Table 12. In confined spaces, such as alcoves, clearances shall not be so reduced.

(e) When a domestic type incinerator that is refractory lined or insulated with heat insulating material is encased in common brick not less than 4 inches in thickness, the clearances may be reduced to 6 inches at the sides and rear, and the clearance at the top may be reduced to 24 inches provided that the construction using combustible material above the charging door and within 48 inches is protected with No. 28 manufacturer's standard gage sheet metal spaced out 1 inch, or equivalent protection.

4.16.2 Mounting:

(a) Listed incinerators specifically listed for installation on combustible floors may be so installed.

(b) Unlisted incinerators, except as provided in 4.16.2(c) and 4.16.2(d), shall be mounted on the ground or on floors of fire-resistive construction with noncombustible flooring or surface finish and with no combustible material against the underside thereof, or on fire-resistive slabs or arches having no combustible material against the underside thereof. Such construction shall extend not less than 12 inches beyond the incinerator base on all sides except at the front or side where ashes are removed where it shall extend not less than 18 inches beyond the incinerator.

(c) Unlisted incinerators may be mounted on floors other than as specified in 4.16.2(b), provided the incinerator is so arranged that flame or hot gases do not come in contact with its base and, further, provided the floor under the incinerator is protected with hollow masonry not less than 4 inches thickness, covered with sheet metal of not less than No. 24 manufacturer's standard gage. Such masonry course shall be laid with ends unsealed and joints matched in such a way as to provide a free circulation of air from side to side through the masonry. The floor for 18 inches beyond the front of the incinerator or side where ashes are removed and 12 inches beyond all other sides of the incinerator shall be protected with not less than ¼ inch asbestos millboard covered with sheet metal of not less than No. 24 manufacturer's standard gage or with protection equivalent thereto.

(d) Unlisted incinerators which are set on legs that provide not less than 4 inches open space under the base of the appliance may be mounted on floors other than as specified in paragraph 4.16.2(b), provided the appliance is such that flame or hot gases do not come in contact with its base, and further provided the floor under the appliance is protected with asbestos millboard not less than ¼ inch thick covered with sheet metal of not less than No. 24 manufacturer's standard gage. The above specified floor protection shall extend not less than 18 inches beyond the front of the incinerator or side where ashes are removed and 12 inches beyond all other sides of the incinerator.

4.16.3 Draft Hood Prohibited:

Draft hoods shall not be installed in the vent connector of an incinerator.

4.16.4 Vent Connector Clearance:

Vent connectors shall have at least 18 inches clearance from combustible material and shall not pass through combustible walls unless guarded at the point of passage as specified in 5.8.14.*

4.16.5 Vent Connector Material:

The vent connector from an incinerator to a chimney shall be galvanized steel of a thickness at least No. 24 manufacturer's standard gage or of material having equivalent or superior heat and corrosion resistant properties, and the joints shall be secured by sheet metal screws.

4.17 REFRIGERATORS

4.17.1 Clearance:

Refrigerators shall be provided with adequate clearances for ventilation at the top and back. They shall be installed in accordance with the manufacturer's instructions. If such instructions are not available, at least 2 inches shall be provided between the back of the refrigerator and the wall and at least a 12-inch clearance above the top.

4.17.2 Venting or Ventilating Kits Approved for Use with a Refrigerator:

If an accessory kit is used for conveying air for burner combustion or unit cooling to the refrigerator from areas outside the room in which it is located, or for conveying combustion products diluted with air containing waste heat from the refrigerator to areas outside the room in which it is located, the kit shall be installed in accordance with the refrigerator manufacturer's instructions.

4.18 HOT PLATES AND LAUNDRY STOVES

(a) Listed domestic hot plates and laundry stoves installed on combustible surfaces shall be set on their own legs or bases. They shall be installed with minimum horizontal clearances of 6 inches from combustible material.

(b) Unlisted domestic hot plates and laundry stoves shall be nstalled with horizontal clearances to combustible material of not less than 12 inches. Combustible surfaces under unlisted domestic hot plates and laundry stoves shall be protected in an approved manner.†

(c) The vertical distance between tops of all domestic hot plates and laundry stoves and combustible material shall be at least 30 inches.

4.19 HOTEL AND RESTAURANT RANGES, DEEP FAT FRYERS AND UNIT BROILERS

4.19.1 Clearance for Listed Appliances:

Listed hotel and restaurant ranges, deep fat fryers and unit broilers shall be installed at least 6 inches from combustible material except that at least 2 inches clearance shall be maintained between the flue box or draft hood and combustible material. Hotel and restaurant ranges, deep fat fryers and unit broilers listed for installation at lesser clearances may be installed in accordance with their listing and the manufacturer's instructions. Appliances designed and marked "For use only in fire-resistive locations" shall not be installed elsewhere.

4.19.2 Clearance for Unlisted Appliances:

(a) Unlisted hotel and restaurant ranges, deep fat fryers and unit broilers, except as provided in 4.19.2(b) and (c) shall be in-

---

[* Given in Chap. 3 of this Section.—EDITOR]

† For details of protection refer to the NBFU Code for the Installation of Heat Producing Appliances, available from the National Board of Fire Underwriters, 85 John Street, New York, N. Y. 10038.

stalled to provide a clearance to combustible material of not less than 18 inches at the sides and rear of the appliance and from the vent connector and not less than 48 inches above the cooking top and at the front of the appliance.

(b) Unlisted hotel and restaurant ranges, deep fat fryers and unit broilers may be installed in rooms, but not in confined spaces such as alcoves with reduced clearances to combustible material, provided the combustible material or the appliance is protected as described in Table 12.

(c) Unlisted hotel and restaurant ranges, deep fat fryers and unit broilers may be installed in rooms, but not in confined spaces such as alcoves, with reduced clearance of 6 inches to combustible material, provided the wall or combustible material is protected by sheet metal of not less than No. 26 manufacturer's standard gage, fastened with noncombustible spacers that are spaced at not less than 2-foot vertical and horizontal intervals to provide a clearance of 1½ inches from such wall or material. Such protection shall extend at least 12 inches beyond the back, side, top or any other part of the appliance and the space between the sheet metal and wall or combustible material shall be open on both sides and top and bottom to permit circulation of air.

4.19.3 Mounting on Combustible Floor:

(a) Listed hotel and restaurant ranges, deep fat fryers and unit broilers listed specifically for installation on floors constructed of combustible material may be mounted on combustible floors.

(b) Listed floor mounted hotel and restaurant ranges, deep fat fryers and unit broilers that are designed and marked "For use only in fire-resistive locations" shall be mounted on floors of fire-resistive construction with noncombustible flooring and surface finish and with no combustible material against the underside thereof, or on fire-resistive slabs or arches having no combustible material against the underside thereof. Such construction shall in all cases extend not less than 12 inches beyond the appliance on all sides.

(c) Hotel and restaurant ranges, deep fat fryers, and unit broilers, which are not listed for mounting on a combustible floor shall be mounted in accordance with 4.19.3(b) or be mounted in accordance with one of the following:

1. When the appliance is set on legs which provide not less than 18 inches open space under the base of the appliance, or where it has no burners and no portion of any oven or broiler within 18 inches of the floor, it may be mounted on a combustible floor without special floor protection, provided there is at least one sheet metal baffle between the burner and the floor.

2. When the appliance is set on legs which provide not less than 8 inches open space under the base of the appliance, it may be mounted on combustible floors provided the floor under the appliance is protected with not less than ⅜ inch asbestos millboard covered with sheet metal of not less than No. 24 manufacturer's standard gage. The above specified floor protection shall extend not less than 6 inches beyond the appliance on all sides.

3. When the appliance is set on legs which provide not less than 4 inches under the base of the appliance, it may be mounted on combustible floors, provided the floor under the appliance is protected with hollow masonry not less than 4 inches in thickness covered with sheet metal of not less than No. 24 manufacturer's standard gage. Such masonry courses shall be laid with ends unsealed and joints matched in such a way as to provide for free circulation of air through the masonry.

4. When the appliance does not have legs at least 4 inches high, it may be mounted on combustible floors, provided the floor under the appliance is protected by two courses of 4-inch hollow clay tile or equivalent with courses laid at right angles and with ends unsealed and joints matched in such a way as to provide for free circulation of air through such masonry courses and covered with steel plate not less than $\frac{3}{16}$ inch in thickness.

4.19.4 Combustible Material Adjacent to Cooking Top:

Any portion of combustible material adjacent to a cooking top section of a hotel or restaurant range, even though certified for close-to-wall installation, which is not shielded from the wall by a high shelf, warming closet, etc., shall be protected as specified in 4.19.2 for a distance of at least 2 feet above the surface of the cooking top.

4.19.5 Install Level:

All hotel and restaurant ranges, deep fat fryers, and unit broilers shall be installed level on a firm foundation.

4.19.6 Ventilation:

Adequate means shall be provided to properly ventilate the space in which hotel and restaurant equipment is installed to permit proper combustion of the gas. When exhaust fans are used for ventilation, special precautions may be required to avoid interference with the operation of the equipment.

4.20 COUNTER APPLIANCES

4.20.1 Vertical Clearance:

A vertical distance of not less than 48 inches shall be provided between the top of all commercial hot plates and griddles and combustible material.

4.20.2 Clearance for Listed Appliances:

Listed counter appliances such as commercial hot plates and griddles, food and dish warmers, and coffee brewers and urns, when installed on combustible surfaces shall be on their own bases or legs, and shall be installed with a minimum horizontal clearance of 6 inches from combustible material except that at least a 2-inch clearance shall be maintained between the flue box or draft hood and combustible material. Counter appliances listed for installation at lesser clearances may be installed in accordance with their listing and the manufacturer's instructions.

4.20.3 Clearance for Unlisted Appliances:

Unlisted commercial hot plates and griddles shall be installed with a horizontal clearance from combustible material of not less than 18 inches. Unlisted gas counter appliances such as coffee brewers and urns, waffle bakers and hot water immersion sterilizers shall be installed with a horizontal clearance from combustible material of not less than 12 inches. Gas counter appliances may be installed with reduced clearances to combustible material provided the combustible material is protected as described in Table 12. Unlisted food and dish warmers shall be installed with a horizontal clearance from combustible material of not less than 6 inches.

4.20.4 Mounting of Unlisted Appliances:

Unlisted counter appliances shall not be set on combustible material unless they have legs which provide not less than 4 inches of open space below the burners, and the combustible surface is protected with asbestos millboard at least ¼ inch thick covered with sheet metal of not less than No. 28 manufacturer's standard gage or with equivalent protection.

4.21 PORTABLE BAKING AND ROASTING OVENS

4.21.1 Clearance for Listed Appliances:

Listed portable baking and roasting ovens shall be installed at least 6 inches from combustible material, except that at least a 2-inch clearance shall be maintained between the flue box or draft hood and combustible material. Portable baking and roasting ovens listed for installation at lesser clearances may be installed in accordance with their listing and the manufacturer's instructions. Appliances designed and marked "For use only in fire-resistive locations" shall not be installed elsewhere.

4.21.2 Mountings for Listed Appliances:

Portable baking and roasting ovens that are listed specifically for installation on a floor constructed of combustible material may be mounted in accordance with their listing.

4.21.3 Clearance for Unlisted Appliances:

(a) Unlisted portable baking and roasting ovens except as provided in 4.21.3(b) and (c) shall be installed to provide a clearance to combustible material of not less than 18 inches at the sides and rear of the appliance and from the vent connector and not less than 48 inches at the front of the appliance.

(b) Unlisted portable baking and roasting ovens may be installed in rooms, but not in confined spaces such as alcoves or closets, with reduced clearance to combustible material, provided the combustible material or the appliance is protected as described in Table 12.

(c) Unlisted portable baking and roasting ovens may be installed in rooms, but not in confined spaces such as alcoves, with reduced clearance of 6 inches to combustible material, provided the wall or combustible material is protected by sheet metal of not less than No. 26 manufacturer's standard gage, fastened with noncombustible spacers that are spaced at not less than 2-foot vertical and horizontal intervals to provide a clearance of 1½ inches from such wall or material. Such protection shall extend at least 12 inches beyond the back, side, top or other part of the appliance and the space between the sheet metal and wall or combustible material shall be open on both sides and top and bottom to permit circulation of air.

4.21.4 Mounting of Unlisted Appliances:

Unlisted portable baking and roasting ovens shall be mounted in an approved manner.*

4.22 AIR CONDITIONING APPLIANCES

4.22.1 Independent Gas Piping:

Gas piping shall be separate and direct from the meter, or service regulator when a meter is not provided, to the air conditioning appliance, unless gas piping of ample capacity exists or is installed. An existing line serving the heating appliance may be satisfactory when heating and cooling appliances are not operated simultaneously. (See 2.4.†)

4.22.2 Connection of Gas Engine–Powered Air Conditioners:

To protect against the effects of normal vibration in service, gas engines shall not be rigidly connected to the gas supply piping.

4.22.3 Manual Main Shutoff Valves:

When a complete shutoff type automatic pilot system is not utilized, a manual main shutoff valve shall be provided ahead of all controls except the manual pilot gas valve.

When a complete shutoff type automatic pilot system is utilized, a manual main shutoff valve shall be provided ahead of all controls.

A union connection shall be provided downstream from the manual main shutoff valve to permit removal of the controls.

4.22.4 Clearances for Indoor Installation:

(a) Listed air conditioning appliances installed in rooms which are large in comparison with the size of the appliance, shall be installed with clearances not less than specified in Line I of Table 11 except as provided in 4.22.4(a) 1, 2 and 3.

1. Air conditioning appliances listed for installation at lesser clearances than specified in Table 11 may be installed in accordance with their listing and the manufacturer's instructions.

2. Air conditioning appliances listed for installation at greater clearances than specified in Table 11 shall be installed in accordance with their listing and the manufacturer's instructions unless protected as specified in 4.22.4(a)3. However, when clearances are specified to provide access for service, they shall not be reduced.

3. Air conditioning appliances may be installed in rooms, but not in confined spaces such as alcoves and closets, with reduced clearances to combustible material provided the combustible material or the appliance is protected as described in Table 12. However, when clearances are necessary or specified to provide access for service, they shall not be reduced.

(b) Air conditioning appliances shall not be installed in confined spaces such as alcoves and closets unless they have been specifically listed for such installation and are installed in accordance with their listing. The installation clearances for air conditioning appliances in confined spaces shall not be reduced by the protection methods described in Table 12.

When the plenum for an air conditioner which includes provisions for heating air is adjacent to combustible material, the clearance shall be measured to the surface of the plaster or other noncombustible finish when the clearance specified is 2 inches or less.

The clearance to these appliances shall not interfere with the requirements for combustion air, draft hood clearance and relief, and accessibility for servicing. (See 3.3.1 and 3.4.)

(c) Unlisted air conditioning appliances shall be installed with clearances from combustible material of not less than 18 inches above the appliance and at sides, front and rear, and 9 inches from projecting flue box or draft hood.

4.22.5 Erection and Mounting:

An air conditioning appliance shall be erected in accordance with the manufacturer's instructions. Unless the appliance is listed for installation on a combustible surface such as a floor or roof, or the surface is protected in an approved manner,* it shall be installed on a surface of fire-resistive construction with noncombustible material and surface finish and with no combustible material against the underside thereof.

4.22.6 Connection of Flow and Return Piping:

The method of connecting the flow and return piping on air conditioning appliances which provide heated or chilled fluid shall be in accordance with the manufacturer's recommendations to facilitate a positive, balanced and unobstructed flow through the system.

4.22.7 Cooling Towers:

A cooling tower used in conjunction with an air conditioning appliance shall be installed in accordance with the manufacturer's installation instructions. The cooling tower shall be provided with a direct connection to a water supply through an individual control valve. A means by which the tower may be flushed or drained shall be provided.

4.22.8 Plenum Chambers and Air Ducts:

A plenum chamber supplied as a part of an air conditioning appliance shall be installed in accordance with the manufacturer's instructions. When a plenum chamber is not supplied with the appliance, any fabrication and installation instructions provided by the manufacturer shall be followed. The method of connecting supply and return ducts shall facilitate proper circulation of air.‡

When the air conditioner is installed within a confined space, the air circulated by the appliance shall be handled by ducts which are sealed to the casing of the appliance and which separate the circulating air from the combustion and ventilation air.

4.22.9 Refrigeration Coils:

(See 4.7.11 and 4.7.12.)

4.22.10 Switches in Electrical Supply Line:

Means for interrupting the electrical supply to the air conditioning appliance and to its associated cooling tower (if supplied

---

* For details of protection refer to the NBFU Code for the Installation of Heat Producing Appliances, available from the National Board of Fire Underwriters, 85 John Street, New York, N. Y. 10038.

[†Not given in this Handbook but subject is covered in Chap. 2 of this Section.—EDITOR]

‡ Reference may be made to the Standard for the Installation of Air Conditioning and Ventilating Systems of Other Than Residence Type, NFPA No. 90A, Standard for the Installation of Residence Type Warm Air Heating and Air Conditioning Systems, NFPA No. 90B, available from the National Fire Protection Association, 60 Batterymarch Street, Boston, Mass. 02110, and to the Design and Installation Manuals of the National Warm Air Heating and Air Conditioning Association, 640 Engineers Building, Cleveland, Ohio 44114.

and installed in a location remote from the air conditioner) shall be provided within sight of and not over 50 feet from the air conditioner and cooling tower.

4.23 ILLUMINATING APPLIANCES

4.23.1 Clearances:

Listed illuminating appliances shall be installed in accordance with their listings and the manufacturer's instructions.

Unlisted enclosed illuminating appliances installed outdoors shall be installed with clearances from combustible material of not less than 12 inches.

Unlisted enclosed illuminating appliances installed indoors shall be installed with clearances from combustible material of not less than 18 inches.

Unlisted open-flame illuminating appliances shall be installed only at locations and with clearances from combustible material acceptable to the authority having jurisdiction.

4.23.2 Mounting on Buildings:

Illuminating appliances designed for wall or ceiling mounting shall be securely attached to substantial structures in such a manner that they are not dependent on the gas piping for support.

4.23.3 Mounting on Posts:

Illuminating appliances designed for post mounting shall be securely and rigidly attached to a post.

Posts shall be rigidly mounted. The strength and rigidity of posts shall be at least equivalent to that of a 2½ inch diameter post constructed of 0.064 inch thick (No. 14 gage) steel.

Drain openings should be provided near the base of posts when there is a possibility of water collecting inside them.

4.23.4 Gas Pressure Regulators:

When a gas appliance pressure regulator is not supplied with an illuminating appliance and the service line is not equipped with a service pressure regulator, it is recommended that an appliance pressure regulator be installed in the line to the illuminating appliance. For multiple installations, one regulator of adequate capacity may be used to serve a number of illuminating appliances.

INDUSTRIAL APPLIANCES

General installation requirements for industrial equipment not covered by ASA Z21.30 or any class of equipment which operates at gas pressures above 0.5 psig are contained in ASA Z83.1.

For installation of specific equipment, reference should be made to the following standards:

NFPA No. 37	Combustion Engines and Gas Turbines
NFPA No. 82	Rubbish Handling Incinerators (Parts II and III)
NFPA No. 85A	Explosion Prevention in Natural Gas-Fired Watertube Boiler-Furnaces with One Burner
NFPA No. 86A	Ovens and Furnaces
NFPA No. 211	Chimneys, Fireplaces, and Venting Systems
NBFU	Code for the Installation of Heat Producing Appliances (Section 8)

Chapter 2

Piping in Buildings

by C. George Segeler, Tony Cramer, and K. R. Knapp

RESIDENTIAL AND COMMERCIAL BUILDINGS

Piping refers to pipe, tubing, and appliance connectors, together with necessary fittings, joints, and valves, which convey gas from the outlet side of the meter set assembly or the outlet of the service regulator (when there is no meter) to the inlet connection of appliances. Pipe refers to rigid pipe; tubing refers to semirigid tubing; and appliance connectors refer to A. G. A.-listed semirigid or flexible connectors.[1,8]

The *American Standard—Installation of Gas Appliances and Gas Piping,* ASA Z21.30[2] (also available as NFPA Standard No. 54),[3] has been incorporated into the four U. S. model building codes and into FHA requirements. This standard is also part of most U. S. and Canadian utility company specifications as well as over 3000 U. S. local building codes and ordinances. Standards covering gas piping in mobile homes are also available.[4]

The following extract from ASA Z21.30[2] covers its scope as well as some of the more significant points to be observed:

1.1 SCOPE

1.1.1 Applicability:

This standard applies to the design, fabrication, installation, tests and operation of appliance and piping systems for fuel gases such as natural gas, manufactured gas, undiluted liquefied petroleum gases, liquefied petroleum gas-air mixtures, or mixtures of any of these gases, as follows:

(a) Low pressure (not in excess of ½ pound per square inch or 14 inches water column) domestic and commercial piping systems extending from the outlet of the meter set assembly, or the outlet of the service regulator when a meter is not provided, to the inlet connections of appliances.

(b) The installation and operation of domestic and commercial appliances supplied at pressures of ½ pound per square inch or less.

1.1.2 Nonapplicability:

This standard does not apply to:

(a) Gas piping systems for industrial installations at any pressure or any other gas piping system operating at pressures greater than ½ pound per square inch. For piping in such installations refer to ASME Code for Pressure Piping, Section 2 of ASA B31.1–1955 and Addenda B31.1a–1961.

(b) Gas equipment supplied through piping systems covered in 1.1.2(a), and

(c) Gas equipment designed and installed for specific manufacturing, production, processing and power generating applications.

1.2 QUALIFIED INSTALLING AGENCY

Installation and replacement of gas piping or gas appliances and repair of gas appliances shall be performed only by a qualified installing agency. By the term "qualified installing agency" is meant any individual, firm, corporation or company which either in person or through a representative is engaged in and is responsible for the installation or replacement of gas piping on the outlet side of the meter or of the service regulator when a meter is not provided, or the connection, installation or repair of gas appliances, and who is experienced in such work, familiar with all precautions required, and has complied with all the requirements of the authority having jurisdiction.

* * *

APPENDIX A

Work on Gas Supply System

This appendix applies only to work on gas supply systems ahead of the outlet of the meter set assembly or the service regulator when a meter is not provided.

Serving Gas Supplier's Main

No person, unless in the employ of or having permission from the serving gas supplier, shall open or make connections with a gas main.

Service Gas Piping

No person, unless in the employ of or having permission from the serving gas supplier, shall repair, alter, open or make connections to the service gas piping, or do any other work on the parts of the gas supply system up to the meter set assembly, or the service regulator when a meter is not provided.

Meter or Service Regulator When a Meter Is Not Provided

No person, unless in the employ of or having permission from the serving gas supplier, shall disconnect the inlet of the gas meter, or service regulator when a meter is not provided, nor move such meter or regulator. A gas fitter or plumber may disconnect the outlet of such a meter or regulator from the house piping only when necessary. He shall remake the joint at the meter or service regulator outlet when a meter is not provided, carefully replacing all insulating fittings or insulating parts of such fittings, and shall leave the gas turned off at the meter or regulator unless the serving gas supplier's rules require or allow deviation from this procedure.

Notify Serving Gas Supplier of Any Repairs Needed

In case any work done by a gas fitter or plumber discloses the need for repairs or alterations on any part of the gas supply system, the serving gas supplier shall be notified promptly of this fact.

Table 12-1a Maximum Capacity of Pipe in Cubic Feet of Gas per Hour[2]

(based on a pressure drop of **0.3 in. w.c.** and 0.6 sp gr gas at pressures up to 0.5 psig in schedule-40 pipe)

IPS, nom., in.	Pipe length, ft													
	10	20	30	40	50	60	70	80	90	100	125	150	175	200
½	132	92	73	63	56	50	46	43	40	38	34	31	28	26
¾	278	190	152	130	115	105	96	90	84	79	72	64	59	55
1	520	350	285	245	215	195	180	170	160	150	130	120	110	100
1¼	1,050	730	590	500	440	400	370	350	320	305	275	250	225	210
1½	1,600	1,100	890	760	670	610	560	530	490	460	410	380	350	320
2	3,050	2,100	1650	1450	1270	1150	1050	990	930	870	780	710	650	610
2½	4,800	3,300	2700	2300	2000	1850	1700	1600	1500	1400	1250	1130	1050	980
3	8,500	5,900	4700	4100	3600	3250	3000	2800	2600	2500	2200	2000	1850	1700
4	17,500	12,000	9700	8300	7400	6800	6200	5800	5400	5100	4500	4100	3800	3500

Notify Serving Gas Supplier of Any Leaks

If gas is leaking from any part of the gas supply system, a gas fitter or plumber not in the employ of the serving gas supplier may make necessary repairs and shall promptly notify the serving gas supplier.

* * *

Design Considerations

The piping installation must supply the probable maximum gas demand without undue pressure loss. Depending on local practices, the allowable pressure loss between the meter and each appliance is generally selected at 0.3 or 0.5 in. w.c. A common operating pressure for residential and commercial piping is 7.0 in. w.c.

Maximum Flow Requirements. A reasonably close estimate can usually be made of the gas requirements of the appliances to be supplied by the piping system. If the make and model number of the individual appliances are known, their input ratings can be found on nameplates, in catalogs, or in the *A. G. A. Directory of Approved Appliances.* Input ratings for domestic appliances may be approximated from Table 9-24.

Diversity of loading should be considered when sizing pipe for other than individual residences; see Tables 9-25 thru 9-28 and Figs. 9-55a thru 9-57b in this regard.

Sizing Pipe or Tubing. To determine the size of each section of pipe or tubing in the system within the ranges specified in Tables 12-1a, 12-1b, 12-2, 12-4, or 12-5, proceed as follows:

1. Calculate the gas demand in cubic feet per hour at each piping system outlet (in MBtu per hour for undiluted LP-gases).

2. Measure the length of piping from the gas meter or service regulator (when a meter is not provided) to the most remote outlet in the building.

3. Using the appropriate table, select the column showing that distance or the next longer distance if the table does not give the exact length. This is the only distance used in determining the size of any section of gas piping. If the specific gravity factor is to be applied, the values in the selected column of Tables 12-1a, 12-1b, and 12-2 are multiplied by the appropriate factor from Table 12-3.

Note: Use the selected column to locate *all* gas demand figures for this particular system of piping.

4. Starting at the most remote outlet, find in the column the gas demand for that outlet. If the exact figure of demand is not shown, choose the next larger figure in the column.

5. Opposite this demand figure, in the first column at the left, read the correct size of gas piping.

6. Proceed in a similar manner for each outlet and each section of gas piping. For each section of piping, determine the *total* gas demand supplied by that section.

7. When it is anticipated that the gas supplied to the piping system will be changed from undiluted LP-gas to utility gas of a lower heating value, size the piping system to provide sufficient capacity for future use with that utility gas.

Supply piping should not be smaller than ½-in. pipe or ⅜-in. tubing. The capacity tables include an allowance for an ordinary number of fittings. In large, complex installations it may be desirable to add some "equivalent" length to compensate for fittings (Table 12-9).

A Two-Pound Domestic House Piping Specification. Although house systems at above 0.5 psig are not covered at present by any national codes, the Minneapolis Gas Com-

Table 12-1b Maximum Capacity of Pipe in Cubic Feet of Gas per Hour[2]

(based on a pressure drop of **0.5 in. w.c.** and 0.6 sp gr gas at pressures up to 0.5 psig in schedule-40 pipe)

IPS, nom., in.	Pipe length, ft													
	10	20	30	40	50	60	70	80	90	100	125	150	175	200
½	175	120	97	82	73	66	61	57	53	50	44	40	37	35
¾	360	250	200	170	151	138	125	118	110	103	93	84	77	72
1	680	465	375	320	285	260	240	220	205	195	175	160	145	135
1¼	1,400	950	770	660	580	530	490	460	430	400	360	325	300	280
1½	2,100	1,460	1,180	990	900	810	750	690	650	620	550	500	460	430
2	3,950	2,750	2,200	1,900	1680	1520	1400	1300	1220	1150	1020	950	850	800
2½	6,300	4,350	3,520	3,000	2650	2400	2250	2050	1950	1850	1650	1500	1370	1280
3	11,000	7,700	6,250	5,300	4750	4300	3900	3700	3450	3250	2950	2650	2450	2280
4	23,000	15,800	12,800	10,900	9700	8800	8100	7500	7200	6700	6000	5500	5000	4600

Table 12-2 Maximum Capacity of Semirigid Tubing in Cubic Feet of Gas per Hour[2]

(based on a pressure drop of **0.3 in. w.c.** and 0.6 sp gr gas at pressures up to 0.5 psig)

OD,	Tubing length, ft									
in.	10	20	30	40	50	60	70	80	90	100
⅜	19	12	10	9	...	...	...	...	...	...
½	45	30	24	20	18	17	15	14	13	12
⅝	97	64	52	44	38	35	32	30	28	26
¾	161	105	88	71	64	59	54	50	46	44
⅞	245	169	135	114	97	91	80	75	71	67

Table 12-3 Specific Gravity Factors for Pipe and Tubing Sizing[2]

(use with Tables 12-1a, 12-1b, and 12-2)

Sp gr	Factor	Sp gr	Factor	Sp gr	Factor
0.35	1.31	0 75	0.90	1.40	0.66
.40	1.23	.80	.87	1.50	.63
.45	1.16	.85	.84	1.60	.61
.50	1.10	0.90	.82	1.70	.59
.55	1.04	1.00	.78	1.80	.58
.60	1.00	1.10	.74	1.90	.56
.65	0.96	1.20	.71	2.00	.55
0.70	0.93	1.30	0.68	2.10	0.54

pany has made over one hundred such installations to date (1964). Use of 2.0 psig at the gas meter outlet permits the use of prefabricated assemblies of small-diameter (⅜ thru ⅝ in. OD), semirigid, types K and L copper tubing. The safety aspects of such systems have been demonstrated.[5] Minneapolis Gas Company's specifications follow:

1. The installation shall satisfy ASA Z21.30[2] with the modifications and additions noted hereafter.
2. The tubing system is designed for a 1.5-psi pressure drop between the meter and the by-pass regulator that further reduces pressure to 6.0 in. w.c. for central furnace use. The piping or tubing between this regulator and the furnace is designed for a 1.0-in. w.c. drop. Table 12-6 gives tubing capacity for both of these pressure drops. The remainder of the tubing system, which has an inlet pressure of 0.5 psig, is designed for a pressure drop of 0.12 psi to the other appliances.
3. The 2.0 psig–to–6.0 in w.c. regulator (a) is rated for up to 5.0 psig operating pressure, (b) maintains reduced pressure at *no-flow* conditions, (c) has the capacity to supply its rated load, (d) either is vented by a No. 80 orifice or has a ball check-type vent limiter, (e) must be installed according to manufacturer's instructions and so that it cannot be concealed.
4. All heating and water heating appliances rated for inches water column pressure operation must have an appliance regulator for reducing pressure from 6.0 in. w.c. to the rated operating pressure.

Table 12-5 Capacity of Semirigid Tubing in Thousands of Btu per Hour of Undiluted LP-Gases[2]

(based on a pressure drop of **0.50 in. w.c.** at pressures up to 0.5 psig)

OD,	Tubing length, ft									
in.	10	20	30	40	50	60	70	80	90	100
⅜	39	26	21	19	...	...	...	...	...	...
½	92	62	50	41	37	35	31	29	27	26
⅝	199	131	107	90	79	72	67	62	59	55
¾	329	216	181	145	131	121	112	104	95	90
⅞	501	346	277	233	198	187	164	155	146	138

5. Piping carrying pressure in excess of 14 in. w.c. must be clearly marked with the pressure the pipe is carrying at intervals as close to 6 ft as possible.
6. All piping carrying gas at pressures above 14 in. w.c. shall be semirigid copper tubing.
7. When copper tubing runs in a wall, flexible metal conduit shall be used where the tubing runs thru the plate* spaces. In the wall, the conduit must extend at least 18 in. on either side of the plate space.
8. A ⅜-in.-OD, spring-loaded, brass-core cock with ⅜-in. flare × pipe thread connections may be used ahead of the regulator in lieu of an appliance shutoff valve.
9. All piping must be run parallel or perpendicular to joists. Tubing run parallel to a joist shall be fastened to the center of the vertical face. Tubing run perpendicular to a joist shall be fastened to the underside of the joist, preferably close to water pipes, conduit, duct work, or center beams. Perpendicular runs may also be routed thru the center of joists.
10. All tubing shall be supported at not more than 4-ft intervals except for vertical drops to appliances.
11. All free-standing range connections shall be made with a 360° coil (12- to 18-in. diam) of tubing.
12. No drips or pipe grading is required. Connections may be taken off the bottom of horizontal runs.
13. All fittings in a copper system shall be flared or sweat with 1100 F solder.

Advantages of Higher Utilization Pressures.[10,11] Besides reducing the size of house piping and permitting its prefabrication, certain advantages of higher pressures at appliances are claimed. These include:

1. Improving appliance performance, since the increased kinetic energy of the gas can be used to inspirate more primary air, reduce combustion space, and improve flame characteristics.
2. Increasing capacity of existing house piping.

* The wood pieces at the top and bottom of the wall studs to which the studs are nailed. The conduit protects the tubing going thru holes in these pieces. Otherwise, the confined tubing might be pierced by a driven nail.

Table 12-4 Maximum Capacity of Pipe in Thousands of Btu per Hour of Undiluted LP-Gases[2]

(based on a pressure drop of **0.5 in. w.c.**)

IPS, nom.,	Pipe length, ft											
in.	10	20	30	40	50	60	70	80	90	100	125	150
½	275	189	152	129	114	103	96	89	83	78	69	63
¾	567	393	315	267	237	217	196	185	173	162	146	132
1	1071	732	590	504	448	409	378	346	322	307	275	252
1¼	2205	1496	1212	1039	913	834	771	724	677	630	567	511
1½	3307	2299	1858	1559	1417	1275	1181	1086	1023	976	866	787
2	6221	4331	3465	2992	2646	2394	2205	2047	1921	1811	1606	1496

Table 12-6 Capacity of Semirigid Tubing in a 2-Psig System in Cubic Feet per Hour

(based on a 0.69 sp gr gas)*

For a 1.5-psi pressure drop				For a 1.0-in. w.c. pressure drop			
	Tubing OD, in.				Tubing OD, in.		
Length, ft	0.375	0.500	0.625	Length, ft	0.375	0.500	0.625
	Capacity, cu ft per hr				Capacity, cu ft per hr		
5	400	890	1620	2	98	218	395
10	285	635	1160	4	70	156	281
15	230	510	930	6	56	124	226
20	200	445	810	8	47	104	190
30	165	367	670	10	40	89	162
40	145	323	570	13	33	73	133
50	125	270	505	16	28	62	113
60	115	255	465	19	24	53	97
70	107	238	430	22	21	47	85
80	100	223	405	26	19	42	77
90	95	211	385	30	17	38	69
100	90	200	365	34	15	33	60
115	84	187	340	38	14	31	56
130	80	178	325	42	13	29	52
150	74	164	300	50	11	24	44

* For other specific gravities, G, multiply given capacities by $(0.69/G)^{0.5}$.

3. Using existing LP-gas piping for natural gas.

4. Complementing the higher distribution pressures in use and those planned for the future.

5. Allowing increased meter pressure drops, subject to limitations arising from increased valve wear.

Meters for Higher Utilization Pressure.[10] Domestic sizes of iron, aluminum, and welded or bolted steel meters are generally built for working pressures of 5 psig max (in some sizes, 10 psig). Compensated meter indexes are available to accommodate different base and metering pressures. Generally, meter manufacturers must be consulted for capacity and corresponding pressure drop data for applications involving more than 0.5 or 2.0 in. w.c. differential, and inlet pressures above "inches water column" and below the maximum working pressure. Manufacturers' literature in 1965 did not generally include these data.

Pressure regulator considerations are also discussed in Reference 10.

Flow Formulas and Tables

Empirical formulas have been applied to building piping for many years, in spite of known limitations, because they furnish practical answers. The **Pole** formula (Eq. 1 of Sec. 9, Chap. 6) has long been used both for setting up flow tables and for designing flow slide rules; its principal limitation is that there is no compensation for change in flow resistance with diameter. The low-pressure **Spitzglass** formula (Eq. 2 of Sec. 9, Chap. 6) provides for a varying friction factor.

Polyflo Formula. Applicable to pressures under 1.5 psig, the **Polyflo computer**[6] uses Eq. **1** to closely approximate results obtained using the low-pressure **Spitzglass** formula. Equation **1** is the basis of Tables 12-1a and 12-1b.

$$Q_h = 2313D^{2.623} (h/CL)^{0.541} \quad (1)$$

where:

Q_h = flow rate at 60 F and 30 in. Hg, cu ft per hr
D = inside diameter, in.
h = pressure drop, in. w.c.
C = factor for viscosity, density, and temperature—0.595 for 0.6 sp gr paraffin base gas at 60 F
L = length of pipe, ft

Effect of Building Height

Relationships involved in the effect of building height are covered by Eq. **1** of Sec. 9, Chap. 2 and its related text. Table 12-7 gives multipliers by which the piping capacities shown in Table 12-1a may be increased, if this effect is to be taken into account.

Table 12-7 Pipe Capacity Multipliers for Applying the Pressure Gain Due to Elevation

(based on 0.60 sp gr gas; may be applied in Table 12-1a)

Building height, ft	Allowable pressure drop, in. w.c.	Capacity multiplier*
0	0.30	1.00
20	.42†	1.18
40	.54	1.34
60	.66	1.48
80	.78	1.61
100	0.90	1.73
120	1.02	1.84
140	1.14	1.95

* Assumes that capacity varies as the square root of the pressure drop.

† Calculated using Eq. **1** of Sec. 9, Chap. 2.

Materials

The following extract from ASA Z21.30[2] applies to piping materials and related topics:

2.6 ACCEPTABLE PIPING MATERIALS

2.6.1 Piping Material:

(a) *Pipe.* Gas pipe shall be steel or wrought-iron pipe complying with the American Standard for Wrought-Steel and Wrought-Iron Pipe, ASA B36.10–1959.* Threaded copper, brass, or aluminum alloy pipe in iron pipe sizes may be used with gases not corrosive to such material except that aluminum alloy pipe shall not be used in exterior locations, or underground, or where it is in contact with masonry, plaster, or insulation, or is subject to repeated corrosive wettings. Aluminum alloy pipe shall comply with specification ASTM B-241 (except that the use of alloy 5456 is prohibited) and shall be suitably marked at each end of each length indicating compliance with ASTM specifications.†

(b) *Tubing.* When acceptable to the serving gas supplier, seamless copper, aluminum alloy, or steel tubing may be used with gases not corrosive to such material. Copper tubing shall be of standard type K or L, or equivalent, complying with specification ASTM B88–62 and having a minimum wall thickness for each tubing size in compliance with ASTM specifications.† Aluminum alloy tubing shall be of standard Type A or B, or equivalent, complying with specification ASTM B-318–62, having a minimum wall thickness for each tubing size, and being suitably marked every 18 inches in compliance with ASTM specifications.† Aluminum alloy tubing shall not be used in exterior loca-

* Available from the American Standards Association, Inc., 10 East 40th Street, New York, New York, 10016.

† Available from American Society for Testing and Materials, 1916 Race St., Philadelphia, Pa. 19103.

tions, or underground, or where it is in contact with masonry, plaster, or insulation, or is subject to repeated corrosive wettings.

(c) *Piping Joints and Fittings.* Pipe joints may be screwed, flanged or welded, and nonferrous pipe may also be soldered or brazed with material having a melting point in excess of 1,000° F. Tubing joints shall either be made with approved flared gas tubing fittings, or be soldered or brazed with a material having a melting point in excess of 1,000° F. Compression type tubing fittings shall not be used for this purpose.

Fittings (except stopcocks or valves) shall be malleable iron or steel when used with steel or wrought-iron pipe, and shall be copper or brass when used with copper or brass pipe or tubing, and shall be aluminum alloy when used with aluminum alloy pipe or tubing. When approved by the authority having jurisdiction, special fittings may be used to connect steel or wrought-iron pipe. Cast iron fittings in sizes 6 inches and larger may be used to connect steel and wrought-iron pipe when approved by the authority having jurisdiction.

2.6.2 Workmanship and Defects:

Gas pipe or tubing and fittings shall be clear and free from cutting burrs and defects in structure or threading and shall be thoroughly brushed, and chip and scale blown.

Defects in pipe or tubing or fittings shall not be repaired. When defective pipe, tubing or fittings are located in a system the defective material shall be replaced.

2.6.3 Pipe Coating:

When in contact with material exerting a corrosive action, piping and fittings coated with a corrosion resisting material shall be used.

2.6.4 Use of Old Piping Material:

Gas pipe, tubing, fittings, and valves removed from any existing installation shall not be again used until they have been thoroughly cleaned, inspected and ascertained to be equivalent to new material.

2.6.5 Joint Compounds:

Joint compounds (pipe dope) shall be applied sparingly and only to the male threads of pipe joints. Such compounds shall be resistant to the action of liquefied petroleum gases.

2.7 PIPE THREADS

2.7.1 Specifications for Pipe Threads:

Pipe and fitting threads shall comply with the American Standard for Pipe Threads (Except Dryseal), B2.1–1960.*

2.7.2 Damaged Threads:

Pipe with threads which are stripped, chipped, corroded, or otherwise damaged shall not be used.

2.7.3 Number of Threads:

Pipe shall be threaded in accordance with Table 6.

Table 6
Specifications for Threading Pipe

Iron pipe size (inches)	Approximate length of threaded portion (inches)	Approximate No. of threads to be cut
½	¾	10
¾	¾	10
1	⅞	10
1¼	1	11
1½	1	11
2	1	11
2½	1½	12
3	1½	12
4	1⅝	13

Installation

The following extract from ASA Z21.30[2] covers installation details:

2.8 CONCEALED PIPING IN BUILDINGS

2.8.1 Minimum Size:

No gas pipe smaller than standard ½ inch iron pipe size shall be used in any concealed location.

2.8.2 Piping in Partitions:

Concealed gas piping should be located in hollow rather than solid partitions. Tubing shall not be run inside walls or partitions unless protected against physical damage. This rule does not apply to tubing which passes through walls or partitions.

2.8.3 Piping in Floors:

(a) Except as provided in 2.8.3(b), gas piping in solid floors such as concrete shall be laid in channels in the floor suitably covered to permit access to the piping with a minimum of damage to the building. When piping in floor channels may be exposed to excessive moisture or corrosive substances, it shall be suitably protected.

(b) When approved by the authority having jurisdiction and acceptable to the serving gas supplier, gas piping may be embedded in concrete floor slabs constructed with portland cement. Piping shall be surrounded with a minimum of 1½ inches of concrete and shall not be in physical contact with other metallic structures such as reinforcing rods or electrical neutral conductors. When piping may be subject to corrosion at point of entry into concrete slab, it shall be suitably protected from corrosion. Piping shall not be embedded in concrete slabs containing quick-set additives or cinder aggregate.

2.8.4 Connections in Original Installations:

When installing gas piping which is to be concealed, unions, tubing fittings, running threads, right and left couplings, bushings, and swing joints made by combinations of fittings shall not be used.

2.8.5 Reconnections:

When necessary to insert fittings in gas pipe which has been installed in a concealed location, the pipe may be reconnected by use of a ground joint union with the nut center-punched to prevent loosening by vibration. Reconnection of tubing in a concealed location is prohibited.

2.9 PIPING UNDERGROUND

2.9.1 Protection of Piping:

Piping shall be buried a sufficient depth or covered in a manner so as to protect the piping from physical damage.

2.9.2 Protection Against Corrosion:

(a) Gas piping in contact with earth or other material which may corrode the piping, shall be protected against corrosion in an approved manner. When dissimilar metals are joined underground, an insulated coupling shall be used. Piping shall not be laid in contact with cinders.

(b) Underground piping for manufactured gas shall be one size larger than that specified by Table 2A† or Table 2B,‡ as designated by the serving gas supplier, but in no case less than 1¼ inch.

2.9.3 Piping Through Foundation Wall:

Underground gas piping, when installed below grade through the outer foundation or basement wall of a building, shall be

* Available from the American Standards Association, Inc., 10 East 40th Street, New York, New York, 10016.

[† Table 12-1a in this chapter.—Editor]
[‡ Table 12-1b in this chapter.—Editor]

either encased in a sleeve or otherwise protected against corrosion. The piping or sleeve shall be sealed at the foundation or basement wall to prevent entry of gas or water.

2.9.4 Piping Underground Beneath Buildings:

When the installation of gas piping underground beneath buildings is unavoidable, the piping shall be encased in a conduit. The conduit shall extend into a normally usable and accessible portion of the building and, at the point where the conduit terminates in the building, the space between the conduit and the gas piping shall be sealed to prevent the possible entrance of any gas leakage. The conduit shall extend at least 4 inches outside the building, be vented above grade to the outside and be installed in a way as to prevent the entrance of water.

2.10 INSTALLATION OF PIPING

Drips, grading, protection from freezing, and branch pipe connections, as provided for in 2.10.2, 2.10.4, 2.10.7, and 2.10.14(a), shall apply only when other than dry gas is distributed and climatic conditions make such provisions necessary.

2.10.1 Building Structure:

The building structure shall not be weakened by the installation of any gas piping. Before any beams or joists are cut or notched, special permission should be obtained from the authority having jurisdiction.

2.10.2 Gas Piping to be Graded:

All gas piping shall be graded not less than ¼ inch in 15 feet to prevent traps. All horizontal lines shall grade to risers and from the risers to the meter, or to service regulator when a meter is not provided, or to the appliance.

2.10.3 Piping Supports:

(a) Gas piping in buildings shall be supported with pipe hooks, metal pipe straps, bands or hangers, suitable for the size of piping, and of adequate strength and quality, and located at proper intervals so that the piping cannot be moved accidentally from the installed position. Gas piping shall not be supported by other piping.
(b) Spacing of supports in gas piping installations shall not be greater than shown in Table 7.

Table 7
Support of Piping

Size of pipe, in.	Support spacing, ft	Size of tubing, in. O.D.	Support spacing, ft
½	6	½	4
¾ or 1	8	⅝ or ¾	6
1¼ or larger (horizontal)	10	⅞ or 1	8
1¼ or larger (vertical)	Every floor level		

2.10.4 Protect against Freezing:

Gas piping shall be protected against freezing temperatures When piping must be exposed to wide ranges or sudden changes in temperatures, special care shall be taken to prevent stoppages.

2.10.5 Overhanging Rooms:

When there are overhanging kitchens or other rooms built beyond foundation walls, in which gas appliances are installed, care shall be taken to avoid placing the gas piping where it will be exposed to low temperatures (40° F or below for manufactured gas) or to extreme changes of temperatures. In such cases the gas piping shall be brought up inside the building proper and run around the sides of the room, in the most practical manner.

2.10.6 Do Not Bend Pipe:

Gas pipe shall not be bent. Fittings shall be used when making turns in gas pipe.

2.10.7 Provide Drips Where Necessary:

A drip shall be provided at any point in the line of pipe where condensate may collect. When condensation is excessive, a drip should be provided at the outlet of the meter. This drip should be so installed as to constitute a trap wherein an accumulation of condensate will shut off the flow of gas before it will run back into the meter.

2.10.8 Location and Size of Drips:

All drips shall be installed only in such locations that they will be readily accessible to permit cleaning or emptying. A drip shall not be located where the condensate is likely to freeze. The size of any drip used shall be determined by the capacity and the exposure of the gas piping which drains to it and in accordance with recommendations of the serving gas supplier.

2.10.9 Use Tee:

If dirt or other foreign material is a problem, a tee fitting with the bottom outlet plugged or capped shall be used at the bottom of any pipe riser (see Fig. 1).

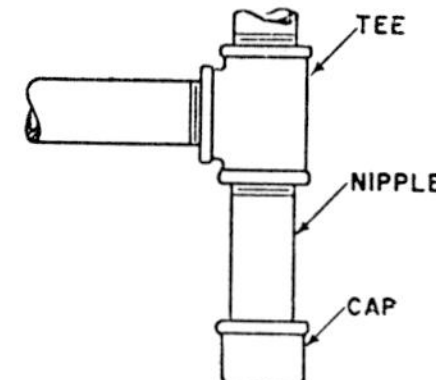

Fig. 1 Suggested method of installing tee.

2.10.10 Avoid Clothes Chutes, etc.:

Gas piping inside any building shall not be run in or through an air duct, clothes chute, chimney or gas vent, ventilating duct, dumb waiter, or elevator shaft.

2.10.11 Cap All Outlets:

(a) Each outlet, including a valve or cock outlet, shall be securely closed gastight with a threaded plug or cap immediately after installation and shall be left closed until an appliance is connected thereto. Likewise, when an appliance is disconnected from an outlet and the outlet is not to be used again immediately, it shall be securely closed gastight. The outlet shall not be closed with tin caps, wooden plugs, corks, or by other improvised methods.
(b) The above provision does not prohibit the normal use of a listed quick-disconnect device.

2.10.12 Location of Outlets:

The unthreaded portion of gas piping outlets shall extend at least one inch through finished ceilings and walls, and when extending through floors shall be not less than 2 inches above them. The outlet fitting or the piping shall be securely fastened. Outlets shall not be placed behind doors. Outlets shall be far enough from floors, walls and ceilings to permit the use of proper wrenches without straining, bending or damaging the piping.

2.10.13 Prohibited Devices:

No device shall be placed inside the gas piping or fittings that will reduce the cross-sectional area or otherwise obstruct the free flow of gas.

2.10.14 Branch Pipe Connection:

(a) All branch outlet pipes shall be taken from the top or sides of horizontal lines and not from the bottom.
(b) When a branch outlet is placed on a main supply line before it is known what size of pipe will be connected to it, the outlet shall be of the same size as the line which supplies it.

2.10.15 Electrical Bonding and Grounding:

(a) A gas piping system within a building shall be electrically continuous and bonded to any grounding electrode, as defined by the National Electrical Code, ASA C1–1962 (NFPA No. 70).*

(b) Underground gas service piping shall not be used as a grounding electrode except when it is electrically continuous uncoated metallic piping, and its use as a grounding electrode is acceptable both to the serving gas supplier and to the authority having jurisdiction, since gas piping systems are often constructed with insulating bushings or joints, or are of coated or nonmetallic piping.

2.11 GAS SHUTOFF VALVES

2.11.1 Accessibility of Gas Valves:

Main gas shutoff valves controlling several gas piping systems shall be placed an adequate distance from each other so they will be easily accessible for operation and shall be installed so as to be protected from physical damage. It is recommended that they be plainly marked with a metal tag attached by the installing agency so that the gas piping systems supplied through them can be readily identified. It is advisable to place a shutoff valve at every point where safety, convenience of operation, and maintenance demands.

2.11.2 Shutoff Valves for Multiple House Lines:

(a) In multiple tenant buildings supplied through a master meter or one service regulator when a meter is not provided, or where meters or service regulators are not readily accessible from the appliance location, an individual shutoff valve for each apartment, or for each separate house line, shall be provided at a convenient point of general accessibility.

(b) In a common system serving a number of individual buildings, shutoff valves shall be installed at each building.

2.12 TEST OF PIPING FOR TIGHTNESS

Before any system of gas piping is finally put in service, it shall be carefully tested to assure that it is gas tight. Where any part of the system is to be enclosed or concealed, this test should precede the work of closing in. To test for tightness, the piping may be filled with the fuel gas, air or inert gas, but not with any other gas or liquid. OXYGEN SHALL NEVER BE USED.

(a) Before appliances are connected, piping systems shall stand a pressure of at least six inches mercury or three pounds gage for a period of not less than ten minutes without showing any drop in pressure. Pressure shall be measured with a mercury manometer or slope gage, or an equivalent device so calibrated as to be read in increments of not greater than one-tenth pound. The source of pressure shall be isolated before the pressure tests are made.

(b) Systems for undiluted liquefied petroleum gases shall stand the pressure test in accordance with 2.12.(a), or, when appliances are connected to the piping system, shall stand a pressure of not less than ten inches water column for a period of not less than ten minutes without showing any drop in pressure. Pressure shall be measured with a water manometer or an equivalent device so calibrated as to be read in increments of not greater than one-tenth inch water column. The source of pressure shall be isolated before the pressure tests are made.

2.13 LEAKAGE CHECK AFTER GAS TURN ON

2.13.1 Close All Gas Outlets:

Before turning gas under pressure into any piping, all openings from which gas can escape shall be closed.

2.13.2 Check for Leakage:

Immediately after turning on the gas, the piping system shall be checked by one of the following methods to ascertain that no gas is escaping:

(a) *Checking for Leakage Using the Gas Meter*

Immediately prior to the test it should be determined that the meter is in operating condition and has not been bypassed.

Checking for leakage can be done by carefully watching the test dial of the meter to determine whether gas is passing through the meter. To assist in observing any movement of the test hand, wet a small piece of paper and paste its edge directly over the center line of the hand as soon as the gas is turned on. Allow five minutes for a one-half foot dial and proportionately longer for a larger dial in checking for gas flow. This observation should be made with the test hand on the upstroke.

In case careful observation of the test hand for a sufficient length of time reveals no movement, the piping shall be purged and a small gas burner turned on and lighted and the hand of the test dial again observed. If the dial hand moves (as it should), it will show that the meter is operating properly. If the test hand does not move or register flow of gas through the meter to the small burner, the meter is defective and the gas should be shut off and the serving gas supplier notified.

(b) *Checking for Leakage Not Using a Meter*

This can be done by attaching to an appliance orifice a manometer or equivalent device calibrated so that it can be read in increments of 0.1 inch water column, and momentarily turning on the gas supply and observing the gaging device for pressure drop with the gas supply shut off. No discernible drop in pressure shall occur during a period of 3 minutes.

(c) *When Leakage is Indicated*

If the meter test hand moves, or a pressure drop on the gage is noted, all appliances or outlets supplied through the system shall be examined to see if they are shut off and do not leak. If they are found tight there is a leak in the piping system. The gas supply shall be shut off until the necessary repairs have been made, after which the test specified in 2.13.2(a) or (b) shall be repeated.

2.14 PURGING

2.14.1 Purging All Gas Piping:

(a) After piping has been checked, all gas piping shall be fully purged. A suggested method for purging the gas piping to an appliance is to disconnect the pilot piping at the outlet of the pilot valve. Piping shall not be purged into the combustion chamber of an appliance.

(b) The open end of piping systems being purged shall not discharge into confined spaces or areas where there are sources of ignition unless precautions are taken to perform this operation in a safe manner by ventilation of the space, control of purging rate, and elimination of all hazardous conditions.

2.14.2 Light Pilots:

After the gas piping has been sufficiently purged, all appliances shall be purged and the pilots lighted. The installing agency shall assure itself that all piping and appliances are fully purged before leaving the premises.

Appliance Connections to Building Piping

The following extract from Z21.30[2] should be noted:

3.5.1 Connecting Gas Appliances:

Gas appliances shall be connected by:

(a) Rigid pipe, or,

(b) Semi-rigid tubing extensions of a tubing piping system, or,

(c) Listed appliance connectors† that are in the same room as the appliance, or,

---

* Available from the National Fire Protection Association, 60 Batterymarch St., Boston, Mass. 02110 in pamphlet form and in the National Fire Codes, Volume 5. Also available from the American Standards Association, Inc., 10 East 40th St., New York, N. Y., 10016.

[† Semirigid and flexible metal connectors are intended for convenience in connecting an appliance. They are not intended to provide mobility for frequent movement, e.g., for cleaning behind a range. However, other listed flexible connectors[8] are available that provide for appliance mobility, especially when used with listed quick-disconnect devices.[9]—EDITOR]

Table 12-8a Capacity of Pipe at Less Than 1.0-psig Inlet Pressure in Cubic Feet of Gas per Hour[7]
(based on a pressure drop of 0.5 in. w.c.* and 0.6 sp gr gas in schedule-40 pipe)

IPS, nom., in.	Equivalent pipe length, ft										
	50	100	150	200	250	300	400	500	1000	1500	2000
1	244	173	141	122	109	99	86	77	54	44	38
1¼	537	380	310	268	240	219	189	169	119	97	84
1½	832	588	480	416	372	339	294	263	185	151	131
2	1,680	1,188	970	840	751	685	594	531	375	306	265
2½	2,754	1,952	1,591	1,379	1,232	1,123	974	869	617	504	436
3	5,018	3,549	2,896	2,509	2,244	2,047	1,774	1,587	1,121	915	793
4	10,510	7,410	6,020	5,170	4,640	4,480	3,660	3,340	2,360	1,910	1,660
5	19,110	13,480	10,960	9,410	8,440	8,150	6,660	6,070	4,290	3,480	3,020
6	31,140	21,960	17,860	15,320	13,760	13,280	10,860	9,890	7,000	5,670	4,920
8	63,310	44,740	36,380	31,220	28,020	27,040	22,120	20,150	14,250	11,550	10,030
10	113,020	79,720	64,830	55,630	49,940	48,180	39,420	35,920	25,400	20,580	17,870
12	177,450	125,180	101,790	87,350	78,400	75,650	61,900	56,400	39,890	32,320	28,060

* Based on **Spitzglass** formula for pressures of less than 1 psig. For pressure drop of 0.3 in. w.c. multiply tabulated values by 0.775 or refer to Table 12-1a, which gives capacities up to 200 ft of length based on pressure drop of 0.30 in. w.c.

(d) Semi-rigid tubing in lengths up to 6 feet that are in the same room as the appliance. When acceptable to the serving gas supplier, greater lengths may be used and need not be connected to an outlet in the same room as the appliance.

The connector or tubing shall be installed so as to be protected against physical damage.

Aluminum alloy tubing and connectors shall not be used in exterior locations nor in interior locations where they are in contact with masonry, plaster, or insulation or are subject to repeated corrosive wettings.

3.5.2 Appliance Shutoff Valves:

Any appliance connected to a piping system supplying two or more appliances should have an accessible manual shut off valve installed upstream of the union or connector and within 6 feet of the appliance it serves.

3.5.3 Use of Gas Hose:

The connection of an appliance with any type of gas hose is prohibited, except when used with laboratory, shop or ironing equipment that requires mobility during operation. Such connections shall have the shutoff or stopcock installed at the connection to the building piping. When gas hose is used, it shall be of the minimum practical length, but not to exceed 6 feet, and shall not extend from one room to another nor pass through any walls, partitions, ceilings or floors. Under no circumstances shall gas hose be concealed from view or used in a concealed location. Only listed gas hose shall be used. Listed gas hose shall be used only in accordance with its listing. Gas hose shall not be used where it is likely to be subject to excessive temperatures (above 125° F).

INDUSTRIAL BUILDINGS

Industrial applications may require gas at any pressure. Other nonresidential installations, particularly certain types of large boilers, may require gas pressures above 0.5 psig, which is the usual residential maximum.[2]

Since industrial gas demands are often large, substantial savings can be realized by using high gas pressures and large pressure drops. At the burners, pressure regulators may be used to provide the appropriate gas pressures. Apart from possible economies, gas burning equipment may require a considerable pressure drop to permit regulators and control valves to operate.

Design Considerations[7]

The design factors in industrial gas piping include:

1. Required pipe capacity—to meet demands and pressures at the various points of use (Tables 12-1a, 12-1b, 12-2, 12-4, and 12-5 for gas pressure up to 0.5 psig; Tables 12-8a thru 12-8g for pressures from below 1.0 psig to 50.0 psig). Coincidence factors, e.g., Table 9-28, may be applied.
2. Allowable pressure drop—generally 0.5 in. w.c. for low pressure systems and 10 per cent of initial supply pressure for systems operating at over 0.5 psig.
3. Equivalent length of pipe—valves and fittings are taken into account by adding their resistance, expressed in equiva-

Table 12-8b Capacity of Pipe at 1.0-psig Inlet Pressure in Cubic Feet of Gas per Hour[7]
(based on a pressure drop of 0.1 psig and 0.6 sp gr gas in schedule-40 pipe)

IPS, nom., in.	Equivalent pipe length, ft									
	50	100	150	200	300	400	500	1000	1500	2000
1	740	520	430	370	300	260	230	170	130	120
1¼	1,540	1,090	890	760	630	540	490	350	280	250
1½	2,330	1,650	1,350	1,160	960	830	740	530	420	380
2	4,550	3,210	2,640	2,260	1,870	1,610	1,440	1,040	830	750
2½	7,330	5,180	4,250	3,650	3,020	2,600	2,320	1,690	1,340	1,200
3	13,100	9,260	7,600	6,520	5,400	4,660	4,160	3,020	2,400	2,160
3½	19,320	13,650	11,210	9,610	7,960	6,870	6,130	4,450	3,540	3,180
4	26,980	19,070	15,650	13,430	11,120	9,590	8,560	6,220	4,940	4,440
5	49,340	34,870	28,620	24,550	20,330	17,550	15,660	11,370	9,030	8,130
6	80,560	56,940	46,740	40,090	33,210	28,650	25,580	18,570	14,760	13,280

Table 12-8c Capacity of Pipe at 2.0-psig Inlet Pressure in Cubic Feet of Gas per Hour[7]
(based on a pressure drop of 0.2 psig and 0.6 sp gr gas in schedule-40 pipe)

IPS, nom., in.	Equivalent pipe length, ft									
	50	100	150	200	300	400	500	1000	1500	2000
1	1,080	760	620	540	440	380	340	240	190	170
1¼	2,250	1,590	1,300	1,120	910	790	710	500	410	350
1½	3,410	2,410	1,970	1,700	1,390	1,200	1,070	760	620	530
2	6,640	4,700	3,840	3,310	2,700	2,350	2,090	1,480	1,210	1,040
2½	10,700	7,580	6,190	5,340	4,360	3,790	3,380	2,390	1,960	1,690
3	19,120	13,540	11,060	9,540	7,790	6,770	6,040	4,280	3,500	3,020
3½	28,200	19,970	16,310	14,070	11,490	9,980	8,900	6,310	5,160	4,450
4	39,380	27,890	22,780	19,650	16,040	13,940	12,440	8,810	7,210	6,220
5	72,010	50,990	41,650	35,930	29,300	25,490	22,740	16,120	13,180	11,370
6	117,580	83,270	68,010	58,670	47,900	41,630	37,140	26,320	21,520	18,570

Table 12-8d Capacity of Pipe at 5.0-psig Inlet Pressure in Cubic Feet of Gas per Hour[7]
(based on a pressure drop of 0.5 psig and 0.6 sp gr gas in schedule-40 pipe)

IPS, nom., in.	Equivalent pipe length, ft									
	50	100	150	200	300	400	500	1000	1500	2000
1	1,860	1,320	1,070	930	760	660	590	410	340	290
1¼	3,870	2,740	2,240	1,930	1,580	1,370	1,220	860	700	610
1½	5,860	4,140	3,390	2,930	2,390	2,080	1,850	1,310	1,060	930
2	11,420	8,070	6,600	5,710	4,660	4,050	3,610	2,550	2,080	1,810
2½	18,400	13,010	10,640	9,200	7,510	6,530	5,820	4,110	3,350	2,920
3	32,860	23,240	19,000	16,430	13,410	11,660	10,390	7,340	5,990	5,220
3½	48,480	34,280	28,030	24,240	19,780	17,200	15,330	10,820	8,840	7,690
4	67,700	47,880	39,140	33,850	27,630	24,020	21,410	15,120	12,340	10,750
5	123,790	87,540	71,570	61,890	50,530	43,920	39,160	27,640	22,570	19,660
6	202,138	142,950	116,870	101,060	82,500	71,720	63,940	45,140	36,860	32,100

Table 12-8e Capacity of Pipe at 10.0-psig Inlet Pressure in Cubic Feet of Gas per Hour[7]
(based on a pressure drop of 1.0 psig and 0.6 sp gr gas in schedule-40 pipe)

IPS, nom., in.	Equivalent pipe length, ft									
	50	100	150	200	300	400	500	1000	1500	2000
1	2,930	2,070	1,690	1,470	1,190	1,030	920	650	530	460
1¼	6,090	4,330	3,520	3,050	2,490	2,150	1,920	1,360	1,110	960
1½	9,210	6,530	5,330	4,620	3,760	3,260	2,910	2,060	1,680	1,460
2	17,940	12,720	10,380	9,000	7,330	6,360	5,680	4,020	3,280	2,840
2½	28,920	20,500	16,730	14,510	11,820	10,250	9,150	6,480	5,290	4,580
3	51,650	36,610	29,880	25,920	21,110	18,300	16,340	11,570	9,450	8,190
3½	76,180	53,990	44,070	38,240	31,140	26,990	24,110	17,070	13,950	12,080
4	106,400	75,410	61,550	53,410	43,500	37,700	33,670	23,850	19,480	16,870
5	194,540	137,890	112,550	97,650	79,540	68,940	61,570	43,600	35,620	30,860
6	317,650	225,150	183,770	159,450	129,870	112,560	100,540	71,200	58,160	50,390

Table 12-8f Capacity of Pipe at 20.0-psig Inlet Pressure in Cubic Feet of Gas per Hour[7]
(based on a pressure drop of 2.0 psig and 0.6 sp gr gas in schedule-40 pipe)

IPS, nom., in.	Equivalent pipe length, ft									
	50	100	150	200	300	400	500	1000	1500	2000
1	4,900	3,470	2,810	2,450	2,000	1,730	1,550	1,070	890	770
1¼	10,190	7,210	5,840	5,090	4,160	3,600	3,220	2,230	1,860	1,610
1½	15,420	10,900	8,830	7,710	6,290	5,450	4,870	3,370	2,810	2,440
2	30,030	21,230	17,190	15,010	12,260	10,610	9,490	6,570	5,480	4,760
2½	48,390	34,220	27,710	24,190	19,750	17,110	15,290	10,590	8,830	7,670
3	86,420	61,110	49,490	43,190	35,280	30,550	27,310	18,910	15,770	13,690
3½	127,480	90,130	73,000	63,710	52,040	45,070	40,280	27,900	23,270	20,200
4	178,040	125,880	101,950	88,980	72,680	62,940	56,260	38,960	32,500	28,210
5	325,530	230,170	186,410	162,700	132,890	115,080	102,870	71,240	59,420	51,590
6	531,530	375,820	304,370	265,660	216,990	187,910	167,980	116,330	97,030	84,240

Table 12-8g Capacity of Pipe at 50.0-psig Inlet Pressure in Cubic Feet of Gas per Hour[7]

(based on a pressure drop of 5.0 psig and 0.6 sp gr gas in schedule-40 pipe)

IPS, nom., in.	Equivalent pipe size, ft									
	50	100	150	200	300	400	500	1000	1500	200
1	10,530	7,450	6,090	5,150	4,350	3,790	3,330	2,350	1,920	1,650
1¼	21,880	15,490	12,650	10,700	9,050	7,870	6,920	4,890	3,990	3,430
1½	33,110	23,430	19,130	16,190	13,690	11,190	10,470	7,410	6,040	5,190
2	64,450	45,610	37,250	31,530	26,660	23,190	20,400	14,420	11,770	10,110
2½	103,870	73,510	60,040	50,820	42,960	37,370	32,870	23,240	18,970	16,300
3	185,490	131,270	107,220	90,750	76,720	66,730	58,700	41,510	33,870	29,100
3½	273,600	193,620	158,140	133,850	113,170	98,430	86,590	61,230	49,970	42,930
4	382,110	270,420	220,870	186,940	158,050	137,480	120,930	85,510	69,780	59,960
5	698,660	494,430	403,840	341,800	288,980	251,360	221,110	156,360	127,600	109,630
6	1,140,780	807,320	659,400	558,110	471,860	410,430	361,040	255,310	208,340	179,010

lent feet of pipe, to the measured length of pipe (Table 12-9). Usually, no allowance need be made for a reasonable number of fittings.

Example: A 300-ft long, 6-in.-diam piping system has three standard elbows and two gate valves. Total equivalent length is: 300 + (3 × 15.2) + (2 × 3.54) = 352.7 ft.

4. Specific gravity—multipliers (Table 12-3) may be used to correct capacity tables for specific gravities other than 0.6.

Safety Requirements and Precautions

Reference should be made to Appendix A of ASA Z21.30 (given earlier in this chapter) for precautions relating to work on those portions of the piping system that belong to the gas company supplying the premises.

Because industrial gas systems often supply gas to many points of use, all affected points of consumption should be notified before the supply of gas is shut off, except under emergency conditions. Each burner valve and pilot valve should be shut off and the meter test hand observed to prove that no gas is passing, before a gas meter valve is shut. When turning on a meter valve, check the test dial hand to be certain that all valves are closed and that there are no piping leaks; only then is it permissible to leave the gas supply on.

Normally, all additions or extensions to existing piping systems are made with the gas shut off ahead of the point of work. With trained and experienced crews, "hot" taps may be made, observing the precautions in ASA B31.8–1963, Par. 841.28 (extracted in Sec. 8, Chap. 2).

When supplying gas to equipment that may force air, oxygen, or stand-by gases into gas piping, back-pressure protection should be provided. Double-diaphragm zero governors or regulators require no further protection unless connected directly to compressed air or oxygen at pressures of 5.0 psig or more. Suitable protective devices include: (1) check valves; (2) three-way cocks (of the type that completely closes one side before starting to open the other side); and (3) reverse-flow indicators controlling positive shutoff valves. Similarly, in the case of gas engines, gas compressors, etc., which may reduce supply pressures suddenly, it is advisable to use low-pressure shutoff valves—mechanical, diaphragm-operated, or electrically operated.

The following extracts, from the proposed ASA Z83.1,[7] cover some of the more significant points to be observed:

3.10 Low-Pressure Protection. A suitable protective device preferably of the manual reset type shall be installed in the supply piping as close to the utilization equipment as practical if the operation of the equipment is such (i.e., gas compressors) that it may produce a vacuum or even a dangerous reduction in gas pressure in the supply piping. Such devices include but are not limited to low-pressure shutoff valves of the following types:

a. Mechanical
b. Diaphragm-operated
c. Electrically operated

3.11 Grounding. Each aboveground portion of a gas piping system shall be electrically continuous and connected to a suitable ground electrode meeting the requirements below.

Gas piping shall not serve as a system ground or as an equipment ground, or as a part of such grounding systems.

Gas piping may be grounded to:

a. A continuous metallic underground water piping system.
b. An effectively grounded metal frame of a building.
c. Local metallic underground tanks or structures.
d. "Made" grounding electrodes meeting the *National Electrical Code, ASA C1.*

Industrial Gas Piping Systems

Basic design recommendations for industrial gas piping systems are covered in the proposed ASA Z83.1. This standard covers the installation of gas piping, both in buildings and between buildings, from the outlet of the consumer's meter set assembly (or point of delivery) to the inlet connection of the equipment. It covers gas piping systems for all industrial installations regardless of the gas pressure and for all other installations operating at pressures greater than ½ psig. The following extracts from the standard are pertinent:

4.1 Selection of Materials.

4.1.1 General. The following sets out basic information regarding selection of materials for piping systems. For more detailed information or for unusual applications, see Section 2 of the American Standard Code for Pressure Piping, B31.1–1955.

4.1.2 Steel Pipe. Standard weight or ASA Schedule 40 steel pipe shall be acceptable for gas pressures up to 125 psig. For pressures in excess of 125 psig, pipe shall be selected in accordance with 214 Section 2 American Standard Code for Pressure Piping, ASA B31.1–1955.

4.1.3 Cast Iron Pipe. Cast iron pipe may be used for underground piping for pressures up to 50 psig. Cast iron pipe shall be at least of six-inch size and be equipped with standardized mechanical joints and made in accordance with American Standard Specification for Cast Iron Pipe Centrifugally Cast in Metal

Molds for Gas, A21.7–1962 or American Standard Specifications for Cast Iron Pipe Centrifugally Cast in Sand-Lined Molds for Gas, A21.9–1962. For provisions covering installations of Cast Iron Pipe, refer to 842.2* of American Standard Gas Transmission and Distribution Piping Systems, B31.8–1963.

4.1.4 Copper or Brass Pipe and Tubing. The use of copper or brass pipe or tubing shall comply with 849.4† of American Standard Gas Transmission and Distribution Piping Systems, B31.8–1963. Since gases containing more than an average of 0.3 grains of hydrogen sulfide per 100 standard cubic feet cause copper and its alloys to be unsuitable for gas piping, the gas supplier shall be consulted before using these materials.

4.1.5 Piping Joints and Fittings.

a. Pipe joints may be screwed, flanged or welded, and nonferrous pipe may also be soldered or brazed with material having a melting point in excess of 1,000 F. Tubing joints shall either be made with approved flared gas tubing fittings, or be soldered or brazed with a material having a melting point in excess of 1,000 F. Compression-type tubing fittings shall not be used for this purpose.

b. Fittings (except stopcocks or valves) shall be malleable iron or steel when used with steel or wrought-iron pipe, and shall be copper or brass when used with copper or brass pipe or tubing, and shall be aluminum alloy when used with aluminum alloy pipe or tubing. Compression or gland type fittings may be used to connect steel or wrought-iron pipe if adequately braced so that neither the gas pressure nor external physical damage will force the joint apart. Cast iron fittings in sizes 6 inches and larger may be used to connect steel and wrought-iron pipe.

c. Reducing couplings are preferred to bushings. Cast iron or plastic bushings shall not be used. Any bushing used shall reduce at least two pipe sizes.

4.1.6 Valves. Shutoff valves shall be selected giving consideration to pressure drop, service involved, emergency use and reliability. Full opening valves cause minimum pressure drop. When plug valves are used in sizes larger than 2 inches, or for pressures greater than ½ psig, they shall be of the lubricated type.

4.1.7 Gaskets. Material for gaskets shall be capable of withstanding the design pressure of the piping system, the chemical constituents of the gas conducted through the piping systems, and of maintaining its physical and chemical properties at the design temperature and pressure. Gaskets shall be made of metal having a melting point of over 1,000 F or shall be confined within an assembly having a melting point of over 1,000 F. Aluminum "O" rings and spiral wound metal gaskets are also acceptable. When a flange is opened, the gasket shall be replaced.

4.1.8 Used Materials. Pipe, fittings, valves, or other materials removed from an existing installation shall not be used again unless they have been thoroughly cleaned, inspected, and ascertained to be adequate for the service intended.

4.1.9 Other Materials. Material not covered by the standards or specifications listed herein shall be investigated, or tested after installation, to demonstrate that it is safe and suitable for the proposed service, and, in addition, shall be recommended for that service by the manufacturer.

4.1.10 Joint Compounds. Joint compounds (pipe dope) shall be resistant to the action of liquefied petroleum gas or to any other chemical constituents of the gases to be conducted through the piping. Consult the gas supplier regarding the use of auxiliary or substitute gases during peak loads.

4.2 Pipe Threads.

4.2.1 Specifications for Pipe Threads. Pipe and fitting threads shall comply with the American Standard for Pipe Threads, B2.1–1960. Basic dimensions for American standard taper pipe thread are given in Table C10, Appendix C.‡

[* Given in Sec. 9, Chap. 7.—EDITOR]
[† Given in Sec. 9, Chap. 3.—EDITOR]
[‡ See Table 1-11 in this Handbook.—EDITOR]

Table 12-9 Equivalent Resistances of Bends, Fittings, and Valves in Feet of Schedule-40 Straight Pipe*

		Screwed fittings†			90° Welding elbows and smooth bends‡							Miter elbows§ (No. of miters)					Welding tees		Valves (screwed, flanged, or welded)			
IPS, nom., in.	ID, in.	45° ell	90° ell	180° close-return bends	Tee	R/d =1	R/d = 1½	R/d = 2	R/d = 4	R/d = 6	R/d = 8	45° (1)	60° (1)	90° (1)	90° (2)	90° (3)	Forged	Miter§	Gate	Globe	Angle	Swing check
					(R—radius of elbow or bend; d—inside diameter)																	
½	0.622	0.73	1.55	3.47	3.10	0.83	0.62	0.47	0.36	0.47	0.62	0.78	1.55	3.10	1.04	0.78	2.33	3.10	0.36	17.3	8.65	4.32
¾	0.824	0.96	2.06	4.60	4.12	1.10	0.82	0.62	0.48	0.62	0.82	1.03	2.06	4.12	1.37	1.03	3.09	4.12	0.48	22.9	11.4	5.72
1	1.049	1.22	2.62	5.82	5.24	1.40	1.05	0.79	0.61	0.79	1.05	1.31	2.62	5.24	1.75	1.31	3.93	5.24	0.61	29.1	14.6	7.27
1¼	1.380	1.61	3.45	7.66	6.90	1.84	1.38	1.03	0.81	1.03	1.38	1.72	3.45	6.90	2.30	1.72	5.17	6.90	0.81	38.3	19.1	9.58
1½	1.610	1.88	4.02	8.95	8.04	2.14	1.61	1.21	0.94	1.21	1.61	2.01	4.02	8.04	2.68	2.01	6.04	8.04	0.94	44.7	22.4	11.2
2	2.067	2.41	5.17	11.5	10.3	2.76	2.07	1.55	1.21	1.55	2.07	2.58	5.17	10.3	3.45	2.58	7.75	10.3	1.21	57.4	28.7	14.4
2½	2.469	2.88	6.16	13.7	12.3	3.29	2.47	1.85	1.44	1.85	2.47	3.08	6.16	12.3	4.11	3.08	9.25	12.3	1.44	68.5	34.3	17.1
3	3.068	3.58	7.67	17.1	15.3	4.09	3.07	2.30	1.79	2.30	3.07	3.84	7.67	15.3	5.11	3.84	11.5	15.3	1.79	85.2	42.6	21.3
4	4.026	4.70	10.1	22.4	20.2	5.37	4.03	3.02	2.35	3.02	4.03	5.04	10.1	20.2	6.71	5.04	15.1	20.2	2.35	112	56.0	28.0
5	5.047	5.88	12.6	28.0	25.2	6.72	5.05	3.78	2.94	3.78	5.05	6.30	12.6	25.2	8.40	6.30	18.9	25.2	2.94	140	70.0	35.0
6	6.065	7.07	15.2	33.8	30.4	8.09	6.07	4.55	3.54	4.55	6.07	7.58	15.2	30.4	10.1	7.58	22.8	30.4	3.54	168	84.1	42.1
8	7.981	9.31	20.0	44.6	40.0	10.6	7.98	5.98	4.65	5.98	7.98	9.97	20.0	40.0	13.3	9.97	29.9	40.0	4.65	222	111	55.5
10	10.02	11.7	25.0	55.7	50.0	13.3	10.0	7.51	5.85	7.51	10.0	12.5	25.0	50.0	16.7	12.5	37.6	50.0	5.85	278	139	69.5
12	11.94	13.9	29.8	66.3	59.6	15.9	11.9	8.95	6.96	8.95	11.9	14.9	29.8	59.6	19.9	14.9	44.8	59.6	6.96	332	166	83.0
14	13.13	15.3	32.8	73.0	65.6	17.5	13.1	9.85	7.65	9.85	13.1	16.4	32.8	65.6	21.9	16.4	49.2	65.6	7.65	364	182	91.0
16	15.00	17.5	37.5	83.5	75.0	20.0	15.0	11.2	8.75	11.2	15.0	18.8	37.5	75.0	25.0	18.8	56.2	75.0	8.75	417	208	104
18	16.88	19.7	42.1	93.8	84.2	22.5	16.9	12.7	9.85	12.7	16.9	21.1	42.1	84.2	28.1	21.1	63.2	84.2	9.85	469	234	117
20	18.81	22.0	47.0	105	94.0	25.1	18.8	14.1	11.0	14.1	18.8	23.5	47.0	94.0	31.4	23.5	70.6	94.0	11.0	522	261	131
24	22.63	26.4	56.6	126	113	30.2	22.6	17.0	13.2	17.0	22.6	28.3	56.6	113	37.8	28.3	85.0	113	13.2	629	314	157

* Values for welded fittings are for conditions where bore is not obstructed by weld spatter or backing rings; if appreciably obstructed, use values for screwed fittings.
† Flanged fittings have three-fourths the resistance of screwed elbows and tees.
‡ Data give the extra resistance due to curvature alone, to which should be added the full length of travel.
§ Small-sized socket-welding fittings are equivalent to miter elbows and miter tees.

4.2.2 Damaged Threads. Pipe with threads which are stripped, chipped, corroded, or otherwise damaged shall not be used. If a weld opens during the operation of cutting or threading, that portion of the pipe shall not be used.

4.3 Entry Piping. Gas piping shall, where practical, enter building aboveground and remain in an aboveground and ventilated location.

4.4 Underground Piping. Piping shall not be buried underground inside of buildings unless installed in a casing, except as provided in 4.7.3. The casing shall extend into a normally usable and accessible portion of the building. At the point where the casing terminates in the building, the space between the casing and the gas piping shall be sealed to prevent the possible entrance of any leakage. The casing shall be vented to a safe location above grade with the vent installed in such a way as to prohibit the entrance of water to the casing.

Underground piping when installed below grade through the outer foundation wall of a building shall be encased in a protective pipe of larger diameter. The casing shall be filled with a suitable sealing compound to prevent the entry of gas or water.

Welded joints or compression-type fittings should be used in all underground piping. For provisions covering the fabrication of joints for underground piping, see 4.1.5.

Where soil conditions are unstable and settling of piping or foundation walls may occur, adequate measures shall be provided to prevent excessive stressing of the piping.

Gas piping in contact with earth or other material which may corrode the piping, shall be protected against corrosion in an approved manner. Piping shall not be laid in contact with cinders.

4.5 Aboveground Piping. Outdoor gas piping in industrial plant yards installed aboveground shall be securely supported and located where it will be protected from physical damage.

Gas piping shall not be installed in crawl spaces or unfrequented basement spaces unless adequate ventilation is provided. Piping shall be protected against cor[illegible]n.

4.6 Piping Inside Buildings. When gas piping is installed in buildings, the building structure shall not be weakened by such installation. Before any structural members are cut or notched, special permission shall be obtained from the authority having jurisdiction. Concealed piping shall be avoided where practical.

4.7 Concealed Piping.

4.7.1 Minimum Size. No gas pipe smaller than standard $\frac{1}{2}$ inch iron pipe shall be used in any concealed location.

4.7.2 Piping in Partitions. Concealed gas piping should be located in hollow, rather than solid, partitions.

4.7.3 Piping in Floors. Gas piping in solid floors such as concrete shall be laid in channels in the floor, suitably ventilated, drained and covered to permit access to the piping. When piping in floor channels may be exposed to excessive moisture or corrosive substances, it shall be suitably protected.

4.7.4 Connections in Original Installations. When installing gas piping which is to be concealed, the following connections shall not be used: unions; tubing fittings; running threads; right and left couplings; bushings; and swing joints made by combinations of fittings.

4.7.5 Reconnections. When necessary to insert fittings in gas pipe which has been installed in a concealed location, the pipe may be reconnected by welding, flanges or the use of a ground joint union with the nut center-punched to prevent loosening by vibration. Reconnection of tubing in a concealed location is prohibited.

4.8 Corrosion Control. Corrosion protection shall be provided for all underground piping by any method or combination of methods where investigation indicates that such protection is needed. Consult the gas supplier for procedures covering protection of underground piping.

4.9 Piping Subjected to Low Temperatures. Piping, which carries gas containing moisture, shall be protected against freezing and the formation of hydrates. Such piping, if installed aboveground, shall be properly insulated. Consult the gas supplier for recommendations concerning protection required.

4.10 Drains and Drips. The following provisions for drips, drains, and gradings shall apply only when other than dry gas is supplied and climatic conditions make such provisions necessary.

a. The pitch of the piping shall be no less than $\frac{1}{4}$ inch in 15 feet; and there shall be no traps in the piping.

b. Drips shall be installed only where they can be readily emptied or cleaned.

c. Drips shall not be located where the condensate is likely to freeze.

d. The size of any drip shall not be smaller than the diameter of the pipe to which it is attached.

e. Where expedient for the removal of condensate and scale, a "T" with its bottom opening extended and capped should be located at each low point in the line.

4.11 Hangers, Supports, and Anchors. Piping and equipment shall be supported in a substantial and workmanlike manner, so as to prevent or damp out excessive vibration, and shall be anchored sufficiently to prevent undue strains on connected equipment.

Supports, hangers, and anchors shall be so installed as not to interfere with the free expansion and contraction of the piping between anchors. Suitable spring hangers or sway bracing shall be provided where necessary. All parts of the supporting equipment shall be designed and installed so that they will not be disengaged by movement of the supported piping.

Structural supports or anchors may be welded directly to the pipe, although it is preferable to support the pipe by a member which completely encircles it.

4.12 Expansion and Flexibility. Piping systems shall be designed to have sufficient flexibility to prevent thermal expansion or contraction from causing excessive stresses in the piping material, excessive bending or unusual loads at joints, or undesirable forces or moments at points of connections to equipment or at anchorage or guide points. Formal calculations or model tests shall be required only where reasonable doubt exists as to the adequate flexibility of the system.

Flexibility shall be provided by the use of bends, loops, or offsets; or provision shall be made to absorb thermal changes by the use of expansion joints or couplings of the slip type, by the use of expansion joints of the bellows type, or by the use of "ball" or "swivel" joints. If expansion joints are used, anchors or ties of sufficient strength and rigidity shall be installed to provide for end forces due to fluid pressure and other causes.

Pipe alignment guides shall be used with expansion joints according to the recommended practice of the joint manufacturer.

4.13 Pipe Bends. Pipe bends shall be made in such a way that they are free from buckling, cracks, or other evidence of physical damage. In addition, the gas carrying capacity of the pipe shall not be reduced.

4.14 Outlets. Each outlet, including a valve or cock outlet, shall be securely closed gastight with a threaded plug or cap immediately after installation and shall be left closed until the gas equipment is connected thereto. Likewise, when the gas equipment is disconnected from an outlet and the outlet is not to be used again immediately, it shall be securely closed gastight.

4.15 Special Local Conditions. Where local conditions include earthquake, tornado, unstable ground or flood hazards, special consideration shall be given to increased strength and flexibility of piping supports and connections. Local utility and insurance engineers can be of assistance.

4.16 Testing.

4.16.1 General.

a. Prior to acceptance and initial operation, installed piping shall be pressure tested to assure tightness.

b. In the event repairs or additions are made following the pressure test, the affected piping shall be retested, except that in the case of minor repairs or additions retest may be omitted, when precautionary measures are taken to assure sound construction.

c. Because it is sometimes necessary to divide a piping system into test sections and install test heads, connecting piping, and other necessary appurtenances for testing, it is not required that the tie-in sections of pipe be pressure tested. Tie-in connections, however, shall be tested with soap suds after gas has been introduced and the pressure has been increased sufficiently to give some indications should leaks exist.

d. The test procedure used shall be capable of disclosing all leaks in the section being tested and shall be selected after giving due consideration to the volumetric content of the section and to its location.

e. A piping system may be tested as a complete unit or in sections as the construction progresses. *Under no circumstances shall a valve in a line be used as a bulkhead between gas in one section of the piping system and air or water in an adjacent section.*

f. Regulator and valve assemblies fabricated independently of the piping system in which they are to be installed may be tested at the time of fabrication.

4.16.2 Test Fluid. The test fluid used shall be air, water, or inert gas (e.g., nitrogen, carbon dioxide) provided proper means are used for removing all of the test fluid.

4.16.3 Test Preparation.

a. Whenever possible all joints, including welds, are to be left uninsulated and exposed for examination during the test. If a joint has been previously tested in accordance with this standard, it may be insulated or covered.

b. Piping shall be provided with additional temporary supports, if necessary to support the weight of the test liquid.

c. Expansion joints shall be provided with temporary restraint if required for the additional pressure load under test.

d. Equipment which is not to be included in the test shall be either disconnected from the piping or isolated by blinds.

e. If a pressure test is to be maintained for a period of time and the test liquid in the system is subject to thermal expansion, precautions shall be taken to avoid excessive pressure.

f. All testing of piping systems shall be done with due regard for the safety of employees and the public during the test. Bulkheads, anchorage, and bracing suitably designed to resist test pressures shall be installed if necessary.

4.16.4 Test Pressure.

a. The test pressure to be used shall be no less than 1½ times the proposed maximum working pressure, but not less than 3 psig, irrespective of design pressure. Where the piping system is connected to equipment or appliances having components designed for operating pressures of less than the test pressure, such equipment or appliances shall be disconnected or the components temporarily removed during the test period.

b. Test duration shall be long enough to determine if there are any leaks but not less than one-half hour for each 500 cubic feet of pipe volume or fraction thereof, except that when testing a system having a volume less than 10 cubic feet, the test duration may be reduced to 10 minutes.

4.16.5 Test Records. Records shall be made of each piping installation during the testing. A certification shall be made that all piping has been pressure tested as required by this standard.

4.16.6 Detection of Leaks and Defects.

a. Any reduction of test pressures as indicated by pressure gages shall be deemed to indicate the presence of a leak unless such reduction can be readily attributed to some other cause. While subjected to test pressures the piping system shall be visually examined for signs of leakage or other defects. If air or inert gas is used, all exposed joints (flanged, threaded, welded, etc.) shall be checked by means of soap bubble test.

b. The piping system shall withstand the test pressure specified, without showing any evidence of leakage or other defects. If leakage or other defects appear, the affected portion of the piping system shall be repaired or replaced and retested afterwards.

4.17. Purging or Clearing. In cases where gas in a piping system is to be cleared with air, a slug of inert gas should be introduced to prevent the formation of an explosive mixture at the interface between gas and air. Nitrogen or carbon dioxide can be used for this purpose.

Suitable precautions shall be taken whenever a piping system which is full of air is cleared with a combustible gas. The gas and air shall be cleared to a safe location outside the building.

4.18 Turning on Gas. Before gas is turned into a system of new gas piping, or back into an existing system after being shut off, the entire system shall be checked to determine that there are no open fittings or ends and that all valves at outlets and equipment are closed.

After the piping system has been tested and determined to be free of leakage, the gas may be turned on and the service established.

Immediately after turning the gas into the piping system, the system shall be checked in accordance with 3.6* to determine that no gas is escaping.

5. INSTALLATION OF CONSUMER-OWNED GAS METERS AND REGULATORS

5.1 Installation of Consumer-Owned Gas Meters.

Note: For special meter piping see Appendix B.*

5.1.1 Capacity. Gas meters shall be properly selected for the maximum expected pressure and permissible pressure drop.

5.1.2 Proper Location.

a. Gas meters shall be located in ventilated spaces readily accessible for examination, reading, replacement or necessary maintenance.

b. Gas meters shall not be placed where they will be subjected to damage, such as adjacent to a driveway, in public passages, halls, coal bins, or where they will be subject to excessive corrosion or vibration.

c. Gas meters shall be located at least three feet from sources of ignition.

d. Gas meters shall not be located where they will be subjected to excessive temperatures or sudden extreme changes in temperature. Meter manufacturers will furnish information regarding safe temperature limits.

5.1.3 Supports. Gas meters shall be securely supported and connected to the piping so as not to exert a strain on the meters.

5.1.4 Meter Protection. Meters shall be protected against over pressuring, back pressure and vacuum. See 3.10, 5.2.1 and 5.2.4.

5.2 Installation of Consumer-Owned Gas Pressure Regulators.

5.2.1 Where Required. When the gas supply pressure is higher than that at which the gas equipment is designed to operate or varies beyond the design pressure limits of the equipment, a gas pressure regulator shall be installed.

5.2.2 Regulator Protection. Regulators shall be protected against tampering and physical damage.

5.2.3 Venting. An adequately sized independent vent to a safe point outside the building shall be provided when the regulator at its location is such that a ruptured diaphragm will cause a hazard. A method of preventing water, insects, or foreign materials from entering the vent shall be provided. Under no circumstances shall a regulator be vented to the gas equipment flue or exhaust system.

5.2.4 Over-Pressure Protection Devices. When failure of the pressure regulator could produce downstream pressures which might result in hazardous conditions or damage to equipment, the downstream piping system shall be provided with adequate overpressure protection such as pressure relieving devices, monitoring regulators, and automatic shutoffs.

5.2.5 By-pass Piping. Suitably valved by-passes should be placed around gas pressure regulators where continuity of service is imperative.

* * *

6.9 Connection [of Equipment] to Building Gas Piping.

6.9.1 Rigid and Semi-Rigid Connectors.

a. Gas equipment, except equipment which requires mobility for operation or cleaning, equipment subject to vibration,

[* Not given in this Handbook.—Editor]

or equipment used in more than one location, shall be connected to the building gas piping system with rigid piping of proper size and material. When installation conditions require, minimum lengths of semi-rigid piping may be used. [Refer to 4.1 for proper pipe materials.]

b. Connections to building gas piping should be made at the top or sides of horizontal lines.

c. All connections shall be protected from physical or thermal damage.

6.9.2 Flexible Connections.

a. Where flexible connections are used, they shall be of the minimum practical length and shall not extend from one room to another nor pass through any walls, partitions, ceilings, or floors. Flexible connections shall not be used in any concealed location. They shall be protected against physical or thermal damage and shall be provided with gas shutoff valves, in readily accessible locations in rigid piping upstream from the flexible connections.

b. Gas equipment subject to vibrations may be connected to the building piping system by the use of all metal flexible connectors suitable for the service required.

c. Gas equipment requiring mobility may be connected to the rigid piping by the use of swivel joints or couplings which are suitable for the service required. Where swivel joints or couplings are used, only the minimum number required shall be installed.

d. Portable gas equipment or gas equipment requiring mobility or subject to vibration may be connected to the building gas piping system by the use of flexible hose (not all metal) suitable and safe for the conditions under which it may be used.

REFERENCES

1. A.G.A. Laboratories. *American Standard Listing Requirements for Metal Connectors for Gas Appliances.* (ASA Z21.24–periodically revised) Cleveland, Ohio.
2. ——. *American Standard Installation of Gas Appliances and Gas Piping.* (ASA Z21.30–1964) New York, 1964.
3. Natl. Fire Protect. Assoc. *Standard for the Installation of Gas Appliances and Gas Piping.* (NFPA 54) Boston, Mass., 1964.
4. Mobile Homes Manufacturers Assoc. *American Standard Installation of Plumbing, Heating, and Electrical Systems in Mobile Homes.* (ASA A119.1–1963) New York, 1963.
5. Hansen, D. F. "Why Not Higher Pressures for House Piping?" *Gas Age* 130: 24–7, July 1963.
6. Polyflo flow computer, copyright 1947, by H. J. Smith and B. C. Shebeko.
7. A.G.A. Laboratories. *Installation of Gas Piping and Gas Equipment on Industrial and Other Non-Residential Premises Not Covered by ASA Z21.30.* (Proposed) (ASA Z83.1) Cleveland, Ohio, 1964.
8. ——. *American Standard Listing Requirements for Flexible Connectors of Other Than All-Metal Construction for Gas Appliances.* (ASA Z21.45-periodically revised) Cleveland, Ohio.
9. ——. *American Standard Listing Requirements for Quick-Disconnect Devices for Use With Gas Fuel.* (ASA Z21.41-periodically revised) Cleveland, Ohio.
10. Loughran, P. H. "Effects on Domestic Metering of Higher Utilization Pressures." *A. G. A. Operating Sect. Proc.* 1964: 64-D-35.
11. "The Case for Higher Utilization Pressures." *Am. Gas J.* 191:17+, Dec. 1964.

Chapter 3
Venting and Air Supply

by Richard L. Stone and C. George Segeler

INTRODUCTION

Venting is the removal of combustion products or flue gases to the outdoors thru a system of piping, ducts, flues, vents, chimneys, or stacks especially designed for that purpose. For a gravity gas vent the sole energy source is the sensible heat above room temperature in the gases delivered to its vent collar or draft hood outlet by an appliance that is required to be vented. Vents attached to appliances use this heat energy to produce flow.

Measurement of volume flow rate or weight flow rate in a vent system is the most useful indication of its adequacy. Static draft pressure measurements are useful if evaluation is made of the resistance in both the air supply and the vent systems. An adequate air supply without appreciable flow resistance is assumed in vent sizing recommendations.

Practically every U. S. residential, commercial, and industrial building has one or more chimneys or vent pipes for conducting flue gases from a fuel appliance to the outside air. The number of vents from appliances for all fuels is conservatively estimated to be at least 50,000,000. Probably 30,000,000 flues serve that many gas customers and have done so satisfactorily, without attention, for many years. The few unsatisfactory installations[2] can generally be traced to a failure to observe published venting standards or to an overloading of existing venting systems. Occasionally, venting trouble is caused by an insufficient number of air openings for entrance of combustion and dilution air or by the creation of a negative pressure in a building by an exhaust fan or a vented clothes dryer. Negative pressures may even completely reverse vent system flow.

A vent or chimney is a very desirable adjunct to a house, serving to remove household odors and combustion gases and to induce fresh air. In case of faulty appliance operation, a vent provides a margin of safety. One cu ft of natural gas requires approximately 10 cu ft of air for complete combustion. Usually 2 to 5 cu ft of excess air are needed to complete the combustion reaction, resulting in the formation of 1.0 cu ft CO_2, 2.0 cu ft water vapor, 8.0 cu ft N_2, as well as excess air.

As gas usage for house heating increased, it became apparent that proper venting of gas appliances involved considerations different from those for coal or oil appliances. For example, chimneys for gas appliances could be smaller but might require linings because of sulfate deposits from the prior fuel used. Different materials and construction of vent pipe were also possible. One of the first to attempt to establish formulas for calculating size, height, and capacity for gas vents was F. Wills.[3] T. H. Gilbert[4] extended Wills' work. In 1946 the A.G.A. Domestic Gas Research Committee established a research project[2,5] at Purdue University on venting direct-fired gas heaters when no chimney connections are available. In 1949 this Committee established a research project at the A.G.A. Laboratories for a general study of venting.[6–21] In 1952 Kinkead mathematically described[22] the simultaneous effects of flue gas heat losses and venting flow rate. Other studies[23,24] have been published.

Disposition of Appliance Flue Products

The following extract from Z21.30[1] should be noted:

5.1.1 Appliances Required to be Vented:

Appliances of the following types shall be provided with venting systems or other means for removing the flue gases to the outside atmosphere.

(a) Steam and hot water boilers, warm air furnaces, floor furnaces, and wall furnaces.

(b) Unit heaters and duct furnaces.

(c) Incinerators.

(d) Water heaters with inputs over 5,000 Btu per hour, except as provided under 5.1.2 (f) and (g).

(e) Built-in domestic cooking units listed and marked only as vented units.

(f) Room heaters listed only for vented use. Room heaters listed as "vented and unvented" units may be installed either vented or unvented (see 4.6.1 and 4.6.2†).

(g) Type 2 clothes dryers (see 4.15.2 and 4.15.5†).

(h) Appliances equipped with gas conversion burners.

(i) Other listed appliances which have draft hoods supplied by the appliance manufacturer.

(j) Unlisted appliances, except as provided under 5.1.2 (l).

5.1.2 Appliances Not Required to be Vented:

(a) Listed ranges.

(b) Built-in domestic cooking units listed and marked as unvented units.

(c) Listed hot plates and listed laundry stoves.

(d) Listed Type 1 clothes dryers (see 4.15.2†).

★(e) Listed water heaters with inputs not over 5,000 Btu per hour.

★(f) Automatically controlled instantaneous water heaters which supply water to a single faucet which is attached to and made a part of the appliance (see 4.5.1†).

★(g) A single listed booster type (automatic instantaneous) water heater when designed and used solely for the sanitizing rinse requirements of a National Sanitation Foundation Class 1, 2 or 3 dishwashing machine, provided that the input is limited to 50,000 Btu per hour, the storage capacity is limited to 12.5 gal-

[† Given in Chap. 1 of this Section.—Editor]

lons, and the heater is installed in a commercial kitchen having a mechanical exhaust system.

★(h) Listed refrigerators.
★(i) Counter appliances.
★(j) Room heaters listed for unvented use (see 4.6.1 and 4.6.2†).
★(k) Other appliances listed for unvented use and not provided with flue collars.
★(l) Specialized equipment of limited input such as laboratory burners or gas lights.

When any or all of the appliances starred above (★) are installed so that the aggregate input rating exceeds 30 Btu per hour per cubic foot of room or space in which they are installed, one or more of them shall be provided with a venting system or other approved means for removing the vent gases to the outside atmosphere so that the aggregate input rating of the remaining unvented appliances does not exceed the 30 Btu per hour per cubic foot figure. When the room or space in which they are installed is directly connected to another room or space by a doorway, archway, or other opening of comparable size, which cannot be closed, the volume of such adjacent room or space may be included in the calculations.

FLUE GAS CONDUCTOR TYPES

Descriptions of the types of passageways for conveying flue gases to the outdoors follow:

Masonry and Site-constructed Chimneys. Construction of these chimneys is of masonry, reinforced concrete, or metal. Their performance characteristics have been published.[25–27] When such chimneys, originally constructed for coal or oil appliances, are used for venting gas appliances, the need for chimney liners varies according to locality. In addition to consideration of deposits, factors to be taken into account depend on climate, local chimney construction practices, and experience. In those areas where condensate problems are encountered in unlined chimneys, flue liners should be used with gas heating. Numerous chimney lining materials have been used by various cities (Table 12-10); however, a tile lining at the time of chimney construction is preferred.

Table 12-10 Materials Used As Flue Liners[28]

Material	No. of cities	No. of gas meters	No. of central house heaters
Tile	8	2,646,943	726,694
Stainless steel	3	2,620,000	206,200
Aluminum*	3	391,138	237,262
Lead-coated copper*	1	320,000	40,000
Plastic-coated aluminum*	1	66,836	22,289
Vitroliner	2	540,000	220,000
Transite*	1	42,500	800
Others	2	157,300	42,500

* Not suitable for venting domestic incinerators.

Factory-built or Prefabricated Chimneys. These chimneys are listed by Underwriters' Laboratories for use where flue gas temperatures do not exceed 1000 F continuously and do not exceed 1400 F for infrequent brief periods. These vents are satisfactory for all domestic gas appliances, including incinerators. Chimneys of this type include all-masonry and all-metal designs and combinations of metal and insulation filling.

Type B Gas Vents. These vents, listed by Underwriters' Laboratories, consist of piping of noncombustible, corrosion-resistant material of sufficient thickness, cross-sectional area, and heat insulating quality to avoid excessive temperatures on adjacent combustible material when installed with specified clearances. This type of vent is suitable for venting gas appliances having flue gas temperatures as high as 550 F. Materials used include clay tile, asbestos-cement, aluminum, and steel. All-metal types employ double walls enclosing an air space.

[† Given in Chap. 1 of this Section.—EDITOR]

Type B-W Gas Vents. These vents, of double-wall construction similar to Type B vents but oval in section, are designed *specifically* for installation in walls for venting A.G.A.-approved vented wall furnaces (recessed heaters) and are listed by Underwriter's Laboratories for that purpose only. Accessory parts provide for spacing, fire stopping, and flue collar or header plate attachment.

Type C Vents. These vents are constructed of single-wall sheet copper of not less than No. 24 U. S. gage, or galvanized iron of not less than No. 20 U. S. gage, or other approved corrosion-resistant material.

SPECIAL VENTING METHODS

The following special methods of venting may be used:

Use of Sealed Combustion Appliances. Listed sealed combustion appliances may be vented directly thru an outside wall. These units are designed for integral venting, and draw combustion air from an outside opening adjacent to the vent outlet. If the two openings are in the form of concentric pipe outlets, the vent system is referred to as balanced. If the outlets are at different elevations in the wall, the system is referred to as unbalanced. Both systems have been successfully applied to recessed heaters, water heaters, clothes dryers, etc.

Ventilating Hoods and Exhaust Systems. This method may be used to vent commercial cooking appliances (see Chap. 15) and much industrial equipment satisfactorily. Natural draft exhaust ducts are sometimes used but a mechanical exhaust system is more common.

Venting thru Louvers or Roof Ventilators. Application of this venting method is in high-ceiling industrial buildings, such as those of sawtooth roof construction. One example is the venting of flue products from unvented industrial radiant gas heaters mounted from 15 to 30 ft above the area to be heated.

Power Venting. Mechanically assisted venting may be used where natural draft is inadequate, as when horizontal runs from appliance to chimney are long, chimneys are undersized, or outside wall terminals are necessary. Three types of power venting are in use: (1) a blower in a flueway inducing flue gases from the appliance and discharging outdoors or to the chimney; (2) a power burner supplying air under sufficient pressure to operate against static resistance of 0.5 in. w. c. min; and (3) an exterior blower discharging air thru a venturi section of flue pipe so as to aspirate flue gases from the appliance and force them out the chimney or vent terminal.

VENTING ACCESSORIES

Draft Hoods

Gas appliances designed for use with draft hoods‡ (Fig. 12-1) are intended to operate without additional chimney

‡ In testing for compliance with ASA Standards, such appliances are *not* connected to a flue pipe; i.e., their venting depends solely on provisions incorporated within the appliance.

Vertical draft hood dimensions, in.

A	B	C	D	E	F	G	H	I	J	K	L	M
3	5.5	7.0	0.7	3.8	2.5	4.4	3.0	1.5	2.3	1.5	3.2	0.7
4	7.2	9.5	1.0	5.0	3.5	6.0	4.0	2.0	3.0	2.0	4.5	1.0
5	9.4	10.8	1.5	5.3	4.0	8.0	5.0	2.3	3.5	2.4	4.9	0.9
6	11.5	12.0	1.9	5.6	4.5	9.8	6.0	2.5	4.0	2.7	5.3	0.8
7	13.5	13.9	2.3	6.4	5.3	11.6	7.0	2.9	4.6	3.1	6.2	0.9
8	15.5	15.8	2.7	7.1	6.0	13.4	8.0	3.2	5.3	3.5	7.0	1.0
9	17.5	17.5	3.1	7.7	6.7	15.2	9.0	3.5	5.8	4.0	7.7	1.0
10	19.7	18.8	3.6	7.9	7.3	17.2	10.0	3.8	6.2	4.3	8.3	1.0
11	22.2	20.7	4.3	8.4	8.0	19.6	11.0	4.1	6.6	4.6	9.5	1.5
12	24.7	22.2	5.0	8.7	8.5	22.0	12.0	4.4	7.0	5.0	10.2	1.7

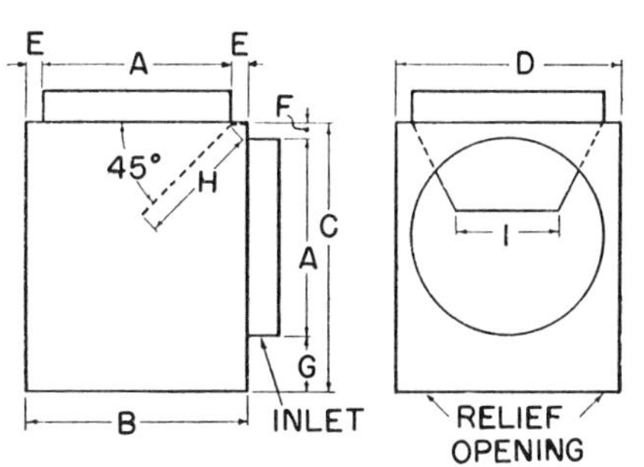

Horizontal draft hood dimensions, in.

A	B	C	D	E	F	G	H	I	J	K	L
3	6	5	9 7/8	1 1/2	2 1/2	1 9/16	3 1/2	3 3/4	1 3/8	2 1/2	4 3/4
4	8	6 3/4	11 5/8	2	3 3/8	2 1/8	4 5/8	5	1 7/8	3 3/8	4 3/4
5	10	8 3/8	13 1/4	2 1/2	4 3/16	2 9/16	5 7/8	6 1/4	2 3/8	4 3/16	4 3/4
6	12	10	15	3	5	3 1/8	7	7 1/2	2 7/8	5	4 3/4
7	14	11 3/4	16 3/4	3 1/2	5 7/8	3 11/16	8 1/8	8 3/4	3 3/8	5 7/8	4 3/4
8	16	13 3/8	18 3/8	4	6 11/16	4 1/8	9 3/8	10	3 7/8	6 11/16	4 3/4
9	18	15	20 1/8	4 1/2	7 1/2	4 11/16	10 1/2	11 1/4	4 3/8	7 1/2	4 3/4
10	20	16 3/4	21 3/4	5	8 3/8	5 1/8	11 5/8	12 1/2	4 7/8	8 3/8	4 3/4
11	22	18 3/8	23 1/2	5 1/2	9 3/16	5 11/16	12 3/4	13 3/4	5 3/8	9 3/16	4 3/4
12	24	20	25 1/4	6	10	6 1/4	14	15	5 7/8	10	4 3/4

Horizontal-to-vertical draft hood dimensions, in.

A	B	C	D	E	F	G	H	I
3	4	4 1/4	4	1/2	1/2	3/4	2	1 1/4
4	5	5 1/2	5	1/2	1/2	1	2 11/16	1 5/8
5	6	6 3/4	6	1/2	1/2	1 1/4	3 5/16	2
6	7	8	7	1/2	1/2	1 1/2	4	2 3/8
7	8	9 1/4	8	1/2	1/2	1 3/4	4 11/16	2 3/4
8	9	10 1/2	9	1/2	1/2	2	5 5/16	3 1/8
9	10	11 3/4	10	1/2	1/2	2 1/4	6	3 1/2
10	11	13	11	1/2	1/2	2 1/2	6 11/16	3 7/8
11	12	14 1/4	12	1/2	1/2	2 3/4	7 5/16	4 1/4
12	13	15 1/2	13	1/2	1/2	3	8	4 5/8

Fig. 12-1 Suggested general dimensions for three types of draft hoods.[1] It should not be construed that these are the only designs that may be used. A hood of any other design that will meet the American Standard Listing Requirements for Draft Hoods, Z21.12–1937, should be satisfactory within the limits of performance specified.

draft. The functions of the draft hood are: (1) to break the chimney draft by introducing dilution air; (2) to relieve downdrafts and prevent them from entering the appliance; and (3) to vent the appliance if the flue is blocked. Certain gas equipment requires chimney draft for its proper functioning. In such cases barometric dampers (Fig. 12-2) are used to control draft.

The following extract from Z21.30[1] should be noted:

3.2.1 Requirements:

(a) Every vented appliance, except incinerators, dual oven type combination ranges, sealed combustion system appliances and units designed for power burners or for forced venting, shall be installed with a draft hood. The draft hood supplied with or forming a part of listed vented appliances shall be installed without alteration, exactly as furnished and specified by the appliance manufacturer. If a draft hood is not supplied by the appliance manufacturer when one is required, it shall be supplied by the installing agency and be of a listed or approved type, and in the absence of other instructions shall be the same size as the appliance flue collar. When a draft hood is required with a conversion burner, it shall be of a listed or approved type supplied by the installing agency or as recommended by the manufacturer.

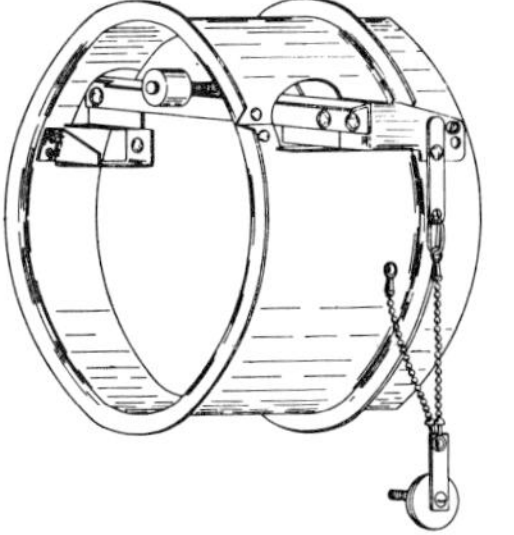

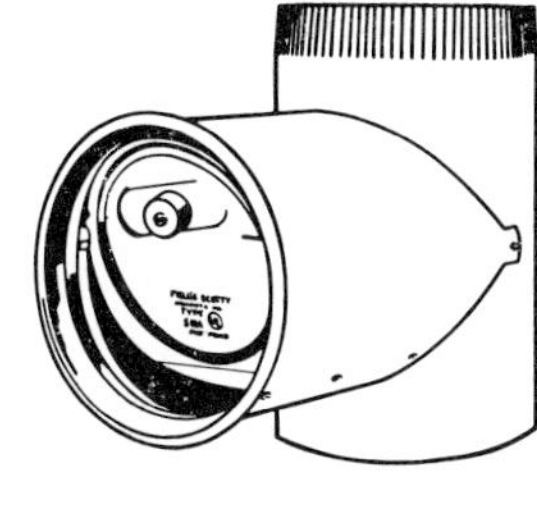

Fig. 12-2 Examples of barometric controls: (left) combination single-acting or double-acting control; (right) single-acting 6 in. incinerator control preset at 0.03 in. w.c. (Field Control Div., H. D. Conkey & Co.)

(b) When the installer determines that a draft hood of special design is needed or preferable for a particular installation, advice of the manufacturer, the serving gas supplier or authority having jurisdiction shall be secured. (For suggested general dimensions of draft hood, see Figs. 2, 3, and 4.*)

[* See Fig. 12-1.—EDITOR]

3.2.2 Installation:

The draft hood shall be in the same room as the combustion air opening of the appliance. In no case shall a draft hood be installed in a false ceiling, in a different room, or in any manner that will permit a difference in pressure between the draft hood relief opening and the combustion air supply. The draft hood supplied for gas conversion burners shall be so located that the burner is capable of safe and efficient operation.

3.2.3 Positioning:

A draft hood shall be installed in the position for which it was designed with reference to the horizontal and vertical planes and shall be located so that the relief opening is not obstructed by any part of the appliance or adjacent construction. The appliance and its draft hood shall be located so that the relief opening is accessible for checking vent operation.

Suggested dimensions of draft hoods are shown in Fig. 12-1. References 18 and 19 give general design data.

Barometric Draft Controls

Appliances with large inputs (over 400 MBtu per hr) are normally installed with barometric controls (Fig. 12-2). Draft hoods larger than 12 in. in diameter become too large for practical use. Some appliances, such as cast iron sectional boilers, use multiple draft hoods in the larger sizes. Draft control is provided by a balanced butterfly gate on a tee

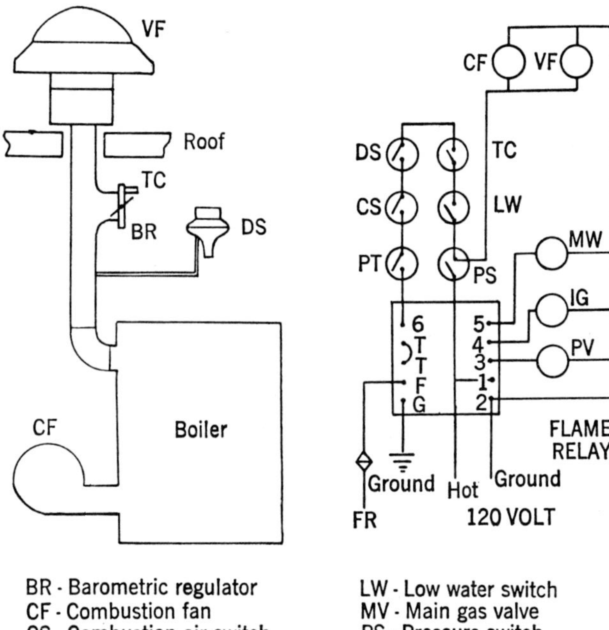

BR - Barometric regulator
CF - Combustion fan
CS - Combustion air switch
DS - Diaphragm switch
FR - Flame rod
IG - Ignition

LW - Low water switch
MV - Main gas valve
PS - Pressure switch
PT - Purge timer
TC - Thermal cut out
VF - Vent fan

Fig. 12-3 Power venting of gas boiler with power burner and accompanying wiring diagram.[29]

fitting, located outside the flue gas flow. The gate may control updraft only (single-acting) or control updraft and relieve downdraft (double-acting). Protection against flue stoppage or downdraft may be obtained thru safety switches, mounted on barometric dampers (such as those shown at the left in Fig. 12-2), which cause the burner to shut off. There are two types of safety switches. One is a thermal switch, also called a spill switch (e.g., TC in Fig. 12-3), which will open, shutting off the burner, as a result of a hot downdraft of sufficient duration, and then close when the proper lower temperature is reached; for example, a switch may be set to shut off the burner when the temperature reaches 160 F and to allow burner operation when the temperature drops to 120 F or below. The second type of safety switch is a mercury switch–delay relay combination. This switch is mounted on the gate so that certain gate positions cause the switch to shut off the burner after a specific time lag. The delay feature prevents momentary instabilities from causing the burner to be shut off.

Vent or Chimney Termination

Vent caps or cowls may be used on gas vents to exclude rain and minimize wind effects; Underwriters' Laboratories–listed designs are available. Figure 12-4 shows an A-type design,

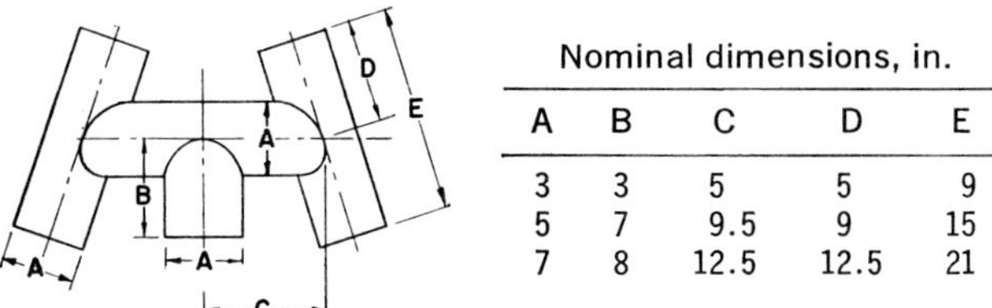

Nominal dimensions, in.				
A	B	C	D	E
3	3	5	5	9
5	7	9.5	9	15
7	8	12.5	12.5	21

Fig. 12-4 A-type cowl.

with dimensions for various sizes. Internal joints in this cowl must be smooth and full size to avoid excessive pressure loss. Figure 12-5 compares the A design with the open top and conical cap designs.

The following extract from Z21.30[1] should be noted:

5.4 TYPE B AND TYPE BW GAS VENTS

5.4.2 Gas Vent Termination:*

(a) Gas vents installed with mechanical exhausters may be terminated not less than 12 inches above the highest point where they pass through a roof surface. (See 5.9.2)

Gas vents installed with listed caps shall terminate in accordance with the terms of the vent cap's listing.

Gas vents installed with approved terminal devices shall be installed in accordance with the terms of their approval.

Gas vents installed without listed caps, approved terminal devices or mechanical exhausters shall extend at least 2 feet above the highest point where they pass through a roof of a building and at least 2 feet higher than any portion of a building within 10 feet.

(b) Type B gas vents shall not terminate less than 5 feet in vertical height above the highest connected appliance draft hood outlet or flue collar.

(c) Type BW gas vents serving a vented wall furnace shall not terminate less than 12 feet in vertical height above the bottom of the heater.

(d) Type B and Type BW gas vents shall terminate in an approved cap or roof assembly with a venting capacity not less than that of the gas vent. The cap or roof assembly shall be of a design to prevent rain and debris from entering the gas vent.

* * *

5.5 MASONRY, METAL, AND FACTORY-BUILT CHIMNEYS

* * *

5.5.2 Termination:

(a) Chimneys shall extend at least 3 feet above the highest point where they pass through the roof of a building and at least 2 feet higher than any portion of any building within 10 feet.

[* See Fig. 12-6.—EDITOR]

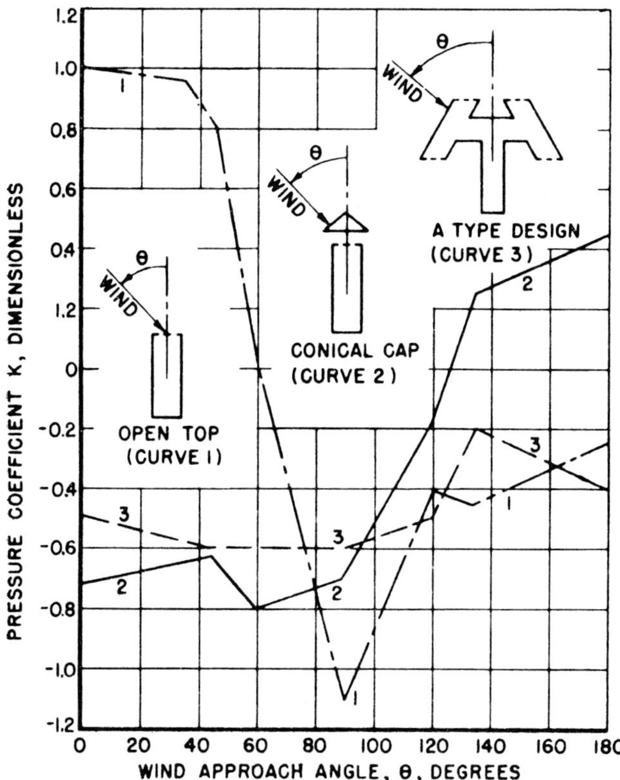

Fig. 12-5 Wind performance characteristics of three vent terminal or vent cowl designs.[30] A negative coefficient, K, indicates that the action of the wind on the vent cowl creates a negative static pressure inside the vent pipe, thus assisting flow. A positive coefficient indicates creation of a positive static pressure in the vent pipe and thus flow opposition.

$$K = \frac{SP_r - SP_w}{VP_w}$$

where:

SP_r = vent pipe static pressure
SP_w = static pressure of the approaching wind stream
VP_w = velocity pressure of the approaching wind stream
(All in consistent dimensions)

Note: The vent pipe velocity pressure is zero for the characteristic curve since the inlet opening of the pipe is capped.

(b) Chimneys shall extend at least 5 feet above the highest connected appliance draft hood outlet or flue collar.

* * *

5.6 SINGLE-WALL METAL PIPE

5.6.1 Installation With Gas Appliances Permitted by 5.3.5(a):*

* * *

(b) Single-wall metal pipe used to vent gas appliances shall comply with the installation provisions of 5.4.2 (a), (b) and (d).

* * *

5.9 SPECIAL VENTING ARRANGEMENTS

5.9.1 Appliances with Sealed Combustion Systems:

* * *

(b) Vent terminals of sealed combustion system appliances shall be located not less than 9 inches from any building opening. A sealed combustion system appliance may be installed in a building opening, such as a window.

5.9.2 Venting System Exhausters:

* * *

(c) The exit terminals of exhauster equipped gas venting systems shall be located not less than 9 inches from any building opening nor less than 2 feet from an adjacent building, and not less than 7 feet above grade when located adjacent to public walkways.

[* Excludes indoor incinerators, appliances which may be converted to the use of solid or liquid fuels, and vented wall furnaces listed for use with Type BW gas vents.—EDITOR]

Fig. 12-6 Vent cowl locations on flat or pitched roofs.[30]

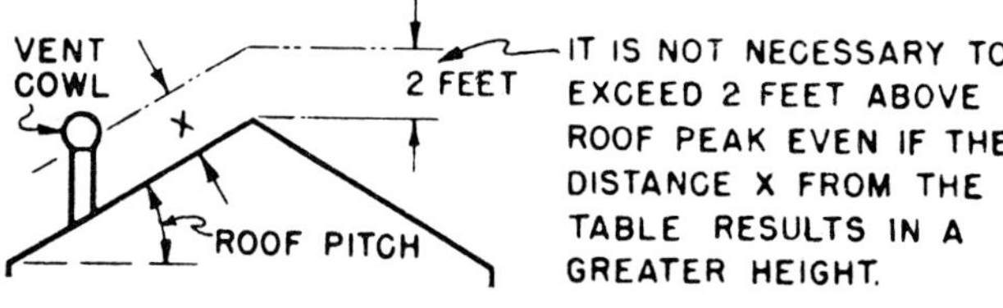

Total length of vent system plus equiv. length of elbows* in terms of vent pipe diameters, dimensionless	Vent pipe appliance heat input loading, MBtu per hr-sq in. of vent pipe cross-sectional area: Up to 3	4	5	6 and up
	Minimum distance, X, between vent cowl opening and roof, ft			
Flat roof to 15° roof pitch				
10	3.0	2.5	2.5	1.5
30–100	2.5	2.0	1.5	1.5
150	2.0	1.5	1.5	1.5
30° roof pitch				
10	5.5	4.0	2.5	2.0
30	5.0	3.5	2.0	2.0
60	4.5	3.0	2.0	2.0
100	4.0	2.5	2.0	2.0
150	3.5	2.0	2.0	2.0

* 90° elbow—20 pipe diameters; 45° elbow—10 pipe diameters.

VENT CAPACITY CALCULATIONS FOR NATURAL DRAFT

A general approach to vent design follows:

1. Draft is due to the weight or density difference between a column of hot flue gases and an equivalent column of outside air.
2. Flue gas flow, produced by the draft, is reduced by friction and by the need to provide initial acceleration to the stream.
3. Flue gas flow is further reduced because of the heat absorption and heat loss of the system.
4. The effects of altitude must be considered.

Design Equations

A semi-empirical equation, Eq. **1**, for calculating vent capacity has been derived from the more exact equations derived by Kinkead[22] to permit more rapid determination of vent capacity; Eq. 1 includes the major variables that affect operation of gravity vents from appliances equipped with draft hoods. With appropriate values assigned to the independent

Table 12-11 Factors in Vent Capacity Calculations Used in Eq. 1

Symbol	Definitions	Units	Span of applicability
I	Maximum vent capacity, expressed as gas appliance heat input	MBtu per hr	All gas-fired appliances having draft hoods
A	Cross-sectional flow area of vent pipe	Sq in.	7 sq in. and up
P	Barometric pressure	In. Hg	Sea level (29.92 in.) up to 15,000 ft
E	Sensible heat fraction in flue gases	Dimensionless	Central furnaces, boilers, unit heaters, water heaters, $E = 0.12$; wall and floor furnaces, room heaters, $E = 0.17$ (exact value may be used; see Eq. **2**)
t_2	Temperature rise above ambient of gases leaving draft hood or entering vent after dilution	°F	For E = 0.12, use $t_2 = 300°$; for E = 0.17, use $t_2 = 350°$ (rise may be found as: t_2 = flue gas rise/dilution ratio)
T	Ambient temperature	°F abs	Generally taken as 520 F (exact value may be used)
H	Total vent height from draft hood to top	Ft	4 to 100+ ft, or as required
R	Total vent system flow resistance	Dimensionless (velocity heads)	Vertical vent, $R = 2.0 + 0.4H/D$*; vent with lateral, $R = 3.5 + 0.4\,(H + L)/D$* (see table 12-12 for components of various arrangements)
D	Vent diameter	In.	For shapes other than round, use D = perimeter/π
L	Horizontal length of lateral connector or breeching	Ft	For $U = 1.2$, $L \leq 3$ ft per in. diam; for $U = 0.6$, $L \leq 8$ ft per in. diam; in no case should L be more than H
U	Overall coefficient of heat transfer from gases thru vent pipe	Btu/hr-°F-sq ft of inside area	$U = 0.0$ to 1.3 (see Table 12-13)

* For metal vents where $D > 24$ in.

parameters (Table 12-11), Eq. **1** may be applied to any size vent of any common material with reasonably accurate results; Eq. **4** is a further simplified version of Eq. **1**.

$$I = \left[\frac{0.067APt_2^{1.5}}{ET^{0.5}(t_2 + T)}\left(\frac{H}{R}\right)^{0.5}\right]\left[1 - 0.039\,\frac{U^{1.5}\,(H + 4L)}{D^2}\right] \qquad \textbf{(1)}$$

Terms are defined in Table 12-11.

Sensible heat fraction in the flue products, E, for natural gas, can be found by Eq. **2**, with constants based on the characteristics of a wide sampling of typical fuel and flue product analyses.

Table 12-12 Flow Resistances for Gas Venting Components
[design values used in determining R (Table 12-11)]

Item or fitting	Resistance, velocity heads: Observed	Resistance, velocity heads: Design
Acceleration head (**must be included in all resistance summations**)	...	1.0
Draft hood	0.5–2.5	1.0
Elbow, 90°	0.5–1.35	0.75
Elbow, 45°	...	0.3
Round 90° tee, in individual vent or chimney breeching	0.75–1.5	1.4
Vent caps, tops, and cowls*	<0.0–>4.0	0.5
Round-to-oval tee fitting	2.5	2.5
Piping losses		
Metal, or asbestos-cement	$0.5(L + H)/D$† for $D \leq 24$ in.	
	$0.4(L + H)/D$† for $D > 24$ in.	
Clay tile, masonry, block	$0.8(L + H)/D$†	
Increaser: A_1 = inlet area, A_2 = outlet area (velocity head at A_2)	$\left(\frac{A_2 - A_1}{A_1}\right)^2$	

* Use of tops with resistances over 1.0 velocity head can seriously reduce vent capacity below tabulated values. Caps can be designed with diffuser sections to produce negative resistance for recovery of the acceleration head.

† Definitions of terms and unit given in Table 12-11.

$$E = 0.00072\left(0.15 + \frac{11}{CO_2\%}\right)c_p t_1 \qquad \textbf{(2)}$$

where:

c_p = specific heat at constant pressure = 0.255 Btu per lb-°F

t_1 = appliance outlet flue gas temperature rise = rise above ambient about 430°F

Example: At 8 per cent CO_2 and an appliance outlet temperature rise, t_1, of 430°F, $E = 0.00072\,(0.15 + 11/8)\,0.255 \times 430 = 0.12$.

To protect against draft hood spillage, the vent flow created must be of sufficient magnitude to induce adequate dilution air flow thru the relief opening in the hood. Dilution air should constitute not less than 40 to 50 per cent of the vent gases.

A heat balance for the flue gases in a draft hood for sensible heat content above ambient level is given by Eq. **3**:

$$W_1 c_{p_1} t_1 = W_2 c_{p_2} t_2 \qquad \textbf{(3)}$$

where:

W = gas weight
c_p = gas specific heat
t = gas temperature rise
1 and 2 = subscripts at entrance and exit, respectively, of draft hood

Equation **3** considers the dilution factor as W_2/W_1 or t_1/t_2, neglecting slight radiation losses and specific heat change. Thus, assuming a dilution factor of 1.4, $t_1 = 1.4t_2$. It is then necessary to choose a characteristic value of outlet flue gas temperature for the appliance, t_1.

For typical approved gas appliances, Eq. **4**, a simplified version of Eq. **1**, can be used for vents and stacks 6.0 in. in diameter and larger by making the following substitutions:

$P = 29.92$ in. Hg; $E = 0.12$; $t_2 = 300$ F; $T = 520$ F; and $D^2 = 4A/\pi$.

$$I = 4.65 \left(\frac{H}{R}\right)^{0.5} [A - 0.031U^{1.5}(H + 4L)] \qquad \textbf{(4)}$$

where:

I = sea level heat input to gas appliance, MBtu per hr
A = cross-sectional flow area of vent, sq in.
H = total vent height from draft hood to top, ft
R = total flow resistance, velocity heads (Table 12-12)
U = overall heat transfer coefficient, Btu/hr-°F-sq ft of inside area (Table 12-13)
L = horizontal length of lateral connector, ft

Example: Find the venting capacity of a steel stack; D = 40 in. diam (area = A = 1256 sq in.), H = 60 ft high, with a breeching L = 10 ft), serving a large efficient gas boiler (E = 0.12).

Flow resistances from Table 12-12:

Acceleration head	1.0
Draft hood	1.0
Tee used in breeching	1.4
70 ft piping = $0.4(H + L)/D$ =	0.7
Total resistance = R =	4.1

From Table 12-13, $U = 1.3$, $L = 10$ ft. Substituting in Eq. **4**:

$$I = 4.65 \left(\frac{60}{4.1}\right)^{0.5} [1256 - 0.031 \times 1.3^{1.5}(60 + 40)]$$
$$= 4.65 \times 3.82 \times (1256 - 4.6) = 22{,}200 \text{ MBtu per hr}$$

Table 12-13 Overall Heat Transfer Coefficients, U, of Various Chimneys and Vents

(design values used in Eqs. **1** and **4**)

Material	U, Btu/hr-°F-sq ft of inside area: Observed	Design	Remarks
Industrial steel stacks	...	1.3	Under wet wind
Clay or iron sewer pipe	1.3–1.4	1.3	Used as single-wall material
Asbestos-cement gas vent	0.72–1.42	1.2	Tested per UL Std 441
Black or painted steel stove pipe	...	1.2	Comparable to weathered galvanized steel
Single-wall galvanized steel	0.31–1.38	1.0	Depends on surface condition and exposure
Single-wall unpainted pure aluminum	...	1.0	No. 1100 or other bright surface aluminum alloy
Brick chimney, tile lined	0.5–1.0	1.0	Residential construction as used for gas appliances
Double-wall gas vent, ¼ in. air space	0.15–0.86	0.6	Galvanized steel outer pipe, pure aluminum inner pipe; tested per UL Std 441
Double-wall gas vent, ½ in. air space	0.37–1.04	0.5	
Insulated prefabricated chimney	...	0.3	Solid insulation meets UL Std 103

Note that heat losses reduce capacity by less than 0.5 per cent even with an uninsulated stack. If a barometric control had been used, the the capacity shown would be slightly conservative, depending on the draft setting.

The input for zero draft with a barometric draft control may be found for this example by reducing the total flow resistance by one velocity head, making R = 3.1. Then the input with a barometric draft control set at very low draft and with no draft hood, would be, from Eq. **4**:

$$I = 4.65 \left(\frac{60}{3.1}\right)^{0.5} (1256 - 4.6) = 25{,}600 \text{ MBtu per hr}$$

In addition to Eqs. **1** and **4**, a number of others[19] have been formulated for calculating vent capacities. These equations were used to develop many of the capacity tables presented in this chapter.

DESIGN OF VENTS FOR A SINGLE NATURAL DRAFT APPLIANCE

Venting arrangements for single natural draft appliances are shown in Fig. 12-7. Table 12-14, which gives sizing recommendations for Type B double-wall vents, is based on the factors of Eq. **1** and was computed using the equations of Ref. 22. Tables 12-15a, 12-15b, and 12-16a, covering other vent materials, are the results of extensive research.[22]

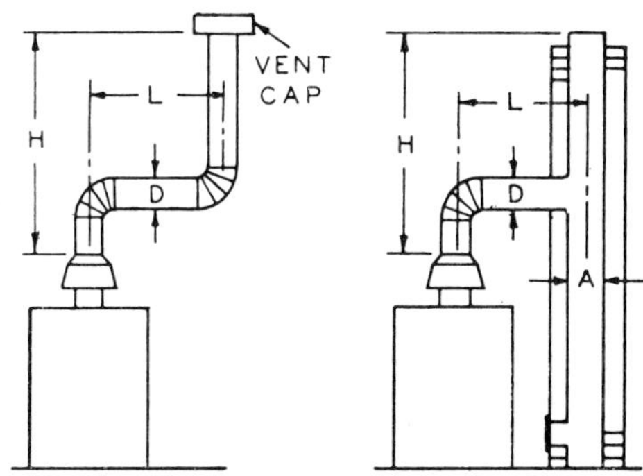

Fig. 12-7 Venting a single natural draft appliance: (left) double-wall or asbestos-cement Type B vents or single-wall metal vents (see Tables 12-14, 12-15a, and 12-15b); (right) masonry chimney (see Table 12-16a).[1]

Notes:[1]

1. If the vent size determined from Tables 12-14 thru 12-16a is less than the size of the draft hood, the smaller vent may be used as long as the vent height, H, is at least 10 ft.
2. Vents for draft hoods 12 in. in diameter or less should not be reduced more than one size (12 in. to 10 in. is a one-size reduction). For larger gas-burning equipment, reductions of more than two sizes are not recommended (24 in. to 20 in. is a two-size reduction).
3. Regardless of the vent size shown, do not connect a 4 in. draft hood to a 3 in. vent.
4. The 0 lateral length, L, applies only to a straight vertical vent attached to a top outlet draft hood.

Sizing of Vents for a Single Appliance

For sizing purposes use Tables 12-14 thru 12-16a. Follow down the column headed by the selected vent diame-

Table 12-14 Capacity of Type B Double-Wall Vents Serving a Single Draft Hood-Equipped Appliance[1]

(see Fig. 12-7 for an explanation of the terms and qualifications on the use of this table)

Total vent height, H, ft	Lateral length, L, ft	Vent diameter, D, in.													
		3	4	5	6	7	8	10	12	14	16	18	20	22	24
		Maximum appliance input rating at sea level,* I, MBtu per hr													
6	0	46	86	141	205	285	370	570	850	1170	1530	1960	2430	2950	3520
	2	36	67	105	157	217	285	455	650	890	1170	1480	1850	2220	2670
	6	32	61	100	149	205	273	435	630	870	1150	1470	1820	2210	2650
	12	28	55	91	137	190	255	406	610	840	1110	1430	1795	2180	2600
8	0	50	94	155	235	320	415	660	970	1320	1740	2220	2750	3360	4010
	2	40	75	120	180	247	322	515	745	1020	1340	1700	2110	2560	3050
	8	35	66	109	165	227	303	490	720	1000	1320	1670	2070	2530	3030
	16	28	58	96	148	206	281	458	685	950	1260	1600	2035	2470	2960
10	0	53	100	166	255	345	450	720	1060	1450	1925	2450	3050	3710	4450
	2	42	81	129	195	273	355	560	850	1130	1480	1890	2340	2840	3390
	10	36	70	115	175	245	330	525	795	1080	1430	1840	2280	2780	3340
	20	NR	60	100	154	217	300	486	735	1030	1360	1780	2230	2720	3250
15	0	58	112	187	285	390	525	840	1240	1720	2270	2900	3620	4410	5300
	2	48	93	150	225	316	414	675	985	1350	1770	2260	2800	3410	4080
	15	37	76	128	198	275	373	610	905	1250	1675	2150	2700	3300	3980
	30	NR	60	107	169	243	328	553	845	1180	1550	2050	2620	3210	3840
20	0	61	119	202	307	430	575	930	1350	1900	2520	3250	4060	4980	6000
	2	51	100	166	249	346	470	755	1100	1520	2000	2570	3200	3910	4700
	10	44	89	150	228	321	443	710	1045	1460	1940	2500	3130	3830	4600
	20	35	78	134	206	295	410	665	990	1390	1880	2430	3050	3760	4550
	30	NR	68	120	186	273	380	626	945	1270	1700	2330	2980	3650	4390
30	0	64	128	220	336	475	650	1060	1550	2170	2920	3770	4750	5850	7060
	2	56	112	185	280	394	535	865	1310	1800	2380	3050	3810	4650	5600
	20	NR	90	154	237	343	473	784	1185	1650	2200	2870	3650	4480	5310
	40	NR	NR	NR	200	298	415	705	1075	1520	2060	2700	3480	4270	5140
40	0	66	132	228	353	500	685	1140	1730	2400	3230	4180	5270	6500	7860
	2	59	118	198	298	420	579	960	1420	2000	2660	3420	4300	5260	6320
	20	NR	96	167	261	377	516	860	1310	1830	2460	3200	4050	5000	6070
	40	NR	NR	NR	223	333	460	785	1205	1710	2310	3020	3840	4780	5820
60	0	NR	136	236	373	535	730	1250	1920	2700	3650	4740	6000	7380	9000
	2	NR	125	213	330	470	650	1060	1605	2250	3020	3920	4960	6130	7400
	30	NR	NR	170	275	397	555	930	1440	2050	2780	3640	4700	5730	7000
	60	NR	NR	NR	NR	334	475	830	1285	1870	2560	3380	4330	5420	6600
80	0	NR	NR	239	384	550	755	1290	2020	2880	3900	5100	6450	8000	9750
	2	NR	NR	217	350	495	683	1145	1740	2460	3320	4310	5450	6740	8200
	40	NR	NR	NR	275	404	570	980	1515	2180	2980	3920	5000	6270	7650
	80	NR	NR	NR	NR	NR	NR	850	1420	2000	2750	3640	4680	5850	7200
100	0	NR	NR	NR	400	560	770	1310	2050	2950	4050	5300	6700	8600	10300
	2	NR	NR	NR	375	510	700	1170	1820	2550	3500	4600	5800	7200	8800
	50	NR	NR	NR	NR	405	575	1000	1550	2250	3100	4050	5300	6600	8100
	100	NR	NR	NR	NR	NR	NR	870	1430	2050	2850	3750	4900	6100	7500

NR—not recommended; vent may be liable to spillage and/or condensation.
* Use sea level input capacity when calculating vent size for high-altitude installation.

ter to the row indicating the vent height and lateral length desired; the maximum appliance input is found at the intersection.

A vent pipe diameter, selected from Tables 12-14 thru 12-16a, equal to or slightly greater than that of the draft hood outlet should satisfactorily vent single gas appliances (excluding masonry chimneys). For single-appliance vent systems that have laterals and flow turns, increasing the vent diameter to one size larger than that of the draft hood outlet tends to increase the venting capacity because of lessened resistance. If practical, it is better to relocate the appliance to permit use of shorter laterals and fewer elbows.

Condensation. Figure 12-8 shows condensation limit curves for single-appliance vents. Based on a dew point of 120 F, the upper curve represents the limit for a vent system with condensation-free inside surfaces. The lower curve represents the limit for bulk condensation–free flue gases. The maximum input values in Tables 12-15a, 12-15b, and 12-16a, marked by a dagger, fall between these curves.

Whether continuous condensation will occur between the two curves depends largely on the individual locality. In cold climates the limit will be close to the upper curve, in warm climates close to the lower one. Selection of either curve or one in between should be based on local experiences. All vent pipe located outside the building should be multiplied by a factor of 2 in calculating the total vent length (abscissas in Fig. 12-8), to obtain reasonable and conservative results.

For each vent connector size, the minimum recommended internal chimney area and diameter are given in the last two lines of Table 12-16a. Dimensions of masonry chimney liners

Table 12-15a Capacity of Single-Wall or Type B Asbestos Vents Serving a Single Draft Hood–Equipped Appliance[1,21]

(see Fig. 12-7 for an explanation of the terms and qualifications on the use of this table)

Total vent height, H, ft	Lateral length, L, ft	Vent diameter, D, in. 3	4	5	6	7	8	10	12
		Maximum appliance input rating at sea level,* I, MBtu per hr							
6	0	39	70	116	170	232	312	500	750
	2	31	55	94	141	194	260	415	620
	5	28	51	88	128	177	242	390	600
8	0	42	76	126	185	252	340	542	815
	2	32	61	102	154	210	284	451	680
	5	29	56	95	141	194	264	430	648
	10	24†	49	86	131	180	250	406	625
10	0	45	84	138	202	279	372	606	912
	2	35	67	111	168	233	311	505	760
	5	32	61	104	153	215	289	480	724
	10	27†	54	94	143	200	274	455	700
	15	NR	46†	84	130	186	258	432	666
15	0	49	91	151	223	312	420	684	1040
	2	39	72	122	186	260	350	570	865
	5	35†	67	110	170	240	325	540	825
	10	30†	58†	103	158	223	308	514	795
	15	NR	50†	93†	144	207	291	488	760
	20	NR	NR	82†	132†	195	273	466	726
20	0	53†	101	163	252	342	470	770	1190
	2	42†	80	136	210	286	392	641	990
	5	38†	74†	123	192	264	364	610	945
	10	32†	65†	115†	178	246	345	571	910
	15	NR	55†	104†	163	228	326	550	870
	20	NR	NR	91†	149†	214†	306	525	832
30	0	56†	108†	183	276	384	529	878	1370
	2	44†	84†	148†	230	320	441	730	1140
	5	NR	78†	137†	210	296	410	694	1080
	10	NR	68†	125†	196†	274	388	656	1050
	15	NR	NR	113†	177†	258†	366	625	1000
	20	NR	NR	99†	163†	240†	344	596	960
	30	NR	NR	NR	NR	192†	295†	540	890
50	0	NR	120†	210†	310†	443†	590	980	1550
	2	NR	95†	171†	260†	370†	492	820	1290
	5	NR	NR	159†	234†	342†	474	780	1230
	10	NR	NR	146†	221†	318†	456†	730	1190
	15	NR	NR	NR	200†	292†	407†	705	1130
	20	NR	NR	NR	185†	276†	384†	670†	1080
	30	NR	NR	NR	NR	222†	330†	605†	1010

NR—not recommended; vent may be liable to spillage and/or condensation.
* Use sea level input capacity when calculating vent size for high-altitude installation.
† Possibility of continuous condensation, depending on locality; consult local utility and/or local codes.

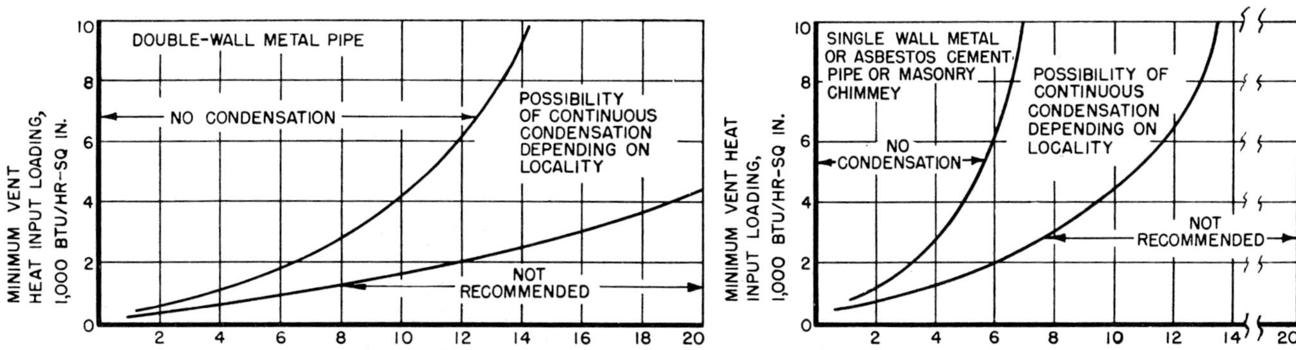

[TOTAL FEET (INSIDE BUILDING) OF EITHER VENT PIPE OR VENT CONNECTOR PIPE AND CHIMNEY + TWICE THE TOTAL FEET (OUTSIDE BUILDING) OF EITHER VENT PIPE OR CHIMNEY] / DIAMETER IN INCHES OF EITHER THE VENT OR INSIDE OF CHIMNEY

Fig. 12-8 Single-appliance vent system condensation limits.[21]

Table 12-15b Capacity of Single-Wall or Type B Asbestos Vents Serving a Single Draft Hood–Equipped Appliance

(see Fig. 12-7 for explanation of the terms and qualifications on the use of this table)

Total vent height, *H*, ft	Lateral length, *L*, ft	Vent diameter, *D*, in. 14	16	18	20	22	24
		Maximum appliance input rating at sea level, *I*, MBtu per hr					
8	0	1160	1520	...	...	...	...
	2*	970	1270	...	...	...	...
	10	900	1190	...	...	...	...
10	0	1320	1750	2160	2940	...	...
	2*	1100	1460	1900	2450	...	...
	10	1020	1370	1800	2340	...	...
15	0	1450	1990	2580	3240	4020	4910
	2*	1210	1660	2150	2700	3350	4100
	10	1120	1560	2040	2580	3210	3960
	15	1090	1510	1970	2500	3120	3850
20	0	1680	2290	3000	3780	4680	5750
	2*	1400	1910	2500	3150	3900	4800
	10	1300	1800	2370	3000	3740	4630
	20	1200	1660	2210	2830	3540	4430
30	0	1920	2630	3420	4320	5400	6850
	2*	1600	2190	2850	3600	4500	5700
	10	1490	2060	2700	3440	4320	5500
	20	1370	1910	2520	3230	4100	5260
	30	1290	1800	2410	3080	3780	4950
50	0	2200	3000	3900	5000	6240	7800
	2*	1830	2500	3250	4170	5200	6500
	10	1700	2350	3080	4000	5000	6260
	20	1570	2170	2870	3760	4740	6000
	30	1470	2060	2740	3560	4550	5850
80	0	2520	3400	4450	5550	6950	8640
	2*	2100	2830	3700	4700	5800	7200
	10	1950	2660	3510	4500	5560	6950
	20	1800†	2460	3270	4220	5280	6650
	30	1690†	2330†	3120	4000	5080	6460
	40	1570†	2210†	2980†	3850	4810	6110

* Lateral length of 2 ft or 2 pipe diameters, whichever is larger.
† Possibility of continuous condensation, depending on locality; consult local utility and/or local codes.

and their circular equivalents are given in Table 12-16b. For rectangular chimney flue passageways, equivalent chimney diameter may be estimated.

Masonry Chimneys. Additional considerations for venting a single appliance follow:

1. Masonry chimneys will rarely be specifically constructed for the venting of individual gas appliances of rated inputs less than 80 MBtu per hour (vent connector sizes less than 5 in.).

2. When chimneys have sectional liners, some slight mismatch of the liner or protrusion at the joints may occur, causing increased flow resistance. Thus, the chimney diameter is usually specified one size larger than the vent connector.

3. Locating the entire vertical rise of a chimney outside a building is not recommended in cold climates. The extent of such exposure depends on experiences in that locality.

General Design Factors

Vent tables for single appliances apply to systems using a draft hood with a top outlet and a maximum of (1) two 90° elbows, or (2) one 90° elbow and a tee, or (3) one 90° elbow and two 45° elbows, or (4) four 45° elbows. If additional turns are needed, a 90° elbow may be considered the equivalent of 20 pipe diameters of lateral or horizontal vent connector pipe. A draft hood with a side outlet is considered equivalent to a top outlet opening type plus a 90° elbow.

Metal chimneys are included in the designation of the single-wall metal or asbestos-cement vent pipe. When designing for such chimneys, 7 to 10 in. in diameter for domestic installations, the vent connector diameter and lateral length should be used for calculating the maximum allowable input and total length limits. For a vent connector, 5 in. or larger in diameter, vertical chimney size will not greatly affect the maximum allowable input to the vent connector if the chimney size is at least as large as the vent connector.

Insulated factory-built chimneys may be sized according to double-wall metal vent pipe specifications (Table 12-14) if the vent connector is to be of double-wall pipe. If such a chimney is used with a single-wall connector, the double-wall vent tables may be used provided the permissible lateral length is reduced by 50 per cent. The same procedure may be followed for vent systems composed of double-wall vent pipe with single-wall connectors. Reasonably smooth and gastight piping junctions are assumed in all cases.

DESIGN OF MULTIPLE VENTS FOR NATURAL DRAFT APPLIANCES

A multiple vent may be defined as a system for venting two or more appliances installed on one level thru a common conduit. For such a system, operation of only one appliance, usually that having the smallest input, produces the most critical phase of vent performance. In addition, with only one appliance operating, dilution air enters the common vent thru the draft hood relief opening of the other appliance(s), lowering the flue gas temperature. Multiple-appliance venting, accordingly, requires a somewhat different design approach from that for single vents.

Lateral Vent Connectors. Figure 12-9 shows, for both single- and multiple-vent systems, typical trends relating to the increase in vent height required to offset an increase in vent connector lateral length. The lower curve of Fig. 12-9 shows that for single-appliance vent connector lengths up to 40 pipe diameters, the required increase in vent height is almost proportional to the increase in lateral length. Compensation for frictional losses in a lateral vent connector is thus possible by an increase in the vent height. The upper curve, applying to multiple venting systems with only the smallest appliance operating, shows that increases in lateral length beyond approximately 20 pipe diameters require considerable increases in common vent height to compensate for the additional frictional losses. Entry of dilution air at the draft hood of the appliance not in operation is responsible. With only the larger appliance operating, the resulting performance curve would fall between the two shown.

From these curves it may be concluded that short vent connector laterals are necessary for satisfactory operation of a multiple-appliance vent system. Flow resistance of the lateral may best be compensated for by increasing the vent connector diameter, not the vent height.

Vertical Vent Connectors. Figure 12-9 also shows the effects of vent connector vertical rise on the performance of single- and multiple-appliance vent systems. The lower curve indicates that for single-appliance vents a draft reduction due to a *decrease* in vertical rise may be compensated for by an equal or slightly larger increase in the common vent height. For multiple-appliance vents, however, a much greater increase in common vent height would be needed; dilution air at the draft hood of the appliance *not in operation*

Table 12-16a Capacity of Masonry Chimneys Serving a Single Draft Hood–Equipped Appliance[1,21]

(see Fig. 12-7 for explanation of the terms and qualifications on the use of this table)

Total vent height, H, ft	Lateral length, L, ft	Vent connector diameter, D, in. (To be used with chimney areas not less than those given at bottom)							
		3	4	5	6	7	8	10	12
		Maximum appliance input rating at sea level,* I, MBtu per hr							
6	2	28	52	86	130	180	247	400	580
	5	25†	48	81	118	164	230	375	560
8	2	29	55	93	145	197	265	445	650
	5	26†	51	87	133	182	246	422	638
	10	22†	44†	79	123	169	233	400	598
10	2	31	61	102	161	220	297	490	722
	5	28†	56	95	147	203	276	465	710
	10	24†	49†	86	137	189	261	441	665
	15	NR	42†	79†	125	175	246	421	634
15	2	35†	67	113	178	249	335	560	840
	5	32†	61	106	163	230	312	531	825
	10	27†	54†	96	151	214	294	504	774
	15	NR	46†	87†	138	198	278	481	738
	20	NR	NR	73†	128†	184	261	459	706
20	2	38†	73	123	200	273	374	625	950
	5	35†	67†	115	183	252	348	594	930
	10	NR	59†	105†	170	235	330	562	875
	15	NR	NR	95†	156	217	311	536	835
	20	NR	NR	80†	144†	202	292	510	800
30	2	41†	81†	136	215	302	420	715	1110
	5	NR	75†	127†	196	279	391	680	1090
	10	NR	66†	113†	182†	260	370	644	1020
	15	NR	NR	105†	168†	240†	349	615	975
	20	NR	NR	88†	155†	223†	327	585	932
	30	NR	NR	NR	NR	182†	281†	544	865
50	2	NR	91†	160†	250	350†	475	810	1240
	5	NR	NR	149†	228†	321†	442	770	1220
	10	NR	NR	136†	212†	301†	420†	728	1140
	15	NR	NR	124†	195†	278†	395†	695	1090
	20	NR	NR	NR	180†	258†	370†	660†	1040
	30	NR	NR	NR	NR	NR	318†	610†	970
Chimney									
Inside area, A, sq in. min		**19**	**19**	**28**	**38**	**50**	**63**	**95**	**132**
ID, equivalent nearest in.		5	5	6	7	8	9	11	13

NR—not recommended; vent may be liable to spillage and/or condensation.

* Use sea level input capacity when calculating vent size for high-altitude installation.

† Possibility of continuous condensation, depending on locality; consult local utility and/or local codes.

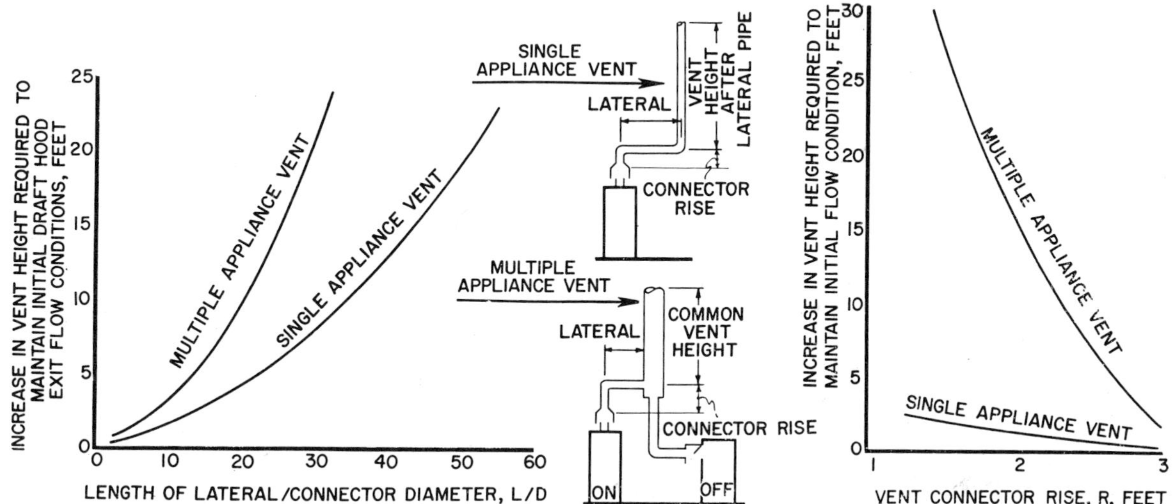

Fig. 12-9 Vent height increase required to overcome the resistance of changes in vent connectors in typical single- and multiple-appliance vent systems: (left) increases in lateral connector length; (right) decreases in connector rise.[21]

Table 12-16b Masonry Chimney Liner Dimensions with Circular Equivalents*,[1]

Nominal liner size, in.	Inside dimensions of liner, in.	Inside diameter or equivalent diameter, in.	Equivalent area, sq in.
4 × 8	2½ × 6½	4	12.3
		5	19.6
		6	28.3
		7	38.3
8 × 8	6¾ × 6¾	7.4	42.7
		8	50.3
8 × 12	6½ × 10½	9	63.6
		10	78.5
12 × 12	9¾ × 9¾	10.4	83.3
		11	95.0
12 × 16	9½ × 13½	11.8	107.5
		12	113.0
		14	153.9
16 × 16	13¼ × 13¼	14.5	162.9
		15	176.7
16 × 20	13 × 17	16.2	206.1
		18	254.4
20 × 20	16¾ × 16¾	18.2	260.2
		20	314.1
20 × 24	16½ × 20½	20.1	314.2
		22	380.1
24 × 24	20¼ × 20¼	22.1	380.1
		24	452.3
24 × 28	20¼ × 24¼	24.1	456.2
28 × 28	24¼ × 24¼	26.4	543.3
		27	572.5
30 × 30	25½ × 25½	27.9	607.0
		30	706.8
30 × 36	25½ × 31½	31	749.9
		33	855.3
36 × 36	31½ × 31½	34.4	929.4
		36	1017.9

* When liner sizes differ dimensionally from those listed in the table, equivalent diameters may be determined from published tables for square and rectangular ducts of equivalent carrying capacity or by other engineering methods.

is mainly responsible. From this curve it may be noted that 1 ft of extra connector rise is as effective as 10 to 20 ft of added common vent height in a multiple-appliance vent.

Sizing of Systems

Venting arrangements are shown in Fig. 12-10. Table 12-17 gives sizing recommendations for Type B vent systems of individual vent connectors and common vents—both double-wall piping.

Definitions (see Figs. 12-9 and 12-10):

Total vent height, H, is the vertical distance from the *highest* draft hood outlet to top of vent (also called *least total height*).

Connector rise, R, is the vertical distance from the draft hood outlet to the point of juncture of the next connector.

A *common vent* is the portion of the venting system above the lowest interconnection (termed a *vertical,* or V, type when vertical; otherwise termed a *lateral,* or L, type). A common vent with a manifold or offset at the base is regarded as an L type.

Using Part A of Table 12-17, determine the vent connector size as follows: Entering at the left, follow the horizontal line thru the selected system vent height, H, and individual appliance connector rise, R, to the column showing the input rating of the individual appliance to be vented. The vent connector diameter, D, is found at the head of that column. Regardless of tabulated size, the vent connector diameter of a multiple appliance vent should never be smaller than the draft hood outlet size.

Using Part B of Table 12-17, determine the common vent size as follows: Entering at the left, follow the horizontal line thru the foregoing selection of H to the column showing the *sum* of the individual appliance input ratings. The diameter of the common vent is found at the head of that column. Regardless of tabulated size, the common vent must always be at least as large as the largest connector. If two connectors are the same size, the common vent must be one size larger or more. Interconnection fittings must be the same size as the common vent.

Multiple-vent design data prepared from extensive research[21] are given for single-wall metal and Type B asbestos-cement (Table 12-18) and masonry materials (Table 12-19). Each table is in two parts; the upper part pertains to the vent connector, the lower to the common vent. Use of these tables is the same as given for Table 12-17.

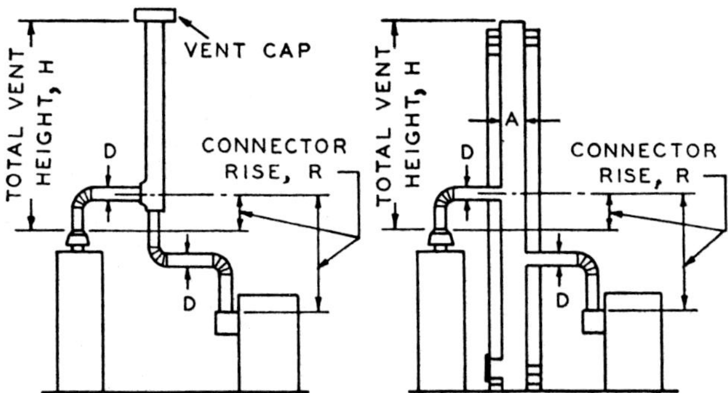

Fig. 12-10 Venting multiple natural draft appliances: (left) double-wall or asbestos-cement Type B vents or single-wall metal vents (see Tables 12-17 and 12-18); (right) masonry chimney (see Table 12-19).[1]

Notes:

1. Maximum vent connector length is $1^1/_2$ ft for every inch of connector diameter. Greater lengths require increase in size, rise, or total vent height to obtain full capacity.
2. Each 90° turn in excess of the first two reduces the connector capacity by 10 per cent.
3. Each 90° turn in the common vent reduces capacity by 10 per cent.
4. Where possible, locate vent closer to or directly over smaller appliance connector.
5. Connectors must be equal to or larger than draft hood outlets.
6. If both connectors are the same size, the common vent must be at least one size larger, regardless of tabulated capacity.
7. The common vent must be equal to or larger than the largest connector.
8. Interconnection fittings must be the same size as the common vent.
9. Use sea level input rating when calculating vent size for high-altitude installation.

Table 12-17 Capacity of Type B Double-Wall Vents Serving Two or More Appliances[1]

(see Fig. 12-10 for an explanation of the terms and qualifications on the use of this table)

A. Vent Connector

Total vent height, H, ft	Connector rise, R, ft	Vent connector diameter, D, in.													
		3	4	5	6	7	8	10	12	14	16	18	20	22	24
		Maximum appliance input rating, at sea level, I, MBtu per hr													
6	1	26	46	72	104	142	185	289	416	577	755	955	1180	1425	1700
	2	31	55	86	124	168	220	345	496	653	853	1080	1335	1610	1920
	3	35	62	96	139	189	248	386	556	740	967	1225	1510	1830	2180
8	1	27	48	76	109	148	194	303	439	601	805	1015	1255	1520	1810
	2	32	57	90	129	175	230	358	516	696	910	1150	1420	1720	2050
	3	36	64	101	145	198	258	402	580	790	1030	1305	1610	1950	2320
10	1	28	50	78	113	154	200	314	452	642	840	1060	1310	1585	1890
	2	33	59	93	134	182	238	372	536	730	955	1205	1490	1800	2150
	3	37	67	104	150	205	268	417	600	827	1080	1370	1690	2040	2430
15	1	30	53	83	120	163	214	333	480	697	910	1150	1420	1720	2050
	2	35	63	99	142	193	253	394	568	790	1030	1305	1610	1950	2320
	3	40	71	111	160	218	286	444	640	898	1175	1485	1835	2220	2640
20	1	31	56	87	125	171	224	347	500	740	965	1225	1510	1830	2190
	2	37	66	104	149	202	265	414	596	840	1095	1385	1710	2070	2470
	3	42	74	116	168	228	300	466	672	952	1245	1575	1945	2350	2800
30	1	33	59	93	134	182	238	372	536	805	1050	1330	1645	1990	2370
	2	39	70	110	158	215	282	439	632	910	1190	1500	1855	2240	2670
	3	44	79	124	178	242	317	494	712	1035	1350	1710	2110	2550	3040
40	1	35	62	97	140	190	248	389	560	850	1110	1405	1735	2100	2500
	2	41	73	115	166	225	295	461	665	964	1260	1590	1965	2380	2830
	3	46	83	129	187	253	331	520	748	1100	1435	1820	2240	2710	3230
60 to 100	1	37	66	104	150	204	266	417	600	926	1210	1530	1890	2280	2720
	2	44	79	123	178	242	316	494	712	1050	1370	1740	2150	2590	3090
	3	50	89	138	200	272	355	555	800	1198	1565	1980	2450	2960	3520

B. Common Vent

Total vent height, H, ft	Common vent diameter, in.													
	3	4	5	6	7	8	10	12	14	16	18	20	22	24
	Combined appliance input rating, I_{total}, MBtu per hr													
6	...	65	103	147	200	260	410	588	815	1065	1345	1660	1970	2390
8	...	73	114	163	223	290	465	652	912	1190	1510	1860	2200	2680
10	...	79	124	178	242	315	495	712	995	1300	1645	2030	2400	2920
15	...	91	144	206	280	365	565	825	1158	1510	1910	2360	2790	3400
20	...	102	160	229	310	405	640	916	1290	1690	2140	2640	3120	3800
30	...	118	185	266	360	470	740	1025	1525	1990	2520	3110	3680	4480
40	...	131	203	295	405	525	820	1180	1715	2240	2830	3500	4150	5050
60	...	NR	224	324	440	575	900	1380	2010	2620	3320	4100	4850	5900
80	...	NR	NR	344	468	610	955	1540	2250	2930	3710	4590	5420	6600
100	...	NR	NR	NR	479	625	975	1670	2450	3200	4050	5000	5920	7200

NR—not recommended.

General Design Factors

If a factory-built chimney and single-wall vent connector are used, vent connector capacity tables for single-wall pipe should be used together with those for double-wall common vents. The same procedure applies for double-wall pipe common vents with single-wall pipe vent connectors. Reasonably smooth and gastight junctions of the two materials is assumed in all cases.

A system using a masonry chimney should always be designed so that the *effective* area of the chimney is not less than the area of the largest vent connector plus 50 per cent of the areas of additional vent connectors,[1] or designed using Parts A and B of Table 12-19.

Vent connector capacity tables apply to connectors with not more than two 90° turns (or the equivalent) including the final turn into the common vent, or one 90° turn and a draft hood with side outlet. If additional 90° elbows are needed, the vent connector rise should be calculated as if the appliance input were increased by 10 per cent per elbow; otherwise, the next larger connector size should be used.

Heat and frictional losses limit maximum vent connector lengths. Maximum lengths, for approximately a 20 per cent maximum flue gas temperature drop thru a single-wall metal or Type B asbestos-cement connector, are 1.5 ft per inch of connector diameter and 3.0 ft per inch for Type B double-wall vent connectors. The common vent height should also exceed that of the vent connector lateral. Where possible, the vent

Table 12-18 Capacity of a Single-Wall Metal or Type B Asbestos-Cement Vent Serving Two or More Appliances[1]

(see Fig. 12-10 for an explanation of the terms and qualifications on the use of this table)

A. Vent Connector Capacity

Total vent height, H, ft	Connector rise, R, ft	Vent connector diameter, D, in.					
		3	4	5	6	7	8
		Maximum appliance input rating at sea level, I, MBtu per hr					
6 to 8	1	21	40	68	102	146	205
	2	28	53	86	124	178	235
	3	34	61	98	147	204	275
15	1	23	44	77	117	179	240
	2	30	56	92	134	194	265
	3	35	64	102	155	216	298
30 and up	1	25	49	84	129	190	270
	2	31	58	97	145	211	295
	3	36	68	107	164	232	321

B. Common Vent Capacity

Total vent height, H, ft	Common vent diameter, in.						
	4	5	6	7	8	10	12
	Combined appliance input rating, I_{total}, MBtu per hr						
6	48	78	111	155	205	320	NR
8	55	89	128	175	234	365	505
10	59	95	136	190	250	395	560
15	71	115	168	228	305	480	690
20	80	129	186	260	340	550	790
30	NR	147	215	300	400	650	940
50	NR	NR	NR	360	490	810	1190

NR—not recommended.

Table 12-19 Capacity of a Masonry Chimney Serving Two or More Appliances[1]

(see Fig. 12-10 for explanation of the terms and qualifications on the use of this table)

A. Vent Connector Capacity

Total vent height, H, ft	Connector rise, R, ft	Vent connector diameter, D, in.					
		3	4	5	6	7	8
		Maximum appliance input rating at sea level, I, MBtu per hr					
6 to 8	1	21	39	66	100	140	200
	2	28	52	84	123	172	231
	3	34	61	97	142	202	269
15	1	23	43	73	112	171	225
	2	30	54	88	132	189	256
	3	34	63	101	151	213	289
30 and up	1	24	47	80	124	183	250
	2	31	57	93	142	205	282
	3	35	65	105	160	229	312

B. Common Chimney Capacity

Total vent height, H, ft	Minimum internal area of chimney, A,* sq in.					
	19	28	38	50	78	113
	Combined appliance input rating, I_{total}, MBtu per hr					
6	45	71	102	142	245	NR
8	52	81	118	162	277	405
10	56	89	129	175	300	450
15	66	105	150	210	360	540
20	74	120	170	240	415	640
30	NR	135	195	275	490	740
50	NR	NR	NR	325	600	910

NR—not recommended.
* Dimensions of masonry chimneys and their circular equivalents are given in Table 12-16b.

should be closer to, or directly over, the *smaller* appliance connector.

Since elbows are a principal source of flow resistance in vent connectors, an equal number of turns in each such connector may be important for "balance," particularly with the larger appliances of a multiple-venting system. Note that a draft hood with a side outlet is the equivalent resistance of one with a top outlet and a 90° elbow.

Common vent capacities in Tables 12-17, 12-18, and 12-19 assume no turns between the last connector junction and the vent terminal. For each 90° elbow or each pair of 45° elbows in this part of the common vent, the maximum allowable total vent input shown in these tables should be reduced by 10 per cent.

Flow resistance of the fittings and the relatively low common vent flue gas temperature offset any advantages of multiple-appliance vent systems with common vent heights (*not* total vent heights) under 5 ft; i.e., single-appliance vents are preferred at these levels.

Vent Manifolds

Two or more appliances located on the same level may be vented into a common horizontal duct or manifold (Fig. 12-11). Manifolds either may be of constant diameter equal to

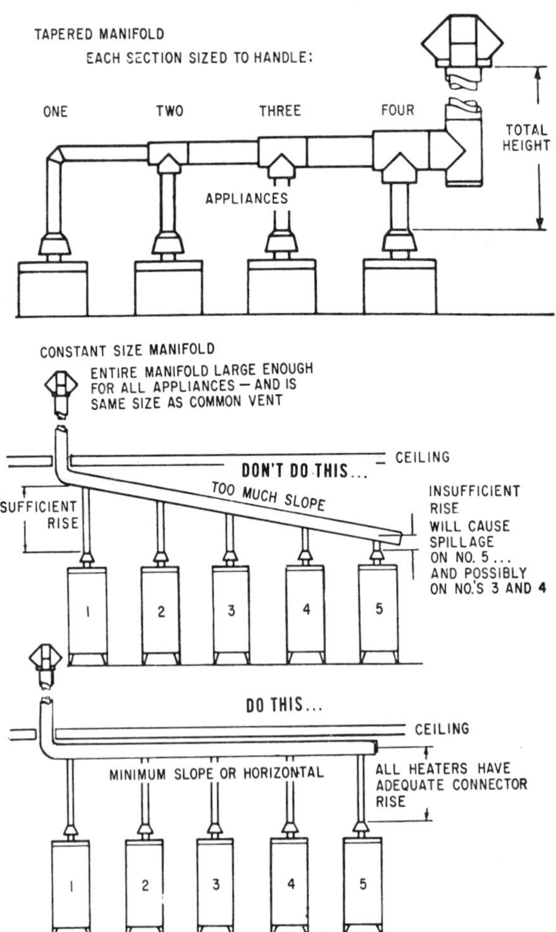

Fig. 12-11 Manifold designs: (upper) tapered manifold (center and lower) constant size manifold arrangements. Note the effect of the insufficient connector rise. (William Wallace Co.)

that of the common vertical vent or may be tapered to accommodate the increased volume of flue gases from each appliance connected. Either design may be used with little difference in venting performance. For best results, shop construction of manifolds is desirable, particularly when they are large or complicated.

DESIGN OF MULTISTORY COMMON VENTS

Multistory common vent systems are used to vent appliances located on different stories; their vent connectors enter the common vent at different levels. Appliances so vented are individually controlled, so that the vent must function properly with any one appliance or any combination of appliances in operation. The operation of a single appliance on a multistory stack constitutes the most critical operating condition.

Joint Action of Connector and Common Vent. One design (and calculation technique) for a multistory double-wall common vent is shown in Fig. 12-12. As with simple combined vents, the vent connectors are designed according to the parameters of Part A of Table 12-17. Each section of vertical common vent is sized, using Part B of Table 12-17, to be large enough to carry the accumulated total input discharging into it.

The vent connector from the lowest appliance to the common vent is readily designed as an individual vent terminating at the first interconnection. The second lowest appliance is considered to have a combined vent terminating at the second interconnection, and so forth. The top floor appliance has a total vent height measured to the outlet of the common vent. All appliances at intermediate levels have a total vent height equal to their connector rise plus the rise between interconnections. Using these heights with appliance inputs permits the designer to choose appropriate sizes of connectors and common vents. Such a system has no limiting height provided that common vent size is increased as required; however, in some cases it may prove more economical to divide a tall building into separate upper and lower systems to avoid very large vents.

Self-Venting Connectors. In the design shown in Fig. 12-13, the constant size vertical common vent is not required to produce any draft. This additional safety factor is desirable for multistory vents serving such appliances as room heaters and wall furnaces. Generally, this more conservative design is specified for appliances that must be located in occupied spaces and may be dependent on infiltration for air supply.

The self-venting connectors that enter the common vent should be designed as individual vents using the methods and factors given under individual appliance vent design. The common vent should be of constant diameter and sized for the total input of all connected appliances, using as total height the shortest distance between successive connections. A trap or cleanout should be provided at the base of the common vent.

European Experiences with Multistory Common Vents

Much postwar European construction has consisted of multistory buildings. Common flues for the venting of gas

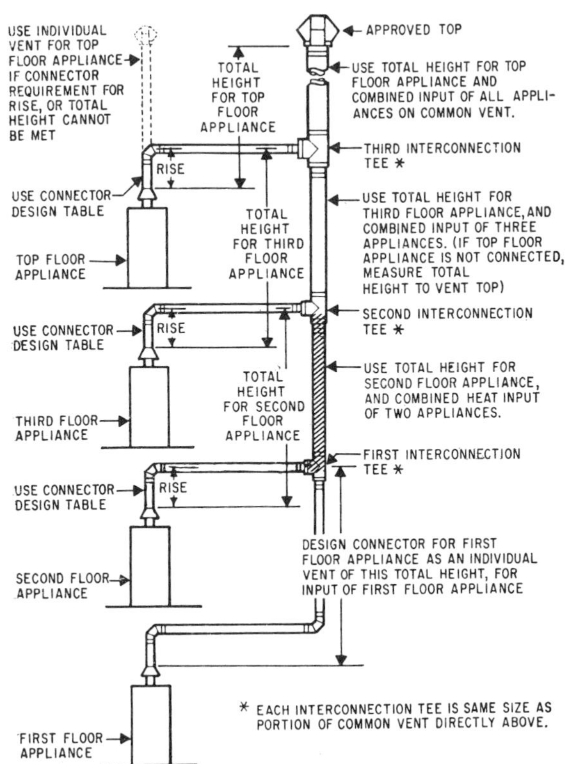

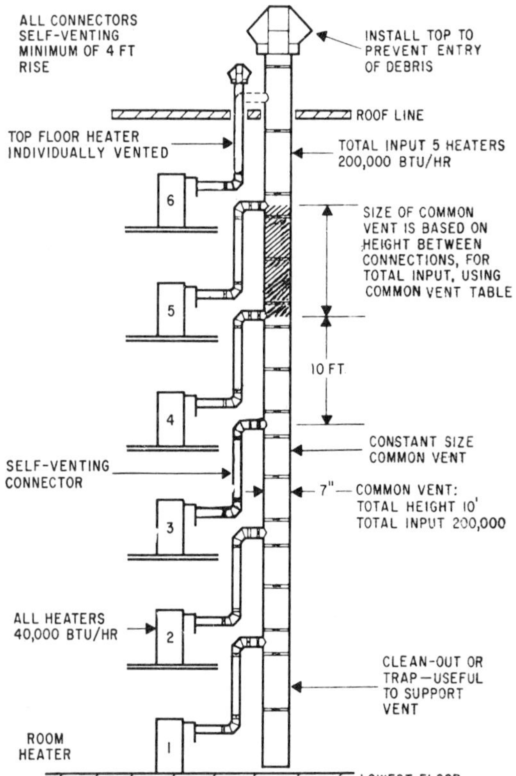

Fig. 12-12 (left) Multistory common vent design in which both the vent connector rise and the common vent effect draft. (William Wallace Co.)

Fig. 12-13 (right) Multistory common vent design in which the vent connectors are completely self-venting. (William Wallace Co.)

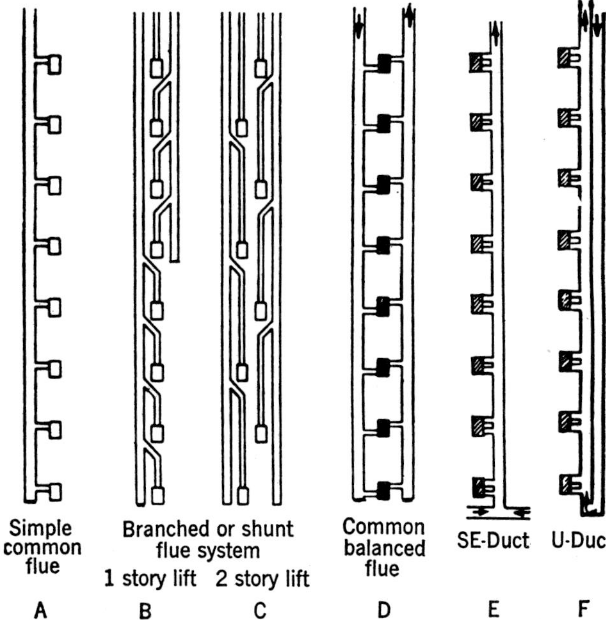

Fig. 12-14 European types of common flue systems.[31,32] Conventional appliances shown unshaded (A, B, and C); appliances with sealed combustion chambers shown shaded (D, E, and F).

appliances in the apartments of such buildings have been found to offer substantial savings of space and labor, stronger building structure, and improved appearance. The common flue systems shown in Fig. 12-14 have been investigated. Principal attention was devoted to the types labeled A, B, and C. The installation of such flues is usually controlled by local building regulations and, when gas appliances are involved, by local gas regulations as well; both sets of regulations vary considerably in different countries. The branched or shunt flue, B, has been used extensively in Belgium, France, and Holland. The greater height of vent connectors in the B system compared with those in the A system has been found to provide better draft control with less likelihood of venting troubles.

Thousands of common flues were installed for venting gas appliances with sealed combustion systems (D, E, and F) during the years 1957 to 1964. Most of these were SE-ducts* (E). More than 2000 SE-ducts were serving about 30,000 appliances in the United Kingdom in 1964. There appears to be no limit to the number of appliances that can be handled on a single flue, provided that it is large enough.

Balancing the D arrangement requires less resistance to combustion air flow for appliances at the upper levels than at the lower levels. The F system is not affected by wind, since the two vent terminals are in close proximity. The E system has two horizontal air intake ducts, not necessarily in a straight line, to overcome unfavorable wind directions. A single air intake has also performed satisfactorily provided that its inlet was located in an area freely and permanently ventilated from both sides of a building. The appliances at the upper levels in both the E and F systems use slightly vitiated air. Two per cent CO_2 may be taken as the vitiation limit for such appliances.[31] The relationship between the CO_2 in the gas stream and the ultimate CO_2 of the gas burned, as well as the means of neutralizing the effects of changes in duct velocity, has been reported.[33] Duct velocity may vary the air flow thru, and thus the efficiency of, an appliance.

* Registered trademark.

Sizing SE-ducts and U-ducts.[32] Empirical Eq. **5** determines the minimum internal cross-sectional area of a masonry SE-duct or U-duct when used with space heating gas appliances that may have little or no diversity of use. This equation is approximated in Fig. 12-15.

$$A = KQ\left[N\left(1 + \frac{0.08N}{A^{0.5}}\right)\right]^{0.5} \qquad (5)$$

where:

A = *minimum* internal cross-sectional area of vertical portion of duct, sq ft
K = 0.0055 for SE-ducts; 0.0069 for *each* U-duct leg
Q = rated input per story, assuming each floor has same input, MBtu per hr
N = number of stories in building

Note that the foregoing equation is implicit for A and thus requires trial and error solutions.

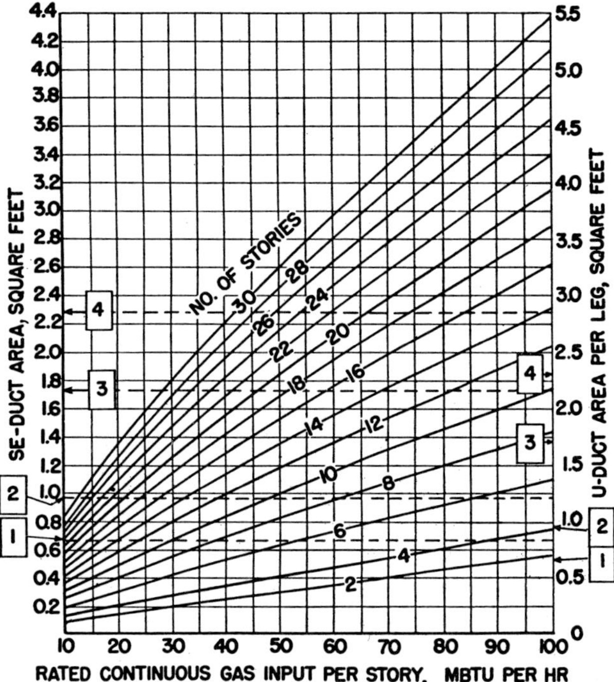

Fig. 12-15 Internal SE-duct and U-duct areas for continuously operating individual appliances in multistory buildings. Numbers in rectangles are the block size given below.[32]

Dimensions of Standard British SE-Duct and U-Duct Blocks

Size	Cross section, in. Internal	External	Weight, lb
No. 1	8 × 12	10 × 14	60
No. 2	9 × 15.5	11 × 17.5	70
No. 3	13 × 19	15.5 × 21.5	110
No. 4	15 × 22	18 × 25	166

All blocks are 18 in. long; material is precast concrete, which either may be made of acid-resistant cement or may be lined with an acid-resistant material.

Joints should be of an acid-resistant material.

A 3-in. layer of masonry is usually specified to protect and insulate the duct assembly. The gas piping may be embedded in this covering.

Example: Sealed combustion space heating appliances, each with an input of 25 MBtu per hr, are to be located in 34 apartments, one above another and back to back in a 17-story apartment building. Find the minimum required SE-duct and U-duct sizes.

From Fig. 12-15, for a 50 MBtu per hr input per floor on 17 floors:

	SE-duct	*U-duct*
Internal cross-sectional area of block, sq ft	1.6	2.0 per leg
Corresponding block size	No. 3	No. 4, per leg

The cross-sectional area required, including a 3-in. layer of masonry: SE-duct, 21.5 × 27.5 in.; U-duct, 31 × 42 in., which contains both of the vertical legs.

POWER VENTING

Many special venting problems may be solved by power venting.[34] Such problems usually involve gas appliances with draft hoods, designed to operate under natural draft, or those designed to operate at some minimum draft provided thru barometric draft controls. Conditions responsible for venting difficulties include the presence of adjacent high buildings that produce very erratic draft performance, absence of chimneys sufficiently high to produce adequate draft, negative pressure around the vented appliance due to building exhaust fans, and need to terminate the vent thru the sidewall of a building.

Fans found acceptable for power venting include the *mushroom wall or roof type,* with enclosed motor, and the *squirrel cage scroll housing type,* with its impeller mounted on the motor shaft. The mushroom type (requiring inlet connections only) is installed outdoors at the discharge point. The scroll housing type usually is installed indoors with its outlet conneeted to the discharge duct work. A weathertight scroll type may be installed on the roof if desired.

The fan selected should have the capacity to handle the dilution air drawn in thru the draft hood relief opening or barometric control as well as the combustion products. Equation **6** has been found by experience to be sufficiently accurate.

$$Q = \frac{1.2ACDT}{60} \qquad \textbf{(6)}$$

where:

Q = fan capacity against 0.375 in. w.c. external static pressure, SCFM

A = air required for combustion per cu ft of gas, (10 cu ft per 1000 Btu gas), cu ft (also, 1.2 applies when 20 per cent excess combustion air is used)

C = gas input, cfh

D = dilution factor, (2 is reasonable with draft hood in use; otherwise, 1.25 to 1.5)

T = factor for correcting vent gas volume to temperature condition at fan inlet, (normally 1.25)

Equation **6** may be used for single appliances vented to individual exhaust fans by ducts up to 10 ft long. For longer runs and for duct manifolds venting several heaters thru one fan, double the capacity obtained by using Eq. **6**. Experience with power venting of appliances with draft hoods indicates that the flue gas temperature at the fan does not exceed 200 F. No motor failures with standard fans have been encountered.

Ducts for power venting should be designed similarly to those for handling heating or cooling air (e.g., see Table 12-97). Runs from more than one appliance should be designed for equal pressure loss in each section between appliances, according to the quantity of diluted products handled, as given by Eq. **6**. Total pressure loss in the ductwork should not exceed 0.1 in w.c.

Since power vent ducts are classified by building codes as vent connectors, required clearances from combustible construction should be provided according to the listing of the duct material used.

The draft on each appliance connected to a power vent should be regulated to prevent excess draft, whether or not a draft hood is used. For appliances equipped with a draft hood, a single-acting barometric control may be installed at least 1 ft downstream from the draft hood. The draft may be considered satisfactory when the flame of a match held at various points around the draft hood relief opening is drawn inward with the appliance operating at full rating. When equipment is intended to operate without a draft hood, draft control by means of a double-acting barometric damper is suggested. Adjustment for proper combustion may be made using a flue gas analyzer.

Power venting requires provisions for shutting off gas flow in case of exhaust system failure. Most safety means sense actual draft in the duct work as a condition of permitting appliance operation. One type of control for this purpose is an adjustable diaphragm-operated switch, sensitive to a draft of 0.05 in. w.c. Another is a sail switch operated by the impact pressure of air or combustion products flowing in the system. Thermal cutouts (Fig. 12-3) are also used.

Usual practice is for the primary control of the appliance to cause only the power vent exhauster to start. When the necessary negative pressure is developed, the safety switch closes the circuit to the basic appliance controls to permit normal operation. When the primary control is satisfied, it stops the exhauster. The ensuing loss of negative pressure causes the safety switch to open the basic control circuit, shutting off the appliance. Figure 12-3 shows a diagram of a power vent system and its wiring.

VENT INSTALLATION PRACTICE

Standards for venting gas appliances are presented in Z21.30.[1] Major installation features commonly encountered are proper sizing, required clearances to combustible materials, and means of interconnection when serving either multiple gas appliances or a gas appliance and one using other fuel. The following extract from Z21.30, covering some of the more important provisions, should be noted:

5.4 TYPE B AND TYPE BW GAS VENTS

5.4.1 Listing:

Type B and Type BW gas vents shall be installed in accordance with their listings and the manufacturer's instructions. Type B and Type BW gas vents may be used for single story or multistory installations when so listed. Type BW gas vents shall have a listed capacity not less than that of the listed vented wall furnaces to which they are connected.

* * *

5.5 MASONRY, METAL AND FACTORY-BUILT CHIMNEYS

5.5.1 Listing or Construction:

(a) Factory-built chimneys shall be installed in accordance with their listings and the manufacturer's instructions.

(b) Masonry or metal chimneys shall be built and installed in accordance with nationally recognized building codes.*

* * *

5.6 SINGLE-WALL METAL PIPE

5.6.1 Installation With Gas Appliances Permitted by 5.3.5(a):†

(a) Single-wall metal pipe shall be constructed of sheet copper not less than No. 24 B&S gage or galvanized sheet steel not less than No. 20 galvanized sheet gage or other approved noncombustible corrosion-resistant material.

(b) Single-wall metal pipe used to vent gas appliances shall comply with the installation provisions of 5.4.2 (a), (b) and (d).‡

(c) Single-wall metal pipe shall be used only for runs directly from the space in which the appliance is located through the roof or exterior wall to the outer air.

(d) Single-wall metal pipe shall not originate in any unoccupied attic or concealed space, and shall not pass through any attic, inside wall, concealed space, or through any floor.

(e) When a single-wall metal pipe passes through an exterior wall constructed of combustible material, it shall be guarded at the point of passage by a method described in 5.8.14.

(f) When a single-wall metal pipe passes through a roof constructed of combustible material it shall be guarded at the point of passage by a method described in 5.8.14 or by a noncombustible nonventilating thimble not less than 4 inches larger in diameter than the vent pipe and extending not less than 18 inches above and 6 inches below the roof with the annular space open at the bottom and closed only at the top.

* * *

5.8 VENT CONNECTORS

* * *

5.8.4 Clearance:

Minimum clearances from vent connectors to combustible material shall be in accordance with Table 13. The clearances from vent connectors to combustible materials may be reduced when the combustible material is protected as specified in Table 12.§

When vent connectors must pass through walls or partitions of combustible material, a thimble shall be used and installed in accordance with one of the methods outlined in 5.8.14.

* * *

5.8.14 Use of Thimbles

(a) When passing through combustible walls or partitions, vent connectors built of listed Type B gas vent material shall be installed so that the clearances required by the listing are maintained.

(b) Vent connectors made of single-wall metal pipe shall not pass through any combustible walls unless they are guarded at the point of passage by ventilated metal thimbles not smaller than the following:

For listed appliances, except incinerators—4 inches larger in diameter than the vent connector, unless there is a run of not less than 6 feet of vent connector in the open, between the draft hood outlet and the thimble, in which case the thimble may be 2 inches larger in diameter than the vent connector.

For unlisted appliances having draft hoods—6 inches larger in diameter than the vent connector.

For incinerators and unlisted appliances without draft hoods—12 inches larger in diameter than the vent connector.

* Article X of the National Building Code of The National Board of Fire Underwriters, 85 John St., New York, N. Y. 10038, or the Standard for Chimneys, Fireplaces and Vent Systems, NFPA No. 211 of the National Fire Protection Association, 60 Batterymarch St., Boston, Mass. 02110, are among such nationally recognized codes and standards.

[† Excludes indoor incinerators, appliances which may be converted to the use of solid or liquid fuels, and vented wall furnaces listed for use with Type BW gas vents.—EDITOR]

[‡ See page 12/46.—EDITOR]

[§ Given in Chap. 1 of this Section.—EDITOR]

Table 13
Vent Connector Clearances for Gas Appliances

Appliance	Minimum distance from combustible material: Listed Type B gas vent material	Vent connectors of other than Type B material
Listed boiler	As listed	6 in.
Listed warm air furnace	As listed	6 in.
Listed water heater	As listed	6 in.
Listed room heater	As listed	6 in.
Listed floor furnace	As listed	6 in.
Listed floor furnace	As listed	6 in.
Listed incinerator	Not permitted	18 in.
Listed conversion burner (with draft hood)	6 in.	9 in.
Unlisted appliances—having draft hoods	6 in.	9 in.
Unlisted appliances without draft hoods	Not permitted	18 in.

(c) In lieu of thimble protection, all combustible material in the wall shall be cut away from the vent connector a sufficient distance to provide the clearance required from such vent connector to combustible material. Any material used to close up such opening shall be noncombustible.

Venting Incinerators to Chimneys Serving Other Appliances. Incinerators should be connected to masonry or factory-built chimneys, except when installed in locations such as open sheds, breezeways, or carports. In such locations metal pipe of not less than #20 U. S. gage galvanized steel or other equivalent noncombustible corrosion-resistant material may be used when fully exposed. A vent connector from an incinerator should be of not less than #24 U. S. gage galvanized steel, and should be as short as possible. No draft hood is permitted. A barometric damper may be used.

Preferably, an incinerator should be connected to a separate chimney. When a chimney that serves other gas equipment must be used, the vent connector from the incinerator should enter the chimney below any other vent connector.

Interconnections. These are gas vent connections to chimneys serving appliances that use other fuels. Sufficient space is sometimes not available to permit chimney entrance, at different levels, of a gas appliance vent connector and a connector from an appliance using solid or liquid fuels. In such cases a suitably designed Y-type fitting may prove satisfactory. It should be of adequate size to accommodate the combined flue gases from both appliances and be located as close as possible to the chimney. The vent connector of the gas appliance should enter the side outlet of the fitting.

The following extract from Z21.30[1] should be noted:

5.5.5 Chimneys Serving Appliances Burning Other Fuels

An automatically controlled gas appliance connected to a chimney which also serves equipment for the combustion of solid or liquid fuel shall be equipped with an automatic pilot. A gas appliance vent connector and a flue connector from an appliance burning another fuel may be connected into the same chimney through separate openings, or both may be connected through a single opening if joined by a suitable fitting located as close as practical to the chimney. If two or more openings are provided into one chimney, they should be at different levels.

VENT DESIGN FOR LARGE GAS APPLIANCES

Draft requirements of large gas appliances, such as furnaces, ovens, boilers, and related units, depend on whether a draft hood is provided. An appliance with a draft hood needs no draft except that produced within its flue passages and combustion chamber. An appliance without a draft hood may require the aid of a chimney or stack to support burning at the required input rate, to maintain desired internal pressure, and to ensure adequate combustion air.

Large Appliances Requiring Supplementary Draft

The following design methods were developed[35] from studies of theoretical relationships governing chimney and gas vent operation. The objective was to develop methods of tabulating important factors affecting draft and capacity in common gas industry terms. Equation **7,** used in computing these factors, is related to those given in Refs. 22, 26, and 36. It shows that available draft results from the difference between "theoretical" draft and energy losses due to flow:

$$S = \frac{0.255P}{T_2}\left[H\left(\frac{T_2}{T_1}-1\right) - \frac{\left(F + 0.03\,\frac{L}{D}\right)}{64.4} \times \left(\frac{2.62I\left(0.15+\frac{11}{CO_2}\right)\frac{T_2}{T_1}}{A \times 10^3}\right)^2\right] \quad (7)$$

where:

S = static draft available for appliance and burner requirements (usually between 0.0 and 0.4 in. w.c.), in. w.c.
P = barometric pressure, in. Hg
T_2 = average temperature of gases in vertical portion of vent, °F abs
H = height of chimney outlet above point of draft measurement, ft
T_1 = ambient temperature of outdoor air, °F abs
F = fitting losses, velocity heads
L = length of all piping in system (connector plus stack) from point of draft measurement to top of vent, ft
D = connector or breeching ID, whichever is the smaller (generally equal), *feet*
I = maximum vent capacity, expressed as gas appliance heat input, MBtu per hr
CO_2 = CO_2 in chimney (dry basis), per cent
A = cross-sectional flow area of chimney, square *feet*

Vent Design. Calculations for gas equipment needing supplementary draft may be performed as follows:

1. From required draft based on expected flue gas temperature and equipment characteristics such as efficiency, over-fire static pressure, firing rate, steam pressure or temperature, choose a stack height to yield a theoretical draft (Table 12-20) greater than that needed for maximum available static draft, S. Other considerations, e.g., building height, fume and gas dispersal, location of adjacent structures, may require greater height than that theoretically called for.

2. From vent operating conditions, choose a diameter at which the sum of the energy losses (Tables 12-21 and 12-22) *plus* the available draft, S, is less than theoretical draft by a margin sufficient to ensure proper operation under all flow conditions (Table 12-20).

Application of Tables 12-20 thru 12-22 requires trial-and-error calculation procedures. With the help of some auxiliary tables, Tables 12-23 thru 12-25, effects of operating variables can be estimated, thus reducing the number of trials.

Design criteria and an explanation of the various factors involved follow; terms are defined under Eq. **7.**

1. **Friction.** This is represented by the ratio L/D. The greater the system L/D ratio, the greater the energy losses. Values given for systems having a connector include effects of energy losses in a tee. When connector and stack sizes differ, the smaller size is used.

2. **Maximum Appliance Input Loading.** This is represented by the ratio I/A, except that A is in square *inches*

Table 12-20 Theoretical or Potential Static Draft Pressure in a Venting System, in. w.c.

$\left[\frac{0.255P}{T_2} H\left(\frac{T_2}{T_1}-1\right)\right]$ in Eq. **7**

Height of stack above level of draft measurement, H, ft	Average temperature rise above 60 F of gas in vertical part of chimney,* °F							
	150	200	300	400	500	600	750	1000
5	0.017	0.020	0.027	0.032	0.036	0.039	0.043	0.049
10	.033	.041	.054	.064	.072	.079	.087	.097
15	.049	.061	.081	.096	.108	.118	.130	.145
20	.066	.082	.11	.13	.14	.16	.17	.19
25	.083	.10	.13	.16	.18	.20	.22	.24
30	.10	.12	.16	.19	.22	.24	.26	.29
40	.13	.16	.22	.26	.29	.32	.95	.39
60	.20	.25	.32	.38	.43	.47	.52	.58
80	.26	.33	.44	.51	.58	.63	.70	.78
100	.33	.41	.54	.64	0.72	0.79	0.87	0.97
150	.50	.61	0.81	0.96	1.08	1.18	1.30	1.45
200	.66	0.82	1.07	1.28	1.44	1.58	1.74	1.94
300	0.99	1.23	1.61	1.92	2.16	2.37	2.61	2.91

* Corresponds to $T_2 - (460 + 60)$ as T_2 is defined under Eq. **7.**

Table 12-21 Static Draft Pressure Losses in Venting System With Four Per Cent CO_2 in the Flue Gas,* in. w.c.

Avg temp rise above 60 F of gas in vert. part of stack,† °F	Max appliance input loading, MBtu per hr-sq in. of vent or chimney area	L/D: System length to diameter ratio, feet of stack (plus connector if used) per foot of diameter							
		Vertical system—no connector‡							
		45	50	60	70	80	100	120	140
		System using connector‡							
		5	10	20	30	40	60	80	100
150	2	0.002	0.002	0.003	0.003	0.004	0.005	0.005	0.006
	3	.005	.005	.006	.007	.008	.010	.012	.014
	4	.009	.010	.011	.013	.015	.018	.022	.025
	6	.020	.022	.025	.029	.033	.041	.048	.056
	8	.035	.039	.045	.052	.059	.073	.086	.100
	10	.055	.060	.070	.081	.092	.113	.134	.155
	15	.123	.135	.159	.183	.207	.254	.301	.349
	20	.219	.240	.283	.326	.368	.453	.536	.622
	25	.343	.376	.443	.510	.576	0.710	0.840	0.975
	30	.490	.540	.634	.730	.825	1.02	1.20	1.40
300	2	.003	.003	.003	.004	.004	0.006	0.007	0.008
	3	.006	.007	.008	.009	.010	.012	.015	.017
	4	.011	.012	.014	.016	.018	.022	.026	.030
	6	.024	.026	.031	.036	.040	.049	.059	.068
	8	.043	.047	.055	.063	.072	.088	.105	.122
	10	.067	.073	.086	.099	.112	.137	.163	.190
	15	.151	.165	.195	.224	.253	.311	.379	.428
	20	.268	.294	.347	.400	.451	.555	0.656	0.761
	25	.420	.460	.542	.624	0.705	0.870	1.03	1.19
	30	.600	.660	.775	.894	1.01	1.25	1.47	1.71
500	2	.003	.004	.004	.005	0.006	0.007	0.008	0.010
	3	.008	.008	.010	.011	.013	.015	.018	.021
	4	.013	.015	.017	.020	.022	.027	.033	.038
	6	.030	.033	.038	.044	.050	.062	.073	.085
	8	.055	.058	.069	.079	.089	.110	.130	.152
	10	.083	.091	.107	.123	.139	.171	.203	.236
	15	.187	.206	.242	.279	.315	.387	.458	.531
	20	.334	.365	.431	.496	.560	0.690	0.816	0.946
	25	.522	.572	.675	0.776	0.877	1.08	1.28	1.48
	30	.746	.822	.965	1.11	1.26	1.55	1.83	2.12
750	2	.004	.005	.005	0.006	0.007	0.009	0.010	0.012
	3	.009	.010	.012	.014	.016	.019	.023	.026
	4	.017	.018	.021	.024	.028	.034	.040	.047
	6	.037	.041	.048	.055	.062	.077	.091	.106
	8	.067	.073	.085	.098	.111	.136	.162	.189
	10	.104	.113	.133	.153	.173	.212	.252	.294
	15	.233	.256	.302	.347	.382	.481	0.570	0.661
	20	.415	.455	.537	.618	0.698	0.860	1.02	1.18
	25	.650	0.713	0.840	0.968	1.09	1.35	1.59	1.85
	30	.930	1.03	1.20	1.38	1.57	1.94	2.28	2.64
1000	2	.005	0.005	0.006	0.007	0.008	0.010	0.012	0.014
	3	.011	.012	.014	.016	.019	.023	.027	.032
	4	.020	.022	.026	.029	.033	.041	.048	.056
	6	.045	.049	.057	.066	.074	.092	.109	.127
	8	.080	.087	.102	.117	.133	.163	.194	.226
	10	.124	.135	.159	.183	.207	.254	.302	.352
	15	.279	.307	.361	.416	.470	0.577	0.683	0.791
	20	.497	.545	0.642	0.740	0.835	1.03	1.22	1.41
	25	0.779	0.854	1.01	1.16	1.31	1.61	1.91	2.21
	30	1.11	1.23	1.43	1.66	1.87	2.32	2.73	3.16

* Applicable only if there is no draft hood; a barometric draft control may be used.

† Corresponds to $T_2 - (460 + 60)$ as T_2 is defined under Eq. **7;** see Table 12-23 for rises not tabulated here.

‡ The difference of 40 in the L/D ratio between a vertical system without a connector and a system with a connector is due to the flow resistance of the 90° turn required between the connector and the vertical stack.

rather than square feet. This ratio is frequently used as a basis for gas chimney design and serves to relate heat input to flow velocity. When connector and stack area differ, the smaller should be used.

3. **Average Gas Temperature in Vertical Stack.** Tables 12-21 and 12-22 show temperature rises of 150° to 1000° F. Energy losses, as shown by Eq. **7,** increase with temperature because of greater flow velocity. Precise work requires compensating average temperature, T_2, for cooling according to Table 12-23. For vent design, ambient temperature, T_1,

Table 12-22 Static Draft Pressure Losses in Venting System with Eight Per Cent CO_2 in the Flue Gas,* in. w.c.

Avg temp rise above 60 F at gas in vert. part of stack,† °F	Max appliance input loading, MBtu per hr-sq in. of vent or chimney area	L/D: System length to diameter ratio, feet of stack (plus connector if used) per foot of diameter							
		Vertical system—no connector‡							
		45	50	60	70	80	100	120	140
		System using connector‡							
		5	10	20	30	40	60	80	100
150	2	0.001	0.001	0.001	0.001	0.001	0.001	0.001	0.002
	3	.001	.002	.002	.002	.002	.003	.003	.004
	4	.002	.003	.003	.004	.004	.005	.006	.007
	6	.005	.006	.007	.008	.009	.011	.013	.015
	8	.010	.011	.012	.014	.016	.020	.024	.028
	10	.015	.017	.019	.022	.025	.031	.037	.043
	15	.034	.037	.044	.050	.057	.070	.083	.096
	20	.061	.066	.078	.090	.101	.125	.148	.171
	25	.094	.104	.122	.140	.159	.195	.232	.270
	30	.136	.149	.175	.201	.227	.280	.332	.384
300	2	.001	.001	.001	.001	.001	.002	.002	.002
	3	.002	.002	.002	.002	.003	.003	.004	.005
	4	.003	.003	.004	.004	.005	.006	.007	.008
	6	.007	.007	.009	.010	.011	.014	.016	.019
	8	.012	.013	.015	.018	.020	.025	.029	.034
	10	.019	.020	.024	.027	.031	.038	.045	.052
	15	.042	.046	.054	.062	.070	.086	.102	.110
	20	.074	.081	.096	.110	.124	.153	.182	.209
	25	.116	.127	.150	.171	.195	.239	.284	.320
	30	.166	.182	.208	.246	.278	.342	.406	.470
500	2	.001	.001	.001	.001	.002	.002	.002	.003
	3	.002	.002	.003	.004	.004	.004	.005	.006
	4	.004	.004	.005	.006	.006	.008	.009	.011
	6	.008	.009	.011	.012	.014	.017	.020	.023
	8	.015	.016	.019	.022	.025	.031	.036	.042
	10	.023	.025	.030	.034	.039	.047	.056	.065
	15	.052	.057	.067	.077	.087	.107	.127	.146
	20	.092	.101	.119	.137	.155	.190	.226	.260
	25	.145	.158	.186	.213	.242	.297	.354	.408
	30	.215	.235	.277	.317	.360	.443	.526	.610
750	2	.001	.001	.001	.002	.002	.002	.003	.003
	3	.003	.003	.003	.004	.004	.005	.006	.007
	4	.005	.005	.006	.007	.008	.009	.011	.013
	6	.010	.011	.013	.015	.017	.021	.025	.029
	8	.018	.020	.024	.027	.030	.038	.045	.052
	10	.029	.031	.037	.042	.048	.059	.070	.081
	15	.064	.070	.083	.095	.118	.133	.158	.182
	20	.114	.125	.148	.170	.192	.236	.281	.324
	25	.180	.196	.231	.265	.300	.370	.440	.508
	30	.256	.281	.322	.381	.430	.530	.628	.730
1000	2	.001	.002	.002	.002	.002	.003	.003	.004
	3	.003	.003	.004	.005	.005	.006	.008	.009
	4	.006	.006	.007	.008	.009	.011	.014	.016
	6	.012	.014	.016	.018	.021	.025	.030	.035
	8	.022	.024	.028	.033	.037	.045	.054	.062
	10	.034	.038	.044	.051	.057	.071	.084	.097
	15	.077	.084	.099	.114	.129	.159	.189	.212
	20	.127	.150	.177	.204	.230	.283	.336	.390
	25	.215	.235	.277	.317	.360	.443	.526	.610
	30	0.307	0.337	0.386	0.456	0.515	0.634	0.753	0.870

* Applicable only if there is no draft hood; a barometric draft control may be used.

† Corresponds to $T_2 - (460 + 60)$ as T_2 is defined under Eq. **7**; see Table 12-23 for rises not tabulated here.

‡ The difference of 40 in the L/D ratio between a vertical system without a connector and a system with a connector is due to the flow resistance of the 90° turn required between the connector and the vertical stack.

is chosen as (60 + 460) °F abs. Boiler room temperature does not influence stack operation. If the design is for the 60 F ambient, any drop in outdoor temperature will improve stack action because of the greater difference between the stack and the ambient temperatures. Choice of a stack temperature lower than that actually expected yields a conservative design, with extra overload capacity.

4. **Gas Composition or CO_2 Concentration (Dry Basis).** Flue gas flow velocity for a given heat input increases as the CO_2 decreases, thereby increasing the energy losses at higher

amounts of excess air. Table 12-21 for 4 per cent CO_2 and Table 12-22 for 8 per cent, represent two common operating conditions required to produce draft.

Table 12-23 Effect of Chimney Gas Temperature on Relative Draft Losses

(based on 300° F rise in chimney above 60 F ambient)

Avg rise,* °F	Relative draft loss	Avg rise,* °F	Relative draft loss
50	0.70	700	1.49
100	.76	800	1.61
200	0.88	900	1.73
300	**1.00**	1000	1.85
400	1.12	1200	2.10
500	1.24	1500	2.46
600	1.37	2000	3.08

*Corresponds to $T_2 - (460 + 60)$ as T_2 is defined under Eq. **7**.

Cooling of gases in the connector, breeching, and vertical section may be estimated as (1) steel connector and stack—4 per cent per 10 diameters; (2) masonry chimney—1 per cent per 10 diameters.

For operation at other CO_2 values, Table 12-24 can be used to correct the losses in Table 12-21 for 4 per cent CO_2. Thus, with 12 per cent CO_2 during theoretical combustion of natural gas (no excess air), Table 12-24 shows the draft loss to be 0.137 that given for 4 per cent CO_2, owing to lowered flue gas velocity.

Table 12-24 Effect of CO_2 Concentration upon Relative Draft Losses

CO_2 in flue gases, %	Relative draft loss	CO_2 in flue gases, %	Relative draft loss
1	14.8	7	0.352
2	3.8	8	.279
3	1.75	9	.224
4	**1.00**	10	.186
5	0.655	11	.157
6	0.455	12	0.137

Table 12-25 Altitude Correction Factor for Theoretical Draft and for Draft Losses

Altitude, ft	Relative draft loss	Altitude, ft	Relative draft loss
Sea level	1.00	6,000	0.81
2000	0.93	8,000	.75
4000	0.86	10,000	0.70

Large Appliances Equipped with Draft Hoods

Equations **1** and **4** may be used for designing vents for large appliances equipped with draft hoods. When double-wall piping is the vent material, capacities may be obtained from Table 12-14. For vent sizes larger than 24 in., capacity may be considered as directly proportional to area, or D^2.

AIR SUPPLY

Proper appliance operation requires an adequate supply of air. Air is needed for combustion (including excess air in the combustion chamber). Air is also required for draft hoods or barometric dampers; in certain circumstances, provisions must also be made to cool the space surrounding the appliance by ventilation. American Standard Z21.30 and corresponding standards for industrial equipment and boilers provide for such adequate air supply.

One cu ft of 1000 Btu natural gas theoretically requires 10 cu ft of air for combustion. To provide an adequate margin most gas appliances are designed to operate with about *15 cu ft of air for combustion per 1000 Btu of input.* Vented gas appliances using draft hoods induce the flow of an equal volume of air thru the skirt of the hood. Therefore, the total volume of air required for combustion and draft hood dilution is 30 cu ft per MBtu of gas input.

An air supply exceeding that for combustion and dilution is indicated by the ratio of the room volume* to total appliance volume. Values of 12 and 16 for furnaces and boilers, respectively, are minimums—below these values additional air is required to maintain safe ambient temperature adjacent to the appliance.

Domestic Appliances

Extracts from Z21.30[1] will be found in Chap. 1 of this Section. Important features to be noted are covered in paragraphs 3.4.2 thru 3.4.6.

Heating and Power Boilers and Similar Equipment

Requirements for combustion air supply, including dilution, for large boilers (over 400 MBtu per hr)—extracted from the proposed ASA Z83.3[38]—follow:

1.2 Combustion Air Supply and Ventilation.

1.2.1 General.

a. Positive means for supplying an ample amount of outside air to permit complete combustion of the gas shall be provided. Automatic or manually adjustable control devices for outside air intake shall be interlocked with the burner.

b. To determine combustion air requirements at the boiler, under standard atmospheric conditions (60 F and 30 in. Hg), the following minimum factors apply:

(a) For boiler with draft hood—3.0 cu ft per 100 Btu input;

(b) For boilers using barometric dampers—1.5 cu ft per 100 Btu input; and,

(c) For boilers directly connected to chimney without neutralizing air openings—1.2 cu ft per 100 Btu input.

c. Where a boiler is located in an inside room or space, air supply shall be provided through ducts or openings leading to the outside air.

d. Openings to the outside shall be unobstructed and screens, if used, shall have a minimum of ¼ in. mesh.

e. Where a room or space in which a boiler is installed is ventilated by mechanical means, air sufficient to replace that exhausted and consumed by combustion shall be supplied from a safe, uncontaminated source. The means for ventilation shall in no case cause an unsafe pressure condition in the boiler room.

f. In addition to the combustion air required, sufficient air shall be supplied to the boiler room to make the room safe for occupancy and proper operation of equipment.

1.2.2 Boilers Equipped with Draft Hoods. The effective cross-sectional area of the permanent outside air opening(s) to the boiler room shall be large enough to supply the air required in the boiler room. For supplying combustion air, the area of the opening shall be of a size at least equal to the boiler(s) breeching but not less

* Ceiling heights above 8 ft are figured as 8 ft.

than one sq in. of free area per 5,000 Btu per hr input (= approximately to 1.4. sq ft per million Btu), except as noted in 1.2.5.

1.2.3 Boilers with Barometric Dampers. The effective cross-sectional area of the permanent outside air opening(s) to the boiler room shall be large enough to supply the air required in the boiler room. For supplying combustion air, the area of the opening shall be of a size at least equal to the boiler(s) breeching but not less than one sq in. of free area per 14,000 Btu per hr input (= approximately to 0.5 sq ft per million Btu), except as noted in 1.2.5.

1.2.4 Boilers Directly Connected to Chimney without Neutralizing Air Openings. The effective cross-sectional area of the permanent outside air opening(s) to the boiler room shall be large enough to supply the air required in the boiler room. For supplying combustion air, the area of the opening shall be of a size at least equal to the boiler(s) breeching but not less than one sq in. of free area per 17,500 Btu per hr input (= approximately to 0.4 sq ft per million Btu), except as noted in 1.2.5.

1.2.5 Exceptions.

a. Ducts to Boiler Room—Where air ducts are used, the increase in friction due to length and shape must be considered in determining the proper area.

b. Forced Air Supply to Boiler Room—Where mechanical means for boiler room air supply are used, the size of the duct or opening may be reduced.

Industrial occupancies in general are covered by the following extract from the proposed Z83.1.[37] It is consistent with the foregoing material on large boilers except that it does not apply to the direct connection of boilers to chimneys *without neutralizing air openings.*

6.10 Air Requirements.

6.10.1 General. The air requirements of the gas equipment to be installed and the air supply in the building in which the equipment is to be installed shall be checked to determine that sufficient air is available. If normal air infiltration is inadequate, sufficient make-up air shall be supplied to prevent any possibility of creating a partial vacuum in the building.

> **Note:** Suitable precautions should be taken to assure that the air supply will be clean. When necessary, make-up air should be heated.

6.10.2 Combustion Air. Complete combustion of gas requires approximately one cu ft of air, at standard conditions, for each 100 Btu of fuel burned, but additional air for proper burner operation (atmospheric burner) may be required. When the building space in which gas equipment is installed does not have adequate air infiltration to assure proper combustion, one or more permanent openings to the out-of-doors or to spaces freely communicating to the out-of-doors shall be required. Such openings shall have a minimum free area of one sq in. for every 5000 Btu per hour for equipment with draft hoods. For other equipment an opening or openings having a minimum free area of 0.5 sq ft per 1,000,000 Btu/hr should be provided.

6.10.3 Process Air. Sufficient process air shall be provided as required for: cooling equipment or material, controlling dew point, heating, drying, oxidation or dilution, safety exhaust, draft hood operation, odor control, and compressors.

6.10.4 Ventilation Air. Sufficient air shall be supplied for ventilation including all air required for comfort and proper working conditions for personnel.

In building where large quantities of combustion and ventilation or process air are exhausted, provision should be made for a sufficient supply of fresh uncontaminated make-up air, warmed if necessary to the proper temperature.[39] Good practice is to provide about 5 to 10 per cent more make-up air than the amount exhausted.

REFERENCES

1. A.G.A. Laboratories. *American Standard Installation of Gas Appliances and Gas Piping,* 4th ed. (ASA Z21.30–1964) Cleveland, Ohio, 1964.
2. Blome, C. E., and Bray, J. L. *Research in Venting Direct Gas Heaters When No Chimney Connections Are Available.* (Research Ser. 103) Lafayette, Ind., Purdue Univ., Engineering Experiment Station, 1948.
3. Wills, F. "Flues and Chimneys for Gas Burning Equipment." *P.C.G.A. Proc.* 23: 169–81, 1932; and "Domestic Vents and Flues and Boiler Stacks." *P.C.G.A. Proc.* 26: 113–6, 1935.
4. Gilbert, T. H. "Venting of Domestic Gas Appliances." *P.C.G.A. Proc.* 33: 83–8, 1942.
5. Hite, S. C., and Bray, J. L. *Research in Venting Direct Gas Heaters When No Chimney Connections Are Available (Report No. 2).* (Research Ser. 105) Lafayette, Ind., Purdue Univ., Engineering Experiment Station, 1948.
6. Kirk, W. B. *Venting of Gas Appliances, Aeration and Humidity Control.* (Domestic Research and Utilization Conf., 1949) New York, A.G.A., 1949.
7. Milener, E. D. *What Is This We Hear About VIC?* (Domestic Gas Research and Utilization Conf., 1950) New York, A.G.A., 1950.
8. Davis, K. T. *Venting: a By-Product Disposal System.* (Research and Utilization Conf., 1952) New York, A.G.A., 1952.
9. Hampel, T. E. *Progress in Venting Research.* (Research and Utilization Conf., 1955) New York, A.G.A., 1955.
10. Kirk, W. B., and Perry, E. H., Jr. *Draft Control Methods and Equipment.* (Research and Utilization Conf., 1956) New York, A.G.A., 1956.
11. ——. *Draft Hood Design Manual.* (Research and Utilization Conf., 1956) New York, A.G.A., 1956.
12. A.G.A. Laboratories. *Effects of Confined Space Installation on Central Gas Space Heating Equipment Performance.* (Research Bull. 53) Cleveland, Ohio, 1949.
13. Stone, R. L., and Kirk, W. B. *Combustion and Ventilation Air Supply to Gas Equipment in Small Rooms.* (Research Bull. 67) Cleveland, Ohio, A.G.A. Laboratories, 1954.
14. Hampel, T. E., and Stone, R. L. *Literature Review and Design Studies of Gas Appliance Venting Systems.* (Research Bull. 68) Cleveland, Ohio, A.G.A. Laboratories, 1955.
15. Reed, H. L. *A Field Survey of Gas Appliance Venting Conditions.* (Research Rept. 1243) Cleveland, Ohio, A.G.A. Laboratories, 1956.
16. Perry, E. H., Jr. *A Field Survey of Gas Appliance Venting Conditions, Part II.* (Research Rept. 1267) Cleveland, Ohio, A.G.A. Laboratories, 1957.
17. ——. *Principles of Draft Hood Operation and Design.* (Research Bull. 74) Cleveland, Ohio, A.G.A. Laboratories, 1956.
18. ——. *Draft Hood Design Manual.* (Research Rept. 1261) Cleveland, Ohio, A.G.A. Laboratories, 1956.
19. ——. *Research in Fundamentals of Gas Vent System Design.* (Research Rept. 1298A) Cleveland, Ohio, A.G.A. Laboratories, 1960.
20. ——. *Gas Vent Design.* (Research Rept. 1300) New York, A.G.A. Laboratories, 1959.
21. ——. *Gas Vent Tables.* (Research Rept. 1319) New York, A.G.A. Laboratories, 1960.
22. Kinkead, A. "Operating Characteristics of a Gas Vent." *P.C.G.A. Proc.* 43: 89–93, 1952; and "A Scientific Approach to Gas Venting." *Gas Age* 110: 33+, Aug. 14, 1952.
23. Stone, R. L. *Practical Gas Vent Design.* (Research and Utilization Conf., 1955) New York, A.G.A., 1955.
24. William Wallace Co. *Metalbestos Gas Vent Tables and Handbook.* Belmont, Calif., 1961.
25. Dill, R. S., and others. "Observed Performance of Some Experimental Chimneys." *Heating-Piping* 14: 252–9, April 1942.
26. Schmitt, L. B. and Engdahl, R. B. "Performance of Residential Chimneys." *Heating, Piping Air Conditioning* 20: 111–8, Nov. 1948.
27. Mitchell, N. D. "Fire Hazard Tests With Masonry Chimneys." *Natl. Fire Protect. Assoc. Quart.* 43: 117–34, Oct. 1949.

28. Drake, D. L. "Chimney Liners." *A.G.A. Proc.* 1952: 425–8.
29. Segeler, C. G. "Gas Heating." *Air Conditioning, Heating Venting* 59: GC33–53, March 1962.
30. Perry, E. H. *Effective Vent Terminal Design and Location.* (Research and Utilization Conf., 1962) New York, A.G.A., 1962.
31. Institution of Gas Engineers. *Investigation on Unconventional Flues.* (Internatl. Gas Conf. IGU/12–58) Brussels, Belgium, I.G.U., 1958.
32. Ringler, M. D., and Cullinane, H. "Multistory Venting With SE-Ducts." *Air Conditioning, Heating Venting* 61: 55–61, Sept. 1964.
33. Stehn, Von W. "Neuartige Methoden der Abgasabführung." *Gas-Wasserfach* 104: 65–70, Jan. 18, 1963.
34. Jones, J. P. "Power Venting of Gas Appliances." *Air Conditioning, Heating Venting* 57: 53–6, May 1960; 150, July 1960.
35. Stone, R. L. "Chimney and Stack Design for Gas-Fired Equipment." *Heating, Piping Air Conditioning* 29: 143–8, Oct. 1957.
36. Am. Soc. of Heating, Refrigerating and Air-Conditioning Engineers. *Heating, Ventilating, Air Conditioning Guide,* 38th ed., p. 532. New York, 1960.
37. A.G.A. Laboratories. *Proposed American Standard for Installation of Gas Piping and Gas Equipment on Industrial Premises and Other Premises.* (ASA Z83.1–Jan. 1964, 4th Draft) Cleveland, Ohio, 1964.
38. ——. *Proposed American Standard Requirements for Gas Equipment in Large Boilers.* (ASA Z83.3–Jan. 1964, 3rd Draft) Cleveland, Ohio, 1964.
39. A.G.A. Chemical Process Com. "A Complete Guide to Direct Gas-Fired Heating of Process Makeup Air." *Actual Specifying Engineer* 6: 88–93, Dec. 1961; 7: 78–84, Feb.; 72+, March 1962.

Chapter 4

Appliance Controls and Accessories

by E. J. Horton and F. E. Vandaveer

Table 12-26 lists those gas appliance accessories for which ASA Standards were in effect in 1964. The A.G.A. listing symbol is shown in Table 12-30.

Table 12-26 ASA Listing Requirements for Gas Appliance Accessories

Gas Hose for Portable Gas Appliances, Z21.2
Draft Hoods, Z21.12
Manually Operated Gas Valves, Z21.15
Domestic Gas Conversion Burners, Z21.17
Domestic Gas Appliance Pressure Regulators, Z21.18
Automatic Burner Ignition and Safety Shutoff Devices, Z21.20
Automatic Valves for Gas Appliances, Z21.21
Relief Valves and Automatic Gas Shutoff Devices for Hot Water Supply Systems, Z21.22
Gas Appliance Thermostats, Z21.23
Metal Connectors for Gas Appliances, Z21.24
Gum Protective Devices, Z21.35
Gas Conversion Burners for Domestic Ranges, Z21.39
Quick-Disconnect Devices for Use with Gas Fuel, Z21.41

The control standards given in Table 12-26 are primarily those used for domestic or commercial equipment. Controls for other equipment, particularly those used in industrial applications and for larger boilers, are available with certification by Underwriters' Laboratories or Factory Mutual Engineering Division. Published listings may be purchased (e.g., *Gas and Oil Equipment List*[1]).

An excellent source for recommendations on the application and operation of such industrial protection equipment is the *Handbook of Industrial Loss Prevention.*[2] Reference may also be made to NFPA Standards 37, 51, 85A, 86A, and 211.[3–7]

The A.G.A. Directory of Approved Appliances and Listed Accessories[8] shows the capacities of each listed control model.

TEMPERATURE CONTROLS

Temperature controls of thermostats complying with ASA Z21.23[9] are listed by A.G.A.[8] under the following four classifications: (1) integral gaseous immersion type; (2) integral liquid immersion type; (3) integral surface contact type; and (4) electric switch type. The first three types are gas appliance thermostats (Table 12-27) in which the thermal elements and thermostatic valve are an integral unit. The last type includes devices that electrically sense changes in temperature and then control, by means of separate components, the flow of gas to the burner(s) to maintain selected temperatures. The names of the three integral types are indicative of the manner in which they are applied.

Capacities. The three integral types of thermostats are rated and listed[8] in Btu per hr of 1000 Btu per cu ft natural gas at 1.0 in. w.c. pressure drop. These types may be either **graduating** (valve motion is approximately proportional to the effective motion of the thermal element induced by temperature change) or **snap-acting** (valve travels instantly from the closed to the open position).

For capacities on other gases, apply the following factors:

Btu per cu ft	<800	800–949	>950	2500
Factor	0.516	0.765	1.000	1.62

Thermostats for Gaseous Immersion

Domestic Range Oven Thermostats. These gaseous immersion thermostats may be either the hydraulic or the electric switch type.

Hydraulic Type. Hydraulic controls are generally designed with nearly all the fluid in a heat-sensing bulb. Expansion and

Table 12-27 Recommended Dial Markings and Calibration Points for Gas Appliance Thermostats[9]

Appliance	Dial markings, °F Min	Dial markings, °F Max	Tolerance*	Calibration point, °F
Water heaters	100	180	...	140
Range top burner and griddle controls				
Throttling-to-off type	150	400	†	‡
Throttling-to-by-pass type	200	400	†	‡
Range ovens (domestic and hotel)				
Cooking	250	550	±10	400
Cooking and warming	140	550	±5§	200
Bake ovens (commercial)	250	550	±10	400
Clothes dryers	100	200	±10	175
Room heaters**	55	95	±5	75
Deep fat fryers	200	400	...	300
Refrigerators**	††	††	±5	...

* At both maximum and minimum numbered temperature dial settings.

† ±8 per cent of the change in dial setting between the calibration point and the highest numbered dial setting.

‡ Lowest degree number on dial.

§ ±5°F for minimum dial setting; ±15°F at 400 F.

** Except for electric switch-type thermostats.

†† Graduated for identification—temperatures not shown.

contraction of this fluid operates the gas valve by means of a bellows or diaphragm. One end of a capillary tube is brazed or welded to a bulb; the other end of the tube is joined to a bellows or diaphragm. Since high ambient temperatures at the capillary or valve body may affect the operation of the thermostat and result in lower oven temperatures, capillaries are short and pass thru low-temperature areas. Valve bodies are kept cool as possible. To counteract high ambient temperature effects, "compensation" is usually provided. In some thermostats, a bimetal counteracts the fluid expansion in the bellows diaphragm; in others, the diaphragm itself acts to do so.

Properly located in an oven, a bulb will neither interfere with loading nor reduce oven volume. Because oven temperatures are checked by a thermometer at the oven center, a bulb must be placed so that changes in temperature at the oven center are practically the same as at the bulb, although the actual temperatures may differ. The thermostat, therefore, is calibrated to take this into account, and its dial reading is adjusted to agree with the temperature at the oven center.

Electric Switch Type. Several modifications of this type are used. The following is illustrative of this type of control: The temperature sensor incorporates a thermistor (resistance increases with rising oven temperatures) that operates a "hot wire" gas valve thru a relay. The system (Fig. 12-16a) provides compensation for voltage changes so as to maintain calibration.

Oven temperature controls may also include an electrically operated automatic gas valve and an automatic pilot (safety shutoff device). The pilot is generally operated by a thermopile. Usage is with a clock, a timer, and a roasting thermostat.

Oven Temperature Control Systems. Domestic ranges incorporate either one of the following systems.

Standard Systems. In these systems, the heat control includes a by-pass feature that maintains a minimum flame at the burner when the thermostat setting is satisfied. This flame limits the lowest controlled operating temperature of the oven to about 250 F. In most of these controls, gas flow is throttled or modulated, but full on-off control may also be used.

Low-Temperature Systems. Such systems (Fig. 12-16b) maintain controlled oven temperatures as low as 140 F. The three basic parts of such systems are a thermostat, a flame-responsive automatic pilot valve, and a pilot—one of many pilot variations.

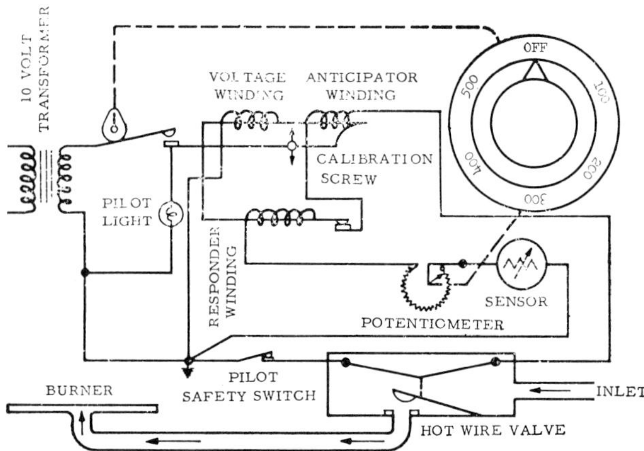

Fig. 12-16a Electric switch oven control system.

Above approximately 325 F, the oven thermostat operates in the same way as in the standard systems. Below 325 F, the heater pilot and, in turn, the oven burner shut off completely when the thermostat setting is satisfied, since the main burner by-pass flame tends to override low-temperature settings. When the temperature drops to a point at which more heat is called for, gas again flows to the heater pilot (or burner pilot), which is promptly relighted by the constant-burning pilot flame (or stand-by pilot). The flame of the heater pilot opens the automatic valve to the main burner.

Calibration of Range Thermostats. Various methods are used, depending on the class of equipment involved. Two examples illustrate the procedures.

1. *Top Burner Thermostats.* Fill a 2- to 3-qt, flat-bottomed aluminum pan with water to a point about 2 in. below the top. Heat the pan for 10 to 25 min. Hold an accurate mercury thermometer in the water without touching the bottom of the pan. (Alternatively, a thermocouple connected to a millivoltmeter may be used to advantage because of the short response time as compared with liquid-filled thermometers.) The temperature of water should be in the range prescribed. One manufacturer specifies 5°F maximum variation for the 200 F setting. Since the range of dial marking on thermostats varies from 150 to 400 F, calibration at two temperatures may be desirable.

2. *Oven Thermostats.* Place a thermometer or a thermocouple connected to a millivoltmeter in the oven. Allow the oven to heat for about 10 min. Then read the temperature. In ovens without glass doors the temperature should be read quickly enough to avoid cooling of the thermometer. Two temperature settings between 200 and 450 F may be used for calibration. If the temperature reading falls within ±10° F of the dial setting, the thermostat is satisfactory. For a thermostat with a "keep warm" feature, a tolerance of ±15°F is satisfactory at 400 F and ±5°F at the minimum dial setting.

Room Heater Thermostats. These thermostats are usually located on the gas line close to, or as a part of, the heater. There are three general types: (1) bellows (vapor filled), with or without auxiliary bulb and capillary; (2) combination of the first type with automatic pilot; (3) combination of the first type with cock, thermostat, and automatic pilot.

The room heater thermostat shown at the left in Fig. 12-17 employs a vapor-filled bellows operating a graduating gas valve. Both graduating and snap-acting types are available, as are combinations of both or combinations of either or both with automatic pilots. These thermostats are usually located close to the floor and near enough to the heater to be affected by the cool air flowing toward it. The right-hand portion of Fig. 12-17 is illustrative of a type of thermostat employing both graduating and snap-acting features.

Some combination designs use a sensing bulb, capillary tube, and bellows. The sensor can be placed in the return air stream, remote from the control body. In recessed heaters and room heaters of other kinds, the appliance manufacturer locates the bulb and control body. A thermostatic device designed to shut off gas in the event of spillage from the draft hood relief opening is sometimes provided on room heaters. Shutoff may be accomplished by locating the sensing element near the draft hood skirt in a position in which it would be subject to any rise in temperature produced by hot spillage gases.

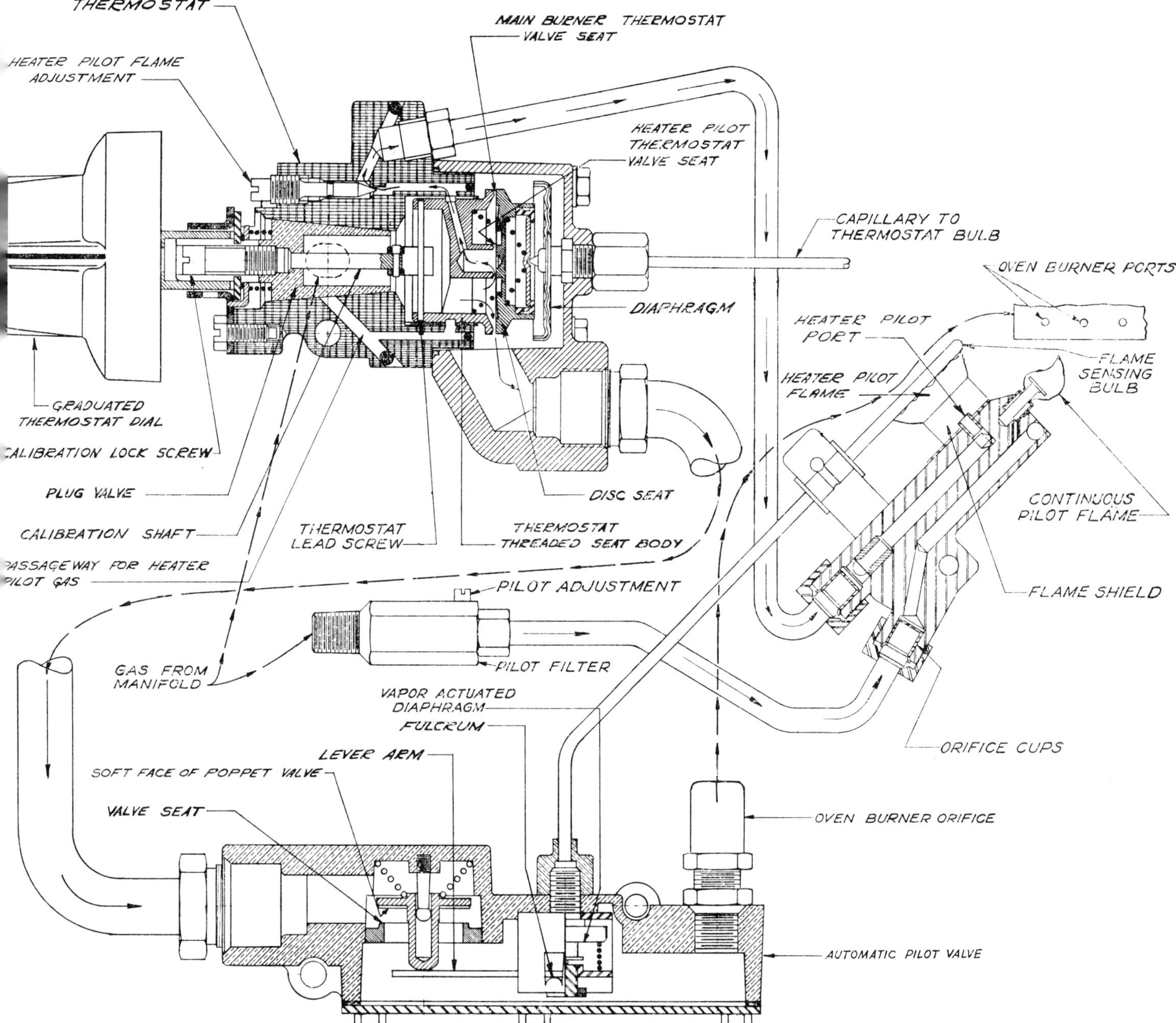

Fig. 12-16b Harper all-gas low-temperature oven control system. (All-Temp, Harper-Wyman Co.)

Room Thermostats for Central Heating Appliances. Most of these are wall-mounted thermostats of the electric switch type and are snap-acting, opening or closing an electric circuit to open or close an automatic valve that controls gas flow. They may also actuate fans. Thermostats generally operate on house voltage or house voltage stepped down by a transformer to 20 to 25 volts, or on self-generated voltages (thermopiles producing less than one volt, Fig. 12-19c) and have a bimetallic sensing element. Bellows types are also used. Contact resistance must be kept to a minimum with thermopile voltages. Use of mercury switches eliminates the possibility of dust accumulating on contacts.

There are many designs and sizes. The essential parts of a simple design are shown in Fig. 12-18. Three residential heating control systems are shown in Figs. 12-19a thru 12-19c.

Day-night thermostats are either of the twin-element or of the single-element type, and may be set manually or by an electric clock. Heating-cooling thermostats incorporate two switches, one for the heating and one for the cooling equipment, with a single sensing element. The differential between heating and cooling may be adjustable or fixed; thermostats with a changeover switch are also available.

To reduce the system lag and the operating differential, many thermostats contain an added feature called the *heat anticipator*, which introduces a small amount of artificial heat

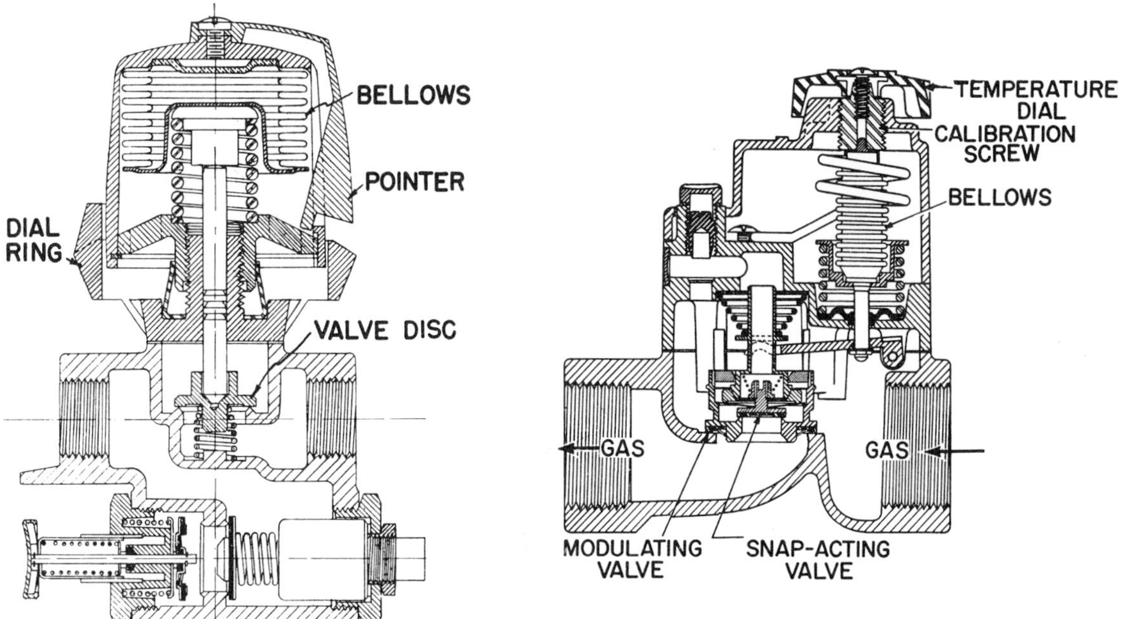

Fig. 12-17 (left) Room heater thermostat and automatic pilot (Robertshaw 2 EC graduating type); (right) thermostat with graduating and snap-acting features—sensing element not shown. (Honeywell, Inc.)

into the room thermostat internally and adjacent to the bimetal, actuating the bimetal at a rate faster than the normal rate of rise in the room air temperature. This minimizes room overheating. Artificial heating is obtained by a small resistor placed within the control and connected in the electric control circuit in such manner that the heater may be switched in or out of the circuit in relation to the position of the thermostat contacts. Depending on the resistors used, anticipators may be fixed or variable; voltage and cycle anticipators are variants of the fixed anticipator.

Some thermostats are equipped with an outdoor reset that automatically advances the thermostat setting when the outdoor temperature drops unduly. For each 10° F drop in outdoor temperature, the thermostat is automatically set up 1°F.

For best results thermostat location is important. Normal location is in the living room (an alternate location is in the dining room) at a level about 5 ft above the floor; on an inside wall away from any outside wall; in natural air circulation when the heating or cooling system is off; in the path of circulating air when the heating or cooling system is on, preferably in the return air path; and at a point between wall studs. Locations to be avoided include those behind doors, furniture, or draperies; in cold drafts (as from an open door); near concealed

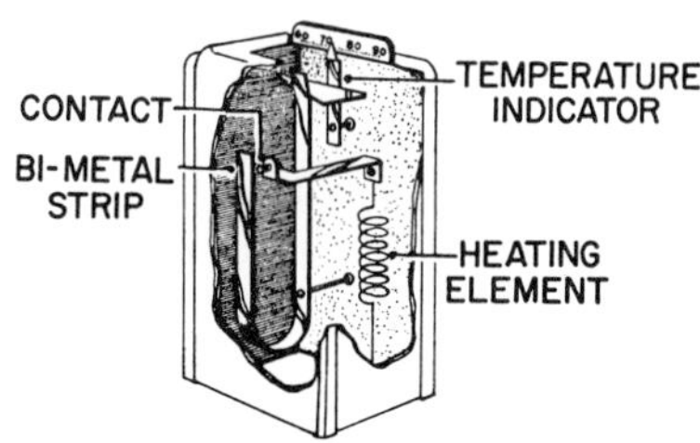

Fig. 12-18 Elements of a room thermostat. Contacts may be metal as shown or a mercury switch. Heat anticipating element (shown as heating element) shortens the on-off cycles.

pipes or ducts; in the direct rays of sun or fireplace; and near discharge air outlets, lamps, radios, television sets, or other heat sources.

Outdoor Thermostats.[29] These thermostats are usually used to control heating of large buildings where a single thermostat will not serve adequately. Location of the thermostat, heat anticipation, proper balancing of the heating system, and sensing duty are determining factors for proper selection of the control. With a standard type of thermostat the outdoor control can be used to act as a compensator giving closer regulation to room temperatures by following changes in outdoor temperatures.

Two-Stage or Modulating-Type Thermostats.[29] These devices vary the gas input to the burners, depending upon the temperature drop in the room. The first contact will

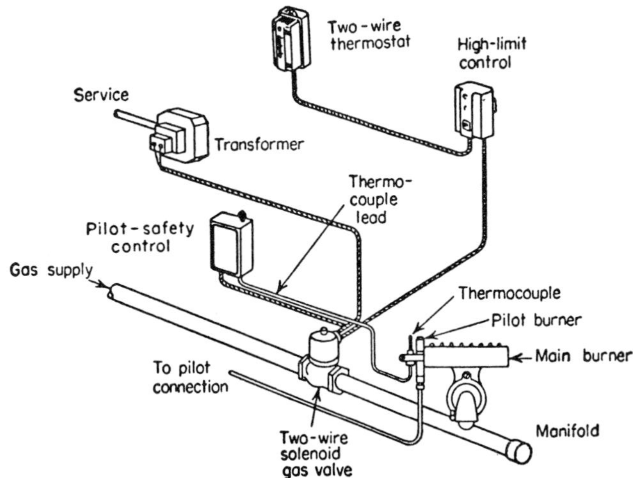

Fig. 12-19a Simple control system for gas-fired steam or gravity warm air or hot water heating system.[30]

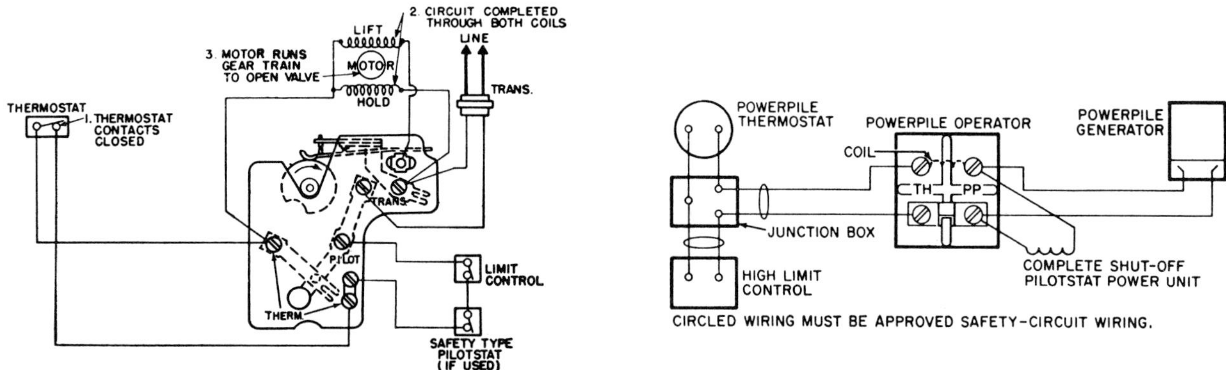

Fig. 12-19b (left) Heating control system using a low-voltage motorized gas valve (detailed in Fig. 12-38). The valve opening operation is shown. (Honeywell, Inc.)
Fig. 12-19c (right) Typical wiring diagram of a thermostat operated by a thermopile (750 mv). Thermopile elements generally consist of long, thin (e.g., $^1/_{64} \times {}^1/_{16}$ in.), flat wire strips to facilitate rapid response to heating and cooling. (Honeywell, Inc.)

bring the burners on thru a modulating-type main control valve at the lowest input. As the temperature continues to drop, the gas input rate is increased, in steps, until the room temperature begins to rise.

Return Air and Return Hot Water Temperature Thermostats.[29] These units sense the temperature of the air or water coming from the occupied spaces. This type of thermostat is usually a liquid- or gas-filled bulb that is placed in the return air duct or return water pipe so that it can accurately sense the temperature in the return of the heating system.

Clothes Dryer Thermostats. Clothes dryer controls usually sense exhaust air temperature. Most dryers use a revolving drum with a blower or fan to blow air thru tumbling wet clothes. The temperature-sensing element of the control is located downstream in the air exhaust that carries moisture from the clothes, thus keeping the element temperature well below the set temperature. As the clothes dry, the exhaust air temperature increases until it reaches the set temperature, at which time gas is shut off. The motor may continue to run until shut off by the timer.

On some dryers, the motor is under the control of an electric snap-acting type of thermostat with two switches. One switch shuts off the gas thru a solenoid valve; the other controls the motor directly, opening its circuit soon after the gas is shut off.

Commercial Cooking Equipment Thermostats. These thermostats require heavier construction and greater capacity than domestic controls in order to endure extensive daily operation, often at high ambient temperatures. Temperatures governed run from 100 to 200 F for a dry food warmer up to 600 or 700 F for a pizza oven. Graduating thermostats are most often used, but where large quantities of gas are required, diaphragm controls (on-off) are employed.

Thermostats for Liquid Immersion

These thermostats may be used on gas appliances such as water heaters, commercial cooking equipment, and industrial equipment.

Water Heater Thermostats. These units consist of a heat-sensing element immersed in the water being heated (or a surface contact-type strapped to the tank) and a gas valve or switch. The sensing element is usually a rod and tube type with a snap-acting valve. Some rod and tube, bellows, and diaphragm applications employ a graduating control of gas flow. Of the three American Standards for water heaters,[11] Z21.10.1, Z21.10.2, and Z21.10.3, only the first requires a thermostat; thermostats, when furnished, should satisfy Z21.23.[9]

The rod and tube type of thermostat, either graduating or snap-acting, usually employs a copper tube with a high coefficient of expansion; the rod is usually Invar, selected for its low coefficient of expansion. Heat applied to the tube causes it to expand, thereby pushing or pulling the rod inside it. To obtain snap-action, a "lost-motion" device is utilized to build up the travel and force necessary to cause the valve disk to snap to the open position. This may be accomplished thru a set of levers, a "clicker," or toggles and permanent magnets. Because copper starts to lose its tensile strength at about 300 F and scales at about 700 F, its use for thermostat tubes should be limited to comparatively low temperatures.

Demand for controls that have maximum utility and incorporate safety devices has led to development of combination designs. Features that may be incorporated in, or used separately with, the thermostat are an automatic pilot, "safe lighting," gas cock, pilot gas filter, pressure regulator, and main and pilot gas adjustment means.

Operation. A typical **rod and tube** thermostat is shown in Fig. 12-20. The heat-sensitive portion, immersed in the water, consists of an Invar rod mounted with one end fixed inside a copper tube. The expanding or contracting movement of the tube, transmitted by the rod, actuates the device. This movement is transmitted to a "clicker" assembly, which amplifies it and provides snap-action. The amplified movement acts to open the thermostat valve. The snap-action principle produces either full opening or full closure of the valve.

The "clicker" is a concave disk that is caused to snap thru at its center. Movement of the Invar rod is applied thru a circular fulcrum plunger near the outer edge of the clicker against another circular fulcrum at this outer edge. Movement of the clicker center is amplified by means of a circular button, mounted thru this center, which pushes a pair of levers against the end of the valve stem.

In the simple thermostat described, different temperature settings are obtained by moving the lever attached to the Invar rod, to lengthen or shorten the rod by screwing it into the outer end of the copper tube. Thus, the temperature is changed, whereupon the "clicker" snaps thru and the valve closes.

The most common type of automatic pilot used in combination controls is actuated by an electromagnet incorporated in the

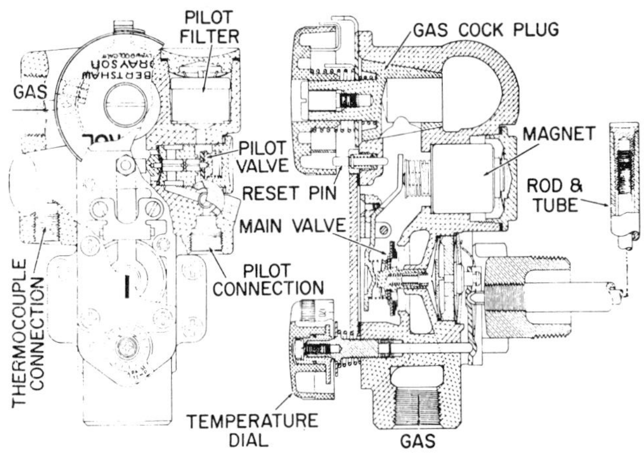

Fig. 12-20 Typical rod and tube snap-acting water heater thermostat.

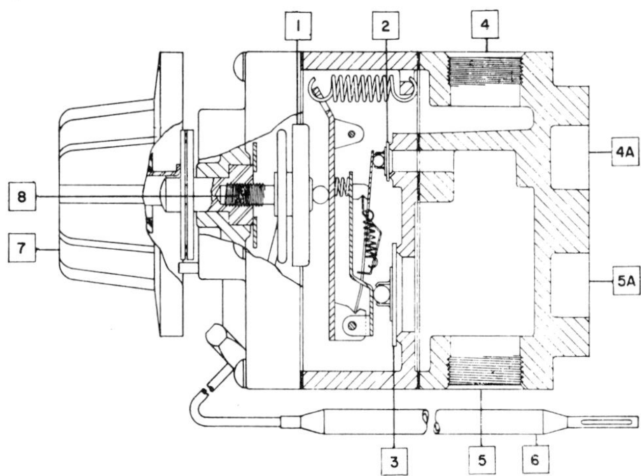

Fig. 12-21 Typical commercial cooking thermostat—combination snap-acting and graduating. (Robertshaw Control Co.)

1. Steel diaphragm
2. Snap disk
3. Throttle disk
4. Gas inlet
4A. Gas inlet
5. Gas outlet
5A. Gas outlet
6. Sensing element
7. Dial
8. Adjusting screw

thermostat body. Power is derived from a thermocouple mounted close to, and receiving its energizing heat from, the pilot flame. The current so induced is sufficient to hold the keeper against the magnet, after they are placed in contact by means of the reset button provided. This opens the valve, allowing gas to flow thru the control normally. If the flame heating the thermocouple is extinguished, however, the current drops to zero as the thermocouple cools. Magnetic attraction to the keeper is then reduced, allowing the spring to pull it away from the magnet and act to close the main gas valve. In the 100-per cent shutoff type (commonly used with LP-gases), pilot gas will also be shut off at the same time.

In a common application, the automatic pilot and the thermostat actuate the same valve. This is accomplished by means of a relief spring within the thermostat valve stem that allows the "clicker" assembly to operate without damage to itself and without opening the valve while the automatic pilot acts to hold it closed. With this arrangement, the automatic pilot can act to close the valve irrespective of thermostatic action.

"Safe Lighting." This feature is found on most combination thermostats with automatic pilots. It permits lighting the pilot flame while the keeper is being held against the magnet, without gas flow to the main burner. The control shown in Fig. 12-20 provides safe lighting since the automatic pilot can be set only with the gas cock in the pilot position and, consequently, no gas can pass to the main burner.

For convenience, economy, and integrated action, the gas shutoff valve may also be made a part of the thermostat. This valve is so arranged that the main gas may be shut off and only the pilot gas allowed to flow during lighting.

Commercial Cooking Equipment Thermostats. Proper temperatures must be maintained in deep fat fryers to avoid "fat taste" and "soggy" food. Since a rapid drop in temperature occurs when a load is put into a fryer, its thermostat must turn on full gas flow as quickly as possible. Figure 12-21 shows a combination snap-acting and graduating thermostat. Sterilizers, steam tables, bains-marie, and coffee urns generally use graduating type controls; dial readings usually range from 100 to 200 F.

Thermostats for Surface Contact

Contact thermostats may employ a hollow disk, straight tube, or coiled tube as the thermal element. In other respects, they are substantially the same as immersion-type units. Electrical resistance means may also be used as the thermal element.

The sensing element is held tightly against a heated storage vessel, container, or plate and reacts to their surface temperature changes. Heat transfer to the sensing element is mainly by conduction. Typical applications are on water heaters, domestic range top burners, commercial cooking appliances, and industrial equipment.

Domestic Range Top Burner Thermostats. Burner-with-a-Brain, Thermo-Set, Thermal Eye, Tem-Trol and Toptrol are some of the names applied to gas range top burner thermostats. Active development started in 1955, although griddle burner applications had been used earlier. The sensing element is usually a liquid-filled tube attached to a metal cap held firmly against the pan bottom by a spring.

Figure 12-22 shows a Toptrol top burner thermostat with accompanying sensing head that incorporates a burner adjustment feature known as Flame Adjust. Figure 12-23 shows a Flame Selector top burner control, which combines graduating thermostatic control with easy maximum burner rate selection. There are two basic types. In the by-pass type, the burner output is reduced thermostatically to a minimum rate set by adjusting the by-pass. In the tower type, the thermostat re-

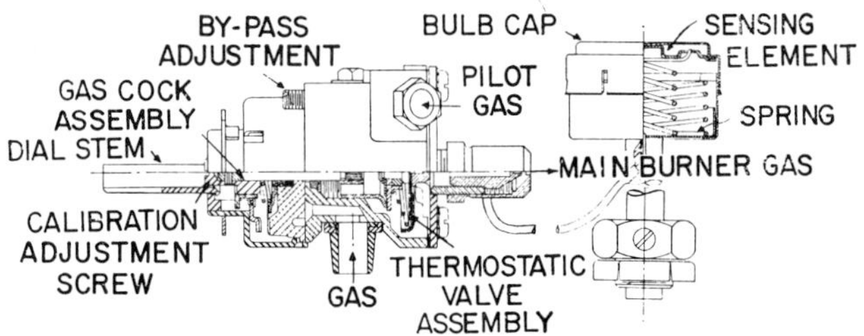

Fig. 12-22 Toptrol top burner thermostat. Temperature range: 150 to 425 F; rated 29 MBtu per hr of natural gas. (Wilcolator Co.)

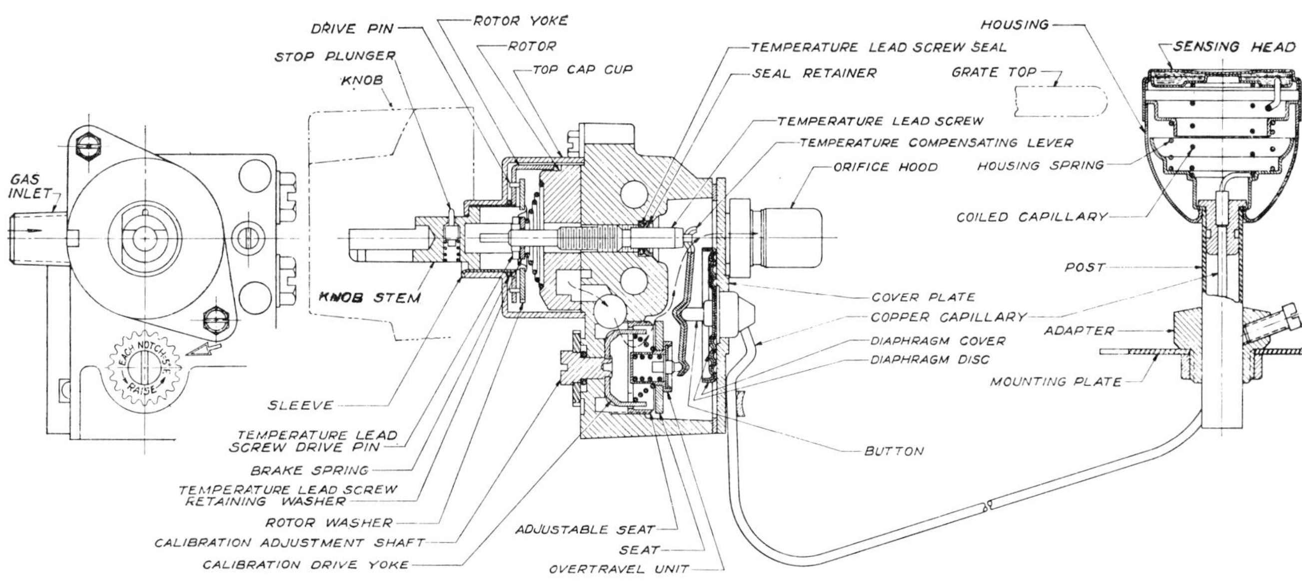

Fig. 12-23 Flame Selector top burner thermostat. (Harper-Wyman Co.)

duces the burner according to need—including a full-off position. A small pilot flame on a pilot tower then provides reignition when the thermostat reopens.

IGNITION AND SHUTOFF DEVICES

Automatic ignition at main burners on appliances is usually accomplished by a gas pilot burner flame. For automatic operation by a thermostat, as on automatic water heaters and central heating appliances, the gas pilot flame must be sized—on a properly located pilot—so as to ignite the main burners; otherwise, the valve controlling the main burner flow will not open. Where burners are visible when turned on, as on gas range tops, standing pilots with flash tubes to each burner, minimum flame pilots at each burner, or electric coil ignition devices may be used for ignition.

ASA Standards[12]

Automatic burner ignition and safety shutoff devices (sometimes called **automatic pilots**) are used (1) to ignite and reignite gas at the main burner(s) and (2) to prevent the flow of unburned gas from the main burner(s) or from the pilot(s) and main burner(s) when the source of ignition is inoperative or not operating to ensure safe ignition.

Classification by Function. Three classes follow:

Class I. Standing Pilot System, Manually Lighted. This type of system consists of a manually lighted pilot that burns constantly, regardless of main burner operation. The pilot is proved by a **flame-responsive element** that will respond to a constant-burning pilot in a Class IA, IB, or IC device and to a separate intermittently operated pilot igniting from a constant-burning pilot in a Class ID device (see Table 12-28).

Class II. Electrically Lighted, Recycling Proved Pilot System. This type of system consists of an electrically ignited pilot that may be automatically lighted each time the burner operates. The pilot is proved by a flame-responsive element. The system will act to reignite the pilot automatically after pilot outage from any cause. The pilot burns continuously during the period the main burner is firing.

Class III. Electrically Ignited, Recycling Direct Ignition System. This type of system consists of an ignition means and a flame-responsive element that operates each time the burner is fired and acts to reignite upon flame failure from any cause. On ignition failure, the system will lock out to prevent the flow of gas to the burner.

Classification by Ignition System Timing. Table 12-28 gives the maximum timing for the foregoing three classes. The ignition system should effectively ignite the gas at the test burner within 4 sec.

Flame-Responsive Elements

Materials. Research[13] has clearly shown the necessity of selecting the best combination of materials for pilot construction and assembly to avoid formation of deposit. Aluminum appears the most suitable material for orifices and supply tubing in pilots for operation on all gases at temperatures below 1000 F; above 1000 F, brass SAE 72 or nickel-free stainless steel may be used for operation on straight natural gas containing no air or sulfur. On other gases, deposits may be expected at temperatures over 1000 F regardless of the material used.

For automatic pilots, the thermal element is customarily a bimetal (strip, or rod and tube), single metal unevenly heated, hydraulic element, or thermocouple.

Bimetal. A bimetal strip type of automatic pilot is shown in Fig. 12-24. Studies[14] of automatic pilot operation have indicated that bimetal material selection considerations include temperature of operation, loads, application, maximum spot temperatures, and method of application. Elements must be heat treated after any forming or bending in order to relieve fabrication strains, particularly the bond between the metals.

The actuating element in a **rod and tube** type unit is a tube closed at one end and secured at the other to the valve body. The tube possesses the necessary coefficient of

Table 12-28 Ignition Systems, Maximum Timing[12]

Class	Description	Lockout timing* (Seconds)	Recycling element response time† (Seconds)	Flame failure shutoff time‡ (Seconds)
IA	Manually lighted standing pilot system			150
IB	Manually lighted standing pilot system			90
IC	Manually lighted standing pilot system			4
ID	Manually lighted standing pilot system, intermittent,§ proved pilot			90
IIA	Electrically lighted, recycling proved pilot system			150
IIB	Electrically lighted, recycling proved pilot system			90
IIC	Electrically lighted, recycling proved pilot system			4
IIIA	Electrically ignited, recycling direct ignition type	120	30	
IIIB	Electrically ignited, recycling direct ignition type	60	15	
IIIC	Electrically ignited, recycling direct ignition type	15	0.8	

* The period of time between the initial ignition trial and lockout by the ignition system. Lockout is defined as the condition that prevents flow of gas to the main burner(s) until normal operating conditions are restored by manual operation.

† Class IIIA, B, C devices act to reignite the main burner after flame failure. Recycling element response time is the elapsed period between flame failure and the start of the reignition cycle. Lockout or shutdown of the main burner will occur only if ignition means fails to ignite main burner gas before expiration of lockout timing period.

‡ The period required by the control to actuate or de-energize the main burner fuel shutoff device after flame extinction.

§ Intermittently operated pilot ignited by a constant-burning pilot.

expansion and resistance to corrosion. Inside the tube is a rod with a very low coefficient of expansion.

Single Metal, Unevenly Heated. The element is usually made of a corrosion-resistant material in the shape of a bar, tube, strip, or horseshoe. Pilot flame impingement on one side or at some point along its length causes twisting, bending, or other distortion, and this action operates an automatic pilot device (Fig. 12-25).

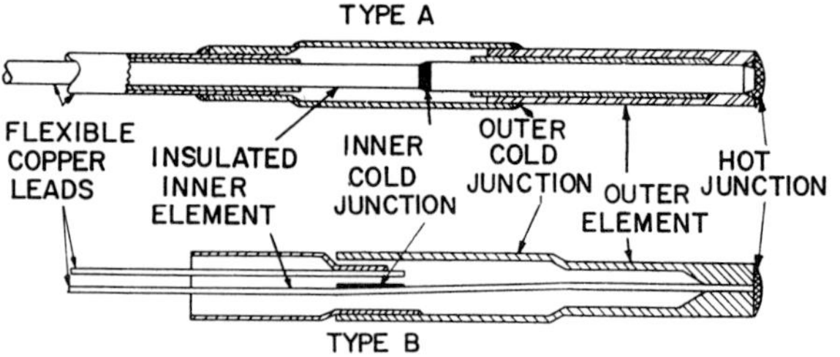

Fig. 12-26 Typical thermocouple elements for an automatic pilot.

Hydraulic Element. The fluid-filled type of element for automatic pilots, especially the one using mercury vapor, has had some successful applications. No fluids known at present have all the properties required by high ambient and spot temperatures. It is difficult to design a fail-safe device using a hydraulic element.

Thermocouple with Electromagnet (Thermoelectric). The most widely used type of automatic pilot is actuated by a small electromagnet energized by self-generated current. The current is generated by the temperature differential produced between the hot and cold junctions of a thermocouple (Fig. 12-26) heated at one end (hot junction) by a pilot flame.

A widely used thermoelectric valve, shown in Fig. 12-27, employs an electromagnetic hood assembly and concentric thermocouple lead as the power unit. The thermocouple lead is attached to the hood assembly by a nut, thus completing a circuit thru the coils of wire wound around the magnet frame. The thermocouple is so placed in the pilot flame as to heat about $\frac{1}{2}$ in. of the couple at the hot junction, while the cold junctions remain substantially colder. Electricity thus generated flows to the electromagnet as long as the thermocouple is properly heated by the pilot flame, but ceases to flow if the pilot flame goes out.

Within the hood assembly lies a circular armature attached to an outside valve disk. The magnet is energized by the thermocouple and the armature is placed against the magnet faces by a reset assembly. Contact is maintained and the connected pilot burner valve is held open as long as the thermocouple

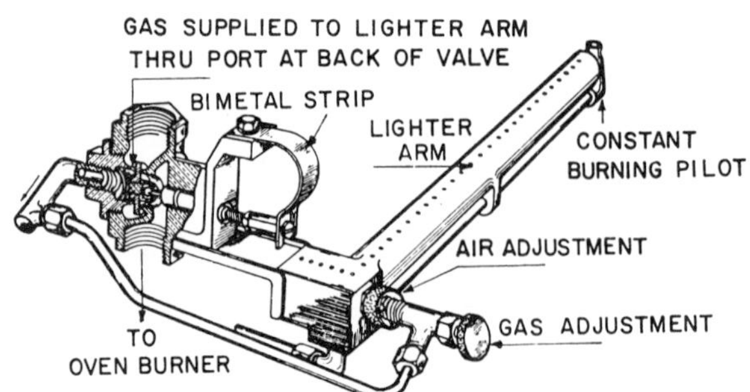

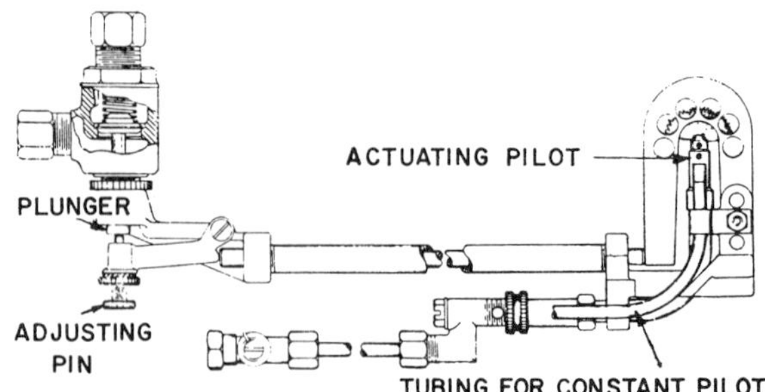

Fig. 12-24 (left) Bimetal strip actuated automatic burner ignition and safety shutoff device (Patrol Valve Type A). Valve shown in position assumed when bimetal is cold; no gas can pass to oven burner. Spring is holding valve against seat; valve is pushed to left when bimetal is heated, allowing gas to go to oven burner.

Fig. 12-25 (right) Single-metal horseshoe shape automatic pilot (Bryant Powerstat oven igniter).

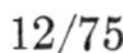

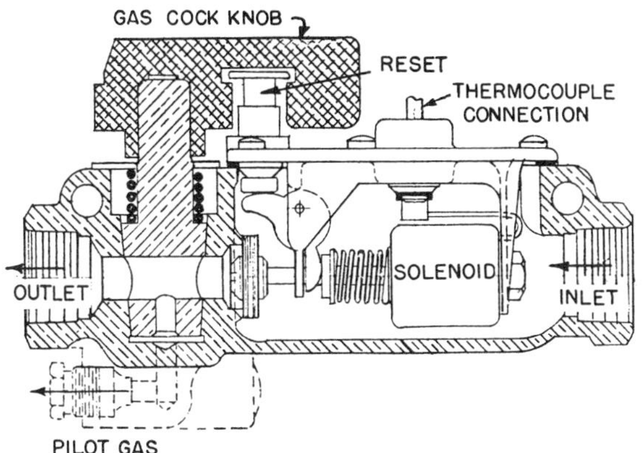

Fig. 12-27 Thermolectric automatic pilot or safety shutoff device. (Baso Div. Penn Controls Co.)

generates current. In the model illustrated, this contact can be accomplished only when the handle is in the "pilot" position. In other models, the reset assembly lies directly opposite the hood assembly. When the reset stem is pushed in, a flow interrupter disk on the reset stem or other means must[12] prevent gas from reaching the *main* burner until the pilot burner is lighted and the electromagnet is holding the armature.

Approximately 30 sec are required to generate sufficient emf to retain the armature. In all designs, if the pilot flame is lost or gas pressure reduces the flame appreciably: (1) the emf is not maintained; (2) the keeper is released; and (3) a spring closes the *main* gas valve. The closing of main gas only or of both main and pilot gas can be effected.

Radiation and Conductivity. Many commercial and industrial combustion safety circuits employ units that sense the radiation in gas flames or their conductivity. In the flame rectification type of sensor, the flame rectifies a-c signals. This signal, in turn, triggers an electronic relay.

Other Ignition Devices

When an appliance has a number of burners, they may be ignited by flash tubes extending from one (single-point ignition) or more than one pilot flame; by very small pilot flames at each burner; or by electric coil or spark ignition of a gas pilot that lights the main burner. For concealed burners,

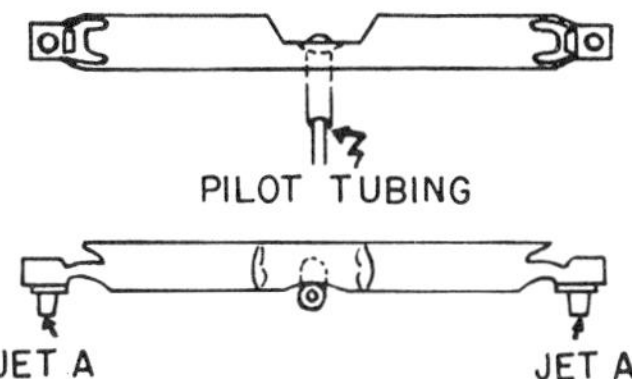

Fig. 12-29 Flash tube ignition device for two burners. (Hardwick Stove Co.)

as on ovens or broilers, and for thermostatically controlled burners, the igniting device must be an automatic pilot (except for top burners on ranges). A combination thermocouple and pilot burner is shown in Fig. 12-28.

Flash Tubes. One to six burners may be lighted by one standing pilot thru separate flash tubes to each burner. When a burner is turned on, raw gas is discharged thru a port into the flash tube; air is injected with the gas. Flame flashes back thru the tube, igniting the burner gas; ignition is almost immediate. In the design shown in Fig. 12-29, jet A fastens into a corresponding opening in the burner. Gas travels up thru this projection and discharges horizontally into the flash tube.

Research[15] in automatic flash tube and pilot ignition of oven burners has stimulated commercial applications of such ignition systems to range ovens and broilers.

Individual Needle Pilot Flames. Ignition of range top burners by the use of needle pilot flames of about 50 Btu per hr constantly burning at each burner has increased greatly since their introduction in 1954.[16] Advantages claimed include low fuel consumption, less heat in the kitchen, and elimination of range top hot spots. Two general designs, shown in Fig. 12-30, are used, with numerous variations in construction and application. All pilots are nonaerated. Manufacturers' instructions for installation should be carefully followed.

Electrical Pilot Igniters. Many gas appliances employ electrical means for ignition of pilot gas. Gas ranges and clothes dryers use such means for direct ignition of the main burners. Figure 12-31 shows the ignition system for an automatic range. Figures 12-96a and 12-96b show ignition systems for clothes dryers. In the case of central heating equipment, however, 1964 ASA requirements permitted electrical ignition of the pilot gas only. Electrical ignition of main burners on approved central heating gas

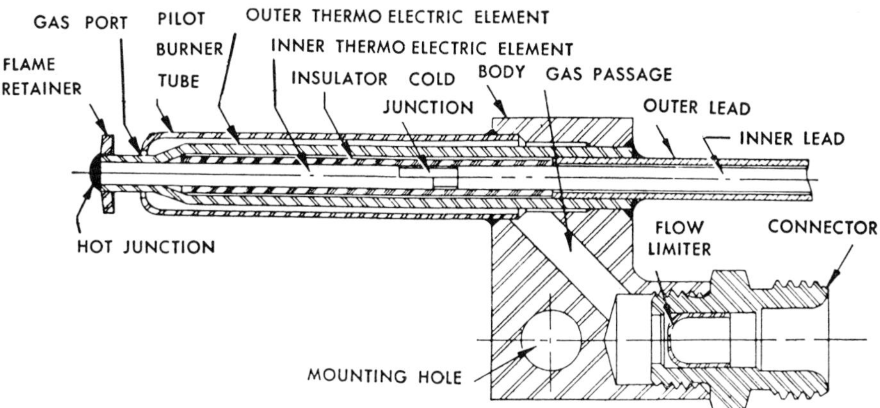

Fig. 12-28 Mini-Pilot—combination thermocouple and pilot burner. (Grayson Div., Robertshaw Controls Co.)

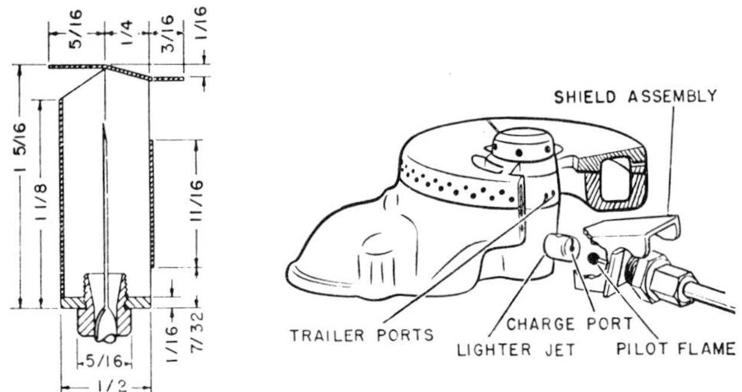

Fig. 12-30 (left) Low-Btu hypodermic needle pilot and shield design. (A.G.A. Lab.)[16] (right) Micro-Lite pilot and range top burner arrangement. (Harper-Wyman Co.)

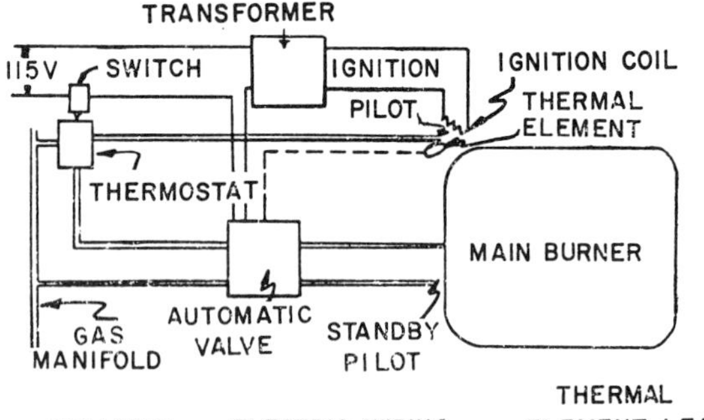

Fig. 12-31 Typical glow coil ignition system for an automatic range.[31]

appliances was under study by an ASA Requirements Committee in 1964.

Available research data[17] cover performance of hot wire coil ignition of both catalytic and noncatalytic materials. Factors in effective ignition include coil surface area, coil temperature, and air-gas mixture velocity. In general the larger the wire size and the coil spacing, the higher the minimum temperature for ignition. A wire coil of Chromel A (80 per cent Ni, 20 per cent Cr), a noncatalytic material, has been found to require about 50 per cent more electric power to raise its temperature to the point necessary to ignite natural gas than does an equivalent coil of platinum, a material possessing catalytic properties.

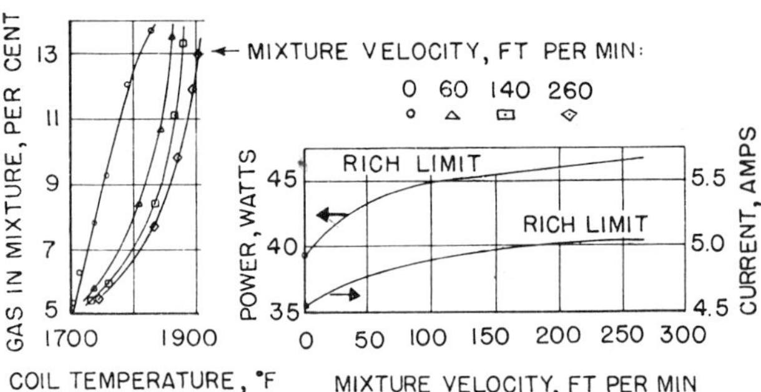

Fig. 12-33 Effect of natural gas–air mixture velocity on minimum ignition temperature and coil power consumption.[17] Chromel A wire coil made of 25 turns of 24-gage wire; coil length ≅ 1.0 in., coil diam = 1/8 in. ID.

Figure 12-32 shows the minimum coil temperatures necessary for ignition using a coil of Chromel A wire and platinum wire. Note that, in regard to the curves for platinum, for a mixture of 16 per cent manufactured gas in air, the coil need attain only a temperature (in air) of 1280 F prior to ignition, since on exposure to the air-gas mixture the coil temperature increases to 2080 F by means of catalysis and effectively produces ignition without an increase in electrical energy input. Effects of air-gas mixture velocity are illustrated in Fig. 12-33.

Figure 12-34 shows the minimum voltage required to create a spark in a range of natural gas-air mixtures. Recent research,[18] coordinated with results of a field survey of ignition conditions, has yielded extensive data on electric spark igniter systems. These include a low-energy unit powered by a thermoelectrically chargeable low-voltage battery. Piezoelectric and hot wire igniters have also been under study.

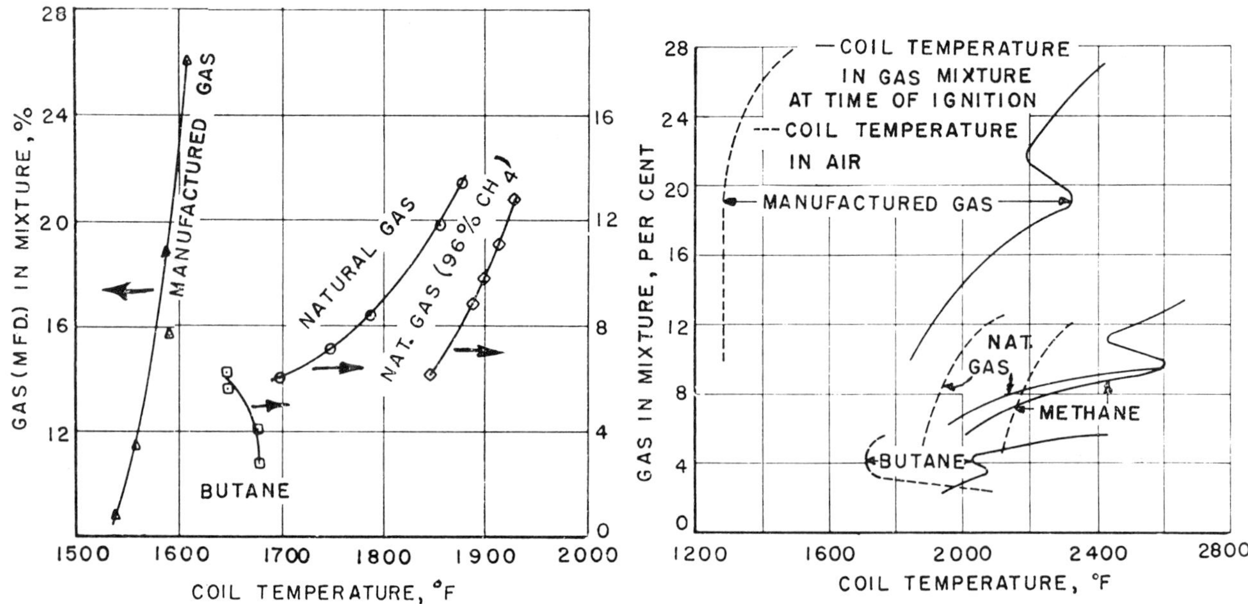

Fig. 12-32 Minimum coil temperatures required for ignition of various gas-air mixtures (at zero velocity).[17] Coils made of 10 turns of 30-gage wire; coil length ≅ 0.25 in., coil diam = 1/8 in. ID. Left—Chromel A wire; right—platinum wire.

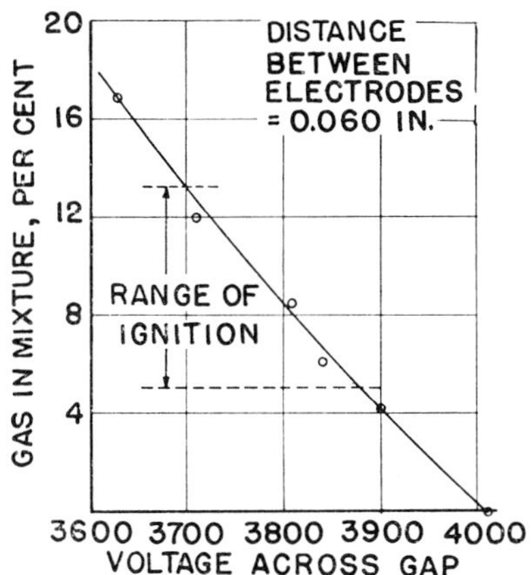

Fig. 12-34 Minimum voltages required for breakdown of natural gas–air mixtures; dotted lines define range of ignitibility.[17]

GAS APPLIANCE PRESSURE REGULATORS

These regulators, of the "inches-to-inches" type, are designed to adjust the gas pressure at appliance orifices and controls so as to maintain uniform flow. Such a regulator usually reduces normal house pressure to between 3.5 to 5.0 in. w.c., depending on its setting. If both heating value and specific gravity remain reasonably constant, a regulator will control pressure closely and maintain a uniform gas input to the appliance.

ASA requirements specify regulators as standard equipment on all central heating gas appliances. Regulators are required on an increasing number of other types of appliances; in many cases they are supplied as standard equipment. Advantages include: (1) elimination of the need for adjustment of burner orifices, thus also permitting the use of fixed orifices—excellent air entrainment devices; (2) better burner performance, because flame fluctuation is eliminated; and (3) better pilot performance and ignition. Figure 9-38a shows the basic parts of an appliance pressure regulator. Such regulators have many design variations, but generally they are spring loaded. The spring is contained in the regulator housing. Table 12-29 gives the characteristics of appliance regulators.

Table 12-29 General Characteristics of Appliance Pressure Regulators*

Type	Connection sizes, in.	Pressure range, in. w.c. Inlet	Outlet	Capacity range, 0.6 sp gr, cu ft per hr
Pilot	⅛ - ¼	18	0 - 12	0.1 - 37
Main burner and pilot	⅜ - ¾	15 - 28	2 - 12	0.1 - 111
Main burner	⅜ - 4	15 - 28	2 - 12	5 - 3725

* For 1964 ASA Standards on gas pressure regulators, see adjacent column; or see the current Z21.18.

Venting of Pressure Regulators.[19] Units requiring access to the atmosphere for successful operation must be equipped with vent piping leading to the outer air or into the combustion chamber adjacent to a constant-burning pilot, unless constructed or equipped with a **vent limiting means** (as directed by ASA Z21.18) to limit the escape of gas from the vent opening in the event of diaphragm failure.

If vent lines terminate in the outer air, means must be provided to prevent water, insects, and other foreign matter from entering the piping. When vents enter a combustion chamber, the vent termination should be positioned so that any escaping gas will be readily ignited from the pilot flame and will not adversely affect the operation of the thermal element of the automatic pilot. For manufactured gas a flame arrester in the vent piping may also be necessary.

ASA Standards[20]

Gas pressure regulators, listed by A. G. A.,[8] (1) are of the inches-to-inches type; (2) may either be individual or be in combination with other controls; (3) are designed for use on individual appliances; (4) are designed for inlet pressures up to 14.0 in. w.c. and outlet pressures up to 7.0 in. w.c.; (5) can withstand 2.0 psig without leakage or deformation; and (6) may be either adjustable or nonadjustable.

Pressure drop capacity is the flow rate thru a wide open regulator, at a pressure drop of 0.3 in. w.c. for an individual regulator or 1.0 in w.c. for a regulator in combination with other controls. The minimum capacity of a unit designed to control **pilot flow** is 0.15 cu ft per hr of 1000 Btu per cu ft (0.64 sp gr) gas. Use of **vent limiting means** limits the flow from the regulator vent, in case of diaphragm rupture, to 1.0 cu ft per hr of 0.6 sp gr gas (undiluted LP-gas—0.5 cu ft per hr of 1.53 sp gr gas at 11.0 in. w.c.); vent connections *not* incorporating a limiting means are threaded for connection to a vent line.

Capacity. Individual units are rated (**pressure drop capacity**) and listed[8] in Btu per hr of 1000 Btu per cu ft of 0.64 sp gr natural gas at 0.3 in. w.c. pressure drop. Two types of applications are rated: (1) main burner only—both maximum and minimum* capacity; (2) main burner and pilot—maximum and 120 Btu per hr minimum.

For capacities on other gases, apply the following factors:

Btu per cu ft	<800	800–949	>950	2500
Factor	0.516	0.765	1.000	1.62

GAS FLOW CONTROL ACCESSORIES

Manual Control Valves

Manual control valves or cocks are used on all appliances to shut off or turn on gas flow positively or to control it. Usually, a main manual control valve is located ahead of all controls (except possibly the pilot gas) to shut off or turn on gas to the appliance. In addition, main burners may have individual valves, as on a gas range. Appliance burner valves are usually of the plug type. Pilot, burner, and main control valves are listed by A. G. A.[8] Figures 12-35 and 12-36 illustrate various types of top burner valves for gas ranges.

Automatic Valves for Gas Appliances

Definition.[21] An automatic or semiautomatic valve consists essentially of a valve and operator that control the gas supply to the burner(s) during operation of an appliance. The operator may be actuated, directly or indirectly,

* Normal appliance input ratings, *not* modulated appliance input ratings.

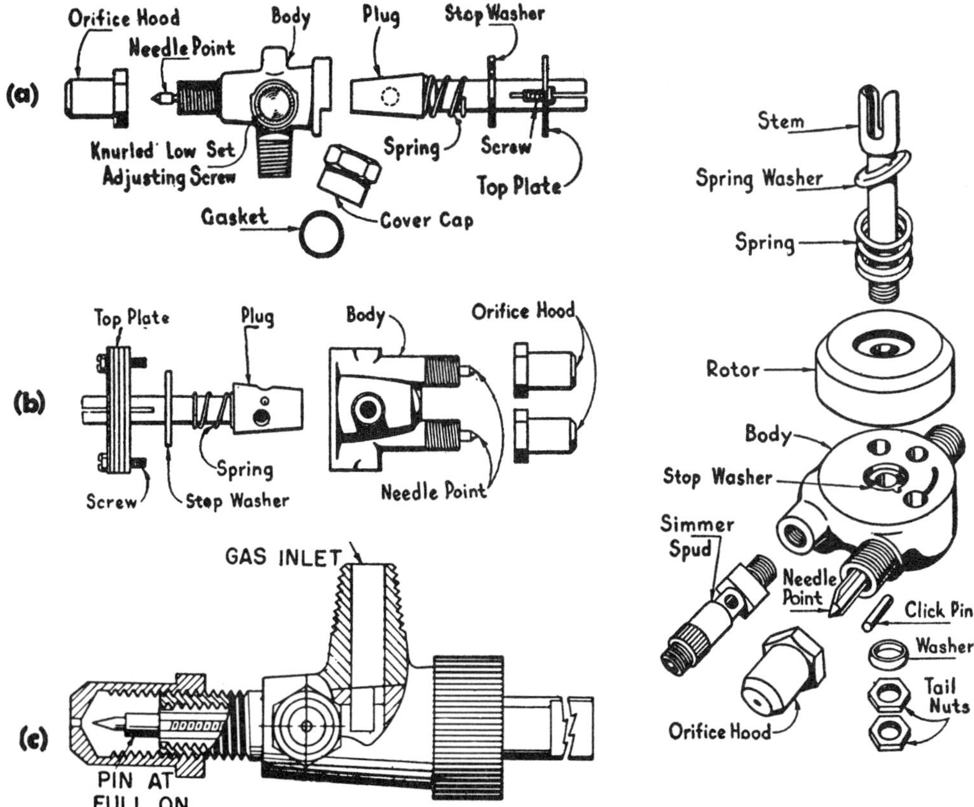

Fig. 12-35 (left) Top burner valves for gas ranges: (a) Hi-Lo single duty plug type; (b) double duty plug type; (c) traveling pin type.
Fig. 12-36 (right) Top burner rotor valve for gas ranges.

by application of gas pressure on a flexible diaphragm, by electrical means (e.g., a thermostat), by mechanical means (e.g., a limit control), or by other means. These valves are listed by A. G. A.[8] The requirements of ASA Z21.21 do *not* apply to domestic gas appliance pressure regulators; *self-contained* water heater, range, or room heater thermostats; self-contained safety shutoff devices; or self-contained automatic gas shutoff valves for hot water supply systems.

Automatic Valve Operating Characteristics.[21] These characteristics are:

Quick opening—requires 3 sec or less from initial flow to the full rated capacity.

Slow opening—opens with a gradually increasing flow rate (without stepping) from initial flow to the full open position in more than 3 sec.

Step opening—retards briefly the opening action at some designed level below full rated capacity.

Multiple stage—permits continuous operation at various selected flow rates up to rated capacity.

Modulating—has an automatically variable flow rate from minimum to maximum as specified by the manufacturer.

Quick closing—closes in less than 5 sec after acting to close (energized or de-energized).

Slow closing—requires 5 sec or more to close after acting to close (energized or de-energized).

Combination Gas Valves. These assemblies (listed by A. G. A.)[8] perform two or more different functions in a single unit. Functions include, but are not limited to, the following: (1) manual or automatic main shutoff; (2) safety shutoff switch; (3) pilot safety shutoff, without mechanical interrupter; (4) complete safety shutoff, with mechanical interrupter; (5) complete safety shutoff, with integral manual main shutoff valve providing separate control of both pilot gas and main burner gas; and (6) gas appliance pressure regulation.

Electrical-Type Valves.[21] These are designed for **line voltage** ($>$30 volts, $\not>$600 volts), **low voltage** ($\not>$30 volts, which is supplied by a primary battery, standard Class 2 transformer, or a transformer and fixed-impedance combination that meets the performance requirements of a Class 2 transformer), or **millivoltage** (supplied by a thermoelectric source). These valves may be installed in conformance with the NEC;[22] tolerances from rated voltage are +10 per cent, −15 per cent.

Four General Types of Automatic Valves. These are:

Solenoid or Magnetic. This type (Fig. 12-37) is opened by an electromagnet and closed automatically (e.g., by weight and spring) when current is broken or secondary controls cause the valve to close. Some controls require manual resetting after shutdown; others are completely automatic. A solenoid valve must be installed in a vertical position in a horizontal line unless specifically designed for location elsewhere.

Motorized. This type of valve may be opened by a motor-driven mechanism. In one commonly used unit (Fig. 12-38), the lever that lifts the valve may be used to stall the motor when the valve is fully opened. An interruption of current flow to the motor permits a spring to close the valve.

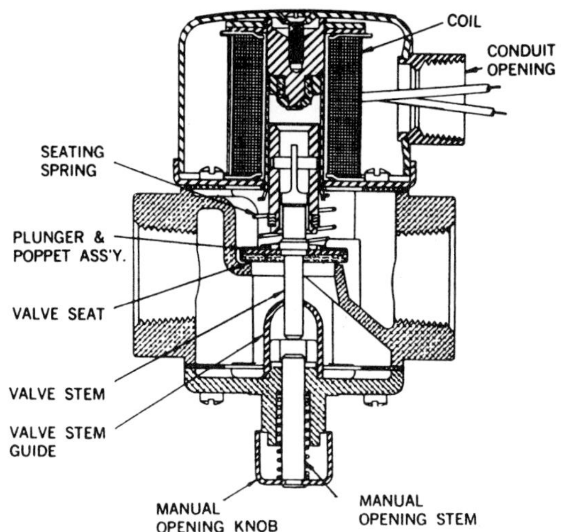

Fig. 12-37 Typical solenoid automatic gas valve. (Grayson Div., Robertshaw-Controls Co.)

Diaphragm. The typical operating principle for this type of valve is shown in Fig. 12-39. If the diaphragm is the only gas seal, means should be provided to vent the control gas. In the event of diaphragm rupture, leakage to the atmosphere is limited to 0.5 cu ft per hr of 1.53 sp gr gas at 11.0 in. w.c.

Heated Element and Permanent Magnet. When the thermostat calls for heat, the heater coil in this type of valve is energized. In one design (Fig. 12-40) this coil heats a bimetal, which slowly bends upward, applying a lifting force to the armature. When this force overcomes the downward pull of the magnet, the valve snaps open. To close, current is interrupted; the heater cools and the bimetal bends downward. The armature, nearing the magnet, is suddenly pulled down with a snap.

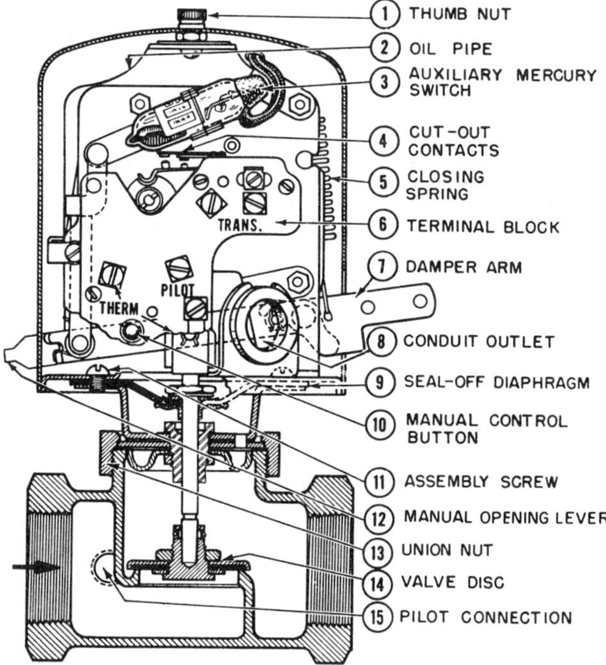

Fig. 12-38 Typical motorized automatic gas valve. (Honeywell, Inc.) See Fig. 12-19b for its use in a system.

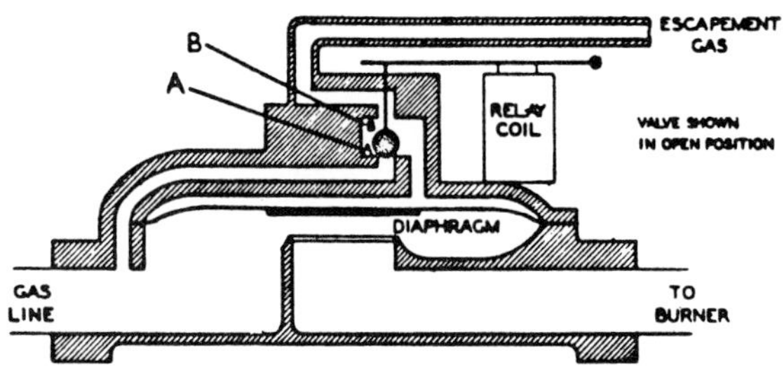

Fig. 12-39 Diaphragm valve. (White-Rogers)
Operation. If the system safety pilot is burning properly and the thermostat is calling for heat, the circuit is completed to the relay coil of the diaphragm gas valve. The armature of the relay is pulled down, closing port A and opening port B. This allows gas to escape from the top of the diaphragm, thru escapement port B. The gas is vented either to the atmosphere or within the valve body. After the gas from the top of the diaphragm has bled off, the pressure of the gas coming from the line forces the diaphragm upward, opening the main port of the valve. The gas now flows to the main burner, which operates until the thermostat is satisfied.

Timer- and Clock-Controlled Valves

Timer-operated gas valves are usually opened manually when the timer is turned on and closed automatically when the set time is reached. The clock mechanisms usually have 2-hr or 4-hr limits; intermediate settings may be made. Applications include incinerators and clothes dryers. Clock-operated valves (e.g., on range ovens) turn on at any set time in a 12-hr period and turn off after any set interval.

Temperature Limit Controls and Fan Controls

On warm air furnaces a temperature limit control is installed or supplied for installation in accordance with the manufacturer's instructions. The sensing element is in representative air temperature. If the air temperature exceeds certain preset temperatures (see page 12/94 and Table 12-40), a circuit is broken and the gas valve is closed, thus preventing an excessive temperature rise in the circulating air. The sensing element is either a metal helix or a liquid-filled element. A fan control to turn on the fan at a selected temperature (60 to 170 F) is usually combined with a limit control.

Similar limit controls are used for hot water and steam boilers to prevent excessive temperature rise or steam pressure by shutting off gas flow.

Aquastats. In hot water boilers an aquastat serves as a temperature limit control. This is a form of thermostat

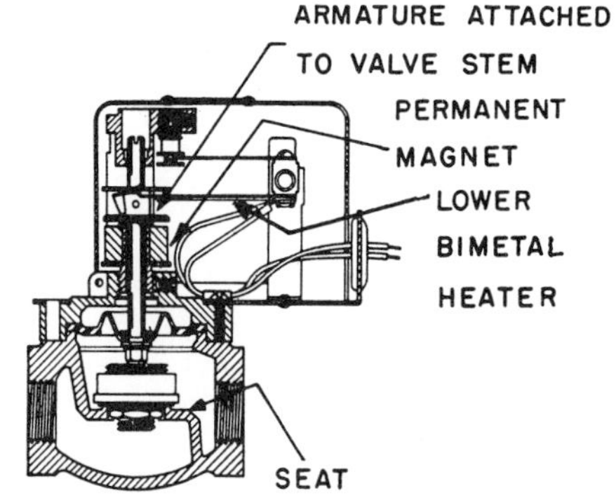

Fig. 12-40 Bimetal–permanent magnet automatic valve.

used to maintain the water temperature in hot water heating system boilers between selected limits, to provide the desired water temperature. Heating requirements for the home generally govern the temperature set on hot water boilers. The temperature range generally lies between 150 and 190 F. The sensing element of the aquastat may be a bimetal, usually helical in shape, gas-filled or liquid-filled bulb enclosed in a watertight tube that is immersed in the water of the boiler. The common practice is to have the room thermostat control and operate the heating water circulating pump and have the aquastat control and operate the gas burner controls of the boiler.

Relief and Automatic Gas Shutoff Valves

Temperature and pressure relief valves and automatic gas shutoff valves are used on gas water heaters to relieve excess pressure or temperature. Such valves are listed by A. G. A.[8]

Pressure relief valves are usually located on the cold water inlet line. They are small spring-loaded valves set to open at a selected pressure, usually 125 psig or above, to allow water to escape.

Temperature relief valves are located near the tank top in the hottest water zone. They are small valves controlled by a fuse plug or expanding rod designed to open at a set temperature, usually above 190 F, and thus allow water to escape. Temperature and pressure relief valves are frequently combined in one casing.

An **automatic gas shutoff valve** operates to prevent a gas burner from heating the contents of a water heater to an excessive temperature if the thermostat should fail to shut off properly. A common type utilizes a thermal element consisting of a metal fuse or a chemical cartridge, located within the lower end of a tube immersed in the heater tank. The fuse or cartridge melts when the tank water reaches the required temperature, allowing the valve to close by spring tension and shut off the gas. Another type designed for surface contact operates as an electrical contactor.

GUM AND DUST FILTERS FOR PILOT LINES

Gum, the result of interaction of nitric oxide, oxygen, and unsaturated hydrocarbons, may be formed in manufactured gas[23] in sufficient quantity to cause pilot outages. Gum particles may vary in size from 0.0001 to 20μ. The gum filters commonly used were developed by the United Gas Improvement Co. These filters (Fig. 12-41) are made under license by several companies and are listed by A. G. A.[8]

In 1964, the A. G. A. Directory of Approved Gas Appliances[8] showed units with capacities at a pressure drop of 0.5 in. w.c., ranging from 0.33 to 13.50 cu ft per hr of air. The gas capacity of a filter tested on air may be calculated by the following equation:[10]

$$Q_{gas} = 182\ Q_{air}/V$$

where:

Q_{gas} = gas capacity, cu ft per hr
Q_{air} = rated air capacity, cu ft per hr
V = viscosity of gas, micropoises—natural, 129; manufactured, 147; mixed, 136; propane, 78; butane, 75; LP-gas–air mixtures, as follows:

Btu/cu ft	1400	800	525
V, μ poises	133	154	164

Filters are primarily installed ahead of pilots, pilot control equipment, and very sensitive gas controls. Research reports on gum, dust, and rust filtering are available.[24,25] Devices to protect against gum are mandatory on appliances approved by A. G. A. for use on manufactured gas. The efficiency of the filter depends upon the size and arrangement of the filaments in the filter pad and the degree to which they are compressed. Thick pad construction provides depth and space in which filtered material may disperse without forming impervious layers. These filters have given good service, with no clogging or pilot outages, on manufactured gas for over 11 years when the gas nitric oxide content has been well controlled.

These filters have also been found satisfactory for removal of dust in natural gas systems. The gum filter, used as a dust filter in pilot lines on some natural gas systems, has removed dust down to the 1-μ level.

Dust filters for pilots, on water heaters approved by A. G. A. for use on *natural* gas, have been mandatory since January 1955. Different types of dust filters for pilot lines have been used with varying success. One company has had good results by using ordinary smoking pipe cleaners for this purpose. Felt pads, wool, glass wool, and hair felt, in varying sizes and shapes to fit the pilot or line, have been used as satisfactory emergency expedients by different companies. None of these makeshift items are the answer to good filtering, since they must be accurately fitted to prevent by-passing. Good filtering material must have a mixture of very fine fibers as small as the particles filtered and must also have a depth characteristic to prevent formation of an impervious layer that would stop gas flow.

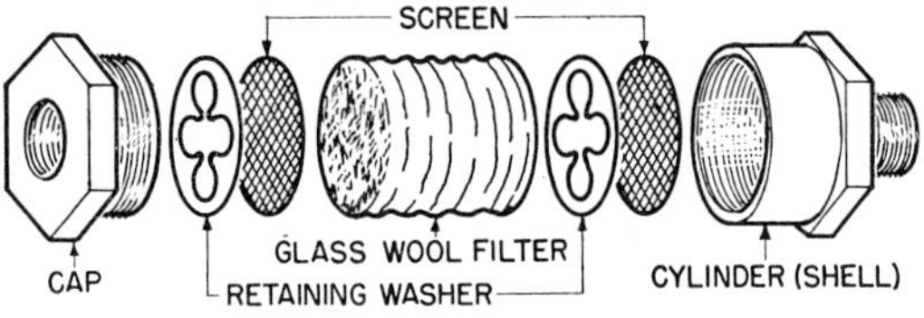

Fig. 12-41 Typical gum filter (exploded view).

CONNECTORS AND QUICK-DISCONNECT DEVICES

Gas appliance **connectors** may be used to join appliances to building piping. Metal appliance connectors (up to 1⅛ in. OD) listed by A. G. A.[8] are specifically approved[19] for such service. Composed of either semirigid metal tubing or flexible all-metal tubing, these connectors have a fitting at each end provided with standard taper pipe threads for connection to a gas appliance and to house piping.[26] Nominal overall length cannot exceed 6 ft. The range of capacities of flexible metal units is from 11 MBtu per hr (¼ in. ID × 6 ft long) to 368 MBtu per hr (1 in. ID × 2 ft long), at a 0.2 in. w.c. pressure drop. Flexible connectors of other than all-metal construction are also listed by A. G. A.[8] Their construction is intended to overcome the problems caused by repeated flexure or sharp bending of the all-metal types.

Installation considerations include the following: (1) The gas outlet must be in the same room as the appliance. (2) Connectors should not be kinked, twisted, or torqued. (3) Connectors are not designed for continuous movement.

A **quick-disconnect**[27] device is a hand-operated fitting (no tools required) for connecting and disconnecting an appliance or an appliance connector to a gas supply. These devices have an automatic means for shutting off the gas when they are disconnected. Mating parts are held together either by a positive locking system or by a means requiring a straight pull of at least 15 lb to disconnect. Capacities at a 0.3-in. w.c. pressure drop are as follows:

Inlet, in.	¼	⅜	½	¾	1
MBtu/hr	30	50	75	115	175

An available report[28] covers the development of two quick-disconnect devices at the A. G. A. Laboratories.

REFERENCES

1. Underwriters' Laboratories, Inc. *Gas and Oil Equipment List.* Chicago, Ill. (annual).
2. Associated Factory Mutual Fire Ins. Companies. *Handbook of Industrial Loss Prevention.* New York, McGraw-Hill, 1959.
3. Natl. Fire Protection Assoc. *Combustion Engines and Gas Turbines.* (NFPA 37) Boston, Mass., 1963.
4. ——. *Standard for the Installation and Operation of Gas Systems for Welding and Cutting.* (NFPA 51) Boston, Mass., 1961.
5. ——. *Explosion Prevention, Boiler-Furnaces, Gas-Fired Watertube (Single Burner).* (NFPA 85A) Boston, Mass., 1964.
6. ——. *Standard for Ovens and Furnaces.* (NFPA 86A) Boston, Mass., 1963.
7. ——. *Chimneys, Fireplaces, Venting Systems.* (NFPA 211) Boston, Mass., 1964.
8. A. G. A. Laboratories. *Directory, Approved Appliances, Listed Accessories.* Cleveland, Ohio (semiannual).
9. ——. *American Standard Listing Requirements for Gas Appliance Thermostats.* (ASA Z21.23–1963) Cleveland, Ohio, 1963.
10. ——. *American Standard Listing Requirements for Gas Filters on Appliances.* (ASA Z21.35–1964) Cleveland, Ohio, 1964.
11. ——. *American Standard Approval Requirements for Gas Water Heaters.* (ASA Z21.10.1, Z21.10.2, Z21.10.3–1962) Cleveland, Ohio, 1962.
12. ——. *American Standard Listing Requirements for Automatic Burner Ignition and Safety Shutoff Devices.* (ASA Z21.20–1963) Cleveland, Ohio, 1963.
13. Griffiths, J. C. *Research in Pilot Burner Design, Construction and Performance: Second Bulletin.* (Research Bull. 69) Cleveland, Ohio, A. G. A. Laboratories, 1955.
14. A. G. A. Laboratories. *A Study of Bimetallic Thermal Elements.* (Research Bull. 42) Cleveland, Ohio, 1947.
15. ——. *Automatic Flash Tube and Pilot Ignition of Oven and Broiler Burners on Manufactured Gas.* (Research Bull. 17) Cleveland, Ohio, 1943.
16. Speidel, P. A., and Hammaker, F. G., Jr. *Research on New Designs for Range Top Burners, Pilots, and Valves.* (Research Bull. 76) Cleveland, Ohio, A. G. A. Laboratories, 1957.
17. A. G. A. Laboratories. *Electric Ignition of Gases.* (Research Bull. 28) Cleveland, Ohio, 1944.
18. Weber, E. J. *Development of Electrical Ignition Systems for Domestic Gas Appliances.* (Research Rept. 1341) Cleveland, Ohio, A. G. A. Laboratories, 1961.
19. A. G. A. Laboratories. *American Standard Installation of Gas Appliances and Gas Piping.* (ASA Z21.30–1964) Cleveland, Ohio, 1964.
20. ——. *American Standard Listing Requirements for Domestic Gas Appliance Pressure Regulators.* (ASA Z21.18–1956; Addenda Z21.18a–1960) Cleveland, Ohio, 1956–60.
21. ——. *American Standard Listing Requirements for Automatic Valves for Gas Appliances.* (ASA Z21.21–1963) Cleveland, Ohio, 1963.
22. Natl. Fire Protection Assoc. *National Electrical Code.* (NFPA 70) Boston, Mass., 1962.
23. Ward, A. L., and others. "Gum Deposits in Gas Distribution Systems." *Ind. Eng. Chem.* 24: 969–77, Sept.; 1238–47, Nov. 1932; 25: 1224–34, Nov. 1933; 26: 947–55, Sept.; 1028–38, Oct. 1934; 27: 1180–90, Oct. 1935.
24. Vandaveer, F. E. "Gum Protective Devices for Gas Appliances." *A. G. A. Proc.* 1943: 270–81.
25. Little, A. D., Inc. *Research in Gum, Dust and Rust Filtering.* New York, A. G. A., 1947.
26. A. G. A. Laboratories. *American Standard Listing Requirements for Metal Connectors for Gas Appliances.* (ASA Z21.24–1963) Cleveland, Ohio, 1963.
27. ——. *American Standards Listing Requirements for Quick-Disconnect Devices for Use With Gas Fuel.* (ASA Z21.41–1962) Cleveland, Ohio, 1962.
28. Hollowell, G. T. *Investigation and Development of Quick-Connect Couplings for Use With Gas Fuel.* (Research Rept. 1365) Cleveland, Ohio, A. G. A. Laboratories, 1963.
29. Gas Appliance Engineers Society. *Gas Appliance Engineers Handbook.* Cleveland, Ohio, 1964, p. 13A–3.
30. Haines, J. E. *Automatic Control of Heating and Air Conditioning.* 2nd ed. New York, McGraw-Hill, 1961.
31. Conner, R. M., and others. *Automatic Range Ignition Field Test Program.* New York, A.G.A., 1953.

Chapter 5
Gas Comfort Heating and Cooling

by Henry Cullinane, Keith T. Davis, C. J. Mathieson, C. George Segeler, F. E. Vandaveer, and James Yund

INTRODUCTION

Air conditioning is the process of treating air to simultaneously control its temperature, humidity, cleanliness, and distribution. The factors involved in year-round air conditioning are: (1) heating, (2) humidifying, (3) cooling, (4) dehumidifying, (5) cleaning, and (6) circulating. The combination of heating, humidifying, cleaning, and circulating is termed *comfort heating and winter air conditioning.* The combination of cooling, dehumidifying, cleaning, and circulating is termed *comfort cooling and summer air conditioning.*

Classifications, Ratings, and Installation Codes for Comfort Heating Appliances

A.G.A. approval of comfort heating gas appliances covers self-contained units in a wide range of capacities, sizes, and uses (Table 12–30). Thousands of models, together with their input and output ratings, are listed in a Directory of Approved Appliances.[1]

Basic standards covering the installation of gas equipment are available as follows:

Z21.30–1964	Gas Appliances and Gas Piping*
NFPA No. 54	Gas Appliances and Gas Piping*
Z21.8–1958	Domestic Gas Conversion Burners
NFPA No. 90A	Nonresidence Air Conditioning and Ventilating Systems
NFPA No. 90B	Residence Warm Air Heating
ASA Z83.1	Gas Piping and Gas Equipment on Industrial and Other Premises (expected in 1966)
ASA Z83.3	Gas Equipment in Large Boilers (expected in 1966)
NFPA No. 85A	Natural Gas-Fired Watertube Furnaces

Furnaces, boilers, and cooling equipment certified by A.G.A. as complying with the appropriate American Standard Approval Requirements must have nameplates showing input and output; efficiencies are given in Table 12–30.

Rating codes for heating boilers have also been issued by various organizations. A number of categories of boilers have been designated, with appropriate rating codes for each.

* Same contents.

Steel Boiler Institute Code.[2] The SBI Code, requiring that construction meet ASME construction standards, covers steel firebox boilers, Scotch-type boilers, packaged boilers, heating boilers, and boiler-burner units. Packaged boilers are built and guaranteed by one supplier who furnishes and assumes responsibility for the components of the assembled units. Boiler-burner units are built and guaranteed by the boiler supplier, the burner supplier, or jointly by both.

The code rates boilers only in standardized sizes. There are no SBI ratings for intermediate sizes. If a boiler producing 2000 lb per hr of steam meets all requirements of the code, the rating will not be approved for more than 1863 lb per hr of steam output, which is the next smaller standardized size.

Fig. 12-42 (left) Steel Boiler Institute symbol.

Fig. 12-43 (middle) Institute of Boiler and Radiator Manufacturers symbol.

Fig. 12-44 (right) Mechanical Contractors Association symbol.

The SBI Code (see Fig. 12–42 for its symbol) comprises four sections, called Tables, as follows:

Table 1 is a dimensional code for commercial steel boilers with inputs up to 161 gph of oil, or the equivalent in other fuels, in which minimum furnace volume and furnace height are specified. This table requires 6⅔ sq ft of heating surface per boiler hp; ratings do not have to be confirmed by test. A boiler hp is defined as 33,475 Btu per hr. The manufacturer assumes no responsibility for the installation, the draft provided, the firing equipment, or any other items not specifically listed in his catalog as standard equipment. Table 1 boilers have natural draft firing.

Table 2 is a dimensional code for boilers used in residential and small commercial applications, with inputs from 1 to 17 gph of oil, or the equivalent in other fuels. The rating basis is

Table 12-30 General Classifications and Sizes of Approved Heating Appliances and Listed Conversion Burners

APPROVAL SEAL

LISTING SYMBOL

(The A.G.A. Approval Seal is granted only to complete gas appliances; the Listing Symbol is the certification mark for gas appliance accessories.)

Appliance	Range of input ratings,* MBtu/hr	Min efficiency, % of input
Boilers	45–5500	80
Central furnaces, forced air and gravity	50–480	80
Decorative gas appliances, vented	17–28	
Duct furnaces	60–2500	80
Floor furnaces		
Forced air	70–85	75
Gravity	30–90	70
Infra-red radiant heaters, unvented	15–90	†
Room heaters		
Sealed combustion system baseboard furnaces	10–15	70
Unvented circulators	5–50	†
Vented circulators	5–125	70‡
Coal basket	28	
Gas logs	12–42	
Radiant heaters	12–30	†
Wall heaters, unvented, open flame radiant type	4–25	†
Fireplace inserts	18–34	
Vented overhead heaters	28.8–150	70
Unvented overhead heaters	12–50	35§
Unit heaters	25–5000	80
Wall furnaces		
Sealed combustion system	5–85	70
Vented	25–75	70
Conversion burners**	45–400	

* Maximum and minimum values as of November 1964.[1]

† Since all of the heat of unvented units enters the premises, efficiencies are approximately 90 per cent; the latent heat is not normally recoverable.

‡ Over 20,000 Btu per hr input; if less, 65 per cent.

§ Radiant efficiency.

** A listed accessory, not an approved appliance.

more liberal than for Table 1; 5½ sq ft of heating surface per boiler hp are permitted. This basis is proved by tests outlined by the SBI and conducted by the manufacturer. The test data on the smallest and largest sizes in a series of boilers are submitted to the SBI and these data are then extended to cover other sizes. The boilers covered by this table are test-rated as boilers only and may be furnished with or without burners and with or without mechanical draft equipment.

Table 3 covers the same range of ratings as Table 2, but without any dimensional limitations. The tests are the same as those conducted on Table 2 boilers and are submitted by the manufacturer. Table 3 covers test-rated boilers only.

Table 4 covers the range of large commercial boiler sizes of Table 1 but with more liberal ratings for equivalent dimensional limitations. Forced draft firing and Scotch boiler design are in this category. Table 4 boilers have 5 sq ft of heating surface per boiler hp. The table covers standardized sizes of boilers, boiler-burner units, and packaged boilers. No tests are conducted.

Ratings. When rating steel boilers, the gross output is the standardized output for boilers operating within all the limits stipulated in the SBI Code. The SBI gross output rating corresponds to the A.G.A. output rating. For Table 1, 2, and 4 boilers, output is specified in forms of heat output per square foot of heating surface; for Table 3 boilers, output is determined by test. Flue gas temperatures are limited to 600 F at gross output in Table 2 boilers. The overall efficiency of the boiler and burner must not be less than 70 per cent. In calculating the SBI firing rate, however, a uniform efficiency of 75 per cent is used. Therefore, an SBI Table 2 or Table 3 boiler with a gross output of 99 MBtu per hr has an SBI firing rate of 132 MBtu per hr, or 1.0 gph of light oil. To make allowances for piping and pickup requirements, the SBI *net* rating has been devised—25 per cent less than the listed gross output.

Institute of Boiler and Radiator Manufacturers Code.[3] The I=B=R code covers low-pressure cast iron boilers. Tests are conducted at a laboratory of the manufacturer's choice and the results are submitted to a committee of the I=B=R for review and approval. (Figure 12–43 shows the Institute's symbol.) Tests have been established for hand-fired anthracite and the results projected to obtain bituminous coal ratings. Other tests have been established for automatically fired oil boilers and the results projected for stoker-fired anthracite and bituminous coal. For A.G.A.-approved boilers, no tests are conducted. For gas-designed boilers not approved by A.G.A., the oil-fired test ratings are used when outputs are over 300,000 Btu per hr; below that, tests have been established for boilers not otherwise A.G.A.-approved.

Tests are conducted on the largest and smallest boilers of a series and the ratings are extended to intermediate sizes. In the case of automatically fired oil boilers, flue gas temperatures are limited to 600 F while the *gross I=B=R output* must be at least 70 per cent of the input for boilers using light oil and 75 per cent for boilers using heavy oil. The *net* I=B=R rating is normally 25 per cent less than the gross I=B=R output to make allowances for piping losses and to make it possible to pick up the heating system more rapidly after night setback or after a prolonged shutdown. The net ratings, however, may be increased 12 per cent when less than normal pickup and piping losses are encountered, as on a series loop baseboard and radiant heating installations. Heating systems using compact gas-fired boilers in conjunction with low water content baseboard and short supply and return piping proved the 25 per cent reserve too large and led to the use of the 12 per cent increase in net rating.

American Boiler Manufacturers Association Ratings. The Packaged Firetube Section of ABMA[4] publishes ratings of steam or hot water firetube boilers. This Section of the Association consists of 11 boiler manufacturers. The ratings are based on the ASME Test Code for Steam Generating Units, modified as follows: (1) The test is conducted at not less than 100 per cent of rated capacity. (2) The 2-hr test is preceded by operation for 1 hr at equilibrium conditions. (3) Low-pressure steam boilers are tested at 10 psig, high pressure steam boilers at 100 psig. A packaged steam or hot water firetube boiler is defined as a modified Scotch-type boiler unit, engineered, built, and fire-tested before shipment; it is guaranteed in material, workmanship, and performance by one firm, with one manufacturer furnishing and assuming responsibility for all components in the assembled unit. Components include burner, boiler, controls, and all auxiliary equipment.

Ratings are obtained by sending test data to a Testing and Rating Committee, which establishes conformity with ABMA standards of output and efficiency, among other things. Gross ratings are published, based on efficiencies of not less than 80 per cent when burning oil and not less than 75 per cent when burning gas. Net load ratings are based upon a maximum allowance of 25 per cent for piping and pickup, of which 15 per cent is the allowance for piping only. The ABMA Code does not specify or limit boiler heating surface.

Mechanical Contractors Association of America.[5] This Association (see Fig. 12–44 for its symbol) has established a rating system that conforms with A.G.A. output ratings, I=B=R net-load ratings, and SBI net-load ratings. For boilers not rated by one of these methods, test procedures have been established to determine the net-load ratings. Unrated boilers are considered by a committee of MCA which assigns a value based on submitted test data that have been developed by tests certified as having been made in accordance with the SBI Code, I=B=R Code, MCA Testing and Rating Codes for Boiler-Burner Units, or any other approved test procedure. Boilers must be constructed in accordance with the rules for construction of low-pressure boilers under Section IV, ASME Boiler Construction Code.

RESIDENTIAL HEATING

American Standard Approval Requirements

Under sponsorship of A.G.A, basic American Standards for safe operation, substantial and durable construction, and acceptable performance have been developed for the equipment given in Table 12–30. Following are some of the test conditions and other major provisions under which an appliance prototype model is tested at the A.G.A. Laboratories.

Combustion Tests. At rated inputs, gas heating equipment must demonstrate complete combustion, with carbon monoxide concentration in an air-free sample of combustion products never exceeding two parts in 10,000 for room heaters, and four parts in 10,000 for the other vented appliances, as tested by an infra-red analyzer. Room heaters specifically approved for unvented operation must pass similar tests in a closed room, with no addition of air.

Venting Limitations. All residential comfort heating gas appliances are required to be vented, except room heaters of not more than 50,000 Btu per hour input approved for unvented operation.

Flue Gas Temperature and Draft Tests.[7] Flue gas temperatures measured upstream of draft hoods on boilers, furnaces, duct furnaces, and unit heaters are not permitted to exceed 550 F. On floor furnaces, wall furnaces, and room heaters, flue gas temperatures are limited to 550 F in the vent pipe downstream of the draft hood. Tests are also conducted to ensure that downdraft pressures do not affect main burners or pilot burners and that combustion products do not spill from the relief openings of a draft hood. Separate draft tests are also conducted on appliances equipped with power burners.

Safety Control Devices. Automatic (safety) pilots are required to shut off main burner gas flow within 2.5 min or less following pilot flame failure. On LP-gas and LP-gas–air fuels, they must also shut off gas flow to the pilot burner. Automatic pilots are mandatory on all boilers, furnaces, duct furnaces, unit heaters, and conversion burners. Such pilots are optional on floor furnaces, wall furnaces, and room heaters; when thermostatic controls are provided on these units, automatic pilots are mandatory.

Sizing Heating Equipment

Heating equipment is sized to provide the desired temperatures at given design conditions. The first step is to evaluate heat losses and gains in order to determine the required equipment output (and indirectly its input). This requires, in effect, solution of Eq. **1**. The heat required for infiltrating air must be added to the result.

Outdoor Design Temperatures. Tabulations of typical outdoor design temperatures for selected cities are available (see Table 12–47). For walls below grade level, outdoor design temperature should be based on ground temperature.

Indoor Design Temperatures. Table 12–31a covers various applications. These temperatures are normally maintained at the "breathing line," 5 ft above the floor. For residences

Table 12-31a Winter Indoor Dry-Bulb Temperatures Usually Specified*[6]

Type of building	°F	Type of building	°F
Schools		Theaters	
Classroom	72–74	Seating space	68–72
Assembly hall	68–72	Lounge room	68–72
Gymnasium	55–65	Toilet	68
Toilet and bath	70	Hotels	
Wardrobe and locker	65–68	Bedroom and bath	75
Kitchen	66	Dining room	72
Dining and lunch	65–70	Kitchen and laundry	66
Playroom	60–65	Ballroom	65–68
Hospitals		Toilets and service	68
Private rooms	72–74	Homes	73–75
Private (surgical)	70–80	Stores	65–68
Operating rooms	70–95	Public buildings	72–74
Wards	72–74	Factories and machine shops	60–65
Kitchen and laundry	66		
Toilet	68	Foundries and boiler shops	50–60
Bathroom	70–80		
Warm air baths	120	Paint shops	80
Steam baths	110		

* The most comfortable dry-bulb temperature to be maintained depends on the relative humidity and air motion. These three factors considered together constitute what is termed the effective temperature. When relative humidity is not controlled separately, optimum dry-bulb temperature for comfort will be slightly higher than shown.

Table 12-31b Approximate Temperature Differentials between Breathing Level and Ceiling, Applicable to Certain Types of Heating Systems*[6]

Ceiling height, ft	Breathing level temperature (5 ft above floor), °F									
	60	65	70	72	74	76	78	80	85	90
10	3.0	3.3	3.5	3.6	3.7	3.8	3.9	4.0	4.3	4.5
11	3.6	3.9	4.2	4.3	4.4	4.6	4.7	4.8	5.1	5.4
12	4.2	4.6	4.9	5.0	5.2	5.3	5.5	5.6	6.0	6.3
13	4.8	5.2	5.6	5.8	5.9	6.1	6.2	6.4	6.8	7.2
14	5.4	5.9	6.3	6.5	6.7	6.8	7.0	7.2	7.7	8.1
15	6.0	6.5	7.0	7.2	7.4	7.6	7.8	8.0	8.5	9.0
16	6.1	6.6	7.1	7.3	7.5	7.7	7.9	8.1	8.6	9.1
17	6.2	6.7	7.2	7.4	7.6	7.8	8.0	8.2	8.7	9.2
18	6.3	6.8	7.3	7.5	7.7	7.9	8.1	8.3	8.8	9.3
19	6.4	6.9	7.4	7.6	7.8	8.0	8.2	8.4	8.9	9.4
20	6.5	7.0	7.5	7.7	7.9	8.1	8.3	8.5	9.0	9.5
25	7.0	7.5	8.0	8.2	8.4	8.6	8.8	9.0	9.5	10.0
30	7.5	8.0	8.5	8.7	8.9	9.1	9.3	9.5	10.0	10.5
35	8.0	8.5	9.0	9.2	9.4	9.6	9.8	10.0	10.5	11.0
40	8.5	9.0	9.5	9.7	9.9	10.1	10.3	10.5	11.0	11.5
45	9.0	9.5	10.0	10.2	10.4	10.6	10.8	11.0	11.5	12.0
50	9.5	10.0	10.5	10.7	10.9	11.1	11.3	11.5	12.0	12.5

* The figures in this table are based on an increase of 1 per cent per foot of height above the breathing level (5 ft) up to 15 ft and an increase of 0.1° F for each foot above 15 ft. This table is generally applicable to forced air types of heating systems. For direct radiation or gravity warm air, increase values 50 to 100 per cent.

Table 12-32a Overall Coefficients of Heat Transmission, U, for Building Materials

(U values are in Btu/hr-sq ft-°F temp diff; R values are in hr-sq ft-°F temp diff/Btu)

(based on 0 F air temperature and 15 mph wind velocity; see also Table 2-93 for k data)

Calculated U values given here provide convenient data for heating estimates. The values are for steady-state heat transfer. Workmanship is an increasingly important factor, particularly when insulation reduces calculated U values to low levels. To adjust U values for added insulation between framing, etc., see Table 12-32b.)

A. Frame Walls

				Type of sheathing		
				Plywood, 5⁄16 in.	Wood, 25⁄32 in., and bldg paper	Insulating board, 25⁄32 in.
Exterior material	(Avg R)	Interior material	(R)	U	U	U
Wood siding, shingles, or panels	0.85	Gypsum board, 3⁄8 in.	0.32	0.30	0.25	0.20
		Gypsum lath and ½ in. plaster (light wt agg)	.64	.27	.23	.19
		Gypsum lath and ½ in. plaster (sand)	.41	.29	.24	.19
		Metal lath and ¾ in. plaster (sand)	0.13	.31	.26	.21
		Insulating board and ½ in. plaster (sand)	1.52	.22	.19	.16
		Wood lath and ½ in. plaster (sand)	0.40	.29	.24	.19
Face brick veneer	0.45	Gypsum lath and ½ in. plaster	0.41	.32	.27	.21
		Insulating board and ½ in. plaster	1.52	.24	.21	.17
Wood shingles over insulating backer board (5⁄16 in.)	1.42	Gypsum lath and ½ in. plaster (sand)	0.41	.25	.21	.18
		Metal lath and ¾ in. plaster (sand)	.13	.27	.23	.18
Asphalt insulating siding		Wood lath and ½ in. plaster (sand)	.40	.25	.21	.18
Asbestos cement siding		Gypsum lath and ½ in. plaster (sand)	.41	.36	.29	.22
Stucco	0.19	Metal lath and ¾ in. plaster (sand)	.13	.40	.31	.24
Asphalt roll siding		Wood lath and ½ in. plaster (sand)	0.40	0.36	0.29	0.22

Table 12-32a (Continued)

B. Masonry Walls

Exterior material	(*R*)	Interior Finish: ⅝ in. plaster (sand agg) on wall ($R = 0.11$) *U*	Gypsum lath and ½ in. plaster (sand agg) on furring ($R = 0.41$) *U*	Insulating board and ½ in. plaster (sand agg) on furring ($R = 1.52$) *U*	Wood lath and ½ in. plaster (sand agg) ($R = 0.40$) *U*
Brick, 8 in., face and common	1.24	0.45	0.29	0.22	0.29
Brick, 12 in., face and common	2.04	.33	.23	.19	.23
Brick, 8 in., common only	1.60	.39	.26	.20	.26
Brick, 12 in., common only	2.40	.30	.22	.17	.22
Stone, 12 in., lime and sand	0.96	.52	.31	.23	.31
Hollow clay tile					
8 in.	1.85	.36	.25	.19	.25
12 in.	2.50	.29	.21	.17	.21
Poured concrete					
8 in., 80 lb/cu ft	3.20	.24	.18	.15	.18
12 in., 80 lb/cu ft	4.80	.17	.14	.12	.14
Poured concrete					
8 in., 140 lb/cu ft	0.64	.63	.35	.25	.35
12 in., 140 lb/cu ft	0.96	.52	.31	.23	.31
Concrete block					
8 in., gravel agg	1.11	.48	.30	.22	.30
12 in., gravel agg	1.28	.45	.28	.22	.29
8 in., lt wt agg	2.00	.34	.24	.19	.24
12 in., lt wt agg	2.27	.31	.22	.18	.22
Face brick and 8 in. concrete block, (sand agg)	1.55	.39	.26	.20	.26
Common brick and 8 in. concrete block, (sand agg)	1.83	.35	.24	.19	.24
Face brick, cavity, and 4 in. concrete block, (gravel agg)	1.15	.32	.23	.18	.23
Common brick, cavity, and 4 in. concrete block, (gravel agg)	1.47	0.29	0.21	0.17	0.21

C. Frame Construction Ceilings and Floors

Ceiling material	Upward heat flow: No floor *U*	Upward heat flow: No floor, and 3 in. insulation *U*	Upward heat flow: Wood subfloor under wood floor *U*	Downward heat flow: No floor *U*	Downward heat flow: No floor, and 3 in. insulation *U*	Downward heat flow: Wood subfloor under wood floor *U*
None	...	...	0.34	...	...	0.28
Gypsum lath and ½ in. plaster (sand)	0.61	0.078	.24	0.44	0.075	.20
Metal lath and ⅜ in. plaster (sand)	.74	.081	.26	.51	.076	.21
Acoustical tile, ½ in. and gypsum board	.37	.073	.19	.30	.069	.17
Wood lath and ½ in. plaster (sand)	0.62	0.079	.24	0.45	0.075	.20

D. Pitched Roofs with Various Rafter Space Treatments

Ceiling (applied directly to roof rafters)	Upward heat flow, Unventilated, not to be further insulated: Asphalt shingles and bldg. paper, 5⁄16 in. Plywood	Asphalt shingles and bldg. paper, 25⁄32 in. Wood sheath	Asbestos cement shingles, paper, and wood sheathing	Insulated	Downward heat flow, Unventilated, not to be further insulated: Asphalt shingles and bldg. paper, 5⁄16 in. Plywood	Asphalt shingles and bldg. paper, 25⁄32 in. Wood sheath	Asbestos cement shingles, paper, and wood sheathing	Insulated
None	0.57	0.44	0.53	0.66	0.51	0.40	0.48	0.56
Gypsum lath and ½ in. plaster (sand)	.33	.28	.31	.52	.29	.25	.28	.45
Insulating board lath and ½ in. plaster (sand)	.24	.21	.23	.33	.22	.20	.22	.30
Wood lath and ½ in. plaster (sand)	0.33	0.28	0.31	0.52	0.29	0.26	0.28	0.46

Table 12-32b Coefficients of Heat Transmission, U, with Added Insulation or Air Spaces to Uninsulated Construction

(Btu/hr-sq ft-°F temp diff)

U, without insulation	Fibrous insulation			One added air space with effective emissivity		
	1 in.	2 in.	3 in.	0.82	0.20	0.05
	A. Walls					
0.60	0.186	0.110	0.078	0.630	0.412	0.341
.40	.161	.101	.074	.409	.299	.258
.30	.142	.093	.069	.300	.234	.207
0.20	0.115	0.081	0.062	0.200	0.165	0.149
	B. Floors, downward heat flow					
0.60	0.186	0.110	0.078	0.600	0.236	0.111
.40	.161	.101	.074	.400	.189	.100
.30	.142	.093	.069	.300	.162	.091
.20	.115	.081	.062	.200	.128	.078
0.10	0.073	0.058	0.047	0.100	0.076	0.054
	C. Ceilings, upward heat flow					
0.40	0.161	0.101	0.074	0.389	0.300	0.270
.30	.142	.093	.069	.292	.237	.215
0.20	0.115	0.081	0.062	0.195	0.173	0.154
	D. Ceilings, downward heat flow					
0.30	0.142	0.093	0.069	0.300	0.175	0.095
0.20	0.115	0.081	0.062	0.200	0.135	0.081

Table 12-32c Coefficients of Heat Transmission, U, for Glass Windows

(Btu/hr-sq ft-°F temp diff)

	Single glass	Double glass	Window w/storm sash
Wood sash, 80 per cent glass area	1.02	0.52	0.48
Wood sash, 60 per cent glass area	0.91	.49	.43
Metal sash	1.13	0.66	0.53

Table 12-32d Coefficients of Heat Transmission, U, for Doors

(Btu/hr-sq ft-°F temp diff)

Nominal thickness	Without storm door	With metal storm door
1½ in. wood	0.49	0.33
2 in. wood	0.43	0.29

Table 12-33a Surface Resistances

(R_1 and R_n in Eq. 2, hr-sq ft-°F temp diff/Btu)

Surface position	Direction of heat flow	Average surface emissivity, e*		
		0.90	0.20	0.05
Still air				
Horizontal	Upward	0.61	1.10	1.32
Sloping 45°	Upward	.62	1.14	1.37
Vertical	Horizontal	.68	1.35	1.70
Sloping 45°	Downward	.76	1.67	2.22
Horizontal	Downward	.92	2.70	4.55
Moving air				
15 mph, winter	...	.17†	...	...
7½ mph, summer	...	0.25†	...	...

* See Table 12-33b.

† For all surface positions and flow directions.

Table 12-33b Effective Emissivity, E, of Air Spaces

$$\frac{1}{E} = \frac{1}{e_1} + \frac{1}{e_2} - 1$$

where: e_1 = average emissivity of one surface
e_2 = average emissivity of second surface

	Average surface emissivity, e	Eff emissivity, E, of air space: When $e_1 = e$, $e_2 = 0.90$	Eff emissivity, E, of air space: When $e_1 = e_2 = e$
Building materials: wood, paper, glass, masonry, paints, etc.	0.90	0.82	0.82
Steel, bright, galvanized	.25	.24	.15
Aluminum-coated paper	.20	.20	.11
Aluminum foil, bright	0.05	0.05	0.03

Table 12-33c Resistances, R, of Air Spaces
(hr-sq ft-°F temp diff/Btu)

Space position	Direction of heat flow	Temp, °F Mean	Temp, °F Temp diff	¾ in. thickness, E = 0.05	¾ in. thickness, E = 0.20	¾ in. thickness, E = 0.82	4 in. thickness, E = 0.05	4 in. thickness, E = 0.20	4 in. thickness, E = 0.82
Horizontal	Upward	50	30	1.67	1.37	0.78	2.06	1.62	0.85
		50	10	2.23	1.71	0.87	2.73	1.99	0.94
		0	20	1.79	1.52	0.93	2.22	1.81	1.03
		0	10	2.16	1.78	1.02	2.67	2.11	1.12
Sloping 45°	Upward	50	30	1.95	1.54	0.83	2.22	1.71	0.88
		50	10	2.78	2.02	0.94	3.00	2.13	0.96
		0	20	2.27	1.74	1.01	2.42	1.95	1.08
		0	10	2.71	2.13	1.13	2.97	2.49	1.17
Sloping 45°	Downward	50	30	3.27	2.27	1.01	3.39	2.33	1.02
		50	10	3.57	2.40	1.02	4.41	2.75	1.08
Vertical	Horizontal	50	30	2.80	2.04	0.96	2.62	1.94	0.94
		50	10	3.48	2.36	1.01	3.45	2.34	1.01
Horizontal	Downward	90	*	3.25	2.08	0.84	8.08	3.38	0.99
		50	*	3.55	2.39	1.02	8.94	4.02	1.23

* Substantially independent of temperature difference.

and structures having ceiling heights under 10 ft, the small temperature differential between the breathing line and ceiling may be neglected. For structures having higher ceilings, the temperature differentials given in Table 12–31b should be used. They should be used, however, with considerable discretion since there are systems where low temperature differentials exist. Table 12–31b can also be used to determine floor temperatures.

Factors influencing the selection of other design temperatures include the amount of glass area, infiltration, amount of installed insulation, and fuel cost in the area.

Heat Transmission thru Building Sections. The equation for heat transfer is:

$$Q = UA\Delta t \tag{1}$$

where: Q = Btu per hr
U = overall coefficient of heat transfer (Tables 12–32a thru 12–32d), Btu per hr-sq ft-°F
A = area involved, sq ft
Δt = difference between indoor and outdoor design temperatures, °F

Calculation of Overall Coefficient, U. When Tables 12–32a thru 12–32d do not furnish the U value for a given construction, it can be calculated by simply adding* appropriate R values selected from Tables 12–33a, 12–33c, and 12–33d. R data are the resistances of specific portions of the heat flow paths.

The relationships are shown in Eqs. **2** and **3**:

$$R_{\text{total}} = R_1 + R_2 + R_3 + \ldots + R_n \tag{2}$$

where:
R_1 = surface resistance given in Table 12–33a
R_n = surface resistance given in Table 12–33a
$R_2, R_3, \ldots, R_{n-1}$ = resistances of other portions of heat flow path given in Tables 12–33c and 12–33d

$$U = 1/R_{\text{total}} \tag{3}$$

Example: Find the U value for a frame wall consisting of ½ × 8 in. lapped wood siding, building paper, 25⁄32 in. wood sheathing, 2 × 4 in. studs on 16 in. centers, and gypsum plaster on wood lath.

Solution: Enter Table 12–33a at the column showing a surface emissivity, e = 0.90 (from Table 12–33b) and at moving air = 15 mph. This yields R_1 for outside surface = 0.17. From Table 12–33d, R_2 for lapped wood siding = 0.81, R_3 for building paper = 0.06, R_4 for wood sheathing = 0.98.

Table 12–33b must be used to establish the effective emissivity of the air space for R_5 (for usual frame walls, E = 0.82). Then for a 4 in. nominal space, the right hand part of Table 12–33c yields R_5 = 0.94 for vertical walls at a mean temperature of 50 F with a 30°F temperature difference.

From Table 12–33d, R_6 for gypsum sand plaster on wood lath = 0.40. From Table 12–33a, R_7 for inside surface = 0.68.

Thus: $R_{\text{total}} = R_1 + R_2 + R_3 + R_4 + R_5 + R_6 + R_7$
$= 0.17 + 0.81 + 0.06 + 0.98 + 0.94 + 0.40 + 0.68$
$= 4.04$

and $U = 1/R = 0.248$

Infiltration. The heat requirement to meet the infiltration load may be obtained using Eq. **4**.

$$Q_i = 0.018q\Delta t \tag{4}$$

* For complex heat flow problems such as parallel plus series paths, reference may be made to Chap. 24. *ASHRAE Guide and Data Book (1965).*

where:

Q_i = heat required because of infiltration, Btu per hr
0.018 = 0.240 (specific heat of air) × 0.075 (density of air)
q = volume of outdoor air entering building, cfh
Δt = difference between indoor and outdoor design temperatures, °F

The *infiltration rate, q,* of outdoor air can be based on either leakage thru building construction cracks (*crack method*—see Table 12–34a for windows) or on the number of air changes per hour (*air change method*—see Table 12–34b).

Latent Heat Loss. When moisture is added to the air for winter comfort, the heat required to evaporate the added water vapor is given by Eq. **5.**

$$Q_l = qd(W_i - W_o)h_{fg} \qquad (5)$$

where:

Q_l = heat required to increase moisture content of air, Btu per hr
q = volume of outdoor air heated, cfh
d = density of air at the indoor temperature (may be taken as 0.075), lb per cu ft
W_i = humidity ratio of indoor air, lb per lb of dry air
W_o = humidity ratio of outdoor air, lb per lb of dry air
h_{fg} = latent heat of vapor at W_i (may be taken as 1060), Btu per lb

Substituting the values for d and h_{fg} in Eq. **5** yields Eq. **5a:**

$$Q_l = 79.5q(W_i - W_o) \qquad (5a)$$

Special Considerations. Because of differences in construction, living habits, and other conditions, various other factors affect the total heat loss. Ignoring such factors results in fuel estimates that reflect maximum conditions and not the more realistic average conditions. Some of these factors follow:

Heat Loss from Slab Floor. This on-grade situation is covered by Table 12–35 and Eq. **6.**

$$H_f = FP\Delta t \qquad (6)$$

where:

H_f = heat loss of slab floor, Btu per hr
F = heat loss coefficient (0.81 for a floor with no edge insulation, 0.55 for one with edge insulation), Btu per hr-ft-°F
P = perimeter or exposed edge of floor, ft
Δt = difference between indoor and outdoor design temperatures, °F

Intermittent Heating. Churches, auditoriums, and other intermittently heated buildings may require an increase in the size of the heating plant to ensure reasonable heatup time. The additional capacity could vary from one-third to two-thirds of the calculated size of a heating plant specified for continuous duty.

Appliance Contribution toward Heating.[7] One method has been proposed to show the appliance contribution toward heating. All or portions of the heat inputs of various appliances may be taken into account. See the referenced publication for a detailed analysis.

Table 12-33d Resistances, R*, of Certain Building Materials

(hr-sq ft-°F temp diff/Btu; at a mean temperature of 75 F except as noted)

Material	Description	Resistance, R
Building boards	Asbestos cement board, 1/8 in.	0.03
	Gypsum or plaster board, 1/2 in.	0.45
	Plywood, per inch thickness	1.25
	Sheathing (impregnated or coated), 25/32 in.	2.06
	Wood fiber (hardboard type), 1/4 in.	0.18
Building paper	Vapor, permeable felt	.06
	Vapor, seal, two layers mopped 15 lb felt	0.12
	Vapor, seal plastic film	0
Flooring	Carpet and fibrous pad	2.08
	Carpet and rubber pad	1.23
	Terrazzo, 1 in.	0.08
	Tile—rubber, vinyl, etc.	.05
	Wood subfloor, 25/32 in.	.98†
	Wood, hardwood finish, 3/4 in.	0.68†
Insulation	Mineral wool (blanket and batt), per inch thickness	3.70
Insulation, loose fill	Mineral wool (glass, slag, or rock), per inch thickness	3.33 4.00‡
	Vermiculite	2.08 2.27‡
Roof insulation	Preformed for use above deck	
	1/2 in. thick (approx)	1.39
	1 in. thick (approx)	2.78
	2 in. thick (approx)	5.26
	3 in. thick (approx)	8.33
Masonry concretes	Cement mortar per inch thickness	0.20†
	Lightweight aggregates per inch thickness	
	Density 100 lb per cu ft	.28†
	Density 40 lb per cu ft	.86†
Masonry materials	Brick	
	Common, per inch thickness	.20
	Face	.11
	Clay tile	
	One cell deep, 3 in.	0.80
	One cell deep, 4 in.	1.11
	Two cells deep, 6 in.	1.52
	Two cells deep, 8 in.	1.85
	Concrete blocks, three oval core	
	Sand and gravel, 8 in.	1.11
	Sand and gravel, 12 in.	1.28
	Cinder aggregate, 8 in.	1.72
	Cinder aggregate, 12 in.	1.89
	Lightweight aggregate, 8 in.	2.00
	Lightweight aggregate, 12 in.	2.27
Plastering materials	Cement plaster, sand aggregate 3/4 in.	0.15
	Gypsum plaster, sand aggregate	
	On wood lath	.40
	On metal lath	.10
Roofing	Asbestos cement shingles	.21
	Asphalt roll roofing	.15
	Built-up roofing	.33
Siding materials	Asbestos cement shingles	.21
	Wood shingles, 16 in., 7 1/2 in. exposure	.87
	Asbestos cement siding, 1/4 in. lapped	.21
	Wood drop siding, 1 × 8 in.	.79
	Wood bevel siding, 1/2 × 8 in. lapped	.81
Woods	Fir, pine, and other softwoods	
	25/32 in.	0.98
	2 5/8 in.	3.28
	3 5/8 in.	4.55

* Manufacturer's data to be used for specific products.
† No mean temperature specified.
‡ Mean temperature = 30 F.

Table 12-34a Infiltration thru Windows[9]
(cfh per ft of crack[a]—multiply by number of feet of crack to obtain q for Eq. **4**)

Type of window	Remarks	Wind velocity, miles per hour					
		5	10	15	20	25	30
Double-hung wood sash windows (unlocked)	Around frame in masonry wall						
	Not calked[b]	3	8	14	20	27	35
	Calked[b]	1	2	3	4	5	6
	Around frame in wood frame construction[b]	2	6	11	17	23	30
	Total for average window, 1/16 in. crack and 3/64 in. clearance.[c] Includes wood frame leakage[d]						
	Nonweatherstripped	7	21	39	59	80	104
	Weatherstripped[d]	4	13	24	36	49	63
	Total for poorly fitted window, 3/32 in. crack and 3/32 in. clearance.[e] Includes wood frame leakage[d]						
	Nonweatherstripped	27	69	111	154	199	249
	Weatherstripped[d]	6	19	34	51	71	92
Double-hung metal windows[f]	Nonweatherstripped						
	Locked	20	45	70	96	125	154
	Unlocked	20	47	74	104	137	170
	Weatherstripped, unlocked	6	19	32	46	60	76
Rolled section steel sash windows[k]	Industrial pivoted, 1/16 in. crack[g]	52	108	176	244	304	372
	Architectural projected						
	1/32 in. crack[h]	15	36	62	86	112	139
	3/64 in. crack[h]	20	52	88	116	152	182
	Residential casement						
	1/64 in. crack[i]	6	18	33	47	60	74
	1/32 in. crack[i]	14	32	52	76	100	128
	Heavy casement section, projected						
	1/64 in. crack[j]	3	10	18	26	36	48
	1/32 in. crack[j]	8	24	38	54	72	92
Hollow metal, vertically pivoted window[f]		30	88	145	186	221	242

[a] The values given in this table, with the exception of those for double-hung and hollow metal windows, are 20 per cent less than test values to allow for the building up of pressure in rooms. These values are based on test data.

[b] The values given for frame leakage are per foot of sash perimeter, as determined for double-hung wood windows. Some of the frame leakage in masonry walls originates in the brick wall itself and cannot be prevented by calking. For the additional reason that calking is not done perfectly and deteriorates with time, it is considered advisable to choose the masonry frame leakage values for calked frames as the average determined by the calked and noncalked tests.

[c] The fit of the average double-hung wood window was determined as 1/16 in. crack and 3/64 in. clearance by measurements on approximately 600 windows under heating season conditions.

[d] The values given are the totals for the window opening per foot of sash perimeter, and include frame leakage and so-called "elsewhere leakage." The frame leakage values included are for wood frame construction, but also apply to masonry construction, assuming a 50-per cent efficiency of frame calking.

[e] A 3/32 in. crack and clearance represent a window more poorly fitted than average.

[f] Windows tested in place in building so that no reduction from test values is necessary; see footnote a.

[g] Industrial pivoted window generally used in industrial buildings—ventilators horizontally pivoted at center or slightly above, lower part swinging out.

[h] Architecturally projected made of same sections as industrial pivoted, except that outside framing member is heavier and there are refinements in weathering and hardware. Used in semimonumental buildings such as schools. Ventilators swing in or out and are balanced on side arms. A 1/32 in. crack is obtainable in the best practice of manufacture and installation; a 3/64 in. crack is considered to represent average practice.

[i] Of same design and section shapes as so-called "heavy section casement," but of lighter weight. A 1/64 in. crack is obtainable in the best practice of manufacture and installation; a 1/32 in. crack is considered to represent average practice.

[j] Made of heavy sections. Ventilators swing in or out and stay set at any degree of opening. A 1/64 in. crack is obtainable in the best practice of manufacture and installation; a 1/32 in. crack is considered to represent average practice. Known as "intermediate windows" by steel window manufacturers.

[k] With reasonable care in installation, leakage at contacts where windows are attached to steel framework and at mullions is negligible. With a 3/64 in. crack, indicative of poor installation, leakage at contact with steel framework is about one-third, and at mullions about one-sixth of that given for industrial pivoted windows in the table.

Insulation

Extensive studies of all factors determining economic thicknesses are warranted only for large installations.

Minimum Property Standards of FHA.[8] These standards require the following in one- and two-family living units (other than those electrically heated):

1. A total structural heat loss not exceeding 50 Btu per hr-sq ft of the total floor area of the space heated.

2. A maximum coefficient of heat transfer, U, in Btu per hr-sq ft-°F for ceilings below unheated space of 0.06 for ceilings with heating panels, 0.15 for ceilings without heating panels.

3. A total heat loss thru all exterior walls, doors, windows, etc., not exceeding 30 Btu per hr-sq ft of total floor area of the space heated.

4. A heat loss thru floors over unheated basements, crawl spaces, breezeways, and garages not exceeding 15 Btu per hr-sq ft of floor area.

Table 12-34b Infiltration to Residences*[9]

(in terms of air changes under average conditions, exclusive of air provided for ventilation, q†)

Kind of room having volume V, cu ft	q, cfh
No windows or exterior doors	0.5 V
Windows or exterior doors on one side	1.0 V
Windows or exterior doors on two sides	1.5 V
Windows or exterior doors on three sides	2.0 V
Entrance halls	2.0 V

* The infiltration allowance for an entire building is usually taken as the sum of the individual room infiltration allowances.
† For rooms with weatherstripped windows or with storm sash, use two-thirds of these values.

5. In areas of 2800 degree days and less, a heat loss from heated or unheated concrete slab-on-grade floors not exceeding 5 Btu per hr-sq ft of floor area achieved by the use of perimeter insulation.

For localities where the minimum temperature is 0 F, with the seasonal average temperature between 35 and 40 F, the usual thickness of building insulation is about 2 in.[10] Experience shows that the most productive use of insulation occurs within a U value range of 0.05 to 0.15 for buildings equipped only with heating systems, and from 0.05 to 0.10 for air-conditioned buildings.[11] However, it is good practice to *design* the heating plant on the basis of a *minimum* U value of 0.10 so that the system is not sized so small that a long warming-up period is required.[10]

Owning and Operating Costs. The amount of insulation or level of performance specified for *each building section*, i.e., ceilings, walls, windows, etc., should produce the lowest owning and operating costs to the owner.

The optimum thermal resistance of the building section will depend upon the following factors: (1) location, i.e., the number of degree days; (2) local fuel cost for usable energy; (3) installed insulation costs, including first cost and cost per incremental unit of thermal resistance of insulation; (4) amount of insulation; and (5) period selected for amortization and price of money.

Example: A plot of the interrelated factors for insulating ceilings is shown in Fig. 12-45. Either material having an R_1 or R_2 resistance value could be recommended, depending upon how much the insulation resistance is above or below the ideal, R_i.

Vapor Barriers. Excessive humidity is common in tightly constructed homes. Vapor barriers applied between the plaster and studding of outside walls prevent

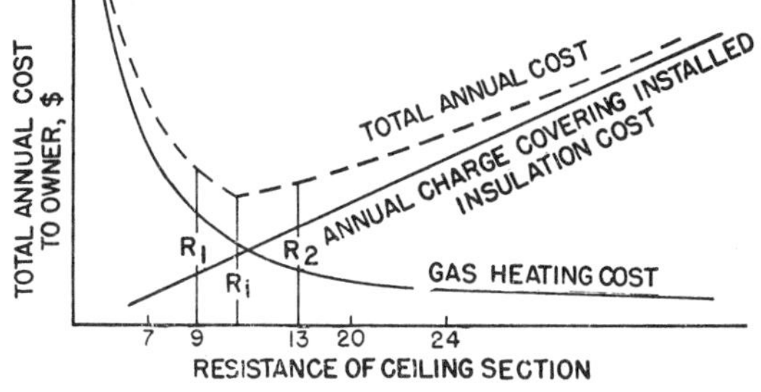

Fig. 12-45 Determining the optimum insulation resistance.

Table 12-35 Heat Loss of Concrete Floors at or Near Grade Level[6]

(Btu/hr-ft of exposed edge)

Outdoor design temp, °F min	Edge insulation thickness and configuration: 2 in.*	1 in.	1 in.	None†
−30	50	55	60	75
−20	45	50	55	65
−10	40	45	50	60

* Recommended.
† Not recommended; shown for comparison only.

migration of indoor vapor. This vapor would otherwise condense on cold insulating material within the wall and eventually lead to paint blistering or dry rot in wood construction. Such barriers, combined with tight storm windows or weatherstripping, tend not only to keep humidity from escaping but even to reduce normal air infiltration to lower levels than desirable. This could interfere with good venting. In such a case, consideration should be given to the installation of an outdoor air intake to the return air side of the furnace or to some unoccupied building space.

Sources of water vapor, including unvented gas appliances, have been evaluated by research at Purdue University (Table 12-36). See Fig. 12-46 for a chart depicting moisture condensation on windows at various temperatures and humidities.

Table 12-36 Home Sources of Water Vapor in Air

	Water vapor, lb
Laundry drying indoors, family of four, weekly	26.7
Cooking, three meals for seven days for four people	33.3
Shower baths, per shower	0.5
People, per person per day	3.0

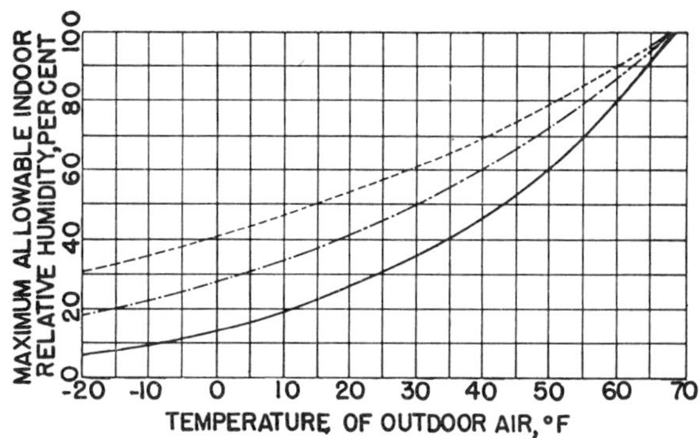

Fig. 12-46 Curves for limiting values of indoor relative humidity. Above these limits, further condensation appears on window panes. Based on room temperature of 69 F and wind velocity of 15 mph.

- - - - - - Double sash, 1½-in. air space
—·—·— Double glazed, ⅛-in. air space
——— Single glazed

Application and Installation

A unit must be sized to meet the maximum load as determined by a heat loss survey. Factors to be considered include amount of glass area, insulation, and unit locations. For these and other considerations—e.g., provision for higher heat losses due to wind velocity, occasional very low outside temperatures, and possible lower system operating efficiency—equipment is seldom sized merely to meet the sum of the losses. A safety or judgment factor is usually applied.

Electric heating contractors often do not include such safety factors. Electric heat systems are sized equal to or sometimes of lower capacity than the computed heat loss. This practice is of doubtful value since ultimate comfort is ostensibly sacrificed to offset high costs.

The I=B=R and the National Warm Air Heating and Air-Conditioning Association have developed piping and pickup factors and multipliers to be applied to the total heat loss to determine unit size needed (Table 12–37).

Installation Standards. Installation requirements for gas heating appliances are given in ASA Z21.30[12] (included in Chap. 1 of this Section). Such units must be erected in accordance with the manufacturer's instructions and, in general, on a floor of fire-resistive construction with noncombustible flooring and surface finish and without combustible material on its underside. Alternatively, mounting on fire-resistive slabs or arches that have no combustible material on their underside is permitted. The exceptions are those A.G.A.-listed[1] appliances specifically approved for installation on a combustible floor. A floor not of fire-resistive construction can be protected and considered fire-resistive when protected according to the NBFU Code for Installation of Heat Producing Appliances.[13]

Clearances. Unless approved for other clearances, use the values in Table 11 of Z21.30,[12] given in Chap. 1 of this Section. NFPA No. 90B[14] gives corresponding data for oil and electricity. Clearances may be reduced as shown in Table 12 of Z21.30,[12] given in Chap. 1 of this Section. Heating units should not be installed in a confined space such as an alcove or closet unless approved for such an installation. Installation in a confined space should then be made in accordance with the approval and the manufacturer's instructions.

Table 12-37 Selection of Gas Boiler and Gravity Warm Air Furnace Sizes

(all values in MBtu per hr)

(1) Structure heat loss	(2) = 1.33* × (1) A.G.A. gross or min bonnet output	(3) = (2)/0.80 Required input: Boiler	(4) = (2)/0.75 Required input: Furnace
25	33	42	44
40	53	67	71
60	80	100	107
80	107	133	143
120	160	200	214
140	187	233	250
160	213	267	284
200	267	333	356

* 1.33 = factor to cover piping losses and to ensure adequate pickup.

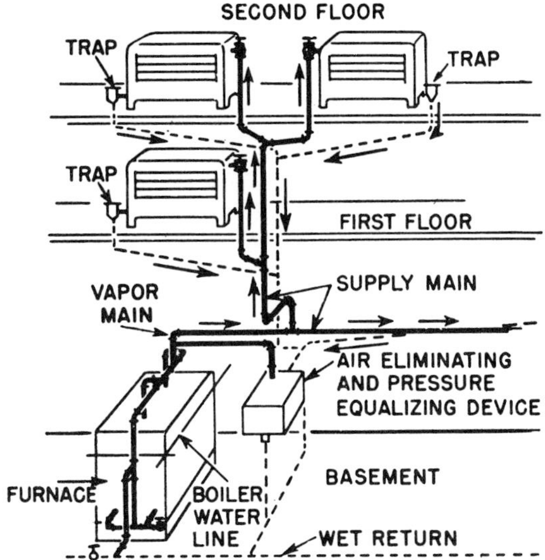

Fig. 12-47 Piping arrangement for two-pipe vapor system.

Vent connector clearances for various gas appliances are given in Table 13 of Z21.30,[12] given in Chap. 3 of this Section. These clearances may also be reduced as shown in Table 12 of Z21.30,[12] given in Chap. 1 of this Section.

Steam Boilers. ASA Z21.13[15] requires that steam boilers conform to ASME construction requirements for 15 psig pressure. Required accessories for a boiler include a main manual shutoff valve, a pilot valve, an automatic pilot, a main automatic control valve, a gas pressure regulator (except for LP-gas), a low-water cutoff, a pressure actuated shutoff device limiting maximum steam pressure to 15 psig, and an ASME safety valve.

Sizing of Boilers. Table 12–37 shows the boiler outputs and inputs required for residential structure heat losses. The gross output rating of a gas steam boiler is defined by ASA Z21.13 as 80 per cent of its input rating.

Radiators, Convectors, Baseboard and Finned-Tube Units. These types of heat-distributing units are used in steam and hot water heating systems. By definition, one sq ft of *steam radiation* emits 240 Btu per hour and one sq ft of *water radiation* emits 150 Btu per hour. To determine the size of a radiator or a convector for a given space, multiply the heat loss of the space in Btu per hour by the applicable factor from Table 12–38 or 12–39. Select a heat distribution unit having an equivalent Btu per hour rating.

When a heating boiler is used for domestic hot water needs, see Chap. 6 of this Section for pertinent information.

Steam Heating Systems. These systems are classified according to (1) piping arrangement, (2) pressure or vacuum systems, (3) condensate return system, or (4) combinations of the three.

Piping arrangement may be either a one-pipe or a two-pipe system, depending on whether or not separate lines are employed for condensate flow. Such arrangement may also be up-flow or down-flow, depending on the direction of steam flow in the risers.

High-pressure systems operate at above 15 psig (usually from 30 to 150 psig) and are normally used in large industrial buildings. With high pressure, returns may be

Table 12-38 Correction Factors for Direct Cast Iron Radiators and Convectors[18]

Steam press (approx)		Heating medium (steam or water) temp, °F	Direct cast iron radiators							Convectors						
			Room temperature, °F							Inlet air temperature, °F						
Gage vacuum, in. Hg	Psia		80	75	70	65	60	55	50	80	75	70	65	60	55	50
22.4	3.7	150	2.58	2.36	2.17	2.00	1.86	1.73	1.62	3.14	2.83	2.57	2.35	2.15	1.98	1.84
20.3	4.7	160	2.17	2.00	1.86	1.73	1.62	1.52	1.44	2.57	2.35	2.15	1.98	1.84	1.71	1.59
17.7	6.0	170	1.86	1.73	1.62	1.52	1.44	1.35	1.28	2.15	1.98	1.84	1.71	1.59	1.49	1.40
14.6	7.5	180	1.62	1.52	1.44	1.35	1.28	1.21	1.15	1.84	1.71	1.59	1.49	1.40	1.32	1.24
10.9	9.3	190	1.44	1.35	1.28	1.21	1.15	1.10	1.05	1.59	1.49	1.40	1.32	1.24	1.17	1.11
6.5	11.5	200	1.28	1.21	1.15	1.10	1.05	1.00	0.96	1.40	1.32	1.24	1.17	1.11	1.05	1.00
Psig																
1	15.6	215	1.10	1.05	1.00	0.96	0.92	0.88	0.85	1.17	1.11	1.05	1.00	0.95	0.91	0.87
6	21	230	0.96	0.92	0.88	.85	.81	.78	.76	1.00	0.95	0.91	0.87	.83	.79	.76
15	30	250	.81	.78	.76	.73	.70	.68	.66	0.83	.79	.76	.73	.70	.68	.65
27	42	270	.70	.68	.66	.64	.62	.60	.58	.70	.68	.65	.63	.60	.58	.56
52	67	300	0.58	0.57	0.55	0.53	0.52	0.51	0.49	0.56	0.54	0.53	0.51	0.49	0.48	0.47

located above the heating unit if a modulating steam valve is not used.

In **low-pressure systems** (from 0 to 15 psig), the air valves normally used do not contain check disks and hence cannot operate under a vacuum. Low-pressure systems are not capable of holding heat as steam rate diminishes.

A **vapor system** (Fig. 12-47) is a two-pipe system operating at low pressure (usually below 1.0 psig). It operates under both vacuum and low-pressure conditions without the use of a vacuum pump. Steam traps are installed on the return side of the radiator, and air is eliminated thru a single air valve at the end of the return mains. Modulation can be obtained by setting the inlet valve.

Vacuum systems employ a vacuum pump, which increases the pressure difference between the supply system and the return system. Rapid circulation is obtained and radiators may be located below the water level of the boiler. An air vent is located on the vacuum pump unit to eliminate air from the system.

Condensate return may use *mechanical means* or *gravity*. When condensate cannot be returned by gravity and either traps or pumps must be used, the system is a mechanical return system. Whenever possible, gravity flow to a receiver and to a pump should be used.

Steam Piping.[16] Factors in sizing include:

1. The initial pressure and the allowable pressure drop between the source and the return.
2. The maximum allowable steam velocity for quiet and dependable operation.
3. The equivalent length of run from the steam source to the farthest heating unit.
4. The direction of flow of the condensate relative to the steam.

See Refs. 16 and 17 for pertinent material in designing piping and steam distribution systems.

If condensate is first drained to an atmospheric receiver, full boiler pressure may be utilized to overcome friction effects. In a simple gravity return system, the boiler water line must be at such an elevation below the farthest distant trap or air eliminator as to at least equal the frictional head generated by the maximum rate of steam and condensate flow. Thus, a system having a 2.0-psi drop thru the piping would require a normal boiler water line at least 56 in. below

Table 12-39 Correction Factors for Finned-Tube and Baseboard Radiation[18]

Steam press (approx)		Heating medium (steam or water) temp, °F	Finned-tube units							Baseboard units						
			Inlet air temperature, °F							Inlet air temperature, °F						
Gage vacuum, in. Hg	Psia		80	75	70	65	60	55	50	80	75	70	65	60	55	50
22.4	3.7	150	2.80	2.50	2.20	1.95	1.81	1.67	1.54	2.86	2.61	2.38	2.20	2.03	1.89	1.76
20.3	4.7	160	2.34	2.14	1.94	1.75	1.62	1.50	1.37	2.38	2.20	2.03	1.89	1.76	1.64	1.56
17.7	6.0	170	2.01	1.86	1.70	1.56	1.46	1.36	1.26	2.03	1.89	1.76	1.64	1.54	1.44	1.38
14.6	7.5	180	1.76	1.65	1.53	1.42	1.32	1.24	1.15	1.76	1.64	1.55	1.44	1.38	1.29	1.23
10.9	9.3	190	1.53	1.45	1.36	1.28	1.20	1.12	1.03	1.54	1.44	1.37	1.29	1.22	1.16	1.09
6.5	11.5	200	1.38	1.31	1.24	1.16	1.09	1.02	0.95	1.38	1.29	1.23	1.16	1.09	1.05	1.00
Psig																
1	15.6	215	1.17	1.12	1.06	1.00	0.95	0.90	.85	1.16	1.10	1.05	1.00	0.95	0.92	0.88
6	21	230	1.00	0.97	0.93	0.90	.86	.82	.78	1.00	0.96	0.92	0.88	.84	.81	.77
15	30	250	0.88	.85	.82	.78	.75	.71	.67	0.86	.82	.79	.76	.73	.70	.68
27	42	270	.75	.73	.70	.68	.65	.62	.59	.73	.70	.68	.66	.63	.61	.59
52	67	300	0.62	0.60	0.58	0.56	0.54	0.52	0.50	0.58	0.57	0.55	0.53	0.52	0.51	0.49

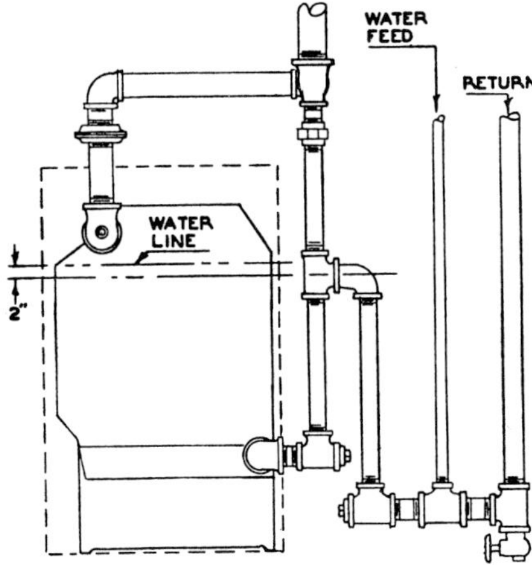

Fig. 12-48 Hartford loop return connection for steam boiler.

the lowest trap, air vent valve, or horizontal portion of a dry return.

Header piping on steam boilers should be so arranged as to drop out entrained particles of moisture and return them directly to the boiler instead of carrying them into the mains and radiator runouts. To keep steam pressure at the top of the boiler from forcing water out of its bottom into the return piping, check the valves or install a Hartford loop connection (Fig. 12–48) in the condensate return.

Hot Water Boilers. ASA Z21.13[15] requires that water boilers meet ASME construction requirements for a minimum pressure of 30 psig. Required equipment includes a pressure or altitude gage, a water temperature thermometer, a manual main gas valve, a pilot valve, a gas pressure regulator, an automatic gas control valve, an automatic pilot, a high temperature safety device that limits boiler water temperature to 250 F, and an ASME relief valve.

Sizing. Table 12–37 may be used for sizing a residential hot water boiler for structural heat losses. Radiators and convectors may be sized using Tables 12–38 and 12–39. For information concerning heat output from ceiling coils, pipe banks on wall, or other types of radiation, see Ref. 18 or the manufacturer.

Hot Water Space Heating Systems. These systems are classified according to (1) piping arrangement—one-pipe or two-pipe, and (2) circulation—gravity or forced. Low-temperature systems have supply water temperatures less than 250 F. Because of greater pressure head maintained by use of a pump, smaller pipe sizes can be used in a forced circulation system. The disadvantage common to all one-pipe hot water systems is that the water temperature becomes successively lower as it reaches the radiators farther away from the boiler. To minimize corrosion possibilities and to reduce the amount of air that may collect in radiation, the water in the system should not be changed. Periodically all radiators and high points in the system should be purged of air.

Advantages of forced hot water systems include:

1. Piping may be run at any pitch or level.
2. Corrosion is minimized, since most of the air in the system is eliminated.
3. Uniform heat distribution occurs during the warm-up period.

System Design. Most low-temperature systems are based on a 20°F drop for the heating medium; this differential is used to determine the water flow rate needed to meet maximum heat loss conditions. For design conditions, allow 10 MBtu per hr for a drop of 20°F in water flowing at 1.0 gpm. Selection of radiation may be based on design supply water temperature and temperature drop thru the unit. A common supply water temperature is 210 F, with a range of 180 to 250 F.

Pipe sizing is based on the average friction loss of the system as affected by the type of pump used, as well as the economics of small pipe size and high pump head vs. large pipe size and low pump head. To reduce the possibility of noise, velocities are normally limited to a maximum of 4.0 fps for pipe sizes less than 2 in. For pipe sizes larger than 6 in., velocities up to 10 fps are frequently used. Average friction drops of 1.0 to 4.0 ft of water per 100 ft of pipe are used in sizing.[19] For details of sizing of hot water systems see Refs. 20, 21, and 22.

Gravity and Forced Air Furnaces. ASA Z21.47[23] requires that gravity and forced air furnaces be equipped or supplied with a manual main gas valve, a pilot valve, a gas pressure regulator, an automatic gas control valve, and an automatic pilot.

Outlet Air Temperature Limit Control. This is an automatic recycling device that should function independently of the operating temperature control. The recycling device has a fixed stop that limits the outlet air temperature of gravity-type furnaces to 250 F for standard clearance and 200 F for reduced clearance. Table 12–40 gives air temperatures for forced air furnaces.

Table 12-40 Temperature Limits for Forced Air Furnaces

Range of air temp rise thru furnace, °F	Max permissible outlet air temp, °F
20–50	150
45–75	175
70–100	200 for reduced clearance
100–130	250 for standard clearance*

* Except downflow furnaces on combustible floors and horizontal furnaces, which are limited to 200 F.

Sizing. The residential gas furnace size is based on the total heat loss of the structure to be heated, the type of duct system, and the allowable pressure and temperature drops in the system. The required input for a gravity furnace and for a forced air furnace may be obtained from Tables 12–37 and 12–41, respectively.

Table 12-41 Selection of Forced Air Furnace Sizes[25]
(all values in MBtu per hr)

Structure heat loss	Min bonnet output	Required input
15	18	23
25	29	36
35	41	51
40	46	58
50	58	73
60	69	87
80	92	115
100	115	144
120	138	173

Table 12-42 Recommended and Maximum Duct Velocities for Conventional Systems,[24] fpm

Designation	Residences		Schools, theatres, public buildings		Industrial buildings	
	Recommended	Max	Recommended	Max	Recommended	Max
Air intakes*	500	800	500	900	500	1200
Filters*	250	300	300	350	350	350
Heating coils*	450	500	500	600	600	700
Air washers	500	500	500	500	500	500
Fan outlets	1000–1600	1700	1300–2000	1500–2200	1600–2400	1700–2800
Main ducts	700–900	800–1200	1000–1300	1100–1600	1200–1800	1300–2200
Branch ducts	600	700–1000	600–900	800–1300	800–1000	1000–1800
Branch risers	500	650–800	600–700	800–1200	800	1000–1600

* These velocities are for total face area, not the net free area; other velocities in table are for net free area.

Types of warm air systems include perimeter-loop, perimeter-radial, perimeter extended-plenum, inside-wall delivery, ceiling-panel, and combinations of the foregoing.

Duct Leader Capacity. Capacities of basement leader pipe for gravity systems, in Btu per hr-sq in. of pipe cross section, follow: For first floor, 111; for second floor, 167; for third floor 200. Second and third floor ducts may have areas 70 per cent of those of the basement leader pipe.

Duct Design. Among the widely used duct design methods are: (1) velocity reduction, (2) equal friction, and (3) static regain. Knowing the air requirement in cfm and the allowable pressure drop permits use of appropriate air friction charts[24] for duct sizing.

Air Handling Capacity. Use Eq. **7** to determine the amount of air handled by a duct system.

$$V = Q/(1.08\Delta t) \quad \textbf{(7)}$$

where:

V = air handling requirement, cfm
Q = minimum bonnet output, Table 12–37 or 12–41, Btu per hr
Δt = temperature rise thru the furnace, °F

For summer cooling the general practice is to handle 400 cfm per ton of air conditioning, which corresponds to a differential of 50° to 60°F. Since heating is generally based on a 100°F temperature rise thru the furnace, twice as much air will be required for cooling as for heating. When both the bonnet output of the furnace and the tons of refrigerating capacity are known, the ducts should be sized for the cycle requiring the greatest air capacity.

Designs are normally based on furnaces that are capable of delivering their rated air volumes against a total pressure of 0.20 in. w.c. external to the furnace. This may be distributed on the basis of three-fourths (0.15 in. w.c.) on the supply side and one-fourth (0.05 in. w.c.) on the return side. Table 12–42 gives recommended values for duct velocities. Velocities should be kept low for quiet operation and for reduction of the required fan horsepower. It should be remembered that duct size increases with decreasing velocities.

Duct design procedures are available.[25] The installation of residential warm air heating and air conditioning systems is covered by NFPA 90B.[14] Table 12–43 gives the minimum requirements for the construction of supply ducts; for exceptions, see NFPA 90B.

Filters used in forced air heating and cooling can generally be classified as dry air filters, viscous impingement-type filters, and electronic air cleaners. They are rated on weight efficiency of dust concentration retained[26] or by a dust spot method.[27] A listing of various applications of filters based on their efficiency is available.[28]

Air Motion. The movement of air within the occupied zone of the conditioned space eliminates stratification and ensures good temperature distribution. A velocity of 25 fpm is desirable; to avoid drafts, velocity should not exceed 40 fpm. A minimum of 4.5 air changes per hour should be provided in terms of warm air movement, exclusive of infiltration thru windows and doors, but more changes are preferable. Desirable face velocity varies with the register height from the floor. A maximum of 300 fpm is recommended for low registers and a minimum of 600 fpm for high registers.

Floor Furnaces. These completely self-contained appliances are suspended from the floor of the space heated (Fig. 12-49a). They take combustion air from outside this space. Means are provided for observing flames and lighting the appliance from the heated space.

Equipment required by ASA Z21.48[29] includes a pilot control valve, a manual control valve, and a pressure regulator. When automatic temperature control means are provided, an automatic pilot must also be included.

Types. Both the gravity and fan types are listed for direct mounting on wooden floors with casings in contact with supporting floor joists. Unless a unit is specifically approved for the connection of auxiliary warm air or return air ducts, additional duct work must not be used, since it can seriously change casing temperature distribution.

Because of the temperature of delivered air and radiation from them, floor furnaces should not be placed in passageways. A distance of at least 15 in. from two adjoining sides of the floor grille to the walls should be provided.[12] Doors should not open over furnace grilles nor should draperies or furniture be within 12 in. of the register.

Table 12-43 Minimum Construction Requirements for Supply Ducts[14]

Diam or width, in.	Nominal thick. in.	Equiv galvanized sheet gage	Approx aluminum B & S gage	Min wt tin plate, lb per base box
Round ducts and enclosed rectangular ducts				
14 or less	0.016	30	26	135
Over 14	0.019	28	24	...
Exposed rectangular ducts				
14 or less	0.019	28	24	...
Over 14	0.022	26	23	...

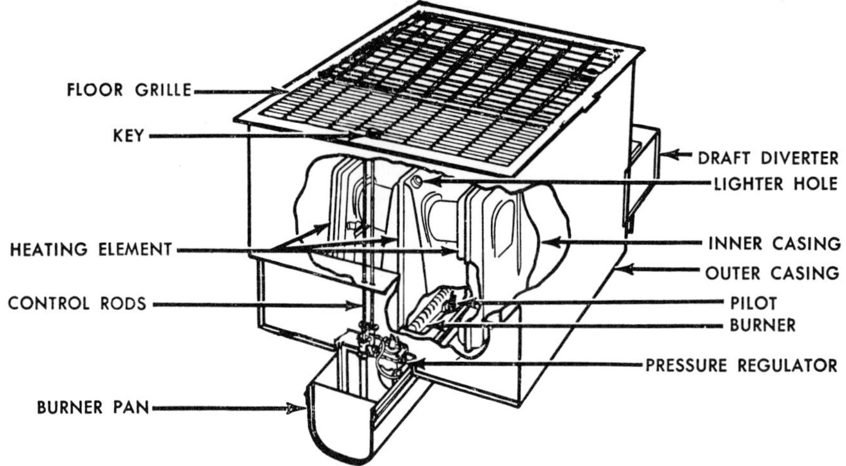

Fig. 12-49a Typical floor furnace.

Vented Wall Heaters. These are self-contained vented appliances complete with grilles or equivalent, designed for incorporation in or permanent attachment to a wall, floor, ceiling, or partition. They furnish heated air directly to the space to be heated thru openings in the casing (Fig. 12–49b). Such appliances must not be provided with duct extensions beyond the vertical and horizontal limits of the casing proper, except that boots for horizontal extension thru walls of nominal thickness (not over 10 in.) may be permitted. These boots must be supplied by the manufacturer as an integral part of the appliance and tested as such.

Recessed heater equipment must include a manual gas valve, a pilot valve, a gas pressure regulator, and an automatic pilot.[30] Many also have automatic temperature controls. A high-temperature limit control and a fan operating device are also required on fan-type units.

Room Heaters. A room heater is a self-contained, free-standing appliance, intended for installation in the space heated and not for duct connection. Air heating appliances covered by other ASA Standards are excluded. A room heater may be vented or unvented, but unvented room heaters are limited to a normal input rating of 50,000 Btu per hour.

Vented Installations. Units installed in sleeping quarters for use of transients, as in hotels, auto courts, and motels, as well as in institutions such as homes for the aged, sanatoriums, convalescent homes, and orphanages, must be of the vented type.

The aggregate input ratings of unvented appliances should not exceed 30 Btu per hr-cu ft of the room volume or space in which they are installed.

Room heaters are subjected to closed room tests under reduced oxygen content in the air and increased flue gas content, in addition to tests in an open room, to ensure that they will operate under these severe conditions. Heater rating plates indicate design for vented or unvented operation. All vented units must be equipped with an automatic pilot. Table 10 of Z21.30,[12] given in Chap. 1 of this Section, specifies the minimum clearances for listed room heaters unless the unit is approved for smaller clearances.

Sealed Combustion Appliances. Such appliances take air for combustion from the outside and return all combustion products to the outside (Fig. 12-50). Numerous appliances with this feature have been approved under ASA requirements for conventional appliances. Among approved sealed combustion appliances are central warm air furnaces, boilers, vented wall furnaces,[32] baseboard heaters, room heaters, unit heaters, and water heaters. These appliances may be either installed individually thru an outside wall (Fig. 12–50) or connected to a flue system (Fig. 12–14).

Individual Unit Installations. Vent terminals shall be

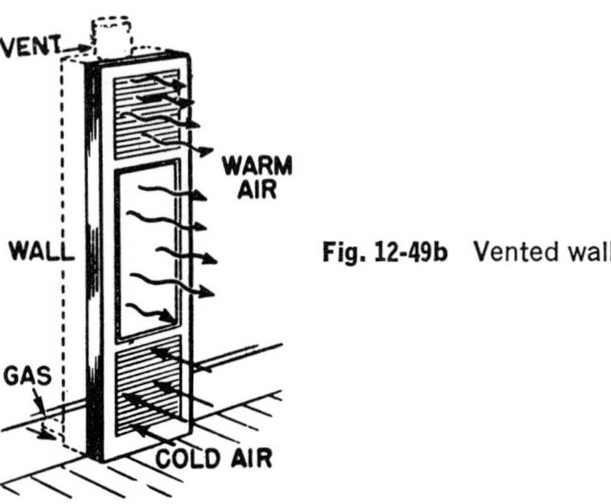

Fig. 12-49b Vented wall heater.

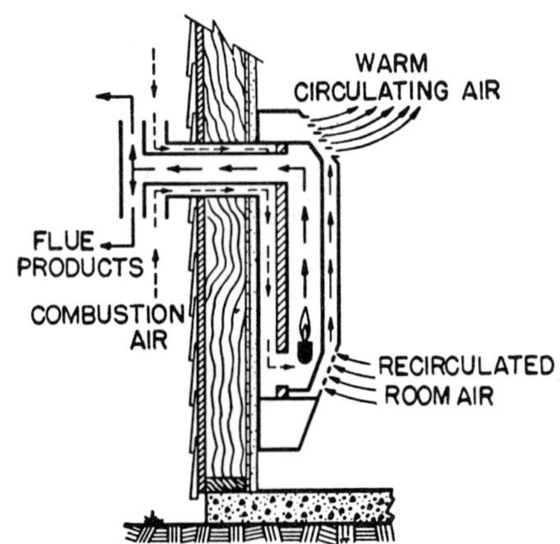

Fig. 12-50 Flow diagram of sealed-combustion-chamber heater.[31] Note that the combustion air supply and the vent outlet are in the same pressure zone.

located at least 12 in. from a building opening and a minimum of 7 ft above grade (to avoid physical contact) when located adjacent to public walkways.[12] Tests have shown that combustion products dilute rapidly and "cannot be considered to represent any hazard to human health."[33] Even with an induced air flow of 3.5 mph (perpendicular or parallel winds of 5.0 and 10.0 mph produced less severe conditions), the combustion products drawn into an open window (normally closed during the heating season) contained insignificant quantities of flue gas constituents.

Domestic Gas Conversion Burners. These burners supply gaseous fuel to an appliance originally designed to use another fuel. Atmospheric burners may be either inshot or upshot types (Fig. 12–118) with drilled-port, slotted-port, or single-port burner heads. For installations in furnaces or boilers with restricted flue passages, power burners may be used (Figs. 12–126 and 12–127). Generally, upshot burners should be used in converting coal equipment since the flame is spread better in what is normally a bigger combustion chamber. Inshot burners normally replace gun-type oil burners.

ASA Z21.17[34] requires that a domestic gas conversion burner be equipped with a manual shutoff valve, a pilot shutoff valve, a gas pressure regulator (except for LP-gas), an automatic control valve, an automatic pilot, and an air duct box or other means for controlling combustion air and locating the burner in the firebox. Other necessary safety controls should be added when the burner is installed. Satisfactory operation of a conversion burner depends primarily on the skill of the installer and also on the performance of both the converted appliance and the entire system.

The required size of the conversion burner can be determined from (1) the input or manufacturer's rating of the existing unit or (2) the heat loss of the building.

To determine gas input, based on the nozzle input of an existing oil-fired heating unit, use the efficiencies given in Table 12–44, which apply in northern areas. Table 2–35 gives heating values of fuel oils. Residential fuel oil is normally Grade No. 2 with an average 140 MBtu per gal heating value.

Table 12-44 Service Efficiency of Heating Equipment with Various Fuels

(residential systems)

Type of fuel burning unit	Efficiency, per cent
Gas, designed unit	70–80
Gas, conversion unit	60–80
Oil, designed unit	65–80
Oil, conversion unit	60–80
Bituminous coal, hand-fired with controls	50–65
Bituminous coal, stoker-fired	50–70
Anthracite, hand-fired with controls	60–80
Anthracite, stoker-fired	60–80
Coke, hand-fired with controls	60–80

For small coal *boilers*, output capacities are based on a hand-fired input rate of approximately 100 MBtu per hr-sq ft of grate area. In most instances gas conversion burner inputs may be reduced below coal firing rates because of increased efficiency.

Capacities of residential types of coal *furnaces* are based on hand-firing at inputs of 70,000 to 100,000 Btu per hr-sq ft of grate area. When ratios of heating surface to grate area are above 20 to 1, and for commercial units, recommended firing rates may be higher.

Installation. The burner should fit the space in the firebox in such a way that the heat generated can be efficiently absorbed by the heating surfaces. In general, the input rating should be between 40,000 and 75,000 Btu per hr-sq ft of grate area of the heating plant.

A draft hood or draft diverter is used on installations up to 400 MBtu per hr input. The draft hood should be installed at the highest practical point in the flue connection. Larger inputs may require double-acting barometric dampers.

The size of the flue connector should be as listed in Table 12–45 for other than revertible or diving flue furnaces and boilers. The flue pipe may also be reduced in area by use of a neutral pressure adjustment.[35]

Table 12-45 Draft Hood and Flue Pipe Sizes for Gas Conversion Burners in Updraft Coal Furnaces and Boilers

(not more than 6500 Btu per sq in. of flue area)

Input, MBtu per hr	Draft hood and flue pipe size, in.
Up to 120	5
120–180	6
180–250	7
250–320	8
320–410	9

Note: If the flue pipe exceeds 10 ft in length or contains more than two elbows, use the next larger size pipe and draft hood.

Combustion air adjustment is described in ASA Z21.8.[35] After the burner has operated for 15 min, a flue gas analysis should be made; CO_2 readings and stack temperature should be within the recommended limits of the local gas company.

Estimating Residential Heating Consumption

The two widely used methods for determining residential heating consumption are the *calculated heat loss method* and the *degree day method.* It should be emphasized that these methods are used for residential heating estimations only; their application to commercial or industrial installations may produce misleading results.

Calculated Heat Loss Method. This method uses Eq. **8.**

$$F = \frac{XN}{E} \tag{8}$$

where:

F = quantity of fuel required (normally for a full heating season), Btu

X = average heat requirement rate for the period under consideration, from Eq. **9**, Btu per hr

N = number of heating hours in the estimate period (for an Oct. 1–May 1 heating season, 212 days × 24 hr per day = 5088 hr)

E = utilization efficiency over the estimate period (usually a heating season), from Table 12–44, decimal

$$X = \frac{H.L.(t_i - t_o)}{\Delta t} \tag{9}$$

where:

$H.L.$ = the calculated heat loss based on *design* conditions, Btu per hr

t_i = average indoor temperature maintained during estimate period, from Table 12–31a, °F

t_o = average outdoor temperature thru estimate period, from Table 12–46, °F

Δt = difference between indoor and outdoor *design* temperatures (see Table 12–47 for outdoor *design* temperatures), °F

Substituting Eq. **9** into Eq. **8** yields Eq. **10**:

$$F = \frac{H.L.(t_i - t_o)N}{E\Delta t} \qquad \textbf{(10)}$$

All terms are defined under Eqs. **8** and **9**.

Table 12-46 Average Monthly and Yearly Degree Days for Cities in the United States and Canada*†‡[36]

(base = 65 F)

State or province	Station	Years	No. of seasons	Avg winter temp§	July	Aug.	Sept.	Oct.	Nov.	Dec.	Jan.	Feb.	Mar.	Apr.	May	June	Yearly total
Ala.	Anniston—A			52.3	0	0	17	118	438	614	614	485	381	128	25	0	2,820
	Birmingham—A			51.8	0	0	13	123	396	598	623	491	378	128	30	0	2,780
	Mobile—A				0	0	0	28	219	376	416	304	222	47	0	0	1,612
	Mobile—C			58.9	0	0	0	23	198	357	412	290	209	40	0	0	1,529
	Montgomery—A				0	0	0	69	304	491	517	388	288	80	0	0	2,137
	Montgomery—C			56.4	0	0	0	55	267	458	483	360	265	66	0	0	1,954
Ariz.	Flagstaff—A			35.9	49	78	243	586	876	1135	1231	1014	949	687	465	212	7,525
	Phoenix—A			59.5	0	0	0	22	223	400	474	309	196	74	0	0	1,698
	Phoenix—C			57.7	0	0	0	13	182	360	425	275	175	62	0	0	1,492
	Yuma—A			62.5	0	0	0	0	105	259	318	167	88	14	0	0	951
Ark.	Bentonville	06/07–40/41	35		1	1	38	216	516	810	879	716	519	247	86	7	4,036
	Fort Smith—A				0	0	9	131	435	698	775	571	418	127	24	0	3,188
	Little Rock—A				0	0	10	110	405	654	719	543	401	122	18	0	2,982
Calif.	Eureka—C			49.3	267	248	264	335	411	508	552	465	493	432	375	282	4,632
	Fresno—A				0	0	0	86	345	580	629	400	304	145	43	0	2,532
	Independence	98/99–40/41	43		0	0	28	216	512	778	799	619	477	267	120	18	3,834
	Los Angeles—A				31	22	56	87	200	301	378	305	273	185	121	56	2,015
	Los Angeles—C			59.3	0	0	17	41	140	253	328	244	212	129	68	19	1,451
	Needles	17/18–38/39	22		0	0	0	19	217	416	447	243	124	26	3	0	1,495
	Point Reyes	98/99–40/41	43		350	336	263	282	317	425	467	406	437	413	415	363	4,474
	Red Bluff—A			52.9	0	0	0	59	319	564	617	423	336	177	51	0	2,546
	Sacramento—A				0	0	22	98	357	595	642	428	348	222	103	7	2,822
	Sacramento—C			53.0	0	0	17	75	321	567	614	403	317	196	85	5	2,600
	San Diego—A				11	7	24	52	147	255	317	247	223	151	97	43	1,574
	San Francisco—A				144	136	101	174	318	487	530	398	378	327	264	164	3,421
	San Francisco—C			54.2	189	177	110	128	237	406	462	336	317	279	248	180	3,069
	San Jose—C			53.5	7	11	26	97	270	450	487	342	308	229	137	46	2,410
Colo.	Denver—A			37.0	5	11	120	425	771	1032	1125	924	843	525	286	65	6,132
	Denver—C			37.9	0	5	103	385	711	958	1042	854	797	492	266	60	5,673
	Durango	04/05–40/41	37		25	37	201	535	861	1204	1271	1002	859	615	394	139	7,143
	Grand Junction—A			39.9	0	0	36	333	792	1132	1271	924	738	402	145	23	5,796
	Leadville	07/08–40/41	34		280	332	509	841	1139	1413	1470	1285	1245	990	740	434	10,678
	Pueblo—A				0	0	74	383	771	1051	1104	865	775	456	203	27	5,709
Conn.	Hartford—A				0	14	101	384	699	1082	1178	1050	871	528	201	31	6,139
	New Haven—A				0	18	93	363	663	1026	1113	1005	865	567	261	52	6,026
D. C.	Washington—A				0	0	37	237	519	837	893	781	619	323	87	0	4,333
	Washington—C			43.4	0	0	32	231	510	831	884	770	606	314	80	0	4,258
Fla.	Apalachicola—C			60.9	0	0	0	17	154	304	352	263	184	33	0	0	1,307
	Jacksonville—A			60.6	0	0	0	16	148	309	331	247	169	23	0	0	1,243
	Jacksonville—C			62.0	0	0	0	11	129	276	303	226	154	14	0	0	1,113
	Key West—A			72.5	0	0	0	0	0	22	34	25	8	0	0	0	89
	Key West—C			73.1	0	0	0	0	0	18	28	24	7	0	0	0	77
	Miami—A				0	0	0	0	8	52	58	48	12	0	0	0	178
	Miami—C			71.4	0	0	0	0	5	48	57	48	15	0	0	0	173
	Pensacola—C			59.7	0	0	0	18	177	334	383	275	203	45	0	0	1,435
	Tampa—A				0	0	0	0	60	163	201	148	102	0	0	0	674
Ga.	Atlanta—A				0	0	8	110	393	614	632	512	404	133	20	0	2,826
	Atlanta—C				0	0	8	107	387	611	632	515	392	135	24	0	2,811
	Augusta—A				0	0	0	59	282	494	521	412	308	62	0	0	2,138
	Macon—A				0	0	0	63	280	481	497	391	275	62	0	0	2,049
	Savannah—A			56.7	0	0	0	38	225	412	424	330	238	43	0	0	1,710
	Thomasville	05/06–40/41	36		0	0	2	48	208	361	359	299	178	52	5	1	1,513
Idaho	Boise—A			39.8	0	0	135	389	762	1054	1169	868	719	453	249	92	5,890
	Lewiston—A				0	0	133	406	747	961	1060	815	663	408	222	68	5,483
	Pocatello—A			35.0	0	0	183	487	873	1184	1333	1022	880	561	317	136	6,976
Ill.	Cairo—C			46.4	0	0	28	161	492	784	856	683	523	182	47	0	3,756
	Chicago—A			35.1	0	0	90	350	765	1147	1243	1053	868	507	229	58	6,310
	Peoria—A			37.3	0	11	86	339	759	1128	1240	1028	828	435	192	41	6,087
	Springfield—A			37.7	0	6	83	315	723	1066	1166	958	769	404	171	32	5,693
	Springfield—C			39.8	0	0	56	259	666	1017	1116	907	713	350	127	14	5,225
Ind.	Evansville—A			45.1	0	0	59	215	570	871	939	770	589	251	90	6	4,360
	Fort Wayne—A			37.6	0	17	107	377	759	1122	1200	1036	874	516	226	53	6,287
	Indianapolis—A			39.0	0	0	79	306	705	1051	1122	938	772	432	176	30	5,611
	Indianapolis—C			39.6	0	0	59	247	642	986	1051	893	725	375	140	16	5,134
	Royal Center	18/19–31/32	14		11	19	116	373	740	1104	1239	976	860	502	245	54	6,239
	Terre Haute—A				0	5	77	295	681	1023	1107	913	725	371	145	24	5,366
Iowa	Charles City—C			31.2	17	30	151	444	912	1352	1494	1240	1001	537	256	70	7,504
	Davenport—C			37.0	0	7	79	320	756	1147	1262	1044	834	432	175	35	6,091

Table 12-46 (Continued)

State or province	Station	Years	No. of sea-sons	Avg winter temp§	July	Aug.	Sept.	Oct.	Nov.	Dec.	Jan.	Feb.	Mar.	Apr.	May	June	Yearly total
Iowa	Des Moines—A			35.4	5	12	99	355	798	1203	1330	1092	868	438	201	45	6,446
	Des Moines—C			36.4	0	6	89	346	777	1178	1308	1072	849	425	183	41	6,274
	Dubuque—A			34.6	8	28	149	444	882	1290	1414	1187	983	543	267	76	7,271
	Keokuk	98/99–41/42	44	39.3	1	3	71	303	680	1077	1191	1025	761	397	136	18	5,663
	Sioux City—A				8	17	128	405	885	1290	1423	1170	930	474	228	54	7,012
Kan.	Concordia—C			40.7	0	0	55	277	687	1029	1144	899	725	341	146	20	5,323
	Dodge City—A				0	0	40	262	669	980	1076	840	694	347	135	15	5,058
	Iola	05/06–40/41	36		0	1	40	236	579	930	1026	817	599	282	98	8	4,616
	Topeka—A				0	8	59	271	672	1017	1125	885	694	326	137	15	5,209
	Topeka—C			42.1	0	0	42	242	630	977	1088	851	669	295	112	13	4,919
	Wichita—A				0	0	32	219	597	915	1023	778	619	280	101	7	4,571
Ky.	Louisville—A				0	0	51	232	579	871	933	778	611	285	94	5	4,439
	Louisville—C			45.1	0	0	41	206	549	849	911	762	605	270	86	0	4,279
	Lexington—A				0	0	56	259	636	933	1008	854	710	368	140	15	4,979
La.	New Orleans—A			60.6	0	0	0	7	169	308	364	248	190	31	0	0	1,317
	New Orleans—C			61.6	0	0	0	5	141	283	341	223	163	19	0	0	1,175
	Shreveport—A				0	0	0	53	305	490	550	386	272	61	0	0	2,117
Me.	Eastport—C			31.5	141	136	261	521	798	1206	1333	1201	1063	774	524	288	8,246
	Greenville	07/08–40/41 42/43–45/46	38		69	113	315	643	1012	1464	1625	1443	1251	842	468	194	9,439
	Portland—A			33.0	15	56	199	515	825	1237	1373	1218	1039	693	394	117	7,681
Md.	Baltimore—A			44.1	0	0	50	278	582	908	955	840	676	378	115	5	4,787
	Baltimore—C			44.3	0	0	29	207	489	812	880	776	611	326	73	0	4,203
Mass.	Boston—A				0	7	77	315	618	998	1113	1002	849	534	236	42	5,791
	Fitchburg	98/99–40/41	43		12	29	144	432	774	1139	1240	1137	940	572	254	70	6,743
	Nantucket—A				22	34	111	372	615	924	1020	949	880	642	394	139	6,102
Mich.	Alpena—C			29.6	50	85	215	530	864	1218	1358	1263	1156	762	437	135	8,073
	Detroit, Willow Run—A				0	10	96	393	759	1125	1231	1089	915	552	244	55	6,469
	Detroit City—A				0	8	96	381	747	1101	1203	1072	927	558	251	60	6,404
	Escanaba—C			27.3	62	95	247	555	933	1321	1473	1327	1203	804	471	166	8,657
	Grand Rapids—A				14	29	144	462	822	1169	1287	1154	1008	606	301	79	7,075
	Grand Rapids—C			36.0	0	20	105	394	756	1107	1215	1086	939	546	248	58	6,474
	Houghton	00/01–40/41 42/43–45/46	45		70	94	268	582	965	1355	1535	1421	1251	820	474	195	9,030
	Lansing—A				13	33	140	455	813	1175	1277	1142	986	591	287	70	6,982
	Ludington	12/13–40/41	29		41	55	182	472	794	1135	1271	1183	1056	698	418	153	7,458
	Marquette—C			28.3	69	87	236	543	933	1299	1435	1291	1181	789	477	189	8,529
	Sault Ste. Marie—A			26.0	109	126	298	639	1005	1398	1587	1442	1302	846	499	224	9,475
Minn.	Duluth—A				56	91	298	651	1140	1606	1758	1512	1327	846	474	178	9,937
	Duluth—C			24.3	66	91	277	614	1092	1550	1696	1448	1252	801	487	200	9,574
	Minneapolis—A				8	17	157	459	960	1414	1562	1310	1057	570	259	80	7,853
	Moorhead	98/99–40/41	43		20	47	240	607	1105	1609	1815	1555	1225	679	327	98	9,327
	Saint Paul—A				12	21	154	459	951	1401	1553	1305	1051	564	256	77	7,804
Miss.	Corinth	09/10–40/41	32		0	1	13	142	418	669	696	570	396	149	32	1	3,087
	Meridian—A				0	0	0	90	338	528	561	413	309	85	9	0	2,333
	Vicksburg—C			56.8	0	0	0	51	268	456	507	374	273	71	0	0	2,000
Mo.	Columbia—A			41.1	0	6	62	262	654	989	1091	876	698	326	135	14	5,113
	Hannibal	98/99–40/41	43		1	3	66	288	652	1037	1139	980	710	374	128	15	5,393
	Kansas City—A				0	0	44	240	621	970	1085	851	666	292	111	8	4,888
	Saint Louis—A			42.3	0	0	45	233	600	927	1017	820	648	297	101	11	4,699
	Saint Louis—C			43.6	0	0	38	202	570	893	983	792	620	270	94	7	4,469
	Springfield—A				0	8	61	249	615	908	1001	790	632	295	118	16	4,693
Mont.	Billings—A			34.9	8	20	194	497	876	1172	1305	1089	958	564	304	119	7,106
	Havre—C			28.4	20	38	270	564	1023	1383	1513	1291	1076	597	313	125	8,213
	Helena—A				36	66	320	617	999	1311	1469	1165	1017	654	399	197	8,250
	Helena—C			31.6	51	78	359	598	969	1215	1438	1114	992	660	427	225	8,126
	Kalispell—A			31.6	47	83	326	639	990	1249	1386	1120	970	639	391	215	8,055
	Miles City—A			27.6	6	11	187	525	966	1373	1516	1257	1048	570	285	106	7,850
	Missoula—A				22	57	292	623	993	1283	1414	1100	939	609	365	176	7,873
Neb.	Drexel	15/16–25/26	11		4	6	95	405	788	1271	1353	1096	843	493	219	38	6,611
	Lincoln—A			35.6	0	12	82	340	774	1144	1271	1030	822	401	190	38	6,104
	Lincoln—C			37.0	0	7	79	310	741	1113	1240	1000	794	377	172	32	5,865
	North Platte—A				7	11	120	425	846	1172	1271	1016	887	489	243	59	6,546
	Omaha—A				0	5	88	331	783	1166	1302	1058	831	389	175	32	6,160
	Valentine—C			33.6	11	10	145	461	891	1212	1361	1100	970	543	288	83	7,075
Nev.	Reno—A				27	61	165	443	744	986	1048	804	756	519	318	165	6,035
	Tonopah				0	5	96	422	723	995	1082	860	763	504	272	91	5,813
	Winnemucca—A			38.0	0	17	180	508	822	1085	1153	854	794	546	299	111	6,369
N. H.	Concord—A				11	57	192	527	849	1271	1392	1226	1029	660	316	82	7,612
N. J.	Atlantic City—C			42.3	0	0	29	230	507	831	905	829	729	468	189	24	4,741
	Cape May	98/99–31/32	34		1	2	38	221	527	852	936	876	737	459	188	33	4,870
	Newark—A				0	0	47	301	603	961	1039	932	760	450	148	11	5,252
	Sandy Hook	15/16–40/41	26	41.2	1	2	40	268	579	921	1016	973	833	499	206	31	5,369
	Trenton—C			42.0	0	0	55	285	582	930	1004	904	735	429	133	11	5,068
N. M.	Albuquerque—A				0	0	10	218	630	899	970	714	589	289	70	0	4,389
	Roswell—A			49.1	0	0	8	156	501	750	787	566	443	185	28	0	3,424
	Santa Fe	98/99–45/46	43		12	15	129	451	772	1071	1094	892	786	544	297	60	6,123
N. Y.	Albany—A				0	24	139	443	780	1197	1318	1179	989	597	246	50	6,962
	Albany—C			35.2	0	6	98	388	708	1113	1234	1103	905	531	202	31	6,319
	Binghamton—A				16	63	192	518	834	1228	1342	1215	1051	672	318	88	7,537
	Binghamton—C			34.7	0	36	141	428	735	1113	1218	1100	927	570	240	48	6,556
	Buffalo—A				16	30	122	433	753	1116	1225	1128	992	636	315	72	6,838
	Canton	06/07–45/46	40	29.5	27	61	219	550	898	1368	1516	1385	1139	695	340	107	8,305
	Ithaca	99/00–42/43	44	34.9	17	40	156	451	770	1129	1236	1156	978	606	292	83	6,914

Table 12-46 Average Monthly and Yearly Degree Days for Cities in the United States and Canada*†‡[36] (Continued)

State or province	Station	Years	No. of sea-sons	Avg winter temp§	July	Aug.	Sept.	Oct.	Nov.	Dec.	Jan.	Feb.	Mar.	Apr.	May	June	Yearly total
N. Y.	New York, La Guardia—A				0	0	28	250	546	908	992	907	760	447	141	10	4,989
	New York—C			41.1	0	0	39	263	561	908	995	904	753	456	153	18	5,050
	New York, Central Park Obs				0	0	31	250	552	902	1001	910	747	435	130	7	4,965
	Oswego—C			34.4	20	39	139	430	738	1132	1249	1134	995	654	355	90	6,975
	Rochester—A				9	34	133	440	759	1141	1249	1148	992	615	289	54	6,863
	Syracuse—A				0	29	117	396	714	1113	1225	1117	955	570	247	37	6,520
N. C.	Asheville—C			46.1	0	0	50	262	552	769	794	678	572	285	105	5	4,072
	Charlotte—A				0	0	7	147	438	682	704	577	449	172	29	0	3,205
	Hatteras—C				0	0	0	63	244	481	527	487	394	171	25	0	2,392
	Manteo	04/05–28/29	25		0	0	7	113	358	595	642	594	469	249	75	7	3,109
	Raleigh—A				0	0	16	149	438	701	732	613	477	202	41	0	3,369
	Raleigh—C			50.0	0	0	10	118	387	651	691	577	440	172	29	0	3,075
	Wilmington—A			54.6	0	0	0	73	288	508	533	463	347	104	7	0	2,323
N. D.	Bismarck—A			22.9	29	37	227	598	1098	1535	1730	1464	1187	657	355	116	9,033
	Devils Lake—C			21.7	47	61	276	654	1197	1668	1866	1576	1314	750	394	137	9,940
	Grand Forks	12/13–40/41 42/43–45/46	33		32	60	274	663	1160	1681	1895	1608	1298	718	359	123	9,871
	Williston—C			24.5	29	42	261	605	1101	1528	1705	1442	1194	663	360	138	9,068
Ohio	Cincinnati—A				0	6	77	295	648	973	1029	871	732	392	149	23	5,195
	Cincinnati—C			43.0	0	0	42	222	567	880	942	812	645	314	108	0	4,532
	Cincinnati, Abbe Obs				0	0	56	263	612	930	989	846	682	347	132	13	4,870
	Cleveland—A				0	10	75	340	699	1057	1132	1019	874	531	223	46	6,006
	Cleveland—C			37.2	0	9	60	311	636	995	1101	977	846	510	223	49	5,717
	Columbus—A			38.2	0	8	69	337	693	1032	1094	946	781	444	180	31	5,615
	Columbus—C			40.4	0	0	59	299	654	983	1051	907	741	408	153	22	5,277
	Dayton—A				0	5	74	324	693	1032	1094	941	781	435	179	39	5,597
	Sandusky—C			38.0	0	0	66	327	684	1039	1122	997	853	513	217	41	5,859
	Toledo—A			35.7	0	12	102	387	756	1119	1197	1056	905	555	245	60	6,394
Okla.	Broken Arrow	18/19–30/31	13		0	0	28	169	513	805	881	646	506	212	61	5	3,826
	Oklahoma City—A				0	0	14	154	480	769	865	650	490	182	40	3	644
	Oklahoma City—C			47.9	0	0	12	149	459	747	843	630	472	169	38	0	3,519
Ore.	Baker—C			35.2	25	47	255	518	852	1138	1268	972	837	591	384	200	7,087
	Medford—A				0	0	77	326	624	822	862	627	552	381	207	69	4,547
	Portland—A			44.3	25	22	116	319	585	750	856	658	570	396	242	93	4,632
	Portland—C			46.1	13	14	85	280	534	701	791	594	515	347	199	70	4,143
	Roseburg—C			46.7	14	10	98	288	531	694	744	563	508	366	223	83	4,122
Pa.	Erie—C			37.3	0	17	76	352	672	1020	1128	1039	911	573	273	55	6,116
	Harrisburg—A				0	0	69	308	630	964	1051	921	750	423	128	14	5,258
	Philadelphia—A			41.4	0	0	47	269	573	902	986	879	704	402	104	0	4,866
	Philadelphia—C			42.7	0	0	33	219	516	856	933	837	667	369	93	0	4,523
	Pittsburg, Allegheny—A				0	6	78	336	678	1004	1073	955	784	447	167	27	5,555
	Pittsburgh, Greater Pittsburgh—A			38.7	0	20	94	377	720	1057	1116	986	818	486	195	36	5,905
	Pittsburgh—C			40.9	0	0	56	298	612	924	992	879	735	402	137	13	5,048
	Reading—C			41.2	0	5	57	285	588	936	1017	902	725	411	123	11	5,060
	Scranton—C			37.7	0	18	115	389	693	1057	1141	1028	859	516	196	35	6,047
R.I.	Block Island—A			40.1	6	21	88	330	591	927	1026	955	865	603	335	96	5,843
	Narragansett Pier	98/99–17/18	20		1	26	121	366	691	1012	1113	1074	916	622	342	113	6,397
	Providence—A				0	26	107	381	672	1035	1125	1019	874	570	258	58	6,125
	Providence—C			37.5	0	7	68	330	624	986	1076	972	809	507	197	31	5,607
S.C.	Charleston—A			55.0	0	0	0	52	270	456	472	379	281	63	0	0	1,973
	Charleston—C			57.4	0	0	0	34	214	410	445	363	260	43	0	0	1,769
	Columbia—A				0	0	0	82	338	558	566	468	340	83	0	0	2,435
	Columbia—C			54.4	0	0	0	76	308	524	538	443	318	77	0	0	2,284
	Due West	21/22–31/32	11		0	0	9	142	393	594	651	491	411	158	39	2	2,890
	Greenville—A			49.2	0	0	10	131	411	648	673	552	442	161	32	0	3,060
S. D.	Huron—A				10	16	149	472	975	1407	1597	1327	1032	558	279	80	7,902
	Pierre	98/99–40/41 42/43–45/46	47		4	11	136	438	887	1317	1460	1253	971	516	238	52	7,283
	Rapid City—A				32	24	193	500	891	1218	1361	1151	1045	615	357	148	7,535
Tenn.	Chattanooga—A			47.8	0	0	24	169	477	710	725	588	467	179	45	0	3,384
	Knoxville—A			46.8	0	0	33	179	498	744	760	630	500	196	50	0	3,590
	Memphis—A			50.0	0	0	17	126	432	673	725	574	427	139	24	0	3,137
	Memphis—C			51.1	0	0	13	98	392	639	716	574	423	131	20	0	3,006
	Nashville—A				0	0	22	154	471	725	778	636	498	186	43	0	3,513
Texas	Abilene—A				0	0	5	98	350	595	673	479	344	113	0	0	2,657
	Amarillo—A				0	0	37	240	594	859	921	711	586	298	99	0	4,345
	Austin—A				0	0	0	30	214	402	484	322	211	50	0	0	1,713
	Brownsville—A				0	0	0	0	59	159	219	106	74	0	0	0	617
	Corpus Christi—A				0	0	0	0	113	252	330	192	118	6	0	0	1,011
	Dallas—A				0	0	0	53	299	518	607	432	288	75	0	0	2,272
	Del Rio—A				0	0	0	26	188	371	419	235	147	21	0	0	1,407
	El Paso—A				0	0	0	70	390	626	670	445	330	110	0	0	2,641
	Fort Worth—A				0	0	0	58	299	533	622	446	308	90	5	0	2,361
	Fort Worth, Amon Carter Field				0	0	0	57	299	524	619	432	326	81	0	0	2,338
	Galveston—A				0	0	0	0	132	286	362	249	176	28	0	0	1,233
	Galveston—C			61.7	0	0	0	0	131	271	356	247	176	30	0	0	1,211
	Houston—A			60.1	0	0	0	7	181	321	394	265	184	36	0	0	1,388
	Houston—C			61.0	0	0	0	0	162	303	378	240	166	27	0	0	1,276
	Palestine—C			57.1	0	0	0	45	260	440	531	368	265	71	0	0	1,980
	Port Arthur—A				0	0	0	20	218	349	406	274	211	39	0	0	1,517
	Port Arthur—C			61.2	0	0	0	8	170	315	381	258	181	27	0	0	1,340

Table 12-46 (Continued)

State or province	Station	Years	No. of sea-sons	Avg winter temp§	July	Aug.	Sept.	Oct.	Nov.	Dec.	Jan.	Feb.	Mar.	Apr.	May	June	Yearly total
Texas	San Antonio—A				0	0	0	25	201	374	462	293	190	34	0	0	1,579
	Taylor	01/02–40/41	40		0	0	2	56	234	462	494	375	214	64	8	0	1,909
Utah	Modena	00/01–45/46	46	36.3	6	11	156	499	832	1142	1190	944	816	567	338	97	6,598
	Salt Lake City—A			38.3	0	0	88	381	771	1039	1194	885	741	453	233	81	5,866
	Salt Lake City—C			40.0	0	0	61	330	714	995	1119	857	701	414	208	64	5,463
Vt.	Burlington—A				19	47	172	521	858	1308	1460	1313	1107	681	307	72	7,865
	Northfield	98/99–42/43	45		62	112	283	602	947	1389	1524	1384	1176	754	405	166	8,804
Va.	Cape Henry—C			49.2	0	0	0	120	366	648	698	636	512	267	60	0	3,307
	Lynchburg—A				0	0	49	236	531	809	846	722	584	289	82	5	4,153
	Norfolk—A				0	0	9	152	408	688	729	644	500	265	59	0	3,454
	Norfolk—C			49.3	0	0	5	118	354	636	679	602	464	220	41	0	3,119
	Richmond—A				0	0	33	210	498	791	828	708	550	271	66	0	3,955
	Richmond—C			47.0	0	0	31	181	456	750	787	675	529	254	57	0	3,720
	Wytheville	02/03–40/41	39		7	13	82	352	662	916	945	836	677	410	168	35	5,103
Wash.	North Head L.H. Reservation			46.4	239	205	234	341	486	636	704	585	598	492	406	285	5,211
	Seattle—C			46.3	49	45	134	329	540	679	753	602	558	396	246	107	4,438
	Seattle-Tacoma—A			45.1	75	70	192	412	633	781	862	675	636	477	307	155	5,275
	Spokane—A				17	28	205	508	879	1113	1243	988	834	561	330	146	6,852
	Tacoma—C			44.9	66	62	177	375	579	719	797	636	595	435	282	143	4,866
	Tatoosh Island—C			45.4	295	288	315	406	528	648	713	610	629	525	437	330	5,724
	Walla Walla—C				0	0	93	308	675	890	1023	748	564	338	171	38	4,848
	Yakima—A				0	7	150	446	807	1066	1181	862	660	408	205	53	5,845
W. Va.	Elkins—A			39.4	9	31	122	412	726	995	1017	910	797	477	224	53	5,773
	Parkersburg—C			42.9	0	0	56	272	600	896	949	826	672	347	119	13	4,750
Wis.	Green Bay—A			29.8	32	58	183	515	945	1392	1516	1336	1132	696	347	107	8,259
	La Crosse—A			30.5	11	20	152	447	921	1380	1528	1280	1035	552	250	74	7,650
	Madison—A				13	31	150	459	891	1302	1423	1207	1008	579	272	82	7,417
	Madison—C			31.4	10	30	137	419	864	1287	1417	1207	1011	573	266	79	7,300
	Milwaukee—A			29.0	20	32	134	428	831	1218	1336	1142	983	621	351	109	7,205
	Milwaukee—C			33.4	11	24	112	397	795	1184	1302	1117	961	606	335	100	6,944
	Wausau	15/16–40/41	26		26	58	216	568	982	1427	1594	1381	1147	680	315	100	8,494
Wyo.	Cheyenne—A				33	39	241	577	897	1125	1225	1044	1029	717	462	173	7,562
	Lander—A				7	23	244	632	1050	1383	1494	1179	1045	687	396	163	8,303
	Yellowstone Park	04/05–40/41	37		125	173	424	759	1079	1386	1464	1252	1165	841	603	334	9,605
Alta.	Calgary	1921–50	30		110	170	410	710	1110	1430	1530	1350	1200	770	460	270	9,520
	Edmonton	1921–50	30		90	180	440	750	1220	1660	1780	1520	1290	760	410	220	10,320
	Lethbridge	1921–50	30		60	100	350	620	1030	1330	1450	1290	1120	690	400	210	8,650
	Medicine Hat	1921–50	30		20	50	300	600	1070	1440	1590	1380	1130	620	320	130	8,650
B. C.	Prince George	1921–50	30		180	220	440	730	1100	1450	1560	1290	1070	730	470	260	9,500
	Prince Rupert	1921–50	30		270	240	340	510	680	860	910	810	790	650	500	350	6,910
	Vancouver	1921–45	25		60	60	200	410	620	790	850	710	650	460	290	130	5,230
	Victoria	1921–50	30		160	150	230	410	600	730	800	660	620	470	350	230	5,410
Man.	Brandon	1921–50	30		60	100	350	730	1290	1810	2010	1730	1440	820	420	170	10,930
	Churchill	1926–50	25		310	370	650	1110	1750	2300	2520	2270	2150	1590	1100	690	16,810
	Winnipeg	1921–50	30		40	70	300	690	1250	1770	2000	1710	1440	810	400	150	10,630
N. B.	Fredericton	1921–50	30		50	70	250	600	940	1410	1570	1410	1180	780	420	150	8,830
	Moncton	1939–50	12		40	80	240	580	890	1340	1510	1340	1180	830	460	210	8,700
	Saint John	1921–50	30		110	110	250	530	830	1250	1400	1270	1100	780	500	250	8,380
Nfld.	Corner Brook	1933–50	18		90	140	320	640	890	1230	1410	1360	1240	900	640	350	9,210
	Gander	1937–50	14		130	160	320	660	920	1230	1430	1320	1270	970	650	380	9,440
	Goose Bay	1941–51	10		130	220	440	840	1220	1740	2020	1710	1530	1110	770	410	12,140
	St. John's		19		170	160	320	580	820	1100	1270	1230	1160	910	680	380	8,780
N. W. T.	Aklavik	1926–50	25		280	460	800	1400	2040	2500	2580	2310	2280	1690	1050	480	17,870
	Fort Norman		25		170	350	700	1220	1940	2460	2550	2190	2040	1390	730	280	16,020
N. S.	Halifax	1921–50	30		50	70	180	470	740	1120	1260	1180	1050	760	480	210	7,570
	Sydney	1921–50	30		60	80	220	510	780	1130	1310	1280	1160	850	570	270	8,220
	Yarmouth	1921–50	30		110	120	230	480	720	1040	1180	1100	1010	750	510	270	7,520
Ont.	Fort William	1921–50	30		90	120	350	690	1140	1590	1780	1550	1360	890	550	240	10,350
	Hamilton		21		10	10	120	430	760	1130	1270	1160	1000	620	320	60	6,890
	Kapuskasing	1921–50	30		110	170	410	780	1270	1780	2060	1770	1570	1030	600	240	11,790
	Kingston	1921–50	30		30	40	160	500	820	1250	1420	1290	1110	710	380	100	7,810
	London	1921–50	30		20	40	150	490	840	1200	1320	1210	1040	650	330	90	7,380
	Ottawa	1921–50	30		30	60	210	590	960	1480	1640	1480	1230	730	340	80	8,830
	Toronto	1921–50	30		20	30	140	460	770	1130	1260	1160	1020	640	320	70	7,020
P. E. I.	Charlottetown	1921–50	30		40	60	200	510	820	1230	1430	1340	1180	840	510	220	8,380
Que.	Arvida	1931–50	20		60	110	330	690	1100	1670	1880	1660	1410	880	480	170	10,440
	Montreal	1921–50	30		10	40	180	530	890	1370	1540	1370	1150	700	300	50	8,130
	Quebec City	1921–50	30		20	70	250	610	990	1470	1640	1460	1250	810	400	100	9,070
	Sherbrooke	1921–50	30		20	70	240	590	920	1400	1560	1410	1190	750	370	90	8,610
Sask.	Prince Albert	1921–50	30		70	140	410	780	1350	1870	2060	1750	1500	850	440	210	11,430
	Regina	1921–50	30		70	110	370	750	1290	1740	1940	1680	1420	790	420	190	10,770
	Saskatoon	1921–50	30		60	110	380	760	1320	1790	1990	1710	1440	800	420	180	10,960
	Swift Current	1921–50	30		50	90	340	680	1170	1550	1710	1490	1260	730	400	190	9,660
Y. T.	Dawson	1921–50	30		170	320	660	1170	1890	2410	2510	2160	1830	1100	570	250	15,040

* Data for U. S. cities are from a publication of the U. S. Weather Bureau (*Monthly Normal Temperatures, Precipitation and Degree Days,* 1954) for the period 1921 to 1950, inclusive. Those U. S. cities for which years are given in Col. 3 are not listed in the above publication; the data were computed by the U. S. Weather Bureau in 1946 and 1947 in accordance with the requirements of the National Joint Committee on Weather Statistics.

† Data for airport stations, A, and city stations, C, are both given where available.

‡ Data for Canadian cities were computed by the Meteorological Division, Department of Transport from normal monthly mean temperatures, adjusted for the summer months by a method described by H. C. S. Thom ("The Rational Relationship between Heating Degree Days and Temperature," *Monthly Weather Review*, 82, Jan. 1954).

§ For period October to April, inclusive.

Table 12-47 Outside Design Conditions for the United States and Canada, °F[37]

State or province and city	Winter DB	Summer DB	Summer WB
Alabama			
Anniston	10	95	78
Birmingham	10	95	78
Gadsden	10	95	78
Mobile	20	90	80
Montgomery	20	95	78
Tuscaloosa	10	95	78
Alaska			
Anchorage	−24	—	—
Barrow	−48	—	—
Bethel	−43	—	—
Cordova	−13	—	—
Fairbanks	−57	—	—
Juneau	−5	—	—
Ketchikan	4	—	—
Kodiak	4	—	—
Kotzebue	−46	—	—
Nome	−36	—	—
Seward	−4	—	—
Sitka	2	—	—
Arizona			
Bisbee	30	100	72
Flagstaff	−5	85	61
Globe	30	105	76
Nogales	30	105	72
Phoenix	35	105	76
Tucson	30	100	72
Winslow	−5	95	65
Yuma	40	110	78
Arkansas			
Bentonville	0	95	76
Fort Smith	5	95	76
Hot Springs	10	95	78
Little Rock	10	95	78
Pine Bluff	10	95	78
Texarkana	10	100	78
California			
Bakersfield	30	105	70
El Centro	35	110	78
Eureka	30	90	65
Fresno	30	105	74
Long Beach	35	90	70
Los Angeles	40	90	70
Montague	15	95	70
Needles	25	115	—
Oakland	30	80	65
Pasadena	40	95	70
Red Bluff	15	100	70
Sacramento	30	100	72
San Bernardino	30	105	72
San Diego	45	80	68
San Francisco	35	80	65
San Jose	40	90	70
Colorado			
Boulder	−15	95	64
Colorado Springs	−10	95	65
Denver	−10	95	64
Durango	−5	95	65
Fort Collins	−15	95	65
Grand Junction	−5	95	65
Leadville	−10	95	64
Pueblo	−15	95	65
Connecticut			
Bridgeport	0	85	75
Hartford	0	90	75
New Haven	0	85	75
New London	5	85	75
Norwalk	0	85	75
Torrington	0	90	75
Waterbury	0	90	75
Delaware			
Dover	10	90	78
Milford	10	90	78
Wilmington	5	90	78
Dist. of Columbia			
Washington	10	90	78
Florida			
Apalachicola	25	95	80
Fort Myers	40	95	78
Gainesville	30	95	78
Jacksonville	30	95	78
Key West	55	100	78
Miami	45	90	79
Orlando	35	90	78
Pensacola	25	95	78
Tallahassee	25	95	78
Tampa	35	95	78
Georgia			
Athens	10	95	76
Atlanta	10	95	76
Augusta	20	100	76
Brunswick	25	95	78
Columbus	20	100	76
Macon	20	95	78
Rome	10	95	76
Savannah	25	95	78
Way Cross	25	95	78
Idaho			
Boise	−10	95	65
Idaho Falls	−15	90	65
Lewiston	−10	95	65
Pocatello	−15	90	65
Twin Falls	−15	95	65
Illinois			
Aurora	−10	95	75
Bloomington	−10	95	76
Cairo	0	100	78
Champaign	−10	95	77
Chicago	−10	95	75
Danville	−10	95	77
Decatur	−10	95	77
Elgin	−15	95	78
Joliet	−10	95	76
Moline	−10	95	76
Peoria	−15	95	76
Rockford	−15	95	78
Rock Island	−10	95	76
Springfield	−10	95	77
Urbana	−10	95	77
Indiana			
Elkhart	−10	95	75
Evansville	−5	95	78
Fort Wayne	−5	95	75
Indianapolis	−10	95	76
Lafayette	−10	95	76
South Bend	−10	95	75
Terre Haute	−5	95	78
Iowa			
Burlington	−10	95	78
Cedar Rapids	−15	95	78
Charles City	−20	95	75
Clinton	−15	95	78
Council Bluffs	−15	100	78
Davenport	−10	95	78
Des Moines	−15	95	78
Dubuque	−15	95	78
Fort Dodge	−15	95	78
Keokuk	−15	95	78
Marshalltown	−15	95	78
Sioux City	−15	95	78
Waterloo	−15	95	78
Kansas			
Atchison	−10	100	76
Concordia	−10	95	78
Dodge City	−10	95	78
Iola	−5	100	75
Leavenworth	−10	100	76
Salina	−10	100	78
Topeka	−10	100	78
Wichita	−5	100	75
Kentucky			
Bowling Green	0	95	78
Frankfort	0	95	78
Hopkinsville	0	95	78
Lexington	0	95	78
Louisville	0	95	78
Owensboro	0	95	78
Shelbyville	0	95	78
Louisiana			
Alexandria	20	95	78
Baton Rouge	20	95	80
New Orleans	25	95	80
Shreveport	15	95	78
Maine			
Augusta	−15	85	73
Bangor	−20	85	73
Bar Harbor	−10	85	73
Belfast	−10	85	73
Eastport	−10	85	70
Lewiston	−10	85	73
Millinocket	−15	85	73
Orono	−20	85	70
Portland	−10	85	73
Presque Isle	−20	85	73
Rumford	−15	85	73
Maryland			
Annapolis	10	90	78
Baltimore	10	90	78
Cambridge	10	90	78
Cumberland	0	90	75
Frederick	5	90	78
Frostburg	−5	90	75
Salisbury	10	90	78
Massachusetts			
Amherst	−5	90	75
Boston	0	85	74
Fall River	0	85	75
Fitchburg	−5	90	75
Framingham	−5	85	75
Lawrence	−5	85	74
Lowell	−5	85	74
Nantucket	0	85	75
New Bedford	0	85	75
Pittsfield	−10	90	75
Plymouth	0	85	75
Springfield	−5	90	75
Worcester	−5	90	75
Michigan			
Alpena	−10	90	75
Ann Arbor	−5	90	75
Big Rapids	−5	90	75

Table 12-47 (Continued)

State or province and city	Winter DB	Summer DB	Summer WB
Michigan (cont)			
Cadillac	−10	90	75
Calumet	−20	80	73
Detroit	−5	90	75
Escanaba	−20	85	74
Flint	−10	90	75
Grand Haven	−5	90	75
Grand Rapids	−5	90	74
Houghton	−20	80	73
Kalamazoo	−5	90	75
Lansing	−10	90	75
Ludington	−5	90	75
Marquette	−15	80	73
Muskegon	−5	90	74
Port Huron	−10	90	75
Saginaw	−10	90	75
Sault Ste. Marie	−20	80	71
Minnesota			
Alexandria	−25	85	74
Duluth	−25	80	71
Minneapolis	−25	90	76
Moorhead	−30	95	75
St. Cloud	−25	90	76
St. Paul	−25	90	75
Mississippi			
Biloxi	25	90	80
Columbus	10	95	78
Corinth	5	95	78
Hattiesburg	20	95	80
Jackson	15	95	78
Meridian	15	95	79
Natchez	15	95	78
Vicksburg	15	95	78
Missouri			
Columbia	−10	100	78
Hannibal	−10	95	77
Kansas City	−10	100	76
Kirksville	−10	95	78
St. Joseph	−10	100	76
St. Louis	−5	95	78
Springfield	−5	100	77
Montana			
Anaconda	−30	85	59
Billings	−30	90	66
Butte	−30	85	59
Great Falls	−40	90	63
Havre	−40	95	70
Helena	−40	90	63
Kalispell	−30	90	63
Miles City	−35	95	69
Missoula	−30	90	63
Nebraska			
Grand Island	−15	100	75
Hastings	−15	100	75
Lincoln	−15	95	78
Norfolk	−15	95	78
North Platte	−15	100	73
Omaha	−15	100	78
Valentine	−20	95	78
York	−15	95	78
Nevada			
Elko	−10	95	63
Las Vegas	10	110	71
Reno	5	95	65
Tonopah	5	90	63
Winnemucca	−10	95	65
New Hampshire			
Berlin	−15	85	73
Claremont	−15	85	73
Concord	−10	85	73
Franklin	−15	85	73
Hanover	−15	85	73
Keene	−10	85	73
Manchester	−10	85	74
Nashua	−10	85	74
Portsmouth	−5	85	74
New Jersey			
Asbury Park	5	90	78
Atlantic City	10	90	78
Bayonne	0	90	75
Belvidere	0	90	75
Bloomfield	0	90	75
Bridgeton	5	90	78
Camden	5	90	78
East Orange	0	90	75
Elizabeth	0	90	75
Jersey City	0	90	75
Newark	0	90	76
New Brunswick	5	90	75
Paterson	0	90	75
Phillipsburg	0	90	75
Trenton	0	90	78
New Mexico			
Albuquerque	10	95	65
El Morro	0	85	65
Raton	−5	95	65
Roswell	5	100	71
Santa Fe	5	90	65
Tucumcari	5	95	70
New York			
Albany	−10	90	74
Auburn	−10	90	74
Binghamton	−5	90	72
Buffalo	−5	85	73
Canton	−20	85	73
Cortland	−10	90	74
Elmira	−5	90	73
Glens Falls	−15	90	73
Ithaca	−5	90	73
Jamestown	−5	90	74
Lake Placid	−15	90	73
New York	5	90	76
Niagara Falls	−5	85	73
Ogdensburg	−20	85	73
Oneonta	−10	90	73
Oswego	−5	90	74
Port Jervis	0	90	75
Rochester	−5	90	74
Schenectady	−10	90	74
Syracuse	−10	90	74
Watertown	−15	85	73
North Carolina			
Asheville	5	90	75
Charlotte	15	95	78
Greensboro	10	90	76
Hatteras	20	90	80
New Bern	20	95	78
Raleigh	15	95	78
Salisbury	10	90	78
Wilmington	20	90	81
Winston-Salem	10	90	76
North Dakota			
Bismarck	−30	95	73
Devils Lake	−30	90	70
Dickinson	−30	95	70
Fargo	−30	95	75
Grand Forks	−30	90	72
Jamestown	−30	95	73
Minot	−35	90	71
Pembina	−35	90	73
Williston	−35	90	73
Ohio			
Akron	−5	90	75
Cincinnati	−5	95	78
Cleveland	−5	90	75
Columbus	−5	90	76
Dayton	−5	90	76
Lima	−5	90	75
Marion	−5	90	75
Sandusky	−5	90	75
Toledo	−5	90	75
Warren	−5	90	75
Youngstown	−5	90	75
Oklahoma			
Ardmore	5	100	78
Bartlesville	−5	100	77
Guthrie	0	100	77
Muskogee	0	95	79
Oklahoma City	0	100	77
Tulsa	0	100	77
Waynoka	−5	105	75
Oregon			
Arlington	5	95	68
Baker	−15	90	66
Eugene	15	90	68
Medford	20	95	68
Pendleton	−10	90	66
Portland	10	85	68
Roseburg	20	90	66
Salem	15	90	68
Wamic	0	90	66
Pennsylvania			
Altoona	−5	90	75
Bethlehem	0	90	75
Coatesville	5	90	75
Erie	−5	85	74
Harrisburg	5	90	75
New Castle	−5	90	75
Oil City	−5	90	75
Philadelphia	5	90	78
Pittsburgh	−5	90	75
Reading	5	90	75
Scranton	0	90	75
Warren	−5	90	75
Williamsport	−5	90	74
York	5	90	75
Rhode Island			
Block Island	5	85	75
Bristol	0	90	75
Kingston	0	85	75
Pawtucket	0	90	75
Providence	0	90	75
South Carolina			
Charleston	20	90	80
Columbia	20	95	78
Florence	20	95	79
Greenville	10	95	75
Spartanburg	10	95	78

Table 12-47 Outside Design Conditions for the United States and Canada, °F[37] (Continued)

State or province and city	Winter DB	Summer DB	Summer WB
South Dakota			
Aberdeen	−25	95	75
Huron	−20	100	75
Pierre	−20	95	73
Rapid City	−20	95	70
Sioux Falls	−20	95	75
Watertown	−25	95	73
Tennessee			
Chattanooga	10	95	76
Jackson	5	95	78
Johnson City	0	95	78
Knoxville	5	95	75
Memphis	5	95	78
Nashville	5	95	78
Texas			
Abilene	5	95	74
Amarillo	0	95	72
Austin	15	100	78
Brownsville	30	95	80
Corpus Christi	25	95	80
Dallas	10	100	78
Del Rio	20	100	78
El Paso	20	100	69
Fort Worth	10	100	78
Galveston	25	95	80
Houston	20	95	80
Palestine	10	100	78
Port Arthur	20	95	80
San Antonio	20	100	78
Waco	10	100	78
Utah			
Logan	−10	95	65
Milford	−5	95	66
Ogden	−5	90	65
Salt Lake City	0	95	65
Vermont			
Bennington	−10	90	73
Burlington	−15	90	73
Montpelier	−20	90	73
Newport	−20	85	73
Northfield	−20	90	73
Rutland	−15	90	73
Virginia			
Cape Henry	15	90	78
Charlottesville	10	90	78
Danville	10	90	78
Lynchburg	10	90	76
Norfolk	15	90	78
Petersburg	10	90	78
Richmond	10	90	78
Roanoke	5	90	76
Wytheville	5	90	76
Washington			
Aberdeen	20	85	64
Bellingham	10	80	65
Everett	15	80	65
North Head	20	80	65
Olympia	15	80	64
Seattle	15	80	65
Spokane	−15	80	65
Tacoma	15	80	64
Tatoosh Island	20	80	65
Walla Walla	−10	90	65
Wenatchee	−10	90	65
Yakima	−5	90	67
West Virginia			
Bluefield	0	95	75
Charleston	0	90	75
Elkins	−5	90	73
Fairmont	0	90	75
Huntington	0	90	76
Martinsburg	0	90	75
Parkersburg	0	90	75
Wheeling	−5	90	75
Wisconsin			
Ashland	−25	80	71
Beloit	−15	95	78
Eau Claire	−20	90	75
Green Bay	−20	90	73
La Crosse	−20	95	75
Madison	−20	90	75
Milwaukee	−15	90	75
Oshkosh	−20	90	75
Sheboygan	−20	90	75
Wyoming			
Casper	−25	90	62
Cheyenne	−20	90	62
Lander	−30	90	65
Sheridan	−30	90	65
Yellowstone Park	−35	85	62
Alberta			
Banff	−30	—	—
Camrose	−35	—	—
Calgary	−30	90	66
Cardston	−30	—	—
Edmonton	−35	90	68
Grande Prairie	−40	—	—
Hanna	−35	—	—
Jasper	−30	—	—
Lethbridge	−30	—	—
Lloydminster	−40	—	—
McMurray	−40	—	—
Medicine Hat	−35	90	65
Red Deer	−35	—	—
Taber	−35	—	—
Wetaskiwin	−35	—	—
British Columbia			
Chilliwack	5	—	—
Courtenay	10	—	—
Dawson Creek	−40	—	—
Estevan Point	15	—	—
Fort Nelson	−40	—	—
Hope	0	—	—
Kamloops	−20	—	—
Kimberly	−25	—	—
Lytton	−5	—	—
Nanaimo	10	—	—
Nelson	−10	—	—
Penticton	−5	—	—
Port Alberni	10	—	—
Prince George	−30	—	—
Prince Rupert	10	—	—
Princeton	−15	—	—
Revelstoke	−25	—	—
Trail	−10	—	—
Vancouver	10	80	67
Vernon	−15	—	—
Victoria	15	—	—
Westview	10	—	—
Labrador			
Goose Bay	−25	—	—
Manitoba			
Boissevain	−35	—	—
Brandon	−30	—	—
Churchill	−40	—	—
Dauphin	−35	—	—
Flin Flon	−40	—	—
Minnedosa	−35	—	—
Neepawa	−35	—	—
La Prairie	−30	—	—
Swan River	−35	—	—
The Pas	−40	—	—
Winnipeg	−30	90	71
New Brunswick			
Bathurst	−10	—	—
Campbellton	−10	—	—
Chatham	−10	—	—
Edmunston	−15	—	—
Fredericton	−5	90	75
Moncton	−10	—	—
Saint John	−5	80	67
Woodstock	−15	—	—
Newfoundland			
Corner Brook	0	—	—
Gander	−5	—	—
Grand Falls	−5	—	—
St. John's	0	—	—
Northwest Territories			
Aklavik	−45	—	—
Fort Norman	−40	—	—
Frobisher	−50	—	—
Resolute	−40	—	—
Yellowknife	−50	—	—
Nova Scotia			
Bridgewater	0	—	—
Dartmouth	0	—	—
Halifax—C*	5	80	75
Halifax—A*	0	80	75
Kentville	0	—	—
New Glasgow	0	—	—
Spring Hill	−5	—	—
Sydney	0	85	67
Truro	0	—	—
Yarmouth	5	—	—
Ontario			
Bancroft	−20	—	—
Barrie	−5	—	—
Belleville	−10	—	—
Brampton	−5	—	—
Brantford	−5	—	—
Brockville	−15	—	—
Chatham	0	—	—
Cobourg	−10	—	—
Collingwood	0	—	—
Cornwall	−15	—	—
Ear Falls	−35	—	—
Fort Frances	−30	—	—
Fort William	−25	85	70
Galt	−5	—	—
Geraldton	−35	—	—
Goderich	0	—	—
Guelph	−5	—	—
Hamilton	0	—	—
Haileybury	−25	—	—
Hanover	−5	—	—
Huntsville	−15	—	—

Table 12-47 (Continued)

State or province and city	Winter DB	Summer DB	Summer WB
Ontario (cont)			
Kapuskasing	−30	—	—
Kenora	−35	—	—
Kingston	−10	—	—
Kirkland Lake	−25	—	—
Kitchener	−5	—	—
Lindsay	−15	—	—
London	0	—	—
Moonsonee	−35	—	—
Newmarket	−5	—	—
Niagara Falls	0	—	—
North Bay	−20	85	69
Orillia	−10	—	—
Oshawa	−5	—	—
Ottawa	−15	90	75
Owen Sound	0	—	—
Parry Sound	−15	—	—
Pembroke	−20	—	—
Peterborough	−10	—	—
Port Arthur	−25	—	—
Port Colborne	0	—	—
Renfrew	−20	—	—
St. Catharines	0	—	—
St. Thomas	0	—	—
Sarnia	0	—	—
Sault Ste. Marie	−10	85	75
Simcoe	0	—	—
Sioux Lookout	−35	—	—
Smith Falls	−15	—	—
Stratford	−5	—	—
Sudbury	−20	—	—
Timmins	−25	—	—
Toronto—C*	0	90	75
Toronto—A*	−5	90	75
Trenton	−10	—	—
Walkerton	−5	—	—
Welland	0	—	—
Windsor	0	95	74
Woodstock	−5	—	—
Prince Edward Island			
Charlottetown	−5	80	69
Summerside	−5	—	—
Quebec			
Amos	−25	—	—
Arvida	−20	—	—
Asbestos	−15	—	—
Chibougamau	−30	—	—
Chicoutimi	−20	—	—
Dorval	−10	—	—
Drummondville	−15	—	—
Fort Chimo	−40	—	—
Gaspe	−10	—	—
Granby	−15	—	—
Harrington Harbour	−15	—	—
Joliette	−15	—	—
Knob Lake	−40	—	—
Lac Megantic	−15	—	—
La Tuque	−25	—	—
Magog	−15	—	—
Mont Joli	−10	—	—
Mount Laurier	−20	—	—
Montreal—C*	−10	90	75
Montreal—A*	−10	90	75
Noranda	−25	—	—
Port Harrison	−40	—	—
Quebec	−15	85	75
Rimouski	−10	—	—
Riviere Du Loup	−10	—	—
Rouyn	−25	—	—
Ste. Agathe	−15	—	—
St. Hyacinthe	−15	—	—
St. Jerome	−15	—	—
St. Johns	−15	—	—
Seven Islands	−20	—	—
Shawinigan Falls	−15	—	—
Sherbrooke	−15	85	—
Sorel	−15	—	—
Thetford Mines	−15	—	—
Three Rivers	−15	—	—
Val D'Or	−25	—	—
Valley Field	−15	—	—
Victoriaville	−15	—	—
Saskatchewan			
Biggar	−35	—	—
Estevau	−35	—	—
Humbot	−40	—	—
Moose Jaw	−35	—	—
Moosomin	−35	—	—
Nipawin	−40	—	—
North Battleford	−35	—	—
Prince Albert	−40	—	—
Regina	−35	90	71
Saskatoon	−40	90	70
Shaunavon	−35	—	—
Swift Current	−35	—	—
Uranium City	−45	—	—
Weyburn	−35	—	—
Yorkton	−35	—	—
Yukon Territory			
Dawson	−55	—	—
Whitehorse	−45	—	—

* C—city; A—airport.

The calculation of X over the period under consideration is given only approximately by Eq. **9**, when the value of $H.L.$ used is the calculated maximum load or design heating load for the building. This is the value commonly used since it is usually available from the heating system design calculations. But this procedure will usually lead to values of X that are too high. In the first place, the design wind speed on which such calculations are based is usually higher than the average wind speed for the period so that infiltration losses may be overestimated. In addition, there will usually be many sources of heat not taken into account in the estimation of maximum heat demand. These sources may actually contribute to the heating and thus may decrease the fuel or energy required for heating without necessarily affecting the maximum heat demand. The more important sources are solar radiation, people, lights, and equipment.

Improved estimates of fuel requirements will result if $H.L.$ can be recalculated so as to be more representative of *average* conditions over the heating period rather than the *maximum* conditions used in maximum heating load estimation.

There is a practical limit to the refinement in calculations which should be attempted, since there will always be some factors that can only be approximately known. One way of treating internal heat gains has been proposed by the National Bureau of Standards.[7] The weather, for example, which very largely determines heat losses, will normally be described in only approximate terms of average values of a limited number of weather elements. There is also a limit to the accuracy of prediction of the weather elements over the period of interest, which is always in the future. For these and other reasons, including such unpredictable factors as the way in which windows and doors will be opened and closed, the determination of $H.L.$ or of X must always remain an estimate only.

The assumption of an October 1–May 1 heating season is reasonably accurate in the well-populated New York–Chicago zone, but heating may be required over a longer period for points as far north as Minneapolis and over a shorter period south of Washington, D.C. Consequently, values taken from Table 12–46 may have to be adjusted accordingly.

When data on degree days are available, these may be used in the calculation of average outdoor temperature, t_o, for use in Eq. **9** or **10**.

Example: Find t_o for use in Eq. **9** for a city not listed in Table 12-46. Given: 4400 degree days, 212 heating days.

$$t_o = 65 - (4400/212) = 44.2 \text{ F}$$

Example: Estimate the fuel that will be used per year for heating a residence with gas, assuming a calculated hourly heat loss, $H.L.$, of 60,000 Btu based on −10 F. The design temperatures are −10 F and 72 F. The normal heating season is 210 days, and the average outdoor temperature during the heating season, t_o, is 35.1 F. The utilization efficiency is 70 per cent. The heating plant is thermostatically controlled, and an indoor temperature, t_i, of 65 F is maintained from 11 p.m. to 7 a.m.

Solution: The average indoor temperature, t_i (for Eq. **10**), is:

$$t_i = \frac{(72 \times 16) + (65 \times 8)}{24} = 69.7 \text{ F}$$

Substituting in Eq. **10**, the seasonal fuel requirement is:

$$F = \frac{60{,}000(69.7 - 35.1)(24 \times 210)}{0.70[72 - (-10)]} = 182 \text{ MMBtu}$$

It should be noted that savings from night setback may not result as calculated.

It may be noted that a portion of Eq. **10**, $(t_i - t_o)N$, when divided by 24, corresponds exactly to the degree days for the period, °D, provided that t_i = 65 F, the usual basis for calculating degree days. Substituting this relationship in Eq. **10** yields Eq. **10a**:

$$F = \frac{(H.L.)(°D) \times 24}{E\Delta t} \qquad \textbf{(10a)}$$

All terms are defined under Eqs. **8** and **9**.

Since this substitution is in accordance with the basis for degree-day calculations, assuming that t_i = 65 F, some allowance is thus made for the fact that F will frequently be overestimated when $H.L.$ is taken as the maximum or design heat load.

Degree-Day Method. This method is based on consumption data which have been taken from residences in operation, and the results computed on a degree-day basis. While this method may not be as theoretically precise as the calculated heat loss method, it is considered by many to be of more practical value.

The amount of heat required in a residence depends upon the outdoor temperature if other variables are eliminated. Theoretically it is proportional to the difference between the outdoor and indoor temperatures. The A.G.A. determined from records in the heating of residences that the gas consumption varies directly as the degree days, or as the difference between 65 F and the mean outdoor temperature. In other words, on a day when the mean temperature was 20° below 65 F, twice as much gas was consumed as on a day when the temperature was 10° below 65 F. For any one day, when the mean temperature is less than 65 F, there are as many degree days as there are degrees difference in temperature between the mean temperature for the day and 65 F. Degree days may be calculated on other than the 65 F base when the calculations are for use mainly for warehouse and other industrial spaces in which temperatures to be maintained are considerably below the 68 to 72 F range.

Studies made by the National District Heating Association of the metered steam consumption of 163 buildings located in 22 different cities and served with steam from a district heating company substantiate the approximate correctness of the 65 F base.

Table 12–46 lists the average number of degree days that have occurred over a long period of years, by months, and the yearly totals for various cities in the United States and Canada. The number of degree days for U. S. cities was calculated by taking the difference between 65 F and the daily mean temperature, computed as half the total of the daily maximum and the daily minimum temperatures. The monthly averages were obtained by adding daily degree days for each month each year and dividing by the number of days in the month, then totaling the respective calendar monthly averages for the number of years indicated and dividing by the number of years. The total or long term yearly average degree-day value is the summation of the 12 monthly averages. Degree days for Canadian cities were supplied by the Canadian Meteorological Division of the Department of Transport. They were computed from the mean temperature normals on record for the various stations; for the method used, see the footnotes to Table 12–46.

Any attempt to apply the degree-day method of estimating fuel consumption for less than one month would be of very little value. It should be noted that this method of calculation is based on a long-term average and cannot be expected to coincide with any single year in calculating fuel requirement. Individual yearly degree-day calculations may vary as much as 20 per cent above and below the long-term average.

If the degree days occurring each day are totaled for a reasonably long period, the fuel consumption during that period compared with another period may be assumed to be in direct proportion to the number of degree days in the two periods. Consequently, for a given installation the fuel consumption can be calculated in terms of fuel used per degree day for any sufficiently long period and compared with similar ratios for other periods to determine the relative operating efficiencies with the outdoor temperature variable eliminated.

Such results should be used with some reservation since it is possible to have wide variations, for example, as between early and late winter periods.

It has been proposed that recent home construction using heavy levels of insulation and the increasing use of appliances within the home help effect small heat transmission losses and considerable heat gains, respectively. These phenomena, which help to justify the reduction of the degree-day base to below 65 F, necessitate the use of a correction factor on the reported degree days to make allowance for the increased usefulness of heat gains within the home.

Computation and Application. The probable fuel consumption by the degree-day method is given by Eq. **11**:

$$F = (U.C.F.) \times N_b \times (°D) \times C_f \qquad \textbf{(11)}$$

where:

F = fuel consumption for the estimate period
$U.C.F.$ = unit consumption factor or quantity of fuel used per (degree day) (*building load unit*), from Table 12–48a
N_b = number of *building load units* (when available, use calculated hourly heat loss instead of actual amount of radiation installed)
$°D$ = number of degree days for the estimate period
C_f = temperature-correction factor, from Table 12–48b

Table 12-48a Unit Consumption Factors, *U.C.F.*

(operating data based on 0 F outdoor temperature, 70 F indoor temperature)

Fuel and units	Utilization efficiency, E, per cent (see Table 12-44)		
	60	70	80
	Unit consumption per degree day per 1000 Btu per hr design heat loss		
Gas, therms*	0.00572	0.00490	0.00429
Oil, gal†	.00405	.00347	.00304
Coal, lb‡	0.0476	0.0408	0.0357

* One therm equals 100,000 Btu.
† Based on a heating value of 141,000 Btu per gal.
‡ Based on a heating value of 12,000 Btu per lb.

Values of N_b depend on the particular residence for which the estimate is being prepared and must be found by surveying plans, by observation, or by measurement of the residence. The quantity of fuel used per degree day in a given heating plant can be reduced to a unit basis in terms of quantity of fuel or steam per degree day per thousand Btu hourly heat loss at design conditions. Since unit consumption factors are based on a design temperature difference of 70°F, the correction factor, C_f, must be applied for any other temperature difference (see Table 12–48b).

The calculated heat loss or the heating capacity of the installed radiation may be used as the building load unit in Eq. **11**. The use of the heating capacity of the installed radiation is of questionable value when referring to heat-transfer surfaces used in warm air furnace or central air-conditioning systems. Where steam or hot water radiation is already installed, care should be exercised when using the installed radiation as the basis for estimating, since actual installed radiation may differ considerably from the exact radiation requirements. In view of all these considerations, it is believed that the unit based on *thousands of Btu of hourly calculated heat loss for the design hour* is probably the most desirable.

Although the relationship among estimated fuel consumption, design heat loss, design temperature difference, and degree days is nonlinear, factors for gas in Table 12–48a, corrected if necessary according to Table 12–48b, are satisfactory for 3500 to 6500 degree days per heating season. In regions with fewer than 3500 degree days, the unit gas consumption is higher than given; with more than 6500 degree days, the unit gas consumption is less than given. Ten per cent addition or deduction, respectively, is recommended in these cases. This table cannot be used for industrial buildings in which low inside temperatures are maintained.

Table 12-48b Correction Factors, C_f, for Outdoor Design Temperatures*

(for use in Eq. **11**)

Outdoor design temp, °F	−20	−10	0	+10	+20
Correction factor, C_f	0.778	0.875	1.000	1.167	1.400

* These multipliers are required to correct for design heat losses in warm areas that are calculated on other than a 0° to 70°F temperature difference. For example, the same size home located in the North would have a lower design heat loss in the South. If the degree days happened to be the same in the North and South during a particular year, the estimated annual fuel consumption would not turn out to be the same unless the correction factor, C_f, was applied to compensate for differences in calculating the heat loss.

For utilization efficiencies other than those given in Table 12–48a, interpolation or extrapolation is necessary.

Statements have appeared from time to time questioning the choice of seasonal gas heating efficiencies in the range of 70 to 80 per cent, although these figures have long been included in recommendations in the *ASHRAE Guide*. In an unpublished report* of the University of Illinois, based on seven research houses, the indicated "seasonal efficiency" ranged between 60 and 90 per cent, with values between 70 and 75 per cent occurring most frequently. Since these results were obtained in research houses, the results do not necessarily reflect all field conditions. However, where field data are available on seasonal fuel consumption, F, that value may be inserted into Eq. **10a** to calculate the value of E.

Using data secured for groups of houses in Elizabeth, N.J., Goldsboro, N.C., Washington, D.C., and Butte, Mont., the fuel consumption was calculated by means of Eq. **10a**, using an efficiency of 75 per cent. This resulted in a minimum error in the calculated value of F when compared with the actual fuel consumption, F, measured in the above groups of houses. Although this procedure is not technically an exact one, it appears to be sufficiently accurate to substantiate the use of 75 per cent for the seasonal efficiency of gas-fired systems.

Example: Estimate the gas (for a gas-designed heating system) required to heat a residence located in Chicago, Ill. The heating season has 6310 degree days (from Table 12–46). The design heat loss of the house is 60,000 Btu per hr based on design temperatures of −10 F outdoors and 70 F indoors.

Solution: Table 12–44 indicates that the expected efficiency of utilization will be between 70 and 80 per cent. In this problem utilization efficiency of 75 per cent is assumed. From Table 12–48a the gas consumption for a design outdoor temperature of 0 F is interpolated to be 0.00460 therm per (degree day) (1000 Btu per hr design heat loss). From Table 12–48b the correction factor is 0.875 for −10 F outdoor design temperature. Hence, 0.875 × 0.00460 = 0.00403.

Substituting in Eq. **11**:

$$F = 0.00460 \times 60 \times 6310 \times 0.875 = 1525 \text{ therms}$$

The A.G.A. Method. The degree-day method has been in use since the 1930s, but the basic data had not been verified in light of changing home construction materials and designs. Therefore, A.G.A. secured the actual meter records (both gas and electric in order to study total energy input to houses) from random-selected groups of similar houses. There were more than 150 homes located in different climatic areas of the U.S., ranging in outdoor design temperature from +30 to −30 F. The groups of houses were large enough to satisfy statistical requirements for accuracy. Monthly data for each house were obtained for a minimum of two years and for as long as seven years.

* W. S. Harris, and others. *Estimating Energy Reqvirements for Residential Heating,* 1965.

One study in a Southern city was extended to include a group of resistance-heated houses and a group of houses heated with heat pumps. Separate meters were used to measure the heating energy usage. A surprising sidelight on living habits resulted from the study. Accepting the claim that the coefficient of performance of a heat pump system (with strip heaters) would be approximately 2.0, one might have expected the usage of electricity in the houses with heat pumps to be about one-half that required for the houses heated by resistance units. The study showed, however, that there was only a 16 per cent difference in favor of the heat pumps. It is believed that this points up the saving attitude of people living in houses with individual thermostats, which enable the home owner to reduce temperatures whenever a room is not in use.

One city in New Jersey used separate gas meters for gas heating usage, thus providing another source of data for direct evaluation of major nonheating heat gains.

Two procedures were used to evaluate the heating usage where only total usage was metered. One was to use the metered data for the nonheating months as the basis for the determination. The other was to use the results from the two areas where separate meters were available to relate heating usage to total energy usage. These served as a mutual check.

Added to the A.G.A. field data were energy consumption data reported from another 210 homes (see Table 12–155c). This material covered periods of one to seven years.

Additional data on electric usage were available from published research studies and field tests which appeared in electric industry trade papers. This information was obtained from 80 electrically heated homes over a period of two years. These data were principally for small groups or for individual houses.

K, the Experience Factor. All of the foregoing results were plotted as K vs. degree days. K was obtained directly from meter readings in occupied houses. It is essential to keep in mind that the span of K values for the individual houses in any one group departs ±25 per cent from the curve plotted thru average values. This wide variation even with identical houses reflects the many differences among people with respect to room temperatures, opening windows and doors, and activities in the home.

The span of heating results obtained in these studies of large groups of houses of closely similar design deserves close attention on the part of all interested in making system or energy comparisons (e.g., Table 12–155c). Many such studies have been reported on pairs of houses. The difference in living habits appears to be so great that tests run on just two houses equipped with different heating systems or different energy sources are likely to be valid only by chance. Reasonably accurate results require the inclusion of at least 20 and preferably 35 houses in each system or energy source being studied.

The experience factor, K, the calculated heat loss of the residence (Btu per hr), and the design temperature difference are used to obtain the annual heating consumption as follows:

Annual heating consumption of fuel or energy, Btu =

$$\frac{(H.L.) \times K}{\Delta t} \qquad \textbf{(12)}$$

where:

$H.L.$ = the calculated heat loss of the houses based on design conditions, Btu per hr

Δt = difference between indoor and outdoor *design* temperatures (see Table 12–47 for outdoor design temperatures), °F

K = experience factor, a function of the area's degree days

The K′ Factor. The experience factor, K, has been multiplied by the cost of gas for heating at 90 cents per 1000 cu ft of 1000 Btu gas* and plotted in Fig. 12–51a as K'_{gas}. This same analysis, applied to reported electric consumption data, was plotted in Fig. 12–51a as K'_{electric} using an average cost of 1.5 cents per kwh.

Thus, the annual cost of heating may be calculated by Eq. **12a**:

$$\text{Annual heating cost, \$} = \frac{(H.L.) \times K'}{\Delta t} \qquad \textbf{(12a)}$$

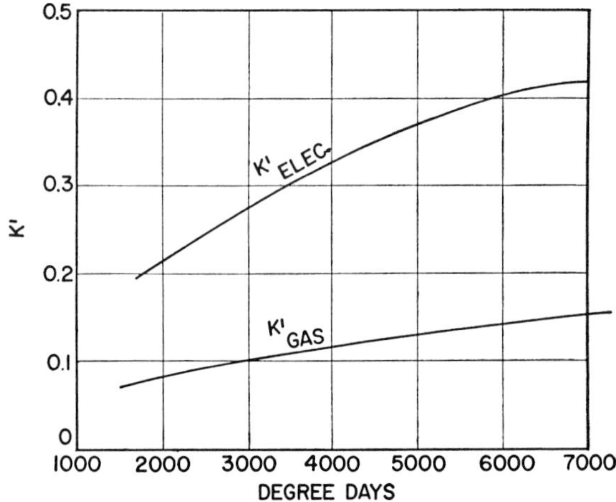

Fig. 12-51a K' for gas heating and electric resistance heating (substitute in Eq. **12a**). Based on an average gas cost of 90¢ per 1000 cu ft of 1000 Btu gas and an average electric cost of 1.5¢ per kwh.

Example: Assume homes to be built in an area with 5000 degree days. The indoor design temperature is 70 F and the outdoor design temperature is 0 F. The calculated heat loss of the gas heated house is 80,000 Btu per hr. More insulation will be used with electric heating, lowering the heat loss to approximately 60,000 Btu per hr. What is the anticipated cost of heating with gas and with electric resistance units?

From Fig. 12–51a:

$$K'_{\text{gas}} \text{ at 5000 degree days} = 0.13$$

$$K'_{\text{electric}} \text{ at 5000 degree days} = 0.37$$

Substituting in Eq. **12a**:

$$\text{Gas heat, annual cost} = \frac{0.13 \times 80{,}000}{(70 - 0)} = \$149$$

$$\text{Electric heat, annual cost} = \frac{0.37 \times 60{,}000}{(70 - 0)} = \$317$$

* National average cost of gas for residential heating as of 1963.

The values of $K'_{electric}$ are relatively lower than would be expected from calculations using assumed efficiencies of 100 per cent for resistance heating and 80 per cent for gas heating. The values as stated are based on actual data from occupied houses. To account for the wide spread between the curves, heating practices must be examined. The curves would be close to the theoretical levels were it not for the differences between electric resistance heating and heating with conventional systems using gas. Two major factors are believed to be involved.

The first factor arises from the use of individual room thermostats with resistance heating. Rooms not in immediate use may be kept at temperatures below 70 F. Thus, the heat loss of the house is less than that calculated for sizing the heating system. With conventional heating, e.g., forced warm air, the entire house is kept at uniform temperatures.

The second factor is the influence of the heavier insulation used with electric heat. This insulation provides a heat sink that apparently serves to permit fuller utilization of internal heat gains and helps to overcome the effects of constantly changing outside temperatures.

In the territory served by the Northern Illinois Gas Co. (approximately 6300 degree days), a field study[38] of a group of houses originally designed for and heated by resistance heating and then changed to gas heating permits verification of these two factors. These heavily insulated houses yielded heating energy usage of 1.31 Btu of gas to 1.00 Btu of electricity. Expressed as a seasonal heating efficiency and taking electricity as 100 per cent, this ratio makes the gas heating efficiency 76 per cent, which is in line with generally accepted values. Referring to Fig. 12–51a, K'_{gas} for 6300 degree days is 0.15, *but if houses insulated as for electric heating had been used as the basis for the curve,* then K'_{gas} would have read 0.09. This, in turn, would be reflected in a substantially lower estimated operating cost.

With the A.G.A. method, estimates can be prepared which can combine any degree of accuracy with any *probability of satisfaction.*[39] Because the curves give average values for K', annual fuel consumption estimates that are prepared using this K' will result in one-half of the bills being bigger than the estimate and the other half of the bills smaller than the estimate. Saying this in different words, there is a 50 per cent probability of satisfaction, that is, 50 per cent of the customers will be satisfied with their bills in view of the estimates given them.

The method works as follows: In any locality, the use of the average K' will result in a probability of satisfaction of 50 per cent. This means that if 1000 estimates of $100 were prepared, 500 satisfied and 500 dissatisfied customers could be expected. It may be assumed that the customer degree of dissatisfaction is related to the degree to which the bill exceeds the estimate. Only when the bill exceeds the customer's *tolerance for error* can complaints be expected. For our purposes, however, this is very important—it tells us that a bill higher than the estimate may well be acceptable. If there is a way to judge or predict the customer's tolerance for error, this can be tied to a multiplier for K' and can be set by the utility in each area according to its needs. Figure 12–51b shows the relationship among the probability of satisfaction, the tolerance for error, and the multiplier for K'.

Figure 12–51b permits determination of the multiplier for

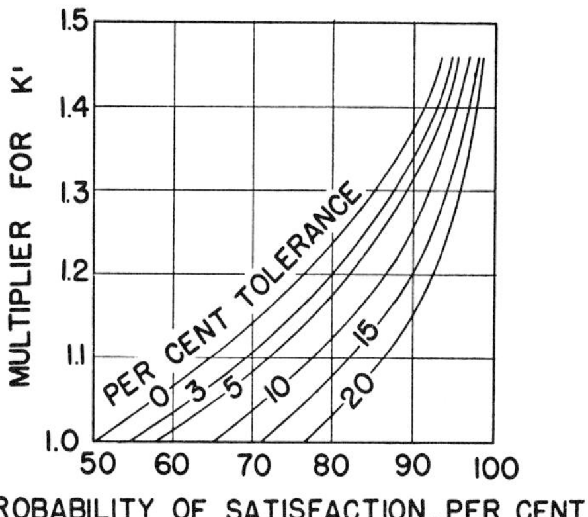

Fig. 12-51b Multiplier for customer satisfaction in estimating utility bills by Eq. **12a**. K' is given in Fig. 12-51a.

K' to be used in any predetermined conditions. Using the foregoing 1000 estimates of $100, the following examples will demonstrate the use of the figure.

Example: Probability of satisfaction desired = 80 per cent; tolerance for error over estimate = 3 per cent (tight). New estimate = $100 × 1.20 = $120; i.e., if the estimates had been issued as $120, probably 80 per cent of the customers would have been satisfied.

Example: Probability of satisfaction desired = 80 per cent; tolerance for error over estimate = 10 per cent (reasonable). New estimate = $100 × 1.12 = $112.

The local utility can thus combine the required probability of satisfaction and the tolerance for error so as to be able to take maximum advantage of the competitive situation.

For another analysis of the various types of residential heating systems, see Fig. 12–52. These data were obtained by a large East coast utility company from a study of numerous case histories.

COMMERCIAL AND INDUSTRIAL SPACE HEATING

No exact line of demarcation exists between the terms "commercial" and "industrial"; space heating equipment cannot specifically be classified as one or the other. Boilers, as an example, may be used for space heating alone or for the

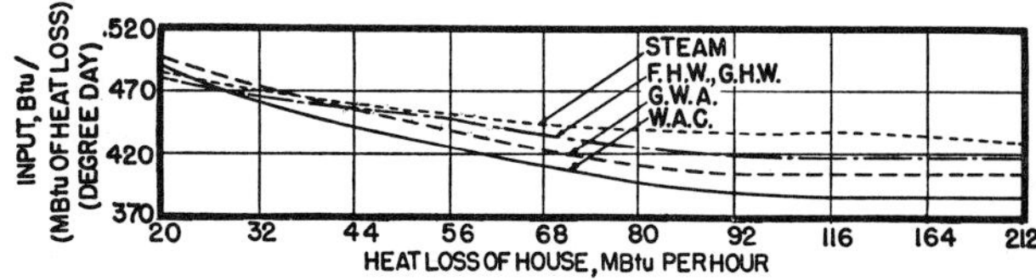

Fig. 12-52 Comparison of residential gas heating inputs for various types of installations.

Type of installation	Code	Number of cases studied: Conversion burner	Gas design	Total
Steam	Steam	465	135	600
Warm air, central	W.A.C.	17	398	415
Gravity, hot water	G.H.W.	311	155	466
Gravity, warm air	G.W.A.	5	119	124
Forced, hot water	F.H.W.	16	26	42
		814	833	1647

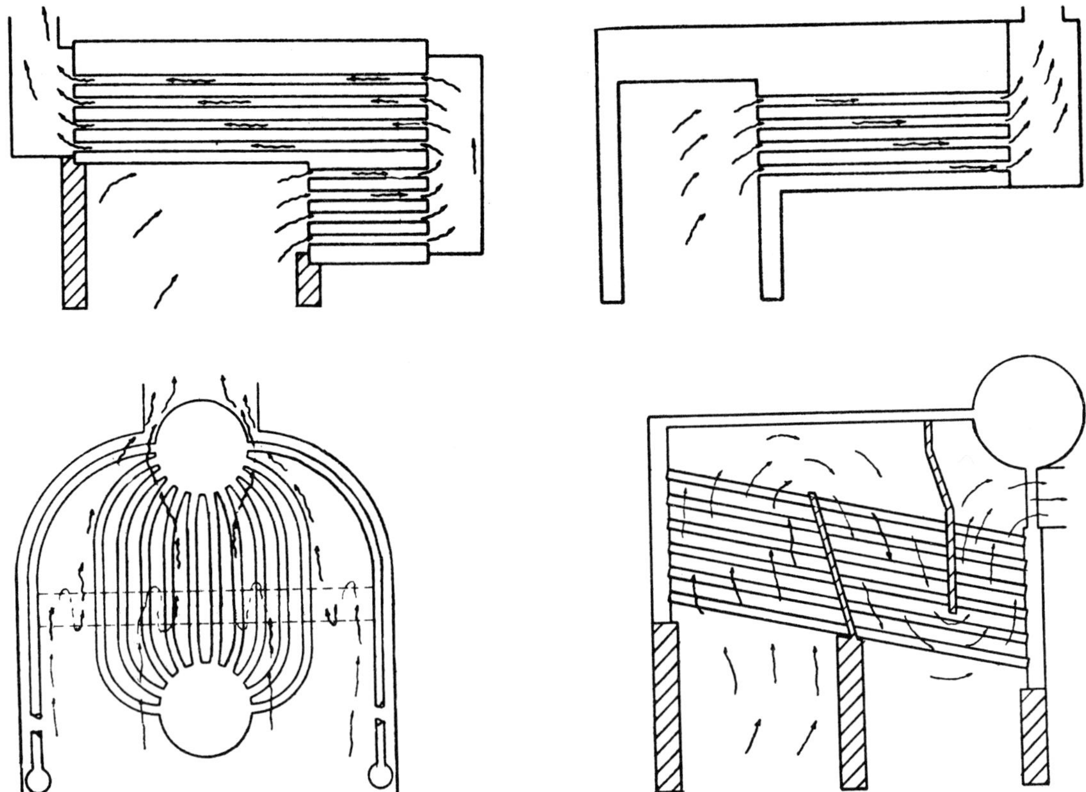

Fig. 12-53a (upper left) Horizontal, two-pass, firetube, externally fired boiler.
Fig. 12-53b (upper right) Locomotive-type, low or high pressure, horizontal, single-pass, firetube, internally fired boiler.
Fig. 12-53c (lower left) Inclined, bent watertube, two-drum, low head, internally fired boiler.
Fig. 12-53d (lower right) Inclined, straight watertube, cross drum, sectional header, externally fired boiler.

combined purpose of space heating and of furnishing steam for some process use. When gas engines or turbines are used, heat recovery equipment may serve in place of conventional boilers.

Boilers

Heating boilers, as listed in the A.G.A. Directory of Approved Appliances,[1] are generally of sectional cast iron construction, although there are many nonsectional steel boilers. Each section of a sectional boiler is of limited capacity (400,000 Btu per hr or less) and each has its own control. (See pages 12/92 and 12/94.) For large capacity, single combustion-chamber steel boilers, manufacturers supply electronic, fast-acting controls in place of the types used on sectional boilers. The size range of these listed boilers (Jan. 1965) is from 18,000 Btu per hr input (hot water central heating boiler) to 6,300,000 Btu per hr input.

Classification of Boilers.[40] The two basic types of gas-fired heating and process boilers or generators are firetube and watertube.

The *firetube boiler* is a closed shell or vessel in which a fluid, usually water, is heated or vaporized by the application of heat. The products of combustion pass thru straight tubes surrounded by water. The firetube boiler is usually limited in size to 600 boiler hp, i.e., about 20,000 lb of steam per hr, and working pressures not exceeding 250 psig. Firetube boiler size is limited by shell diameter and working pressure, which determine the plate thickness. In general, these boilers are well suited to gas and liquid firing. Some firetube boiler types are horizontal return tubular (Fig. 12–53a), locomotive (Fig. 12–53b), horizontal firebox, refractory lined firebox, vertical, and Scotch marine (Fig. 12–180).

The *watertube boiler* is the reverse of the firetube boiler, with the hot gases contacting the outside surfaces of the tubes and the boiler water and steam inside the tubes. The watertube boiler has the advantage of being able to withstand much higher pressure since its tubes are relatively small in diameter. Generally speaking, watertube boilers are not used in the smaller size range where firetube boilers are practical. Some watertube boiler types are bent tube (Fig. 12–53c), horizontal, sectional header, box header, cross drum (Fig. 12–53d), longitudinal drum, and low head. Another type is the high-temperature, high-pressure water generator. This is a forced circulation water heater that has the advantage of a closed system, which minimizes corrosion and water treatment problems.

The *packaged steam generator* may be a watertube or firetube type; the latter predominates. It is usually skid mounted, requiring no special foundations other than a level, noncombustible floor of adequate strength. Trim includes pressure gages, safety valves, low-water cutoffs, water level controls, water column with tricocks and blowdown connections, injectors, main steam stop valves, blowdown valves, and feed water stop and check valves. Gas controls include a radiant jet type of burner with automatic electric ignition of the gas pilot and electronic supervision of pilot and main burner flame ignition period; a blower or an eductor fan for combustion air and flue gas venting; a motor-operated gas valve closing automatically in the event of power, combustion air, or flame failure, or low-water condition; a manual shutoff valve; a burner and flame failure safeguard; and a control panel.

Gas-fired packaged units usually provide three or four passes (Figs. 12–54a and 12–54b) of flue gas travel thru the generator. Combustion gases may be either forced thru the tubes, the fan handling cool air, or drawn thru, the fan handling hot gases. By decreasing the cross-sectional area of each pass, high flue gas velocity and, therefore, high heat transfer is maintained.

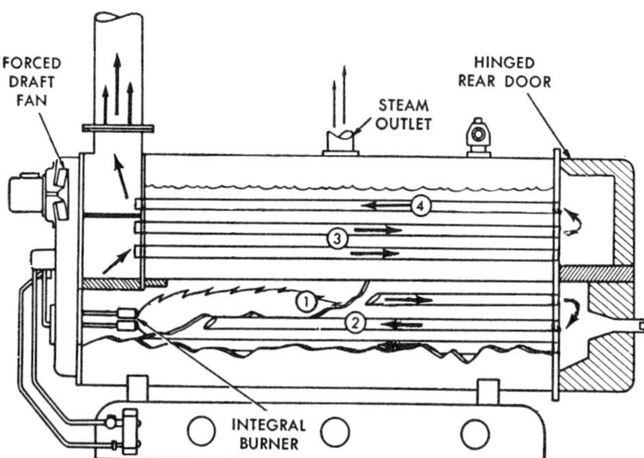

Fig. 12-54a A four-pass packaged steam generator. Circled numbers indicate passes. (Cleaver Brooks)

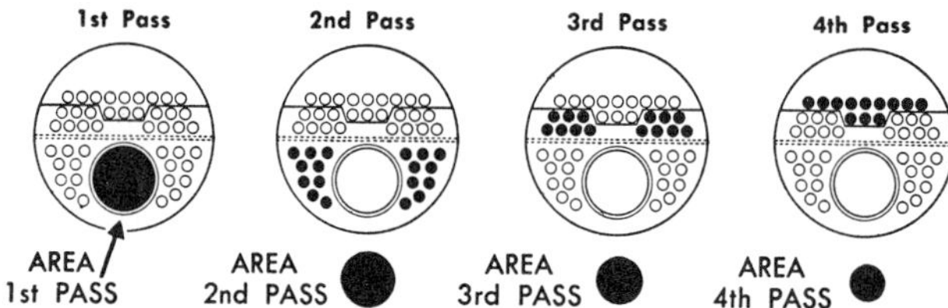

Fig. 12-54b Location and relative size of each of four passes of flue gases thru generator.

Application of Boilers.[41] The size or type of boiler that can be used for space heating ranges from cast iron boilers for small applications, thru firetube boilers for middle-size jobs, to watertube boilers for very large heating jobs.

Watertube boilers seldom operate at pressures as low as 10 or 15 psig. There are several reasons. First, if the heating load is so great that a watertube unit must be used, the steam lines are long and the line loss, that is, the pressure drop, is too large and the required pipe size is too great. Second, the circulation in a watertube boiler suffers as the pressure is reduced, with the result that the boiler output is reduced. Third, with relatively low steam and water temperature, flue gas condensation is encountered on the tubes near the stack outlet.

Since a packaged firetube boiler is fired under pressure, there is no infiltration of cold air into the furnace. In a conventional unit, negative pressures exist in the furnace and, therefore, infiltration occurs thru the furnace walls. At low loads this represents a substantial increase in the percentage of excess air. The packaged boiler is completely insulated and the furnace is entirely submerged or water cooled, reducing radiation from the boiler to a very minor amount. The integrated design of burner and boiler makes it possible, thru close control of fuel-air ratios, to maintain high combustion efficiencies over a wide range of load.

Selection of Boilers.[42] Information that may be required for boiler selection includes client or owner preference, location, steam conditions, loads, and fuels available.

For space heating and industrial processing,[43] steam temperature is generally the most important consideration. Steam pressure is the prime consideration in some industrial applications. Typical examples are steam blasting for cleaning, atomization of fuel in oil burners, and processes requiring humidification in depth, e.g. adding moisture to textile felt or expansion of polystyrene beads. Prime movers, either reciprocating or rotary, require close control of pressure and temperature.

Unit Size. Generator size can be approximated by the quantity of steam required, but fitting a new boiler into an existing space may pose problems. Often a boiler may be selected to fit the space at hand. Sometimes the overall load is split among several small units located where space is available. In this case, steam is either discharged into a common header or fed directly to the steam-consuming unit with little or no connection between units.

In general, firetube boilers meet needs up to about 25,000 lb per hr capacity. Applications of the packaged and shop-assembled watertube boiler overlap many of those of the firetube boiler, as well as of the small and medium-size, field-erected watertube boiler. Natural-circulation, field-erected steam generators span a wide range of applications.

Examples of standardization may be found at most pressures. In the low-pressure range, boilers with a design pressure of about 160 psig are generally selected for a working pressure between 140 and 150 psig. Above this range, 225 to 250 psig is the next working pressure range favored. For what might be called industrial boilers, design pressures of 450, 650, and 900 psig are widely adopted with steam (saturated and superheated) temperatures up to 950 F. It is mutually advantageous to maker and user to select standard designs where possible.

In commercially available standard steam generators the following pressure-capacity relationships were widespread, but not exclusive, in 1964. Boilers of 1450 psig had capacities up to 750,000 lb per hr; 2000 psig, about 1 million lb per hr; 2400 psig, up to 2 million lb per hr; and 3500 psig, above 2 million lb per hr.

Load Analysis.[42] Economic evaluation of alternatives for replacement or modernization of existing boiler plants is simplified by availability of statistics on past load magnitude, frequency, and duration, as well as statistics on fuel and operating costs. For new projects, load data must be computed and assembled. Care should be taken to avoid erroneous pyramiding of loads. Only those loads that will actually be concurrent should be used to establish maximum demand.

For the determination of monthly and annual loads for a project which, in addition to transmission loads, includes loads due to make-up air heating, steam driven auxiliaries, steam-powered air conditioning, and process requirements, it is suggested that the magnitude, frequency, and duration of each of these loads be determined separately and that internal heat gains, such as those from lights, machine tools, or process equipment, be subtracted from the sum of these loads at concurrent times.

The average hourly steam demand for transmission and ventilation load components during a 24-hr day for a specific month can be computed using outdoor temperature intensity-frequency-duration data available from the U.S. Weather Bureau. Computer techniques can frequently be applied economically to this type of load determination. Average daily boiler loads for a specific month can be estimated by graphical integration of load intensity-frequency-duration curves that take into consideration average daily variations in make-up air, internal load, and process load. From these data, annual load duration curves can be drawn for each load component for initial and future loads. Annual load factors for each load component can be determined from annual load duration curves, which are valuable in sizing boiler plant components and in economic analyses.

Seasonal Efficiencies. The operating characteristics of larger boilers, primarily firetube boilers such as the Scotch marine type, are different from those of equipment used for residential heating. Table 12–49 shows what might be expected in such equipment.

Table 12-49 Seasonal Efficiency of Large Boilers

Fuel	Efficiency, %
Natural gas	72–82*
No. 5 fuel oil	70–83
Hand-fired coal (over 100 boiler hp)	55†
Stoker-fired coal (up to 100 boiler hp)	65†
Stoker-fired coal (over 100 boiler hp)	70†

* A.G.A. Monthly, Oct. 1964, p. 16.
† Data recommended by U.S. Air Force Guide, Feb. 1958.

Installation of Boilers. Provisions of ASA Z21.30[12] (included in Chap. 1 of this Section) should be followed. Boilers should be erected according to the manufacturer's instructions—in general, on a floor of fire-resistive construction, with noncombustible flooring and surface finish and no combustible material on its underside. Mounting on fire-resistive slabs or arches with no combustible material on their undersides is permissible. Exceptions are those boilers specifically listed for installation on combustible floors. A floor not of fire-resistive construction can be considered fire resistive when protected according to the NBFU Code for Installation of Heat Producing Appliances.[13]

Clearances. Boilers installed in rooms large in comparison with the size of the boiler should be installed in accordance with the data in Table 11 of Z21.30, given in Chap. 1 of this Section. Clearance to the boiler front should be at least 18 in., more if needed for servicing. When combustible material is properly protected, reduced clearances as given in Table 12 of Z21.30[12] (see Chap. 1 of this Section) are permissible.

When American Standard Z83.3[44] becomes available (probably in 1966), its provisions should be followed.

Installation of Gas Burners for Boilers. Combustion chamber volumes for converted brick set boilers using multi-tube and register-type burners usually are from 0.8 to 2.5 cu ft per developed boiler hp. Horizontally baffled or rear-fired, vertically baffled boilers may be operated with the smaller combustion chamber volumes. In some three- and four-drum boilers, combustion volumes should be large; 1.5 to 2.5 cu ft per developed horsepower are not excessive.

Changes Required for Conversion. Stokers or grates should usually be removed. Rear or front ignition arches or other devices used with coal should also be removed.

After cleaning, a new floor should be installed and the space formerly occupied by the bridge wall and stoker bricked in. Usually, in making these changes part of the side wall is lost and must be replaced. Some of the older, vertically baffled, straight-tube boilers of the horizontal type will be found to have only a 4½ in. firebrick facing of the side walls, the remainder being common brick. In such cases the common brick should be torn out sufficiently to install a 9-in. firebrick wall to protect the old one. An existing wall in good condition can be faced with an additional course of firebrick.

Boilers with concrete foundations, ash hoppers, or dust collectors require protection against furnace heat. This may be provided by a ventilated floor of 6-in. hollow tile covered with a course of insulation and two courses of firebrick. Good practice is to place ¼ in. asbestos millboard between the tile and the brick to prevent air infiltration. In some installations part or all of the combustion air has been drawn thru the hollow floor. Usually the hollow tile is left open at both ends, permitting air to circulate. When use of hollow tile has been impossible, a solid floor consisting of 5 in. each of insulating brick, insulating refractory brick, and fire clay brick has proved satisfactory if the underside of the concrete has been ventilated.

When stokers and grates must be left intact, they should be protected from excessive heat. Undesirable air infiltration can best be minimized by limiting the number of cracks, using asbestos board sheets or asbestos paper rolls under the brick.

With chain grate stokers the opening to the ash hopper at the rear wall may be closed and the grate protected by covering it with 2.5 in. of insulating brick and 5 in. of fire clay brick. Coverings of 9 to 12 in. of ash have been used, with occasional additions of new ash.

Underfed stokers may be protected by filling the feeding mechanism with sand or cinders before laying the firebrick. Spreader-type stokers may be protected in much the same manner as chain grate stokers.

Firetube boilers such as horizontal return-tubular (Fig. 12–53a) or Scotch marine (Fig. 12–180) generally have large flues into which thin steel helical restrictors may be inserted to improve scrubbing action between the hot gases and the tube surface. Use of such restrictors is customarily limited to the last 5 or 6 ft of tube length unless ample draft is available.

Horizontal straight-tube, front-fired, vertically baffled boilers generally need no baffle changes. When changing to rear firing, good practice is to cut off the front vertical baffle up to the third row of tubes and insert on the second row a horizontal baffle extending from the front baffle to the rear boiler wall. The bottom of the rear vertical baffle then should be cut off to avoid too much loss of draft. Horizontally baffled boilers usually are low set and have C-tile on the bottom row of tubes to eliminate smoke. It is customary in converting to gas to remove the C-tile and install T-tile on the second row of tubes.

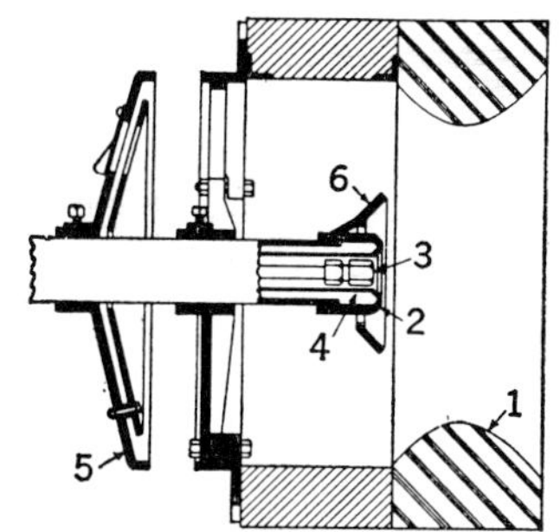

REGISTER TYPE BURNER WITH GAS ORIFICES IN CENTER OF AIR STREAM

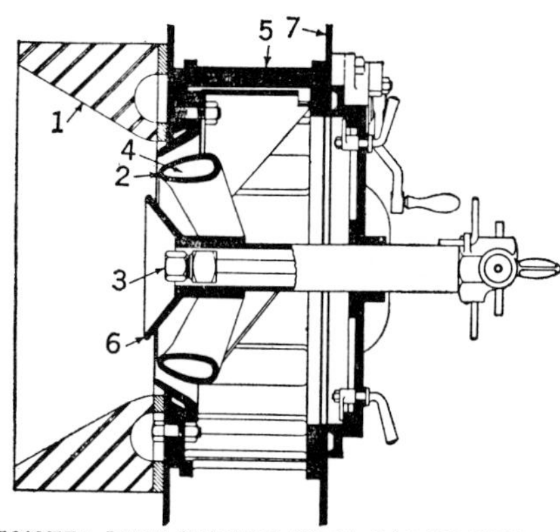

REGISTER TYPE BURNER WITH GAS MANIFOLD DIVIDING THE AIR STREAM

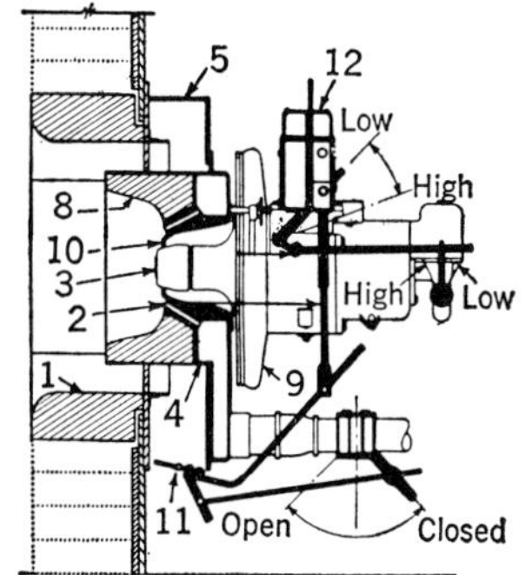

LOW PRESSURE REGISTER BURNER

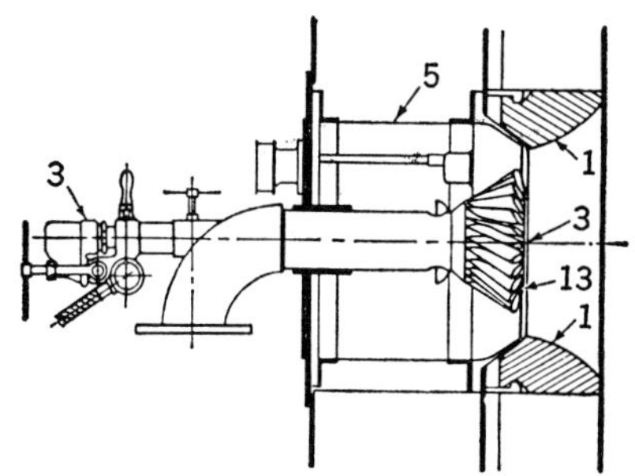

LOW PRESSURE REGISTER TYPE GAS BURNER

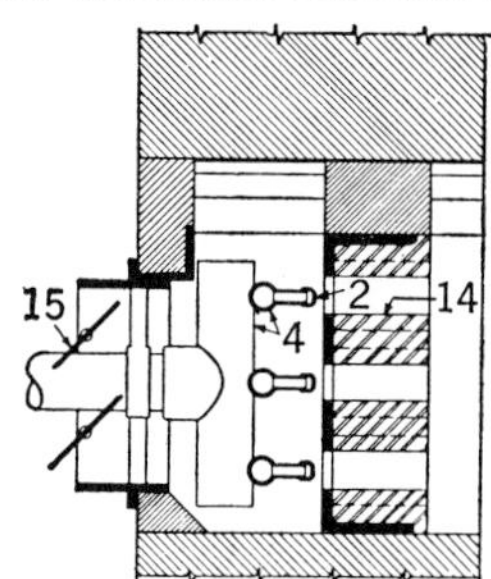

MULTITUBE BURNER WITH GAS ORIFICES AT CENTER OF CROSS SHAPED TUBES

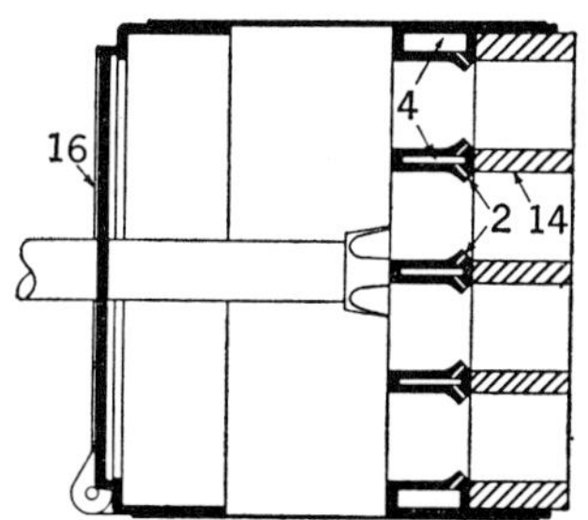

MULTITUBE BURNER WITH GAS ORIFICES AT PERIPHERY OF ROUND TUBES

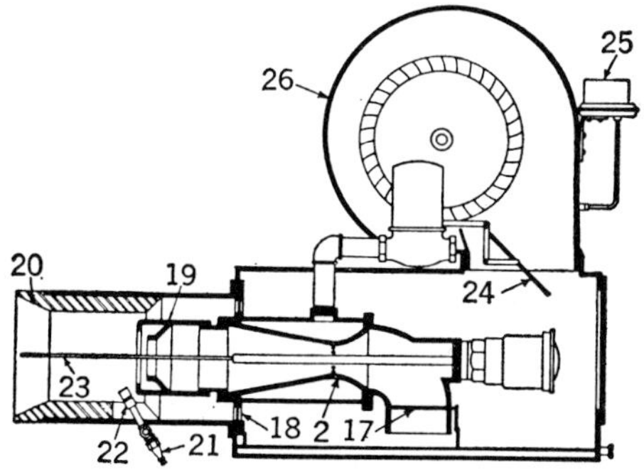

PRE-MIXING BURNER WITH ELECTRIC IGNITION AND FLAME PROTECTION

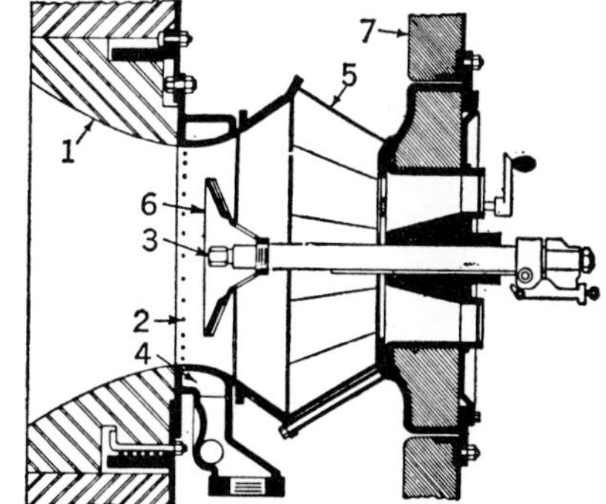

REGISTER TYPE BURNER WITH GAS ORIFICES AT PERIPHERY OF AIR STREAM

1. Refractory tuyere block
2. Gas orifices
3. Oil burner
4. Gas manifold
5. Air register
6. Diffuser cone
7. Outer windbox wall
8. Burner block (Nose tile)
9. Primary air blower
10. Primary air supply
11. Secondary air damper
12. Modulating motor
13. Gas nozzles
14. Refractory tubes
15. Air louver
16. Air door
17. Primary air control
18. Secondary air control
19. Flame retention nozzle
20. Refractory tunnel
21. Spark plug
22. Pilot burner
23. Flame electrode
24. Air damper
25. Air pressure switch
26. Air blower

Fig. 12-55 Types of large conversion burners.[45]

Burner Selection. Figure 12–55 shows several types of burners for converting large boilers to gas firing. A.G.A.-listed conversion burners are limited to input ratings of 400,000 Btu per hour. Such listed burners are sometimes used in multiples, principally in heating boilers, where overloads are not excessive.

In selecting the burner, the load characteristics must first be analyzed. The burner must be capable of meeting the maximum load, which may include both heating and process requirements. Piping losses and pickup load must be taken into account. Minimum load conditions must be satisfactorily provided for, as by throttling or cutting burner units out of service.

With a satisfactory multiple burner selection made for maximum and minimum load conditions, it should be determined whether prolonged intermediate loads can be carried without frequent burner turn-on and shutoff. If not, burners of various sizes should be chosen to give the necessary overlap of capacity.

Example: A 750-hp boiler is equipped with three 250-hp burners, each having a turndown ratio of two to one. To handle a fluctuating load of 200 to 300 hp during the night would require one burner part of the night and two burners for the remainder. This would require cutting a burner in and out of service as the load crossed 250 hp, the maximum capacity of one burner and the minimum of two. Operating conditions could be improved by selecting one 350 hp burner and two burners of 200 hp each.

Boilers of the same rating may require combustion chambers of different lengths. Mounting a burner in a combustion chamber too short for it will result in flame impingement on the rear wall or tubes. A solution might be to use large burners operated at low capacity. A forced draft burner operating with high draft loss might have a flame sufficiently shortened to be acceptable. Based on available information, horsepower per burner tube may well be limited to 0.5 d, d being the distance in feet from burner to bridge wall or tube bank.

When questions of burner application arise, the boiler manufacturer should be consulted regarding recommended method of firing, and the burner manufacturer regarding flame geometry and travel. For specific types of boilers, the following may be observed.

Internally fired boilers present a different problem from those with refractory wall furnaces. Designed primarily for heating and to be fired with coal on a grate, these boilers have small combustion chambers and cold metal walls. Because of their short fireboxes, gas flames must be short to avoid undue chilling by contact with cold surfaces before combustion is completed. A premix type burner best meets the requirements for short flames and is preferable for firing these boilers. With such a burner, 1 boiler hp can be developed per 0.5 cu ft of combustion chamber volume.

Scotch marine boilers not provided with an extension furnace require burner systems that produce maximum turbulence and as much radiant heat as practical to make maximum use of the Morrison tube surface.

Application of burners to *brick-set boilers* depends on the type of boiler, its method of baffling, the rating to be developed, and other details of the specific installation.

Horizontal return tubular boilers (Fig. 12–53a) are best fired from the front, either over the grates, or preferably with grates and bridge wall removed.

Straight-tube, vertically baffled boilers may be either front or rear fired, depending on the operating rate. If front fired, the relation between the probable loads and the amount of exposed *black surface* should be checked. The heat input should not exceed about 7 hp per sq ft (250,000 Btu per hour) of exposed tube surface. *Black surface* is defined as that portion of the heating surface which "sees" the flame or the hot refractory furnace walls. *Black surface area* (square feet) is the product of the exposed tube length and the furnace width, plus the projected area of any water wall tubes in the furnace. It is frequently necessary to move the bridge wall and change the vertical baffling to expose the required black surface.

Watertube boilers operating at high ratings are preferably fired from the rear. Exposure of the entire tube bank to radiant furnace heat is then possible.

Three- or four-drum boilers may be fired from either the front or the rear, preferably from the rear since this permits better utilization of combustion chamber space and longer flame travel without striking cold surfaces. Side firing may occasionally be necessary because physical limitations preclude front or rear firing. For a narrow boiler it may be necessary to use more burners than with front or rear firing, to avoid flames striking the opposite wall.

Burner Application. Burners may be placed on two different levels where forced draft is used. This may be done with natural draft, provided the difference in levels is not too great. Operation has been found satisfactory with two rows of burners spaced 2 ft apart, center to center. In other cases with 5 ft spacing using natural draft, difficulty has been encountered in adjusting the two sets of burners owing to a difference in draft conditions. The upper burners showed a smaller air–gas ratio than the lower. To compensate, draft loss thru the lower burners was increased by closing their air louvers.

Safety Control Systems. Standards for safety control systems are published by many agencies, e.g., FIA, FM, and UL, but requirements differ markedly. NFPA Standards 31, 85, and 86, which have an important influence on equipment choice, have their own requirements and recommendations. A new American Standard, Z21.52, sponsored by A.G.A., will become effective January 1, 1966, covering gas-fired, single-firebox boilers; many of the requirements and recommendations in this Standard differ from those already in existence. It is also expected that by 1966 American Standard Z83.3 will be available. This Standard, sponsored by A.G.A., will cover the installation of gas burners in all firetube boilers and those watertube boilers smaller than the watertube boilers now covered by NFPA Standard 85.

Under the conditions existing in 1965, reference to all pertinent publications is necessary in order to evaluate the degree to which safeguards are desirable in any specific case. For example, NFPA Standard 85 requires eight air changes of the boiler only, at 70 per cent air flow, in contrast to the FIA requirement of four air changes of the boiler and breeching, at 60 per cent air flow. Other relatively minor differences can readily be located. These differences could largely be reconciled without lowering overall safety.

Unit Heaters

Unit heaters are self-contained, automatically controlled, vented gas-burning appliances, used for heating the nonresidential space in which they are installed. They may be classified according to (1) their method of location, suspended, or floor (or platform) mounted; or (2) the static pressure at which they deliver heated air.

Propeller Fan Type. This type is designed mainly for suspended installation. It is best operated against no static pressure, with full face opening available for air delivery. Duct work should not be used.

Location of propeller fan heaters is governed by the individual application. They are usually placed so as to circulate heated air along outside walls; or they are located on outside walls so as to pick up cold air, heat it, and force it toward the center of the area or where heat is most needed. The upper portion of Fig. 12–56a illustrates a typical installation.

Suspended-Blower Type. Usage is much the same as for the propeller fan type. Since blowers handle larger quantities of air under pressure, duct work may be used if desirable, increasing materially the scope of application.

Blower-type unit heaters are particularly effective in heating buildings where the heaters must be suspended more than 12 ft above the floor. Heat distribution ducts may be attached for heating several rooms by one unit. Return ducts can be used to pick up cold air off the floor, as for overcoming drafty conditions. The lower portion of Fig. 12–56a shows a typical installation.

Suspended Heavy-Duty Blower Type. Heaters of this type with inputs up to 5 MMBtu per hr serve locations such as airplane hangars, mills, and factories, where heaters must be suspended because of code requirements or the need for space conservation. High-velocity nozzles and special plenums permit directing automatically controlled heat down to any point in a large area and in any direction.

Floor-Mounted Vertical Blower Type. This type is widely employed in industrial heating. Another important use is for heating office suites and special enclosures when separation from the main heating system is desired. Delivered air may easily be filtered if necessary. Common applications include heating spaces such as candy plants, dress factories, and other occupancies where cleanliness is a prime requirement.

In larger areas, e.g. assembly floors, heaters are usually placed around the outside walls and deliver air toward the center. Return air is drawn from the floor at the heater base.

Floor-Mounted Heavy-Duty Blower Type. These unit heaters may be (1) equipped with high velocity nozzles but without connection to duct work or (2) designed for duct work. Figure 12-56b shows typical applications.

Nozzle-equipped heaters may deliver heated air at a level about 8 ft above the floor. Nozzles are rotatable. Both horizontal and vertical louvers are usually obtainable. Placement of the heater is optional, depending on heating requirements of the building and good practice fundamentals. Generally, location is either around the outside walls, with discharge toward the center, or down the center of the area, with discharge in both directions.

Such systems are ideal for heating very large spaces, particularly in industrial occupancies where reasonably high exit air temperatures are desirable to provide sufficient delivered heat at long distances.

Sizing Unit Heaters. To provide a satisfactory margin

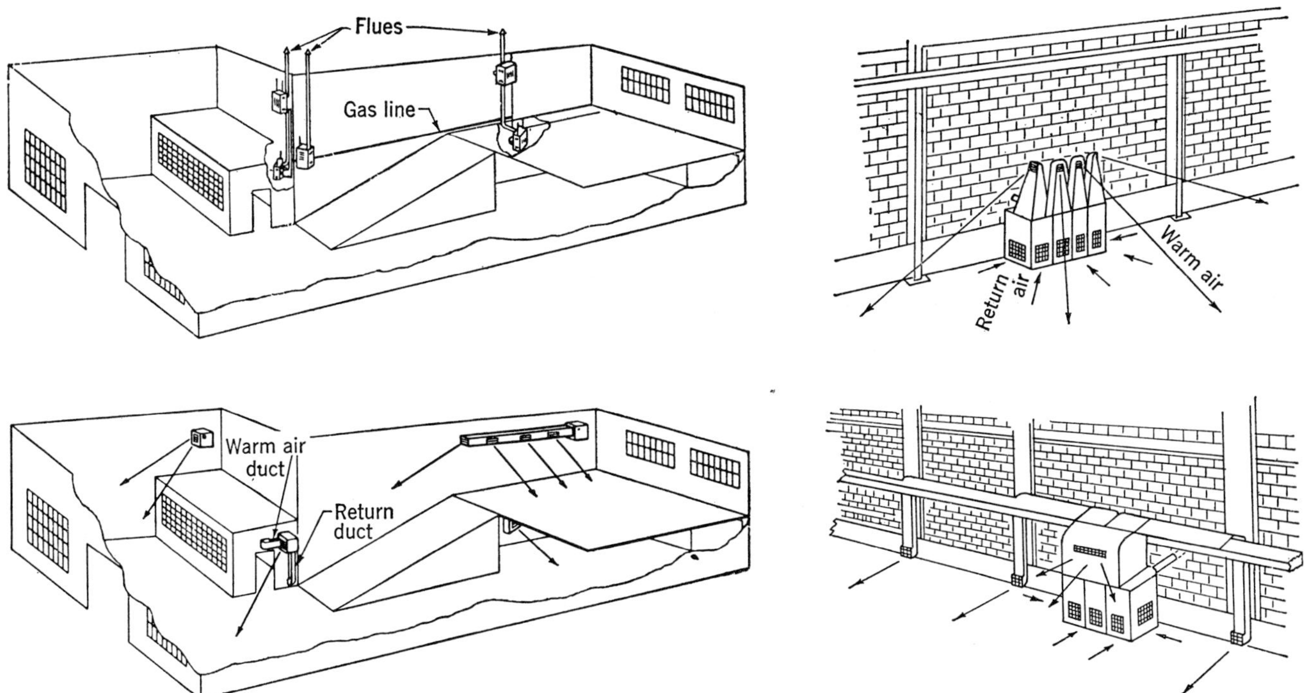

Fig. 12-56a (left) Industrial application of suspended gas unit heaters:[45] (upper) use of suspended propeller-type units; (lower) use of suspended blower-type units.

Fig. 12-56b (right) Industrial application of heavy-duty floor mounted unit heaters:[45] (upper) discharge thru high-velocity nozzles; (lower) duct-type installation.

of safety in severe weather and necessary pickup capacity, a unit heater should have an output at least 10 per cent larger than the heat loss of the structure to be heated. Because unit heater output is based on 80 per cent of the input, the input required should be about 1.4 times the heat loss (i.e., heat loss $\times$ 1.1 $\div$ 0.8). When temperature is reduced at night or for intermittent heating of cold areas, a larger allowance may be needed for adequate pickup.

Unit heaters for industrial occupancies may need to provide heated air for process use, for heating cold stock, and for other miscellaneous purposes in addition to heating the structure. All purposes should be considered in determining the heat loss.

Type of occupancy may influence heat loss calculations. For example, in shop heating, where a working temperature of 60 F may be adequate, unit heaters should be sized accordingly. Unnecessary on-off operation may then be avoided and steadier circulation maintained than would be the case if the unit heater were designed to provide normal room temperatures.

Installation of Unit Heaters. Suspended units should be placed as low as headroom requirements permit. High mounting makes it more difficult to force heated air down to the working plane and increases the floor-ceiling temperature differential.

Provisions of paragraphs 3.1.7, 3.1.8, and 4.13 of ASA Z21.30[12] should be observed when applicable. These requirements are included in Chap. 1 of this Section.

Aircraft Hangar Installations. Unit heaters approved as suitable for use in aircraft hangars[46] must be installed at least 10 ft above the upper surface of wings or of engine enclosures of the highest aircraft which may be housed in the hangar. (The measure should be made from the wing or engine enclosure, whichever is higher from the floor, to the bottom of the heater.) In shops, offices, and other sections of aircraft hangars communicating with aircraft storage or servicing areas, unit heaters must be installed not less than 8 ft above the floor. Suspended or elevated heaters must be so located in all spaces of aircraft hangars that they shall not be subject to injury by aircraft, cranes, movable scaffolding, or other objects. Provision should be made to ensure accessibility to suspended heaters for recurrent maintenance purposes.

Air Flow and Circulation. The direction of heated air from a unit heater should generally be toward the location where the most uniform temperature is desired, usually a space occupied by people not in motion. Exposure of occupants to direct blasts of heated air should be avoided. An air stream should not be directed against any exposed wall or other cold surface near the heater.

Multiple units heating a large area may be spaced so that air will be circulated completely around its inside walls. In small rooms, for good circulation from one unit heater, accepted practice is to place the unit near an unexposed wall with the heated air directed toward the exposed wall surfaces but not striking them. Existence of any obstructions such as partitions or material displays that might interfere with heated air distribution should be fully considered.

Thermostats controlling unit heaters should be located on an unexposed wall or column at about eye level. If such locations are not suitable or available, insulated mounting on an exposed wall may be necessary. Warm air from a heater should not strike the thermostat directly. Each heater may be individually controlled or several may be operated by a single thermostat.

Duct Furnaces

A duct furnace is an appliance normally installed in distribution ducts of an air-conditioning system to supply warm air for heating and depending for air circulation on a blower not furnished as an integral part.

A.G.A.-listed (Table 12–30) duct furnaces are available. Such units are tested to ensure that temperatures of combustible materials will not be excessive at points 6 in. from the sides and top, and on the floor at clearances of zero or 2 to 6 in., as specified by the manufacturer. For floor clearance greater than zero, the manufacturer must provide a noncombustible stand or legs. Duct furnaces designed and approved only for suspension must carry a permanent plate giving installation instructions and stating the required floor clearance. Where draft hoods extend above the casing top, a 6-in. clearance must be maintained above a hood designed for horizontal connection and above a standard 90° elbow when designed for a vertical outlet.

Circulating air should not be taken from the same enclosure in which the duct furnace is located.

Duct furnaces used in conjunction with a refrigeration system should not be located downstream from the refrigeration coil unless specifically approved for such installation.

The provisions of paragraph 4.10 of Z21.30[12] should be observed where applicable. These requirements are included in Chap. 1 of this Section.

Enclosed Furnaces

Enclosed furnaces are designed primarily for heating places of public assembly such as school rooms, churches, and auditoriums. Incorporating an integral enclosure around the entire unit, the furnaces take only outside air for combustion. Insulation ensures quiet operation and precludes condensation.

These furnaces provide automatic room ventilation, blending outside air and recirculating air. They may be used against static pressures up to 0.5 in. w.c. Normal air temperature rise is from 70° to 105°F.

Enclosed furnaces are A.G.A.-listed for installation with zero clearance to combustible materials on the floor and one vertical side. At the manufacturer's option they may be approved additionally with zero clearance at one or both ends. Surfaces exposed to air in the heated space may not exceed 60°F above room temperature.

Installations in Schools. These classroom installations rapidly bring the environment to a comfort level. Uniform temperatures and even distribution of the room air are maintained at any required ventilation rate. Since the enclosed furnace is a combination heating and ventilating unit, a separate classroom ventilation system is not used. The savings offered include:

1. No requirement for a space or a building to house a central heating plant.
2. Reduction in fuel costs—a heating unit enclosed in the area to be heated is a highly efficient arrangement.

There are, for example, no piping losses to be taken into account. Also note that rooms not in use may be maintained at a lower temperature than other classrooms.

3. Reduction in the cost of mechanical contracting, e.g., no distribution ducts.

4. Ease of adding units to heat any classrooms that are constructed at a later date.

The A.G.A. Laboratories test enclosed furnaces according to the special provisions included in ASA Z21.47.[23]

Building Code Recognition. Twenty-one states permit classroom installation of enclosed furnaces. Another five states permit them with certain restrictions (mainly a function of the number of classrooms).

Figure 12–56c shows the general arrangement of an enclosed furnace, together with diagrams of its installation.

Infra-red Radiant Heaters

Gas-fired heaters of this type are applicable to a wide variety of heating problems. They have proved successful in many different occupancies such as factories, warehouses, garages, shopping centers, bus stations, race tracks, and outdoor needs. An infra-red unit may be approved by A.G.A. (Table 12–30) as an unvented overhead heater, or as a vented overhead heater when tested under the room heater requirements, or as an unvented infra-red radiant heater intended for heating of nonresidential spaces when tested under the unvented infra-red radiant heater requirements. Unvented overhead heaters approved under room heater requirements are limited in size to an input of 50,000 Btu per hr, but there is

Table 12-50 Heat Input of Infra-red Heaters

(values given here are for the comfort of an individual who is spot heated; values based on Raber and Hutchison comfort equation and a surface area of the clothed body of 19.5 sq ft)

Air and unheated surface temp, °F	Direct radiation heat transfer reqd for comfort, Btu per hr	Infra-red heater input required for various heater radiant efficiencies,* Btu per hr-sq ft		
		40	50	60
40	1190	161	129	107
45	995	134	108	90
50	800	108	87	72
55	585	79	63	53
60	410	56	45	37
65	210	28	23	19

* Radiant efficiency or the ratio of useful radiant energy output to total energy output.

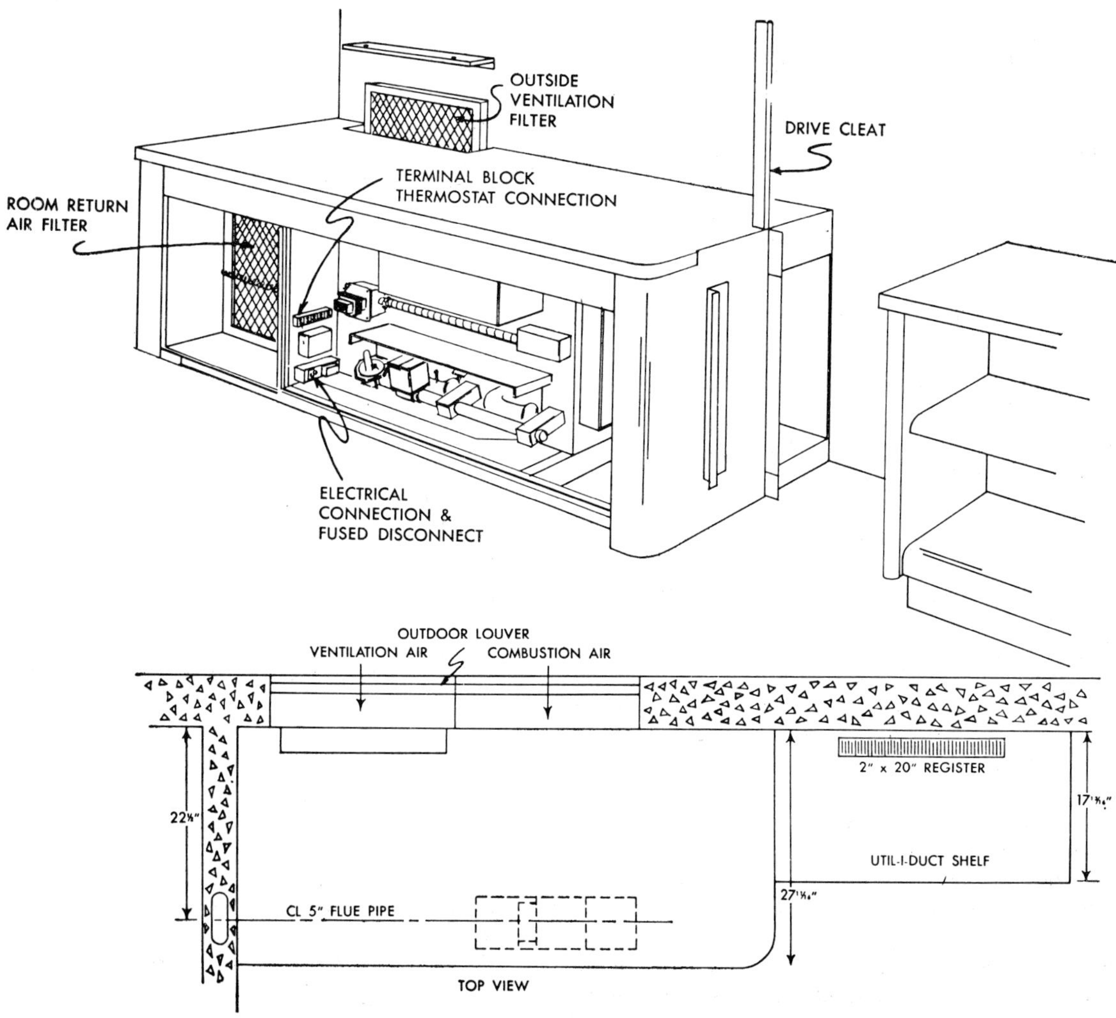

Fig. 12-56c Enclosed furnace installation diagrams. Front louvered return air panel and access trim panel not shown. (Norman ITT)

Below, unvented ceramic type; at right, vented metallic type. Mounted overhead, they can be used in systems designed for partial or full area heating.

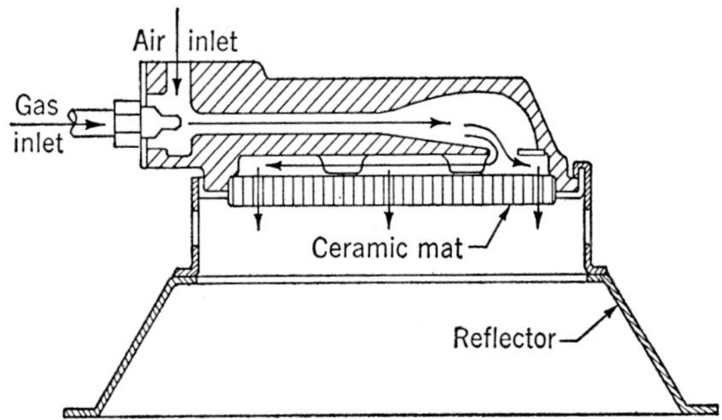

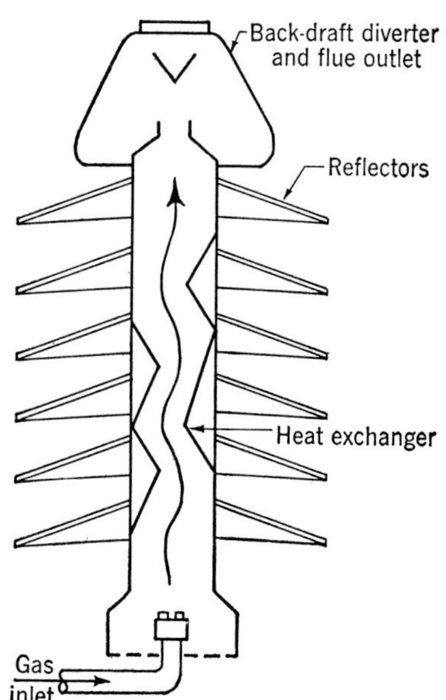

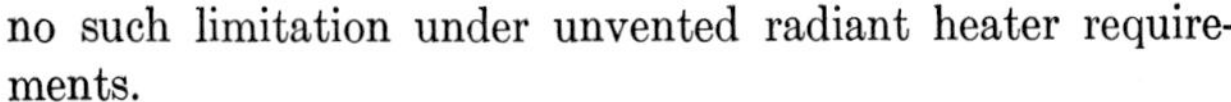

Fig. 12-56d Two types of gas-fired radiant (infra-red) heaters.[45]

no such limitation under unvented radiant heater requirements.

Unvented overhead heaters are usually installed in high-bay buildings well above breathing level. Vents are usually provided in the roof for discharge of combustion products. A minimum roof vent opening of 50 sq in. per 100,000 Btu per hour input is generally desirable but may vary, depending on local regulations.

Figure 12–56d shows two types of overhead heaters.

Application.[47] High-intensity infra-red units may be used in three ways: (1) total space heating, in which an entire plant is heated; (2) area heating, in which only a small area within a plant is heated; and (3) spot heating, in which each work station is treated as a separate heating problem. Some of the factors that should be considered in determining the approach to be taken include the occupant density; the nature of the work stations; the amount of infiltration or the quantity of ventilation required; and the overall level of temperature required to safeguard the product, to prevent freeze-ups of water services, etc.

Total Space Heating.[47,48] The infra-red heater is suited for heating large spaces. Many such systems are in operation.

From 30 to 60 per cent of the input to gas infra-red heaters is radiated to the floor of the heated space. The radiation is absorbed in varying quantities by the floor, tools, materials, and intervening structural members. A small part of it may be absorbed by water vapor, carbon dioxide, and dust particles in the air. For typical mounting heights in well-ventilated plants, the loss of radiation due to absorption of materials in the air is small and may be neglected.

All of the surfaces upon which the radiation is directed become warm and contribute to the overall comfort of the heated space. These surfaces warm the air by convection heat transfer and warm other surfaces by reradiation. The heat that is lost from the unit by convection or with the products of combustion from unvented heaters rises to warm the underside of the roof to take care of its heat loss. Limited temperature data[49] published indicate that the floor level air and the floor surface temperatures are higher than those at higher levels. Thus, the system operates, for comfort purposes, much as a radiant floor panel, with one exception—individuals receive greater quantities of direct radiation from the heaters. This serves to offset body heat losses by convection in areas of high air motion and radiation losses to cold exterior surfaces. This "reserve heating" capacity increases as the heat loss per square foot of floor of the structure and thus the installed heater input increases. At high rates of heat loss, the horizontal component of radiation can approach values used for spot heating.

This relationship is shown in Fig. 12–57, in which the heater input rates (Btu per hr-sq ft of floor) and equal air and unheated surface temperatures have been plotted (solid lines) for various heater radiant efficiencies.

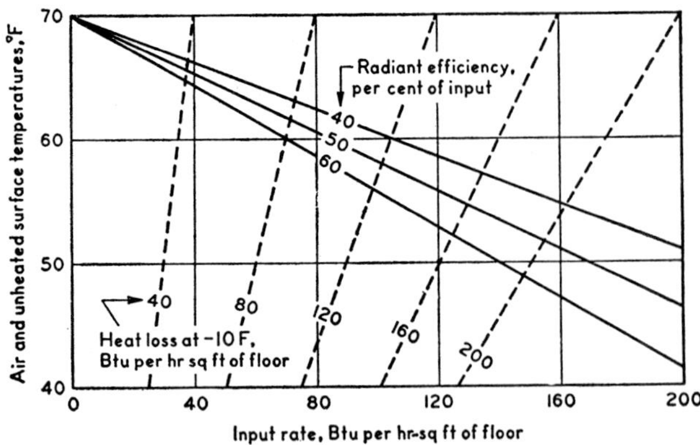

Fig. 12-57 Relationship of heat input and air temperature for comfort (shown as solid lines).[47] Dashed lines represent temperature maintained in space at various rates of heat input for given computed rates of heat loss at 70 F indoors and —10 F outdoors. Intersection of solid and dashed lines for certain efficiency and heat loss indicates minimum heater capacity that can be expected to provide comfort. See text for further discussion.

Example: Enter Fig. 12-57 at the left, at an air temperature of 50 F, move to the right to the 60-per cent efficiency curve, and read an input rate of 140 Btu per hr-sq ft as the comfort requirement. These data assume that the individual will be irradiated from at least two sides, that is, that two heaters having input rates of 70 Btu per hr-sq ft, installed 180° apart, will direct radiation on the worker. These curves are based upon the data given in Table 12–50 and are subject to the same limitations.

Superimposed upon these curves is another set of curves (dash lines) which give the relationship between the installed heat input rate and the air temperature that will be maintained at various rates of calculated heat loss. It is assumed that design conditions are 70 F indoors and −10 F outdoors. In an actual case, the relationship among the heat input, calculated heat loss, and air temperature should take into account the internal heat load. This would in effect decrease the reduction in temperature for a given reduction in input.

The intersection of the appropriate heat loss curve and the temperature input curve at the radiant efficiency of the units installed gives the minimum capacity of radiation that would be required for comfort heating of an entire plant.

Example: In a building with an estimated heat loss of 160 Btu per hr-sq ft of floor (at an 80°F indoor-outdoor temperature difference) and a heater efficiency of 60 per cent, the balance point occurs at an air temperature of 52 F and an input of about 125 Btu per hr-sq ft.

The radiation is assumed to be horizontal (and therefore perpendicular to the major portion of the surface area of the individual) and the individual to be irradiated from two sides. In practice, the radiation enters the living zone at angles ranging from about 45° to 90° from the horizontal. Thus, the horizontal component of radiation is less than that taken into account in the curves for a given heater input. A higher input rate and a correspondingly higher control temperature will generally be required for comfort in practice.

Keeping in mind the limitations of the curves in Fig. 12–57, the following qualitative observations can be made:

1. As the heat loss per square foot of floor area increases, a reduction of heater input can probably be made. High heat losses per square foot, however, will be associated with light construction, high infiltration rates, or large wall-to-floor area ratios. In such cases, the "reserve heating" offered by the difference between the direct radiation received by the body and that required when an air temperature of 70 F is automatically maintained will minimize the effects of drafts and cold wall surfaces.

2. The closer the installed input is to the minimum input indicated by the curves, the greater the dependence on direct radiation for comfort. Thus, the infra-red units would have to be controlled in such a way that a portion of them operated continuously to provide just the required amount of direct radiation at any given outdoor temperature. Greater emphasis must also be placed in design on locating the units to provide uniform radiant intensities throughout the heated space.

Again bearing in mind the limitations of the curves in Fig. 12–57, consider the foregoing example (160 Btu per hr-sq ft heat loss, 125 Btu per hr-sq ft heat input). An air temperature of 70 F will be held until the outdoor temperature drops to about 10 F. Further reduction in outdoor temperature will cause the indoor temperature to drop. At −10 F outdoors, the indoor temperature would be about 52 F; however, the heaters will operate continuously below 10 F and the direct radiation gains to the workers plus the reradiation from floor, tools, etc., will provide adequate comfort.

Another consideration in reducing the air temperature, and thus the installed heat input rate, is the decrease in the temperature of the floor, material handled, tools, etc. This decrease will tend to eliminate one of the principal subjective advantages that have led to the acceptance of high-intensity infra-red heating—the warmth of everything touched.

Distribution of Units.[47] Application begins with a conventional heat loss calculation for the space to be heated, to determine the number of units required or the total input. Ventilation air is needed to ensure removal of combustion products from unvented heaters. If it exceeds that supplied by infiltration or for other purposes, it must be included in the estimated heat loss.

The efficiency of the unvented gas radiant heaters may be taken as 90 per cent, the vented types as 80 per cent.

Locating Units. It is desirable to consider the perimeter heat losses and the fact that an individual in a perimeter location will more likely be subjected to drafts due to infiltration and will lose more heat by radiation to the cold surfaces. Thus, where possible, units should be located between the individual and the perimeter.

Part of the infra-red energy must reach the floor near the perimeter of the space. If partitions, stored products, or other opaque structures are interposed between the radiant heater and the floor so that the floor is not heated, cold pockets of discomfort will result.

Some general rules to follow in locating units are:

1. Locate at least one row of heaters arranged for uniform distribution along each exposed wall to offset radiant losses in this area and to compensate for infiltration and heat transmission thru the wall and floor.

2. Treat the areas near doors that will be opened periodically in order to counteract large amounts of infiltration as a spot heating problem so that the direct radiation gains to the workers will effectively offset the reduced air temperature.

3. Locate the remainder of units in a pattern that will achieve uniform distribution of radiation in spaces normally occupied.

4. Avoid locating units where structural members and machinery will cast cold shadows on the subject. The infra-red waves will be absorbed or reflected so that they do not reach the subject.

5. Watch placement so that combustible materials and sensitive equipment are not overheated by either direct radiant heat transfer or by contact with the heater. Shielding with aluminum sheets will provide protection from radiation.

6. When heaters emit a large portion of their radiant energy in the visible and near infra-red regions, avoid direct irradiation of glass surfaces. Glass will transmit about 75 per cent of the incident radiation at these wave lengths. In the far infra-red region, only 5 to 10 per cent are transmitted.

Heating Limited Areas.[47] When work stations are confined to a portion of a large space and heating is not required elsewhere, the area concept of heating is applied. The infra-red system for a small area in a plant would be designed in the same way as a spot heating installation. Units should be located to provide uniform radiation throughout the space at intensities consistent with the minimum temperature that will be maintained.

By varying the amount of radiation installed, it is possible to maintain different levels of comfort in different portions of a large space without resorting to partitions. Comfort depends upon the intensity of radiation provided in the occupied zone since field experience indicates that the air temperatures in the heated and unheated portions of the space will tend to equalize by natural convection. Data given in Table 12–50 include the heat input rates required per square foot of floor for providing comfort in limited areas of a plant.

For either spot or area heating an estimate must be made of the minimum air temperature that will be maintained if adequate radiation is to be installed. This requires an estimate of the heat loss of the structure and knowledge of the internal heat gains arising from the processes involved. Solutions of some spot and area heating problems must of necessity be empirical because of the difficulty of estimating heat losses when only small areas within a structure are to be heated. Even so, such designs can be checked to determine within reasonable limits whether they will be adequate, by applying the principles outlined for spot heating.

Spot Heating.[47] When the occupant density is low and work stations are relatively static, spot heating will provide the most economical solution, both from a first cost and an operation cost standpoint. An extremely high infiltration rate in conjunction with limited work stations also suggests spot heating.

In spot heating applications, it is desirable to provide radiation to two sides of the individual when this is practicable. The major amount of the radiation should preferably be directed below the individual's head. The input rates required to provide the necessary direct radiation for comfort are given in Table 12–50.

The values in Table 12–50 are conservative since they apply to lightly clothed, sedentary individuals. The heat production of working individuals and thus their required rates of heat loss will be greater, so that less direct radiation will be required. Heavy clothing will usually be worn which makes the values even more conservative. Offsetting these heat gains is the increased loss from the worker due to convection heat transfer in areas of high air motion. The values of Table 12–50 apply for relatively low air motion (25 to 30 fpm); the higher air velocities that may be experienced in practice will increase the heat loss.

The values of Table 12–50 assume that the radiation is directed perpendicularly to the surface of the body and that the body is uniformly irradiated.

In many instances, the nature of the work will prevent irradiation from both sides. In such cases, it should be remembered that all surfaces irradiated become warm and reradiate or reflect energy to the individual. This effect plus the warm floor will tend to offset the disadvantages of providing radiation from only one side.

Ventilation and Infiltration Air.[48] Ventilation is defined as the process of supplying and removing air by natural or mechanical means to and from any space. Natural ventilation can be obtained by air movement thru walls or around windows (commonly called infiltration), thru open sash, or by vent flues thru the roof. Mechanical ventilation is accomplished by blowers or fans forcing or drawing air thru the space. There must be sufficient air to keep the concentration of combustion products at tolerable levels and to prevent condensation.

With powered ventilation the rating or capacity of the unit is usually known. Nevertheless, the amount of air exhausted must be replaced by an equal amount of air. Therefore, inlet openings must be provided if air is exhausted from a space. By the same token, if air is delivered to a space, outlet openings must be provided. The greater the inlet area, the slower the movement of air thru the inlet. Cold air spots can be eliminated if inlets are small, well distributed around the periphery, and away from personnel. This may mean entry at high or eave levels. A velocity of 200 to 400 fpm at the opening is satisfactory in most cases.

The temperature differential between indoors and outdoors causes a stack effect that produces "gravity" ventilation thru vent stacks or openings high in the building. This will be true only if there is a sufficient area of inlet openings at an elevation lower than the outlet openings. The amount of air that will be exhausted by these openings can be approximated by Eq. **13**:

$$V = 9.4A[h(t_1 - t_2)]^{0.5} \tag{13}$$

where: V = volume of air exhausted, cfm
A = free area of outlets, sq ft
h = height from inlet to outlet, ft
t_1 = avg temperature of indoor air in height h, °F
t_2 = outdoor air temperature, °F

Example: Using Eq. **13,** determine the volume of air exhausted by a 2-ft-diam opening located 20 ft from the floor. Inlet air is provided by cracks around windows and doors. Average height of cracks is 5 ft above the floor. Average temperature between 5 ft and 20 ft is 65 F. The outside temperature is 10 F.

Substituting in Eq. **13**:

$$V = 9.4\pi(1)^2 \times [(20 - 5)(65 - 10)]^{0.5} = 848$$

The amount of air that will infiltrate thru cracks around doors and windows and thru walls can be ascertained from Table 12–34a. In a building with four exposed walls, only half of the total crack is to be used in computing the heat load. With three exposed walls use the wall with the most crack but never less than half the total crack for the three walls. For two exposed walls use the wall with the most crack.

The heat required to bring the temperature of the infiltration air from the outside design temperature to the inside temperature is given by Eq. **4**.

Example: 4000 cfh of air will infiltrate around a door; 65 F inside temperature and 0 F outside temperature.

Substituting in Eq. **4**, the heat required to warm the air to 65 F will be:

$$Q_i = 0.018 \times 4000 \times (65 - 0) = 4680 \text{ Btu per hr}$$

Example: The volume of a building is 1,000,000 cu ft; 50 F inside temperature and 0 F outside temperature; two air changes per hour are expected.

Substituting in Eq. **4**, the heat required to warm the air to 50 F will be:

$$Q_i = 0.018 \times (2 \times 10^6) \times (50 - 0) = 1.8 \text{ MMBtu per hr}$$

In most buildings heated with warm air, there is a definite temperature differential between the breathing level and the roof; see Table 12–31b. To obtain an accurate heat loss in a warm air system, these temperature differentials must be included in determining heat loss of walls and roofs, and heat loss due to air changes.

Dilution of Combustion Products.[48] Suggested maximum or allowable working levels (in the breathing zone) for CO and CO_2 are given in Table 1–30. With units mounted overhead and with efficient exhaust systems, exhaust products will never be present at breathing levels.

Use Eq. **14** to compute the minimum air circulation rate required to keep the building atmosphere at a tolerable level.

$$Q_A = \frac{Q_c \times 10^6}{(MAC) - (SAC)} \qquad \textbf{(14)}$$

where:

Q_A = quantity of air circulated, cfm
Q_c = rate of generation of contaminant, cfm
(MAC) = maximum allowable contaminant concentration, ppm by volume
(SAC) = concentration of contaminant in supply air; 300 to 600 ppm for CO_2, zero for CO; ppm by volume.

Operating Costs.[47] Only limited data are available to estimate differences in the operating cost of radiant heating and conventional systems. In some reported cases, radiant system operating cost was considerably less than estimates for other types of heating. In others, the costs were about the same or somewhat higher than those for conventional systems. This agrees with experience with conventional low-temperature panels. Lower costs are likely to occur when the air temperature is reduced and comfort maintained by increasing the mean radiant temperature, thus also lowering the cost of warming infiltration air.

An extensive survey[48] covering various size buildings has indicated that heat loss can be reduced by 15 per cent if the perimeter method is used. If for some reason heat cannot be directed toward the slab perimeter and wall or if the products of combustion are immediately exhausted, this percentage reduction does not hold true. As a result, the total heat loss may be required to provide the inside design condition.

The reasons for the 15 per cent apparent reduction in heat loss can be explained by one or more of the following points:

1. There is a virtual absence of temperature difference from floor to ceiling.
2. Practically no movement of air is required to circulate heat.
3. The heat is directed to the floor and released at moderate temperatures.
4. About 90 per cent of the heat generated in the combustion of the gas is released in the building enclosure.
5. Flexibility permits placing the heat where it is needed. With the proper number of thermostats, heat can be supplied at the moment it is needed.
6. Because radiant heat warms personnel, floors, equipment, and materials directly, comfort can be achieved at air temperatures lower than those necessary with warm air systems.

Maintenance.[47] The maintenance required to ensure proper operation of infra-red units will depend to some extent upon the nature of the manufacturing processes carried out in the heated space. In general, the most frequent maintenance operation consists of cleaning the reflector surfaces to ensure maximum efficiency of the units. The frequency with which the reflecting surfaces must be cleaned varies with the type of operation carried out in the heated space.

When operations produce dust or other solid contaminants, the burners and the pilots may have to be cleaned occasionally to remove solids deposited from the combustion air. An annual inspection at the time the pilots are lighted in the fall is a good routine practice. When cleaning the combustion components, procedures recommended by the manufacturer should be followed to avoid damaging the burners or radiating elements.

Make-up Air Heating

The elements of much of the air exhaust load of an industrial plant are self-evident, e.g., those discharged thru ducts directly to the outside from enclosures at the source of air contamination. Numerous other less obvious ventilation requirements also exist. Among them are combustion air for burners for heating and processing equipment, recirculating air necessary under safety or insurance requirements, and air losses thru loading dock doors.

Creation of negative pressure within a building due to inadequate air admission may result in various problems such as back drafts in flues and ventilators, excessive drafts thru doors and windows, and dispersal of contaminants throughout the building. These conditions may be corrected by introduction of sufficient make-up air to replace that lost from the building. Provision of 5 to 10 per cent more volume of make-up air than the total air exhausted is considered good practice.

Make-up air passes thru the building only once; make-up air heating should not be confused with space heating. Make-up air temperatures are usually maintained at approximately 65 to 80 F, the same as the desired comfort level for the space heating system in the building. Once the delivery air temperature is determined, the total heat input required to maintain it is a function of outdoor temperature.

Because the outdoor temperature varies from hour to hour during the day, and from month to month during the year, the heating equipment for make-up air must throttle smoothly to maintain the desired delivery temperature. Variations of more than 5°F in temperature of the air delivered are usually unacceptable, and the minimum temperature rise required from a make-up air system should be within this tolerance. The maximum temperature rise is determined by the lowest outside temperature that may be expected during the winter months.

Types of Make-up Air Heating Systems. These systems fall into two broad classifications: (1) directly compensating systems and (2) area ventilation systems. In many cases both types of systems may be used in the same building. One or more directly compensating make-up air heaters will be used for the processes or areas of maximum exhaust, while a central general area ventilation system will provide for the various smaller ventilation needs.

Directly Compensating Systems. These systems deliver their heated air into the immediate vicinity of the equipment

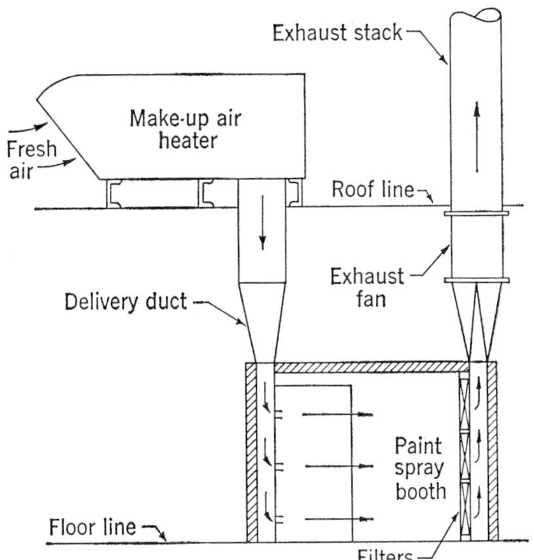

Fig. 12-58a Directly compensating system supplying heated make-up air for paint spray booths.[45]

or process being exhausted, directly replacing the air removed from that point. Make-up air heaters for paint spray booths shown in Fig. 12–58a are typical of these systems.

Area Ventilation Systems. Systems of this type introduce their heated air (usually thru a duct system) so as to blanket entire areas or buildings to replace air being exhausted at many points (Figs. 12–58b and 12–58c). Such systems, properly designed, often can eliminate borderline areas of contamination which would be difficult to exhaust directly. They work well in foundries, for example, where there is general ventilation but where direct exhaust would be prohibitively expensive.

Heat Input Calculations. To determine the make-up air heating requirements for a given system, it is necessary to know the amount of air to be delivered, its delivery temperature, and the maximum expected outdoor temperature.

A fan delivering heated make-up air into the building usually is downstream from the burner. Thus, it handles heated and expanded air. Since this air is at about 70 F and at very low pressure, it is customary to disregard correction factors applying to fans handling air at other than standard conditions.

For direct-fired (unvented) make-up air heating, efficiency is taken as 90 per cent since all the heat except the latent portion is absorbed by the air passing over the burner.

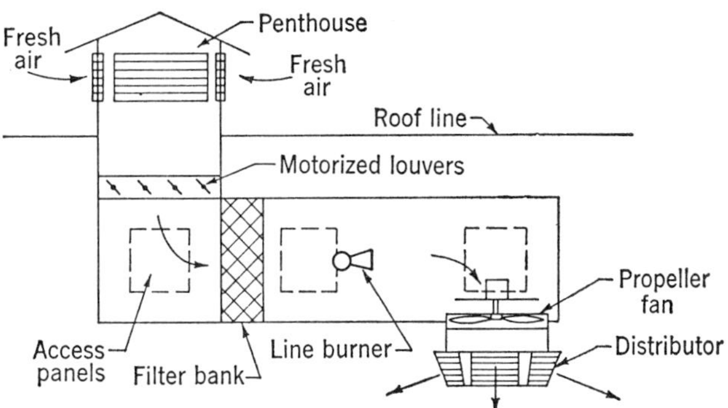

Fig. 12-58c Direct-fired make-up air system with penthouse intake and propeller fan, which discharges thru a distributor.[45]

The required Btu per hour output size of a unit can be determined by use of Eq. **4**. The Btu outputs per hour required for 0 to 50 F, 0 to 60 F, 0 to 70 F, and 0 to 80 F rises are given for various cfm requirements in Table 12–51.

Table 12-51 Heat Required to Increase Air Temperature Output,* MBtu per hr[50]

Air flow, cfm	Temperature rise			
	50°F	60°F	70°F	80°F
1,000	54	65	76	86
1,500	81	97	114	130
2,000	108	130	151	173
2,500	135	162	188	216
3,000	162	195	227	259
3,500	189	227	265	302
4,000	216	259	303	346
5,000	270	324	378	432
10,000	540	648	756	864

* For vented units, if input Btu per hour ratings are used, multiply above figures by 1.25.

Installation of Area Ventilation Systems. Locate the system (Fig. 12-58b) close to an outside wall so that duct runs can be kept short. Filters may be provided in the inlet duct when conditions warrant their use. An automatic gravity damper opens when the make-up air blower starts, to prevent cold outside air from chilling the heat exchanger and creating drafts when the blower is not operating.

The blower should be sized to handle all the make-up air flow load plus any static (resistance) developed in the make-up air system (inlet duct, filters, distribution ducts, etc.). System pressure loss is usually low and can be handled by a relatively small motor run at slow speeds.

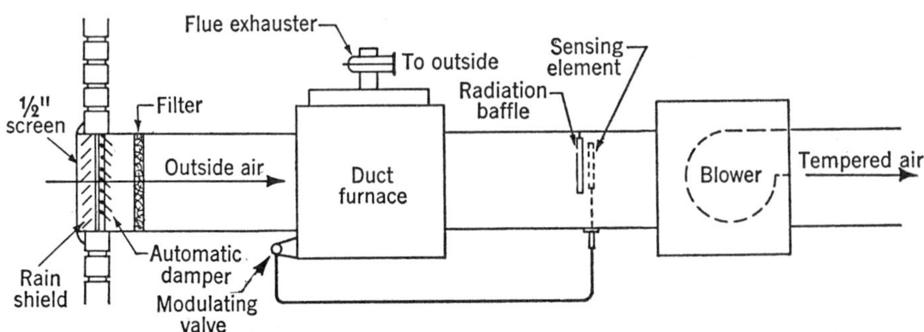

Fig. 12-58b Indirect type of make-up air heating system, consisting of a vented gas-fired duct furnace, blower, and filter section.

Outside air flowing over the heat exchanger during the colder seasons will cause condensate to form inside the heating tubes. When entering air temperature will be below 40 F, the use of stainless steel heat exchangers is recommended. Because corrosion is also promoted by the burners cycling on and off when the entering air temperatures are less than 40 F, a modulating control that will vary the height of the flame and maintain a constant discharge temperature is necessary.

Low firing rates and cold entering air chill flue gases below the temperature required for gravity venting. Therefore, make-up air systems should have a power-driven flue gas exhauster.

The sensing element of the modulating gas valve should be placed in the discharge air stream so that it senses the average discharge air temperature. It should be shielded by a baffle from radiant heat, as illustrated in Fig. 12–58b. A room thermostat is not used.

Make-up air should be discharged into the building at or slightly below room temperature to prevent interference with the normal heating system.

Air Intakes. Make-up air should be clean and fresh, and taken from an outside source. The most common means of introducing it are thru existing window openings, openings in the building wall, roof penthouses, or hooded stacks extending above the roof. In all cases the following precautions should be observed:

1. Make sure that the air inlet is not too close to stacks, chimneys, or exhaust outlets from which products discharged from the building can enter with incoming make-up air.
2. Direct the air inlet in such a way that any outside contaminants will not be blown into it by prevailing winds.
3. Avoid air inlet openings just above roof level where incoming air may draw in dust and dirt off the roof.
4. Provide sufficient inlet air to avoid high air velocities that may entrain dirt.
5. Screen the inlet, using mesh small enough to keep out birds, rodents, etc., and large enough not to become clogged with dirt.
6. Provide louvers or a 45° downward elbow to prevent direct entrance of rain or snow.
7. If extended shutdowns of the system are probable when the rest of the plant is operating, a set of automatic louvers should be installed to shut off air intake entirely.

Direct Gas-Fired Make-up Air Systems. These systems are those in which the flue products mix directly with the air being heated (e.g. Fig. 12–58c). Two factors make the direct gas-fired make-up heaters safe and feasible.[51] These are: (1) high rate of dilution of the flue gases by the make-up air stream; and (2) constant air flow with no recirculation. The dilution rate can be examined from the viewpoints of explosion or fire hazard, and air contamination.

Standard electronic, fast-acting combustion safeguards together with interlocks are a part of every heater. They are used for a check of all associated equipment. Assuming, nevertheless, total failure of the safety equipment resulting in a full, unignited fuel flow, no fire or explosion hazard exists because the fuel is highly overdiluted and thus becomes nonflammable immediately upon leaving the burner.

For example, assuming the extreme case of a high temperature rise heater of 110°F, 2 Btu are allotted per cu ft of make-up air at full burner capacity. Approximately 100 Btu per cu ft are required for complete combustion; the lower flammability limit requires approximately 50 Btu per cu ft. It is apparent that the safety factor is 25 to 1 at the greatest extreme, over and above the safety effects of the combustion safeguard system. A perfectly safe demonstration can be made by circumventing the safety controls to open the main fuel valve without ignition. Ignition can be applied at any time thereafter—even hours later—without harm.

Only two contaminants warrant serious consideration. These are carbon dioxide and carbon monoxide.

Carbon Dioxide. In small quantities CO_2 is a harmless gas. The U.S. Public Health Service states that a concentration of 5000 ppm is safe for continuous exposure (see Table 1–30). Assuming again the extreme case of a high temperature rise heater of 110°F, at full input the maximum CO_2 generation will run less than 2000 ppm. This represents a safety factor of 2.5 to 1 under the worst possible condition. Variations due to lesser input or variations in combustion can only reduce CO_2 generation and consequently can only increase the safety factor.

Carbon Monoxide. Although CO is not a product of complete combustion, it can occur where insufficient combustion air is furnished or where combustion has been arrested before being brought to completion.

In an effort to determine possible concentrations of CO in the air flow under conditions of extreme maladjustment of fuel–air ratio, exhaustive tests of make-up air heating were conducted under closely simulated field operating conditions.[52] The tests were made using a burner with sufficient maximum capacity to give a 200°F rise in air temperature (more than twice the normal size burner for the volume of air flow involved). The air supply was sampled at a number of points downstream of the burner with a precision CO detector. The amount of CO detected in the air flow when the burner was properly adjusted was not measurable. When the burner was in extreme misadjustment, resulting in a strong odor of aldehydes, the maximum amount of CO detected in the air flow was less than 0.001 per cent, or less than 10 ppm.

The U. S. Public Health Service publishes the following information regarding the effects of CO (also see Fig. 6–24):

100 ppm—safe for continuous exposure
200 ppm—slight effect after six hours
400 ppm—headache after three hours
900 ppm—headache and nausea after one hour
1000 ppm—death on long exposure
1500 ppm—death after one hour

Most codes written for the control of industrial and commercial breathing atmospheres specify that CO concentration shall not exceed 100 ppm, considered safe for continuous exposure (Table 1–30). At a maximum heater output of 10 ppm, it is evident that there is no CO hazard; a safety factor of at least 10 to 1 will exist at worst. The discharge air of the heater is purer than the air in many outdoor areas, e.g., a busy expressway.

A further safety consideration is that incomplete combustion of fuel gas gives off the distinctive odor of aldehydes long before CO concentration becomes serious. Thus, while

CO is odorless, the aldehyde odor would give sufficient warning to occupants of the building that there is trouble.

Water Vapor. The outside air drawn into a make-up air system in winter is quite dry, becoming drier the colder it gets. Even saturated air at 0 F holds only 0.0008 lb of water per lb of dry air.

Reference to standard psychometric charts will show that raising the temperature of this saturated air from 0 F to 70 F will drop its relative humidity from 100 per cent to about 6 per cent. Thus, the addition of the water vapor in the products of combustion actually is beneficial in partially restoring humidity to the air during extremely cold weather.

From Table 12-52 it will be seen that the 6 per cent relative humidity of 0 F saturated air heated to 70 F will be increased to about 10 per cent because of the addition of the

Table 12-52 Effects of the Water Vapor in the Products of Combustion

(an example)

	Assuming saturated incoming air at	
	0 F	65 F
(a) Humidity ratio, H_2O per lb of dry air	0.0008 lb	0.0132 lb
(b) Relative humidity	100%	100%
(c) Heating this air to 70 F will reduce the relative humidity to	6%	84%
(d) Amount of natural gas* required to heat 1 lb of incoming air to 70 F	0.017 cu ft	0.0013 cu ft
(e) Water added in products of combustion, per lb of incoming air		
(H_2O produced is 2.246† lb per lb of CH_4. CH_4 weighs 0.04242 lb per cu ft; 2.246 × 0.04242 = 0.0954 lb H_2O produced per cu ft of CH_4 consumed in the heating)		
With 0 F air: 0.017 × 0.0954 =	0.0016 lb	
With 65 F air: 0.0013 × 0.0954 =		0.00012 lb
(f) Total H_2O per lb dry air after heating (a + e) =	0.0024 lb	0.01332 lb
(g) Corresponding relative humidity	15%	84½%

* 1000 Btu per cu ft.
† See Table 2-89.

water in the products of combustion from direct gas-fired burners. Assuming the unlikely situation of 65 F saturated air (column 2 of Table 12-52), heating will reduce the relative humidity to 84 per cent, with the water from direct gas-fired combustion bringing it up only to 84.5 per cent.

The conclusion, therefore, is that the relative humidity of air delivered into a building thru a direct gas-fired make-up heating system will always be less than the relative humidity of the outside air, even when the water of combustion is added. In very cold weather the water of combustion can be a valuable humidifying agent.

Burner Types. Types in use include the partial premix, full premix, and raw gas.

Partial Premix Burners. These employ fully premixed fuel and air at low firing rates. Combustion occurs beneath the ignition rails (see Fig. 12-59,a). Rich mixtures form as the firing rate is increased. The high-velocity air from the make-up air stream creates air jets thru the openings in the mixing plates (Fig. 12-59,a). Each air jet creates a vacuum pocket on the adjacent flat surface of the mixing plate. This pocket draws in the rich mixture for mixing with the air and subsequent ignition.

For severe service, alloy steel ignition rails and stainless steel mixing plates are used. Turndown ratios of 25 to 1 or 30 to 1 are readily available. For example, one commercially available partial premix burner has a minimum input of 15,000 Btu per hr and a maximum input of 450,000 Btu per hr, with air velocities that range from 2000 to 3000 fpm. With turbulent, uneven air flow or with propane or butane (other than for use as a stand-by fuel) the minimum capacity will be 23,000 Btu per hr.

Full Premix Burners. These units, with turndown ranges from 5 to 1 to 7 to 1, have limited application. Because of the small turndown range, dual-burner arrangements provide the full range of throttling required. The burners are generally specified in two separately controlled sections so that the total heat release may be divided in ratios of ¾ to ¼ or ⅔ to ⅓. Delivery temperatures can vary widely, necessitating a sophisticated control system.

Raw Gas Burners. These burners (Fig. 12-59,b) are similar to the partial premix burners except that the "burner" is primarily a distributing manifold for undiluted fuel gas. The diverging perforated combustion baffles shown are designed to take all air for combustion from the make-up air stream. This burner design is capable of a 20 to 1 or 25 to 1 turndown range, with air velocities in the range of 2600 to 3000 fpm.

The raw gas burner of the type shown in Fig. 12-59,c makes use of a gas volume-operated, free-floating piston. As the air temperature leaving the heater decreases, a modulating gas flow control valve opens. The gas pressure under the free-floating piston in the vertical distribution header causes the piston to rise. Thus, the number of orifices in operation depends on the amount of gas flowing thru the control valve in direct response to the leaving air temperature. The design is based on a maximum temperature rise of 90°F; temperature control within 3.6°F is obtained by use of 25 orifices. All air for combustion comes from the make-up air stream. This burner is used in a packaged make-up air heater that has outputs of 1 MM to 5 MMBtu per hr with outlet air velocities ranging approximately from 2000 to 2100 fpm.

Selection of Type of Burner.[52] In choosing the type of gas combustion system for use in the heating of process make-up air, the following basic requirements should be noted:

1. Heat must be introduced into the air stream in a manner that effects a uniformity of temperature.
2. The system must be able to throttle smoothly over a total range of turndown that can be as large as 20 or 25 to 1.
3. The burner must be able to produce complete combustion of the fuel throughout the turndown range.
4. Flame retention and stability must be maintained in high-velocity flow of air.
5. The system should be as simple and as economical in first cost and operating cost as feasible, consistent with top quality performance.

For simplicity and low initial and operating cost, the wide-range line burner system is preferred. Its tolerance for a large range of fuel-air ratio adjustability makes it completely

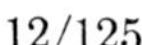

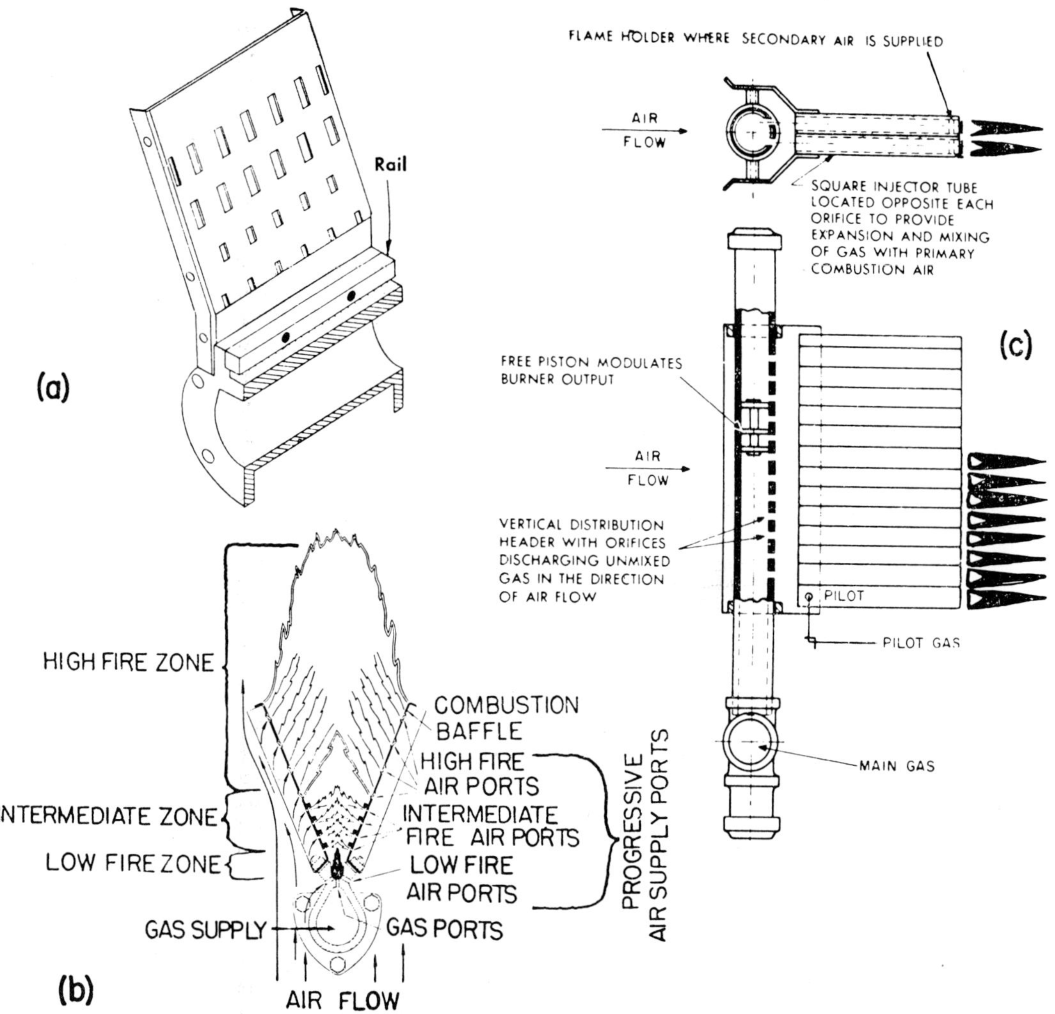

Fig. 12-59 Burners for make-up air heating: (a) section thru partial premix burner showing mixing plate and ignition rail (Maxon Premix Burner Co.); (b) flame pattern of a raw gas burner; (c) raw gas burner operated by a free-floating piston. (Westinghouse Electric Corp.)

insensitive to minor changes in gas pressure or other operating conditions.

The burner manufacturer's installation instructions should be followed, especially with regard to the velocity of air flow across the burner.

Calculation of Air Velocity at Burner.[52]

$$v = Vf_t/A \qquad (15)$$

where: v = flow velocity across burner, fpm
V = air flow rate across fan, cfm
$f_t = 460 + t_o/460 + t_i$
where: t_o = outdoor air temperature, °F
t_i = air temperature *at downstream fan*, °F
A = net cross-sectional area of duct at burner, sq ft

Example: A burner with an effective frontal area of 10 sq ft is installed in a 6 × 10 ft duct (cross section). A downstream fan passes 100,000 cfm at 70 F. The outdoor air temperature is −10 F. Using Eq. **15**, find the flow velocity, v, across the burner.

$$v = 100{,}000 \times \frac{460 - 10}{460 + 70} \Big/ [(6 \times 10) - 10] = 1700 \text{ fpm}$$

In the event that this velocity is more than that recommended by the burner manufacturer, the duct should be increased in size accordingly. On the other hand, if a higher velocity is desired to give optimum burner system performance, it can be secured by installing a profile plate at the burner location. A profile plate is a sheet metal plate or plates installed at right angles to the air flow and in the plane of the face of the burner so as to reduce the cross-sectional area at the burner.

Control System.[51] In spite of the inherent safety of make-up air heating systems, the control system is applied according to standard practices for the combustion of fuel gases. The controls perform three fundamental functions: (1) ignition and flame supervision; (2) safety interlocks; and (3) temperature control.

Ignition and flame supervision are carried out by the flame safeguard relay, which is usually an instantaneous electronic type using a flame rectification rod for flame detection. The relay may be a simple one providing for spark ignition, and pilot and main flame sequencing, or it may be a full programming type providing a more elaborate sequence, including preignition and postignition purges.

The control system will also carry interlocking arrangements to prevent start-up or cause shutdown if a faulty condition develops. Commonly, these arrangements will include: (1) a blower starter interlock; (2) an air flow proving device; and (3) an excess temperature limit. All three devices prove blower operation in one manner or another so that failure of any one will not cancel proof of air flow. Other auxiliary interlocks may be supplied, such as a damper limit switch, gas pressure switches, and an outdoor temperature thermostat to turn on the heating system.

Outlet Air Temperature Control. This device is by necessity of the modulating type. Since the outdoor temperature is infinitely variable, the temperature control must likewise be variable.

For smaller heaters the control may be a direct acting type of valve. Electrically or pneumatically operated types can be installed on any size heater. The modulating control may be of the reset type, which will automatically raise the outlet temperature a few degrees as the outdoor air grows colder.

Low Temperature Cutout. This device, which shuts down the blowers in freezing weather if for any reason the burner fails to function, protects against the possibility of water piping inadvertently bursting in an unoccupied building.

Fuel Requirements of Commercial and Industrial Buildings[53]

Because of the wide range of operating conditions, gas consumption cannot be determined as readily and as accurately for commercial and industrial buildings as for domestic heating.

An approximate method of calculating a season's fuel requirements is to arrive at an average hourly temperature and use this as a basis for a degree-day computation.

Example:

Temp, °F	×	hr per day	×	days per week	=	deg-hr per week
70	×	10	×	5	=	3,500
60	×	14	×	5	=	4,200
50	×	24	×	2	=	2,400
						10,100

$$\frac{10{,}100 \text{ deg-hr per week}}{168 \text{ hr per week}} = 60 \text{ F avg}$$

$$\frac{\text{Btu loss per hour (design basis)} \times 24}{\text{Design temperature difference}} = \text{Btu loss per °D} \quad \textbf{(16)}$$

$$\frac{\text{Btu loss per °D} \times \text{°D per season (at average temp carried)}}{\text{Efficiency of heating system}} = \text{Btu required per season} \quad \textbf{(16a)}$$

The number of degree-days per season at the average temperature for the locality can be taken from Weather Bureau data. It is suggested that 5 degrees be added to the actual night temperature in the foregoing calculation to allow for the additional pickup load in the morning heating period.

RESIDENTIAL AND COMMERCIAL COOLING

Summer air-conditioning equipment using natural gas as the primary source of energy can be divided into two categories. In the first, gas is used *directly* in the equipment and the energy released by the burning gas operates the unit; examples are direct-fired absorption or adsorption dehumidification units, and gas engine or gas turbine driven units. In the second category, or *indirect* application, gas is used in boilers to generate steam or hot water, which in turn operates the air-conditioning equipment. Residential equipment falls in the first category.

Gas-operated air-conditioning refrigeration units are available (1965) in capacity ranges from 3 to 4500 tons. Larger units can be made, but physical size limitations keep them from being commercially feasible or necessary, since multiple-unit installations will meet specifications where a single unit of the required tonnage is unavailable.

The following sizes are available for the indicated systems:

System	Size, tons
Absorption	3–1000
Gas engine	
Reciprocating compressor	5–375
Centrifugal compressor	85–750
Gas or steam turbine, centrifugal compressor	200–3000+

Basic American Standards cover absorption,[54] gas-engine-powered,[55] and combination gas-fired[23] furnaces with an electrically driven compression air-conditioning unit in a common casing. Required equipment includes a main control valve, an automatic valve, an automatic pilot, a gas pressure regulator, a means for limiting refrigerant pressure, and provisions for venting. The testing requirements applicable to the electrically driven compression air-conditioning unit also include UL465, Central Cooling Air Conditioners and ASA B9.1, Safety Codes for Mechanical Refrigeration.

Summer air-conditioning cycles may be classified as: (1) absorption refrigeration; (2) vapor compression refrigeration; and (3) dehumidification.

A comparison of the vapor compression and the absorption cycles shows that both vaporize and condense a refrigerant but that the former uses a compressor unit and the latter an absorber-generator unit for producing the necessary pressure difference between vaporizing and condensing levels. Both cycles require energy for operation—mechanical energy for the vapor compression cycle and heat energy for generation in the absorption cycle.

Dehumidification Cycles. The open absorption and open adsorption cycles are closely related and are known as dehumidification cycles. They differ from all other refrigeration cycles in that they bring the air to be conditioned into direct contact with the sorbent.

Absorption Systems

Gas-fired absorption systems are of two types, those using lithium bromide and those using ammonia. The former are applicable wherever chilled water in the range of 40 F (min) and above is required. The ammonia systems can, where required, maintain an evaporator temperature as low as 32 F.

The systems are fired with gas, either directly or indirectly thru the medium of steam or hot water. The major advantages of absorption systems are summarized in Table 12–53a.

Table 12-53a Major Advantages of Absorption Systems[45]

1. Used with low-pressure steam (2 to 14 psig). Higher steam pressures require the use of a pressure-reducing valve.
2. No major moving parts in the refrigeration system. Easy to operate and maintain, and has long life expectancy.
3. Completely automatic.
4. Most flexible. Automatic controls give smooth throttling characteristics throughout entire range from 0 to 100 per cent of capacity.
5. Rapid response to load changes.
6. Economical to operate. Steam input decreases with drop in load. Uses 20 lb or less of steam per ton-hr at full load.
7. Quiet and free of vibration.
8. Not injured by overloads. All systems include full safety control against damage.
9. Available in a wide range of sizes.

*10. Compact and light in weight.
*11. No special foundations required. Easy to install in either new or old structures.
*12. Particularly adaptable to roof top installation.
*13. High-vacuum operation eliminates need for insulating chiller.
*14. A safe and inexpensive refrigerant water. The permanent absorbent is a nontoxic, nonflammable salt.

* Lithium bromide systems only.

Lithium Bromide–Water Cycle. Water is the refrigerant and lithium bromide is the absorbent. The absolute pressure in the generator and condenser is about 50 to 60 mm Hg and that in the absorber and cooling coil is 6 to 9 mm Hg. Operation is under a vacuum at all times. Performance data are given in Fig. 12–60a.

As shown in Fig. 12–60b, the spent (dilute) solution is heated in the generator to a temperature of about 230 F and the hot regenerated (concentrated) solution is raised by means of vapor lift to the separator, where the refrigerant vapor and the regenerated solution are separated by baffles. Approximately 5 per cent of the water is boiled off from the solution. After the refrigerant (water vapor) leaves the condenser, it flows thru a restriction that creates a pressure barrier to separate the slightly higher absolute pressure in the condenser from the lower pressure in the cooling coil. The refrigerant (water) vaporizes upon entering the cooling coil because of the lower absolute pressure. In other words, the high vacuum in the cooling coil lowers the boiling temperature of water sufficiently to produce refrigeration; the heat of evaporation for the refrigerant is extracted from air passing over the coil, and cooling and dehumidifying are accomplished.

Since lithium bromide in either dry or solution form has a very strong affinity for water vapor, the cool regenerated solution in the absorber absorbs the refrigerant vapor. The resultant spent solution drains back thru the heat exchanger to the generator, and the cycle is repeated.

The heat exchanger conserves heat in the cycle by using the hot regenerated solution from the separator to preheat the spent solution from the absorber. Cooling water is passed first thru the absorber to remove the heat of absorption of the refrigerant (water) into the solution and then thru the condenser. This cooling water may be supplied by a cooling tower or other suitable sources.

The equipment may consist of either two shells or a single shell. In the former, the generator and condenser are located in the upper shell (high-pressure shell) and the evaporator and absorber in the lower shell. In the single-shell machine, the shell is divided in a similar manner.

The smaller size units (3 to 25 tons) operate on the basis of natural circulation; on the larger units pumps are used to reduce the physical size of the equipment. A purge unit removes any noncondensable gas that might accumulate by means of a vacuum pump, an eductor, or a palladium cell.

Double-Effect Absorption Chiller. This chiller (Fig. 12–60c) operates on a modification of the lithium bromide cycle. After the lithium bromide is heated in the "first-effect"

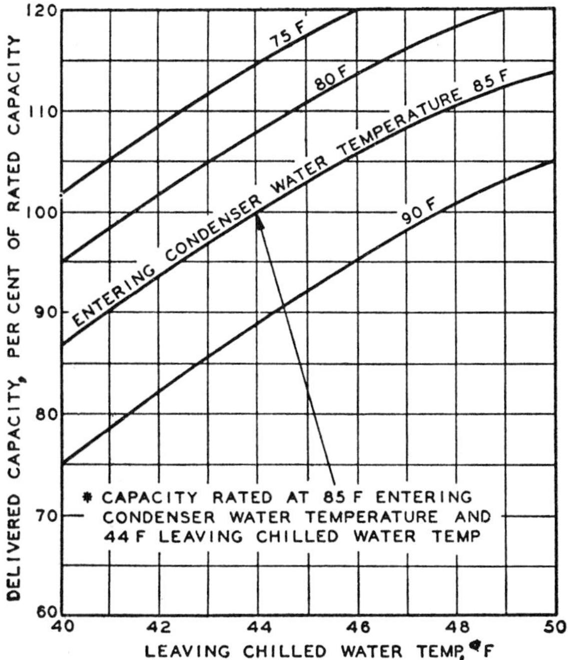

Fig. 12-60a Performance characteristics of lithium bromide cycle water chiller.[56]

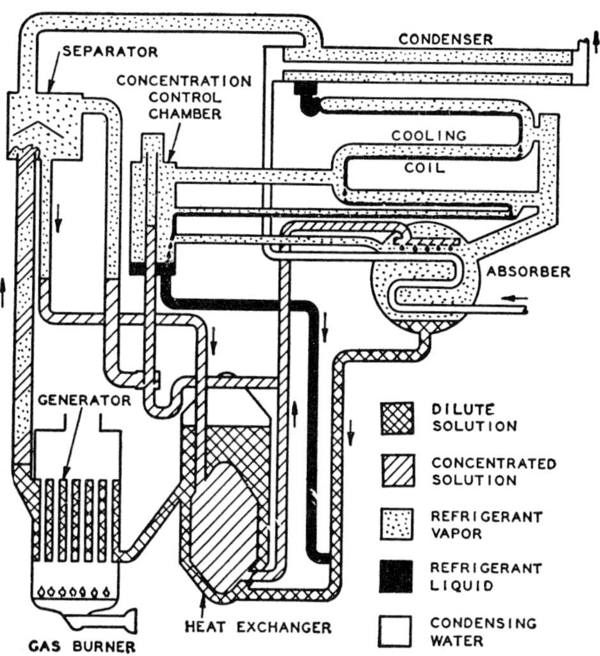

Fig. 12-60b Direct-fired lithium bromide cycle.[56]

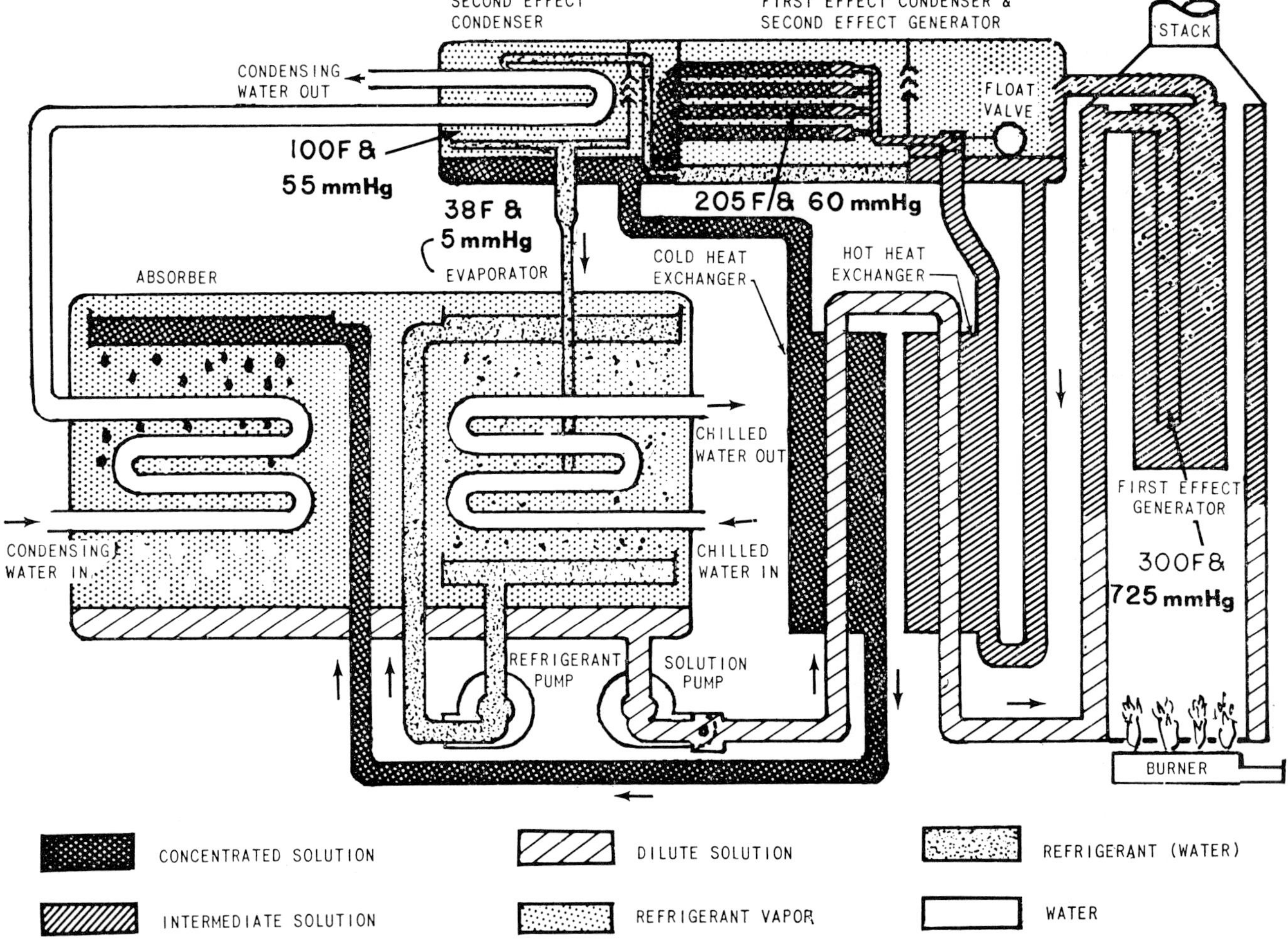

Fig. 12-60c Double-effect absorption chiller.

generator, the concentrated lithium bromide passes thru a coil which serves as a "second-effect" generator.

Since pressure in the second-effect generator is lower than that in the first, the solution boils once more but at a lower temperature. This second coil is heated by the condensing vapor from the first distillation step, increasing the efficiency of the cycle. Vapor from the second-effect generator is condensed on a water-cooled condenser coil. The fully concentrated lithium bromide solution then acts as the absorbent for the refrigerant vapor. The combined condensate from both generators is delivered to the evaporator, and the resultant refrigerant is pumped over the chiller coil, where it cools the coil by evaporation. Vapor from the evaporation is absorbed into the concentrated lithium bromide solution. The dilute solution is then pumped back to the first-effect generator to start thru the cycle again.

Ammonia-Water Cycle. Thermodynamically the ammonia-water absorption cycle (Fig. 12-60d) is identical to the lithium bromide–water cycle. In the ammonia-water cycle, however, water plays a different role—it is the absorbent. Ammonia is the refrigerant and consequently temperatures below 32 F can be reached. The other principal difference between the two absorption cycles is that the ammonia-water cycle operates in a pressure range comparable to that of compression systems. Table 12–53b shows typical cycle conditions.

Although the major components of all absorption systems are the same, the characteristics of specific refrigerant-absorbent combinations may require auxiliary devices in order to improve the operating efficiency of the cycle. In the ammonia-water cycle, the volatility of water under the generator conditions is the major consideration.

The proportion of water in the ammonia-water vapor leaving the generator depends on the composition and temperature of the liquid in the generator. In a simple cycle, no provision is made for removing water vapor from the refrigerant vapor before it enters the condenser. The water vapor becomes liquid in passing thru the condenser on its way to the evaporator (chiller) and, being almost nonvolatile under evaporator conditions, accumulates as evaporation proceeds, thus raising the boiling point of the refrigerant. To keep such an evaporator operating, the accumulated water would have to be purged to the absorber, either continuously or periodically, as a liquid of high ammonia content.

Two devices commonly used to reduce the amount of water vapor reaching the condenser and evaporator in the actual ammonia absorption cycle are the analyzer and the rectifier. The analyzer consists of a chamber thru which the

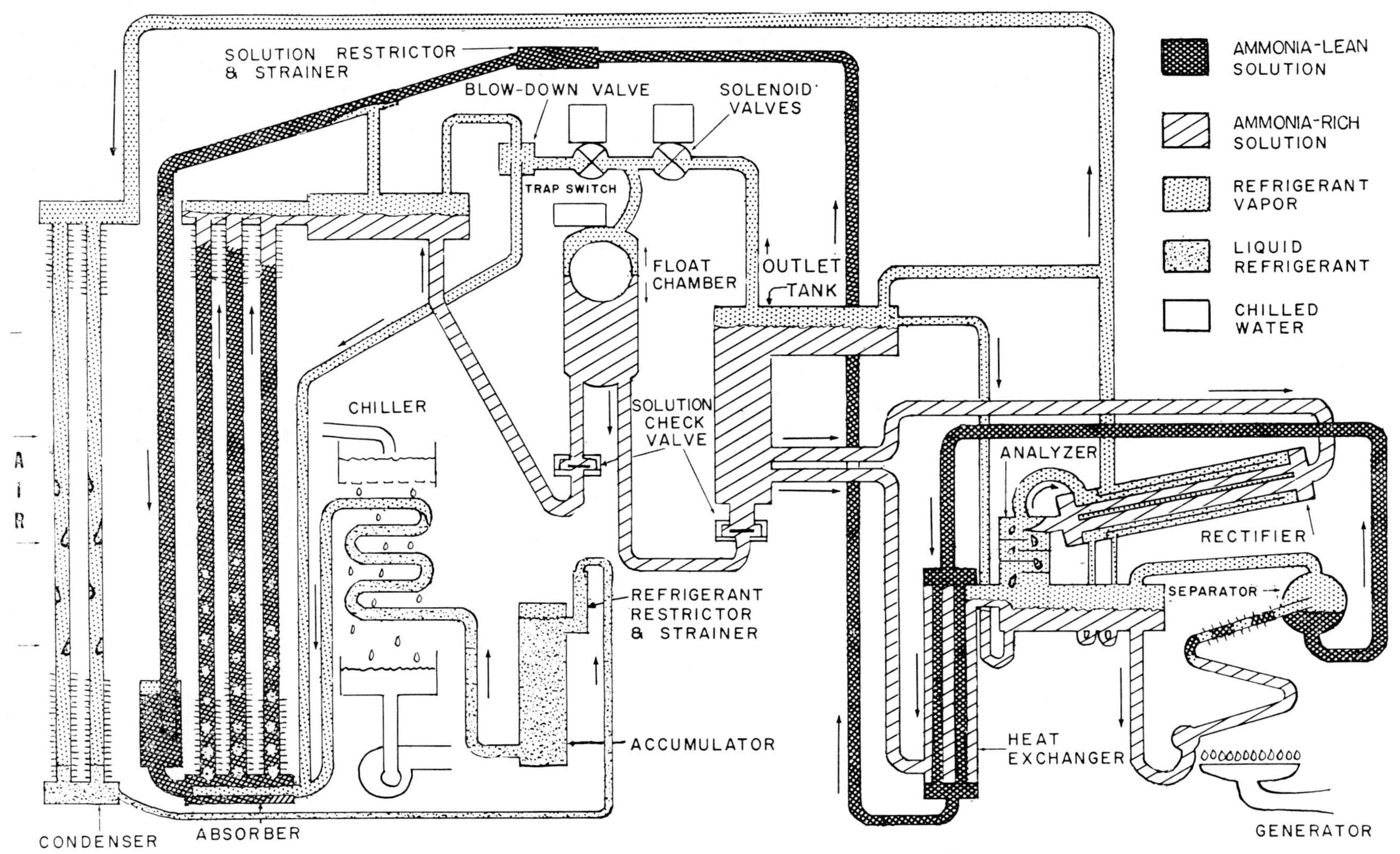

Fig. 12-60d Direct-fired ammonia-water cycle.[57]

Table 12-53b Cycle Conditions for an Ammonia-Water Refrigeration Machine[57]

(entering air at 95 F dry bulb and 75 F wet bulb; chilled water output at 44 F, 2.5 gal per min-ton)

Point in cycle	Press., psia	Temp, °F	Concentration: Liquid, % H_2O	Concentration: Vapor, % NH_3	Calc. weight flow, lb/min-ton
Ammonia-lean solution leaving generator	286.4	305	78.0	76.5*	2.150
Ammonia-lean solution leaving heat exchanger	286.4	170	78.0	...	2.150
Vapor leaving analyzer	286.4	257	...	91.0*	...
Vapor leaving rectifier	286.4	175	...	99.2*	0.488
Liquid leaving condenser	286.4	110	0.8*	...	0.448
Vapor leaving chiller	69.0	52	...	100.0*	...
Liquid residue leaving chiller	69.0	52	23.0*	...	...
Ammonia-rich solution leaving absorber	69.0	137	63.6	...	2.640
Ammonia-rich solution leaving trap	286.4	137	63.6	...	2.640
Ammonia-rich solution leaving rectifier	286.4	245	63.6	93.4*	...

* Values read from property tables.

vapors leaving the generator pass in counterflow contact with the ammonia-rich solution flowing to the generator. A bubble column, packing, or a series of baffle plates may be used to make the vapor-liquid contact more effective. As the vapor passes upward thru the analyzer, it is enriched with ammonia, as well as cooled, and the solution is heated. Thus, the vapor going to the condenser is cooler and richer in ammonia, and the heat input required at the generator is decreased. The vapor flows from the analyzer into the rectifier. The rectifier is a heat exchanger set before the condenser (or it may be the inlet to the condenser) arranged in a manner to enable its condensate to drain to the analyzer or the generator. The warm refrigerant vapor from the analyzer is cooled by the ammonia-rich solution flowing to the generator.

To permit flow from the comparatively low-pressure inlet tank to the high-pressure outlet tank, a float assembly is alternately "equalized" thru the opening and closing of low-side and high-side solenoid valves.

As in the case of the lithium bromide–water cycle, it is universal practice to conserve heat in this cycle by the use of a heat exchanger, which uses the hot ammonia-lean solution from the generator to preheat the ammonia-rich solution from the absorber. This improves the cycle efficiency by reducing the heat input required. It also reduces the cooling water requirements. Still further improvements in the efficiency of the operating cycle may be obtained by subcooling the liquid ammonia from the condenser with the cool refrigerant vapor from the evaporator. This increases the refrigerating capacity of the refrigerant at no cost except that of slightly warmer cooling water leaving the absorber.

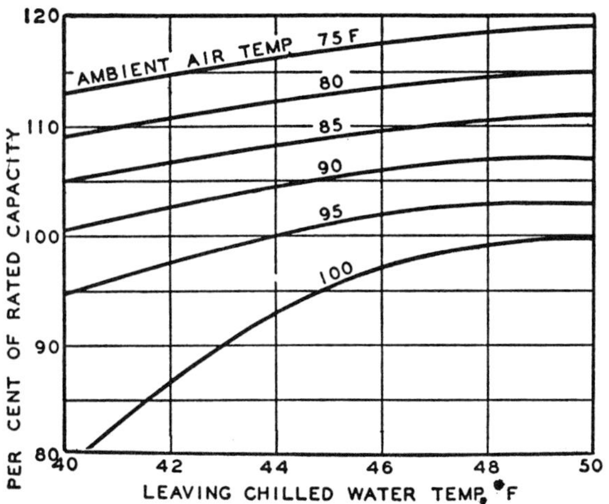

Fig. 12-60e Performance characteristics of an ammonia-water cycle water chiller.[56]

Not all the auxiliary equipment and processes described are used in every application. Consideration must be given to the efficiency improvement vs. the cost of auxiliaries. There is a combination which will yield the best economic balance of initial costs to operating expenses.

Coefficient of Performance. In the absorption cycle, the coefficient of performance (*COP*) is the ratio of the heat absorbed in the evaporator to the heat absorbed in the generator, and is given by Eq. **17**.

$$COP = \frac{T_1(T_4 - T_3)}{T_4(T_2 - T_1)} \tag{17}$$

where: T_1 = evaporator temperature, °R
T_2 = condenser temperature, °R
T_3 = absorber temperature, °R
T_4 = generator temperature, °R

Effects of Variables on Equipment Capacity. A decrease in condenser water temperature increases the capacity of a unit (see Fig. 12-60a). Figure 12-60e shows the effect of decreasing ambient air temperature on an air-cooled unit.

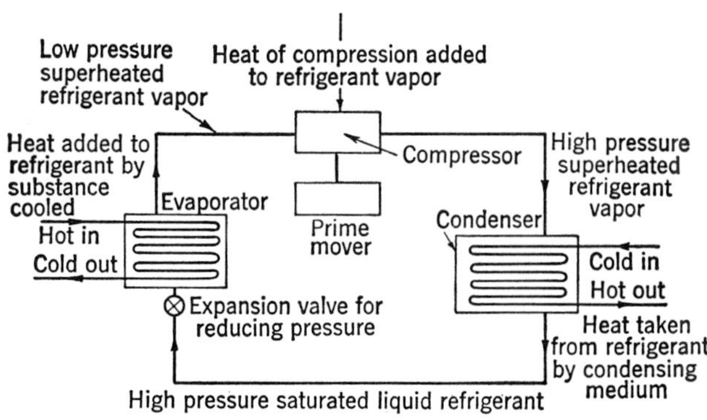

Fig. 12-61 Schematic of a vapor compression cycle.

Condensing water may be supplied from wells, lakes, or cooling towers; however, water treatment is desirable to eliminate corrosion or excessive scaling. Condensing water must be maintained above 75 F;[45] this is usually accomplished with a three-way blending valve that allows the water to by-pass a cooling tower until 75 F is reached. Cooling towers should be sized to dissipate the heat of activation as well as the heat taken from the area to be cooled.

Figure 12–60a shows the effect of changing the temperature of the water leaving the chiller. For example, increasing the temperature of the leaving chilled water from 44.0 to 47.8 F increases the capacity for handling the sensible cooling load by 10 per cent for a fixed condenser water temperature of 85 F. The capacity of the unit for meeting dehumidification requirements, however, will be decreased.

Controls. The devices which may be used include:

1. A low temperature cutout to prevent low evaporator temperatures.
2. A liquid level switch that will stop the unit when the refrigerant level is too low or too high in the unit.
3. A cooling water pressure switch to ensure cooling water supply flow.
4. A chilled water flow switch to stop the unit when the chilled water flow is reduced below design limits.
5. A high-pressure cutoff switch on ammonia-water units to limit pressure.
6. A high-temperature limit switch on a direct-fired unit to limit temperatures in the generator and thus prevent dissociation or other changes in the refrigerant, solvent, or absorbent.

Vapor Compression Refrigeration Cycle

In the vapor compression cycle (Fig. 12–61) a mechanical compressor is used to pump the refrigerant vapor from the evaporator to the condenser. Vapor from the evaporator is compressed and delivered to the condenser, where it is liquefied. The heat of vaporization is absorbed by the condenser cooling medium—water or air at normal temperatures.

Liquefied refrigerant collects in the bottom of the condenser or in a separate receiver. From there it is fed back to the evaporator thru suitable control valves, and the cycle is repeated. The compressor can be powered by a natural gas engine, a natural gas turbine, or a steam turbine. Steam for the steam turbine can be generated by natural gas.

Engine-Driven Reciprocating Systems. Factory-assembled, engine-driven reciprocating compressor systems are available in sizes from 5 to 150 tons; larger units are field assembled. Advantages of these systems include a low operating cost and a high degree of flexibility in some models, effected by combining speed variation with cylinder unloading. Units can achieve capacity control by modulation of engine speed to approximately 50 per cent of rated speed; for further capacity reduction, the compressor may be unloaded in increments.

Systems can be used with remote direct expansion coils located in air handlers or ductwork, and are adaptable to air-cooled condensing, evaporative condensing, or water tower applications.

Advantages. These systems operate on approximately 8 to 13 cu ft of 1000-Btu natural gas per ton-hr (see Fig. 12–221a). The cooling tower pump can be driven by the engine. Engine jacket water and exhaust heat can be recovered and used to provide boiler water preheat or used as reheat on a dehumidification cycle at no additional cost for energy.

Most engine-driven units are equipped with a control that provides for idle speed starting and a mechanical override control on the compressor's inherent cylinder loading mechanism. This results in gradual cylinder loading at idle speed and then gradual increase in engine speed over a period of 2 or 3 min. The net effect on the system is to provide very low suction line velocity and gradual suction pressure reduction (and consequent slow opening of the thermal expansion valve) during start-up while the expansion valve remote bulb is cooling down from room temperature to suction line temperature. In other words, the flow control valve is given time to gain control at flow before the compressor begins to pump at full capacity.

Close humidity control is also effected. The typical engine-compressor unit varies both speed and cylinder loading, cycling on and off infrequently. Since the speed and pumping rate, rather than the suction pressure, change when the load varies, the cooling temperature is held constant. The result is the very close humidity control from 5 thru 75 tons on a typical 75-ton unit.

Another advantage claimed for the gas engine is that compressors last longer when driven by variable-speed engines, since they operate for a large portion of the time at 1000 to 1200 rpm instead of cycling on and off at a fixed speed. Also, compressor life is extended by the virtual elimination of refrigerant liquid floodback, which tends to dilute the compressor lubricating oil and damage plates and bearings.

Design Considerations. The following are some of the considerations that should be observed when designing any engine-driven system:

1. Exhaust lines must be vented out of doors or into a flue. Therefore, adequate provision must also be made for combustion air flow into the equipment room.
2. The allowable noise level for heating and air-conditioning equipment varies from one sone (conference room) to 12 sones (tabulating areas).[58] One sone is defined as equal to 1000 cps tone at 40 dbC (decibel C scale). When commercial grade mufflers are installed, engine-driven compressors fall well within this range. Also see Fig. 12–232.
3. Suction line piping risers should be selected so as to give 1500 fpm gas velocity at minimum load. Dual risers should be used where coil sections exceed 10 tons and piping must rise when leaving the evaporator coil.
4. Forced draft cooling towers should be oversized 30 per cent in areas where the design wet bulb temperature is 78 F or less. In areas with a higher design temperature, towers should be oversized 40 per cent.[59] The exact heat rejection for a particular engine should be obtained from its manufacturer for use in sizing the tower.
5. Evaporator coils for over 10 tons capacity should be split in order to permit operation at light loads. On larger single-section coils, pilot-operated thermal expansion valves should be specified.
6. Power wiring recommended is: three No. 10 AWG wires with 30 amp fused disconnect for 208 to 230 volts, a-c, three-phase or single-phase circuits.

Engine-Driven Centrifugal Units. Completely engi-

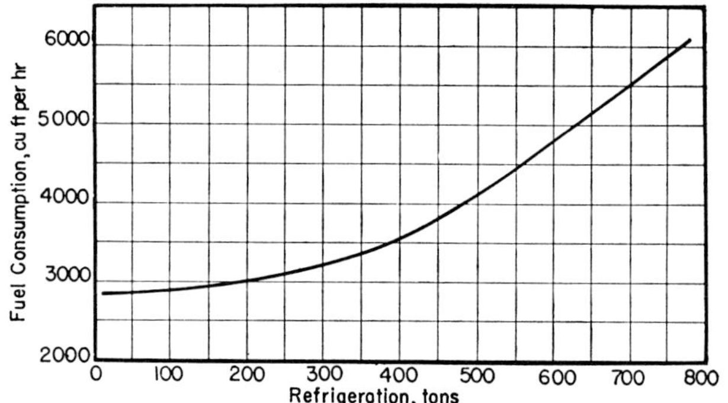

Fig. 12-62 Fuel consumption of a specific natural gas engine driving a 750-ton centrifugal compressor operating under varying load conditions.[60] Heating value of fuel is equal to 1030 Btu per cu ft (HHV).

neered, "packaged" water chilling systems are available in sizes from 85 to 750 tons. These units may be equipped with manual or automatic start-stop systems and engine speed controls. Above 100 or 125 tons, it is generally advisable to consider use of centrifugal and absorption machines.

The centrifugal compressor is driven thru a speed increaser. Engine speed control provides economical operation over a wide output range. In Fig. 12–62, for example, the following relationship may be seen:

Output, per cent	100	33
Fuel, MBtu per ton-hr	8.0	12.8

Heat Pump Drives. An additional economic gain may be had by operating a gas engine-driven air-conditioning unit as a heat pump to provide heat during the winter months. Using the same equipment for both heating and cooling reduces capital investment. A gas engine drive for heat pump operation also makes it possible to recover rejected heat from the engine exhaust gas and jacket water.

A heat pump can be defined as a thermodynamic cycle that will heat or cool as required. Any substance at a temperature above −459.7 F contains some heat. The requirement is to remove the heat from a low-temperature substance and elevate it to a usable temperature level. The gas engine-driven heat pump transfers Btu by removing them from a heat source and delivering them at a higher temperature to the area requiring the heat. When planning to use an air source for Btu, three items must be considered: (1) the maximum and minimum outdoor temperatures encountered; (2) the hours at which these temperatures occur; and (3) the total number of hours at each temperature.

A heat pump system is for the most part made up of standard components. Variations in systems come from the manner in which the components are arranged and controlled.

Water-to-Water Heat Pump.[61] This unit consists of a reciprocating or centrifugal water chiller with a water-cooled condenser; controls are modified for heat pump service.

Figure 12–63a shows the water flow diagram for a water-to-water heat pump supplying *either heating or cooling.* For the sake of clarity, the refrigerant flow is not shown, but it would be the same as that on any chiller with a water-cooled condenser. Since the refrigerant flow is never reversed, the chiller is always a chiller and the condenser always a condenser. The water flow, however, is reversed.

When *heating* is required, the circulating water from the coil(s) passes thru the condenser, where it absorbs the heat rejected by the compressor. At the same time, the refrigeration load is absorbed by the water circulating thru the chiller. This water can be the source water, if it is clean enough, or an intermediate water supply, which transfers the refrigeration load to the source water by means of the auxiliary water-to-water heat exchanger. For this operation, all valves will have ports b and c open.

When *cooling* is required, the circulating water from the coil(s) passes thru the chiller, where it absorbs the required refrigeration load. The heat rejected by the compressor is transferred directly to the source water or indirectly to it by means of the intermediate water supply and auxiliary water-to-water heat exchanger. For this operation, all valves will have ports a and b open.

Figure 12–63b shows the water flow diagram for a water-to-water heat pump for *simultaneous heating and cooling.* This cycle is basically the same as that shown on Fig. 12–63a except that an extra coil has been added. The coil is used in the same circuit as the auxiliary water-to-water heat exchanger. Any auxiliary heating or cooling required would be supplied by the extra coil; otherwise, that load would be absorbed by the source water. When the greatest demand is for heating, ports b and c of the valves would be open; when the greatest demand is for cooling, ports a and b of the valves would be open.

Air Source Heat Pump.[61] Both air-to-air and air-to-water are available. Figure 12–63c shows the actual coefficients of performance for a gas engine-driven air-to-water heat pump operated by the Washington Gas Light Co.

Both systems can also be designed for simultaneous heating and cooling. In an air source heat pump, the air is the sink

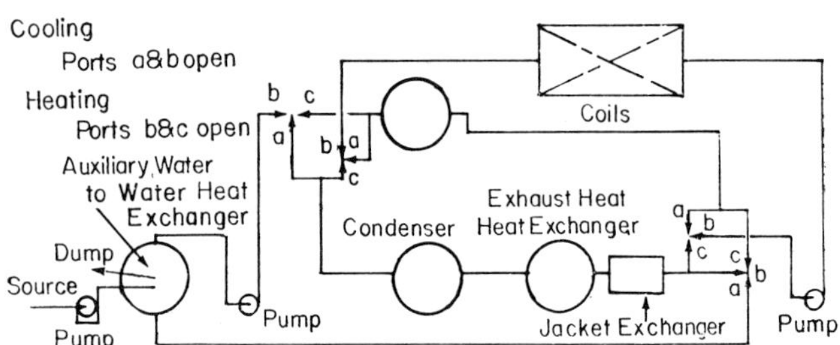

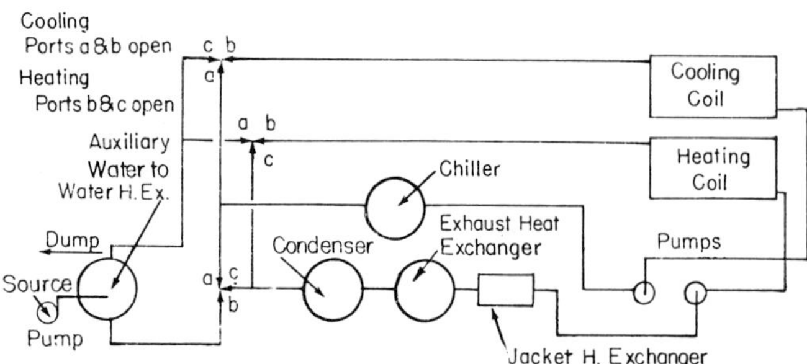

Fig. 12-63a (left) Water flow diagram for a water-to-water heat pump arranged for alternate heating and cooling operation.[61]
Fig. 12-63b (right) Water flow diagram for a water-to-water heat pump arranged for simultaneous heating and cooling operation.[61]

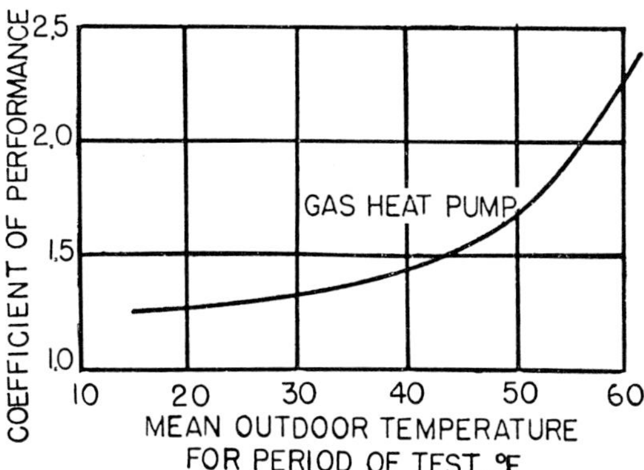

Fig. 12-63c Gas heat pump coefficient of performance vs. outdoor temperature.

or condenser in the summer and the source or the evaporator in the winter. In an air-to-air heat pump, the conditioned space is the source or evaporator in the summer and the sink or condenser in the winter. By reversing source and sink, the conditioned space will be heated or cooled as required.

Many types of heat pump systems have been designed over the years. Three major systems are (1) direct expansion; (2) flooded; and (3) force feed.

Ground Source Heat Pump. On peak demand days there will be more heat available from the earth than from the air but less than from well water at the same location. If ground coils are used, either a direct expansion or a secondary refrigeration system may be used. The maintenance of buried coils is a major disadvantage.

Another disadvantage is the variation of ground temperatures in a relatively small area. Even if test borings are made that give accurate temperatures, the fact that these temperatures will vary somewhat with changes in ground moisture content makes the coil design uncertain.

Simultaneous Heating and Cooling. In some industrial and commercial buildings, the room exposures, room arrangements, and variation in occupancy and use require that simultaneous heating and cooling be available throughout the year. The heat pump offers a means of transferring heat from those portions of the building requiring cooling to other areas requiring heating.

The operating cost of such a system is often favorable since heat removed from areas requiring cooling will be made available directly to those requiring heating. In a conventional heating and cooling system, the mechanical cooling system would be operated for certain portions of the building and the boiler for other areas. Throughout most of the spring and fall seasons the amount of heat required is small, and the boiler would operate at a low load factor.

Heat Balance. Table 12–54 shows the total energy available from a typical engine-driven heat pump operating at one brake hp per ton.

Load Requirements. Summer load calculations present no special problem. The winter load, however, requires a careful study of all internal heat gains and losses. Winter load calculations must consider the following:

1. Exhaust air is one of the largest potential heat losses of any building. Using an air source heat pump, a maximum of heat may be recovered from this air before it is exhausted.

Table 12-54 Heat Balance and Coefficient of Performance for Engine-Driven Heat Pumps[61]

(all heat rates in Btu per ton-hr)

	Heat from refrigeration condenser only	Heat from refrigeration condenser and jacket water	Heat from refrigeration condenser, jacket water, and exhaust gas
Total heat input to engine	10,000	10,000	10,000
Cooler heat rejection to condenser (from building load)	12,000	12,000	12,000
Heat of compression	2,545	2,545	2,545
Heat from engine jacket water heater	...	2,500	2,500
Heat from exhaust gas heater	...	...	3,000
Total heat to heating circuit	14,545	17,045	20,045
Coefficient of performance	1.45	1.70	2.0

Many methods can be employed to achieve this result. For example, a coil can be installed in both the fresh air intake and the exhaust air discharge. The two coils can be connected by means of a pump, and an antifreeze solution can be pumped thru them. Considerable heat can be recovered by cooling the exhaust air and preheating the fresh air, with only a small power expenditure required to drive the centrifugal pump.

2. Infiltration-exfiltration losses should be kept to a minimum.

3. Recovery of heat from all other available sources in the building should be evaluated. Computers and other machinery in office buildings and industrial plant machinery are important sources of heat in heat pump installations. Recovery of heat from waste water should also not be forgotten since hot water use in such a building or plant may be large.

Upon completion of the heat balance study of the building, the winter load should be expressed in the following form:

Btu required: during the day—occupied
during the night—unoccupied

These two requirements should be specified at various outdoor temperatures and a graph plotted from the lowest winter design temperature to indoor design temperature.

When the preliminary studies of both cooling and heating loads have been completed, the type of equipment that will be required to meet these loads is determined. In most cases, the summer tonnage will indicate whether the heat pump will require reciprocating or centrifugal compressors. For example, if the summer load is 150 tons, centrifugal compressors cannot be considered if a heat pump is to be installed, since only Refrigerants 12 and 22 fulfill heat pump temperature and heat requirements and no centrifugal unit of such small tonnage utilizing these refrigerants is economically feasible.

Gas Turbine-Driven Centrifugal Units. The gas turbine is suitable for a variety of industrial and commercial air-conditioning installations. It fits a wide range of tonnage requirements when paired with steam turbine or absorption equipment (Fig. 12–64a). Such combinations are competitive in first cost with comparable equipment and, in addition,

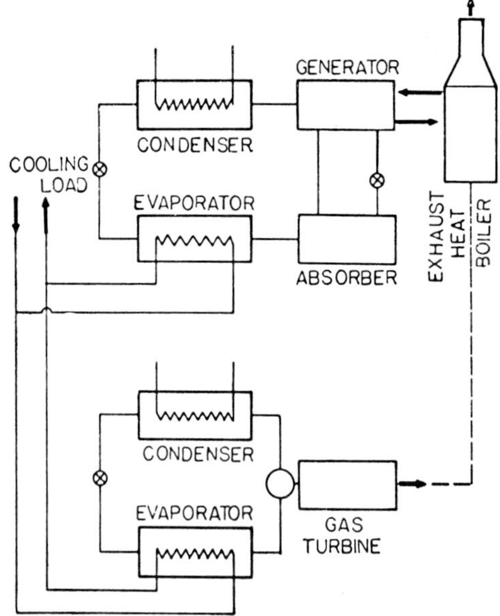

Fig. 12-64a Gas turbine-driven centrifugal compressor-absorption system for air conditioning.

possess the ability to provide heating and cooling simultaneously when needed.

In general, the shaft power output of the gas turbine drives a centrifugal compressor directly. An absorption chiller or a steam turbine-driven centrifugal unit, of approximately half the tonnage of the gas turbine-driven centrifugal unit, is operated by steam generated in an exhaust heat recovery unit. Such a system permits maximum use of the input energy thru modulation of both the centrifugal and absorption units. Furthermore, when heating is required during the winter, it is possible for a sizable air-conditioning load to continue to exist, since the integrated system permits both loads to be met.

Process heating and cooling loads can be met in the same way. The exhaust heat is generally economically recoverable down to within 75°F of the temperature of the water or steam in the exhaust heat unit. For 15 psig steam this can vary from 7,000 to 11,000 Btu per hr per centrifugal output horsepower; for 125 psig steam, the recovery may be two-thirds of this range. Combined centrifugal absorption units can be designed for a heat rate of less than 11,000 Btu per ton-hr.[62] Figure 12–64b compares the use of a centrifugal unit with the use of the combination of a centrifugal unit and an absorption system.

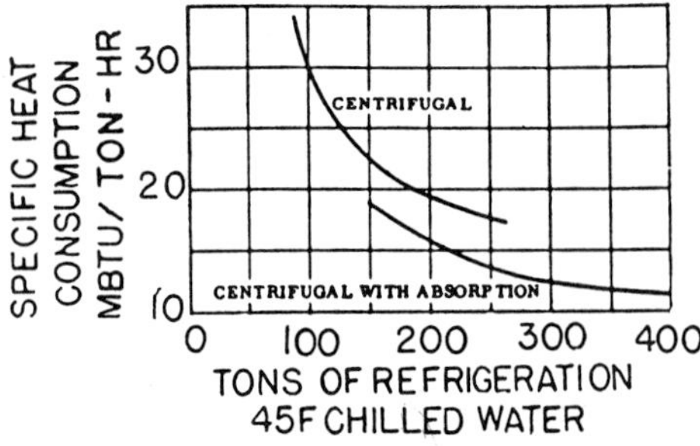

Fig. 12-64b Performance of centrifugal chiller with absorption system.

In applying hot water to an absorption machine, the inlet hot water temperature to the generator tubes must be within the range of 225 to 400 F.[63] The outlet temperature will then range between 215 and 240 F. Should the hot water temperature exceed 400 F, a low-pressure steam converter must be put into the line leading to the absorption machine, resulting in a standard steam-operated installation; or a water-to-water heat exchanger may be used to lower temperatures to the absorption machine below 400 F.

Steam Turbine-Driven Centrifugal Units. Centrifugal compressors are well suited to direct steam turbine drive because of their high operating speed. Installations ranging from 400 to 3000 tons per unit are common.

High-pressure steam is needed with most standard types of turbines used with centrifugal compressors. High-pressure systems are in favor in most areas to meet the steam requirements of hospitals, hotels, many large commercial and public buildings, and industry. Commonly, pressures of 70 psig and above are used. When low-pressure steam is available, absorption refrigeration equipment can be employed.

Steam consumption depends on: (1) heat content of the supply steam, which in turn depends on pressure and quality, whether dry or superheated; (2) turbine exhaust pressure; and (3) turbine efficiency.

Types of Steam Turbines. Units used for driving centrifugal compressors can be grouped into three types: (1) condensing, (2) noncondensing, and (3) variable-pressure bleeder.

1. *Condensing-Type Turbine.* This is the most commonly used steam turbine. It operates with steam at almost any pressure, is available in a wide range of speeds and capacities, and is relatively easy to install and operate.

The turbine exhausts to a steam condenser, usually at a vacuum of 26 in. Hg. With a multistage turbine operating at an efficiency of 60 per cent and using 125 psig dry steam entering (26 in. Hg vacuum exhaust), steam consumption would be 14.7 lb per hp-hr. At about 1.0 hp per ton, about 15 lb of steam per ton-hr would be used at full load.

The consumption of a condensing turbine ranges from 12 to 25 lb of steam per ton-hr. Gas consumption at 80 per cent boiler efficiency averages 20 cu ft per ton-hr of 1000 Btu gas.

Condensing water is passed in series, first thru the refrigeration system condenser and then thru the steam condenser, usually at a rate of 3 gpm per ton, in a cooling tower application.

Steam condensate return piping, including insulation, is required. Condensing water and steam condensate return pumps can be driven by a steam turbine to improve the load factor.

2. *Noncondensing-Type Turbine.* This turbine acts as a reducing valve for high-pressure (e.g., 125 psig) steam. It is a popular selection where low-pressure exhaust steam (e.g., 12 psig) serves applications such as process equipment and absorption machines for additional water chilling (e.g., turbine topping absorption). In these cases, the turbine fits into the heat balance system of the industrial plant or commercial building with little cost for its motive energy.

3. *Variable-Pressure Bleeder-Type Turbine.* This is, in effect, a combination of the condensing and noncondensing types and is used where the steam source may be variable. Higher initial costs are normally offset by the applicability of this turbine to the varying steam supply situations that

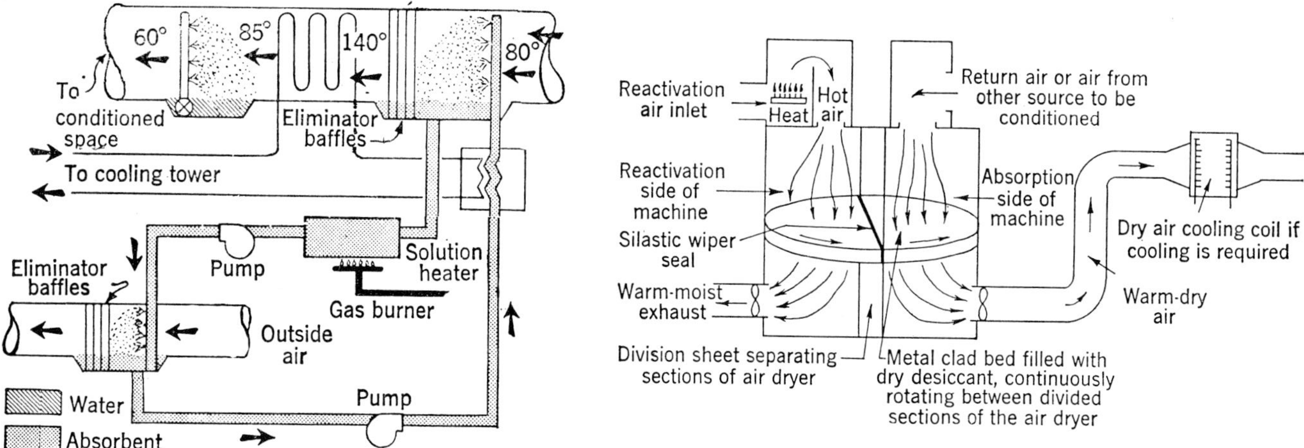

Fig. 12-65a (left) Wet absorption dehumidifier.[45]
Fig. 12-65b (right) Bryant dehumidifier with flat bed, dry desiccant, with sensible dry air cooler. The process is one of adsorption.

frequently exist in industry, hotels, hospitals, and other types of large-scale projects.

Advantages. Turbines are rugged, durable, dependable, and simple to operate and maintain. High speed, flexibility, and ease of turbine speed control and compactness make direct drive practical. Centrifugal action of both compressor and turbine results in minimum vibration. Relatively small floor space is required; floor loading is light. Turbines have good partial load performance, down to approximately 10 per cent of normal load, and it is easy to control the output. A wide range of overload can also be accommodated.

Application. There is no absolute point at which use of a steam turbine should be considered. The turbine has found its greatest application on air-conditioning systems above 500 tons. Below this size, however, many have been installed and have proved economical.[45]

Centrifugal and Absorption Combinations. The combination of a steam turbine-driven centrifugal compressor and an absorption unit may be the optimum choice in large buildings requiring at least 400 to 500 tons of cooling. For such systems the following conditions are required:

1. The availability of high-pressure steam.
2. A fairly constant need for refrigeration, whether at partial or full load.
3. A need that from 25 to 40 per cent of the system's total capacity be met by the absorption unit at all times (below 25 per cent capacity, the specific power requirements of the centrifugal machine rises rapidly).
4. A need for more than one refrigeration unit.

With a single-stage, back-pressure steam turbine operating at 125 psig and exhausting at 13 psig, the approximate steam rate is 40 lb per ton-hr. This steam rate applied to the absorption unit produces about 2 tons of refrigeration. Thus, the proper ratio of refrigeration capacity is 2 tons absorption per ton of centrifugal. The full-load steam rate is approximately 13.33 lb per hr per ton. At low loads the steam turbine is shut down and the absorption unit handles the load.

Steam Jet Cycle. Steam jet pumps handle large volumes of gases such as water vapor for specialized industrial and process cooling (generally 30 to 500 tons). They have high overload capacity, relatively light weight, and no moving parts.

In common practice, steam jet refrigeration systems produce chilled water temperatures between 35 and 70 F. Systems consist of an evaporator, one or more booster ejectors, a barometric or surface condenser, and a two-stage ejector air pump. They operate on high-pressure steam, 100 psig and above. The steam leaving the nozzle reaches a supersonic velocity between 3500 and 4500 fps. Steam consumption is a function of steam pressure, condenser temperature, and evaporator temperature. The consumption decreases with decreasing condenser temperature, with increasing chilled water temperature, and with increasing steam pressure. Pressures below 60 psig are uneconomical regardless of steam cost.

Dehumidification Cycles

In the *open adsorption* process the sorbent does not change, either physically or chemically. Some sorbents used in this process are activated alumina, activated bauxites, and silica gel.

In the *absorption action* the sorbent material changes physically, chemically, or both. Commonly used sorbents in this process are solid calcium chloride; solutions of lithium chloride, calcium chloride, and lithium bromide; and the ethylene glycols.

In both the absorption and adsorption cycles the sorbent is continuously reactivated for reuse in the cycle. Reactivation is by heat, which can be supplied by natural gas.

Wet-Type Unit. This air dryer (Fig. 12-65a) consists of a spray coil contactor in which a solution of lithium chloride is sprayed on a cooling coil as air is being drawn thru it. Moisture is removed by the lithium chloride solution, and the heat of absorption is removed by the cooling agent in the coil. The diluted solution drains to a sump in the base of the contactor. The quantities of drained solution pumped to a reconcentrator depend on the amount of moisture load absorbed.

The reconcentrator consists essentially of another spray coil contactor in which the lithium chloride solution is heated

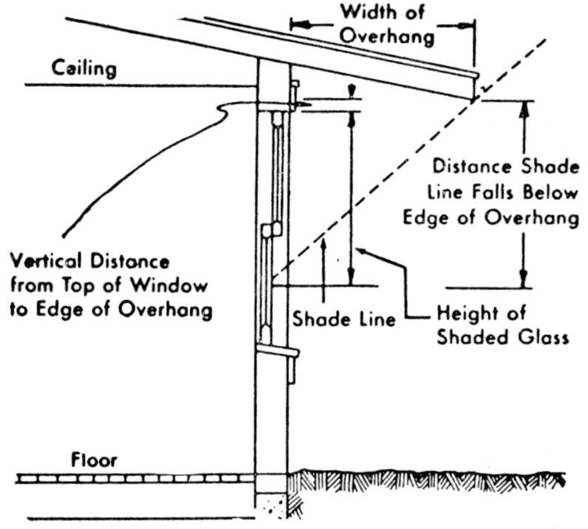

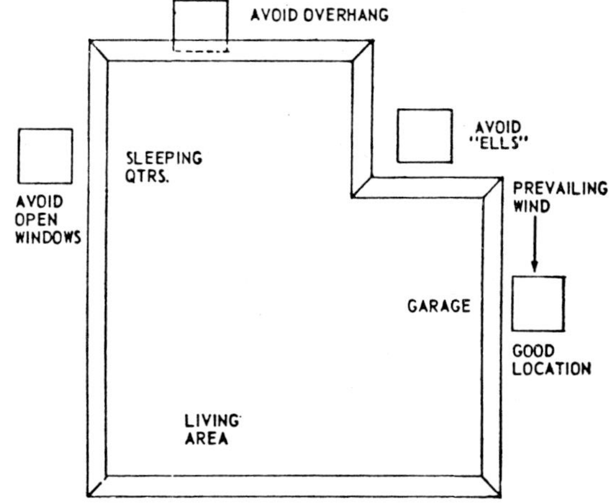

Fig. 12-66a (left) Location of shade line from roof overhang.[37]
Fig. 12-66b (right) Evaluation of alternative outdoor condenser locations.[69]

by steam in the contactor coil. As the solution is heated and air is passed across the coil, moisture is drawn from the solution and the air bearing this moisture is exhausted. The reconcentrated lithium chloride solution is then ready for reuse in the absorption contactor.

The wet system is generally used for drying from high-moisture levels. The equivalent coil surface temperature that would be used to accomplish the drying effect with a straight refrigeration system lies between 45 and 65 F.

This equipment is applicable to comfort or industrial applications, where wide variations occur in either the internal sensible or latent heat loads and where it is desirable to maintain control of the relative humidity within close limits.

Dry-Type Unit. This air dryer (Fig. 12-65b) consists essentially of one or more rotating flat beds. The beds, which are filled with a dry desiccant such as silica gel in granular form, rotate continuously between two compartments of the machine. In one compartment, air is drawn in and passes thru the bed of desiccant, where the air is dried to levels often as low as one grain per pound of dry air. A fan then draws the air from the machine and sends it into the distribution system, which may contain facilities for controlling the dry bulb temperature of this dry air. As the desiccant bed absorbs moisture, it rotates into the other compartment of the machine, where it is reactivated.

Reactivation is accomplished by a separate stream of air drawn into a heating chamber, heated either by the direct combustion of gas or by steam coils, and passed thru the desiccant bed in a path parallel to the path of the dry air. This heated air stream removes moisture from the desiccant, and a fan discharges the warm, moist air as exhaust.

Another type of unit employs two separate chambers charged with desiccant; while one chamber is drying air, the other is regenerated. Periodically the air streams are reversed.

Dry-type desiccant machines are generally applied in the lower humidity ranges, for example, in applications with a requirement for an apparatus dew point lower than 40 F.

Residential System Design

Heat Gain Calculations. In designing a year-round residential air-conditioning system, heat loss and heat gain must be calculated. The larger of the air volumes required for heating and cooling determines the size of the distribution system. When studying existing installations, be certain that they comply with the foregoing rule.[37]

Because the systems in most residences are operated 24 hr per day, calculations should reflect the thermal flywheel effects of structural components and furnishings within the structure. Since the cooling load is largely affected by conditions outside the house and only a few days each season are design days, a partial load situation exists most of the time.

The loads on residential cooling systems are primarily those imposed by heat flow thru structural components and by air leakage or ventilation. Internal loads, particularly those imposed by occupants and lights, are small in comparison with those experienced in commercial or industrial installations; nevertheless, they should be taken into account in determining the size of the unit.

Cooling Load Determination. Six main components of heat gain are involved. It is necessary to segregate and evaluate the amount of dry heat (sensible heat) and moisture load (latent heat) for each. A general classification of all sources of heat gain includes:

	Sensible heat	Latent heat
External loads		
Normal transmission	Yes	None
Sun heat	Yes	None
Internal loads		
People	Yes	Yes
Lights	Yes	None
Appliances	Yes	Yes (in most cases)
Outside air load	Yes	Yes (in most cases)

All load components are evaluated in terms of Btu per hour. The sum of the sensible and latent heat loads, the total heat load, divided by 12,000, gives the *tons of refrigeration* capacity required.

The first step in sizing cooling equipment is to make a heat gain calculation by use of Eq. **18**:

$$Q = UA\Delta t \tag{18}$$

where:

Q = heat gain, Btu per hr

U = overall coefficient of heat transfer for cooling, as follows:

roofs (Table 12–32a, "downward heat flow" in Parts C, D, and E), Btu per hr-sq ft-°F

walls (Table 12–32a, Parts A and B may be used, since there is very little difference between summer and winter wall transmission coefficients), Btu per hr-sq ft-°F

windows (Table 12–55b gives $U\Delta t$ products; apply Table 12–55c if appropriate), Btu per hr-sq ft

infiltration and ventilation (Table 12–55d gives $U\Delta t$ products), Btu per hr-sq ft

A = area involved, sq ft

Δt = equivalent temperature difference, from Table 12–55a (interpolation permitted), °F

Solar radiation, absorptivity of the structure, air and building surface temperatures, and the heat transfer conductance between the air and building surface form a complex interrelationship that makes it difficult to evaluate the heat gain of a residence. The equivalent temperature difference accounts for this interrelationship of factors.

In order to determine the effect of these factors on the heat flow into a residence, the sol-air temperature is used. The *sol-air temperature* is the temperature of the outdoor air which, in the absence of all radiation exchanges, would give the same rate of heat entry into the surface as exists with the actual combination of incident solar radiation, radiant energy exchange with the sky and other outdoor surroundings, and convective heat exchange with the outdoor air.[64] Temperature differentials based on sol-air temperatures and an indoor design temperature of 75 F have been determined by analytical procedures[65] and are known as *equivalent temperature differences*. They have been interpreted for residential structures[66] (Table 12–55a).

Heat Gain by Transmission. The sensible heat gain thru wall, floor, and ceiling areas of each room is calculated using Eq. **1**. Outdoor design temperatures are given in Table 12–47. Equipment selection and system design should be based upon a room-by-room calculation.

Table 12–55b can be used to determine *heat gains thru windows*. These data take account of both shaded and unshaded portions of the window.

When *permanent shading devices* such as wide roof overhangs are used, their effects must be considered separately. The area of *unshaded glass* in a window protected by permanent shading can be determined by the location of the line of shadow (*shade line*), due to the overhang, on the window. Shade line factors are given in Table 12–55c. The shade line factor multiplied by the width of overhang will determine the distance the shadow line falls below the edge of the overhang (see Fig. 12–66a). The area of glass above this distance is shaded; the remainder is unshaded. The shaded portion of the glass is considered to be north-facing glass and the heat gain thru it can be obtained by the use of Table 12–55b. The heat gain thru the unshaded portion is calculated by using the appropriate factor in Table 12–55b. Since windows facing northeast and northwest are not effectively protected by

Table 12-55a Equivalent Temperature Differences, °F[64]
(based on an indoor design temperature of 75 F)

	Outdoor design temperature, °F											
	85		90			95			100		105	110
	Outdoor daily temperature range,* °F											
	L	M	L	M	H	L	M	H	M	H	H	H
Walls and doors												
Frame and veneer-on-frame	17.6	13.6	22.6	18.6	13.6	27.6	23.6	18.6	28.6	23.6	28.6	33.6
Masonry walls, 8-in. block or brick	10.3	6.3	15.3	11.3	6.3	20.3	16.3	11.3	21.3	16.3	21.3	26.3
Partitions, frame	9.0	5.0	14.0	10.0	5.0	19.0	15.0	10.0	20.0	15.0	20.0	25.0
masonry	2.5	0	7.5	3.5	0	12.5	8.5	3.5	13.5	8.5	13.5	18.5
Wood doors	17.6	13.6	22.6	18.6	13.6	27.6	23.6	18.6	28.6	23.6	28.6	33.6
Ceilings and roofs†												
Ceilings under naturally vented attic or vented flat roof—dark	38.0	34.0	43.0	39.0	34.0	48.0	44.0	39.0	49.0	44.0	49.0	54.0
—light	30.0	26.0	35.0	31.0	26.0	40.0	36.0	31.0	41.0	36.0	41.0	46.0
Built-up roof, no ceiling—dark	38.0	34.0	43.0	39.0	34.0	48.0	44.0	39.0	49.0	44.0	49.0	54.0
—light	30.0	26.0	35.0	31.0	26.0	40.0	36.0	31.0	41.0	36.0	41.0	46.0
Ceilings under unconditioned rooms	9.0	5.0	14.0	10.0	5.0	19.0	15.0	10.0	20.0	15.0	20.0	25.0
Floors												
Over unconditioned rooms	9.0	5.0	14.0	10.0	5.0	19.0	15.0	10.0	20.0	15.0	20.0	25.0
Over basement, enclosed crawl space or concrete slab on ground	0	0	0	0	0	0	0	0	0	0	0	0
Over open crawl space	9.0	5.0	14.0	10.0	5.0	19.0	15.0	10.0	20.0	15.0	20.0	25.0

* Outdoor daily temperature range—L (low) calculation value: 12°F, applicable range: less than 15°F; M (medium) calculation value: 20°F, applicable range: 15 to 25°F; H (high) calculation value: 30°F, applicable range: more than 25°F.

† Factors for light-colored roofs calculated using procedure of footnote for Table 8, Chap. 27, 1965 ASHRAE Guide. For roofs in shade, 8-hr avg = 11°F. At a 90 F outdoor design temperature and a medium daily range, equivalent temperature difference for light-colored roof = 11 + (0.71)(39 − 11) = 31°F.

Table 12-55b Sensible Cooling Load thru Windows Due to Transmitted and Absorbed Solar Energy and Air-to-Air Temperature Difference,[64] Btu per hr-sq ft

(based on an indoor design temperature of 75 F)

Direction window faces	Outdoor design temperature, °F																	
	Regular single glass						Regular double glass						Heat absorbing double glass					
	85	90	95	100	105	110	85	90	95	100	105	110	85	90	95	100	105	110
No awnings or inside shading																		
North	23	27	31	35	38	44	19	21	24	26	28	30	12	14	17	19	21	23
NE and NW	56	60	64	68	71	77	46	48	51	53	55	57	27	29	32	34	36	38
East and West	81	85	89	93	96	102	68	70	73	75	77	79	42	44	47	49	51	53
SE and SW	70	74	78	82	85	91	59	61	64	66	68	70	35	37	40	42	44	46
South	40	44	48	52	55	61	33	35	38	40	42	44	19	21	24	26	28	30
Draperies or venetian blinds																		
North	15	19	23	27	30	36	12	14	17	19	21	23	9	11	14	16	18	20
NE and NW	32	36	40	44	47	53	27	29	32	34	36	38	20	22	25	27	29	31
East and West	48	52	56	60	63	69	42	44	47	49	51	53	30	32	35	37	39	41
SE and SW	40	44	48	52	55	61	35	37	40	42	44	46	24	26	29	31	33	35
South	23	27	31	35	38	44	20	22	25	27	29	31	15	17	20	22	24	26
Roller shades half-drawn																		
North	18	22	26	30	33	39	15	17	20	22	24	26	10	12	15	17	19	21
NE and NW	40	44	48	52	55	61	38	40	43	45	47	49	24	26	29	31	33	35
East and West	61	65	69	73	76	82	54	56	59	61	63	65	35	37	40	42	44	46
SE and SW	52	56	60	64	67	73	46	48	51	53	55	57	30	32	35	37	39	41
South	29	33	37	41	44	50	27	29	32	34	36	38	18	20	23	25	27	29
Awnings																		
North	20	24	28	32	35	41	13	15	18	20	22	24	10	12	15	17	19	21
NE and NW	21	25	29	33	36	42	14	16	19	21	23	25	11	13	16	18	20	22
East and West	22	26	30	34	37	43	14	16	19	21	23	25	12	14	17	19	21	23
SE and SW	21	25	29	33	36	42	14	16	19	21	23	25	11	13	16	18	20	22
South	21	24	28	32	35	41	13	15	18	20	22	24	11	13	16	18	20	22

Table 12-55c Shade Line Factors for Windows[64]

Direction window faces	Latitude, degrees						
	25	30	35	40	45	50	55
East	0.8	0.8	0.8	0.8	0.8	0.8	0.8
SE	1.9	1.6	1.4	1.3	1.1	1.0	0.9
South	10.1	5.4	3.6	2.6	2.0	1.7	1.4
SW	1.9	1.6	1.4	1.3	1.1	1.0	0.9
West	0.8	0.8	0.8	0.8	0.8	0.8	0.8

Note: Distance shadow line falls below the edge of the overhang equals shade line factor multiplied by width of overhang (see Fig. 12-66a). Values are averages for the 5 hr of greatest solar intensity on Aug. 1.

roof overhangs, in most cases no credit should be taken for their shading.

Infiltration and Ventilation. Natural air infiltration into a house depends upon the area of the house exposed to the wind. Table 12–55d lists the factors that are used in determining *infiltration.* They are based upon a leakage rate of one-half air change per hour. If mechanical ventilation is used, the factors in Table 12–55d are used to determine the heat gain based on the amount of air (cfm) introduced.

Occupant Load. The *heat release per occupant* in a residence is usually assumed to be 300 Btu per hr of sensible heat. Unless otherwise stated, a rule of thumb is to assume that there will be approximately twice as many occupants as the residence has bedrooms, never assuming fewer than three occupants. This load is divided among the living quarters of the residence.

Table 12-55d Sensible Cooling Load Due to Infiltration and Ventilation[64]

	Outdoor design temperature, °F					
	85	90	95	100	105	110
Infiltration, Btu per hr-sq ft of gross exposed wall area	0.7	1.1	1.5	1.9	2.2	2.6
Mechanical ventilation, Btu per hr-cfm	11.0	16.0	22.0	27.0	32.0	38.0

The *heat gain from appliances* will in most cases be considered only for the kitchen. The *ASHRAE Guide*[64] recommends adding 1200 Btu per hr to the calculated heat gain of the kitchen.

An allowance is also made for *duct heat gain.* The following percentage increase, suggested by FHA, can be applied to the sum of the heat gain thru the building construction, the sensible heat gain from outside air, the sensible heat gain from people, and the sensible heat gain from appliances:

	Per cent
Ducts with 1-in. insulation in attic	15
Ducts with 2-in. insulation in attic	10
Ducts in slab floors; insulated (1-in.) ducts in enclosed crawl space, unconditioned basement, or furred spaces	5

Table 12-55e Capacity Multiplier for Selection of Air-Conditioning Units[37]

Air-cooled units

Outside design conditions		Desired inside temp swing,* °F		
Temp, °F	Daily range†	6.0	4.5	3.0
90	M	0.69	0.83	0.97
	L	0.71	0.84	0.98
95	H	0.74	0.88	1.02
	M	0.75	0.89	1.03
	L	0.77	0.90	1.04
100	H	0.81	0.95	1.08
	M	0.82	0.96	1.09
105	H	0.86	1.01	1.16
	M	0.87	1.02	1.17
110	H	0.92	1.07	1.22

Evaporatively cooled units

Outside design conditions		Outside design wet bulb temp, °F														
Temp, °F	Daily range†	65	70	75	78	80	65	70	75	78	80	65	70	75	78	80
		(6°F	inside	temp	swing)		(4.5°F	inside	temp	swing)		(3°F	inside	temp	swing)	
90	M, L	0.74	0.75	0.77	0.78	0.79	0.89	0.90	0.92	0.93	0.94	1.03	1.05	1.07	1.08	1.09
95	H, M, L	0.79	0.80	0.81	0.82	0.83	0.93	0.94	0.96	0.97	0.98	1.07	1.09	1.11	1.12	1.13
100	H, M	...	0.84	0.85	0.87	0.88	...	0.99	1.00	1.02	1.03	...	1.12	1.14	1.15	1.16
105	H, M	...	0.87	0.88	0.89	0.90	...	1.02	1.03	1.05	1.06	...	1.17	1.19	1.20	1.21
110	H	...	...	0.92	0.93	0.94	...	...	1.08	1.09	1.10	...	...	1.23	1.24	1.25

Water-cooled units

Outside design conditions		Leaving water temp, °F														
Temp, °F	Daily range†	90	95	100	105	110	90	95	100	105	110	90	95	100	105	110
		(6°F	inside	temp	swing)		(4.5°F	inside	temp	swing)		(3°F	inside	temp	swing)	
90	M, L	0.74	0.77	0.81	0.83	0.86	0.88	0.93	0.97	1.00	1.03	1.03	1.08	1.12	1.16	1.20
95	H, M, L	0.78	0.82	0.85	0.88	0.91	0.93	0.97	1.01	1.05	1.08	1.07	1.12	1.17	1.21	1.25
100	H, M	0.82	0.87	0.91	0.94	0.97	0.97	1.02	1.06	1.09	1.12	1.10	1.16	1.20	1.24	1.28
105	H, M	0.86	0.90	0.94	0.97	1.00	1.00	1.05	1.10	1.14	1.17	1.15	1.20	1.25	1.30	1.34
110	H	0.89	0.94	0.98	1.01	1.04	1.04	1.09	1.14	1.18	1.22	1.18	1.24	1.30	1.34	1.38

Note: This table is to be used according to the relationship:
(calculated heat gain) × (capacity multiplier) = (equipment standard ARI capacity rating)
ARI—Air Conditioning and Refrigeration Institute
* Thermostat must be set number of degrees below 78 F as degree swing desired.
† For explanation of H, L, and M, see footnotes to Table 12-55a.

A more accurate procedure for calculating duct heat gain can be found in NWAHACA Manual J.[37] This procedure takes into account the duct length; for example, the 15 per cent increase for ducts with 1 in. insulation located in attics applies to ducts that are from 16 to 25 ft long. A smaller increase is made for shorter duct runs.

The total *sensible heat load* will be the sum of the sensible heat gains of all rooms. The *latent heat load* is usually estimated as 30 per cent of the total sensible heat load; experience has shown that the *total heat load* is 1.3 times the sensible heat load. Because latent load will depend upon weather conditions and family living habits, the use of an average factor to determine total heat load causes negligible errors in most geographical locations.

Sizing Equipment. The calculated heat gain is used to determine the equipment capacity required to maintain the inside temperature within limits desired. Table 12–55e lists factors for determining the equipment capacity required to maintain inside temperature within a 3.0°, 4.5°, or 6.0°F temperature swing. This temperature swing, the maximum indoor temperature minus the temperature set at the thermostat, gives a practical degree of control.

As an example, many people will start to "feel uncomfortable" at 78 F. If the equipment is sized on the basis of a capacity multiplier obtained from Table 12–55e for a temperature swing of 6°F, the thermostat must be set at 72 F (78 − 6 = 72) to maintain the indoor temperature below this limit.

Selection of Distribution System. Factors considered include system performance characteristics, indoor design temperature, climate, and the type of structure and occupancy. A 3°F maximum difference between any two rooms is the generally accepted comfort standard.

Table 12–56 lists the various systems used in residential summer air conditioning. A full discussion of these systems citing the advantages and disadvantages of each is contained in NWAHACA Manual G[67] and I=B=R Installation Guide No. 900.[68]

Supply grilles must be located so that no objectionable

Table 12-56 Types of Residential Distribution Systems

1. Warm air perimeter
   a. Perimeter loop
   b. Perimeter radial
2. Horizontal-flow duct
3. Inside wall
   a. High outlets
   b. Low outlets
4. Ceiling outlet
   a. Center placed diffuser
   b. Perimeter placed diffuser
5. Crawl space plenum
6. Chilled water—fan convector units

drafts result within the space below the 6 ft level, termed the "occupied zone." Introduction of supply air at a lower level is acceptable provided the occupant does not normally move within two or three feet of its outlet and that grille velocity and type are such that supply air and room air are well mixed in the occupied zone. Sufficient air circulation must be maintained to provide even room temperatures throughout the day as the heat load on various rooms changes.

Outdoor* Installation. An absorption air-conditioning unit may be installed outdoors when the unit has been approved for this service. Pertinent tests may include operation under a simulated rainstorm; pilot ignition and combustion tests when the unit is exposed to a 10 mph wind; and pilot and main burner operation and main burner ignition when the unit is exposed to a 40 mph wind.[54]

The equipment should be installed in such a manner that combustion products cannot enter fresh air intakes of a building. When the condenser is located outdoors, the unit should be clear of any roof overhang; if it is necessary to place the unit within the overhang line, a 6 ft clearance between the fan discharge and overhang is required. A clearance of 2 ft on all sides of the unit ensures adequate air supply and access for service. Figure 12–66b evaluates outdoor condenser locations.

The installation of the air-conditioning system should meet the requirements of NFPA 90B, Standard for the Installation of Residence Type Warm Air Heating and Air Conditioning Systems.

Commercial System Design

Heat Gain Calculations. Calculating summer cooling loads for *commercial and industrial* applications involves indoor and outdoor design conditions, heat loads within the space, air quantities, and apparatus dew point.

Calculation procedures deal with *instantaneous* rates of heat gain and thus considerable judgment is essential to effect successful design. There may be an appreciable difference between the net instantaneous rate of heat gain and the total cooling load at any instant.

Chapter 27 of the 1965 *ASHRAE Guide*[64] presents the methods and data used in calculating the cooling loads from which the required capacity of the equipment can be determined. The major factors in this determination are:

1. Heat transmission thru walls, roofs, and floor.
2. Solar radiation.
3. Heat gain from occupants, lights, and motors and other equipment.
4. Amount of outside air infiltrating and introduced into the conditioned area.

Procedures for calculating factors 1, 2, and 4 are given in the *ASHRAE Guide*. Data for use in determining factor 3 are available.[70]

Distribution Systems. The three types of distribution systems are: (1) all air, (2) all water, and (3) mixed air and water. Table 12–57 lists most of these systems.

Table 12-57 Types of Distribution Systems in Commercial Buildings[71]

1. All air
   Single duct
   Double duct
   Multizone
2. All water
   Zoned water (cooled or heated)
   Water (or direct expansion) and steam
3. Mixed air and water
   Fan coil unit
   Induction unit
   Radiant ceiling

System Selection.[71] Experience, the particulars of the rooms to be conditioned, the quality of results considered appropriate, the relative importance of initial vs. operating costs, whether the space can be modified to accommodate a desirable system or not—these and many other considerations lead to the selection of the distribution system to be used in any given case.

A few important factors follow:

1. In a multistory office building, space requirements and cost practically prohibit one central fan room. This may not be true concerning central fan rooms serving one or two floors or only a few floors in separate wings.

2. Peripheral systems are convenient and economical for long, narrow buildings or wings where nearly all usable space is within 15 to 20 ft of windows, especially if the space is greatly subdivided into small private rooms and particularly if there are large windows.

3. Among peripheral systems, the induction type of unit reduces the scatter of equipment. The fan type of unit allows primary air to be delivered separately where this is convenient. For high buildings, the overall costs are often somewhat lower with induction units.

4. Where large interior spaces are to be conditioned, horizontal duct systems are essential unless the building layout will permit the running of vertical primary air ducts and secondary water piping, systematically spaced at interior partitions and columns. This is a very rare situation. Four categories of such systems are:

a. "Conventional," with supply ducts handling 100 per cent and return ducts about 85 per cent of computed total air, based on sensible load and a permissible differential of perhaps 20°F.

b. High-velocity systems and "mixing" outlets, with supply ducts handling about 65 per cent of the computed

* For indoor installation, see Chap. 1 of this Section.

Table 12-58a Average Hourly Temperature Occurrences in Chicago[72]
(bearing on air-conditioning refrigeration operation; based on U. S. Weather Bureau records for 1951 thru 1956)

Temp occurrences	Monthly average hours									Yearly avg hr
	Mar	Apr	May	June	July	Aug	Sept	Oct	Nov	
55 F and above	40	237	480	675	743	741	625	372	100	4013
60 F and above	14	167	365	624	733	714	520	260	53	3449
65 F and above	*	103	268	548	680	652	379	173	21	2824
70 F and above	*	62	171	432	558	500	248	92	*	2063

* Less than 10.

total air (at 30°F differential or more) thru ducts having about one-fourth the cross section of the ducts used in a conventional system and return ducts handling about 55 per cent.

c. Double-duct, high-velocity systems with cold duct sized for 100 per cent, hot duct for about 50 per cent, and return duct for about 85 per cent of computed total air.

d. A modification of the second category (b), which is sometimes useful in alterations or special cases, in which a high-velocity, horizontal primary duct delivers only the outside air (20 to 25 per cent of the required total air) to peripheral-type (or larger) secondary units appropriately distributed in the interior spaces (and connected with either vertical or horizontal secondary water piping).

5. Radiant cooling ceilings (with either horizontal or vertical water piping) in conjunction with a much reduced air supply system of either conventional or high-velocity horizontal distribution. These may be considered in any space where their appearance and other details are acceptable, and where window sizes are not so great as to make them impracticable or too costly. They have special application where any of the following apply:

a. Any enclosures under windows or at outside walls and columns are objectionable.

b. Very dry climates which require little or no dehumidification and have low dew points. Water temperatures for radiant ceilings may then be lowered to give more sensible cooling, but will still be high enough to give significant savings in cost of refrigerating equipment or in quantity of condensing water needed.

c. Well water at 50 to 55 F is available in such ample quantities as to be useful both for conditioning of primary air supply and for cooling the circulating water for the ceiling panels.

Energy Requirements. In estimating the energy consumption for commercial summer air conditioning, both the air-conditioning unit and its auxiliaries must be taken into account. The estimated energy consumption for the air-conditioning unit can be arrived at as follows:

1. Determine the energy input for various load levels and the corresponding number of operating hours for the different types of equipment under consideration.

2. Determine the number of hours that the air-conditioning equipment operates at different outdoor temperature levels; i.e., establish the air-conditioning season.

For example, Table 12–58a lists the total number of monthly hours for which air conditioning may be required (according to Table 12–58a). Relate the type of occupancy to the total number of hours that air conditioning is required (according to the last column of Table 12–58b).

3. For the total number of hours of operation found in Step 2, determine the number of hours that the air-conditioning equipment will operate at the various load levels (according to the load columns of Table 12–58b). Such load data can be obtained from similar existing installations. Table 12–58b lists the results of a metered study conducted on several buildings in the Chicago area; Table 12–58c shows the physical characteristics of these buildings.

4. Obtain the sum of Step 3 times Step 1 to arrive at the estimated energy consumption for summer air conditioning.

Table 12-58b Air-Conditioning Hours at Various Temperature and Load Levels for Selected Types of Buildings[72]
(a metered study; see Table 12-58c for building characteristics)

Type of building	Operating schedule		Temp at which air cond is required, ° F min	Operating hr				Total
	Day/week	Hour/day		¼ load	½ load	¾ load	Full load	
Apartment	7	24	55	710	1713	1409	181	4013
Hospital	7	24	60	610	1472	1212	155	3449
Hotel	7	24	65	499	1205	992	128	2824
			70	365	881	724	93	2063
Office	5½	12	55	418	684	404	71	1577
			60	359	588	347	61	1355
Department store	6	12	55	26	901	655	138	1720
			60	22	775	563	118	1478
College	7	13	60	495	811	478	84	1868
Library	7	13	65	405	664	392	69	1530
			70	296	485	286	50	1117

Table 12-58c Characteristics of Buildings Covered in Table 12-58b[72]

A. Building data

Apartment	
Floors	26
First (machine room and garage)	
Second (garage)	
Third (storage)	
Fourth to 25th (88 apartments)	
26th (heating plant)	
Floor area (apartments), sq ft	98,700
Building volume (living space), cu ft	807,000
Air distribution system	
Fan coil units	
Hotel	
Guest rooms (nine floors)	221
Total floor area, sq ft	76,950
Building volume, cu ft	705,000
Office	
Floors	22
Floor area, gross sq ft	667,000
Department store	
Floors (including basement)	4
Total floor area, sq ft	160,000
Building volume, cu ft	1,920,000
Lighting, watts per sq ft	6

B. Air-conditioning system data

Building	Installed capacity, tons	Max load attained, tons	Demand factor, per cent
Apartment	170	126.5	74.5
Hotel	100	69.0	69.0
Office	1400	1000.0	71.5
Department store	500	416.0	83.2

REFERENCES

1. A.G.A. Laboratories. *Directory, Approved Appliances, Listed Accessories.* Cleveland, Ohio (semiannual).
2. Inst. Boiler and Radiator Manufacturers. *SBI Testing and Rating Code for Steel Boilers,* 9th ed. New York, 1963.
3. ——. *I=B=R Testing and Rating Code for Low-Pressure Cast Iron Heating Boilers,* 6th ed. + amendments. New York, 1958–64.
4. Am. Boiler Manufacturers Assoc. *Packaged Firetube Boiler Ratings for Heating Boilers,* 3rd ed. Newark N. J., 1962.
5. Mech. Contractors Assoc. of Am., 666 Third Avenue, Suite 1464, New York, N. Y., 10017.
6. "Heating Load." (In: Am. Soc. Heating, Refrig. Air-cond. Engrs. *ASHRAE Guide and Data Book,* chap. 26. New York, 1965).
7. Achenbach, P. R., and others. *Analysis of Electric Energy Usage in Air Force Houses Equipped With Air-to-Air Heat Pumps.* (Natl. Bur. Std. Monograph 51) Washington, G. P. O., 1962.
8. U.S. Federal Housing Admin. *Minimum Property Standards for One and Two Living Units.* Washington, G. P. O. (loose-leaf service).
9. "Infiltration and Ventilation." (In: Am. Soc. Heating, Refrig. Air-cond. Engrs. *ASHRAE Guide and Data Book,* chap. 25, New York, 1965).
10. Baumeister, T., ed. *Marks' Mechanical Engineers' Handbook,* 6th ed., p. 12–81. New York, McGraw-Hill, 1958.
11. Owens-Corning Fiberglas Corp. *Economics of Sensible Heat Control.* (Pub. 5–IN–2460) New York, 1963.
12. A. G. A. Laboratories. *American Standard Installation of Gas Appliances and Gas Piping.* (ASA Z21.30–1964) Cleveland, Ohio, 1964.
13. Natl. Board of Fire Underwriters. *Code for the Installation of Heat Producing Appliances.* New York, 1955.
14. Natl. Fire Protect. Assoc. *Residence Warm Air Heating.* (NFPA 90B) Boston, 1963.
15. A.G.A. Laboratories. *American Standard Approval Requirements for Gas-Fired Steam and Hot Water Boilers.* (ASA Z21.13) Cleveland, Ohio, 1964.
16. "Steam Heating Systems." (In: Am. Soc. Heating, Refrig. Air-cond. Engrs. *ASHRAE Guide and Data Book,* chap. 8. New York, 1964.)
17. Inst. Boiler and Radiator Manufacturers. *Piping Guide, Residential Heating Systems.* (No. 700) New York, 1962.
18. "Radiators, Convectors, Baseboard and Finned-Tube Units." (In: Am. Soc. Heating, Refrig. Air-cond. Engrs. *ASHRAE Guide and Data Book,* chap. 55. New York, 1965.)
19. "Basic Water System Design." (In: Am. Soc. Heating, Refrig. Air-cond. Engrs. *ASHRAE Guide and Data Book,* chap. 10. New York, 1964.)
20. Inst. Boiler and Radiator Manufacturers. *One Pipe Forced Circulation Hot Water Heating Systems,* 2nd ed. (Installation Guide 100) New York, 1950.
21. ——. *Two Pipe Reverse Return Gravity Hot Water Heating System.* (Installation Guide 4) New York, 1950.
22. ——. *Baseboard Heating Systems,* 2nd ed. (Installation Guide 5) New York, 1950.
23. A.G.A. Laboratories. *American Standard Approval Requirements for Gas-Fired Gravity and Forced Air Central Furnaces.* (ASA Z21.47) Cleveland, Ohio 1964.
24. "Air Duct Design." (In: Am. Soc. Heating, Refrig. Air-cond. Engrs. *ASHRAE Guide and Data Book,* chap. 31. New York, 1965.)
25. Natl. Warm Air Heating Air Cond. Assoc. *Equipment Selection and System Design Procedures.* (Manual K) Cleveland, Ohio, 1964.
26. Am. Soc. Heating and Ventilating Engrs. "ASHVE Standard Code for Testing and Rating Air Cleaning Devices Used in General Ventilation Work." *ASHVE Trans.* 39:228, 1933.
27. Dill, R. S. "A Test Method for Air Filters." *ASHVE Trans.* 44: 379, 1938.
28. "Air Cleaners." (In: Am. Soc. Heating, Refrig. Air-cond. Engrs. *ASHRAE Guide and Data Book,* chap. 36. New York, 1965.)
29. A.G.A. Laboratories. *American Standard Approval Requirements for Gas-Fired Gravity and Fan Type Floor Furnaces.* (ASA Z21.48) Cleveland, Ohio, 1964.
30. ——. *American Standard Approval Requirements for Gas-Fired Gravity and Fan Type Vented Wall Furnaces.* (ASA Z21.49) Cleveland, Ohio, 1964.
31. Kirk, W. B., and DeWerth, D. W. *Past Decade of Comfort Heating and Cooling With Gas.* (ASME Paper 62–FUS–1) New York, ASME, 1962.
32. A.G.A. Laboratories. *American Standard Approval Requirements for Gas-Fired Gravity and Fan Type Sealed Combustion System Wall Furnaces.* (ASA Z21.44) Cleveland, Ohio, 1963.
33. ——. *Diffusion of Combustion Products Leaving Vent Terminals of Sealed Combustion System Appliances.* (Report 1357A) Cleveland, Ohio, 1963.
34. ——. *American Standard Listing Requirements for Domestic Gas Conversion Burners.* (ASA Z21.17) Cleveland, Ohio, 1962.
35. ——. *American Standard Installation of Domestic Gas Conversion Burners.* (ASA Z21.8) Cleveland, Ohio, 1958.
36. "Estimating Fuel or Energy Consumption for Space Heating." (In: Am. Soc. Heating, Refrig. Air-cond. Engrs. *ASHRAE Guide and Data Book,* chap. 16. New York, 1964.)
37. Natl. Warm Air Heating Air Cond. Assoc. *Load Calculation.* (Manual J) Cleveland, Ohio, 1964.
38. Burgart, H. F. *How Big Big Bill?* (Research and Utilization Conf., 1963) New York, A.G.A., 1963.
39. Sarno, W. R., *Gas Heating Estimates: the New Way.* (Research and Utilization Conf., 1961) New York, A.G.A., 1961.
40. Simpson, J. H., Jr. *Gas Firing of Heating and Process Boilers.* (Reprint 80) Cleveland, Ohio, North Am. Manufacturing Co., n. d.
41. Jones, N. D. "The Packaged Firetube Boiler." (In: A.G.A. Industrial and Commercial Gas Sect. *Industrial Gas School.* New York, 1957.)

42. Tallafuss, W. J. "Industrial Steam Boiler Plants and Piping." *ASHRAE J.* 6: 39–44, July 1964.
43. Bender, R. J., ed. "Steam Generation." *Power* 108: S2–48, June 1964.
44. A.G.A. Laboratories. *American Standard Requirements for Gas Equipment in Large Boilers.* (ASA Z83.3—Jan. 1964, 3rd draft) Cleveland, Ohio, 1964.
45. A.G.A. *Gas Comfort Heating/Cooling 1962.* (Reprint: *Air Conditioning, Heating Ventilating* 59: GC 1–80, March 1962.)
46. Natl. Fire Protect. Assoc. *Standard on Aircraft Hangars.* (NFPA 409) Boston, 1962.
47. "How to Apply High Intensity Infrared Heaters for Comfort Heating." *Heating, Piping Air Conditioning* 33: 232, Jan. 1961.
48. Hupp Corp. Perfection Div. *Perfection Schwank Gas Infra-Red Heating Manual.* Cleveland, Ohio, 1962.
49. Miller, H. I., Jr. "Designed Conditions in Direct Fired Radiant Heating." *A.M.A. Arch. Ind. Health* 19: 312–19, 1959.
50. Hama, G. M. "How to Size Make-Up Air Heaters." *Air Eng.* 4: 33, Sept. 1962.
51. Zavodny, S. W. "How to Use Direct-Fired Gas Make-Up Air Systems." *Air Eng.* 5: 26+, Nov. 1963.
52. "Direct Gas-Fired Heating of Process Make-Up Air." *Actual Specifying Engr.* (Reprint: vol. 6, No. 6, 1961; vol. 7, No. 2 and No. 3, 1962.)
53. A.G.A. *Comfort Heating.* New York, 1938.
54. A.G.A. Laboratories. *American Standard Approval Requirements for Gas-Fired Absorption Summer Air Conditioning Appliances.* (ASA Z21.40.1) Cleveland, Ohio, 1961.
55. ——. *American Standard Approval Requirements for Gas Engine-Powered Summer Air Conditioning Appliances.* (ASA Z21.40.2) Cleveland, Ohio, 1961.
56. "Absorption Air-Conditioning and Refrigeration Equipment." (In: Am. Soc. Heating, Refrig. Air-cond. Engrs. *ASHRAE Guide and Data Book,* chap. 40, New York, 1965.)
57. Merrick, R. H., and English, R. A. "An Air-Cooled Absorption Cycle." *ASHRAE J.* 2: 39–41, Aug. 1960.
58. Hartwein, C. E. "Natural Gas Engines—Prime Movers for Air Conditioning Compressors." (In: A.G.A. Industrial and Commercial Gas Sect. *Gas Air Conditioning Sales School.* New York, 1960.)
59. Hall, N. K. "Gas Engine Drive for Air Conditioning Systems." *Actual Specifying Engr.* 6, No. 2: *Gas Comprehensive,* p. 67, Aug. 1961.
60. Conrad, J. C. "Natural Gas Engine Drive for Centrifugal Compressors." *Heating, Piping Air Conditioning* 34: 111–16, Dec. 1962.
61. Japhet, R. E. "How to Heat/How to Cool With a Gas Powered Heat Pump." *Actual Specifying Engr.* 11: 64+, March 1964.
62. Apitz, C. R. "Application of Gas Turbines to Large Tonnage Air Conditioning." (In: A.G.A. Industrial and Commercial Gas Sect. *Gas Air Conditioning Sales School.* New York, 1960.)
63. Kuhen, G. O. *Popular Heat Sources for Absorption Refrigeration Machines.* (63–OGP–12) New York, ASME, 1963.
64. "Air-Conditioning Cooling Load." (In: Am. Soc. Heating, Refrig. Air-cond. Engrs. *ASHRAE Guide and Data Book,* chap. 27. New York, 1965.)
65. Steward, J. P. "Solar Heat Gain Thru Walls and Roof for Cooling Load Calculation." *ASHVE Trans.* 54: 361, 1948.
66. Natl. Warm Air Heating Air Cond. Assoc. *The All Industry Procedure for Heat Gain Calculations.* (Bulletin, May 1961) Cleveland, Ohio, 1961.
67. ——. *Selection of Distribution System.* (Manual G) Cleveland, Ohio, 1963.
68. Inst. Boiler and Radiator Manufacturers. *Chilled Water Cooling Systems.* (Installation Guide 900) New York, 1957.
69. Bryant Manufacturing Co. *Gas-Fired Air Cooled Summer Air Conditioner.* (EH 36–450–2) Indianapolis, Ind. 1959.
70. McPartland, J. F. *Electrical Systems for Power and Light.* New York, McGraw-Hill, 1964.
71. Jaros, A. L., Jr. "Heating-Cooling Distribution: Which System Best for Your Job?" *Actual Specifying Engr.* 5: 90–111, Jan. 1961.
72. "How to Estimate Energy Requirements for Big Building Air Conditioning." *Heating, Piping Air Conditioning* 32 157–64, Jan. 1960.

Chapter 6

Gas Water Heating

by J. W. Farren, E. A. Jahn, K. R. Knapp, and D. D. Williams

Gas water heaters are available in many types and capacities to meet the hot water requirements of any residential or commercial application. Boilers for building heating systems are discussed in Chap. 5.

DIRECT GAS-FIRED WATER HEATERS

A. G. A. tests representative samples of water heaters for conformance with the appropriate American Standards.[1–3] Units thus approved are listed,[4] with their respective input rates and recovery capacities, by the A. G. A. Laboratories. Installation standards are available in ASA Z21.30[5] and in the similar NFPA Standard No. 54.[6]

Automatic Storage Heaters

In automatic storage heaters, the vertical storage tank, insulation, gas burner, and automatic controls are combined in a single unit. These heaters can be obtained with storage tank capacities varying from 10 to 100 gal. Recovery capacities normally range from 3.8 to over 200 gal per hr, based on a 100° F rise in temperature. However, at least one A. G. A. model is rated as high as 756 gal per hr. The smaller sizes, with storage capacities of 20 to 40 gal and inputs up to 50 MBtu per hr are used in private residences. Sizes over 50 MBtu per hr—tested under ASA Z21.10.3[3]—are used in commercial and industrial applications either individually or in parallel.

Automatic storage heaters with attractive outer jackets are designed to be installed in kitchens, playrooms, and utility rooms. Units of table-top height may be installed in line with kitchen cabinets, ranges, etc. Heaters with inputs up to 5000 Btu per hr do not require a vent connection.

The heat in the combustion products is transferred to the stored water thru the tank bottom and either an internal or an external flue (Fig. 12-67).

Circulating Tank Heaters

These heaters are commonly referred to as **side-arm** or **tank** heaters when input ratings are below 50,000 Btu per hr. (See "Coil or Fin Heaters" for larger circulating tank heaters.) The small side-arm heater is used for domestic hot water in conjunction with a 20- to 40-gal storage tank (Fig. 12-68). The heater contains cast elements or coils of ¾ to 1 in. diam copper tubing thru which the supply water circulates. These heaters may be either manually or automatically controlled. The performance of the latter type is similar to that of the automatic storage heater; however, if the tank is not insulated, operating costs will be higher. Few side-arm heaters have been installed in recent years.

Coil or Fin Heaters (Circulating Tank Type)

For large homes, apartment houses, institutions, restaurants, clubs, and other applications requiring more water than conventional types of heaters can supply, a large volume water heater is recommended. Such a heater is the coil heater, which resembles an instantaneous heater but which is equipped with a series of parallel coils connected to manifolds, shown at the left in Fig. 12–69. The burner is controlled by a thermostat inserted into a separate insulated horizontal storage tank.

Multiple installations of this type are practically unlimited in their range of application as far as delivery of hot water is concerned. Individual units have recovery rates from 15 to 640 gal per hr thru a 100°F temperature rise.

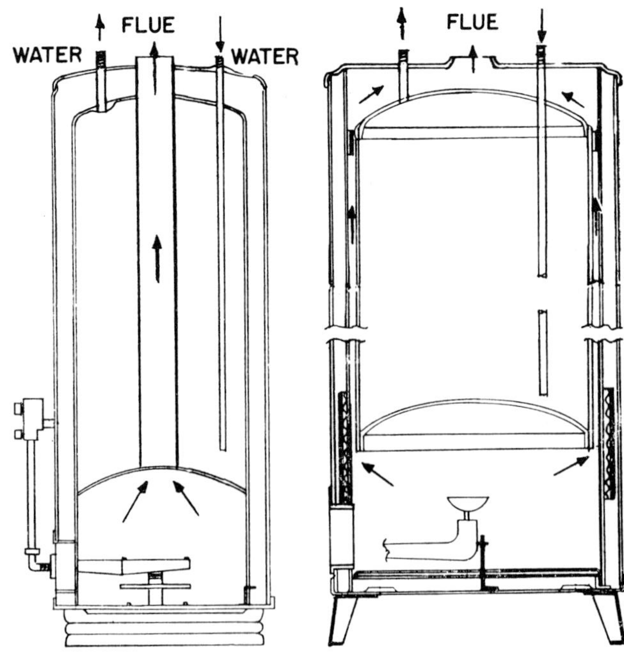

Fig. 12-67 Section thru a typical automatic storage heater: (left) internal flue type; (right) external flue type.

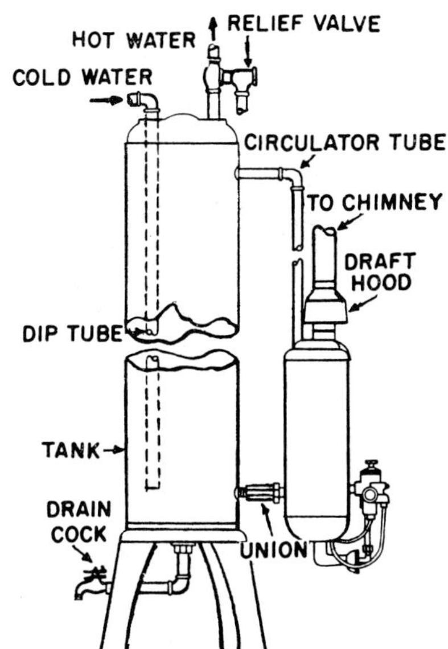

Fig. 12-68 Side-arm, automatic, circulating water heater connected to a vertical storage tank.

Multi-Flue Heaters

Multi-flue heaters bridge the gap between a coil heater with a large external storage tank and an instantaneous heater without a storage tank. This kind of heater (Fig. 12-69, right) is similar in design to a storage-type heater, but the ratio of gas input or recovery rate to storage tank capacity is much greater—5000 Btu (or more) per hour per gallon of storage. Storage capacity may range from 20 to 100 gal and recovery rates from 50 to 300 gal per hr.

As a result of the high input ratios, these heaters are designed with multiple flues thru the tank to provide the necessary heat transfer surface. The heaters are intended for large volume installations and are suitable substitutes for coil-type or instantaneous heaters. Although the multi-flue heater can be used as a complete water heating system by itself, it is frequently advisable to use it in conjunction with an auxiliary storage tank. The most economical ratio between storage capacity and recovery capacity will depend upon the particular hot water demand requirements.

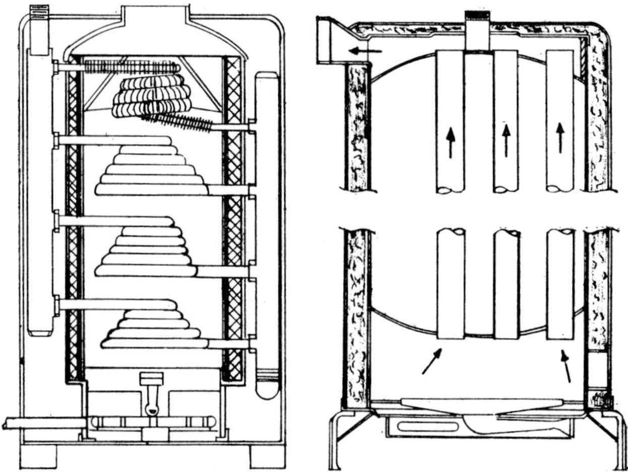

Fig. 12-69 (left) Section thru a coil-type circulating water heater; (right) section thru a typical multi-flue heater.

Instantaneous Heaters

The automatic instantaneous or continuous flow type of water heater is a self-contained unit in which the cold water is heated as it flows thru the heater. It may be a copper coil type or a tank type with a relatively small volume of stored water in proportion to the input rating. The burner is controlled by means of water pressure drop, water flow, or a thermostat in the heater. These heaters are recommended for applications in which nearly constant flow rates are encountered over long periods of time, making hot water storage of little value. Such uses are found in certain industrial applications, car washing, some photo process plants, etc. Capacities range up to 9.43 gal per min at 100°F rise, or 566 gal per hr.

INDIRECT WATER HEATERS

These heaters utilize steam or hot water, generated in cast iron or steel boilers, as the heating medium. The following systems are those most commonly employed with gas-fired boilers.

Boiler with Auxiliary Storage Tank

In one type of system, shown in Fig. 12-70(a), a gas-fired steam boiler supplies steam to a coil in the hot water storage tank or to a heat exchanger mounted on the boiler. The tank, coil, and boiler are available in any desired capacity. It is particularly necessary with this type of equipment to study the available space and the allowable floor loadings. Figure 12-70(b) shows a similar system using a water coil in the steam boiler. Circulation between the coil and tank may be by gravity or forced circulation. In both systems consideration should be given to the use of corrosion-resistant storage tanks.

Boiler with Submerged Coils for Instantaneous Heating

The entire output of the steam boiler in this type of heater is used for water heating (Fig. 12-71). The gas input to the steam boiler is selected to match the maximum hourly hot water demand. The water heating coil has still greater capacity, being adequate to meet the requirements of the maximum minute—1 sq ft of copper coil area for each 15 MBtu per hr to be transmitted is usually specified.

Although no storage tank is provided, sudden heavy demands of brief duration may be met by the hot water stored within the boiler and coil. When used without a storage tank, these units should be considered as *gallon-per-minute heaters.* Care should be given to sizing, particularly when the unit is used to provide both 140 and 180 F water. Coil and heater sizing should be based on the maximum total gpm draw of both general purpose and 180 F rinse water.

TWO-TEMPERATURE WATER HEATING SYSTEMS

For installations such as those in commercial kitchens where hot water at two different temperatures is needed, any of the previously described water heating equipment may be used, either singly or in combination. There are five basic arrangements for providing both 140 and 180 F water.

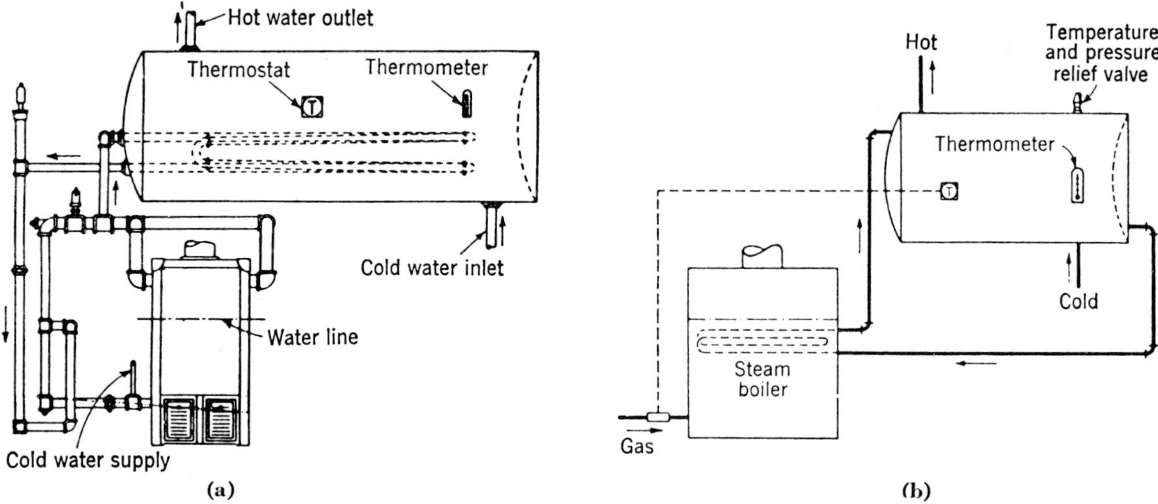

Fig. 12-70 Indirect water heating systems using boiler and auxiliary storage tank: (a) steam from boiler flows thru coils of storage tank: (b) water flows thru coils in boiler to storage tank.

Separate Heaters System. This arrangement (Fig. 12-72) is logical when an existing heater has sufficient 140 F water capacity but is unable to carry any additional load. Cold water is heated to 180 F in a separate heater.

Booster Heater System. This system (Fig. 12-73) may be used if the 140 F water supply is capable of carrying an additional load. The booster heater raises part of the primary 140 F supply to 180 F.

All 180 F System. Water is heated and stored at 180 F; 140 F water is obtained by means of a mixing valve (Fig. 12-74).

Boiler with Two Submerged Coils. This system (Fig. 12-71) meets demands for both 140 and 180 F water. Since water is often stored in the boiler at temperatures well above 180 F, it is customary to use a separate mixing valve for each hot water service. Each coil must be sized to handle the maximum flow rate imposed on it. The boiler must have a combination of input and/or stored hot water capacity to handle the largest volumes normally encountered during peak load draws for both temperatures of hot water.

Dual-Temperature System (Patented). The heater is connected to a storage tank used to store 140 F water (Fig. 12-75). The same heater also serves as a booster to meet the 180 F water demand by raising the temperature of the stored water.

WATER HEATER CORROSION

Steel storage tanks are designed to be corrosion resistant; they may be galvanized or lined with copper, glass, or concrete. Tanks of stainless steel as well as ones made of copper, aluminum, Monel, and other nonferrous alloys are also available. Requisites of corrosion-resistance, strength, and economics dictate the selection of the tank, as does the type of usage to which the water heater will be put.

Sand and similar materials carried in the water system sometimes deposit in storage tanks. If *hard* waters are heated above 150 to 160 F, solid deposits occur; this is called *liming.* Thin deposits of these solids, called *scaling,* may act as corrosion barriers. Thick coatings, however, may close off the flow thru tubing or cause localized hot spots in the tank bottom or the tank area where heat is applied. Such overheated areas are subjected to higher temperatures and to potentially more severe corrosion conditions. It is commonly recommended that tanks and heat exchangers periodically be "blown down" or drained in order to clean out undesirable deposits. Manufacturers of coil heaters recommend preventive maintenance (e.g., acid treatments) for removing lime from heater coils. *Soft* waters (60 to 120 ppm of solids) are responsible for more corrosion problems than hard waters (121 to 180 ppm), but very hard waters (above 180 ppm) accelerate liming problems. The output of water softening equipment may be controlled to provide the desired hardness.

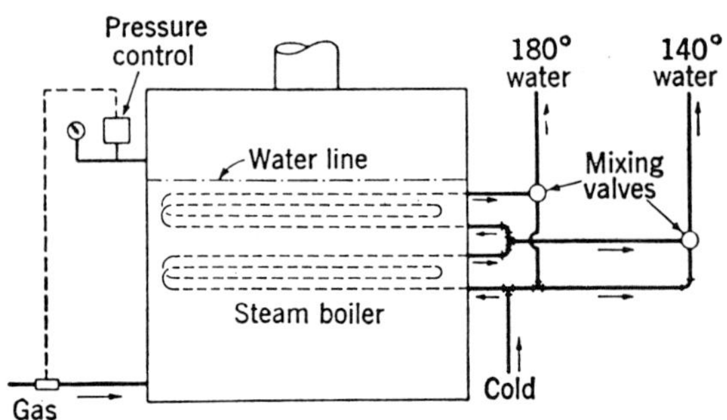

Fig. 12-71 Steam boiler used solely for instantaneous water heating and supplying water at both 140 and 180 F.

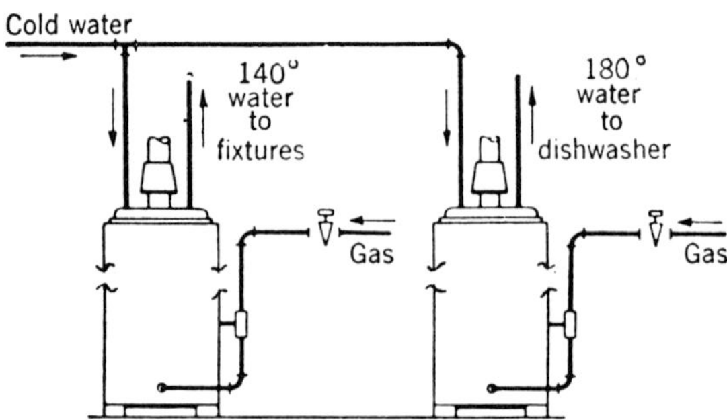

Fig. 12-72 Two-temperature water heating system using separate heaters.

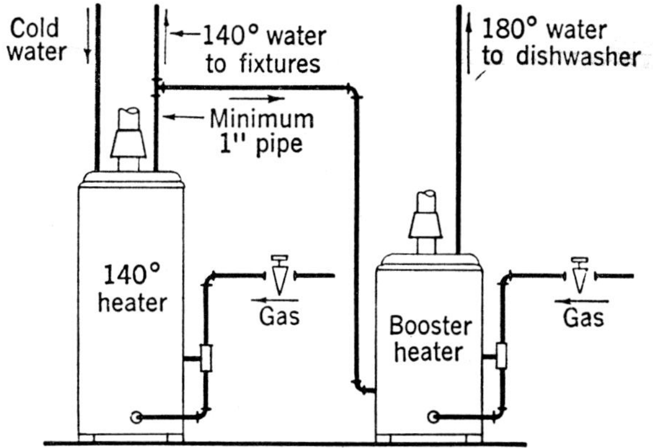

Fig. 12-73 Booster heater which raises 140 F water from one heater to 180 F for rinsing.

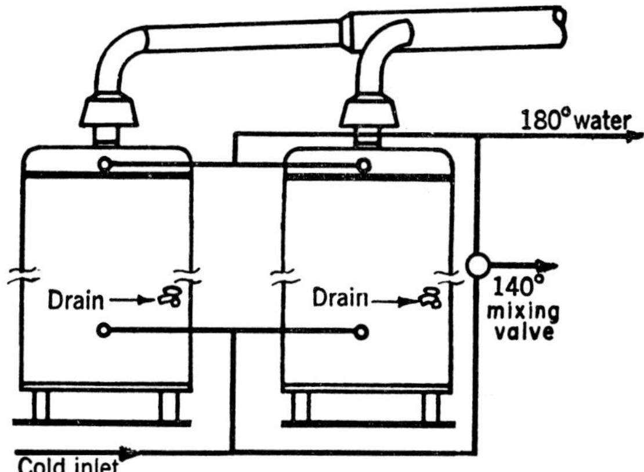

Fig. 12-74 Multiple heaters supplying 180 F water directly and 140 F water by means of a mixing valve.

Contact between dissimilar metals should be avoided in the water surfaces, since this combination produces galvanic couples; insulated pipe couplings prevent this effect. **Cathodic protection** may also be used to prevent corrosion. Usually this is accomplished by inserting a sacrificial magnesium anode in the tank; magnesium is more active than other metals employed in the construction of water heaters (Table 11-1).

Since studies show low corrosion rates for most waters at temperatures below 140 F, it is desirable to keep the water temperature no higher than necessary in tanks made of galvanized or lined common steel. Gas inputs to water heaters have no appreciable effect on the corrosion rate, provided that metal temperatures are not significantly increased. Fast rates of hot water withdrawal tend to increase turbulence in the tank, and more oxygen, brought in by the water, contacts the tank interior, accelerating corrosion.

WATER HEATER SELECTION AND SIZING

Listing of Approved Direct Gas-Fired Water Heaters

Units tested under American Standards[1-3] and listed[4] by the A. G. A. Laboratories are as follows:

Type A. Miscellaneous, including circulating tank, instantaneous, uninsulated, underfired.

Type B. Underfired, insulated automatic storage heaters.

Type C. Type B units with one or more flat sides, tested for installation flush to wall.

All listings[4] include **recovery** capacities in U. S. gallons raised 100°F per min (instantaneous heaters) or per hr (storage or circulating heaters); nominal storage vessel capacities of assembled heaters are also shown. Based on 70 per cent efficiency and a 100°F rise:

Recovery, gal per min = Btu per hr/71,400
Recovery, gal per hr = Btu per hr/1190

Domestic Water Heating

Direct Gas-Fired Water Heaters. Automatic units of the fast recovery type (inputs of 1000 Btu per hr or more per gal of storage) are generally used for residential service. When hot water requirements are quite low, low-recovery heaters are sometimes installed; small circulating tank heaters and instantaneous heaters are also used occasionally. FHA sizing standards for one- and two-family units are shown in Table 12-59a.

Table 12-59a FHA Minimum Direct Gas-Fired Water Heater Sizes for One- and Two-Family Units[7]

	Number of bathrooms									
	1–1½			2–2½			3–3½			
Number of bedrooms	2	3	4	3	4	5	3	4	5	6
Storage tank capacity, gal	30	30	40	40	40	50	40	50	50	50
Input rating, MBtu per hr	30	30	30	33	33	35	33	35	35	35

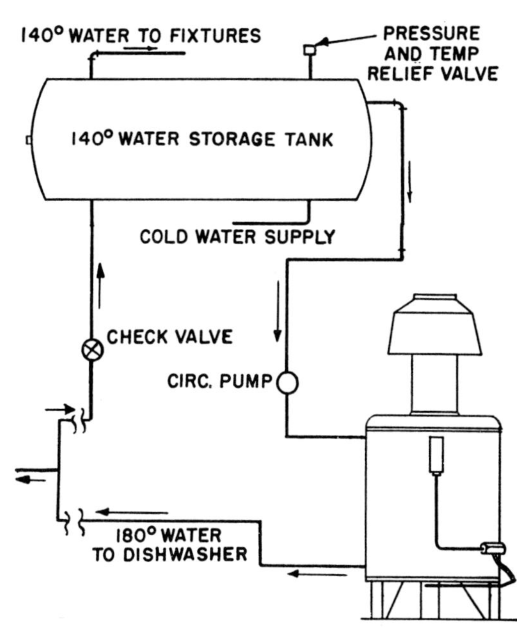

Fig. 12-75 Two-temperature system with 180 F water supplied directly from the heater; 140 F water for fixtures is drawn off the top of the storage tank. (Patented, A. O. Smith Co.)

Besides a wide range of combinations of heat input and storage capacity, models are available with one or more of the following features: (1) noncorrosive or corrosion-resistant tanks; (2) listed for water storage up to 180 F; (3) mixing valves for two-temperature water service; (4) automatic variation of input ratings to meet different draw-off rates; (5) sealed combustion design requiring no vertical vent; and (6) table-top height for installation in or under kitchen work counters.

Automatic cycle laundry washers add severe peak hot water demands to family hot water needs. Water heaters sized according to Table 12-59a will furnish the hot water used by cycle washers, but the drain imposed by some of the popular makes (20 to 27 gal per wash) will leave little reserve for a second and third washer cycle. By using slightly larger or faster water heaters, superior hot water service can be provided. A. G. A. recommends that the heater for the average family be capable of supplying 81 gal of hot water during a 2-hr period. Such heaters will take care of repeated family washing cycles and other hot water uses that may occur during the laundering period.

Table 12-59b furnishes the sizing information for water heaters for one-bath homes in which repeated use of cycle washers must be considered; note that incoming water temperatures affect sizing for heavy loads. Add 10 gal of tank capacity to Table 12-59b values for each additional bath in homes having four or more bedrooms.

Table 12-59b A. G. A. Recommended Gas Water Heater Size for Average Family with One Bath and Automatic Cycle Washer

Inlet water temp, °F	Tank capacity, gal	Input rating, MBtu/hr
40	30	45
	40	42
50	30	40
	40	35
60	30	35
	40	30

Daily Hot Water Usage. The U. S. Department of Agriculture Technical Bulletin No. 1073 (1953) used 38 gal per day per family as the basis for water heater tests. A 1959 survey by electric utilities indicated 41 gal per day based on the readings from 60,000 meters. A 1962 survey of 23 large gas utility companies by the A. G. A. Committee on Comparison of Competitive Services (CCCS) showed an average gas consumption for water heating of 255 therms per year, or a hot water usage of 44.5 gal per day (based on the average service efficiency established by the University of Illinois). A water heater performance study conducted by the Building Research Advisory Board (BRAB) for the FHA in 1956 recommended 55 gal per day as the mean minimum quantity of hot water required in low-income American homes, with an increase of 7 per cent for each $1000 of income above $3000. The foregoing data vary from 38 to over 55 gal per day and therefore substantiate the commonly used gas industry practice of *estimating water heating fuel costs* based on an average usage of 50 gal per day or, if an automatic cycle washer is used, 75 gal per day.

Peak Demand. In 1956 BRAB also concluded that the peak water heater load should be estimated on the basis of two automatic cycle washer loads in 2 hr, or 46 gal *plus* 7 gal for other simultaneous uses—a total of 53 gal in the peak 2-hour periods. A comprehensive study (1958) of hot water demands in the home by CCCS also determined that use of an automatic clothes washer represented the peak hot water demand. The Committee recommended that water heaters be *sized* to supply three consecutive washer cycles of 25 gal each in 2 hr plus 6 gal for other uses, or 81 gal of hot water for the 2-hr peak demand.

Comparative Monthly Fuel Usage. Service efficiency tests of automatic storage water heaters using various fuels have been made by the University of Illinois and A. G. A. Laboratories. The tests were based on typical daily withdrawal schedules and include stand-by losses. Table 12-60 shows the relative monthly gas, oil, and electric fuel usages based on the service efficiencies obtained.

Table 12-60 Gas, Oil, and Electric Water Heater Performance Based on Usage of 50 and 100 Gal per Day, 100° F Rise*

Water usage	Avg service efficiency, %	Monthly fuel consumption
1500 gal per mo		
Gas heater	52.4	23.9 therms
Oil heater	44.2	20.5 gal
Electric heater	84.0	435 kwh
3000 gal per mo		
Gas heater	62.0	40.4 therms
Oil heater	50.4	36.2 gal
Electric heater	90.6	810 kwh

* Based on University of Illinois tests—Bulletin 436, "Investigation of the Performance of Automatic Storage-Type Gas and Electric Domestic Water Heaters." Oil data from A. G. A. Laboratories tests (Report No. 1361A) using University of Illinois draw schedules.

Indirect Water Heaters. Hydronic systems (i.e., steam boilers or hot water boilers) equipped with indirect water heaters that either have instantaneous coils or have slow-recovery coils connected to storage tanks are widely used in the Northeastern states. FHA sizing standards for one- and two-family units are shown in Table 12-61.

Table 12-61 FHA Minimum Indirect Water Heater Sizes for One- and Two-Family Units[7]

Number of bathrooms	1–1½		2–2½		3–3½	
Number of bedrooms	2–3	4	3–4	5	3	4–6
With tank; boiler water at 180 F:*						
I-W-H†-rated, gal in 3 hr, 100°F rise	40	66	66	66	66	66
Manufacturer-rated, gal in 3 hr, 100°F rise	49	75	75	75	75	75
Tank capacity, gal	66	66	66	82	66	82
Tankless; boiler water at 200 F:*						
I-W-H rated, gpm, 100°F rise	2.75	3.25	3.25	3.75	3.25	3.75
Manufacturer-rated, draw in 5 min, 100°F rise	15	25	25	35	25	35

* Boiler-connected internal or external type.
† Indirect Water Heater Testing and Rating Code of Institute of Boiler and Radiator Manufacturers.

When instantaneous coils are used in typical one-family homes, the boiler and coil should be sized to deliver at least 15 gal of hot water in five minutes at a minimum temperature of 130 F, with the boiler water temperature at 180 F. The quantity and temperature of hot water during any draw-off is limited to the heat supplied to the boiler by the burner during the draw plus the stored heat in the boiler water before the draw. The time the burner is in operation during any draw is equal to the length of the draw-off period *minus* the time delay before the aquastat starts the burner. Usable heat in the stored boiler water is limited to that above the required minimum outlet temperature of the drawn water. Figure 12-76 shows performance data for a typical unit.

In general, when slow recovery coils are used with storage tanks, no additional allowance need be made in boiler sizing beyond that needed for the heating load. The pickup and piping loss allowances used in boiler sizing are more than adequate to take care of the additional water heating load during the winter months.

Internal tankless coils are not recommended in hard water areas except when provision is made to treat the water supply.

The Institute of Boiler and Radiator Manufacturers tests and rates indirect water heating coils in accordance with its Indirect Water Heater (I-W-H) Testing and Rating Code. A tankless coil is tested when installed in the particular boiler in which it is to be used. The I-W-H ratings are in gallons per minute (100°F rise) for tankless coils and in gallons in three hours (100°F rise) for indirect water heaters connected to a storage tank.

Large Volume Water Heating

Gas-fired water heating equipment meets the needs for large or small volumes of water at temperatures up to 180 F for sanitizing rinses in dishwashing. Either direct-fired water heaters or indirect water heaters connected to a gas boiler may be used. Table 12-62 shows the maximum daily hot water requirements for several types of commercial installations.

Sizing. The usual practice in determining inputs and storage tank sizes for the lowest installed cost consistent with reasonable operating cost follows: The recovery rate (or input) is based on *dividing* the estimated number of gallons of hot water usage over a period of 24 hr by the maximum number of hours considered good practice for the heater to operate. For sizing purposes, ten hours of daily operation is considered good practice for a copper coil water heater or an underfired storage water heater. For indirect systems using either cast iron or steel boilers, continuous operation, if necessary, is considered good practice; the conservative custom, however, is to base the hourly recovery on 20 hr of operation. Knowing the recovery per *heater-operating hour* and the temperature rise, the input to the heater is readily calculated. The storage tank capacity is determined by *subtracting* the recovery rate during the period of the maximum peak demand from the use of water during that period. The result represents the amount of water that must be available from storage to make up the deficit during the peak. Because of stratification and mixing in the tank, only 70 per cent of the tank capacity is *usable* hot water; hence, the tank must be oversized by at least 43 per cent.

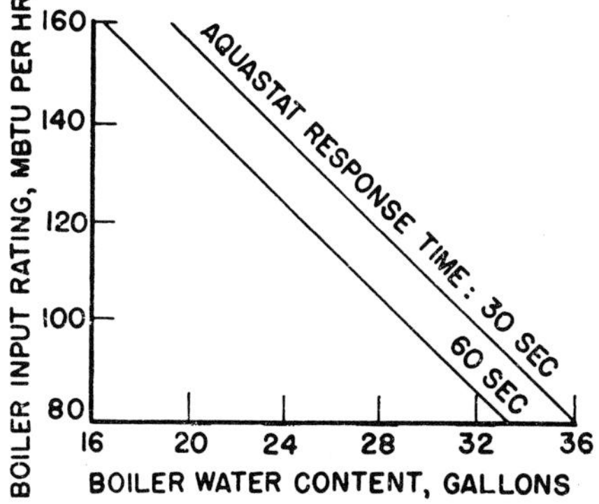

Fig. 12-76 Gas input vs. boiler water volume required for an instantaneous coil, rated at 3 gpm, supplying 15 gal of 100°F rise water in five minutes.[8] Boiler water temperature dropped from 180 to 150 F.

Table 12-62 Maximum Requirements for 140 F Water in Various Commercial Establishments[9]

Establishment	Requirement, gal per 24 hr
Hotels and restaurants	
Public basins (or 0.5 gal per use)	150
Public basins with attendant	200
Slop sink for cleaning purposes	30
Room with basin	10
Room with basin and bath or shower	
Transient	40
Resident (male)	40
Resident (female)	70
Resident (mixed)	60
2 Rooms, bath and/or shower	80
3 Rooms, bath and/or shower	100
Public bath	150
Public shower	200
Garages	
Approx 50 gal of 90 F water per car washed	
Hospitals	
Approx 125 gal per bed on max day, 80 gal on avg day	
Kitchens (exclusive of dishwashers)	Gal per filling
Pot sinks	
Small	20
Medium	30
Large	40

Instantaneous gas water heaters that have no storage must be sized to meet the maximum hot water demand at the desired temperature rise, even if this demand may exist for only a few minutes. When an instantaneous, indirect water heater is installed in a boiler, the required gas input may be reduced by the amount of available heat in the stored boiler water. This stored heat is an important factor when the indirect coil surface is ample and circulation over the coil is active.

When making these calculations, it is necessary to select an appropriate thermal efficiency for the water heating system to be used. A. G. A.-approved water heaters are rated at 70 per cent and boilers at 80 per cent. Actual system efficiencies will not usually reach these levels, however, because of piping losses, stand-by losses, type of hot water demand, type of system, equipment size, and other variables. A well-designed system can operate at 60 per cent efficiency if the piping and stand-by losses are held to a minimum. On the other hand, a space-heating boiler used to provide steam for water heating during the summer months may operate at 20 per cent efficiency or less if the storage tank is not insulated and the

Table 12-63a Rinse Water (180 F) for Dishwashers
(pressure at dishwashers assumed to be 20 psig)

Type and size of dishwasher	Flow rate,[10] gpm	Heater requirements, gal per hr: No internal storage	Heater requirements, gal per hr: Sufficient internal storage to meet gpm flow rate*
Door type			
16 × 16 in. rack	6.94	416	69
18 × 18 in. rack	8.67	520	87
20 × 20 in. rack	10.4	624	104
Undercounter type	5	300	70
Conveyor type			
Single tank	6.94	416	416
Multiple tank (dishes flat)	5.78	347	347
Multiple tank (dishes inclined)	4.62	277	277
Silver washers	7	420	45
Utensil washers	8	480	75
Make-up water	2.31	139	139

*Dishwasher operation at 100 per cent mechanical capacity.

water heating load is small in relation to the boiler size. The efficiency during winter operation will naturally be much better.

Food Service Establishments

Two water temperatures are needed in food service establishments. In a restaurant, sanitizing to completely acceptable levels is usually achieved by exposing the washed dish to a 180 F water rinse for ten seconds. In addition, an ample supply of general purpose hot water (usually 140–150 F) is required for the wash cycle of dishwashers, personnel use, pot washing, general cleaning, and other uses.

The National Sanitation Foundation has established minimum requirements[10] for the construction, sizing, and installation of gas water heaters for food service establishments that use dishwashing machines. Many water heater manufacturers construct heaters complying with these requirements and many local health authorities require conformance when sizing and supplying heaters for public eating establishments.

Sizing.[11] Equation **1** may be used to determine the Btu per hr input required to heat water at 20 psig:

$$H = 0.012WT \quad (1)$$

where:

H = gas input, MBtu per hr
W = water requirement from Tables 12-63a (180 F water) or Table 12-63b (140 F water), *less* the *available* (Eq. 2) portion of storage volume, gal per hr
T = temperature rise, °F

Available hot water from storage for systems with more than 100 gal storage capacity may be calculated from Eq. **2**:

Available hot water from storage, gal per hr = 0.7 × *capacity of tank* in gal/*duration of peak hot water demand*, hr **(2)**

This *peak demand* will usually coincide with the dishwashing period during and after the main meal and may last as long as three hours or more. Any hour in which the dishwasher is used at 70 per cent or more of mechanical capacity should be considered a peak hour. If the peak demand period lasts four hours or more, the heater recovery rate should be sized to equal the peak hour demand and the storage tank sized to meet heavy, short draws within the peak hour.

The gas input rating to automatic storage water heaters with storage capacities up to 100 gal should also be sized on the basis of the maximum hourly hot water demand, since the small quantity of hot water available from storage may be needed to meet heavy draws of short duration.

Sanitizing Rinse. Both the National Sanitation Foundation Standard No. 5[10] recommendations for hot water rinse demand and Table 12-63a are based on the use of 100 per cent of operating capacity. However, it is generally recognized by industry, and stated by NSF Standard No. 5, that 70 per cent of operating rinse capacity is all that is normally attained except for rackless-type conveyor machines.

Booster Water Heaters. Approximately 70 per cent of all dishwashing machines are of the single-tank rack type that meet the NSF[10] requirements of a six quart intermittent rinse of about ten seconds. For these machines a booster water heater of 50 MBtu per hr or less with nominal (up to 12.5 gal) storage capacity meets this requirement based on a 140 F inlet water temperature. Since the practical maximum usage of such dishwashers is 70 per cent of the theoretical mechanical capacity of these machines, the actual input to the water heater is also about 70 per cent of the rated input. The booster operates only during dishwashing periods—when the kitchen exhaust fan is also customarily

Table 12-63b General Purpose Hot Water (140 F) for Various Kitchen Uses

Application	Consumption, gal per hr
Vegetable sink	45
Single-compartment sink	30
Double-compartment sink	60
Triple-compartment sink	90
Prescraper (open type)	180
Preflush	
Hand operated	45
Closed-type	240
Recirculating preflush	40
Bar sink	30
Lavatory	5

running. Assuming three 1.5-hr dishwashing periods at 70 per cent input, the daily gas consumption would be 158 MBtu.

ASA Z21.30–1964 permits these booster heaters to be installed unvented, since the kitchen exhaust system will provide adequate ventilation for the capacity relationship involved. An electrical interlock between the exhaust fan and the heater gas valve may be specified but is not required.

Water Heater Sizing Checklists. To size a water heating system correctly, it is necessary to know: (1) types and sizes of dishwashers used; (2) required quantity of general purpose hot water (manufacturers' catalogs should be consulted to determine the initial fill requirements of the wash tanks); (3) duration of peak hot water demand period; (4) inlet water temperature; (5) type and capacity of existing water heating system; and (6) type of water heating system desired.

Example I: Size a new cafeteria water heating system. Maximum dishwasher usage at 70 per cent of capacity, 2 consecutive hr; supply water temperature, 60 F; hot water requirements as follows:

No. of installed facilities	General purpose, gal per hr at 140 F (from Table 12-63b)	Dishwasher at 100% capacity, gal per hr at 180 F (from Table 12-63a)
1 Vegetable sink	45	...
5 Lavatories	25	...
1 Prescraper	180	...
1 Two-tank (20 gal each) conveyor dishwasher with make-up (dishes inclined)	...	277
Initial fill of tanks	40	...
Make-up water	...	139
1 Utensil washer	...	75
Initial fill of tank	85	...
Total requirements	375 gal per hr	491 gal per hr at 0.7* of capacity ≅ 350 gal per hr

* Based on the NSF recommendation for probable operating capacity for rack-type dishwashers.

These hot water demands can be met by many systems. The three solutions given here are not necessarily the "best" ones.

Solution A: Using separate self-contained storage heater(s), with sufficient storage capacity to meet the gpm flow rate:

1. Gas input, H_1, for 180 F water, substituting in Eq. **1**:

$$H_1 = 0.012WT = 0.012 \times (350 - 0) \times (180 - 60) = 504 \text{ MBtu per hr}$$

Heater(s) with this input may now be selected from manufacturers' catalogs.

2. Similarly, H_2, for the 140 F water:

$$H_2 = 0.012WT = 0.012 \times (375 - 0) \times (140 - 60) = 360 \text{ MBtu per hr}$$

Heater(s) with this input may now be selected from manufacturers' catalogs.

Solution B: Using instantaneous-type heater(s), having no internal storage, to supply both 180 F water and, thru a mixing valve, 140 F water. Sizing is dependent upon the gpm flow rate of the 180 F rinse water. The hourly 180 F water consumption based on the gpm flow rate can be obtained from Table 12-63a, the hourly consumption of 140 F water, from Table 12-63b:

1. For the 180 F water:

Use	180 F rinse water use based on gpm flow rate, gal per hr
Dishwasher	277
Make-up	139
Utensil washer	480
Total	896 of 180 F water (use 900 gal per hr)

2. Gas input, H_1, for 180 F water, substituting in Eq. **1**:

$$H_1 = 0.012\,WT = 0.012 \times (900 - 0) \times (180 - 60) = 1296 \text{ MBtu per hr}$$

3. Similarly, H_2, for 140 F water = 360 MBtu per hr (from *Solution A*).

4. Total gas input = $H_1 + H_2$ = 1656 MBtu per hr.

Heater(s) with this input may now be selected from manufacturers' catalogs.

Solution C: Using (1) a booster heater having sufficient storage capacity to meet the gpm flow demand to supply the 180 F rinse water requirements from a 140 F supply; and (2) a heater used in conjunction with a 500 gal external storage tank to supply 140 F water for both general purposes and the booster heater:

1. Gas input, H_1, to the booster heater, substituting in Eq. **1**:

$$H_1 = 0.012WT = 0.012 \times (350 - 0) \times (180 - 140) = 168 \text{ MBtu per hr}$$

A booster heater with this input may now be selected from manufacturers' catalogs.

2. Available 140 F water from storage = 0.7 × 500 gal/2 hr = 175 gal per hr (from Eq. **2**).

3. Gas input, H_2 for 140 F water, substituting in Eq. **1**:

$$H_2 = 0.012WT = 0.012 \times (350 + 375 - 175) \times (140 - 60) = 528 \text{ MBtu per hr}$$

4. Total gas input = $H_1 + H_2$ = 168 + 528 = 696 MBtu per hr.

A heater with this input may now be selected from manufacturers' catalogs. In solutions using storage tanks, it is assumed that the water in these tanks will have been brought up to temperature before the peak dishwashing period and that sufficient time will have elapsed before the next peak period for the water in the storage tank to recover its required heat content.

Example II: Size a booster water heater to supply 180 F water for a door-type dishwasher to handle 16 × 16 in. racks. The existing hot water system can supply sufficient 140 F general purpose water to meet all requirements, including this booster heater.

Solution A: Using an instantaneous (booster) heater having no storage capacity:

1. Since the heater has no 180 F storage, it must be sized to supply 180 F water at a flow rate of 6.94 gpm or 416 gal per hr (from Table 12-63a).
2. Gas input, H, for 180 F water, substituting in Eq. **1**:

$$H = 0.012WT = 0.012 \times (416 - 0) \times (180 - 140) = 200 \text{ MBtu per hr}$$

A heater with this input may now be selected from manufacturers' catalogs.

Solution B: Using an instantaneous (booster) heater having sufficient 180 F storage capacity to meet the gpm flow rate for the 10 sec rinse cycle:

1. Since the heater has sufficient internal storage, it requires 69 gal per hr (from Table 12-63a) at 100 per cent capacity or $0.7 \times 69 = 48.3$ gal per hr at 70 per cent of capacity.
2. Gas input, H, for the 180 F water, substituting in Eq. **1**:

$$H = 0.012WT = 0.012 \times (48.3 - 0) \times (180 - 140) = 23 \text{ MBtu per hr}$$

A heater with this input may now be selected from manufacturers' catalogs.

Mass Housing

The hot water supply in motels, hotels, military barracks, dormitories, etc. (but not apartment buildings) is included in this category. Hotels will usually have a three-hour peak load, while in motels this peak will usually last only two hours, with about the same volume of hot water used per person. Barracks, dormitories, and fraternity houses are usually found to have a peak load of one hour. The peak load in housing usually results from shower and/or tub baths.

In some types of housing, such as fraternity houses or dormitories, all occupants may take showers within a very short period of time. In this case, the peak load is best determined by the number of shower heads, the rate of flow per head, and by estimating the length of time that the showers will be on.

Shower Heads. The rate of flow will vary depending on the type, size, and water pressure. At 40-psi water pressure, available shower heads have nominal flow rates from about 2.5 to 10 gpm; about half of this is hot water. In multiple shower installations, flow control valves are recommended, since they reduce the flow rate and maintain it regardless of fluctuations in water pressure. Even though the flow rate may be reduced by as much as 50 per cent, the spray pattern of the shower head is not affected. Flow control valves come in three sizes: 4.0, 2.6, and 1.6 gpm. If the manufacturer's flow rate for a shower head is not available and a flow control valve is not used, the following will serve as a guide for sizing the water heater:

Shower head size:	Small	Medium	Large
Flow rate, gpm:	2–3	4–6	7–9

Other uses for hot water may also exist. Motels and hotels may have a restaurant; all types of housing may have coin-operated laundry machines; and sometimes a swimming pool may be heated by the water heater. If these loads fall during the peak draw, they must be added to the shower and bathing load.

Apartment Houses

For multiple dwellings (four or more apartments), A. G. A. has developed recommendations[12] for selecting water heaters. The combinations of storage capacity and heater inputs obtained from Table 12-64 (which includes diversity considerations) should meet the peak shower and tub bathing periods, as well as any concurrent use of washing machines. Where other uses of hot water are present, additional system capacity may have to be provided.

Table 12-64 Gas Input Required for Water Heating per Apartment, Btu per hr[12]

(based on 100° F temp rise and 40 psi water inlet pressure)*

Hot water stored in tank per apt,† gal	Heavy use‡	Avg use‡	Light use‡
	4 to 9 apts (avg use based on 37 gal per apt in 3 hr)		
5	16,700	13,300	10,000
10	14,900	11,900	8,900
15	13,100	10,500	7,900
20	11,500	9,200	6,900
25	9,500	7,600	5,700
30	7,900	6,300	4,700
Immersion coil boiler	(Currently available equipment oversized for this market)		
	10 to 19 apts (avg use based on 35 gal per apt in 3 hr)		
5	15,600	12,500	9,400
10	13,900	11,100	8,300
15	12,200	9,800	7,400
20	10,400	8,300	6.200
25	8,600	6,900	5,200
30	7,000	5,600	4,200
Immersion coil boiler§	17,400	13,900	10,400
	Over 19 apts (avg use based on 32 gal per apt in 3 hr)		
5	14,100	11,300	8,500
8	13,100	10,500	7,900
10	12,400	9,900	7,400
15	10,800	8,600	6,400
20	9,000	7,200	5,400
25	7,300	5,800	4,400
30	5,500	4,400	3,300
Immersion coil boiler§	15,900	12,700	9,900

* For 90°F temp rise, multiply required Btu input by 0.9; for 80°F temp rise, multiply required Btu input by 0.8.

† Recommended hot water storage per apt:

No. of apts:	4–6	7–10	11–19	Over 19
Storage, gal:	20	15	10	8

‡ Heavy use: Over six persons per apt, or more than one bathroom in each apt, or more than one washing machine per five apts.

Avg use: Three to five persons per apt.

Light use: One to two persons per apt.

§ The coil should be sized as follows for a continuous 5-min draw period:

No. of apts:	10–19	20–29	30–39	40–49	50–59	60–79	Over 79
Gpm per apt:	1.2	1.0	0.84	0.67	0.6	0.53	0.5

Sizing. To determine the required Btu per hr input to a water heater using Table 12-64, ascertain: (1) number of apartments; (2) type of water use; and (3) gallons of storage capacity per apartment. Then, enter the appropriate section of Table 12-64 with the storage capacity per apartment and directly obtain the required Btu input per apartment in the column corresponding to type of water use. If an existing storage tank is to be used, its size will fix the storage capacity per apartment. For new installations a footnote to the table suggests appropriate storage per apartment, subject to space limitations and system designer preferences.

Swimming Pools

The desirable water temperature for residential and commercial swimming pools is 78 F. To maintain this temperature usually requires a water heating system that will raise the temperature of circulated water so as to compensate for any colder fresh water entering the pool and for any heat losses to the ground and air.

Many commercial swimming pool installations include shower facilities and sometimes a steam room. It is usual practice to use independent heating equipment for these purposes—a standard water heater for the showers and a small boiler or central steam supply for the steam room.

Special water heating equipment suitable for heating pools is available from most manufacturers of gas-fired water heaters or hot water boilers. Some manufacturers offer units that are factory packaged with all necessary operating controls, including a pool temperature control and a water by-pass, which prevents condensation. The water heating system is usually installed in the normal circulation system of the pool at a point between the treatment stage and the pool. Indirect systems using hot water piping embedded in the walls or floor of the pool may also be used and, since pool water does not pass thru the heater, corrosion, scaling, and condensation problems are greatly reduced. However, the initial cost of this system is relatively high and direct heating systems are the most common.

In addition to the safety controls normally used, pool heaters should be equipped with a pool temperature control and a water pressure or flow safety switch. The temperature control is usually installed in the return line from the pool to the heater, preferably at the inlet to the heater. The pressure or flow switch is mounted either in the heater inlet or in the outlet, depending on the manufacturer's instructions. Its purpose is to protect against no flow or inadequate water flow thru the heater.

A number of methods for sizing pool water heating systems are available. Some are rule-of-thumb and are based simply on pool area or water volume. Others are more complicated and include many factors, such as heat loss to the ground and to the air from the water surface. Unless unusual conditions exist for a particular job, A. G. A. recommends[13] sizing the equipment on the basis of the amount of heat necessary to raise the temperature of the volume of water the desired number of degrees in a specified time.

One of the most important considerations in calculating the Btu per hr input of the gas water heater is whether the pool will be used intermittently or continually.

Intermittent Use. Some pools are used only periodically, for example, on weekends. For the most economical operation, therefore, the heater should be on only just before and during the period of pool use. Thus, a high-Btu heater will be desirable as a means of reducing the heatup time to a practical maximum—say, 24 hr.

Continual Use during a Season. Continual use requires the heating of the pool water over a maximum temperature rise only once in a season. It is conventional to allow 48 hr to reach the desired pool temperature. When longer pickup periods are used in sizing, the heater selected may be inadequate to meet the heat loss of the pool. Therefore, the heater capacity should be checked against an assumed pool surface heat loss of 12 Btu per hr-sq ft-°F temperature difference between the desired pool temperature and the average air temperature during the coldest month of pool use. The resultant heat loss must be divided by the heater efficiency, 0.7, to obtain the required heater input rating.

Sizing Pool Heaters. The A. G. A.-recommended method[13] is as follows:

1. Obtain pool water capacity in gal; if unknown, multiply length (in ft) × width (in ft) × 5.5 ft (assumed average depth) × 7.5 gal/cu ft.
2. Determine the desired heatup time, hr.
3. Determine the required pool water temperature rise, °F.
4. Use Table 12-65 to determine the required gas heater input.

Example: Size a heater for a pool 25 ft × 60 ft × 5.5 ft for intermittent use. Average temperature during coldest operating month is 50 F; water temperature specified is 80 F.

Table 12-65 Gas Input Required for Swimming Pool Heaters, MBtu per hr

(based upon water temp rise of 25° F* and max wind velocity of 15 mph)

Pool heatup time, hr	Pool water capacity, thousands of gal													
	5	10	15	20	25	30	35	40	45	50	75	100	150	200
8	186	371	556	742	926	1110	1300	1482	1670	1860	2780	3710	5560	7420
12	124	247	371	494	618	742	865	990	1110	1240	1860	2470	3710	4940
16	93	186	278	371	463	556	648	742	833	926	1390	1860	2780	3710
24	62	124	186	247	309	371	433	494	556	618	926	1240	1860	2470
36	41	83	124	165	206	247	288	330	371	412	618	830	1240	1650
48	31	62	93	124	155	186	216	247	278	310	463	618	926	1240

* Temp rise correction factors to be applied to pool water capacity **before** entering table:

Temp rise, °F:	10	15	20	25	30	35
Factor:	0.4	0.6	0.8	1.0	1.2	1.4

Solution:

1. Capacity of pool = 25 ft × 60 ft × 5.5 ft (assumed average depth) × 7.5 gal/cu ft = 62,000 gal.
2. Assumed heatup period for intermittent use is 24 hr.
3. Pool water temperature rise = 80° F − 50° F = 30° F.
4. Pool capacity corrected for 30° F temperature rise = 62,000 gal × 1.2 (from Table 12-65) = 74,400 gal. Thus, the heater input (in the 75,000 gal column for a 24 hr heatup period) is 926 MBtu per hr.

Heat Loss Check:

1. Surface area of pool = 25 × 60 ft = 1500 sq ft.
2. Temperature differential is 30° F.
3. Average heat loss = 12 Btu per hr-sq ft-°F × 1500 sq ft × 30° F = 540 MBtu per hr.
4. Input needed (assuming 70 per cent heater efficiency) = 540/0.7 = 771 MBtu per hr.

Thus, the heater selected (926 MBtu per hr input) satisfies the heat loss requirements.

Coin-Operated Laundries

The automatic washing machines used in these centers require water in the 140 to 160 F range. Any direct or indirect water heaters may be used; usually some water storage capacity is necessary to meet short peak draw periods when a number of machines are operating simultaneously on the fill cycle. If no storage is provided, the required gas input may be excessive in places where a large number of machines are installed.

Since the hot water requirements vary considerably for different models of washing machines, manufacturer's data should be initially obtained for the length of operating cycle in minutes and for gallons of hot water per machine for both initial fill and the remainder of the cycle.

Sizing. The A. G. A.-recommended procedure[14] that follows meets both the total maximum hourly usage of hot water and the hot water demand for short periods of high draw when many machines are simultaneously operating on the initial fill cycle.

1. *Maximum Hourly Usage.* Usually based on *simultaneous* operation of all machines:

$$G_t = 60nG_c/t_c \tag{3}$$

where:

G_t = total maximum hot water use, gal per hr
n = number of installed machines
G_c = hot water per machine, gal per cycle
t_c = time for one cycle = machine cycle time plus 10 min for loading and unloading, min

2. *Short Periods of High Draw.* Best supplied from boiler water heat or hot water storage. Initial fill portion of a cycle may require from 9 to 16 gal per machine, depending on the model. Suggested coincidence factors for simultaneous start-up of machines:

No. of machines:	1–11	12–24	25–35	Over 35
Factor:	1.0	0.8	0.6	0.5

Example: Size a water heater for 25 washing machines, each using 20 gal of 150 F water per 20 min machine cycle; the fill position of the cycle requires 15 gal.

1. Maximum hourly use, substituting in Eq. **3**: $G_t = 60nG_c/t_c = 60 \times 25 \times 20/(20 + 10) = 1000$ gal per hr
2. Demand for short periods of high draw = 15 gal/machine × 25 machines × 0.6 coincidence = 225 gal. Since the available hot water from storage does not exceed 70 per cent of the tank capacity, the required tank capacity = 225/0.7 = 321 gal.
3. Thus, the water heater should have a recovery rate of 1000 gal per hr and a storage tank with a capacity of at least 321 gal. The required heater input can be obtained from Eq. **1**.
4. If an immersion coil boiler without a storage tank is used, the recovery capacity must also be 1000 gal per hr, but the combination of heat input and heat in the boiler water must be sufficient to meet the simultaneous fill demand of 225 gal. To check this, the time for the fill portion of the cycle must be obtained for the particular washing machine installed. For example, if the fill lasts 3.0 min, the peak flow capacity of the immersion coil boiler has to be 225 gal/3.0 min, or 75 gpm for three minutes. The peak flow ratings for short draws can usually be obtained directly from the boiler manufacturer's literature.

Schools

The hot water used in the **cafeteria** is generally about 70 per cent of the amount normally required in a commercial restaurant serving adults; the necessary amount can be estimated as discussed in the material under Food Service Establishments. Where NSF sizing[10] is required, Standard No. 5 should be followed. For the **showers,** the hot water needs should be determined from the number of shower heads, the water flow rate of the shower heads, and the length of time needed to accommodate the largest athletic contest (largest number of men to use the showers). The sizing techniques for mass housing may be applied.

The shower and cafeteria loads will not ordinarily run concurrently. Each should be determined separately, and the larger load should determine the size of the water heater(s) and tank. Where feasible, the same water heating system may be used for both needs, but provision must be made for the 180 F sanitizing rinse for the dishwashers. Where the distance between the two points of use is great, separate water heating systems should be used.

A separate water heating system may be specified for the **swimming pool.**

Ready-Mixed Concrete

In cold weather, ready-mixed concrete plants need hot water for mixing. If cold water is used, the concrete may freeze before it sets and thus be ruined. Use of hot water in cold ambients also shortens hardening time and improves quality. With cold aggregate, in cold weather, water at about 150 F is normally used. When the water temperature is too high, some of the cement will "flash-set." Operators like to pour mixes at about 70 F.

Thirty gallons of hot water per cubic yard of concrete mixed is generally used for sizing. Table 12-66 indicates a common method of sizing water systems for ready-mixed concrete plants. Since hot water is dumped into the mix as fast as possible, ample hot water storage is required to handle each loading. If storage is not used, very large heat exchangers must be employed for the high draw rate.

Table 12-66 Sizing Water Heating Systems for Ready-Mixed Concrete Plants*

(gas input and storage tank capacity required to supply 150 F water, 40 F inlet)

Time interval between loadings, min	Truck capacity 6 cu yd	7.5 cu yd	9 cu yd	11 cu yd
	Input, MBtu per hr (with 430 gal storage)	(with 490 gal storage)	(with 560 gal storage)	(with 640 gal storage)
50	458	527	596	687
35	612	790	792	915
25	785	900	1020	1175
10	1375	1580	1790	2060
5	1830	2100	2380	2740
0	2760	3150	3580	4120

* Based on 10 min loading time per truck. Water usage also includes filling of 120 gal truck-mounted storage tank; this hot water is used for cleaning the truck container after unloading the concrete.

Part of the heat needed in the mix may be supplied by heating the aggregate bin. This may be done by installing pipe coils in the walls and/or sides of the bin and circulating hot water thru the coils. By using warm aggregate, the required temperature of the mixing water is lower. Warm aggregate will also flow more easily from bins than frozen material.

Jets of steam are sometimes used for thawing large chunks of aggregate to facilitate handling and mixing the material. If hot water is used for the thawing, it is difficult to maintain the proper aggregate–water ratio. A small gas-fired steam boiler will easily supply this "spot steam."

Industrial Plants

Hot water uses include the cafeteria, showers, lavatories, laundries, gravity sprinkler tanks, and many industrial processes. If the same hot water supply system is used only for the cafeteria, employee cleanup, laundry, and small miscellaneous uses, the water heater can generally be sized to meet the employee cleanup load, since it is usually heaviest and, coming at the end of the day shift, is not concurrent with the other uses. The other loads, however, should be checked against this capacity. The cafeteria load is estimated as for any other restaurant.

Employee Cleanup. The load is composed of one or more of the following: wash troughs or standard lavatories; Bradley wash fountains; and showers. The hot water requirements for employees using standard wash fixtures may be estimated at one gallon per clerical and light industrial employee and two gallons per heavy industrial worker. The hot water demand for *Bradley wash fountains* is based on full flow for the entire cleanup period (Table 12-67). The **shower load** depends on the flow rate of the shower heads, the total number of showers, and the period of use, as in mass housing installations (Table 12-67).

Water heaters used to prevent freezing in gravity sprinkler tanks should be part of a separate system. Load depends on tank heat loss, tank capacity, and winter design temperature.

Table 12-67 Hot Water Use for Bradley Wash Fountains and Showers

Wash fountains: Type	140 F water required* in 10 min period, gal	Showers: Flow rate, gpm	140 F water required† for 15 min period, gal
36 in. circular	40	3	29
36 in. semicircular	22	4	39
54 in. circular	66	5	48.7
54 in. semicircular	40	6	58

*Based on 110 F wash water and 40 F cold water at average flow rates.

† Based on 105 F shower water and 40 F cold water.

Process hot water volume and temperature vary with the specific processes. If this load coincides with the shower or restaurant load, the system must be sized to reflect the total demand. Separate systems may also be used, depending on the size of the various loads and the distance between loads. Information on various process hot water applications is available.[15]

WATER HEATER INSTALLATION

The size of the cold water supply line to a water heater should never be smaller than the connection on the heater. If the heater is equipped with water-actuated controls, care must be taken in piping design to ensure the required water pressure at the heater so that the controls will function properly.

The circulating water lines connecting a water heater to an external storage tank should be sized to conform to the water connections on the heater, unless otherwise specified by the manufacturer. All storage tanks must be equipped with a drain valve at the bottom of the tank to permit removal of sediment or complete emptying. External storage tanks connected to a circulating water heater should be provided with a device to prevent siphoning off the stored water. One satisfactory method is to install a cold water tube with a hole near the top (Fig. 12-77).

When multiple storage-type heaters are used, the outlet water connections must be manifolded together and connected with piping of equivalent pressure drop in order to

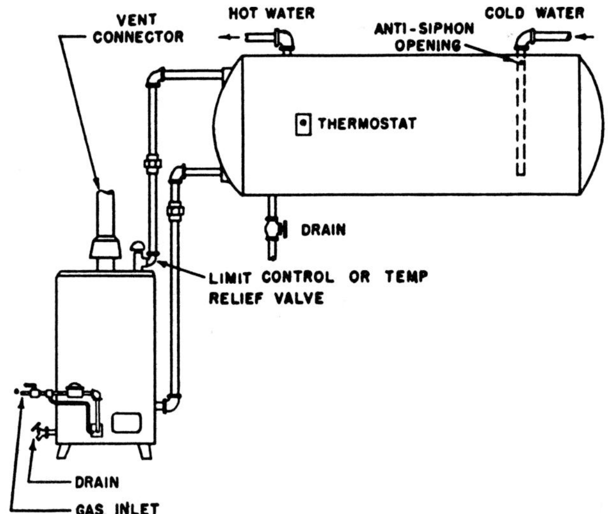

Fig. 12-77 Suggested location for anti-siphon opening in cold water inlet.

allow each heater to handle its proportion of the load. If this is not done, one or more of the heaters will operate excessively, and, in addition, the system efficiency and capacity will be reduced. An overdraw on one heater will also cause a drop in the water temperature.

If a return recirculating line is used in conjunction with a circulating pump, an adjustable flow control should be used to produce the desired temperature in the return water line. Too much circulation may also cause improper heater operation. In gravity systems, flow control should be unnecessary if the return line is properly sized.

Gas water heaters approved under ASA requirements[1–3] are equipped with the necessary safety pilots and other controls for automatic operation of the heater. Nevertheless, additional safety devices must be supplied and/or installed in the hot water supply system at the time of the heater's installation. These devices are designed to prevent excessive water temperatures and pressures in a malfunctioning system.

Three types of listed[4] protective devices[16] may be specified to prevent the build-up of unsafe water conditions in storage tank and system:

1. A high-limit energy cutoff designed to prevent stored water temperatures in excess of 210 F, by stopping the flow of gas to the main burner.
2. A spillage-type temperature relief valve designed to prevent stored water temperatures in excess of 210 F. This valve should have a steam discharge capacity in excess of the input rating of the water heater.
3. A pressure relief valve with a discharge capacity sufficient to relieve the incremental expansion of stored water caused by the input rating of the water heater. This valve may be of the thermal expansion type (comply with ASA Z21.22) or steam rated type (comply with the ASME Code[17]).

Some difference of opinion exists among various groups and regulatory agencies as to which of the three kinds of protective devices should be used and how they should be installed. It is essential, therefore, to check and comply with the manufacturer's instructions and applicable local codes in each individual case. In the absence of such instructions and codes, the following recommendations may be used as a guide.

1. Water heating systems should be equipped with both a temperature protective device and a pressure relief valve of the types previously described. These devices may be separate or in combination. Sometimes pressure relief valves are not required by local codes, but their use is always required in closed systems in which water meters, check valves, or pressure-reducing valves are installed in the cold water supply line.
2. High-limit gas cutoff devices or temperature relief valves should be installed so that the temperature-sensitive element is immersed in the hottest stored water. For underfired heaters with integral tanks the recommended location is within the top six inches of the tank (Fig. 12-78). The same recommendation applies to combination temperature and pressure relief valves.
3. If a separate pressure relief valve is used, the best location is usually the cold water inlet line because cold water tends to be less corrosive than hot water. In some hard water areas where scale formation is a problem, the hot water line from the heater may be the best location. The choice of location depends on local experience. Under no conditions should a check valve or shutoff valve be installed between the pressure relief valve and the tank. Unless local codes specify otherwise, it is suggested that pressure relief valves be factory-set at 125 psig.

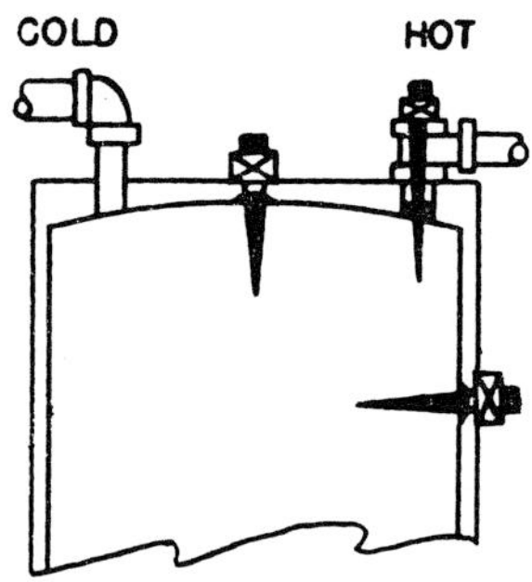

Fig. 12-78 Three alternative locations for an immersion-type temperature-sensitive relief valve on a heater.

REFERENCES

1. A. G. A. Laboratories. *American Standard Approval Requirements for Gas Water Heaters: Volume I, Automatic Storage Type.* (ASA Z21.10.1–periodically revised) Cleveland, Ohio.
2. ——. *American Standard Approval Requirements for Gas Water Heaters: Volume II, Side-Arm Type.* (ASA Z21.10.2–periodically revised) Cleveland, Ohio.
3. ——. *American Standard Approval Requirements for Gas Water Heaters: Volume III, Circulating Tank, Instantaneous and Large Automatic Storage Type.* (ASA Z21.10.3–periodically revised) Cleveland, Ohio.
4. ——. *Directory, Approved Appliances Listed Accessories.* Cleveland, Ohio, semiannual.
5. ——. *American Standard Installation of Gas Appliances and Gas Piping.* (ASA Z21.30–periodically revised) Cleveland, Ohio.
6. Natl. Fire Protect. Assoc. *Standard for the Installation of Gas Appliances and Gas Piping.* (NFPA 54–periodically revised) Boston.
7. U. S. Federal Housing Admin. *Minimum Property Standards for One and Two Living Units.* Washington, D. C., 1964.
8. Harris, W. S. and Hill, L. L. *Performance of Three Types of Indirect Water Heaters.* (Univ. Illinois Bull. Eng. Exp. Sta. 432) Urbana, Ill., 1955.
9. A. G. A. *Enough Hot Water ... Hot Enough.* New York, 1950.
10. Natl. Sanitation Found. *Commercial Gas Fired and Electrically Heated Hot Water Generating Equipment.* (Standard 5) Ann Arbor, Mich., 1960.
11. A. G. A. *Enough Hot Water ... Hot Enough.* New York, 1959. (Reprint of "Water Heating for Commercial Kitchens." *Air Cond., Heating and Ventilating* 56: 70–83, May 1959.)
12. A. G. A. Industrial and Commercial Gas Section. *Sizing Water Heaters for Multiple Dwellings.* (Information Letter 130) New York, 1962.
13. ——. "Sizing and Equipment Data for Specifying Swimming Pool Heaters." *Actual Specifying Eng.* 4: 72+, Aug. 1960.
14. ——. *Water Heating Applications in Coin Operated Laundries.* (Information Letter 127) New York, 1962.
15. "Water Heater Special." *Actual Specifying Eng.* 8:H1–50, Dec. 1962.
16. A. G. A. Laboratories. *American Standard Listing Requirements for Relief Valves and Automatic Gas Shutoff Devices for Hot Water Supply Systems.* (ASA Z21.22–periodically revised) Cleveland, Ohio.
17. ASME *Rules for Construction of Low-Pressure Heating Boilers.* (ASME Boiler and Pressure Vessel Code Sect. IV) New York, 1962.

Chapter 7
Domestic Gas Cooking Appliances

by K. R. Knapp, E. A. Jahn, and Evelyn M. Kafka

The most commonly used of all gas appliances is the domestic range. Sales[1] of ranges have averaged more than two million a year over the past ten years, and the total number in daily use in the U. S. in 1963 was estimated by A.G.A. to be 35,275,000. This figure includes built-in ovens but does not include ranges in trailers, boats, summer homes, or camps.

TYPES OF COOKING APPLIANCES

Cooking appliances may be classified in various ways. The system used in this Chapter follows ASA Standards[2] and divides the appliances into two classes: (1) free-standing ranges and (2) built-in cooking units.

Free-Standing Ranges

A free-standing range is self-supporting and does not require any special provision in the building structure for its installation. By ASA definition,[2] it is a self-contained gas burning appliance designed for domestic cooking purposes. It has a top section and an oven section; a broiling section is optional. Models are available in widths of 20, 24, 30, 36, and 40 in., with an occasional odd size in between.

The most popular arrangement of top burners is called the divided top—two burners on each side of a center work area. In some models the work area cover conceals a griddle that may be converted to a fifth burner. Figure 12–79 shows these and several other burner arrangements.

Flush-to-Wall Range. A flush-to-wall range is designed to be installed with its back and side walls in direct contact with combustible construction. This is a distinct advantage where space is a problem. The so-called **slide-in** model is particularly popular because it has a built-in appearance without requiring the additional space necessary for separate built-in units. Installation costs are considerably lower for slide-ins since they require no expensive cabinetry. They simply slide into the space normally occupied by a base cabinet.

Bungalow Range. In addition to a gas oven and a conventional open-top burner section, a bungalow range has a gas, solid, or liquid fuel section designed for space heating and for heating a solid-top section. An all-gas design, known as a **domestic gas room heater type,** has a gas oven, gas top burners, and a separate gas-fueled space heater section.

Built-in Units

Built-in cooking appliances are designed to be recessed in, placed on, or attached to counters, cabinets, walls, or partitions. These appliances are separately installed, individual range components that provide top or surface cooking, oven cooking, broiling, or any combination of the three. The last category includes those **slide-in** types of free-standing ranges that are tested under Volume II of the Approval Requirements[2] which covers built-in units. (See Fig. 12-80.)

Surface Cooking Sections. Surface sections are units of two, four, or six top burners that may be installed either in or on a counter top. They are available in a variety of arrangements and are sometimes combined with a griddle or an open-top hearth broiler. Some typical arrangements are shown in Fig. 12-79.

Ovens and Broilers. Built-in ovens are generally designed with a broiling compartment below the baking compartment and are intended for installation in a cabinet

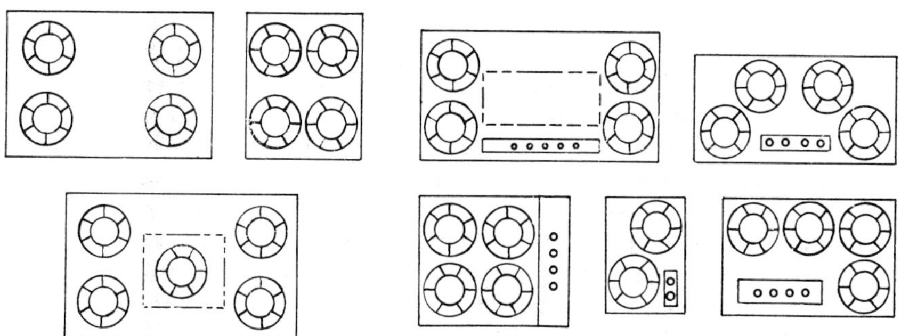

Fig. 12-79 Burner arrangements for free-standing and built-in gas ranges.

or wall. The average model is 24 in. wide although some models are 30 in. wide. Heights vary from 35 to 48 in.

Other models have only a baking section and no broiling compartment. Such units are considerably smaller and are designed to be installed on a counter top, although they may also be recessed into a cabinet or wall.

Fig. 12-80 Built-in wall oven and slide-in model range.

Built-in broilers are available as individual wall units as well as in combination with baking ovens. Wall models are all of the over-fired type.

Several manufacturers offer **open-top hearth broilers** that are underfired and have a bed of ceramic "coals" above the gas burner. The food, on a grid above these glowing "coals," is exposed to more intense heat than from a charcoal fire. While these are essentially outdoor appliances, some models have been approved under ASA Standards[2] for indoor use. Indoor units are designed for counter top installation and require the use of an exhaust hood vented to the outside.

Many outdoor broilers are portable, either connected to LP-gas cylinders or joined to house piping by means of "listed" flexible connectors and **quick-disconnect** couplings. Other models are permanently installed in brick or stone fireplaces or locations of similar construction.

In preparation (1964) are proposed revisions to the domestic range approval requirements that will cover outdoor broilers. A major objective is to provide for protection of the unit and its components from various weather conditions.

Eye-Level Consoles

Also referred to as a **high-oven range,** the eye-level console is a one-piece unit in which the oven is elevated above the cooking top. There is a wide variety of oven and broiler arrangements offered in 30, 36, and 40 in. sizes. Some typical styles are illustrated in Fig. 12-81.

Not all of these ranges stand directly on the floor. When there is no lower oven, the unit is mounted on a base cabinet or on the wall.

RANGE SELECTION

Some of the points to consider in selecting a range are:

1. Evaluation of the following factors in deciding on one of the three basic types: space limitations, relative installation costs, amount of remodeling or cabinet work required, and the possibility that the range might have to be moved at a future date.

2. Surface, oven, and broiler capacity needed for size of family and amount of entertaining.

3. Construction features desired, such as acid-resistant outside finish; one-piece top and back splasher, with rounded corners for easy cleaning; adequate provision for spillovers; rust- and corrosion-resistant burner heads, burner boxes, and liners of other compartments and drawers; and one-piece oven liners.

4. Convenience features evaluated in terms of their contribution to easier meal preparation and better home management. Among the more outstanding features are: top burner thermostats, programmed oven control systems, automatic meat thermometers, automatic clock controls, interval timers, rotisseries, and built-in griddles.

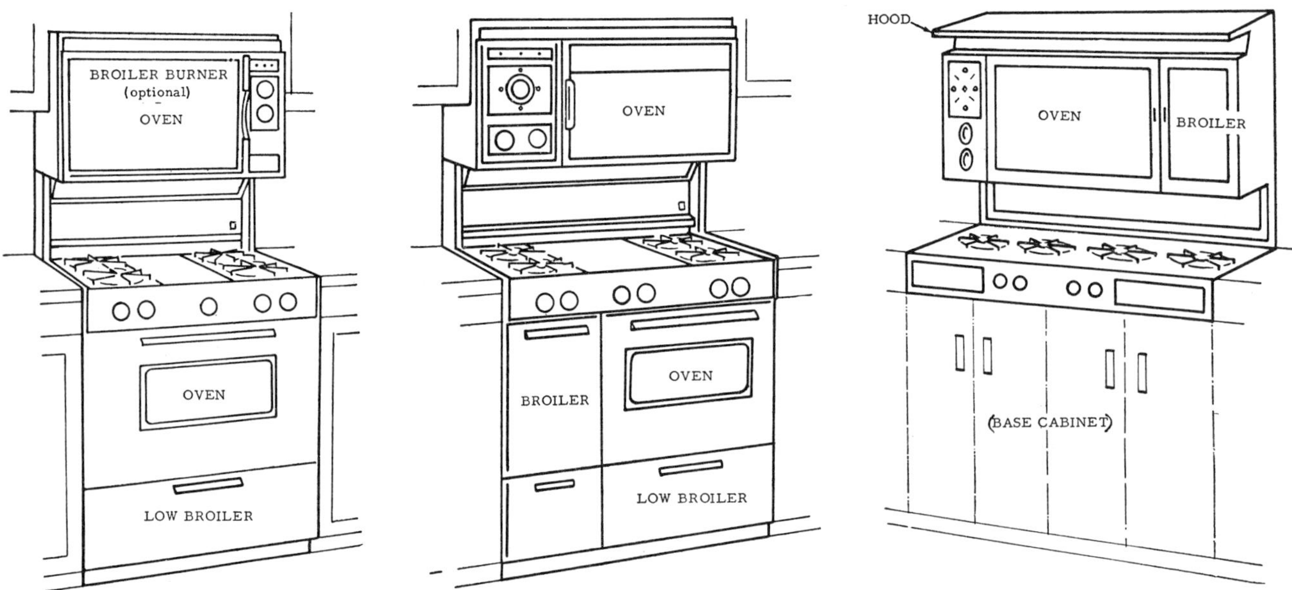

Fig. 12-81 Eye-level console ranges: (left) 30 in. model; (center) 36 in. model; (right) 40 in. model mounted on base cabinet.

DESIGN FACTORS

Thru the years extensive A.G.A. research studies have been devoted to gas ranges and their applications. Valuable data are presented in more than 20 published bulletins and reports covering various phases of gas cooking research. References 3 thru 13 are of special importance since the data contained in them have greatly influenced the design of contemporary gas ranges and contributed to improved performance and durability.

Burners

Gas range burners are generally designed for input ratings at normal gas pressure as follows:

	Btu per hour
Regular top burner—natural, manufactured, and mixed gases	9,000
Regular top burner—LP-gas	7,000 to 8,000
Giant top burner—natural, manufactured, and mixed gases	12,000
Giant top burner—LP-gas	9,000 to 10,000
High-speed top burner	13,000 to 20,000
Simmer burner section	1,200 (min)
Griddle burner	7,000 to 10,000
Oven burner	18,000 to 25,000, (about 10,000 Btu/ cu ft of oven volume)
Broiler burner	12,000 to 18,000
Constant burning pilots	175 (max)

Pressure Variations. All of the foregoing burners will operate satisfactorily and maintain stable flames on gas pressure variations between 0.5 and 1.25 of the normal adjustment pressure. However, any pressure variation causes a change in input by a factor of (new pressure/adjusted pressure)$^{0.5}$. For example, a burner adjusted for 9000 Btu at normal pressure would provide $9000 \times 1.12 = 10{,}100$ Btu per hour at 1.25 times normal pressure. At 0.5 normal pressure, the same burner would operate at 6360 Btu per hour.

Turndown. Gas range burners are required to perform satisfactorily at turndown rates of one-fifth normal rated input with no change in adjustment. As a result, by turning the control valve, a 9000 Btu per hr top burner can provide any desired heat input between 9000 and 1800 Btu per hr and a simmer burner between 1200 and 240 Btu per hr. Many contemporary top burners are capable of much lower turndown rates.

Shape and Material. Most contemporary top burners have round heads varying in diameter from ¾ in. to about 3 in., with either drilled ports or slotted ports that are horizontal or directed slightly upward. On the larger burners and those with a simmer section, some ports are located around a center opening. This opening permits passage of air for combustion of gas on the outer ring of ports. Burner head material may be aluminum, enameled cast iron, or stainless steel. If the burner has a simmer section there are two mixing tubes; if not, one. These tubes are usually of enameled cast iron or pressed steel tubing and are fitted with fixed or adjustable primary air openings. Burner parts and tubes must have melting points above 950 F.

Oven and broiler burners, usually of cast iron or steel, vary in shape and size to fit the oven compartment and must have melting points above 1450 F. Generally these burners are of rectangular, loop, bar, or tee shape, with drilled or slotted ports. Better broiling performance with these burners has been achieved by putting wire screen or ceramics adjacent to the burners. The increased radiant heat results from impingement of the flame on the radiant materials.

Recently, radiant broiling has been improved thru the development and use of atmospheric infra-red burners. This type of burner generally has a porous refractory, a drilled port refractory, or a metallic screen or mesh burner face. Air and gas mixtures pass thru the face, burn just above the surface and heat it to incandescence. The operating temperature span of these atmospheric burners is approximately from 1400 to 1650 F. At these temperatures they radiate about 70 per cent of their energy within the wave length span from 2 to 6 microns. The high percentage of infra-red energy available from these burners results in improved food quality and flavor. In addition, preheating and broiling times are substantially reduced.

Manual Valves and Orifices

Range valves for each individual burner are commonly of the plug and barrel type, although rotor valves (Fig. 12-36) are sometimes used for top burners. Plug and barrel valves are usually constructed of cast or forged brass and may be of the single- or double-duty type (Fig. 12-35, a and b). The latter is used with dual-throat burners having a separate center simmer burner head and hence a greater range of input rates.

The outlets of range valves are equipped with an orifice fitting (spud or hood) that regulates the maximum gas flow to the burners. The opening in the orifice fitting may be fixed, adjustable, or universal. Gas inputs between full-on and off are obtained by rotating the knob attached to the valve stem. Some valves have "click" settings, which permit quick and reproducible input changes to preset adjustments.

The latest advancement in burner valve design utilizes a traveling pin (Fig. 12-35, c) that controls burner input by moving a threaded needle in or out of the orifice to restrict the area open to gas flow. Since the gas pressure behind the orifice remains constant, the air injection qualities of the burner also remain constant throughout the range of input rates. The use of this valve also gives a greater valve handle rotation for improved control of burner heat.

Ignition Devices

On ranges complying with ASA Standards,[2] automatic top burner ignition has been mandatory since 1954 and automatic ignition of ovens and broilers since 1959.

Top burner ignition is usually accomplished either from a standing pilot thru flash tubes to the burners or by individual small pilots (Fig. 12-30) at each burner. Maximum gas input to all pilots is limited[2] to 175 Btu per hr, and a safety shutoff device is not required. Some needle pilots burn as little as 50 Btu per hr. Oven and broiler ignition utilizes an automatic pilot, which shuts off the gas supply to the burners if the ignition pilot is inoperable. Gas shutoff may be to the main burner only or to both the main and pilot burners (100 per cent shutoff). Figure 12-27 shows a thermoelectric pilot assembly that is typical of the kind most commonly used.

Electric ignition devices for both top and oven sections have had limited application to date, but recent research and development work by A.G.A.[14] and manufacturers on systems using spark and glow coils indicates more widespread use and acceptance in the future.

Thermostatic Controls and Timers

Thermostatic oven controls have been mandatory on A.G.A.-approved ranges for many years. The basic control consists of a temperature sensor, which reacts to the temperature in the oven, and an automatic valve, which responds to the sensor and controls the flow of gas to the oven burner. Until recently, nearly all thermostatic oven controls were of the modulating type; these regulated the heat input between the maximum rate and the rate available from a minimum stable flame on the oven burner. The latter rate is called the "by-pass rate"; the minimum gas input is dependent on the requirements for a stable flame that will not extinguish. Most oven burners have approximately a 10 to 1 ratio between the maximum and minimum rates. The advantage of modulating controls is that they provide a uniform temperature at the control setting. However, since all modern range ovens are well insulated, the use of a minimum flame prevents maintenance of oven temperatures much below 300 F.

"Start-and-Stop" Oven Control System. When used in conjunction with an automatic pilot and an electric clock or timer, the thermostatic oven control provides fully automatic oven cooking to the degree that the user is required only to preset the time cooking is to start and the desired length of the cooking period. Called "start-and-stop" programmed cooking, this system is limited to cooking at one temperature and does not provide defrost temperatures at the start nor holding temperatures when the cooking process is completed. After being automatically turned off, the oven cools toward room temperature. If not removed at the "stop" time, the food in the oven will also cool and may at the same time overcook because of residual oven heat. These shortcomings have led to the development of low-temperature oven control systems.

Low-Temperature Oven Controls. While automatic temperature control and programmed cooking were available with the modulating systems, further progress was made in providing lower oven temperatures.

Control systems were modified to obtain temperatures in the oven that could be used for defrosting food, holding food at serving temperatures, and even warming dishes. The control system, the oven construction, and the oven burner design all play a part in making possible the maintenance of an oven temperature below 250 F.

One revision of the control system to obtain the low oven temperatures consisted of modifying the previous modulating control system to cycle after the minimum by-pass rate was reached. This made possible much lower temperatures without the need to change the oven or oven burner design.

Usually the change from modulating operation to cycling operation was made when the oven temperature was approximately 325 F, permitting oven operation at temperatures down to room temperature. However, a temperature of 170 F for defrosting and holding has been generally accepted.

Several systems were developed that used the method of cycling the main burner on and off, including all-gas systems not affected by electrical supply failures, as well as those that depend entirely upon electricity for the operation of the control system. One point of difference is that the gas system (Fig. 12-16b) employs an automatic pilot consisting of a standing pilot which is used to light a second, larger pilot (called a heater pilot) when the system is operated. The heat from the heater pilot actuates a main burner gas flow control valve by means of a mercury-filled tube. The electric system uses an automatic pilot that has only one pilot burner, which stays lighted at all times. In case of pilot outage the circuit to an electric valve is opened by means of a heat-sensitive switch and the valve closes to prevent gas flow to the main burner. Another difference is that the thermostat valve in the gas system controls the flow of gas to the heater pilot, whereas the electric system thermostat controls the gas flow to the main burner.

Programmed Oven Control Systems. All the low-temperature systems that operate on the cycling principle offer a "hold" feature for use in programmed cooking. That is, each of the gas and electric cycling systems provides a means for changing from the original oven temperature to a second oven temperature automatically during the programmed operation of the oven. The usual procedure is to change the thermostat from a "cook" setting to a "hold-for-serving" setting. The change in the thermostat setting is accomplished by clock-controlled timer motors that mechanically reset the thermostat for the hold temperature or by switching from the original thermostat to a fixed-setting thermostat or equivalent.

This form of programmed cooking differs from that of the earlier "start-and-stop" programming. In the newer system the programming controls the cooking procedures with respect to both temperatures and cooking time. When combined with appropriate timing devices and accessories, any one of the following 12 cooking programs is possible:

Two-cycle programs

1. Delay—Cook
2. Defrost—Cook
3. Cook—Off
4. Cook—Hold

Three-cycle programs

1. Delay—Cook—Off
2. Delay—Cook—Hold
3. Defrost—Cook—Off
4. Defrost—Cook—Hold

Four-cycle programs

1. Delay—Cook—Cool—Off
2. Delay—Cook—Cool—Hold
3. Defrost—Cook—Cool—Off
4. Defrost—Cook—Cool—Hold

The "cook-and-hold" systems provide the user with a convenient way to cook foods and then hold them at serving temperatures for extended periods of time. Systems of this type require that the cooking must be started when the timer is set; that is, they will not start at a preselected time. After the oven has operated the required period, the clock activates the means for changing to holding temperature, which is held until the system is reset to manual operation.

All of the low-temperature systems that do not modulate the flow of gas to the oven burner can be provided with the "cook-and-hold" feature. Oven temperature, however, does not instantly change from the "cook" temperature to the "hold" temperature when the thermostat setting is changed. The time required for an oven to cool to the holding temperature will vary, depending on insulation thickness and air flow thru the oven. To prevent overcooking, some systems anticipate the time of temperature change and automatically allow for the cooling period; other systems do not and the user must make allowance when setting cooking time. A few systems have accomplished rapid cooling thru the use of an oven fan that comes on at the end of the cooking cycle and stops when the holding temperature is reached.

In order to make available a system with a delayed start, that is, without the cook's being present, the "start" components from the "start-and-stop" systems were incorporated into the "cook-and-hold" systems. This combination provides "delay-cook-and-hold" programming and can be used with any system that has the "cook-and-hold" feature if an extra valve is provided to prevent gas from flowing until the clock triggers the start switch. Electric systems used on gas ranges do not require the extra valve since the clock can be made to control the electrical valve already used in the system.

The gas systems and electrically controlled systems that provide low-temperature oven control perform the same functions and provide the same features. The choice of one system over the other is more often a matter of the sales features available with the control system and of personal preference rather than of operation or performance.

Top Burner Temperature Control. Range top sections are also equipped with thermostatically controlled top burners popularly known as the Burner-with-a-Brain.

The Burner-with-a-Brain is an automatic temperature control system used in conjunction with a top section burner. The system senses the temperature of the cooking vessel bottoms, and when the selected temperature is reached, the control system maintains that temperature by one of various methods. The control system modulates the flow of fuel to provide the correct amount of heat; cycles the flow of fuel to maintain the selected temperature; or operates with a combination of these methods. Those controls that modulate the flow of fuel will either modulate to a minimum size flame or cycle the gas on and off at the minimum flame. In the latter case a small burner, called a tower burner, stays on and provides ignition for the main burner as soon as the flow of gas is re-established to the main burner by the thermostat.

To compensate for the variables introduced by the difference in materials and size of cooking vessels and the difference in the foods being prepared, a manual valve was added to the automatic control (Fig. 12-23) that permitted the user to adjust the flame size to match the utensil being used.

As with any automatic control system, accurate and reliable response is dependent upon proper application of the temperature sensor. A clean sensing element, good contact with the cooking vessel, a cooking vessel made of a material that will conduct heat quickly, and optimum maximum flame size will provide the most satisfactory operation.

General Construction

Many changes have recently been made in range design. These include:

1. Better cabinet insulation, resulting in cooler kitchens.
2. Increased use of porcelain enamel and stainless steel for better appearance, easier cleaning, and longer life.
3. Elimination of joints, rough edges, and other places where dirt or grease may collect, thru use of larger stamped pieces of metal with rounded edges and curved joints.
4. Use of aeration pans around top burners for directing air flow and catching spillovers.
5. Improved grate design for better and more uniform heat transfer to pans, for better support of pans, and for longer life without discoloration.[15]
6. Smokeless broiling pans made of stainless steel or enameled steel for easier cleaning and handling.
7. Safe automatic ignition of all burners, thereby eliminating delayed or improper ignition during manual lighting.
8. Deflectors on oven outlets to prevent discoloration or greasy deposits on walls behind ranges.
9. Noiseless, easy-to-use oven racks and supports.
10. Increased body strength.

INSTALLATION OF DOMESTIC COOKING APPLIANCES

Every cooking appliance should be a listed model, evidenced by display of the symbol of a nationally recognized testing agency. It should be ascertained that the appliance is designed for use with the gas to which it will be connected. When it is to be converted from use with the gas specified on its rating plate to a different gas, the supplying gas company should be consulted.

The appliance should be so located as to be readily accessible for operation and servicing. Its installation should be such that its operation does not create a hazard to persons or property. Facilities for ventilation should permit satisfactory combustion under normal usage conditions. General requirements for gas piping are covered in Chap. 2 and specific installation requirements for ranges, open-top broilers, built-in sections, and hot plates in Chap. 1.

Venting

Listed domestic gas ranges and hot plates do not require flue connections. Such appliances have been proved by test under severe conditions to comply with rigid safety standards. Furthermore, it is reliably estimated that more than 35 million domestic gas ranges are in satisfactory daily use in the United States without flue connections. Recognizing these conditions, American Standards covering installation of gas appliances[16] have for many years specifically exempted those that are listed from the necessity of flue connections. The same standards, however, require that open-top broilers be installed under a metal ventilating hood since large quantities of heat and smoke may be produced when cooking some foods. Paragraph 4.4.3 of ASA Z21.30, given in Chap. 1 of this Section, covers hood requirements.

Kitchen Ventilation

Regardless of the kind of fuel used, cooking of food produces heat, moisture, and cooking odors. Temperature and humidity may rise beyond the point of comfort, and undesirable odors may spread to other parts of the house. To overcome such conditions satisfactorily, means in addition to natural ventilation may be necessary.

Best estimates are that about 80 per cent of domestic cooking is done on the range top. Cooking vapors from grease and fat are released in addition to substantial amounts of water vapor from boiling operations. For efficient removal of the heat and vapors it is necessary to contain and exhaust them to the outdoors before they escape into the kitchen.

Several kitchen ventilation methods are available that employ a small blower, usually 300 to 400 cfm capacity, to exhaust the cooking by-products to the outside. Tests[4] of these systems have shown them capable of removing at least 50 per cent of the heat produced by a gas range and in some cases as much as 70 per cent. Practically all water vapor formed during cooking can be removed and cooking odors confined to the kitchen or diluted to an acceptable extent in adjacent rooms. Effective operation depends on locating the inlet of the exhaust system immediately behind or above the range.

Two kitchen ventilation systems found by test to have good performance possibilities are shown in Figs. 12-82a and 12-82b.

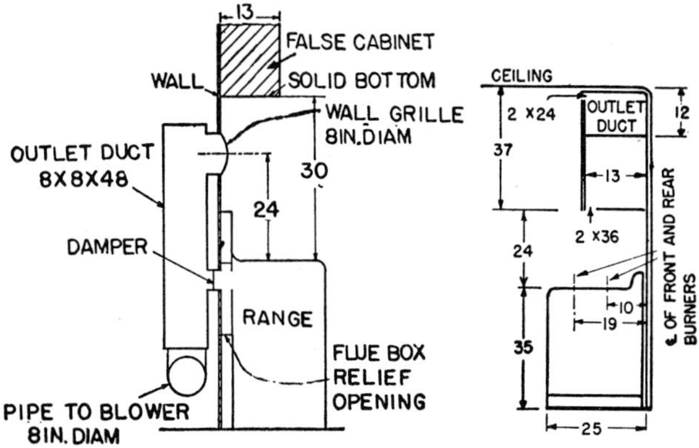

Fig. 12-82a (left) General arrangement of Rochester ventilating system.[4]
Fig. 12-82b (right) Cabinet hood with ceiling outlet.[4] All dimensions in inches.

Both provide more efficient heat removal from the rear burners than from the front.

Studies[7] have also been made of the accumulation of combustible deposits such as dust, lint, and grease in kitchen ventilating systems. Tests showed that such deposits tend to be more concentrated at the air intake and in the first few feet of ductwork. Apparently no fire hazard was created by deposited dust, lint, and grease even when they were subjected to higher temperatures than normally encountered in domestic range operation.

EFFICIENCY

Efficiency of ranges is a term having technical significance only with respect to top surface cooking. Ovens and broilers are normally large in comparison with the heat loads imposed on them; this is simply because ovens and broilers are required to take care of occasional large items (roasts or turkeys). The major day-to-day use, however, is with relatively small loads for the available oven volume.

Data collected as far back as 1949,[17] illustrating the relationship of inputs to typical efficiencies for top cooking and related items, are reasonably up-to-date. Information on electric and gas ranges (Tables 12-68, 12-69, and 12-70) may

Table 12-68 A Comparison of Domestic Top Cooking Efficiencies with Gas and Electricity

(using a 9 in. OD, 8 qt aluminum pan, 7½ in. high, with a 5.0 lb water load)

	Efficiency, per cent	
Input, Btu/hr	Gas burner rating	Electric element rating
	9 MBtu/hr	**1250 watts**
1,500	38	75
2,000	46	72
4,000	58	72
6,000	59	(Limit:
8,000	55	4080 Btu/hr
9,000	51	at 236 volts)
	12 MBtu/hr	**2100 watts**
2,000	45	80
4,000	54	76
6,000	52	71
8,000	50	(Limit:
10,000	48	6900 Btu/hr
12,000	46	at 236 volts)

Table 12-69 A Comparison of Domestic Cooking Efficiencies for Various Utensils, Using Gas and Electric Surface Units

(2.0 lb water load in all tests)

Pan type	Gas, rated 9 MBtu/hr	Electric, rated 1250 watts	Gas, rated 12 MBtu/hr	Electric, rated 2100 watts
7¼ in. aluminum	41%	57%	35%	43%
6½ in. enamel	38	62	31	41
6⅞ in. stainless steel, copper bottom	38	55	32	39
6½ in. stainless steel	32	53	29	37
6⅞ in. Pyrex	39%	62%	31%	45%

Table 12-70 Relative Heating Speeds and Energy Consumptions, Gas vs. Electric Range Tops[18]

(data given are averages obtained by testing a number of gas and electric ranges)

	Relative speed to boiling, min		Relative use of energy, Btu required	
Utensils	Gas ranges	Elec. ranges	Gas ranges	Elec. ranges
Pyrex coffee maker, using giant burners	5.92 (29% faster)	7.62	1184 (3% more)	1146
Pyrex coffee maker, using regular burners	6.36 (68% faster)	10.68	954 (13% more)	846
Pyrex saucepan, using giant burners	7.65 (3% faster)	7.89	1535 (30% more)	1183
Pyrex saucepan, using regular burners	8.90 (27% faster)	11.28	1336 (49% more)	897
Aluminum saucepan, using giant burners	8.42 (3% slower)	8.16	1683 (36% more)	1233
Aluminum saucepan, using regular burners	10.00 (2% slower)	9.78	1500 (93% more)	775
Aluminum utility pot, using giant burners	8.77 (2% faster)	8.98	1755 (29% more)	1364
Aluminum utility pot, using regular burners	10.93 (48% faster)	16.18	1640 (28% more)	1285

be of interest. Values are for cold-start tests. Hot-start tests would yield somewhat higher electric range top efficiencies since a relatively large amount of "stored heat" is retained by electric heating elements. Gas range burners would benefit only about one per cent because of small heat storage.

REFERENCES

1. A.G.A. Bur. of Statistics. *Gas Facts.* New York (annual).
2. A.G.A. Laboratories. *American Standard Approval Requirements for Domestic Gas Ranges, Volume I* (ASA Z21.1); *Volume II* (ASA Z21.2). Cleveland, Ohio (periodically revised).
3. ——. *Domestic Gas Range Research.* (Bull. 7) Cleveland, Ohio, 1936.
4. ——. *Study of Various Methods of Kitchen Ventilation.* (Research Bull. 40) Cleveland, Ohio, 1946.
5. ——. *Research in Fundamentals of Design Features Affecting Oven Performance.* (Research Bull. 47) Cleveland, Ohio, 1948).
6. ——. *Research in Broiler Design.* (Research Bull. 48) Cleveland, Ohio, 1948.
7. ——. *Study of Accumulation of Combustible Deposits in Kitchen Ventilating Systems.* (Research Bull. 49) Cleveland, Ohio, 1948.
8. Speidel, P. A., and Hammaker, F. G., Jr. *Research on New Designs for Range Top Burners, Pilots, and Valves.* (Research Bull. 76) Cleveland, Ohio, A. G. A. Laboratories, 1957.
9. Hampel, T. E. *Study of Experimental Range Top Burner Systems.* (Research Bull. 88) Cleveland, Ohio, A. G. A. Laboratories, 1961.
10. Nead, L. A., and Honaker, B. G., Jr. *Evaluation of the Various Methods for Attaining Cooler Domestic Gas Kitchens.* (Research Bull. 89) Cleveland, Ohio, A. G. A. Laboratories, 1961.
11. Kirk, W. B., and Komora, L. J. *New Concepts for Gas Oven Cookery and for Gas Broiling.* (Research Bull. 94) Cleveland, Ohio, A. G. A. Laboratories, 1963.
12. Pountney, C. H., Jr. *Study of Prevention of High Surface Temperatures around and on Oven and Broiler Doors of Domestic Gas Ranges.* (Research Rept. 1218) Cleveland, Ohio, A. G. A. Laboratories, 1954.
13. Speidel, P. A. *Performance Characteristics of Gas Ovens.* (Research Rept. 1264) Cleveland, Ohio, A. G. A. Laboratories, 1956.
14. Weber, E. J. *Development of Electrical Ignition Systems for Domestic Gas Appliances.* (Research Rept. 1341) Cleveland, Ohio, A. G. A. Laboratories, 1962.
15. Spinell, D. M., and Wallace, J. F. *Improvement of Domestic Gas Range Grates; Part 1: Industry Survey, Literature Analysis, and Laboratory Evaluation of Base Metals and Finishes.* (Research Rept. 1337) Cleveland, Ohio, A. G. A. Laboratories, 1961.
16. A. G. A. Laboratories. *American Standard Installation of Gas Appliances and Gas Piping.* (ASA Z21.30–periodically revised) Cleveland, Ohio, 1964.
17. ——. *Comparative Laboratory Tests of Gas and Electric Ranges, Part I–Medium Price Ranges.* (Research Rept. 1132) Cleveland, Ohio, 1949.
18. Geltz, C. F., and Segeler, C. George. "Correction and Rebuttal." *A. G. A. Monthly* 45: 6–7, Sept. 1963.

Chapter 8
Domestic Gas Refrigerators

by B. A. Phillips

Gas-fired household refrigerators first came into prominence in the United States during the 1920s. The earliest models, operating on an intermittent-absorption principle, included cold reservoirs in the form of brine tanks as a means of avoiding large cyclic temperature variations within the cabinet. The continuously operating, single-pressure absorption system invented in Sweden in 1922 by Baltzar von Platen and Carl Munters[1] was introduced into this country in 1926 by Servel, Inc., and A. B. Electrolux of Sweden. This system soon became the basic gas-refrigeration system. Other systems have been developed and a limited number produced, but as yet no other has had an appreciable commercial success.[2-4]

The condensers in early Servel models were water cooled. In 1933 an air-cooled model was introduced, and by 1935 all were air cooled. The features of no moving parts, silence, minimum service, and low operating cost made the gas refrigerator popular during the 1930s and 1940s, but during the 1950s sales declined drastically. In 1958, the Whirlpool Corporation purchased the Servel gas-refrigeration business, started production of the RCA Whirlpool models, and established a renewed research and engineering program. In 1960 Whirlpool introduced 13- and 14-cu ft refrigerators featuring many improvements over previous designs. The Norge Sales Corporation also entered the field in 1960, with an 11-cu ft model of improved performance. Smaller size refrigerators are sold by Norcold, Inc. Other models are imported from Europe. Well over 4,000,000 gas refrigerators had been sold in the United States up to 1960.

Abroad, particularly in Europe, absorption refrigerators are produced by numerous manufacturers. Many are electrically heated rather than gas heated owing to limited gas supplies. Recent European production has been high. Foreign refrigerators are smaller than American models; their average size is about 3 cu ft.

THE ABSORPTION SYSTEM

The most common and practical methods of refrigeration depend on the evaporation of a liquid at a low temperature; the heat of evaporation is extracted from the refrigerated space. Since the liquid must be replenished as it evaporates, refrigeration machines are fundamentally closed-cycle systems for recovering the vapor, liquefying it, and returning the liquid to the evaporator.

In *compression* systems, the recovery cycle starts with the compression of the vapor to a high pressure. At this pressure and a correspondingly high saturation temperature, the vapor is delivered to a condenser, where its heat is transferred to a cooling fluid, usually air or water, causing the vapor to condense. In *absorption* systems, the high solubility of the refrigerant in the absorbent serves as the means for compressing the vapor into a small volume, that is, the volume of the absorbent. Subsequently, the refrigerant is boiled out of the solution at a high pressure, liquefied in a condenser, and returned to the evaporator.

The basic **two-pressure absorption system,** or Carré system (Fig. 12-83), from which the Platen-Munters system was developed, consists of four major components. These are the evaporator and absorber, which make up the low-pressure side, and the generator and condenser, which make up the high-pressure side. A pump forces absorbent solution from the absorber to the generator. Flow-limiting devices control the flow of absorbent liquid from the generator back to the absorber and the flow of refrigerant liquid from the condenser to the evaporator.

The **single-pressure**, or Platen-Munters, system embodies the same four basic components but eliminates the flow-limiting devices and replaces the pump with a liquid circulator. These improvements are made possible by filling the absorber and evaporator with hydrogen to a total pressure which exactly equals that in the generator and condenser and by utilizing gravity for circulation of the fluids.[5-7] This system (Fig. 12-84) is charged with an ammonia-water solution containing approximately 30 per cent ammonia by weight and with hydrogen to a total pressure of about 300 psia. The solution fills the lower portions of the unit to a predetermined level at the bottom of the absorber, as shown in the figure. All system parts are interconnected and under equal pressure.

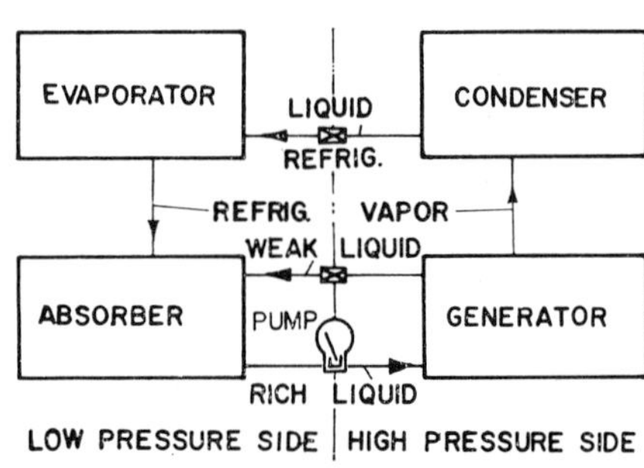

Fig. 12-83 Carré two-pressure absorption refrigeration system.

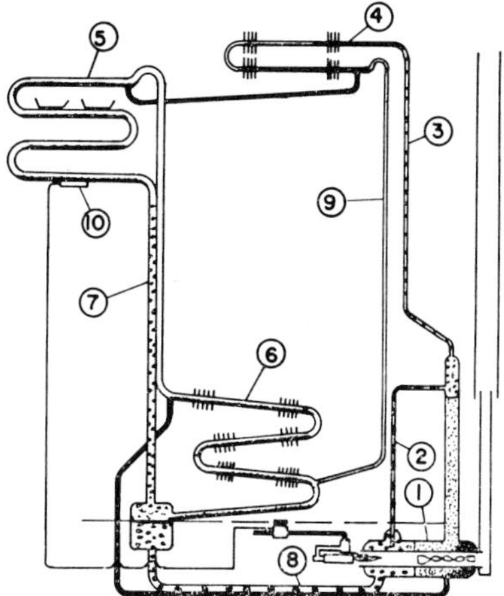

Fig. 12-84 Platen-Munters single-pressure absorption refrigeration system.

Operation starts when the generator (1) is heated by a gas flame. The generator has two purposes: first, to boil off the ammonia from the ammonia-water solution; second, to circulate the solution by means of a vapor-lift tube (2). This tube acts similarly to the lift tube in a coffee percolator, utilizing the vapor bubbles to raise slugs of liquid to a higher level. From that level the liquid, further stripped of ammonia when passing thru the second chamber of the boiler, flows by gravity thru a liquid heat exchanger (8) into the top of the absorber (6).

The ammonia-vapor that is boiled out of the solution in the generator rises thru the rectifier tube (3) into the condenser (4), displacing the hydrogen and pushing it ahead. As the hydrogen is thus compressed, the total pressure rises, perhaps to above 400 psia in rooms at high temperatures. In the condenser tubes the ammonia vapor liquefies as it gives up its heat to the surrounding air. The resulting liquid refrigerant flows thru the drain tube to the evaporator (5). A pressure-equalizing tube (9), leading from the end of the condenser to the absorber, ensures that the pressure always remains equalized throughout the system. This tube also serves as a vent for the ammonia vapor if the condenser is not properly cooled by air circulation and as a vent for any hydrogen present in this part of the system.

In the evaporator (5) the liquid refrigerant comes into contact with hydrogen containing only a minor amount of ammonia vapor. The liquid ammonia evaporates into the gaseous space, absorbing heat and thus producing a cooling effect. The temperature attained is determined primarily by the low pressure of ammonia vapor in the hydrogen. Dalton's law expresses the basic principle involved: A mixture of gases has a pressure equal to the sum of the pressures exerted by each gas; each gas exerts its partial pressure and in other ways acts as though it alone occupied the space. In principle, therefore, the liquid ammonia evaporates into the gas space as though no hydrogen were present. Theoretically, the temperature attained depends only on the partial pressure of ammonia vapor in the gaseous space. In practice, however, the hydrogen molecules impede evaporation, and other factors act to raise the evaporating temperature. Nevertheless, a close approximation to theoretical value can be attained. Evaporator temperatures corresponding to 15 to 20 psia pressure may readily be reached even though the actual total pressure may be over 400 psia. The temperature-sensing element is shown as (10).

The lowest temperature occurs at only one spot in the evaporator. As the ammonia evaporates into the gas, its partial pressure rises and the temperature at which evaporation takes place also rises. The temperature of the evaporator tube therefore changes along its length. A difference of 20° to 40° F may exist from one end to the other of an evaporator comprising both a freezer section and a fresh-food coil.

Another result of the evaporation of the liquid into the gas mixture is evident. Since ammonia is much heavier than hydrogen, the density of the gas mixture increases as ammonia evaporates into it. The final heavy gas flows downward from the evaporator toward the absorber, which should be located as far below the evaporator as possible. Entering the absorber, the ammonia portion of the gas mixture begins to dissolve in the absorbent solution (water weak in ammonia); the hydrogen dissolves only to a minor extent.

The gas flows thru the absorber countercurrent to the weak solution, which enters at the top. The ammonia vapor dissolves in the solution throughout the length of the absorber, the heat of absorption being dissipated by the cooling fins. Because of the countercurrent flow, the gas continuously encounters liquid that is weaker in ammonia, until, at the top of the absorber, it contacts the solution arriving from the generator, where its concentration had been reduced as much as practicable, generally to a level of 12 to 15 per cent of ammonia by weight. The hydrogen leaving the absorber is therefore as free of ammonia as it is practicable to make it. This gas, rising from the absorber, is much lighter than the mixture of ammonia vapor and hydrogen descending from the evaporator. The force resulting from the unbalanced weights of the two gases produces the continuous flow of gas that, in its downward travel, transports the refrigerant vapor from the evaporator to the absorber.

The gas rising from the absorber is 20° to 30° F warmer than room temperature, whereas that descending from the evaporator is cold, generally below 32 F. These two streams are passed thru a countercurrent gas heat exchanger (7), where the refrigerating effect of the cold gas is recovered by precooling the stream from the absorber. The minimum temperature and the cooling capacity attainable by the refrigerator are determined to a great extent by the completeness with which this gas heat exchange is accomplished. In the average unit, the heat transferred in the gas heat exchanger may equal the total refrigerating capacity of the unit.

In the absorber the absorbent solution becomes enriched in ammonia as it descends, reaching the storage chamber at the bottom of the absorber at its maximum operating concentration, 25 per cent or more. From this chamber it flows by gravity thru the liquid heat exchanger (8), where it is preheated by heat transfer from the weak liquid, to the generator to repeat the cycle.

The operating cycle incorporates three semi-independent fluid circuits that must work in unison. One is the hydrogen circuit, thru the evaporator, gas heat exchanger, and absorber, whose rate of circulation is regulated in part by the height of the circuit and by the resistance to gas flow that is incorporated in the various components. This resistance is utilized to improve heat transfer and mass transfer in the evaporator, gas heat exchanger, and absorber. The second circuit is the liquid absorbent circuit, which incorporates the absorber, the generator, and the liquid heat exchanger. The circulation rate of the liquid absorbent is regulated primarily by the pump-tube design, and by the distribution of the heat input between the two chambers of the boiler to control the supply of vapor to the pump tube. The third circuit is the ammonia circuit, which includes all parts of the unit. The entire unit operation is summarized in this circuit. In the evaporator, the ammonia evaporates into the hydrogen at a low partial pressure. In the absorber, the ammonia is separated from the hydrogen by absorption into water. In the generator, the ammonia is separated from the water by distillation. Freed of hydrogen and water, the ammonia flows to the condenser at full system pressure, condenses at the high saturation temperature corresponding to the system pressure, and flows as a liquid to the evaporator. The major factors that determine the rate of ammonia circulation are the net heat input to the generator, the concentration of ammonia in the solution, the efficiency of the generator–rectifier–liquid heat exchanger system, and the rate of circulation of the absorbent solution.

Optimum performance depends on proper matching of the operation of the three circuits. Since the heat from the burner is the total source of energy, the burner must be accurately matched to the unit to attain peak performance. The operating components must also be well matched at partial inputs since refrigerators operate more at partial than at full input.

Operating Fluids

Although other absorbents and refrigerants have been considered, the only practical pair for air-cooled household refrigeration is ammonia and water. Intensive searches for better combinations have been conducted in the past,[8–15] all with negative results. A few new pairs have been found useful for air conditioning purposes.[16] Studies are continuing but are directed primarily toward air conditioning.

The unique set of properties necessary in a good absorbent-refrigerant combination starts with a strong affinity between the refrigerant and the absorbent. This characteristic is best demonstrated by the *depression* of the vapor pressure of the solutions. The diagonal straight line in Fig. 12-85 represents the vapor pressure of normal solutions that follow Raoult's law. The lower curve represents the actual vapor pressures of ammonia-water solutions. The reduction of the vapor pressure is indicated by the vertical distances between these curves. At the lower concentrations the depression is well over 50 per cent.

An effective practical test for affinity is the measurement of the temperature rise that results when the two liquids under investigation are mixed.[17] Although some fluid combinations achieve results comparable with those achieved by ammonia water in this test, they fail in other respects. There are additional important requirements. The refrigerant must have a high heat of vaporization and its boiling point must be low compared with that of the absorbent. Both of these materials and their solutions must have low specific heats, low viscosities, and low freezing points. Both must have high stability and be relatively inert to materials of construction. Further requirements relating to operating pressures, freedom from hazards, costs, etc., are discussed by Buffington.[17]

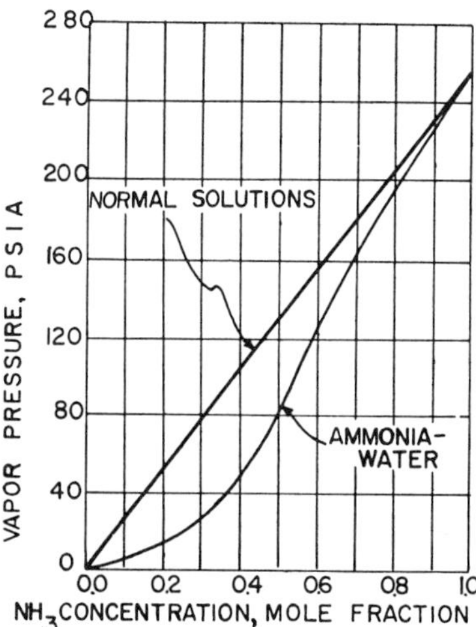

Fig. 12-85 Vapor pressure of NH_3-H_2O solutions at about 115 F.

All known absorbent-refrigerant combinations fail in one or more specific requirements, but success or failure depends more on the sum total of the properties than on any single one. A pair that satisfies the primary characteristics is next evaluated from a thermodynamic viewpoint by calculating its performance under operating conditions. Complete system calculations, including equipment cost, operating efficiency, and applicability of the equipment, become the final criteria that determine the relative value of an absorbent-refrigerant combination.[14,17] In some cases, actual field experience may determine the worth of the combination.

In the single-pressure system a third component is required, the so-called inert gas. Its characteristics should include minimum density and viscosity; high thermal conductivity; low heat capacity; low solubility in the absorbent; low cost; inertness toward the absorbent, refrigerant, and materials of construction; freedom from hazards. For gas refrigerators, hydrogen has the best total combination of properties. Helium is considerably less effective.

BASIC THERMODYNAMICS OF ABSORPTION SYSTEMS

The absorption refrigeration system combines within itself both a refrigerating machine and a heat engine. Thermodynamically, refrigeration is a process by which heat, collected at a low temperature, is raised to a high temperature and there dissipated. Work is required to raise the heat from a low temperature level to a high one. A heat engine produces work by absorbing heat at a high temperature, converting part of it into work, rejecting the rest at a low temperature.

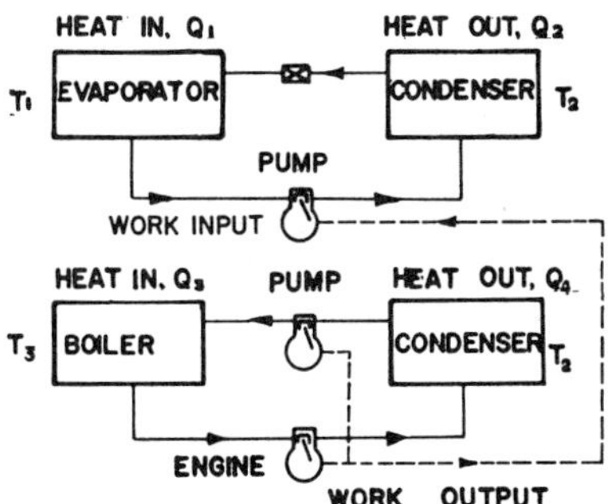

Fig. 12-86 Refrigerator-heat engine cycle.

A mechanical compression system (Fig. 12-86) shows the basic thermodynamics of these processes. The refrigerating machine consists of a compressor, condenser, liquid flow control, and evaporator. The heat engine includes the boiler, steam engine or turbine, condenser, and condensate pump. Work from the engine or turbine shaft is transferred to the compressor either directly or thru an electric system comprising a generator, transmission circuit, and mo

The minimum theoretical work required to o frigeration machine can be determined thermody as by a Carnot cycle analysis (Eq. 1).

$$W_{min} = Q_1 \frac{T_2 - T_1}{T_1}$$

where:

W_{min} = minimum work required
Q_1 = heat absorbed at the low (evaporator) temperature, T_1
T_1 = evaporator temperature, °R
T_2 = heat dissipation (condenser) temperature, °R

$$Q_2 = Q_1 + W_{min} = \text{heat dissipated}$$

A similar expression, Eq. **2**, represents the maximum work obtainable from a given quantity of heat by means of a heat engine.

$$W_{max} = Q_3 \frac{T_3 - T_2}{T_3} \quad (2)$$

where: Q_3 = heat required at a high temperature, T_3
T_3 = heat input temperature, °R
T_2 = heat dissipation temperature, °R

$$Q_4 = Q_3 - W_{max} = \text{heat dissipated}$$

Equating these two expressions for the work involved yields Eq. **3**.

$$Q_1 = Q_3 \frac{T_1}{T_3}\left(\frac{T_3 - T_2}{T_2 - T_1}\right) \quad (3)$$

This basic equation, applicable to all refrigeration systems that utilize heat as their prime source of energy, states that Q_1, the maximum quantity of refrigeration obtainable from a quantity of heat, Q_3, is ultimately dependent only on the temperatures involved in the cycles. Since Eq. **3** has no work term, W, the possibility that refrigeration can be obtained directly from heat is indicated. This thermodynamic possibility instigated the long search that resulted in the invention of Platen-Munters.[5]

A corollary to Eq. **3** is that the heat input to the system, $Q_1 + Q_3$, must equal the heat output, $Q_2 + Q_4$. In the example used here, all the heat output occurs at T_2. In the absorption system both Q_2 and Q_4 are rejected at the refrigerator. In compression systems Q_4 is dissipated at the electric power plant.

The **coefficient of performance,** *COP*, is defined as the ratio of the refrigeration capacity, Q_1, to the heat input, Q_3, required to produce that capacity. Equation **4,** a form of Eq. **3,** represents the maximum theoretical coefficient of performance for a given set of temperatures.

$$COP = \frac{Q_1}{Q_3} = \frac{T_1\,(T_3 - T_2)}{T_3\,(T_2 - T_1)} \quad (4)$$

The true efficiency of an actual system may be obtained by dividing the actual Q_1'/Q_3' by the theoretical Q_1/Q_3. This step, however, is rarely taken. It is common practice merely to judge the actual coefficient, Q_1'/Q_3', in relation to other well-known *COP*s. It is also common in gas refrigeration practice to use as Q_3' the total or *gross* heat energy in the fuel used (the *net* Btu input might be more logical). A practical *COP* of this nature is very useful but to be comparable must be measured under strictly defined conditions. For true comparisons these conditions must be equivalent in all cases.

Similarly, practical *COP*s of other refrigeration systems are not always defined in the same way. Those for compression machines are commonly defined as Q_1/W, where W is the electrical work input into the motor. This *COP* is related to Eq. **5**, a form of Eq. **1**.

$$\frac{Q_1}{W} = \frac{T_1}{T_2 - T_1} \quad (5)$$

Since this *COP* does *not* refer to the heat input at the power plant, it is numerically a much larger quantity than the *COP* in gas refrigeration. An interesting comparison can be made by carrying the calculations back to the power plant—overall efficiencies for gas absorption and motor-driven systems may be about the same. Measured *COP*s for absorption refrigerators have been reported over a range from 0.17 to nearly 0.50. This wide variation appears to be due not only to the state of development but also to differences in conditions and methods of measurement. The smaller coefficients generally refer to tests run at high ambient and at low evaporator temperatures, and are based on the net useful refrigeration and on the total heating value of the gas input. The higher coefficients appear to refer to a total refrigeration effect in cooler ambients and to be related to a net heat input to the generator.

Absorption System Calculations

The foregoing analysis defines theoretical refrigeration performance in the terms of the thermodynamically important temperatures, all temperatures being considered independent variables. No considerations of the operating cycle or fluids are involved. When an operating cycle and the fluids are chosen, other limiting conditions are placed on the system. The operating temperatures become interrelated by the cycle and by the properties of the fluids. As a result, the thermodynamic properties of the fluids must be known or calculated. These properties include the vapor pressures of the refrigerant, the absorbent, and their solutions; the compositions of the equilibrium vapors; and the heat contents or enthalpies of the pure components, their solutions, and the vapors. Freezing points may be important if they restrict the possible cycles. For single-pressure systems, the enthalpy and composition of the hydrogen-ammonia mixtures are also needed.

For preliminary studies many data may be calculated or estimated from fundamental properties such as boiling points, specific heats, heats of formation, mixing, vaporization, etc., but for serious evaluation and design work, accurate experimental measurements are required. Thermodynamic properties of the ammonia-water system have been measured by a number of investigators.[18–22] A variety of correlations of these data has appeared;[23,24] one of the more useful is by Scatchard et al.[25] These correlations are limited by the 250-psi limit of the basic data; however, refrigerators built since 1959 have operating pressures above 400 psi. Recent determinations[26] have extended vapor pressure data to over 600 psia, and an A.G.A. research project has obtained data on the thermodynamic properties up to 500 psia. These data have been used to prepare a new Mollier chart for ammonia-water mixtures at pressures up to 500 psia.[41]

With such data, many methods of calculation may be utilized in the development of an absorption system.[7] Basic to all systems are cycle and fluid analyses to establish the most suitable or desirable operating conditions, and material and enthalpy balances to establish a complete First Law analysis. A Second Law analysis may also be helpful at times. All such calculations are made most readily and the cycles visualized best when the data are plotted in suitable chart form.

Pressure-Temperature-Composition Diagrams

For the absorption system, pressure-temperature-composition (P-T-X) diagrams and enthalpy-composition diagrams are most helpful. A number of useful ammonia-water P-T-X diagrams have been published.[27–29] An especially convenient form is shown as Fig. 12-87. This extrapolated chart is based

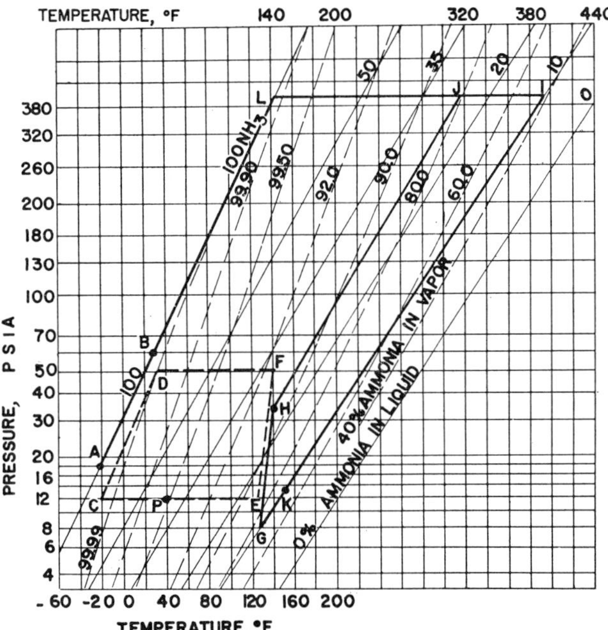

Fig. 12-87 Saturation vapor pressures of NH_3–H_2O solutions.

on the Dühring Relation. Relatively straight lines resulted when the saturation pressures were plotted against a linear temperature scale, and the pressure scale was found to be reasonably well expanded in both the high- and low-pressure ranges.[30] In Fig. 12-87 the saturated vapor pressures of the pure components and of their solutions are shown as solid lines. The dotted lines represent the compositions of the equilibrium vapors.

Example. A typical single-pressure refrigerator circuit is outlined on Fig. 12-87. Since a P-T-X diagram depicts only saturation conditions, most of this circuit is superimposed on the coordinates rather than plotted on the chart. Only the 400-psi operations are at saturation. Strictly speaking, some small errors may be expected, but these are generally outweighed by the advantages of a complete plot.

The evaporator temperatures range from −20 to +30 F; the absorber temperatures from 130 to 140 F, and the condenser temperature is 143 F. Although the major part of the diagram is drawn assuming equilibrium conditions, pressure differentials in the evaporator and absorber have been indicated to illustrate the application of the necessary driving forces. The evaporator temperatures are located along the pure ammonia line from A to B. The pressure at which ammonia evaporates to produce the temperature of −20 F is then read on the ordinate scale as 18 psia. The points C and D, representing the partial pressure of ammonia in the weak and rich gases, are located appreciably below the evaporating pressures. The vertical distances between A and C and between B and D represent the pressure differential that must exist for evaporation to proceed rapidly. A similar differential of partial pressure to vapor pressure is also indicated for the absorber. The conditions at the weak and rich end of the absorber are designated by the points G and H for the liquid and points E and F for the gas.

Transfer of the rich and weak gases between the evaporator and absorber, with heat exchange, is shown as occurring along constant partial pressure lines DF and EC. The liquid streams between the absorber and the boiler follow the constant composition lines HJ and IG. Point K on line IG indicates the probable limit of cooling of the weak liquid by heat transfer to the rich liquid. Further cooling from K to G occurs in the finned weak liquid precooler. The various processes in the boiler and analyser occur along the line JI, with rectification of the boiler vapors following along line JL. The condensation of the purified ammonia vapor takes place at point L. The cooling of the liquid refrigerant is represented by line LBA. Use of the P-T-X chart in this manner for detailed development of a proposed cycle readily establishes many important design requirements or necessary cycle modifications.

Material Balances

After the desired cycle has been established on a P-T-X chart (e.g., Fig. 12-87), material balances are made to determine the relative quantities of fluid flowing in each system component. Such balances, as well as heat balances, are generally calculated on the basis of one unit of refrigerant. For absorption systems a suitable unit is 1 lb of ammonia evaporating in the evaporator. All other flow rates throughout the system are then related to this unit. The hourly refrigerating capacity in Btu per hour defines the rate of refrigerant flow in pounds per hour.

When a material balance is made around the absorber, Eqs. **6** and **7** are readily derived:

$$L_w = \frac{1 - C_r}{C_r - C_w} \tag{6}$$

$$G_w = \frac{P_t - P_{pr}}{P_{pr} - P_{pw}} \tag{7}$$

where:

L_w = related weak liquid circulation, lb weak liquid per lb NH_3
G_w = relative weak gas circulation, moles weak gas per mole NH_3
C_r = fraction of NH_3 in rich liquid
C_w = fraction of NH_3 in weak liquid
P_t = total pressure, abs
P_{pr} = partial pressure of NH_3 in rich gas, abs
P_{pw} = partial pressure of NH_3 in weak gas, abs

The relative circulation of rich liquid is $L_r = L_w + 1$, and the rich gas circulation rate is $G_r = G_w + 1$.

It is often useful to express the gas circulation in pounds of weak gas per pound of ammonia. The relation in that case is shown by Eq. **8**:

$$R_{wg} = \frac{1 - F_r}{F_r - F_w} \tag{8}$$

where:

R_{wg} = relative weak gas circulation, lb weak gas per lb NH_3
F_r = wt fraction of NH_3 in rich gas
F_w = wt fraction of NH_3 in weak gas

Material balances around other components are also very helpful. For example, a balance around the rectifier determines the rate of condensate flow from the rectifier, and one around the gas heat exchanger accounts for the effect of water vapor in the weak gas leaving the absorber.

Enthalpy-Composition Diagrams

With the aid of the flow quantities determined by material balances, heat balances are then calculated around each component and around the entire system. Occasionally, it may be necessary to solve material balance and heat balance equations simultaneously.

Enthalpy-composition diagrams are very convenient for calculations of heat balances.[31] For ease in studying the complete circuit, Fig. 12-88 shows the ammonia-water and

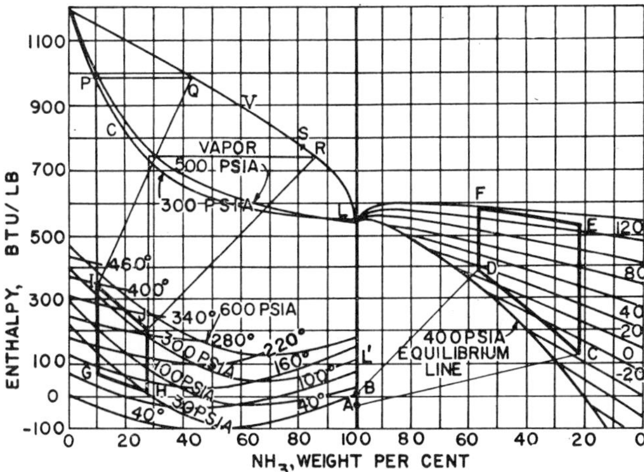

Fig. 12-88 Enthalpy-composition diagrams: (left) NH_3–H_2O mixtures; (right) NH_3–H_2 mixtures at 400 psia total pressure.

ammonia-hydrogen enthalpy diagrams placed back-to-back with a common axis of 100 per cent ammonia. The liquid ammonia-water lines on the left side of the figure are saturation lines, but no significant errors occur in utilizing them also for the unsaturated sections of the Platen-Munters unit. Vapor lines V and C are also saturation lines, calculated as ideal gas mixtures. In this unit the ammonia-water vapors occur saturated.

Two sets of vapor lines are shown. Line V represents the enthalpies of vapors having compositions as read from the horizontal axis (the enthalpies at 300 thru 500 psia fall very close together and are here represented as a single line). The lines C are used primarily as *construction lines*, representing the enthalpies of those saturated vapors in equilibrium with solutions of the composition corresponding to the horizontal axis. For example, a 10-per cent ammonia solution at 400 psia (point I) is in equilibrium with a vapor having an enthalpy represented by point P. The composition and enthalpy of that vapor are represented by point Q.

The ammonia-hydrogen gas lines on the right side of the figure are of necessity for a single pressure and are based on ideal mixtures. By judicious use, and the addition of adjacent saturation curves, a chart of this type can be used for other pressures close to the base pressure. Thus, relatively few charts can be used to cover the total range of operating pressures.

The refrigerator circuit shown in Fig. 12-87 is also superimposed on Fig. 12-88, the letters retaining their original significance. The ammonia-water solution in the absorber is designated by line GH; line IJ represents the solution undergoing boiling in the generator. The vapor resulting from the boiling of the solution falls along line QR; the vapor in equilibrium with the weakest liquid being at Q and that in equilibrium with the rich liquid being at R. An ideal boiler and analyzer combination would enrich the weaker vapor along line QR until all vapor leaving the analyzer would be represented by R. In practice this is not achieved. The vapor leaving the analyzer is weaker and has an enthalpy higher than R, and may be represented by some point, S. Removal of water to produce pure ammonia vapor proceeds along line SL′, L′ representing the vapor enthalpy at the condenser. Upon condensation, the enthalpy represented by the distance LL′ is given off. The liquid ammonia is subsequently cooled from L′ to B, evaporation occurring along line AB.

For the gas portion of the circuit (the right half of Fig. 12-88), line CD represents the gas conditions in the evaporator and line EF those in the absorber, with gas heat exchange occurring along lines DF and CE.

Besides being useful in depicting the circuit and in picking out enthalpy values for calculations, the enthalpy-composition diagram can be used to calculate graphically all the heat quantities involved in the circuit. Complete heat balances can be calculated in a few minutes.[31] An analysis of this kind also directly portrays the problems that need be solved for better performance and high efficiency. Explanation of these graphical analyses is too extensive for inclusion here. Fundamentals of graphical calculations on enthalpy-composition diagrams may be found in the available literature.[31–34] Partial summaries in English are available.[12,35]

Though not absolutely essential, a Second Law analysis may be very helpful in establishing the presence or location of serious Second Law losses. This analysis can best be made as an availability balance around the circuit[36] using the data provided by Scatchard, et al.[25]

P-T-X analyses and material and heat balances may be reinforced when needed by other detailed studies of components using similar or related techniques.[7] These calculations provide the quantitative operating specifications that the system must fulfill. The system and its components are then designed to the specifications according to best engineering practice. However, since the heat and mass transfer rates occurring in gas refrigerators are often of a magnitude uncommon to other related operations, reference to previous designs is very helpful. Final development by building and testing prototypes is always necessary because details must be worked out to an accuracy greater than that which can be obtained by calculations based on current (1964) theories.

ABSORPTION UNITS AND COMPONENTS

Absorption units for gas refrigerators are fabricated of low-carbon steel tubing, which is unaffected by ammonia. Corrosion of the steel by the water in the solution is prevented by the use of sodium chromate as an inhibitor in concentrations of between 1 and 2 per cent. Experience has shown that protection for twenty years or more is attainable when the units have been rigorously cleaned internally and are free of foreign materials, such as oils, greases, chlorides, copper, etc., which may accelerate corrosion or react with the chromate.

Seamless tubing is commonly used, but welded tubing is now proving satisfactory. The ASA has specified that the entire system must withstand pressures of five times the maximum working pressure.[37] Welded joints not only must be leaktight but also must have adequate strength to withstand handling and transportation stresses. Acetylene hand welding is used for the majority of joints; automatic arc welding is used whenever practical.

To obtain the lowest temperatures and rapid evaporation rates, the tubing for the evaporator may be serrated internally (Servel patent) with circumferential grooves. Liquid ammonia flowing along the bottom of the tube is drawn up into these serrations by capillary action, wetting the entire internal surface. Electrolux, Norge, and others use a fine, soft iron wire screen to form a similar capillary surface.

To take maximum advantage of gravity effects, absorption units are normally built to the full height available within the cabinet dimensions. The condenser, absorber, and rectifier are mounted on the back of the cabinet in an enclosed space thru which cooling air rises. The condenser is mounted horizontally at the top of this space so that the liquid ammonia outlet may be as high above the evaporator as possible. The refrigerant drain from the condenser to the evaporator forms a U-tube with unequal legs. The *down* leg must be at least one and one-half *times* the length of the *up* leg because liquid ammonia, with a specific gravity of about 0.6, is only two-thirds as heavy as the solution that may be present in the tube on start-up. The drain trap should be made as deep as possible and should operate at the limit of the 1.5:1 gravity ratio so that the liquid refrigerant tube may be cooled by contact with the gas heat exchanger.

Figure 12-89 shows the refrigerating unit of a typical pre-1960 11-cu ft refrigerator. This unit utilizes two condenser drain traps. The liquid inlet to the upper coil of the freezer evaporator is so high that the 1.5 to 1 ratio could not be maintained. The liquid for the upper plate is therefore precooled only by evaporation into the rich gas leaving the high-temperature finned evaporator.

The evaporator shown in Fig. 12-89 is divided into a freezer coil and a fresh-food compartment coil. Because both coils involve considerable lengths of tubing in horizontal planes, the refrigerators must be carefully leveled when installed. Otherwise, a low spot may occur in the evaporator coils. If such a low spot fills with liquid ammonia, gas circulation can be stopped completely.

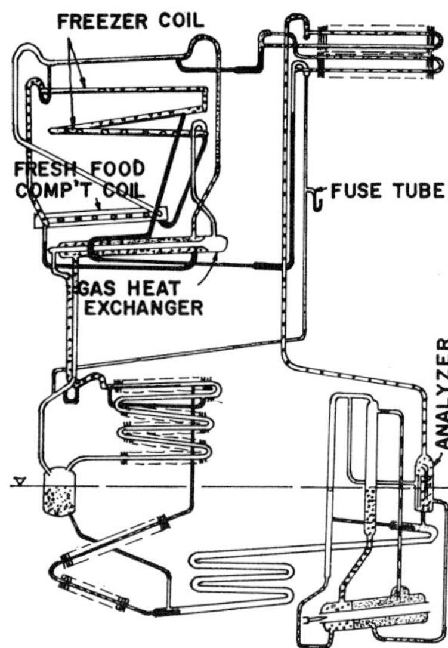

Fig. 12-89 Schematic diagram of 1959 11-cu ft refrigerator unit.

Aluminum plates are normally used to form an evaporator box that subsequently becomes the inner liner of the freezer compartment. The freezer evaporator coils are attached to the top and bottom plates by an aluminum brazing process, the metallurgical bond providing a good thermal conduction path from plates to tubes. During brazing, the aluminum alloy covers the steel tubing completely, providing corrosion protection.

The finned coil that cools the fresh-food compartment is located beneath the freezer box. Steel fins normally are used and the whole assembly is galvanized as a unit to provide good thermal conductivity as well as corrosion protection. A louvered drip tray under the coil catches condensate and defrost water. The finned coil is automatically defrosted daily by means of an electric heater nested into its fins. Automatic defrosting of box-type freezer evaporators is not used. Such defrosting has been proved impractical, owing to drippage and refreezing of water on frozen foods.

The gas heat exchanger extends horizontally just below the finned high-temperature coil. This is the highest level at which it can be located and still allow excess refrigerant to drain from the evaporator. Vertical heat exchangers have frequently been used, but the horizontal types have the advantage of increasing overall efficiency by precooling the liquid refrigerant. The rich gas from the evaporator has a higher heat capacity than the weak gas since its mass is more than 50 per cent greater. Hence, the weak gas cannot utilize all the cooling effect available in the rich gas. By attaching the liquid ammonia line to the gas heat exchanger, the exchanger becomes unbalanced in the opposite direction. The weak gas and liquid ammonia together have a higher heat capacity than the rich gas. All the cooling effect remaining in the rich gas can therefore be utilized, with a net gain in total refrigerating capacity.

The vertical spacing between the evaporator and absorber, which establishes the gravity force circulating the gas, should be as great as possible. However, other factors such as cabinet shape and evaporator design often determine this dimension.

In the absorber, the liquid must expose a large surface area to the gas, and the surface must be continuously stirred and replenished. The required liquid characteristics are obtained by an adequate length of absorber tubing and by insertion of closely spaced disks within the tubes. The center-hole disks shown in Fig. 12-90 form dams, which enlarge the liquid surface and produce turbulence as the liquid spills over each dam. The disks also produce turbulence in the gas stream.

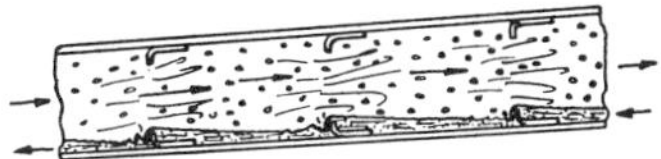

Fig. 12-90 Section of absorber tube showing gas and liquid flow.

The storage chamber at the bottom of the absorber serves a dual purpose. It provides a large liquid surface so that production variations in charge volume do not greatly affect the liquid level or the operation of the generator and the analyzer. By storing considerable solution, the chamber also serves as a source and supply of inhibitor for long unit life.

Figure 12-89 shows, as part of the generator, an analyzer in the form of a small chamber in which vapor from the boiler contacts the entering rich liquid. Intimate contact is accomplished by causing the vapor to pump the liquid thru a short tube. The liquid recirculates between the upper and lower analyzer chambers. When ammonia solution is boiled, both water and ammonia vapors are released. As seen in Fig. 12-87, the vapor from weak solution may contain as much as 30 to 40 per cent water, which must be removed before the ammonia enters the condenser at the 99.7 per cent or higher purity required to obtain good evaporator temperatures. Removal of this water vapor is accomplished partly in the analyzer and partly in the rectifier tube. In the analyzer the heat in the water vapor is recovered, but in the normal rectifier the heat of condensing the water vapor is lost. Minimizing this loss is important; it is done by improving the efficiency of the analyzer and by maintaining good vapor and liquid equilibrium in the rectifier. It can also be accomplished by a triple heat exchanger that would transfer this heat to the rich liquid, but high cost and difficult design problems have limited the use of triple exchangers.

A fuse plug is welded to the unit so that the charge may be safely vented in case of fire. To comply with American Standards[37] the plug must melt below 280 F and provide an outlet $\frac{1}{16}$ in. in diameter or more to allow the charge to vent at a safe rate.

Components other than those shown in Fig. 12-89 have been used at various times to provide increased efficiency or added features. Among the more common items are pressure control vessels, internal defrost systems, and concentration controls.

REFRIGERATOR CABINETS

The refrigerator cabinet (Fig. 12-91) is primarily an insulated box in which food may be stored or water frozen. The optimum storage temperatures and humidities of foods vary greatly. Frozen foods should be kept at 0 F or below. At 0 F, however, ice cream is generally too hard; its optimum serving temperature is about 5 to 8 F. The majority of fresh foods keep well at 35 to 40 F. Meats, fish, and poultry should be stored just above their freezing point, which may be below 32 F, and melons, for example, store best at 50 to 55 F. To suit these requirements with a minimum of compromise, refrigerators, whether with one door or two, generally have two main compartments, one for frozen foods, the other for fresh—with temperatures adjustable to suit the user's needs. The fresh-food compartment may also include separate containers for vegetables, fruits, meats, butter, and cheese. Overall convenience in food storing and ease of access are of major importance in design.[38]

The freezer compartment generally provides for storing specialized items like frozen juice concentrates, as well as for general frozen storage and ice-making. Gas refrigerators have long featured the automatic ice maker for maintaining a supply of ice cubes. The ice maker successively fills the ice tray or cube mold with water, freezes it, frees the cubes by melting a thin surface film, and mechanically moves them to a temporary location for refreezing the surface film. The cubes then are dumped into a pan. These operations are repeated until the pan is filled to a predetermined level, at which time they stop. The operations are restarted automatically as the cubes are used.

To maintain proper cabinet temperatures without excessive losses, 3 to 4 in. of insulation have been used in the walls and doors. Many insulating materials have been utilized: cork, mineral wool, Styrofoam, fiberglass, and polyurethane foam. The maintenance of almost constant cabinet temperatures in varying ambients is an important feature of food preservation. Norge, for example, claims food compartment temperatures that do not vary more than $\pm 1°$ F (measured by bare thermocouples) while the unit is cycling (with doors closed, no load) in ambients from 70 to 100 F. Even when unimportant to the total load, localized heat leaks may cause exterior "sweating" during humid weather. If insulation cannot eliminate such cold spots, electric heating wires are attached to the inside of the outer surface to raise spot temperatures to the ambient.

All exterior cabinet joints are sealed against leakage of room air or humidity into the insulation space. Water vapor condensing or freezing in this space increases heat leakage and may permanently affect some insulations. Sealing materials such as hot asphalt are used at all joints or seams exposed to room air. Openings and cracks from the interior into the insulation space may be allowed—or provided—so that moisture from the insulation may move to the evaporator surfaces. When closed, the doors are sealed against leakage by suitable gaskets and latches. To comply with American Standards,[37] the latching means must permit opening the doors from the interior by application of a force of not over 15 lb. Soft gaskets and improved latches have been developed; the use of magnetic gaskets is increasing.

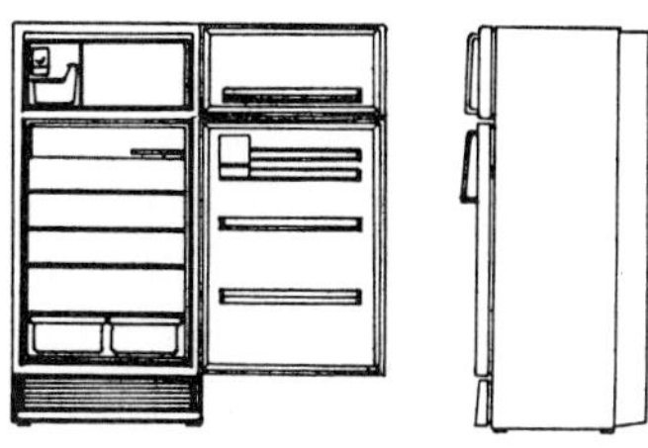

Fig. 12-91 Two-door gas refrigerator cabinet.

A gas refrigerator cabinet includes a machine compartment at the bottom and rear. The 5- to 6-in. rear space, housing the condenser, rectifier, absorber, and parts of the generator, serves as a chimney for the air that cools the absorber and condenser. The generator flue and the dilution flue also are located in this space. The space under the cabinet is utilized for the burner and controls and for portions of the generator. All items that may require adjustment or servicing are generally made accessible from the front.

Air for cooling the absorber and condenser enters at the cabinet back and thru the machine compartment. Cooling air, rising thru the condenser and absorber, may gradually deposit lint and dust on the fins, eventually restricting air flow and limiting the refrigerating performance. Normal performance is restored by cleaning the fins. Use of a deflector grill surmounting the flue prevents the direct contact of warm air and flue gases with the adjacent walls and minimizes the smudging that results when a warm air stream contacts a cooler surface.

Installation

Gas refrigerators must be carefully installed to the specifications of the manufacturer and in agreement with the provisions of local codes and ASA Z21.30.[39] Of particular importance are provisions regarding gas piping and connections, spacing of the refrigerator from adjacent walls and cabinets, leveling, setting of inputs, and checking of the controls and all components of the combustion system.

COMBUSTION AND HEAT TRANSFER SYSTEM

In 110 F or 115 F rooms, most gas refrigerators have a sharp peak of performance at the optimum heat input to the generator. For practical field use the maximum variation in the gas input setting is normally specified as ±5.0 per cent. In laboratory hot rooms a change of 2 per cent in input will generally show measurable change from the peak performance. In cooler rooms the peak is broader. To have the correct net input to the generator, other controlling factors besides gas input must be considered. The primary and secondary air must be correct, the combustion must be completed within the combustion space, and the flame must be properly located. The fire tube must be clean and free from deposits, and the convection transfer flue baffle must be properly located.

Typical Combustion System. As shown in Fig. 12-92a, the boiler (1) of the generator has a central fire tube surrounded by the boiling liquid in the annular space. The flame of the burner (2) extends into the fire tube, which has a flue baffle (3) located at its rear. The flue (4) discharges into the dilution flue (5).

Gas Refrigerator Burners

These burners (Fig. 12-92b) perform reliably over a range of inputs from 750 to about 3400 Btu per hr in unvented appliances under conditions of dust, dirt, lint, and grease. Burners must meet rigid ASA performance requirements[37] for completeness of combustion, good flame characteristics, and flexibility of operation. Different burners may be used for manufactured, natural, mixed and LP-gases. Since the main and pilot flames burn from the same ports and the mixing tube length is limited, a turbulator is used ahead of the orifice to ensure good gas–air mixing at low inputs. Thirty-six sizes of precision orifices are used to cover various gases and pressures.

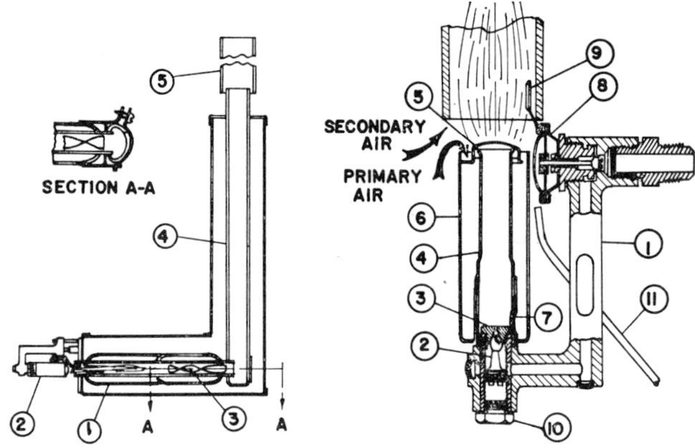

Fig. 12-92a (left) Combustion and heat transfer system for a gas refrigerator.

1. Boiler
2. Burner
3. Flue baffle
4. Flue
5. Dilution flue

Fig. 12-92b (right) Typical gas refrigerator burner.

1. Burner body casting
2. Turbulator
3. Orifice
4. Mixing or venturi tube
5. Burner cap
6. Dust tube and shutter assembly
7. Primary air ports
8. Safety shutoff valve (automatic pilot)
9. Heat conductor
10. Manometer connection
11. Lighter tube

Operation. The dust tube and air shutter assembly (6) shown in Fig. 12-92b is designed to clean the primary air of dust and lint by forcing the air to make a sharp turn to enter the dust tube. Both primary and secondary air flow together around the burner toward the entrance to the fire tube. The secondary air, about three-fourths of the total, flows directly into the fire tube. Lint and dust particles, having appreciable inertia, also continue into the fire tube and are incinerated. Only minute particles are capable of following the primary air into the dust tube. To ensure that these particles do not lodge in the burner, the primary air openings and the burner ports are made as large as the burning characteristics of the gases permit.

The shutoff valve (8) is a bimetal disk valve that must be heated, by means of a temporary flame from the lighter tube (11), to open. Once open, the valve is kept open by heat conducted to it from the flame by the aluminum heat conductor (9). If the flame goes out, the valve cools and shuts off the gas.

For proper location of the flame in the fire tube and for best operation of the incinerating feature, the burner must be accurately spaced from the end of this tube and located axially so that the flame is centered in the tube and surrounded by a blanket of secondary air. The air prevents the flame from impinging on the tube. Burner brackets to accomplish these objectives are generally designed to fit around the fire tube and mount either on the insulation box or on the generator proper. Generally, by tightening a single mounting screw, the burner is automatically aligned with the fire tube by the bracket pins. Spacing of the burner to the specifications of the particular model is accomplished manually.

When all components are set according to specifications, the gas refrigerator combustion system is highly reliable. Under normal conditions, modern refrigerators operate for years without change in flue performance. Improper servicing or flue stoppage, however, may cause difficulties. Except in early models, flue stoppages are almost always due to improper service or foreign objects.

Flue System

The flue system consists of the fire tube, flue baffle, lower flue, and dilution flue. The primary purpose of the flue baffle is to increase heat transfer in the convection section of the generator. The baffle may also be used to control excess air. After combustion is completed, the flue gases are cooled rapidly by accelerating convection transfer with the baffle. At maximum input, the exit flue gas temperature is 500 to 600 F, with combustion efficiencies from 70 to 75 per cent.

The flue gases then enter the vertical lower flue. In horizontal generators the lower flue is the sole source of the draft used to supply the combustion air. In vertical generators most of the flue effect may be developed in the fire tube. In both types the minimum possible flue height may produce excess draft. The flue baffle then provides the restriction required to limit the excess air, and also improves heat transfer.

As a result of early experiences with condensation from flue gases of high sulfur content, lower flues are ceramic, with ample cross section, and are located within the generator insulation. Formerly, external flues, at minimum input, might cool to the dew point of high-sulfur flue products, with resulting corrosion of susceptible materials. Insulating the flue and using noncorrosive materials has eliminated these difficulties.

Gases leaving the lower flue enter a dilution flue and leave the refrigerator at temperatures less than 90°F above the ambient. The dilution flue is sized to produce the draft for entraining the additional volumes of air to cool flue products. This flue is constructed of materials unaffected by sulfur compounds.

CONTROLS

Gas Controls

The arrangement of gas controls is shown in Fig. 12-93. The regulator is adjusted to the proper pressure for the burner orifice used, using a manometer connected at the base of the burner. The minimum gas input is set by means of an adjustment screw in a by-pass in the thermostat. This minimum input, generally 750 Btu per hr, should be adjusted accurately. If too high, it can cause excess refrigeration in cool rooms, and in some models it may prevent complete defrosting. If too low, the flame may go out because of drafts or inadequate heating of the heat conductor of the safety valve.

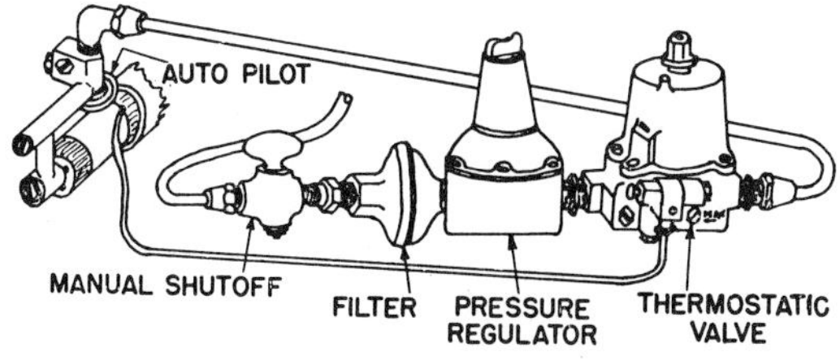

Fig. 12-93 Gas controls for gas refrigerator.

The thermostat, primarily a diaphragm-actuated gas valve responsive to the temperature of the refrigerated space, includes a minimum input bypass system and a manual pushbutton valve supplying the burner lighter tube. Some thermostats have also included a maximum input adjustment.

The thermostat control bulb (item 10 in Fig. 12-84) is generally attached to a suitable evaporator tube. In two-compartment models (Fig. 12-89) the refrigerant and the weak gas flow thru the freezer coil first and then thru the finned lower coil. Since evaporating temperatures in the lower coil depend not only on the local load but also on the upstream load, i.e., the evaporation that has occurred upstream, high loads in either compartment result in a rise in the temperature of the lower coil. The bulb, when properly located on the lower coil, thus controls the gas input in sensitive response to the requirements of both compartments. The thermostat is commonly a modulating control and tends to maintain a continuous flow of refrigerant thru the freezer and to adjust the refrigerant overflow from the freezer to suit the needs of the fresh-food compartment.

Defrost Controls

Defrosting the evaporators of low-cost or older model refrigerators is accomplished manually by turning the thermostat dial to a continuous-minimum input setting until defrosting is complete. Normal operation is then restored manually (or automatically if the thermostat includes that feature). In higher priced models, rapid defrost is commonly accomplished electrically; heat may be applied directly to the evaporator tubes or it may vaporize stored refrigerant, which transfers the heat throughout the length of the evaporator.

With pushbutton defrost control, the defrost period is started manually. The control then automatically shuts off the heater when defrosting is complete. With fully automatic controls both the time and duration of the defrost period are clock-controlled. The defrost period is generally set to occur at 2:00 A.M. The resulting defrost water is evaporated in an aluminum container that is attached to the rectifier. Fully automatic defrosting is generally applicable only to the food compartment coils, which collect the majority of the frost. Unattended automatic defrosting is not suited to the defrosting of the box-type evaporators of the freezer section, owing to the dripping and refreezing of water on the frozen foods.

Developments Since 1960[40]

Polyurethane Insulation. Use of this material in board form and foamed-in-place form has permitted reduced wall thicknesses, for about a 20-per cent increase in usable storage volume. The *no-frost* feature has been facilitated by burying the evaporator coil in the cabinet insulation.

Forced Interior Air Circulation. Use of a fan in both the freezer and refrigeration cabinets has permitted an additional 10° to 12° F of cooling. Units without this feature are generally called "static" models.

Forced Exterior Air Circulation. A fan with downward air flow cools the condenser and absorber and supplies the draft for combustion. After evaporating the water in the defrost pan, the air is vented thru the bottom grille. Smudging of ceilings and walls is thus eliminated. The fan has also eliminated the need for installation clearances at the cabinet sides and back and greatly simplified built-in installations. Clearances for air circulation above the cabinet have been reduced to four inches.

The improved refrigerating unit is shown in Fig. 12-94. Major innovations are the evaporator–gas heat exchanger system, which operates at much lower evaporator temperatures, and the high-efficiency, low-input generator. The evaporator–gas heat exchanger is a continuous tubular system folded into three sections. The two upper sections are finned, serving as evaporators. An inner gas heat exchanger assembly conducts the weak gas thru the lower heat exchanger tube and also thru the two upper evaporator tubes. Resulting evaporator temperatures may run 20° F colder than those in previous models. Sloping the evaporator tubing has eliminated the need for accurate leveling of the unit.

Defrosting of the top tube, i.e., the freezer evaporator, is accomplished once every 24 hr by means of hot refrigerant vapor and liquid pumped up from the storage chamber by an electric heater. The liquid refrigerant fills a trap that virtually stops gas circulation. The ammonia vapor can then build up to a high partial pressure and condense throughout the evaporator, melting the frost. The food compartment coil is defrosted during each thermostat cycle. During the off period of the cycle, the minimum gas input supplies sufficient refrigerant for the freezer tube but none for the food compartment. Continued air flow over the finned coil melts the frost. The on cycle starts only after all the frost is melted.

The generator of the new unit is smaller and requires 17 per cent less gas input than the one on the previous 11-cu ft model. The single-chamber boiler of the new model provides very uniform pumping. Higher efficiency is obtained by a bubble-cap analyzer tower that ensures good contacting of the liquid and vapor and by recovery of excess sensible heat in the weak liquid thru heat exchange with the analyzer.

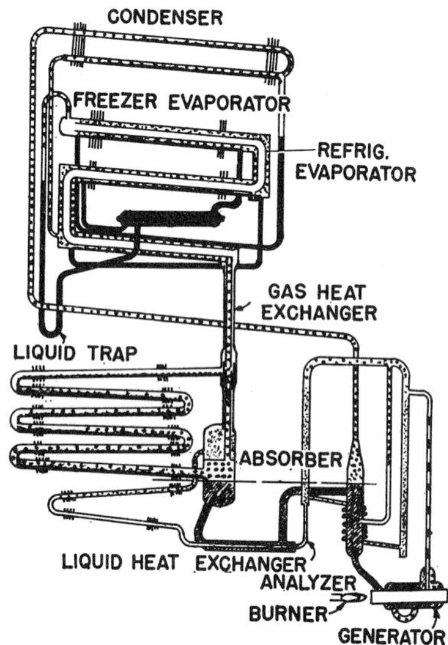

Fig. 12-94 1960 Whirpool No-Frost unit.

The new unit is about 70 lb lighter and has 35 fewer welds than the former one, and its reliability and low-service incidence are now well established. A Klixon gas valve attached to the bracket closes on overheating at the onset of malcombustion and shuts off the gas.

The gas input control valve, with built-in pressure regulator and filter, provides three input levels: maximum for full capacity; minimum for supplying refrigeration to the freezer only; and stand-by pilot for power failure. All three inputs are set automatically, by interrelated orifices, when the maximum input is set with the pressure regulator. The orifices for the burner and the gas valve are supplied as a set, to be used together for proper operation. The gas valve is actuated electrically by a cyclic-defrost thermostat, which senses the food compartment coil temperature. The cut-on temperature of the thermostat is constant; set slightly above freezing, it assures that all frost is melted before the input is turned back on maximum. The cutoff temperature is adjusted by the housewife when she sets the control to the desired refrigerator temperature.

Further new developments occurred in 1962. Norge marketed a 14-cu ft frost-free model, utilizing fans in the freezer, in the food compartment, and in the machine compartment, with an all-foam-insulated cabinet. This model includes a thermoelectric gas safety control as a gas shutoff in the event of flame outage. An electric gas valve that stays on at full input during power failures is used. The burner and control assembly is self-aligning to the rear-mounted vertical generator. Norge also displayed a 10-cu ft frost-free freezer utilizing fans internally and externally.

Whirlpool has developed a new line of units and cabinets. The units, reduced 40 per cent in weight and 60 per cent in the number of joints over the 1960–61 14-cu ft models, are suited to a complete line of refrigerators. Economies in cabinet design were also effected. Fans are eliminated from all models with the exception of a freezer fan in the No-Frost refrigerators. Mechanical gas controls are used so that the refrigerators operate under thermostatic control irrespective of power failure. In the No-Frost models stoppage of the freezer fan by power failure affects only the freezer; its temperature may rise to that of pre-1960 two-door gas refrigerators.

REFERENCES

1. Platen, B. von, and Munters, C. G. *Refrigeration.* (U. S. Patent 1,609,334) Washington, D. C., G.P.O., 1926.
2. Browne, R. "Absorption Refrigeration: A General History and Explanation." *Refrig. J.* 3: 292+, Jan. 1950.
3. Taylor, R. S. "Heat Operated Refrigerating Machines of the Absorption Type." *Refrig. Eng.* 17: 136–43, May 1929.
4. ——. "Heat Operated Absorption Units." *Refrig. Eng.* 49: 188–93, Mar. 1945.
5. Platen, B. von, and Munters, C. G. "Production of Low Temperatures." *Refrig. Eng.* 12: 142–8, Nov. 1925.
6. Ashby, C. T. "Absorption Refrigeration." (In: Am. Soc. Refrig. Engrs. *Air Conditioning, Refrigerating Data Book, Design Volume,* 10th ed., chap. 5. New York, 1957.)
7. Niebergall, W. *Sorptions-Kältemaschinen.* (Plank, R. *Handbuch der Kaltetechnik,* vol. 7) Berlin, Springer, 1959.
8. Hainsworth, W. R. "Refrigerants and Absorbents. *Refrig. Eng.* 48: 97+, Aug.-Sept. 1944.
9. Sherwood, T. K., and others. "Absorbent-Refrigerant Combinations." *Servel Report,* Jan. 1943.
10. Zellhoefer, G. F. "Solubility of Halogenated Hydrocarbon Refrigerants in Organic Solvents." *Ind. Eng. Chem.* 29: 548–51, May 1937.

11. ——, and others. "Hydrogen Bonds Involving the C–H Link; the Solubility of Haloforms in Donor Solvents." *J. Am. Chem. Soc.* 60: 1337–43, June 1938.
12. Ellington, R. T., and others. *The Absorption Cooling Process: a Critical Literature Review.* (Res. Bull. 14) Chicago, I.G.T., 1957.
13. Mastrangelo, S. V. R. "Solubility of Some Chlorofluorohydrocarbons in Tetraethylene Glyco Dimethyl Ether." *ASHRAE J.* 1: 64–8, Oct. 1959.
14. Eisman, B. J., Jr. "Why Refrigerant 22 Should Be Favored for Absorption Refrigeration." *ASHRAE J.* 1: 45–50, Dec. 1959.
15. Albright, L. F., and others. "Solubility of Refrigerants 11, 21 and 22 in Organic Solvents Containing an Oxygen Atom." *Trans. ASHRAE* 66: 423–33, 1960.
16. Ashby, C. T. and Whitlow, E. P. "Absorption and Steam Jet Units." (In: Am. Soc. Refrig. Engrs. *Air Conditioning, Refrigerating Data Book, Design Volume,* 9th ed., chap. 23. New York, 1955.)
17. Buffington, R. M. "Qualitative Requirements for Absorbent-Refrigerant Combinations." *Refrig. Eng.* 57: 343+, April 1949.
18. Wucherer, J. "Messung von Druck, Temperatur und Zusammensetzung der Flüssigen und Dampfförmigen Phase von Ammoniak-Wassergemischen in Sättigungszustand." *Z. Ges. Kälte Ind.* 39: 97–104, June ; 136–40, July 1932.
19. Zinner, K. "Wärmetönung beim Mischen von Ammoniak und Wasser." *Z. Ges. Kälte Ind.* 41: 21–9, Feb. 1934.
20. *Tables of Thermodynamic Properties of Ammonia.* (Natl. Bur. Std. Circ. 142) Washington, D. C., G.P.O., 1923.
21. Din, F., ed. *Thermodynamic Functions of Gases, Volume 1: Ammonia, Carbon Dioxide and Carbon Monoxide.* London, Butterworths Scientific Pub., 1956.
22. Keenan, J. H., and Keyes, F. G. *Thermodynamic Properties of Steam.* New York, Wiley, 1936.
23. Jennings, B. H., and Shannon, F. P. "Thermodynamics of Absorption Refrigeration." *Refrig. Eng.* 35: 333–6, May 1938.
24. Stickney, A. B. "New Tables and Chart for Ammonia Solutions." *Refrig. Eng.* 30, Oct. 1935 supplements.
25. Scatchard, G., and others. "Thermodynamic Properties—Saturated Liquid and Vapor of Ammonia-Water Mixtures." *Refrig. Eng.* 53: 413–19, May 1947.
26. Backstrom, M., and Pierre, B. "Total Vapor Pressure in Bar Over Ammonia-Water Solutions." *Tidsskr. Kiltek.* 18: 89, Aug. 1959.
27. Kohloss, F. H., and Scott, G. L. "Equilibrium Properties of Aqua-Ammonia in Chart Form." *Refrig. Eng.* 58: 970, Oct. 1950.
28. Bulkley, W. L., and Swartz, R. H. "Temperature-Pressure Concentration Chart for Ammonia-Water Solutions." *Refrig. Eng.* 59: 660–2, July 1951.
29. Tandberg, J., and Widell, N. "Diagram för Ammoniak-Vatten-Lösningar." *Tek. Tidskr.: Mekanik Suppl.* 67: 117–20, Sept. 18, 1937.
30. Buffington, R. M., and Phillips, B. A. *Servel Ammonia Charts.* Evansville, Ind., Servel, Inc., 1948.
31. Merkel, F., and Bosnjakovic, F. *Diagramme und Tabellen zur Berechnung der Absorptions-Kältemaschinen.* Berlin, Springer, 1929.
32. Dodge, B. F. *Chemical Engineering Thermodynamics.* New York, McGraw-Hill, 1944.
33. Mickley, H. S., and others. *Applied Mathematics in Chemical Engineering,* 2nd ed. New York, McGraw-Hill, 1957.
34. Bosnjakovic, F. *Technische Thermodynamik.* Dresden, Steinkopff, 1935–37. 2 vol.
35. Ruhemann, M. "Ammonia Absorption Machine." *Trans. Inst. Chem. Engrs.* 25: 143+, 1947.
36. Sherwood, T. K. "Availability Balance on Electrolux Cycle." *Servel Reports,* Oct. 1938.
37. A. G. A. Laboratories. *American Standard Approval Requirements for Refrigerators Using Gas Fuel.* (ASA Z21.19–1961) Cleveland, Ohio, 1961.
38. McCloy, G. S. "Household Refrigerators and Freezers." (In: Am. Soc. Refrig. Engrs. *Air Conditioning, Refrigerating Data Book, Design Volume,* 10th ed., chap. 28. New York, 1957.)
39. A. G. A. Laboratories. *American Standard Installation of Gas Appliances and Gas Piping.* (ASA Z21.30–periodically revised) Cleveland, Ohio.
40. Phillips, B. A. *Recent Developments in Gas Refrigeration.* (A. G. A. Res. and Util. Conf. 1963) Cleveland, Ohio, A. G. A., 1963.
41. Macriss, R. A., and others. *Physical and Thermodynamic Properties of Ammonia-Water Mixtures.* (Res. Bull. 34) Chicago, I.G.T., 1964.

Chapter 9

Direct-Fired Gas Clothes Dryers

by C. S. O'Neil, E. H. Smith, and E. A. Jahn

Most gas clothes dryers built before World War II consisted essentially of a large sheet-metal enclosure housing a dryer chamber equipped with collapsible clothes racks. The chamber was heated, either directly or indirectly, by a gas burner at its base. Postwar domestic clothes dryers of the tumbler type are among the major gas appliance developments of the period. About 4.1 million domestic gas clothes dryers were in use, and 539,000 units were reported sold, in 1963.[1]

Types of Clothes Dryers.[2,3] *Type 1* is primarily used in family living environments. *Type 2* is installed in commercial establishments and may be operated by the public or by an attendant. Both types are packaged units, using gas at up to 0.5 psig; either may be coin-operated.

Type 1 dryers have dry weight capacities up to about 9 lb and *Type 2* commercial dryers up to 50 lb. The following listing may be used to estimate the size of a dry load:

	Dry weight, lb
1 single bedsheet (muslin)	1.50
1 double bedsheet (muslin)	2.00
1 single bedsheet (percale)	1.25
1 double bedsheet (percale)	1.50
1 Turkish towel (avg wgt)	0.50
4 to 5 Turkish hand towels	1.00
15 Turkish washcloths	1.00
5 to 6 dish towels	1.00
12 linen napkins	0.75
3 housedresses	2.00
2 men's shirts	1.00
3 yd cotton material (avg wgt)	1.00
2 yd heavy cotton fabric (corduroy)	1.00

The U. S. Department of Agriculture reports[4] that the tumbling action in a clothes dryer causes no greater loss of fabric strength than other drying methods. Another report[5] states that the thermostat setting—high, medium, or low—made little difference in graying, yellowing, loss of bursting strength, or dimensional change in fabrics.

Servicing. A continuing A.G.A. program makes up-to-date procedures available.[6]

GENERAL DESIGN AND OPERATION

Wet clothes are tumbled in a cylindrical basket or drum that rotates on a horizontal axis at about 50 rpm. On some dryers the entire periphery of the drum is perforated, allowing the heated air to enter directly. On others, the perforated or screened area may be either on the drum front or rear head section. The drum revolves in a full casing or a partial casing that has an opening thru which the heated air is directed into the drum, circulated thru the clothes, and then exhausted thru another opening located opposite the inlet.

The burner and combustion chamber are housed within the cabinet structure. Products of combustion along with make-up air are drawn into the drum and circulated thru the clothes. As evaporation takes place, the moisture-laden air is removed by means of a fan or blower. The exhaust air, discharged from the drum and drum case, is filtered thru a screen that traps any lint picked up from the clothes.

Internal exhaust and venting arrangements of a typical dryer are shown in Fig. 12-95a.

Performance

Table 12-71a presents performance data of a laboratory test[7] using a contemporary gas clothes dryer. The dryer had air flowing into the front of the drum and out the center, at the rear of the drum (axial flow). The dryer was run as

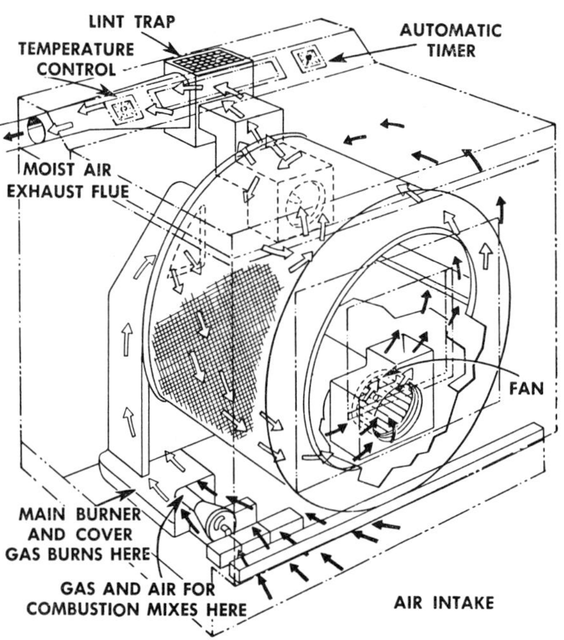

Fig. 12-95a Typical gas-fired clothes dryer.

Table 12-71a Effect of Heat Input and Air Flow Rate on Drying Performance[7]
(axial flow dryer)

Air flow, cfm	Heat input rate, Btu per hr	Time of run, min	Gas consumed, Btu	Water evap, lb†	Avg rate of evap, lb of water per min	Evap factor, lb of water per 1000 Btu of gas	Performance factor, gas Btu per lb of water evap	Exhaust temp, °F	Ambient humidity ratio, lb of water per lb of dry air	Ambient temp, °F
76*	17,100	50.0	14,200	9.00	0.18	0.635	1580	125	0.0190	92.0
43	18,100	50.0	15,100	9.00	.18	.595	1680	134	.0170	97.5
93	19,000	42.6	13,500	9.00	.21	.665	1500	114	.0185	93.0
74	8,700	90.0	13,100	9.00	.10	.685	1450	91	.0085	81.0
42	8,600	96.0	13,700	9.00	.09	.660	1520	104	.0080	87.0
101	9,000	81.0	12,100	8.75	.11	.725	1300	86	.0080	81.0
105	30,000	27.6	13,700	8.50	0.31	0.620	1615	122	0.0130	83.5

* Run as received at rated input. The dryer was a representative sample of an axial flow dryer. A heat input rate of 18,000 Btu per hour was used in 1960 by 39.5 per cent of the approved domestic gas clothes dryers.
† 9.0-pound load of Indianhead cloth, 100 per cent saturated.

received, as well as with variations in heat input rate and blower air throughput.

The change in drying efficiency (performance factor) in these runs indicates that the efficiency is improved by decreasing the heat input rate while maintaining all other conditions at constant values. If the blower air throughput is increased with no change in gas input rate, the drying efficiency is also improved.

It is significant, however, that both air flow rate and heat input rate can be increased to attain a much shorter drying cycle with very little loss in drying efficiency, as seen by comparing the first and last runs. Note that the first run was for the dryer as received from the manufacturer and required 1580 Btu of gas per lb of water evaporated.

Figure 12-95b, taken from the same report,[7] shows the relationship between exhaust temperature and cloth moisture content during a drying cycle. Figure 12-95c indicates the heat input distribution during each period of the cycle.

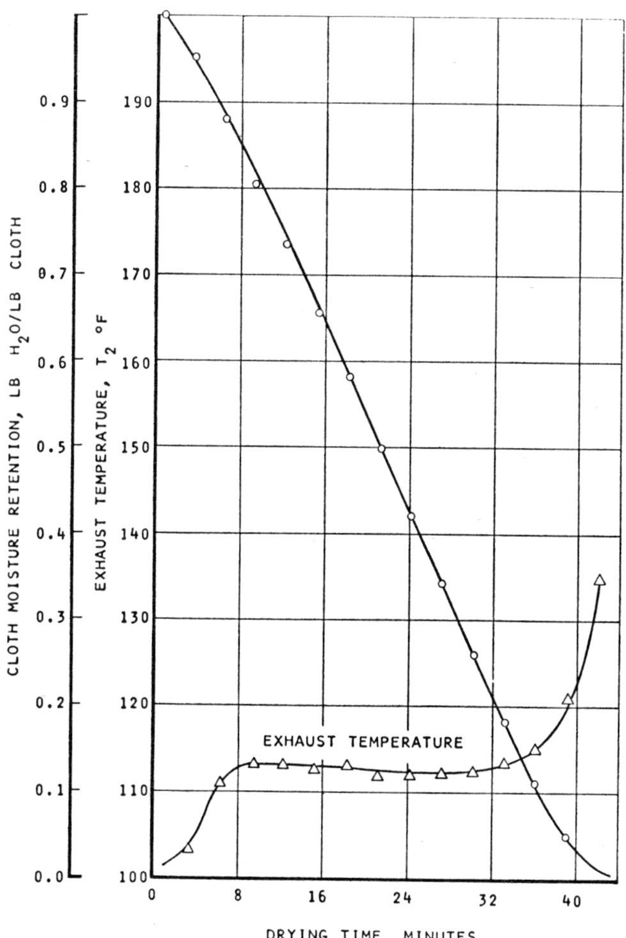

Fig. 12-95b Operating characteristics of a gas clothes dryer.[7]

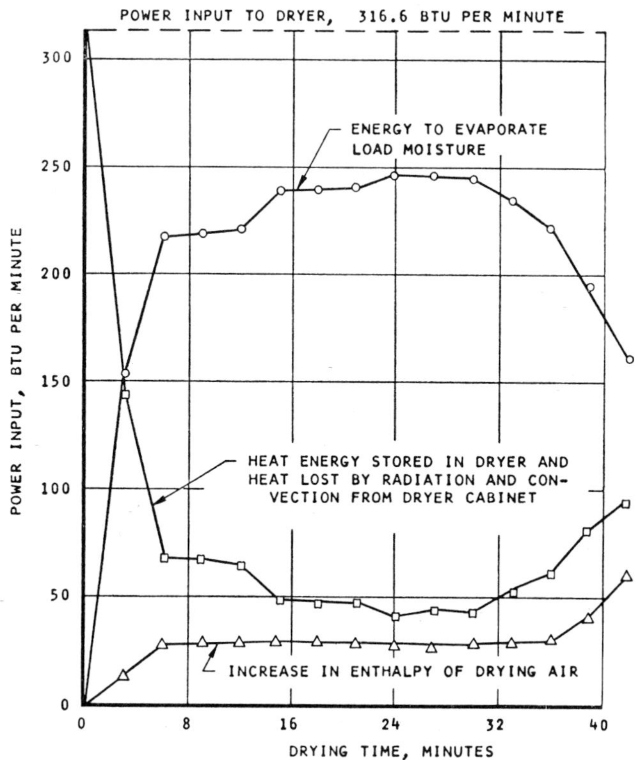

Fig. 12-95c Performance evaluation of a gas clothes dryer.[7]

Table 12-71b Distribution of Heat Input to Domestic Clothes Dryers and Measured Temperatures for Various Conditions*

Condition	Ambient		Energy input distribution, %†			Load temp, °F	Exhaust temp, °F	
	Temp, °F	Moisture, grains/lb air	To evap moisture	To heat air	For heatup and other losses		Dry bulb	Wet bulb
Cold start	46	20	47.0	43.0	13.3	79	98	79.0
Hot start	90	48	69.5	26.6	9.0	97	120	93.5
9-lb test load	45	20	49.4	39.4	9.9	80	94	79.0
Reduced air flow	45	26	47.2	35.2	13.9	89	111	90.0
Hot and wet ambient	110	241	69.2	23.2	13.2	112	137	112.0
Reduced heat input	110	93	75.2	11.0	11.9	97	120	97.0
4-lb nylon load‡	80	78	41.0	27.0	32.8	91	110	91.0

* 1961 Whirlpool Corp. research data.
† Does not add up to 100 because of error of measurement.
‡ Includes 37.5 per cent moisture by weight.

Table 12-71b also shows the energy input distribution to a dryer, but under various operating and ambient air conditions. The close correlation between the load temperature and the wet-bulb temperature of the drying air indicates that the heat is transferred primarily by convection;[8] conduction and radiation are negligible. This conclusion has also been confirmed by A.G.A. research.[7]

Fuel Consumption. Many independent tests (see page 12/342 in Chap. 22) of domestic dryers using normal drying loads have indicated an average gas consumption of 10,000 Btu per load.

	Test A	*Test B**	*Test C*	*Test D*
Avg consumption per load, MBtu	9.6	14.5	10	9.95
Avg drying time per load, min	...	72	29	33

* Heavy load of sheets and terry cloth towels.

ASA Approval Requirements.[2,3] Tests covering satisfactory combustion under normal conditions and with the exhaust outlet blocked are of principal importance. Rigid provisions for operation of burners, pilots, and igniters must also be met, with the dryer equipped with an exhaust duct at least 14 ft long so as to simulate conditions of actual use.

The temperature of the air and flue products leaving the exhaust outlet is limited to 200 F, with the control at its maximum setting. The load temperature is limited to 240 F. Air or combustion products discharged from any cabinet opening are kept from exceeding 250 F by a temperature-limiting device. The temperature rise on floors and on adjacent back and side walls must not exceed room temperature by more than 90°F after extended operation of a dryer installed with the clearance to walls as specified by its manufacturer. No evidence of scorching or color change of the drying charge is permitted.

A means for exhaust duct connection or other means for moisture and lint disposal is required. A lint trap or other device for minimizing discharge of lint from the dryer exhaust must be provided.

DESIGN FEATURES

Burners

Atmospheric burners are widely used in domestic dryers. Nearly all have fixed orifices in both pilot and main burner and therefore must be correctly sized to obtain the specified heat input for the fuel conditions encountered. Input rates range from about 18,000 to 37,000 Btu per hr for *Type 1* dryers and up to 130,000 for *Type 2* dryers.[9] A pressure regulator is mandatory[2,3] except for units using LP-gas.

Ignition Systems

Automatic ignition of listed gas dryers, which is mandatory, may be accomplished with a constant burning automatic gas pilot; a means for direct ignition (usually an electric spark) of main burner gas, together with a main burner flame-sensing device; or a recycling automatic gas pilot and glow coil igniter.

Units installed for multiple family or public use must be equipped with approved automatic pilots. Some dryers have a convenient means for manually igniting the pilot burner; others are equipped with automatic pilot igniters. The latter may effect automatic ignition by use of a hot-wire igniter or a high-intensity electric spark. The gas consumption of constant burning gas pilots averages about 4 therms per month.

Manual Ignition of Pilots. Dryers with this feature are equipped with an automatic shutoff valve that has a manually operated pilot by-pass and cocking mechanism. This mechanism is activated by the operator and held until the pilot is ignited manually and its flame impingement on a thermocouple generates sufficient current to hold the valve open electrically. With the valve held open, gas can flow freely to the next control in the main burner line, usually an electric solenoid wired in series with a thermostat and, sometimes, a time switch. Modifications to this basic arrangement include a door switch in the solenoid line to shut off the main burner when the loading door is opened, or additional safety thermostats or fusible links for overtemperature control. These additional controls may be placed in either the thermocouple line or the solenoid line.

Hot-Wire Automatic Ignition. With this arrangement, there are two independently operated solenoid valves in the manifold system. The first-stage valve opens when current is supplied either thru a manually operated switch or a time switch; gas flows only to the pilot burner. Simultaneously, the resistance wire mounted adjacent to the pilot burner is energized thru a stepdown transformer. The heated wire ignites the pilot, whose flame impinges on a mercury-filled bulb and vaporizes the mercury. Expansion of the mercury vapor operates a double-throw switch by means of a bellows or diaphragm, breaking the circuit to the igniter coil and closing the circuit to the second, or main burner solenoid valve, either directly or thru the time switch.

Some dryers using hot-wire ignition also incorporate a *warp switch,* which is a bimetal thermostat calibrated to open when current is applied thru its resistance heater circuit for from 3 to 6 min. The first, or pilot solenoid circuit, and the igniter circuit are connected to this switch. In the event of pilot ignition failure due to gas stoppage, low gas pressure, or mechanical failure, the circuits will be broken as the switch cycles. Temperature controls and additional safety controls for this system can be wired in series with the pilot solenoid, with the main burner solenoid, or with both, effecting complete or partial shutoff.

Dryer operation with a hot-wire automatic pilot igniter system (Fig. 12-96a) is as follows:[6] With the dryer in the idle position, the starting switch (usually the timer) is open and gas is shut off at the pilot valve (Diagram 1). When the starting switch is closed or the timer is turned to the on position, the pilot valve is opened, admitting gas to the pilot burner and the cycling valve. The igniter coil is energized and ignites the pilot (Diagram 2). The pilot sensing bulb is heated by the pilot flame, causing the mercury in the bulb to vaporize. The increased mercury vapor pressure extends the diaphragm in the pilot device, a single-pole, double-throw switch, moving it from its "cold" to its "hot" position. The cycling valve is then actuated to admit gas to the main burner, which is ignited by the pilot flame. The ignition coil is de-energized (Diagram 3).

During normal operation, the thermostat causes the main burner to cycle as necessary to maintain the desired temperatures. At the end of the drying cycle, the main burner is shut off, although the drum continues to rotate for a short period. In some dryers, both the pilot and cycling valves close at the end of the drying period; in others, only the cycling valve closes. When the pilot valve is de-energized, the pilot flame goes out and the pilot sensing bulb cools. The mercury vapor condenses, actuating the pilot device switch, which returns to its "cold" position. The dryer is then completely shut down.

If, for any reason, the pilot flame is not established soon after the starting switch is closed, the warp switch opens the pilot valve circuit, causing the pilot valve to close (Diagram 4). If the pilot flames goes out, condensation of the mercury vapor in the pilot sensing bulb actuates the pilot device switch, which returns to its "cold" position, causing the cycling valve to close. The igniter coil is then re-energized in an attempt to re-establish the pilot flame (Diagram 2). If the pilot flame is not re-established, the warp switch opens the circuit to both the pilot and cycling valves, shutting off all gas.

Spark Automatic Ignition. This method consists of an automotive-type spark plug mounted adjacent to the pilot, with the plug electrodes directly in the path of the pilot gas flow. An automatic shutoff valve controls the pilot gas. On some dryers, manual turning of a single control knob simultaneously operates the cocking button of the pilot shutoff valve and an electric switch, causing current to flow to a transformer that provides current to the spark plug. When the control knob is released after the length of time required to heat the thermocouple, the electric switch is opened and current to the spark plug is interrupted. A solenoid valve, downstream from the automatic shutoff valve and controlling the main burner gas, is energized when the operator starts the dryer with the main control or time switch. At the end of the drying cycle, the solenoid valve shuts off the main burner.

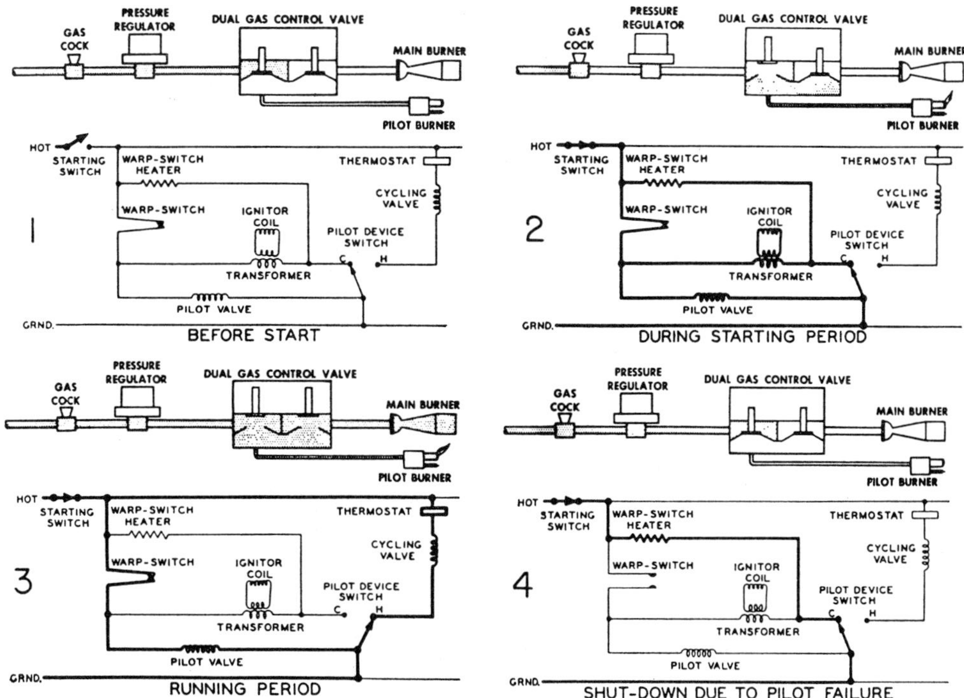

Fig 12-96a Operating cycle diagram for a hot-wire automatic pilot ignition system on a clothes dryer.[6]

The pilot burner remains lighted until a gas cock in the manifold system is closed manually.

SIGI Automatic Ignition System. This is an example of a spark system that does not use a pilot burner. The magnetic vibrator (Fig. 12-96b) is energized thru the flame detector switch and the warp switch, which is normally closed. The safety relay is energized in series with the heating coil and equalizing resistor, thru the flame detector switch. The relay closes switch Z, completing the circuit to the gas solenoid valve. The gas flows to the burner and is ignited by the vibrator. The burner heats up the sensor bar of the igniter assembly, causing the flame detector switch to open. This action opens the circuit to the magnetic vibrator, warp switch, heaters, and relay coil. However, the relay coil circuit is now completed thru the lockout resistor.

If flame is not established at the burner, the warp switch heating coil will cause the warp switch contacts to open in 30 to 75 sec, opening the circuit to the magnetic vibrator and shutting off gas flow at the solenoid valve. The warp switch continues to heat and remains open until the power is shut off by a timer or other means.

Dryer Cycle Controls

Timer switches, temperature controllers (thermostats), a combination of these two, and moisture-sensing devices are used to control the degree of drying.

Timer Switch. The timer switch mechanism is essentially the same on all dryers using this control; there may be variations in the operating sequence of the electrical components, depending on the make of dryer. The timer mechanism actuates two sets of contacts. One set is connected in series with the main motor, the other with the solenoid valve circuit and, on some models, electric ignition. When the timer is set by the operator for the drying period desired, both sets of contacts are closed and the machine starts. On most dryers, about five or six minutes before the end of the operating period, the contacts that control the solenoid circuit open, shutting off either the main burner gas or the entire supply. The motor control contacts may remain closed for the remainder of the operating cycle, allowing continued tumbling of the clothes with the heat off so as to cool them to a comfortable temperature for handling on removal.

Most dryers with a timer control also have a thermostatically operated temperature control. The thermostat

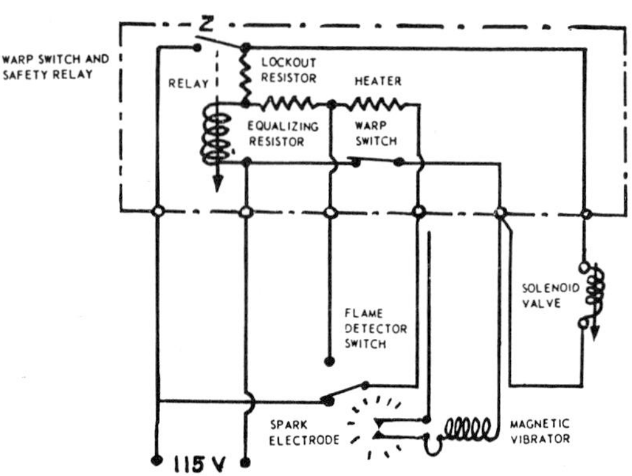

Fig. 12-96b SIGI automatic ignition system. (Hamilton Mfg. Co.)

may be adjustable, permitting the operator to select temperature settings from high to low as indicated by the control knob. The thermostat is wired in series with the solenoid valve that controls the main burner gas. The temperature-sensitive element may be inside the drum casing or in the exhaust air stream.

Thermostat. This control uses a heat-operated switch with adjustments that enable the operator to select the degree of dryness desired. The heat-sensitive element, or thermostat bulb, is located within the dryer, where it is subject to the temperature of the exhaust air. When drying has progressed to a point at which the air contains little moisture, the air temperature increases sharply. The thermostat detects this increased temperature and operates to open the solenoid circuit. Current to the motor is maintained until the exhaust air temperature drops approximately 40° F. The reverse movement of the thermostat bellows then opens the motor circuit, shutting off the dryer.

When this control is used, the reset mechanism usually incorporates an interlocking arrangement, allowing the dryer to be reset at a lower temperature than that for the previous run so that it may be restarted immediately.

Moisture-Sensing Devices. Available on some dryers, these sense the moisture content of the dryer load and shut down the dryer at the desired degree of dampness. A moisture-sensing element has been reported on that is not influenced by factors such as atmospheric humidity, temperature, time, and load size.[10] The clothes contact the element and the moisture present changes the electrical resistance of the device.

Other Safety Controls

Various other devices within the dryer protect against failure of one or more of its operating controls. Application and location of these devices differ considerably in individual dryers.

Safety Pilot Thermocouple Switch or Fusible Link. On a system using the Baso-type safety pilot, a thermostatically operated temperature switch or fusible link may be placed in series with the thermocouple lead. If overheating occurs owing to failure of the main control thermostat or some other malfunction of the main burner controls, the thermocouple circuit will be broken. The safety pilot valve will then close, shutting off both pilot and main burner gas.

Solenoid Valve Circuit Switch or Fusible Link. On dryers using electric solenoids for gas control, a safety switch or fusible link is placed in series with one or the other of the solenoids, or both, bringing about a partial or complete shutoff of the gas supply. A dryer equipped with a fusible link will remain inoperative after the link melts until it is replaced. If a cycling type of safety switch is used, the cycling temperature differential is usually so great that the operator notices the difference in operation after the safety switch has cycled. In some applications, even though a Baso-type safety pilot valve is used, the safety switch or fusible link is in series with the main burner solenoid. The main burner then may cycle, the pilot remaining on.

Flue Thermostat. In addition to a conventional safety switch or fusible link, some dryers have a flue thermostat located on the flue or purge stack. This thermostat may be in series with one or both solenoids, thus providing complete or partial shutoff. Shutoff will occur as a result of excessive exhaust restriction, fan stoppage, or some other malfunction that causes the heated air and combustion products to purge thru the exhaust duct or purge stack instead of passing into the drying chamber.

CLOTHES DRYER INSTALLATION

Although the following discussion pertains specifically to clothes dryers, it should be remembered that general requirements[11] for gas piping and appliance clearances must also be observed.

It is generally desirable to direct moisture-laden air and products of combustion from *Type 1* units to the outside air. *Type 2* dryers must be so installed. Manufacturers' instructions regarding duct size and length should be followed closely for satisfactory results. The exhaust duct should never be connected to a vent connector, gas vent, or chimney. A volume of replacement air to the room, equal to the clothes dryer exhaust products, should be provided for *Type 2* dryers, either mechanically or thru a free area, opening to the outdoors, of one sq in. for each 1000 Btu of gas input to the dryers(s). Commercial units may require about 500 to 750 cfm per dryer.[12] Without replacement air, flue gases from relief openings of draft hoods of other appliances may be drawn into the room.

Halogenated Hydrocarbons. Such vapors, from aerosol spray cans in the home and dry cleaning agents (e.g., perchlorethylene) in commercial cleaning establishments, react at temperatures as low as 500 F to form acid gases that are a hazard to health, corrosive to equipment, and damaging to fabrics (discoloring) and painted surfaces. Mitigating measures include keeping such vapors away from hot surfaces, ventilating areas where such vapors may be released, avoiding the introduction of contaminated make-up air, and connecting combustion air inlets directly to an uncontaminated air supply.

REFERENCES

1. American Home Laundry Manufacturers Assoc.
2. A.G.A. Laboratories. *American Standard Approval Requirements for Gas Clothes Dryers, Volume I: Type 1 Clothes Dryers.* (ASA Z21.5.1–1962) Cleveland, Ohio, 1962.
3. ——. *American Standard Approval Requirements for Gas Clothes Dryers, Volume II: Type 2 Clothes Dryers.* (ASA Z21.5.2–1962) Cleveland, Ohio, 1962.
4. U. S. Dept. of Agriculture. *Home Laundering: The Equipment and the Job.* (Home and Garden Bull. No. 101) Washington, D. C., 1964.
5. Ross, E. S., and others. *Automatic Clothes Dryers—Their Performance and Effect on Certain Fabric Properties.* (U. S. Dept. of Agriculture Home Econ. Research Rept. 6) Washington, D. C., G.P.O., 1958.
6. A.G.A. *Gas Appliance Service Manual, Volume 1: Clothes Dryers.* 4th ed. plus supplements. New York, 1963.
7. A.G.A. Laboratories. *Investigation of Domestic Gas Clothes Dryers.* (Research Bull. 100) Cleveland, Ohio, 1964.
8. Treybal, R. E. *Mass-Transfer Operations.* New York, McGraw-Hill, 1955, p. 543.
9. A.G.A. Laboratories. *Directory, Approved Appliances, Listed Accessories.* Cleveland, Ohio, semi-annual.
10. Maas, J. *A Direct Sensing Control System.* (A.G.A. Research and Util. Conf., 1961) Cleveland, Ohio, A.G.A., 1961.
11. A.G.A. Laboratories. *American Standard Installation of Gas Appliances and Gas Piping.* (ASA Z21.30–periodically revised) Cleveland, Ohio.
12. A.G.A. *Water Heating Applications in Coin Operated Laundries.* (Information Letter No. 127) New York, 1962.

Chapter 10
Gas Incinerators

by F. E. Vandaveer and M. Khan

TYPES OF INCINERATORS

I.I.A.* Definitions.[1] An *incinerator* is a combustion apparatus in which solid, semisolid, or gaseous combustible wastes are ignited and burned to CO_2 and water vapor, the solid residues of the process containing little or no combustible material. In the *primary combustion chamber*, ignition and burning of the waste occurs. In the *secondary combustion chamber*, the unburned combustible materials from the primary chamber are completely burned.

Gas-fired incinerators are generally classified as: (1) domestic, (2) commercial, (3) industrial, and (4) flue-fed. An alternate classification system is given in Table 12-72.

Note that many so-called "incinerators" are merely containers or primary combustion chambers in which combustibles may be burned with no provision for burning smoke and odor or for retaining the fly ash.

TYPES AND PRODUCTION RATES OF WASTES

Annual U. S. waste production per capita ranges from 587 to 1575 lb of refuse (garbage) and 173 to 573 lb of rubbish. Table 12-73 shows the annual refuse collected in 12 American cities; Table 12-74 gives four analyses of municipal refuse. A method to make a chemical and physical analysis of refuse is available.[2] The make-up of residential waste on a daily per person basis is generally taken as 0.6 lb of food waste plus an equal weight of dry combustible material for a total of 1.2 lb. Thus, an average family of four has a waste output of about five pounds per day.[3]

The types of waste which may be burned in incinerators are classified in Table 12-75. Estimated daily production rates of wastes and their average densities and volumes are given in Table 12-76.

AIR POLLUTION CONTROL LAWS

Principally a legislative right of the state, air pollution control is usually delegated to local communities. A survey (1962) of 327 communities with a population of 50,000 or more indicated that of the 166 respondents, 107 did have air pollution control regulations, while 59 did not.[4] While their

Table 12-72 I.I.A. Classification of Incinerators[1]

Class	Type of waste*	Fed	Capacity or burning area	Remarks
I	1 or 2	Direct	Up to 5 cu ft storage or 25 lb/hr burning	Portable, packaged
IA	1 or 2	Direct	5 to 15 cu ft primary chamber or 25 to under 100 lb/hr burning	Portable, packaged or job assembled
II†	1 or 2	Flue‡	Over 2 sq ft area	...
IIA†	1 or 2	Flue§	Over 2 sq ft area	...
III	1 or 2	Direct	100 lb/hr and over	...
IV	3	Direct	75 lb/hr and over	...
V	Municipal	...	...	...
VI	4	Direct	...	Crematory and pathological use
VII	5 or 6	Direct	...	...

* Determined by analysis of many samples; used in computing heat release, burning rate, velocity, and other details of incinerator design.

† Not recommended for industrial wastes.

‡ Single flue for both charging waste and exhausting effluent.

§ Two flues, one for charging and one for exhaust.

* Incinerator Institute of America.

Table 12-73 Total Refuse Collected in 12 United States Cities, 1957–1958*,[2]

City	Pounds per capita per year	City	Pounds per capita per year
St. Petersburg, Fla.	1690	Seattle, Wash.	1370
Los Angeles, Calif.	1677	Omaha, Nebraska	1370
Washington, D. C.	1638	New York, N. Y.	1325
Chandler, Arizona	1587	Atlanta, Georgia	1252
Garden City, N. Y.	1438	Cincinnati, Ohio	1103
Hartford, Conn.	1430	San Francisco, Calif.†	794

* Includes refuse actually collected, regardless of the method of collection or disposal. It does not include automobile bodies, cinders, and bones and fats. Since they are usually salvaged before collection and sold, they cannot be considered wastes. Apartment house incinerator ash is calculated in terms of equivalent amounts of refuse.

† San Francisco is set apart from the other 11 cities because it is the only one in which the householder is charged for refuse collection on the basis of quantity. The method of charging is an incentive for the householder to dispose of as much refuse as possible on his premises. Thus, the figure for the amount collected is much lower than it is in other cities.

Table 12-74 Composition and Analyses of an Average Municipal Refuse[14]

	Per cent of total refuse	Proximate analysis,* "as received" basis, weight per cent				Ultimate analysis, dry basis, weight per cent								Btu per lb‡	
		Moisture	Volatile matter	Fixed carbon	Non-comb.†	Carbon	Total H_2	Available H_2	O_2	N_2	S	Non-comb.†	Ratio C:(H)	Dry basis	Dry, ash-free basis
Rubbish, 64%															
Paper, mixed	42.0	10.24	75.94	8.44	5.38	43.41	5.82	(0.28)	44.32	0.25	0.20	6.00	155	7572	8055
Wood and bark	2.4	20.00	67.89	11.31	0.80	50.46	5.97	(0.672)	42.37	0.15	0.05	1.00	75	8613	8700
Grass	4.0	65.00	...	...	2.37	43.33	6.04	(0.83)	41.68	2.15	0.05	6.75	52	7693	8250
Brush	1.5	40.00	...	...	5.00	42.52	5.90	(0.75)	41.20	2.00	0.05	8.33	56.7	7900	8600
Greens	1.5	62.00	26.74	6.32	4.94	40.31	5.64	(0.77)	39.00	2.00	0.05	13.00	52.4	7077	8135
Leaves, ripe	5.0	50.00	...	...	4.10	40.50	5.95	(0.31)	45.10	0.20	0.05	8.20	131	7069	7700
Leather	0.3	10.00	68.46	12.44	9.10	60.00	8.00	(6.56)	11.50	10.00	0.40	10.10	9.1	8850	9850
Rubber	0.6	1.20	83.98	4.94	9.88	77.65	10.35	(10.35)	...	...	2.0	10.00	7.5	11330	12600
Plastics	0.7	2.00	...	...	10.00	60.00	7.20	(4.40)	22.60	...	...	10.20	13.6	14368	16000
Oils, paints	0.8	0.00	...	...	16.30	66.85	9.65	(9.00)	5.20	2.00	...	16.30	7.43	13400	16000
Linoleum	0.1	2.10	64.50	6.60	26.80	48.06	5.34	(3.00)	18.70	0.10	0.40	27.40	16	8310	11450
Rags	0.6	10.00	84.34	3.46	2.20	55.00	6.60	(2.70)	31.20	4.62	0.13	2.45	20.4	7652	7844
Sweepings, Street	3.0	20.00	54.00	6.00	20.00	34.70	4.76	(0.36)	35.20	0.14	0.20	25.00	96	6000	8000
Dirt, Household	1.0	3.20	20.54	6.26	70.00	20.62	2.57	(2.07)	4.00	0.50	0.01	72.30	10	3790	13650
Unclassified	0.5	4.00	...	...	60.00	16.60	2.45	(0.166)	18.35	0.05	0.05	62.50	100	3000	8000
Food wastes, 12%															
Garbage	10.0	72.00	20.26	3.26	4.48	44.99	6.43	(2.845)	28.76	3.30	0.52	16.00	15.8	8484	10100
Fats	2.0	0.00	...	...	0	76.70	12.10	(10.70)	11.20	0	0	0	7.2	16700	16700
Noncombustibles, 24%															
Metallics	8.0	3.00	0.5	0.5	96.0	0.76	0.04	(0.02)	0.2	...	...	99.0	51	124	12000
Glass and ceramics	6.0	2.00	0.4	0.4	97.2	0.56	0.03	(0.02)	0.11	...	...	99.3	34	65	8000
Ashes	10.0	10.00	2.68	24.12	63.2	28.0	0.5	(0.40)	0.8	...	0.5	70.2	70	4172	14000
	100.0														

Organic analysis of composite, %	
Moisture	20.73
Cellulose, sugar, starch	46.63
Lipids (fats, oils, waxes)	4.50
Protein, 6.25N	2.06
Other organic (plastics)	1.15
Ash, metal, glass, etc.	24.93
	100.00

Analysis of composite refuse, as received basis										Btu per lb	
Moisture	Carbon	Total H_2	Available H_2	O_2	N_2	S	Non-comb.	Ratio C:(H)	Btu per lb	Dry basis	Mean, ash-free basis
20.73	28.00	3.50	(0.71)	22.35	0.33	0.16	24.93	39.4	4917	6203	9048

* Based on ASTM methods of analysis of coal and coke, as adapted for refuse.
† Noncombustibles—ash, metal, glass, and ceramics.
‡ Lower heating value will be considerably lower, depending on moisture content.

Table 12-75 Types of Wastes Which May Be Incinerated[1]

Waste						
Type	Description	Principal components and sources	Approximate composition, % by wt	Moisture content,* % by wt	Incombustible solids, %*	Btu/lb of refuse as fired*
1	Rubbish	Combustible waste, paper, cartons, rags, wood scraps, floor sweepings; domestic, commercial, industrial	Rubbish, 100; garbage up to 20	25	10	6500
2	Refuse	Rubbish, garbage; residential	Rubbish, 50; garbage, 50	50	7	4300
3	Garbage	Animal and vegetable; restaurants, hotels, markets, institutional, commercial, clubs	Garbage, 100; rubbish, up to 50	70	5	2500
4	Animal solids and organic	Carcasses, organs, solid organic wastes; hospitals, laboratories, abattoirs, animal pounds, and similar sources	Animal and human, 100	85	5	1000
5	Gaseous, liquid, or semiliquid	Industrial processes	Variable	†	‡	‡
6	Semisolid and solid	Combustibles requiring hearth, retort, or grate burning equipment	Variable	†	‡	‡

* Determined by analysis of many samples; used in computing heat release, burning rate, velocity, and other details of incinerator design.
† Dependent on predominant components.
‡ Variable, according to waste surveys.

primary purpose is to maintain clear air in the community, local codes and ordinances vary considerably in detail and complexity. The more definite ordinances establish emission and ambient air standards which make engineering decisions on air pollution control possible before the plants or equipment are built. Open burning contributes the highest pollution level per pound of refuse burned. It is prohibited in many communities. Air pollution legislation to date (1964) has been carefully worded to avoid the interpretation that adherence to the standards will also prevent a nuisance; thus, a process which emits air pollutants can still be held at common law for damages even though it complies with the local air pollution ordinance. The engineer must decide the ground level concentration that will be acceptable and produce no nuisance. If the air pollutant emitted is odorous, the proper criterion is for the concentration of the pollutant to be below the odor threshold at ground level outside the plant property. For example, hydrogen sulfide has an odor threshold concentration of 0.1 parts per million—concentrations below this level will not be detected as an odor.[5] The threshold limit for SO_2 is 5 ppm for 8 hours duration during a 24 hour period (Table 1-30). This concentration should be lower for exposure round the clock. *Sulfur-bearing auxiliary fuel should be avoided.*

Almost all air pollution control laws permit installation of incinerators complying with local laws; however, some communities require approval of the incinerator installation by local authorities.

PRINCIPLES OF INCINERATION[3,6,7]

An ideal incinerator turns combustible refuse into CO_2 and H_2O. It has been found that all combustible constituents of refuse can be adequately burned at 1400–1800 F. Higher temperatures accelerate oxidation of hydrocarbons and combustibles. However, excessively high temperature may dissociate CO_2 to CO unless sufficient excess O_2 is present to shift combustion equilibrium to maximum CO_2.

Btu value of refuse is supplemented by auxiliary gas to raise the temperature of products of combustion to 1200–1800 F or higher. Secondary chamber should provide additional heat and residence time, and thorough mixing.

The following basically applies to domestic and commercial incinerators:

1. The nonhomogeneity of incinerator charges precludes any exact determination of auxiliary heat requirements. It was found that the modified smokeless-odorless domestic design considered at the A.G.A. Laboratories required about 4000 Btu *per pound* of refuse for the primary burner plus 10,000 Btu per lb for the secondary burner.[3] Thus, 42 MBtu per hr were required to burn three pounds of refuse per hour. A later A.G.A. study[8] showed that increasing the gas input to contemporary (originally about 34 MBtu per hr gas input for 2.0 cu ft of capacity) incinerators by 10,000 Btu per hr decreased particulate matter and uncarbonized residue in the ash by as much as 87 and 55 per cent, respectively; the time required for incineration decreased; thus, the gas consumption per incinerator charge does not necessarily increase.

2. The direction of flow in the settling chamber of a commercial incinerator should be turned at right angles at least twice, with reduction of flow velocity below 8.0 fps.

3. A barometric damper is necessary, with an opening thru the flap area at least equal to the breeching or chimney area to control draft.

4. Inlet air openings of an incinerator exposed to wind should be protected; for example, it may be protected with a sheet metal box which has an elbow directed downward.

5. A flame safeguard on large, and an automatic pilot on small, units should be provided for each gas burner, to shut off gas in case of flame outage.

6. Incinerator secondary chamber temperature should be about 1400 to 1800 F. This temperature should be achieved after a reasonable period following ignition.

7. Total combustion air required is the sum of air for burning the refuse and the auxiliary gas fuel, plus excess combustion air. Total air requirements, allowing for 100 per cent excess air, may be computed by providing 20 cu ft of air for each 1000 Btu of refuse and auxiliary fuel to be consumed.

Example: *Refuse Combustion Air.* If 100 lb per hr of Type 3 waste (2500 Btu per lb, from Table12-75) is to be incinerated, the *refuse* combustion air required is:

$$20 \frac{\text{cu ft air}}{\text{MBtu waste}} \times 2.5 \frac{\text{MBtu}}{\text{lb waste}} \times 100 \frac{\text{lb waste}}{\text{hr}} = 5000 \frac{\text{cu ft air}}{\text{hr}}$$

Auxiliary Fuel Combustion Air. If 300 MBtu per hr of auxiliary fuel is required to incinerate this waste, the combustion air required is:

$$20 \times 300 = 6000 \text{ cu ft air per hr}$$

Total combustion air required = 5000 + 6000 = 11,000 cu ft per hr.

Table 12-76 Estimated Waste Production Rates*

Type of installation	Daily rate	Type of waste†
Apartment house	2 lb per person	1 and 2
Cafeteria	½ to ¾ lb per meal served	1, 2, and 3
City	2 to 4 lb per capita	
Club	2 lb per person	1, 2, and 3
Department store	1 lb per 25 sq ft of floor space	1‡
Hospital	7 to 8 lb per bed	1, 2, 3, and 4
Hotel, class	Per guest room *plus* per meal	
First	2 lb + 2 lb	1, 2, and 3
Medium	1.5 lb + 1.5 lb	
Institution	3 lb per person	1, 2, and 3
Market (produce, meat, fish, etc.)	1 lb per 10 to 25 sq ft of sales area	2 and 3
Office building	1 lb per 100 sq ft floor area	1
Restaurant, first class	1.5 lb per meal served	1, 2, and 3
School	8 lb per classroom plus 0.7 lb per student if cafeteria is used	1 and 3
Warehouse, department, or chain store	1 lb per 35 to 50 sq ft of floor area	1
Warehouse, produce	1 lb per 25 sq ft of floor area	1, 2, and 3
Factory	Same as warehouse plus trade waste	1, 2, and 3

Average waste density, lb per cu ft		Container volume conversions, cu ft	
Loose paper	4	Garbage can,§ 16 × 22	2.0
Dry rubbish, office building	7	18 × 24	3.6
Garbage, 70% wet, 30% dry	45	Bushel	1.25
Pathological	55	Barrel, U. S. std	4
Sawdust	10	55 gal drum	7
Sewage sludge	60	1 gal	0.134
Water	62.4		

* These estimates are subject to variations as influenced by local and special conditions; individual surveys should be made wherever possible.

† Determined by analysis of many samples; used in computing heat release, burning rate, velocity, and other details of incinerator design.

‡ Type 2 or 3 where restaurant is in operation.

§ Diameter × height, in.

The foregoing calculation provides only for 100 per cent excess air. If 200 or 300 per cent excess air is required, multiply total combustion air (in foregoing calculation) by 1.5 or 2, respectively. Local ordinances throughout the U. S. require excess combustion air in amounts ranging from 100 to 300 per cent.

DOMESTIC GAS-FIRED INCINERATORS

ASA requirements[9] cover domestic gas-fired incinerators of the smokeless-odorless type; the 1957 edition of this Standard was the first to do so. Requirements for units for outdoor installations were prepared later. A comparable Underwriters' Laboratories standard is also available.[10]

A typical incinerator is shown in Fig. 12-97. Note the primary chamber for incineration and the secondary chamber for burning smoke and odorous constituents and collecting fly ash.

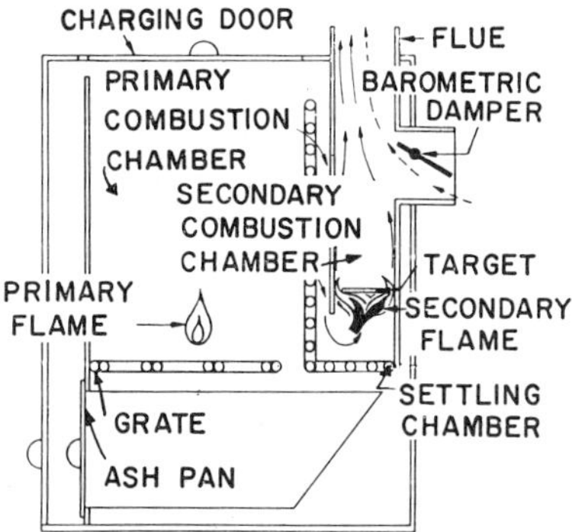

Fig. 12-97 Features of a typical smokeless-odorless domestic incinerator.[8]

Gas burner arrangements used include: (1) a burner firing horizontally thru a tube, part of the flames or heat issuing thru holes in this tube into the primary chamber but most of them continuing to the back chamber for burning smoke and odorous constituents; (2) a horizontal tube thru the primary chamber with no openings in it, all heating and ignition being by radiation thru the tube; this tube is bent upward in the secondary chamber, and a metal baffle above it diverts its heat and flame to mix with the smoke and odorous constituents for ignition and burning; (3) a burner across the back of the base of the unit, with part of the flames directed at an angle to the primary chamber for incineration, and the major portion directed upward into the secondary chamber for ignition and burning of smoke and odorous constituents; and (4) a power burner directing a flame at the top of the pile of garbage.

ASA Requirements

The 1962 incinerator approval requirements[9] include provisions as follows:

1. Automatic pilots, pilot filters, pressure regulators, timers, draft control, valve, and large ash receptacle are mandatory.

2. Gaseous combustion must be complete—not over 0.04 per cent CO in air-free sample.

3. A garbage test charge duplicates average household refuse.

4. Incinerating effectiveness is measured over a four-day operating period with only six ounces of uncarbonized material per bushel of capacity permitted at its end.

a. Smoke must not exceed No. 4 Bacharach filter paper test for more than ten consecutive minutes.

b. Fly ash must not exceed 0.3 grain per cu ft of flue gases corrected to 500 F and 50 per cent excess air.

c. Odor must be no greater than that of two sheets of burning newspaper. No smoke or odor may escape into test room.

5. During the fire hazard test, a charge of 1 lb of shredded paper per bushel is burned intermittently at 8 minute intervals. Wall and floor temperatures are limited to less than 90° F rise above room temperature; flue gas temperature must not exceed 1400 F; and smoke, odor, and fly ash produced during this test must meet the same limitations as in 4 above.

Installation[11,12]

Many cities have ordinances covering the installation of incinerators.

Chimneys. Unless the incinerator is installed outdoors, a masonry chimney* or an approved factory built chimney is required, built in accordance with recognized codes or standards. Type "B" vents are not acceptable.

In many localities in the South and West outdoor installations are made where the exhaust may be an exposed metal pipe (at least No. 20 U. S. gage).

The incinerator exhaust pipe should preferably enter the chimney below other vents sharing the chimney (Fig. 12-98) but above the extreme bottom of the chimney itself.

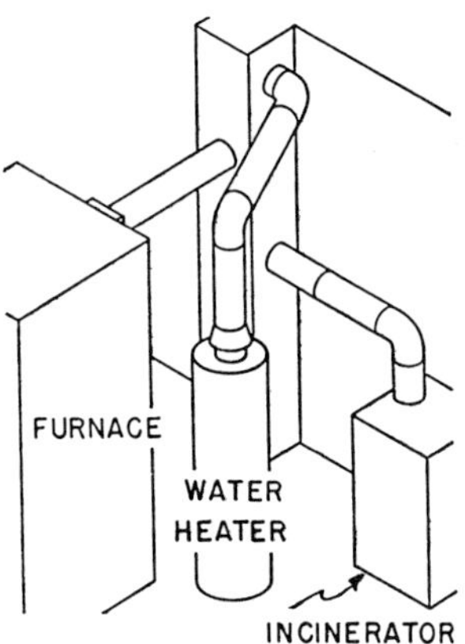

Fig. 12-98 Venting incinerator directly into chimney flue in combination with other gas appliances.[11] Alternatively, a common chimney side may be used; or a common vent connector with its intersection close to the chimney, and the included angle of its Y connector not exceeding 45°. Note that a draft hood is not used with an incinerator.

Draft may be controlled by a barometric damper, or, for greater control, a guillotine damper. In a test,[8] the variation in input rate from 28,000 to 48,000 Btu, interchange of standard A.G.A. garbage[9] with shredded paper, variation in static pressure from 0.0125 to 0.0300 in. w.c. and the combination of above did not affect the air flow thru a domestic incinerator.

* Tile linings or steel linings may be required.

If a ventilating fan or clothes dryer draws air from the incinerator room, such action must not create a downdraft in the incinerator (may pull flue gases out of the incinerator).

Manufacturer's installation and operating instructions should be followed.

Operation of Domestic Incinerators

With each incinerator the manufacturer supplies operating instructions, including air adjustment, pressure regulator setting, lighting of pilot, and operation of controls.

The incinerator is charged with Types 1 and 2 waste to a given level. The timer is turned on and set for the time necessary to burn the charge completely. This timer automatically turns the gas on, then turns it off after the time needed has elapsed. Most charges of one wastebasket full of refuse require from one to two hours' operation. A full incinerator load, including wet garbage, may require three to four hours.

Incinerator operation is improved by: (1) wrapping garbage in small packages, using several sheets of newspaper; (2) charging packages with accumulated paper and cardboard into incinerator daily; (3) not charging noncombustibles (tin cans, bottles, aluminum foil); (4) shaking ash grate daily and keeping primary chamber compartment reasonably clean; and (5) removing ashes and clearing the grate weekly or often enough to prevent build-up of ashes above ash receptacle sides or in primary chamber.

Ash Disposal. Analyses of the ashes remaining from domestic incinerator tests have not been reported to date, but they are undoubtedly similar to that of municipal incinerators,[13] as given in Table 12-77a. Much of the ash from domestic incinerators is light and easily blown around. Part of the material collected from dumping of grates or poking thru them is noncombustible and generally not desirable for mixing with garden or yard soil.

Flue Gas Constituents, Including Trace Effluents

Over 99.96 per cent of the volume of incinerator flue gases is composed of CO_2, water vapor, N_2, and O_2. Table 12-77b gives the trace constituents in flue gases from "new," "old," and research prototypes of domestic gas-fired models, as well as in flue gases from municipal incinerators, industrial

Table 12-77a Chemical Analysis of Municipal Incinerator Fly Ash[14]

	Collected	Emitted
Organic	3.3%	18.2%
Inorganic	96.7	81.8
	100.0%	100.0%
Components of inorganic portion		
Silicon as SiO_2	49.5%	36.3%
Aluminum as Al_2O_3	22.9	25.7
Iron as Fe_2O_3	6.3	7.1
Calcium as CaO	8.8	8.8
Magnesium as MgO	2.2	2.8
Sodium as Na_2O		
Potassium as K_2O	6.0	10.4
Titanium as TiO_2	1.3	0.9
Sulfur as SO_3	3.0	8.0
	100.0%	100.0%

Table 12-77b Trace Effluents in Stack Emissions from Domestic Gas-Fired, Municipal, and Commercial Incinerators, and Backyard Trash Burners, Compared with Gas- and Oil-Fired Heating Units and Automobile Exhaust[19]

Source	Aldehydes (as formaldehyde)		Nitrogen oxides (as nitrogen dioxide)		Organic acids (as acetic acid)		Ammonia		Hydrocarbons (as hexane)		Sulfur oxides (as sulfur dioxide)		Carbon monoxide, ppm	Particulate matter (including tarry organic materials), grains per cu ft at		Smoke compliance with ASA requirements[9]
	ppm	lb/ton	ppm	lb/ton	ppm	lb/ton	ppm	lb/ton	ppm	lb/ton	ppm	lb/ton		STP (actual)	500 F, 50% excess air	
Domestic gas-fired incinerators																
A. G. A. prototype, shredded paper	8–21	0.9–2.3	6–13	1.0–2.2	...	...	<5	...	...	...	...	...	100	0.017–0.019	0.023–0.028	Yes
A. G. A. prototype, ASA domestic wastes	8	0.8	15	2.1	7	1.8	<5	...	0.7	0.3	2	...	200–400	.005–.018	.006–.026	Yes
New mfrs.' units, shredded paper	4–67	0.7–15.9	2–7	0.3–2.6	...	...	<5	...	...	...	...	...	200–400	.013–.095	.030–.222	Yes
New mfrs.' units, ASA domestic wastes	25–40	...	2–5	...	...	...	<5	...	...	...	...	...	200–1000	.006–.012	.011–.026	Yes
Older units, shredded paper	24–48	0.3–0.4	5	0.6–0.8	...	...	5	...	...	...	...	...	...	.039–.132	.084–.282	No
Older units, ASA domestic wastes	5–30	5–6	1–3	0.6–2	17	6.6	5	...	4.7	2.5	...	...	...	.019–.097	.122–.526	No
Municipal incinerators																
Glendale, Calif. with scrubber	1–9	...	24–58	...	...	...	...	...	...	None	...	...	<1000	.035–.060	.12–.16	...
Glendale, Calif. without scrubber	1–22	...	58–92	...	...	...	...	...	...	None	...	...	<1000–3000	.128–.347	0.14–0.66	...
Other units																
Single chamber	...	0.03–2.7	...	3.9–4.6	...	2.0–3.9	...	0.33–0.5	...	Nil	...	1.4–2.3	...	0.75	...	...
Backyard (Battelle), paper & trimmings	760	29	<1.5	<0.1	...	...	65	1.8	...	...	...	...	5500–27,000	...	...	...
Backyard, 6 cu ft, paper	49	2.1	7	0.5	18	1.5	4	0.1	...	...	34	1.2	...	...	...	...
Industrial heating units*																
Large gas-fired	49	2	215	14	30	2.5	0.6	...	...	...	4	0.3	...	...	...	...
Large oil-fired	61	2.4	390	26	365	30	0.6	...	...	...	750	60	...	...	...	...
Automobile exhaust																
Accelerating	1369	41	4180	190	...	...	...	...	410	30	...	...	<100	...	...	...
Cruising	264	8	1606	75	...	...	...	...	354	28	...	...	4000	...	...	...

*See Tables 2-90 and 2-91 for additional data.

incinerators, gas- and oil-fired heating units, and in automobile exhaust gas.

Certified smokeless-odorless incinerators produce considerably smaller traces of aldehydes, organic acids, saturated hydrocarbons, and particulate matter than older domestic incinerators. Slightly more nitrogen oxides are produced, because of higher operating temperatures, but they are still extremely low. Compared with municipal incinerators, the new smokeless-odorless units have lower particulate emission, about the same aldehyde and nitrogen oxide emission, and slightly higher organic acid and hydrocarbon emission.

Economics of On-Site Domestic Incineration

In municipal refuse systems, collection cost and disposal cost constitute about 80 and 20 per cent, respectively, of system cost.[32] On-site gas incinerators, which reduce refuse to noncombustibles, effect a substantial decrease in the municipal collection cost, and are therefore desirable.

Generally, the refuse generated per capita and the cost of collection advance at a more rapid rate than population growth. Data for Saskatoon, Saskatchewan, Canada, illustrate these trends (base year is 1948):[33]

	Increase, per cent		
Year	Collection cost	Refuse generated per capita	Population
1953	40	26	21
1958	165	105	69

That city now has an effective, municipally subsidized on-site domestic incinerator program.

Other communities have ordinances calling for on-site incineration. For example, a community of 8000 in the Cleveland, Ohio area passed its ordinance in 1955. Refuse collection and disposal cost data attributable to this move follow.[34]

	Year		
	1961	1963	1965
Estimated cost without ordinance, $	17,000	32,000	37,000
Actual and forecasted cost with ordinance, $	9,000	16,000	18,000

Other cities have considered on-site incineration programs for one-family dwellings. A 1960 study in Milwaukee showed that refuse collection and disposal costs had increased 400 per cent in a 15-yr period and that $15 million could be saved directly over a 10-year period with an on-site incineration program.

COMMERCIAL AND INDUSTRIAL GAS-FIRED INCINERATORS

Table 12-72 summarizes the I.I.A. incinerator classification system. Incinerator manufacturers supply custom-built units to meet special needs and to comply with local codes. Frequently, if the waste contains a high percentage of dry combustibles, no auxiliary fuel (gas) burners are provided. Incinerators without these burners will emit smoke, odor, and fly ash, particularly during start-up and cooling down periods when the secondary chamber is below 1200 F.

Most commercial and industrial units consist of a primary and a secondary chamber (Fig. 12-99). The gas burner in the primary chamber burns the refuse. Power burners are recommended; the input of auxiliary fuel varies according to the type of waste incinerated, as shown below in a classification based on the I.I.A. Standard for inputs of 100,000 Btu or more:[2]

Type of waste:	2	3	4
Fuel input, Btu/lb:	1500	3000	8000

The secondary burner incinerates polynuclear hydrocarbons and combustible solids escaping the primary chamber.

Burning paper, wood, rubbish, tar paper, grass clippings, garbage, and other combustible refuse in a unit similar to that illustrated in Fig. 12-99 showed that a gas input of 180 MBtu per hr to the primary chamber and 720 MBtu per hr to the secondary chamber gave best operation. Preheating time to reach 1200 F in the secondary chamber was about 20 minutes. About 175 lb of municipal refuse per hr or 400 lb of paper and cardboard per hr can be incinerated without smoke, odor, or fly ash. The barometric damper was set to maintain about 0.05 in. updraft in the primary and secondary chambers. The settling chamber collected 4.95 grains of fly ash per lb of charge or 0.033 grain per cu ft of flue gas. Gas consumption varied from 4.6 cu ft to 3.24 cu ft of gas per lb of refuse incinerated for 400 to 2400 lb per day charges.

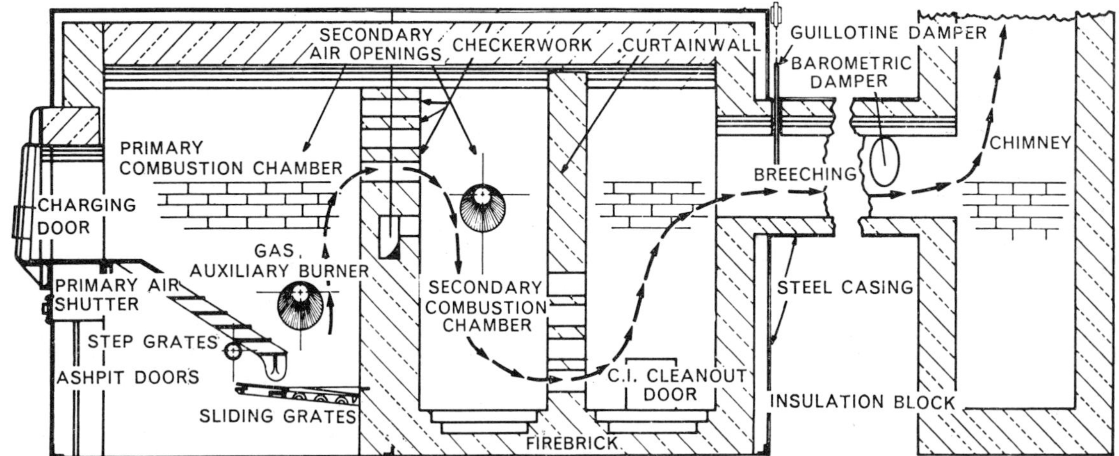

Fig. 12-99 Commercial gas incinerator nomenclature.[15]

There appears to be a lack of agreement among the various local requirements for incinerator test standards, procedures, and instrumentation. These nonuniformities were studied[15] for Class III incinerators in terms of the New York City, Los Angeles County, and I.I.A. requirements. Conclusions drawn follow:

1. The nature of the refuse and the operating procedure will affect incinerator performance so materially that a standard test charge, procedure, and instrumentation are called for.

2. There is considerable latitude in incinerator design to meet a specified performance.

3. The I.I.A. design standards will safely meet the limitations required by most American cities to ensure proper performance.

ASA Requirements (Commercial)

Requirements proposed[16] (1964) for gas-fired Class IA incinerators (Table 12-72) include the following major provisions:

1. Manufacturer should specify the static updraft and the location at which it is to be measured, and provide means to facilitate this measurement.

2. Incinerator must be connected to a chimney capable of withstanding temperatures of 2000 F. Manufacturers are permitted to specify necessary clearance to combustible construction and to mark approved clearances on incinerator.

3. Main burners shall be equipped with pilot burner(s) and a safety shutoff device or devices. In the event of a flame or ignition failure, complete shutoff of gas must be provided.

4. Requirements also cover fly ash and particulate matter.

Installation

Clearances of 36 in. at sides, 48 in. at top, and 8 ft in front of incinerator are recommended.[17] Incineration room should be provided with enough fresh air. The air, in cfm, required *divided by* 1000 will give the area of inlet duct in square feet.[15] Surface temperature of adjacent combustible structure in industrial installations[18] using gas at above 0.5 psig shall be below 160 F. Temperature of unprotected combustibles on installations operating at 0.5 psig or lower shall not be more than 90° F *above ambient room temperature.*[19] Structural members of a building shall not pass thru an incinerator with a temperature in excess of 500 F and they shall not be part of an incinerator unless protected against a deterioration in strength, expansion, and ignition.[18] Minimum clearance to combustible construction is specified for Class III incinerators.[16]

Venting. Chimneys 12 in. diam or less shall be made of No. 16 U. S. gage steel lined with 2.5 in. fire brick; those larger than 12 in. diam shall be made of 12 gage steel lined with 4.5 in. firebrick clay—reinforced concrete shall also be lined with firebrick.[20] A chimney shall extend 4 ft above the highest point in a sloping roof, 8 ft above flat roofs and 2 ft above the highest point of the building within 20 ft. Local ordinances should also be consulted.

Suitable factory-built chimneys designed for high-temperature flues are also recommended. Automatic draft control is recommended to control combustion. Use of automatic controls to control draft, burner operation, flue gas temperature, firing time, and scrubbing of flue gas is increasing.

Automatic Systems. A typical incinerator automated to eliminate all human handling of refuse and ashes in a 35 story apartment house is described below.[21] Another system is shown in Fig. 12-102(a).

Charging. Waste material is deposited into the three charging hoppers located at every floor level. The material then drops thru stainless steel chutes. At the base of each chute is the first of a series of pneumatically operated gates and devices to control, measure, and regulate the flow of materials to the incinerator. When the waste builds up to a predetermined level, a photocell puts the charging cycle into operation:

1. The first gate opens, permits the load to drop into a measuring bin, and closes.

2. The second gate opens, permits the load to drop into a ram pit, and closes.

3. The guillotine-type charging door of the incinerator opens.

4. The ram extends, pushes the charge into the combustion chamber, and retracts.

5. The incinerator door closes and the burning cycle starts.

The charging cycle is protected against jamming or malfunction by a fully instrumented electropneumatic system. The sequence is such that if a jam or malfunction should occur, the system will repeatedly attempt to clear itself while actuating an audible and visible alarm. The system or any part can also be easily and readily energized by a series of manual control switches.

Firing. The burning cycle is also automated to the point of being self-initiating in the event of heat build-up, as would occur from smoldering waste or accidental ignition. This cycle is tied into the same control system used for the charging cycle, which provides thermostatically controlled induced draft and flue gas scrubbing and maintains the proper operating temperatures throughout. Temperatures are maintained by two 450 MBtu per hr power gas burners on each of the three incinerators.

The flue gas scrubbers use both the impingement and water spray methods; the water is recirculated for economy.

Residue Removal. A conveyor system automatically removes the indestructible solids from the incinerator and deposits them *directly* into trucks at curbside.

Performance

With proper air–fuel ratio, sufficient mixing of reactants, adequate residence time, high temperatures (1400 to 1800 F), and low velocities, performance will be satisfactory. Stack temperatures not in excess of 1800 F for 15 minutes or above 2000 F at any time are recommended.[16] The burning rate is determined either by loading the incinerator to its full capacity or by loading a fixed percentage at regular intervals during the burning period. Shaking of grates is not recommended during the burning period. Local particulate and ash emission requirements vary considerably; however, 0.85 lb of particulate matter per 1000 lb of flue gases is a widely accepted figure. Smoke, odor, and fly ash at start and shutdown are eliminated in Alleghany County by requiring automatic control. A timing device lights the burner at the base of the chimney 15 minutes prior to the lighting of the ignition burner and permits it to remain lighted 15 minutes after the ignition burner shuts off.

Dust Collectors. Particles 10 microns and over are generally classified as dust; between 0.1 and 10 microns, clouds; between 0 001 and 0.1 micron, smoke. The terminal velocities of settling particles are given as follows:

2000 micron particle or above: $v = k_1\sqrt{sd}$
1000–100 micron particle: $v = k_2\sqrt{s^{2/3}d}$
50–2 micron particle: $v = k_3\sqrt{sd^2}$

where: v = terminal velocity, ft per min
d = particle diam, microns
s = specific gravity of gas

For particles falling in air, constants k are listed below:

For irregularly shaped particles, $k_1 = 28$, $k_2 = 0.51$, and $k_3 = 0.0039$.
For spheres, $k_1 = 48$, $k_2 = 0.81$, and $k_3 = 0.0059$.

Since the terminal particle velocity is limited, dynamic particle collectors are more compact than gravity dust collectors. In a dynamic dust collector, the direction of the gas stream is changed rapidly. The dust particle momentum carries the particles to the baffle, where they are collected. Also see Figs. 1–6a and 1–6b.

Sizing and Operation

For a general guide in sizing commercial incinerators, consult data supplied by Table 12-72 which establish various classes of incinerators according to type of waste and burner capacity.

Special Applications

Grease Incinerator. Figure 12-100 shows a prototype of a commercial kitchen grease incinerator which can dispose of grease and oil from kitchen exhaust. Effective grease vapor removal is necessary to avoid the deposit of grease particles in exhaust ducts, a fire hazard as well as a frequent cleaning problem. When hot baffles are used to incinerate grease, excess air should be limited, to avoid a cooling effect.

Incineration of Fumes. A multifaceted commercial and industrial application of incinerators is the disposal of noisome or toxic fumes and air-pollutant by-products of industrial processes. Equipment is usually custom designed to perform specific tasks and to meet the requirements of pertinent local laws and ordinances. Successful efforts to eliminate objectionable industrial odors,[23,24] halogenated hydrocarbon wastes,[25] and cyanide wastes[26] have been reported.

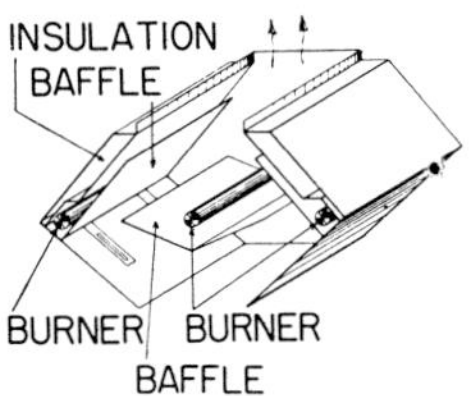

Fig. 12-100 Commercial kitchen grease incinerator (three burner impingement type).[22]

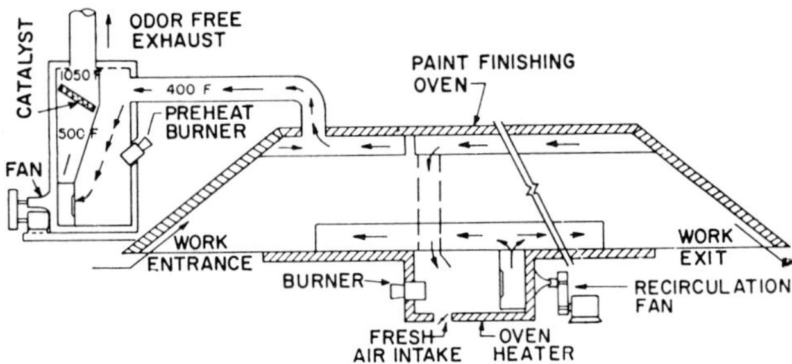

Fig. 12-101 Catalyst system independent of oven heater for air pollution control.[27]

A widely used method of correcting and controlling organic and combustible types of air pollution emission is the application of *catalytic combustion* systems. Figure 12-101 shows a catalyst system independent of oven heater for air pollution control.

After burner sizing, based on 15–30 ft per sec effluent air velocity (at the after burner exit temperature of 1100–1500 F) and a residence time of 0.25–0.5 sec, has been found to give good results.[28] Length of combustion chamber equals effluent velocity *times* residence time. Length of combustion chamber for unusual vapors can be calculated by selecting suitable flue temperature and residence at required effluent velocity.

Incinerating Toilets. The Destroilet disposes of liquids and solids by way of a gas power burner (25 MBtu per hr). It measured 15 × 18 × 24 in. and weighs about 80 lb. This unit is capable of disposing of ¼ lb of solid matter in 16 min and 4 oz of liquid in about 6 min; the resultant negligible quantity of ash is inert. Approval was granted under domestic incinerator requirements[29] for installation in other than public facilities.

Crematory. Gas consumption varies from 1000 to 4500 cu ft per cremation, depending on design and continuous or noncontinuous operation. Cremation time varies from 1.5 to 4 hours per cremation.

Miscellaneous Applications. Incinerators are used in *hospitals* for burning dressings, amputations, and experimental animals. In *banks, offices,* and *warehouses,* incinerators eliminate pilferage thru garbage and burn confidential papers. Gas incinerators burn auto bodies and control air pollution. In manufacturing plants and junk yards valuable metallic parts are separated after combustibles are burned.

FLUE-FED INCINERATORS

Recent designs use separate flues for charging and venting. Special charging arrangements are used in some incineration systems—Fig. 12-102(a); these are *not* termed "flue-fed" incinerators. In some schemes, Fig. 12-102(b), the same stack or flue is used to deliver refuse to the combustion chamber and for venting. Charging doors are generally installed at each floor. Unless an auxiliary fuel is used to burn smoke and odorous components, a flue-fed unit usually produces smoke, odor, and fly ash. Fig. 12-103 shows a typical gas burner for an incinerator; note the automatic spark ignition system that uses a mercury-filled capillary to sense the pilot (20–30 sec response time).

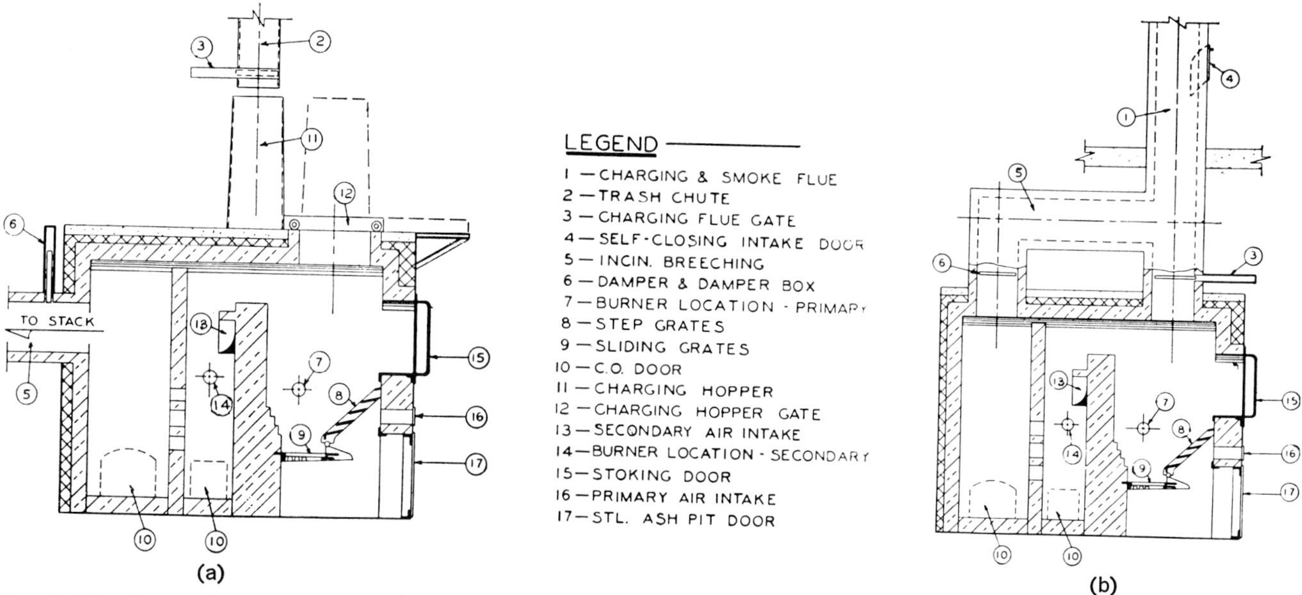

Fig. 12-102 Multichamber incineration systems for high-rise buildings:[29] (a) Chut-O-Matic automatic hopper-fed design; (b) single flue with by-pass design (Joseph Goder, Inc.)

When the charge is ignited, combustion takes place at the surface, where air from under the grates and thru air ports is readily available. The generated heat dehydrates the charge and drives out volatiles, some of which may escape unburned if the temperature in the chamber is not high enough to ignite them. Severe combustion with high temperature may inspirate more air than needed, causing low-temperature combustion with destructive distillation. A draft regulator may be used. Damper may be sized approximately 40 per cent greater than the theoretical size.[30] A damper which opens at 0.2 in. w.c. was reported satisfactory for multiple-chamber operation.

An analysis of design attitudes for flue-fed incinerators[31] showed that power burners were preferred. Seventy-one per cent of the air pollution control officials contacted were for the use of flue-fed incinerators in schools, hospitals, hotels, and other commercial places.

Settling Chambers. A luminous flame gas burner may be installed in a settling chamber on the roof to prevent smoke and odor emission.[7] A luminous flame burner is used to get the proper flame shape and size, ensuring that all gases rising from the refuse material will pass thru the flame and be ignited. A turboblower supplies combustion air. This avoids hot first-floor hoppers and controls smoke even in case of flue stoppage or burning in the flue. The burner is adjustable for various capacity ratings and is equipped with approved flame safety devices, air and gas failure switches, and automatic gas–electric ignition.

Specifications. The following summarizes the requirements of a major municipal housing authority: (1) powered gas burner installed in combustion chamber; 155 MBtu per hr per 100 rooms of 100 people in the building; minimum turndown ratio, 5 to 1; electronic flame failure protection; and (2) cycle control; seven day program timer for adjustment at 15 min intervals. Cycling and safety features include locking (and proving) hopper doors, operating flue gate, sensing and monitoring draft, controlling air blower, igniting gas, shutdown of burner and blower if fire door is opened, tolerance of 5 to 8 sec draft interruptions, and combustion chamber temperature limited to 1600 F.

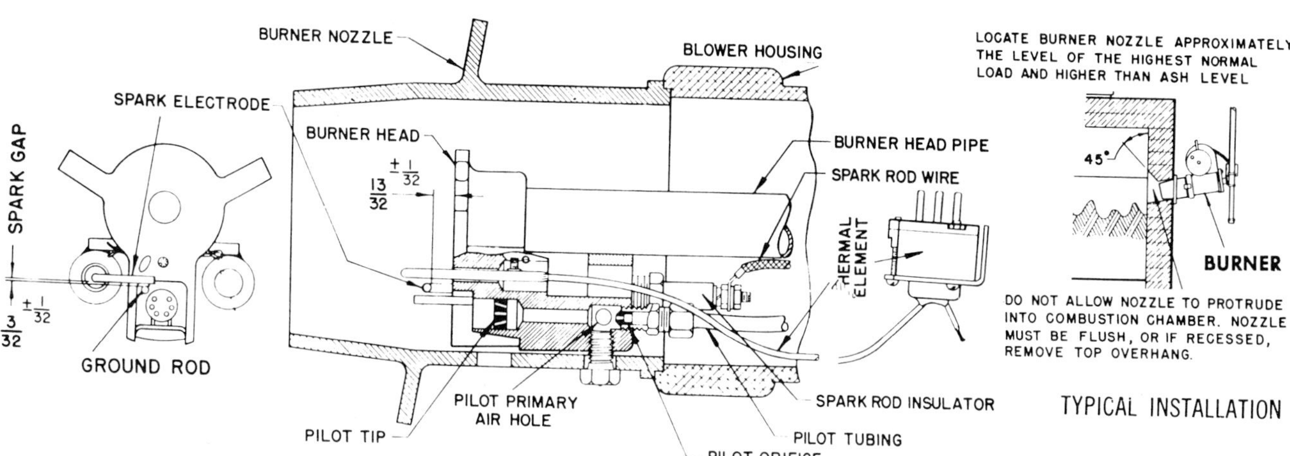

Fig. 12-103 Automatic spark-ignited gas burner for commercial and industrial incinerators. The pilot sensor is connected to a remote diaphragm-type snap switch which operates the main gas valve. (Mid-Continent Metal Products Co.)

REFERENCES

1. Incinerator Institute of America. *I. I. A. Incinerator Standards.* New York, 1963.
2. Am. Public Works Assoc. *Municipal Refuse Disposal.* Chicago, Public Administration Service, 1961.
3. Skipworth, D. W., and others. *Design of Domestic Gas-Fired Incinerators for Elimination of Smoke, Odors, and Fly Ash.* (Res. Bull. 78) Cleveland, Ohio, A. G. A. Laboratories, 1958.
4. U. S. Public Health Service. *Digest of State Air Pollution Laws.* Washington, D. C., G.P.O., 1963.
5. Salzenstein, M. A. "Review of Air Pollution Legislation." *Actual Specifying Eng.* 10: 78+, Oct. 1963.
6. Marble, G. E. *Case Histories of Commercial Incinerator Installations.* (A. G. A. Sales Conference on Industrial and Commercial Gas) New York, A. G. A., 1954.
7. Reed, R. J. and Truitt, S. M. "Selecting Incinerator Smoke and Odor Burners." *Air Repair* 4: No. 3, Nov. 1954.
8. Houry, E. and Kain, H. W. *Principles of Design of Smokeless-Odorless Incinerators for Maximum Performance.* (Res. Bull. 93) Cleveland, Ohio, A. G. A. Laboratories, 1962.
9. A. G. A. Laboratories. *American Standard Approval Requirements for Domestic Gas-Fired Incinerators.* (ASA Z21.6–1962 and Z21.6a-1963) Cleveland, Ohio, 1962 and 1963.
10. *Standards for Safety, Domestic-Type Incinerators.* (Underwriters' Lab. UL Bull. Res. 791). Chicago, 1963.
11. GAMA. *Recommended Practices for the Installation of the Domestic Gas-Fired Incinerator,* 2nd ed. New York, 1959.
12. East Ohio Gas Co. *Recommended Practices for Installation of Domestic Gas-Fired Incinerators.* Cleveland, Ohio, 1955.
13. *Municipal Incineration* (Univ. Calif. Sanitary Eng. Res. Lab. Tech. Bull. 5), p. 17, 56, and 71. Richmond, Calif., 1951.
14. Kaiser, E. R. "Refuse Composition and Flue-Gas Analysis from Municipal Incinerators." *Natl. Incinerator Conf., (ASME) Proc.* 1964.
15. Goder, R. "Incineration and Air Pollution Standards." *Actual Specifying Eng.* 10: 83+, Sept. 1963.
16. A. G. A. Laboratories. *Proposed American Standard Approval Requirements for Commercial Gas-Fired Incinerators.* Cleveland, Ohio, 1963–4.
17. Natl. Fire Protection Assoc. *Standard on Incinerators.* (NFPA 82) Boston, 1960; *Standards for Rubbish Handling and Incinerators.* (NFPA 82A) Boston, 1948.
18. A. G. A. Laboratories. *Installation of Gas Piping and Gas Equipment on Industrial and Other Non-Residential Premises Not Covered by ASA Z21.30.* Cleveland, Ohio, 1964. (Proposed) (ASA Z83.1).
19. Hein, G. M. and Engdahl, R. B. *A Study of Effluents from Domestic Gas-Fired Incinerators.* New York, A. G. A. 1959.
20. Natl. Fire Protection Assoc. *Standard for Chimneys, Flues and Vents.* (NFPA 211) Boston, 1961.
21. "Hi-Rise Gas Incinerators Have Brain." *A. G. A. Monthly* 46: 16, April 1964.
22. Marn, W. L. *A Study of Methods of Grease Removal from Commercial Kitchen Exhaust Air.* (Res. Rept. 1323) Cleveland, Ohio, A. G. A. Laboratories, 1960.
23. Baggs, J. W. "Objectionable Roofing Plant Odors Eliminated." *Ind. Gas* 28: 7+, Feb. 1950.
24. Helme, W. "Hall Printing Co. Solves the Smoke Problem." *Ind. Gas* 26: 5+, Jan. 1948.
25. Haggin, J. H. S. "Disposing of Halogenated Hydrocarbon Waste." *Ind. Eng. Chem.* 56: 10+, Jan. 1964.
26. "New Way to Incinerate Cyanide Wastes." *Air Eng.* 4: 36+, Feb. 1962.
27. Natl. Fire Protection Assoc. *Standard for Ovens and Furnaces.* (NFPA 86A) Boston, 1963.
28. Krenz, W. B., and others. "Control of Solvent Losses in Los Angeles County." (Paper 57–43) *Air Pollution Control Assoc. Proc.* 1957.
29. Goder, R. "Various Types of Incinerators for High Rises." *Actual Specifying Eng.* 11: 80–3, June 1964.
30. MacKnight, R. J., and others. "Controlling Flue-Fed Incinerator." *Air Pollution Control Assoc. J.* 10: 103+, April 1960.
31. Sterling, M. "Attitudes on Design of Flue-Fed Incinerators." *Air Pollution Control Assoc. J.* 10: 110+, April 1960.
32. Sarsfield, J. R. *The Residential Incinerator—Why Bother.* (A. G. A. Res. and Util. Conf., 1962) Cleveland, Ohio, A. G. A., 1962.
33. *A Case for "On-the-Site" Disposal of Household Wastes.* Regina, Saskatchewan, Saskatchewan Power Corporation, n.d.
34. East Ohio Gas Co. *Residential Incineration and Its Benefit to Your Community.* Cleveland, Ohio, 1964.

Chapter 11
Gas Illumination

by F. E. Vandaveer

Starting about 1957, gas lamps have again come into rapidly increasing use for outdoor lighting. Principal residential lighting applications include yards, driveways, patios, porches, play areas, and swimming pools. Among commercial applications are streets, shopping centers, air fields, hotels, and restaurants. Signs, incorporating gaslight burners, are often used in these applications. Most gas lamps employ incandescent mantles of the upright type originally developed by Welsbach for street light use (for further information, see Ref. 1); inverted mantles are also used to some degree. The number of outdoor gas lamps in use in January 1964 totaled about 895,000 of which approximately 9 per cent were nonresidential installations. These lamps were produced by some 20 companies, a few of them making indoor lamps as well.

LIGHT UNITS AND STANDARDS

The **international candle,** a unit of luminous intensity, is the light emitted by 5.0 sq mm of platinum at the temperature of solidification.

The **standard candle,** a unit of luminous intensity, is the light emitted by a ⅞ in. sperm candle burning at the rate of 120 grains per hr.

Candlela, formerly candle, is the intensity of light given out by a standard candle, i.e., the unit of luminous intensity in a given direction.

The **lumen** is a unit of luminous flux. One lumen is intercepted by a one square foot surface, all points of which are one foot from a uniform point source of one international candle; i.e., one candlepower source delivers 4π lumens to a surface enclosing it. One lumen at the wave length of maximum visibility (0.556 μ) equals 0.00161 watts. However, the number of lumens actually produced by a watt is considerably less. For example, a 60 w incandescent bulb is initially rated to produce about 14 lumens per watt.

A **foot-candle,** the direct illumination on a surface everywhere one foot from a uniform point source of one international candle, equals one **lumen** per square foot. Illumination levels are usually specified in terms of foot-candles.

Illuminating Efficiency of Gaslights. Units incorporating a mantle shall produce an average light output of *at least* one candlepower for each 65 Btu per hr of input rating.[2] The candlepower equals the illumination intensity measured in foot-candles, *FC*, *times* the *square* of the distance in feet between the light source and the light meter.

Example. A corrected reading of 2.0 foot-candles would be measured at a point 5.0 ft from an *ideal* 50 candlepower source; $4\pi \times 50$ candlepower = 628 lumens would be the resulting ideal total light output. To obtain 50 candlepower and to satisfy the foregoing efficiency requirement, the *maximum* gas input would be 50 candlepower × 65 Btu per hr-candlepower = 3250 Btu per hr (compare with actual efficiencies in last column of Table 12-78).

Table 12-79a shows efficiencies of illuminating devices; Table 12-78 gives performance data for three mantle arrange-

Table 12-78 Light Intensity, Output, and Efficiencies of Various Mantle Arrangements[3]

Arrangement and type of mantle(s)	Operated at: Total input, Btu/hr (1)	Operated at: Optimum primary aeration (2)	Overall light intensity, mean spherical candlepower (3)	Total light output, lumens (4)*	Lighting efficiency, lumens per MBtu/hr (5)†	Lighting efficiency, lumens per Watt (6)‡	Illuminating efficiency, candlepower per 65 Btu/hr** (7)§
One A†† upright	2200	74%	34.50	434	197	0.673	1.0
Two No. 222‡‡ inverted	2200	70%	67.75	851	387	1.320	2.0
Three No. 222‡‡ inverted	3300	70%	86.90	1090	330	1.130	1.7

* (4) = 4π × (3).
† (5) = (4) × 1000/(1).
‡ (6) = (4) × 3.413/(1).
§ (7) = (3) × 65/(1).
** 1.0 is minimum, per ASA Z21.42-1963.
†† Has a coated single wire top support and a single stitch weave.
‡‡ Representative of most inverted mantles.

Table 12-79a Efficiencies of Various Illuminating Devices
(see Table 12-78 for mantles)

Lamp	Rating or specification	Lumens/watt input
Acetylene	1.0 liter/hr	0.67
Arc, electric high intensity	150 amp bare arc	18.5
Gas burner, open flame	Bray high pressure	0.22
Tungsten filament, 120 v, gas filled, frosted	40 watt (100 hr life)	11.7
	60 watt (1000 hr life)	13.9
	100 watt (750 hr life)	16.2
Fluorescent lamp, standard	20 watt daylight	38.0
	40 watt daylight	45.0

ments. Note that the light output of a gas lamp with a type A upright mantle (434 lumens, from Table 12-78) is about the same as that of a 40 w incandescent bulb (11.7 lumens per watt, from Table 12-79a, *times* 40 watts = 468 lumens). Similarly, the arrangement of two No. 222 inverted mantles has about the same light output as a 60 w bulb. Figure 12-104 gives the vertical light intensities for a single upright mantle and a two-cluster inverted mantle. In the span of normal operating temperatures, light output varies as the tenth power of mantle brightness temperature, °F.

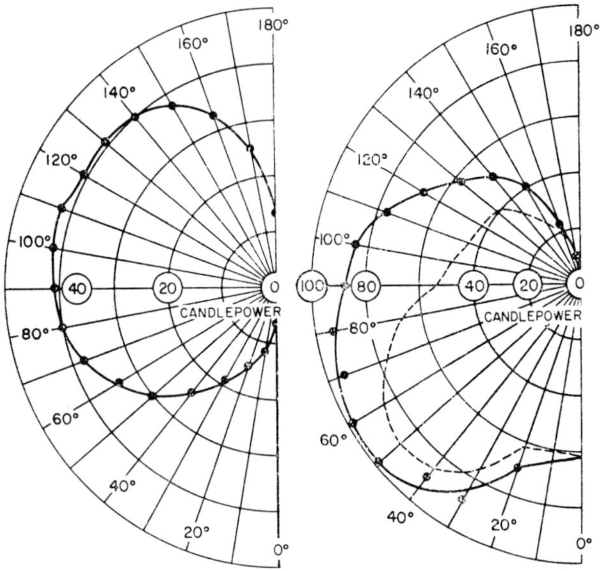

Fig. 12-104 Vertical light intensities,* in candlepower, for mantles operated with optimum adjustment at total gas inputs of 2200 Btu per hr: (left) a single type A upright mantle—view unobstructed by mantle support bracket; (right) a two-cluster No. 222 inverted mantle (solid line—both mantles foremost, dotted line—one mantle foremost.)[3] Table 12-78 (column 4) gives integration of these data; no significance should be attributed to the relative areas enclosed by curves in polar coordinates.

* Note difference in scale; 180 deg shown—the other 180 deg are symmetrical.

Gaslight Standards[2]

First issued in 1963, these cover the construction and performance of gas-fired illuminating appliances intended for either outdoor or indoor (including mobile homes) installation for attachment to fixed gas piping systems. Open flame illuminating appliances with inputs above 5000 Btu per hr are not covered. Requirements included cover flashback, combustion performance, illuminating efficiency (see last column of Table 12-78), and the temperatures of walls and ceilings adjacent to units (not greater than 90 F above room temperature). Furthermore, units designed for outdoor installation must be protected against insect entry, rainstorms, and winds (ignition at 10 mph, not extinguished at 40 mph). Pressure regulators and automatic valves may be incorporated.

LAMPS WITH INCANDESCENT MANTLES

These units use primary aerated burners. Typical upright mantle-type gas lamps are shown in Fig. 12-105; inverted mantle arrangements are shown in Fig. 12-106a. Gas requirements for various mantle–light combinations are shown in Table 12-79b.

The largest gaslight in the world (1960) has a 13,000 candlepower output produced by 78 mantles receiving 148 cfh of gas.[4]

Table 12-79b Gas Burning Capacities of Various Gas Lamps*
(cubic feet per hour)

Lamp name	Mantle	Natural	Mfd.	Propane
Cabildo, Flair, Heritage, Watchman	Upright No. 813	2.6	7.0	0.65
Doorman, Sentry	Upright No. 813S	1.9	5.4	.47
Welsbach street inverted burner	Inverted No. 844	1.0	2.0	0.33

* Contributed by Francis Werring, The Welsbach Corp., Philadelphia, Pa.

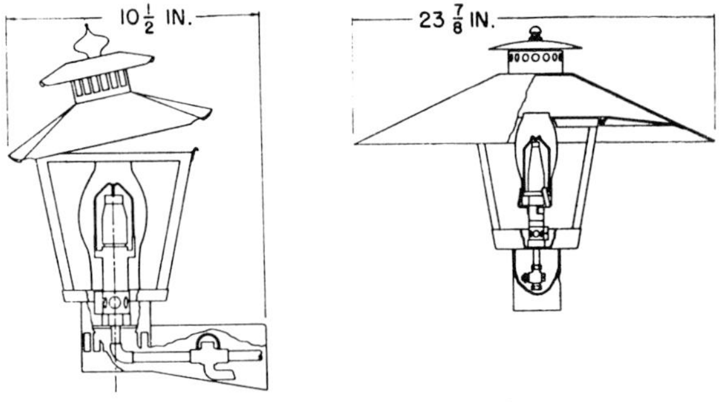

Fig. 12-105 Two upright mantle-type gaslights.

Automatic Controls. To turn gaslights on or off, clocks (Fig. 12-106a) or photocells may be used. Clocks for street lighting may be of the spring-wound, eight-day, 24-hour dial type. Some clocks have self-adjusting astronomical dials. Usual schedules are to turn gaslights on one-half hour after sundown and off one-half hour before sunrise. A photocell, actuated by daylight, may be used to turn gaslights on or off.

One scheme, developed by the Boston Gas Company, eliminates the need for an outside source of electric power to operate a gaslight control valve. A buried battery, consisting of a magnesium anode and an iron cathode, energizes a normally open gas valve. (This battery is similar to arrangements used extensively to provide protective current to steel pipelines, thereby reducing corrosion.) A light-sensitive cell and transistor, installed in series with the gas valve, causes the gas valve to close during the daytime. The results indicate that this arrangement produces an economical automatic control unit with a life expectancy of more than nine years before battery renewal. The battery is buried at the base of the gaslight post.

Street Lighting. For all-night service, about two to three mantles per lamp per year are required; for electric lights, about four bulbs per year are needed. Prices in 1962 were about $1.00 for the large upright mantles and 60 cents for the inverted mantles.

Table 12-80 gives illumination recommendations for various roadway classifications. These data assume compliance with ASA recommendations on light distribution, mounting height, spacing, and size of lamp.[5]

Airport Runway Lights. Gaslights have been used at many airports.[6] Installation suggestions made by the Arkla Air Conditioning Co. include: (1) mantle height about 21 in. from the ground; (2) flexible construction of assembly to minimize damage to colliding aircraft—e.g., lamp post not embedded in concrete; about four feet of 3/8 in. copper tubing coiled in cone under light; (3) welding of a 3/8 in. pipe-to-flame union adapter concentric to a 1/8 in. drilled hole in gas main at each light—this additional orifice limits gas flow to atmosphere in the event of a break in the copper tubing to a light; (4) use of a photoelectric cell to close a solenoid valve which controls gas flow to a string of lights, a by-pass supplying lights with a minimum flame; (5) separate shutoff valve and regulator under each cone housing; (6) consultation of both the U. S. Dept. of Commerce and the FAA for information on airport lighting.

Floodlights. Incandescent gaslights with outputs of up to 10,000 candlepower are available for "all-weather" use; these employ LP-gas at 14 psig.

Mantle Construction. A fabric such as rayon, silk, cotton ramie, or viscose is woven into fabric tubing of the desired stitch and impregnated with a solution of thorium nitrate, cerium nitrate, beryllium nitrate, aluminum nitrate, and magnesium nitrate. Exact formulas used are not generally disclosed, but one large maker uses 99.25 per cent thorium nitrate and 0.75 per cent cerium nitrate. After impregnation, the knitted tubing is cut into short lengths and attached to individual rings or mountings.

Soft mantles are sold unburned and must be shaped on the burner, a familiar task for those who own gasoline lanterns. The Welsbach Corp. uses soft mantles on street light inverted burners only in two cities in Massachusetts.

Hard mantles are preshaped and preshrunk during the burning operation of their manufacture, which burns out the fabric, the remaining ash consisting of oxides of the impregnating metals. A collodion coating is then applied to strengthen the burned mantle for handling and shipping. Because of the shrinkage in burning, almost two-thirds of original size, upright mantles must be of the hard variety.

Mantle Performance

Incandescence of a gas mantle is *mainly* dependent on an exacting cerium content, and on flame temperature and flame velocity. The type of gas, injection pressure, and burner design may contribute to the candlepower obtained. There is no known change in the chemical content of contemporary mantles which would improve their lighting efficiency. The weave of the mantle is well established for each type of gas and is not a factor in mantle life. The single stitch weave gives the same light output as the double stitch; however, light from a double stitch mantle may appear more yellow to the eye. Experiments at Armour Research Institute with heavier ceramic mantles did not prove them to be as efficient in light qualities as the traditional mantle.

Primary Air Adjustments. Each gas input rate has an optimum per cent primary air for maximum light output[3] (single type A upright mantle):

Candlepower at 3 ft	45	34	27	16	12
Gas input, MBtu/hr	2.6	2.2	1.6	1.5	1.3
Primary air, %	78	74	68	55	45

Port Loadings. For 1962 designs:[3]

Mantle type	Upright	Inverted
MBtu per hr-sq in. of open port area	4.8–5.5	9.2–14.4
Corresponding primary air needed, %	70–74	62–70
Press. at D.M.S. 68 to 72 orifices, in. w.c.	3–5	5–8

Table 12-80 Recommendations for Roadway Illumination*,[5]

(lumens per square foot)

Roadways (other than Expressways or Freeways)			
Roadway classification	Area classification: Downtown	Intermediate	Outlying and rural
Major	2.0	1.2	0.9
Collector	1.2	0.9	0.6
Local or minor	0.9	0.6	0.2†

Expressways and Freeways‡	
Classification	Expressways
Continuous urban	1.4§
Continuous rural	1.0
Interchange urban	2.0
Interchange rural	1.4

* The average horizontal foot-candles recommended represent average illumination on the roadway pavement, straight and level roadway areas, and areas having minor curves and grades when the illuminating source is at its lowest output and when the luminaire is in its dirtiest condition. Intersecting, converging, and diverging roadway areas at grade require higher illumination. The lowest foot-candle value at any point on the pavement should not be less than one-third the average value. The only exception to this requirement applies to residential roadways, where the lowest foot-candle value at any point may be as low as one-sixth the average value.

† Residential.

‡ The value of 0.6 foot-candle for freeways is being continued from the 1953 **American Standard Practice for Street and Highway Lighting** pending the results of research currently being carried on under the supervision of the Highway Research Board.

§ 2.0 foot-candles in downtown area.

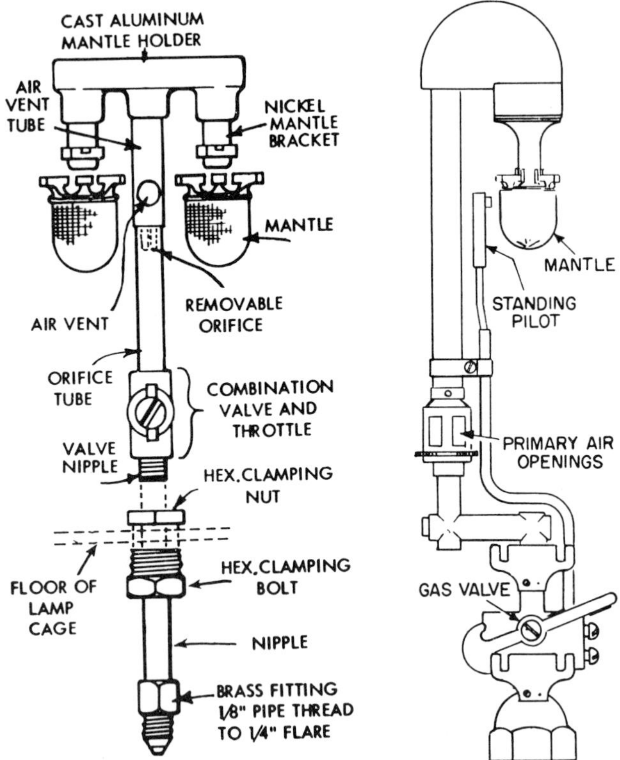

Fig. 12-106a Inverted mantle arrangements: (left) details—Charmglow Products; (right) clock-operated unit—gas valve accommodates a clock mechanism.

Inverted burners (Fig. 12-106a) are generally more efficient than **upright burners** (Fig. 12-106b) because of superheating of the gas–air mixture in the burner head. Also, because the flame burns back on itself, more heat is concentrated on the mantle fabric, and thus more light is evolved. On the other hand, the combustion quality of units with upright mantles is superior to that of units using inverted mantles.[3]

A properly designed clear glass chimney (about 3 in. diam) surrounding an upright mantle (not used with inverted mantles) will aid in directing the hot gases upward along the mantle surface, thereby increasing light output thru the chimney as much as 35 per cent above that of a bare mantle

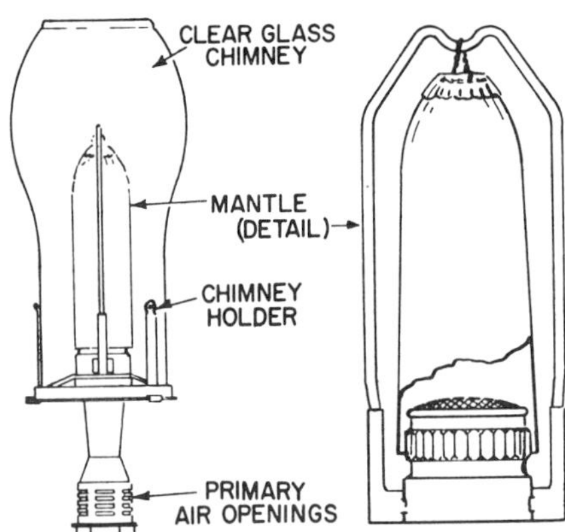

Fig. 12-106b Welsbach No. 813 burner for manually operated lights: (left) assembly; (right) mantle details, half size; wire screen base acts as burner ports.

with no chimney. Clear glass globes are used because the soft mellow light of the gas mantle needs no further diffusion. White opal domes may be used to produce a low-intensity upward light to dispel darkness further and to afford a police lighting of house fronts. An eight to nine foot mounting height is ideal for tree-lined streets.

Fogging. This refers to the deposit of a milky gray-white film on gaslight components. These deposits, which should not be confused with the momentary water vapor condensation immediately following gas lamp ignition, give chimneys and other glass components a frosted appearance. The most effective way of controlling fog deposits on gaslight glass components is to minimize the amount of impingement of combustion products onto the glass. Reference 3 covers lamp design factors directed at reducing both fog particle deposits and susceptibility of units to wind effects. In these regards, cylindrical chimneys were found to be superior to traditional globe designs.

Upright Mantle Life.[7] A 1962 study which measured the resistance of mantles to natural gas flame exposure and mechanical shock showed that the average service life of mantles—four months to a year—may be increased many times over without modifying the lamps now in service. The improvement was achieved by substituting a molded ceramic loop for the metal suspension wire used in most upright mantles, and by better manufacturing control of mantle fabric sintering.

Because a mantle is composed of ceramic threads of low mass and fine diameter, it is extremely resistant to **thermal shock.** In some commercial sign applications, the fuel input has been rapidly cycled from high to low millions of times without apparent detrimental effect to the mantle fabric. On the other hand, ceramic threads are delicate and can be torn readily during installation and lighting.

Outdoor Gas Lamp Installation

The usual practice is to connect the gas supply to an outdoor lamp from a point downstream from the customer's meter. In a number of special situations, as in housing developments, unmeasured gas may be supplied at a flat rate. Some gas lamps include a meter and/or service regulator in their bases.[8]

Copper tubing, ⅜ in. OD, is commonly used for gas supply lines to outdoor lamps (tinned, where gas contains sulfur). When steel pipe is used, its size may vary from ½ to 1½ in. Connection to the existing house piping may be made with standard fittings or by the use of a saddle where conditions permit.

Figure 12-107 shows a typical method of installation of the gas supply line to a post-mounted outdoor lamp, as used by one large natural gas utility company.

OPEN FLAME GASLIGHTS

Open flame gaslights have no mantles and use nonprimary aerated burners of either the fishtail or single port type. They are used mainly for unique ornamental effects rather than for their lighting ability. Fuel gases containing illuminants are better for this purpose than natural gas.

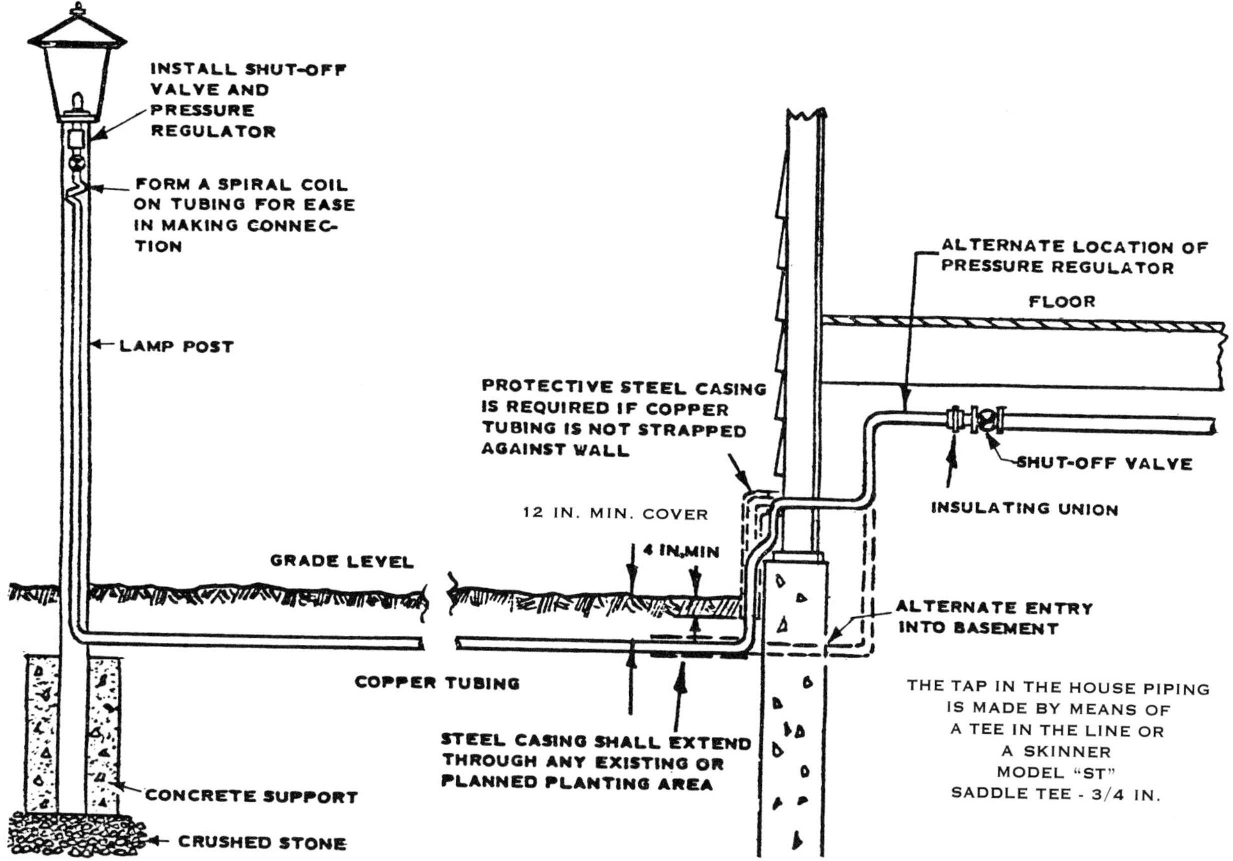

Fig. 12-107 Typical method of installation of gas supply line to outdoor lamp from inside meter location.

Notes:

1. If distance from house line piping connection to lamp post is 30' or less, use ¼ in. copper tubing. If distance is greater than 30' or if multiple lamp installation is being made, use larger size copper tubing, size depending on distance and load.
2. No soldered fittings are permitted.
3. Steel piping may be used in lieu of copper tubing but shall be no smaller than ¾ in. standard weight pipe. (Steel piping as small as ⅛ in. has been successfully applied.)
4. Use an all-brass or brass-to-steel insulating union. If an all-steel union is used, it should be wrapped with polyvinyl tape or equivalent or coated with a protective material.
5. All copper tubing must be reamed out where cut.

Fishtail Burners

These nonprimary aerated burners yield a fan-shaped or fishtail-shaped flame. The head can either be slotted or have two ports angled so that their respective gas streams impinge. Such burners may be installed in gaslight assemblies if the maximum height of the flame is controlled and flame impingement is avoided (to minimize carbon deposits). Lamps using these burners are available as table models with LP-gas containers in the base or as post-mounted units for outdoor use.

Single Port Burners

Two types of gaslights incorporate single port burners.

Torch Lights. Also known as *Luau* torches, these units are often shaped like frustums of right circular cones. It was claimed that a unit 11.25 in. high, with a maximum diameter of 8 in. and a gas input of 12.5 MBtu per hr at 11 in. w.c., produced a 9 to 12 in. flame length and withstood rain and a 50 mph wind. It was separately reported[3] that a natural gas input of 20 MBtu per hr produced only 15 candlepower (contrast with incandescent mantle data in Table 12-78).

Targets or flame spreaders, vanes to redirect drafts, and wire screens or porous plugs to coalesce and harden flames are used to improve flame stability and wind resistance.

Table Models. These lamps generally operate on LP-gas from a small container concealed in the base. The burner simulates a candle flame about 1 in. high in a miniature model of an outdoor gaslight. Other table types resemble small candles set in a base with a glass chimney. Still others are designed as 12 or 15 in. tapered candles that burn butane stored in the candles themselves.

GAS ORNAMENTALS

Many unique decorative arrangements featuring open gas flames have been custom designed, see Fig. 12-108; others use Luau torches, which are commercially available. Many designs combine water sprays and gas flames. Both low and high pressure gas (20 psig) have been applied.[9,10]

The four general types of "ornamentals" or "spectaculars" (besides torch lights) follow:

Lakes. These large installations generally use multiple

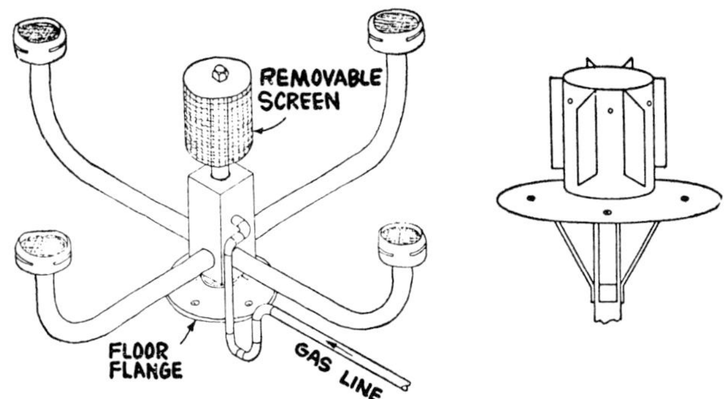

Fig. 12-108 Two burner designs for decorative gas and water displays. (left) Underwater gas burner arrangement for "pools of fire." Constant burning pilot in center is ½ inch above water surface; burners on arms are ½ inch below water surface. (Roman Fountains, Van Nuys, Calif.) (right) Constant burning pilot burner for "aqua torch." Gas is ignited at the six 60-gage pilot holes on the circumference of the tube. The protective fins maintain flame stability.

gas outlets, either slightly submerged below or elevated above the water surface. Variation of the gas flow rates by mechanical means produces changes in flame configuration. Indoor installations, in high-roofed spacious enclosures and malls, have been made.

Fountains. These attractions are located indoors or outdoors. The more elaborate oudoor designs use mechanical means to produce variations in both water sprays and gas flames. The latter are usually controlled by varying the gas flow rate. The effectiveness of many unusual displays depends on the size and design of the fountain bowl complementing a voluminous gas flame pattern.

Submerged gas outlets are generally used. After the initial ignition of the gas on the surface of the water, the flame is self-sustaining. Use of a small standing safety pilot is recommended. (See left side of Fig. 12-108.)

While a 1964 report on "dancing" fountains does not mention gas flames, it may be pertinent for its discussion of the electronic control of light, water sprays, and music.[11]

Volcanoes. Here an open gas flame pours out thru crevice openings or thru the open top of a man-made simulated mineral formation; sometimes cascading waterfalls are incorporated in the display. Replicas of dragons spewing intermittent flames are an interesting variation.

Flame Caldrons. Copper "wool" or spinnings are placed within large, and generally colorful, ornate bowls. The porous material on top of a round plate serves to diffuse and spread the flame. This plate expands the cross section of the gas stream issuing from the open pipe burner.

Burner Design

Generally, the burners used are of the nonaerated type. Drilled pipe burners are frequently specified and shaped to conform to the flame pattern desired. Pipe burners tapped for fixed or adjustable "hood" jets are also used. Brass pipe and jets reduce corrosion problems.

Figure 12-121 suggests port spacing and gas input for non-coalescing flames up to 16 in. high. Coalescence, however, is often a desirable phenomenon. It leads to the upward extension of the flames. Any deficiency of oxygen at the various levels of the flame promotes random variations in flame configuration. Where spreader plates are used, open end pipe or a drilled pipe cap with single or multiple ports is simple and effective.

Outdoor installations require both rain shields and wind shields. The former protect exposed burners and help to diffuse the flame while the latter help to maintain the contour stability of the flame. Flame retention type (Sticktite) nozzles are sometimes used.

Gas Pressure and Input

Burners for indoor lakes and fountains, installed at or above water level, generally operate at about 6 in. w.c. On the other hand, if gas outlets are located 8 to 12 in. *below* water level, at least 1.0 psig gas pressure is desirable for flame stability as well as to overcome the hydrostatic head in the gas line. A check valve installed in the gas supply prevents the gas line from filling with water in the event of a sizable gas pressure drop. To allow for the hydrostatic head and the pressure drop thru gas piping, fittings, and valves, a conservative design would limit the height of the water level above the check valve to not more than one half of the gas pressure in inches w.c.; for example, 14 inches with 1 psig available gas pressure.

Outdoor fountains and lakes with burners above water level also operate at about 6 in. w.c. but with rain and water shields and with greater gas flow to sustain flame.

Most volcanoes operate at pressures below 0.5 psig. Gas input depends upon size of flame desired and whether flame is open or semishielded.

Flame caldrons generally require gas pressures of 2 to 5 psig and an input of about 350 cfh.

Ignition

By and large, manual ignition is used for outdoor installations. Electric ignition with timers and safety shutoff devices may be specified. Indoor installations should be protected with an automatic safety pilot.

REFERENCES

1. Stotz, L. *History of the Gas Industry.* New York, Stettiner Bros. Press, 1938.
2. A.G.A. Laboratories. *American Standard Approval Requirements for Gas-Fired Illuminating Appliances.* (ASA Z21.42-1963) New York, 1963.
3. Zielinski, R. J. *Gas Light Performance and Design.* (Research Bull. 91) Cleveland, Ohio, A.G.A. Laboratories, 1962.
4. "World's Biggest Gas Light." *Am. Gas J.* 187: 20, Aug. 1960.
5. Illuminating Engineering Soc. *American Standard Practice for Roadway Lighting,* p. 11, Table II. (ASA D12.1-1963) New York, 1964.
6. "Gas Runway Lights Prove Out in Year-Long Test." *Gas* 37: 70-3, Jan. 1961.
7. Mason, D. M., and others. *Investigation of Factors Affecting Mantle Life.* (I.G.T. Tech. Rept. 6) New York, A.G.A., 1962.
8. Vogel, O. B. "Gas Lights—Installation and Service." *A.G.A. Operating Sec. Proc.* 1960: DMC-60-25.
9. "Utilities Mix Gas and Water for Eye-Catching Displays." *Am. Gas J.* 190: 25–9, May 1963.
10. "Flaming Fountains." *Am. Gas J.* 187: 25, Feb. 1960.
11. Beamer, S. "Designing 'Dancing' Fountains." *Consulting Engineer* 22: 82–92, May 1964.

Chapter 12
Gas Burner Design

by E. J. Weber and F. E. Vandaveer

ATMOSPHERIC BURNERS

Practically all domestic and commercial gas appliances and many industrial gas units employ atmospheric gas burners. Undoubtedly, more burners of this type are in use than any other kind. An evaluation of the domestic utilization potential of "unusual" gas burners and combustion processes, both atmospheric and otherwise, is available.[1]

An atmospheric gas burner is used in a low-pressure gas (up to 14 in. w.c.) system; the momentum of the jet of gas entrains, from the atmosphere, a portion or all the air required for combustion. Air premixed with the gas is designated as primary air, and the remainder supplied around the flame as secondary air.

The first American literature on burner design was issued by the National Bureau of Standards.[2,3] Many other contributions have since been published.[4] Notable advancements resulting from research sponsored by A.G.A. have been reported by the A.G.A. Laboratories.[5–19]

General Characteristics

The various parts of a domestic atmospheric gas burner are shown in Fig. 12-109. Modifications in size and shape are made to fit the desired combustion chamber and heating element.

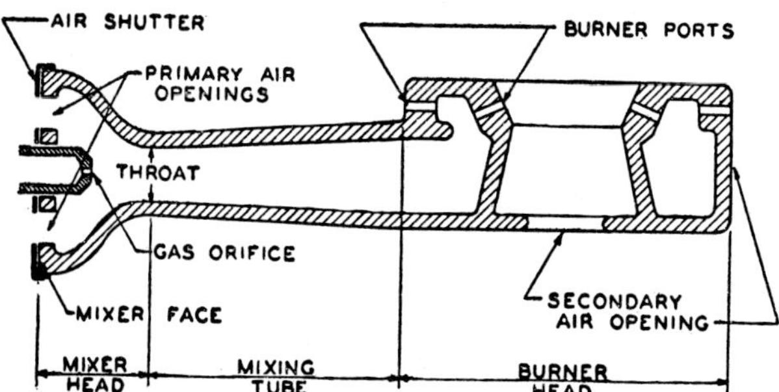

Fig. 12-109 Parts of an atmospheric gas burner.

The purpose of most gas burners is to transform gas into useful heat to be absorbed by an object. Attainment of this purpose involves much more than the burner; the design of the combustion chamber, element to be heated, and flue gas passageways is an inseparable part of the problem.

In general, a burner must have the following characteristics: (1) be controllable over a wide range of turndown without flashback or outage—mainly applicable to gas range and manually controlled room heater burners; (2) provide uniform heat distribution over the area heated; (3) be capable of completely burning the gas; (4) no lifting of flames away from ports; (5) provide ready ignition with flame traveling from port to port over the entire burner rapidly and positively; (6) operate quietly during ignition, burning, and extinction; and (7) substantial construction to withstand severe heating and cooling for the life of the appliance. These requirements must be met under a wide variety of service conditions. Differences in gas composition and changes in pressure and specific gravity should not prevent satisfactory operation.

Materials. Cast iron is frequently used for the entire burner, with either cast iron or heavy gage steel air shutters. A ground coat (single application) of porcelain enamel is generally applied on burners when good appearance and corrosion resistance are desired.

Aluminum burner head caps have also been widely used, particularly on range top burners, because of appearance, light weight, and ease of removal for cleaning. Due to high burner head temperature and design factors of certain burners, aluminum has not given full satisfaction on manufactured gases.

Brass is used for tubular burners with single or multiple rows of ports or slots. It is also used for single port Bunsen or Meker laboratory burners and for tips in jet-type burners. On all brass burners head temperatures must be kept low with ample secondary air in order to reduce oxidation.

Use of **steel** and **stainless steel** burners with welded or seamed joints has increased; porcelain enamel may be applied as a further protection against corrosion.

A **Nichrome** alloy is used to form the port area in ribbon burners.

Ceramic materials of various compositions are used in many types of infra-red burners such as the Schwank and the Cercor.

Molded lava has found favor for pilot burner tips and small burner heads.

All *American Standard Approval Requirements* for domestic gas appliances include sections on burner construction and performance. The principal points differ only slightly in meeting specific requirements of individual appliances.

Burner Heads

Port Design and Arrangement. Because nearly all the foregoing characteristics are dependent on port design and port arrangement, they should be the starting point in designing a burner. Head shape and size, total port area, and port design and arrangement can be determined by knowing the gas input rating and the available combustion space to accommodate the burner head.

Generally recommended maximum input rates per unit port area, as well as maximum port sizes for various gases, are given in Table 12-81. Total port area may be determined by *dividing* the selected input rate per unit port area into the total input rate required. Number of ports is fixed by the total port area with selection of port size. Distribution of ports over the available burner head area determines their spacing and arrangement.

Table 12-81 Maximum Input Rates per Unit Port Area and Maximum Port Sizes

Gas	Maximum input rate, Btu/hr-sq in. of port area	Maximum drilled port size, D.M.S. (Table 1-17)	Maximum slotted port width, in.	
			Coalescing flames	Non-coalescing flames
Natural	20,000	Nos. 30–32	3/32	1/16
Mixed	20,000	Nos. 34–36	5/64	3/64
Manufactured	28,000	Nos. 38–40	5/64	3/64
Butane or propane	18,000	No. 32	3/32	1/16

A universal city gas burner should be designed to have a port loading and a port size between those shown for natural and manufactured gases in Table 12-81. The port loading of a universal burner preferably should approach the recommended value for natural gas, while the port size should be almost as small as that recommended for manufactured gas. Usually, the smaller the individual port, provided total port areas are not reduced beyond recommended limits, the better the burner performance, particularly in resistance of flashback.

Primary Air. The first step in the design of an appliance burner is the selection of the primary air value (Table 12-82). Further requirements are stable flames and performance flexibility when operating at the chosen primary air value. Generally radiant burners use 100 per cent primary air.

Attention to the following design factors reduces primary air requirements: (1) distance from ports to nearest heating or impinging surface; (2) direction of gas flow from ports, particularly to adjacent surfaces upon which flames may impinge; (3) temperature of surface upon which flames may impinge; (4) distribution and spacing of burner ports as affecting mass and height of secondary combustion zone or outer flame mantle, and availability of secondary air; (5) volume and direction of secondary air flow; and (6) direction of venting of flue gases.

Table 12-82 Minimum Primary Air Requirements for Various Types of Appliance Burners

Type of burner	Primary air, per cent of air theoretically required
Range top	55 to 60
Range oven	35 to 40
Water heater	35 to 40
Radiant-type space heater	65
Other heating appliances	As low as 35

Typical performance characteristics of atmospheric gas burners with respect to flame stability are shown in Figs. 12-110a and 12-110b. To obtain stable flames at the chosen primary aeration, the gas input rate per unit of port area must be such that the burner adjustment point is located in the stable flame zone and, preferably, in the usual design area. To obtain flexible performance, the adjustment point in the stable flame zone should be located as far as possible from the three curves defining flashback, yellow tipping, and lifting limits.

Changes in port size and port arrangement will displace the curves in Figs. 12-110a and 12-110b upward or downward. For example, on natural gas:

Gas input, MBtu/hr-sq in.	Primary air, per cent	
	Lifting curve	Yellow tip curve
10	65 to 95	12 to 35
50	45 to 65	24 to 38

Similar variations occur for manufactured gas. Flashback curves also vary, depending on port size and gas composition.

Secondary Air. Secondary air complements primary air; together they constitute the air necessary to complete combustion. While increased primary aeration provides better mixing of air with fuel gas, it can also result in poor combustion by indirectly increasing the amount of recircula-

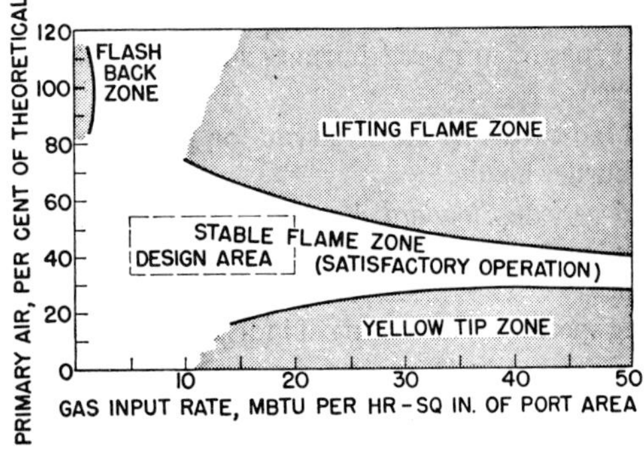

Fig. 12-1'"a Characteristic lifting, yellow tipping, and flashback curves on natural gas. (Burner with No. 36 D.M.S. ports, 1/4 in. spacing, one row.)

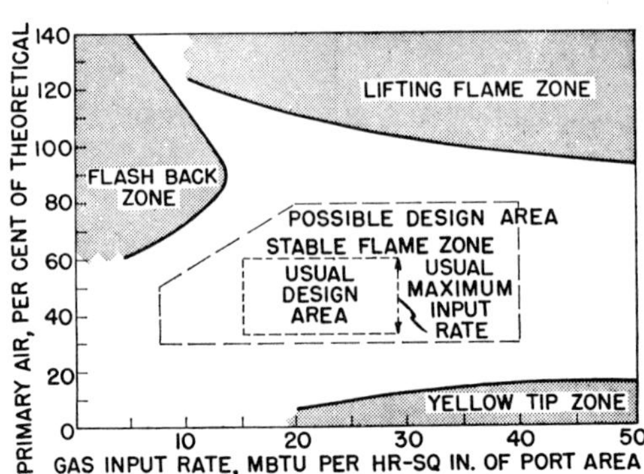

Fig. 12-110b Characteristic lifting, yellow tipping, and flashback curves on manufactured gas. (Burner with No. 36 D.M.S. ports, 1/4 in. spacing, one row.)

tion of combustion products. This recirculation is relatively severe immediately after ignition with a cold start.

Lifting of Flames from Ports. When the velocity of an air–gas mixture normal to the inner cone surfaces of flames on atmospheric burner ports exceeds its burning velocity (Table 2-70), the flame will lift (Table 12-83 and Figs. 12-110a and 12-110b) or blow off the ports to varying degrees. If the velocity is further increased, a point will be reached at which flames will be extinguished. Lifting may result from too much primary air or, with no primary air, from increasing the gas velocity until it exceeds the burning velocity. The point at which the flame starts to leave a single port burner, or at which several flames start to lift from a multiport burner, is known as the **lifting limit.** This characteristic is so definite that it has long been used as a measure of burner performance. It is usually expressed in percentage of primary air at which lifting occurs for a given set of conditions.

Table 12-83 Typical Lifting and Yellow Tip Limits for Various Gases

(for No. 36 D.M.S. single port burner at 25 MBtu/hr-sq in. of port area)

Gas	Heating value, Btu/cu ft	Lifting limit, % primary air	Yellow tip limit, % primary air
Natural	1116	54	22
Natural, high-inert	1021	45	...
Manufactured, high H_2	555	100	14
Manufactured	518	75	...
High-Btu oil	918	92	...
	1276	74	44
Propane	2503	54	42
Butane	3207	51	45

Most gas appliance burners are designed to avoid lifting by regulation of primary air (usually by an air shutter). Except under special conditions, such as when the appliance is designed to use blowing flames, lifting is undesirable. It usually indicates escape of some unburned gas or incomplete combustion. Burner applications usually require fairly definite ranges of primary aeration for good combustion. Therefore, a burner with a low lifting limit may not be able to attain the minimum required aeration without lifting, and incomplete combustion results. Ignition of gas–air mixtures at or above the lifting limit is difficult and flame travel unreliable. Excessive lifting or blowing also decreases efficiency of heat transfer and may produce disturbing noise.

Table 12-84a indicates, for various primary aerations, limiting port loadings *above* which lifting flames will be obtained on a natural gas (1040 Btu per cu ft, 0.62 sp gr) with an arbitrarily selected reference burner. Input multipliers for other gases are also included. Table 12-84b gives input multipliers covering variations in port size, depth, and spacing. The number of rows of ports, with spacing between rows of 1/4 in. or more, has no effect on lifting limits. The data in Tables 12-84a and 12-84b are for a cold burner, in which lifting is most likely. The tendency for lifting decreases as the burner temperature increases.

Example: Determine the limiting burner port loading for lifting flames for propane at 60 per cent primary air. Ports are vertical No. 38 D.M.S., 3/8 in. deep, and spaced 1/8 in. edge to edge.

Table 12-84a Lifting Limits on a Natural Gas Burner and Input Factors for Other Gases

Primary aeration, %	Input, max, MBtu/hr-sq in. of port area	Gas	Input factor
35	45.7	Propane	1.10
40	37.2	Butane	1.04
45	30.2	Natural	1.00
50	24.6	Oil	3.46
55	20.0	Mixed	1.94
60	16.2	Manufactured	6.45
65	13.2		

Note: To anticipate a 50 per cent increase in gas pressure, apply the additional factor 0.82; for a 25 per cent increase, apply the additional factor 0.89.

Table 12-84b Factors for Port Size, Depth, and Spacing to be Applied to the Heat Inputs in Table 12-84a

Size, D.M.S.	Factor	Depth, in.	Factor	Port spacing, port edge to port edge: Spacing, in.	Vertical factor	Horizontal factor
36	1.14	1/8	0.96	1/4	1.00	0.83
38	1.09	1/4	1.00	1/8	1.15	1.06
40	1.05	3/8	1.11	1/16	1.85	1.85
42	1.00					

From Table 12-84a, the limiting port loading of the reference burner operating on the reference gas at 60 per cent primary air is 16.2 MBtu per hr-sq in. of port area × 1.10. From Table 12-84b, the factor for a No. 38 D.M.S. port is 1.09; for the 3/8 in. port depth, 1.11; and for the 1/8 in. spacing of vertical ports, 1.15. The product of these values equals the limiting port loading with respect to lifting flames for the given conditions; i.e., 16.2 × 1.10 × 1.09 × 1.11 × 1.15 = 24.8 MBtu per hr-sq in. of port area. If a 25 per cent increase in gas pressure is anticipated, the limiting port loading should be reduced to 89 per cent of this value (see Note to Table 12-84a).

Various design expedients may be used to increase the limit of port loading by reducing lifting. For example, a ledge placed slightly below horizontal ports will tend to raise the factors in Table 12-84b for these ports to equal those for vertical ports. Lifting limits for horizontal ports may be raised even beyond those for vertical ports by placing the ledge rights at their bottom edge so that the issuing air–gas stream impinges on the ledge. Ports arranged to provide **self-piloting flames** will increase the limiting port loading with respect to lifting: (1) two staggered rows of horizontal ports; (2) a pattern of closely spaced rows; or (3) a small secondary flame at the base of the primary flame. Flame targets or impinging surfaces may also be used.

Data in Table 12-84b on horizontal ports apply to ports drilled in a straight-sided burner head. Flames, issuing from horizontal ports drilled in a circular head or one with curved surfaces, diverge from each other; hence, they have less self-piloting action than flames from a straight-sided burner. Consequently, the lifting limit of such horizontal ports will be somewhat less than indicated by Table 12-84b.

Other factors in the appliance design may reduce the limiting port loading with respect to lifting. For example, recirculation of flue gases into the flame base or a high velocity flow of secondary air across the port lowers the lifting limit considerably.

Ribbon or Slotted Ports. These forms may be considered closely spaced ports of a diameter equal to that of the drill which just enters the port. A square port practically always has a lower lifting limit than the circular port of a diameter equal to the side of the square.

Yellow Tipping. This limit on port loading is indicated in Table 12-83 and Figs. 12-110a and 12-110b. Table 12-85a indicates, for various port loadings, limiting primary aerations *below* which yellow tipped flames will be obtained on a natural gas (1040 Btu per cu ft, 0.62 sp gr), with an arbitrarily selected reference burner; corrections of primary aeration for other gases are also included. Table 12-85b provides corrections of primary aeration to be algebraically added for variations in port size and spacing.

Table 12-85a Yellow Tip Limits of Primary Aeration for Various Natural Gas Port Loadings and Corrections for Other Gases

Input, MBtu/hr-sq in. of port area	Minimum primary aeration, %	Gas	Correction of min primary aeration*
15.0	15.5	Propane	+18.0
20.0	20.0	Butane	+21.0
25.0	22.3	Natural	0.0
30.0	24.0	Oil	+20.0
35.0	25.0	Mixed	−3.0
		Manufactured	−7.0

* To be algebraically added.

Flashback. This limit on port design and arrangement, indicated in Figs. 12-110a and 12-110b, is controlled by the burner design as well as by gas composition, air–gas ratio, and operating temperature. Flashback will occur at increasingly greater port loadings as port size is increased and port depth is decreased (Fig. 12-111).

Manufactured gas may flash back at input rates as high as 9000 to 14,000 Btu per hr-sq in. of port area with the ordinary port sizes. The maximum port loading at which flashback occurs on manufactured gas at 800 F will be about 1.5 times that at room temperature. Port spacing has no effect on flashback at higher burner operating temperatures. At lower temperatures, close port spacing may possibly raise the air–gas mixture temperature.

Flashback on natural or butane gas (Fig. 12-112) is practically never encountered (see also Fig. 12-110a). Even with the large No. 26 ports used for Fig. 12-112, flashback on natural gas occurs at very low rates of 3 to 5 MBtu per hr-sq in. of port area, which are below normal design limits, and at primary air percentages of 85 to 90, which are much above normal. With No. 30 ports, the curves would move to the left and practically vanish from the plot shown.

Flashback limits, as illustrated in Fig. 12-111, define areas of input rate and primary air adjustment which will produce flashback; i.e., flashback will not occur to the right of a curve for the port condition given. A burner is designed so that its adjustment at full input rate is out of the flashback zone. However, a change in gas composition, as in peak shaving, without an appliance adjustment, might shift operation into the flashback zone. It is also important to consider whether equilibrium conditions at a turndown rate will be in this zone. Transient conditions of burner operation are a major cause for flashback.

Port Loading. Design of a burner for one general group of gases, e.g., natural gases, is more complicated than for one specific gas. If a universal burner is not used, an appliance requires different burners for natural, manufactured, and possibly LP-gases. These burners are usually alike except for port size; their design is a compromise. In designing a burner for one general type or group of gases, a certain amount of flexibility must be provided to meet composition variations. For example, natural gases vary between high Btu and low inerts and low Btu and high inerts. Sufficient flexibility must be provided to permit some variation in gas composition without either air shutter or orifice readjustment.

Example (use of gas characteristic limit data in designing a burner port arrangement): An appliance requires an input of 25 MBtu per hr on natural gas; its combustion space accommodates a narrow vertical port burner 16 in. long. The type of appliance burner requires 55 per cent primary air (Table 12-82).

Trial Solution: Assume No. 36 D.M.S. ports, ⅜ in. deep, spaced ¼ in. edge to edge, and arranged in three rows:

From Table 12-84a, port loading at 55 per cent primary air = 20.0 MBtu per hr-sq in. of port area

From Table 12-84a, factor for 50 per cent pressure increase = 0.82

From Table 12-84b, factors for:

No. 36 D.M.S. port = 1.14
⅜ in. port depth = 1.11
¼ in. vertical port spacing = 1.00
Port loading = 20.0 × 1.14 × 1.11 × 1.00 × 0.82 = 20.8 MBtu per hr-sq in.
Total port area = 25.0/20.8 = 1.2 sq in.

Table 12-85b Corrections for Port Size and Spacing to be Algebraically Added to Yellow Tip Primary Aeration Limits in Table 12-85a

		Port spacing, port edge to port edge			No. of rows	
Size, D.M.S.	Correction	Spacing, in.	Vertical port correction	Horizontal port correction	Of ports*	correction
36	+2.0	¼	0.0	0.0	1	0.0
42	0.0	⅛	+5.0	+9.0	2	+6.0
45	−4.0	1/16	+8.0	+14.0	3	+8.5

* The effect of number of rows of ports is independent of row spacing if flames from various rows coalesce.

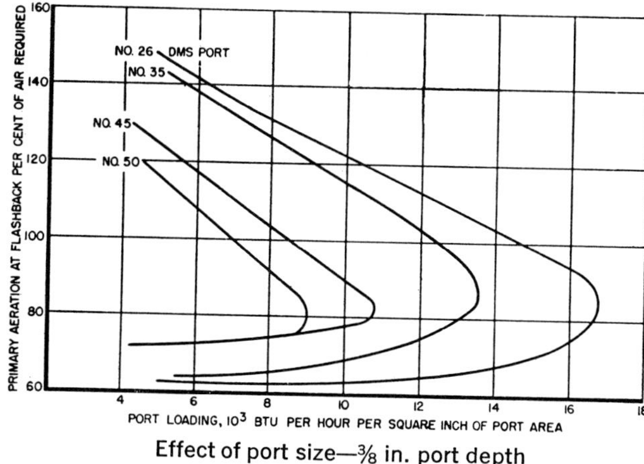

Effect of port size—⅜ in. port depth

PORT DEPTH (INCHES) 1-1/4, 3/4, 3/8, 1/4, 1/8, 1/30

PRIMARY AERATION AT FLASHBACK, PER CENT OF AIR REQUIRED

PORT LOADING, 10^3 BTU PER HOUR PER SQUARE INCH OF PORT AREA

Effect of port depth

Fig. 12-111 Flashback limits of coke oven gas (540 Btu per cu ft, 0.42 sp gr, 800 F).[12,19]

Total number of ports = 1.2/0.00891 = 135
Number of ports per row = 135/3 = 45
Length of burner for ports = (44 × ¼) + (45 × 0.1065) = 15.8 in.

From Table 12-85a, at a port loading of 20.8 MBtu per hr-sq in. of port area:

Yellow tip limit = 20.4 per cent primary air
Adjustment for natural gas = 0.0

From Table 12-85b, adjustment for:

No. 36 D.M.S. port = +2.0
¼ in. vertical port spacing = 0.0
Three rows of ports = +8.5
Yellow tip limit = 30.9 per cent primary air
From Fig. 12-110a, lifting limit = 60 per cent

This trial design would be satisfactory with respect to lifting and yellow tip limits. The span of primary air adjustment between these two limits would be from 31 to 60 per cent primary air with this design. Flashback would not have to be considered for this natural gas burner.

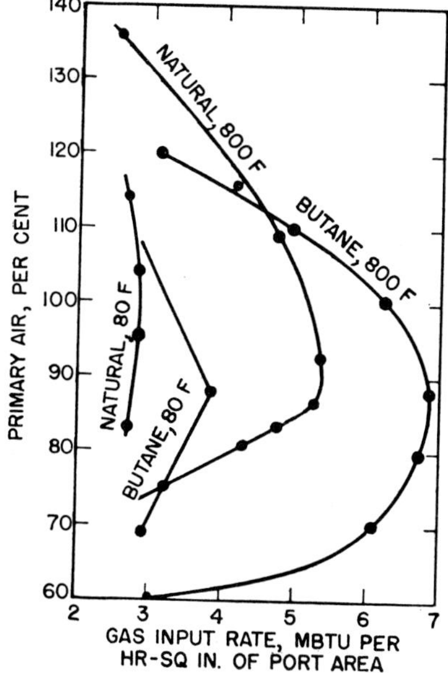

Fig. 12-112 Flashback limit curves for natural and butane gases (Burner with No. 26 D.M.S. ports spaced 0.33 in. edge to edge.)

Burner Flexibility. The span of primary air adjustment between lifting and yellow tip limits is important in considering burner flexibility on peak or substitute gases. Even though these gases may vary somewhat in composition from that on which the burner is originally adjusted, satisfactory performance should be obtained without either air shutter or orifice readjustment. Generally, the greater this span, as shown in Figs. 12-110a and 12-110b, the better the burner's flexibility. Flexibility is greater for a design with lower port loadings. However, port loading cannot be made low enough to permit flashback.

Flexibility also depends on whether the primary aeration selected for design is near the lifting or yellow tip limits. With a substitute gas having a high yellow tip limit and increased input rate on substitution, any selected design primary aeration should be near the lifting limit of the adjustment gas. With a substitute gas having a low lifting limit and decreased input rate on substitution, the selected design primary aeration should be near the yellow tip limit of the adjustment gas.

Any tentative burner design may be examined for flexibility by constructing a flame stability diagram (Fig. 12-110a) for the proposed burner. A lifting limit curve is constructed by plotting the primary aerations in Table 12-84a vs. their corresponding port loadings multiplied by the product of the appropriate factors in Tables 12-84a and 12-84b. A second lifting limit curve for the substitute gas is similarly constructed. A burner in an appliance is subjected to greater flows of air than it would be in an open room. While such flows have no effect on flashback limits, they generally lower burner flows for blowoff and yellow tip limits.[20]

Yellow tip limit curves for both adjustment and substitute gases are similarly constructed, using Table 12-85a and factors from Table 12-85b.

The operating adjustment point for which the burner was designed is next located on the graph by plotting the selected primary aeration vs. the port loading at which the burner is to operate. Flexibility may be evaluated by plotting the relocation of this adjustment point on substitution of one gas for another. For flexibility under the given conditions, the relocated adjustment point must be within the limit curves of the substitute gas.

Relocation of the adjustment point on substitution may be calculated. The port loading with the burner operating on the substituted gas would equal:

$$PL_s = PL_a \left(\frac{H_s}{G_s^{0.5}}\right)\left(\frac{G_a^{0.5}}{H_a}\right) \qquad (1)$$

where: PL = port loading, MBtu per hr-sq. in. of port area
H = heating value of gas, Btu per cu ft
G = specific gravity of gas (air = 1.0)
s and a = subscripts for substitute gas and adjustment gas, respectively

The primary aeration on the substitution would equal:

$$PA_s = PA_a \left(\frac{G_s^{0.5}}{B_s}\right)\left(\frac{B_a}{G_a^{0.5}}\right) \qquad (2)$$

where:

PA = primary aeration, per cent of air required for combustion
B = cu ft of air required for combustion per cu ft of gas
Other terms are defined under Eq. **1**.

PL_a and PA_a in Eqs. **1** and **2** are the selected design values of port loading and primary aeration, respectively, on the adjustment gas. When an appliance is adjusted to these design values on a substitute gas, the substitute and adjustment gas data should be interchanged in these equations.

Calculations to determine burner flexibility are simplified if the compositions of all gases involved are known. Burners adjusted on natural gas may be checked with a 1200 Btu per cu ft propane–air gas as a substitute. This gas is critical for both lifting and yellow tipping.

In considering any gas–air fuel, allowance should be made for the quantity of air it contains. By definition, primary aerations at the various limits correspond to injected primary air or air entering the burner thru the mixer face. On this basis, primary aerations at these limits would be lower for a gas–air fuel than for an all-gas fuel.

Limit curves for straight propane may be converted to propane–air curves as follows: The per cent primary air values at the limits for straight propane are converted to air–gas ratios. By definition, per cent primary air equals 100 R/B, where R equals the air–gas ratio and B equals the air required for the straight gas in cubic feet per cubic foot of gas. The primary air values for the straight gas are then converted to primary air values for the gas–air fuel by the equation:

$$\text{Per cent primary air for gas–air fuel} = 100\,\frac{R - r}{B - r} \qquad (3)$$

where: R = air-gas ratio for straight gas
B = air required for straight gas, cu ft per cu ft of gas
r = ratio of air to gas in the gas–air fuel.

When using gas–air fuel, the value of B in Eq. **2** must be the number of cubic feet of air required for combustion per cubic foot of that fuel.

Minimizing Lint Stoppage of Atmospheric Burner Ports.[21] These deposits of airborne particles on the underside of ports reduce the effective port diameters which, in turn, may reduce primary air injection to a point at which yellow tipping occurs. The three major ways of eliminating dust are: (1) incineration *in situ* (port areas at 500 to 600 F); (2) filtering combustion air; (3) changing the direction and velocity of the combustion air so that entrained dust drops out before it reaches the burner. On the other hand, burners may be designed to tolerate airborne dust for long periods. Alleviating factors include: (1) flexibility with respect to primary air adjustment; and (2) mixer tubes with rough interior surfaces.

Universal Burners. All three operating limits considerably restrict these burner designs. Such burners work within the flashback limits of a fast-burning manufactured gas, the lifting limits of slow-burning natural gas, and the yellow tip limits of oil or LP-gases with high propensities for yellow tipping.

Turndown Rates. The input rate to gas range burners, some water heater burners, and other burners may be manually or thermostatically throttled to low by-pass rates or turndowns. With a high ratio of turndown to full rate as a requirement, the minimum input rate at which combustion is sustained becomes an important design consideration. Within the practical range of port sizes, 30 Btu per hr *per port* is approximately the minimum. This rate may be slightly decreased with very close port spacing, as in ribbon burners. It is also affected by primary aeration and burner head temperature.

Flame Travel. The ability of the flame to move from port to port depends on gas rate, type of gas, port size and spacing, and primary aeration. More difficulty with flame travel usually is encountered at minimum by-pass rates. Flames travel less rapidly with natural or other slow-burning gases than with fast-burning manufactured gases. In general, port spacing (center to center) should not be greater than 9/32 to 3/8 in. where ports are located on a straight line. For ports on arcs,

Radius:	1 in.	2 in. or more
Port spacing:	5/32 to 7/32 in.	3/16 to 1/4 in.

Generally, the maximum spacing applies to No. 36 D.M.S. ports. For turndown rates approaching the minimum, and with the port spacing given, mechanical aids to flame travel, such as a slight ledge over the port row for oven burners or countersunk ports, are helpful.

Burner Head Configuration. Calculations in burner design assume an even distribution of mixture flow to the port area. Therefore, burner head volume should be adequate to assure an even conversion of velocity pressure in the mixer tube to static pressure in the burner head. It is recommended that the head be tapered so that its cross-sectional area at any point is 1.5 to 2.0 times the port area upstream of the point.

Other factors which may cause uneven distribution in the burner head are: (1) burner ports drilled too close to the mixer tube; (2) impingement of the main mixer stream directly against certain ports; and (3) friction in the gas stream passage thru the burner. In some cases, internal baffling in the burner may be needed to provide even distribution.

Burner Head Pressure. Figure 12-113 may be used.

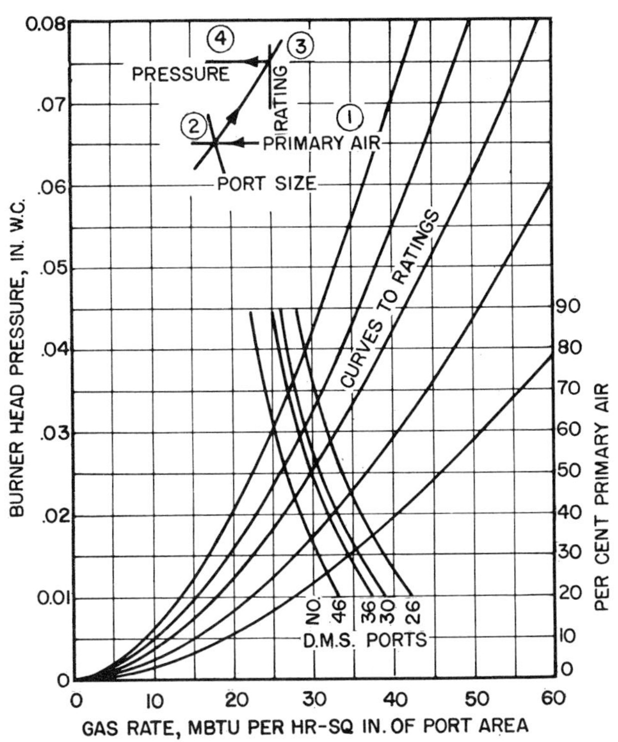

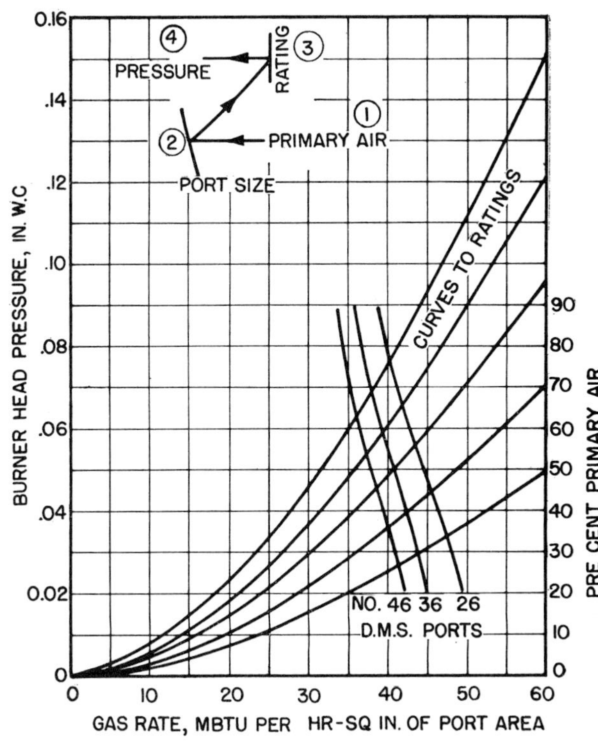

Fig. 12-113 Burner head pressures for natural gas (left) and butane (right) for various port sizes, primary aerations, and gas inputs.

Example: Determine the burner head pressure of a burner with No. 36 ports operating on natural gas at a port loading of 20 MBtu per hr-sq in. of port area at a primary aeration of 60 per cent.

Solution: Follow to the left the horizontal 60 per cent aeration line on the left-hand chart of Fig. 12-113 until it intersects with the curve for the No. 36 D.M.S. port. From this point, trace sloping curve until it intersects the vertical line for a port loading at 20 MBtu per hr-sq in. of port area. The value of this point on the left-hand scale is the burner head pressure, in this case 0.016 in. w.c. If the intersection of a primary air line with a port size curve does not fall on a curve to rating, trace an interpolated curve to the desired port loading.

Burner Head Temperature. This is mainly a function of appliance design. Other factors involved are type of gas, port loading, primary aeration, and burner design. High temperatures adversely influence flashback, noise of extinction, yellow tipping, and primary air injection performance. On the other hand, high temperature aids flame travel and reduces lifting tendencies. Table 12-86 gives burner head temperatures for some appliances.

Table 12-86 Burner Head Temperatures of Various Appliances

Appliance burner	Burner head temp, °F		Air-gas mixture temp, max, °F
	Average	Maximum	
Boiler	525	710	468
Water heater	598	660	590
Floor furnace	685	766	621
Range high broiler	918	982	960
Range oven	835	915	893
Range top	500	675	330
Radiant heater	460	537	416

Flame Height. An impinging inner cone is not conducive to good combustion, nor is an impinging outer mantle of flame. Distribution and spacing of ports on a burner affect the mass and height of the secondary combustion zone and secondary aeration of the flame.

In an initial design, the ports may be arranged and spaced closely to obtain a high degree of flame stability against lifting. However, this spacing may result in such a high outer flame mantle that the first estimate of primary air aeration may need to be raised for good combustion. Flame height should be matched with the available combustion space.

Inner Cone Height. This may be approximated by Eqs. **4a** thru **4c**:

For natural gas:

$$\log H_i = 0.65 - 0.00064PA(dT)^{0.5} \quad \textbf{(4a)}$$

For butane:

$$\log H_i = 0.71 - 0.00068PA(dT)^{0.5} \quad \textbf{(4b)}$$

For manufactured gas:

$$\log H_i = 0.58 - 0.00102PA(dT)^{0.5} \quad \textbf{(4c)}$$

where:

H_i = inner cone height, in. per MBtu per hr *per port*
PA = primary air, per cent of air required for combustion
d = port diameter, in.
T = temperature of air–gas mixture, °F abs

Equations **4a** and **4b** may be used for up to 90 per cent primary air. Equation **4c** applies only up to 70 per cent primary air. The equations do not apply to very close port spacing and low primary aeration, since inner cones may merge together. Note that the inner cone height, H_i, is directly proportional to the input rate *per port*.

Outer Mantle Height. This is the height from port to mantle tip. It may be approximated by Eqs. **5a** thru **5c** for flames in the open. Decrease in secondary aeration increases flame height above the values indicated by these equations.

For natural gas:

$$H_o = 4.46\ sn\ [1 - (0.005PA)^2] \quad \textbf{(5a)}$$

For butane:

$$H_o = 5.2\ sn\ [1 - (0.005PA)^2] \quad \textbf{(5b)}$$

For manufactured gas:

$$H_o = 3.8\ sn\ [1 - (0.005PA)^2] \quad \textbf{(5c)}$$

where:

H_o = outer mantle height, in. per MBtu per hr *per port*
PA = primary air, per cent of air required for combustion
s = factor for port spacing (Table 12-87)
n = number of rows of ports in a straight bar burner

Spacing between rows does not affect outer mantle height if the flames from the rows coalesce. If they do not, the burner should be treated as a single row burner. Usually, the spacing which would prevent coalescing exceeds any practical spacing.

Determination of n (in Eqs. **5a, 5b,** and **5c**) becomes difficult with circular burners. It is possible for flames to coalesce with small diameter burners having only one row of vertical ports. Therefore, with circular burners n may equal at least twice the number of rows of ports.

Because flames from horizontal ports curl upward, their horizontal extent is less than calculated for vertical flames.

Mixing Tubes

In a mixing tube, air entrained by the gas stream issuing from the orifice or spud forms a mixture which enters the burner head (Fig. 12-130). Its design is flexible enough to satisfy a range of primary air values. With the burner port area designed for desired flame characteristics and appliance requirements, the mixing tube may be specified as follows:

1. Throat to port area ratio of about 0.45; the range of 0.3 to 0.6 results in a deviation of about five per cent from maximum air injection.[5]
2. Mixing tube length, throat to burner head, of at least six throat diameters; within limits, greater lengths do not appreciably affect injection, although shorter lengths decrease it.
3. Total included angle between venturi sides of four degrees; primary air injection decreases appreciably with larger angles.
4. The inlet approach of the throat should have a radius of curvature of 3 to 4 in. and an air shutter opening at least 1.25 times the port area; with high port loadings, the air shutter opening should be increased to about twice the port area.
5. The orifice should be located one to two throat diameters from the throat inlet. If the orifice fitting is large relative to the throat size, it should not be located so close to the throat that it restricts flow of air into the mixing tube.

Table 12-87 Effect of Port Spacing on Outer Mantle Heights
(used in Eqs. **5a, 5b,** and **5c**)

Port spacing, edge to edge, in.	Factor for port spacing, s
½	1.10
⅜	1.19
5/16	1.26
¼	1.37
3/16	1.53
⅛	1.77
1/16	2.14

Primary air injection implicit relationships are given by Eq. **6**:

$$r + 1 = \frac{KH^{0.5}(hG_g)^{0.25}}{(PLG_m)^{0.5}} \quad \textbf{(6)}$$

where:

r = primary air–gas ratio, cu ft of air per cu ft of gas
H = heating value of gas, Btu per cu ft
h = gas pressure, in w.c.
G_g = specific gravity of gas (air = 1.0)
G_m = specific gravity of air–gas mixture; see Eq. **7** (air = 1.0)
PL = port loading, Btu per hr-sq in. of port area
K = constant depending on burner design; about 28 at room temperature (see Table 12-88)

The specific gravity of the air–gas mixture, G_m, may be calculated by Eq. **7**:

$$G_m = (r + G_g)/(r + 1) \quad \textbf{(7)}$$

All terms are defined under Eq. 6.

Substituting G_m from Eq. **7** into Eq. **6** results in the quadratic:

$$r^2 + r(G_g + 1) + G_g = K^2H(hG_g)^{0.5}/PL$$

All terms are defined under Eq. **6**.

This may be reduced, without appreciable error, to Eq. **8**:

$$r = \frac{KH^{0.5}(hG_g)^{0.25}}{(PL)^{0.5}} - \frac{G_g + 1}{2} \quad \textbf{(8)}$$

All terms are defined under Eq. **6**.

With a mixing tube designed as outlined above, K may be taken as 28.0 (at "room" temperature)[19] to approximate the air injection obtained with a wide open air shutter; for high temperatures—T, °F abs—multiply Eq. **8** by $(540/T)^{0.5}$, where T is the arithmetic average of the air inlet temperature and the gas mixture temperature at the ports, °F abs. Modifying factors for various port sizes and depths are given in Table 12-88.

Table 12-88 Effects of Port Size and Depth on Primary Air Injection
(multiply K in Eqs. **6** and **8** by factors shown)

Port size, D.M.S.	Port depth		
	⅛ in.	¼ in.	⅜ in.
25	0.977	0.973	0.962
27	.971	.965	.953
29	.963	.955	.942
31	.945	.930	.915
33	.937	.920	.902
35	.935	.915	.896
37	.926	.905	.885
39	.923	.900	.878
41	.918	.892	.865
43	.912	.880	.850
45	0.906	0.870	0.835

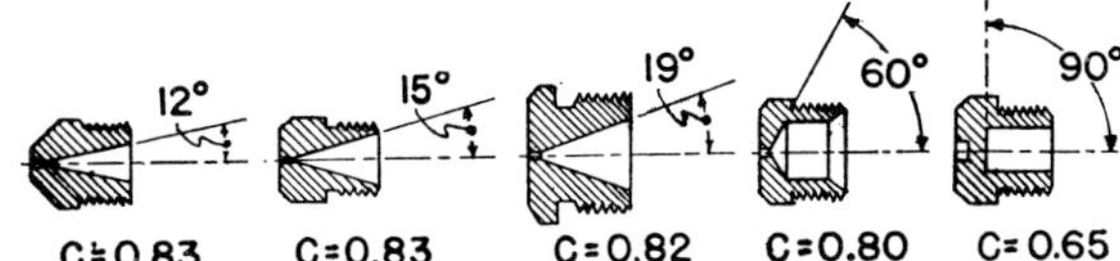

Fig. 12-114 Five types of orifice spuds and their discharge coefficients C for angles of approach shown.

The value of K, and consequently primary air injection, is influenced considerably by the port discharge coefficient, type of flow thru burner passageways, and flow of mixture into the mixer head. All flows should be as smooth as possible. Air flow into the mixing head should be symmetrical around the orifice so as not to disturb the velocity profile of flow thru the mixer tube.

Burner temperature also affects primary air injection as shown in Eq. **9**:

$$PA_t = PA_r \left(\frac{2T_r}{T_t + T_r}\right)^{0.5} \qquad (9)$$

where:

PA_t and PA_r = primary air at temperatures T_t and T_r, respectively, per cent of air required for combustion

T_t and T_r = air–gas mixture and room temperatures, respectively, °F abs

With fast burning gases, the maximum operating primary aeration would be limited by the injection ability of a burner rather than, as with slow burning gases, by its lifting limit. In many instances, the manufactured gas burner is the same as a natural gas burner for the same application, except for port size; injection with manufactured gas may be calculated using the port loading with the manufactured gas port size.

The injection relationships (Eq. 6) indicate that the air–gas ratio is only slightly dependent on gas pressure. Thus, the air–gas ratio at turndown rates does not vary appreciably from that at full rate. However, in many cases it will start decreasing when the gas pressure is reduced to less than 1.0 in. w.c.

A vertical mixing tube, directed upward, may cause an increase in the air–gas ratio at turndown rates because of stack action. On the other hand, with the tube directed downward, the ratio may decrease rapidly with turndown. Stack action of the appliance itself may have the same effect of either increasing or decreasing the air–gas ratio with turndown, depending on the location of the mixer face.

In some instances, it is desirable to have injection fall off rapidly with turndown in order to avoid flashback at very low rates. A straight-sided mixing tube provides this characteristic in the same way as a tube in which the throat is extended about one inch before the sides diverge to form the venturi.

Gas Orifices

Various types of fixed orifices are used in domestic gas appliances, and their discharge coefficients are given in Fig. 12-114. Flow thru these orifices, at sea level, may be expressed by Eq. **10**:

$$Q = 1658.5\,CA(h/G)^{0.5} \qquad (10)$$

where:

Q = gas flow thru orifice at 60 F and 30 in. Hg, cu ft per hr
C = discharge coefficient of orifice (see Fig. 12-114)
A = area of orifice, sq in.
h = gas pressure, in w.c.
G = specific gravity of gas (air = 1.0)

Sizing: (1) Select factor from Table 12-89a for the specific gravity and heating value of the gas. (2) Multiply this factor by the burner input rate in Btu per hour. (3) Locate the resultant product (input rate) in Table 12-89b,

Table 12-89a Orifice Capacity Factor for Heating Value and Specific Gravity of Gas
(basis: 800 Btu per cu ft, 0.6 sp gr gas)

Heating value, Btu/cu ft	Specific gravity (air — 1.0)											
	0.3	0.4	0.5	0.6	0.7	0.8	0.9	1.0	1.1	1.2	1.3	1.4
400	1.41	1.63	1.83	2.00	2.16	2.31	2.45	2.59	2.70	2.83	2.95	3.06
500	1.13	1.31	1.46	1.60	1.73	1.85	1.96	2.07	2.16	2.27	2.36	2.45
525	1.08	1.29	1.39	1.52	1.64	1.76	1.87	1.97	2.06	2.16	2.25	2.33
550	1.03	1.24	1.33	1.45	1.57	1.68	1.78	1.88	1.97	2.06	2.15	2.23
575	0.983	1.14	1.27	1.39	1.50	1.61	1.70	1.80	1.88	1.97	2.05	2.13
600	.942	1.04	1.22	1.33	1.44	1.54	1.63	1.73	1.80	1.89	1.97	2.04
700	.81	0.932	1.04	1.12	1.24	1.32	1.39	1.48	1.54	1.62	1.69	1.75
800	.707	.816	0.912	1.00	1.08	1.15	1.22	1.30	1.35	1.41	1.48	1.53
825	.685	.791	.885	0.970	1.05	1.12	1.19	1.26	1.31	1.37	1.43	1.48
850	.665	.769	.860	.942	1.02	1.09	1.15	1.22	1.27	1.33	1.39	1.44
875	.646	.746	.835	.915	0.987	1.05	1.12	1.19	1.24	1.29	1.35	1.40
900	.628	.725	.811	.890	9.960	1.03	1.09	1.15	1.20	1.26	1.31	1.36
950	.595	.687	.769	.842	0.910	0.972	1.03	1.09	1.14	1.19	1.24	1.29
1000	.565	.653	.730	.800	.864	.924	0.980	1.04	1.08	1.14	1.18	1.22
1025	.551	.636	.712	.781	.842	.900	.955	1.01	1.06	1.10	1.15	1.19
1050	.538	.621	.696	.762	.823	.880	.934	0.985	1.03	1.08	1.12	1.16
1075	.526	.606	.679	.745	.804	.859	.911	.961	1.00	1.05	1.10	1.14
1100	.514	.594	.664	.727	.785	.840	.891	.938	0.982	1.03	1.07	1.11
1125	.502	.580	.649	.711	.767	.820	.871	.920	.961	1.00	1.05	1.09
1150	.491	.567	.635	.696	.751	.802	.852	.899	.940	0.985	1.02	1.06
1175	.481	.555	.621	.681	.735	.786	.834	.878	.920	.965	1.00	1.04
1200	0.417	0.544	0.608	0.666	0.720	0.770	0.816	0.860	0.900	0.945	0.985	1.02

Table 12-89b Orifice Capacities, Btu per Hour

(C, the discharge coefficient = 0.8; 800 Btu per cu ft, 0.6 sp gr gas; see Table 12-89a for corrections to apply for other gases)
Note: Since most contemporary coefficients range from 0.85 to 0.94, enter Table at 0.9 times desired capacity.

Orifice,	Pressure, in. w.c.									
D.M.S.	2	3	4	5	6	7	8	9	10	11
1	78,500	96,400	111,000	124,100	136,000	147,000	157,100	167,000	176,500	185,100
2	73,600	90,500	104,100	116,300	127,200	139,000	147,100	156,400	165,300	173,900
3	68,500	84,100	97,000	108,200	118,500	129,000	137,000	145,500	153,900	161,500
4	66,000	81,000	93,500	104,200	114,100	124,000	132,000	140,100	148,100	155,700
5	63,700	78,300	90,300	100,900	110,200	120,000	127,400	135,400	143,100	150,200
6	62,700	77,000	88,900	99,400	108,900	117,000	125,900	133,900	141,400	148,600
7	60,900	75,000	86,200	96,400	105,500	114,000	122,000	129,800	137,000	144,000
8	59,700	73,400	84,500	94,500	103,500	112,000	119,700	127,000	134,100	141,000
9	58,000	71,200	82,100	91,900	100,500	109,000	116,100	123,500	130,500	137,000
10	56,500	69,400	80,000	89,500	98,000	106,000	113,100	120,100	127,000	133,200
11	55,000	67,500	77,900	87,000	95,300	104,000	110,100	117,100	124,000	130,100
12	53,900	66,100	76,400	85,400	93,500	101,000	108,000	114,900	121,300	127,500
13	51,600	63,400	73,100	81,700	89,500	97,000	103,400	110,000	116,200	122,100
14	50,000	61,400	70,800	79,100	86,600	94,000	100,000	106,100	112,100	118,000
15	48,900	60,000	68,300	76,400	83,600	92,000	96,600	102,800	108,800	114,100
16	47,200	58,000	66,900	74,700	81,900	89,000	94,600	100,500	106,100	111,700
17	45,200	55,500	64,000	71,500	78,300	85,000	90,500	96,200	101,600	106,800
18	43,400	53,300	61,500	68,800	75,400	82,000	87,200	92,700	98,000	103,000
19	41,500	51,000	58,800	65,700	72,000	78,000	83,200	88,500	93,500	98,300
20	39,100	48,000	55,400	62,000	68,000	75,000	78,600	83,600	88,500	93,000
21	38,100	46,700	53,900	60,300	66,100	72,000	76,500	81,400	86,000	90,400
22	37,200	45,600	52,600	58,900	64,500	70,000	74,600	79,400	84,000	88,300
23	35,800	44,000	50,700	56,700	62,100	68,000	71,900	76,500	81,000	85,100
24	34,800	42,700	49,300	55,100	60,400	66,000	69,900	74,400	78,600	82,600
25	33,700	41,400	48,000	53,700	59,000	64,000	68,500	73,000	77,200	81,200
26	32,600	40,100	46,500	52,100	57,200	62,100	66,500	70,800	74,900	78,700
27	31,300	38,500	44,600	50,000	54,900	59,600	63,900	68,000	72,000	75,700
28	29,800	36,600	42,400	47,500	52,100	56,600	60,600	64,500	68,200	71,700
29	27,900	34,300	39,700	44,500	48,800	53,000	56,700	60,400	63,900	67,200
30	25,000	30,800	35,700	40,000	43,900	47,600	51,000	54,300	57,500	60,500
31	21,700	26,700	30,950	34,700	38,100	41,400	44,300	47,100	49,800	52,500
32	20,300	24,950	28,900	32,400	35,600	38,700	41,450	44,100	46,600	49,000
33	19,300	23,700	27,450	30,800	33,800	36,700	39,300	41,800	44,200	46,500
34	18,600	22,800	26,400	29,600	32,500	35,300	37,800	40,300	42,600	44,800
35	18,250	22,400	26,000	29,100	31,900	34,600	37,000	39,400	41,600	43,800
36	17,100	21,000	24,300	27,200	29,800	32,400	34,700	37,000	39,100	41,100
37	16,300	20,000	23,200	26,000	28,500	31,000	33,200	35,300	37,300	39,200
38	15,510	19,080	22,100	24,800	27,200	29,600	31,700	33,800	35,800	37,700
39	14,910	18,300	21,200	23,800	26,100	28,400	30,400	32,400	34,300	36,100
40	14,500	17,800	20,600	23,050	25,250	27,400	29,300	31,200	33,000	34,700
41	13,900	17,100	19,800	22,200	24,350	26,450	28,300	30,150	31,900	33,600
42	13,200	16,200	18,800	21,050	23,100	25,100	26,900	28,650	30,300	31,900
43	11,950	14,700	17,000	19,050	20,900	22,700	24,300	25,900	27,400	28,800
44	11,190	13,730	15,900	17,800	19,500	21,200	22,700	24,200	25,600	27,000
45	10,120	12,450	14,400	16,100	17,650	19,200	20,600	21,950	23,200	24,400
46	9,900	12,180	14,100	15,800	17,350	18,850	20,200	21,500	22,700	23,900
47	9,300	11,410	13,210	14,800	16,200	17,600	18,800	20,000	21,100	22,200
48	8,730	10,710	12,400	13,900	15,250	16,550	17,700	18,850	19,900	20,900
49	8,050	9,860	11,400	12,730	13,930	15,070	16,100	17,080	18,000	18,850
50	7,400	9,060	10,470	11,700	12,820	13,850	14,800	15,700	16,550	17,330
51	6,770	8,300	9,590	10,720	11,730	12,680	13,540	14,370	15,170	15,880
52	6,090	7,460	8,620	9,640	10,550	11,400	12,180	12,900	13,630	14,280
53	5,340	6,550	7,560	8,450	9,250	10,000	10,680	11,330	11,950	12,500
54	4,560	5,590	6,450	7,210	7,900	8,530	9,120	9,670	10,200	10,700
55	4,070	4,990	5,760	6,440	7,050	7,620	8,140	8,640	9,110	9,550
56	3,260	3,990	4,610	5,160	5,650	6,100	6,520	6,910	7,300	7,640
57	2,785	3,420	3,940	4,410	4,830	5,210	5,570	5,900	6,230	6,520
58	2,660	3,260	3,760	4,210	4,610	4,970	5,320	5,640	5,950	6,230
59	2,535	3,110	3,590	4,010	4,390	4,740	5,070	5,370	5,670	5,940
60	2,410	2,950	3,410	3,810	4,170	4,510	4,820	5,110	5,400	5,650
61	2,295	2,815	3,250	3,630	3,980	4,300	4,590	4,865	5,140	5,380
62	2,180	2,670	3,090	3,450	3,780	4,080	4,360	4,620	4,880	5,110
63	2,060	2,525	2,920	3,260	3,570	3,855	4,120	4,360	4,610	4,820
64	1,953	2,395	2,765	3,090	3,380	3,650	3,906	4,140	4,370	4,570
65	1,849	2,265	2,620	2,925	3,200	3,460	3,700	3,920	4,140	4,340
66	1,640	2,010	2,320	2,595	2,840	3,070	3,280	3,470	3,670	3,840
67	1,541	1,890	2,180	2,435	2,670	2,880	3,080	3,270	3,450	3,610
68	1,449	1,775	2,040	2,285	2,505	2,705	2,890	3,670	3,240	3,390
69	1,282	1,573	1,810	2,030	2,220	2,400	2,560	2,720	2,880	3,010
70	1,180	1,450	1,670	1,865	2,040	2,200	2,360	2,500	2,640	2,765

Table 12-90 Rate of Flow of Undiluted Propane and Butane and Their Mixtures thru Orifices[22]

(in Btu per hour, at sea level)

		Propane	Butane
Pressure at orifice = 11 in. w.c.	Heating value, Btu per cu ft	2500	3175
Orifice coefficient = 0.8	Specific gravity	1.53	2.00

For altitudes above 2000 ft, reduce appliance input, if required, by reducing the size of the orifice (Eq. **11**).

Orifice size, in., D.M.S.	Propane	Butane or butane-propane mixures	Orifice size, in., D.M.S.	Propane	Butane or butane-propane mixtures
0.008	445	492	51	31,400	35,000
.009	570	630	50	34,200	38,000
.010	703	778	49	37,200	40,300
.011	845	936	48	40,400	44,700
0.012	1,005	1,110	47	43,000	47,600
80	1,270	1,410	46	45,800	50,700
79	1,470	1,625	45	47,000	52,000
78	1,790	1,980	44	51,600	57,200
77	2,260	2,500	43	55,300	61,300
76	2,790	3,090	42	61,100	67,700
75	3,080	3,410	41	64,400	71,300
74	3,540	3,920	40	67,000	74,200
73	4,020	4,450	39	69,200	76,600
72	4,370	4,840	38	72,000	79,600
71	4,730	5,240	37	75,500	83,600
70	5,490	6,070	36	79,300	87,800
69	5,960	6,600	35	84,500	93,600
68	6,720	7,440	34	86,200	95,300
67	7,150	7,920	33	89,800	99,500
66	7,600	8,420	32	94,000	104,000
65	8,560	9,480	31	100,600	111,500
64	9,050	10,030	30	115,300	127,600
63	9,570	10,600	29	129,500	145,200
62	10,100	11,140	28	137,500	152,500
61	10,600	11,800	27	145,000	160,000
60	11,170	12,300	26	151,000	167,000
59	11,750	13,000	25	156,000	173,000
58	12,300	13,600	24	161,500	179,200
57	12,930	14,300	23	166,000	183,500
56	15,100	16,700	22	172,000	190,700
55	18,850	20,900	21	176,500	195,700
54	21,200	23,400	20	181,100	200,000
53	24,700	27,400	19	193,000	215,000
52	28,200	31,200	18	200,500	222,000

under the desired pressure. (4) Read the orifice size in the left-hand column. See Table 12-90 for propane and butane data.

Example: Find the orifice for a burner rated at 20,000 Btu per hr using natural gas at 7.0 in. w.c., 0.60 sp gr, and 1100 Btu per cu ft heating value. Orifice coefficient is in 0.85 to 0.94 range.

Solution: From Table 12-89a the factor for heating value and specific gravity is 0.727. Multiplying this by 20,000 (the burner rate) gives a product of 14,540. After applying the 0.9 factor noted in the heading to Table 12-89b, enter the 7 in. w.c. column of this table with 13,086 and select a D.M.S. No. 51 orifice.

Fixed orifices always give higher air injection than adjustable ones at rates of 15,000 Btu per hr or less. As the size of an adjustable orifice is made greater than that of a fixed orifice, more adjustment with the needle will be required to obtain the desired rate, with greater difference in injection with the two types. However, for input rates above 15,000 Btu per hr, there is little difference in primary air injection with the two types.

Sizing Orifices at High Altitudes.[23] Ratings for most gas appliances are only for elevations up to 2000 ft. Above this, rated heat input should be reduced four per cent for each 1000 ft *above sea level.* While the heating value of gas decreases with altitude—due to the accompanying pressure decrease—it is still necessary to reduce the size of an appliance orifice when derating is required. Thus, the reduced orifice size offsets the increase in flow *per unit of orifice area* attributable to the decrease in density of the gas. Equation **11** may be used to determine a reduced orifice size for the recommended four per cent input rate reduction. Equation **12,** on the other hand, applies when no reduction in input rate is necessary; that is, when orifice size is increased.

$$A_a = (5.48 - 0.219E)A_s/P_a^{0.5} \quad (11)$$

$$A_a = 5.47A_s/P_a^{0.5} \quad (12)$$

where:

A_a = area of orifice at altitude E, sq in.
E = altitude, thousands of feet
A_s = area of orifice required at sea level, sq in.
P_a = atmospheric pressure at altitude E (see Table 6-7), in. Hg

Air Shutters and Air Control Devices

Five ways to adjust primary air flow readily with a given burner are with: air shutters (Fig. 12-115), angle orifices, extended orifices, pressure reduction orifices, and double orifices (Fig. 12-116). Air shutters are most widely used.

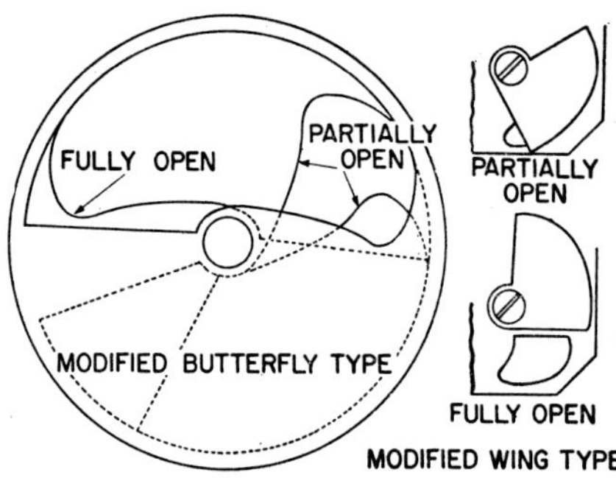

Fig. 12-115 Air shutter design providing approximately circular opening near point of closure.[8]

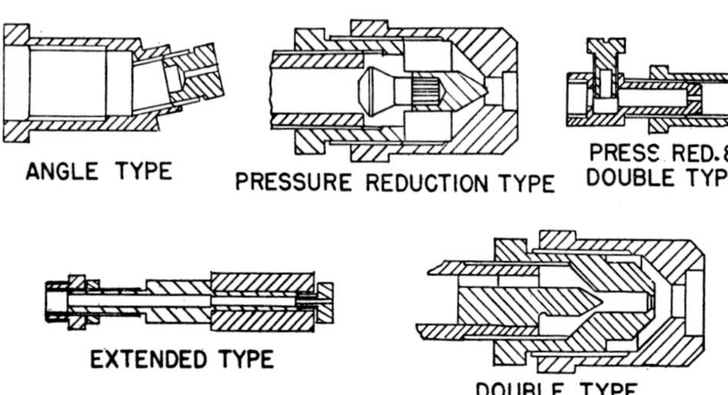

Fig. 12-116 Orifices used to control primary air.[8]

Noise Elimination

Combustion of gas usually is accomplished without noise. Yellow flame burner operation is silent. Bunsen-type burners have the following potential sources of noise.

Flame Noise. Caused by high primary aeration, high velocity of flow thru the ports, turbulence, and poor mixing, such noise may be reduced or eliminated by reducing primary aeration, reducing gas input if possible, increasing port size to reduce velocity, increasing number of ports, or increasing port depth. Fast burning gases have a greater tendency to produce noise.

Orifice Noise. A shrill noise may be caused by an orifice burr, a badly out-of-round hole (improper drilling), or high gas pressure. Elimination of imperfections and reduction in pressure will reduce such noise.

Air Inspiration and Mixing Noise. Extreme roughness of the internal surface of the mixing tube or a metal projection into the air stream may cause inspirational noise. Replacement of mixing tube or smoothing of the surface will correct it. Improper distance between orifice and throat may create sound which can be corrected by changing this distance. Fewer and larger openings thru the air shutter may also reduce inspirational noise.

Ignition Noise. Delayed ignition, which permits accumulation of a flammable air–gas mixture, may produce a concussion. Faulty ignition may result from improper pilot location or from too great a port spacing for rapid flame travel. Poor flame distribution over the port area or a tendency to lifting or blowing with a cold burner may cause delayed ignition and resultant noise.

Flashback of flames thru burner ports may also cause ignition noise. This flashback may be caused by a lean mixture either already present in the burner head or produced when gas flow is initiated. A lean mixture in the head, as a result of short off-on operating cycles, may result from a combination of the stack action of the appliance and the relative burner location in the combustion chamber. This type of flashback on ignition may be controlled by port size and depth and by the air shutter adjustment.

Flashback on ignition may also occur if flames are swept back to the mixture face and ignite the orifice jet.

Noise on Rapid Turndown. Flashback on turndown may occur with manually operated valves which can quickly throttle to a very low rate. Rapid turndown momentarily interrupts primary air injection, so that mixture flow to the ports is reduced greatly. During this interval, flashback is a possibility. It may be controlled by proper selection of port size and depth and by valve design.

Noise of Extinction. Caused by sudden shutoff of the gas supply, it may be considered as flashback at zero gas rate. While more prevalent with manufactured than with natural gas, it may occur on any gas.

Even though adjustment at full input rate may be well outside the flashback zone when gas is shut off, input drops to zero at a more or less rapid rate, depending on burner size. In extreme cases it may require several seconds. During this interval the additional air drawn into the mixer tube produces a lean mixture. This combination of lean mixture and low input rate may be inside the flashback zone. Consequently, flashback results if a small flame bead remains on a few ports to serve as an ignition source.

There is a limiting value of primary air for any given port loading below which flashback on extinction will not occur. This limiting aeration increases as port loading decreases and may be even less than the yellow tip limit on large burners.

Small port size, small mixing tube diameter, and low port loading tend to increase the primary aeration at which a burner may operate without promoting extinction noise. Ledges, either above or below ports, tend to retain flames after gas shutoff and thus increase the tendency for such noise. Secondary air flow and combustion chamber pressures may either increase or decrease this tendency, depending on the relative location of burner and mixer face. Low gas pressure also favors the occurrence of this type of noise. Noise of extinction is usually louder with large burner volumes, higher primary aerations, and fast burning gases. The burner shape is also a factor.

Laboratory Burners

These burners, commonly known as Bunsen burners, exist in wide variety; well-known types are: Meker, Argand, blast, micro, multiflame, and blowtorch. Basically, the Meker type burner uses the venturi configuration to aspirate primary air. The Bunsen type, on the other hand, uses a straight tube for this purpose. Units for any kind of gas and for nearly any kind of flame shape, size, and temperature can be obtained.

The Meker burner (Fig. 12-117a) conforms to recommendations of the National Bureau of Standards and Federal Specification GG-B-817. It has a 10 mm deep grid with a diameter of 40 mm, a venturi mixing tube, adjustable gas orifice, large primary air inlet openings, and large port area. It will burn 10,500 Btu per hr of natural gas.

The Bunsen burner shown in Fig. 12-117b has a flame retainer with eight small pilot jets fed by the regular mixture from the mixing tube, adjustable primary air shutter, and adjustable gas orifice. It will burn 5000 Btu per hr of natural gas. No flame retainer is needed for manufactured gas.

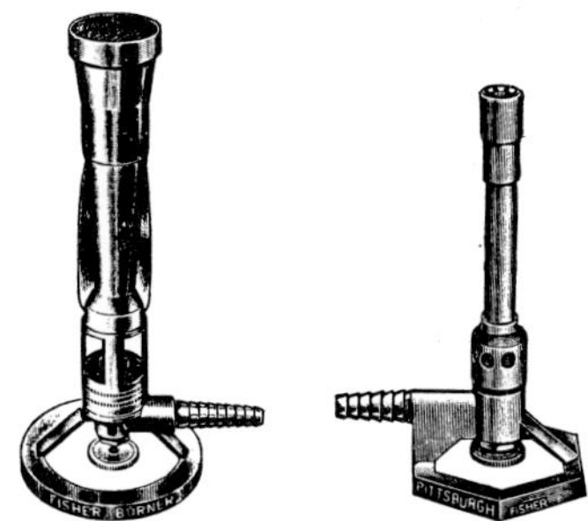

Fig. 12-117a (left) Fisher natural gas laboratory burner (Meker type).
Fig. 12-117b (right) Pittsburgh-Universal Fisher laboratory burner (Bunsen type).

Single Port Burners[16]

These burners (Fig. 12-118) have come into wide use since 1945, primarily for central heating appliances, both gas designed and converted, and for incinerators. Some upshot burners use spiders of various designs in the riser tube to effect better mixture distribution. Conversion burners may be designed for a range of input ratings from 65 to 400 MBtu per hr, or for one input rating for a given appliance. These burners are sometimes known as "target burners," but this term should be restricted to impingement-type burners.

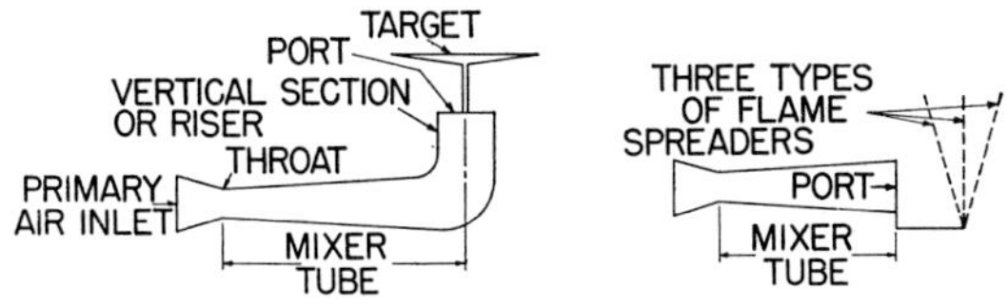

Fig. 12-118 Two basic designs of single port burners: (left) upshot type; (right) inshot type.

Principal advantages of single port burners are simplicity of design and low cost. They have a tendency toward flashback and yellow tipped flames, their area of adjustment for primary air and gas input rate is smaller, and their operation is less flexible on supplementary gases than drilled port burners. Figure 12-119 shows their narrow range of adjustment on natural gas.

Flashback in these burners appears to be a function of flow distribution rather than of port size. Flashback on extinction can be minimized by using a high primary aeration and a high input rate. Lifting flames are minimized by use of flame retention devices or targets and, therefore, are not a problem on these burners.

High primary air injection appears to be the most important design factor for these burners. This is a function of throat loading. Other usual factors of design for good injection, such as mixing tube length, port area, and venturi slope, appear to have little influence here. Suggested operating range is:

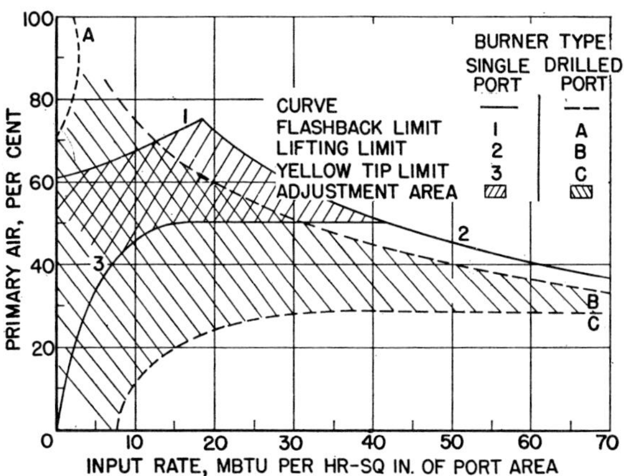

Fig. 12-119 Comparison of burner operating diagrams: single port burner vs. drilled port burner (natural gas at 3.5 in. w.c.).

MBtu per hr-sq in. of throat area	50	100
At per cent primary air	90	65

The riser section of upshot burners should be designed for even flow distribution to minimize its effect on air entrainment. The flame target (or spreader) controls flame shape and distribution. Yellow tipping, flashback, lifting, and primary air entrainment are not affected. Proper spacing can be determined by trial and error.

Impingement Target Burners[17]

The elements are an orifice and a target, each properly sized and spaced to attain desired flame characteristics (Fig. 12-120). Advantages are simplicity, low cost, and ease of servicing. No air shutter, venturi, mixer tube, or burner head is normally use. Disadvantages are noisy operation and limitation to use on natural, butane, and propane gases. On manufactured gas the flame may burn at the orifice with incomplete combustion. These burners are used on a limited scale in water heaters as pilot burners, and experimentally in ranges, water heaters, and furnaces.

Noisy operation can be reduced by: (1) reduction in air entrainment by placement of the orifice relatively close to the impingement target, consistent with complete combustion; (2) operation of the burner in a condition of natural flame retention by proper selection of orifice size, design, and gas pressure; and (3) use of flame stabilization means built into the target design. Flame size is determined by target size. Target temperatures have varied from 750 to 1300 F on natural gas, indicating need for a high-temperature alloy.

Blue Flame, No Primary Air, and No Target.[10] One type of nonluminous burner, without either primary air

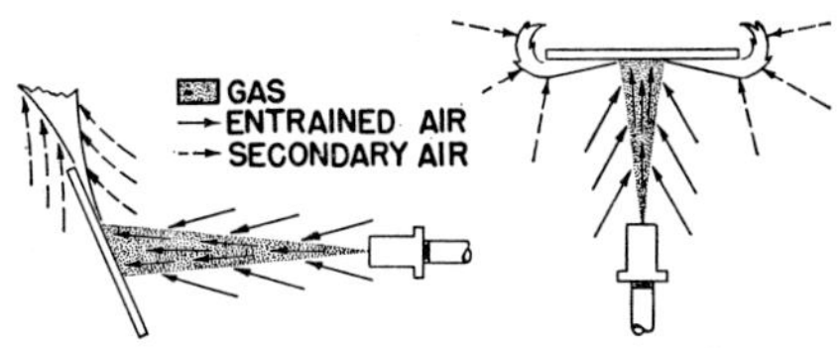

Fig. 12-120 Impingement target burner operation.

or a target, employs a sufficiently low rate per port to give a blue flame on almost any gas without primary aeration. This rate is about 180, 80, and 60 Btu per hr per port, or less, for manufactured, natural, and butane gases, respectively. Port size is not critical, but port spacing must be adequate to prevent flame coalescing. This type of burner is free from lifting, yellow tipping, and flashback. Its disadvantages are a comparatively large number of ports and poor turndown characteristics.

Such "neat" gas burners are commonly used with manufactured gases in England and in some other parts of Europe. A U. S. study reported on the design, construction, and operating factors affecting their performance characteristics.[15] Commercial British designs, such as the Bray type of jet, were studied as well as experimental types.

Luminous Flame Burners

Although most domestic appliance burners are of the Bunsen type, luminous flame-type burners have been employed to a limited extent in water heaters, space heaters, and central heating equipment, usually for manufactured gas operation.

Luminous flames are obtained on most gases when no primary air is admitted or when the gas rate is above the blue flame input. Their height is the same for various port sizes and gases at equal heat inputs. Flame height largely depends on port spacing (Fig. 12-121). Because the upper limit for butane is 475 Btu per hr per port, luminous flame burners on LP-gases are inadvisable.

Burner input varies inversely with burner head temperature; for example, input to a burner with No. 65 D.M.S. ports will be reduced about 35 per cent on heating from room temperature to a head temperature of 600 F. Thus,

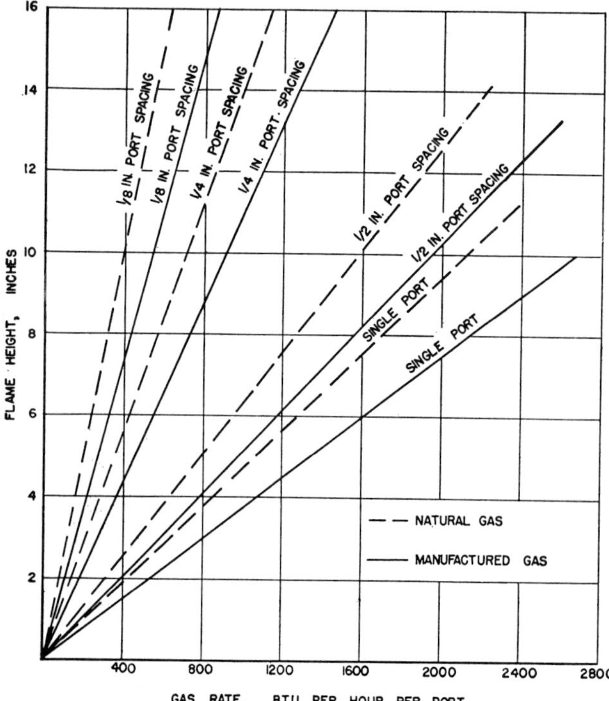

Fig. 12-121 Effects of port spacing on luminous flame height.[10]

design provisions are necessary to assure proper combustion at above normal operating rates.

Effects of port depth on obtainable input may be neglected with larger ports. With smaller ports they become appreciable. Use of shallow ports may prove helpful in obtaining the desired input.

Burners for Small Diameter Tubes[24]

Tubes 0.5 to 1.0 in. in diam and 6.5 to 13.75 ft long may be heated uniformly with elongated flames by means of an atmospheric burner (Figs. 12-122a and 12-122b) and an eductor

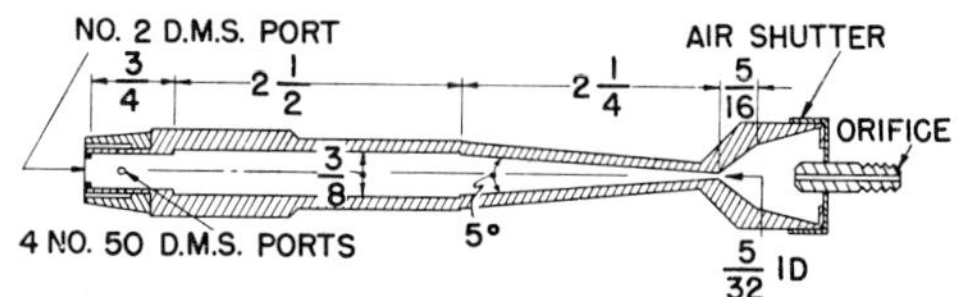

Fig. 12-122a Flame retention burner used to heat a straight ½ in. tube. (Dimensions in inches.)

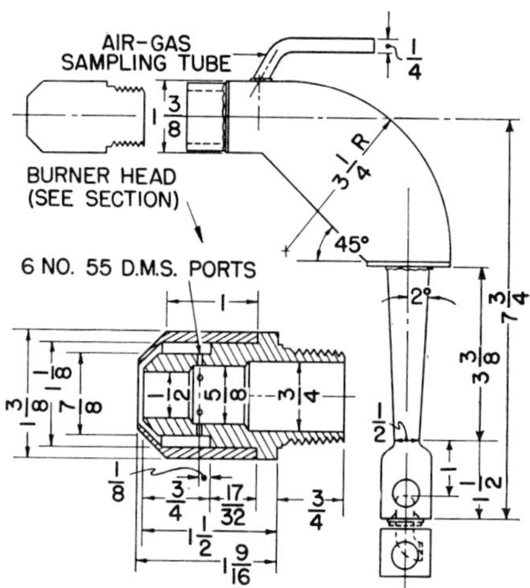

Fig. 12-122b Burner used to heat a straight 1.0 in. tube.

blower, developing from 1.0 to 6.75 in. w.c. vacuum pressure. A ¼ in. tube could not be heated satisfactorily. Maximum inputs of 9000 and 36,000 Btu per hr were applied to the 0.5 and 1.0 in. tubes, respectively, with satisfactory combustion. This amounted to 60 MBtu per hr-sq in. of internal tube cross section area. Port loadings of the flame retention burners were as high as 175 MBtu per hr-sq in. of port area.

PILOT BURNERS

Pilot burners provide the source of main burner ignition. They are constant-burning for ignition of manually operated main burners or of automatically controlled burners for which they actuate automatic pilot devices. On self-energized automatic pilots the pilot flame also heats thermocouples to generate electricity for actuation of automatic valves. Electric ignition by spark or coil, when used on domestic appliances, must ignite the pilot; its flame, in turn, actuates the automatic pilot device and ignites the main burner gas. Most of the following applies to domestic appliances only.

Pilot burners burn a maximum of ten per cent of main burner input. Inputs are usually 50 to 300 Btu per hr on ranges and small appliances, and from 300 to 1000 Btu per hr on larger ones, with a maximum limit of 2000 Btu per hr on conversion burners. Industrial pilot burners are tailored to fit the application, sometimes consuming several thousand Btu per hour. Most pilot burners on domestic appliances are designed to burn constantly, while main burners normally operate intermittently. If pilot outage occurs, there is no means of ignition or of operation of the main gas valve on automatically controlled appliances.

Experience has shown that pilots are remarkably reliable. A constant-burning gas pilot is far more reliable than any other automatic ignition means. However, so many pilots are in use that a high percentage of utility service calls are caused by pilot outage.

Causes of pilot outage are: (1) closure of orifice due to solids or liquids in the gas, corrosion of interior or tubing, and/or gas cracking and deposition of carbon; (2) closure of primary air openings due to lint and dust; (3) plugging of burner ports due to lint, dust, and carbon deposition caused by gas cracking; (4) wind, draft, flue gases, or inadequate air supply; and (5) improper location. Optimum pilot location, to avoid outage due to recirculation of flue products, would be somewhat below port level and as close to the burner as possible.[25]

Many types of industrial equipment, e.g., gas-fired water-tube boiler-furnaces,[30] make use of a spark ignited gas light-off pilot. The **interrupted** type burns during light-off only and shuts off during normal operation of main burner, while the **intermittent** type burns during light-off and while the main burner is firing and shuts off with the main burner.

Nonclogging and Nonlinting Features

Gum or dust filters in the pilot line remove solid and liquid particles which might otherwise clog a pilot orifice. Such filters are required on A.G.A. approved appliances for manufactured gas. They are also widely used on natural gas.

When gas contains oxygen or sulfur, materials resistant to these elements should be used to avoid stoppages from corrosion products. Aluminum will not corrode and is satisfactory at temperatures below 1000 F. Stainless steel may be used above 1000 F. Copper should not be used. Extensive tests[26] show that a very low concentration of oxygen and sulfur in the gas supply reduces clogging troubles; high concentrations increase them.

High temperatures favor the formation of deposits in tubing and orifices due to gas cracking and deposition of carbon; e.g., manufactured gas and butane form deposits with copper at 450 to 550 F, and manufactured gas alone forms deposits with stainless steel at 800 F. Natural gas and butane are not affected up to 1000 F. With aluminum none of these three gases forms deposits up to 1000 F. Aluminum or stainless steel is preferable for parts heated to a high temperature.

The A.G.A. Laboratories developed four "lintproof" automatic pilots (Fig. 12-123): two target types (a and b) and two nonprimary aerated types (c and d). The flames of these pilots were unaffected by 100 hr of operation in an accelerated linting study. By way of comparison, the performance of some contemporary primary aerated pilots was radically affected in 1.5 to 3.5 hr under similar operating conditions. The A.G.A. pilots also resisted steady cross drafts ranging from about 5 to 17 mph, depending on gas input rate, before their flames were extinguished.

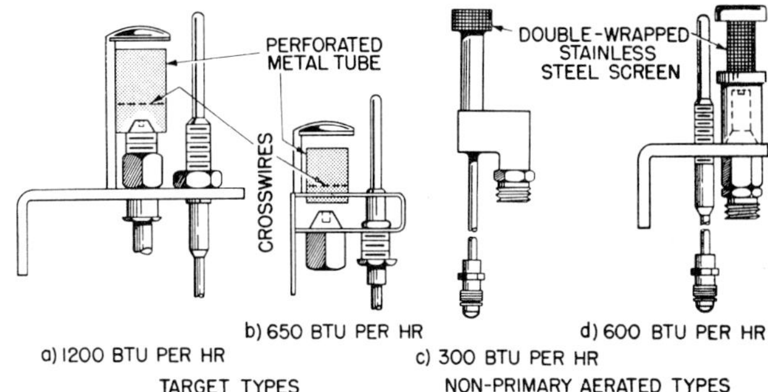

Fig. 12-123 Four A.G.A. lintproof automatic pilots[27] (shown about one-third full size).

Means of eliminating lint and dust troubles include:

1. Fine mesh screens, having considerable air inlet area, tightly fastened to the housing around the primary air openings.
2. A rather large primary air opening: entire bottom of the burner tube is flared and open and placed some distance from orifice fittings.
3. An incineration effect provided by a snorkel type of construction: primary air is drawn down past the flame and any lint present is ignited.
4. A nonprimary aerated burner.

Protection from Drafts and Flue Gases

A pilot flame can be snuffed out by strong winds or drafts unless protected by proper baffling or by special design for increased flame stability. Usually, baffles or protective plates are experimentally located as additional protection. In many ranges and water heaters, sufficient pilot protection is afforded by the enclosed combustion chamber. Recessed ports or metal lips around ports protect the base of the flame and permit greater stability.

If flue gases from a burner compartment below the pilot or from the main burner become too concentrated, they may snuff out the pilot flame. Better location of the pilot or baffling should correct such conditions.

A.G.A. Pilot Requirements

All A.G.A. approval requirements for gas appliances include provisions for pilots and automatic pilots. The major ones follow:

1. Pilots shall be so placed that they can be easily seen and safely lighted.
2. Pilot ports shall be positively positioned with respect to main burner ports and not be subject to accidental displacement.
3. Ignition ports and operating mechanism shall be protected from falling scale and dirt.
4. Tips of aerated pilots must be made of AISI 416 steel (or better material) for heat and corrosion resistance. More than 1.0 per cent nickel alloys are not acceptable.
5. Slotted ports connecting ignition and thermal element ports shall be accurately formed.

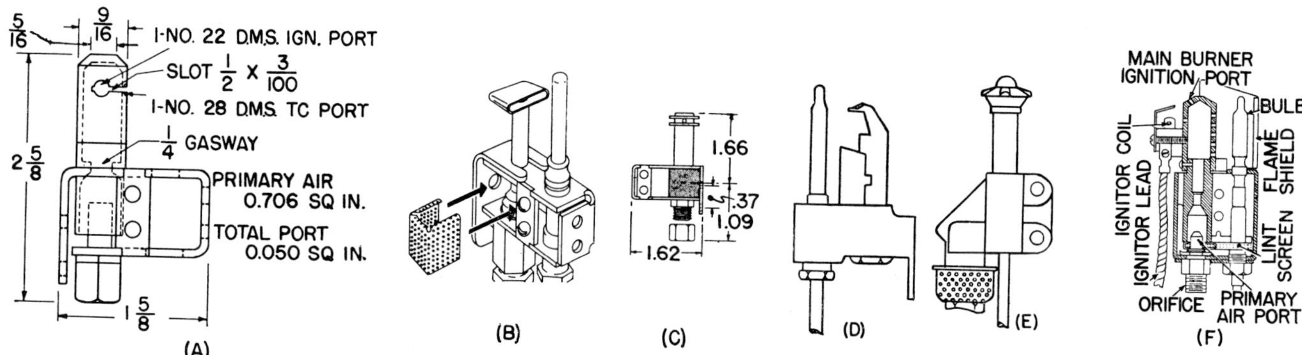

Fig. 12-124 Typical primary aerated pilot burners: (A) 1000 Btu, round stainless steel head type with lint screen, two ports, carryover slot; (B) single port type with venturi throat and lint screen; (C) 1000 Btu, drilled port, lip protected type with lint screen and stainless steel head; (D) 500 Btu, single port, target and snorkel type; (E) 300–500 Btu, lint screen type with imbedded thermocouple for maintaining fairly constant input; (F) assembly with mercury-filled sensing bulb, electric glow-coil ignition, venturi-type mixing, lint screen multiport burner provides for complete envelopment of sensing bulb. (All dimensions in inches.)

6. Ignition of gas at the main burner shall be effected immediately after gas reaches main burner ports, without extinguishment when the main burner is turned on or off.

7. Flames of pilot burners shall not deposit carbon.

8. An automatic pilot must ignite gas at the main burner when the pilot gas supply is reduced to a point just sufficient to keep the automatic valve open.

9. In a multiflame pilot, all pilot ports except those for heating the thermal element are blocked; with reduced pilot gas supply, main burner must ignite immediately.

Primary Aerated Pilot Burners

Typical primary aerated pilot burners are shown in Fig. 12-124. The trend is toward a single port pilot; the same flame heats the thermocouple to hold the automatic valve open and ignites the main burner.

Primary aerated-type pilot burners generally conform to design principles for larger main burners with respect to air injection, lifting, yellow tips, and flashback.[28] Their normal primary air injection varies from 27.4 to 62.2 per cent with natural and from 30.8 to 80.5 per cent with manufactured gas for single orifices. With dual orifices maximum injection is somewhat lower.

Average operating temperature of a pilot burner head is 250 F. As it is increased to 350 F, primary air injection drops about 14 per cent. On further increase to 650 F, air decreases only about 4 per cent more. Cracking of manufactured gas and carbon deposition begins at 500 F.

Variations in pilot burners for heating appliances are:

Orifice, natural gas, No. 72 to 77 D.M.S. (one mfr. uses 0.016 to 0.018 in. diam)

Orifice, mfd. gas, No. 66 to 75 D.M.S. (one mfr. uses 0.020 in. diam)

Primary air openings, two No. 31 D.M.S. or one No. 15 D.M.S.

Burner mixer tube, 1 to 8½ in. long, No. 10 to No. 22 D.M.S. diam made of brass, stainless steel, or iron

Burner ports—size varies greatly with application

Nonprimary Aerated Pilot Burners

Typical nonprimary aerated pilot burners are shown in Fig. 12-125. The advantage of a burner with one port and no primary air inlet is its freedom from linting. If properly designed and protected, the flame is constant and stable. Design features are simple—proper orifice size for input rating and proper size and conformation for flame structure.

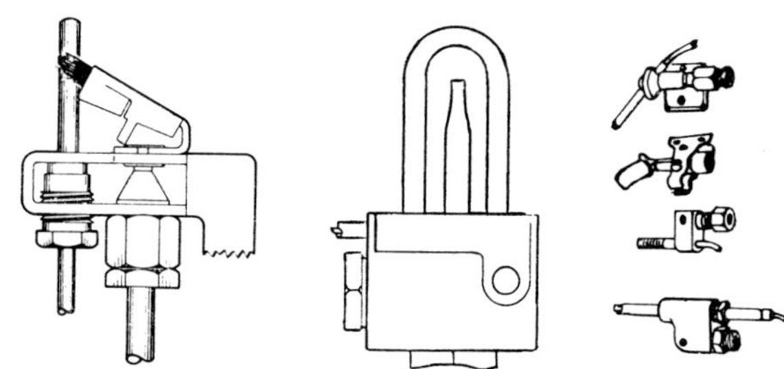

Fig. 12-125 Contemporary nonprimary aerated, nonlinting pilot burners: (left) target type pilot with divided flame; (middle) microflame pilot, low Btu input, blue flame; (right) four design variations of single port pilots.

POWER CONVERSION BURNERS

The pressure of either gas or air (or both) is boosted in a power conversion burner. Four burner types are: (1) forced draft burners—combustion air is supplied by a fan; (2) induced draft burners—depend on draft induced by a fan; (3) premixing burners—all, or nearly all, combustion air is mixed with the gas as primary air; and (4) pressure burners—an air–gas mixture under pressure is supplied, usually at 0.5 to 14 in. w.c.

These burners are used for many purposes, e.g., converting large boilers and appliances originally fired with oil, firing revertible flue furnaces, and various industrial applications. They are usually shipped completely assembled with all controls. Available models cover inputs from 75,000 to 20,000,000 Btu per hour. ASA Listing Requirements[29] are applicable to burners with gas inputs up to 400,000 Btu per hour.

Forced Draft Conversion Burners

Such burners (Fig. 12-126) are available in gas-fired or combination gas- and oil-fired models. They are used as original equipment or conversions for firing boilers ranging from relatively small heating boilers to large watertube proc-

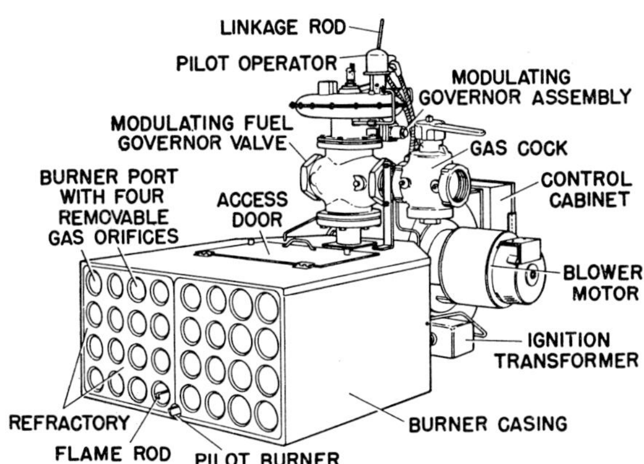

Fig. 12-126 Forced draft conversion burner. (Mettler Div., Eclipse Fuel Engineering Co.)

ess boilers. These burners may also be used on commercial incinerators, ovens, and air-heating applications.

When large boilers are to be fired and the required input exceeds the rated capacities available from a single burner, it is common practice to go to a multiple-burner arrangement. Such an arrangement provides for a great deal of combustion flexibility and a wide turndown range.

The burners are built to operate off–on, high–low–off, or full modulation, and are controlled by steam pressure or by temperature. Low-fire start and fuel–air interlock are supplied as a part of all but the smallest burners. Burners are available to meet FIA and FM requirements; they are also UL listed.

Fan-Assisted Conversion Burners

A typical burner is shown in Fig. 12-127; a description of its numbered components and features follows:

1. Furnace—natural draft here does not affect combustion. Induction of combustion air will not vary appreciably from cold to hot conditions. Air–gas ratio is constant.
2. High heat release rate—increases capacity rating of many boilers.
3. Insulating refractory brick, forming the combustion chamber—rapidly reaches high incandescence after burner starts.
4. In-shot firing system—facilitates burner installation and inspection.
5. Durable refractory tuyeres or ignition tunnels, individual for each mixer—become incandescent and stabilize the flame base.
6. Burner flame ports—take secondary air from plenum chamber thru slotted openings.
7. Cast iron mixers—control primary air–gas mixture ratio.
8. Plenum chamber or wind box—receives combustion air from blower at pressures of from 0.1 to 0.75 in. w.c., depending on size and capacity of burner.
9. Primary combustion air—is inspirated into mixers by combined action of plenum chamber pressure and injector action of gas spuds.
10. Orifices—meter the gas input to each mixer.
11. Burner housing—cast iron for burners up to 1.2 MMBtu per hr capacity, heavy steel for large burners.
12. Gas manifold.
13. Centrifugally operated valve on motor shaft—assures blower operation before main diaphragm gas valve can open.
14. Operating parts of burner—are protected with dirt-tight cover (not shown).
15. Electric transformer and relay—for low voltage (20 volts) room thermostat and limit control circuit.
16. Sound-deadening mounted electric blower motor of 1/40 to 1/4 hp sizes (100,000 to 5,000,000 Btu per hr burner sizes).
17. Controlled source of combustion air—low-speed forward curved blade blower.
18. Main automatic gas shut off valve—is of the diaphragm snap-action type.
19. Main manual gas shutoff cock—is closed while pilot burner is lighted.
20. Gas pressure regulator.
21. Pressure regulator adjustment.
22. Pilot burner—functions as an atmospheric burner during stand-by periods, with minimum gas consumption (3 to 5 cu ft per hr). Pilot flame automatically expands before main gas valve opens, and burns as a blast flame during ignition and burner operation.
23. Electric solenoid valve—controls bleed gas which operates main snap-action diaphragm gas valve.
24. Shield, rigidly attached to main gas manual shutoff cock—prevents access to manual reset on pilot failure automatic shutoff valve unless main gas is manually shut off.
25. Baso thermoelectrically operated valve—for pilot ignition before main diaphragm gas valve can open.
26. Pilot—is lighted here with match; flame is visible from outside.
27. Accessible pilot and safety element assembly—may be removed as a unit without disturbing main burner.

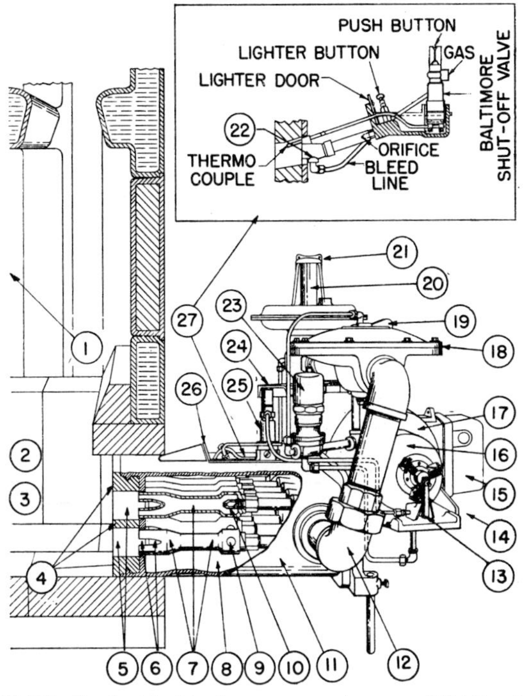

Fig. 12-127 Typical lo-blast power gas burner. (Mid-Continent Metal Products Co.)

Induced Draft Burners

Induced draft burners are used when resistance to flow of flue gases is so great they will not vent properly without an eductor fan. A fan or blower of adequate capacity is located on the outlet of the appliance or vent pipe. Examples are the burners shown in Figs. 12-122a, 12-122b, and 12-159, and wherever unusually long horizontal flue runs are necessary. For this type of operation the fan must be operating both before the gas burner comes on and while it is on. An interlock control, in the form of a pressure or vacuum switch actuated by the induced draft fan, should be provided.

REFERENCES

1. Griffiths, J. C., and others. *New or Unusual Burners and Combustion Processes.* (Research Bull. 96) Cleveland, Ohio, A.G.A. Laboratories, 1963.
2. Berry, W. M., and others. *Design of Atmospheric Gas Burners.* (Tech. Paper 193) Washington, D. C., Bur. of Standards, 1921.
3. Brumbaugh, I. V. "How Natural Gas Burners Can be Improved." *Natl. Gas Assoc. Am. Proc.* 13: 41-126, 1921.
4. Vandaveer, F. E "Atmospheric Gas Burners." (In: Shnidman, L., ed. *Gaseous Fuels*, 2nd ed., Chap. VII. New York, A.G.A., 1954).
5. A.G.A. Laboratories. *Research in Fundamentals of Atmospheric Gas Burner Design.* (Bull. 10) Cleveland, Ohio, 1940.
6. ——. *Fundamentals of Design of Atmospheric Gas Burner Ports.* (Research Bull. 13) Cleveland, Ohio, 1942.
7. ——. *Gas Burners Utilizing All Air for Combustion as Primary Air.* (Research Bull. 20) Cleveland, Ohio, 1944.
8. ——. *Primary Air Control Devices for Atmospheric Gas Burners.* (Research Bull. 22) Cleveland, Ohio, 1944.
9. ——. *Primary Air Injection Characteristics of Atmospheric Gas Burners.* (Research Bull. 26) Cleveland, Ohio, 1944.
10. ——. *Non-Aerated Burners.* (Research Bull. 32) Cleveland, Ohio, 1944.
11. ——. *Primary Air Injection Characteristics of Atmospheric Gas Burners, Part II.* (Research Bull. 37) Cleveland, Ohio, 1945.
12. ——. *Fundamental Data for Design of Totally Aerated Atmospheric Gas Burners.* (Research Bull. 38) Cleveland, Ohio, 1946.
13. Weber, E. J. *Investigation of Primary Air Injection by Use of Variable Chamber Pressures.* (Research Bull. 55) Cleveland, Ohio, A.G.A. Laboratories, 1950.
14. Badger, K. L. *A Study of Small Diameter Gas Glow Tubes.* (Research Bull. 56) Cleveland, Ohio, A.G.A. Laboratories, 1950.
15. Anthony, J. C. *Study of Fundamentals of Design of Non-Primary Aerated Blue Flame Gas Burners.* (Research Bull. 62) Cleveland, Ohio, A.G.A. Laboratories, 1951.
16. Nead, L. A., and Weber, E. J. *A Study of Single Port Gas Burners.* (Research Bull. 72) Cleveland, Ohio, A.G.A. Laboratories, 1956.
17. Griffiths, J. C., and Weber, E. J. *Design and Application of Impingement Target Burners.* (Research Bull. 75) Cleveland, Ohio, A.G.A. Laboratories, 1957.
18. Speidel, P. A., and Hammaker, F. G., Jr. *Research on New Designs for Range Top Burners, Pilots, and Valves.* (Research Bull. 76) Cleveland, Ohio, A.G.A. Laboratories, 1957.
19. Griffiths, J. C., and Weber, E. J. *Influence of Port Design and Gas Composition on Flame Characteristics of Atmospheric Burners.* (Research Bull. 77) Cleveland, Ohio, A. G. A. Laboratories, 1958.
20. Grumer, J., and others. *Relations of Fundamental Flashback Blowoff and Yellow-Tip Limits of Fuel Gas–Air Mixtures to Design Factors of Burners in Gas Appliances.* New York, A.G.A. Comm. on Domestic Gas Research, 1960.
21. Weber, E. J. *Minimizing Lint Stoppage of Atmospheric Gas Burner Ports.* (Research Bull. 79) Cleveland, Ohio, A.G.A. Laboratories, 1960.
22. A.G.A. Laboratories. *American Standard Installation of Gas Appliances and Gas Piping.* (ASA Z21.30-1964) Cleveland, Ohio, 1964.
23. Scherer, R. "Sizing Orifices at High Altitudes." *Gas Age* 116: 32+, Aug. 11, 1955.
24. Hammaker, F. G., Jr., and Hampel, T. E. *Study of Combustion and Heat Transfer Fundamentals in Small Diameter Tubes.* (Research Rept. 1255) Cleveland, Ohio, A.G.A. Laboratories, 1956.
25. Weber, E. J. *Principles of Secondary Aeration of Atmospheric Gas Burners.* (Research Bull. 84) Cleveland, Ohio, A.G.A. Laboratories, 1960.
26. Griffiths, J. C. *Research in Pilot Burner Design, Construction and Performance: Second Bulletin.* (Research Bull. 69) Cleveland, Ohio, A.G.A. Laboratories, 1955.
27. ——. *Development of Gas Pilot Designs for Domestic Gas Appliances.* (Research Rept. 1340) Cleveland, Ohio, A.G.A. Laboratories, 1962.
28. Kane, L. J. *Research in Pilot Burner Design, Construction and Performance: First Bulletin.* (Research Bull. 57) Cleveland, Ohio, A.G.A. Laboratories, 1950.
29. A.G.A. Laboratories. *American Standard Listing Requirements for Domestic Gas Conversion Burners.* (ASA Z21.17-1962) Cleveland, Ohio, 1962.
30. Natl. Fire Protection Assoc. *Standard for Prevention of Furnace Explosions in Natural Gas-Fired Watertube Boiler-Furnaces with One Burner.* (NFPA 85A) Boston, 1964.

Chapter 13
Industrial Combustion Systems

*by D. A. Campbell**

Although industrial gas utilization equipment is as diversified as industry itself, it may be classified according to a small number of categories. Combustion systems differ mainly as to the method of combining oxygen (usually from air) and gas at ratios best suited to produce the heat, temperature, and other process conditions wanted. It is assumed that practically all of the industrial processes discussed herein approach stoichiometric combustion, i.e., perfect combustion. Certain important exceptions, however, will also be evaluated in this chapter. Figure 12-128 shows the efficiency obtainable by proper air–gas mixing; note the significance of process temperature.

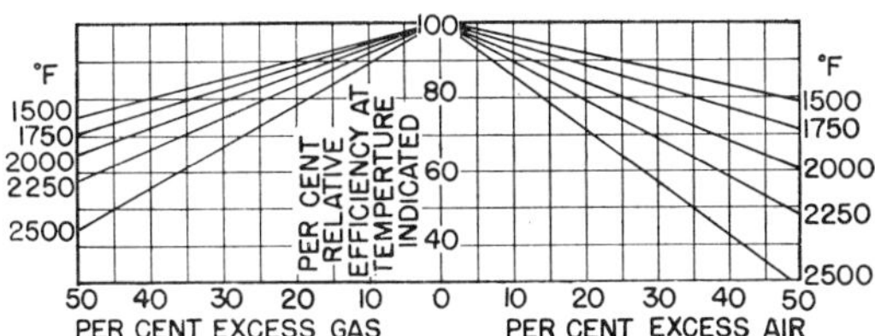

Fig. 12-128 Relative efficiency with excess gas or air. (Surface Combustion Div., Midland-Ross Corp.)

INDUSTRIAL GAS NOMENCLATURE[1]

Gas–Air Proportioning and Mixing Systems for Combustion

1. **Low-Pressure Gas or "Atmospheric" System** (gas pressure less than 0.5 psig or 14 in. w.c.): uses the momentum of a jet of low-pressure gas to entrain from the atmosphere a portion of the air required for combustion.

2. **High-Pressure Gas System** (gas pressure 0.5 psig or higher): uses the momentum of a jet of high-pressure gas to entrain from the atmosphere all, or nearly all, of the air required for combustion.

3. **Low-Pressure Air System** (air pressure up to 5 psig): uses the momentum of a jet of low-pressure air to entrain gas to produce a combustible mixture.

4. **High-Pressure Air System** (air pressure 5 psig or higher): uses the momentum of a jet of high-pressure air to entrain gas, or air and gas, to produce a combustible mixture.

5. **Suction System:** applies suction to a combustion chamber to draw in the air and/or gas necessary to produce the desired combustible mixture.

6. **Two-Valve System:** uses separate controls of air and gas, both of which are under pressure. The valves, controlling the air and gas flows, may or may not be interlocked.

7. **Mechanical System:** proportions air and gas and mechanically compresses the mixture for combustion purposes. A central mixing unit may be used or each individual appliance may have its own mixer.

Mixers and Mixing Devices

1. **Mixer, General:** mixes gas and air in any desired proportion.

2. **Manual Mixer:** requires manual adjustments to maintain the desired air–gas ratio as rates of flow are changed.

3. **Automatic Mixer:** automatically maintains within its rated capacity a substantially constant air–gas ratio at varying rates of flow. All types defined below can be designed to fit this classification.

4. **Gas Jet Mixer:** uses the kinetic energy of a jet of gas issuing from an orifice to entrain all or part of the air required for combustion. Commonly used names of gas jet mixers include: injector, Lojector, venturi mixer, two-stage mixer, inspirator, Hijector, tube mixer, atmospheric mixer, and Bunsen mixer.

5. **Air Jet Mixer:** uses the kinetic energy of a stream of air issuing from an orifice to entrain the gas required for combustion. In some cases, this type of mixer may be designed to entrain some of the air for combustion as well as the gas. Commonly used names of air jet mixers include: low-pressure inspirator, aspirator, Flomixer, Mixjector, mixing tee, low-pressure proportional mixer, and Vari Flame.

6. **Mechanical Mixer:** uses mechanical means to mix gas and air, neglecting entirely any kinetic energy in the gas and air, and compresses the resultant mixture to a pressure suitable for delivery to its point of use. Mixers in this group utilize either a centrifugal fan or some other type of mechanical compressor, with a proportioning device on its intake thru which gas and air are drawn by the fan or compressor suction. The proportioning device may be automatic or may require manual adjustment to maintain

* Deceased.

the desired air–gas ratio as rates of flow are changed. Names of mechanical mixers include: Fan-Mix, industrial carburetor, PreMix, Fantype Mixer, combustion controller, and diluter.

Burners

1. **Burner, General:** releases air–gas mixtures, oxygen–gas mixtures, or air and gas separately into the combustion zone. Industrial gas burners may be classed as atmospheric burners and blast or pressure burners.

2. **Atmospheric Burner:** used in a low-pressure gas or "atmospheric" system which requires secondary air for complete combustion.

3. **Blast Burner:** delivers a combustible mixture under pressure, normally above 0.3 in. w.c., to the combustion zone.

4. **Pressure Burner:** same as blast burner.

5. **Single Port Burner:** has only one discharge opening or port.

6. **Multiport Burner:** has two or more separate discharge openings or ports which may be either flush or raised.

7. **Line Burner:** has a flame that is a continuous "line" from one end to the other. Normally applied to a blast burner.

8. **Pipe Burner:** any type of atmospheric or blast burner made in the form of a tube or pipe, with ports or tips spaced over its length.

9. **Ribbon Burner:** has many small, closely spaced ports, usually made up by pressing corrugated metal ribbons into a slot or an opening of some other shape (Fig. 12-148).

10. **Open Port Burner:** fires across a gap into an opening in the furnace or combustion chamber wall and is not sealed into the wall. Burners of this type include: Torch, Tile, Box, Ventite, and Burnix.

11. **Tunnel Burner:** sealed in a furnace wall; combustion takes place mostly in a refractory tunnel or tuyere which is really part of the burner. Common names for tunnel burners include: Walltite, Impact, Hyperblo, Pyronic, andRefrak.

12. **Flame Retaining Nozzle:** nozzle with built-in features to hold the flame at high mixture pressures. Names of flame retaining nozzles include: Sticktite, Ferrofix, Staylite, and F. R. Nozzle.

13. **Blast Tip:** a small metallic or ceramic burner nozzle made so that flames will not blow away from it, even with high mixture pressures.

14. **Nozzle Mixing Burner:** device in which the gas and air are kept separate until discharged from the burner into the combustion chamber or tunnel. Generally used with low-pressure gas (up to 0.5 psig or 14 in. w.c.) and low-pressure air (up to 5 psig).

15. **Proportional Mixing Burner:** incorporates an automatic mixer and a burner as an integral unit. Trade names include: L. P. Velocity, H. P. Velocity, and Walltite.

16. **Radiant Burner:** transfers a significant part of the combustion heat in the form of radiation from surfaces of various shapes which are usually of refractory material. Trade names include: Red Ray, Duradiant, and Burdette.

17. **Luminous Flame Burner:** discharges nonturbulent parallel strata of air and gas to produce an extended flame of high luminosity.

18. **Ring Burner** (two types):

a. Atmospheric burner made with one or more concentric rings.

b. Burner used in firing boilers; it consists of a perforated vertical gas ring with air admitted generally thru the center of the ring. Combustion air may be supplied by natural, induced, or forced draft.

19. **Multijet Burner:** generally consists of gas manifolds with a large number of jets arranged to fire horizontally thru openings in a vertical refractory plate. These openings are of various shapes: round, square, clover-leafed, etc. Combustion air may be supplied by natural, induced, or forced draft. Complete assemblies combining burner, refractory plate, wind box, blower and controls are generally known as forced draft boiler burners. Trade names include: Lo-Blast, Fan-Mix, Flame King, and Gas Pak.

20. **Enclosed Combustion Burner:** confines the combustion in a small chamber or miniature furnace, and only the high-temperature completely burned gases, in the form of high-velocity jets or streams, are used for heating. Trade names include: Superheat and Zigzag.

21. **Diaphragm Burner:** utilizes a porous refractory diaphragm as the port so that the combustion takes place over the entire area of this refractory diaphragm.

22. **Gas–Oil Burner:** burns gas and oil simultaneously.

23. **Dual Fuel Burner:** burns either gas or oil, but not both together.

24. **Combination Gas and Oil Burner:** burns either gas or oil or both together.

GENERAL CLASSIFICATION OF SYSTEMS

Industrial combustion systems consist of several major parts which may be either in an integral unit or in an assembly. They are: mixing equipment; ratio control equipment or devices; volume control equipment; and burner or combustion equipment.

Assuming that adequate air and gas are available at a constant pressure and temperature, the energy required for their mixing in various systems is obtained by one of the following four distinct methods:

1. Entraining necessary air by means of the kinetic energy of a gas stream issuing from an orifice in a venturi mixer.
2. Entraining necessary gas by means of the kinetic energy of an air stream issuing from an orifice in a mixing device.
3. Supplying all energy by a pump, compressor, or blower, disregarding entirely the kinetic energy of both gas and air.
4. Using the energy from both gas and air streams.

Based on the foregoing energy considerations, industrial combustion systems may be classified as shown in Table 12-91. The general arrangement of the components of the various numbered systems shown are given in Fig. 12-129. Table 12-91 shows that each system has two principal components, the ratio controller and the burner equipment. Installed costs of premix systems and systems in which air and gas are mixed at the point of combustion are comparable; the saving on piping and regulators in the former tends to offset the cost for the automatic carburetor.

System I—Separate Gas and Air Feeds to Combustion Chamber

This simple system (Table 12-91 and Fig. 12-129) is used in many large installations in which high temperatures in the combustion chamber assure continuous ignition of the gas–air mixture. The mixing rate is relatively slow, since the air and gas are admitted separately. Combustion takes place over a large area, and only by using preheated air can high temperatures be achieved. As the gas and air do not mix until they enter the combustion chamber, air may be preheated as much as is permissible and economical. Mechanical means for maintaining air–gas ratios include dampers or valves in the supply lines.

System II—Nozzle Mixing

This simple and very flexible system (Table 12-91 and Fig. 12-129) may be used in practically any type of industrial application, at 1000 to 2800 F, with combustion chamber heat releases up to 250 MBtu per hr-cu ft.

The ratio control may be either mechanical (operating valves in both the air and gas lines) or proportioning (arranging for variations in one fluid to initiate proportional variations in the other). By suitable variations in the burner and control equipment, a nozzle mixing system (400 to 36,000 fpm) can be operated either stoichiometrically or with excess air or excess gas as required, without possibility of flashback. Usually the burner system is sealed so that all air is con-

Table 12-91 General Classifications of Combustion Systems
(see Fig. 12-129)

System	Type of mixers and method of mixing	Ratio control method	Volume control, manual or auto.	Burner types used	Typical furnace or process uses
I Separate gas and air feeds to combustion chamber	None (both gas and air under pressure)	1. Manual: 2-valve 2. Automatic: a. Pressure balance b. Mech. linkage	Separate valves on gas and air lines	. . .	Open-hearth glass tanks, and lime kilns (used in large industrial gas applications)
II Nozzle mixing	None except at point of combustion (both gas and air under pressure)	1. Manual: 2-valve 2. Automatic: a. Pressure balance b. Mech. linkage	Separate valves on gas and air lines	Nozzle mixing with combustion block	Air heaters, heat-treating furnaces, and forge furnace boilers
III Partial premixing requiring secondary air	1. Gas jet mixers; gas at press., air at zero 2. Air jet mixers air at press., gas at zero 3. Mechanical mixers; both air and gas at zero	Manual or automatic	Valve in line of entraining fluid or valve in discharge line from mech. mixer	Open burners: 1. nozzles 2. tips 3. line 4. ribbon	All applications in which excess air is not harmful; heating machines
IV Complete premixing using kinetic energy only	1. Gas jet mixers; gas at press; air at zero 2. Air jet mixers; air at press., gas at zero	Manual (variable) or automatic (fixed)	Valve in line handling entraining fluid	Sealed tunnel burners, some non-tunnel and open burners	Furnaces, kilns, melters, processes and heating machines using open burners, generators, and boilers (units requiring close control of atmosphere in combustion chamber)
V Complete premixing using mechanical mixers	Fans, compressors, pumps, and diluters (both gas and air at atmospheric press.)	Adjustable or fixed over range of operation; manual or automatic	Mech. control of inlet gas and air or valves in mixture line from mixer outlet	Sealed tunnel burners, some nontunnel and open burners	Furnaces, kilns, melters, generators, boilers, and processes and heating machines using open burners (units requiring close control of atmosphere in combustion chamber and/or wide range of operation)
VI Combination using parts of Systems I–V	As needed to fit systems				Special processes mainly

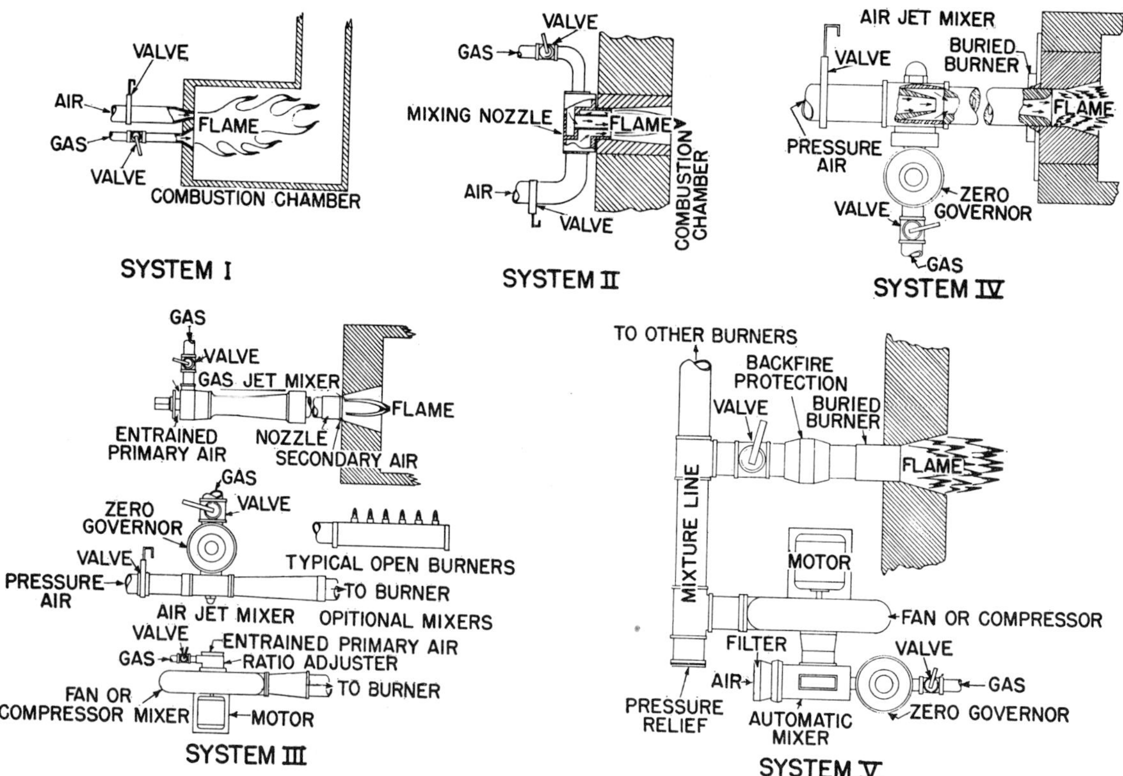

Fig. 12-129 Arrangement of components of Systems I thru V inclusive (see Table 12-91).

trolled by either a manual or an automatic ratio control system.

System III—Partial Premixing Requiring Secondary Air

This simple and very flexible system (Table 12-91 and Fig. 12-129) employs a wide variety of open nozzles or burners of either the single orifice or multitip type. Usually the mixture contains less than half the air required for combustion; the nozzle velocity ranges from 50 to 1200 fpm. Because of the relatively poor mixing of the secondary air with the flame in many types of burners, a large quantity of excess air may be required. Control of secondary air may be either variable or fixed throughout the capacity range. This system is generally used for temperatures of 100 to 1200 F and combustion chamber heat releases up to 35 MBtu per hr-cu ft.

System IV—Complete Premixing by Kinetic Energy

This system (Table 12-91 and Fig. 12-129) is based on the use of burners requiring a complete mixture of gas and air at, or close to, the stoichiometric ratios. Because of its simplicity and reasonably accurate ratio and temperature control, it is the most commonly used system in industry. The absence of secondary air facilitates attainment of relatively high temperatures as well as continuous, close control of a neutral, rich, or lean combustion atmosphere, as desired. Furnace pressure and draft conditions must be considered in relation to the system operation to maintain suitable ratio control.

Both systems III and IV use mixture velocities of 1200 to 36,000 fpm thru nozzles. The applicable temperature range is 1200 to 3000 F and the maximum combustion chamber releases are 250 and 5000 MBtu per hr-cu ft for open and tunnel burners, respectively.

System V—Complete Premixing with Mechanical Mixers

This system (Table 12-91 and Fig. 12-129) produces and burns mixtures at any pressure desired to overcome high combustion chamber pressures or to give a wide range of operation to burner equipment. Energy utilization is high because the mixture is used directly from the mixing machine without any further mixing losses. Because ratio control is accurate and independent of the volume delivered, a number of burners may be supplied independently, and the burners may be turned off and on separately without affecting the ratios in the other burners. This is extremely advantageous in many operations in which heat patterns or a wide range of heat input can only be achieved by shutting off some burners. Suitable protective devices must be installed between the mechanical mixer and the burner equipment so that no damage to the mixer will result from a burner backfire.

System VI—Combinations of Systems I thru V

No hard and fast lines may be drawn among different combustion systems. A combination most suitable for a special process may be evolved by use of parts of two or three systems. For example, all gas required for a process can be mixed with part of the air in a mechanical mixer, and the rich mixture thus secured can be supplied to a nozzle mixing burner in which the balance of the air for complete combustion may be added. Thru various combinations (Table 12-91 and Fig.

12-129), it is possible to change from one gas to another by simply introducing variations to accommodate different air–gas ratios or heat capacities.

RATIO CONTROL BY PREMIXING EQUIPMENT

Gas–air mixtures must be within flammability limits (Figs. 2-24b, 2-25, and 2-26) in order to burn. Some combustion systems depend on using the combustion space as a mixing chamber for the gas and air, which are admitted separately. However, most industrial gas utilization requires a method of mixing prior to combustion which can be regulated and will produce consistently uniform results.

Safety. Air and gas mixtures at or near stoichiometric ratios ignite readily; therefore, burner equipment should be properly designed to minimize flashback. Proper design provides velocities thru the burner orifice(s) safely in excess of the flame propagation rate of the gas. Standard practice is to place a fire check just upstream from each burner or group of burners for flame entrapment in the event of flashback. On long distribution lines provision should be made to relieve excess pressures due to flashback by installing rupture disks or backfire preventers. Many components for carburetor-type combustion systems are approved and listed by Underwriters' Laboratories and by Factory Mutual Engineering Division.

Compressors. Compressors may be of the sliding vane type, rotary positive displacement type, or turbo type. Since the mixtures compressed are flammable, construction should be of nonsparking materials. Positive displacement blowers should be equipped with a pressure controlled by-pass to suction so that variable output can be obtained.

Gas Jet Mixers

The amount of fluid (air) that can be entrained by a given fluid (gas) jet depends on the gas pressure at the orifice, the efficiency of entrainment in the mixers, and the resistance to entrainment at the mixer outlet. As these factors are independent of size, the mixture pressure that can be developed by a given fluid may be estimated by the empirical Eq. 1. Table 12-92a gives maximum mixture pressures for various gases calculated from this equation.

$$P_m = \frac{CE^2GP_g}{(1 + R)(G + R)} \qquad (1)$$

where:

P_m = static mixture pressure, in w.c. (see Table 12-92a)

C = coefficient of conversion of velocity pressure to static pressure in mixer for well-designed mixers:

	C range	C avg
Air entrained by gas	0.85–0.95	0.9
Gas entrained by air	0.60–0.95	0.8

E = coefficient of entrainment, dependent on design (see Table 12-92b)

G = specific gravity of *entraining* fluid (air = 1)

P_g = gage pressure of entraining fluid at orifice, in. w.c.

R = ratio of *entrained* fluid to *entraining* fluid

Table 12-92a Maximum Mixture Pressures with Various Gases (based on entrainment coefficent $E = 1.0$ and velocity conversion coefficient $C = 0.9$)

	Natural gas			Propane	Butane	Coke oven
Btu per cu ft, approx.	950	1050	1250	2500	3200	525
Specific gravity, G	0.6	0.62	0.65	1.52	2.07	0.45
Pressure,* P_g, psig	30	30	30	30	30	10
Air–gas ratio, approx.	9/1	10/1	12/1	25/1	32/1	5/1
Mixture sp gr	0.96	0.975	0.975	1.02	1.03	0.91
Air entrained by gas						
Max possible mixture press. ($E = 1.0$, $C = 0.9$), P_m, in. w.c.	4.65	4.0	2.94	2.75	1.35	4.1
Mixture press., P_m ($C = 0.9$, E from Table 12-92b), in w.c.	2.5	2.2	1.6	1.4	0.7	3.4
Gas entrained by air						
Max possible mixture press.† ($C = 0.8$, E from Table 12-92b), in. w.c.		14.8 avg		16.6	16.9	9.4

* For pressures P_g other than those given, the following mixture data[14] (air entrained by gas) may be linearly interpolated:

	Natural Gas	Water Gas	Mfd Gas
Specific gravity, G	0.65	0.55	0.50
Heating value, Btu/cu ft	1000	310	530
Pressure, P_g, psig	30	18	30
Mixture press., P_m, in. w.c.	3.5	17.8	14.2

Example. The mixture pressure for a 1000 Btu gas at 10 psig = (10/30) × 3.5 = 1.2 in. w.c.

† For well-designed mixers under *ideal* conditions, based on air at 1.0 psig and gas at 0.0 psig; actual mixture pressures may be only 40 per cent of these values; for water gas, 9.8; for mixed gas, 12.0.

Table 12-92b Average Coefficients of Entrainment, E

	Air entrained by gas			Gas entrained by air		
Kind of gas	Air–gas ratio	E	E^2	Gas–air ratio	E	E^2
Natural	10	0.74	0.55	0.10	0.90	0.81
Propane	25	.71	.50	.04	.90	.81
Butane	...	...	...	.031	.90	.81
Coke oven	4	.90	.81	.25	.75	.56
Water gas	5	.85	.72	.20	.80	.64
Mixed (natl. & mfd.)	8	0.82	0.67	0.125	0.85	0.72

Ideally, gas jet mixers should be individually designed for each installation to secure the maximum mixture pressure. Practically, the discharge end must conform with standard pipe sizes and the throat diameter can only be varied slightly in each size mixer.

For low-pressure (atmospheric) mixers, which usually entrain only a portion of the total air required, the throat area varies from 35 to 50 per cent of the discharge pipe area; throats may be used "as cast" without machining. For high pressure (above 0.5 psig), where industrial mixers are designed to entrain all air required for complete combustion, the throat area varies from 20 to 30 per cent of the discharge area. For best results these throats should be accurately machined and centered (Fig. 12-130).

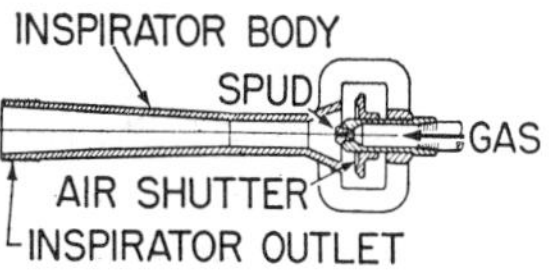

Fig. 12-130 High-pressure gas jet mixer. (Surface Combustion Div., Midland-Ross Corp.)

Table 12-93 Typical Ratio of Burner Area to Orifice Area for Various Gases under Conditions Stated

Kind of gas	Heat value, Btu gross	Approx. air–gas ratio	Approx. ratio of burner discharge area to orifice area*
Natl. gas: Low	976	9.17	214
Med.	1140	10.7	250
High	1245	11.7	286
Propane	2500	25	585
Butane–air mix.	750	6.8	65
Carb. water gas	520	4.37	54
Mixed coke oven and water gas	560	4.8	66
Coke oven gas	595	5.2	100
Mixed natl. & mfd.	800	8.0	159

* Burner coefficient of discharge = 1.0.

The burner discharge area served by a gas jet mixer should be less than that of the outlet pipe, but large enough to handle the volume of gas–air mixture at the pressure developed according to Eq. 1. The coefficient of discharge of the burner discharge area also must be considered. Common practice is to relate the burner discharge area (Table 12-93) to that of the gas orifice (Table 12-94a). Variables such as discharge coefficients and furnace draft must also be evaluated.

Gas orifice capacity (Table 12-94a) varies with the pressure and density of the gas. Table 12-94b gives correction factors for specific gravities other than 0.55.

With a given gas pressure and a fixed orifice area, the throat velocity which can be developed varies with the amount of air entrained. This amount is governed by resistance at the mixture outlet and the air inlet. The resistance may be measured as the static pressure required to push the mixture thru the piping and burner. Increasing the burner discharge area decreases the resistance and permits more air entrainment, increasing the air–gas ratio. A strong furnace draft will do the same. Reducing the burner area, closing the air shutter, or operating against an internal furnace pressure reduces the amount of air entrained and decreases the air–gas ratio. The furnace back pressure may be high enough to upset the operation, requiring capacity reduction or system redesign.

Air Jet Mixers

The simplest type requires separate valves for air and gas control. By linking valves with identical flow characteristics, the gas and air flows can be kept proportional. If not so linked, the valves may be manually operated separately, but any variation in either gas or air pressure will change the air–gas ratio at the mixer outlet.

A proportioning air jet mixer (Fig. 12-131) maintains desired air–gas ratios if the jet and venturi throat are accurately machined and aligned. Gas enters the mixing chamber thru an adjustable orifice which, after setting, may be locked in place. Gas pressure at the adjustable orifice is maintained at zero gage pressure by a sensitive regulator (zero governor) which reduces the normal supply pressure to zero regardless of the flow rate.

In operation, air flow thru the restricted throat of the removable insert aspirates gas maintained at constant pressure at the spud by the governor. The air–gas ratio desired is set by sizing the spud. Once set, an accurate air–gas ratio prevails at all generally used air flows, provided the gas specific gravity and all mechanical conditions remain constant. The air flow can be controlled by a manual or automatic valve in the supply line. This valve is the only volume control that must be operated, gas flowing automatically in proportion at all times.

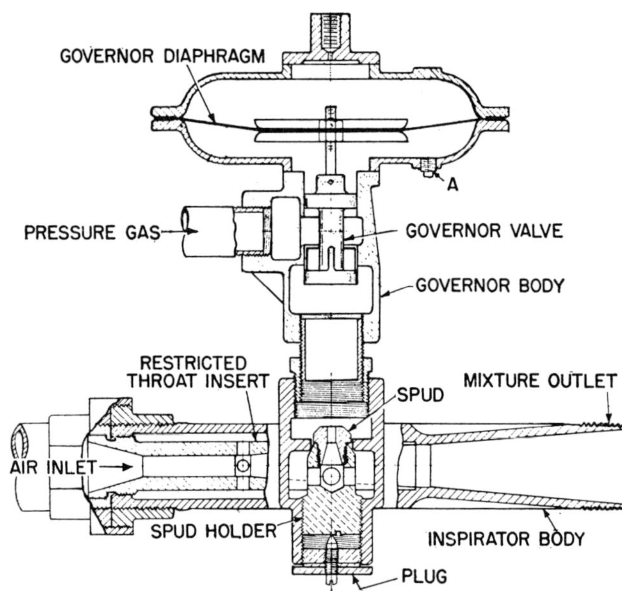

Fig. 12-131 Low-pressure air jet proportional mixer. (Surface Combustion Div., Midland-Ross Corp.)

As the gas flow is determined by the pressure drop between the "zero" chamber supplying the spud and the tee suction or mixing chamber, the governor must maintain "zero" pressure at all times, regardless of flow.

Example. Assumed conditions: air pressure at jet, 25 in. w.c.; suction developed measured in mixing chamber, −9 in. w.c.; measured air passing air jet, 10,000 cfh; measured gas passing spud, 1000 cfh; air–gas ratio, 10 to 1; and pressure at outlet of zero governor, 0 in w.c.

If the air pressure at the jet is reduced to 12.5 in. w.c., the suction will reduce proportionately, becoming −4.5 in. The other new conditions will be: gas pressure, 0 in. w.c.; air pressure drop thru jet = 12.5 + 4.5 = 17 in. w.c.; gas pressure drop into chamber = 0 + 4.5 = 4.5 in. w.c.; air volume = $[(12.5 + 4.5)/(25 + 9)]^{0.5} \times 10{,}000 = 7080$ cfh; gas volume = $[(0 + 4.5)/(0 + 9)]^{0.5} \times 1000 = 708$ cfh; and air–gas ratio = 7080/708 = 10, or 10 to 1.

If the governor should lose its accuracy and develop a pressure at its outlet of 1 in. w.c., the conditions would be: gas pressure drop = 1 + 4.5 = 5.5 in. w.c.; gas volume = $[(1 + 4.5)/(0 + 9)]^{0.5} \times 1000 = 783$ cfh; and air–gas ratio = 7080/783 = 8, or 8 to 1. The flame characteristics would change materially.

If the governor continued to hold 1 in. w.c. outlet pressure and air pressure were further reduced to 1 in. w.c., the other conditions would be: suction = −9/25 = −0.36 in. w.c.; air pressure drop thru jet = gas pressure drop into tee = 1 + 0.36 = 1.36 in. w.c.; air volume = $[(1 + 0.36)/(25 + 9)]^{0.5} \times 10{,}000 = 2000$ cfh; gas volume = $[(1 + 0.36)/(0 + 9)]^{0.5} \times 1000 = 390$ cfh; and air–gas ratio = 2000/390 = 5.1, or 5 to 1.

Table 12-94a Capacities of Orifices in Cubic Feet per Hour*

(based on 1.0 coefficient of discharge and 0.55 specific gravity; see Table 12-94b for specific gravity correction factors)

Diam, in.	MTD size	Diam, in.	Area, sq in.	Pressure, psig																	
				1	2	3	4	5	6	7	8	9	10	12	14	15	16	18	20	25	30
	80	0.0135	0.000143	1.7	2.3	2.9	3.4	3.8	4.1	4.5	4.8	5	5.4	5.8	6.3	6.5	6.7	7.2	7.5	8.4	9.3
	79	.0145	.000165	2.0	2.8	3.4	3.8	4.3	4.8	5.2	5.5	5.8	6.2	6.7	7.3	7.5	7.8	8.2	8.7	9.7	10.8
1/64	...	.0156	.00019	2.2	3.2	3.9	4.5	5.0	5.5	6.0	6.3	6.7	7.2	7.8	8.4	8.7	9.	9.2	10	11.2	12.4
	78	.016	.00020	2.4	3.3	4.1	4.7	5.3	5.8	6.2	6.7	7.1	7.5	8.2	8.8	9.2	9.4	10	10.5	11.8	13
	77	.018	.00025	3.0	4.2	5.1	5.9	6.6	7.2	7.8	8.3	8.8	9.4	10.2	11	11.4	11.8	12.5	13.2	14.8	16.3
	76	.020	.00031	3.7	5.2	6.3	7.3	8.2	9.0	9.7	10.3	11	11.6	12.7	13.7	14.2	14.6	15.5	16.4	18.3	20
	75	.021	.00035	4.1	5.9	7.1	8.3	9.3	10.1	11	11.7	12.4	13.1	14.3	15.4	16	16.5	17.5	18.5	20.6	23
	74	.0225	.00040	4.7	6.7	8.2	9.4	10.6	11.5	12.5	13.3	14.1	15.0	16.4	17.6	18.3	19	20	21	24	26
	73	.024	.00045	5.3	7.5	9.2	10.6	11.8	13.0	14	15	15.9	17.0	18.4	19.7	20.6	21	22.5	24	27	29
	72	.025	.00049	5.8	8.2	10.0	11.5	13.0	14.1	15.3	16.3	17.3	18.4	20	22	22	23	25	26	29	32
	71	.026	.00053	6.3	8.8	10.8	12.5	14.0	15.3	16.5	17.6	18.7	20	22	23	24	25	27	28	31	34
	70	.028	.00062	7.3	10.4	12.5	14.6	16.4	17.9	19.3	20.6	22	23	25	27	28	29	31	33	37	40
	69	.0292	.00067	7.9	11.3	13.6	15.8	17.6	19.3	21	22	24	25	27	29	31	32	34	35	40	44
	68	.030	.00075	8.9	12.5	15.3	17.6	19.8	21.6	23	25	26	28	31	33	34	35	37	40	44	49
1/32	...	.0312	.00076	9.0	12.7	15.5	18	20.0	21.9	24	25	27	28.5	31	33	35	36	38	40	45	50
	67	.032	.00080	9.5	13.4	16.4	19	21.0	23.0	25	27	28	30	33	35	37	38	40	42	47	52
	66	.033	.00086	10.2	14.2	17.6	20	23	25.0	27	29	30	32	35	38	39	40	43	45	50	56
	65	.035	.00096	11.3	16.0	19.6	23	25	28	30	32	34	36	39	42	44	45	48	51	56	62
	64	.036	.00102	12.1	17.0	20.6	24	27	30	32	34	36	38	42	45	47	48	51	54	60	66
	63	.037	.00108	12.7	18	22.0	25	28	31	34	36	38	40	44	47	49	51	54	57	64	70
	62	.038	.00113	13.3	19	23	27	30	33	35	38	40	42	46	50	52	53	57	60	66	74
	61	.039	.00119	14	20	24	28	31	34	37	40	42	45	49	52	54	56	60	63	70	78
	60	.040	.00126	15	21	26	30	33	36	38	42	44	47	53	55	58	59	63	66	74	82
	59	.041	.00132	15.6	22	27	31	35	38	41	44	47	50	54	57	60	62	66	70	78	86
	58	.042	.00138	16.4	23	28	33	36	40	43	46	49	52	56	61	63	65	69	73	81	90
	57	.043	.00145	17.1	24	30	34	38	42	45	48	51	55	59	64	66	68	72	76	85	93
	56	.0465	.00170	20.0	28	35	40	45	49	53	57	60	64	69	75	78	80	85	90	100	110
3/64	...	.0469	.00173	20.4	29	35	41	46	50	54	58	61	65	71	76	79	81	86	91	102	112
	55	.0520	.00210	25	35	43	50	55	60	65	70	74	79	86	92	96	99	105	111	124	135
	54	.0550	.0023	27	38	47	54	61	66	72	77	81	36	94	101	105	108	115	121	136	148
	53	.0595	.0028	33	47	57	66	74	81	87	93	99	105	114	123	128	132	140	148	165	18[illegible]
1/16	...	.0625	.0031	37	52	63	73	82	90	97	103	110	116	127	136	142	146	155	163	183	20[illegible]
	52	.0635	.0032	38	53	65	76	84	92	100	107	113	120	131	141	146	151	160	169	189	205
	51	.0670	.0035	41	58	71	83	92	101	109	117	123	131	143	154	160	165	175	184	207	226
	50	.070	.0038	45	63	78	90	100	110	119	126	134	143	155	167	174	178	190	200	225	245
	49	.073	.0042	50	70	86	99	111	121	131	140	148	158	172	185	192	198	210	222	248	271
	48	.076	.0043	51	72	88	101	113	124	134	143	152	161	176	189	196	203	215	227	254	278
5/64	...	.0781	.0048	57	80	98	113	127	139	150	160	170	180	196	212	220	226	240	253	283	310
	47	.0785	.0049	58	82	100	116	129	142	153	163	173	184	200	216	224	230	245	258	288	316
	46	.081	.0051	60	85	104	120	135	147	159	170	180	192	208	225	233	240	255	269	300	329
	45	.082	.0053	63	89	108	125	140	153	165	176	187	200	217	234	242	250	265	280	312	345
	44	.086	.0058	69	97	118	137	153	168	181	193	205	218	238	256	265	273	290	306	342	372
	43	.089	.0062	73	103	126	146	163	179	194	207	219	233	254	273	283	292	310	328	366	405
	42	.0935	.0069	82	115	141	163	182	199	215	230	244	259	282	304	316	325	345	364	406	440
3/32	...	.0937	.0069	82	115	141	163	182	199	215	230	244	259	282	304	316	325	345	364	406	445
	41	.096	.0072	85	120	147	170	190	208	225	240	254	270	294	318	329	339	360	380	425	465
	40	.098	.0075	88	125	153	177	198	217	234	250	265	282	307	330	342	354	375	396	442	483
	39	.0995	.0078	92	130	159	184	206	225	244	260	276	293	320	344	356	368	390	411	460	503
	38	.1015	.0081	96	135	165	191	214	234	253	270	286	304	331	367	370	382	405	428	478	522
	37	.104	.0085	100	142	173	200	224	246	265	283	300	318	348	374	388	400	425	449	500	547
	36	.1065	.0090	106	150	183	212	238	260	271	300	318	348	370	396	411	425	450	475	530	580
7/64	...	.1093	.0094	111	157	192	222	248	271	293	312	332	354	385	414	430	443	470	496	555	606
	35	.110	.0095	112	159	194	224	250	275	297	316	336	357	388	419	434	448	475	501	560	612
	34	.111	.0097	115	162	198	228	256	280	303	323	343	364	396	428	444	458	485	511	572	625
	33	.113	.0100	118	167	204	236	264	289	312	333	354	376	409	441	457	471	500	528	590	645
	32	.116	.0106	125	177	216	250	280	306	331	353	385	400	434	467	485	500	530	560	625	685
	31	.120	.0113	133	189	230	267	298	326	355	376	400	425	462	497	526	534	565	596	676	728
1/8	...	.125	.0123	145	205	250	290	325	356	384	410	435	462	503	521	563	580	615	650	725	793
	30	.1285	.0130	153	217	265	307	344	376	405	433	460	490	532	574	595	614	650	686	767	840
	29	0.136	0.0145	171	242	296	342	383	419	452	482	513	545	594	640	663	685	725	765	838	935

Table 12-94a Capacities of Orifices in Cubic Feet per Hour* (Continued)

Diam, in.	MTD size	Diam, in.	Area, sq in.	Pressure, psig 1	2	3	4	5	6	7	8	9	10	12	14	15	16	18	20	25	30
	28	0.1405	0.0155	183	259	316	367	410	447	485	516	547	582	635	683	709	730	775	818	915	1,000
9/64	...	.1406	.0156	184	262	318	368	412	450	488	520	552	588	638	687	714	736	780	823	920	1,005
	27	.144	.0163	192	272	332	385	430	470	509	542	576	614	667	718	745	770	815	860	960	1,050
	26	.147	.0174	205	290	353	410	460	502	543	580	615	655	712	765	790	820	870	918	1,025	1,122
	25	.1495	.0175	207	292	357	413	462	505	547	582	620	658	717	770	800	825	875	924	1,030	1,130
	24	.152	.0181	214	302	370	428	478	522	565	602	640	681	742	797	830	853	905	955	1,068	1,170
	23	.154	.0186	219	310	380	440	491	538	580	620	657	700	760	820	850	877	912	980	1,095	1,200
5/32	...	.1562	.0192	226	321	391	443	507	555	600	640	680	721	785	846	879	906	960	1,012	1,130	1,240
	22	.157	.0193	228	322	394	455	510	556	602	642	683	726	790	851	883	910	965	1,014	1,140	1,242
	21	.159	.0198	234	332	405	468	525	572	619	660	700	745	810	872	906	935	988	1,043	1,170	1,278
	20	.161	.0203	240	340	414	480	535	586	634	676	718	763	830	894	928	960	1,015	1,070	1,200	1,310
	19	.166	.0216	255	361	440	510	570	625	674	720	764	814	882	950	987	1,020	1,080	1,140	1,275	1,390
	18	.1695	.0226	267	368	460	534	595	654	705	752	800	850	925	995	1,030	1,065	1,130	1,190	1,330	1,457
11/64	...	.1719	.0232	274	388	473	548	614	670	725	772	820	874	950	1,020	1,060	1,093	1,160	1,223	1,370	1,495
	17	.175	.0235	278	392	480	555	620	678	734	782	830	885	961	1,033	1,075	1,109	1,175	1,240	1,385	1,520
	16	.177	.0246	290	412	502	580	650	710	768	820	870	925	1,005	1,082	1,123	1,160	1,230	1,292	1,450	1,590
	15	.180	.0254	300	425	518	600	670	735	792	845	900	955	1,040	1,120	1,160	1,200	1,270	1,340	1,500	1,640
	14	.182	.0260	307	435	530	615	687	750	810	865	920	980	1,062	1,142	1,190	1,225	1,300	1,360	1,530	1,680
	13	.185	.0269	318	450	548	635	710	776	838	895	950	1,010	1,100	1,182	1,230	1,268	1,345	1,420	1,590	1,732
3/16	...	.1875	.0276	326	460	564	650	728	797	860	920	975	1,040	1,130	1,215	1,260	1,300	1,380	1,465	1,630	1,780
	12	.189	.02805	330	467	573	664	742	810	875	935	990	1,045	1,145	1,238	1,281	1,320	1,403	1,483	1,650	1,810
	11	.191	.02865	342	479	586	675	758	827	893	955	1,012	1,070	1,170	1,262	1,310	1,350	1,430	1,513	1,690	1,850
	10	.1935	.0294	347	490	600	695	775	846	916	980	1,040	1,095	1,200	1,295	1,343	1,382	1,470	1,550	1,732	1,920
	9	.196	.0302	356	504	617	712	795	870	941	1,005	1,070	1,125	1,235	1,330	1,380	1,420	1,510	1,592	1,781	1,945
	8	.199	.0311	366	519	635	734	820	895	970	1,035	1,100	1,160	1,270	1,370	1,420	1,461	1,552	1,643	1,835	2,000
	7	.201	.0316	373	527	646	745	833	910	985	1,050	1,118	1,180	1,290	1,390	1,442	1,485	1,580	1,670	1,865	2,040
13/64	...	.2031	.0324	383	541	664	765	855	935	1,010	1,080	1,145	1,210	1,325	1,430	1,480	1,520	1,620	1,710	1,910	2,090
	6	.204	.0327	386	545	668	771	860	940	1,025	1,090	1,155	1,218	1,335	1,440	1,493	1,535	1,632	1,730	1,930	2,105
	5	.2055	.0332	392	554	678	782	875	960	1,040	1,105	1,175	1,238	1,356	1,460	1,517	1,560	1,660	1,752	1,960	2,140
	4	.209	.0343	404	571	700	810	905	961	1,065	1,142	1,212	1,278	1,400	1,510	1,568	1,610	1,715	1,810	2,020	2,210
	3	.213	.0356	420	583	725	840	940	1,025	1,113	1,185	1,260	1,325	1,455	1,570	1,628	1,675	1,780	1,880	2,100	2,300
7/32	...	.2187	.0376	443	626	766	886	995	1,085	1,178	1,250	1,330	1,400	1,535	1,656	1,720	1,765	1,880	1,985	2,220	2,420
	2	.221	.0384	452	640	785	905	1,011	1,110	1,202	1,280	1,360	1,430	1,570	1,690	1,752	1,805	1,920	2,030	2,265	2,480
	1	.228	.0409	482	680	835	965	1,080	1,180	1,280	1,361	1,448	1,522	1,675	1,800	1,870	1,920	2,045	2,160	2,415	2,640
	A	.234	.0430	506	716	880	1,015	1,135	1,240	1,348	1,430	1,520	1,600	1,755	1,898	1,965	2,020	2,200	2,270	2,540	2,770
15/64	...	.2343	.0431	509	718	883	1,018	1,138	1,270	1,350	1,432	1,521	1,605	1,760	1,900	1,970	2,025	2,205	2,275	2,550	2,780
	B	.238	.0444	524	740	905	1,048	1,170	1,280	1,390	1,480	1,570	1,652	1,810	1,955	2,030	2,090	2,220	2,342	2,620	2,860
	C	.242	.0460	542	766	940	1,082	1,210	1,325	1,440	1,530	1,628	1,713	1,880	2,030	2,100	2,160	2,300	2,430	2,720	2,960
	D	.246	.0475	560	792	970	1,120	1,252	1,370	1,490	1,581	1,680	1,770	1,940	2,095	2,170	2,185	2,375	2,510	2,800	3,060
1/4	E	.250	.0491	579	820	1,000	1,160	1,300	1,418	1,540	1,635	1,736	1,830	2,000	2,160	2,240	2,310	2,455	2,595	2,890	3,180
	F	.257	.0519	610	864	1,060	1,226	1,370	1,500	1,625	1,730	1,835	1,930	2,120	2,290	2,370	2,440	2,590	2,690	3,060	3,240
	G	.261	.0535	630	895	1,095	1,260	1,410	1,540	1,675	1,780	1,890	1,995	2,190	2,360	2,440	2,520	2,670	2,815	3,160	3,450
17/64	...	.2656	.0554	654	925	1,132	1,309	1,465	1,600	1,735	1,848	1,960	2,065	2,260	2,440	2,530	2,600	2,770	2,930	3,270	3,570
	H	.266	.0556	658	930	1,138	1,312	1,470	1,602	1,738	1,850	1,963	2,075	2,275	2,450	2,540	2,610	2,780	2,940	3,280	3,590
	I	.272	.0580	683	965	1,185	1,370	1,530	1,670	1,816	1,931	2,050	2,160	2,370	2,562	2,654	2,730	2,900	3,060	3,420	3,740

	J	.277	.0601	709	1,000	1,230	1,415	1,590	1,732	1,880	2,000	2,120	2,240	2,450	2,646	2,744	2,830	3,000	3,180	3,540	3,880
	K	.281	.0620	730	1,030	1,270	1,461	1,630	1,785	1,940	2,060	2,190	2,310	2,530	2,732	2,834	2,920	3,101	3,270	3,660	4,000
9/32	...	.2812	.0621	735	1,040	1,275	1,462	1,640	1,790	1,945	2,070	2,200	2,320	2,540	2,743	2,840	2,930	3,110	3,280	3,665	4,005
	L	.290	.0660	777	1,095	1,350	1,565	1,740	1,900	2,065	2,200	2,332	2,460	2,700	2,916	3,020	3,100	3,300	3,480	3,890	4,250
	M	.295	.0683	805	1,140	1,395	1,610	1,800	1,970	2,140	2,280	2,417	2,545	2,790	3,013	3,120	3,180	3,410	3,600	4,030	4,400
19/64	...	.2968	.0692	816	1,152	1,412	1,630	1,830	2,000	2,165	2,300	2,450	2,580	2,830	3,056	3,160	3,260	3,460	3,660	4,080	4,460
	N	.302	.0716	840	1,192	1,460	1,690	1,880	2,060	2,240	2,390	2,530	2,670	2,930	3,150	3,270	3,370	3,580	3,860	4,220	4,610
5/16	...	.3125	.0767	905	1,280	1,568	1,790	2,020	2,210	2,400	2,560	2,710	2,860	3,140	3,380	3,500	3,610	3,830	4,050	4,520	5,000
	O	.316	.0784	922	1,310	1,600	1,850	2,070	2,260	2,460	2,610	2,780	2,920	3,210	3,450	3,580	3,690	3,920	4,140	4,620	5,050
	P	.323	.0820	968	1,370	1,680	1,930	2,160	2,360	2,570	2,740	2,900	3,060	3,350	3,620	3,750	3,860	4,100	4,330	4,830	5,300
21/64	...	.3281	.0846	1,000	1,412	1,730	2,000	2,240	2,440	2,650	2,820	2,990	3,160	3,460	3,730	3,870	3,980	4,230	4,463	5,000	5,450
	Q	.332	.0866	1,020	1,445	1,770	2,040	2,290	2,500	2,710	2,880	3,060	3,230	3,540	3,820	3,960	4,070	4,340	4,567	5,100	5,590
	R	.339	.0901	1,060	1,500	1,840	2,130	2,380	2,580	2,820	3,000	3,190	3,360	3,690	3,970	4,120	4,240	4,500	4,760	5,310	5,800
11/32	...	.3437	.0928	1,094	1,550	1,890	2,190	2,450	2,680	2,890	3,100	3,280	3,460	3,800	4,090	4,240	4,370	4,640	4,890	5,570	5,980
	S	.348	.0950	1,121	1,585	1,950	2,240	2,500	2,740	2,960	3,160	3,360	3,550	3,890	4,190	4,340	4,470	4,750	5,010	5,600	6,120
	T	.358	.1005	1,185	1,675	2,060	2,370	2,650	2,900	3,140	3,350	3,560	3,750	4,120	4,440	4,590	4,730	5,020	5,300	5,930	6,480
23/64	...	.3593	.1014	1,200	1,695	2,080	2,400	2,680	2,930	3,160	3,380	3,600	3,795	4,150	4,480	4,640	4,790	5,090	5,350	5,980	6,550
	U	.368	.1063	1,255	1,780	2,170	2,510	2,810	3,070	3,320	3,540	3,770	3,970	4,350	4,700	4,860	5,000	5,420	5,620	6,300	6,850
3/8	...	.375	.1104	1,300	1,840	2,260	2,610	2,920	3,180	3,450	3,680	3,900	4,120	4,520	4,870	5,050	5,200	5,510	5,830	6,520	7,120
	V	.377	.1116	1,315	1,860	2,280	2,640	2,940	3,250	3,480	3,710	3,950	4,160	4,560	4,910	5,100	5,250	5,590	5,900	6,580	7,200
	W	.386	.1170	1,380	1,950	2,390	2,760	3,090	3,380	3,650	3,900	4,140	4,360	4,790	5,150	5,350	5,500	5,850	6,180	6,900	7,550
25/64	...	.3906	.1198	1,412	2,000	2,450	2,830	3,160	3,450	3,740	4,000	4,240	4,470	4,900	5,290	5,480	5,630	5,980	6,320	7,080	7,740
	X	.397	.1236	1,450	2,060	2,520	2,910	3,260	3,560	3,860	4,120	4,370	4,600	5,050	5,440	5,645	5,810	6,180	6,520	7,300	7,960
	Y	.404	.1278	1,505	2,090	2,610	3,020	3,380	3,690	3,990	4,260	4,520	4,770	5,210	5,625	5,840	6,020	6,390	6,750	7,540	8,250
13/32	...	.4062	.1296	1,530	2,160	2,650	3,060	3,420	3,740	4,040	4,320	4,590	4,840	5,300	5,700	5,920	6,100	6,460	6,840	7,650	8,350
	Z	.413	.1340	1,580	2,240	2,740	3,160	3,540	3,870	4,180	4,460	4,740	5,000	5,490	5,900	6,120	6,300	6,700	7,080	7,900	8,650
7/16	...	.4375	.1503	1,775	2,510	3,070	3,550	3,960	4,340	4,700	5,010	5,320	5,600	6,150	6,620	6,870	7,100	7,540	7,950	8,860	9,700
29/64	...	.4531	.1613	1,890	2,690	3,300	3,800	4,250	4,650	5,030	5,380	5,800	6,020	6,600	7,120	7,370	7,600	8,060	8,530	9,530	10,400
15/32	...	.4687	.1726	2,040	2,970	3,520	4,060	4,550	4,970	5,390	5,750	6,100	6,430	7,050	7,600	7,890	8,120	8,610	9,100	10,180	11,120
31/64	...	.4843	.1843	2,170	3,070	3,760	4,350	4,860	5,310	5,750	6,140	6,500	6,870	7,630	8,130	7,970	8,690	9,200	9,750	10,900	11,860
1/2	...	.5000	.1963	2,320	3,280	4,000	4,640	5,180	5,660	6,130	6,550	6,950	7,320	8,050	8,650	8,980	9,250	9,800	10,350	11,600	12,650
33/64	...	.5156	.2088	2,460	3,480	4,260	4,920	5,500	6,030	6,500	6,960	7,300	7,780	8,550	9,200	9,550	9,850	10,420	11,000	12,300	13,500
17/32	...	.5312	.2217	2,610	3,690	4,510	5,220	5,840	6,400	6,900	7,400	7,830	8,250	9,060	9,750	10,100	10,400	11,090	11,700	13,100	14,300
35/64	...	.5468	.2349	2,770	3,920	4,800	5,560	6,200	6,780	7,320	7,500	8,300	8,750	9,600	10,300	10,720	11,100	11,750	12,400	13,850	15,180
9/16	...	.5625	.2485	2,930	4,150	5,080	5,860	6,550	7,160	7,750	8,270	8,780	9,270	10,180	10,950	11,360	11,700	12,400	13,100	14,650	16,000
37/64	...	.5781	.2625	3,100	4,370	5,360	6,200	7,200	7,560	8,190	8,750	9,280	9,800	10,720	10,550	12,000	12,310	13,100	14,850	15,500	16,800
19/32	...	.5937	.2769	3,270	4,410	5,650	6,540	7,300	8,000	8,640	9,230	9,800	10,310	11,320	12,200	12,650	13,000	13,820	14,600	16,300	17,850
39/64	...	.6093	.2916	3,440	4,850	5,950	6,900	7,700	8,400	9,100	9,700	10,300	10,850	11,900	12,820	13,300	13,700	14,550	15,400	17,200	18,800
5/8	...	.625	.3068	3,720	5,200	6,160	7,240	8,100	8,850	9,570	10,210	10,820	11,420	12,550	13,500	14,000	14,400	15,320	16,200	18,100	19,800
41/64	...	.6506	.3223	3,800	5,370	6,570	7,600	8,500	9,300	10,050	10,750	10,400	12,000	13,200	13,750	14,700	15,150	16,100	17,000	19,000	20,800
21/32	...	.6562	.3382	4,000	5,640	6,900	7,980	8,940	9,750	10,550	11,300	11,950	12,600	13,850	14,900	15,400	15,900	16,900	17,850	19,900	21,850
43/64	...	.6718	.3545	4,200	5,900	7,240	8,350	9,350	10,250	11,050	11,700	12,520	13,200	14,500	15,600	16,200	16,700	17,700	18,700	20,900	22,900
11/16	...	.6875	.3712	4,370	6,160	7,560	8,750	9,800	10,700	11,580	12,400	13,100	13,800	15,200	16,300	16,900	17,450	18,500	19,600	21,900	23,900
45/64	...	.7031	.3883	4,600	6,500	7,950	9,150	10,250	11,200	12,100	12,950	13,750	14,500	15,900	17,100	17,700	18,300	19,400	20,500	22,900	25,100
23/32	...	.7187	.4057	4,800	6,750	8,300	9,550	10,707	11,700	12,680	13,500	14,350	15,100	16,600	17,850	18,500	19,100	20,300	21,450	23,900	26,200
47/64	...	.7343	.4236	5,000	7,060	8,650	10,000	11,150	12,200	13,200	14,100	14,950	15,800	17,300	18,650	19,500	19,900	21,100	22,317	25,000	27,300
3/4	...	.750	.4418	5,230	7,400	9,050	10,400	11,670	12,700	13,800	14,700	15,600	16,500	18,000	19,400	20,200	20,800	22,000	23,220	26,000	28,440
49/64	...	.7656	.4604	5,430	7,650	9,360	10,900	12,200	13,300	14,400	15,350	16,300	17,100	18,800	20,300	21,000	21,700	23,000	24,300	27,200	29,700
25/32	...	.7812	.4794	5,680	8,000	9,800	11,300	12,620	13,800	14,900	15,900	16,900	17,900	19,600	21,100	21,900	22,700	23,900	25,300	28,200	30,900
51/64	...	.7968	.4987	5,900	8,300	10,200	11,791	13,150	14,300	15,450	16,600	17,600	18,600	20,400	22,000	22,800	23,500	24,800	26,900	29,400	32,100
13/16	...	.8125	.5185	6,110	8,650	10,600	12,200	13,680	14,950	16,150	17,300	18,300	19,300	21,200	22,800	23,600	24,400	25,900	27,400	30,600	33,400
53/64	...	.8281	.5386	6,350	8,950	10,950	12,700	14,200	15,500	16,740	17,900	19,000	20,000	22,000	23,700	24,600	25,300	26,900	28,400	31,800	34,700
27/32	...	.8337	.5591	6,600	9,300	11,400	13,200	14,720	16,100	17,400	18,650	19,750	20,800	22,900	24,600	25,500	26,300	28,000	29,500	33,000	36,100
55/64	...	.8593	.5800	6,840	9,650	11,820	13,700	15,030	16,600	18,100	19,300	20,500	21,600	23,800	25,100	26,500	27,300	29,000	30,600	34,200	37,400
7/8	...	0.875	0.6013	7,120	10,000	12,300	14,200	15,850	17,300	18,750	20,000	21,300	22,500	24,600	26,500	27,500	28,300	30,000	31,800	35,400	38,800

* Calculated, based on: a constant × coefficient of discharge × orifice area × (pressure drop/specific gravity)$^{0.5}$.

Table 12-94b Specific Gravity Correction Factors* for Use with Table 12-94a

Sp gr	Factor	Sp gr	Factor	Sp gr	Factor
0.36	1.24	0.54	1.01	0.85	0.805
.38	1.20	.56	0.988	.90	.783
.40	1.17	.58	.973	0.95	.762
.42	1.14	.60	.956	1.00	.741
.44	1.12	.62	.942	1.20	.678
.46	1.09	.65	.920	1.40	.626
.48	1.07	.70	.887	1.52	.600
.50	1.05	.75	.857	1.80	.552
0.52	1.03	0.80	0.829	2.00	0.524

* Based on the formula $(0.55/\text{specific gravity})^{0.5}$.

If the governor functioned properly and maintained zero pressure at its outlet, the conditions on turndown would be: air pressure = 1 in. w.c.; gas pressure = 0; suction = $-9/25 = -0.36$ in. w.c.; air pressure drop thru jet = $1 + 0.36 = 1.36$ in. w.c.; gas pressure drop into chamber = $0 + 0.36 = 0.36$ in. w.c.; air volume = $[(1 + 0.36)/(25 + 9)]^{0.5} \times 10{,}000 = 2000$ cfh; gas volume = $[(0 + 0.36)/(0 + 9)]^{0.5} \times 1000 = 200$ cfh; and air–gas ratio = 2000/200 = 10, or 10 to 1.

Because air jet proportional mixer operation depends on velocity, it is essential that no changes be made in the flow conditions between the mixer and the burner(s) supplied (Fig. 12-132). A valve is sometimes used in the discharge line of a mixer to protect against flow of hot furnace gases during off conditions. Such a valve must be operated either wide open or fully closed and not to throttle or control the flow.

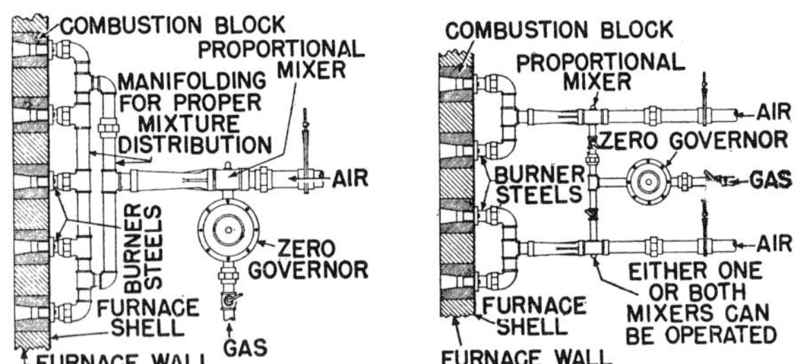

Fig. 12-132 Plan views of two typical installations for supplying multiple burners thru one or more proportional mixers using a zero governor.

Figure 12-133 shows some of the many arrangements of jets, venturi throats, mixing chambers, and governors used.

Mixer Sizing. Established practice for natural gas is to size mixers, air jets, and burners so that the maximum

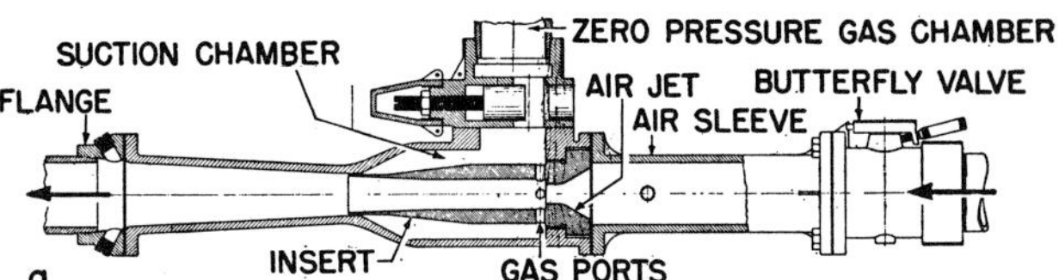

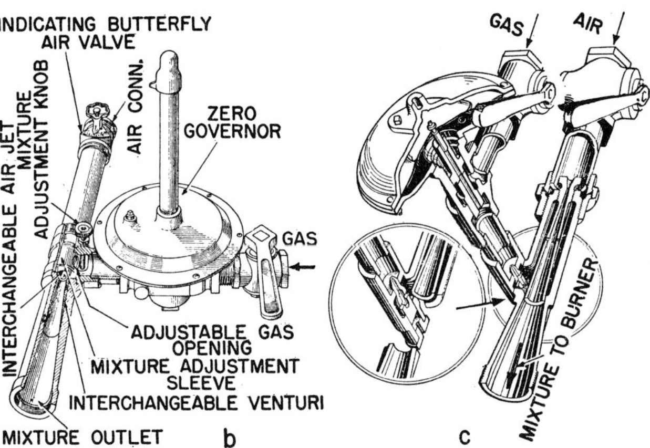

Fig. 12-133 Typical low-pressure proportional mixers: (a) Flomixer (Pyronics, Inc.); (b) mixer (Hauck Mfg. Co.); (c) inspirator with plug-type balanced governor (Surface Combustion Div., Midland-Ross Corp.).

mixture pressure will be about 30 to 35 per cent of that of the incoming air. This will vary with different specific gravities and air–gas ratios, from as low as 20 per cent to as high as 50 per cent. The exact mixture pressure that can be developed in a given installation depends on many variables, including ratio of air jet area to burner discharge area, piping resistance, air–gas ratios, specific gravity of mixture, and design of mixer equipment. Hence, the exact mixture pressure that may be developed cannot be predicted in advance.

Mixture pressure in proportional mixers in which the volume of entraining fluid is greater than that of the entrained fluid may be calculated from Eq. 1.

Several types of air jet mixers provide for a certain amount of adjustment in the mixer to meet variations not previously accounted for in external piping resistance or burner discharge areas (Figs. 12-134a and 12-134b).

Air jet mixers are made to fit standard pipe sizes from 0.5 to 8.0 in. In general, it is bad practice to carry air–gas mixtures of more than three to four million Btu per hour from a mixing device of this type, due to the possibility of backfire*

* See p. 2/82 for pertinent detonation data.

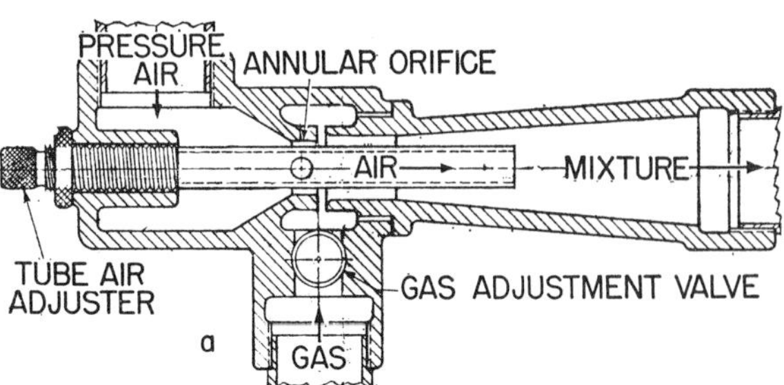

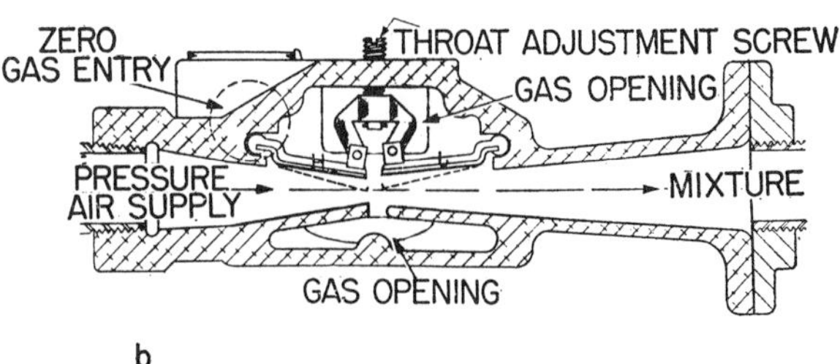

Fig. 12-134a (left) Low-pressure adjustable mixer (Multi-Flomixer—Pyronics, Inc). When the center tube is screwed back via the tube air adjuster, more air is passed, which in turn draws in more gas and increases mixture pressure.
Fig. 12-134b (right) Low-pressure adjustable mixer (Vari-Set—Eclipse Fuel Engineering Co). Throat areas, which are rectangular, can be simultaneously changed via the yoke and throat adjustment screw shown; enlarged areas produce leaner mixtures.

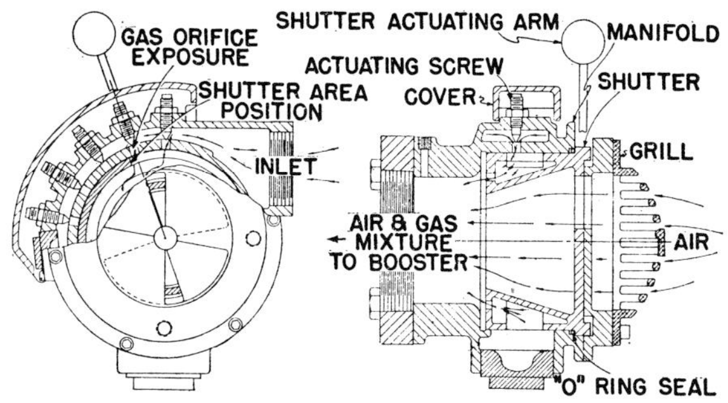

Fig. 12-135a Mechanical mixing with manual ratio adjustment. (Eclipse Engineering Co.)

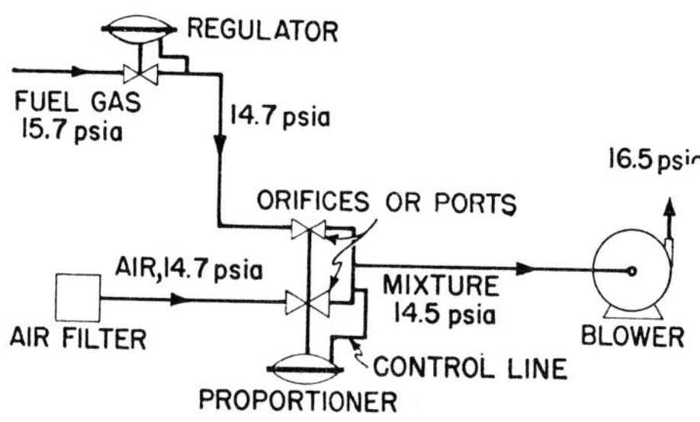

Fig. 12-136 Diagram of flow thru automatic air-gas proportioning combustion system. Absolute pressures are shown to avoid consideration of positive and negative pressures.

into the manifold, which might cause serious damage and shut down the unit. It is usually better to use multiple mixers so that a manifold does not carry over three to four million Btu per hour.

Mechanical Mixers

Mechanical Mixers with Manual Ratio Adjustment. These are mechanically operated but manually adjusted mixers, employing gas and air orifices (Figs. 12-135a and 135b). Flows of gas and air are controlled by varying the setting of the gas and air valves and by interlocking them so as to keep the flows in proportion. Since the flow depends on the pressure drop, fixed pressure conditions must be maintained for accurate proportioning throughout the range.

The usual practice is to have the air valve nonadjustable throughout its range and the other valve equipped with several adjustment points, so that gas flow can be contoured to follow the air flow. Thru these adjustments, the air–gas ratios may be varied at different points in the range to give, for example, a rich mixture at low fire and a lean mixture at high fire. Manual setting is required for each increment of opening of the gas and air valves because the pressure drop varies at different flows and the coefficient of discharge may vary widely. The performance of these mixers is quite satisfactory when proper manual settings are made at each point over the range.

Volume control may be either manual or automatic. It is not desirable to place any variable orifices or valves in the

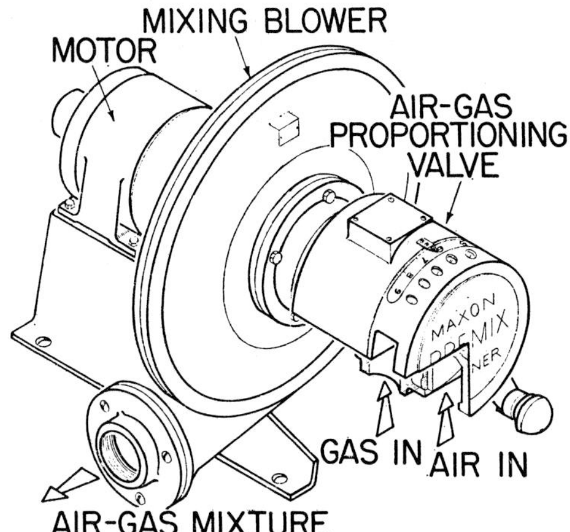

Fig. 12-135b Mechanical mixing with manual ratio adjustment. (Maxon Premix Burner Co.)

discharge line of this type of mechanical mixer or to vary the burner discharge area because all volume control is handled at the mixer inlet.

Mechanical Mixers with Automatic Proportioning Valves. Mixers of this type include automatic devices which maintain flows in an exact ratio at all flow rates by variable restrictions in the gas and air lines (Fig. 12-136). However, variations in the gas pressure regulator outlet pressure will cause variations in the air–gas mixture ratio. Discharge pressures of 1.0 to 3.0 psig are customary, to permit piping the mixture to points of application. Any restriction beyond the compressor will have no effect on the air–gas ratio, but will reduce the mixture flow. Provision must be made, therefore, for sufficient pressure to offset the pressure drop thru any throttling controls.

Operation of this type of mixer depends on a *constant pressure drop* thru air and gas orifices that have similar or identical characteristics; the incremental area changes in these orifices are automatically maintained in ratio. Therefore, it is only necessary to adjust for the proportion desired at one point in the range. A weighted or spring-loaded diaphragm and suitable linkage are used to provide the adjusting force and to position the orifices properly. The operation of a proportioning mechanism closely parallels that of a gas regulator.

Advantages of automatic mechanical proportioning include: (1) high heat input in a limited combustion space; (2) simple and accurate temperature control; (3) throttling of the mixture to control temperature without upsetting mixture ratio; (4) exceptionally wide turndown without ratio change—assures constant flame characteristics, maximum combustion efficiency, and uniformity of combustion products; (5) on–off, manual or automatic, burner(s) control which does not influence other burners in the system—the proportioning mechanism adjusts to demand; (6) after the initial ratio setting, accurate proportioning in accordance with the ratio setting—maintained for long periods; (7) uniform temperatures maintained in heat applications requiring many small burners in banks; and (8) high mixture pressures which permit high flame velocities when desirable.

Three basic mechanical arrangements of automatic proportioning valves used are: (1) the piston type; (2) the cone type; and (3) the slide type.

Piston-Type Proportioning Valve (Fig. 12-137). Vertical motion imparted by the mixing valve diaphragm automatically controls volume. Rotation of the sleeve outside the piston changes the area ratio of the ports governing the air and gas flows. To maintain uniform air and gas pressure conditions,

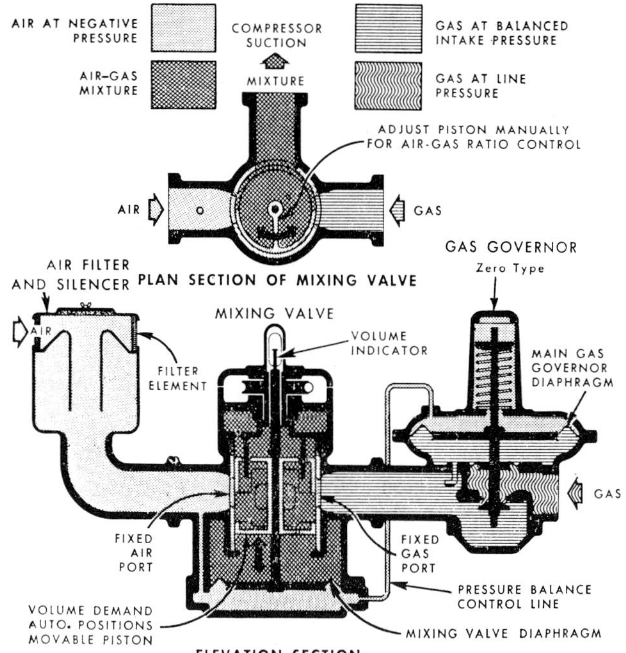

Fig. 12-137 Piston-type automatic proportioning valve. (Selas Corp. of America)

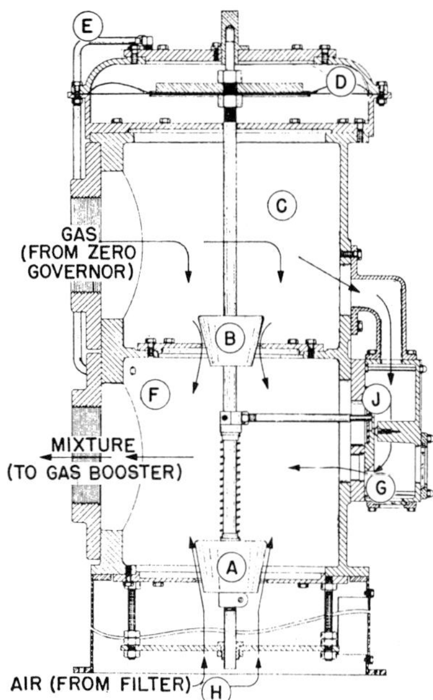

Fig. 12-138 Cone-type proportioning valve. (Eclipse Fuel Engineering Co.)

the zero gas governor is cross-connected thru a pressure balance control line to the air inlet. This compensates for varying resistance of air filters, air flow meters, piping, etc.

Cone- or Plug-Type Proportioning Valve (Fig. 12-138). Primary proportioning is obtained by the ratio and contour of the plugs.

Operation: The chamber above diaphragm (D) is connected by pressure line (E) to mixture chamber (F). As the booster or compressor draws the mixture from (F), the suction also raises (D), and simultaneously opens air cone (A) and gas cone (B), allowing air and gas to pass from the air inlet (H) and the zero gas chamber (C), respectively.

Normally, the resistance of the moving parts is such that a suction of about 5.0 in. w.c. is required; the plugs (A) and (B) rise until this suction is achieved. A separate opening leading from the chamber (C) thru adjustable opening (G) permits adding gas to enrich the mixture. This adjustment is divided into 16 increments, allowing a very fine adjustment at any desired point. Valve plate (J) controlling opening (G) moves up and down with valves (A) and (B). By the maintenance of a zero gas pressure in chamber (C) and zero air pressure at inlet (H), the resulting air–gas ratio of the mixture will be very accurately controlled over the entire range of operation.

Slide-Type Proportioning Valve (Fig. 12-139). Separate rectangular ports for gas and air are located in the valve seat. The slide valve, which has a single port shorter in length than the total length of the two ports in the fixed plate, is positioned lengthwise (manually) to expose gas and air port areas in the proper ratio. A weighted diaphragm and bell crank linkage automatically positions the slide valve crosswise according to demand.

As greater demand for fuel at (A) draws mixture from (B), the suction also raises diaphragm (C). This lifts stem (D), which pivots yoke lever (E) to move slide valve (F), increasing the size of the valve passage and reducing the suction at (B) until the diaphragm is balanced.

Capacity. Mixing machines with automatic proportioning valves which maintain air–gas ratios within one per cent throughout a capacity range of 250 to 180,000 cfh of mixture have been made. These are particularly applicable to processes requiring exact air–gas mixtures for a number of parallel burners, each of which may be turned off and on or throttled individually.

AIR–GAS RATIO CONTROL WITHOUT PREMIXING—NOZZLE MIXING BURNERS

In combustion systems I and II (Table 12-91 and Fig. 12-129), the gas and air in correct proportions are mixed at the point of combustion.

Mechanical Ratio Controllers

A basic way to control the ratio is to operate valves in the air and gas lines manually or mechanically. Two valves, identical in flow characteristics, can be linked together and operated by one lever or mechanism (Fig. 12-140). For good results, a constant pressure is necessary at the inlet of each. Such ratio controls are satisfactory for many applications and their cost is relatively low.

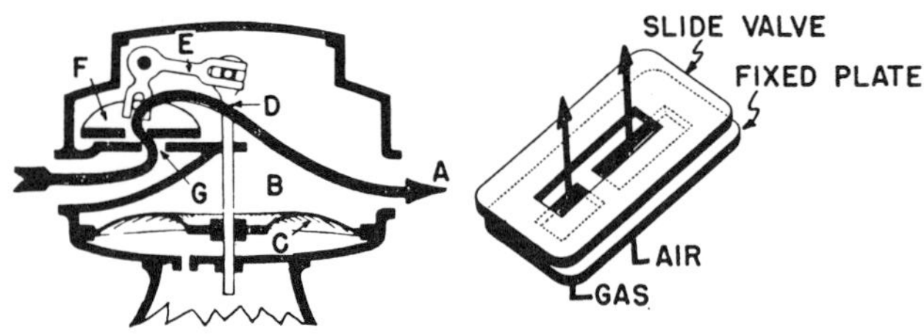

Fig. 12-139 Arrangement of slide-type proportioning valve. Slide valve arrangement shown in detail on right. (Kemp Mfg. Co.)

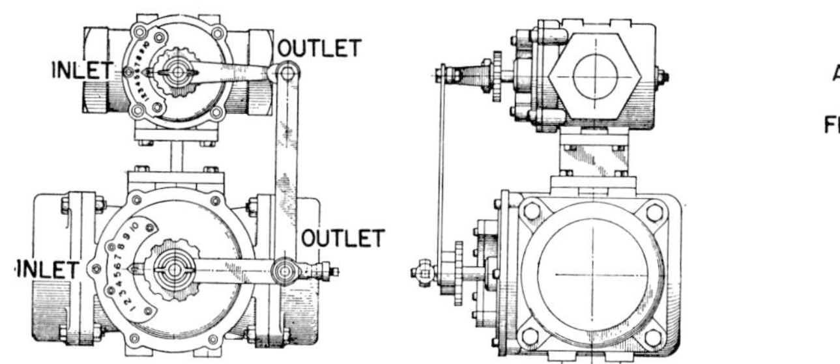

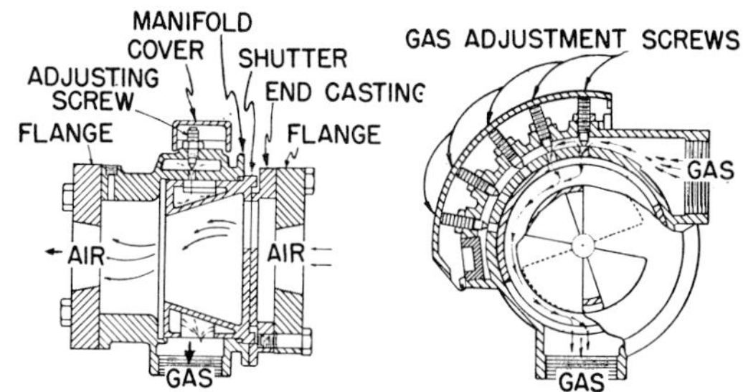

Fig. 12-140 Two automatic two-valve arrangements: left-hand arrangement—Hauck Mfg. Co.; right-hand arrangement—Eclipse Fuel Engineering Co.

Ratio Control by Balancing Flows or Pressures

These devices automatically control the flow of one fluid thru flow variations of another. They fall into two general classes: hydraulic balancing devices and balanced governors.

Hydraulic Balancing Devices. These balance hydraulically measured pressure differentials across fixed orifices to maintain air–gas flow ratios. A typical device (Fig. 12-141, left) has two diaphragms (A) and (B) which are controlled by the differential pressures at points (C) and (D) in the gas and air lines, respectively. These diaphragms position the jet pipe (G) with respect to the two ports (H), while the operating cylinder (E), which is activated by the jet fluid, positions the air damper to maintain the air–gas ratio set by manual adjustment of the ratio slider (F).

Another hydraulic device is the jet relay (Fig. 12–141, right). Here, high-pressure fluid flows thru a pivoted hydraulic nozzle, and the resulting jet stream is directed at two adjacent ports which are connected to the ends of a double-acting hydraulic cylinder. As the jet stream is pivoted a few degrees, from full impingement on one port to full impingement on the other, the piston speed is varied from full speed in one direction thru zero to full speed in the opposite direction. The reversal is entirely free of hydraulic shock.

Balanced Governors. The so-called "cross-connected governor control" (Fig. 12-142a) for air–gas ratio is widely used for nozzle mixing burners. The device uses variations in pressure on the discharge side of an air control valve to move a diaphragm-operated gas valve. Such a device can maintain accurate ratio control over the entire range of air pressure available.

Operation: Gas passes thru the zero governor, adjusting valve (A), and the nozzle mixing burner. Air from a blower or some other source also enters the nozzle mixing burner. The cross-connection runs from the air line thru fitting (B) into the governor top at (C), where a gage should be installed for setting. With (C) open to the atmosphere, the governor delivers zero pressure gas only and there is no flow into the burner. However, with (C) closed and air pressure acting above the diaphragm, the governor will deliver a corresponding pressure as long as the imposed air pressure is not greater than the inlet gas pressure at (D).

Note: If a higher air pressure is imposed on the governor than that of the incoming gas, the governor will be held wide open until the air pressure is reduced below that of the gas.

Air pressure above the diaphragm is varied by adjusting fitting (B); for example, if the air pressure is 1.0 psig, 4.0 in. w.c. may be imposed above the diaphragm by adjustment of (B). As the air pressure is changed by means of the air valve, the gas pressure will remain in proportion.

The mixture ratio may be adjusted at valve (A) for any flame condition desired at the burner nozzle. The air–gas ratio then will be maintained throughout the range of operation. To turn the burner up or down it is necessary only to vary the air valve setting; an automatic air valve, operated by furnace temperature, may be used.

The operation just described is based on a negligible furnace pressure or draft. Should the furnace pressure vary widely for any reason, compensation is necessary to add or subtract the furnace pressure from the governor balancing line so as to maintain the ratio desired (see, for example, Fig. 12-142b). The gas and air pressure must always be sufficiently higher than the maximum back pressure in the furnace.

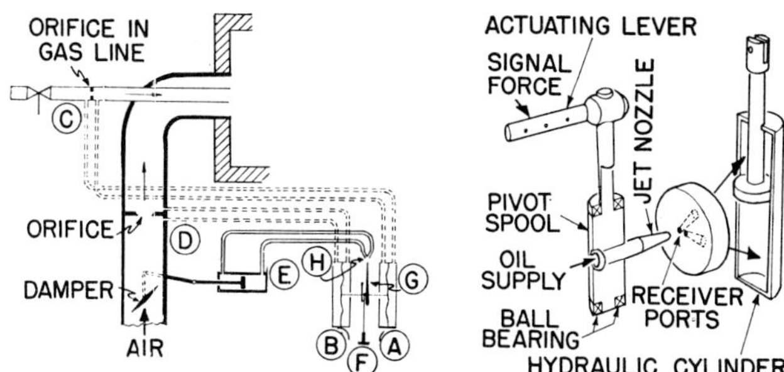

Fig. 12-141 Hydraulic balancing devices: (left) Askania-Werke ratio regulator; (right) jet relay—North American Mfg. Co.

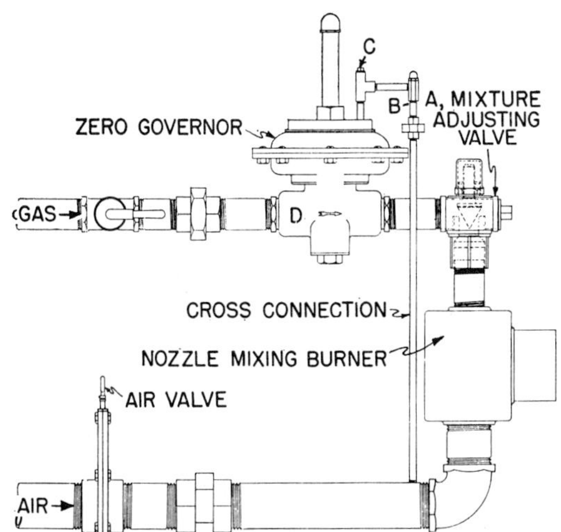

Fig. 12-142a Nozzle mixing proportionator or cross-connected controls. (Eclipse Fuel Engineering Co.)

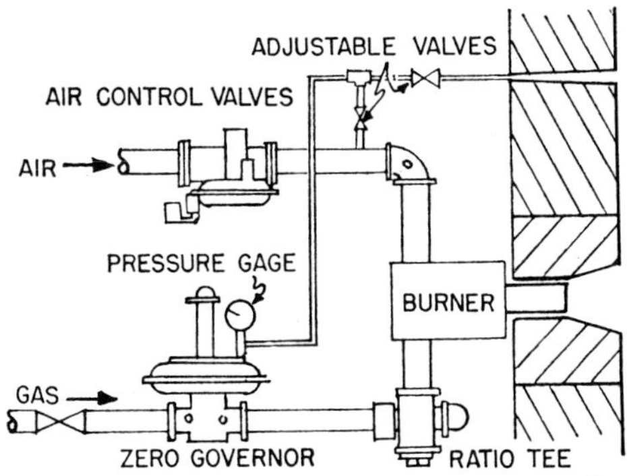

Fig. 12-142b Cross-connected control arrangement for variable furnace pressure (for use where furnace pressure is not atmospheric and gas pressure is lower than air pressure).

INDUSTRIAL GAS BURNERS

In terms of this chapter, an industrial gas burner requires a supply of air at a pressure of 0.25 in. w.c. min or utilizes gas at sufficiently high gas pressure to entrain air to give a mixture pressure of 0.25 in. w.c. min. The major features of industrial burners include: flame retention; maximum capacity; operating range; reasonable life; ease of maintenance; and simplicity of installation.

Industrial gas burners may be broadly classified as: (1) open burners; or (2) buried or sealed burners. Many special burners are also utilized for particular purposes. Such burners are generally custom built; they are not covered in this chapter.

Open Burners

Ring Burners. A drilled port ring-type industrial atmospheric burner is shown in Fig. 12-143.

Single Nozzle Burners. These burners (Fig. 12-144) are of the single orifice type with some form of ignition ring. Usually they have a large capacity relative to nozzle size; they are often used where torch-like flames are required. They may be fired thru the sides of furnaces or underneath tanks, or used as torches for other purposes.

Multitip and Multiport Burners. These are pipe burners using a number of nozzles or tips. The tips may be manifolded to a low-pressure inspirator; for example, such an arrangement may be used where low capacity with uniform heat distribution is required. Figure 12-145 shows two mounting methods.

Another type, a drilled pipe burner, is shown in Fig. 12-146. A $\frac{5}{32}$ in. diam port will burn about 1.0 cu ft of natural gas per hr. Ports should be on $\frac{1}{2}$ to $\frac{3}{4}$ in. centers, depending upon the flame distribution required. For best results, the burner may be equipped with a venturi mixer but will also operate with an ordinary mixer, with sufficient distance between the orifice and the first port. Pipe burners should not be over six feet long.

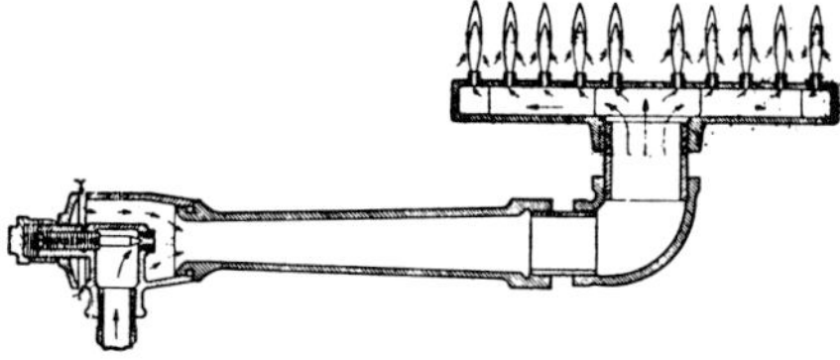

Fig. 12-143 Drilled port ring-type industrial atmospheric burner

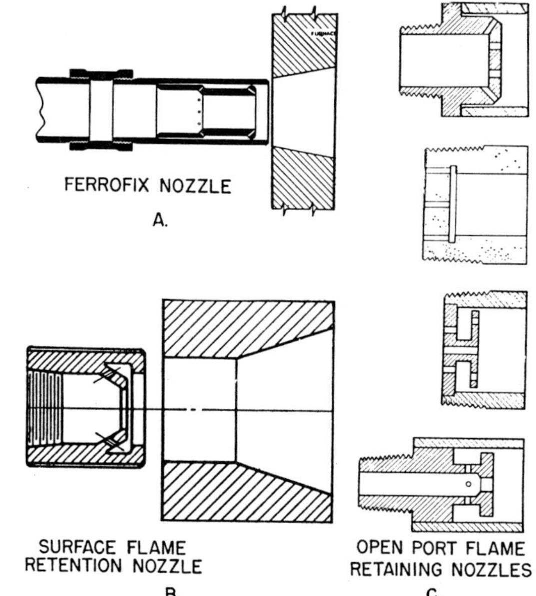

Fig. 12-144 Types of single-nozzle burners.

Fig. 12-145 Two lava tip burner mounting arrangements: (left) tips in extension fittings; (right) tips in street ells. The lighter pipe shown is favored for large capacity and long pipe burners.

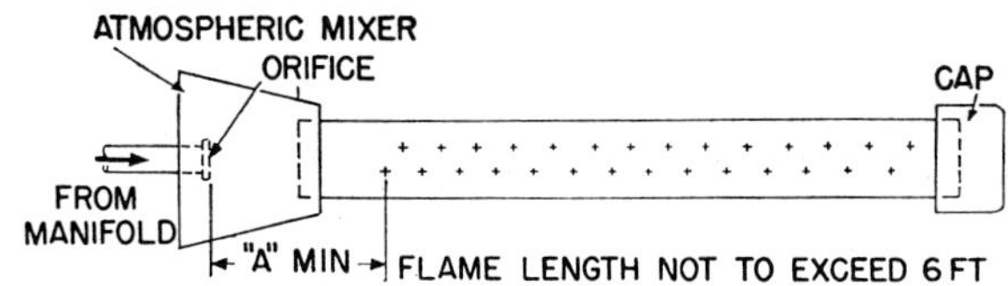

Pipe size, in.	A, in.	Number of $\frac{5}{32}$-in. ports Allowable	Recommend
3/4	8½	28	21
1	10⅞	45	34
1¼	12½	78	58
1½	14½	106	80
2	17½	174	131
2½	20½	250	188
3	25	385	290

Fig. 12-146 Drilled pipe industrial atmospheric burner with double row of ports.

Line Burners. This design (Fig. 12-147) provides a line of flame over the entire burner length; various configurations of continuous flame can be produced by means of elbows, tees, and crosses as shown. Line burners are much used in air streams (velocities up to 2500 fpm) to heat large volumes of air.

Ribbon Burners. These are similar to line burners, except that the flames are formed by a series of ribbons (see detail in Fig. 12-148) so installed in pipes as to give continuous flames. Made in widths of from one up to six ribbons, they are very desirable for providing uniform heating over a long length.

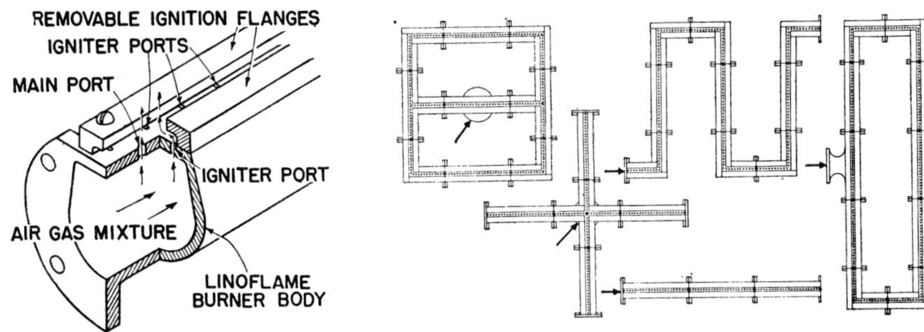

Fig. 12-147 (left) Line burner detail (Maxon Premix Burner Co.); (right) various burner assemblies to provide desired flame configurations— arrows show air-gas mixture inlets (Eclipse Fuel Engineering Co.).

Enclosed Combustion Burners. In this type of burner, Fig. 12-149(a), approximately stoichiometric combustion takes place in a small furnace-like enclosure, and the resulting hot gases (little loss by radiation) are utilized for high-temperature heating. The high-temperature combustion products leave the burner at high velocities. A chief advantage of these burners is that they are capable of fast local temperature build-up.

Pulse Combustion.[5] One unit of this type uses a burner assembly with a resonant burner chamber and rotary port-type valve. Gas flows to the burner continuously, but is interrupted during the brief burning interval by spark ignition. This type of burner can release large quantities of heat in small volumes.

Pulse jet combustion has been employed in a domestic boiler (Lucas-Rotax Pulsamatic sealed combustion boiler). The principle has also been experimentally applied to a water heater[6] and other domestic appliances.[7] Studies were underway in 1963 at the A.G.A. Laboratories to apply such burners to warm air furnaces and industrial and commercial burners.

Radiant Cup Burners. These, Fig. 12-149(b), may be used both as open and as enclosed burners. Combustion is arranged to scrub the surface of the inside of a cup so that a large portion of the heat is transferred as radiant energy; i.e., a considerable area of highly radiant refractory is developed. Air–gas mixture supplied thru (1) issues under refractory cup (2), burns on surface (3), and develops high-temperature radiation.

Trough Burners. These are similar to the aforementioned radiant units. The entire trough is scrubbed by the flame to develop a radiation throughout its full length. Such burners are commonly utilized for paper drawing and textile work.

Infra-Red Generators. These have relatively large radiant areas, usually flat (Fig. 12-150), but sometimes cup shaped as in (b) of Fig. 12-149. To develop a uniformly radiant surface, a gas–air mixture is forced thru a porous or perforated plate and burned over the entire surface, above which a secondary or diffusing screen may be placed. The primary plate may be metal or ceramic. The secondary, or diffusing, screen is generally heat-resisting stainless steel. Burner heads are usually arranged so that they may be banked close together to simplify safe ignition.

When used for process heating, open radiant burners, supplied with gas and air mixture at several inches water column pressure, develop surface temperatures up to about 1800 F. Input capacities of 60,000 to 100,000 Btu per hr-sq ft of radiant surface are easily achieved. Some enclosed radiant

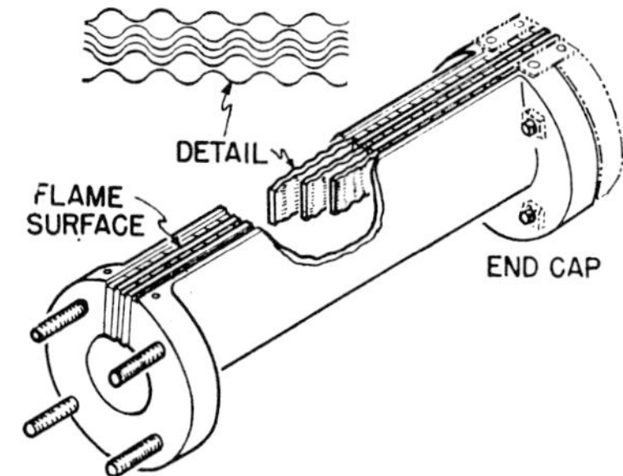

Fig. 12-148 Ribbon burner. (Flynn Burner Co.)

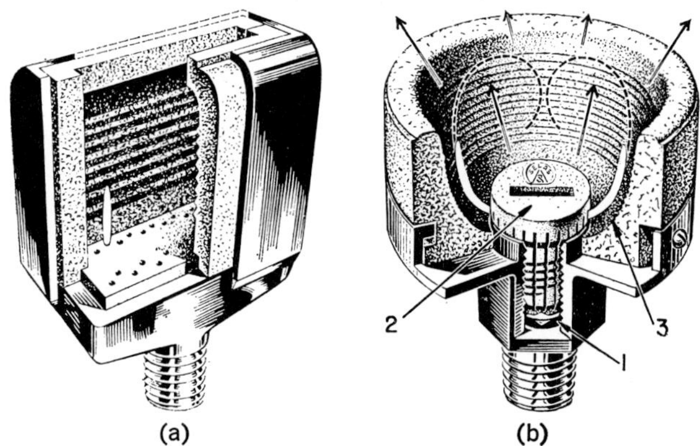

Fig. 12-149 (a) Enclosed combustion burner—combustion product velocity may exceed 1000 fps (b) radiant cup burner—surface temperature at 2000 to 3000 F. (Selas Corp. of America)

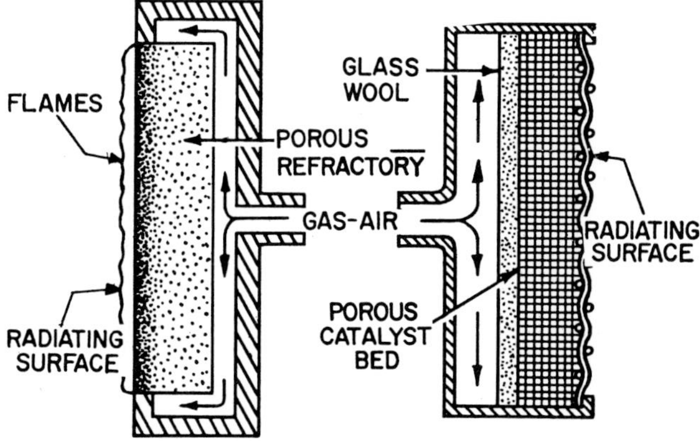

Fig. 12-150 Two types of gas-fired infra-red burners:[8,9] (left) porous refractory type; (right) catalytic oxidation type.

burners can develop up to 3000 F with input capacitities as high as 1.0 MMBtu per hr-sq ft. (See Tables 2-99a and 2-99b). Individual flanged sections may be bolted together.

One catalytic version of an infra-red heater, the Therm-A-Tron, operates at about 650 F. It has been molded to fit around a pipe in a meter house application, where it prevents the formation of ice.[10]

Infra-Red Energy Generated by Radiant Gas Burners. Radiation characteristics of atmospheric, powered, and catalytic gas-fired infra-red generators and those of various types of electric generators have been developed and tabulated.[9] See Tables 2-99a and 2-99b for data from this source. These data show how much total normal energy is emitted by the different types of radiant heat sources at different temperatures and the spectral distribution of this energy in the infra-red spectrum from 1.4 to 16.0 microns. A **gas infra-red radiation factor** (GIR) was developed for gas burners. This GIR factor takes flue gas radiation into account, since it adds considerably to incandescent gas burner surface radiation.

The heat radiated by a radiant gas burner comes from two sources: the hot burner surface and the hot flue gases adjacent to the burner surface. A typical spectral emission curve for a gas-fired burner is shown by curve 1 of Fig. 12-151. This curve results from curves 2* and 3 of this figure, which are emission curves for hot flue gases and the hot burner surface, respectively. The shaded portion of curve 1 represents the amount of emitted energy contributed by gas radiation. This gas radiation amounts to 5 to 18 per cent of a radiant burner's total normal radiation. Spectral radiation curves for two burners at various temperatures are shown in Figs. 12-152a and 12-152b. These figures correspond to the first and eighth burners, respectively, in Table 2-99a.

* Comparable to Fig. 2-60.

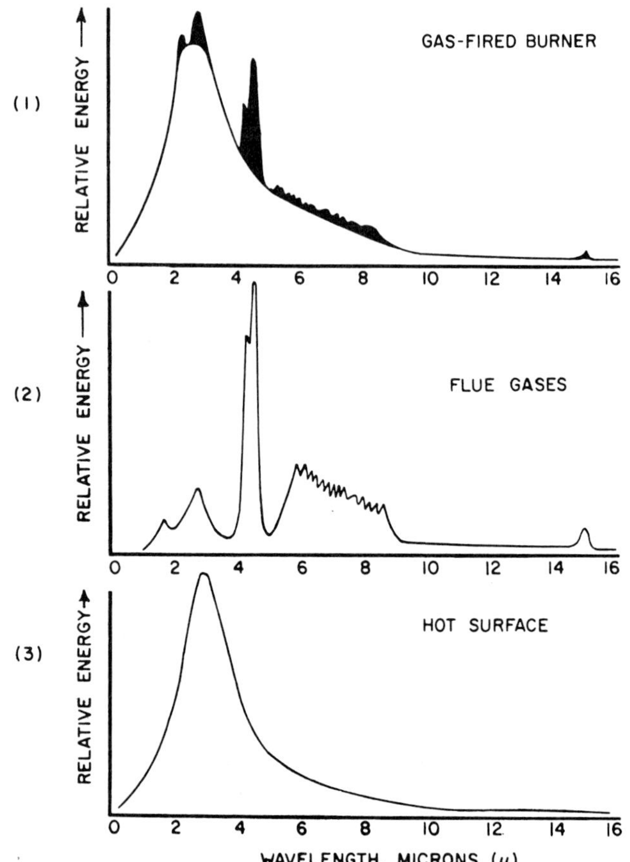

Fig. 12-151 Spectral radiation curves: (1) gas-fired infra-red burner; (2) flue gases; (3) hot surface.[9]

Note that H_2O and CO_2, both major constituents of flue gases, are absorbers and emitters of appreciable amounts of thermal radiation only in narrow bands of wave length[11] (curve 2 of Fig. 12-151):

H_2O: 2.24–3.27, 4.8–8.5, 12–25

CO_2: 2.36–3.02, 4.01–4.80, 12.5–16.5

Figures 2-61 and 2-62 may be used to determine radiation from flue gases, since other than a surface phenomenon is involved.

Example:[9] A burner of the type shown in Fig. 12-152a with a 1 ft × 1 ft ceramic tile face (A_b = 1.0 sq ft) operating at t_b = 1675 F (curve 3 of Fig. 12-152a) has been selected to heat a 1 ft × 1 ft surface (A_1 = 1.0 sq ft), having an emissivity e_1 = 0.75, to t_1 = 340 F. Find the radiant heat transfer rate q_r, assuming that the transfer takes place in an enclosure and that the burner and work are L = 1.0 ft apart. What portion of the total normal radiation is this?

$$q_r = 0.1713 A_b \langle F \rangle \left[\left(\frac{t_b + 460}{100} \right)^4 - \left(\frac{t_1 + 460}{100} \right)^4 \right]$$

where:

q_r = net rate of direct radiant heat transfer, Btu per hr
A_b = area of radiant burner, sq ft
$\langle F \rangle$ = geometric view factor:
in enclosures—

$$= \frac{1}{1/\bar{F} + [(1/\text{GIR}) - 1] + A_b/A_1[(1/e_1) - 1]}$$

in the open—

$$= \frac{1}{(1/e_b e_1 F_A) - [(1/e_b) - 1][(1/e_1) - 1]F_A}$$

$\bar{F}$ = geometric configuration-reflection factor (Fig. 2-57; also see Table 2-97), dimensionless
GIR = gas infra-red radiation factor, essentially the burner emissivity factor which takes gas radiation into account
F_A = geometric configuration factor (Fig. 2-57; also see Table 2-97), dimensionless

Substituting in the equation:

$\bar{F}$ = 0.53 (from curve 6 in Fig. 2-57 at a side/L ratio of 1)
GIR = 0.84 (from caption of Fig. 12-152a)

$$\langle F \rangle = \frac{1}{1/0.53 + [(1/0.84) - 1] + 1.0/1.0[(1/0.75) - 1]} = 0.415$$

$$q_r = 0.1713 \times 1.0 \times 0.415 \left[\left(\frac{1675 + 460}{100} \right)^4 - \left(\frac{340 + 460}{100} \right)^4 \right]$$

$$= 14{,}500 \text{ Btu per hr}$$

Radiant heat transferred, per cent of total = q_r/total normal radiation (from caption of Fig. 12-152a)

= 14,500/29,920 = 0.485 or 48.5 per cent.

Physiological Effects of Infra-Red Energy on Eye and Skin.[9] The lens of the eye will focus infra-red energy onto the retina. If the iris does not have time to contract and the flux density

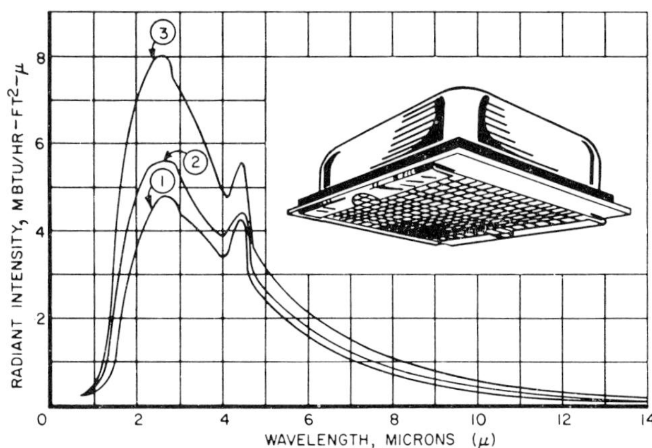

Curve	Red brightness temp, °F	Total normal radiation, Btu per hr-sq ft	GIR factor
1	1420	18,400	0.85
2	1500	22,420	.87
3	1675	29,920	0.84

Fig. 12-152a Spectral radiation curves for an atmospheric gas burner.[9] Perforated refractory radiating surface operating temperatures: with reradiating screen, 1420–1675 F; without reradiating screen, 1400–1490 F (low-medium temperature applications).

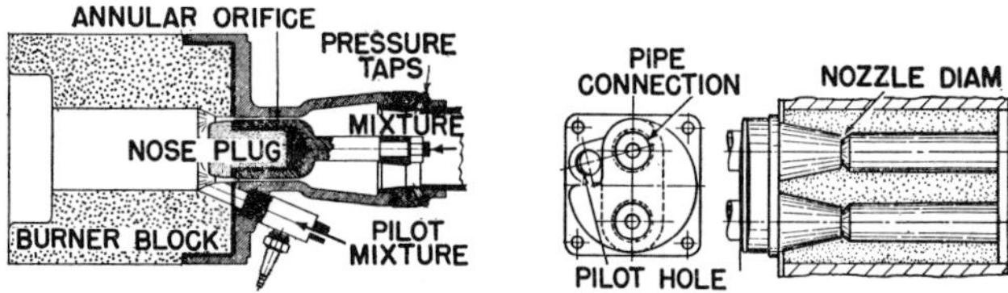

Fig. 12-153 Typical short nonluminous flame burners: (left) Pyronic tunnel burner—Pyronics, Inc.; (right) twin-nozzle tunnel burner—Surface Combustion Div., Midland-Ross Corp.

of this energy is high enough, damage to the eye will result. However, studies have shown that short wave length infra-red energy (1.0–2.0 microns) can cause cataracts of the eye, whereas the longer wave length energy (2.0 microns and longer) will not. Cataracts have been observed in occupations such as glass-blowing and foundry work. For all practical purposes, gas-fired generators produce energy in the longer wave length category.

Skin absorption of infra-red energy is also somewhat wave length selective in nature. In this case, however, the only effect is a rise in temperature of the skin made evident by a feeling of

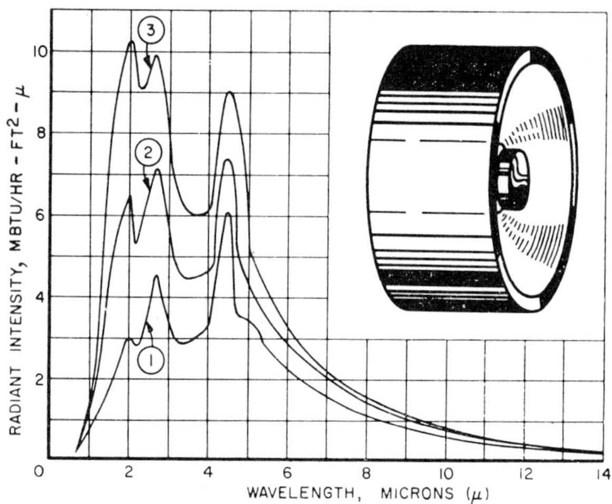

Curve	Red brightness temp, °F	Total normal radiation, Btu per hr-sq ft	GIR factor
1	1500	22,250	0.87
2	1800	32,700	.73
3	2000	44,400	0.69

Fig. 12-152b Spectral radiation curves for a pressure gas burner.[9] Refractory radiating surface operating temperature range is 1500–2000 F (medium-high temperature applications).

warmth. The absorptivity of the skin for energy longer than 2.0 microns is between 0.97 and 0.98, while the absorptivity of energy shorter than 2.0 microns is about 0.75. It was shown further that the longer wave length energy penetrates the skin about 0.0008 inches and in shorter wave length, about 0.0012 inches. This work also showed that the skin is less sensitive to short exposures of short wave length energy than it is to the longer wave length energy. This means that a smaller amount of long wave length energy is required to produce a sensation of comfort.

Buried or Sealed Burners

Industrial gas burners operating in tight enclosures and installed in furnace walls without any secondary air openings are called buried or sealed burners. Any mixture supplied them must be within the combustible range of the gas. Operation is usually at a point close to the stoichiometric ratio. Buried burners are made in many sizes and arrangements to supply the variety in flame geometry and characteristics desired for individual applications.

Tunnel Burners. The majority of buried or sealed burners may be classified as tunnel burners, because a

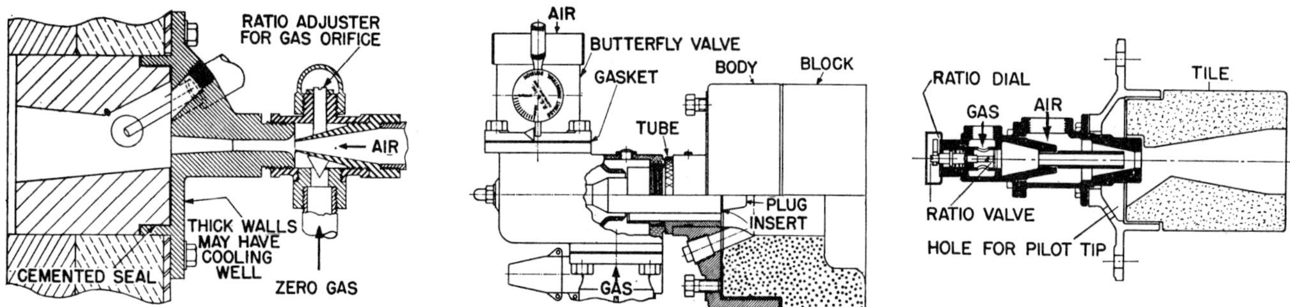

Fig. 12-154 Short nonluminous flame burner-mixer arrangements: (left) tunnel burner combined with air jet mixer—Eclipse Fuel Engineering Co.; (center) Mixjector burner—Pyronics, Inc.; (right) integral tunnel burner and mixer—North American Mfg. Co.

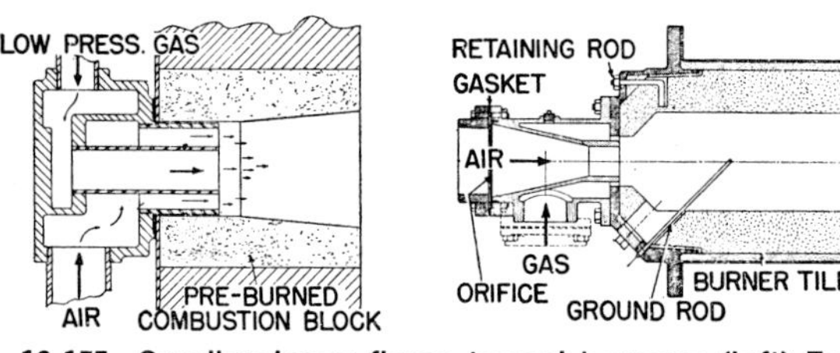

Fig. 12-155 Semiluminous flame tunnel burners: (left) Eclipse Fuel Engineering Co.; (right) North American Mfg. Co.

nozzle supplies the mixture to a combustion tunnel, where combustion takes place. The tunnel, in general, diverges from the nozzle to permit expansion of the mixture as it burns and to form an enclosed space for completion of combustion without dilution of furnace gases or too much temperature loss by radiation from the flame. Most tunnels consist of refractory material which accelerates the combustion rate, permitting the development of very high temperatures. The flue gases issue in a torch-like stream. Since their high velocity produces a turbulence in the furnace enclosure, tunnel burners, properly applied, provide very uniform furnace temperatures. By varying the method and location of mixing, as well as the nozzle velocity, flame pattern can be modified over a wide range. Flame patterns may be generally classified as: (1) short, nonluminous; (2) long, semiluminous; and (3) long, highly luminous.

Short, Nonluminous Flame Burners. The better the mixing of the gas and air supplied and the greater the turbulence in the combustion tunnel, the shorter the flame and the higher the temperatures in the combustion tunnel will be. High mixture pressures accelerate this process. Unless the heat is dissipated rapidly, high-temperature refractory is necessary for the combustion tunnel. Figure 12-153 shows typical short, nonluminous flame burners.

Tunnel burners producing short, nonluminous flames may be either of the premix or nozzle mixing type. The premix type may utilize a mixing device remote from the burner(s). Otherwise the mixture may be produced by either a gas jet or an air jet mixer incorporated into the burner itself. Such an arrangement provides a compact combustion system which operates with a minimum loss of velocity pressure in the air or mixture stream. Some burner-mixer combinations are shown in Fig. 12-154.

Combination burner–mixer units provide means of adjusting burner air–gas ratios to fit the requirements of each furnace part and of operating all burners thru one valve that varies the gas or air. On small furnaces it is usually more convenient to operate premix burners by one separate mixing device so that a single adjustment takes care of the entire furnace.

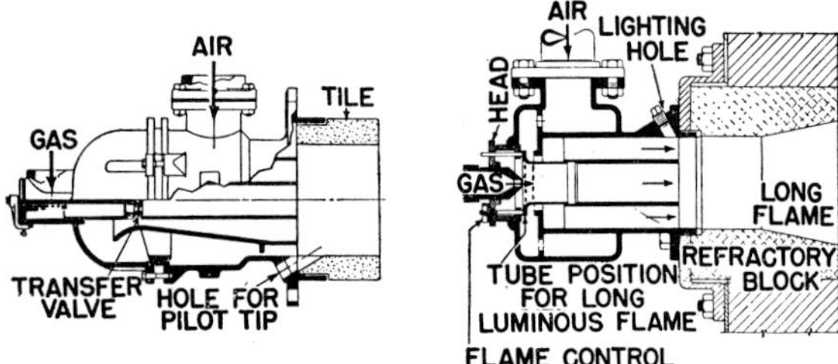

Fig. 12-156 Luminous flame tunnel burners: (left) North American Mfg. Co.; (right) Tate Jones Burner Co. Rotation of micrometer head shown determines initial air–gas ratio by varying orifice area; the annular orifice is indicated by dotted lines.

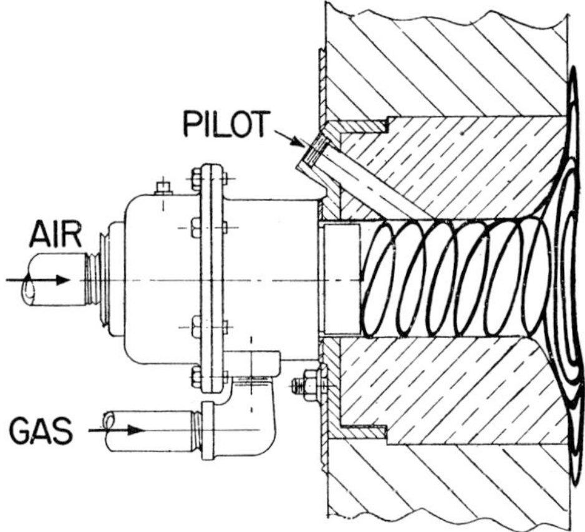

Fig. 12-157 Spiroflame burner. (Eclipse Fuel Engineering Co.)

Semiluminous Flame Burners (Fig. 12-155). By use of a nozzle mixing burner and relatively low gas and air pressures to produce low exit velocity into the combustion tunnel, the resulting flame is lengthened, and evidence of incomplete combustion appears. Such a flame, characterized as semiluminous, tends to radiate heat over a considerable length but, because it is largely nonluminous, heats more by convection than by radiation.

Luminous Flame Burners (Fig. 12-156). These nozzle mixing burners, which operate in the laminar flow range in the combustion tunnel and furnace enclosure, achieve highly luminous flames (as long as 30 ft). By maintenance of laminar flow throughout the length, combustion takes place on the interfaces of the gas and air streams. The resulting breakdown of gaseous hydrocarbons releases free carbon, which is quickly heated to incandescence and forms a mass of radiant flame. Since the flame is so long and so highly radiant, the theoretical flame temperature cannot be reached; however, a high percentage of the combustion energy can be recovered in the form of radiation from the flame. This type of burner is particularly suitable to large, long furnaces where uniformity of temperature is most desirable and extreme flame temperatures are not required.

Flat Flame Burners. These tunnel burners are arranged to reduce the velocity of the combustion products leaving the tunnel. Either a premix or nozzle mix type may be used. By shaping the tunnel and controlling the gas and air, the flame may be made to flow out of the tunnel sides with only negligible forward velocity. Some burners direct the flow radially, others spirally (Fig. 12-157), with a resulting wide expansion of highly heated radiant surface which can be placed very close to the material to be heated without flame impingement.

Radiant Tube Burners (Fig. 12-158). These nozzle mixing burners are especially designed to provide complete combustion in long, small-diameter tubes, and to develop uniform heating throughout their length. In general, rather low velocities are desirable in these tubes to obtain such uniformity of heating. When properly designed and supplied with gas and air at relatively low pressure, these burners are quiet and efficient. By using radiant tubes, combustion products can be kept out of the furnace chamber, which may contain special heat-treating atmospheres.

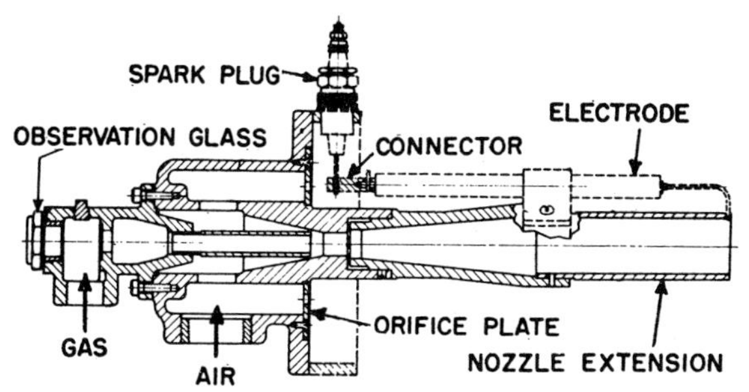

Fig. 12-158 Sealed-in radiant-tube heating burner. (North American Mfg. Co.)

Another design using an "eductor" is shown in Fig. 12-159. The eductor employs air at 1 to 1.5 psig to create a vacuum at the outlet end of the radiant tube. In this way, combustion products are completely withdrawn from the tube. Should a leak develop in the tube, the negative pressure within it prevents the entrance of combustion products into the furnace.

Basic design factors:

1. A rule of thumb for tube capacity: 25,000 Btu per square inch of cross-sectional area for quiet operation and fairly uniform heating over the entire tube length.
2. Radiation from the tube surface to the furnace will vary with the temperature differential between these two, the tube material, and the rate of circulation over the tube. Table 12-95 shows the approximate radiation and input required per square foot of outside tube surface for various furnace and tube temperatures.
3. See Eq. **49** of Section 2, Chapter 6.

Forced circulation around a tube increases the radiation rate as much as ten per cent.

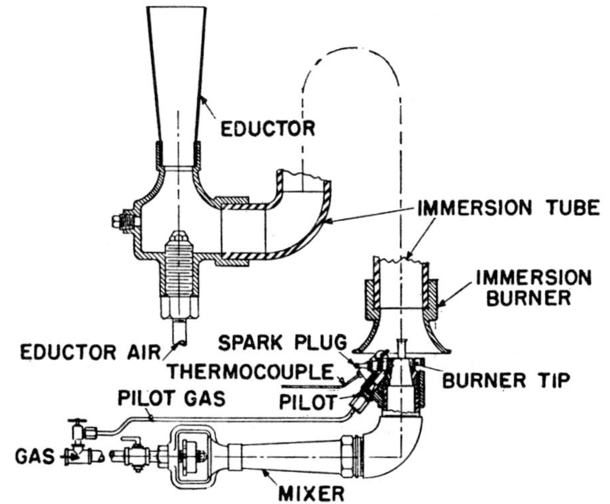

Fig. 12-159 Radiant-tube burner with eductor. (Surface Combustion Div., Midland-Ross Corp.)

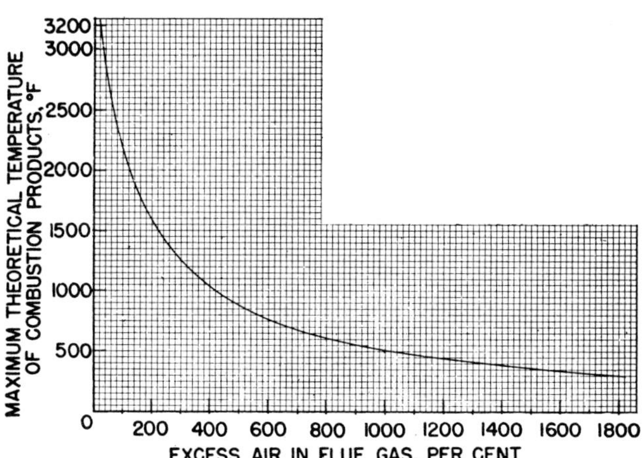

Fig. 12-160 Variation in approximate maximum theoretical flame temperature with excess air in flue gas for the complete combustion of natural gas. (Eclipse Fuel Engineering Co.)

Excess Air Burners. Any wide departure from stoichiometric ratios usually causes combustion either to occur outside the tunnel or to cease. The maximum variation which can be maintained by most tunnel burners is about ten per cent either way. However, burners of special design will maintain complete combustion even though a large amount of excess air is fed into the tunnel to dilute the flue products and reduce their temperature (Fig. 12-160).

Flue products from these excess air, or tempered air, burners (Fig. 12-161) are entirely free of unburned gas or aldehydes which normally result from quenching flames with too much air. In effect, this type of burner can be used as a direct-fired air heater for those processes in which the combustion products in the air to be heated will have no deleterious effect on the materials treated.

Excess air burners are available for a wide range of mixtures; ranges from ten per cent excess *gas* to 2000 per cent

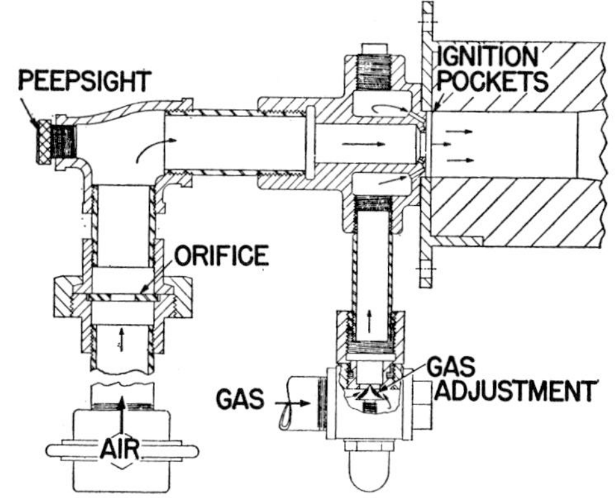

Fig. 12-161 Excess air burner. (Eclipse Fuel Engineering Co.)

Table 12-95 Approximate Radiation from Tube Surfaces and Heat Input Required

Outside tube temperature, °F	1,200	1,400	1,500	1,600	1,700	1,800	1,850	1,950
Furnace temperature, °F	1,000	1,000	1,200	1,400	1,500	1,600	1,750	1,850
Radiation, Btu/hr-sq ft tube surface	5,400	12,850	12,070	10,000	11,500	13,200	7,400	8,360
Gas input,* Btu/hr	7,500	19,200	18,600	16,100	19,000	22,300	13,300	15,500
Combustion efficiency,† %	72	67	65	62	60	57	56	54

* Required for tube temperatures given.
† Based on flue temperature equal to tube temperature (shown) and stoichiometric air–gas ratio.

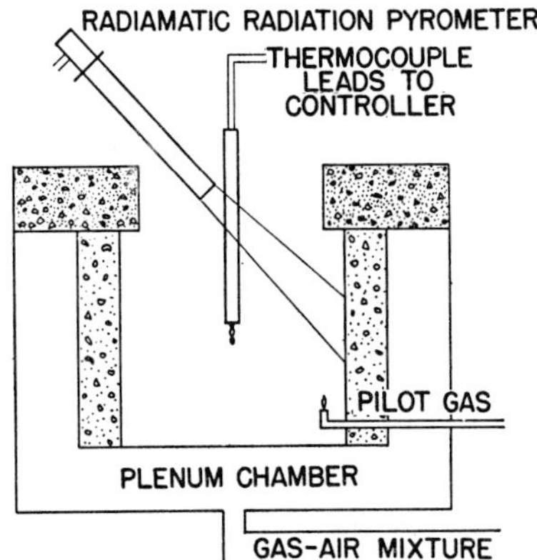

Fig. 12-162 Luminous wall furnace. (A. F. Holden Co.)

excess *air* are in common use. The air curve in Fig. 2-18 may be used to determine the heat content of air at various temperatures.

Radiant Luminous Walls (Fig. 12-162). While not strictly burners, radiant luminous walls serve the same purpose in a combustion system. They consist of a porous refractory furnace wall or section. The exterior is enclosed in a gastight chamber to which a controlled mixture of gas and air is supplied. As this mixture diffuses thru the wall or diaphragm it is ignited on the furnace side, producing—over the entire wall surface—a uniform radiant panel. There is little forward velocity. Since the combustion products are in the furnace chamber, radiant walls can only be used where such products are not harmful to the material heated.

Submerged Combustion. This method, which involves heat transfer without an exchanger, is often used to heat or evaporate corrosive or scale-forming solutions. Gas and air are mixed at a burner plate and ignited by a pilot flame, both located beneath the liquid surface (Fig. 12-163). The combustion products enter the solution thru exhaust ports.

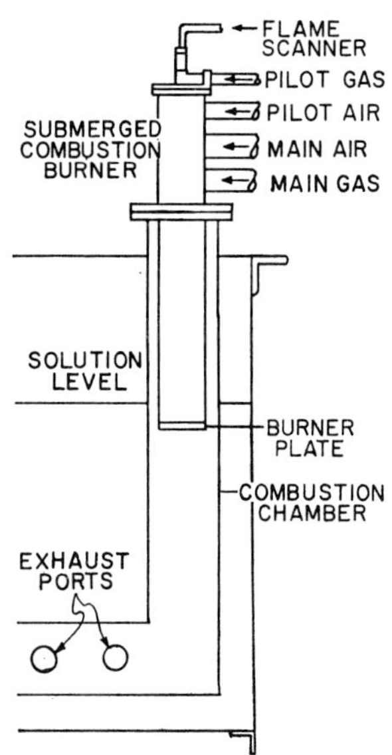

Fig. 12-163 Submerged combustion burner arrangement. (Submerged Combustion, Inc.)

Softening Hard Water.[12] In one municipal water system, the water hardeners, calcium bicarbonate and magnesium bicarbonate, are precipitated out of solution by calcium oxide. However, any excess of the latter causes, in turn, water hardness; CO_2 generated by burning natural gas precipitates excess CaO as $CaCO_3$. The combustion system mixes gas (25 psig) and air (15 psig) about 10 ft below the water level. The 10 psig mixture is burned and the resultant CO_2 issues from $\frac{3}{32}$ in. holes in submerged piping.

Accessory Equipment

Mounting Flanges. These are made in a variety of shapes and sizes to permit installing either buried or open burners in furnace walls. Some seal the burner tightly in the wall and hold the combustion block incorporating the combustion tunnel. Others merely mount on the furnace

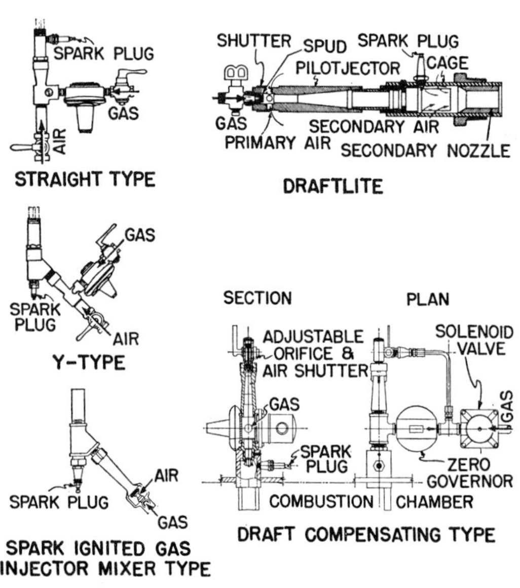

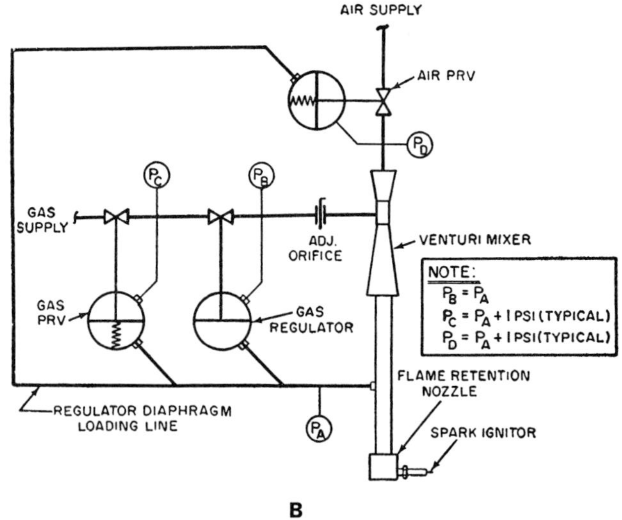

Fig. 12-164 Industrial pilot burner assemblies. **A.** Typical industrial pilot burners. The Draftlite is made by Pyronics, Inc. and the draft compensating unit by Eclipse Fuel Engineering Co. In the latter unit, gas is supplied to the injector, which entrains primary air; increased draft opens the zero governor, allowing more gas to flow. **B.** Wide range carbureting pilot for large boilers and furnaces.[15] This system readily lends itself to preadjustment.

exterior and locate the nozzle. Many mounting flanges incorporate openings for pilots, flame failure devices, peepsights, and other openings required. Some flanges are cast integral with the burner nozzles.

Combustion Blocks. Preburned refractory blocks, incorporating combustion tunnels, may be shaped for particular burners. The combustion tunnel may also be molded directly in the wall, using a mandrel and plastic refractory materials. Preburned blocks are favored because they are simple to install and replace. Materials used must be capable of withstanding thermal shock and temperatures developed in the combustion tunnel. Where combustion tunnel temperature approaches theoretical flame temperature, materials like Sillimanite or Mullite may be specified. For ordinary furnaces, below 2200 F, a good quality fire clay block is generally used. Block size is determined by that of the burner, the necessary length of the combustion tunnel, and the thickness needed for a good operating life.

Pilots. Small ignition burners or pilots are used on many industrial gas burners; they provide a simple means of lighting main burners. Pilot fuel may be premixed, supplied by gas jet or air jet mixers, or nozzle mixed (Fig. 12-164). Pilots frequently incorporate a method of ignition, such as spark plugs or hot wires. When used with flame sensing equipment, pilots can be proved to be in operation before the main burner can be turned on. Because of the increasing trend to automatic equipment, automatic lighting of pilots is a requirement of many installations.

Flame Sensing Devices. These are used to make sure that a pilot flame exists at a location suitable to ignite the mixture immediately after it leaves the main burner. Flame failure control equipment must therefore be installed at a point in the pilot flame at which it confirms that the pilot will immediately ignite the main flame. Two sensing devices may be installed, one on the pilot and one on the main burners, since it may be desirable to extinguish the pilot after it has performed its lighting function. Flame sensing devices include rods conducting electric current thru the flame itself to ground, heat sensing devices which must "sight" on the flame at a certain point, and visual electronic devices operated by the ultra-violet waves produced by the flame.

Peepsights. While not essential to industrial burner systems, peepsights are often desirable for checking the operation of either the pilot or main burner, the flame sensing device, and other equipment otherwise not visible. Many industrial gas burners are equipped with peepsights for sighting the junction of the pilot flame with the main burner.

General Burner Design Data

Flame Retention. Methods for obtaining flame retention account for many differences in burner design. The objective is to localize the flame on the nozzle and prevent it from blowing away and eventually being diluted by air or combustion products to a point of extinguishment. This is accomplished in two general ways. On open burners, the most common way is by means of an ignition ring which is shielded from the velocity effect of the main mixture and acts continually to light it. On buried burners, the flame is enclosed in a combustion tunnel, and continuous ignition is maintained by turbulence around the edges of the main stream and continuous contact with hot refractory walls. An ignition ring is not used on buried burners, since it tends to overheat the burner nozzle if a cooling medium is not applied.

Open Burner Flame Retention (Fig. 12-165a). The pilot combustion chamber around the main nozzle is fed from the main mixture stream by the series of small weepholes. The mixture passing thru these weepholes loses velocity in the chamber and forms a continuous ring of low-velocity flame. This flame is protected from the action of secondary air by the outer rim of the burner. The velocity effect of the main stream flowing thru the nozzle tends to "suck in" the pilot flame, continuously igniting the main flame.

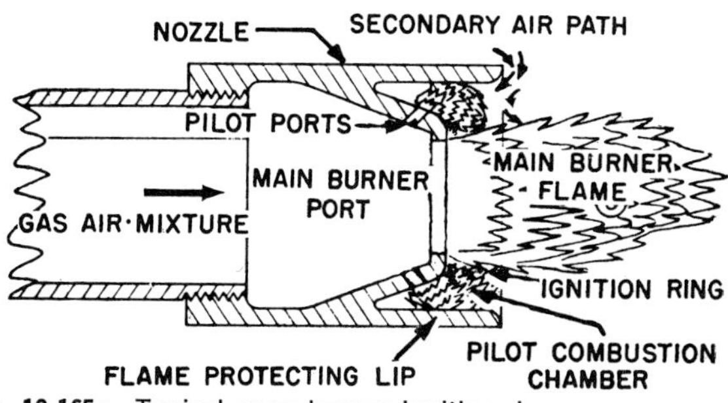

Fig. 12-165a Typical open-burner ignition ring.

Properly designed open nozzles will hold flames even when operated with mixture pressures as high as 1.0 psig. In general, open nozzles operate better with slightly rich mixtures which, while using the air drawn over the burner to cool it, are not overaerated.

Buried or Sealed Burner Flame Retention. Here retention is maintained by a ring of slow-moving mixture which peels off from the main stream as the latter issues into the combustion tunnel. This separation is caused by the slight suction formed near the nozzle by venturi action. The refractory forming the combustion tunnel becomes incandescent and further acts to speed up the combustion process. Both the tunnel shape and the fast expansion of burning gases prevent infiltration of the flame by furnace gases.

Generally, the shape of the combustion tunnel should follow the expansion of the burning mixture; tunnel length should permit sufficiently complete combustion. Combustion tunnels that are too short or diverge too rapidly prevent the functioning of the ignition ring. The results are unstable flames and the tendency for combustion to be extinguished or to take place beyond the tunnel unless the entire furnace first becomes incandescent.

A gas–air mixture in the combustible range will ignite quickly near the end of the burner if the combustion tunnel is designed to operate with the burner, and a pilot or ignition flame is present. This is true regardless of the rate of flow. However, it is generally recommended that burners be ignited at low fire to prevent blowing out the pilot flame.

Combustion Tunnel Design. Tunnel shapes vary widely in diameter-to-length ratios. In general, the parameters of Table 12-96 apply.

If the tunnel length, L (Fig. 12-165b), is too short, the flame may tend to go out or to be difficult to ignite; if too long, so much heat may be retained in the tunnel that its walls may melt, making the burner inoperative. If the divergence angle, A, is too great, the flame may lose contact with the walls and

Table 12-96 Design Data for Combustion Tunnels for Buried Burners
(based on stoichiometric mixtures)

Type of burner	Fast burning gases, 400 to 700 Btu per cu ft				Slow burning gases, 900 to 3200 Btu per cu ft			
	L/D*	Angle A,* deg	Max exit* velocity at 2600 F, fpm	Heat release, MMBtu/hr-cu ft for tunnel volume	L/D*	Angle A,* deg	Max exit* velocity at 2600 F, fpm	Heat release, MMBtu/hr-cu ft of tunnel volume
Premixed, 8 in. w.c. mixture press.	1 to 2.5	10 to 16	10,000 to 12,000	15 to 20	2 to 4.5	10 to 14	8,000 to 10,000	10 to 15
Nozzle mixing, 8 oz. air press.	1.5 to 3	10 to 14	9,000 to 11,000	15 to 20	2.5 to 5	10 to 14	7,000 to 9,000	10 to 15
Semiluminous, 6 oz. air press.	1 to 2.5	10 to 12	5,000 to 7,000	10 to 12	2 to 4	10 to 12	4,000 to 6,000	9 to 11
Luminous or radiant, 4 oz. air press.	1 to 2	10 to 12	3,000 to 4,000	6 to 8	1.5 to 3	10 to 12	2,000 to 4,000	4 to 6

* See Fig 12-165b.

blow out of the tunnel, particularly on high fire. Conversely, if this angle is too small, the tunnel may act as a continuation of the burner, causing an unstable flame which burns outside the tunnel. Refractory tunnels are subjected to very severe conditions. Therefore, great care must be taken to select material that will withstand the thermal stock on lighting and will not melt or distort under severe full fire conditions.

Capacity. It is desirable to obtain as much output as possible from a burner, consistent with its use. All industrial gas burners, regardless of their shape or type, are actually orifices which control the flow of a gas–air mixture into the combustion zone. The mixture should be stoichiometric for ideal conditions.

A formula to determine capacity follows:

$$Q = 1658.5KA\left(\frac{h}{G}\right)^{0.5} \quad (2)$$

where: Q = *mixture* flow thru orifice, cu ft per hr
K = discharge coefficient of burner
A = area of burner orifice, sq in.
h = pressure drop thru burner, in. w.c.
G = specific gravity of *mixture* (air = 1)

Example: Mixture pressure = 8.0 in w.c.; furnace pressure = 0.5 in. w.c.; specific gravity of gas = 0.62; coefficient of discharge K = 0.85. Determine the orifice area of a burner to pass 900 cfh of natural gas requiring an air–gas ratio of 9.74 to 1.

Solution: The mixture volume to be supplied will be 900 cfh gas plus (900 × 9.74) cfh air, or a total of 900 + 8760 = 9660 cfh.

Specific gravity of mixture =

$$\frac{(1.0 \times 9.74) + (0.62 \times 1)}{10.74} = 0.97$$

Substituting in Eq. **2**:

$$9660 = 1658.5 \times 0.85 \times A\left(\frac{8.0 - 0.5}{0.97}\right)^{0.5}$$

$$A = 2.5 \text{ sq in.}$$

The coefficient of discharge of a burner, K, has to be experimentally determined unless available from its manufacturer. In general, **ribbon burners** have very low K values because of their multiplicity of passageways and the tortuous travel that the mixture must take. **Venturi throat-type burners** appear to have K values greater than unity (using the critical dimension of the throat as the burner area). The velocity thru the throat is so high that it creates a negative static pressure which acts to increase the pressure drop as applied in Eq. **2**. If, however, the area at the burner **discharge** is used (where the velocity pressure has been converted back to static pressure), the coefficient of discharge is somewhat less than unity.

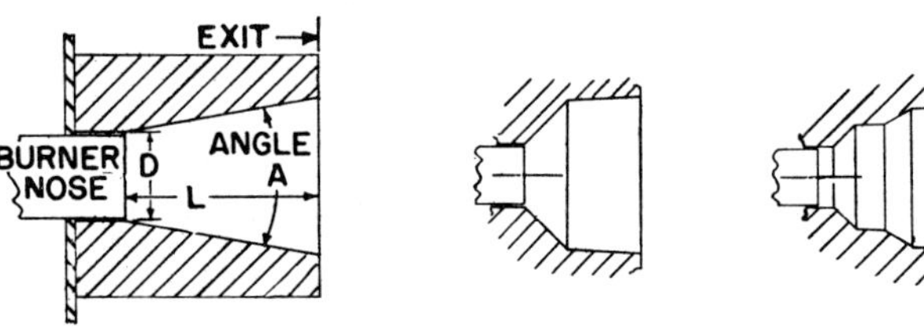

Fig. 12-165b Some shape variations of combustion tunnels. Cavity sections are basically frustums of right circular cones.

Operating Range. Sometimes called the **turndown range,** it indicates the upper and lower capacity limits in a given installation, either on the basis of pressure, or in cubic feet or Btu per hour. The *upper capacity limit,* or the upper limit of useful mixture pressure, depends on the maximum mixture pressure available or the capacity at which the flame leaves the burner nozzle and becomes unstable. The *lower capacity limit,* or turndown point of mixture pressure, is determined by the mixture pressure at which the burner backfires into the mixture line. Note that capacity varies as the square root of pressure (Eq. **2**).

In general, to prevent backfire, a minimum mixture pressure of at least 0.2 in. should be maintained on open burners and about 0.3 to 0.4 in. on buried burners. Under some conditions these figures will be much higher. Some burners, notably some types of ribbon burners, will operate with a mixture pressures as low as 0.05 in. w.c.

GENERAL APPLICATION DATA

Several usage factors must be considered in selecting industrial burners:

1. The shape or geometry of the flame; e.g., straight, torch-like flames vs. widespread, flat surface flames.

2. A large burner vs. a battery of small burners; note that the turndown range on one of a group of burners may be insufficient to meet process demands. Two or more manifolds may be required, one being shut off entirely at times.

3. The capacity of the system must be adequate for producing the temperature needed and for meeting all other anticipated demands.

4. The type of combustion selected must be correct for the furnace volume; e.g., a luminous flame which requires considerable length would not be practical in a small rectangular furnace.

5. The operating cycle of the burner and its control equipment must be suitable for all process demands.

Piping for Gas, Air, and Mixtures

Combustion equipment rating must be based on gas and air pressures available at the point of mixing or, for premixed burners, on mixture pressures available at the burner. Proper sizing of piping and controls is essential to avoid excessive pressure drops; normally, the lengths of pipe between the combustion system components are so short that friction losses in straight piping *per se* are negligible. However, *dynamic pressure losses*, which occur at every fitting or turn, may seriously affect the available burner pressure. These dynamic losses vary as the square of fluid velocity.

Sizing Air and Mixture Piping. Air piping should deliver air to proportional mixers or burners at a practically uniform pressure, regardless of load fluctuations. In combustion work, piping runs are usually short (under 50 ft) but often have many bends. By assuming that one velocity head is lost or dissipated at each change of direction, and using a pipe size that gives a very low velocity head, other losses can be disregarded. A general rule for selecting pipe is to use the size (Table 12-97) for a given volume that gives a velocity head, P_v, less than 0.3 in. w.c. For air, this is equivalent to a velocity of about 2200 fpm.

Table 12-97 shows velocities, V, and velocity heads, P_v, for air flow thru 1 to 10 in. pipe. This table can be used for mixtures as well as for air with little error, since most air–gas mixtures have specific gravities near unity.

Example: Initial pressure at blower outlet = 15 in. w.c.; system length = 15 ft of piping (including four 90° bends); capacity = 10,000 cfh air max; outlet pressure ≐ 13 in. w.c. min at load end of pipe. Determine the pipe size required.

Solution: Table 12-97 shows P_v for 1 in. pipe is 58.1 in. w.c. for 10,000 cfh, or four times the available pressure. Thus, 1 in. pipe is too small. Similarly, sizes thru 3 in. are also too small. P_v for 4 in. pipe is 0.23 in. w.c. Loss due to four bends will be 4×0.23 or 0.92 in. w.c. The resulting delivery pressure is 14.08 in. at full load, neglecting friction which is negligible in short runs with the velocity pressure below 0.3 in.

By eliminating one of four turns, it would be possible to use 3 in. pipe, since the P_v for 3 in. pipe is 0.72 in. w.c., which results in a delivery pressure of 12.84 in. w.c. (slightly less than required, but probably sufficient). The pressure variation from no load to full load would then be 2.16 in. w.c.

Piping should be of ample size to conserve pressure. By using sufficiently large pipe and few turns, high-pressure blowers are unnecessary and considerable initial and operating cost is saved.

Factors Affecting Btu Liberation in a Given Space

The amount of heat that can be liberated in a given space depends upon the amount of gas that can be burned completely in it. Assuming a stoichiometric mixture, to avoid either unburned gas or excess air, the factors determining the amount of gas burned are the pressure maintained in the enclosure and the amount of gas–air mixture discharged against this pressure. There seems to be no limit on the pressure at which combustion can be carried out after it has started in an enclosure, providing a stoichiometric mixture is maintained. The enclosure must be tight in order for its flue gas opening to be properly controlled; its material must be capable of withstanding the temperatures evolved.

The combustion system must be able to pass the gas–air mixture thru the burner at a pressure substantially above that in the enclosure in order to prevent possibility of backfiring. By operating at high mixture pressures or high air and gas pressures with nozzle mixing burners, in tight enclosures sufficiently insulated and with minimum flue openings, temperatures approaching the theoretical flame temperature of the gas may be attained. The amount of heat liberated per cubic foot of enclosure under such conditions may easily reach millions of Btu per hour; 10 to 15 MMBtu per cu ft of enclosure are quite common.

For ordinary industrial combustion applications, mixture pressures of 8.0 to 9.0 in. w.c. against enclosure pressures of 2.0 to 3.0 in. w.c. are quite common. These require air pressures of about 1.0 psig and gas pressures of about 4 in. w.c. Heat releases up to 500 MBtu per cu ft are quite common and can easily be exceeded if the process requires it and the cost of higher pressure equipment is not excessive.

Combustion System Selection

No hard and fast line separates the various types of combustion systems. It may be necessary to modify or combine several systems to get the results desired. Evaluation of the following will be helpful in determining the general system to be used and in selecting the specific sizes and types of burners, mixers, and controls.

Work Load in Btu. Before considering the exact size and type of combustion equipment, the amount of fuel necessary to accomplish the work desired must be known. It is usually best to base calculations on the maximum conditions, i.e., the heaviest load on the furnace or heating application at any given hour. The heat needed to maintain the required furnace temperature must be added to that used by the product or process. The former heating requirement or "holding requirement" also applies to stand-by equipment.

Available Heat. The available heat produced by combustion depends on the flue loss at the operating temperature. If the flue gases leaves a furnace at an excessive temperature, they leave only a fraction of the available heat for useful work. For example, the available heat per cubic foot in a high-inert natural gas (see Fig. 2-38) with a flue temperature of 1500 F (and 12 per cent CO_2) is $1000 - 420 = 580$ Btu; at 2600 F, this is reduced to 300 Btu

Table 12-97 Velocity, V, and Velocity Heads, P_v,* for Air thru Standard Pipe

	Nominal pipe size																			
	1 in.		1¼ in.		1½ in.		2 in.		2½ in.		3 in.		4 in.		6 in.		8 in.		10 in.	
Flow rate, cfh	V, fpm	P_v, in. w.c.	V, fpm	P_v, in. w.c.	V, fpm	P_v, in. w.c.	V, fpm	P_v, in. w.c.	V, fpm	P_v, in. w.c.	V, fpm	P_v, in. w.c.	V, fpm	P_v, in. w.c.	V, fpm	P_v, in. w.c.	V, fpm	P_v, in. w.c.	V, fpm	P_v, in. w.c.
1,000	3,055	0.58	1,603.9	0.1604	1,193.2	0.088	763.6	0.036												
1,500	4,582	1.3	2,450.9	.36	1,790	.2	1,145.5	.082												
2,000	6,109	2.32	3,208	0.64	2,386	.354	1,527	.145												
2,500	7,636	3.63	4,010	1	2,983	.55	1,909	.23	1,222	0.09										
3,000	9,164	5.23	4,812	1.44	3,579	0.798	2,291	.33	1,466	.134	1,018	0.065								
4,000	12,218	9.3	6,416	2.57	4,773	1.42	3,055	.58	1,955	.24	1,358	.115								
5,000	15,273	14.5	8,020	4.01	5,966	2.22	3,818	0.91	2,444	.37	1,697	.18								
6,000	18,327	20.9	9,624	5.77	7,159	3.19	4,582	1.31	1,932	.53	2,036	.26	1,145	0.082						
7,000	21,382	28.5	11,228	7.86	8,352	4.35	5,345	1.78	3,421	.73	2,376	.35	1,336	.111						
8,000	24,436	37.2	12,832	10.26	9,545	5.68	6,109	2.33	3,910	0.95	2,715	.46	1,527	.145						
9,000	27,491	47.1	14,435	12.99	10,739	7.18	6,872	2.94	4,398	1.20	3,055	.58	1,718	.184						
10,000	30,545	58.1	16,039	16.04	11,932	8.87	7,636	3.63	4,887	1.49	3,394	.72	1,909	.23						
11,000	33,600	70.3	17,643	19.41	13,125	10.73	8,400	4.4	5,376	1.79	3,733	0.87	2,100	.27						
12,000	36,665	83.7	19,247	23.09	14,318	12.77	9,164	5.2	5,864	2.14	4,073	1.03	2,291	.33						
13,000	39,709	98.2	20,851	27.1	15,511	14.99	9,927	6.1	6,353	2.51	4,412	1.21	2,482	.38						
14,000	42,764	113	22,455	31.2	16,705	17.39	10,691	7.1	6,842	2.92	4,752	1.41	2,673	.45	1,187	0.092				
15,000	45,818	130	24,059	36	17,898	19.96	11,455	8.17	7,331	3.35	5,091	1.62	2,864	.51	1,273	.101				
16,000	48,873	148	25,663	41	19,091	22.71	12,218	9.3	7,819	3.81	5,430	1.84	3,055	.58	1,357	.115				
18,000	54,982	188	28,871	52	21,478	28.75	13,745	11.8	8,796	4.82	6,109	2.33	3,436	.74	1,527	.145				
20,000	61,091	232	32,079	64	23,864	35.48	15,272	14.5	9,774	5.95	6,788	2.87	3,818	0.91	1,697	.18				
25,000			40,099	100	29,830	55.44	19,091	22.7	12,218	9.31	8,485	4.49	4,773	1.42	2,121	.28	1,193	0.088		
30,000			48,118	144	35,795	79.8	22,909	33	14,661	13.39	10,182	6.5	5,727	2.04	2,545	.40	1,432	.127		
40,000			64,158	259	47,727	142	30,545	58	19,548	23.8	13,576	11.49	7,636	3.63	3,394	0.72	1,909	.227	1,222	0.093
50,000			80,197	401	59,659	222	38,182	90.8	24,435	37.2	16,970	17.9	9,545	5.68	4,242	1.12	2,386	.354	1,527	.145
60,000					71,591	319	45,818	131	29,322	53.6	20,364	25.85	11,455	8.17	5,091	1.62	2,864	.509	1,833	.21
70,000					83,523	435	53,454	178	34,209	73	23,758	35.18	13,353	11.1	5,939	2.18	3,341	.694	2,138	.28
80,000					95,454	568	61,091	233	39,096	95	27,152	45.96	15,272	14.5	6,788	2.87	3,818	0.906	2,443	.37
90,000					107,386	718	68,727	294	43,985	120	30,545	58	17,182	18.4	7,636	3.63	4,296	1.15	2,749	.47
100,000							76,364	363	48,872	149	33,939	72	19,091	22.7	8,485	4.49	4,773	1.42	3,055	.58
120,000							91,636	523	58,646	214	40,728	103	22,909	33	10,182	6.5	5,727	2.04	3,666	0.84
140,000									68,420	292	47,516	141	26,727	45	11,878	9.2	6,681	2.78	4,276	1.14
160,000											54,304	184	30,545	58	13,574	11.5	7,635	3.63	4,887	1.49
180,000											61,092	233	34,363	74	15,270	14.5	8,589	4.59	5,496	1.88
200,000											68,000	287	38,200	96	17,000	18	9,550	5.6	6,100	2.3
220,000											75,000	349	42,000	104	18,700	21.5	10,500	6.8	6,720	2.8
240,000											81,500	410	46,000	115	20,400	25	11,400	8.0	7,450	3.4
260,000											88,500	485	49,500	123	22,000	30	12,400	9.5	7,950	3.9

* Based on the formula:

$P_v = (V/4005)^2$

where: P_v = velocity head, in. w.c.

V = velocity = (flow rate in cfh × 2.4)/inner pipe area in sq. in., fpm

Basis of Burner Selection. The ratio of the "holding requirement" of the heating application to its maximum requirement has much to do with determining the type and arrangement of the burners. Because of the physical limitations of handling gas and air, each type of combustion system has a definite, limited range of operation. With a given air pressure and a nozzle mixing type of combustion, a useful operating range with the maximum as much as eight times the minimum is possible. In a proportional mixing type of system with 1.0 psig air, the useful range is normally limited to four to one.

For example, if calculations indicate a range of fifteen to one between maximum and holding requirements, a double burner arrangement may be specified. One set operates as holding burners; the second set is turned on under full load conditions. The **turndown** requirements greatly influence the size of burners selected. Also, the heat pattern of the application must be considered. With only a single burner, the range might be adequate but the heat pattern would be so nonuniform that performance could only be poor.

Natural Gas for Cutting Torches.[2] First the preheat flame raises the cutting path temperature on ferrous material—just ahead of the cutting jet—to 1600 F. At this temperature, the accompanying cutting jet of pure oxygen will rapidly oxidize the flame path. The advantages of using natural gas over acetylene for preheating ferrous materials include: (1) substantially equal preheat times, with less tendency to flash back and "pop"; (2) no sooting and lighter colored flame—easier on the eyes; (3) usually less expensive, particularly for thicknesses up to 6 in.; for greater thicknesses, the higher flame intensity of acetylene becomes more significant; and (4) costs attributable to handling acetylene tanks or the additional insurance premium for acetylene piping are absent.

Both low pressure—universal, aspirating (7.0 in. w.c. to 1.0 psig)—and high pressure (8.0 to 10.0 psig) equipment are available; the only advantage of the latter is that it preheats somewhat faster.

Suggested Combustion Requirements:[3] Preheat gas velocity, 1000 fps; preheat O_2-to-natural gas ratio, 2:1 (reduce to 1.5:1 once cut is started, 1:1 for 10 in. stock thickness and over); and sufficient heat liberation, 65 cfh of natural gas.

A comparison of the pertinent operating characteristics of four gases used to obtain preheating temperature in oxygen cutting follows:[2]

Gas	Btu/ cu ft	O_2 required for combustion/ cu ft	O_2 premixed flame temp, °F	Gas, cu ft/ equivalent Btu
Acetylene	1440	2.50	6300	1.00
Manufactured	550	1.09	4900	2.62
Natural	1100	2.16	5200	1.31
Propane	2400	5.00	5200	0.60

A standard covering the installation and operation of gas systems for welding and cutting is available.[4]

MAINTENANCE

Planned maintenance, involving periodic inspection of gas burning equipment, helps to ensure economy and safety of operation. The details of scheduled maintenance procedures vary according to (1) the type or part of the equipment and (2) the nature of the service in terms of plant cleanliness and temperature of the air supply.

Table 12-98 presents an overall maintenance program for industrial combustion systems.

Premix Equipment. The device that mixes the air and gas in exact ratios requires regular attention to ensure proper functioning. The proportioning valves must operate smoothly in order to respond to the small changes in fuel requirements that are caused by turning a burner on or off.

Care should be directed to maintaining the necessary film of lubricant between the working surfaces of the valve. Avoid overgreasing, which may cause a build-up of grease on the accurately machined valve parts and thereby alter the mix ratio. Oil all moving parts as recommended by the manufacturer. Lubricate the motor that drives the compressor according to the manufacturer's instructions.

Protective Equipment. Many of these controls have operating diaphragms. Most recently made diaphragms consist of synthetic materials that are impervious to the type of gas to be handled. Usually, they are also resistant to the effects of heat up to 150 F. Older installations may have leather diaphragms, which will dry up and become stiff when subjected to heat for a long time. This will result in valves being held in a fixed position, causing poor operation of the burners.

The basic requirements for keeping diaphragms in good operating condition follow:

1. Keep them as far from heat as possible.
2. Protect controls by asbestos shields where necessary.
3. Oil the diaphragm (if leather) with neat's-foot oil at regular intervals (varying from one month in hot locations to one year in cool locations).

The fire checks and flame arresters require little maintenance. At regular intervals clean any metal gauzes in the flame arresters, as indicated by the manufacturer. Replacement fuses or disks should be kept on hand for ready use in the event that safety devices operate. Premix machines are usually equipped with filters on the air intake. The filters must be regularly cleaned or replaced.

IGNITION PILOTS IN LARGE BOILERS AND FURNACES[15]

The pilot, which is the source of ignition for a main burner, must provide sufficient energy to ignite the main fuel stream quickly and positively.

The time required to establish conditions for ignition at the ignition source point is subject to many variables, such as burner design, furnace configuration, primary aeration, gas concentration gradients between the pilot and main gas flow, and convection currents. In view of the many possible combinations of these variables, data from isolated experiments are useful for reference only.

Increasing the pilot flow rate in effect increases the number of possible ignition sources. The probability that one of these sources is located at a point where conditions for ignition would be quickly established is accordingly increased. On this basis, more effective ignition might be obtained by in-

Table 12-98 Suggested Maintenance and Number of Inspections per Year for Gas Burning Equipment[13]

Type of equipment	Parts of equipment that require maintenance	Maintenance required	Service condition*		
			I	I	III
Atmospheric, high- or low-pressure gas	Mixers and venturi injectors	Brush or scrape clean inside; readjust air-mixer shutter; check orifice alignment.	1	4	12
	Burner heads or nozzles	Clean; line up with firing opening; replace or repair if burned or damaged.	1	4	12
	Pilot burners	Clean and readjust to produce satisfactory flame properly located to light main burners; tighten mounting brackets.	3	6	12
	Pressure regulators	Gage outlet pressure at various loads. If not reasonably constant, repair or replace. If regulator does not hold low loads, clean seat. Be sure breather hole is open.	2	4	6
Blast	Blowers	Oil or grease bearings; check for impeller slippage on shaft; clean inlet guard, filter, and inside of case.	2	4	12
	Air–gas mixers	Clean all air passages. Check air–gas ratio.	1	3	6
	Burners, open	Check for proper location in opening; replace nozzle if burned or damaged.	12	12	12
	Burners, closed	Check for tight joint with refractory; check combustion tunnel for smoothness and correct shape; replace nozzle or refractory if damaged or burned.	12	12	12
	Pressure regulators	See pressure regulators above.	2	4	6
	Zero governors	Same as for pressure regulator. It is much more sensitive. Should not be used as shutoff valve unless specially adapted for that purpose. A true zero governor will always leak gas at no load.	2	4	6
Control & protective	Motor and solenoid valves	Check for sticking; oil as required; if leaking, repair or replace.	4	4	6
	Ignition devices	Check for clean spark points, correct spark gap, high-tension leakage, grounds, proper placement for lighting pilots; confirm that pilot flames immediately, and lights main flame.	12	12	12
	Auto pilots	Shut off pilot and make sure automatic feature functions within specified time limit.	12	12	12
	Wiring connections	Check for grounds, open or loose contacts, overloaded wire, sticking relays, defective insulation.	4	4	6
	Electronic equipment	Follow manufacturers' instructions.	12	12	12
	Gas cocks	Assure easy turning, be sure cocks can be turned easily; lubricate if required; check for leakage in open and closed positions; assure access for shutoff in emergency.	12	12	12
Auxiliaries	Heat surfaces	Remove deposits; patch holes or replace part; if burnouts occur, change firing method or combustion adjustment.	2	4	12
	Secondary air openings	Keep unobstructed; adjust to minimize overventilation; be sure secondary air is getting to all parts of the gas flame.	12	12	12
	Dampers	Same as above; be sure damper is not sticking.	12	12	12
	Flue pipes & building flues	Remove any flue pipe blockage; be sure joints are tight and no holes have developed; be sure there is no sagging in pipe and that roof extensions have not blown down.	6	6	6

* I: cleaned air and controlled temperature.
II: average air and temperature conditions.
III: lint, fibers, oil, or soot in air; variable temperatures.

creasing pilot flame dimension either vertically or horizontally, depending on the relative point location where conditions for ignition are established in the shortest time.

An air purge of four or more furnace volumes to clear the furnace prior to pilot ignition is almost universally accepted as a precaution against accidents. Often means are also provided to monitor pilot piping, since an idle furnace can fill with gas from a leaking pilot valve as well as from a leaking main gas valve.

Pilot Sizing

Selecting a pilot size that will reduce the inevitable ignition delay to a reasonable amount is a difficult task. One of the few published experiments[16] shows that a 50-per cent increase in pilot flame height gives a 50-per cent reduction in contact time required to ignite a methane-and-air mixture. Although a "weak" pilot may ignite a main burner, ignition will be delayed more than with a "strong" pilot. The trend of many installations has been to larger pilots, which reduce the ignition puff by reducing the ignition delay on the main burner.

As a guide, pilots may be sized to have a minimum capacity in Btu per hour of 5 per cent of the firing rate of the main burner at the time of light-off. For example, if a 10 MMBtu per hr capacity main burner is lighted off at a low fire rate of 2 MMBtu per hr, the minimum pilot size should be 100 MBtu per hr.

In no case should a pilot burner exceed 400 MBtu per hr capacity at light-off if it is spark ignited. This is the generally accepted level above which the oil or gas burner requires a proven pilot of its own. Moreover, no pilot trial-for-ignition period should exceed 10 sec. The belief that longer times are needed when valves are not located immediately adjacent to the pilot burner is not valid, provided that the piping is not vastly oversized.

Normal pipeline gas velocities run in the order of 60 fps. If a pilot (or main burner) does not light in 10 sec, there is no safe reason to try any longer. The limits of a maximum of 400 MBtu per hr capacity at light-off and a 10-sec trial-for-ignition result in a maximum accumulation of about 1100 Btu, or one cu ft of natural gas in case of a misfire. The universally required purge after a misfire can easily clear this away.

For **ribbon type burners** this method of sizing is not satisfactory. A substitute rule is to provide a pilot equal in size to the light-off capacity of one lineal ft of the ribbon burner. As an added precaution, no pilot should be more than 10 lineal ft from the farthest point of ribbon burner that it is igniting. This restriction is necessary to ensure light-off of the entire ribbon burner within 10 sec.

The foregoing guides for capacity are empirical. Nevertheless, they are consistent with good practice and offer the purchaser of equipment a positive specification.

Pilot Fuels

Industrial pilots may use either gas or light oil. Most manual torches consist of asbestos rope or cloth bound to a metal holder and saturated in kerosene. Such a torch has a good heat release rate when properly constructed. The trend has been to fixed, automatically ignited pilots burning either natural or LP-gas.

The reasons for this trend follow:

1. Investment is lower since either natural gas is already available as the main fuel or propane can be readily purchased in rented containers. Light oil pilots require storage tanks, pumps, and filters, which represent both space and investment.
2. Maintenance is easier with gas pilots since no mechanical equipment is involved other than valves, which are common to either fuel. Gas pilots need no pumps, stuffing boxes, filters, or other items requiring mechanical maintenance.
3. Dirt and carbon formations can cause difficulties in the small nozzles used for oil pilots. Passages for gas are larger and the control of air-fuel ratios is inherently easier.
4. For spark-ignited pilots gas pilots are preferred because gas-air mixtures ignite more easily than oil-air mixtures.

Spark Ignition

Commercial ignition transformers commonly range in secondary voltage from 5000 to 12,000 volts. Rated currents vary but are generally approximately 20 milliamperes.

Voltage and Current Requirements. The energy required for ignition is uniformly low if the spark gap exceeds the quenching distance.[17] For methane the energy requirement is less than 1 millijoule if the spark gap exceeds 0.075 in., which is the quenching distance. Since a 5000-volt, 20-milliampere transformer has an energy equivalent of 100 joules per sec, this would appear to show a large reserve of energy to do the igniting. It was also shown, however, that the velocity of the air-gas mixture past the spark increases the amount of energy required to a much higher level.[18] These data show a curve for propane at 3 in. Hg abs approaching a required 80 millijoules at a velocity of 54 fps. On this basis a transformer capacity of 5000 volts and 20 milliamperes is ample from an energy viewpoint but must be regarded as a minimum for the ignition of a *single* pilot. In fact, the frequent failure of a single transformer to ignite several pilots simultaneously bears out this conclusion. Data[19] for voltage required to produce a spark between two 2-cm spheres cast serious doubt on the adequacy of a 5000-volt transformer. In particular, 5000 volts is barely adequate for the minimum gap spacing of 0.075 in., and 10,000 volts is adequate for a spacing of only 0.153 in. Even if needle point electrodes approximately double these spacing values, it is not good engineering to design without any safety factor. Thus, a 10,000-volt, 20-milliampere transformer with a grounded secondary winding, sparking a single spark plug or electrodes with spacing of 0.125 in., represents good design.

The spark ignition of oil-air mixtures is a more difficult problem and is one of the primary reasons for specifying gas-air pilots. More ignition energy must be supplied, since the oil must be volatilized by the spark. Energy values of 250 millijoules per sec are common for light oil ignition spark systems, and the spacing of the electrodes is quite critical. Because direct insertion of the electrodes in the oil spray results in quenching, careful placement and shaping of the electrode tips is necessary to achieve a spark that originates outside the atomized oil spray but will penetrate it sufficiently to ignite it.

An interesting comparison with the minimum pilot size requirement of 5 per cent may be made with spark igniters. In a manner of speaking, the spark is a pilot for a pilot. The 10,000-volt, 20-milliampere transformer represents 5 per cent of a 13,652 Btu per hr pilot. The apparent inconsistency of using it for gas pilots of a size up to 400,000 Btu per hr is justified when it is realized that the probable increased ignition delay is not serious with a pilot of this size or smaller.

Spark Location. Locating the spark electrode outside the furnace ensures that both the wiring and electrode are protected from heat and are accessible. Unfortunately, most such igniters incorporate a flame retainer at the spark igniter. With this design most of the pilot flame is within the pilot tube and not at the pilot tube tip, where it would be effective in lighting the main burner. Where furnace temperatures are not excessive, the spark plug may be located at the pilot tip, as shown in part **A** (straight type) of Fig. 12-164. In an alternate design, an extended electrode bridges the distance between the spark plug mounted at the exterior of the furnace and the electrode tip at the inside wall of the furnace. Ceramic insulating spacers prevent grounding of the electrode lead. Corona discharges may be a problem. The arrangement provides for accessibility, protection from furnace heat, and the retention of all wiring outside the furnace enclosure.

Carbureted Blast-Type Igniters. Use of these units offers a means of compensating for changes in furnace and/or ambient conditions. Preadjustment of pilots is facilitated by connecting both air and gas pressure regulators with feedback to their diaphragms (part **B** of Fig. 12-164).

COMBUSTION NOISE[20]

Industrial gas combustion is sometimes accompanied by noise. Analysis of data obtained at eight industrial locations indicates that the portion of the combustion noise generated in the higher frequency bands (300 cps and up) tends to be

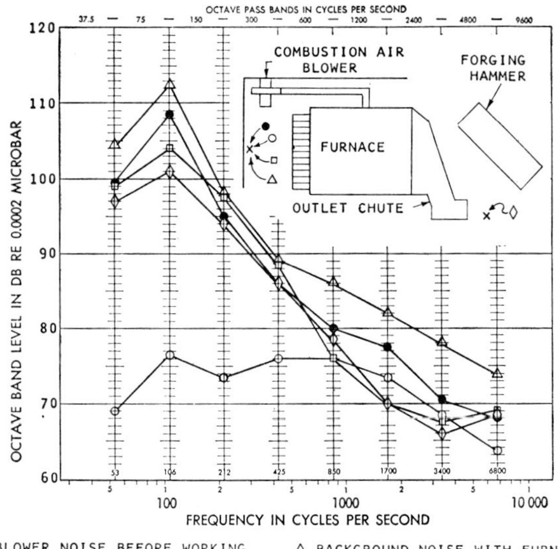

Fig. 12-166 Octave band noise analysis for forging furnace.[20]

muffled by the burner environment. Noise output peak levels were found to be predominantly in the 75 to 150 and 150 to 300 cps bands.

Figure 12-166 shows the tape-recorded octave band analysis for a forging furnace. Isolated operation, blower operation, and background noises are shown separately. Note that the burners contributed relatively little to the overall noise level created by forging hammers, fork trucks, etc., despite the fact that the burner peak noise level was above 100 db in the lower bands. It also appears that the air issuing from the burner without presence of flame is noisier at 850 cps than is the aerated flame itself.

Other units investigated in this study include high- and low-pressure rotary asphalt kilns, a reverberatory furnace, an aluminum melting crucible, and 50- and 80-hp package boilers.

The cited reference also contains background data on sound technology and acoustical terminology.

REFERENCES

1. Campbell, D. A. *Industrial Gas Nomenclature.* (Inform. Letter 44) New York, A.G.A. Industrial and Commercial Gas Section, 1951.
2. Reuter, J. S., and White, E. B. *Natural Gas for Cutting Torches.* (Ind. Data Bull. 169) (File B-10) San Francisco, P.C.G.A., 1957.
3. Anthes, C. C. "Recent Developments in Oxy-Fuel-Gas Cutting." *Welding J.* 39: 1022-7, Oct. 1960.
4. Natl. Fire Protection Assoc. *Standard for the Installation and Operation of Gas Systems for Welding and Cutting.* (NFPA 51) Boston, 1961.
5. Turin, J. J., and Huebler, J. *Advanced Studies in the Combustion of Industrial Gases: Interim Report.* New York, A.G.A. Committee on Ind. & Commercial Gas Research, 1950.
6. Griffiths, J. C. *Some New or Unusual Methods for Heating Water with Gas.* (Research Bull. 97) Cleveland, Ohio, A.G.A. Laboratories, 1963.
7. ——, and others. *New or Unusual Burners and Combustion Processes.* (Research Bull. 96) Cleveland, Ohio, A.G.A. Laboratories, 1963.
8. De Werth, D. W. *Literature Review of Infra-Red Energy Produced with Gas Burners.* (Research Bull. 83) Cleveland, Ohio, A.G.A. Laboratories, 1960.
9. ——. *A Study of Infra-Red Energy Generated by Radiant Gas Burners.* (Research Bull. 92) Cleveland, Ohio, A.G.A. Laboratories, 1962.
10. Kridner, K. "Infra-Red: Colorado Uses It to Heat Field Meter Houses." *Gas* 38: 95–6, July 1962.
11. Kreith, F. *Principles of Heat Transfer,* p. 212. Scranton, Pa., International Textbook Co., 1958.
12. "Natural Gas Makes Hard Water Soft." *Gas* 39: 103–4, Mar. 1963.
13. Campbell, D. A. "How to Keep Gas-Fired Equipment in First-Class Condition." *Factory* 107: 135-9, July 1949.
14. A.G.A. *Combustion,* 3rd ed. New York, 1938.
15. Monroe, E. S., Jr. "Ignition of Burners in Industrial Furnaces." *Combustion* 36: 44–7, Sept. 1964.
16. Morgan, J. D. *Principles of Ignition.* London, Pitman, 1942.
17. Lewis, B., and Von Elbe, G. "Ignition and Flame Stabilization in Gases." *Trans. ASME* 70: 307–16, May 1948.
18. Swett, C. C., Jr. "Energies to Ignite Propane-Air Mixtures in Pressure Range of 2 to 4 Inches Mercury Absolute." [In: U. S. Natl. Advisory Committee for Aeronautics. *Spark Ignition of Flowing Gases,* Part I. (RM E9E17) Washington, D. C., 1949.]
19. Peek, F. W. *Dielectric Phenomena in High Voltage Engineering,* New York, McGraw-Hill, 1929.
20. Westberg, F. W. *Combustion Noise Evaluation of Selected Field and Laboratory Industrial Burners and Environments.* (Research Rept. 1358A) Cleveland, Ohio, A. G. A. Laboratories, 1964.

Chapter 14

Interchangeability of Fuel Gases

by E. J. Weber

Large winter gas demands make it necessary in some areas to mix sendout gases or to substitute one type of gas for another during these peaks. The use of natural gas in locations remote from supply sources has required stand-by gas production facilities and/or underground or pipe storage. Accordingly, appliance burners may have to perform satisfactorily without readjustment on fuel gases that vary considerably in their combustion characteristics from the gas on which initial adjustment was made.

The range of acceptable variations in gas compositions which may be used for peak loads will be determined by the least flexible appliances served. It is usually necessary for each gas utility company to analyze its situation independently of the practices of other companies.

The lack of an explicit definition of interchangeability of gases exemplifies its complexity. A typical definition has described it as the possibility of using a substitute gas for that usually distributed to customers, without interfering with the operation of their appliances. Unless the gases are completely interchangeable, some degree of interference must be tolerated. However, the degree which might be considered acceptable will vary from one gas company to another and will be influenced by a number of factors, including the type of burner adjustment for a given utility and the tolerance of its consumers under peak load conditions.

There has been extensive research for ways to predict the interchangeability of one gas with another. *It was early recognized that correlation of heating values and specific gravities alone was insufficient, and that a third factor, namely, flame characteristics, which depend on chemical composition, must also be included. However, formulas and indexes based only on these two factors are widely used, apparently because of their simplicity. Results* cannot be considered as reliable as those determined by actual tests on representative appliances.*

In some studies, flame characteristics have been defined by the tendencies for lifting, yellow tipping, and flashback. Much work has been devoted to combining the factors of heating value, specific gravity, and flame characteristics into equations or charts for predicting interchangeability. Applications of such methods have met with varying degrees of success, depending on the selection of limiting values for the indexes of interchangeability.

* Results may be satisfactory locally, where a limited range of burner types is in use. This is not generally the case in larger industrial cities.

To date, there has not been any success in predicting the probability of occurrence, after substitution, of extinction or ignition noise, flashback on rapid turndown, unsatisfactory combustion, faulty automatic ignition performance with either thermal elements or flash tubes, and unfamiliar gas odors. The number of customer complaints in these regards are a practical criterion of successful interchangeability.

The most satisfactory method of determining interchangeability is to operate, in a laboratory, some ten or more appliances representative of the most critical types served. Interchangeability will be indicated by their performance on proposed substitute gases after initial adjustment on the base load gas.

FUNDAMENTALS OF INTERCHANGEABILITY†

Two gases may be regarded as interchangeable if flame characteristics are satisfactory after substitution of one gas for another. A flame which does not lift, yellow tip, or flash back is considered satisfactory in this frame of reference. A set of flame limiting conditions, i.e., a point within the stable flame zone, as functions of primary aeration and gas input rate may be established for any given aerated burner; see Fig. 12-110a. These flame limits will vary with the composition of the gas supplied.

Assume that the primary air and gas input rate are within the flame limits for a burner and the base gas. With a *substitute* gas, the primary air and gas input rate may change, depending on the relative specific gravities and heating values of the two gases involved. However, for the substitute gas to be *interchangeable* with the base gas, the base settings of primary air and gas input rate must be within the flame limits of the substitute gas.

Changes in gas input rate (or port loading) and primary aeration with substitution of a gas for the base or adjustment gas may be calculated from Eqs. **1** and **2** of Sec. 12, Chap. 12. Figure 12-167 illustrates these principles. It presents typical flame limit curves of a reference burner on an adjustment gas (subscript *a*) and a substitute gas (subscript *s*):

1. Curve L_s is the lifting limit curve for the substitute gas. L_s is above L_a; hence, the substitute gas is faster burning than the adjustment gas.
2. Curve Y_s is the yellow tipping limit for the substitute gas; curve Y_a is the yellow tip limit of the adjustment gas.

† For atmospheric low-pressure burners.

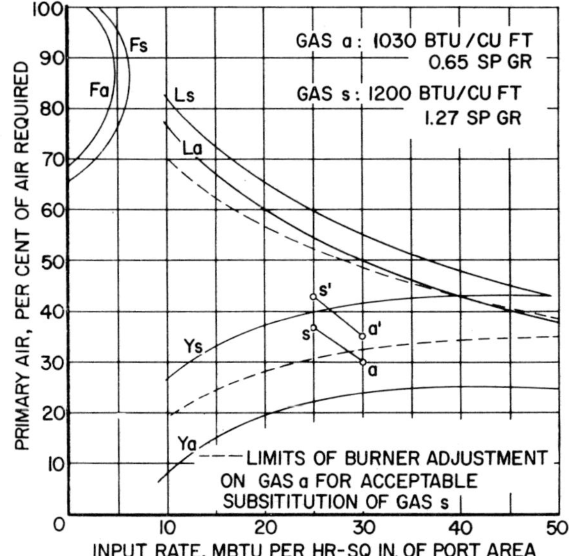

Fig. 12-167 A reference burner performance diagram for adjustment gas a and substitute gas s.

Since Y_a is below Y_s, the adjustment gas has less yellow tipping constituents.

3. Curves F_s and F_a show flashback limits of the substitute and adjustment gases, respectively. The substitute gas is the more susceptible to flashback, as indicated by curve F_s extending farther to the right.

Satisfactory and stable flames will, therefore, result with the adjustment gas for any burner adjustment of primary air and input rate within the zone bounded by the limit curves L_a, Y_a, and F_a. However, after initial satisfactory adjustment on gas a, substitution of gas s will cause a shift in primary aeration and input rate, which must now be within the limit curves L_s, Y_s, and F_s for acceptable interchange.

Examples. Gas a has a heating value of 1030 Btu per cu ft and a specific gravity of 0.65. Gas s has a heating value of 1200 Btu per cu ft and a specific gravity of 1.27. Substituting in Eq. **1** of Sec. 12, Chap. 12:

$$PL_s = PL_a(H_s/G_s^{0.5})(G_a/H_a^{0.5}) = PL_a(1200/1.27^{0.5})(0.65^{0.5}/1030) = 0.833PL_a$$

In other words, the input rate or port loading with gas s would be 0.833 *times* that with gas a.

The air per cubic foot of gas required for combustion for gas a is 9.7 cu ft, and for gas s, 11.0 cu ft. Substituting in Eq. **2** of Sec. 12, Chap. 12:

$$PA_s = PA_a(G_s^{0.5}/B_s)(B_a/G_a^{0.5}) = PA_a(1.27^{0.5}/11.0)(9.7/0.65^{0.5}) = 1.23PA_a$$

In other words, the primary aeration with gas s, in terms of per cent of air required would be 1.23 *times* that with gas a.

Figure 12-167 also determines the limits of adjustment on gas a which will also satisfy substitute gas s. Equations may be used to calculate adjustment points for gas a which will shift to adjustment points on curves L_s, Y_s, and F_s with substitution of gas s.

For example, a point on curve L_s is selected at the input rate, PL_s, of 20 MBtu per hour and the primary aeration, PA_s, of 65 per cent primary air. From the relations $PL_a = PL_s/0.833$ and $PA_a = PA_s/1.23$ calculated in the foregoing, the adjustment on gas a which would give the selected adjustment point on L_s after substitution of gas s would be at an input rate of 24 MBtu per hr and a primary aeration of 52.8 per cent primary air. Similarly, a second point on curve L_s at 30 MBtu per hr and 55 per cent primary air results from substitute of gas s from an adjustment on gas a at 36 MBtu per hr and 44.7 per cent primary air.

These adjustments on gas a fall on the upper dotted line. This dotted line was determined by selecting a series of points on L_s and, in the foregoing manner, calculating the adjustments on gas a which would give the selected points on substitute gas s. Thus, *this dotted line is the upper limit of adjustment on gas a for satisfactory substitution of gas s.* Any adjustment on gas a above the dotted line will produce lifting flames when gas s is substituted.

Similarly, the lower flame limit for adjustment on gas a for satisfactory substitution of gas s may be determined. Points on curve Y_s were used to locate the lower dotted line in Fig. 12-167.

Flashback curves may also be treated in the same manner. In this case, they are outside the limits of practical burner adjustments.

Figure 12-167 shows that gas s may be substituted for gas a if the initial adjustments on gas a are made within the dotted lines. It also indicates what changes in adjustment may be necessary to make gas s interchangeable. For example, if a burner were adjusted on gas a at point a, the resulting adjustment after substitution of gas s would be at point s. Since point s is below Y_s, flames would be yellow tipped. For satisfactory performance, increased primary aeration would be necessary so as to locate the *initial adjustment* at point a'. The resulting adjustment on substitution would then be at point s' above Y_s.

These examples also show that interchangeability of gases cannot be predicted entirely from their characteristics. Note, too, that the type of burner adjustments made greatly influences interchangeability. If typical adjustments for a given situation fell below the lower dotted line in Fig. 12-167, gas s could not be considered a satisfactory substitute unless readjustments were made, but if, on the other hand, typical adjustments fell entirely between the dotted lines, it could be considered a satisfactory substitute.

Figure 12-167 also shows that interchangeability of gases is not always reversible. From its curves, any adjustment between the limits of L_s and Y_s would be satisfactory on gas s with initial adjustment made on it. Substitution of gas a for gas s then would result in adjustments somewhere between the dotted lines. Because these lines are entirely within the limits of L_a and Y_a, gas a could be satisfactorily substituted for gas s. However, as previously shown, gas s could not be substituted for gas a if initial adjustments on gas a were below the lower dotted line or above the upper dotted line.

While not evident from Fig. 12-167, other seemingly contradictory conditions may be clarified by applying the fundamentals involved. Suppose flashback troubles are encountered with a substitute gas high in hydrogen. It might be assumed that they could be eliminated by replacing some of the hydrogen with carbon monoxide or inerts to lower the burning speed. Under some conditions, such an alteration of the substitute gas might actually intensify flashback troubles. In this case, carbon monoxide or inerts replacing hydrogen would increase specific gravity without a corresponding increase in heating value. Compared with the high hydrogen substitute gas, the altered substitute gas would result in a lower input rate and a leaner mixture, thus increasing flashback tendency.

The performance diagram of Fig. 12-167 is for a particular reference burner. Analysis of an interchangeability problem is more accurate and complete if it includes performance diagrams of several critical burners on the line, since burner design influences interchangeability. If the design of the reference burner were modified to raise the lifting limits appreciably without affecting the yellow tip limits, the burner would have greater flexibility and, hence, would tolerate a greater variation in substitute gas composition.

Since the abscissa of Fig. 12-167 denotes input *per square inch of port area* rather than simply input rates, the diagram is applicable to a wide variety of burners having the same pattern of port design and arrangement as the reference burner, although varying in the relative proportions of their component parts. Similar performance diagrams of a few critical burners would, therefore, provide a fairly accurate analysis of a system's interchangeability problem.

A.G.A. INTERCHANGEABILITY INDEXES

"C" Factor

An extensive A.G.A. research project[1] on interchangeability of gases involved some 250 different gas mixtures. The formula developed for calculating what was known as the *index of change in appliance performance, C*, is given in Eq. **1**:

$$C = \frac{H_s B_s G_a}{H_a B_a G_s} + \left(\frac{H_a B_a}{5000 E_a F_a} - \frac{H_s B_s}{5000 E_s F_s} \right) \quad (1)$$

where:

C = index of change in appliance performance
H = heating value of gas, Btu per cu ft
B = air theoretically required for combustion, cu ft per cu ft of gas
G = specific gravity of gas (air = 1.0)
E = heat content of theoretical products of combustion from 60 to 1600 F, Fig. 2-18, Btu per cu ft
F = summation of products of the mole fraction and a constant resolving each combustible constituent to the basis of equivalent free hydrogen
a and s = adjustment gas and substitute gas, respectively

For satisfactory performance on the substitute gas of appliances adjusted on the adjustment gas, C should be between 0.85 and 1.15. C values greater than unity indicate a trend toward softer flames and incomplete combustion. For values less than unity the tendency is toward flashback. As Eq. **1** was developed using manufactured gases of less than 800 Btu per cu ft, it makes no provision for lifting or yellow tips. Therefore, if these characteristics of a substitute and adjustment gas are different, this equation may not be adequate for determining interchangeability.

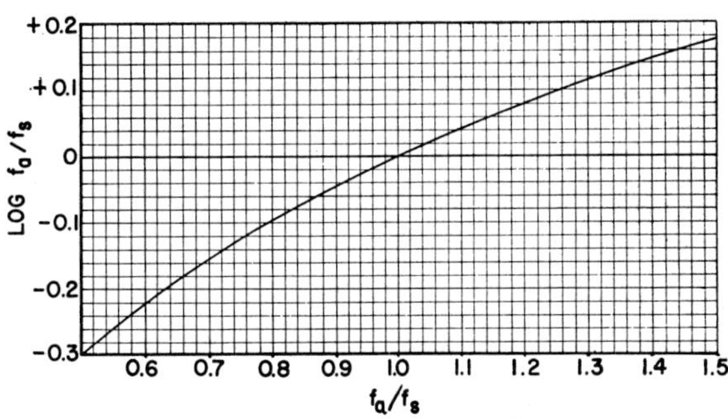

Fig. 12-168 Logarithms of various values of f_a/f_s. (Used in Eq. **2**.)

The C factor proved of great value for calculating interchangeability for gases of 800 Btu per cu ft or less.

Lifting, Flashback, and Yellow Tip Indexes

A.G.A. research[2] showed how gases of over 800 Btu per cu ft could supplement or be substituted for natural or high-Btu mixed gases. This study yielded equations for calculating the three flame limits in terms of indexes to indicate possibilities of lifting, flashback, or yellow tips when substituting various high-Btu gases for natural gases. These developments were believed applicable to all base load gases, but verification by further laboratory study was felt necessary. This later A.G.A. research on interchangeability of low heating value gases, using contemporary burners, was presented in four extensive reports;[3–6] a summary of these is available.[7]

Equations and limits follow for the three interchangeability indexes: lifting, I_L; flashback, I_F; and yellow tips, I_Y.

$$I_L = \frac{K_a}{(f_a a_s / f_s a_a)\,[K_s - \log(f_a/f_s)]^*} \leq 1.0 \text{ (limit)} \quad (2)$$

$$I_F = (K_s f_s / K_a f_a)(H_s/1000)^{0.5} \leq 1.18 \text{ (limit)} \quad (3)$$

$$I_Y = (f_s a_a Y_a / f_a a_s Y_s) \geq 1.0 \text{ (limit)} \quad (4)$$

where:

K = lifting limit constant, Eq. **6b**
f = $1000G^{0.5}/H$ = primary air factor
G = specific gravity of gas (air = 1.0)
H = gross heating value of gas, Btu per cu ft
a = $100B/H$ = air theoretically required for complete combustion, cu ft per 100 Btu of gas
B = air theoretically required for complete combustion, Table 2-63, cu ft per cu ft of gas
Y = maximum primary air on yellow tip limit curve, Eq. **7**, per cent

s and a = substitute gas and adjustment gas, respectively

Lifting Index, I_L. I_L is the ratio of the resultant primary aeration after substitution to the primary aeration at the lifting limit of the substitute gas at the resultant input rate. Derivation of Eq. **2** is based on lifting limit curves (see Fig. 12-167) obtained with the *A.G.A. precision burner*. It is assumed that a burner is adjusted on a base gas at an input rate of 10 MBtu per hr-sq in. of port area and at a primary aeration equal to the lifting limit of this gas at

* Log (f_a/f_s) is given in Fig. 12-168.

this input rate. With introduction of the substitute gas, both input rate and primary aeration usually change from their initial values.

If I_L is less than 1.0, the resultant primary aeration is less than the primary aeration at the lifting limit of the substitute gas. This gas would, therefore, be interchangeable with the base gas insofar as the lifting flame characteristic is concerned. On the other hand, if I_L is greater than 1.0, the resultant primary aeration is above the *theoretical* lifting limit of the substitute gas—lifting flames would result with substitution. Practically, however, the limiting value of I_L is usually somewhat greater than 1.0. It must be determined from experience, since it is dependent on local conditions. Two factors allow the limiting value of I_L to be somewhat greater than unity. The first is that a burner is adjusted on the base gas at an input rate of 10 MBtu per hr-sq in. of port area, an unusually low input rate. Also, the actual value of I_L will vary, depending on the port loading of any given burner on the line. The second factor is that, experimentally, the burner is adjusted on the base gas right at its lifting limit. Actually, burners are adjusted to primary aerations below the lifting limit to an extent varying with local conditions and the prevailing adjustment practice. These factors permit a practical limiting value of I_L greater than 1.0 to an amount depending on local conditions.

An additional factor regarding this limiting value is that I_L is specific to the *A.G.A. precision burner*, and only relative insofar as other burners are concerned. I_L, in accordance with its definition, may be determined graphically from lifting limit curves (see Fig. 12-167). To obtain a relative index, curves of any suitable reference burner may be used.

The lifting limit constant, K, for Eqs. **2** and **3** originates from empirical Eq. **5** of the lifting limit curve:

$$K = \log PL + 0.016PA \qquad (5)$$

where:

PL = port loading, 10 MBtu per hr-sq in. of port area
PA = primary aeration, per cent of air required for theoretical combustion

K for any given gas indicates the relative lifting tendency of that gas, although it is also a function of burner design. The K values of simple gases may be averaged to obtain the value for a mixture, K_{mix}. In the first study[2] from which I_L originated, K values of simple gases were averaged on a per cent by weight basis. A later study[7] indicated that more accurate results could be obtained if the K values of the constituent gases were averaged on the basis of the per cent, P, each uses of the theoretical air required for complete combustion of the mixture. This may be done by means of Eq. **6a**:

$$K_{mix} = (K_1B_1P_1 + K_2B_2P_2 + \ldots)/100B_{mix} \qquad (6a)$$

where:

B = air theoretically required for complete combustion of constituent gas, cu ft per cu of gas

K may also be determined from Eq. **6b**; note that the general term K_nB_n in Eq. **6a** may be replaced by F from Eq. **6b** to get K_{mix} on a per cent by volume basis.

$$K = F/G \qquad (6b)$$

where: F is given on Fig. 12-169
G = specific gravity

Flashback Index, I_F. Derivation of Eq. **3** is entirely empirical. This is because characteristic flashback limit curves can be obtained for very few high-Btu gases with the *A.G.A. precision burner* and also because these curves are much influenced by burner design and burner head temperatures. The recommended maximum limiting value of I_F for satisfactory interchange, based on actual test substitution of various supplementary gases *for natural gases only*, is 1.18. It is probable that separate indexes are required to fully cover flashback on extinction, on ignition, and at equilibrium.

A study of factors affecting flashback indicated that its tendency varied with the percentage of primary air in the burner head. Maximum ignition velocities usually were attained at high primary air percentages, so flashback was most likely at high primary aerations. Typical flashback limit curves also showed flashback to occur only at low port loadings. Thus, substitution of a gas with a primary air factor higher than that of the adjustment gas aggravated flash-

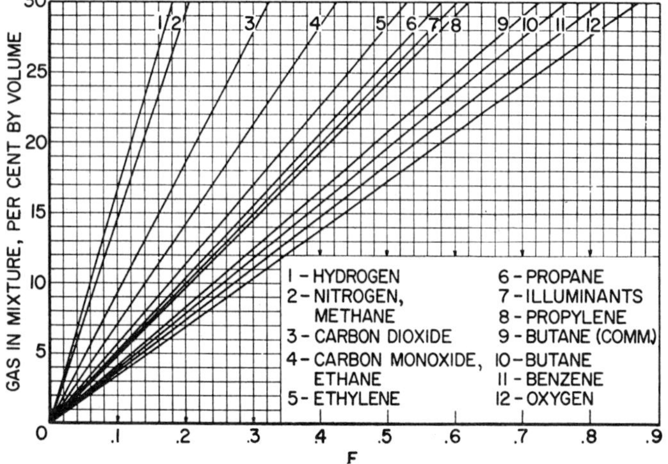

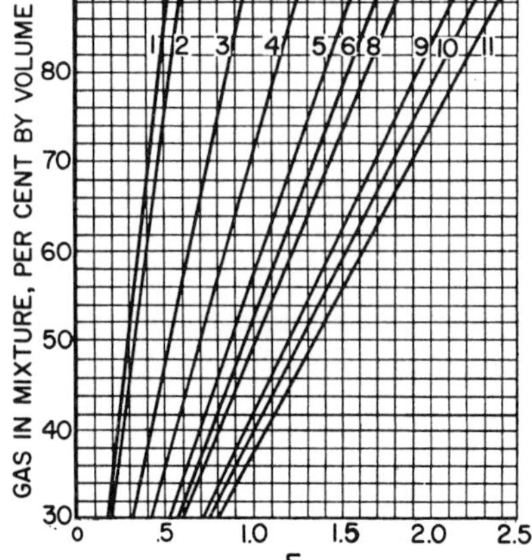

Fig. 12-169 Graphs for calculating F values of gases.[2] (Used in Eq. **6b**.)

back tendencies in two ways; it decreased input rate and, consequently, increased the percentage of primary air in the burner head.

Experiments with some gases indicated that I_F might be expressed as a ratio between K values and primary air factors of the adjustment and substitute gases. However, further tests showed that flashback took place with butane–air and other high-gravity mixtures, although their satisfactory interchange was indicated by use of an equation based on K and f factors. A possible explanation of such flashback was that uneven mixing and unequal distribution occurred with the heavier gases, causing some ports to flash back. To reflect this condition, the heating value of the substitute gas was included in Eq. **3**; I_F values so calculated showed better agreement with test results.

Yellow Tip Index, I_Y. I_Y is the ratio of the primary aeration after substitution to the maximum primary aeration on the yellow tip limit curve of the substitute gas. Derivation of Eq. **4** is also based on lifting limit curves (see Fig. 12-167). It is assumed that a burner is adjusted on a base gas at a primary aeration equal to the maximum primary aeration on the yellow tip limit curve of the base gas. This initial primary aeration value usually changes with substitution. No provision for input rate is made in the equation for I_Y because the yellow tip limit curve is horizontal over a considerable range of input rates.

If I_Y is greater than 1.0, the resulting primary aeration with substitution will be above the yellow tip limit curve and yellow tips will *not* occur with the substitute gas; i.e., the gases are interchangeable. With I_Y less than 1.0, the opposite is true and the substitute gas is theoretically not interchangeable.

As with the lifting index, I_L, the theoretical minimum limiting value of I_Y is 1.0. For reasons generally similar to those discussed in connection with I_L values *above* unity, the practical limiting value of I_Y may be somewhat less than 1.0, depending on local conditions of burner adjustments.

The yellow tip limit in maximum percentage of primary air, Y, for Eq. **4** may be calculated by Eq. **7**:

$$Y = \frac{T_1V_1 + T_2V_2 + \ldots}{B + 7C - 26.3\,O_2} \qquad \textbf{(7)}$$

where:

T = air required to eliminate yellow tips, cu ft per cu ft of gas, Table 12-99

V = combustible gas constituent, per cent by volume

B = air theoretically required for complete combustion, cu ft per cu ft of gas

C = total volume of N_2 and CO_2 in mixture, fraction or decimal

Application of I_L, I_F, and I_Y. As previously explained, it is sometimes practical to exceed theoretical limits (Eqs. **2**, **3**, and **4**), depending on local conditions. Therefore, specific limits for use in an interchangeability problem should be established after a careful analysis of the entire situation. If all three indexes for a given substitution are found to be within the limits established, the substitute gas should be fully interchangeable with the base gas. If one of the calculated indexes is outside the established limit, the substitute gas may be expected to give unsatisfactory performance on interchange. However, it may be considered as a supplemental gas satisfactory for mixing, up to some limit, with the base gas for peak demands. Comparison of the sets of indexes for various mixtures will permit selecting of the mixture with the highest percentage of the supplemental gas for satisfactory interchange.

Table 12-99 Yellow Tip Constant, T, for Various Simple Gases[2]

Gas	T	Gas	T
Hydrogen	0.0	Ethylene	8.7
Carbon monoxide	0.0	Propylene	13.0
Methane	2.18	Acetylene	17.4
Ethane	5.8	Benzene	52.0
Propane	9.8	Illuminants†	19.53
Butane (comm.)*	15.3	Oxygen	−4.76
Butane (pure)	16.85		

* Contains 22 per cent propane.
† Contain 75 per cent ethylene and 25 per cent benzene.

Extensive tests have shown that I_L and I_Y indexes give accurate predictions of lifting and yellow tips in about 90 per cent of the test cases. They also indicated that these indexes may be used with almost complete accuracy to predict the performance of an individual appliance (including the effects of the burner and its adjustment) on a substitute gas, by establishing experimentally the limiting indexes for that appliance. Limits so established for one appliance are not likely to be applicable to all others in a system, since they might include many different types of burners.

The I_F index as calculated by Eq. **4** was not found to be fully reliable in predicting the probability of flashback. It appears that two equations may be needed, one covering flashback during normal operation, the other on ignition or extinction. So far, attempts to modify Eq. **4** to permit better correlation with actual flashback performance have not been successful.

OTHER INTERCHANGEABILITY INDEXES

Knoy Formulas

The assumption that the burner head mixture of primary air and gas has a heating value of 175 Btu per cubic foot when an atmospheric burner is properly adjusted yielded[8–11] Eq. **8**, which is plotted in Fig. 12-170. For good interchangeability, the substitute gas determined from this chart should have a C value equal to that of the adjustment gas. A deviation of ± 5 per cent may be tolerated in some cases, or even ± 10 per cent in special gases. Actually, variations in adjustments of appliance burners cause some deviation from the mixture heating value of 175 Btu per cu ft. A C factor for a substitute gas above that of the adjustment gas indicates softer flames.

$$C = (H - 175)/G^{0.5} \qquad \textbf{(8)}$$

where:

C = a constant
H = heating value of gas, Btu per cu ft
G = specific gravity of gas (air = 1.0)

Figure 12-170 and Eq. **8** have the advantages of simplicity and ready availability of the data required. These are important factors in meeting emergencies with substitute gases

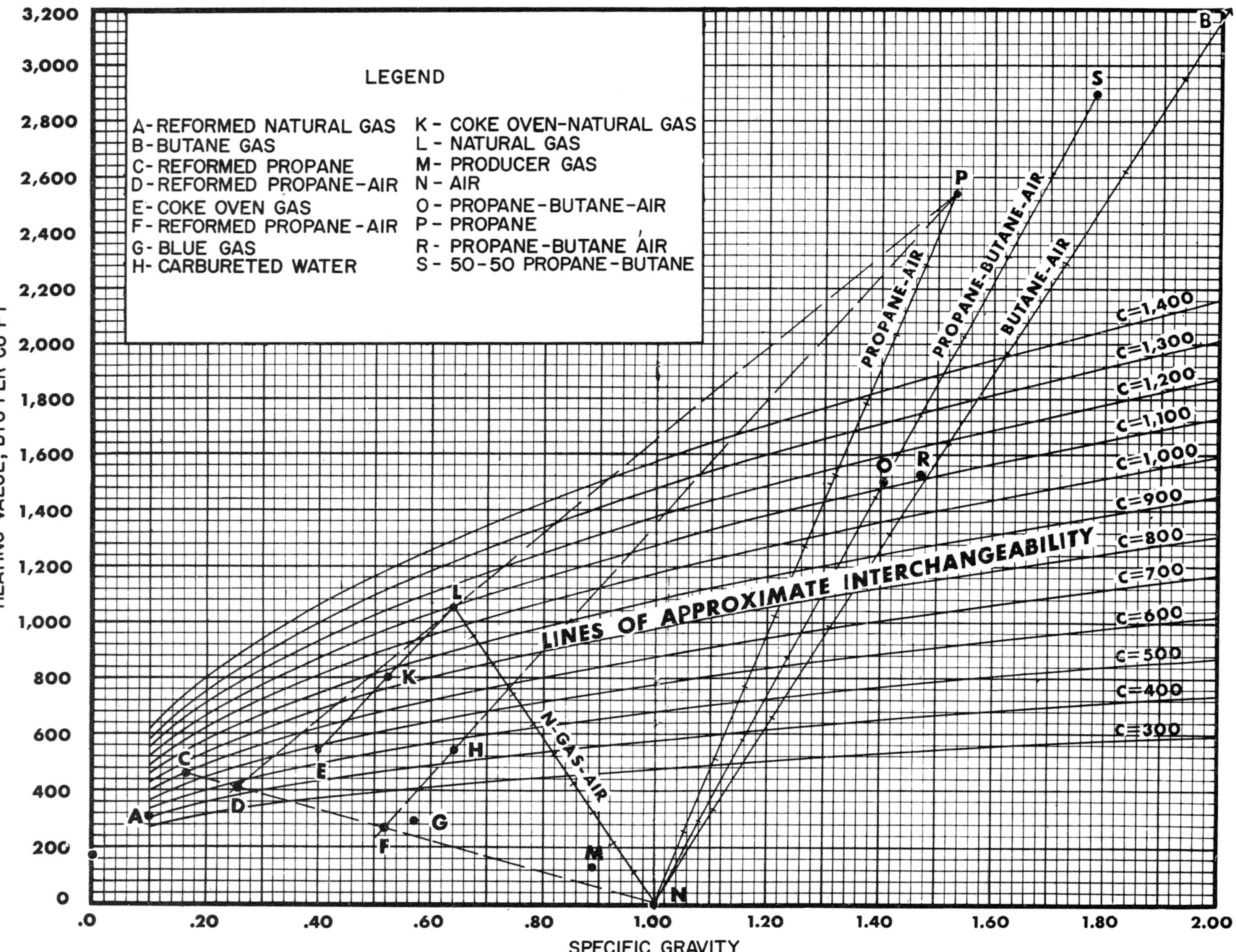

Fig. 12-170 Knoy interchangeability chart.

of diverse properties in proportions varying so rapidly that only heating value and specific gravity can be known at all times.

Both Eq. **8** and Fig. 12-170 are based on heating value and specific gravity relations and provide no means for evaluating burning characteristics of a substitute gas. Without duly allowing for such characteristics, misleading results might be obtained. Criteria have been described by Knoy[11] which, used in connection with Fig. 12-170 and some test burners at the plant, will ensure as good substitute gas performance as is possible with available equipment and materials.

Examples Using Fig. 12-170.

1. Find the proportions of gases corresponding to points C and N in a mixture F on the dashed line joining these points.

From the bottom scale the following specific gravities may be read: gas C, 0.165; gas F, 0.52; gas N, 1.00.

By the law of inverse ratios, proportions of C and N in mixture F are:

$$\text{Gas } C = \frac{1.00 - 0.52}{1.00 - 0.165} = 0.58$$

$$\text{Gas } N = \frac{0.52 - 0.165}{1.00 - 0.165} = 0.42$$

2. What proportions of propane, gas P; reformed propane, gas C; and air, N, are needed to give a mixture of the same heating value and specific gravity as gas L?

Draw line PL, extending it to intersect CN at D. D, being on line CN, represents a mixture of gas C and gas N. The proportions of C and N in D and, in turn, the proportions of D and P in L, may be determined as in the foregoing example. Note that gas D may be mixed with gas P to give the same heating value and specific gravity as gas L, since L is on line PD.

3. Substitute gas interchangeability may be determined by using the "lines of approximate interchangeability" on Fig. 12-170. For example, determine the heating value and specific gravity of gas H enriched by propane, gas P, to be approximately interchangeable with natural gas L.

Follow interchangeability line C = 1100, which passes thru L, to its intersection with line PH. This intersection corresponds to a heating value of 1250 Btu per cu ft and a specific gravity of 0.95. Proportions of gas H and P may be determined by the law of inverse ratios (as in the first example).

Using the heating value of the burner head mixture as a criterion of interchangeability, Eq. **9**, also developed by Knoy, may be employed.

$$P_s = \frac{H_s G_a^{0.5}}{G_a^{0.5} + \left(\frac{H_a - P_a}{P_a}\right) G_s^{0.5}} \tag{9}$$

where:

P = heating value of burner head mixture (usually assumed to be 175 Btu per cu ft for the adjustment gas)
H = heating value of gas, Btu per cu ft
G = specific gravity of gas (air = 1.0)
a and s = adjustment gas and substitute gas, respectively

For interchangeability, P_s should be within ±15 of the assumed P_a of 175 with the adjustment gas.

Weaver Indexes[12]

These indexes are expressions which give an approximation of the relative tendencies of a substitute gas to give unsatisfactory results when supplied without appliance readjustment. The indexes were derived partly from theory and partly empirically from former research.[1–7]

The effects of a change in gas composition on heat output and on primary air changes accompanying a substitute gas are given by Eqs. **10** and **11**, respectively.

$$J_H = \frac{H_s}{H_a}\left(\frac{G_a}{G_s}\right)^{0.5} \tag{10}$$

$$J_A = \frac{B_s}{B_a}\left(\frac{G_a}{G_s}\right)^{0.5} \tag{11}$$

where:

J_H = effect of substitute gas on heat output
H = heating value of gas, Btu per cu ft
G = specific gravity of gas (air = 1.0)
J_A = change in primary air with substitute gas
B = air theoretically required for complete combustion, cu ft per cu ft of gas
a and s = adjustment gas and substitute gas, respectively

J_A provides not only an accurate measure of the relative conditions of primary air supply for burning two gases but of secondary air also. Consequently, it is a measure of the possibility of incomplete combustion on those appliances in which flame impingement is not involved and secondary air is closely adjusted to increase thermal efficiency.

Formulas for the four proposed indexes are:

$$J_L = J_A \frac{v_s}{v_a}\left(\frac{100 - Q_s}{100 - Q_a}\right) \tag{12}$$

$$J_F = \frac{v_s}{v_a} - 1.4 J_A + 0.4 \tag{13}$$

$$J_Y = J_A + \frac{N_s - N_a}{110} - 1 \tag{14}$$

$$J_I = J_A - 0.366 \frac{R_s}{R_a} - 0.634 \tag{15}$$

where:

J_L = lifting index
v = flame speed, Table 2-70, fps
Q = O_2 in gas, per cent
J_F = flashback index
J_Y = yellow tip index
N = number of carbon atoms easily liberated by combustion per 100 molecules of gas; saturated hydrocarbons are assumed to have one carbon atom per molecule
J_I = relative tendency of the two gases to burn incompletely, i.e., to liberate CO
R = ratio of number of hydrogen atoms in the gas to number of carbon atoms in the hydrocarbons only

a and s = adjustment gas and substitute gas, respectively

Other terms are defined under Eq. **11.**

When two gases are fully interchangeable with regard to lifting, $J_L = 1.0$. When $J_F = 0$, there is no difference in flashback tendency. Similarly, when $J_Y = 0$, there is no difference in yellow flame tendency. J_I represents a correction of J_A and reflects effects of flame impingement in some appliances; when $J_I = 0$, there is no change in this tendency.

Weaver's study[12] includes detailed comparison with the I_L, I_F, and I_Y indexes developed by A.G.A. research.[3–6] It also indicates that where the gas supply is changed on a large number of appliances in service, any uncertainty as to the accuracy with which either set of indexes represents the relative properties of the gases is small compared with the uncertainties involved in the initial appliance adjustments.

Wobbe Index

In British practice, the heating value, H, of a gas *divided* by the square root of its gravity, $G^{0.5}$, is defined as the Wobbe number.[13] This number is proportional to the heat input to a burner at constant pressure. It is widely used in Europe, together with a measured or calculated flame speed factor, for calculating interchangeability. Note that the relationship, $H/G^{0.5}$, which is also known as the **heat input factor,** is included in some of the interchangeability formulas given in this chapter.

BUREAU OF MINES INTERCHANGEABILITY STUDIES

Extensive interchangeability data have been developed[14–17] on the basis of the theory of flame propagation, which claims that a flame is stabilized on a burner port at points in the gas stream at which normal burning velocity is equal to the local linear velocity of the unburned gas stream. Points of such equality generally exist near the boundary of the stream. Flame stability is, therefore, dependent on boundary conditions.

Lifting and Flashback

Flame lifting or blowoff will occur when the slope of the flow velocity distribution profile or boundary velocity gradient at the stream boundary reaches a critical value. This critical boundary velocity gradient varies with the air–gas ratio of the mixture burned and with the gas composition. The same also applies to the occurrence of flashback.

For laminar flow, the boundary velocity gradient may be expressed by Eq. **16:**

$$g = 1.27V/R^3 \qquad \textbf{(16)}$$

where:

g = boundary velocity gradient, per sec
V = volume rate of flow, cu in. per sec
R = burner port radius, in.

Critical boundary velocity gradients for blowoff, g_B, are obtained with values of V corresponding to flows at which the flames just blow off the port. Similarly, at values of V just causing flashback, critical boundary velocity gradients for flashback, g_F, are obtained.

By plotting critical boundary velocity gradients g_B and g_F vs. mixture composition, blowoff and flashback limit curves may be obtained; Fig. 12-171 shows such curves for methane. Mixture composition is expressed in terms of gas concentration as a *fraction of stoichiometric*. For example, since there is 9.46 per cent methane in a stoichiometric mixture (one that has the primary aeration theoretically required for combustion), 12.0 per cent CH_4 in a mixture is the equivalent of a stoichiometric fraction of 12.0/9.46, or 1.27.

Yellow Tipping

Each gas has a minimum characteristic gas–air ratio at which yellow appears in the flame. The corresponding gas concentration, as a fraction of stoichiometric, is termed the constant yellow tip limit, F_c.[17] When secondary air diffuses into the entire flame, the gas–air ratio in the flame is leaner than that in the burner, and the apparent yellow tip limit for the burner and gas becomes richer. The corresponding gas–air concentration, as a fraction of stoichiometric, is termed the nonconstant yellow tip limit, F_y.

Figure 12-171 shows yellow tip limit curves for methane, with critical boundary velocity gradients plotted against gas concentration. Calculations necessary for plotting yellow tip limit curves similar to those in Fig. 12-171 are given by Grumer and others.[17]

Application to Interchangeability

Flame limit curves like those in Fig. 12-171 are treated in the same manner as those discussed in Fig. 12-167. Interchangeability is indicated by whether or not the zone of adjustment on substitution lies within the region of stable flames as established by the flame limit curves of the substitute gas. If the resultant zone of adjustment lies entirely within this stable flame region, no question of interchangeability exists. On the other hand, if the resultant zone of adjustment lies only partially in the stable flame region, interchangeability depends on the type of adjustment made on the base gas. In this case, laboratory tests of critical burners adjusted close to the limit curves should be made. In any event, the relative positions of the curves will indicate the type of readjustment required for interchangeability.

The nature of the flame limit curves of the Bureau of Mines is such that it rules out the effects of burner design. In other words, only the relative burning characteristics of the fuel gases are involved; hence, the need for a reference burner is eliminated.

As with other methods, judgment is necessary as to the effect of the type of adjustment on the appliances served.

STUDIES OF SUBSTITUTES FOR NATURAL GAS

Research has been conducted by I.G.T. on interchangeability of several specific substitute gases with natural gas. The following results are presented under the substitute gas employed.

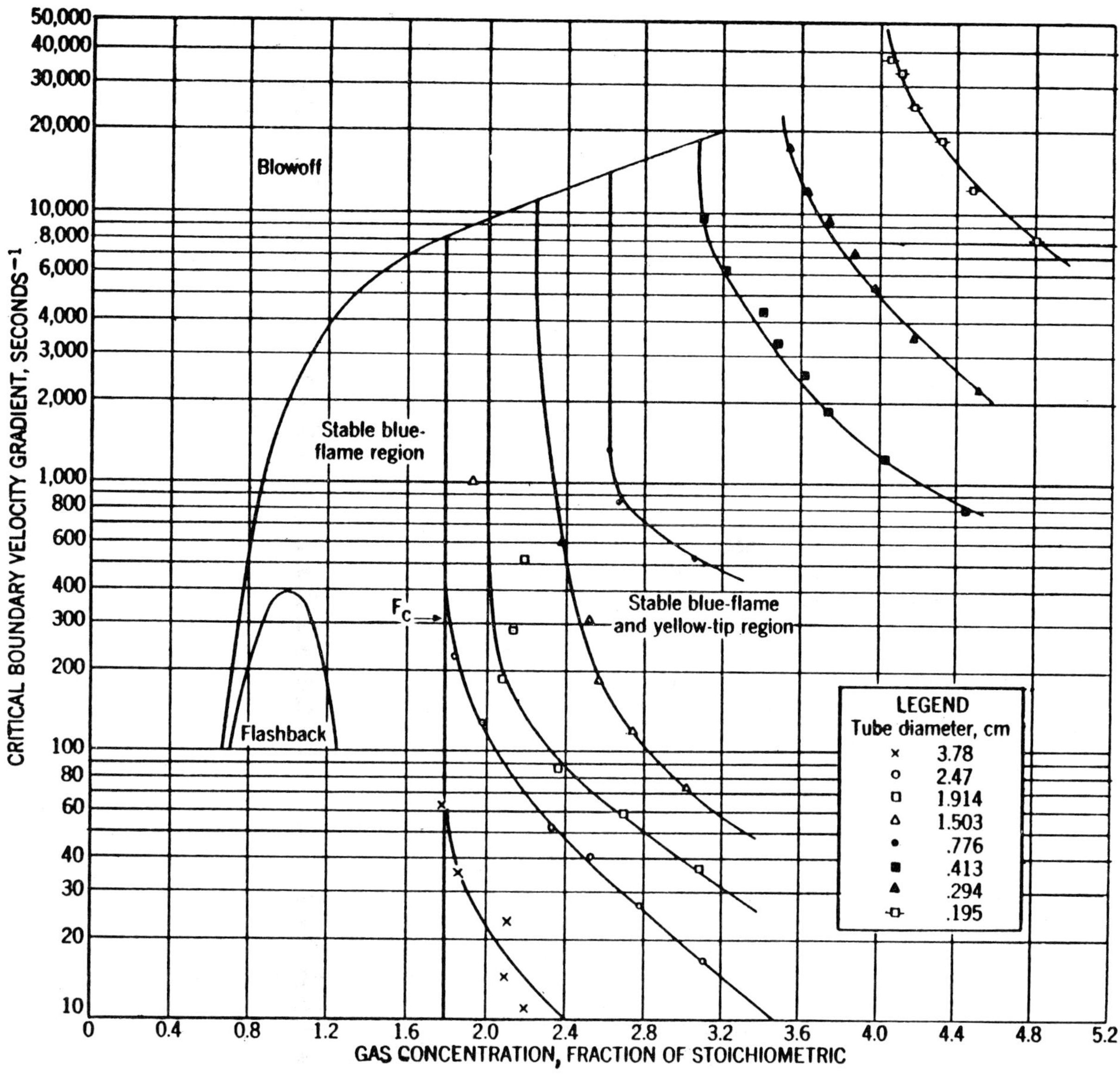

Fig. 12-171 Flame stability diagram for methane showing blowoff, flashback, and the effects of port size on yellow tipping.[17] (Tubes of diameters shown used as burners.)

High-Btu Oil Gas[18]

Tests were made using typical scrubbed high-Btu oil gas with true (inert-free) heating values ranging from 900 to 1500 Btu per cu ft to replace high-methane natural gas in a group of critical contemporary appliances and burners (Fig. 12-172). Substitution was made under two test conditions: (1) dilution with a synthetic blast–blow-run gas (20 per cent CO_2, 80 per cent N_2) to about 1000 Btu per cu ft (that of the base natural gas); and (2) dilution with the foregoing inert gas to heating values and specific gravities that would maintain the rated burner heat inputs. Yellow tipping, flashback, and blowoff tendencies were determined at three primary aerations. These corresponded to moderately hard (+2, A.G.A. Code, Table 12-100), normal (0, A.G.A. Code), and moderately soft (−2, A.G.A. Code) adjustments on the base natural gas.

Interchangeability was reported in terms of the maximum per cent oil gas in the oil gas–base natural gas mixture which gave satisfactory combustion performance on five of the critical test burners and on all test appliances. These substitutability values for oil gases, adjusted to constant heating value and constant heat input, are related to the true (diluent-free) oil gas heating values and thus provide a simple method for estimating the performance of high-Btu oil gases introduced into a natural gas system.

Ignition delay periods were observed for all range top burners with automatic ignition. When oil gases were substituted at constant *heat input*, such delay periods averaged 73 per cent of those with base natural gas. For oil gases substituted at constant *heating value*, delays averaged 86 per cent of those with the base gas.

On 1350 Btu oil gas diluted with inerts to constant heat input, pilot operation was unsatisfactory due to heavy carbon deposition on pilots after 25 hr. A similar 1350 Btu true oil gas, diluted with 65 per cent high-methane natural gas to produce a substitutable mixture with regard to yellow tipping and flashback, showed only slight carbon deposition after 100 hr.

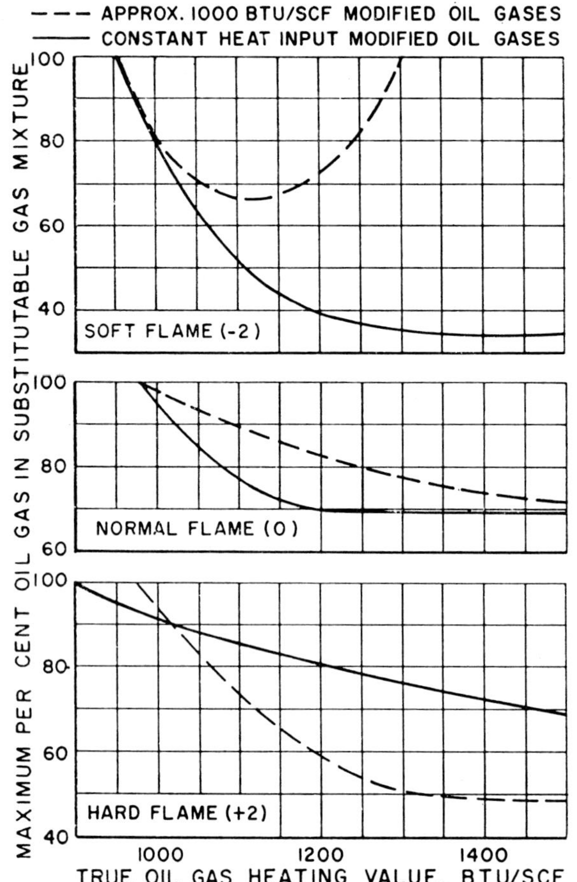

Fig. 12-172 Substitution of scrubbed oil gases, modified to constant heat input and constant heating value, for high-methane natural gas.[18] (Critical burner tests.) Numbers in parentheses defined in Table 12-100.

Table 12-100 A.G.A. Code for Describing Flame Characteristics

Code	Flame description
+5	Flames lift from ports with no flame on 25 per cent or more of the ports.
+4	Flames tend to lift from ports, but become stable after short period of operation.
+3	Short inner cone, flames may be noisy.
+2	Inner cones distinct and pointed.
+1	Inner cones and tips distinct.
0	Inner cones rounded, soft tips.
−1	Inner cones visible, very soft tips.
−2	Faint inner cones.
−3	Inner cones broken at top, lazy wavering flames.
−4	Slight yellow streaming in the outer mantles or yellow fringes on tops of inner cones. Flames deposit no soot on impingement.
−5	Distinct yellow in outer mantles or large volumes of luminous yellow tips on inner cones. Flames deposit soot on impingement.

Such deposits burned off readily when natural gas was supplied again. Pilot performance did not appear to introduce additional limitations on the proportion of oil gas which may be added to natural gas for brief periods.

Operating Experiences. Emergency substitution[19] of high-Btu oil gas has been made by an Eastern utility during interruption of its normal supply of 1040 Btu per cu ft high-methane natural gas. Oil gas, produced at about 1200 Btu per cu ft, was reduced by compression to 1130–1150 Btu per cu ft and 0.84 sp gr. This gas was supplied to 10,000 consumers for 30 hr until normal natural gas service could be restored. The substitution resulted in 180 complaints—all of yellow tipping. These occurred on appliances which had been converted about five years previously from 537 Btu manufactured gas to natural gas at critical adjustments.

In describing this experience, the author emphasizes that no claim is made that the oil gas used can be considered a complete substitute for high-methane natural gas.

In other instances this utility has replaced 85 per cent of its natural gas supply to some 30,000 customers for 12-hr periods without complaints.

Propane-Enriched Test Gases

Interchangeability studies[20] of propane-enriched test gases covered mixtures ranging from about 1000 Btu per cu ft to about 100 per cent relative *heat input factor* to the base natural gases (Table 12-101). Approximate *mixture* heating values corresponding to a constant heat input factor were about as follows:

	Btu per cu ft
Propane–hydrogen	1000
Propane–low-gravity reformed gas	1200
Propane–high-gravity reformed gas	1300
Propane–air	1500+

The performances of propane-enriched test gas and base gas mixtures containing CO, N_2, CO_2, or air were compared both at constant heating value and various heat inputs and at constant heat input and various heating values, since heat input and heating value of high-methane natural gas can be duplicated simultaneously only with hydrocarbon–hydrogen mixtures.

Table 12-101 presents data using each of the four propane-enriched gas mixtures tabulated in the foregoing in terms of substitutability (i.e., the maximum per cent of test gas in the test gas–base gas mixture) for high-methane natural gas on five critical burners mounted in the open. Flame adjustments were for soft, medium, and hard flames (−2 to +2 in Table 12-100). The data of Table 12-101 are arranged in order of increasing heating value.

Further tests were made on a group of 13 critical and contemporary appliances used in previous oil gas interchangeability studies (Fig. 12-172). The range of adjustments on the base gas was the same as that on the five burners mounted in the open.

For the adjustment range used, yellow tipping and transient flashback proved the limiting conditions on burners and appliances. Lifting and blowoff occurred with propane–air mixtures in some instances.

High hydrogen content test gases gave better combustion performance than propane–air gases near 100 per cent relative heat input. Best overall substitutability of propane-enriched reformed gases was about 80 per cent at 90 per cent relative heat input on a group of seven critical burners mounted in the open. Mixtures found substitutable on the

Table 12-101 Critical Burner Tests on Various Propane-Enriched Test Gases[20]

Test gas: Composition* C_3H_8	N_2	CO_2	CO	H_2	CH_4	C_2H_6	O_2	Heating value, Btu/SCF	Specific gravity (air=1)	Relative heat input to burner,††%	Substitutability, %: Flame adjustment† −2	0	+2
Propane–Reformed Gas (high specific gravity)													
24.7	3.7	5.5	12.9	45.5	1.3	2.2	...	960	0.762	83.7 b‡	100	62§	62§
24.7	7.3	3.2	16.6	40.7	1.5	2.0	...	968	.789	82.8 b	100	75§	60§
30.3	16.9	0.2	16.9	29.5	0.3	2.1	...	1050	.914	84.0 b	92**	62§	62§
35.9	9.7	4.2	9.3	31.2	7.2	1.2	...	1132	.942	86.6 a‡	78**	92§	78§
31.2	2.4	5.2	10.8	36.4	4.8	2.9	...	1170	.855	96.2 b	78**	92§	70§
39.3	...	4.2	11.4	35.6	3.2	0.6	...	1290	0.921	102.2 b	66**	81**	70§
Propane–Reformed Gas (low specific gravity)													
27.7	2.2	1.1	15.8	48.1	3.8	0.4	...	971	0.713	85.2 a	100	92§	70§
27.1	4.3	...	17.5	46.6	...	1.6	...	1018	.701	92.3 b	78**	78§	...
32.4	10.4	1.4	...	50.5	4.7	0.6	...	1029	.730	89.3 a	92**	100	78§
37.1	...	0.2	14.9	47.8	...	...	...	1133	.778	95.4 a	...	78**	78§
36.8	2.5	0.1	14.7	40.2	0.8	0.7	...	1196	.827	99.8 b	36**	44**	55§
42.5	...	...	18.5	39.0	..	...	...	1258	0.856	103.3 b	20**	32**	55§
Propane–Air													
31.7	44.6	0.3	...	1.2	2.8	1.9	11.6	1008	1.180	70.4 b	92§	48§	48§
40.9	40.8	...	...	...	0.7	1.2	11.0	1192	1.230	81.6 b	92**	78§	78§
45.7	31.9	...	...	5.0	0.5	3.1	5.7	1413	1.293	94.5 b	55**	78§	78§
61.6	30.4	...	...	..	...	...	8.0	1528	1.326	98.4 a	25**	25**	62**
Propane–Hydrogen													
32.5	5.4	...	...	61.4	0.4	0.2	...	997	0.591	98.5 b	70**	92§	62§
30.9	...	...	...	68.4	0.4	0.2	...	1011	.546	103.9 b	70**	85**	62§
28.1	0.9	1.4	...	63.5	3.6	1.1	...	1027	0.595	98.8 a	55**	63**	62§

* Remainder consists of unsaturated and higher saturated hydrocarbons. † See Table 12-100.

‡ Base gases:

	CH_4	C_2H_6	C_3H_8+	N_2	CO_2	Btu/SCF	sp gr
a	94.0	3.3	1.5	0.6	0.6	1049	0.605
b	96.6	1.6	...	1.7	0.1	993	0.570

§ Limiting condition: flashback on turndown. ** Limiting condition: yellow tipping.

†† Relative heat input compared to base gas adjustment $= \frac{H_T}{S_T^{0.5}} \times \frac{S_B^{0.5}}{H_B} \times 100$, where H = heating value, S = specific gravity, and subscripts T and B are for the test and base gases, respectively.

critical burners were also generally satisfactory at equivalent adjustments on the critical appliances.

Propane–air mixtures showed best overall substitutability of 80 per cent on the same base gas adjustments, but at about 85 per cent relative heat input. Such substitutability corresponded to a 1200 Btu heating value and 1.2 specific gravity for the hydrogen-rich carrier gas mixtures. On approaching 100 per cent relative heat input, most of the propane-enriched reformed or process hydrogen carrier gases maintained a substitutability level of about 60 per cent. The level for propane–air gases dropped to 25 per cent due to their greater yellow tipping tendency at a heating value of about 1500 Btu per cu ft. Overall substitutabilities of about 60 per cent at full relative heat input were obtained with propane–hydrogen mixtures of only 1000 Btu and 0.6 specific gravity.

When substituting propane–carrier gas mixtures for high-methane natural gas, better combustion performance near rated heat inputs appears indicated at substantially lower heating value and gravity levels; high hydrogen content gases such as catalytically reformed hydrocarbons or process hydrogen are used rather than air.

TEMPORARY GAS SERVICE EQUIPMENT[25]

A trailer-mounted LP-gas–air plant is available to provide peak shaving and temporary gas service to consumers disconnected from a normal gas supply. One mixture in actual use has a heating value of about 1300 Btu per cu ft and a 1.3 sp gr; it is substantially interchangeable with a 1000 Btu per cu ft air-modified natural gas where appliances have been initially adjusted with a soft setting. For a discussion of a larger portable propane-air plant, see page 5/54.

Applications of Unit

Besides temporary service for disconnections arising from construction maintenance and other causes, the unit may also be used for local peak shaving and for increasing the effectiveness of Hortonsphere storage.

Equipment Specifications

The UL-approved direct-fired vaporizer supplies propane vapor, which is mixed with air from an 85-cfm-capacity compressor. A 110-volt a-c supply is required for the latter. The vaporizer can deliver 9000 cfh of 1350 Btu gas. A liquid pump may be used to complement the vapor pressure that forces the propane into the vaporizer.

Substitution Example. About 3000 homes in a natural gas base load system were switched over to LP-gas-air for five hours during the summer; there were no complaints. The 20 house heating units run on the LP-gas-air mixture operated satisfactorily. If limited feedstock supply necessitates decreasing the gas heating value to 1200 Btu, complaints may be expected from users of hard-set and critical appliances.

Table 12-102 Classification of Test Burners[21]

Class	Observations	Measurements	Name of burner(s)
1	Start of yellow tipping of flames	Per cent of opening of primary air shutter	A.G.A. precision; Rochester
2	Flashback*	Per cent of opening of primary air shutter	Ott; Hawes; Burna
3	Change in inner cone height at constant primary aeration	Inner cone height of flame	Hofsass; Grebel–Velter
4	Inner cone height fixed	Per cent of opening of primary air shutter	Czako–Schaack; A.T.B., Rochester; A.G.A. precision
5	Same as 4	Gas pressure, air pressure, and mixture pressure	C_k burner
6	Flame temperature	Temperature	Flamdicator
7	Start of cup-shaped flames	Amount of air supplied	T.A.R. (total air requirement)
8	Constant luminous flame†	Gas pressure	A.G.A.; Gaz de France; luminous test burner

* See Fig. 12-173(c).
† See Fig. 12-173(d).

FLAME CHARACTERISTICS INDICATORS (TEST BURNERS)

Many of these burners (Table 12-102) have been used: (1) to determine the interchangeability of a substitute gas rapidly without appliance tests and calculations, and (2) to control plant mixing operations so as to maintain acceptable appliance performance. Atmospheric burners are usually employed with primary air or gas control devices operating over arbitrary scales. Some characteristics of the flame, such as its flashback or yellow tip points or the adjustment to produce a definite cone height, serve as the indicating means to differentiate between two gases. Results are expressed in empirical units. Test burners can be used only after their calibration in terms of gases similar in composition and characteristics to those to be compared. Accordingly, the range of usefulness of these burners is limited.

Successful use of various test burners has been reported within set limits for certain types of gas mixtures after correlation with appliance servicing experiences. Full advance knowledge was available of the particular types of appliances served and their adjustments. Test burners have not been recommended for use where new or untried substitute gases are involved or where the types of appliances and their adjustments are unknown. Studies reported in Reference 7 indicate that no single test burner index reading could be used to predict fully satisfactory performance of all appliances on interchange of a substitute gas with a known adjustment gas. However, readings of certain burners, e.g., the Rochester test burner, could be correlated with types of unsatisfactory burner performance in about 90 per cent of the cases once an average limiting index value was established.

A.G.A. Precision Test Burner

Both gas and primary air are independently controlled and metered in this precisely machined burner; see Fig. 12-173(a). Temperature of the burner head is measured by a thermocouple, permitting determination of each type of limit curve at a constant temperature. The temperature maintained was 225 F for lifting and flashback curves and 300 F for yellow tip curves.

Data obtained with this burner were translated into factors for use in Eqs. **2, 3,** and **4** for calculating interchangeability. Such data, expressed in the form of lifting, flashback, and yellow tip limit curves, are not direct indications of gas interchangeability. Thus, application of this burner differs from that of others in which index readings indicate performance characteristics of substitute gases.

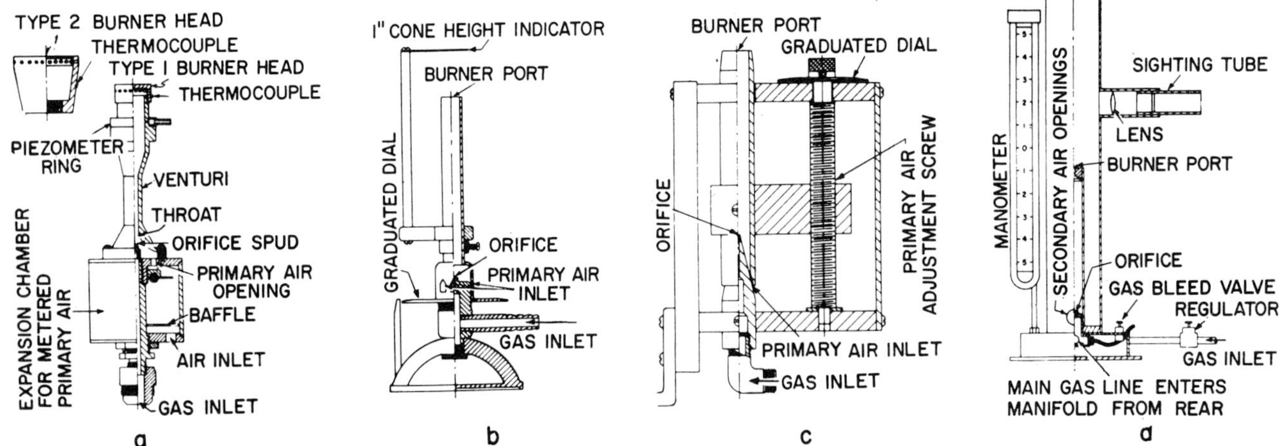

Fig. 12-173 Four types of test burners:[7] (a) A.G.A. precision; (b) Rochester; (c) Modified Swedish flashback; (d) A.G.A. luminous flame.

Rochester Test Burner

This Bunsen-type burner, Fig. 12-173(b), employs an air shutter with a graduated scale to indicate the extent of shutter opening and a fixed gas orifice. Means are also provided for measuring the height of the inner cone.

Scale readings of air shutter openings when the flame is at the points of lifting, flashback, and yellow tipping are taken as indexes of burner performance for the gas employed. Comparison of these indexes with those for a base adjustment gas permits direct indication of burner performance on a substitute gas.

Data have been obtained with this burner on numerous substitute gases for manufactured gas and natural gas. These data show that it may be used successfully for measurement of flame characteristics and for determining interchangeability of substitute gases for both manufactured and natural gases. Several additional test burners of this general type and their application are discussed in Refs. 7 and 22.

Thermal Conductivity Gas Analyzer

This type of flame characteristics indicator differs from those previously described in that its index reading is determined from a sample of unburned gas rather than from the flame. Such a device was included in studies reported in Reference 7. Gases with high thermal conductivity readings do not give trouble from lifting, while those with low readings produced lifting. The degree of correlation of these readings with appliance performance varied somewhat, depending on the particular adjustment gas.

A.G.A. Air-Gas Ratio Analyzer

Unsatisfactory appliance performance on a substitute gas may be due to the nature of the existing adjustment. Consideration was given, therefore, to developing a device for assisting a serviceman in making proper appliance adjustments to fit the circumstances. Measurements of primary air and input rate were involved. As input was readily determinable by means already available, efforts were concentrated on finding a means of primary air measurements thru thermal conductivity of the air–gas mixture.

An air–gas ratio analyzer was developed thru A.G.A. sponsored research. Several models were built and field-tested by operating utilities. Primary aeration was indirectly measured by indication of relative thermal conductivities of air–gas mixtures. These instruments were found to be fully satisfactory as laboratory means for studying interchangeability problems and burner performance characteristics. However, they were not considered practical for field use to indicate burner adjustment, mainly because of the complexity of the operation and the time required, as well as the need of many preliminary laboratory data to establish a base for field use. Construction and operating details of this analyzer are presented in Reference 13.

INTERCHANGEABILITY IN INDUSTRIAL GAS EQUIPMENT

From the standpoint of interchangeability, there are several differences between industrial and domestic gas appliances. Domestic customers are supplied on a firm basis; thus, gas available to them must burn properly at the set burner adjustment.

Industrial customers, on the other hand, frequently are supplied on an interruptible basis, permitting the gas supply to be shut off when conditions require it. Furthermore, industrial equipment is controlled by instruments or is under the supervision of experienced operators qualified to make any changes necessary to permit the equipment to continue to operate satisfactorily. When the utility gas supply is on an interruptible basis, either another fuel may be used or the affected equipment may be shut down until the normal gas supply is restored. Responsibility for the availability and use of such a stand-by fuel rests with the individual industrial customer.

Equipment for mixing LP-gas and air in closely controlled proportions is often used to provide stand-by fuel for industrial plants when the base gas supply is interrupted; see, for example, Fig. 5-30a.

When a propane–air mixture is supplied which is interchangeable on domestic appliances with the base natural gas normally served, it has been the experience of a large utility operating in the Middle West that practically no trouble is encountered on industrial equipment. Operating variations were so minor as to be undetectable by the average industrial customer.

Atmospheric Generators

These generators, both endothermic and exothermic, may continue to operate when change is made to a substitute gas. However, since their operation depends on the composition of the fuel supplied, any change in the hydrogen-to-carbon ratio will be reflected in the composition of the atmosphere produced. Problems are most apparent in applications in which the atmosphere must be controlled within close tolerances, such as those required by rigid heat treating specifications. A readjustment of the gas–air ratio to the generator is often necessary. Devices may be used which will detect a variation in the atmosphere generated and make a compensating adjustment.

Glass Working Flames and Similar Applications

Lamps, ampoules, lenses, and other articles requiring precise application of heat from a gas flame will be affected by changes in gas composition, heating value, or specific gravity. Effects of such changes appear in the displacement of the maximum temperature area in the flame; for example, a natural gas flame cone is 50 per cent longer than the comparable cone on manufactured gas.[23] It was indicated[24] that the specific flame intensity equation (Eq. **3a**, Sec. 2, Chap. 5) may be applied to correct conditions produced by a gas change. Necessary steps may range from relatively simple adjustment of pressure to more involved procedures requiring oxygen or hydrogen additions. Burner redesign has been used sometimes to get maximum flame intensity on substitution.

STABILIZATION

The term "stabilization" as used in natural gas service refers to the processes for slightly modifying a gas, generally involving an adjustment of its heating value. Modifications may be desirable to make the delivered product comform to contract levels or to maintain a uniform product when differing natural gases are obtained from two or more sources. This may be accomplished with relatively simple mixing equipment obtainable from several manufacturers. The addition may be air, CO_2, water vapor, or LP-gases, depending on the objective desired and the circumstances involved.

REFERENCES

1. A.G.A. Testing Laboratories. *Mixed Gas Research Investigation.* (Research Repts. 597, 645, 689) Cleveland, Ohio, 1930–32.
2. ——. *Interchangeability of Other Fuel Gases With Natural Gases.* (Research Bull. 36) Cleveland, Ohio, 1946.
3. ——. *Interchangeability of Other Fuel Gases With Coke Oven Gas.* (Research Rept. 1106A) Cleveland, Ohio, 1948.
4. ——. *Interchangeability of Other Fuel Gases With Carburetted Water Gas.* (Research Rept. 1106B) Cleveland, Ohio, 1948.
5. ——. *Interchangeability of Other Fuel Gases With a Mixed Coke Oven–Carburetted Water Gas.* (Research Rept. 1106C) Cleveland, Ohio, 1949.
6. ——. *Interchangeability of Carburetted Water Gases With a Mixed Natural Gas–Blue Gas–Cracked Natural Gas–Producer Gas.* (Research Rept. 1106D) Cleveland, Ohio, 1949.
7. ——. *Interchangeability of Various Fuel Gases With Manufactured Gases: Summary Bulletin.* (Research Bull. 60) Cleveland, Ohio, 1950.
8. Knoy, M. F. "Combustion Experiments With Liquefied Petroleum Gases." *Gas* 17: 14–9, June 1941.
9. ——. "Experiences With Butane–Air Standby." *Gas* 18: 20–2, July 1942.
10. ——. "Master Interchangeability Chart." *Gas* 23: 46–52, June 1947.
11. ——. "Graphic Approach to the Problem of Interchangeability." *A. G. A. Proc.* 1953: 938–47.
12. Weaver, E. R. *Formulas and Graphs for Representing the Interchangeability of Fuel Gases.* (RP 2193) Washington, D. C., Natl. Bur. of Standards, 1951. Also in *J. Res. Natl. Bur. Std.* 46, Jan.–June 1951.
13. Weber, E. J. *Study of Possible Burner Adjustment Indicator Devices.* (Research Rept. 1214) Cleveland, Ohio, A. G. A. Laboratories, 1954.
14. Grumer, J. *Study of Combustive Characteristics of Fuel Gases.* (Proj. PDC-3, Interim Research Rept. 1) (Bur. of Mines Proj. PX3-102, Rept. 3223) New York, A. G. A., 1951.
15. —— and others. "Predicting Interchangeability of Fuel Gases." *Ind. Eng. Chem.* 44: 1554, July 1952.
16. Grumer, J. and Harris, M. E. "Flame Stability Limits of Methane, Hydrogen and Carbon Monoxide Mixtures." *Ind. Eng. Chem.* 44: 1547, July 1952.
17. Grumer, J. and others. *Fundamental Flashback, Blowoff, and Yellow-Tip Limits of Fuel Gas–Air Mixtures.* (R. I. 5225) Washington, D. C., Bur. of Mines, 1956.
18. Searight, E. F. and others. *Interchangeability of High-Btu Oil Gases and Natural Gas.* (Research Bull. 24) Chicago, I. G. T., 1956.
19. Voelker, J. G. "Emergency Substitution of 100% High Btu Oil Gas for Natural Gas." (CEP-55-9) *A. G. A. Proc.* 1955: 847–54.
20. Ellington, R. T. and others. "Substitutability of Propane-Enriched Carrier Gases for Natural Gas." *Am. Gas J.* 185: 35–42, June 1958.
21. Roux, A. "Principal Types of Gas Burners." *J. Usines Gaz* 74: 216–20, 1950.
22. Sackmann, von W. "Aufbau und Eigenschaften der Wichtigsten Pröfbrennertypen." *Gaswärme* 8, 1959.
23. Richardson, H. K. "Burners for High Speed Glass Forming." *Ceram Ind.* 52: 89–90, March 1949.
24. ——. "Burners for Lamp Working." *Ind. Gas* 27: 8+, April 1949.
25. Cook, R. H., and Deutsch, I. "Portable Unit Provides Temporary Service—No Interruption to the Customer." *A.G.A. Operating Sect. Proc.* 1964: 64-D-204.

Chapter 15
Commercial Kitchen Gas Appliances

by E. A. Jahn

Gas cooking equipment is designed for a specific operation. Therefore it is not possible merely to specify a battery of ranges on the basis that the range is a universal appliance and can perform all types of cooking by use of appropriate pots, pans, or kettles. Specialized or functional appliances can do the various cooking jobs faster, better, and more economically—especially in large volume. Generally, the larger the cooking volume, the greater the need for specific, functional appliances of the various types covered in this chapter. ASA Approval Requirements exist for all of the appliances and those models that comply are A.G.A. approved. In addition, those conforming to National Sanitation Foundation Standard No. 4 carry the NSF approval seal.

RANGES

The two basic styles of ranges used in commercial kitchens are commonly known as "heavy duty" and "restaurant" or "cafe" ranges.

Heavy Duty Ranges

Both in construction and in cooking capacity, heavy duty ranges are ideal for commercial kitchen use. They are usually made in unit sections which can be joined to form a battery. These ranges are suited for severe service conditions—used for long hours, cooking large amounts of food, rough handling, and heavy cooking utensils. Large restaurants, hospitals, army camps, and other places in which large volume cooking is regularly done require such equipment.

A number of manufacturers have introduced models of "super heavy duty" ranges designed to meet the continuous heavy requirements of the hotel or other large-volume kitchen. These ranges offer superior performance and reliability and conform with NSF standards for ease of cleaning and sanitation. Some of the features requested by the food service industry and incorporated in this appliance design include:

a) High-speed open top burners (up to 35,000 Btu per hr with a blower).

b) Insulated burner compartments to reduce surface and valve handle temperatures.

c) Recessed controls and handles to prevent breakage.

d) Leak-proof oven bottoms to protect burners and controls.

e) Warp-resistant plates for hot tops with outer gutter to collect grease and spillovers.

f) Sealed space between ranges to permit individual gas connection and prevent grease and food accumulation.

g) Front combustion air openings, better insulation, and double wall flue risers to allow flush-to-wall mounting.

Heavy duty range sections may have a solid hot top, a set of open top burners (grate top), or a solid griddle for frying. Beneath the top is usually an oven suitable for roasting and baking. Sometimes a *skeleton* range with shelves or a storage cabinet is supplied instead, or the top section is installed on legs or on a wall bracket. Other appliances, such as broilers, salamanders, roasting ovens, and deep fat fryers, are built to the same dimensional standards as the ranges so that they may be incorporated into the range battery.

Spreader plates are available to provide additional working surface between range sections. Extensions, usually half the width of a range section, may be installed at either side of a section or between two sections. They are available with hot, open, or fry tops.

Individual heavy duty range sections are between 29 and 37 in. wide, and from 34 to 42 in. deep, depending on make and model. When a manufacturer makes two sizes, the smaller has less cooking top area, but usually the same size oven and burners and similar construction. See Table 12-103 for typical dimensions and gas inputs.

Table 12-103 Typical Dimensions and Gas Inputs for Heavy Duty Ranges

	Dimensions, in.*							Input, MBtu/hr	
	Overall		Cooking top†		Oven			Top† section	Oven section
Type of range	*W*	*D*	*W*	*D*	*W*	*D*	*H*		
Extra heavy duty									
Hot top	32	42	32	39	25	28	15	60	42
Uniform hot top	32	42	32	39	25	28	15	70	42
Open top	32	42	32	39	25	28	15	60	42
Fry top	32	42	30	32	25	28	15	70	42
Heavy duty									
Hot top	32	35	32	32	25	28	15	50	42
Uniform hot top	32	35	32	32	25	28	15	50	42
Open top	32	35	32	32	25	28	15	58	42
Fry top	32	35	30	26	25	28	15	50	42

* W = width, D = depth, H = height.

† High-speed top sections, with inputs up to 70 per cent greater than these typical values, are available from a few manufacturers.

Each section or complete range may be equipped with a single or a double high shelf, an elevated broiler, or a stub flue riser. Ranges are sometimes placed back to back and connected to a concealed flue. This arrangement without back splashers is usually required for "island" batteries.

Most heavy duty ranges are available in stainless steel when a particularly long lasting, attractive, and easily cleaned finish is desired. Standard models usually have black or gray japan finish. Enameled finishes in four pastel shades may also be obtained.

Hot Top Ranges. Intended for heavy duty continuous cooking, they have a heavy cast-iron (or alloy) top to support pots and pans. The entire top area can be used.

Several burner arrangements are used on hot tops. A common one consists of a series of concentric rings beneath the center of the top. Use of one or more of these rings gives a hot spot in the center of the top with gradually decreasing temperatures toward its edges. For fast heating up and *hard* boiling, the center of the top is used; and for slower and gentler cooking processes, its sides and back. A second type uses a row of burners under the front of the top, giving the highest temperature there, and less heat toward the rear. Another arrangement is the uniform hot top heated by a series of bar burners which produce a substantially uniform temperature over the entire surface. For longer life and freedom from warping, tops made of alloy castings are available.

Open Top Ranges. These have grates to hold vessels directly over the burners. While continuous capacity is somewhat less than that of a hot top, an open top is superior for short order work. Full cooking heat is available instantly, and each burner may be shut off when not required.

Fry Top Ranges. The entire top is one piece, heated from beneath. Its surface is not intended for use with pots but as a griddle for frying. Provision is made to carry excess grease to a suitable receptacle. A fry top is sometimes combined with a broiler underneath to make a griddle and broiler section.

Restaurant or Cafe Ranges

These ranges are lighter in construction than heavy duty types. They are not usually intended for battery use, each being a complete unit. The name *restaurant* does not mean that all restaurants use these ranges, nor that they are suited only for restaurant use. Most restaurants, particularly larger ones, require heavy duty equipment. Restaurant ranges, however, are well suited for smaller establishments stressing short order cooking, and for intermittent use, such as in service pantries, diet kitchens, lodges, clubs, churches, large homes, and similar locations. Burners are often of lower capacity and ovens are usually smaller.

Restaurant ranges vary in size from those of domestic types up to 7½ ft long. The most popular models are available in seven basic styles (Table 12-104), with many variations obtainable depending on the top section type desired. Any two front and back open burners may be replaced by either a solid hot or fry top section. Many of these ranges are equipped with a combination griddle and broiler; the griddle forms the top surface of the broiler and is heated by the broiler burners. The griddles are available either flush with the cooking top or raised to locate the broiler at cooking top level. Ranges employing the latter type of griddle usually have two ovens.

Luncheonette Ranges. These are commonly used in drugstores, snack bars, small coffee shops, etc., for very light duty cooking. They are light equipment, similar in some respects to domestic ranges.

BAKING AND ROASTING OVENS

Gas baking and roasting ovens are available in many varieties and sizes. Their selection depends on the quantity and type of food to be prepared.

Range Ovens

Most heavy duty and restaurant ranges are equipped with roasting ovens (up to 130 lb capacity) which may also be used for baking operations. Similar ovens are also available as separate units, usually constructed in pairs, one over the other. They are designed for installation in batteries with heavy duty ranges and are known as *double-deck range-type ovens.* Internal dimensions and inputs follow:

Type	*Height*	*Volume*	*Input, MBtu/hr*
Heavy duty	14–15 in.	5.4–6.3 cu ft	42
Restaurant	13–14 in.	2.9–5.5 cu ft	30

Table 12-104 Typical Dimensions and Gas Inputs for Restaurant-Type Ranges[1]

	Dimensions, in.*									Input, MBtu per hr			
	Overall		Burner section		Oven			Broiler-griddle					
Type of range	*W*	*D*	*W*	*D*	*W*	*D*	*H*	*W*	*D*	Top burners	Oven(s)	Broiler-griddle	Entire range
Single oven model													
6 open burners	36	30	36	28	24	22	14	...	...	84	30	...	114
Single oven w/broiler-griddle													
4 open burners	38	30	26	28	26	22	14	12	26	56	30	30	116
6 open burners	60	30	36	28	24	22	14	24	26	84	30	42	156
Double oven model													
10 open burners	60	30	60	28	24	22	14	...	...	140	60	...	200
Double oven w/broiler-griddle													
6 open burners	60	30	36	28	24	22	14	24	26	84	60	42	186
8 open burners	70	30	46	28	24	22	14	14	26	112	60	42	214
10 open burners	90	30	65	38	26	22	14	25	26	140	60	42	242

* W = width, D = depth, H = height.

Portable Baking and Roasting Ovens

These may be direct- or indirect-fired ovens, constructed in sections so that baking and roasting decks may be combined in any desired arrangement. They are portable in that they are factory shipped as a complete unit and require a minimum of assembly on the job. For best performance each deck of a sectional oven is equipped with its own burner and thermostat, but some models (cabinet ovens) use one burner and control for two or more decks. Except for special application requiring tile hearths, the entire oven is constructed of steel to provide quick heating and low initial cost. These ovens are frequently used in large kitchens, commissaries, baking departments of institutions, and small commercial bakeries. Table 12-105 shows typical dimensions and capacities.

Table 12-105 Capacities and Dimensions of Portable Baking and Roasting Ovens[1]

	Baking deck	Roasting deck
Overall width, in.	47 to 60	47 to 60
Overall depth, in.	28 to 40	28 to 40
Interior height, in.	7	12
Hearth area, sq ft	4.8 to 9.3	4.8 to 9.3
Capacity:		
10 in. pie pans	6 to 12	...
20 × 22 in. roast pans	...	1 to 2
Input, MBtu per hr	20 to 40	22 to 50

Muffle-Type Ovens. These indirect deck ovens have dimensions and capacities similar to those in Table 12-105. They differ somewhat from the conventional direct-fired design, as combustion products do not enter the baking chamber. The heated oven bottom sets up secondary convection air currents which circulate thru the oven compartment. This design is suited for baking processes requiring a high degree of product uniformity and direct injection of steam for moisture control.

Forced Circulation Cabinet Ovens. These indirect ovens represent the most recent design for baking and roasting in the commercial kitchen. They differ from the conventional deck oven in that the rapid circulation of heated air permits the food to be cooked on multiple open racks rather than on a solid hearth. As a result the full volume of the oven space can be utilized and production capacity greatly increased. The high-velocity forced air circulation within the oven is accomplished by means of a motor-driven fan unit. This is possible because the oven is of the muffle type and the combustion products pass around the oven chamber and do not enter the cooking compartment. The rapid air circulation also ensures even temperature distribution to all parts of the oven and to the food products. The high heat input permits effective reconstitution of frozen foods, while the use of an "on" and "off" type of thermostat makes possible low-temperature meat roasting at 190 to 225 F.

Pizza Ovens

The pizza oven is an adaptation of a direct-fired single-deck type, with two important differences. Most time and temperature recipes for pizza call for a very *fast* oven which requires a special high-temperature thermostat with a working range up to 750 F. Also, the hearth temperature must be low compared with the baking temperature; hence, the deck is usually constructed of Marinite or other insulating material.

Ovens are available with hearth areas of from 6 to 12 sq ft and capacities of two to six 16 in. pies per loading. High recovery rates are essential, since the baking period is only a matter of minutes and thus the door is constantly opened for loading and unloading. As a result, higher input burners are required; a 6 sq ft oven is rated at 48 MBtu per hr and a 12 sq ft oven at 70 MBtu per hr.

Revolving Tray or Reel Ovens

The chief characteristic of these ovens is the use of flat trays suspended between two spiders rotating in an arrangement similar to a ferris wheel. The entire assembly is usually housed in a semidirect gas-fired chamber, although some are direct- or indirect-fired. The food is loaded on the trays as they appear opposite the door opening. Provision is made to prevent escape of a hot blast of air when opening door for loading and unloading. For many years these ovens were primarily used in very large commissaries and bakeries, but today small six-pan units, having a depth of only 3.5 ft, are available for smaller restaurants and institutions. Mechanical ovens of this type can be used to cook complete meals, including vegetables, entrees, and dessert. They hold up to 350 lb of meat or fowl and 48 loaves of bread. When desired, steam injection can be provided.

Rotary Ovens

In principle, these ovens are similar to the revolving tray ovens, except that the hearth rotates about a vertical instead of a horizontal axis. The hearth is generally heated from below by gas burners; heat may also be provided at the top. The walls are of firebrick. Some ovens have two hearths on the same axis, the shaft being raised or lowered to bring either to the level of the doors. Ovens of this type are most frequently used for large-volume pie and bread baking. They are seldom found in individual restaurants. Capacities may range from 300 to 800 lb per hr.

Peel Ovens

This name is derived from the long paddle or peel used to load and unload the baked goods. Older "built-in" brick ovens were usually coal burning, but many have been successfully converted to gas. Stock model ovens equipped with gas burners are available. Construction is of firebrick and angle iron, with a large stationary hearth, all completely enclosed in insulation and an outer jacket. A steam-tight baking chamber is provided to accommodate baked goods requiring steam injection. Models suitable for retail bakeries may have hearth areas of 25 to 60 sq ft and baking capacities of 80 to 200 one-pound bread loaves per loading. Larger models with hearth areas up to 120 sq ft are also available.

Traveling Ovens

Ovens of this type are confined almost entirely to large-volume industrial food production, such as bread baking. The food is carried along on a continuous conveyor thru the length of the oven, which is heated by multiple gas burners.

BROILERS

The distinguishing characteristic of broilers is that cooking is primarily by radiant heat instead of by conduction or convection. This radiant heat is generally provided by ceramic radiants adjacent to the broiler burners. Quantity of broiled products and available floor space are usually the two most important considerations in choice of broiling equipment.

Table 12-106 gives inputs and broiling areas of various types commonly used in restaurants and institutions.

Table 12-106 Typical Broiler Grid Areas and Gas Inputs[1]

Type of broiler	Broiling area per section, sq ft	Gas input, MBtu per hr
Unit broiler	3.3 to 5.0	70 to 100
Combination broiler & griddle	2.3 to 3.3	35 to 55
Salamander (heavy duty)	1.6 to 2.8	30 to 44
Underfired or hearth broiler	1.2 to 2.7	40 to 52
Rotisserie	2 to 7 spits	30 to 90

Unit or Heavy Duty Broilers

Also referred to as "hotel broilers," these are designed for large-volume production and for installation in a heavy duty range battery. They are the same width as the range sections and have a large grid area and powerful gas burners for fast, continuous broiling.

Two types are available: one is the same height as the range tops and equipped with a standard high shelf; the other is integral with an overhead oven heated by the burners in the broiling compartment below. The overhead oven can serve as a warming compartment, a precook chamber, or a finishing oven, thereby greatly increasing the broiler capacity.

The oven volume may vary from 3.2 to 5.8 cu ft. The burners are always equipped with ceramic radiants to provide the intense, uniform heat required. The burners may be located in the center with ports angled toward the outer edges, along the sides and firing toward the center, or spaced evenly across the broiling compartment top.

For high production capacity, these broilers are also available equipped with the infra-red Schwank burners. Since these burners do not require any secondary air, the broilers may be mounted one on top of the other. Another advantage of the Schwank burner is that it preheats to operating temperature in 90 seconds and it can therefore be turned off when not in use.

Experience has shown that a unit broiler can medium broil 75 to 100 lb of meat or fish per hour. Factors influencing the actual capacity include the thickness of cut, type and quality of food, and degree of *doneness* desired.

Combination Broiler and Griddle

This type is most commonly used in small kitchens where space is limited and volume production is not required. Some are designed for use in range batteries, while others are separate free-standing units.

One set of burners supplies heat for the broiler and for the frying griddle which forms the top. Because of this limitation, this type is not recommended for simultaneous large-volume frying and broiling loads. Griddle area is approximately 4.0 sq ft. Ceramic radiants improve broiling performance but adversely affect that of the griddle.

Combination broilers and griddles also comprise part of some restaurant-type ranges (Table 12-104).

Salamander or Elevated Broilers

This type is a miniature broiler mounted above the top of a heavy duty range (usually closed or hot top) or above a spreader plate as part of the back shelf assembly. It has many of the features found in a heavy duty broiler, including ceramic radiants, but a grid area of only 1.6 to 2.8 sq ft. Although its chief advantage is that it requires no floor space, it is sometimes mounted in pairs on special stands or legs to provide additional broiling capacity in a specific location.

A salamander broiler is frequently used as an auxiliary for preparation of short orders during off-peak hours when other equipment is shut down. In smaller restaurants which serve limited quantities of broiled food, the salamander can often handle the entire broiling load.

Hearth or Underfired Broilers

Chunks of irregularly sized ceramic or refractory form a radiant bed above the gas burners. A heavy cast-iron grate which holds the food is located horizontally above this heat source. Since some of the juice from the meat drips directly onto the hot radiants and burns, the resulting smoke and flame impart a typical charcoal flavor and appearance to the meat. Restaurants specializing in charcoal broiled food have found that these broilers produce equally palatable food without the labor and dirt problems associated with charcoal fires. The units are usually located in a window or other prominent location, since the broiling operation has much eye appeal for the customer.

Most manufacturers produce multiple broiler sections of any desired length and broiling area. Because of the large volume of smoke given off during the cooking process, installation must be made under an efficient exhaust hood. Flame control is accomplished by such methods as grooved or tilting grids and a directed stream of air across the surface of the ceramics.

Rotisseries

These broilers, designed for restaurants specializing in rotisserie roasting (barbecuing), are usually placed in the restaurant in a prominent, eye-appealing manner. They are essentially vertical broilers, heated by refractory-type gas burners located in the rear. From two to seven mechanically rotated spits hold the meat or fowl in position while it is slowly broiled on all sides. Units are available in window, back-bar, and floor models. They are also popular for on-the-premises cooking of fowl in retail poultry markets. The broiling capacity is four whole chickens per spit, or 28 chickens per loading for the seven-spit model.

DEEP FAT FRYERS

A deep fat fryer consists of a deep kettle containing cooking oil or fat, heated by thermostatically controlled burners. The tube type is heated by gas flames directed into large seamless immersion tubes running thru the lower part of the kettle and welded securely in place. The underfired type is heated by burners located directly below a V-shaped kettle with the flame and hot flue gases contacting its bottom and sides.

In both types, provision is made for crumbs and other sediment to collect in the kettle bottom in a relatively cool spot or "cold zone." This prevents charring the sediment and damaging the food flavor. At the lowest point a drain valve is provided for periodic oil straining and removal.

Proper selection and use of the frying equipment can greatly reduce, and sometimes eliminate, the need to discard cooking fat. Means of keeping fat losses to a minimum include:

1. Use of modern, high-input fryers.
2. Sizing appliance to meet peak-hour demand exactly.
3. Maintenance of thermostat calibration.
4. Reduction of fat temperature when equipment is not in use.
5. Use of more smaller fryers to meet intermittent peaks rather than fewer large fryers, since the ratio of fat capacity to production can be very high for underloaded large fryers.
6. Frequent filtering of oil.

Conventional restaurant-type deep fat fryers vary from a 10 × 10 in. kettle to 24 × 24 in. Gas inputs range from 25 to 186 MBtu per hour and fat capacities from 15 to 130 lb. The most popular floor model has a kettle which measures approximately 14 × 14 in. and a fat capacity of about 30 lb. Its production capacity will vary greatly, depending upon input, which ranges from 30 to 120 MBtu per hr for different manufacturers. Since 600 Btu are required to fry 1.0 lb of raw potatoes and fryer efficiency may be 50 per cent, the produicty capacity can be estimated by dividing the gas input bon1200. Hence, the 120 MBtu input model can produce up to 100 lb of potatoes per hr. Counter, free-standing, and built-in models are available, as well as models with removable fry kettles.

Another type of fryer is semiautomatic and has a relatively high production rate. It is about 68 in. long and 16 in. deep, and equipped with a screw-type conveyor. Its main feature is that it produces continuously instead of by single load operation. Each portion is inserted as ordered and automatically discharged when cooked. Different foods starting to cook at different times all fry simultaneously. Fat capacity is 140 lb and gas input 165 MBtu per hr.

Some fryers have indicating lights which show when the pilot light is burning, when the fryer is preheating, and when the fryer is up to temperature. Fryers so equipped generally have an additional indicating light which shows if the high limit control (a safety device preventing the fryer from overheating) is operating, and batch timers to control the production of the fried foods.

Pressure Fryers. These are similar to the conventional fryers except in one important respect. The fry kettle is equipped with a tightly sealed lid which permits the moisture given off in the cooking process to build up 9 to 14 psig of steam pressure within the kettle. The fat temperature is held at about 300 to 310 F for all types of food, and cooking time (even for chicken parts) does not exceed seven minutes. Because of the pressure in the kettle, less moisture is removed from the food and the finished product is moist inside but crisp on the outside. The units are furnished with batch timers, pressure control valves, relief valves, indicating lights, and other accessories.

STEAM COOKING APPLIANCES

Two popular types of steam cooking appliances in commercial kitchens are: (1) the steam-jacketed kettle (stock kettle) and (2) the compartment steamer (vegetable steamer or pressure cooker). Steam at or above atmospheric pressure is the heat source, although in jacketed kettles it does not come into contact with the food as it does in compartment steamers.

For a limited number of steam-heated appliances, it is advantageous to use equipment with an integral gas-fired steam generator. For a larger number all steam can be supplied by a gas-fired boiler at a close or remote location.

Steam-Jacketed Kettles

Capacities range from 10 to 150 gal for stationary kettles and from one quart to 80 gal for other types. The choice of size is made in terms of specific foods to be prepared and volume and speed required. Kettles larger than 40 gal capacity are used for products having a high liquid content that are easy to stir and require little attention. Shallow kettles are suitable for browning meat or roasting. The cooking temperature depends on the steam pressure and can be varied from 215 F at 0.9 psig to 298 F at 50 psig pressure.

The differences in models relate to depth, steam jacketing, (either full or two-thirds jacketed); mounting (on pedestal, legs, or wall), tilt or stationary type, and method of supplying steam (external or self-generated).

The *deep jacketed type* is cylindrical and has a one- or two-piece hinged cover. From two-thirds to the entire kettle area, excluding the top, is jacketed with steam. It is equipped with valves for steam inlet, outlet and safety, a drawoff line, and a faucet. Mounting may be made with a tubular leg stand, a pedestal, or a wall bracket. The height recommended for the top rim is 36 to 42 in. above the floor. For institutions, sizes up to 120 gal capacity are popular, but for average kitchens, smaller kettles are preferable. The flexibility obtainable with two smaller kettles rather than a single large one is the determining factor. In estimating steam requirements for stock kettles, common practice is to provide from 0.5 to 0.8 boiler hp for each 10 gal capacity.

The *shallow, or full jacketed type*, is similar in construction but has a greater diameter for a given capacity, and the steam jacket covers the full height of the kettle. Its shallow construction permits successful cooking of large quantities of poultry and other delicate meats, whereas the weight of too many layers would damage the product on the bottom.

The *trunnion, or tilting type*, has a pouring lip, is mounted on trunnions, and is equipped with either a hand-operated or a worm-and-gear tilting device for easy unloading. Small models in 1 to 40 quart capacity are designed for table top installation. When a number of these kettles are mounted in a group, they can be installed on a special table fitted with a gas-fired steam generator sized to meet their entire steam demands.

Larger floor-mounted models of 10 to 80 gal capacity may be either deep or shallow jacketed. The tilting mechanism may be locked in any desired position.

The *direct gas-fired unit* is comparable in construction to the deep jacketed type except that the jacket contains a small quantity of water and requires no steam connection. A gas burner under the kettle generates jacket steam at the desired rate and pressure. This steam is regenerated as it condenses. The unit is equipped with a water sight gage, safety valve, low water cutoff, and pressure control.

Another design of this type contains distilled water in a hermetically sealed jacket. Under normal conditions, this water lasts the life of the kettle and eliminates the need for a sight gage, safety valve, and low water cutoff. The jacket is equipped with a fusible plug which melts at about 300 F, and protects against pressures above 50 psig.

Capacities of these kettles are from 10 to 120 gal, with inputs of from 40 to 300 MBtu per hr.

Compartment Steamers

These appliances differ from stock kettles in that the steam comes into direct contact with the food cooked. This is accomplished by introducing steam of up to 15 psig into a chamber equipped with sealed doors and racks to hold the baskets (perforated or solid) containing the food. Designs are available for use with an external steam supply or with an integral gas-fired generator having the necessary controls. Such a generator may be obtained with extra capacity to supply steam to other equipment. Since the steam comes into direct contact with the food, it is important that water softeners and other chemicals not be used in the boiler water.

Food capacities of these steamers are from 1 to 3 bushels or 30 to 100 lb per compartment. They may carry various sized trays or baskets to accommodate the type of food cooked. The smallest single compartment models are suitable for serving up to 50 and the largest four compartment units up to 1000 meals per hour.

Small counter-type steamers hold three 12 × 20 × 2½ in. cafeteria pans and may be direct gas-fired. Another two compartment model of twice this capacity holds six full-sized pans and handles 60 to 70 lb of food per hour. They are usually constructed of stainless steel or aluminum and may be equipped with automatic water feed, temperature control, automatic timer, and other safety devices. Gas input to the single compartment steamer is about 25 MBtu per hr. Designs are available with a cabinet base containing a 20 or 30 gal steam kettle or a smaller capacity tilting kettle. Such combinations have a gas boiler in the base, sized to supply all cooking units. Input is about 90 MBtu per hr.

Larger, floor-type compartment steamers may use either external or self-generated steam. There are three compartment sizes: 24 in. wide by 24 in. deep, 24 in. wide by 26 in. deep, and 29 in. wide by 21 in. deep. Compartment heights may be 13 in., 14 in., or 16½ in. Two to six standard pans are accommodated, depending on size and type.

Direct gas-fired models may have two or three compartments with a capacity of 100 lb of food per compartment. The steam boiler gas input is about 90 MBtu per hr. These steamers are also available in combination with stock kettles. In this case the gas input to the steam boiler in the base cabinet may be 140 to 180 MBtu per hr, depending on the number and size of kettles.

Units using externally generated steam may have two, three, or four compartments and a food cooking capacity of 100 lb per compartment. To assure ample reserve capacity, the steam boiler should prove 1.5 boiler hp per compartment. These compartment steamers are availale in combination with 20, 30, or 40 gal stock kettles.

COUNTER APPLIANCES

Counter appliances may be used for many cooking operations. Some are designed for installation with other appliances so as to form a complete matching line. Counter appliances may be installed on back-bar counters, on serving tables, or on individual cabinet or leg stands.

Coffee Urns

Standard gas-fired coffee urns range in size from 1 to 80 gal coffee capacity. Larger units are availale on special order. The urn consists of an underfired water tank which holds a ceramic, glass, Monel, or stainless steel coffee container. Hot water, supplied from its own water jacket or from an outside source, is poured over the ground coffee, contained in a fabric bag or perforated metal container which fits into the top of the urn. The coffee container is surrounded by hot water, keeping the brewed coffee at the proper serving temperature, usually 175 to 185 F.

To meet large demands, one or more coffee urns may be connected in a battery with a water urn of adequate capacity to supply all hot water required for brewing. A more compact variation is to have twin coffee urns immersed in a single large water tank.

Most conventional round coffee urns do not use flueways up the side walls, the only heat transfer surface being the tank bottom. For this reason gas input for reasonable thermal efficiency is limited to about 140 Btu per sq in. of bottom area. Inputs to conventional small urns are about as follows:

Size, gal	3	5	8
Input, MBtu per hr	10	15	20

A three-unit battery with two 5 gal coffee urns and one 10 gal water boiler can deliver up to 20 gal (400 cups) per hr. Larger or smaller sizes can deliver proportional amounts. The coffee capacity may also be increased by connecting the urn battery to the 180 F water supply for the dishwasher, providing the water is free of rust and chemicals.

Automatic Coffee Makers. Somewhat similar in operation to conventional coffee urns, these units have many quality control refinements which ensure correct water temperature, water pressure, and filtering time. The only manual operations required are the placement of the extractor containing the ground coffee in the machine and the pressing of a start button. From there on the brewing process begins and continues automatically until the coffee is ready to serve.

One available model is rated at 400 cups per hr, while another type with twin coffee compartments holds 6 or 12 gal of coffee and 8 or 12 gal of water, depending on the size selected.

Glass Coffee Makers

These units consist of a gas brewing stove with open burners and a number of glass coffee decanters, each having two globular glass containers or bowls of about one-half gal capacity. The upper bowl has a long open neck which fits thru the opening of the lower one with a tight seal. Ground coffee is placed on a filter in the upper bowl and water in the lower one. When the water boils, steam pressure forces it up thru the neck of the upper bowl. When most of the water is in the upper bowl, gas is shut off and the brewed coffee passes thru the filter and returns to the lower bowl. It is served directly from the lower bowl.

Sizes range from the smallest unit, having a single burner stove and one decanter, to a terraced stove with eight burners and a complete set of decanters. Burners are often of the dual type with a large burner for brewing the coffee and a small simmer section for keeping it at serving temperature. On some larger stoves, part of the burners are designed for warming only, the others for use either for brewing or warming.

Burners are individually controlled. Their "high heat" settings are about 5000 Btu per hour. "Low heat" or "keep warm" burners have inputs as low as 400 Btu per hour.

Griddles

Griddles are used for short order frying. They consist of a one-piece polished steel or cast-iron plate, usually with a raised edge and a grease trough to drain off any excess grease produced during frying. Multiple burners under the plate may be of the "bar" or "loop" type, thermostatically controlled if desired to maintain a constant frying plate temperature.

Standard griddles range in size from 7 × 14 in. (0.7 sq ft) up to 60 × 24 in. Larger sizes can be made up on special order.

The minimum recommended gas input to griddles is 10 MBtu per hr-sq ft of plate area. Where high production capacity is required, input should approach 16 MBtu per hr-sq ft of cooking surface.

An average gas griddle which measures 18 × 42 in. (5.3 sq ft) is capable of producing hourly about 120 orders of griddle cakes or 600 hamburgers or frankfurters.

Counter Broilers

Several different designs of broilers suit varying requirements. One has a single set of burners with a broiler grid below. A second and more common type is similar in construction, except that a griddle plate over the burners forms the top of the unit. Still another type has an additional set of gas burners below the broiler grid so as to cook both sides of the food at the same time. Usually, the gas burners at the broiler top are provided with radiant elements of ceramic, Nichrome wire, or similar material in order to increase the amount of radiant heat directed toward the broiler grid. Schwank radiant burners are also used.

Counter broilers are available with grid areas from 1.4 to 5.2 sq ft. Their inputs are usually between 10 and 15 MBtu per hr-sq ft of broiler grid area.

Hot Plates

Hot plates normally have from one to eight open top burners, with larger units obtainable on special order. The "ring" or "star" type of burner is generally used, its input varying from 6 to 15 MBtu per hr. Grates above the burners hold the cooking utensils. Each burner may have an individual grate, or the entire top may consist of bar-type grate sections.

Toasters

The popular gas toaster consists of a motor-driven conveyor that carries the bread thru a thermostatically controlled toasting compartment and automatically discharges it into a serving box at the front. Inside the insulated toasting compartment are ceramic radiants on both sides of the conveyor. Gas flames directed on these radiants produce the high degree of radiant heat necessary for proper toasting. Similar toasters are made for toasting rolls, buns, and sandwiches.

Two available sizes have maximum capacities of 360 and 720 slices of bread per hour; corresponding gas inputs are 12 and 22 MBtu per hr.

FOOD HOLDING AND WARMING EQUIPMENT

In addition to the gas appliances required in a commercial kitchen for food preparation and cooking, equipment is necessary for holding the finished food at proper serving temperature. This temperature varies for different foods and must be accurately controlled if flavor is to be retained and overcooking prevented. A number of appliances meeting these requirements are available.

Steam Tables

A steam table consists of a shallow water tank, up to 10 in. deep, with a cover which has cutouts sized to accommodate standard meat trays, vegetable crocks, gravy boats, and roll warmers. The tank is partially filled with 3 in. or more of water. Its cover supports the food insets in or slightly above the water bath usually maintained at 180 to 190 F. Thermostatic controls are desirable.

Tables are obtainable in any desired length; above 9 ft they are generally custom-made. Small counter units are also available with one to four insets. Bases may be fully open, partially closed, or fully enclosed with sliding doors. When closed, the compartment may be used as a dish warmer.

Steam tables are almost always heated by urn-type burners spaced about 2 ft apart beneath the water tank, its entire bottom area acting as the heat transfer surface. Typical inputs are about 2500 Btu per hr-sq ft of bottom area. If cold water is piped to the steam table, an input of 5000 Btu per sq ft is recommended to reduce the heating up period.

Bains-Marie

These units are similar to and serve the same function as steam tables. The essential difference is that they do not have a top cover to support the food insets. Instead, the food dishes are placed on a perforated rack near the bottom of the water tank and well below the water level. Bains-marie are mainly used for the chef's convenience in preparing foods and are, therefore, frequently located in or near the cook's table.

Gas inputs may vary as for steam tables from 2500 to 5000 Btu per hr-sq ft of bottom area.

Dry Food Warmers

Dry food warmers differ from steam tables in that warm air is the heating medium. In addition, the unit is divided into sections, each having its individual burner and, when desired, its own thermostatic control. The burner is located under a circular opening in each compartment bottom, the heat striking a baffle plate for even distribution.

Since each compartment has a separate burner, each type of food stored may be held at its most suitable temperature; for example, mashed potatoes at 125 F, soup at 180 F, and rolls at 110 F. The input to each burner for a typical compartment is 3500 to 4000 Btu per hour.

KITCHEN VENTILATION

Installation of an exhaust fan has long been accepted practice for removing heat, odors, and grease-laden vapors produced in cooking. Fan capacity often used to be determined by rule-of-thumb methods without consideration of effects on the heating or cooling system required for other areas. Also, no provision was made for admitting fresh air to replace that withdrawn. With all windows in the kitchen closed, as in winter, an exhaust fan creates a negative pressure in its area, with the following adverse results: (1) reduced fan capacity; (2) improper venting of water or space heaters or other individually vented gas appliances; and (3) a rush of cold air into the dining area whenever entrance doors are opened.

More recently, due to the installation of summer and winter air-conditioning in most restaurants, kitchen ventilation system design has become a major criterion for assuring low operating costs for mechanical systems. The objectives of a ventilation system are to provide clean fresh air for the dining room and kitchen, to maintain comfortable air temperatures and humidities, and to exhaust effectively all odors, moisture, and grease vapors so that they will not discolor kitchen surfaces and permeate the dining area.

Sizing of Hood Exhaust Fans

Lacking any completely acceptable engineering formula, exhaust fan capacities are usually calculated by a simple rule-of-thumb method that experience has shown will result in fan sizes reasonably satisfactory for the average kitchen. This method is based on maintaining a constant air velocity across the entire area of the hood opening. To ensure effective removal of the heavy grease-laden cooking vapors, the air velocities shown in Table 12-107 are recommended by many local and national codes.

To prevent excessive temperature or humidity in a kitchen that is *not* air-conditioned, it is good practice to provide 20 to 30 air changes per hour within the kitchen. Designs based on Table 12-107 should result in at least 20 to 30 air changes per hour. This practice is followed by the Veterans Administration and many independent designers, and is recommended by NFPA.[3]

Table 12-107 Recommended Hood Face Velocities[2]

No. of exposed hood sides	Air velocity across face of hood, fpm
4 (central hood)	150
3 (wall hung)	100
2 (corner hung)	85
1 (aprons on three sides)	85

In 1962, A.G.A. published[4] detailed design criteria for sizing commercial kitchen ventilation systems. These are applicable to all types of kitchens and combinations of cooking appliances. Field experience with these data were not available at this writing.

Effects of Kitchen Exhaust on Dining Room Air-Conditioning

Proper sizing of the exhaust system will not provide satisfactory kitchen ventilation unless replacement air is provided in sufficient quantity. This air must be distributed so as to prevent drafts and condition for comfort.

To balance the ventilation system, replacement fresh air must be introduced to both the kitchen and dining areas in such proportions that a slight negative pressure will exist in the kitchen. This is necessary to assure a moderate movement of air from the dining area into the kitchen, thus preventing cooking odors from entering the dining room.

If the restaurant is not air-conditioned, it is common practice to draw most, if not all, of the replacement air from the dining area. If air-conditioned, however, it is desirable to reduce the amount of air taken from the dining area so as to avoid too heavy a load on the air-conditioning equipment; it is usual to supply half the replacement air from the dining area and half directly from outdoors. In no case, however, should the quantity drawn from the dining area be less than that required for proper ventilation. In the absence of specific ventilation requirements in local codes, a fresh air supply of 15 cfm per person is a good figure for restaurant occupancy.

This ventilation air represents a basic, necessary load on the air-conditioning system, since an equivalent amount of fresh air must simultaneously be introduced into the dining area. This air must first pass thru the air-conditioning apparatus and be cooled, dehumidified, and filtered to maintain the restaurant in the desired comfort zone.

The only additional load on dining room air-conditioning chargeable to kitchen ventilation is any air volume drawn from the dining room over and above that required to supply fresh air for the human load. Such extra volume depends on the exhaust system design *and is in no way related to the type of fuel used in the cooking appliances.* As previously stated, the kitchen exhaust system is sized (see Table 12-107) so as to maintain sufficient air velocity at the appliances to remove heat, odors, and cooking vapors effectively. Appliance fuel consumption does not enter into the calculations. Hence, the type of cooking fuel has no effect on the size or operating cost of equipment used to condition the air in the dining area.

Table 12-108 Estimated Heat Gains from Restaurant Appliances Located in Air-Conditioned Area[2,5]

			Manufacturer's input rating			Heat gain, Btu/hr			
	Overall dim., in.*				Probable maximum	Without hood			With hood,
Appliance	$W \times D \times H$	Miscellaneous data	Watts or boiler hp	Btu/hr	input, Btu/hr	Sensible	Latent	Total	all sensible
		Gas Burning Counter Type Appliances							
Broiler–griddle	31 × 20 × 18			36,000	18.000	11,700	6,300	18,000	3,600
Coffee brewer									
Per burner		With "warm" position		5,000	2,500	1,750	750	2,500	500
Water heater burner		With storage tank		11,000	5,500	3,850	1,650	5,500	1,100
Coffee urn	12 diam	3 gal		10,000	5,000	3,500	1,500	5,000	1,000
	14 diam	5 gal		15,000	7,500	5,250	2,250	7,500	1,500
	25 wide	8 gal twin		20,000	10,000	7,000	3,000	10,000	2,000
Deep fat fryer	14 × 21 × 15	15 lb of fat		30,000	15,000	7,500	7,500	15,000	3,000
Dry food warmer		Per sq ft of top		1,400	700	560	140	700	140
Griddle, frying		Per sq ft of top		15,000	7,500	4,900	2,600	7,500	1,500
Short order stove		Per burner, open grates		10,000	5,000	3,200	1,800	5,000	1,000
Steam table		Per sq ft of top		2,500	1,250	750	500	1,250	250
Toaster, continuous									
360 slices/hr	19 × 16 × 30	Two slices wide		12,000	6,000	3,600	2,400	6,000	1,200
720 slices/hr	24 × 16 × 30	Four slices wide		20,000	10,000	6,000	4,000	10,000	2,000
		Gas Burning Floor-Mounted Type Appliances							
Broiler, unit	24 × 26 grid	Same burner heats oven		70,000	35,000	Exhaust hood required	Exhaust hood required	Exhaust hood required	7,000
Deep fat fryer	14 in. kettle	32 lb of fat		65,000	32,500				6,500
	18 in. kettle	56 lb of fat		100,000	50,000				10,000
Oven, deck, per sq ft of hearth area		Same for 7 & 12 in. high decks		4,000	2,000				400
Oven, roasting	32 × 32 × 60	Two ovens, 24 × 28 × 15		80,000	40,000				8,000
Range, heavy duty	32 × 42 × 33								
Top section		32 in. W × 39 in. D		64,000	32,000				6,400
Oven		25 in. × 28 in. × 15 in.		40,000	20,000				4,000
Range, Jr., heavy duty	31 × 35 × 33								
Top section		31 in. W × 32 in. D		45,000	22,500				4,500
Oven		24 in. × 28 in. × 15 in.		35,000	17,500				3,500
Range, restaurant type									
Per 2 burner section		12 in. W × 28 in. D		24,000	12,000				2,400
Per oven		24 in. × 22 in. × 14 in.		30,000	15,000				3,000
Per broiler–griddle		24 in. W × 26 in. D		35,000	17,500				3,500
		Electric Counter Type Appliances							
Coffee brewer									
Per burner			625	2,130	1,000	770	230	1,000	340
Per warmer burner			160	545	300	230	70	300	90
Automatic 240 cups/hr	27 × 21 × 22	Four burners & water heater	5,000	17,000	8,500	6,500	2,000	8,500	2,700
Coffee urn		3 gal	2,000	6,800	3,400	2,550	850	3,400	1,000
		5 gal	3,000	10,200	5,100	3,850	1,250	5,100	1,600
		8 gal twin	4,000	13,600	6,800	5,200	1,600	6,800	2,100
Deep fat fryer	13 × 22 × 10	14 lb of fat	5,500	18,750	9,400	2,800	6,600	9,400	3,000
	16 × 22 × 10	21 lb of fat	8,000	27,300	13,700	4,100	9,600	13,700	4,300
Dry food warmer		Per sq ft of top	240	820	400	320	80	400	130
Egg boiler	10 × 13 × 25	Two cups	1,100	3,750	1,900	1,140	760	1,900	600
Griddle, frying		Per sq ft of top	2,700	9,200	4,600	3,000	1,600	4,600	1,500
Griddle–grill (w/top and bottom grids)	18 × 20 × 13	200 sq in. each	6,000	20,400	10,200	6,600	3,600	10,200	3,200
Hotplate	18 × 20 × 13	Two heating units	5,200	17,700	8,900	5,300	3,600	8,900	2,800
Roaster	18 × 20 × 13		1,650	5,620	2,800	1,700	1,100	2,800	900
Roll warmer	18 × 20 × 13		1,650	5,620	2,800	2,600	200	2,800	900
Toaster, continuous									
360 slices/hr	15 × 15 × 28	Two slices wide	2,200	7,500	3,700	1,960	1,740	3,700	1,200
720 slices/hr	20 × 15 × 28	Four slices wide	3,000	10,200	5,100	2,700	2,400	5,100	1,600
Toaster, pop-up	12 × 11 × 9	Four slice	2,450	8,350	4,200	2,230	1,970	4,200	1,300
Waffle iron	18 × 20 × 13	Two grids	1,650	5,620	2,800	1,680	1,120	2,800	900

Table 12-108 Estimated Heat Gains from Restaurant Appliances Located in Air-Conditioned Area (Continued)

Appliance	Overall dim., in.*	Miscellaneous data	Manufacturer's input rating		Probable maximum input, Btu/hr	Heat gain, Btu/hr			
						Without hood			With hood, all sensible
	$W \times D \times H$		Watts or boiler hp	Btu/hr		Sensible	Latent	Total	
Electric Floor Mounted Type Appliances									
Broiler									
No oven		23 in. × 25 in. grid	12,000	40,900	20,500				6,500
With oven		23 in. × 27 in. × 12 in. oven	18,000	61,400	30,700				9,800
Deep fat fryer									
28 lb of fat	20 × 38 × 36	14 in. × 15 in. kettle	12,200	40,900	20,500	Exhaust hood required	Exhaust hood required	Exhaust hood required	6,500
60 lb of fat	24 × 36 × 36	20 in. × 20 in. kettle	18,000	61,400	30,700				9,800
Oven, baking, per sq ft of hearth		Compartment 8 in. high	500	1,700	850				270
Oven, roasting, per sq ft of hearth		Compartment 12 in. high	900	3,070	1,500				490
Range, heavy duty	36 × 36 × 36								
Top section			15,000	51,100	25,600				8,200
Oven			6,000	20,400	10,200				3,200
Range, medium duty	30 × 32 × 36								
Top section			8,000	27,300	13,600				4,300
Oven			3,600	12,300	6,200				1,900
Range, light duty	30 × 29 × 36								
Top section			6,600	22,500	11,200				3,600
Oven			3,000	10,200	5,100				1,600
Steam Heated Appliances									
Coffee urn		3 gal	0.2	6,600	3,300	2,180	1,120	3,300	1,000
		5 gal	0.3	10,000	5,000	3,300	1,700	5,000	1,600
		8 gal twin	0.4	13,200	6,600	4,350	2,250	6,600	2,100
Steam table per sq ft of top		With insets	0.05	1,650	825	500	325	825	260
Bain-marie per sq ft of top		Open tank	0.10	3,300	1,650	825	825	1,650	520
Oyster steamer			0.5	16,500	8,250	5,000	3,250	8,250	2,600
Steam kettles per gal capacity		Jacketed type	0.06	2,000	1,000	600	400	1,000	320
Compartment steamer per compartment	24 × 25 × 12 compartment	Floor mounted	1.2	40,000	20,000	12,000	8,000	20,000	6,400
Compartment steamer		3 pans, 12 × 20 × 2½; single counter unit	0.5	16,500	8,250	5,000	3,250	8,250	2,600
Plate warmer per cu ft			0.05	1,650	825	550	275	825	260

* W = width, D = depth, H = height.

Notes: 1. Heat gains from cooking appliances located in the conditioned area but not included in the table should be estimated as follows: (a) obtain "probable maximum input" by multiplying the manufacturer's hourly input rating by the usage factor of 0.50; (b) for installations without an exhaust hood the estimated latent heat gain is 34 per cent of the "probable maximum input" and the sensible heat gain is 66 per cent; (c) for installations under an exhaust hood the estimated heat gain is all sensible heat and can be calculated from Eqs. 1 and 2 in the text.

2. Heat gains from cooking appliances not located in the conditioned area are not considered in cooling load calculations.

3. For poorly designed or undersized exhaust systems the heat gains in the last column should be doubled and half of the increase assumed to be latent heat.

Effects of Hooded Appliances on Air-Conditioned Kitchens

If the area in which the cooking appliances are located is to be air-conditioned, all hoods over the appliances must be properly designed and the exhaust fan capacity must satisfy Table 12-107 or other recognized design criteria.[2,4]

The cooling load attributable to the exhaust fan is that required to condition a quantity of replacement air equal to the exhaust fan capacity *minus* the volume of fresh air supplied to the dining room for ventilation and exhausted thru the kitchen. Furthermore, the cooling load due to the air exhausted is independent of the type of cooking fuel, since the type of fuel does not affect the cfm rating of the exhaust fan or the exhaust system design. The only effect the choice of fuel for cooking has on the room cooling load is that resulting from any difference in the heat radiated to the kitchen by the respective appliances.

Field studies of electric and gas commercial kitchens have repeatedly shown that 1.0 Btu of electricity is equivalent to 1.6 Btu of gas for cooking purposes. The added heat input with gas appliances compensates for the difference between their cooking efficiencies and those of comparable electric appliances. This additional heat is not absorbed by the appliance, but is contained in the products of combustion exhausted thru its flue outlets.

If gas cooking appliances are connected to a flue, all heat in the combustion products will be directly vented outdoors and never enter the kitchen. If they are not so connected, installation must be under an exhaust hood which will evacuate all flue gases and prevent them from heating the kitchen.

Accordingly, effects of gas and electric cooking appliances on the kitchen cooling load will be *identical*[4,5] for equipment of the same size operating at the same cooking temperatures. For practical purposes, all heat released to the room, mainly by radiation from hot appliance surfaces, will be sensible. Convected heat and latent heat in the cooking vapors are removed at the source by a properly designed exhaust hood. A conservative estimate of the maximum heat released into the kitchen by radiation is 32 per cent of the *maximum hourly input* to the appliance.

The magnitude of this maximum hourly input can be estimated as 0.5 *times* the *total rated manufacturers' input* of the appliances. This *usage factor* is based on numerous demand studies in both large and small commercial kitchens by gas and electric utilities. The studies showed the diversity of cooking appliance use and the effects of thermostatic controls during peak 15 or 30 minute periods. Therefore, the maximum hourly heat gain for *electric appliances* can be determined from Eq. **1**.

$$\underset{\text{(Btu/hr)}}{\text{Heat gain}} = \underset{\text{(Btu/hr)}}{\text{rated input}} \times \underset{0.50}{\text{usage factor}} \times \underset{(0.32)}{\text{radiant heat factor}}$$

or

$$\underset{\text{(Btu/hr)}}{\text{Hooded electric appliance heat gain}} = \underset{\text{(Btu/hr)}}{\text{rated input}} \times 0.16 \qquad \textbf{(1)}$$

As previously described, the higher input ratings of gas over electric cooking appliances compensate for the difference in their heating efficiencies. It is, therefore, necessary to modify Eq. **1** for use with gas appliances. Dividing the right side of Eq. **1** by the *flue loss factor*, 1.6, corrects for the heat contained in the completely exhausted combustion products. The heat gain for gas appliances is thus given by Eq. **2**.

$$\underset{\text{(Btu/hr)}}{\text{Hooded gas appliance heat gain}} = \underset{\text{(Btu/hr)}}{\text{rated input}} \times 0.10 \qquad \textbf{(2)}$$

Table 12-108 lists average inputs for typical gas and electric cooking appliances. Its last column shows the probable sensible heat gain when an exhaust hood is used.

Effects of Unhooded Appliances on Air-Conditioned Kitchens

If any cooking appliance is installed without a vent or an exhaust hood, all of its heat input will eventually escape into the kitchen. While this may be advantageous during the heating season, such installations should be avoided (especially for high-input appliances) if the kitchen is to be air-conditioned. Nevertheless, it is sometimes expedient to install, unhooded serving urns, coffee warmers, bun and plate warmers or similar low-input equipment.

In this case, the same *usage factor*, 0.50, must be applied to the rated appliance input to determine the probable maximum input released to the room as sensible and latent heat. This applies to both gas and electric appliances, assuming the gas equipment is unvented.

The problem in estimating heat gain for unhooded appliances is not in calculating the total heat, but rather in establishing what percentage of it is sensible and what percentage latent heat. Based on knowledge of the type of cooking operation, fuel used, amount of moisture in the flue gases, available test data and other factors, probable sensible and latent heat ratios were calculated for various cooking appliances and applied to the "probable maximum input" column in Table 12-108. On some devices, such as a frying griddle, where the ratio of sensible to latent heat varies greatly with the type of food being cooked, the sensible heat load was assumed to be 66 per cent and the latent heat load 34 per cent of the total.

The "without hood" columns of Table 12-108 reflect these calculations and assumptions, and are a good guide for estimating heat gains from unhooded and unvented counter cooking appliances. Their magnitude indicates why it is impractical to permit unhooded appliances in an air-conditioned kitchen, since they result in higher initial cost for air-conditioning equipment and greatly increased operating costs.

REFERENCES

1. Schneider, N. F. and others. *Commercial Kitchens.* New York, A.G.A., 1962.
2. Jahn, E. A. *Commercial Kitchen Ventilation.* New York, A.G.A., 1960.
3. Nat. Fire Protection Assn. *Standard for Ventilation of Restaurant Cooking Equipment.* (NFPA 96) Boston, 1961.
4. Marn, W. L. *Commercial Gas Kitchen Ventilation Studies.* (Research Bull. 90) Cleveland, Ohio, A.G.A. Laboratories, 1962.
5. "Air-Conditioning Cooling Load." (In: Am. Soc. of Heating, Refrigerating and Air-Conditioning Engineers. *ASHRAE Guide and Data Book*, chap. 26. New York, 1963.)

Chapter 16

Other Commercial Gas Appliances

by E. A. Jahn, C. J. Mathieson, and F. E. Vandaveer

INTRODUCTION

Lines of demarcation between commercial and industrial appliances may vary among gas companies; certain commercial appliances may be classified as industrial and vice versa. Some companies classify any appliance as commercial if used in a retail store or shop, theater or sports arena, office or office building, church, school, hospital, or other public institution.

Besides the building heating, cooling, water heating, cooking, and incineration equipment discussed in preceding chapters, commercial categories and their corresponding applications for gas include the following:

Bakeries—process steam, proof ovens, pan washers, bake ovens, and confectioners' stoves.

Barber and beauty shops—sterilizers, towel steamers, and hair dryers.

Dairies—process steam, pasteurizers, and can and bottle washers.

Dry cleaners, tailor shops, and laundries—process steam, dryers, presses, finishers, and hand irons.

Garages and auto repair—process steam, car washers, degreasing tanks, torches, and tire molds.

Greenhouses—soil sterilizers, heaters, and atmosphere generators.

Hat renovators—process steam, blockers, and dryers.

Hospitals—process steam, sterilizers, volume cooking, and laundry equipment.

Meat processors—process steam, rendering kettles, bake ovens, and pan washers.

Printing shops—dryers.

Schools—laboratory burners, shop equipment, and home-economics appliances.

At the end of 1963 commercial gas customers in the United States numbered 2,770,300 or 7.64 per cent of the total.[1] Commercial gas consumption in 1963 was 11,648,700,000 therms, 10.73 per cent of all gas sold by gas utilities and pipelines. Revenue was $928,261,000, or 13.73 per cent of total revenue.

PROCESS STEAM AND HOT WATER[2–10]

Many of the appliances found in commercial establishments utilize steam or hot water either as the heat source or otherwise in the process. The required steam supply is usually high pressure (over 10 psig) and is sometimes furnished from the building heating plant but, more frequently, from separate process steam generators. Use of the latter is generally more economical and satisfactory.

In comparison with the steam requirements for many industrial processes, the steam consumption of typical commercial appliances and processes is relatively low. The bulk of the commercial market can be handled by steam generators rated from ¾ to 20 boiler hp with working pressures from 10 to 100 psig. Boilers with higher output and/or working pressures up to 500 psig are readily available if needed.

By ASME definition a boiler horsepower is the capacity to evaporate 34.5 pounds of water from feed water at 212 F to steam at the same temperature in 1 hour. The absorption of heat equals 33,479 Btu per hr (33,500 commonly used). At 70-per cent boiler efficiency, the required heat input per boiler hp would be approximately 48,000 Btu per hr.

Use Table 12-109a to estimate steam boiler capacity under specific conditions more accurately. Note that with excessively low feed water temperature a boiler generating steam at reasonably high pressure can have an actual steam output of only 80 per cent, or even less than that indicated in published tables which are "from and at 212 F" figures.

Table 12-109a Fraction of Boiler Rating Under Operating Conditions[23]

Feed water temperature, °F	Boiler operating pressure, psig						
	10	25	50	100	150	200	250
	Per cent of "from and at 212 F" rating						
40	84	83	82	82	81	81	81
60	86	85	84	83	83	82	82
80	87	86	85	85	84	84	84
100	89	88	87	86	86	85	85
120	90	89	89	88	88	87	87
140	92	91	90	89	89	89	89
160	94	93	92	91	91	91	91
180	95	95	94	93	93	92	92
200	97	97	96	95	94	94	94

Gas-Fired Steam Boilers

Since the steam requirements for commercial appliances and processes vary widely, it is difficult to outline all the desirable features that an "ideal" gas boiler should incorporate. However, a satisfactory process steam boiler should meet the following general requirements:

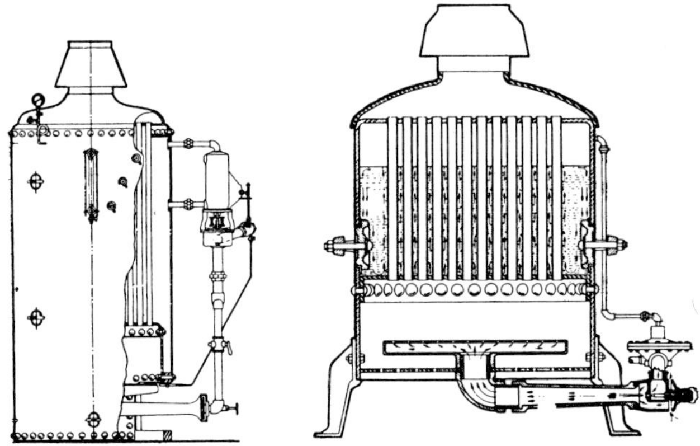

Fig. 12-174 Vertical firetube boilers: (left) water-leg type; (right) low water line design.

1. Simple design, easily understood by boiler operator.
2. Readily accessible heating surfaces for ease in cleaning out mud and scale.
3. Sufficient steam and water capacity to prevent fluctuations in steam pressure or water level.
4. A design that provides a constant and thorough circulation of water thru the boiler.
5. Adequate water surface for the disengagement of steam without foaming.
6. Sufficient heat transfer surface to maintain good efficiency when operating at maximum gas input.
7. Simple accessories and controls, compatible with safe operation and the degree of automatic operation desired.
8. Conformance to ASME Boiler Code and listing by Underwriters' Laboratories or other approved testing agency.

Experience has shown that high pressure gas-fired steel boilers satisfactorily meet these requirements and the process steam demands of the commercial market. Steel boilers may be firetube or watertube. In the smaller sizes, the watertube boiler has a comparatively small water-carrying capacity but generates steam quickly. It is best suited for intermittent steam service. The firetube design has a much greater water capacity and is therefore suitable for heavy-duty and continuous operation. For this reason most of the gas designed boilers used for commercial process steam are the firetube type.

Vertical Firetube Boilers. Gas-fired boilers of the vertical firetube type have proved highly successful and dependable for commercial applications in sizes from a fractional horsepower upward. They consist of the boiler shell, upper and lower tube sheets, and small-diameter boiler tubes. The tubes and lower crown sheet are the basic heat transfer surfaces, but some designs also pass the hot gases between the shell and an outer jacket. Approximately 8 to 15 sq ft of water-backed heating surface per boiler horsepower are used.

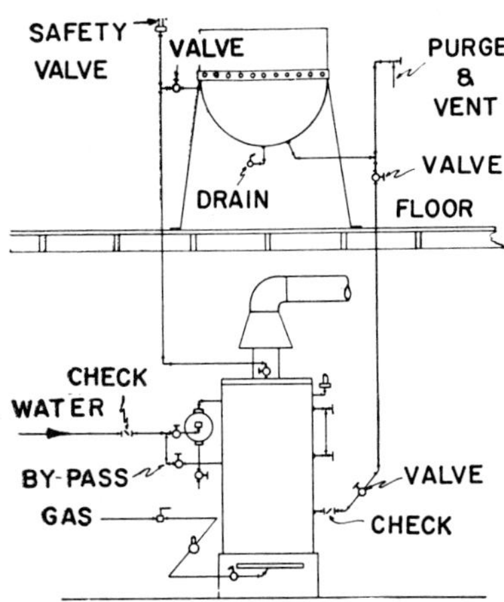

Fig. 12-176 Gravity return system with boiler on lower floor.

Atmospheric gas burners have been used very successfully with vertical firetube boilers. In units of this type the burners are located in a firebox under the lower crown sheet. In some cases, as shown at the left in Fig. 12–174, the firebox is enclosed in a water leg that also serves as a settlement chamber for mud and sediment. Figure 12-174 also shows (right) a typical boiler without a water leg. The short boiler tubes used in this design provide a low water line. Such boilers may therefore be installed on the same floor level as the steam-using equipment, utilizing gravity return of condensate so that pumps or traps are not needed. Figures 12-175 and 12-176 show typical gravity return installations, and Fig. 12-177 shows a forced return system.

In addition to the basic boiler trim of try cocks, water gage, steam gage, blowdown valve, and safety valve, the manufacturers of vertical firetube boilers can supply any required combination of controls and accessories. Depending upon the type of installation and degree of automaticity desired, these may include a safety pilot, low-water cutoff, modulating or on–off gas controller, or feed water pump and regulator with or without condensate return assembly.

Horizontal Return Tubular (HRT) Boilers. Where floor space is available, modified HRT boilers may be used in place of vertical firetube boilers. Common standard sizes range from 3 to 50 boiler hp and working pressures up to 125 psig.

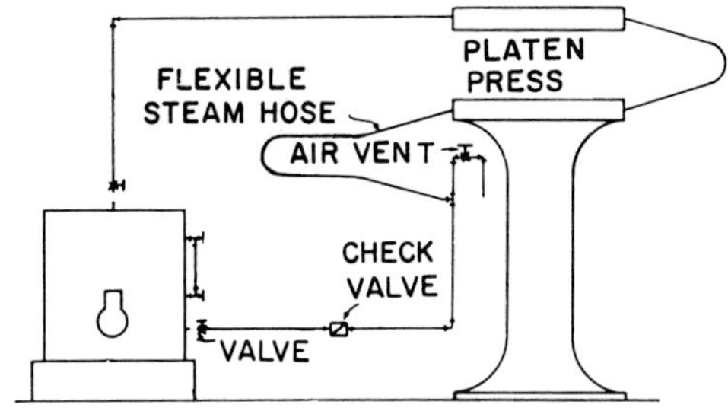

Fig. 12-175 Gravity return system for platen press.

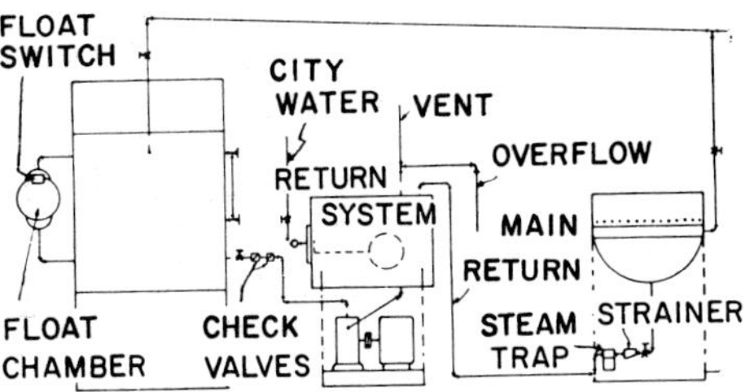

Fig. 12-177 Forced return system using traps and return pump—kettles on same floor level as boiler.

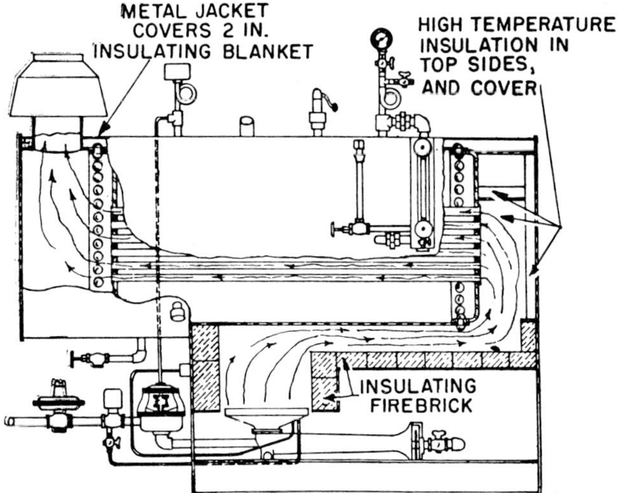

Fig. 12-178 Horizontal return tubular steam boiler.

The typical HRT boiler (Fig. 12-178), factory-assembled, including gas burner, firebox, base, insulation, and basic operating controls, is essentially a two-pass design. The hot combustion products, leaving the firebox, pass on the underside of the boiler shell, upward thru an insulated breeching, and then thru fire tubes located below the water line to the flue collection box, from which they are vented. Water-backed heat transfer surface may vary from 5 to 10 sq ft per boiler hp.

Either atmospheric gas burners and draft hoods (Fig. 12-178) or power burners and draft regulators may be used. Manufacturers can supply any desired combination of auxiliary equipment and controls for semi- or fully-automatic boiler operation. Since the HRT boiler design does not provide a water leg to collect sediment, periodic cleaning of the interior shell bottom is required to maintain efficient heat transfer. Removable cover plates, at each end of the boiler, provide direct access to handhole openings and boiler tubes and thus facilitate cleaning operations.

Vertical Tubeless Boilers. This design is available for small process steam applications in the 1.5- to 8-boiler hp range, with working pressures up to 100 psig. Figure 12-179 (left) shows the simplest type, which is basically a shell within a shell, similar to the conventional externally heated storage water heater. Welded to the outside of the inner shell are U-shaped heat conductors that increase heat transfer and baffle the flow of hot gases. The outer shell is made of heavy boiler plate and insulated from the exterior metal jacket.

An atmospheric gas burner is located under the lower crown sheet. Combustion products scrub this sheet, pass up the annular flueway, over the heat conductors, and out the top of the boiler. Because no water leg is provided, scale and sediment can collect on the lower crown sheet. A large opening for cleaning the boiler interior is located in the top head.

Figure 12-179 (right) shows an internally fired design that incorporates a water leg. Heat transfer is increased by U-shaped metal heat conductors welded to the entire inside of the furnace and to the upper crown sheet. The boiler is bottom-fired by atmospheric gas burners. The combustion products are deflected by a two-section cylindrical baffle to the heat conductors, over the crown sheet, inside the baffle, and out the side flue outlet. The entire heating surface is below the water line. Handhole openings are provided to permit cleaning the boiler.

Horizontal "Package" Boilers. These boilers, increasingly popular for high-pressure steam generation in the medium- to high-capacity range, are available from 12 to 500 or more boiler hp, with working pressures up to 350 psig. The boilers, usually of the firetube Scotch marine type, are completely assembled and pretested at the factory. They are shipped on skids that serve as the boiler base; installation requires only gas, steam, water, electric, and vent connections.

Figure 12-180 illustrates the typical two-pass Scotch marine design incorporating a relatively large firing tube and a battery of small return tubes that are all totally submerged in the boiler water. Three- and four-pass designs are also

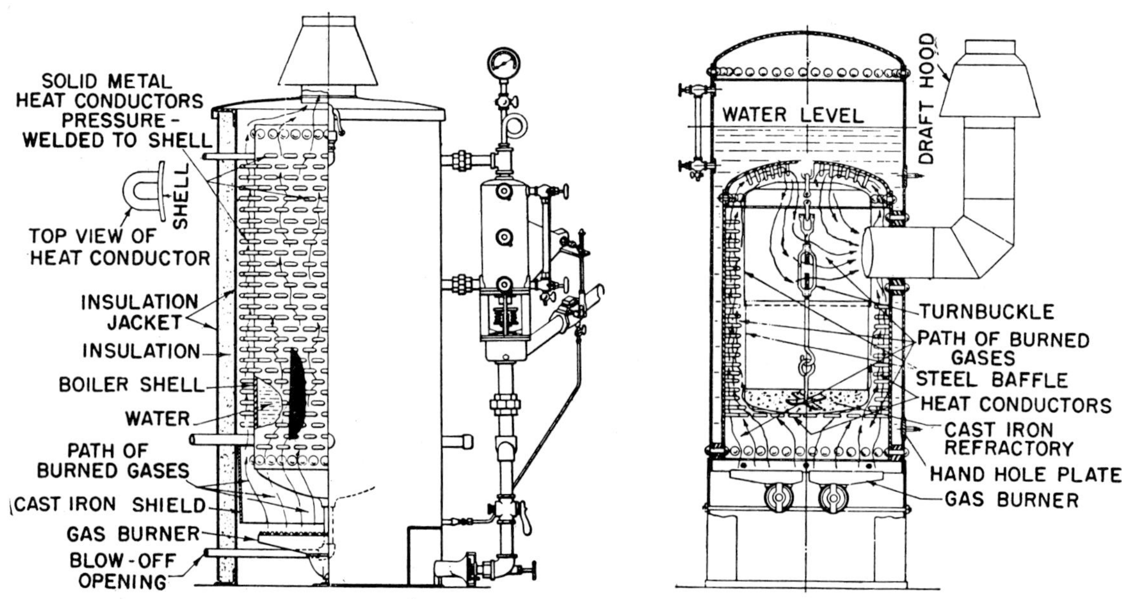

Fig. 12-179 Two types of vertical tubeless boilers: (left) tub type; (right) water-leg type.

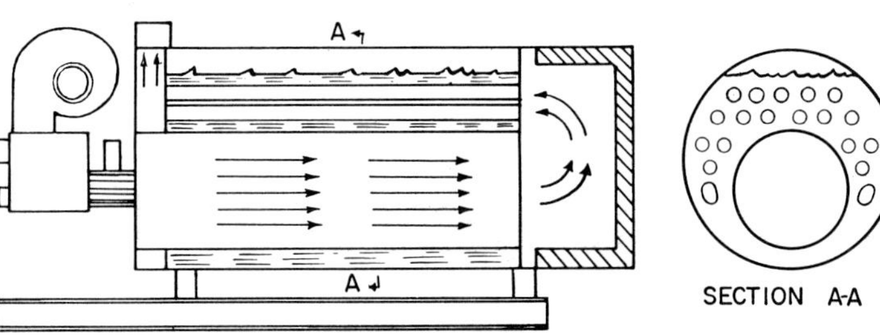

Fig. 12-180 Typical two-pass Scotch marine boiler.

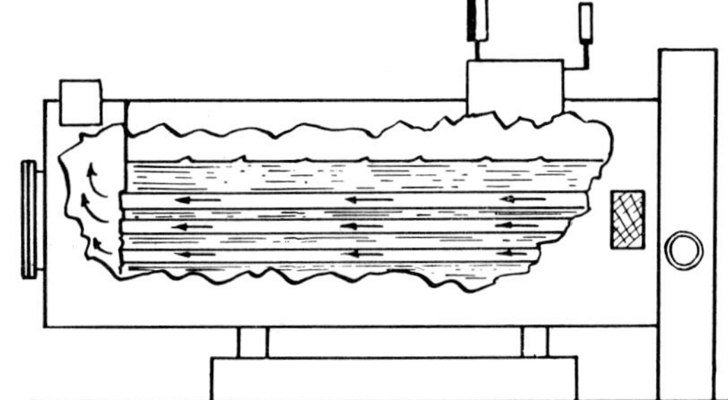

Fig. 12-181 Single-pass, horizontal immersion-fired steam boiler.

available. All designs provide large heating surface, steam space, combustion area, and water storage. The space below the firing tube acts as a settling chamber for sludge and is readily accessible for cleaning. Smaller sizes may be equipped with atmospheric gas burners, but standard burners are usually blast-type gas or gas–oil combination units.

Immersion Fired Boilers. A horizontal package unit is shown in Fig. 12-181. It is available in sizes from 10 to 500 boiler hp and for either low- or high-pressure steam. The single-pass design shown consists of multiple small-diameter firing tubes running the full length of the boiler shell; all tubes are below the water line. Each tube is provided with its own blast-type gas burner; combustion takes place within the tube.

The smaller size immersion boilers (10 to 70 boiler hp) are of two-pass design (Fig. 12-182) and utilize return tubes, larger in diameter than the firing tubes and located below them, to carry the combustion products back to the front of the boiler for venting. The water-backed heating surface is approximately 7.2 sq ft per boiler hp.

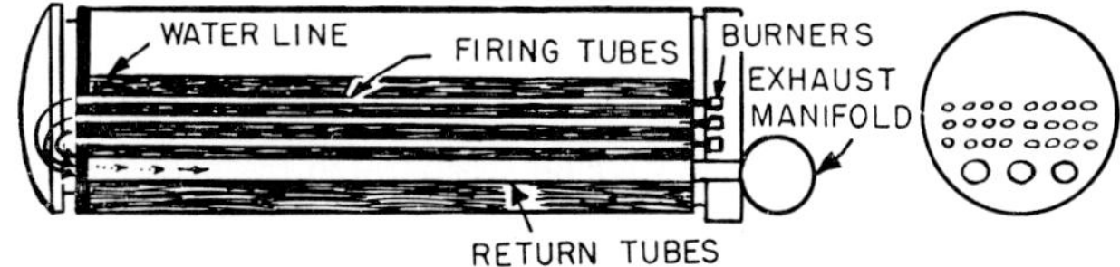

Fig. 12-182 Two-pass, horizontal immersion-fired steam boiler.

Because each tube is fired independently, both designs provide uniform heat distribution and effective use of available heating surface. The relatively cool space at the bottom of the boiler shell below the tubes collects sludge and sediment.

Forced-Circulation Water Coil Boiler. One type of water-tube steam generator that has been successfully used for process steam applications is shown in Fig. 12-183. The boiler, with accessories, including forced-draft gas burner, pump, accumulator, valves, controls, and piping are factory assembled on a common base. The only component shown that is not an integral part of the boiler is the storage tank, or feed water hot well.

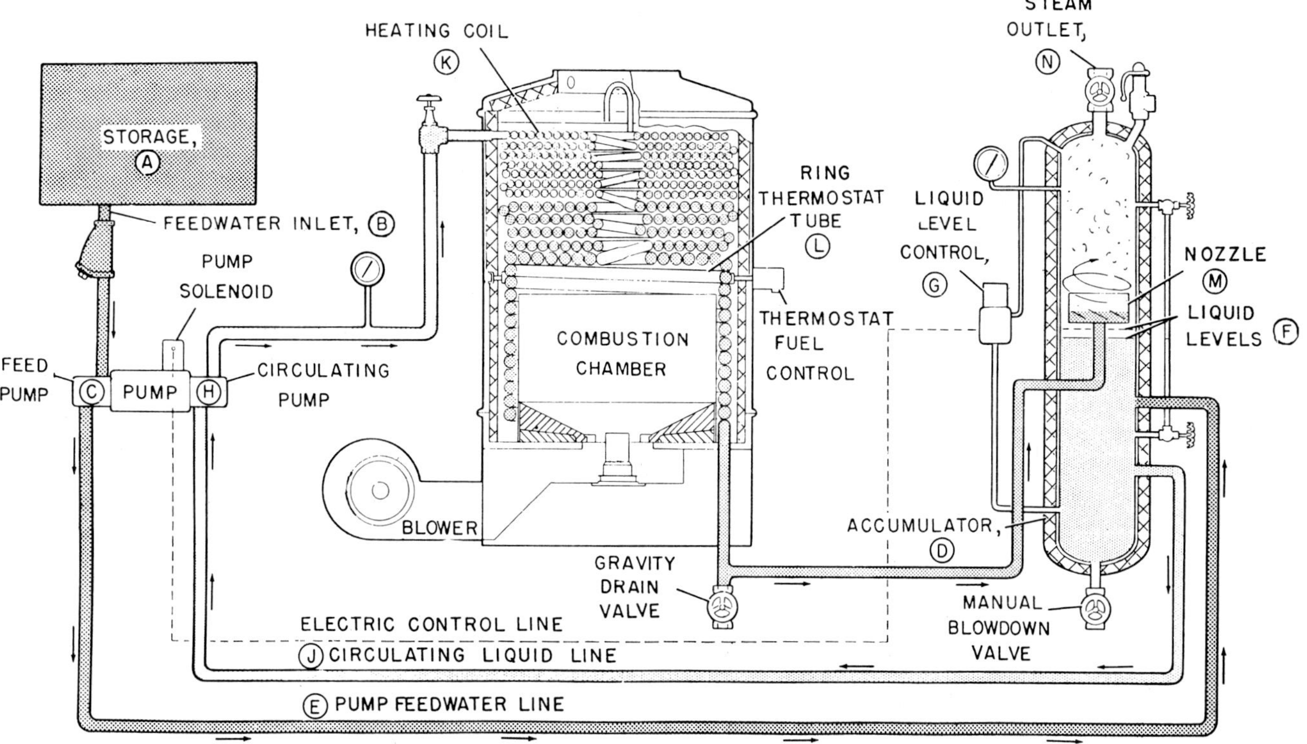

Fig. 12-183 Forced-circulation, water coil steam generator. (Clayton Mfg. Co.)

Operation. Referring to Fig. 12-183, supply water and returned condensate are held in storage (A) until they enter the feed water inlet (B) and flow to the feed pump (C), which forces them thru the feed water line (E) to the accumulator (D). The feed water pumping is started and stopped by the liquid level control (G) in response to the low and high liquid levels (F). As water is evaporated, the liquid level changes.

The circulating liquid is drawn from the accumulator (D) thru the line (J) by the circulating pump (H) and pumped to the heating coil (K). The liquid under forced circulation passes down thru the spiral single-tube heating coil in counterflow to the combustion gases that are passing over the outside of the tube. Leaving the spiral generating section, the fluid passes thru the ring thermostat tube (L), which permits temperature limit control, and is delivered thru the helically wound water wall tube to the separating nozzle (M) in the accumulator (D). The dry steam is separated by means of the centrifugal action of the nozzle (M) and delivered thru the steam outlet (N). The surplus liquid goes to the lower section of the accumulator.

One advantage of this forced-circulation and feed system is that the rate of water flow thru the boiler can be closely controlled to match the rate of steam output required. The system also incorporates a manual and automatic blowdown feature that serves to reduce the amount of dissolved solids and sediment in the circulating water passing thru the boiler coil.

Available sizes for this type of boiler range from 15 to 175 boiler hp. Standard pressures are 65 to 200 psig; higher or lower pressure models may be specified.

Dowtherm Generators

Dowtherm generators are basically designed for the high temperature–low pressure process heating applications that are generally found in the industrial, rather than the commercial field. Advantages claimed for these units as compared with direct-firing units are:

1. No local overheating of the process load.
2. Close temperature control.
3. Uniform temperature throughout the load.
4. Process vessel life lengthened—less carbonization.
5. Heat transmission rates to work may be increased.
6. Fire hazard when handling flammables minimized.
7. High temperatures; low pressures.
8. Possibility of heating various process vessels to different temperatures with the same Dowtherm vaporizer.

Dowtherm boilers, usually called vaporizers or liquid-phase heaters, may be vertical or horizontal units of firetube or liquid-tube design. Gas inputs range from 30,000 to several million Btu per hr.

Vapor-phase systems are ideal where high temperatures are required and where a temperature gradient is not desirable. A liquid-phase system is usually used for the lower temperature range and in cases where a temperature gradient can be permitted. A platen press with complex passages is an example of equipment in which liquid might be used instead of vapor. A heating and cooling cycle could be worked out.

Dowtherm "A" is a mixture of diphenyl $(C_6H_5)_2$ and diphenyl oxide $(C_6H_5)_2O$. It freezes at 53.6 F and boils at 495.8 F, at atmospheric pressure. While steam is excellent for process temperatures up to 350 or 400 F, Dowtherm is more practical for temperatures above that level. For example, saturated steam at 700 F requires a pressure of over 3100 psig; at the same temperature Dowtherm vapor is at only 95 psig pressure.

The most popular Dowtherm vaporizer is probably the vertical firetube type. This unit is similar in appearance and design to the vertical steam boiler shown in Fig. 12-174 and also uses atmospheric gas burners. The design provides good natural circulation that prevents breakdown of the Dowtherm resulting from localized overheating. When gravity return of condensate is practical, a completely closed system is possible without the use of pumps, traps, or other moving parts.

Applications for a Dowtherm vaporizer exist in the chemical, food, paint and varnish, fats and oils, and petroleum industries, among others.

Hot-Water Generators

Hot water at the desired temperature and volume for commercial processes can be obtained from gas-fired water heaters and water heating systems (Figs. 12-67 thru 12-75). In addition, direct high-pressure steam injection may be employed as well as various heaters of specialized designs.

Direct Immersion Heater. This type of heater may be similar in construction to the horizontal, single-pass, immersion-fired boiler shown in Fig. 12-181 or it may be of vertical design. Storage capacity of horizontal units ranges from 100 to 1000 gal, and input ranges from 100 to 5000 MBtu per hr; auxiliary storage tanks may be used for extra capacity. Vertical heaters store up to 600 gal and use gas inputs up to 500 MBtu per hr.

Indirect Immersion Heater. This package water heater (Fig. 12-184) consists of a power gas burner firing into a horizontal tube containing a finned tube bundle. A separate heat transfer fluid, usually pure water, is pumped thru the finned bundle and then to the water heating bundle located below the firetube in the shell or storage tank. The heat transfer fluid, which is continuously circulated at a controlled velocity, is the basic source of heat transfer to the stored water, even though the firing tube is immersed in the water and therefore serves as additional heat transfer surface. The heat transfer fluid is circulated at a maximum temperature of 250 F and the closed system is operated under pressure to prevent boiling.

Corrosion and scale formation are eliminated or minimized because no make-up water, and hence no oxygen, is brought into the circulating fluid system. The metal temperature of the service water side of the water heating bundle averages about 160 F.

Indirect immersion heaters are available with gas inputs from 390 to 2215 MBtu per hr and with storage capacities from 260 to 4010 gal. The storage tanks may be lined with copper or cement.

Direct Flame Heater. This heater (Fig. 12-185) is suitable for many processes requiring large quantities of water at temperatures up to 200 F. Available sizes range from 5 to 50 MMBtu per hr and use gas at 3 to 4 in. w.c.

Referring to Fig. 12-185, the burner fires into a water-jacketed combustion chamber and the combustion gases are

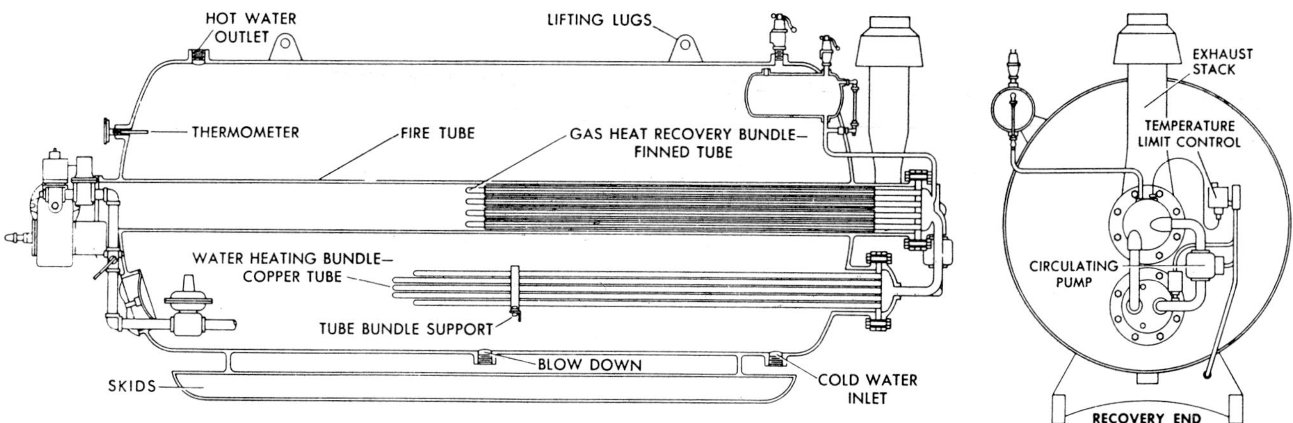

Fig. 12-184 Indirect immersion-fired water heater. (Patterson-Kelley Co., Inc.)

subsequently quenched and washed by recirculated water sprays. The resulting pure steam and flue gases travel up the heat recovery stack as cold process water flows down. The heated process water from the recovery stack is further heated by steam from the combustion chamber. The hot water then passes thru the degasser, where carbon dioxide and oxygen are removed before the water enters the process lines. Since the cleansing spray water is recirculated and kept separate from the process water, the latter comes in contact only with pure steam and wasted flue gases.

Thermal efficiency approaching 99 per cent is claimed, since the temperature of the saturated flue gases leaving the recovery stack is under 100 F.

Process Steam Uses

The potential uses for process steam in the commercial field are varied and numerous. The following listing illustrates some of the more common applications and appliances that can be effectively served by small and medium-sized gas-fired steam boilers:

Sterilizers	Garment presses	Can washing
Water heating	Puffers and shapers	Disinfectors
Dishwashers	Steam irons	Evaporators
Proof boxes	Dry cleaners	Pasteurizers
Food warmers	Blocking machines	Steam rooms
Coffee urns	Clothes dryers	Solution heating
Vegetable steamers	Tar removal	Matrix tables
Stock kettles	Degreasing	Tire molds
Mangles	Bottle washers	Vulcanizing

Steam Requirements and Boiler Sizing

In commercial applications process steam may be utilized directly and indirectly. Some applications use a combination of both. The direct method, commonly referred to as "live steam" or "open end jet," may utilize the sensible as well as the latent heat in the steam. With the indirect method, only the latent heat is utilized in the heat transfer cycle and all of the sensible heat is returned to the boiler in the form of con-

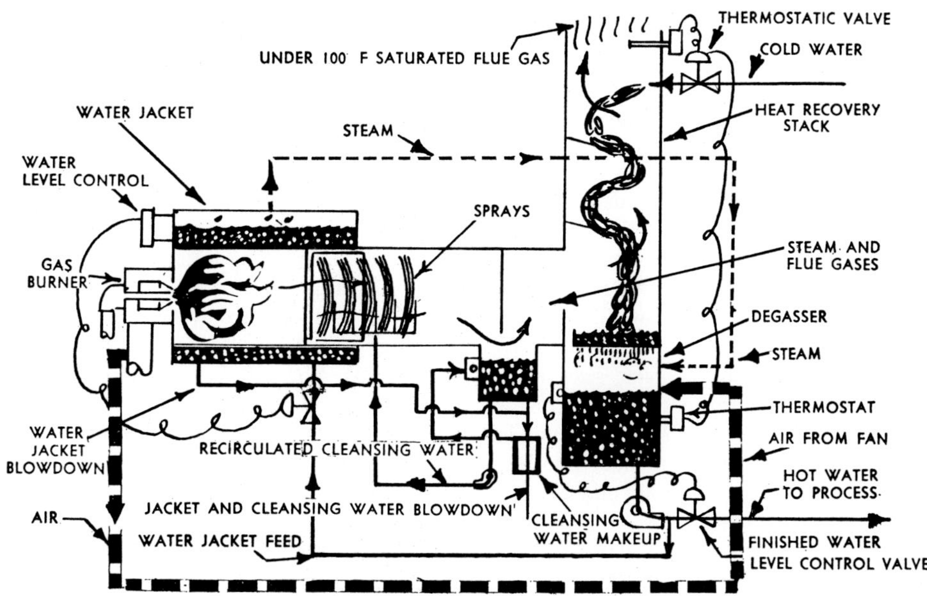

Fig. 12-185 APCO flame water heater. (Patented, Applied Engineering Co.)

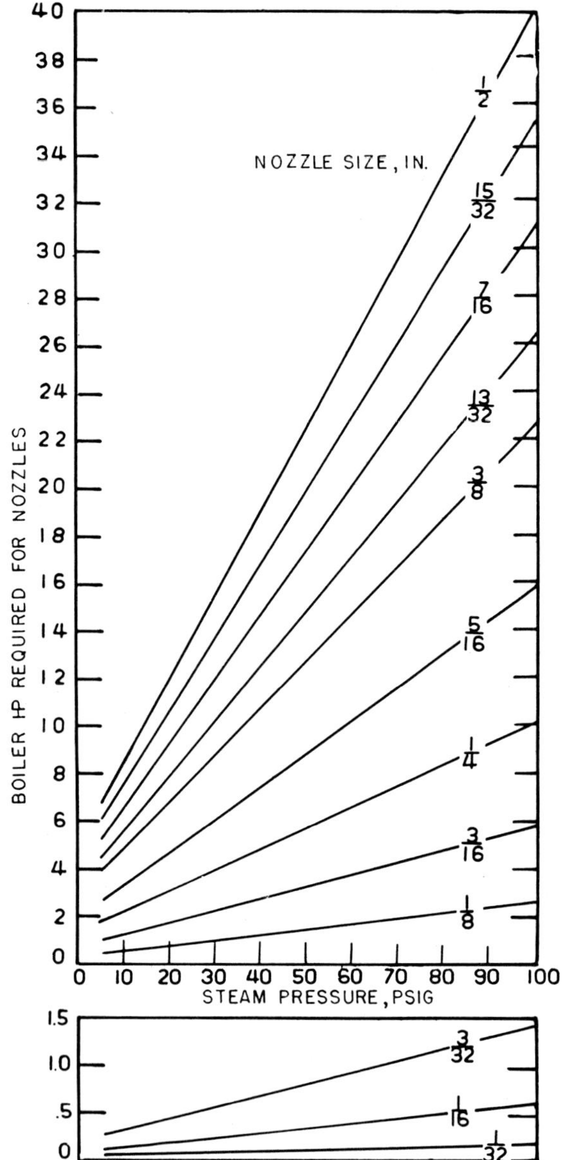

Fig. 12-186a Boiler horsepower requirements for steam nozzles.

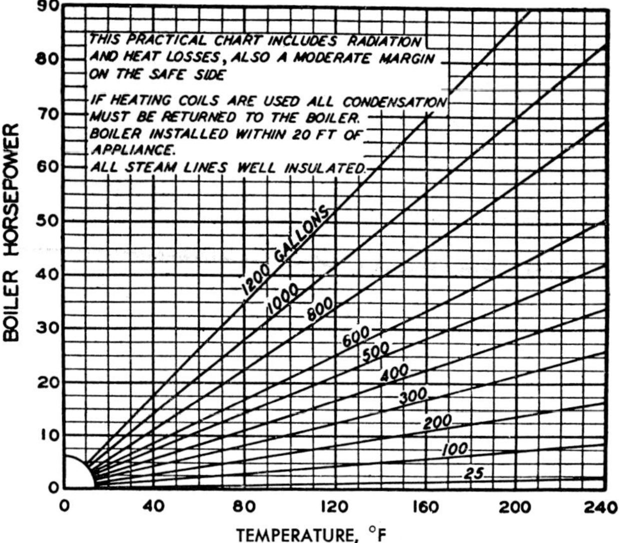

Fig. 12-186b Boiler horsepower requirements for heating water or solutions.

densate. For economy of operation and of boiler maintenance, the latter method is preferred.

The operating pressure of the boiler depends primarily upon the temperature required in the process. The higher the steam pressure the smaller the surface area required, because the heat transfer is more rapid. However, the higher the pressure the lower the operating efficiency of the boiler will be.

There are few commercial steam applications that present any particular problems in determining boiler sizes since many of the appliances using steam are standardized. Process boiler manufacturers and manufacturers of steam-consuming appliances can supply steam ratings, by model number, of the more common appliances. Table 12-109b may be used as a guide if exact steam ratings are not known.

Where open end jet or live steam is used, it may be necessary to determine the amount of steam flowing thru a jet or nozzle. Figure 12-186a indicates the approximate steam flow thru nozzles when discharging to the atmosphere, into cabinets, against a product, or in other processes with little or no back pressure at the outlet. For slots or nozzles of other shapes or sizes, a rough estimate can be obtained from Napier's equation:

$$W = 0.0143\,Ap$$

where: W = steam flow, lb per sec
A = area of opening, sq in.
p = steam pressure at opening, psia

Some commercial processes involve the heating of water or aqueous solutions in tanks[13] or containers. In most cases the boiler horsepower required is determined by the heat necessary to raise the solution temperature in the given time. Boiler size may be estimated from Fig. 12-186b. Alternatively, curve B of Fig. 2-83, together with Eq. **51** of Sec. 2, Chap. 6, may be used to estimate the heat output required. If the allowable heat-up period is long and the tank radiation and evaporation losses high, the heat required to maintain solution temperature may be the determining factor in boiler size. See Table 12-110 for tank heat losses. Note also Table 2-113.

On every job an attempt should be made to determine the *actual* steam requirements as accurately as possible since two identical appliances can have substantially different steam requirements, depending on the usage and type of installation. The following factors must be considered:

1. **Heating time.** If 2 hr can be allowed to attain the desired temperature, the required boiler size can usually be cut in half. However, it is not usually safe to allow more than 3 to 4 hr without careful calculation of line and other radiation losses.

2. **Appliance usage.** Although it is normally good practice to size the boiler for the maximum rated steam consumption of the equipment it serves, a smaller boiler may be warranted if there is a known diversity in appliance use or if it is definitely established that the appliance will be operated only at some level below rated capacity.

3. **Future requirements.** Every effort should be made to determine future steam needs. Consideration should be

Table 12-109b Approximate Steam Requirements for Typical Commercial Processes[2–6, 10–12]

Process	Steam, lb per hr or per unit	Boiler hp	Range of steam pressure, psig
Bakeries, retail			
Dough room trough, 8 ft long	4.6	0.13	
Proof box, 500 cu ft capacity	34.5	1.00	
Ovens, peel- or Dutch-type			
White bread, per 120 sq ft	34.5	1.00	
Rye bread, per 50 sq ft	34.5	1.00	
Ovens, revolving			
8–10-bun pans	34.5	1.00	
12–18-bun pans	69.0	2.00	5 to 25
18–28-bun pans	103.5	3.00	
Ovens, rotary			
Single deck, 14-bun pans	34.5	1.00	
28-bun pans	69.0	2.00	
32-bun pans	103.5	3.00	
Double deck, 18-bun pans	69.0	2.00	
28-bun pans	103.5	3.00	
Bottle washing			
Soft drink, beer, 100 per min	340.5	10.00	10
Milk, quarts, 12 per min	69.0	2.00	10
General, 1000 lb per hr	34.5	1.00	10
Creameries and dairies			
Creamery cans, 3 per min	340.5	10.00	10
Pasteurizer, 100 gal in 20 min	258.5	7.50	10
Hospital equipment			
Stills, per 100 gal water	121.0	3.50	
Sterilizers			
Bed pan	3.5	0.10	
Dressing, per 10-in. length	8.5	.25	60 to 100
Instrument, per 100 cu in.	3.5	.10	
Water, per 10 gal	6.9	.20	
Disinfecting cabinets, per 10 cu ft	27.8	.80	
Laundry equipment			
Vacuum stills, per 10 gal	20.7	0.60	
Spotting board, trouser stretcher	34.5	1.00	
Steam irons	5.5	0.16	
Flatwork ironers			
48 in. × 120 in., 1 cylinder	278.0	8.00	
48 in. × 120 in., 2 cylinder	345.0	10.00	
4 roll, 100 to 120 in.	242.0	7.00	
6 roll, 100 to 120 in.	380.0	11.00	
8 roll, 100 to 120 in.	518.0	15.00	70 to 100
Shirt equipment			
Single cuff, neckband, each	8.5	0.25	
Double sleeve	17.3	0.50	
Body	34.5	1.00	
Bosom	51.8	1.50	
Laundry presses, per 10-in. length	8.5	0.25	
Deodorizer	69.0	2.00	
Restaurant equipment			
Stock kettles, per 10 gal	17.3	0.50	
Coffee urns, per gal	3.5	.10	
Steam tables, per sq ft	1.7	.05	
Plate warmer, per cu ft	1.7	0.05	
Vegetable steamers, per compartment	34.5	1.00	
Clam, lobster, potato steamers	34.5	1.00	
Egg boilers, 3 compartments	17.3	0.50	10 to 75
Oyster pots	17.3	.50	
Bain marie, per sq ft	3.5	.10	
Warming ovens, per cu ft	1.7	0.05	
Jets for sinks, avg use	34.5	1.00	
Silver burnisher	69.0	2.00	
Dishwasher, tub, 2 compartments	69.0	2.00	
Mechanical (use manufacturer's rating)		...	
Tailors and cleaners			
Steam-head utility press			
28 in.	34.5	1.00	
43 in.	58.7	1.70	
Hot-head skirt press			
28 in.	34.5	1.00	
42.5 in.	58.7	1.70	
Dress finisher, manikin type	69.0	2.00	
Overcoat shaper, manikin type	69.0	2.00	
Jacket finisher, manikin type	43.2	1.25	
Puff irons, 2 per table	25.9	0.75	75
Dry cleaning tumbler	69.0	2.00	
Synthetic solvent cleaning unit, 30 lb capacity	25.9	0.75	
Petroleum solvent cleaning unit			
Single, no vacuum still (42 lb)	103.5	3.00	
Single, vacuum still (42 lb)	190.0	5.50	
Double, no vacuum still (84 lb)	207.0	6.00	
Double, vacuum still (84 lb)	311.0	9.00	

given to the possible addition of new appliances and/or the use of existing equipment at an increased production rate.

4. **Multiple boilers.** For some critical operations, the use of two or more boilers to meet the steam requirements may be desirable so that the forced shutdown of one unit will not completely halt production. Multiple units also provide greater flexibility in meeting variable steam demands.

5. **Line losses.** For steam economy the boiler should always be located as close as possible to the steam-using appliances. If remotely located, the steam and condensate return lines should be insulated. A rule-of-thumb method of estimating radiation losses from uninsulated piping is to allow 20 per cent of the total appliance steam demand. To calculate this loss with greater accuracy, Table 12-111 may be used.

Table 12-110 Solution Tank Heat Losses[13]

(Btu per sq ft-hr)

	Heat loss from surface			Heat loss thru tank walls			
					Insulation		
Water temp, °F	Evapo-ration loss	Radia-tion loss	Total loss	Bare steel walls	1 in.	2 in.	3 in.
90	80	50	130	50	12	6	4
100	160	70	230	70	15	8	6
110	240	90	330	90	19	10	7
120	360	110	470	110	23	12	9
130	480	135	615	135	27	14	10
140	660	160	820	160	31	17	12
150	860	180	1040	180	34	18	13
160	1100	210	1310	210	38	21	15
170	1380	235	1615	235	42	23	16
180	1740	260	2000	260	46	25	17
190	2160	290	2450	290	50	27	19
200	2680	320	3000	320	53	29	20
210	3240	350	3590	350	57	31	22
220	4000	380	4380	380	61	33	23

MISCELLANEOUS GAS USES

In addition to process steam and hot water, there are numerous commercial heat applications where gas appliances can be utilized successfully.

Torches

Among the most common small appliances are the gas–air and the gas–oxygen torches found in practically every shop in which metal is shaped or fabricated. Torches are used principally for the following purposes:

1. **Soldering**—the joining of metal parts by means of a low-melting-point alloy. The torch melts the solder, heating the edges to be joined. Solder irons (or coppers) are also used for this purpose. A flux is used to "wet" the surfaces and permit the easy flow of solder to the joint. Torches for soldering usually use low-pressure air, although atmospheric torches are sometimes found.

2. **Brazing**—the joining of similar or dissimilar metals by means of copper, brass, or silver alloys which melt at a temperature appreciably below the melting temperature of the metals to be joined. Brazing is a higher temperature operation than soldering; oxygen instead of air is generally used for combustion, thus increasing the rate of burning and raising the flame temperature. With compressed gas (10 to

Table 12-111 Pipe Lengths Equivalent to 1.0 Boiler Hp[2]

(34.5 lb steam per hour from and at 212 F)

Nominal pipe diam, in.	Outside surface area of pipe, sq ft/lin ft	Internal flow area of pipe, sq in.	Pipe length with heat loss equiv. to 1.0 boiler hp, lin ft		Steam from and at 212 F condensed per ft pipe, lb	
			Bare	Covered 1½ in. insul.	Bare	Covered 1½ in. insul.
¾	0.275	0.614	260	1000	13.5	3.5
1	.346	0.945	200	820	17.5	4.3
1¼	.434	1.638	160	660	22.0	5.3
1½	.494	2.222	140	620	25.0	5.7
2	.622	3.654	110	560	31.8	6.3
2½	0.753	5.450	90	440	38.9	8.0

20 psig), results are even better. Natural, manufactured, or LP-gas can be made to perform comparably to acetylene, which is more costly. Brazing material, a rod or wire, is either fed by hand or placed on the joint, sometimes in specially prepared shapes. The flame must be reducing or neutral, never oxidizing. Torches for this type of work are made in many varieties and sizes. They are convenient, easy to handle, and permit uniform heating of the work.

3. **Cutting**—the severing of ferrous metals by means of an oxygen cutting jet. The cutting operation consists of two phases, preheating and cutting. Oxygen is mixed with fuel gas for the preheating flame; pure oxygen accomplishes the cutting. Oxygen cutting torches generally consist of a handle, tubes, a mixer, and a tip. The mixers and tips are available for use with low-pressure fuel gas (7.0 in. w.c. to 1.0 psig) as well as high-pressure fuel gas (8–10 psig).

4. **Battery burning**—the joining of plate lugs to the terminal straps of lead cell batteries. In this process the material is lead or hardened lead containing 6 per cent antimony. The terminal strap socket is placed over the plate lugs and a small oxygen–gas flame applied until the metal begins to run. A rod of lead is then melted over the joint, forming a clean, smooth assembly. On large batteries connectors are also burned on to the terminals in the same manner. For best results the gas should be at 12 to 15 psig and the metal surfaces must be extremely clean. No flux is used.

Degreasing Tanks

Solvent degreasing equipment has largely taken the place of solution cleaning tanks in plating plants and other plants where metal parts must be completely clean before a final finish is applied. These degreasing tanks use a solvent such as trichlorethylene with a stabilizer; most such solvents boil at about 180 F. Heat can be applied with steam or gas. Vapor degreasing consists of lowering the work into the vapor formed over the boiling solvent. The vapor level is controlled by cooling coils or water jackets around the upper part of the tank. When cold work reaches the vapor, condensation takes place, and the condensate bathes the parts, carrying off any grease or oil. The work is removed when it reaches the temperature of the vapor. If a more thorough cleaning is required, the work can be subjected to a spray of hot liquid solvent.

Hoods should be installed over tanks since high-temperature mediums break down halogenated hydrocarbons into toxic (e.g., phosgene) and corrosive products. Paint or enamel finishes being baked nearby can be discolored. These finishes are sensitive to the most minute quantities of solvent vapor.

Candy Cooking

Although steam in large quantities plays an important part in preparation of confectionery, some products require the direct application of higher heats and shorter heating periods than can usually be obtained with steam. If certain white mixtures, for instance, are not brought to a boil quickly, discoloration will result. Thus, the flexibility of gas, which provides fast or slow heat and controllable inputs, is of great advantage in a candy plant. For rapid heating of kettles and cylindrical containers, high-input gas confectioners' stoves are commonly used. These may be fired with ring-type atmospheric burners or with low-pressure blast burners.

Bread and Cake Baking

Bread Making. Two kinds of dough are used to make bread. For straight dough, all ingredients are mixed at one time for 10 to 20 min. For sponge dough, about 60 per cent of the ingredients is mixed for 5 to 10 min. In either case, dough-filled troughs are set in a warm (80 F), humid (75 per cent relative humidity) fermentation or dough room. After rising for 3 to 5 hr, sponge dough is returned to the mixer and the remaining ingredients are added.

The dough is then cut into units, called billets, in the divider and shaped in the rounder to seal the outside pores and unify the mass. Each billet is formed in the molder and placed in a pan. The billets are sent to a proof box, where the dough is permitted to rise and fill the pan. (In the proof box the yeast is very active as a result of the 95 F temperature and the 85 per cent relative humidity.) The dough is then baked in the oven for 25 to 40 min at a temperature of 425 to 475 F. Baking kills the yeast and stops fermentation.

The interior temperature of the bread as it leaves the oven is about 210 F. Prior to slicing or wrapping, it is cooled to 80 to 90 F in the cooler. About 10 per cent of the weight of the dough is lost by water evaporation in the oven.

Certain types of bread, such as rye, require a high glaze, which is produced by the introduction of water, usually in the form of steam, into the oven. The formation of crust is a process of hydrolysis of the surface starch (into gum and sugar) followed by caramelization.

Gas consumption for baking bread in modern traveling ovens amounts to approximately 525 Btu per lb of finished product. The heat absorbed per lb of bread and pan is 238 Btu.

Cake Baking. Eggs or chemicals such as baking powder are used as leavening agents in cake dough. To be light, dough must be made spongy, full of bubbles of entrapped air, carbon dioxide, or steam. The froth of beaten eggs traps air; sponge and angel food cakes require no other aerating agent. In other types of cakes, chemicals release carbon dioxide and water vapor, which cause the dough to rise.

Gas consumed for sweet goods amounts to approximately 790 Btu per lb. In **cracker baking** (e.g., soda crackers) gas consumption is approximately 940 Btu per lb.

Traveling Ovens

There are two types of traveling ovens, the **tray traveler** and the **tunnel traveler.**

In tray traveler ovens, trays are supported from conveying chains that travel thru the baking chamber. The trays may make a single or a double lap thru the oven before they return to the starting point where both loading and unloading take place. Capacity ranges up to about 3000 lb an hour for a variety of baked goods. In newer ovens, trays have been stabilized and unloading made automatic. In the tunnel traveler the product makes a single pass thru a long baking chamber on a continuous conveyor. The conveyor may be one of several types: plate, wire mesh, rod, or steel band. These ovens vary widely in length and capacity but are often between 60 and 110 ft and produce up to 7500 lb an hour. The wire mesh and band ovens are used in cracker, cake, and pretzel baking, the other types for cake and bread. Early ovens were directly fired with ribbon burners.

Pan Washers

Pan washers (Fig. 12-187) for large bakeries use gas for water heating and for drying. Conveyors move the pans under spray nozzles for washing and rinsing, then into a drying chamber. Water is heated to about 160 F (lower wash temperatures, 120 F, may be used for some utensils such as angel cake pans). Wash and rinse tank sizes may be about 500 gallons each. Immersion heating with blast combustion equipment is popular.

Drying compartments may be heated directly by pipe burners or by recirculated hot air thru nozzles to help blow

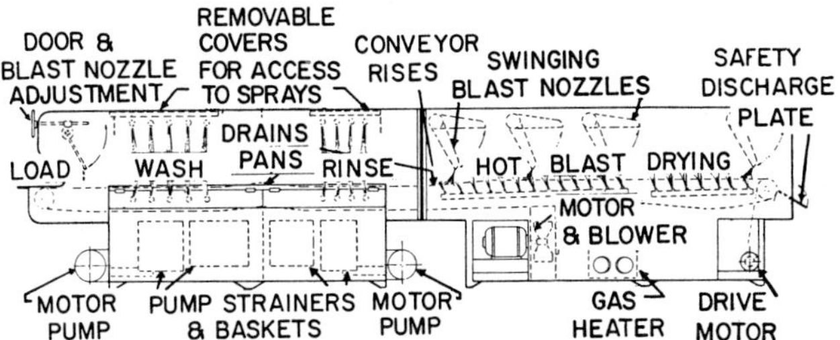

Fig. 12-187 Tunnel-type pan washer—operation of sprays and adjustable dryer nozzles.

water off the pans. Temperatures of about 250 F are usual. (When pan greasers are used, absolute dryness is essential.) Washers may be about 36 ft long with 20-ft drying chambers and conveyor speeds of 4 to 12 fpm.

Doughnut Cookers

Baking powder doughnuts are produced in batch-type and automatic conveyor-type equipment. The batch-type doughnut cooker consists of the common 24 × 34 in. deep fat fryer, in which the doughnuts are handled on screens. Automatic conveyor-type cookers drop the doughnuts into the hot oil bath. The doughnuts are cooked on one side, moved to a midpoint by a conveyor while floating on the oil, and then automatically turned over. Baking is completed in the last half of the machine. Yeast doughnuts and crullers are generally baked in batch-type appliances. Heat required to cook doughnuts by gas is approximately 1200 Btu per dozen.[14]

Deep Fat Fryers

Gas-fired deep fat frying equipment is utilized in the manufacture of potato chips, julienne or shoestring potatoes, doughnuts, twisted corn meal strips, and corn chips such as Fritos. Batch-type equipment is often used, but for larger production the continuous cooker is used. Both immersion- and bottom-firing are employed. Cookers for as much as 500 pounds of potato chips (or for about 200 pounds of Fritos) are available. An input of about 5000 Btu is required to produce one pound of chips or Fritos in the batch-type fryer, and 6000 Btu in the continuous-type. Because of the high water content of potatoes, the weight ratio of raw potato to chip is about 4 to 1. Fast recovery is needed to maintain the 360 F temperature necessary to prevent soggy products.

Food Preservation[15]

Most foods are subject to deterioration when exposed to air at normal room temperature. Deterioration is due to a wide range of causes, including bacteriological action, oxidation, and dehydration. Products containing edible fats, aldehydes, alcohols, amino acids, or unsaturated hydrocarbons in the presence of oxygen can develop bitterness, oiliness, or rancidity. Deterioration is commonly controlled by refrigeration. In many cases an inert gas atmosphere will preserve food products without refrigeration or increase the effectiveness of refrigeration. Some applications of inert gas and the gases used are given in Table 12-112.

Atmosphere generators for the production of inert gases are available in sizes using from 100 to several thousand cubic feet of natural gas per hour. Such generators are also used for many other purposes where special atmospheres are needed, for example, metal processing. The exothermic generator is the type used almost exclusively in the food processing field since the gases produced are noncombustible and inert to most foods. Natural, manufactured, or LP-gas may be used as the fuel.

Effect of Gases on Color of Meat. The effect of gases on the color of meats exposed to them in sealed glass containers has been studied in the laboratory. Results are shown in Table 12-113.

Table 12-112 Applications of Inert Gas in Food Processing[15]

Application	Typical inert gas used: 100% N_2	100% CO_2	88% N_2 + 12% CO_2
Storing and handling dehydrated alfalfa	...	...	x
Storing and handling grains	...	...	x
Storing and handling fresh vegetables*	x	x	x
Packaging meat†	x	x	x
Making, storing, and packaging vegetable oils	x	...	...
Storing and shipping milk	...	...	x
Canning peanuts, walnuts, etc.	x	...	x
Canning soluble coffee, tea, etc.	x	x	x
Canning fruit juices	x	...	...
Bottling soft drinks (noncarbonated)	x	...	...
Bottling beer	...	x	...

* Since oxygen is the offender, it is often desirable to operate the generator slightly on the rich side, thereby reducing the possibility of oxygen being in the generated inert gas.
† Varies with product.

Nitrogen and carbon dioxide may be used to preserve the color of meat, in addition to refrigeration, processing, and packaging methods. Gases that darken meat rather rapidly are nitric oxide, oxygen, formaldehyde, sulfur dioxide, ammonia, carbon tetrachloride, and hydrogen sulfide. Gases that tend to turn meat gray or white (depending on the meat) are formaldehyde, ammonia, sulfur dioxide, and ethyl mercaptan.

Greenhouse Atmospheres[16,17]

Direct-fired natural gas heaters will not only heat the greenhouse, but will also increase its CO_2 content, thereby improving the conditions for photosynthesis which is the basis of plant growth.

The carbon dioxide level in the open air is about 300 parts per million; at that level photosynthesis just balances respiration. In enclosed greenhouse atmospheres this balance may be disturbed, especially in the winter when ventilators are closed. The CO_2 level may then fall below that of the outside air, decreasing plant growth rate even in bright sunlight. Precisely because greenhouse air is largely contained, however, the CO_2 balance can be altered favorably. Extra CO_2 benefits plant growth, increases yields, and improves quality; it can also compensate partially for a lack of sunlight. Experiments have shown that yields of greenhouse-grown vegetables can be increased by 25 to 100 per cent; for example, in one test, lettuce and tomatoes were found to increase in weight at the following rates in CO_2-enriched atmospheres:

	Carbon dioxide (ppm): 125–500 (normal)	800–2000 (above normal)	Increase, per cent
Lettuce, lb/head	0.41	0.83	>100
Tomatoes, lb/plant	18.9	29.5	> 50

Extra CO_2 also promotes earlier flowering and fruiting (e.g., in the tomato and cucumber). However, it must be kept in mind that the changed CO_2 level implies oth er changes in crop culture and management.

To supply CO_2 naturally, i.e., by ventilation, requires between two and ten air changes per hour depending on plant size, resulting, at the higher rate, in a heat loss more than half the total. When enriching with CO_2, ventilation should be reduced to retain the added CO_2. Using direct-fired forced-

Table 12-113 Effect of Various Gases on Color of Meat

Gas	Exposure, hr	Lamb	Veal	Ham	Beef	Pork
Natural gas	24	Red, firm; no change	Red, firm; no change	No change	Red, firm; no change	Red, firm; no change
Oxygen	24	Slight darkening	No change	No change	Darker red than original	Darker red than original
Nitrogen	18	Red, firm	No change	No change, better color than original	No change; better color	No change; better color
Flue gases from natural gas	24	No change	Lighter red	Lighter red	Darker red	No change
Carbon dioxide	6	No change	No change	Slightly darker	Slightly darker	No change
Carbon dioxide	72	Darker red	Light red; O.K.	Slightly darker	Slightly darker	Lighter red, O.K.
Carbon monoxide	24	Brighter red; firm	No change	No change	Brighter red	Brighter red
Hydrogen	18	Slightly lighter red	Lighter red; good appearance	No change; good appearance	Slightly lighter red	Lighter red
Formaldehyde	8	Grey; putrid	Grey; putrid	Almost white	Black; putrid	Dark; slimy
Sulfur dioxide	16	Dark red; dried, fat shriveled	White-bleached; putrid	Lighter color; dried out	Darker red; greasy	White-bleached; greasy
Nitric oxide, concentrated, plus air	18	Darkened; putrid, yellow (green at first)	Dark, soggy, putrid, yellow (green at first)	Slightly darker	Black; $\frac{3}{16}$-in. penetration; putrid, yellow (green at first)	Verk dark; putrid, yellow (green at first)
Hydrogen sulfide	5	Greenish black, putrid	Light green, putrid	Edges green or brown	Very dark, black, green, putrid	Light green, putrid, greasy
Ethyl mercaptan	24	Dirty brown	Almost white	Almost white	Dark red, brown	Dark red, brown
Ammonia	72	Quite dark	Nearly white	Nearly white	Dark, putrid	Light red, greasy
Carbon tetrachloride	24	Dark red	Lighter red; greasy	Lighter red, almost grey	Dark red, almost black	Lighter red, greyish, greasy

circulation gas heaters, the following results are achieved: (1) an optimum concentration of CO_2 can be maintained; (2) heating requirements can be minimized thru the reduced ventilation rate; and (3) air circulation is provided, which benefits greenhouse crops, since turbulence maintains a higher CO_2 level at leaf surfaces.

The following is a determination of the rate of natural gas combustion for a required CO_2 level.

Example. Find the natural gas input required to maintain a one-part-per-thousand carbon dioxide–air ratio in a greenhouse with a volume of 60,000 cu ft and a construction infiltration rate of one air change per hour.

$$CO_2 \text{ required} = \frac{1 \text{ part}}{1000 \text{ parts}} \times 60{,}000 \text{ cu ft} \times 1\,\frac{\text{change}}{\text{hour}} = 60\,\frac{\text{cu ft}}{\text{hour}}$$

Since the combustion of 1 cu ft of natural gas produces 1 cu ft of CO_2, 60 cu ft of natural gas must be burned per hour. The accompanying release of 60,000 Btu per hr helps to overcome normal heat losses in cold weather.

In practice, local considerations, especially the variations in outdoor temperatures and weather conditions, substantially affect the balance among heating requirements, CO_2 produced, and ventilation. Special attention must be given to the selection of fuel so that no toxic levels of trace combustion products are exhausted into the greenhouse. Tolerances of tomato plants, which are particularly sensitive to toxic gases, are listed in Table 12-114.

Table 12-114 Tolerance of Tomato Plants for Various Gases[18]

Gas	Tolerances, ppm	Gas	Tolerances, ppm
Ethylene	0.1	Propylene	50
Hydrogen fluoride	0.1	Ethyl mercaptan	50
Sulfur dioxide	0.5	Hydrogen sulfide	50
Hydrogen chloride	10	Hydrogen cyanide	100
Ammonia	10	Carbon monoxide	500
Nitrogen oxide	25	Carbon dioxide	20,000
Acetylene	50	Butylene	50,000

Fruit and Vegetable Storage[19]

When fruits and vegetables are harvested and placed in storage, ripening and subsequent spoilage take place. These normal physiological changes necessitate consumption in a relatively short period of time unless the changes can be retarded.

In recent years control of the atmosphere of the storage warehouse has been found to substantially decrease the spoilage rate of fruits and vegetables. Controlled atmosphere as used in food storage refers to the process whereby the oxygen and carbon dioxide in the air are *reduced* to such proportions as to substantially extend storage life. This is the reverse of the procedure used in hothouses in which the atmosphere is supplemented with carbon dioxide to increase the rate of growth and production of plants and vegetables.

Conventional Method of Atmosphere Control. The conventional method for controlling atmosphere in storage warehouses has been used in the apple industry for many years. The following data refer to the storage of apples but the procedure described would be that for any fruit or vegetable. In the conventional method the storage warehouse must be tightly constructed so as to permit a maximum of 0.02 air change per day. In this situation, the conversion of oxygen to carbon dioxide by the fruit is retarded once the initial oxygen level is reduced or "pulled down" to the acceptable limits of 2 to 5 per cent. The excess CO_2 produced is removed by a scrubber until reduced to 1.5 to 5 per cent. The initial oxygen pulldown takes 3 to 4 weeks. Maximum storage time for apples maintained in this manner is 7 months.

Tectrol Method. The Whirlpool Corporation markets a controlled atmosphere service* that has received a degree of acceptance in the apple industry. The heart of their patented system is the mixing and burning unit, which combines the oxygen from air with natural gas or propane in a catalytic combustion unit, producing carbon dioxide and water. Any substantial formation of oxides of nitrogen, a major producer of adverse effects on food, is avoided, since combustion temperature does not exceed about 2000 F (1600 F nominal).[20] The nitrogen passes thru the system to the storage area. If desired, water may be removed by cooling the combustion products. The CO_2 level is controlled by a scrubber and by-pass valve arrangement.

Using the Tectrol system, the allowable leakage rate for the warehouse is one-half air change per hour. This affords the user a sizable initial cost saving over the conventional method, which requires much tighter construction. The oxygen pulldown rate is comparatively rapid (usually three days), providing another important advantage. The storage warehouse may be entered for short periods of time, allowing addition or removal of produce as dictated by market conditions.

Storage atmospheres with oxygen reduced to such an extent (2 to 5 per cent vs. 21 per cent for air) will not support human life. Extreme caution must be exercised by personnel entering the storage areas. The use of oxygen-breathing equipment is required; the "buddy" system for workers is urged.

Smokehouses

Gas is widely used for smoking meats such as ham, bacon, butts (pork tenderloin), tongue, and bologna in rooms or cabinets in a variety of sizes. The smoking process conditions meat and fish to enhance taste and aroma as well as to preserve. Creosote vapors distilled from burning sawdust penetrate the meat. Hickory is thought to be best, although other hardwoods are suitable. Excessive free carbon is to be avoided, since it produces smudge on the surface of the meat.

Older smokehouses were an integral part of the building, constructed of hollow tile or masonry and lined with glazed tile. Figure 12-188 illustrates a smokehouse using atmospheric pipe burners located under a sawdust pan. Newer houses have been built with stainless steel panels backed with 2 or 3 in. of insulation; external smoke generators are used.

The most recent development is the "air-conditioned"[21,22] or recirculating type of smokehouse. Air is heated in a heater located outside the house proper. The smoke is picked up in the air stream and both air and smoke are distributed thru the house. The gases are then recirculated. Temperature is carefully controlled and, in some installations, humidity control has been added. Operating temperatures between 110

*A comparable service for greenhouses is available.

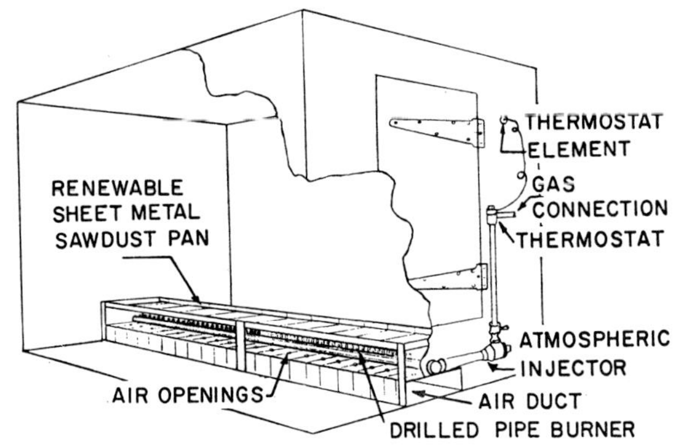

Fig. 12-188 Older style smokehouse, using atmospheric pipe burner and sawdust pan.

and 190 F are used. Frankfurters are smoked 1½ hours; bologna, 3 hours; hams, 12 hours or more. Other than mackerel, which requires a heavy, long smoking period, most fish need only a short time in a smokehouse.

Smoking procedures vary somewhat with the operator. The operator must strike a heat balance between the removal of moisture, penetration of the creosote vapors, and regard for the softening of fatty substances. Good air and smoke circulation with temperature variations not over 5 to 7° F are desirable.

Gas consumption may vary from 200 to 600 Btu per lb of meat. Fish smoking and drying generally require somewhat less gas.

Coffee Roasting

Roasters may be of either the batch or continuous type. Older batch types use a premixed blast of air and gas blown into the center of the roasting cylinder. Helical flanges on the interior of the perforated cylinder and a baffle over the flame keep the coffee revolving about the flame but prevent contact. These roasters are capable of roasting a batch of 500 lb of coffee in about 17 min. Efficiency is claimed to be 70 per cent.

Low-temperature roasting was the next step in roaster evolution. In older roasters high temperatures (2000 F) of the heating medium are essential to penetrate the gaseous film surrounding each bean. When this protecting film is thinned by contact with other beans, exposure to excessive temperatures can break down the oils and scorch the surface; also, chaff is ignited and forms soot. In the newer roasters the insulating air film is washed uniformly thin by a very large volume of high-velocity, heated gases. Heat transfer is then accomplished as rapidly as before but at lower temperatures (800 to 900 F). Chaff is not burned; oil, waxes, and fibers are not broken down; and the center of the bean is more thoroughly roasted. Fuel is burned in a separate chamber and the hot gases are recirculated after passing thru the roaster and chaff collector. Roasting time is about 17 min for a 500-lb batch. The four-bag size is the most widely used and produces roughly 1500 lb an hour.

The most recent development is the continuous roaster; models with capacities up to 9000 lb per hr of green coffee are available. The coffee is advanced thru a single, long, perforated cylinder by means of helical flanges. Hot gases are drawn thru the coffee at about 500 F, passed thru a cyclone collector, and recirculated. The roasting period may be as short as 5 min; the coffee then passes thru an air lock into the cooling section of the cylinder, where it is brought to room temperature. Another style comprises several perforated, horizontal cylinders, staggered one above another. Coffee passes thru one cylinder after the other for roasting and is cooled in the last cylinder.

Depending on the age of the beans, gas consumption in batch-type roasters may vary from 360 to 470 Btu per pound of green coffee.[14] Continuous roasters may require a somewhat higher rate.

Printing

Gas is used in the printing industry for melting stereotype, linotype, electrotype, monotype, and foundry type alloys of lead, tin, and antimony. Melting points of these alloys vary from 475 to 605 F. They are easily melted in pots with atmospheric gas burners or power burners. Gas is also used to heat air for drying ink in continuous dryers after the printing operation.

Sterilizers

Sterilizers are widely used in hospitals, laboratories, and doctors' and dentists' offices. Many varieties and sizes are available using gas or steam heat. Gas-heated vertical and horizontal steam pressure sterilizers, autoclaves, and hot-air sterilizers are available in many sizes and shapes and are described in chemical and hospital supply catalogues.

REFERENCES

1. A. G. A. Bureau of Statistics. *Gas Facts.* New York (annual).
2. Reiter, F. M. "Design Data for Service Steam." *Air Cond., Heating & Vent.* 52: 92–8, Feb. 1958.
3. Olsen, O. M. "Steam Generation." (In: A. G. A. Industrial and Commercial Gas Section. *1958 A. G. A. Commercial Gas School.* New York, 1958).
4. Magnuson, E. E. "Steam Generation." (In: A. G. A. Industrial and Commercial Gas Section. *1956 A. G. A. Commercial Gas School.* New York, 1956).
5. Leudemann, A. V. "Steam Generation." (In: A. G. A. Industrial and Commercial Gas Section. *1954 A. G. A. Commercial Gas School.* New York, 1954).
6. ——. "Steam Requirements of Industrial Processes." (In: A. G. A. Industrial and Commercial Gas Section. *1953 A. G. A. Industrial Gas School.* New York, 1953).
7. Baker, J. K. "Gas Designed Steam Boilers." (In: A. G. A. Industrial and Commercial Gas Section. *1953 A. G. A. Industrial Gas School.* New York, 1953).
8. Wierum, C. "Miscellaneous Small Industrial Processes." (In: A. G. A. Industrial and Commercial Gas Section. *1953 A. G. A. Industrial Gas School.* New York, 1953).
9. Craig, P. W. "Types of Gas-Designed Boilers, Ratings and Capacities." (In: A. G. A. Industrial and Commercial Gas Section. *1949 A. G. A. Industrial Gas School.* New York, 1949).
10. A. G. A. Industrial Gas Section. *Steam Boilers.* (Industrial Gas Series) New York, 1927.
11. A. G. A. Industrial and Commercial Gas Section. *Gas Fuel in Retail Tailor Shops.* (Inform. Ltr 36) New York, 1950.
12. ——. *Steam for the Retail Bakery.* (Inform. Ltr 67) New York, 1954.

13. Dewey, M. J. "Heating Liquids in Tanks." (In: A. G. A. Industrial and Commercial Gas Section. *1953 A. G. A. Industrial Gas School*. New York, 1953).
14. Harris, E. L. "Specialized Large Volume Food Processing." (In: A. G. A. Industrial and Commercial Gas Section. *1949 A. G. A. Industrial Gas School*. New York, 1949).
15. A. G. A. Industrial and Commercial Gas Section. *Application of Inert Gas in the Preservation and Protection of Food Products*. (Inform. Ltr 91) New York, 1958.
16. Wittwer, S. H. and Robb, W. "CO_2 Does Increase Yield and Quality." *Am. Vegetable Grower* 11: 9–11, Nov. 1963.
17. "Utilizing Facts." *A. G. A. Monthly* 46: 16, April 1964.
18. Wittwer, S. H. and Robb, W. "Carbon Dioxide Enrichment of Greenhouse Atmospheres for Vegetable Crop Production." *Econ. Botany* 18: 34–56, Jan.-Mar. 1964.
19. Lannert, J. W. *A New Device for Controlling Atmospheres for Fruit and Vegetable Storage*. Chicago, Am. Soc. of Agric. Engrs., 1963.
20. Bedrosian, K., and others. *Apparatus for Preserving Animal and Plant Materials*. (U. S. Pat. 3,102,778) Washington, D. C., G.P.O., 1963.
21. Moran, S. G. *Air Conditioned Smokehouses*. (Inform. Ltr 18) New York, A. G. A. Industrial and Commercial Gas Section, 1948.
22. Gehnrich, H. *Air Conditioned Smoke Houses*. (1948 A. G. A. Sales Conference on Industrial and Commercial Gas) New York, 1948.
23. Distelhorst, S. D. "Estimating Industrial/Commercial Steam Needs." *Air Cond., Heating & Vent.* 61: 73–84, Nov. 1964.

Chapter 17

Prepared Atmospheres

by F. E. Vandaveer and C. George Segeler

GENERAL TYPES AND SOURCES

Fuel gases and products of their complete or incomplete combustion are used industrially for preparing atmospheres to surround, protect, or react with a product in various processing operations.

Protective gases are used to minimize or prevent an undesirable reaction such as oxidation or decarburization. For instance, annealing, brazing, hardening, normalizing, and sintering are performed in protective atmospheres; and apples may be stored in a preserving atmosphere. **Reactive gases** chemically change a product, usually its surface, as in carburizing or nitriding steels. The addition of hydrogen to vegetable oils is another illustration. Atmospheres may be reactive in some processes and protective in others. For example, water vapor, which is undesirable when treating steel, may be used in annealing copper. **Inert gases,** another type of protective atmosphere, are used as purges to prevent fire and explosion hazards in the paint and varnish, oil, gas, and chemical industries. Canned foods may be improved and vitamin loss diminished by displacing air with inert gas between the canned product and the container top.

Atmosphere Definitions. Atmosphere types are categorized as follows:

Atmosphere	O_2, per cent	$CO + H_2$, per cent
Oxidizing[1]	Above 0.05	Below 0.05
Neutral* (or inert)	Below 0.05	Up to 4.0
Reducing	Below 0.05	Above 0.05

These definitions have limited significance since the effect of the atmosphere on the product processed is the major consideration. In this regard, it should be noted that an atmosphere may oxidize one metal, reduce another, and remain neutral toward a third. Also, an atmosphere may oxidize a metal at one temperature but reduce it at another temperature. In heat treating operations, furnaces must have well-constructed jackets and tight-fitting doors to exclude air and confine protective atmospheres.

The major sources of prepared atmospheres are natural gas, propane, and ammonia. Other sources are natural gas and steam, coke-oven gas, butane, charcoal, dissociated ammonia, and cracked methanol. Typical composition of these atmospheres as generated is given in Table 12-115, part I. Natural gas or butane will not crack without external heat in mixtures containing less than 60 per cent of the air required for perfect combustion. For lesser amounts of air, heat was applied, with the results shown. Part II of Table 12-115 is a recalculation of part I after moisture was removed to the dew point at 40 F (0.825 per cent water by volume) by a water cooler, contact scrubber, or dehydrator. Part III is another recalculation of part I, with the moisture removed to 40 F, the dew point, and all the carbon dioxide removed.

Tests made during incomplete combustion of gases[1] identified and determined small amounts of aldehydes, organic acids, alcohols, and combined nitrogen. The total maximum volume obtained of all these intermediate products was 0.01816 per cent from burning natural gas, 0.00499 per cent from coke-oven gas, and 0.11956 per cent from butane.

Characteristics of commercially prepared atmospheres are given in Table 12-116. Types of atmospheres suitable for heat treatment of various materials are given in Table 12-117.

* Gases not reactive with the product.

ATMOSPHERE GENERATORS[5]

Manufacturers of atmosphere generators have been listed by the A.G.A.[3] A proposed standard[6] for generators was being prepared in 1965.

Hydrocarbon Generators—Exothermic Base

Exothermic gas is produced in a generator (Fig. 12-189) burning natural gas, propane, coke-oven gas, or butane with less air than required for perfect combustion, to form a wet effluent gas, largely carbon dioxide and nitrogen. By reducing

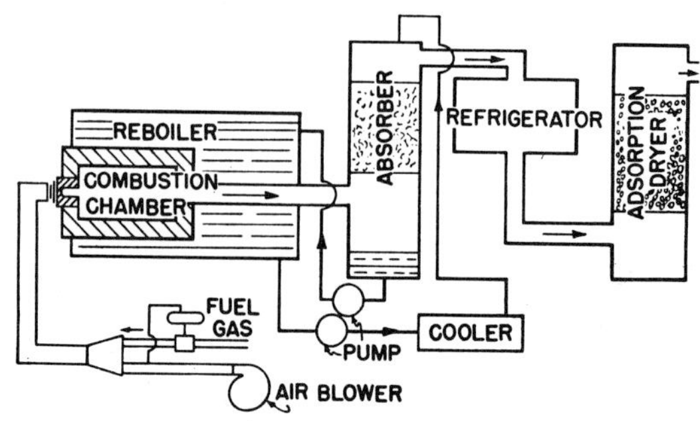

Fig. 12-189 Exothermic generator with CO_2 and moisture removal auxiliaries. (Surface Combustion Div., Midland-Ross Corp.)

the proportion of air, a richer exothermic atmosphere is produced. Exothermic atmospheres are widely applied in bright heat treatment of low-carbon steels, bright annealing of copper, bright brazing of steel where decarburization is not a factor, silver brazing of nonferrous metals, and sintering of nonferrous and low-carbon steel powders.

Exothermically produced gases contain appreciable quantities of water, which almost always must be removed. First-stage removal is by surface condensers or by direct-contact water spray. If necessary, refrigeration or adsorption dryers may be used to obtain complete removal. Characteristics of the material processed determine the extent of drying.

Sulfur, when present in hydrocarbon gases, must be removed for many atmosphere applications. Sulfur dioxide may be removed by caustic scrubbing or adsorption; hydrogen sulfide may be removed by adsorption in iron oxide, adsorption by activated carbon, or solution in organic liquids. Organic sulfur, if present in the fuel gas, may be catalytically converted to H_2S or adsorbed by activated carbon.

Removal of CO_2, if desirable, may be accomplished by (1) absorption in solutions of alkali carbonates, ethanolamines, alkali salts, and weak organic acids; (2) adsorption on special dessiccants (e.g., solid synthetic zeolites); or (3) passage thru an externally heated charcoal bed.

Maintaining an exothermic atmosphere of consistent composition presupposes relatively constant fuel gas characteristics, continuously correct air–gas ratios, adequate combustion chamber temperatures, and combustion chamber dimensions adequate for the quantities produced. Many inconsistencies are traceable to too low reaction temperatures or to such rapid gas flows that time is insufficient for reactions to be completed. Combustion chamber temperatures between 2000 and 2450 F help to ensure production of stable atmospheres. At lower temperatures, thermal decomposition with heavy sooting may occur. At higher ones, a hard carbon coke is among the products of methane decomposition. These effects are particularly noticeable with low air–to–gas ratios.

The generalized composition and properties of the gases produced in exothermic gas generators are shown in Fig. 12-190. Traces of oxygen may be eliminated by low-temperature combustion of the effluent gas on a suitable catalyst bed, provided that enough hydrogen is present or is added to complete the combustion reaction.

Hydrocarbon Generators—Endothermic Base

Endothermic gas is produced by mixing hydrocarbon gases with about 28 per cent of the air required for complete combustion and passing the mixture thru an externally heated retort at about 1800 F (Fig. 12-191). By the use of this heat, a process called cracking, a very dry gas is formed, with a composition of about 41.5 per cent N_2, 38 per cent H_2, 20 per cent CO, and 0.5 per cent CH_4.

Endothermic gas costs roughly twice as much as exothermic gas. The composition and **carbon potential** (ability to carburize steel) of endothermic gas may be easily controlled by dew point measurement. The gas is useful for annealing, normalizing, hardening, brazing, carburizing, and sintering of medium- and high-carbon steel and alloy steels.

Table 12-115 Typical Composition of Atmospheres from Different Sources[2]

Source	Percent of air needed for perfect combustion	External heat, °F	Product gas composition, volume %																				
			(I) As generated							(II) With moisture removed to 40 F dew point							(III) With moisture removed to 40 F dew point and all CO_2 removed						
			CO_2	O_2	CO	H_2	Hydrocarbons	H_2O	N_2	CO_2	H_2	CO	H_2	Hydrocarbons	H_2O	N_2	CO_2	O_2	CO	H_2	Hydrocarbons	H_2O	N_2
Natural gas	100	None	9.8	0	0	0	0	18.4	71.8	11.9	0	0	0	0	0.8	87.3	0	0	0	0	0	0.9	99.1
	95	None	9.3	0	1.0	0.9	0	18.1	70.7	11.3	0	1.2	1.1	0	.8	85.6	0	0	1.4	1.2	0	.9	96.5
	60	None	4.4	0	8.3	8.1	0.4	15.8	62.0	5.2	0	9.8	9.5	0.4	.8	73.0	0	0	10.3	10.0	0.5	.8	77.1
	40	1800	3.1	0	14.3	24.3	.1	3.6	53.8	3.2	0	14.7	25.0	.1	.8	55.4	0	0	15.2	25.8	.1	.8	57.3
	20	1800	0.1	0	21.0	40.0	0.2	0.3	34.5	0.1	0	21.0	40.0	0.2	.3	34.5	0.1	0	21.0	40.0	0.2	.3	34.5
Natural gas and steam*	0	1800	1.1	0	24.8	71.2	2.0	...	0.9	1.1	0	24.8	71.2	2.0	...	0.9	0	0	25.0	72.1	2.0	...	0.9
Coke-oven gas	100	None	9.3	0	0	0	0	20.8	69.9	11.7	0	0	0	0	.8	87.5	0	0	0	0	0	.9	99.1
	95	None	8.8	0	0.8	1.0	0	20.7	68.7	11.0	0	1.0	1.3	0	.8	85.9	0	0	1.1	1.5	0	.9	96.5
	50	None	3.4	0	10.0	13.1	0.5	18.3	53.9	4.1	0	12.2	15.9	0.5	.8	65.4	0	0	12.7	16.6	0.5	.8	68.2
	15	1800	1.2	0	21.0	47.2	2.2	...	28.4	1.2	0	21.0	47.2	2.2	...	28.4	0	0	21.3	47.8	2.2	...	28.7
Butane	100	None	11.9	0	0	0	0	15.8	72.8	13.9	0	0	0	0	.8	85.3	0.	0	0	0	0	.9	99.1
	95	None	11.0	0	1.0	0.4	0	15.1	72.5	12.9	0	1.2	0.5	0	.8	84.6	0	0	1.4	0.6	0	.9	97.1
	60	None	5.1	0	10.6	8.7	0.9	11.5	63.2	5.7	0	11.9	9.8	1.0	.8	70.8	0	0	12.6	10.4	1.1	.8	75.1
	20	1800	0	0	23.4	33.6	0.4	...	39.7	0	0	23.4	33.6	0.4	...	39.7	0	0	23.4	33.6	0.4	...	39.7
Charcoal producer	...	2300	0	0	34.7	2.5	0	...	62.8	0	0	34.7	2.5	0	...	62.8	0	0	34.7	2.5	0	...	62.8
Dissociated ammonia	...	...	0	0	0	75.0	0	0	25.0	0	0	0	75.0	0	0	25.0	0	0	0	75.0	0	0	25.0
Cracked methanol	0	1400	0.2	0	31.8	65.5	2.5	...	0	0.2	0	31.8	65.5	2.5	...	0	0.2	0	31.8	65.5	2.5	...	0

* 28.5 cu cm H_2O per cu ft of gas.

Table 12-116 Characteristics of Various Prepared Atmospheres[3,4]

Class No.	Method of Preparation	Input, Btu per cu ft of gen. atm	N_2	CO	CO_2	H_2	CH_4	Dew point, °F	Safety qualities
	Exothermic base—burn gas–air mixture (may be followed by cooling or dehydration to desired dew point)								
101	Exothermic base with lean mixture	120	86.8	1.5	10.5	1.2	...	*	Noncombustible; toxic
102	Exothermic base with rich mixture	155	71.5	10.5	5.0	12.5	0.5	*	Combustible; toxic
103	Class 101 prepared in furnace								
104	Class 102 prepared in furnace								
105	Class 101 followed by passage thru incandescent charcoal		77.8	20.1	...	2.1	...		Combustible; toxic
106	Class 102 followed by passage thru incandescent charcoal		67.3	19.3	...	12.9	0.5		Combustible; toxic
112	Class 102 plus burned mixture of Cl_2, hydrocarbon gas, and air								
113	Class 101 with sulfur removed								
114	Class 102 with sulfur removed								
116	Class 102 carrying lithium vapor (red-line cartridge)								
118	Class 102 carrying lithium vapor (blue-line cartridge)								
120	Class 102 prepared directly in furnace with lithium vapor added								
	Prepared nitrogen base—CO_2 and H_2O removed from 100 series gas								
201	Prepared N_2 base with lean mixture	135	97.1	1.7	...	1.2	...	−40	Noncombustible; inert; toxic
202	Prepared N_2 base with rich mixture	160	75.3	11.0	...	13.2	0.5	−40	Combustible; toxic
207	Class 201 plus raw hydrocarbon enriching gas								
208	Class 202 plus raw hydrocarbon fuel gas								
213	Class 201 with sulfur and odors removed								
214	Class 202 with sulfur and odors removed								
223	Class 201 plus steam with catalyst to convert CO		96.9	...	...	3.0	...	−40	Noncombustible; toxic
224	Class 202 plus steam with catalyst to convert CO		89.9	...	...	10.0	...	−40	Combustible; toxic
	Endothermic base—react gas-air mixture in catalyst-filled externally heated retort								
301	Endothermic base partially reacted followed by quick cooling to eliminate breakdown of CO to C and CO_2	190†	45.1	19.6	0.4	34.6	0.1	+50	Combustible; toxic
302	Endothermic base completely reacted and cooled as in Class 301	200†	39.8	20.7	...	38.7	0.5	0 to −5	Combustible; toxic
305	Class 301 followed by passage thru incandescent charcoal								
307	Class 301 plus raw hydrocarbon enriching gas								
308	Class 302 plus raw hydrocarbon enriching gas								
309	Class 301 plus raw hydrocarbon enriching gas and raw NH_3								
310	Class 302 plus raw hydrocarbon fuel gas and raw NH_3								
315	Class 301 carrying lithium vapor (white-line cartridge)								
323	Gas–air–steam mixture with catalyst to convert CH_4	260	5.0	21.4	3.0	65.6	5.0	*	Combustible; toxic
325	Gas–air–steam mixture with catalyst to convert CO and CO_2 removal		Balance	0.05 to 1.0	0.05 to 2.0	50.0 to 99.6	0.0 to 0.4		
	Charcoal base—pass air thru incandescent charcoal‡								
402	Charcoal base	12.5 lb charcoal	64.1	34.7	...	1.2	...	−20	Combustible; toxic
408	Class 402 plus raw hydrocarbon enriching gas								
410	Class 402 plus raw hydrocarbon enriching gas and raw NH_3								
421	Air plus NH_3 and benzol passed thru incandescent charcoal without external heating	16.0 lb charcoal	63.0	33.5	1.0	2.0	0.5	−10	Combustible; toxic

Table 12-116 (Continued)

Exothermic–endothermic base—blend combustion products of air–gas mixture with natural gas and react in catalyst-filled externally heated retort									
501	Exothermic-endothermic lean mixture	120	63.0	17.0	...	20.0	...	−70	Combustible; toxic
Ammonia base									
600	NH_3 raw								
601	NH_3 dissociated in externally heated chamber	23.5 lb NH_3	25.0	...	...	75.0	...	−60	Combustible
621	Class 601 almost completely burned followed by cooling or dehydration to desired dew point	13.7 lb NH_3	99.0	...	...	1.0	...	*	Noncombustible; inert
622	Class 601 partially burned followed by cooling or dehydration to desired dew point	14.9 lb NH_3	80.0	...	...	20.0	...	*	Combustible

* Dew point corresponds to room temperature using tap water cooling; may be reduced to +40 F by refrigeration or —50 F by adsorbent towers.

† Plus 250 Btu per cu ft for heating gas.

‡ Vertical retort internally heated, prepared gas drawn off at maximum temperature zone, undesirable constituents of green charcoal eliminated by venting small portion of the gas formed at top.

Ammonia-Base Generators

The ammonia dissociator (Fig. 12-192) carries out, in the presence of catalysts, at a temperature of 1600 to 1800 F, the endothermic reaction $2NH_3 = N_2 + 3H_2$. The dissociated gas is pure and dry, and contains less than 0.1 per cent free ammonia, which is commonly removed by passage thru an adsorbent.

Dissociated ammonia may be partially burned to produce an atmosphere ranging from 0.25 to 25 per cent H_2 with the remainder N_2. Such an atmosphere contains no free ammonia and, because of its low hydrogen content, is less hazardous. It is saturated at about 10° F higher than the cooling water, and for many applications must be dried further. Any mixture of N_2 and H_2 with less than 6 per cent H_2 will remain outside the explosive limit regardless of the amount of air added (Fig. 2-24b). This characteristic is used to advantage to provide safe atmospheres for purging and similar operations.

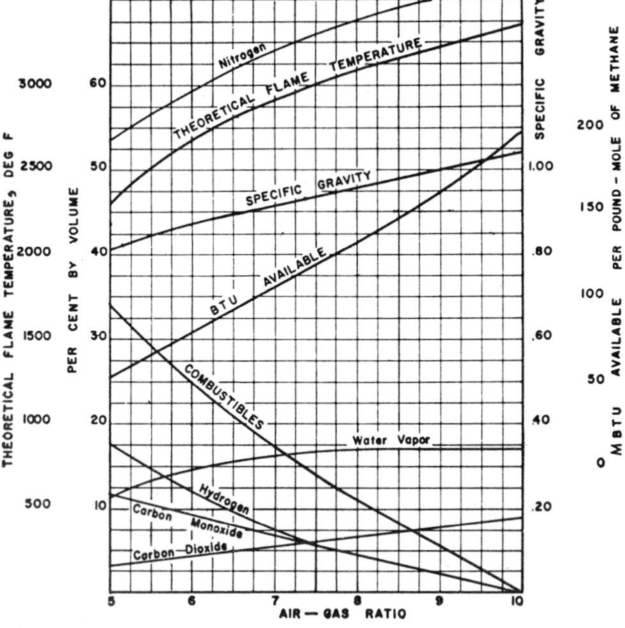

Fig. 12-190 Analyses of gases and other thermal factors for a range of exothermic air–gas ratios.[7]

The specific field of application of dissociated ammonia is the bright heat treatment of stainless steel. In many applications where pure dry hydrogen is suitable, dissociated ammonia may also be utilized.

Charcoal-Base Generators

Atmospheres may be produced by the partial combustion of charcoal in chambers (Fig. 12-193) in which temperature control is established by (1) air input, (2) external heating of the combustion chamber, and (3) incorporation of the charcoal retort within the furnace which uses the atmosphere. Carbon monoxide concentration in the generated atmosphere depends on the temperature of the carbon bed at the point from which the gases are withdrawn. Variations from 34 per cent CO and 66 per cent N_2 to 20 per cent CO_2 and 80 per cent N_2 may be obtained. The high CO concentrations are obtainable at temperatures near 1800 F, the high CO_2 near 900 F.

Water vapor present in the charcoal or in the combustion air reacts to form H_2 and CO, so that small quantities of H_2 are always present. At high generator temperatures, the water vapor concentration in charcoal is very low. Typical dew points (about −10 F) are so low that further drying is seldom necessary. By incorporating the charcoal retort within the furnace, the atmosphere leaves the generator at the furnace temperature, at which its composition is close to the required equilibrium.

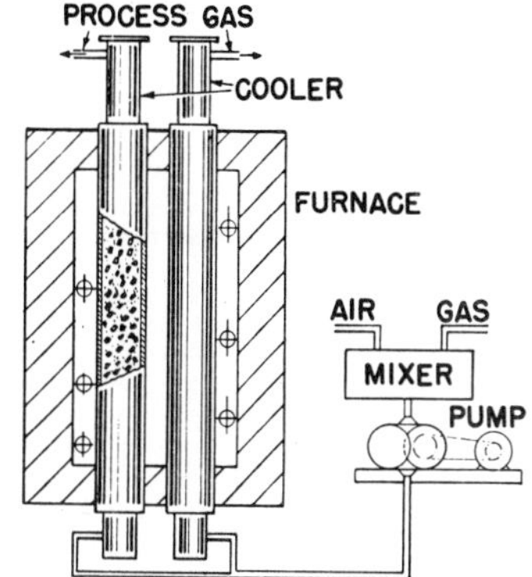

Fig. 12-191 Schematic flow diagram of endothermic generator. (Surface Combustion Div., Midland-Ross Corp.)

Table 12-117 Prepared Atmospheres Suitable

Material	Process	Desired surface	Temp range, °F	Cycle*
	Annealing and Normalizing			
Steel				
0.20 C	Anneal	Bright	1200–1350	Long
0.20–0.60 C	Anneal, no decarb.	Bright	1200–1450	Long
0.20–0.60 C	Anneal, no decarb.	Clean	1200–1450	Long
0.20–0.60 C	Anneal, no decarb.	Bright		Under ½ hr
To 0.60 C	Normalizing	Bright		Long
0.60 C and up	Anneal, no decarb.	Bright	1200–1450	Long
Alloy 0.20 C and up	Anneal, no decarb.	Bright or clean	1300–1600	Long
0.60 C and up, and alloy	Normalizing			
High-speed, incl. Mo	Anneal, no decarb.	Bright or clean	1400–1600	Long
Stainless steel Ni–Cr	Anneal	Bright	1800–2100	Short or long
High-Si steel	Anneal	Clean	1900–2000	Long
Ferrous metals	Patenting			
Si–Cu alloys	Anneal	Bright		
Copper	Anneal†	Bright	400–1200	Short or long
Various brasses	Anneal†	Clean	800–1350	Short or long
Nickel	Anneal	Bright		
Cu–Ni alloys	Anneal†	Bright	800–1400	Short or long
Monel	Anneal	Discolored, no scale	1600	3–4 min at temp
Monel	Anneal	Bright	1600	3–4 min at temp
	Hardening and Tempering			
Steel				
0.20–0.60 C	Hardening, no decarb.	Bright or clean	1400–1600	Under ½ hr
0.20–0.60 C	Hardening, no decarb.	Clean	1400–1600	Under ½ hr
0.60 C and up	Hardening, no decarb.	Bright or clean	1400–1800	Short
0.60 C and up	Hardening, no decarb.	Clean	1400–1800	
Alloy 0.20 C and up	Hardening, no decarb.	Bright or clean	1400–1800	Short
Alloy 0.20 C and up	Hardening, no decarb.	Clean	1400–1800	Short
Air hardening	Hardening	Bright or clean		
High-speed, incl. Mo	Hardening, no decarb.	Bright or clean	1800–2400	Short
Tool, high C, high Cr	Hardening, no decarb.	Bright		
Tool, high C, high Cr	Hardening	Clean		
Stainless steel	Temper or draw	Bright		
All other ferrous metals	Temper or draw	Bright or clean	400–1200	Short
Ferrous metals	Gas quenching	Bright		
	Forging			
Steel	Forging	Clean	1900–2400	Short
	Steel Surface Chemistry			
Steel				
To 0.20 C	Carburizing	High C	1500–1900	Long
0.20–0.60 C	Carburizing	High C	1500–1900	
Alloy 0.20 C and up	Carburizing	High C	1500–1900	
	Cyaniding	High C, high N_2		
	Nitriding	High N_2		
	Skin recovery	Normal C		
	Gas pickling	Virgin metal		
	Automatic Brazing and Soldering			
Steel				
To 0.20 C	Cu brazing	Bright	2050	Short
0.20 C and up	Cu brazing, no decarb.	Bright	2050	Short
Alloy 0.20 C and up	Cu brazing, no decarb.	Bright	2050	Short
High C, high Cr	Cu brazing	Bright	2050	Short
Stainless steel	Cu brazing	Bright	2050	Short
Copper or brass	Ph–Cu brazing	Bright or clean	1500–1600	Short
Copper or brass	Silver soldering	Bright or clean	1500–1600	Short
	Powder Metallurgy			
Metal powders	Sintering and other processes	Slightly reducing		
		Highly reducing		
Powders				
Up to 0.40 C	Sintering			
Over 0.40 C	Sintering			
Steel powder				
To 0.20 C	Carburizing			
0.20–0.60 C	Carburizing			
Steel powder alloy 0.20 C and up	Carburizing			
	Miscellaneous Uses			
Metal melting	Surface protection of bath			
Metal melting	Magnesium			
Tinning	Surface protection of bath			
Paint manufacturing	Inert gas			
	Varnish cooking			
Chemical plants	Inert gas			
	Agitation during cooking			
	Blanketing combustible liquids and powders			
	Reactivating catalysts			
	Purging stills, holders, and pipelines			
Food processing	Protecting food product reactions, purging containers, etc.			
General	Inert atmosphere for fire extinguishing			
	Replacement of bottled nitrogen			
	Storage of materials in an oxygen-free atmosphere			
Fuel use	Glass working			
	Peak-shaving device			

X—Most commonly used; O—can be used; . . .—not recommended.
* Short cycles are less than 2 hr, long ones over 2 hr.
† 102, 114, 116, 301, 325, and 501 can be used only if metal is fully deoxidized.

for Heat Treatment of Various Materials[3,4]

Class No. of suitable atmospheres

101	102	103	104	105	106	112	113	114	116	118	201	202	207	208	213	214	223	301	302	307	308	309	310	315	323	325	402	408	410	421	501	600	601	621	622
...	O	...	...	...	...	...	...	...	O	...	X	...	...	...	...	...	O	O	...	...	...	...	...	...	...	O	...	...	...	...	O	...	...	...	...
...	...	...	...	...	...	...	...	...	...	...	O	O	O	...	...	...	...	X	O	O	O	...	...	...	...	O	...	...	...	...	O	...	O	O	O
...	...	...	...	...	...	...	...	...	...	...	O	O	X	...	...	...	...	X	O	O	O	...	...	...	...	O	...	...	...	...	O	...	O	O	O
...	...	...	...	...	...	...	...	...	...	...	O	...	...	O	...	...	...	X	O	O	O	...	...	...	...	O	...	...	...	...	O	...	O	O	O
...	...	...	...	...	...	...	...	...	...	...	X	O	X	...	...	...	...	...	O	...	...	...	...	...	...	O	...	...	...	...	O	...	O	O	O
...	...	...	...	...	...	...	...	...	...	...	O	O	X	O	...	...	...	X	O	O	O	...	...	...	...	O	...	...	...	...	O	...	O	O	O
...	...	...	...	...	...	...	...	...	...	...	O	...	X	O	...	...	...	X	O	O	O	...	...	...	...	O	...	...	...	...	O	...	O	O	O
...	...	...	...	...	...	...	...	...	O	...	O	O	X	...	...	...	...	X	O	O	O	...	...	...	...	O	...	...	...	O	O	...	O	O	O
...	...	...	...	...	...	...	...	...	...	...	O	...	...	...	...	...	...	X	O	...	O	...	...	...	...	O	X	...	...	O	O	...	O	O	O
...	...	...	...	...	...	...	...	...	...	...	...	...	...	...	...	...	...	...	...	...	...	...	...	...	...	O	...	...	...	...	...	...	X	O	O
...	X	...	...	...	...	...	...	...	...	...	X	...	...	...	...	...	O	...	...	...	...	...	...	...	...	O	...	...	...	...	...	...	O	O	...
...	O	X	...	...	...	...	...	...	...	...	O	...	...	...	...	...	O	O	...	...	...	...	...	...	...	...	...	...	...	...	O	...	...	...	...
...	...	...	...	...	...	...	O	O	...	...	O	...	...	...	...	...	...	...	...	...	...	...	...	...	...	O	...	...	...	...	...	...	O	O	X
X	O	O	...	...	...	...	O	O	O	...	O	...	...	...	O	...	O	O	...	...	...	...	...	...	...	O	...	...	...	...	O	...	...	O	...
X	O	O	...	...	...	...	O	O	O	...	O	...	...	...	...	...	O	O	...	...	...	...	...	...	...	O	...	...	...	...	O	...	...	O	...
X	O	O	...	...	...	...	...	...	...	...	O	...	...	...	...	...	O	O	...	...	...	...	...	...	...	O	...	...	...	...	O	...	O	O	O
X	O	O	...	...	...	...	O	O	O	...	O	...	...	...	...	...	O	O	...	...	...	...	...	...	...	O	...	...	...	...	O	...	O	O	O
X	O	X	...	...	...	...	...	...	...	...	O	...	...	...	...	...	O	O	...	...	...	...	...	...	...	O	...	...	...	...	O	...	O	O	O
X	O	O	...	...	...	...	...	...	...	...	O	...	...	...	...	...	O	O	...	...	...	...	...	...	...	O	...	...	...	...	O	...	O	O	O
...	...	...	...	...	...	...	...	...	...	...	O	O	O	...	...	...	...	O	X	O	O	...	...	...	...	O	O	...	...	O	O	...	...	O	O
...	...	...	...	...	...	...	...	...	...	...	O	O	O	...	...	...	...	O	X	O	O	...	...	...	...	O	O	...	...	O	O	...	...	O	O
...	...	...	...	...	...	...	...	...	...	...	O	O	O	...	...	...	...	...	X	...	O	...	...	...	...	O	O	...	...	O	O	...	...	O	O
...	...	...	...	...	...	...	...	...	...	...	O	O	O	...	...	...	...	...	X	...	O	...	...	...	...	O	O	...	...	O	O	...	...	O	O
...	...	...	...	...	...	...	...	...	...	...	O	O	O	...	...	...	...	...	X	...	O	...	...	...	...	O	O	...	...	O	O	...	...	O	O
...	...	...	...	...	...	...	...	...	...	...	O	...	O	...	...	...	...	...	X	...	O	...	...	...	...	O	O	...	...	O	O	...	...	O	O
...	...	...	...	...	...	...	...	...	...	...	O	O	O	...	...	...	...	...	X	...	...	...	...	...	...	O	O	...	...	O	O	...	...	O	O
...	...	...	...	...	...	...	...	...	...	...	...	...	O	...	...	...	...	...	X	...	...	...	...	...	...	...	X	...	...	O	O	...	...	...	...
...	...	...	...	...	...	...	...	...	...	...	...	...	...	...	...	...	...	...	X	...	O	...	...	...	...	O	...	...	...	...	O	...	O	O	O
...	...	...	...	...	...	...	...	...	...	...	O	...	O	...	...	...	...	...	X	...	O	...	...	...	...	O	O	...	...	O	O	...	O	O	O
...	...	...	...	...	...	...	...	...	...	...	...	...	...	...	...	...	...	...	...	...	...	...	...	...	...	O	...	...	...	...	...	...	X	O	O
...	X	...	...	...	...	...	...	...	...	...	O	...	...	...	...	...	...	O	...	...	...	...	...	...	...	...	...	...	...	...	...	...	...	...	...
...	X	...	...	...	...	...	...	...	...	...	O	...	...	...	...	...	...	O	O	...	...	...	...	...	...	...	...	...	...	...	O	...	...	...	...
...	O	...	...	...	...	...	...	...	...	...	X	...	...	...	...	...	...	O	O	...	...	...	...	...	...	...	...	...	...	...	O	...	...	...	...
...	...	...	...	...	...	...	...	...	...	...	...	...	O	...	...	...	...	...	O	O	X	...	...	...	...	...	...	O	...	...	O	...	...	...	...
...	...	...	...	...	...	...	...	...	...	...	...	...	O	...	...	...	...	...	O	...	X	...	...	...	...	...	...	O	...	...	O	...	...	...	...
...	...	...	...	...	...	...	...	...	...	...	...	...	O	...	...	...	...	...	O	O	X	...	...	...	...	...	...	O	...	...	O	...	...	...	...
...	...	...	...	...	...	...	...	...	...	...	...	...	...	...	...	...	...	...	...	...	...	O	X	...	...	...	...	...	O	...	...	...	...	...	...
...	...	...	...	...	...	...	...	...	...	...	...	...	...	...	...	...	...	...	...	...	...	...	...	...	...	...	...	...	...	...	...	X	...	...	...
...	...	...	...	...	...	...	...	...	...	...	...	...	O	...	...	...	...	O	X	...	X	...	...	...	...	...	...	...	...	...	O	...	...	...	...
...	...	...	...	...	...	X	...	...	...	...	...	...	...	...	...	...	...	...	...	...	...	...	...	...	...	...	...	...	...	...	...	...	...	...	...
...	X	...	...	...	...	...	...	...	...	...	...	...	...	...	...	...	...	O	O	...	...	...	...	...	...	O	...	...	...	O	O	...	O	O	O
...	...	...	...	...	...	...	...	...	...	...	...	...	O	...	...	...	...	O	X	...	O	...	...	...	...	...	...	...	...	O	O	...	...	...	...
...	...	...	...	...	...	...	...	...	...	...	...	...	O	...	...	...	...	O	X	...	O	...	...	...	...	...	...	...	...	O	O	...	...	...	...
...	...	...	...	...	...	...	...	...	...	...	...	...	...	...	...	...	...	...	O	...	...	...	...	...	...	O	...	...	...	...	O	...	X	O	O
...	...	...	...	...	...	...	...	...	...	...	...	...	...	...	...	...	...	...	...	...	...	...	...	...	...	O	...	...	...	...	...	...	X	O	O
X	O	...	...	...	...	...	...	...	...	...	...	...	...	...	O	...	...	O	...	...	...	...	...	...	...	O	...	...	...	...	O	...	...	O	...
X	X	...	...	...	...	...	...	...	...	...	...	...	...	...	O	...	...	O	...	...	...	...	...	...	...	O	...	...	...	...	O	...	...	O	...
...	X	...	...	...	...	...	...	...	...	...	...	...	...	...	...	...	...	...	...	...	...	...	...	...	...	...	...	...	...	...	...	...	...	O	O
...	O	...	...	...	...	...	...	...	...	...	...	O	...	...	...	...	...	O	X	...	...	...	...	...	...	O	O	...	...	...	O	...	X	O	O
...	...	...	...	...	...	...	...	...	...	...	...	O	X	...	...	...	...	O	X	...	...	...	...	...	...	O	O	...	...	...	O	...	O	O	O
...	...	...	...	...	...	...	...	...	...	...	...	...	...	...	...	...	...	...	O	...	O	...	...	...	...	O	...	...	...	...	O	...	X	O	O
...	...	...	...	...	...	...	...	...	...	...	...	...	O	...	...	...	...	O	O	O	X	...	...	O	...	...	...	O	...	...	O	...	...	...	...
...	...	...	...	...	...	...	...	...	...	...	...	...	O	...	...	...	...	...	...	...	X	...	...	O	...	...	...	O	...	...	O	...	...	...	...
...	...	...	...	...	...	...	...	...	...	...	...	...	O	...	...	...	...	...	...	...	X	...	...	O	...	...	...	O	...	...	O	...	...	...	...
...	O	X	O	...	...	...	...	...	...	...	O	...	...	...	...	...	...	...	...	...	...	...	...	...	...	O	...	...	...	...	...	...	...	O	...
...	...	...	...	...	...	...	...	...	...	...	X	...	...	...	...	...	O	...	...	...	...	...	...	...	...	...	...	...	...	...	...	...	...	...	...
O	...	...	...	...	...	...	...	...	O	...	O	...	...	...	...	...	...	...	...	...	...	...	...	...	...	...	...	...	...	...	...	...	...	...	...
X	...	...	...	...	...	...	...	...	O	...	X	...	...	...	O	...	...	...	...	...	...	...	...	...	...	...	...	...	...	...	...	...	...	O	...
X	...	...	...	...	...	...	...	...	...	...	...	...	...	...	...	...	...	...	...	...	...	...	...	...	...	...	...	...	...	...	...	...	...	...	...
X	...	...	...	...	...	...	...	...	O	...	X	...	...	...	O	...	...	...	...	...	...	...	...	...	...	...	...	...	...	...	...	...	...	O	...
X	...	...	...	...	...	...	...	...	...	...	...	...	...	...	...	...	...	...	...	...	...	...	...	...	...	...	...	...	...	...	...	...	...	...	...
X	...	...	...	...	...	...	...	...	...	...	...	...	...	...	...	...	...	...	...	...	...	...	...	...	...	...	...	...	...	...	...	...	...	...	...
...	...	...	...	...	...	...	...	...	...	...	X	...	...	...	...	...	...	...	...	...	...	...	...	...	...	...	...	...	...	...	...	...	...	...	...
X	...	...	...	...	...	...	...	...	...	...	...	...	...	...	...	...	...	...	...	...	...	...	...	...	...	...	...	...	...	...	...	...	...	...	...
...	...	...	...	...	...	...	...	...	...	...	...	...	...	...	X	X	...	...	...	...	...	...	...	...	...	...	...	...	...	...	...	...	...	...	...
...	...	...	...	...	...	...	...	...	...	...	O	...	...	...	...	...	...	...	...	...	...	...	...	...	...	...	...	...	...	...	...	...	...	X	...
...	...	...	...	...	...	...	...	...	...	...	X	...	...	...	O	...	...	...	...	...	...	...	...	...	...	...	...	...	...	...	...	...	...	O	...
X	...	...	...	...	...	...	...	...	...	...	...	...	...	...	...	...	...	...	...	...	...	...	...	...	...	...	...	...	...	...	...	...	...	...	...
...	...	...	...	...	...	...	...	...	...	...	...	...	...	...	...	...	...	...	...	...	...	...	...	X	...	...	...	...	...	...	...	...	...	...	...
...	...	...	...	...	...	...	...	...	...	...	...	...	...	...	...	...	...	...	...	...	...	...	...	X	...	...	...	...	...	...	...	...	...	...	...

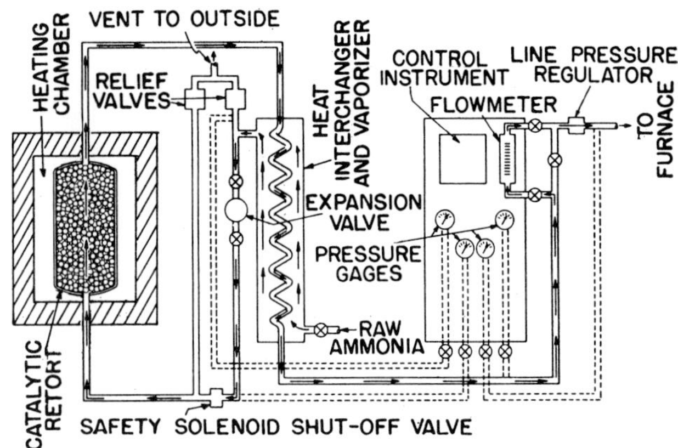

Fig. 12-192 Schematic flow diagram of ammonia dissociator. (Lindberg Engineering Co.)

Charcoal generators externally heated with a fuel gas may be supplied with the dried combustion products of the burned fuel gas, largely CO_2, CO, and N_2, with small quantities of water vapor. With the charcoal held at a sufficiently high temperature, CO_2 is reduced to CO and water to H_2. A typical atmosphere thus produced is very dry, containing about 20 per cent CO, 2 per cent H_2, and the balance N_2.

All charcoal-gas generators are inherently of the batch type since fresh charcoal must be supplied at intervals. Production of a satisfactory atmosphere may be difficult because of charcoal impurities. To obtain a consistent atmosphere composition, both the flow rate thru, and the temperature of, the charcoal bed must be controlled.

Charcoal-gas generators have largely been supplanted by endothermic gas generators.

Steam Atmospheres[9]

Furnaces are available in which air (up to temperatures of 650 to 750 F) is used, followed by a steam atmosphere. After heat treatment is complete, air cooling or quenching in soluble oil occurs; with some nonferrous materials a water quench may be used. The use of steam atmospheres is applicable in processes in which steel parts are to be blued or an oxide coating (controlled as to depth) is permissible or desired.

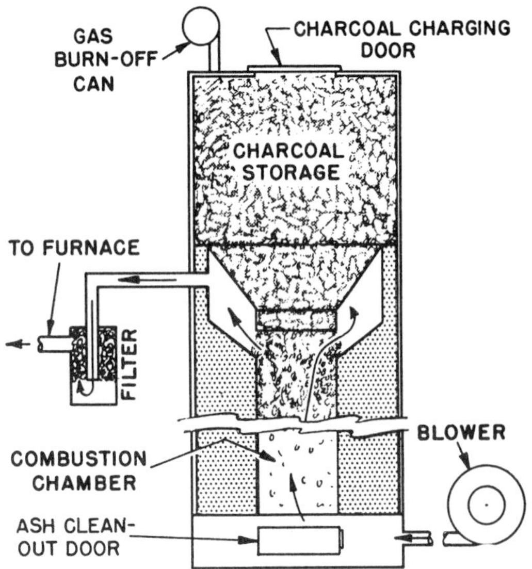

Fig. 12-193 Charcoal-gas generator. A flowmeter and valve are usually installed between the blower and the generator. (Lindberg Engineering Co.)

A wider range of applications appears possible with nonferrous alloys, including aluminum.

Protective Equipment for Generators

Protective devices shut down generators in the event of utility (gas, air, power) or mechanical failure; manual resetting is required. Abnormally low or high gas pressures cause gas supply shutdown. Low gas pressure also shuts down the reaction air supply. A vent is provided to dispose of unwanted generated atmosphere, which may be burned (e.g., by catalytic combustion) before release. Other means of protection include automatic fire checks, visible and/or audible alarms, temperature control of reactant flows (endothermic generators), and gas-analyzing devices.

Effect of Atmospheres—Electrically Heated Furnaces

The effect of atmospheres on various types of electric heating elements needs consideration. Table 12-118 provides guide lines; note that element temperatures are higher than furnace control temperatures. The reactive atmospheres require lower power input density when elements are operating near their maximum temperature levels.

If contamination (particularly from zinc, lead, and sulfur)* is liable to occur, resistors lower in nickel than the normal 80 nickel–20 chromium should be used. In air furnaces, this type of alloy should be replaced by iron-chromium–aluminum alloys when the temperature range is from 2100 to 2350 F. Above 2350 F, silicon carbide is normally used, preferably in air furnaces or in slightly reducing atmospheres.

Platinum resistors have been used in smaller furnaces. Platinum should not be exposed to reducing atmospheres.

APPLICATIONS

Tables 12-116 and 12-117 show typical analyses, applications, and surfaces achieved by various commercial atmospheres.

Equilibrium conditions are not normally achieved in special atmosphere furnaces. The equilibrium constants for five common gas reactions occurring in furnaces, given in Table 12-119, indicate the extent of these reactions at various temperatures.

Steel

Heat treatment of steel may subject its surface to oxidation or decarburization. Oxidation produces an undesirable surface finish which must be cleaned; decarburization produces surface weakness.

To prevent **oxidation,** oxidizing gases, e.g., O_2, CO_2, and H_2O, must be removed as far as practicable and replaced with neutral or reducing gases. Reduction is promoted in the presence of H_2 and CO. Nitrogen is relatively inactive in

* Zinc from zinc stearate used as lubricant; lead from work such as sintered powder compacts; sulfur from cutting oils or special atmospheres.

Table 12-118 Comparative Life of Various Heating Elements in Different Furnace Atmospheres*,[8]

(temperatures are element temperatures, not furnace temperatures)

Element material	Oxidizing air	Reducing-type H_2 or 501	Reducing-type 102 or 202	Reducing-type 301 or 402	Carburizing-type 307 or 309	Reducing or oxidizing with sulfur	Reducing with lead or zinc	Vacuum
Nickel–Chromium and Nickel–Chromium–Iron Alloys								
80 Ni–20 Cr	Good to 2100 F	Good to 2150 F	Fair to 2100 F	Fair to 1850 F	NR†	NR	NR	Good to 2100 F
60 Ni–16 Cr–24 Fe	Good to 1850 F	Good to 1850 F	Good to fair to 1850 F	Fair to poor to 1700 F	NR	NR	NR	...
35 Ni–20 Cr–45 Fe	Good to 1700 F	Good to 1700 F	Good to fair to 1700 F	Fair to poor to 1600 F	NR	Fair to 1700 F	Fair to 1700 F	...
Iron-Chromium–Aluminum Alloys‡								
Fe–23 Cr–4.5 Al–1 Co	Good to 2100 F	Fair to poor to 2100 F§	NR	NR	NR	Fair in oxidizing atmosphere	NR	...
37 Cr–7.5 Al–remainder Fe	Good to 2400 F	Fair to poor to 2350 F§	NR	NR	NR	Fair in oxidizing atmosphere	NR	...
Pure Metals								
Molybdenum	NR**	Good to 3000 F	NR	NR	NR	NR	NR	Good to 3000 F
Platinum	Good to 2550 F	NR	NR	NR	NR	NR	NR	...
Tantalum	NR	NR	NR	NR	NR	NR	NR	Good to 4500 F
Tungsten	NR	Good to 4500 F††	NR	NR	NR	NR	NR	Good to 3000 F
Nonmetallic Heating Element Materials								
Silicon carbide	Good to 2900 F	Fair to poor to 2200 F	Fair to 2500 F	Fair to 2500 F	NR	Good to 2500 F	Good to 2500 F	NR
Graphite	NR	Fair to 4500 F	NR	Fair to 4500 F	Fair to poor to 4500 F	Fair to 4500 F in reducing	Fair to 4500 F	...
Cermets	Good to 2900 F	...	...	...	...	...	...	...

NR—Not recommended.
* Inert atmosphere of argon or helium can be used on all materials. Nitrogen recommended only for the nickel-chromium group.
† Special 80 Ni–20 Cr resistors with ceramic protective coatings designated for low voltage (8 to 16 volts) can be used successfully.
‡ Cobalt additions of 0.50 to 2.0% are used by several manufacturers.
§ Must be oxidized first.
** Special molybdenum heating elements with $MoSi_2$ coating can be used in oxidizing atmospheres.
†† Good with pure hydrogen.

Table 12-119 Equilibrium Constants, K_p, for Selected Reactions

(see Table 3-36 for data on 14 reactions; Table 3-42 gives an application of equilibria data)

Temperature °C	Temperature °F	$\frac{(H_2)^*}{(H_2O)}$	$\frac{(CO)}{(CO_2)}$	$\frac{(H_2)(CO_2)}{(H_2O)(CO)}$	$\frac{(CO)^2}{(CO_2)}$	$\frac{(H_2)^2}{(CH_4)}$
400	752	9.12†	0.74†	12.3	9×10^{-5}	5.66×10^{-2}
450	842	6.38†	.86†	7.38	7.3×10^{-4}	0.164
500	932	4.68†	0.96†	4.88	4.7×10^{-3}	.422
550	1022	3.53†	1.03†	3.45	0.023	0.997
600	1112	2.99	1.17	2.55	.096	2.09
650	1202	2.65	1.35	1.96	0.343	3.92
700	1292	2.38	1.53	1.56	1.06	7.16
750	1382	2.17	1.72	1.27	2.96	12.3
800	1472	2.00	1.90	1.05	7.48	20.1
850	1562	1.84	2.07	0.891	17.46	31.8
900	1652	1.72	2.24	.765	37.76	48.3
950	1742	1.61	2.41	.668	76.70	71.0
1000	1832	1.51	2.57	.589	146.5	102.4
1050	1922	1.44	2.72	.527	264.0	141.2
1100	2012	1.37	2.88	.474	463.4	192.0
1150	2102	1.31	3.03	.433	767.4	256.0
1200	2192	1.26	3.21	.395	1244	335.0
1250	2282	1.22	3.36	.363	1945	431.5
1300	2372	1.18	3.49	0.339	2951	547.0

* (H_2), (H_2O), etc., are the partial pressures.
† Stable oxide at this temperature is Fe_3O_4 not FeO; the equilibrium constants so marked are not so accurate as the others.

Table 12-120 Carbon Potential of High-Purity Hydrogen[16]

(impurities: 0.02% CO, 0.01% CH_4, and 0.002% CO_2)

Water volume, %	Dew point, °F	Carbon potential, %
0.0003	−100	0.13
.0015	−80	.09
.0056	−60	.045
.01	−50	.03
.02	−40	.02
0.056	−20	0.009

molecular form and may be considered as having no effect on oxidation or reduction.

To prevent **decarburization,** decarburizing gases, e.g., O_2, CO_2, and H_2O, must be removed as far as practicable and a suitable carbon potential maintained in the atmosphere to nullify the tendency of carbon to leave steel. This potential is maintained by control of the CO and CH_4 content in the generated atmosphere or by additions of CH_4 or higher hydrocarbons to it. Hydrogen may decarburize under some conditions and both H_2 and N_2 will not prevent decarburization. Carbon monoxide, CH_4, and higher hydrocarbons have a carburizing effect.

With steel of low carbon content and at relatively low temperatures, decarburization is seldom a problem; as bright a finish as desired can readily be attained. As the carbon content of steel goes up, less attention is paid to brightness and more to surface hardness. Freedom from decarburization is more desirable and more difficult to attain with the higher carbon content of the steel.

Carbon monoxide is an especially useful atmosphere gas because it is both reducing and carburizing. Unfortunately, it is difficult to generate free from other gases such as CO_2, H_2, H_2O, and N_2.

Oxidation–reduction and carburizing–decarburizing reactions are both reversible. Reaction rates vary with temperature, composition of atmosphere and of steel, presence of contaminants in atmosphere or steel, composition of furnace walls, and catalytic effect of furnace walls or steel surface processed.

The dew point–temperature relationship vs. the carbon potential for high-purity hydrogen is given in Table 12-120.

One important operation is **bright annealing** of low-carbon steels at temperatures of 1200 to 1250 F. This process can be carried on in an atmosphere of partially burned gas with a CO_2-to-CO ratio of 0.5 or less, provided that little moisture is present. At these temperatures and with this low carbon content, carburizing or decarburizing effects are negligible. Even effects of sulfur in relatively large concentrations may be neglected. This atmosphere, however, tends to deposit carbon and is somewhat oxidizing to steel at temperatures below 700 F. Thus, where extremely clean metal surfaces are required, an H_2–N_2 atmosphere containing no CO_2 or H_2O is employed.

Bright normalizing of low-carbon steels of temperatures up to 1750 F requires an initially dry gas that is low in, or free of, CO_2.

Normalizing of high-carbon steels at temperatures above 1340 F must be carried on in atmospheres of a very low dew point and almost free of H_2. Traces of CH_4 help to establish nondecarburizing conditions, which are difficult to obtain with control of the CO_2-to-CO ratio only.

Hardening medium- or high-carbon steels without decarburizing requires a dry, CO_2-free atmosphere of N_2, CO, and H_2, or N_2 and CO. Small additions of CH_4 may be necessary to compensate for the decarburizing tendencies of the other gases. Hardening without oxidation requires delivery of the work to the quench tank while surrounded by the atmosphere.

Gas Carburizing. Carburizing steel parts to provide hard surfaces while leaving interiors tough and ductile has long been practiced. The process involves both carbon absorption and subsequent diffusion. Although the following discussion deals with gas carburizing, liquid salt bath and pack (solid) carburizing are also practiced.

To provide control of carbon potential in gas carburizing, combinations of a carrier gas (generally an endothermic generator gas) and propane or natural gas (as principal carbon sources) are used. Absorption speeds are influenced by temperature and carbon potential. Temperatures used are 1675 F or higher, with excessive grain growth, effect on furnace fixtures, and parts distortion as limiting factors. Development work at temperatures up to 2000 F has been completed.[10] Carburizing can be controlled by monitoring the dew point of the furnace atmosphere.

Uniform carburizing is maintained by circulating the hydrocarbon atmosphere. The small amounts of carbon required are easily provided by relatively small flows of gas. Forced circulation also helps to prevent excess soot deposit. The gas inlet should be located so as to prevent overrich local gas concentration with resultant excess sooting.[11]

Alloy Steels. Effects of alloy additions on oxidation and decarburization tendencies of steel underlie the selection of suitable protective atmospheres. Additions of nickel and chromium individually or together improve resistance to oxidation. Silicon alone has little effect but in the presence of nickel and chromium it reduces the extent of scaling considerably. Effects of molybdenum, manganese, and cobalt, in concentrations generally used and in the presence of other strong antioxidants such as chromium, are very small. At temperatures up to 1650 F, vanadium increases resistance to scaling except in steels of high nickel content.

Nickel–chromium alloys are particularly vulnerable to attack by gases containing sulfur. Destructiveness is mainly intergranular in nature and leads to embrittlement. Steels containing only nickel as an alloying element may be bright annealed in a desulfurized nonoxidizing atmosphere.

Direct-fired scale-free heating became practical in 1962 with the introduction of firing techniques using "recuperative" atmosphere burners[12,13] for temperatures up to 2375 F. Completely scale-free heat treatment of certain stainless steels[14] requires the use of tightly constructed furnaces supplied with dried atmospheres of hydrogen or dissociated ammonia.

For **precipitation hardening,** various carefully selected atmospheres are required. Air is satisfactory for some (e.g., PH 17-7 and PH 15-7 molybdenum); others may be handled in dry hydrogen,[15,16] etc. In some instances, vacuum furnaces are needed to ensure completely scale-free results.

Copper and Its Alloys

Annealing of copper for reworking and for control of final temper is an important operation. Temperatures range from 700 to 1200 F. Heating and cooling rates are relatively unimportant since copper contains no hardening agent.

Much industrial copper is impure—a mixture of copper and cuprous oxide. Hydrogen may deoxidize the surface of such a metal and may diffuse into its interior to form water vapor at internal pressures sufficient to rupture the metal and cause what is termed hydrogen embrittlement.

Deoxidation by CO is controlled by the rate of diffusion of cuprous oxide to the metal surface and is independent of the CO concentration in the atmosphere. The deoxidation rate by H_2 is controlled by its rate of diffusion into the metal and is partially dependent on the H_2 concentration in the atmosphere.

Free oxygen will oxidize copper and is therefore an undesirable constituent of a protective atmosphere. Nitrogen, inert to copper, is a satisfactory constituent. Carbon dioxide does not oxidize copper at temperatures up to 1100 F. Hydrogen sulfide attacks copper at elevated temperatures to form a black film of copper sulfide removable only by treatment in a dilute alkali cyanide solution. Sulfur dioxide is practically inert at ordinary annealing temperatures but in the presence of H_2 may form H_2S. Carbon monoxide and hydrogen are both effective reducing gases but must be limited in quantity because of the possibility of embrittlement.

Sulfur present in minute quantities in the original fuel is usually oxidized to SO_2. Large concentrations of SO_2 may be absorbed by lime or adsorbed on desiccants. Sulfur present as H_2S may be removed by iron oxide absorption or activated carbon adsorption. A frequent source of sulfur is the lubricant present on copper. Counterflow of annealing metal and atmosphere may minimize the problem in continuous furnaces. Complete degreasing of the copper is often advocated as the only real solution.

Steam is frequently used as a protective atmosphere. The major objection to it or to water vapor in other gases is the water staining that frequently occurs at the cooling end of the furnace.

In commercial practice an atmosphere containing 0.5 to 1.0 per cent H_2, 1 to 2 per cent CO, with the remainder CO_2 and N_2, cooled to reduce moisture content so that water will not condense on the cooling copper, makes a satisfactory bright annealing atmosphere. Carbon monoxide and hydrogen are active gases that reduce any undesirable surface oxides but are present in insufficient quantity to promote hydrogen embrittlement. They provide the necessary blanket of inert gases.

Copper alloys containing nickel or as much as 15 per cent zinc or tin behave quite similarly to copper. Bright annealing can be carried on in the same atmosphere.

Nickel and Its Alloys

Although nickel is not usually heat treated, effects of atmospheres on nickel or its alloys are of general interest because of the many uses nickel alloys have in muffles, containers, wire belts, and heating elements in the heat treating industry.

Nickel oxidizes in air at a much slower rate than copper or iron. Its oxidation rate is increased by the presence of small quantities of iron, manganese or silicon. Nickel–iron alloys composed of 25 per cent nickel and 75 per cent iron are slightly more resistant to oxidation than pure nickel or iron. Nickel–chromium alloys of up to 30 per cent chromium have a very high resistance to oxidation in air.

Nickel and its alloys are subject to intergranular disintegration and embrittlement under alternately oxidizing and reducing atmospheres. In steam, at temperatures above 840 F, there is a tendency toward embrittlement. Both SO_2 and H_2S attack nickel at elevated temperatures, particularly in reducing atmospheres, and may cause embrittlement.

Nickel and nickel–copper alloys may be successfully bright annealed in sulfur-free atmospheres containing N_2, H_2, CO_2, and CO, dried to a low dew point. Nickel–zinc alloys are difficult to bright anneal.

Aluminum and Magnesium

Heat treatment of aluminum and magnesium has long been carried on in an atmosphere of air. Aluminum can be heat treated in atmospheres containing N_2, O_2, CO_2, and small quantities of water vapor. Both aluminum and magnesium can be successfully heat treated in direct-fired furnaces using only combustion products and dilution air as the atmosphere.

Precious Metals

Silver and its alloys are highly sensitive to staining by gases containing sulfur. Steam is frequently used as a silver annealing atmosphere, as are dissociated burned ammonia and almost completely burned city or natural gas freed of sulfur. At temperatures above 1380 F, water vapor may dissociate and the H_2 and O_2 may diffuse into the metal, recombining internally to form water again and thereby embrittling the metal. Since the solubility of O_2 in silver and the diffusion rate of O_2 thru hot silver are high, silver will be damaged if heat treated first under oxidizing conditions, then in hydrogen.

Gold of high purity apparently requires no annealing, but its various alloys are frequently annealed. A reducing atmosphere which will avoid oxidation of the base metals is generally used. Gold–zinc alloys may be difficult to anneal for the same reason that zinc-rich brasses are—volatilization of zinc and consequent surface roughening at relatively low temperatures.

Miscellaneous Applications

Paint and varnish manufacturers maintain an inert gas atmosphere above their products in storage to prevent skin formations. In synthetic-resin manufacture, inert gases are bubbled thru heated (about 300 F), closed reaction kettles to increase heat transfer; the inert gas blanket facilitates the removal of the liberated water vapor. Carbon monoxide may be used as a tracer in natural gas fields in repressuring and recycling operations. Oxygen-free atmospheres are used to preserve in-process synthetic crystalline vitamin A. An inert gas, low in O_2, is similarly used in making synthetic fibers such as rayon and nylon.

ATMOSPHERE CONTROL

When atmosphere composition changes occur, they should be discovered quickly. A complete chemical analysis of an atmosphere can be made within a few minutes by chromatography. If the chromatograph packing materials yield analyses on a *dry* basis, instruments to measure water vapor content are used.

Each gas has individual physical properties. The properties most easily measured are specific gravity and thermal conductivity. Somewhat more difficult to measure are magnetic susceptibility (attraction to a magnetic field) and absorption of infra-red light. In an endothermic generator, measurement of dew point is sufficient to determine carbon potential, which is usually all that is required. It is possible to absorb CO_2 and measure the volume of the remaining gas. Instruments are also available to measure the increase in temperature caused by catalytic combustion of small samples of atmosphere with measured quantities of oxygen or hydrogen, thus indicating the presence of combustibles or oxygen.

Table 12-121 gives some measurable quantities of commonly encountered gases. Table 12-115 shows the products of partial combustion of natural gas with varying proportions of air. In an exothermic generator, measurement of either specific gravity or thermal conductivity quickly shows the approximate composition. A spot check with a chemical absorbent for CO_2 or CO can be used for verification. Both specific gravity and thermal conductivity readings can be obtained continuously with standard instruments. Any variations in readings may be observed quickly and the causes remedied.

Table 12-121 Specialized Properties of Gases Useful in Monitoring of Atmospheres

Gas	Volume magnetic suscep-tibility, 10^{-10} emu	Ratio of thermal conduc-tivities, gas-to-air, at 32 F	Wave lengths of max absorption of infra-red, microns	Specific gravity
CO_2	−8.83	0.585	4.2–4.4	1.520
CO	−4.50	0.958	4.7–4.8	0.968
CH_4	−5.08	1.31	3.3–3.5 7.6–7.8	0.554
O_2	+139	1.044	None	1.105
N_2	−4.97	0.996	None	0.970
H_2	−1.665	7.10	None	0.0696

Many operations can be carried on in a broad range of atmospheres, provided that minute quantities of contaminants such as sulfur are absent. No control may be necessary other than occasional checks for the contaminating material. Atmosphere generators as purchased are usually equipped only with flow gages on air and fuel lines. These gages measure momentum of the fluid stream and are accurate only for one specific gravity. Under peak load conditions fuel gas specific gravity may vary, and the flow gage readings will then be inaccurate. Readings from a specific gravity indicator in the fuel supply line can be used to make corrections. Since the completeness of reaction in the combustion or cracking chamber is a function of temperature, a suitable temperature indicator as part of the control system is desirable.

Table 12-122 may be used as a checklist in establishing a quality control of protective atmospheres.

Table 12-122 Sources of Impurities in Protective Atmospheres[17]

Oxygen may come from:
1. Air leakage into furnace chamber.
2. Air carried into furnace with work.
3. Free oxygen in generated atmosphere.
4. Oxygen in atmosphere from cooling water (direct-contact cooling methods).

Water vapor may come from:
1. Oxide on work entering furnace (in contact with a reducing gas).
2. Surfaces of parts entering furnace.
3. Atmosphere introduced into furnace chamber.
4. Reformation of atmosphere by heat (water-gas reaction).
5. Leakage from water-jacketed members (cooling chamber, door fronts, etc.).

Carbon dioxide may come from:
1. Atmosphere as generated.
2. Air leakage if atmosphere contains CO_2.
3. Reformation of atmosphere by heat (water-gas reaction).

Sulfur may come from:
1. Fuel gas used in atmosphere producer.
2. Air burned with fuel gas.
3. Lubricants on surface of work being treated.
4. Brickwork and insulation of furnace.
5. Enclosed atmosphere oil quench chamber.

Carbon monoxide may come from:
1. Atmosphere as generated.
2. Vapors from oil on work.

Methane may come from:
1. Atmosphere as generated.
2. Lubricants on work.
3. Deliberate addition to atmosphere after generation.

Soot may come from:
1. Improper atmosphere (e.g., excessive carbon potentials in carburizing).
2. Vapors from oil on work.

REFERENCES

1. A.G.A. Laboratories. *Combustion of Gas With Limited Air Supply* (Res. Bull. 15) Cleveland, Ohio, 1942.
2. Shnidman, L. *Gaseous Fuels*, 2nd ed., chap. 9, New York, A.G.A., 1954.
3. Seizert, D. K. *Prepared Atmospheres*. (Inform. Letter 104) New York, A.G.A. Ind. & Comm. Gas Section, 1959.

4. Eeles, C. C., and Shriner, M. E. *Prepared Atmospheres.* (Inform. Letter 9) New York, A.G.A. Ind. & Comm. Gas Section, 1946.
5. Lanning, E. S., Jr. "Industrial Uses of Prepared Atmospheres." (In: *Industrial Gas School Lectures.* New York, A.G.A., 1955.)
6. A.G.A. *Proposed American Standard for Gas Atmosphere Generators: 4th Draft.* (ASA Z83.2–1965) New York, 1965.
7. Bayer, E. C. "Installation and Operation of Gas Atmosphere Generators." *Ind. Heating* 27: 517+, March 1960.
8. Am. Soc. for Metals. *Metals Handbook,* 8th ed., vol. 1, "Atmospheres," p. 622.
9. Spangler, F. L. "Steam Atmosphere Heat Treating." *Steel Process. Conversion* 44: 35–7, Jan. 1958.
10. Ipsen, H. N. "High Temperature Carburizing." *Metal Treat.* 10: 4–8, Jan.-Feb. 1959.
11. Am. Soc. for Metals. "Gas Carburizing." *Metal Progr.* 68: 132–43, Aug. 15, 1955.
12. Turner, C. A., Jr. "Scale-Free Heating in Direct Gas-Fired Furnaces." *Ind. Heating* 30: 72–6, Jan. 1963.
13. Demoulin, A. C. "Scale-Free Heating of Steel." *Metal Progr.* 84: 77–80, Aug. 1963.
14. Cole, S. W. "Heat Treating Stainless Steels." *Ind. Heating* 26: 908+, May 1959.
15. Chang, W. H. "A Dew Point-Temperature Diagram for Metal-Metal Oxide Equilibria in Hydrogen Atmospheres." *Welding Res.* [suppl. to *Welding J.* (N. Y.)] 35: 622s–4s, Dec. 1956.
16. Ruediger, B. A. "Bright Treating Stainless Steels in Protective Atmosphere Furnaces." *Ind. Heating* 28: 2184–96, Nov. 1961.
17. Hotchkiss, A. G., and Webber, H. M. *Protective Atmospheres.* New York, Wiley, 1953.

Chapter 18

Industrial Gas Ovens and Dryers

by F. E. Vandaveer and C. George Segeler

OVEN APPLICATIONS AND GENERAL TYPES

Oven walls, roof, floor, doors, etc., are fabricated of insulated steel panels of a type and quality consistent with the operating temperature. Oven temperatures range from 120 to 1200 F. Mild or aluminized steels are used up to 900 F (at low temperatures, self-supporting sheets of resilient insulating material) and alloy steels for higher temperatures. Frequently ovens are identified by their use, such as core baking, japanning, heat treating, etc.

Industrial furnaces, in contrast with industrial ovens, are constructed essentially of brick of a type and quality consistent with the temperature employed, usually above 1000 F.

Table 12-123 gives the NFPA oven classification system. Table 12–124 classifies ovens according to temperature and process requirements.

The combustion systems for ovens are of two types:

1. *Direct-fired.* Combustion products contact the work regardless of whether the combustion equipment is inside or outside the oven. Burners may be atmospheric, premix, or infra-red.
2. *Indirect.* Combustion products do not contact the work but are kept separate from the oven atmosphere by means of a heat exchanger or an intermediate fluid such as steam or hot water. Burners may be atmospheric or premix.

The location of the heat source may be internal (equipment lies within the oven walls) or external (equipment is located outside the oven walls).

OVEN DESIGN AND OPERATING FACTORS

Ventilation[1,2]

Basic factors in ventilation of ovens are (1) type of oven, (2) nature of the product, (3) heating required, and (4) heating method and fuel used. In most ovens the main volume of heated gases is recirculated thru or past the heat source. In the absence of hazardous solvent vapors, minimal fresh air displacement is required for meeting combustion air and process needs. Thermal efficiency is thus promoted by minimizing the quantity of hot gases exhausted. What is more important, better temperature control and uniformity throughout the oven chamber and faster heat transfer are achieved by active, forced recirculation.

No arbitrary rule can be set up for rate of recirculation. It may vary from 4 to 100 or more complete air changes per minute. The volume of air moved is but one consideration in the design of a good recirculating system. The nature of the work processed and its method of handling must be considered in relation to the manner of air delivery to the working space.

Besides volume, the location, direction, velocity, and temperature of air delivery must be considered. Provision should be made for carrying the heat far enough in the direction of discharge to provide uniformity throughout the working space.

Under certain conditions recirculation may not be desirable or advisable. In those cases the air, before admission

Table 12-123 Oven and Furnace Classification
(pressures and temperatures given are approximations)

Class	Pressure	Operating temperature	Remarks
A*	1 atm	≥ 700 F†	Fuel and/or product flammable
B	1 atm	> 700 F	...
C	...	...	Flammable special atmosphere
D	Ovens only, flammable vapor, solvent recovery with or without recirculation		

* NFPA Std 86A[1] provides rules covering this class.
† ≤ 900 F for food processing.

Table 12-124 Temperatures for Various Oven Applications

Approximate range	Process
110 to 180 F	Drying rugs, chemicals, and powders; curing rubber; baking finishes on vacuum metalized plastics
180 to 350 F	Drying; baking finishes on vacuum metalized metals and on metal parts (cabinets, partitions, shelving, etc.); baking insulating varnishes on transformers and electric windings; curing plastics and adhesives, etc.
350 to 550 F	Precipitation heat treatment or aging of aluminum or magnesium; relief annealing brass; tempering carbon steel parts such as springs; core and mold baking, baking or curing abrasives and silicon varnishes, fusing organosols and plastisols
550 to 1200 F	Solution heat treatment of aluminum and magnesium; stress relieving weldments; drawing certain steels after heat treatment in furnaces; preheating forging dies; annealing aluminum foil

to the processing space, must be heated to the desired temperature, properly distributed throughout the oven, and then discharged.

Baking of finishes on metal parts requires various ventilation rates. Certain black japans can be baked with a high ratio of recirculated air to make-up air, while lacquers and synthetics require larger amounts of fresh make-up air. An oven should only be used to handle the finish for which it is designed. If more than one type of finish is to be processed, fans and duct work must be arranged to provide a wide range of recirculation and/or fresh make-up air.

Arrangement. Ventilation is entirely independent of and in addition to any recirculation within the oven enclosure and should be arranged in an oven enclosure in such a way that there are no zones in which circulation does not take place. Consideration must be given to the proportioning of fresh air and recirculated air inlets and exhaust outlets, so that maximum dilution is obtained at points of maximum solvent evaporation. Consideration must be given also to the specific gravity of the solvent vapor and the vapors from the fuel used. The vapors of all volatile solvents and thinners commonly used in finishing materials are heavier than air and bottom ventilation is of prime importance. Fuel gases are generally lighter than air, LP-gases heavier than air.

Types of Ventilation. Forced mechanical ventilation shall be used where flammable volatiles or toxic fumes are given off by the product. However, subject to the approval of the authority having jurisdiction, **natural ventilation** may be used in ovens having only the fuel hazard provided the burner–mixer design is such that all air necessary for complete combustion of the fuel is reliably obtained by means independent from and not adversely affected by the natural draft.[1]

Rate of Ventilation.

Continuous Process Ovens. The controlled ventilation rate R given by Eq. **1** will maintain a particular solvent at 25 per cent of its lower explosive limit (L.E.L). An R value of 10,000 covers most solvents.

$$R = \frac{444G_l\,(100 - \text{L.E.L.})}{G_v\,(\text{L.E.L.})} \times \frac{t + 460}{530} \quad (1)$$

where: R = ventilation air rate at t °F, cu ft per gal of solvent evaporated

G_l = specific gravity of solvent *liquid* (water = 1)

G_v = specific gravity of solvent vapor (air = 1)

L.E.L. = lower flammability limit of solvent (Table 2–63), per cent by volume

t = make-up air temperature, °F

Batch Process Ovens. A ventilation rate of 380 cfm of 70 F air per gallon of flammable volatiles in the batch usually satisfies the maintenance of 25 per cent of the solvent's L.E.L. Often the maximum evaporation rate is established by tests. Correct for make-up air temperatures other than 70 F by using the last factor in Eq. **1**. Ventilation for ovens operating above 250 F should also be increased by a multiplier of 1.4.

In certain processes, such as baking varnished coils, only a small portion of the impregnating liquid will be found on the coil surface. If ventilation were provided at the rates mentioned above, uneconomical operation would result. An approach to such a problem is to conduct a test on a small batch of coils and determine the rate at which the solvent is actually evaporated. The required rate of ventilation is calculated from the maximum rate of solvent evaporation.

General. Any *duct damper* in a ventilating system (intake and exhaust) shall pass the volume required for safer ventilation—*in the closed position.* Gravity-retained *explosion vents* in all ovens must be provided in a ratio of not less than one square foot of vent for each 15 cu ft of oven volume. Control devices or interlocks shall be provided to assure that the ventilation is in operation and to provide for preventilation (*purging*) of ovens before heating systems can be turned on so as to dissipate accumulated vapors in an unventilated oven. They must shut down the heating system and any conveyor used in the event of failure of the ventilation system. If natural draft ovens are permitted, they shall be provided with fixed openings of at least one square foot for each 15 cu ft of oven volume or shall have doors or panels providing the same ratio and so interlocked that they must be fully open before the heating system can be operated.

Heating

The method of heating depends on the process involved. When the finish to be baked or material to be cured emits volatiles, the oven heating system should introduce sufficient air to prevent the concentration of volatile vapors from reaching the L.E.L. When oven space is at a premium, external-type heating systems, with heaters located on top or under the oven or in "dead" spaces nearby, are desirable.

When higher oven temperatures (500 to 1200 F) are required, direct-fired systems with large-capacity recirculating blowers and minimum fresh air requirements are preferable. Indirect systems are used when products of combustion adversely affect the color, taste, surface finish, metallurgy, or chemical composition of the product.

When zoned ovens are needed, external-type heating systems, which can usually be inexpensively segmented along the oven length, are recommended rather than internal systems.

When direct–external heating systems are used, the heater should be on the intake side of the blower. This arrangement aids combustion system control, prevents back pressure on the burners, and facilitates locating fresh air entry points.

When small ovens and intermittent usage are indicated, internal-type designs should be selected. Most drying processes require ovens with liberal fresh air supplies and exhaust provisions.

Zoning.[3] Continuous ovens are usually zoned. For example, in enamel finishing there should be a heating zone in which the work is brought up to temperature and volatiles are exhausted, holding zones to provide baking, a finishing zone for hardening (in some cases), and a cooling zone for further handling and processing.

Both ends of an oven should be provided with large-capacity fans. One at the intake prevents volatile-laden air from spilling out into the plant. The other at the exhaust end provides air for cooling. A separate fan or fans should be furnished to ventilate the oven proper; such a fan also serves to purge the oven before ignition.

Fouling.[3] The oven atmosphere should not harm the finish. Discoloration may be caused by uneven heat distribution which, in turn, may be the result of poor air circulation.

Proper oven atmosphere varies according to the material cured and type of surface desired. If the material contains oxidizing oils, provision must be made for ample oxygen. The percentage of oxygen in the oven atmosphere has been found to influence the speed of curing. The exact amount of recirculation possible should be determined for the particular finish used. Fresh air admission must be adequate to carry combustible thinner vapors away safely. Care should be exercised to have an excess of O_2 and no CO in the oven atmosphere. Concentration of CO_2 should not be high enough to cause acid catalysis on the surface. Normally the fresh air introduced for solvent removal supplies sufficient O_2 and prevents air fouling.

Precautions must be taken against introduction of water vapor, halogenated hydrocarbons, and dust. During shutdowns, finishing oven interiors should be protected from dust.

Humidity.[3] Water vapor produced by gas combustion in a direct-fired oven will not create a drying problem. For example, when nitrocellulose lacquer is dried at 250 F with 50 per cent recirculation (make-up air initially at 70 F and 70 per cent relative humidity), the relative humidity in the oven will be only 9 per cent. This figure includes moisture in combustion air and moisture formed by combustion. A considerable quantity of moisture might be introduced accidentally, but this would be an exceptional case. In curing baked wood finishes at schedules of one hour at 150 F, it is desirable to add humidity to the oven and maintain it at between 40 and 45 per cent relative humidity.

Combustion Safeguards. Combustion safeguards include flame conductivity, flame rectification, and ultraviolet sensing types. The thermal type is rarely used. To ensure safe oven operation, the pilot flame and "flame rod" must be properly installed and located so that no delayed lighting of the main burner or burners occurs.[4]

Approved types of combustion safeguards are available together with pilots or electric ignition systems for oven use. In order to permit automatic control of the main burner "trial-for-ignition period," the main burner is monitored. In other cases, both the gas pilot and the main burner are simultaneously monitored. When two adjacent burners in a multiburner oven heating system will positively light from each other, a single combustion safeguard serving both burners is sometimes used.

Ovens heated by explosion-resistant radiant heating tubes are not generally required to be equipped with combustion safeguards. Some types of direct-fired multiburner ovens with continuous ignition systems are of such design that the installation of combustion safeguards is not considered very practical. However, grouping of burners should permit their application.

Example: A batch-loading oven process-heats 1000 lb of sheet steel parts with 1.0 gal of solvents in the coating in 30 min to 350 F and then maintains it at that temperature. Aside from the radiation losses, which are usually a small portion of the total heat needed, requirements for heating up the steel parts and for supplying the proper amount of ventilation are.

$$\text{Heating steel:}\quad 1000\text{ lb} \times 40\,\frac{\text{Btu}}{\text{lb}^*} \times \frac{1}{0.5\text{ hr}} = 80\text{ MBtu per hr}$$

$$\text{Ventilation:}\quad 380\,\frac{\text{cu ft}}{\text{min}} \times 5\,\frac{\text{Btu}}{\text{cu ft}^\dagger} \times 1.4 \times 60\,\frac{\text{min}}{\text{hr}} = 160\text{ MBtu per hr}$$

In calculating the heat absorbed by the product, consideration must be given to the nature of the load. In the foregoing example, it has been assumed that the temperature rise of 280°F in the steel will take place within 30 min. This would undoubtedly be true if the load consisted of fairly thin sheet steel parts. However, in the baking of varnish-impregnated electrical transformers or coils, for example, the product bulk is a major factor. While the varnish on the surface of the windings can be baked in a comparatively short time, the varnish in the center of the winding cannot be baked unless the temperature of the entire winding reaches the operating temperature. Obviously, the load must remain in the oven for a sufficient length of time to permit penetration of the heat to all parts of the load. (See Fig. 2-70.) This is known as "soak time" and applies to all types of loads which have large cross sections.

On all except small ovens, a purge timer should be employed to ensure purging before ignition. Usually three to four complete oven changes will suffice. A door switch ensuring open doors before lighting up is recommended on small to medium ovens.

Catalytic combustion-type heaters may be used to burn fuel–air or fume–air mixtures for release of heat to an oven. Alternatively, these heaters may be installed in the oven exhaust stream both to conserve oven fuel and to oxidize combustible air pollution emissions. A 500°F maximum temperature differential across the catalyst bed is recommended. Discharge should be below 40 per cent of the L.E.L. if an interlocked continuous gas–air mixture analyzer is used; otherwise, 25 per cent of the L.E.L. is maximum.

DRYERS FOR LIQUIDS AND SOLIDS

Drying is a process for driving off moisture thru proper application, direct or indirect, of heat.[5] A dryer or apparatus employed for this purpose differs from a dehydrator or a calciner. A dehydrator extracts moisture by mechanical means, while a calciner drives off volatiles, etc., by roasting at high temperatures. Sometimes drying and roasting may be combined. For further discussion of drying liquids and solids see Refs. 6 and 7.

Dryers may be classified in several different ways as to methods of product handling and of heat transfer employed. The three methods of product handling are batch, semicontinuous, and continuous. Foods containing less than eight per cent water are practically immune to bacterial action.

* See Fig. 2-21.
† See Fig. 2-18.

Spray Dryers[8,9]

Spray dryers are in use by the food, chemical, and other industries for many continuous and semiautomatic operations. The material to be dried is generally a slurry or solution. It is forced under pressures thru nozzles or high-speed atomizers at the top of the drying chamber. Pressures of 100 to 10,000 psig have been used with various nozzles from 0.01 to 0.15 in. diam. In small dryers, where pressures under 60 psig are used, two fluid nozzles mixing an atomizing stream of air or steam at about 10 psig pressure produce more uniform particle distribution.

If a high-speed centrifugal atomizer is used, a thin sheet of the solution of slurry is projected and instantly broken up on striking the air by the combination of surface tension and impact at the periphery of the rapidly rotating disk. The resulting foglike mist easily floats on the air currents.

Air is the heating medium generally used but, in special processes in which there is danger of chemical reaction or explosion, heat may be applied indirectly by an inert medium such as nitrogen. Drying chambers are of two general designs, conical and flat-bottom.

Temperatures range from 200 to 1400 F, with special construction required for higher temperatures. Control equipment is necessary to maintain constant inlet temperatures of the heating medium. Recording thermometers are required in the walls and outlet air duct. Direct gas-fired drying of chemical and food products has been found to be very satisfactory, since there is no contamination. Gas firing also allows finer control over the drying unit.

Drying Milk.[10] It is claimed that natural gas, which is the only fuel that can be direct-fired, spray dries skim milk with 95 per cent efficiency. The dryer heat requirement is a function of the solids content of the feed.

Making Instant Coffee.[10] Following an extraction process in which about 32 per cent of the weight of the ground roasted beans is dissolved in water, the coffee liquid, containing 60 to 70 per cent water, is pumped to an atomizer at pressures up to 2000 psig. Direct gas-fired heating is used. The heated atmosphere contacts the product at 475 F and leaves at 245 F.

Other Products. In general, any slurry, emulsion, or solution that can be pumped can be spray-dried. Some chemicals that have been dried successfully are:

Amino acids	Latex	Soap and detergents
Animal blood	Pigments	Sodium chromite
Chlorophyl	Salts	Tannery wastes
Flavorings	Stearates	Penicillin wastes

Dryers for Textiles and Similar Materials

Gas is used for drying clothes, rugs, curtains, dry-cleaned garments, textiles, paper, and related products. Direct gas-fired tumbler dryers, similar to but larger than domestic clothes dryers, are used in laundries and dry cleaning establishments. They are made in various sizes—capacities up to 200 lb dry weight and gas inputs up to 1,280,000 Btu per hr.

Room dryers are used for drying rugs, carpets, and large fabric pieces which may be hung in the space. A room may be any desired size needed to accommodate the size and amount of materials to be dried. For 9 × 12 rugs a room 16 ft high × 14 ft wide × 15 ft long has been used. Warm air, forced into the room near the ceiling, is uniformly distributed, and removed (0.5 air change per min) near the floor level by means of exhaust fans. A 9 × 12 ft rug holds about 27 lb of water. A well-constructed room with adequate warm air circulation will dry rug materials in 3 to 5 hr.

For continuous drying of rolls of cloth or paper, steam drums are commonly used for paper and steam cans for textiles. Paper or textile is passed over the heated drum or can. Infra-red burners are often used for this purpose, either alone or to augment drum drying. Burner heat is directed as needed on one or both sides of the material.

Draw-type cabinet dryers are frequently used for curtains and blankets. They are equipped with stretcher frames for material up to 75 in. × 90 in. or 75 in. × 190 in. if wrapped around the frame. Models have two, three, or four compartments and gas inputs up to 335 MBtu per hr. Capacity per compartment is four blankets or six pairs of curtains per hr.

Conveyor Dryers

In conveyor-type dryers the material to be dried is passed thru a succession of drying zones.

The flash conveyor dryer is very similar to the spray dryer in that the material is floated on a current of preheated air from an air heater and separators are used to free the product from the drying medium. Air at temperatures as high as 1200 F is introduced into a falling flow of wet solids, which are carried at velocities of up to 4000 fpm to a hammer mill for grinding. The fine particles leave the mill thru a duct small enough to maintain carrying velocities, and enter a cyclone separator. Separated material can be dropped into a stream of hot fresh air for final drying, after which it is again separated in a second collector. Air from the second stage is reheated for use in the first stage for greater economy.

A *tower dryer* depends on gravity alone to carry the material from one inclined baffle to another in the path of an ascending current of hot air.

Other conveyor dryers depend on mechanical means for transporting materials thru the drying zones. The belt conveyor dryer usually employs a woven wire mesh thru which drying air passes. To save space, a multipass arrangement may be installed. A belt formed of hinged perforated metal shelves or vibrating screens may be used, and the material may be subjected to zones of varying heat and humidity. The belt conveyor dryer lends itself well to recirculation of part of the air after it has been reheated, which increases thermal efficiency.

The screw conveyor dryer is simply a heated trough thru which material is moved by means of a screw- or paddle-type agitator while the heat is supplied from a jacket. A countercurrent of air or inert gas may be passed thru the dryer to assist in water vapor removal. Oxides, carbonates, lava, and ores may be dried in conveyor dryers of the screw or belt type.

The festoon or roll dryer uses the material itself as a belt which travels over rollers thru a tunnel-like chamber concurrent to the warm air. Chemically treated papers and sheet plastic are examples of products thus dried.

The endless chain dryer has hooks suspended from overhead links and carries the product thru a tunnel in which warm air is circulated.

Kettle

Jacketed kettles with double-motion agitators like the Dopp kettle may also be used for drying solids. The outside agitator has scrapers which keep the wall clean. Horizontal arms move in a direction opposite to that of a set of paddle arms, causing the material to break up more readily and mix more thoroughly without forming large lumps. Kettles used in this way become kettle dryers.

Rotary

Basically, a rotary dryer is a revolving cylinder inclined to move drying material to the discharge end. Both small laboratory and large capacity production units are used. Cylinders for the latter are often 8 to 10 ft in diam and 40 to 50 ft long. The main cylinder is usually inclined to the horizontal and either the shell or the agitators revolve to advance the wet material from the upper to the discharge end. Rate of food, speed of rotation, and volume and temperature of heated air or gases are varied to dry the product before discharge. The use of longitudinal shelves or lifts on the shell interior improves efficiency as the material is carried farther up the sides of the shell and then dumped into the air stream. Air speeds vary from 2 to 15 fps. The air direction is generally countercurrent to the flow of the material but may be concurrent in some cases in which heat-sensitive material is being dried. Double shells are also used for heat-sensitive materials. The hot gases enter the inner shell at the feed end and return thru the center section after temperatures are reduced.

There are several modifications using steam tubes or flues for high-temperature gases when heat transfer is by radiation and convection. With the heat sensitivity of the material duly considered, the more thorough the exposure of wet surfaces to the heat, the shorter the dryer need be and, consequently, the lower its initial cost. Moisture reduction will be from about 1 to 12 per cent, with fuel requirements of 1800 to 3000 Btu per lb of water evaporated. Oxides, salts, and ores are materials typically dried in such equipment.

Drum

Drum dryers or flakers are steam cans used in paper and textile drying. Wet material is applied to the rotating drums and scraped off after drying is completed. Two drums may be rotated toward each other and the slurry fed into the trough thus formed. Coating thickness is governed by the clearance between the drums. Drum dryers may be operated under vacuum. Tanning extract, ores, and oxides are dried at atmospheric pressures. Sodium ferrocyanide crystals, albumen, and similar heat-sensitive materials are dried under vacuum.

Pan Dryers

Atmospheric. The atmospheric pan dryer has a jacketed round pan in which a slowly revolving stirrer exposes fresh product surfaces. Dried material is discharged by opening a gate in the side of the pan.

The pan dryer is a small-batch machine. It may be used first to evaporate a solution to crystallizing strength and then as a crystallizer by sending cold water instead of steam into the jacket. The effect of the stirrer during crystallization is to promote formation of small, uniform crystals rather than large crystals. The mother liquor is then dried off and the crystals dried in the same apparatus. For the production of "hypo" (sodium thiosulfate) in rice size, for example, the pan dryer transformed into a crystallizer gives excellent results. Similarly, it can be used for carrying out a number of other operations in succession without transfer of material.

Vacuum. The vacuum pan dryer consists of a round pan surmounted by an airtight hood and fitted with a slow-moving stirrer. The pan is jacketed and strong enough to take steam at 100 psig. For drying a wet solid, the vapor would be drawn out by a suction pump, condensed, and saved if desired. Stirring speeds drying by exposing fresh surfaces. A discharge door is at the side of the dryer.

Vacuum dryers possess several advantages over atmospheric types: (1) danger of contamination by dirt in the drying air is avoided as well as chemical change in the substance by atmospheric oxygen; (2) valuable vapors such as those of organic solvents may be easily recovered; and (3) working temperature is also lower, and this is the controlling factor for certain pharmaceuticals, extracts, and similar substances.

Disadvantages of vacuum drying are the higher initial and operating costs and limited production. Thus, the vacuum dryer is seldom used if any other dryer will do.

As long as sufficient moisture is present to wet the surface of the material completely, product temperature will be low and controlled by the chamber vacuum. As soon as the moisture drops below the critical point, the temperature of the material increases. Finally, when the material is dry throughout, its temperature closely approaches that of the heating medium which, therefore, must be kept below the critical temperature.

Tunnel and Compartment Dryers

The tunnel type, also known as a truck or kiln dryer, is very similar to a compartment dryer, which may also be called a cabinet, chamber, or tray dryer. All consist of a chamber, closet, or room with heat supplied by heating tubes or a direct-firing source. Extent of recirculation depends on the application and dryer length involved. Generally, warping, checking, and cracking are best prevented by introducing heat at the wet end of the tunnel.

Direct-fired air heaters or radiant burners have been used in these dryers to dry magnesia forms, pigments, dye stuffs, pharmaceuticals, and dye intermediates.

Another similar type is the vacuum-operated shelf dryer in which heat is supplied from hollow shelves via water, steam, or other media. A wide range of temperatures may be used. The cabinet may be circular and provided with rotating arms or the material may be moved from one shelf to another alternately, with central and then circumferential discharge. Vacuum is generally used for substances that give off toxic or valuable vapors and/or are decomposed by heating at atmospheric pressure. Pharmaceuticals, dyes, and anhydrous bisulfide of soda are dried in this type of dryer.

Characteristics of Materials Dried

A major consideration in a drying problem is the relation of the material dried to the method of drying. Consideration should be given to the economical soundness of mechanically separating as much water as practicable before drying. Centrifuging, pressing, filtration, or other mechanical means may be used. Table 12-125 shows typical temperature ranges for common baking and drying operations.

Table 12-125 Baking and Drying Temperatures, °F

Baking			
Cores	300 to 500	Lithographing	250 to 350
Japan, synthetic enamels, varnishes	250 to 400	Wire coatings, coils, armatures	250 to 350
Drying			
Chemicals	100 to 400	Lumber	70 to 150
Citrus peels	250	Macaroni	110 to 150
Clothes	100 to 180	Mattresses	100 to 180
Eggs	200 to 280	Nuts	90 to 140
Enamels (incl. porcelain)	250 to 350	Painted materials	250 to 350
		Paper	100 to 200
Feathers	150 to 180	Photo papers and films	90 to 110
Fruit	70 to 100	Rice	105
Furs	150 to 190	Rugs	120 to 180
Hair	150 to 190	Sand	200 to 450
Hay	130 to 190	Seed corns	85 to 120
Hops	120 to 180	Steel wire	300
Ladles	300 to 400	Sweet potatoes	140 to 220
Leather goods	100 to 200	Textiles	100 to 180

Physical and chemical properties of materials govern the type of dryer. The following should be ascertained:

1. "Free moisture content" of the material fed into dryer.
2. Specific heat of the material.
3. Specific gravity of the dry material.
4. Quantity of free moisture allowable in material at dryer discharge.
5. Quantity of material fed into dryer per hour.
6. Reaction of material as free moisture is driven off.
7. Whether material will be injured or affected by combustion products.
8. Maximum temperature to which material may be heated without danger of ignition and without undesirable chemical or physical changes.
9. Absence of halogenated hydrocarbon vapors in heated spaces or near hot surfaces.

Matching Material to Process. Four classifications follow:

1. *High temperature, contact with combustion products.* Construction materials such as sand, trap rock, gravel, clays, bitumastics, etc. The simplest and most commonly used dryer for these materials is a rotating drum with built-in agitators. Heat is applied at one end, usually the one at which materials enter; stack or fan is at the other end. Precautions must be taken to maintain exhaust temperatures high enough to avoid recondensation. A disadvantage of this type is its high fuel cost. Another is the difficulty in obtaining a low final moisture content when the heater air flow and material flow are parallel.

2. *High temperature, no contact with combustion products.* Materials such as china clay, fuller's earth, talc rock, and kaolin can be dried economically by indirect air heaters or radiant tubes. Another method is to pass the materials thru a drum encased in brickwork, with heat applied to the drum exterior.

3. *Temperatures below 400 F, in contact with combustion products.* Materials such as sawdust, gypsum, wood chips, fertilizer materials, various chemicals, and lignite can be dried in the single-drum bricked-in type of dryer, but the operation is uneconomical, and has some attendant dangers. The more practical method is to use a direct type of dryer providing two passes of the flue products inside one revolving drum. Many foods such as skim milk, instant coffee, and eggs are dried by direct contact with the products of combustion of natural gas.[10] The products of combustion do not enter the processed material, since the vapor pressure of the liquid evaporated is rising during drying, which tends to prevent any contamination. See Table 1-25 and, in particular, its cautionary footnote.

4. *Varied temperature, cannot be agitated.* This group comprises the largest and most varied number of materials dried. Basically, it contains those which cannot be agitated because of their physical properties. Examples are ceramic wares, dyed products, extremely sticky products, knitted products, some food products, and very finely divided materials.

Fluidized Beds. Here combustion products or heated air is "bubbled" thru a column containing the product. The product layer expands until the pressure drop thru the bed exactly balances the weight per unit area of the solid particles in the bed, whereupon the bed of solid particles behaves like a liquid.[11] A high rate of uniform heat transfer results from this intimate intermingling. Products treated[12] using natural gas include limestone, shale, foundry sand, blast furnace slag, phosphate rock,[13] chemicals, and food products.

Curing Concrete Blocks.[14] Curing at elevated temperatures shortens the normal curing period of three or four weeks to one day. After a minimum one-hour waiting period (usual upper limit: three hours), the blocks are heated from 100 F at the rate of 28°F per hour for about 3.5 hr.

Thermal Efficiency

The thermal efficiency of a dryer is the ratio of the theoretical heat required for drying to the total heat supplied.

Efficiency of dryers for materials in foregoing Groups 3 and 4 can be regarded as about 60 per cent—and as high as 78 per cent for certain dryers in Group 3. Dryers for Group 2 materials range in efficiency from 25 to 67 per cent, while those for Group 1 materials run as follows:

1. A single-shell dryer of good design: 45 to 60 per cent.
2. A bricked-in direct heater: 50 to 65 per cent.
3. A rotary drum, direct forced: 70 to 80 per cent.

Heat losses can be assumed to be as follows, depending upon the design of the dryer: radiation, 4 to 30 per cent; flue gas, 7 to 40 per cent; absorbed by dried material, 3 to 25 per cent.

Theoretical Heat Required. Use Eq. 2, which assumes that the water is heated from 60 to 212 F and evaporated at that temperature. It does not take into account anything but drying of the material. Heating the air required for drying is not considered.

$$H = 1120(2000W/100 - W) + 2000S(212° - 60°) \quad (2)$$

where: H = Btu required per ton of dry material
W = moisture in material, weight per cent
S = specific heat of material, Btu per lb-°F

Continuous circulation of warm dry air over or thru moist material to be dried provides an efficient drying medium. If the moisture-laden air is passed over cooling coils to remove moisture, reheated to its drying temperature, and recirculated over the material, the system efficiency increases markedly.

To use air as the drying medium it is necessary to know: (1) the required moisture removal rate; and (2) the volume of air that will carry that moisture at the temperature at which it leaves the material. Since the warm moist air leaving the material may not be fully saturated, due allowance should be made.

Air Volume. Calculate air flow by using psychrometric charts or Eq. 3.

$$N(H_2 - H_1) = S(W_1 - W_2) \quad (3)$$

where: N = pounds of dry air passing thru dryer per unit of time
H = humidity ratio of air, pounds of water vapor per pound of dry air
S = pounds of material dried per unit of time
W = pounds of water per pound of dry stock
Subscripts 1 and 2 are the dryer entrance and exit conditions, respectively.

Total Heat Requirement. Evaluate the following four terms: (1) sensible heat loss in effluent air; (2) latent heat of vaporization of the water plus heat required to raise it to the temperature of vaporization; (3) heat to raise the temperature of the stock plus the water remaining unevaporated in it; and (4) total radiation and conduction losses.

Controls normally consist of a thermostat in a main return duct to maintain a constant dry bulb temperature by controlling heat input and a wet bulb controller in the return circulating duct, controlling dampers to regulate amounts of make-up air, exhaust air, and recirculated air.

Dry Rate Periods

Ample supplies of air are essential for good dryer performance. Most dryers permit a wide latitude in the selection of temperature, humidity, and air flow rates. In studying drying, it is customary to consider two stages: (1) when the capacity of the hot air to remove water is the limiting factor, and (2) when it is the moisture flow out of the material that limits performance. These are called, respectively, the constant drying and falling rate stages. Finally, there is a moisture content in the material being dried which is in equilibrium with the drying air. This moisture percentage (regain percentage) is related to the specific circumstances involved.

The behavior of water in materials is complex. Some of it may be absorbed or held in crevices, in the cell structure, or otherwise be unable to develop the expected vapor pressure at the surface of the material dried. However, the basis of most dryer calculations is the unbound moisture which does not exert full vapor pressure.[15]

Air flow rates expressed in cubic feet per minute are convenient terms used with duct and outlet velocities, fan capacities, etc. However, their use in dryer calculations may be misleading and one must take care not to neglect temperature and humidity changes. Calculation on the basis of grains of H_2O per lb of dry air obviates such problems. This unit is independent of variations in the dry bulb temperature. Tables 12-126 and 12-127 give the moisture pick-up capacities of air at constant dry bulb and constant dew points, respectively.

The *constant rate drying period*[16] can be evaluated by Eq. 4.

$$\frac{dw}{d\theta_c} = \frac{h_t A(t_a - t_s)}{H} = k_g A(p_s - p_a) \quad (4)$$

where:

$\frac{dw}{d\theta_c}$ = constant drying rate, lb of water per hr-lb of bone-dry material
h_t = total heat transfer coefficient, Btu per hr-sq ft-°F
A = area of heat transfer and evaporation, sq ft per lb of bone-dry material
H = enthalpy of evaporation at t_s, Btu per lb
k_g = mass transfer coefficient, lb per hr-sq ft-atm
t_a = air temperature, °F
t_s = temperature of surface of evaporation, °F
p_s = vapor pressure of water at t_s, atm
p_a = partial pressure of water vapor in air, atm

Table 12-126 Theoretical Moisture Pick-up of Air at Constant Dry Bulb Temperatures

Dew point, °F	Air temp, dry bulb, °F	Moisture pick-up, grains per lb of dry air
0	150	120
50	150	110
100	150	70
0	200	180
50	200	170
100	200	150
150	200	100
0	260	250*
50	260	250
100	260	230
150	260	220
180	260	300
0	375	410
50	375	410
100	375	410
150	375	450
170	375	600

* Above 260 F dry bulb, an increase in recirculation should result in heat economy and greater pick-up. Drying speed will not be impeded. Below 260 F dry bulb, sacrifice in heat economy must be balanced against a gain in drying speed, or vice versa.

When $h_t = h_c$, the coefficient of heat transfer by convection only, then t_s under equilibrium conditions becomes t_w, the wet bulb temperature of the air, and p_s is the vapor pressure at this temperature. If heat is also supplied by radiation, then h_t is the sum $(h_c + h_r)$, where h_r is the radiation coefficient and h_c is the convection coefficient, and t_s becomes higher than the wet bulb temperature. A similar result occurs

Table 12-127 Theoretical Moisture Pick-up of Air at Constant Dew Points

Dew point, °F	Air temp, dry bulb, °F	Moisture pick-up, grains per lb of dry air
0	100	60
0	200	180
0	300	290
0	400	440
0	500	570
100	150	50
100	200	150
100	300	290
100	400	440
100	500	590
180	200	60
180	300	370
180	400	670

when heat reaches the surface of evaporation by convection and conduction.

Effect of Air Velocity. The principal effect of air velocity is on h_c and k_g, since the rate of transfer of heat and mass in the constant rate period depends mainly on the rate of diffusion of heat and vapor thru the air film at the surface of the solid, and air velocity is the chief factor affecting the thickness of this film. The influence of direction of air flow on the heat transfer coefficient h_c and on the corresponding drying rate $(dw/d\theta)_c$ is shown in Table 12-128.

Effect of Temperature or Humidity. Temperature or humidity enters the drying rate equation as a driving force across the air film. The wet bulb depression, which is the difference between the dry bulb temperature and the wet bulb temperature, is directly proportional to the drying rate in this period.

When we deal with heat transfer coefficients the wet bulb depression is the driving force, whereas with mass transfer coefficients the driving force is expressed in terms of humidity or vapor pressure differential.

Table 12-128 Convection Heat Transfer, h_c, and Drying Rates, $(dw/d\theta)_c$, for Constant Rate Period[16]

Direction of air flow	h_c*	$(dw/d\theta)_c$*
Parallel to plane surfaces	$0.0128G^{0.8}$	$\frac{0.0128G^{0.8}A}{H}(t_a - t_w)$
Perpendicular to plane surfaces	$0.37G^{0.37}$	$\frac{0.37G^{0.37}A}{H}(t_a - t_w)$
Thru circulation (for Reynolds number > 300)	$\frac{0.37G^{0.59}c_s}{D_p^{0.41}}$	$\frac{0.37c_s aG^{0.59}\Delta t_m}{\rho_s H D_p^{0.41}}$

* Where:

h_c = convection heat transfer coefficient, Btu per hr-sq ft-°F
G = mass velocity of dry air, lb per hr-sq ft
t_w = wet bulb temperature of drying air, °F
a = drying area of bed volume, sq ft per cu ft
ρ_s = bulk density of dry granular bed, lb per cu ft
D_p = average diameter of particle, ft
Δt_m = logarithmic mean difference between the air temperature entering and leaving the bed and the wet bulb temperature, °F
c_s = humid heat, Btu per °F-lb of dry air
Other terms are defined under Eq. **4**.

The *falling rate drying period*[16] can be evaluated by Eq. **5** or **6**. When the surface is dry or at the equilibrium moisture content and the solid has a uniform initial moisture distribution, Eq. **5** holds for relatively large values of time θ; and when $(w - w_e)/(w_0 - w_e) < 0.6$:

$$\frac{dw}{d\theta} = -\frac{\pi^2\delta}{4L^2}(w - w_e) \tag{5}$$

where:

w = moisture content on dry basis, at any time θ, lb water per lb dry material
w_e = moisture content at equilibrium with external conditions, lb water per lb dry material
w_0 = moisture content at start of diffusional period, lb water per lb dry material
δ = liquid diffusivity, sq ft per hr
L = one-half material thickness, ft

Equation **5** is restricted to a slab-shaped solid, the length of which is large compared with the thickness.

For some materials the drying time in the falling rate period varies directly with the thickness. When this occurs the falling rate can be expressed with *fair* accuracy by Eq. **6**:

$$\left(\frac{dw}{d\theta}\right)_f = -\frac{(dw/d\theta)_c}{(w_c - w_e)}(w - w_e) \tag{6}$$

where:

$(dw/d\theta)_c$ = constant drying rate (Eq. **4** and Table 12-128), lb per hr-lb dry material
$(dw/d\theta)_f$ = falling rate, lb water per hr-lb dry material
w_c = critical moisture content, lb per lb dry material

Approximate Classification of Materials Most Likely to Obey Eqs. 5 and 6. *Materials Obeying Eq.* **5**.

1. Single-phase solid systems such as soap, gelatin, and glue.
2. Wood and similar solids below the fiber saturation point.
3. Last stages of drying starches, textiles, paper, clay, hydrophilic solids, and other materials when bound water is being removed.

Materials Obeying Eq. **6**.

1. Coarse granular solids, such as sand, paint pigments, and minerals.
2. Materials in which moisture flow occurs at concentrations above the equilibrium moisture content at atmospheric saturation or above the fiber saturation point.

Moisture Regain. Following drying, materials pick up moisture from ambient air (see Table 12-129).

DRYING OF GASES[17]

Gases may be dried by (1) absorption; (2) adsorption; (3) compression; (4) cooling; or (5) a combination of compression and cooling.

The mechanical drying methods, compression and cooling, are used in large-scale operations. They are generally more expensive than those employing desiccants and are used when compression of the gas is a necessary step in the operation or when its cooling is required.

Absorption

Either spray chambers or packed columns may be used with liquid desiccants in continuous absorption processes. Desiccants used in spray chambers are organic liquids such as glycerol, or aqueous solutions of salts such as calcium or lithium chloride. Organic liquids are also used in packed columns with countercurrent flow, as well as sulfuric or phosphoric acids.

General properties of liquid and soluble desiccants are given in Table 12-130.

Adsorption

Solid desiccants such as activated alumina and silica gel are generally used in intermittent processes which require periodic interruption for regeneration of the desiccant. Continuous regeneration is effected if the desiccant bed is rotated thru two compartments isolated by a seal; drying occurs in one compartment, regeneration in the other.

Table 12-131 lists general properties of solid desiccants; see also Figs. 4-51 and 4-53.

Compression

By compression to a partial pressure of water vapor greater than the saturation pressure, condensation of liquid water may be effected. Gases may be dried by compression since the humidity of saturation, H_s, decreases with total pressure according to Eq. **7**:

$$H_s = [p_s/(P - p_s)](M_v/M_g) \quad (7)$$

where:

H_s = saturation humidity, lb vapor per lb dry gas
p_s = saturation pressure of the vapor at ambient temperature, abs
P = total pressure, abs
M_v = molecular weight of vapor
M_g = molecular weight of gas

Saturation humidity, H_s, values for water vapor in air, as related to total pressures, P, at 68 F follow:

P, atm	*H_s, lb water/lb dry air*
1	0.0147
2	.0072
5	.0029
10	.00144
50	.000287
200	0.000072

When initially unsaturated gas is compressed, it must become saturated at some point in the compression before condensation begins. If compression is adiabatic, the air becomes heated and must be cooled in an aftercooler to obtain condensation and complete benefit from the compression. The air leaving the aftercooler may contain water as a fog or mist which must be removed; additional drying by solid desiccants may follow.

Table 12-129 Moisture Regain of Some Common Materials
(moisture content in per cent of dry weight)

Material	Relative humidity, per cent 20	40	60	80
Cotton cloth	3.7	5.2	6.8	10.0
Wool skein	7.0	10.8	14.9	19.9
Silk skein	5.5	8.0	10.2	14.3
Rayon skein	5.7	7.9	10.8	14.2
Paper, newsprint	3.2	4.7	6.1	8.7
Paper, kraft	4.6	6.6	8.9	12.6
Leather	8.5	13.6	18.3	24.0
Wood (average)	4.4	7.6	11.3	17.5

In general, compression cannot compete as to cost with desiccants when the latter can effect the required degree of dryness. Equipment, power, and cooling water costs are generally excessive. Drying by compression is usually not used unless the gas must be compressed as part of its later treatment.

Cooling

Surface condensers or water sprays may be used for cooling below the dew point. When air is cooled well below its dew point by such means as brine, ammonia, Freon expansion coils, or solid CO_2, much of its water vapor can be condensed and the air humidity greatly reduced (see column 10 of Table 1-27). The usual practice in cooling gases to low temperatures is to put them in contact with refrigerated surfaces such as pipe or finned tubes. However, the freezing which results when they are cooled to low humidities may require the frequent shutdown of coil sections for defrosting. Some types of coolers permit continuous defrosting.

Gas leaving a cooling coil may be somewhat less than saturated at its dry bulb temperature. Its dew point tends to approach the coil surface temperature because of the heat and mass transfer at the boundary between the gas and the coil.

CURING PROCESSES[3]

Finishing materials such as japans, lacquers, paints, and synthetic enamels are cured by evaporation, oxidation, and/or polymerization. For example, refrigerator enamels with urea-type bases and alkyd plasticizers dry by all three methods—evaporation of thinner, polymerization of the urea, and oxidation of the alkyds.

Evaporation

Although evaporation is only a part of the curing mechanism in most finishes, a few dry and harden solely by evaporation. Two good examples are shellac and nitrocellulose lacquer. Evaporation rate increases with temperature. If evaporation is too rapid, the film is cooled below the dew point of the surrounding air; this causes moisture to condense and dull the finish, a condition known as blushing. Adequate provision for safe removal and disposal of flammable solvent vapors must be provided.

Oxidation

Curing by oxidation is a process in which oxygen in the curing atmosphere helps to bring about the formation of the

Table 12-130 Properties of Liquid and Soluble Desiccants for Gas Drying

Desiccant	Relative humidity at room temp economically obtainable, %	Useful temp range, °F	Concentration range, %	Toxicity	Corrosiveness	Stability	Principal applications	Remarks
Calcium chloride (aqueous solution)	20–25	90–120	40–50 as solution, sometimes used as a dissolving solid	Nontoxic	Noncorrosive	Stable	Drying city gas	Same as for $LiCl_2$ solution
Diethylene glycol and triethylene glycol	5–10	60–110	70–95	...	Noncorrosive	Stable	Drying natural gas (Fig. 4-48)	High boiling points eliminate need for elaborate refluxing equip, regeneration at 300 F (DEG) removes all but trace of water without loss of desiccant
Glycerol	30–40 with 70–80 conc., 5 with anhydrous	70–100	70–80	Nontoxic	Noncorrosive	Oxidizes and decomposes at high temperatures	Drying manufactured gas	Regenerated under vacuum; low heat requirements; at concentrations as low as 50 to 60%, glycerol can still absorb considerable water
Lithium chloride (aqueous solution)	10–20	70–100	Usual, 40–45; however, 30 has good capacity	Nontoxic	Noncorrosive except on Mg alloys; corrosive at regen. temp	Stable at boiling temperatures	Air conditioning, industrial processes, low temp vacuum drying	Aftercoolers to remove heat of condensation and solution, solution cooler to control inlet solution temp, hydrometer control to regulate concentration during regeneration
Phosphoric acid	5–20	60–130	80–95	Toxic	Corrosive	Stable	Laboratory, especially the anhydride	Corrosive, toxic nature makes it im practical for industrial drying; regeneration does not have a fuming problem
Sodium and potassium hydroxides	10–20	85–120	Sat. sol; also used as dissolving solids	Toxic	Corrosive	Stable	Liquid air manufacture, compressed gases, gas plants	High heats of solution and corrosiveness preclude use in ordinary installations; frequently used to remove CO_2 and H_2O simultaneously
Sulfuric acid	5–20	70–120	60–70	Toxic	Corrosive	Stable	Miscellaneous gas drying	Is being displaced by other desiccants because of handling hazards; where its use is feasible, it is highly efficient

Table 12-131 General Properties of Solid Desiccants

	Silica gel	Activated alumina	Activated carbon	Fuller's earth
Mesh size for gas drying	3 to 8, or 16-16	...	...	1
Specific heat, Btu/lb-°F	0.2	0.24	0.2	...
Bulk density, lb/cu ft	40 to 45	50	28.34	35
Reactivation temp, °F	300 to 350	350 to 600	220 to 240*	Up to 800
No. of activations without deterioration	Unlimited	...	...	...
Residual moisture content, %	5	...	...	6 to 8
Heat to remove water (commercial practice), Btu/lb	2000 to 2500	...	...	500
Water adsorption, % of its weight	20	...	...	...

* Special up to 550 F.

finish film. In direct-fired ovens, presence of sufficient oxygen must be assured either thru excess air in the air–gas mixture or by the addition of fresh air to the oven atmosphere.

Vehicles used in materials which cure by oxidation are oils such as linseed oil. Formerly, absorption of oxygen by paint films containing oils was very slow and the curing cycle was a matter of weeks. Addition of so-called dryers was found to reduce the drying time to a few days. Later, synthetic resins and harder oils came into use and reduced the time of curing in air to a matter of hours. Air drying still takes several hours with oxidizing finishes, but can be reduced to one-half hour by the application of heat.

Polymerization

Polymerization is a curing mechanism in which simple molecules of an organic substance unite to form complex organic molecules. All components necessary to form a film are present in the material as it is applied. However, there are some polymerization reactions of the "condensation" type in which water or alcohol is liberated during the curing of the film. Because no film needs to be added, it is not necessary to provide a specified atmosphere as in the case of oxidation.

In curing paint by evaporation and oxidation, heat serves merely to shorten the time cycle. However, in polymerization, heat serves as a catalyst. Polymers cured at elevated temperatures are quick-drying, water-resistant, more durable, and more abrasion-resistant than the same finishes cured at low temperatures.

Sometimes acids are used as polymerization catalysts. The quantity of acid needed is very small. It is so small that on some occasions an atmosphere high in CO_2 and H_2O has produced enough carbonic acid to cause the painted surface to dry so rapidly that it has microscopic surface wrinkling, giving it a dull finish.

REFERENCES

1. Natl. Fire Protection Assoc. *Standard for Ovens and Furnaces.* (NFPA 86A) Boston, 1963.
2 Associated Factory Mutual Fire Insurance Co. *Industrial Ovens and Dryers; Safe Installation and Operation.* (Loss Prevention Bull. 14.15) Norwood, Mass., 1957.
3. Humphrys, L. M., and Olinger, S. T. *Drying Processes, Atmospheres and Oven Equipment for Curing Industrial Finishes.* (Information Letter 15) New York, A.G A. Industrial and Commercial Gas Section, 1947.
4. Gehnrich, H. "Convection Oven Applications." (In: A.G.A. *Industrial Gas School Lectures.* New York, 1959.)
5 Mack, E. S. *Dryers and Drying.* (Information Letter 61) New York, A.G.A. Industrial and Commercial Gas Section, 1954.
6. Bagnoli, E. "Drying of Solids." (In: Perry, J. H., ed. *Chemical Engineers' Handbook,* 4th ed., p. 15/32–15/50. New York, McGraw-Hill, 1961.)
7. Marks, L. S., ed. *Mechanical Engineers' Handbook,* 5th ed., p. 1651–7. New York, McGraw-Hill, 1951.
8. Sims, W. S. "Gas Fuel in Chemical Industry." (In: A.G.A. *Industrial Gas School Lectures.* New York, 1955.)
9. Ellington, R. T. *Spray Drying.* (Information Letter 101) New York, A.G.A. Industrial and Commercial Gas Section, 1959.
10 Canner, T. R., and Cole, C. B. *Spray Drying of Milk, Coffee and Eggs.* (Information Letter 108) New York, A.G.A Industrial and Commercial Gas Section, 1959.
11. A.G.A. Industrial and Commercial Gas Section. *Fluidized Bed: Heating, Roasting, Calcinating, Drying.* (Information Letter 134) New York, 1963.
12. Priestley, R. J. "Where Fluidized Solids Stand Today." *Chem. Eng.* 69: 125–32, July 9, 1962.
13. Guffey, A. A. "Processing Calcined Phosphate Rock." *Chem. Eng. Progr.* 58: 91–3, Oct. 1962.
14. Ellington, R. T., and Mack, E. S. *High Temperature Curing of Concrete Block.* (Information Letter 77) New York, A.G.A. Industrial and Commercial Gas Section, 1955.
15. Victor, V. P. "Air Recirculation in Drying of Unbound Moisture" *Chem. Eng.* 52: 105–9, July 1945.
16. Am. Soc. of Heating, Refrig. and Air-Cond. Engrs. *ASHRAE Guide and Data Book, 1962, Applications.* New York, 1962.
17. Marshall, W. R., and Friedman, S. J. "Drying of Gases." (In: Perry, J. K , ed. *Chemical Engineers' Handbook,* 3rd ed., p. 877–84. New York, McGraw-Hill, 1950.)

Chapter 19
Industrial Furnaces

by A. H. Koch, H. C. Weller, R. C. Le May, C. George Segeler, W. Wirt Young, and K. R. Knapp

INTRODUCTION

Furnaces are used for processing metals and various nonmetallic materials at temperatures of about 1000 F and higher. When such operation involves lower temperatures, the term *oven* may be employed (see Table 12–123). The term *kiln* is applied to a furnace (e.g., ceramic kilns) if the heating operation first raises the temperature to a point where either gas or moisture is given off, then increases it to a higher final value. Typical industrial heating operations with approximate temperatures required are given in Table 12-132. The vapor pressures of various metallic elements are shown in Fig. 12-194. The thermal conductivities of various furnace materials are shown in Fig. 12-195.

Industrial gas furnaces are most satisfactorily classified according to the particular purpose for which they are designed and used. Those most commonly encountered are discussed later in this chapter. Figure 12-196 shows some of the many types of furnaces and their respective burner locations.

Furnaces may be broadly divided into two general types, depending on the manner in which the heated material is handled. In the intermittent, or batch, type, the material is placed in a specific location in the furnace and remains there until heating is completed. The interior temperature of the furnace is practically constant during the heating period. In the continuous type, the material moves with respect to the furnace during its heating. Motion is accomplished in several ways. The material may pass over a stationary hearth; the hearth itself may move, the material remaining stationary above it; or both the material and hearth may move.

Direct Furnaces

In direct furnaces, products of combustion come into direct contact with the material to be heated. The actual location where combustion takes place and the manner of handling the products vary considerably. The following subdivisions have been established:

Direct-fired—combustion takes place in the heating chamber itself. Such a furnace may also be termed an oven or box type.

Underfired—combustion takes place under the hearth and sweeps up into the heating chamber.

Sidefired—combustion takes place in a chamber at one side of the heating space and products pass over a bridge wall into this space.

Overfired—combustion takes place above the heating space and products enter it thru a perforated arch.

Reverberatory—combustion takes place at some distance from the hearth and products are deflected onto it by an arched or sloping roof.

Recuperative—incoming air for combustion is heated by an indirect exchange with flue gases. Both air and flue gases flow without interruptions and always in the same direction over the heat-exchanging surfaces and thru the furnace.

Regenerative—incoming air for combustion is heated by passing it thru a heat exchanger of refractory material, generally checker-brick. The air is then mixed with the fuel, and the mixture is ignited as it passes into the furnace. The combustion products give up part of their heat to the furnace and to the material being processed. Before passing out the flue, the products of combustion pass thru an alternate set of checker-bricks, heating them up. The direction of air and gas flow is reversed periodically and the checker-bricks are alternately either heating up the combustion air or being heated by the flue gases.

Scale-Free Heating. Essentially, the direct gas-fired furnaces involved, both the batch and the continuous designs, are self-contained atmosphere generators. The furnaces maintain a protective atmosphere of partially burned fuel gases around the work in process.[4] The oxidizing constituents, CO_2 and H_2O, are suppressed, while the concentrations of CO and H_2 are favored.

One process report[5] noted the following points: (1) Reaching 2375 F with reducing conditions, e.g., 50 per cent of air required for complete combustion, requires either preheating air to a high temperature, 1650 F, or preheating both air and gas simultaneously to a lower temperature, 1100 F.* (2) Mild steel can be heated up to 1200 F (stainless steel up to 1550 F) under oxidizing conditions, but near the oxidizing-reducing limit, without exceeding an oxidation rate of 0.001 mm per hr. (3) The reduction in radiant heat transfer attributable to the lower concentrations of CO_2 and H_2O in reducing atmospheres is more than compensated for by the radiation from the fine particles of incandescent carbon forming when hydrocarbons are cracked.

* The latter technique was found to be more practical.

Table 12-132 Temperatures for Various Industrial Operations
(highest temperature during heating process)

Heating process or subsequent operation	Temp, °F
Annealing	
Aluminum	750
Brass	1000
Cold-rolled strip	1250–1400
Copper	1150
German silver	1200
Glass	1150
High-carbon steel	1500
Low-carbon steel	1350
Malleable iron, long cycle	1600
Malleable iron, short cycle	1800
Manganese-steel castings	1900
Nickel or Monel wire or sheets	1470
Steel castings	1650
Stainless steel (400 series)	1425–1460
Stainless steel (300 series)	1820–1900
Bar and pack heating, stainless steel	1900
Bisque-firing porcelain	2250
Blueing	500
Box-annealing steel sheets	1600
Burning firebrick	2400–2700
Burning Portland cement	2600
Calcining limestone	2500
Calorizing (baking in aluminum powder)	1700
Carburizing	1750
Cracking petroleum	750
Cyaniding	1800
Decorating porcelain	1400
Enameling, wet process	1200
Forging steel	2150–2250
Glazing porcelain	1830
Glost-firing porcelain	2050
Hardening high-speed steel	2200
Heating	
Aluminum for rolling	850
Brass for rolling	1450
Copper for rolling	1600
Sheet, bars	1700
Sheet steel for pressing	1920
Spring steel for rolling	2000
Steel blooms and billets for rolling; also rivets	2275
Steel for drop-forging or die-pressing	2370
Tool steel for rolling	1900
Heat-treating medium-carbon steel	1550
Hot-dip tinning	500
Melting	
Chromium steel	3250
Glass	2850
Steel	3050
Nitriding steel	950
Normalizing	
Sheet steel	1750
Stainless steel	1700–2000
Steel pipes	1650
Pack-heating sheet steel	1750
Patenting wire	700
Rolling stainless steel	1750–2400
Rolling steel	2100–2350
Strain-relieving	500
Tempering high-speed steel	630
Tempering high-speed steel in oil	500
Vitreous enameling castings	1850
Vitreous enameling sheet steel	1600
Welding steel tubes from preformed skelp	2550

Indirect Furnaces

These furnaces are used when combustion products should not come into contact with the material to be heated, since such contact is injurious to it. Enclosed muffle, radiant tube or sealed crucible types of furnaces may be used. In some cases when protection against high temperature is of greater importance than protection against furnace atmosphere, the muffle roof may be omitted, giving a semimuffle.

FORGES

Forging equipment generally consists of a stationary anvil, or die, and a movable hammer, or head. The piece to be forged is usually placed on the anvil and the hammer is forced against it in a series of sudden blows (hammer forging) or as a single slowly applied force (press forging). Pinch rolls are also considered forging equipment.

Forging Temperatures and Heating Capacity

Stock Temperature. Metals most commonly hot-forged and temperatures to which they may be heated follow:[6b]

Aluminum	680–880 F	Nickel	1600–2300 F
Brass	1200–1500 F	Nickel silver	1350–1500 F
Bronze	1300–1600 F	Steel	See Table 12-133
Copper	1400–1600 F	Steel, stainless	See Table 12-134
Magnesium	650–850 F		

The actual temperature to which the stock should be heated for forging depends on its chemical composition and the degree of metal reduction required.

Table 12-133 Maximum Forging Temperatures for Steel
(excluding stainless steels)

Carbon, per cent	Temp, °F: Carbon steels	Alloy steels
0.10	2400	2350
.20	2375	2300
.30	2350	2250
.40	2350	2250
.50	2300	2200
.60	2250	2175
.70	2225	...
0.90	2150	...
1.10	2075	...
1.50	1900	...

Heating-Chamber Temperature. When heating steel for forging, scale and decarburization should be kept at a minimum. This can best be accomplished if the metal is under furnace heat for the shortest possible time. The temperature to which the stock must be heated can vary somewhat. The greater the temperature difference between heat source and stock, the greater the quantity of heat transmitted. Accepted practice is to operate the furnace at a considerably higher temperature than that to which the stock is to be heated. The difference is termed *temperature differential*, or *head*.

In the conventional manually operated forge furnace, the temperature head is usually 100 to 200 F above the forging temperature. When operating at a high head, the stock must be removed as soon as the desired temperature is attained, to

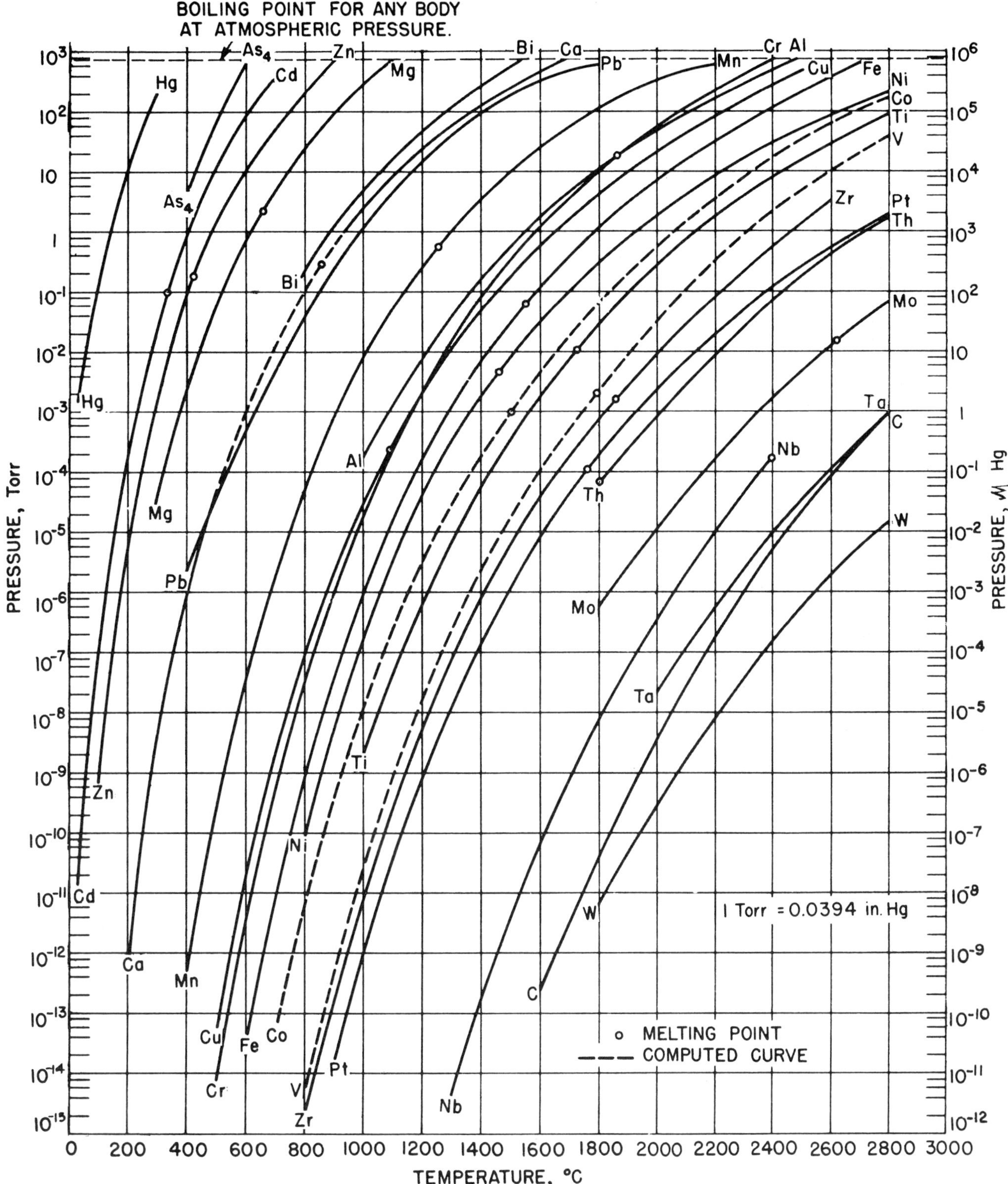

Fig. 12-194 Vapor pressures of metallic elements. (S. A. Heurtey)

prevent surface damage. The allowable head depends upon the thickness of the stock and the rate at which it can absorb heat. The thicker its cross section, the lower the temperature head should be. Up to 1 in. thickness the maximum temperature head will, in most cases, produce no harmful results. For metal thicknesses above 3 in. or for materials that absorb heat slowly, the head should be maintained at a minimum.

High-speed or *rapid heating* in connection with forging are terms used to describe heating systems that provide for the greatest heat absorption by the stock in the shortest time. These processes materially reduce the scaling, decarburization,

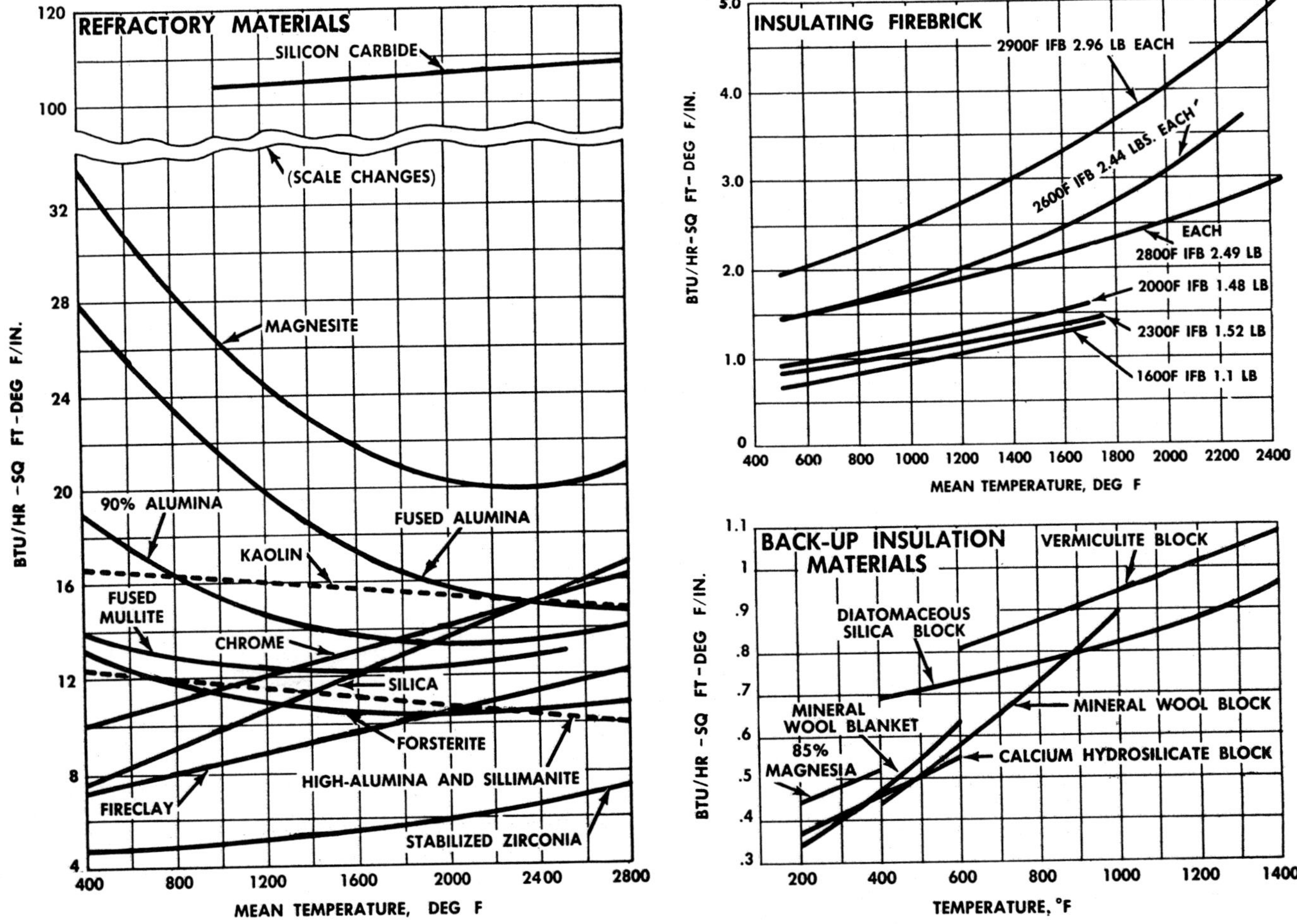

Fig. 12-195 Thermal conductivities, k, of various refractories (brick and castables) and insulating materials in air.[1] When a furnace gas other than air is involved, the following equation may be used to take account of the atmosphere penetration of porous brick:

$$k_{bg} = k_{ba} - Pk_a + Pk_g$$

where: k = thermal conductivity, Btu/hr-sq ft-°F/in.
P = brick porosity, decimal fraction
bg = brick in gas
ba = brick in air (from above charts)
a = air (from Table 2-95)
g = gas (from Table 2-95)

Alternatively, curves are available enabling direct determination of thermal conductivity of refractories when the thermal conductivity of the ambient gas is known.[2]

and grain growth of the stock; greatly diminish the floor space required; and considerably increase the production rates over those of conventional forges (Table 12-140c).

Heating Capacity. The normal heating capacity of oven and slot forges is based on an empirical heating rate, expressed in *pounds per hour per square foot of hearth area* for various types of metals as follows:

Metal	*Hearth area, lb per hr-sq ft*
Steel	80 to 100
Copper and brass	70 to 80
Aluminum and magnesium	25 to 30

If operation is at a higher heating rate, overheating of the forge and excessive maintenance usually result.

Heating Time

How long the piece of stock must remain in a forge depends on (1) the kind of material; (2) the thickness of stock; (3) the temperature required; and (4) the furnace temperature. Figure 12-197 gives the estimated times it takes heat to penetrate a round steel rod at various furnace and rod temperatures. For square rods the time is assumed to be the same as for a round rod of a diameter equal to the side of the square. For rectangular rods the time is assumed to be the same as for a round rod of a diameter equal to the smaller side.

Types of Forges

The forge suitable for a given application depends largely on the type and dimensions of the stock to be forged and on

Table 12-134 Forging Temperatures for Stainless Steels

AISI type No.	Composition, per cent			Forging temperatures, °F		
	Carbon	Chromium	Nickel	Preheat	Initial	Finish
			Ferritic chromium steels			
430	0.12 max	14.0–16.0	...	1400	1900–2050	1300–1450
442	.35 max	18.0–23.0	...	1400	1900–2050	1300–1450
446	0.35 max	23.0–27.0	...	1400	1900–2050	1300–1450
			Martensitic chromium steels			
403	0.15 max	11.5–13.0	...	1500	2050–2150	1600–1700
410	.15 max	10.0–14.0	...	1500	2050–2150	1600–1700
416	0.15 max	12.0–14.0	...	1500	2050–2150	1600–1700
420	Over 0.15	12.0–14.0	...	1500	2000–2100	1600–1700
440	0.50–0.70	14.0–18.0	...	1500	1950–2050	1600–1700
440	.75–0.95	14.0–18.0	...	1500	1950–2050	1600–1700
440	0.90–1.05	14.0–18.0	...	1500	1950–2050	1600–1700
			Austenitic chromium-nickel steels			
302	0.08–0.20	17.0–20.0	8.0–10.0	1500	2150–2200	1600–1700
303	0.20 max	17.0–20.0	8.0–10.0	1500	2150–2200	1600–1700
304	.08 max	17.0–20.0	8.0–10.0	1500	2150–2200	1600–1700
309	.20 max	22.0–26.0	12.0–15.0	1500	2100–2150	1700
310	.25 max	24.0–26.0	19.0–22.0	1500	2100–2150	1700
311	.25 max	19.0–21.0	24.0–26.0	1500	2100–2150	1700
321	.10 max	17.0–20.0	7.0–11.0	1500	2150–2200	.600–1700
325	.25 max	7.0–10.0	19.0–23.0	1500	2000–2050	1600–1700
327	0.10 max	17.0–20.0	8.0–12.0	1500	2150–2200	1600–1700

the portion to be heated. Four types of forges are generally used: furnace, slot, rapid heating, and continuous.

Furnace Forge. For large pieces that must be heated all over, a furnace, or oven, forge is usually the most advantageous. Built in the form of a box with a door at one end and a solid hearth, it is exclusively overfired. Burners located in the side walls are directed to fire across the product.

Slot Forge. The conventional slot forge has a box-shaped heating chamber and a slot opening in one of its longer walls, just above the hearth. The solid hearth is overfired with burners in either the end or the side walls. In a forge with a heating chamber considerably longer than it is wide, the burners are located in the rear wall; otherwise, they are located in the end walls. Overfiring is necessary because molten slag, formed when metals are heated to high temperatures, collects on the hearth and will readily flow into any openings below its level. Therefore, burners are not located at or below hearth level. Table 12-135 gives dimensions and operating characteristics of some common sizes of slot forge equipped with clear flame gas burners.

Rapid Heating. This type (Fig. 12-198) combines generation of a large quantity of heat in a small combustion space with exposure of the heated stock to large radiating surfaces. This design uses the inner surface of the refractory lining as a radiant burner face, transferring heat at high rates to exposed surfaces of the charge. Resulting high velocities and centrifugal action cause the burning gases to hug the refractory surfaces and spread out along them, completing combustion with extreme rapidity. Seven case histories are available.[7]

Barrel types incorporate a cylindrical heating chamber section. Each section consists of a refractory-lined steel shell, equipped with a series of burner ports on relatively close centers. These forges are used primarily for continuous production of round billets from 2 to 8 in. (or 1 to 8 in. square). The split-barrel type is used for end-heating of forging stock.

Table 12-135 Dimensions and Operating Characteristics of Typical Slot-Type Forges

Furnace dimensions, in.							Operating characteristics			
Heating chamber			Working opening		Overall		Heating from cold to 2300 F		Gas to hold at 2300 F, MBtu/hr	Max gas capacity, MBtu/hr
W	× H	× L	W	× H	L	× W	MBtu	Time, hr		
17	9.5	6	12	14.5	42	53	170	1.0	56	170
17	11.5	18	12	14.5	72	55	380	1.0	81	380
23	12	9	18	14.5	52	58	310	1.0	92	310
23	15	24	18	14.5	84	56	700	1.0	135	700
29	16.5	13.5	24	15	58	62	560	1.0	154	560
29	15	24	24	15	90	56	880	1.0	175	880
35	16.5	13.5	30	15.5	62	62	840	1.25	196	670
35	16.5	24.9	30	15.5	96	67	1375	1.25	235	1060
41	16.5	13.5	36	16	65	64	1060	1.25	245	850
41	16.5	24.9	36	16	102	66	1620	1.25	290	1300

W—width, H—height, L—length

Fig. 12-196 Typical designs of various types of furnaces.[3] (Hauck Mfg. Co.)

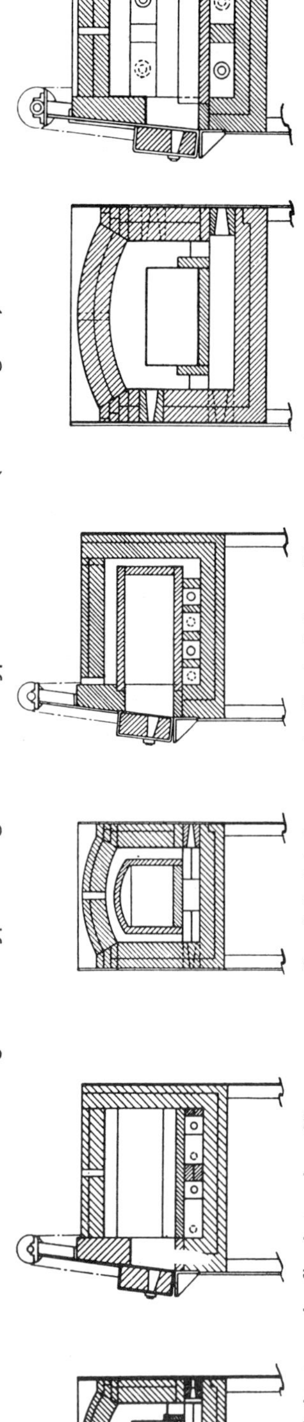

Small batch type, under-fired, semi-muffle, heat-treating furnace.

Small batch type, under-fired, full muffle, heat-treating furnace.

Large batch type, under- and over-fired, semi-muffle, heat-treating furnace.

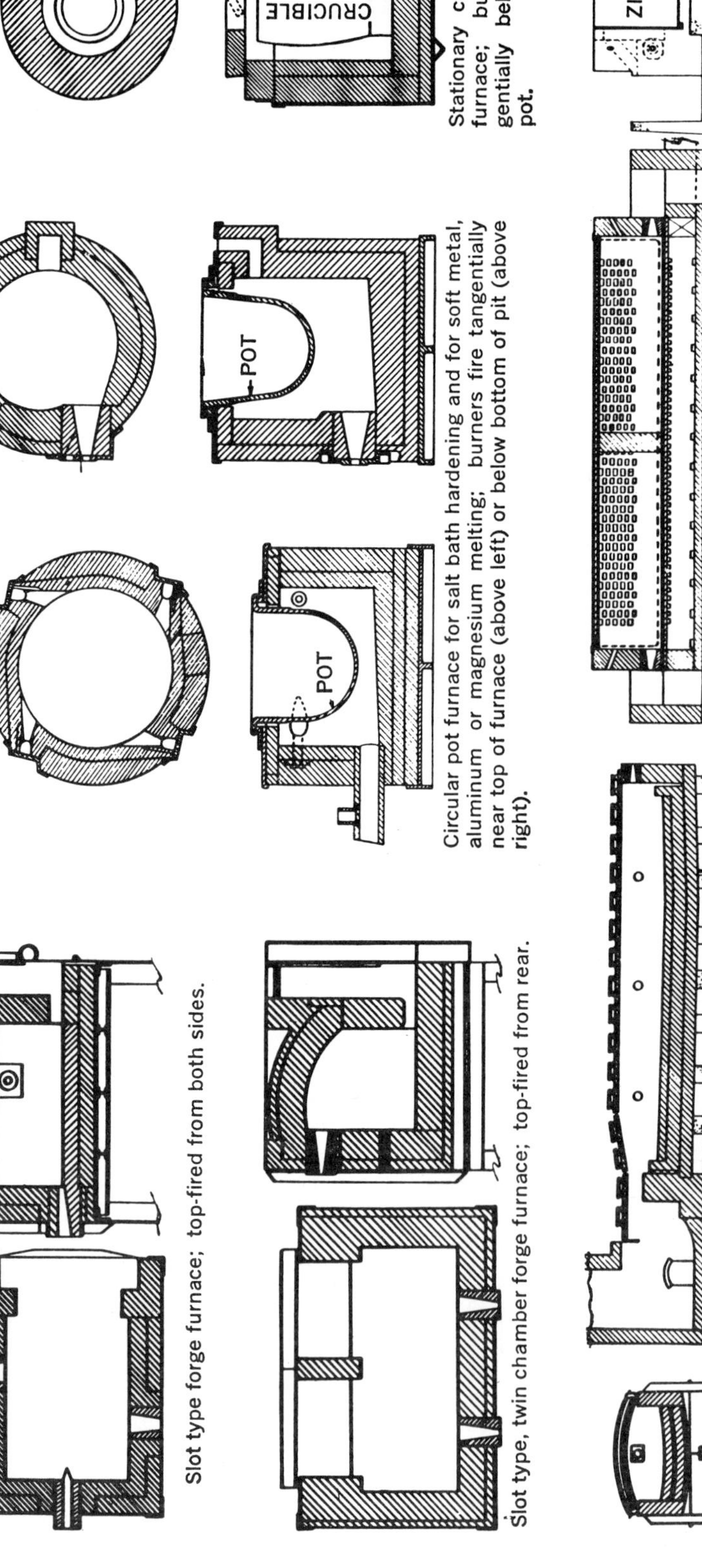

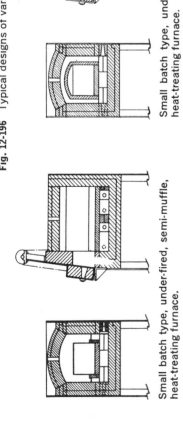

Slot type forge furnace; top-fired from both sides.

Circular pot furnace for salt bath hardening and for soft metal, aluminum or magnesium melting; burners fire tangentially near top of furnace (above left) or below bottom of pit (above right).

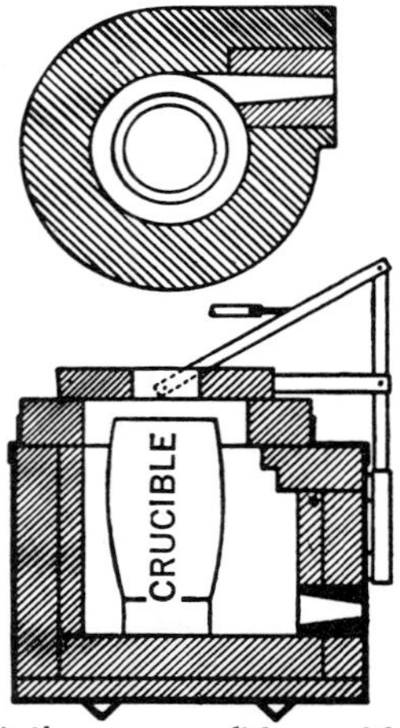

Stationary crucible melting furnace; burner fires tangentially below bottom of pot.

Slot type, twin chamber forge furnace; top-fired from rear.

Reverberatory furnace for malleable iron melting.

End-fired galvanizing kettle.

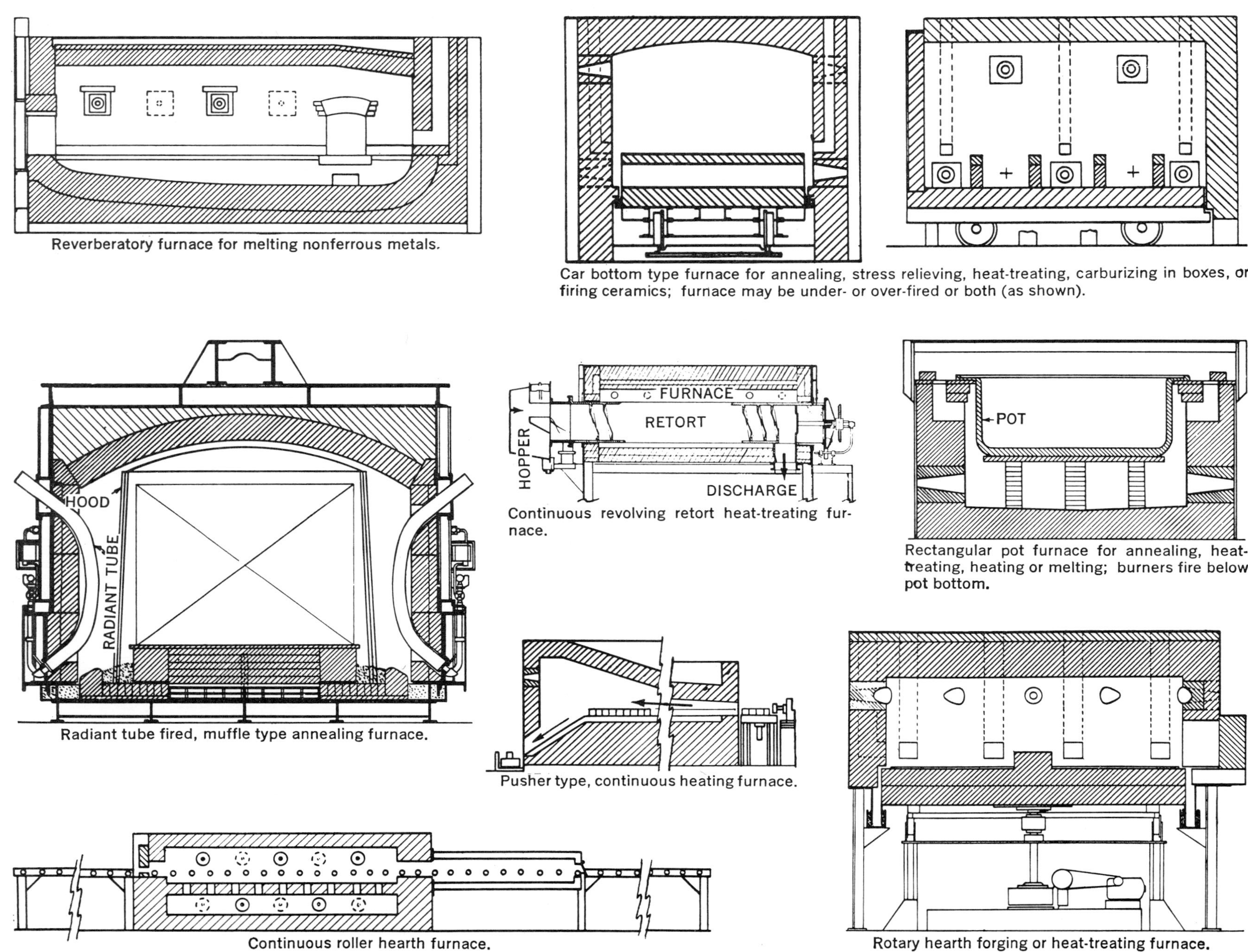

Reverberatory furnace for melting nonferrous metals.

Car bottom type furnace for annealing, stress relieving, heat-treating, carburizing in boxes, or firing ceramics; furnace may be under- or over-fired or both (as shown).

Radiant tube fired, muffle type annealing furnace.

Continuous revolving retort heat-treating furnace.

Rectangular pot furnace for annealing, heat-treating, heating or melting; burners fire below pot bottom.

Pusher type, continuous heating furnace.

Continuous roller hearth furnace.

Rotary hearth forging or heat-treating furnace.

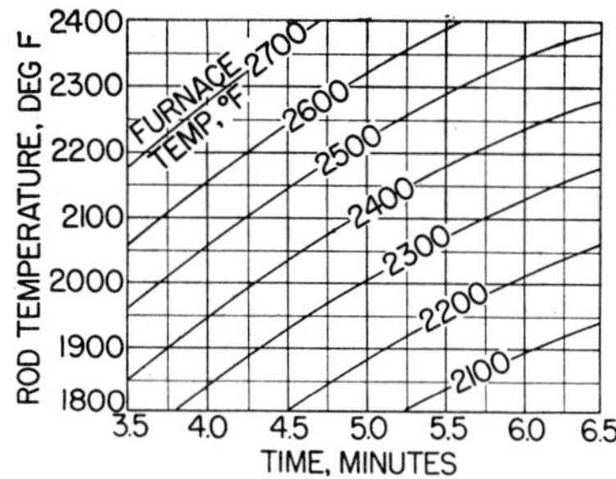

Fig. 12-197 Approximate time required to raise round steel rod to temperature with various furnace temperatures—*per inch of diameter* for rods up to 3 in. Example: heating a 2-in. rod to 2000 F in a 2300 F furnace requires about 2 × 5 min = 10 min.

The line-burner type is suitable for heating any portion of a piece of forging stock.

Continuous. For applications requiring large volume production, forges are equipped with mechanical systems (e.g., moving hearths, rails, and chains) that move heated stock at a constant rate. Some systems are built directly into the forge. Others are applied to its outside structure. One type, frequently used when the entire billet is to be heated, is the power-driven *rotary hearth* with a circular furnace casing. The stock is introduced thru a door opening in the casing and removed thru the same or an adjacent opening after reaching the required temperature while making one revolution.

Another type is the *pusher*, which has an opening at both ends and has rails on the hearth that extend thru the furnace. The stock, which must have flat sides or ends, rides on these rails and is pushed, piece against piece, by a pusher mechanism at the charging end. A variation of the pusher type is applicable to stock that must be end-heated. The rails in this case are mounted in front of a slot-type forge, the portion of the stock to be heated extending into the heating chamber. The unheated portion rides on the rails. If it is not heavier than the heated portion, holddown rails must be provided.

To end-heat stock which has a contour that precludes pushing, a series of continuous parallel chains may be mounted on the front of a slot-type furnace with open ends to convey the charge thru.

Combustion Systems

Most small gas forges are equipped with tunnel-type burners. A number of variations of this type of burner are given in Figs. 12-153 thru 12-157.

Clear Flame. A gas-air mixture under pressure is discharged into a tunnel, where it burns rapidly and almost completely before it enters the heating chamber, giving what is commonly known as clear-flame combustion. The

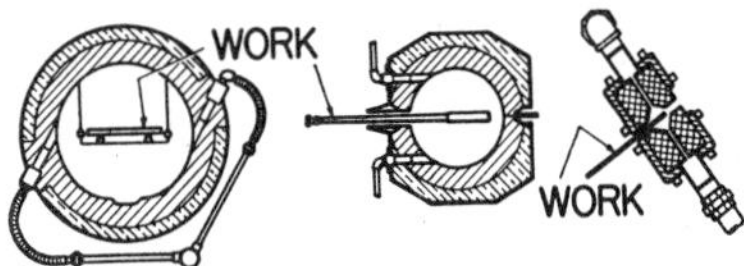

Fig. 12-198 Three rapid heating forge designs: (left) barrel; (center) split barrel; (right) line burner. (Surface Combustion Div. of Midland-Ross Corp.)

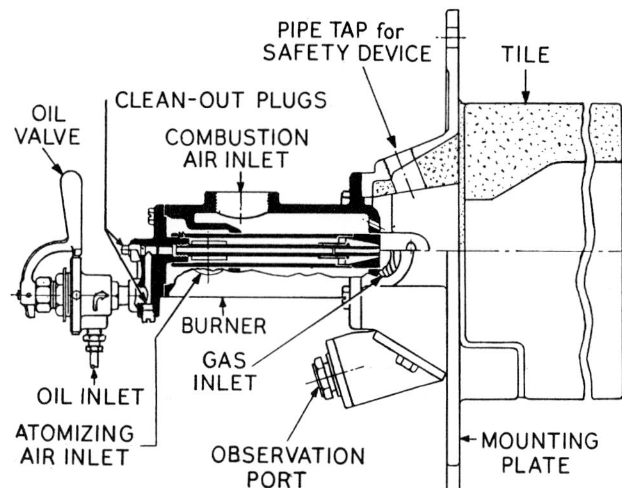

Fig. 12-199 Combination gas and oil burner. (North American Mfg. Co.)

mixture is automatically proportioned over the entire operating range by an inspirator. A single valve controls input.

Burner adjustment is usually "rich," with a slight deficiency of air, to keep scaling of the stock at a minimum. A relatively light scale is formed and adheres tightly to the metal surface. Free-falling scale can be produced by oxidizing conditions; this may be desirable from the point of view of forging-die life.

Luminous Flame. Luminous flame has the advantage of radiating considerable heat from the flame itself, and the ability to operate at a higher degree of *richness*. Sufficient combustion space must be provided, both in cross-sectional area and in length, to prevent flame turbulence before combustion is completed. The usual practice is to provide at least 1 cu ft of combustion space for each 50 MBtu per hr of heat input.

Combined Gas and Oil Burners. There are three kinds of these burners: (1) the type designed to burn gas and oil simultaneously; (2) the dual-fuel type, designed to burn either gas or oil, but not simultaneously (gas is used normally, oil during peaks); (3) the combination—designed to burn either gas or oil, or both together. A closed combination (sealed-in) forge burner with separate air atomizing oil and gas inlets is shown in Fig. 12-199. These burners should provide a means of mixing gas with primary air to ensure the rapid combustion necessary to atomize the oil, as well as a means of introducing additional air required for combustion. Because a luminous flame is required, combustion space must be adequate to prevent turbulence from destroying the flame pattern before combustion is completed. One cubic foot of combustion space should be provided for each 50 MBtu per hr of heat input.

One gal of No. 5, or heavier, oil per 280 MBtu of gas burned gives an atmosphere that will produce a thin, loose, easily removable scale. Increased heat radiation from the flame accomplishes maximum heating capacity with a minimum forging temperature. Heat generated by the rapid burning of the gas accelerates the gasification, cracking, and combustion of the oil, producing in turn a luminous flame and a very desirable forging atmosphere.

Temperature Controls

Forges are usually operated with a head temperature, i.e., a considerable differential between furnace and product temperatures. Accurate temperature control of product is, therefore, rather difficult with common types of controls that sense only furnace temperature. In many cases, operation depends entirely on the operator's skill in manual control and in removing the stock before it is overheated.

Some batch-type forges have high-limit temperature controls to prevent damage by overheating during stand-by periods when no cold stock is charged. Forges with continuous systems can be controlled by varying the speed of the stock's passage thru the heating chamber and, consequently, the stock temperature at discharge. In forges so operated, the temperature at the discharge point governs the temperature control. The most common type of temperature control has a suitably protected thermocouple element located near the arch in the center of a batch-type forge. On continuous forges, the thermocouple is located near the point of stock discharge. Another practical type of control uses a radiation pyrometer as the sensor. It may be located outside the heating chamber, thus avoiding the problem of high-temperature deterioration faced by other types of controls.

In batch-type forges the radiant element is sighted on a target, located at the end of a tube that extends down thru the furnace arch near the heating chamber center. In the continuous type, the element can be sighted directly on the piece of stock next to be discharged. In such an application the radiation type of control has the added advantage of controlling the actual temperature of the stock as it is delivered to the forge.

Forge Selection

The size and type of forge furnace for a given operation depend on the following:

1. **Kind of metal charged**—establishes operating temperature.
2. **Thickness of stock**—determines time it must spend in forge. For steel, see Fig. 12-197.
3. **Production rate required**—the number of forgings required per hour, multiplied by the time, in hours, stock must spend in furnace, indicates the number of pieces its hearth must accommodate. In slot-type forges, or in those into which a cold piece of stock is inserted as each heated one is removed, good practice provides for three or four additional pieces. A gap can then be left between the heated and the newly charged stock, to prevent cooling of the former by radiation to the latter.
4. **Dimensions of stock**—the maximum area of the portion of the stock resting on, or occupying space horizontally above, the hearth plus an allowance for clearance between pieces (usually ¼ to ¾ in.) should be calculated. This area, multiplied by the number of pieces to be accommodated, gives the hearth area required.
5. **Weight of stock**—the total number of pieces going thru the furnace, multiplied by the weight of the portions to be heated, expressed in pounds per hour, will determine whether the calculated hearth area is sufficient to keep the heating rate within allowable limits. For steel, this should not exceed 80 to 100 lb per hr-sq ft of hearth area.
6. **Heating chamber height**—usually established by the space required to install burner equipment without flame impingement on the charge, and to provide sufficient combustion space

Heating Requirements

The gas input required by a forge furnace in active production (after it has reached operating temperature) must be sufficient to supply the heat absorbed by the product and the radiation losses. Input may be calculated from Eq. **1**:

$$Q = (R + M)T/A \tag{1}$$

where:

Q = gas input, Btu per hr
R = radiation losses from forge, Btu per hr
M = heat absorbed by metal worked, Btu per hr
T = gross heating value of gas, Btu per cu ft
A = available heat in gas, T *minus* heat content in combustion products from Figs. 2-36 thru 2-42, Btu per cu ft

Figures 2-21 to 2-23 show the heat content of several common metals at varying temperatures. Similar graphs are available for steels which show the effects of different percentages of carbon and common alloying metals.

Radiation losses are determined by the temperature of the forge, its refractory construction, and areas of walls (including those of arch and hearth) and openings. All openings are assumed to lose heat by radiation to a black body.

Estimating Daily Gas Consumption. The heat required during the forge's inactive and active periods must be considered in determining the total daily gas consumption. Heat losses during a stand-by period consist of those due to radiation and heat carried away by the flue gases. In addition, if the forge is operated only one or two shifts per day, the fuel required to bring it up to temperature following shutdown should be included.

The daily fuel requirement of a forge may be estimated by Eq. **2**:

$$C = [K(R + M) + SR]\frac{T}{A} + Qh \tag{2}$$

where:

C = gas consumption, Btu per day
K = production time, hr per day
S = stand-by time, hr per day
h = estimated time to bring forge to temperature after shutdown; depends on temperature of operation and hr per day it must be maintained, hr

Other terms are defined under Eq. **1**.

FERROUS HEAT TREATING

The addition of carbon greatly affects the properties of iron. Steel contains less than 1.7 per cent carbon; cast iron contains more. The iron-carbon equilibrium diagram (Fig. 12-200) shows that the changes in the constitution of alloys with temperature are the basis for the heat treatment of plain carbon steels.

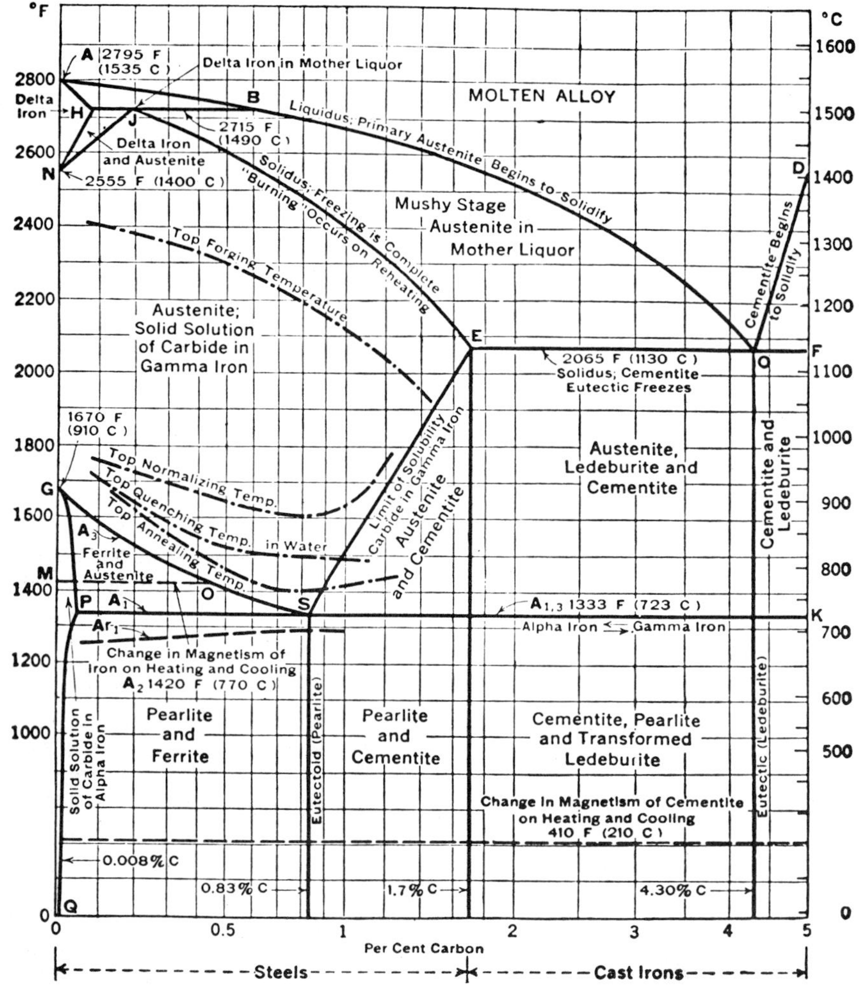

Fig. 12-200 Iron-carbon equilibrium diagram.

The presence of common alloying elements such as chromium, manganese, molybdenum, nickel, and others materially affects the metal's response to heat treatment and its resulting physical characteristics. Controlled additions of such elements are made to produce a steel of definite required properties. Identification of standard alloy types may be determined by the type and quantity of such alloy additions, as shown by Table 12-136.

"S" Curves. Plots of temperature and hardness (vertical) against time (horizontal, on a logarithmic scale) indicate the transformation of various steel alloys. Diagrams for the plain carbon steels are quite similar, but those for alloy steels vary widely.

Critical temperatures, defined as the temperatures at which phase changes take place in iron-carbon alloys during heating and cooling, are also affected by the presence of alloying additions. Consequently, selection of heat treating cycles is also influenced, as Table 12-137 indicates.

Steel possesses a wide range of physical properties obtainable by changes in chemical composition and by suitable heat treatments. Heat-treating processes involve the controlled heating of the alloy, with minimum thermal stresses, to a temperature determined by its critical temperature, followed by an equalizing or soaking period. Then the steel is uniformly cooled by quenching in the medium suited to produce the desired structure. Finally, the steel is reheated to a definite temperature, modifying the structure as quenched, to obtain the required physical characteristics.

Long practice has given generally recognized names to numerous separate phases of the basic heat-treating process and its subsequent variations. The most common are:

Annealing, or heating and cooling of a metal to induce softness or to alter physical and mechanical properties. The objective is usually to recrystallize the metal completely by heating it above its critical point while minimizing grain growth. Specific methods are bright, full, isothermal, process annealing, graphitizing, malleableizing (cast iron), and spheroidizing.

Hardening, or heating and cooling of a metal to increase its hardness, either throughout or on the surface only. Types of surface hardening, known as **case hardening,** follow:

Carburizing, or addition of carbon to the surface by heating in contact with carbonaceous material.

Cyaniding, or simultaneous addition of carbon and nitrogen by heating in a molten cyanide salt.

Nitriding, or addition of nitrogen by exposure to nitrogen or nitrogenous material.

Gas cyaniding (carbo-nitriding), or nitriding thru use of a gaseous atmosphere.

Table 12-136 Basic Numerals for Various Types of Steel*
(SAE numbering system)†

Type of steel	Numerals‡
Carbon	1xxx
Plain carbon	10xx
Free cutting (screw stock)	11xx
Manganese	13xx
Nickel	2xxx
3.50% Ni	23xx
5.00% Ni	25xx
Nickel-chromium	3xxx
1.25% Ni–0.60% Cr	31xx
1.75% Ni–1.00% Cr	32xx
3.50% Ni–1.50% Cr	33xx
3.00% Ni–0.80% Cr	34xx
Corrosion- and heat-resistant	30xxx
Molybdenum	4xxx
Chromium	41xx
Chromium-nickel	43xx
Nickel	46xx and 48xx
Chromium	5xxx
Low-chromium	51xx
Medium-chromium	52xxx
Corrosion- and heat-resistant	51xxx
Chromium-vanadium	6xxx
Silicon-manganese	9xxx

* From Surface Combustion Div. of Midland-Ross Corp.

† The AISI numbers are the same as the SAE numbers except for an additional prefix indicating method of manufacture. These prefixes are: A, open hearth alloy steel; B, acid Bessemer carbon steel; C, basic open hearth carbon steel; CB, either B or C; D, acid open hearth carbon steel; E, electric furnace alloy steel.

‡ The first digit indicates the type to which the steel belongs; for example 1 indicates a carbon steel, 2 a nickel steel. For simple alloy steels the second digit generally indicates the approximate percentage of the predominant alloying element. Usually the last two or three digits indicate the average carbon content in "points," or hundredths of 1 per cent. Thus, 2340 indicates a nickel steel of approximately 3.0 per cent nickel and 0.40 per cent carbon.

Interrupted quenching, or quenching from hardening temperatures at a critical-rate by immersion in hot oil or salt to obtain the physical characteristics desired.

Normalizing, or heating of steel above the transformation range, followed by air cooling.

Rapid (high-speed) heating, or heating steel under a high-temperature head, giving differences up to 2000 F between furnace and work at discharge. Practical applications of rapid-heating rates (above 100 lb per hr-sq ft of hearth area) include processes for forging, annealing, brazing, hardening, stress relieving, and tempering.

Stress relieving, or heating of steel to a suitable temperature held sufficiently long to relieve residual stresses as from casting, quenching, cold-working, or welding.

Tempering, or reheating quench-hardened or normalized steel to a temperature below the transformation range, followed by cooling.

Furnace Design

The two basic furnace designs are (1) the batch, or "in-and-out," and (2) the continuous furnace. The batch type with special automatic handling equipment is sometimes called *intermittent* or semicontinuous. Table 12-138 shows variations of the two basic types.

Material Handling System. The system used may influence the choice of furnace. Overhead handling systems work well with a battery of pit furnaces. Industrial fork trucks, transfer cars, and other over-the-floor equipment as well as gravity and power conveyors work well with horizontal hearth or oven-type furnaces. This equipment is mainly utilized for material handling during charging and discharging.

Material movement within continuous and semicontinuous furnaces can be accomplished in several ways (Table 12-138), depending on the type of part and the volume of production required.

Furnace Selection and Sizing

The required volume and rate of production principally govern the selection of furnace type, but the desired product and processing also have considerable influence on the design utilized. Processing includes heat-treating and meeting any special requirements such as appearance, subsequent machining operation, and intended use of part; these requirements often indicate use of a special furnace atmosphere.

Example.[8] Batch annealing of steel strip may require 25 to 35 hr to heat the entire charge to 1250 F, plus a 15-hr soaking period to obtain temperature uniformity. At this point recrystallization is complete and grain growth commences. The product is then cooled for as long as 60 hr. On the other hand, the cycle for continuous strip annealing at 650 fpm may take less than 2 min, broken down as follows:

	Seconds
Preheating to 400 F	7.0
Heating from 400 to 1320 F	3.5
Cooling from 1320 to 800 F	50.5
Cooling from 800 to 400 F	55.0

Pot furnaces are often used where production rates vary, surface oxidation must be prevented, and rapid uniform or selective heating is desired. Salt "drag out" or loss must be considered as a cost item under full protection. With special overhead handling facilities continuous production can be achieved.

Determination of the most practical heat transfer medium is of major importance in selection of a furnace type. Selection is also greatly influenced by the factors mentioned previously. Air and furnace atmospheres are the most common media, but wall radiation plays the principal part in heat transfer. The usual media for heating by conduction are oil, lead, and salt, as used in pot furnaces. The particular heat transfer medium selected influences in turn the type of burner equipment and the method of firing.

Selection of a furnace of the proper size involves three main factors: (1) heating rate; (2) time that charge must spend in furnace; and (3) load pattern, dictated by size and contour of pieces composing charge. Furnace size for a given application depends on the type and size of parts to be treated, on production rates required, and, to some extent, on plant layout and possible future expansion. Proper sizing is essential for best operating performance. Too large a unit may require excessive fuel consumption; one too small may have a shortened service life resulting from operation under overload conditions.

Table 12-137 Approximate Critical Temperatures for Various Steels, °F[6a]

	On slow heating*			On slow cooling†		
No.	Ac_1	Ac_2	Ac_3	Ar_3	Ar_2	Ar_1
Carbon steels						
1010	1350	1405	1605	1570	1400	1255
1015	1355	1410	1585	1545	1395	1265
1020	1355	1410	1570	1535	1395	1260
1025	1355	1405	1545	1515	1405	1255
1030	1350	1405	1495	1465	1405	1250
1035	1345	...	1475	1455	1395	1275
1040	1340	...	1455	1415	...	1275
X1040	1340	...	1450	1340	...	1270
1045	1340	...	1450	1405	...	1275
X1045	1335	...	1420	1330	...	1270
1050	1340	...	1425	1390	...	1275
X1050	1335	...	1400	1330	...	1270
1055	1340	...	1425	1390	...	1275
X1055	1335	...	1400	1330	...	1270
1060	1340	...	1410	1370	...	1275
1065	1340	...	1385	1345	...	1285
X1065	1335	...	1380	1330	...	1280
1070	1345	...	1370	1340	...	1280
1075	1350	...	1365	1340	...	...
1080	...	...	1360	1285	...	...
1090	...	...	1360	1285	...	...
1095	...	...	1360	1290	...	...
10150‡	...	...	1355	1290	...	...
Free-cutting steels						
1112	1355	1410	1590	1545	1395	1265
1120	1355	1405	1550	1510	1400	1255
X1315	1345	1420	1520	1495	1370	1245
X1330	1320	1400	1490	1380	1360	1240
X1335	1315	1390	1420	1360	1340	1220
X1340	1310	...	1400	1340	...	1210
Manganese steels						
T1330	1325	...	1480	1340	...	1160
T1335	1315	...	1460	1340	...	1165
T1340	1315	...	1435	1310	...	1160
T1345	1315	...	1410	1300	...	1160
T1350	1310	...	1400	1255	...	1105
T1360‡	1305	...	1405	1200	...	1095
Nickel steels						
2015	1375	1475	1575	1450	1400	1215
2115	1345	1455	1525	1475	1380	1195
2315	1300	1350	1440	1350	1260	1100
2320	1285	1345	1420	1235	1160	920
2330	1275	1315	1400	1180	...	1050
2335	1275	...	1375	1180	...	1050
2340	1280	...	1360	1180	...	1060
2345	1280	...	1350	1180	...	1060
2350	1280	...	1340	1180	...	1070
2515	1250	1335	1420	1220	1140	825
2520‡	1240	1340	1390	1175	1025	825

	On slow heating*			On slow cooling†		
No.	Ac_1	Ac_2	Ac_3	Ar_3	Ar_2	Ar_1
Nickel-chromium steels						
3115	1355	1400	1500	1470	1380	1240
3120	1350	1400	1480	1455	1380	1230
3125	1350	1395	1465	1400	1380	1220
3130	1345	1380	1460	1360	...	1220
3135	1340	...	1445	1300	...	1220
3140	1355	...	1415	1295	...	1220
X3140	1350	...	1430	1300	...	1240
3145	1355	...	1395	1295	...	1220
3150	1355	...	1380	1275	...	1215
3215	1350	1410	1465	1415	1350	1240
3220	1350	1415	1460	1405	1355	1240
3230	1340	...	1435	1395	...	1240
3240	1335	...	1425	1280	...	1240
3245	1345	...	1400	1270	...	1225
3250	1340	...	1375	1255	...	1200
3312	1330	1370	1435	1240	...	1160
3325	1335	1365	1400	1230	...	1160
3330‡	1320	1360	1380	1225	...	1145
3335	1310	...	1360	1200	...	1100
3340	1290	...	1380	1180	...	1100
3415	1330	1370	1425	1340	1300	1220
3435	1290	...	1380	1200	...	1150
3450	1290	...	1360	1200	...	1100
Molybdenum steels						
4130	1395	1435	1485	1405	1395	1280
X4130	1395	1435	1480	1405	...	1250
4135	1395	1440	1475	1380	1360	1280
4140	1380	...	1460	1370	...	1280
4150	1365	...	1395	1355	...	1280
4340	1350	...	1425	1220	...	725
4345	1345	...	1415	1200	...	725
4615	1335	1400	1485	1400	1320	1200
4620	1335	...	1470	1390	...	1175
4640	1320	...	1430	1300	...	1125
4650‡	1315	...	1410	1260	...	1125
4815	1300	...	1440	1310	...	800
4820	1300	...	1440	1260	...	760
Chromium steels						
5120§	1410	1460	1540	1470	1420	1295
5140	1370	...	1440	1345	...	1280
5150§	1330	...	1420	1280	...	1220
52100	1340	...	1415	1315	...	1280
Chromium-vanadium steels						
6115	1420	1460	1550	1450	1380	1300
6120§	1410	1460	1545	1440	1380	1300
6125§	1400	1440	1490	1390	1360	1295
6130	1390	1440	1485	1370	1340	1285
6135§	1390	...	1480	1370	...	1280
6140	1390	...	1455	1375	...	1295
6145	1390	...	1450	1375	...	1290
6150	1385	...	1450	1375	...	1270
6195	1370	...	1425	1360	...	1300
Tungsten steels						
7260	1360	...	1430	1370	...	1310
Silicon-manganese steels						
9255	1400	...	1500	1380	...	1320
9260	1400	...	1500	1380	...	1315

* Ac_1—lower transformation temperatures; Ac_2—magnetic points; Ac_3—upper transformation temperatures.
† Ar_1—lower transformation temperatures; Ar_2—magnetic points; Ar_3—upper transformation temperatures.
‡ Not a standard SAE steel.
§ The critical point determinations for these steels were obtained on a Leeds and Northrup transformation apparatus.

Table 12-138 Typical Basic Furnace Designs

Batch	Continuous
Horizontal hearth (oven)	Pusher—piece-to-piece, pans or trays on roller rails, gravity, reciprocating "dog"
Vertical hearth (pit)	Belt—mesh, cast link, flight, slat
Car bottom	Chain—hearth level, overhead, vertical
Rectangular cover	Roller hearth
	Disk hearth
Circular cover	Walking beam
	Rotating—hearth, casing, drum or retort
Pot type	Miscellaneous

Heating Requirements

Furnace heating requirements are similar to those for forges; i.e., Eqs. **1** and **2** may be used.

Heat removed by the charge, Eq. **1,** includes that needed for heating both the material under treatment and any fixtures or conveyors, if used. It is given by the formula:

$$M = (m + t + c)f \tag{3}$$

where:

M = total heat removed by work, fixtures, and conveyors, Btu per hr
m = weight of work treated, lb per hr
t = weight of trays and fixtures, lb per hr
c = weight of conveyors, lb per hr
f = heat input factor according to material and temperature, Btu per lb

Losses from radiation are made up of heat radiated from the furnace top, hearth, sides, and ends, and from any furnace openings (black body). Each may be calculated using the formula:

$$R = Af_r \tag{4}$$

where:

R = radiation loss, Btu per hr
A = exposed area, sq ft
f_r = heat input factor according to type of wall or opening and temperature (Table 12-139), Btu per hr-sq ft

Heat Input to Furnaces. The sum of the heat requirements of the maximum hourly furnace load *plus* the heat required to maintain the furnace temperature furnish the basis for establishing the maximum hourly heat requirement under steady-state conditions. These data may be used to select the required burner inputs and turndown range when the furnace operating time is long. Compensating increases may be needed when the furnace is operating only for limited periods. These increases will seem to increase the "apparent" radiation losses from the furnace.

In any event, the sum of the heat requirements must be translated into heat input or cubic feet of gas per hour. To do this, the available heat from the burning gas is calculated or read from a combustion chart. For example, if a furnace is operating at 2200 F flue gas exit temperature, then the available heat is under 400 Btu per cu ft of high-inert natural gas burned (1000 Btu gas, Fig. 2-38):

$$\frac{\text{Max hour heat requirement per hr}}{\text{Available heat per cu ft of gas burned}} = \text{gas input, cu ft per hr}$$

The **holding gas input** is determined in the same way, by *dividing* the heat required to maintain the furnace (unloaded) by the available heat. The quotient is the holding input required. The **turndown ratio** is the ratio of the total gas input to the holding input.

The characteristics of various mixing and burner systems will suggest how these requirements may best be met. In general, for a given air pressure and a nozzle-mixing combustion system, a useful operating range of 8:1 is practical. With proportional mixing systems and air at 1.0 psig the useful range will normally be 4:1. If the requirements are 15:1, a double set of burners may be required—one set used for holding and the other, a larger set, used when the furnace is loaded. The heat pattern and temperature ranges of the furnace must also be taken into account in burner selection and sizing.

In determining the total burner capacity needed by a furnace, a safety factor should be applied according to Eq. **5**, particularly if the working temperature must be reached rapidly or the charge must be heated in minimum time.

$$G = KH \tag{5}$$

where:

G = total burner capacity, Btu per hr
H = maximum heat input, Btu per hr
K = safety factor, as follows:

Rapid heating	$K = 1.50$
High furnace temperature (1800–2500 F)	$K = 1.33$
Medium furnace temperature (1400–1800 F)	$K = 1.25$
Low furnace temperature (500–1400 F)	$K = 1.15$

Tables 12-140a thru 12-140d give examples of heating steel, aluminum, and copper in various furnaces. Heatup from cold, heating preheated stock, high-speed heating, and forced convection examples are included.

The rate of heat transfer by conduction from gases or from physical contact with the furnace is very low compared with other methods of transfer. Some authors tend to minimize the effects of convective heat transfer at temperatures above 2000 F. This is erroneous, since furnaces can be designed for high convection at high temperatures. Under theoretically ideal conditions it may be reasonable to expect the mechanism of heat transfer to reduce heating times approximately 20 per cent under pure radiation, in contradiction to the commonly used design procedure. The importance of convection becomes more pronounced as the temperature is reduced, becoming equal to radiation at about 1350 F.[10] Table 12-141a compares the time required for 2-in. plate to reach a core temperature of 2000 F by radiation, by convection, and by both mechanisms together. Table 12-141b gives the data for other thicknesses of plate in a 2800 F furnace.

Note that it is possible to decrease the furnace operating temperature by a combination of radiation and convection. One practical aspect of such a decrease is the increased life of the refractory. One such radiation-convection furnace is a

Table 12-139 Heat Losses from Average Furnace Walls, f_r*

(Btu per hr-sq ft of internal wall area)

Temp, °F	Wall construction													
	4.5A	4.5A 2.5B	4.5A 4.5B	4.5A 9B	9A	9A 2.5B	9A 4.5B	9A 9B	2.5B 4.5C	4.5B 4.5C	9C	2.5B 9C	4.5B 9C	Black body
200	174	70	46	26	97	53	35	21	50	38	55	34	25	175
400	546	220	143	81	306	165	110	66	158	119	174	108	77	750
600	918	370	241	137	514	278	185	111	266	200	292	181	130	1,900
800	1,327	535	348	198	744	401	268	161	385	289	423	262	187	3,900
1000	1,786	720	468	266	1001	540	360	216	518	389	569	353	252	7,000
1200	2,294	925	601	342	1286	694	463	278	666	500	731	453	324	11,700
1400	2,852	1150	748	426	1599	863	575	345	828	621	909	564	403	18,900
1600	3,472	1400	910	518	1946	1050	700	420	1008	756	1106	686	490	29,100
1800	4,191	1690	1099	625	2349	1268	845	507	1217	913	1335	828	592	42,800
2000	5,010	2020	1313	747	2808	1515	1010	606	1454	1091	1596	990	707	60,100
2200	5,927	2390	1554	884	3322	1793	1195	717	1721	1291	1888	1171	837	81,000
2400	6,944	2800	1820	1036	3892	2100	1400	840	2016	1512	2212	1372	980	107,000
2600	8,060	3250	2113	1203	4518	2438	1625	975	2340	1755	2568	1593	1138	139,000
2800	9,275	3740	2431	1384	5199	2805	1870	1122	2693	2020	2955	1833	1309	178,000
3000	10,590	4270	2776	1580	5935	3203	2135	1281	3074	2306	3373	2092	1495	225,000

A—firebrick, B—insulation, C—insulated firebrick.
* From surface Combustion Div. of Midland-Ross Corp.

cylinder, tangentially fired so as to have the major portion of the combustion occur while the gases are scrubbing the furnace walls.[10]

The value of h_c, the convective heat transfer coefficient, has been found to be independent of the temperature difference between the charge surface and the furnace atmosphere at a constant atmosphere temperature.[11] The heating of steel in a recirculating furnace from room temperature up to temperatures of 1200 F has been studied. Figure 12-201a shows the variation of the convection coefficient with furnace temperature and velocity. These experimental data were based on the flow of air rather than a combustion atmosphere. Air was used as the heat transfer fluid in order to avoid the complication of radiation from combustion gases; hence, the figure should not be used for radiating gases.

Figure 12-201b shows the variation in overall heat transfer coefficient as a function of furnace temperature and surface temperature of charge.[10] Note that the heat coefficient increases with the charge surface temperature. Actually, however, less heat is transferred as the product temperature rises, because the temperature differential used in calculating the heat transfer decreases more rapidly than the coefficient of heat transfer increases. The convection coefficient is constant at 15.

Figure 12-202 illustrates the radiation sources in a furnace. Note that the radiation attributed to the gas is less than 5 per cent.

Sizing Flue Openings in Furnaces. This depends on many factors discussed in texts on furnace design. Some rough indications may be useful:

- Oven-type furnaces—1.0 sq in. for each 30,000 Btu per hr input
- Pot-type furnaces—1.0 sq in. for each 15,000 Btu per hr input
- Atmospheric-burner equipment—1.0 sq in. for each 5000 to 7500 Btu per hr input, depending on temperatures and stack effect.

Larger areas may be needed, for example, when high-pressure (i.e., 20 psig), one-valve control firing is used, and where the burner manifold may show about 2 in. w.c. pressure. If necessary, the area can be temporarily reduced when starting the furnace. Furnaces using both oil stand-by and gas (interruptible) may require larger flue areas to permit clean burning of oil, i.e., to take care of the excess air required.

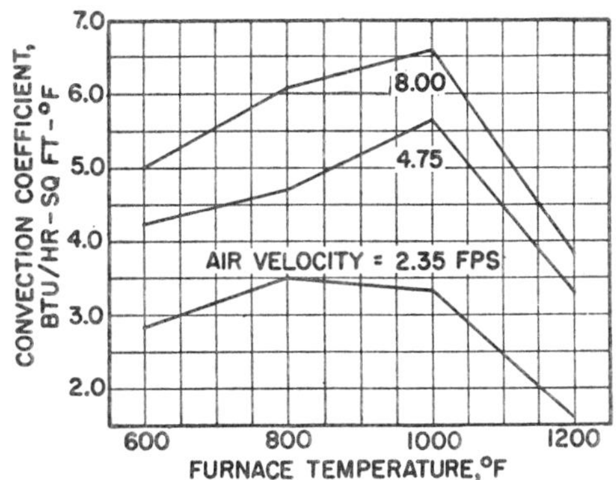

Fig. 12-201a Convection coefficient vs. furnace temperature for selected air velocities.[11]

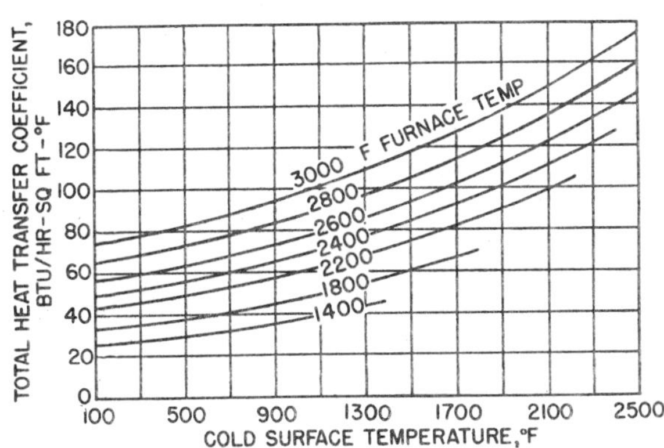

Fig. 12-201b Overall heat transfer coefficient as a function of furnace temperature and work-surface temperature.[10]

Table 12-140a Heating Cold Steel in Open Fired Furnaces (28 Examples)[9]

	Hearth, width × length, ft	Production, lb per hr	Heating rate, lb per hr-sq ft	Pieces per hr	Size pieces	No. pieces in furnace	Area exposed to heat in furnace, A, sq ft	Absorption, Btu per hr-sq ft	Mean furnace temp, t_1, °F	Final steel temp, °F	Mean steel temp, t_2, °F	Mean temp diff., °F	Coefficient, Btu per hr-sq ft-°F	Ratio t_2/t_1	Refractory area, A_R, sq ft	Ratio A/A_R	$e'F_A$	Emmissivity, e'	F_A
Pusher forging	4 × 27	4,500	41.6	88	2½ in. sq × 28 in. lg	90	175	9,780	2280	2300	1560	720	13.6	0.685	350	0.500	0.134	0.85	0.158
Pusher forging	4 × 27	8,960	83.0	176	2½ in. sq. × 28 in. lg	90	175	19,500	2330	2300	1475	855	22.9	.632	350	.500	.235	.85	.275
Groove pusher forging	4.5 × 20	3,420	38.0	76	4¼ in. sq. × 9 in. lg	160	127	9,450	2200	2200	1440	760	12.4	.655	350	.362	.142	.85	.167
Batch forging	3 × 7.5	1,800	80.0	60	2 in. diam × 35 in. lg	25	29.5	21,400	2250	2200	1420	830	25.8	.630	100	.295	.289	.85	.340
Batch rolling	7.5 × 12.5	3,160	33.7	12	4 in. sq × 57 in. lg	33	197	5,290	1840	2150	1310	530	10.0	.708	234	.840	.159	.85	.188
Car anneal	14.5 × 30	3,040	6.9	...	20 in. diam × 72 in. lg	17	572	1,430	1120	1700	970	150	9.6	.865	1320	.432	.336	.65	.517
Rotary forge	8.3 diam	1,890	44.0	60	10 in. × 4½ in. × 2½ in.	42	40.5	17,000	2400	2300	1500	900	18.9	.625	188	.214	.176	.85	.207
Rotary normalize	29 diam	3,220	5.8	134	36 in. × 96 in. sheets	6	144	6,800	2030	1900	1250	780	8.7	.615	1200	.120	.126	.70	.180
Groove pusher hardening	2 × 13	850	32.7	116	3 in. diam rollers	100	51.5	3,950	1430	1500	1010	320	12.4	.713	107	.482	.238	.60	.398
Pusher shoes hardening	6.5 × 24	3,400	21.8	88	Axles	100	412	1,740	1350	1380	1160	190	9.1	.858	526	.785	.227	.65	.350
Walking beam normalizing	4.3 × 16	2,040	29.3	40	Axles	60	114	4,640	1555	1600	1105	450	10.4	.711	280	.408	.230	.65	.353
Pusher shoes drawing	6.5 × 23	3,650	24.5	88	Axles	96	425	1,110	995	1020	810	185	6.1	.815	506	.840	.258	.65	.396
Walking beam drawing	4.3 × 20	2,040	23.5	40	Axles	74	142	1,400	800	800	560	240	5.8	.700	344	.552	.368	.65	.565
Car anneal	7 × 47	1,925	5.9	...	20 in. diam × 45 ft lg	3	708	600	1290	1350	1135	155	3.8	.876	1160	.518	.106	.65	.163
Strip heating	2.5 × 30	4,000	53.3	...	22-in.-wide strip	1	100	10,900	1700	1800	1120	680	16.0	.660	475	.211	.380	.65	.585
Pipe normalize rolldown	37 × 40	22,500	15.2	18	9⅝ in. OD × 33 ft lg	46	2880	2,110	1400	1700	1110	290	7.3	.792	4000	.720	.185	.70	.265
Pusher heating blooms	14.5 × 60	50,000	58.0	19	12 in. sq × 66 in. lg	120	1560	10,800	1900	2100	1250	650	16.7	.658	2485	.488	.261	.85	.308
Pusher anneal	24 × 8	3,036	15.8	1.33	21 ft × 3 ft × 4 in. box	3	378	2,160	1550	1550	1080	470	4.6	.697	630	.600	.107	.70	.153
Chain conveyor hardening	32 × 10	2,930	9.2	60	¾ in. diam × 30 ft lg	50	295	2,390	1287	1550	925	362	6.6	.718	1018	.290	.200	.65	.308
Chain conveyor drawing	32 × 10	1,220	3.8	25	¾ in. diam × 30 ft lg	50	295	535	857	1020	636	221	2.4	.743	1018	.290	.137	.65	.211
Batch forging	2 × 5	253	25.3	325	1¼ in. diam × 2¼ in. lg	300	23.6	3,220	1950	1950	1460	490	6.6	.750	41	.575	.087	.80	.110
Pusher heating blooms	20 × 69	100,000	73.0	20	9 in. sq × 18 ft lg	92	2594	13,100	1850	2100	1250	600	21.8	.675	3830	.610	.358	.85	.422
Wire patenting	6 × 60	3,870	10.8	...	0.312 in. wire	16 wires	79	13,300	1750	1600	835	915	14.5	.476	820	0.097	.340	.65	.525
Roof slot drawing	8.2 × 44.5	13,000	36.5	15	38 in. diam wheels	90	2160	785	1000	1000	800	200	3.9	.800	1226	1.760	.165	.65	.255
Roller hearth strip normalize	5 × 60	10,000	33.3	2	26-in.-wide strip	2	520	5,200	1650	1700	1220	430	12.1	.740	1200	0.433	.234	.65	.360
Roller hearth hardening	5 × 42.3	7,500	34.7	300	24 in. diam disks	42	265	6,800	1590	1600	975	615	11.0	.612	800	.331	.270	.65	.415
Roller hearth drawing	5 × 60	7,500	25.0	300	24 in. diam disks	60	378	2,600	985	950	685	300	8.7	.696	1120	.336	.343	.65	.530
Strip anneal	4.7 × 188	10,000	11.4	...	32 in. wide × .028 in. gage	1	1000	2,450	1620	1600	1385	235	10.4	0.855	3300	0.300	0.174	0.50	0.348

NONFERROUS HEAT TREATING

General specifications for heat-treating furnaces, fuel consumption, and heating rates of nonferrous metals are similar to those for ferrous metals. Applications vary, depending on the physical properties of the particular metal (Table 1-1) or alloy undergoing treatment.

Copper and Its Alloys

Hot-Working. Copper alloys (Table 1-8) are quite plastic within a certain temperature range above the recrystallization point but below the melting point. Within this range they lend themselves to much more extensive working with less power expenditure than when cold. They are usually hot-worked in the early stages of fabrication when heavy reduction can be made. Such operations include hot-rolling of cakes and bars for sheet and wire; extension for rods, tubes, and shapes; piercing for tubes; and hot-forging or pressing for special shapes.

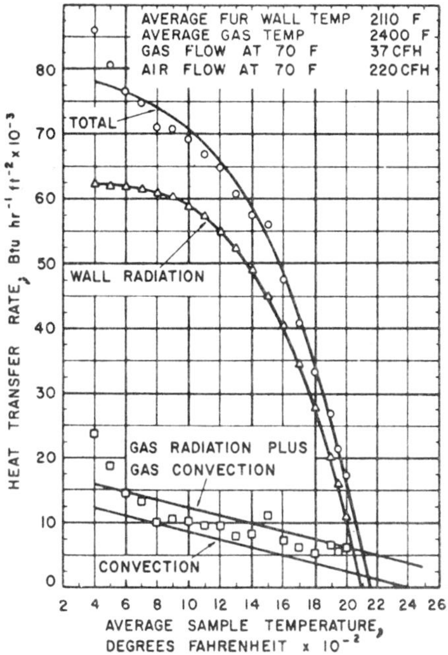

Fig. 12-202 Heat transfer in a gas-fired furnace.[12]

Table 12-142 gives hot-working and annealing temperatures for most common copper alloys. Forging is generally done at temperatures about 100° F higher than the hot-working temperatures of this table.

Annealing. Wire, tube or shell drawing, sheet rolling, and other cold-working operations cause strain and distortion in the grain structure of copper alloys, with accompanying increase in hardness and decrease in ductility. A point is reached in progressive cold-working at which further deformation cannot be made economically or without structural damage to the metal. It must, therefore, be annealed by heating to a temperature where the change in grain size will restore ductility for further cold-working.

Several factors affect the quality of the annealing and its cost. Furnace atmosphere is of special importance. Practically all copper and most of its alloys high in copper are bright annealed, requiring a furnace atmosphere free of oxygen and sulfur.

Thermal Stress Relief. Metals, including some copper alloys, subjected to nonuniform plastic deformation below the recrystallization temperature, may develop residual stresses. This results in a "season cracking" failure. Susceptibility increases with zinc content, especially at 20 per cent or more zinc.

In thermal stress relief it is considered best practice to use temperatures as low and times as long as practicable. Since the operation is performed at temperatures usually between 300 and 650 F, little temperature head can be carried in the furnace without danger of some overloading of the work. Furnaces providing forced convection at high velocity are well suited for stress-relieving operations.

Sintering.[13] Heating of powdered metal (Table 12-143) to effect bonding of contacting particle surfaces without melting is known as sintering. Pressures up to 200,000 psi may be used. In **pressureless sintering,** processing temperature is below the melting point of the main component of the compact but above the melting point of minor constituents.

Muffle-type or radiant tube-type furnaces are used for sintering. The former are preferred for low production; the latter are normally used with atmospheres that are not highly pure. In high production, continuous sintering furnaces are generally used, e.g., for temperatures of 1470 to 2100 F. Pusher, roller hearth, and mesh-belt conveyor types are the means for handling the product. Protective atmosphere is retained in the muffle while the work is passed thru the furnace. The compacts are brought from room temperature to maximum sintering temperature and back to room temperature (in a cooling zone) in the presence of a controlled atmosphere. Controlled atmospheres prevent oxidation at sintering temperatures and reduce existing oxides. Nitrogen will not reduce existing oxides; hydrogen decarburizes and is therefore not used where carbon content must be controlled.

The majority of powder metallurgy parts are made of either copper-base or iron-base materials. There is, however, an infinite variety of metal powders, alloys, and mixtures used. Increasing use is being made of stainless steel, titanium, nickel, beryllium, chromium, and other refractory and exotic metals. Some metals, such as tungsten, molybdenum, and beryllium, can be worked only thru powder metallurgy.

Aluminum and Its Alloys

Melters for Aluminum and Its Alloys.[14] Large reverberatory furnaces are primarily employed to alloy, melt down, or break down large volumes of all kinds of metal, from primary ingots to dirty, oily scrap. Metal thus melted will invariably be handled, processed, remelted, or held subsequently in other furnaces before being cast. The quality of the finished cast product will be affected by the subsequent melting, holding, or casting procedures. This does not mean that careless handling or melting technique can be allowed simply because ingots or hot metal processed in a reverberatory breakdown furnace will later be subjected

Table 12-140b Heating Preheated Steel in Open Fired Furnaces (6 Examples)[9]

	Hearth, width × length, ft	Production, lb per hr	Heating rate, lb per hr-sq ft	Pieces per hr	Size pieces	No. pieces in furnace	Area exposed to heat in furnace, A sq ft	Absorption, Btu per hr-sq ft	Mean furnace temp, t_1, °F	Final steel temp, °F	Mean steel temp, t_2, °F	Mean temp diff., °F	Coefficient, Btu per hr-sq ft-°F	Ratio t_2/t_1	Refractory area, A_R sq ft,	Ratio A/A_R	$e'F_A$	Emmissivity, e'	F_A
Batch forging 800 F preheat	5.7 × 10	1,480	26.0	1	44 in. diam ingot	1	130	2,950	2100	2200	1934	166	18.0	0.920	223	0.582	0.158	0.85	0.186
Pipe anneal 1000 F preheat	20 × 37	50,000	67.2	30	20 in. diam × 18 ft lg	16	1508	5,000	1660	1720	1350	310	16.7	.820	1980	.760	.295	.50	.599
Axle harden 100 F preheat	4.3 × 12.5	2,040	38.2	40	Axles	40	89	2,520	1550	1580	1420	130	19.3	.920	238	.319	.345	.65	.530
Disk anneal 1000 F preheat	4.2 × 15	2,460	39.2	378	16 in. diam disks	32	88	3,070	1500	1525	1350	150	20.5	.900	220	.400	.427	.70	.610
Crank normalize 1200 F preheat	4.5 × 21.5	5,510	57.0	120	Cranks	96	165	2,330	1600	1600	1515	85	27.4	.950	402	.411	.430	.70	.614
Batch forge 1700 F preheat	2.5 × 4	818	81.8	156	2 in. × 2 in. slugs	25	12.5	8,500	2450	2450	2280	170	50.0	0.930	46	0.318	0.300	0.85	0.352

Table 12-140c Heating Steel in High-Speed Furnaces (14 Examples)[9]

(for heating from cold)*

	Hearth, width × length, ft	Production, lb per hr	Heating rate, lb per hr-sq ft	Pieces per hr	Size pieces	No. pieces in furnace	Area exposed to heat in furnace, A, sq ft	Absorption, Btu per hr-sq ft	Mean furnace temp, t_1, °F	Final steel temp, °F	Mean steel temp, t_2, °F	Mean temp diff., °F	Coefficient, Btu per hr-sq ft-°F	Ratio t_2/t_1	Refractory area, A_R, sq ft	Ratio A/A_R	$e'F_A$	Emmissivity, e'	F_A
Skelp heating, continuous	2.5 × 140	48,000	137	...	11½ in. × 0.200 in.	1	269	64,200	2450	2250	1200	1250	51.3	0.490	1400	0.192	0.545	0.70	0.775
Skelp heating, continuous	2.5 × 154	60,000	155	...	15¾ in. × 0.388 in.	1	404	53,500	2450	2250	1200	1250	42.7	.490	1540	.261	.454	.70	.650
Tube anneal, continuous	2.5 × 10	3,000	120	3.2	10 in. diam × 15 ft lg	0.67	26.2	27,500	1810	1450	1000	810	34.0	.553	100	.262	.625	.80	.780
Billet heating	2.6 × 85	40,000	175	...	3¾ in. diam	2 rows	176	79,500	2450	2200	1150	1300	61.1	.470	880	.200	.660	.85	.775
Pusher forging	1.75 × 9.5 diam	3,400	204	336	1 in. × 4 in. × 9 in.	30	14.1	75,000	2600	2300	1180	1420	58.5	.455	52	.272	.511	.85	.602
End-heating, continuous	1.25 × 1.25	475	304	180	1 in. diam × 12 in. lg	7	1.87	75,000	2400	1900	1050	1350	55.5	.438	3	.625	.655	.80	.818
End-heating, batch	1.5 × 4	870	145	194	2¼ in. diam × 9 in. lg	17	7.5	34,800	2300	2000	1200	1100	31.6	.522	12	.500	.362	.70	.520
End-heating, continuous	1.5 × 1.5	416	185	173	⅞ in. diam × 13½ in. lg	8	2.1	70,000	2600	2230	1200	1400	50.0	.462	5	.052	.480	.70	.685
End-heating, continuous	1.5 × 1.5	287	128	167	¾ in. diam × 15 in. lg	8	2.0	50,000	2600	2220	1200	1400	35.6	.462	5	.050	.345	.70	.493
End-heating, continuous	1.5 × 1.5	211	94	167	⅝ in. diam × 16 in. lg	8	1.8	42,000	2600	2240	1200	1400	30.0	.462	5	.044	.287	.70	.410
Tube hardening	2.5 × 32.5	10,800	134	7.6†	7 in. OD × 26 #	1	59.8	43,500	2245	1600	1185	1060	41.0	.528	325	.184	.510	.70	.730
Tube hardening	2.5 × 32.5	10,800	134	5.0†	9⅝ in. OD × 40 #	1	82.0	31,600	2195	1600	1420	775	41.0	.645	325	.252	.448	.70	.640
Tube drawing	2.5 × 32.5	10,800	134	7.6†	7 in. OD × 26 #	1	59.8	25,400	1610	1100	700	910	27.8	.435	325	.183	.660	.70	.940
Tube reheating*	1.5 diam	2,400	232	60	4.1 in. diam × 25 ft lg	1	74	47,000	2360	1890	1650	700	67.2	0.702	324	0.228	0.590	0.75	0.785

* Tube reheating is for heating from 1300 F preheat. † Ft per hr.

Table 12-140d Heating Copper and Aluminum in Forced Convection Furnaces[9]

	Hearth, width × length, ft	Production, lb per hr	Heating rate, lb per hr-sq ft	Pieces per hr	Size pieces	No. pieces in furnace	Area exposed to heat in furnace, A, sq ft	Absorption, Btu per hr-sq ft	Mean furnace temp, t_1, °F	Final steel temp °F	Mean steel temp, t_2, °F	Mean temp diff., °F	Coefficient, Btu per hr-sq ft-°F	Ratio t_2/t_1	Refractory area, A_R, sq ft	Ratio A/A_R	$e'F_A$	Emmissivity, e'	F_A
Copper																			
Billet heating	5.5 × 30	15,000	91.0	60	4 in. sq. × 54 in. lg	90	291	8,300	1700	1600	1050	650	12.8	0.620	580	0.500	0.249	0.75	0.330
Coil annealing	4 × 17	2,000	29.5	22	Coils of ⅜ in. rod	37	370	460	825	825	625	200	2.3	.760	304	1.220	.078	.60	.130
Aluminum																			
Billet heating	4 × 22	2,500	28.3	250	4 in. diam × 9 in. lg	375	295	1,860	900	925	600	300	6.2	.667	384	0.770	.090	.20	.450
Billet heating	7.5 × 15	2,500	22.2	50	6 in. diam × 24 in. lg	50	175	3,140	1000	925	650	350	8.9	.650	405	.432	.095	.20	.470
Ingot heating	3.75 × 42	4,560	28.9	16	11 in. diam × 24 in. lg	42	345	3,310	1150	1100	750	400	8.3	.655	683	.505	.160	.30	.530
Strip heating with rider strip	6.5 × 66	9,920	23.1	...	0.25 in. × 72 in. strip	1	792	2,624	1000	900	660	340	7.8	.660	1368	.580	.045	.08	.560
Strip heating with rider strip	6.5 × 66	4,000	9.3	...	0.010 in. × 72 in. strip	1	792	1,060	900	900	760	140	7.6	.840	1368	0.580	.045	.08	.560
Coil annealing, batch	12.5 × 27	22,500	66.5	2.25	48 in. diam coils	18	2520	1,520	800	700	510	290	5.3	.640	1300	1.930	.045	.08	.560
Pit furnace ingots	9.5 × 17	5,700	35.2	0.6	16 in. × 48 in. × 12½ ft lg	15	1820	750	1000	1000	700	300	2.5	0.700	1065	1.710	0.160	0.30	0.530

Table 12-141a Time Required to Heat 2-In. Plate to 2000 F, Min[10]

Furnace temp, °F	Radiation and convection	Radiation	Convection	Per cent time saved by convection
3000	5.60	6.46	42.1	15.3
2800	6.97	8.27	45.9	18.6
2600	9.02	10.83	52.1	20.0
2400	11.98	14.82	62.5	23.7

Table 12-141b Time Required to Heat Plate to 2000 F in 2800 F Furnace, Min[10]

Plate thickness, in.	Radiation and convection	Radiation	Convection	Per cent time saved by convection
0.5	1.61	1.97	8.8	22.4
1.0	3.33	3.97	20.6	19.2
4.0	15.34	17.62	119.2	14.8

Table 12-142 Hot-Working and Annealing Temperatures for Copper and Wrought Copper Alloys[6b]

(composition given in Table 1-8)

	Hot-working temp, °F	Annealing temp, °F
Copper	1300–1650	700–1200
Gilding metal	1300–1650	800–1450
Commercial bronze	1400–1600	800–1450
Red brass	1450–1650	800–1350
Low brass	1450–1650	800–1300
Cartridge brass	1350–1550	800–1300
Yellow brass	*	800–1300
Muntz metal	1150–1450	800–1100
Leaded commercial bronze	*	800–1200
Low-leaded brass (tube)	*	800–1200
Low-leaded brass	*	800–1300
Medium-leaded brass	*	800–1200
High-leaded brass (tube)	*	800–1200
High-leaded brass	*	800–1100
Extra-high-leaded brass	*	800–1100
Free-cutting brass	1300–1450	800–1100
Leaded Muntz metal	1150–1450	800–1100
Free-cutting Muntz metal	1150–1450	800–1100
Forging brass	1200–1500	800–1100
Architectural bronze	1200–1400	800–1100
Admiralty brass	1200–1500	800–1100
Naval brass	1200–1450	800–1100
Leaded Naval brass	1200–1450	800–1100
Manganese bronze	1250–1450	800–1100
Aluminum brass	1450–1550	800–1100
Phosphor bronze, 10%	*	900–1250
Phosphor bronze, 1.25%	1700–2000	1200–1600
Cupronickel, 30%	1700–2000	1200–1600
Nickel silver, 18% (A)	*	1100–1500
Nickel silver, 18% (B)	*	1100–1400
High-silicon bronze (A)	1300–1650	900–1300
Low-silicon bronze (B)	1300–1650	900–1250

* Usually hot-extruded after casting but further hot-working is uncommon.

to remelting or holding operations in other furnaces. An inherent disadvantage attributed by some to the reverberatory-type furnace for aluminum melting is the large ratio of exposed metal surface to the mass of the metal bath, facilitating gas absorption from direct flue gas and flame contact. The open-flame rapid melters capable of breaking down large batches in 8 to 30 min have led to different attitudes toward the subject of gas absorption, principally because of the great reduction in the time the molten bath is exposed to the products of combustion.

Microporosity in die castings is generally not the result of hydrogen but of air trapped in the dies. The rapid freezing and chilling of die castings would of itself tend to keep dissolved hydrogen in solution, just as in the case of permanent-mold casting. The gain in density and in other physical properties of castings produced by vacuum die casting offers strong proof of these points.

Aluminum, however, is not a "well-behaved" metal in the melting process and has certain characteristics that must be taken into consideration in order to produce a satisfactory, dense, and strong end product. At elevated temperatures aluminum will readily combine to form oxides. During melting, the amount of oxide, etc. formed increases with the temperature. Agitation of molten aluminum also accelerates the formation of oxides, etc. Since the specific gravity of aluminum alloys is very close to that of the oxides, agitation of molten metal should be avoided, or at least held to a minimum, according to most authorities.

The melt in a reverberatory furnace should be less susceptible to the formation of oxides than that in open-pot furnaces or low-frequency induction melters, in which the metal is exposed to the atmosphere. The atmosphere within the reverberatory furnace can be set with an Orsat and regulated so as to offer some means of protection.

An inherent characteristic of the low-frequency electric induction melting furnace is the agitation that occurs as current is applied and released. This agitation, besides exposing more molten metal to the atmosphere, can also hold oxides and refractory particles in suspension. The comparatively still bath of gas-fired furnaces will permit oxides to settle out or to be brought to the surface during degassing operations, when they become trapped in the surface dross and can be easily skimmed off.

Hydrogen in aluminum sand castings (and in other kinds of castings to a lesser extent) results initially from the breakdown of H_2O in the melting furnace where contact with molten aluminum is possible. Water may be introduced from many sources, including hydrogen-forming materials carried into the melter on the surface of the cold charge metal, and from the moisture produced by fuel combustion. Whether H_2 contamination of the metal will occur depends on more than just the presence of H_2O. The furnace atmosphere composition, the burner flame characteristics, and the temperature affect the problem. Above 1400 F, H_2 solubility problems increase markedly.

Pinhole porosity in sand castings is not infrequently the result of hydrogen contamination of the molten metal because of the negligible solubility of hydrogen in solid aluminum that has been slowly cooled. Removal of dissolved hydrogen consists of bubbling a flushing gas thru the molten metal, inducing an artificial boil, which is followed by the release of dissolved gas. Dry nitrogen and chlorine are the most effective flushing gases. When using chlorine, which is both toxic and corrosive, certain safeguards must be employed.

Table 12-143 Sintering Temperature and Time

(in high-heat chamber)

Material	Temp, °F	Time, min	Material	Temp, °F	Time, min
Bronze	1400–1600	10–20	Stainless steel	2000–2350	30–60
Copper	1550–1650	12–45			
Brass	1550–1650	10–45	Alnico magnets	2200–2375	120–150
Iron, iron-graphite	1850–2100	8–45			
			Tungsten carbide	2600–2700	20–30
Nickel	1850–2100	30–45			

Degassing with chlorine has the disadvantage of rapidly removing the magnesium in an aluminum alloy. This depletion may be compensated for by adding magnesium to the melt.

Numerous suitable fluxes are also commonly used. Most fluxes, however, are hygroscopic and should be properly stored, kept dry, and applied in the prescribed manner. Their indiscriminate use is as likely to cause trouble as to cure it. Degassing may be accomplished in a separate holding pot or furnace by allowing the molten metal to cool down slowly, below the pouring temperature, and then rapidly reheating it to the pouring temperature. Preheating metal to approximately 900 F prior to charging it into a molten bath should help to eliminate gas that results from surface moisture or any hydrogen-forming materials.

Nitrogen Degassing Procedure.[32] It is essential that the fluxing tube be preheated just prior to use to ensure that it is moisture-free since it has a tendency to pick up moisture from the atmosphere when standing at room temperature. The temperature at which the operation is conducted is also important. In the case of aluminum alloys, degassing is begun when the temperature is 50°F below the pouring temperature, whereas with **copper-base alloys** the temperature is 50°F above the pouring temperature.

There seems to be a difference of opinion on the amount and the rate of nitrogen to be injected. The suggested amount of nitrogen ranges from 1 to 3 cu ft per 100 lb molten aluminum. The rate of injection ranges from an amount just sufficient to cause a gently rolling action on the surface to an amount that will cause the molten metal to agitate as rapidly as possible without resulting in excessive splashing. The slower rate of admission is associated with the lower amount of gas, and vice versa. The action of nitrogen in degassing is mechanical; with high rates of injection, more gas is required to obtain the desired results than when the injection is slower.

The rate of injection of nitrogen is also related to the quantity of metal to be treated. For a 150-lb batch a rate of 20 cu ft per hr at 2 psi and 1 to 1.25 cu ft per 100 lb metal will provide a gentle rolling action. With larger amounts of metal the rate may be increased. In any case it is recommended that the metal be allowed to rest for 15 to 20 min after treatment; during that period the metal is gradually brought up to the pouring temperature.

With **copper-base alloys** the suggested minimum amount of nitrogen is about 0.4 cu ft per 100 lb of alloy, except in the case of brasses containing 10 per cent or more Zn, when the nitrogen is reduced to 0.1 cu ft. The reason for the lower quantity is that Zn thru volatilization functions as a degasifier, and excessive injection of nitrogen will cause a greater evolution and loss of Zn. In general, treatment of copper-base alloys is similar to that for aluminum alloys.

In treatment of both types of alloys, the injection or fluxing tube should be pushed close to the bottom of the crucible and preferably moved slowly about to promote intimate contact between gas and metal. The possibility that thoroughly degassed metal may result in internal shrinkage problems unless adequate gating and risering systems are employed should be noted.

Types of Aluminum Melters. Many variants of open-flame furnaces of the potless type, used for aluminum melting, are broadly classified as reverberatory furnaces. They include large-volume barrel-type furnaces for aluminum as well as for brass melting, adaptable to many types of plants, including smelters. These furnaces are also well suited for aluminum melting in jobbing foundries. Crucible furnaces, particularly those that do not permit the products of combustion to contact the work, are used in many sand foundries.

The term "reverberatory" is used loosely in respect to large-volume barrel-type furnaces. A reverberatory furnace is usually defined as a furnace with a low roof so inclined that the flame, in passing to the chimney, is reflected down on the hearth where the charged material can be heated.

There are five types of open-flame hearth melting furnaces.

1. Large Stationary Reverberatories for Volume Melting. These units are used principally in smelting and refining plants to produce primary and secondary alloys for the aluminum trade. Large producers of die castings also employ such furnaces for melting and alloying scrap or returns. Usually, these furnaces consist almost entirely of firebrick, held together by stays and tie rods. A charging door or a skimming door, or both, and a tap hole for metal discharge are provided. Capacity may range as high as 30 tons of aluminum.

2. Combination Breakdown and Holding Reverberatories. Both single- and multiple-well types are used. The single-well dip-out type is the most common open-flame aluminum furnace. Sizes range up to a capacity of about 800 lb of aluminum. Large producers of castings frequently use such a furnace as a hot-charged holding unit for die- and permanent-casting operation, with molten metal supplied by a breakdown melter. Occasionally, this furnace is used for both melting and holding purposes. Several pouring stations may be accommodated by units (up to 4000-lb capacity) equipped with multiple dip-out wells. If all stations are operated continuously, use of such a furnace may require very low fuel consumption per pound of metal cast. Other advantages are savings in space and reduction in man-hours needed to transfer hot metal from a separate breakdown furnace. Small combination breakdown and holding furnaces are well adapted to the needs of small- and medium-sized die-casting plants.

3. Stack- or Hopper-charged Large-volume Reverberatories. These furnaces are generally used where large volumes of melted metal are required for hot-transfer operations. Rapid melters break down metal in 8 to 30 min, depending on size. Fast melting is accomplished by using burners of relatively high input, and by charging cold metal into the flue or hopper with recovery of considerable heat from combustion products. Standard factory-made stack- or hopper-charged furnaces are available in sizes of 600 to 5000 lb capacity. Types include stationary; manually tilted; mechanically tilted; and mechanically tilted and/or charged.

4. Two-chamber Combination Dry Hearth Melting and Holding Reverberatories. Most formerly objectionable characteristics of open-flame melters have been overcome in these furnaces, which perform both melting and holding functions. A firebrick wall divides the furnace near its center into two chambers, an opening between them permitting passage of molten metal from the melting to the holding compartment. The melting chamber embodies a "dry" or "sweat" hearth on which metal for melting is placed. The hearth slopes downward toward the center wall opening. The holding section is a refractory basin form holding the metal bath, placed below the melting chamber so that the top level of the bath will be below the melting hearth. This general type of furnace is produced by several manufacturers. Some are factory built and delivered as complete units. Others of large capacity, up to 50 tons of metal, are erected in the field.

5. Continuous Aluminum Melters. These melters, capable of supplying a steady stream of molten aluminum, are particularly suited for continuous casting. Characteristics claimed include: (1) adaptability to automatic feeding; (2) steady stream of metal for automatic pouring; (3) lower fuel consumption (1900 Btu per lb of aluminum melted); (4) cooler surroundings; (5) close metallurgical control (same casting and ingot analysis); (6) practically no gas absorption; (7) fast alloy changeover; and (8) wide high-efficiency range.

Factors in Melting. Oxide formation increases with temperature. Agitation of molten metal should be avoided, since the densities of oxides and alloys are about equal. Furnace atmosphere, burner flame characteristics, and purity of product charged should be controlled, to avoid liberation of hydrogen (readily absorbed by aluminum). Molten aluminum tends to decompose water vapor from the atmosphere above the molten surface. An unbroken oxide envelope surrounding the metal affords considerable protection against gas absorption.

Fuel Usage. Fuel consumption for aluminum melting varies with the type, size, and operating temperature of the furnace and with the alloy. Data[14] range from 1900 Btu per lb for a continuous melter to 2500 Btu per lb for sand-casting foundry practice. Holding furnaces are said to use about 1500 Btu per lb. In the case of permanent-mold or die-casting operations in which alloys of lower melting points may be used, 2400 Btu per lb is suggested, with holding operations correspondingly reduced.

Heat-Treating Furnaces. Large volumes of circulating air in convection furnaces are generally used for heat treating of aluminum and its alloys in both cast and wrought forms. For special applications salt baths are sometimes employed. The objective of heat-treating is to dissolve as many of the eutectic particles as possible as a solid solution in the core of aluminum grains. Later, the hardening particles are precipitated out of solution at controlled temperatures in order to disperse fine alloy particles widely. Wrought all~ seldom require more than 1 hr soaking, while cast alloys n need up to 24 hr.

Certain differences exist between cast- and wrought-aluminum alloys. Rolling breaks up and scatters particles of undissolved alloying constituents thru the matrix of the alloy. The solution time for rolled stock is therefore materially less than for cast alloys. However, rolling does not dissolve all eutectic particles; some remain unaltered in general size, shape, and distribution, regardless of later heat-treating.

Calculation of the size required in an aluminum heat-treating furnace to deliver a given output can be carried out by using a modified exponential heat transfer equation involving the Biot modulus (Fig. 2-70). The method has been described in detail by Marsh.[15]

Annealing. Strain hardening, resulting from cold-working of aluminum alloys, may be removed by heating to permit recrystallization. Recommended annealing temperatures are given in Table 12-144a. The rate of cooling from these temperatures is not important if the maximum temperature limit has been observed. However, if any part of the work has been heated above this point, slow cooling to about 500 F is desirable.

Solution Heat Treatment. When heat-treatable aluminum alloys solidify in production of castings or ingots, an appreciable part of the hardening elements segregate as relatively large distinct particles. These particles are eutectic mixtures, the melting point of which is well below that of the alloy and just above the optimum solution heat-treating temperature. This melting point is the basis for requiring temperature uniformity and close control during heat-treating, to avoid ruining the alloy by incipient melting or blistering.

Solution heat treatment is the first step toward redistribution of the hardening elements. By holding them at temperatures (930 to 970 F) that are specific for each alloy, the eutectic particles are absorbed without melting. The temperature is held sufficiently long for solution and diffusion to occur so as to produce a homogeneous solid solution with very little undissolved eutectic. The alloy is then rapidly cooled or quenched so that the hardening elements do not precipitate from solid solution during cooling. This results in an unstable condition at room temperature. Forming or straightening the alloy to correct any distortion due to heat-treating starts a precipitation of its constituents that can be accelerated by proper heating until final tempering is achieved.

Precipitation Heat Treatment. Most aluminum alloys will age-harden at room temperature for a long period, attaining higher tensile and yield strengths. Natural precipitation or age-hardening occurs satisfactorily with some wrought alloys at room temperature, starting about an hour after quenching. Common practice is to accelerate aging by heating to 250 to 400 F.

Tables 12-144a and 12-144b show times and temperatures for both solution and precipitation heat treatment of wrought- and cast-aluminum alloys, respectively.

Table 12-144a Recommended Conditions for Heat Treating Wrought-Aluminum Alloys[16]

(see Table 1-6 for compositions)

Alloy No.	Former designation	Annealing treatment: Metal temp, °F	Annealing treatment: Heating time, hr	Solution treatment,[a,b] metal temp, °F	Precipitation treatment: Metal temp, °F[c]	Precipitation treatment: Heating time, hr[d]
1100	2S	650	[e]	...	...	...
3003	3S	775	[e]	...	...	...
3004	4S	755	[e]	...	...	...
2014	14S[f]	775[g]	2–3	940	340	8–12
2017	17S	775[g]	2–3	940	...	...
2117	A17S	775[g]	2–3	940	...	...
2218	18S	775[g]	2–3	940	340	8–12
					375[h]	11–13
2024	24S	775[g]	2–3	920	375[h]	8–10
2025	25S	775[g]	2–3	960	340	8–12
4032	32S	775[g]	2–3	950	340	8–12
4043	43S	650	[e]	...	...	...
5151	A51S	775[g]	2–3	960	340	8–12
5052	52S	650	[e]	...	...	...
6053	53S	775[g]	2–3	970	320	16–20
					350	6–10
5056	56S	650	[e]	...	...	...
					320	16–20
6061	61S	775[g]	2–3	970	350	6–10
					450	1–2
6063	63S	...	...	...	350	3–5
				970	350	6–10
7072	72S	650	[e]	...	...	...
7075	75S[f]	775[i]	2–3	870[j]	250[k]	22–26

[a] Heating time varies with product, type of furnace, and size of load. For sheets, heat-treated in a molten salt bath, time may range from 10 min for thin to 60 min for thick material. Several hours may be required in air furnaces. A minimum of 4 hr is suggested for average forgings.

[b] Material should be quenched from solution heat-treating temperature as rapidly as possible. Quenching in a large tank of cold water is preferred. Bulky sections, such as large forgings, are usually quenched in water at 140 to 212 F, to minimize quenching strains.

[c] Specified temperature should be attained by all portions of the load as rapidly as possible and closely maintained during recommended time. Available gas furnaces can maintain variations within ±10° F.

[d] Rate of cooling from precipitation heat treatment is unimportant.

[e] Time in the furnace need not be longer than necessary to bring all parts of load to annealing temperature. Cooling rate is unimportant.

[f] Alclad sheet is heat treated under the same conditions as core alloy. Shortest heat-treatment time consistent with obtaining required properties should be used, repeated reheat treatments avoided.

[g] This treatment is intended to remove the effect of heat treatment and includes cooling at a rate of about 50° F per hr from the annealing temperature to 500 F. Rate of subsequent cooling is unimportant. The treatment recommended for 2S can be used to remove cold-work effects or partially to remove heat-treatment effect if a fully annealed material is not required.

[h] Cold-working subsequent to solution heat treatment and prior to precipitation treatment is necessary to obtain required properties.

[i] Should be followed by heating at 450 F for about 6 hr if material is to be stored for an extended period before use.

[j] Sheet may also be heat treated at higher temperatures (to 925 F), if desired.

[k] A two-stage treatment may be used for sheet—4 to 6 hr at 210 F, followed by 8 to 10 hr at 315 F.

Magnesium and Its Alloys

Some magnesium alloys readily lend themselves to solution and precipitation heat treatments. Most common are Dow Metal C and Dow Metal H. Table 12-145 shows treatment conditions for both.

The need for extreme uniformity requires the use of forced convection furnaces with very accurate temperature control. Both limit and program controls should be used. Lack of care in uniform and controlled temperature may result in severe fires. It has been demonstrated that direct gas-fired external recirculating furnaces for magnesium alloy treatment can maintain a furnace atmosphere of almost constant CO_2 content. This is apparently the result of the balance of heat input with furnace and flue gas losses. In long-time solution heat treatment, equilibrium is attained with about 2.5 per cent CO_2. In one case, a furnace maintained 2.3 to 2.5 per cent CO_2 throughout a 36-hr test period. Provided that accurate temperature uniformity is maintained at proper levels, no eutectic melting will occur, thus affording the best protection against fire. Specific concentration of inhibitor gases such as CO_2 and others used as protection against incipient fires are known to

Table 12-144b Recommended Conditions for Heat Treating Aluminum Alloy Castings[16]
(see Table 1-6 for compositions)

Alloy and final temper	Sand castings				Permanent mold castings			
	Solution heat treatment*		Precipitation heat treatment		Solution heat treatment*		Precipitation heat treatment	
	Temp, °F †	Time, hr	Temp, °F†	Time, hr	Temp, °F†	Time, hr	Temp, °F†	Time, hr
122-T2	...	...	600	2–4	...	...	...	...
122-T52	...	...	...	...	...	...	310	5–7
122-T551	...	...	...	...	...	...	340	18–22
122-T61	950	12	310	10–12	...	...	...	...
122-T65	...	...	...	...	950	8	340	7–9
A132-T551	...	...	...	...	...	...	340	14–18
A132-T65	...	...	...	...	960	8	340	12–16
D132-T5	...	...	...	...	...	...	400	7–9
142-T21	...	...	650	2–4	...	...	...	...
142-T571	...	...	...	...	...	...	340	40–48
142-T61	...	...	...	...	960	4	400	3–5
142-T77*	970	6	650	1–3	...	...	...	...
195-T4	960	12	...	...	...	...	...	...
195-T6	960	12	310	3–5	...	...	...	...
195-T62	960	12	310	12–16	...	...	...	...
B195-T4	...	...	...	...	950	8	...	...
B195-T6	...	...	...	...	950	8	310	5–7
B195-T7	...	...	...	...	950	8	500	4–6
319-T6	940	12	310	2–5	940	8	310	2–5
355-T51	...	...	440	7–9	...	...	440	7–9
355-T6	980	12	310	3–5	980	8	310	3–5
355-T61	980	12	310	8–10	...	...	...	...
355-T62	...	...	...	...	980	8	340	14–18
355-T7	980	12	440	7–9	980	8	440	7–9
355-T71	980	12	475	4–6	980	8	475	4–6
356-T51	...	...	440	7–9	...	...	...	...
356-T6	1000	12	310	2–5	1000	8	310	3–5
356-T7	1000	12	440	7–9	1000	8	440	7–9
356-T71	1000	12	475	2–4	...	...	...	...

* Solution heat treatment is followed by quenching in water (150 to 212 F) except for 142-T77 sand castings, which are quenched in still air. A boiling-water quench is recommended, since it minimizes quenching stresses and distortion.

† Nominal metal temperature should be maintained as closely as possible during the recommended time. Available gas furnaces can maintain variations within ±10° F.

be desirable. Fires in direct-fired gas furnaces have been notably few, contrasted with those in furnaces using other fuels and an SO_2 atmosphere. An SO_2 atmosphere is not required in gas furnaces.

Nickel and Its Alloys

In hot-working, almost all nickel alloys are first forged to blooms and then hot-rolled. Temperatures range from 2100 to 2300 F, depending on the particular alloy.

Annealing. Many nickel products are bright-annealed by controlled furnace atmospheres. Since sulfur attacks nickel, an essentially sulfur-free fuel such as natural gas is desirable for direct furnace firing. Table 12-146 shows annealing temperatures for most commonly used nickel alloys. Stress relief annealing, desirable on some nickel products, is generally done in convection furnaces at about 600 F.

Age-Hardening. Several nickel alloys show appreciable response to age-hardening. Some Monels are usually hardened around 1100 F. Cooling rates are relatively unimportant. The A and B *Hastelloy* combinations are age-hardened at about 1400 F, the C at 1600 F.

Table 12-145 Typical Heat Treating Magnesium Alloys
(cooled in air outside furnace)

Alloy	Temp, °F	Time, hr
Solution treatment		
Dow metal H	730	12
Dow metal C	760	20
Alloys without zinc	780	20
Precipitation treatment		
All alloys	350	12–20

Lead, Tin, and Zinc Annealing

Both tin and zinc are used extensively as coatings for steel in tin-plating and galvanizing, respectively. Annealing of tin plate is considered as a ferrous heat-treating process. Zinc may be preheated before rolling. It is sometimes annealed, usually on drum-type annealers, at about 500 F.

Titanium

Possessing a very high melting point, titanium is used for many high-temperature applications. Temperatures of 1800 to 1900 F are used for hot-working, 1500 F for annealing.

Table 12-146 Annealing Temperature for Various Nickel Alloys[6b]

	Composition, per cent										Annealing temp, °F
	Ni	Cu	Fe	Al	Mn	Cr	Mo	W	Si	C	
Nickel	99+	...	...	...	...	...	...	...	...	...	1500–1700
"L" nickel	99+	...	...	...	...	...	...	...	...	...	1500–1700
"D" nickel	95	...	...	...	5	...	...	...	...	...	1600–1800
"Z" nickel	94	...	...	4.5	...	...	...	...	...	...	1600–1900
Monel	67	30	1.5	...	1	...	...	...	...	...	1600–1800
"K" Monel	66	30	...	3	...	...	...	...	...	...	1600–1800
"KR" Monel	66	30	...	3	...	...	...	...	...	0.25	1600–1800
"S" Monel	63	30	2.0	...	0.5	...	...	...	4.0	0.1	...
80 Ni–20 Cr	80	...	...	...	...	20	...	...	...	...	1800–2000
Inconel	80	...	5	...	...	15	...	...	...	...	1800–2000
60 Ni–15 Cr	60	...	25	...	...	15	...	...	...	...	1600–2000
Illium G	58	6	6	...	1.2	22	6	...	0.6	...	...
Hastelloy A	56	...	22	...	...	...	22	...	...	...	2100–2150
Hastelloy B	62	...	6	...	...	...	32	...	...	...	2100–2150
Hastelloy C	53	...	6	...	...	17	19	5	...	...	2200–2250
Hastelloy D	85	3	...	...	...	...	...	...	10	...	...

Precious Metals

Platinum, silver, and gold are the primary precious metals fabricated in quantity. Besides melting, the only heat-treating process applied to them is annealing. In practically all cases, bright annealing is done, using a special furnace atmosphere.

HIGH-SPEED HEATING FURNACES

In this discussion, high-speed heating refers to the acceleration of heat processes by appropriate use of high thermal heads or high-velocity combustion products. Maximum acceleration of heating rates is usually obtained by equipment that minimizes both heat storage capacity and heat losses. Benefits claimed for high-speed heating include: (1) improved product quality; (2) shorter heating times with corresponding production increases; (3) selective differential or patterned heating, when desired; (4) integration of heating with production lines; (5) reduced space requirements; and (6) reduced overall cost.

Adoption of high-speed heating methods has permitted abandonment of many old rule-of-thumb heating rates. Many high-speed processes produce lower overall production costs despite their lower thermal efficiencies (via savings in labor and other nonfuel costs). Several typical applications are given in Table 12-147 with performance data from operating installations. Some high-speed heating processes do not use conventional furnace structures.

Rapid heating often produces superior or special product qualities. Fast heating of metals, for example, often combines advantages of improved forming characteristics with significant reduction in surface oxidation and undesirable grain growth. Frequently, preferential heating of one section or area of a work piece or surface is desirable; this can often be

Table 12-147 Typical High-Speed Heating Processes

Process	Material	Max working temp, °F	Type of firing	Heating time or speed	Remarks
Continuous fusion of electrolytically deposited tin plate	Tin on steel	470 ± 5	Radiant	1500 fpm	Purpose: to smooth plated surface and eliminate its porosity.
Drying printing ink on continuous web	Ink on coated paper	Ink 275 Paper 300	Radiant	½ sec 500–1200 fpm	Accomplished without discoloring or off-setting.
Baking enamel (polymerization)	Enamel on thin brass shells	350	Radiant	1 min	Special enamel formulations.
Tooth-hardening automatic starter ring gears (14.6 in. OD × ½ in. face)	Carbon steel	1600	Enclosed combustion	32 sec	Produces good transition from hardened surface to tough core.
Annealing and baking coatings on fluorescent lamps	Glass	1000	Radiant	1.5 min in lehr	Also serves as straightening operation.
Singeing textiles	Piled fabrics	Varies	Enclosed combustion	85 yd/min single pass	Uniformity is quite critical.
Reheating large tubes prior to redrawing	Steel	1800	Radiant	To 240 fpm	Replaces batch with continuous heating at mill speeds.
Heating centers of 1½ in. square bars for piercing	Steel	2200	Radiant	8 min (300/hr)	Steel flows more readily after high-speed heating.
Drying ¼ in. filter cake	Alumina powder	450	Radiant	10 min	Permits much shorter unit than convection drying.
Continuous anneal of thin metal strip	Steel	1250	Radiant	~1000 fpm	Strip emerges "bright."

done successfully by local high-speed heating before thermal conduction disperses the heat to other areas where it is unwanted.

High-speed heating is usually best applied to separated work pieces, rather than to work handled in bulk, since with such heating each individual work piece is exposed directly, and most of the time equally, to the heat source. The heat "soak" (extended time at temperature) normally employed with conventional heating to dissipate work piece temperature differentials is not employed; with proper work arrangement it is unnecessary and may result in overheating the work. With bulk material, high-speed heating is often undesirable because overheating of the nearest pieces of the load may be expected before shielded parts reach desired temperature.

Products to be heated by accelerated methods must be able to tolerate high heat-transfer rates without deterioration. Alloys are sometimes reformulated to permit safe, rapid temperature response. To prevent both overheating and underheating, heat transfer must be more closely regulated thru more uniform combustion, work exposure, and time control than is required with conventional methods.

High-Speed Burners and Heating Chambers

High-speed heating requires small heating chambers, relatively high heat release, and containment materials with relatively low heat storage and conductivity. High-heat transfer may be accomplished by suitable use of one or more of the following burner types: (1) radiant (incandescent) ceramic burners [Fig. 12-149(b), Tables 2-99a and 2-99b]; (2) highly incandescent burners developing luminosity thru delayed gas-air mixing; (3) high-velocity jet type, enclosed combustion, or "superheat" burners [Fig. 12-149(a)]; (4) elevated temperature burners employing oxygen enrichment or preheated air.

Substantial reductions in heating space are possible because combustion is normally completed within burner cavities. Since radiant burners normally operate without appreciable "torching," they may be placed very close to objects that flame impingement would damage. On the other hand, hot blasts from high-velocity enclosed combustion burners are often played directly against the more durable work pieces. In special cases blast nozzles are shaped to the contours of the work.

Elevated flame temperatures are secured by preheating or oxygen enrichment. Table 12-148 shows the approximate effects of preheating combustion air and of replacing part or all with oxygen (see also Fig. 2-11b). Results vary with fuel gas analyses (Tables 2-69 and 2-92). Although preheating air is quite effective in increasing flame temperatures, preheating oxygen is hardly justified. Because gas preheat temperatures are limited by the danger of thermal decomposition (or preignition, if premixed), fuel preheating is infrequently used.

Comparative Heating Rates

Figure 12-203a shows comparative heating speeds for a 1-in.-diam steel bar at various radiant chamber temperatures. Because radiant heat transfer varies as the difference of the fourth powers of source and receiver temperatures, much greater transfer rates occur in the high-temperature ranges.

Table 12-148 Effect of Preheat on Theoretical Flame Temperatures, °F

Preheat, air or oxygen	Stoichiometric air	Stoichiometric oxygen
None	3550	4970
500	3750	5010
1000	3925	5040
1500	4100	5070
2000	4250	5100

Note: For example, preheating combustion air 1000° F raises the theoretical flame temperature of natural gas from 3550 to 3925 F.

Figure 12-203b shows the effects of hot gas velocities on transfer rates to round steel bars. The high velocities shown are usually attained with an enclosed combustion burner [Fig. 12-149(a)] in close proximity to work surfaces.

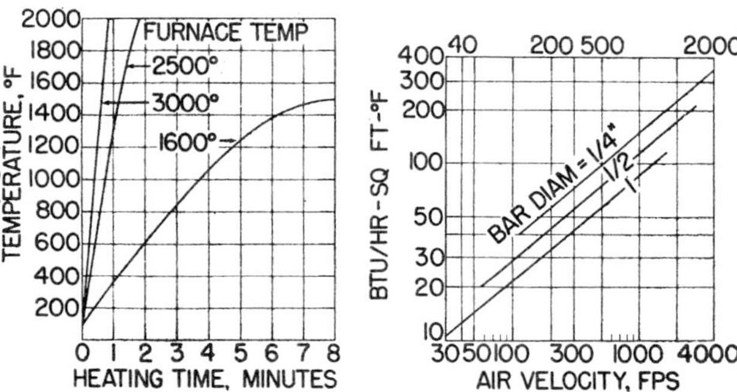

Fig. 12-203a (left) Heating time for 1-in.-diam steel bar at various furnace temperatures.

Fig. 12-203b (right) Convection heat transfer from 3000 F air stream over steel bars. (Selas Corp. of America)

IRON AND STEEL MELTING FURNACES

Open Hearth Furnaces

The steel industry employs open hearth furnaces extensively for production of medium- to high-grade steel. Unheated steel scrap and pig iron form the charge, with limestone or lime as a flux. Table 12-149 gives the heat balance for a typical open hearth furnace. A procedure for the thermal design of open hearth regenerators is available.[17]

Figure 12-204 shows an open hearth furnace. A 150-ton-capacity unit has hearth dimensions of about 15 by 40 ft, with a bath depth of 15 to 24 in.; furnace width is 21 ft; furnace vertical cross-sectional area is 137 sq ft; combustion

Table 12-149 Heat Balance for Typical Open Hearth Furnace[18]
(all in per cent)

Inputs		Outputs			
Fuel	100	Useful output	28.5	Heating to steam	30.0
Preheated air	40	Radiation loss	37.5	Heating air for combustion	40.0
Exothermic reactions	13	Stock loss	17.0		

volume is 6300 cu ft. The charge is introduced thru refractory-lined doors in the front wall. Finished steel and slag are removed thru a tap hole in the back wall. Inlet ports for fuel and air are located at the ends, 46 ft apart. Regenerators are placed below the furnace level and extend under the charge floor.

The tonnage in a "heat" (batch) is determined largely by the area of the bath surface, altered somewhat by the bath depth, which may vary from 15 to 44 in. An open hearth at the Fairless Works of U. S. Steel was designed with the following parameters in mind:[19]

1. Unit loading of molten steel at 350 lb per cu ft to compensate for the turbulent condition of the bath during refining.
2. Maximum hearth bank slope of 36° to coincide with the angle of repose of raw dolomite.
3. Uptake area not less than 0.75 sq ft per MMBtu per hr average firing rate, which is 180 MMBtu per hr maximum and 120 MMBtu per hr average in an open hearth 68 ft × 19 ft.

Heat sent to the regenerators is 67.3 MMBtu per hr. Of this amount, 41 MMBtu per hr are recovered and returned to the furnace, indicating 61 per cent recovery in the regenerators.

Natural gas is generally used in open hearths in combination with heavy oils and tars (or pitch) for atomizing, as a supplemental fuel, in reaction-type jet burners, and in oxygen-gas roof lances. In some steel plants there has been a steady increase in the use of natural gas from year to year. Advantages include the flexibility and ease of operation, as well as the economy resulting from practically sulfur-free gas fuel, when compared with expensive low-sulfur oils ($<0.7\%$ sulfur, for example).

In a specific open hearth the following data were collected in late 1959:

The 8-hr firing cycle is divided into four parts, although the total Btu input rate is practically constant.

1. Melting down: This period of about 3.0 to 3.5 hr calls for 100 MMBtu per hr of gas and 50 MMBtu per hr of oil.

2. Hot metal oxygen boil: This period of about 1 hr is characterized by the use of oxygen for decarburizing the steel, since this furnishes considerable energy within the bath itself. Fuel requirements are down to 50 MMBtu per hr of gas and 50 MMBtu per hr of oil. Operators stress that flames having high kinetic energy are desired in order to "open up" the slag cover for efficient heating.

3. Lime boil: This period of about 1.5 hr uses gas at 50 MMBtu per hr, but the oil heat rate is increased slightly.

4. Metal refining: This final period completes the 8-hr cycle. During this time the metal temperature is checked repeatedly until the pour point, 2850 F, is reached. The fuel rate in this period calls for 50 MMBtu per hr of gas and 70 to 80 MMBtu per hr of oil.

Overall performance has averaged 2.8 MMBtu per ton of steel produced, although as little as 2.5 MMBtu per ton of steel has been obtained. This performance was obtained in open hearths that normally yield 345 tons of steel per heat. Occasional heats as high as 375 tons or even 400 tons have been tried. Thus far, no complete heat run on natural gas has been attempted. Operators feel that the luminosity of oil flames is necessary. There is also a strong opinion that a great deal of turbulence is necessary (i.e., momentum of the flue gas streams) in order to open up the bath for good heat penetration thru the slag layer. It was also noted that the slag cover did not present an opaque layer with respect to roof radiation into the steel.

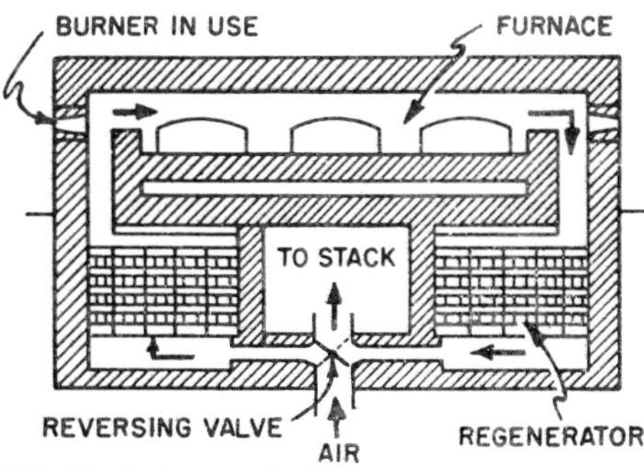

Fig. 12-204 Typical open hearth steel furnace with regenerators.[3] (Hauck Mfg. Co.)

The range of open hearth steel furnace capacities is generally between 25 to 500 tons per heat. Time required to complete a heat varies from 9 to 13 hr, using a charge containing equal weights of cold steel scrap and molten iron.

Combustion takes place with a sluggish flame directed downward to impinge on the bath surface; slow combustion occurs throughout the greater part of the furnace length. An example of fuel results is as follows: Burning fuel tar with 35 per cent excess air, fuel consumption in the furnace (Fig. 12-204) will be 3450 lb per hr and combustion air 12,820 cfm. Gas and wall temperatures will be 3130 F and 2920 F, respectively, and heat of combustion 49 MMBtu per hr. Since incoming preheated air returns about 41 MMBtu per hr, total heat throughput is 90 MMBtu per hr. The total heat-transfer coefficient is 25.8 Btu per hr-sq ft-°F of temperature difference, equivalent to 5420 Btu per hr-sq ft. About 98.5 per cent of this heat transfer is due to radiation.

Atomizing Oil. Data for various methods of atomizing oil are shown in Fig. 12-205. Open hearth practice requires long luminous flames that result from the slow entrainment of preheated air into the fuel jet. The momen-

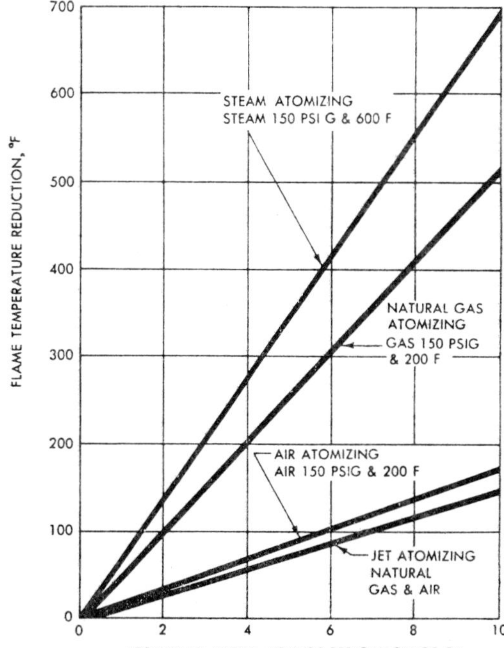

Fig. 12-205 Curves showing relationship between atomizing agent and reduction in flame temperature.[20] Furnace combustion air at 1800 F and 10 per cent excess.

tum and velocity of the stream depend on the conversion of pressure (typically 150 psig and up) to kinetic energy at the burner. For example, if 8 lb of gas replaces 8 lb of steam for atomizing oil, the replacement of the water vapor raises the flame temperature about 140° F (Fig. 12-205), which in turn increases the heat transfer rate and heat input. Some operators have increased production by readjusting combustion air flow to compensate, although increased overall fuel consumption may occur. Shorter regenerator reversal cycles are then required so as to prevent excessively high roof temperatures and accelerated firebrick deterioration. When high-pressure gas is not supplied to the steel plant, the cost of gas compression will approximate the cost of steam. Nevertheless, the lower sulfur content of gas ultimately benefits the steel.

The jet-type combustors operate at 30 to 50 psig (only 50 to 75 psig gas pressure required) and make use of the high-temperature products of combustion to atomize oil at approximately 900 F (compared with steam atomization at 350 F). Note that this type of firing will markedly increase the open hearth flame temperature, as shown in Fig. 12-205.

Advantages of Using Natural Gas.[21] Using natural gas in open hearth furnaces has the following advantages:

1. Substituting gas for steam as the oil-atomizing agent (where the flame luminosity attributable to oil is desired) increases flame temperature as much as 200° F; resultant reduction in meltdown time increases production at least 5 per cent. Note that while steam absorbs heat, gas adds heat.

2. Replacing steam with high-pressure gas reduces flue gas volume, permitting greater heat input rates to the furnace. Increase in production may approach 20 per cent.

3. High-pressure (150 psig) gas atomization permits the control of flame direction necessary to avoid overheating the furnace roof and its subsequent early burnout.

4. Use of high-pressure gas eliminates the need for steam or pressure air (4–6 lb steam or 120–150 cu ft of air per gal of oil). Where gas must be compressed by the customer, the compression cost will about equal the steam cost.

5. Fuel oil adds sulfur to the product; natural gas does not.

Reverberatory Furnaces

(as applied to cast-iron and ductile-iron melting)

Many foundries require facilities for melting smaller batches of metals of varying composition than the cupola can economically handle. Several types of internally fired tilting reverberatory furnaces are particularly applicable. Typically, the charge is loaded thru the stack, the molten metal collecting on the hearth, where its temperature is raised to the pouring temperature.

These furnaces, relatively "clean" as compared with the air pollution effect of the cupola-type melter, are compact in design and easy to reline. Capacities range in general from 400 to 10,000 lb. Gas usage, according to various manufacturers, ranges from 2400 to 6000 Btu per lb. (A number of jobs using 4000 to 4500 Btu per lb have been reported in the literature.) The time required to melt a charge and raise it to the pouring temperature varies from 20 to 60 min, after the initial melt. One type of "rapid melter" is produced for charges of 400, 550, 1000, and 2000 lb per melt. (Fig. 12-206 shows a furnace for a charge of 550 lb.) Acid brick linings of

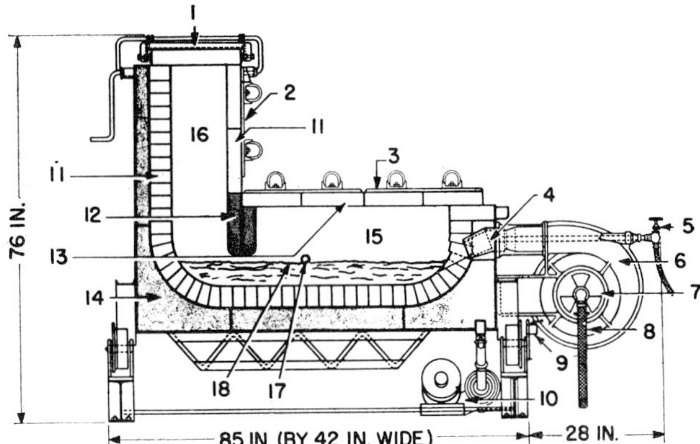

Fig. 12-206 Reda series 550 reverberatory furnace.[22] (Reda Pump Co., Bartlesville, Okla.)

1.	Swing stack cover	10.	Tilting mechanism
2.	One-piece stack front	11.	Firebrick
3.	Hearth cover	12.	Crystolon baffles
4.	Gas burner	13.	Firebrick tile
5.	Burner tunnel	14.	Insulation
6.	Blower	15.	Bath area
7.	Air intake	16.	Charge
8.	Gas line	17.	Tap hole
9.	Tilt control	18.	Bath level

standard insulating and firebrick shapes are used wherever possible.

Under proper operating conditions, these furnaces should yield about two heats per hour at a tapping temperature of 2750 F. Furnace efficiency improves markedly as the number of heats is increased. A preheating period of 30 to 45 min is allowed, with minimum heat input maintained during most of it. Charging is then done thru the stack. Tapping should be done immediately on reaching desired temperature. The furnace is capable of producing temperatures above 2900 F, but serious oxidation losses may occur unless strongly reducing atmospheres are employed.

Two furnaces of a different type (1200 lb and 2000 lb capacity) in the rapid-melt class, located in a small cast-iron foundry, may be described as follows: Both are of the tilting type. Tuyeres are located at one end, and the exhaust chamber, which also serves as a retaining hopper, is at the other. The pouring spout is central. The hearth is concave, 6 in. at the spout and running to a feather edge at both ends. Burners are of the proportional mix blast type with a capacity of 4000 cu ft per hr of natural gas, firing into rammed tunnels. Preheating time before charging is about 30 min; after standing over a weekend, preheating may require 1 hr. Pouring temperature is 2700 F. Natural gas consumption of 4.8 cu ft per lb of metal melted during an 8 hr shift is reported.

A third type of melter uses a combustion-oxidation system to raise temperatures and reduce sulfur content, and to produce grain refinement required for ductile irons and high alloys. The operation involves an acid slag to reduce and neutralize iron oxide and to retain carbon and silicon. This melter uses 6000 to 8000 cu ft of gas per ton.

Direct Gas-Fired Cupola Air Preheating[23]

This preheating results in many economies and reduces smoke and fumes. The case history of a No. 8 Whiting cupola, comparing hot blast with cold blast, showed (1) increase in iron melt capacity—14 per cent; (2) decrease in coke used—20 per cent; (3) increase in iron temperature at cupola spout—125° F; (4) fuel savings—64 cents per ton; (5) decrease in oxidation of product, with concomitant savings in alloying changes and more uniformity of product; (6) increase in refractory life—30 to 50 per cent; and (7) minimizing of slag clogging around tuyeres.

Cupola Melting with Supplementary Natural Gas

In a cupola there are two separate zones of reaction: the combustion-oxidation zone, where the carbon is burned to CO_2, and the reducing zone, where some of the CO_2 is reduced by the incandescent coke to CO. The need for excess O_2, to effect complete combustion in a given furnace, increases with higher flue gas temperature. If O_2 from an outside source were added to the high-temperature flue products, dissociation of O_2 would be repressed, and some of the CO would combine with the extra O_2 to form additional CO_2. Note that dissociation increases exponentially with temperature.

The composition of the natural gas combustion products favors the formation of cupola gas mixtures that bring about complete combustion within the cupola. This increase in the combustion ratio (i.e., higher ratio of CO_2 to CO in the stack gas) results in (1) a smaller heat loss owing to the reduced quantity of unburned CO in the stack gas; (2) increased thermal efficiency of the cupola; and (3) higher temperatures in the cupola. Thus, the height of the melting zone increases. In this larger zone of "superheating," metal melts fast.

Installation and Operation. Basic data follow:

1. The burner input may be calculated from the total Btu heat balance of the coke charge (the coke used in the initial bed should not be included). *As much as 50 per cent of the charged coke can be replaced with natural gas.*

2. Burner location is the most important factor.

   a. The burners should be placed between the tuyere openings, not in or directly above them.

   b. The burners should be located 10 to 15 in. above the tuyeres.[31]

   c. One burner less than the total number of tuyeres should be used. The burner directly over the pouring spout should be left out.

3. The burner should be of the nozzle-mixing type with a short, stable flame capable of operating against the cupola back pressure. The burner tile must be simple in design to facilitate minor repair.

4. The combustion air must come from a separate blower and never from the tuyere air supply.

   a. A constant air-to-gas ratio of 9:1 must be maintained thru the burners.

   b. The burner air pressure and gas pressure must exceed the cupola back pressure; 0.5 psig is suggested in order to give good turndown.

   c. Control of the air thru the tuyeres is important.

   d. Safety controls should meet the requirements of the authority having jurisdiction.

5. Higher stack temperature, which incinerates smoke particles, provides marked improvement in terms of reduced air pollution.

6. When producing malleable iron, it is possible to replace more coke with natural gas since carbon pickup is not a factor.

In an installation at the Ohio Products Co., Orville, Ohio, production was increased and coke consumption was decreased by using natural gas, as follows:

Production, tons per hr	3.6	5.0
Fuel, coke, lb per ton	266	167
natural gas, cu ft per ton	...	470*

Cupola data: 48 in. shell with 30 in. ID lining; six tuyeres centered 23 in. from cupola bottom; five gas burners, each using 600 cu ft per hr of gas, centered 34.5 in. from bottom (and 8 in. below top of burned-in coke bed); bottom of charging door located 18.5 ft from bottom of cupola; air required for burners and tuyeres is 450 cfm (at 1.5 psig) and 1200 cfm at 1.0 psig, respectively; hearth pressure is 0.75 psig; continuous tap into the forehearth.

When *no* natural gas is used, air is generally supplied at the tuyere level at a rate of 0.18 lb per min-sq in. of cross sectional area. When natural gas is used, the gas and its accompanying combustion air result in sufficient hot gas mass flow to reduce the tuyere air requirement to 0.12–0.15 lb per min-sq in.[31]

Metal charge, pounds:

Pig iron	140	Steel scrap	80
Remelt	240	Limestone	20
C.I. briquettes	140	Manganese	1

Product analysis:

Component	C	Si	S	P	Mn
Per cent	3.34	2.08	0.12	0.096	0.50

Miscellaneous: Pour temperature—2720 F; tensile stress—31,000 psi; chills, 32nd—7 (Inco test).

NONFERROUS METAL MELTING FURNACES[24]

The most important nonferrous metals in commercial melting and processing are aluminum, copper, zinc, and magnesium, and their various alloys. Melting points of metallic elements are given in Table 1-1. Heat required for melting operations varies, depending on the kind of metal (Figs. 2-21 thru 2-23), its pouring temperature, and the type and size of furnace.

Melting furnaces may be classified generally into two types, pot and reverberatory. Both types have many variations, such as manual tilt, mechanical tilt, hydraulic tilt, nose pour, stationary nose pour, melting, holding, and dip-out.

Pot-type furnaces accommodate standard sizes of crucibles, or iron pots, or special iron pots as used in coating operations. Crucibles are standardized by number. Capacity is indicated by the product of this number and a factor depending on the particular metal. For aluminum this factor is 1, for brass or other copper alloys it is 3; thus, the brass capacity of a No. 80 crucible is 240 lb. Factors are available for other metals.

Brass Melting

The term "brass melting," as used loosely in the industry, includes high-grade bronzes, alloys of copper and zinc, and alloys of copper, lead, and tin. A more appropriate term is "copper-bearing alloys"; common compositions are given in Table 1-8.

* Reference 31 recommends 300 to 1000 SCF per ton of iron.

Formerly, brass melting for foundry practice was a crude operation. Its general objective was to bring a crucible filled with metal to the required temperature as rapidly as possible, then remove the crucible from the furnace and pour its contents. Various attempts were made to develop furnaces that would reduce metal loss and melting requirements. High input and rapid melting usually reduced metal loss but increased melting costs. Vapor pressure data are given in Fig. 12-194. Data on the subject of metal loss and the effect of atmospheres on molten copper-bearing alloys are available.[25] Lower inputs and close combustion control resulted in lower unit melting costs and longer crucible life but usually higher metal loss. To speed up melting it was found necessary to use reverberatory-type furnaces with considerably higher inputs per pound of metal capacity. The term "reverberatory" is used loosely as it applies to open flame or open-hearth furnaces of the rapid-melting type. These furnaces represent modern applications or adaptations of the reverberatory principle. Rapid melters are becoming more and more popular in foundries where melting capacity is sufficient to warrant their use.

In recent years some appliance manufacturers specializing in metal-melting equipment, particularly Reda, Stroman, and Eclipse, have developed standard reverberatory-type furnaces suitable for use in medium- and large-sized foundries. Principal advantages claimed for these furnaces are: (1) greater production per unit; (2) minimum floor space per unit of melting capacity; and (3) rapid-melting feature conducive to better metallurgical control and lower metal loss since alloying constituents of low boiling point are exposed to high temperature for a minimum time.

Standard factory-built and field-erected reverberatory melters, now available in a wide range of sizes and designs, can be classified as stationary type; tilt type; and skip hoist-charged, both stationary and tilt type. The method of burner application varies somewhat. Usually, burners are placed at one end of the furnace and fired diagonally downward toward the melter bottom and against or just over the metal bath. Combustion products vent at the opposite end thru a combination hopper and flue. The contact of hot combustion gases with the incoming metal utilizes the heat otherwise wasted. Degassing methods are similar to those used for aluminum.

CERAMIC AND GLASS FURNACES

Ceramic Kilns

Ceramic ware made by potteries and the heavy clay industry is formed from natural inorganic materials. The plasticity necessary for molding or shaping is developed by mixing with water.

Stages of Firing Ceramic Bodies. The three general stages follow:

1. *Dehydration*, or "water smoking," is the period for eliminating all mechanically contained water not removed by drying, as well as the hygroscopic moisture absorbed from the air during setting. Progress is most rapid when heated air in large volume is used as a moisture-carrying vehicle. Dehydration of hygroscopic moisture starts at about 800 F and is completed at about 1300 F.

2. *Oxidation* is the period for removal of volatile constituents and hardening of the product. This stage usually immediately follows dehydration and is inseparable from it. Removal of volatile matter is a function of time and temperature, and often requires excess air. Oxidation is usually completed at about 1650 F. Since CO_2 is given off by the product, combustion controls that reflect CO_2 readings would be invalid.[26]

3. *Vitrification* is the final stage, when soaking and finishing temperatures are attained and thermal uniformity is established. Vitrification occurs in two steps, incipient vitrification, during which the clay grains begin to soften, and complete vitrification, which renders the product impervious. Various clay types have different vitrification ranges.

Types of Kilns Used. *Intermittent, or periodic, kilns* may be of the updraft, downdraft, muffle, or electric type. *Continuous kilns* may be of the straight car tunnel type, or they may have circular moving or roller-type hearths.

Fuels Used. Fuel gas is used chiefly in the pottery industry for all types of continuous and intermittent kilns; oil and coal for brick, sewer tile, and other heavy clay ware. Electric or fuel gas-fired radiant tubes are commonly specified for decorating and other critical firing operations.

Typical Btu requirements for various types of ware are:

1. Chinaware in continuous kilns:
   Bisque firing, 2300 F, 15 to 20 MBtu per doz
   Glost firing, 2150 F, 20 to 30 MBtu per doz
   Decorating, 1350 F, 10 to 14 MBtu per doz
2. Wall tile in intermittent downdraft kilns:
   10,000 sq ft of tile per setting
3. Face brick in rectangular downdraft kilns:
   150,000 bricks per kiln, 8 to 10 MMBtu per thousand bricks
4. Paving brick in continuous car-type kilns:
   7 MBtu per thousand bricks
5. Glazed wall bricks in continuous car-type kilns:
   8.5 MBtu per thousand bricks

Burner Equipment and Gas Requirements. Recommendations follow:

1. *Intermittent kilns* generally use atmospheric or low-pressure tile burners or similar hand-operated burners with atmospheric mixers to facilitate individual adjustment during the firing stages.

Gas input and pressure vary for each firing stage, with highest rates at the firing-off period. As a rule, a series of small burners is used for each firing port; the number of burners used and the pressure are increased with firing time and temperature reached. Gas consumption is a factor of kiln size; pressure may vary from 4 to 12 oz. The drying of ware comprises the major use for waste heat.

2. *Continuous kilns* generally use low-pressure inspirator-type burners (3 to 4 oz gas and 16 oz air pressure) or radiant-type burners (premixed gas and air), all suitable for full automatic control in predetermined kiln sections.

The size of the continuous kiln, the firing temperature, and the type of ware determine the hourly gas input. With automatic temperature controls, gas pressure for inspirators remains constant (4 to 6 oz for low-pressure or 10 to 20 psig for high-pressure equipment). Typical examples for natural gas-fired kilns follow:

Type of ware	Kiln length, ft	Temp, °F	Gas input, MMBtu/hr
China, bisque	320	Cone 9*	6.0
glost	280	Cone 5*	7.0
decorating	160	1350	2.0
Face brick	375	...	8.5
Radiant backwalls	150	2210	4.2
Steatite bodies	80	2700	1.0

In continuous kilns, exhaust gases, usually recirculated in the preheating zone of the kiln, ensure uniform heating of the incoming ware. The remainder of the waste heat supplies dryers or other related equipment. Humidity controls can be applied to waste gases where necessary.

Glass Tanks and Lehrs

Glass is defined as an inorganic substance in a condition which is continuous with, and analogous to, the liquid state of that substance, but which, as a result of having been cooled, has attained so high a degree of viscosity as to be for all practical purposes rigid. The principal glass products are bottles and windowpanes. About 75 per cent of all U. S. glass is in the form of containers and other hollow ware made by similar processes. Fibrous glass manufacturing uses natural gas almost exclusively.[27] Raw materials for glass include silica, with limestone and soda ash as principal fluxes and small amounts of other fluxes or modifiers.

Types of Plants. The glass industry can conveniently be divided into three major parts, depending upon the end product:

1. Hand Plants. These use pot furnaces and day tanks to produce hand-blown, pressed, or semiautomatic machine-made glass such as tableware, bar accessories, lamp bases, optical glass, and tubing.

2. Machine Plants. These use continuous-melting tanks, fully automatic blowing, and pressing machinery to produce chiefly glass containers, bottles, vials, glass blocks, lamp and television bulbs, and other items on a continuous operating schedule.

3. Flat Glass Plants. These use pot and continuous-tank furnaces to produce plate, window, rolled, ribbed, wire and ornamental glass on a continuous operating basis.

Definitions. Glass melting is a thermal process by which raw materials are fused into glass without any undissolved batch material. **Annealing** involves the removal of internal stresses caused by the blowing, pressing, drawing, or forming operations performed on the glass. **Tempering,** also known as toughening or case hardening, requires heating glass nearly to its softening point, then cooling it rapidly and uniformly by means of air jets, or a salt or oil bath, to obtain a uniform compression pattern. **Safety glass** may be either tempered glass or two or more layers of glass laminated with a plastic to prevent shattering when breakage occurs. **Bending or convexing** of flat glass is accomplished by heating it nearly to its softening point in suitable molds that form it to the desired curvatures.

Glass Melting Furnaces. Three types are used:

1. Pot Furnaces. These intermittently operating units are used mainly in hand glass and optical glass plants. Thermal efficiency is low compared with continuous units. Careful attention during all stages of melting is necessary. One or more pots, each pot holding between one and two tons of glass, are arranged around the periphery of a circular furnace structure and fired by mounting the burners along the bottom at the back, directing the flame upward thru a wall and over the pots to the flue openings near the front. Large circular pot furnaces may contain 10 to 20 pots, fired by a single large burner underneath and in the center of the furnace bench, or floor. Cyclic heat interchangers (regenerators) are used, alternately receiving heat from combustion products and preheating the combustion air. The regenerative pot furnace cycle is usually 20 min. Atmospheric-type inspirators with pressure requirements from 8 to 16 oz are commonly used; the hourly consumption may vary from 1.0 to 10.0 MMBtu, according to the number of pots in the furnace.

2. Day Tanks. These intermittently operating units are batch-charged at the end of the working day. The charge melts overnight and is consumed the next day. Used in hand plants, usually for special glasses, the tanks have a capacity ranging from 2 to 10 tons. The method of firing is usually direct, but some are equipped with a recuperator or a regenerator, especially for high-temperature glasses such as borosilicate. Premix-type burner applications are favored, using one or more burner tips for the direction of flame. Gas pressure is 4 to 6 oz; air pressure is 16 oz. Gas input, which is a function of the tank capacity and kind of glass melted, is about 12 MMBtu per ton of glass.

3. Continuous Tanks. These are used in manufacturing glass containers, sheet glass, and in all machine-operated plants. Capacities range from 10 to 125 tons for container plants to 1400 tons for plate and window glass plants. Regenerators or recuperators are used as heat exchangers between combustion products and air.

Continuous tanks may be fired either from end or from side ports. Firing equipment consists of an ordinary piece of swedged pipe thru which gas, at 2.0 to 5.0 psig, is introduced above the regenerator checkers and mixed with the preheated combustion air to produce a highly luminous flame. Periods between regenerator reversals are usually 20 to 25 min. Between 35 and 45 per cent of the waste heat is recovered. Typical operating data on natural gas fired continuous tanks follow:

Melting area, sq ft	250	520
Bottle glass made, tons/hr	2 (amber)	3 (clear)

Heat Treatment of Glass. Three types of furnaces are used:

1. Annealing Furnaces or Lehrs. These are usually of the continuous-belt conveyor type, although small batch-type ovens are used for annealing special glasses such as optical lenses on a controlled time-temperature cycle. Any glass annealing cycle consists of three distinct periods: (1) *soaking* at the maximum annealing temperature; (2) *initial cooling* thru the critical range; and (3) subsequent *faster cooling* to room temperature.

Lehrs vary from 2 to 12 ft effective width, with lengths up to 175 ft; plate and window glass lehrs run from 250 to 600 ft long. Combustion equipment on annealing lehrs includes low-pressure inspirators (4 to 6 oz gas and 16 oz air

* Seger cones; see Table 6-18 for some examples.

pressure); four to six burners are commonly used with automatic temperature and safety controls. The tunnel atmosphere is usually recirculated.

Examples of natural gas-fired annealing lehrs follow:

Product	Belt width, ft	Tunnel length, ft	Cycle, min	Gas input, MBtu/hr
Bottles	4	100	90	350
Table ware	10	60	120	750
Lighting globes	10	60	150	800
Pyrex ware	10	90	45	800

2. Decorating, Reannealing, and Bending Lehrs. The design of these lehrs usually follows that of the straight annealing lehrs, with the addition of a preheating zone that is about one-third of the overall lehr length. The preheating zone varies in length from 60 to 375 ft since the glass is charged at room temperature. Operating temperatures vary from 900 F for reannealing, 1050 to 1150 F for decorating, and up to 1350 F for bending or convexing glass with subsequent annealing.

For combustion equipment similar to that used on annealing lehrs, recirculating the tunnel atmosphere for uniformity is advisable. For fuel gases containing more than 4.0 grains of sulfur per 100 cu ft, full muffle or radiant tube construction is necessary to prevent contamination of the ware.

Examples of this type of lehr follow:

Operation and product	Belt width, ft	Tunnel length, ft	Cycle, min	Gas input, MMBtu/hr
Decorating glass tumblers	8	110	45	1.8
Convexing lighting ware	6	90	90	0.9
Convexing front windshields	9*	180	36	2.6

3. Tempering Kilns or Furnaces. Most of these are of the roller hearth kiln type. Glass is charged cold, heated to between 1150 and 1350 F, and then rapidly quenched with air jets. For curved automobile glass, ring molds are used for convexing; sheets remain on the mold during air tempering. For flat glass, an overhead conveyor system is preferable, with sheets in a vertical position during heating and quenching. The furnace, prior to tempering or air quenching, is heated by radiant tube elements placed in close proximity to both glass surfaces. Gas at 6 to 8 oz and combustion air at atmospheric pressure are drawn into tubes by exhaust fans. Gas consumption for this critical operation varies with the size of the heating furnace. As a rule, each radiant tube should be capable of burning from 150 to 200 MBtu per hr to ensure uniformity of heat release. Fully automatic control is used.

Furnace Atmospheres. During all *melting and refining* operations, an oxidizing atmosphere with an excess of air is necessary. One exception is the melting of certain glasses (lead, ruby, etc.), when a controlled reducing atmosphere is beneficial because of the high affinity of some ingredients for oxygen. A slight pressure must be maintained to ensure uniformity across the melting and refining areas. Preferably, gaseous fuels are used but oil and electricity may be applied as booster fuels in some sections of the tank and feeder forehearth.

The preparation, maintenance, and control required for producer gas makes it too costly an alternative. In *annealing, decorating, bending,* and other heat-treating operations, oxidizing atmospheres are essential. Protecting muffles such as radiant tubes may be used with fuels containing undesirable trace constituents, in order to prevent contamination of the furnace atmosphere. For example, sulfur oxides cause a "blooming" effect on sodium oxide in glass.

Electronic Tube Making.[28] There are many applications of natural gas in this art. Some processes still require the use of oxygen and hydrogen.

Porcelain Enameling.[29] In the U. S., 82 per cent of the enamel furnaces are of the muffle type, using gas and/or oil; 8 per cent use gas-fired radiant tubes; 10 per cent use electricity.

Fire Polishing of Glassware.[30] The removal of dangerous cutting edges, unsightly mold and grinding marks, cracks and fissures, and dull surface finish is effected by fire polishing. The use of natural gas in equipment designed for manufactured gas often requires the addition of hydrogen or oxygen to increase flame velocity.

* Rollers.

REFERENCES

1. J. D. McCullough, Refractories Div., Babcock & Wilcox Co.
2. Young, R. C., and others. "Effect of Various Temperatures on Thermal Conductance of Refractories." *J. Am. Ceram. Soc.* 47: 205–10, May 21, 1964.
3. Hauck Manufacturing Co. *Hauck Industrial Combustion Data.* New York, 1953.
4. Turner, C. A., Jr. "Scale-Free Heating in Direct Gas-Fired Furnaces." *Ind. Heating* 30: 72–6, Jan. 1963.
5. Demoulin, A. C. "Scale-Free Heating of Steel." *Metal Progr.* 84: 77–80, Aug. 1963.
6. (a) Am. Soc. for Metals. *Metals Handbook,* 1939 ed., p. 475. Cleveland, Ohio, 1939.
   (b) Am. Soc. for Metals. *Metals Handbook,* 1948 ed., p. 1032. Novelty, Ohio, 1948.
7. Thurston, A. M. *Rapid Heating of Steel for Hot Forming.* (Information Letter 95) New York, A.G.A. Industrial & Commercial Gas Section, 1958.
8. Ocean, E. J.. and Kiehle, C. B. *Continuous Strip Annealing of Steel.* (Information Letter 85) New York, A.G.A. Industrial & Commercial Gas Section, 1957.
9. Mawhinney, M. H "Heat Transfer in Industrial Heating Furnaces." *Ind. Heating* 22: 2011+, Oct. 1955; 23: 54+, Jan. 1956, 292+, Feb. 1956.
10. Huebler, J. "Radiative and Convective Heat Transfer Rates Pertaining to Heating Processes." *Ind. Eng. Chem.* 40: 1094–8, June 1948.
11. Sinnott, M. J., and Siebert, C. A. "Heat Transfer in a Recirculating Furnace." *Ind. Eng. Chem.* 40: 1039–44, June 1948.
12. Genna, S. J., and others. "Heat Transfer in Gas-Fired Furnace." *Am. Soc Mech. Engrs. Trans.* 76: 527–36, May 1954.
13. Kulas, E. F. *Powder Metallurgy and Natural Gas.* (Information Letter 133) New York, A.G.A. Industrial & Commercial Gas Section, 1963.
14. Parker, S. C. *Reverberatory & Other Large Furnaces for Melting Aluminum Including Melting Problems & Practices.* (Information Letter 103) New York, A.G.A. Industrial & Commercial Gas Section. 1959.

15. Marsh, K. *Heat Transfer to Aluminum.* Pittsburgh, Pa., Furnace Div., Aluminum Co. of Am.
16. Aluminum Co of Am. *Alcoa Aluminum and Its Alloys.* Pittsburgh, Pa., 1942.
17. Hlinka, J. W., and others. "AISE Manual for Thermal Design of Regenerators" *Iron & Steel Engr.* 38: 59–76, Aug. 1961.
18 United States Steel Co. *Making, Shaping and Treating of Steel,* 6th ed. Pittsburgh, Pa., 1951.
19. Burch, C. J. "Decade of Open Hearth Furnace Development." *Iron Steel Engr.* 39: 97–102, Oct. 1962.
20. Gardner, E. R. "Use of Natural Gas in Open Hearths." *Am. Gas J.* 190: 29–34, March 1963.
21. Thompson, V. B. *Use of Natural Gas in Open Hearth Furnaces.* (Information Letter 119) New York, A.G.A. Industrial & Commercial Gas Section, 1960.
22. Michaels, S. J. *Reverberatory Iron Melting.* (Information Letter 116) New York, A.G.A. Industrial & Commercial Gas Section, 1960.
23. Apthorp, H. *Hot Blast for the Foundry Cupola.* (Information Letter 120) New York, A.G.A. Industrial & Commercial Gas Section, 1961.
24. Parker, S. C. *Brass Melting and Brass Melting Practices.* (Information letter 102) New York, A.G.A. Industrial & Commercial Gas Section, 1959.
25. Upthegrove, C., and Herzig, A. J. *Effect of the Products of Combustion on the Shrinkage of Metal in the Brass Industry.* (Engr. Research Bull. 22) Ann Arbor, Mich., Dept. of Engr. Research, Univ. of Michigan, 1931.
26. Moran, S. G. *Gas for Glass Melting.* (Information Letter 75) New York, A.G.A. Industrial & Commercial Gas Section, 1956.
27. McEvoy, R. J. *Fibrous Glass Industry and Natural Gas.* (Information Letter 81) New York, A.G.A. Industrial & Commercial Gas Section, 1957.
28. Schaefer, F. C. *Gas Uses in the Manufacture of Electronic Tubes.* (Information Letter 135) New York, A.G.A. Industrial & Commercial Gas Section, 1963.
29. Hartwein, C. E. *Vitreous Enameling.* (Information Letter 78) New York, A.G.A. Industrial & Commercial Gas Section, 1956.
30. LeMay, R. C. *Fire Polishing of Glassware.* (Information Letter 47) New York, A.G.A. Industrial & Commercial Gas Section, 1952.
31. Stone, A.J., and others. *Use of Natural Gas in Foundry Cupolas* (A.G.A. Project IG-27) Columbus, Ohio, Battelle Memorial Institute, 1965.
32. "Degassing Nonferrous Alloys," *Foundry,* 93: 68–9, July 1965.

Chapter 20

Kilns for Lime, Cement, and Other Products

by C. George Segeler

LIME

Loosely defined, lime means finely divided limestone ($CaCO_3$ and dolomitic limestone, $CaMgCO_3$) as well as the burned products derived from it. The dictionary, on the other hand, defines lime as calcined limestone, i.e., CaO (quicklime); but hydrated, or slaked, lime—$Ca(OH)_2$—is also included in this definition. In each case there are dolomitic, i.e., high-magnesium, counterparts. Table 12-150 gives typical lime analyses.

Table 12-150 A Typical Range of Lime Analyses, Per Cent

	High-calcium type	High-magnesium type
CaO	93.3–98.0	55.5–57.5
MgO	0.3– 2.5	37.6–40.8
SiO_2	.2– 1.5	0.1– 1.5
Fe_2O_3	.1– 0.4	.1– 0.4
Al_2O_3	.1– .5	.1– .5
H_2O	.1– 0.9	.1– 0.9
CO_2	0.4– 1.5	0.4– 1.5

Before calcining, limestone averages around 98 per cent carbonate rock, with the remainder chiefly silica, alumina, and iron oxides. Nearly 50 per cent of the weight of the original limestone is driven off as CO_2 during calcining. A small percentage, called "core," may resist calcining. The amount of core is one index of the effectiveness of the kiln; from 0.5 to 3.5 per cent core is generally acceptable.

Although limestone begins to break down at its surface at approximately 1480 F, practical lime kilns operate at levels as high as 2600 F at points of maximum temperature. The performance of the kiln can be evaluated from studies of the exhaust gas analysis.[1]

The specific heat of limestone is approximately 0.27 Btu per lb, and the heat required to drive off CO_2 is approximately 1400 Btu per lb. Heats of reaction are given in Table 12-151.

Table 12-151 Heats of Reaction, ΔH (at 25 C)

Reaction	Heat, Btu/lb-mole
$CaCO_3 \leftrightharpoons CaO + CO_2$	+78,000
$MgCO_3 \leftrightharpoons MgO + CO_2$	+52,000
$CaO + H_2O \leftrightharpoons Ca(OH)_2$	−27,500
$MgO + H_2O \leftrightharpoons Mg(OH)_2$	−18,000

Heat requirements for calcining given here are approximations; they vary with the temperatures in the kilns, the type of rock, etc. Additional heat is required for preheating and "postheating" to the desired finishing temperatures. These additions bring the approximate heat required to about 2500 Btu per lb for dead burned dolomite and 2200 Btu per lb for lime. Shrinkage in the stone varies with the intensity of calcination. Soft burn results in a rather porous lime with little shrinkage; hard burn results in a denser product but with greater shrinkage. Detailed data will be found in the references.[2,3] The time required for calcining is roughly proportional to the 1.7 power of the diameter of the stone.

Types of Kilns

Rotary Kilns. These kilns are inclined, lined steel cylinders, sometimes with larger diameters for the final part of their lengths, that are rotated 60 to 150 rpm so as to produce peripheral speeds of 25 to 50 fpm. Countercurrent firing is used with most kilns. Fuel consumption runs 6 to 8 MMBtu per ton of lime produced.

Since thermal efficiency of rotary kilns is not high, considerable attention has been paid to means for improving their performance and recovering heat. Exhaust heat boilers, coolers, stone preheaters, heat exchanger cross systems, etc., have been applied. Figure 12-207 shows two rotary kilns, equipped with preheater and coolers, at the Gibsonburg Lime Products Co. in Gibsonburg, Ohio. Kiln No. 1 is 8 × 208 ft and rotates at 60 rpm. Lime discharges at 2000 F into the rotary cooler. The kiln is fired by natural gas.

Rotary kilns have also been used in lime recovery; an example is the installation at the Mead Corp. plant in Chillicothe, Ohio.[5] The charge is a slurry (35 per cent $CaCO_3$ and 65 per cent H_2O), which is converted back to CaO for reuse in paper mill operation. Natural gas is fired at 15 psig thru Cohen burners.

Dolomite calcination with natural gas firing is being done in rotary kilns at a plant in Thornton, Illinois.

Rotary kiln performance may be evaluated by reference to procedures outlined in a book by Azbe.[6] Good performance depends on instrumentation,[7] heat salvage, and insulation.

Fluidized Vertical Lime Kilns. Calcination of limestone and dolomite in fluidized beds of stone has been successfully demonstrated[8] (Fig. 12-208). Gas (or oil) is injected into the fluidized bed[9] of the calcining compartment. The bed serves to mix gas and air and to maintain ignition.

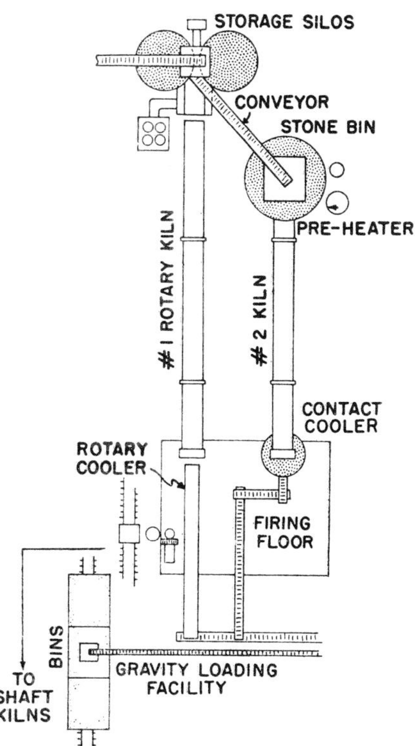

Fig. 12-207 Rotary kilns for processing dolomitic limestone.[4] The No. 1 kiln is fired with natural gas using a Kennedy Van Saun burner.

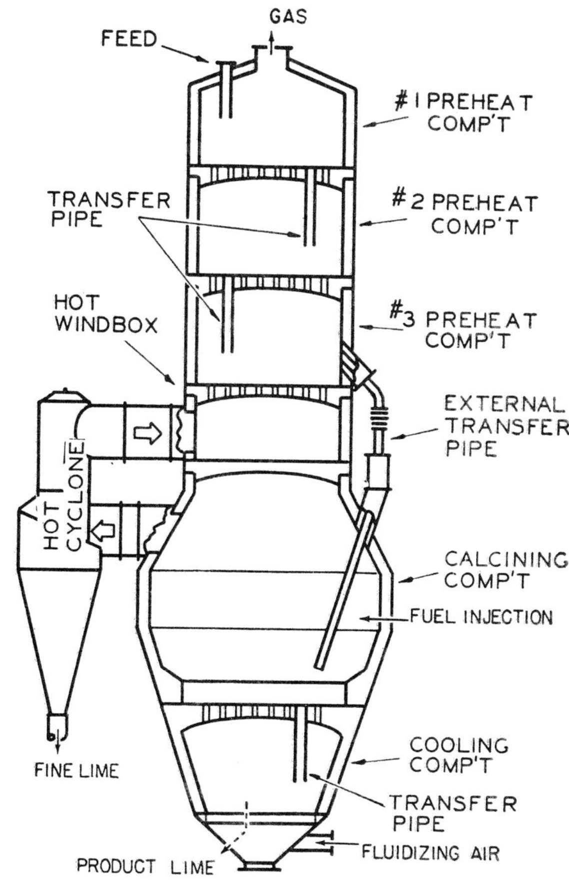

Fig. 12-208 Fluidized-solids calciner for converting limestone to lime.[9] (Dorr-Oliver Inc.)

Calcining proceeds at rather low temperatures, thus minimizing the formation of silicates and other impurities. The size of crystals in the stone limits the application of this process. Small crystals may lead to excess dust losses. Plants of this kind are operated in Adams, Massachusetts, and elsewhere.

Vertical Kilns. Vertical shaft lime kilns date well back into history. Some are adapted for mixed feed (coal and stone charged together); others introduce fuel thru a separate combustion chamber (Dutch oven) or thru burners near the bottom of the shaft. Fuel consumption runs 4.5 to 6 MMBtu per ton of lime produced.

The flexibility of natural gas firing and the high temperatures attainable by preheating combustion air raises the question: What is the optimum temperature in the kiln? Within limits, higher temperature reduces calcining time. The effect on refractory lining is one limit; the loss of porosity of stone under high temperature and the pressure of the load is another. Large-diameter stone introduces additional heat transfer considerations, since the CO_2 moving toward the surface absorbs heat and retards the temperature rise at the center of the stone. The presence of impurities (SiO_2, Al_2O_3, etc.) in limestone may cause the formation of certain hard silicates. Lower firing temperatures are used to prevent these formations, especially with large-diameter stones.

Smaller stones offer advantages in shortening calcining time, but the pressure drop thru the kiln increases. Since a variety of stone sizes is usually handled, it is necessary to adjust operations so as to prevent overburning or underburning. Field tests of kilns indicate that the resistance to gas flow (i.e., pressure drop) is proportional to V^a where V is the gas flow rate and a is between 1.86 and 1.90. When using charts based on the assumption that resistance is proportional to V^2, this difference should be kept in mind. However, the actual behavior of stones in a kiln introduces other flow differences that may have even greater effect. These include channeling, irregular packing, the effects due to the cleavage of stones, dust accumulation, and bridging.

Natural gas applications to vertical shaft kilns involve the use of various means for ensuring proper flame spread. Two arrangements are illustrated:

1. Figure 12-209 (left) shows the Chemstone Corp. lime kilns in Strausberg, Virginia. Design is based on a production rate of 2 tons per sq ft-day, with a gas usage of less than 5.0 MMBtu per ton of lime. Pressure drop is indicated by an induced draft of 8.5 in. w.c. The combustion arrangement has four quadrants, each served with two groups of four burner ports. One leads out from the center and the other from the periphery of the charge. Gas pressure is 60 psig.

2. Figure 12-209 (right) shows the vertical kilns at the Ashtabula, Ohio plant of the Union Carbide Metals Corp. It has been reported that these kilns have from three to six *times* the capacity of former conventional-shaft kilns of equal size. The kilns are 11 ft square internally. Production rates over 500 tons per day have been maintained. Heat distribution is secured by balancing the gas burned in the upper and lower burner beams and by controlling the air flow. Gas usage is from 4.5 to 5.0 MMBtu per ton of lime. Figure 12-210 shows the details of the burner system.

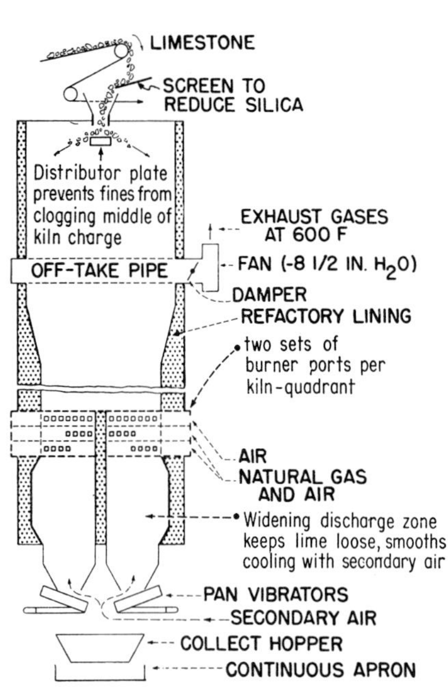

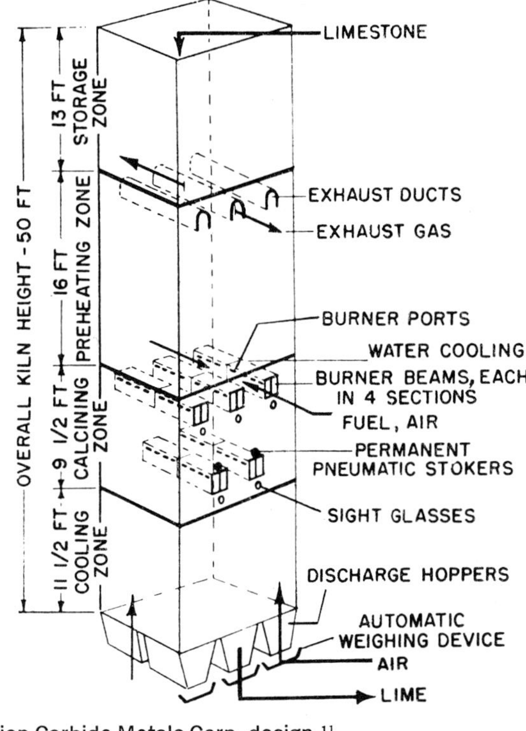

Fig. 12-209 Vertical lime kilns: (left) Chemstone Corp. design;[10] (right) Union Carbide Metals Corp. design.[11]

CEMENT

Cement kilns are inclined rotary kilns (Fig. 12-211) similar to the rotary type of lime kiln. Approximate capacities are shown in Table 12-152. The average fuel consumption per barrel of cement is: gas, 1179 cu ft; oil, 0.1900 bbl; coal, 99.9 lb.[12] The largest kiln reported (1959) was built by a Belgian firm for use in Siberia. Each of two kilns was 574 ft long and 19¾ ft in diameter.

Table 12-152 Approximate Capacities of Rotary Cement Kilns

		Capacity, bbl/day		
Diam, ft	Length, ft	Dry process	Wet process	Approximate horsepower
7	100	400	250	10–20
8	125	650	450	15–25
9	150	1000	750	25–50
9	200	1200	850	30–60
10	240	1450	1100	50–75
11	300	2200	1900	75–125
12	400	3700	3200	150–200
12	500	4600	4000	200–300

An example of successful natural gas firing is found in the Tulsa, Oklahoma plant of the Dewey Portland Cement Co.[13] The kiln is 12 ft in diameter and 425 ft long, with a design capacity of 4000 bbl per day, requiring 3.5 MMCF of natural gas. The drying of the raw material also depends on natural gas, used in a Todd heater burning 25 MCFH.

Wet Process. This process begins by using heat to drive off water—about 25 per cent of the Btu is used for this function. The dried material, at about 212 F, then passes thru the heating zone, where the temperature rises to about 1500 F and calcination starts. The calcining zone operates between 1650 and 1700 F. In the reaction zone, at 2700 to 2900 F, the process is completed.

An explanation of how the various individual cement components react during setting and curing is available.[14] Such information can be used to fix the kiln reaction temperatures in order to make the desired product.

Dry Process. This process, which differs from the wet process in the preparation and feeding of raw materials, lends itself better to economical heat salvage from the exhaust gases (higher exhaust temperatures). Although "dry" raw materials are used, these may contain considerable moisture (from 15 to 20 per cent in limestone shale and clay being common). Therefore, dryers are also needed.

Since large volumes of air are involved, fuel-gas usage is influenced by changes in air temperature and air humidity. These plus wind also affect stack draft. Appropriate means for adjustment of air–fuel ratio and control of air movement are suggested.

Natural gas (in comparison with coal) does not furnish any iron or ash, which may be of some value in cement operations.

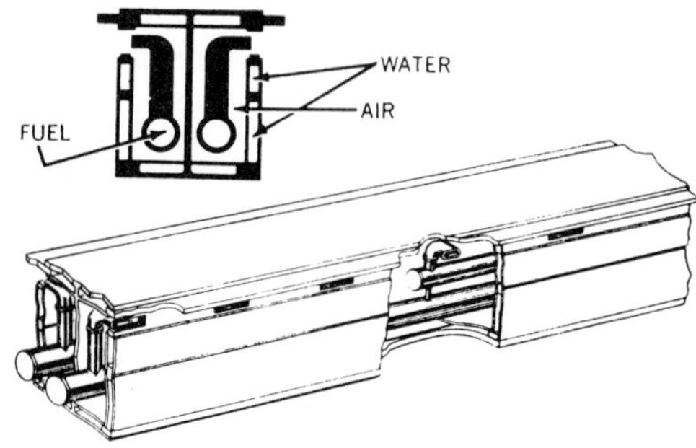

Fig. 12-210 Burner beam detail for the Union Carbide Metals Corp. vertical lime kiln[11] shown in Fig. 12-209 (right).

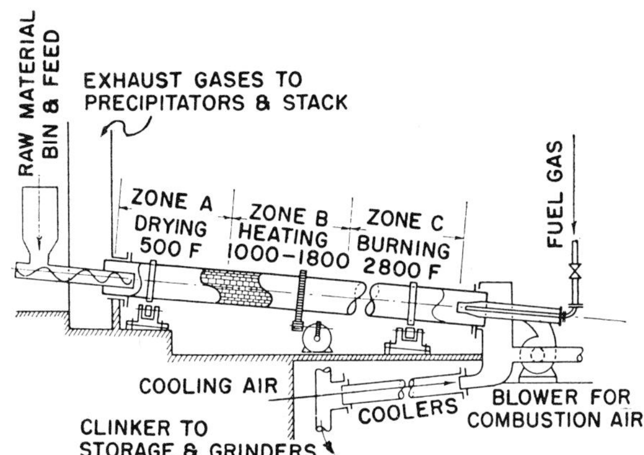

Fig. 12-211 Rotary cement kiln.

On the other hand, natural gas does not add sulfur. Normally, clay provides all of the elements that may be needed to make cement.

The oxygen content of the kiln exhaust gases is held at low levels in order to maintain good economy (0.5 per cent is typical with natural gas firing). With coal firing, however, a reported O_2 content of about 2 per cent may be needed to hold down CO losses.

OTHER KILN PRODUCTS

Many other materials besides lime, dolomite, and cement are pyroprocessed in rotary and vertical shaft kilns. Each material poses its own requirements as to time, temperature, air movement, etc. Kilns for magnesite (periclase, MgO) require the highest possible temperatures, close to 4000 F, which are reached by preheating air thru countercurrent material flow. As in the case of forging, dual-fuel use, with some oil added to the gas, increases flame radiation. On the other end of the temperature scale are phosphate rock kilns, which need low temperatures to burn out organic materials present.

REFERENCES

1. Weisz, W. H. "Kiln Performance Charted from Studies of Gas Analysis." *Rock Prod.* 55: 88+, March, 1952.
2. Azbe, V. J. "Specific Heat and Heat of Calcination." *Rock Prod.* 54: 122–3, Jan. 1951.
3. Murray, J. A. "Specific Heat Data for Evaluation of Lime Kiln Performance." *Rock Prod.* 50: 148+, Aug. 1947.
4. "Visit to Gibsonburg Lime Products Plant." *Nonmetal. Minerals Processing* 3: 15–16, Dec. 1962.
5. Rohr, H. R. "Lime-Recovery Kiln Reduces Costs, Eliminates Waste-Disposal Problems." *Ind. Gas* 39: 10–1, Jan. 1961..
6. Azbe, V. J. *The Rotary Kiln: Its Performance, Evaluation and Development.* St. Louis, Mo., Azbe Publications, 1954.
7. Blackman, A. A. "Automatic Temperature Control of Rotary Kilns." *Rock Prod.* 46: 72+, Aug. 1943.
8. "Use of FluoSolid Kilns Increasing." *Chem. Eng. News* 41: 56–7, Oct. 21, 1963.
9. Zenz, F. A. and Othmer, D. F. *Fluidization and Fluid-Particle Systems.* New York, Reinhold, 1960.
10. "Revamped Kilns Smooth Lime Burning." *Chem. Eng.* 65: 64+, July 28, 1958.
11. Chase, H. "New Lime Kilns Multiply Output and Boost Economy." *Ind. Heating* 30: 2166–76, Nov. 1963.
12. U. S. Bur. of Mines. *Minerals Yearbook, 1963, Vol. 1: Metals and Minerals (Except Fuels)* Washington, D. C., G.P.O., 1965.
13. "New Stackless Plant for Dewey Portland." *Nonmetal. Minerals Processing* 2: 21–4, July 1961.
14. Brunauer, S. and Copeland, L. E. "Chemistry of Concrete." *Sci. Am.* 210: 80–92, April 1964.

Chapter 21

Blast Furnace Fuel Injection*

by J. C. Agarwal

Crude iron ores consist mainly of iron oxides plus other impurities such as silica, alumina, and sulfur. Iron oxides are reduced to iron in the blast furnace by the carbon monoxide formed from the coke. The impure (pig) iron thus formed is subsequently refined into steel.

The largest single conversion cost item in the manufacture of iron is coke. One method of decreasing the consumption of coke significantly is the injection of other fuels into the tuyere zone. A tuyere is the nozzle thru which air is blown into the blast furnace. The advent of fuel injection into blast furnace operation has had a far-reaching impact on the process of making iron. The status of fuel injection in American industry is shown in Table 12-153. To date, the results from various plants indicate that even though the reported coke savings appear to vary greatly, the actual coke savings are approximately the same in all cases after assigning proper credits for higher blast temperature, blast humidity, and burden changes, and for hot metal quality.

Table 12-153 Fuel Injection in American Iron and Steel Industry (1962)

	Natural gas	Oil	Oil and natural gas	Coke-oven gas	Coal
No. of furnaces*	30	5	2	1	1
Fuel injected, lb/ton hot metal	50–100	100–150	100–150	60–70	200–300
Coke saved, lb/ton hot metal	75–100	120–170	125–175	80–100	200–350

* See Addendum to this Chapter for December 1964 data.

The reaction of coke with air to form CO and H_2 in the blast furnace produces 19,620 Btu per lb mole of carbon at a tuyere temperature of 3400 F and a blast air temperature of 1800 F. Under the same conditions at the tuyeres, the following reactants produce endothermic incomplete-combustion reactions:

	Btu/lb mole of carbon
Natural gas with air	+79,325
Coke with moisture	+75,040
Oil ($C_{15}H_{32}$) with air	+21,135
Coal with air	+8,335

Therefore, it can be seen that the injection of fuels into the blast furnace will produce an endothermic effect.

The primary reason that the partial combustion of coke releases heat while the partial combustion of coal, for example, consumes heat is that the coke is preheated to very high temperatures before it reaches the tuyere zone. On the other hand, coal is injected at low temperatures so as to prevent its sticking in the injection system. The endothermic effect is also related to the heat of formation, which for methane is −32,198 Btu per lb mole as compared with −4800 for fuel oil and −2800 for coal.

In early trials of fuel injection, no attempt was made to compensate for the localized endothermic effect of natural gas injection at tuyeres by increasing the hot blast temperature or decreasing the blast humidity, at least temporarily. Thus, these trials failed, because the furnace gradually cooled.

COMPENSATING FOR FUEL INJECTION

The three main procedures available for compensating for the endothermic effect of fuel injection and thus maintaining thermal balance in the furnace are: (1) removing moisture from the blast; (2) increasing the hot blast temperature; and (3) injecting oxygen instead of air. In the first method, for example, the heat that would have been consumed by the moisture if it had not been removed can now be consumed by the endothermic reactions caused by the injected fuels. In this manner the heat balance around the tuyere zone is essentially unchanged, and the furnace continues to operate as efficiently as when coke alone was charged.

Reducing Blast Humidity

It is assumed that the blast humidity can be decreased to the ambient level of 7.0 grains per cu ft, permitting the removal (at most) of approximately 100 lb of water from the blast per ton of hot metal. The amounts of the various fuels that can be injected when this amount of moisture is removed, while maintaining maximum efficiency, are given in Table 12-154. Because the heats of reaction of natural gas, fuel oil, and coal are different and decrease with respect to an increasing carbon–to–hydrogen ratio, more coal can be injected than fuel oil, and more fuel oil than natural gas, for the same degree of moisture removed from blast air.

Oxygen Injection

At the tuyeres, the chilling effect that results from the endothermic reaction can be compensated for by injecting oxygen into the tuyeres, thereby permitting more fuel to be

* Extracted from J. C. Agarwal, "Blast-Furnace Fuel Injection," *Mechanical Engineering* (November 1963), 52–5.

Table 12-154 Economics of Blast Furnace Fuel Injection When Removing 100 lb of Water per Ton of Hot Metal*

	Natural gas	Fuel oil	Coal
Coke saved, lb per lb of fuel injected	1.59	1.28	1.16
lb per ton of iron	108	180	231
Fuel injected, lb per ton of iron	68†	140	198

* Based on blast humidity of 7.0 grains per cu ft.
† Equivalent to 1600 SCF.

injected at a given condition. Two factors, however, should be recognized in the injection of oxygen: (1) the solids–to–gas ratio in the blast furnace; (2) the cost of the oxygen injected into the blast furnace.

When oxygen is injected into the blast furnace, the rate of production will go up in the furnace. The production rate rises 50 per cent when the oxygen concentration in the blast increases to 32 per cent. However, the gas-to-solids ratio in the blast furnace stack will go down, because the nitrogen associated with the oxygen injected is no longer present; therefore, the moles of bosh gas generated decrease very rapidly. This decreases the ratio of the amount of gases to the amount of solids in the stack and also decreases the temperature in the upper half of the stack. As a result, the temperature profile in the stack becomes very steep. Beyond a certain point, this lowering of temperature in the upper half of the stack results in improper reduction of the burden in the stack, thereby nullifying any beneficial effect of the injection of oxygen. In other words, there are conditions that determine the point beyond which oxygen cannot be profitably injected into the blast furnace without making drastic chemical changes in the furnace operation.

Another consideration is the expense of the oxygen. To make an incremental ton of hot metal in blast furnaces requires half a ton of oxygen. Even though injection of oxygen permits higher rates of fuel injection resulting in greater coke savings and increased production rates, careful analyses will have to be made to justify its use.

A major advantage of oxygen is that higher production rates can be obtained without the use of high top pressure, eliminating the maintenance problems associated with ultra-high top pressure (30 psig) operation.

Fuel Preheating

From time to time there have been proposals for preheating the fuel injected into the blast furnace so that more fuel can be injected. Preheating of coal above 400 F is difficult because of the release of volatile matter and the increased tendency to coke. Preheating of oil above approximately 250 to 300 F is not recommended because of the possible coking of oil in the pipes. It is quite practical to preheat natural gas by means of relatively inexpensive tube heaters, although the preheat temperature must be limited to 1000 F so as to preclude carbon deposition.

Preheating natural gas to 1000 F makes it possible to increase the natural gas input to the furnace by 10 per cent. If heat is added to the furnace in this manner, but the amount of input is not increased, the hot blast temperature need not be increased by 10°F for every per cent of natural gas preheated to 1000 F. These choices are possible because of the decrease in endothermic heat effects of natural gas injection at the tuyeres. However, preheating natural gas to 1000 F requires the expenditure of both capital and fuel. To determine the economic benefits of maximum natural gas injection, all factors, including the cost of preheat fuel and the attendant capital costs, must be considered.

COKE SAVINGS

The total possible savings in coke consumption resulting from the removal of moisture from the blast and the injection of various fuels are given in Table 12-154. Larger quantities of these fuels can be injected than those shown in the table. However, the utilization of additional amounts will be less efficient than the maximum amounts listed, because the tuyere-zone flame temperature will decrease and additional carbon will have to be burned to compensate for the temperature deficit. Where the price differential between the fuel injected and the coke is large, slightly lower efficiency of utilization will still yield economic benefits.

No fuel can be successfully injected when the flame temperature at the tuyeres becomes so low that the blast furnace becomes inoperable. Since natural gas has the greatest chilling effect on the furnace and coal the least, the amount of coal that can be injected into the blast furnace before inefficient or inoperable flame temperatures are reached is considerably greater than the amount of natural gas.

PRODUCTION CONSIDERATIONS

The production rate of the blast furnace is directly dependent upon (1) the amount of heat released and (2) the temperature at which this heat is released. Accordingly, the production rate of a blast furnace will decrease if the tuyere-zone temperature is decreased. This temperature reduction may be caused by injecting blast air or fuels containing excessive amounts of moisture. Fuel injection *per se* does not increase blast furnace production rate. Production rate increases with increased hot blast temperature or decreased blast humidity; both are generally used with fuel injection.

Figure 12-212 shows the effect of both gas and oil injection on the production rate as compared with the rate without fuel injection. For any specific hot blast temperature, the production rates are lower when fuel is injected; however, fuel injection permits the use of a higher hot blast temperature and thus results in higher than normal production rates. Excessively high oil injection rates lower the production rate.

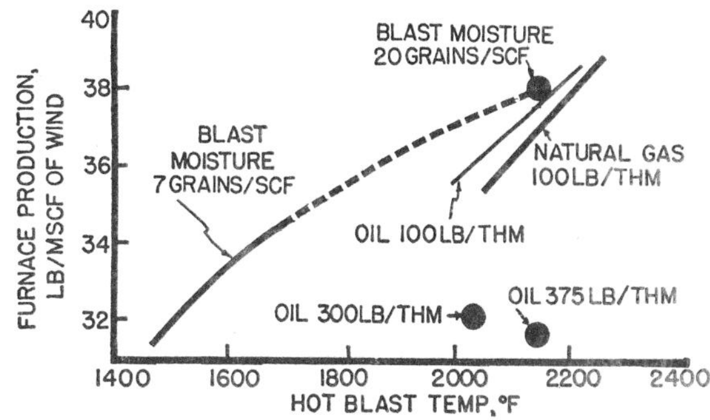

Fig. 12-212 Effect of both oil and gas injection on blast furnace production rate, compared with no injection.

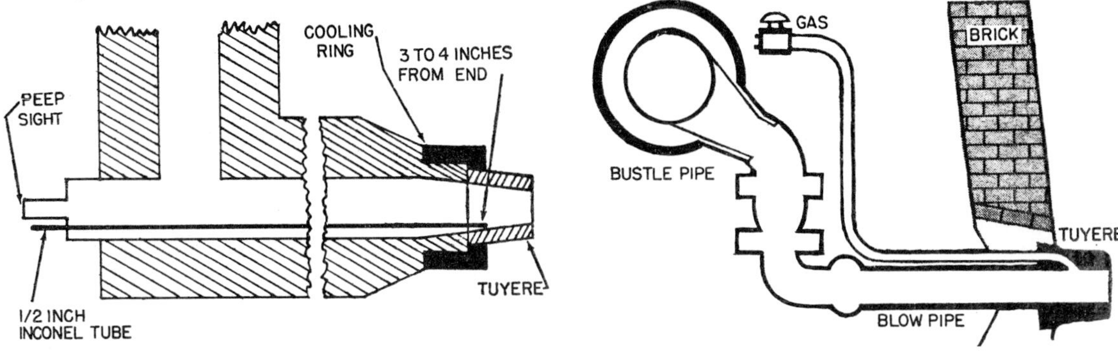

Fig. 12-213 Two ways of injecting natural gas into the blast furnace: (left) by an Inconel tube connected at the bottom of the blow pipe—this method has also been used to inject oil; (right) by a channel cast into the tuyere, connected to the blow pipe.

Natural Gas Injection

Natural gas is the most commonly used fuel injectant in the American iron and steel industry (Table 12-153). In most commercial furnaces, natural gas is injected by the two methods shown in Fig. 12-213.

Fuel Injection and Fuel Balance

One of the important economic factors in the blast furnace process is the Btu value of the top gas and the total Btu obtained per ton of hot metal from the top gas. Generally, the Btu value of the top gas is between 90 to 100 Btu per cu ft. Even though the coke rate is decreased considerably, the heating value in the top gas does not decline appreciably. The burners fueled by top gas, produced by furnaces using injected fuel, should be adjusted to compensate for the higher H_2 content of such gas compared with "normal" top gas.

The total Btu in the top gas per ton of hot metal, however, does decrease from 12 million Btu per ton of hot metal to less than 8 million Btu per ton of hot metal as the coke rate declines. Approximately 25 to 30 per cent of the top gas is used at the blast furnace to preheat the blast to approximately 1500 F. If the blast air is heated to a higher temperature, as is likely with fuel injection, the top-gas usage around the blast furnace will increase to as high as 50 per cent. This increased usage at the furnace may create a fuel deficiency at the boiler house, a condition that will have to be rectified by supplementary fuel. Similarly, in an integrated plant, it is necessary to take the loss of coke-oven gas not produced in the coke plant into account to make a proper fuel balance in the plant.

Sulfur Control in Hot Metal

The various fuels injected into the blast furnace have different sulfur contents. Natural gas has essentially no sulfur; fuel oil contains from 0.5 to 3.5 per cent; and coal may contain as much as 1.5 per cent. Depending upon the amount of fuel injected and the amount of sulfur in it, the sulfur load in the blast furnace will decrease or increase in proportion to the sulfur content of the coke that the fuel displaces. When injecting natural gas, the blast furnace sulfur-control problem should become easier, because the coke that will be displaced by natural gas contains far more sulfur than the natural gas injected. With coal injection, the sulfur input may be about what it is with coke (or even more, depending on the coal). Normally, in the blast furnace approximately 6 to 12 lb of sulfur is contained in each ton of hot metal charged. When fuel oil containing 0.5 to 3.5 per cent sulfur is injected, the sulfur input will go up, making the control of the sulfur in the hot metal difficult.

Results on an experimental blast furnace with oil injection have demonstrated that it should be possible to inject a fuel with a high sulfur content by either increasing the basicity of the slag to 1.50 or more, or increasing the slag volume. The partition ratio (defined as per cent sulfur in slag divided by per cent sulfur in hot metal) is less than 85 in all cases. Note that in commercial practice values of approximately 90 for partition ratio are attained with a sulfur content of 2.0 per cent in the slag. Therefore, it appears that adequate control of sulfur content in the hot metal can be achieved if the slag volume is large enough and the total sulfur input low enough so as not to require the partition ratio to exceed approximately 90.

Gas Density and Pressure Drop

With natural gas injection, it has been claimed, the furnace behaves more smoothly, the burden movement is more uniform, and the number of checks on the furnace is decreased markedly. It should be recognized that the injection of various fuels introduces a substantial quantity of H_2 into the blast furnace. Because H_2 is approximately 0.07 as light as the other gases present in the blast furnace, even a small amount of H_2 in the blast furnace gas tends to decrease the density of the gas and thereby decrease the pressure drop within the furnace. A decrease in the pressure drop in the furnace when the furnace is being driven at the same wind rate should result in a smoother movement of the processing materials; it should also result in less flue dust being blown out of the furnace. The presence of H_2 in the high-temperature zone of the furnace tends to increase the rate of reduction so that the furnace performs in a smoother manner.

Iron Content of Stack Solids

The injection of various fuels into the furnace facilitates the removal of coke from the burden. When these fuels are injected, the iron-oxide concentration increases in the stack of the furnace, permitting more efficient use of the reducing gases. The density of the solid burden increases with the increased iron-oxide content. Increased density permits higher pressure drop across the stack without excessive fluidization of the dust or hindrance of the smooth movement of the burden.

ADDENDUM

[The following material, added just prior to the printing of the Handbook, was not prepared by Dr. J. C. Agarwal, author of the Chapter. This addendum is based in part on an unpublished research project carried out by Jack Huebler and A. R. Khan at the Institute of Gas Technology for the Con-Gas Service Corporation and in part on a gas industry blast furnace symposium.— EDITOR]

Huebler and Khan's findings may be summarized as follows:

Natural gas burns (that is, to $CO + H_2$) rapidly and smoothly—no soot in the stack indicates burning rather than cracking. Natural gas can be injected in large quantities.

The coke savings per unit of natural gas can be established at different values, depending on what credit is given to which variable involved. At least two variables must be changed in a blast furnace to maintain balance.

The blast temperature must be increased in order to maintain steady conditions when natural gas is added.

One or two per cent of natural gas can be added when the furnace bottom is running relatively too hot, without adding heat to the furnace by means of the preheating operations.

If the blast temperature is increased, the most economical step to keep the blast furnace in balance is to add steam. The net effect is a moderate savings in coke.

The main advantages of natural gas injection follow:

1. Injections up to 12 per cent of the blast have actually been achieved. When substantial amounts of coal or oil have been injected, stack problems have been encountered.

2. Compared with coal or oil, natural gas furnishes large amounts of hydrogen. When a blast furnace is driven toward maximum production rates, the "heat line" is driven upward in the furnace. If the reduction rate of the ore is slow compared with the rising heat line, the oxides may slag with silica, melt, run down, and then be reduced to iron, which may solidify (because its melting point is higher than that of the slag).

The higher H_2 content of natural gas is of considerable value: It reduces iron about four times as fast as CO can; it can penetrate more rapidly into the iron oxides because of the low molecular weight; and it stops the adverse effects of the rising heat line. These effects are quite aside from coke savings.

3. The equipment required for natural gas injection is simple and is easy to operate.

4. Natural gas contains no sulfur or ash. This is an important plus factor, although it has not proved possible to pinpoint the results with respect to sulfur partition by slag. The relatively large amounts of silica in the ore and even in the coke are removed in the form of slag produced by the use of dolomite and limestone. The slag also serves to remove most of the sulfur, at least reducing it to acceptable levels in the iron. If ore or pellets with less than 2 per cent silica should become available, the low sulfur content of natural gas would play an important role.

5. Natural gas is of real value in reducing air pollution problems. With natural gas, less slippage of burden will occur, thus sharply reducing the heavy dust load attributed to that cause, particularly with high production rates.

P. L. Woolf* suggests four principal limitations in connection with fuel injection. These govern smooth furnace performance. The first two regulate the minimum quantity of blast additives; the latter two set the maximum rate of injection.

1. An upper combustion zone temperature above which the furnace operation will get sticky and hang.

2. A minimum quantity of reducing gas per ton of metal below which metal quality cannot be maintained.

3. A lower combustion zone temperature below which metal quality cannot be maintained.

4. Sufficient coke in the burden to maintain satisfactory permeability.

A summary of the use and benefits of fuel saving additives follows:

Steam permits high blast temperature, substantial gains in productivity, and moderate coke economy.

Natural gas permits high blast temperature and gains in productivity similar to those obtained by moisture, but with high coke savings.

Oil permits high blast temperature and productivity about equal to that obtained with moisture and gas injection, but with lower coke rates than with gas. Compared with gas, however, the higher quantity of oil used may more than offset the saving. Oil may be injected without increase in blast temperature, and this leads to very low coke rates. The relative prices of oil and coke would determine the economics of fuel use.

Calculations based on the Bureau of Mines experiment indicate that coal will not permit as large an increase in blast temperature as the other additives. Productivity would therefore be restricted to highly volatile coals. Coal used at constant blast temperature would usually result in a substantial economy. Furthermore, although the blast temperature may be limited, it may still be more than the furnace can safely tolerate.

The Bureau of Mines reported that as of December 1964† fuel injection in 77 blast furnaces in the U. S. was divided as follows: 52 use natural gas injection; 15 fuel oil; 4 coke oven gas; 2 coal tar; 3 gas and oil; and 1 uses gas, oil, and tar with oxygen.

Discussion at the blast furnace symposium also included the following topics:

Prereducing ores at mine sites is also a factor in the changing technology of iron production. This would affect the potential use of natural gas at blast furnaces.

Another potential use of gas in the blast furnace is to apply it to gas turbine-driven air blowers, which in a typical case may require approximately 25,000 hp. The turbine exhaust can be recovered by a suitable boiler that in turn furnishes steam to a steam turbine (about 8000 hp) on the same shaft. The combination of the two turbines would develop the desired power.

---

* Of the U. S. Bureau of Mines. Unpublished address at the A.G.A. Blast Furnace Symposium, Pittsburgh, Pa., Dec. 1964.

† See Table 12-153 for 1962 data

Chapter 22

Fuel Comparisons

by F. E. Vandaveer and E. A. Jahn

AREAS OF FUEL COMPETITION

Competition in the selection of fuels varies geographically, depending on nearness to point of usage and availability, and varies also among the heating processes. In some areas there is little competition, in others one fuel predominates, and in still others several fuels share the market.

Coal is widely used for the generation of steam and electricity, for other industrial process heating, and for house and large building heating. Its use is confined largely to the North, East, and Midwestern sections of the United States, where the major coal fields are located. Relatively small amounts of coal, however, are used in those sections for cooking or water heating. Very little coal is used in the Southwest and Pacific Coast states for any purposes.

Coke made from coal is now used primarily for metallurgical and industrial purposes. Very little is used for house heating.

Oil (petroleum) is used nationally for gasoline, diesel oil, kerosene, and heating oil production. Its major uses are for power generation in automobiles, trucks, railroad engines, and ships; steam-electric generation; other industrial process heating; and, in some areas of the country, for house and water heating.

In 1965, oil shale and tar sands were not being used commercially in the United States and Canada, although some large pilot plants have been operated to make **shale oil** in limited amounts. These potential sources for oil and for substitute natural gas by hydrogenation of the oil are so large that they will undoubtedly be used in the future. It is conceivable that shale oil may become competitive with oil from crude petroleum.

District steam is usually made from coal in large generating stations and supplied to business districts of large Northern and Eastern cities in the United States. Steam under pressure is supplied to large buildings for heating and air-conditioning purposes.

Electricity is available in nearly all urban and most rural districts in the United States. Its primary use is for lighting and power, but its use is being extended for domestic purposes into cooking, water heating, air conditioning, and heating, and for industrial purposes into these same uses as well as into heating and processing.

Nuclear heat is beginning to be used to generate steam to produce electricity. Such use will doubtless increase. Predictions of installed electric generating capacity in 1980 vary from 50 to 250 million kw.

Solar heat has long been used on a limited scale for heating water in the South and Southwest. Many of the new building designs take advantage of this heat for supplementary space heating. Methods for its storage are being studied. There is the possibility that solar heat will come into greater use in some areas.

Table 12-155a Actual Cost of Supplying Natural Gas and of Making Electricity from Coal[1]

(based on 1957 annual company reports of a large Pennsylvania electric company and a large Northeastern natural gas company)

Natural gas company		Electric company	
Operating revenues	$281,000,000	Operating revenues	$98,681,000
Gas sold, MCF	472,119,000	Capacity of system, kw	1,226,000
		Kwh per yr sold	6.1×10^9
		Avg capacity used, kw (57% of total)	697,000
Plant investment	$685,624,000	Plant investment (about $180,000,000 for generator facilities at $150 per kw)	$464,000,000
Avg daily gas sold, MCFD	1,300,000		
Plant investment		Plant investment	
$ per MCFD	526	$ per kw system capacity	380
		$ per kw avg capacity	665
$ per MMBtu-day	526	$ per MMBtu-day (57% load factor)	8,120
Avg residential rate, $ per MCF	0.693		
Avg cost, $ per MCF	0.330	Total system, ¢ per kwh	1.61
Cost of distribution, storage, etc., $ per MCFD	0.363	Avg residential rate cost, ¢ per kwh	2.87

Energy Sources and Reserves

Data on recoverable U. S. fuel resources are given in Table 2-1. Geological evidence indicates that large, undiscovered resources exist. Since the development of these potentialities is mainly an economic decision, it is reasonable to assume that the next two decades will witness no great changes in the availability of the fossil fuels. Furthermore, additional supplies of gas could be provided from oil shale, tar sand gasification, or coal.

PRODUCTION COST OF SUBSTITUTE NATURAL GAS COMPARED WITH ELECTRICITY

Production of substitute natural gas from coal by methanation of synthesis gas or hydrogenation of coal (Fig. 3-68) or a combination of these methods should be possible at less cost than that for making electricity from coal. Comparative data given in Tables 12-155a and 12-155b show that on a Btu basis electricity would cost 3.09 times as much as substitute natural gas and that plant investment would be from 5.3 to 10.1 times as much. The efficiency of the operation is shown to be nearly double in favor of the gas-making process.

Table 12-155b Comparative Costs of Making Substitute Natural Gas from Coal and Generating Electricity from Coal[1]

(using best available data on manufacture of substitute natural gas and data from Table 12-155a)

Investment costs	Substitute natural gas	Electricity
Calculated cost of substitute natural gas plant, $ per MMBtu-day	$1,000	...
Present plant investment, $ per MMBtu-day	526	...
Total plant investment, $ per MMBtu-day	$1,526	...
$ per kw installed	...	$ 380
$ per kw, avg (57% load factor)	...	665
$ per kw, avg (30% load factor for house heating)	...	$1,270
$ per MMBtu-day	$1,526*	$8,120–15,500*
Thermal efficiency, %	50–75	35–40
Cost to residential customer	$1.36†/MCF	1.5¢‡/kwh (proposed by company)
Cost to residential customer per MMBtu	$1.36§	$4.40§

* Ratio in favor of gas made from coal, 5.3 and 10.1 to 1.

† Estimated cost of making substitute natural gas from coal used was $1.00 per MCF; present figures vary from $0.55 to $1.20. Distribution cost of 36¢ from Table 12-155a was added to $1.00 to obtain $1.36.

‡ This is 1.37¢ per kwh below present residential rate and 0.11¢ per kwh below rate for entire system shown in Table 12-155a.

§ Ratio in favor of gas made from coal, 3.09 to 1.

FUEL AND ENERGY COMPUTATIONS

In competitive situations it may be necessary to provide the prospective purchaser with an estimate of the comparative operating cost using gas and any other fuel or form of energy under consideration. If the estimate is to be accurate, authentic fuel equivalents must be applied that are applicable to the appliance or operation involved.

Energy Ratios or Replacement Factors

Development of an acceptable fuel equivalent begins with consideration of the heat contents of the fuels themselves. For example, if both natural gas and electricity were utilized at 100 per cent efficiency, 1000 cu ft of natural gas (1000 Btu per cu ft) and 293 kwh of electricity (3413 Btu per kwh) would both provide 1 million Btu of useful heat, and be equivalent. It should be kept in mind that in the U. S. the gross, or higher, heating value is used in these comparisons. Fuels, however, develop different efficiencies and thus a comparison among them cannot be made strictly on the basis of their Btu content. Their comparative performance in commercially available equipment under actual operating conditions is the most desirable yardstick, but laboratory tests of thermal efficiency may in some instances be the only available comparison. An *energy ratio, ER* (also called a *fuel replacement factor*), is the number of Btu of one fuel equivalent to one Btu of another fuel supplying the same amount of useful heat. Stating the energy ratio as an equation:

$$ER = \frac{\text{Amount of fuel A used} \times \text{heat content of fuel A}}{\text{Amount of fuel B used} \times \text{heat content of fuel B}}$$

Establishing a fair energy ratio requires carefully controlled laboratory testing under simulated field conditions or, preferably, a large number of actual field comparisons of similar installations. Even more valid is a field comparison of similar appliances in which one fuel replaces another, the operating conditions remaining the same before and after the fuel change. For many years all of these methods have been used successfully. When impartially and competently supervised, studies have yielded unusually similar results.

Table 12-155c lists source material for reliable energy ratios for domestic and commercial applications. These data are reflected in Table 12-155d, which indicates the most frequently used and accepted energy ratios applied to competitive fuel situations. Variations from these ratios will naturally be encountered. Whenever possible, local data adequately documented are highly desirable.

Use of Energy Ratios

In competitive situations the use of an acceptable and eliable energy ratio makes fuel cost comparisons easy to establish. As an example, suppose that an electric bill of $45.00 for commercial cooking is estimated at an electric rate (energy plus demand) of 1.5¢ per kwh. At that rate such a bill would result from an electric consumption of 3000 kwh, which equals 10,240,000 Btu. The energy ratio of gas to electricity for this use is given as 1.60:1 in Table 12-155d. The amount of gas required will be 10,240,000 Btu × 1.60, or 16,400,000 Btu. Assuming the use of a natural gas (1000 Btu per cu ft), 16,400 cu ft of gas would be used as the basis for figuring the monthly cost.

Working the problem the other way, the procedure is as follows: Suppose the gas bill is $24.60, with a gas rate of $1.50 per MCF of natural gas (1000 Btu per cu ft). This would buy 16,400 cu ft of gas, or 16,400,000 Btu. If the energy ratio of gas to electricity is assumed to be 1.60, the gas Btu must be divided by this figure, yielding a quotient of 10,240,000 Btu. This figure can be converted into kilowatt hours by dividing by 3413 (Table 1-41), giving 3000 kwh.

Table 12-155c Fuel Comparisons, References and Data

Domestic cooking

1. Beveridge, E., and McCracken, E. C. *Comparative Utilization of Energy by Household Electric and Liquefied Petroleum Gas Ranges, Refrigerators, and Water Heaters.* (U. S. Dept. of Agric. Tech. Bull. 1073) Washington, G.P.O., 1953.

Laboratory test of LP-gas vs. electricity confirms an energy ratio of 2:1 for cooking.

2. Prettyman, B. "The Facts About Gas Efficiency." *L-P Gas* 9: 25+, Nov. 1949.

Energy ratio of gas to electricity is reported as 2:1.

3. A.G.A. Laboratories. *Comparative Laboratory Tests of Gas and Electric Ranges.* (Research Rept. No. 1132) Cleveland, Ohio, 1949.

Covers two years of laboratory tests. Overall energy ratio of gas to electricity was 1.99:1.

4. Carpenter, J. W. "Prove Favorable Gas Cooking Ratio." *A.G.A. Monthly* 30: 9+, Feb. 1948.

Field study by the Long Island Lighting Co. of large comparable groups of private homes using gas and electric ranges to cook identical menus. Resulted in a reported 2.03:1 energy ratio.

Commercial cooking

1. A.G.A. *Gas and Electric Consumption in Two College Cafeterias.* New York, 1950.

A 6-month study of two cafeterias at Southern Methodist University, one electric and one gas equipped. Results were 2035 Btu per meal for the gas kitchen vs. 1262 Btu per meal for the electric kitchen, or an energy ratio of 1.61:1. The energy ratios by types of appliance were: ranges and griddles, 1.25:1; deep fat fryers, 2.06:1; ovens, 2.47:1.

2. A.G.A. *A Comparison of Gas and Electric Usage for Commercial Cooking.* New York, 1948.

Field study involving new electric and gas appliances in the kitchen of Alfonso's Restaurant in Washington, D.C., for a period of 26 months. Consumption data for electric and gas equipment for a comparable number of meals yielded a gas-to-electric energy ratio of 1.37:1.

3. A.G.A. *Comparative Test on Gas and Electric Kitchens in Two Welfare Association Cafeterias in Washington, D.C.* New York, 1933.

An extended field test of two cafeteria kitchens operated by the Welfare and Recreational Association of Public Buildings and Grounds in Washington, D.C. Consumption data per pound of food cooked in the Mall Cafeteria (gas) vs. the Internal Revenue Cafeteria (electric) yielded an energy ratio of 1.65:1.

Domestic clothes dryers

1. "Clothes Dryers." *Consumer Bulletin* 48: 6–12, Oct. 1964; "Gas Clothes Dryers." *Consumer Bulletin* 48: 27–30, Nov. 1964.

Laboratory test of 13 gas and 12 electric dryers with regular or normal loads averaged 9950 Btu per load for gas vs. 10,441 Btu per load for electric, or an energy ratio of 1:1.05 in favor of gas. Average drying time (using the automatic cycle where available) was 33 mins for the gas dryers and 46 mins for the 240-volt electric dryers.

2. "Clothes Dryers—Electric and Gas." *Consumer Bulletin* 44: 10–7, Oct. 1961.

Laboratory test of 10 gas and 13 electric dryers using regular or normal loads averaged 10,000 Btu per load for gas vs. 9360 Btu per load for electric, or an energy ratio of 1.07:1.

3. *Automatic Clothes Dryers—Their Performance and Effect on Certain Fabric Properties.* (U. S. Dept. of Agriculture—Home Economics Research Rept. No. 6) Washington, G.P.O., 1958.

Two gas dryers tested averaged 14,500 Btu per load vs. 14,700 Btu per load for three electric dryers when using heavy loads of sheets and towels, for an energy ratio of 1.00:1.01 in favor of gas.

4. Weaver, E. K., and Thomas, M. E. "Automatic Clothes Dryers." *Ohio Farm and Home Research* 37: 58–61 July–Aug. 1952.

Fuel consumption averaged 9600 Btu per load for a gas dryer vs. 9200 Btu per load for an electric dryer, for an energy ratio of 1.04:1.

Domestic water heating

1. Consolidated Edison Co. *Field Tests of Oil and Gas Residential Space Heating.* New York, 1964.

A 15-month study conducted by Consolidated Edison Co. of New York; report on file with A.G.A. As part of this test, fuel consumption for domestic water heating was segregated from heating consumption. The results for 81 separate gas water heaters vs. 58 oil boilers equipped with built-in heat exchanger coils showed an overall annual fuel ratio of 10 therms of gas equivalent to 12.5 gal of fuel oil. This is an oil-to-gas energy ratio of 1.75:1.

2. Grier, R. F., and Jahn, E. A. *Performance Characteristics of Oil-Fired Water Heaters.* New York, A.G.A., 1964.

Comparison of the A.G.A. Laboratories service efficiency test for a high-input (1 gph) oil water heater and the University of Illinois service efficiencies for gas and electric water heaters yielded the following results:

100° F rise water usage	Service efficiency, %	Monthly fuel consumption
1500 gal per mo		
Gas	54.1	23.1 therms
Oil	44.2	20.5 gal
Electric	85.6	430.0 kwh
3000 gal per mo		
Gas	63.7	39.2 therms
Oil	50.4	36.2 gal
Electric	92.2	795.0 kwh

3. Hebrank, E. F. *Investigation of the Performance of Automatic Storage-type Gas and Electric Domestic Water Heaters.* (Univ. of Ill. Bull. No. 436) Urbana, Ill., 1956.

Controlled laboratory test of six representative gas water heaters and seven electric heaters under typical household draw-off schedules resulted in the following gas-to-electric energy ratios for a 90° F rise:

50 gals per day, 1.60:1
100 gals per day, 1.46:1
150 gals per day, 1.41:1

4. Beveridge, E., and McCracken, E. C. *Comparative Utilization of Energy by Household Electric and Liquefied Petroleum Gas Ranges, Refrigerators, and Water Heaters.* (U. S. Dept. of Agric. Tech. Bull. 1073) Washington, G.P.O., 1953.

Laboratory test of four gas and four electric water heaters financed by NEMA indicated gas-to-electric energy ratios between 1.58:1 and 1.81:1 for 78 gal per day draw-off.

Domestic refrigeration

1. Private memoranda to A.G.A. from two manufacturers of both gas and electric refrigerators, 1960.

Laboratory tests by each manufacturer of comparable 13-cu ft model gas and electric refrigerators resulted in a combined average monthly energy usage of 16.5 therms for gas models (including electric consumption) and 5.23 therms for electric units, or a gas-to-electric energy ratio of 3.16:1. Exclusive of the electric consumption of the gas models, the energy ratio was 2.8:1.

Central residential heating (gas vs. oil)

1. Consolidated Edison Co. *Field Tests of Oil and Gas Residential Space Heating.* New York, 1964.

A 15-month study; report on file with A.G.A. The results of this test were based on actual usage data from 100 gas heated

Table 12-155c Fuel Comparisons, References and Data (Continued)

homes and 82 oil heated homes in the New York City area. The test established an overall average relationship of 10 therms of gas equivalent to 8.48 gal of No. 2 fuel oil. This is an oil-to-gas energy ratio of 1.22:1 for a normal heating season of 4871 degree-days.

2. Long Island Lighting Co. *Space Heating Study*. New York, 1957.

An 8-month field study; report on file with A.G.A. This test was conducted for a full heating season on 50 gas heated homes and 44 oil heated homes. It reported that 1.74 Btu of oil were required to obtain 1 Btu of effective heating, whereas only 1.22 Btu of gas provided 1 Btu of effective heating. This is an oil-to-gas energy ratio of 1.42:1 in a heating season of approximately 5300 degree-days.

3. Paterson, R. E., and Wilcox, R. S. "Is Oil Heat More Efficient?" *Fuel Oil and Oil Heat* 16: 66–8, March 1957.

An investigation by The California Research Corp., an affiliate of Standard Oil of California, of gas and oil usage in 35 homes in Salt Lake City, Utah, that had been converted to gas heat. This test claims an oil-to-gas energy ratio of 0.945:1. Serious doubt as to the validity of this test remains, since it indicated that oil consumption was reduced to 0 therms per day on days having 5 degree-days or less, while the gas consumption was 2.8 therms per day when there were no degree-days. Furthermore, the substitutions did not always involve the same type of heating system. In a number of instances, gas floor furnaces were installed. These are not central heating systems and should not have been included in the study.

4. McGuckin, J. "Philadelphia Spikes Oil Heat Claims" *A.G.A. Monthly* 37: 6+, April 1955.

A comparison of 64 gas heated and 64 oil heated homes for a full heating season. This test indicated that 527 Btu of oil were required for each degree-day and each 1000 Btu per hr of heat loss, whereas only 362 Btu of gas satisfied the same requirements. This is an oil-to-gas energy ratio of 1.458:1 for a heating season of approximately 4000 degree-days.

5. Carroll, A. L. *Field Tests of Oil and Gas Residential Space Heating*. (ASME Paper No. 52-SA-21) Cincinnati, Ohio, ASME, 1952.

This test covers the fuel usage of 51 oil heated homes and 56 gas heated homes in the New York City area for an entire heating season. The overall results showed that 22.14 therms of gas or 31.5 therms of oil were required per 1000 Btu of heat loss. This is an oil-to-gas energy ratio of 1.42:1 for a normal heating season of 5090 degree-days.

6. "Relative Efficiency of Gas and Other Fuels" (In: A.G.A. *Comfort Heating*, p. 51–2, Sect. 38. New York, 1938).

Field study of a large group of homes in Rochester, N. Y., which were converted from oil to gas heat in 1937. The average oil-to-gas energy ratio in all cases was 1.42:1.

7. Sherman, R. A., and Cross, R. C. "Efficiencies and Costs of Various Fuels in Domestic Heating." *Heating & Ventilating* 34: 63–4, March 1937.

Field study, conducted by Battelle Memorial Institute in Columbus, Ohio on behalf of Bituminous Coal Research Inc. in 1936–37; included 30 test homes. It recommended efficiencies of 60 per cent for oil conversion burners and 70 per cent for gas conversion burners. This is an oil-to-gas energy ratio of 1.17:1.

Central residential heating (gas vs. electric)

1. Segeler, C. G. *Energy Losses in Space Heating, Air Conditioning and Water Heating*. (World Power Conf. Paper No. 117) Lausanne, Switzerland, 1964.

Evaluation of estimating methods that makes use of recorded energy usage data for heating from over 150 U. S. homes located in different climatic areas (outdoor design temperature areas ranging from −30 to +30 F). Added to these field data are published energy consumption data from another 210 homes. From an analysis of the comparative experience factor discussed in the paper (called K'), the energy ratios for gas to electric heating range from 1.88:1 in a 3000-degree-day area to 1.68:1 in a 7000-degree-day area. These data were obtained from electrically heated homes that had high levels of insulation and individual baseboard units and from gas heated homes that had minimum amounts of insulation and central warm air heating units.

This same paper compares homes with heat pumps and homes with individual electric baseboard units, and reports that the baseboard heated homes require 16 per cent more energy. This relationship applied to gas heated homes with a central warm air heating system and homes with heat pumps results in a gas-to-electric energy ratio of 2.18:1 for the 3000-degree-day area and 1.95:1 for the 7000-degree-day area.

2. Burgart, H. F. *How Big, Big Bill?* (Research and Utilization Paper No. 154/DR–22) Cleveland, Ohio, A.G.A. Laboratories, 1963.

Metered data from six homes that were heated by individual baseboard electric units and were converted to a central gas heating system. These homes all had the heavy insulation required for electric heat; the occupants remained the same. In these essentially identical homes the gas-to-electric energy ratio was 1.31:1.

School heating

1. Rudoy, W., and Walukas, D. *An Economic Study of Heating and Ventilating Systems for School Buildings*. Pittsburgh, Pa., Univ. of Pittsburgh, 1960.

Study of fuel consumption in nine gas heated schools and ten electrically heated schools conducted by the University of Pittsburgh. An analysis showed that gas heated buildings similar in construction to those designed for electric heat would require 10 cu ft per degree-day per 1000 sq ft of floor area, while the electrically heated schools actually required 2.0 kwh per degree-day per 1000 sq ft. This usage establishes an energy ratio of gas to electric of 1.52:1.

2. Sarno, W. R. "Gas Wins Top Marks In School Heating." *A.G.A. Monthly* 42: 12+, June 1960.

Review of a comparative test, conducted by Prof. Paul Mohn of the University of Buffalo, of two functionally identical schools built from the same single-story floor plan. One school was heated by electric unit ventilators and the other by a gas-fired, forced hot water system. For the school year 1958–59 the gas-to-electric energy ratio was 1.44:1.

3. Hill, F. W. "Gas vs. Electric Heating at Shingle Creek School." *The Minnesota Engineer*, p. 6–11, Dec. 1964.

Comparison of two identical classroom cluster units, one heated by electric resistance unit ventilators and the other by a central gas-fired boiler with individual unit ventilators. For the school year Sept. 7, 1959 to May 27, 1960, gas consumption, including auxiliary power, was 505 MMBtu, while electric heating consumption, including auxiliaries, was 384 MMBtu. This is an electric-to-gas energy ratio of 1:1.32. After the first school year, insulation was added to the electrically heated school, causing the electric heating energy usage to decrease; no changes were made in the gas heated school. The electric-to-gas energy ratio for the school year 1960–61 was 1:1.68; for 1961–62, 1:1.75.

In comparing gas and electric heating, Dr. Hill found: (1) no significant differences in room cleanliness, in paint deterioration, or in janitorial and custodial expense; (2) some initial paint scorching of furniture when cabinets were placed too close to baseboard electric resistance heating units; (3) a greater variation in the "overall room heating comfort" with electric units, suggesting more critical thermostat calibration and location for electric heating.

Table 12-155d Gas-to-Electric and Gas-to-Oil Energy Ratios

Use	Gas-to-electric	Gas-to-oil
Domestic cooking	2.00:1	...
Domestic storage water heating		
50 gal per day	1.60:1	0.80:1
100 gal per day	1.46:1	0.77:1
150 gal per day	1.41:1	NA
Domestic water heating		
Separate gas vs. oil boiler coil	...	0.57:1
Domestic refrigeration	3.16:1*	...
Domestic clothes drying	1.00:1	...
Commercial cooking		
All appliances	1.60:1	...
Ranges and griddles	1.25:1	...
Deep fat fryers	2.06:1	...
Ovens	2.47:1	...
Central residential heating	(See discussion of factors affecting energy usage and energy ratios in Table 12-155c and in Chap. 5 of this Section.)	
Oil furnaces and boilers		
Electric resistance		
Electric heat pump		
Residential summer air-conditioning		
Gas absorption		
Gas engine-driven		

NA—not available.
* Includes electric consumption of gas models.

These examples of the use of energy ratios merely demonstrate the mechanics involved. The real significance is that they reflect the values of various fuels to the customer based on heat content and efficiency of application. Similar comparisons for residential gas uses can be readily made by applying the energy ratios from Table 12-155d and the estimated annual gas usages given in Table 12-155e.

FUEL EFFICIENCY

In this country the principal fuels used for the production of energy (heat and power) are coal, oil, gas, and electricity. While electricity is not normally classified as a fuel, it is generated mainly by fossil fuels, although hydro or nuclear power is sometimes used. Conversion of the heat energy in fuels to electricity is therefore one important factor affecting the overall efficient use of our natural fuel resources.

Delivered Efficiency to Consumer

Delivered efficiency for various fuels has been summarized by Ayres and Scarlott[2] (Table 12-156a). Delivered efficiency to the consumer of coal, petroleum, and natural gas is about the same, whereas the refined or altered products of heating

Table 12-155e Annual Estimated Gas Usage for Residential Appliances (Therms)

(data reported by 21 gas utility companies to A.G.A. Committee on Comparison of Competitive Services, Oct. 31, 1962)

Reporting company and geographical location	Range	Refrigerator	Clothes dryer (elec pilot)	Clothes dryer (gas pilot)	Incinerator	Water heater	Yard light
A, East North Central	89	120	63	120	133	300	170
B, East North Central	104	152	...	92	195	278	180
C, East North Central	120	180	...	100	180	280	216
D, South Atlantic	87	192	34	...	303	216	173
E, East North Central	96	192	48	...	...	300	...
F, West South Central	96	185	48	100	120	270	170
G, Pacific	124	95	47	...	...	233	...
H, South Atlantic	108	120	34	...	...	336	...
I, New England	96	180	36	86	48	384	61
J, West North Central	96	156	...	72	...	287	...
K, West South Central	153	148	28	74	144	261	...
L, Middle Atlantic	76	115	...	102	81	292	96
M, Middle Atlantic	...	180	...	...	...	...	170
N, Middle Atlantic	74	...	...	...	...	192	...
O,* West Coast	115	100	44	...	...	210	180
Total	2,124	2,715	646	746	1,204	5,099	2,496
Avg	106	136	43	93	150	255	167

* Seven-company average.

Table 12-156a Efficiency of Fuels Delivered to Consumer

(efficiency of utilization not included)

Fuel	Efficiency of production,[2] %	×	Efficiency of generation or refining, %	×	Efficiency of transmission,[2] %	=	Efficiency delivered to consumer, %
Coal	96	×	...	×	97	=	93
Petroleum	96	×	...	×	97	=	93
Natural gas	96	×	...	×	97	=	93
Heating oil	96	×	87	×	97	=	81
Manufactured gas	31–46	×	...	×	95	=	37
Electricity (coal)	96–97	×	22	×	80	=	17

Table 12-156b Estimated Average Utilization Efficiency of Domestic Heating and Water Heating Equipment Using Different Fuels

Fuel	Equipment	Est eff of utilization, %
Coal (bituminous)	Central heating, hand fired[3]	45.0
	Central heating, stoker fired[3]	55.0
	Water heating, pot stove, 50 gal per day[4]	14.5
Oil	Central heating (84% of gas efficiency)[5]	63.0
	Storage water heater, 50 gal per day[6]	44.2
	Storage water heater, 100 gal per day[6]	50.4
Gas	Central heating[3]	75.0
	Room heater, unvented*	91.0†
	Room heater, vented*	75.0
	Storage water heater, 50 gal per day[7]	54.1
	Storage water heater, 100 gal per day[7]	63.7
Electricity	Central heating, resistance*	95.0
	Central heating, heat pump (air to air)[8]	226.0
	Room heaters*	100.0
	Storage water heater, 50 gal per day[7]	85.6
	Storage water heater, 100 gal per day[7]	92.2

* Estimated.

† This is a minimum figure and assumes adequate building ventilation to prevent condensation of the water vapor in the products of combustion.

oil, manufactured gas, and electricity have delivered efficiencies of 81, 37, and 17 per cent, respectively.

Estimated Utilization Efficiency

The efficiency of heating equipment to convert delivered fuel to useful heat energy in service, as contrasted to laboratory test efficiency at full, steady operation, varies according to the way the appliances are operated and the type of heating system. Table 12-156b lists typical utilization efficiencies for various domestic heating and water heating applications.

Coal in a hand-fired domestic furnace usually yields rather low efficiency because of smoke and flue gas losses, coal falling into the ash pit, and difficulty in controlling heat loss in excess air in the flue gas. Stoker operation is much better. Large automatic stoker-fed steam boilers have been perfected to burn coal very efficiently, in some cases more so than gas or oil in similar installations owing to less heat loss in the water vapor produced by coal.

Oil furnaces for house heating are also relatively low in efficiency, 84 per cent of that for gas furnaces, as demonstrated by field tests.[5] This may be due to relatively higher flue gas temperatures and excess air, poor combustion, poor efficiency during fall and spring months, loss during ignition, and poor adjustment of input. Large steam boilers designed for oil are highly efficient compared with many conversion units used for domestic heating.

The efficiency of **gas** house heating furnaces and large boilers is relatively high. The operating efficiency of water heaters is quite high, but lower than that of furnaces, as a result of stand-by losses and intermittent operation. The more hot water used the higher the efficiency, as shown in Table 12-156b.

Electricity for house heating is the most efficient if only utilization efficiency is considered. But this is a deceptive value since the customer does not receive this efficiency of total fuel utilization (see Table 12-156c). Heating by electricity is a very inefficient overall use of energy (16.1 per cent for resistance heating with electricity made from coal).

Ultimate Fuel Efficiency

Table 12-156c gives the ultimate fuel efficiency of various fuels, taking into consideration efficiencies of production, transmission, refining, generation, and utilization. The much higher ultimate efficiency of gas over that of all other fuels is readily apparent. To determine the comparative value of fuels to the consumer, other factors must be considered: maintenance and repairs, fuel handling and control, initial cost.

APPLIANCE AND INSTALLATION COSTS

Domestic Appliances

Table 12-156d indicates the relative cost of various domestic appliances for the year 1964. Costs are based on factory and dollar shipments and therefore do not represent the retail price to the consumer. Assuming equal markups, however, the cost relationship would be the same.

Table 12-156c Estimated Ultimate Fuel Efficiency
(based on data from Tables 12-156a and 12-156b)

Fuel and use	Delivered efficiency, %	×	Utilization efficiency, %	=	Ultimate efficiency, %
Coal (bituminous)					
Central heating, hand fired	93.0	×	45.0	=	41.8
Central heating, stoker fired	93.0	×	55.0	=	51.1
Water heating, pot stove	93.0	×	14.5	=	13.5
Oil					
Central heating	81.0	×	63.0	=	51.0
Water heating, 100 gal per day	81.0	×	50.4	=	40.8
Natural gas					
Central heating	93.0	×	75.0	=	69.7
Water heating, 100 gal per day	93.0	×	63.7	=	59.2
Electricity					
Central heating, resistance	17.0	×	95.0	=	16.1
Central heating, heat pump	17.0	×	226.0	=	38.4
Water heating, 100 gal per day	17.0	×	92.2	=	15.6

Table 12-156d Average Factory Cost of Ranges, Water Heaters, and Warm Air Furnaces*

(1964 data)

Appliance	Total units shipped	Total cost, $ (thousands)	Avg cost per unit, $
Ranges (free standing)			
Electric	1,150,000	171,891	149.00
Gas	1,458,000	134,607	92.25
Water heaters (storage)			
Electric	1,000,000	44,700	44.70
Gas	2,724,000	110,867	40.80
Furnaces (central)			
Gas	1,163,541	145,590	125.00
Oil	254,169	50,415	198.00
Solid fuel	11,723	2,673	228.00

* Based on statistics prepared by the Institute of Appliance Manufacturers.

In 1955 A.G.A. reported[9] on the average retail price for the most popular model gas and electric ranges selling in 88 representative areas of the United States. The survey revealed that the average price of gas ranges was $218.13 as against $270.41 for electric models. The same report indicated that retail prices for water heaters with glass lined tanks and comparable hot water delivery were $134.57 for gas and $173.76 for electric models.

Table 12-156e may be used as a guide to help estimate the relative cost of owning (exclusive of yearly fuel costs) domestic central heating systems operated on various fuels. These figures are national averages as of June 1963 and are applicable only to low-cost housing. It should also be noted that the annual expense figures for gas would be substantially lower in areas where the gas utility provides free or below-cost service and/or maintenance.

Commercial Cooking Appliances

In 1962 A.G.A. undertook a comprehensive study to determine the comparative cost of installing gas and electric cooking appliances in a typical commercial restaurant. A member of the International Society of Food Service Consultants and a reputable firm of consulting engineers were retained to select comparable appliances, determine the kitchen layout, draw up complete electrical and plumbing plans, and write detailed installation specifications. Itemized bids for labor and material were obtained from electrical and plumbing contractors in 15 representative U. S. cities.

An analysis of the bids[10] showed that the average installation cost of electric appliances was $1700 higher than comparable gas equipment, for an installation cost ratio of 5.9:1 in favor of gas. Table 12-156f lists the gas and electric appliances provided for the "typical" restaurant, and Table 12-156g summarizes the individual contractor bids in each of the fifteen cities.

Table 12-156e Comparative Initial and Annual Costs for Central Heating Systems*

Heating system	Fuel used	Initial cost, $	Maintenance, replacement, and repairs, $/yr
Forced warm air system (complete with ducts, electrical connections, and 275 gal oil storage tank or coal bin)	Gas	441–531	20.53–24.23
	Oil	621–693	31.11–33.89
	Coal	546–626	29.34–32.80
Hot water system (complete with pump, electrical connections, and 275 gal oil storage tank or coal bin)	Gas	781	32.77
	Oil	1070	49.81
	Coal	881	48.10

* Based on average national data issued by the Public Housing Administration (Bulletin No. LR-11) as a guide in the selection of utilities for low-cost public housing with individual systems for each dwelling unit.

Table 12-156f Commercial Cooking Appliances Specified to Obtain Installation Costs Shown in Table 12-156g

		Total input	
Quantity	Description of items	Gas, MBtu	Electric, kw/v
3	Ranges	286	68.0/220
2	Fryers (floor)	130	24.0/220
2	Broilers	200	24.0/220
1	Dishwasher	60	10.0/220
1	Steam table	45	6.0/220
1	Coffee urn (twin)	28	8.0/220
1	Coffee warmer unit	12	1.07/110
1	Fryer (counter)	30	4.5/220
1	Hot plate	16	2.5/220
1	Griddle (36 in.)	58	12.0/220
1	Hot food table	10.5	1.4/220
1	Open hearth broiler	96	10.0/220

FUEL HANDLING COSTS

Because gas is piped and electricity is wired to the appliance, handling costs for these fuels are negligible. Oil and coal are delivered to the consumer by truck or other conveyance, with the cost of handling usually included in that of the fuel. Maintenance of an oil storage tank and a coal storage bin is at the customer's expense. Oil may flow by gravity to the burner or be pumped to it, depending upon the burner; pumping represents an extra cost. Oil burners must be cleaned and serviced periodically. Stoker operating costs or the salary of a fireman are additional costs for coal. Ashes from coal-fired equipment must be removed and hauled away.

Both coal and oil furnaces and the electric generating plant, either coal or oil fired, are sources of air pollution. Special apparatus may be required to remove the pollutants.

RELATIVE FUEL PRICES

The national average prices of gas and electricity for various types of service have remained quite constant for many years. Table 12-157a notes the year-to-year changes for the five-year period 1960 thru 1964. Gas prices for each class of service have declined each year since 1962, electric prices since 1960 for residential service, since 1962 for small commercial service, and since 1961 for large commercial service.

The residential retail price indexes compiled by the U. S. Bureau of Labor Statistics show a similar trend when related to 1957–1959 as base 100. Table 12-157b includes the 1960–1964 price indexes for gas, oil, and electricity and indicates only minor year-to-year fluctuations during this five-year period.

Table 12-156g Installation Costs and Cost Ratio for Installing Comparable Gas and Electric Commercial Cooking Appliances in 15 U. S. Cities[10]

(see Table 12-156f for list of appliances)

Contractor estimates	Mont-gomery, Ala.	Charles-ton, W. Va.	Tulsa, Okla.	Nashville, Tenn.	Chicago, Ill.	Dayton, Ohio	Portland, Ore.	Detroit, Mich.
Bid 1, all-electric kitchen								
Material	...	$3425.00	*	$2214.89	*	$2706.00	$2055.00	$3825.00
Labor	...	2472.00	*	1966.59	*	1694.00	1525.00	3275.00
Total	†	$5897.00	$4131.63	$4181.48	$5136.00	$4400.00	$3580.00	$7100.00
Bid 2, basic electric system								
Material	...	$1969.00	*	$1187.29	*	$1371.00	$1030.00	$1845.00
Labor	...	1667.00	*	1282.69	*	1096.00	690.00	2255.00
Total	†	$3636.00	$2065.55	$2469.98	$3705.00	$2467.00	$1720.00	$4100.00
Installation cost of electric cooking equipment	$2153.00	$2261.00	$2066.08	$1711.50	$1431.00	$1933.00	$1860.00	$3000.00
Bid 3, installation cost of gas cooking equipment								
Material	$ 104.00	$ 180.00	*	$ 183.56	$ 113.43	$ 139.00	*	$ 135.56
Labor	180.00	170.00	*	157.50	178.20	206.50	*	299.14
Total	$ 284.00	$ 350.00	$ 522.00	$ 341.06	$ 291.63	$345.50	$ 406.00	$ 434.70
Increased cost to install electric kitchen	$1869.00	$1911.00	$1544.08	$1370.44	$1139.37	$1587.50	$1454.00	$2565.30
Installation cost ratio	**7.6:1**	**6.5:1**	**4:1**	**5:1**	**4.9:1**	**5.6:1**	**4.6:1**	**6.9:1**

Contractor estimates	Boston, Mass.	Jackson, Miss.	Atlanta, Ga.	Buffalo, N. Y.	Seattle, Wash.	Cleveland, Ohio	Dallas, Tex.
Bid 1, all-electric kitchen							
Material	$3146.00	$3043.00	$3000.00	$2402.05	*	...	$2149.19
Labor	1972.00 85.00 (permit)	1869.00	1600.00	2128.00	*	...	1141.58
Total	$5203.00	$4912.00	$4600.00	$4530.05	$4275.00	†	$3290.77
Bid 2, basic electric system							
Material	$1397.00	$1839.00	$1440.00	$1225.50	*	...	$ 931.93
Labor	1127.00 44.50 (permit)	1143.00	1000.00	1178.00	*	...	590.56
Total	$2568.50	$2982.00	$2440.00	$2403.50	$1750.00	†	$1522.49
Installation cost of electric cooking equipment	$2634.50	$1930.00	$2160.00	$2126.55	$2525.00	$1640.00	$1768.28
Bid 3, installation cost of gas cooking equipment							
Material	$ 375.80	$ 101.07	*	$ 120.00	*	$ 104.00	*
Labor	479.42	66.00	*	273.00	*	336.00	*
Total	$ 855.22	$ 167.07	$ 384.04	$ 393.00	$ 800.00	$ 440.00	$ 175.00
Increased cost to install electric kitchen	$1779.28	$1762.93	$1775.96	$1733.55	$1725.00	$1200.00	$1593.28
Installation cost ratio	**3.1:1**	**11.6:1**	**5.6:1**	**5.4:1**	**3.2:1**	**3.7:1**	**10:1**

* Contractor estimates did not itemize material and labor costs.
† Contractor estimated electric cooking equipment installation only.

Table 12-157a National Average Gas and Electric Prices By Class of Service, 1960–1964

Class of service	Cents per therm or kwh				
	1960	1961	1962	1963	1964*
Gas					
Residential	9.97	10.17	10.19	10.16	10.08
Commercial	7.87	7.99	8.00	8.01	7.88
Industrial	3.32	3.47	3.52	3.50	3.48
Electric					
Residential	2.47	2.45	2.41	2.37	2.31
Commercial and industrial					
Small	2.46	2.35	2.37	2.28	2.19
Large	0.97	0.97	0.96	0.93	0.91

Source: A.G.A. and EEI annual statistics.
* Preliminary.

Table 12-157b Average Annual Indexes of Residential Retail Prices of Gas, Oil, and Electricity

(1957–1959 = 100)

Use	1960	1961	1962	1963	1964
Gas (heating)	112.9	113.6	112.8	112.7	113.0
Gas (other uses)	109.9	111.8	112.2	112.4	112.8
Electricity (excluding heating)	102.7	103.0	103.0	103.0	102.5
Oil No. 2 (heating)	97.2	101.1	101.2	103.3	101.0

Source: U. S. Bureau of Labor Statistics.

COST COMPARISON OF FUELS

For accurate comparison of fuels it is necessary either to conduct tests on the competing appliances on the site of operation or to accept authoritative tests by others. In making such tests or calculations the following information is needed:

1. Cost of the fuels for the amounts to be used at the particular location. Fuel costs vary widely from one area to another.

2. Either the amount of each fuel required for the same heating operation per month or per year, or the utilization efficiency or energy ratio data established for the appliances involved.

Costs of domestic cooking and water heating with gas and electricity in 70 cities in the United States have been calculated[9] from authoritative test data. Similar calculations[11] for commercial cooking equipment in cafeteria kitchens are available for 28 cities.

A 1957 study[8] of multiple-dwelling housing projects resulted in operating cost comparisons for gas and electric utility service in various combinations. Included were the calculated costs for operating a 1000-unit housing project using gas for cooking, water heating, space heating, and summer cooling as against an all-electric operation including heat pumps for space heating and summer cooling. The costs for the gas projects included the electric consumption for lighting, small appliances, and refrigeration. *Except for one city, where the all-electric costs were 5.4 per cent less, all the others show an advantage of 3.5 to 198.3 per cent in favor of gas usage.*

NATURAL GAS VS. RESIDUAL OILS FOR BOILER FIRING

Besides establishing relative performance factors, comparisons of major fuel-using equipment are normally made by including evaluations of all related factors.[15] A number of the engineering aspects are presented here; see Ref. 15 for details, including a sample problem covering a 25,000 lb per hr steam boiler operated at 60- and 90-per cent load.

Combustion Efficiency

Manufacturers' catalog data covering large boilers generally indicate that the efficiency of their boilers fired with oil is 4 per cent higher than when fired with gas. This difference in theoretical efficiency is based on the higher combined hydrogen content of natural gas.

In actual practice, the difference in theoretical efficiency is usually offset by some practical problems in utilizing oil. Perfect atomization does not occur, and poor atomization results in long flames, with the burning partially in the liquid state producing soot accumulations and lower efficiency.

Atomization fluctuates with oil pressure, temperature, and viscosity. To overcome this, more excess air is used for a comparable oil-fired installation, lowering the overall efficiency in proportion to the excess air requirement. In addition, there are flame property variations caused by fluctuations in fuel properties, foreign matter, and reduction of orifice size by carbonization. These operating variations lead to the elimination of the theoretical efficiency difference. For these reasons, oil and gas firing efficiencies are considered equal. Moreover, gas has the added advantage of having consistent properties, thus eliminating the necessity of continually checking the burner setting.

Figures 12-214a and 12-214b show the heat lost in combustion products of boilers using No. 2 and No. 6 fuel oils.

Example: Find the theoretical efficiency of a boiler using No. 6 oil. There is 10.7 per cent CO_2 in the 600 F flue gas.

Solution:

1. Sensible heat portion of the flue loss = 16.0 per cent (from Fig. 12-214a).
2. Latent heat portion of flue loss attributable to the fuel = 6.7 per cent.
3. Latent heat loss attributable to air = 0.36 per cent.
4. Add 1.50 per cent for radiation and unaccounted-for losses to items 1 thru 3 to give a total loss of 24.56 per cent.
5. The theoretical efficiency is 100 − 24.56 = 75.44 per cent; i.e., 0.7544 × 18,300 Btu per lb (HHV).

Viscosity and Its Effect on Combustion Efficiency. Viscosity of oil is a measure of the relative ease or difficulty with which it flows. It is measured by the time, in seconds, that a standard amount of oil takes to flow thru a standard orifice. In the U. S. the usual standard is the Saybolt Universal (SSU) or for highly viscous oils the Saybolt Furol (SSF). Because viscosity changes with temperature (Fig. 1-8), the tests are made at a standard temperature of 100 F for SSU and 122 F for SSF. How closely the viscosity of oil at the burner must be controlled depends on the atomizing method. Each type of burner can atomize a certain maximum viscosity oil.

Steam Atomizing Burner. A well-designed unit will handle oil at 250–300 SSU if the steam is dry and has sufficient pressure and temperature to break up the oil. The smaller the droplet issuing from the burner tip, the greater the surface area available to air, and the better will be the combustion efficiency.

Mechanical Atomizing Burner. The recommended viscosity is 150 SSU. Low-pressure air atomizing burners are very sensitive to viscosity. However, they can normally operate at a viscosity of 100 SSU.

Fuel oil viscosity must be maintained at proper levels so that the oil can be pumped thru the system and atomized thoroughly by the burner. To obtain the proper viscosity for good atomization, final heating is usually done in steam and electrical preheaters connected in series ahead of the burner.

The combustion efficiency of highly viscous fuel oils depends on proper fuel preparation so that the oil will be delivered to the combustion chamber sufficiently atomized and at the right temperature.

Residual Oil. In the following discussion residual oil is considered to be No. 5 and No. 6 grade or the comparable grades known, respectively, as PS 300 and PS 400 on the Pacific Coast (Tables 2-36 and 2-37 give fuel oil specifications). The sulfur content of U.S. fuels is given in Table 12-158a.

PS 300 or No. 5 fuel oil usually requires preheating for pumping and combustion. The maximum viscosity for this oil is 40 SSF. Water content may be as high as 1 per cent. The heat content is about 150,000 Btu per gal.

PS 400 or No. 6 fuel oil is a highly viscous oil that requires preheating to 180 F or more to circulate and maintain proper combustion and atomization. Permissible viscosity of this

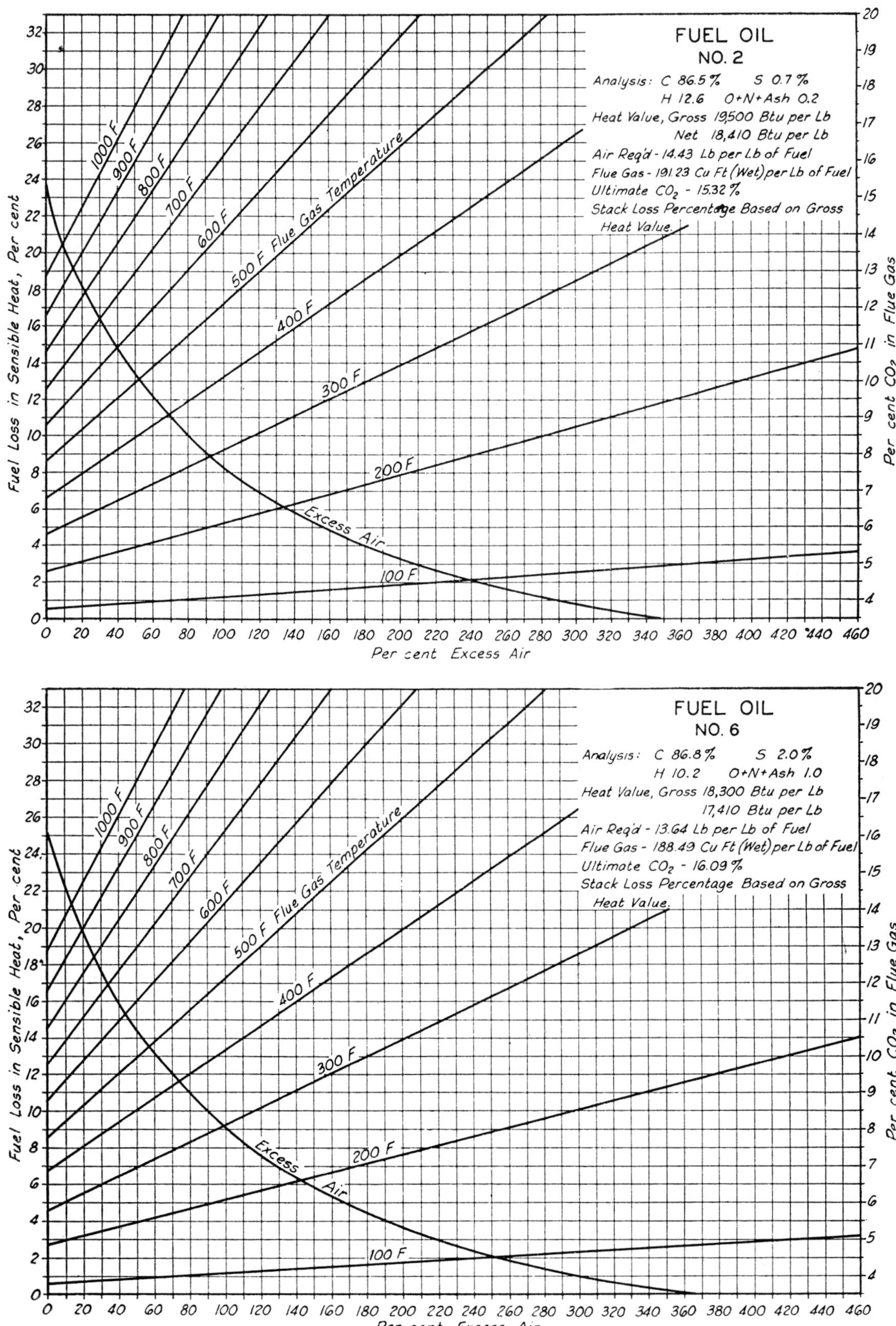

Fig. 12-214a Sensible heat lost in combustion products of boilers using a No. 2 (upper diagram) and a No. 6 (lower diagram) fuel oil.[12] Figure 12-214b shows latent heat losses. Text gives illustrative example.

Table 12-158a Sulfur Content of U. S. Fuel Oils in 1963[13]

(per cent by weight)

(these data cover 386 samples, from 114 refineries, manufactured by 37 companies)

	Eastern region		Southern region		Central region		Rocky Mountain region		Western region	
Oil grade	Min	Max	Min	Max	Min	Max	Min	Max	Min	Max
1	0.01	0.28	0.01	0.19	0.004	0.75	0.002	0.40	0.002	0.43
2	0.029	0.65	0.03	0.62	0.029	1.20	0.05	0.90	0.07	0.78
4	0.22	2.00	0.22	1.40	0.22	2.97	1.18	2.97	1.18	2.97
5 (light)	0.35	1.83	1.76	2.01	0.60	3.5	1.95	3.5	2.81	3.5
5 (heavy)	0.62	3.12	0.52	3.03	0.30	3.12	1.1	2.82	0.83	2.34
6	0.49	3.45	0.56	3.20	0.60	4.52	0.33	4.52	0.76	4.0

oil can range from a low of 45 SSF to a high of 300 SSF viscosity. Water content can be as high as 2 per cent. The heat content is about 153,000 Btu per gal.

Fuel oils do not have standard specifications as to gravity, Btu content, sulfur content, water content, metallic content, etc. *Care must be taken to establish certain minimum and maximum limits when ordering.*

The residual oils on the market in 1965 are by-products of modern refining techniques. They are priced to compete with coal or other low-cost fuels. To upgrade the quality of residual oil would mean pricing it out of the market, which might not be economically justifiable. Most crude oils, particularly imports, are high in impurities such as sulfates and vanadium, which are concentrated in the residuals.

Slagging and Corrosion Considerations. High-temperature, high-pressure steam boilers burning residual oils have difficulty because of slag deposits and corrosion stemming from the sulfur and vanadium in the oil. High-temperature corrosion is mainly a problem with boilers operating at and above 1000 F. However, "cold-end" boiler equipment such as air heaters, economizers, duct work, and stacks are also attacked by the sulfuric acid created by sulfur in the oil. This low-temperature corrosion can affect a boiler of any size, from central station units down to those used in small commercial and industrial plants.

Slagging has an adverse effect on heat transfer surfaces and disrupts gas flow patterns. It usually collects on tubes in high-temperature zones. Residual oil ash deposits are likely to produce a hard slag that is extremely difficult to remove. Beneath this slag the alkali sulfates and vanadium are in

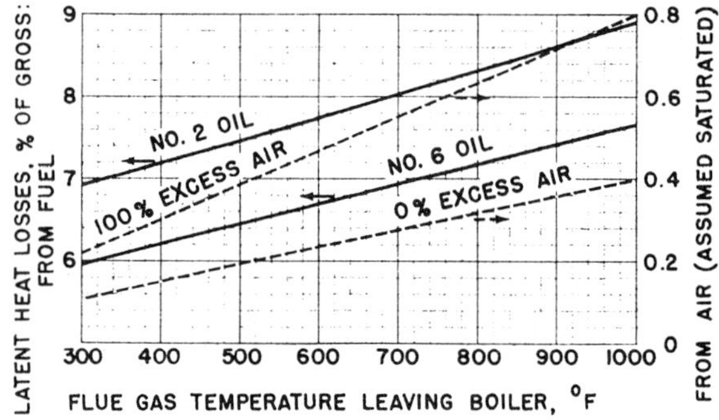

Fig. 12-214b Latent heat content in combustion products from No. 2 and No. 6 fuel oils. The fuel oil compositions and sensible heat losses are given in Fig. 12-214a.

intimate contact with the metal, and corrosive action takes place. As the temperature increases so does the corrosive action.

No. 4 Oil. This light residual fuel oil has a heating value of about 145,000 Btu per gal. It is produced either as a straight distillate or as a blend of No. 2 and No. 6 oils (see Table 2-37). The maximum viscosity of No. 4 oil is 15 SSF. Sulfur content can be as high as 2.05 per cent by weight, and the water and sediment content up to 0.5 per cent by volume. Normally, No. 4 oil does not require preheating, but preheating is practiced when blends contain No. 6 oil.

Viscosity Fluctuations. The majority of the problems associated with burning No. 4 oil is the result of fluctuations in the viscosity. These fluctuations are directly attributable to variations in the proportion of No. 2 and No. 6 oils in the blended product. One result of improper blending of No. 4 oil is line freeze-up in the cold weather. When this occurs, flow ceases until the line is thawed out.

Changes in viscosity cause improper or incomplete atomization of the oil. This in turn results in soot, smoke, carbon deposition on the heat transfer surfaces of the boiler, and, in general, in inefficient and uneconomical operation. Not the least of the problems of incomplete atomization is air pollution.

Air Quality Standards.[14] Growing numbers of governmental agencies, domestic and foreign, are considering or

Table 12-158b Oxides of Sulfur Emitted in Combustion

(estimated as SO_2)

Fuel	Lb per MMBtu	Equivalent
Natural gas	0.0086	6 grains per 100 cu ft*
Fuel oil		
No. 2	0.46	0.88 lb per 100 lb of oil†
No. 4	1.61	3.06 lb per 100 lb of oil†
No. 6	2.36	4.48 lb per 100 lb of oil†
Coal	1.4	40 lb per ton of coal‡

* Based on a relatively high sulfur content of 3 grains per 100 cu ft.

† Based on averages of data in Table 12-158a.

‡ Chambers, L. A. "Where Does Air Pollution Come From?" In: *Proceedings—National Conference on Air Pollution.* (U. S. Dept. of Health, Education, and Welfare) Washington, G.P.O., 1959. Dr. Chambers is the Director of Research, Los Angeles County Air Pollution Control District.

have adopted and are enforcing standards for a number of air pollutants. Comparison of the sulfur dioxide produced on burning common fuels is shown in Table 12-158b.

California, for example, has set 0.3 part of SO_2 per million parts of ambient air (by volume, averaged over an 8-hr period) as the "adverse" level. New York State standardized on 0.15 ppm SO_2 averaged over 24 hr for high-density areas and 0.1 ppm for low-density areas.

The U.S.S.R. reportedly limits SO_2 pollution to 0.05 ppm over a 24-hr period. Manitoba (Canada), Poland, West Germany, and others have set levels about the same as California.

Storage of Oil. Sludge should be removed from time to time. Its composition follows:

1. *Emulsions*, found on tank bottoms, which are a mixture of heavy fuel oil and water.
2. Heavy *insoluble compounds* plus some *coke* and carbon from high cracking of residuals.
3. *Oxidation products* of these insoluble solids due to reaction with the oxygen in the air.
4. *Precipitates* formed by blending of residual and distillate oils.
5. *Dirt* and *rust* scale.

Sludge can clog lines and burner tips, producing undesirable flame characteristics and inefficient combustion.

Cost of Oil Supply Systems

In making a cost analysis between gas and oil, estimate the following "extra" costs, which are generally based on a useful life of 25 years for fuel oil equipment: (1) pumping and atomizing of oil; (2) burners (*extra* cost for oil averages 15 per cent); (3) maintenance of oil equipment; (4) depreciation (*extra* cost for oil); (5) interest on oil equipment investment (*extra* cost for oil); (6) heating of oil; and (7) steam atomization.

Owning and Operating—Fixed Costs[15]

Capital Equipment Costs. These costs are incurred to make the installed plant ready for operation. They include the price of the boiler, burners, pipe, tanks, pumps, and heaters, as well as all transportation and the labor and materials necessary for unloading, setting, and installing the boiler and its accessories.

Oil Burners. Steam atomizing units require dry steam at 30 psig and up. They are generally the least expensive to install. Combustion air must be mixed with the oil after it has been atomized by the steam, producing a longer flame, slower combustion, and a limited range of capacity, particularly at low firing rates. The amount of steam required to atomize one gal of oil will vary with the operator and the boiler. The best operators under good conditions will use 2 to 4 lb of steam per gal of oil. Some older installations may require up to 8 lb per gal. Manufacturers' recommendations are usually conservative. Some manufacturers suggest using 1 per cent of the steam output, but 2 per cent is more realistic.

High-pressure oil burners require at least 30 psig compressed air. Table 12-159 gives the air requirements for different pressures. These burners may be used when excess

Table 12-159 Approximate High-Pressure Air Requirements for Oil Atomization

Atomizing air pressure at burner, psig	Air required for atomization, % of total combustion air*	Atomizing air, cfm per gph of oil
40	13	3.25
60	11	2.75
80	9	2.25
100	7	1.75
125	6	1.50
150	5	1.25

(Hauck Manufacturing Co.)
* For combustion, about 1500 cu ft are required per gal of oil.

compressed air is available at the plant. If air compressors must be purchased for air supply, increased cost normally precludes the use of these burners. Operating costs may be estimated from the horsepower requirements given in Table 12-160.

Table 12-160 Brake Horsepower Required by Air Compressors
(per 100 cu ft of free air actually delivered)

Altitude, ft	Single-stage, psig			Two-stage, psig			
	60	80	100	60	80	100	125
0	16.3	19.5	22.1	14.7	17.1	19.1	21.3
1000	16.1	19.2	21.7	14.5	16.8	18.7	20.9
2000	15.9	18.9	21.3	14.3	16.5	18.4	20.5
3000	15.7	18.6	20.9	14.0	16.1	18.0	20.0
4000	15.4	18.2	20.6	13.8	15.8	17.7	19.6
5000	15.2	17.9	20.3	13.5	15.5	17.3	19.2

(Hauck Manufacturing Co.)

Low-pressure oil burners require from 1 to 5 psig compressed air, which is usually furnished by a blower. The air required for atomization may amount to as much as 50 per cent of the total combustion air. Table 12-161 shows approximate air requirements. For operating cost calculations, requirements can be found in blower manufacturers' literature.

Table 12-161 Approximate Low-Pressure Air Requirements for Oil Atomization

Atomizing air pressure, oz	Air required for atomization, % of total combustion air*	
	With balanced furnace pressure	With moderate draft
8	66	52
16	58	42
24	52	37
32	50	33
40	40	30
48	30	25

(Hauck Manufacturing Co.)
* For combustion, about 1500 cu ft are required per gal of oil.

Mechanical atomizing burners atomize oil by forcing it thru an orifice at 75 to 300 psig. This is regarded as one of the most economical methods of atomizing. The higher pressure pumps, heater equipment, and piping will add considerably to the initial cost of the installation. Since the capacity can be reduced only by changing orifices or reducing pressure, the degree of atomization and efficiency will vary with the firing rate.

Rotary cup burners atomize oil with a high-speed rotating cup, breaking the oil up by centrifugal force and entraining it in a stream of low-pressure air. A single motor usually drives the shaft, blower, and oil pump.

Comparison of Atomizing Methods and Burner Costs. Table 12-162 shows the quantity of atomizing air and power requirements for five types of burners. Figure 12-215 compares burner equipment costs for oils and gas.

Table 12-162 Air and Power Requirements for Five Types of Oil Burners

Burner type	Atomizing agent per gph	Power requirement per gph
Low-pressure air	1–15 cfm	0.1 hp
High-pressure air	2½–3 cfm	0.4 hp
Steam atomizing	2–8 lb	0.12 boiler hp
Mechanical atomizing	...	0.32 hp
Rotary cup	...	0.05 hp

(Hauck Manufacturing Co.)

Table 12-163 Recommended Fuel Oil Temperatures for Pumping and Atomizing

(see Fig. 1-8 for viscosity-temperature relationships)

		Recommended temp to which oil should be preheated for proper atomization, °F		
Viscosity, SSU at 100 F	Recommended pumping temp to obtain 1500 SSU, °F	For low-pressure air burners to obtain 90 SSU	For high-pressure air burners to obtain 150–180 SSU	For high-pressure steam burners to obtain 180–200 SSU
4000	120	210	185	175
3000	115	205	180	170
2500	110	200	175	170
2000	105	195	170	165
1500	100	190	160	155
1000	90	180	150	145
800	85	175	145	140
600	80	170	140	135
400	70	155	130	125
300	60	145	120	115
200	50	135	105	100
150	40	125	95	90
100	25	105	80	75

(Hauck Manufacturing Co.)

Table 12-164 Aboveground Horizontal Oil Tank Space Requirements

Tank capacity, gal	Approx area required by tank,* sq ft	Tank capacity, gal	Approx area required by tank,* sq ft
1000	80	6,000	200
1500	85	7,500	240
2000	110	10,000	250
3000	135	12,000	280
4000	165	15,000	330
5000	200	20,000	410

* Includes the area between tank and building.

Oil Pumps and Heaters. In most large installations it is advisable to provide two pumps and two heat exchangers so that repair of any malfunction in one set may be made without interruption of the oil supply to the boiler.

Oil Storage Tank. For No. 5 and No. 6 oils, a tank suction heater must be provided to heat the oil to pumping viscosity (Table 12-163).

Aboveground storage tanks may tie up space as indicated in Table 12-164.

Locate aboveground tanks according to the following:

Tank (horizontal) capacity, M gal	1–12	15–20
Distance to nearest building, ft	10	15

Underground tanks do not restrict the use of land except for the construction of additional buildings; storage cost data are given in Table 12-165.

Amortization of Fixed Capital Costs. Depreciation of the boiler plant serves several functions. The total installed cost of the boiler plant less its salvage value is divided over the estimated life of the equipment. This yearly cost is charged against the boiler plant and represents the lessening in value of the equipment thru its use. To a business, depreciation can also be a source of income.

Figure 12-216 shows a typical piping arrangement for a complete oil-fired installation. The complexity of the arrange-

Table 12-165 Estimated Underground Oil Storage Installed Tank Costs

(1963 data)

Tank capacity, gal	Cost, $*	Tank capacity, gal	Cost, $*
1,000	600	20,000	4,200
2,000	900	25,000	5,000
3,000	1100	30,000	7,000
5,000	1800	50,000	8,700
10,000	3100	75,000	11,000
15,000	3500	100,000	16,000

* These estimates include tank, connections, concrete foundations, piping, and preheater in tank.

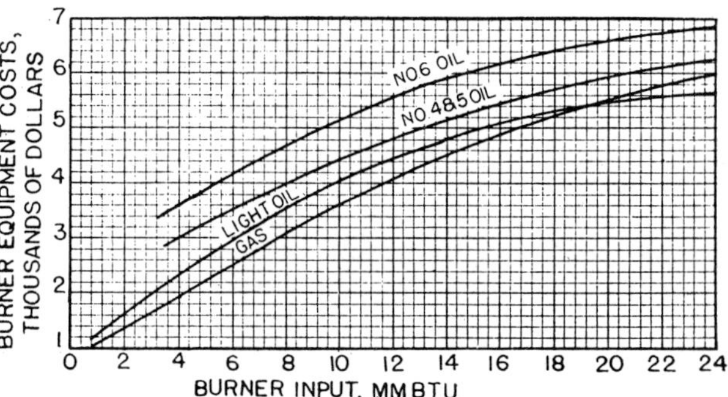

Fig. 12-215 Burner equipment costs for various oils and gas. Prices are for UL-approved installations. FM and FIA requirements will add from $50 to $1200, depending on the size of the burner (1963 data).

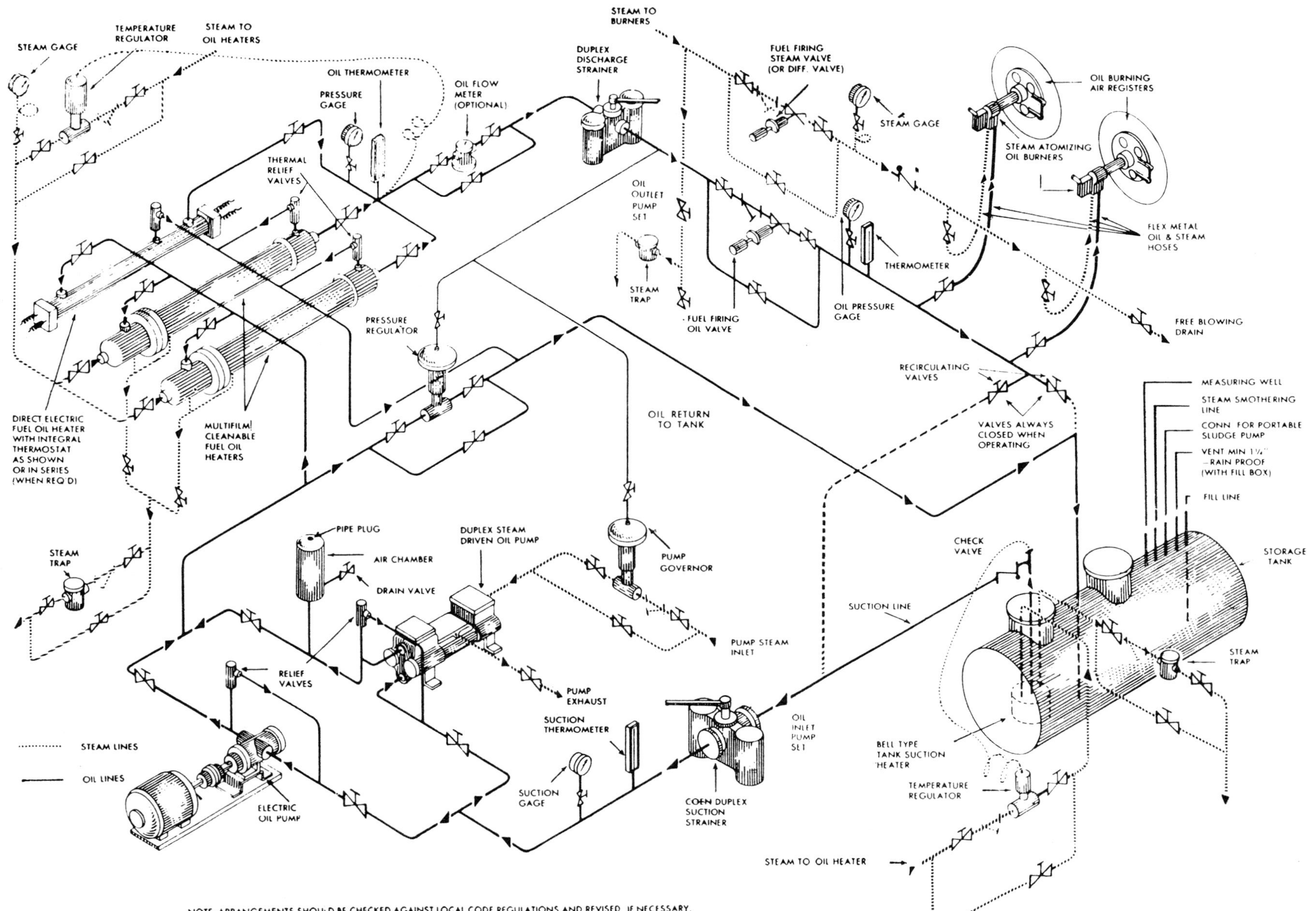

Fig. 12-216 Typical piping arrangement for oil-fired installation. (Coen Co.)

ment clearly indicates the reasons for higher investment and subsequent higher amortization costs for oil-fired boilers.

Maintenance and Parts Replacement. Table 12-166 gives estimates prepared by Frank R. Valvoda and Associates, Oak Park, Ill., of annual maintenance costs for oil-firing equipment. Maintenance for any one system could deviate appreciably from these estimates in the event of such unpredictable circumstances as equipment failure or clogging of fuel lines.

Table 12-166 Annual Maintenance Costs for Oil-Firing Equipment

Oil input, MBtu per hr	Cost, $	Oil input, MMBtu per hr	Cost, $*
1,600	350	20	900
2,560	400	24	1030
4,470	450	28	1150
6,300	600	32	1270
12,750	700	36	1390

* Extrapolated from first half of table.

Owning and Operating—Variable Costs

Oil. Oil costs fluctuate from time to time depending on demand. Prices generally are for delivered oil. When railroad tank car lots are quoted FOB refinery, transportation costs must be added.

If the oil arrives by tank car, steam may be required to heat the oil to pumping fluidity. See Table 12-163 for the pumping temperature required.

Costs of sludge and leakage, oil inventory, and additives should be taken into account.

NATURAL GAS VS. COAL FOR LARGE BOILERS

A sample problem[15] is available to illustrate comparisons of capital equipment costs and owning and operating costs. The example involves a watertube boiler capable of producing 25,000 lb of steam per hour at 200 psig with 50° F superheat; a 60 per cent annual load is assumed. The superheater, an integral part of the boiler, is of the radiant type. A combination gas-oil system and a coal-to-gas conversion are also considered.

REFERENCES

1. Benson, H. Consolidated Natural Gas Co., Pittsburgh, Pa., 1959. Private communication.
2. Ayres, E., and Scarlott, C. A. *Energy Sources—The Wealth of the World.* New York, McGraw-Hill, 1952.
3. Sherman, R. A., and Cross, R. C. "Heat Losses and Efficiencies of Fuels in Residential Heating." *Heating, Piping Air Conditioning* 9: 53–64, Jan. 1937.
4. *Final Report of Comparative Water Heater Tests Conducted at Massachusetts Institute of Technology.* New York, A.G.A., 1935.
5. *Field Tests of Oil and Gas Residential Space Heating.* New York, Consolidated Edison Co., 1964.
6. Grier, R. F., and Jahn, E. A. *Performance Characteristics of Oil-Fired Water Heaters.* New York, A.G.A., 1964.
7. Hebrank, E. F. *Investigation of the Performance of Automatic Storage-type Gas and Electric Domestic Water Heaters.* (Univ. of Ill. Eng. Exp. Sta. Bull. 436) Urbana, Ill., 1956.
8. Brandt, C. A. *Criteria for Determining Costs of Gas and Electric Service in Military and Public Housing Projects.* Washington, D.C., H. Zinder and Assoc., Inc., 1957.
9. Segeler, C. G., and Kafka, E. M. *Comparative Total Costs of Gas and Electricity for Cooking and Water Heating in Residences.* New York, A.G.A. General Management Section, 1955.
10. *Commercial Cooking Installation Cost Comparison.* (Information Letter 132) New York, A.G.A. Industrial and Commercial Gas Section, 1962.
11. *What's The Score?—Costs of Gas versus Electricity in Commercial Kitchens.* New York, A.G.A., 1952.
12. Strock, C., ed. *Handbook of Air Conditioning, Heating, and Ventilating.* New York, Industrial Press, 1964, p. 3–19.
13. Blade, O. C. *Burner Fuel Oils, 1963.* (U.S. Bur. Mines, Petrol. Prod. Surv. 31) Washington, D.C., G.P.O., 1963.
14. Stern, A. C. "Summary of Existing Air Pollution Standards." *J. Air Pollution Control Assoc.* 14: 5–15, Jan. 1964.
15. *Sales Idea Book: Gas vs. Oil.* New York, A.G.A. Industrial and Commercial Gas Section, 1963.

Chapter 23

Gas Total Energy Systems

by Marvin D. Ringler

Total energy systems are those in which electric power is generated on customer premises by means of prime movers (engines or turbines), with exhaust heat (and jacket heat) recovered and used for space heating, water heating, absorption air conditioning, and/or process applications.

On-site power generation is not a new concept. The 1950's, however, saw the start of a gas industry effort to increase the use of gas reciprocating engines and turbines (and steam turbines supplied by gas-fired boilers) for both on-site power generation and other shaft power applications, together with heat recovery systems. Table 12-167a provides data on 58 gas-fired, one oil-fired, and two gas- and oil-fired systems.

THE ENERGY SYSTEMS CONCEPT

In its simplest definition, total energy is energy conversion at the point of use. Energy is not purchased—only the fuel for a prime mover and perhaps for supplementary firing is purchased. The efficiency of such systems is a function of the degree to which the prime mover's heat and shaft outputs are utilized.

On-site energy systems are combinations of mechanical, thermal, and electrical devices (Fig. 12-217a) that must operate adequately as systems if customers are to be satisfied. Thus, each component in such systems must be capable of both meeting the specifications written for that particular component and performing satisfactorily within the system.

Principles of Total Energy

1. Direct fired heat is the most efficient source of heat supply, unless the prime mover used has a higher overall efficiency than a boiler.
2. A total energy system is the most efficient source of energy supply when both electricity and heat are required and when the ratio of electric demand to heat demand is equal to or less than the ratio of the prime mover's shaft to recoverable heat outputs (see the S/R columns in Tables 12-167b and 12-167c).
3. It may pay to operate a prime mover solely for electric power when the electric and gas rate differential is great enough to compensate for the difference in efficiency and cost of energy conversion.

Table 12-167a Analyses of Characteristics of 61 On-Site Generation Systems[1]

(all data in per cent of 61)

Base load fuel	
Gas	95
Oil	2
Gas and oil	3
Prime mover type	
Gas engine	48
Gas turbine	39
Miscellaneous*	13
Exhaust heat recovered	87
Type of application	
Industrial plants	62
Office buildings	20
Miscellaneous†	18
Generating capacity	
75 to 290 kw	26
300 to 500 kw	23
600 to 930 kw	25
1000 to 8000 kw	23
12,000 to 18,075 kw	3
Provision for stand-by	
On-site	64
Utility connection	8
On-site and utility connection	3
None	15
Not reported	10
Lighting frequency	
60 cps	87
420 or 428 cps	10
840 cps	3

* Steam turbines, dual-fuel engines, and various prime mover combinations.

† About equal distribution among apartments, shopping centers, schools, universities, hotels, etc.

Coincidence of Demands

Prime movers produce power and exhaust heat simultaneously. Since neither can be stored economically, each must be used as it is produced. The power-to-heat ratio will be different for any two prime movers, and it will vary for each prime mover under different conditions of load. Therefore, the more closely the power and heat outputs of the prime mover match the energy demands of the building as they vary over time, the greater the efficiency of the system. Additional heat output may be readily obtained by supplementary firing in the exhaust heat recovery unit.

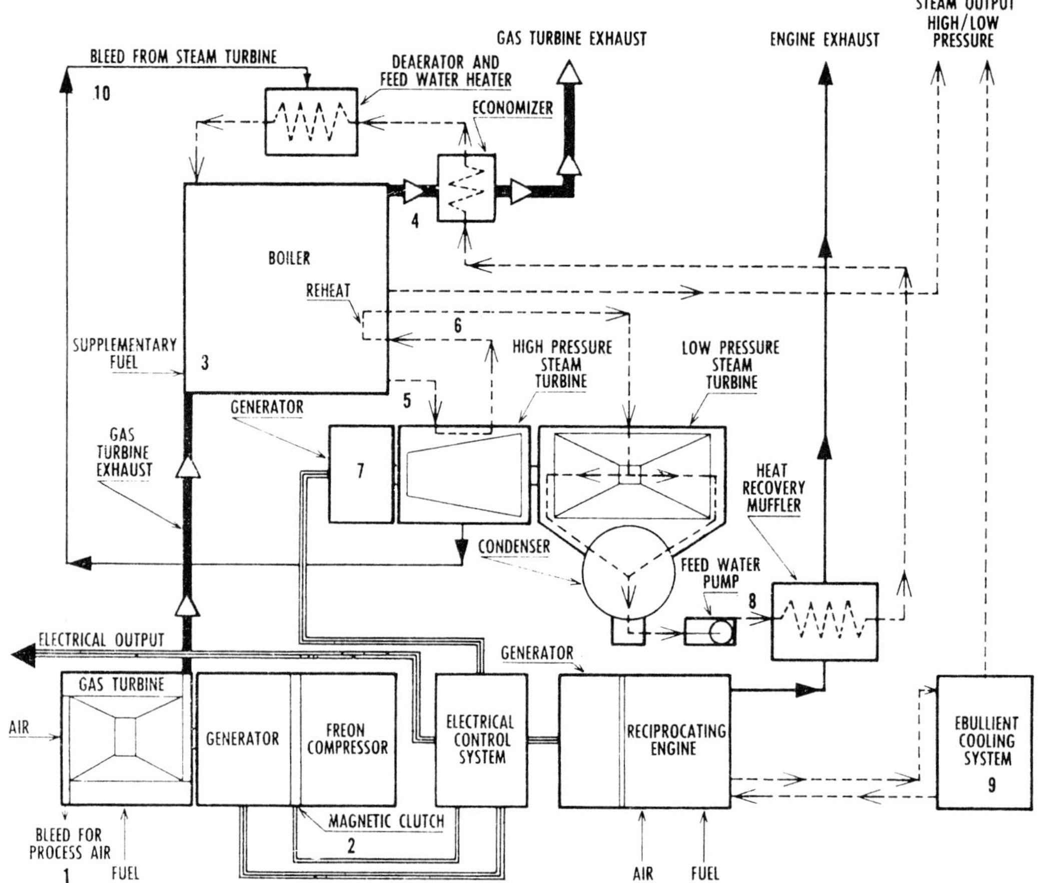

Fig. 12-217a Hypothetical total energy system. This arrangement shows a number of possibilities of on-site energy systems. It is unlikely that all of the components shown would be included in any one system.

The system shown would work as follows: (1) Pressurized air is bled from gas turbine compressor for process use. (2) Gas turbine drives both generator and Freon compressor for refrigeration (magnetic clutch releases refrigeration compressor load when generator requires full output of turbine). (3) Gas turbine exhaust heat is largely consumed in boiler, with steam deficiencies compensated for by supplemental firing. (4) After passing thru boiler, exhaust gas enters economizer for final heat recovery before passing out thru stack. (5) High-pressure steam from boiler drives high-pressure steam turbine and provides plant with its high-pressure steam supply. (6) Steam from discharge of high-pressure turbine returns to boiler for reheat and enters divided-flow, low-pressure steam turbine. (7) Both steam turbines drive one generator. (8) Low-pressure turbine exhaust enters condenser; condensate is pumped thru heat recovery muffler of reciprocating engine, thru economizer and feed water heater, and back to boiler. (9) Heat of engine jacket water is recovered by ebullient cooling system, which provides low-pressure steam for process use. (10) Steam bled from high-pressure steam turbine boosts temperature in feed water heater.

Transportation of Thermal Energy

Heat, unlike electrical energy, cannot be transported economically for any significant distance without substantial loss. In the case of district steam systems, heat is transported only on a limited scale. In hot and chilled water systems served by central plants, transport distances are also relatively short. Conventional energy systems supply heat from thermal converters located at the point of utilization or on-site. These devices supply heat independent of any other energy supply and deliver it as needed, thru the building's heat distribution system, to points at which demand occurs.

ON-SITE FEASIBILITY ECONOMICS

On-site energy systems generally require a greater capital investment than the conventional electrical and mechanical systems. Feasibility studies compare this incremental investment cost with the operating economies of on-site installation. The most desirable systems are those in which energy demand and energy output attain a high degree of coincidence with respect to both quantity and form of energy. The resultant composite of operating efficiencies is translated into financial terms.

Total energy is likely to cost more to install but less to operate than a conventional system. Management often views the initial cost differential as an investment and the operating saving as a return upon that investment. To win management approval, the profitability of the total energy investment must compare favorably with the profitability of other investment opportunities available to the company. To enable management to make this evaluation, the engineer must provide precise and complete data on installation and operating cost differentials. Illustrative examples are available for a school[2] and an office building.[3]

Table 12-167b Typical Reciprocating Gas Engine Inputs and Recoverable* Outputs[6]

Shaft load, per cent	75 kw; C = 12:1; 1800 rpm; NA; gas press. = 6 in. w.c.				175 kw; C = 10:1; 1200 rpm; TA; gas press. = 15 psig				225 kw; C = 10:1; 1200 rpm; TA; gas press. = 15 psig				450 kw; C = 10:1, 1200 rpm; TA; gas press. = 15 psig			
	Fuel rate (HHV), MMBtu per hr	Outputs: S (shaft) kw	Outputs: R (heat) MMBtu per hr	S/R†	Fuel rate (HHV), MMBtu per hr	Outputs: S (shaft) kw	Outputs: R (heat) MMBtu per hr	S/R†	Fuel rate (HHV), MMBtu per hr	Outputs: S (shaft) kw	Outputs: R (heat) MMBtu per hr	S/R†	Fuel rate (HHV), MMBtu per hr	Outputs: S (shaft) kw	Outputs: R (heat) MMBtu kw	S/R†
100	0.999	75.0	0.504	1/2.0	2.18	175	0.842	1/1.4	2.75	225	1.062	1/1.4	5.50	450	2.13	1/1.4
90	.930	67.5	.459	1/2.0	2.03	158	.759	1/1.4	2.51	203	0.943	1/1.4	5.04	405	1.90	1/1.4
80	.863	60.0	.414	1/2.0	1.86	140	.675	1/1.4	2.35	180	.853	1/1.4	4.67	360	1.69	1/1.4
70	.797	52.5	.372	1/2.1	1.72	122	.603	1/1.4	2.13	158	.745	1/1.4	4.23	315	1.48	1/1.4
60	.730	45.0	.331	1/2.2	1.56	105	.528	1/1.5	1.92	135	.645	1/1.4	3.80	270	1.28	1/1.4
50	0.661	37.5	0.296	1/2.3	1.38	88	0.450	1/1.5	1.71	113	0.555	1/1.4	3.35	225	1.09	1/1.4

C—compression ratio; NA—naturally aspirated; TA—turbocharged.
* Assumes a 315 F stack temperature.
† This means of expressing outputs facilitates matching prime movers to load profiles; S/R ratios are dimensionless, i.e., both S and R are in terms of either MMBtu per hr or kw.

Table 12-167c Typical Gas Turbine Inputs and Recoverable* Outputs[6]

Shaft load, per cent	150 kw; C = 3.4:1; 30,000 rpm; gas press. = 100 psig				200 kw; C = 4.1:1; 35,000 rpm; gas press. = 100 psig				300 kw; C = 6:1, 43,500 rpm; gas press. = 150 psig				900 kw; C = 4:1, 6000 rpm; gas press. = 100 psig			
	Fuel rate (HHV), MMBtu per hr	Outputs: S (shaft) kw	Outputs: R (heat) MMBtu per hr	S/R†	Fuel rate (HHV), MMBtu per hr	Outputs: S (shaft) kw	Outputs: R (heat) MMBtu per hr	S/R†	Fuel rate (HHV), MMBtu per hr	Outputs: S (shaft) kw	Outputs: R (heat) MMBtu per hr	S/R†	Fuel rate (HHV) MMBtu per hr	Outputs: S (shaft) kw	Outputs: R (heat) MMBu per hr	S/R†
100	5.78	150	2.80	1/5.5	7.40	200	3.68	1/5.4	6.32	300	2.60	1/2.5	22.6	900	11.2	1/3.7
90	5.51	135	2.78	1/6.0	6.90	180	3.37	1/5.5	5.70	270	2.40	1/2.6	21.2	810	10.3	1/3.8
80	5.28	120	2.75	1/6.7	6.45	160	3.06	1/5.6	5.45	240	2.10	1/2.6	19.8	720	9.5	1/3.9
70	5.04	105	2.70	1/7.5	6.00	140	2.74	1/5.7	5.04	210	2.00	1/2.8	18.5	630	8.7	1/4.1
60	4.80	90	2.60	1/8.5	5.60	120	2.43	1/5.9	4.55	180	1.80	1/2.9	17.2	540	7.9	1/4.4
50	4.55	75	2.40	1/9.3	5.21	100	2.12	1/6.2	4.25	150	1.60	1/3.1	15.8	450	7.1	1/4.7

C—compression ratio.
* Assumes a 315 F stack temperature.
† This means of expressing outputs facilitates matching prime movers to load profiles; S/R ratios are dimensionless, i.e., both S and R are in terms of either MMBtu per hr or kw.

Feasibility Screening Procedure

There is no way in which the economic feasibility of total energy systems can be predicted on a general basis. Each potential application must be examined individually if a reliable answer is to be obtained regarding its feasibility. It is possible, however, to screen prospective total energy installations economically and reliably. One such procedure was developed thru an A.G.A. research project.[4,5]

Total energy system prospects deemed favorable by an appropriate screening procedure should be analyzed in detail to determine whether the investment may be justified in terms of the prospective customer's own financial criteria.

Bases of Comparison

Reliability. This enters into feasibility analysis only if the proposed system requires a provision for reliability that will affect its installation or operating cost. Stand-by generating capacity in a conventional electrical system and interruptible fuel supply in an on-site system are illustrations of the ways in which the reliability requirement can affect cost and hence must be calculated as part of a feasibility analysis.

Installation Cost. This includes all direct and indirect costs of installing a system. Direct costs are the costs of purchasing and installing all components of the system itself. Indirect costs are the costs of purchasing and installing all other mechanical or electrical components of the building affected by the system, as well as all costs of construction not common to both systems. Indirect costs are often difficult to identify and to calculate, but they must be known for each system if the analysis is to be valid.

Operating Cost. This includes all costs of running each system on an annual basis, e.g., all fuel or purchased energy expenditures, service, maintenance, operating labor, insurance, cost of space occupied by system components, and cost of money used to purchase or finance acquisition of the system.

As in the estimating of installation cost, use of a particular system may also have indirect cost effects upon operation. These may be due to changes in building or process operation because of the characteristics of the energy system, and must be considered part of the operating cost.

Example: The "compact" school, with its relatively low construction costs, is not feasible without air conditioning. Yet, in many communities, air conditioning itself is not feasible for school construction on the basis of operating costs. A total energy system, on the other hand, may result in practical total annual and operating costs that permit innovations in school design.

Annual Owning and Operating Cost. When both total installation cost and total operating cost are known for each system, the systems may be compared in terms of total annual owning and operating cost. To arrive at this figure, the cost of repaying the installation cost of each system over the number of years of its anticipated useful life is determined in accordance with the owner's accounting procedure. This yields an annual owning cost that, when added to the annual operating cost, provides an expression of the system's total cost in a single, annual figure.

The system that offers the lower annual owning and operating cost is the more feasible. Nevertheless, the fact that a total energy system and a conventional energy system do not enter the analysis on an equal footing may make it necessary to use the owning and operating cost of the conventional energy system as the base line.

When all of these costs have been accurately calculated, it becomes possible to report to management: (1) the dollar investment required to install a total energy system, and (2) the operating savings (or rate of return) that this investment will produce. Once these data have been supplied, management is able to subject the total energy proposal to financial analysis.

Discounted Cash Flow Analysis

One favored method of financial analysis is the discounted cash flow study in which alternate investments are compared in terms of their "true" rates of profitability.

Operating savings (income) are discounted by applying a present worth factor, which reflects the present worth of money to be earned in the future. This is determined by standard financial tables calculated for varying rates of return or interest (e.g., Table 1-52). The cash flow, discounted to reflect the year in which it occurs, will equal the cash required to make the investment if the correct rate of return is chosen (see example on page 1/48). Finally, tax considerations must be applied. These include taxes payable on energy "profits" and any increase in property taxes due to increased valuation of the total energy plant. Depreciation and investment credits are beneficial tax effects.

Note that a discounted cash flow analysis requires the establishment of a fixed analysis period. The length of this period is influenced by anticipated equipment life, "payouts" generally accepted in the particular industry, and the market for capital investment.

ENERGY SYSTEM DESIGN

All predictable and potential demands of the projected or existing facility should be analyzed and modified to make maximum use of the probable outputs of a total energy system. The prime movers generally used in total energy systems are: (1) gas turbines, (2) gas engines, and (3) steam turbines. Tables 12-167b and 12-167c give design data for some gas engines and gas turbines, respectively. For most on-site total energy systems, the size range from 250 to about 1000 hp generally complements the energy system market. Overall efficiency, as determined by the difference in the absolute temperature of the air entering the first-stage turbine wheel or engine manifold and the absolute temperature of the exhaust gases leaving the final heat recovery unit, can run as high as 85 per cent.

Gas Turbines. These are available in sizes ranging from 100 to over 100,000 hp.

Gas Engines. Small, high-speed units are available in sizes from 15 to 900 hp; they run at speeds from 2400 to 12,000 rpm.

Steam Turbines. The available size range is similar to that of the gas turbine. The necessity for a boiler makes initial investment higher; however, this can often be justified by other advantages such as the interchangeability of many kinds of fuel and a need for high pressure steam.

Combining Prime Movers. Energy requirements can frequently be met most efficiently thru use of more than one prime mover. For example, it may prove feasible to use a steam turbine as part of the prime mover complement, operating it on steam generated by the gas turbine's exhaust heat and recovering shaft power from the steam turbine. Similarly, maximum economy may sometimes be gained by including a gas engine within the prime mover complement, for stand-by, peaking, and light load operation.

Modular System. Satisfying a capacity requirement by means of a number of equally sized prime movers (e.g., four 500-kw turbines to meet a 1500-kw demand) offers all the advantages associated with standardization such as lower unit costs and simplified maintenance.

Determining Energy Demands

Every energy system must be designed to meet the total of all energy requirements. These occur in two forms, heat and work. *Heat demands* are those energy requirements that can be satisfied directly by heat or indirectly by its effects. *Work demands* are those that can be satisfied electrically or mechanically. Mechanical work may be provided by a turning shaft or compressed air. Mechanical drives are limited almost exclusively to large and readily identifiable units. Miscellaneous drives are usually electric.

Heat demand and electric demand share several important characteristics. The *consolidated demand curve* for each type of demand is composed of individual demands, usually a considerable number. The consolidated electric demand curve of a building will include all its lighting and electric power demands. The consolidated heat demand curve of a building will include all its individual heat demands.

It is necessary to plot curves of energy demands showing the magnitude of the load for each hour of the year (energy units may be shown most conveniently as millions of Btu). These curves are called *load profiles* (Fig. 12-217b). A curve representing heat recoverable from the prime mover for each condition of load on the shaft (see Tables 12-167b and 12-167c) is then superimposed on the load profile. This will indicate when supplementary firing may be needed to compensate for deficiencies in the prime mover's heat output and will also show those periods when excess heat is available. Accurate collection of energy demand data is a major aspect of total energy feasibility analysis.

Commercial Structures. Here energy demands are largely restricted to power for lighting, appliance, and auxiliary use and to heat for space heating, water heating, and absorption air-conditioning systems. To calculate these demands, it is necessary to measure electric demands as well as the heating and air-conditioning demands of the structure. These are functions of ambient temperature and structural characteristics as they affect heat losses and gains. Demand must be anticipated hourly thru the year.

Industrial Structures. Although there are likely to be many different kinds of energy demands (depending upon the particular process in which the plant is engaged), most of these may be predicted. It is also generally true that the wider the range of energy demands, the greater the oppor-

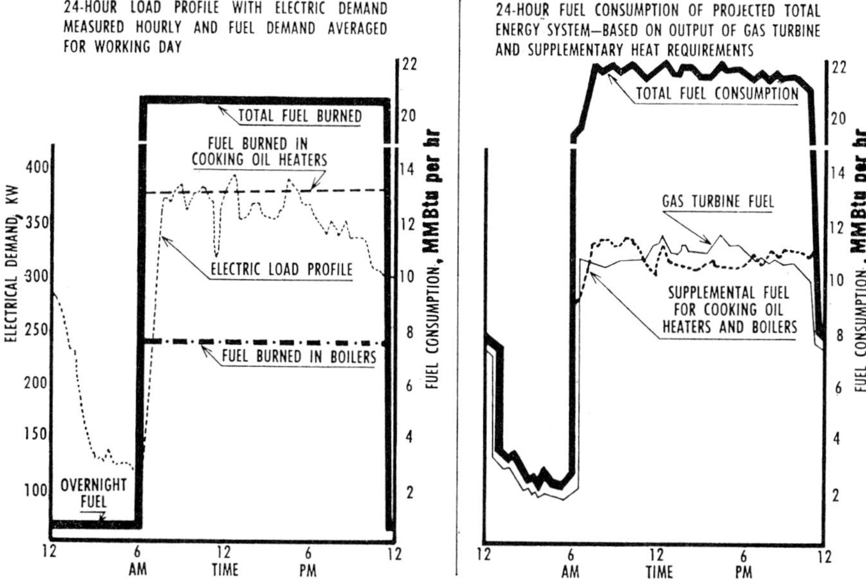

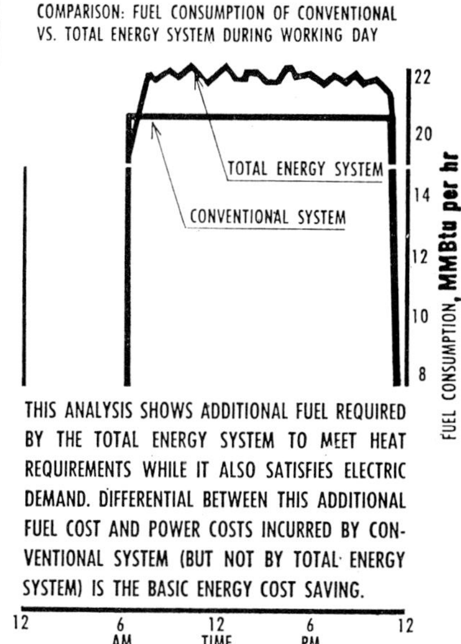

Fig. 12-217b Load profile and fuel use analysis for typical food processing plant.

tunities for maximum utilization of energy output—and the higher the efficiency of the total energy system.

Shaft Power Considerations

The prime mover represents a major cost element in the total energy package. Cost of the prime mover varies with its size, and size is a function of the shaft power demand it must satisfy.

As a general rule, since most of the shaft output is likely to be employed in driving an electric generator (although it may also be used for direct drive, hydraulic power, or pneumatic power), the energy system should be designed to minimize electric demand, for example, substituting gas engines for electric motors wherever possible and using gas ranges instead of electric ranges.

Conventional on-site prime movers operate at thermal efficiencies of 10 to 38 per cent (Table 12-168a).

High-Frequency Fluorescent Lighting

In installations with appreciable fluorescent lighting requirements, high-frequency power offers economies. Power generated on-site can be in a variety of frequencies without the attendant losses and expenses of the rotary or static converter equipment required to produce high frequencies from a standard 60-cps electric utility system.

A summary of the advantages claimed for fluorescent lighting of 420 cps over 60 cps follows: (1) as much as 35 per cent more lumens per system watt for 40-watt tube systems; (2) about 3 to 7 per cent more lumens per tube; (3) longer tube life; (4) less tube deterioration; (5) reduction of power consumption and ballast weight at the fixtures, a 420-cps capacitor ballast weighing 0.2 lb and a 60-cps commercial ballast 7.8 lb; (6) reduced air-conditioning tonnage because of reduced heat gain (by virtue of item 1); (7) "whiter," steadier light with less "strobe" effect.

Table 12-168a Thermal Efficiencies of Various Prime Movers, Per Cent

Output, kw	Piston engine (1200 rpm)		Gas turbine		Steam turbine*	
	Full load	Half load	Full load	Half load	Full load	Half load
75	28.9	20.2	...	...	...	...
175	29.6	23.6	13	10	...	...
225	31.0	24.7	15	12	...	...
300	31.0	25.2	18	15	...	...
450	31.0	25.2	18	15	...	...
750	...	...	18	15	34	30
900	...	...	18	15	36	32
1100	...	...	16	14	38	34

* Includes boiler efficiency of 80 per cent.

EXHAUST HEAT RECOVERY

Exhaust heat may be used to make steam or used directly for drying or other processes. The steam thus produced will provide space heating, hot water, and absorption refrigeration, which may supply air conditioning and process refrigeration. Heat recovery systems generally involve equipment specifically tailored for the job, although conventional fire-tube boilers are sometimes used.

Exhaust heat recovery from reciprocating engines is usually accomplished by **ebullient cooling** of the jacket water and use of a muffler type of exhaust heat recovery unit.

Total energy packages consisting of an air intake filter and cooler, prime mover, generator, exhaust heat recovery unit, jacket heat recovery unit, and exhaust noise attenuator are available.

Table 12-168b gives a typical heat balance for a four cycle, naturally aspirated gas engine. Of the five categories shown in the table, only jacket water and the recoverable exhaust heat outputs are usually considered in heat recovery applications. Table 12-168c gives typical heat recovery data for a variety of engines. Table 12-168d gives the temperature levels normally required for various heat recovery applications.

Table 12-168b Typical Heat Balance for a Four-Cycle, Naturally Aspirated Gas Engine[7]

(Btu per hr)

Heat output	Load, per cent 40	70	100
Shaft power	1020	1790	2550
Jacket water	830*	1890	2950
Exhaust, recoverable	550†	1080	1600
unrecoverable	850	1230	1600
Lubrication oil	180	240	300
Radiation and other	750	800	850
Total	4180	7030	9850

* Zero at 20 per cent of full load.
† Zero at 10 per cent of full load.

Ebullient Systems. These pressurized reciprocating engine cooling–heat recovery systems are also termed "high-temperature" or Vapor-Phase since they operate, with a few degrees of temperature differential, in the range from 212 to 250 F (sometimes as high as 270 F).

Engine builders have approved various conventionally cooled models for ebullient cooling application. The supposition of engine "hot spots," i.e., inadequately cooled areas in the engine, appears groundless since many installations, both large (including municipal power) and small, have not reported operating difficulties. Azeotropic antifreezes should be used to ensure constant coolant composition in boiling.

The system components generally displaced by ebullient cooling include the radiator, the belt-driven fan, and the gear-driven water pump.

Operation. Increasing quantities of steam are formed as the coolant ascends in the engine jacket and riser leading to the steam separator. The correspondingly decreasing head on the coolant-steam column permits increased steam formation. The resultant change in coolant density facilitates natural circulation in the system.

In the separator, the steam (at about 15 psig) is removed from the coolant and sent on as the energy source to heating and absorption cooling equipment. The steam-free water from the separator is returned to the engine, as is the condensate from the heating and/or cooling equipment.

Table 12-168d Temperature Levels Normally Required for Various Heating Applications[7]

Application	Temp, °F
Absorption refrigeration machines	225–245
Space heating (radiation converters)	215–250
Water heating (domestic)	215–250
Process heating	150–250
Evaporators (water)	190–250
Residual fuel heating	212–330
Auxiliary power producers, with steam turbines or binary expanders	230–350

Example: The jacket cooling water has engine inlet and outlet temperatures of 246 and 250 F, respectively. The pressure in the jacket is 15.1 psig. Atmospheric pressure is 14.7 psia. Find the steam-forming rate necessary to reject 3000 Btu per hp-hr to a steam separator.

Latent heat of steam at 250 F and 29.8 psia = 945.5 Btu per lb (from steam tables)

Steam forming rate = 3000 Btu per hp-hr/945.5 Btu per lb = 3.17 lb per hr at a pressure somewhat under 15 psig

Heat from Exhaust Gases. In many engines, e.g., turbocharged models, exhaust heat rejection considerably exceeds jacket water rejection.

Generally, gas engine exhaust temperatures run from 700 to 1000 F (Table 12-168e). Cooling exhaust gases to a minimum of approximately 300 F precludes condensation in the exhaust line. Thus, about one-half to three-quarters of the sensible heat in the exhaust may be considered recoverable.

Economics of exhaust heat boiler design generally limits the temperature differential between exhaust gas and generated steam to 100°F minimum. Therefore, in low-pressure steam boilers, gas temperature can be reduced to 300 or 350 F; the corresponding final exhaust temperature range in high pressure steam boilers is 400 to 500 F. Table 12-168d gives application data.

Exhaust Heat Boilers. Most of these make low-pressure steam or hot water. They develop approximately one boiler hp per 30 sq ft of heating surface. Boilers that incorporate silencers are available.

Vertical exhaust boilers may automatically vary the boiler water level; this in turn determines steam output (Fig.

Table 12-168c Typical Heat Recovery Rates of Various Types of Engines[7]

(Btu per bhp-hr)

(based on 15 psig steam pressure and 100 F ambient)

Type of engine	Fuel input at rated load	Heat recoverable at rated load and bmep: Jacket water, Air-cooled manifold	Jacket water, Water-cooled manifold	Exhaust unit
Two-cycle				
Mechanical supercharged gas	8,300	1750	2200	1,200
Naturally aspirated gas	12,000	3350	3900	1,500
Blower charged diesel	8,200	1650	1950	1,100
Four-cycle				
Naturally aspirated gas	8,500	1900	2350	1,250
Naturally aspirated diesel	8,500	1900	2350	1,250
Turbocharged diesel	7,300	1100	1350	1,200
Turbocharged dual-fuel	6,500	950	1100	1,000
Gas turbine				
Simple cycle	18,200			12,750
Regenerative cycle	13,000			7,500

Table 12-168e Approximate Full Load Exhaust Mass Flows and Temperatures for Various Types of Engines[7]

Type of engine	Mass flow lb/bhp-hr	Temp, °F
Two-cycle		
Blower charged gas	16	700
Turbocharged gas	14	800
Blower charged diesel	18	600
Turbocharged diesel	16	650
Four-cycle		
Naturally aspirated gas	9	1200
Turbocharged gas	10	1050
Naturally aspirated diesel	12	750
Turbocharged diesel	13	850
Gas turbines, nonregenerative	60	1050

12-218a). The ASME Code does not permit such operation in horizontal exhaust heat boilers such as that shown in Fig. 12-218b.

Heat Recovery from Turbines. The cost of the heat recovery equipment depends upon its design effectiveness; the design pressure rating; and whether or not it includes economizers, superheater, supplementary firing, and exhaust gas by-pass controls. The incremental cost of additional heat transfer surface added to the recovery boiler may be small compared with the incremental value of the added steam production, since the additional steam is obtained at no additional fuel consumption.

Supplementary firing can be used in conjunction with any heat recovery unit. The materials used generally limit the gas temperature to 1100 or 1200 F maximum.

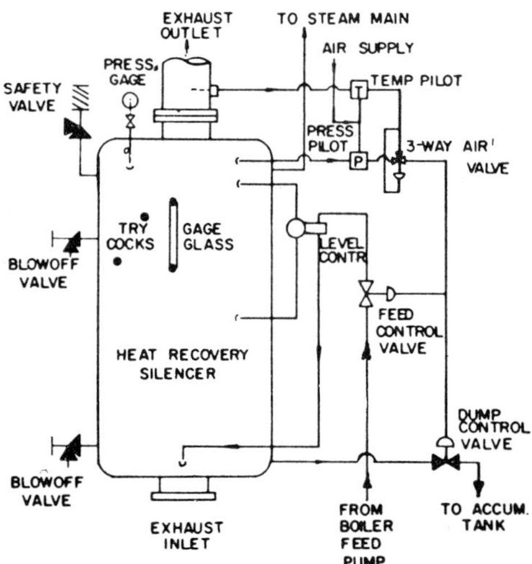

Fig. 12-218a Vertical exhaust heat recovery boiler-silencer with variable water level control system. When steam production begins to exceed the demand, the feed control valve begins to close, throttling the feed supply. Concurrently, the dump valve begins to open, thus lowering the water level in the boiler. These valves reach an equilibrium position that maintains a level in the boiler to match the steam demand. This system can be fitted with an overriding exhaust temperature controller that will regulate the boiler output to maintain a preset minimum exhaust temperature at the outlet. (AMF Beaird, Inc.)

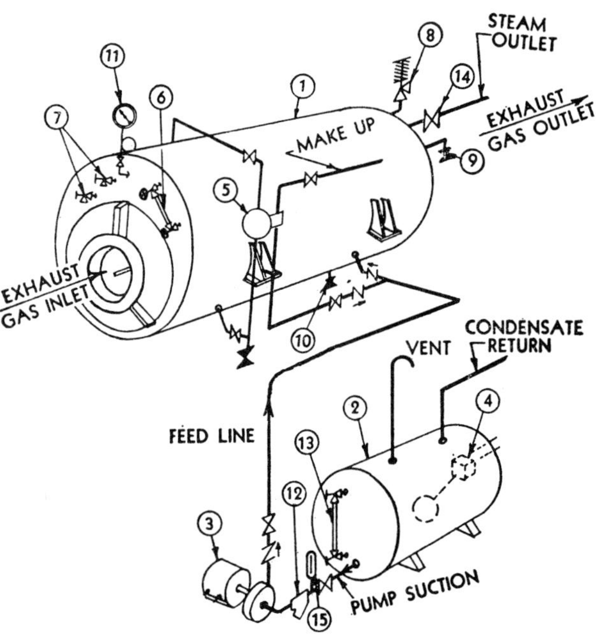

Fig. 12-218b Horizontal heat recovery boiler-silencer arrangement. (AMF Beaird, Inc.)

1. Heat recovery silencer
2. Condensate tank
3. Boiler feed pump
4. Condensate float switch
5. Make-up water feeder
6. Boiler water gage
7. Try cocks
8. Safety valve
9. Surface blowoff
10. Bottom blowoff
11. Steam pressure gage
12. Suction strainer
13. Condensate water gage
14. Stop valve
15. Thermometer

Low-pressure steam units or water heaters are either of the watertube or of the firetube type; ordinarily, high-pressure recovery units are of the watertube type. The watertube unit lends itself most universally to all forms of heat recovery applications and, because of its inherent compactness, is most compatible with the total energy concept.

Figure 12-218c shows the heat transfer relationships in an exhaust heat boiler. The figure also indicates the phenomenon known as the "pinch point" that occurs in high-pressure recovery units with an economizer section. This is the point in the water circuit at which boiling begins at the start of the

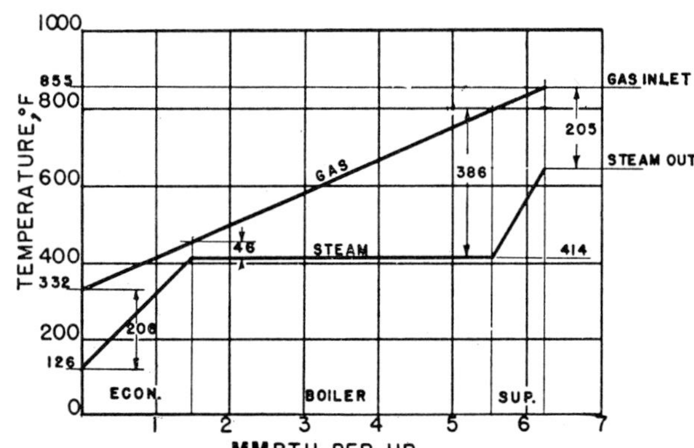

Fig. 12-218c Temperature and heat transfer distribution in a typical exhaust heat recovery boiler for a turbine.[8] Turbine gas flow, 13 pps (pounds per second); steam production, 5000 lb per hr at 275 psig.

latent heating or evaporating section. In the diagram it is 46 F. This is the closest approach of water temperature to gas temperature in the entire boiler circuit. The pinch point temperature spread becomes smaller as steam pressure is increased or an attempt is made to reduce boiler exhaust gas temperature. Because the pinch point is a determining factor in the logarithmic mean temperature difference in both the economizer section and the evaporating section, it is also a determining factor of the amount of heating surface in each section as well as the cost of the entire heat recovery unit. This fact favors the selection of a heat exchanger section having a gas passage area that diminishes in the direction of gas flow.

The superheater section may consist either of a tube assembly surrounding the evaporating section or of a grid section located directly in the turbine exhaust stream and connected to the steam separator steam outlet.

Low-pressure boilers are not influenced by the pinch point effect. In these boilers, the heating surface determining factor is only the logarithmic mean temperature difference across the evaporating surfaces and the gas mass velocity.

Exhaust heat recovery water heaters are comparatively simple. They can be the once-through multiple circuit type, with turbine gas by-pass control as a function of water temperature. The heating surface design principles are essentially the same as those discussed in the foregoing, except that in many cases the water pressure drop thru the heater determines the size and number of parallel circuits.

Systems having gas turbines providing base load with reciprocating engine stand-by can be integrated with the recovery equipment described by exhausting the reciprocating engine directly into the boiler inlet plenum, then recirculating boiler water from the separator reservoir or drum thru the engine jacket and back to the steam separator or steam drum. A separate recirculating pump can be used for this purpose, or sufficient static head on the engine jackets will ensure positive natural circulation. This latter method is known as ebullient cooling. Either method will maintain the engine temperature at a constant level.

Recovery Equipment Codes. Jacket water steam separators and exhaust recovery exchangers may be built under the ASME Unfired Pressure Vessels Code[9] unless stand-by firing directly into the exhaust exchanger is desired. In the latter case, the unit should be built to the appropriate fired boiler code. Exhaust exchangers operated above 50 psig are considered to be in the unfired steam boiler classification. Many cities and some states may require the approval of the National Board of Boiler and Pressure Vessel Inspectors, the so-called National Board stamp, particularly if any auxiliary firing is involved. If the exhaust units are to be operated dry, coding must be according to the maximum expected exhaust temperature.

Requirements for licensing of operating personnel should be considered when the plant operating pressure is selected.

The ASME Low-Pressure Heating Boilers Code[10] covers boilers at design pressures up to 15 psig. The ASME Power Boilers Code[11] covers boilers for operating pressures above 15 psig steam; the higher construction costs associated with boilers of this design, as well as the lower exhaust heat conversion attainable, often preclude the use of exhaust-heat boilers of this type.

REFERENCES

1. Marks, R. H. "Total Energy Approach Scores Impressive Gains." *Power* 108: S34–5, Oct. 1964.
2. "Determining Feasibility of Onsite Energy Systems." *Heating, Piping Air Conditioning* 36: 159–74, Nov. 1964; 37: 161–76, March 1965.
3. Wright, W. F. "Estimating Gas Turbine Profitability." (In: A.G.A. *Gas Turbine Manual*, chap. 8. New York, Industrial Press, 1965). Also in: *Air Conditioning, Heating Ventilating* 61: 65–74, March 1964.
4. Bunch, H. M., and others. *A Method of Estimating the Economic Feasibility of a Gas Fired Total Energy Package.* New York, A.G.A., 1964.
5. Martin, R. "Briefcase Computer Makes TEP Study in an Hour." *Am. Gas J.* 192: 55–62, April 1965.
6. *How to Make Total Energy Move.* (A.G.A. Total Energy School) New York, A.G.A., 1964.
7. Baker, M. L. *Systems for Extracting and Utilizing Engine Rejected Heat.* (ASME Paper 63–OGP–6) New York, ASME, 1963.
8. Boyen, J. L. *Heat Recovery Problems in Total Energy Systems.* (ASME Paper 65–GTP–2) New York, ASME, 1965.
9. ASME. *Rules for Construction of Unfired Pressure Vessels.* (ASME Boiler and Pressure Vessel Code, sect. VIII) New York, 1962.
10. ——. *Rules for Construction of Low-Pressure Heating Boilers.* (ASME Boiler and Pressure Vessel Code, sect. IV) New York, 1962.
11. ——. *Rules for Construction of Power Boilers.* (ASME Boiler and Pressure Vessel Code, sect. I) New York, 1962.

Chapter 24

Gas Internal Combustion Engines (Reciprocating)

by Marvin D. Ringler and F. E. Vandaveer

The *Diesel and Gas Engine Catalog*[1] lists and describes available engines and auxiliary equipment.

GAS-FUELED ENGINES

Gas-fueled engines may use either a spark-ignited or a diesel cycle; in the latter case there is a need for "pilot" fuel oil (5 to 7 per cent of the total Btu input) to initiate the combustion process. Both types of engines can use all types of gaseous fuels. The diesel engine can, in addition, use normal diesel fuels interchangeably with gaseous fuels without interruption of service.

Table 12-169 gives some results of a 1964 survey covering about 300 U. S. and Canadian reciprocating engine installations.

Table 12-169 Analysis of Engine Installations[2]
(all data in per cent of number surveyed)

Principal use	
Generating power	
Prime	35
Stand-by only	25
Peaking	3
Driving equipment	
Compressors	19
Pumps	18
Type of engine	
Full diesel	49
Spark-ignition gas	38
Dual-fuel	13
Operating cycle	
Two-cycle	35
Four-cycle	65
Engine horsepower	
300–499	18
500–699	20
700–899	11
900–1999	13
2000–3999	31
4000–9999	7
Action	
Single-acting	89
Opposed piston	11
Air supply	
Supercharged	55
Not supercharged	45

Natural Gas as an Engine Fuel

Natural gas has been used as an engine fuel in the U. S. for over 45 years.

Fuel Gas Systems. Systems range from extremely simple, manually controlled direct hookups (e.g., oil field engines) to engineered, automatic, precisely metered municipal power generating installations.

Gaseous Fuel Advantages. These are:

1. Has higher octane ratings than gasoline, with resultant antiknock protection—thus greater compression ratios may be used without preignition or detonation.
2. Requires only a simple carburetion system.
3. Mixes thoroughly with air, burning more completely, quietly, and smoothly than liquid fuel-air mixtures
4. Utility gases are generally uniform in composition and quality,* with consistent specific gravity and Btu, in contrast to the possible deterioration of stored fuels.
5. Permits engines to be started and stopped easily.
6. Permits a fuel system designed for natural gas to be readily switched over to propane stand-by, provided that engine timing is retarded at the higher compression ratios (12 and 10.5 to 1) to avoid detonation due to preignition.
7. Has practically no offensive odor associated with the fuel or its combustion.

Service Aspects. Advantages of natural gas over liquid fuel follow:

1. Gas does not collect in the combustion chamber as an unburned liquid to find its way past the piston rings to dilute the lubricating oil.
2. Less sludge accumulates in the lubricating oil because the higher combustion temperature of natural gas prevents the condensation of any water vapor in the combustion products.
3. Fewer filter and oil changes are required because natural gas burns completely, leaving little or no fuel soot.
4. For the spark-ignited engine, internal starting loads are lower and less critical with respect to starting accessories.
5. Spark ignition is less critical than its diesel compression pressure equivalent.
6. There are few upper-cylinder lubrication problems since gaseous fuel does not wash down the upper-cylinder walls on cold start-ups.

* That is, there are negligible sulfur compounds, water, and other foreign matter. This does not preclude possible rust or scale forming in the piping to a given engine (in the absence of suitable filters).

7. There is no tetraethyl lead to foul spark plugs.

8. Gas engines require only a simple fuel-filtering system.

9. Engine speed modulation is relatively easy, providing good operating economy at partial loads.

Gas Pressure. Applicable code requirements[3–5] must be accommodated. The design pressure range should be maintained at the inlet to the engine regulator (when present) or the carburetor. Pressure *below* this range results in excessive engine exhaust temperature, which damages engine exhaust valves and the flexible exhaust connector, besides derating the engine. Pressure *above* that accommodated by the engine regulator results in rich air-fuel mixtures, particularly at low loads. Such mixtures cause surging and rough operation.

Heating Value. Fuel consumption data may be reported in terms of either *higher heating value*, HHV, or *lower heating value*, LHV. The former is preferred by the gas utility industry, and it is the basis for most gaseous fuel considerations. Most natural gases have an HHV/LHV factor of 1.11.* For gaseous fuels in general, however, this factor may range from 1.00 to 1.15. For fuel oils this factor ranges from 1.05 (heavy oils) to 1.07 (light oils);[6] the HHV is customarily used for oil (including pilot oil in dual-fuel engines).

Many manufacturers' engine power ratings are based on an LHV of 1000 Btu per cu ft. Manufacturers sometimes suggest derating engine output about 2 per cent per 100 Btu per cu ft decrease in fuel heating value below this base. The following example illustrates the theoretical considerations involved.

Example: Based on data in Tables 2-59 and 2-63:

Gas constituents, % by vol			Gas LHV, Btu/cu ft	Cu ft/cu ft of gas		Net Btu/cu ft of comb prod
CH_4	C_2H_6	N_2		Air	Products	
87.5	12.5	0.0	1000	10.42	11.42	87.5
78.8	11.2	10.0	900	9.38	10.38	86.7

Thus, (87.5 − 86.7) 100/86.7 = 0.92 per cent less energy evolved by a specific 900 Btu gas, compared with a specific 1000 Btu gas (LHV) going thru the same engine. This reduction in energy content of the gas is reflected in a reduction of about 2 per cent in engine power output.

Types of Gas-Fired Internal Combustion Engines

Engines may be classified in several ways: the arrangement of cylinders or pistons, type of fuel burned, general design features such as cooling means and cycle used, and cycle of operation. The arrangement of cylinders or pistons is usually obvious from the appearance. They can be arranged "in-line," in the form of a "V," in a circle ("radial"), or in other less familiar forms such as "W," opposed piston, or "delta."

Classification by Type of Fuel Burned. There are three types:

1. A *gas* engine uses spark plugs as the means of ignition and burns 100 per cent gas, e.g., natural gas, LP-gas, or LP-gas–air. This type of engine is similar to the ordinary gasoline engine in automobiles.

2. A *dual-fuel* engine (also known as a gas-diesel engine) burns a small amount of "pilot" fuel oil, which is ignited by the heat of compression and in turn ignites the balance of the fuel, which is gas. If the gas supply is interrupted, the dual-fuel engine can burn 100 per cent oil without any changes and without stopping the engine. Thus, gas can be purchased on an interruptible basis (available from many utility companies) usually at a lower price. When operating chiefly on gas, a dual-fuel engine consumes about 3 to 12 per cent pilot oil, depending upon the design. The gas fuel may be either compressed in the cylinders with the air or compressed separately and injected into the combustion chamber when the fuel oil is added. Since the combustion system of a dual-fuel engine is rather complex, it may not readily handle fluctuating or low loads. These engines are best applied to large, well-maintained systems handling continuous loads at relatively constant speed.

3. The straight oil engine, or *diesel* engine, burns 100 per cent fuel oil, which is ignited by the heat resulting from the compression of the combustion air.

Classification by Cooling Means. There are two types:

1. *Small air-cooled engines* (1.5 to 60 hp, 1200 to 3600 rpm). These self-contained engines are relatively light, compact, and readily started.

2. *Water-cooled engines* (8 to over 1100 hp, 40 to 6000 cu in. engine displacement). The larger engines generally run at 700 to 1300 rpm, smaller ones at 1200 to 2100 rpm.

Classification by Thermodynamic Cycle—Otto vs. Diesel Cycle. Merits of the Otto cycle compared with the diesel cycle are: (1) higher output per cubic inch or displacement; (2) better power-to-weight ratio; (3) smoother running, especially at idle; (4) less expense to build. The thermal efficiencies for these cycles are shown in Fig. 12-219a.

Classification by Cycle of Operation. Engines may be classified by the cycle of events taking place in their cylinders. In a **four-cycle engine** four piston strokes (i.e., two crankshaft revolutions) complete a cycle: (1) intake of gas-air mixture; (2) compression of mixture; (3) expansion of ignited mixture for power; (4) exhaust. Figure 12-219b shows the relationship between the opening and closing of the exhaust and intake valves on the one hand and the position of the piston on the other. In a **two-cycle engine** two piston strokes (i.e., one crankshaft revolution) complete a cycle, the first stroke comprising the foregoing first two strokes and the second stroke comprising the foregoing last two.

Two-cycle gas engines have long been associated with applications requiring only a relatively narrow span of torque and control. The use of mixture controls, however, has broadened both the torque and speed ranges so that smooth and efficient operation is possible at any normal torque-speed combination above one-half rated torque and one-half rated speed (both blower-scavenged and turbocharged units). A bmep of 40 psig is a recommended minimum.[8]

Since four-cycle engines tolerate comparatively high vacuum pressures, they may be operated smoothly down to no-load.

Terms for Power. Table 12-170 presents some basic power definitions.

Direction of Rotation. Direction of engine rotation is

* A check of the data given in Table 2-9 indicates that 1.11 applies to the natural gases distributed in all 48 cities, with a tolerance of +0.00, −0.01.

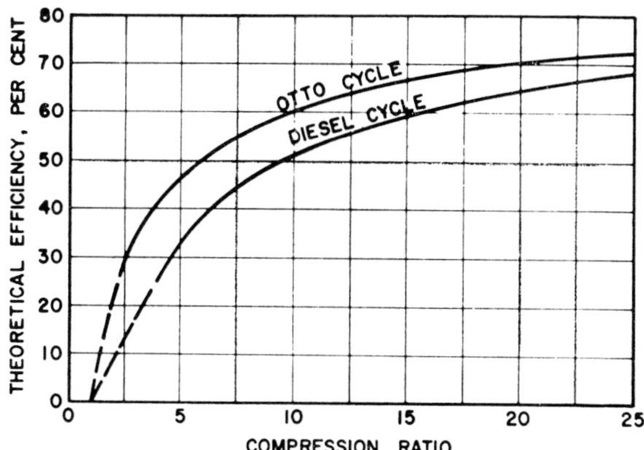

Fig. 12-219a Theoretical efficiencies of Otto and diesel cycles.[7]

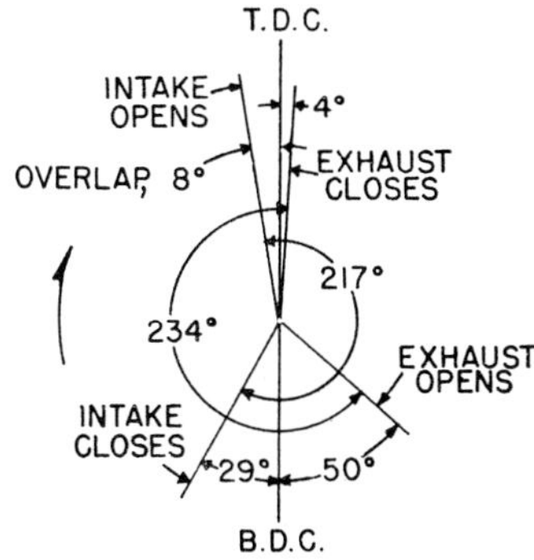

Fig. 12-219b Valve timing diagram. View from flywheel end of small naturally aspirated engine. (Continental Motors Co.)

determined as *clockwise* or *counterclockwise* when viewed from the power takeoff (flywheel) end of the engine.

Engine Rating and Atmospheric Variables. A commonly used rating system relates to type of duty. *Maximum* rating is defined under Eq. 1. *Intermittent* rating bases vary with the manufacturer's definition, ranging from an engine's ability to carry a given load for less than 1 hr to the capability of carrying such a load for 12 hr per year. In general the engine idles or operates at substantially reduced load at frequent intervals. *Continuous* rating is defined as the engine operating 24 hr per day with little or no variation in load or speed. Rating curves are generally available for the latter two types. Other designations, e.g., stand-by, are sometimes used. The *load factor* is the ratio of the average load to the maximum capability of the engine, and is usually expressed in per cent.

The Diesel Engine Manufacturers Association, composed of most of the large engine manufacturers, rates engines on the basis of being about to carry a load continuously and an overload of 10 per cent for at least 2 hr in any 24 hr. This "standard" is based on an ambient temperature not exceeding 90 F and an altitude not exceeding 1500 ft, unless otherwise stated.

The power required for the various engine accessories, such as cooling fans, accounts for the difference between *gross* and *net* maximum horsepower. Cooling fans generally account for a 6 per cent difference on small engines and 3 per cent on large ones. Some manufacturers recommend that 75 or 80 per cent of *net* maximum horsepower be specified for continuous duty and 90 per cent for intermittent duty. The American Petroleum Institute recommends derating maximum bhp by 35 per cent for continuous duty.

Atmospheric variables included in Eq. 1 may be used to match prime movers to loads at various ambients.

Table 12-170 Power Terms and Definitions[9]

Term	Definition
Measured power output, mechanical	Uncorrected power delivered at the coupling or power takeoff connection of any of the engine assemblies defined in the referenced code, as determined by a brake or other approved power measuring device, expressed as brake horsepower.
Net power output, mechanical	Horsepower output determined from the measured power output, mechanical, by application of corrections, charges, and credits as prescribed in the referenced code.
Measured power output, electrical	Uncorrected power delivered at the generator terminals of any of the engine driven generator units defined in the referenced code, as determined by electrical measurements, expressed in kilowatts.
Net power output, electrical	Kilowatt output determined from the measured power output, electrical, by application of corrections, charges, and credits as prescribed in the referenced code.
Rated power output (continuous rating)	Net full-load power output, either mechanical or electrical, stated or guaranteed at rated speed under specified operating conditions and on the basis of continuous operation.
Intermittent rating	Net power output, either mechanical or electrical, stated or guaranteed under specified speed and operating conditions on the basis of stated load variations and corresponding time periods at stated load conditions.
Overload rating	Net power output, either mechanical or electrical, stated or guaranteed in horsepower or kilowatts, or as a percentage, over and above the continuous rating; produced at rated speed under specified operating conditions and maintained continuously for a stated period of time.
Peak horsepower capacity	Maximum horsepower output that engine, when provided with minimum of accessories required for its operation or "stripped," will develop under specified speed and operating conditions for a period of not less than 1 min.
Indicated horsepower	Horsepower exerted by the working media in the cylinders of an engine assembly as obtained by calculations based on the indicated mean effective pressure or by summation of the net power output, mechanical, and the total power losses, including friction horsepower.

$$\text{uhp} = \text{max hp} \times \text{rating* factor} \quad (1)$$

where:

uhp = usable horsepower

max hp = horsepower rating under ideal conditions, based on individual manufacturer's laboratory performance data (usually a dynamometer test), corrected in accordance with manufacturer's specifications to, for example:
60 F and 29.92 in. Hg, *or*
80 F and 1000 ft elevation (NEMA),[10] *or*
85 F and 500 ft elevation (SAE),[11] *or*
90 F and 1500 ft elevation (DEMA)[12]

rating factor = [100 − (% altitude correction + % temperature correction + heating value correction + % reserve)]/100

where:

altitude correction = 3 per cent per 1000 ft above a specified level for naturally aspirated engines; 2 per cent per 1000 ft for turbocharged engines

temperature correction = 1 per cent per 10°F rise above a specified base temperature for air intake

heating value correction—see "Heating Value" on page 12/364

reserve = a percentage allowance (safety factor) to permit design output under unforseen operating conditions that would reduce output, such as dusty environment, poor maintenance, higher ambient temperature, lowered cooling efficiency. Recommended values follow:

Application	*Reserve, %*
Air conditioning and refrigeration	See Table 12-173
Irrigation pumping	20
Oil field service	35

Example: A naturally aspirated gas engine has a maximum performance laboratory rating of 100 hp at standard NEMA conditions (1000 ft elevation and 80 F). Using Eq. 1 and Table 12-173, find the usable horsepower for refrigeration service at 2000 ft elevation and 100 F.

$$\text{Usable hp} = 100 \times [100 - (3 + 2 + 0 + 14)]/100 = 81 \text{ hp}$$

Engine Size. Physical size does not reflect power in any simple rule-of-thumb relationship. In general, *per horsepower*, low-speed engines are larger than high-speed ones.

Fuel Economy Curves. These characteristic fuel consumption curves give results of various types of laboratory tests. *Brake specific fuel consumption data* should be obtained from engine manufacturers for the loads and speeds pertinent to applications under consideration. In the absence of any data, 8 to 10 MBtu per hp-hr may be used in estimating. These values correspond to 32 and 25 per cent thermal efficiency, respectively.† *Heat recovery* appreciably increases the overall thermal efficiency of an engine-driven system.

Fuel Consumption Guarantees. Manufacturers' guarantees generally are in terms of the lower heating value of gas, Btu per net bhp at one-half, three-quarters, and full load. Any liquid fuel, including pilot oil in dual-fuel engines, is reported in terms of higher heating value.

Altitude and temperature are specified as noted for Eq. **1**; engine jacket water and lubricating oil temperatures should be within ±10°F of values specified in engine builder's bid.

Shop and/or field tests[9] may be specified to establish both fuel consumption and horsepower capacity.

Substituting Engines for Motors. Many electric motors can develop 125 per cent of their rating for 1 or 2 hr periods and 150 per cent for shorter periods. When sizing an engine to replace a motor, these factors, as well as atmospheric variables, should be taken into account whenever the requirements of the driven load are not fully known.

Piston and Engine Speeds. Average piston velocity involves a sliding motion and is therefore a truer measure of wear than engine crankshaft speed, N (in rpm). The relationship between piston and engine speed is expressed by:

$$\text{Piston speed, fpm} = \frac{N}{6} \times \text{stroke, in.}$$

Early gas engines were generally rated at 75 psi bmep and 900 fpm (two-cycle) or 1200 fpm (four-cycle) for long life at continuous operation. More recently improvements in engine design, metallurgy, manufacture, and lubricants have resulted in many manufacturers rating their engines at piston speeds up to 1600 fpm. Even higher piston speeds are sometimes used in stand-by and lightly loaded engines.

Figure 12-220 shows that the higher the engine output at a specific speed (rpm), the lower the specific fuel consumption. When engines speeds are categorized, the following relationship generally applies:[14] slow, below 700 rpm; medium, 700 to 1500 rpm; high, above 1500 rpm.

Speed Modulation. In applications where gas engine speed may be varied (e.g., chillers for air conditioning), essentially constant bmep can be maintained by throttling engine output from full-load at rated speed to half-load at approximately one-half rated speed. Figure 12-221a illustrates the concept for a compressor unit of 100 tons refrigeration capacity.

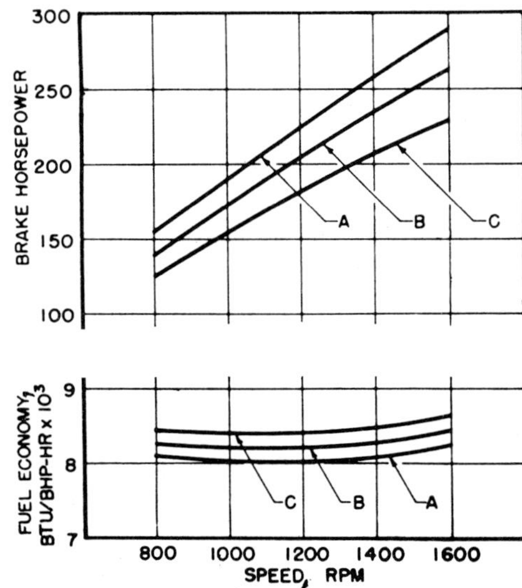

Fig. 12-220 Gas engine performance curves.[7] Ratings: A—maximum; B—intermittent; C—continuous.

* Sometimes called derating.

† Thermal efficiency may range from less than 20 to more than 40 per cent.

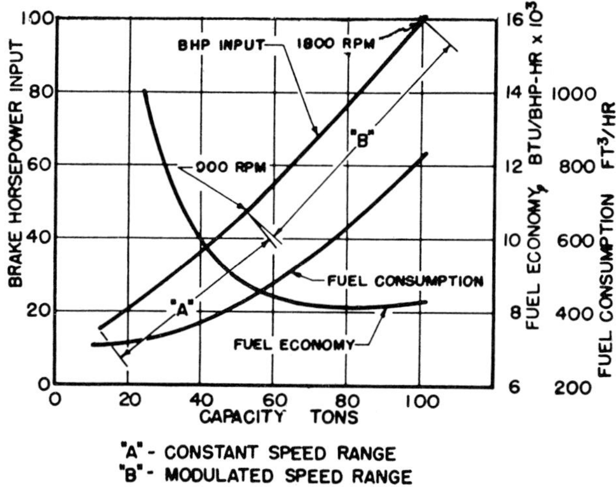

Fig. 12-221a Performance curves for a 100 ton capacity, gas engine-driven reciprocating chiller unit.[7]

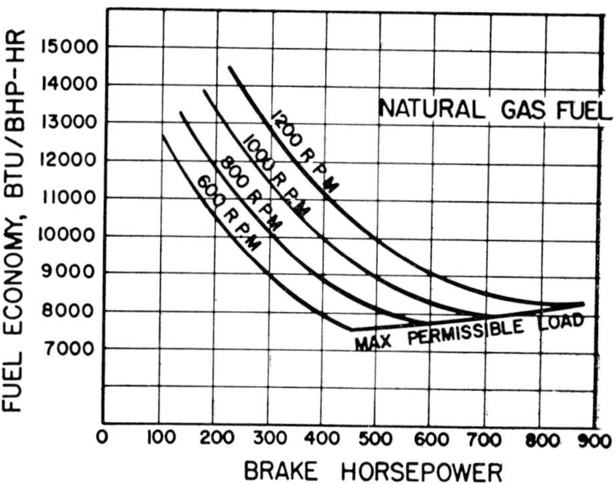

Fig. 12-221b Part load fuel economy curves.[7]

Figure 12-221b shows how specific fuel input at partial load varies at a number of engine speeds.

Manifold Vacuum. This pressure is frequently used to estimate gas engine loading when the horsepower being used by the driven unit is unknown; to aid in tuning; and to balance loads between engines. Generally, for four-cycle engines (not supercharged) operating at full load, the carburetor throttle is wide open, and the vacuum, measured by a vacuum gage, is at a minimum of zero to 2 in. Hg (zero in. vacuum corresponds to atmospheric pressure). At zero load, the throttle is almost closed and the intake manifold pressure runs about 20 in. Hg vacuum. For partial loads linear interpolation must be made; e.g., 10 in. Hg vacuum corresponds to about 50 per cent load.

When a supercharger is used, the pressure in the intake manifold rises to above atmospheric pressure at high loadings (Fig. 12-222). A manifold pressure gage that measures *absolute* pressure may be used to indicate pressures both above and below atmospheric.

Compression Ratio. Theoretically, the compression ratio, C, is the sum of the piston displacement, PD, plus the clearance volume V_c, *divided* by the clearance volume; actually, C is less than this nominal value owing to late valve openings. For early natural gas and propane engines the compression ratio was about 6:1. In 1962, it was 12:1 for dry* natural gas, naturally aspirated engines and 10.5:1 for propane stand-by in nonlugging applications. The 10.5:1 ratio, when used on dry natural gas or sewage gas, gives about a 12 per cent increase in either fuel economy (or power) over the 7.5:1 ratio, which is suitable for wet natural gas, mixed gases, or butane.

Engine efficiency varies with compression ratio (Fig. 12-223). As shown in the foregoing paragraph, however, fuels limit the ratios.

Detonation. Detonation is the major fuel-interchangeability consideration. It is characterized by violent, sudden

* Less than 5.0 per cent butane (sometimes less than 5.0 per cent propane is specified instead).

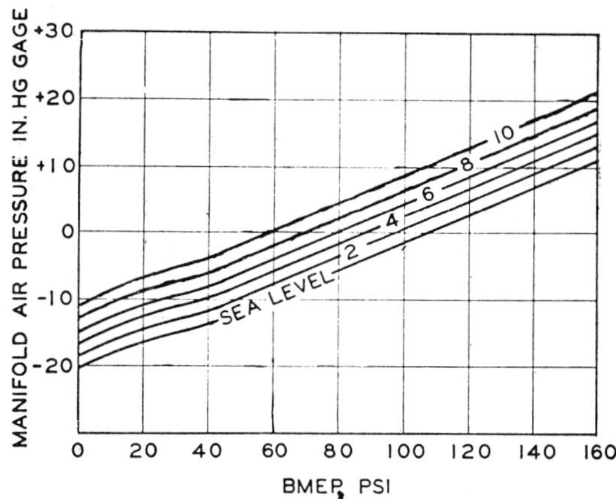

Fig. 12-222 Manifold air pressure vs. bmep at altitudes from sea level to 10,000 ft[15] for a typical turbocharged and aftercooled gas engine. Turbocharging in conformance with the altitude curves in above sea level installations permits development of sea level output at elevations shown; i.e., sea level air-fuel charge density is maintained at all altitudes.

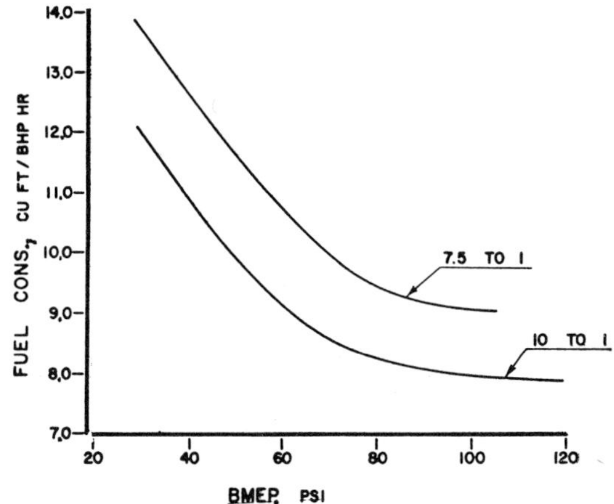

Fig. 12-223 Effect of compression ratio on fuel consumption at part throttle (for a 5.75 in. × 8 in., 4 stroke, 1200 rpm naturally aspirated gas engine).[15]

Table 12-171 Typical Relationships of Compression Ratio, Ignition Timing, and Fuels for Gas Engines

Engine type*	Compression ratio	Ignition timing,†	Fuel gas
NA	12:1	27.5	Dry‡ natural
NA	10.5:1	20	Dry natural
TA	10.5:1	12.5	Dry natural
NA	10.5:1	10	Propane§
NA	7:1	30	Dry natural-propane
TA	7:1	30	Dry natural
NA	7:1	20	Wet** natural-butane

(Caterpillar Tractor Co.)
* NA—naturally aspirated; TA—turbocharged aftercooled.
† Degrees before top dead center.
‡ Less than 5 per cent butane and heavier gases.
§ Stand-by.
** Field gas or well gas.

rises in combustion pressure that produce audible knocks. Proper gas engine design and operation minimize the occurrence of these knocks. Knocking, a postignition phenomenon, is generally said to be caused when an advancing flame front compresses a final portion of an unburned charge above its ignition temperature.[17] Four factors influence knocking. These are fuel characteristics, compression ratio, ignition timing, and mixture temperature. Examples of relationships among the first three factors are given in Table 12-171.

Fuel Characteristics. The knocking tendency varies with the size of the hydrocarbon fuel molecule. Thus, for example, methane has a lesser tendency to knock than butane. Optimum spark advance varies inversely with burning velocity (Table 2-70). Excessive spark advance permits a portion of the air-gas mixture to burn too much before the piston reaches the top-dead-center position. This premature burning may cause detonation.

Fuels containing H_2, the fastest burning fuel listed in Table 2-70, are particularly prone to detonate. One engine manufacturer limits H_2 to 12 per cent in a mixed gas (1000 Btu per cu ft) containing H_2 and natural gas. Another manufacturer's naturally aspirated, high-compression engines reportedly use mixtures of natural gas and 30 per cent H_2.

Compression Ratio. Detonation tendency increases with increases in compression ratio because of the increased temperatures reached during the compression stroke. In addition, there is a decrease in dilution of the charge by exhaust products; the exhaust products tend to retard detonation. Table 12-172 gives critical compression ratios, C_c, at which incipient detonation occurs, and "modified" octane numbers for various gases.

Table 12-172 Critical Compression Ratios, C_c, and Modified Octane Ratings for Various Engine Fuels

Fuel	Critical compression ratio	Octane rating
Methane	15:1	~130
Ethane	14:1	103
Propane	12:1	99.6
iso-Butane	...	98.4
n-Butane	6.4:1	91.6
Pentane	3.8:1	See Table 2-40
Propylene	8.4:1	81.0

Standardized apparatus and techniques for determining octane numbers of gaseous fuels are available.[18-21] A maximum of 120 octane can be handled in these tests. Although ratings on mixtures of saturated hydrocarbon gases can be calculated with acceptable accuracy, the presence of unsaturates or of antiknock additives makes an engine test necessary.[7]

Table 12-172 shows that butane cannot be used in high-compression ratio engines. Engines with 10:1 compression ratios may tolerate fuels containing up to approximately 5.0 per cent by volume of hydrocarbons heavier than propane. Severe detonation occurs above this limit.* Propane, a satisfactory fuel for 7:1 ratios, can sometimes be used, if *pure* or at least 95 per cent pure, in 10:1 compression ratio engines. These engines may be either naturally aspirated or turbocharged aftercooled.

Ignition Timing and Mixture Temperature. Ignition timing controls the portion of the charge burned before and immediately after the top-dead-center position of the piston. This setting is usually a compromise between maximum engine output and optimum fuel economy on one side (large advance) and minimum detonation on the other (small advance). Accommodation of sudden load accelerations and major variations in fuel supply (e.g., propane stand-by for natural gas) are other considerations. Note that no spark advance would be necessary if the combustion of a cylinder charge could be instantaneous.

For each fuel (Table 12-171) and each mixture temperature (see the example in Fig. 12-224), there is a relationship between ignition timing and bmep in a specific engine.

Supercharging. This process involves supplying a gas-air† mixture to, and retaining it in, the cylinders at above atmospheric pressure prior to the compression stroke. Supercharging constitutes a major means of increasing engine horsepower. For example, the **bmep** of a naturally aspirated engine may be increased from 120 to 175 psig.

The two general methods of supercharging are: (1) **turbocharging**, using engine exhaust gases to drive a turbine that in turn powers an air blower (Fig. 12-225); and (2) using an air blower that is driven directly by the engine. The net brake horsepower rating of the engine should take account of the power to the blower.

Scavenging, which is always used in two-cycle engines and sometimes in four-cycle engines, forces combustion products out of engine cylinders during the exhaust portion of a cycle and displaces them with fresh air. The process differs from turbocharging with air in that the latter includes provision for retaining the fresh air furnished at above atmospheric pressure. Means for scavenging include turbocharging and compressing air either at the underside of the pistons or in special crankcase chambers.

* Use of propane HD5 (Table 2-17) in combination with tetramethyl lead (TML) raises the octane number of this fuel from a minimum of 95 to over 100. An additive concentration of about 4 cu cm of TML (costing about one cent) per gallon of propane was suggested by one engine manufacturer. The addition is generally made by suppliers under carefully controlled conditions since TML is very toxic.

† Air alone when fuel under pressure is added to the cylinders separately.

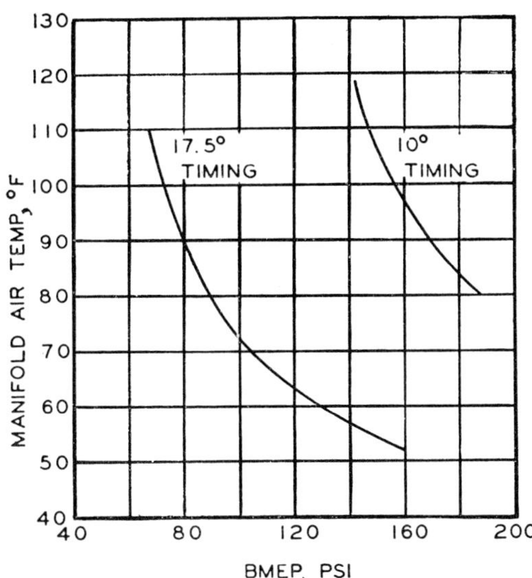

Timing for the detonation limit curves is in °BTDC (before top dead center)

Fig. 12-224 Manifold air temperature vs. bmep for a typical turbocharged, aftercooled gas engine.[15] Compression ratio, 10:1; fuel, dry natural gas with propane stand-by; engine rpm, 1200; displacement, 1473 cu in.; jacket water temperature, 190 F; water temperature to aftercooler is 20° to 30°F lower than manifold temperature. The air-fuel mixture temperature is practically equal to the manifold air temperature. (Caterpillar Tractor Co.)
Example: Determine the aftercooler water temperature to permit a bmep of 157 psi using propane (95 per cent pure). Locate 157 on the horizontal scale; project this point vertically to some point safely under the 10° timing curve, then horizontally to the left to read a manifold air temperature of about 90 F. The cooling water temperature is 90 — 30 = 60 F.

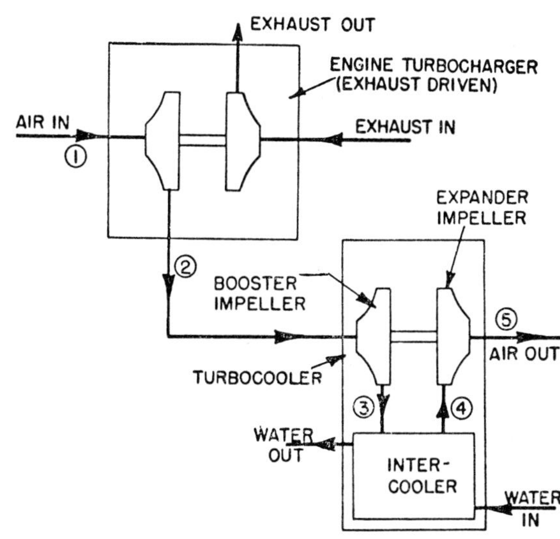

Fig. 12-225 Turbocharger and intercooler operation. Filtered air enters the turbocharger at (1) and is compressed by the exhaust gas-driven blower. From the turbocharger (2) the heated pressurized air is piped to the turbocooler booster impeller to undergo a second stage of compression (not in all systems). The expander and booster impellers are fixed on a common shaft and are driven by the flow of air from the intercooler. Leaving the booster impeller at (3), the air is at a temperature and pressure level considerably higher than at (2). The air is then substantially cooled by the air-to-water intercooler and enters the expander impeller at (4). Since the air pressure at (4) is greater than the engine requirements, the air is expanded at this point to enter the engine intake manifold (5) at a temperature and pressure ideal for good combustion. Expanding the air (reducing the pressure) by using it to drive the expander impeller further reduces the temperature of the air. The air out of the turbocooler (5) may be regulated by opening and closing nozzles that direct the air to the expander impeller at (4). A hydraulic actuator, operated by the engine manifold pressure regulator, may be used to position the nozzles to give the desired manifold pressure at all loads and speeds. Best performance from the turbocooler is obtained when the pressure drop across the unit is at its maximum. (Cooper-Bessemer Corp.)

Turbocharging. The horsepower delivered by a turbocharged, high-compression (10.5:1) engine will be as much as 35 per cent greater than a naturally aspirated, low-compression (7.5:1) engine having the *same* displacement. The former, however, may require larger bearings and crankshafts, as well as more exacting valve and cooling jacket designs. The initial cost *per horsepower* for a turbocharged engine may approach that of a naturally aspirated engine rated at the same horsepower. Turbocharged engines should be loaded to at least 75 per cent of rating to achieve their advantage over naturally aspirated units. Turbocharging is also a means of reducing fuel consumption (see the captions of the three curves in Figs. 12-228 thru 12-230).

Turbocharging may be used to effect torque control for lugging applications and other applications in which engines are subject to severe load surges. For example, in electric power generation, large, instantaneous load changes may have to be accommodated.

A typical drooping torque curve for an engine with a *free-running* turbocharger is shown as curve 1 in Fig. 12-226. Curve 2 of this figure shows the "rising" torque curve effected by controlling turbocharger speed by means of the turbocharger outlet pressure. This pressure acts on an exhaust bypass valve or "waste gate." The size of the turbocharger nozzle ring used with such a pressure control permits the development of an adequate air supply even if only light loads are applied to the engine.

Torque. An engine will have a definite relationship between torque and horsepower at a particular speed:

$$\text{Torque, lb-in.} = 63{,}025 \text{ hp/rpm}$$

Many applications are not critical with respect to torque. Centrifugal pumps, propellers, generators, and fans are examples of **nonlugging** applications that may be directly connected (i.e., no clutch) to engines. Such applications operate at approximately constant speed and load or under conditions in which the load drops rapidly with reductions in speed (curve 1 of Fig. 12-226). On the other hand, lugging applications call for engines with pronounced torque buildup at certain speeds (e.g., curve 2 of Fig. 12-226, results from a turbocharger design).

Lugging applications include piston pumps, compressors, drill rigs, and, in particular, positive displacement machines in which speed modulation is required. Appropriate **couplings** between the engine and the load facilitate the handling of lugging applications.

Brake Mean Effective Pressure. This term, signifying the average pressure on the piston during its working stroke, reflects the intensity of effort exerted by an engine

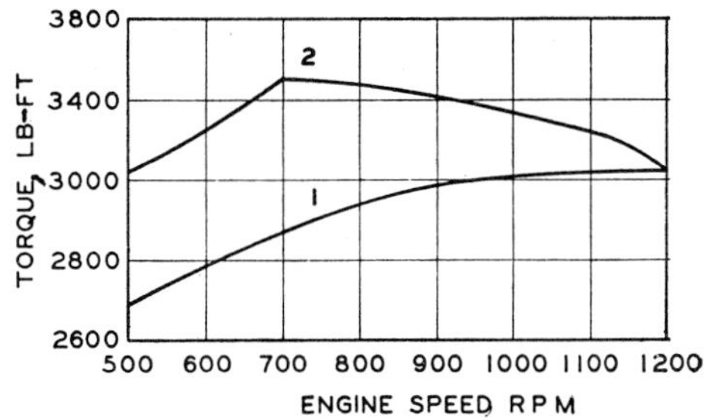

Fig. 12-226 Torque curves for a 700 hp turbocharged gas engine: curve 1, free-running turbocharger; curve 2, pressure-controlled turbocharger.[15]

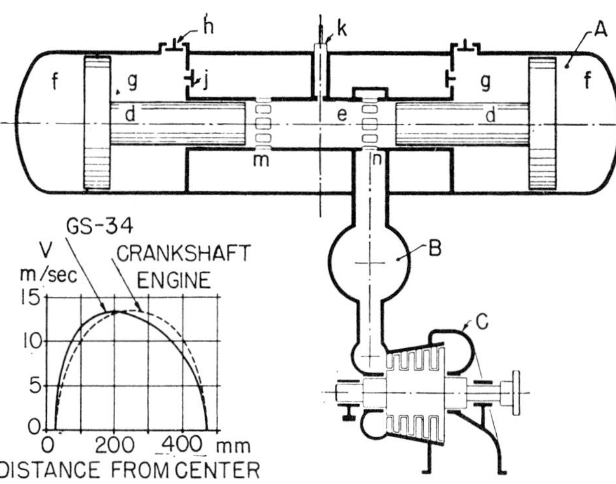

Fig. 12-227 Free piston engine (S.I.G.M.A.). (upper right) Free piston gasifier driving expansion turbine: (A) gasifier, (B) gas receiver, (C) expansion turbine, (d) piston, (e) diesel cylinder, (f) cushion cylinder, (g) compressor cylinder, (h) intake valve, (j) discharge valve, (k) fuel injector, (m) scavenge ports, (n) exhaust ports.
(lower left) Comparison of piston speed vs. stroke for a free piston (GS-34) and a conventional crankshaft engine having the same piston speed (differences exaggerated for emphasis). Free piston speed is higher near inner dead points and lower near outer dead points, leaving less time for combustion and more time for scavenging.

operating at a given speed under a given brake horsepower. Equation **2** covers single-acting* engines. The *indicated* or actual cylinder pressures are used in Eq. **28** of Sec. 8, Chap. 3. Figures 8–48 and 8–49 show engine indicator cards.

$$\text{bmep} = \frac{C \times \text{bhp}}{D^2 l N n} \qquad (2)$$

where:

bmep = brake mean effective pressure, psig
C = 1,000,000 for a four-cycle engine; 500,000 for a two-cycle engine
D = bore diameter, in.
l = piston stroke, in.
N = number of cylinders
n = engine speed, rpm

Example: Find the bmep of a four-cycle, six-cylinder gas engine with a 3.75 in. bore, a 4-in. stroke, driving a 50-hp load at 1400 rpm. Substituting in Eq. **2**:

$$\text{bmep} = \frac{1{,}000{,}000 \times 50}{(3.75)^2 \times 4 \times 6 \times 1400} = 106 \text{ psig}$$

Converting Diesels. Many manufacturers supply conversion kits for their own units. Some kit fabricators do not make engines; others market converted engines.

Changing a diesel engine to a spark-ignition gas engine may involve: (1) replacing the fuel injections by a magneto drive; (2) replacing the precombustion chambers by spark plug adapters; (3) mounting transformer coils, one per cylinder, as close as practical to the spark plugs (low-tension ignition system); (4) replacing air intake manifolds by manifolds flanged for carburetors; (5) adding a cylinder plate to reduce the compression ratio as necessary; (6) changing the shape of the piston face.

Free-Piston Engine. Designed and developed by S.I.G.M.A. in France in 1950, these two-cycle diesel engines function mainly as gasifiers. A standard model (Fig. 12-227) produces 1300 gas hp, which corresponds to 1100 shaft hp at a gas turbine. Thermal efficiencies are: driving gas and air compressor, above 40 per cent; driving expander power unit, 36 per cent. Pressure ratios are: power cylinder compression, 15:1; compression prior to firing, approximately 24:1 (firing pressure 1200 psi max); scavenging, approximately 2.4:1.

* Combustion takes place on one end of cylinder only.

The cyclic speed is basically fixed by the masses of moving parts. It varies from 600 cps at full load to 240 cps at no-load.[23] In 1961, free-piston engines totaling about 500,000 shaft hp were reportedly in operation in 40 countries, burning practically every kind of liquid fuel. Apparently the unit is insensitive to octane considerations. Trial operation on natural gas was also reported satisfactory.

Early in 1965, S.I.G.M.A. reported the installation of their free-piston, oil-fueled units at Mobil Oil Co. and Belridge Oil Co. facilities in California.

Expansion Reciprocating Engines.[24] Such engines are mainly used for cryogenic applications (to about −320 F), e.g., oxygen for steel mills, low-temperature chemical processes, the space program.

Operation. Relatively high-pressure air or gas expanded in an engine drives a piston and is cooled in the process. About 42 Btu are removed per shaft horsepower developed. Available units, developing as much as 600 hp, handle flows ranging from 100 to 10,000 SCFM. Throughput control at a given pressure is effected by varying the cutoff point and/or the engine speed. The conversion efficiency of heat energy to shaft work ranges from 65 to 85 per cent. A 5:1 pressure ratio and an inlet pressure of 3000 psig are recommended. Outlet temperatures of −450 F have been handled satisfactorily.

ENGINE APPLICATIONS

The National Engine Use Council publishes *Energy System Reports*[25] on specific applications of gas engines.

Figure 12-221b is indicative of the energy savings involved in varying engine speed to meet different output requirements. Figure 8-46 compares the fuel consumption and thermal efficiency at part load for seven types of prime movers.

Energy Cost Comparisons

Breakeven. Figures 12-228 thru 12-230 show breakeven charts of fuel costs for gas engines vs. electric motor drives, gas engines vs. diesel drives, and gas engines vs. gasoline and LP-gas drives.

Minimum Owning and Operating Costs

Considerable data are available based on the experiences of one utility company that has a comprehensive gas engine program for irrigation pumping.[27] The results of this program for a constant load application should be evaluated within this context. Details of the program are available.[28]

Electrical Generation

The majority of engines in stationary service drive a-c generators directly. Installations for generation service range from a few horsepower to the 475,000 hp generated by radial gas engines at the Chalmette (La.) Works of the Kaiser Aluminum and Chemical Corp. Many engines are used in public works power generation.[32]

Naturally Aspirated vs. Turbocharged Engine-Alternators. A speed-governing comparison between two four-cycle gas engine-driven systems (rate 880 hp at 514 rpm) of these types showed that comparable performances were achieved.[29] Test criteria were satisfied for limited steady-state speed deviation (± 0.33 per cent) and speed recovery from transient loadings.[30]

Sizing. Electrical demand meters, which are often included in metering systems, give accurate estimates of peak requirements. In the absence of such data, diversity factors are available for a wide variety of electrical loads.[31]

D-C Generation. This application is served to great advantage by gas engines because utility d-c power is not generally available and it is necessary to purchase rectifiers or motor-generator sets to change alternating current to direct current. Direct-current requirements, however, are limited mainly to certain electrolytic processes, such as the reduction of alumina to aluminum; the production of chlorine, caustic soda, and metallic sodium; and the electrorefining of copper and other metals.

Air Conditioning and Refrigeration

A basic advantage of an engine drive for this service is the ability to modulate (follow the load) without going off and on.

Variables used in calculating the usable horsepower requirement (Eq. 1) include: compressor displacement, suction and condensing pressures, refrigerant used, and compressor speed. These factors should be ascertained from compressor manufacturer data.

Several manufacturers package gas engine-driven chiller units. These are cross-instrumented and controlled for maximum effectiveness. Equipment is often factory tested or "authorized" for individual applications. Many completely automated units (capacities from 3 to about 150 tons) use reciprocating compressors. Above 150 tons, completely automated, built-up, engine-driven centrifugal compressors may be specified. At loads of about 1000 hp, gas turbine drives are usually a favored alternative.

Sizing. Eq. 1 used with Table 12-173 gives a method of rating engines for air-conditioning and refrigeration.

Speed Modulation. Figure 12-221a illustrates the fuel economy effected by varying prime mover speed with load (reciprocating compressor) until the machine is operating at about half-capacity. Below this level, the load is reduced at

Fuel Cost Breakeven Charts

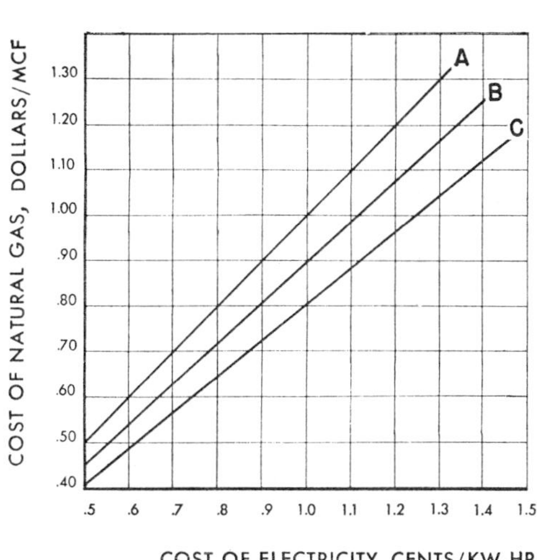

Fig. 12-228 Three gas engines vs. electric motor drives.

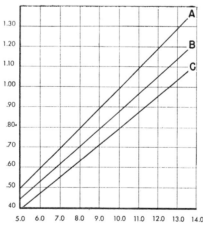

COST OF DIESEL, CENTS/GALLON

Fig. 12-229 Three gas engines vs. diesel drives.

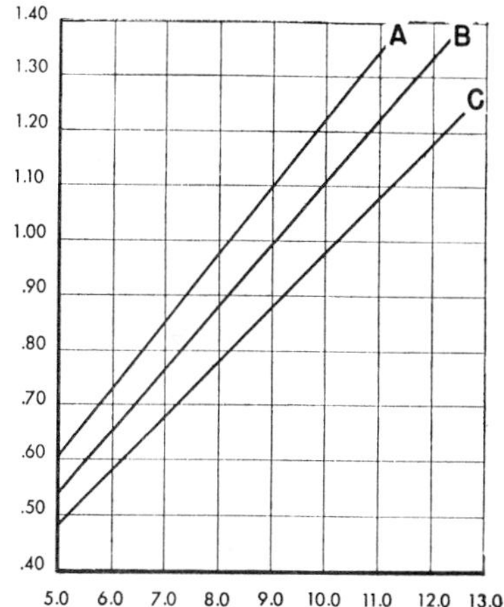

Fig. 12-230 Three gas engines vs. gasoline and LP-gas drives.

Basis of charts: 1000 Btu per cu ft gas; squirrel cage induction motor running fully loaded at 1800 rpm; 92 per cent efficiency. Cost of electricity should be adjusted for demand and/or power factor charges.

Fuel	Heating value, Btu/gal	Consumption, Btu/hp-hr
Diesel oil	144,000	11,800
Gasoline	130,000	13,000
LP-gas	92,000	9,200

Curves	Compression ratio	Gas, consumption, Btu/hp-hr
A—Turbocharged	10.5:1	8,100
B—Naturally aspirated	10.5:1	9,200
C—Naturally aspirated	7.5:1	10,250

Table 12-173 Minimum Gas Engine Reserves for Air Conditioning and Refrigeration, Per Cent[7]

Altitude, ft	Naturally aspirated		Turbocharged aftercooled	
	Air conditioning	Refrigeration	Air conditioning	Refrigeration
Sea level	15	20	20	30
1,000	12	17	18	28
2,000	10	14	16	26
3,000	10	11	14	24
4,000	10	10	12	22
5,000	10	10	10	20
10,000	10	10	10	10

essentially constant engine speed by unloading compressor cylinders.

The frequent operation of the system at low engine idling speed may require the use of an auxiliary oil pump for the compressor.

Fuel Economy. The fuel economy curve in Fig. 12-221a shows how constant-speed operation under declining load conditions reduces bmep, in turn increasing specific fuel consumption.

Reduced Wear. The reduced wear on both engines and compressors running at below rated speed, compared with rated speed operation, constitutes an additional advantage.

Load Tracking. Constant speed machinery, e.g., conventional electric motor drives, must unload cylinders in large increments to meet partial loads. Gas engine-driven compressors can track load variations by means of speed modulation and compressor cylinder unloading down to about 25 per cent of rated output. Thus, the gas fueled units "stay on the line" and track load (e.g., temperature and humidity) much better than electric motor-driven machinery.

Centrifugal Compressors.[33] Many of these compressors operate at about six times the speed of the gas engines used. Compressor speeds up to 14,000 rpm have been used. To effect the best compromise between equipment (engine, couplings, and transmission) first cost and maintenance cost, engine speeds between 900 and 1200 rpm are generally used. Completely automatic start-stop centrifugal compressor installations are frequently made.

Speed Modulation.[34] The following relationship applies:

Rated operating speed, per cent	100	75
Centrifugal compressor output, per cent	100	30

A vane-type control that regulates the flow of gas to the compressor is usually used to effect any further reductions in output.

Heat Pump Drives. Advantages of gas engine drives for this service include: (1) the ability of the engine to operate at the optimum speeds for the low head requirements of summer cooling and the high heads needed for winter heating; (2) the engine heat recovery from the jacket and exhaust that make a possible overall thermal efficiency of about 75 per cent (vs. about 30 per cent for electric motor or steam turbine installations). The combination of recovered engine heat and heat pump effect results in a heating COP approaching 2.0 (see Table 12-54).

ASA Approval Requirements. American Standard Z21.-40.2[35] is available covering the construction and performance of gas engine-driven summer air-conditioning appliances. The Standard applies only to automatically operated "assemblies of matched components." Units designed for indoor and outdoor installation are included.

Dual-Service Engine Application

Thru the use of magnetic or pneumatic clutches, rapid release of a given engine application may be accomplished. An electric generator already spinning on the gas engine shaft can then pick up the electric load. In 1965, such systems were in use for normal air-conditioning compressor drives in the Morris Cafritz Memorial Hospital in Washington, D. C., and the Mt. Sinai Hospital in Minneapolis, Minn. Such arrangements conserve space and reduce capital costs.

Water Pumping

Applications include irrigation, transmission of water to cities and its subsequent distribution, transfer of storm water from catch basins to drainage systems and other flood projects, sewage transport, and injection of high-pressure salt water into oil formations to increase oil production and control land subsidence.

Gas Engine vs. Electric Motor. Unlike an electric motor, gas engine speed can be varied to match pump speed and power requirement. In wells, for example, both capacity and water level may vary considerably over the life of the well.

A study of overall system efficiencies of 122 electric motor-operated wells in Colorado showed that 3 had a system efficiency above 65 per cent, 76 below 50 per cent, and the remainder between these extremes. Low efficiencies are due to the mismatching of the pumping bowl to the water requirement and lift (matching may have been proper at time of installation) and the inability of the fixed motor speed to compensate for this mismatching during well life or even an irrigation season.

Horsepower Required to Pump Water. To size the gas engine, proceed as follows:

1. Determine the dynamic head, H, in feet of water, by summing its four components:

   a. *Vertical distance from surface of water,* either in the well while the pump is operating or, if an open stream or pond, to the horizontal centerline of the engine drive-shaft.

   b. *Vertical distance from engine drive shaft,* either to the free water discharge point or to the surface of the stored water for underwater discharge.

   c. *Friction head loss,* which is the flow resistance of the pipe length, L (in feet), and associated fittings, F, in items a and b. The friction head loss varies with the flow rate and size of piping used (Table 12-174); i.e., [$L \times$ (value from Table 12-174)/100] + F, per Table 12-9.

   d. *Product of the minimum outlet water pressure required times 2.31.*

2. Determine the horsepower required, using Eq. **3**:

$$\text{Horsepower required} = \frac{8.34\,GH}{33{,}000\,E} \qquad (3)$$

Table 12-174 Friction Head Loss for Water Flowing in Pipe*

Gal per min	Friction head in ft for each 100 ft in length of pipe for sizes of clean iron pipe† indicated in **boldface type**					
	¾ in.	**1 in.**	**1½ in.**	**2 in.**		
5	7.59	1.93	0.27	0.092	**3 in.**	
10	29.90	10.26	1.08	.277	0.046	**3½ in.**
20	115.92	28.29	3.81	0.97	.138	0.069
30		63.25	8.62	2.09	.30	.138
	4 in.					
40	0.1383	110.40	14.99	3.68	.53	.254
50	.208	**5 in.**	23.00	5.61	0.80	.393
60	.3	0.1156	32.95	8.88	1.155	.555
80	.581	.185	58.45	14.55	2.08	0.948
100	0.763	.277	**6 in.**	21.75	3.01	1.478
150	1.59	0.578	0.231	48.76	6.55	3.12
200	2.82	0.972	.393	**8 in.**	11.54	5.50
250	4.37	1.504	.601	0.162	17.84	8.55
300	6.15	2.15	0.85	.208	25.76	11.63
350	8.44	2.914	1.15	.2775		16.4
400	10.92	3.72	1.50	.37	**10 in.**	21.36
450	13.88	4.62	1.87	.462	0.1618	
						12 in.
500	17.16	5.55	2.22	0.5785	.204	0.0925
750	**14 in.**	11.28	5.11	1.224	.416	.1843
1,000	0.1432		8.98	2.17	0.74	.304
1,250	.22	**15 in.**	13.86	3.378	1.132	.462
1,500	.306	0.22	**18 in.**	4.84	1.618	0.67
2,000	.541	.38	0.195	8.70	2.842	1.132
2,500	0.8365	.575	.295	**20 in.**	4.43	1.78
3,000	1.19	0.79	.405	0.25	6.30	2.567
3,500	1.61	1.045	.53	.325	8.52	4.18
4,000	2.10	1.445	.68	.415	**26 in.**	4.45
4,500	2.65	1.67	0.84	.51	0.14	5.55
5,000	3.3	2.03	1.02	.61	.22	6.65
6,000	4.8	2.9	1.425	0.87	.241	**28 in.**
7,500	**30 in.**	4.4	2.175	1.3	.363	0.252
10,000	0.315		3.67	2.245	0.62	0.43

(International Harvester Co.)

* For fittings, see Table 12-9.

† For iron pipe 15 years old multiply tabulated head loss by 1.4; for pipe 25 years old by 1.7.

where:

8.34 = weight of a U. S. gallon of fresh water at about 56 F

G = rate of water flow required, gpm

H = dynamic head as determined in step 1, ft of water

33,000 = ft-lb per minute-hp

E = pump efficiency, per cent*/100

The value from Eq. **3** must be equal to or less than that from Eq. **1**.

A comprehensive utility company specification is available for spark-ignited natural gas engines and accessories to drive deep well, turbine-type, 50 to 600 bhp irrigation pumps.[37] This company maintains its engines at customer sites and sells horsepower-hours measured at the engine drive shaft. The Foxboro Co., manufacturer of an instrument that measures horsepower, claims a ±1 per cent hp-hr indication at 80 F, ±1.5 per cent from 32 to 130 F, with their 48-in.-long (between faces of couplings), 160-lb (less couplings) weatherproof unit.

Specifying Pumps. A wide variety of designs is in use. A comprehensive reference is available.[36]

Water Requirement. About 450 gpm are generally required for 70 to 100 acres; the same amount of water will irrigate up to 160 acres with favorable soil conditions and diversified crops. An irrigation season generally lasts about four months. The water is pumped 24 hr per day for this period. An **acre-foot** of water is the amount of water required to cover one acre, one foot deep; i.e., 43,560 cu ft or 326,000 gal. A **miner's inch** is 9 gpm in California; several states use 12 gpm.

Wells. Irrigation wells range as deep as 2000 ft. Often water enters a well from more than one stratum. Pumping levels are usually at a maximum of 500 ft. Well capacities range up to 6000 gpm.

Oil Field Applications[38]

Oil field operators expect a minimum of four years or 35,000 hr between overhauls. Proper assignment of load factors and speed factors helps to achieve this objective. Equation **1** may be used to establish usable horsepower from maximum engine horsepower.

In general, for oil well pumping:

$$\text{Average bhp} = \frac{\text{bbl in 24 hr} \times \text{pump depth, ft}}{68{,}000}$$

The downtime records of several major oil producers indicate a negligible difference in this regard between gas engines and electric motors.

Gas Engine vs. Electric Motor. Items sometimes neglected in estimating electric energy costs are energy charge, line charge, demand charge, and fuel adjustment clauses.

The operation of lease automatic custody transfer units need not preclude the use of gas engines for other field applications or for generating the electricity required to run the unit. Comparison of gas and electric proposals should include installation, maintenance, and repair costs for electric motors.

Gas fuel cost should be reduced by 12.5 per cent from the wellhead price to reflect the fact that gas used on a lease is not usually subject to the commonly specified 12.5 per cent royalty rate to the landowner.

Injection and Repressuring. Engine exhaust for these gas field purposes may be treated to reduce any NO present to N_2. The Tenex system employs catalytic reduction of the products of slightly rich gas-air mixtures. Exhaust from four-cycle, naturally aspirated engines is the preferred medium since this catalysis is effective in the absence of free oxygen.

API Recommended Practices. Suggestions are available[39] covering installation, including hazard considerations; daily, weekly, and monthly maintenance; and operating troubles. Information generally applicable to reciprocating internal combustion engines, as well as data particularly applicable to oil field applications, is included. An API Standard[40] covers test methods for rating internal combustion engines and unit-type radiator coolers for oil field service.

* Nominally 75 per cent, the product of the following component efficiencies (in per cent): suction and strainer (97), bowl (81), column (97), pump head discharge (98).

Sawmills

Intermittent sawmill power requirements vary with size and skill of crews, size of logs, sizes of lumber being cut, and species and hardness of the wood. Approximately 6 to 7 hp are required per 1000 board feet of daily capacity.

Circular Saws. The approximate power requirement of a circular saw follows:

$$\text{hp} = \text{No. of teeth} \times \text{saw rpm}/360$$

Cotton Gins[41]

Continuous horsepower requirements for gins with complete cleaning and drying equipment (for mechanically harvested or snapped cotton) follow:

Gin size	Required hp
2–70, 2–80	145
3–70, 3–80	220
4–70, 4–80	290
5–70, 5–80	360

About 38 per cent of these values are required for basic gins made up of feeders, distributors, separators, and bailers for hand picked cotton.

Installation Recommendations. Radiation cooling units, if used, should be well screened from lint. A prescreener should be used on the engine air cleaner.

Foundations[42]

Loadings for outdoor mountings should not exceed the soil bearing capacity. Installations within buildings require consideration of space allotment, flotation, vibration isolation, and service accessibility.

Types of Mountings. Five basic mounting arrangements follow:

1. *Flexible*—permits engine to move very freely. These mountings are suitable when the engine connects directly to a driven load, i.e., no clutch, external drive shaft, or belt is used. Examples include generators, pumps, and compressors mounted on engine structures, with the engine flexibly supported.
2. *Semiflexible*—mobile installations that operate while the vehicle is moving.
3. *Rigid*—heavy duty mobile installations, e.g., earth moving equipment.
4. *Subbase*—ranging from a large concrete mass (or cast iron frame) isolated by a layer of cork to a slight, pressed steel frame isolated by rubber or spring supports (Fig. 12-231). The mounting base holds the engine true and imposes no loads of its own. Isolation means that engine vibration is not transferred to the driven load or building structure. Often the driven unit shares the base with the driver.
5. *Isolation*—a somewhat more sophisticated version of subbase designs.

Isolating Materials. A layer of processed natural cork, placed between the engine foundation and the subsoil and between the sides of the engine foundation and the surrounding building foundation, constitutes effective engine foundation system isolation. Alternatively, wet gravel or sand may be substituted for cork.

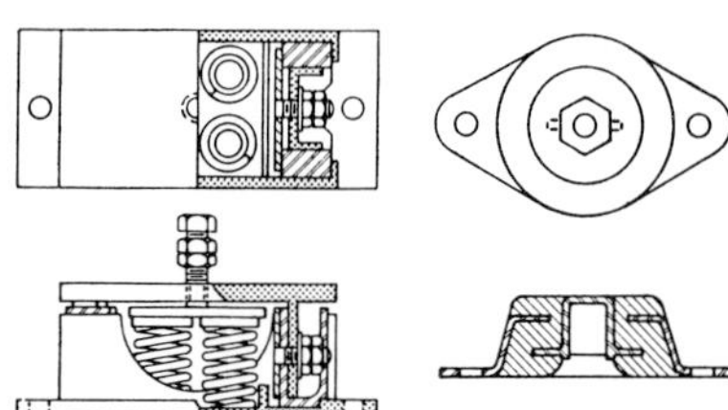

Fig. 12-231 Two types of vibration isolators having special provisions for resisting side loads.[43]

Linear vibration, which is at a minimum in most engine designs, may require the addition of pilings under the foundation where soil bearing capacity is poor. If a foundation rests freely on the subsoil, an elastic system is formed. In such systems the foundation must be nonresonant with any unbalanced forces in the engine. The frequency and magnitude of unbalanced forces in the prime mover-driven load arrangement (e.g., unbalanced assemblies rotating at high speeds), specified by the manufacturer, establishes to a considerable degree what the natural frequency of the installation must be to avoid resonance. In the absence of a foundation specification or suitable experience, a person competent in this area should be consulted, since a poor foundation design can be detrimental to an installation.

In general, linear vibration amplitude varies inversely with system speed. The selection of vibration isolators must take this phenomenon into account, especially where speed modulation is involved.

Torsional Vibration and Critical Speed. These complex phenomena cannot readily be summarized. The combination of a particular engine and load should generally be approved by the manufacturers concerned. A torsional vibration analysis will establish whether resonance exists (or is approached) between any periodic force impulses in the system and the natural frequency of the system. Normally, no harmful (above 5000 psi) torsional vibration stresses should occur within ±10 per cent of rated speed.[26] Such analysis may be performed by the engine manufacturer, the driven equipment manufacturer, or an independent torsional analyst.

Torsional problems are sometimes encountered in the following applications:[14] generator sets, refrigeration units, drilling rigs, air compressors, large impeller pumps, and remotely mounted transmissions. The operating range of variable speed units should be kept as free of vibrations as possible; a specific speed at which harmful stresses occur must be avoided. Any system alterations should reflect consideration of torsional vibration and critical speed.

Specifying Foundations. A checklist for specifying foundations follows:

1. Observe the engine manufacturer's recommendations.
2. Use the manufacturer's certified prints for engine mounting dimensions and outline.
3. Establish the bearing capacity of the soil (use Table 5-20 for reference only).
4. Grout the engine package to the foundation using a nonshrink material (e.g., an epoxy resin with plasticized base) for uniform bearing load distribution.

5. Isolate the foundation from floor and building foundations.
6. Check resonant frequencies if the foundation is to be set on rock.
7. Place vibration isolators between skid and block (for skid-mounted units).
8. Consider foundations with hollow spaces at the four corners[42] that will permit subsequent filling, if necessary, to change the natural frequency of the installation.

Types of Foundations.[42] There are two types. The first is the *conventional concrete* type, which is poured directly on the ground, with or without piling; a frequency ratio of less than 0.5 (preferably under 0.3) is used. (The frequency ratio is equal to the frequency of the exciting force divided by the natural frequency of the engine-foundation-ground system.) The second type is *resiliently mounted concrete*, with the concrete block resting on springs, rubber cork, or felt; a frequency ratio of 2 to 3 is common.

The natural frequency, f_x, in cps, for the conventional concrete type in the vertical direction is:*

$$f_x = 0.028 \frac{(KA)^{0.5}}{w + w_s}$$

where: K = subgrade modulus (from soil lab), lb per cu ft
A = area of foundation contact plane, sq ft
w = weight of engine plus concrete foundation, lb
w_s = weight of soil that participates in vibration, lb

It has been suggested[33] that a concrete base should be designed for one to two times the weight of its load (engines, accessories, driven load, exchangers). Lighter designs may be considered where space is limited if vibration can create no problems or if vibration isolators are provided.

Grouting. A recommended method follows:[14]

1. Level the engine with wedges and jackscrews. A maximum of 0.001 in. crankcase distortion is permitted.[44]
2. Tighten anchor bolts to about one-half the torque required.
3. Pump the grout in under the engine or subbase.
4. After the grout has hardened, complete the tightening of the anchor bolts.

Grouting and anchoring of engines to foundations are not always necessary. Many engines for deep well irrigation pumping rest on their foundations only.

Air Requirements

Combustion Air. Considerations include:

1. Two to 5 cfm per bhp, depending on type, design, and size. Two-cycle units consume about 40 per cent more air than four-cycle units; natural aspiration requires more air than turbocharging.
2. Avoiding heated air since power output varies with the factor $[(t_r + 460)/(t + 460)]^{0.5}$, where t_r is the temperature at which the engine is rated and t is the engine air intake temperature, both in °F.
3. Locating intake remote from sources of air contamination.
4. Filtering. Properly sized air cleaners should be installed so that they can be readily inspected (indicators are available) and maintained.

Air Handling Systems. Equipment may include exhaust fans, louvers, shutters, bird screens, and air filters. It is recommended that the total static pressure opposing the fan be 0.35 in. w.g. max.

Shutters and Louvers. These restrictions, which are either manually or motor operated, control the quantity of intake or exhaust air. Vent area should be increased 25 to 50 per cent to account for them. Thermostatically controlled shutters regulate air flow to maintain desired temperature range. In cold climates, louvers should be closed when the engine is shut down, to help maintain engine ambient temperature at a safe level.

Air Cleaners. Clean air minimizes cylinder wear and piston ring fouling. About 90 per cent of valve, piston ring, and cylinder wall wear is due to dust.[45] Both dry and wet cleaners are used. Wet cleaners, if oversized or operated at below their capacity, are generally inefficient; if too small, the resultant oil pullover reduces filter life. Filters may also serve as flame arresters.

Table 12-175a Ventilation Air for Engine Equipment Room (cfm per horsepower)

Engine room air temp rise, exhaust *minus* inlet	Muffler and exhaust pipe insulated or enclosed in ventilated duct; manifold water-cooled	Muffler and exhaust pipe not insulated	Engine, air- or radiator-cooled; heat discharged in engine room*
10°F	140	280	550
20°F	70	140	280
30°F	50	90	180

* Not recommended for continuous operation.

Ventilation Air. In general, sufficient ventilation should be provided to offset heat losses from engine, driven equipment, muffler, and exhaust piping, and to protect against all fuel supply leaks except rupture of the supply line.

Equation 4 may be used to calculate ventilation air requirements:[33]

$$V = UA_T(t_1 - t_2)/1.08(t_2 - t_3) \tag{4}$$

where:

V = ventilation air required, cfm
U = 2 Btu per hr-sq ft-°F
A_T = total engine surface, 2[h(l + w) + lw],† *plus* insulated exhaust piping area, sq ft
t_1 = mean engine surface temperature (180 F is typical), °F
t_2 = engine room temperature (generally between 110 and 120 F), °F
t_3 = maximum outside air temperature (generally between 90 and 105 F), °F
$(t_2 - t_3)$ = 20°F when t_3 = 90 F; 10°F when t_3 = 120 F (generally)

One utility company uses Table 12-175a for minimum ventilation air requirements. This company recommends

* Other modes of vibration should also be considered.

† Cubical dimensions: h—height, l—length, and w—width.

using a ventilated metal hood over the engine. The hood is vented, preferably using an induced draft fan (2 SCFM of air per cu ft per hr of gas input), thru a sleeve surrounding the exhaust pipe. A slight positive pressure is maintained in the engine room.

Engine Cooling

An engine converts fuel to shaft power and heat. Means of dissipating this heat include: (1) jacket water system; (2) exhaust gas, which includes latent heat; (3) lubrication and piston cooling oil; (4) turbocharger and air intercooler; and (5) radiation from engine surfaces.

The manner and amount of heat rejection vary with the type, size, and make of engine, as well as with the extent of engine loading.

Installation. Proceed as follows:

1. Provide a fresh air entrance at least as large as the radiator face; increase the size of the air entrance by 25 to 50 per cent if protective louvers impede air flow.
2. Use auxiliary means, e.g., a hydraulic actuator, to open the louvers blocking the heated air exit since cooling fan pressure is insufficient for this purpose.
3. Control jacket water temperature by operation of the louvers in lieu of a by-pass arrangement.
4. Position the engine so that the face of the radiator is in a direct line with an air exit leeward to the prevailing wind.
5. Provide an easily removable shroud so that exhaust air cannot re-enter the radiator.
6. Separate the units in a multiple installation to avoid air recirculation among them.

The low static heads achieved by propeller-type engine fans preclude their use with long ducts.

Directing the radiator cooling air over the engine promotes good circulation around the latter, permitting it to run somewhat cooler than for air flows in the opposite direction.

Water Cooling. In most indoor installations, heat pump systems, and where noise tolerance is low, heat in the engine coolant is removed by heat exchange with a separate water system. The water in this latter system, often called raw water, may or may not be recirculated. For example, an installation intended for brief stand-by operation is more likely to pass water to waste. Recirculated water is cooled in cooling towers. Water is added to make up for evaporation.

Since the engine coolant travels in a closed loop, it is usually circulated on the shell side of the exchanger. A minimum fouling factor of 0.002 should be assigned to the tube side.

Operating Temperatures. Water jacket outlet and inlet temperature ranges of 175 to 190 F and 165 to 175 F, respectively, are generally recommended. These temperatures are maintained by one or more thermostats that act to by-pass water as required. A 10° to 15°F temperature rise is generally accompanied by a circulating water rate of about 0.5 to 0.7 gpm per engine hp.

Installation. Proceed as follows:

1. Size water piping according to the engine manufacturer's recommendations (also see Table 12-174).
2. Avoid restrictions in the water pump inlet line.
3. Never connect piping rigidly to engine.
4. Provide shutoffs to facilitate maintenance.

Exhaust Systems

Engine exhaust must be safely conveyed from the engine thru piping and any auxiliary equipment to the atmosphere, within an allowable pressure drop and noise level. Limiting **back pressures,** which vary with engine design, run from 2 to 25 in. w.c. (about 6 and 12 in. w.c., respectively, for low- and high-speed engines). Adverse effects of excessive pressure drops include power loss, poor fuel economy, and excessive valve temperatures, all of which result in shortened service life and jacket water overheating.

General Installation Recommendations. These include: (1) locating a high-temperature* flexible connection between engine and exhaust piping; (2) adequately supporting the exhaust system weight downstream of the connector; (3) minimizing the distance between the silencer and the engine; (4) checking for a 30° to 45° tail pipe angle in order to reduce turbulence; (5) specifying tail pipe length (in the absence of other criteria) in *odd* multiples of $12.5T_e^{0.5}/P$, where T_e = temperature of exhaust gas, °R, and P = exhaust frequency, pulses† per sec; (6) specifying an engine-to-silencer pipe length that is 25 per cent of the tail pipe length; (7) maintaining a separate exhaust for each engine to reduce the possibility of backfire on an ignition failure; (8) favoring individual silencers; (9) preventing excessive heat radiation from exhaust piping by means of a ventilated sleeve surrounding the pipe or high-temperature insulation; (10) using welded tube turns with a radius of at least four diameters; (11) specifying a muffler with an outer casing at least equal in strength to that of the exhaust pipe; and (12) providing for thermal expansion in exhaust piping, about 0.09 in. per ft of length.

Flexible Connections. Design and installation features follow:

1. Material. Convoluted steel (Grade 321 stainless steel is recommended) or a strip-wound element, copper or asbestos packed. Stainless steel is favored for interior installations.
2. Location. Principal imposed motion (vibration) should be at right angles to the connector axis.
3. Assembly. Do not stretch or compress the connector (not an expansion joint); secure without bends, offsets, or twisting (use of floating flanges is recommended).
4. Anchor. Rigidly secure exhaust pipe immediately downstream of connector in line with downstream pipe.

Exhaust Piping. Wrought iron or steel pipe of standard weight may be joined by fittings of malleable cast iron. See Table 12-9 for the equivalent resistances of fittings and bends. This table used in conjunction with Table 12-175b determines the exhaust pipe size. The exhaust pipe should be at least as large as the engine exhaust connection.

---

* Exhaust gas temperature does not normally exceed 1000 F; however, 1400 F may be reached for short periods.

† Rpm/120 and rpm/60 for four-stroke and two-stroke engines, respectively, where rpm = engine speed for V-engines with two exhaust manifolds. A second but less desirable exhaust arrangement is a Y-connection with branches entering the single pipe at about 60°; a T-connection should never be used, since the pulses of one bank would interfere with the pulses from the other.

Table 12-175b Minimum Exhaust Pipe Diameter (in Inches) to Limit Engine Exhaust Back Pressure to 8 in. w.c.

Horsepower	Equivalent length of exhaust pipe, ft			
	25	50	75	100
25	2	2	3	3
50	2½	3	3	3½
75	3	3½	3½	4
100	3½	4	4	5
200	4	5	5	6
400	6	6	8	8
600	6	8	8	8
800	8	8	10	10

(Washington Gas Light Co.)

Moisture Considerations. Condensation, which is a problem particularly with long exhaust lines, should be trapped and drained before it can flow back into the engine manifolds. Otherwise, a portion of the gallon of water formed per Mcf of gas burned would run back into the engine during shutdown, corroding the engine and fouling the valves and rings. Hydraulic "lock" is another possibility. Open condensate drains should not be used within buildings.

Turbocharged Engines. A turbocharger reduces exhaust noise level, eliminating the need for mufflers in many installations. When noise restrictions are exceptionally strict, an appropriate muffler can be used to obtain a residential level of silence.

Operation Noise Levels

Installation of gas engine-driven machines indoors, where the background noise level is high, usually requires no special provisions for noise attenuation. Installations in more sensitive areas may be isolated and/or receive sound treatment. Figure 12-232 compares the noise level of an electric motor and that of a gas engine, both loaded and unloaded.

Silencers or mufflers greatly attenuate engine noise. Shell and tube cooling systems are quieter than air-cooled radiators.

Exhaust heat recovery boilers may include provisions for silencing. Designs that can be operated dry eliminate the need for the by-pass arrangement, often otherwise necessary, with separate silencers and boilers.

Basic attenuation suggestions include: (1) turning air intake and exhaust openings away (*up* in most cases) from the potential listener; (2) limiting blade tip speed, if forced draft air cooling is used, to 12,000 fpm for industrial applications, 10,000 for commercial, and 8000 for critical locations; (3) considering the fan shroud and plenum between blades and coils acoustically; (4) isolating (or covering) moving parts, including the unit, from its shelter (where used); (5) properly selecting the gas meter and regulator(s) to prevent "singing."

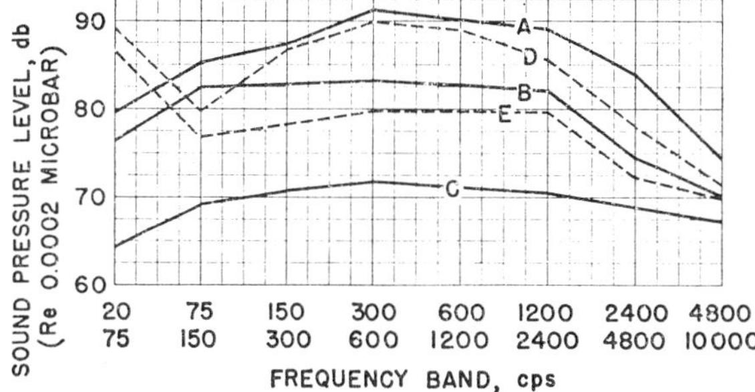

Fig. 12-232 A comparison of the actual sound pressure levels for gas engine-driven and electric motor-driven 100 ton reciprocating units.[46] As explained in the text, sound levels expressed in decibels are not directly additive. Gas engine curves: A, 1800 rpm, full load; B, 1800 rpm, no load; C, 1200 rpm, no load. Electric drive curves: D, full load; E, no load; both 1800 rpm.

Further attenuation means are: (1) insulating the intake and exhaust manifolds with sound-absorbing materials; (2) mounting the unit, particularly a smaller engine, on vibration isolators, thereby reducing foundation vibration; (3) installing a barrier, preferably insulated, between the prime mover and the listener (often a cement block enclosure suffices); (4) enclosing the unit with a cover of absorbing material, since attenuation varies with insulating mass; (5) locating the unit in a building constructed of massive materials, paying particular attention to the acoustics of the ventilating system and doors.

Sound Intensity.[47] Airborne sound, a variation in atmospheric pressure, is measured in decibels. The base of this scale, 0 db, corresponds to a sound pressure of 0.0002 microbar (where 1.0 atm = 1.0×10^6 microbars), which is considered the lowest sound audible to a person with good hearing in a quiet location. Table 12-176a describes various levels of sound intensity. Note that a reduction of 3.0 db will halve a sound level since logarithmic relationships are involved. The determination of an acceptable sound level should take account of the background noise level—the noise permitted in a steel mill exceeds that acceptable in a school.

Table 12-176a Various Levels of Sound Intensity

Decibels*	Relative* energy	Typical sound level
170	10^{17}	Large ram jet engine
90	10^{9}	Heavy city traffic, fire siren at 75 ft
40	10^{4}	Public library, quiet office
20	10^{2}	A whisper
0	1	Threshold of hearing

* Decibels = 10 log (relative energy).

Combining Sound Levels.[47] Noise levels, when expressed in decibels, cannot be added directly. Thus, two machines that individually create a noise level of 40 db result in a 43 db level when run together. This result is calculated as follows: The relative energy for one 40 db unit = 10^4 (from Table 12-176a). The relative energy for two 40 db units = 2×10^4, which is the equivalent of about 43 db.

The noise level chargeable to a new noise source in an existing background level, L_1, may be determined by measuring L_2, which is the ambient noise level with the operation of the new noise source, as follows: (a) if $L_2 - L_1 < 3$ db, the new noise is less than the original background noise; (b) if $L_2 - L_1 > 10$ db, the effect of the original background noise is negligible; and (c) between the foregoing conditions, i.e., 3 db $< (L_2 - L_1) <$ 10 db, the noise level caused by a new source is equal to $L_2 - C$, where:

$L_2 - L_1$	3	4	5	6	7	8	9	10
C	3	2.2	1.6	1.2	1	0.8	0.6	0.4

Engines vs. Motors. The overall noise level of a *fully loaded* gas engine-driven machine and a corresponding electric motor-driven unit is shown in Fig. 12-232. A comparison of curves C and E of this figure is indicative of the fact that under partial loading gas engine-driven systems generate less noise than motor-driven systems.

Residential Exhaust Mufflers. Used for critical installations, these units are preferably located *both* out of doors and near the engine. Reportedly, these mufflers can attenuate engine noise so that it is barely audible (30 ft from the exhaust outlet in one case).[48] Separations over 30 ft may result in "pipe bang" at maximum load. Nominal exhaust pipe diameters are about twice the gas supply piping diameter, whereas muffler diameter is about three to four times exhaust diameter.

Suggested means for routing the exhaust pipe between an interior engine installation and a roof-mounted muffler are: (1) thru an existing unused flue in a chimney (or one serving gas appliances only); (2) thru an exterior fireproof wall, with provision for condensate drip to the vertical run; (3) thru the roof provided that a galvanized thimble with flanges having an annular clearance of 4 to 5 in. is used. A clearance of 1 to 2 in. between the flue terminal and the rain cap on the pipe permits the venting of the flue. Only one engine should be connected to an exhaust line. A clearance of 30 in. between the muffler and roof is common. Vent passages and chimneys should be checked for resonance.

When interior mufflers are unavoidable, the practice in foregoing item 3 should be followed, except that the inside of the muffler portion of the flue should be insulated. Flue runs exceeding 25 ft may have to be power vented.

Maintenance

Maintenance, which may be said to be insurance on the investment in an engine, provides continuous and economical engine operation.

Periodic Servicing. The following items should be replaced or restored on an elapsed-time or hours-run basis:

1. *Ignition system.* The life of parts is predictable, and parts costs are relatively insignificant.
2. *Lubricating oil.* Tests are made to determine the safe-use period. Changes on small engines are usually on an hours-run basis.
3. *Oil and fuel filters.* The service period is determined by operating inspections; reasons for any decreaes in life should be determined.
4. *Cooling water.* The need for change should be based on tests.

Careful record should be kept of all servicing done; check-lists should be used for this purpose.

Scheduling. There are three related goals: minimum downtime, maximum engine-part life, and prevention of premature equipment failure. Preventive maintenance on the basis of operating inspection and periodic servicing, which show gradual (e.g., week-to-week) changes in engine condition, is preferred to periodic maintenance and parts replacement. Contaminated or other unfavorable atmospheres may in themselves determine the maintenance program.

Preventive Maintenance. Reportedly, "99% of all [engine] failures are preceded by certain signs, conditions, or indications that these parts were going to fail."[49] A preventive maintenance program, which should be used whenever economically feasible, is based on the analysis of such indicators. It minimizes downtime, avoids dismantling engines for unnecessary inspections, and realizes the maximum life of engine parts.

A preventive maintenance program should include inspections for: (1) leaks (this is a visual inspection facilitated by engine cleanliness); (2) abnormal sounds or smells; (3) unaccountable speed changes; (4) condition of fuel and lubricating oil filters; (5) water and lubricating oil temperatures; (6) individual cylinder compression pressures, useful in indicating blow-by indicator; (7) changes in valve tappet clearance, which indicate the extent of wear in the valve system; (8) lubricating oil condition, including viscosity, acidity, foreign matter (including water), deposits, and consumption rate.

Maintenance Contracts. These are offered by engine manufacturers, service agencies and sometimes by gas utilities. Besides operating inspections and periodic servicing, contracts can call for: (1) reasonable time limit for emergency repairs, with parts furnished by the contractor; (2) confining of work on the engine to the contracting agency; (3) maintenance charge to be a function of fuel consumed,* engine size, and either hours run (Table 12-176b) or lapsed time; (4) contract renewals, specifying engine overhaul at a guaranteed maximum price per cylinder; (5) guaranteed fuel rate and/or horsepower output, stating the conditions and methods of measuring and verifying the measurements; (6) access to the engine at all times; (7) visual inspection by the owner; (8) shutdown of the engine by the owner in event of abnormal performance; (9) limiting of agency responsibility to the engine and its accessories, excluding driven equipment.

Table 12-176b Gas Engine Maintenance Charge Schedule
(various vendors' fees per running hour; reported by a gas utility company in 1962)

Naturally aspirated		Turbocharged	
Size, hp	¢/hr	Size, hp	¢/hr
75–125	25	180–280	38
125–200	30	200–320	42
175–225	34	220–350	48
230–300	35	300–425	50
300–475	45	400–650	60

Component Failures. One analysis[50] of internal combustion engine failures has indicated that two categories account for 58 per cent of all troubles:

1. Bearing failures, 30 per cent. Of these, 43 per cent are on connecting rods, 40 per cent on the main crankshaft, 9 per cent on other engine parts, and 8 per cent on auxiliaries (e.g., turbocharger).
2. Pistons and rings; cylinders and liners and heads, 28 per cent. Of these, 70 per cent involved the overheating, scuffing, and scoring of pistons and cylinder liners.

Lubrication. Manufacturers' recommendations should be followed. In general, both the crankcase oil and oil filter element(s) should be changed at least once every six months. One gas utility recommends three-month periods.

* For example, in 1962, eight cents per Mcf (including small parts, excluding lubricating oil).

Consumption. Since no crankcase oil dilution occurs in a gas engine and since there is no build-up of residues in the crankcase to compensate for oil consumption (such as might occur in a gasoline or diesel engine) it may appear that a natural gas engine uses more lubricating oil than an equivalent liquid fueled engine. However, the extended oil and filter change periods usually result in better oil economy for the natural gas engine. Use of an oil level regulator reduces oil consumption and makes for more effective lubrication.

The following factors influence the length of oil change periods of a particular engine:

1. *Type of oil.* Superior lubricants (Series 3), Supplement I oils, and oils meeting specification MIL-L-2104A are the three basic categories. The service life ratios for these three oil types run about 4:2:1, respectively.

2. *Oil reservoir size.* Engines may be furnished with oil tanks in a variety of sizes. The oil change period varies roughly with tank size. The filter change period, however, may remain fairly constant.

3. *Properties.* Many engine oils contain **detergent** additives that reduce the possibility of oil-oxidation products, fuel soot, resins, and other materials from settling out of the oil and depositing on engine parts. Such detergents have the ability to remove some already deposited material. Engine tests devised by the Caterpillar Tractor Co. are used to define the detergency levels for the three types of oil. The combustion products of **ashless, detergent lubricating oils** are gaseous and hence leave no deposits on valves and pistons. Oxidation inhibitor additives prolong lubrication life.

The *total base number* (TBN) of an oil is a measure of its alkalinity. This index is mainly significant when fuels containing high levels of sulfur are used. The acid content of the combustion products of such fuels require neutralization.

Viscosity, the most important single property of a fuel oil, determines the ease with which an engine may be started in cold weather. Manufacturers' viscosity recommendations reflect design clearances. Viscosity of engine oils is usually defined by SAE numbers or in units of seconds Saybolt universal at specified temperatures. The relationship between these two systems at 210 F follows:

SAE	20	30	40	50
SSU	45–58	58–70	70–85	85–110

Field test kits, available from some engine oil suppliers, generally contain instruments for four tests: viscosity; acidity; presence and quantity of insolubles; and presence and quantity of water. Such kits have been found particularly useful in measuring effectiveness of filtration and in setting periods between drains.

Effect of Rich Mixture on Lubricating Oil.[51] When combustible material is in the exhaust gas, not only fuel is wasted but incomplete combustion permits fuel carbon, or soot, to form. The term "black stack," or dark exhaust gas, is used to describe this condition. Some of this carbon will contribute to piston deposits, ring groove filling, crankcase deposits, more rapid filter plugging, and depletion of detergent, if used. An analysis of crankcase deposits and filter debris in such cases will generally show about 50 per cent of these deposits to be insoluble in benzene. Benzene is soluble with the oil and dissolves oxidation products, leaving fuel carbon, dirt, and traces of engine wear metals. When the benzene insolubles are ashed, about 90 per cent of its weight, fuel carbon, is burned away, leaving only the metal oxides.

Lubricant Nitration.[52] Nitrogen unites with oxygen in engine cylinders to form nitrogen oxide (**nitrogen fixation**) that blows by piston rings and combines with lubrication oil (lubricant nitration) on cylinder walls. These phenomena occur in four-cycle engines; in two-cycle engines, combustion products are exhausted directly and therefore do not easily get into engine oil.

Resultant varnishlike deposits on cylinders, pistons, and rings interfere with lubrication, cause scoring of cylinder walls and piston skirts, increase ring wear, and make rings stick.

Factors other than lubricant nitration, particularly oil oxidation, make oil changes necessary before lubricant nitration is effective. Use of lubricating oils specifically designed for gas engines, in accordance with their manufacturers' recommendations, will avoid any serious nitration effects. If desired, infra-red analyses may be used periodically to evaluate an oil with regard to nitrogen fixation.

If crankcase oil temperature is considered normal at 120–160 F, oil life is halved for every 20°F continuous oil temperature above normal.[13]

Crankcase Breathers. These units reduce contamination of oil by drawing blow-by gases out of the crankcase. The gases are filtered and returned to the intake manifold, where they lubricate the intake valve stem and top of the cylinder. The use of breathers also eliminates condensation under valve covers and helps to prevent spark plug fouling if the engine is frequently idled.

Engine Analyzers. Spring-loaded, piston-type indicators and timing lights are generally adequate for balancing the load among cylinders and ignition timing. The use of electronic engine analyzers permits greater accuracy in these operations. These instrument systems[53,54] convert vibrations, pressures, and actions in the ignition system to electric impulses. The impulse patterns corresponding to the functions under study are projected on the screen of a cathode ray tube. A comparison of these patterns with known standard patterns indicates any malfunctions.

CONTROLS AND ACCESSORIES

More often than not malfunctions in prime mover systems are attributable to a poor selection of a "minor" component. These components are of two kinds: (1) devices that shut down an engine to protect it against mechanical damage, e.g., a shutoff activated by low oil pressure; and (2) devices that interrupt the gas fuel supply.

Codes and Ordinances. The safety requirements set forth by the authority having jurisdiction over the installation must be met. Beyond these basic considerations, however, competent engineering judgment is necessary. The local codes (e.g., zoning ordinances) involved are usually based on state,[55] regional,[56] and national[57] codes. Gas codes covering service at all pressures are available.[3,4]

Explosionproof Controls. NFPA Standard 37, the combustion engine and gas turbine installation standard,[5] indicates the environment and applications under which prime mover installations are considered to be in hazardous locations. The *National Electrical Code*[58] specifies the type of equipment and wiring to be used in certain hazardous locations; however, the Code does *not* define these hazardous locations.

The usual gas engine installation does not require explosion-proof considerations.

Line-Type Gas Pressure Regulator

Turbocharged (and aftercooled) engines, as well as many naturally aspirated units, are equipped with line regulators designed to control the gas pressure to the engine regulator according to the following schedule:

Line regulator	Turbocharged engine	Naturally aspirated engine
Inlet*	14–20 psig	2–50 psig
Outlet†	12–15 psig‡	7–10 in. w.c.

The same regulators, both line and engine, used on naturally aspirated gas engines may be used on turbocharged equipment. The pressure sensed by the engine regulator is the boost pressure in the turbocharger outlet (Fig. 12-225). In other words, regulators used on naturally aspirated engines are vented to the atmosphere (in the carburetor), while regulators on turbocharged equipment are vented to the turbocharger outlet.

Line-type gas pressure regulators are commonly known as service regulators (and occasionally as field regulators). They are located just upstream of the engine regulator§ and ensure that the required pressure range exists at the inlet to the latter control. Although this intermediate regulation does *not* constitute a safety device, it does permit initial regulation (by the gas utility at meter inlet) at a higher outlet pressure, thus affording an extra "cushion" of gas between the line regulator and the meter for both full gas flow at engine start-ups and delivery to any future branches from the same supply line. The engine manufacturer will specify the size, type, orifice size, and other regulator characteristics based on the anticipated gas pressure range.

* Overall ranges, not the variation for individual installations.

† Also inlet to engine regulator.

‡ Turbocharger boost plus 7 to 10 in. w.c., as sensed by the balance line.

§ Sometimes a remote location is specified; check authorities having jurisdiction.

Engine-Type Gas Pressure Regulator

This engine-mounted pressure regulator, also called a carburetor regulator (and sometimes a secondary regulator or a "B" regulator), controls the fuel pressure to the carburetor. Regulator construction may vary with the fuel passed. The unit is similar to a **zero governor.**

Operation. (See Fig. 12-233). At rest, main valve (13) is closed, and carburetor inlet pressure is exerted on both sides of lower diaphragm (14). When the engine is started, the engine manifold vacuum is communicated to the underside of diaphragm (3). The resultant downward movement of this diaphragm opens pilot valve (7), which in turn reduces the pressure above diaphragm (14) and permits main valve (13) to open and pass gas to the carburetor.

The gas flow relieves some of the engine manifold vacuum, as communicated thru passage (11), which partly closes main valve (13). The net result is a slight vacuum at the regulator outlet (e.g., 0.19 in. w.c. vacuum). Shutting down the engine relieves the vacuum, thus shutting the main valve.

Idling. Here the carburetor throttle plate (17) of Fig. 12-234 is nearly closed; thus, insufficient vacuum is applied at outlet (12) of Fig. 12-233. In this event, vacuum

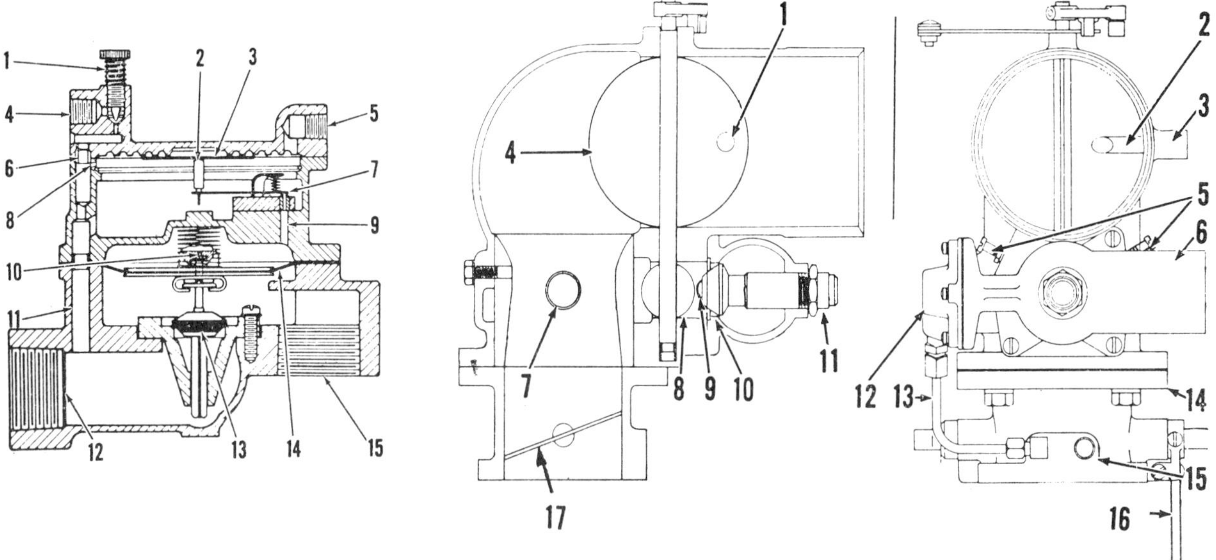

Fig. 12-233 (left) Carburetor regulator: (1) idle adjustment screw, (2) pushrod, (3) diaphragm, (4) idle tube connection, (5) balance line connection, (6) passage, (7) valve, (8) orifice, (9) passage, (10) orifice, (11) passage, (12) outlet, (13) valve, (14) diaphragm, (15) inlet. (Caterpillar Tractor Co.)

Fig. 12-234 (center and right) Gas engine carburetor: (1) air orifice, (2) Pitot tube, (3) balance line connection, (4) air choke disk, (5) starting gas adjustment, (6) gas inlet, (7) venturi nozzle (venturis with various throat dimensions may be substituted), (8) gas choke disk, (9) by-pass opening, (10) main fuel orifice, (11) main gas adjustment screw, (12) economizer, (13) economizer connection line (not on turbocharged engines), (14) flange, (15) idle gas connection, (16) throttle lever, (17) throttle plate. (Caterpillar Tractor Co.)

from the engine side of the carburetor throttle (see item 15 in Fig. 12-234) is communicated to idle tube connection (4) of Fig. 12-233, where it is in turn applied to the underside of diaphragm (3). Part of the fuel also flows to the carburetor thru connection (4).

Balance Line. A connector joining the balance line connection (5) of Fig. 12-233 (the regulator) to the balance line connection (3) of Fig. 12-234 (the carburetor) automatically reduces the flow of gas in correct proportion to any reduction in air flow. Air cleaner restriction is a possible cause of reduced air flow.

Air-Fuel Control

The quantity flow of air-fuel mixtures in definite ratios must be controlled under all of the load and speed conditions required of engines.

Air-Fuel Ratios. High-rated, naturally aspirated, spark-ignited engines require closely controlled air-fuel ratios. The following air rate data generally apply:

Air rate, lb/min-hp (approx)	Result
Under 0.15	Detonation
0.15 to 0.18	Satisfactory
Above 0.18	Misfiring

Excessively lean mixtures cause valve burning, excessive oil consumption, and engine overheating. Engines using pilot oil ignition can run at air rates above 0.18 lb per min-hp without misfiring. Air rates may vary with changes in compression ratio, valve timing, and ambient conditions.

A study[59] of piston ring and cylinder wall wear as related to fuel-air ratios indicated the following: (1) Engines burn lubrication when there is up to 4 per cent O_2 in the exhaust. (2) From 4.0 to 9.0 per cent O_2 in the exhaust (preferably 6.0 to 7.0) permits a good oil film on the cylinder wall and thus results in minimum wear. (3) Exhaust O_2 exceeding 10 per cent causes detrimental backfiring effects.

Exhaust Gas Analyzers. These instruments help to control fuel-air mixtures for minimum downtime, reduced maintenance cost, optimum fuel economy, and high thermal efficiency. Engines tuned by ear may exhaust excessive combustibles caused by overly rich fuel mixtures.

Exhaust gas analyzers have been applied to gas engine-driven transmission pipeline compressors.[51] In two-cycle gas engines, where the scavenging air mingles with the exhaust gases, analyzers show about 10 per cent O_2. However, a trapped exhaust gas sample will generally show no combustibles. Exhaust from turbocharged gas engines will generally contain from 9 to 12 per cent O_2; a trapped sample will contain from 7 to 9 per cent O_2.

The four-cycle, low-speed, naturally aspirated gas engine benefits most from the results indicated by the exhaust gas analyzer. It has been determined that the best operating range on fuel mixtures is effected by *either* 1.0 *or* 5.0 per cent O_2 in the exhaust gas. Under these conditions all the fuel has been burned. Some four-cycle naturally aspirated engines, however, will not carry an overload with 5 per cent O_2 in the exhaust. In this case 1.0 per cent O_2 operation is recommended.

Proper Mixture Settings. The following adjustment procedures are recommended:[8]

1. *High-compression, turbocharged four-cycle gas engine.* Optimum air manifold pressure for any speed, torque, and air temperature is found by reducing the air charging pressure until incipient detonation is encountered and then raising the charging pressure slightly to clear detonation.

2. *Low-compression, turbocharged four-cycle gas engine.* This engine has no "built-in" index as does the high-compression four-cycle engine. It is necessary to obtain performance data and determine manifold pressures that will permit good operation consistent with acceptable fuel consumption, firing pressures, and exhaust temperatures.

3. *Turbocharged two-cycle gas engine.* Optimum manifold settings at higher torques can be determined in the same manner as for the high-compression four-cycle engine. At three-quarters torque and below, the satisfactory operating band is characterized by obvious cylinder misfiring for excess air manifold pressure and detonation or possible floodout due to insufficient air manifold pressure.

4. *Mechanical blower-scavenged two-cycle gas engine.* Optimum manifold pressure at any torque can be determined in the same manner as for the two-cycle turbocharged engine at three-quarters torque and below.

Control Devices. Means of effecting air-fuel control follow:

1. *Mechanical mixer.* This combines the desired quantities of gas and air in a set proportion by opening of its gas and air ports, mechanically actuated by the governor.

2. *Piston valve control.* Mixing is accomplished by the vacuum created in the intake by piston movement. This vacuum actuates a piston valve that opens up gas and air ports in the mixer. The gas flow is adjustable to obtain the best mixture. Quantity flow of mixture is controlled by a governor-actuated butterfly valve (Fig. 12-235).

3. *Dual-fuel control.* The gas admission arrangement (one per cylinder) for one such system is shown in Fig. 12-236. Leaving the admission valve, the gas passes into the air intake elbow. The mixture then enters the cylinder head and goes thru the intake valve after the exhaust valve has closed. A regular diesel fuel injection system is used for pilot oil and 100 per cent diesel operation. The air-fuel ratio regulator is controlled by a governor thru a Select-O-Matic control mecha-

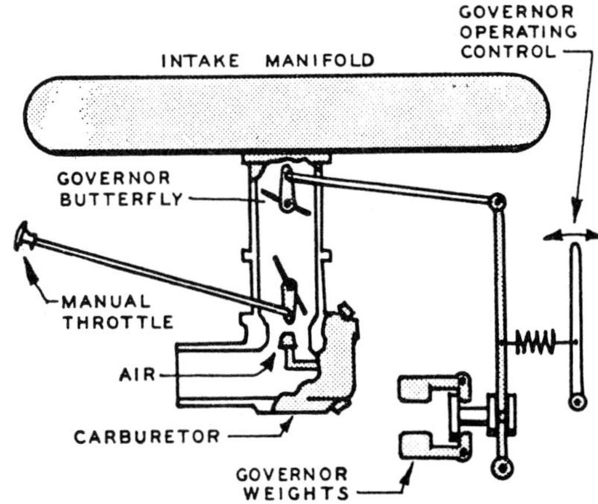

Fig. 12-235 Fuel control for a spark-ignition gas engine.[42] Alternatively, both the manual throttle and the governor may be connected to the same butterfly. The weights of the centrifugal governor are rotated by the engine.

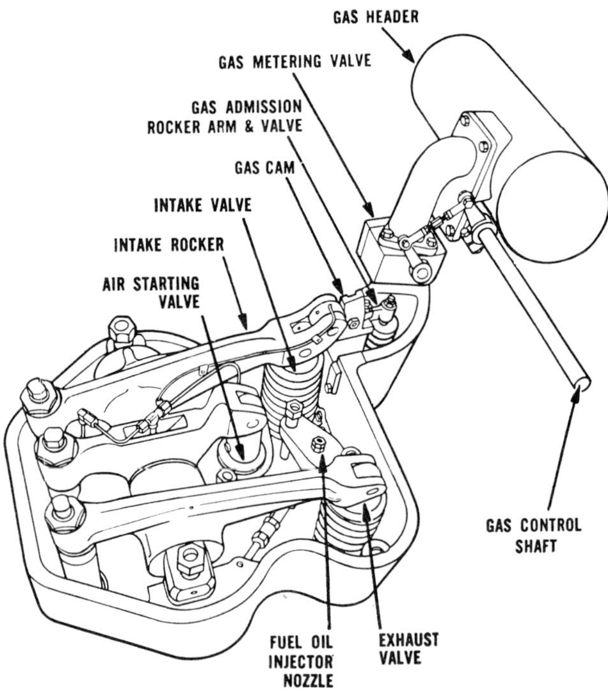

Fig. 12-236 Gas flow control valve for a dual-fuel engine. (Enterprise Engine and Machining Co.)

nism, which sets the engine either on diesel or on one of an infinite number of gas-to-oil ratios (19:1 max) for dual-fuel operation.

4. *Carburetor.* In this venturi-type device, the quantity of mixture is controlled by a governor-actuated butterfly valve (Fig. 12-235). This type of air fuel control has no moving parts other than the butterfly valve.

The motivating force in naturally aspirated engines is the vacuum created by the intake strokes of the pistons; turbocharged engines, on the other hand, supply the additional energy of pressurized air and pressurized fuel. Updraft and downdraft indicate that the carburetor is, respectively, below and above the intake manifold.

Gas Carburetors. These are simpler than their gasoline counterparts. Fuel enters already in a gaseous state. Much of the gas regulating and metering function is done by the engine regulator.* A greater volume flow than the equivalent in liquid fuel calls for larger passages, which are both less subject to plugging and less critical dimensionally, so that finesse in adjustments is unnecessary. A single gas carburetor size is often used with a number of different engine sizes by changing the venturi (7, Fig. 12-234) to match the engine displacement and horsepower since air aspiration varies with venturi size.

Operation. (See Fig. 12-234). During start-up, air choke (4) and gas choke (8) are completely closed; orifice (1) in choke (4), however, passes sufficient air for starting. The starting gas goes thru main fuel orifice (10) and by-passes the opening blocked by choke (8) via opening (9). This system passes enough gas and air to run the engine past idling speed, at which point the main air-fuel orifices replace those used during start-up. Economizer (12) varies the air-fuel ratio with engine load, i.e., provides lean mixtures at part throttle. At low manifold vacuum (below about 4 in. Hg), on the other hand, the economizer supplies richer mixtures both to increase power input and to reduce exhaust temperature. It is often desirable to deactivate economizers on constant-load applications to eliminate unnecessary maintenance.

Spud Type. Here a gasoline carburetor on a small engine is modified for gas service. A connector from the outlet of an engine carburetor regulator (e.g., Fig. 12-233) is secured thru a drilled and tapped hole into the venturi area of the carburetor casting. An adjustable cutoff valve is usually installed between the flexible hose and tube portions of this connector. Gasoline remains as either an alternate fuel or the starting fuel, or both. Its use as the starting fuel is recommended if hand cranking is used.

Gas-Gasoline Type. Here a gas carburetor is modified for gasoline service by adding a float bowl and liquid feed components. A gasoline line shutoff valve and float lock should be provided.

* The engine regulator and carburetor may be in a single unit.

Engine Starters

Sizing. The "breakaway" torque of the driven unit and gearing, if any, must be specified. Sometimes the WR^2 values and a speed-torque curve for these components must also be considered. Usually, however, the WR^2 values of rotating engine parts, principally the flywheel, make the driven load and gear inertia values almost negligible in determining acceleration time.

Ignition

An electrical system or pilot oil ignition may be used. Electrical systems are either low tension (make-and-break) or high tension (jump spark).

Low-Tension Ignition. The relatively low-voltage (90 to 200 volts) electrical energy that is developed in a low-tension magneto winding, ignition or spark coil, or pulse generator is boosted to 20,000 volts by a transformer coil, either one per cylinder bank or one per cylinder. In the latter case, the coil is mounted close to or unitized with the spark plug (Fig. 12-237). In designs with one coil per cylinder, high voltage is handled only in the short lead from the coil to the spark plug.

Advantages of low-tension ignition include: (1) reduction of capacitance, one of the causes of short spark plug life; (2) increase in breaker point life since low voltage is handled at this point; (3) minimizing of the corona effect; (4) increase in spark plug life since shielding is unnecessary (shielding prevents radio interference from high-tension leads); (5) use of

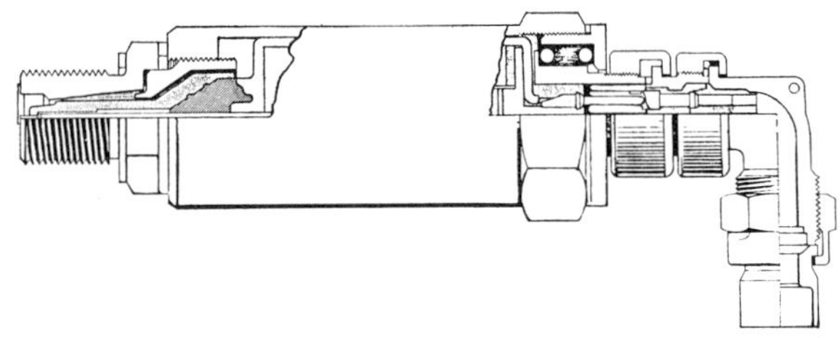

Fig. 12-237 Spark plug and transformer coil assembly. (Scintilla Division, Bendix Corp.)

transistorized circuitry, which is essentially maintenance-free, although not foolproof per se.

Advances in discharge circuitry and transformer design have speeded up sparking use time to less than 5 μsec. Most ignition systems in operation in 1964 had rise times between 25 and 200 μsec.

Pulse Generators. These components of low-tension systems operate without a distributor and without breakers at normal operating speeds from 75 to 1000 rpm. Starting ignition at very low speeds, e.g. 15 rpm, requires breakers. In effect, the breakers *temporarily* convert a pulse generator into a low-tension magneto. The breakers retract automatically and remain inoperative once normal operating speed has been reached.

The only moving parts are an Alnico magnet rotor and its supporting bearings. The turning rotor induces a steep wave front pulse in each of the precisely positioned generating coils in the stator. These coils are in effect a distributor.

The relatively low-voltage, high-current pulses are fed thru low tension cables to the primary winding of high-tension ignition transformers mounted at or near each engine cylinder.

High-Tension Ignition. In these systems, high-voltage sparks are created by transformers somewhat remote from the spark plugs. Either batteries or magnetos (Fig. 12-238) may be used.

Pilot Oil Ignition. Table 12-177 compares spark ignition and compression (pilot oil) ignition, mainly based on one manufacturer's test data. The heat added by pilot oil should run 4 to 7 per cent of the total heat input in a properly maintained installation.[60] Sometimes, however, increased pilot oil consumption is substituted for proper engine tuning.

In **surface ignition engines** ignition is either solely or partially the result of hot surfaces such as an uncooled tube, bulb, or plate, or an electric resistance; sometimes spark ignition is included within the limits of this definition. Hot surfaces in compression ignition engines help to effect combustion at relatively low pressures (about 100 to 300 psi).

Electronic Ignition. In one system (Fig. 12-239)[61] 115 volts ac is stepped up to 240 volts ac thru a transformer and then rectified to direct current by means of silicon rectifiers. The d-c voltage charges a capacitor and supplies bias voltage of −7 volts to the grid and 240 volts to the anode

Tabe 12-177 Ignition System Comparison, Spark vs. Compression[60]

Features	Comparisons
Carburetion	CI uses a simpler control since it tolerates a greater variation in air rate per horsepower.
Interruptible gas	CI systems can be switched automatically to straight diesel operation during periods when gas is unavailable.
Fuel handling	The involvement of an additional fuel (oil) in CI generally justifies SI, except for interruptible gas supply.
System cost	SI systems are more economical.
Exhaust temperature	Somewhat lower with CI since leaner air-fuel mixtures can be tolerated without misfiring (940 F with CI and 975 F with SI in a 15.5- × 22-in. cylinder).
Maintenance	Nozzles are more complicated than spark plugs, but the latter require more frequent adjustment or replacement.
Fuel economy	A 0.5 per cent advantage claimed for CI over SI; however, the higher cost of diesel fuel negates this performance economy.
Energy for ignition	Energy input per cylinder, joules per cycle cylinder: CI, 5000 to 10,000; SI, 0.32 for 4μf capacitor discharge and 0.04 for 0.5μf discharge (generally small and large sparks tend to work equally well); CI fires throughout a volume rather than at a point or two.
Timing	Longer delay, which permits better mixing, with CI (see Energy for ignition).

CI—compression or pilot oil ignition, gas-diesel engine; SI—spark ignition.

of the thyratron tube. When the engine flywheel magnet passes the ignition pulse coil, it generates a voltage pulse in the coil and drives the tube grid positive with respect to the zero potential cathode. This allows the tube to conduct electrons momentarily, in turn allowing the charged capacitor to discharge its energy thru the primary of the ignition coil so as to induce a 27,000-volt potential in the ignition secondary to fire the plugs. Proper timing is effected by the relative positioning of the pulse coil and the flywheel magnet. There are no breaker points in the circuit to erode, and the high-tension voltage does not vary, because the initial current thru the primary is always constant.

Spark Plugs.[62] The following data are based on spark plug service in large gas engines. Normal plug tip temperatures range from 900 to 1200 F. Gases with high sulfur

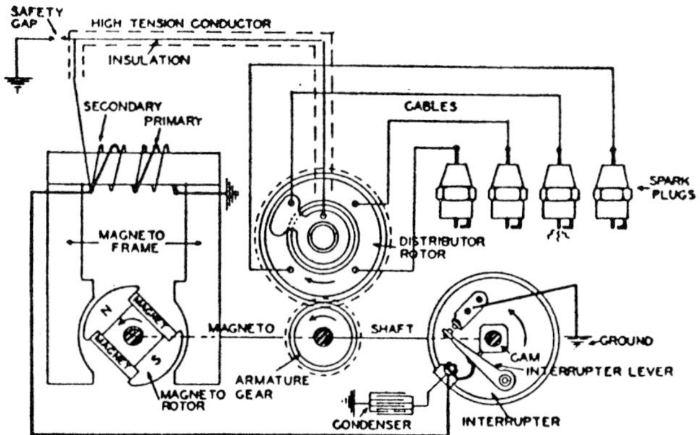

Fig. 12-238 Rotating magnet magneto–high-tension ignition system.

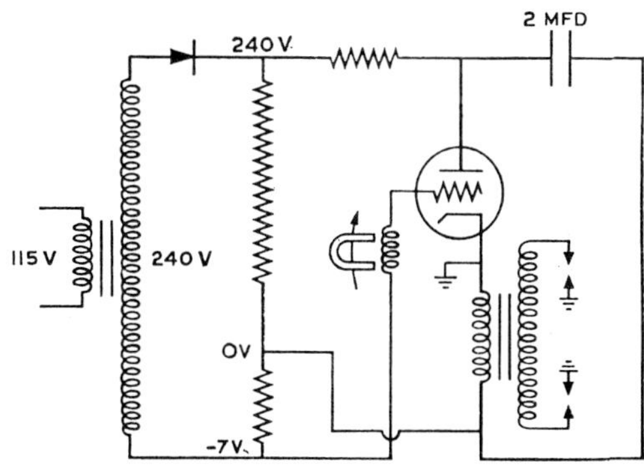

Fig. 12-239 Electronic ignition system used on a 15 hp gas engine.[61]

content require hotter plugs. Above 1600 F, electrodes erode rapidly.

The predominant tip configuration is the four-prong aircraft electrode design. Extended electrodes, often used with sour gases, reduce the possibility of deposits bridging and hence shorting the spark gaps. Heat dissipation of electrodes may be increased by use of high-conductivity cored centers.

The electrical erosion rate of electrodes is a critical parameter. Platinum may erode less than certain nickel alloys; the cost of the former generally dictates servicing such electrodes for reuse. Folded steel gaskets are favored over copper ones.

Shielding a standard (unshielded) ignition system reduces voltage output about 35 per cent. Similarly, painting or grounding a plug lead wire reduces ignition power output about 15 to 25 per cent. The required spark plug firing voltage is about 20 per cent less with a negative center electrode, compared with reverse polarity.

Governors

A governor senses speed (and sometimes load), either directly or indirectly, and acts by means of linkages to control the flow of gas and air thru engine carburetors or other fuel-metering devices to maintain a desired speed. Speed control extends engine life by minimizing forces on engine parts, permits automatic throttle response without operator attention, and prevents destructive overspeeding.

Use of a separate overspeed device (e.g., a maximum speed type of governor, sometimes called an overspeed stop) prevents runaway in the event of a failure that renders the governor inoperative.

Both constant and variable engine speed controls are available. For *constant speed,* the governor is set at a fixed position. These positions may be manually reset.

The adjustment of *variable speed* governors may accommodate speeds below one-half of rated values. Electric generator drives require adjustment only for synchronizing purposes and load balancing. Many use a governor with an air motor speed changer. The air motor operates linkage arms to adjust the set point of the governor. An electric engine speed modulator effects similar control. When operating under variable speed control, the manual governor should be set at slightly above design speed to serve as an overspeed safety control.

Characteristics. Figure 12-240, part A, shows a typical response curve for an industrial engine governor and the terms generally used to describe such curves. Many governors use the speed droop inherent in the operation of a governor mechanism to achieve **stability.**

Droop may be defined as the change in governor speed corresponding to the used portion of the angular movement of the governor output shaft (or servomotor stroke). It is generally expressed as a percentage of governor speed when the governor output shaft is at the end of its travel in the direction to increase fuel.[26]

Engine speed regulation, per cent $= (N_0 - N_r) \times 100/N_r$

where: N_0 = no-load steady-state engine speed, rpm
N_r = rated steady-state engine speed, rpm

Stability. As shown in Fig. 12-240, part B, poor stability means that the governor takes an excessive period of time to cease unstable oscillations and return to a steady-state condition following a disturbance. Note that the illustrations for fair and good stability reflect both the amplitude of the oscillation and the time over which the oscillations extend.

Drift. A stable speed oscillation of small, constant amplitude from a fixed mean value and long period may occur continuously even with a constant load (Fig. 12-240, part C). Such oscillations tend to stabilize the system. The extent of this variation is rarely serious.

Specifications. Basically, the control specification and the required actuating torque (or other form of work) for the gas flow control determine governor selection. A 1.5:1 relationship between stalled work capacity and required work is suggested by one governor manufacturer.

A test code covering speed-governing systems (except emergency or overspeed governors) for internal combustion engine-generator units is available.[63] This code may be used to determine the functional and performance characteristics of industrial and utility type governors specified in AIEE No. 606.[30]

Transient speed variations resulting from load changes can be minimized on some applications by using only part of the available angular movement of the governor output shaft.

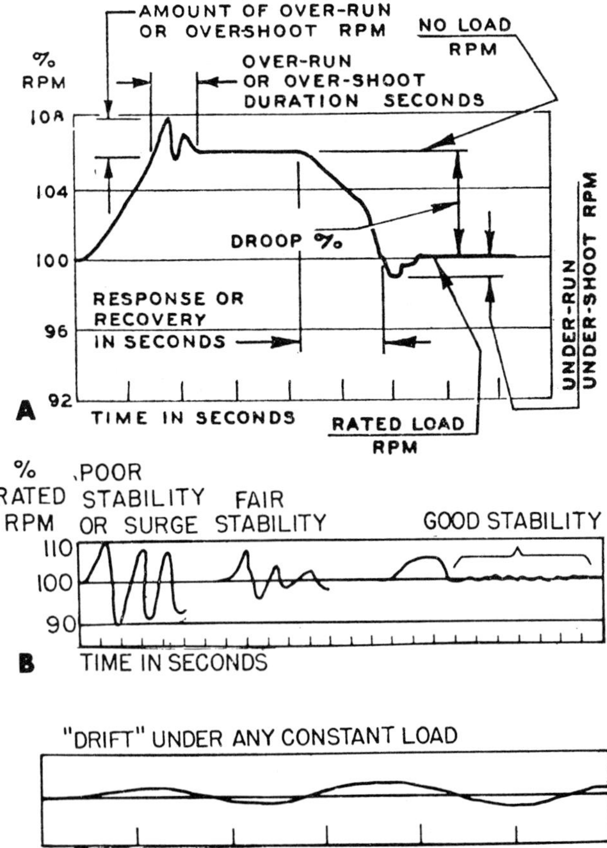

Fig. 12-240 Governing engine speed—terms and concepts:[14] **(A)** Cycle showing changes from steady-state conditions. Sequence: starts at steady-state loading, sudden reduction to no-load, sudden increase in loading, ends at steady-state loading. **(B)** Three stability conditions. **(C)** Drift at steady-state conditions.

For example, as little as 19 of a total of 30 degrees may be used in one hydraulic governor design.

Types. Either nonisochronous or isochronous (i.e., constant-speed, zero speed regulation) governors may be specified. The nonisochronous may be either centrifugally or relay powered; the isochronous are always relay powered. Descriptions of commonly used types of constant-speed gas engine governors follow.[43]

Centrifugal. This is the most frequently used industrial engine governor. It makes use of the centrifugal ball head rotating about an axis at a speed proportional to the engine speed (Fig. 12-235).

Hydraulic. Mechanical governors are used to control a servo or relay mechanism that applies power from another source to the actual fuel control. This power can be electrical, pneumatic, or hydraulic; hydraulic is favored. A hydraulic pump built into the governor gives enough physical power to control the engine. The governor merely moves a small and almost friction-free servo valve to direct the hydraulic flow properly. Some units can be adjusted for isochronous operations over the entire engine load range. Booster servomotors may be used to overcome the lack of oil pressure on engine start-up.

Electrical-Hydraulic. On some applications, e.g., when an engine is used to drive an a-c generator, it may be very important to maintain engine speed, and hence electrical frequency, with great precision. (Esterline Angus charts may be used for accurate plots of frequency vs. per cent of load.) A governor that responds to speed changes may not suffice, because some speed change must take place before the governor initiates corrective action. By using either a load-sensing or a frequency-sensing system (or both in combination) in conjunction with a hydraulic governor, it is possible to initiate corrective fuel compensation for load changes before the engine has time to react significantly to these changes.

Maximum Speed. These units may be directly coupled to throttle a butterfly valve by an overriding arm and permit the user to manage the throttle unless speed exceeds a preset maximum. When the manual throttle is wide open, the engine is "on the governor" and will attempt to maintain governed speed, minus a normal drop of about 10 per cent on most simple installations.

Tail Shaft. These governors sense the speed on the output side of nonpositive drives (e.g., a torque converter).

Instantaneous Load Changes. Abrupt changes, e.g., from 100 per cent of rated output to light or no-load, may tend to cause momentary overspeeding. Such overspeeding, generally expressed as a percentage of the steady-state speed at the time of load change, is beyond the control of a governor since the condition is caused by the fuel already in the manifold and in some of the cylinders. In most cases "step" unloading can overcome this deficiency.

Safety Controls and Considerations

The following devices are "standard" for most gas engines: a low lubrication pressure switch; a high jacket water temperature cutout; and high and low gas pressure cutouts. Other safety controls used include an engine speed governor, an ignition current failure shutdown (battery-type ignition only), and the safety devices associated with a driven machine.

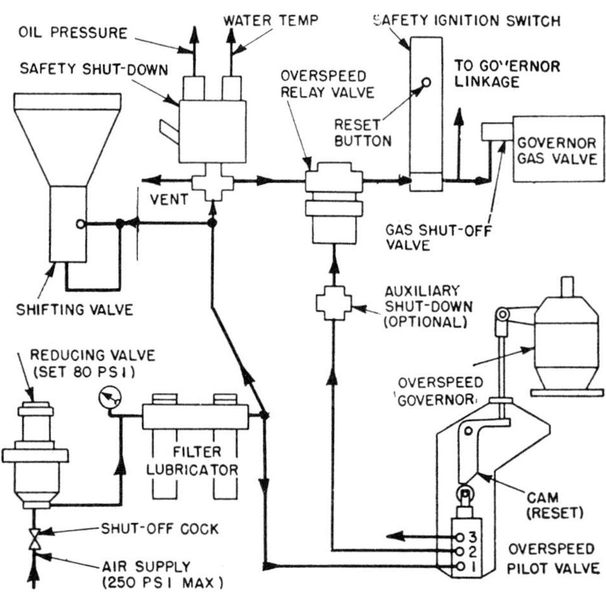

Fig. 12-241a Fulton-Sylphon shutdown control system. (Cooper-Bessemer Corp.)

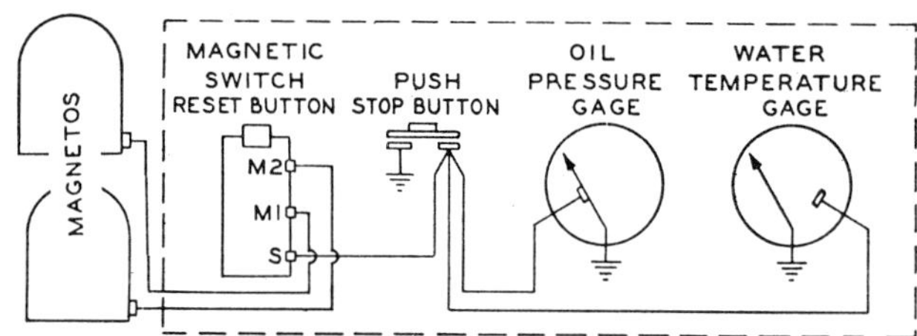

Fig. 12-241b Safety shutoff control wiring diagram. (Caterpillar Tractor Co.)

Figure 12-241a shows a shutdown control system for a large (2500 to 5500 hp) engine.

These devices shut the engine down (see Fig. 12-241b) to protect it against mechanical damage—*they do not necessarily shut off the gas fuel supply unless specifically arranged to do so.* For example, a control may stop one engine and simultaneously energize the cranking circuit of its stand-by.

Gas Leakage Prevention. There are two methods of avoiding gas leakage due to engine regulator failure. The first is the use of a solenoid shutdown valve with a positive cutoff, installed either upstream or downstream of the engine regulator. The second method is the use of a sealed combustion system that carries any leakage gas directly to the outdoors, i.e., all combustion air is ducted to the engine directly from the outdoors.*

Crankcase Explosion Relief Valve. This device is occasionally specified, particularly for larger engines. When this valve is used, the Chowning Regulator Corp. recommends (1) a minimum of 1.5 sq in. of relief area per cu ft of crankcase volume; and (2) a flame trap consisting of an internal oil-wetted screen that has an area at least twice that of the fully open valve area.

* The possibility of gas leakage thru the crankcase breather vent should, however, be considered.

REFERENCES

1. *Diesel and Gas Engine Catalog.* Milwaukee, Wisc., annual.
2. "Energy System Design Survey." *Power* 108: S1–55, Oct. 1964.
3. A.G.A. Laboratories. *American Standard Installation of Gas Appliances and Gas Piping.* (ASA Z21.30) Cleveland, Ohio, 1964.
4. ——. *Proposed American Standard for Installation of Consumer-Owned Gas Piping and Gas Equipment.* (ASA Z83.1–4th draft) Cleveland, Ohio, 1964.
5. Natl. Fire Protection Assoc. *Combustion Engines and Gas Turbines.* (NFPA 37) Boston, Mass., 1963.
6. ASME. *1960–61 Report on Oil and Gas Engine Power Costs.* New York, 1962.
7. D'Amour, R. A. *Natural Gas Engines as Prime Movers for Air Conditioning.* (SAE Paper 876A) New York, Soc. of Automotive Engineers, 1964.
8. Fellows, F. H., and others. *Controls Extend Stable and Efficient Operating Ranges of Gas Engines.* (ASME Paper 62-PET-5) New York, ASME, 1962.
9. ASME. *Internal Combustion Engines.* (Power Test Code 17) New York, 1957.
10. Natl. Electrical Manufacturers Assoc., 155 East 44th St., New York, N. Y. 10017.
11. Soc. of Automotive Engineers, 485 Lexington Ave., New York, N. Y. 10017.
12. Diesel Engine Manufacturers Assoc., 122 East 42nd St., New York, N. Y. 10017.
13. Yelton, F. L. *Lubricating Oils as Developed for Crankcase and Power Cylinders of Natural-Gas Engines.* (ASME Paper 65–OGP-8) New York, ASME, 1965.
14. Internal Combustion Engine Inst. *Installation Practices for Internal Combustion Engines.* Chicago, Ill., 1962.
15. Gill, Jack. "Developments in Medium Speed Turbocharged Gas Engines." *Diesel Gas Eng. Prog.* 30: 44–7, Oct. 1964.
16. Henderson, R. D., and Hallinan, J. C. *Development of High-Compression-Ratio Gas Engine.* (ASME Paper 59–OGP-5) New York, ASME, 1959.
17. Bullard, H. P. "Engine Operating Characteristics." *Actual Specifying Engr.* 8: G118+, Aug. 1962.
18. Am. Soc. for Testing and Materials. "Tentative Method of Test for Knock Characteristics of Motor Fuels Above 100 Octane Number by the Motor Method." (Standard D1948–63T) *1964 Book of ASTM Standards,* pt. 18. Philadelphia, Pa., 1964.
19. ——. "Tentative Method of Test for Knock Characteristics of Motor Fuels Above 100 Octane Number by the Research Method." (Standard D1656–63T) *1964 Book of ASTM Standards,* pt. 18. Philadelphia, Pa. 1964.
20. ——. "Standard Method of Test for Knock Characteristics of Motor Fuels of 100 Octane Number and Below by the Motor Method." (Standard D357–63) *1964 Book of ASTM Standards,* pt. 18. Philadelphia, Pa., 1964.
21. ——. "Standard Method of Test for Knock Characteristics of Motor Fuels of 100 Octane Number and Below by the Research Method." (Standard D908–63) *1964 Book of ASTM Standards,* pt. 18. Philadelphia, Pa., 1964.
22. Hedrick, H. L. "Gas Engine Application." *Actual Specifying Engr.* 8: G98, Aug. 1962.
23. Barthalon, M. M. *Free Piston Engine for All Fuels—the World-Wide Sigma Experience.* (ASME Paper 61–WA–63) New York, ASME, 1961.
24. "Expansion Engines Gaining New and Wider Applications." *Diesel Gas Eng. Progr.* 30: 30–1, Aug. 1964.
25. Natl. Engine Use Council. *Engine System Reports* (series) Chicago, Ill.
26. Diesel Engine Manufacturers Assoc. *Standard Practices for Low and Medium Speed Stationary Diesel and Gas Engines.* Washington, D. C., 1958.
27. Yates, C. T. "Arizona Public Service Provides a New Look." (In: Natl. Engine Use Council. *Proc.* St. Charles, Ill., 1963).
28. Owens, J. R. "Dollars and Sense of Gas Engines for Irrigation." (PCGA Gas Prime Mover and Air Conditioning Symposium, 1962.) (In: A.G.A. Direct Gas Fired Prime Mover Committee. *Natural Gas Fueled Reciprocating Engine Utility Information Portfolio,* B502. New York, 1962).
29. Kauffmann, W. M. "Turbocharged Gas Engine Control Smoothes A.C. Drive Regulation." *Pipe Line News* 35: 51–2, April, 1963.
30. Am. Inst. of Electrical Engineers. *Recommended Specifications for Speed-Governing of Internal Combustion Engine-Generator Units.* (AIEE 606) New York, 1959.
31. McPartland, J. F. *Electrical Systems for Power and Light.* New York, McGraw-Hill, 1964.
32. Monroe, J. M. "Public Works Power Generation." (In: Natl. Engine Use Council. *Proc.,* p. 27–32. St. Charles, Ill., 1964.)
33. Stryon, J. S. "Specifications and Operating Data for Gas Engine Driven Refrigerating Machines." *Actual Specifying Engr.* 11: 70+, March 1964.
34. Gill, J. H. "Using Natural Gas Engines As Air Conditioning Prime Movers." *Heating, Piping Air Conditioning* 33: 152–6, Nov. 1961.
35. A.G.A. Laboratories. *American Standard Approval Requirements for Gas Engine-Powered Summer Air Conditioning Appliances.* (ASA Z21.40.2) Cleveland, Ohio, 1961.
36. Farm Implement News. *Pump Engineering.* Chicago, Ill.
37. Jorgensen, E. M. *Operation of Natural Gas Engines by Arizona Public Service Company.* (SAE Paper 554B) New York, Soc. of Automotive Engineers, 1962.
38. Sahlen, L. O. "Engine Power vs. Electric Power in the Oil Fields." (In: Natl. Engine Use Council. *Proc.,* p. 1–26. St. Charles, Ill., 1964.)
39. Am. Petrol. Inst. *API Recommended Practice for Installation, Maintenance, and Operation of Internal-Combustion Engines.* (API RP7C–11F) New York, 1960.
40. ——. *Specification for Internal-Combustion Engines and Unit-Type Radiator Coolers for Oil-Field Service.* (API 7B–11C) New York, 1956.
41. Intern. Harvester Co. *Power to Gin Cotton.* (IS27) Chicago, Ill.
42. Newcomb, W. K. "Principles of Foundation Design for Engines and Compressors." *ASME Trans.* 73: 307–18, 1951.
43. Waukesha Motor Co. *Waukesha and Climax Installation Manual.* (Form 1846) Waukesha, Ill., 1961.
44. Mallow, J. E. "Preventive Maintenance Practices for Reciprocating Gas Engines." *Pipe Line News* 36: 27–31, Nov. 1964.
45. Calloghan, J. P. "Designing Vent Systems for I-C Engine Room." *Power* 108: 80–3, Sept. 1964.
46. Gamze, M. "A Consulting Engineer's View of Engine-Driven Air Conditioning." (In: Natl. Engine Use Council. *Proc.,* p. 75–90. St. Charles, Ill., 1964.)
47. Peterson, A. P. G., and Gross, E. E., Jr. *Handbook of Noise Measurement,* 4th ed. West Concord, Mass., General Radio Co., 1960.
48. Conrad, J. C. "Natural Gas Engine Drive for Centrifugal Compressors." *Heating, Piping Air Conditioning* 34: 111–6, Dec. 1962.
49. Caldwell, J. H. *Practical Preventive Maintenance for Gas Engines,* 2nd ed. Mount Vernon, Ohio, Kokosing Press, 1951.
50. Shoephoester, K. F. "Internal Combustion Engines Do Fail." *The Locomotive* (Hartford Steam Boiler Inspection and Insurance Co., Hartford, Conn.), July 1962.
51. Mastin, R. G. "Cut a Gas Engine's Operating Costs $3,400 a Year." *Pipe Line Ind.* 19: 49–56, Nov. 1963.
52. Schnack, D. D. "Control of Lubricant Nitration." *Diesel Gas Eng. Progr.* 29: 46–7, Dec. 1963.
53. Gabriles, G. A. "Application and Results of the Use of Engine Analyzers." (In: Oklahoma Univ. *Gas Compressor Engines Short Course,* chap. II. Norman, Okla., 1961.)
54. Southard, D. F. "Electronic Engine Analyzing." (In: Oklahoma Univ. *Gas Compressor Engines Short Course,* sect. E. Norman, Okla., 1962.)

55. New York State Div. of Housing. *State Building Construction Code*. Albany, N. Y., periodically revised.
56. Southern Building Code Congress. *Southern Standard Building Code*, rev. ed. Birmingham, Ala., 1961.
57. Natl. Board of Fire Underwriters. *National Building Code*. New York, 1955–63.
58. Natl. Fire Protection Assoc. *National Electrical Code*. (NFPA 70) Boston, 1962.
59. Smith, C. J. "Getting the Most from Gas Engine Lubricants." *Diesel Gas Eng. Progr*. 28: 44, Nov. 1962.
60. Ulrey, L. S. "Gas Engine Ignition Today." *Gas Age* 131: 19–23, Sept. 1964.
61. "Compact Engine-Compressor Package for Domestic Air Conditioning." *Diesel Gas Eng. Progr*. 28: 24–5, April 1962.
62. Nielsen, R. E. "Spark Plugs in Large Stationary Gas Engines." *Gas* 39: 100–2, Jan. 1963.
63. ASME. *Speed-Governing Systems for Internal Combustion Engine-Generator Units*. (Power Test Code 26) New York, 1962.

Chapter 25

Gas Turbines

by W. Roger Sarno

INTRODUCTION

The gas turbine prime mover combines many of the better features of other engines with advantages of its own. These include low unit weight per horsepower, an absence of reciprocating parts and their inherent unbalanced forces, and the ability to use a wide variety of fuels.

Turbine applications can be divided into two basic groups. The first comprises those applications in which a specific feature of the turbine is of decisive importance; e.g., short time interval from cold start to full load operation; ability to operate without water at the site (a completely "dry" installation); or operating speed closely matched to the speed of the driven equipment, thus eliminating expensive gearing. The second group comprises those applications in which the turbine is evaluated as "just another engine" against competitive equipment.

There are a number of turbines installed in *total energy (TE) systems*, providing power and thermal energy for use on the customer's premises. Although the current (1965) economic position of the gas turbine can often compare favorably with that of the reciprocating engine, it is possible to further improve the competitive position of the turbine by the use of heat recovery devices.

Engineering-economy considerations are overriding in any analysis, but it is interesting to note from the viewpoint of conservation of resources that the conversion of the energy content of the fuel to more useful forms of power and heat energy can easily exceed 65 per cent in the TE system, whereas the effective conversion of fuel energy to electricity does not exceed 40 per cent even in the very largest utility generating stations.

The portions of this chapter that deal with accessories, system components, installation requirements, and maintenance and overhaul are adapted from the *Gas Turbine Manual* written by the A.G.A. Prime Mover and Large Tonnage Air Conditioning Sales and Promotion Committee, and published by The Industrial Press in 1965.

GLOSSARY

Availability. The ratio of the time the unit is in use to the total time, or

$$\frac{\text{(total installed hours)} - \text{(planned and forced outages)}}{\text{(total installed hours)}}$$

Bleed (air) turbine. A gas turbine capable of supplying compressed air, which is bled from the engine compressor (also called an *extraction gas turbine*).

Compressor. The mechanical component of a gas turbine in which the pressure of the working medium (gas or air) is increased.

Axial flow type. A compressor in which the flow takes place in an axial direction essentially parallel to the compressor shaft.

Centrifugal type. A compressor in which centrifugal force causes radial flow outward from the compressor shaft.

Combustor. The mechanical component of the gas turbine in which fuel is burned to increase the temperature of the working medium.

Direction of rotation. For a gas turbine, the clockwise or counterclockwise rotation, determined by looking at the face of the gas turbine output shaft coupling.

Effectiveness. The per cent of available heat that is recovered in a regenerator or recuperator, or

$$\frac{\text{air temperature rise}}{\text{available temperature difference}}$$

Exhaust heat recovery. The process of extracting heat from the working medium leaving the gas turbine and transferring it to a second fluid stream or to a product.

Gas producer (gasifier). A gas turbine that produces hot discharge gases and no mechanical output power.

Gas turbine. A rotary prime mover in which a gaseous working medium, usually air, is compressed, heated, and expanded to produce useful power.

Heat consumption. The quantity of heat used per unit of time under specified conditions. It is expressed in Btu per hour based on the higher heating value of the fuel.

Heat rate. The unit heat consumption of the gas turbine. It is expressed in Btu per horsepower-hour or Btu per kilowatt-hour based on the higher heating value of the fuel.

Intercooler. A heat exchanger located between two compressor stages to reduce the air temperature entering the high-pressure compressor stage and thereby reduce the power to drive the compressor.

Normal operating conditions. The following normal operating conditions are used to determine ratings and performance: inlet temperature = 80 F; inlet pressure = exhaust pressure = 14.17 psia = 28.86 in. Hg. The inlet conditions shall be measured at the engine inlet flange; the exhaust condition shall be measured at the engine exhaust flange with simple cycle operation and at the regenerator exhaust flange with regenerative cycle operation.

Open cycle. A cycle in which the working medium enters the gas turbine from the atmosphere and discharges to the atmosphere.

Rated power. There are two classifications: normal rated power and site rated power.

Normal rated power. The stated power of the gas turbine when it is operated under the conditions of 80 F and 14.17 psia at the inlet and discharges to 14.17 psia.

Site rated power. The stated power of the gas turbine when it is operated under specified conditions of compressor inlet temperature, compressor inlet pressure, and gas turbine exhaust pressure. It is measured at or is referred to the output shaft of the gas turbine or to the generator terminals.

Rated speed. The speed of a designated shaft at which rated power is developed.

Regenerative cycle. A cycle in which the working medium passes successively thru the compressor, regenerator (or recuperator), combustor, turbine(s), and regenerator (or recuperator).

Recuperator. A heat exchanger having the exhaust gas and the combustion air streams separated by a thin wall (usually of metal) thru which heat is transferred by conduction.

Regenerator. A heat exchanger having a matrix that is alternately exposed to the exhaust gas stream and the compressed air stream, soaking up heat from the former and rejecting it to the latter.

Reheater. A combustor located between two turbine stages to increase the temperature of the working fluid and the power available from it.

Reliability. The ratio of the time the unit is in use to the time planned for the unit to be in use, or

$$\frac{(\text{total installed hours} - \text{planned outages}) - (\text{forced outages})}{(\text{total installed hours} - \text{planned outages})}$$

Simple cycle. A cycle in which the working medium passes successively thru the compressor(s), combustor, and turbine(s) only.

Single-shaft gas turbine. A turbine in which all the rotating components are mechanically coupled together.

Specific fuel consumption. The quantity of a stated fuel used per unit of work under specified conditions. It is expressed in pounds per horsepower-hour or pounds per kilowatt-hour for liquid fuel operation or in standard cubic feet per horsepower-hour or per kilowatt-hour for gaseous fuel operation. (For fuel consumption in terms of Btu per unit output normally recommended, see *Heat rate.*)

Turbine. The mechanical component of the gas turbine in which the energy of a working medium is converted to mechanical energy by kinetic action on a rotary element.

Axial type. A turbine in which the flow takes place in an axial direction essentially parallel to the turbine shaft.

Radial type. A turbine in which the flow is essentially radially inward toward the turbine shaft.

Turbine thermal efficiency. The ratio of shaft output power (heat equivalent) to the rate at which heat is supplied to the turbine. It is expressed as a percentage based on the higher heating value of the fuel.

Two-shaft gas turbine. A turbine in which the rotating components are arranged on two separate shafts. The shaft connected to the load is the power, or output, shaft; the other shaft is the compressor, or gas producer, shaft.

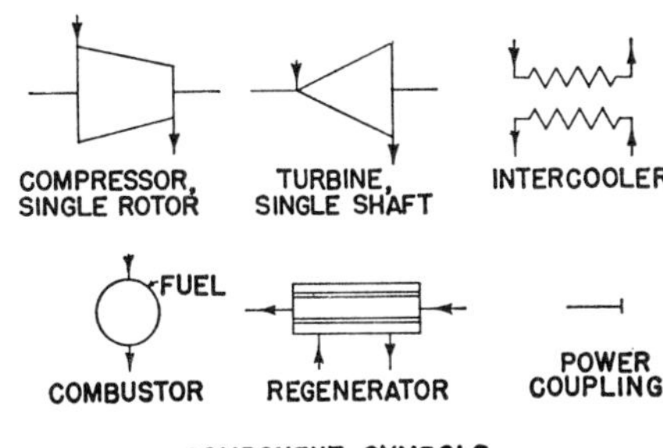

COMPONENT SYMBOLS

THERMODYNAMICS

Reduced to its basic elements, the gas turbine consists of a compressor, a combustor, and a turbine. Together, the compressor and combustor produce a high-energy gas stream that is expanded in the turbine, producing useful power. The process is a continuous flow cycle as contrasted with the essentially "batch" or nonflow processes of the internal combustion and diesel cycles.

Ideal Cycles

Simple Cycle. The ideal cycle which describes the operation of the gas turbine is the *Brayton cycle* (called the *Joule cycle* in Europe). In this cycle there are two constant entropy processes (compression and expansion) and two constant pressure processes (heating the working fluid stream, then cooling it). In the actual simple cycle, the exhaust gases from the turbine are discharged to the atmosphere, and there is no cooling process; the total energy content of the exhaust

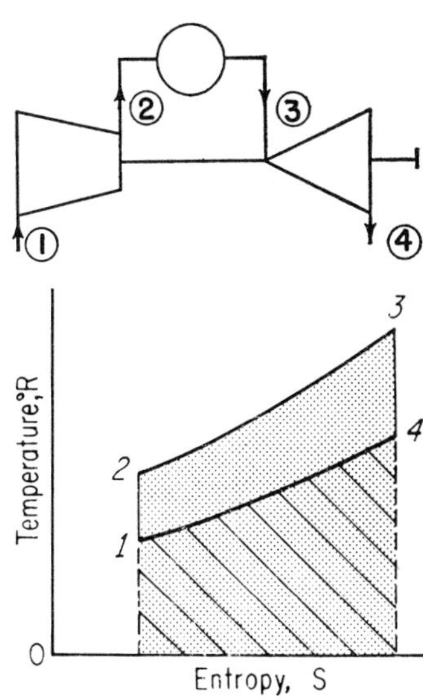

Fig. 12-242 Schematic and T-S diagrams of ideal simple cycl (T-S diagram reproduced by permission from *Power*, Dec. 19€

gases is dissipated. Fig. 12-242 shows the schematic and temperature-entropy (T-S) diagrams of the ideal simple cycle. The areas in the T-S diagram show the quantities of heat transferred during the cycle. The heat input from the fuel is shown by the shaded area under 2-3; the heat rejected in the exhaust gases is shown by the crosshatched area under 4-1. The difference between these two areas (bounded by 1-2-3-4-1) represents the available work output of the cycle (or, more accurately, the heat equivalent of the work output).

Net Work. The net work output of the ideal Brayton cycle is given by Eq. **1**:

$$W_{net} = c_p T_1 (P_r^{(k-1)/k} - 1)\left(\frac{T_3/T_1}{P_r^{(k-1)/k}} - 1\right) \tag{1}$$

where:

W_{net} = the net work output of the cycle, Btu of air flow
c_p = specific heat of air at T_1 and at constant pressure, Btu per lb-F
T_1 = compressor inlet temperature (ambient air temperature), °R
P_r = pressure ratio of the cycle
k = ratio of specific heats of air, approximately 1.4
T_3 = turbine inlet temperature, °R

The net work output is zero at the limits when $P_r = 1$ and $P_r = (T_3/T_1)^{k/(k-1)}$. Between these limits, the net work output reaches a maximum at the pressure ratio defined by Eq. **2**:

$$P_r' = (T_3/T_1)^{k/2(k-1)} \tag{2}$$

where:

P_r' = pressure ratio for maximum net output work between fixed temperature limits T_1 and T_3

Other terms are defined under Eq. **1**.

Cycle Efficiency. The efficiency, η, of the simple ideal Brayton cycle is given by Eq. **3**:

$$\eta = 1 - \frac{1}{P_r^{(k-1)/k}} \tag{3}$$

All terms are defined under Eq. **1**.

Equation **3** shows that the efficiency of the ideal simple Brayton cycle is independent of cycle temperatures and increases with an increasing pressure ratio.

The net work output and the cycle efficiency are plotted against P_r in Fig. 12-243. Note that the point of maximum net work output is at a relatively low pressure ratio compared with the point of maximum efficiency.

Modifications to Simple Cycle. Several modifications can be made to the simple Brayton cycle to improve its performance. These include *regeneration, intercooling,* and *reheating*, or a combination of the three, called the *compound cycle*.

Regenerative Cycle. Regeneration is the recovery of heat from the exhaust gas stream by the compressed air before it flows to the combustor. The efficiency, η, of the ideal regenerative cycle is given by Eq. **4**:

$$\eta = 1 - \frac{P_r^{(k-1)/k}}{(T_3/T_1)} \tag{4}$$

All terms are defined under Eq. **1**.

Here the cycle efficiency decreases as the pressure ratio increases (exactly opposite to the basic Brayton cycle) and is, furthermore, dependent on the temperature ratio. Regeneration improves the efficiency of the cycle up to the pressure ratio at which the temperature of the air leaving the compressor equals the temperature of the exhaust gases leaving the turbine; this pressure ratio can be calculated from Eq. **2**. A plot of regenerative cycle efficiency vs. pressure ratio is shown in Fig. 12-244.

Figure 12-245 shows schematic and T-S diagrams of a regenerative cycle. In the T-S diagram the net output (area 1-2-3-4-1) remains the same as in the simple cycle, but the rejected heat in the exhaust stream is reduced from that of the simple cycle. This is seen in the figure, which shows that the heat recovered from the exhaust stream (area under 4-b)

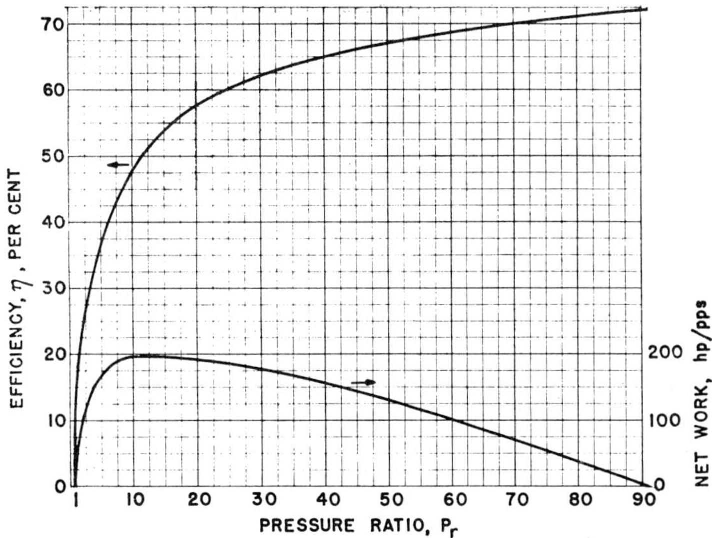

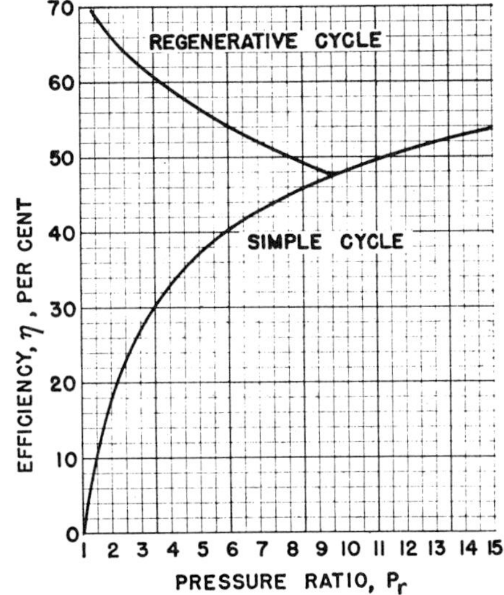

Fig. 12-243 (left) Net work and cycle efficiency vs. pressure ratio in ideal simple cycle (pps = pounds per second).
Fig. 12-244 (right) Cycle efficiency vs. pressure ratio in the ideal regenerative cycle; $t_1 = 80$ F, $t_3 = 1500$ F; $\eta_c = \eta_t = 1.0$.

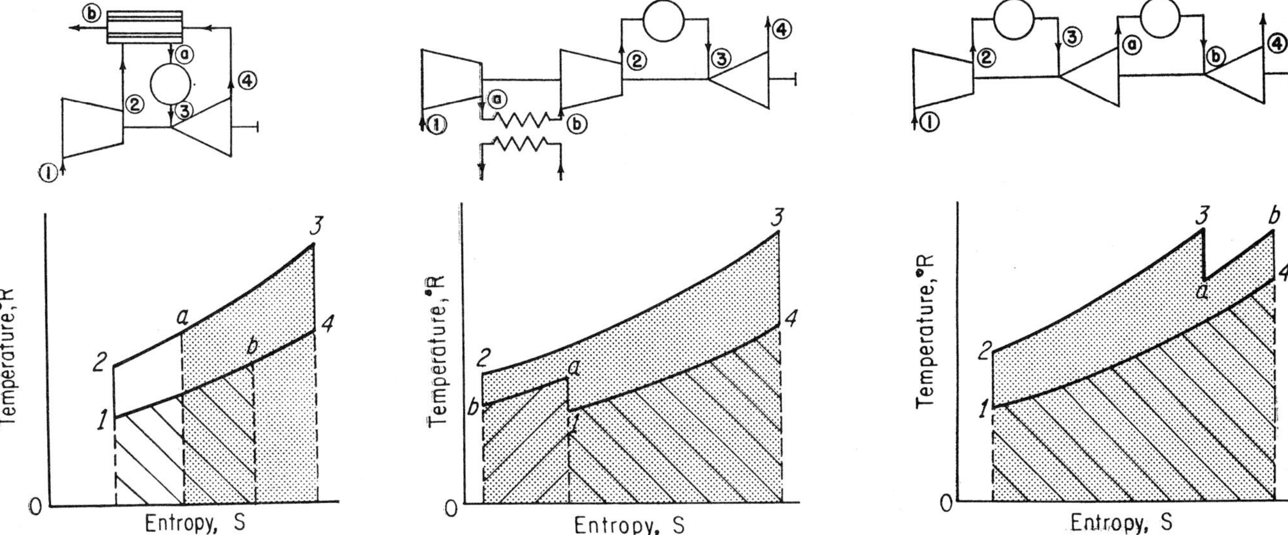

Fig. 12-245 (left) Schematic and T-S diagrams of ideal regenerative cycle.
Fig. 12-246 (center) Schematic and T-S diagrams of ideal intercooled cycle.
Fig. 12-247 (right) Schematic and T-S diagrams of ideal reheat cycle. (T-S diagrams reproduced by permission from *Power*, Dec. 1963.)

equals the heat added to the compressed air stream (area under 2-a). The heat rejected (crosshatched area under b-1) and the heat input (shaded area under a-3) are significantly less than in the simple cycle.

Intercooled Cycle. Intercooling is the cooling of compressed air between stages of compression. Used alone, it offers an increased net work output from the cycle, but it decreases the efficiency. If regeneration is added along with the intercooling, both the efficiency and the net work are improved over the simple cycle. Figure 12-246 shows the schematic and T-S diagrams of an intercooled cycle.

Reheat Cycle. Reheating requires a second combustor between expansion stages. It has the same effects as intercooling, but they are not as pronounced. Used alone, reheating provides an increased net work output at a decrease in cycle efficiency. If regeneration is added, however, both the net work and the efficiency are increased over the simple cycle. Figure 12-247 shows the schematic and T-S diagrams for the reheat cycle.

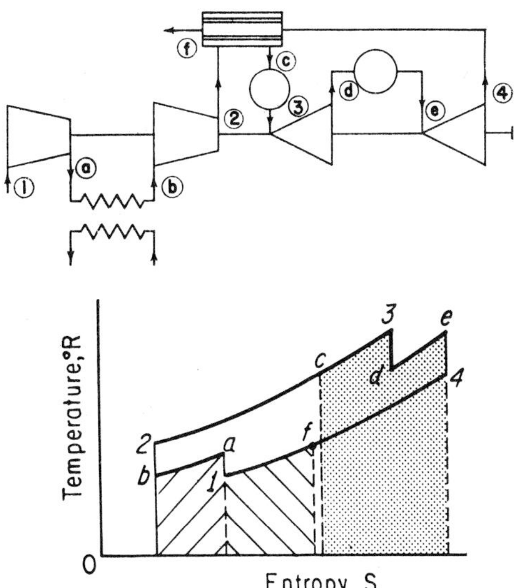

Fig. 12-248 Schematic and T-S diagrams of ideal compound cycle. (T-S diagram reproduced by permission from *Power*, Dec. 1963.)

Compound Cycle. Regeneration, intercooling, and reheating can all be used together in the *compound* cycle. Figure 12-248 shows the schematic and T-S diagrams. In this cycle, the net work output is represented by the area 1-a-b-2-c-3-d-e-4-f-1; the heat input by the shaded area under c-3-d-e; the heat rejected in the intercooler by the crosshatched area under a-b; and the heat rejected in the exhaust stream by the crosshatched area under f-1. The compound cycle, at the higher pressure ratios, achieves the highest efficiency of any cycle; however, it is apt to be found only in the largest sizes because of the amount and complexity of the additional equipment and controls.

The thermal efficiencies of the various ideal cycles are compared in Fig. 12-249. It can be seen that the intercooled regenerative cycle is the most efficient at the lower pressure ratios, but that the compound cycle surpasses it at all normal pressure ratios. The important point is that the ideal regenerative cycle is only *slightly less efficient* than the ideal intercooled regenerative cycle for a substantial range of pressure ratios, suggesting that additional cycle complexity may not be warranted.

Actual Cycles

The work output and efficiency of any actual cycle are less than those of the corresponding ideal cycle. This is due primarily to the effects of component inefficiencies and secondarily to system losses such as pressure losses within the machine, mechanical (friction) losses, changes in the specific heat of air and gases. In the case of regenerative, intercooled, or compound cycles, these secondary losses also include pressure losses in the external portions of the power plant. It is the inherent irreversibility of the compression and expansion processes that causes the largest portion of the deviation from the ideal, setting the pattern of the deviation. The

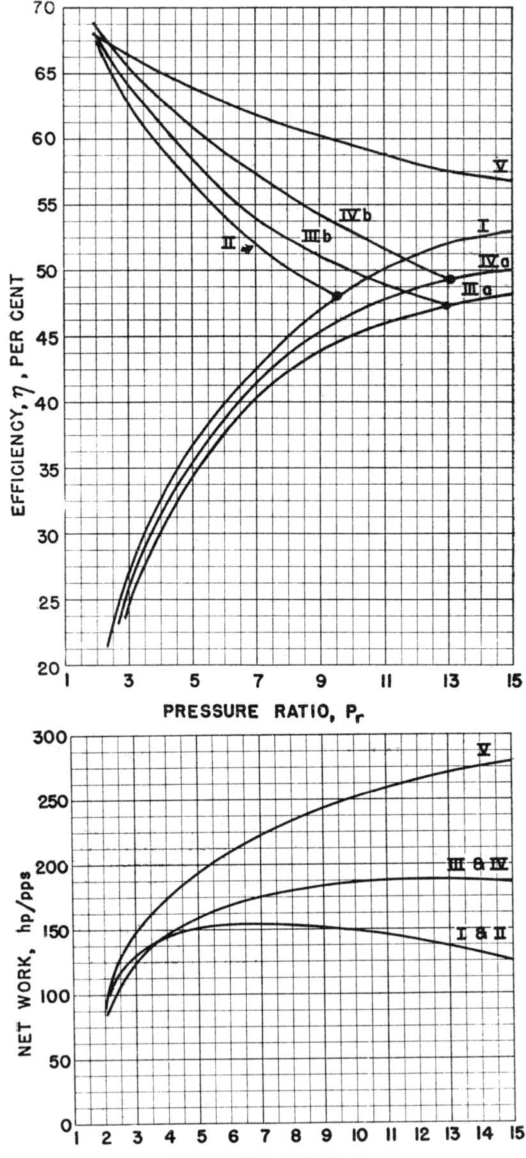

Fig. 12-249 Comparison of net work and cycle efficiency of ideal cycles. Key to curves: (I) simple cycle; (II) regenerative cycle; (IIIa) reheat cycle; (IIIb) reheat cycle with regeneration; (IVa) intercooled cycle; (IVb) intercooled cycle with regeneration; (V) compound cycle.

secondary, or system, losses only modify the pattern in a quantitative manner.

In the following paragraphs, seven variations from a fixed basic simple cycle are explored and their effects shown. A representative set of values was assumed and was used as the "standard" simple cycle against which all the variations were compared. The figures and curves should be interpreted as indicating the effect of variations of the listed items, rather than as giving absolute results.

Ambient Air Temperature and Pressure. Varying ambient conditions affect the gas turbine to a greater degree than they do any other internal combustion engine. Their effect on the turbine is felt chiefly thru their effect on the density of the inlet air. Higher-than-rated inlet temperatures and lower-than-rated inlet pressures reduce the density of

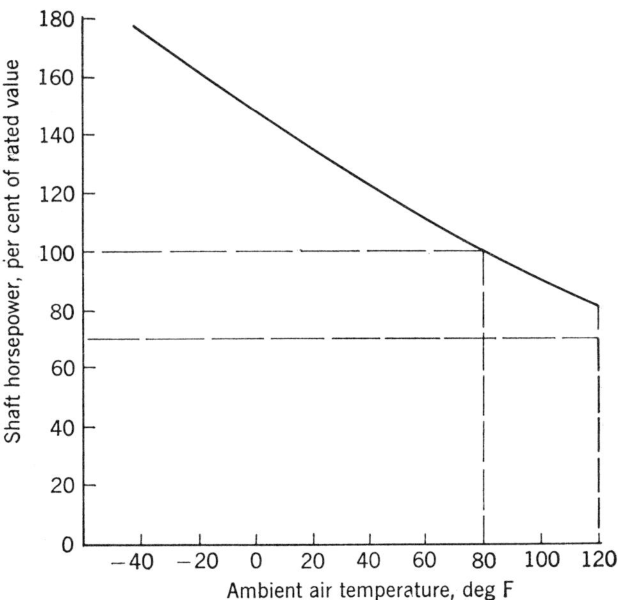

Fig. 12-250 Typical variation of performance with ambient temperature.

the inlet air and consequently the mass flow thru the engine. This reduced mass flow results in a reduced power output. The higher-than-rated ambient air temperatures also reduce the efficiency of the cycle; variations in ambient pressure, however, have virtually no effect on the cycle efficiency.

Figure 12-250 shows the typical variation of performance of a gas turbine as a function of ambient air temperatures. Figure 12-251 shows the typical correction factor curve for the

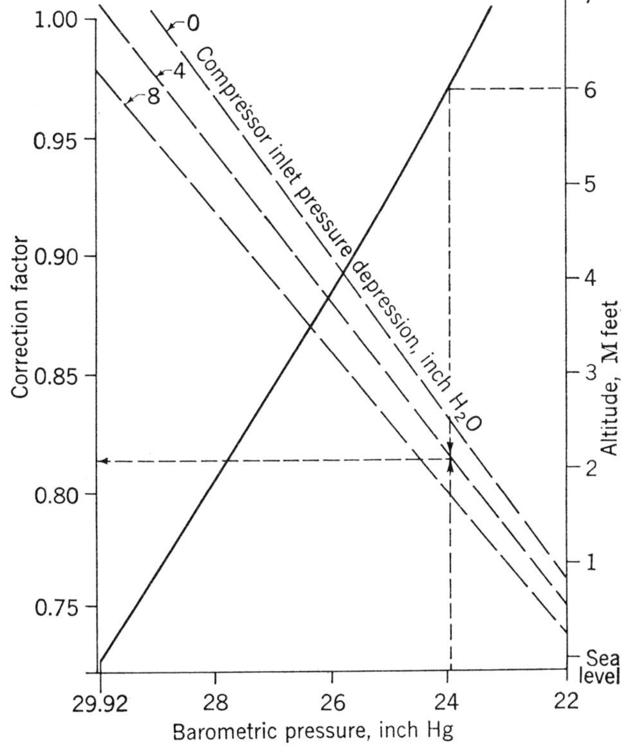

Fig. 12-251 Typical variation in power with ambient pressure.

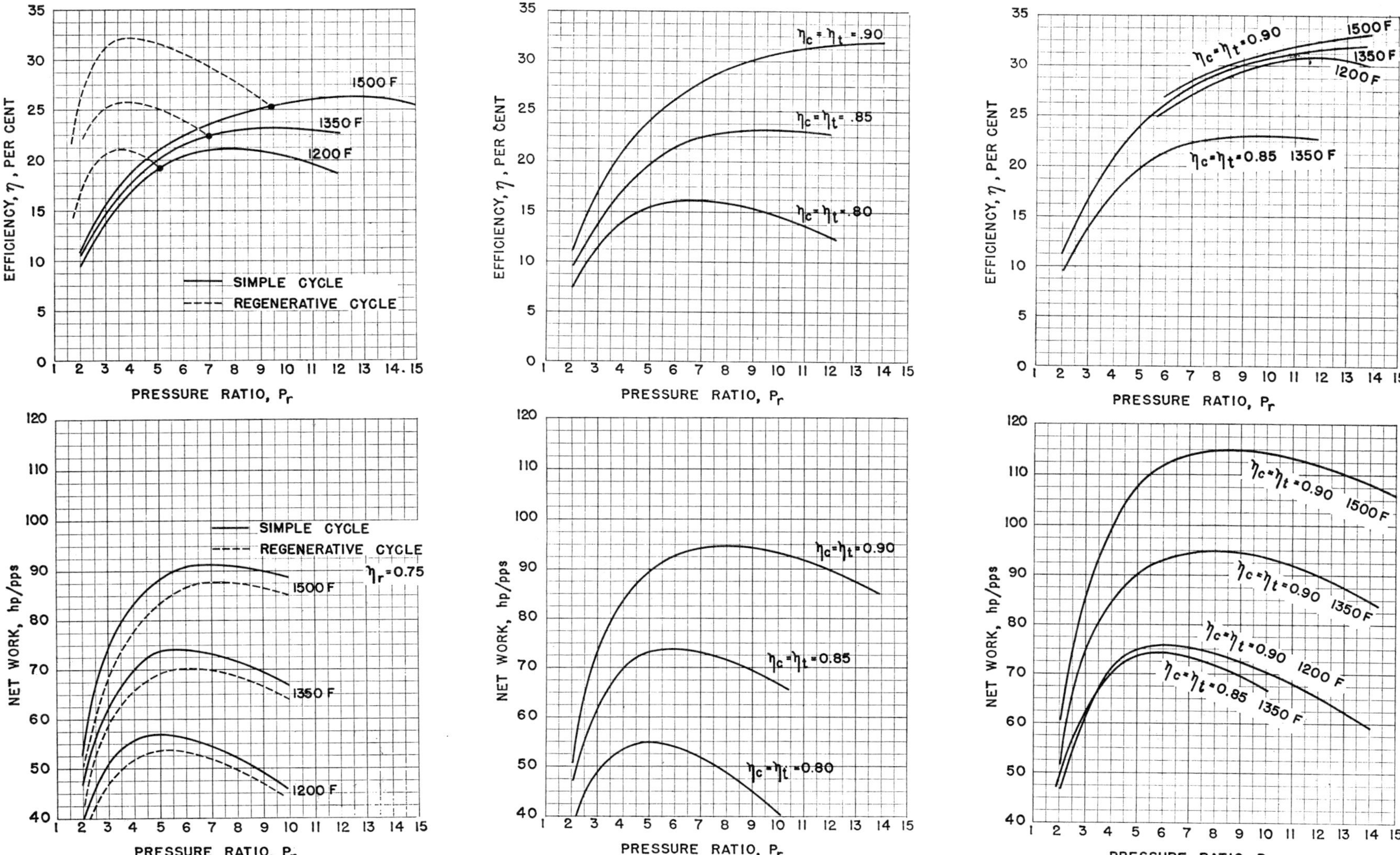

Fig. 12-252 (left) Effect of turbine inlet temperature on net work and cycle efficiency of actual simple and regenerative cycles.
Fig. 12-253 (center) Effect of compressor and turbine efficiency on net work and cycle efficiency of the actual simple cycle. Turbine inlet temperature $t_3 = 1350$ F.
Fig. 12-254 (right) Combined effect of high component efficiencies and turbine inlet temperatures.

turbine power output as a function of ambient air pressure, as well as the effect of inlet pressure depression resulting from inlet screens, filters, ducting, etc.

Turbine Inlet Temperatures. Increasing the temperature of the gas stream at the turbine inlet increases both the work output and the efficiency of the cycle, and also increases the pressure ratios at which the maximum values for the output and efficiency occur. An increased turbine inlet temperature has a greater effect on turbine output than on efficiency, resulting in a persistent drive to ever higher turbine inlet temperatures. Figure 12-252 shows the effect of turbine inlet temperature for the simple and regenerative cycles.

Compressor and Turbine Efficiencies. Irreversibility in compression and expansion manifests itself by a rise in the temperature of the air or gas stream; i.e., the compressor delivery temperature and the turbine exhaust temperature are higher than isentropic conditions would dictate. The extent of this irreversibility is measured by the ratio of ideal work to actual work.

For the compressor:

$$\eta_c = \frac{W_{\text{ideal}}}{W_{\text{actual}}} = \frac{\Delta h_{\text{ideal}}}{\Delta h_{\text{actual}}} = \frac{\bar{c}_p \Delta T_{\text{ideal}}}{\bar{c}_p \Delta T_{\text{actual}}} = \frac{\Delta T_{\text{ideal}}}{\Delta T_{\text{actual}}} \quad (5)$$

where:
η_c = isentropic efficiency of the compressor
ΔT_{ideal} = temperature rise thru the compressor, from gas laws
ΔT_{actual} = temperature rise thru the compressor as measured
$\bar{c}_p$ = average specific heat during compression

For the turbine:

$$\eta_t = \frac{W_{\text{actual}}}{W_{\text{ideal}}} = \frac{\Delta h_{\text{actual}}}{\Delta h_{\text{ideal}}} = \frac{\bar{c}_p \Delta T_{\text{actual}}}{\bar{c}_p \Delta T_{\text{ideal}}} = \frac{\Delta T_{\text{actual}}}{\Delta T_{\text{ideal}}} \quad (6)$$

where:
η_t = isentropic efficiency of the turbine
ΔT_{ideal} = temperature drop thru the turbine, from gas laws
ΔT_{actual} = temperature drop thru the turbine, as measured
$\bar{c}_p$ = average specific heat during expansion

Reduced component efficiencies reduce the net output and efficiency decisively. For a given turbine inlet temperature, the pressure ratio at which the maximum output and efficiency occur is also reduced; the maximum efficiency occurs at a higher pressure ratio than does the maximum work output.

Figure 12-253 shows the effect of different compressor and turbine efficiencies on the simple and regenerative cycles. Note that high pressure ratios are not necessary when low efficiency components are used, because the maximum values of both net output and cycle efficiency occur at relatively low pressure ratios.

Figure 12-254 shows the combined effect of turbine inlet temperatures with component efficiencies. Even with fairly low turbine inlet temperatures, useful performance can be obtained by using high component efficiencies. This approach allows the use of less expensive turbine materials.

Regenerator Effectiveness. In the ideal regenerative cycle (with 100 per cent regeneration) the efficiency is highest at P_r = 1. The efficiency remains higher than that of the simple cycle for pressure ratios up to that calculated from Eq. 2. A regenerator having an effectiveness of 1.0 would have an infinite heat transfer surface, but for practical reasons, the effectiveness of actual regenerators does not greatly exceed 0.75. In exchange for improving the efficiency of the cycle, the regenerator reduces the turbine output because of pressure losses in the regenerator. Usually, the more effective the regenerator, the greater the pressure loss thru it; therefore, the more highly regenerated cycles are more efficient but have greater specific power output losses. Figure 12-255 shows the efficiency and work output of the regenerative cycle for different values of regenerator effectiveness.

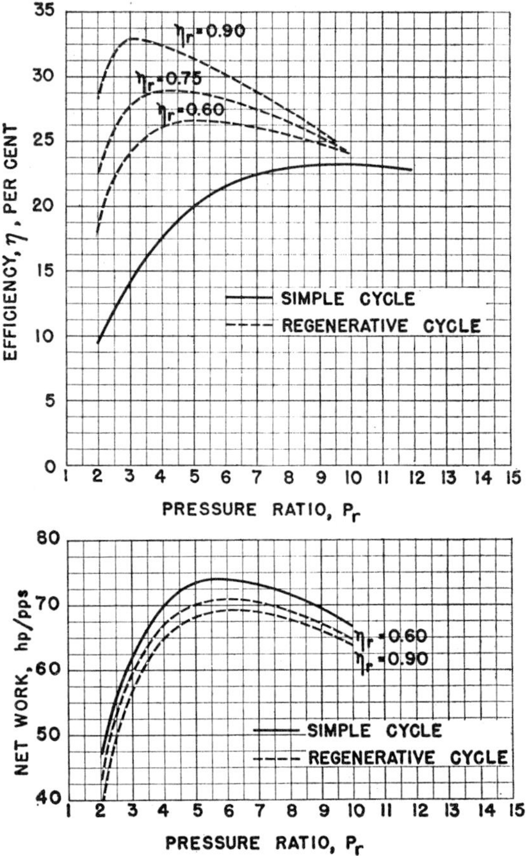

Fig. 12-255 Effect of varying regenerator effectiveness on net work and cycle efficiency. Turbine inlet temperature t_3 = 1350 F.

Intercooling. In the ideal intercooled cycle, the work output is increased at the cost of reduced efficiency. In the actual cycle, the net work increase is retained and the efficiency is increased. The efficiency is only slightly less than the simple cycle at low pressure ratios and crosses at the higher pressure ratios.

The reason for this is that there has been a beneficial tradeoff of saving in compressor work for additional input since compressor inefficiency makes it more desirable to save compressor work than to reduce input. As compressor efficiency approaches 100 per cent, there is less to gain and the cycle performance approaches that of the ideal. This increase in cycle efficiency is not great nor necessarily sought after. The

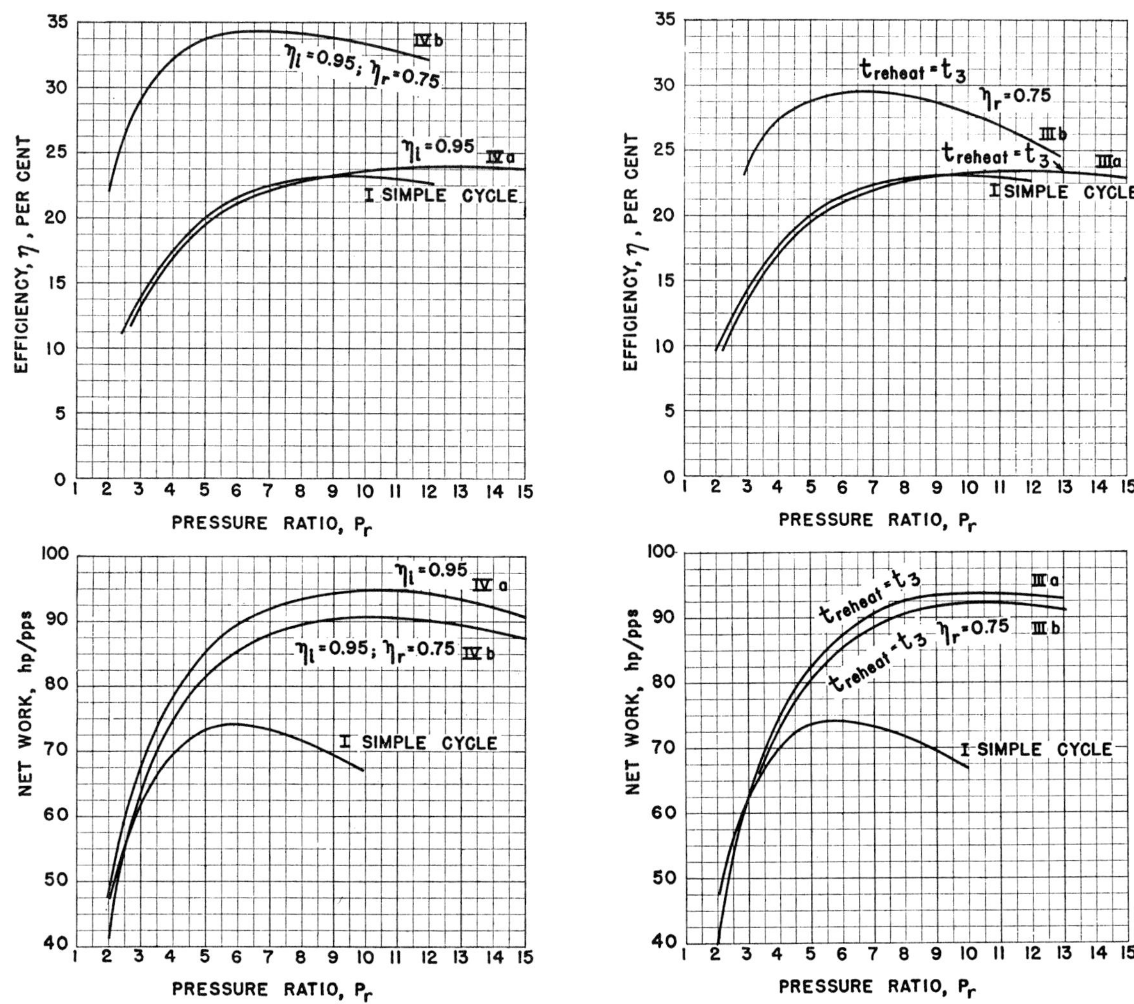

Fig. 12-256 (left) Effect of intercooling on net work and cycle efficiency. Turbine inlet temperature $t_3 = 1350$ F. See Fig. 12-249 for curve identification key.
Fig. 12-257 (right) Effect of reheating on net work and cycle efficiency. Turbine inlet temperature $t_3 = 1350$ F.

chief purpose of intercooling is to increase the work output per pound of mass flow, thus permitting smaller machinery. With intercooling, the pressure ratio must be high to achieve the peak efficiency level. Figure 12-256 shows the efficiency and work output of the intercooled cycle with and without regeneration.

The *intercooled-regenerative* combination gives one of the most effective turbine cycles. Intercooling gives a significant increase in output, while regeneration provides the dual benefit of an increased efficiency and the reduced pressure ratio at which maximum efficiency occurs. As will be seen later, the intercooled-regenerative cycle has the highest efficiency of any cycle at the lower pressure ratios, and at the higher pressure ratios its efficiency is exceeded only by that of the compound cycle.

Intercooling, however attractive, does eliminate a unique feature of the gas turbine: without intercooling it is the only prime mover which can be used without water in all sizes.

Reheat. In the ideal reheat cycle, the same effects are noted as for intercooling, but not to the same extent. In the actual cycle, the net work increase is retained as in the intercooled cycle and the efficiency is improved, closely following the curve for the simple cycle and again crossing it at the higher pressure ratios. As with intercooling, the chief purpose of reheating is to increase the work output per pound of mass flow, permitting smaller machinery. With reheat, the pressure ratio must be high to obtain the efficiency inherent in the cycle. Figure 12-257 shows the efficiency and work output of the reheat cycle, with and without regeneration. Comparing the actual reheat and intercooled cycles, the reheat-regenerative cycle gives results similar to those of the intercooled-regenerative cycle, though not as pronounced. The reheat cycle is not widely used because of combustor design problems, control difficulties, and the high temperature problems that are brought to the physically larger low-pressure stage.

The Compound Cycle. The compound cycle incorporates every input-reducing and output-increasing device so far discussed; it represents an approach to the ideal Ericsson cycle, which is equal in efficiency to the Carnot cycle. The compound cycle has a rather flat efficiency curve over a wide range of pressure ratios and a power output level far higher than any other cycle (Fig. 12-258). The compound cycle also incorporates the disadvantages of all input-reducing and output-increasing devices: it requires water, requires extra care in the design of the reheat combustor and its controls,

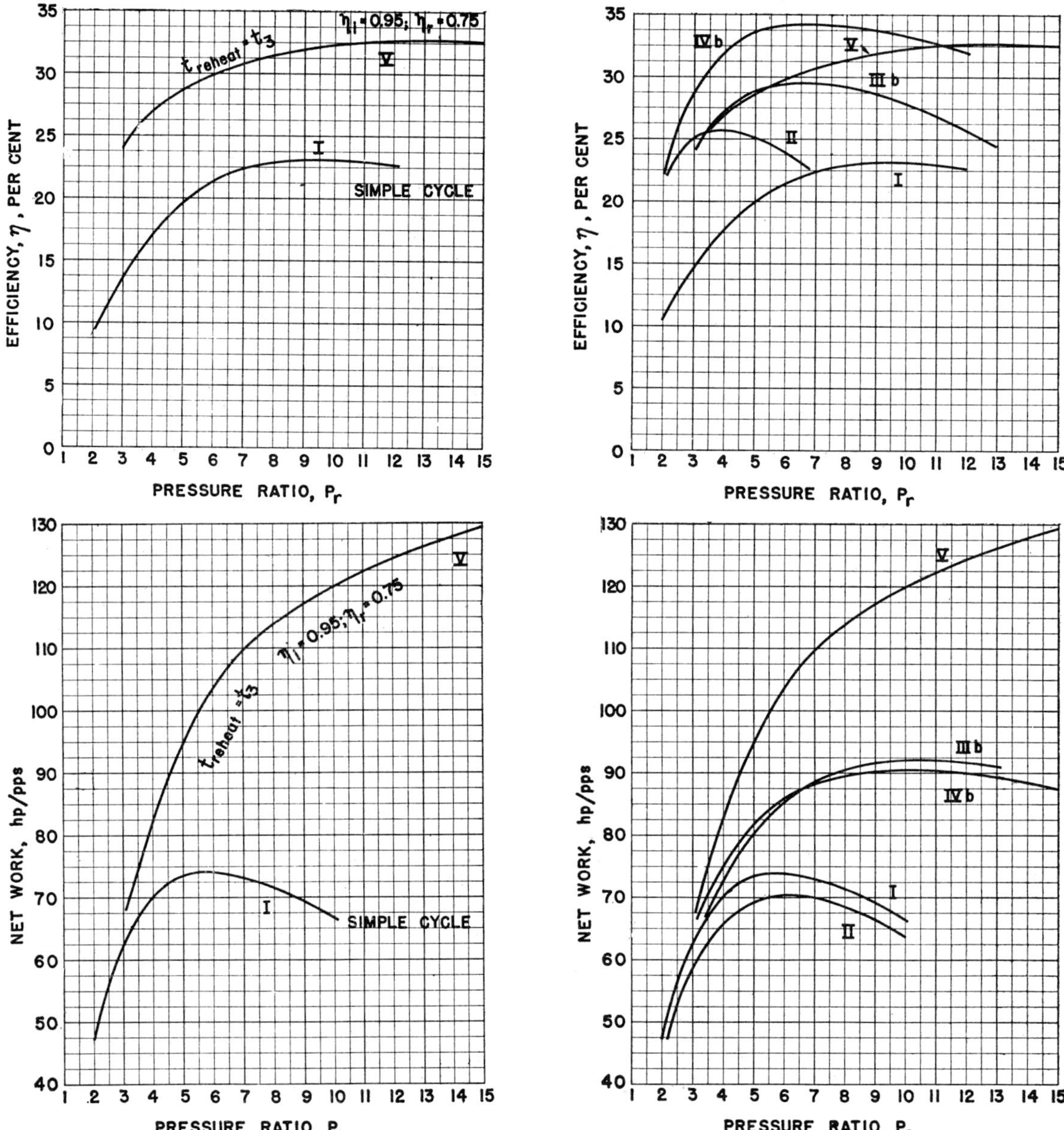

Fig. 12-258 (left) Net work and cycle efficiency of the compound cycle. Turbine inlet temperature $t_3 = 1350$ F.
Fig. 12-259 (right) Comparison of the cycle efficiency and net work of several actual cycles. Key to curves in Fig. 12-249.

and is closer to the conventional steam plant in its space requirements than it is to the simple cycle.

Summary. For comparative purposes, the efficiency curves for all of the actual cycles are shown in Fig. 12-259. As in the ideal cycles, the intercooled-regenerative cycle is most efficient at the lower pressure ratios, and the compound cycle at the higher ones. Also, the regenerative cycle is only slightly less efficient than the intercooled-regenerative cycle, and that peak efficiency occurs at low-pressure ratios where simpler, more rugged equipment might be utilized.

TYPES OF GAS TURBINES

Open Type

All of the working fluid (air and products of combustion) passes thru the machine only once; the intake and discharge are open to the atmosphere. It is the most commonly used type because of its high specific power/weight ratio, its relatively simple control requirements, and its ability to operate without water. Figures 12-242 thru 12-259 reflect open cycle turbines.

Closed Type

As shown in Fig. 12-260, the working fluid (air, N_2, CO_2, and He have all been used) is sealed into the unit and is continuously recycled. The working fluid does not come into contact with combustion products, the heat being externally supplied, transferred thru the walls of a closed heater. The advantages of the closed cycle are: (1) lack of fouling, attack, or erosion of the compressor and turbine blades and the consequent maintenance of design efficiency levels; (2) control of the density of the working medium, giving the ability to operate over a wide load range with nearly constant efficiency while keeping constant temperatures throughout the unit, minimizing thermal shock effects; and (3) the ability to use a wide range of fuels, including low-pressure natural gas. The disadvantages of the cyle are: (1) cost, complexity, and total size of plant; (2) high "thermal inertia," resulting in relatively slow response to load changes; and (3) the temperature limitations of the air heater. The closed cycle is not in general use; it is found only in special applications with high power levels.

Semiclosed Type

This cycle (Fig. 12-261) is an attempt to combine the best features of both the open and closed cycles. It retains the closed cycle's ability to operate at substantially constant efficiency over a wide range of loads thru control of working fluid density. Further, the combustor is the compact, high-pressure device of the open cycle, eliminating the large, highly stressed air heater of the closed cycle. The disadvantages of the semiclosed cycle are the plant complexity and the buildup of combustion products in the working fluid as a result of recycling. If the cycle is arranged to keep the combustion products out of a number of the compressors and turbines, then the advantage of the compact combustor is lost because the large, closed cycle type of air heater must be used.

COMPONENTS

Compressors

The compressor is the first of the basic components in the air stream of the cycle. Two types of compressors are in general use: centrifugal and axial flow. Generally, the higher horsepower rated turbines (above 1000 hp) use multistage axial flow compressors; the smaller turbines (300 hp or less) use centrifugal compressors. The turbines in the intermediate size range utilize various combinations of axial flow and centrifugal compressors. The air flow rate thru the compressor ranges from 8.5 to 20 cfm per output hp of the turbine.

Characteristics. The centrifugal compressor can efficiently develop pressure ratios as high as 6:1 in a single stage, although the usual upper limit is less than 5:1. The centrifugal compressor impeller compared to its axial counterpart is simpler, more compact, more rugged, cheaper

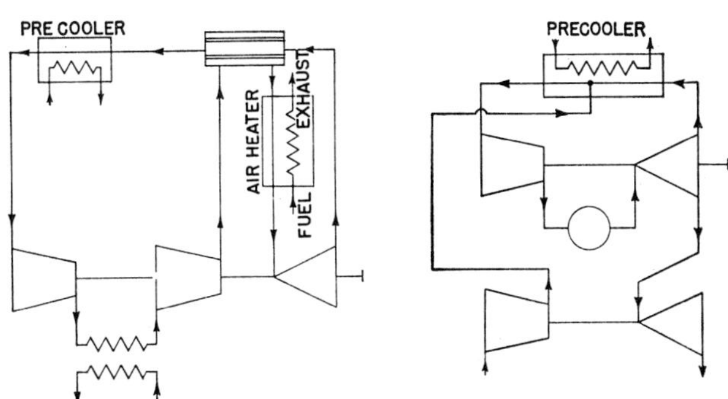

Fig. 12-260 (left) Schematic of closed cycle turbine.
Fig. 12-261 (right) Schematic of semiclosed cycle turbine.

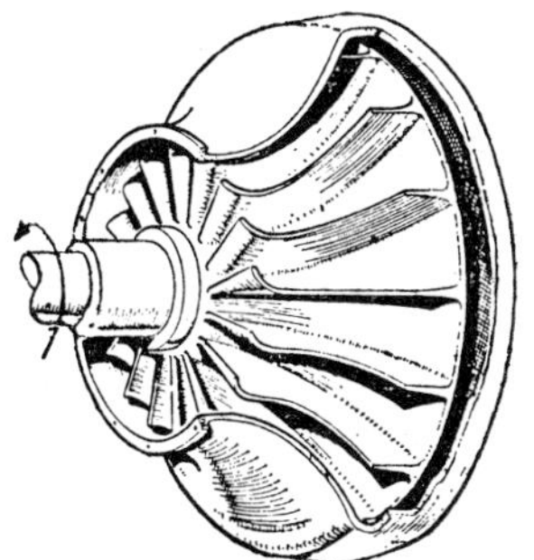

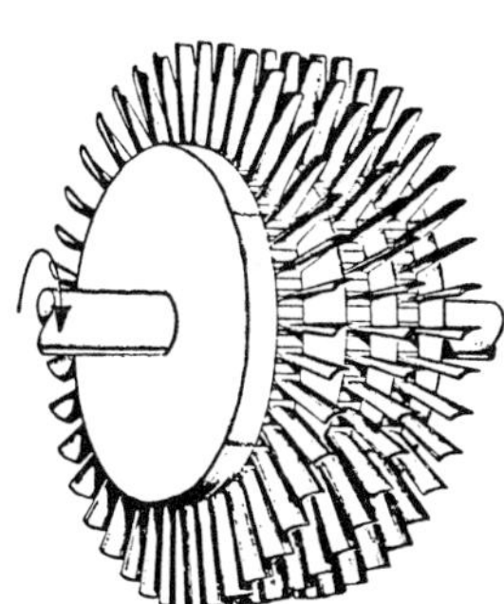

Fig. 12-262 (left) Typical centrifugal compressor.
Fig. 12-263 (right) Typical multistage axial flow compressor.

to produce, less sensitive to fouling or to particle ingestion, operable over a wider range, and less efficient than the axial flow compressor. The axial flow compressor develops a stage pressure ratio of 1.2:1 to 1.4:1, requiring the use of multiple stages. The axial flow compressor has a greater air handling capacity, higher efficiency, more sensitivity to fouling or to erosion by ingested particles, a smaller overall diameter for a given air flow, a narrower operating range, and a higher cost of manufacture than the centrifugal compressor. Figure 12-262 shows a typical centrifugal compressor; Fig. 12-263 shows a typical multistage axial flow compressor.

Arrangements. Compressors may be of single- or dual-rotor design. In the single rotor, all the compressor stages are on one shaft. In the dual rotor, there are two shafts independently driven by separate turbines; the two shafts rotate at different speeds, the low-pressure spool generally operating at lower speeds than the high-pressure spool. This arrangement allows each spool to operate at its most effective speed and simplifies compressor-turbine matching problems.

Compressor Surge. Compressors are designed for a specific air flow and pressure combination. The centrifugal compressor has a broader range of operation than the axial flow unit. This is seen in Fig. 12-264, which superimposes characteristics of the two types at the same design point.

If, at any given speed, the flow is progressively restricted in some manner, the compressor will tend to deliver progres-

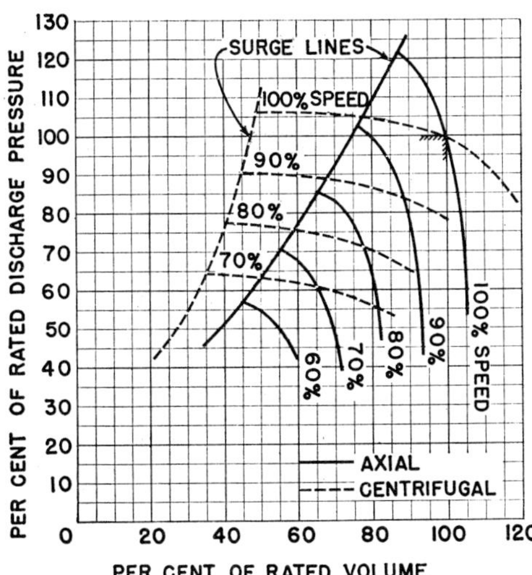

Fig. 12-264 Typical compressor maps: axial flow and centrifugal types.

sively higher pressure until the discharge pressure reaches the maximum of which the compressor is capable at the speed, as indicated by the surge line. If the restriction is removed, the flow and pressure return to normal. If the flow is restricted further, the discharge pressure of which the compressor is capable at that speed decreases and the compressor is not able to deliver against the existing downstream pressure. The flow stops or may even reverse. Then the compressor attempts to deliver air again. The discharge pressure and flow rate build up to the point at which the compressor cannot deliver against the downstream pressure, and again flow stops or reverses. This phenomenon, repeated at high frequency, is called *surging*. It can be extremely severe and damaging owing to shock loads and high-frequency vibrations. At the least, it will be a mild instability, with a fluctuating compressor discharge pressure and turbine output.

Surging can be caused by a sudden demand for power that would cause the fuel flow to increase abruptly, increasing the volume flow thru the turbine and causing it to "choke." This would then initiate the sequence of events described. To prevent this, acceleration-limiting devices are usually built into the governor mechanism.

Some gas turbines go thru potential surge conditions during start-up. Surge-free start-ups can be achieved by bleeding compressed air to the atmosphere to keep the flow thru the compressor above the danger level minimum. This bleeding, or "blowing-off," is done thru automatic valves at an appropriate point in the compressor. Because bleeding is a power wasting process, it is usually done at an early point in the compression process when little work has been done on the air. Another method for eliminating surging on start-up is the use of variable (adjustable) inlet guide vanes ahead of the compressor to effectively change the basic characteristics of the first stages.

Combustors

The combustor is the second basic element in the air stream of the cycle. It must operate with widely varying fuel flow rates to satisfy several criteria: (1) complete combustion of fuel; (2) minimum loss of pressure; (3) no deposit-forming or fouling products of combustion; (4) rapid and stable ignition; (5) uniform temperature and velocity distribution at the turbine inlet.

Air Distribution. In the open cycle turbine, the air flow rate is determined by the dilution air required to cool the gas stream from flame temperatures (approximately 3500 F) to the permissible turbine inlet temperature; this amounts to a total air-fuel ratio of 50 to 100 lb of air per lb of fuel. It is necessary to operate the primary zone of the combustor at the stoichiometric air-fuel ratio to develop the highest flame temperatures and rate of flame propagation. The common hydrocarbon fuels have a stoichiometric air-fuel ratio (by weight) of about 15:1. Thus only a small portion of the total air flow can be admitted to the primary zone; the remainder of the air is admitted, as dilution air, after combustion is complete.

Combustion Efficiency. Combustion efficiency is defined as the ratio of actual heat delivered to the air stream to the heat equivalent of the fuel burned. Combustion efficiencies as high as 99 per cent are common in gas turbines. Poor combustion efficiencies can result from dilution air being admitted too soon, resulting in quenching or chilling of the flame and exhausting of unburned combustibles from the turbine.

Arrangements. Gas turbines may use single or multiple combustors. Aircraft turbines almost universally use multiple combustors. Industrial turbines usually have a single combustor because of pressure drop and outlet temperature distribution, as well as for better operating control and ease of maintenance. Figure 12-265 shows several combustor arrangements. Combustors that perform satisfactorily on liquid fuels generally operate without difficulties on gaseous fuels. All-metal construction is almost universally used.

Fuel Injection. Liquid fuels must be injected into the combustor in as fine a mist as possible to permit rapid evaporation and mixing. An atomizer or nozzle is generally used. Some liquid-fueled turbines use vaporizing injection. In this method, fuel is metered into a tube surrounded by hot gases. The fuel is vaporized and flows to the combustion chamber.

Gaseous fuels need only to be delivered to and dispersed within the combustor, eliminating much of the complexity and cost of liquid fuel injectors. Because of the combustion characteristics of gases and their generally nonluminous flames, it is easier to obtain a more uniform temperature in the gases leaving the combustor.

Ignition. Combustion is commonly initiated by an ignition unit that is energized only during the starting cycle. The ignition unit is an igniter plug supplied from a high-energy capacitor discharge unit, providing the large, hot spark required by the turbine.

As soon as normal combustion is established, the igniter is de-energized and in some cases partially retracted to protect it from sustained high operating temperatures. The stabilized flame front is self-sustaining in the primary zone of the combustor. For extra assurance, multiple combustor arrangements have flame tubes between combustors to provide continuous ignition during operation.

Turbines

The final basic element in the gas turbine engine is the turbine itself. The high-energy gas stream from the combustor(s) is expanded thru the turbine, producing the power to drive the compressor and provide net output. The most important item in the gas turbine is the turbine wheel. It carries the most highly stressed parts operating at high temperatures, its

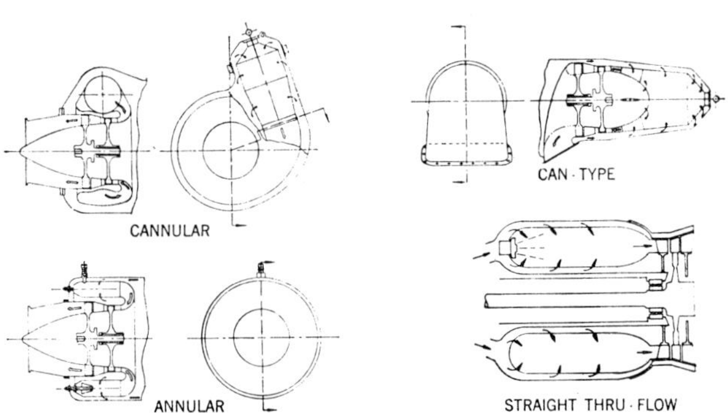

Fig. 12-265 Typical combustor arrangements.

temperature limitations are the criteria for the maximum cycle temperature, and its efficiency is more important to the efficiency of the machine than that of the compressor because it operates at a higher power level.

Characteristics. There are two types of turbines in general use: *radial inflow* (sometimes called *centripetal*) and *axial flow*. The characteristics and application of these types are similar to those of the corresponding types of compressors. The radial inflow turbine is used chiefly in the smaller output ratings (under 400 hp). Where the expansion ratio does not exceed 4:1, the radial inflow turbine offers acceptable efficiency levels with compact, rugged, one-piece construction. The chief limitation of this type is the allowable design stresses at the high temperatures to which the entire turbine disk is exposed, since the entire disk is the gas flow passage.

The axial flow turbine is used in nearly all medium- and large-sized turbines because of its higher efficiencies. The stage expansion ratio for an axial flow turbine is lower than for a radial inflow turbine, but it is a much simpler affair to add stages to an axial flow turbine than to a radial inflow unit. Only the blades themselves "see" the high-temperature gases, permitting the use of more ordinary alloys for the main disk, with the very costly, exotic alloys required only in the blades.

Air-cooling of the turbine rotor is widely used because the relatively small loss of air involved permits a relatively large saving to be made in turbine rotor costs.

Arrangements. Turbines, like compressors, may be of single or multirotor design. The number and arrangement of turbine shafts is one of the major classification criteria; see page 12/400.

Regenerators

A regenerator is a heat exchanger that transfers heat from the turbine exhaust gas to the compressor discharge air. The heat so transferred reduces the requirement for fuel input. Regenerators are an important part of any gas turbine plant designed for high efficiency operation. Regenerators must withstand rapid and large temperature changes while operating under full compressor discharge pressure. They must have a minimal pressure drop, particularly on the exhaust side.

Types of Regenerators. Regenerators are of two types: *recuperative* and *regenerative*. The recuperative heat exchanger has the air and gas streams separated by a thin wall (usually of metal) thru which heat is transferred by conduction. The regenerative heat exchanger has a matrix (of corrosion-resistant metals or of ceramic) that is exposed alternately to the turbine exhaust stream and the compressor discharge air stream, picking up heat from the former and rejecting it to the latter. Since the recuperative heat exchanger has the advantage of being a stationary device with absolute separation of the exhaust gas and air streams, with no leakage of compressed air, it is the type of regenerator in almost universal use.

Effectiveness. The effectiveness of a heat exchanger is a function of heat transfer surface, flow arrangements, and material of construction. The effectiveness of recuperative heat exchangers (recuperator) is primarily dependent upon the amount of heat transfer surface. Figure 12-266 shows the nominal amount of surface required as a function of recuperator effectiveness. As this curve suggests, practical considerations tend to place a limit on the effectiveness of

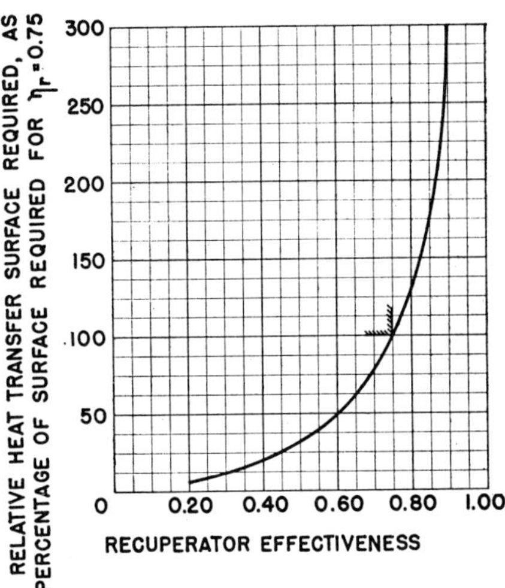

Fig. 12-266 Heat transfer surface required for various recuperator effectiveness levels.

the recuperator. As the effectiveness increases, its incremental effect on the specific fuel consumption decreases; the cost of high effectiveness may never be recovered. In addition, the maximum benefit that can be obtained from the recuperator decreases as the pressure ratio increases because of the approach of the compressor discharge temperature to the turbine exhaust temperature.

Regenerative heat exchangers are more compact and can achieve a higher effectiveness than the recuperative heat exchanger. The chief disadvantage of the former is the loss of air resulting from leakage past the seals of the matrix sections.

Intercoolers

Intercoolers are special types of heat exchangers and have the same general limitations on their desirability (i.e., size, weight, cost, and pressure drop); in addition, they require a source of water.

Intercoolers are generally of shell-and-tube construction. Because the water side has a much higher heat transfer coefficient than the air side, the air side resistance is the controlling factor, and effective use can be made of fins or other secondary heat transfer surface. Very high levels of effectiveness can be achieved even in a single-pass arrangement.

Reheaters

Reheaters are combustors, located between stages of expansion. Their use has so far been restricted to experimental units because of added cycle complexity and cost, although automotive turbines have included reheating to maximize the power output.

CHARACTERISTICS OF TURBINES

Aircraft turbine manufacturers normally rate their units at sea level pressure (29.92 in. Hg abs) and an inlet air temperature of 59 F. Standard industrial ratings are at an altitude of 1000 ft (28.86 in. Hg abs) and 80 F inlet air temperature. These ratings are based on no compressor inlet or turbine discharge losses measured at the engine inlet and exhaust flanges. *Normal rated power* (the net maximum continuous output of the engine) and *site rated power* are defined on page 12/389.

Gas turbine cycles and types may be arranged to provide shaft power, high pressure air, or thrust, either individually or in combinations.

Shaft Power Turbines

The gas turbine used for its shaft power output can be assembled in single- or two-shaft arrangements. The simplest arrangement is that of the single-shaft turbine in which all the rotating components are on one shaft; the two-shaft turbine adds an independent shaft and a turbine that develops the output power. In the latter arrangement, the section of the turbine having the compressor, combustor, and compressor-drive turbine is called the *gas producer section;* the section having the independent output turbine is called the *free turbine*, or *power section.*

Speed vs. Load Characteristics. The choice between single- and two-shaft engines is dependent upon the characteristics of the load to be driven. The single-shaft turbine is basically best adapted to constant-speed applications such as generators; the two-shaft turbine is more suitable where the flexibility of its speed-load characteristic is of greater importance, as in pump and compressor drives. The speed-load characteristic of the single-shaft turbine is very "stiff"; i.e., the speeds tends to remain essentially constant even with wide load variations. This results from the fact that the external load represents only a minor portion of the total power being developed, the compressor absorbing three to four times the rated power output of the machine. Thus, the turbine is always at 65 to 75 per cent of full capacity regardless of the external load. In the two-shaft turbine, the variable speed gas producer section will provide mass flow thru the power turbine sufficient to maintain the load. The speed of the power output section is independent of the gas producer section; in the event of load change, there is a lag as the gas producer section responds to the new condition. The fuel controls for a single-shaft turbine tend to be simpler and less costly than those of a two-shaft turbine.

Torque. The output torque of the single-shaft turbine increases with increasing engine speed because the mass flow and gas temperature to the turbine increase as the speed increases. The output torque of the two-shaft machine, on the contrary, increases as the speed decreases, because the gas producer section does not change its speed with the output shaft. The total energy available to the power section is constant (see Fig. 12-267). If an increase in load torque is felt, the turbine will slow down to develop the additional torque. This torque availability increases to the stall or breakaway point of the power section when the torque is at the maximum (that is, when the torque is equivalent to the locked rotor torque of an electric motor). An overload on the single-shaft engine may cause it to lose power completely (stall), whereas the two-shaft turbine will continue to carry load at a reduced speed. Subsequent clearing of the overload will allow the two-shaft turbine to resume normal speed operation, but the single-shaft engine will have to be restarted.

Figure 12-268 shows characteristic performance curves for simple open-cycle engines of both single- and two-shaft arrangements.

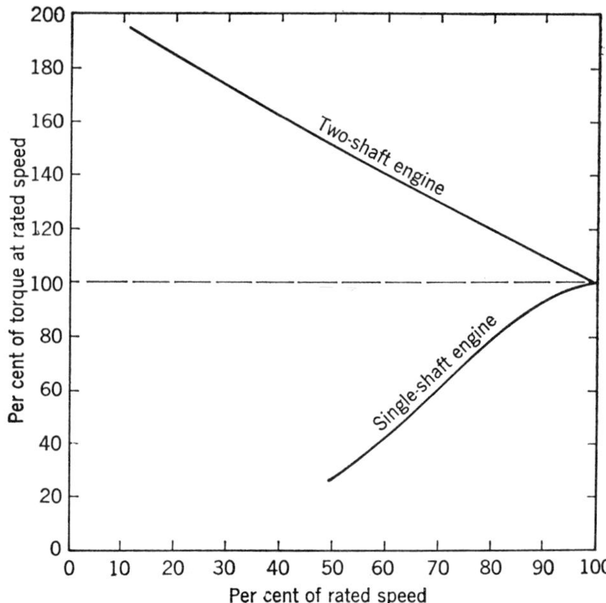

Fig. 12-267 Comparison of torque characteristics for single- and two-shaft turbines.

Part Load Efficiencies. The part load efficiencies of single- and two-shaft turbines vary substantially. In the single-shaft machine, the air mass is constant at all loads. Any reduction in the fuel rate in response to a load reduction will cause an immediate reduction in the turbine inlet temperature, with a corresponding loss of efficiency. The addition of a *regenerator* to the cycle does not substantially alter this effect, although it does cause it to occur at higher levels of efficiency. Furthermore, the regenerator has the effect of building a lag into the response of the single-shaft turbine. The two-shaft turbine can maintain design turbine inlet temperatures (and consequently operating efficiencies) over a considerable load range, because the air flow can be varied. Again, a regenerator does not alter the effect, only the efficiency level at which it occurs. The regenerator also intensifies the lag inherent in response to the two-shaft turbine, particularly its tendency to overspeed when the external load is dropped completely and suddenly. Suitable safeguards, however, can be built into the load to prevent this.

Starting Requirements. For any given application, the starting torque and power requirements will be greater for a single-shaft turbine than for a two-shaft turbine because the starter of the former has to accelerate the combined rotating mass of the engine and the driven unit, whereas the starter of the latter has to accelerate only the gas producer section of the engine. This is significant both from the view of choosing the type of starter and from the view of preventing hot or false starts. Starter power is required for relatively short periods of time (30 to 120 seconds), with the torque requirement increasing as the start proceeds until light-off occurs. At light-off the engine cannot sustain itself, and the starter must continue to provide assistance so that the engine can complete the start within time and temperature limits. Starting power may be as high as 5 per cent of the power rating of the turbine in the case of a single-shaft turbine and 2 per cent of the power rating of a two-shaft turbine.

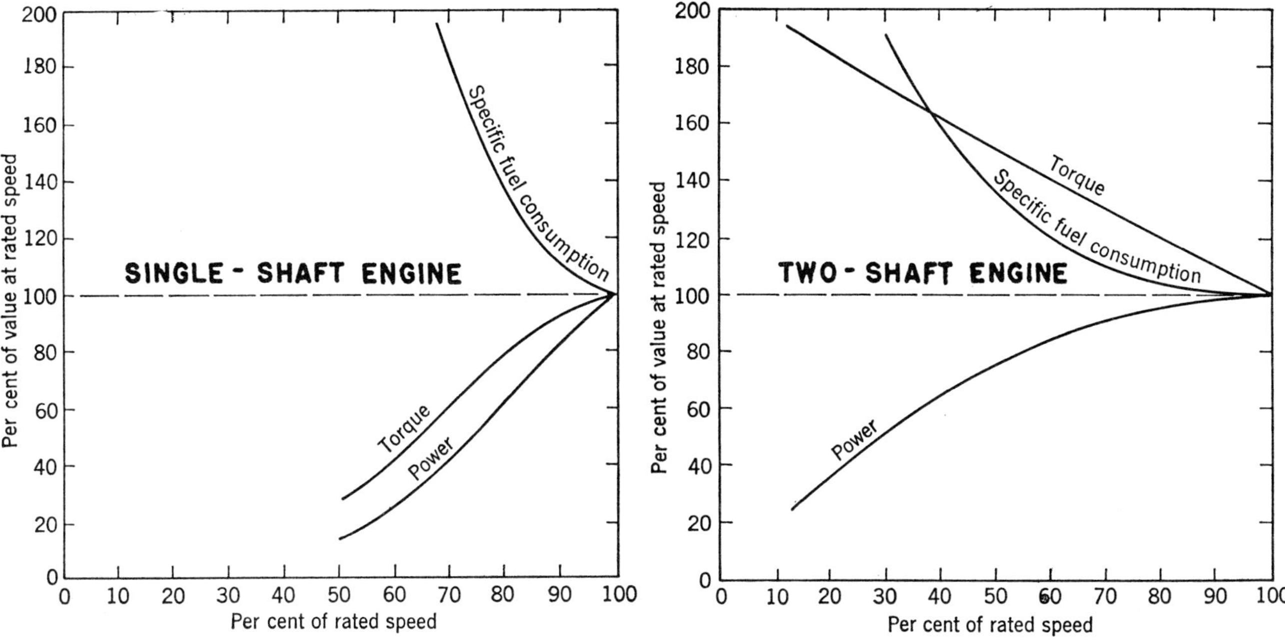

Fig. 12-268 Characteristic performance curves for simple open cycle engines.

Provisions to unload the driven equipment may have to be made in applications of single-shaft turbines.

The starter must also be capable of turning the gas generator whenever it becomes necessary to purge the unit. This purging capability may be more rigorous than the start capability.

High-Pressure Air Supply Turbines

Many small gas turbines, particularly under 200 hp in size, are designed to provide their entire useful output in the form of compressed air. These are commonly referred to as "bleed air" units. The gas turbine compressor is oversized to permit the bleeding of the desired quantity of air from a point between the compresser and the combustor. The advantage of this practice, as opposed to driving an external compressor, is that it results in lower cost of equipment with an efficiency at least equal to that of a turbine and external compressor combination. This type of engine was specifically developed to provide small, lightweight air compressors for driving starters for large aircraft turbines.

Thrust Output Turbines

Most aircraft turbines are designed to produce thrust. In the strictest sense, there is no direct industrial application for such engines; however, the engine is widely used as the gas producer section in large, built-up shaft power output turbines. The stream gas discharged from the gas producer section has not been completely expanded to atmospheric pressure and thus has energy that can be converted into power. In the thrust configuration, this stream is accelerated thru a nozzle, providing the required thrust; in the application of the engine to shaft power production, a free turbine stage is substituted for the nozzle. This free turbine performs the same task as the free turbine in the two-shaft arrangement described previously.

The chief advantages of this arrangement stem from the complete independence of the gas producer and free turbine sections (in contrast with a conventional two-shaft engine, in which the output shaft is on the same frame as the gas producer section), permitting gas producer sections to be changed without disturbing the free power turbine or the driven equipment. This greatly simplifies the operation and ensures retention of the alignment of the power section and the load.

The shaft power output from this type of turbine is found in three primary applications: gas transmission line compressor drives, electric generator drives, and marine propulsion units. In those cases in which the power requirement exceeds that available from a single gas producer, several turbines may be arranged in parallel, either thru the gas stream or thru a common gear drive. Such arrangements permit one or more of the turbines to be removed for servicing without shutting down the entire operation. As many as 10 turbines have been clustered to provide as much as 121 megawatts. Figure 12-269 shows several possible multiple turbine arrangements.

FUELS

Gaseous, liquid, or solid fuels can be used in a gas turbine, gaseous and liquid being the most prevalent. Solid fuel usage is still experimental in open-cycle turbines, although solid fuels have been satisfactorily used in closed-cycle units. Neither octane nor cetane rating is a relevant turbine fuel consideration.

Gaseous Fuels

In the open-cycle turbine, where products of combustion directly contact the turbine blades, gaseous fuels are almost ideal in many respects. They require no vaporization; they are simple to meter accurately; they require a negligible amount of combustor and fuel nozzle maintenance; they have no solid particles to erode turbine blades; and they burn more evenly, giving a more uniform temperature in the gas stream going to the turbine. Natural gas, LP-gas, refinery gas, and blast furnace gas have all been used satisfactorily.

Natural Gas. Natural gas, as distributed throughout the United States, varies somewhat in composition depending upon the source of supply, although in all cases by far the largest component is methane. Turbine performance guarantee is dependent upon gas analysis, which should be obtained from the local supplying utility in all cases. Nominal natural gas analyses for 48 U. S. cities are given in Table 2-9.

Natural gas is conventionally described and sold in terms of the higher, or gross, heating value. The gas turbine, however, utilizes only the lower, or net, heating value in developing power. The difference may vary from 8 to 10 per cent. To convert efficiency values from an LHV basis to an HHV basis, multiply by 0.91; to convert heat rates from an LHV basis to an HHV basis, multiply by 1.1.

Natural gas distributed by utility companies averages less than 100 ppm sulfur, and as such is an insignificant contributor to air pollution. (On the same basis, oils and coal contribute 50 to 250 times as much sulfur per Btu delivered.) Unprocessed (field) natural gases containing high sulfur may be used in oil and gas field operations; special provisions may be required to use this gas satisfactorily, since its combustion products are corrosive.

LP-Gas. For propane and butane the combustion characteristics of importance to the turbine are very similar to those of natural gas, and can be used in similar fashion. Both propane and butane have higher HHV than natural gas, and this permits smaller fuel gas piping and valving. LP-gases may also be blended with air to produce a mixture that is completely compatible and interchangeable with natural gas in the same burner/nozzle setup. Dual-fuel systems, employing natural gas and distillate fuels or LP-gas and capable of rapid changeover, are available from several turbine manufacturers. Such installations permit the purchase of natural gas at interruptible rates, enhancing the economics of the turbine.

LP-gas can be stored and pumped as a liquid; NFPA Standard 58[1] provides guidelines for such installations. It is more efficient to pump liquid propane and vaporize it with auxiliary heat than to vaporize it and boost the pressure of the vapor.

Refinery Gas and Blast Furnace Gas. The use of such by-product combustible gases can improve process economics; generally, they are unsuitable for use in reciprocating engines or boiler furnaces because of low calorific values

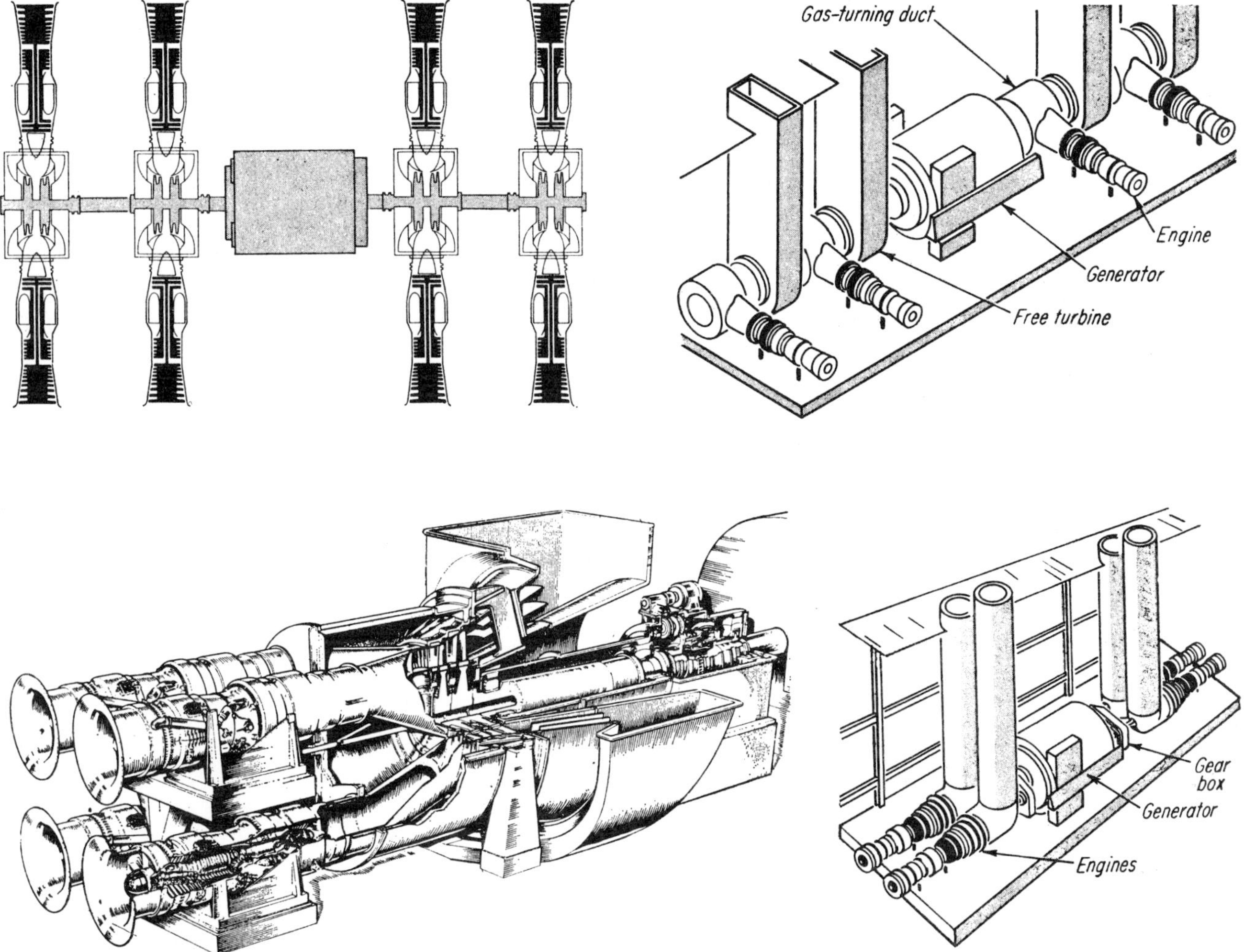

Fig. 12-269 Multiple turbine arrangements. (Reproduced by permission from *Power*, Dec. 1963.)

or other combustion characteristics. These gases may require pretreatment, cooling, and/or cleaning; they have been successfully used in several installations.

Liquid Fuels

Most gas turbines in use are liquid fueled. Mobile power applications—aircraft, marine, and automotive—use liquid fuel. To date, with the exception of aircraft fuels, there does not exist any "gas turbine fuel" similar in scope to "diesel fuel" or "automobile gasoline." Gas turbines have successfully used kerosene, gasoline, alcohol, diesel oil, No. 2 fuel oil, and residual oils. Their versatility in this respect provides an advantage not possessed by reciprocating engines.

Aviation Turbine Fuels. Three types of aviation turbine fuels are widely available. Jet A and Jet A-1 are relatively high flash point, kerosene-type fuels having a lesser volatility than Jet B, which has a relatively wide boiling range. Jet A-1 has special low temperature characteristics for flight operations not necessary for ground applications. For all practical purposes, the fuels are compatible and interchangeable in current gas turbines.

Diesel Oil and No. 2 Oil (Distillate). These oils are widely available and familiar fuels. Dual-fuel burners can accommodate them without interruption to service. Like the aviation-type fuels, they can be used directly with no pretreatment or preheating required for satisfactory operation.

Residual Oils. Heavy, residual oils for large stationary power applications offer reduced fuel costs. These oils, however, are not desirable fuels because their sulfur and vanadium contents create greater maintenance problems, causing corrosion and undesirable deposits in the turbine's hot parts. These oils can be satisfactorily pretreated to minimize the effects of the vanadium, but doing so reduces their economic attractiveness. If the residual oils also have a high ash content, serious erosion may result.

Heavy oils can be burned satisfactorily only if they are adequately preheated to reduce their viscosity. To facilitate the atomization of these heavy oils, several techniques are used, including air injection, steam injection, and very high pressure "return flow" nozzles. Because heavy residual oils are less volatile than the lighter fuels, they are slower burning; they require a larger combustor so that the fuel is in the combustion zone for a longer period of time. It is extremely important to prevent flame impingement on the combustor since the flames of residual oils tend to be particularly corrosive.

Storage and Handling of Liquid Fuels. Figure 12-216 shows a fuel oil handling system. Aviation fuels and distillates are easy to pump into the engine fuel system; residual oils generally require preheating in order to reduce the viscosity to a reasonable level. The gum forming tendencies of liquid fuels generally increases as the oil becomes heavier, and must be considered in the design of the storage and handling system.

Products of Combustion. Liquid fuels, as a group, contain more sulfur than gaseous fuels. This could cause air pollution problems, particularly with the heavy residual oils. If the exhaust gases are to be used in any industrial process, the nature of the exhaust gases and the degree to which they are subject to unacceptable variations become important and may bar their use in applications in which product appearance is important, such as brick and tile drying or food processing.

Solid Fuels

Several gas turbines are in operation on coal and peat, but in the open cycle they must be considered experimental. Erosion by ash significantly limits the life of the turbine blades since ash deposits can build up in stationary portions such as ducts and nozzle passages.

Actual combustion of solid fuels is different from that of gaseous or liquid fuels because solid fuels do not vaporize. Even in regions of great turbulence they require much larger combustor volumes. The equipment required to satisfactorily use solid fuels adds greatly to the complexity and cost of the system, especially when the closed cycle is used to circumvent the erosion problem.

ACCESSORIES

Gas turbine accessories are those parts of the engine not directly involved in the production and transmission of the engine's net power output. They perform auxiliary services essential to the production, control, and transmission of that power output.

Fuel Systems

Fuel systems consist of filters, piping, control valves, regulators, manifolds, and, if necessary, fuel booster compressors.

Gaseous Fuel Systems. The pressure required to inject gaseous fuels into the combustor is essentially compressor discharge pressure plus the pressure losses caused by the control system and manifolds plus a small nozzle pressure differential to facilitate dispersion of the fuel in the combustor. The actual pressure requirement for a given pressure ratio engine will vary among engine manufacturers as a result of differing design parameters.

Gas Booster. If the gas supply pressure is not high enough to be used directly, a gas booster compressor must be provided. The most efficient way to drive this booster is by direct power takeoff from the turbine, although it may be more convenient to use a separate drive. The power required to compress the gas to the necessary pressure depends on available gas supply pressure, arrangement of the controls, turbine pressure ratio, turbine efficiency, booster efficiency, and calorific value of the fuel gas. From 5 to 10 per cent of the total power output of the turbine may be absorbed by the booster.

The booster may require some form of cooling, particularly if it is staged. This cooling may be by air or water.

When individual booster compressors are direct-driven by the turbine, a supply of pressurized gas is needed for engine start-up. Possible methods of providing this start-up fuel are: LP-gas, stored high-pressure gas, and manifolding of all the boosters so that any engine can obtain pressurized gas from any operating booster. In this situation, the control system must be able to shift from the start-up fuel source to the regular fuel source when the turbine-booster combin[illegible] becomes self-sustaining. If individual boosters a[illegible] trically driven, or if compressors are set up in a booster "station," there is no problem because the boosters are tied into a common electrical network fed by all the generators.

Liquid Fuel Systems. The prime requirement for a liquid fuel system is that it develop sufficient pressure to atomize the fuel at the nozzle in the combustor. The pressure requirement for atomization varies with fuel viscosity and surface tension. Even though some liquid fuel systems require pressures as high as 1000 psi, pump power requirements are low, in the range of 0.5 to 0.75 per cent of turbine output. The fuel pump is often an integral part of the control system. Liquid fuels must be filtered.

Dual-Fuel Systems. The primary fuel for a dual-fuel engine is usually a gaseous fuel; the use of a secondary fuel that can be stored on the premises is most often dictated by an interruptible gas supply contract or by an absolute requirement for continuous power. This secondary fuel is generally a liquid fuel. The fuel and combustion system can be arranged so that transfer from the primary to the secondary fuel can be accomplished under load. The system to do this may be complex, since it essentially involves two complete fuel systems, each having its fuel transfer and control devices continuously in operation, and a combination nozzle that can be used for either fuel.

When the transfer is to be automatic, a sensor for failing primary fuel supply pressure is required, as well as primary fuel storage volume (in the gas piping or other reservoir) sufficient to keep the gas pressure up until the secondary fuel reaches the combustor. If interruption to service is permissible or if it can be scheduled, the turbine is taken off the line and stopped. The fuel nozzles are changed, the governor reconnected or switched over to the secondary fuel, and the engine restarted and restored to service. Depending upon the engine design, this changeover can be done in a period ranging from 30 min to 8 hr. In either case, automatic or manual changeover, no change is necessary to the combustors or to the remainder of the engine.

Starting Controls

The sequence of events required to start, run, load, and shut down a gas turbine can be controlled manually but is usually handled automatically. Engines of over 1000 hp are ordinarily supplied with a means to furnish oil pressure to the bearings before the cranking cycle is started. After cranking has started and a sufficient speed has been reached, the ignition system is energized and fuel is admitted to the combustor to start combustion. The starter continues to assist the engine, usually until it attains 40 to 60 per cent of full speed, beyond which the engine is capable of continuing the start under its own power. As a rule, ignition is continued until the end of starter assist when the engine is then self-sustaining. In the event of a fast start (on the order of 20 sec or less to idle speed), it is good practice to add approximately 5 sec of holding, with fuel off, at or near light-off speed for purging purposes before ignition is supplied.

Acceleration and Shutdown Controls

Acceleration after light-off is controlled by an acceleration limiter that schedules fuel to provide an acceleration rate within turbine design limits. The speed governor assumes

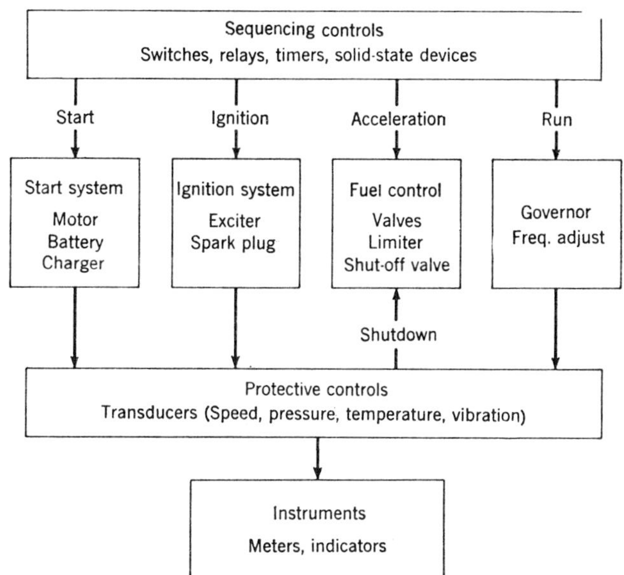

Fig. 12-270 Block diagram of typical gas turbine engine control system.

control near full speed conditions and provides fuel corrections as needed to maintain turbine speed under varying load. A shutdown signal supplied either manually or by any one of several protective devices will cause the fuel valve to be closed, thus stopping the engine. The components of such an elemental control system are shown in Fig. 12-270.

Optional Controls

The gas turbine application determines the degree of control system complexity. Engineering design of complex systems is sufficiently developed to provide nearly any function desired. As a rule it is more difficult and expensive to satisfy extremes of performance than extremes of complexity. For instance, a requirement for 0.1 per cent frequency regulation in a driven alternator may be more difficult to obtain reliably than the control necessary to operate several alternators in parallel.

Most sequencing controls of an automatic nature are accomplished by electromechanical devices, such as relays, contactors, stepping switches, and timers. Depending upon the application, purely electronic equipment, such as magnetic amplifiers, semiconductor devices, or electron tubes, may be used to provide complete or partial control of gas turbine engines and their driven alternators.

Governors

The operation of a gas turbine engine generally requires automatic fuel regulation for both starting and normal operation. Both functions may be performed by the governor, or a separate acceleration limiter may be used for starting.

A typical fuel control or governing system is shown in Fig. 12-271; it represents a minimum fuel control system for a natural gas-burning turbine engine. A two-position shutoff valve admits the fuel to the pressure regulator, which meters the fuel as required. The engine is started and accelerated to governed speed with the acceleration fuel flow scheduled as

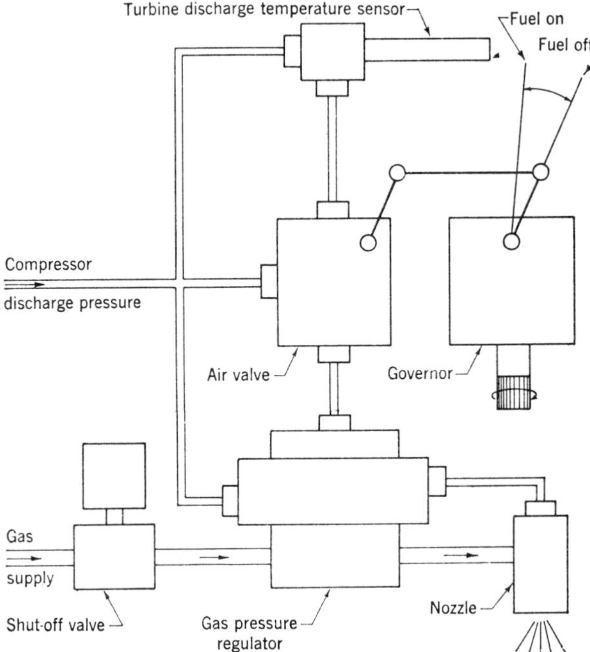

Fig. 12-271 Typical fuel control or governing system.

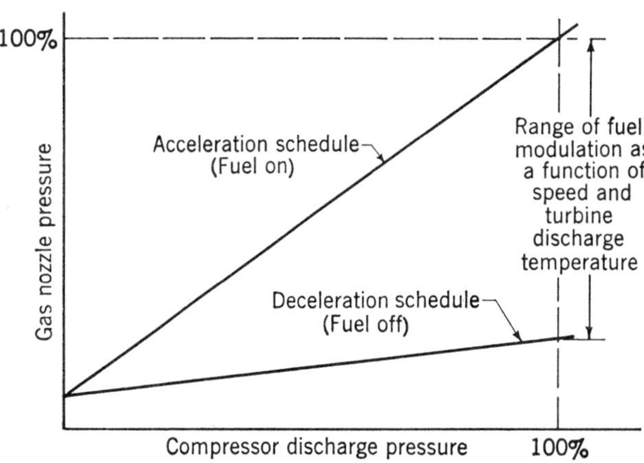

Fig. 12-272 Sample fuel schedule for gas turbine.

indicated in Fig. 12-272. The regulator schedules the pressure at the gas nozzle to a fixed ratio of the engine compressor discharge pressure. The gas nozzle area constitutes the metering area, thus fixing the fuel schedule. Either the speed governor or the turbine discharge temperature sensor can lower the schedule in order to regulate speed or to limit the turbine discharge temperature to the preset value corresponding with full load. The lower limit of the scheduled gas nozzle pressure allows the engine to decelerate under no-load conditions but is maintained at a level sufficient to prevent flame-out under any operating conditions.

This basic system can be altered or added to as required. The governor itself can be any one of several commercially available types of governors such as a simple droop type, an isochronous type, or a load sensing governor.

The system can be applied either to single- or two-shaft engines. When used with a two-shaft engine, the addition of a power turbine governor to control power turbine speed will be required to regulate fuel flow.

Protective Devices

Certain overload conditions or malfunctions of a gas turbine engine can be monitored to prevent possible destruction or damage to the engine. To provide this protection, certain basic protective devices are generally included with all gas turbine engines. The limits at which such devices operate may vary with engine application, their operation usually resulting in an engine shutdown and an alarm indication. Normal practice is to maintain alarm indication and lockout to prevent restart until a deliberate alarm reset has been made. Devices are usually provided to protect against the following conditions:

1. *Engine overspeed.* The control device provides a means of detecting speeds approaching safety limits of the engine.
2. *Temperature.* Turbine inlet temperature is the physical quantity for which protection is desired, but exhaust temperatures are more easily measured and are directly related to inlet temperatures.
3. *Low lubrication oil pressure.*
4. *High lubrication oil temperature.*
5. *Flame detection.* Two conditions exist in which it is important to know that combustion is proceeding normally in order to prevent damage to the engine. These conditions are *light-off* and *flame-out.*

   a. *Light-off.* In starting, the igniter is energized and then the fuel valve is opened. If the fuel fails to ignite, an explosive amount of fuel may soon accumulate and present a hazard. A light-off detection method will usually be provided that shuts off the fuel supply if combustion does not occur within a very short time after fuel is admitted.

   b. *Flame-out.* If a loss of combustion occurs during turbine engine operation, the same hazardous condition will arise as in failure to light off and the fuel supply must be shut off quickly. Although the engine will lose power immediately, rotational energy will continue to provide fuel pressure for a short period.
6. *Vibration.* Any abnormal operation or malfunction that causes an unbalance in the rotating parts of a turbine engine or unstable combustion will cause a vibration level higher than normal for the engine.
7. *Fuel underpressure.* Low fuel pressure, either momentary or sporadic, could result in erratic operation and hazardous conditions. A pressure-operated switch is usually included in the protective controls to shut down the gas turbine in the event that fuel pressure falls below a safe level.

Instrumentation

Instrumentation to provide the following information is usually included with a gas turbine engine for installation and checkout purposes.

Speed. The usual engine speed indication is provided by a tachometer generator and indicator set using a small permanent magnet driven by the engine.

Inlet Temperatures. Knowledge of inlet temperature frequently is of use during operation and monitoring of an engine but *exhaust* gas temperatures are usually measured. Such knowledge is essential for manual throttle operation.

Event Count and Running Hours. The engine manufacturer usually includes one or two devices for counting the number of starts and the total engine operating hours to aid in establishing warranty periods and to provide a reliable record for maintenance schedules.

Lubrication System

The lubrication system lubricates and cools bearings and gears, whether the engine uses journal bearings throughout, antifriction bearings throughout, or a mixture of the two. Large engines with heavy rotating components on sleeve bearings require that lubrication pumps be driven separately from the engine. This is to provide the full lubrication pressure from start of rotation to standstill after shutdown. Smaller engines, particularly those with antifriction bearings, use a single engine-driven lubrication pump. Some large engines use a combination of engine-driven and separate electric motor-driven or steam turbine-driven lubrication pumps. Oil reservoirs can be integrated with the engine or can be entirely separate. A common separate reservoir for several engines can be used. The latter arrangement is not to be recommended owing to the possibility of transfer of contaminants from one engine to all of the others sharing the lubrication system.

In cold surroundings, an oil prestart heater is useful. Depending upon engine design, a separate oil cooler may or may not be required. If a water-cooled oil cooler is used, the oil side pressure should be higher than the water side pressure and the cooler arranged so that the engine will shut down because of low oil pressure or level in the event of cooler oil leakage. Use of one of the many commercially available oil filters is an economical insurance measure. On larger installations, parallel filter systems should be used, with separate pressure drop measurement and shutoff provisions made to detect the need for filter service and to permit it under load.

Starter Systems

Starter systems accelerate the turbine to a speed sufficient to complete the start on its own power within reasonable time and temperature limits.

Electric. An electric starter system consists of a motor and a connect-disconnect arrangement. The power supply can be a battery pack or a connection to the main electrical distribution system.

Motor. The high starting torque of a d-c motor generally makes it the starter of choice; the a-c motor is usually heavier for the same torque.

Motor-Generator. A d-c motor requires a d-c power source to drive it, usually a battery. It is common to use the motor as a generator after the engine starts, to recharge the battery and for other purposes. This usage generally requires a larger motor than one used only for starting, since the motor-generator must be rated for continuous duty, whereas a starter is subject only to intermittent duty. The motor-generator also requires a voltage regulating system.

Batteries and Chargers. Both lead-acid and nickel-cadmium batteries, usually 28 volts, are used for gas turbine applications. Lead-acid batteries are usually specified because of lower cost. Where temperatures below approximately −30 F are encountered, nickel-cadmium batteries are considered superior.

Engine. Reciprocating engines are frequently used to crank large gas turbine engines, especially those used in remote locations. This arrangement requires use of a torque converter or other hydraulic transmission to supply the necessary high torque to the cranked engine from standstill until it has reached a self-sustaining speed.

High-Pressure Impingement. Gas turbine engines can be started by high-pressure air supplied to a special set of nozzles. The nozzle arrangement causes the air to impinge upon the turbine blades, setting the turbine and compressor in motion and developing sufficient rotational speed to effect a start. The simplicity of the starting system is at the expense of some complication of the engine turbine. High-pressure impingement also requires a greater amount of compressed air energy to accomplish a start than would be required by a more efficient reciprocating or air motor.

Hydraulic. A hydraulic starter is used primarily for flexibility. For example, a single power supply such as a small reciprocating engine, an electric motor, or a running gas turbine engine can supply pressure to a central system from which any of several engines might be started by hydraulic start motors. In this respect, hydraulic starters resemble electricity for starting purposes. The hydraulic starting system, however, is more flexible than the electric system. By use of reservoirs, variable displacement motors, or pump motors, the torque can be held virtually constant from initiation of cranking to assist speed, or it can be tailored to fit the power available to the cranking torque requirement.

Hand Crank. Gas turbines of 50 to 100 hp may be hand cranked. For example, a 50 hp engine might be hand cranked by a sturdy man at an ambient temperature as low as −25 F, while 0 F is about the limit for one man hand cranking a 100 hp output engine. Higher output power engines can be started by stored manual energy, as in a flywheel or in a hydraulic reservoir.

Compressed Air or Gas. Usable sources are high-pressure natural gas, plant compressed air, or high-pressure air bled from either an intermediate stage or a final stage of an operating gas turbine. Bottled gas, such as nitrogen, and steam may also be used. All of these may be used in a small reciprocating or rotary starting motor or in a direct injection system. Shutoff and (usually) pressure control valves are required in all instances.

SYSTEM COMPONENTS

System components are those items of equipment necessary for operation of the power plant but not included as a part of the turbine assembly.

Intake Ducting. The intake duct opening should be protected by weather louvers to protect against the entry of solid water and freezing rain. Inlet ducting will be required to mate with the intake filter, cooler, or silencer, as needed, and with the engine intake. Although thermal expansion problems in the intake system are not severe, an expansion joint should be provided at the engine intake and perhaps elsewhere, depending on the length and complexity of the system. Expansion bellows of the rubberized fabric type will usually be adequate.

Because a 1 per cent loss in intake pressure will result in a 2.5 per cent loss in engine power, inlet depression must be kept to a minimum; a typical recommendation is a maximum depression of 2 in. w.c. for rated power. Careful design of the intake system is important, particularly where duct velocities are over approximately 100 ft per sec and dynamic head is a significant fraction of total pressure. Where a 90° change in direction of the air stream is required, vaned corners will provide minimum loss, approximately 13 per cent of the dynamic head. An elbow in which the bend radius of the center line is three times the diameter, D, of the duct or pipe (i.e., $3D$ elbow) will give a loss of about 15 per cent of dynamic head; a $2D$ elbow, 20 per cent.

Sudden increases in duct cross-section in total area or in either longitudinal plane should be avoided. If it is necessary to discharge air from a relatively low velocity volume such as the plenum behind an intake filter, a bellmouthed entry to the duct will give minimum loss. A convergent section of 30° included angle may also be used or a convergent section of 90° included angle with a fairing of generous radius at the entrance to the duct.

Intake Air Filters. The necessity for filtering gas turbine intake air is not in some respects as acute as for filtering piston engine intake air.

The following summarizes gas turbine foreign body ingestion hazards and the means of preventing damage:

Type of foreign object	Engine damage	Prevention
Small hard objects (gravel, hail)	Compressor blade nicking, bending, or fracture	Screen
Finely divided abrasives (sand and dust)	Compressor and turbine blade erosion	Dust filter
Large blanketing objects (birds, paper, rags)	Compressor blade jamming, and fracture or air passage blockage	Periodic filtering, inspection, and cleaning
Insects	Compressor fouling	Periodic filtering, inspection, and cleaning
Liquid water	Flame-out, possible compressor blade fracture	Antisplash louvers
Oily vapor	Compressor fouling	Routine cleaning

The compressor sustains the greatest damage from foreign particle ingestion because it is the first piece of equipment encountered by the foreign object. The resulting impingement force on the compressor rotating at high speed can result in considerable damage.

Centrifugal compressors are less susceptible to damage than axial compressors because the centrifugal compressor impeller is more massive, with relatively few blades of heavy section.

At installations where the turbine uses dust-laden air, more elaborate air filtering must be considered. Research has shown that air-borne dust, if sufficiently concentrated and of sufficient particle size, will cause destructive erosion of compressors and turbines. These research results can be summarized as follows:

1. Dust particles 2 to 3 microns and smaller in diameter have little effect on turbine components other than polishing of the compressor.
2. The larger the dust particles, the more rapid is the erosion of turbine components.
3. Total turbine component erosion is a direct function of the total weight of ingested dust.

Air-borne dust survey research reported in the *ASHRAE Guide* shows the outdoor air dust concentrations, grams per MMCF, measured in three cities to be: Minneapolis, 2.06; Pittsburgh, 2.32; Louisville, 3.42.

Dust particle size distribution was found to be as follows:

Distribution, %	99.99	88	65	50
Size, microns	<25	<10	<5	<3

As a general precaution against dust erosion, the turbine inlet should be located as far from dust generating sources as possible. Whenever air-borne dust can be present in any magnitude either on a continuous basis or from periodic dust storms, an intake air filtering system is required.

Types of Filters. Two general types of filter are used: the *extended area filter* and the *inertial separator*. The extended area filter is an array of filter panels of a suitable filter medium thru which all of the intake air passes. A dry type of filter is preferable because of the possibility of carry-over from oil-wetted surfaces with subsequent compressor fouling. The filter medium should be capable of removing most of the dust larger than 2 to 3 microns. Clean filter pressure drop must be about 0.5 in. w.c. to allow an operating margin. Arrangements must be provided to monitor the pressure drop so that the filter can be cleaned or replaced when necessary. The extended area filter has the advantages of low cost and simplicity, but it is bulky and requires periodic servicing. It is preferred in dusty locations.

Two effective inertial separators are the dust louver type and the cyclone, or centrifugal, type (Fig. 8-57). Both types use scavenging air to remove the dust that has separated from the inlet air. A scavenging blower, usually electric motor-driven, is required with an air flow of 10 per cent of the inlet flow. Inertial filters are thus self-cleaning, but there must be provisions for storage and cleaning out at the ultimate dust collection point. Pressure drop does not increase with use. Inertial separators have the advantage of small size and they are self-cleaning; but they require a scavenging air system and will probably be more expensive installed than extended area filters.

Intake Air Cooling. Cooling of intake air during periods of high ambient temperature is beneficial. Cooling can be done by mechanical refrigeration or by water evaporation.

A water evaporation cooler consists of a set of extended area air filters and an array of nozzles to spray water onto the filter's upstream surface. When the intake air passes over the wet surfaces, its temperature will be reduced as it evaporates water. The cooler should be placed as close as possible to the entrance of the intake air ducting system so that the system handles cooled air and so that the maximum time for vaporization of entrained water is available. Where an intake air filter is used, the cooler should be placed immediately downstream of the filter. Some precautions are needed to prevent water carry-over into the compressor. Air velocity thru the cooler panels should be kept below 250 fpm to avoid droplet entrainment. Air should not be cooled to within less than 10°F of the wet bulb temperature, to preclude subsequent water dropout and accumulation on the duct walls. Such accumulated water might eventually separate from the duct and enter the compressor as a damaging slug.

Centrifugal compressor turbines have been operated with a water cooling system consisting only of a spray nozzle spraying directly into the compressor inlet. The chief disadvantage of this simple system is that any dissolved solids in the water will deposit on the compressor blades when the water evaporates and eventually will affect engine performance adversely.

Intake Sound Attenuation. The most objectionable sound generated by a gas turbine comes largely from the intake. In many installations acoustic treatment of the gas turbine enclosure or power plant walls will be required. Intake filters and intake coolers do not provide significant attenuation.

A commercial intake silencer of the acoustic splitter vane or cylindrical type will usually be required. It should be specified on the basis of allowable pressure drop, normally from 0.5 to 1 in. w.c., at design point air weight flow, and on the basis of desired sound attenuation. A typical silencer has an attenuation of 16 db in the third octave band (75 to 150 cps), increasing to 40 db in the sixth octave band (1200 to 2400 cps). Higher attenuations can be achieved by increasing passage length and, to maintain the same pressure drop, by increasing flow area.

If the intake system has a right-angled bend, acoustic treatment of the duct walls may provide sufficient attenuation, but the acoustic performance of such an arrangement is difficult to predict with any degree of certainty. Special care will be required to ensure that bits of acoustic material and miscellaneous items of hardware will not work loose and be drawn thru the engine.

Anti-Icing System. Under atmospheric conditions of low temperatures and high relative humidity, icing of the gas turbine inlet can occur. Ice will build up when air impinging on metal surfaces is below 40 F, causing the inlet flow area to decrease, decreasing the air mass flow, and causing turbine inlet temperatures to rise. Another serious hazard is that accumulated ice may break off in chunks and be drawn into the engine, causing severe damage to the compressor.

Ice can be prevented from forming by supplying heat to those inlet surfaces upon which the cold air stream impinges. Electrical resistance heating elements are a convenient method of supplying that heat. A simpler method of preventing ice formation is to heat the entire inlet air supply to more than 40 F; this method is attractive because it brings the inlet air temperature closer to design conditions, which is the point of maximum engine efficiency. Although lower-than-design air temperatures allow a given engine to develop greater-than-design power, the efficiency is adversely affected.

If icing conditions are expected, automatic initiation of the anti-icing system is recommended. This can be keyed to ice detectors or, more simply, to inlet air temperatures. Sufficient overlap into nonicing conditions should be provided to ensure prevention of ice formation, which is preferable to melting the ice after it has begun to form. An anti-icing turn-on control temperature of 40 F is satisfactory.

External Lubrication Systems. The lubrication of a gas turbine is normally provided by a gear-driven pump supplied with oil from a tank built into the base of the engine. For starting and for cooling on shutdown, as well as for emergency use, an auxiliary, externally driven pump may be required. The external system would include a pressure regulating valve, liquid level gage, filter(s), and associated valves. An external oil cooler would also be required. An oil-to-water cooler is most economical in first cost and is recommended in locations where a reliable source of low cost water is available. Control will normally be by a thermostatically operated by-pass at the inlet to the cooler.

Oil-to-air coolers are more expensive than the oil-to-water type, particularly if high ambient air temperatures are to be accommodated. If their size and cost become excessive, consideration should be given to the use of a high-temperature oil. The temperature can be controlled in many ways, including variable fan speed and modulating dampers. Both of these methods are superior to by-passing the oil around the cooler since they save fan power.

Output Gearbox. An inherent characteristic of the gas turbine is high rotational speed, about 6000 rpm for large machines to 40,000 rpm or more for small engines. The engine manufacturer furnishes reduction gearing as an integral part of his engine to provide usable output shaft speeds. The general engine lubrication system serves these gears and their bearings. Since speed-matching of the turbine wheel to its load is important, most turbine manufacturers offer a variety or output reduction gear ratios.

Exhaust Heat Recovery Equipment. The economics of gas turbine installations almost always dictate that the large amount of heat in the exhaust stream be recovered. This recovery can be made in the form of regeneration (described previously), or by the production of steam or hot water (or both), or directly by a process heater.

The gas turbine is sensitive to exhaust back pressure. The specification for any exhaust gas heat exchanger must therefore establish the hot gas side allowable pressure drop. A target value for this pressure drop is between 3 and 5 in. w.c. When additional heat is supplied to the boiler by direct firing there may be a higher pressure drop in the boiler.

The exhaust gas temperature and flow depend on the engine model and the load at which it is operating. This information is available in the form of curves from the engine manufacturer. As a general approximation, rated load exhaust temperature may be taken as 1000 F and rated load exhaust flow as 8.5 to 20 standard cfm per horsepower.

Outlet By-Pass and Dampers. Use of exhaust heat recovery equipment requires a control system to regulate the quantity of heat delivered by the heat exchanger. A common method involves the use of an exhaust duct by-passing the heat exchanger to permit partial use of the engine exhaust flow when heat demand is low. A control system modulates the by-pass damper valve in accordance with the temperature of the heated product.

An alternative heat control system is one that passes all of the engine exhaust thru the heat exchanger and regulates the flow of the cold medium thru the heat exchanger or regulates the boiler water level in accordance with the heat demand.

Exhaust Ducting. Exhaust ducting may lead directly to the stack or thru a heat exchanger, exhaust heat boiler, silencer, or other unit, or combination of units before entering the stack. Further requirements for exhaust duct installation are given in NFPA Standard No. 37.[2]

Up to a gas temperature of approximately 900 F, mild steel exhaust ducting will be adequate for most installations. Above this temperature, specialized low-alloy steels should be considered to minimize corrosion. Expansion joints in exhaust ducting must accommodate considerable thermal expansion in the ducting; these joints are highly stressed in the hot end of the engine. They should be fabricated from an appropriate stainless steel. When they are located in a high velocity section of the exhaust ducting or when the engine will burn a fuel that could produce deposits interfering with the free flexing of the bellows, the joints may be fitted with an inner sleeve. Where the ducting is insulated, the expansion joint should also be fitted with an outer sleeve.

The previous comments on the aerodynamic design of intake ducting are generally applicable to exhaust ducting. Since 1 per cent reduction in turbine back pressure will cause an increase in available power of about 1.5 per cent, the ducting and stack must be designed to gradually reduce the gas velocity for maximum recovery of dynamic head. Losses in the intake and exhaust ducting have less effect on the performance of gas turbines with high pressure ratios. A conical diffuser of (for usual lengths) about 7° included angle provides the minimum pressure loss.

For practical reasons, rectangular diffusers are often preferred. When two walls of a rectangular diffuser are parallel, the diverging walls should have an angle between 8° and 10°. When all walls of a rectangular diffuser are diverging, an included angle of 6° is appropriate. Where space limitations dictate, a rectangular diffuser of higher outside angle may be provided with a system of splitters to provide a number of passages, each one of which has the optimum internal angles.

Some engines require external piping for carrying compressor delivery air to the high-pressure side of a recuperative heat exchanger and for carrying the heated air back to the engine combustion system inlet. This ducting is generally commercial steel pipe of a gage appropriate to the pressure and temperatures. Bends should preferably be long radius, and duct lengths kept as short as possible.

Bellows expansion joints are a special problem, owing to the pneumatic ram effect of the air at compressor delivery pressure. There must be special restraint to prevent "flip out" of a section of pipe between two bellows and to eliminate the very substantial loads on the engine and recuperator casings. Where engine and recuperator flanges have intersecting centerlines and permit an arc-shaped piping layout, a three-joint layout using hinge-pinned expansion bellows resolves pneumatic and dynamic loads to a hoop tension of low value and permits relative motion of the engine and recuperator flanges in the plane of the pipe centerline.

In some installations, accessibility to the engine itself is improved by arranging the ductwork to enter and leave the engine from below. This is important with large sizes where heavy lifting equipment is required for removal of casings for inspection.

Outlet Sound Attenuation. Attenuation of gas turbine exhaust noise may also be required to satisfy noise restriction requirements. Exhaust noise is of low frequency, about 300 cps. Exhaust heat recovery equipment functions very well as an exhaust noise attenuator, either eliminating the need for or reducing the size of a separate silencer.

Exhaust silencers are also commercially available. They may be of the detuner type or be generally similar to intake silencers in configuration. Because of the low-frequency nature of the sound and because of the elevated temperatures, exhaust silencers are comparatively expensive. In noncritical locations a carefully designed simple stack arrangement may provide adequate silencing, since jet shear, which is the principal component of noise from a high-velocity exhaust system and varies as the eighth power of jet velocity, will be insignificant if the exhaust system provides controlled diffusion down to exhaust velocities of approximately 50 fps.

Operational Controls. Gas turbine plants are well suited to automatic and semiautomatic operation. An **annunciator panel** will generally be provided that will indicate the source of potential trouble or the reason for a shutdown. Once the difficulty is corrected, the same panel will indicate that the gas turbine is ready for restart. The actual starting operation will often consist of selecting the appropriate fuel and pushing a start button and, once the engine is self-sustaining, setting the required speed. In the case of electric generator drives, the generators can then be synchronized, if necessary, and loaded.

In the case of a multi-unit gas turbine generator installation, it will be advantageous to install automatic load sensing equipment for adding sets as required to maintain optimum loading. Such a system should have provision for interchanging the lead or master unit to equalize wear on the various sets.

Instrumentation of the equipment in a gas turbine plant should be arranged so that indications of difficulties are brought to the attention of the operator at one point, perhaps in an office adjacent to the machinery room. The control equipment may also be centralized, or may be located close to the units it controls. Certain operations such as speed setting may be controlled from both a local engine control panel and an electrical control center.

INSTALLATION REQUIREMENTS

The installation requirements for gas turbines include many features peculiar to the specific engine or installation. A number of common requirements exist, fundamental to all industrial gas turbines.

Machine Mount. Generally, industrial gas turbines are supplied by the manufacturer complete with an integral base that supports the basic engine and the various accessories required for the particular installation, such as starting motor, accessories gearbox, engine-driven fuel and lubrication pumps, auxiliary fuel and lubrication pumps, and reduction gearbox if required. In some instances the base may be extended to support the driven component (e.g., pump or generator). The base should be designed to preclude pockets of flammable fuel or lubricating oil vapor. The installation site must be adequately safeguarded against similar hazards. If engines are to be installed in a foundation pit of any kind, forced ventilation of the pit should be provided.

Foundations. One of the advantages of the industrial gas turbine is the virtually vibration-free operation of the unit. From an installation standpoint this results in two distinct advantages: (1) There is no need to provide isolation between the machine mount and the foundation to prevent the transfer of vibration, although some manufacturers do recommend isolation pads or mounts in some installations. (2) Massive foundations capable of withstanding reciprocating loads are unnecessary. Gas turbines lend themselves particularly to roof installation.

Depending on the weight of the equipment, the inertia of rotating parts (WR^2), and the characteristics of the connected equipment (including the braking characteristics of the generator under short circuit conditions, where it forms the main load), a load of 1.2 times the dead weight of the equipment may suffice in design of the foundations.

Fuel System. Although vibration isolation of fuel supply lines to a gas turbine is not generally a problem, it should be considered in all cases. The installation should conform to local codes and NFPA Standard No. 37.[2]

The diameter of the lines will depend on the type of fuel being carried, the permissible pressure losses in the lines, and their lengths. Tables 12-8a thru 12-8g may be used to size gas piping.

As a convenience for periodic performance checks on the gas turbine, a meter of the pressure differential type may be installed on the gas supply line to each engine. If it is desired to check liquid fuel performance on a dual-fuel engine, the liquid fuel meter installed will normally be of the rotometer or turbine type.

Instrumentation. The instrumentation associated with a gas turbine ordinarily indicates the following: (1) rpm of power turbine; (2) rpm of gas producer, if separate; (3) exhaust temperature; (4) lubricating oil pressure; (5) fuel supply pressure; (6) oil cooler temperature, in and out; and (7) oil tank level.

Other indications that might be read on instruments include: (1) air inlet depression; (2) oil filter differential pressure; (3) fuel filter differential pressure; (4) fuel tank quantity indicator; (5) fuel flow meter; (6) scavenge oil temperature; (7) flame-out detection; (8) engine vibration; and (9) fuel flow rate.

As a matter of convenience, some data from the driven equipment, such as generator current, voltage, kva, and frequency, may appear on the engine control panel. The most important items, such as output rpm and exhaust temperature, may be continuously recorded or may be read on a regular schedule along with other significant indications, and recorded manually for future reference.

Many of the indications do not need to be shown numerically, since what matters is only whether certain limits are exceeded. Warning lights and/or audible alarms are adequate in these cases. Automatic control devices to prevent damage to equipment are usually incorporated to provide shutdown of the gas turbine under conditions of excessive speed, excess exhaust temperature, or lack of oil pressure.

Certain instruments may require electrical power for their operation. In addition, continuous operation of the gas turbine may depend on a continuous supply of electrical power. This may be direct current, with a battery bank and charger, or alternating current, with some means of maintaining it in the event of failure of the main power source. Even engines with completely pneumatic control systems require electrical power for the ignition circuit during engine start.

Water System. Water is not required for the operation of a gas turbine alone, but it is often needed by equipment associated with one, such as an evaporative air inlet cooler, a lubricating oil cooler, a gas compressor intercooler, or an exhaust heat boiler. The boiler will require its own feedwater pumps; the remaining equipment can be supplied from the mains. Piping should conform to ASA B31.1.[3]

Insulation. Thermal insulation of turbine and combustion system casings, exhaust ducts, heat exchanger ducting, and boilers is required for personnel protection and avoidance of excessive heat release to the building. Thermal insulation is also effective in reducing sound transmission. Special consideration should be given to the insulation of engine casings so that component removal for inspection is not made difficult. In low-velocity sections of exhaust ducting, thermal insulation may be installed in the inside rather than the outside of the duct to provide absorption of high-frequency noises that would otherwise be radiated up the stack.

Inside heated buildings, insulation of the intake system will avoid condensation and frost formation.

Ventilation. Ventilation in a gas turbine power plant will be required to remove heat radiated from the equipment and ducting. This may require as many as 10 air changes per hour. Hot surfaces within the building will normally be insulated to provide a skin temperature of 150 F maximum, and heat release can be calculated using standard formulas. Heat release from a gas turbine with short duct lengths may be as low as 3 Btu per hp-hr. Heat release from a gas turbine and connected waste heat boiler or recuperator located within the building may be 300 Btu per hp-hr, requiring appropriate air change capacity. Consideration must also be given to the connected load. A generator has a full-load efficiency of about 96 per cent, and the remaining 4 per cent of energy input is converted to heat. If the generator is "self-cooled" from building air, the heat release adds an additional 100 Btu per hp-hr to the heat load.

Internal ventilation of the gas turbine is handled automatically by the engine itself during regular operation. The gaseous fuel and lubricating oil vapor that may be present inside the engine at rest should be purged by motoring the engine prior to ignition and the introduction of fresh fuel. When equipment such as an exhaust heat boiler is installed in the exhaust from the turbine, it may be advantageous to use a separate blower to reduce the purging time and save energy, since the gas turbine is not particularly effective as a fan at normal motoring speeds.

Accessibility. Space should be provided for access to a gas turbine for maintenance. This should include clearance for removal of parts as they are withdrawn and a place to put them while further work is being done. This problem will be made more difficult by overhead ductwork and piping, but will be simplified by adequate lifting facilities.

Although a moderately large opening may be required to pass a gas turbine to its mounting position, replacing the whole unit in service should never be necessary. Permanent passages and doorways need be large enough only to accommodate the largest individual part. In gas turbine-driven electrical generator sets, the parts of the generator are likely to be larger and heavier than those of the gas turbine.

Electrical Services. Electrical services to a gas turbine are always required for ignition, and usually for instrumentation, control, and starting. The supply may come from batteries, an independent auxiliary generator, a gas turbine-driven generator, or more often a combination of these.

MAINTENANCE AND OVERHAUL

Factors Affecting Maintenance and Overhaul

The maintenance and overhaul requirements of the gas turbine engine, as for any type engine, are determined by the design of the turbine and by the requirements of the individual installation.

The design goal for the compact or adapted aircraft type of gas turbine engine necessitates compromises with respect to endurance life. Consequently, this engine generally requires more frequent and sophisticated inspection, maintanence, and overhaul. This disadvantage, however, is in part offset by separable, easily handled subassemblies. These features permit the changing of engine components or a complete engine replacement in a few hours. The larger industrial engines are generally serviced and overhauled at the facility where they are located. Overhaul usually requires shutdown of operations for a short period of time.

Duty Cycle. Hot section parts of the gas turbine engine are designed to operate over a wide temperature range. Inspection and maintenance requirements are, however, increased as thermal shock resulting from extreme load fluctuations is experienced.

Frequency of Start-Up. The more often a gas turbine engine starts and stops, the more often the engine components are subjected to thermal cycles. Sometimes an engine start-up may produce temperatures higher than the normal operating temperatures. This so-called "hot start" produces abnormal thermal stresses in the engine components.

Turbine Inlet Temperature. Gas turbine engines are designed to operate at high-temperature levels to gain maximum efficiency. The use of high-temperature alloys in recent years has greatly increased the life expectancy and performance of hot section components. Hot section components include such items as the combustion chambers, transition liners, turbine nozzles, and blades. Overtemperature peaks and extended operation above the maximum recommended temperatures increase the requirements for inspection and repair for these components.

Overtemperature Recording. Thermal shock occurs to a degree during every engine start and every load change. It cannot be eliminated but it can be controlled by strict attention to the reduction of peak starting temperature. High thermal shocks are the responsibility of the operator or the automatic control equipment. Recording of excessively hot gas stream temperatures and prompt inspection following such malfunctions may prevent costly repairs.

Environment. The general environment of a gas turbine engine installation is not a critical factor if proper maintenance procedures are followed. The quality of the engine inlet air, however, is important. Air flow thru the engine is necessary for combustion, internal cooling, and mass flow thru the turbine for the development of power. Contaminants in the engine air supply such as dust and smoke can be deposited in the engine and reduce engine efficiency. Abrasive particles that may pass thru the engine tend to scar or erode rotor blades. Foreign chemicals in the air supply may be deposited on compressor components and may cause corrosion or pitting of internal engine parts, especially the hot section components.

Dirty compressor blading can cause loss of air flow and power as well as excessive starting temperature. The contamination slowly collects on compressor blades and vanes usually as the result of engine environment, but a malfunctioning component such as a front oil seal may also be the cause. Contamination can readily be detected because of increasing peak starting temperatures and loss of output power.

Filters are usually placed in the compressor inlet air ducting to remove dust, smoke, and other common contaminants, as well as larger objects that may cause serious damage to gas turbines.

Inspection

Routine inspections may include the examination of external lines, connections, and engine mounts. Attention should be given to signs of engine vibration that may have caused visual damage. The gas turbine normally is not subject to the low-frequency vibration expected in a reciprocating-type engine; however, in case of damage to compressor or turbine blades by a foreign object, a very high frequency vibration may result. This type of vibration has a very small amplitude but can be very damaging. It is prudent to maintain a continuous or frequent check for vibration with suitable engine instrumentation.

Usually, the first stage of compressor blading can easily be inspected for damage or dust accumulation. Certain industrial atmospheres may form an oily dust deposit that may cause a loss of power and an increase in fuel consumption. Blade erosion from sand particles may also be found.

The last stage of the turbine blading should be examined for possible damage. Its condition may be indicative of the condition of other stages.

Combustion chambers, turbine inlet vanes, and fuel nozzles should be examined if turbine temperatures are above normal. This may be caused by dirt in a fuel nozzle or turbine blading deterioration from overtemperature operation. It is important that a defective fuel nozzle be replaced promptly; serious damage can result from continuous overtemperature operation.

The engine may be rotated very slowly without fuel flow, to examine for unnatural noises from the main bearings or excessive interference and rubbing of air seals.

Troubleshooting

Typically, troubleshooting of gas turbines is performed as described in the following paragraphs.

Power Loss. Loss of power and high fuel consumption can be produced by deposits on compressor blading. These can be readily cleaned off by introducing a cleaning material (usually ground walnut shells) into the compressor inlet while the engine is running. Damaged turbine blading can cause similar problems. Low turbine inlet temperature and low fuel flow may indicate a defective fuel control.

Vibration. High-frequency vibration, which is usually

at the same frequency as engine rotation, is checked with an instrument. Vibration may be caused by an excessive rubbing condition between a large-diameter rotor and seal or between blade tips and a shroud. Vibration also results from out-of-balance conditions, which usually come from damaged blading in the turbine or compressor.

Turbine Inlet Temperature. Since the gas turbine is sensitive to excessive turbine inlet temperature, it may be necessary on occasion to record the temperature during a starting cycle or to check the circumferential temperature distribution to determine if one nozzle or one combustion chamber is malfunctioning. The fuel control system may also require examination.

Failure to Start. A defective fuel control or lack of ignition are the usual reasons underlying a failure to start. Most engines have two spark plugs for ignition even though there may be a greater number of combustion chambers. Only one spark plug is required to start the combustion and the flame then travels thru crossover tubes to all combustors.

Repair

Generally, major components and accessories are readily accessible; many designs are also adaptable to sectionalized repair techniques. Therefore, the repair can frequently be accomplished by replacing components or removing them to a repair or overhaul shop. On some engines it is possible to remove a portion of the casing and replace damaged blades at the installation site if only a limited number of blades are damaged. This is usually done by replacement of each blade with another blade having the same "moment-balance."

Major assembly or component repairs are specialized operations and must be performed according to the engine manufacturer's specification.

Maintenance

A routine maintenance schedule is determined largely by experience. This experience factor is dependent largely on the duty cycle of the installation.

A general engine maintenance schedule follows:

1. Clean air, oil, and fuel filters.
2. Make electrical continuity check on magnetic oil drain plugs. This will determine whether any foreign metallic material has been picked up from the oil.
3. Check turbine exhaust temperature thermocouples for accuracy.
4. Make electrical check of electronic control components.
5. Test igniter plugs.
6. Clean compressor blading with walnut shell treatment, as determined by experience.
7. Check for engine vibration with appropriate instrumentation if preliminary examination shows any sign of excessive vibration.
8. Flow-test fuel nozzles to ensure proper flow rate and spray pattern.

Compressor Cleaning. Long-time operation of a gas turbine may result in a gradual buildup of an adhesive oily deposit on the compressor components. The black deposit that is slightly sticky to the touch reflects the pollution present in metropolitan atmospheres. Engine performance gradually deteriorates as the deposit builds up; this deterioration is detectable as an increase in exhaust gas temperature over that of the new engine exhaust temperature for a given load. Clean the compressor according to the manufacturer's instructions when the exhaust temperature has risen to the *condemning limit.*

An accepted cleaning method consists of feeding an abrasive such as ground walnut shells into the compressor inlet of a running engine. The air inlet duct or plenum should contain access panels immediately adjacent to the compressor inlet to permit introduction of the cleaning compound and inspection of the compressor.

Overhaul

Overhaul of gas turbines includes such operations as the following:

1. Precision balance of all high speed rotor assemblies.
2. Inspection for cracks in all highly stressed parts such as turbine and compressor blades and wheels, engine shafting, gears, welded joints, sheet metal assemblies, etc. Magnetic parts may be examined by magnetic particle inspection. Turbines have many nonmagnetic, stainless steel, high-temperature materials which are examined by fluorescent penetrant inspection with ultraviolet light.
3. Examination of large-diameter labyrinth seals for excessive rubbing. These may require rework to remove burrs that have been raised from normal light rubbing.
4. A thorough check of electronic equipment usually used in fuel control systems to ensure proper control of turbine inlet temperature.

Factory Unit Exchange and Repair

Turbine engine production has reached a level where factory maintenance and overhaul services are common practice. These services provide the user with the technique approaching mass production overhaul and an inventory of spare parts. Production-type facilities enable the manufacturer to restore the engine to "new condition" in a minimum of time. Thorough inspection and production-testing by experienced personnel provide prompt, reliable repair and overhaul, obviating the requirement for specialized equipment, a large inventory of spare parts, and highly skilled repair personnel.

APPLICATIONS

Gas turbines have been successfully used in four major categories of applications where a specific characteristic of the gas turbine is of decisive importance. These four categories are: marine, motive power, electric generation, and industrial uses. Only the latter two categories are discussed here.

Electric Generation

Gas turbines are generally used in peaking applications and for hydroelectric stand-bys. For these purposes, turbines have been built in sizes up to 55,000 kw. A number of rail-mounted and flatbed truck-mounted mobile plants have been built for emergency service; these turbines have been built in sizes up to 6200 kw. For **peak load** service, the prime consideration is first cost since the units are not run long enough to make fuel costs significant. The units can be remotely controlled, need not be located at established plant sites, and can thus be located near the load center. These installations can be made with modified aircraft-type or industrially designed turbines. These large units have thermal efficiencies as high as 25 per cent without regeneration. The quick-start-to-full-load feature of the gas turbine also reduces the spinning reserve requirements of the generating system.

Industrial Uses

Gas Pipeline Compressors. Gas turbine compressor drives have been installed in sizes ranging from 1000 to 10,500 hp. Gas from the pipeline is generally the fuel. These are usually simple open-cycle units. The first cost and maintenance costs of the turbine-compressor station are usually sufficiently lower than those of a reciprocating engine compressor station to outweigh the efficiency difference. Added advantages of the gas turbine in this application are its ability to run without water and the ease with which it can be adapted to remote control.

Petroleum. The gas turbine finds application in oil fields for several purposes: oil well pressure maintenance programs, oil well cementing, and fracturing. One installation for well pressure maintenance in Venezuela uses 12 gas turbine-driven compressors, each having a maximum rating of 8000 hp, to compress 350 MMSCFD to 1950 psia for injection into the wells. Cementing equipment, mounted on trucks, is widely used, employing turbines in sizes up to 1100 hp. Fracturing of subsurface rock formations is also possible with gas turbine-driven reciprocating pumping equipment. Pressures as high as 10,000 psi are required. The light weight of the turbine, allowing it to be brought directly to the job site, is its major feature.

Steel. In steel production, gas turbines, burning blast furnace gases, are used to drive blast furnace compressors. An unusual feature of this application is that the volume of gases going to the power turbine is *less* than the volume of air and fuel gas supplied to the combustors because of the CO content of the blast furnace gas. These gas turbines are a form of bleed-air turbine, and have been built to deliver as much as 94,000 cfm at 22 psig.

Petrochemical. Gas turbines are found in many process applications, where they are successfully able to burn by-product gases that could not otherwise be used because of their low calorific value. These units do not generally require sophistication; moderate efficiencies and temperatures can produce economic advantages. A well-known use of the gas turbine is in the Houdry process. Electrical power and steam from exhaust heat recovery equipment are also produced in these applications.

Air Separation Plants. Large quantities of air are handled in this work, for which the gas turbine-driven centrifugal or axial compressor is highly suitable. In addition, exhaust heat from the turbine, recovered in the form of steam, can be used to drive electric generators, and other compression and refrigeration equipment.

Drying Operations. Brick-making operations, users of large amounts of low-temperature heat, can utilize the exhaust of a gas turbine directly, while the shaft power drives an electrical generator supplying the entire plant service for various conveyors. Similarly, a hay drier uses exhaust heat to dry the hay, while the electric power is used to operate the mechanical aspects of the drying operation.

Total Energy Systems. A number of these exist in which the turbine drives a generator for lighting, elevators, and miscellaneous motors and machines. Recovered exhaust heat in the form of steam is used to operate absorption air-conditioning units and provide winter heating. See Chap. 23 of this Section for information concerning total energy systems.

REFERENCES

1. Natl. Fire Protect. Assoc. *Standard for the Storage and Handling of Liquefied Petroleum Gases.* (NFPA 58) Boston, 1963.
2. ——. *Standard for the Installation and Use of Combustion Engines and Gas Turbines.* (NFPA 37) Boston, 1963.
3. ASME. *American Standard Code for Pressure Piping.* (ASA B31.1–1955) New York, 1955.

SECTION 13

CHEMICALS FROM NATURAL GAS

Frank H. Dotterweich, *Section Chairman,* Texas College of Arts and Industries, Kingsville, Tex.

The following authors assisted in the preparation of this Section:

Keith Buell, Phillips Petroleum Co., Bartlesville, Okla.

R. A. Cattell, U. S. Bureau of Mines, Washington, D. C.

Gordon Kiddoo, National Research Corp., Cambridge, Mass.

W. A. Kohlhoff (deceased), Portland Gas and Coke Co., Portland, Ore.

John J. McKetta, Jr., University of Texas, Austin, Tex.

ACKNOWLEDGMENTS

The authors wish to make acknowledgments to the following for their contributions to this Section:

H. O. Ervin, Research Engineer, Portland Gas and Coke Co., Portland, Oregon; **James Q. Wood,** Chemical Engineering Div., Phillips Petroleum Co., Bartlesville, Oklahoma; **Richard Gooding, Henry P. Wheeler, Jr.,** and **C. C. Anderson,** Petroleum and Natural Gas Branch, U. S. Bureau of Mines, Washington, D. C.; **M. W. Mayer,** Esso Engineering Dept., Standard Oil Development Co., Linden, New Jersey; **L. B. Pope,** Managing Editor, *Chemical Engineering,* New York, N. Y.; **Wynkoop Kiersted, Jr.,** Div. Manager, Process Div., Texaco Development Corp., New York, N. Y.; **Norman C. Updegraff,** Assistant Technical Director, Girdler Corp., Louisville, Kentucky; **A. W. Sprague,** Supt., Lion Chemical Co., El Dorado, Arkansas; **O. O. Wilson,** Editor, *The Oil and Gas Journal,* Tulsa, Oklahoma; **A. R. Powell,** Associate Manager, Research Dept., Koppers Co., Inc., Pittsburgh, Pennsylvania; **Arch L. Foster,** Editor, Refining and Gas Processing, *Petroleum Engineer,* Dallas, Texas; **Thomas C. Ponder,** Petrochemicals Editor, *Petroleum Refiner,* Houston, Texas; **A. F. Dyer,** Phillips Petroleum Co., Bartlesville, Oklahoma.

CONTENTS

Chapter 8—Production of Butadiene

Chapter 9—Production of Helium

Tables

Figures

Chapter 1
Introduction

by Frank H. Dotterweich

This Section covers the basic products made from natural gas. A discussion of the end products which are in turn made from these chemicals is beyond the scope of this Handbook. Many processes are patented; reference therefore should be made to lists of licensors for additional information.

On the basis of modern extraction and separation methods, the components of natural gas can be grouped as follows: (1) nonhydrocarbon material; (2) dry gas (methane, ethane, some propane); (3) liquefied petroleum gas (propane, *iso*-butane, and *n*-butane; and (4) natural gasoline (*iso*-butane thru normal heptane). This grouping is dictated in part by the demand for the components for use as fuel and as chemicals for solvents, for conversion to organic chemicals, or for production of carbon black.

During World War II, many new processes and products were developed that utilized components of natural gas as the starting raw material. The processes by which these components, chiefly methane, ethane, propane, and the butanes, could be shattered or rearranged included thermal, electrical, catalytic, and photosynthetic methods and allowed addition of elements such as oxygen, chlorine, and nitrogen.

CONVERSION METHODS APPLICABLE TO NATURAL GAS

The following 13 basic methods are directly or indirectly applicable for the conversion of natural gas to basic chemical raw materials:[1]

1. **Decomposition** (thermal, catalytic, electric)—splitting of the hydrocarbon molecule into smaller molecules or into carbon and hydrogen by heat alone (pyrolysis), with the aid of catalysts, or by electric discharge. It is generally accompanied, especially in pyrolysis, by recombination of some of the products into new compounds.

$$C_2H_6 \xrightarrow[\text{electric arc}]{\text{heat or}} H_2 + C_2H_4$$

$$C_2H_6 \xrightarrow[\text{electric arc}]{\text{heat or}} 2H_2 + C_2H_2$$

$$C_2H_6 \xrightarrow[\text{decomposition}]{\text{maximum}} 3H_2 + 2C$$

2. **Oxidation** (thermal, catalytic)—reaction of the hydrocarbon molecule with oxygen, air, or oxygen-containing compounds, activated by heat or catalyst, whereby oxygen is introduced into the hydrocarbon molecule or the molecule is changed to carbon monoxide and hydrogen, carbon dioxide and hydrogen, or finally to carbon dioxide and water.

$$2C_2H_6 + O_2 \xrightarrow[\text{pressure}]{\text{heat and}} \underset{\text{ethyl alcohol}}{2C_2H_5OH}$$

$$C_2H_6 + 4H_2O \xrightarrow[\text{catalyst}]{\text{heat and}} 7H_2 + 2CO_2$$

$$C_2H_6 \xrightarrow[\text{oxidation}]{\text{maximum}} 3H_2O + 2CO_2$$

$$C_2H_6 + O_2 \xrightarrow[\text{oxidation}]{\text{partial}} 3H_2 + 2CO$$

3. **Halogenation** (thermal, photolytic, catalytic)—reaction of the hydrocarbon molecule with a halogen (fluorine, chlorine, bromine, iodine), activated by heat, light, or catalyst, whereby one or more halogen atoms replaces an equivalent number of hydrogen atoms.

$$C_2H_6 + Cl_2 \xrightarrow[\text{or catalyst}]{\text{heat, light}} \underset{\text{ethyl chloride}}{C_2H_5Cl} + HCl$$

4. **Nitration** (thermal, vapor phase)—reaction of the hydrocarbon molecule with nitric acid, accelerated by heat and pressure, whereby a nitrogroup, NO_2, replaces a hydrogen atom.

$$C_2H_6 + \underset{\text{nitric acid}}{HNO_3} \xrightarrow[\text{phase}]{\text{vapor}} \underset{\text{nitroethane}}{C_2H_5NO_2} + H_2O$$

5. **Sulfurization**—reaction of the hydrocarbon molecule with sulfur or hydrogen sulfide to form sulfur-containing compounds such as organic sulfides, mercaptans, disulfides, or thiophenes.

$$C_2H_6 + S \xrightarrow{\text{catalyst}} \underset{\text{ethyl mercaptan}}{C_2H_5SH}$$

6. **Desulfurization** (catalytic)—removal of the sulfur atom from a sulfur–carbon–hydrogen molecule to form a sulfur-free molecule (activated by catalyst).

$$\underset{\text{ethyl mercaptan}}{C_2H_5SH} + H_2 \xrightarrow[\text{and heat}]{\text{catalyst}} C_2H_6 + H_2S$$

7. **Hydrogenation** (thermal catalytic, catalytic)—addition of hydrogen atoms to a hydrocarbon molecule, activated by heat and catalyst or catalyst alone to produce

one or more saturated molecules. Hydrogenation is termed (a) "destructive" when original saturated molecule is "cracked" to form more than one subsequently hydrogenated smaller molecule or (b) "nondestructive" when no cracking occurs.

(a) *Destructive:*

$$\underset{\text{pentane}}{C_5H_{12}} + H_2 \xrightarrow[\text{catalyst}]{\text{heat and}} C_3H_8 + C_2H_6$$

(b) *Nondestructive:*

$$\underset{\text{thylene}}{C_2H_4} + H_2 \xrightarrow{\text{catalyst}} C_2H_6$$

8. **Dehydrogenation** (thermal, catalytic)—a form of controlled decomposition whereby hydrogen atoms are removed from hydrocarbon molecules to form less saturated molecules. Dehydrogenation is termed (a) "destructive" when the original molecule is *cracked* to form more than one smaller molecule, or (b) "nondestructive" when no breaking of carbon–carbon bonds occur.

(*a*) *Destructive:*

$$\underset{\text{pentane}}{C_5H_{12}} \xrightarrow[\text{catalyst}]{\text{heat or}} \underset{\text{propylene}}{C_3H_6} + \underset{\text{ethylene}}{C_2H_4} + H_2$$

(*b*) *Nondestructive:*

$$C_2H_6 \xrightarrow{\text{heat}} \underset{\text{ethylene}}{C_2H_4} + H_2$$

9. **Alkylation** (thermal, catalytic)—chemical union of an alkyl radical to a hydrocarbon molecule. It is used particularly to designate combination of an olefin and an *iso*-paraffin or aromatic under high temperature conditions or in the presence of a catalyst.

$$\underset{\text{ethylene}}{C_2H_4} + \underset{\textit{iso}\text{-butane}}{C_4H_{10}} \xrightarrow{\text{heat}} \underset{\substack{\text{neohexane}\\ \text{(2,2-dimethylbutane)}}}{C_6H_{14}}$$

$$\underset{\text{ethylene}}{C_2H_4} + \underset{\textit{iso}\text{-butane}}{C_4H_{10}} \xrightarrow{\text{catalyst}} \underset{\substack{\text{diisopropyl}\\ \text{(2,3-dimethylbutane)}}}{C_6H_{14}}$$

$$\underset{\text{butylene}}{C_4H_8} + \underset{\textit{iso}\text{-butane}}{C_4H_{10}} \xrightarrow{\text{catalyst}} \underset{\substack{\text{isomeric}\\ \text{octanes (alkylate)}}}{C_8H_{18}}$$

$$\underset{\text{benzene}}{C_6H_6} + \underset{\text{propylene}}{C_3H_6} \xrightarrow{\text{catalyst}} \underset{\substack{\text{isopropyl}\\ \text{benzene (cumene)}}}{C_6H_5CH(CH_3)_2}$$

10. **Polymerization** (thermal, catalytic)—combination of small molecules or monomers into chain molecules or polymers of higher molecular weight.

(*a*) *Selective:*

$$\underset{\text{ethylene}}{nC_2H_4} \rightarrow \underset{\text{polyethylene}}{(C_2H_4)_n}$$

$$\underset{\text{propylene}}{C_3H_6} \rightarrow \underset{\text{propylene tetramer}}{C_{12}H_{24}}$$

(b) *Nonselective:*

$$\underset{\text{propylene}}{C_3H_6} + \underset{\text{amylene}}{C_5H_{10}} \rightarrow \underset{\text{mixed octylenes}}{C_8H_{16}}$$

11. **Isomerization** (thermal, catalytic)—transformation of the molecular structure of a hydrocarbon molecule without changing its empirical composition or molecular weight.

$$\underset{\text{butane}}{CH_3CH_2CH_2CH_3} \xrightarrow{\text{catalyst}} \underset{\textit{iso}\text{-butane}}{CH_3{-}\overset{\overset{CH_3}{|}}{C}H{-}CH_3}$$

12. **Cyclization** and **Aromatization** (thermal, catalytic) —conversion of paraffinic or olefinic hydrocarbon molecules to cyclic and aromatic molecules. It is accompanied by dehydrogenation and by prior polymerization in some cases, as when gaseous hydrocarbons are used as raw material.

$$\underset{\text{hexane}}{C_6H_{14}} \xrightarrow{\text{catalyst}} \underset{\text{cyclohexane}}{C_6H_{12}} + H_2$$

$$\underset{\text{cyclohexane}}{C_6H_{12}} \xrightarrow{\text{catalyst}} \underset{\text{benzene}}{C_6H_6} + 3H_2$$

Table 13-1 Hydrocarbon Raw Material Requirements and Primary Product Output of Petrochemicals by End-Use Groups in U. S.[3]—1965 to 1975

(millions of pounds annually)

End-Use group	1965 H.C.	1965 P.P.	1970 H.C.	1970 P.P.	1975 H.C.	1975 P.P.
1. Synthetic fibers	2,010	2,100	2,330	2,470	2,540	2,740
2. Synthetic rubbers	2,590	2,960	2,790	3,180	3,130	3,510
3. Plastics and plasticizers	10,440	14,460	13,920	19,320	17,540	24,670
4. Protective coatings	1,830	1,870	2,180	2,230	2,530	2,590
5. Automotive chemicals	1,670	2,890	1,820	3,150	1,930	3,400
6. Nitrogen products	2,830	5,370	3,430	6,500	4,040	7,670
7. Detergents	860	940	970	1,060	1,020	1,100
8. Miscellaneous	4,850	6,880	5,420	7,700	6,050	8,500
Hydrocarbons to products and losses	27,080		32,860		38,780	
Liquid hydrocarbons degraded to fuel	9,200		11,840		14,450	
Total hydrocarbons consumed	36,280		44,700		53,230	
Primary product output		37,470		45,610		54,180

H.C.—.ydrocarbon raw materials.
P.P.—Primary petrochemical products.

13. **Hydrocarbon Synthesis**—formation of hydrocarbons from organic or inorganic materials identical with those from petroleum by synthetic processes such as the Fischer-Tropsch synthesis or action of water on calcium carbide.

$$\underset{\text{synthesis gas}}{nCO + (2n + 1)H_2} \xrightarrow[\text{reaction (catalytic)}]{\text{Fischer-Tropsch}} \underset{\substack{\text{straight chain}\\ \text{paraffin}\\ \text{hydrocarbons}}}{C_nH_{2n+2}} + nH_2O$$

$$\underset{\text{calcium carbide}}{CaC_2} + 2H_2O \rightarrow \underset{\text{calcium hydroxide}}{Ca(OH)_2} + \underset{\text{acetylene}}{C_2H_2}$$

Table 13-2 Classified Petroleum and Natural Gas Requirements for Petrochemicals in U. S.[3]—1965 to 1975

(millions of pounds annually)

	1965	1970	1975
Natural gas:*			
Use other than ethylene†	8,100	9,680	11,150
Ethylene from natural gas	2,500	2,880	3,120
Total natural gas	10,600	12,560	14,270
Petroleum:†			
Ethylene:			
To product and losses	3,230	4,020	4,940
End products—fuel gases	760	890	1,020
Total consumed	3,990	4,910	5,960
Propylene	2,280	2,580	2,820
Butylenes and butanes:			
To product and losses	2,450	2,680	3,040
End products—fuel gases	420	390	380
Total consumed	2,870	3,070	3,420
Pentenes, pentanes, and higher	740	900	1,060
Aromatics:			
To product and losses	7,780	10,120	12,650
End products—fuel gases	8,020	10,560	13,050
Total consumed	15,800	20,680	25,700
Summary:			
Natural gas consumed	10,600	12,560	14,270
Petroleum to product and loss	16,480	20,300	24,510
Total to product and loss	27,080	32,860	38,780
Petroleum degraded to fuel gas	9,200	11,840	14,450
Total hydrocarbons consumed	36,280	44,700	53,230

* Cumulative requirements for both existing and new petrochemicals production.

† Includes some higher paraffins than methane and ethane.

HYDROCARBON RAW MATERIAL REQUIREMENTS FROM NATURAL GAS[2]

The period following World War II indicated that a readjustment and stabilization in the petrochemical industry using natural gas as a raw material was imminent. Tables 13-1 thru 13-3 show the estimated industry requirements for the period 1965 to 1975.

Table 13-3 Investment Requirements* For Petrochemicals By End-Use Groups in U. S.[3]—1965 to 1975

(figures in millions of dollars for 5-yr periods ending in the years indicated)

End-Use group	1965	1970	1975
1. Synthetic fibers	34.7	46.6	26.7
2. Synthetic rubbers	114.2	94.4	113.9
3. Plastics	519.0	556.1	572.8
4. Surface coatings	36.3	37.1	38.3
5. Automotive chemicals	81.9	77.5	69.2
6. Nitrogen products	121.5	122.4	123.4
7. Detergents	8.0	4.8	6.8
8. Miscellaneous	116.1	115.4	103.7
Total	1,031.7	1,054.3	1,054.8

* Investments are for June 1950 construction costs, and include on-site, off-site and utilities requirements. The investment coverage is about 89 per cent for primary products and ethylene. No investments are included for any end products except synthetic rubber.

REFERENCES

1. Smith, H. M., and Holliman, W. C. *Utilization of Natural Gas for Chemical Products.* (Bureau of Mines I.C. 7347) Washington, D. C. 1947.
2. U. S. President's Materials Policy Commission. *Resources for Freedom, Vol. IV: Promise of Technology.* Washington, D. C., G.P.O., 1952.
3. "Forecasts for Petroleum Chemicals." *Petrol. Process.* 7: 1289–93, Sept. 1952. Also in: *Oil Gas J.* 51: 102–5, Sept. 1; 114+, Sept. 8; 98+, Sept. 15, 1952.

Chapter 2

Production of Synthesis Gases and Hydrogen

by Frank H. Dotterweich

Natural gas hydrocarbons will react with steam over catalysts at elevated temperatures to produce hydrogen, carbon, and oxides of carbon. With the aid of catalysts the hydrocarbon may be almost completely converted to oxides of carbon and hydrogen. The two principal reactions which occur with the paraffin hydrocarbons are as follows:

$$C_nH_{(2n+2)} + nH_2O \rightleftharpoons nCO + (2n + 1)H_2 \qquad (1)$$

$$C_nH_{(2n+2)} + 2nH_2O \rightleftharpoons nCO_2 + (3n + 1)H_2 \qquad (2)$$

These reactions are both endothermic. Reaction **2** is considered to be the summation of Reaction **1** and the "water gas reaction," which is:

$$nCO + nH_2O \rightleftharpoons nCO_2 + nH_2 \qquad (3)$$

At high temperatures, natural gas hydrocarbons can be partially oxidized with oxygen to produce hydrogen and oxides of carbon. The reaction is exothermic and no catalyst is required.

HYDROGEN–CARBON MONOXIDE MIXTURES

Since the gas mixture resulting from the steam–hydrocarbon reaction consists of carbon monoxide carbon dioxide, and hydrogen, plus any unchanged methane, it is possible to produce mixtures of carbon monoxide and hydrogen[1] which are suitable for use in various synthesis operations, such as the methanol and the Oxo process and Fischer-Tropsch synthesis. To obtain the high $CO:H_2$ ratios required in many of these synthesis gas mixtures, it is necessary to introduce carbon dioxide into the reformer furnace to replace part or all of the steam used to react with the hydrocarbon. The following reaction is involved:

$$C_nH_{(2n+2)} + nCO_2 \rightleftharpoons 2nCO + (n + 1)H_2 \qquad (4)$$

Under many conditions the reactions can be carried out so that there will be only a small amount of carbon dioxide in the product gas. If complete removal is required, this can be accomplished by scrubbing with suitable reagents.

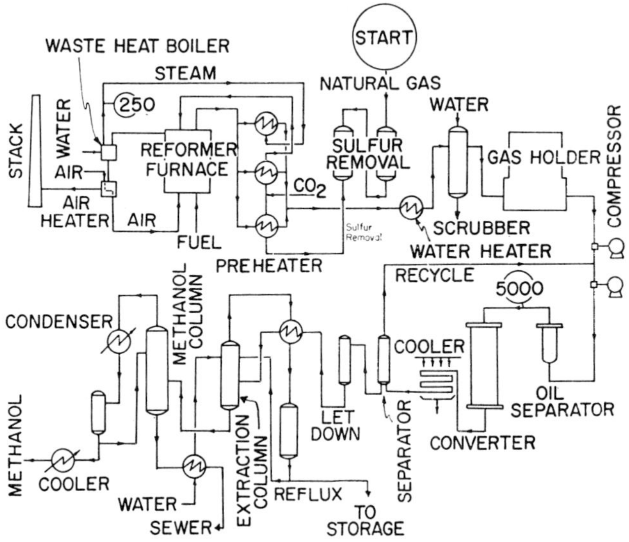

Fig. 13-1 Process for methanol production.[3] (Commercial Solvents Corp.; copyright, *Petrol. Refiner*)

Table 13-4 Production of Various Synthesis Gas Mixtures from Natural Gas[2]

	Inlet Flows, SCF/hr				Inlet gas ratios		Product gas, SCF/hr					Composition*		
	Run No.	Natl. gas	CO_2	Steam	CO_2: natl. gas	H_2O: natl. gas	Meas.	Calc.	H_2:CO ratio	H_2	Product gas, CO	CO_2	CH_4	N_2
High H_2:CO ratio	1	1260	1330	2445	1.06	1.94	5460	5660	1.97	65.0	33.0	0	1.4	0.6
	2	612	478	1265	0.78	2.07	2760	2750	2.00	65.6	32.8	0	0.4	1.2
	3	99	328	852	3.31	8.60	477	444	1.98	65.2	32.8	0	0.2	1.8
Medium H_2:CO ratio	4	835	1740	1345	2.08	1.61	3710	3750	1.04	51.1	48.9	0	0	0
	5	1650	3420	1570	2.08	0.95	7450	7400	0.93	47.7	51.2	0.4	0.5	0.2
Low H_2:CO ratio	6	629	1990	0	3.17	...	2830	2820	0.44	30.5	69.2	...	...	0.3
	7	619	4360	0	7.05	...	2790	2780	0.21	17.5	82.2	0.1	Trace	0.2
	8	292	4320	0	14.80	...	1580	1310	0.10	7.5	73.9	18.6	...	...

*Run No. 4 analysis by precision Orsat; others by mass spectrometer.

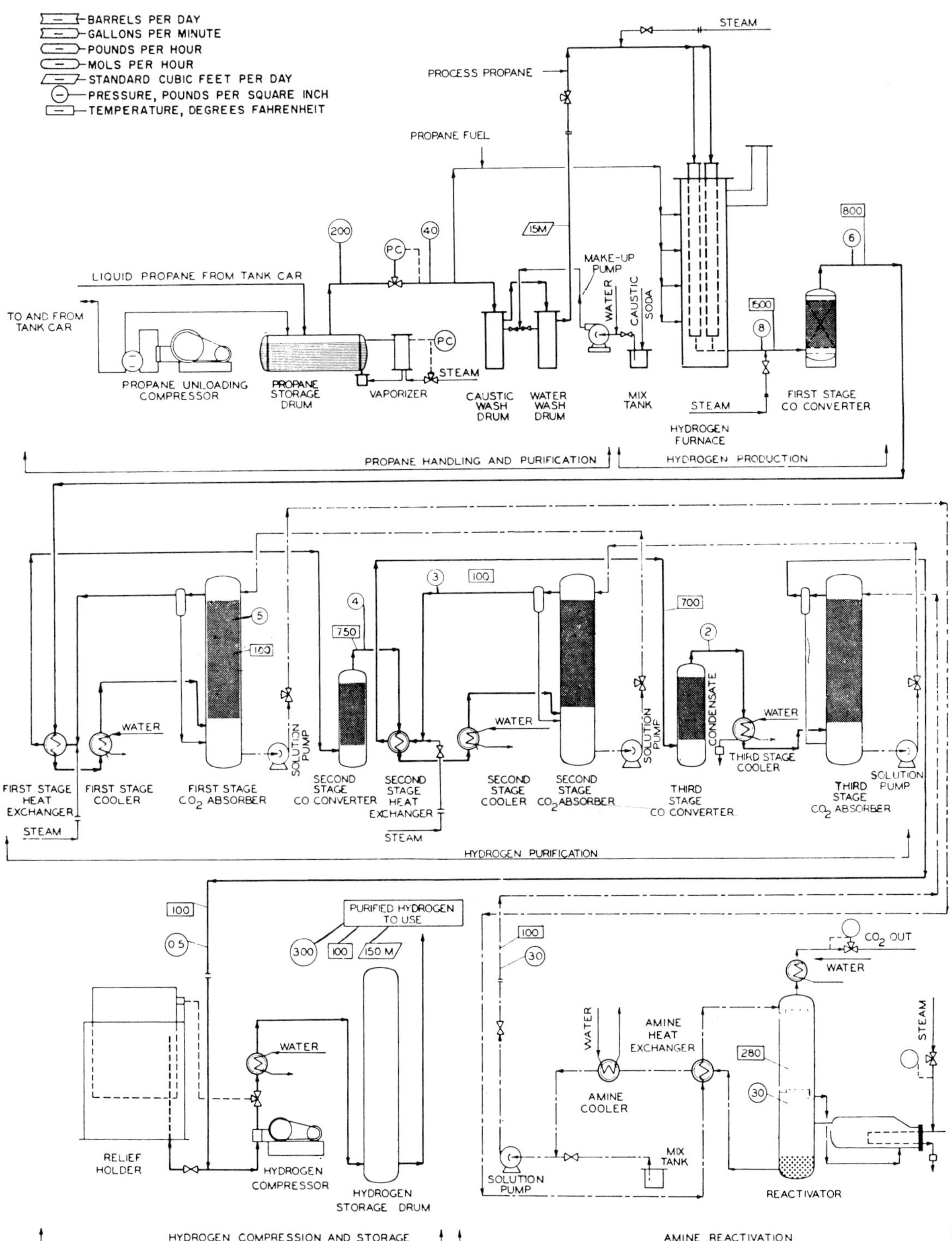

Fig. 13-2 Girdler Hygirtol process for hydrogen production.[2]

Table 13-4 presents data on production of synthesis gas mixtures of three hydrogen–carbon monoxide ratios.

Methanol from Synthesis Gas

In the production of methanol (methyl alcohol), natural gas is reformed to synthesis gas, preferably two parts hydrogen to one of carbon monoxide. These are then combined to form methanol. It may also be formed when carbon dioxide is combined with hydrogen. This method results in a diluted product requiring excess hydrogen but has its advantages in industrial production. Using methane as the basic raw material, the following reactions occur:

$$CH_4 + H_2O \rightarrow 3H_2 + CO$$

$$CH_4 + 2H_2O \rightarrow 4H_2 + CO_2$$

$$CO + 2H_2O \rightarrow CH_3OH$$

$$CO_2 + 3H_2 \rightarrow CH_3OH + H_2O$$

When methane is primarily the raw material, the hydrogen ratio is high. It may be reduced by adding heavier hydrocarbon gases to the feed or adding carbon dioxide from an outside source.

The reaction between hydrogen and carbon monoxide to form methanol is highly exothermic, while that between carbon dioxide and hydrogen is less so, producing about one-half the heat. Use of carbon dioxide and hydrogen reduces the tendency toward "hot spots" in the conversion unit and allows the synthesis to proceed under more favorable conditions.

Figure 13-1 is a diagram of the process for production of methanol. Various other processes are in commercial operation.[4]

Natural gas feedstock is purified of sulfur, preheated, mixed with carbon dioxide and steam, and then passed downward thru the reformer furnace. The reformed gas is quenched, passed thru a holder, and then compressed to 5000 psig. Entrained oil is removed in oil traps and the main outlet gas stream introduced to the converter. The effluent gas is cooled and flows to a separator. The condensed product from the separator flows to the low-pressure letdown and separator drum. The liquid product from the letdown drum flows to crude methanol storage. Impurities are then removed by extractive distillation.

The converter operates at 5000 psig and 400 C with about 12 to 15 per cent conversion of feed gas to methanol per pass.

HYDROGEN

Production of hydrogen from natural gas on a large scale was initiated in 1931. Hydrogen has been used in the petroleum industry for hydrogenation of various petroleum materials. Its second large-scale utilization was for production of synthetic ammonia. This process has subsequently been applied on both a large and small scale, in several branches of both the chemical and food industries.

Most hydrogen for ammonia production is manufactured either by partial oxidation of natural gas or by steam reforming of natural gas to a mixture of hydrogen and oxides of carbon. In partial oxidation of natural gas to produce hydrogen and oxides of carbon, additional yield results from the water–gas shift reaction, during which most of the carbon monoxide is converted into CO_2 and H_2. The CO_2 is removed by alkaline scrubbing, either with an amine solution or a regenerative caustic solution. The hydrogen-rich gas is refrigerated to low temperatures and purified by fractionation. The final liquefaction of hydrogen occurs at temperatures below −450 F. In addition to the usual heat of liquefaction, which has to be removed, there is a high heat of transition involved (resulting from a catalyzed change in the hydrogen molecule).

High-Purity Hydrogen

A typical plant[2] for production of high-purity hydrogen from natural gas, propane, or butane will consist of the following five sections: (1) hydrocarbon handling and purification; (2) hydrogen production; (3) hydrogen purification; (4) amine reactivation; and (5) hydrogen compression and storage.

Figure 13-2 shows the flow diagram for Girdler's Hygirtol hydrogen manufacturing plant using propane gas. A plant of this type will produce hydrogen sufficiently pure for all applications (for example, hydrogenation of edible oils) except those few specialized uses, such as the heat treating of vacuum tube electrodes, in which traces of carbon compounds cannot be tolerated. Hygirtol plant hydrogen can, however, be purified further so as to be suitable for this use also.

Since commercial natural gas, propane, and butane may contain traces of sulfur compounds, either as impurities or as added odorants, their desulfurization is necessary before they can be utilized for hydrogen production. A caustic wash and a water wash may be employed to remove traces of mercaptans that may be present in propane or butane from natural gasoline. When materials obtained from refinery operations which may contain impurities such as carbonyl sulfide are handled, a catalytic process should be employed for treating the feedstock.

The hydrogen production section of the plant (Fig. 13-2) consists of the hydrogen furnace itself, containing vertical alloy steel tubes filled with a supported nickel catalyst thru which the hydrocarbon–steam mixture is passed at a temperature of about 1400 to 1600 F. At the furnace outlet, steam is added to the product gas stream to cool it to about 700 F before it enters the first-stage converter (iron oxide, water–gas shift catalyst), where most of the carbon monoxide is converted into CO_2 and H_2. The hot product gases from the converter after cooling enter the first-stage carbon dioxide absorber, in which CO_2 is scrubbed out (35 lb per MCF of H_2 produced) with aqueous monoethanolamine solution. The product gas leaving the absorber is mixed with steam, reheated to 700 F, and then passed thru the second-stage carbon monoxide converter. Methanation is sometimes used in place of a third stage of conversion, as shown in Fig. 13-2.

Process material and utility requirements for producing hydrogen of the analysis given in a typical Hygirtol plant are approximately as follows:

Material	Quantity per MCF of H_2	H_2 Analysis	
		Component	Mole per cent
Propane feed	2.75 gal	CO_2	0.01
Natural gas feed	250 cu ft	CO	0.01
Heat in fuel (gas or oil)	250 MBtu	CH_4	0.10
Steam	380 lb	N_2*	0.01
Cooling water (30°F rise)	1600 gal	O_2	0.00
Power	2 kwh	H_2	99.87
		Total	100.00

* By difference.

REFERENCES

1. Clark, E. L., and others. "Synthesis Gas Production." *Chem. Eng. Progr.* 45: 651–4, Nov. 1949.
2. Read, R. M., and Eriksen, A. "Hydrogen and Synthesis Gas Production." *Gas* 24: 53–6, Oct. 1948.
3. "Methanol." *Petrol. Refiner* 38: 267, Nov. 1959.
4. "Petrochemical Handbook Issue." *Petrol. Refiner* 40, No. 11, Nov. 1961.

Chapter 3
Production of Synthetic Ammonia and Ammonia Products

by Frank H. Dotterweich

SYNTHETIC AMMONIA

Figure 13-3 shows the process flow for an intermediate-pressure synthetic ammonia plant using natural gas as a source of hydrogen. Natural gas often contains a relatively high percentage of nitrogen which is of advantage for production of ammonia. About one-half of the natural gas used is required by the process itself and by the gas engine compressors. This gas is scrubbed with monoethanolamine to remove H_2S. The balance of the natural gas is used as boiler and reforming furnace fuel. It may not be necessary to remove H_2S from this gas. Roughly 33.5 MCF of natural gas is required per ton of ammonia.

Organic sulfur is removed by zinc oxide catalyst, after preheating to about 750 F. Steam similarly preheated is then mixed with the reaction gas in a 1:1 weight ratio. The mixture passes down thru the stainless steel tubes of the primary reform furnaces (temperature about 1300 F), which contain nickel oxide in pellet form. The gas reforming reactions are mainly:

$$CH_4 + H_2O \rightleftharpoons CO + 3H_2$$

$$CH_4 + 2H_2O \rightleftharpoons CO + 4H_2$$

In addition, some carbon is formed by the following reaction, requiring periodic catalyst steaming:

$$CH_4 \rightleftharpoons C + 2H_2$$

The gas then passes to the secondary reformers, where air is introduced to furnish the required amount of nitrogen. These reformers, also containing nickel oxide catalyst, serve to clean up residual methane and to use up the oxygen (from the air) by combustion of part of the gases, the temperature rising to 1830 F. The secondary reformers are brick-lined horizontal cylindrical vessels, with the catalyst supported on grids.

Waste heat boilers then reduce the temperature to 950 F. The gases next pass thru CO converters, where steam again is added and the water–gas shift reaction $CO + H_2O \rightleftharpoons CO_2 + H_2$ takes place.

The reformed gas is quenched to atmospheric temperature by direct spray water coolers and stored. Its approximate composition follows:

Component:	H_2	N_2	CH_4 + inerts	CO	CO_2
Mole per cent:	60.0	20.0	0.2	2.5	16.3

The reformed gas is drawn from the holder by gas engine-driven compressors and compressed in three stages to about 250 psig. At this pressure it is passed thru absorbers (packed towers) countercurrent to aqueous monoethanolamine solution, which reduces its CO_2 content to about 0.5 per cent. The amine solution is regenerated by heating at low pressure, and substantially pure CO_2 is vented.

The process gas pressure is raised in two additional stages to 1800 psig. At this pressure it flows through packed tower scrubbers countercurrent to a cold (32 F) aqueous ammoniacal solution of copper ammonium formate for substantially complete absorption of CO and partial removal of residual CO_2. The copper ammonium formate is regenerated by reducing the pressure, then heating and blowing with air to oxidize part of the reduced copper back to the cupric form. The rest of the CO_2 is then absorbed in aqueous ammonia in a separate scrubber. The process gas now contains, besides N_2 and H_2, only the contaminants methane, argon, and other inerts.

After a sixth stage of compression to 5400 psig, the process gas is blended with recirculated unconverted gas containing part of the synthesized ammonia. These gases pass thru heat exchangers for precooling and for final deep cooling to remove ammonia. The blended gas then goes to the combined converter-heat exchangers, where synthesis of ammonia over granular iron catalyst in tubular "baskets" takes place at about 930 F.

Under these conditions the conversion per pass is about 12 per cent and the overall hydrogen efficiency is about 85 per cent. Purge (so-called voluntary purge) facilities are provided for the gas entering the converters to keep methane, argon, and other inerts low. This reduces the efficiency of the process.

The greater part of the ammonia produced per pass over the synthetic catalyst is condensed out at 5000 psig by water-cooled condensers; some plants operate at 10,000 psig. This, together with the remainder of the ammonia, condensed out in the indirect, ammonia-cooled condensers, is removed from the process in lower pressure letdown tanks as liquid. It flows to lower pressure storage spheres, which are refrigerated by taking ammonia vapor from the top of the liquid and then compressing and condensing it. Noncondensable gases are purged from this compression system. From the spheres, liquid ammonia may be pumped to tank cars or used in other ways.

The conventional method for producing synthetic ammonia as described above is undergoing modification. In one development, the catalytic steam–methane process takes

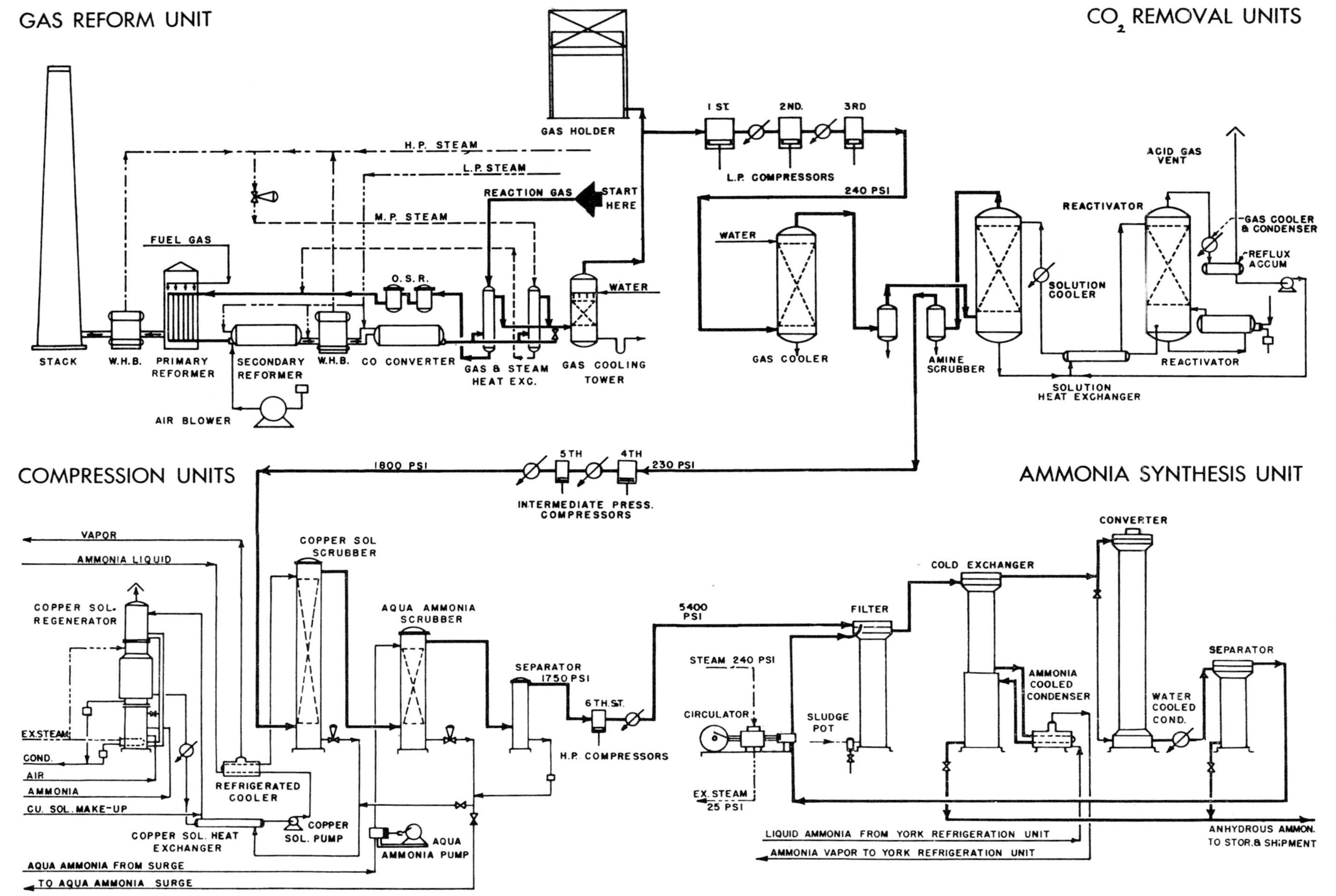

Fig. 13-3 Cactus synthetic ammonia plant, Etter, Tex.[1]

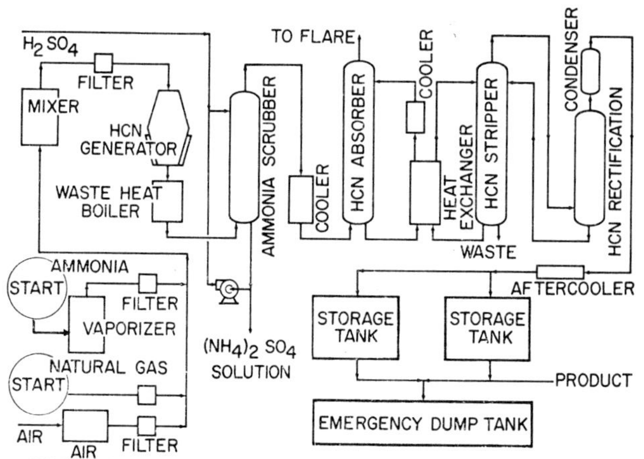

Fig. 13-4 Andrussow process for hydrogen cyanide production. (Girdler Corp.; copyright, *Petrol. Refiner.*)

place at elevated pressure so that savings in compression costs can be realized.[2] Another modification employs partial oxidation of natural gas feed.[3] A further development uses liquid nitrogen (nitrogen wash) to absorb the impurities from hydrogen, thus eliminating copper ammonium formate scrubbing.

Manufacturing cost data for ammonia made from natural gas and a report[4] of the processes in use in 1964 indicate that, at a production rate of 400 tons of ammonia per day, manufacturing cost would be $30.00 per short ton, with natural gas at $0.40 per MMBtu. About 40 per cent of this cost is the cost of natural gas.

Approximately 31,000 cu ft of natural gas are required per ton of ammonia produced when electric motor compression is used, and the process requires 625 kwh of electric power. When gas compression is used, the natural gas requirements rise to 33,500 cu ft per ton of ammonia, and the electric power requirements drop to 55 kwh.

In January 1964, the capacity of U. S. anhydrous ammonia plants was 7.6 million tons per year; 81.2 per cent of this production was based on the use of natural gas as the raw material.

Significant advances have taken place in the development of the partial oxidation process.[5] Reactions are employed to produce hydrogen and carbon monoxide with either oxygen or oxygen-enriched air. Pressures as high as 32 atm have generally been used, but at least one variation of the process has operated at a pressure as high as 82 atm.

Reference 4 furnishes detailed breakdowns of the manufacturing costs with various daily capacities.

Products Derived from Ammonia

Hydrogen Cyanide. Production may be by the Andrussow process, in which methane, air, and ammonia react as follows:

$$2CH_4 + 2NH_3 + 3O_2 \rightarrow 2HCN \pm 6H_2O$$

In industrial operations (Fig. 13-4) the reaction is carried out at about 2000 F and 1.0 atm over a platinum catalyst. The raw materials are mixed and enter a reactor where a portion of the natural gas reacts with the oxygen of the air to supply heat for the foregoing endothermic reaction. The products from the reactor are scrubbed with an acid solution to recover unreacted ammonia. The gases then pass to an absorber where the hydrogen cyanide is removed by water and the unreacted methane, air, and products of combustion are vented. The hydrogen cyanide is removed from the water solution in a stripper; the anhydrous product is a stable liquid since it is handled and stored under refrigeration. The yield of hydrogen cyanide is reported at about 75 per cent of the theoretical yield, based on the quantity of the ammonia raw material used.

REFERENCES

1. McCullough, G. W., and others. "Fertilizer from Petroleum." *Petrol. Process.* 6: 380–7, Apr. 1951.
2. Reidel, J. C. "Natural Gas to Ammonia: a Look at Round 2 of NH_3 Expansion." *Oil Gas J.* 52: 86+, Mar. 8, 1954.
3. ——. "Unique Ammonia Plant Opened." *Oil Gas J.* 52: 60–2, Feb. 8, 1954.
4. Strelzoff, S. "Ammonia Manufacturing Processes Completely Change in 10 Years." *Oil Gas J.* A63: 76+, Jan. 11, 1965.
5. Strelzoff, S., and Pan, L. C. "Synthetic Ammonia." (In: McKetta, J. J., Jr., ed. *Advances in Petroleum Chemistry and Refining*, vol. 9, chap. 7. New York, Interscience, 1964.)

Chapter 4
Production of Ethylene and Major Derivatives

by Keith Buell

Ethylene ranks first in importance as a raw material to the petrochemical industry. Its commercial production is generally achieved by: (1) recovery from petroleum refinery gases; (2) thermal cracking of light hydrocarbons, principally ethane and propane; or (3) a combination of (1) and (2). Recovery of ethylene, in general, is by low-temperature absorption and fractionation, at moderate to high pressures.

THERMAL CRACKING OF LIGHT HYDROCARBONS

Chemistry of Cracking Ethane and Propane

When propane–ethane mixtures are cracked at low pressures, high temperatures, and relatively short contact times, the major products resulting are hydrogen, methane, ethylene, and propylene:

$$C_2H_6 \rightarrow H_2 + C_2H_4 \quad (1)$$

$$C_3H_8 \rightarrow CH_4 + C_2H_4 \quad (2)$$

$$C_3H_8 \rightarrow H_2 + C_3H_6 \quad (3)$$

The relationship between reaction time and the percentage of propane or ethane cracked at a given fixed temperature is expressed by the equation:

$$k_1 = \frac{1}{t} \ln \left(\frac{100}{100 - x} \right) \quad (4)$$

where: k_1 = cracking velocity constant per sec

t = time, sec

x = lb propane or ethane cracked per 100 lb of propane or ethane in feed

The cracking velocity (reaction rate) constant, k_1, varies with temperature and reactant used.

For **propane:**[1]

$$\log_{10} k_1 = 13.11 - \frac{24{,}650}{460 + °F} \quad (5)$$

for the temperature range of 1050 to 2000 F.

For **ethane:**

$$\log_{10} k_1 = 14.02 - \frac{27{,}423}{460 + °F} \quad (6)$$

for the temperature range of 1400 to 2000 F.

Table 13-5 gives values of cracking velocity constants for propane and ethane at various temperatures. It shows that appreciably higher temperatures are required for cracking ethane.

Secondary reactions following the primary decomposition of propane and ethane involve not only decomposition of the primary reaction products, but also thermal polymerization of the olefinic products. These olefins polymerize to yield cyclic olefins, which in turn dehydrogenate to aromatics. These further condense and dehydrogenate under progressively more severe conditions, yielding products such as tars and coke. Both thermal polymerization reactions and those of primary cracking are affected by time, temperature, and pressure.

Table 13-5 Cracking Velocity Constants for Propane and Ethane[2]

	Cracking velocity constant, k_1	
Temp, °F	Propane	Ethane
1300	0.13	...
1350	0.31	...
1400	0.72	0.19
1450	1.60	0.46
1500	3.41	1.07

Primary pyrolysis reactions are endothermic and relatively rapid, whereas those of secondary polymerization and condensation are moderately exothermic and relatively slow. Under the operating conditions of the propane–ethane cracking process, the extent of such secondary reactions is not great. It is very important that the cracking reaction be conducted under conditions of as short a residence time and as low a pressure as possible, since the greater the time and pressure, the greater the extent of thermal polymerization. The secondary reactions should be suppressed to prevent consumption of ethylene thru formation of heavy oils, tars, and coke. Production of these undesirable by-products also imposes certain operating difficulties.

ETHYLENE PRODUCTION PROCESSES

Phillips Process[2]

A propane-rich stream containing considerable ethane is cracked thermally in a tube still* to produce ethylene at tem-

* Processes with radiant heaters and cracking units have also been used.

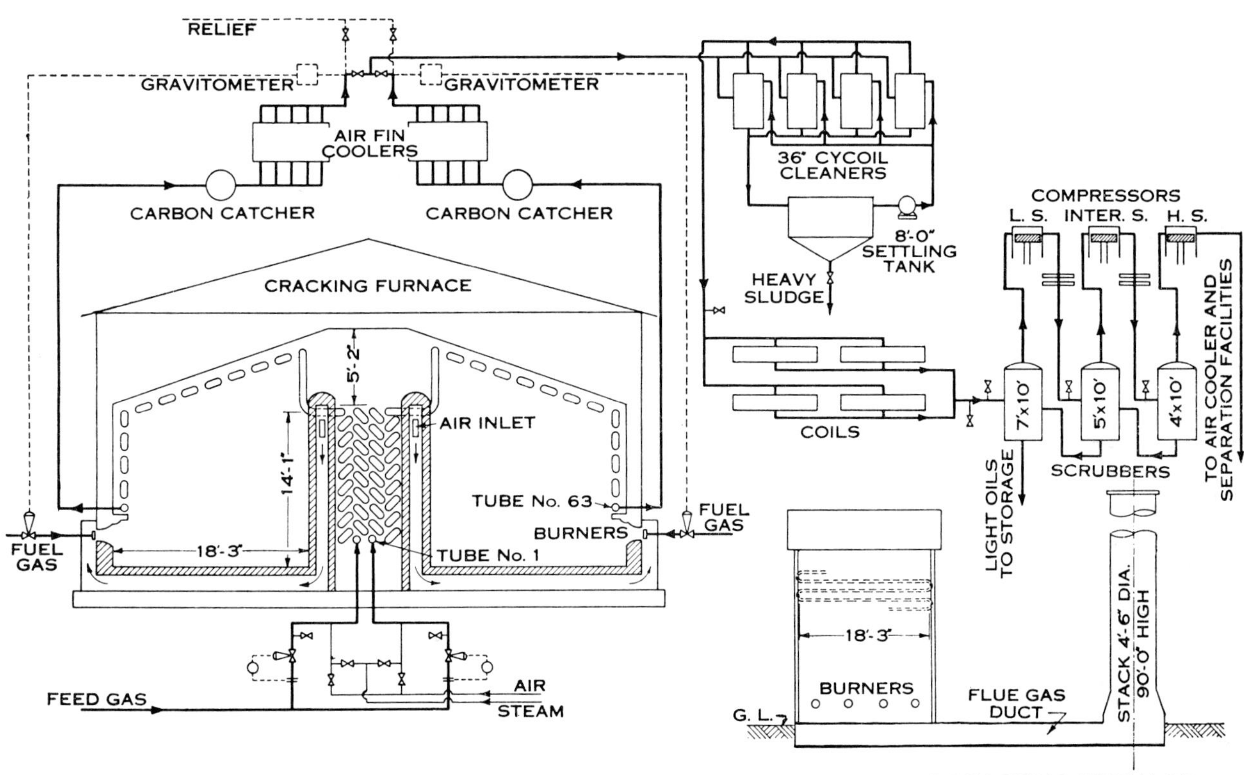

Fig. 13-5 Thermal cracking furnace arrangement for ethylene production from propane–ethane mixtures.[2]

Table 13-6 Cracking Furnace Feed Volume and Composition[2]

(feed to one-half of one furnace)

Component	Pounds	Wt %	Moles	Mole %	Cu ft*
Hydrogen	83	0.88	42	1.17	15,918
Methane	12,333	11.53	771	21.53	292,209
Ethylene	5,052	4.72	180	5.03	68,220
Ethane	52,100	48.63	1,737	48.50	658,323
Propylene	3,250	3.04	77	2.15	29,183
Propane	33,600	31.40	763	21.31	289,177
C_4	637	0.60	11	0.31	4,169
Total per day	107,055	100.00	3,581	100.00	1,357,199
Total per hour					56,550
Average molecular weight: 29.9					
Specific gravity* = 1.032 (air = 1.000)					

* At 14.7 psia and 60 F.

Table 13-7 Cracking Furnace Effluent Gas Volume and Composition[2]

(effluent from one-half of one furnace)

Component	Pounds	Wt %	Moles	Mole %	Cu ft*
Hydrogen	2,619	2.45	1,308	24.18	495,732
Methane	25,996	24.28	1,625	30.04	615,875
Acetylene	531	0.50	20	0.37	7,580
Ethylene	38,365	35.84	1,370	25.32	519,230
Ethane	21,181	19.78	706	13.05	267,574
Propylene	8,747	8.17	208	3.84	78,832
Propane	4,454	4.16	101	1.87	38,279
Butylene	1,430	1.34	25	0.46	9,475
Butane	687	0.64	12	0.22	4,548
C_5+	3,045	2.84	35	0.65	13,265
Total per day	107,055	100.00	5,410	100.00	2,050,390
Total per hour					85,433
Average molecular weight: 19.78					
Specific gravity* = 0.682 (air = 1.000)					

* At 14.7 psia and 60 F.

Table 13-8 Summary of Cracking Furnace Design Factors[2]

	Three furnaces	One-half furnace
Feed, lb/day	642,327	107,055
Steam temperatures, °F		
Inlet convection section	100	100
Inlet radiant section	1,290	1,290
Outlet radiant section	1,500	1,500
Stream pressures, psig		
Inlet convection section	44	44
Inlet radiant section	20.6	20.6
Outlet radiant section	10	10
Heat requirements, MBtu/hr		
Reaction heat	18,500	3,080
Sensible heat	30,000	5,000
Total	48,500	8,080
Heat distribution in furnace, MBtu/hr		
Radiant coil:		
Reaction heat	18,500	3,080
Sensible heat	6,000	1,000
Convection coil sensible heat	24,000	4,000
Radiation loss from radiant section	2,910	485
Radiation loss from convection section	1,940	323
Lost in stack gases	43,650	7,275
Total	97,000	16,163
Fuel gas requirements,* cu ft/hr at 14.7 psia and 60 F	100,000	16,700
Furnace temperatures, °F		
Flame	2,250	2,250
Bridge wall	1,670	1,670
Bottom of stack	1,090	1,090

	Radiant coil	Conv. coil	Radiant coil	Conv. coil
Furnace tubes:				
Surface required, sq ft	4083	4080	681	680
Size, in.	5.5 × ¼	4 × ¼	5.5 × ¼	4 × ¼
Length, ft	20	20	20	20
Eff. heating length, ft	18¼	18¼	18¼	18¼
Number:				
Preheat	38.1	216	6.35	36†
Reaction	117.9	...	19.65	...
Total	156.0	216	26.00	36†
Alloy	25-20	18-8‡	25-20	18-8‡

	Radiant coil: Reaction section	Radiant coil: Preheat section	Conv. coil	Total
Pressure drop, psi	8.75	1.83	23.35	33.93
Residence time, sec	1.37	0.74	4.47	6.58

Three stacks, 4½ ft diam, 90 ft high (one on each furnace)

* Net heating value, 970 Btu/cu ft.

† In furnace as erected, 37 convection tubes were actually installed.

‡ Two top rows of tubes in convection section 25-20, remainder 18-8.

peratures between 1400 and 1500 F. The ethylene is purified and used in manufacture of aviation gasoline (or in other processes).

Cracking Furnace Design. A furnace must accomplish the twin objectives of preheating the feedstream to cracking temperature and of furnishing the heat necessary for the desired degree of cracking quickly. The heat of reaction must be added to the gas at as low a pressure as possible. Thus, the cracking coil must be designed for a low-pressure drop. To keep residence time and cracking temperature to a minimum, gas velocities must be high in the last portion of the cracking coil.

The cracking furnace design (Fig. 13-5) discussed here was specified for the cracking of 642,327 lb per day, that is, 8,143,000 cu ft/day at standard conditions (14.7 psia and 60 F) of a propane–ethane mixture. Three identical furnaces connected in parallel were determined to be the most practical design for the charge rate desired. Each furnace has two parallel convection and radiant coils, thus dividing the total designed furnace feed into six equal parts.

For certain economies in design, each furnace has two separately fired combustion chambers housing radiant coils, and a common convection section housing two convection coils. Horizontal gas burners along each side wall of the combustion chamber supply heat. Flue gases flow from the combustion chambers over the bridgewalls and thence downward into the convection section, leaving thru an underground duct to a stack.

The designed feed composition and volume charged to one-half of one furnace are shown in Table 13-6. A reaction temperature of 1500 F was selected for design purposes. Table 13-7 shows the cracked gas yield and composition calculated from previous work, and Table 13-8 summarizes the major design factors.

The furnace design was purposely conservative. Considerable flexibility in the composition of the feed that could be handled was desired. At a fixed feed rate, heat requirements vary, depending upon the temperature of cracking (i.e., depth of cracking) as well as the feed composition. Since higher temperatures are required to crack ethane than propane at comparable conversion levels, a feedstream analysis was chosen containing the greatest percentage of ethane ever expected to be charged.

Any sulfur compounds present in the feed will crack and carry over into the product gas. If sulfur is undesirable in the process in which the ethylene, or propylene, will be utilized, the impurity is generally removed from the feed stock before charging. To prevent "freeze-ups," the cracked gas must contain absolutely no water if low-temperature (refrigerated) fractional distillation processes are used for ethylene recovery or purification.

Operating Data. The effects of cracking depth on the composition and yield of the cracked gases, and the operability of the furnaces using a feed stock representative of that normally available, are summarized in Table 13-9. These data were used to plot the feeds and products obtained per 100 lb of charge stock at various cracking depths (Fig. 13-6). As the temperature was increased, the specific gravity of the cracked gas progressively decreased. At the same time the propane conversion rate (pounds of propane destroyed per 100 lb of propane in the feed) progressively increased (Fig. 13-6).

From a standpoint of maximum ethylene yield from the feed described, it would be desirable to operate the cracking furnaces at a cracked gas specific gravity of about 0.725 (Fig. 13-6). However, from an operating standpoint, difficulties from increased tar, coke, and carbon formation make such an extremely deep cracking rate most undesirable. If coke deposition progresses too far, not only will the tubes be overheated, causing serious damage, but they may rupture because of excessive back pressure due to coke build-up. The deeper cracking rates will also cause plugging of air fin coolers and lines to the carbon scrubbers with large amounts of coke, carbon, and tarry material. To avoid such plugging, the cracking furnaces are normally operated at about 0.80 to 0.85 cracked gas specific gravity.

At this cracking depth the effluent gas contains only rather small quantities of a "fluffy" finely divided carbon, along with small amounts of tar and coke which are easily removed periodically. Removal is ordinarily accomplished by shutting off the feed and passing an air and steam mixture thru the tube coil until the coke, etc., is burned away. One side of one furnace can be cleaned while the other is operating; thus, with three furnaces in operation, the total cracked gas stream is only decreased one-sixth during burnout operations. Normally each coil is cleaned about every two weeks.

Propane or heavier hydrocarbon feeds are employed when propylene is desired. No commercial installation is reported[3] for the preferential production of ethylene by catalytic breakdown of higher hydrocarbons. However, catalysts have been used for dehydrogenation of propane and butane to the corresponding olefins.

A lower operating temperature is necessary when cracking propane than when cracking ethane for ethylene production. When propylene and ethylene are required, recycle propylene from ethylene production may be added to propane and cracking continued in a separate furnace for optimum time–temperature conditions for the propylene production.

The foregoing illustrates the principles and variables which are important in thermal cracking for ethylene. Several other types of cracking units have been used successfully, including the radiant heaters of Selas Corp. Rapid growth of the ethylene industry has encouraged significant developments in furnace design for specific feed stocks and in higher capacity, operating economy, and materials of construction.

The recovery and purification of ethylene will not be described here. However, it should be pointed out that this part of the process is very costly in a commercial ethylene plant. Many process variations have been used to minimize the investment and energy demands of the purification process. Equipment developments in the fields of compression, fractionation, and low-temperature heat exchange have contributed greatly to making ethylene the most important petrochemical raw material.

Pebble Heater Process

The raw materials in this process (Fig. 13-7) are primarily methane, ethane, and propane. Depending on processing conditions, i.e., reaction time, pressure, and temperatures, the product is primarily ethylene with other unsaturated hydrocarbons, predominantly olefinic. This process allows for attainment of temperatures beyond those in the conventional tube furnace.

In the reactor the hydrocarbon feed contacts downward-moving hot refractory balls (pebbles). The processed hydrocarbons leaving the reactor are quickly cooled by direct water spray. The cooled pebbles leave the reactor at a controlled rate thru the engaging pot and travel upward in a stream

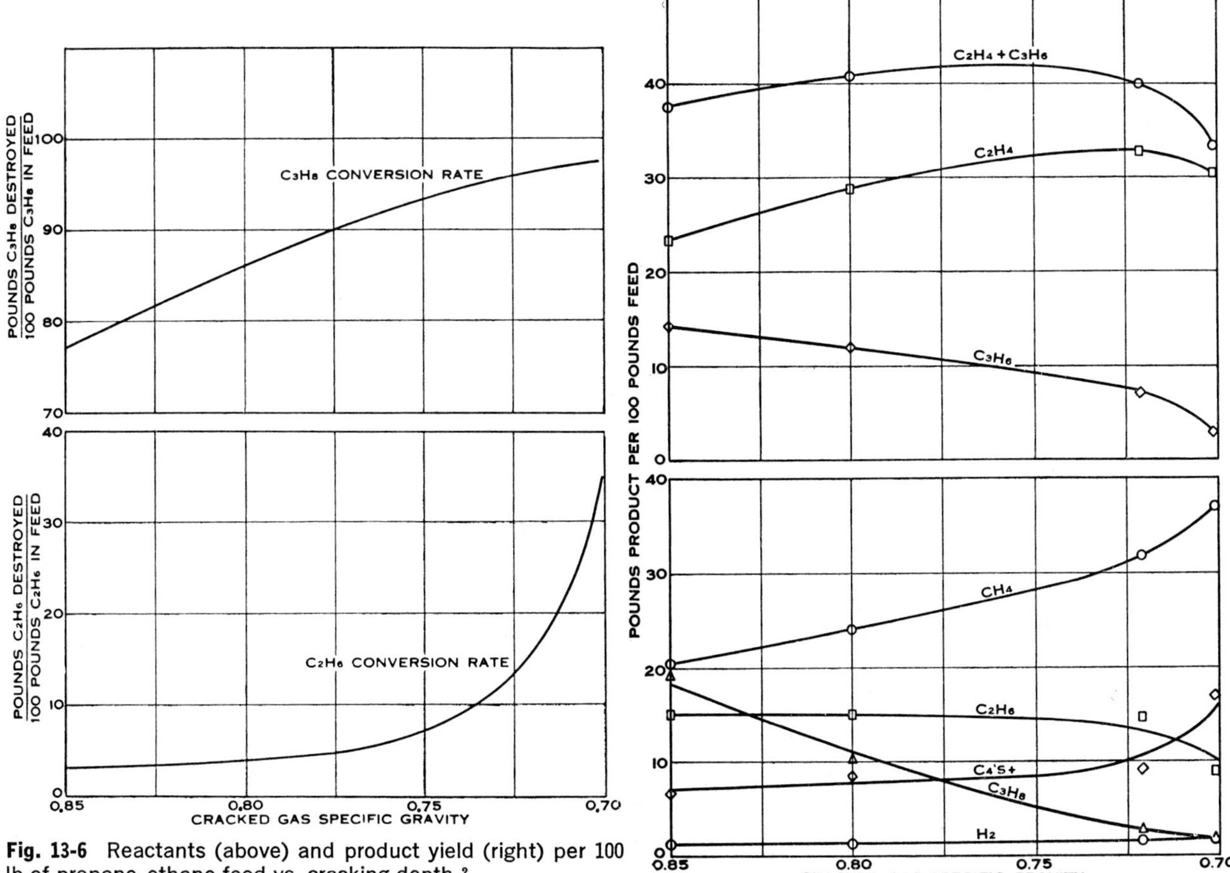

Fig. 13-6 Reactants (above) and product yield (right) per 100 lb of propane–ethane feed vs. cracking depth.[2]

Table 13-9 Propane–Ethane Cracking Furnace Test Runs
(all data for one-half furnace)

	Test number							
Operating conditions:	**Run 1**		**Run 2**		**Run 3**		**Run 4**	
Pressures, psig*								
Furnace inlet	34.0		35.4		37.8		39.4	
Furnace outlet	...		3.6		4.1		4.2	
Temperatures, °F								
Furnace inlet	65		72		85		87	
Outlet convection section	...		1300		1330		1350	
Outlet radiant section†	1230		1253		1298		1319	
Calculated reaction temp‡	1407		1428		1464		1471	
Flow rate (metered), MSCFD	1034		1037		1025		1022	
Product composition:	**Wt %**	**Mole %**	**Wt %**	**Mole %**	**Wt %**	**Mole %**	**Wt %**	**Mole %**
Hydrogen	1.22	14.83	1.35	15.45	1.62	16.79	1.85	18.57
Methane	20.43	31.29	24.03	34.65	31.87	41.43	37.02	46.81
Ethylene	23.44	20.54	28.81	23.78	32.73	24.34	30.31	21.93
Ethane	14.92	12.19	15.05	11.58	14.72	10.21	8.96	6.04
Propylene	14.15	8.27	11.94	6.57	7.14	3.54	2.97	1.43
Propane	19.11	10.65	10.25	5.38	2.77	1.31	1.91	0.88
C_{4S}	2.98	1.26	3.27	1.30	2.68	0.96	5.27	1.84
$C_{5S}+$	3.75	0.97	5.30	1.29	6.47	1.42	11.71	2.50
Total	100.00	100.00	100.00	100.00	100.00	100.00	100.00	100.00
Unsaturates (bromine Orsat)	33.4		34.7		34.0		29.0	
Specific gravity (air = 1.000)	0.850		0.800		0.721		0.701	
Flow rate, MSCFD	1630		1745		1915		1949	
M lb/day	106		106		105		104	
Calculated residence time, sec								
Convection section	5.35		5.45		5.66		5.78	
Radiant preheat section	0.90		0.92		0.95		0.98	
Radiant reaction section	1.82		1.76		1.69		1.70	
Total time	8.07		8.13		8.30		8.46	
Conversions (smoothed values from curves):								
lb C_2H_6 destroyed per 100 lb C_2H_6 feed	3.1		3.8		15.2		34.3	
lb C_3H_8 destroyed per 100 lb C_3H_8 feed	77.0		86.1		96.5		97.6	
lb $C_2H_6 + C_3H_8$ destroyed per 100 lb $C_2H_6 + C_3H_8$ feed	65.1		72.4		83.7		86.6	
Efficiencies (smoothed values from curves):								
lb $C_2H_4 + C_3H_6$ produced per 100 lb $C_2H_6 + C_3H_8$ destroyed	60.4		58.8		49.9		40.1	
Yields:								
lb $C_2H_4 + C_3H_6$ produced per 100 lb $C_2H_6 + C_3H_8$ feed	39.3		42.5		41.7		34.8	

* Barometric pressure, 13.2 psia.
† Fourth tube from outlet of furnace.
‡ Reaction temperatures calculated using propane cracking velocity constants.

Feed composition for all four runs:	Wt %	Mole %
Methane	3.66	8.86
Ethane	15.69	20.26
Propane	79.98	70.43
C_{4S}	0.67	0.45
	100.00	100.00

Specific gravity for all four runs = 1.343 (air = 1.000).

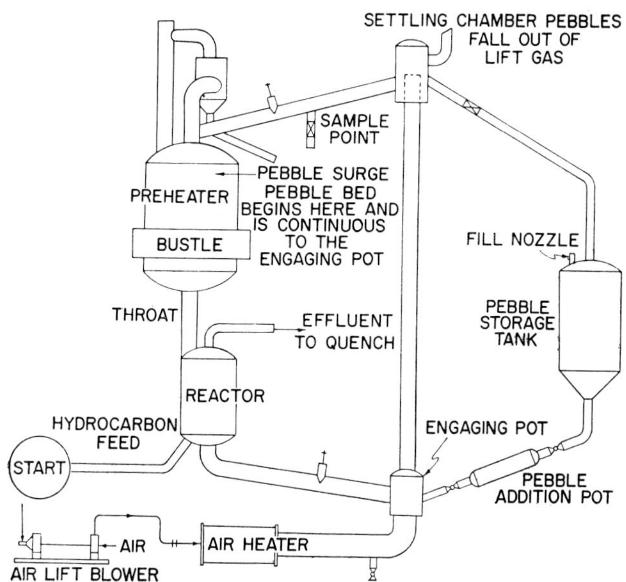

Fig. 13-7 Process for ethylene production with pebble heater. (Phillips Petroleum Co.; copyright, *Petrol. Refiner.*)

of heated air. They settle out and run down to the preheater. As they move downward they are heated by the hot combustion gases in the bustle-type combustion chamber surrounding the preheater. The hot pebbles pass thru a throat into the reactor where the raw material is processed. A pressure balance is maintained across the throat to prevent undesirable mixing of the combustion gases in the preheater and the effluent from the reactor. The Phillips Petroleum Co.[4] has installed a 30 MMBtu per hr commercial plant.

Typical feed stocks and reactor effluent compositions follow:

Feed	Mole %	Reactor effluent	Mole %
Methane	30.0	Hydrogen	32.0
Ethane	61.0	Methane	29.0
Propane	6.0	Acetylene	1.0
Other	3.0	Ethylene	25.0
		Ethane	11.0
	100.0	Heavy unsat.	2.0
			100.0

REFERENCES

1. Phillips Petroleum Co. Research Dept. Unpublished data.
2. Buell, C. K., and Weber, L. J. "Ethylene Production by Thermal Cracking of Propane-Ethane Mixtures." *Colo. School Mines Quart.* 45: 59–89, Apr. 1950. Also in: *Petrol. Process.* 5: 266–72, Mar.; 387–91, Apr. 1950.
3. Sherwood, P. W. "Production of Ethylene from Petroleum Sources." *Petrol. Refiner* 30: 220+, Sept. 1951.
4. "Petrochemical Process Handbook." *Petrol. Refiner* 32, No. 11, Nov. 1953.

Chapter 5
Production of Acetylene

by Frank H. Dotterweich

The chemical reactions for making acetylene from hydrocarbons are endothermic; e.g., for methane (from Table 3-36):

$$2CH_4 \rightarrow C_2H_2 + 3H_2 + 161{,}840 \text{ Btu}$$

Three methods may be used to supply this heat of reaction: (1) electric arc or spark; (2) heat from an auxiliary fuel as in a regenerative furnace; and (3) partial combustion of some of the feed gas with oxygen. The fundamental principle underlying all such methods of cracking hydrocarbons to acetylene is to raise the gas quickly to the required temperature and then to quench the product after a short reaction period.[1]

ELECTRIC METHODS

Two methods which may be termed the German and Schoch processes have been reported for production of acetylene from methane.

German Process

In this process[2,3] as operated in Huls, Germany, 24,000 MCF of natural gas were cracked daily. Cracking was done at high temperature and normal pressure in an electric arc with rapid quenching to retard polymerization of the low molecular weight hydrocarbons formed. Two useful grades of carbon black were removed and the acetylene produced was absorbed by scrubbing with water under 18 atm. Hydrogen and ethylene were also recovered from the arc gas by low-temperature fractionation and the residual gas recycled to the arc. The mean gas temperature was estimated at 2900 F with 7000 v applied to the converter. The reaction time of greater than 0.01 sec was controlled by the length of the arc. The flame tube life was from 250 to 300 hr. On leaving the reactor the gas was quenched with water to 284 F.

The average analysis of the outlet gas using natural gas as feed was:

Product	Volume, %
Acetylene	13.3
Hydrogen	46.0
Ethylene	1.0
Sat. hydrocarbons	27.8
Others	11.9
Total	100.0

Recoveries of acetylene and carbon black per MCF of natural gas were 2.74 and 2.04 lb, respectively.

Schoch Process[4]

Figure 13-8 is a flow diagram of a single stage electric discharge chamber for production of acetylene from natural gas and heavier hydrocarbons. In the commercial process the gas passes thru six reaction chambers in series, each having three arcs in parallel. Three impellers introduce the feed, each acting as an electrode with the opposite electrode mounted outside each opening. Recycling the gas limits the chamber temperature to 550 F. By-product carbon black is removed as shown. Commercial acceptability of this process at present is retarded[5] by the relatively complicated electric arc process and high power consumption.

Table 13-10 Analyses of Product Gases from Various Feed Gases[4]
(in mole per cent)

Feed	H_2	CH_4	C_2H_2	C_2H_4	C_2H_6	Methyl acetylene	C_3H_8	Di-acetylene	Vinyl acetylene	C_4H_{10}
Methane	39.3	48.9	10.0	1.0	...	0.2	...	0.4	0.2	...
	47.2	38.7	12.0	1.2	...	.2	...	.5	.2	...
	55.1	28.3	14.0	1.4	...	.3	...	.6	.3	...
Ethane	31.1	2.3	10.0	2.3	5.33	.3	...	.5	.2	...
	37.3	2.7	12.0	2.7	44.1	.3	...	.6	.3	...
	43.5	3.0	14.0	3.0	35.1	.4	...	.7	.3	...
Propane	32.8	3.2	12.0	3.2	1.5	.3	46.1	.6	.3	...
	38.3	3.8	14.0	3.6	1.7	.4	37.2	.7	.3	...
	43.8	4.4	16.0	3.9	2.0	.4	28.3	.8	.4	...
Butane	30.6	3.4	12.0	3.5	1.5	.3	...	.6	.3	47.8
	35.7	4.2	14.0	3.9	1.7	.4	...	.7	.3	39.1
	40.8	5.0	16.0	4.3	2.0	0.4	...	0.8	0.4	30.3

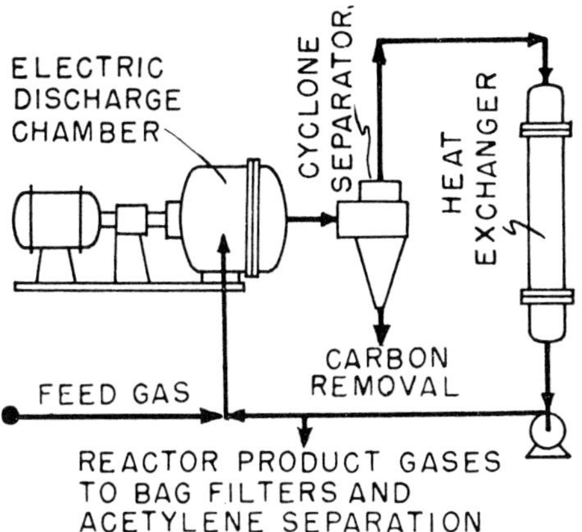

Fig. 13-8 Flow diagram of single stage electric discharge chamber for production of acetylene.

Table 13-10 gives the analysis of product gases from various feeds. About 1.6 lb of natural gas or 2.15 lb of liquid hydrocarbons are required per lb of acetylene produced.

PYROLITIC METHODS

Attempts have been made in pilot plant operations to supply the heat for reaction by: (1) indirect heat exchangers or (2) regenerative furnaces. Extremely high flame temperatures are required in indirect heat exchange systems, with resulting excessive strain on their materials. Accordingly, this method has given way to pilot plant studies of the regenerative furnace; e.g., the Wulff process (U.S.) and the vacuum operated method (Ruhrchemie, A. G., Germany).

Wulff Process

This process employs a regenerative furnace consisting of three masses of ceramic checkers of special design forming numerous small circular passages (Fig. 13-9). Operation is on a four cycle phase controlled by a cycle timer.

First, air is drawn into one end of the furnace and preheated. Fuel enters the first combustion chamber between the first and second ceramic masses and heats the second and third masses. The feed, diluted with steam, enters the opposite end of the furnace and is preheated, cracked, and quenched by passing thru the three masses in series. The third step introduces air in the same direction as that of the previous cracking phase, fuel now being added in a second combustion chamber between the second and third masses. The fourth phase is introduction of feed and steam in the opposite direction to that of the second heating phase.

For maximum production, furnace operation is at reduced pressure. The cracked gas passes to a quench system to condense most of the dilution steam and any liquid products formed during cracking. The acetylene content is recovered and purified in a selective solvent system using dimethylformamide. Over 50 per cent of C_2 and heavier feeds are converted to acetylene plus ethylene. The process is not sensitive to feed composition. Table 13-11 shows operating results in earlier pilot plant studies using propane as feed.

Vacuum Pyrolitic Method[6]

This process (Ruhrchemie, A. G., Germany) is intermittent with alternate one-minute reaction and heating periods in a

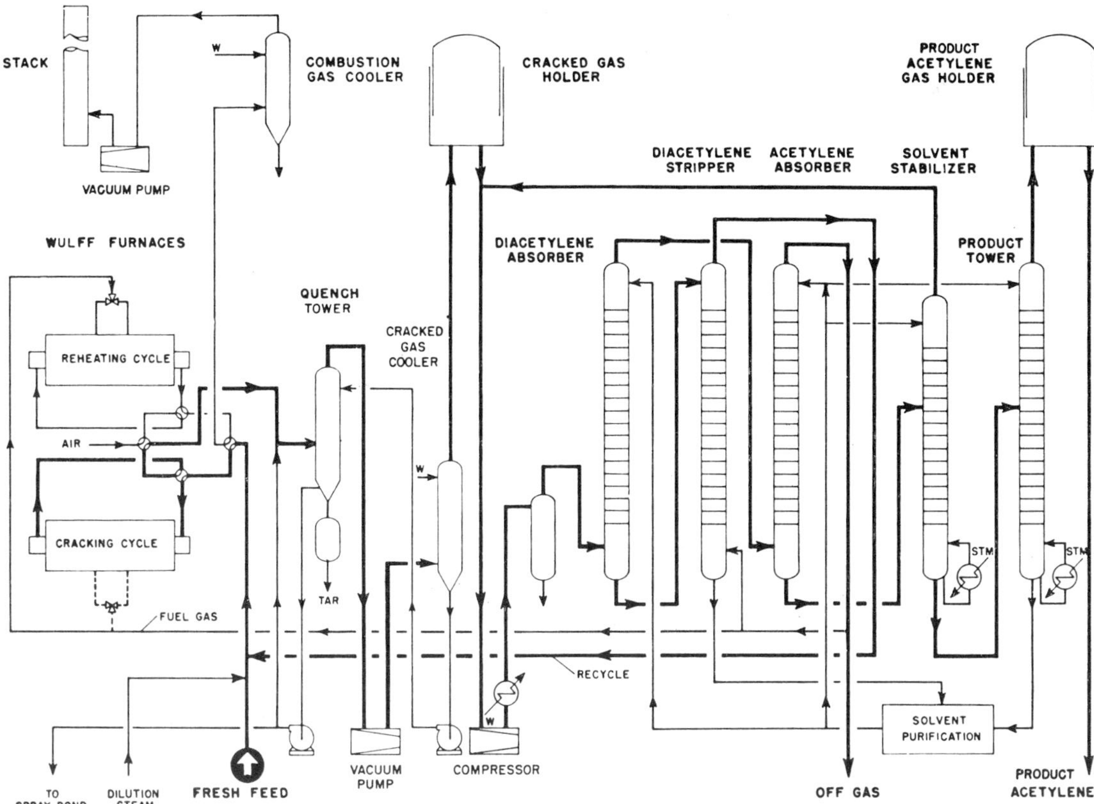

Fig. 13-9 Wulff process for acetylene production. (Copyright, *Petrol. Refiner.*)

Table 13-11 Pyrolysis of Propane

	Run 1	Run 2	Run 3	Run 4	Run 5	Run 6	Run 7	Run 8	Run 9
Temperature, °C	900	950	1000	1050	1075	1100	1190	1250	1325
Steam dilution	6.5	6.5	6.5	6.5	6.5	6.5	6.5	6.5	6.5
Expansion	1.29	1.51	1.64	1.90	2.14	2.28	2.77	3.11	3.66
Carbon balance, per cent	100	100	99.6	98.6	98.0	97.6	96.1	92.8	91.5
Yield, $C_2H_2 + C_2H_4$, per cent	10.65	20.3	25.5	37.9	47.5	51.8	61.3	60.2	57.6
Product, volume per cent									
Carbon dioxide	0.1	0.1	0.1	0.1	0.1	0.2	0.2	0.4	0.7
Acetylene	0.4	1.1	2.4	4.0	6.2	7.5	13.0	15.8	15.3
Ethylene	12.0	19.4	20.9	25.8	27.1	26.5	20.3	13.3	8.3
Propylene	7.0	8.4	7.7	6.7	5.3	4.6	1.4	0.6	0.0
Benzene	0.8	0.9	1.0	0.7	0.9	0.8	1.2	1.0	0.7
Oxygen	0.1	0.1	0.1	0.1	0.1	0.1	0.1	0.0	0.1
Hydrogen	11.3	16.1	18.9	23.4	27.3	30.0	39.0	45.7	51.7
Carbon monoxide	0.1	0.1	0.1	0.4	0.5	0.8	2.0	3.2	7.2
Methane	...	...	...	6.3	15.5	17.6	21.1	19.9	15.7
Ethane	22.0	33.7	40.0	32.3	16.7	11.6	1.4	0.0	0.0
Propane	46.1	20.1	8.8	...	0.0	0.0	0.0	0.0	0.0

refractory furnace. Optimum reaction time at 2600 F and 3.0 in. Hg vacuum pressure is 0.01 sec with a rapid quench. Heating is conducted at atmospheric pressure. Table 13-12 gives the products of such a pyrolysis.

Table 13-12 Product Gas from Pyrolysis of Methane Feed
(99 per cent pure)

Product	Mole %
Methane	15.7
Acetylene	9.8
Diacetylene	0.3
Hydrogen	70.6
Nitrogen	<3.1*
Oxygen	<0.1
Carbon dioxide	0.4
Total	100.0

[* N_2 probably comes mainly from air infiltration.—Editor.]

Disadvantages of this process are high power requirements for vacuum cracking and the higher compression needed to bring the cracked gas from 3.0 in. Hg vacuum to 160 psi for purification.

OXIDATION PROCESSES

In these processes, heat for acetylene formation is supplied by burning a portion of the feed gas with oxygen. Using methane at 2300 F,

$$2CH_4 + O_2 \rightarrow 2CO + 4H_2 - 30{,}732 \text{ Btu}$$

$$2CH_4 \rightarrow C_2H_2 + 3H_2 + 161{,}840 \text{ Btu (from Table 3-36)}$$

Carbon dioxide and water also are formed as dictated by the water gas equilibrium.

The **Sachse oxidation process** using oxygen for partial combustion was developed commercially at Oppau, Germany.[6] In this process, preheated oxygen and natural gas, heated separately to about 950 F, enter a special burner (Fig. 13–10) developed primarily for the uniform mixing and flow of the reactants in a molar ratio of 0.65 oxygen to 1.00 methane. The cracking reaction takes place at approximately 2700 F at atmospheric pressure with a reaction time of about 0.01 sec. The reactants are immediately quenched with a water spray. Yield of acetylene based on natural gas is 30 to 35 per cent. Composition of a typical outlet gas is:

Component	Mole %	Component	Mole %
Acetylene	8–9	Hydrogen	56
Carbon dioxide	3–4	Diacetylene	0.15
Methane	6–7	Others	0.35
Carbon monoxide	24–26		

Carbon formation is about 1.25 lb per MCF of feed gas.

The oxidation process is limited to natural gas feeds and requires 95 per cent oxygen. Although the oxygen plant requires a substantial additional investment, this is partly balanced by low energy costs and a by-product gas suitable for Fischer-Tropsch or Oxo synthesis.

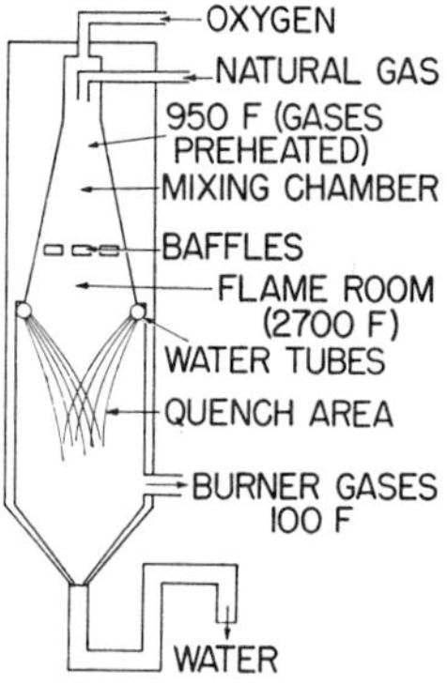

Fig. 13-10 Partial oxidation reactor of Sachse process.

An oxidation process using air and one employing 95 per cent oxygen are very similar. With air, the preheat temperature is higher and more dilute effluent gas, partly nitrogen, results. Other oxidation processes[7] include those of Societé Belge de L'Azote, Hoechst, Montecantini, Phillips, Kellogg, Chemical Construction Corp., and Hydrocarbon Research Inc. These processes are adapted to the use of a variety of hydrocarbon feed stocks and to the production of ethylene as well as acetylene if desired.

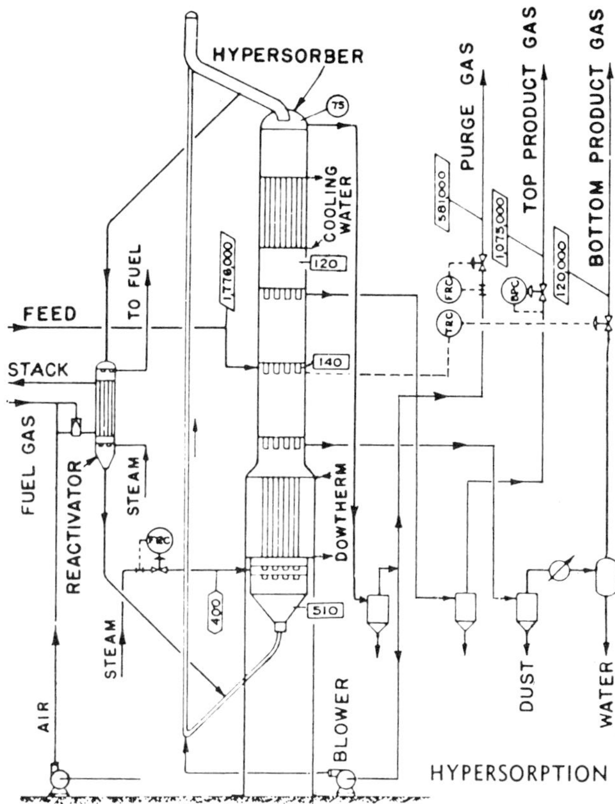

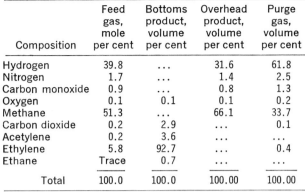

Fig. 13-11 Hypersorption purification process. (Copyright, *Petrol. Refiner.*)

Purification

Purification of the cracked gases is carried out in complex solvent recovery systems. Compounds to be removed include higher acetylenes, such as methyl and vinyl acetylenes, and diacetylene. Solvents employed include acetone, trimethyl urea, acetonyl acetone, and dimethylformamide. Hypersorption[8] may also be used (Fig. 13–11, Table 13–13). Purification is one of the most important steps in these processes because its cost factor is of great importance in producing acetylene of such purity as to allow it to be competitive with that from calcium carbide.

Table 13-14 sets forth power and raw material requirements and production costs for various acetylene production processes.

Table 13-13 Application of Hypersorption to the Recovery of Ethylene from Air Oxidation of Natural Gas[8]

Composition	Feed gas, mole per cent	Bottoms product, volume per cent	Overhead product, volume per cent	Purge gas, volume per cent
Hydrogen	39.8	...	31.6	61.8
Nitrogen	1.7	...	1.4	2.5
Carbon monoxide	0.9	...	0.8	1.3
Oxygen	0.1	0.1	0.1	0.2
Methane	51.3	...	66.1	33.7
Carbon dioxide	0.2	2.9	...	0.1
Acetylene	0.2	3.6	...	...
Ethylene	5.8	92.7	...	0.4
Ethane	Trace	0.7	...	...
Total	100.0	100.0	100.00	100.00

Table 13-14 Power and Raw Material Requirements for Production of Synthesis Acetylene from Light Hydrocarbons[1]

(for plant production rate of 1000 lb per hr acetylene)

Process	Power requirements, kwh		Raw material needs		
	For 98% C_2H_2	For dilute C_2H_2	Methane, MCF	Oxygen, lb	Gas credit, MMBtu
Electric arc	5.80	5.00	0.065	...	0.048
Pyrolytic (vacuum)	2.95	1.17	.118	...	.062
Oxidation	1.01*	0.06	0.104	5.7	0.051

* Production of 70 per cent acetylene.

REFERENCES

1. Sherwood, P. W. "Production of Ethylene from Petroleum Sources." *Petrol. Refiner* 30: 220–5, Sept. 1951.
2. Morrow, G. M. *Arc Process for Acetylene Production.* (U. S. Navy, Tech. Rept. 115–45) Washington, D. C.
3. Pettyjohn, E. S. "German Use of Natural Gas." *A.G.A. Natl. Gas Dept. Proc.* 1946: 33–44.
4. Daniels, L. S. "Acetylene Manufactured by Electric Discharge in Schoch Process." *Petrol. Refiner* 29: 221–4, Sept. 1950.
5. "Petrochemical Process Handbook." *Petrol. Refiner* 32, No. 11, Nov. 1953.
6. Kornfeld, J. A. *Natural Gas Economics.* Dallas, Transportation Pr., 1949.
7. "Petrochemical Handbook Issue." *Petrol. Refiner* 40, No. 11, Nov. 1961.
8. Berg, C. "Hypersorption in Modern Gas Processing Plants." *Petrol. Refiner* 30: 241–6, Sept. 1951. Also in: *Chem. Eng. Progr.* 47: 585–90, Nov. 1951.

Chapter 6
Production of Miscellaneous Chemicals

by Frank H. Dotterweich

CARBON BLACK

Common varieties of carbon black and processes for their production[1,2] are discussed briefly in this Chapter. In each case the product name is indicative of the process used for its manufacture.

Channel Process. Natural gas is burned in thousands of small flames in a "hothouse." Only secondary air is supplied, in amounts insufficient to obtain complete combustion. The flames impinge on steel channels which move slowly across stationary scrapers to remove the deposited carbon black. This falls into conveyors, which move it to grit separators. The carbon black is blown thru pneumatic separators, some of which are merely inefficient cyclones. The heavy, gritty particles are thrown against the walls and fall to the bottom outlet of the cyclone. The light, fluffy black remains suspended in the gas stream and passes out thru the central discharge tube. It subsequently is removed from the gas stream by more efficient cyclone separators. The black is fed to micropulverizers, then to pellet mills.

Carbon black may be pelleted by either a dry or a wet process. In the dry process it is tumbled with other pellets in long drums until it is properly pelleted. In the wet process the carbon black is mixed with water in high-speed pug mills to form pellets which must then be dried. After pelleting, the carbon black is stored in bins from which it may be bagged or loaded in bulk into hopper cars for shipment.[3] Figure 13-12a shows the channel black process. Analysis shows approximately 95 per cent total carbon content.

Furnace Process. Oil, gas, or a mixture of the two is burned in a furnace with a deficiency of air. The design of the furnace, the burner system, and the operating conditions are critical factors in controlling the qualities of the product. After the carbon black laden gases leave the furnace they are cooled to about 450 F by quenching and atmospheric cooling. In most carbon black plants the black particles are agglomerated in an electrostatic precipitator. Cyclone collectors then remove the bulk of the carbon black. Frequently the tail gases are treated by an auxiliary collector such as a bag filter.

After the carbon black is collected, it is conveyed to micropulverizers, pellet mills, and storage tanks. From these tanks it may be loaded into hopper cars for bulk shipment or packaged in paper bags. In general, the process is similar to the channel black process after the black is collected.[3] Figure 13-12b shows two furnace black processes.

Thermal Black. The Thermatomic method for producing thermal black may best be described as one of thermal decomposition. The furnaces are packed with firebrick nearly to the stack. The bricks, regularly spaced with openings, are heated by admitting the natural gas–air mixture under pressure in the bottom, where the mixture is ignited, the flames rising through the bricks to the stack. When the operating temperature reaches between 2200 and 2600 F, the heating cycle is stopped and natural gas is admitted at the top of the furnace, descending thru the hot bricks, where it is cracked. The product gases are cooled by water spray and pass to the

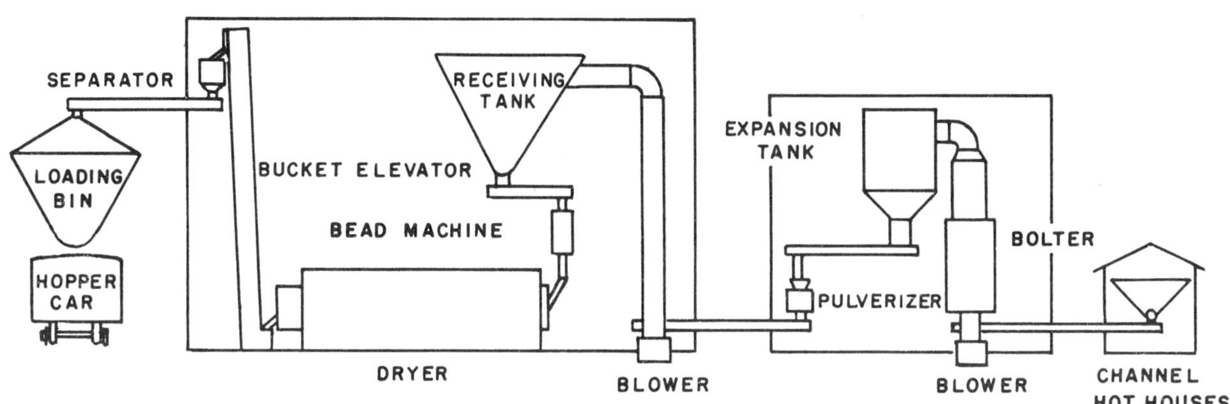

Fig. 13-12a Channel black process of manufacturing carbon black.[4]

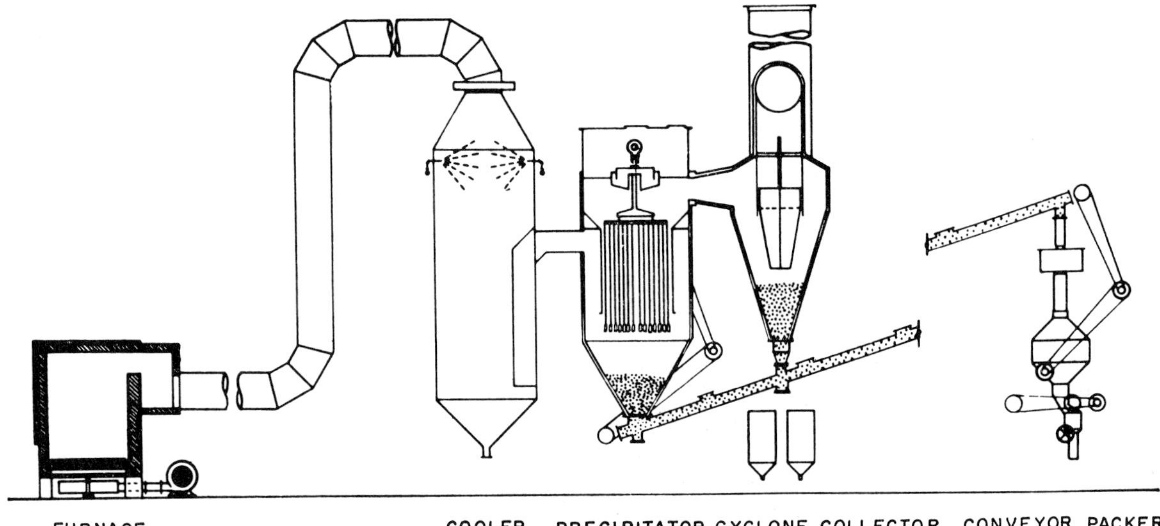

Fig. 13-12b Furnace black production processes:[4] (top) Gastex process, and (bottom) Furnex process.

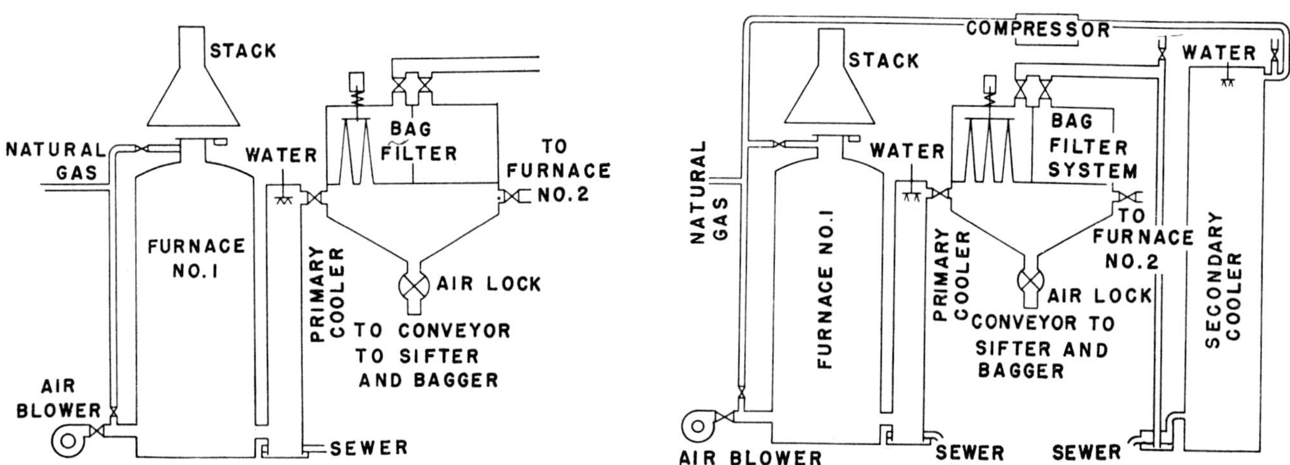

Fig. 13-12c Systems for thermal black production:[4] (left) Thermax process; (right) P-33 process.

bag filter system. The exit gas in the Thermax process (Fig. 13-12c) is composed of about 85 per cent hydrogen, 5 per cent methane, and small quantities of carbon dioxide, carbon monoxide, illuminants, and nitrogen.

The P-33 system (Fig. 13-12c) is generally similar to the Thermax process, except that it uses natural gas diluted with at least twice its volume of an inert. This permits production of a carbon of much smaller particle size. A larger collector system is required to handle the increased volume of gas to be filtered. Also, the cooling water system must be adequate to handle the larger quantities of gas for cooling from 2200 to about 240 F.

Thermal decomposition methods were first used in the production of hydrogen for ammonia synthesis with carbon black as a by-product. In 1922 commercial production of carbon black by this method was begun.

Acetylene Black. This is produced by thermal decomposition of acetylene feed stock. Since acetylene decomposes spontaneously and exothermally at about 800 C, the furnace only needs to be heated initially for acetylene fed into it to be cracked continuously to carbon and hydrogen. Carbon black is collected from the exit gases, which may first be burned to convert the hydrogen to water.[3]

Properties

The channel, furnace, and thermal blacks vary in fixed carbon content from 80 to almost 100 per cent, in relative color index from low to very high, in tinting strength from medium to very high, in particle size from 10 to 270 millimicrons, and in relative oil absorption from medium to very high.

AROMATIC CHEMICALS

Aromatic chemicals, such as benzene, toluene, and xylenes, once produced almost entirely as by-products of coke-oven operations, are now produced in ever-increasing quantities from petroleum and natural gas ingredients. In 1958, 52 per cent of the benzene, 82 per cent of the toluene, and 92 per cent of the xylene production was from petroleum and natural gas.

Benzene, toluene, and the xylenes are produced by catalytic reforming[5] of the carbon-six, seven, and eight raw material fractions, respectively. The potential capacity for petroleum aromatics in the United States is large, because aromatic production is possible in any refinery with a catalytic reformer. The additional investment required to produce aromatics from a light naphtha feed stock is relatively small.

For production of aromatics the **platforming process** developed about 1948 has been employed by independent refiners, while processes utilizing a platinum catalyst are used extensively by large refiners. All these processes are unusual because some molecules are hydrogenated while others are dehydrogenated.

Reactions for aromatic production may be as follows:

1. Naphthene dehydrogenation

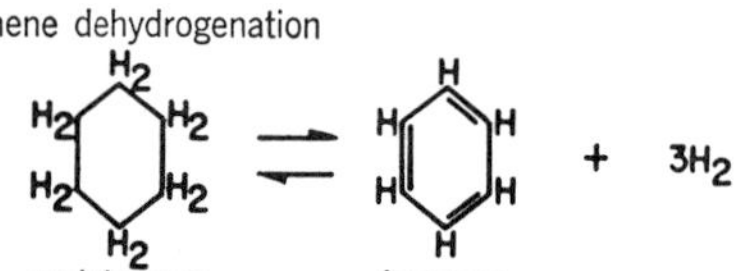

cyclohexane ⇌ benzene + $3H_2$

2. Dehydrocyclization of paraffins

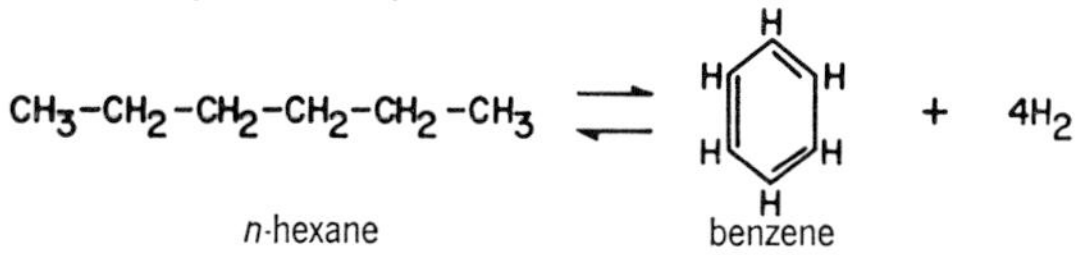

$CH_3-CH_2-CH_2-CH_2-CH_2-CH_3$ ⇌ benzene + $4H_2$

n-hexane ⇌ benzene

3. Naphthene dehydroisomerization

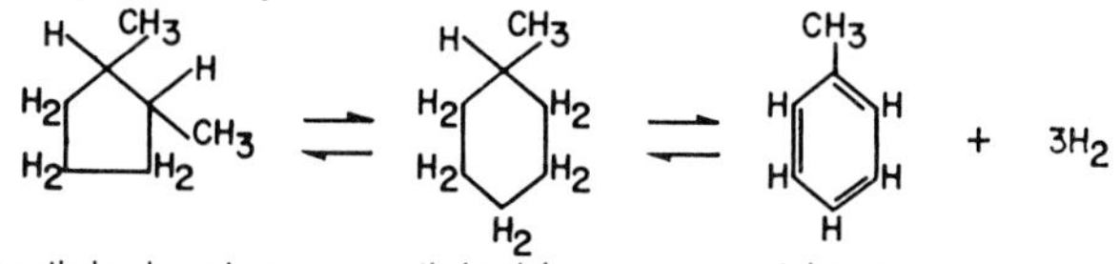

1,2-dimethylcyclopentane ⇌ methylcyclohexane ⇌ toluene + $3H_2$

4. Hydrodealkylation

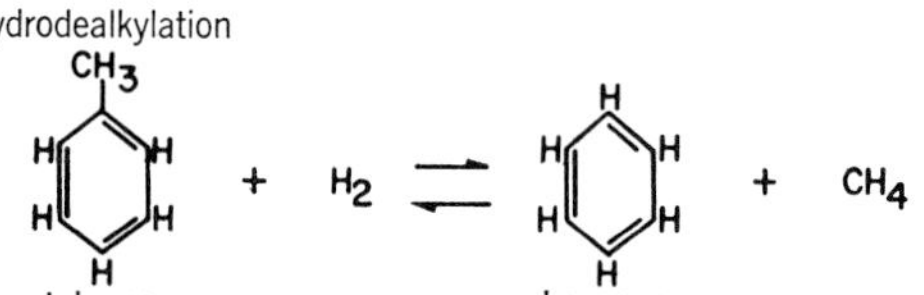

toluene + H_2 ⇌ benzene + CH_4

The great flexibility of refinery operations permits catalytic reformers to operate on a blocked basis; a portion of the schedule is for aromatic production, with an easy switch to motor fuel production to round out the schedule. Additional flexibility in aromatic production allows for varying the respective amounts and specifications of the benzene, toluene, and xylene production. Thus, hydrodealkylation is used to convert excess toluene to benzene, and the benzene may then be hydrogenated to cyclohexane.

ANHYDROUS HYDROGEN CHLORIDE

Hydrogen produced from natural gas may be reacted with chlorine in the **Hooker process** to produce anhydrous hydrogen chloride. The basic reaction is:

$$H_2 + Cl_2 \rightarrow 2HCl$$

Figure 13-13 is a flow sheet of the **Hooker process,** in which dry gaseous chlorine and hydrogen at a minimum of 5 psig react in the water cooled burner vessels. Essentially, all the chlorine and 97 per cent or more of the hydrogen are reacted. A typical analysis of the finished gas (per cent by volume) follows:

Hydrogen chloride	99.00
Chlorine	None
Hydrogen	0.64
Nitrogen	0.20
Water vapor	0.10
Carbon dioxide	0.05
Carbon monoxide	0.01
Total	100.00

As the hydrogen and chlorine must be metered carefully to avoid the formation of explosive mixtures, specifically adapted control devices are employed, making the plant essentially automatic in operation. The plant is completely equipped with safety devices to prevent troubles from hydrogen, chlorine, or water failures.

Following the production of hydrogen chloride the gases are cooled and any traces of chlorine removed. Drying may be done by washing with concentrated sulfuric acid.

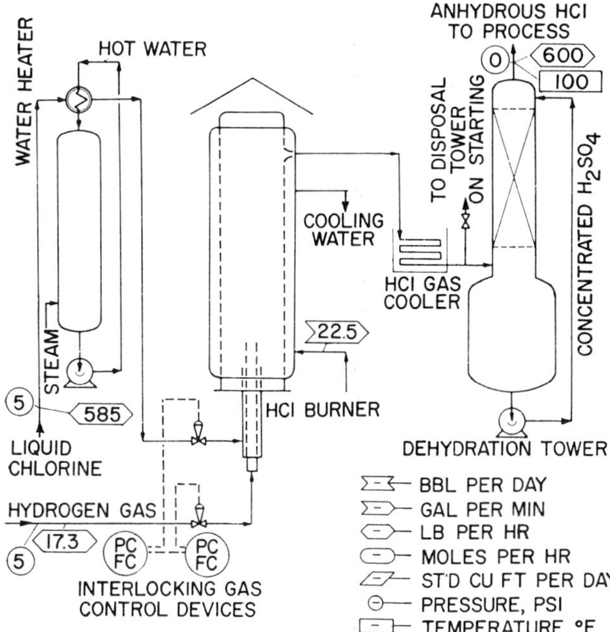

Fig. 13-13 Process for production of anhydrous hydrogen chloride. (Girdler Corp.)

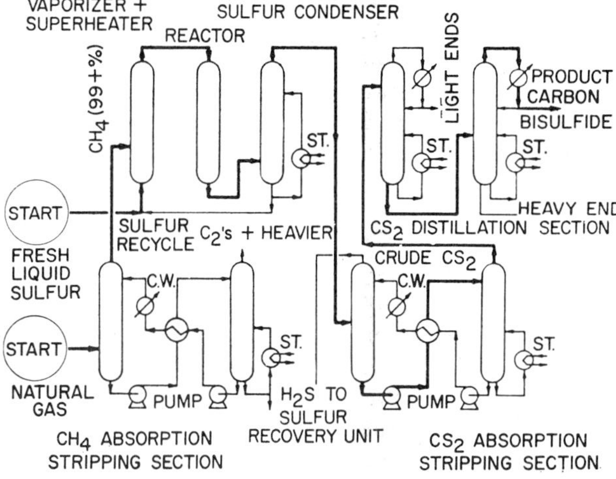

Fig. 13-14 Carbon disulfide process.[6]

Raw material requirements for producing one ton of anhydrous hydrogen chloride are approximately 1950 lb of chlorine and 10,800 cu ft of hydrogen.

CARBON DISULFIDE[6]

The carbon disulfide process (Fig. 13-14) uses methane of high purity (99 per cent or more). Liquid sulfur is combined with recycle sulfur in an insulated tank held at 250 to 270 F by steam coils. The liquid is transferred continuously to a furnace provided with high chrome steel coils, where sulfur is vaporized and mixed with methane and the mixture is superheated to between 1200 and 1300 F. The superheated vapors pass thru a reactor in which CS_2 is formed according to the reaction:

$$CH_4 + 4S(\text{vapor}) \rightleftharpoons CS_2 + 2H_2S$$

The reactor is a vertical high chrome steel vessel filled with a synthetic or artificial clay catalyst operated adiabatically between 15 and 25 psig and at a vapor space velocity of approximately 600 SCF per hr per cu ft of catalyst. With five to ten per cent excess sulfur, 85 per cent of the methane is utilized.

After cooling, unreacted sulfur is absorbed in liquid sulfur and CS_2 is separated by selective mineral oil absorption and stripped off the oil in the distillation section of the stripper.

REFERENCES

1. U. S. President's Materials Policy Commission. *Resources for Freedom, Vol. IV: Promise of Technology.* Washington D. C., 1952.
2. Gallie, J. F. "Carbon Black." *Petrol. Refiner* 23: 97–108, Mar.; 115–24, Apr. 1944.
3. Studebaker, M. L. *Carbon Black—a Survey for Rubber Compounders.* Akron, Ohio, Phillips Chemical Co., Philblack Sales Div.
4. Campbell, A. W. "Manufacture, Properties and Uses of Carbon Black." *Rubber Age* 50: 21–7, Oct. 1941.
5. "39% Benzene Yields from a C_6 Straight-Run Fraction." *Petrol. Process.* 6: 249–50, Mar. 1951.
6. Thacker, C. M. "Carbon Bisulfide." *Petrol, Refiner* 40: 228, Nov. 1961.

Chapter 7
Production of Liquid Fuels

by Frank H. Dotterweich

An analysis of premium motor fuel production is indicative of the importance of natural gas-derived liquid hydrocarbons. Only one-quarter the amount of steel and one-half the capital expenditures are required for the development of 1000 bbl of total recoverable premium motor fuel materials from natural gas condensate fields as compared with major oil fields.[1] For the efficient use of butanes, processes such as the diisopropyl process and the HF and H_2SO_4 alkylation processes are of importance.

METHODS OF HYDROCARBON SEPARATION

Methods for extraction and separation of the heavier hydrocarbons from natural gas streams include hypersorption, stage separation, adsorption, and absorption processes. Absorption is most widely used. See Figs. 5-4 thru 5-6.

FEED STOCK PREPARATION

Processes for the manufacture of liquefied hydrocarbon fractions from lighter hydrocarbons require feeds like ethylene, propylene, butene, and *iso*-butane. When sufficient quantities of *iso*-butane or *iso*-pentane are not available the liquefied hydrocarbon fractions may be formed from normal butane or pentane, respectively.

Isomerization

Figures 13-15 and 13-16 show Shell Development Co. processes for isomerization of butane and pentane. The processes developed by the Standard Oil Company of Indiana are shown in Fig. 13-17. The Butamer and Penex units developed by the Universal Oil Products Co. utilize a solid platinum-containing catalyst for isomerization of normal butene and pentane, respectively.

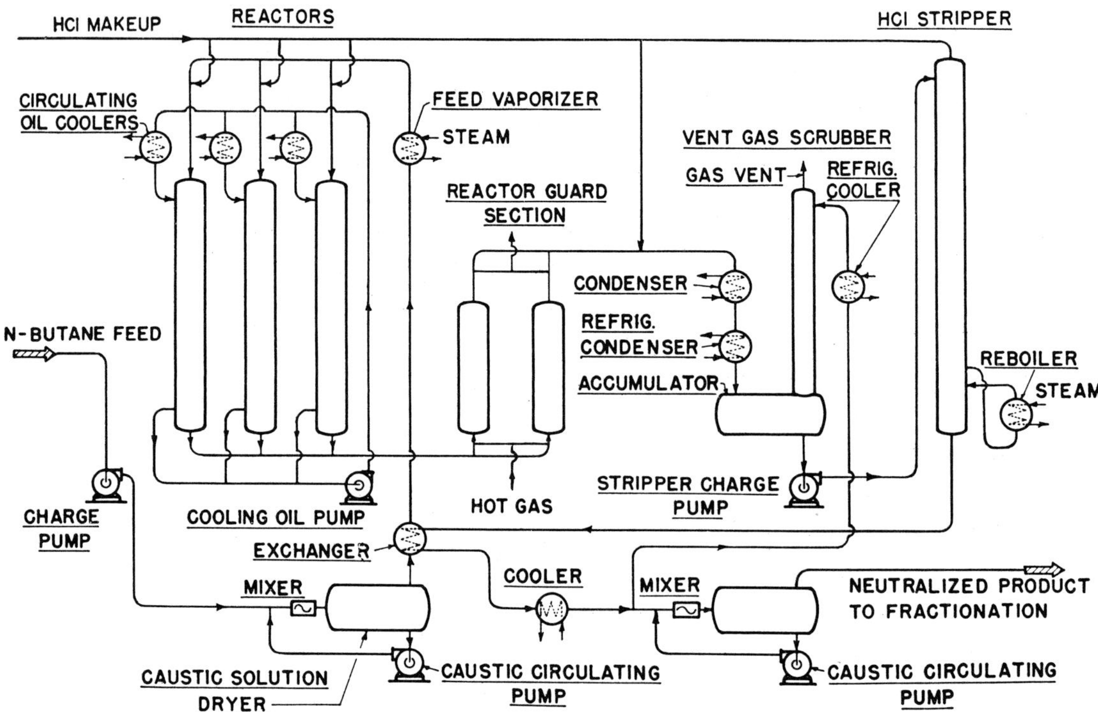

Fig. 13-15 Butane isomerization process: 1000 bbl per day of *iso*-butane. (Shell Development Co.; copyright, *Petrol. Refiner.*)

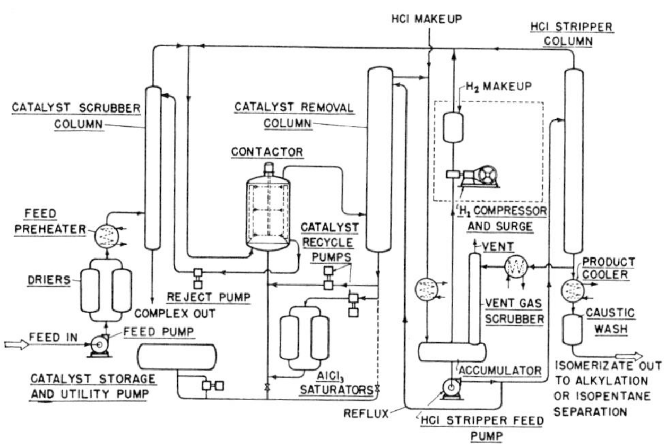

Fig. 13-16 Liquid phase paraffin isomerization process: 1000 bbl per day of *iso*-butane or *iso*-pentane. (Shell Development Co.; copyright, *Petrol. Refiner.*)

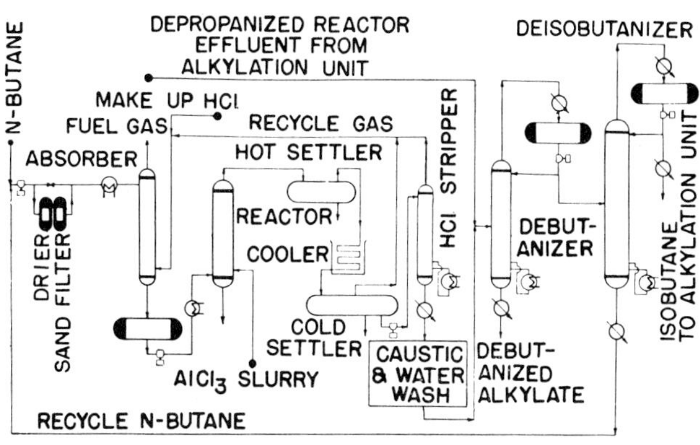

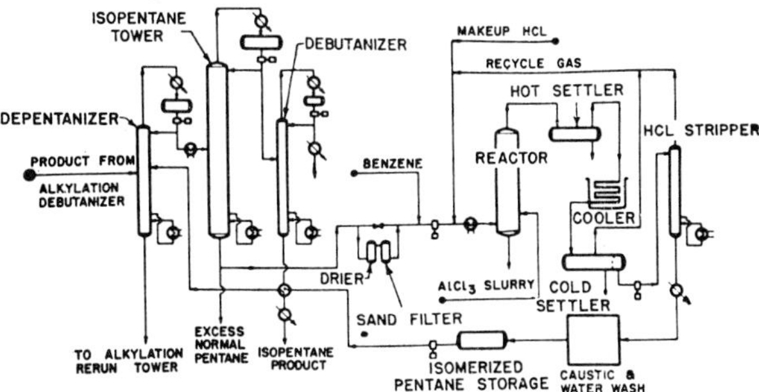

Fig. 13-17 Butane and pentane isomerization processes. (Standard Oil Co., Indiana; copyright, *Petrol. Refiner.*)

Dehydrogenation

Dehydrogenation of butane to butene may be accomplished by various processes, such as the production of butadiene. Figure 13-18 illustrates the Universal Oil Products Co. process. The reaction is carried out over a catalyst at 1000 to 1100 F and about 10 to 50 psig. Gas from the catalyst tube is washed free of tar or coke and compressed to 100 to 150 psig before entering the separation system. The small quantity of carbon produced must be burned off the catalyst periodically. This is accomplished without interruption to processing by reactors installed in pairs, one dehydrogenating while the other reactivates. A conversion of 30 per cent per pass with recycle operation gives ultimate yields of 75 to 80 bbl of butene per 100 bbl of butane.

Butane is also dehydrogenated to butene as the first step of the Phillips Petroleum Co. butadiene process (Fig. 13-26).

PROCESSES FOR LIQUID FUEL PRODUCTION

Hydrofluoric and Sulfuric Acid Alkylation

Anhydrous hydrofluoric acid is the catalyst used in Fig. 13-19 to promote the union of *iso*-paraffins with olefins.[2] The most desirable feed stock for manufacture of high octane premium motor fuel is a mixture of *iso*-butane and butylenes. Propylene and amylenes may also be used as the olefin feed. The alkylate produced from these olefins has a slightly lower octane number than that produced from butylenes. With the use of *iso*-butane–butylene feed under good operating conditions, a product having an ASTM clear octane number of 94 to 96 is obtained; under selected conditions, the product may have a number as high as 97. Operation is at low pressure at about 80 to 115 psig with a very short contact time. The yield of motor fuel alkylate amounts to 100 per cent of total alkylate. An HF alkylation plant is complete in that it possesses its own acid regenerating system. There are no disposal prob-

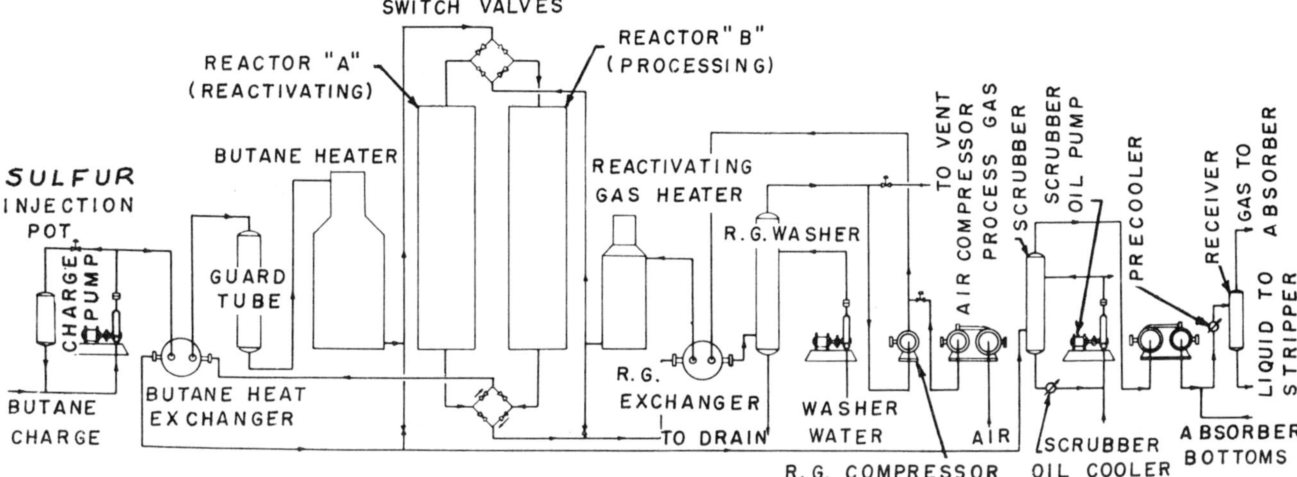

Fig. 13-18 Catalytic dehydrogenation process. (Universal Oil Products Co., copyright, *Petrol. Refiner.*)

lems of any consequence and no partially spent acid to be recovered for other uses. Table 13-15 gives operating data.

An HF alkylation plant is comparable in cost to a sulfuric acid alkylation plant of the same capacity. However, if the sulfuric acid plant includes an acid regenerating system, it will be more expensive than the HF unit. An HF alkylation plant can be operated without refrigeration, an added attraction over sulfuric alkylation.

Feed stock, dehydrated by a suitable drying agent, combines with a recycled *iso*-butane stream to enter the reactor (Fig. 13-19). The recycled and regenerated acid also enters the reactor. From the reactor the combined hydrocarbon–acid stream is transferred to an acid settling tank for separation of acid from hydrocarbons during a brief settling time. The acid

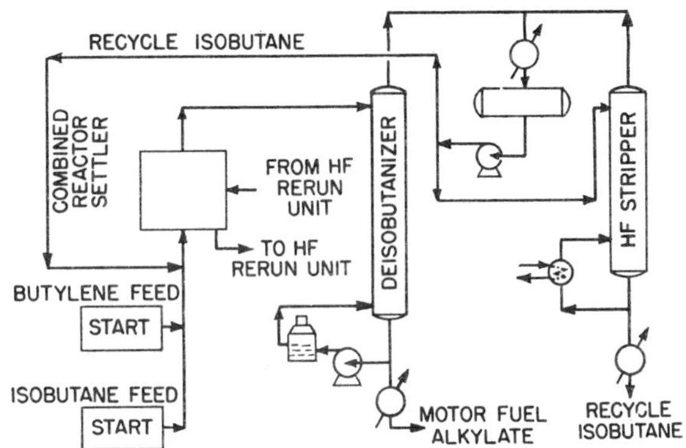

Fig. 13-19 Perco HF alkylation process. (Phillips Petroleum Co.)

from this tank is recycled to the reactor and the hydrocarbon effluent stream is fed to the deisobutanizer, the first unit in the fractionation system. Part of the overhead is depropanized and this, combined with the remainder of the deisobutanizer overhead, constitutes the recycle stream introduced with fresh feed into the reactor. The deisobutanizer bottoms are debutanized to yield butane and motor fuel alkylate.

An acid purification unit is provided for removal of acid soluble oils and water. The feed to this purification unit is bled off the acid recirculation system.

Figure 13-20 is a flow diagram for the HF and sulfuric acid alkylation processes. Typical sulfuric acid alkylation results are given in Table 13-16, which is based on the performance of units constructed for aviation alkylate and modified for motor fuel operation.

Reactor design has progressed with advances in reaction kinetics. Acid pumping requirements have been eliminated and total alkylate quality has greatly improved.

Table 13-15 Typical Operation of Phillips HF Alkylation Unit[3]

Charge to unit, bbl per day:	
Propylene	300
Butylene	600
Amylenes	100
Total olefins	1000
iso-Butane	1250
n-Butane	100
Total feed	1350
Yield, bbl per day:	
Motor fuel alkylate	1770
Propane	30
Butane	100
Acid-soluble oils	2
Motor alkylate yield, bbl per bbl olefin	1.77
***iso*-Butane consumption, bbl per bbl olefin**	1.25
Acid catalyst:	
Titratable acidity, wt per cent	89.6
Oils plus solids, wt per cent	3.0
Water, wt per cent	1.6
Typical properties of motor fuel alkylate:	
Gravity, °API	70
ASTM distillation, °F:	
Initial bp	120
10 per cent	185
50 per cent	212
90 per cent	260
End point	360
Reid vapor pressure, psia	4.2
Octane number, motor clear	93.5
+ 3 cu cm TEL	106
Octane number, research clear	94
+ 3 cu cm TEL	105
Utilities:	
Steam, lb per hr (150 psig)	5000
Electricity, kwh	120
Water circulation, gpm	3000
Water, make-up, gpm	100
Fuel gas, MMBtu/hr	30
Chemicals, lb per day:	
Caustic soda	100
Hydrofluoric acid	175
Labor:	
Operation, man-hours per day	48
Maintenance, man-hours per day	8

Table 13-16 Typical H_2SO_4 Alkylation Results

	Butylene	Propylene	Amylene
Alkylate, volume per cent on olefin	170	170	160
iso-Butane, volume per cent on olefin	110	125	110–120
Acid consumption, lb/gal of alkylate	0.7	3	1–2
ASTM octane rating of alkylate	94	89	90
Reid vapor pressure of alkylate, psia	3–4	5	2
End point of alkylate, °F	350	350	350

Production of *neo*-Hexane

Figure 13-21 shows a commercial noncatalytic alkylation process. The plant has two units: (1) the gas cracking unit by which the olefin feed stock is prepared and (2) the alkylation unit. Alternatively, the olefin feed stock may be secured from any source. In this process (Fig. 13-21) the olefins are caused to react directly with paraffins as follows:

$$C_2H_4 + i\text{-}C_4H_{10} \rightarrow \text{2,2-dimethylbutane}(\textit{neo}\text{-hexane})$$

A mixture of ethane and propane is subjected to low-pressure cracking to produce ethylene which is absorbed in *iso*-butane cooled to about −30 F. This ethylene–*iso*-butane mixture from the absorption tower passes thru a pump, where the pressure is increased to about 4000 to 5000 psi. At this point the ethylene absorbed in *iso*-butane is ready for injection into the alkylation furnace.

In the alkylation operation, *iso*-butane is pumped at about 4000 to 5000 psi thru a preheat coil in the alkylation furnace and preheated to about 950 F. It then passes to the alkylation coil, where the ethylene–*iso*-butane mixture described above

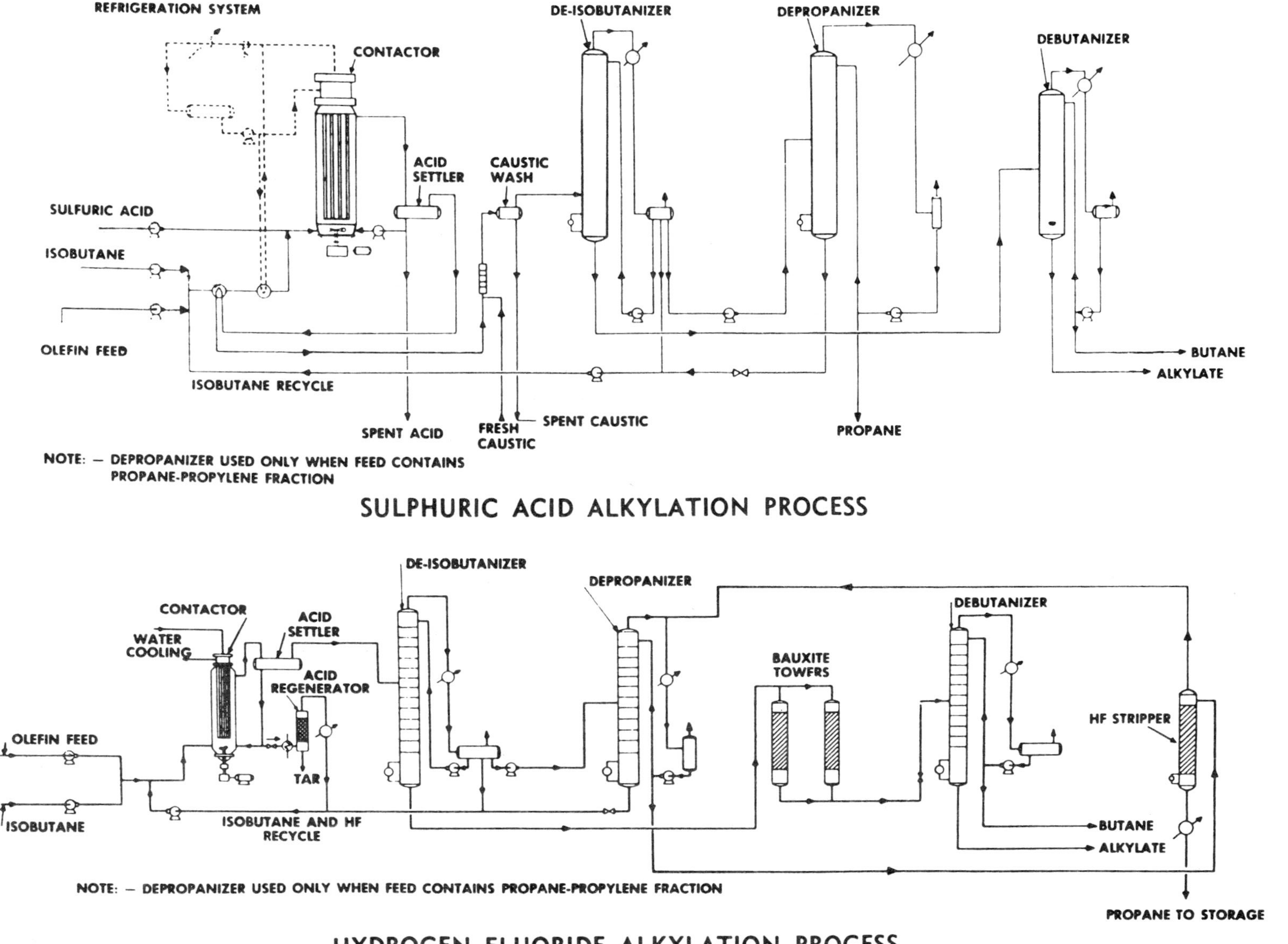

Fig. 13-20 H_2SO_4 and HF alkylation processes. (Universal Oil Products Co.; copyright, *Petrol. Refiner.*)

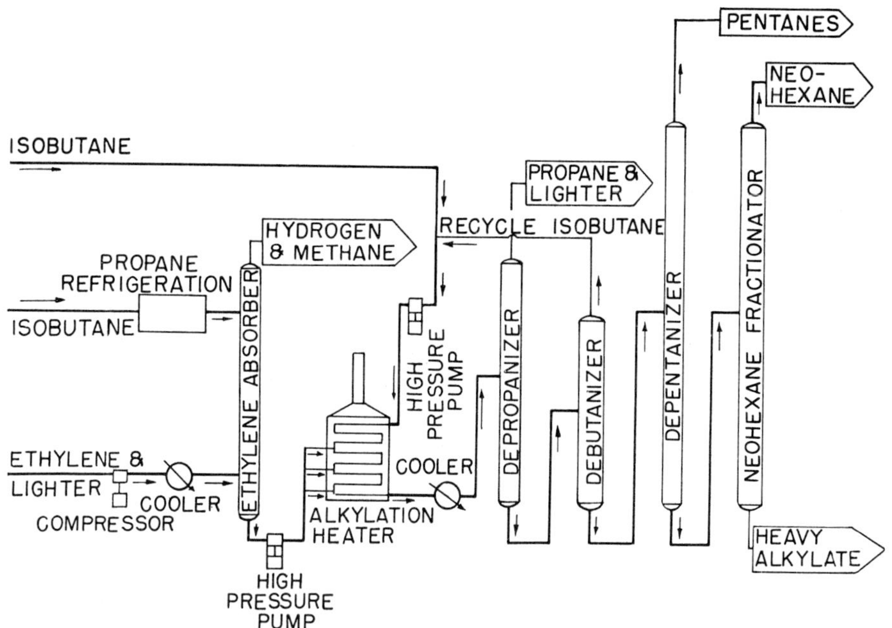

Fig. 13-21 Phillips noncatalytic alkylation process for *neo*-hexane.[4]

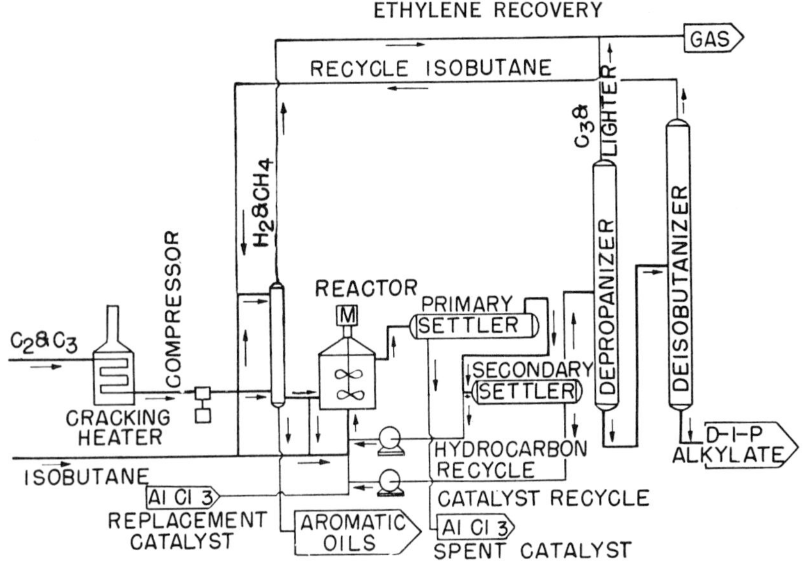

Fig. 13-22 Phillips diisopropyl process.[5]

is added to the heated *iso*-butane stream at numerous injection points. An *iso*-butane:ethylene ratio of about 9:1 in the reaction zone is maintained. The reaction mixture leaving the alkylation furnace flows thru a pressure reduction valve and cooler and enters the depropanizer column. Propane and lighter materials are removed overhead and the bottoms pass directly into the debutanizer column. Here the *iso*-butane is taken overhead and recycled to the *iso*-butane preheater coil. The bottoms from the debutanizer, containing pentanes and heavier fractions, are pumped to the depentanizer, and an overhead product of pentanes is removed. The bottoms are then sent to the *neo*-hexane fractionator, the overhead from which is *neo*-hexane. The bottoms from the fractionator constitute a high grade motor fuel.

Neo-hexane has an 84.9° API gravity; a 9.5 Reid vapor pressure; a 121 F boiling point; and a 95 ASTM octane number. Because of these properties and its excellent blending properties and response to tetraethyl lead, it can be used to make aviation gasoline on a commercial scale with an octane number as high as 115 or more.

The yield of liquid products, including all hydrocarbons in the boiling range of gasoline as well as *neo*-hexane, is approximately 60 to 70 per cent by weight of the net ethane–propane and *iso*-butane consumption during decomposition and alkylation.

This process is not limited to *neo*-hexane production. By using different paraffin and/or olefin feeds, many different motor fuel blending stocks may be produced. When only motor fuels are desired some of the fractionating steps used in feed stock preparation and in *neo*-hexane separation may be eliminated.

Production of Diisopropyl (2,3-Dimethylbutane)

The first manufacture of diisopropyl on a large scale was begun by the Phillips Petroleum Co.[5] in 1944, which used a process (Fig. 13-22) based on alkylation with ethylene and *iso*-

butane as feed stocks. Raw materials are ethane, propane, and *iso*-butane. Ethane and propane are cracked at low pressure to produce ethylene. Ethylene and *iso*-butane react as shown below in the presence of an aluminum chloride catalyst to yield diisopropyl alkylate. This alkylate contains a high percentage of diisopropyl and may be employed either in motor fuels or in aviation gasoline.

$$C_2H_4 + i\text{-}C_4H_{10} \rightarrow \text{2,3-dimethylbutane (diisopropyl)}$$

Laboratory studies show (Table 13-17) that the yield of total alkylate per pound of olefin reacted varies with the olefin composition and the molal ratio of *iso*-paraffin to olefin. For the feed composition shown, the yield ranges from 2.7 to 2.8 lb per lb of olefin and 1.5 to 1.6 lb per lb of *iso*-butane consumed. The total alkylate contained 67 wt per cent diisopropyl under favorable conditions. The yield of total alkylate per pound of aluminum chloride catalyst ranged from about 8 to 35 gal per lb, the consumption of catalyst being highest at low ethylene conversions.

Table 13-18 gives the antiknock blending characteristics of pure diisopropyl.

Catalytic Polymerization

The Universal Oil Products Co. catalytic polymerization processes (Fig. 13-23) were the first of the many catalytic processes employed in the refining industry. They consist of contacting an olefin-containing feed stream with the catalyst at 350 to 475 F and 150 to 1200 psig. Suitable cooling is required to remove the heat of the exothermic reaction.

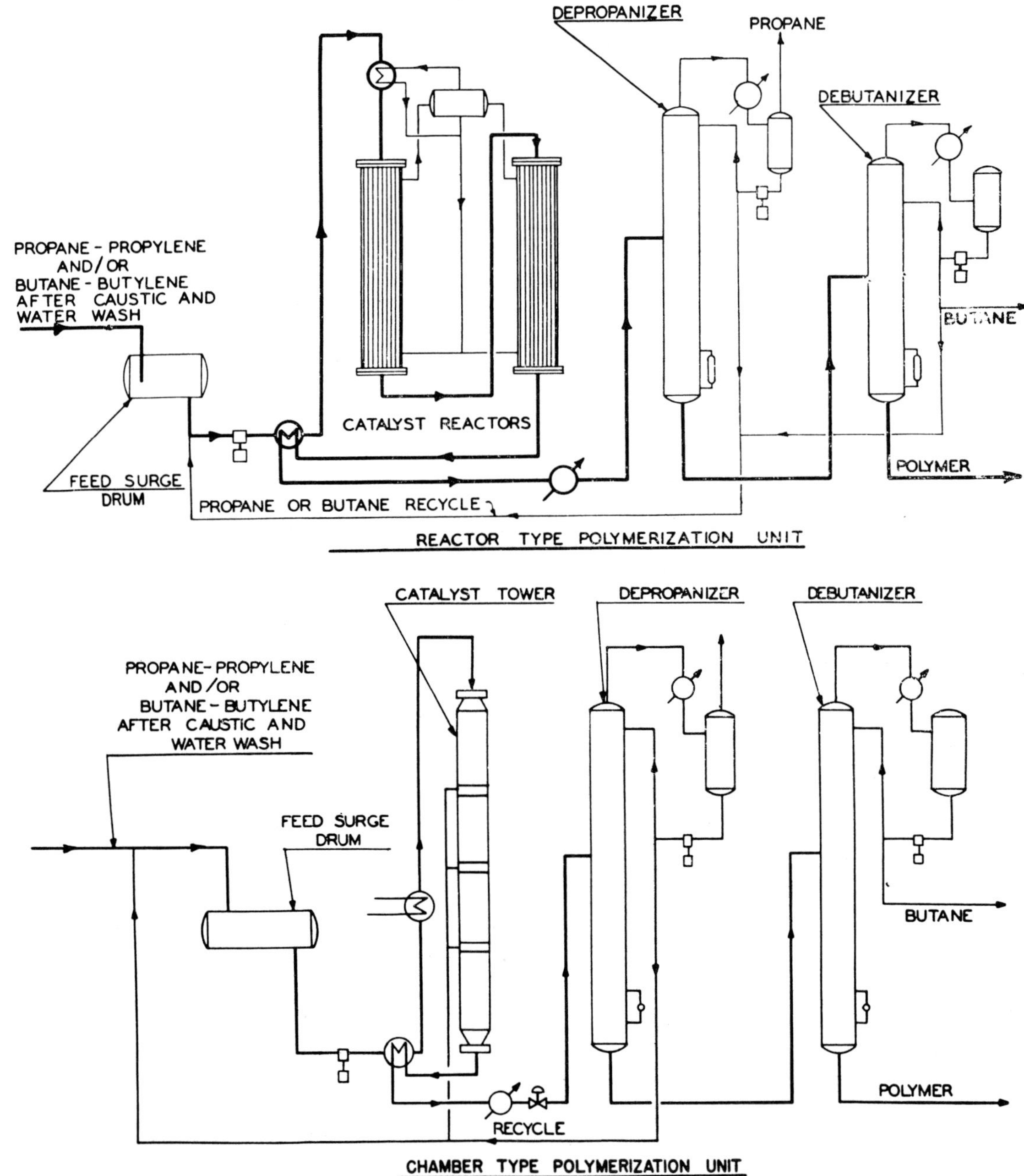

Fig. 13-23 Catalytic polymerization processes. (Universal Oil Products Co.; copyright, *Petrol. Refiner.*)

Table 13-17 Laboratory Ethylene-*iso*-Butane Alkylation Data[5]

(at 130 F and 300 psi; reactor volume, 560 ml)

Flow rate, lb/hr	1.17	1.34	1.29	1.20
Composition of fresh feed, wt per cent:				
Ethylene	6.9	6.45	6.17	6.80
Propylene*	1.1	1.10	1.00	1.05
Propane*	9.0	1.57	8.91	9.10
iso-Butane	75.95	83.50	76.80	75.80
n-Butane*	7.0	7.30	6.80	7.20
HCl	0.05	0.08	0.32	0.05
	100.00	100.00	100.00	100.00
Mole ratio:				
i-C_4H_{10}:olefin	4.8	5.4	5.5	4.9
Catalyst:				
ml in reactor	71	90	141	232
Viscosity†	260	600	400	60
Charges, lb:				
Fresh feed	93.51	124.0	156.3	238.2
$AlCl_3$	0.252	0.215	0.259	0.538

Products, lb:				
C_4 and lighter	59.39	88.27	105.8	177.4
Alkylate	11.82	20.12	24.8	44.9
Recovery, per cent	76.20	87.40	83.5	93.3
Conversions, per cent:				
Ethylene	68	84	92.8	89.2
Propylene	100	100	100	100
Alkylate yields:				
Per lb $AlCl_3$	8.20	16.40	16.80	14.60
Per lb olefin‡	2.62	2.72	2.75	2.79
Efficiency§	1.83	2.15	2.19	2.17
Composition of alkylate, wt per cent:				
Pentanes	3.1	1.9	2.4	3.6
neo-Hexane	0.2	0.6	0.7	0.6
Diisopropyl	58.1	66.5	67.8	66.1
Other hexanes**	1.7	2.8	3.6	4.3
Heptanes	13.5	11.6	10.2	11.2
Octanes	14.4	11.0	8.5	10.0
Heavier	9.0	5.6	6.8	4.2
	100.0	100.0	100.0	100.0

* These components are included to simulate commercial plant feed stock.
† Centistokes at 100 F.
‡ Corrected for low recoveries.
§ Pounds diisopropyl per pound ethylene reacted.
** Mainly 2-methylpentane, although infra-red absorption also shows 3-methylpentane and *n*-hexane to be present.

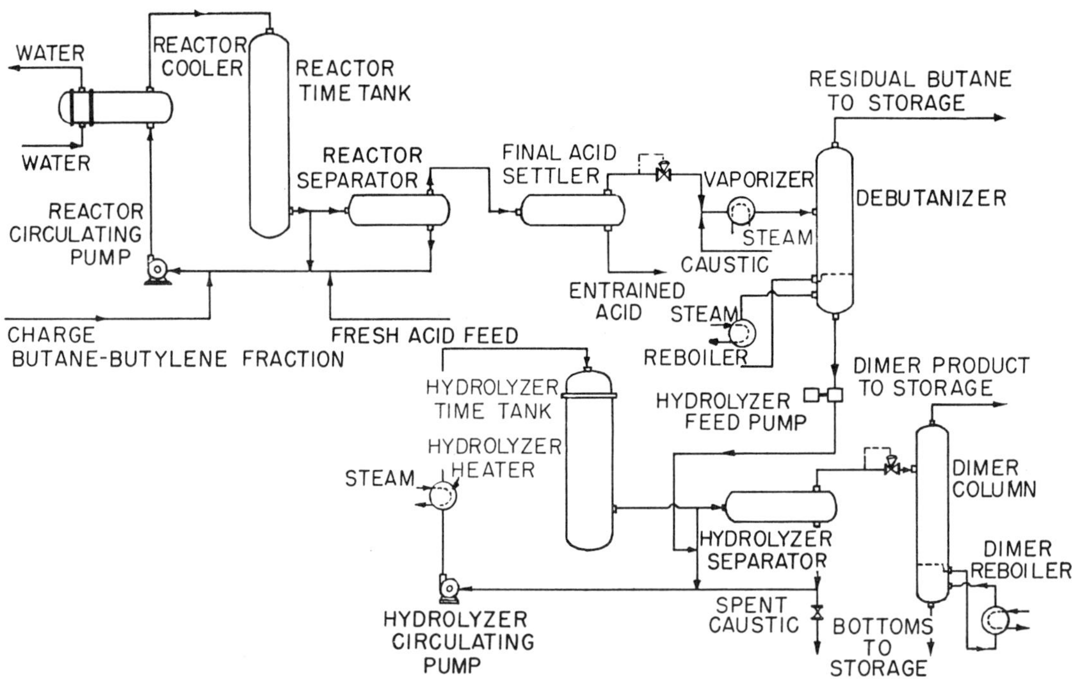

Fig. 13-24 Hot acid polymerization process. (Shell Development Co.; copyright, *Petrol. Refiner.*)

Two types of units are now in use. Table 13-19 gives properties of the polymer formed. In the more common chamber type the catalyst is contained in one or more vertical chambers, each with multiple catalyst beds 5 to 10 ft thick. In the reactor type, the catalyst is loaded in vertical small diameter tubes surrounded by a cooling medium.

Figure 13-24 shows the Shell Development Co. hot acid polymerization process.

Other Polymerization and Alkylation Processes

A plant for polymerization of propylene and butylene may be modified without difficulty for alkylation of propylene and benzene to form isopropyl benzene. The latter was used as a rich mixture ingredient in aviation gasoline during World War II, and later as the starting material in the synthesis of phenol and acetone.

Table 13-18 Antiknock Blending Characteristics of Pure Diisopropyl

Composition, per cent												
	Reference fuel blends		Aviation base stock No. 1	Reid vapor press.	Gravity, °API	Evap. temperature, °F‡					ASTM aviation oct. No. with 4 ml TEL§/gal	Super-charged rating with 4 ml TEL§/gal, % power
Diisopropyl	83% S,* 17% M†	93% S,* 7% M†				Initial bp	10%	50%	90%	End point		
100	...	...	...	7.35	80.7	134	135	135	136	136**	113.2	...
...	100	...	...	1.95	69.8	194	205	213	238	371	104.1	101.1
...	...	100	...	1.90	71.1	199	207	210	218	339	111.8	122.6
...	...	...	100	6.75	70.4	119	142	165	198	217**	94.5	100.0
10	90	...	...	2.75	70.9	176	193	208	238	368	105.4	108.4
30	70	...	...	3.65	73.0	156	171	193	226	353	108.2	124.5
50	50	...	...	4.95	75.0	145	157	175	222	343	110.3	141.5
10	...	90	...	2.50	72.3	180	194	206	217	339	112.0	130.9
20	...	80	...	3.15	73.1	166	182	199	215	331	112.2	139.8
40	...	60	...	4.25	75.1	150	164	183	215	311	112.2	152.0
10	...	...	90	6.75	71.5	120	140	161	193	212**	96.5	106.4
30	...	...	70	6.85	73.3	125	141	155	188	215**	100.3	119.1
50	...	...	50	6.90	75.4	128	139	148	178	207**	103.0	134.8

* Secondary reference fuel for 2,2,4-trimethylpentane.
† Secondary reference fuel for normal heptane.
‡ Corrected to 29.92 in. Hg.
§ Tetraethyl lead.
** Dry point.

Table 13-19 Catalytic Polymer Properties[3]

(Universal Oil Products Co. catalytic polymerization process)

	Type of plants		
	Chamber	Reactor	Reactor
Pressure, psig	500	1000	1000
Temperature, °F	370–465	400	450
Feed stock	Cracking plant stabilizer net overhead liquid	Butane-butylene from refinery gas concentration system	Propane-propylene from refinery gas concentration system
Polymer inspection:			
Gravity	64.0	64.3	62.4
Initial boiling point	96	78	144
10 per cent	156	152	204
30 per cent	210	238	244
50 per cent	235	258	266
70 per cent	279	...	284
90 per cent	366	379	330
End point	430	416	402
Recovery, per cent	97	95	...
Reid vapor pressure, psia	9	11	4
Average polymer properties:			
Bromine No.	130–150		
Induction period:			
Clear	15 min		
+ 0.025 per cent UOP* No. 1	500–550		
+ 0.005 per cent UOP* No. 4	450–500		
Copper dish gum	25 mg		
ASTM octane No. clear		82.5	
ASTM octane No. + 3 cu cm TEL		85.0	
Research octane No. clear		97.0	
Research octane No. + 3 cu cm TEL		100.00	
Polymer blending values:			
5–10 per cent blends in average straight run gasoline			110–120
5–10 per cent blends in thermal cracked gasoline			95–100

* Universal Oil Products Co.

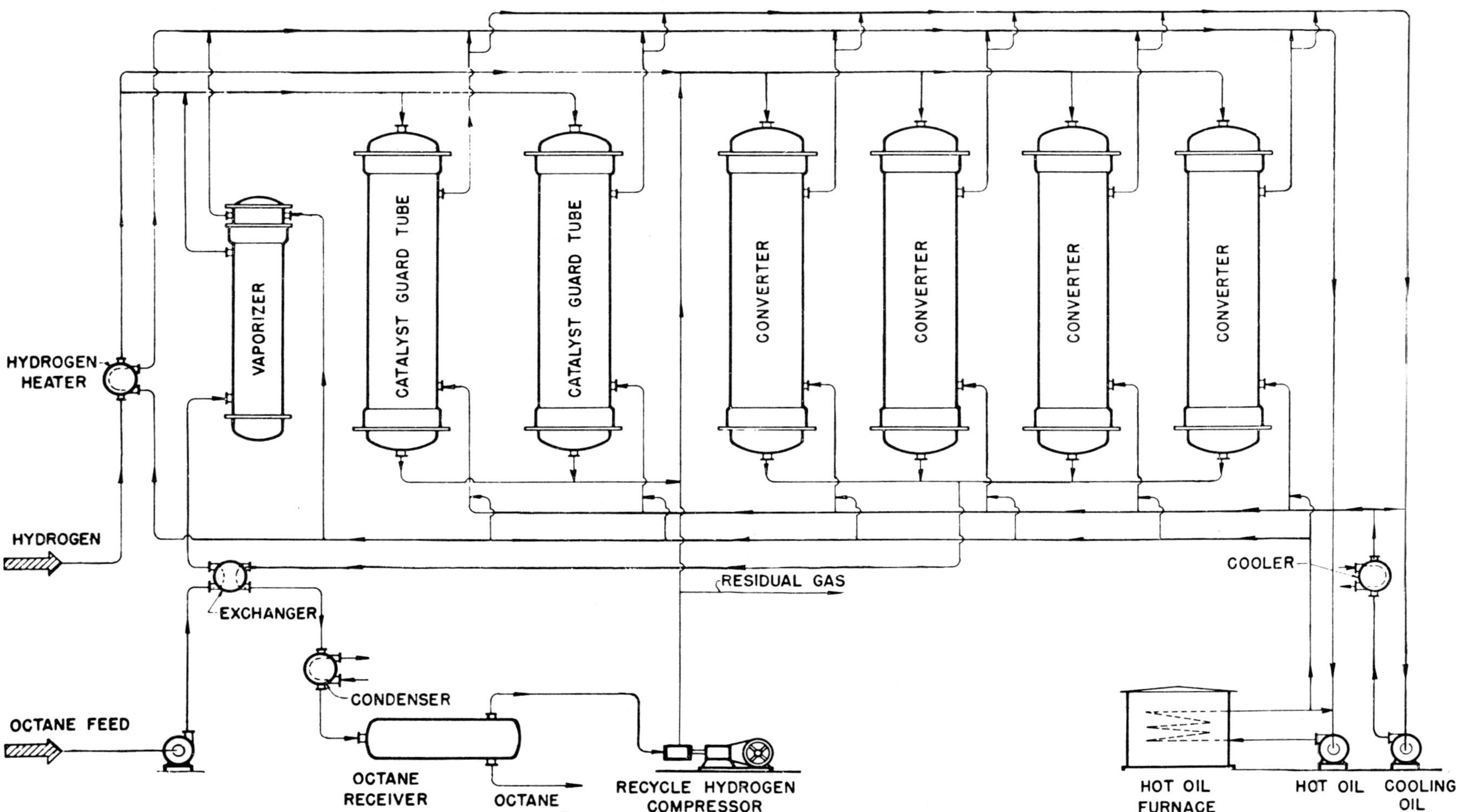

Fig. 13-25 Catalytic hydrogenation process. (Shell Development Co.; copyright, *Petrol. Refiner.*)

Hydrogenation. High- and low-pressure hydrogenation processes are used in refinery operations.[6] The catalytic process developed by the Shell Development Co. has been successful for hydrogenation of octenes produced by polymerization of butanes (Fig. 13-25). Octane volumetric yield is 104 per cent of the octene charge.

Production of Gasoline

The gas synthesis process with natural gas as the raw material may become an important source of liquid hydrocarbons.[7]

The Fischer-Tropsch process exists today in three major variations: (1) low-pressure and medium-pressure synthesis with cobalt catalysts; (2) medium-pressure synthesis with iron catalysts; and (3) high-pressure synthesis ("isosynthesis") with thoria catalysts.

The first two approaches have achieved commercial significance; the third is not economically promising. Wartime German gas synthesis was based entirely on the first process. The higher gasoline quality and greater production of valuable organic chemicals possible with iron catalysts tend to make the second process more attractive for American conditions. The gas synthesis demonstration plant of the U. S. Bureau of Mines at Louisiana, Missouri was based on the use of iron catalyst. Various other companies (Phillips, Texas, Esso Standard) have also pursued active pilot plant research along the same lines.

The underlying reaction is the hydrogenation of carbon monoxide to form hydrocarbons, steam, and carbon dioxide as follows:

$$n\mathrm{CO} + 2n\mathrm{H_2} \rightarrow (\mathrm{CH_2})_n + n\mathrm{H_2O} \qquad (1)$$

$$2n\mathrm{CO} + n\mathrm{H_2} \rightarrow (\mathrm{CH_2})_n + n\mathrm{CO_2} \qquad (2)$$

Ratios of H_2:CO larger than 2:1 are required for most iron-catalyzed gas syntheses, appreciable quantities of CO_2 being formed according to Eq. **2**. Exact synthesis gas composition depends here on the method by which the catalyst has been prepared; it may be as high as 8:1.

The major engineering problem in the design of Fischer-Tropsch reactors is effective removal of the large amounts of heat of reaction from the temperature-sensitive synthesis system. A great American contribution has been the development of fluidized operation in connection with iron-catalyzed gas synthesis. This, combined with the recent commercial process of producing high-volume, low-cost oxygen for partial oxidation of the natural gas at high pressures for the synthesis gas reaction, is the key to the development of the Fischer-Tropsch process in the United States.

REFERENCES

1. Dotterweich, F. H. "Intrinsic Value of Natural Gas—National Asset." *Am. Gas J.* 165: 11+, Dec. 1946.
2. Phillips Petroleum Co. *Hydrofluoric Acid Alkylation.* Bartlesville, Okla., 1964.
3. "Process Handbook Edition." *Petrol. Refiner* 27, No. 9.2, Sept. 1948 (Sec. 2).
4. Alden, R. C. "Neohexane." *N.G.A.A. Proc.* 19: 39–44, 1940.
5. —— and others. "Story of Diisopropyl." *Oil Gas J.* 44: 70+, Feb. 9, 1946.
6. "Gas Conversion Processes." *Petrol. Refiner* 28: 184+, Sept. 1949.
7. Sherwood, Peter W. "Petrochemicals from Water Gas." *Petrol. Engr.* July 1952.

Chapter 8
Production of Butadiene

by Frank H. Dotterweich

PHILLIPS PETROLEUM CO. BUTADIENE PROCESS

The Phillips Petroleum Co. butadiene process (Fig. 13-26) uses a two-step catalytic normal butane dehydrogenation unit. The feed stock, essentially pure *n*-butane and recycle stock, flows to a heater, where its temperature is raised to about 1100 to 1150 F at about 25 psia. The preheated butane flows to a chamber containing chrome–alumina catalyst in small pellets, where the dehydrogenation reaction takes place. Of a feed stock of 98 per cent or more *n*-butane dried over bauxite, about 30 to 32 per cent may be converted to butylenes per pass. The effluent is cooled, compressed to 200 psi, and sent to a vapor recovery system. Suitable fractionation and extraction steps (Fig. 13-27) follow in which the unconverted *n*-butane is separated from the butylenes. The *n*-butane is recycled and the butylenes pass to the second-stage dehydrogenation unit.[1]

This second stage consists of converting the butylenes to butadiene. The butylene feed diluted with steam is heated in a preheater to about 1200 F at close to atmospheric pressure (low pressure favors butadiene formation). The hot mixture then flows to the catalyst chamber, where the butylenes are dehydrogenated to butadiene. The effluent passes thru a cooler into a vapor recovery system and from there to fractionation and extraction systems (Fig. 13-28), where butadiene is recovered in high purity. Unconverted normal butylenes are recycled.

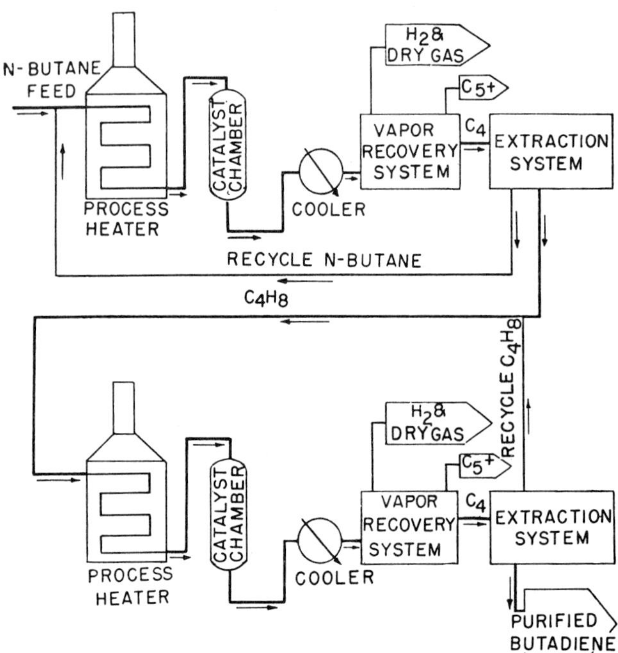

Fig. 13-26 Phillips two-step butadiene process.

STANDARD OIL DEVELOPMENT CO. BUTADIENE PROCESS

This process (Fig. 13-29) utilizes two single fixed-bed reactors, with equal cycle times on reaction and regeneration. Butene from storage is preheated to not over 1100 F, and diluent steam is superheated to about 1300 F. Jet-type mixers contact the two streams directly over the reactor-catalyst bed. A 25 psi pressure drop across the mixing nozzles ensures adequate mixing energy.

Reaction products leaving the catalyst bed are quenched to below 1000 F and further cooled in a waste heat boiler and a series of quench towers. A condensate containing the bulk of the C_4 hydrocarbons is finally charged to a fractionator for stabilization by separation of a light overhead stream, and to a rerun tower where a polymer bottoms stream is removed. The crude butadiene is condensed as an overhead product with other C_4 hydrocarbons. The mixture constitutes the

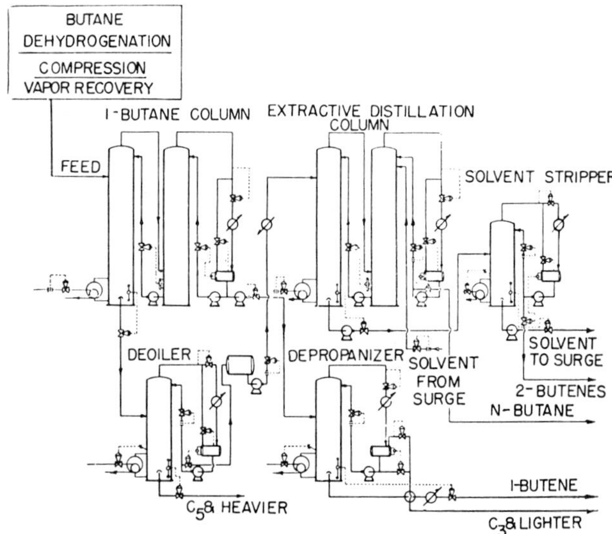

Fig. 13-27 Phillips process for *n*-butylene purification.

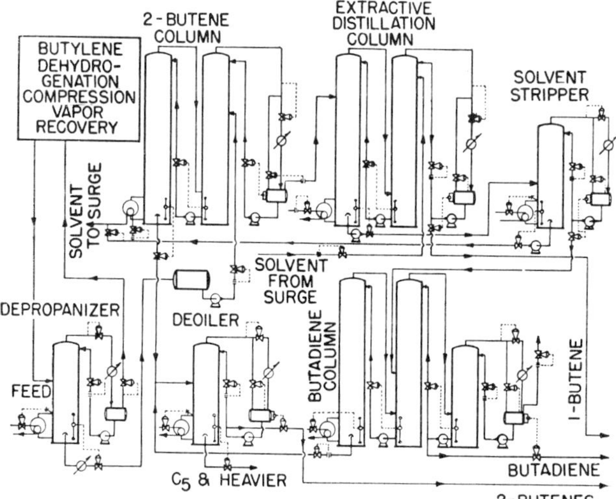

Fig. 13-28 Phillips process for butadiene purification.

feed to a subsequent purification step, when commercial butadiene is produced by extraction with cuprous salt solutions or by other means. Feed may be butane or butylene.

The on-stream period on each reactor is followed by a vacuum purge to remove hydrocarbons and then by a regeneration period using preheated air. The reactor is purged by steam prior to the following on-stream period. In a three-reactor group, one reactor receives hydrocarbons while the second receives air and the third undergoes valve changes and purging.

The normal butane fresh feed and the butane–butene recycle stock from the butadiene recovery plant are pumped to an accumulator and thence thru a feed-quench oil exchanger and thru a heater before they enter the reactor at about 1150 F. The reactor effluent, after partial quenching in a line exchanger and further cooling in the quench tower, passes to a knockout drum prior to the first stage of the four-stage compressors. The overhead from the last drum is compressed and then passes to an absorber and stripper for recovery of C_4 and heavier materials. The effluent from the stripper passes to the butadiene recovery and purification unit.

Major process variables are temperature, pressure, space rate, and time on stream. For butadiene operation, average temperature is 1100 F, pressure is 5.0 in. Hg abs, space rate is 2.2–2.7 cu ft of feed per cu ft of catalyst per hr, and time on stream is 8–10 min. For butene production from *n*-butane, the pressure is increased to 10 psig with other conditions unchanged.

A typical charge stock would yield the following product:

Material	Charge, wt %	Product, wt %*	Yield, wt %†
C_3 and lighter	2.2	6.2	23.4
iso-Butane	0.9	1.8	5.3
iso-Butene	2.1	2.5	2.3
n-Butene	25.0	25.0	...
n-Butane	67.9	50.8	...
Butadiene	1.0	10.5	55.5
C_5+	0.9	1.7	4.7
Coke	...	1.5	8.8
	100.0	100.0	100.0

* Of total feed.
† Of butane fresh feed (ultimate).

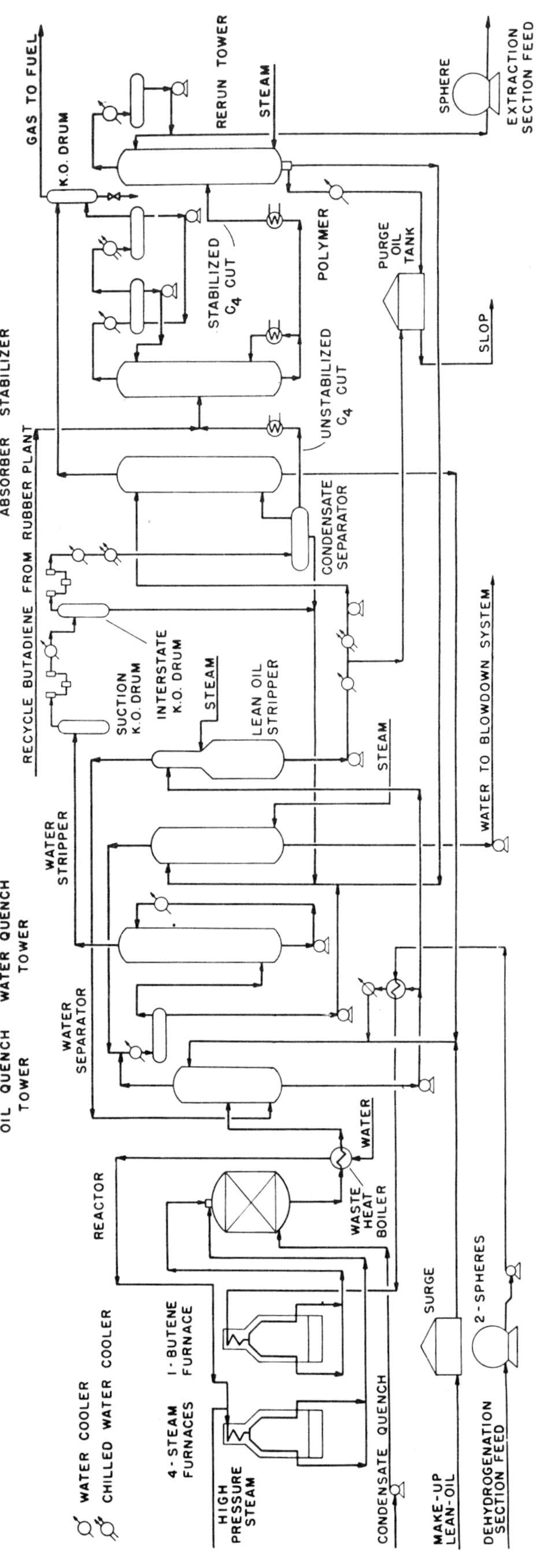

Fig. 13-29 Butadiene production process. (Standard Oil Co. of Indiana.)

HOUDRY DEHYDROGENATION PROCESS

This fixed-bed process is cyclic and adiabatic. Operating conditions are so chosen that the heat required for reaction during the on-stream period is slightly greater than that supplied by combustion of the coke deposit during the regeneration period. The heat deficiency and the desired operating temperature are controlled by adjustment of reactant and air temperatures, and by use during regeneration of a greater amount of air than combustion requires. The air equalizes temperature differences in a large size bed due to heat losses from various causes.

Inert material of high heat capacity, mixed with the active catalyst pellets, acts as a heat-storage medium, absorbing heat liberated by combustion of the coke deposit during regeneration with a consequent increase in temperature of the catalyst bed. This heat is released to the reacting hydrocarbons during dehydrogenation. Heat exchange between the material, catalyst pellets, and reactants takes place with small temperature differences.

With on-stream periods between 7 and 15 min, temperature variation during the cycle is usually less than 50°F, thus providing the optimum temperature range necessary for high conversions. Coke deposition on the catalyst is thereby limited, ensuring high catalyst activity.

Production of Butadiene

Figure 13-30 shows the Houdry process. Following preheating to about 1150 F, the fresh feed, consisting mainly of *n*-butane, and the butane–butene recycle stock from the butadiene recovery unit pass over a chrome–alumina type of catalyst in the reactors. The effluent after quenching enters a knockout drum prior to the first stage of the four-stage compressors (details not shown). Liquid from the first four knockout drums returns to the quench system. The overhead from the last drum passes to a conventional absorber and stripper for recovery of C_4 and heavier materials. The effluent from the stripper passes to the butadiene recovery and purification unit.

The on-stream period of each reactor is followed by a vacuum purge to remove hydrocarbons. Regeneration follows with air, preheated by direct combustion of fuel in the air stream. After regeneration the reactor is purged by means of a steam ejector prior to the following on-stream period. Thus, in a three-reactor group, one reactor receives hydrocarbons while the second receives air and the third undergoes valve changes and purging operations.

Major process variables are temperature, pressure, space rate, and time on stream. For butadiene operation, average temperature is 1125 F, pressure is 5.0 in. Hg abs, space rate is 1.0–1.5 volume feed per volume catalyst per hour, and time on stream is 8–10 min. Table 13-20 shows typical yields.

Production of Mono-Olefins

For production of butenes and *iso*-butene from the corresponding normal and *iso*-butane, the reaction is conducted at or slightly above atmospheric pressure. The space rate is adjusted to balance coke formation with the necessary heat of reaction.

The flow sheet is substantially the same as for the butadiene unit. Line sizes and compressor requirements are smaller for the mono-olefin operation, as the reactor pressure is maintained slightly above atmospheric. Approximately 1 to 2

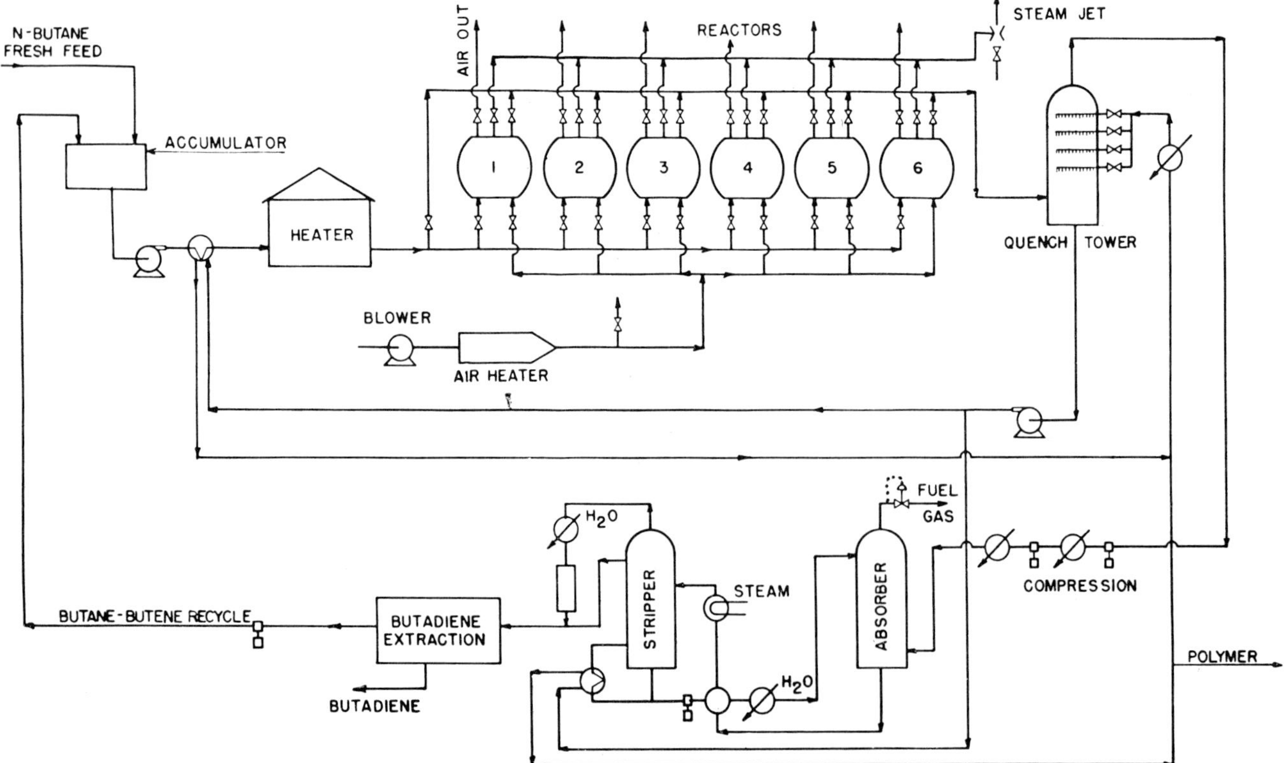

Fig. 13-30 Houdry dehydrogenation process.

per cent of butadiene is produced at the higher operating pressure.

If additional butadiene as well as butenes is desired, the process is operated under reduced pressure and the desired quantity of butenes is separated from the recycle stream before the remaining butane–butene is recycled. Table 13-21 shows typical yields for this operation.

Table 13-20 Production of Butadiene by Houdry Process

	Fresh feed, wt %	Recycle, wt %	Total feed, wt %	Reactor product, wt %	Ultimate yield, wt % butane fresh feed
Hydrogen	...	...	...	1.9	10.1
Methane	...	...	...	1.0	5.3
C_2	...	...	...	1.1	5.9
C_3	...	...	...	1.1	5.9
iso-Butane	0.5	*	*	*	...
iso-Butene	...	*	*	*	...
n-Butene	...	41.4	33.6	33.6	...
n-Butane	99.5	57.9	65.8	47.0	...
Butadiene	...	0.7	0.6	11.1	55.8
C_5+	...	...	...	0.5	2.7
Coke	...	...	...	2.7	14.3
Total	100.0	100.0	100.0	100.0	100.0

* Included in *n*-C_4 fractions.

Table 13-21 Production of Butene and Butadiene by Houdry Process

	Fresh feed, wt %	Recycle, wt %	Total feed, wt %	Reactor product, wt %	Ultimate yield, wt % butane fresh feed
Hydrogen	...	...	...	2.1	7.4
Methane	...	...	...	1.4	4.9
C_2	...	...	...	1.5	5.3
C_3	...	...	...	1.6	5.6
iso-Butane	0.5	*	*	*	...
iso-Butene	...	*	*	*	...
n-Butenes	...	35.6	25.5	34.9	33.1
Butadiene	...	0.7	0.5	9.8	32.7
n-Butane	99.5	63.7	74.0	45.6	...
C_5+	...	...	...	0.5	1.8
Coke	...	...	...	2.6	9.2
Total	100.0	100.0	100.0	100.0	100.0

* Included in *n*-C_4 fraction.

Normal operating conditions for producing a major quantity of mono-olefins and a minor quantity of diolefins are: temperature, 1125 F; pressure, 10 psig; space rate, 1.5–2.0 cu ft of feed per cu ft of catalyst per hr; and on-stream time, 8–10 min. Typical yields for producing mainly butene from butane are given in Table 13-22.

Table 13-22 Production of Butene by Houdry Process

Material	Charge, wt %	Product, wt %	Ultimate yield at high conversion, wt % *n*-butane fresh feed
Hydrogen	...	1.6	3.4
Methane	...	2.4	5.1
C_2	...	2.7	5.8
C_3	...	3.4	7.2
iso-Butane	0.5	0.3	Trace
iso-Butene	...	0.2	Trace
n-Butene	...	32.0	68.2
n-Butane	99.5	52.6	...
Butadiene	...	2.0	4.3
C_5+	...	0.6	1.3
Coke	...	2.2	4.7
Total	100.0	100.0	100.0

Note: Ultimate yield of butene would be increased at lower conversion and at lower pressure.

Process Applications

Commercial production		Pilot plant investigation	
Product	Source	Product	Source
Butadiene*	*n*-Butane or butane–butene	Propane	Propane
		iso-Butene	*iso*-Butane
Butene	*n*-Butane	Isoprene	*iso*-Pentane
		Styrene	Ethyl benzene
		Methyl styrene	Isopropyl benzene

* Only single-step process.

The Standard Oil Company of California unit at El Segundo, Calif., has a capacity of 18,000 tons of butadiene per year. Started in 1943, it has been operated to produce butadiene and a butene–butadiene mixture.

REFERENCE

1. Phillips Petroleum Co. *Furfural Extractive Distillation for Separation and Purification of C_4 Hydrocarbons.* (Bull. 184) Bartlesville, Okla.

Chapter 9
Production of Helium

by Frank H. Dotterweich

At ordinary temperatures and pressures this light, odorless, tasteless, colorless gas* conforms closely to the laws of an ideal gas. Naturally occurring helium is virtually all He_4, but contains a trace of the stable isotope He_3. Unstable isotopes He_5 and He_6 have been listed, but the existence of He_5 is questioned by some scientists. He_6 is radioactive, with a half-life of 0.8 sec.[1,2]

Helium has a negative Joule-Thomson effect (heats upon free expansion) except at very low temperatures.

RESOURCES

Helium is estimated at one part in 185,000 to 200,000 parts of air at the earth's surface.[3-6] The only known raw material from which helium can be extracted economically in large quantities is helium-bearing natural gas. Usually natural gases of higher helium contents are found in fields that lie over buried granite ridges, such as the deeply buried Amarillo Mountains of the Texas Panhandle and the Nemaha Ridge of Kansas, and in fields closely associated with igneous intrusions, such as the Rattlesnake field of San Juan County, New Mexico.

Helium resources of the United States are contained principally in four helium-bearing natural gas fields (per cent helium content): (1) the Hugoton field of Kansas, Oklahoma, and Texas (0.46); (2) the Panhandle field of Texas (0.48); (3) the Keyes field of Oklahoma; and (4) the Greenwood field of Kansas (0.6). According to a 1959 estimate these fields contain about 125 billion cubic feet of recoverable helium. In some areas, helium contents of nearly one per cent are found.

Major helium-bearing natural gas fields in the United States as of June, 1950, are given in Table 2-14.

PRODUCTION

Production facilities are centered in southwestern United States, at Amarillo and Exell, Texas; Keyes, Oklahoma; Otis, Kansas; and Shiprock, New Mexico, where relatively large volumes of helium-bearing natural gas are available. Average helium content (volume per cent) of gas processed in U. S. Government plants follows: Amarillo (1.8); Exell (0.9); Keyes (2.0); Otis (1.0); and Shiprock (5.7). Table 13-23 gives helium production by U. S. Government plants for the period 1921–1962, inclusive. The cumulative production for this period was 5,104,894 MCF.

The Bureau of Mines helium plants at Exell, Texas; Keyes, Oklahoma; and Otis, Kansas extract helium from natural gas produced by privately owned companies.

Table 13-23 Helium Production in the United States*,[7]

Year	Production, MCF
1921–28 (avg)	5,761
1929–42 (avg)	11,776
1943–49 (avg)	83,545
1950–54 (avg)	137,957
1955	220,711
1956	243,880
1957	291,457
1958	334,175
1959	476,892
1960	642,033
1961	727,103
1962†	683,082

* 1961 and 1962 data from personal communication.
† Includes 2.2 MMCF purchased by the Bureau of Mines from private plants under the helium conservation program.

Separation and Purification

Helium is separated from natural gas by the gas being cooled to a temperature below the liquefaction point of its ordinary constituents but above that of helium. First, the gas must be treated chemically to remove small quantities of carbon dioxide and water vapor that would solidify in the low-temperature equipment. Helium may also be extracted from gases by mass diffusion methods.

Liberal Plant.[8] The National Helium Corp. plant at Liberal, Kansas, is the world's largest cryogenics plant (Fig. 13-31). It was designed to recover 95 per cent of the helium in 850 MMSCFD natural gas containing about 0.42 per cent He. The design product, 5.8 MMSCFD crude He, contains 65 per cent He; the remainder is mostly N_2.

Exell Plant. The natural gas enters the Bureau of Mines Exell (Amarillo, Tex.) plant at about 450 psi. It passes first thru a bubble tower in contact with a solution of monoethanolamine and diethylene glycol to remove water vapor, carbon dioxide, and hydrogen sulfide, and next thru activated bauxite to remove remaining water vapor.

After this preliminary treatment the natural gas, still at about 450 psi, enters a heat exchanger in which its temperature

* See Table 1-1 for physical properties.

is reduced to about −238 F by the cold outgoing residue gas. The partly liquefied natural gas is throttled from this heat exchanger into a separator (operated at about 225 psi) and cooled with vapor from liquid nitrogen. Gaseous crude helium is drawn from the top of the separator. The liquefied portion of the natural gas (residue gas) is withdrawn from the bottom to a pressure of about 200 psi and run back thru the heat exchanger. There it chills the incoming gas and is vaporized and warmed to approximately atmospheric temperature. It then passes into a pipeline for transportation to fuel gas markets (except when the residue gas consists largely of nitrogen and is discharged to the atmosphere).

The helium extraction process increases the heating value of the natural gas by removing all the helium and carbon dioxide and part of the nitrogen. It also sweetens gas initially containing hydrogen sulfide.

Some refrigeration is supplied by expansion of the gas from its initial pressure of 450 psi to the discharge pressure of about 200 psi. The rest of the refrigeration, responsible for the lowest temperature, is supplied by an auxiliary nitrogen refrigeration cycle. Any nitrogen extracted from the natural gas in the process is compressed in three stages to 600 psi, cooled with water, and brought to its liquefaction temperature in a Claude-type expansion engine cycle.

The crude product drawn off in the first stage consists of about 70 per cent helium and 30 per cent nitrogen. This mixture is compressed in four stages to 2700 psi, cooled with water, dried with activated alumina, and passed to another low-temperature processing unit. Here, without substantial reduction in pressure, consecutive phase separations produce first 98.2 per cent, then 99.5 per cent helium. This is followed by final purification thru activated charcoal.

In the primary step, under high pressure and a temperature of about −320 F produced by a bath of liquid nitrogen boiling at about atmospheric pressure, most of the nitrogen in the crude helium condenses to a liquid. The condensed nitrogen is withdrawn and used as make-up for the nitrogen refrigeration cycle. The helium is drawn off as a gas and passes to the intermediate step, where additional nitrogen is condensed at a temperature of about −340 F produced by a bath of liquid nitrogen boiling at subatmospheric pressure. Finally, the helium passes thru one of a pair of purifiers containing activated charcoal cooled by liquid nitrogen boiling at approximately atmospheric pressure. The charcoal absorbs virtually all the remaining impurities and yields helium 99.995 per cent pure (Grade A). One purifier is reactivated by warming while the other is being used for this final purification. The helium passes from the charcoal purifier, is warmed to atmospheric temperature thru heat exchangers, and passes directly into high-pressure containers for storage or shipment.

Significant reductions in compressor horsepower required for helium extraction can be made by reducing the throttling pressures of the helium-bearing natural gas stream from the 600 to 70 psig range used at the Exell plant to a range of 450 to 200 psig. This is accomplished by the addition of heat exchanger surface to the gas liquefaction section of the crude helium extraction units.

The crude helium separation cycle is the principal part of the process, since the purification step involves comparatively small volumes because of the low helium content of the processed natural gas. The Exell plant separation cycle is shown on a theoretical pressure–enthalpy chart (Fig. 13-32) for a typical natural gas processed for helium extraction.

The crude helium extraction cycle is represented in Fig. 13-32 by the dotted path ABCDEG. Starting at point A, corresponding to 600 psia and 80 F, path AB represents the cooling and liquefaction of the gas, which takes place during passage thru the gas interchanger, with transfer of heat to cold outgoing gas from which helium has been extracted. Path BC represents throttling of the nearly completely liquefied gas into the crude helium separator operated at 250 psia and approximately −230 F. Path CD represents the cooling of the liquefied natural gas by cold nitrogen vapors passing thru the separator vessel during crude helium separation. Path DE corresponds to throttling of the liquefied gas from

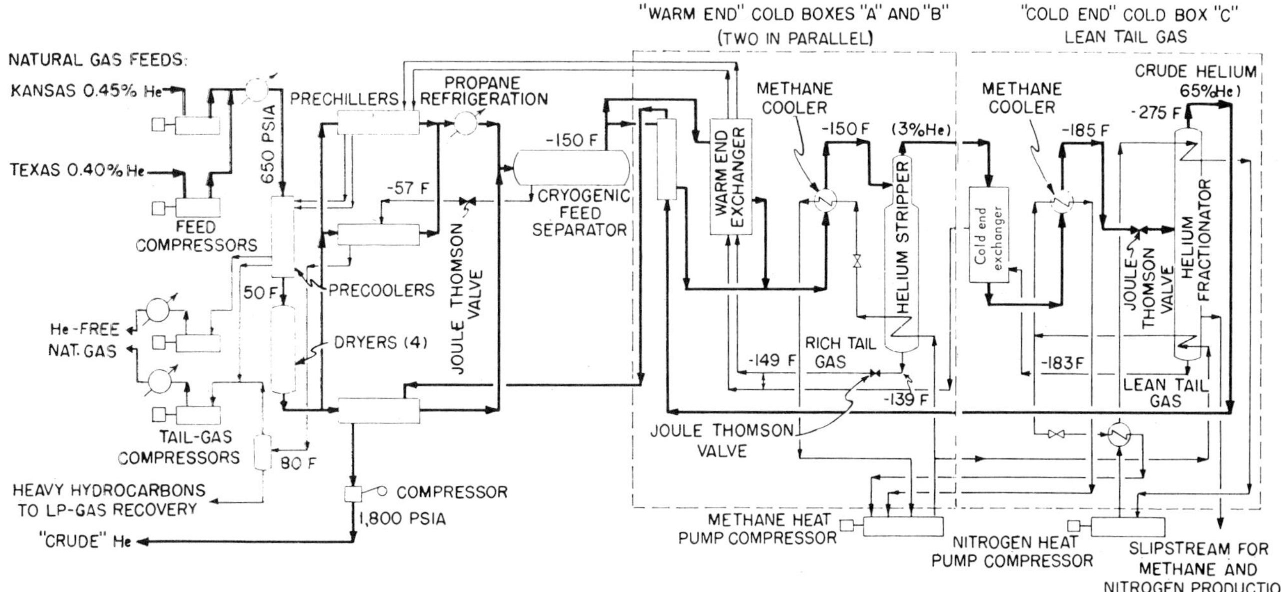

Fig. 13-31 Process flow diagram for National Helium Corp. helium plant at Liberal, Kan.[8]

the separator to about 70 psia. Path EG represents the warming of the processed gas stream to 80 F in the gas interchanger by the high-pressure incoming gas traversing path AB.

In actual practice, the low-pressure processed gas will not warm to the temperature of point G but to a somewhat lower temperature, depending upon the amount of refrigeration supplied by nitrogen cooling, the amount required to offset heat infiltration, and the efficiency of the heat exchangers. Figure 13-32 takes no account of the comparatively small amount of refrigeration lost to the cycle in the removed crude helium or of the small change in composition and heat capacity of the processed gas due to removal of crude helium.

The total amount of refrigeration theoretically supplied to the cycle is that represented by C–D (about 6 Btu) plus G–F (about 12 Btu), making a total of approximately 18 Btu per lb of gas processed. Under favorable conditions, the crude helium separation cycle can be operated without nitrogen cooling in the crude helium separator, thus reducing the refrigeration requirements to about 12 Btu per lb of natural gas processed. The throttling range used is dictated primarily by two requirements: (1) the establishment of sufficient temperature differential to accommodate the necessary heat exchange between incoming and outgoing gas streams; and (2) provision of sufficient differential heat capacity between incoming and outgoing gas streams for refrigeration to offset heat infiltration and losses. Experience indicates that the refrigeration is required principally to accommodate heat exchange, that is, to allow for heat exchanger inefficiencies.

Helium is soluble in liquefied natural gas at the pressures and temperatures existing in the separating chamber of the helium extraction cycle. A principal consideration in the process is to effect separation under conditions that will permit a high degree of helium recovery by minimizing helium solubility and loss in the liquefied gas. Figure 13-33 shows the helium content of the gas and liquid phases at various indicated temperatures and pressures, as computed for a typical natural gas processed for extraction of helium of the composition indicated. Crude helium separation is normally effected at pressures of 200 to 225 psi and temperatures of −235 F to −245 F to provide 70 per cent or more helium in the gas phase while keeping solubility to about 0.12 per cent in the liquefied gas. This permits helium recovery of 88 per cent or greater when the initial helium content of the gas is only about one per cent.

The production equipment at the Exell plant consists of modules, each with a rated processing capacity of 10 MMCF of natural gas per 24 hr. This plant has processed more than 100 MMCF (about 2300 tons) of natural gas in 24 hr. Although the 100 MMCF gas contains less than one per cent helium, the plant has produced more than 20 MMCF of helium in a month.

The entire process for extraction and purification is continuous. The residence time of the natural gas, from entry into the equipment to discharge into the pipeline, is less than one minute.[9–11]

The helium produced at the Bureau of Mines helium plants is 99.995 per cent pure and is designated "Grade A." The major impurity is neon.

A hydrogen analysis apparatus has been developed by the Bureau of Mines which is capable of detecting less than 0.005

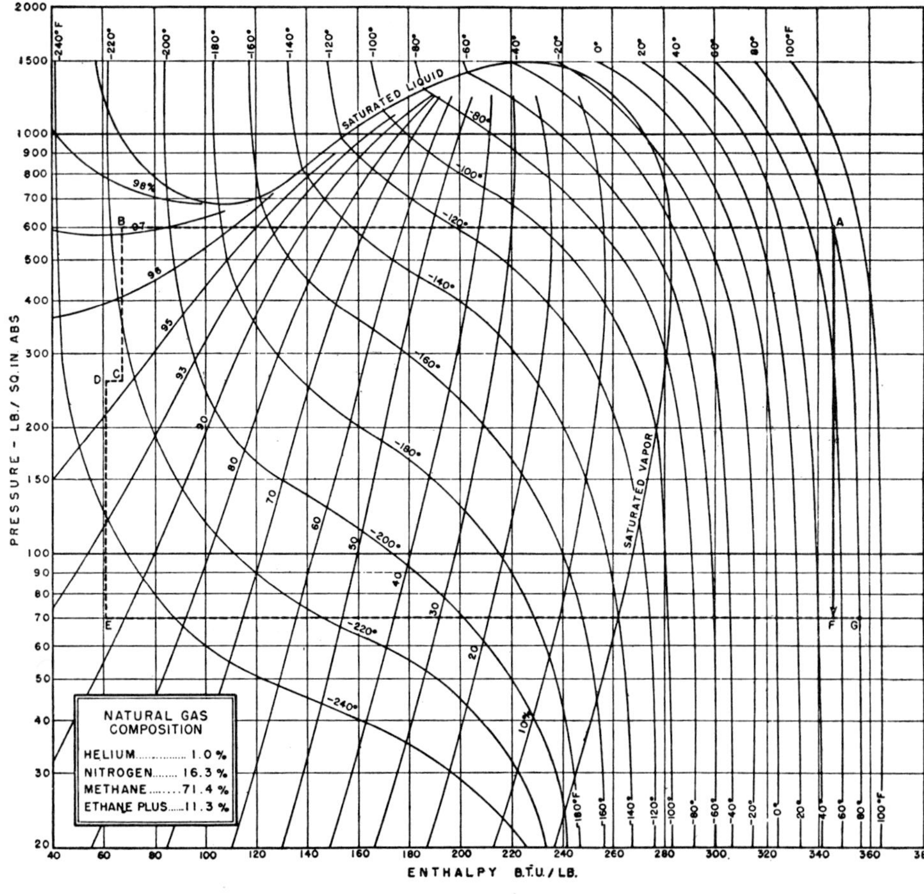

Fig. 13-32 Crude helium separation cycle represented on pressure–enthalpy diagram for typical helium-bearing natural gas. (U. S. Bureau of Mines data, courtesy *Chem. Eng. Progr.*)

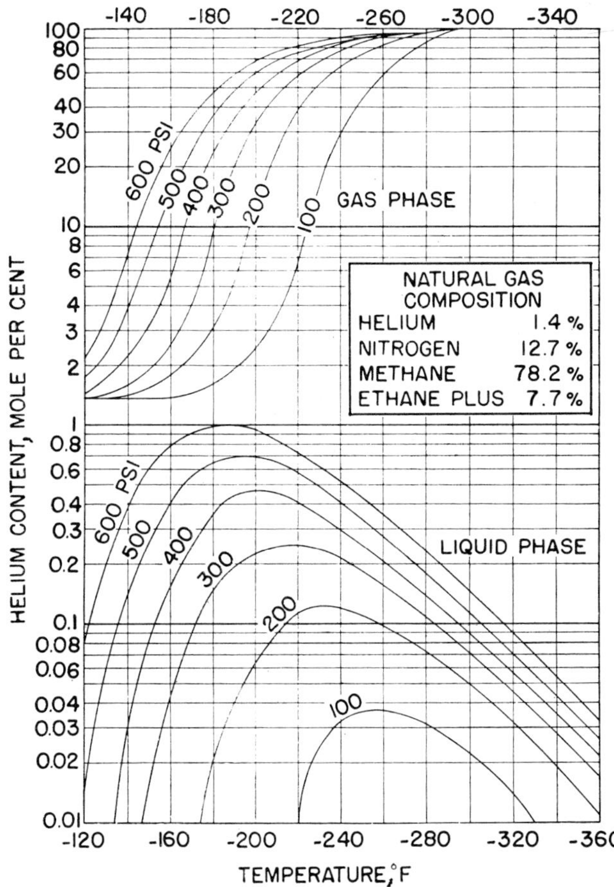

Fig. 13-33 Helium content of gas and liquid phases of typical helium-bearing natural gas.[9] (U. S. Bureau of Mines data, courtesy *Chem. Eng. Progr.*)

per cent hydrogen in helium. This apparatus is used as an operating guide in producing "Grade A" helium.

STORAGE AND SHIPMENT

Most helium plants have provisions for storage in limited quantities at 1800–2000 psi. Storage containers consist of 10 in. heavy-wall pipe in 80 ft lengths, suitably manifolded for convenience in filling and discharging.

Most helium shipments are made in special railroad tank cars, each having 28 or 30 steel cylinders mounted lengthwise. The cylinders accommodate shipping pressures of 2000–3600 psi and cars have capacity of up to about 300 MCF of helium measured at standard pressure and temperature. Helium is also shipped at similar pressures in standard compressed gas cylinders and automotive trailers. Trailer construction is similar to that of the tank cars.

AVAILABILITY AND COSTS

Helium to meet future increases in demand can be obtained by constructing additional extraction plants. However, known resources of helium-bearing natural gas are being depleted at rates such that they could not supply predicted demand beyond about 1985. Legislative authority to accommodate a long-range helium conservation program was obtained when the President approved the Helium Act Amendments of 1960 (Public Law 86-777). The program proposed by the Department of the Interior is aimed at saving for future use helium that would otherwise be wasted in helium-bearing natural gases consumed as fuel.

Congress authorized the program in 1961. All five plants authorized under the program have been built. In addition to these there are five government-owned plants.

Production from the privately owned plants during the 22-year contract period will be about 62.5 billion cu ft, supplemented by 15 billion cu ft from the government plants. This will be enough to meet projected demand until 1985, at which time there will still be 41.5 billion cu ft in storage.

The helium plants in the conservation program were built by:[8]

Cities Service Helex Corp. (Cities Service Co.) at Ulysses, Kansas, producing 610 MMCF per year for sale at $11.78 per MCF.

Northern Helex Co. (Northern Natural Gas Co.) at Bushton, Kansas, producing 675 MMCF per year for sale at $11.24 per MCF.

National Helium Corp. at Liberal, Kansas, producing 1053 MMCF per year for sale at $11.78 per MCF.

Phillips Petroleum Co., at Dumas and Sherman County, Texas, two plants producing a total of 788 MMCF per year for sale at $10.30 per MCF.

The only helium privately produced in the United States outside these facilities comes from the Kerr-McGee Oil Industries, Inc., plant in Arizona. Its capacity output of 65 MMCF per year of "Grade A" helium is sold to commercial distributors, mainly on the West Coast.

The conservation program is designed to be self-liquidating.

By an act of Congress in 1937, provisions were made for the sale to private parties of helium not needed for government use. The government's 1963 price was $35.00 per MCF for commercial, scientific, and medical use.

REFERENCES

1. Cattell, R. A., and Wheeler, H. P., Jr. "Helium." *Encyclopedia of Chemical Technology* vol. 7, p. 398–408. New York, Wiley, 1951.
2. ——. "Growing Demand Is Seen for Helium." *Petrol. Refiner* 30: 91–4, Mar. 1951.
3. Keesom, W. H. *Helium.* Amsterdam, Elsevier, 1942.
4. Rogers, G. S. *Helium-Bearing Natural Gas.* (U. S. Geol. Survey, Prof., Paper 121) Washington, D. C., 1921.
5. Stewart, A. *About Helium.* (Bur. of Mines I.C. 6745) Washington, D. C., 1933.
6. Anderson, C. C., and Hinson, H. H. *Helium-Bearing Natural Gases of the United States.* (Bur. of Mines Bull. 486) Washington, D. C., 1951. Suppl.: Boone, W. J., Jr. *Analyses and Analytical Methods.* (Bur. of Mines Bull. 576) Washington, D. C., 1958.
7. Lipper, H. W., and Wilcox, Q. L. "Helium." In: U. S. Bur. of Mines. *Mineral Facts and Problems,* p. 383–91. Washington, D. C., 1960.
8. Kinney, G. T. "World's Biggest Helium Plant Opens." *Oil Gas J.* p. 54, Sept. 30, 1963.
9. Mullins, P. V. "Helium Production Process." *Chem. Eng. Progr.* 44: 567–72, July 1948.
10. Cattell, R. A. "Helium—the Wonder Gas." *Sci. Monthly* 69: 222–8, Oct. 1949.
11. Frost, E. M., Jr. *Improved Apparatus and Procedure for the Determination of Helium in Natural Gas.* (U. S. Bur. of Mines R.I. 3899) Washington, D. C., 1946.

SECTION 14

PURGING*

G. Russell King (deceased), *Section Chairman,* Philadelphia Electric Co., Philadelphia, Pa.
H. S. Carpenter (deceased), Public Service Electric and Gas Co., Newark, N. J.
Mark G. Eilers, (retired), Rochester Gas and Electric Corp., Rochester, N. Y.
Hugh E. Ferguson (deceased), The Peoples Gas Light and Coke Co., Chicago, Ill.
P. W. Geldard, The Consumers' Gas Co. of Toronto, Toronto, Ont., Canada
J. W. Penney, Boston Gas Co., Jamaica Plain, Mass.
W. E. Russell, Baltimore Gas and Electric Co., Baltimore, Md.
Charles J. Smith, Consolidated Edison Co. of New York, Inc., New York, N. Y.
F. E. Vandaveer (retired), Con-Gas Service Corp., Cleveland, Ohio
Jesse S. Yeaw (deceased), Rochester Gas and Electric Corp., Rochester, N. Y.

The authors of this Section chose to be identified with the Section as a whole rather than with individual chapters.

CONTENTS

* Material in this Section is abstracted from "Purging Principles and Practice," published by the American Gas Association in 1954. This publication may be consulted for additional information.

Tables

Figures

Chapter 1

General

INTRODUCTION

When the combustible gas content of a chamber is directly replaced by air, a mixture within the *flammable limits* of the gas and air will be formed and remain in the chamber during part of the replacement operation. Similar mixtures within the flammable limits will also be formed if the air in a chamber is directly replaced by a combustible gas. Although ignition of such mixtures can be avoided, their formation should be prevented. When the content of an enclosed space is changed from combustible gas to air or vice versa, formation of flammable mixtures should be avoided by a procedure known as "purging."

Purging consists of interposing an inert substance between the two media being interchanged. For example, when a piece of equipment is taken out of service for inspection or repair, the combustible gas content is replaced by an inert gas, which is, in turn, replaced by air. Inert gases most commonly used for purging are carbon dioxide, nitrogen, and mixtures of the two produced by carefully controlled combustion of fuel gases, oil, or gasoline. Steam and water are also used as intervening media.

The purging of boilers, furnaces, ovens, and atmospheric generators before starting the burner lighting procedure, to ascertain that no combustible mixture is present, is not specifically covered in this section.

TERMINOLOGY

Words and expressions commonly used in purging procedures are defined below:

Aeration: Provision of a constant supply of air by mechanical means.

Blanking: Insertion of a solid metal plate across a main at fitting flanges.

Clearing: Act of replacing the atmosphere by direct displacement by another medium so rapidly that there is a minimum of mixing, thus reducing to a minimum the duration of any explosive mixture. Venting takes place into a large, open area, high above ground (eight feet or more), with means of ignition excluded.

Combustible: Capable of being ignited and rapidly oxidized when mixed with proper proportions of air.

Concentration: Per cent by volume unless otherwise noted.

Dilution: A form of purging in which replacement of one substance by another is accomplished with appreciable mixing.

Displacement: A form of purging in which replacement of one substance by another is accomplished without appreciable mixing.

End point: Attainment of such a *concentration* (see definition) of inert substance in the closed system being purged that subsequent admission of air if purging out of service, or admission of gas or vapor if purging into service, will not result in formation of an explosive mixture.

Explosive limits: See *flammable limits.*

Explosive mixtures: Flammable gas and air mixtures between the upper and lower flammable limits.

Flammable: See *combustible.*

Flammable limits: The lowest (lower limit) and highest (upper limit) concentrations of a specific gas or vapor in mixture with air that can be ignited at ordinary temperature and pressure. See Table 2-63 and Figs. 2-24b, 2-25, and 2-26. (Synonymous terms: explosive limits, limits of inflammability, and limits of flame propagation.)

Holding purge: Replacement of the normal combustible content of a closed system by inerts which are maintained during alterations or repairs and then replaced by the normal combustible.

House line: Gas piping beyond the outlet of the consumer's meter.

Inert gas: A gas, noncombustible and incapable of supporting combustion, which contains less than two per cent oxygen and combustible constituents of less than 50 per cent of the lower explosive limit of the combustible being purged.

Isolation: Disconnection from all other equipment or mains of a chamber or space to be purged.

Pig: A cylindrical or barrel-shaped body, slightly smaller than the inside of a main, which is moved thru the main by gas or air introduced behind it. A purging pig separates gas and air in travel thru a long main. It must be nonabrasive and covered with nonsparking material, such as rubber, leather, or wood.

Purging into service: Replacement of the air in a closed system by an inert substance, which is in turn replaced by combustible gas, vapor, or liquid.

Purging out of service: Replacement of the normal combustible content of a closed system by an inert substance, which is in turn replaced by air.

Ventilation: Procedure in which doors, manholes, or valves are opened to permit ingress of air by natural circulation to replace inert gas contents.

FACTORS AFFECTING MECHANICAL PURGING OPERATIONS

Replacing one gas by another in an enclosed space or chamber is accomplished by means of two distinct actions: (1) *displacement;* and (2) *dilution* or mixing. In a purging effected entirely by displacement, the gas or air originally present is pushed out of the escape vents by the entering inerts, with little or no mixing. Thus, the quantity of inert gas required for a purging by displacement approximates the quantity of gas or air to be replaced. Quite frequently certain conditions, such as the chamber size or shape and the nature of the gases, cause the inert gas to mix with the original contents, so that the purging tends to proceed by dilution. To purge by dilution or mixing requires a volume of inerts possibly four or five times that of the free content of the purged chamber, because increasing amounts of inert gas will be lost from the escape vents in mixture with the original contents as the purging proceeds. Most purging operations are combinations of displacement and dilution actions. In actual practice, it is impossible to avoid some mixing of the inert gas with the combustible gas or air replaced. Generally, though, the less the mixing or dilution, the more efficient the purging is.

Causes of Dilution or Mixing

The more important causes of mixing during a purging follow. Failure to recognize their importance may result in a purging operation of 80–85 per cent dilution and only 15–20 per cent displacement.

Area and Time of Contact. The area of contact between the inert gas and the original contents depends upon the size, shape, and internal construction of the chamber being purged. Ordinarily, little can be done to limit the mixing that results from this factor.

When a tall narrow tower is purged, the area of contact between the gases is small compared with their volumes, so that mixing is limited and the volume of inerts used need not be much greater than that of gas or air to be cleared out. By contrast, crowns of storage holders are flat or concave shallow domes, with heights less than one-tenth their diameters. It is impossible to avoid large areas of contact in such chambers; therefore, it is usually necessary to use at least 1.8 to 2.0 volumes of inerts per volume of free space in purging them.

When a main is purged, the area of contact may be so small that little mixing will occur. Advantage can be taken of this condition to use a quantity of inert gas that is only a fraction of the volume of combustible gas or air to be replaced. It is possible to introduce merely enough inert gas to form a "slug" or piston between the original gas (or air) content and the entering air (or gas). This slug and the original gas or air ahead of it are pushed thru the section being purged by air or gas introduced after the slug.

The duration of contact between the surfaces of the inert gas and the gas or air should be as short as possible, to minimize mixing by diffusion. Interruptions, variations, and too low an inert gas input rate should be avoided.

Input Velocities. The velocity of entrance of the inert gas materially affects the nature of the purging. In general, the inert gas inlet of containers other than gas mains should be as large as practical, so that the input velocity will not exceed 2 or 3 ft per sec, thus keeping agitation or stirring of the chamber contents at a minimum. If the inert gas input connection is small compared with the rate of input, the velocity of the inert gas may carry it completely across the chamber, resulting in thorough mixing.

Densities of Gases. The relative densities of the inert gas and the medium being purged have important effects on the mechanics of purging. The specific gravity of carbon dioxide is 1.5; that of CO_2–N_2 mixtures produced by controlled combustion is about 1.05. These gravities are sufficiently greater than those of usual city gases (0.4–0.7) for such inert gases to tend to stratify and remain on the bottom of a chamber filled with lighter gas. Therefore, when purging a light gas from a chamber, it is preferable to vent it from the chamber top by admitting the heavier inert gas at the base. Conversely, in putting equipment back into service, when heavy inert gas is to be replaced by a light gas the latter should be introduced at the chamber top and the heavier inerts vented from the bottom.

Introducing a fourth atmosphere in some cases will facilitate purging of a tank containing vapors appreciably heavier than the inert gas available. Thus, heavy butane, propane, or benzol vapors can be displaced downward and out thru bottom connections by natural or manufactured gas first, this gas then displaced thru top vents by inert gas, and the inert gas finally replaced by air.

Temperature Effects. It is desirable to keep the temperature of inert gas entering a large chamber as low as practicable to minimize the possibility of setting up any "thermal currents" and to maintain as great a gravity differential as possible.

The contraction in gas volume resulting from decrease in temperature should be given attention. A positive, though slight, pressure should be maintained within a chamber being purged.

It may be impractical to control temperatures and avoid thermal currents when the chamber being purged contains deposited solids or liquids, such as naphthalene or tar deposits, oils, or solvents. These or other materials which will volatilize and give off combustible vapors with relatively small increases in temperature above atmospheric should be heated enough before or during purging so that no further volatilization of combustible vapors after purging can occur on admission of air. This may be accomplished by: (1) steaming of the chamber or system before an inert gas purging; (2) purging with steam as the inert gas; or (3) admitting the inerts at an elevated temperature (150–180 F), saturated with water vapor.

ORGANIZATION OF PURGING OPERATION

Good organization, planning, and preparation with full agreement of all concerned are essential for a successful purging project. The following factors must be decided upon:

1. Equipment to be purged, and how it should be separated.
2. Inerts to be used, how obtainable, and how introduced and vented.
3. Method for testing completeness of the purging and end point to be attained.
4. Selection and assignment of a responsible supervisor and operating personnel.

5. Preparation of a written "procedure," detailing the sequence of all operations related to the purging, including the time of performance and estimated duration.

Responsibility and authority for the purging operations should be vested in a person who is familiar with the properties and nature of the materials involved and the construction and function of the equipment to be purged. He should be capable of deciding how the purging should be done and of judging whether it is proceeding satisfactorily and when it is properly completed. He should be able to detect any hazards and to decide how best to overcome any difficulties that might arise. He should plan and discuss the schedule of the entire operation with other personnel.

Selection of the time for performing the purging may be affected by many factors not directly related to the operation itself, such as: demands and loads, availability of personnel to perform the repair work for which the purging is undertaken, and weather conditions.

It is desirable to start a purge at a time that will permit completion of the introduction of inerts and removal from or return to service during daylight hours.

When more than one unit or piece of equipment is involved, the purging should be broken down into several successive operations, with their sequence definitely decided upon and their timing carefully calculated and scheduled. Each successive part of a large scale operation may well be considered a separate purging.

It is important to set down the decisions reached in a written "procedure" which is definite and detailed as to consecutive steps. Sample procedures in Chap. 4 illustrate the details necessary to ensure, for instance, that no valve is left open when it should be closed, or vice versa.

The heads of the various departments that might be affected should then review the purging procedure, so that any misunderstandings or conflicts of thought can be reconciled before, not during, the actual purging operation.

After executive approval of the written procedure and the signature of supervisory personnel concerned, the purging supervisor may proceed with the selection of those required to assist in the operation. All should then be instructed together in the work to be done. Each man should understand what he is to do and its importance in relation to the work others must perform.

Those selected to aid in the purging operation should have definite responsibilities. For example, one man may be made responsible for the production and continuity of supply of inerts; a second, who has analytical and chemical testing training, responsible for the testing of the contents in or escaping from the purged chambers. These men should concentrate all their attention on their indicated duties and should not be expected to perform any other tasks. As many other men as are deemed necessary should be assigned for the general purging operations.

SELECTION OF INERT MEDIUM

Proper selection of a purging medium depends principally upon the relative merits and availability of various inerts, the contents and construction of the equipment to be purged, the overall purpose of its purging, and the method to be used.

In general, equipment may be purged with any of the commonly used purging media; namely, carbon dioxide, nitrogen, purging machine products of combustion, automotive exhaust gas, steam, or water. The advantages and disadvantages of each are discussed in Chap. 3.

ISOLATION OF EQUIPMENT

Whenever any equipment space is to be purged, it is essential for it to be isolated from the rest of the system either by blocking or by severing all its connections. Two distinct but related objectives are thus accomplished: (1) prevention of any inleakage of vapors (or air) or any outleakage of inerts during the purging; and (2) prevention of any inleakage of vapors after the purging, when the equipment is out of service for inspection, repair, or demolition. If possible, the measures adopted for isolating the space for purging out of service should also provide the desired post-purging isolation. If bags, liquid seals, or similar means of isolation must be used while purging out of service, they should be replaced by permanent means of isolation, like blanks or gaps, before the purging inerts are replaced by air. Similarly, when purging into service, one should remove any such temporary methods of isolation and make all final connections after the air has been replaced with inerts and before the gas is introduced.

Methods by which a space may be isolated from adjoining equipment, piping, or mains follow (in descending order of general reliability):

1. **Actual disconnection,** or breaking of the physical continuity of a pipe or connection, by removal of a fitting or spacer piece is recognized as the most dependable method of isolation, because it makes inleakage of gases impossible.

2. A **blank** or a sheet metal plate inserted between the flanges in a pipe can be considered a very effective method of isolation. Its use for isolation purposes is generally restricted to locations where: (a) it is possible to force flanges apart to receive it; (b) the flanges can subsequently be regasketed without difficulty or made gastight without gaskets; and (c) valving or disconnection cannot be used.

3. The use of **valves** already located in the system is the simplest and easiest method of isolating the section or equipment to be purged, but it is not recommended unless two valves can be closed and the piping between them either removed or vented. Frequently valves may be in such a condition or position that they cannot be made gastight merely by closing them. Determination of whether or not a particular valve can be used should be made *before* starting the purging operation.

Plug valves or lubricated cocks may also provide dependable isolation for the purging operation if they are properly conditioned in advance.

Gate valves of either the heavy duty or ordinary type in some cases will provide satisfactory isolation for the purging operation. Because of deposits or erosion, the disks of ordinary gate valves may not seat well enough to be gastight.

Valve gates may be sealed with a liquid, such as water, oil, or tar. For such sealing, two connections are required; one for liquid admission, the other for overflow after the proper depth is obtained.

Flow valves, including check valves, regulators, pressure

controllers, and similar types of flow equipment, cannot provide the degree of shutoff required for isolation during or after a purge alone. With these, a more effective type of shutoff should be used.

4. Water sealing is often an effective method of isolation for mains and plant equipment. Several considerations that should be taken into account are: (a) the weight of water might damage foundations; (b) the depth of seal, indicated at all times, should be about twice the pressure against it; (c) an ample supply of water should be available, with provision for its quick removal, if necessary.

5. Bags and stoppers should not be used alone for effecting isolation for purging, except in low-pressure (inches of water) mains. They should not be used for post-purging isolation.

Bags and stoppers are frequently used to prevent gas that might leak past a valve from entering the equipment being purged, the intervening space being vented to the atmosphere thru a 2- or 3-in. pipe extending to a safe height. The stopper is placed toward the source of the inleakage, and another stopper or bag is placed behind it.

On low-pressure mains up to eight inches in diameter some companies use one stopper or bag where no open flame cutting or welding is employed. On larger lines two stoppers or bags are used.

To ensure that isolation is complete when purging out of service, reduce the pressure in the system to barely above atmospheric. Any rise in pressure over a period of time related to the size of the space can then be noted with a water manometer. A pressure rise of 4.0-in. w.c. indicates an inleakage equivalent to 1.0 per cent of the volume of the space. The source of any leakage should be found and eliminated before an attempt is made to start purging. Soap suds (or detergent) or linseed oil should be used in searching for leaks.

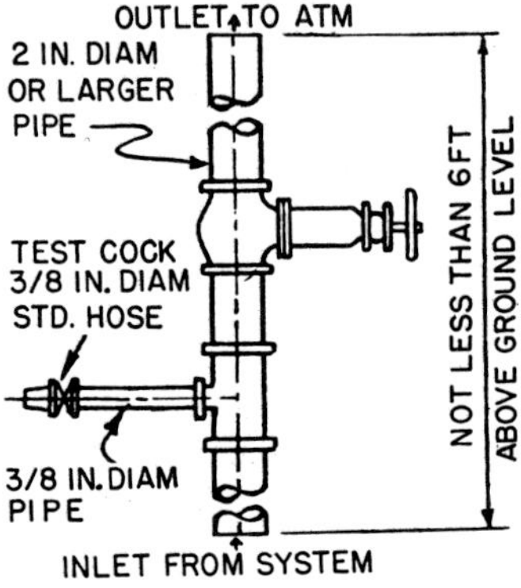

Fig. 14-1 Suggested purge pipe.

PURGE OUTLETS AND THEIR PIPING

Permanent purge outlets should be installed on equipment, including all plant apparatus and piping, during their installation or construction. In general, insufficient consideration is given to such outlets at that time. Their subsequent installation is far more expensive and troublesome than immediate installation.

Vent Pipes. These carry purged gas from the apparatus being purged to a point from which it can diffuse into the air without hazard to persons and property. For outside equipment a vent pipe long enough (six to ten feet) to carry the gases above the heads of workmen usually suffices (Fig. 14-1). Equipment within buildings may require longer vent pipes to the outside. Mains in congested areas may require longer vent pipes to discharge purge gases above nearby buildings, or to some other point. It may even be necessary to install several vent pipes to purge a piece of equipment completely. No traps should be permitted to remain unpurged.

Vent pipe size is important. In general, the total cross-sectional area of all vents in operation at one time should be less than that of all pipes admitting the purging medium to the equipment.

The rate of flame travel in mixtures of most commercial gases with air, in tubes 1 or 2 in. in diameter, ranges from about 2 to 7 ft per sec. Thus, a minimum exit velocity of 10 to 12 ft per sec should minimize hazards at vent points.

Table 14-1 lists the pressure drop in inches of water for various hourly flows.

Table 14-1 Pressure Loss for Gas Flow thru Purge Vent Pipes 10 Ft Long*

(inches of water)

Gas flow, MCFH	Nominal inside pipe diameter				
	1 in.	$1^1/_2$ in.	2 in.	3 in.	4 in.
2.0	6.30	0.60	0.12	...	...
2.5	9.84	0.96	.24	...	...
3.0	14.16	1.38	.36	0.06	...
3.5	19.26	1.86	.48	.06	...
4.0	25.20	2.46	.60	.06	...
4.5	31.86	3.12	.72	.06	...
5.0	39.36	3.84	0.90	.12	...
6.0	56.70	5.52	1.32	.18	...
7.0	77.16	7.50	1.80	.24	0.06
8.0	...	9.78	2.34	.30	.06
9.0	...	12.36	3.00	.36	.06
10.0	...	15.30	3.66	.42	.12
15.0	...	34.38	8.28	0.96	.24
20.0	...	61.14	14.70	1.74	.42
25.0	...	...	22.98	2.70	.66
30.0	...	...	33.12	3.90	0.96
35.0	...	...	45.12	5.34	1.26
40.0	...	...	58.92	6.96	1.68
50.0	...	...	...	10.86	2.58

* Based upon the gas flow formula

$$Q = C\left(\frac{d^5 \Delta p}{SL}\right)^{0.5}$$

where: Q = gas flow, cfh
d = internal diameter of pipe, in.
Δp = pressure drop thru pipe, in. w.c.
S = specific gravity of gas = 0.60
L = length of pipe, yards
C = gas flow constant:

Diam., in.	1	$1^1/_2$	2	3	4
Constant	1000	1000	1200	1300	1350

This table may be used in selecting pipe sizes, with the following corrections for different gas specific gravities:

1. The pressure loss varies directly as the square of the quantity of gas flowing.

2. The pressure loss varies directly as the specific gravity of the gas.

3. To obtain the flow velocities, V, in feet per second from the cubic feet per hour, Q, for the various pipe sizes with area, a, in square inches, apply the formula $V = 0.04\, Q/a$.

SOURCES OF IGNITION

Sources of ignition should be eliminated from the proximity of all purging operations. They include:

1. Flames from open lights, pilot lights, matches, welding torches or blowtorches, cigarette lighters, boilers, water heaters, burning materials, incinerators, and fireworks.

2. Sparks and arcs from locomotives, chimneys, static electricity, short circuits, faulty flashlights, lightning, cutting or welding operations, and solids traveling at high velocity in mains.

3. Such heated materials as glowing metals, cinders, and filaments.

Static electricity is one of the most difficult sources of ignition to control. It is generated by friction, such as from slipping belts and pulleys, and by passage of solids, liquids, or gases at high velocity thru small openings. Therefore, every piece of equipment should be "grounded" with a heavy bond wire.

Before a main is severed or disconnected, a bond wire should be attached at two points that provide a connection across the proposed break.

In purging from combustible gas to air, especially with old piping, it should be kept in mind that purging only removes gaseous or volatile materials. Solid combustible material may remain in the lines after purging is completed. Autoignition can then take place as soon as an adequate air supply is available. Deposits of iron sulfide and other easily oxidizable materials can provide centers for autoignition. Special care should be exercised after purging such piping before the piping is entered or disassembled.

Chapter 2

General Practices, Control Methods, and Instrumentation

END POINT OF INERT GAS ADDITION

One method of establishing the end point of purging is to continue the operation until no combustible gas is present in the equipment being purged out of service or until no oxygen is present in that being purged into service.

Another method is to determine whether or not the gas mixture involved is: (1) flammable or (2) likely to become flammable during the course of operations. The relationships among constituents of the three-component system (flammable gas, atmospheric air, and inert gases) may be represented on triangular or rectangular coordinates. A rectangular plot for methane is given in Fig. 14-2.

The horizontal axis *XH* of Fig. 14-2 represents natural gas concentration, the vertical axes *XV* and *XV′* show concentration of atmospheric air and oxygen, respectively, and the diagonal axis *VH* illustrates concentration of the inert gases CO_2 and N_2. Point *V* denotes 100 per cent air or 21 per cent O_2, zero per cent natural gas, and zero per cent inert gases. Point *O* represents 100 per cent inert gases, zero per cent air

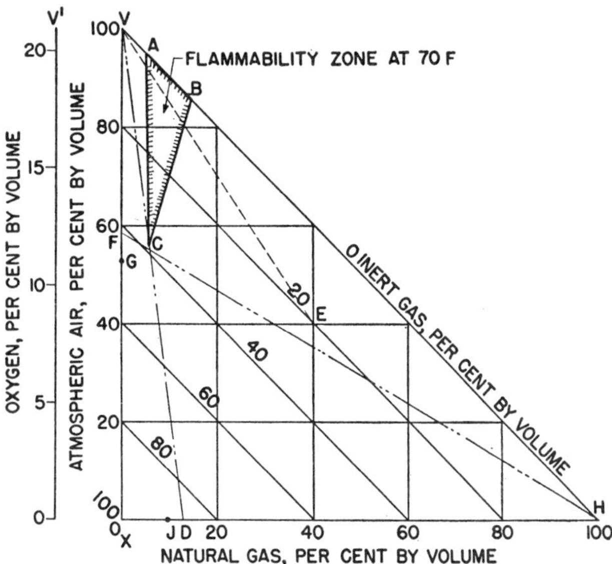

Fig. 14-2 Flammability end point diagram for the purging of natural gas at 70 F. (At 1300 F, the flammability zone area is more than twice as large as that at 70 F. The A and B coordinates are 3.1 and 19.6 per cent natural gas, respectively, and the areas are approximately similar.) To see the effects of composition changes, move in a straight line from any point toward: (1) V, when adding air; (2) X, when adding inert gases; (3) H, when adding natural gas.

(or O_2), and zero per cent natural gas. Point *H* denotes 100 per cent natural gas, zero per cent air or oxygen, and zero per cent inert gases. Therefore, line *VH* represents all possible concentrations of air and natural gas and zero inerts; all possible mixtures of natural gas, air, and inert gases are included within the area *XVH*. Points *A* and *B* on *VH* represent the lower and upper flammable limits of natural gas in air, respectively.

As inert gas is mixed with natural gas and air in the flammable range, other mixtures are formed which have different lower and upper flammable limits. These new limiting mixtures are represented by the lines *AC* and *BC*. As more inert gas is added, *AC* and *BC* converge at *C*. No mixture of natural gas which contains less than the amount of air represented at point *C* is flammable within itself, but all mixtures within *ABC* are within the flammable limits and must be avoided for safe purging practice.

Mixtures within the area *DCBH* are above the flammable limits, but will become flammable when air is added. Thus, on Fig. 14-2, a mixture containing 40 per cent air, 40 per cent natural gas, and 20 per cent inerts (point *E*) is not flammable. If air is added to this mixture, its composition will vary along the line *EV*, and as it enters the area *ABC*, the mixture becomes flammable.

Similarly, all mixtures within the area *VACF* are below flammability limits but will become flammable if natural gas is added, since they will then enter the area *ABC*.

Mixtures indicated by points in the area *XDCF* are not only nonflammable, but they *cannot* be made flammable by adding either natural gas or air.

PURGING EQUIPMENT INTO SERVICE

The operation of safely purging air from a container subsequently to be filled with natural gas may be indicated on Fig. 14-2. As inert gas is added, the air concentration drops along ordinate *VX* to any point *G* below *F*. Subsequent addition of natural gas causes the mixture composition to change along line *GH* (not shown), which crosses no part of the flammable zone *ABC*. In the example shown in Fig. 14-2, inert gas should be added until the purged atmosphere contains at least 42 per cent inerts, thereby reducing the air content in the purged atmosphere to 58 per cent, an oxygen concentration of about 12 per cent. (Table 14-2 makes more precise recommendations.)

To render a given combustible air mixture nonflammable, it is desirable to know what percentages of inert gases are

Table 14-2 Inert Gas End Points for Purging into Service Using Carbon Dioxide or Nitrogen

Combustible	Per cent required to render mixtures nonflammable[1] CO_2	N_2*	Purging end points with 20% safety factor CO_2	N_2*
Hydrogen	57	71	66	77
Carbon monoxide	41	58	53	66
Methane	23	36	38	49
Ethane	32	44	46	55
Propane	29	42	43	54
n-Butane	28	40	42	52
iso-Butane	26	40	41	52
Pentane	28	42	42	54
Hexane	28	41	42	53
Gasoline	29	43	43	55
Ethylene	40	49	52	59
Propylene	29	42	43	54
Cyclopropane	30	41	44	53
Butadiene	35	48	48	49
Benzene	31	44	44	55

* Nitrogen percentages do not include nitrogen of the air in the mixtures.

required. Table 14-2 gives the values for a number of combustibles investigated by the U. S. Bureau of Mines. To ensure safety, a purging should be continued to a point at least 20 per cent beyond the flammability limit, as shown by the inert gas end points tabulated at the right. Where combustion products are employed, the CO_2 percentage may be used as a measure of the inert gas present, but the percentage of the end point for N_2 should be taken for safe purging control.

Sometimes it is more convenient to determine the oxygen content of the purged gases. In purging into service, inert gas is added until the oxygen concentration of the container contents decreases to the point where no mixture with the combustible gas would be flammable. Table 14-3 gives these data. Suggested purging end point data with a 20 per cent safety factor are given at the right in terms of per cent of oxygen for the purging of containers in preparation to receive the various combustibles shown.

Table 14-3 Oxygen End Points for Purging into Service Using Carbon Dioxide or Nitrogen

Combustible	Per cent of oxygen below which no mixture is flammable[1] CO_2	N_2	Purging end points with 20% safety factor CO_2	N_2
Hydrogen	5.9	5.0	4.7	4.0
Carbon monoxide	5.9	5.6	4.7	4.5
Methane	14.6	12.1	11.7	9.7
Ethane	13.4	11.0	10.7	8.8
Propane	14.3	11.4	11.4	9.1
n-Butane	14.5	12.1	11.6	9.7
iso-Butane	14.8	12.0	11.8	9.6
Pentane	14.4	12.1	11.5	9.7
Hexane	14.5	11.9	11.6	9.5
Gasoline	14.4	11.6	11.5	9.3
Ethylene	11.7	10.0	9.4	8.0
Propylene	14.1	11.5	11.3	9.2
Cyclopropane	13.9	11.7	11.1	9.4
Butadiene	13.1	10.4	10.5	8.3
Benzene	13.9	11.2	11.1	9.0

PURGING EQUIPMENT OUT OF SERVICE

The operation of purging natural gas from a container to be filled subsequently with air may also be indicated on Fig. 14-2. As inert gas is added, the natural gas concentration decreases from point H (at the left) along abscissa HX to a point J beyond D (per safety factors in Table 14-3). Subsequent addition of air results in a change in the mixture composition along line JV (not shown), which crosses no part of flammable zone ABC. In the example shown in Fig. 14-2, at least 88 per cent of the natural gas should be replaced by inert gas when the container is purged out of service. (Tables 14-4 and 14-5 make more precise recommendations.)

Table 14-4 Inert Gas End Points for Purging out of Service Using Carbon Dioxide or Nitrogen

Combustible	Per cent required to render mixtures nonflammable when air is added in any amount CO_2	N_2	Purging end points with 20% safety factor CO_2	N_2
Hydrogen	91	95	93	96
Carbon monoxide	68	81	74	85
Methane	77	86	82	89
Ethane	88	93	91	95
Propane	89	94	91	95
n-Butane	91	95	93	96
iso-Butane	91	95	93	96
Pentane	96	97	97	98
Hexane	96	97	97	98
Gasoline	93	96	95	97
Ethylene	90	94	92	95
Propylene	94	96	95	97
Benzene	93	96	95	97

Table 14-5 Combustible Gas End Points for Purging out of Service Using Carbon Dioxide or Nitrogen

Combustible	Per cent of combustible below which no mixture is flammable when air is added in any amount CO_2	N_2	Purging end points with 20% safety factor CO_2	N_2
Hydrogen	9	5	7	4
Carbon monoxide	32	19	26	15
Methane	23	14	18	11
Ethane	12	7	9	5
Propane	11	6	9	5
n-Butane	9	5	9	4
iso-Butane	9	5	7	4
Pentane	4	3	3	2
Hexane	4	3	3	2
Gasoline	7	4	5	3
Ethylene	10	6	8	5
Propylene	6	4	5	3
Benzene	7	4	5	3

To render a given combustible nonflammable if air is added to it, the required percentages of inert gases should be added. Table 14-4 gives the data for a number of combustibles investigated by the U. S. Bureau of Mines. Inert gas end points at least 20 per cent beyond the flammable limit are given at the right. Where combustion products are employed, the CO_2 percentage may be used as a measure of the inert gas present, but that of the end point for N_2 should be taken for safe purging control.

Sometimes a more convenient control method is to determine the combustible content of the purged gases. In purging out of service, inert gas is added until the combustible gas concentration of the container contents is decreased to the point where no mixture with any amount of air would be flammable. Table 14-5 gives such data. Suggested purging end point data with a 20 per cent safety factor are given at the right in terms of the percentage of combustible in a mixture which will remain nonflammable regardless of the amount of air which may be added to it.

HOLDING PURGE

A holding purge is similar to the purging of equipment out of service except that an inert atmosphere is maintained and is not replaced at once by air. Alterations or repairs can sometimes be made safely on closed systems under such conditions, after which combustible gas is readmitted and the equipment is returned to service.

Figure 14-2 may also be applied to a holding purge for natural gas. Natural gas concentration decreases during purging from point H (at the left) along abscissa HX to a point J beyond D (per safety factors in Tables 14-4 and 14-5). Combustible gas may then be readmitted at any time, the composition of the mixture changing from J along XH until H is reached and the equipment is returned to service.

GENERAL PRECAUTIONS

In any purging operation, it is a good rule to purge too much rather than too little. When there is any doubt, the purging should be conducted as though the container were to be purged of hydrogen or to be prepared for its admission. Hydrogen, which is present in most manufactured gases, has the widest known explosive limit range of the common fuel gas constituents. Natural gas, however, does not contain any free hydrogen.

After a purging operation has been conducted according to a safe procedure and brought to a satisfactory end point, the purged atmospheres must be closely rechecked, so that condensates, residues, leaks, or some other condition will not create a dangerous condition within the container later. Due consideration should be given to the possible presence of substances which, because of chemical reactions, may produce combustible elements or cause spontaneous combustion.

SAMPLING

Control of purging operations is based on periodic sampling and testing of the gas mixture discharged from equipment during its purging. Three general types of gases are involved: (1) inert gas; (2) air or, more specifically, oxygen; and (3) flammable gas.

Representative samples of gas mixtures of adequate and convenient size are required for subsequent analysis, to show the extent of variations in such mixtures in different parts of the system and their relation to performance, health, and safety. Samples must be representative to be fully satisfactory. Precautions should be observed to assure that they are not contaminated or altered by any agent which might affect the representativeness or quality of the sample.

For satisfactory results, sampling points should be chosen with care and located close to the desired reaction or process. A sufficient number of sampling points should be established to furnish all necessary information for purging control. Information secured at one or more of the original points may not be pertinent, and additional or substitute sampling points may be found necessary during purging. There should be no hesitation in justifiable changing of sampling locations.

Sample tube connections should be of the correct size and as short as possible. They may be of rubber (except for LP-gas), glass, copper, iron, plastic, or any other convenient material which will not allow any adulteration, contamination, or loss of the sample.

An adequate sample may be obtained thru simple connections in places where the gas is well mixed, as in purge vents or small mains. Sampling connections which extend only thru the wall or shell of large containers or mains are generally not satisfactory. They should extend far enough inside to prevent possible surface condensates from entering the sample tube. In large mains the sampling tube should extend inside from one-third to one-half the main diameter.

GAS ANALYSIS AND DETECTION

The Orsat gas analysis apparatus is widely used for chemical analysis of samples from purging operations. Concentrations of carbon dioxide, oxygen, and carbon monoxide are determined by their successive selective absorption in chemical solutions. If a more complete analysis is required, other absorption pipettes and combustion tubes may be included to remove additional constituents. Chromatographs and infra-red analyzers may also be used.

Combustible Gas Indicators

These instruments indicate the presence of combustible gases without identifying them. They are used chiefly in purging equipment out of service. They are most useful up to the lower explosive limit. An advantage in using them is that results are available at once on passage of the sample, so that numerous tests can be made quickly.

These indicators should be calibrated for the gases to be tested. They are not sensitive enough to detect concentrations below about 0.2 per cent of combustible. They carry flame arresters as standard equipment. However, when large concentrations of hydrogen or acetylene are tested, a small sample should be withdrawn into a container. This sample should then be tested with the combustible gas indicator at a distance from the purge site. Activated charcoal filters should not be used during purging operations.

Pauling Oxygen Analyzer

Oxygen is one of the three general constituents most significant in purging control. The Pauling oxygen analyzer is based upon the magnetic susceptibility of gases. It is practically specific for oxygen in any gas mixture ordinarily encountered in the gas industry. It may be used for determining the oxygen content of the container mixture when purging equipment either into or out of service.

Specific Gravity Indicators

The change in specific gravity of the gas mixture during a purging operation can be used to indicate purging progress. Gas is drawn thru the instrument at a high rate of flow. The indication is instantaneous. Self-contained indicators are available. These units operate on either 120 v a-c or 6 v d-c, and can run continuously for at least 5 hr on one charging.

TESTS FOR GASES HAZARDOUS TO HEALTH

Purging containers either into or out of service may involve handling gases which are injurious to health. Since one of the principal objectives of purging is to remove equipment from service for repairs, tests should be made of the contents of such purged containers to make sure that their atmospheres are safe and will remain so for repairmen.

Inert Gases

Adequate vents to carry excess inerts (CO_2 and N_2) outside the containers should be provided. Absence of such vents may allow the oxygen content to fall low enough to cause oxygen starvation or smothering.

When the oxygen of the air is decreased (to about 16 per cent), the breathing rate increases, the pulse rate accelerates, and the ability to think clearly diminishes. Constant indications of oxygen content are advisable to warn of oxygen deficiencies (Table 14-6).

Carbon dioxide is odorless and nontoxic in small quantities, but it acts as a respiratory stimulant. Above six per cent concentration, physical impairments are experienced, such as headache, drowsiness, and general nervousness.

Flue or Exhaust Gases and Purging Machine Gas

When used as inerts, these gases contain small percentages of carbon monoxide. Instruments capable of determining CO in concentrations as low as 0.002 per cent should be used to analyze the contents of a space where men may work. Physiological effects of carbon monoxide (Table 6-4 and Fig. 6-24) and instruments for its determination are available; see Fig. 6-27. Table 1-30 gives the threshold limits for a number of gases and vapors which may be encountered.

Table 14-6 Physiological Effects of Oxygen Deficiency

O_2, vol %	Effect
20.99	Normal air supply.
17.5	Flame lamp extinguished. Atmosphere can be breathed and work done without ill effect for several hours.
13.0	Acetylene flame extinguished. Work difficult; considerably increased rate of breathing; lips blue; nausea and headache.
8 to 11	Loss of consciousness on exertion.
Below 6	Unconsciousness and death if adequate supply not quickly restored.

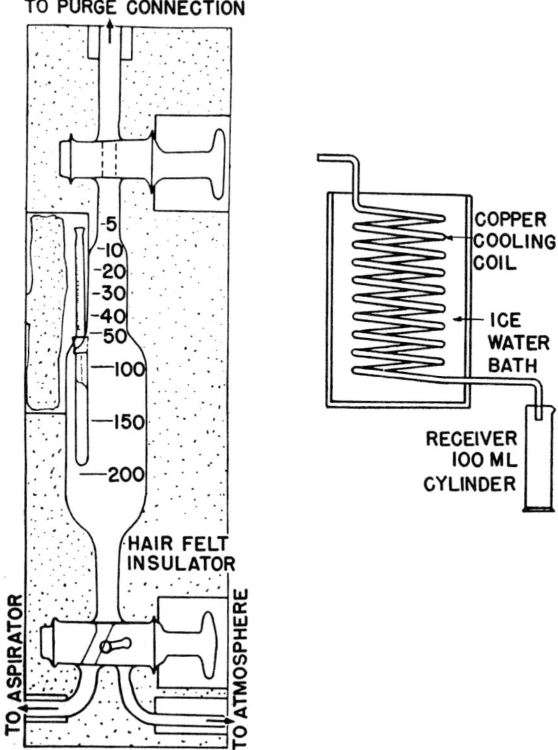

Fig. 14-3 (left) Steam purge indicator.
Fig. 14-4 (right) Condensing coil for steam purge testing with volatile combustibles present.

Steam Purging Indicators. Steam, as a purging medium, has the advantage of usually being available in quantity. It is well suited to operations where volatile combustibles are present. The pressure at which it is supplied should not exceed the design pressure of the equipment to be purged. The operation must be continuous, to avoid drawing in air by steam condensation.

The steam purge indicator (Fig. 14-3) consists of an insulated Pyrex glass bulb of 400-ml capacity with a graduated small diameter neck at its upper end and suitable cocks at each end. One leg of the three-way bottom cock is connected to a water-filled leveling bottle.

The bulb is filled with water to displace all air and then connected to the sampling line at the purging vent. Purge gas is drawn in by allowing the water to drain into a bottle. Then the cocks are manipulated so that purge gas is allowed to flow thru the indicator until temperature equilibrium is reached. By closing the top cock and reversing the bottom, water from the bottle will be drawn in by the condensing steam. The per cent of gas or air remaining will be indicated by reading the water level in the graduated neck.

When volatile combustibles are present, a second operation is necessary. The purge gas is passed thru an adequately cooled condensing coil (Fig. 14-4), and the condensate is collected in a 100-ml cylinder. The oil layer may then be measured. Condensate temperature should not exceed 70 F.

REFERENCE

1. Jones, G. W., and Scott, G. S. *Extinction of Isobutane Flames by Carbon Dioxide and Nitrogen.* (U. S. Bur. Mines Rept. Invest. 4095) Washington, D. C., 1947.

Chapter 3
Purging Media

Inert substances used in purging operations should (1) not be combustible; (2) not support combustion; (3) contain less than two per cent oxygen; and (4) have a combustible content of less than 50 per cent of the lower explosive limit of the gas to be purged.

For convenience, these inert substances for purging may be divided into first, second, and third class media.

FIRST CLASS MEDIA

Carbon Dioxide and Nitrogen

Both CO_2 and N_2 as commercially prepared are satisfactory for purging equipment of practically all sizes and types. Their general advantages for such operations include: (1) constant quality; (2) availability on completion of connections without waiting until purging machines can be put into operation; (3) ease of transportation and connection; and (4) reasonable cost.

Relative characteristics of these gases for purging operations are:

1. CO_2 is ideal for purging low points (as in mains) because it is heavier than all usual utility gases, except LP-gas. It will produce stratification if used correctly; this may be advantageous in one case and disadvantageous in another. Because of its higher specific gravity, it causes diffusion less readily than N_2; thus, less CO_2 than N_2 is generally required.
2. CO_2 depresses the explosive limit range of combustible gas mixtures to a greater extent than N_2.
3. CO_2 is more soluble in water than N_2. Resulting increase in water acidity and corrosiveness and formation of carbonates may be factors to consider where contact between the inert and water is prolonged.
4. Both CO_2 and N_2 cause the gas temperature to drop because of pressure release. CO_2 may freeze solid if it is not handled properly; N_2 will not.

Carbon Dioxide. As a *liquid,* CO_2 is available in cylinders for small purging operations and in tank trucks or tank cars for large volumes. As a *solid,* it is available in blocks, as "dry ice."

Cylinders. A unit contains 50 lb of liquid CO_2 at a pressure of about 850 psig. This forms 440 cu ft of gas on release. Both the latent heat of evaporation and some sensible heat must be supplied. Cold CO_2, for example, would not vaporize oils or liquid butane (below 32 F).

Heat is supplied either by the surrounding atmosphere or by artificial means (steam or water bath and open flames should not be used). Rapid evaporation may reduce the temperature of the liquid to the point at which it will freeze, thus restricting or stopping flow. To aid in preventing this stoppage, evaporation preferably should take place outside the cylinder. This may be accomplished by using either an upright siphon tube cylinder or an inverted standard cylinder, in which case the liquid CO_2 is against the outlet valve and orifice.

A siphon tube cylinder carries a tube extending to its bottom and connected to the gas outlet valve. With the cylinder in a normal upright position and the outlet valve open, gas pressure in the cylinder forces the liquid up thru the siphon tube and out thru the outlet orifice. Then, evaporation occurs in the exterior piping.

The high supply pressure of CO_2 may be utilized in clearing LP-gas lines of liquid in certain cases. For instance, if it is advisable to force liquid LP-gas back into the storage tanks instead of evaporating it into a holder or distribution system, CO_2 may be connected to the piping to force the liquid from it.

Manifold connections between a CO_2 cylinder and the equipment to be purged may either by purchased or field assembled. A commercially produced hose manifold should be metallic, at least $\frac{3}{8}$-in. diam, covered with sheet rubber and strong cotton duck. Strong connecting fittings should be attached at each end. The field-assembled manifold can be of extra heavy pipe and fittings, of at least 1-in. diam ($1\frac{1}{4}$ in. or $1\frac{1}{2}$ in. is recommended). Heavy rubber hose is sometimes used in cylinder connections. Wire ground or jumper wires must then be used so that the entire system, including cylinders, is grounded. Connections must be electrically sound without dependence on rivets, threaded connections, etc.

Tank Cars or Tank Trucks. These are used to transport large volumes of CO_2. CO_2 can be handled in much the same manner as LP-gas. Transport units are available in several sizes, including 3–5 tons (51–85 MCF expanded) trucks, 8–10 tons (136–170 MCF expanded) semitrailers, and 24 tons (408 MCF expanded) railroad tank cars.

Stationary Storage Units. These are available in capacities ranging from 750 lb (6375 cu ft expanded) to 125 tons (2125 MCF expanded) under working pressures of 300 to 325 psig. To permit prolonged storage at lower pressures, refrigerated, heavily insulated pressure vessels are used.

The equipment required for using liquid CO_2 from the transport unit consists of means of vaporizing the liquid and of regulating gas flow for purging. Ordinarily, the vaporizer is on the delivery truck. Heat required is obtained from the truck's exhaust or water cooling system. The supplier may furnish a unit vaporizer and the purchaser the heating me-

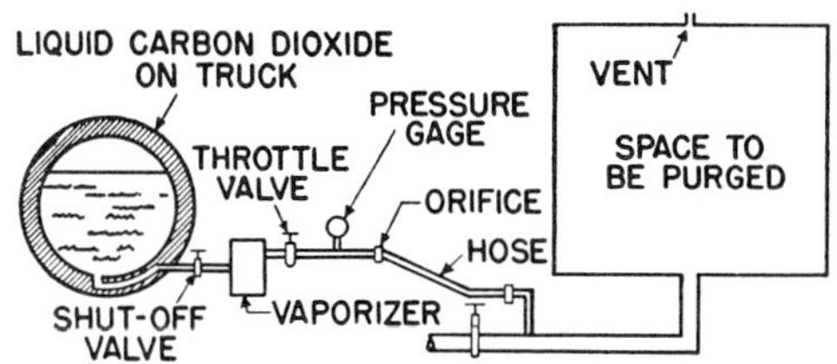

Fig. 14-5 Purging with carbon dioxide.

dium. The rate of flow to the equipment can be controlled by a pressure gage, a valve, and a calibrated orifice; Fig. 14-5 shows a typical assembly.

Solid Carbon Dioxide. Dry ice is a satisfactory source of inert gas for purging operations. Well-insulated gastight boxes (Fig. 14-6) provide storage without appreciable loss. Solid CO_2 can be changed into a gas in a "converter." The converter may be fabricated commercially (low and high pressure) or by the local utility (low pressure only).

POUNDS CAPACITY	BLOCK CAPACITY	LIDS	THICKNESS INSULATION	A	B	C
50	1	1	4 1/2	21 1/2	21 1/2	23 1/2
100	2	1	4 1/2	21 1/2	32	23 1/2
200	4	1	4 1/2	21 1/2	32	33 1/2
400	8	1	4 1/2	31 1/2	31 1/2	33 1/2
600	12	1	4 1/2	31 1/2	42	33 1/2
900	18	1	5	32 1/2	43 1/2	44
1350	27	1	5	43 1/2	43 1/2	44
1800	36	1	5	44	54	44
2250	45	2	5	44	64	44
3000	60	2	5	54	64	44
3600	72	2	5	54	75	44

HEAVY DUTY HINGES
AIRCRAFT STITCH RIVETING
STRUCTURAL ALUMINUM MOULDING FRAME
ALUMINUM SHEET ON PROTEKWOOD
LID
LIFT BAR
A
B
C

Fig. 14-6 Dry ice storage box.

High-pressure converters for producing liquid and/or gaseous CO_2 should comply with code regulations for tanks and unfired pressure vessels.[1] Their design working pressure is about 1400 psig, and the hydrostatic test pressure is 3000 psig. Converters range in diameter from 8 to 20 in. and in length from 6 to 18 ft. They can be purchased or rented in capacities ranging from 150 lb (1200 expanded cu ft) to 1000 lb (8000 expanded cu ft). For larger flows, single units can be manifolded.

To operate, place dry ice in the converter and close its outlets. The dry ice passes into its gaseous phase until the vapor pressure has risen to the point at which CO_2 liquid is formed. It remains liquid as long as this pressure continues, and either liquid or gaseous CO_2 may be withdrawn. One of the chief advantages of a high-pressure converter is that dry ice may be charged into it a day or two before the purging job, during which period the heat gain completely liquefies the dry ice.

These units can also be used as low-pressure converters. Except in special cases of gas apparatus purging, it should not be necessary to allow the pressure to rise to that for the liquid phase, since low-pressure gas should be satisfactory.

Utility-made low-pressure converters for converting solid to low-pressure gaseous CO_2 may be constructed of pipe. Figure 14-7 shows the design used by one gas utility. It may be made of 12- to 36-in. diam pipe, in lengths up to 15 ft. At least one charging opening "A" must be provided; more if desired. If gate valves are not available, charging openings can be blanked with any material strong enough to withstand the low operating pressure. To protect against excessive pressure, a water seal is recommended.

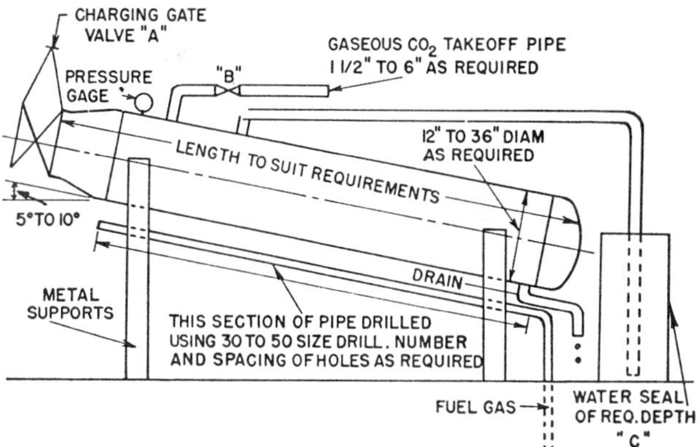

Fig. 14-7 Low-pressure dry ice converter.

With the design shown, a 20-in. diam converter 10 ft long will produce gaseous CO_2 at a rate of between 2 and 6 MCF per hour, depending on heat transfer. The size of converter that may be constructed seems to have no practical limit, and capacities of 20 to 50 MCF per hour can easily be attained. A minimum of about 100,000 Btu of heating gas is needed per MCF of gaseous CO_2 produced per hour. The gas burner, consisting simply of a section of 1-in. or 2-in. pipe drilled as shown in Fig. 14-7, supplies enough heat for complete gasification. Such a burner has been used satisfactorily with fuel gas pressures up to 20 psig. The addition of water in the converter nearly doubles the heat transfer to the dry ice.

Nitrogen. As a *gas,* it is available in cylinders for small volumes; as a *liquid,* it is available at slightly above atmospheric pressure and below critical temperature (−233 F) in tank cars and tank trucks for large volumes.

Cylinders. The most common size contains about 16 to 17 lb of gas at 2200 psig (about 220 expanded cu ft). Nitrogen may or may not flow freely from cylinders. If it is "oil pumped" and moisture free, no freezing will result on expansion; if "water pumped" and not moisture free, the water may (rarely) freeze on expansion and close off the connections.

Since nitrogen is not a liquid in the cylinder, the cylinder may be held in any convenient position. Connections between it and the equipment to be purged should be heavy duty, because cylinder pressures run as high as 2200–2350 psig. Commercially manufactured tubing is recommended.

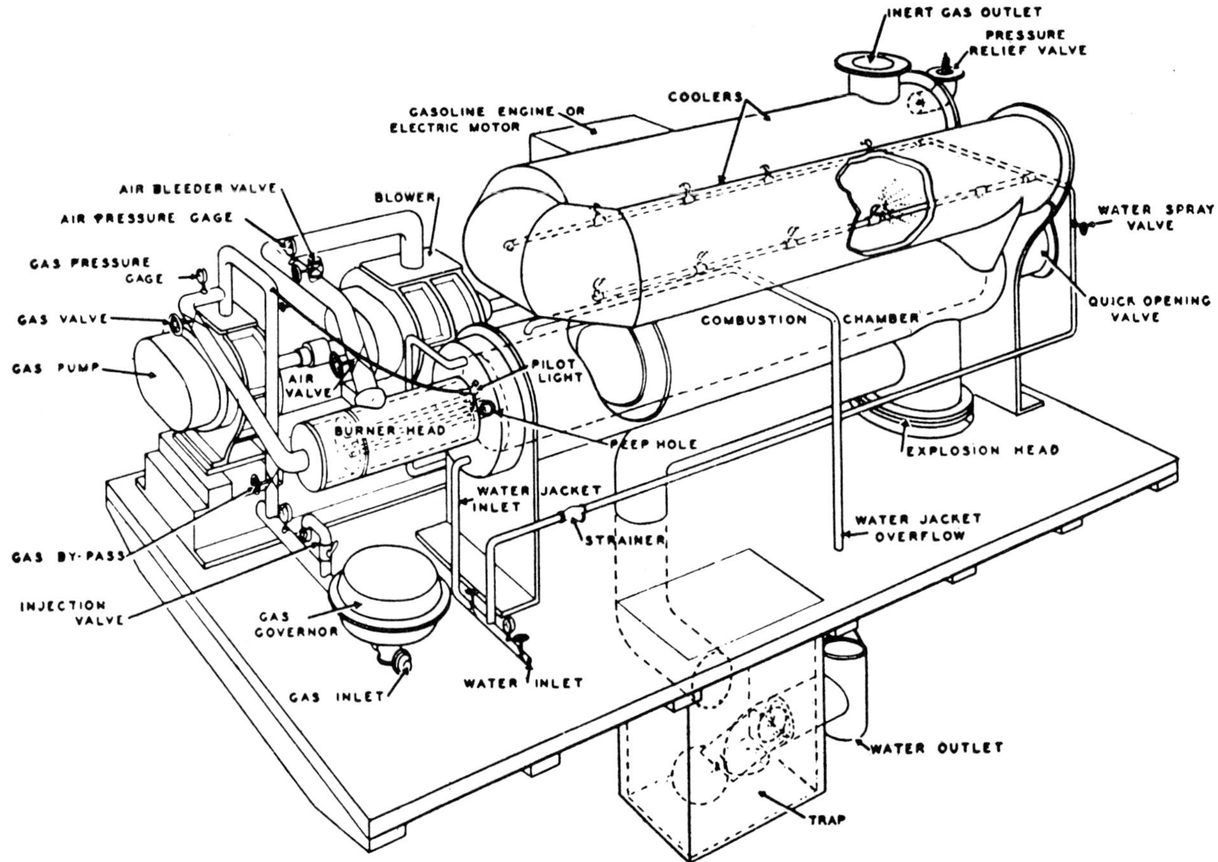

Fig. 14-8 Phantom view of Harrison inert gas generator mounted on platform.

If a pressure regulator is used on the cylinder, connections may be made with standard pipe fittings or tubing.

Truck Transports. Truck transports are available in several sizes, the common one having an equivalent capacity of 100 MCF at standard conditions. Pumps on the trucks will deliver at rates between 10 and 20 MCF per hour at temperatures controlled as desired between 20 and 100 F. These units are usually obtainable with a few days' notice, and may be shipped to any point within one day's traveling time. Shipment by railroad tank car is practical to distances within one week's traveling time.

Stationary Storage Units. These may be installed at plants for any desired capacity. They are constructed of multiple special heavy wall cylinders 35 or 40 ft in length, manifolded together.

Purging Machines or Inert Gas Generators

Within these (see, for example, Fig. 14-8), controlled combustion results in an inert gas mixture of CO_2 and N_2. After practically perfect combustion of fuel gases or oil is effected, the combustion products are cooled to 135 to 150 F (depending principally on the temperature of the cooling water). Most machines yield mixtures of CO_2 and N_2 containing less than 0.5 per cent of either O_2 or CO. Capacities range from 1 to 50 MCF per hour. Units may be powered by electric motors or by gasoline or gas engines.

Mobile units may be used for holder and main purging work. Stationary equipment is usually used for purging oxide boxes and other chemical process equipment and for generation of special metallurgical atmospheres.

Operation requires an ample supply of fuel and clean cooling water (170 to 300 gal per MCF of inert gas per minute), and a drain for the used water. An operator should be stationed at the inert gas producer at all times to control the water flow and to make necessary adjustments of the quality of the products formed.

Water Gas Sets

These are adaptable for generating inerts only when there are certain types of large scale purging operations of gas plant apparatus directly connected to the set by large piping, and when large quantities of inerts must be generated at a rapid rate. Using water gas sets is indicated when purging an entire plant for dismantling, when purging the foul main system of a plant for new construction or repair, or when purging a gas holder.

Production of inerts in a water gas set consists primarily of continuously blowing the generator with simultaneous and carefully controlled combustion in the carburetor to produce blow gases containing practically no combustibles and only a small quantity of excess oxygen. With proper control, the inerts produced will contain less than two per cent oxygen and combustibles in concentration less than 50 per cent of the lower flammable limit.

The time that the blowing can be continued before the machine reaches a dangerous temperature varies from a few minutes to approximately one hour, depending upon the blowing rate. For longer periods, steam can be used with the generator air, and/or water sprays can be installed in the take-off pipe from the superheater or in the carburetor to protect against overheating.

Coke is the usual fuel. The volume of inerts produced in a number of actual purgings has averaged about 1.9 *times* that of primary air used for generator blast.

Steam

Steam has been found to be a useful purging medium in gas plants where its high temperature is not objectionable. At pressures near atmospheric, one pound of steam occupies about 26.5 cu ft. When steam is admitted rapidly into a space containing flammable gas or air, it expands quickly into a large, relatively solid "slug" of steam which tends to push ahead of it any gas or air present. This is particularly true in apparatus of small diameter; it is not true in large diameter equipment, such as gas holders.

The comparatively high temperatures of steam atmospheres volatilize any light oils, benzol, naphthalene, tar, and other combustible material that will volatilize slightly above ambient temperature. High temperatures keep them volatile, and they are vented with the steam. The heating effect tends to soften and melt any tarry deposits in the apparatus, so that they may run off with the condensing steam. Drains should be examined frequently to prevent stoppages by cooling of the draining mixture.

Steam can be used in any equipment not damaged by higher temperatures. This includes steel scrubber towers, water gas machines and associated equipment, producer gas machines, tubular and direct contact condensers, and relatively short lengths of piping (except cast iron).

Equipment not generally recommended for steam purging includes: piping with Bitumastic or other asphalt coating; cast iron equipment; machines with close clearances, such as boosters, exhausters, or compressors; equipment with large and effective condensing surfaces such as station wet meters or purifier boxes; and tanks used for gasoline or oil storage under atmospheric pressure, with an air space on top, since static discharges may cause an explosion.

To provide for the possibility of interruption of purging operations, there should be suitable connections made allowing admission of the proper media into the equipment to overcome effects of the contraction caused by cooling of its contents. If air or an inert medium is being purged, air or an inert gas must be admitted. If flammable gas is being purged, either the flammable gas or an inert gas must be admitted.

SECOND CLASS MEDIA

Stack Gas

Stack gas is defined as waste or exhaust products of combustion after a fuel gas has served its purpose in the apparatus in which it forms stack gas. Thus, at any point in the boiler breeching or stack, products of combustion are stack gas. Its usual sources in a gas plant are the steam boilers and the coke oven battery.

The principal advantage of stack gas for purging is its availability in gas plants.

The principal disadvantages of using stack gas include:

1. Possible high equipment costs. Equipment includes piping and exhausters on compressors for transferring the stack gas from its source to the apparatus being purged.
2. Lack of assurance that a constant volume of stack gas will be produced during purging. With boilers, stack gas volume depends upon the fuel fired to meet steam requirements. In coke oven battery operation, the "pause" at the reversal of direction of flow of the air–gas mixture to the oven burners shuts off the stack gas supply. Such interruptions can interfere with purging operations.
3. Considerable variance in quality. Among other things, charging coal or opening fire doors changes quality. Most fuels require ample excess oxygen, which is undesirable in an inert. Adjusting to give products with less than two per cent oxygen results in considerable loss of efficiency of the heating equipment. However, coke oven batteries can be adjusted to give two to three per cent oxygen in the stack gas.
4. The necessity for tight, leakproof boiler and battery flue settings to prevent excess oxygen.
5. The fact that drawing off stack gas by the exhauster may affect heating equipment efficiency by unbalancing the pressures thru the setting.
6. The need to clean and cool stack gas before injection into the apparatus to be purged. Fly ash, or soot, and high temperatures may be harmful.

Prime Mover Exhaust Gas

Exhaust gas from automotive engines is suitable for purging operations under certain conditions; the equipment required is flexible, portable, and usually readily available. Gasoline-driven air compressors are better suited for purging purposes than automobile engines because they may be operated under load, thus consuming more fuel and producing a larger volume of inert gas in a given time. An automobile engine may be damaged if it is run at no load (especially above idling speed) for long periods.

Any type of automotive equipment usually requires carburetor adjustments. Normally, carburetors are adjusted to a mixture "richer" than is suitable for purging operations. They should be readjusted by an experienced mechanic to give minimum CO (at the most, 4.0 per cent). The volume of inert gas produced by an automotive internal combustion engine is from 2500 to 5000 cfh for the sizes generally used by gas utility companies. It may be estimated from the normal gasoline consumption rate, which is known from normal usage. A moderate figure would be 1000 cu ft of inert gas per gallon of gasoline; however, a good engine and good adjustment may yield as high as 1200 to 1400 cu ft per gal. Where comparatively large volumes of inert gas are required, two or more engines may be manifolded together.

For air compressors, a load of not less than 40 psig is recommended. The back pressure on the engine should not exceed 1.0 psig. The spark should be operated at the "advanced" side of the neutral position to minimize engine heating.

The quality of the exhaust gas should be tested at not less than half-hour intervals. Readjustment should be made immediately as required.

It is recommended that the inert gas be passed thru a

wash box before entering the equipment to be purged, the wash box serving as a seal to prevent backflow. The wash box will also cool the hot exhaust gases emerging from the exhaust pipe at 600 to 900 F. This temperature is usually too high for injection directly into equipment and should be reduced to 200 F or less. Alternatively, sufficient lengths of pipe to allow air cooling or a spray cooler may be used.

The exhaust gases are saturated with water. When water vapor is objectionable or may affect the subsequent operation of LP-gas equipment and apparatus, exhaust gases should be dried before their admission to the equipment. Exhaust gas (at temperatures up to 200 F) may also be used, instead of steam, to remove volatile oils.

THIRD CLASS MEDIA

Water

Water as used in purging operations is not a purging medium, but a "displacing medium."

Some advantages of water are:

1. Water tends to fill every space in the equipment, so long as vents are properly provided to prevent pockets of gas from being trapped within the apparatus. Connections should be provided to dispose of the overflow water and oils.
2. Hot water (over 160 F) tends to "top" light oils within the equipment, so that the volatile fractions of any oily matter will have been removed on completion of purging.
3. Water may be used to inspirate the final contents of the purged apparatus as it is drained.

Principal disadvantages of using water are:

1. Water is heavy. Before using it, it must be determined whether the equipment and foundations can stand the weight and pressure when filled.
2. Proper disposal of used water, in accord with local laws, must be provided. The water may contain oils and tars, which are not usually permitted to be run to city drains.
3. Hydrates may be formed with natural gas under high pressures (Figs. 4-43 and 4-44). Careful consideration should be given before using water in natural gas equipment.

If necessary, dry the equipment after its drainage by entering it and drying it with mops and cloths, or by passing dried gas, air, or inert gas (whichever is the proper medium). If these expedients are not possible or practical, a dehydrator may be connected into the main gas flow line to absorb the water that is picked up by the flammable gas.

REFERENCE

1. A.S.M.E. *Rules for Construction of Unfired Pressure Vessels.* New York, 1962.

Chapter 4
Typical Purging Procedures

INTRODUCTION

Procedures presented here are believed sufficiently representative to enable an operator to apply them to his own particular situation.

A purging procedure begins with:

1. Careful planning and preparation, as previously discussed.
2. Selection of the inert medium most adaptable to the operation.
3. Estimation of the required quantity of inert medium.
4. If a major operation, preparation of a complete written procedure.

The written procedure should include all pertinent information concerning the work to be done, and should not be limited to the various steps in the purging operation. It should include the company's name, the equipment to be purged, its location, the reason for purging, the total system to be affected (whether or not all of it will be purged), the estimated duration of the work, a short *Summary of Operations* covering the important steps to be carried out, and signatures of all who planned the work and of those who reviewed and approved the proposed plan. Finally, the itemized and detailed sequence of steps in the purging operation should be stated, including both the preliminary work and the actual purging. A suggested form for noting the preliminary information follows.

Sample of Preliminary Data

Company: Longshore Gas Co.
Location: Ocean Ave. (between 1st and 9th Ave.).
Equipment to be Purged: 6400 ft of new 12 in. main.
Reason for Purging: New 12 in. main is to be put into service.
System Affected: Present main on Ocean Ave. from valve at 14th Ave. Pressure will be reduced in main supplying 14 customers.
Estimated Duration of Work: Eight hours.
Summary of Operations: Reduce pressure in old main; purge new main with nitrogen from cylinders; install stoppers at end of old main; connect old and new main; remove stoppers and gas out* new main; gas out services on new main; return pressure to normal; check customers on old main.
Approval Signatures: (Enough lines for the planners, reviewers, and others involved).
Purging Procedure: Itemized steps covering both preliminaries and the actual operation to be given in sequence.

Suggested forms are given later for inclusion under the *Purging Procedure* part of the written form. Experience has proved the value of taking time to prepare a thorough procedure before the actual purging and to enumerate the steps in detail, particularly when operations like numerous valve changes are involved.

* Fill with flammable gas.

PURGING GAS PLANT APPARATUS

This apparatus includes aftercoolers, condensers, tar precipitators, scrubbers, purifiers, station meters, exhausters, compressors, work mains, oil tanks, wash boxes, gasoline and drip oil tanks, tar tanks, barges, drip oil trucks, coal gas ovens, and LP-gas equipment. Each of these units is basically a "container" with inlet and outlet connections for storage or handling of a combustible gas or material.

The volumetric capacity of different units, even of the same type, may vary considerably (Table 14-7).

Table 14-7 Approximate Volumes of Gas Plant Apparatus

Apparatus	Volume, cu ft
Aftercoolers	300–1,000
Condensers	1,500–2,760
Cottrell tar precipitators	200–700
Exhausters	100–220
Meters	300–1,000
Purifiers	1,000–30,000
Compressors, boosters	175–250

Many municipalities have ordinances governing the handling of tanks containing volatile oils such as gasoline. These should be investigated before purging is attempted. Rigid regulations have also been adopted for control of gas hazards on tank ships, tank barges, and similar craft. Before purging such equipment, consult the pertinent regulations.[1–3]

Removing Apparatus from Service (Gas to Air)

Preliminary Preparations. Figure 14-9 is involved in many of the following steps:

1. *Study the apparatus.* Determine its internal construction and its external connections and decide just what must be isolated for inerting. Prepare a written procedure and have it approved by the persons in charge.

2. *Determine the steps for isolating the apparatus.* If this is to be done by closing valves (such as *1* and *2* in the figure), advance preparation should be made to seal their gates with liquid under sufficient head to exceed the pressure on either side. In some cases, the valves may be vented between the gates.

3. *Select the inert medium to be used.* Except when there is positive displacement, the required volume of inerts will be at least three times that of the apparatus. If light oils which vaporize are present, the required inert volume will be still greater, and inerting will be hastened by increasing the temperature to 130–160 F. For scrubbers with wood grids saturated with oil or when quantities of naphthalene are present, steam may be used, provided its high temperatures do not set up harmful stresses. Hot water may also be used. Steam alone is not recommended for inerting large oil tanks. Apparatus cannot be inerted by filling with water unless both it and its supports have been designed to withstand the additional weight. Inerts should be introduced at inert gas inlet or at bottom of apparatus if introduction directly into its inlet is impractical.

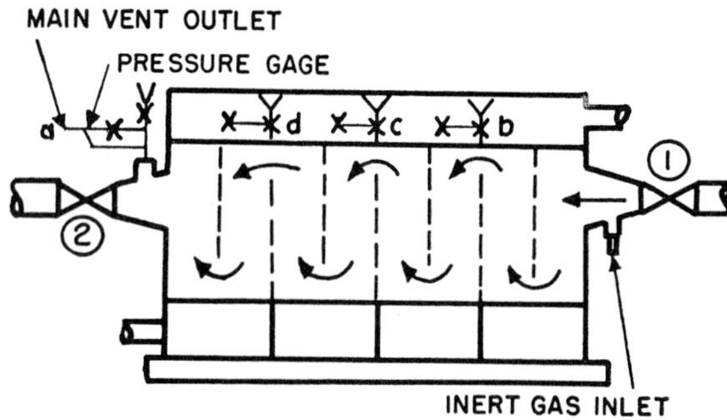

Fig. 14-9 Gas plant condenser.

4. *Install vent and sampling cocks.* Vents for discharging the purged gas from the apparatus and cocks for sampling the purged gas should be installed where necessary. Main vent *a* should be large enough to discharge the purged gas as rapidly as the inerts are admitted. It should be located just inside outlet valve *2*. For apexes or pockets which may not be readily inerted, additional vents and test cocks, such as at points *b*, *c*, and *d*, should be installed.

5. *Install a pressure gage.* One gage should be installed on the inert gas apparatus or at its point of admittance to the purged equipment. Another may be provided at the outlet of the purged equipment. Both gages should be readily readable.

6. *Have competent gas testers.* Qualified test personnel should be available throughout the purging operation. They should be equipped with the necessary testing instruments in good working condition.

7. *Isolate.* All auxiliary sources of combustible material should be physically disconnected from the apparatus before purging is started.

Purging Operation.

1. Close and seal inlet and outlet valves *1* and *2*, isolating the apparatus from all sources of combustibles. Also close and seal all other valves that may be connected to combustible sources.

2. Open one vent and reduce the pressure to approximately 1 in. w.c., close the vent, and watch the gage at least five minutes for any pressure increase. Such an increase indicates leakage into the apparatus which must be stopped.

3. If the pressure remains constant, open the vents and admit the inert gas immediately. Continue its admission until samples taken at each vent prove to be in the safe range on the particular instrument used for end point determination. Note Tables 14-4 and 14-5 in this regard.

4. When the desired end point is reached, stop admission of inerts and open the unit to the atmosphere (unless a "holding purge" is desirable).

5. Permanently isolate the unit if it is to be out of service for some time or maintain the seals on the valves if the job is short and does not require permanent isolation.

Post-Purging Care.

1. If the apparatus cannot be opened to the air, provision should be made to maintain positive pressure with inerts.

2. Apparatus which has been opened should be tested periodically for vaporizing of oils present. If ventilation by natural draft is not adequate to maintain a safe concentration of these vapors, provision should be made for forced ventilation.

3. Tests should be made for toxic vapors or gases before workmen are allowed to enter the apparatus.

4. If there is danger of spontaneous combustion while the apparatus is open (as in purifier boxes with fouled oxide), a fire hose or suitable extinguishers should be instantly available.

5. If "hot work" like cutting or welding is to be done, all combustible material should be removed from within the apparatus. If the "hot work" is on its exterior, it may be more convenient and advisable to maintain a "holding purge."

Placing Apparatus into Service (Air to Gas)

Preliminary Preparations. Same as for removing from service.

Purging Operation. Figure 14-9 is involved in many of the following steps:

1. Reconnect any pipe connections removed for permanent isolation and reestablish all temporary isolation such as sealed valves.

2. Close all openings other than vents.

3. Introduce the inert gas and continue until samples taken at the various test cocks show a satisfactory end point.

4. Shut off the inert gas supply.

5. Admit gas by opening discharge valve *2* a few turns (this usually admits gas at the highest part of the apparatus) until samples taken at the vents are combustible. Place an additional vent where the inerts were introduced.

6. Close the vents.

7. Open outlet valve *2* wide, and when it is desirable to return the apparatus to service, open inlet valve *1*.

Post-Purging Care.

1. If the apparatus is not returned to service and it is not desirable to admit flammable gas, a positive pressure should be maintained with inerts. If it has been gassed out, positive pressure should be maintained with flammable gas.

2. If any valves, normally closed, were sealed during the purging, they should be opened enough to release the sealing

material or otherwise drained. Neglecting to do this may cause serious trouble, particularly if a valve sealed with water is exposed to freezing.

PURGING LP-GAS EQUIPMENT

Any of the standard purging media, as discussed in Chap. 3, may be used. All LP-gas installations should have permanent purging connections as part of their original construction. These should include openings at both ends of storage tanks and at other suitable locations for admission of purge media and for venting.

A detailed written purging procedure, specifying every operation, should be prepared, approved, and signed.

All personnel involved in LP-gas purging operations should understand the characteristics and behavior of this gas.

Preliminary Preparations

All LP-gas in the liquid state should be removed before purging is started. Since this liquid evaporates comparatively slowly, due allowance must be made when the equipment has been reduced to atmospheric pressure. When pressure is released, part of the liquid evaporates and removal of the heat of evaporation cools the remainder to a point at which its vapor pressure is low and complete vaporization is delayed. This condition may exist for a considerable time before enough heat is absorbed to complete vaporization. Evaporation will be hastened by the use of steam on the bottom (exterior) of a storage tank or on a trapped section of pipe, or by passing a noncondensable like natural or manufactured gas thru the equipment for at least 24 hr before purging.

Equipment to be purged should have all pipe connections physically disconnected and capped or otherwise tightly closed.

LP-gas should not be vented to the atmosphere if other disposal means are available.

If it is impractical to use the LP-gas, it should be burned at the outlet of a water seal (a 30 or 55 gal drum of water is a simple seal) or in a burner of the flame retention type. This seal or burner should be at least 50 ft from the nearest tank, with a valve in the vent line to control the flame size and purging rate. A permanent flare (Fig. 14-10) is another method of disposal.

If venting to the atmosphere is the only resort, a windy day or bright sunny day with ground temperature higher than air temperature is desirable. Wind or convection currents will then tend to minimize settling and ground travel of LP-gas. An overcast, foggy, or rainy day with little air movement is unfavorable because under those conditions the gas will remain in heavy concentration close to the ground. The vent pipe outlet should not be less than eight feet above the apparatus. It should not be near any door or window, but should be extended to a safe position.

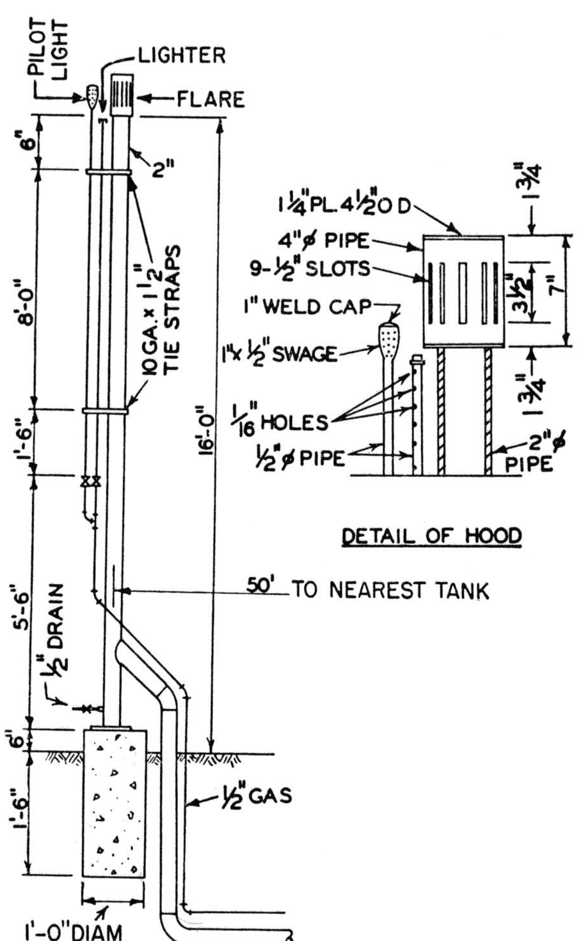

Fig. 14-10 Permanent LP-gas flare. (Manufacturers Light and Heat Co.)

Inerting End Points

The object of purging LP-gas equipment is to prevent the formation of an explosive mixture at any time while passing from air to LP-gas or vice versa. It is necessary to know the limits of flammability of propane, air, and carbon dioxide mixtures. These are shown in Fig. 14-11. For example, if air is added to the mixture *G* (11 per cent propane and 89 per

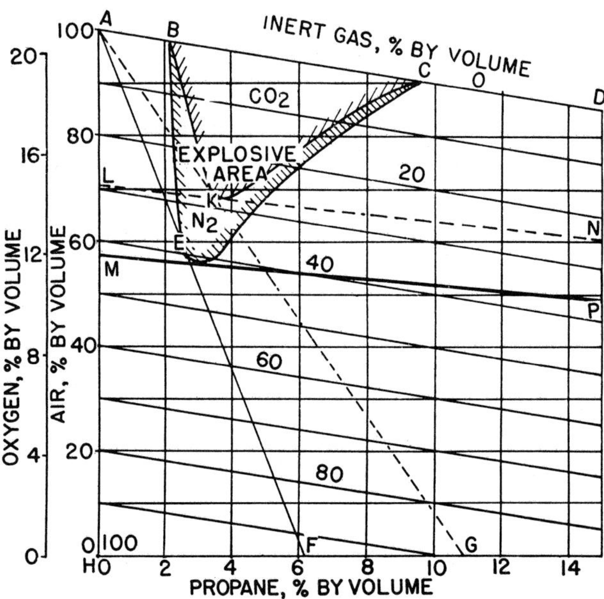

Fig. 14-11 Flammability end point diagram for purging of propane with carbon dioxide or nitrogen.

cent CO_2), the composition change to 100 per cent air is represented by passing from point *G* to point *A*. The explosive area is avoided. The inerting end point is therefore 11 per cent propane (or 89 per cent carbon dioxide) when purging out of service.

When equipment containing air is to be *purged into service* using CO_2, the inerting end point for propane (including a 20 per cent safety factor) is reached when CO_2 is 43 per cent or greater; if N_2 is used, when N_2 is 54 per cent or greater (Table 14-2). If the end point is determined by oxygen analysis, then the end point (including a 20 per cent safety factor) is reached when O_2 is 11.4 per cent or less using CO_2, and when O_2 is 9.1 per cent or less using N_2 (Table 14-3).

When equipment containing propane is *purged out of service* using CO_2, the end point (including 20 per cent safety factor) is reached when CO_2 is 91 per cent or greater; if N_2 is used, when N_2 is 95 per cent or greater (Table 14-4).

End point determinations may be made as previously outlined. Rubber tubing should not be used. Neoprene tubing and nonsparking probes are suggested.

Precautions

1. The evaporation of LP-gas cools the remaining liquid. Contact with this cold liquid can destroy skin tissue. Treat as for a burn. Goggles and gloves should be worn and first-aid supplies for burns should be available.

2. LP-gas both as a liquid and as a gas can cling to clothing for a long time, especially in cold weather. On entering a warm room, gas is released and a hazardous condition results.

3. Expansion of LP-gas liquid with temperature rise in a closed system can cause high pressures. The pressure of the vapor in equilibrium with the liquid depends upon the liquid temperature, increasing rapidly with temperature rise.

4. Precautions should be taken to prevent ignition of escaping vapors. Open flames and other sources of ignition should be prohibited and equipment should be grounded to prevent static sparks. The use of nonsparking tools is recommended.

5. Compressed air or "air movers" should be provided to ventilate equipment in which men will work. Gas masks should also be available.

PURGING GAS HOLDERS

The following procedures apply to all types of gas storage holders (*except LP-gas storage tanks* for which consult Purging Principles and Practice[4]).

Preliminary Preparations

In addition to the general preliminary preparations listed in Chap. 1, the following items are essential to holder purging:

1. Whenever practical, perform and complete the entire purging operation in daylight. If a long purge is necessary and interruptions cannot be tolerated, it is advisable to start at midday, using the night hours for admitting the required amount of purge gas and completing the operation in daylight.

2. An adequate and reliable supply of inert gas should be available. It may be estimated as at least two volumes for each holder volume purged.

3. Holders previously in service may contain deposits of volatile oils or oil emulsion, which should be removed before purging. In a **water-seal holder,** use the permanent or a portable skimmer to remove the oil and emulsion before initially deflating the holder. In a **waterless-type holder** (M.A.N., Klonne, or Wiggins), after the piston has been deflated to a height of about three feet, introduce water onto the holder bottom by a connection thru the piston to float off the oil to the lowest holder connection or to a sealed condensate drain. Drain **pressure holders** of any oil accumulation before reducing the holder pressure.

4. The inert gas supply for a water-seal or waterless holder can be connected to any convenient holder connection. For a pressure holder it should be connected at or near the holder bottom. If a water-seal holder has seal bonnet (Livesey seal) covers for the standpipes, provide the seal bonnets with full-size by-passes; if such an installation is not convenient, provide an inert gas connection to the holder crown. In the latter case, provide a vent in each seal bonnet and an inert gas supply connection at each holder connection, to permit purging the standpipes and seal bonnets. Provide test cocks on all vents. For horizontal cylindrical pressure holders, also provide either permanent or temporary test connections for sampling across the holder cross section between the inert connection and the vents.

5. Connect a water gage to the holder to indicate pressure.

6. Provide readily accessible valves adjacent to the holder in its inlet and outlet flammable gas lines.

7. Before severing the last flammable gas connection of a water-seal or waterless holder, deflate the holder to a height of about six inches above the landing beams. For a pressure holder, reduce the gas pressure to about six inches of water.

8. Provide facilities for displacing the inert gas and ventilating the holder after purging has been completed.

9. Caution workmen that both inert and purge gases are suffocating and toxic, and should not be inhaled.

10. Take precautions against all sources of ignition (open flames, welding and burning, smoking, or electrical equipment) in the immediate vicinity of the holder. Purging should not be started during an electrical storm.

Removing a Holder Containing Flammable Gas from Service

The procedures on page 14/21 apply after the foregoing preliminaries have been completed. For vent locations refer to Figs. 14-12 thru 14-17.

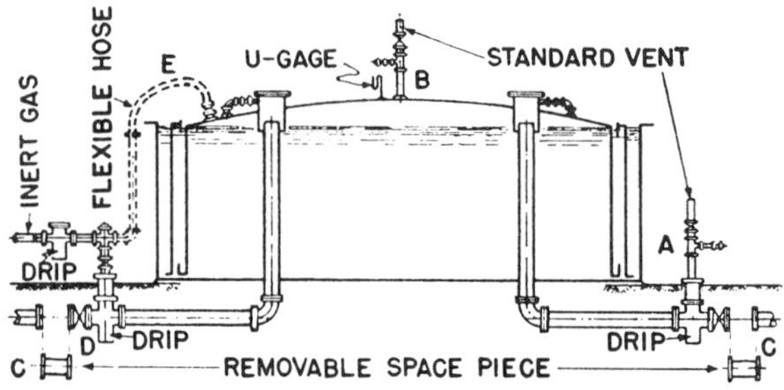

Fig. 14-12 Purging connections on a water-seal holder.

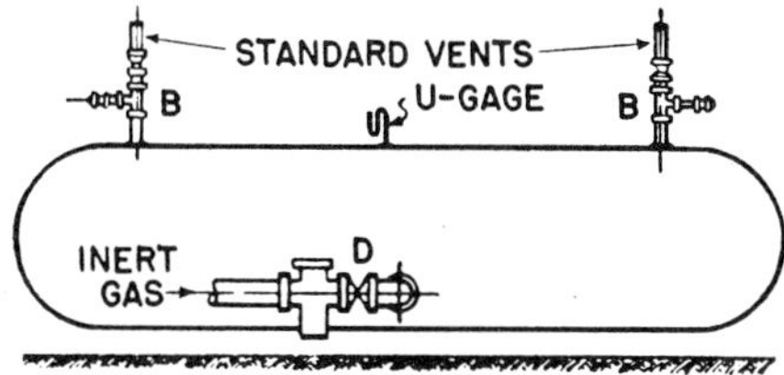

Fig. 14-13 Purging connections on a cylindrical pressure tank.

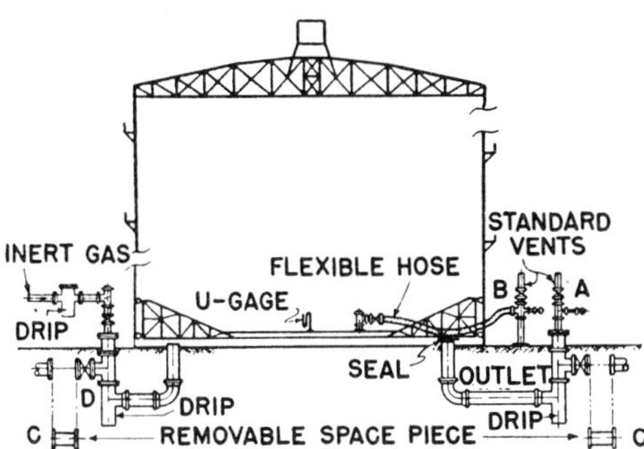

Fig. 14-14 Purging connections on an M.A.N. holder (not to scale).

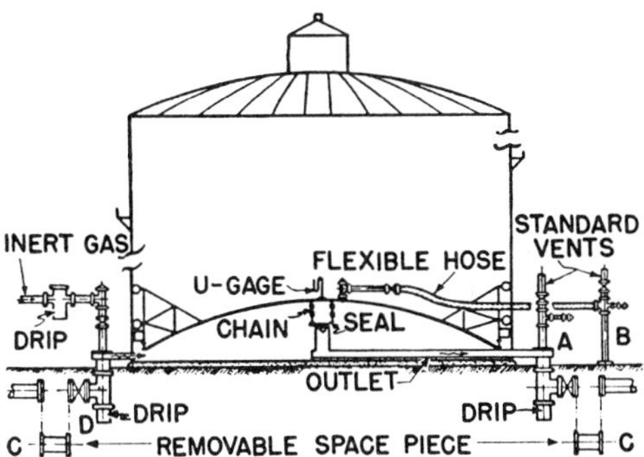

Fig. 14-15 Purging connections on a Klonne holder (not to scale).

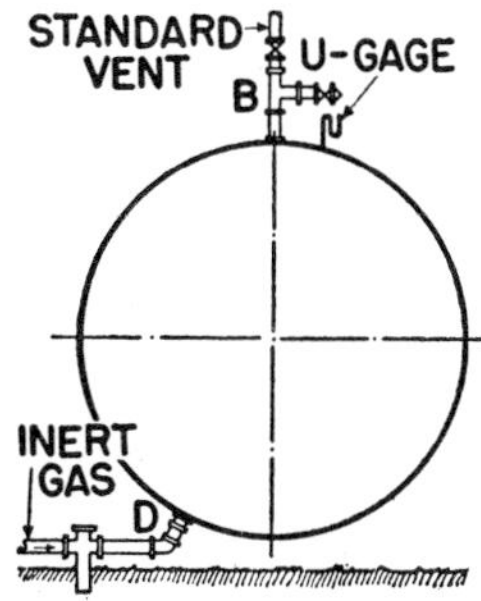

Fig. 14-16 Purging connections on a spherical tank or Hortonsphere.

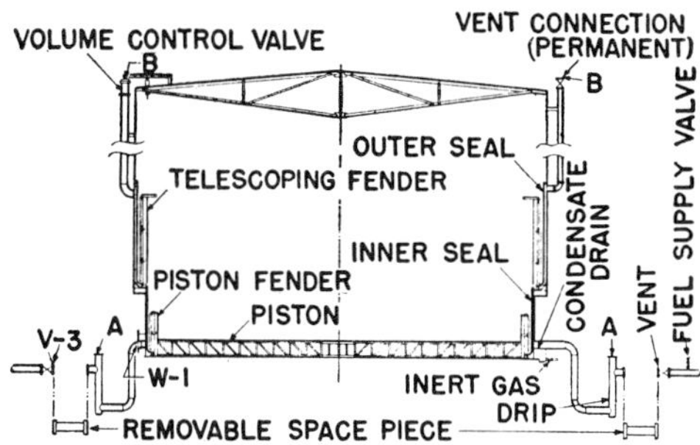

Fig. 14-17 Purging connections on a Wiggins dry-seal holder (not to scale).

1. Flammable gas remaining in a holder should be allowed to escape thru vent *B*. With all vents *A* closed and vents *B* open, admission of inert gas is started while a slight pressure still remains in the holder. On a water-seal holder the seal bonnet by-passes, if provided, should be opened. Maintain a minimum holder pressure of one inch of water throughout the purging operation.

2. During the purging period, concurrent admission of steam with the inert gas promotes removal of any volatile oils which remain. However, steam should not be used in purging a **Wiggins holder,** because of the possibility of diaphragm damage. Stop stream flow into a holder when all purge samples (by a combustible gas indicator) show 85 per cent or less of the lower explosive limit concentration. Inert gas supply to the holder should be continued for a time to compensate for shrinkage of contents upon cooling.

3. When the holder contents have been satisfactorily purged to 85 per cent or less of the lower explosive limit, vents *B* are closed, vents *A* opened (except for pressure holders), and holder drips pumped, if water sealed. Purging of the holder connections, standpipes, and seal bonnets thru vents *A* or thru those in the seal bonnet covers, as the case may be, is continued until samples at these vents are shown by test to be of a safe composition. All vents should then be closed, and the holder should be inflated about two feet with inert gas. Care should be taken to purge out the drips of a water-seal holder and to purge out the tar seal tanks, tar risers, and outer annular chamber for receiving sealing tar on the bottom of a waterless holder.

4. When a holder has been withdrawn from service for inspection or repairs, the inert gas should be displaced by air, and ventilation maintained by a blower or compressor. All manholes in the crown of a water-seal holder should be opened immediately, and the seal bonnet covers removed. Also, it is advisable to remove several of the crown plates. The piston of a waterless holder should be landed and all manholes opened.

Placing a Holder Containing Air into Flammable Gas Service

1. Close all manholes and replace any seal bonnet covers and crown sheets. Admit inert gas to displace air in the holder standpipes and connections until purge gas samples from all vents show an oxygen content below five per cent by volume (see Table 14-3). Close all vents and inflate the water-seal or waterless holder with inert gas to a height of about two feet,

or, for a pressure holder, maintain a positive pressure, preparatory to reconnecting the flammable gas lines. After they are reconnected, purge air from these sections with inert gas, and pump out the holder drips if water sealed.

2. Displace the inert gas from the holder and connections by admitting flammable gas thru one holder valve *D* and allowing it to purge thru vents *A* of a water-seal or waterless holder or vents *B* of a pressure holder until the purge gas samples give satisfactory analysis. Close the seal bonnet by-passes of a water-seal holder after the inert gas has been displaced with flammable gas. Before placing the tar circulating pumps for a waterless holder in service, purge the tar seal tanks, tar risers, and outer annular chamber for receiving sealing tar on the holder bottom with inert gas.

3. Notify those in authority that the holder is physically connected to the distribution system, and is ready for service.

Purging an LP-Gas Cavern.[5,6] This operation has been accomplished thru a novel method employing combustion. In this process, the cavern atmosphere is drawn out, combined with LP-gas, burned to reduce the oxygen content below the lower flammability limit, and returned to the cavern as the purging medium. The usually remote locations of caverns make this method more economical than others for the operation.

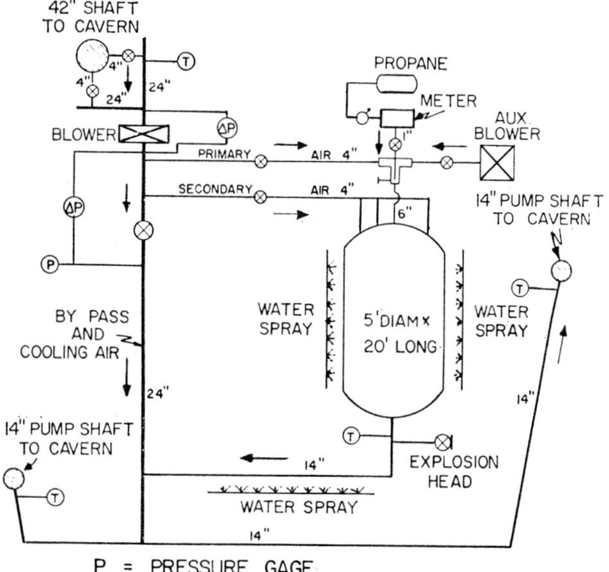

Fig. 14-18 Purging an LP-gas cavern into service.[5]

The arrangements tried for inducing combustion within the cavern included placing burners inside the cavern, which proved unsatisfactory. The best procedure appears to be one where combustion is carried out in an aboveground chamber, which is connected to the cavern thru a closed system that utilizes the cavern atmosphere as combustion air and forces the combustion products into the cavern, using the circulating equipment on hand.

Apparatus (Fig. 14-18) includes: a 4500 SCFM motor-driven blower; specially designed combustion chamber, 5 ft in diameter, 20 ft long, with water sprays and burner (6-in. diam, with 1-in. spud, drilled to ¼-in. hole); auxiliary blower; explosion head; meter; and piping and valves. Fuel choice depends upon type of LP-gas stored (e.g., propane vapor for propane).

A standard procedure for a five-day purge of a one MMCF cavern was: (1) establish proper flow rate and pressure thru apparatus before lighting; (2) light burner, opening special valve to atmosphere to supply initial combustion air; (3) begin at low fuel rate, gradually increasing to about 300 cfh; (4) for first two days, use cavern atmosphere as primary and secondary air; (5) during other three days, admit some atmospheric air as primary, continuing cavern air as secondary; (6) at end of purge, seal cavern openings, and displace inert purge gas with stored LP-gas.

The operation used 30 MCF of propane to produce an atmosphere containing 7.6 per cent CO_2, 9.8 per cent O_2, and no CO. As a precaution during the warmer daylight hours, purge gas stream temperature was held below 150 F, thru cooling with by-pass air. Instruments were read every 30 min, and Orsat readings were taken every hour.

PURGING, BLEEDING, AND CLEARING OF GAS PIPELINES

For industry standards on purging operations reference should be made to Section 841.285* of ASA B31.8-1963, Gas Transmission and Distribution Piping Systems. Pressure testing of pipelines with air, gas, or water as covered in Sections 841.41 to 841.417* of that code are also pertinent.

It was reported[7] that air has been used as a purging medium in 4-in. diam or smaller pipes. Further, a length of 40 ft of flammable mixture would have to be produced to develop hazardous conditions. In repeated tests using air for purging, it was found that a length of only eight feet or less of flammable mixture was produced by interfacial turbulence between air and natural gas. However, further tests showed that the length of flammable mixture produced by air purging increased rapidly with pipe size. Therefore, the use of inert gases as a slug to separate gas from air has become common practice in purging larger size pipe.

The methods and practices outlined here are selected examples for the gas industry, but they should not be construed as compulsory in any case. Their choice or modification is optional with utility management, depending upon local conditions.

The following safety precautions should be observed, in addition to those given in Chaps. 1 and 2:

1. Company regulations should be observed in separating or reconnecting a line and locating air compressors and trucks (away from the site and upwind).

2. Four "rough" methods of estimating purging progress which may be used in addition to, but not as substitutes for, the instruments listed in Chap. 2 include: (a) observation of natural gas waves in bright sunlight; (b) gas odors; (c) time for purging a line; and (d) the bag test.

3. Disposal of expelled gas in urban areas varies with circumstances. Some methods which have been used include burning the gas using a double safety fire screen, venting it from very high pipe extensions or lamp posts, and transferring it by portable compressor to a line remaining in service.

* Extracted in Sec. 8, Chap. 2.

4. Clearing volatile oil, gasoline, tar vapors, or naphthalene from a pipeline may be accomplished by one or more of the following steps: (a) continue air flow until the vapor content is below an acceptable level; (b) locate the liquid and drain it out of the line; (c) add steam to the purge and, if possible, raise the line temperature to 150–160 F to volatilize the material.

Bleeding House Lines in Restoring Service

In bleeding house lines containing air or an air–gas mixture adequate precautions must be taken to avoid:

1. Liberation of enough gas to form an explosive mixture with air within the appliance or in the room.
2. Escape of enough unburned gas thru a defective safety pilot valve to cause an explosion or flashing of flame outside the appliance. An explosion within house lines two inches or less in diameter resulting from flashback is not likely to rupture the pipe.
3. Liberation of enough gas to cause odor or a toxic condition.

Follow company instructions to ascertain that the line is gastight and in turning on gas at the meter. Supplementary procedures follow:

Open the valve of an open burner on a small gas appliance and apply a match until ignition occurs. Another method is to disconnect the pilot line at the pilot valve outlet and ignite. If a large appliance such as a funace or boiler must be used, a lighted match or taper must be applied continuously to the burner until ignition occurs. To minimize any concussion at the instant gas appears in the burner, its air shutter may be closed temporarily. Any flashback of the flame into the burner is not likely to pass the burner orifice with propagation of flame back into the piping.

House lines larger than two inches in diameter and large meters should be bled to the outside air thru a hose or pipe extending thru a door or window.

Clearing Service Lines to Restore Service

For clearing normal sized service lines (3/4 and 1 inch) containing air or an air–gas mixture, the flammable gas itself is satisfactory. Precautions must be taken, however, to avoid liberating sufficient gas to form an explosive mixture, objectionable odors, or a toxic condition within the space where the purge outlet is located. Large service lines should be considered as mains when clearing or purging.

The following procedure is recommended:

After the line is tested for tightness according to company regulations with the meter lock cock closed, admit gas to the line. Loosen the meter outlet union and connect a hose (low-pressure systems) or rigid piping (high-pressure systems) exiting outdoors at a safe point. Open the meter cock and purge until a gas odor appears at vent exit. Close the meter cock and substitute the house piping for the venting arrangement. The remainder of the line and the house line may be cleared thru a burner or pilot valve, as outlined previously.

Purging Pipelines

All purging operations as defined herein involve the use of inert gas to separate gas and air so that an explosive mixture will not be formed in the pipeline if proper procedures are followed. Selecting the method depends upon local conditions. In some situations, purging is essential; in others it would be undesirable, and clearing should be specified. Any inert gas may be used for purging pipelines. Nitrogen, carbon dioxide, purging machine products, and engine exhaust gas are most commonly used. Water or steam may be used where conditions permit. Several purging methods follow:

1. Fill the line completely with inert gas; the capacity of various sized pipes can be calculated from Table 1-9 or 8-7. Inerts to the extent of 150 per cent or more of the total volume of the line, added rapidly, will ensure complete filling, but the vent gases should be analyzed with suitable instruments to make sure that the line has been completely filled with inerts.

Table 14-8 Length of Slug of Nitrogen* Required for Various Pipe Sizes and Lengths†
(feet at purging velocity of 200 lineal pipe feet or higher)

Pipe length, thousand ft		Pipe diameter nominal, in.												
Over	Up to	4	6	8	10	12	14	16	18	20	22	24	26	30
0	1	39	59	78	98	117	137	156	176	195	215	234	253	294
1	2	46	69	91	115	137	160	182	205	228	250	273	295	344
2	4	60	90	120	150	180	210	240	270	300	329	360	390	452
4	6	73	109	146	183	220	256	292	329	365	402	432	475	551
6	8	87	131	174	218	262	305	348	392	435	478	522	565	656
8	10	101	151	202	252	304	353	404	455	505	556	606	656	762
10	15	136	204	272	340	408	476	544	612	680	748	816	885	1026
15	20	171	256	342	428	514	598	684	770	855	940	1026	1110	1291
20	25	206	309	412	515	617	720	824	928	1030	1131	1236	1340	1555
25	30	241	361	482	605	722	845	964	1085	1210	1325	1446	1570	1820
30	40	311	467	622	778	935	1090	1244	1400	1555	1710	1866	2021	2350
40	50	381	571	762	953	1150	1340	1524	1710	1880	2100	2286	2480	2880
50	60	451	676	902	1130	1360	1580	1804	2030	2260	2480	2706	2930	3410
60	70	521	784	1042	1320	1570	1830	2084	2350	2610	2870	3126	3400	3930
70	80	591	889	1182	1480	1780	2070	2364	2670	2980	3250	3546	3860	4460

* Same length for carbon dioxide; for exhaust gas add 20 per cent.

† Calculated from Table 3 in reference 7 for 8-in. diam pipe. For example, for 10-in. pipe, $10/8 \times 78 = 98$. Slug length has been doubled for safety.

Table 14-9 Volume of Slug of Nitrogen* Required for Various Pipe Sizes and Lengths†
(cubic feet at purging velocity of 200 lineal pipe feet or higher‡)

Pipe length, thousand ft		Pipe diameter nominal, in.												
Over	Up to	4	6	8	10	12	14	16	18	20	22	24	26	30
0	1	7	24	54	107	184	274	430	588	806	1,070	1,390	1,768	2,722
1	2	8	28	73	126	216	320	480	686	940	1,250	1,624	2,060	3,190
2	4	11	36	83	165	282	420	632	902	1,240	1,640	2,142	2,720	4,180
4	6	13	44	103	200	348	512	784	1118	1,540	2,000	2,580	3,300	5,100
6	8	15	52	123	236	418	610	936	1334	1,840	2,380	3,120	3,930	6,080
8	10	18	60	143	270	484	706	1088	1550	2,040	2,790	3,620	4,580	7,060
10	15	24	82	188	372	640	952	1430	2046	2,800	3,720	4,860	6,160	9,520
15	20	30	102	238	468	808	1196	1800	2580	3,540	4,680	6,120	7,720	12,000
20	25	36	124	288	528	976	1440	2170	3014	4,280	5,640	7,400	9,320	14,400
25	30	42	146	338	588	1144	1690	2540	3448	5,020	6,600	8,620	10,960	16,900
30	40	55	186	432	1042	1470	2180	3280	4680	6,420	8,520	11,120	14,100	21,800
40	50	67	228	530	1234	1820	2680	4000	5720	7,960	10,440	13,680	17,300	26,800
50	60	80	270	630	1426	2170	3160	4720	6760	9,500	12,320	16,200	20,500	31,700
60	70	92	312	730	1618	2520	3660	5440	7900	11,040	14,300	18,700	23,620	36,500
70	80	101	354	830	1812	2870	4140	6160	8940	12,580	16,200	21,220	27,000	41,400

* Same volumes for carbon dioxide; for exhaust gases add 20 per cent.
† Calculated from Table 14-8: length of slug × cubic feet of pipe volume per linear foot × 100% factor of safety.
‡ Number of large cylinders of nitrogen required can be obtained by dividing the volume required by 200.

2. Separate air and gas by means of a slug of inert gas. Such a slug must be long enough to prevent the air and gas meeting and mixing during the purging. A delay of approximately three minutes between addition of the inert gas and the following air or gas may destroy the slug. Lengths of inert slug required for various sizes and lengths of pipe are given in Table 14-8. Corresponding volumes of such slugs are given in Table 14-9. A gravity meter may be used to check the presence of gas.

3. Put two spheres or pigs in the line and separate them with a slug of inert gas (nitrogen or carbon dioxide). The slug of inert gas contained between the spheres is then pushed thru the line at about 20 mph with natural gas—forcing air out of the line via a series of blowoffs. A 150-mile length of 36-in. pipe was successfully purged in this way.[8]

The data in Table 14–9 have generally been accepted in the gas industry. Table 14–10 compares some of these recommendations to the less conservative recommendations of one gas utility company. Although there is fair agreement for very short lengths of pipeline, Table 14–10 indicates considerably smaller amounts of nitrogen for long lengths. The data in Table 14–10 reflect test results from some different sizes and lengths of line. The referenced report[10] gives test results for slug velocities ranging from 100 to 200 fpm.

The report notes:

> The data may be used with confidence with the precaution that each individual spot where welding or cutting is to be done should be checked with a combustible gas indicator before work begins. Particular attention should be given to the top of the pipeline when it has been purged to air and to the bottom of the pipeline when it has been purged to gas. No practical method of purging can guarantee that pockets of gas-air mixture may not be trapped in drip legs, valves, or fittings.
>
> In addition, the following significant facts have been established by this investigation:
>
> (1) Purge velocity is extremely important. Velocities less than 100 feet per minute in large pipelines allow stratification between heavier and lighter gases.
>
> (2) Reynolds' number is not a satisfactory criterion for purge velocity. Large pipelines require greater purge velocities than smaller pipelines which is exactly opposed to what might be expected from using Reynolds' number as an index for required purge velocity.
>
> (3) The amount of nitrogen necessary to purge short lengths (500 feet or less) of large-diameter pipelines satisfactorily at practical purge velocities exceeds the volume of the line.
>
> (4) The amount of nitrogen required to purge a large-diameter pipeline satisfactorily is much less than previously supposed.
>
> (5) Changes in horizontal or vertical direction because of ells or return bends do not tend to destroy the nitrogen slug.

The data in this report are based upon a slug of pure nitrogen that approaches zero feet in length at the end of the pipeline. A factor of safety results from the fact that a mixture of 85 per cent or more of nitrogen with natural gas cannot be made to burn regardless of the amount of air present, as shown in Fig. 14–2. Accordingly, the effective and safe length of noncombustible slug is the length of any pure nitrogen plus the length of mixture including more than 85 per cent nitrogen. This nitrogen-rich mixture varied from about one to several hundred linear feet, depending upon the purge velocities and quantities of nitrogen used in the field test. Because it is difficult to measure injected volumes accurately and an extra amount of nitrogen increases the cost of a purge only slightly, the Southern California Gas Co. arbitrarily

Table 14–10 Comparison of Purging Recommendations of One Gas Utility Company with Some Data from Table 14–9[10]
(cubic feet of slug of nitrogen at slug velocity of 200 fpm)

Length of pipeline, ft	20-in. pipeline		30-in. pipeline	
	Test	Table 14–9	Test	Table 14–9
500	340	806	2,100	2,722
1,000	400	873*	2,500	2,956
2,000	480	1,090	3,000	3,685
5,000	570	1,540	3,600	5,100
10,000	630	2,420	4,000	8,290
20,000	690	3,910	4,500	13,200

* From Table 14–9 for 20 in. pipe: [806 (from 0 to 1000 ft of pipe) + 940 (from 1000 to 2000 ft of pipe)]/2 = 873.

adds an additional amount of nitrogen so that a slug of pure nitrogen 100 feet long may be expected at the end of a purge.

Purging Gas Distribution Mains and Large Services

Involved here may be putting a new or reconstructed line containing air into gas service or taking a gas line out of service for repair, replacement, or abandonment. In any case, an interface of air–gas mixture within the flammable limits will probably be present unless inert gas is interposed. At some time during a pipeline purging, combustible gas is discharged. It is, therefore, essential (except in special cases where the gas is deliberately lighted) to avoid its accidental ignition as by sparks, static electricity, cigarettes, or any flames.

When taking a low-pressure main out of service, decision whether to purge with an inert gas usually rests with the job supervisor. Experience in street work is a requisite for safe practice. In decidirg on the purging procedure, it must first be determined that the line to be purged is completely isolated from all others.

The longer the line and the larger its diameter, the greater the advisability of purging with inert gas becomes. The minimum length of the line which should be purged cannot be definitely stated. The larger the pipe diameter and the greater the volume of gas contained, the greater the chance of ignition, with resulting possibility of explosion, is. As a very general example, consideration should be given to purging with inert gas when 500 ft or more of 4 in. or larger diameter pipe is taken out of natural gas service, or when 2 in. or larger diameter pipe is removed from manufactured gas service.

The following procedures may be used when it is necessary to cut into a "live" main at a location where there is no effective shutoff valve: (1) the installation of proprietary devices such as Mueller Co. Line Stopper fittings—purging may be involved here; (2) use of two canvas or rubber (depending on pressure) inflated stoppers—the volume between them may be purged; (3) positive shutdown of gas at upstream block valve and purging downstream of this valve; (4) use of stoppers set to permit a reduced flow rate of gas in the pipe sufficient to preclude in-leakage of air—the gas is ignited at the end of the pipe.

Purging Transmission Lines

When operations indicate that purging is desirable on transmission lines, the considerations previously described for gas distribution mains apply.

Clearing Natural Gas Distribution and Transmission Lines

Clearing has been widely used throughout the natural gas industry for many years, particularly for transmission lines and high-pressure lines not passing thru populated areas. It has also been used for short lengths of low-pressure mains. No instances have been reported to date where proper clearing procedures have resulted in explosions and rupture of the pipeline. The ruptures that have been recorded seem to have been due to factors other than clearing or purging operations.

Gas introduced directly and rapidly into a line containing air will form only a short section of gas–air mixture within the

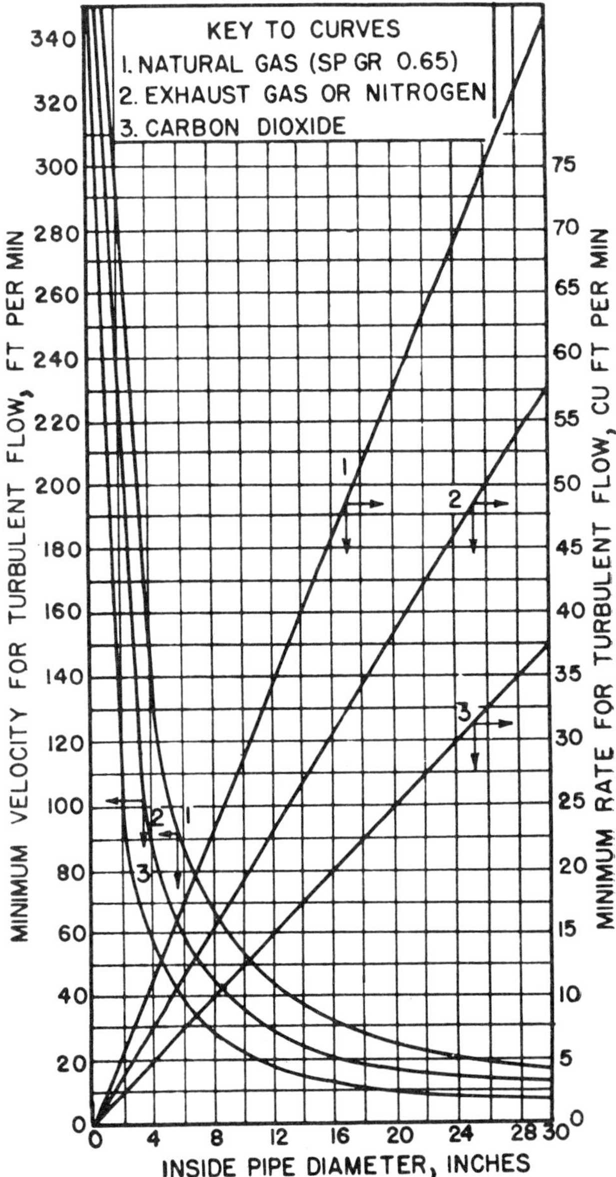

Fig. 14-19 Calculated minimum velocities and flow rates for turbulent flow in pipelines.

flammable range. In Fig. 14-19, the minimum velocity for turbulent flow and the minimum flow rate for various sized pipe are given. A velocity of 20 lin ft per min or greater will cause turbulent flow in 30-in. diam pipe, while a flow of about 95 ft per min is required in 6-in. diam pipe. Therefore, flow rates of 100 lin ft per min would be satisfactory for a minimum of mixing in all sizes. To be doubly sure of a minimum of mixing, a flow velocity of 200 lin ft per min or greater is recommended. The mixing that may take place will vary with the length and size of line. On a 10-in. diam line 25,000 ft long, the zone of possible mixing would be less than 500 ft in length, and on a 20-in. diam line this zone would be less than 1000 ft. Since the line is completely enclosed during this clearing operation, and since the vent line extends eight to nine feet aboveground, the only possibility of ignition would be from friction of solid particles inside the line or from static electricity. A slow clearing of less than 100 lin ft per min should be avoided, because it may permit more mixing and

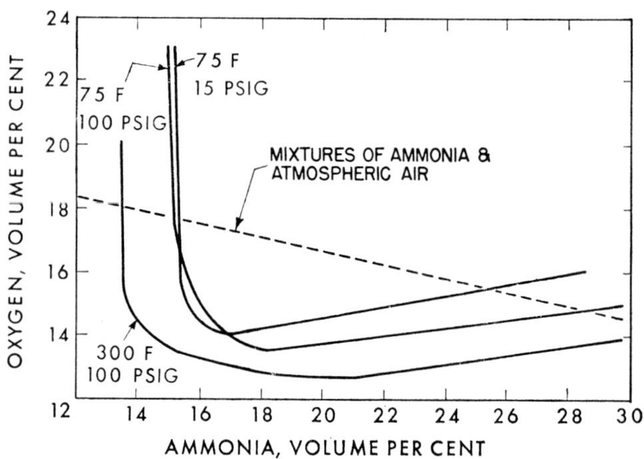

Fig. 14-20 Flammability limits of ammonia-oxygen-nitrogen mixtures[9]

stratification of air and gas. Velocities as high as 1300 lin ft per min have been used. The greater the velocity, the greater the turbulence is, with less chance of creating a long section of air–gas mixture.

AMMONIA[9]

Ammonia burns readily in oxygen; in air, however, a flame-stabilizing hot surface is necessary to maintain combustion. Figure 14-20 gives the flammability limits of ammonia. Note that the intersections of the dotted line and the 75 F–15 psig curve approximate the lower and upper flammability limits of ammonia in air (Table 2-63). The minimum ignition energy for ammonia is 680 millijoules (compare with 0.4 for *n*-hexane).

REFERENCES

1. Natl. Fire Protection Assoc. *Standard for the Control of Gas Hazards on Vessels to be Repaired.* (NFPA 306) Boston, 1962.
2. U. S. Coast Guard. *Manual for Safe Handling of Inflammable and Combustible Liquids.* (Ed. ser. 1)(CG 174) Washington, D. C., 1951.
3. Natl. Fire Protection Assoc. *Flammable Liquids Code.* (NFPA 30) Boston, 1962.
4. A.G.A. *Purging Principles and Practice.* New York, 1954.
5. Schreiner, W. J. "Purging Storage Caverns by Internal Combustion." *Gas* 38: 61–3, June 1962.
6. Ormston, R. H. "Testing, Volumetric Determination and Purging of Mined LPG Caverns." *A.G.A. Operating Sec. Proc.* 1962: CEP-62-3.
7. Henderson, E. "Combustible Gas Mixtures in Pipe Lines." *P.C.G.A. Proc.* 32: 98–111, 1941. Also partially in *Gas* 17: 23–9, Sept. 1941.
8. "How P.G. & E. purged a 292 mile line with only seven welds required." *Gas Age* 129: 8, 9, June 7, 1962.
9. Buckley, W. L., and Husa, H. W. "Combustion Properties of Ammonia." *Chem. Eng. Progr.* 58: 81–4, Feb. 1962.
10. Cortelyou, J. T., and Curtis, J. M. "Safe and Economical Purging Practices." *A.G.A. Operating Sect. Proc.* 1964: 64-T-165.

SECTION 15

ELECTRICAL APPLICATIONS IN THE GAS INDUSTRY

C. W. Nessel, *Section Chairman*, Honeywell, Inc., Minneapolis Minn.
Raymond P. Flagg, Chapter 5, Honeywell, Inc., Minneapolis, Minn.
H. G. Nafe, Chapter 3, Century Electric Company, St. Louis, Mo.
R. W. Pashby, Chapter 5, Honeywell, Inc., Freeport, Ill.
Sigward A. Stavnes, Chapters 1, 2, and 4, Honeywell, Inc., Minneapolis, Minn.
C. E. Swanson, Chapter 5, Honeywell, Inc., Minneapolis, Minn.
F. E. Vandaveer (retired), Chapter 5, Con-Gas Service Corp., Cleveland, Ohio

CONTENTS

Chapter 5—Electric and Thermoelectric Gas Appliance Controls and Unconventional Electrical Systems

Tables

Figures

Chapter 1

Electrical Terminology and Measurements

by Sigward A. Stavnes

TERMINOLOGY

Conduction. Conduction, in electricity, is the flow of electrons between two points that occurs when a difference in potential is established between them. This flow is called *current*. In insulators and dielectrics current is negligible; in conductors (e.g., metals) current is considerable and continuous so long as the source of the potential remains continuous. The source of potential is called the *electromotive force* (*emf*).

Direction of Emf. Electric current is a flow of negatively charged particles, called electrons, from the negative terminal of a battery to the positive terminal. In U. S. practice, however, it is customary to regard current as flowing from the positive to the negative terminal of the battery.

Electrical Current (Ampere). The unit measure of a quantity of electricity is the coulomb. The rate of flow of electricity is given in coulombs per second, a term seldom used in practice. A shorter term, ampere, or one coulomb per second, is generally used.

There are two kinds of current: *direct current,* which flows always in the same direction; and *alternating current,* which reverses direction periodically.

Frequency. The periodic directional changes of alternating current are called *cycles*. The number of cycles per second is termed frequency. The two common frequencies in general use in the United States are 60 cps and 25 cps. Lately, higher frequencies for use in high-frequency lighting have been introduced.

The relationship between the rotational speed of an a-c generator and the number of poles on its field winding, and the frequency of the voltage produced is expressed by:

$$\text{Frequency, cps} = \text{No. of poles} \times \text{rpm}/120$$

The following tabulation illustrates this relationship:

60 cycles		420 cycles	
Poles	Rpm	Poles	Rpm
2	3600	2	25,200
4	1800	4	12,600
6	1200	14	3,600
10	720		

Potential Difference (Voltage). A potential difference of one joule per coulomb is called a volt. Thus, if one joule of work is required to move one coulomb of electricity between two points, the difference of potential is one volt.

Electrical Resistance (Ohm). Opposition to the flow of current is called electrical resistance. The unit of resistance is the volt per ampere. A resistance of one volt per ampere is called one ohm.

The resistance, R, of a conductor is given by the equation:

$$R = \frac{\rho L}{A} \qquad \textbf{(1)}$$

where: ρ = resistivity of the conductor
L = length of the conductor
A = cross-sectional area

The resistivity, ρ, varies with different metals and, in general, increases with the temperature of the metal.

The amount of current flowing thru a circuit is equal to the emf, in volts, *divided* by the resistance, in ohms, provided that the temperature is held constant. This relationship, known as **Ohm's law,** may be expressed as:

$$I = E/R \qquad \textbf{(2)}$$

where: I = current, amp
E = emf, volts
R = resistance, ohms

Instantaneous values of this relationship are expressed by i and e.

Internal Resistance of Source of Emf. The measured voltage of a battery in a closed circuit will be somewhat less than in an open circuit. Thus, it can be seen that the emf of the battery depends to some extent upon the current it is furnishing. As current increases, the emf of the source decreases, owing to a potential drop in the source itself across its internal resistance.

Capacitance. Capacitance in an electrical circuit is analogous to elasticity in a mechanical circuit. As mechanical energy is stored in an elastic spring, electrical energy is stored in a capacitor,* which consists of two conductors separated by an insulating material or dielectric.

As shown in Fig. 15-1, when a capacitor is connected in a circuit and an emf is applied, there will be a momentary flow of electricity when switch A is closed as electrons leave the positive plate of the capacitor and build up on the negative plate. The capacitor will continue to charge and current will

* Capacitors were previously called condensers.

flow until the potential difference across the capacitor is equal and opposite to that of the battery. When the capacitor has been charged, switch A can be opened and the capacitor will still retain its charge. When switch B is closed, current will flow thru the circuit until the capacitor is discharged.

The quantity of electricity stored by a capacitor is expressed by:

$$Q = CE \tag{3}$$

where: Q = quantity of electricity, coulombs
C = capacitance, farads
E = impressed emf, volts

When a potential difference of 1 volt occurs across a capacitor charged with 1 coulomb, the capacitance is 1 farad. The farad, a very large unit, is seldom used in practice. The microfarad (μf), or 10^{-6} farads, is more common.

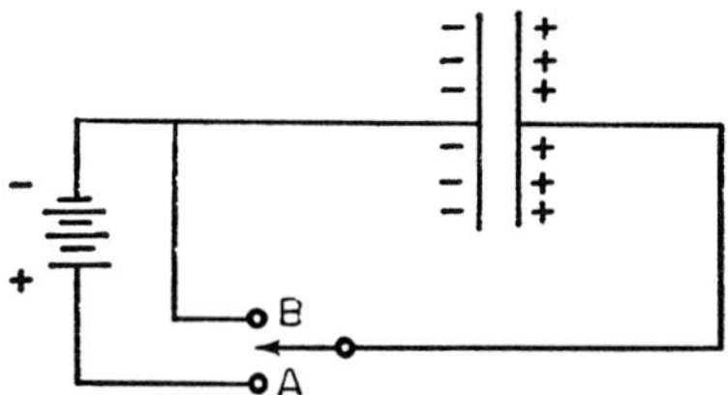

Fig. 15-1 Capacitor—charging and discharging.

Capacitors in Series. When two or more capacitors are connected in series, the voltage across them is equal to that of the the battery. Therefore, $E = E_1 + E_2 + E_3$, where E_1, E_2, and E_3 are the voltages across the respective capacitors. Under action of the battery voltage, E, a charge, Q, is sent into all the capacitors since they are in series.

Dividing by Q from Eq. **3**:

$$\frac{1}{C} = \frac{1}{C_1} + \frac{1}{C_2} + \frac{1}{C_3} \tag{4}$$

Capacitors in Parallel. Connecting capacitors in parallel has the same effect as increasing the number of plates—the capacitance of the circuit is increased. With three capacitors in parallel across a supply voltage, E, the charge is: $Q = Q_1 + Q_2 + Q_3$, and the voltage across each capacitor is equal to the supply voltage, E.

Dividing by E from Eq. **3**:

$$C = C_1 + C_2 + C_3 \tag{5}$$

Effect of Capacitance in Circuits. Illustrating a capacitive circuit with resistance in series, Fig. 15-2a shows a time plot of e_R, e_c, and i, with the switch closed at time t_1. There is no initial charge on the capacitor. From $t = 0$ to $t = t_1$, $e_c = 0$, $e_r = 0$, and $i = 0$. At $t = t_1$, i jumps immediately to a value of e_R/R. In this case, because of resistance R, time is required for the capacitor to charge. As a result, voltage e_c equals zero and e_R must equal E, yielding $i = e_R/R = E/R$. As e_c increases, e_R decreases ($e_R = E - e_c$) and i decreases.

As t approaches infinity, e and i both approach zero. With capacitance constant, an increase in resistance will decrease the rate at which the capacitor charges. With resistance constant, an increase in capacitance will also decrease the rate at which the capacitor charges. This relationship is expressed as the time constant, T, of the resistor-capacitor combination, or:

$$T = RC \tag{6}$$

where T is in seconds, R in ohms and C in farads. The time constant, T, is the time required for the voltage e_c to build up to 63 per cent of its final value. In a decreasing curve such as e_R or i, it is the time required for the curve to decrease 63 per cent from its maximum value.

Figure 15-2b shows a circuit that allows a capacitor to charge when switch 1 is closed and allows it to discharge when switch 1 is open and switch 2 is closed. The figure also shows a plot of e_c, i_c, and i_2 for the condition when switch 1 is closed at t_1 and opened at t_2, switch 2 is closed at t_3 where $t_1 < t_2 < t_3$, and the initial charge on the capacitor is zero.

Resistance-capacitance circuits are useful when time delays are necessary for switching.

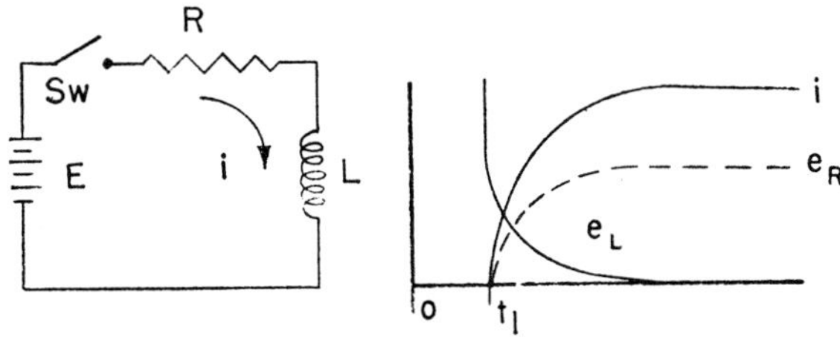

Fig. 15-3 Circuit with inductance in series and its response.

Inductance. A magnetic field surrounds every conductor carrying electric current. (A permanent magnet produces the same effect.) As current increases, energy is added to the magnetic field; as current decreases, magnetic energy is returned to the electric circuit. Since the energy in the magnetic field depends on the current, one cannot change without the other. The magnetic field cannot change instantaneously but requires time.

An emf applied to a circuit attempts to start a current. The current establishes a magnetic field that in turn reacts

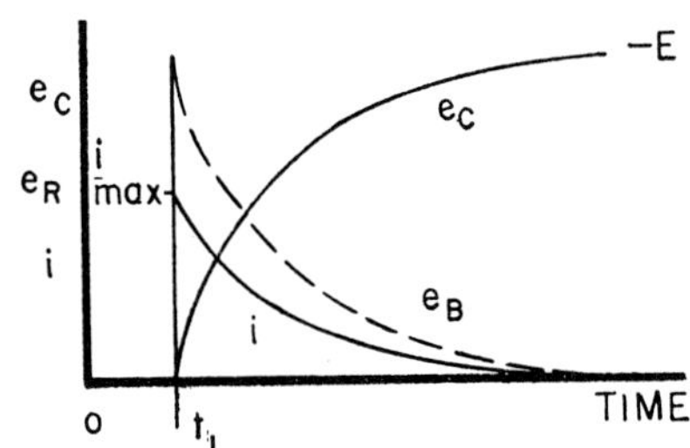

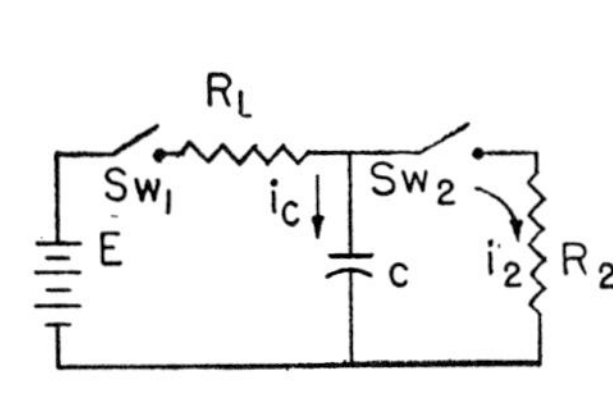

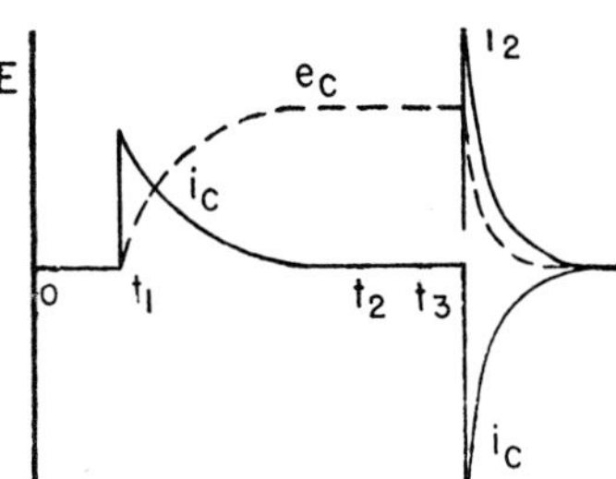

Fig. 15-2a (left) Time plot of e_c, e_R and i, with capacitance and resistance in circuit.
Fig. 15-2b (center and right) Capacitor switching circuit and its transient response.

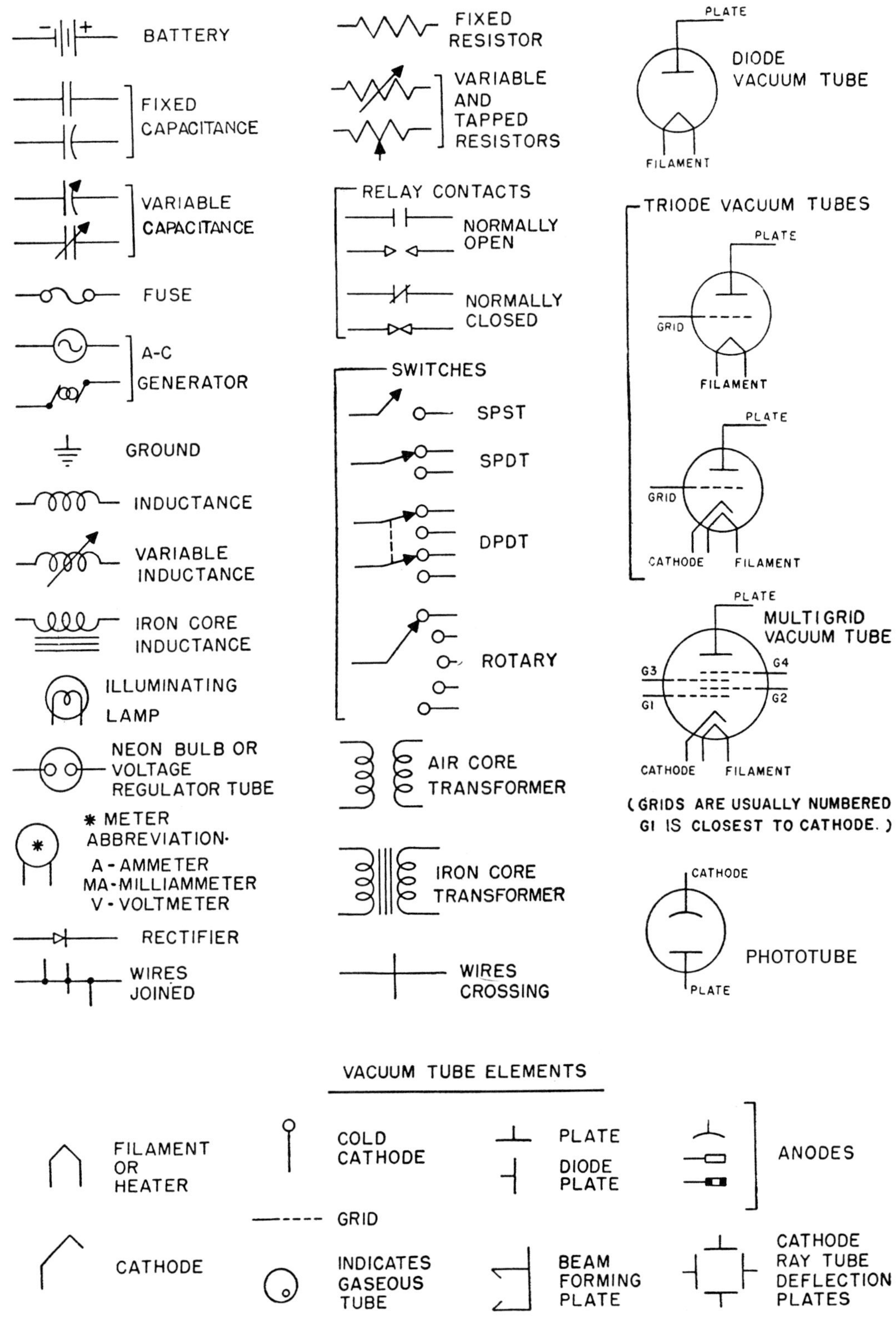

Fig. 15-4 Schematic symbols for electrical equipment.

upon the circuit so that part of the circuit emf must be used to overcome the magnetic field reactions. A mechanical analogy is the case of pushing a movable object. Part of the force exerted must overcome friction, while the rest goes into kinetic energy. The longer the force is applied, the faster the object moves until all energy imparted to it is used to overcome friction, at which time the speed is constant. If the force is suddenly removed, the object does not stop immediately, but coasts to a gradual stop as its energy is dissipated.

Resistance and capacitance must always be considered in an inductance circuit. Figure 15-3 shows a practical circuit in which R is the total resistance, including that of battery and coil, and L is the coil inductance. The capacitance, C, is negligible compared with R and L. The current eventually reaches a final value after the switch is closed at t_1. Since the current is zero from t_0 to t_1, there is no voltage drop across R; therefore, e_L, the voltage across the inductance, must equal E, the applied voltage. The rate of change of current with respect to time equals e_L/L; thus, at t_1 the rate of change equals E/L since $e_L = E$. As current starts to flow, $e_L = E - e_R$. Finally, current builds up to such a value that $e_R = E$. When e_l is zero, the magnetic field is complete and the applied emf is used to overcome resistance. The maximum value of current is $I = e_R/R = E/R$.

The time constant of the curves in a resistance-inductance circuit is:

$$T = L/R \tag{7}$$

where T is in seconds, L in henrys, and R in ohms.

Power. Power is the rate of doing work. The common unit of power is the *horsepower* (*hp*). To convert horsepower to other units of measurement, see Table 1-41.

Electrical Work. Total work is equal to power *multiplied* by time. The common electrical unit of work is the *kilo-watt-hour* (*kwh*). To convert kilowatt-hours to other units of measurement, see Table 1-43.

Electrical Power. The power required, in *watts*, to keep electricity flowing against a certain resistance equals the flow, in coulombs per second (i.e., amperes), *times* the pressure, in volts, or:

$$W = EI \tag{8}$$

Since $E = IR$,

$$W = I^2R \tag{9}$$

Schematic Symbols

Figure 15-4 shows the symbols for various items of common electrical equipment.

SIMPLE ELECTRICAL MEASUREMENTS

Resistance Measurements

Wheatstone Bridge. A resistance bridge is a convenient and accurate piece of test equipment for laboratory or shop use in measuring resistance. Figure 15-5a shows the circuit employed in a Wheatstone bridge. When the bridge is in balance (current i_g thru galvanometer G is zero), the current thru resistor X is the same as that thru resistor P, and points a and b are at the same potential. The voltage drop across P equals that across Q. If P, Q, and R are known, the resistance of X can be determined.

Portable bridges are commercially available. In such bridges, P/Q is the ratio, and R (the variable resistance) is used for balancing. Adjustments are usually made by means of dials.

Voltmeter-Ammeter. The voltmeter-ammeter method may also be used to measure resistance. Figure 15-5b shows two methods of connection. In the diagram at the left, the ammeter measures the sum of the current flowing thru the resistor and the voltmeter. If the resistance of the voltmeter is high compared with that of R, the current thru it will

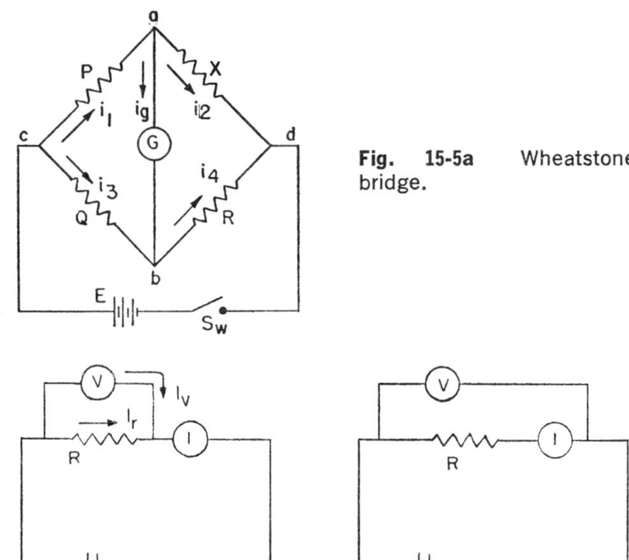

Fig. 15-5a Wheatstone bridge.

Fig. 15-5b Voltmeter-ammeter connections for resistance measurement.

be small and the current the ammeter indicates is considered to be that thru R. Then by Ohm's law, $R = V/I$. A more exact value of R also may be obtained. From Kirchhoff's first law, $I = I_R + I_v$, where I_R = current thru R and I_v = current thru the voltmeter. Since $I_v = V/R_v$, where R_v = the resistance of the voltmeter, substitution gives $R = V/(I - I_v)$.

When R is larger, the voltmeter connection shown at the right in Fig. 15-5b may be used. Here the ammeter measures only the current thru R, but the voltmeter measures the voltage drop across R and the ammeter in series. Thus, $R + R_a = V/I$, where R_a = the resistance of the ammeter. This correction is easy to make.

Current and Voltage Measurement

The *portable ammeter* is an accurate instrument for current measurement. It may be inserted into the circuit at the point where the current is to be measured. Caution should be exercised never to allow a larger current to flow thru an ammeter than it is intended to carry. In d-c circuits it is important to have the correct polarity on the meter; a momentary contact may be made to see if the polarity is correct and the ammeter large enough before connection is made. An a-c ammeter must be used on a-c circuits. Polarity need not be considered. The ammeter's resistance must be low, to minimize error.

A *shunt* is a known value of resistance often used to measure large amounts of current. The voltage drop across it is proportional to the current. Measurement of this drop by a sensitive voltmeter will indicate the shunt current.

A *voltmeter* is used to measure the difference in potential between two points. Current thru the voltmeter is proportional to the voltage between its points of connection. When using a voltmeter, care must be taken to see that its resistance is large compared with that of the circuit. For best results in measuring voltage to within 0.1 per cent of its true value, a voltmeter having a resistance about 1000 times that of the circuit is necessary.

Meter manufacturers produce *multimeters,* which measure voltage, current, or resistance on a single multiscale indicating instrument. These instruments are useful for making field measurements.

Power Measurement

The power dissipated in a resistor at any instant is expressed by the equation:

$$P = i^2R$$

where P is the instantaneous value of power, i the instantaneous value of current, and R the resistance of the circuit. Since $e = iR$, the instantaneous value of power can also be written as $P = ei$.

A *wattmeter* is commonly used to find the average power dissipated in a circuit. A combination ammeter and voltmeter, the wattmeter multiplies the instantaneous values of current and voltage and gives an average value of the product, whether the power is constant or varies periodically. The product of voltmeter and ammeter readings is d-c power.

In a-c circuits, current and voltage are usually sinusoidal

Table 15-1 Allowable Current-Carrying Capacities, in Amperes, of Insulated Copper Conductors*

(not more than three conductors in raceway or cable or direct burial; based on 86 F room temperature)

Size AWG MCM	Rubber Type R Type RW — Type RU Type RUW (14-2) — Type RH-RW† — Thermoplastic Type T Type TW	Rubber Type RH — RUH (14-2) — Type RH-RW† — Type RHW Thermoplastic Type THW — THWN	Paper — Thermoplastic asbestos Type TA — Thermoplastic Type TBS — Silicone Type SA — Var-Cam Type V — Asbestos Var-Cam Type AVB — MI Cable — RHH‡	Asbestos Var-Cam Type AVA Type AVL	Impregnated asbestos Type AI (14-8) Type AIA	Asbestos Type A (14-8) Type AA
14	15	15	25	30	30	30
12	20	20	30	35	40	40
10	30	30	40	45	50	55
8	40	45	50	60	65	70
6	55	65	70	80	85	95
4	70	85	90	105	115	120
3	80	100	105	120	130	145
2	95	115	120	135	145	165
1	110	130	140	160	170	190
0	125	150	155	190	200	225
00	145	175	185	215	230	250
000	165	200	210	245	265	285
0000	195	230	235	275	310	340
250	215	255	270	315	335	...
300	240	285	300	345	380	...
350	260	310	325	390	420	...
400	280	335	360	420	450	...
500	320	380	405	470	500	...
600	355	420	455	525	545	...
700	385	460	490	560	600	...
750	400	475	500	580	620	...
800	410	490	515	600	640	...
900	435	520	555	...	...	...
1000	455	545	585	680	730	...
1250	495	590	645	...	...	...
1500	520	625	700	785	...	...
1750	545	650	735	...	...	...
2000	560	665	775	840	...	...

* From *National Fire Codes, Volume 5, Electrical,* Boston, Mass., Natl. Fire Protection Assoc., 1964.

† In wet locations, capacities are those in column 2; in dry locations, those in column 3.

‡ Capacities for Type RHH conductors for sizes AWG 14, 12, and 10 are the same as designated for Type RH conductors in this table.

and may or may not be in phase. If in phase, their product is equal to the power; if not in phase, a wattmeter is generally used for measuring the power. A wattmeter reads average instantaneous power.

Figure 15-6 shows wattmeter connections in a single-phase circuit.

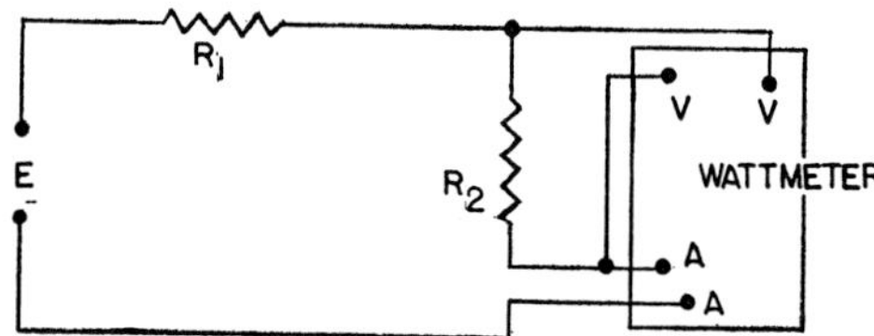

Fig. 15-6 Wattmeter connections for measurement of power dissipated in R_2.

The product of current and voltage in an a-c circuit is called the *apparent power*. The unit for apparent power is the volt-ampere. The actual power is expressed by $W = VI \cos \theta$, where θ is the phase angle between current and voltage.

The *power factor* of a circuit is the ratio of actual power (watts) to apparent power (volt-amperes) and is equal to the cosine of the phase angle of the circuit:

$$pf = \frac{\textit{actual power}}{\textit{apparent power}} = \frac{\textit{watts}}{\textit{volts} \times \textit{amperes}} = \frac{kw}{kva} = \frac{R}{Z}$$

The *reactive power* in this circuit equals the sine of the power factor angle *times* the apparent power. The actual power and apparent power are ordinarily expressed in kilowatts and kilovolt-amperes, respectively, and the reactive power in kilovars.

Devices used in a-c circuits (e.g., transformers) are rated in volt-amperes, or voltage and current, as well as in watts. Transformers are rated according to the number of volt-amperes they can supply.

Impedance, Inductance, and Capacitance Measurement

The impedance of a circuit or circuit component is given by the equation $Z = E/I$, where Z is the impedance of the circuit, E the voltage drop across the impedance, and I the current in the circuit. The relationships among impedance, resistance, power factor angle θ, inductance, and capacitance are given in Fig. 15-7; all except the angle θ are measured in ohms.

If the resistance of the component is known, its inductive reactance, X_L, can be calculated when E and I are measured.

Impedance. In general, impedance, Z, is equal to the joint effect of resistance, R, inductance, X_L, and capacitance, X_C, which are present in an a-c circuit, and may be expressed:

$$Z = \sqrt{R^2 + (X_L - X_C)^2}$$

where Z, R, X_L, and X_C are in ohms.

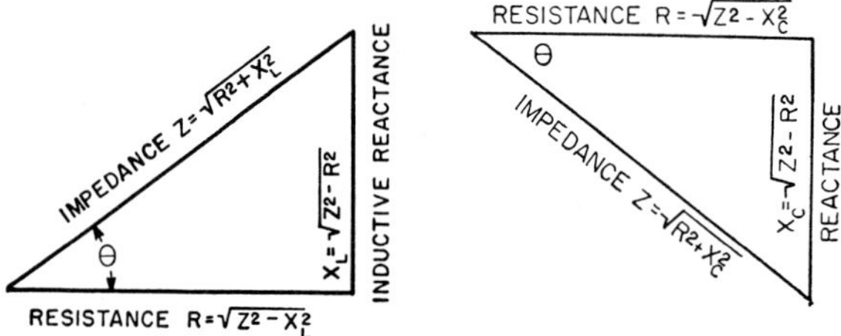

Fig. 15-7 Impedance triangles: (left) inductive circuit; (right) capacitive circuit. $X_L = 2\pi fL$ and $X_c = 1/2\pi fC$, where f = frequency, cps; L = inductance, henrys; C = capacitance, farads.

CURRENT-CARRYING CAPACITY OF CONDUCTORS

Table 15-1 lists the allowable current-carrying capacities of copper conductors. For further information consult the latest edition of the *National Electrical Code*.

Chapter 2
Electrical and Magnetic Circuits

by Sigward A. Stavnes

ELECTRICAL CIRCUITS

Circuits with constant parameters, which are common in general practice, are called *linear circuits*. Determining the response of such circuits to arbitrarily impressed potential and currents leads to a set of linear differential equations set up by **Kirchhoff's laws**:*

1. The algebraic sum of the currents at any junction point in an electric circuit is zero.
2. The algebraic sum of the electromotive forces and potential drops around any closed path or mesh of an electric circuit is zero.

DIRECT-CURRENT CIRCUITS

Series Circuits

The characteristics of series circuits are: (1) the current everywhere is the same; (2) the algebraic sum of the emfs equals the sum of the voltage drops (Kirchhoff's second law). Because only one value of current flows, it may be measured by an ammeter inserted at any point.

The total resistance of a series circuit is equal to the sum of its separate resistances. For a circuit with three resistances, using Kirchhoff's second law:

$$E = IR_1 + IR_2 + IR_3 = I(R_1 + R_2 + R_3)$$

If a single resistance, R_s, replaces the three separate resistances such that the current remains the same, then:

$$E = IR_T$$

Substituting and simplifying:

$$R_T = R_1 + R_2 + R_3 \quad \textbf{(1)}$$

Parallel Circuits

The characteristics of parallel circuits are: (1) the voltage across a parallel group is the same for all its components; (2) the algebraic sum of the currents flowing into any junction is zero (Kirchhoff's first law).

In Fig. 15-8 the voltage across R_1, R_2, and R_3 is the same. Adding the voltages about one loop gives $E - E_1 = 0$ (Kirchhoff's second law). Since $E_1 = I_1R_1$, this equation becomes $E = I_1R_1$. Similarly, for the other loops, $E = I_2R_2$ and $E = I_3R_3$.

If the parallel group shown is replaced by R_T, then $E = I_TR_T$. From Kirchhoff's first law:

$$I_T = I_1 + I_2 + I_3 \quad \textbf{(2)}$$

Hence, by substitution:

$$E/R_T = E/R_1 = E/R_2 = E/R_3$$

or

$$1/R_T = 1/R_1 + 1/R_2 + 1/R_3 \quad \textbf{(3)}$$

Direct-Current Networks

One or more voltage sources with several resistances may be so connected as to form a network. Figure 15-9 represents a simple network.

Assuming currents to flow as shown, applying Kirchhoff's first law at point a:

$$I_1 - I_2 - I_3 = 0$$

Adding the voltages about the two loops, applying Kirchhoff's second law:

$$10 - 2I_1 - 8I_1 - 2I_2 = 0$$

$$20 - 4I_3 + 2I_2 - 6I_3 = 0$$

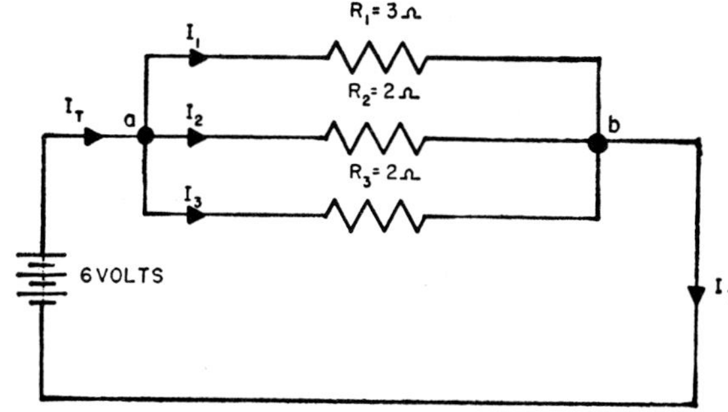

Fig. 15-8 Parallel circuit. $R_T = R_{ab} = 0.75$ ohm (by Eq. 3).

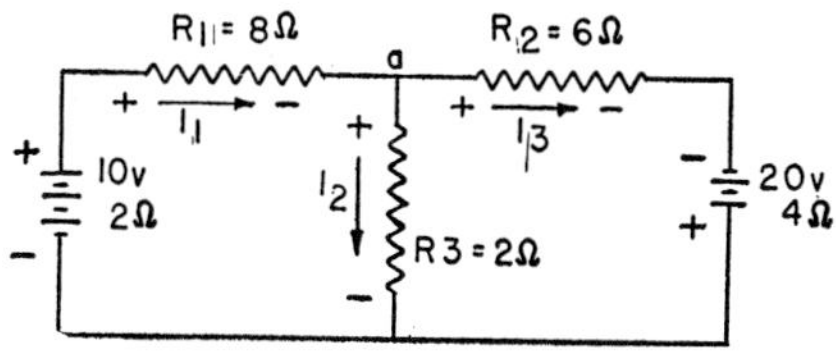

Fig. 15-9 D-C network.

* Comparable to the Hardy Cross method of gas network analysis.

Simultaneous solution of these three equations yields the following values:

$I_1 = 8/7$ amp
$I_2 = -5/7$ amp (showing that the original assumption regarding the direction of current flow thru R_3 was the reverse of the actual flow)
$I_3 = 13/7$ amp

Superposition

Superposition is a useful principle in the solution of linear circuits with two or more sources of emf. The principle states that current in any branch of a circuit is the algebraic sum of currents produced by each source acting alone, assuming the emf of the other sources is reduced to zero. The internal resistance of each source must be left in the circuit when the currents from other sources are considered.

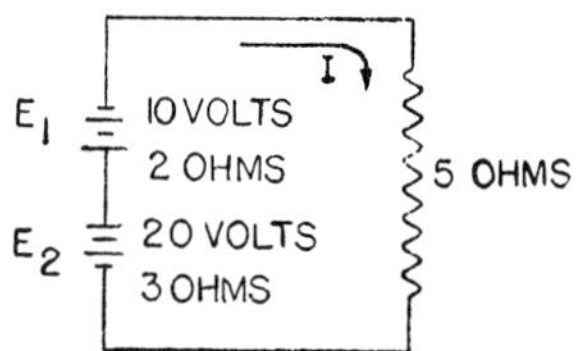

Fig. 15-10 Series circuit with two voltage sources, illustrating superposition principle.

To illustrate, if E_1 in Fig. 15-10 has its voltage reduced to zero, the current that is due to E_2 will be 20 *divided* by 10, or 2 amp flowing in the direction shown. If E_2 is reduced to zero, then the current due to E_1 will be 10 volts *divided* by 10 ohms, or 1 amp. The algebraic sum of the currents with both batteries in the circuit is 1 amp.

Superposition, which permits reducing the solution of a complex problem to that of several simpler ones, is valuable in circuits in which a-c and d-c sources exist simultaneously.

MAGNETS AND MAGNETIC CIRCUITS

Magnets may be classified as either **permanent magnets** or **electromagnets.** A hard steel bar when magnetized becomes a permanent magnet and tends to retain its magnetism under normal conditions for a long period unless subjected to heat or jarring. Soft iron is easily magnetized when subjected to a magnetic flux, but upon removal of this flux only slight residual magnetism is retained. Permanent magnets are steel or alloys with cobalt, nickel, etc. Electromagnets are generally soft iron.

All magnets establish a field. The imaginary lines of force that constitute this field are generally called *magnetic flux lines.* The electromagnetic unit for measurement of flux density is the *gauss.*

Magnetic Circuits

Magnetic flux is analogous to electric current. As electricity is caused to flow in an electric circuit, so magnetic flux can be established in a magnetic circuit. **Flux density,** B, or flux lines per unit area, may be expressed as $B = \phi/A$, where ϕ is the total flux and A the cross-sectional area. The flux density depends on the strength of the magnet and the material of the circuit.

The property of a magnetic circuit analogous to electrical resistance is called **reluctance,** R. Another property of a magnetic circuit similar to voltage is called **magnetomotive force,** M. For a complete magnetic circuit the total flux may be expressed as:

$$\phi = M/R \tag{4}$$

Although magnetic and electrical circuits are in many ways analogous, they differ in at least two aspects of major importance. First, a magnetic circuit can never be entirely open but must exist at all times near a magnet. (A magnetic circuit is more nearly analogous to an electrical circuit under water that is completed thru the water when the continuity of its metal conductor is broken. The current strength then is decreased but the circuit is never entirely broken.) Second, flux is not quite like current, since current is rate of flow, whereas flux is more nearly a state of the medium.

Electromagnets

A magnetic field exists around all conductors carrying current. The "right-hand rule" indicates the relationship between the current flow and direction of the field. With the conductor grasped in the right hand, the thumb points in the direction of current flow and the fingers point in the direction of concentric flux lines.

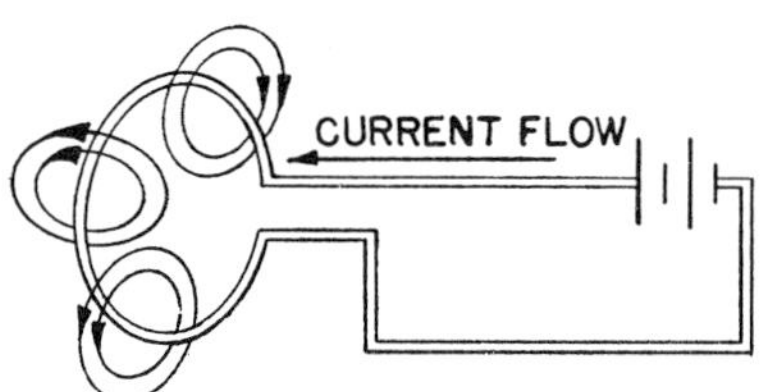

Fig. 15-11 Magnetic field about a conductor in the form of a loop.

If the electrical conductor is in the form of a loop, the groups of lines forming loops for every unit of its length can be imagined as arranging themselves as shown in Fig. 15-11. Flux lines enter the center of the loop from the bottom and leave thru the top. Because of crowding in the center, the field intensity increases and the flux paths become ellipses. If two loops are present instead of one, the intensity of the field doubles provided the current remains the same. The magnetomotive force of an electromagnet can be calculated and, if the circuit reluctance is known, the total flux determined. The total flux can be increased by use of a low-reluctance material such as soft iron.

The term **permeability,** μ, is ordinarily used in place of reluctance when discussing electromagnets. Permeability, which is inversely proportional to reluctance (and analogous to electrical conductivity), is the ratio of the magnetic conductivity of a substance to that of air. Iron and its alloys have high permeabilities, which vary with the flux density of the metal. As this density increases, permeability decreases until it is the same as air at the saturation point. Other magnetic materials behave similarly.

Reluctance is expressed by:

$$R = l/\mu A \tag{5}$$

where: l = length of magnetic material, in.
A = cross-sectional area, sq in.
μ = permeability of the magnetic material

Since $\phi = BA$, $M = NI$, and $R = l/\mu A$, substitution in Eq. 4 gives:

$$BA = \frac{\mu NIA}{l} \quad \text{or} \quad B = \frac{\mu NI}{l}$$

Since NI/l (i.e., ampere turns per unit length of magnetic path) = H (i.e., field intensity):

$$\mu = B/H \tag{6}$$

Figure 15-12a is a B-H curve for one grade of iron; H, magnetizing force per inch of magnetic path, is plotted against B, flux density. Such curves are different for various materials. From the B-H curve the permeability can be obtained, using Eq. 6. To solve for total flux in a complete magnetic circuit, the magnetizing force and the total reluctance of the circuit must be known. Because reluctance depends upon flux density, it is necessary to assume a total flux and, from the physical size of the circuit and the B-H curve, to calculate the total magnetizing force for the assumed flux. If the answer is not the same as the magnetizing force applied, another trial-and-error solution must be attempted.

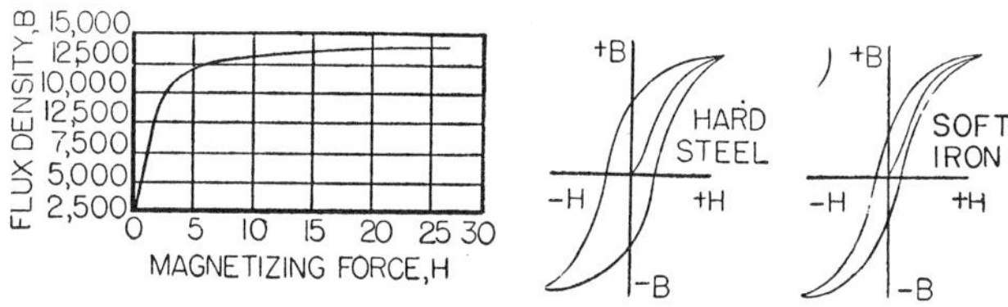

Fig. 15-12a (left) B-H curve for iron.
Fig. 15-12b (right) Hysteresis loops for soft iron and steel.

Simple Solenoid and Plunger.[1] The pull of a simple solenoid on a plunger depends on ampere-turns, shape and size of the coil and plunger, and induction. For practical purposes the maximum pull, P, is:

$$P = \frac{CA(NI)}{L_c} \tag{7}$$

where:

C = pull in pounds per sq in.-amp turn per inch length of coil
A = cross-sectional area of plunger, sq in.
N = number of turns
I = current in the coil, amp
L_c = length of the coil, in.

For a saturated plunger, C varies between 9×10^{-3} and 10.5×10^{-3}.

Hysteresis

Curves similar to those in Fig. 15-12b are called hysteresis loops. They represent relationships between flux density and field intensity when a piece of iron is subjected to an increasing magnetizing force until saturation is reached. At saturation, the force is decreased to zero and similarly established in the opposite direction. The cycle is then repeated.

Figure 15-12b shows that the iron does not return to its original magnetic condition after it has reached saturation. The figure further shows that when the magnetizing force returns to zero, the iron retains some magnetization. If the force was insufficient to saturate the iron, a smaller hysteresis loop will exist.

The area inside the hysteresis loop represents the energy lost during one cycle. This loss is important in a-c circuits. To restore the iron to its original nonmagnetized condition, the magnetizing force per cycle must be gradually decreased to zero. This principle is used in demagnetizing a watch.

ALTERNATING-CURRENT CIRCUITS

In d-c circuits the current either is constant or reaches a steady value after a brief transient state. The current then remains constant until a change occurs in the emf or in the circuit elements. The effects of inductance and capacitance are brief.

An alternating emf is cyclic and induces a cyclic current. Since the current never becomes constant, inductance and capacitance exert a continuous influence. Figure 15-13 shows four kinds of alternating emf waves.

Sinusoidal Variations

In practice, the emf of an a-c generator is nearly sinusoidal (C of Fig. 15-13). A nonsinusoidal wave can be reduced to a series of sine waves.

The equation describing a sinusoidal variation of a quantity, y, as a function of angle θ can be written $y = A \sin \theta$, where A is the wave amplitude. θ is usually measured in radians. If θ is a linear function of time, $\theta = \omega t$, where ω is angular velocity in radians per second. As the radius, A, rotates thru one revolution, θ increases by 2π radians, and y passes thru a complete cycle.

At the end of T seconds (time required for one cycle), θ has increased by 2π radians; hence.

$$\omega T = 2\pi$$

Since, by definition, frequency $f = 1/T$, substitution gives:

$$\omega = 2\pi f$$

Also, y may be expressed as $A \sin (2\pi ft)$. If, when $t = 0$, y is not equal to zero, then:

$$y = A \sin (2\pi ft + \phi) = A \sin (\omega t + \phi)$$

where ϕ is a constant.

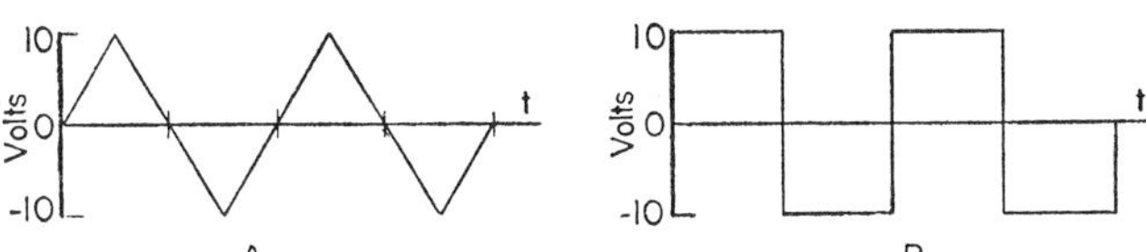

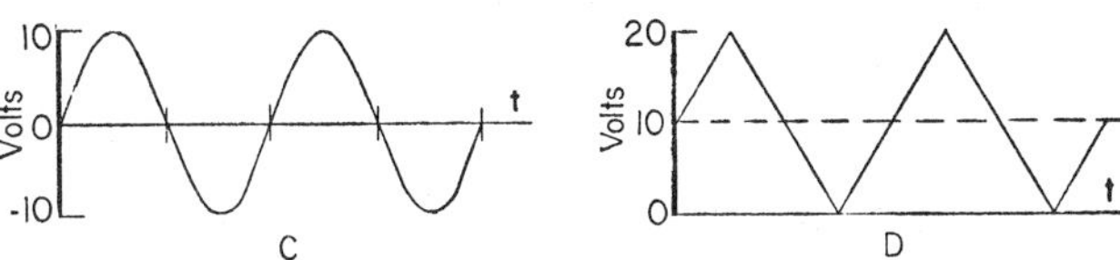

Fig. 15-13 Alternating emf waves: A—triangular wave; B—square wave; C—sine wave; D—triangular wave superimposed on 10 v d-c.

Pure Resistance

In the circuit shown in Fig. 15-14a current at any instant is given as $i = e/R$, where e = instantaneous voltage, i = instantaneous current, and R = constant resistance. With R constant, the current is proportional to the voltage. The current curve is similar to the voltage curve and is in phase with it. If e is a voltage sine wave, its division by a constant resistance, R, gives i as a current sine wave. The right side of Fig. 15-14a shows a current-voltage relationship with respect to time.

Instantaneous power, p, delivered to the resistor in the circuit shown in Fig. 15-14a is:

$$p = i^2R = RI_m{}^2 \sin^2 \omega t \quad \textbf{(8)}$$

where I_m = maximum value of current, i.

Effective Values of Sinusoidal Voltages and Currents

In an a-c circuit, since the current flow keeps changing cyclically, the measurement of the rate of flow is expressed in amperage *equivalent* to the amperage of a d-c circuit, where flow is constant. This equivalency is based on the heating effect produced by the two different currents and is called the *effective* or *rms* (root-mean-square) value, I_{eff}. Thus:

$$I_{\text{eff}} = I_{\text{rms}} = 0.707\ I_{\text{max}} \quad \textbf{(9a)}$$

$$E_{\text{eff}} = E_{\text{rms}} = 0.707\ E_{\text{max}} \quad \textbf{(9b)}$$

When alternating current and voltage values are given, effective values are meant unless otherwise specified. Most measuring instruments indicate effective values.

Pure Inductance

A back emf is developed whenever a current changes in an inductor. This back emf is an alternating voltage of the same frequency as the applied voltage, has a maximum value of $L\omega I_m$, and lags the current by the value $\pi/2$. The applied voltage required to overcome the back emf and force the current thru the inductor equals the back emf in magnitude but differs by 180° in phase. Since E_m equals ωLI_m, in terms of effective values $E = \omega LI$ or $E/I = \omega L$, where L = inductance in henrys.

A pure inductance does not exist. Since all coils wound with wire have some resistance, the impedance of the inductor is a resistance plus a reactance.

Capacitance

The left side of Fig. 15-14b shows a pure capacitor connected to an a-c source. Applied voltage $e = E_m \sin \omega t$ is chosen as the reference voltage. The charge, q, at any instant, t_1, is given by $q = Ce = CE_m \sin \omega t$. Differentiating with respect to t gives $dq/dt = i = CE_m\, \omega \cos \omega t$. The current is sinusoidal, has a maximum value of ωCE_m, and leads the applied voltage by 90°. In terms of effective values, $I = \omega CE$ or:

$$E/I = 1/\omega C \quad \textbf{(10)}$$

$1/\omega C$ is called the capacitive reactance, X_c, in ohms.

Impedance

If the current thru an a-c circuit component and the voltage across it are known, the equation $I = E/Z$ applies, where Z is the circuit impedance, consisting of resistance, inductive reactance, and capacitive reactance, or any combination of them.

The same current flows thru all components of a circuit with resistance, inductance, and capacitance in series with a a-c generator. The total impedance, Z_t, of such a circuit is the sum of the impedances of its individual components, or:

$$Z_t = Z_1 + Z_2 + Z_3 + \ldots \quad \textbf{(11a)}$$

This is similar to the expression for total resistance in d-c circuits.

In an a-c circuit with three impedances in parallel, the voltage across each is the applied voltage. The individual impedance may be replaced by one impedance, Z_T. The following relationship applies:

$$\frac{1}{Z_T} = \frac{1}{Z_1} + \frac{1}{Z_2} + \frac{1}{Z_3} \quad \textbf{(11b)}$$

This is similar to the equation for total resistance of several parallel resistors in d-c circuits.

Power in Alternating-Current Circuits

In circuits containing more inductive reactance than capacitive reactance, i.e., $2\pi fL > \frac{1}{2}\pi fC$, current lags behind voltage. Conversely, in circuits that are predominantly capacitative, current leads voltage.

The actual or effective power, P, in a circuit equals the product of the effective voltage and the portion of the current that is in phase with the voltage.

The **power factor** is the cosine of the angle by which voltage and current are out of phase with each other:

$$P = aEI \cos \theta/1000 \quad \textbf{(12)}$$

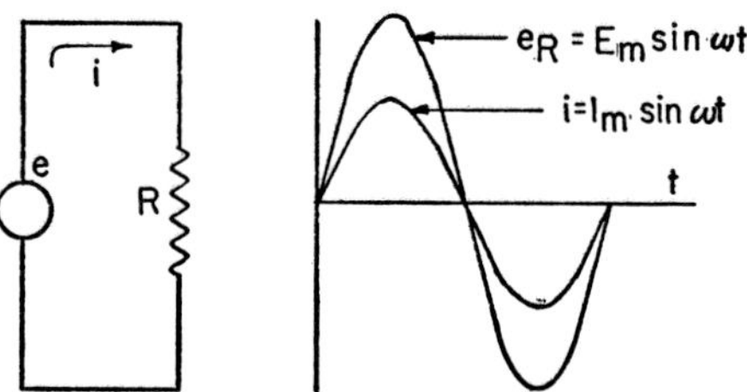

Fig. 15-14a Alternating-current circuit with pure resistance.

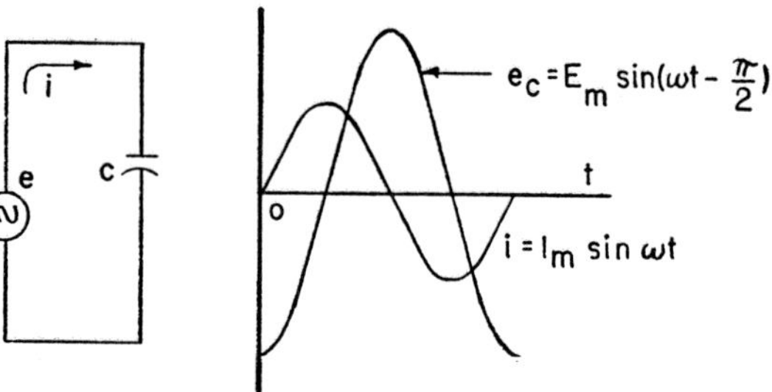

Fig. 15-14b Pure capacitor with a-c source.

where:

a = system factor (1 for single-phase circuits, 2 for balanced two-phase circuits, 1.73 for balanced three-phase circuits)
P = actual or effective power, kw
E = effective or rms voltage between phases or lines, volts
I = effective or rms current in each phase, amp
θ = power factor angle
$\cos\theta$ = power factor = actual power/apparent power,* = kw/kva

Transformers

A transformer consists of two or more separate coils insulated electrically but coupled by means of magnetic flux. An alternating current in the primary coil produces a magnetic flux that links the turns of the secondary. The changing flux linking the secondary coil induces an alternating emf in it. When the secondary circuit is closed, its current also produces a changing flux in the core, opposing that produced by the primary current. The reduction of flux thru the primary because of the presence of current in the secondary lowers the primary back emf. Primary current thus increases proportionally with the secondary current.

If the same flux is assumed to link all coils (as is usually true), the voltage generated in each turn will be the same. With N_1 turns in the primary, the primary back emf, which almost equals the applied voltage, will equal the voltage induced per turn e_t *times* the number of primary turns, or $E_1 = e_tN_1$. In like manner, in the secondary, $E_2 = e_tN_2$. By division:

$$\frac{E_1}{E_2} = \frac{N_1}{E_2} \qquad (13)$$

In most applications, the exact theory gives results little different from this simple analysis, which omits the resistance of transformer coils and core losses that are due to hysteresis and eddy currents.

* Sin θ = reactive power/apparent power = kvars/kva.

REFERENCE

1. Kinnard, I. F. and Goss, J. H. "Electromagnets." (In: Pender, H., and Del Mar, W. A., eds. *Electrical Engineers Handbook: Electric Power*. New York, Wiley, 1949.)

Chapter 3
Electric Motors

by H. G. Nafe

GENERAL USE AND SELECTION[1,2]

Electric motors of all sizes and types are used in the gas industry. Over 12 million motors provide power for blowers on central heating gas furnaces, and over 4.1 million are used to rotate drums on gas clothes dryers. Many more drive fans on unit heaters; pumps on hot water circulating systems; and equipment on gas conversion and industrial burners and on various gas, air, and water systems in processing plants. Gas-fueled prime movers are suitable for many of these applications.

Basic considerations in motor selection and application are as follows:

1. Electrical characteristics
   a. Starting torque requirements
   b. Breakdown torque requirements
   c. Starting current limitations
   d. Power factor and efficiency
   e. Power supply (voltage, number of phases, frequency; see Table 15-2)
2. Protection (open, dripproof, splashproof, totally enclosed, etc.)
3. Ambient temperature
4. Maximum winding temperature rise
5. Number of starts, stops, and reversals required
6. Possible unusual overload
7. Altitude at point of installation

Table 15-2 Recommended Motor Horsepower Range for Various Distribution Voltages

Voltage	Horsepower
Direct current	
115	0–30 (max)
230	0–200 (max)
550-600	½ and upward
Alternating current, one-phase	
110-115-120	0–1½
220-230-240	0–10
440-550	5–10*
Alternating current, two- and three-phase	
110-115-120	0–15
208-220-230-240	0–200
440-550	0–500
2200-2300	40 and upward
4000	75 and upward
6600	400 and upward

* Not recommended.

Further consideration may be given to the merits of direct current vs alternating current when deciding on the type of motor best suited for a particular application.* The following will serve as a guide:

Advantages of Direct Current

Motors—possibility of inherent, or automatic or adjustable speed variation.
Generators—self-contained, easy to operate, parallel.
Unity power factor

Advantages of Alternating Current

Motors and generators—simple, light, rugged (no commutators or brushes); little maintenance required.
Controls—simple, low-cost, light.
Cables and windings—reduction in cost and weight because of higher voltages.

ALTERNATING-CURRENT MOTORS

Alternating-current motors are divided into two main classifications, polyphase and single phase, according to their type of power supply. Each classification may be further subdivided by type of motor winding, as shown in Table 15-3a.

Polyphase Motors

Many types and designs of polyphase motors have been developed for various applications. Motors of this kind are used to drive compressors, fans, and pumps in gas plants.

Squirrel Cage Motors. This type is the simplest and most rugged of all motors, lowest in cost, highest in efficiency, and easiest to maintain. Because these motors obtain their starting torque from their rotating magnetic fields, an inherent characteristic of polyphase circuits, they require no auxiliary devices such as starting switches, brush mechanisms, or commutators.

A squirrel cage stator contains a distributed polyphase winding. The rotor consists of slotted laminations with uninsulated copper or aluminum bars, short-circuited at both ends.

* The Ward Leonard system is an important composite of a-c and d-c machinery. It is often used for elevators, steel rolling, printing, and other applications where good speed response, fast braking, and reversing ability are important considerations. An a-c motor drives a d-c generator that in turn drives a d-c motor connected to the load.

Table 15-3a General Classification of A-C Motors[3]

	Type	Speed characteristics	Full voltage: Starting torque	Full voltage: Starting current	Horsepower range
		Constant-speed drives			
Polyphase	Squirrel cage general purpose Design A	Constant	Normal, 1–2.5 times*	High, 6–8 times	All
	Squirrel cage Design B	Constant	Normal, 1–2.5 times*	Normal, 5–6 times	Medium, small
	Squirrel cage Design C	Constant	High, 2–2.5 times*	Normal, 5–6 times	Medium, small
	Squirrel cage Design F	Constant	Low, 1.25 times	Low, 4 times	Medium, large
	Wound rotor	Constant or variable	High, 1–2.5 times (with secondary control)	Low, 1–3 times	All
	Synchronous high speed	Exactly constant	Normal, 0.75–1.75 times	Normal, 5–7 times	Medium, large
	Synchronous low speed	Exactly constant	Low, 0.3–0.4 times	Low, 3–4 times	Medium, large
Single-phase	Two-value capacitor	Constant	High	Normal	Small
	Permanent split capacitor	Constant	Low	Normal	Fractional
	Capacitor start	Constant	Moderate	Normal	Small, fractional
	Repulsion-induction	Constant	High	Normal	Medium, small
	Split-phase	Constant and adjustable	Normal	Normal	Fractional
		Adjustable-speed drives			
Polyphase	Squirrel cage high slip, transformer adjustment	Variable	Normal	Normal	Medium, small
	Squirrel cage separate winding or regrouped poles	Constant multispeed	Normal or high	Normal or low	All
	Wound rotor	Variable	High (with secondary control)	Low	All
Single-phase	Repulsion	Variable	High	Normal	Low, fractional
	Capacitor low torque tapped winding	Variable two-speed	Low	Normal	Fractional
	Capacitor low torque transformer adjustment	Variable	Low	Low	Fractional
	Split-phase regrouped poles	Constant	Normal	Normal	Fractional

* Torque depends on horsepower rating and speed. See NEMA Standard MG1–1963 on Motors and Generators.

Squirrel cage motors are available in several standardized electrical designs. Each design is defined by NEMA[1] in terms of horsepower, speed, torque, current, and slip as follows:

Breakdown torque of a motor is the maximum torque that it will develop with rated voltage applied at rated frequency, without an abrupt drop in speed.

Full-load torque of a motor is the torque necessary to produce its rated horsepower at full-load speed.

Locked rotor current of a motor is the steady-state current taken from the line with the rotor locked and rated voltage applied at rated frequency.

Locked rotor torque of a motor is the minimum torque that it will develop at rest for all angular positions of the rotor, with rated voltage applied at rated frequency.

Power factor of an a-c motor is the ratio of the kilowatt input to the kilovolt-ampere input, usually expressed in per cent.

Pull-up torque of an a-c motor is the minimum torque developed during the period of acceleration from rest to the speed at which the breakdown torque occurs. For motors not having a definite breakdown torque, the pull-up torque is the minimum developed up to rated speed.

Slip of a motor is the relation of synchronous speed to full-load speed. It can be calculated with the formula $S = (N_s - N)/N_s$, where S = slip, N_s = synchronous speed, and N = full-load speed. Slip is usually expressed in per cent.

Squirrel cage polyphase motors are divided by NEMA into five classes, Designs A, B, C, D, and F. The stator windings of all classes are similar, but their rotor designs vary widely. The depth and shape of rotor slots as well as the resistance of the rotor bar and rotor and ring conductors are changed to obtain the desired performance characteristics. Figure 15-15 shows typical speed-torque curves.

Design A motors, used only where their special characteristics are required, provide normal starting torque at starting current in excess of Design B motors. The rotor design of Design A motors provides higher breakdown torque than Design B or C motors by allowing locked rotor current to increase.

Design B motors are the most widely used squirrel cage motors. They can withstand full-voltage starting and have a slip at rated load of less than 5 per cent. Because they develop locked rotor and breakdown torques adequate for general application, they are widely used on centrifugal pumps, fans, and compressors equipped with unloaders. These motors will accelerate to full speed any load that the motor can start. Their locked rotor currents should not exceed those shown in Table 15-3b. Starting current of Design B motors in sizes from 30 to 200 hp should not exceed 14.5 amp per hp at 220 volts.

Design C motors are constructed for applications requiring higher starting torques than those furnished by Design B. The Design C motors employ double-cage rotors; the inner cage has low resistance and high reactance, the outer cage high resistance and low reactance. During starting, most of the current is carried in the outer cage, enabling the high starting torque to be developed. As the motor accelerates, the current

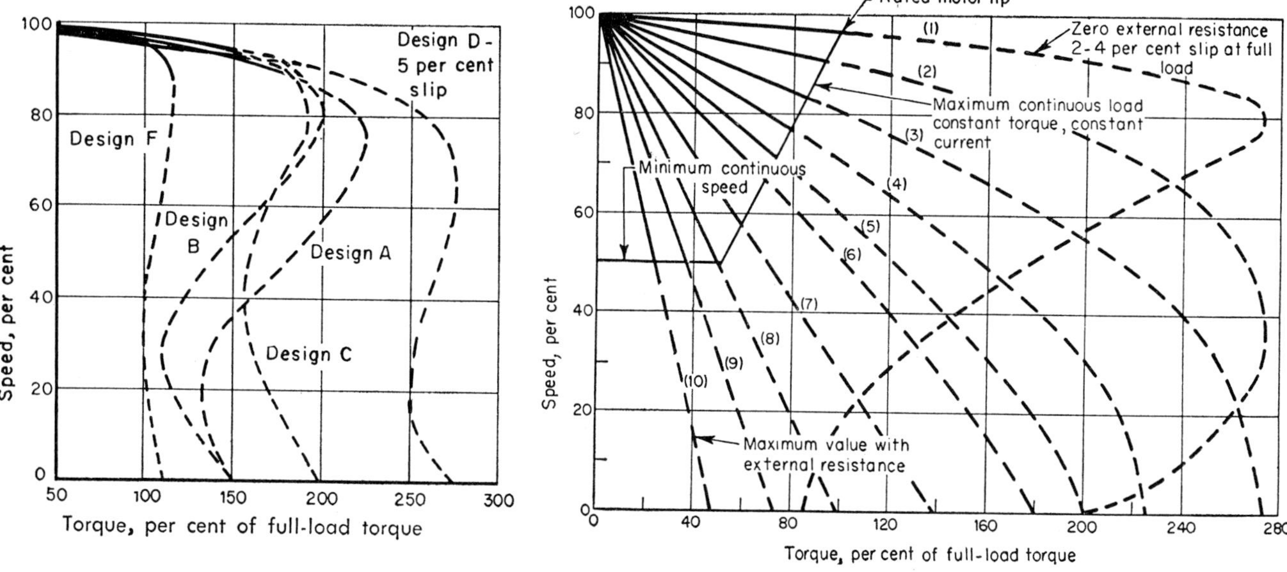

Fig. 15-15 (left) Per cent synchronous speed-torque curves for squirrel cage induction motors (30 to 50 hp range).[2]

Fig. 15-16 (right) Family of per cent synchronous speed-torque curves for a typical wound rotor motor. Various values of resistance are placed in series with the rotor windings to change the motor characteristics.[2] The locked rotor current drawn by a rotor with zero external resistance, curve (1), is the maximum value.

shifts to the low-resistance inner cage. The breakdown torque of Design C motors is slightly less than that of Design B. The Design C motors withstand full-voltage starting and have locked rotor currents that are the same as those of Design B. Design C motors, commonly known as high starting torque, normal starting current motors, are used on refrigeration compressors, reciprocating pumps, and air and gas compressors that are not unloaded during starting.

Design D motors will handle loads requiring high starting and accelerating torques, such as punch presses, grinders, crushers with flywheels, and high-inertia loads. These motors have a higher slip than do Designs B and C, usually about 7 to 13 per cent at full load. Their maximum torque is usually developed at zero speed (locked rotor), as compared with 75 to 85 per cent of synchronous speed for Design B.

Design F motors have low starting current and starting torque, and may be used where limited locked rotor current is required. Often they are used to meet certain power company regulations as to allowable motor starting current, thus avoiding the high-cost motor control necessary with other designs. Generally these motors are available only in sizes of 30 to 200 hp, a range whose excessive starting current, 9 amp per hp at 220 volts, frequently causes difficulty. To obtain a motor with this low starting current, starting and breakdown torques must be low. Because of their limited torques, Design F motors should be used only on applications that start easily and can be brought up to speed rapidly. These motors are often satisfactory for blowers, fans, and compressors that start unloaded.

Table 15-3b Locked Rotor Current of Polyphase Motors
(three-phase, 60 cycles, 220 volts)

Horsepower	Amp/hp	Rotor current, kva/hp
	Designs B, C, and D	
1 and less	24	10.0–11.19
1½	35	9.0–9.99
2	45	8.0–8.99
3	60	7.1–7.99
5	90	6.3–7.09
7½	120	5.6–6.29
10	150	5.6–6.29
15	220	5.0–5.59
20	290	5.0–5.59
25	365	5.0–5.59
30–200	14.5	5.0–5.59
	Design F	
30–200	9	0–3.14

Multispeed Motors. Multispeed squirrel cage motors are useful on air and gas handling equipment when it is desirable to control air or gas movement by adjusting the fan speed. Two, three, or four different speeds are obtainable.

Two-speed squirrel cage motors have either one or two stator windings. Single-winding motors, which obtain a second speed by reconnection of the stator poles, produce speeds only in the ratio of two to one. For other ratios two windings must be used.

Three types of multispeed squirrel cage motors are available: constant torque, variable torque, and constant horsepower. Most are built with two windings, since two speeds produced by one winding must be in a ratio of two to one. Three-speed, three-winding motors are not usually recommended for reasons of space because large frame sizes would be required at high cost. Only a limited number of speed combinations are available in four-speed motors, usually consisting of two windings each of a reconnectable type.

Constant-horsepower multispeed motors, which produce the same horsepower on all speeds, are used for torque requirements that increase as the speed decreases. Torque changes inversely as the speed. Machine tools, winches, etc., require motors of this type if two-speed operation is necessary.

Constant-torque motors produce the same starting and breakdown torque at all speeds. Horsepower is proportional to torque and varies directly as the speed. Multispeed motors of this type are used on applications such as reciprocating compressors and positive pressure blowers, where the torque required is about the same regardless of speed.

Variable-torque motors develop starting and breakdown torques proportional to the speed. Horsepower developed varies as the square of the speed. These motors are regularly used for fans, blowers, and other devices in which the horsepower required decreases as rapidly as the square of the speed.

For any given horsepower rating, speed, and combination of speeds, a constant-horsepower motor will be large, comparable in size to a single-speed motor of the same horsepower operating at its lowest speed. Constant-horsepower motors are therefore considerably more expensive than either constant- or variable-torque designs. A multispeed polyphase unit represents the most economical means of obtaining an adjustable speed a-c motor.

Control Equipment. Squirrel cage motors are designed for full-voltage, across-the-line starting. Most starters for these motors consist of a simple three-pole switch or contactor, arranged to connect the stator windings directly to the distribution system. The starters usually include protection against motor overloads.

Both manual and magnetic starters are available. Manual across-the-line starters are used with small motors of about 7½ hp and less. Magnetic starters have many advantages over manual starters, such as remote control, low-voltage protection, or low-voltage release.

Low-voltage protection is used when restarting is governed by machine operators. Such protection disconnects the motor so that the control will not return to the start position until the operator is ready.

Low-voltage release disconnects the motor in case of a power failure and automatically restarts it when power is again available. Magnetic starters with low-voltage release are often specified on applications where continuous duty is important or where motors are remotely located and no operator is available to them. A typical application is its use with a ventilating fan.

Current Limiting Starters. Squirrel cage motors often require some type of current limiting starter to meet the current inrush specifications of the power company. Standard across-the-line magnetic or manual starters will not suffice. Reduced-voltage starters are often necessary, with one or more steps of starting resistance or inductance to keep the starting current within established limits. Such starters may be required to limit the torque imposed on driven equipment in order to prevent possible damage to it thru the locked rotor torque developed at rated voltage.

The most common reduced-voltage starters are the primary resistor and autotransformer types. The former makes use of resistors in series with the stator winding. Line current is thus reduced in direct proportion to voltage reduction and starting torque in proportion to the square of this reduction. This type is usually recommended when low first cost is more important than high starting torque.

Autotransformer starters or compensators provide higher starting torque than resistor-type starters for the same given reduction in starting current. With an autotransformer, line current varies as the square of the transformer voltage. Since starting torque varies as the square of the voltage across the motor winding, higher starting torques are possible for a given reduction in starting current than with resistor starters. For the highest possible starting torque with the minimum starting current, an autotransformer starter may be used and its additional cost justified.

Reduced-voltage starters, either manually or magnetically operated, are available. The manual type includes the necessary resistors or autotransformers, along with a switching mechanism and an external operating handle. When starting the motor, the resistors or autotransformers are connected in series with the stator winding. After acceleration, the operating lever is moved to the run position, connecting the motor directly across the line.

For automatic control, magnetic reduced-voltage starters should be selected. Pushbuttons replace the operating handle and contactors change connections between the motor and the reduced-voltage starter to obtain the desired starting sequence. Magnetically operated starters, which are considerably more expensive than the manually operated type, are available for starting squirrel cage polyphase motors of 5 hp and larger.

Increment and Part Winding Starting. Any polyphase motor with a two-section winding can be used on increment starting if the proper motor leads are available. Such a motor energizes one section of the winding at the time the motor circuit is closed. After a set time, the remaining section is placed in the circuit by the use of a starter consisting of two sets of contacts and a timer. One set of contacts is closed, energizing part of the motor winding. After a time delay relay operates, the other set is closed, completing the circuit to the full motor winding. Thru this combination, 48 per cent of the locked rotor torque may be obtained while drawing only 60 per cent of the full-voltage locked rotor current.

Most power companies impose limitations on the current drawn during starting. These restrictions may cover only the maximum increment of such current. A motor and starter combination of the increment type will provide compliance, giving a lower cost installation and avoiding an expensive motor control.

When starting current limitations are extremely severe, increment starters with resistors limiting current inrush may be connected in series with the stator winding. A time delay relay short-circuits these resistors after a set interval, and the motor operates again.

Three-step increment starters are more expensive than conventional increment starters since resistances and additional relays are needed. The low starting torque available may be insufficient to start the driven equipment on reduced-voltage steps.

Synchronous Motors. Polyphase synchronous motors and squirrel cage induction motors have similar stators, but their rotors are different. Synchronous motors employ a salient pole rotor with windings energized by a d-c source thru collectors or slip rings on its shaft. This source may be a distribution system but more often is a direct-connected d-c exciter, or a motor-generator set.

Synchronous motor speed is independent of the load and is determined by the power frequency and the number of field poles on the rotor. Synchronous motors are not self-starting.

When power is applied to the stator windings with the rotor stationary, the rotating magnetic field of the primary revolves at synchronous speed. The rotor poles are attracted first in one direction and then in the other, resulting in zero torque. To obtain starting torque, amortisseur starting windings are incorporated in each pole face. This combined with a short-circuiting field winding gives the effect of a squirrel cage rotor and results in a starting and accelerating torque to start the motor and bring it up to nearly synchronous speed. At about 95 per cent of synchronous speed, the d-c field excitation is applied with the short circuit on the field winding open, permitting the motor to operate at constant speed.

The power factor of a synchronous motor can be controlled by varying the field excitation. Unity power factor is usually obtained at normal excitation and *leading* power factor (current leads voltage) with stronger excitation.

Synchronous motors are often installed to increase the overall plant power factor. In such cases a motor rated 80 per cent *leading* power factor would normally be selected. Such a motor represents an increased investment over one with unity power factor, but its additional cost is often justified by the resultant power saving, particularly in plants with low power factor loads. Synchronous motors not only improve the power factor but attempt to act as voltage regulators, keeping line current at a minimum on low voltage and at maximum on high voltage.

In addition to their constant speed, synchronous motors may have several advantages over induction motors. These are:

1. Initial cost is lower for 15 hp or larger units, with speeds below 500 rpm; also lower for 500 hp and larger units, with operating speeds from 500 to 900 rpm.
2. Power factor in the average plant may be improved.
3. Efficiency of low-speed synchronous motors is higher than that of squirrel cage motors of the same horsepower and speed.

It is especially desirable to drive large air or gas compressors with synchronous motors. Such compressors are designed for constant speed, and their output efficiencies are often affected if they are driven at a different speed.

Synchronous motors, even though they require somewhat complex starting equipment and d-c power for field excitation, usually provide the most sensible and economical method of keeping the power factor high and the power cost low. The pull-in torque necessary for satisfactory operation should be thoroughly studied, since it depends upon the inertia of the load that must be accelerated to the synchronous speed of the motor. For reciprocating compressors (equipped with unloaders) the motor is usually designed for a pull-in torque of about 50 per cent of full-load torque. For other applications it may be as high as 150 per cent of full-load torque.

Control Equipment. Starting equipment similar to that used to connect the stator winding of a squirrel cage polyphase motor is used to connect a synchronous motor to the line. The same primary starting methods apply.

Synchronous motor starters, however, must perform several additional functions. Excitation must be applied to the motor field at the proper time and the motor synchronized at close to synchronous speed. Provision must also be made for the power factor adjustment.

Most synchronous motors are started across the line with full-voltage, full-magnetic starters. Semimagnetic starters may be used with manual primary circuit control and automatic magnetic excitation control, but their application is limited.

Self-Synchronous Drive. Many applications require two or more machines to be kept in exact step so that one follows the other at the same operating speed. In such cases, the secondary windings of two identical wound rotor motors are connected together, with the primaries energized from the same power source.

The motors will not rotate when so connected but will act as two transformers with primaries and secondaries in parallel. The two rotors will maintain the same phase relationship with their respective primaries and therefore stay in exact step. When one is rotated, the other follows with the exact rotation of the first. This will occur for continuous rotation as well as for only a fraction of a turn. Such combinations are widely used, for example, to provide remote indications of liquid levels and to position valves.

A complicated mechanical drive can often be replaced with a properly designed self-synchronized drive. A few copper wires then replace heavy line shafting and complicated mechanical parts. The more complicated the drive, the more pronounced will be the advantages of the electrical system. Such drives have become increasingly popular in processing plants.

Wound Rotor Motors. Wound rotor motors are generally used when speed must be adjustable with only alternating current available or when high starting torques and low starting currents are required. Such a motor has stator windings similar to those of squirrel cage motors but the rotor has an insulated winding with leads connected to collector rings. Carbon brushes riding on these rings can be connected to an external resistance that is cut out with each successive step of control.

When speed adjustment is required, controllers can be used for variations over a range of about two to one. Greater speed reductions give unstable operation and cannot be recommended. On any point of adjustment speed varies with any change in load. By proper controller setting, the initial starting current can be reduced to full-load values. Figure 15-16 shows typical torque-speed and load current–speed curves for full and reduced voltage by insertion of different secondary resistances.

Wound rotor motors require the proper controller resistors for adjustable speed operation. The type used depends on the application, since constant-torque loads require different resistors from variable-torque loads.

Control Equipment. The simplest and cheapest method of controlling a wound rotor motor is by a manual drum controller. Such a controller has primary contacts for connecting the stator winding to the power supply and has several secondary contacts for controlling the resistance inserted into the motor secondary. This resistance may be cut out in steps to bring the motor up to normal speed. For speed control continuous-rated resistors must be used in the secondary circuit. Short-time-rated resistors for starting and accelerating the motor to full load can be left in the secondary circuit only temporarily.

Fully magnetic wound rotor motor controllers consist of a panel with primary and secondary contactors controlled by pushbuttons. When starting, a pushbutton closes a magnetic contactor and connects the motor stator to the line, at the same time energizing the accelerating relays. These relays close the secondary contactors in a predetermined sequence so as to cut out the secondary resistances in steps, bringing the motor up to its rated speed. Because such a control is very elaborate and expensive, it is seldom used.

Single-Phase Motors

A large percentage of motors used on gas appliances is of the single-phase type, which requires auxiliary devices for starting, since the current in a single-phase circuit does not provide a rotating magnetic field. These motors are more complicated than polyphase motors, more susceptible to trouble, and their power factors and efficiencies are lower. As a result, single-phase motors are used only in smaller sizes or where polyphase power is not available.

Split-Phase Motors. The split-phase motor is the most common fractional horsepower, single-phase induction motor. It is lowest in cost and can handle numerous applications.

Split-phase motors depend for starting torque on a phase or auxiliary winding consisting of a fine wire of relatively high resistance and hence a high power factor. The main winding of heavier wire has a lower resistance and higher reactance. The resulting displacement between currents in the main and phase windings develops a rotating magnetic field. When the motor reaches a predetermined speed, approximately 70 per cent of full-load speed, a centrifugal device opens a switch in the phase winding circuit. The motor then draws current in the main winding only and operates as a single-phase squirrel cage induction motor (Fig. 15-17).

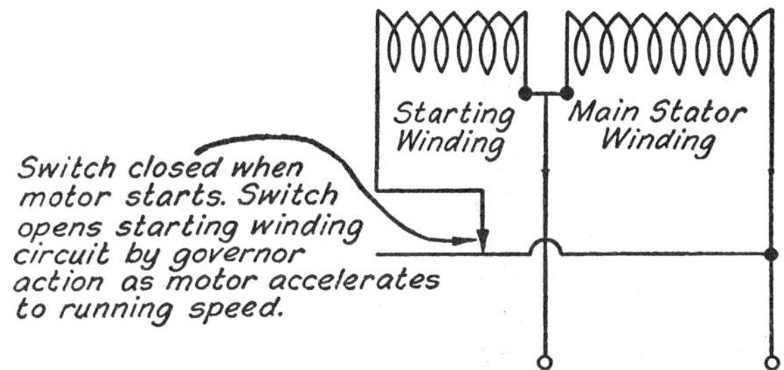

Fig. 15-17 Diagram of a split-phase motor circuit.

Figure 15-18 shows speed-torque curves of an average fractional horsepower, split-phase motor. Locked rotor torque is usually about 150 per cent and pull-up torque about 135 per cent of full-load torque. Breakdown torque is about double full-load torque. The torque required for a given load must be carefully considered; if the motor is not able to come up to speed rapidly, the phase winding will remain in the circuit too long, with resulting damage, or trip out of the safety protector, if one is provided.

The inertia of the load should always be considered when selecting motors. Split-phase motors should be used only on applications that are easy to start and can be brought up to speed rapidly; they are mainly used on fans and blowers. Because they have a low starting torque and draw a relatively high starting current, their use is usually limited to sizes of 1/3 hp and less.

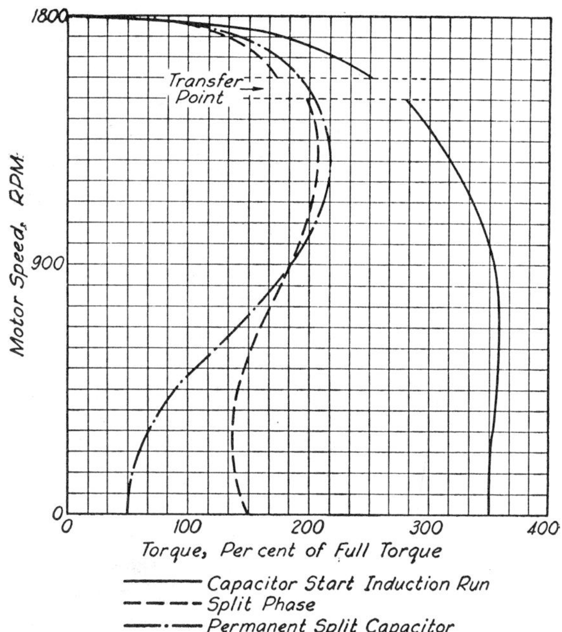

Fig. 15-18 Speed-torque curves of fractional horsepower, single-phase motors.

The type of motor thus far described is known as a *general purpose split-phase motor*. Another, called a *special service split-phase motor*, is sometimes used on heating and ventilating equipment, but is not ordinarily recommended for applications requiring frequent starting. Special service motors have running currents high enough to produce light flicker. They are normally used on washing machines, ironers, and equipment requiring infrequent starting.

General purpose split-phase motors of 1/3 hp and less are ideal for handling gas furnace blowers. The torque characteristics of these motors enable smooth and easy starting. Starting and accelerating noises are at a minimum as full speed is attained.

Capacitor Start, Induction Run Motors. Development of a reliable low-cost electrolytic capacitor made the capacitor start, induction run motor commercially successful. This type of motor, now widely used, embodies a capacitor in series with the phase winding. Displacement between currents in the main and phase windings is increased, permitting increased starting and pull-up torques. Usually, a centrifugal device opens a switch in the phase circuit at a predetermined speed to disconnect the starting winding and capacitor.

The speed-torque curves (Fig. 15-18) of an average fractional horsepower motor indicate the torque improvement possible with capacitor start motors. Because of their high starting torque, they are ideal for driving reciprocating compressors and other hard-starting loads. Their increased pull-up torque enables these motors to bring connected loads up to speed more rapidly than split-phase motors. Therefore, capacitor start motors are often used to drive heavy fans or blowers or equipment requiring additional torque for rapid acceleration.

All fractional horsepower motors of this type are designed for starting currents not exceeding those listed in Table 15-4.

Table 15-4 Locked Rotor Current of Fractional Horsepower (Single-Phase) Motors

Horsepower rating	Locked rotor current, amp	
	115 volts	230 volts
⅙ and smaller	20	10.0
¼	23	11.5
⅓	31	15.5
½	45	22.5
¾	61	30.5

Permanent Split Capacitor Motors. These motors have been widely used for many years on unit heaters and unit coolers; recently they have been used for a variety of other applications. Quantity production, along with improvements in design and manufacturing techniques, has made them relatively inexpensive. They are ideal for many fractional horsepower, direct-connected fan and blower usages.

Both the main and phase windings of these motors are designed to remain in the circuit at all times. Starting switches, governors, brushes, brush holders, etc., are thus eliminated, giving a simpler and more trouble-free motor design. Because the starting torque developed (Fig. 15-18) is necessarily low, usually about 40 to 50 per cent of full-load torque, these motors are not recommended for belted applications or other applications requiring considerable starting or pull-up torque.

The operating speed of these motors can be adjusted in several ways. The main windings may be in series or in parallel, giving two-speed operation with proper controls and connections. Tapped windings and the use of autotransformers or variacs are two additional means that permit multispeed operation. For satisfactory speed reduction, the motor should be fully loaded at its top speed.

A motor of this type is ideal for gas or steam unit heater applications for the following reasons: (1) high efficiency and high power factor; (2) smooth acceleration from standstill, eliminating the "noisy start" often encountered on split-phase motors; (3) elimination of the starting switch and other starting mechanisms.

Repulsion Start, Induction Run Motors. This is one of the most popular types of single-phase motors, with a very large field of application. Sizes from ⅙ to 10 hp are generally available.

This type, which starts as a repulsion motor and runs as an induction motor, has a distributed stator winding and a wound rotor similar to the armature of a d-c motor. The rotor winding is connected to a commutator. With the brushes short-circuited and set correctly, a very high starting torque can be developed. A centrifugal device connects all commutator bars together at a predetermined speed, thereby converting the armature into a squirrel cage rotor. This arrangement permits the high starting torque of the repulsion motor to be obtained, together with the constant-speed characteristics of the squirrel cage type at rated speed.

These motors have the highest starting efficiency of any single-phase type, and are capable of a higher starting torque than other single-phase types described—features that should be considered when the driven load is hard to start or when low voltage is likely.

Effect of Voltage Variation. The importance of operating an induction motor at or near design voltage is well known. Since its locked rotor torque and breakdown torque vary as the square of the applied voltage, difficulty due to low voltage may occur. Motor torque may be reduced to the point at which the motor will not start or operate the driven equipment properly. Table 15-5 outlines effects of both high and low voltage on induction motor performance.

Integral Horsepower Single-Phase Motors. Most motors larger than 5 hp on process equipment applications are polyphase. Single-phase motors are regularly used in sizes up to 5 hp when polyphase power is unavailable; these motors can be obtained in larger sizes. The two most popular types of integral horsepower single-phase motors are the capacitor start, induction run and the capacitor start, capacitor run.

Capacitor Start, Induction Run. These motors are built regularly in sizes up to 5 hp. Their construction and operating characteristics are basically the same as those of fractional horsepower units of this type described on page 15/19. In some sizes and speeds centrifugal switches are not satisfactory and voltage sensitive relays are employed.

The following torque (in per cent of full-load torque) may be expected from these motors: 250 to 300 per cent starting torque; 125 per cent pull-up torque; 220 to 235 per cent breakdown torque.

Speed-torque curves are fundamentally the same as for fractional horsepower motors.

Capacitor Start, Capacitor Run. Motors of this type, which use capacitors in both the starting and running circuits, have the following advantages: (1) higher efficiency and power factor; (2) lower full-load current; and (3) lower locked rotor current.

Most designs use electrolytic capacitors in the phase circuit in parallel with oil-filled capacitors. When a predetermined speed is reached, the electrolytic capacitors are removed from the line, but the oil-filled ones remain.

Table 15-5 Effects of High and Low Voltage on Induction Motor Performance

	Starting and max torque	Full-load speed	Full-load current	Starting current	Temp. change, full load	Full-load efficiency	Full-load power factor
110% rated voltage	Increases 21%	Increases 1%	Decreases 7%	Increases 10 to 12%	Decreases 3° or 4°C	Increases ½ to 1 percentage point	Decreases 3 points*
90% rated voltage	Decreases 19%	Decreases 1.5%	Increases 11%	Decreases 10 to 12%	Increases 6° or 7°C	Decreases 2 percentage points	Increases 1 point

* For example from 0.80 to 0.77.

In some sizes these motors are necessarily more expensive than those of the capacitor start, induction run type because of the addition of the oil-filled capacitors and the necessity of adjusting the motor windings.

INHERENT MOTOR PROTECTIVE DEVICES

Many motors, especially single-phase types, are protected against overheating by built-in devices that utilize a bimetallic disk to open the circuit when the motor overheats. Figure 15-19 is a diagram of a popular type of inherent motor protector and its connection to a single-phase motor circuit.

The protector, mounted where its ambient temperature bears a direct relationship to that of the motor, is connected in series with it. Protection is thus given against overloading, high ambient temperature, poor motor ventilation, stalled rotor conditions, etc. Both automatic and manual types of these protectors are available. On oil burners and stokers manual protectors are usually required. When such a protector breaks the circuit, the motor will not restart until the protector is reset. Since the motor is prevented from cycling, an investigation of the cause of shutdown can be made before the motor is restarted.

Most single-phase motors on heating equipment are equipped with protectors. Polyphase motors as large as 7½ hp may also be equipped with protectors, and then have the same type of protection as single-phase motors. Protectors prevent excessive burnouts if the motor becomes overloaded or is subjected to unusual ambient temperature.

The degree of motor protection or enclosure required varies widely. Many problems can be satisfactorily solved by use of standard, open, dripproof motors, constructed so that dripping water and falling particles cannot readily enter them.

Corrosive agents such as acids or alkalies, in the form of fumes, gases, etc., make a greater degree of motor protection necessary. Sometimes only an extra dip and bake on the motor windings is required, but often chemical service motors must be used. Such motors can be either totally enclosed, nonventilated ones or totally enclosed, fan cooled ones. Their materials are not easily attacked by corrosive agents; their frames, end bells, and terminal boxes are usually cast iron. Totally enclosed, fan cooled designs require bronze or plastic fans, which are not affected by the chemicals present. Totally enclosed motors offer the maximum degree of protection and should be used when dust, metallic particles, and liquid spray are present.

Applications* which require explosionproof equipment may need motors designed to prevent an explosion within the motor from igniting the surrounding explosive atmosphere. There is a practical limit to the size to which they can be economically constructed. Motors as large as 1250 hp, at 3600 rpm, or motors of approximately 900 hp, at 1800 rpm, have been built with Underwriters' Laboratories approval.

It is sometimes possible to separate a motor from a hazardous atmosphere by a fire wall. A totally enclosed motor may also be filled with an inert gas under slight pressure, to prevent entrance of any hazardous atmosphere. Heat losses then are removed by water thru a heat exchanger. Such construction, used regularly in very large motors in chemical plants, may be suitable for explosionproof service.

Explosionproof starters employing an air break are enclosed in cast iron boxes with wide flanges and several holddown bolts to provide the required flame pass, as covered in Underwriters' specification for explosionproof apparatus. These starters may have their apparatus submerged in oil and be housed in steel tanks of approved design with sealed conduit connections.

This type of control equipment is far more expensive than standard controls. Whenever possible, it is located outside an explosive atmosphere or in a special space ventilated with uncontaminated air.

In instances when a motor is to be installed in an area that is hazardous only under abnormal and infrequent conditions, standard, open, squirrel cage motors are often used without any special enclosure because of the large resultant saving. Motor location should be carefully checked for this possibility. Wound rotor, synchronous, and d-c motors have slip rings or commutators that are a constant source of sparking. Special enclosures around the slip rings must be provided for wound rotor and synchronous motors, and complete enclosures for d-c motors.

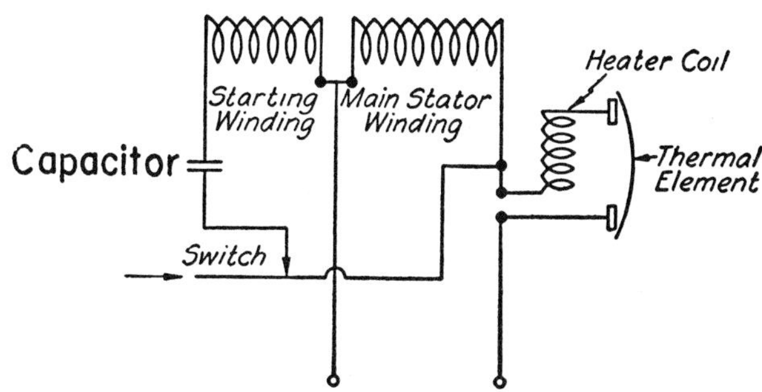

Fig. 15-19 Diagram of single-phase motor circuit with inherent protector.

When motors must be installed in outdoor locations where they will be subjected to very severe weather, totally enclosed, fan cooled motors and weather resistant control equipment may be used. Outdoor applications often can be handled satisfactorily with splashproof motors that give adequate protection against weather. These motors are designed so that drops of liquid or solid particles falling on the motor, or traveling toward it in a straight line at any angle not greater than 100 degrees from vertical, cannot enter. Since the splashproof design is cheaper than a totally enclosed, fan cooled motor, the possibility of the use of the former for an outdoor application should always be studied.

Careful attention should be given to adequate ventilation of any area in which a motor is installed. High ambient temperatures often prevail in small rooms or cabinets, owing to heat losses from the motors. These areas should always be well ventilated with clean, safe air.

It is sometimes advantageous to construct a motor so that the heated air discharged from it can be exhausted to the outside thru ducts. In some cases, motors are constructed so that air is taken in and exhausted at a remote point.

* For additional information see Chap. 5, "Special Occupancies," *National Electrical Code*, current edition.

FRAME SIZES

Motor mounting dimensions have been standardized. All motors of a given NEMA frame size are interchangeable as to bolt holes, shaft diameter, height, length, and certain other dimensions. Conforming manufacturers have established frames for motors up to 125 hp, 1800 rpm, open construction, built in the NEMA 445U frame.

This standardization is particularly helpful when many motors are installed in a plant and a replacement stock is maintained. Integral horsepower motors have recently been rerated and new NEMA frame sizes worked out.* Most motor manufacturers offer motors in the new rerated frames. These rerated motors are smaller and lighter in weight because of the more effective use of material and improved ventilating methods. Former standards for temperature rise, service factor, and allowable voltage variation have been maintained.

BEARINGS

Antifriction bearings are rapidly becoming the accepted standard on small and medium-size motors of average speeds. When the peripheral velocity increases (as with large-shaft diameters and high operating speeds), sleeve bearings are usually recommended for good performance and quiet operation. Bearings and couplings must be carefully selected so as to prevent imposing excessive thrust on the motor bearings. This selection is often critical in certain pump applications, where the pump impeller might place an excessive thrust load on the bearings. When large high-speed motors with sleeve bearings are used on pump applications, precautions should be taken to ensure proper field performance.

Ball bearings will handle more thrust load than sleeve bearings. It is advisable to check with the motor manufacturer whenever such a load is imposed, to make sure that the bearings are not excessively overloaded.

REVERSAL OF SHAFT ROTATION[4]

When starting new commercial equipment or replacing a motor, the shaft may rotate the wrong way. Repulsion-induction motors may be reversed by shifting the location of their brushes. To reverse a three-phase motor, any one of the three line wires may be interchanged until proper rotation is attained. Direction of rotation on a split-phase motor is reversed by switching the wires that provide current for either its running or starting winding. A capacitor start, induction run motor has its rotation reversed in a similar manner.

DIRECT-CURRENT MOTORS

Direct-current motors and direct-current generators are structurally the same. A given unit may be used as either motor or generator, although its characteristics will be slightly better for one method of operation.

Except for fractional horsepower sizes, starting of d-c motors requires a rheostat in the armature circuit, since the armature resistance is inherently low. The rheostat is usually built into a starting box.

* See latest edition of NEMA Standard on Motors and Generators.

Shunt Motors

Shunt motors, the most common d-c motors, give nearly constant speed (within 5 per cent) from no-load to full-load. Speeds of commercial motors may be increased nearly 25 per cent by field circuit (rheostat) control; for small motors 50 per cent speed range is available. Figure 15-20 shows characteristic curves.

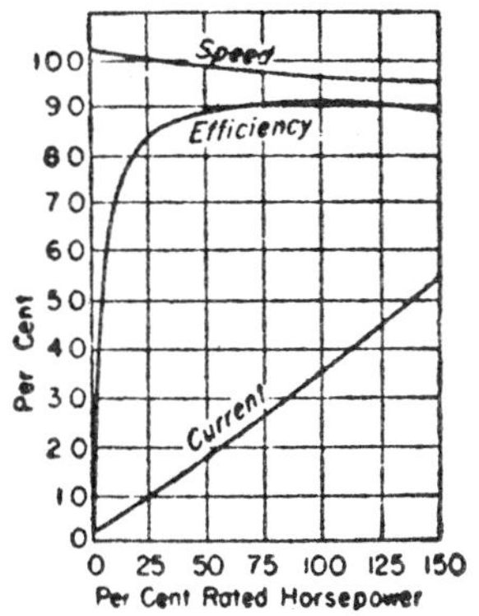

Fig. 15-20 Typical characteristics of shunt motor.

Adjustable speed shunt motors are standard for 2 to 1, 3 to 1, and 4 to 1 speed ranges, at higher prices. For greater speed range, means other than a field rheostat are required. One means is to shift the armature axially, giving a flux change thru it. In a shunt motor the commutating poles are offset from the center line of the main poles so that, with the weak main field and hence greater armature reaction, the effect of the commutating poles is increased, giving good commutation over a large speed range. One of the best arrangements for increased speed range and good speed regulation provides independent voltage supplies for the field and armature. With constant field voltage the motor speed will vary directly with the voltage across the armature. This arrangement requires a d-c generator for each motor, a common practice for applications requiring wide speed ranges, such as in rolling mills and elevators.

Armature circuit resistance may be used to reduce the motor speed with the load. The losses are relatively large, the regulation poor, and the speed decrease at light load negligible. For a given speed decrease, the rheostat loss and the input loss are in the same proportion. Potentiometer control is convenient for reducing speed from small motors when the accompanying losses are not excessive.

Series Motors

Series motors are used for loads requiring heavy starting torque when variable speed is permissible. The armature current is also the field current; hence, the torque is approximately proportional to the current squared. The speed varies nearly inversely as the load current. Because of its excessive speed at no-load, a series motor should not be belt connected.

Compound Motors

Characteristics of shunt and series motors are combined in compound wound motors. The starting torque is good, with speed regulation halfway between that of shunt and series motors. Compound motors are often used with a flywheel for heavy duty of short duration, as on a punch press.

To reverse the direction of rotation of a d-c motor, it is necessary to reverse the current thru the armature circuit with respect to the field circuits. Commutating poles are a part of the armature circuit and should always be connected to the armature in the same way. Any series-field turns must be connected so that current thru them is not reversed with respect to that in the shunt field.

TIMING MOTORS[5]

There are three basic types of timing motors: a-c, d-c, and stepper. The stepper is an a-c motor in design and a d-c motor in function. The available power supply generally determines the choice of motor, but devices that permit converting direct current to alternating current allow selection to be based on the merit of a design to the application in question.

Alternating-Current Type

Alternating-current timing motors, the most widely used timing motors, are of two kinds: hysteresis and synchronous induction.

Hysteresis Motor. This is a synchronous motor without salient poles and without d-c excitation. It may be two-phase and therefore reversible, or shaded-pole single-phase and therefore unidirectional. Starting torque is approximately 100 per cent of full-load torque.

These motors have been built in sizes varying from as large as 1/4 hp to as small as 1 1/2 μhp, and operate at frequencies from 25 to 2000 cps. A motor designed to operate at a specific frequency should not be used at any other frequency.

Hysteresis motors are useful for repeat-cycle timers and elapsed-time indicators.

Synchronous Induction Motor. This uses a permanent-magnet (usually barium ferrite) rather than an electromagnetic rotor, and may be a shaded-pole, split- or two-phase type. Most of these motors operate over a frequency range of from less than 1 to 400 cps, and have the advantage of a relatively high torque.

If these motors are operated at a subsynchronous speed, they will stall, a characteristic useful in designing system trouble warnings.

Direct-Current Type

Direct-current timing motors are reversible, have high starting torque, and can have their speed varied by variation of the applied voltage. High-inertia loads do not cause starting difficulties for these motors. The efficiency of these motors ranges from 30 to 60 per cent compared with as low as 5 per cent for some synchronous types. Some type of governor must be used on d-c motors when speed tolerances may not exceed 10 per cent.

Two common types of d-c timing motors are the printed circuit motor and the transistorized d-c motor.

Printed Circuit Motor. This consists of a plastic disk with the conductor photoetched directly on it. The brushes bear directly on the conductors, making a separate commutator unnecessary. The winding is limited by the number of effective turns that can be printed. Advantages of this type of motor include low armature inertia, low armature reactance, smooth torque, and high-temperature operation.

Transistorized D-C Motor. This is a motor without commutator and brushes. It consists of a permanent-magnet rotor between two field coils. When current is applied, a circuit is completed thru one of its coils, starting the rotor. In the starting phase both coils act as running coils and the rotor receives four impulses per revolution. When the rotor reaches a speed of approximately 1000 rpm, centrifugal action on the starter mechanism causes the contact pins to move inward toward the rotor axis and out of contact with the brushes. The transistors now operate. One becomes conducting and nonconducting in phase with the rotation of the rotor; the other is biased nonconducting when the first conducts. As a result, a series of pulses in phase with the rotor is applied to the running coil.

Stepper Motors

Stepper motors respond to random d-c input pulses as well as to alternating current. Primarily pulse counting devices, they are used, in sophisticated applications, to position switches, synchros, and potentiometers. These motors are easily reversible and therefore useful when rapid reset is required, but this application precludes the use of normal reset clutching mechanisms because of the inherent shock or vibration.

Use of the stepper motor is practical today for timing control. In many instances these motors can replace both a-c and d-c timing motors.

Stepper motors are adaptable to a varying frequency. Synchronous motors, on the other hand, are suitable only for a single, fixed frequency.

REFERENCES

1. Natl. Elec. Manufacturers Assoc. *Motors and Generators.* (MG1–1963) New York, 1963.
2. "Electric Motors Reference Issue." *Machine Design* 36, No. 7, March 19, 1964.
3. "Motors and Motor Protection." (In: Am. Soc. of Heating, Refrigerating and Air-Cond. Engineers. *ASHRAE Guide and Data Book 1963,* Chap. 49. New York, 1963).
4. Hastings, G. T. "Electric Motors Used in Gas Heating." *Gas Heat* 4: 34+, Oct. 1953.
5. "Electric Motor Book Issue." *Machine Design* 33: Sec. 2, Dec. 21, 1961.

Chapter 4
Electronic Equipment

by Sigward A. Stavnes

ELECTRONIC EMISSION

Electronics may be defined as a study of the motion of electrons in a field of force. Electronic devices are many and varied; mainly vacuum tubes and gaseous tubes are discussed here:

Four general methods of liberating electrons from surfaces are:

1. *Field emission.* Sometimes called high field emission, this refers to the liberation of electrons from a surface by the attraction of a very intense electric field of such a polarity as to make the surface negative to the space around it. Both a high voltage and a sharply pointed surface are required.
2. *Photoelectric emission.* This applies to those electrons emitted from a metallic surface on which electromagnetic radiation falls.
3. *Secondary emission.* This is the emission of electrons from a surface bombarded by a beam of fast-moving particles. It occurs when a particle with high kinetic energy is suddenly stopped by a surface.
4. *Thermionic emission.* This is the liberation of free electrons from a hot surface, depending on its material and temperature.

Thermionic Cathode Materials

A surface acting as a source of electrons is called a *cathode.* Cathode types are (1) pure tungsten, (2) thoriated tungsten, and (3) oxide-coated.

A *pure tungsten* cathode is used in large vacuum tubes in which bombardment is likely by positive ions. Of high melting point, such a cathode can be operated at about 2500 K without damage. Filaments are sturdy and long-lived. Tungsten requires considerable heat energy for a given amount of emission.

Thoriated tungsten or tungsten with a small amount of imbedded thorium oxide is used in many medium-size tubes. The oxide alters the nature of the surface so that less energy is required by electrons for their escape. Thoriated tungsten emits more electrons than pure tungsten, other conditions being the same. Since thoriated tungsten may be damaged by positive ion bombardment, it is not used in tubes with very high voltage gradients.

Oxide-coated cathodes, used mostly on small tubes, consist of a layer of barium oxide or strontium oxide deposited on a core, usually of a nickel alloy. These cathodes furnish a large supply of electrons at a fairly low temperature but may be destroyed by positive ion bombardment.

Cathode Structure

The cathode must be heated to its proper temperature to liberate free electrons. Directly or indirectly heated cathodes are in use.

A *directly heated cathode* or filament is a resistance that reaches its proper temperature under the recommended voltage. The filament is the electron emitter. A directly heated filament is well suited to d-c operation. An a-c filament may develop a temperature fluctuation that causes emission to fluctuate.

An *indirectly heated cathode* is formed as an oxide-coated cylinder. A resistance wire is inside but insulated from the cylinder, which it heats. Either alternating or direct current may be used. The coating acts as a source for the free electrons.

DIODES AND THEIR APPLICATION

A diode is a tube with an electron source and a collector to which free electrons can travel. The tube consists of an evacuated envelope containing a *cathode* and a *collector* electrode called the *anode,* or *plate.* The electrons emitted by the heated cathode move toward the plate, causing a current flow. If the potentials of the cathode and plate are reversed and the plate acquires a negative potential, electrons are no longer attracted to the plate and the current flow stops. As a circuit element, a diode behaves like an automatic switch that opens whenever the plate potential becomes negative relative to the cathode.

Diode Characteristics

Diode current is a function of heater temperature and potential difference between the anode and the cathode. Cathode current follows **Child's law,** $i = \alpha V^{3/2}$, over a wide range. The limiting current depends on the cathode temperature.

Both the positive voltage of the plate and neighboring electrons act on the electrons leaving the cathode so that an electron concentration termed *space charge* occurs around the cathode. This charge tends to repel other electrons escaping from the cathode surface, thus controlling their flow. An increase in accelerating voltage causes electron current to increase until the space charge becomes intense enough to limit the current again.

Use of Diodes as Rectifiers

A rectifier is a device that transforms an alternating current into a direct current. *Crystals, semiconductors, nonlinear resistors,* and *diodes* are such devices.

Figure 15-21a shows a diode used as a full-wave rectifier. Equal resistances R_1 and R_2 form a low-resistance voltage divider. When A is positive to B, the plate of tube T_1 is positive to its cathode and that of T_2 is negative to its cathode. Tube T_1 conducts and T_2 does not; current flows thru the load, R_L, from T_1. When B is positive to A, tube T_2 conducts and T_1 is cut off; current flows thru the load, R_L, from T_2. In either case, current flows thru the load in the same direction.

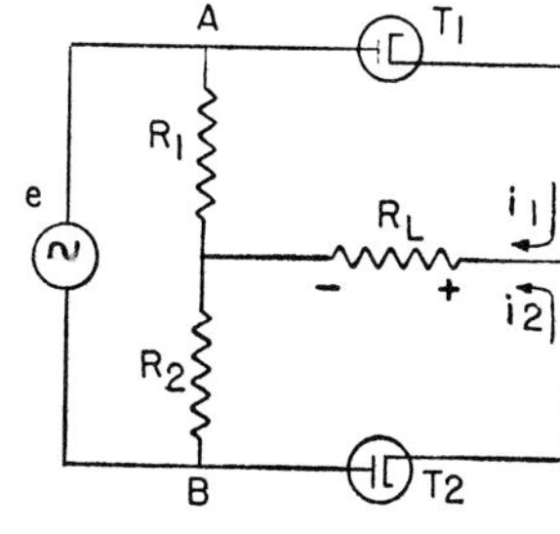

Fig. 15-21a Full-wave rectifier.

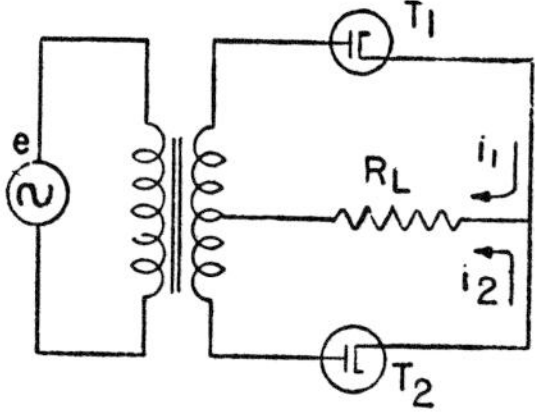

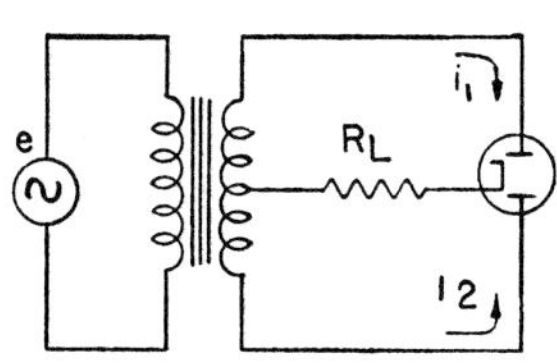

Fig. 15-21b Practical full-wave rectifier.

Figure 15-21b shows a practical full-wave rectifier circuit. A transformer used for flexibility also acts as a voltage divider. T_1 and T_2 can be contained in the same envelope (as shown at the right in Fig. 15-21b) when the two plates have a common cathode. Double diodes as shown save one envelope and the power to one heater. They are applicable to low-voltage rectifiers, in which no extreme potential difference exists at any time between the two plates.

Filters

Undesirable pulsations, or ripple, in a rectified current can be reduced by *smoothing filters.* The effect of ripple is generally expressed by a *ripple factor,* which is defined as the ratio of the effective value of the a-c components of voltage (or current) to the average voltage (or current), both at the load. The larger the factor, the larger is the departure from pure direct current.

Smoothing filters are used in most electronic circuits, ordinarily between the rectifier and the load. Figure 15-22 shows three types. The *capacitor input filter* supplies a higher d-c voltage to the load than does the *choke input* type with the same rectifier, but requires high peak currents from the rectifier.

If more filtering is needed, several sections of filters can be used in succession. Because inductors are expensive, resistors may be used in some cases.

In some industrial controls, the filter consists of a section known as an *RC filter,* shown in (c) of Fig. 15-22. The capacitor energy increases as the voltage rises; the energy is fed back to the circuit as the voltage falls. The capacitor thus tends to maintain a constant voltage across the load.

Voltage Doublers

A transformer is used to step the voltage up or down before rectification. A rectified voltage approximately double that of a conventional rectifier may be obtained by connecting a circuit as shown at the right in Fig. 15-23. The circuit is then called a voltage doubler because the potential difference between A and B is twice the peak voltage of the a-c source.

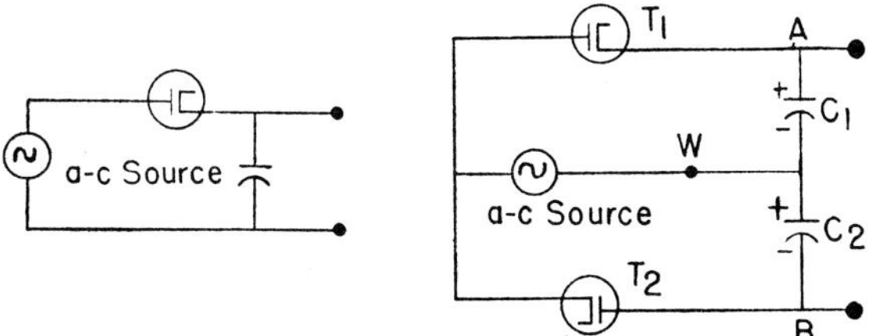

Fig. 15-23 (left) Conventional rectifier and (right) voltage doubler circuits.

TRIODES

Space charge in the region of the cathode plays an important role in determining the current thru a tube. In diodes a large change in the anode-to-cathode potential is required to change this field distribution appreciably. This field distribution can be changed, however, by insertion of a third electrode in the region close to the cathode, thus making a three-element tube, or a triode. Open-mesh wire is used as the electrode, which is generally called a *grid.* The grid in the electron stream absorbs very few electrons because it is kept negative to the cathode.

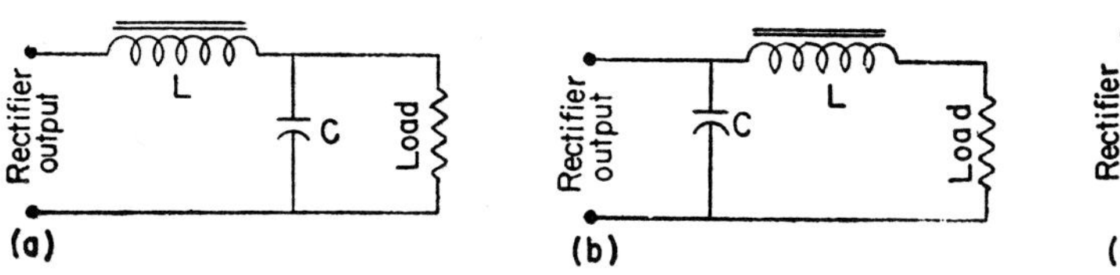

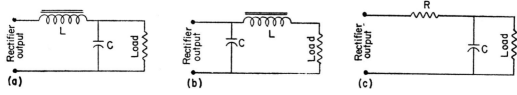

Fig. 15-22 Types of filters: (a) choke input (L section); (b) capacitor input; (c) *RC*.

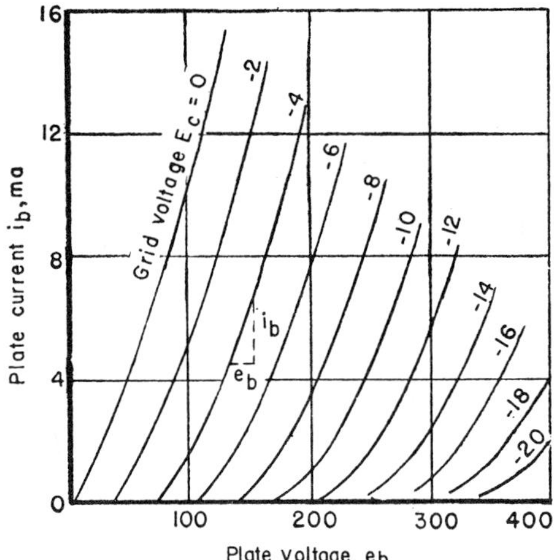

Fig. 15-24 Plate characteristics of a triode.

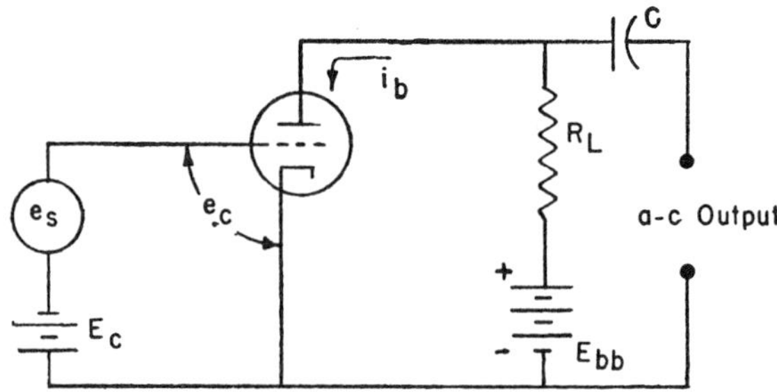

Fig. 15-25 Triode amplifier for a-c signals.

Static Characteristic Curves

Plate current in a triode may be varied by varying the plate voltage and holding the grid voltage constant, or by varying the grid voltage and holding the plate voltage constant.

Plate characteristic curves (Fig. 15-24) are obtained by plotting plate current, i_b, against plate voltage, e_b, for various constant grid voltages. For each plate voltage there is a grid voltage, called the *cutoff voltage*, that reduces plate current to a negligible value.

Amplifiers

The triode is most commonly used as an amplifier of a-c signals.

In the setup shown in Fig. 15-25, the alternating voltage from the signal source alternately makes the grid more and then less negative. Assuming that the grid voltage, e_c, is always negative and never reaches cutoff, the signal shifts the operating point continuously along the load line. The plate current, alternately increasing and decreasing, is fluctuating and unidirectional and may be considered as alternating current superimposed on direct current.

The ratio of output voltage to signal voltage is the *amplifier gain*. It can be found by dividing either their peak or effective values.

MULTIELECTRODE TUBES

Tubes with multiple electrodes, e.g., *tetrodes* and *pentodes*, are available; their characteristics are given in tube manuals. Many other types of tubes exist and specialized tubes have been developed, such as twin triodes or twin diode-triodes. Consult tube manuals for their characteristics and manufacturers' catalogs for specific uses.

MULTISTAGE AMPLIFIERS

Signal inputs to electronic amplifiers often are a few millivolts or less. A gain of 10^6 may be required to obtain a usable voltage from them. If the output of a one-tube amplifier is used as a signal for a second amplifier, the gain is much increased, giving the effect of a two-stage amplifier. Generally, three or four stages in cascade are the practical limit for stable operation.

OSCILLATORS

An important function of tubes is to generate alternating voltages from direct voltage.

Electronic oscillators are used for induction and dielectric heating, instrumentation devices, communications, and in cases where rotating a-c generators are not suitable. Electronic oscillators have no moving parts, are light and small, and offer a large range and easy adjustment of frequency. Their amplitude of oscillation may be controlled by the addition of energy at the proper time. If this amplitude is constant, the energy added just equals that lost.

Figure 15-26 shows common types of oscillator circuits.

Whether a circuit is oscillating may not always be obvious. An oscilloscope connected across a part of the oscillator a-c circuit will show whether oscillation is occurring. If the oscillation is in the audible range, a head set loudspeaker connected across an a-c circuit element should give a rough indication of the frequency.

GAS-FILLED TUBES

In manufacturing high-vacuum tubes, as much air as possible is removed from the envelope. If too much air remains, tube characteristics change considerably. Some low-vacuum tubes may be designed to contain a gas instead of air. The characteristics of these tubes are very different.

Figure 15-27 shows voltage-current characteristics with the voltage applied to the plate of a gas-filled thermionic diode. In the voltage range 0b, the current is space-charge limited. At b, it suddenly increases greatly with a drastic change in the type of conduction. With this sudden increase in current, the tube is said to "fire." After firing, it has a nearly constant voltage across it.

Operating Precautions

Gas tubes, which should be used more carefully than vacuum tubes, must have enough external circuit resistance to limit currents to safe values. Heater current must be maintained close to its recommended value. If too high, the heater may burn out; if too low, a voltage increase may occur with possible cathode damage.

Normal cathode emission should be reached before the plate voltage is applied; otherwise the increased voltage drop across the tube results in cathode damage. A delay of from 15 to 60 sec is required between the application of heater supply voltage and the energizing of the plate circuit. Some

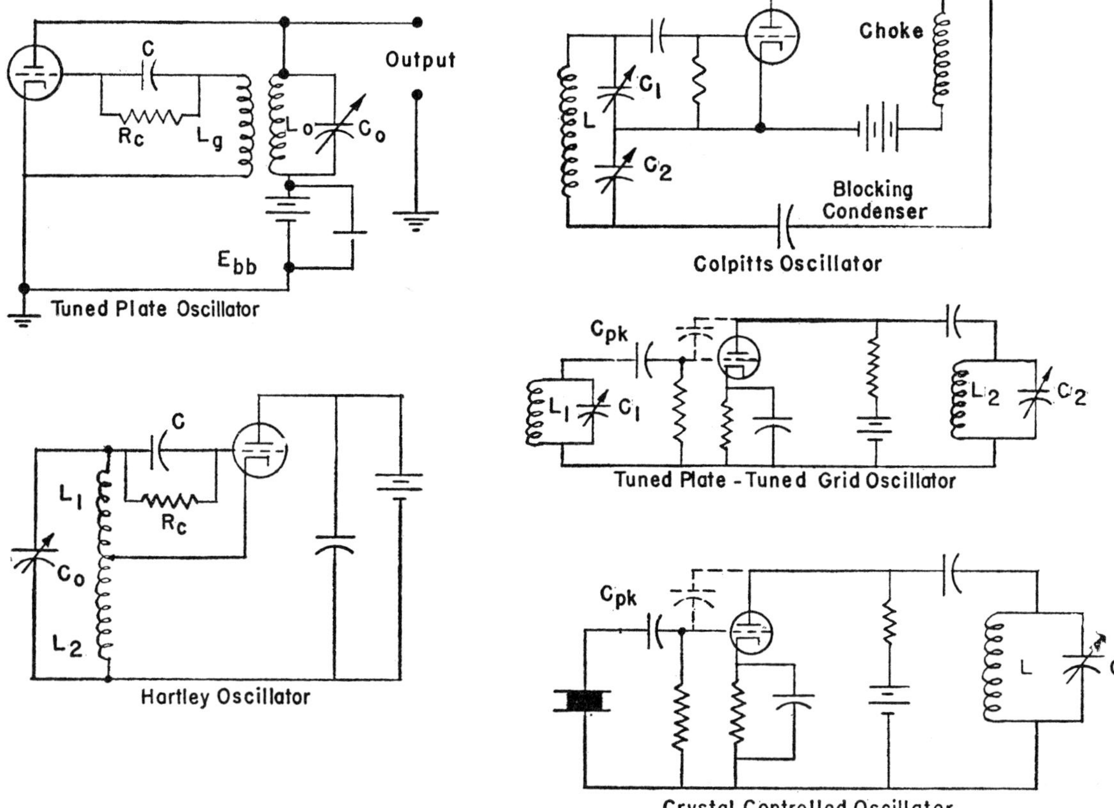

Fig. 15-26 Common types of oscillator circuits.

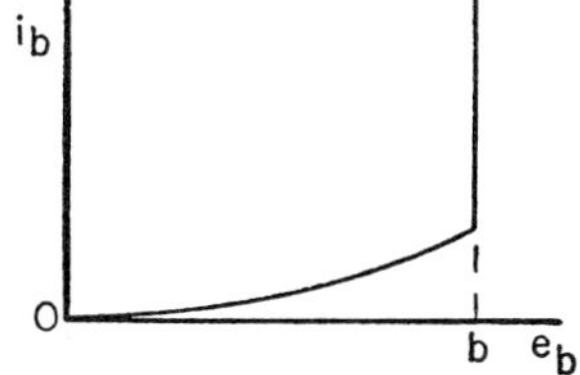

Fig. 15-27 Current-voltage characteristics of a thermionic gas-filled diode.

gas tubes may need a longer delay. Heater voltage should not be removed while the plate supply is still connected.

Cathodes of Gas-Filled Tubes

Compared with the cathodes of high-vacuum tubes, thermionic cathodes of gas tubes are generally large physically. Utilizing the large current capacity of gas tubes requires high cathode emission. In oxide-coated cathodes of spiral ribbon form, the adjacent layers act as heat shields and reduce the heat lost by radiation, thus increasing the emission efficiency.

Heat-shielded cathodes are found in gas tubes whose current is not space-charge limited after the tube has fired. Some gas tubes have cold cathodes from which emission is obtained by secondary action. Firing (breakdown) potentials of such tubes, much higher than those of thermionic cathode tubes, depend on the kind of gas in the tube and its pressure. Cold cathode or glow tubes are used as light sources, voltage regulators, rectifiers, protective devices, etc.

The gas diode acts as a very efficient switch of low resistance and high current carrying capacity. Thus it is a very efficient rectifier. A gas tube may be considered a low-voltage, high-current device as compared with a vacuum tube, which is a high-voltage, low-current device.

Voltage Regulator Tubes

A definite voltage is necessary for firing a cold-cathode diode. After firing, the tube current varies over a considerable range, with potential nearly constant. A nearly constant voltage can then be maintained across a load resistor in parallel. A diode of this type, designed to stabilize voltage, is called a voltage regulator; the voltage across the tube may be held close to the rated voltage when the tube current is within a certain range.

Gas-Filled Triodes

Gas-filled triodes are known as *thyratrons*. Their operation is shown by a firing curve rather than the conventional static characteristics of vacuum tubes. Such a curve is obtained by plotting plate voltage against the critical grid voltage just sufficient to prevent firing.

Mercury vapor tubes have firing curves that vary with the temperature of the gas in the tube.

Regulated Power Supplies

Rectified power supplies for most high-grade instruments are regulated to prevent fluctuations in the operating points of tubes. A combination of vacuum and gas tubes is generally employed.

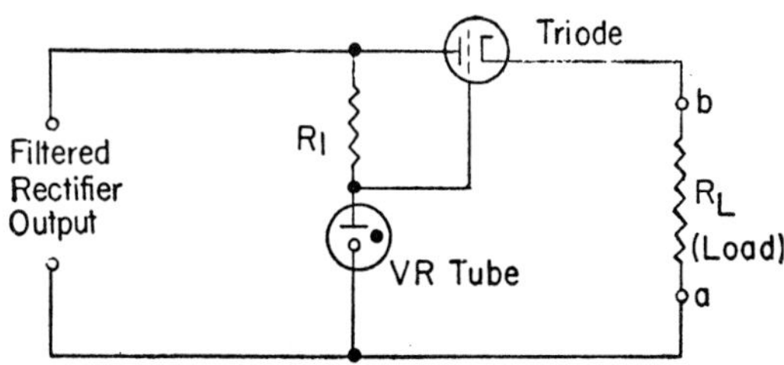

Fig. 15-28 Simple stabilized power supply.

Figure 15-28 shows a triode in series with a load to maintain a constant or nearly constant voltage across the load resistor, despite variations in rectifier voltage or load resistance. A triode is used as the variable resistor, its resistance being varied with a change in plate voltage or grid voltage. Proper design of components can make the increase in tube plate voltage almost equal to the rise of rectifier voltage, load voltage remaining fairly constant.

PHOTOELECTRIC DEVICES

Photoelectric cells are classified according to how their output is made available in the circuit. Cell sensitivity is usually given in terms of current per unit of radiant power striking the element surface, generally as microamperes per microwatt of incident power at a specified color or wave length of light. Radiant energy wave length must be specified, since emission efficiency changes with wave length. Tubes may be more or less sensitive to certain wave lengths, depending on their cathode material.

Photoemissive Cells

This type of cell consists of a cathode, coated with a light-sensitive material such as cesium oxide, and an anode. The cathode is usually a semicylindrical surface, receiving incident light on its concave face. The anode consists of a straight wire, wire ring, or metal plate mounted so as to cast no shadow on the cathode. The anode and cathode are assembled in a glass or quartz bulb. A light beam striking the cathode surface imparts enough energy to some electrons in the cathode to allow them to escape and be attracted to the positive plate. A small voltage applied between the cathode and the anode causes a small current to flow. As voltage increases, current rises very rapidly to saturation.

Introduction of a small amount of gas into the envelope of a photoemissive cell allows ionization to occur as electrons move from cathode to anode, the current increasing with the increase in anode voltage. Current must be limited, since it may become so large that positive-ion bombardment may damage the cathode. The presence of ionized gas molecules raises the current for a given amount of light. The increase in sensitivity of a gas-filled cell over that of a vacuum cell is called *gas amplification*. Attempts to increase it more than tenfold usually result in ionization so intense that a glow discharge is visible.

Photovoltaic Cells

These cells generate an emf when radiant energy strikes their surface. This emf is applied to an external circuit, and the current flow from it may be made directly proportional to the incident light energy. Cells consist of oxides, sulfides, etc. coated on metal plates. They need not be vacuum enclosed. and they may be liquid or dry. Such cells are commonly employed in light meters used in photography. Photovoltaic cells, either wet or dry, act as batteries that take light energy and transform it into electrical energy. They are sensitive to temperature and lose their sensitivity permanently if operated above 55 C.

Owing to physical construction, photovolatic cells have a fairly high internal capacity, which acts as a shunt with the cell and effectively short-circuits any alternating current developed by the use of modulated light. Even at a 60-cycle variation of the light, cell response is less than half the output for a steady light source.

Photoconductive Cells

Electrical resistance of these cells varies as a function of the light energy received. The cell, consisting of a thin coating of selenium between two electrodes on an insulating plate, acts as a variable resistance in an electrical circuit. Cell resistance changes slowly, showing as a time lag between change in light and change in current. To avoid erratic dark current and time lag, photoemissive and photovoltaic cells are usually preferred for commercial devices.

INDUSTRIAL ELECTRONIC CIRCUITS

Elementary Diagrams

An understanding of the fundamental tube actions and the operation of basic circuits is a necessary preliminary to a knowledge of electronic circuits. The basic circuits previously discussed are only a few of the many encountered. A large part of a complex industrial electronic circuit often consists of a combination or variation of the basic circuits.

A circuit as presented in a wiring diagram is usually difficult to study. The first step is to resolve it into its elementary form. For simplification, tube-heaters are omitted unless the heater (filament) circuit is part of the control circuit.

For the elementary circuit a sheet about twice the size needed for the diagram should be used. Circuits should be drawn in orderly sequence. Input to the electronic circuit usually is at the left and power input at the bottom right. Starting at the left with the input signal, each circuit is added in turn as the control signal is amplified or converted. All components should be properly labeled. Their values are often clues to their physical action.

One or more redrawings may be necessary for a clear elementary diagram for study. Manufacturers of electronic equipment generally include in their instructions clear elementary drawings for the most efficient servicing and maintenance.

Signal Tracing

An instruction book or description of the circuit operation, if available, should be studied. Otherwise some idea is usually available of what the input signal will be and what output results are desired. It is usually best to start at the signal input. If it becomes difficult to continue, a stop may be made at the last point where operation is clear. Then starting at the

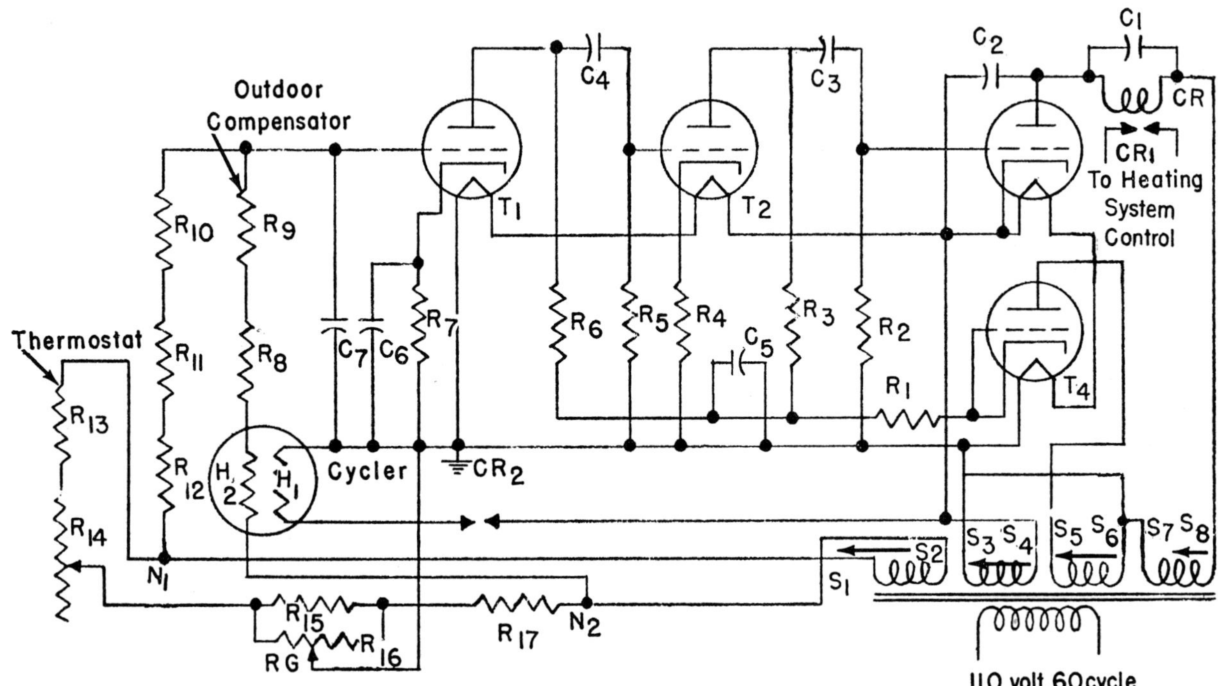

Fig. 15-29a Electronic circuit for control of heating system.

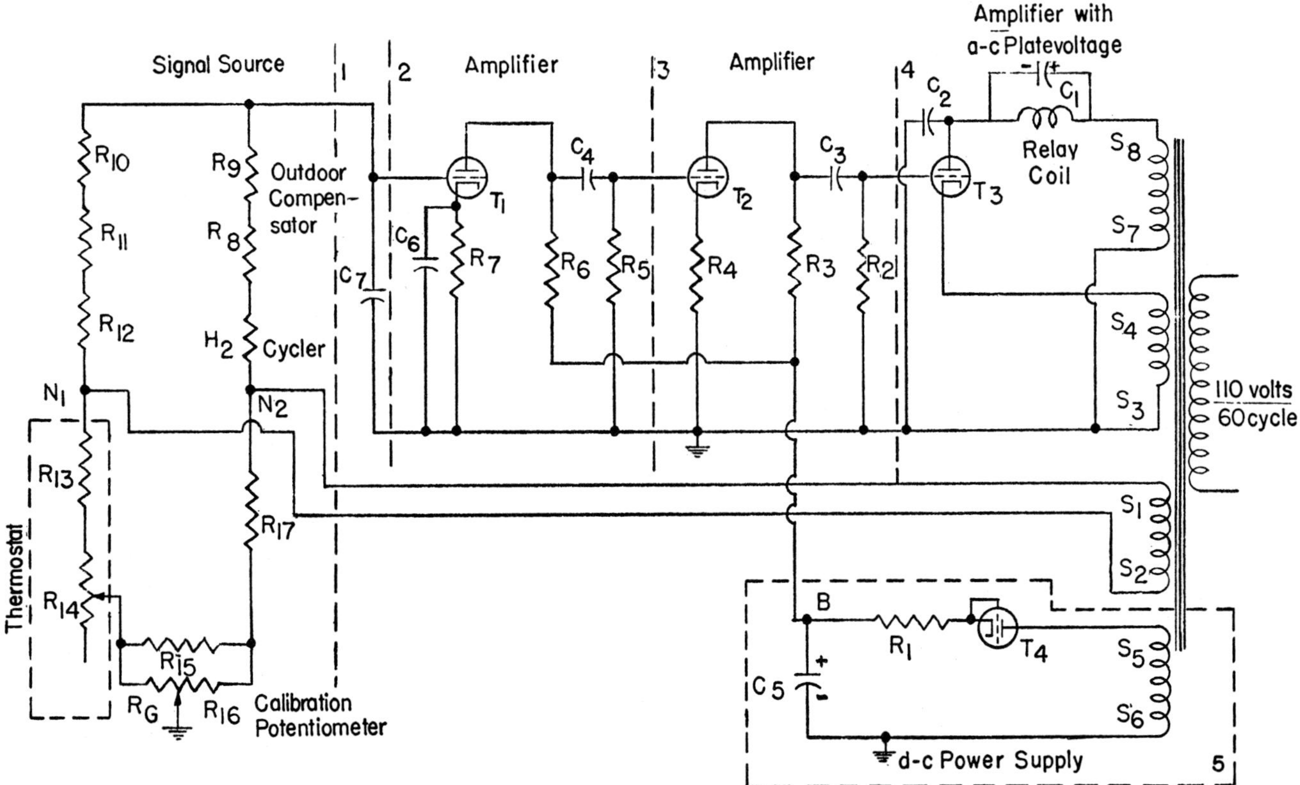

Fig. 15-29b Separation of basic circuits of electronic heating system control of Fig. 15-29a.

output, the circuit may be back-traced and the unknown portion narrowed down to solution.

Generally the complete elementary diagram can be divided into a number of basic circuits in series and parallel combinations. Each may be represented by a block, and the blocks connected to show the path of the signal and the effect of each basic circuit on it. Another plan is to separate each basic circuit by dotted lines, with the function of each circuit written between them. The first, or *block diagram,* method shows at a glance what is happening to the signal as it travels from input to output. The second method is more detailed, permitting determination of the function of a single component.

The Honeywell Moduflow circuit shown in Fig. 15-29a may be used as an example of signal tracing by both methods.

The signal input appears obscure in this circuit for controlling temperature in a building. To govern the heat input, the signal must be supplied by some sort of thermostat. Inspection shows R_{13} and R_{14} in series combination, labeled "Thermostat." When the temperature of a resistance changes, the resistance must change. Therefore the change in resistance of resistors R_{13} and R_{14} that is due to temperature is the signal source.

The basic circuit then is determined in which R_{13} and R_{14} are components. The thermostat is drawn in on the left side of Fig. 15-29b, at N_1 downward to the dot next to the wiper arm of R_{14}. Resistors R_{12}, R_{11}, and R_{10} are then drawn from N_1 to the junction dot below "Signal Source." Downward from that dot, resistors R_9, R_8 and H_2 are sketched in, all in series from the dot to junction N_2. The leads from one of the secondary windings of the transformer are seen to connect across the junctions N_1 and N_2, so that S_1 is connected to N_2 and S_2 to N_1. From N_2, R_{17} is drawn and then the parallel combination of resistor R_{15} and potentiometer R_{16}. The section to the left of dotted line 1 now appears as a Wheatstone bridge, with a source of emf supplied by the secondary winding S_1–S_2. If balanced, zero voltage appears between the junction dot below "Signal Source" and ground. If a change occurs in any of the bridge resistors, an imbalance results, and a voltage appears between that junction and ground.

Next are sketched the grid, plate, and cathode of T_1, connecting the grid to the junction below "Signal Source." The cathode is connected to R_7 and C_6 in parallel. R_7 and C_6 are recognized as a method used for obtaining grid bias, termed *self-bias.* From grid to ground there is a capacitor which is not the normal input to a tube. These components are drawn in with the plate (anode) of T_1 connected to the grid of T_2, thru coupling capacitor C_4, and to R_6. The cathode of T_2 is connected to ground thru R_4. The grid, besides being connected to T_1, is connected to ground thru R_5. Then R_4 and R_5 are drawn in. The plate of T_2 is connected to T_3, thru capacitor C_3, and to R_3. Next, C_3, R_3 and T_3 are drawn in. The grid of T_3 is also connected to ground thru R_2.

It will now be seen that tube T_1, with its associated circuits between dotted lines 2 and 3, is an amplifier. Tube T_2, with its associated circuits between dotted lines 3 and 4, is a similar amplifier. The two amplifiers differ in their cathode circuits. The cathode resistor of tube T_2 has no by-pass capacitor; as a result, degeneration, which acts to stabilize the amplifier, occurs.

The cathode of T_3 is connected to lead S_4 of a secondary winding of the transformer. The other end of the winding, lead S_3, is connected to ground. This secondary winding S_3–S_4 is the emf supplied to the filaments. As now sketched, the bias of T_3 is the a-c voltage feeding the filaments. The plate of T_3 is connected to a relay coil in parallel with capacitor C_1. The other end of the coil and capacitor combination is connected to secondary winding S_8 of the transformer; side S_7 is connected to ground.

Inspection of tube T_3 and its associated circuits shows that the bias is an a-c voltage from secondary winding S_3–S_4 and that the voltage supplied to the plate of the tube is from secondary winding S_7–S_8. Cathode and plate voltages with respect to ground are seen to be in phase. An a-c plate supply means that the tube conducts only during the positive half of the cycle. When the plate voltage becomes positive, the tube conducts, energizing the relay coil and charging capacitor C_1. When the plate voltage is negative, the tube will not conduct, and the tube acts like an open switch. When the tube is not conducting, capacitor C_1 partially discharges thru the relay coil, keeping the relay energized until the plate again becomes positive. If the grid signal is still of proper value, C_1 will be charged to a value that will keep the relay energized on the negative excursion of the plate voltage.

A plate voltage has not yet been supplied to tubes T_1 and T_2. Plate resistors R_6 and R_3 of T_1 and T_2, respectively, connect to capacitor C_5 and resistor R_1. C_5 connects to ground and R_1 connects to the cathode of T_4. Secondary winding S_5–S_6 is connected across the plate of T_4 and ground. This basic circuit shown enclosed is the familiar half-wave rectifier.

Fig. 15-30 Block diagram for circuit of Fig. 15-29a.

To construct a block diagram, the bridge may be called the source of signal and represented by the extreme left block of Fig. 15-30. The signal then feeds thru the two stages of amplification into the amplifier having an a-c plate voltage. The proper signal will energize a relay in the plate circuit of the last amplifier.

Returning to Fig. 15-29b, it is known that the cathode and plate voltages are in phase. The signal to the grid of T_3 will have the same frequency but may or may not have the same phase as the plate voltage. If the signal to the grid of T_3 is large enough and in phase with the plate voltage, the relay will be energized. If out of phase, the tube is cut off on the positive excursion of the plate voltage and will not conduct on the negative excursion. The conclusion must be drawn that if T_3 conducts only on an in-phase signal, then the signal from the bridge (signal source) must be unbalanced in the manner that yields an in-phase signal at the grid of T_3.

Working backward from T_3 and knowing that its grid voltage and plate voltage must be in phase for tube conduction, the input to T_2 is found to be 180° out of phase with the output of T_2 owing to phase reversal thru the tube. The output of T_2 is the input to T_3. The input to T_1 must be 180° out of phase with its output (the input to T_2). Therefore, T_1 must have an input in phase with the plate voltage of T_3 if the relay is to be energized.

Chapter 5

Electric and Thermoelectric Gas Appliance Controls and Unconventional Electrical Systems

by Raymond P. Flagg, R. W. Pashby, C. E. Swanson, and F. E. Vandaveer

Ignition devices (Table 12-28), automatic valves (Fig. 12-37), and relays are examples of controls for domestic gas appliances that are often operated on alternating current.

The most important d-c controls for domestic gas appliances are of the thermoelectric type.

Major control devices not dependent on electricity for operation are thermostats of the type used on ranges and water heaters, gas ignition pilots, gas pressure regulators, relief valves, draft controls, and manual control valves.

THERMOELECTRIC CONTROLS FOR DOMESTIC GAS APPLIANCES

Thermoelectricity was first used in the gas industry to operate automatic pilot devices to prevent the flow of unburned gas. Such use is claimed to date back nearly to 1900. Its most significant advance was the development of the Baso safety pilot in 1934.

Operation. Heating a thermocouple by a gas pilot flame generates electricity, which energizes an automatic pilot device (Fig. 12-28) that in turn permits gas to flow to the main burner. If the pilot is extinguished, the current ceases and the automatic pilot device will shut off. Table 6-15 gives temperature-millivolt relations for various thermocouples.

Materials. Requirements of thermocouples or thermopiles for automatic pilots limit usable thermoelectric materials to those with the following properties: ability to generate sufficient thermoelectric power; ability to operate satisfactorily at temperatures encountered; power output stability throughout life; reproducible characteristics; and ease of fabrication.

Currently used materials capable of withstanding the temperatures reached in gas appliances have efficiencies of one per cent or less under ideal conditions. Conversion of heat into electrical energy depends on the thermoelectric power of the couple, the thermal and electrical conductivities of its materials, and the temperature difference between its hot and cold junctions.

Since thermocouples and thermopiles are usually listed as being capable of withstanding hot junction temperatures between 1100 and 1700 F in furnace applications, satisfactory performance at high ambient temperatures is important (see also Table 6-16). Under such conditions thermoelectric materials must have:

1. Resistance to excessive oxidizing or reducing atmospheres.
2. Resistance to furnace gases and fumes.
3. Resistance to preferential oxidation and/or reduction of alloying elements.
4. A melting point above the operating temperature.

A number of thermoelectric materials have high thermoelectric power. Some are difficult to fabricate with uniform thermoelectric output, and thus lack reproducibility. Others with high thermoelectric power are difficult to bond together and may have high junction resistances, which reduce power. Still others have relatively low melting points. Taken together, these requirements eliminate all but a few thermoelectric materials from use with gas pilots. Typical materials used for positive elements are Chromel P and stainless steel. Constantan is usually used for the negative element.

Thermocouple-Operated Systems

Since the power output of a thermocouple is limited, it is used primarily in automatic pilots.

Thermocouples. Figure 12-26 shows the basic design of a common type of thermocouple. Various models differ primarily in size and mounting features. Since the hot junction reaches temperatures of 1700 F, the exterior shell is frequently made of stainless steel. The constantan round wire in the center is the negative element. It is heliarc welded to the stainless steel cylinder at the hot junction. The cylinder both gives corrosion-resistant protection for the constantan and forms the positive element. The cold junctions are at the junction of the stainless steel to the brass mounting rod and at that of the constantan to the copper wire. Temperatures of these junctions are seldom identical.

The voltage produced is approximately 75 per cent of that produced by a Chromel P–constantan couple with similar hot and cold junction temperatures. On the other hand, the Chromel P–constantan couple must have a protective covering because both elements lack corrosion resistance in the flame, and a more complex assembly would be necessary. The heat loss due to such a protective shell would somewhat reduce the increased voltage obtainable from the Chromel P–constantan couple. Thus, the stainless steel–constantan couple has distinct advantages and is more commonly used.

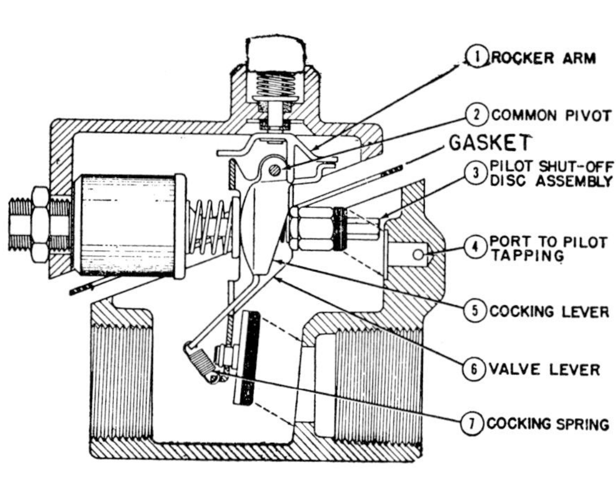

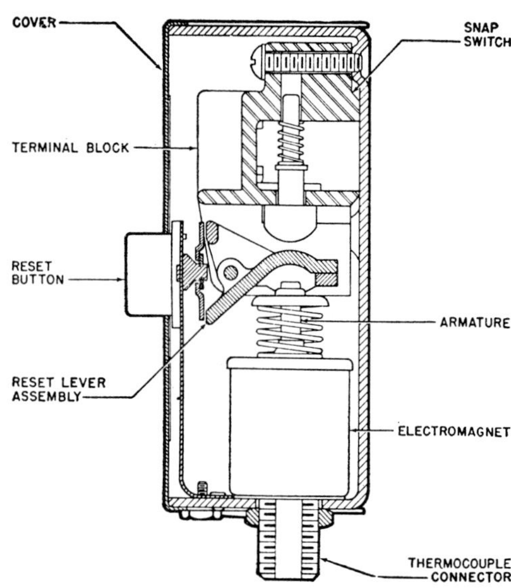

Fig. 15-31a (left) Valve-type automatic pilot device. Reset button is shown on top.
Fig. 15-31b (right) Electric switch-type automatic pilot device.

Some increase in power may be obtained by increasing the distance between hot and cold junctions, but at a sacrifice in safety shutdown timing. The spacing shown in Fig. 12-26 is a compromise between maximum power output under normal operating conditions and fast equalization of hot and cold junction temperatures on pilot flame failure for fast safety shutdown.

A principal disadvantage of the thermocouple is its low power output. Heavier leads, which decrease electrical resistance, permit improvement. The main objections to them are reduced flexibility and higher cost.

Automatic Pilot Devices. Electromagnetic units usually employed with thermocouples are the *valve type* and the *switch type*.

Figure 15-31a shows a valve type. With a normal pilot flame, the power generated by an appropriate pilot burner–thermocouple combination holds the valve disk open, permitting gas flow to the main burner and pilot. If the pilot becomes extinguished, the thermocouple output drops off and the valve disk snaps shut, shutting off all gas flow. To relight the pilot, a reset button must first be depressed. After the thermocouple has been heated, its output will be sufficient to "hold in" the pilotstat. The reset button may then be released. An automatic safe-light feature incorporates an integral means of blocking main burner flow during relighting of the pilot.

Figure 15-31b shows a manual reset electric switch model. It interrupts the electrical circuit to prevent opening of the control valve if the pilot flame becomes extinguished. It incorporates a safe-light feature whereby the switch is mechanically held open when the reset button is depressed during lighting of the pilot. On release, the switch closes the electrical circuit only if the pilot flame condition is normal.

Thermopile-Operated Systems

Self-C‿nerating Electric Control Systems. A system of this type (Fig. 12-19c) consists essentially of a thermopile-pilot burner, a main gas control valve, and a thermostat. The thermopile-pilot burner performs the dual function of igniting main burner gas and of supplying power to operate the main gas control valve, which is also controlled by the thermostat. Sufficient power may be generated to permit incorporation of a heat anticipation feature within the thermostat if desired.

When a self-generating electric control system is used, the main gas control valve closes if the pilot flame fails, thus eliminating the need for a switch model automatic pilot device. Since this type of system does not depend on outside power, failure of such power does not cause shutdown. On the other hand, long electrical leads to the thermostat may cause excessive resistance, so that a slight reduction in pilot flame results in generation of insufficient voltage for system operation.

Thermopiles and Pilot Burners. Power necessary to operate the system shown in Fig. 12-19c is obtained from a thermopile, which differs from the thermocouple previously discussed. Large in size for greater power, the thermopile is composed of numerous thermocouples attached in series for increased voltage, which is desirable to reduce line loss and the effects of contact resistance. Even a small increase in voltage may reduce contact resistance appreciably in thermostats and limit switches. The elements are Chromel P–constantan wire. Since neither substance has satisfactory life when exposed to a gas flame, the thermopile must be encased in a protective shell of stainless steel. Element insulation is accomplished by mechanical spacing with magnesium oxide, mica, or ceramic separators. The thermopile shown has a glass fill; its pilot burner is of the secondary aerated type.

Valves. Power available for operation of the main burner control valve is about 10 milliwatts or less. Consequently, a diaphragm-type valve is utilized, operated by a pilot valve that in turn is controlled by a sensitive relay. Such a valve is shown in Fig. 12-39; it operates when the pilot burner flame is burning properly and the thermostat calls for heat.

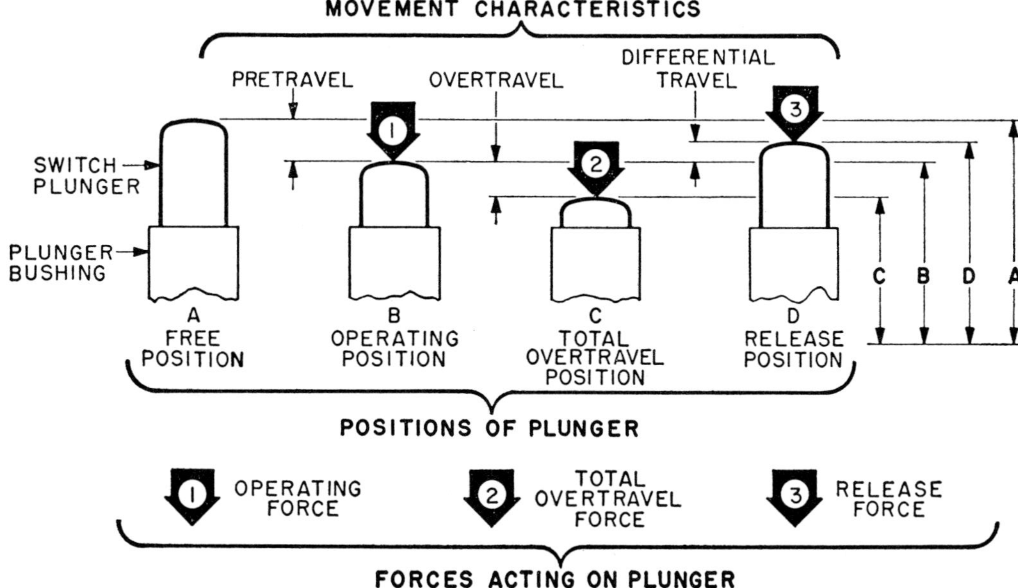

Fig. 15-32a Movement characteristics and descriptive terms of precision snap-acting switches.

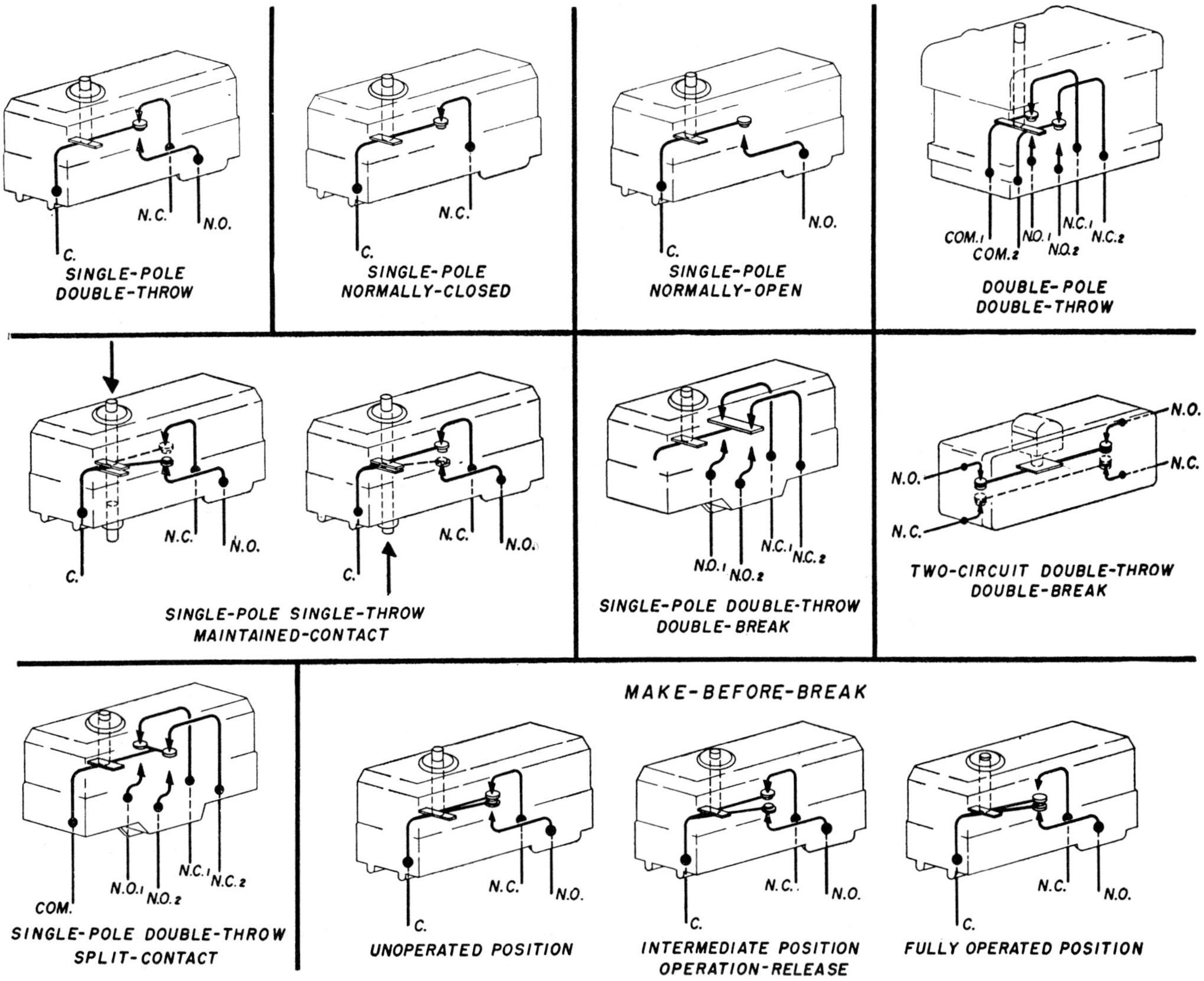

Fig. 15-32b Circuit variations of precision snap-acting switches.

ELECTRIC SWITCHES

Precision Snap-Acting Switches

The term "precision snap-acting switch" is applied to small switches with snap mechanisms that operate with low differential travel and differential force. Their air gap or contact separation may be as small as 0.006 in. or as large as ⅛ in. For their size, the switches have high electrical ratings, up to 20 amp at 480 volts ac and 2 hp at 230 volts ac.

Characteristics. The simple design of precision snap-acting switches has led to their wide application when long life and accurately repeated characteristics are required. Figure 15-32a illustrates the terms associated with these characteristics. One of the most important is *movement differential.* In precision snap-acting switches, movement differential is very low, usually 0.0002 to 0.005 in. but sometimes higher. Another important feature is the high degree of *repeat accuracy,* approximately ±0.001 in; on special types of light electrical loads, ±0.0001 in. is achieved. Other features responsible for the popularity of this type switch are its *small size, dependability,* and *high electrical rating.* Most such switches are listed by Underwriters' Laboratories.

Electrical circuits are becoming more and more complex, and their requirements can no longer be met completely by simple on-off switches. Figure 15-32b illustrates the variety of circuits available in precision snap-acting switches.

Actuators. In precision snap-acting switches, travel of the switch spring beyond the point at which the switches are snapped or operated is generally limited to about 0.005 to 0.030 in. Overtravel mechanisms are used for a greater motion after snap and when the operating force is high enough to damage the switch. Some mechanisms are built into spring-loaded plungers so that, after the switch snaps, the plunger can be depressed 1⁄16 to ¼ in. farther without harming the switch. Similar results are achieved by the use of external leaf springs. When a switch must be operated with its plunger perpendicular to a sliding cam, a leaf spring may be used between these parts to prevent excessive side thrust on the plunger or damage to it. Alternatively, a roller lever or a roller-leaf combination may be used.

Enclosures. Enclosures are used around precision snap-acting switches to prevent mechanical damage; to keep dirt, oil, and moisture away; and to protect terminals. Enclosures are usually aluminum or zinc die castings. Some are sealed with gaskets, O-rings, and flexible diaphragms to prevent entry of oil or moisture. Special housings for hazardous locations are available.

Uses. Precision snap-acting switches are widely used in instruments, controls, machine tools, aircraft, electrical and electronic military devices, and vending and packaging machines. Major uses in the heating, ventilating, and gas industries are in thermostats and controls of all kinds, in instruments, and on remotely controlled valves.

Means by which a switch may be thermostatically actuated include: (1) gas- or vapor-filled bellows or wafers; (2) expansion of a liquid either in a bellows or in a bulb connected to it thru capillary tubing; (3) a bimetal; and (4) a rod and tube combination of materials having different coefficients of expansion.

Thousands of different types of precision snap-acting switches are manufactured. They have been extensively used in combination control instruments employed in large power plants and industrial heating installations. Explosionproof types have also been used in oil and gas pipelines as limit switches, interlocks, and indicators, and on hand-operated, pneumatically operated, and motor-operated valves.

Mercury Switches

Mercury switches (Fig. 15-33) are simple and widely used (e.g., in room thermostats). They consist of two or more electrodes, a glass or metal enclosure called a tube, and mercury as the contact-making medium. In their manufacture, the tubes are evacuated and filled with hydrogen before sealing. Some mercury switches have electrodes that contact the mercury, the electrical circuit being interrupted between them and the mercury. Others make mercury-to-mercury contact; the arc is broken over a ceramic barrier for the highest current ratings and across the glass enclosure for lower ratings. Metal-enclosed mercury switches may use the enclosure as one of the electrodes.

Fig. 15-33 Typical mercury switch (SPST type).

Mercury switches are rated up to 50 amp (occasionally higher) at 120 volts ac.

Simplicity is the basis for the low cost of the mercury switch and for its ease of application. It is inherently a low-force switch, requiring for its operation only a force that tilts it sufficiently to make the mercury roll from one end of the tube to the other. Because electrical contact is made and broken within a hermetically sealed enclosure, mercury switches are vaporproof, waterproof, and oilproof when supplied with suitable leads. They can be used in moist and corrosive atmospheres without fear of failure owing to surrounding conditions. Most mercury switches are designed so that the mercury rolls rapidly, accomplishing rapid make and break of the electrical circuit. Protection for glass enclosures can be provided by a line of metal enclosures into which the glass tubes are sealed with a potting compound.

Minimum temperature for standard mercury switches is −35 F, but special winterized units can be used below −65 F. Some types can be used at temperatures as high as 300 F. Most mercury switches operate with a differential angle of 3 to 12°. Some low angle designs require as little as a 0.25° change in angle for their operation.

CONTROLS FOR COMMERCIAL AND INDUSTRIAL GAS BURNERS

Systems for commercial and industrial gas burner control may consist of a number of different devices, each providing a portion of the overall operation desired. No single control is capable of providing all required functions. A complete control system should provide the following:

1. Means of starting and stopping the burner (operating control).

2. Checks to ensure that regulating devices are in proper position and other conditions are satisfactory prior to burner start-up and during its firing cycle (interlocks).

3. Starting of the burner in a safe sequence and flame supervision during operation (programming flame safeguard controls).

4. Regulation of the burner firing rate (firing rate or combustion controls).

5. Protection against excessive pressures, temperatures, or low water conditions (limit controls).

6. Instantaneous on-off control of pilot and main fuel supply (valves).

7. Regulation of furnace draft for proper combustion (draft regulators).

8. Prevention of electrical overloads from endangering or damaging the burner or wiring (electrical overload releases and fuses).

9. Regulation of the boiler water level on steam boilers (water level controls).

The extent to which a control system will include all these functions will depend upon the burner size and type as well as on local regulations.

Individual Controls and Functions

Operating Controls. Such a control is an automatic temperature- or pressure-sensing device designed to start the burner when heat is demanded and to stop it when the demand has been satisfied. This control may also be a start-stop station that allows the operator to manually start and stop the burner when desired.

Burner Interlocks. Burner interlock controls are designed to prove that conditions for combustion are established and the burner is ready to be fired; and that the burner can continue to be fired safely.

Interlocks consist of electrical switches which indicate that a satisfactory starting or operating condition exists. Starting interlocks may be provided to indicate the availability of adequate gas or combustion air pressure. They may also be used to indicate that the burner firing rate positioner is in the low-fire position or that the uptake damper is in the start burner position. Running interlocks indicate that fuel and air pressures are maintained within operating limits. Selection of interlocks depends largely upon burner size and method of operation required.

Programming Flame Safeguards. These consist of a programming control and a flame detector. The programming control initiates the safe start-up of the burner when heat is demanded, providing power to the burner motor (if used), pilot valve, ignition, main fuel valve, and firing rate controls. A thermally operated safety switch is an integral part. Its function is to limit the trial-for-ignition period on starting. In the event of an ignition failure on starting or flame failure during the running cycle, the safety switch trips out and must be manually reset prior to a restart. Some programming controls may also provide for preignition and postignition purges and a timed ignition cycle.

The flame detector proving pilot and/or main flame is interlocked with the programming control and continues its normal sequence. On the running cycle, in the event of flame failure, the main fuel supply is immediately cut off and safety shutdown results. Provision may also be made on some controls for switching from the pilot-sensing device to the main flame-sensing device at the end of the trial-for-ignition cycle. Several different sequences are available; selection depends on burner type and size, sequencing requirements, and needs of the installation. The following typical sequences may be used:

Basic Sequence for Atmospheric Natural Draft Gas Burners (Fig. 15-34)

1. When heat is demanded, the pilot gas valves and the ignition transformer are energized simultaneously.
2. On proof of pilot, the main gas valve is permitted to open and the ignition transformer is de-energized.
3. When heat demand is satisfied, the pilot and main gas valve are de-energized.
4. If the pilot is not proved within the trial-for-ignition period, safety shutdown results and a manual reset is required before the burner can be restarted.

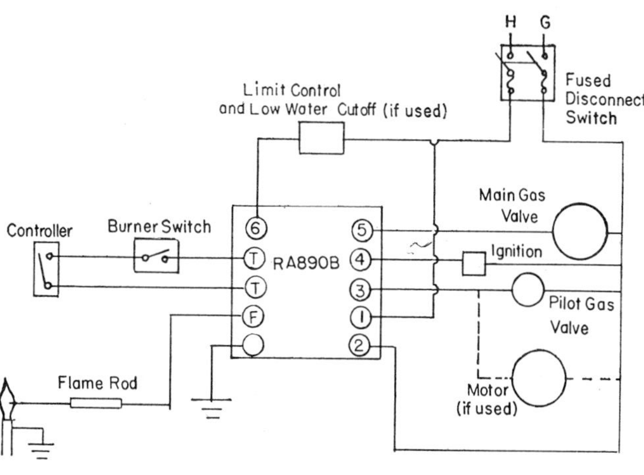

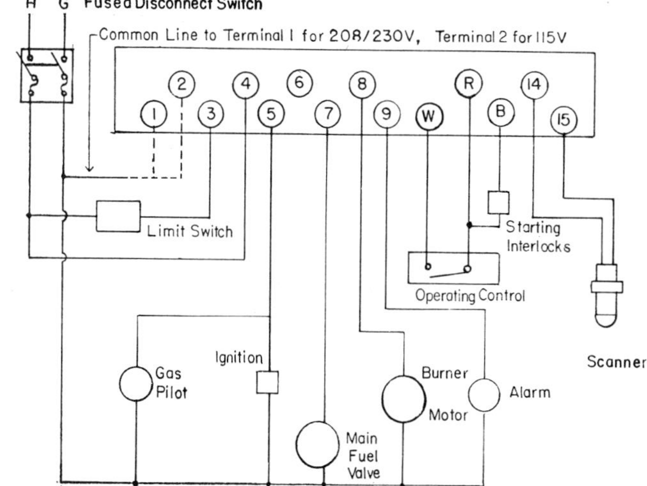

Fig. 15-34 (left) Typical control system wiring diagram for an atmospheric-type burner (natural or forced draft). Pilot is established and proved before main fuel valve can be energized.

Fig. 15-35 (right) Typical control system wiring diagram for a forced draft burner requiring a preignition purge, postignition purge, and timed ignition, in addition to proving the pilot before the main fuel valve can be opened.

5. In the event of flame failure, the main gas valve is immediately cut off and the ignition is returned to re-establish the gas pilot.

6. If the pilot is not re-established, the pilot and ignition are de-energized on safety shutdown. A manual reset is then required to restart the burner.

Alternate Sequence for Forced Draft Burners (Figs. 15-34 and 15-35)

1. If a preignition purge is required, the burner motor is energized 15 to 30 sec before the ignition and pilot valve can be energized. After the firing cycle, the burner motor continues to run for 15 sec to provide a postignition purge (Fig. 15-35).

2. Without a preignition purge, the burner motor may be started at the same time as the ignition and pilot valve.

Foregoing items 2 thru 6 apply.

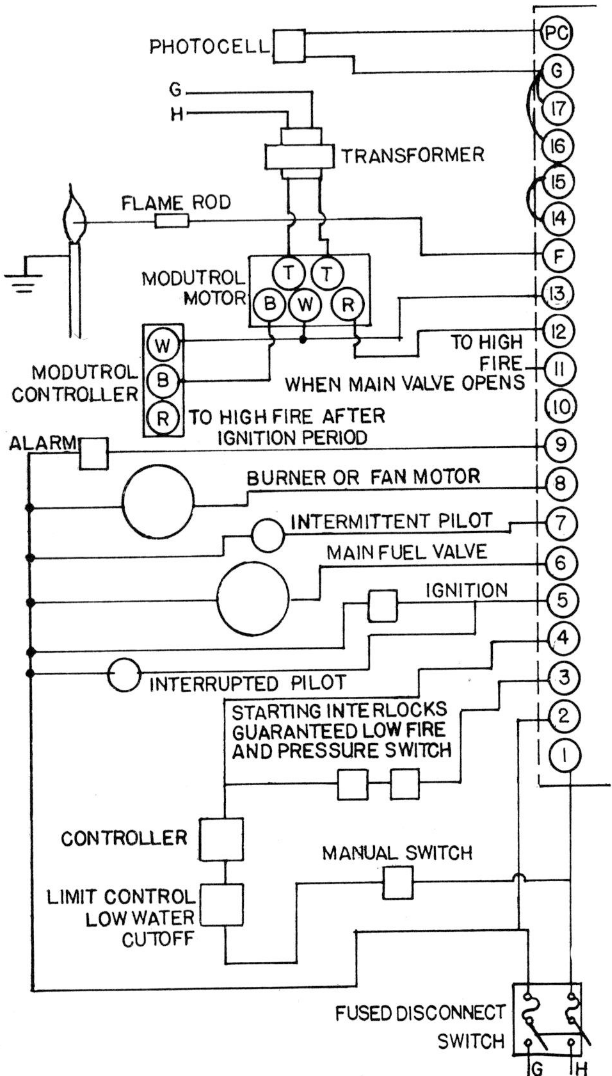

Fig. 15-36 Typical wiring diagram for a burner control system with preignition purge, postignition purge, and ignition timing, as well as a modulating firing rate control with a guaranteed low-fire start. The pilot is proved by a flame rod, the main flame by a photoconductive cell. Jumpers: G-17, photocell test; G-16, 30-sec preignition purge (no jumper for 5-sec preignition purge); 14–15, 15-sec ignition (no jumper for 60-sec ignition).

Complete Sequence for Induced or Forced Draft Mechanical Gas Burners (Fig. 15-36)

1. When heat is demanded, the draft control is energized, driving the damper to the starting position.

2. Burner motor and/or induced draft fan are energized for preignition purge.

3. Following the preignition purge cycle, the pilot valve and ignition are energized. After the pilot has been established and proved, the main fuel valve can be opened.

4. The firing rate control may go to the high fire at the same time the main fuel valve opens or at the end of the ignition timing.

5. At the end of the ignition period, the ignition is de-energized. The pilot valve may or may not be de-energized, as required.

6. At the end of the ignition timing period, flame detection is transferred to main flame proving only.

7. On a call for less heat, the fuel valves are de-energized and the firing rate control starts immediately to low-fire. Burner and/or draft motor continue until the end of the postignition purge.

8. In the event of failure to prove the pilot, the main fuel valve cannot be opened, and safety shutdown results.

9. In the event of main flame ignition failure or failure during the operating cycle, the main fuel valve is de-energized and the programming flame safeguard shuts down on safety, followed by a postignition purge period.

Flame Detection. Available means of flame detection based on response timing may be divided into two categories, thermal and electronic.

In gas-fired watertube boiler furnaces, rated at 10,000 lb of steam per hr and up, the time interval between loss of flame and stopping of fuel must not exceed 2 sec.[1] In smaller units of this type and in other types of gas-fired boilers having inputs over 400 MBtu per hr, the response time of the primary safety control must be not more than 4 sec.[2] A pilot flame establishing period of 15 sec is specified for expanding, intermittent, and interrupted pilots.

Thermal Means. Flame-sensing by this means involves proving the pilot only thru a bimetallic element or a thermocouple. Neither is normally used in conjunction with a programming-type relay. The switching mechanism is normally in series with the main fuel valve for a constant-pilot application. Response of the bimetal or thermocouple is relatively slow, since reaction to a change in the flame condition usually requires a minimum of 30 sec. Both prove the pilot only at the source and not at the point where it is capable of safely igniting the main burner.

The thermocouple type of flame-sensing utilizes two dissimilar metals. The hot junction is immersed in the pilot flame. The difference between hot and cold junction temperatures generates a current sufficient to hold in or, in some cases, pull in an electric switch or valve that is connected with the main gas supply valve. The bimetal sensing element type is an assembly on which the pilot flame impinges. The mechanism either establishes an electrical contact in series with the main gas supply or operates a mechanical valve in this supply line.

Electronic Means. An electronic flame safeguard causes closing of the main gas safety shutoff valve within 2 to 4 sec after flame failure. Some safeguards operate in less than one second; however, for elimination of nuisance shutdowns, longer periods are usual. Incorporated in the flame safeguard panel is a limited trial-for-ignition period, anywhere from 15 to 45 sec. Certain governing agencies require the time limit to be set at not more than 15 sec. At the end of the trial period, the control locks out, leaving the unit in an off position; in some cases, the control repeats the light off procedure before locking out. After lockout, the unit must be manually reset before it can be used.

Electronic flame-sensing utilizes either the electrical characteristics of the flame or its spectral radiation as a signal source.[24] Response to flame presence or absence occurs within 4 sec. Flame-sensing units employed are classified by the means of flame-proving. One is the *flame rod,* which depends on current-carrying ability and can be used to check a gas pilot and/or the main flame. Flame-sensing devices, rods or scanners, are normally located so that they sense the junction between the pilot and main flame, thus proving the pilot at a point where it is capable of igniting the main flame. The advantage of a flame-sensing device, therefore, is its ability to pinpont the location of the flame being proved.

Flame-proving may also be accomplished by means of photocells. Two types now available may be used to supervise the gas pilot and/or the main flame. Both respond to the invisible radiant energy from the flame.

A *lead sulfide photocell* responds (reduced resistance to current flow) to the invisible radiant energy from the infra-red portion of the spectrum present in yellow and blue flames. Flame pulsations (generally between 5 and 25 cps) cause the resistance of the lead sulfide element to change. If such pulsations are within certain frequency limits, the flame safeguard system will respond, indicating flame presence. Since hot refractory also gives off infra-red radiations, it too must be eliminated from the range of the sensor. With air motion, a shimmer can result that may cause either an overriding effect or an afterburner hold-in.

An *ultraviolet photocell* responds only to the invisible radiant energy in the ultraviolet portion of the spectrum. Such energy striking the photocell results in a pulsating d-c signal that can be used with the rectification systems. Ultraviolet photocells are not affected by an adjacent hot refractory because they do not respond to its portion of the spectrum.

Both types of photocells view an area but cannot distinguish depth. For that reason care should be taken in their application to see that they will prove a pilot capable of igniting the main flame. Either type of cell may be used to prove the main oil flame on a combination gas-oil burner.

Pilot Turndown and Hot Refractory Tests. A pilot turndown test is essential on every application to ensure that the flame-sensing means used (including location) will always prove a pilot capable of safe main burner ignition. To conduct this test, the pilot should be turned down to the point at which it just barely holds in the flame relay of the programming control. The remaining pilot then must provide safe light-off of the burner.

One additional test should be performed whenever a photocell is used, to make sure that it does not respond improperly to shimmer from a hot refractory. After the refractories have been brought to the maximum firing temperature, the fuel supply should be turned off. The lead sulfide cell should then immediately detect the flame failure. Any shimmer due to a hot refractory should neither cause the flame-sensing means to be held in improperly if flame failure occurs nor override the effect of a signal from a pilot. If either condition exists, suitable changes should be made to correct it.

Firing Rate on Combustion Controls. Many commercial and industrial burners have a means of matching the firing rate to the load demand. This involves changing from high or low fire or modulating between them as the pressure or temperature of the sensing element of the control builds up toward its cutoff point. This may be done either electrically or pneumatically.

At times the burner may be started on low fire and so held until the burner stabilizes and draft is established before the firing rate control is permitted to take over. Regardless of the means of such control, a guaranteed low-fire start provides a smoother light-off and allows the burner to stabilize before going to the firing rate demanded by the load.

Limit Controls. Limit controls and safety controls provide a check and establish maximum limits beyond which the burner should not be allowed to operate. Such a control may respond to steam pressure or to air or water temperature. It is normally set to shut off the burner should conditions exceed those required by the operating control. Steam boilers require a low-water cutoff to prevent burner operation if the water level drops to an unsafe level.

Valves. Valves are required for final shutoffs on both pilot and main gas lines. Their type depends upon the service and the burner characteristics. Valves may be divided into the following basic types:

1. *Solenoid Valves.* These valves (Fig. 12-37) provide quick opening and closing, are acceptable for final shutoff service in sizes of $\frac{3}{4}$ in. and smaller. Normally, for pilot supply service, they may be used on main gas lines where their quick-opening characteristics are not objectionable.

2. *Motorized Valves.* Electrically or hydraulically operated, these valves (Fig. 12-38) normally provide relatively slow-opening and quick-closing characteristics. Their opening timing is not adjustable. Provision may be made for either direct-acting or reverse-acting secondary air arm operation. These valves are acceptable for final safety shutoff service.

3. *Diaphragm Valves.* These are usually operated by the available gas pressure; some larger models may be air or steam operated. Provision is made for adjustable opening timing, according to the burner operating characteristics, and for relatively fast closing timing, within 5 sec. These valves (Fig. 12-39) are acceptable as final safety shutoff valves in most areas. They also provide for secondary air operation, either direct- or reverse-acting.

4. *Manually Opened Safety Shutoff Valves.* These valves can be manually opened only when power is available. With power off, they trip free for fast closure. They are normally used on semiautomatic and manually fired installations.

5. *Burner Input Control Valves.* Some localities require these valves in addition to the final safety shutoff valve. They may or may not provide for final tight shutoff and may be of a motorized butterfly type for high-low or modulating service. Local regulations should be checked to be sure that valves selected will be acceptable for the requirements of the installation.

Draft Regulators. Draft regulators are normally utilized on commercial burners in place of draft diverters for maintaining a constant overfire draft to ensure stable firing and to increase overall efficiency. There are two basic types:

1. *Barometric dampers* (Fig. 12-2) are generally installed in the breaching between boiler and chimney. They are balanced,

closing when chimney draft is low and opening when it is high. When the dampers open, air from the boiler room is allowed to enter the breaching, thereby maintaining a constant draft in the combustion chamber. Barometric dampers are generally satisfactory for constant firing rate burners and may be used on some high-low burners.

2. *Motorized damper positioners* regulate the uptake damper in the boiler breaching. They are designed to hold the damper in a nearly closed position with the burner off; to open it wide during light-off; and to modulate the uptake damper while the burner is firing, to maintain constant overfire draft. The damper actuator has an integral switch that prevents the burner from starting with the damper closed. For multiple installations or for modulating control, *mechanical damper positioners* are normally used to maintain greater accuracy in overfire draft conditions so as to increase efficiency.

Electrical Load Releases and Fuses. Electrical load releases and fuses are a part of the overall control system used to disconnect power in the event of short circuits or any electrical overload that could endanger or damage the equipment. Small burner motors generally have overload releases built in. Larger motors require external overload releases, which are generally included in their magnetic starters. Fuses should be provided as specified by local regulations.

Water Level Controls. Steam boilers require water level controls for added safety against low-water conditions. There are two basic types:

1. *Float type* water level controls are installed either directly in the boiler or in the adjacent float chamber. As the water level falls, the float lowers and thru linkage operates a switch to start the water feed pump or energize the feed water electrical solenoid valve for low-pressure water. On low-pressure boilers only, falling water level may directly open the mechanical feed water valve. If the water level continues to fall, another switch, actuated by the linkage, shuts off the burner before a hazardous water level is reached.

2. *Probe type* controls provide the same function as the float type. They operate on the principle of conduction of a small current from the probe thru the water to hold the relay energized. As the water level falls, contact is broken, current flow stops, and the relay is de-energized. Relay contacts provide switching as required.

Definitions

Boiler, Automatically Lighted. A boiler on which the burner is automatically started on a call for heat by the controller in its normal sequence. Provision is normally made for proof of pilot before the main fuel supply can be turned on.

Boiler, Semiautomatic. A boiler on which the burner is normally started by means of a manual switch, with provision for remote ignition of a pilot that when automatically proved allows the main valve to be opened either automatically or manually.

Boiler, Manually Lighted. A boiler on which the pilot is normally a manual torch that should be proved as capable of igniting the main burner before the main fuel supply can be turned on.

Flame Safeguard. A device that will automatically shut off the gas supply to the main burner or a group of burners when the means of ignition is not proved or when flame failure occurs on the main burner or on one or more of a group of burners.

Supervised Flame. A flame the presence or absence of which is detected by a flame safeguard.

Proved Pilot. A pilot flame supervised by a flame safeguard that senses the presence of the pilot flame prior to permitting the main burner fuel safety shutoff valve to open.

Flame-Sensing Element of a Flame Safeguard. A flame rod, a photoemissive cell, or a photoconductive cell.

Flame Rod. An electrode extending into the gas flame being supervised. Owing to the conductive property of the ionized gases in the flame, the electrode causes current to flow thru the rod to the flame safeguard circuit of the programming control.

Photoconductive Cell. A device the internal resistance of which changes on exposure to pulsations of either a luminous or nonluminous flame, thus permitting current flow to the flame safeguard control circuit. Suitable for both gas and oil flames.

Photoemissive Cell. A device that increases the emission of electrons on exposure to luminous flames, thus permitting current flow to the flame safeguard control circuit when a flame is present. Generally not suitable for nonluminous gas flames. A specialized form (ultraviolet cell) permits current flow to the flame safeguard when ultraviolet energy from the flame is present. This type is suitable for gas or oil flames.

Trial-for-Ignition Period. That period of time when the programming flame safeguard permits the main burner fuel valve to be opened before the main flame-sensing device is required to detect the main flame. If not detected within the required time, all fuel is shut off and locked out immediately with no further ignition attempt.

STATIC ELECTRICITY

Static electricity, or electricity that is standing still as contrasted to that which is dynamic or flowing, is generated in dry weather, below 55 per cent relative humidity, by the contact and separation of materials under a wide variety of conditions. These conditions include walking across rugs, sliding across automobile seats, movement of automobiles and trucks on highways, flow of liquids or solids thru pipes, moving belts or pulleys, flow of liquid into a tank, and moving parts of machinery. Static electricity may also be developed between clouds and between earth and clouds.

Discharge of static electricity by a spark may create a hazard by igniting flammable or explosive mixtures. In other instances, it may interfere with the handling of materials, because of attraction or repulsion, or in the operation of machinery. Lightning is the principal manifestation of static electricity. Its effect on striking pipelines and buildings is well known. The potential between cloud and ground, just prior to a stroke of lightning has been estimated from 10^8 to 10^9 volts and from 10,000 to 114,000 amp. Potentials up to 40,000 volts have been measured between a gasoline truck body and ground.

Various methods and procedures[3–5] can be used either to prevent the static charge from getting so high that it will discharge in the form of a spark or to drain the charge continuously. These are:

1. Ground the affected machinery, pipes, or equipment. This must be done adequately and properly. When grounding, it is not enough to ground only stationary parts of equipment; moving parts must also be grounded, generally thru a brush rubbing on the shaft.

2. For belting, a grounded metal comb may be used, so placed that its teeth project toward the belt over its full width. Special hygroscopic belt dressings may also be used.

3. Where possible to do so, humidity should be kept above 55 per cent in rooms or above 65 per cent in some machines.

4. Static neutralizers may be used in some cases to supply a liberal amount of positive and negative ions in the air between installed neutralizer and electrified stock. Commercial static neutralizers are of three types: high voltage, radioactive, and induction.

5. Paper coming from a printing press may have a high negative charge in spots and a positive charge in others. Heating with a gas flame for drying the ink also reduces the static electricity.

Detailed data on static electricity, processes in which it may be a hazard, instruments for its detection, and methods for prevention of its accumulation are given in Ref. 6. Included also is a code for protection against lightning for buildings and for structures containing flammable liquids and gases.

THERMOELECTRIC GENERATION—SEEBECK EFFECT

Applications

Direct-current electric power can be generated without moving parts by passing heat thru solid materials of requisite thermoelectric properties. This so-called *Seebeck effect* named after its discoverer manifests itself as the "development of an electrical potential or voltage across the heated junction of two dissimilar conductors or semiconductors."* The voltage developed is called a Seebeck emf. A Seebeck generator is a series of thermocouples (also known as a thermopile).

Present gas industry applications of the Seebeck effect include operation of electric controls such as thermostats, automatic pilots, automatic valves, and generators for cathodic protection of pipelines and instrumentation purposes. Provided generator efficiency is increased and more favorable thermoelectric materials are found, potential applications include:

1. Cathodic protection of water heater tanks.
2. Operation of domestic heating appliance blowers, making gas appliances completely independent of utility power lines.
3. Operation of cooling equipment by passing current generated thru another series of thermocouples to cause cooling (**Peltier effect**).
4. Emergency electrical supply for use during power failures.
5. Exhaust-heat recovery (e.g., from prime movers).

* In 1964, elements such as tellurium and germanium, which have a small energy gap between their respective conduction and valence bands, were called *semiconductors;* the term *semimetal* was restricted to elements such as bismuth and antimony, whose conduction and valence bands overlap.[11]

Seebeck generator efficiencies (conversion of heat to direct current) of 4 to 6 per cent are attainable.[7] It was reported early in 1965 that a General Atomics unit, costing $23 per watt, achieved an overall efficiency of 4 per cent. Units available from the Minnesota Mining and Manufacturing Co., costing from $20 to $40 per watt, have an overall efficiency of 2.7 per cent. French Petroleum Institute units, weighing 45 to 150 lb per kw, operate at 2 to 10 per cent efficiency. Westinghouse[8] has developed a 100-watt thermoelectric generator operating on a propane flame using indium arsenide at 850 F as the thermoelectric material. A 5000-watt (equal to 6.7 hp) unit is being made, large enough to operate good-sized motor blowers.

Thermoelectric Materials

The best properties for maximum efficiency of thermoelectric materials are: (1) high thermoelectric voltage or Seebeck emf; (2) low electrical resistance; and (3) low thermal conductivity. Full-conducting metals represent unsatisfactory thermoelectric materials because of low Seebeck emf and excessive thermal conductivities. Compounds of metals with nonmetals, although better because of high Seebeck emf and lower thermal conductivities, are generally still unsatisfactory because of high electrical resistance. Nevertheless, some high-temperature ceramics and ionic crystals such as silver iodide and silver bromide appear promising. The most promising materials recently developed are intermetallic compounds of heavy elements (compounds or alloyed mixtures of two or more metals) such as tellurium selenide, lead telluride, bismuth telluride, and manganese telluride, activated with small concentrations of positive and negative conducting promoters. Other materials under study for use in generating electricity

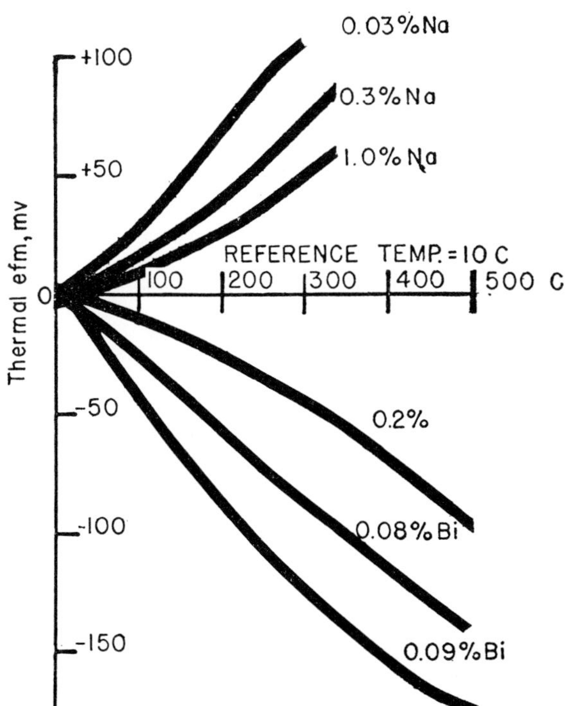

Fig. 15-37a Thermal emf for specific lead telluride alloys, indicating temperature and voltage ranges possible with various compositions.

directly from heat include cobalt silicide, germanium telluride, gallium phosphide, gallium arsenide, and silver antimony telluride.

Data are available covering a number of materials.[9]

The function of positive and negative conducting promoters is to improve the electrical conductivity and to increase the Seebeck emf. The combination of a positive and a negative conducting element as the two legs of a thermoelectric generator yields the highest Seebeck emf and the highest potential efficiency of conversion of heat into electricity of any thermoelectric device developed thus far. This combination also yields the best thermoelectric coefficients of performance for cooling and heating by the Peltier effect. Promoted intermetallic thermoelectric elements are now relatively costly since strict manufacturing control is needed in the alloying and promoter compounding steps. These elements cannot generally be operated at much above 800 to 1000 F or above the melting points of the alloys; lead telluride is restricted to 1000–1200 F.

Complex relationships exist among thermal emf, resistivity, and temperature in a semiconductor thermocouple.[10] Composition for positive and negative thermocouple elements is governed by the designed function of the thermocouple and its temperature range in service. Figure 15-37a shows thermal emfs taken with respect to copper for several specific lead telluride alloys. When positive and negative lead telluride elements are used in a single thermocouple, the thermal emf developed is the sum of the emfs for the two components or the ordinate differential between two curves. The electrical resistivity of these materials exhibits a large positive temperature coefficient at ordinary temperatures. At elevated temperatures this coefficient becomes negative.

High-voltage sensitivity and maximum power delivery to a matched load cannot be obtained with a single choice of positive and negative compositions. High-voltage sensitivity is obtained with elements "doped" with small amounts of sodium and bismuth. High matched load power is obtained with elements doped with relatively higher amounts of these additives. The data for thermocouples designed for maximum power delivery to a matched load illustrate how the selection of composition depends on the temperature ranges to be encountered.

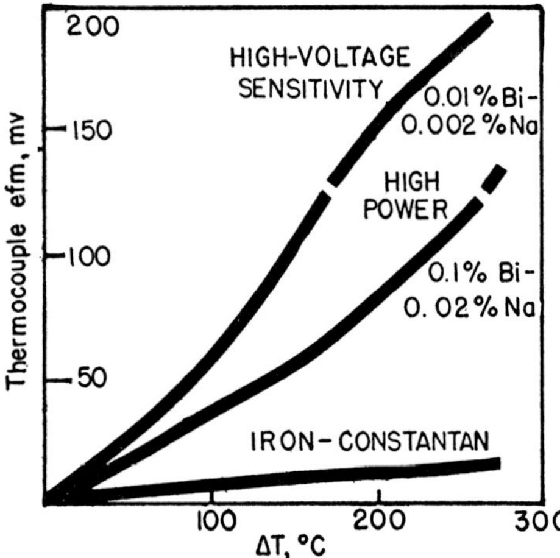

Fig. 15-37b Voltage sensitivity of semimetals vs. conventional iron-constantan thermocouple.

Figure 15-37b compares the voltage sensitivity of semimetal materials with that of a conventional thermocouple alloy (iron-constantan). As shown, semimetal high-voltage sensitivity and high-power compositions achieve as much as ten times the output of an iron-constantan thermocouple.

A practical Seebeck generator requires many individual thermoelectric elements connected electrically in series to provide a usable output voltage. A single set of elements produces, for example, about 0.04 volt for every 100° C temperature difference between hot and cold junctions. At a maximum temperature difference of 600° C, each unit would yield approximately 0.24 volt, about 460 sets of elements being required to develop an open circuit of 110 volts. When power is drawn from the Seebeck generator, approximately one-half of the voltage appears over the external load (in a matched load situation only); the other half is required to overcome the internal resistances. Therefore, nearly 1000 sets of elements would be required to develop 110 volts under actual load conditions.

A possible arrangement of multiple single-stage thermoelectric elements connected in series is shown in Fig. 15-38.

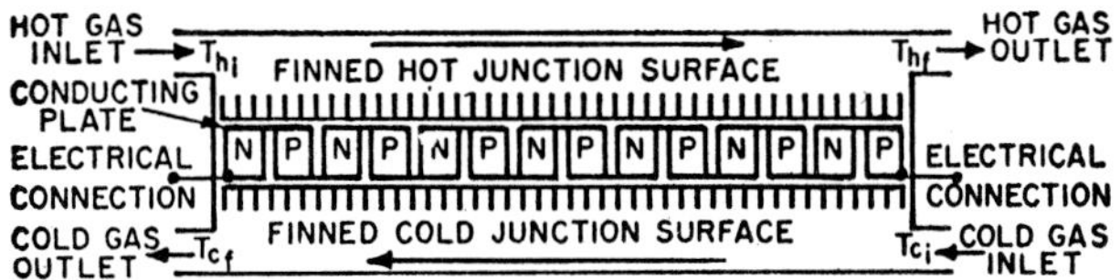

Fig. 15-38 Multiple thermoelectric element Seebeck generator.[7]

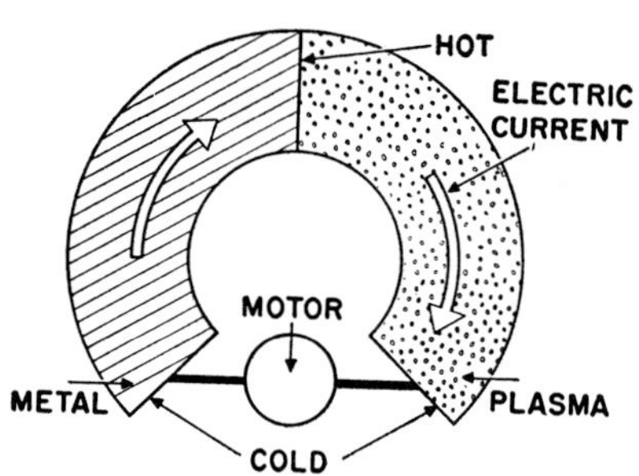

Fig. 15-39 Plasma thermocouple. One metal and a gas replace the two metals used in normal thermocouples.

PLASMA THERMOCOUPLES

A plasma thermocouple (Fig. 15-39) with a metal-plasma junction is theoretically capable of converting heat into electricity with an efficiency of 30 per cent; 5 per cent has been achieved. The plasma is generated when heat ionizes cesium vapor in a cell at a pressure of 0.1 to 2 mm Hg. Cesium vaporizes at a low temperature (670 C) and has the lowest ionization potential (3.88 electron volts) of any element. Cesium cost over $500 per lb in 1963. Possible substitutes for cesium are rubidium, potassium, sodium, and lithium.

Thermionic Converter Engines—Edison Effect

The principle of operation of thermionic converter engines[12] is based on the phenomenon of electronic emission from metals at high temperatures (Edison effect). Their basic feature is the partial conversion of kinetic energy of electrons emitted from a metallic surface into electrical potential energy. The process is similar to that in a steam engine in which the heated steam gives out its kinetic energy and drives a turbine. Unlike steam, however, electrons possess an electrical charge; consequently, the volume filled with them exhibits a space charge, which inhibits their further emission. The main technological problem in thermionic converters is the reduction of this space charge.

A thermionic converter can be considered as a thermocouple in which an evacuated space has been substituted for one of the conductors. In Fig. 15-40, R_1 and R_2 represent two parallel metallic plates separated by an evacuated gas of width y. R_1 is maintained at a relatively high temperature, T_1, and R_2 at a lower temperature, T_2. Because of temperature difference, the plates will emit electrons at different rates and hence a potential difference will be established between them.

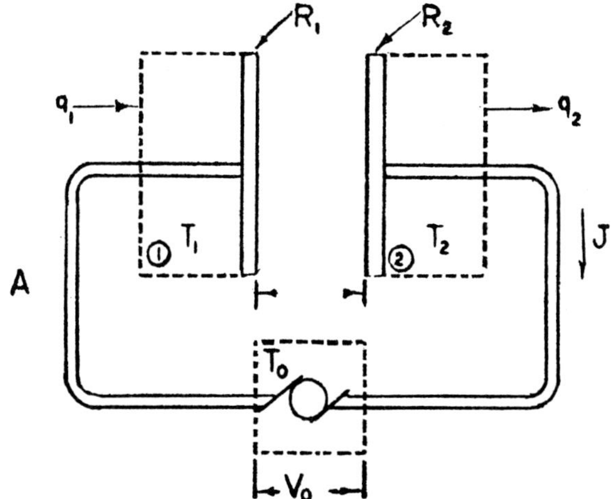

Fig. 15-40 Principle of thermoelectron engine.

If now the two plates are externally connected by conductor A to a load, a current will flow and the potential difference between the two plates will adjust itself to that necessary to sustain this current. Analysis of the thermionic converter shows that as the load resistance is increased, the output voltage across the load will increase and the current thru it will decrease.

Power output obtainable from a thermionic converter operating across a fixed temperature difference will depend primarily on the electron affinity (work function) of the emitting and collecting plates. It can be shown, however, that if the electron affinity of the emitting hot plate is below a certain limit, the power output from the converter will depend only on the electron affinity of the collector. A second important factor affecting the power output of a thermionic engine is the value of a potential barrier, resulting from the space charge of the emitted electrons. A space charge barrier of one volt would result in a power output of the device 20,000 times less than the maximum power output obtainable without a space charge.

The efficiency of a thermionic converter is proportional to the power output divided by the heat that must be supplied to the hot plate to maintain the high temperature (see Table 15-6). The heat input to the plate can be separated into two parts. The first, called *electron cooling*, is associated with the energy carried along with the electrons flowing from the hot to the cold plate. The second part is the *heat loss* between the hot and cold plates that is due to thermal conduction and thermal radiation. For a vacuum converter, conduction losses are negligible, and the radiation heat transfer between the two plates will depend on their thermal emissivities. To obtain high values of thermal efficiencies, the electrode plates must have low thermal emissivity values.

Three types of thermionic engines have been analyzed extensively. These devices differ only in the method used to reduce the space charge. Basically they operate as previously described. The types are:

1. The *cesium diode*, in which the space charge is reduced by means of positive ions.
2. The *thermoelectron engine*, in which the space charge is reduced by means of a crossed electric and magnetic field.
3. The *close-spaced diode*, in which the space charge is reduced by means of close spacing.

Many design configurations have been studied using a close-spaced diode in combination with a variety of heat sources. Any fuel may be used as long as it is consumed at sufficiently high temperatures to heat the emitting electrode. Nuclear fission and isotopes have been considered and models are being built around them; the fossil fuels, which are readily available, show the most promise for future use. Problems of combustion are under examination and combustors and combustion systems are being developed.

Table 15-6 Performance and Objectives of Thermionic Converters

(I.G.T. January 1965)

Manufacturer	Temp, °C	Power density	Efficiency
RCA Model A–1192	1200	1 w/sq cm, 6w output	5%
	1700	24 w/sq cm	19.7%
Goal	...	6–10 w/sq cm	
Thermo Electron Engineering Corp. (10¢/w estimated)			
Experimental	1760	56 w/sq cm	
Goal	...	50 w/sq cm	20–40 w/lb
Best life (255 hr)	1250–1365	1.25–4 w/sq cm	
Allison Div., General Motors	1600 1900	2.87 w/sq cm 12.5 w/sq cm	
French Petroleum Institute	...	...	2–30 lb/kw, 11–17%

In one proposed design, emitting surfaces in the form of fins are mounted on a hollow cylinder. Finlike collector electrodes are mounted in a bellows-type container. The two electrodes are separated by ceramic spacers. A gaseous fuel may be consumed within the hollow structure, permitting very compact design, or hot flue gas may be passed thru it. Various types of combustors and combustion reactors utilizing the honeycomb structure of catalytic ceramic are in preliminary stages of design. These in conjunction with regenerative heaters may be inserted within the hollow cylinder to reduce heat losses and increase efficiency. One converter of this design is 14 in. long, 2⅜ in. in diameter, and produces 800 watts at an efficiency of 10 per cent.

The economics of the diode are interesting. Tungsten cathodes cost approximately $10 per pound and the same amount for processing. These cathodes constitute a small portion of the total weight of the converter. The remainder is composed of common ceramics and metals at perhaps a tenth of the cost per pound.

THERMOELECTRIC COOLING—PELTIER EFFECT

Peltier discovered, in 1934, that when a direct current was passed thru two conductors or dissimilar metals, heat was absorbed at one junction and given off at the other. Refrigeration or cooling based on this effect may be considered as using electron gas as the refrigerant instead of the conventional Freons.

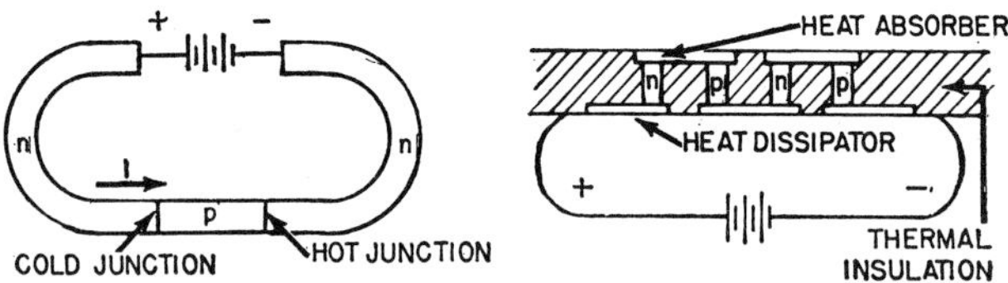

Fig. 15-41a (left) Illustration of Peltier effect. Electrons move counterclockwise.
Fig. 15-41b (right) Arrangement of thermoelectric materials for refrigeration.

In Fig. 15-41a the two dissimilar thermoelectric materials are represented by n and p. The n-type, with negative thermoelectric power, has an excess of electrons. Current is carried by the electrons. The p-type has positive thermoelectric power. It appears that the current, I, in the circuit[13] is carried by the holes (an electron deficiency), or positive charges. I equals the electron current plus the hole current. Since the current passing into the cold junction requires heat to be supplied, the electrons take on energy from the materials at this junction. At the hot junction work is done, and this work plus the energy picked up at the cold junction is given up as heat. Thus, the cold junction may be considered an expansion valve and evaporator, the hot junction, a compressor and condenser.

Thermoelectric materials arranged as in Fig. 15-41a would not make a practical refrigeration device. They may be rearranged, however, as shown in Fig. 15-41b. Here all cold surfaces are placed on one side and all hot surfaces on the other, with thermal insulation between them.

Thermoelectric cooling using the Peltier principle is recognized as having many potential uses for refrigeration,[14] but it is generally conceded that great improvements must be made in the materials before thermoelectric cooling is ready to serve conventional refrigeration and air conditioning applications. Such improvement appears to be at least several years away, since a figure of merit Z of at least 8 to $10 \times 10^{-3}/°C$ is said to be necessary for practical use, whereas materials readily available today have values of only 2 to $2.5 \times 10^{-3}/°C$.

For laboratory and scientific instruments, electronic component cooling, and small table appliances of small capacity in which cost may be less critical, thermoelectric cooling can be successfully used even in its present state of low capacity and relatively high cost. This is illustrated by a temperature-controlled chamber (Fig. 15-42a) for determining specific gravity of liquids, developed by Whirlpool Corp.

The heat-dissipating media consist of a set of lightweight aluminum fins for each thermoelectric module, with a small fan to move air over them. The Peltier couples are assembled in modular form (Fig. 15-42b). A total of 60 couples is used, divided into 4 modules of 15 couples each. The shape of each module matches the outside surface of the cooling chamber and the ends of the dissipating fins so that maximum surface contact and good heat exchange are provided.

Each module consists of 15 p-type and 15 n-type bismuth-telluride thermoelements, each approximately $\frac{1}{4}$ in. in di-

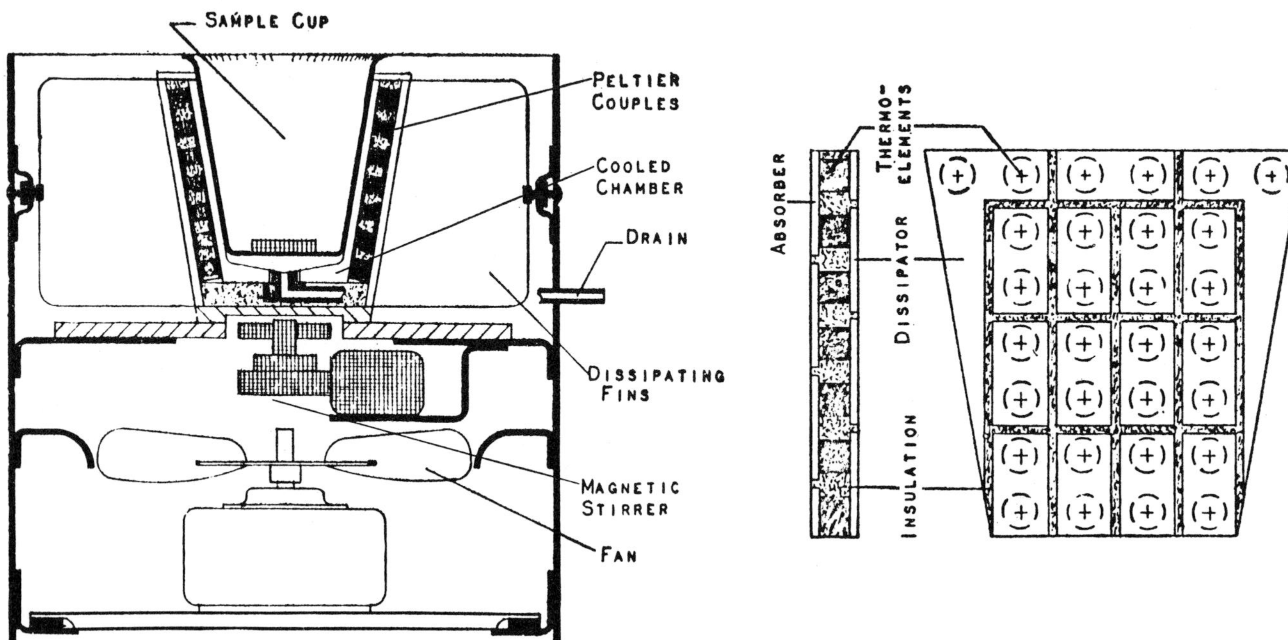

Fig. 15-42a (left) Temperature-controlled chamber using Peltier couples.
Fig. 15-42b (right) Construction of couple banks (end and plan views).

ameter and 1/4 in. long. The elements, alternate p-type and n-type, are soldered to flat copper absorbers and dissipators to form a series circuit. The space between the elements is then filled with foam-in-place insulation, which separates the cold and hot sides of the couple bank and strengthens the whole assembly.

The d-c power supply for this unit is obtained from a 115-volt ac supply thru a transformer, rectifier, choke, and filter to 22.5 amp, 10 volts, with less than 5 per cent ripple.[15] Such d-c supply could also be provided by thermoelectric generation without the above conversion equipment.

Additional data on thermoelectric cooling are given in Refs. 16, 17, 18, and 19.

FUEL CELLS

Fuel cells are d-c generators in which energy is obtained directly from the chemical reaction between an oxidant and a fuel that are continuously fed into the cell. A fuel cell differs from a battery in that the electrodes are inert; that is, they do not react and are therefore not consumed. Some cells use consumable electrodes.[20]

Classification

Fuel cells may be classified as direct or indirect. In the *direct cell,* the fuel and oxidant are fed directly to the electrodes. In the *indirect cell,* the fuel, usually a hydrocarbon, is reformed into a more readily reactive fuel, hydrogen. A third type, a *pseudofuel cell,* is the regenerable fuel cell. With this type cell the fuel is regenerated by thermal or other means to produce a closed cycle system subject to Carnot limitations. In the strictest sense this is more of a Carnot engine than a fuel cell.

Either the direct or the indirect fuel cell can be further subclassified as (1) very high temperature, 1000 C solid oxide; (2) high temperature, above 500 C, molten salt; (3) intermediate temperature, 200 to 400 C, Bacon alkaline electrolyte; (4) low temperature, less than 150 C, acid or alkaline, immobilized electrolyte, free electrolyte or ion exchange membrane.

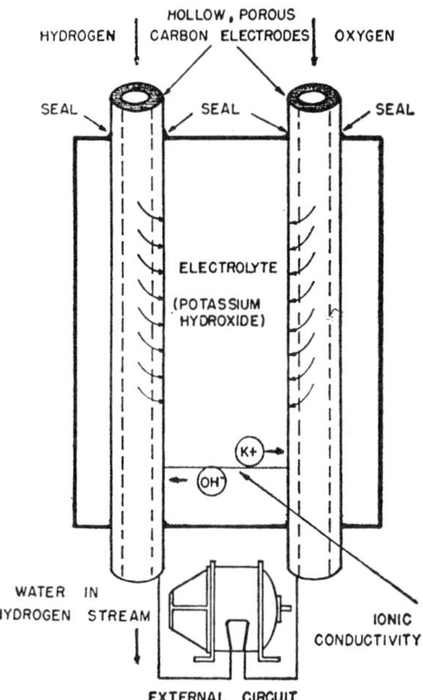

Fig. 15-43 Hydrogen-oxygen fuel cell.

Performance and Operation

Performance. Data are given in Table 15-7.

Operation. Figure 15-43 shows a fuel cell using hydrogen and oxygen gas as fuel. Hydrogen gas enters one of the hollow electrodes (anode), which is both porous and electrically conductive, and diffuses into the potassium hydroxide electrolyte-electrode interface. Water is thus formed by the oxidation of the hydrogen and an electron freed electrochemically. This electron enters the electrical system, passes thru the external load circuit, and returns to the cell at a second electrode (cathode), similar to the first and having oxygen gas supplied to it. At the oxygen electrode, the oxygen is reduced by the reaction between the oxygen, the electrolyte, and the electron.

A fuel cell designed to burn methane would operate along analogous principles. Natural gas is one of the cheapest sources of hydrogen. Experimental fuel cells have operated with efficiencies as high as 80 per cent in contrast to the close to 40 per cent efficiency of a superior central generating station. Research is in progress on the use of natural gas as fuel, with air as oxidant. In some schemes the natural gas is reformed (together with water) and the resultant products, H_2, CO_2, CO, CH_4, and H_2O, are charged to the cell. In other approaches only the resultant H_2 is used as fuel.

Table 15-7 Fuel Cell Performance and Economics
(I.G.T., January 1965)

Source	Cost, $/kw	Energy density w/lb	Energy density kw/cu ft	Current density, amp/sq ft	Efficiency, %
Special report* to the President of the U. S.	71,000	18	0.85	...	27
Special report,† French Petroleum Institute	~1,500	3	0.25	...	...
Battelle Institute,‡ Frankfurt					
Allis Chalmers	...	20	1.47	...	...
Shell Oil	...	40	4.7	...	...
General Electric	...	40	...	...	...
Union Carbide	...	13	1.1	...	...
Institute of Defense§ analysis	300–500	...	...	...	...
ASEA**	††	11	0.37	200	35

* Special survey conducted for the Executive Office of the President, Office of Science and Technology, June 1963—based on external reforming and fuel cell.

† Report from *Institut Français du Pétrole,* September 1964—based on methanol.

‡ Report from Battelle Institute, Frankfurt, April 1964—based on hydrogen-air.

§ Estimated for 1970.

** *Allmanna Svenska Elektriska Aktiebolaget Journal,* Sweden, 1964—data based on hydrocarbon-oxygen cell with external reformer included.

†† $20,000 to $200,000/kw in 1964, $200 in 1974, $20 in 1984.

A series of fuel cell reports is available. These reports collect and review unclassified (from U. S. security viewpoint) and nonproprietary information.[21–23]

REFERENCES

1. Natl. Fire Protection Assoc. *Explosion Prevention Boiler-Furnaces, Gas-Fired Watertube (Single Burner)* (NFPA 85A). Boston, 1964.
2. A.G.A. Laboratories. *American Standard Requirements for Gas Equipment in Large Boilers.* (ASA Z83.3–Jan. 1964, 3rd draft) Cleveland, Ohio, 1964.
3. Knowlton, A. E., ed. *Standard Handbook for Electrical Engineers,* 9th ed. New York, McGraw-Hill, 1957.
4. Silsbee, F. B. *Static Electricity.* (Natl. Bur. of Standards C438) Washington, G.P.O., 1942.
5. Howard, J. C. "Static Electricity in the Petroleum Industry." *Elec. Eng.* 77: 610–14, July 1958.
6. Natl. Fire Protection Assoc. *National Fire Codes, Volume 5: Electrical.* Boston, 1964.
7. Von Fredersdorff, C. G. "New Methods of Generating Electricity from Gas Sources." *A.G.A. Operating Sect. Proc.* 1960: CEP–60–12.
8. "Thermoelectricity Moves Up." *Chem. Eng. News* 37: 36, June 29, 1959.
9. Davisson, J. W., and Pasternak, J. *Status Report on Thermoelectricity.* (NRL Memorandum Rept. 1241) Washington, U. S. Naval Research Laboratory, 1962.
10. Fritts, R. W., "High Voltage Semimetal Thermocouples." *Electronic Design* 6: 28–31, May 28, 1958.
11. Wolfe, R. "Magnetothermoelectricity." *Sci. Am.* 210: 70–82, June 1964.
12. Hatsopoulos, G. N., and Welsh, J. A. *Present Developments in Thermionic Conversion.* (Research & Utilization Conf., 1959) New York, A.G.A., 1959.
13. Eichhorn, R. L. "Thermoelectric Refrigeration." *Refrig. Eng.* 66: 31–5, June 1958.
14. Danielson, W. R. "Temperature-Controlled Chamber Using Thermoelectric Cooling." *Refrig. Eng.* 67: 30–3, Feb. 1959.
15. Heinicke, J. B. "Design and Performance of Thermoelectric Refrigerator." *Refrig. Eng.* 67: 34–6, Feb. 1959.
16. Staebler, L. A. "Primer of Thermoelectric Refrigeration." *ASHRAE J.* 1: 60–5, Aug. 1959.
17. Joffe, A. F. *Semiconductor Thermoelements and Thermoelectric Cooling.* London, Infosearch, Ltd., 1957.
18. Hannay, N. B., ed. *Semiconductors.* New York, Reinhold, 1959.
19. Goldsmid, H. J. *Thermoelectric Refrigeration.* New York, Plenum Press, 1964.
20. McCormick, J. E. *Fuel Cell Systems.* (U. S. Air Force RADC–TN–60–118; U. S. Office of Tech. Services PB 161972) Washington, Dept. of Commerce, 1960.
21. Stein, B. R. *Status Report on Fuel Cells.* (U. S. Army Research Office, ARO Rept. 1; U. S. Office of Tech. Services PB 151804) Washington, Dept. of Commerce, 1959.
22. ——, and Cohn, E. M. *Second Status Report on Fuel Cells.* (U. S. Army Research Office, ARO Rept. 2; U. S. Office of Tech. Services PB 171155) Washington, Dept. of Commerce, 1960.
23. Hunger, H. H., and others. *Third Status Report on Fuel Cells.* (U. S. Army Signal Research and Development Laboratory; U. S. Office of Tech. Services AD 286686) Washington, Dept. of Commerce, 1962.
24. Piatt, W. R., "Flame Safety on Gas Burners." *Gas* 41: 56–61, Feb. 1965.

INDEX

In page references, numbers before slashes are section numbers; thus, 8/20 is page 20 in Section 8. In table and figure references, numbers before dashes are section numbers; to find their page numbers, see the Contents at the beginning of those sections. References in italics are to material extracted from other sources and include the page in this Handbook on which they are located.